国家出版基金项目
NATIONAL PUBLICATION FOUNDATION

A Dictionary of Seed Plant Names

Vol. 3 In Latin, Chinese and English (P-Z)

种子植物名称

卷3 拉汉英名称 (P-Z)

(279357-418831)

尚衍重 编著

中国林业出版社

图书在版编目(CIP)数据

种子植物名称. 卷3,拉汉英名称. P - Z / 尚衍重编著.
—北京:中国林业出版社,2012.6

ISBN 978 - 7 - 5038 - 6653 - 1

Ⅰ.①种… Ⅱ.①尚… Ⅲ.①种子植物 - 专有名称 - 拉丁语、
汉语、英语 Ⅳ.①Q949.4 - 61

中国版本图书馆 CIP 数据核字(2012)第 138914 号

中国林业出版社·自然保护图书出版中心

出 版 人: 金 旻
策划编辑: 温 晋
责任编辑: 刘家玲 温 晋 周军见 李 敏

出版 中国林业出版社(100009 北京市西城区刘海胡同7号)
      网址 http://lycb. forestry. gov. cn
      E-mail wildlife_cfph@163. com 电话 010 - 83225836
发行 中国林业出版社
      营销电话:(010)83284650 83227566
印刷 北京中科印刷有限公司
版次 2012 年 6 月第 1 版
印次 2012 年 6 月第 1 次
开本 889mm×1194mm 1/16
印张 124.75
字数 10240 千字
印数 1~2000 册
定价 750.00 元

279357 Pabellonia Quezada et Martic. = Leucocoryne Lindl. ■☆

279358 Pabstia Garay(1973);帕布兰属(帕勃兰属)■☆

279359 Pabstia jugosa(Lindl.)Garay;帕布兰■☆

279360 Pabstia triptera(Rolfe)Garay;三翅帕布兰■☆

279361 Pabstia viridis(Lindl.)Garay;绿帕布兰■☆

279362 Pabstia viridis(Lindl.)Garay var. parvifolia(Hoehne)Garay;小叶绿帕布兰■☆

279363 Pabstiella Brieger et Senghas = Pleurothallis R. Br. ■☆

279364 Pachea Pourr. ex Steud. = Crypsis Aiton(保留属名)■

279365 Pachea Steud. = Crypsis Aiton(保留属名)■

279366 Pachecoa Standl. et Steyerm. (1943);棱柱豆属■☆

279367 Pachecoa prismatica(Sessé et Moc.)Schub.;棱柱豆■☆

279368 Pacheya Scop. = Coussarea Aubl. ●☆

279369 Pachidendron Haw. = Aloe L. ●■

279370 Pachidendron africanum(Mill.)Haw. = Aloe africana Mill. ●☆

279371 Pachidendron africanum (Mill.) Haw. var. angustum Haw. = Aloe africana Mill. ●☆

279372 Pachidendron africanum (Mill.) Haw. var. latum Haw. = Aloe africana Mill. ●☆

279373 Pachidendron angustifolium(Haw.)Haw. = Aloe africana Mill. ●☆

279374 Pachidendron ferox(Mill.)Haw. = Aloe ferox Mill. ●☆

279375 Pachidendron pseudoferox(Salm-Dyck)Haw. = Aloe ferox Mill. ●☆

279376 Pachidendron supralaeve(Haw.)Haw. = Aloe ferox Mill. ●☆

279377 Pachila Raf. = Erucaria Gaertn. ■☆

279378 Pachiloma Raf. = Polytaenia DC. ■☆

279379 Pachiphillum La Llave et Lex. = Pachyphyllum Kunth ■☆

279380 Pachira Aubl. (1775);瓜栗属(巴拿马栗属,中美木棉属); Pachira ●

279381 Pachira aquatica Aubl.;水瓜栗(瓜栗,墨西哥瓜栗); Good Luck Money Tree,Guiana-chestnut,Guinea Chestnut,Provision Tree, Shaving Brush Tree,Water Pachira ●

279382 Pachira glabra Pasq.;光瓜栗●☆

279383 Pachira insignis(Sw.)Sw. ex Sav.;野瓜栗;Wild Chestnut ●☆

279384 Pachira insignis Sav. = Pachira insignis(Sw.)Sw. ex Sav. ●☆

279385 Pachira longifolia Hook. = Pachira macrocarpa (Schltdl. et Champ.) Walp. ●

279386 Pachira lukayense (De Wild. et T. Durand) Bakh. = Bombax lukayense De Wild. et T. Durand ●☆

279387 Pachira macrocarpa(Schltdl. et Champ.)Walp.;瓜栗(巴拉巴栗,巴拿马栗,大果瓜栗,大果木棉,大果木棉树,马拉巴栗,南洋土豆,中美木棉); Large-fruit Pachira, Large-fruited Malabar-chestnut,Pachira-nut,Sapotòn ●

279388 Pachira macrocarpa (Schltdl. et Champ.) Walp. = Pachira aquatica Aubl. ●

279389 Pachira oleaginea Decne. = Pachira glabra Pasq. ●☆

279390 Pachira sessilis Benth.;无梗瓜栗●☆

279391 Pachistima Raf. = Paxistima Raf. ●☆

279392 Pachites Lindl. (1835);厚兰属■☆

279393 Pachites bodkinii Bolus;博德金厚兰■☆

279394 Pachyacris Schltr. = Xysmalobium R. Br. ■☆

279395 Pachyaeris Schltr. ex Bullock = Xysmalobium R. Br. ■☆

279396 Pachyandra Post et Kuntze = Pachysandra Michx. ●■

279397 Pachyanthus A. Rich. (1846);粗花野牡丹属●☆

279398 Pachyanthus Post et Kuntze = Pachysanthus C. Presl ■☆

279399 Pachyanthus Post et Kuntze = Rudgea Salisb. ■☆

279400 Pachyanthus angustifolius Griseb.;窄叶粗花野牡丹●☆

279401 Pachyanthus cordifolius Cogn.;心叶粗花野牡丹●☆

279402 Pachyanthus glaber Cogn.;光粗花野牡丹●☆

279403 Pachyanthus longifolius Jenn.;长叶粗花野牡丹●☆

279404 Pachyanthus monocephalus(Urb.)Borhidi;单头粗花野牡丹●☆

279405 Pachyanthus reticulatus Britton et P. Wilson;网脉粗花野牡丹●☆

279406 Pachycalyx Klotzsch = Simocheilus Klotzsch ●☆

279407 Pachycalyx glaber(Thunb.) Klotzsch = Erica inaequalis(N. E. Br.) E. G. H. Oliv. ●☆

279408 Pachycalyx hispidus Klotzsch = Erica inaequalis(N. E. Br.) E. G. H. Oliv. ●☆

279409 Pachycalyx inaequalis Klotzsch = Erica inaequalis (N. E. Br.) E. G. H. Oliv. ●☆

279410 Pachycalyx pubescens Klotzsch = Erica inaequalis (N. E. Br.) E. G. H. Oliv. ●☆

279411 Pachycarpus E. Mey. (1838);大果萝藦属■●☆

279412 Pachycarpus albens E. Mey. = Asclepias albens (E. Mey.) Schltr. ■☆

279413 Pachycarpus appendiculatus E. Mey.;附属物大果萝藦■☆

279414 Pachycarpus asperifolius Meisn.;糙叶大果萝藦■☆

279415 Pachycarpus bisacculatus(Oliv.)Goyder;双囊大果萝藦■☆

279416 Pachycarpus bullockii Cavaco = Pachycarpus bisacculatus (Oliv.) Goyder ■☆

279417 Pachycarpus campanulatus(Harv.)N. E. Br.;风铃草状大果萝藦■☆

279418 Pachycarpus campanulatus (Harv.) N. E. Br. var. sutherlandii N. E. Br.;萨瑟兰大果萝藦■☆

279419 Pachycarpus chirindensis(S. Moore)Goyder;奇林达大果萝藦■☆

279420 Pachycarpus concolor E. Mey.;同色大果萝藦●☆

279421 Pachycarpus coronarius E. Mey.;饰冠大果萝藦●☆

279422 Pachycarpus crispus(P. J. Bergius)E. Mey. = Asclepias crispa P. J. Bergius ■☆

279423 Pachycarpus dealbatus E. Mey.;白色大果萝藦■☆

279424 Pachycarpus decorus N. E. Br.;装饰大果萝藦■☆

279425 Pachycarpus distinctus(N. E. Br.)Bullock;离生大果萝藦■☆

279426 Pachycarpus eximius(Schltr.)Bullock;优异大果萝藦■☆

279427 Pachycarpus firmus(N. E. Br.)Goyder;坚硬大果萝藦■☆

279428 Pachycarpus fulvus (N. E. Br.) Bullock = Asclepias dregeana Schltr. ■☆

279429 Pachycarpus galpinii(Schltr.)N. E. Br.;盖尔大果萝藦■☆

279430 Pachycarpus gerrardii (Harv.) N. E. Br. = Pachycarpus campanulatus(Harv.) N. E. Br. var. sutherlandii N. E. Br. ■☆

279431 Pachycarpus goetzei(K. Schum.)Bullock;格兹大果萝藦■☆

279432 Pachycarpus gomphocarpoides E. Mey. = Xysmalobium gomphocarpoides(E. Mey.) D. Dietr. ■☆

279433 Pachycarpus graminifolius Wild = Trachycalymma graminifolium (Wild) Goyder ●☆

279434 Pachycarpus grandiflorus(L. f.)E. Mey.;大花大果萝藦■☆

279435 Pachycarpus grandiflorus (L. f.) E. Mey. subsp. tomentosus (Schltr.) Goyder;毛大花大果萝藦●☆

279436 Pachycarpus grandiflorus(L. f.)E. Mey. var. chrysanthus N. E. Br. = Pachycarpus grandiflorus(L. f.)E. Mey. ●☆

279437 Pachycarpus grandiflorus(L. f.)E. Mey. var. elatocarinatus N. E. Br. = Pachycarpus grandiflorus(L. f.)E. Mey. ●☆

279438 Pachycarpus grandiflorus (L. f.) E. Mey. var. tomentosus (Schltr.) N. E. Br. = Pachycarpus grandiflorus (L. f.) E. Mey. subsp. tomentosus(Schltr.) Goyder ●☆

279439 Pachycarpus grantii(Oliv.)Bullock;格兰特大果萝藦●☆

279440 Pachycarpus humilis E. Mey. = Asclepias humilis (E. Mey.) Schltr. ■☆

279441 Pachycarpus inconstans N. E. Br. = Pachycarpus asperifolius Meisn. ■☆

279442 Pachycarpus insignis (Schltr.) N. E. Br. = Pachycarpus transvaalensis(Schltr.) N. E. Br. ■☆

279443 Pachycarpus lebomboensis D. M. N. Sm. ;莱邦博大果萝藦●☆

279444 Pachycarpus ligulatus E. Mey. = Pachycarpus dealbatus E. Mey. ■☆

279445 Pachycarpus linearis(E. Mey.) N. E. Br. ;线状大果萝藦●☆

279446 Pachycarpus lineolatus(Decne.)Bullock;线条大果萝藦■☆

279447 Pachycarpus mackenii(Harv.) N. E. Br. ;马肯大果萝藦■☆

279448 Pachycarpus macrochilus(Schltr.)N. E. Br. ;大唇大果萝藦■☆

279449 Pachycarpus marginatus E. Mey. = Woodia mucronata(Thunb.) N. E. Br. ■☆

279450 Pachycarpus mildbraedii Bullock = Pachycarpus lineolatus (Decne.)Bullock ■☆

279451 Pachycarpus natalensis N. E. Br. ;纳塔尔大果萝藦■☆

279452 Pachycarpus orbicularis E. Mey. = Xysmalobium orbiculare(E. Mey.) D. Dietr. ■☆

279453 Pachycarpus pachyglossus Goyder;粗舌大果萝藦■☆

279454 Pachycarpus petherickianus(Oliv.)Goyder;彼瑟大果萝藦■☆

279455 Pachycarpus plicatus N. E. Br. ;折叠大果萝藦■☆

279456 Pachycarpus reflectens E. Mey. ;骤折大果萝藦■☆

279457 Pachycarpus rhinophyllus(K. Schum.) N. E. Br. = Pachycarpus concolor E. Mey. ●☆

279458 Pachycarpus richardsiae Goyder;理查兹大果萝藦■☆

279459 Pachycarpus rigidus E. Mey. ;硬大果萝藦■☆

279460 Pachycarpus rigidus E. Mey. var. tridens? = Pachycarpus rigidus E. Mey. ■☆

279461 Pachycarpus robustus(A. Rich.)Bullock;粗壮大果萝藦■☆

279462 Pachycarpus rostratus N. E. Br. ;喙状大果萝藦■☆

279463 Pachycarpus scaber(Harv.) N. E. Br. ;粗糙大果萝藦■☆

279464 Pachycarpus schinzianus(Schltr.) N. E. Br. ;欣兹大果萝藦■☆

279465 Pachycarpus schumannii Chiov. = Pachycarpus petherickianus (Oliv.)Goyder ■☆

279466 Pachycarpus schweinfurthii (N. E. Br.)Bullock = Pachycarpus lineolatus(Decne.)Bullock ■☆

279467 Pachycarpus spurius(N. E. Br.)Bullock;长舌大果萝藦■☆

279468 Pachycarpus stelliceps N. E. Br. ;星头大果萝藦■☆

279469 Pachycarpus stenoglossus(E. Mey.) N. E. Br. ;窄头大果萝藦■☆

279470 Pachycarpus suaveolens(Schltr.)Nicholas et Goyder;芳香大果萝藦■☆

279471 Pachycarpus transvaalensis(Schltr.) N. E. Br. ;德兰士瓦大果萝藦■☆

279472 Pachycarpus validus (Schltr.) N. E. Br. = Pachycarpus asperifolius Meisn. ■☆

279473 Pachycarpus vexillaris E. Mey. ;旗瓣大果萝藦■☆

279474 Pachycarpus vexillaris E. Mey. var. stenoglossus? = Pachycarpus stenoglossus(E. Mey.) N. E. Br. ■☆

279475 Pachycarpus viridiflorus E. Mey. = Asclepias dregeana Schltr. ■☆

279476 Pachycentria Blume (1831);厚距花属（大蕊野牡丹属）; Pachycentria ●

279477 Pachycentria constricta Merr. ;缢缩厚距花●☆

279478 Pachycentria fengii S. Y. Hu = Medinilla fengii(S. Y. Hu)C. Y. Wu et C. Chen ●

279479 Pachycentria formosana Hayata;厚距花(红果野牡丹,台湾大蕊野牡丹,台湾厚距花,台湾酸脚杆);Formosan Pachycentria, Taiwan Pachycentria ●

279480 Pachycentria formosana Hayata = Medinilla fengii(S. Y. Hu)C. Y. Wu et C. Chen ●

279481 Pachycentron Pomel = Centaurea L. (保留属名)●■

279482 Pachycereus(A. Berger)Britton et Rose(1909);摩天柱属(武伦柱属);Pachycereus ●

279483 Pachycereus Britton et Rose = Pachycereus(A. Berger)Britton et Rose ●

279484 Pachycereus grandis Rose;巨摩天柱(丹羽太郎)●☆

279485 Pachycereus hollianus(F. A. C. Weber) Buxb. ;霍尔摩天柱, Acompes, Baboso ●☆

279486 Pachycereus marginatus(DC.)Britton et Rose;白云角;Central Mexico Organ Pipe, Jarritos, Margined Pachycereus, Organo ●☆

279487 Pachycereus militaris(Audot) D. R. Hunt;金毛翁●☆

279488 Pachycereus orcuttii(K. Brandegee)Britton et Rose;飞云阁; Orcutt Pachycereus ●☆

279489 Pachycereus pecten-aboriginum(Engelm.)Britton et Rose;摩天柱(土人节柱)●

279490 Pachycereus pringlei (S. Watson)Britton et Rose Britton et Rose;多刺摩天柱(武伦柱);Elephant Cactus, Giant Cardon, Giant Mexican Cereus, Mexican Giant Cactus ●

279491 Pachycereus pringlei Britton et Rose = Pachycereus pringlei(S. Watson)Britton et Rose Britton et Rose ●

279492 Pachycereus schottii(Engelm.)D. R. Hunt;上帝阁(大风柱); Senita, Whisker Cactus Old-man-cactus ●☆

279493 Pachycereus weberi(Coult.)Backeb. ;武卫柱;Candelabro ●☆

279494 Pachychaeta Sch. Bip. ex Baker = Ophryosporus Meyen ■●☆

279495 Pachychilus Blume = Pachystoma Blume ■

279496 Pachychilus chinensis(Lindl.)Blume = Pachystoma pubescens Blume ■

279497 Pachychilus chinensis Blume = Pachystoma pubescens Blume ■

279498 Pachychilus pubescens(Blume)Blume = Pachystoma pubescens Blume ■

279499 Pachychlaena Post et Kuntze = Pachylaena D. Don ex Hook. et Arn. ●☆

279500 Pachychlamys Dyer ex Ridl. = Shorea Roxb. ex C. F. Gaertn. ●

279501 Pachyclado Hook. f. (1864);粗枝芥属■☆

279502 Pachyclado novae-zelandiae Hook. f. ;粗枝芥■☆

279503 Pachycormus Coville = Pachycormus Coville ex Standl. ●☆

279504 Pachycormus Coville ex Standl. (1923);粗茎木属●☆

279505 Pachycormus discolor(Benth.)Coville ex Standl. ;粗茎木;Baja Elephant Tree, Copalquin, Elephant Tree, Torote Blanco ●☆

279506 Pachycornia Hook. f. (1880);盐角木属●☆

279507 Pachycornia triandra(F. Muell.)J. M. Black. ;盐角木●☆

279508 Pachycornus Willis = Pachycormus Coville ex Standl. ●☆

279509 Pachyctenium Maire et Pamp. (1936);厚栉芹属☆

279510 Pachyctenium mirabile Maire et Pamp. ;厚栉芹☆

279511 Pachycymbium L. C. Leach(1978);粗冠萝藦属■☆

279512 Pachycymbium abayense(M. G. Gilbert) M. G. Gilbert;阿拜粗冠萝藦■☆

279513 Pachycymbium baldratii (A. C. White et B. Sloane) M. G. Gilbert;巴尔粗冠萝藦■☆

279514 Pachycymbium baldratii(A. C. White et B. Sloane) M. G. Gilbert subsp. subterraneum(E. A. Bruce et P. R. O. Bally) M. G. Gilbert = Orbea subterranea(E. A. Bruce et P. R. O. Bally) Bruyns ■☆

279515 Pachycymbium carnosum(Stent) L. C. Leach = Orbea carnosa (Stent)Bruyns ■☆

279516 Pachycymbium circes (M. G. Gilbert) M. G. Gilbert = Orbea circes(M. G. Gilbert)Bruyns ■☆

279517 Pachycymbium commutatum (A. Berger) M. G. Gilbert = Orbea sprengeri(Schweinf.)Bruyns subsp. commutata(A. Berger)Bruyns ■☆

279518 Pachycymbium decaisneanum(Lem.) M. G. Gilbert = Orbea decaisneana(Lehm.)Bruyns ■☆

279519 Pachycymbium decaisneanum (Lem.) M. G. Gilbert subsp. hesperidum (Maire) J. -P. Lebrun et Stork = Orbea decaisneana (Lehm.)Bruyns ■☆

279520　Pachycymbium denboefii（Lavranos）M. G. Gilbert = Orbea denboefii（Lavranos）Bruyns ■☆

279521　Pachycymbium distinctum（E. A. Bruce）M. G. Gilbert = Orbea distincta（E. A. Bruce）Bruyns ■☆

279522　Pachycymbium dummeri（N. E. Br.）M. G. Gilbert = Orbea dummeri（N. E. Br.）Bruyns ■☆

279523　Pachycymbium gemugofanum（M. G. Gilbert）M. G. Gilbert = Orbea gemugofana（M. G. Gilbert）Bruyns ■☆

279524　Pachycymbium gilbertii（Plowes）M. G. Gilbert = Orbea gilbertii（Plowes）Bruyns ■☆

279525　Pachycymbium huernioides（P. R. O. Bally）M. G. Gilbert = Orbea huernioides（P. R. O. Bally）Bruyns ■☆

279526　Pachycymbium keithii（R. A. Dyer）L. C. Leach = Orbea carnosa（Stent）Bruyns subsp. keithii（R. A. Dyer）Bruyns ■☆

279527　Pachycymbium kochii（Lavranos）M. G. Gilbert = Orbea sacculata（N. E. Br.）Bruyns ■☆

279528　Pachycymbium laikipiense M. G. Gilbert = Orbea laikipiensis（M. G. Gilbert）Bruyns ■☆

279529　Pachycymbium lancasteri Lavranos = Orbea carnosa（Stent）Bruyns subsp. keithii（R. A. Dyer）Bruyns ■☆

279530　Pachycymbium laticoronum（M. G. Gilbert）M. G. Gilbert = Orbea laticorona（M. G. Gilbert）Bruyns ■☆

279531　Pachycymbium lugardii（N. E. Br.）M. G. Gilbert = Orbea lugardii（N. E. Br.）Bruyns ■☆

279532　Pachycymbium miscellum（N. E. Br.）M. G. Gilbert = Orbea miscella（N. E. Br.）Meve ■☆

279533　Pachycymbium rogersii（L. Bolus）M. G. Gilbert = Orbea rogersii（L. Bolus）Bruyns ■☆

279534　Pachycymbium sacculatum（N. E. Br.）M. G. Gilbert = Orbea sacculata（N. E. Br.）Bruyns ■☆

279535　Pachycymbium schweinfurthii（A. Berger）M. G. Gilbert = Orbea schweinfurthii（A. Berger）Bruyns ■☆

279536　Pachycymbium semitubiflorum（L. E. Newton）M. G. Gilbert = Orbea semitubiflora（L. E. Newton）Bruyns ■☆

279537　Pachycymbium sprengeri（N. E. Br.）M. G. Gilbert = Orbea sprengeri（Schweinf.）Bruyns ■☆

279538　Pachycymbium sprengeri（N. E. Br.）M. G. Gilbert subsp. foetidum（M. G. Gilbert）M. G. Gilbert = Orbea sprengeri（Schweinf.）Bruyns subsp. foetida（M. G. Gilbert）Bruyns ■☆

279539　Pachycymbium sprengeri（N. E. Br.）M. G. Gilbert subsp. ogadense（M. G. Gilbert）M. G. Gilbert = Orbea sprengeri（Schweinf.）Bruyns subsp. ogadensis（M. G. Gilbert）Bruyns ■☆

279540　Pachycymbium tubiforme（E. A. Bruce et P. R. O. Bally）M. G. Gilbert = Orbea tubiformis（E. A. Bruce et P. R. O. Bally）Bruyns ■☆

279541　Pachycymbium ubomboense（I. Verd.）M. G. Gilbert = Australluma ubomboensis（I. Verd.）Bruyns ■☆

279542　Pachycymbium vibratilis（E. A. Bruce et P. R. O. Bally）M. G. Gilbert = Orbea vibratilis（E. A. Bruce et P. R. O. Bally）Bruyns ■☆

279543　Pachycymbium wilsonii（P. R. O. Bally）M. G. Gilbert = Orbea wilsonii（P. R. O. Bally）Bruyns ■☆

279544　Pachydendron Dumort. = Aloe L. ●■

279545　Pachydendron Dumort. = Pachidendron Haw. ●■

279546　Pachyderis Cass. = Pteronia L.（保留属名）●☆

279547　Pachyderma Blume = Olea L. ●

279548　Pachydesmia Gleason = Miconia Ruiz et Pav.（保留属名）●☆

279549　Pachydiscus Gilg et Schltr. = Periomphale Baill. ●☆

279550　Pachydiscus Gilg et Schltr. = Wittsteinia F. Muell. ●☆

279551　Pachyelasma Harms（1913）；厚腺苏木属（厚腔苏木属）●☆

279552　Pachyelasma tessmannii（Harms）Harms；厚腺苏木（厚腔苏木）●☆

279553　Pachyelasma tessmannii Harms = Pachelasma tessmannii（Harms）Harms ●☆

279554　Pachyglossum Decne. = Oxypetalum R. Br.（保留属名）●■☆

279555　Pachygone Miers ex Hook. f. et Thomson = Pachygone Miers ●

279556　Pachygone Miers（1851）；粉绿藤属；Pachygone, Palegreenvine ●

279557　Pachygone dasycarpa Kurz；粗果粉绿藤●☆

279558　Pachygone ovata（Poir.）Miers ex Hook. f. et Thomson；卵叶粉绿藤；Ovate-leaf Pachygone ●☆

279559　Pachygone pubescens Benth.；柔毛粉绿藤；Pubescent Pachygone ●☆

279560　Pachygone sinica Diels；粉绿藤（广西粉绿藤，华粉绿藤）；Chinese Pachygone, Palegreenvine ●

279561　Pachygone valida Diels = Limaciopsis valida（Diels）H. S. Lo ●

279562　Pachygone yunnanensis H. S. Lo；滇粉绿藤；Yunnan Pachygone, Yunnan Palegreenvine ●

279563　Pachygraphea Post et Kuntze = Aspalathus L. ●☆

279564　Pachygraphea Post et Kuntze = Pachyraphea C. Presl ●☆

279565　Pachylaena D. Don ex Hook. et Arn.（1835）；厚被菊属●☆

279566　Pachylaena atriplicifolia D. Don ex Hook. et Arn.；厚被菊●☆

279567　Pachylaena elegans Phil.；雅致厚被菊●☆

279568　Pachylaena rosea I. M. Johnst.；粉红厚被菊●☆

279569　Pachylarnax Dandy（1927）；厚壁木属●

279570　Pachylarnax pleiocarpa Dandy；厚壁木●☆

279571　Pachylarnax sinica（Y. W. Law）N. H. Xia et C. Y. Wu = Manglietiastrum sinicum Y. W. Law ●◇

279572　Pachylecythis Ledoux = Lecythis Loefl. ●☆

279573　Pachylepis Brongn. = Parolinia Endl. ●☆

279574　Pachylepis Brongn. = Widdringtonia Endl. ●☆

279575　Pachylepis Less. = Crepis L. ■

279576　Pachylobium（Benth.）Willis = Dioclea Kunth ■☆

279577　Pachylobus G. Don = Dacryodes Vahl ●☆

279578　Pachylobus albiflorus Guillaumin = Dacryodes klaineana（Pierre）H. J. Lam ●☆

279579　Pachylobus balsamifera（Oliv.）Guillaumin = Santiria trimera（Oliv.）Aubrév. ●☆

279580　Pachylobus barteri Engl. = Dacryodes klaineana（Pierre）H. J. Lam ●☆

279581　Pachylobus buettneri（Engl.）Engl. = Dacryodes buettneri（Engl.）H. J. Lam ●☆

279582　Pachylobus deliciosus（A. Chev. ex Hutch. et Dalziel）Pellegr. = Dacryodes klaineana（Pierre）H. J. Lam ●☆

279583　Pachylobus edulis G. Don = Dacryodes edulis（G. Don）H. J. Lam ●☆

279584　Pachylobus ezigo Pierre = Dacryodes buettneri（Engl.）H. J. Lam ●☆

279585　Pachylobus ferrugineus A. Chev. ex Pellegr. = Dacryodes heterotricha（Pellegr.）H. J. Lam ●☆

279586　Pachylobus gossweileri Exell = Dacryodes pubescens（Vermoesen）H. J. Lam ●☆

279587　Pachylobus heterotrichus Pellegr. = Dacryodes heterotricha（Pellegr.）H. J. Lam ●☆

279588　Pachylobus klaineanus（Pierre）Engl. = Dacryodes klaineana（Pierre）H. J. Lam ●☆

279589　Pachylobus letestui Pellegr. = Dacryodes letestui（Pellegr.）H. J. Lam ●☆

279590　Pachylobus macrophyllus（Oliv.）Engl. = Dacryodes macrophylla（Oliv.）H. J. Lam ●☆

279591　Pachylobus mayumbensis Exell = Santiria trimera（Oliv.）Aubrév. ●☆

279592　Pachylobus osika Guillaumin = Dacryodes osika（Guillaumin）H. J. Lam ●☆

279593 Pachylobus paniculatus Hoyle = Dacryodes klaineana(Pierre)H. J. Lam ●☆

279594 Pachylobus pubescens Vermoesen = Dacryodes pubescens (Vermoesen) H. J. Lam ●☆

279595 Pachylobus saphu(Engl.)Engl. = Dacryodes edulis(G. Don) H. J. Lam ●☆

279596 Pachylobus zenkeri Engl. = Dacryodes klaineana(Pierre)H. J. Lam ●☆

279597 Pachyloma DC.(1828);厚缘野牡丹属●☆

279598 Pachyloma DC. = Comolia DC. ●☆

279599 Pachyloma Post et Kuntze = Pachiloma Raf. ■☆

279600 Pachyloma Post et Kuntze = Polytaenia DC. ●☆

279601 Pachyloma Spach = Hericinia Fourr. ■

279602 Pachyloma Spach = Pfundia Opiz ex Nevski ■

279603 Pachyloma Spach = Ranunculus L. ■

279604 Pachyloma coriaceum DC.;厚缘野牡丹●☆

279605 Pachyloma nanum Wurdack;矮厚缘野牡丹●☆

279606 Pachyloma setosum Wurdack;刚毛厚缘野牡丹●☆

279607 Pachylophus Spach = Oenothera L. ●■

279608 Pachymeria Benth. = Meriania Sw.(保留属名)●☆

279609 Pachymitra Nees = Rhynchospora Vahl(保留属名)■

279610 Pachymitus O. E. Schulz(1924);粗线芥属■☆

279611 Pachymitus cardaminoides(F. Muell.)O. E. Schulz;粗线芥■☆

279612 Pachyne Salisb. = Phaius Lour. ■

279613 Pachynema R. Br. ex DC.(1817);粗蕊花属(粗丝木属)●☆

279614 Pachynema R. Br. ex DC. = Hibbertia Andréws ●☆

279615 Pachynema complanatum R. Br. ex DC.;粗蕊花●■

279616 Pachynema diffusum Craven et Dunlop;铺散粗蕊花●☆

279617 Pachyneurum Bunge(1839);厚脉荠属■☆

279618 Pachyneurum grandiflorum Bunge;厚脉荠■☆

279619 Pachynocarpus Hook. f. = Vatica L. ■

279620 Pachypharynx Aellen = Atriplex L. ■●

279621 Pachyphragma(DC.)N. Busch = Thlaspi L. ■

279622 Pachyphragma(DC.)Rchb.(1841);厚隔芥属■☆

279623 Pachyphragma(DC.)Rchb. = Thlaspi L. ■

279624 Pachyphragma Rchb. = Pachyphragma(DC.)Rchb. ■☆

279625 Pachyphragma macrophyllum(Hoffm.)N. Busch;厚隔芥; Caucasian Penny-cress ■☆

279626 Pachyphyllum Kunth(1816);厚叶兰属■☆

279627 Pachyphyllum falcifolium Rchb. f.;镰叶厚叶兰■☆

279628 Pachyphyllum gracillimum C. Schweinf.;纤细厚叶兰■☆

279629 Pachyphyllum mexicanum Dressler et Hágsater;墨西哥厚叶兰■☆

279630 Pachyphyllum micranthum Schltr.;小花厚叶兰■☆

279631 Pachyphyllum minus Schltr.;小厚叶兰■☆

279632 Pachyphyllum parvifolium Lindl.;小叶厚叶兰■☆

279633 Pachyphyllum uniflorum La Llave et Lex.;单花厚叶兰■☆

279634 Pachyphytum Link = Pachyphytum Link,Klotzsch et Otto ●☆

279635 Pachyphytum Link,Klotzsch et Otto(1841);厚叶草属(肥天属,厚叶属);Moonstones ●☆

279636 Pachyphytum bracteosum Link,Klotzsch et Otto;多苞厚叶草●☆

279637 Pachyphytum compactum Rose;千代田之松●☆

279638 Pachyphytum hookeri A. Berger;群雀厚叶草(群雀,群雀草)●☆

279639 Pachyphytum oviferum J. A. Purpus;厚叶草(星美人); Moonstones,Sugared-almond Plum ●☆

279640 Pachyphytum viride E. Walther;绿厚叶草●☆

279641 Pachyplectron Schltr.(1906);粗距兰属■☆

279642 Pachyplectron aphyllum T. Hashim.;无叶粗距兰■☆

279643 Pachyplectron arifolium Schltr.;粗距兰●☆

279644 Pachypleurum Ledeb.(1829);厚棱芹属(厚肋芹属); Thickribcelery ■

279645 Pachypleurum alpinum Ledeb.;高山厚棱芹;Alpine Thickribcelery ■

279646 Pachypleurum condensatum(L.)Korovin = Libanotis condensata Crantz ■

279647 Pachypleurum condensatum Korovin = Libanotis condensata Crantz ■

279648 Pachypleurum dolichostylum(Schischk.)Korovin ex Kamelin = Ligusticum mucronatum(Schrenk)Leute ■

279649 Pachypleurum lhasanum Hung T. Chang et R. H. Shan;拉萨厚棱芹;Lasa Thickribcelery ■

279650 Pachypleurum mucronatum(Schrenk)Schischk. = Ligusticum mucronatum(Schrenk)Leute ■

279651 Pachypleurum muliense R. H. Shan et F. T. Pu;木里厚棱芹; Muli Thickribcelery ■

279652 Pachypleurum nyalamense Hung T. Chang et R. H. Shan;聂拉木厚棱芹;Nielamu Thickribcelery ■

279653 Pachypleurum xizangense Hung T. Chang et R. H. Shan;西藏厚棱芹;Tibet Thickribcelery,Xizang Thickribcelery ■

279654 Pachypodanthium Engl. et Diels(1900);粗柄花属(厚足属)●☆

279655 Pachypodanthium barteri(Benth.)Hutch. et Dalziel = Duguetia barteri(Benth.)Chatrou ●☆

279656 Pachypodanthium confine Engl. et Diels;亲缘粗柄花●☆

279657 Pachypodanthium confine Engl. et Diels = Duguetia confinis (Engl. et Diels)Chatrou ●☆

279658 Pachypodanthium confine Engl. et Diels var. sargosii(R. E. Fr.) Le Thomas = Duguetia confinis(Engl. et Diels)Chatrou ●☆

279659 Pachypodanthium gossweileri Exell et Mendonca = Uvaria scabrida Oliv. ●☆

279660 Pachypodanthium sargosii R. E. Fr. = Duguetia confinis(Engl. et Diels)Chatrou ●☆

279661 Pachypodanthium simiarum Exell et Mendonca = Letestudoxa bella Pellegr. ●☆

279662 Pachypodanthium staudtii(Engl. et Diels)Engl. et Diels = Duguetia staudtii(Engl. et Diels)Chatrou ●☆

279663 Pachypodanthium staudtii(Engl. et Diels)Engl. et Diels var. letestui Pellegr. = Duguetia barteri(Benth.)Chatrou ●☆

279664 Pachypodanthium staudtii Engl. et Diels;施氏粗柄花●☆

279665 Pachypodanthium tessmannii R. E. Fr. = Duguetia barteri (Benth.)Chatrou ●☆

279666 Pachypodium Lindl.(1830);棒棰树属(棒锤树属,粗根属,瓶干树属);Pachypodium ●☆

279667 Pachypodium Nutt. = Thelypodium Endl. ■☆

279668 Pachypodium Nutt. ex Torr. et A. Gray = Thelypodium Endl. + Pleurophragma Rydb. ■

279669 Pachypodium Nutt. ex Torr. et A. Gray = Thelypodium Endl. ■☆

279670 Pachypodium Webb et Berthel. = Sisymbrium L. ■

279671 Pachypodium Webb et Berthel. = Tonguea Endl. ■

279672 Pachypodium ambongense Poiss.;安博棒棰树●☆

279673 Pachypodium baroni Constance et Bois;赤花棒棰树(巴氏棒锤树,赤花粗根);Baron Pachypodium ●☆

279674 Pachypodium bispinosum(L. f.)A. DC.;双刺棒棰树●☆

279675 Pachypodium brevicaule Baker;短茎棒棰树(短茎粗根)●☆

279676 Pachypodium densiflorum Baker;密花棒棰树(密花粗根)●☆

279677 Pachypodium erysimoides(Desf.)Webb et Berthel. = Sisymbrium erysimoides Desf. ■☆

279678 Pachypodium giganteum Engl. = Pachypodium lealii Welw. ●☆

279679 Pachypodium griquense L. Bolus = Pachypodium succulentum (Jacq.)Sweet ●☆

279680 Pachypodium jasminiflorum L. Bolus = Pachypodium succulentum(Jacq.)Sweet ●☆

279681 Pachypodium lamerei Drake;拉默棒棰树（草原之星，长叶瓶干树，锦堂，拉默粗根）;Lamer Pachypodium,Madagascar Palm ●☆

279682 Pachypodium lealii Welw.;巨棒棰树（巨粗根）●☆

279683 Pachypodium marginatum?;墨西哥棒棰树（墨西哥粗根）; Mexican Fence Post Cactus ●☆

279684 Pachypodium mikea Lüthy;马岛棒棰树●☆

279685 Pachypodium namaquanum(Laxm. ex Harv.)Welw.;纳马夸棒棰树（埃氏棒棰树，埃氏粗根，棒槌树）;Club Foot,Elephant's Trunk,Hale-men ●☆

279686 Pachypodium namaquanum (Laxm. ex Harv.) Welw. var. cristatum Hort.;光堂冠●☆

279687 Pachypodium namaquanum (Wyley ex Harv.) Welw. = Pachypodium namaquanum(Laxm. ex Harv.)Welw. ●☆

279688 Pachypodium namaquanum Welw. = Pachypodium namaquanum (Laxm. ex Harv.)Welw. ●☆

279689 Pachypodium rosulatum Baker;莲座叶棒棰树（莲座叶粗根）●☆

279690 Pachypodium saundersii N. E. Br.;簇刺棒棰树（簇刺粗根）; Star of the Lundi ●☆

279691 Pachypodium succulcntum(L. f.)Sweet;南非棒棰树（南非粗根）●☆

279692 Pachypodium succulentum(Jacq.)Sweet;多肉棒棰树（多肉瓶干树）●☆

279693 Pachypodium tomentosum G. Don = Pachypodium succulentum (Jacq.)Sweet ●☆

279694 Pachyptera DC. = Mansoa DC. ●☆

279695 Pachyptera DC. ex Meisn. = Mansoa DC. ●☆

279696 Pachypteris Kar. et Kir. = Isatis L. ■

279697 Pachypteris Kar. et Kir. = Pachypterygium Bunge ■

279698 Pachypteris densiflora(Bunge)Parsa = Pachypterygium multicaule (Kar. et Kir.)Bunge ■

279699 Pachypteris lamprocarpa (Bunge) Parsa = Pachypterygium multicaule(Kar. et Kir.)Bunge ■

279700 Pachypteris multicaulis Kar. et Kir. = Pachypterygium multicaule (Kar. et Kir.)Bunge ■

279701 Pachypteris persica (Boiss.) Parsa = Pachypterygium brevipes Bunge ■

279702 Pachypterygium Bunge = Isatis L. ■

279703 Pachypterygium Bunge (1845);厚壁荠属（厚翅荠属）; Pachypterygium ■

279704 Pachypterygium brevipes Bunge;短梗厚壁荠; Breviped Pachypterygium ■

279705 Pachypterygium brevipes Bunge = Isatis brevipes(Bunge)Jafri ■

279706 Pachypterygium brevipes Bunge var. persicum Boiss. = Pachypterygium brevipes Bunge ■

279707 Pachypterygium densiflorum Bunge;密花厚壁荠■

279708 Pachypterygium densiflorum Bunge = Pachypterygium multicaule (Kar. et Kir.)Bunge ■

279709 Pachypterygium echinatum Jarm. = Pachypterygium multicaule (Kar. et Kir.)Bunge ■

279710 Pachypterygium heterotrichum Bunge = Isatis brevipes (Bunge) Jafri ■

279711 Pachypterygium heterotrichum Bunge = Pachypterygium brevipes Bunge ■

279712 Pachypterygium lamprocarpum Bunge = Pachypterygium multicaule(Kar. et Kir.)Bunge ■

279713 Pachypterygium microcarpum Gilli = Pachypterygium multicaule (Kar. et Kir.)Bunge ■

279714 Pachypterygium multicaule (Kar. et Kir.) Bunge;厚壁荠; Manystem Pachypterygium ■

279715 Pachypterygium praemontanum Jarm. = Pachypterygium multicaule (Kar. et Kir.)Bunge ■

279716 Pachypterygium ramosum Jarm. = Pachypterygium multicaule (Kar. et Kir.)Bunge ■

279717 Pachypterygium ramosum Jarm. ex Pavlov = Pachypterygium multicaule(Kar. et Kir.)Bunge ■

279718 Pachyra A. St. -Hil. et Naudin = Pachira Aubl. ●

279719 Pachyraphea C. Presl = Aspalathus L. ●☆

279720 Pachyraphea propinqua (E. Mey.) C. Presl = Aspalathus triquetra Thunb. ●☆

279721 Pachyraphea triquetra (Thunb.) C. Presl = Aspalathus triquetra Thunb. ●☆

279722 Pachyraphea zeyheriana C. Presl = Aspalathus triquetra Thunb. ●☆

279723 Pachyrhiza B. D. Jacks. = Pachyrhizus Rich. ex DC.（保留属名）■

279724 Pachyrhizanthe(Schltr.)Nakai = Cymbidium Sw. ■

279725 Pachyrhizanthe aphyllum(Ames et Schltr.)Nakai = Cymbidium macrorrhizum Lindl. ■

279726 Pachyrhizanthe macrorhizon (Lindl.) Nakai = Cymbidium macrorrhizum Lindl. ■

279727 Pachyrhizanthe nipponicum (Franch. et Sav.) Nakai = Cymbidium macrorhizon Lindl. ■

279728 Pachyrhizus Rich. ex DC.（1825）（保留属名）;豆薯属（沙薯属）;Yambean,Yam Bean ■

279729 Pachyrhizus angulatus A. Rich. ex DC.;窄豆薯;Wayaka Yam Bean ■☆

279730 Pachyrhizus angulatus Rich. = Pachyrhizus angulatus A. Rich. ex DC. ■☆

279731 Pachyrhizus angulatus Rich. = Pachyrhizus erosus(L.)Urb. ■

279732 Pachyrhizus erosus(L.)Urb.;豆薯（草瓜茹，地瓜，地萝卜，番葛，葛瓜，葛薯，凉瓜，凉薯，六缚，贫人果，沙葛，土瓜，土萝卜）; Jicama,Wayaka Yambean,Yam Bean ■

279733 Pachyrhizus orbicularis Welw. ex Baker = Neorautanenia mitis (A. Rich.)Verdc. ■☆

279734 Pachyrhizus palmatilobus Benth. et Hook. f.;掌裂豆薯;Jicana ■☆

279735 Pachyrhizus thunbergianus Siebold et Zucc. = Pueraria lobata (Willd.)Ohwi ●■

279736 Pachyrhizus trilobus (Lour.) DC. = Pueraria lobata (Willd.) Ohwi subsp. thomsonii(Benth.)H. Ohashi et Tateishi ●

279737 Pachyrhizus tuberosus (Lam.) Spreng.;西印度豆薯;Ajipo, Manioc Bean,Yam Bean ■

279738 Pachyrhynchus DC.（1838）;粗喙菊属■☆

279739 Pachyrhynchus DC. = Lucilia Cass. ■☆

279740 Pachyrhynchus xeranthemoides DC.;粗喙菊■☆

279741 Pachyrrhizos Spreng. = Pachyrhizus Rich. ex DC.（保留属名）■

279742 Pachysa D. Don = Erica L. ●☆

279743 Pachysandra Michx.（1803）;板凳果属（板凳草属，粉蕊黄杨属，富贵草属，吉祥草属，三角咪属）;Alleghany Spurge, Benchfruit,Pachysandra,Spurge ●■

279744 Pachysandra axillaris Franch.;板凳果（白金三角咪，草本叶上花，顶蕊三角咪，粉蕊黄杨，黄芩，金丝矮陀，金丝矮陀螺，金丝矮陀陀，千年矮，三果吉祥草，三角咪，三角咪草，山板凳，山板凳果，宿柱三角咪，万年青，小清喉，腋花板凳果，腋花富贵草，腋花三角咪）;Benchfruit,China Spurge Pachysandra,Chinese Pachysandra ●■

279745 Pachysandra axillaris Franch. f. kouytchensis H. Lév. = Pachysandra axillaris Franch. var. stylosa(Dunn)M. Cheng ■

279746 Pachysandra axillaris Franch. subsp. stylosa (Dunn) Boufford et

Q. Y. Xiang = Pachysandra axillaris Franch. var. stylosa (Dunn) M. Cheng ■

279747　Pachysandra axillaris Franch. var. glaberrima (Hand. -Mazz.) C. Y. Wu；光叶板凳果（板凳果）；Glabrescent Benchfruit, Glabrescent Pachysandra ■

279748　Pachysandra axillaris Franch. var. stylosa (Dunn) M. Cheng；多毛板凳果（毛叶板凳果，三角咪，山板凳，宿柱三角咪）；Style Benchfruit, Style Pachysandra ■

279749　Pachysandra axillaris Franch. var. tricarpa Hayata = Pachysandra axillaris Franch. ■

279750　Pachysandra bodinieri H. Lév.；毛叶板凳果；Bodinier Benchfruit, Bodinier Pachysandra ■

279751　Pachysandra bodinieri H. Lév. = Pachysandra axillaris Franch. var. stylosa (Dunn) M. Cheng ■

279752　Pachysandra bodinieri H. Lév. var. stylosa (Dunn) M. Cheng = Pachysandra axillaris Franch. var. glaberrima (Hand. -Mazz.) C. Y. Wu ■

279753　Pachysandra mairei H. Lév. = Sarcococca hookeriana Baill. var. digyna Franch. ●

279754　Pachysandra procumbens Michx.；伏卧板凳果；Allegheny Spurge, Mountain Spurge ■☆

279755　Pachysandra stylosa Dunn = Pachysandra axillaris Franch. var. glaberrima (Hand. -Mazz.) C. Y. Wu ■

279756　Pachysandra stylosa Dunn = Pachysandra axillaris Franch. var. stylosa (Dunn) M. Cheng ■

279757　Pachysandra terminalis Siebold et Zucc.；顶花板凳果（长青草，顶蕊三角咪，粉蕊黄杨，富贵草，黄秋连，黄秧连，吉祥草，捆仙绳，上天梯，石莲藤，四季青，孙儿茶，土桔梗，雪山林，雪山苓，转筋草）；Japan Benchfruit, Japanese Pachysandra, Japanese Spurge, Terminalflower Pachysandra ■

279758　Pachysandra terminalis Siebold et Zucc. 'Variegata'；花叶顶花板凳果 ■☆

279759　Pachysandra terminalis Siebold et Zucc. f. subrhombea? = Pachysandra terminalis Siebold et Zucc. ■

279760　Pachysandraceae J. Agardh = Buxaceae Dumort.（保留科名）●■

279761　Pachysandraceae J. Agardh = Paeoniaceae Raf.（保留科名）■●

279762　Pachysandraceae J. Agardh；板凳果科 ●■

279763　Pachysanthus C. Presl = Rudgea Salisb. ■☆

279764　Pachysolen Phil. = Nolana L. ex L. f. ●☆

279765　Pachystachys Nees(1847)；麒麟吐珠属（红珊瑚属，厚穗爵床属，黄苞花属，金苞花属）；Pachystachys ●

279766　Pachystachys albiflora Rizzini；白花麒麟吐珠 ●☆

279767　Pachystachys cardinalis? = Pachystachys coccinea Nees ●☆

279768　Pachystachys coccinea Nees；红珊瑚（绯红厚穗爵床）；Cardinal's Guard, Sally-my-handsome, Scarlet Jacobinia ●☆

279769　Pachystachys lutea Nees；金苞花（黄虾花，金包银）；Golden Candles, Golden Shrimp Plant, Lollilop Plant, Shrimp Plant, Yellow-bract Pachystachys ●■

279770　Pachystachys rosea Wassh.；粉红麒麟吐珠 ●☆

279771　Pachystachys spicata (Ruiz et Pav.) Wassh.；厚穗麒麟吐珠（厚穗爵床）；Cardinal's-guard ●☆

279772　Pachystegia Cheeseman = Olearia Moench（保留属名）●☆

279773　Pachystegia Cheeseman(1925)；厚冠菊属 ●☆

279774　Pachystegia insigne Cheeseman；厚冠菊 ■☆

279775　Pachystela Pierre = Pachystela Pierre ex Radlk. ●☆

279776　Pachystela Pierre ex Baill. = Synsepalum (A. DC.) Daniell ●☆

279777　Pachystela Pierre ex Radlk. (1899)；粗柱山榄属 ●☆

279778　Pachystela Pierre ex Radlk. = Synsepalum (A. DC.) Daniell ●☆

279779　Pachystela albida A. Chev. = Synsepalum pobeguinianum (Pierre ex Lecomte) Aké Assi et L. Gaut. ●☆

279780　Pachystela antunesii (Engl.) Lecomte = Englerophytum magalismontanum (Sond.) T. D. Penn. ●☆

279781　Pachystela argentea A. Chev. = Synsepalum msolo (Engl.) T. D. Penn. ●☆

279782　Pachystela argyrophylla (Hiern) Lecomte = Englerophytum magalismontanum (Sond.) T. D. Penn. ●☆

279783　Pachystela batesii A. Chev. = Synsepalum batesii (A. Chev.) Aubrév. et Pellegr. ●☆

279784　Pachystela bequaertii De Wild. = Synsepalum msolo (Engl.) T. D. Penn. ●☆

279785　Pachystela brevipes (Baker) Engl. = Synsepalum brevipes (Baker) T. D. Penn. ●☆

279786　Pachystela buluensis (Greves) Aubrév. et Pellegr.；布卢粗柱山榄 ●☆

279787　Pachystela cinerea (Engl.) Radlk. = Synsepalum brevipes (Baker) T. D. Penn. ●☆

279788　Pachystela cinerea (Engl.) Radlk. var. batangensis (C. H. Wright) Engl. = Synsepalum brevipes (Baker) T. D. Penn. ●☆

279789　Pachystela cinerea (Engl.) Radlk. var. cuneata (Radlk.) Engl. = Synsepalum brevipes (Baker) T. D. Penn. ●☆

279790　Pachystela cinerea (Engl.) Radlk. var. ogowensis Engl. = Synsepalum brevipes (Baker) T. D. Penn. ●☆

279791　Pachystela cinerea (Engl.) Radlk. var. undulata Engl. = Synsepalum brevipes (Baker) T. D. Penn. ●☆

279792　Pachystela conferta Radlk. = Synsepalum brevipes (Baker) T. D. Penn. ●☆

279793　Pachystela gossweileri Engl.；戈斯粗柱山榄 ●☆

279794　Pachystela laurentii (De Wild.) C. M. Evrard；洛朗粗柱山榄 ●☆

279795　Pachystela lenticellosa Radlk. = Synsepalum brevipes (Baker) T. D. Penn. ●☆

279796　Pachystela liberica Engl. = Synsepalum brevipes (Baker) T. D. Penn. ●☆

279797　Pachystela longepedicellata (De Wild.) J. Léonard = Zeyherella longepedicellata (De Wild.) Aubrév. et Pellegr. ●☆

279798　Pachystela longistyla (Baker) Radlk. = Synsepalum brevipes (Baker) T. D. Penn. ●☆

279799　Pachystela macrocarpa A. Chev. = Synsepalum brevipes (Baker) T. D. Penn. ●☆

279800　Pachystela magalismontana (Sond.) Lecomte = Englerophytum magalismontanum (Sond.) T. D. Penn. ●☆

279801　Pachystela micrantha (A. Chev.) Hutch. et Dalziel = Synsepalum afzelii (Engl.) T. D. Penn. ●☆

279802　Pachystela mildbraedii Engl.；米尔德粗柱山榄 ●☆

279803　Pachystela msolo (Engl.) Engl. = Synsepalum msolo (Engl.) T. D. Penn. ●☆

279804　Pachystela ovatostipulata De Wild.；卵托叶粗柱山榄 ●☆

279805　Pachystela pobeguiniana Pierre ex Dubard = Synsepalum pobeguinianum (Pierre ex Lecomte) Aké Assi et L. Gaut. ●☆

279806　Pachystela pobeguiniana Pierre ex Lecomte = Synsepalum pobeguinianum (Pierre ex Lecomte) Aké Assi et L. Gaut. ●☆

279807　Pachystela robusta Engl.；粗柱山榄 ●☆

279808　Pachystela seretii De Wild. = Synsepalum seretii (De Wild.) T. D. Penn. ●☆

279809　Pachystela subverticillata E. A. Bruce；轮生粗柱山榄 ●☆

279810　Pachystela subverticillata E. A. Bruce = Synsepalum subverticillatum (E. A. Bruce) T. D. Penn. ●☆

279811　Pachystele Schltr. (1923)；粗柱兰属 ■☆

279812　Pachystele Schltr. = Scaphyglottis Poepp. et Endl.（保留属名）■☆

279813　Pachystele jimenezii (Schltr.) Schltr.；粗柱兰 ■☆

279814　Pachystelis Rauschert = Pachystele Schltr. ■☆

279815　Pachystelma Brandegee = Matelea Aubl. ●☆

279816　Pachystemon Blume = Macaranga Thouars ●

279817　Pachystemon Blume. (1826);粗蕊大戟属●☆

279818　Pachystemon populifolius Miq. ;杨叶粗蕊大戟●☆

279819　Pachystemon trilobus Blume;粗蕊大戟●☆

279820　Pachystigma Hochst. (1842);厚柱头木属●☆

279821　Pachystigma Hook. = Peltostigma Walp. ●☆

279822　Pachystigma Meisn. = Paxistima Raf. ●☆

279823　Pachystigma Raf. = Paxistima Raf. ●☆

279824　Pachystigma albosetulosum Verdc. ;白刚毛厚柱头木●☆

279825　Pachystigma ancylanthum (Schweinf.) K. Schum. = Fadogia ancylantha Schweinf. ●☆

279826　Pachystigma bowkeri Robyns;鲍克厚柱头木●☆

279827　Pachystigma burttii Verdc. ;伯特厚柱头木●☆

279828　Pachystigma burttii Verdc. subsp. hirtiflorum Verdc. ;毛花厚柱头木●☆

279829　Pachystigma caffrum (Sim) Robyns = Pachystigma macrocalyx (Sond.) Robyns ●☆

279830　Pachystigma coeruleum Robyns;青蓝厚柱头木●☆

279831　Pachystigma cymosum Robyns = Pachystigma venosum Hochst. ●☆

279832　Pachystigma decussatum K. Schum. = Rytigynia decussata (K. Schum.) Robyns ■☆

279833　Pachystigma gillettii (Tennant) Verdc. ;吉莱特厚柱头木●☆

279834　Pachystigma humilis Bews = Pygmaeothamnus chamaedendrum (Kuntze) Robyns ●☆

279835　Pachystigma kenyense Verdc. = Pachystigma gillettii (Tennant) Verdc. ●☆

279836　Pachystigma lasianthum Sond. = Lagynias lasiantha (Sond.) Bullock ■☆

279837　Pachystigma latifolium Sond. ;宽叶厚柱头木●☆

279838　Pachystigma loranthifolium(K. Schum.) Verdc. ;桑寄生厚柱头木●☆

279839　Pachystigma loranthifolium (K. Schum.) Verdc. subsp. salaense Verdc. ;萨拉厚柱头木●☆

279840　Pachystigma macrocalyx(Sond.) Robyns;大萼厚柱头木●☆

279841　Pachystigma macrophyllum Farr = Paxistima macrophylla Farr ●☆

279842　Pachystigma micropyren Verdc. ;小果厚柱头木●☆

279843　Pachystigma myrsinites (Pursh) Raf. = Paxistima myrsinites (Pursh) Raf. ●☆

279844　Pachystigma obovatum (N. E. Br.) Eyles = Fadogia ancylantha Schweinf. ●☆

279845　Pachystigma pygmaeum(Schltr.) Robyns;矮小厚柱头木●☆

279846　Pachystigma rhodesianum (S. Moore) Robyns = Pachystigma pygmaeum(Schltr.) Robyns ●☆

279847　Pachystigma schumannianum(Robyns) Bridson et Verdc. ;舒曼厚柱头木●☆

279848　Pachystigma schumannianum(Robyns) Bridson et Verdc. subsp. mucronulatum(Robyns) Bridson et Verdc. ;微凸舒曼厚柱头木●☆

279849　Pachystigma triflorum Robyns;三花厚柱头木●☆

279850　Pachystigma venosum Hochst. ;厚柱头木●☆

279851　Pachystigma zeyheri Sond. = Pygmaeothamnus zeyheri (Sond.) Robyns ●☆

279852　Pachystoma Blume(1825) ;粉口兰属;Pachystoma ■

279853　Pachystoma brevilabium Schltr. = Pachystoma pubescens Blume ■

279854　Pachystoma chinense(Lindl.) Rchb. f. = Pachystoma pubescens Blume ■

279855　Pachystoma chinense (Lindl.) Rchb. f. var. formosanum (Schltr.) S. S. Ying = Pachystoma pubescens Blume ■

279856　Pachystoma formosanum Rchb. f. = Pachystoma pubescens Blume ■

279857　Pachystoma formosanum Schltr. = Pachystoma pubescens Blume ■

279858　Pachystoma ludaoense S. C. Chen et Y. B. Luo;绿岛粉口兰(短毛美冠兰,毛芋兰) ;Ludao Pachystoma ■

279859　Pachystoma pubescens Blume;粉口兰(台湾粉口兰) ;Pachystoma, Pubescent Pachystoma ■

279860　Pachystoma rothschildianum (O'Brien) Sander = Ancistrochilus rothschildianus O'Brien ■☆

279861　Pachystoma thomsonianum Rchb. f. = Ancistrochilus thomsonianus (Rchb. f.) Rolfe ■☆

279862　Pachystorna brevilabium Schltr. = Pachystoma pubescens Blume ■

279863　Pachystorna formosanum Schltr. = Pachystoma pubescens Blume ■

279864　Pachystrobilus Bremek. = Strobilanthes Blume ●■

279865　Pachystroma(Klotzsch) Müll. Arg. = Pachystroma Müll. Arg. ☆

279866　Pachystroma Müll. Arg. (1865) ;厚垫大戟属☆

279867　Pachystroma ilicifolium Müll. Arg. ;厚垫大戟☆

279868　Pachystylidium Pax et K. Hoffm. (1919) ;小粗柱大戟属☆

279869　Pachystylidium hirsutum Pax et K. Hoffm. ;小粗柱大戟☆

279870　Pachystylis Blume = Pachystoma Blume ■

279871　Pachystylum(DC.) Eckl. et Zeyh. = Heliophila Burm. f. ex L. ●■☆

279872　Pachystylum Eckl. et Zeyh. = Heliophila Burm. f. ex L. ●■☆

279873　Pachystylus K. Schum. (1889) ;粗柱茜属●☆

279874　Pachystylus guelcherianus K. Schum. ;粗柱茜●☆

279875　Pachythamnus(R. M. King et H. Rob.) R. M. King et H. Rob. (1972) ;肉皮菊属●☆

279876　Pachythamnus(R. M. King et H. Rob.) R. M. King et H. Rob. = Eupatorium L. ■●

279877　Pachythamnus crassirameus(Rob.)R. King et H. Rob. ;肉皮菊●☆

279878　Pachythelia Steetz = Epaltes Cass. ■

279879　Pachythrix Hook. f. = Pleurophyllum Hook. f. ■☆

279880　Pachytrophe Bureau = Streblus Lour. ●

279881　Pachyurus Post et Kuntze = Calocephalus R. Br. ■●☆

279882　Pachyurus Post et Kuntze = Pachysurus Steetz ■●☆

279883　Pacifigeron G. L. Nesom(1994) ;大洋蓬属●☆

279884　Pacifigeron rapensis(F. Br.) G. L. Nesom;大洋蓬■☆

279885　Packera Á. Löve et D. Löve(1976) ;金千里光属(蛮鬼塔属) ; Ragwort ■☆

279886　Packera anonyma(A. W. Wood) W. A. Weber et Á. Löve;斯莫尔金千里光;Small's Ragwort ■☆

279887　Packera antennariifolia(Britton) W. A. Weber et Á. Löve;瘠地金千里光;Shalebarren Ragwort ■☆

279888　Packera aurea(L.) Á. Löve et D. Löve;金千里光(黄蛮鬼塔) ; Golden Groundsel, Golden Ragwort, Heart-leaved Groundsel, Squaw Weed ■☆

279889　Packera bernardina(Greene) W. A. Weber et Á. Löve;圣贝尔纳多金千里光;San Bernardino Ragwort ■☆

279890　Packera bolanderi(A. Gray) W. A. Weber et Á. Löve;鲍氏金千里光;Bolander's Ragwort ■☆

279891　Packera bolanderi (A. Gray) W. A. Weber et Á. Löve var. harfordii(Greenm.) Trock et T. M. Barkley;哈福德金千里光; Harford's Ragwort ■☆

279892　Packera breweri(Burtt Davy) W. A. Weber et Á. Löve;布鲁尔金千里光;Brewer's Ragwort ■☆

279893　Packera buekii Trock et T. M. Barkley = Packera subnuda (DC.) Trock et T. M. Barkley ■☆

279894　Packera cana(Hook.) W. A. Weber et Á. Löve;浅灰金千里光; Woolly Groundsel ■☆

279895　Packera cardamine(Greene) W. A. Weber et Á. Löve;苦金千里光;Bittercress Ragwort ■☆

279896　Packera castoreus(S. L. Welsh) Kartesz;乳金千里光●☆

279897　Packera clevelandii(Greene) W. A. Weber et Á. Löve;克利夫兰金千里光;Cleveland's Ragwort ■☆

279898　Packera crocata(Rydb.) W. A. Weber et Á. Löve;番红花金千里光;Saffron Ragwort ■☆

279899　Packera cymbalaria(Pursh) Á. Löve et D. Löve;木犀叶金千里光(矮北极金千里光,木犀叶千里光);Dwarf Arctic Ragwort ■

279900　Packera cymbalaria (Pursh) Á. Löve et D. Löve = Senecio resedifolius Less. ■

279901　Packera cynthioides(Greene) W. A. Weber et Á. Löve;赖特金千里光;White Mountain Ragwort ■☆

279902　Packera debilis(Nutt.) W. A. Weber et Á. Löve;柔弱金千里光;Weak Groundsel ■☆

279903　Packera dimorphophylla(Greene) W. A. Weber et Á. Löve;二型叶金千里光;Splitleaf Groundsel ■☆

279904　Packera dimorphophylla(Greene) W. A. Weber et Á. Löve var. intermedia(T. M. Barkley) Trock et T. M. Barkley;全缘二型叶金千里光■☆

279905　Packera dimorphophylla(Greene) W. A. Weber et Á. Löve var. paysonii(T. M. Barkley) Trock et T. M. Barkley;佩森金千里光;Payson's Groundsel ■☆

279906　Packera eurycephala(Torr. et A. Gray) W. A. Weber et Á. Löve;宽头金千里光;Widehead Groundsel ■☆

279907　Packera eurycephala(Torr. et A. Gray) W. A. Weber et Á. Löve var. lewisrosei (J. T. Howell) J. F. Bain;莱维金千里光;Lewis' Groundsel ■☆

279908　Packera fendleri(A. Gray) W. A. Weber et Á. Löve;芬德勒金千里光;Fendler's Ragwort ■☆

279909　Packera fernaldii Greenm. = Packera cymbalaria(Pursh) Á. Löve et D. Löve ■

279910　Packera flettii(Wiegand) W. A. Weber et Á. Löve;弗莱特金千里光;Flett's Ragwort ■☆

279911　Packera franciscana(Greene) W. A. Weber et Á. Löve;旧金山金千里光;San Francisco Peaks Ragwort ■☆

279912　Packera ganderi(T. M. Barkley et R. M. Beauch.) W. A. Weber et Á. Löve;甘德金千里光;Gander's Ragwort ■☆

279913　Packera glabella (Poir.) C. Jeffrey;卡罗林金千里光;Butterweed ■☆

279914　Packera greenei(A. Gray) W. A. Weber et Á. Löve;格林金千里光;Flame Ragwort ■☆

279915　Packera hartiana(A. Heller) W. A. Weber et Á. Löve;哈特金千里光;Hart's Ragwort ■☆

279916　Packera hesperia(Greene) W. A. Weber et Á. Löve;西部金千里光;Western Ragwort ■☆

279917　Packera hyperborealis(Greenm.) Á. Löve et D. Löve;北部金千里光;Northern Groundsel ■☆

279918　Packera indecora(Greene) Á. Löve et D. Löve;雅致金千里光;Elegant Groundsel, Plains Ragwort, Rayless Ragwort, Taller Discoid Groundsel ■☆

279919　Packera ionophylla (Greene) W. A. Weber et Á. Löve;紫罗兰金千里光;Tehachapi Ragwort ■☆

279920　Packera layneae(Greene) W. A. Weber et Á. Löve;莱氏金千里光;Layne's Ragwort ■☆

279921　Packera macounii(Greene) W. A. Weber et Á. Löve;锡斯基尤金千里光;Siskiyou Mountain Ragwort ■☆

279922　Packera malmstenii(S. F. Blake ex Tidestr.) Kartesz;马尔金千里光;Podunk Ragwort ■☆

279923　Packera millefolia(Torr. et A. Gray) W. A. Weber et Á. Löve;皮特蒙特金千里光;Piedmont Ragwort ■☆

279924　Packera moresbiensis (Calder et R. L. Taylor) J. F. Bain = Packera subnuda (DC.) Trock et T. M. Barkley var. moresbiensis (Calder et R. L. Taylor) Trock ☆

279925　Packera multilobata(Torr. et A. Gray) W. A. Weber et Á. Löve;多裂金千里光;Lobeleaf Groundsel ■☆

279926　Packera musiniensis(S. L. Welsh) Trock;穆地金千里光■☆

279927　Packera neomexicana(A. Gray) W. A. Weber et Á. Löve;新墨西哥金千里光;New Mexico Groundsel ■☆

279928　Packera neomexicana (A. Gray) W. A. Weber et Á. Löve var. mutabilis(Greene) W. A. Weber et Á. Löve;易变新墨西哥金千里光■☆

279929　Packera neomexicana (A. Gray) W. A. Weber et Á. Löve var. toumeyi(Greene) Trock et T. M. Barkley;图米金千里光;Toumey's Groundsel ■☆

279930　Packera obovata(Muhl. ex Willd.) W. A. Weber et Á. Löve;倒卵叶金千里光;Roundleaf Groundsel, Squaw Weed ■☆

279931　Packera ogotorukensis(Packer) Á. Löve et D. Löve;奥地金千里光;Ogotoruk Creek Ragwort ■☆

279932　Packera ovina (Greene) J. F. Bain = Packera subnuda (DC.) Trock et T. M. Barkley ■☆

279933　Packera ovinus Greene = Packera subnuda(DC.) Trock et T. M. Barkley ■☆

279934　Packera pauciflora(Pursh) Á. Löve et D. Löve;少花金千里光;Alpine Groundsel ■☆

279935　Packera paupercula (Michx.) Á. Löve et D. Löve;北方甜金千里光; Balsam Groundsel, Balsam Ragwort, Northern Meadow Groundsel, Northern Ragwort ■☆

279936　Packera plattensis(Nutt.) W. A. Weber et Á. Löve;草原金千里光;Platte Groundsel, Prairie Groundsel, Prairie Ragwort ■☆

279937　Packera porteri (Greene) C. Jeffrey;波特金千里光;Porter's Groundsel ■☆

279938　Packera prionophyllus Greene = Packera hartiana(A. Heller) W. A. Weber et Á. Löve ■☆

279939　Packera pseudaurea(Rydb.) W. A. Weber et Á. Löve;拟金千里光;False-gold Groundsel, Golden Ragwort, Heart-leaved Groundsel, Squaw Weed, Western Golden Ragwort, Western Heart-leaved Groundsel ■☆

279940　Packera pseudaurea (Rydb.) W. A. Weber et Á. Löve var. flavula(Greene) Trock et T. M. Barkley;浅黄金千里光■☆

279941　Packera pseudaurea (Rydb.) W. A. Weber et Á. Löve var. semicordata(Mack. et Bush) Trock et T. M. Barkley;半心形金千里光;False-gold Groundsel, Heart-leaved Groundsel, Western Golden Ragwort ■☆

279942　Packera quaerens (Greene) W. A. Weber et Á. Löve = Packera hartiana(A. Heller) W. A. Weber et Á. Löve ■☆

279943　Packera quaerens Greene = Packera hartiana(A. Heller) W. A. Weber et Á. Löve ■☆

279944　Packera quercetorum (Greene) C. Jeffrey;栎林金千里光;Oak Creek Ragwort ■☆

279945　Packera resedifolia (Less.) Á. Löve et W. A. Weber = Packera cymbalaria(Pursh) Á. Löve et D. Löve ■

279946　Packera resedifolius Less. = Packera cymbalaria(Pursh) Á. Löve et D. Löve ■

279947　Packera sanguisorboides(Rydb.) W. A. Weber et Á. Löve;地榆金千里光;Burnet Ragwort ■☆

279948　Packera schweinitziana(Nutt.) W. A. Weber et Á. Löve;施氏金千里光;Schweinitz's Ragwort ■☆

279949　Packera spellenbergii (T. M. Barkley) C. Jeffrey;斯佩金千里光;Carrizo Creek Ragwort ■☆

279950　Packera streptanthifolia(Greene) W. A. Weber et Á. Löve;落基山金千里光;Rocky Mountain Groundsel ■☆

279951　Packera streptanthifolia(Greene) W. A. Weber et Á. Löve var. borealis(Torr. et A. Gray) Trock = Packera streptanthifolia (Greene) W. A. Weber et Á. Löve ■☆

279952　Packera subnuda(DC.) Trock et T. M. Barkley;裂叶金千里光;

Cleftleaf Groundsel ■☆

279953　Packera subnuda（DC.）Trock et T. M. Barkley var. moresbiensis（Calder et R. L. Taylor）Trock;莫尔斯比金千里光■☆

279954　Packera tampicana（DC.）C. Jeffrey;大草原金千里光;Great Plains Ragwort ■☆

279955　Packera texensis O'Kennon et Trock;得州金千里光■☆

279956　Packera tomentosa（Michx.）C. Jeffrey;毛金千里光;Groundsel, Woolly Ragwort ■☆

279957　Packera tridenticulata（Rydb.）W. A. Weber et Á. Löve;三齿金千里光;Threetooth Ragwort ■☆

279958　Packera werneriifolia（A. Gray）W. A. Weber et Á. Löve;沃纳菊叶金千里光;Hoary Groundsel ■☆

279959　Pacoseroca Adans. = Amomum L.（废弃属名）■

279960　Pacoseroca Adans. = Amomum Roxb.（保留属名）■

279961　Pacoseroca Adans. = Zingiber Mill.（保留属名）■

279962　Pacouria Aubl.（废弃属名）= Landolphia P. Beauv.（保留属名）●☆

279963　Pacouria amoena（Hua）Pichon = Ancylobotrys amoena Hua ●☆

279964　Pacouria angustifolia（K. Schum. ex Engl.）Kuntze = Ancylobotrys petersiana（Klotzsch）Pierre ●☆

279965　Pacouria capensis（Oliv.）S. Moore = Ancylobotrys capensis（Oliv.）Pichon ●☆

279966　Pacouria crassifolia（K. Schum.）Hiern = Ancylobotrys scandens（Schumach. et Thonn.）Pichon ●☆

279967　Pacouria dubia（Lassia）Pichon = Saba comorensis（Bojer ex A. DC.）Pichon ●☆

279968　Pacouria dulcis（Sabine）Roberty = Landolphia dulcis（Sabine）Pichon ●☆

279969　Pacouria echinata（A. Chev.）Pichon = Ancylobotrys scandens（Schumach. et Thonn.）Pichon ●☆

279970　Pacouria florida（Benth.）Hiern = Saba comorensis（Bojer ex A. DC.）Pichon ●☆

279971　Pacouria lucida（K. Schum.）Kuntze = Dictyophleba lucida（K. Schum.）Pierre ●☆

279972　Pacouria owariensis（P. Beauv.）Hiern = Landolphia owariensis P. Beauv. ●☆

279973　Pacouria parvifolia（K. Schum.）Hiern = Landolphia parvifolia K. Schum. ●☆

279974　Pacouria petersiana（Klotzsch）S. Moore = Ancylobotrys petersiana（Klotzsch）Pierre ●☆

279975　Pacouria petersiana（Klotzsch）S. Moore var. schweinfurthiana（Hallier f.）S. Moore = Ancylobotrys amoena Hua ●☆

279976　Pacouria pyriformis（Pierre）Pichon = Ancylobotrys pyriformis Pierre ●☆

279977　Pacouria robusta（Pierre）Pichon = Ancylobotrys pyriformis Pierre ●☆

279978　Pacouria scandens（Schumach. et Thonn.）Pichon = Ancylobotrys scandens（Schumach. et Thonn.）Pichon ●☆

279979　Pacouria scandens（Schumach. et Thonn.）Pichon var. floribunda（Pellegr.）Pichon = Ancylobotrys scandens（Schumach. et Thonn.）Pichon ●☆

279980　Pacouriaceae Martinov = Apocynaceae Juss.（保留科名）●■

279981　Pacourina Aubl.（1775）;水红菊属■☆

279982　Pacourina edulis Aubl.;水红菊■☆

279983　Pacourinopsis Cass. = Pacourina Aubl. ■☆

279984　Pacurina Raf. = Messerschmidia L. ex Hebenstr. ●■

279985　Padbruggea Miq.（1855）;异鸡血藤属●☆

279986　Padbruggea Miq. = Callerya Endl. ●■

279987　Padbruggea dasyphylla Miq.;异鸡血藤●☆

279988　Padbruggea filipes（Dunn）Craib = Afgekia filipes（Dunn）R. Geesink ●◇

279989　Padbruggea filipes（Dunn）Craib = Whitfordiodendron filipes（Dunn）Dunn ●◇

279990　Padellus Vassilcz. = Cerasus Mill. ●

279991　Padellus Vassilcz. = Prunus L. ●

279992　Padellus mahaleb（L.）Vassilcz. = Cerasus mahaleb（L.）Mill. ●

279993　Padellus maximowiczii（Rupr.）Eremin et Yushev = Cerasus maximowiczii（Rupr.）Kom. ●

279994　Padia Moritzi = Oryza L. ■

279995　Padia Zoll. et Moritzi = Oryza L. ■

279996　Padostemon Griff. = Podostemum Michx. ■☆

279997　Padota Adans. = Marrubium L. ■

279998　Padus Mill.（1754）;稠李属;Bird Cherry, Birdcherry, Cherry ●

279999　Padus Mill. = Prunus L. ●

280000　Padus acrophylla C. K. Schneid. = Padus grayana（Maxim.）C. K. Schneid. ●

280001　Padus acrophylla Schneid. = Padus grayana（Maxim.）C. K. Schneid. ●

280002　Padus asiatica Kom. = Padus avium Mill. var. asiatica（Kom.）T. C. Ku et B. M. Barthol. ●

280003　Padus avium Mill.;稠李（臭耳子,臭李子）;Asiatic Bird Cherry, Bird Cherry, Black Dogwood, Black Merry, Cluster Cherry, Cluster-cherries, Eckberry, Eggberry, Europe Birdcherry, European Bird Cherry, European Bird-cherry, Fool's Cherry, Hack, Hackberry, Hacker, Hackwood, Hagberry, Hawkberry, Heckberry, Hedgeberry, Hegberry, Hicberry, Hog Cherry, Hog-berry, Jackwood, Mayday Tree, Mazar-tree, Polstead Cherry, Wild Cluster Cherry, Wild Clusterberry ●

280004　Padus avium Mill. var. asiatica（Kom.）T. C. Ku et B. M. Barthol.;北亚稠李（亚洲稠李）;Asia Birdcherry ●

280005　Padus avium Mill. var. pubescens（Regel et Tiling）T. C. Ku et B. M. Barthol. = Padus avium Mill. ●

280006　Padus avium Mill. var. pubescens（Regel et Tiling）T. C. Ku et B. M. Barthol. = Prunus padus L. var. pubescens Regel et Tiling ●

280007　Padus avium Mill. var. pubescens（Regel et Tiling）T. C. Ku et B. M. Barthol.;毛叶稠李（稠梨子,稠李,臭李子,多毛稠李,樱额,樱额梨）;Hairyleaf Bird Cherry, Pubescent Bird Cherry ●

280008　Padus beijingensis Y. L. Han et C. Y. Yang = Padus avium Mill. var. pubescens（Regel et Tiling）T. C. Ku et B. M. Barthol. ●

280009　Padus brachypoda（Batalin）C. K. Schneid.;短梗稠李（短柄稠李,短柄椆木,短柄樱桃）;Shortstalk Bird Cherry, Shortstalk Birdcherry, Short-stalked Bird Cherry ●

280010　Padus brachypoda（Batalin）C. K. Schneid. = Padus obtusata（Koehne）Te T. Yu et T. C. Ku ●

280011　Padus brachypoda（Batalin）C. K. Schneid. var. eglandulosa（W. C. Cheng）H. J. Wang = Padus brachypoda（Batalin）C. K. Schneid. ●

280012　Padus brachypoda（Batalin）C. K. Schneid. var. microdonta（Koehne）Te T. Yu et T. C. Ku = Prunus brachypoda Batalin var. microdonta Koehne ●

280013　Padus brachypoda（Batalin）C. K. Schneid. var. microdonta（Koehne）Te T. Yu et T. C. Ku;细齿短柄稠李;Thintooth Bird Cherry ●

280014　Padus brachypoda（Batalin）C. K. Schneid. var. pubigera C. K. Schneid. = Padus obtusata（Koehne）Te T. Yu et T. C. Ku ●

280015　Padus brunnescens Te T. Yu et T. C. Ku;褐毛稠李;Brownhairy Bird Cherry, Brown-hairy Bird Cherry, Brownhairy Birdcherry ●

280016　Padus buergeriana（Miq.）Te T. Yu et T. C. Ku;椆木（布氏稠李,高山小白樱,华东稠李,椆木稠李,椆木樱,犬樱）;Buerger Bird Cherry, Buerger Birdcherry, Racemosa Cherry, Undulate Bird Cherry ●

280017　Padus buergeriana（Miq.）Te T. Yu et T. C. Ku = Prunus

buergeriana Miq. ●

280018　Padus cornuta(Wall. ex Royle)Carrière;光尊稠李(喜马拉雅稠李);Himalayas Birdcherry,Horny Bird Cherry,Smooth Calyx Bird Cherry,Smooth-calyx Bird Cherry ●

280019　Padus cornuta(Wall. ex Royle)Carrière = Prunus cornuta(Wall. ex Royle)Steud. ●

280020　Padus cornuta(Wall. ex Royle)Carrière var. glabra Fritsch ex C. K. Schneid. = Padus cornuta(Wall. ex Royle)Carrière ●

280021　Padus cornuta(Wall. ex Royle)Carrière var. typica Schneid. = Padus cornuta(Wall. ex Royle)Carrière ●

280022　Padus germanica Borkh. = Padus avium Mill. ●

280023　Padus grayana (Maxim.) C. K. Schneid. ;灰叶稠李(灰毛稠李);Gray Bird Cherry, Gray Chokecherry, Gray Choke-cherry, Japanese Bird Cherry ●

280024　Padus grayana (Maxim.) C. K. Schneid. = Prunus grayana Maxim. ●

280025　Padus integrifolia Te T. Yu et T. C. Ku;全缘叶稠李(全缘光尊稠李);Entireleaf Bird Cherry, Entireleaf Birdcherry, Entire-leaved Bird Cherry ●

280026　Padus integrifolia Te T. Yu et T. C. Ku = Prunus cornuta(Wall. ex Royle)Steud. var. integrifolia Te T. Yu ●

280027　Padus laxiflora(Koehne)T. C. Ku;疏花稠李(疏花樱,兴山樱桃);Loose-flower Cherry ●

280028　Padus maackii(Rupr.)Kom. ;斑叶稠李(斑叶樱,披针形斑叶稠李,山桃,山桃稠李);Amur Bird Cherry, Amur Birdcherry, Amur Choke Cherry, Amur Chokecherry, Amur Choke-cherry Manchurian Bird-cherry,Maack Laurel Cherry,Manchurian Cherry ●

280029　Padus maackii(Rupr.)Kom. = Prunus maackii Rupr. ●

280030　Padus maackii(Rupr.)Kom. f. lanceolata Te T. Yu et T. C. Ku;披针形斑叶稠李;Lanceolate Birdcherry ●

280031　Padus maackii(Rupr.)Kom. f. lanceolata Te T. Yu et T. C. Ku = Padus maackii(Rupr.)Kom. ●

280032　Padus mahaleb(L.)Borkh. = Cerasus mahaleb(L.)Mill. ●

280033　Padus mahaleb Borkh. = Cerasus mahaleb(L.)Mill. ●

280034　Padus maximowiczii (Rupr.) Sokoloff = Cerasus maximowiczii (Rupr.)Kom. ●

280035　Padus nana(Du Roi)M. Roem. = Prunus virginiana L. ●☆

280036　Padus napaulensis(Ser.)C. K. Schneid. ;粗梗稠李(尼泊尔稠李);Nepal Bird Cherry, Nepal Chokecherry ●

280037　Padus napaulensis(Ser.)C. K. Schneid. = Prunus napaulensis (Ser.)Steud. ●

280038　Padus napaulensis(Ser.)C. K. Schneid. var. sericea(Batalin) C. K. Schneid. = Padus wilsonii C. K. Schneid. ●

280039　Padus napaulensis(Ser.)C. K. Schneid. var. typica Schneid. = Padus napaulensis(Ser.)C. K. Schneid. ●

280040　Padus obtusata(Koehne)Te T. Yu et T. C. Ku;细齿稠李(塔山樱,台湾稠梨,台湾稠李);Obtuse Bird Cherry, Vaniot Bird Cherry ●

280041　Padus obtusata(Koehne)Te T. Yu et T. C. Ku = Prunus obtusata Koehne ●

280042　Padus perulata (Koehne) Te T. Yu et T. C. Ku;宿鳞稠李;Perules Bird Cherry, Scale Bird Cherry, Scale Birdcherry ●

280043　Padus perulata (Koehne)Te T. Yu et T. C. Ku = Prunus perulata Koehne ●

280044　Padus racemosa(Lam.) Gilib. ;欧稠李;Bird Cherry, European Bird Cherry ●

280045　Padus racemosa(Lam.) Gilib. = Padus avium Mill. ●

280046　Padus racemosa(Lam.) Gilib. = Prunus padus L. ●

280047　Padus racemosa(Lam.) Gilib. var. asiatica (Kom.) Te T. Yu et T. C. Ku = Padus avium Mill. var. asiatica(Kom.)T. C. Ku et B. M. Barthol. ●

280048　Padus racemosa(Lam.) Gilib. var. pubescens (Regel et Tiling) C. K. Schneid. = Padus avium Mill. ●

280049　Padus racemosa(Lam.) Gilib. var. pubescens (Regel et Tiling) C. K. Schneid. = Prunus padus L. var. pubescens Regel et Tiling ●

280050　Padus racemosa(Lam.) Gilib. var. pubescens (Regel et Tiling) C. K. Schneid. = Padus avium Mill. var. pubescens (Regel et Tiling) T. C. Ku et B. M. Barthol. ●

280051　Padus serotina Borkh. = Cerasus serotina (Ehrh.)Loisel. ●

280052　Padus serrulata(Lindl.) Sokoloff = Cerasus serrulata(Lindl.)G. Don ex Loudon ●

280053　Padus ssiori(F. Schmidt) C. K. Schneid. ;稀归稠李(日本稠李);Ssior Bird Cherry, Ssior Birdcherry ●

280054　Padus stellipila(Koehne)Te T. Yu et T. C. Ku;星毛稠李(星毛椿木稠李);Starhaiey Bird Cherry, Starhairy Birdcherry, Stellate Bird Cherry ●

280055　Padus stellipila (Koehne) Te T. Yu et T. C. Ku = Prunus buergeriana Miq. var. stellipila(Koehne)Te T. Yu et C. L. Li ●

280056　Padus velutina (Batalin) C. K. Schneid. ;毡毛稠李; Velvety Bird Cherry, Velvety Birdcherry ●

280057　Padus velutina (Batalin) C. K. Schneid. = Prunus velutina Batalin ●

280058　Padus virginiana(L.)Mill. = Prunus virginiana L. ●☆

280059　Padus vulgaris Borkh. = Padus avium Mill. ●

280060　Padus vulgaris Borkh. = Padus racemosa(Lam.)Gilib. ●

280061　Padus wilsonii C. K. Schneid. ;绢毛稠李(绢毛粗梗稠李);E. H. Wilson Bird Cherry, E. H. Wilson Chokecherry, Sericeous Bird Cherry,Wilson Bird Cherry ●

280062　Padus wilsonii C. K. Schneid. = Prunus wilsonii Diels ex Koehne ●

280063　Paederia L. (1767)(保留属名);鸡矢藤属(鸡屎藤属,牛皮冻属);Fever Vine, Fevervine, Paederia ●■

280064　Paederia argentea(A. Rich.) K. Schum. ;银白鸡矢藤●☆

280065　Paederia axilliflora Puff;腋花鸡矢藤●☆

280066　Paederia bodinieri H. Lév. = Gardneria multiflora Makino ●

280067　Paederia bodinieri H. Lév. = Paederia yunnanensis(H. Lév.) Rehder ●

280068　Paederia bojeriana(A. Rich.) Drake;博耶尔鸡矢藤●☆

280069　Paederia bojeriana (A. Rich.) Drake subsp. foetens (Hiern) Verdc. ;臭博耶尔鸡矢藤●☆

280070　Paederia cavaleriei H. Lév. ;耳叶鸡矢藤(臭皮藤,卡氏鸡矢藤,卡氏鸡屎藤,毛鸡屎藤,圆锥鸡矢藤);Cavalerie Fevervine ●

280071　Paederia chinensis Fedde = Paederia scandens(Lour.)Merr. ●

280072　Paederia chinensis Hance = Paederia foetida L. ●

280073　Paederia diffusa(Britton)Standl. ;伸展鸡矢藤●☆

280074　Paederia dunniana H. Lév. = Paederia scandens(Lour.)Merr. ●

280075　Paederia esquirolii H. Lév. = Paederia scandens(Lour.)Merr. ●

280076　Paederia farinosa(Baker) Puff;被粉鸡矢藤●☆

280077　Paederia foetens (Hiern) K. Schum. = Paederia bojeriana (A. Rich.) Drake subsp. foetens(Hiern) Verdc. ●☆

280078　Paederia foetida L. ;臭鸡矢藤 (鸡矢藤,鸡屎藤);Foetid Fevervine,Skunk Vine,Stinkvine ●

280079　Paederia foetida Thunb. = Paederia scandens(Lour.)Merr. ●

280080　Paederia grandidieri Drake;格朗鸡矢藤●☆

280081　Paederia kerrii Craib;南臭皮藤;Kerr Fevervine ●

280082　Paederia lanata Puff;绵毛鸡矢藤●☆

280083　Paederia lanuginosa Wall. ;绒毛鸡矢藤;Lanuginose Fevervine ■●

280084　Paederia laxiflora Merr. ex H. L. Li;疏花鸡矢藤; Laxflower Fevervine ■

280085　Paederia lingun Sweet = Paederia bojeriana(A. Rich.)Drake ●☆

280086　Paederia macrocarpa Wall. = Paederia lanuginosa Wall. ■●

280087　Paederia majungensis Homolle ex Puff;马任加鸡矢藤●☆

280088　Paederia mandrarensis Homolle ex Puff;曼德拉鸡矢藤●☆

280089　Paederia minima J. König ex Retz. = Microcarpaea minima(Jos. König ex Retz.)Merr. ■

280090　Paederia owariensis(P. Beauv.)Spreng. = Landolphia owariensis P. Beauv. ●☆

280091　Paederia pertomentosa Merr. ex H. L. Li;白毛鸡矢藤(广西鸡矢藤, 广西鸡屎藤,鸡屎臭药,毛鸡藤); Guangxi Fevervine, Kwangsi Fevervine ●

280092　Paederia petrophila K. Schum. = Paederia pospischilii K. Schum. ●☆

280093　Paederia pospischilii K. Schum. ;波斯皮希尔鸡矢藤●☆

280094　Paederia praetermissa Puff;奇异鸡矢藤; Neglect Fevervine ■●

280095　Paederia rehderiana Hand. -Mazz. = Paederia yunnanensis (H. Lév.)Rehder ●

280096　Paederia sambiranensis Homolle ex Puff;桑比朗鸡矢藤●☆

280097　Paederia scandens(Lour.)Merr. ;鸡矢藤(斑鸠饭,臭狗藤,臭鸡屎藤,臭藤,臭藤子,臭腥藤,狗屁藤,鸡屎藤,鸡香藤,皆治藤,解暑藤,毛葫芦,母狗藤,牛皮冻,牛皮消,女青,清风藤,却节,仁骨蛇,甜藤,五德藤,五香藤,香藤); China Fevervine, Chinese Fevervine ●

280098　Paederia scandens(Lour.)Merr. = Paederia foetida L. ●

280099　Paederia scandens (Lour.) Merr. f. mairei (H. Lév.) Nakai = Paederia scandens(Lour.)Merr. ●

280100　Paederia scandens(Lour.)Merr. f. rubescens Asai;变红鸡矢藤●☆

280101　Paederia scandens (Lour.) Merr. f. velutina (Nakai) Sugim. = Paederia scandens(Lour.)Merr. var. velutina(Nakai)Nakai ●☆

280102　Paederia scandens(Lour.)Merr. var. longituba(Nakai)H. Hara; 长管鸡矢藤●☆

280103　Paederia scandens (Lour.) Merr. var. mairei (H. Lév.) H. Hara = Paederia scandens(Lour.)Merr. ●

280104　Paederia scandens (Lour.) Merr. var. maritima (Koidz.) H. Hara;海滨鸡矢藤●☆

280105　Paederia scandens(Lour.) Merr. var. maritima(Koidz.) H. Hara f. rubrae-stellaris Konta et S. Matsumoto;红星海滨鸡矢藤●☆

280106　Paederia scandens (Lour.) Merr. var. tomentosa (Blume) Hand. -Mazz. ;毛鸡矢藤(白鸡屎藤,臭茎子,臭皮藤,打屁藤,肺痈藤,毛鸡屎藤,迎风子); Tomentose Fevervine ●

280107　Paederia scandens (Lour.) Merr. var. velutina (Nakai) Nakai;绒毛鸡屎藤●☆

280108　Paederia spectatissima H. Li ex Puff;云贵鸡矢藤(狗屁藤,云桂鸡矢藤); Yunnan-Guizhou Fevervine ■

280109　Paederia stenobotrya Merr. ; 狭序鸡矢藤; Narrow-raceme Fevervine ●

280110　Paederia stenophylla Merr. ; 狭叶鸡矢藤; Narrow-leaved Fevervine ●

280111　Paederia taolagnarensis Razafim. et C. M. Taylor;陶拉纳鲁鸡矢藤●☆

280112　Paederia thouarsiana Baill. ;图氏鸡矢藤■☆

280113　Paederia tomentosa Blume = Paederia cavaleriei H. Lév. ●

280114　Paederia tomentosa Blume = Paederia foetida L. ●

280115　Paederia tomentosa Blume = Paederia scandens (Lour.) Merr. ●

280116　Paederia tomentosa Blume = Paederia scandens(Lour.) Merr. ●

280117　Paederia tomentosa Blume var. glabra Kurz = Paederia scandens (Lour.) Merr. ●

280118　Paederia tomentosa Blume var. mairei H. Lév. = Paederia scandens(Lour.) Merr. ●

280119　Paederia tomentosa Blume var. purpureacaerulea H. Lév. et Vaniot = Paederia yunnanensis(H. Lév.)Rehder ●

280120　Paederia wallichii Hook. f. = Paederia yunnanensis (H. Lév.) Rehder ●

280121　Paederia yunnanensis(H. Lév.)Rehder;云南鸡矢藤(白鸡矢藤, 滇鸡矢藤,狗屁藤,毛叶黄药,云南鸡屎藤); Yunnan Fevervine ●

280122　Paederota L. (1758);亮耳参属;Paederota ■☆

280123　Paederota L. = Veronica L. ■

280124　Paederota ageria L. ;亮耳参■☆

280125　Paederota axillaris Siebold et Zucc. = Veronicastrum axillare (Siebold et Zucc.)T. Yamaz. ■

280126　Paederota bona-spei L. = Diascia capensis(L.)Britten ■☆

280127　Paederota humilis Steph. ex Link = Veronica densiflora Ledeb. ■

280128　Paederota minima J. König ex Retz. = Microcarpaea minima (Jos. König ex Retz.)Merr. ■

280129　Paederota minima Koen. = Microcarpaea minima(Jos. König ex Retz.)Merr. ■

280130　Paederota racemosa Houtt. = Hemimeris racemosa (Houtt.) Merr. ■☆

280131　Paederota villosula Miq. = Veronicastrum villosulum (Miq.) T. Yamaz. ■

280132　Paederotella(Wulf)Kem. -Nath. (1953);小亮耳参属■☆

280133　Paederotella(Wulf)Kem. -Nath. = Veronica L. ■

280134　Paederotella pontica(Rupr.)Kem. -Nath. ;小亮耳参■☆

280135　Paedicalyx Pierre ex Pit. (1922);匙萼木属;Paedicalyx ●

280136　Paedicalyx Pierre ex Pit. = Xanthophytum Reinw. ex Blume ●

280137　Paedicalyx attopevense Pierre ex Pit. = Xanthophytum attopevense(Pierre ex Pit.)H. S. Lo ●

280138　Paedicalyx attopevensis Pierre ex Pit. ; 匙 萼 木; Common Paedicalyx, Paedicalyx ●

280139　Paennaea Meerb. = Penaea L. ●☆

280140　Paenoe Post et Kuntze = Panoe Adans. ●☆

280141　Paenoe Post et Kuntze = Vateria L. ●☆

280142　Paenula Orchard(2005);斗篷菊属☆

280143　Paeonia L. (1753);芍药属(牡丹属);Paeony, Peony ●■

280144　Paeonia × baokangensis Andréws;保康牡丹;Baokang Peony ●

280145　Paeonia × papaveracea Andréws;延安牡丹;Yanan Peony ●

280146　Paeonia abchasica Miscz. ;阿伯哈斯芍药■☆

280147　Paeonia albiflora Pall. ;白花芍药(白术,草芍药,赤芍,将离,离草,犁食,没骨花,芍药,余客); Chinese Peony, Whiteflower Peony, White-flowered Peony ■

280148　Paeonia albiflora Pall. = Paeonia lactiflora Pall. ■

280149　Paeonia albiflora Pall. var. hirta Regel;毛脉白花芍药■

280150　Paeonia albiflora Pall. var. trichocarpa Bunge = Paeonia lactiflora Pall. ■

280151　Paeonia algeriensis Chabert = Paeonia mascula(L.) Mill. subsp. atlantica(Coss.)Greuter et Burdet ■☆

280152　Paeonia altaica K. M. Dai et T. H. Ying;阿尔泰芍药; Altai Peony ■

280153　Paeonia altaica K. M. Dai et T. H. Ying = Paeonia anomala L. ■

280154　Paeonia anomala L. ;新疆芍药(阿尔泰芍药,变叶芍药,块根

芍药,毛实芍药,奇特芍药,西北草芍药,杂芍药,窄叶芍药);
Pubescent-follicled Peony,Sinjiang Peony,Ural Peony,Xinjiang Peony ■

280155　Paeonia anomala L. subsp. veitchii(Lynch) D. Y. Hong et K. Y. Pan;
川芍药(草芍药,赤芍,臭牡丹根,川赤芍,红芍药,木芍药,条赤芍);
Veitch Peony,Veitch Ural Peony ■

280156　Paeonia anomala L. var. intermedia (C. A. Mey.) O. Fedtsch. et B.
Fedtsch. = Paeonia anomala L. ■

280157　Paeonia anomala L. var. intermedia (C. A. Mey.) O. Fedtsch. et B.
Fedtsch. = Paeonia intermedia C. A. Mey. ■

280158　Paeonia anomala L. var. nudicarpa Huth =Paeonia anomala L. ■

280159　Paeonia arborea Donn =Paeonia suffruticosa Andréws ●■

280160　Paeonia arietina G. Anderson;羊角芍药■☆

280161　Paeonia atlantica(Coss.) Trab. = Paeonia mascula(L.) Mill. subsp.
atlantica(Coss.)Greuter et Burdet ■☆

280162　Paeonia baokangensis Z. L. Dai et T. Hong = Paeonia ×
baokangensis Andréws ●

280163　Paeonia beresowskii Kom. = Paeonia anomala L. subsp. veitchii
(Lynch)D. Y. Hong et K. Y. Pan ■

280164　Paeonia biebersteiniana Rupr. ;毕氏芍药■☆

280165　Paeonia bifurcata Schipcz. = Paeonia mairei H. Lév. ■

280166　Paeonia brownii Douglas ex Hook. 布朗芍药;Western Peony ■☆

280167　Paeonia californica Nutt. 加州芍药■☆

280168　Paeonia cambessedesii Willk. ;康氏芍药(巴里阿里芍药)■☆

280169　Paeonia caucasica Schipcz. ;高加索芍药;Caucasian Peony ■☆

280170　Paeonia chinensis Oken = Paeonia suffruticosa Andréws ●■

280171　Paeonia chinensis Vilm. = Paeonia lactiflora Pall. ■

280172　Paeonia corallina Retz. = Paeonia mascula(L.) Mill. ■☆

280173　Paeonia corallina Retz. subsp. atlantica(Coss.) Maire = Paeonia
mascula(L.) Mill. subsp. atlantica(Coss.)Greuter et Burdet ■☆

280174　Paeonia corallina Retz. subsp. coriacea(Boiss.) Maire = Paeonia
mascula(L.) Mill. subsp. coriacea(Boiss.) Malag. ■☆

280175　Paeonia corallina Retz. var. atlantica Coss. = Paeonia mascula
(L.) Mill. subsp. atlantica(Coss.)Greuter et Burdet ■☆

280176　Paeonia corallina Retz. var. coriacea (Boiss.) Coss. = Paeonia
mascula(L.) Mill. subsp. coriacea(Boiss.) Malag. ■☆

280177　Paeonia coriacea Boiss. = Paeonia mascula (L.) Mill. subsp.
coriacea(Boiss.) Malag. ■☆

280178　Paeonia coriacea Boiss. var. maroccana(Pau et Font Quer)Romo
= Paeonia mascula(L.) Mill. subsp. coriacea(Boiss.) Malag. ■☆

280179　Paeonia decomposita Hand. -Mazz. = Paeonia suffruticosa
Andréws ●■

280180　Paeonia decompsonia Hand. -Mazz. ;四川牡丹;Sichuan Peony ●

280181　Paeonia decompsonia Hand. -Mazz. subsp. rotundiloba D. Y.
Hong;圆裂四川牡丹;Round-lobed Sichuan Peony ●

280182　Paeonia delavayi Franch. ;滇牡丹(白药,赤丹皮,赤芍,滇藏
牡丹,野牡丹,云白芍,紫牡丹);Chinese Tree Peony, Delavay
Peony,Maroon Tree Peony,Wild Peony ●

280183　Paeonia delavayi Franch. subsp. angustiloba (Rehder et E. H.
Wilson)B. A. Shen = Paeonia delavayi Franch. ●

280184　Paeonia delavayi Franch. subsp. lutea (Delavay ex Franch.) B.
A. Shen = Paeonia delavayi Franch. ●

280185　Paeonia delavayi Franch. var. alba Bean = Paeonia delavayi
Franch. ●

280186　Paeonia delavayi Franch. var. angustiloba Rehder et E. H.
Wilson;狭叶牡丹(保氏牡丹,裂叶紫牡丹,乌花牡丹,狭裂牡丹,
狭叶野牡丹);Narrowleaf Delavay Peony, Narrowleaf Peony ●

280187　Paeonia delavayi Franch. var. angustiloba Rehder et E. H.

Wilson = Paeonia delavayi Franch. ●

280188　Paeonia delavayi Franch. var. atropurpurea Schipcz. = Paeonia
delavayi Franch. ●

280189　Paeonia delavayi Franch. var. ludlowii Stern et Taylor;大黄牡
丹;Ludlow Peony ●

280190　Paeonia delavayi Franch. var. lutea (Delavay ex Franch.) Finet
et Gagnep. = Paeonia delavayi Franch. ●

280191　Paeonia delavayi Franch. var. lutea (Delavay ex Franch.) Finet
et Gagnep. f. superba Lemoine = Paeonia delavayi Franch. ●

280192　Paeonia delavayi Franch. var. lutea (Delavay ex Franch.) Finet et
Gagnep. ;黄牡丹(野白芍);Yellow Delavay Peony, Yellow Peony ●

280193　Paeonia delavayi subsp. ludlowii (Stern et Taylor) B. A. Shen =
Paeonia ludlowii(Stern et Taylor) D. Y. Hong ●

280194　Paeonia emodi Wall. ex Royle f. glabrata (Hook. f. et Thomson)
H. Hara = Paeonia emodii Wall. ex Royle ■

280195　Paeonia emodi Wall. ex Royle subsp. sterniana (H. R. Fletcher)
Halda = Paeonia sterniana H. R. Fletcher ■

280196　Paeonia emodi Wall. ex Royle var. glabrata Hook. f. et Thomson
= Paeonia emodii Wall. ex Royle ■

280197　Paeonia emodii Wall. ex Royle;多花芍药(块根芍药,喜马拉雅芍
药,喜马牡丹);Himalayan Peony, Himalayas Peony,Peony Rose ■

280198　Paeonia emodii Wall. ex Royle f. glabrata(Hook. f. et Thomson)
H. Hara = Paeonia emodii Wall. ex Royle ■

280199　Paeonia emodii Wall. ex Royle subsp. sterniana(H. R. Fletcher)
Halda = Paeonia sterniana H. R. Fletcher ■

280200　Paeonia emodii Wall. ex Royle var. glabrata Hook. f. et Thomson
= Paeonia emodii Wall. ex Royle ■

280201　Paeonia franchetii Halda = Paeonia delavayi Franch. ●

280202　Paeonia fruticosa Dum. Cours. = Paeonia suffruticosa Andréws ●■

280203　Paeonia handel-mazzettii Halda = Paeonia delavayi Franch. ●

280204　Paeonia hybrida Pall. ;窄叶芍药(杂芍药,赤芍,块根芍药);
Tuberousroot Peony ■

280205　Paeonia hybrida Pall. = Paeonia intermedia C. A. Mey. ■

280206　Paeonia intermedia C. A. Mey. ;块根芍药■

280207　Paeonia japonica(Makino) Miyabe et Takeda;山芍药(日本芍
药);Japan Peony,Japanese Peony ■

280208　Paeonia japonica(Makino) Miyabe et Takeda = Paeonia obovata
Maxim. ■

280209　Paeonia japonica(Makino) Miyabe et Takeda f. hirsuta H. Hara;
毛叶山芍药■☆

280210　Paeonia japonica(Makino) Miyabe et Takeda f. pilosa (Nakai)
W. T. Lee = Paeonia japonica (Makino) Miyabe et Takeda f. hirsuta
H. Hara ■☆

280211　Paeonia japonica(Makino) Miyabe et Takeda var. pilosa Nakai =
Paeonia japonica(Makino) Miyabe et Takeda f. hirsuta H. Hara ■☆

280212　Paeonia jishanensis T. Hong et W. Z. Zhao;金山牡丹(矮牡丹,
樱山牡丹);Jishan Peony ●

280213　Paeonia lactiflora Pall. ;芍药(白芍,白芍药,草芍药,赤芍,赤
芍药,臭牡丹根,川白芍,川白芍,大白芍,殿春客,伏丁,伏贡,冠
芳,杭白芍,毫白芍,红芍药,将离,焦白药,解仓,金芍药,可离,
婪尾春,梨食,离草,没骨花,木芍药,其积,天斗,天魁,艳友,余
容,玉斗,玉魁,中江芍);Chinese Peony, Common Garden Peony,
Common Peony, Herbaceous Peony, Peony,Shaoyao ■

280214　Paeonia lactiflora Pall. var. trichocarpa(Bunge)Stearn;毛果芍
药(赤芍);Hairfruit Peony ■

280215　Paeonia lactiflora Pall. var. trichocarpa(Bunge)Stern. = Paeonia
lactiflora Pall. ■

280216　Paeonia lactiflora Pall. var. villosa M. S. Yan et K. Sun；毛茎芍药；Villose Peony ■

280217　Paeonia lactiflora Pall. var. villosa M. S. Yan et K. Sun = Paeonia lactiflora Pall. ■

280218　Paeonia lactiflora Vilm. var. trichocarpa（Bunge）Stern = Paeonia lactiflora Pall. ■

280219　Paeonia lactiflora Vilm. var. villosa M. S. Yan et K. Sun = Paeonia lactiflora Pall. ■

280220　Paeonia lemoinei Rehder；杂种牡丹●☆

280221　Paeonia ludlowii（Stern et G. Taylor）J. J. Li et D. Z. Chen = Paeonia ludlowii（Stern et Taylor）D. Y. Hong ●

280222　Paeonia ludlowii（Stern et Taylor）D. Y. Hong；大花黄牡丹（大花牡丹）；Lluslow's Peony ●

280223　Paeonia lutea Delavay ex Franch. ；黄牡丹树（白芍，丹皮，野牡丹）；Tree Peony，Yellow Peony，Yellow Tree Peony ●

280224　Paeonia lutea Delavay ex Franch. = Paeonia delavayi Franch. ●

280225　Paeonia lutea Delavay ex Franch. var. ludlowii Stern et G. Taylor = Paeonia ludlowii（Stern et G. Taylor）J. J. Li et D. Z. Chen ●

280226　Paeonia lutea Delavay ex Franch. var. ludlowii Stern et Taylor = Paeonia ludlowii（Stern et Taylor）D. Y. Hong ●

280227　Paeonia macrophylla（Albov）Lomakin；大叶芍药；Largeleaf Peony ■☆

280228　Paeonia macrophylla Lomakin = Paeonia macrophylla（Albov）Lomakin ■☆

280229　Paeonia mairei H. Lév. ；美丽芍药；Maire Peony ■

280230　Paeonia mairei H. Lév. f. oxypetala（Hand. -Mazz.）W. P. Fang = Paeonia mairei H. Lév. ■

280231　Paeonia mascula（L.）Mill. ；光叶羊角芍药；Cabbage Rose，Cheese，Chesses，Coral Peony，English Peony，He Peony，Male Peony，Marmaritan，Nanpie，Peony，Pianet，Piano Rose，Piano-roses，Pie-nanny，Piny，Piony，Posy，Pyanot，Sheep-shearing Rose，Whitsun Ball ■☆

280232　Paeonia mascula（L.）Mill. = Paeonia officinalis Retz. ■☆

280233　Paeonia mascula（L.）Mill. subsp. atlantica（Coss.）Greuter et Burdet；大西洋光叶羊角芍药■☆

280234　Paeonia mascula（L.）Mill. subsp. coriacea（Boiss.）Malag. ；革质光叶羊角芍药■☆

280235　Paeonia mascula（L.）Mill. var. maroccana Pau et Font Quer = Paeonia mascula（L.）Mill. ■☆

280236　Paeonia mlokosewitschi Lomakin；恩洛克氏芍药（高加索芍药）；Lemon Peony，Mlokose Peony ■☆

280237　Paeonia moutan Sims = Paeonia suffruticosa Andréws ●■

280238　Paeonia moutan Sims subsp. atava Brühl = Paeonia rockii（S. G. Haw et Lauener）T. Hong et J. J. Li ●

280239　Paeonia moutan Sims var. papaveracea（Andréws）DC. = Paeonia suffruticosa Andréws ●■

280240　Paeonia moutan Sims var. vitata Van Houtte = Paeonia suffruticosa Andréws ●■

280241　Paeonia obovata Maxim. ；草芍药（赤芍，卵叶芍药，山芍药，野芍药）；Obovate Peony，Obovate-leaved Peony ■

280242　Paeonia obovata Maxim. f. albiflora M. Mizush. ex T. Shimizu；白花草芍药■☆

280243　Paeonia obovata Maxim. f. glabra（Makino）Kitam. = Paeonia obovata Maxim. var. glabra Makino ■

280244　Paeonia obovata Maxim. f. oreogeton（S. Moore）Kitag. = Paeonia obovata Maxim. ■

280245　Paeonia obovata Maxim. f. oreogeton（S. Moore）Kitag. = Paeonia suffruticosa Andréws ●■

280246　Paeonia obovata Maxim. subsp. japonica（Makino）Halda = Paeonia obovata Maxim. ■

280247　Paeonia obovata Maxim. subsp. willmottiae（Stapf）D. Y. Hong et K. Y. Pan；拟草芍药（毛果芍药，毛芍药，毛叶草芍药，毛叶芍药）；Hairyleaf Peony ■

280248　Paeonia obovata Maxim. var. glabra Makino；无毛草芍药■

280249　Paeonia obovata Maxim. var. glabra Makino = Paeonia obovata Maxim. ■

280250　Paeonia obovata Maxim. var. japonica Makino = Paeonia japonica（Makino）Miyabe et Takeda ■

280251　Paeonia obovata Maxim. var. japonica Makino = Paeonia obovata Maxim. ■

280252　Paeonia obovata var. willmottiae（Stapf）Stern = Paeonia obovata Maxim. subsp. willmottiae（Stapf）D. Y. Hong et K. Y. Pan ■

280253　Paeonia officinalis Retz. ；药用芍药（欧洲芍药）；Common Peony，Kingsbloom，Peony，Rose-of-the-mount ■☆

280254　Paeonia officinalis Retz. 'Alba Plena'；白重瓣欧洲芍药■☆

280255　Paeonia officinalis Retz. 'China Rose'；月季欧洲芍药■☆

280256　Paeonia officinalis Retz. 'Rubra Plena'；红重瓣欧洲芍药■☆

280257　Paeonia oreogeton S. Moore = Paeonia obovata Maxim. ■

280258　Paeonia ostii T. Hong et J. X. Zhang；凤丹（扬州牡丹，杨山牡丹，药用牡丹）；Osti Peony，Yangshan Peony ●

280259　Paeonia ostii T. Hong et J. X. Zhang subsp. lishizhenii（B. A. Shen）B. A. Shen = Paeonia ostii T. Hong et J. X. Zhang ●

280260　Paeonia ostii T. Hong et J. X. Zhang var. lishizhenii B. A. Shen = Paeonia ostii T. Hong et J. X. Zhang ●

280261　Paeonia oxypetala Hand. -Mazz. = Paeonia mairei H. Lév. ■

280262　Paeonia papaveracea Andréws = Paeonia suffruticosa Andréws ●■

280263　Paeonia peregrina Mill. ；华丽芍药（欧洲芍药）■☆

280264　Paeonia peregrina Mill. 'Otto Froebel'；阳光华丽芍药■☆

280265　Paeonia potaninii Kom. = Paeonia delavayi Franch. ●

280266　Paeonia potaninii Kom. f. alba（Bean.）Stern. = Paeonia delavayi Franch. ●

280267　Paeonia potaninii Kom. var. trollioides（Stapf ex Stern）Stern；金莲花叶牡丹●☆

280268　Paeonia potaninii Kom. var. trollioides（Stapf ex Stern）Stern = Paeonia delavayi Franch. ●

280269　Paeonia qiui Y. L. Pei et D. Y. Hong；卵叶牡丹；Ovate-leaf Peony，Qiu Peony ●

280270　Paeonia ridleyi Z. L. Dai et T. Hong；红斑牡丹；Ridley Peony ●

280271　Paeonia ridleyi Z. L. Dai et T. Hong = Paeonia qiui Y. L. Pei et D. Y. Hong ●

280272　Paeonia rockii（S. G. Haw et Lauener）T. Hong et J. J. Li；紫斑牡丹；Rock's Peony ●

280273　Paeonia rockii（S. G. Haw et Lauener）T. Hong et J. J. Li = Paeonia suffruticosa Andréws var. papaveracea（Andréws）A. Kern. ●

280274　Paeonia rockii（S. G. Haw et Lauener）T. Hong et J. J. Li ex D. Y. Hong subsp. linyanshanii T. Hong et Osti = Paeonia rockii（S. G. Haw et Lauener）T. Hong et J. J. Li ●

280275　Paeonia rockii（S. G. Haw et Lauener）T. Hong et J. J. Li ex D. Y. Hong subsp. taibaishanica D. Y. Hong = Paeonia rockii（S. G. Haw et Lauener）T. Hong et J. J. Li subsp. taibaishanica D. Y. Hong ●

280276　Paeonia rockii（S. G. Haw et Lauener）T. Hong et J. J. Li ex D. Y. Hong = Paeonia rockii（S. G. Haw et Lauener）T. Hong et J. J. Li ●

280277　Paeonia rockii（S. G. Haw et Lauener）T. Hong et J. J. Li ex D. Y. Hong = Paeonia suffruticosa Andréws var. papaveracea（Andréws）A. Kern. ●

280278　Paeonia rockii（S. G. Haw et Lauener）T. Hong et J. J. Li subsp.

linyanshanii T. Hong et Osti = Paeonia rockii(S. G. Haw et Lauener) T. Hong et J. J. Li ●

280279　Paeonia rockii(S. G. Haw et Lauener) T. Hong et J. J. Li subsp. taibaishanica D. Y. Hong;太白牡丹(太白山紫斑牡丹);Taibaishan Rock's Peony ●

280280　Paeonia russi Biv. var. coriacea (Boiss.) Coss. = Paeonia mascula(L.) Mill. subsp. atlantica(Coss.)Greuter et Burdet ■☆

280281　Paeonia sinensis Steud. = Paeonia lactiflora Pall. ■

280282　Paeonia sinjiangensis K. Y. Pan = Paeonia anomala L. ■

280283　Paeonia smouthii Hort. ;思茂芍药■☆

280284　Paeonia spontanea(Rehder) T. Hong et W. Z. Zhao = Paeonia jishanensis T. Hong et W. Z. Zhao ●

280285　Paeonia sterniana H. R. Fletcher;斯氏白花芍药;White Peony, Whiteflower Peony ■

280286　Paeonia suffruticosa Andrews;牡丹(百两金,丹皮,花王,鹿韭,洛阳花,牡丹花,木芍药,鼠姑,吴牡丹,野牡丹花,云南牡丹);Chinese Tree Peony,Moutan,Moutan Peony,Subshrubby Peony, Suffruticosa Peony,Tree Peony ●■

280287　Paeonia suffruticosa Andrews 'Cardinal Vaughan';卡迪纳尔·沃恩牡丹●☆

280288　Paeonia suffruticosa Andrews 'Godaishu';五大洲牡丹●☆

280289　Paeonia suffruticosa Andrews 'Hanadaijin';丽丹牡丹●☆

280290　Paeonia suffruticosa Andrews 'Hana-kisoi';赛花牡丹●☆

280291　Paeonia suffruticosa Andrews 'Jewel';莲花宝石牡丹●☆

280292　Paeonia suffruticosa Andrews 'Joseph Rock' = Paeonia rockii (S. G. Haw et Lauener) T. Hong et J. J. Li ●

280293　Paeonia suffruticosa Andrews 'Kamadunishiki';卡马达锦锻牡丹●☆

280294　Paeonia suffruticosa Andrews 'Large Globe' = Paeonia suffruticosa Andrews 'Godaishu' ●☆

280295　Paeonia suffruticosa Andrews 'Magnificent' = Paeonia suffruticosa Andrews 'Hanadaijin' ●☆

280296　Paeonia suffruticosa Andrews 'Reine Elizabeth';瑞恩·伊丽莎白牡丹●☆

280297　Paeonia suffruticosa Andrews 'Renkaku';飞鹤牡丹●☆

280298　Paeonia suffruticosa Andrews 'Rock's Variety' = Paeonia rockii (S. G. Haw et Lauener) T. Hong et J. J. Li ●

280299　Paeonia suffruticosa Andrews subsp. atava(Brühl) S. G. Haw et Lauener = Paeonia rockii(S. G. Haw et Lauener) T. Hong et J. J. Li ●

280300　Paeonia suffruticosa Andrews subsp. ostii (T. Hong et J. X. Zhang) Halda = Paeonia ostii T. Hong et J. X. Zhang ●

280301　Paeonia suffruticosa Andrews subsp. rockii S. G. Haw et Lauener = Paeonia rockii(S. G. Haw et Lauener) T. Hong et J. J. Li ●

280302　Paeonia suffruticosa Andrews subsp. spontanea (Rehder) S. G. Haw et Lauener = Paeonia jishanensis T. Hong et W. Z. Zhao ●

280303　Paeonia suffruticosa Andrews subsp. yinpingmudan D. Y. Hong, K. Y. Pan et Zhang W. Xie;银屏牡丹;Yinping Suffruticosa Peony ●

280304　Paeonia suffruticosa Andrews var. hiberniflora Makino;冬花牡丹●☆

280305　Paeonia suffruticosa Andrews var. humei Bailly = Paeonia suffruticosa Andrews ●■

280306　Paeonia suffruticosa Andrews var. jishanensis(T. Hong et W. Z. Zhao) Halda = Paeonia jishanensis T. Hong et W. Z. Zhao ●

280307　Paeonia suffruticosa Andrews var. papaveracea (Andrews) A. Kern. ;矮牡丹(粉牡丹,紫斑牡丹);Poppy Tree Peony ●

280308　Paeonia suffruticosa Andrews var. papaveracea (Andrews) A. Kern. = Paeonia suffruticosa Andrews ●■

280309　Paeonia suffruticosa Andrews var. purpurea Andrews = Paeonia suffruticosa Andrews ●■

280310　Paeonia suffruticosa Andrews var. qiui(Y. L. Pei et D. Y. Hong) Halda = Paeonia qiui Y. L. Pei et D. Y. Hong ●

280311　Paeonia suffruticosa Andrews var. rosea(Lodd.) Bailly = Paeonia suffruticosa Andrews ●■

280312　Paeonia suffruticosa Andrews var. roseosuperba Bailly = Paeonia suffruticosa Andrews ●■

280313　Paeonia suffruticosa Andrews var. rubugoplena Bailly = Paeonia suffruticosa Andrews ●■

280314　Paeonia suffruticosa Andrews var. spontanea Rehder = Paeonia jishanensis T. Hong et W. Z. Zhao ●

280315　Paeonia suffruticosa Andrews var. vittata(Van Houtte) Bailly = Paeonia suffruticosa Andrews ●■

280316　Paeonia suffruticosa subsp. atava(Brühl) S. G. Haw et Lauener = Paeonia rockii(S. G. Haw et Lauener) T. Hong et J. J. Li ●

280317　Paeonia szechuanica W. P. Fang = Paeonia decompsonia Hand. - Mazz. ●

280318　Paeonia tenuifolia L. ;细叶芍药(碎叶芍药,细裂芍药); Fennel-leaved Peony, Fine-leaved Peony, Fringed Peony, Slender-leaved Peony,Thinleaf Peony ■☆

280319　Paeonia tomentosa (Lomakin) N. Busch;茸毛芍药;Tomentose Peony ■☆

280320　Paeonia triternata Pall. ;三回三出芍药;Triternate Peony ■☆

280321　Paeonia trolioides Stapf ex Stern = Paeonia delavayi Franch. ●

280322　Paeonia veitchii Lynch = Paeonia anomala L. subsp. veitchii (Lynch) D. Y. Hong et K. Y. Pan ■

280323　Paeonia veitchii Lynch subsp. altaica(K. M. Dai et T. H. Ying) Halda = Paeonia anomala L. ■

280324　Paeonia veitchii Lynch var. beresowskii (Kom.) Schipcz. = Paeonia anomala L. subsp. veitchii(Lynch) D. Y. Hong et K. Y. Pan ■

280325　Paeonia veitchii Lynch var. leiocarpa W. T. Wang et S. H. Wang;光果赤芍(毛果川赤芍);Glabrousfruit Peony ■

280326　Paeonia veitchii Lynch var. leiocarpa W. T. Wang et S. H. Wang = Paeonia anomala L. subsp. veitchii(Lynch) D. Y. Hong et K. Y. Pan ■

280327　Paeonia veitchii Lynch var. leiocarpa W. T. Wang et S. H. Wang ex K. Y. Pan = Paeonia anomala L. subsp. veitchii (Lynch) D. Y. Hong et K. Y. Pan ■

280328　Paeonia veitchii Lynch var. uniflora K. Y. Pan;单花赤芍; Oneflower Peony ■

280329　Paeonia veitchii Lynch var. uniflora K. Y. Pan = Paeonia anomala L. subsp. veitchii(Lynch) D. Y. Hong et K. Y. Pan ■

280330　Paeonia veitchii Lynch var. woodwardii(Stapf ex Cox) Stern;毛赤芍(毛脉川芍药,毛芍药);Wood-ward Peony ■

280331　Paeonia veitchii Lynch var. woodwardii (Stapf ex Cox) Stern = Paeonia anomala L. subsp. veitchii(Lynch) D. Y. Hong et K. Y. Pan ■

280332　Paeonia vernalis Mandl;春芍药;Spring Peony ■☆

280333　Paeonia willmottiae Stapf = Paeonia obovata Maxim. subsp. willmottiae(Stapf) D. Y. Hong et K. Y. Pan ■

280334　Paeonia wittmanniana Hartwiss ex Lindl. ;紫丝芍药■☆

280335　Paeonia wittmanniana Lindl. = Paeonia obovata Maxim. ■

280336　Paeonia wittmanniana Lindl. = Paeonia wittmanniana Hartwiss ex Lindl. ■☆

280337　Paeonia woodwardii Stapf ex Cox = Paeonia anomala L. subsp. veitchii(Lynch) D. Y. Hong et K. Y. Pan ■

280338　Paeonia yananensis T. Hong et M. R. Li = Paeonia × papaveracea Andrews ●

280339 Paeonia yinpingmudan(D. Y. Hong,K. Y. Pan et Zhang W. Xie) B. A. Shen = Paeonia suffruticosa Andréws subsp. yinpingmudan D. Y. Hong,K. Y. Pan et Zhang W. Xie ●

280340 Paeonia yui W. P. Fang = Paeonia lactiflora Pall. ■

280341 Paeonia yunnanensis W. P. Fang = Paeonia suffruticosa Andréws ●■

280342 Paeoniaceae F. Rudolphi = Paeoniaceae Raf. (保留科名)■●

280343 Paeoniaceae Kunth = Ranunculaceae Juss. (保留科名)●■

280344 Paeoniaceae Raf. (1815)(保留科名);芍药科 ■●

280345 Paepalanthus Kunth(1834)(保留属名);四籽谷精草属 ■☆

280346 Paepalanthus Mart. = Paepalanthus Kunth(保留属名)■☆

280347 Paepalanthus affinis Kunth;近缘四籽谷精草;Country Pellitory,False Pellitory ■☆

280348 Paepalanthus flavidulus (Michx.) Kunth = Syngonanthus flavidulus(Michx.)Ruhland ■☆

280349 Paepalanthus hispidissimus Herzog = Paepalanthus pulvinatus N. E. Br. ■☆

280350 Paepalanthus kanaii Satake;金井四籽谷精草 ■☆

280351 Paepalanthus lamarckii Kunth;拉马克四籽谷精草 ■☆

280352 Paepalanthus poggeanus (Ruhland) Hess = Syngonanthus poggeanus Ruhland ■☆

280353 Paepalanthus pulvinatus N. E. Br. ;叶枕四籽谷精草 ■☆

280354 Paepalanthus schlechteri(Ruhland)J. F. Macbr. = Syngonanthus schlechteri Ruhland ■☆

280355 Paepalanthus sessilis Lecomte = Eriocaulon sessile Meikle ■☆

280356 Paepalanthus wahlbergii Wikstr. ex Körn. = Syngonanthus wahlbergii(Wikstr. ex Körn.)Ruhland ■☆

280357 Paepalanthus welwitschii Rendle = Syngonanthus welwitschii (Rendle)Ruhland ■☆

280358 Paeudobaecharis Cabrera = Baccharis L. (保留属名)●■☆

280359 Pagaea Griseb. = Irlbachia Mart. ■☆

280360 Pagamaeaceae Martinov = Rubiaceae Juss. (保留科名)●■

280361 Pagamea Aubl. (1775);帕加茜属 ●☆

280362 Pagamea coriacea Spruce ex Benth. ;帕加茜 ●☆

280363 Pagameopsis Steyerm. (1965);拟帕加茜属 ●☆

280364 Pagameopsis garryoides(Standl.)Steyerm. ;拟帕加茜 ●☆

280365 Pagapate Sonn. = Sonneratia L. f. (保留属名)●

280366 Pagella Schönl. = Crassula L. ●■☆

280367 Pagerea Pierre ex Laness. = Sageraea Dalzell ●☆

280368 Pageria Juss. = Lapageria Ruiz et Pav. ●☆

280369 Pageria Raf. = Pagesia Raf. ■☆

280370 Pagesia Raf. = Mecardonia Ruiz et Pav. ■☆

280371 Pagetia F. Muell. = Bosistoa F. Muell. ex Benth. ●☆

280372 Pagiantha Markgr. (1935);巴基山马茶属(圆头花属)●☆

280373 Pagiantha Markgr. = Tabernaemontana L. ●

280374 Pagiantha cerifera(Pancher et Sebert)Markgr. ;巴基山马茶(蜡圆头花)●☆

280375 Pagiantha corymbosa (Roxb. ex Wall.) Markgr. = Tabernaemontana corymbosa Roxb. ex Wall. ●

280376 Pagiantha dichotoma (Roxb.) Markgr. = Rejoua dichotoma (Roxb.)Gamble ●

280377 Pagiantha macrocarpa (Jack) Markgr. = Ervatamia macrocarpa (Jack)Merr. ●

280378 Pagiantha pandacaqui (Lam.) Markgr. = Tabernaemontana pandacaqui Lam. ●

280379 Pahudia Miq. = Afzelia Sm. (保留属名)●

280380 Pahudia africana(Sm. ex Pers.)Prain = Afzelia africana Sm. ex Pers. ●☆

280381 Pahudia attenuata(Klotzsch)Prain = Afzelia quanzensis Welw. ●☆

280382 Pahudia bequaertii (De Wild.) Dewit = Afzelia bipindensis Harms ●☆

280383 Pahudia bracteata (Vogel ex Benth.) Prain = Afzelia parviflora (Vahl)Hepper ●☆

280384 Pahudia brieyi(De Wild.)Dewit = Afzelia pachyloba Harms ●☆

280385 Pahudia cochinchinensisi Pierre = Afzelia xylocarpa(Kurz)Craib ●

280386 Pahudia quanzensis(Welw.)Prain = Afzelia quanzensis Welw. ●☆

280387 Pahudia rhomboidea Prain = Afzelia rhomboidea S. Vidal ●

280388 Pahudia xylocarpa Kurz = Afzelia xylocarpa(Kurz)Craib ●

280389 Paillotia Gand. = Erodium L'Hér. ex Aiton ■●

280390 Painteria Britton et Rose = Havardia Small ●☆

280391 Paiva Vell. = Sabicea Aubl. ●☆

280392 Paivaea O. Berg = Campomanesia Ruiz et Pav. ●☆

280393 Paivaea Post et Kuntze = Paiva Vell. ●☆

280394 Paivaea Post et Kuntze = Sabicea Aubl. ●☆

280395 Paivaeusa Welw. = Oldfieldia Benth. et Hook. f. ●☆

280396 Paivaeusa dactylophylla Welw. ex Oliv. = Oldfieldia dactylophylla (Welw. ex Oliv.)J. Léonard ●☆

280397 Paivaeusa orientalis Mildbr. = Oldfieldia somalensis (Chiov.) Milne-Redh. ●☆

280398 Paivaeusa Welwitsch ex Benth. et Hook. f. = Oldfieldia Benth. et Hook. f. ●☆

280399 Paivaeusaceae A. Meeuse = Euphorbiaceae Juss. (保留科名)●■

280400 Paivaeusaceae A. Meeuse = Picrodendraceae Small(保留科名)●☆

280401 Pajanelia DC. (1838);帕亚木属 ●☆

280402 Pajanelia longifolia(Willd.)Schum. ;帕亚木 ●☆

280403 Pakaraimaea Maguire et P. S. Ashton(1977);美洲龙脑香属 ●☆

280404 Pakaraimaea dipterocarpacea Maguire et P. S. Ashton;美洲龙脑香 ●☆

280405 Pala Juss. = Alstonia R. Br. (保留属名)●

280406 Pala scholaris(L.)Roberty = Alstonia congensis Engl. ●☆

280407 Pala scholaris(L.)Roberty = Alstonia scholaris(L.)R. Br. ●

280408 Pala scholaris Roberty = Pala scholaris(L.)Roberty ●

280409 Paladelpha Pichon = Alstonia R. Br. (保留属名)●

280410 Palaeconringia E. H. L. Krause = Erysimum L. ■●

280411 Palaeno Raf. = Campanula L. ■●

280412 Palaeocyanus Dostal = Centaurea L. (保留属名)●■

280413 Palafoxia Lag. (1816);对粉菊属;Palafoxia ■☆

280414 Palafoxia arida B. L. Turner et M. I. Morris;旱地对粉菊 ■☆

280415 Palafoxia arida B. L. Turner et M. I. Morris var. gigantea(M. E. Jones)B. L. Turner et M. I. Morris = Palafoxia arida B. L. Turner et M. I. Morris ■☆

280416 Palafoxia callosa (Nutt.) Torr. et A. Gray;西班牙对粉菊;Spanish Needles ■☆

280417 Palafoxia feayi A. Gray;费氏对粉菊 ■☆

280418 Palafoxia hookeriana Torr. et A. Gray;胡克对粉菊 ■☆

280419 Palafoxia integrifolia(Nutt.)Torr. et A. Gray;全缘叶对粉菊 ■☆

280420 Palafoxia reverchonii(Bush)Cory;勒韦雄对粉菊 ■☆

280421 Palafoxia riograndensis Cory;格朗德对粉菊 ■☆

280422 Palafoxia rosea(Bush)Cory;蔷薇对粉菊 ■☆

280423 Palafoxia rosea(Bush)Cory var. macrolepis(Rydb.)B. L. Turner et M. I. Morris = Palafoxia rosea(Bush)Cory ■☆

280424 Palafoxia sphacelata(Nutt. ex Torr.)Cory;暗点对粉菊 ■☆

280425 Palafoxia texana DC. ;得州对粉菊 ■☆

280426 Palafoxia texana DC. var. ambigua(Shinners)B. L. Turner et M. I. Morris = Palafoxia texana DC. ■☆

280427　Palala Kuntze = Myristica Gronov.（保留属名）●

280428　Palala Rumph. = Horsfieldia Willd. ●

280429　Palala Rumph. ex Kuntze = Myristica Gronov.（保留属名）●

280430　Palamostigma Benth. et Hook. f. = Croton L. ●

280431　Palamostigma Benth. et Hook. f. = Palanostigma Mart. ex Klotzsch ●

280432　Palandra O. F. Cook = Phytelephas Ruiz et Pav. ●☆

280433　Palandra O. F. Cook.（1927）;美柱椰属（拔兰抓属,帕兰德拉象牙椰属,帕兰象牙椰属）●☆

280434　Palandra aequatorialis O. F. Cook;帕兰美柱椰（帕兰象牙椰）; Vegetable Ivory ●☆

280435　Palanostigma Mart. ex Klotzsch = Croton L. ●

280436　Palaoea Kaneh. = Tristiropsis Radlk. ●☆

280437　Palaquium Blanco（1837）;胶木属（大叶山榄属）; Gutta Percha, Nato Tree, Natotree, Palaktree ●

280438　Palaquium amboinense Burck;安汶胶木●☆

280439　Palaquium ellipticum Engl. ;椭圆胶木●☆

280440　Palaquium formosanum Hayata;台湾胶木（大叶山榄,千仔树,马古公,椭圆叶胶木）; Formosa Nato Tree, Formosa Natotree, Hayata Natotree, Taiwan Natotree, Taiwan Palaktree ●

280441　Palaquium gutta（Hook. ）Baill. ;胶木; Gutta-percha Tree, Malay Gutteperch Natotree, Malay Guttapercha Nato Tree, Malay Guttapercha Nato-tree, Malay Palaktree, Taban Merah ●☆

280442　Palaquium hayatae Lam. = Palaquium formosanum Hayata ●

280443　Palaquium hexandrum King et Gamble;六雄蕊胶木●☆

280444　Palaquium javense Burck;爪哇胶木●☆

280445　Palaquium lanceolatum Blanco;披针叶胶木; Lanceolate Palaktree ●☆

280446　Palaquium oblongifolium Burck;马来胶木●☆

280447　Palaquium obovatum Engl. ;倒卵胶木●☆

280448　Palaquium philippense（Pers. ）C. B. Rob. ;菲律宾胶木; Philippine Nato Tree ●☆

280449　Palaquium polyandrum Hayata = Palaquium formosanum Hayata ●

280450　Palaquium rostratum Burck;曲胶木●☆

280451　Palaua Cav.（1785）;帕劳锦葵属■☆

280452　Palaua Ruiz et Pav. = Apatelia DC. ●

280453　Palaua Ruiz et Pav. = Saurauia Willd.（保留属名）●

280454　Palaua concinna I. M. Johnst. ;灰帕劳锦葵■☆

280455　Palaua micrantha Ulbr. ;小花帕劳锦葵■☆

280456　Palaua tomentosa Hochr. ;毛帕劳锦葵■☆

280457　Palaua velutina Ulbr. ;黏帕劳锦葵■☆

280458　Palava Juss. = Palaua Cav. ■☆

280459　Palava Pers. = Palaua Ruiz et Pav. ●

280460　Palava Pers. = Saurauia Willd.（保留属名）●

280461　Palavia Ruiz et Pav. ex Ortega = Calyxhymenia Ortega ■

280462　Palavia Ruiz et Pav. ex Ortega = Mirabilis L. ■

280463　Palavia Schreb. = Palaua Cav. ■☆

280464　Paleaepappus Cabrera（1969）;叉枝菀属●☆

280465　Paleaepappus patagonicus Cabrera;叉枝菀■☆

280466　Paleista Raf. = Eclipta L.（保留属名）■

280467　Palenga Thwaites = Drypetes Vahl ●

280468　Palenga Thwaites = Putranjiva Wall. ●

280469　Palenia Phil. = Heterothalamus Less. ●☆

280470　Paleodicraeia C. Cusset（1973）;糠叉苔草属■☆

280471　Paleodicraeia imbricata（Tul. ）C. Cusset;糠叉苔草■☆

280472　Paleolaria Cass. = Palafoxia Lag. ■☆

280473　Paletuviera Thouars ex DC. = Bruguiera Sav. ●

280474　Paletuvieraceae Lam. ex Kuntze = Rhizophoraceae Pers.（保留科名）●

280475　Paleya Cass. = Crepis L. ■

280476　Palgianthus G. Forst. ex Baill. = Plagianthus J. R. Forst. et G. Forst. ●☆

280477　Paliavana Vand. = Paliavana Vell. ex Vand. ●☆

280478　Paliavana Vell. ex Vand.（1788）;帕里苣苔属（帕拉瓦苣苔属）;Paliavana ●☆

280479　Paliavana gracilis（Mart. ）Chautems;细帕里苣苔●☆

280480　Paliavana lasiantha Wiehler;毛花帕里苣苔●☆

280481　Palicourea Aubl.（1775）;巴茜草属（巴西茜属,帕立茜草属）;Palicourea ●☆

280482　Palicourea densiflora Mart. ;密花巴茜草（密花巴西茜）●☆

280483　Palicourea fendleri Standl. ;芬氏巴茜草（芬氏巴西茜）; Fendler Palicourea ●☆

280484　Palicourea grandiflora（Kunth）Sdandl. ;大花巴茜草（大花巴西茜）●☆

280485　Palicourea guianensis Aubl. ;圭亚那巴茜草（圭亚那帕立茜）●☆

280486　Palicourea rigida Kunth;坚挺巴茜草（坚挺巴西茜）●☆

280487　Palicuria Raf. = Palicourea Aubl. ●☆

280488　Palilia Allam. ex L. = Heliconia L.（保留属名）■

280489　Palimbia Besser = Palimbia Besser ex DC. ■

280490　Palimbia Besser ex DC.（1830）;额尔齐斯芹属■

280491　Palimbia rediviva（Pall. ）Thell. ;额尔齐斯芹■☆

280492　Palina japonica Herm. = Cycas revoluta Thunb. ●◇

280493　Palindan Blanco ex Post et Kuntze = Orania Zipp. ●☆

280494　Palinetes Salisb. = Ammocharis Herb. ■☆

280495　Palinetex Salisb. = Ammocharis Herb. ■☆

280496　Paliris Dumort. = Liparis Rich.（保留属名）■

280497　Palisota Rchb. = Palisota Rchb. ex Endl.（保留属名）■☆

280498　Palisota Rchb. ex Endl.（1836）（保留属名）;浆果鸭跖草属■☆

280499　Palisota alopecurus Pellegr. ;看麦娘浆果鸭跖草■☆

280500　Palisota ambigua（P. Beauv. ）C. B. Clarke;可疑浆果鸭跖草■☆

280501　Palisota barteri Hook. ;巴特浆果鸭跖草■☆

280502　Palisota bicolor Mast. ;双色浆果鸭跖草■☆

280503　Palisota bogneri Brenan;博格纳鸭跖草■☆

280504　Palisota brachythyrsa Mildbr. ;短序鸭跖草■☆

280505　Palisota bracteosa C. B. Clarke;多苞片浆果鸭跖草■☆

280506　Palisota caillei A. Chev. ;卡耶鸭跖草■☆

280507　Palisota congolana Hua = Palisota ambigua（P. Beauv. ）C. B. Clarke ■☆

280508　Palisota flagelliflora Faden;鞭花鸭跖草■☆

280509　Palisota gracilior Mildbr. ;纤细鸭跖草■☆

280510　Palisota hirsuta（Thunb. ）K. Schum. ;粗毛鸭跖草■☆

280511　Palisota lagopus Mildbr. ;兔足鸭跖草■☆

280512　Palisota laurentii De Wild. ;洛朗鸭跖草■☆

280513　Palisota laxiflora C. B. Clarke;疏花浆果鸭跖草■☆

280514　Palisota maclaudii Cornu = Palisota hirsuta（Thunb. ）K. Schum. ■☆

280515　Palisota mannii C. B. Clarke;曼氏浆果鸭跖草■☆

280516　Palisota mannii C. B. Clarke subsp. megalophylla（Mildbr. ）Faden;大叶曼氏浆果鸭跖草■☆

280517　Palisota megalophylla Mildbr. = Palisota mannii C. B. Clarke subsp. megalophylla（Mildbr. ）Faden ■☆

280518　Palisota megalophylla Mildbr. var. robusta? = Palisota mannii C. B. Clarke subsp. megalophylla（Mildbr. ）Faden ■☆

280519　Palisota micrantha K. Schum. ex C. B. Clarke = Palisota ambigua（P. Beauv. ）C. B. Clarke ■☆

280520　Palisota myriantha K. Schum.;多花浆果鸭跖草■☆

280521　Palisota ombrophila K. Schum. ex C. B. Clarke;喜雨浆果鸭跖草■☆

280522　Palisota orientalis K. Schum.;东方浆果鸭跖草■☆

280523　Palisota pedicellata K. Schum.;梗花浆果鸭跖草■☆

280524　Palisota plagiocarpa Hua = Palisota ambigua (P. Beauv.) C. B. Clarke ■☆

280525　Palisota preussiana K. Schum. ex C. B. Clarke;普罗伊斯浆果鸭跖草■☆

280526　Palisota prionostachys C. B. Clarke = Palisota ambigua (P. Beauv.) C. B. Clarke ■☆

280527　Palisota pseudoambigua A. Chev.;疑似浆果鸭跖草■☆

280528　Palisota pynaertii De Wild.;皮那浆果鸭跖草■☆

280529　Palisota schweinfurthii C. B. Clarke;施韦浆果鸭跖草■☆

280530　Palisota staudtii K. Schum.;施陶浆果鸭跖草■☆

280531　Palisota thollonii Hua;托伦浆果鸭跖草■☆

280532　Palisota thyrsiflora Benth. = Palisota ambigua(P. Beauv.) C. B. Clarke ■☆

280533　Palisota thyrsostachya Mildbr.;聚伞浆果鸭跖草■☆

280534　Palisota waibelii Mildbr.;魏贝尔浆果鸭跖草■☆

280535　Palissya Baill. = Necepsia Prain ☆

280536　Palissya Baill. = Neopalissya Pax ☆

280537　Paliuros St. -Lag. = Paliurus Tourn. ex Mill. ●

280538　Paliurus Mill. = Paliurus Tourn. ex Mill. ●

280539　Paliurus Tourn. ex Mill. (1754);马甲子属（铜钱树属）; Cointree,Jerusale Thorn,Paliurus ●

280540　Paliurus aculeatus Lam. = Paliurus australis Gaertn. ●

280541　Paliurus aculeatus Lam. = Paliurus spina-christi Mill. ●

280542　Paliurus aubletia Roem. et Schult. = Paliurus ramosissimus (Lour.) Poir. ●

280543　Paliurus aubletia Schult. = Paliurus ramosissimus(Lour.)Poir. ●

280544　Paliurus australis Gaertn.;刺马甲子●

280545　Paliurus australis Gaertn. = Paliurus hemsleyanus Rehder ●

280546　Paliurus australis Gaertn. = Paliurus spina-christi Mill. ●

280547　Paliurus australis Gaertn. var. orientalis Franch. = Paliurus orientalis(Franch.) Hemsl. ●

280548　Paliurus hemsleyanus Rehder;铜钱树(串树,刺凉子,金钱木,金钱树,马鞍,马鞍秋,鸟不宿,钱串树,摇钱树); Chinese Paliurus,China Cointree ●

280549　Paliurus hirsutus Hemsl.;硬毛马甲子（长梗铜钱树,钩交刺）; Hirsute Cointree, Hirsute Paliurus ●

280550　Paliurus hirsutus Hemsl. var. trichocarpus C. Z. Gao = Paliurus hirsutus Hemsl. ●

280551　Paliurus mairei H. Lév. = Ziziphus mauritiana Lam. ●

280552　Paliurus orientalis(Franch.)Hemsl.;短柄铜钱树（刺楸树,蒙自刺,蒙自铜钱树）; Shortstalk Cointree, Shortstalk Paliurus, Short-stalked Paliurus ●

280553　Paliurus orientalis (Franch.) Hemsl. = Paliurus hemsleyanus Rehder ●

280554　Paliurus perforatus Blanco = Harrisonia perforata(Blanco)Merr. ●

280555　Paliurus ramosissimus(Lour.)Poir.;马甲子(白棘,刺盘子,狗骨芳,棘刺,棘盘子,棘针,芳仔,簕盘子,簕子,萝荸,马鞍刺,牛角刺,企头簕,石刺木,铁篱笆,铜钱树,乌刺仔,雄虎刺); Branchy Cointree, Branchy Paliurus,Thorny Wingnut ●

280556　Paliurus ramosissimus (Lour.) Poir. var. japonica? = Paliurus ramosissimus(Lour.)Poir. ●

280557　Paliurus sinicus C. K. Schneid. = Paliurus orientalis (Franch.) Hemsl. ●

280558　Paliurus sinicus Schneid. = Paliurus orientalis (Franch.) Hemsl. ●

280559　Paliurus spina-christi Mill.;滨枣（刺马甲子）; Christ's Thorn, Christ's-thorn, Christ's-thorn Paliurns, Christthorn Cointree, Christthorn Paliurus,Christ-thorns,Chritton Paliurus, Garland Thorn, Jerusalem Thorn ●

280560　Paliurus tonkinensis Pit. = Paliurus hirsutus Hemsl. ●

280561　Paliurus virgatus D. Don = Paliurus spina-christi Mill. ●

280562　Palladia Lam. = Blakwellia Gaertn. ●■

280563　Palladia Moench = Lysimachia L. ●■

280564　Pallasia Houtt. = Calodendrum Thunb. (保留属名)●☆

280565　Pallasia Klotzsch = Wittmackanthus Kuntze ●☆

280566　Pallasia L. f. = Calligonum L. ●

280567　Pallasia L. f. = Pterococcus Pall. (废弃属名)●☆

280568　Pallasia L'Hér. = Encelia Adans. ●■☆

280569　Pallasia L'Hér. ex Aiton = Crypsis Aiton(保留属名)■

280570　Pallasia Scop. = Crypsis Aiton(保留属名)■

280571　Pallassia Houtt. (废弃属名) = Calodendrum Thunb. (保留属名)●☆

280572　Pallastema Salisb. = Albuca L. ■☆

280573　Pallavia Vell. = Pisonia L. ●

280574　Pallavicinia Cocc. = Alliaria Heist. ex Fabr. ■

280575　Pallavicinia De Not. = Cyphomandra Mart. ex Sendtn. ●■

280576　Pallenis(Cass.) Cass. = Asteriscus Mill. ●■☆

280577　Pallenis(Cass.) Cass. = Pallenis Cass. (保留属名)■●☆

280578　Pallenis Cass. (1822)(保留属名);苍菊属（叶苞菊属）■●☆

280579　Pallenis aurea (Willk.) DC. = Pallenis spinosa (L.) Cass. subsp. aurea(Willk.) Nyman ■☆

280580　Pallenis cuspidata Pomel;骤尖苍菊■☆

280581　Pallenis cuspidata Pomel subsp. canescens(Maire)Greuter;灰尖凸苍菊■☆

280582　Pallenis cyrenaica Alavi;昔兰尼苍菊■☆

280583　Pallenis hierochuntica(Michon)Greuter;耶路苍菊■☆

280584　Pallenis maritima(L.)Greuter;沼泽苍菊（沼泽北非菊）; Gold Coin, Golden Dollar, Mediterranean Beach Daisy ■☆

280585　Pallenis spinosa(L.)Cass. ;多刺苍菊■☆

280586　Pallenis spinosa (L.) Cass. subsp. asteroidea (Viv.) Greuter;星形多刺苍菊■☆

280587　Pallenis spinosa (L.) Cass. subsp. aurea (Willk.) Nyman;黄多刺苍菊■☆

280588　Pallenis spinosa (L.) Cass. subsp. cuspidata (Pomel) Batt. = Pallenis cuspidata Pomel ■☆

280589　Pallenis spinosa (L.) Cass. subsp. maroccana (Aurich et Podlech)Greuter;摩洛哥多刺苍菊■☆

280590　Pallenis spinosa (L.) Cass. var. asteroidea (Viv.) Asch. = Pallenis spinosa(L.)Cass. subsp. asteroidea(Viv.)Greuter ■☆

280591　Pallenis spinosa (L.) Cass. var. aurea Batt. = Pallenis spinosa (L.)Cass. subsp. aurea(Willk.) Nyman ■☆

280592　Pallenis spinosa (L.) Cass. var. canescens Maire = Pallenis cuspidata Pomel subsp. canescens(Maire)Greuter ■☆

280593　Pallenis spinosa (L.) Cass. var. crocea Webb et Heldr. = Pallenis spinosa(L.)Cass. ■☆

280594　Pallenis spinosa (L.) Cass. var. cuspidata (Pomel) Hochr. = Pallenis spinosa(L.)Cass. subsp. aurea(Willk.) Nyman ■☆

280595　Pallenis spinosa (L.) Cass. var. eriophora Burnat = Pallenis spinosa(L.)Cass. ■☆

280596 Pallenis spinosa (L.) Cass. var. pallida Bornm. = Pallenis spinosa(L.) Cass. ■☆

280597 Pallenis spinosa (L.) Cass. var. straminea Maire et Weiller et Wilczek = Pallenis spinosa(L.) Cass. ■☆

280598 Pallenis spinosa (L.) Cass. var. sulfurea Braun-Blanq. et Maire = Pallenis spinosa(L.) Cass. ■☆

280599 Pallenis teknensis(Dobignard et Jacquemoud) Greuter et Jury; 特克纳纳苍菊■☆

280600 Pallenis vietnamensis Ha et Grushv. ;越南苍菊(越南人参)■☆

280601 Palma Mill. = Phoenix L. ●

280602 Palma Plum. ex Mill. = Phoenix L. ●

280603 Palma argentata Jacq. = Coccothrinax argentata (Jacq.) L. H. Bailey ●☆

280604 Palma elata W. Bartram = Roystonea regia(Kunth) O. F. Cook ●

280605 Palma prunifera Mill. = Copernicia prunifera (Mill.) H. E. Moore ●☆

280606 Palmae Adans. = Arecaceae Bercht. et J. Presl(保留科名)●

280607 Palmae Adans. = Palmae Juss. (保留科名)●

280608 Palmae Juss. (1789)(保留科名);棕榈科(槟榔科);Palm Family ●

280609 Palmae Juss. (保留科名) = Arecaceae Bercht. et J. Presl(保留科名)●

280610 Palma-filix Adans. (废弃属名) = Zamia L. ●☆

280611 Palmangis Thouars = Angraecum Bory ■

280612 Palmerella A. Gray = Laurentia Adans. ■☆

280613 Palmerella A. Gray(1876);帕尔桔梗属■☆

280614 Palmerella debilis A. Gray;帕尔桔梗■☆

280615 Palmeria F. Muell. (1864);藤桂属●☆

280616 Palmeria scandens F. Muell. ;藤桂●☆

280617 Palmeroandenbrockia Gibbs = Polyscias J. R. Forst. et G. Forst. ●

280618 Palmerocassia Britton = Senna Mill. ●■

280619 Palmeroeassla Britton = Cassia L. (保留属名)●■

280620 Palmervandenbroeckia L. S. Gibbs. = Polyscias J. R. Forst. et G. Forst. ●

280621 Palmervandenbroekia L. S. Gibbs. = Polyscias J. R. Forst. et G. Forst. ●

280622 Palmia Endl. = Hewittia Wight et Arn. ■

280623 Palmia Endl. = Shutereia Choisy(废弃属名)■

280624 Palmifolia Kuntze = Palma-Filix Adans. (废弃属名)●☆

280625 Palmifolia Kuntze = Zamia L. ●☆

280626 Palmijuncus Kuntze = Calamus L. ●

280627 Palmijuncus Rumph. = Calamus L. ●

280628 Palmijuncus Rumph. ex Kuntze = Calamus L. ●

280629 Palmofilix Post et Kuntze = Palma-Filix Adans. (废弃属名)●☆

280630 Palmofilix Post et Kuntze = Zamia L. ●☆

280631 Palmoglossum Klotzsch ex Rchb. f. = Pleurothallis R. Br. ■☆

280632 Palmolmedia Ducke = Naucleopsis Miq. ●☆

280633 Palmonaria Boiss. = Pulmonaria L. ■

280634 Palmorchis Barb. Rodr. (1877);掌兰属■☆

280635 Palmorchis pubescentis Barb. Rodr. ;掌兰■☆

280636 Palmstruckia Retz. (废弃属名) = Chaenostoma Benth. (保留属名)■☆

280637 Palmstruckia Retz. (废弃属名) = Sutera Roth ■●☆

280638 Palmstruckia Retz. f. = Palmstruckia Retz. (废弃属名)■☆

280639 Palmstruckia Sond. = Thlaspeocarpa C. A. Sm. ■☆

280640 Palmstruckia capensis Sond. = Heliophila suborbicularis Al-Shehbaz et Mummenhoff ■☆

280641 Palmstruckia foetida(Andréws) Retz. = Sutera foetida Roth ■☆

280642 Paloue Aubl. (1775);帕洛豆属■☆

280643 Paloue brasiliensis Ducke;巴西帕洛豆■☆

280644 Palovea Aubl. = Paloue Aubl. ■☆

280645 Palovea Juss. = Paloue Aubl. ■☆

280646 Palovea Raf. = Sabicea Aubl. ●☆

280647 Paloveopsis R. S. Cowan(1957);凹叶豆属■☆

280648 Paloveopsis emarginata R. S. Cowan;凹叶豆■☆

280649 Paltoria Ruiz et Pav. = Ilex L. ●

280650 Paludana Giseke(废弃属名) = Amomum Roxb. (保留属名)■

280651 Paludana Salisb. = Monocaryum(R. Br.) Rchb. ■

280652 Paludana Salisb. = Paludaria Salisb. ■

280653 Paludaria Salisb. = Colchicum L. ■

280654 Paluea Post et Kuntze = Paloue Aubl. ■☆

280655 Palumbina Rchb. f. (1863);洁兰属■☆

280656 Palumbina candida Rchb. f. ;洁兰■☆

280657 Palura(G. Don) Buch. -Ham. ex Miers = Symplocos Jacq. ●

280658 Palura(G. Don) Miers = Symplocos Jacq. ●

280659 Palura Buch. -Ham. ex D. Don = Symplocos Jacq. ●

280660 Palura Buch. -Ham. ex Miers = Symplocos Jacq. ●

280661 Palura argutidens Nakai = Symplocos coreana(H. Lév.) Ohwi ●☆

280662 Palura chinensis(Lour.) Koidz. = Symplocos chinensis(Lour.) Druce ●

280663 Palura chinensis (Lour.) Koidz. = Symplocos paniculata (Thunb.) Miq. ●

280664 Palura chinensis(Lour.) Koidz. = Symplocos sawafutagi Nagam. ●

280665 Palura chinensis (Lour.) Koidz. var. leucocarpa? = Symplocos sawafutagi Nagam. ●

280666 Palura chinensis (Lour.) Koidz. var. pilosa (Nakai) Nakai = Symplocos sawafutagi Nagam. ●

280667 Palura chinensis (Lour.) Koidz. var. pilosa (Nakai) Nakai f. leucocarpa? = Symplocos sawafutagi Nagam. ●

280668 Palura chinensis (Lour.) Koidz. var. pilosa Nakai = Symplocos paniculata(Thunb.) Miq. ●

280669 Palura chinensis Koidz. = Symplocos paniculata(Thunb.) Miq. ●

280670 Palura chinensis Koidz. var. pilosa Nakai = Symplocos paniculata (Thunb.) Miq. ●

280671 Palura chinensis Koidz. var. pubescens? = Symplocos tanakae Matsum. ●☆

280672 Palura ciliata? = Symplocos sawafutagi Nagam. ●

280673 Palura coreana(H. Lév.) Nakai = Symplocos coreana(H. Lév.) Ohwi ●☆

280674 Palura odorata Buch. -Ham. ex D. Don = Symplocos paniculata (Thunb.) Miq. ●

280675 Palura paniculata Nakai = Symplocos paniculata(Thunb.) Miq. ●

280676 Palura paniculata Nakai var. glabra? = Symplocos paniculata (Thunb.) Miq. ●

280677 Palura paniculata Nakai var. leucocarpa? = Symplocos sawafutagi Nagam. ●

280678 Palura paniculata Nakai var. pallida? = Symplocos paniculata (Thunb.) Miq. ●

280679 Palura paniculata Nakai var. pilosa? = Symplocos sawafutagi Nagam. ●

280680 Palura paniculata Nakai var. pubescens? = Symplocos tanakae Matsum. ●☆

280681 Palura sinica Miers = Symplocos paniculata(Thunb.) Miq. ●

280682 Palura tanakana Nakai = Symplocos tanakae Matsum. ●☆

280683　Pamburus Swingle(1916)；全尾木属（攀布鲁木属，攀布鲁属）●☆

280684　Pamburus missionis Swingle；全尾木●☆

280685　Pamea Aubl.（废弃属名）= Buchenavia Eichler（保留属名）●☆

280686　Pamea Aubl.（废弃属名）= Terminalia L.（保留属名）●

280687　Pamianthe Stapf(1933)；鳌瓣花属■☆

280688　Pamianthe peruviana Stapf；鳌瓣花■☆

280689　Pamphalea DC. = Panphalea Lag. ■☆

280690　Pamphalea heterophylla Less.；互叶纤细钝柱菊■☆

280691　Pamphalea maxima Less.；大纤细钝柱菊■☆

280692　Pamphalea smithii Cabrera；史密斯纤细钝柱菊■☆

280693　Pamphilia Mart. = Pamphilia Mart. ex A. DC. ●☆

280694　Pamphilia Mart. ex A. DC.(1844)；全喜香属●☆

280695　Pamphilia aurea Mart.；黄全喜香●☆

280696　Pamphilia styracifolia A. DC.；全喜香●☆

280697　Pamplethantha Bremek.(1940)；全花茜属●☆

280698　Pamplethantha gilletii（De Wild. et T. Durand）Bremek.；吉氏全花茜●☆

280699　Pamplethantha gilletii (De Wild. et T. Durand) Bremek. = Pauridiantha viridiflora(Hiern)Hepper ●☆

280700　Pamplethantha verticillata(De Wild. et T. Durand)Bremek.；全花茜●☆

280701　Pamplethantha verticillata (De Wild. et T. Durand) Bremek. = Pauridiantha verticillata(De Wild. et T. Durand) N. Hallé ●☆

280702　Pamplethantha viridiflora (Hiern) Bremek. = Pauridiantha viridiflora(Hiern)Hepper ●☆

280703　Panacea Mitch. = Panax L. ■

280704　Panactia Cass. = Podolepis Labill.（保留属名）■☆

280705　Panamanthus Kuijt(1991)；巴拿马桑寄生属●☆

280706　Panamanthus panamensis(Rizz.)Kuijt；巴拿马桑寄生●☆

280707　Panargyrum D. Don = Panargyrus Lag. ●☆

280708　Panargyrus Lag. = Nassauvia Comm. ex Juss. ●☆

280709　Panarica Withner et P. A. Harding = Epidendrum L.（保留属名）■☆

280710　Panarica Withner et P. A. Harding(2004)；中美柱瓣兰属■☆

280711　Panax Hill = Opopanax W. D. J. Koch ■☆

280712　Panax L.(1753)；人参属；Ginsen，Ginseng ■

280713　Panax L. = Polyscias J. R. Forst. et G. Forst. ●

280714　Panax aculeatus Aiton = Acanthopanax trifoliatus(L.)Merr. ●

280715　Panax aculeatus Aiton = Eleutherococcus trifoliatus(L.)S. Y. Hu ●

280716　Panax amplifolium Baker = Polyscias amplifolia(Baker)Harms ●☆

280717　Panax armatus Wall. = Aralia armata(Wall. ex G. Don)Seem. ●

280718　Panax armatus Wall. ex G. Don = Aralia armata (Wall. ex G. Don)Seem. ●

280719　Panax atropurpureus (Franch.) Hand. -Mazz. = Aralia atropurpurea Franch. ■

280720　Panax bakeriana Drake = Polyscias multibracteata(Baker)Harms ●☆

280721　Panax bijugus Wall. ex G. Don = Pentapanax fragrans(D. Don) T. D. Ha ●

280722　Panax bipinnatifidus Seem.；疙瘩七（白地瓜，大金线吊葫芦，复羽裂参，花叶三七，黄连三七，鸡腰参，纽子三七，钮三七，钮子七，钮子三七，羽叶七，羽叶三七，羽叶竹节参，珠儿参，珠子参，竹根七）；Featherleaf Sanqi ■

280723　Panax bipinnatifidus Seem. = pseudoginsengvar. bipinnatifidus (Seem.)C. Y. Wu et K. M. Feng ex C. Chou ■

280724　Panax boivinii (Seem.) Benth. et Hook. f. = Polyscias boivinii (Seem.)Bernardi ●☆

280725　Panax cephalobotrys F. Muell. = Cephalaralia cephalobotrys(F. Muell.)Harms ●☆

280726　Panax chapelieri Drake = Polyscias chapelieri(Drake)Harms ex R. Vig. ●☆

280727　Panax confertifolius Baker = Polyscias confertifolia (Baker) Harms ●☆

280728　Panax cussonioides Drake = Polyscias cussonioides (Drake) Bernardi ●

280729　Panax davidii Franch. = Nothopanax davidii(Franch.) Harms ex Diels ●

280730　Panax delavayi Franch. = Nothopanax delavayi(Franch.) Harms ex Diels ●

280731　Panax divaricatus Siebold et Zucc. = Acanthopanax divaricatus (Siebold et Zucc.)Seem. ●☆

280732　Panax ferrugineus Hiern = Polyscias fulva(Hiern)Harms ●☆

280733　Panax finlaysonianus Wall. ex G. Don = Aralia finlaysoniana (Wall. ex G. Don)Seem. ●

280734　Panax floccosa Drake = Polyscias floccosa(Drake)Bernardi ●☆

280735　Panax foliolosus Wall. = Aralia foliolosa(Wall.)Seem. ●

280736　Panax fragrans Roxb. = Heteropanax fragrans(Roxb.)Seem. ●

280737　Panax fragrans Roxb. ex DC. = Heteropanax fragrans (Roxb. ex DC.)Seem. ●

280738　Panax fraxinifolius(Baker)Drake = Polyscias fraxinifolia Harms ●☆

280739　Panax fruticosus L. = Polyscias fruticosa(L.)Harms ●

280740　Panax fulvus Hiern = Polyscias fulva(Hiern)Harms ●☆

280741　Panax ginseng C. A. Mey.；人参（白干参，白杆参，白物，白修参，百尺杵，百济参，棒棰，棒槌，棒锤，参，朝鲜参，朝鲜人参，大力参，地精，东洋参，高丽参，鬼盖，孩儿参，海腴，黄参，黄石，吉林参，金井玉阑，力参，楝参，辽参，辽东参，人身，人葠，人薓，人微，人衔，人御，三七，神草，汤参，土精，血参，羊角参，药用人参，野参，野山参，玉精，御种人参，皱面还丹，子参，紫团参）；Asia Ginsen，Asiatic Ginsen，Asiatic Ginseng，Ginsen ■

280742　Panax gomphophylla Baker = Polyscias amplifolia(Baker)Harms ●☆

280743　Panax grevei Drake = Polyscias boivinii(Seem.)Bernardi ●☆

280744　Panax hildebrandtii Drake = Polyscias nossibensis(Drake)Harms ●☆

280745　Panax horridus Sm. = Oplopanax horridus(Sm.)Miq. ●☆

280746　Panax japonicus(T. Nees)C. A. Mey.；竹节参■

280747　Panax japonicus (T. Nees) C. A. Mey. f. angustatus (Makino) Makino et Nemoto = Panax japonicus (T. Nees) C. A. Mey. var. angustatus(Makino)H. Hara ■☆

280748　Panax japonicus(T. Nees)C. A. Mey. f. dichrocarpus(Makino) Nakai；二色果锐裂人参■☆

280749　Panax japonicus(T. Nees)C. A. Mey. f. incisus Nakai；锐裂人参■☆

280750　Panax japonicus(T. Nees)C. A. Mey. f. xanthocarpus(Makino) Nakai；黄果竹节参■☆

280751　Panax japonicus(T. Nees)C. A. Mey. var. angustatus(Makino) H. Hara；狭竹节参■☆

280752　Panax japonicus(T. Nees)C. A. Mey. var. angustifolius(Burkill) C. Y. Cheng et Y. C. Chu；狭叶竹节参（鸡头七，土三七，土田七，狭叶假人参，野三七）■

280753　Panax japonicus(T. Nees)C. A. Mey. var. angustifolius(Burkill) W. C. Cheng et Y. C. Chu = Panax pseudoginseng Wall. var. angustifolius(Burkill)H. L. Li ■

280754　Panax japonicus (T. Nees) C. A. Mey. var. bipinnatifidus (Seem.)C. Y. Wu et K. M. Feng = Panax bipinnatifidus Seem. ■

280755　Panax japonicus(T. Nees)C. A. Mey. var. bipinnatifidus(Seem.)C.

Y. Wu et K. M. Feng ex C. Chou = Panax bipinnatifidus Seem. ■

280756　Panax japonicus(T. Nees)C. A. Mey. var. major(Burkill)C. Y. Wu et K. M. Feng;珠子参(白地瓜,大金线吊葫芦,大药子,疙瘩七,鸡腰参,扣子七,钮子七,盘七,土三七,野三七,珠儿参,竹节参)●■

280757　Panax japonicus(T. Nees)C. A. Mey. var. major(Burkill)C. Y. Wu et K. M. Feng = Panax pseudoginseng Wall. var. japonicus(C. A. Mey.)G. Hoo et C. J. Tseng ■

280758　Panax japonicus C. A. Mey. = Panax japonicus(T. Nees)C. A. Mey. ■

280759　Panax japonicus C. A. Mey. = Panax pseudoginseng Wall. var. japonicus(C. A. Mey.)G. Hoo et C. J. Tseng ■

280760　Panax lancifolia Drake = Polyscias lancifolia(Drake)R. Vig. ●☆

280761　Panax lantzii Drake = Polyscias lantzii(Drake)Harms ex R. Vig. ●☆

280762　Panax leschenaultii DC. = Pentapanax fragrans(D. Don)T. D. Ha ●

280763　Panax lokobensis Drake = Polyscias nossibensis(Drake)Harms ●☆

280764　Panax major(Burkill)K. C. Ting = Panax pseudoginseng Wall. var. japonicus(C. A. Mey.)G. Hoo et C. J. Tseng ■

280765　Panax major(Burkill)K. C. Ting ex C. P'ei et Y. L. Chou = Panax japonicus(T. Nees)C. A. Mey. var. major(Burkill)C. Y. Wu et K. M. Feng ■

280766　Panax major Burkill = Panax pseudoginseng Wall. var. japonicus(C. A. Mey.)G. Hoo et C. J. Tseng ■

280767　Panax maralia(Roem. et Schult.)Decne. = Polyscias maralia(Roem. et Schult.)Bernardi ●☆

280768　Panax multibracteatum Baker = Polyscias multibracteata(Baker)Harms ●☆

280769　Panax nigericus A. Chev. = Polyscias fulva(Hiern)Harms ●☆

280770　Panax nossibensis Drake = Polyscias nossibensis(Drake)Harms ●☆

280771　Panax notoginseng(Burkill)F. H. Chen ex C. Chow et W. G. Huang;三七(白芷三七,不换金,参三七,春七,滇七,滇三七,冬七,峨眉三七,汉三七,旱三七,猴头三七,假人参,剪口三七,金不换,筋条七,盘龙七,人参三七,三七参,山漆,田七,田漆,田三七,秀丽假人参,血参,羽叶三七,昭参,竹节三七);Pseudoginseng,Sanchi,Sanqi ■

280772　Panax notoginseng(Burkill)F. H. Chen ex C. Y. Wu et K. M. Feng = Panax notoginseng(Burkill)F. H. Chen ex C. Chow et W. G. Huang ■

280773　Panax notoginseng F. H. Chen ex C. Y. Wu et K. M. Feng = Panax pseudoginseng Wall. var. notoginseng(Burkill)G. Hoo et C. J. Tseng ■

280774　Panax oligoscias Drake = Polyscias madagascariensis(Seem.)Harms ●☆

280775　Panax palmatus Roxb. = Euaraliopsis palmata(Roxb.)Hutch. ●☆

280776　Panax pentamerus Baker = Polyscias pentamera(Baker)Harms ●☆

280777　Panax pinnatus A. Rich. = Polyscias farinosa(Delile)Harms ●☆

280778　Panax pseudoginseng Wall. ;假人参(参三七,人参三七,三七);Falseginsen,Sanchi ■

280779　Panax pseudoginseng Wall. subsp. himalaicus H. Hara = Panax bipinnatifidus Seem. ■

280780　Panax pseudoginseng Wall. subsp. himalaicus H. Hara = Panax pseudoginseng Wall. var. japonicus(C. A. Mey.)G. Hoo et C. J. Tseng ■

280781　Panax pseudoginseng Wall. subsp. japonicas(C. A. Mey.)H. Hara = Panax pseudoginseng Wall. var. japonicus(C. A. Mey.)G. Hoo et C. J. Tseng ■

280782　Panax pseudoginseng Wall. subsp. japonicum C. A. Mey. =

Panax pseudoginseng Wall. var. japonicus(C. A. Mey.)G. Hoo et C. J. Tseng ■

280783　Panax pseudoginseng Wall. subsp. japonicus(C. A. Mey.)H. Hara = Panax japonicus(T. Nees)C. A. Mey. ■

280784　Panax pseudoginseng Wall. subsp. japonicus(C. A. Mey.)H. Hara f. incisus(Nakai)H. Hara = Panax japonicus(T. Nees)C. A. Mey. f. incisus Nakai ■☆

280785　Panax pseudoginseng Wall. subsp. japonicus(C. A. Mey.)H. Hara f. xanthocarpus(Makino)H. Hara = Panax japonicus(T. Nees)C. A. Mey. f. xanthocarpus(Makino)Nakai ■☆

280786　Panax pseudoginseng Wall. subsp. japonicus(C. A. Mey.)H. Hara var. angustatus(Makino)H. Hara = Panax japonicus(T. Nees)C. A. Mey. var. angustatus(Makino)H. Hara ■☆

280787　Panax pseudoginseng Wall. subsp. japonicus(T. Nees)H. Hara = Panax japonicus(T. Nees)C. A. Mey. ■

280788　Panax pseudoginseng Wall. var. angustifolius(Burkill)H. L. Li;狭叶假人参(西洋参,狭叶竹节参);Narrowleaf Ginsen ■

280789　Panax pseudoginseng Wall. var. angustifolius(Burkill)H. L. Li = Panax japonicus(T. Nees)C. A. Mey. var. angustifolius(Burkill)C. Y. Cheng et Y. C. Chu ■

280790　Panax pseudoginseng Wall. var. bipinnatifidus(Seem.)H. L. Li = Panax bipinnatifidus Seem. ■

280791　Panax pseudoginseng Wall. var. elegantior(Burkill)G. Hoo et C. J. Tseng;秀丽假人参(秀丽三七,珠子参,竹节三七);Elegant Falseginsen ■

280792　Panax pseudoginseng Wall. var. japonicus(C. A. Mey.)G. Hoo et C. J. Tseng = Panax japonicus(T. Nees)C. A. Mey. ■

280793　Panax pseudoginseng Wall. var. japonicus(C. A. Mey.)G. Hoo et C. J. Tseng;大叶三七(白三七,定风草,峨三七,佛掌七,疙瘩七,鸡头七,节人参,扣子七,罗汉三七,萝卜七,明七,钮子七,盘七,七叶子,人参三七,水三七,甜七,土参,土精,蜈蚣七,狭叶假人参,雪三七,血参,野三七,芋儿七,珠参,珠儿参,珠子参,竹鞭人参,竹鞭三七,竹根七,竹节参,竹节七,竹节人参,竹节三七,竹七);Bigleaf Sanqi,Creeping Ginsen,Japanese Ginsen ■

280794　Panax pseudoginseng Wall. var. japonicus(T. Nees)G. Hoo et C. J. Tseng = Panax japonicus(T. Nees)C. A. Mey. ■

280795　Panax pseudoginseng Wall. var. major(Burkill)H. L. Li = Panax japonicus(T. Nees)C. A. Mey. var. major(Burkill)C. Y. Wu et K. M. Feng ■

280796　Panax pseudoginseng Wall. var. major(Burkill)H. L. Li = Panax pseudoginseng Wall. var. japonicus(C. A. Mey.)G. Hoo et C. J. Tseng ■

280797　Panax pseudoginseng Wall. var. notoginseng(Burkill)G. Hoo et C. J. Tseng = Panax notoginseng(Burkill)F. H. Chen ex C. Chow et W. G. Huang ■

280798　Panax pseudoginseng Wall. var. wangianus(S. C. Sun)G. Hoo et C. J. Tseng = Panax pseudoginseng Wall. var. japonicus(C. A. Mey.)G. Hoo et C. J. Tseng ■

280799　Panax quinquefolius L. ;西洋参(顶光参,粉光西洋参,佛兰参,广东人参,花旗参,毛西参,美国人参,美洲人参,片参,西参,西洋人参,洋参,原皮西洋参,正光洁参,正面参);America Ginsen,American Ginsen,Five Fingers,Five-fingers,Ginseng,Redberry,Tartar-root ■☆

280800　Panax quinquefolius L. var. ginseng(C. A. Mey.)Regel et Maack = Panax ginseng C. A. Mey. ■

280801　Panax quinquefolius L. var. ginseng(C. A. Mey.)Regel et Maack ex Regel = Panax ginseng C. A. Mey. ■

280802   Panax repens Maxim. = Panax japonicus(T. Nees)C. A. Mey. ■

280803   Panax repens Maxim. = Panax pseudoginseng Wall. var. japonicus (C. A. Mey.)G. Hoo et C. J. Tseng ■

280804   Panax rhododendrifolius Griff. = Schefflera rhododendrifolia(Griff.) Frodin ●

280805   Panax ricinifolius Siebold et Zucc. = Kalopanax septemlobus (Thunb.)Nakai ●

280806   Panax schin-seng Nees = Panax ginseng C. A. Mey. ■

280807   Panax schin-seng S. S. Chien = Panax pseudoginseng Wall. var. japonicus(C. A. Mey.)G. Hoo et C. J. Tseng ■

280808   Panax schin-seng T. Nees = Panax ginseng C. A. Mey. ■

280809   Panax schin-seng T. Nees var. japonicus T. Nees = Panax japonicus (T. Nees)C. A. Mey. ■

280810   Panax sessiliflorus Rupr. et Maxim. = Acanthopanax sessiliflorus (Rupr. et Maxim.)Seem. ●

280811   Panax sessiliflorus Rupr. et Maxim. = Eleutherococcus sessilif-lorus(Rupr. et Maxim.)S. Y. Hu ●

280812   Panax spinosus L. f. = Acanthopanax spinosus(L. f.)Miq. ●☆

280813   Panax stipuleanatus H. T. Tsai et K. M. Feng = Panax japonicus(T. Nees)C. A. Mey. var. bipinnatifidus(Seem.)C. Y. Wu et K. M. Feng ex C. Chou ■

280814   Panax stipuleanatus H. T. Tsai et K. M. Feng ex C. Chow;屏边三七 (白三七,土三七,香刺,野三七,竹节七);Pingbian Ginsen ■

280815   Panax stipuleanatus H. T. Tsai et K. M. Feng ex C. Chow = Panax bipinnatifidus Seem. ■

280816   Panax transitorius G. Hoo = Panax pseudoginseng Wall. var. japonicus (C. A. Mey.)G. Hoo et C. J. Tseng ■

280817   Panax trifolius L. ;三叶人参; Dwarf Ginsen, Dwarf Ginseng, Groundnut,Threeleaf Ginsen ■☆

280818   Panax tripinnatus Baker = Polyscias tripinnata Harms ●☆

280819   Panax wangianus S. C. Sun = Panax pseudoginseng Wall. var. japonicus(C. A. Mey.)G. Hoo et C. J. Tseng ■

280820   Panax zanthoxyloides Baker = Polyscias zanthoxyloides(Baker)Harms ●☆

280821   Panax zingiberensis C. Y. Wu et K. M. Feng;姜状三七(鸡蛋七,野三七);Ginger Ginsen,Gingerlike Sanqi ■

280822   Panaxus St. -Lag. = Panax L. ●

280823   Pancalum Ehrh. = Hypericum L. ■●

280824   Panchena Montrouz. = Ixora L. ●

280825   Panchera Post et Kuntze = Ixora L. ●

280826   Panchera Post et Kuntze = Pancheria Montrouz. ●

280827   Pancheria Brongn. et Gris(1862)(保留属名);潘树属●☆

280828   Pancheria Montrouz. = Ixora L. ●

280829   Pancheria calophylla Guillaumin;美叶潘树●☆

280830   Pancheria elegans Brongn. et Gris. ;雅致潘树●☆

280831   Pancheria elliptica Pamp. ;椭圆潘树●☆

280832   Pancheria ferruginea Brongn. et Gris;锈色潘树●☆

280833   Pancheria fusca Schltr. ;褐潘树●☆

280834   Pancheria laevis Vieill. ex Guillaumin;平滑潘树●☆

280835   Pancheria lanceolata Vieill. ex Guillaumin;披针叶潘树●☆

280836   Pancheria lucida Vieill. ex Guillaumin;亮潘树●☆

280837   Pancheria minima J. Bradford;微小潘树●☆

280838   Pancheria obovata Brongn. et Gris;倒卵潘树●☆

280839   Panchezia B. D. Jacks. = Ixora L. ●

280840   Panchezia B. D. Jacks. = Pancheria Montrouz. ●

280841   Panchezia Montrouz. (废弃属名)= Pancheria Brongn. et Gris(保留属名)●☆

280842   Panciatica Picciv. = Cadia Forssk. ●■☆

280843   Panciatica purpura G. Piccioli = Cadia purpurea(G. Piccioli) Aiton ●☆

280844   Pancicia Vis. (1858);潘奇克草属☆

280845   Pancicia Vis. et Schltdl. = Pancicia Vis. ☆

280846   Pancicia serbica Vis. ;潘奇克草☆

280847   Pancovia Fabr. = Potentilla L. ■●

280848   Pancovia Heist. ex Adans. = Pancovia Willd. (保留属名)●☆

280849   Pancovia Heist. ex Fabr. (废弃属名)= Comarum L. ●■

280850   Pancovia Heist. ex Fabr. (废弃属名)= Pancovia Willd. (保留属名)●☆

280851   Pancovia Willd. (1799)(保留属名);潘考夫无患子属●☆

280852   Pancovia delavayi Franch. = Sapindus delavayi (Franch.) Radlk. ●

280853   Pancovia floribunda Pellegr. ;繁花潘考夫无患子●☆

280854   Pancovia golungensis(Hiern)Exell et Mendonca;戈龙潘考夫无患子●☆

280855   Pancovia harmsiana Gilg;哈姆斯潘考夫无患子●☆

280856   Pancovia heckeli Claudel;赫克尔无患子●☆

280857   Pancovia hildebrandtii Gilg;希尔德潘考夫无患子●☆

280858   Pancovia holtzii Gilg ex Radlk. ;霍尔茨潘考夫无患子●☆

280859   Pancovia holtzii Gilg ex Radlk. subsp. faulknerae Verdc. ;福克纳潘考夫无患子●☆

280860   Pancovia laurentii(De Wild.)Gilg ex De Wild. ;洛朗潘考夫无患子●☆

280861   Pancovia le-testui Pellegr. ;勒泰斯蒂潘考夫无患子●☆

280862   Pancovia macrophylla Gilg = Chytranthus punctatus Radlk. ●☆

280863   Pancovia mortehanii De Wild. = Chytranthus mortehanii (De Wild.)de Voldere ex Hauman ●☆

280864   Pancovia pedicellaris Radlk. et Gilg = Pancovia turbinata Radlk. ●☆

280865   Pancovia polyantha Gilg ex Engl. ;多花潘考夫无患子●☆

280866   Pancovia sessiliflora Hutch. et Dalziel;无梗花潘考夫无患子●☆

280867   Pancovia subcuneata Radlk. ;楔形潘考夫无患子●☆

280868   Pancovia thyrsiflora Gilg ex Radlk. ;聚伞潘考夫无患子●☆

280869   Pancovia tomentosa(Kurz)Kurz = Sapindus tomentosus Kurz ●

280870   Pancovia tomentosa Kurz. = Sapindus tomentosus Kurz ●

280871   Pancovia turbinata Radlk. ;陀螺形潘考夫无患子●☆

280872   Pancratiaceae Horan. ;全能花科●

280873   Pancratiaceae Horan. = Amaryllidaceae J. St. -Hil. (保留科名)●■

280874   Pancratiaceae Horan. = Poaceae Barnhart(保留科名)■●

280875   Pancratio-Crinum Herb. ex Steud. = Crinum L. ■

280876   Pancratium Dill. ex L. = Pancratium L. ■

280877   Pancratium L. (1753);全能花属(金钟花属,力药花属); Allroundflower,Pancratium,Sea Lily ■

280878   Pancratium americanum Mill. = Hymenocallis littoralis(Jacq.) Salisb. ■

280879   Pancratium arabicum Sickenb. ;阿拉伯全能花■☆

280880   Pancratium biflorum Roxb. ;全能花;Allroundflower,Twoflower Pancratium ■

280881   Pancratium canariense Ker Gawl. ;加那利全能花■☆

280882   Pancratium centrale(A. Chev.)Traub;刺全能花■☆

280883   Pancratium chapmannii Harv. = Pancratium tenuifolium Hochst. ex A. Rich. ■☆

280884   Pancratium collinum Coss. et Durieu = Pancratium foetidum Pomel ■☆

280885   Pancratium coronarium J. Le Conte = Hymenocallis coronaria(J. Le Conte) Kunth ■☆

280886   Pancratium crassifolium (Herb.) Schult. = Hymenocallis crassifolia Herb. ■☆

280887　Pancratium foetidum Pomel;臭全能花■☆

280888　Pancratium foetidum Pomel var. brachysiphon Maire = Pancratium foetidum Pomel ■☆

280889　Pancratium foetidum Pomel var. oranense Batt. = Pancratium foetidum Pomel ■☆

280890　Pancratium foetidum Pomel var. rifanum Maire = Pancratium foetidum Pomel ■☆

280891　Pancratium foetidum Pomel var. saldense Batt. = Pancratium foetidum Pomel ■☆

280892　Pancratium foetidum Pomel var. tunetanum Batt. = Pancratium foetidum Pomel ■☆

280893　Pancratium fragrans Salisb. ;芳香全能花;Fragrant Pancratium ■☆

280894　Pancratium hirtum A. Chev. = Pancratium tenuifolium Hochst. ex A. Rich. ■☆

280895　Pancratium humile Cav. = Braxireon humile(Cav.) Raf. ■☆

280896　Pancratium illyricum L. ;伊利里亚全能花(带叶全能花,力药花);Corsican Lily, Illyrian Pancratium, Spirit Lily ■☆

280897　Pancratium latifolium Mill. = Hymenocallis latifolia (Mill.) M. Roem. ■☆

280898　Pancratium liriosme Raf. = Hymenocallis liriosme (Raf.) Shinners ■☆

280899　Pancratium littorale Jacq. = Hymenocallis littoralis (Jacq.) Salisb. ■

280900　Pancratium maritimum L. ;海滨全能花(海滨力药花);Mediterranean Lily, Sea Daffodil, Sea Lily, Sea-daffodil, Seadaffodil Pancratium, Sea-side Pancratium Lily ■☆

280901　Pancratium occidentale J. Le Conte = Hymenocallis occidentalis (J. Le Conte) Kunth ■☆

280902　Pancratium parvicoronatum Geerinck;小冠全能花■☆

280903　Pancratium rotatum Ker Gawl. = Hymenocallis rotata (Ker Gawl.) Herb. ■☆

280904　Pancratium saharae Batt. = Pancratium trianthum Herb. ■☆

280905　Pancratium saharae Batt. var. chatinianum? = Pancratium trianthum Herb. ■☆

280906　Pancratium sickenbergeri Asch. et Schweinf. ;西肯贝格全能花■☆

280907　Pancratium sickenbergeri Asch. et Schweinf. var. desertorum Sickenb. = Pancratium sickenbergeri Asch. et Schweinf. ■☆

280908　Pancratium sickenbergeri Asch. et Schweinf. var. littorale Sickenb. = Pancratium sickenbergeri Asch. et Schweinf. ■☆

280909　Pancratium speciosum Salisb. ;美丽全能花;Showy Pancratium, Specious Pancratium ■☆

280910　Pancratium tenuifolium Hochst. ex A. Rich. ;细叶全能花■☆

280911　Pancratium tortuosum Herb. ;扭曲全能花■☆

280912　Pancratium trianthum Herb. ;三花全能花;Threeflower Pancratium ■☆

280913　Pancratium trianthum Herb. var. chatinianum (Batt.) Maire et Weiller = Pancratium trianthum Herb. ■☆

280914　Pancratium trianthum Herb. var. saharae (Coss. ex Batt. et Trab.) Maire = Pancratium trianthum Herb. ■☆

280915　Pancratium zeylanicum L. ;单花全能花;One-flowered Sea-daffodil ■☆

280916　Panctenis Raf. = Aureolaria Raf. ■☆

280917　Panda Pierre(1896);攀打属(盘木属,箫属,油树属)●☆

280918　Panda oleosa Pierre;油树(油盘木,油箫)●☆

280919　Pandaca Noronha ex Thouars = Tabernaemontana L. ●

280920　Pandaca Noronha ex Thouars(1806);山马菜属(潘达加属,山

马茶属)●

280921　Pandaca Thouars = Tabernaemontana L. ●

280922　Pandaca affinis Markgr. = Pentopetia ovalifolia (Costantin et Gallaud)Klack. ■☆

280923　Pandaca anisophylla (Pichon) Markgr. = Tabernaemontana ciliata Pichon ●☆

280924　Pandaca boiteaui Markgr. ;鲍氏山马菜(鲍氏山马茶,潘达加)●☆

280925　Pandaca boiteaui Markgr. = Tabernaemontana mocquerysii Aug. DC. ●☆

280926　Pandaca calcarea (Pichon) Markgr. = Pentopetia ovalifolia (Costantin et Gallaud)Klack. ■☆

280927　Pandaca callosa (Pichon) Markgr. = Tabernaemontana ciliata Pichon ●☆

280928　Pandaca ciliata (Pichon) Markgr. = Tabernaemontana ciliata Pichon ●☆

280929　Pandaca ciliata (Pichon) Markgr. var. lanceolata Markgr. = Tabernaemontana ciliata Pichon ●☆

280930　Pandaca ciliata(Pichon) Markgr. var. sambiranensis Markgr. = Tabernaemontana ciliata Pichon ●☆

280931　Pandaca crassifolia (Pichon) Markgr. = Tabernaemontana crassifolia Pichon ●☆

280932　Pandaca cuneata (Pichon) Markgr. = Tabernaemontana ciliata Pichon ●☆

280933　Pandaca cuneata (Pichon) Markgr. var. exserta Markgr. = Tabernaemontana ciliata Pichon ●☆

280934　Pandaca debrayi Markgr. = Tabernaemontana debrayi(Markgr.) Leeuwenb. ●☆

280935　Pandaca eusepala (Aug. DC.) Markgr. = Tabernaemontana eusepala Aug. DC. ●☆

280936　Pandaca eusepaloides Markgr. = Tabernaemontana eusepaloides (Markgr.) Leeuwenb. ●☆

280937　Pandaca humblotii (Baill.) Markgr. = Tabernaemontana humblotii(Baill.) Pichon ●☆

280938　Pandaca longissima Markgr. = Tabernaemontana ciliata Pichon ●☆

280939　Pandaca longituba (Pichon) Markgr. = Tabernaemontana ciliata Pichon ●☆

280940　Pandaca minutiflora(Pichon)Markgr. ;微花山马菜(小花潘达加,小花山马茶)●☆

280941　Pandaca minutiflora (Pichon) Markgr. = Pentopetia ovalifolia (Costantin et Gallaud)Klack. ■☆

280942　Pandaca minutiflora (Pichon) Markgr. = Tabernaemontana minutiflora Pichon ●☆

280943　Pandaca mocquerysii(Aug. DC.) Markgr. ;莫氏山马菜(莫氏山马茶)●☆

280944　Pandaca mocquerysii(Aug. DC.) Markgr. = Tabernaemontana mocquerysii Aug. DC. ●☆

280945　Pandaca mocquerysii(Aug. DC.) Markgr. var. lancifolia Markgr. = Tabernaemontana mocquerysii Aug. DC. ●☆

280946　Pandaca mocquerysii (Aug. DC.) Markgr. var. parvifolia (Pichon) Markgr. = Tabernaemontana mocquerysii Aug. DC. ●☆

280947　Pandaca mocquerysii(Aug. DC.) Markgr. var. pendula Markgr. ;垂枝莫氏山马菜●☆

280948　Pandaca mocquerysii (Aug. DC.) Markgr. var. pendula Markgr. = Tabernaemontana mocquerysii Aug. DC. ●☆

280949　Pandaca ochrascens(Pichon)Markgr. = Tabernaemontana ciliata Pichon ●☆

280950　Pandaca parvifolia (Pichon) Markgr. = Tabernaemontana mocquerysii

Aug. DC. ●☆

280951　Pandaca pichoniana Markgr. = Tabernaemontana ciliata Pichon ●☆

280952　Pandaca retusa(Lam.)Markgr. ;微凹叶山马菜(微凹叶山马茶)●☆

280953　Pandaca retusa(Lam.)Markgr. = Tabernaemontana retusa(Lam.)Pichon ●☆

280954　Pandaca sambiranensis(Pichon)Markgr. = Tabernaemontana sambiranensis Pichon ●☆

280955　Pandaca speciosa Markgr. = Tabernaemontana humblotii(Baill.)Pichon ●☆

280956　Pandaca stellata(Pichon)Markgr. = Tabernaemontana stellata Pichon ●☆

280957　Pandaca verrucosa Markgr. = Tabernaemontana mocquerysii Aug. DC. ●☆

280958　Pandacastrum Pichon = Tabernaemontana L. ●

280959　Pandacastrum saccharatum Pichon = Tabernaemontana ciliata Pichon ●☆

280960　Pandaceae Engl. et Gilg(1913)(保留科名);攀打科(盘木科,萧科,小盘木科,油树科);Panda Family ●

280961　Pandaceae Pierre = Pandaceae Engl. et Gilg(保留科名)●

280962　Pandamus Raf. = Pandanus L. f. ●■

280963　Pandanaceae R. Br. (1810)(保留科名);露兜树科;Screwpine Family,Screw-pine Family ●■

280964　Pandanophyllum Hassk. = Mapania Aubl. ■

280965　Pandanus L. f. = Keura Forssk. ●■

280966　Pandanus L. f. = Pandanus Parkinson ex Du Roi ●■

280967　Pandanus Parkinson = Pandanus Parkinson ex Du Roi ●■

280968　Pandanus Parkinson ex Du Roi(1773);露兜树属;Pandanus,Screw Pine,Screwpine,Screw-pine ●■

280969　Pandanus Rumph. ex L. f. = Pandanus Parkinson ex Du Roi ●■

280970　Pandanus acanthostylus Martelli;刺柱露兜树●☆

280971　Pandanus akeassii Huynh;阿克斯露兜树●☆

280972　Pandanus alpestris Martelli;高山露兜树●☆

280973　Pandanus alveolatus Huynh;蜂窝露兜树●☆

280974　Pandanus amaryllifolius Roxb. ;香露兜(板蓝香);Amaryllis-leaf Screwpine,Pandan Wangi ●■

280975　Pandanus ambalavaoensis Huynh;安巴拉沃露兜树●☆

280976　Pandanus ambongensis Martelli;安邦露兜树●☆

280977　Pandanus analamazaotrensis Martelli;阿地露兜树●☆

280978　Pandanus andringitrensis Huynh;安德林吉特拉山露兜树●☆

280979　Pandanus androcephalanthos Martelli = Martellidendron andro-cephalanthos(Martelli)Callm. et Chassot ●☆

280980　Pandanus angolensis Huynh;安哥拉露兜树●☆

280981　Pandanus angustifolius Baker = Pandanus dyckioides Baker ●☆

280982　Pandanus aquaticus F. Muell. ;水生露兜树;Aquatic Screwpine ●☆

280983　Pandanus arenicola Huynh;沙生露兜树●☆

280984　Pandanus arenosus Martelli = Pandanus rollotii Martelli ●☆

280985　Pandanus aridus H. St. John;旱生露兜树●☆

280986　Pandanus austrosinensis T. L. Wu;露兜草;Herb Screwpine,South China Screwpine ■

280987　Pandanus austrosinensis T. L. Wu var. longifolius L. Y. Zhou et X. W. Zhong;长叶露兜树(长叶露兜草);Longleaf Herb Screwpine,Longleaf South China Screwpine ●■

280988　Pandanus bakeri Warb. ;贝克露兜树●☆

280989　Pandanus baptisti Hort. ;英国露兜树;New Britain Island Screwpine ●☆

280990　Pandanus bilamellatus Martelli = Pandanus malgassicus Pic. Serm. ●☆

280991　Pandanus bilobatus H. St. John ex Huynh;双裂露兜树●☆

280992　Pandanus bipyramidatus Martelli;双塔露兜树●☆

280993　Pandanus boninensis Warb. ;小笠原露兜树(小笠原露兜);Bonin Pandanus,False Pineaple ●

280994　Pandanus butayei De Wild. ;布塔耶露兜树●☆

280995　Pandanus candelabrum P. Beauv. ;烛台露兜树●☆

280996　Pandanus centrifugalis H. St. John = Pandanus concretus Baker ●☆

280997　Pandanus centrifugalis St. John;离心露兜树●☆

280998　Pandanus cephalotus B. C. Stone;头状露兜树●☆

280999　Pandanus ceratophorus Baker;角状露兜树●☆

281000　Pandanus chevalieri H. St. John ex Huynh;舍瓦利耶露兜树●☆

281001　Pandanus chiliocarpus Stapf;千果露兜树●☆

281002　Pandanus collinus Martelli = Pandanus pseudocollinus Pic. Serm. ●☆

281003　Pandanus columellatus Huynh;小圆柱露兜树●☆

281004　Pandanus columnaris H. St. John;圆柱露兜树●☆

281005　Pandanus comatus Martelli;束毛露兜树●☆

281006　Pandanus concretus Baker;联合露兜树●☆

281007　Pandanus connatus H. St. John;合生露兜树●☆

281008　Pandanus coriaceus Huynh;革质露兜树●☆

281009　Pandanus cruciatus Pic. Serm. = Martellidendron cruciatum(Pic. Serm.)Callm. et Chassot ●☆

281010　Pandanus cuneatus Huynh = Pandanus leptopodus Martelli ●☆

281011　Pandanus cyaneoglaucescens Martelli = Pandanus bakeri Warb. ●☆

281012　Pandanus dauphinensis Martelli;多芬露兜树●☆

281013　Pandanus denudatus Huynh;裸露兜树●☆

281014　Pandanus diffusus Martelli;铺散露兜树●☆

281015　Pandanus djalonensis Huynh;贾隆露兜树●☆

281016　Pandanus dyckioides Baker;雀舌兰露兜树●☆

281017　Pandanus echinops Huynh;刺露兜树●☆

281018　Pandanus embuensis H. St. John = Pandanus kajui Beentje ●☆

281019　Pandanus engleri Warb. = Pandanus rabaiensis Rendle ●☆

281020　Pandanus erectus H. St. John;直立露兜树●☆

281021　Pandanus farakoensis Huynh;法拉科露兜树●☆

281022　Pandanus fascicularis Lam. ;簇生露兜树●☆

281023　Pandanus fascicularis Lam. = Pandanus tectorius Parkinson ex Du Roi ●

281024　Pandanus ferox Huynh = Pandanus guillaumetii B. C. Stone ●☆

281025　Pandanus fibrosus Gagnep. = Pandanus fibrosus Gagnep. ex Humbert ●

281026　Pandanus fibrosus Gagnep. ex Humbert;小露兜●

281027　Pandanus flagellaris B. C. Stone;鞭状露兜树●☆

281028　Pandanus flagellibracteatus Huynh = Pandanus stellatus Martelli ●☆

281029　Pandanus forceps Martelli;箝古子;Tongs Screwpine ●

281030　Pandanus forceps Martelli = Pandanus kaida Kurz ●

281031　Pandanus furcatus Roxb. ;分叉露兜树(分叉露兜,露兜,罗金堆,帕梯,帕梯果,山菠萝,野菠萝);Furcate Screwpine ●

281032　Pandanus gabonensis Huynh;加蓬露兜树●☆

281033　Pandanus globulatus Huynh;小球露兜树●☆

281034　Pandanus goetzei Warb. = Pandanus rabaiensis Rendle ●☆

281035　Pandanus gossweileri H. St. John ex Huynh;戈斯露兜树●☆

281036　Pandanus grallatus B. C. Stone;高腿露兜树●☆

281037　Pandanus graminifolius Miq. ;禾叶露兜树(禾叶露兜);Grassleaf Screwpine,Grassy Screwpine ●

281038　Pandanus gressittii B. C. Stone = Pandanus fibrosus Gagnep. ex Humbert ●

281039　Pandanus gressittii Merr. ex B. C. Stone；小露兜树（小果露兜，小露兜）；Gressitt Screwpine, Small Screwpine ●■

281040　Pandanus guillaumetii B. C. Stone；多刺露兜树●☆

281041　Pandanus guineabissauensis Huynh；几内亚比绍露兜树●☆

281042　Pandanus heddei Warb. = Pandanus kirkii Rendle ●☆

281043　Pandanus hemiacanthus Peter = Pandanus chiliocarpus Stapf ●☆

281044　Pandanus humbertii Laivao, Callm. et Buerki；亨伯特露兜树●☆

281045　Pandanus imerinensis Martelli；伊梅里纳露兜树●☆

281046　Pandanus insolitus Huynh；异常露兜树●☆

281047　Pandanus intricatus Martelli = Pandanus dyckioides Baker ●☆

281048　Pandanus isalicus Huynh；伊萨卢露兜树●☆

281049　Pandanus kaida Kurz；勒古子●

281050　Pandanus kajui Beentje；卡朱露兜树●☆

281051　Pandanus karaka Martelli = Martellidendron karaka (Martelli) Callm. ●☆

281052　Pandanus kariangensis Huynh = Martellidendron kariangense (Huynh) Callm. ●☆

281053　Pandanus kerstingii Warb. ；克斯廷露兜树●☆

281054　Pandanus kirkii Rendle；柯克露兜树●☆

281055　Pandanus kuepferi Callm. , Wohlhauser et Laivao；屈氏露兜树●☆

281056　Pandanus lachaisei Huynh；拉查斯露兜树●☆

281057　Pandanus laferrerei Huynh；拉费雷露兜树●☆

281058　Pandanus latifolius Hassk. = Pandanus amaryllifolius Roxb. ●

281059　Pandanus latiloculatus Huynh；宽囊露兜树●☆

281060　Pandanus latistigmaticus Huynh；宽柱头露兜树●☆

281061　Pandanus laxespicatus Martelli = Pandanus ceratophorus Baker ●☆

281062　Pandanus leonensis Lodd. ex H. Wendl. ；莱昂露兜树●☆

281063　Pandanus leptopodus Martelli；细梗露兜树●☆

281064　Pandanus liberiensis Huynh；利比里亚露兜树●☆

281065　Pandanus linguiformis B. C. Stone；舌状露兜树●☆

281066　Pandanus livingstonianus Rendle；利文斯顿露兜树●☆

281067　Pandanus longecuspidatus Pic. Serm. ；长尖露兜树●☆

281068　Pandanus longipes Perrier ex Martelli；长梗露兜树●☆

281069　Pandanus longissimepedunculatus Martelli；长柄露兜树●☆

281070　Pandanus longistylus Martelli et Pic. Serm. ；长柱露兜树●☆

281071　Pandanus macrophyllus Martelli；大叶露兜树●☆

281072　Pandanus malgassicus Pic. Serm. ；马尔加斯露兜树●☆

281073　Pandanus mammillaris Martelli et Pic. Serm. ；乳突露兜树●☆

281074　Pandanus manamboloensis Huynh；马南布卢露兜树●☆

281075　Pandanus mangokensis Martelli；曼戈基露兜树●☆

281076　Pandanus marojejicus Callm. et Laivao；马鲁杰露兜树●☆

281077　Pandanus maromokotrensis Callm. et Wohlhauser；马鲁穆库特鲁山露兜树●☆

281078　Pandanus masoalensis Laivao et Callm. = Martellidendron masoalense (Laivao et Callm.) Callm. et Chassot ●☆

281079　Pandanus membranaceus Huynh；膜质露兜树●☆

281080　Pandanus microcarpus Perrier = Pandanus saxatilis Martelli ●☆

281081　Pandanus microcephalus Baker；小头露兜树●☆

281082　Pandanus mosambicius H. St. John ex Huynh；莫桑比克露兜树●☆

281083　Pandanus muralis Huynh；厚壁露兜树●☆

281084　Pandanus murira Beentje = Pandanus kajui Beentje ●☆

281085　Pandanus myriocarpus Baker；多果露兜树●☆

281086　Pandanus namakiensis Martelli；纳马基露兜树●☆

281087　Pandanus neoleptopodus Pic. Serm. ；新细梗露兜树●☆

281088　Pandanus oblongicapitellatus Huynh；长圆头露兜树●☆

281089　Pandanus odoratissimus L. f. ；林投露兜树（林投露兜，露兜树，浓香露兜，甜香露兜树，香露兜）；Breadfruit, Chinese Screwpine, Hala Screw Pine, Pandang ●

281090　Pandanus odoratissimus L. f. = Pandanus tectorius Parkinson ex Du Roi ●

281091　Pandanus odoratissimus L. f. = Pandanus tectorius Sol. ex Balf. f. var. sinensis Warb. ●

281092　Pandanus odoratissimus L. f. f. laevis (Warb.) Hatus. ；平滑林投露兜●☆

281093　Pandanus odoratissimus L. f. var. liukiuensis (Warb.) T. Ito = Pandanus odoratissimus L. f. ●

281094　Pandanus odoratissimus L. f. var. sinensis (Warb.) Kaneh. = Pandanus odoretissimus L. f. ●

281095　Pandanus odoratissimus L. f. var. sinensis (Warb.) Kaneh. = Pandanus tectorius Sol. ex Balf. f. var. sinensis Warb. ●

281096　Pandanus odoratissimus L. f. var. sinensis (Warb.) Martelli = Pandanus tectorius Parkinson ex Du Roi ●

281097　Pandanus odorus Ridl. = Pandanus amaryllifolius Roxb. ●

281098　Pandanus oligocarpus Martelli；寡果露兜树●☆

281099　Pandanus oligocephalus Baker；头露兜树●☆

281100　Pandanus parachevalieri Huynh；拟舍瓦利耶露兜树●☆

281101　Pandanus parvicentralis Huynh；小刺露兜树●☆

281102　Pandanus perrieri Martelli；佩里耶露兜树●☆

281103　Pandanus pervilleanus Solms；佩尔露兜树●☆

281104　Pandanus petrosus Martelli；岩生露兜树●☆

281105　Pandanus pichi-sermollii B. C. Stone = Pandanus guillaumetii B. C. Stone ●☆

281106　Pandanus platyphyllus Martelli；宽叶露兜树●☆

281107　Pandanus pluriloculatus H. St. John = Pandanus columnaris H. St. John ●☆

281108　Pandanus princeps B. C. Stone；帝王露兜树●☆

281109　Pandanus problematicus Huynh；普罗露兜树●☆

281110　Pandanus pseudobathiei Pic. Serm. ；假巴西露兜树●☆

281111　Pandanus pseudochevalieri Huynh；假舍瓦利耶露兜树●☆

281112　Pandanus pseudocollinus Pic. Serm. ；假山丘露兜树●☆

281113　Pandanus pulcher Martelli；美丽露兜树●☆

281114　Pandanus pygmaeus Hook. ；矮露兜树（禾叶露兜树）；Dwarf Screwpine ●

281115　Pandanus pygmaeus Thouars；矮小露兜树●☆

281116　Pandanus rabaiensis Rendle；拉巴伊露兜树●☆

281117　Pandanus ramenensis Perrier = Pandanus sambiranensis Martelli ●☆

281118　Pandanus raynalii Huynh；雷纳尔露兜树●☆

281119　Pandanus remotus H. St. John = Pandanus tectorius Parkinson ex Du Roi ●

281120　Pandanus rollotii Martelli；罗洛露兜树●☆

281121　Pandanus sambiranensis Martelli；桑比朗露兜树●☆

281122　Pandanus sanderi Sand. ；桑得露兜树；Timor Screwpine ●☆

281123　Pandanus satabiei Huynh；萨塔比露兜树●☆

281124　Pandanus saxatilis Martelli；岩栖露兜树●☆

281125　Pandanus senegalensis H. St. John ex Huynh；塞内加尔露兜树●☆

281126　Pandanus serrimarginalis H. St. John ex Huynh；齿边露兜树●☆

281127　Pandanus sessilis Bojer = Pandanus chiliocarpus Stapf ●☆

281128　Pandanus sierra-leonensis Huynh；塞拉里昂露兜树●☆

281129　Pandanus sinensis (Warb.) Martelli = Pandanus tectorius Parkinson ex Du Roi ●

281130　Pandanus sparganioides Baker；黑三棱露兜树●☆

281131　Pandanus spicatus H. St. John；穗状露兜树●☆

281132　Pandanus spiralis H. Wendl. ；螺旋状露兜树●☆

281133　Pandanus stellatus Martelli；星状露兜树●☆

281134　Pandanus stuhlmannii Warb. = Pandanus rabaiensis Rendle ●☆

281135　Pandanus sylvicola Huynh;森林露兜树●☆

281136　Pandanus taiwanianus Hosok. = Pandanus odoretissimus L. f. ●

281137　Pandanus tectorius Parkinson = Pandanus tectorius Sol. ex Parkinson ●

281138　Pandanus tectorius Parkinson ex Du Roi;露兜树(假菠萝,假番梨,老锯头,笋菠萝,笋鲁子,箭菠萝,箭古,龙般笋,露兜笋,露兜勒,露兜箭,露兜子,檜筥子,路兜勒,母猪锯,婆锯勒,山波罗,山菠萝,山荫古子,水拖髻,野菠萝,猪母锯);Beach Screw Pine,Ketaki,Pandang,Screwpine,Tahitian Screwpine,Thatch Screw Pine,Thatch Screwpine,Thatch Screw-pine ●

281139　Pandanus tectorius Parkinson ex Du Roi'Veitchii';维奇露兜树●

281140　Pandanus tectorius Parkinson ex Du Roi var. sinensis Warb. = Pandanus tectorius Parkinson ex Du Roi ●

281141　Pandanus tectorius Sol. ex Balf. f. ' Veitchii';花边露兜树;Veitch's Screwpine ●

281142　Pandanus tectorius Sol. ex Balf. f. = Pandanus tectorius Sol. ex Parkinson ●

281143　Pandanus tectorius Sol. ex Balf. f. var. sinensis Warb.;林投;Chinese Screwpine ●

281144　Pandanus tectorius Sol. ex Parkinson = Pandanus tectorius Parkinson ex Du Roi ●

281145　Pandanus tectorius Sol. ex Warb. = Pandanus odoratissimus L. f. ●

281146　Pandanus tectorius Sol. ex Warb. f. laevis (Warb.) Masam. = Pandanus odoratissimus L. f. f. laevis (Warb.) Hatus. ●☆

281147　Pandanus tectorius Sol. ex Warb. var. ferreus Y. Kimura = Pandanus odoratissimus L. f. ●

281148　Pandanus tectorius Sol. ex Warb. var. liukiuensis Warb. = Pandanus odoratissimus L. f. ●

281149　Pandanus tectorius Sol. ex Warb. var. utinensis Masam. = Pandanus odoratissimus L. f. ●

281150　Pandanus tenuiflagellatus Huynh;细鞭露兜树●☆

281151　Pandanus tenuimarginatus Huynh;细边露兜树●☆

281152　Pandanus teuszii Warb.;托兹露兜树●☆

281153　Pandanus thomensis Henriq.;爱岛露兜树●☆

281154　Pandanus tiassaleensis Huynh;蒂亚萨莱露兜树●☆

281155　Pandanus togoensis Warb.;多哥露兜树●☆

281156　Pandanus tolanarensis Huynh;托拉纳尔露兜树●☆

281157　Pandanus tonkinensis Martelli;小果山菠萝;Tonkin Screwpine ●

281158　Pandanus triangularis H. St. John ex Huynh;三角露兜树●☆

281159　Pandanus tsaratananensis Martelli;察拉塔纳纳露兜树●☆

281160　Pandanus ugandaensis H. St. John = Pandanus chiliocarpus Stapf ●☆

281161　Pandanus uliginosus Perrier = Pandanus pulcher Martelli ●☆

281162　Pandanus umbellatus Martelli;小伞露兜树●☆

281163　Pandanus urophyllus Hance;分叉露兜 ●

281164　Pandanus urophyllus Hance = Pandanus furcatus Roxb. ●

281165　Pandanus usaramensis Martelli = Pandanus kirkii Rendle ●☆

281166　Pandanus utilis Bory;扇叶露兜树(红刺露兜,红刺露兜树,江边露兜,麻露兜,有用露兜树);Common Screw Pine, Common Screwpine,Common Screw-pine,Screwpine ●

281167　Pandanus variabilis Martelli;易变露兜树●☆

281168　Pandanus variegatus Miq.;斑叶露兜树■☆

281169　Pandanus veitchii Dallim.;台湾露兜树(斑叶露兜,斑叶露兜树,斑叶树,斑缘露兜,金边露兜,威氏露兜树);Taiwan Screwpine,Veitch Screwpine,Veitch Screw-pine,Veitch's Screwpine ●

281170　Pandanus veitchii Dallim. = Pandanus tectorius Parkinson ex Du Roi' Veitchii' ●

281171　Pandanus warburgii Martelli = Pandanus kirkii Rendle ●☆

281172　Pandanus welwitschii Rendle;韦氏露兜树●☆

281173　Panderia Fisch. et C. A. Mey. (1836);兜藜属(齿兜藜属,潘得藜属);Panderia ■

281174　Panderia piiosa Fisch. et C. A. Mey.;毛兜藜■

281175　Panderia turkestanica Iljin;兜藜;Turkestan Panderia ■

281176　Pandiaka(Moq.) Benth. et Hook. f. (1880);脊被苋属■☆

281177　Pandiaka(Moq.)Hook. f. = Pandiaka(Moq.)Benth. et Hook. f. ■☆

281178　Pandiaka Benth. et Hook. f. = Pandiaka(Moq.)Hook. f. ■☆

281179　Pandiaka andongensis Hiern = Pandiaka rubro-lutea (Lopr.) C. C. Towns. ■☆

281180　Pandiaka andongensis Hiern var. gracilis Suess. = Pandiaka rubro-lutea(Lopr.) C. C. Towns. ■☆

281181　Pandiaka angustifolia(Vahl)Hepper;窄叶脊被苋■☆

281182　Pandiaka aristata Suess. = Pandiaka rubro-lutea (Lopr.) C. C. Towns. ■☆

281183　Pandiaka benthamii (Lopr.) Schinz = Pandiaka angustifolia (Vahl)Hepper ■☆

281184　Pandiaka carsonii(Baker)C. B. Clarke;卡尔脊被苋■☆

281185　Pandiaka carsonii(Baker)C. B. Clarke var. linearifolia Hauman;线叶卡尔脊被苋■☆

281186　Pandiaka confusa C. C. Towns.;混乱脊被苋■☆

281187　Pandiaka cylindrica Hook. f. ex Baker et C. B. Clarke = Pandiaka elegantissima(Schinz)Dandy ■☆

281188　Pandiaka debilis(Baker)Hiern = Pandiaka welwitschii(Schinz)Hiern ■☆

281189　Pandiaka deserti N. E. Br. = Cyathula lanceolata Schinz ■☆

281190　Pandiaka elegantissima(Schinz)Dandy;雅致脊被苋■☆

281191　Pandiaka fasciculata Suess. = Achyranthes fasciculata (Suess.) C. C. Towns. ■☆

281192　Pandiaka glabra(Schinz)Hauman = Pandiaka carsonii(Baker)C. B. Clarke ■☆

281193　Pandiaka heterochiton (Lopr.) C. B. Clarke = Sericocoma heterochiton Lopr. ■☆

281194　Pandiaka heudelotii (Moq.) Hiern = Pandiaka angustifolia (Vahl)Hepper ■☆

281195　Pandiaka incana Suess. et Overkott = Pandiaka ramulosa Hiern ■☆

281196　Pandiaka involucrata(Moq.)B. D. Jacks.;总苞脊被苋■☆

281197　Pandiaka involucrata(Moq.)B. D. Jacks. var. megastachya Moq. = Pandiaka involucrata(Moq.)B. D. Jacks. ■☆

281198　Pandiaka kassneri Suess. = Pandiaka welwitschii(Schinz)Hiern ■☆

281199　Pandiaka lanata(Schinz)Hauman = Pandiaka carsonii(Baker)C. B. Clarke ■☆

281200　Pandiaka lanceolata(Schinz)C. B. Clarke = Cyathula lanceolata Schinz ■☆

281201　Pandiaka lanuginosa(Schinz)Schinz;多毛脊被苋■☆

281202　Pandiaka lindiensis Suess. et Beyerle = Pandiaka rubro-lutea (Lopr.) C. C. Towns. ■☆

281203　Pandiaka longipedunculata Peter = Centemopsis longipedunculata (Peter)C. C. Towns. ■☆

281204　Pandiaka metallorum P. A. Duvign. et Van Bockstal;光泽脊被苋■☆

281205　Pandiaka milnei Suess. et Overkott = Pandiaka carsonii(Baker)C. B. Clarke var. linearifolia Hauman ■☆

281206　Pandiaka oblanceolata (Schinz) C. B. Clarke = Pandiaka elegantissima(Schinz)Dandy ■☆

281207　Pandiaka obovata Suess. = Pandiaka carsonii (Baker) C. B.

Clarke ■☆

281208 Pandiaka polystachya Suess. = Pandiaka ramulosa Hiern ■☆

281209 Pandiaka polystachya Suess. var. incana (Suess. et Overkott) Cavaco = Pandiaka ramulosa Hiern ■☆

281210 Pandiaka porphyrargyrea Suess. et Overkott;银脊被苋 ■☆

281211 Pandiaka ramulosa Hiern;多枝被苋 ■☆

281212 Pandiaka richardsiae Suess.;理查兹脊被苋 ■☆

281213 Pandiaka rubro-lutea(Lopr.) C. C. Towns.;红黄脊被苋 ■☆

281214 Pandiaka schweinfurthii (Schinz) C. B. Clarke = Pandiaka welwitschii(Schinz) Hiern ■☆

281215 Pandiaka schweinfurthii(Schinz) C. B. Clarke var. glabra Schinz = Pandiaka carsonii(Baker) C. B. Clarke ■☆

281216 Pandiaka schweinfurthii C. B. Clarke var. compacta Suess. et Overkott = Pandiaka confusa C. C. Towns. ■☆

281217 Pandiaka schweinfurthii C. B. Clarke var. minor Suess. = Pandiaka confusa C. C. Towns. ■☆

281218 Pandiaka schweinfurthii C. B. Clarke var. parvifolia Suess. et Overkott = Pandiaka confusa C. C. Towns. ■☆

281219 Pandiaka trichinioides Suess.;澳洲苋脊被苋 ■☆

281220 Pandiaka welwitschii(Schinz) Hiern;韦尔脊被苋 ■☆

281221 Pandiaka wildii Suess. = Cyathula lanceolata Schinz ■☆

281222 Pandora Noronha ex Thouars = Rhodolaena Thouars ●☆

281223 Pandorea(Endl.) Spach(1840);粉花凌霄属;Pandorea ●

281224 Pandorea Spach = Pandorea(Endl.) Spach ●

281225 Pandorea jasminoides (Lindl.) K. Schum.;粉花凌霄(大素馨,南天素馨,洋凌霄);Bower Plant, Bower Vine, Bowerplant of Australia, Jasmine Pandorea ●

281226 Pandorea jasminoides K. Schum. ' Variegata Charisma';斑叶素馨 ●☆

281227 Pandorea pandorana(Andréws) Steenis;万佳藤;Pandora Vine, Wonga-wonga Vine ●☆

281228 Pandorea pandorana Steenis = Pandorea pandorana (Andréws) Steenis ●☆

281229 Pandorea ricasoliana (Tanfani) Baill. = Podranea ricasoliana (Tanfani) Sprague ●

281230 Paneguia Raf. = Sisyrinchium L. ■

281231 Paneion Lunell = Poa L. ■

281232 Panel Adans. (废弃属名) = Glycosmis Corrêa(保留属名) ●

281233 Panel Adans. (废弃属名) = Terminalia L. (保留属名) ●

281234 Panemata Raf. = Gymnostachyum Nees ■

281235 Paneroa E. E. Schill. (2008);墨西哥藿香蓟属 ■☆

281236 Paneroa E. E. Schill. = Ageratum L. ■●

281237 Panetos Raf. = Hedyotis L. + Arcytophyllum Willd. ex Schult. et Schult. f. ■☆

281238 Panetos Raf. = Houstonia L. ■☆

281239 Pangiaceae Blume = Pangiaceae Blume ex Endl. ●☆

281240 Pangiaceae Blume ex Endl.;马来刺篱木科(潘近树科) ●☆

281241 Pangiaceae Blume ex Endl. = Achariaceae Harms(保留科名) ●■☆

281242 Pangiaceae Blume ex Endl. = Flacourtiaceae Rich. ex DC. (保留科名) ●

281243 Pangiaceae Endl. = Achariaceae Harms(保留科名) ●■☆

281244 Pangiaceae Endl. = Flacourtiaceae Rich. ex DC. (保留科名) ●

281245 Pangium Reinw. (1823);马来刺篱木属(马来大风子属,潘近树属);Pangium ●☆

281246 Pangium edule Reinw.;马来刺篱木(马来大风子,马来潘近树);Edible Pangium, Food Pangium ●☆

281247 Panhopia Noronha ex Müll. Arg. = Macaranga Thouars ●

281248 Panhopia Noronha ex Müll. Arg. = Panopia Noronha ex Thouars ●

281249 Panicaceae(R. Br.) Herter = Gramineae Juss. (保留科名) ■●

281250 Panicaceae(R. Br.) Herter = Poaceae Barnhart(保留科名) ■●

281251 Panicaceae Bercht. et J. Presl = Gramineae Juss. (保留科名) ■●

281252 Panicaceae Bercht. et J. Presl = Poaceae Barnhart(保留科名) ■●

281253 Panicaceae Herter = Gramineae Juss. (保留科名) ■●

281254 Panicaceae Herter = Poaceae Barnhart(保留科名) ■●

281255 Panicaceae Voigt = Gramineae Juss. (保留科名) ■●

281256 Panicaceae Voigt = Poaceae Barnhart(保留科名) ■●

281257 Panicastrella Moench(废弃属名) = Echinaria Desf. (保留属名) ■☆

281258 Panicularia Fabr. = Glyceria R. Br. (保留属名) ■

281259 Panicularia Fabr. = Poa L. ■

281260 Panicularia Heist. ex Fabr. = Glyceria R. Br. (保留属名) ■

281261 Panicularia Heist. ex Fabr. = Poa L. ■

281262 Panicularia acutiflora(Torr. ex Trin.) Kuntze = Glyceria acutiflora Torr. ex Trin. subsp. japonica(Steud.)T. Koyama et Kawano ■

281263 Panicularia acutiflora(Torr.) Kuntze = Glyceria acutiflora Torr. ex Trin. subsp. japonica(Steud.) T. Koyama et Kawano ■

281264 Panicularia borealis Nash = Glyceria borealis(Nash) Batch. ■☆

281265 Panicularia canadensis (Michx.) Kuntze = Glyceria canadensis (Michx.) Trin. ■☆

281266 Panicularia grandis (S. Watson) Nash = Glyceria grandis S. Watson ■☆

281267 Panicularia laxa Scribn. = Glyceria laxa(Scribn.) Scribn. ■☆

281268 Panicularia nervata (Willd.) Kuntze = Glyceria striata (Lam.) Hitchc. ■☆

281269 Panicularia nervata (Willd.) Kuntze var. parviglumis Scribn. et Merr. = Glyceria striata (Lam.) Hitchc. ■☆

281270 Panicularia pallida(Torr.) Kuntze = Puccinellia pallida(Torr.) R. T. Clausen ■☆

281271 Panicularia septentrionalis (Hitchc.) E. P. Bicknell = Glyceria septentrionalis Hitchc. ■☆

281272 Paniculum Ard. = Oplismenus P. Beauv. (保留属名) ■

281273 Paniculum Ard. = Panicum L. ■

281274 Panicum L. (1753);黍属(稷属);Crab Grass, Millet, Panic Grass, Panicgrass, Panic-grass, Panicum, Switch Grass, Witch Grass, Witchgeass, Witch-grass ■

281275 Panicum abludens Roem. et Schult. = Digitaria abludens (Roem. et Schult.) Veldkamp ■

281276 Panicum abyssinicum Hochst. ex A. Rich. = Digitaria abyssinica (Hochst. ex A. Rich.) Stapf ■☆

281277 Panicum abyssinicum Hochst. ex A. Rich. var. velutinum Chiov. = Digitaria pearsonii Stapf ■☆

281278 Panicum acariferum Trin. = Thysanolaena latifolia (Roxb. ex Hornem.) Honda ■

281279 Panicum acariferum Trin. = Thysanolaena maxima (Roxb.) Kuntze ■

281280 Panicum accrescens Trin. = Cyrtococcum patens(L.) A. Camus var. latifolium(Honda) Ohwi ■

281281 Panicum acroanthum Steud. = Panicum bisulcatum Thunb. ■

281282 Panicum acroanthum Steud. var. brevipedicellatum Hack. = Panicum bisulcatum Thunb. ■

281283 Panicum acuminatifolium Robyns = Panicum sadinii (Vanderyst) Renvoize ■☆

281284 Panicum acuminatissimum Steud. = Oplismenus undulatifolius (Ard.) Roem. et Schult. ■

281285　Panicum acuminatum Sw. = Dichanthelium acuminatum（Sw.）Gould et C. A. Clark ■

281286　Panicum acuminatum Sw. var. columbianum（Scribn.）Lelong = Dichanthelium acuminatum（Sw.）Gould et C. A. Clark subsp. columbianum（Scribn.）Freckmann et Lelong ■☆

281287　Panicum acuminatum Sw. var. columbianum（Scribn.）Lelong = Panicum portoricense Desv. ex Ham. ■☆

281288　Panicum acuminatum Sw. var. fasciculatum（Torr.）Lelong = Dichanthelium acuminatum（Sw.）Gould et C. A. Clark subsp. fasciculatum（Torr.）Freckmann et Lelong ■☆

281289　Panicum acuminatum Sw. var. fasciculatum（Torr.）Lelong = Panicum acuminatum Sw. ■

281290　Panicum acuminatum Sw. var. implicatum（Scribn.）Beetle = Panicum acuminatum Sw. ■

281291　Panicum acuminatum Sw. var. implicatum（Scribn.）C. F. Reed = Dichanthelium acuminatum（Sw.）Gould et C. A. Clark subsp. implicatum（Scribn. ex Nash）Freckmann et Lelong ■☆

281292　Panicum acuminatum Sw. var. lindheimeri（Nash）Beetle = Dichanthelium acuminatum（Sw.）Gould et C. A. Clark subsp. lindheimeri（Nash）Freckmann et Lelong ■☆

281293　Panicum acuminatum Sw. var. unciphyllum（Trin.）Lelong = Panicum acuminatum Sw. ■

281294　Panicum acuminatum Sw. var. villosum（A. Gray）Beetle = Panicum acuminatum Sw. ■

281295　Panicum acutiglumum Steud. = Hymenachne acutigluma（Steud.）Gillies ■

281296　Panicum acutiglumum Steud. = Hymenachne amplexicaulis（Rudge）Nees ■

281297　Panicum acutissimum Peter = Panicum poaeoides Stapf ■☆

281298　Panicum addisonii Nash = Dichanthelium ovale（Elliott）Gould et C. A. Clark subsp. pseudopubescens（Nash）Freckmann et Lelong ■☆

281299　Panicum adenophorum K. Schum. ;腺黍■☆

281300　Panicum adenophyllum Pilg. = Panicum adenophorum K. Schum. ■☆

281301　Panicum adhaerens Forssk. = Setaria verticillata（L.）P. Beauv. ■

281302　Panicum adscendens Kunth = Digitaria ciliaris（Retz.）Koeler ■

281303　Panicum aequinerve Nees;等脉黍■☆

281304　Panicum africanum（P. Beauv.）Poir. = Oplismenus hirtellus（L.）P. Beauv. ■☆

281305　Panicum afrum Mez = Panicum phragmitoides Stapf ■☆

281306　Panicum afzelii Sw. ;阿芙泽尔黍■☆

281307　Panicum agrostideum Salzm. ex Steud. = Ichnanthus pallens（Sw.）Munro ex Benth. ■☆

281308　Panicum agrostoides Spreng. = Panicum rigidulum Bosc ex Nees ■☆

281309　Panicum agrostoides Spreng. var. condensum（Nash）Fernald = Panicum rigidulum Bosc ex Nees ■☆

281310　Panicum agrostoides Spreng. var. ramosius（C. Mohr）Fernald = Panicum rigidulum Bosc ex Nees ■☆

281311　Panicum alabamense Ashe = Dichanthelium ovale（Elliott）Gould et C. A. Clark subsp. pseudopubescens（Nash）Freckmann et Lelong ■☆

281312　Panicum albemarlense Ashe = Dichanthelium acuminatum（Sw.）Gould et C. A. Clark subsp. implicatum（Scribn. ex Nash）Freckmann et Lelong ■☆

281313　Panicum albens（Trin.）Steud. = Isachne albens Trin. ■

281314　Panicum albidulum Steud. = Panicum laetum Kunth ■☆

281315　Panicum albovellereum K. Schum. = Brachiaria dictyoneura（Fig. et De Not.）Stapf ■☆

281316　Panicum alopecuroides L. = Pennisetum alopecuroides（L.）Spreng. ■

281317　Panicum amarum Elliott;苦黍;Short Dune-grass ■☆

281318　Panicum ambiguum Trin. = Urochloa paspaloides J. Presl ex C. Presl ■

281319　Panicum ambositrense A. Camus;安布西特拉黍■☆

281320　Panicum americanum L. = Pennisetum americanum（L.）Leeke ■

281321　Panicum americanum L. = Pennisetum glaucum（L.）R. Br. ■

281322　Panicum ammophilum Steud. = Tricholaena capensis（Licht. ex Roem. et Schult.）Nees subsp. arenaria（Nees）Zizka ■☆

281323　Panicum amoenum Balansa;可爱黍;Lovable Panicgrass ■

281324　Panicum amplexicaule Rudge = Hymenachne amplexicaulis（Rudge）Nees ■

281325　Panicum amplexifolium Hochst. = Brachiaria lata（Schumach.）C. E. Hubb. ■☆

281326　Panicum anabaptistum Steud. ;非洲黍■☆

281327　Panicum anceps Michx. ;喙黍;Beaked Panic Grass ■☆

281328　Panicum andongense Rendle = Brachiaria andongensis（Rendle）Stapf ■☆

281329　Panicum andringitrense A. Camus;安德林吉特拉山黍■☆

281330　Panicum angolense（Rendle）K. Schum. = Digitaria angolensis Rendle ■☆

281331　Panicum angustum Trin. = Sacciolepis indica（L.）Chase ■

281332　Panicum anisotrichum Mez = Brachiaria xantholeuca（Hack.）Stapf ■☆

281333　Panicum annulatum A. Rich. = Eriochloa fatmensis（Hochst. et Steud.）Clayton ■☆

281334　Panicum annulum Ashe = Dichanthelium dichotomum（L.）Gould ■☆

281335　Panicum annulum Ashe = Panicum dichotomum L. ■☆

281336　Panicum antidotale Retz. ;蓝黍;Blue Panicum ■☆

281337　Panicum aparine Steud. = Setaria verticillata（L.）P. Beauv. ■

281338　Panicum aphanoneurum Stapf = Panicum fluviicola Steud. ■☆

281339　Panicum appendiculatum Hack. = Setaria appendiculata（Hack.）Stapf ■☆

281340　Panicum appletonii Stapf = Panicum pinifolium Chiov. ■☆

281341　Panicum aquaticum A. Rich. = Panicum hygrocharis Steud. ■☆

281342　Panicum aquaticum Poir. ;水生稷;Aquatic Panicgrass ■☆

281343　Panicum arabicum Steud. = Paspalidium desertorum（A. Rich.）Stapf ■☆

281344　Panicum arborescens L. = Panicum brevifolium L. ■

281345　Panicum arbusculum Mez;树状黍■☆

281346　Panicum arcuatum R. Br. = Sacciolepis indica（L.）Chase ■

281347　Panicum arenarium Brot. = Panicum repens L. ■

281348　Panicum argyrotrichum Andersson = Digitaria argyrotricha（Andersson）Chiov. ■☆

281349　Panicum aridum Mez = Brachiaria arida（Mez）Stapf ■☆

281350　Panicum aristatum Retz. = Oplismenus compositus（L.）P. Beauv. ■

281351　Panicum aristiferum Peter = Echinochloa haploclada（Stapf）Stapf ■☆

281352　Panicum aristulatum Steud. = Digitaria aristulata（Steud.）Stapf ■☆

281353　Panicum arnottianum Nees ex Steud. = Ottochloa nodosa（Kunth）Dandy ■

281354　Panicum arrectum Hack. ex T. Durand et Schinz = Brachiaria arrecta（Hack. ex T. Durand et Schinz）Stent ■☆

281355　Panicum arundinifolium Schweinf. = Panicum deustum Thunb. ■☆

281356　Panicum asperum Lam. = Setaria verticillata（L.）P. Beauv. ■

281357 Panicum assamicum Hook. f. = Hymenachne assamicum (Hook.) Hitchc. ■

281358 Panicum atrofuscum Hack. = Digitaria atrofusca (Hack.) A. Camus ■☆

281359 Panicum atrosanguineum Hochst. ex A. Rich. ;深褐稷■☆

281360 Panicum atroviolaceum A. Rich. = Echinochloa pyramidalis (Lam.) Hitchc. et Chase ■☆

281361 Panicum aubertii Mez = Alloteropsis paniculata(Benth.)Stapf ■☆

281362 Panicum auritum J. Presl ex Nees;紧序黍;Aurtcled Panicgrass ■

281363 Panicum austroasiaticum Ohwi = Panicum humile Nees ex Steud. ■

281364 Panicum austroasiaticum Ohwi = Panicum walense Mez ■

281365 Panicum bambusiusculum Stapf;竹秆黍■☆

281366 Panicum barbatum Lam. = Setaria barbata(Lam.) Kunth ■☆

281367 Panicum barbifultum Hochst. ex Schltdl. = Oplismenus undulatifolius(Ard.) Roem. et Schult. ■

281368 Panicum barbigerum Bertol. = Setaria sagittifolia (A. Rich.) Walp. ■☆

281369 Panicum barbinode Trin. = Brachiaria mutica(Forssk.)Stapf ■

281370 Panicum barbipedum Hayata;刺髭柄稷(刺髭稷);Bearded-stalk Panicgrass ■☆

281371 Panicum barbipulvinatum Nash;叶枕稷;Barbed Panic-grass ■☆

281372 Panicum barbipulvinatum Nash = Panicum capillare L. ■☆

281373 Panicum barbulatum Michx. = Dichanthelium dichotomum(L.) Gould ■☆

281374 Panicum basisetum Steud. = Setaria barbata(Lam.) Kunth ■☆

281375 Panicum baumannii K. Schum. = Panicum nervatum (Franch.) Stapf ■☆

281376 Panicum beccabunga Rendle = Panicum parvifolium Lam. ■☆

281377 Panicum bechuanense Bremek. et Oberm. ;贝专黍■☆

281378 Panicum bellum Steud. = Isachne pulchella Roth ■

281379 Panicum bengalense Spreng. = Arundinella bengalensis (Spreng.) Druce ■

281380 Panicum benthami Steud. = Alloteropsis paniculata (Benth.) Stapf ■☆

281381 Panicum bergi Arechav. ;贝格黍;Berg's Panicgrass ■☆

281382 Panicum bicknellii Nash = Dichanthelium boreale (Nash) Freckmann ■☆

281383 Panicum bicknellii Nash = Panicum boreale Nash ■☆

281384 Panicum bicknellii Nash var. bushii (Nash) Farw. = Dichanthelium boreale(Nash)Freckmann ■☆

281385 Panicum bicorne (Lam.) Kunth = Digitaria bicornis (Lam.) Roem. et Schult. ■

281386 Panicum bicorne Lam. = Digitaria bicornis (Lam.) Roem. et Schult. ■

281387 Panicum bifalcigerum Stapf = Brachiaria platynota (K. Schum.) Robyns ■☆

281388 Panicum biforme (Wild.) Kunth = Digitaria ciliaris (Retz.) Koeler ■

281389 Panicum biforme Willd. = Digitaria bicornis (Lam.) Roem. et Schult. ■

281390 Panicum bisulcatum Thunb. ;糠稷(糠黍);Chaff Panicgrass, Japanese Panicgrass ■

281391 Panicum bolbodes (Hochst. ex Steud.) Asch. = Urochloa oligotricha(Fig. et De Not.)Henrard ■☆

281392 Panicum boliviense Hack. ;玻利维亚黍;Bolivian Panicgrass ■☆

281393 Panicum bongaense Pilg. = Setaria homonyma(Steud.)Chiov. ■☆

281394 Panicum boreale Nash;北方稷;Boreal Panic-grass,Panic Grass ■☆

281395 Panicum boreale Nash = Dichanthelium boreale (Nash) Freckmann ■☆

281396 Panicum boreale Nash var. michiganense Farw. = Dichanthelium boreale(Nash)Freckmann ■☆

281397 Panicum borzianum Mattei = Urochloa panicoides P. Beauv. ■

281398 Panicum boscii Poir. ; 博斯克黍; Bosc's Panic-grass, Panic Grass ■☆

281399 Panicum boscii Poir. var. molle (Vasey) Hitchc. et Chase = Panicum boscii Poir. ■☆

281400 Panicum bovonei Chiov. = Brachiaria bovonei(Chiov.)Robyns ■☆

281401 Panicum brachiariaeforme Steud. = Echinochloa colona (L.) Link ■

281402 Panicum brachylachmum Steud. = Brachiaria ramosa(L.)Stapf ■

281403 Panicum brachythyrsum Peter = Urochloa reptans(L.) Stapf ■

281404 Panicum brachyurum Hack. = Urochloa brachyura(Hack.)Stapf ■☆

281405 Panicum braunii Steud. = Melinis repens(Willd.) Zizka ■

281406 Panicum brazzae Franch. = Digitaria brazzae (Franch.)Stapf ■☆

281407 Panicum brazzavillense Franch. ;布拉柴维尔黍■☆

281408 Panicum brevifolium L. ;短叶黍;Shortleaf Panicgrass ■

281409 Panicum brevifolium L. var. heterostachyum Peter = Panicum heterostachyum Hack. ■☆

281410 Panicum brevifolium L. var. hirtifolium (Ridl.) Jansen = Panicum brevifolium L. ■

281411 Panicum brevipedicellatum Peter = Echinochloa brevipedicellata (Peter) Clayton ■☆

281412 Panicum brevispicatum Rendle = Brachiaria brevispicata (Rendle) Stapf ■☆

281413 Panicum brizoides Jacq. = Paspalidium flavidum (Retz.) A. Camus ■

281414 Panicum brizoides L. = Echinochloa colona(L.) Link ■

281415 Panicum brodiei H. St. John = Dichanthelium acuminatum(Sw.) Gould et C. A. Clark subsp. fasciculatum (Torr.) Freckmann et Lelong ■☆

281416 Panicum bromoides Lam. = Oplismenus burmannii (Retz.) P. Beauv. ■

281417 Panicum brunonianum Wall. et Griff. = Pseudoraphis brunoniana (Wall. et Griff.) Pilg. ■

281418 Panicum bulawayense Hack. = Urochloa oligotricha (Fig. et De Not.) Henrard ■☆

281419 Panicum bulbodes (Hochst. ex Steud.) A. Chev. = Urochloa oligotricha(Fig. et De Not.) Henrard ■☆

281420 Panicum bulbosum Kunth = Zuloagaea bulbosa(Kunth) Bess ■☆

281421 Panicum bullockii Renvoize;布洛克黍■☆

281422 Panicum burgu A. Chev. = Echinochloa stagnina (Retz.) P. Beauv. ■☆

281423 Panicum burmannii Retz. = Oplismenus burmannii (Retz.) P. Beauv. ■

281424 Panicum bushii Nash = Dichanthelium boreale (Nash) Freckmann ■☆

281425 Panicum busseanum Mez = Melinis nerviglumis(Franch.)Zizka ■☆

281426 Panicum caerulescens Hack. ex Hitchc. = Dichanthelium dichotomum(L.) Gould ■☆

281427 Panicum caesium Nees = Panicum cambogiense Balansa ■

281428 Panicum caffrorum Retz. = Sorghum caffrorum (Thunb.) P. Beauv. ■

281429 Panicum calliphyllum Ashe = Dichanthelium boreale (Nash)

Freckmann ■☆

281430 Panicum calliphyllum Ashe = Panicum boreale Nash ■☆

281431 Panicum callopus Pilg. = Echinochloa callopus(Pilg.)Clayton ■☆

281432 Panicum callosum Hochst. ex A. Rich. ;硬皮黍■☆

281433 Panicum calocarpum Berhaut;美果黍■☆

281434 Panicum calvum Stapf;光秃黍■☆

281435 Panicum cambogiense Balansa = Panicum luzonense J. Presl et C. Presl ■

281436 Panicum canescens Roth ex Roem. et Schult. = Brachiaria ramosa(L.)Stapf ■

281437 Panicum capense Licht. ex Roem. et Schult. = Tricholaena capensis(Licht. ex Roem. et Schult.)Nees ■☆

281438 Panicum capillare L. ;毛线稷(纤枝稷);Capillary Panic-grass, Common Panicgrass, Common Witch Grass, Common Witchgrass, Common Witch-grass, Old Witch Grass, Old-witch Grass, Witch Grass,Witch-grass ■☆

281439 Panicum capillare L. subsp. barbipulvinatum (Nash)Tzvelev = Panicum capillare L. ■☆

281440 Panicum capillare L. var. agreste Gatt. = Panicum capillare L. ■☆

281441 Panicum capillare L. var. barbipulvinatum (Nash)McGregor = Panicum capillare L. ■☆

281442 Panicum capillare L. var. brevifolium Vasey ex Rydb. et Shear = Panicum capillare L. ■☆

281443 Panicum capillare L. var. flexile(Gatt.)Scribn. ■☆

281444 Panicum capillare L. var. occidentale Rydb. = Panicum capillare L. ■☆

281445 Panicum capillare L. var. sylvaticum Torr. = Panicum philadelphicum Bernh. ex Trin. ■☆

281446 Panicum carinatum J. Presl et C. Presl = Cyrtococcum patens (L.)A. Camus ■

281447 Panicum carneovaginatum Renvoize;肉斑黍■☆

281448 Panicum catangense Chiov. = Panicum ecklonii Nees ■☆

281449 Panicum catumbense Rendle = Sacciolepis catumbensis (Rendle)Stapf ■☆

281450 Panicum caucasicum Trin. = Brachiaria eruciformis (Sm.)Griseb. ■

281451 Panicum cauda-ratti Schumach. = Pennisetum polystachion(L.)Schult. ■

281452 Panicum caudiglume Stapf = Panicum delicatulum Fig. et De Not. ■☆

281453 Panicum cenchroides Rich. = Pennisetum polystachion (L.)Schult. ■

281454 Panicum cenchroides Rich. = Pennisetum setosum(Sw.)Rich. ■

281455 Panicum chaetophoron Roem. et Schult. = Cyrtococcum chaetophoron(Roem. et Schult.)Dandy ■☆

281456 Panicum chinense Trin. = Setaria italica(L.)P. Beauv. ■

281457 Panicum chionachne Mez;雪黍■☆

281458 Panicum chlorochloe K. Schum. = Hylebates chlorochloe (K. Schum.)Napper ■☆

281459 Panicum chloroticum Nees ex Trin. ;绿稷;Green Panicgrass ■☆

281460 Panicum chondrachne Steud. = Setaria chondrachne (Steud.)Honda ■

281461 Panicum chromatostigma Pilg. = Panicum repens L. ■

281462 Panicum chusqueoides Hack. = Brachiaria chusqueoides (Hack.)Clayton ■☆

281463 Panicum ciliare Retz. = Digitaria bicornis (Lam.)Roem. et Schult. ■

281464 Panicum ciliare Retz. = Digitaria ciliaris(Retz.)Koeler ■

281465 Panicum ciliocinctum Pilg. = Sacciolepis ciliocincta (Pilg.)Stapf ■☆

281466 Panicum cimicinum(L.)Retz. = Alloteropsis cimicina(L.)Stapf ■

281467 Panicum cinctum Hack. ;围绕黍■☆

281468 Panicum cinereo-vestitum Pilg. = Sacciolepis typhura (Stapf)Stapf ■☆

281469 Panicum cinereo-viride Mez = Panicum anabaptistum Steud. ■☆

281470 Panicum clandestinum L. = Dichanthelium clandestinum (L.)Gould ■☆

281471 Panicum clavipilum Chiov. = Brachiaria clavipila (Chiov.)Robyns ■☆

281472 Panicum claytonii Renvoize;克莱顿黍■☆

281473 Panicum cocoospermum Steud. = Brachiaria villosa(Lam.)A. Camus ■

281474 Panicum cognatissimum Steud. = Brachiaria ramosa(L.)Stapf ■

281475 Panicum cognatum Schult. = Digitaria cognata(Schult.)Pilg. ■☆

281476 Panicum collare Schumach. = Brachiaria falcifera(Trin.)Stapf ■☆

281477 Panicum colonum L. = Echinochloa colona(L.)Link ■

281478 Panicum colonum L. var. bussei Peter = Echinochloa cruspavonis(Kunth)Schult. ■

281479 Panicum coloratum L. ; 光头黍; Buffalo Grass, Coloured Panicgrass,Klinegrass ■

281480 Panicum coloratum L. = Panicum bisulcatum Thunb. ■

281481 Panicum coloratum L. var. makarikariense Gooss. = Panicum coloratum L. ■

281482 Panicum coloratum L. var. minus Chiov. ;小光头黍■☆

281483 Panicum coloratum Walter = Panicum virgatum L. ■

281484 Panicum columbianum Scribn. = Dichanthelium acuminatum (Sw.)Gould et C. A. Clark subsp. columbianum (Scribn.)Freckmann et Lelong ■☆

281485 Panicum columbianum Scribn. = Panicum portoricense Desv. ex Ham. ■☆

281486 Panicum columbianum Scribn. var. commonsianum (Ashe)McNeill et Dore = Dichanthelium ovale(Elliott)Gould et C. A. Clark subsp. pseudopubescens(Nash)Freckmann et Lelong ■☆

281487 Panicum columbianum Scribn. var. oricola (Hitchc. et Chase)Fernald = Dichanthelium acuminatum (Sw.)Gould et C. A. Clark subsp. implicatum(Scribn. ex Nash)Freckmann et Lelong ■☆

281488 Panicum columbianum Scribn. var. thinium Hitchc. et Chase = Dichanthelium acuminatum (Sw.)Gould et C. A. Clark subsp. implicatum(Scribn. ex Nash)Freckmann et Lelong ■☆

281489 Panicum comatum Hochst. ex A. Rich. = Brachiaria comata (Hochst. ex A. Rich.)Stapf ■☆

281490 Panicum commonsianum Ashe = Dichanthelium ovale (Elliott)Gould et C. A. Clark subsp. pseudopubescens (Nash)Freckmann et Lelong ■☆

281491 Panicum commonsianum Ashe var. addisonii (Nash)Fernald = Dichanthelium ovale (Elliott)Gould et C. A. Clark subsp. pseudopubescens(Nash)Freckmann et Lelong ■☆

281492 Panicum commonsianum Ashe var. addisonii (Nash)R. W. Pohl = Dichanthelium ovale (Elliott)Gould et C. A. Clark subsp. pseudopubescens(Nash)Freckmann et Lelong ■☆

281493 Panicum commonsianum Ashe var. euchlamydeum(Shinners)R. W. Pohl = Dichanthelium ovale(Elliott)Gould et C. A. Clark subsp. pseudopubescens(Nash)Freckmann et Lelong ■☆

281494 Panicum commutatum Roxb. var. nodosum (Parl.) Hack. ex T. Durand et Schinz = Digitaria nodosa Parl. ■☆

281495 Panicum commutatum Schult. ;变异黍;Panic Grass ■☆

281496 Panicum commutatum Schult. var. nodosum (Parl.) T. Durand et Schinz = Digitaria nodosa Parl. ■☆

281497 Panicum comorense Mez;科摩罗黍■☆

281498 Panicum compositum L. = Oplismenus compositus (L.) P. Beauv. ■

281499 Panicum compressum Forssk. = Ochthochloa compressa (Forssk.) Hilu ■☆

281500 Panicum conglomeratum L. = Sacciolepis indica(L.) Chase ■

281501 Panicum congoense Franch. ;刚果黍■☆

281502 Panicum contractum Wight et Arn. ex Nees = Sacciolepis indica (L.) Chase ■

281503 Panicum controversum Steud. = Urochloa panicoides P. Beauv. ■

281504 Panicum convolutum P. Beauv. ex Spreng. = Panicum repens L. ■

281505 Panicum cooperi C. E. Hubb. = Panicum hochstetteri Steud. ■☆

281506 Panicum cordatum Büse = Panicum notatum Retz. ■

281507 Panicum corymbiferum Steud. = Panicum deustum Thunb. ■☆

281508 Panicum corymbosum Roxb. = Digitaria setigera Roth ex Roem. et Schult. ■

281509 Panicum crassiapiculatum Merr. = Acroceras munroanum (Balansa) Henrard ■

281510 Panicum crassipes Mez = Panicum coloratum L. ■

281511 Panicum crispum Llanos = Panicum sumatrense Roth ex Roem. et Schult. ■

281512 Panicum cristatellum Keng;冠稷(小冠稷);Crested Panicgrass ■

281513 Panicum cristatum Andersson = Digitaria perrottetii (Kunth) Stapf ■☆

281514 Panicum cruciabile Chase = Panicum cambogiense Balansa ■

281515 Panicum cruciabile Chase = Panicum luzonense J. Presl et C. Presl ■

281516 Panicum cruciatum Nees ex Steud. = Digitaria cruciata(Nees ex Steud.) A. Camus ■

281517 Panicum crus-galli (L.) P. Beauv. var. brevisetum Döll = Echinochloa crus-galli(L.) P. Beauv. var. breviseta (Döll) Podpéra ■

281518 Panicum crus-galli L. = Echinochloa crus-galli(L.) P. Beauv. ■

281519 Panicum crus-galli L. subsp. colonum(L.) Makino et Nemoto = Echinochloa colona(L.) Link ■

281520 Panicum crus-galli L. subsp. colonum Makino et Nemoto = Echinochloa colona(L.) Link ■

281521 Panicum crus-galli L. var. brevisetum Döll = Echinochloa crus-galli(L.) P. Beauv. var. breviseta(Döll) Neilr. ■

281522 Panicum crus-galli L. var. brevisetum Döll = Echinochloa crus-galli(L.) P. Beauv. ■

281523 Panicum crus-galli L. var. colonum (L.) Coss. = Echinochloa colona(L.) Link ■

281524 Panicum crus-galli L. var. frumentacum (Roxb.) Trin. = Echinochloa frumentacea(Roxb.) Link ■

281525 Panicum crus-galli L. var. hispidulum (Retz.) Hack. = Echinochloa hispidula(Retz.) Nees ■

281526 Panicum crus-galli L. var. longisetum Doell. = Echinochloa crus-galli(L.) P. Beauv. ■

281527 Panicum crus-galli L. var. mite Pursh = Echinochloa crus-galli (L.) P. Beauv. var. mitis(Pursh) Peterm. ■

281528 Panicum crus-galli L. var. submuticum Mey. = Echinochloa crus-galli(L.) P. Beauv. var. praticola Ohwi ■

281529 Panicum crus-pavonis(Kunth) Nees = Echinochloa crus-pavonis (Kunth) Schult. ■

281530 Panicum cupressifolium A. Camus;铜花黍■☆

281531 Panicum curvatum L. = Sacciolepis curvata(L.) Chase ■☆

281532 Panicum curviflorum Hornem. ;弯花黍(旱黍草,毛叶黍,弯花稷);Curvateflower Panicgrass, Drought Panicgrass ■

281533 Panicum curviflorum Hornem. var. suishanense (Hayata) Veldkamp; 水社黍 (台岛稷); Suisha Panicgrass, Sunmoon Panicgrass ■

281534 Panicum curviflorum Hornem. var. suishanense (Hayata) Veldkamp = Panicum elegantissimum Hook. f. ■

281535 Panicum cyrtococcoides Napper = Cyrtococcum multinode (Lam.) Clayton ■☆

281536 Panicum dactylon L. = Cynodon dactylon(L.) Pers. ■

281537 Panicum daltonii Parl. = Echinochloa colona(L.) Link ■

281538 Panicum debile Desf. = Digitaria debilis(Desf.) Willd. ■☆

281539 Panicum decaryanum A. Camus;德卡里黍■☆

281540 Panicum decempedale Kuntze = Arundinella decempedalis (Kuntze) Janowski ■

281541 Panicum decompositum Rendle = Panicum paludosum Roxb. ■

281542 Panicum decompositum R. Br. ;多子黍■

281543 Panicum deflexum Schumach. = Brachiaria deflexa (Schumach.) C. E. Hubb. ex Robyns ■☆

281544 Panicum delicatulum Fig. et De Not. ;姣美黍■☆

281545 Panicum denudatum (Link) Kunth = Digitaria stricta Roth ex Roem. et Schult. ■

281546 Panicum denudatum (Link) Kunth = Digitaria stricta Roth ex Roem. et Schult. var. denudata (Link) Henrard ■

281547 Panicum denudatum Kunth = Digitaria denudata Link ■

281548 Panicum depauperatum Muhl. = Dichanthelium depauperatum (Muhl.) Gould ■☆

281549 Panicum depauperatum Muhl. var. involutum (Torr.) A. W. Wood = Dichanthelium depauperatum(Muhl.) Gould ■☆

281550 Panicum depauperatum Muhl. var. psilophyllum Fernald = Dichanthelium depauperatum(Muhl.) Gould ■☆

281551 Panicum desertorum A. Rich. = Paspalidium desertorum (A. Rich.) Stapf ■☆

281552 Panicum despreauxii Steud. = Brachiaria villosa (Lam.) A. Camus ■

281553 Panicum deustum Thunb. ;焦黍■☆

281554 Panicum deustum Thunb. var. eburneum Chiov. = Panicum deustum Thunb. ■☆

281555 Panicum deustum Thunb. var. hirsutum Peter = Panicum deustum Thunb. ■☆

281556 Panicum dewinteri J. G. Anderson;德温特黍■

281557 Panicum diagonale Nees = Digitaria diagonalis(Nees) Stapf ■☆

281558 Panicum diagonale Nees var. glabrescens K. Schum. = Digitaria diagonalis(Nees) Stapf var. uniglumis(Hochst. ex A. Rich.) Pilg. ■☆

281559 Panicum diagonale Nees var. hirsutum De Wild. et T. Durand = Digitaria diagonalis(Nees) Stapf var. hirsuta(De Wild. et T. Durand) Troupin ■☆

281560 Panicum diamesum Steud. = Digitaria nuda Schumach. ■☆

281561 Panicum dichotomiflorum Michx. ;洋野黍;Autumn Millet, Fall Panic Grass, Fall Panicgrass, Fall Panicum, Knee Grass, Zigzag Grass ■

281562 Panicum dichotomiflorum Michx. var. geniculatum (A. W. Wood) Fernald = Panicum dichotomiflorum Michx. ■

281563 Panicum dichotomiflorum Michx. var. imperiorum Fernald =

Panicum dichotomiflorum Michx. ■

281564　Panicum dichotomum Forssk. = Pennisetum divisum（Forssk. ex J. F. Gmel.）Henrard ■☆

281565　Panicum dichotomum L.；二歧野黍；Panic Grass ■☆

281566　Panicum dichotomum L. = Dichanthelium dichotomum（L.）Gould ■☆

281567　Panicum dichotomum L. var. barbulatum（Michx.）A. W. Wood = Dichanthelium dichotomum（L.）Gould ■☆

281568　Panicum dichotomum Michx. var. barbulatum（Michx.）A. W. Wood = Panicum dichotomum L. ■☆

281569　Panicum dichotomum Michx. var. fasciculatum Torr. = Dichanthelium acuminatum（Sw.）Gould et C. A. Clark ■

281570　Panicum dichotomum Michx. var. lanuginosum（Elliott）A. W. Wood = Dichanthelium acuminatum（Sw.）Gould et C. A. Clark ■

281571　Panicum dictyoneurum Fig. et De Not. = Brachiaria dictyoneura（Fig. et De Not.）Stapf ■☆

281572　Panicum dilatatum Steud. = Digitaria setigera Roth ex Roem. et Schult. ■

281573　Panicum dimidiatum L. = Stenotaphrum dimidiatum（L.）Brongn. ■

281574　Panicum dinklagei Mez；丁克黍■☆

281575　Panicum distachyon L. = Brachiaria distachya（L.）Stapf ■☆

281576　Panicum distachyon L. = Urochloa distachya（L.）T. Q. Nguyen ■☆

281577　Panicum distichophylloides Mez = Brachiaria xantholeuca（Hack.）Stapf ■☆

281578　Panicum distichophyllum Trin. = Brachiaria villosa（Lam.）A. Camus ■

281579　Panicum diversinerve Nees = Digitaria diversinervis（Nees）Stapf ■☆

281580　Panicum divisum Forssk. ex J. F. Gmel. = Pennisetum divisum（Forssk. ex J. F. Gmel.）Henrard ■☆

281581　Panicum djalonense A. Chev. = Panicum hymeniochilum Nees ■☆

281582　Panicum doloense Vanderyst；多罗黍■☆

281583　Panicum dorsisetum Hack. ex T. Durand et Schinz = Urochloa trichopus（Hochst.）Stapf ■☆

281584　Panicum dregeanum Nees；德雷黍■☆

281585　Panicum drosocarpum Stapf = Panicum tenellum Lam. ■☆

281586　Panicum eburneum Trin. = Panicum maximum Jacq. ■

281587　Panicum echinochloa T. Durand et Schinz = Echinochloa colona（L.）Link ■

281588　Panicum ecklonii Nees；埃氏黍■☆

281589　Panicum eickii Mez；艾克黍■☆

281590　Panicum elegans Wight et Arn. ex Steud. = Sphaerocaryum malaccense（Trin.）Pilg. ■

281591　Panicum elegantissimum Hook. f.；旱黍草；Elegant Panicgrass ■☆

281592　Panicum elongatum Mez = Melinis nerviglumis（Franch.）Zizka ■☆

281593　Panicum elongatum Pursh var. ramosius C. Mohr = Panicum rigidulum Bosc ex Nees ■☆

281594　Panicum emergens Hochst. = Paspalidium geminatum（Forssk.）Stapf ■

281595　Panicum eminii Mez = Brachiaria eminii（Mez）Robyns ■☆

281596　Panicum endolasion Mez ex Peter = Entolasia imbricata Stapf ■☆

281597　Panicum ephemerum Renvoize；短命黍■☆

281598　Panicum equitans Hochst. ex A. Rich. = Echinochloa colona（L.）Link ■

281599　Panicum erubescens Willd. = Pennisetum polystachion（L.）Schult. ■

281600　Panicum erubescens Willd. = Pennisetum setosum（Sw.）Rich. ■

281601　Panicum eruciforme Sm. = Brachiaria eruciformis（Sm.）Griseb. ■

281602　Panicum eruciformis（Sibth. et Sm.）Rchb. = Brachiaria eruciformis（Sm.）Griseb. ■

281603　Panicum esculentum A. Braun = Echinochloa esculenta（A. Braun）H. Scholz ■

281604　Panicum euchlamydeum Shinners = Dichanthelium ovale（Elliott）Gould et C. A. Clark subsp. pseudopubescens（Nash）Freckmann et Lelong ■☆

281605　Panicum euryphyllum Peter = Urochloa setigera（Retz.）Stapf ■☆

281606　Panicum excurrens Trin. = Setaria plicata（Lam.）T. Cooke ■

281607　Panicum falciferum Trin. = Brachiaria falcifera（Trin.）Stapf ■☆

281608　Panicum fatmense Hochst. et Steud. = Eriochloa fatmensis（Hochst. et Steud.）Clayton ■☆

281609　Panicum fenestratum A. Rich. = Digitaria velutina（Forssk.）P. Beauv. ■☆

281610　Panicum fibrosum Hack. = Digitaria fibrosa（Hack.）Stapf ex Craib ■

281611　Panicum fibrosum Hack. = Digitaria setifolia Stapf ■☆

281612　Panicum figarianum Chiov. = Panicum delicatulum Fig. et De Not. ■☆

281613　Panicum filicaule Stapf = Panicum tenellum Lam. ■☆

281614　Panicum filiculme Hack. = Panicum hymeniochilum Nees ■☆

281615　Panicum filifolium Clayton；线叶黍■☆

281616　Panicum flacciflorum Stapf；柔花黍■☆

281617　Panicum flavidum Retz. = Paspalidium flavidum（Retz.）A. Camus ■

281618　Panicum flexile（Gatt.）Scribn.；纤细黍草；Slender Panic Grass，Wiry Panic Grass，Wiry Panic-grass，Wiry Witch Grass，Wiry Witchgrass ■☆

281619　Panicum flexuosum Retz. = Panicum psilopodium Trin. ■

281620　Panicum flexuosum Retz. = Panicum sumatrense Roth ex Roem. et Schult. ■

281621　Panicum flexuosum Retz. var. glabrum Retz. = Panicum psilopodium Trin. ■

281622　Panicum floridum Royle = Paspalidium flavidum（Retz.）A. Camus ■

281623　Panicum fluitans Retz. = Paspalidium geminatum（Forssk.）Stapf ■

281624　Panicum fluviicola Steud.；河岸黍■☆

281625　Panicum forbesianum Nees ex Steud. = Setaria forbesiana（Nees）Hook. f. ■

281626　Panicum formosanum（Rendle）Makino et Nemoto = Digitaria radicosa（J. Presl et C. Presl）Miq. ■

281627　Panicum formosanum Makino et Nemota = Digitaria radicosa（J. Presl et C. Presl）Miq. ■

281628　Panicum frederici Rendle = Panicum brazzavillense Franch. ■☆

281629　Panicum frederici Rendle var. minor？= Panicum brazzavillense Franch. ■☆

281630　Panicum frumentaceum Roxb. = Echinochloa frumentacea（Roxb.）Link ■

281631　Panicum frutescens Mez；灌木状黍■☆

281632　Panicum fulgens Stapf = Panicum nervatum（Franch.）Stapf ■☆

281633　Panicum fulgens Stapf var. pubescens Robyns = Panicum nervatum（Franch.）Stapf ■☆

281634　Panicum funaense Vanderyst；富纳黍■☆

281635　Panicum fuscoviolaceum Peter = Urochloa echinolaenoides Stapf ■☆

281636　Panicum gabunense Hack. = Acroceras gabunense（Hack.）

Clayton ■☆

281637　Panicum galli Thunb. = Echinochloa stagnina(Retz.)P. Beauv. ■☆

281638　Panicum gattingeri Nash;加氏黍草;Gattinger's Panic-grass ■☆

281639　Panicum gattingeri Nash = Panicum philadelphicum Bernh. ex Trin. ■☆

281640　Panicum gayanum Kunth = Digitaria gayana(Kunth) A. Chev. ex Stapf ■☆

281641　Panicum geminatum Forssk. = Paspalidium geminatum (Forssk.) Stapf ■

281642　Panicum geminatum Forssk. = Paspalidium punctatum (Burm. f.) A. Camus ■

281643　Panicum geniculatum Lam. = Setaria geniculata (Lam.) P. Beauv. ■

281644　Panicum geniculatum Poir. = Setaria parviflora (Poir.) Kerguélen ■

281645　Panicum genuflexum Stapf;膝曲黍■☆

281646　Panicum geometra Chiov. = Brachiaria platynota (K. Schum.) Robyns ■☆

281647　Panicum germanicum Mill. = Setaria italica(L.)P. Beauv. ■

281648　Panicum germanicum Mill. = Setaria italica P. Beauv. var. germanica(Mill.) Schrad. ■

281649　Panicum giganteum Mez = Panicum maximum Jacq. ■

281650　Panicum giganteum Scheele = Panicum virgatum L. ■

281651　Panicum gilvum Launert;淡黄褐黍■☆

281652　Panicum gimmae Fiori = Acroceras macrum Stapf ■☆

281653　Panicum glaberrimum Steud. = Panicum virgatum L. ■

281654　Panicum glabrescens Steud. = Panicum subalbidum Kunth ■☆

281655　Panicum glabrum (Schrad.) Gaudin = Digitaria ischaemum (Schreb.) Schreb. ex Muhl. ■

281656　Panicum glanduliferum K. Schum. ;腺体黍■☆

281657　Panicum glandulopaniculatum Renvoize;腺锥黍■☆

281658　Panicum glandulosum Nees ex Trin. = Pseudechinolaena polystachya(Kunth) Stapf ■

281659　Panicum glaucidulum Peter = Sacciolepis indica(L.) Chase ■

281660　Panicum glaucifolium Hitchc. ;灰绿叶黍■☆

281661　Panicum glaucocladum C. E. Hubb. ;灰绿枝黍■☆

281662　Panicum glaucum L. = Pennisetum americanum(L.) Leeke ■

281663　Panicum glaucum L. = Pennisetum glaucum(L.) R. Br. ■

281664　Panicum glaucum L. = Setaria pallidifusca(Schumach.) Stapf et C. E. Hubb. ■

281665　Panicum globulosum Mez = Panicum nervatum(Franch.)Stapf ■☆

281666　Panicum glomeratum Hack. ex Schinz = Brachiaria glomerata (Stapf) A. Camus ■☆

281667　Panicum glutinoscabrum Fernald = Dichanthelium acuminatum (Sw.) Gould et C. A. Clark subsp. fasciculatum(Torr.) Freckmann et Lelong ■☆

281668　Panicum golae Chiov. = Brachiaria humidicola (Rendle) Schweick. ■☆

281669　Panicum gossypinum A. Rich. = Brachiaria serrata (Thunb.) Stapf ■☆

281670　Panicum gracilicaule Rendle;细茎黍■☆

281671　Panicum graciliculme Napper = Panicum poaeoides Stapf ■☆

281672　Panicum graciliflorum Rendle = Panicum fluviicola Steud. ■☆

281673　Panicum gracillimum K. Schum. = Tricholaena monachne (Trin.)Stapf et C. E. Hubb. ■☆

281674　Panicum gracillimum Mez = Melinis nerviglumis(Franch.)Zizka ■☆

281675　Panicum granulare Lam. = Paspalidium flavidum (Retz.) A. Camus ■

281676　Panicum griffonii Franch. ;格里丰黍■☆

281677　Panicum grossarium L. = Panicum bisulcatum Thunb. ■

281678　Panicum grossum Salisb. = Echinochloa crus-galli (L.) P. Beauv. ■

281679　Panicum hackelii Pilg. = Digitaria abyssinica (Hochst. ex A. Rich.) Stapf ■☆

281680　Panicum hamadense Mez = Brachiaria lata (Schumach.) C. E. Hubb. ■☆

281681　Panicum hanningtonii Stapf;汉宁顿黍■☆

281682　Panicum haplocaulos Pilg. ;单茎黍■☆

281683　Panicum haplocladum Stapf = Echinochloa haploclada (Stapf) Stapf ■☆

281684　Panicum hayatae (Honda) Makino et Nemoto = Digitaria mollicoma(Kunth) Henrard ■

281685　Panicum hayatae (Honda) Makino et Nemoto var. magnum (Honda) Makino et Nemoto = Digitaria mollicoma(Kunth) Henrard ■

281686　Panicum hayatae Makino et Nemoto = Digitaria mollicoma (Kunth) Henrard ■

281687　Panicum helleri Nash = Dichanthelium oligosanthes (Schult.) Gould subsp. scribnerianum(Nash) Freckmann et Lelong ■☆

281688　Panicum helopus Trin. = Urochloa panicoides P. Beauv. ■

281689　Panicum helopus Trin. f. glabrescens K. Schum. = Urochloa panicoides P. Beauv. ■

281690　Panicum helopus Trin. var. glabrescens (K. Schum.) Stapf = Urochloa panicoides P. Beauv. ■

281691　Panicum henryi (Rendle) Makino et Nemoto = Digitaria henryi Rendle ■

281692　Panicum hensii K. Schum. = Acroceras gabunense (Hack.) Clayton ■☆

281693　Panicum hermaphroditum Steud. = Cyrtococcum oxyphyllum (Hochst. ex Steud.) Stapf ■

281694　Panicum heteranthum Link = Pseudechinolaena polystachya (Kunth) Stapf ■

281695　Panicum heteranthum Nees et Meyen = Digitaria heterantha (Hook. f.) Merr. ■

281696　Panicum heteranthum Nees et Meyen ex Nees = Digitaria mollicoma(Kunth) Henrard ■

281697　Panicum heteranthum Nees et Meyen ex Nees var. pachyrhachis Hack. = Digitaria mollicoma(Kunth) Henrard ■

281698　Panicum heterochlamys Peter = Pseudechinolaena polystachya (Kunth) Stapf ■

281699　Panicum heterocraspedum Peter = Brachiaria scalaris Pilg. ■☆

281700　Panicum heterophyllum Bosc ex Nees = Dichanthelium acuminatum (Sw.) Gould et C. A. Clark subsp. columbianum (Scribn.) Freckmann et Lelong ■☆

281701　Panicum heterophyllum Bosc ex Nees var. thinium (Hitchc. et Chase)F. T. Hubb. = Dichanthelium acuminatum (Sw.) Gould et C. A. Clark subsp. implicatum(Scribn. ex Nash) Freckmann et Lelong ■☆

281702　Panicum heterostachyum Hack. ;异穗黍■☆

281703　Panicum heynii Roth = Panicum maximum Jacq. ■

281704　Panicum hians Elliott;张口黍;Panic Grass ■☆

281705　Panicum hians Elliott = Steinchisma hians(Elliott)Nash et Small ■☆

281706　Panicum hirsutissimum Steud. = Panicum maximum Jacq. ■

281707　Panicum hirsutulum Rendle = Panicum griffonii Franch. ■☆

281708　Panicum hirsutum Koen. ex Roxb. = Urochloa panicoides P. Beauv. ■

281709　Panicum hirtellum L. = Oplismenus hirtellus(L.)P. Beauv. ■☆

281710　Panicum hirtifolium Ridl. = Panicum brevifolium L. ■

281711　Panicum hirtiglume H. Scholz;毛颖黍■☆

281712　Panicum hispidissimum Peter = Panicum merkeri Mez ■☆

281713　Panicum hispidissimum Peter var. gracile？ = Panicum merkeri Mez ■☆

281714　Panicum hispidulum Retz. = Echinochloa crus-galli (L.) P. Beauv. ■

281715　Panicum hispidulum Retz. = Echinochloa hispidula(Retz.)Nees ■

281716　Panicum hochstetteranum A. Rich. = Urochloa panicoides P. Beauv. ■

281717　Panicum hochstetteri Steud. ;霍赫黍■☆

281718　Panicum holubii Stapf = Echinochloa pyramidalis (Lam.) Hitchc. et Chase ■☆

281719　Panicum holzii Peter = Urochloa rudis Stapf ■☆

281720　Panicum homblei Robyns;洪布勒黍■☆

281721　Panicum homonymum Steud. = Setaria homonyma (Steud.) Chiov. ■☆

281722　Panicum hookeri Parl. = Echinochloa colona(L.)Link ■

281723　Panicum hordeoides Lam. = Pennisetum hordeoides (Lam.) Steud. ■☆

281724　Panicum hostii M. Bieb. = Echinochloa oryzoides(Ard.)Fritsch ■

281725　Panicum huachucae Ashe = Dichanthelium acuminatum (Sw.) Gould et C. A. Clark subsp. fasciculatum (Torr.) Freckmann et Lelong ■☆

281726　Panicum huachucae Ashe = Dichanthelium acuminatum (Sw.) Gould et C. A. Clark ■

281727　Panicum huachucae Ashe var. fasciculatum(Torr.)F. T. Hubb. = Dichanthelium acuminatum (Sw.) Gould et C. A. Clark subsp. fasciculatum(Torr.)Freckmann et Lelong ■☆

281728　Panicum huachucae Ashe var. fasciculatum(Torr.)F. T. Hubb. = Dichanthelium acuminatum(Sw.)Gould et C. A. Clark ■☆

281729　Panicum huillense Rendle = Sacciolepis spiciformis (Hochst. ex A. Rich.)Stapf ■

281730　Panicum humbertii Robyns = Panicum robynsii A. Camus ■☆

281731　Panicum humidicola Rendle = Brachiaria humidicola (Rendle) Schweick. ■☆

281732　Panicum humifusum (Pers.) Kunth = Digitaria ischaemum (Schreb.)Schreb. ex Muhl. ■

281733　Panicum humile Nees ex Steud. ;南亚稷（矮黍，南亚黍）;S. Asia Panicgrass,South-asian Panicgrass,Southasiatic Witchgrass ■

281734　Panicum humile Nees ex Steud. = Panicum walense Mez ■

281735　Panicum humile Thunb. ex Trin. = Panicum walense Mez ■

281736　Panicum hydaspicum Edgew. = Panicum atrosanguineum Hochst. ex A. Rich. ■☆

281737　Panicum hygrocharis Steud. ;喜湿黍■☆

281738　Panicum hymeniochilum Nees;膜质黍■☆

281739　Panicum hymeniochilum Nees var. glandulosum？ = Panicum hymeniochilum Nees ■☆

281740　Panicum hystrix Steud. ;毛刺黍■☆

281741　Panicum ianthum Stapf = Panicum brazzavillense Franch. ■☆

281742　Panicum imbecille (R. Br.) Trin. = Oplismenus undulatifolius (Ard.)Roem. et Schult. var. imbecillis(R. Br.)Hack. ■

281743　Panicum imbelle Spreng. = Panicum tenellum Lam. ■☆

281744　Panicum imberbe Poir. = Setaria geniculata(Lam.)P. Beauv. ■

281745　Panicum impeditum Launert;累黍■☆

281746　Panicum implicatum Scribn. = Dichanthelium acuminatum

(Sw.) Gould et C. A. Clark subsp. implicatum (Scribn. ex Nash) Freckmann et Lelong ■☆

281747　Panicum implicatum Scribn. = Panicum acuminatum Sw. ■

281748　Panicum inaequilatum Stapf et C. E. Hubb. ;不等黍■☆

281749　Panicum incanum Schumach. = Oplismenus hirtellus (L.) P. Beauv. ■☆

281750　Panicum incomptum Fig. et De Not. ;装饰黍■☆

281751　Panicum incomtum Trin. = Panicum incomtum Trin. ex Nees ■

281752　Panicum incomtum Trin. = Panicum sarmentosum Roxb. ■

281753　Panicum incomtum Trin. ex Nees;藤竹草■

281754　Panicum incrassatum Hochst. = Setaria incrassata (Hochst.) Hack. ■☆

281755　Panicum indicum L. = Sacciolepis indica(L.) Chase ■

281756　Panicum indutum Steud. = Brachiaria comata (Hochst. ex A. Rich.)Stapf ■☆

281757　Panicum infestum Andersson;有害黍■☆

281758　Panicum ingens Peter = Panicum subalbidum Kunth ■☆

281759　Panicum inops Peter = Panicum brazzavillense Franch. ■☆

281760　Panicum insculptum Steud. = Brachiaria lata (Schumach.) C. E. Hubb. ■☆

281761　Panicum insigne Steud. = Melinis repens(Willd.) Zizka subsp. grandiflora(Hochst.)Zizka ■☆

281762　Panicum insulicola Steud. = Hymenachne insulicola(Steud.)L. Liou ■

281763　Panicum insulicola Steud. = Panicum auritum J. Presl ex Nees ■

281764　Panicum intermedium (Roem. et Schult.) Roth = Setaria intermedia Roem. et Schult. ■

281765　Panicum intermedium Hornem. = Setaria intermedia Roem. et Schult. ■

281766　Panicum interruptum (Willd.) Büse = Sacciolepis interrupta (Willd.) Stapf ■

281767　Panicum interruptum Willd. = Sacciolepis interrupta (Willd.) Stapf ■

281768　Panicum inundatum Kunth = Sacciolepis interrupta (Willd.) Stapf ■

281769　Panicum isachne Roem. et Schult. = Brachiaria eruciformis (Sm.)Griseb. ■

281770　Panicum isachne Roth ex Roem. et Schult. = Brachiaria eruciformis(Sm.)Griseb. ■

281771　Panicum isachnoides Vanderyst;柳叶箬黍■☆

281772　Panicum ischaemoides Retz. = Panicum repens L. ■

281773　Panicum ischaemum Schreb. = Digitaria ischaemum (Schreb.) Schreb. ex Muhl. ■

281774　Panicum isolepis Mez;同鳞黍■☆

281775　Panicum italicum L. = Setaria italica(L.)P. Beauv. ■

281776　Panicum italicum L. var. germanicum (Mill.) Koeler = Setaria italica(L.)P. Beauv. ■

281777　Panicum italicum L. var. germanicum (Mill.) Koeler = Setaria italica P. Beauv. var. germanica(Mill.)Schrad. ■

281778　Panicum japonicum Steud. = Oplismenus undulatifolius(Ard.) Roem. et Schult. var. japonicus(Steud.)Koidz. ■

281779　Panicum jardinii Steud. = Cyrtococcum chaetophoron (Roem. et Schult.)Dandy ■☆

281780　Panicum javanicum Poir. = Urochloa oligotricha (Fig. et De Not.)Henrard ■☆

281781　Panicum javanicum Poir. = Urochloa panicoides P. Beauv. ■

281782　Panicum jubatum Fig. et De Not. = Brachiaria jubata(Fig. et De

Not.) Stapf ■☆

281783　Panicum jumentorum Pers. = Panicum maximum Jacq. ■

281784　Panicum juncifolium Stapf = Panicum natalense Hochst. ■☆

281785　Panicum kafuroense K. Schum. = Digitaria abyssinica (Hochst. ex A. Rich.) Stapf ■☆

281786　Panicum kalaharense Mez;卡拉哈利黍■☆

281787　Panicum kasumense Renvoize;卡苏穆黍■☆

281788　Panicum katentaniense Robyns = Panicum pectinatum Rendle ■☆

281789　Panicum kermesinum Mez = Panicum subalbidum Kunth ■☆

281790　Panicum kerstingii Mez = Panicum pansum Rendle ■☆

281791　Panicum khasianum Munro ex Hook. f. ;滇西黍(喀西黍,卡西山黍);Khasi Panicgrass ■

281792　Panicum kisantuense Robyns = Panicum hymeniochilum Nees ■☆

281793　Panicum klingii Mez = Panicum phragmitoides Stapf ■☆

281794　Panicum kotschyanum Steud. = Brachiaria comata (Hochst. ex A. Rich.) Stapf ■☆

281795　Panicum kraussii Steud. = Oplismenus undulatifolius (Ard.) Roem. et Schult. ■

281796　Panicum kunthii Steud. = Panicum virgatum L. ■

281797　Panicum kurzii Hook. f. = Brachiaria kurzii(Hook. f.) A. Camus ■

281798　Panicum lachnanthum Hochst. = Brachiaria lachnantha (Hochst.) Stapf ■

281799　Panicum laetum Kunth;愉悦黍■☆

281800　Panicum laevifolium Hack. ;光叶黍■☆

281801　Panicum laevifolium Hack. = Panicum schinzii Hack. ■☆

281802　Panicum laevifolium Hack. var. contractum Pilg. = Panicum gilvum Launert ■☆

281803　Panicum lanceolatum Retz. = Oplismenus compositus (L.) P. Beauv. ■

281804　Panicum languidum Hitchc. = Dichanthelium acuminatum (Sw.) Gould et C. A. Clark subsp. fasciculatum(Torr.) Freckmann et Lelong ■☆

281805　Panicum lanipes Mez;西方毛梗黍■☆

281806　Panicum lanuginosum Elliott = Dichanthelium acuminatum (Sw.) Gould et C. A. Clark ■

281807　Panicum lanuginosum Elliott = Dichanthelium acuminatum (Sw.) Gould et C. A. Clark subsp. implicatum (Scribn. ex Nash) Freckmann et Lelong ■☆

281808　Panicum lanuginosum Elliott = Dichanthelium acuminatum (Sw.) Gould et C. A. Clark subsp. lindheimeri (Nash) Freckmann et Lelong ■☆

281809　Panicum lanuginosum Elliott = Dichanthelium acuminatum (Sw.) Gould et C. A. Clark subsp. fasciculatum(Torr.) Freckmann et Lelong ■☆

281810　Panicum lanuginosum Elliott = Panicum acuminatum Sw. ■

281811　Panicum lanuginosum Elliott var. praecocius(Hitchc. et Chase) McNeill et Dore = Dichanthelium ovale(Elliott) Gould et C. A. Clark subsp. praecocius(Hitchc. et Chase) Freckmann et Lelong ■☆

281812　Panicum lasiocoleum Engl. = Panicum infestum Andersson ■☆

281813　Panicum lasiopodum Stapf;毛梗黍■☆

281814　Panicum lasiosoma Peter = Panicum nervatum(Franch.)Stapf ■☆

281815　Panicum lassenianum Schmoll = Dichanthelium acuminatum (Sw.) Gould et C. A. Clark subsp. fasciculatum(Torr.) Freckmann et Lelong ■☆

281816　Panicum latifolium Ham. var. majus Hook. f. = Acroceras tonkinense(Balansa) C. E. Hubb. ex Bor ■

281817　Panicum latifolium Ham. var. majus Hook. f. = Neohusnotia tonkinensis(Balansa) A. Camus ■

281818　Panicum latifolium L. ;宽叶黍;Broad-leaved Panic-grass , Panic Grass ■☆

281819　Panicum latifolium L. = Dichanthelium latifolium(L.) Harv. ■☆

281820　Panicum latum Schumach. = Brachiaria lata(Schumach.) C. E. Hubb. ■☆

281821　Panicum laxiflorum Lam. = Dichanthelium laxiflorum (Lam.) Gould ■☆

281822　Panicum laxum Sw. ;疏松黍■☆

281823　Panicum leersioides Hochst. = Brachiaria leersioides (Hochst.) Stapf ■☆

281824　Panicum leibergii (Vasey) Scribn. = Dichanthelium leibergii (Vasey) Freckmann ■☆

281825　Panicum leibergii (Vasey) Scribn. var. baldwinii Lepage = Dichanthelium xanthophysum(A. Gray) Freckmann ■☆

281826　Panicum lelievrei A. Chev. = Echinochloa stagnina (Retz.) P. Beauv. ■☆

281827　Panicum lepidotum Steud. = Isachne globosa(Thunb.) Kuntze ■

281828　Panicum lepidum Chiov. = Panicum delicatulum Fig. et De Not. ■☆

281829　Panicum leptocaulon Trin. = Panicum deustum Thunb. ■☆

281830　Panicum leptorhachis Pilg. = Digitaria leptorhachis(Pilg.)Stapf ■☆

281831　Panicum letouzeyi Renvoize;勒图黍■☆

281832　Panicum leucacranthum K. Schum. = Brachiaria leucacrantha (K. Schum.) Stapf ■☆

281833　Panicum leucanthum A. Rich. = Tricholaena teneriffae (L. f.) Link ■☆

281834　Panicum leucothrix Nash = Dichanthelium acuminatum (Sw.) Gould et C. A. Clark subsp. implicatum(Scribn. ex Nash) Freckmann et Lelong ■☆

281835　Panicum lindenbergianum Nees = Setaria lindenbergiana (Nees) Stapf ■☆

281836　Panicum lindheimeri Nash = Dichanthelium acuminatum (Sw.) Gould et C. A. Clark subsp. lindheimeri(Nash)Freckmann et Lelong ■☆

281837　Panicum lindheimeri Nash = Panicum acuminatum Sw. var. lindheimeri(Nash) Beetle ■☆

281838　Panicum lindheimeri Nash = Panicum acuminatum Sw. ■281839 Panicum lindheimeri Nash var. fasciculatum (Torr.) Fernald = Dichanthelium acuminatum (Sw.) Gould et C. A. Clark subsp. fasciculatum(Torr.)Freckmann et Lelong ■☆

281840　Panicum lindiense Pilg. = Brachiaria lindiensis(Pilg.)Clayton ■☆

281841　Panicum lindleyanum Nees ex Steud. = Panicum tenellum Lam. ■☆

281842　Panicum lineare Krock. = Digitaria ischaemum (Schreb.) Schreb. ex Muhl. ■

281843　Panicum linearifolium Scribn. = Dichanthelium linearifolium (Scribn.) Gould ■☆

281844　Panicum linearifolium Scribn. = Panicum linearifolium Scribn. ex Nash ■☆

281845　Panicum linearifolium Scribn. ex Nash var. werneri (Scribn.) Fernald = Panicum linearifolium Scribn. ex Nash ■☆

281846　Panicum linearifolium Scribn. var. werneri (Scribn.) Fernald = Dichanthelium linearifolium(Scribn.) Gould ■☆

281847　Panicum lineatum Schumach. = Setaria barbata(Lam.) Kunth ■☆

281848　Panicum lineatum Trin. = Panicum sadinii (Vanderyst) Renvoize ■☆

281849　Panicum littorale Sosef;滨海黍■☆

281850　Panicum loliaceum Lam. = Oplismenus hirtellus(L.) P. Beauv. ■☆

281851　Panicum longepetiolatum Pilg. = Setaria longiseta P. Beauv. ■☆

281852　Panicum longicaudum Mez = Melinis ambigua Hack. subsp.

longicauda(Mez) Zizka ■☆

281853　Panicum longiflorum (Retz.) J. F. Gmel. = Digitaria longiflora (Retz.) Pers. ■

281854　Panicum longifolium Schumach. = Setaria longiseta P. Beauv. ■☆

281855　Panicum longiglume H. Peng et L. H. Zhou;长颖黍■

281856　Panicum longiglume H. Peng et L. H. Zhou = Panicum brevifolium L. ■

281857　Panicum longijubatum (Stapf) Stapf = Panicum subalbidum Kunth ■☆

281858　Panicum longiramum Peter = Panicum subalbidum Kunth ■☆

281859　Panicum lukwangulense Pilg. ;卢夸古尔黍■☆

281860　Panicum lutescens Weigel = Pennisetum glaucum(L.) R. Br. ■

281861　Panicum lutescens Weigel = Setaria glauca(L.) P. Beauv. ■

281862　Panicum lutetense K. Schum. = Acroceras zizanioides (Kunth) Dandy ■☆

281863　Panicum luzonense J. Presl et C. Presl;大罗湾草(大罗网草,网脉稷) ;Netveined Panicgrass ■

281864　Panicum macroblepharum Hack. = Digitaria macroblephara (Hack.) Stapf ■☆

281865　Panicum macrocarpon Lecomte ex Torr. = Dichanthelium latifolium(L.) Harv. ■☆

281866　Panicum macrotrichum Steud. = Melinis longiseta (A. Rich.) Zizka ■☆

281867　Panicum madagascariense Spreng. = Tricholaena monachne (Trin.) Stapf et C. E. Hubb. ■☆

281868　Panicum madagascariense Spreng. var. brevispiculum Rendle = Tricholaena monachne(Trin.) Stapf et C. E. Hubb. ■☆

281869　Panicum mahafalense A. Camus;马哈法尔黍■☆

281870　Panicum malaccense Trin. = Sphaerocaryum malaccense(Trin.) Pilg. ■

281871　Panicum malacodes Mez et K. Schum. = Brachiaria malacodes (Mez et K. Schum.) H. Scholz ■☆

281872　Panicum malacon Nash = Dichanthelium ovale (Elliott) Gould et C. A. Clark ■☆

281873　Panicum malacophyllum Nash;软叶黍;Panic Grass ■☆

281874　Panicum mananarense A. Camus;马诺纳拉黍■☆

281875　Panicum mandshuricum Maxim. = Arundinella hirta (Thunb.) Tanaka ■

281876　Panicum mandshuricum Maxim. var. pekinense Maxim. = Arundinella anomala Steud. ■

281877　Panicum mandshuricum Maxim. var. pekinense Maxim. = Arundinella hirta(Thunb.) Tanaka ■

281878　Panicum manicatum Stapf = Panicum massaiense Mez ■☆

281879　Panicum mannii Mez = Panicum monticola Hook. f. ■☆

281880　Panicum manongarivense A. Camus;马诺黍■☆

281881　Panicum margaritiferum(Chiov.) Robyns;珍珠黍■☆

281882　Panicum marlothii Hack. = Brachiaria marlothii(Hack.) Stent ■☆

281883　Panicum marunguense Robyns;马龙古黍■☆

281884　Panicum massaiense Mez;马萨黍■☆

281885　Panicum matsumurae (Hack.) Hack. = Setaria chondrachne (Steud.) Honda ■

281886　Panicum matsumurae Hack. = Setaria chondrachne (Steud.) Honda ■

281887　Panicum maximum Jacq. ; 大黍(羊草) ; Guinea Grass, Guineagrass, Guinea-grass ■

281888　Panicum maximum Jacq. var. hirsutissimum (Steud.) Oliv. = Panicum maximum Jacq. ■

281889　Panicum maximum Jacq. var. trichoglume Robyns = Panicum maximum Jacq. ■

281890　Panicum mayumbense Franch. = Acroceras gabunense(Hack.) Clayton ■☆

281891　Panicum megalanthum Steud. = Rhynchelytrum repens(Willd.) C. E. Hubb. ■

281892　Panicum megaphyllum Steud. = Setaria megaphylla (Steud.) T. Durand et Schinz ■☆

281893　Panicum melananthum L. = Panicum bisulcatum Thunb. ■

281894　Panicum melanospermum Mez;黑籽稷■☆

281895　Panicum melanotylum Hack. = Brachiaria nigropedata (Munro ex Ficalho et Hiern) Stapf ■☆

281896　Panicum melinis Trin. = Melinis minutiflora P. Beauv. ■

281897　Panicum melinis Trin. var. inerme Döll = Melinis minutiflora P. Beauv. ■

281898　Panicum membranaceum Robyns = Panicum gracilicaule Rendle ■☆

281899　Panicum menyharthii Hack. = Panicum deustum Thunb. ■☆

281900　Panicum meridionale Ashe = Dichanthelium acuminatum (Sw.) Gould et C. A. Clark subsp. implicatum(Scribn. ex Nash) Freckmann et Lelong ■☆

281901　Panicum meridionale Ashe = Panicum acuminatum Sw. ■

281902　Panicum meridionale Ashe var. albemarlense (Ashe) Fernald = Dichanthelium acuminatum (Sw.) Gould et C. A. Clark subsp. implicatum(Scribn. ex Nash) Freckmann et Lelong ■☆

281903　Panicum merkeri Mez;梅克尔黍■☆

281904　Panicum meyeranum Nees var. umbratile (Mez) Chiov. = Panicum monticola Hook. f. ■☆

281905　Panicum meyerianum Nees = Eriochloa meyeriana(Nees) Pilg. ■☆

281906　Panicum meyerianum Nees var. grandeglume Stent et J. M. Rattray = Eriochloa meyeriana(Nees) Pilg. subsp. grandiglumis(Stent et J. M. Rattray) Gibbs Russ. ■☆

281907　Panicum microbachne J. Presl = Digitaria microbachne(J. Presl) Henrard ■

281908　Panicum microbachne J. Presl = Digitaria setigera Roth ex Roem. et Schult. ■

281909　Panicum microcarpon Muhl. ex Elliott = Dichanthelium dichotomum(L.) Gould ■☆

281910　Panicum microcarpon Muhl. ex Elliott = Panicum dichotomum L. ■☆

281911　Panicum microcephalum Peter = Panicum flacciflorum Stapf ■☆

281912　Panicum microlemma Pilg. = Panicum comorense Mez ■☆

281913　Panicum microstachyum Lam. = Sacciolepis indica(L.) Chase ■

281914　Panicum microthyrsum Stapf = Panicum mueense Vanderyst ■☆

281915　Panicum milanjianum Rendle = Digitaria milanjiana (Rendle) Stapf ■☆

281916　Panicum miliaceum L. ;稷(白黍,赤黍,赤虾米,刺,大黑黍,红莲米,黄米,黄黍,稷米,穄,穄米,秬,糜穄黍,糜,糜子,糜子米,蘼子,米子,牛黍,芑,秫黍,黍,黍糜,黍米,鸭蹄黍,粢,粢米) ;Broom Millet, Broomcom Millet, Broomcorn, Broom-corn Millet, Broomcorn Panicgrass, Common Millet, French Millet, Hirse, Hog Millet, Millet, Panic Millet, Proso, Proso Millet, Russian Millet, Wild Proso Millet ■

281917　Panicum miliaceum L. subsp. ruderale (Kitag.) Tzvelev;野生稷(野稷) ;Broomcorn Millet, Broom-corn Millet, Wild Panicgrass ■

281918　Panicum miliaceum L. var. effusum Alef. ;散枝稷;Expanded Panicgrass ■☆

281919　Panicum miliaceum L. var. ruderale Kitag. = Panicum miliaceum L. subsp. ruderale(Kitag.) Tzvelev ■

281920   Panicum miliare Lam. = Panicum antidotale Retz. ■☆

281921   Panicum miliiforme C. Presl = Brachiaria subquadripara(Trin.) Hitchc. var. miliiformis(C. Presl) S. L. Chen et Y. X. Jin ■

281922   Panicum miliiforme J. Presl = Urochloa subquadripara(Trin.) R. D. Webster ■

281923   Panicum miliiforme J. Presl et C. Presl = Brachiaria subquadripara(Trin.) Hitchc. var. miliiformis(C. Presl) S. L. Chen et Y. X. Jin ■

281924   Panicum milium Pers. = Panicum miliaceum L. ■

281925   Panicum minus Stapf = Panicum stapfianum Fourc. ■☆

281926   Panicum minutiflorum(P. Beauv.) Raspail = Melinis minutiflora P. Beauv. ■

281927   Panicum minutiflorum Hochst. ex A. Rich. = Digitaria pseudodiagonalis Chiov. ■☆

281928   Panicum minutiflorum Steud. = Digitaria violascens Link ■

281929   Panicum mite Steud. = Eriochloa meyeriana(Nees) Pilg. ■☆

281930   Panicum mitophyllum Pilg. = Panicum congoense Franch. ■☆

281931   Panicum mitopus K. Schum. ;线梗黍■☆

281932   Panicum mixtum Mez = Panicum griffonii Franch. ■☆

281933   Panicum mokaense Mez = Panicum hochstetteri Steud. ■☆

281934   Panicum molle Sw. ;柔软黍;Water Grass ■☆

281935   Panicum mollifolium Link;毛叶黍■☆

281936   Panicum monachne Trin. = Tricholaena monachne(Trin.) Stapf et C. E. Hubb. ■☆

281937   Panicum moninense(Rendle) K. Schum. = Digitaria brazzae(Franch.) Stapf ■☆

281938   Panicum monodactylum Nees = Digitaria monodactyla (Nees) Stapf ■☆

281939   Panicum montanum Roxb. = Panicum notatum Retz. ■

281940   Panicum monticola Hook. f. ;山地黍■☆

281941   Panicum mosambicense Hack. = Urochloa mosambicensis (Hack.) Dandy ■☆

281942   Panicum mucronatum Roth ex Roem. = Paspalidium punctatum (Burm. f.) A. Camus ■

281943   Panicum mucronatum Roth ex Roem. et Schult. = Paspalidium punctatum(Burm. f.) A. Camus ■

281944   Panicum mueense Vanderyst;穆埃黍■☆

281945   Panicum multifolium Peter = Brachiaria arrecta (Hack. ex T. Durand et Schinz) Stent ■☆

281946   Panicum multinode J. Presl et C. Presl = Ottochloa nodosa (Kunth) Dandy ■

281947   Panicum multinode Lam. = Cyrtococcum multinode (Lam.) Clayton ■☆

281948   Panicum multinode Lam. = Ottochloa nodosa(Kunth) Dandy ■

281949   Panicum multisetum Hochst. ex A. Rich. = Oplismenus burmannii(Retz.) P. Beauv. ■

281950   Panicum mundum Fernald = Dichanthelium ovale (Elliott) Gould et C. A. Clark subsp. pseudopubescens(Nash) Freckmann et Lelong ■☆

281951   Panicum munroanum Balansa = Acroceras munroanum(Balansa) Henrard ■

281952   Panicum muricatum Retz. = Cyrtococcum patens(L.) A. Camus var. schmidtii(Hack.) A. Camus ■

281953   Panicum muricatum Retz. = Cyrtococcum patens(L.) A. Camus ■

281954   Panicum muricatum Retz. = Isachne globosa(Thunb.) Kuntze ■

281955   Panicum muticum Forssk. = Brachiaria mutica(Forssk.) Stapf ■

281956   Panicum myosuroides R. Br. = Sacciolepis myosuroides(R. Br.) Chase ex E. G. Camus et A. Camus ■

281957   Panicum myuros (P. Beauv.) Lam. = Sacciolepis indica (L.) Chase ■

281958   Panicum myuros Lam. = Sacciolepis indica(L.) Chase ■

281959   Panicum natalense Hochst. ;纳塔尔黍■☆

281960   Panicum natans Koen. ex Trin. = Panicum paludosum Roxb. ■

281961   Panicum neesianum Wight et Arn. ex Steud. = Panicum tripheron Schult. ■

281962   Panicum neglectum Roem. et Schult. ;忽视黍■☆

281963   Panicum nemorosum Trin. = Pseudechinolaena polystachya (Kunth) Stapf ■

281964   Panicum nervatum(Franch.) Stapf;多脉黍■☆

281965   Panicum nervosum Lam. = Setaria palmifolia(J. König) Stapf ■

281966   Panicum neurodes Schult. = Setaria palmifolia(J. König) Stapf ■

281967   Panicum neurodes Schult. var. blepharoneuron A. Braun = Setaria plicata(Lam.) T. Cooke ■

281968   Panicum nidulans Mez = Brachiaria ramosa(L.) Stapf ■

281969   Panicum nigerense Hitchc. ;尼日利亚黍■☆

281970   Panicum nigrirostre Nees = Setaria nigrirostris (Nees) T. Durand et Schinz ■☆

281971   Panicum nigritianum Hack. = Digitaria leptorhachis(Pilg.) Stapf ■☆

281972   Panicum nigromarginatum Robyns;黑边黍■☆

281973   Panicum nigropedatum Munro ex Ficalho et Hiern = Brachiaria nigropedata(Munro ex Ficalho et Hiern) Stapf ■☆

281974   Panicum nilagiricum Steud. = Brachiaria semiundulata(Hitchc. ex A. Rich.) Stapf ■

281975   Panicum nitens(Rendle) K. Schum. = Digitaria nitens Rendle ■☆

281976   Panicum nitens K. Schum. = Ichnanthus pallens(Sm.) Munro ex Benth. var. major(Nees) Stieber ■

281977   Panicum nitens K. Schum. = Ichnanthus vicinus(F. M. Bailey) Merr. ■

281978   Panicum nitidum Lam. = Dichanthelium dichotomum(L.) Gould ■☆

281979   Panicum nitidum Lam. = Panicum dichotomum L. ■☆

281980   Panicum nodibarbatum Hochst. ex Steud. = Isachne dispar Trin. ■

281981   Panicum nodosum Kunth = Ottochloa nodosa(Kunth) Dandy ■

281982   Panicum nodosum Kunth var. micranthum Balansa = Ottochloa nodosa(Kunth) Dandy var. micrantha (Balansa ex A. Camus) S. L. Chen et S. M. Phillips ■

281983   Panicum nodosum Kunth var. micranthum Balansa ex A. Camus = Ottochloa nodosa (Kunth) Dandy var. micrantha (Balansa ex A. Camus) S. L. Chen et S. M. Phillips ■

281984   Panicum notatum Retz. ;心叶稷(山黍,心叶黍,硬骨草); Cordate Panicgrass, Mountain Panicgrass, Vine Panicgrass ■

281985   Panicum nubicum Fig. et De Not. ;云雾黍■☆

281986   Panicum nudiflorum Renvoize;裸花黍■☆

281987   Panicum nudiglume Hochst. var. major Rendle = Brachiaria grossa Stapf ■☆

281988   Panicum numidianum Lam. = Brachiaria mutica(Forssk.) Stapf ■

281989   Panicum nyanzense K. Schum. = Panicum repens L. ■

281990   Panicum nyassense Napper = Panicum delicatulum Fig. et De Not. ■☆

281991   Panicum obliquum Roth ex Roem. et Schult. = Cyrtococcum patens(L.) A. Camus ■

281992   Panicum obtusiflorum Hochst. ex A. Rich. = Echinochloa rotundiflora Clayton ■☆

281993   Panicum obtusifolium Delile = Paspalidium obtusifolium(Delile) N. D. Simpson ■☆

281994   Panicum obtusifolium Delile var. acutifolium Coss. et Durieu =

Paspalidium obtusifolium(Delile) N. D. Simpson ■☆

281995　Panicum obtusum Kunth;藤黍（钝状黍）;Vine Mesquite, Vine Mesquite Grass ■☆

281996　Panicum occidentale Scribn. = Dichanthelium acuminatum （Sw.) Gould et C. A. Clark subsp. fasciculatum(Torr.) Freckmann et Lelong ■☆

281997　Panicum ogowense Franch. = Acroceras zizanioides (Kunth) Dandy ■☆

281998　Panicum oligobrachiatum Pilg. = Brachiaria oligobrachiata （Pilg.) Henrard ■☆

281999　Panicum oligosanthes Schult. = Dichanthelium oligosanthes （Schult.) Gould ■☆

282000　Panicum oligosanthes Schult. var. helleri (Nash) Fernald = Dichanthelium oligosanthes (Schult.) Gould subsp. scribnerianum （Nash) Freckmann et Lelong ■☆

282001　Panicum oligosanthes Schult. var. helleri (Nash) Fernald = Dichanthelium oligosanthes(Schult.) Gould ■☆

282002　Panicum oligosanthes Schult. var. helleri (Nash) Fernald = Panicum oligosanthes Schult. ■☆

282003　Panicum oligosanthes Schult. var. scribnerianum (Nash) Fernald = Dichanthelium oligosanthes (Schult.) Gould subsp. scribnerianum （Nash) Freckmann et Lelong ■☆

282004　Panicum oligotrichum Fig. et De Not. = Urochloa oligotricha （Fig. et De Not.) Henrard ■☆

282005　Panicum oplismenoides Hack. = Poecilostachys oplismenoides （Hack.) Clayton ■☆

282006　Panicum oricola Hitchc. et Chase = Dichanthelium acuminatum （Sw.) Gould et C. A. Clark subsp. implicatum (Scribn. ex Nash) Freckmann et Lelong ■☆

282007　Panicum orthostachys Mez = Brachiaria orthostachys (Mez) Clayton ■☆

282008　Panicum oryzetorum (A. Chev.) A. Chev. = Echinochloa stagnina(Retz.) P. Beauv. ■☆

282009　Panicum oryzicola Vasinger = Echinochloa oryzicola (Vasinger) Vasinger ■

282010　Panicum oryzicola Vasinger = Echinochloa phyllopogon (Stapf) Stapf ex Koss ■

282011　Panicum oryzoides Ard. = Echinochloa oryzoides(Ard.)Fritsch ■

282012　Panicum ovale Elliott = Dichanthelium ovale (Elliott) Gould et C. A. Clark ■☆

282013　Panicum ovale Elliott var. addisonii (Nash) C. F. Reed = Dichanthelium ovale (Elliott) Gould et C. A. Clark subsp. pseudopubescens(Nash) Freckmann et Lelong ■☆

282014　Panicum ovale Elliott var. pseudopubescens (Nash) Lelong = Dichanthelium ovale (Elliott) Gould et C. A. Clark subsp. pseudopubescens(Nash) Freckmann et Lelong ■☆

282015　Panicum ovale R. Br. = Brachiaria ovalis Stapf ■☆

282016　Panicum ovalifolium Poir. = Panicum brevifolium L. ■

282017　Panicum owenae E. P. Bicknell = Dichanthelium ovale(Elliott) Gould et C. A. Clark subsp. pseudopubescens (Nash) Freckmann et Lelong ■☆

282018　Panicum oxycephalum Peter = Urochloa panicoides P. Beauv. ■

282019　Panicum oxyphyllum Hochst. ex Steud. = Cyrtococcum oxyphyllum(Hochst. ex Steud.) Stapf ■

282020　Panicum pabulare Aitch. = Digitaria nodosa Parl. ■☆

282021　Panicum pachystachys Franch. et Sav. = Setaria viridis(L.) P. Beauv. subsp. pachystachys(Franch. et Sav.) Masam. et Yanagita ■

282022　Panicum pacificum Hitchc. et Chase = Dichanthelium acuminatum(Sw.) Gould et C. A. Clark subsp. fasciculatum(Torr.) Freckmann et Lelong ■☆

282023　Panicum pallens Sw. = Ichnanthus pallens (Sw.) Munro ex Benth. ■☆

282024　Panicum pallens Sw. var. majus Nees = Ichnanthus pallens （Sm.)Munro ex Benth. var. major(Nees)Stieber ■

282025　Panicum pallidefuscum K. Schum. = Setaria pallidifusca （Schumach.)Stapf et C. E. Hubb. ■

282026　Panicum pallidefuscum K. Schum. = Setaria parviflora (Poir.) Kerguélen ■

282027　Panicum pallidefuscum K. Schum. = Setaria pumila (Poir.) Roem. et Schult. ■

282028　Panicum pallidum Peter = Brachiaria ramosa(L.) Stapf ■

282029　Panicum palmifolium J. König = Setaria palmifolia (J. König) Stapf ■

282030　Panicum palmifolium Willd. ex Poir. = Setaria palmifolia (J. König) Stapf ■

282031　Panicum paludosum Roxb. ; 水 生 黍（ 沼 地 黍）; Aquatic Panicgrass, Chesapeake Panicgrass, Marshy Panicgrass ■

282032　Panicum paludosum Roxb. = Panicum dichotomiflorum Michx. ■

282033　Panicum panicoides(P. Beauv.) Hitchc. = Urochloa panicoides P. Beauv. ■

282034　Panicum panicoides Hitchc. = Urochloa panicoides P. Beauv. ■

282035　Panicum paniculiferum Steud. = Setaria palmifolia (J. König) Stapf ■

282036　Panicum pansum Rendle;伸展黍■☆

282037　Panicum paradoxum Roth ex Roem. et Schult. = Digitaria abludens(Roem. et Schult.) Veldkamp ■

282038　Panicum parlatorei Steud. = Digitaria nodosa Parl. ■☆

282039　Panicum parvifolium Lam. ；小叶黍■☆

282040　Panicum parvulum Trin. = Digitaria longiflora (Retz.) Pers. ■

282041　Panicum paspaloides Hayata;花稗;Kodo-millet-like Panicgrass ■☆

282042　Panicum paspaloides Pers. = Paspalidium geminatum (Forssk.) Stapf ■

282043　Panicum patens L. = Cyrtococcum patens(L.) A. Camus ■

282044　Panicum patens L. f. latifolium Honda = Cyrtococcum patens （L.) A. Camus var. latifolium(Honda) Ohwi ■

282045　Panicum patens L. var. latifolium(Honda) Honda = Cyrtococcum patens(L.) A. Camus var. latifolium(Honda) Ohwi ■

282046　Panicum paucinode Stapf;少节黍■☆

282047　Panicum pauperulum Steud. = Brachiaria villosa (Lam.) A. Camus ■

282048　Panicum pectinatum Rendle;篦状黍■☆

282049　Panicum pectinellum Stapf;蓖齿黍■☆

282050　Panicum pedicellare (Trin. ex Hook. f.) Hack. = Digitaria abludens(Roem. et Schult.) Veldkamp ■

282051　Panicum pedicellare Hack. = Digitaria abludens (Roem. et Schult.) Veldkamp ■

282052　Panicum pedicellare Trin. ex Hook. f. = Digitaria abludens （Roem. et Schult.) Veldkamp ■

282053　Panicum pedicellatum Nees ex Duthie = Digitaria abludens （Roem. et Schult.) Veldkamp ■

282054　Panicum pennatum Hochst. = Digitaria pennata (Hochst.) T. Cooke ■☆

282055　Panicum perlaxum Stapf = Panicum aequinerve Nees ■☆

282056　Panicum perlongum Nash = Dichanthelium linearifolium

(Scribn.) Gould ■☆

282057　Panicum perlongum Nash = Dichanthelium perlongum (Nash) Freckmann ■☆

282058　Panicum perlongum Nash = Panicum linearifolium Scribn. ex Nash ■☆

282059　Panicum perrieri A. Camus；佩里耶黍■☆

282060　Panicum perrottetii Kunth = Digitaria perrottetii（Kunth）Stapf ■☆

282061　Panicum peteri Peter = Sacciolepis curvata（L.）Chase ■☆

282062　Panicum peteri Pilg.；彼得黍■☆

282063　Panicum petiveri Trin. = Brachiaria ramosa（L.）Stapf ■

282064　Panicum phaeocarpum Nees；褐果黍■☆

282065　Panicum phalarioides Roem. et Schult. = Sacciolepis indica (L.) Chase ■

282066　Panicum philadelphicum Bernh. = Panicum philadelphicum Bernh. ex Trin. ■☆

282067　Panicum philadelphicum Bernh. ex Trin.；费城黍；Philadelphia Panic Grass，Philadelphia Panic-grass，Philadelphia Witchgrass ■☆

282068　Panicum philadelphicum Bernh. ex Trin. var. tuckermanii （Fernald）Steyerm. et Schmoll = Panicum philadelphicum Bernh. ex Trin. ■☆

282069　Panicum philadelphicum Trin. = Panicum philadelphicum Bernh. ex Trin. ■☆

282070　Panicum philadelphicum Trin. var. tuckermanii（Fernald）Steyerm. et Schmoll = Panicum philadelphicum Bernh. ex Trin. ■☆

282071　Panicum phippsii Renvoize；菲普斯黍■☆

282072　Panicum phragmitoides Stapf；篦笆黍■☆

282073　Panicum phragmitoides Stapf var. lasioneuron? = Panicum phragmitoides Stapf ■☆

282074　Panicum phyllomacrum Steud. = Setaria megaphylla（Steud.）T. Durand et Schinz ■☆

282075　Panicum phyllopogon Stapf = Echinochloa oryzoides (Ard.) Fritsch ■

282076　Panicum phyllopogon Stapf = Echinochloa phyllopogon（Stapf） Stapf ex Koss ■

282077　Panicum pilgerana Scholz = Brachiaria scalaris Pilg. ■☆

282078　Panicum pilgeri Mez；皮尔格黍■☆

282079　Panicum pilgerianum（Schweick.）Clayton；皮氏黍■☆

282080　Panicum pilipes Nees et Arn. ex Büse = Cyrtococcum oxyphyllum （Hochst. ex Steud.）Stapf

282081　Panicum pinifolium Chiov.；松叶黍■☆

282082　Panicum piriferum Chiov. = Digitaria nodosa Parl. ■☆

282083　Panicum plagianthum Stapf = Panicum nervatum（Franch.）Stapf ■☆

282084　Panicum plantagineum Link = Brachiaria plantaginea (Link) Hitchc. ■

282085　Panicum plantagineum Schumach. = Panicum brevifolium L. ■

282086　Panicum platynotum K. Schum. = Brachiaria platynota (K. Schum.）Robyns ■☆

282087　Panicum pleianthum Peter；多花黍■☆

282088　Panicum plicatile Hochst. = Setaria plicatilis（Hochst.）Hack. ex Engl. ■☆

282089　Panicum plicatulum（Michx.）Kuntze = Paspalum plicatulum Michx. ■

282090　Panicum plicatulum Kuntze = Paspalum plicatulum Michx. ■

282091　Panicum plicatum Lam. = Setaria plicata（Lam.）T. Cooke ■

282092　Panicum plicatum Willd. = Setaria palmifolia（J. König）Stapf ■

282093　Panicum poaeoides Stapf；早熟禾黍■☆

282094　Panicum poecilanthum Stapf = Panicum dregeanum Nees ■☆

282095　Panicum poiretianum Schult. = Setaria poiretiana (Schult.) Kunth ■☆

282096　Panicum pole-evansii C. E. Hubb.；埃文斯黍■☆

282097　Panicum polyanthes Schult. = Panicum sphaerocarpon Elliott ■☆

282098　Panicum polygamum Sw. = Panicum maximum Jacq. ■

282099　Panicum polyphyllum Peter；多叶黍■☆

282100　Panicum polystachion L. = Pennisetum polystachion (L.) Schult. ■

282101　Panicum polystachyum（Kunth）K. Schum. = Pseudechinolaena polystachya（Kunth）Stapf ■

282102　Panicum porphyrrhizos Steud.；紫根黍■☆

282103　Panicum porranthum Steud. = Digitaria horizontalis Willd. ■☆

282104　Panicum portoricense Desv. ex Ham.；哥伦比亚黍；Panic Grass ■☆

282105　Panicum praealtum Afzel. ex Sw.；高大黍■☆

282106　Panicum praecocius Hitchc. et Chase = Dichanthelium ovale （Elliott）Gould et C. A. Clark subsp. praecocius（Hitchc. et Chase） Freckmann et Lelong ■☆

282107　Panicum praecocius Hitchc. et Chase = Panicum acuminatum Sw. ■

282108　Panicum praelongum Mez = Panicum porphyrrhizos Steud. ■☆

282109　Panicum praelongum Steud. = Panicum maximum Jacq. ■

282110　Panicum proliferum Hook. f. = Panicum paludosum Roxb. ■

282111　Panicum proliferum Hook. f. var. paludosum（Roxb.）Stapf = Panicum paludosum Roxb. ■

282112　Panicum proliferum Lam. = Panicum paludosum Roxb. ■

282113　Panicum proliferum Lam. var. decompositum（R. Br.）Thell. = Panicum decompositum R. Br. ■

282114　Panicum proliferum Lam. var. longijubatum Stapf = Panicum subalbidum Kunth ■☆

282115　Panicum prolisetum Steud. = Setaria megaphylla (Steud.) T. Durand et Schinz ■☆

282116　Panicum propinquum R. Br. = Digitaria longiflora（Retz.）Pers. ■

282117　Panicum prostratum Lam. = Brachiaria reptans（L.）C. A. Gardner et C. E. Hubb. ■

282118　Panicum prostratum Lam. = Urochloa repans（L.）Stapf ■

282119　Panicum protractum Peter = Panicum trichocladum K. Schum. ■☆

282120　Panicum pruriens Fisch. ex Trin. = Digitaria setigera Roth ex Roem. et Schult. ■

282121　Panicum pruriens Trin. = Digitaria setigera Roth ex Roem. et Schult. ■

282122　Panicum psammophilum Welw. ex Rendle = Brachiaria psammophila（Welw. ex Rendle）Launert ■☆

282123　Panicum pseudoagrostis Steud. = Digitaria perrottetii（Kunth） Stapf ■☆

282124　Panicum pseudodistachyum Hayata = Brachiaria subquadripara （Trin.）Hitchc. ■

282125　Panicum pseudoinfestum Chiov. = Panicum infestum Andersson ■☆

282126　Panicum pseudopubescens Nash = Dichanthelium ovale（Elliott） Gould et C. A. Clark subsp. pseudopubescens（Nash）Freckmann et Lelong ■☆

282127　Panicum pseudoracemosum Renvoize；假总花黍■☆

282128　Panicum psilopodium Trin. = Panicum sumatrense Roth ex Roem. et Schult. ■

282129　Panicum psilopodium Trin. var. coloratum Hook. f. = Panicum psilopodium Trin. ■

282130　Panicum psilopodium Trin. var. coloratum Hook. f. = Panicum sumatrense Roth ex Roem. et Schult. ■

282131　Panicum psilopodium Trin. var. epaleatum Keng ex S. L. Chen,

T. D. Zhuang et X. L. Yang = Panicum sumatrense Roth ex Roem. et Schult. ■

282132　Panicum psilostachyum Hochst. ex T. Durand et Schinz = Digitaria velutina(Forssk.) P. Beauv. ■☆

282133　Panicum puberulum Kunth var. tricostulatum Hack. = Digitaria thouarsiana(Flüggé) A. Camus ■☆

282134　Panicum pubifolium Mez = Brachiaria xantholeuca(Hack.)Stapf ■☆

282135　Panicum pubiglume Stapf = Panicum brazzavillense Franch. ■☆

282136　Panicum pubivaginatum K. Schum. = Panicum deustum Thunb. ■☆

282137　Panicum pumilum Poir. = Setaria glauca (L.) P. Beauv. ■

282138　Panicum pumilum Poir. = Setaria pumila (Poir.) Roem. et Schult. ■

282139　Panicum punctatum (Burm. f.) Stapf ex Ridl. = Paspalidium punctatum(Burm. f.) A. Camus ■

282140　Panicum punctatum Burm. f. = Paspalidium punctatum(Burm. f.) A. Camus ■

282141　Panicum punctatum Burm. f. = Setaria punctata (Burm. f.) Veldkamp ■

282142　Panicum purinisum Bernh. ex Trin. = Panicum virgatum L. ■

282143　Panicum purpuracens Raddi = Brachiaria mutica(Forssk.)Stapf ■

282144　Panicum purpurascens Mez = Panicum fluviicola Steud. ■☆

282145　Panicum purpurascens Raddi = Brachiaria mutica (Forssk.) Stapf ■

282146　Panicum purpurascens Raddi ex Opiz = Brachiaria mutica (Forssk.)Stapf ■

282147　Panicum pusillum Hook. f. ;微小黍■☆

282148　Panicum pusillum Hook. f. var. glabriglumatum Schnell = Panicum glaucocladum C. E. Hubb. ■☆

282149　Panicum pycnocomum Steud. = Setaria viridis (L.) P. Beauv. subsp. pycnocoma(Steud.) Tzvelev ■

282150　Panicum pyramidale Lam. = Echinochloa pyramidalis (Lam.) Hitchc. et Chase ■☆

282151　Panicum quadrifarium Hochst. ex A. Rich. = Echinochloa pyramidalis(Lam.) Hitchc. et Chase ■☆

282152　Panicum radicans Retz. = Cyrtococcum patens(L.) A. Camus ■

282153　Panicum radicosum J. Presl et C. Presl = Digitaria radicosa (J. Presl et C. Presl) Miq. ■

282154　Panicum radicosum Mez = Panicum fluviicola Steud. ■☆

282155　Panicum radula Mez = Panicum merkeri Mez ■☆

282156　Panicum ramosum L. = Brachiaria ramosa(L.) Stapf ■

282157　Panicum raripilum Kunth = Panicum parvifolium Lam. ■☆

282158　Panicum rautanenii Hack. = Brachiaria humidicola (Rendle) Schweick. ■☆

282159　Panicum ravenelii Scribn. et Merr. ;雷夫纳尔黍;Panic Grass ■☆

282160　Panicum regulare Nees = Brachiaria deflexa(Schumach.) C. E. Hubb. ex Robyns ■☆

282161　Panicum reimarioides Andersson = Digitaria debilis (Desf.) Willd. ■☆

282162　Panicum repens L. ;铺地黍（大广草,风台草,枯骨草,马鞭草,藤竹草,硬骨草）;Torpedo Grass,Torpedograss ■

282163　Panicum repens L. var. paludosum (Roxb.) Kuntze = Panicum paludosum Roxb. ■

282164　Panicum repentellum Napper = Panicum hygrocharis Steud. ■☆

282165　Panicum reptans L. = Brachiaria reptans (L.) C. A. Gardner et C. E. Hubb. ■

282166　Panicum reptans L. = Urochloa reptans(L.) Stapf ■

282167　Panicum rescissum Trin. = Setaria longiseta P. Beauv. ■☆

282168　Panicum respiciens (A. Rich.) Hochst. ex Steud. = Setaria verticillata(L.) P. Beauv. ■

282169　Panicum restioideum Franch. = Setaria restioidea (Franch.) Stapf ■☆

282170　Panicum reticulatum Torr. = Panicum cambogiense Balansa ■

282171　Panicum rhachitrichum Hochst. = Setaria barbata(Lam.)Kunth ■☆

282172　Panicum rigens Mez = Sacciolepis rigens (Mez) A. Chev. ■☆

282173　Panicum rigidulum Bosc ex Nees ;红头稷;Redtop Panic Grass, Red-top Panic Grass,Red-top Panicum ■☆

282174　Panicum rigidulum Bosc ex Nees var. elongatum (Pursh) Lelong = Panicum rigidulum Bosc ex Nees ■☆

282175　Panicum rivae Chiov. = Digitaria rivae(Chiov.) Stapf ■☆

282176　Panicum robynsii A. Camus ;罗宾斯黍■☆

282177　Panicum roseum(Nees)Steud. = Melinis repens(Willd.)Zizka ■

282178　Panicum roseum (Nees) Steud. = Rhynchelytrum repens (Willd.) C. E. Hubb. ■

282179　Panicum rothii Spreng. = Urochloa panicoides P. Beauv. ■

282180　Panicum rovumense Pilg. = Eriochloa rovumensis (Pilg.) Clayton ■☆

282181　Panicum rowlandii Stapf = Panicum fluviicola Steud. ■☆

282182　Panicum roxburghii Spreng. = Panicum tripheron Schult. ■

282183　Panicum roxburghii Spreng. = Panicum trypheron Schult. var. suishaense(Hayata) C. C. Hsu ■

282184　Panicum rubiginosum Steud. = Setaria parviflora (Poir.) Kerguélen ■

282185　Panicum rubiginosum Steud. = Setaria pumila (Poir.) Roem. et Schult. ■

282186　Panicum rudimentosum Steud. = Setaria sphacelata (Schumach.) Stapf et C. E. Hubb. ex M. B. Moss ■☆

282187　Panicum ruspolii Chiov. ;鲁斯波利黍■☆

282188　Panicum sabiense Renvoize;萨比黍■☆

282189　Panicum sabulorum Lam. var. thinium (Hitchc. et Chase) C. F. Reed = Dichanthelium acuminatum (Sw.) Gould et C. A. Clark subsp. columbianum(Scribn.) Freckmann et Lelong ■☆

282190　Panicum sadinii(Vanderyst) Renvoize;萨丁黍■☆

282191　Panicum sagittifolium (A. Rich.) Hochst. ex Steud. = Setaria sagittifolia(A. Rich.) Walp. ■☆

282192　Panicum sanguinale L. = Digitaria sanguinalis(L.) Scop. ■

282193　Panicum sanguinale L. var. biforme(Willd.) Hack. ex T. Durand et Schinz = Digitaria ciliaris(Retz.)Koeler ■

282194　Panicum sanguinale L. var. biforme(Willd.) Hack. ex T. Durand et Schinz = Digitaria bicornis(Lam.) Roem. et Schult. ■

282195　Panicum sanguinale L. var. lecardii Pilg. = Digitaria argillacea (Hitchc. et Chase) Fernald ■☆

282196　Panicum sanguinale L. var. microbachne (J. Presl) Hack. = Digitaria setigera Roth ex Roem. et Schult. ■

282197　Panicum sanguinale L. var. microbachne Hack. = Digitaria microbachne(J. Presl) Henrard ■

282198　Panicum sanguinale L. var. timorense (Kunth) Hack. = Digitaria radicosa(J. Presl et C. Presl) Miq. ■

282199　Panicum sapinii (Vanderyst) Robyns = Panicum strictissimum Afzel. ex Sw. ■☆

282200　Panicum sarmentosum Roxb. ;卵花黍（长匍茎黍,藤竹草,亚高山黍草,竹篙草）;Longrunner Panicgrass, Mountain Panicgrass, Vinebamboo Panicgrass ■

282201　Panicum scabriusculum Chapm. ;略糙黍;Wooly Panicum ■☆

282202　Panicum scabrum Lam. = Echinochloa stagnina (Retz.) P.

Beauv. ■☆

282203 Panicum scabrum Lam. subsp. oryzetorum A. Chev. = Echinochloa stagnina(Retz.) P. Beauv. ■☆

282204 Panicum scalare Mez = Brachiaria scalaris Pilg. ■☆

282205 Panicum scalarum Schweinf. = Digitaria abyssinica (Hochst. ex A. Rich.) Stapf ■☆

282206 Panicum scalarum Schweinf. var. elatior Chiov. = Digitaria abyssinica(Hochst. ex A. Rich.) Stapf ■☆

282207 Panicum scandens Mez = Panicum sadinii (Vanderyst)Renvoize ■☆

282208 Panicum schimperanum Hochst. ex A. Rich. = Eriochloa meyeriana(Nees)Pilg. ■☆

282209 Panicum schinzii Hack. ;平滑黍;Transvaal Millet ■☆

282210 Panicum schlechteri Hack. = Panicum hymeniochilum Nees ■☆

282211 Panicum schmidtii Hack. = Cyrtococcum patens (L.) A. Camus var. schmidtii(Hack.) A. Camus ■

282212 Panicum schmidtii Hack. = Cyrtococcum patens(L.)A. Camus ■

282213 Panicum schweinfurthii Hack. ;施韦黍■☆

282214 Panicum scoparium Lam. ;帚黍;Panic Grass ■☆

282215 Panicum scoparium Lam. = Dichanthelium scoparium (Lam.) Gould ■☆

282216 Panicum scoparium Lam. var. leibergii Vasey = Dichanthelium leibergii(Vasey) Freckmann ■☆

282217 Panicum scoparium S. Watson ex Nash = Dichanthelium oligosanthes(Schult.) Gould subsp. scribnerianum(Nash) Freckmann et Lelong ■☆

282218 Panicum scribnerianum Nash = Dichanthelium oligosanthes (Schult.)Gould subsp. scribnerianum(Nash)Freckmann et Lelong ■☆

282219 Panicum scribnerianum Nash var. leibergii (Vasey) Scribn. = Dichanthelium leibergii (Vasey)Freckmann ■☆

282220 Panicum secernendum Hochst. ex Mez = Brachiaria comata (Hochst. ex A. Rich.) Stapf ■☆

282221 Panicum semialatum R. Br. = Alloteropsis semialata (R. Br.) Hitchc. ■

282222 Panicum semialatum R. Br. var. eckloniana (Nees) Hack. ex T. Durand et Schinz = Alloteropsis semialata (R. Br.) Hitchc. var. eckloniana(Nees)C. E. Hubb. ■

282223 Panicum semiundulatum A. Rich. = Brachiaria semiundulata (Hitchc. ex A. Rich.) Stapf ■

282224 Panicum semiundulatum Hitchc. ex A. Rich. = Brachiaria semiundulata(Hitchc. ex A. Rich.) Stapf ■

282225 Panicum semiundulatum Hochst. = Brachiaria semiundulata (Hitchc. ex A. Rich.) Stapf ■

282226 Panicum senegalense Desv. ;塞内加尔黍■☆

282227 Panicum sennii Chiov. = Panicum infestum Andersson ■☆

282228 Panicum serrarium Fig. et De Not. ;齿黍■☆

282229 Panicum serratum (Thunb.) Spreng. = Brachiaria serrata (Thunb.) Stapf ■☆

282230 Panicum serratum (Thunb.) Spreng. var. brachylophum (Stapf) A. Chev. = Brachiaria serrata(Thunb.) Stapf ■☆

282231 Panicum serrifolium Hochst. = Brachiaria serrifolia (Hochst.) Stapf ■☆

282232 Panicum serrulatum Schumach. = Brachiaria villosa (Lam.) A. Camus ■

282233 Panicum seslerioides Rendle = Sacciolepis rigens (Mez) A. Chev. ■☆

282234 Panicum setarioides Peter = Urochloa panicoides P. Beauv. ■

282235 Panicum setigerum P. Beauv. = Cyrtococcum chaetophoron

282236 Panicum setigerum Retz. = Urochloa setigera(Retz.) Stapf ■

282237 Panicum setinsigne Mez = Melinis repens(Willd.) Zizka subsp. grandiflora(Hochst.)Zizka ■☆

282238 Panicum shinyangense Renvoize ;希尼安加黍■☆

282239 Panicum simbense Mez = Panicum trichocladum K. Schum. ■☆

282240 Panicum simpliciusculum Wight et Arn. ex Steud. = Coelachne simpliciuscula(Wight et Arn.) Munro ex Benth. ■

282241 Panicum simulans Smook ;相似黍■☆

282242 Panicum snowdenii C. E. Hubb. = Panicum hymeniochilum Nees ■☆

282243 Panicum sociale Stapf = Panicum laetum Kunth ■☆

282244 Panicum sordidum Thwaites = Pseudoraphis sordida (Thwaites) S. M. Phillips et S. L. Chen ■

282245 Panicum spadiciferum Peter = Echinochloa pyramidalis (Lam.) Hitchc. et Chase ■☆

282246 Panicum sparmannii Mez = Panicum fluviicola Steud. ■☆

282247 Panicum sparsum Schumach. = Panicum maximum Jacq. ■

282248 Panicum sphacelatum (Benth.) Steud. = Melinis repens (Willd.) Zizka ■

282249 Panicum sphacelatum Schumach. = Setaria sphacelata (Schumach.)Stapf et C. E. Hubb. ex M. B. Moss ■☆

282250 Panicum sphaerocarpon Elliott ;球果黍;Panic Grass ■☆

282251 Panicum sphaerocarpon Elliott var. inflatum (Scribn. et J. G. Sm.) Hitchc. et Chase = Panicum sphaerocarpon Elliott ■☆

282252 Panicum spiciforme Hochst. ex A. Rich. = Sacciolepis myosuroides(R. Br.) Chase ex E. G. Camus et A. Camus ■

282253 Panicum spiciforme Hochst. ex A. Rich. = Sacciolepis spiciformis(Hochst. ex A. Rich.) Stapf ■

282254 Panicum spinescens R. Br. = Pseudoraphis spinescens (R. Br.) Vickery ■

282255 Panicum spongiosum Stapf = Panicum funaense Vanderyst ■☆

282256 Panicum sprengelianum Schult. ;野生稷草;Wild Panicgrass ■☆

282257 Panicum squamigerum Pilg. = Sacciolepis catumbensis(Rendle) Stapf ■☆

282258 Panicum stagninum Retz. = Echinochloa stagnina (Retz.) P. Beauv. ■☆

282259 Panicum stapfianum Fourc. ;施塔普夫稷■☆

282260 Panicum steudelianum Domin = Digitaria violascens Link ■

282261 Panicum stigmatisatum Mez = Brachiaria stigmatisata(Mez)Stapf ■☆

282262 Panicum stipitatum Nash = Panicum rigidulum Bosc ex Nees ■☆

282263 Panicum strictissimum Afzel. ex Sw. ;刚直黍■☆

282264 Panicum strictum Pursh var. linearifolium (Scribn.) Farw. = Dichanthelium linearifolium(Scribn.) Gould ■☆

282265 Panicum strictum Pursh var. perlongum (Nash) Farw. = Dichanthelium perlongum(Nash)Freckmann ■☆

282266 Panicum strictum Pursh var. psilophyllum (Fernald) Farw. = Dichanthelium depauperatum(Muhl.) Gould ■☆

282267 Panicum stuhlmanni K. Schum. = Brachiaria comata(Hochst. ex A. Rich.) Stapf ■☆

282268 Panicum subalbidum Kunth ;亚白黍■☆

282269 Panicum subangustum Schumach. = Pennisetum polystachion (L.) Schult. ■

282270 Panicum subflabellatum Stapf ;亚扇黍■☆

282271 Panicum sublaetum Stapf = Panicum gracilicaule Rendle ■☆

282272 Panicum submontanum Hayata = Panicum incomtum Trin. ex Nees ■

282273 Panicum subobliquum Stapf = Panicum brevifolium L. ■

282274 Panicum subpilosum Peter;疏毛黍■☆

282275 Panicum subquadriparum Trin. = Brachiaria subquadripara (Trin.) Hitchc. ■

282276 Panicum subquadriparum Trin. = Urochloa subquadripara (Trin.) R. D. Webster ■

282277 Panicum subrepandum Rendle = Panicum nervatum (Franch.) Stapf ■☆

282278 Panicum subtilissimum Renvoize;纤细黍■☆

282279 Panicum subulifolium Mez = Brachiaria subulifolia (Mez) Clayton ■☆

282280 Panicum subvillosum Ashe = Dichanthelium acuminatum (Sw.) Gould et C. A. Clark subsp. fasciculatum (Torr.) Freckmann et Lelong ■☆

282281 Panicum subvillosum Ashe = Panicum acuminatum Sw. ■

282282 Panicum suishaense Hayata = Panicum curviflorum Hornem. var. suishanense (Hayata) Veldkamp ■

282283 Panicum suishaense Hayata = Panicum curviflorum Hornem. ■

282284 Panicum suishaense Hayata = Panicum elegantissimum Hook. f. ■

282285 Panicum suishaense Hayata = Panicum tripheron Schult. ■

282286 Panicum sumatrense Roth ex Roem. et Schult. ;细柄黍;Barefoot Panicgrass, Little Millet, Sama, Slenderstalk Panicgrass ■

282287 Panicum sumatrense Roth ex Roem. et Schult. subsp. psilopodium de Wet = Panicum sumatrense Roth ex Roem. et Schult. ■

282288 Panicum supervacum C. B. Clarke = Brachiaria ramosa (L.) Stapf ■

282289 Panicum swynnertonii Rendle = Panicum coloratum L. ■

282290 Panicum syzigachne Steud. = Beckmannia syzigachne (Steud.) Fernald ■

282291 Panicum tambacoundense Berhaut = Panicum gracilicaule Rendle ■☆

282292 Panicum tenellum Griff. = Panicum walense Mez ■

282293 Panicum tenellum Lam. ;柔弱黍■☆

282294 Panicum tenellum Roxb. = Panicum tripheron Schult. ■

282295 Panicum teneriffae (L. f.) R. Br. = Tricholaena teneriffae (L. f.) Link ■☆

282296 Panicum teneriffae R. Br. = Tricholaena teneriffae (L. f.) Link ■☆

282297 Panicum tennesseense Ashe = Dichanthelium acuminatum (Sw.) Gould et C. A. Clark subsp. fasciculatum (Torr.) Freckmann et Lelong ■☆

282298 Panicum tenuiflorum R. Br. = Digitaria longiflora (Retz.) Pers. ■☆

282299 Panicum tenuissimum Mart. ex Schrank = Sporobolus tenuissimus (Mart. ex Schrank) Kuntze ■

282300 Panicum tephrosanthum Hack. = Panicum maximum Jacq. ■

282301 Panicum ternatum (Hochst. ex A. Rich.) Hochst. = Digitaria ternata (Hochst. ex A. Rich.) Stapf ex Dyer ■

282302 Panicum ternatum (Hochst. ex A. Rich.) Hochst. ex Steud. = Digitaria ternata (Hochst. ex A. Rich.) Stapf ex Dyer ■

282303 Panicum ternatum Hochst. ex Steud. = Digitaria ternata (Hochst. ex A. Rich.) Stapf ex Dyer ■

282304 Panicum texanum Buckley;得州黍;Colorado Grass, Colorado Millet, Texas Millet ■☆

282305 Panicum texanum Buckley = Urochloa texana (Buckley) R. D. Webster ■☆

282306 Panicum tholloni Franch. = Setaria homonyma (Steud.) Chiov. ■☆

282307 Panicum thwaitesii Hack. = Digitaria thwaitesii (Hack.) Henrard ■

282308 Panicum timorense Kunth = Digitaria radicosa (J. Presl et C. Presl) Miq. ■

282309 Panicum tomentosum Roxb. = Setaria intermedia Roem. et Schult. ■

282310 Panicum tonkinense Balansa = Acroceras tonkinense (Balansa) C. E. Hubb. ex Bor ■

282311 Panicum tonsum (Nees) Steud. = Melinis repens (Willd.) Zizka ■

282312 Panicum transvenulosum Stapf = Panicum monticola Hook. f. ■☆

282313 Panicum trichocladum Hack. ex K. Schum. ;毛枝黍■☆

282314 Panicum trichocladum K. Schum. = Panicum trichocladum Hack. ex K. Schum. ■☆

282315 Panicum trichocladum K. Schum. var. parviflorum Peter = Panicum trichocladum K. Schum. ■☆

282316 Panicum trichoides Sw. ;发枝稷;Hair Panicgrass ■

282317 Panicum trichonode Launert et Renvoize;毛节黍■☆

282318 Panicum trichopodioides Mez = Urochloa setigera (Retz.) Stapf ■

282319 Panicum trichopodon A. Rich. = Urochloa trichopus (Hochst.) Stapf ■☆

282320 Panicum trichopus Hochst. = Urochloa trichopus (Hochst.) Stapf ■☆

282321 Panicum trigonum Retz. = Cyrtococcum trigonum (Retz.) A. Camus ■☆

282322 Panicum tripheron Schult. = Panicum curviflorum Hornem. ■

282323 Panicum tripheron Schult. var. suishanense (Hayata) C. C. Hsu = Panicum curviflorum Hornem. var. suishanense (Hayata) Veldkamp ■

282324 Panicum tristachyoides Trin. = Loudetiopsis tristachyoides (Trin.) Conert ■☆

282325 Panicum trypheron Schult. = Panicum curviflorum Hornem. ■

282326 Panicum trypheron Schult. var. suishaense (Hayata) C. C. Hsu = Panicum curviflorum Hornem. ■

282327 Panicum trypheron Schult. var. suishaense (Hayata) C. C. Hsu = Panicum elegantissimum Hook. f. ■

282328 Panicum trypheron Schult. var. suishaense (Hayata) C. C. Hsu = Panicum tripheron Schult. ■

282329 Panicum tsaratananense A. Camus;察拉塔纳纳黍■☆

282330 Panicum tsugetorum Nash = Dichanthelium acuminatum (Sw.) Gould et C. A. Clark subsp. columbianum (Scribn.) Freckmann et Lelong ■☆

282331 Panicum tuberculiflorum Steud. = Echinochloa villosa (Thunb.) Kunth ■☆

282332 Panicum tuberculiflorum Steud. = Eriochloa villosa (Thunb.) Kunth ■

282333 Panicum tuckermanii Fernald;图氏稷;Tuckerman's Panic-grass ■☆

282334 Panicum tuckermanii Fernald = Panicum philadelphicum Bernh. ex Trin. ■☆

282335 Panicum tunicatum Hack. ;衣稷■☆

282336 Panicum turgidum Forssk. ;膨胀黍■☆

282337 Panicum turritum Thunb. = Sacciolepis interrupta (Willd.) Stapf ■

282338 Panicum tylanthum Hack. = Panicum pusillum Hook. f. ■☆

282339 Panicum typhurum Stapf = Sacciolepis typhura (Stapf) Stapf ■☆

282340 Panicum uliginosum Roth. = Sacciolepis interrupta (Willd.) Stapf ■

282341 Panicum umbellatum Trin. = Brachiaria umbellata (Trin.) Clayton ■☆

282342 Panicum umbratile Mez = Panicum monticola Hook. f. ■☆

282343 Panicum unciphyllum Trin. var. thinium Hitchc. et Chase = Dichanthelium acuminatum (Sw.) Gould et C. A. Clark subsp. columbianum (Scribn.) Freckmann et Lelong ■☆

282344 Panicum undulatifolium Ard. = Oplismenus undulatifolius (Ard.) Roem. et Schult. ■

282345 Panicum unguiculatum Trin. = Panicum deustum Thunb. ■☆

282346 Panicum uniglume Hochst. ex A. Rich. = Digitaria diagonalis (Nees) Stapf var. uniglumis (Hochst. ex A. Rich.) Pilg. ■☆

282347 Panicum urochloa Desv. = Urochloa panicoides P. Beauv. ■

282348 Panicum vagiflorum Stapf = Panicum fluviicola Steud. ■☆

282349 Panicum vaginatum (Sw.) Gren. et Godr. = Paspalum vaginatum Sw. ■

282350 Panicum verruciferum Mez = Panicum gracilicaule Rendle ■☆

282351 Panicum verrucosum Muhl. ; 疣稷; Warty Panicum ■☆

282352 Panicum verticillatum L. = Setaria verticillata (L.) P. Beauv. ■

282353 Panicum verticillatum L. var. ambiguum Guss. = Setaria verticillata (L.) P. Beauv. ■

282354 Panicum vescum R. R. Stewart = Panicum humile Nees ex Steud. ■

282355 Panicum vexillare Peter = Brachiaria humidicola (Rendle) Schweick. ■☆

282356 Panicum viciniflorum Stapf; 邻花黍■☆

282357 Panicum vicinum F. M. Bailey = Ichnanthus pallens (Sm.) Benth. var. major (Nees) Stieber ■

282358 Panicum vicinum F. M. Bailey = Ichnanthus pallens (Sm.) Munro ex Benth. var. major (Nees) Stieber ■

282359 Panicum vicinum F. M. Bailey = Ichnanthus vicinus (F. M. Bailey) Merr. ■

282360 Panicum villosissimum Nash var. pseudopubescens (Nash) Fernald = Dichanthelium ovale (Elliott) Gould et C. A. Clark subsp. pseudopubescens (Nash) Freckmann et Lelong ☆

282361 Panicum villosissum Nash = Panicum acuminatum Sw. ■

282362 Panicum villosum Lam. = Brachiaria villosa (Lam.) A. Camus ■

282363 Panicum violaceum Klein ex Thiele = Isachne dispar Trin. ■

282364 Panicum violaceum Lam. = Pennisetum violaceum (Lam.) Rich. ■☆

282365 Panicum violascens (Link) Kunth = Digitaria violascens Link ■

282366 Panicum virgatum L. ; 柳枝稷; Chinese Fountaingrass, Switch Grass, Switchgrass, Switch-grass, Venezuelan Panicgrass ■

282367 Panicum virgatum L. ' Cloud Nine '; 极乐柳枝稷; Tall Switch Grass ■☆

282368 Panicum virgatum L. ' Rehbraun '; 红叶柳枝稷; Red-leaved Switch Grass ■☆

282369 Panicum virgatum L. 'Warrior'; 武士柳枝稷; Switch Grass ■☆

282370 Panicum virgatum L. var. cubense Griseb. = Panicum virgatum L. ■

282371 Panicum virgatum L. var. obtusum A. W. Wood = Panicum virgatum L. ■

282372 Panicum virgatum Roxb. ex Steud. = Panicum psilopodium Trin. ■

282373 Panicum viride L. = Setaria viridis (L.) P. Beauv. ■

282374 Panicum viride L. var. gigantea Franch. et Sav. = Setaria viridis (L.) P. Beauv. subsp. pycnocoma (Steud.) Tzvelev ■

282375 Panicum viride L. var. majus Gaudin = Setaria viridis (L.) P. Beauv. subsp. pycnocoma (Steud.) Tzvelev ■

282376 Panicum viride L. var. majus Gaudin = Setaria viridis (L.) P. Beauv. var. major (Gaudin) Peterm. ■☆

282377 Panicum viride Steud. = Setaria viridis (L.) P. Beauv. ■

282378 Panicum viridescens Steud. = Setaria viridis (L.) P. Beauv. ■

282379 Panicum viviparum Schumach. = Brachiaria villosa (Lam.) A. Camus ■

282380 Panicum vollesenii Renvoize; 福勒森黍■☆

282381 Panicum volutans J. G. Anderson; 旋卷黍■☆

282382 Panicum walense Mez = Panicum humile Nees ex Steud. ■

282383 Panicum walteri Pursh = Echinochloa walteri (Pursh) A. Heller et Nash ■☆

282384 Panicum warburgii Mez = Cyrtococcum patens (L.) A. Camus ■

282385 Panicum werneri Scribn. = Dichanthelium linearifolium (Scribn.) Gould ■☆

282386 Panicum wiehei Renvoize; 韦厄黍■☆

282387 Panicum wilcoxianum Vasey = Dichanthelium wilcoxianum (Vasey) Freckmann ■☆

282388 Panicum wilcoxianum Vasey var. breitungii B. Boivin = Dichanthelium wilcoxianum (Vasey) Freckmann ■☆

282389 Panicum williamsii Hance = Arundinella anomala Steud. ■

282390 Panicum williamsii Hance = Arundinella hirta (Thunb.) Tanaka ■

282391 Panicum wilmingtonense Ashe = Dichanthelium ovale (Elliott) Gould et C. A. Clark subsp. pseudopubescens (Nash) Freckmann et Lelong ■☆

282392 Panicum xantholeucum Hack. = Brachiaria xantholeuca (Hack.) Stapf ■☆

282393 Panicum xanthophysum A. Gray; 黄稷; Yellow Panic-grass ■☆

282394 Panicum xanthophysum A. Gray = Dichanthelium xanthophysum (A. Gray) Freckmann ■☆

282395 Panicum xanthotrichum Hack. = Digitaria xanthotricha (Hack.) Stapf ■☆

282396 Panicum yadkinense Ashe = Dichanthelium dichotomum (L.) Gould ■☆

282397 Panicum yadkinense Ashe = Panicum dichotomum L. ■☆

282398 Panicum zamba Vanderyst; 桑巴黍■☆

282399 Panicum zambesiense Renvoize; 赞比西稷■☆

282400 Panicum zelayense (Kunth) Steud. = Echinochloa crus-galli (L.) P. Beauv. var. zelayensis (Kunth) Hitchc. ■

282401 Panicum zenkeri K. Schum. = Panicum griffonii Franch. ■☆

282402 Panicum zeyheri Nees = Digitaria velutina (Forssk.) P. Beauv. ■☆

282403 Panicum zeylayense (Kunth) Steud. = Echinochloa crus-galli (L.) P. Beauv. var. zelayensis (Kunth) Hitchc. ■

282404 Panicum zizanioides Kunth = Acroceras zizanioides (Kunth) Dandy ■☆

282405 Panicum zollingeri Steud. = Isachne albens Trin. ■

282406 Paniopsis Raf. = Inula L. ●■

282407 Paniopsis Raf. = Nidorella Cass. ■☆

282408 Panios Adans. = Erigeron L. ■●

282409 Panisea (Lindl.) Lindl. (1854) (保留属名); 曲唇兰属; Panisea ■

282410 Panisea (Lindl.) Steud. = Panisea (Lindl.) Lindl. (保留属名) ■

282411 Panisea Lindl. = Panisea (Lindl.) Lindl. (保留属名) ■

282412 Panisea bia (Kerr) Ts. Tang et F. T. Wang = Panisea tricallosa Rolfe ■

282413 Panisea cavalerei Schltr. ; 平卧曲唇兰 (单叶曲唇兰, 曲唇兰); Prostrate Panisea, Simpleleaf Panisea ■

282414 Panisea demissa (D. Don) Pfitzer; 矮曲唇兰■

282415 Panisea pantlingii (Pfitzer) Schltr. = Panisea tricallosa Rolfe ■

282416 Panisea parviflora (Lindl.) Lindl. = Panisea demissa (D. Don) Pfitzer ■

282417 Panisea tricallosa Rolfe; 曲唇兰 (双叶曲唇兰); Common Panisea ■

282418 Panisea uniflora (Lindl.) Lindl. ; 单花曲唇兰; Monoflower Panisea, Singleflower Panisea ■

282419 Panisea unifolia S. C. Chen = Panisea tricallosa Rolfe ■

282420 Panisea yunnanensis S. C. Chen et Z. H. Tsi; 云南曲唇兰; Yunnan Panisea ■

282421 Panisia Raf. = Cassia L. (保留属名) ●■

282422 Panke Molina = Gunnera L. ■☆

282423 Panke Willd. = Francoa Cav. ■☆

282424 Pankea Oerst. = Gunnera L. ■☆

282425 Pankea Oerst. = Panke Molina ■☆

282426 Panmorphia Luer = Epidendrum L. (保留属名)■☆

282427 Panmorphia Luer(1828);热带火炬兰属■☆

282428 Panninia T. Durand = Fanniinia Harv. ☆

282429 Panoe Adans. = Vateria L. ●☆

282430 Panope Raf. = Lippia L. ●■☆

282431 Panopia Noronha ex Thouars = Macaranga Thouars ●

282432 Panopsis Salisb. = Panopsis Salisb. ex Knight ●☆

282433 Panopsis Salisb. ex Knight(1809);热美山龙眼属●☆

282434 Panopsis ferruginea(Meisn.)Pittier;锈色热美山龙眼●☆

282435 Panopsis rubescens(Pohl)Pittier;红变热美山龙眼●☆

282436 Panopsis suaveolens(Klotz. et H. karst.)Pittier.;甜热美山龙眼●☆

282437 Panoxis Raf. = Hebe Comm. ex Juss. ●☆

282438 Panphalea Lag. (1811);纤细钝柱菊属■☆

282439 Panphalea Lag. = Pamphalea DC. ■☆

282440 Panslowia Wight ex Pfeiff. = Kadsura Kaempf. ex Juss. ●

282441 Panstenum Raf. = Allium L. ■

282442 Panstrepi Raf. = Coryanthes Hook. ■☆

282443 Panstrepis Raf. = Coryanthes Hook. ■☆

282444 Pantacantha Speg. (1902);全刺茄属☆

282445 Pantacantha ameghinoi Speg. ;全刺茄☆

282446 Pantadenia Gagnep. (1925);全腺大戟属☆

282447 Pantadenia chauvetiae(Leandri et Capuron)G. L. Webster;全腺大戟☆

282448 Pantasachne Endl. = Pentasacme Wall. ex Wight et Arn. ■

282449 Pantathera Phil. = Megalachne Steud. ■☆

282450 Panterpa Miers = Arrabidaea DC. ●☆

282451 Panterpa Miers = Petastoma Miers ●☆

282452 Panthocarpa Raf. = Acacia Mill. (保留属名)●■

282453 Pantlingia Prain = Stigmatodactylus Maxim. ex Makino ■

282454 Pantlingia palawensis (Tuyama) Rauschert = Disperis neilgherrensis Wight ■

282455 Pantocsekia Griseb. = Convolvulus L. ■●

282456 Pantocsekia Griseb. ex Pantoc. = Convolvulus L. ■●

282457 Pantorrhynchus Murb. = Trachystoma O. E. Schulz ■☆

282458 Pantorrhynchus maroccanus Murb. = Trachystoma ballii O. E. Schulz ■☆

282459 Panulia(Baill.)Koso-Pol. = Apium L. ■

282460 Panulia Baill. = Apium L. ■

282461 Panurea Spruce ex Benth. = Panurea Spruce ex Benth. et Hook. f. ■☆

282462 Panurea Spruce ex Benth. et Hook. f. (1865);南美长叶豆属■☆

282463 Panurea longifolia Spruce ex Benth. ;南美长叶豆■☆

282464 Panza Salisb. = Narcissus L. ■

282465 Panzera Cothen. = Calodendrum Thunb. (保留属名)●☆

282466 Panzera Cothen. = Pallassia Houtt. (废弃属名)●☆

282467 Panzera Willd. = Eperua Aubl. ●☆

282468 Panzeria J. F. Gmel. = Lycium L. ●

282469 Panzeria Moench = Leonurus L. ■

282470 Panzeria Moench = Panzerina Soják ■

282471 Panzeria alaschanica Kuprian. var. minor C. Y. Wu et H. W. Li = Panzerina alaschanica Kuprian. var. minor C. Y. Wu et H. W. Li ■

282472 Panzeria argyracea Kuprian. = Panzerina lanata(L.)Bunge var. argyracea(Kuprian.)H. W. Li ■

282473 Panzeria canescens Bunge = Panzerina canescens(Bunge)Soják ■

282474 Panzeria kansuensis C. Y. Wu et H. W. Li = Panzerina kansuensis C. Y. Wu et H. W. Li ■

282475 Panzeria parviflora C. Y. Wu et H. W. Li = Panzerina parviflora C. Y. Wu et H. W. Li ■

282476 Panzeria tomentosa Moench = Panzerina lanata(L.)Bunge ■

282477 Panzerina Soják(1982);脓疮草属;Panzerina ■

282478 Panzerina alaschanica Kuprian. ;脓疮草(阿拉善脓疮草,白花益母草,白龙苍,白龙昌菜,白龙穿彩,白龙串彩,白龙疮,白益母草,脓疮草,野芝麻);Alashan Panzerina ■

282479 Panzerina alaschanica Kuprian. var. minor C. Y. Wu et H. W. Li;阿拉善小花脓疮草(白龙昌菜,小花脓疮草);Smallflower Alashan Panzerina ■

282480 Panzerina canescens(Bunge)Soják;灰白脓疮草(灰脓疮草)■

282481 Panzerina kansuensis C. Y. Wu et H. W. Li;甘肃脓疮草;Gansu Panzerina,Kansu Panzerina ■

282482 Panzerina lanata(L.)Bunge;绒毛脓疮草(白龙穿彩)■

282483 Panzerina lanata (L.) Bunge var. alaschanica (Kuprian.) Tschern. = Panzerina alaschanica Kuprian. ■

282484 Panzerina lanata(L.)Bunge var. albescens(Kuprian.)H. W. Li = Panzeria albescens Kuprian. ■

282485 Panzerina lanata(L.)Bunge var. argyracea(Kuprian.)H. W. Li;银白脓疮草■

282486 Panzerina lanata(L.)Bunge var. parviflora(C. Y. Wu et H. W. Li)H. W. Li = Panzerina parviflora C. Y. Wu et H. W. Li ■

282487 Panzerina parviflora C. Y. Wu et H. W. Li;小花脓疮草;Littleflower Panzerina,Smallflower Panzerina ■

282488 Panzhuyuia Z. Y. Zhu = Alocasia(Schott)G. Don(保留属名)■

282489 Panzhuyuia Z. Y. Zhu(1985);繁株芋属■

282490 Panzhuyuia omeiensis Z. Y. Zhu;繁株芋■

282491 Paolia Chiov. = Coffea L. ●

282492 Paolia jasminoides Chiov. = Coffea rhamnifolia(Chiov.)Bridson ●☆

282493 Paoluccia Gand. = Lathyrus L. ■

282494 Papas Opiz = Battata Hill ●■

282495 Papas Opiz = Solanum L. ●■

282496 Papaver L. (1753);罂粟属;Oriental Poppy,Poppy ■

282497 Papaver aculeatum Thunb. ;皮刺罂粟■☆

282498 Papaver agrivagum Jord. = Papaver rhoeas L. ■

282499 Papaver ajanense Popov;阿加罂粟■☆

282500 Papaver alaskanum Hultén = Papaver radicatum Rottb. ex DC. subsp. alaskanum(Hultén)J. P. Anderson ■☆

282501 Papaver alaskanum Hultén var. grandiflorum Hultén = Papaver radicatum Rottb. ex DC. subsp. alaskanum(Hultén)J. P. Anderson ■☆

282502 Papaver alaskanum Hultén var. latilobum Hultén = Papaver radicatum Rottb. ex DC. subsp. alaskanum(Hultén)J. P. Anderson ■☆

282503 Papaver alaskanum Hultén var. macranthum Hultén = Papaver macounii Greene ■☆

282504 Papaver alboroseum Hultén;粉白罂粟■☆

282505 Papaver album Mill. ;白花罂粟■☆

282506 Papaver alpinum L. ;高山罂粟(伯舍罂粟);Alpine Poppy ■☆

282507 Papaver alpinum L. var. croceum Regel = Papaver nudicaule L. ■☆

282508 Papaver ambiguum Popov;可疑罂粟■☆

282509 Papaver amoenum Lindl. = Papaver somniferum L. ■

282510 Papaver amurense (N. Busch) N. Busch ex Tolm. = Papaver nudicaule L. subsp. amurense N. Busch ■

282511 Papaver angustifolium Tolm. ;狭叶罂粟■☆

282512 Papaver anomalum Fedde = Papaver nudicaule L. subsp.

amurense N. Busch ■

282513　Papaver anomalum Fedde = Papaver nudicaule L. var. aquilegioides Fedde ■

282514　Papaver arenarium M. Bieb. ;沙生罂粟■☆

282515　Papaver argemone L. ;大红罂粟（蓟罂粟）; Bastard Wild Poppy, Cock's Head, Eye Poppy, Long Pricklyhead Poppy, Long Prickly-headed Poppy, Long Rough-headed Poppy, Pale Poppy, Prickly Poppy, Rough Long-beaded Poppy, Rough-headed Bastard Poppy, Wind Rose ■☆

282516　Papaver argemone L. var. hybridum (L.) Schmalh. = Papaver hybridum L. ■

282517　Papaver armeniacum Lam. ;亚美尼亚罂粟;Armenia Poppy ■☆

282518　Papaver atlanticum (Ball) Coss. = Papaver atlanticum Ball. ex Coss. ■☆

282519　Papaver atlanticum Ball. ex Coss. ;大西洋罂粟;Atlas Poppy, Poppy ■☆

282520　Papaver belangeri Boiss. ;白氏罂粟■☆

282521　Papaver bipinnatum C. A. Mey. ;双羽罂粟■☆

282522　Papaver bracteatum Lindl. ;苞罂粟（大红罂粟）; Bracted Poppy, Great Poppy, Great Scarlet Poppy ■☆

282523　Papaver burseri Crantz = Papaver alpinum L. ■☆

282524　Papaver californicum A. Gray;加州罂粟; Fire Poppy, Western Poppy ■☆

282525　Papaver canescens Tolm. ;灰毛罂粟(阿尔泰黄罂粟,天山罂粟);Greyhair Poppy ■☆

282526　Papaver caucasicum Henning;高加索罂粟;Caucasia Poppy ■☆

282527　Papaver chelidoniifolium Boiss. et Buhse;白屈菜叶罂粟■☆

282528　Papaver chinense(Regel) Kitag. = Papaver nudicaule L. ■

282529　Papaver collinum Bisch. = Papaver dubium L. ■☆

282530　Papaver commutatum Fisch. ,Mey. et Trautv. ;变化罂粟(梦幻罂粟)■☆

282531　Papaver commutatum Fisch. , Mey. et Trautv. ' Ladybird ' = Papaver commutataum Fisch. ,Mey. et Trautv. ☆

282532　Papaver commutatum Fisch. ,Mey. et Trautv. = Papaver rhoeas L. ■

282533　Papaver conigerum Stocks = Papaver pavoninum Schrenk ■

282534　Papaver cornwallisensis D. Löve;极地罂粟■☆

282535　Papaver crassifolium(Benth.) Greene = Stylomecon heterophylla G. Taylor ■☆

282536　Papaver croceum Ledeb. ;橘黄罂粟(冰岛罂粟,橙黄罂粟,野罂粟);Iceland Poppy,Orange Poppy ■

282537　Papaver croceum Ledeb. = Papaver nudicaule L. ■

282538　Papaver croceum Ledeb. subsp. chinense (Regel) Rändel = Papaver nudicaule L. ■

282539　Papaver cylindricum Cullen;圆柱罂粟;Cylindrical Poppy ■☆

282540　Papaver dalechianum Fedde = Papaver macrostomum Boiss. et Huet ex Boiss. ■☆

282541　Papaver decaisnei Boiss. = Papaver dubium L. ■☆

282542　Papaver decaisnei Elkan = Papaver decaisnei Hochst. et Steud. ex Boiss. ■☆

282543　Papaver decaisnei Hochst. et Steud. ex Boiss. ;南亚罂粟■☆

282544　Papaver denalii Gjaerev. = Papaver mcconellii Hultén ■☆

282545　Papaver divergens Fedde = Papaver macrostomum Boiss. et Huet ex Boiss. ■☆

282546　Papaver dubium L. ;淡红罂粟（可疑罂粟）; Blaver, Blind Eyes, Blindeyes, Cockrose, Cock's Head, Cootamundra Wattle, Corn Poppy, Field Poppy, Headache, Long Smooth-headed Poppy, Longhead Poppy, Long-headed Poppy, Long-pod Poppy, Smooth Long-headed Poppy, Smooth Prickly-headed Poppy, Smooth-headed Bastard Poppy ■☆

282547　Papaver dubium L. subsp. mairei(Batt.) Maire = Papaver mairei Batt. ■☆

282548　Papaver dubium L. var. collinum(Bisch.) = Papaver dubium L. ■☆

282549　Papaver dubium L. var. decaisnei (Hochst. et Steud. ex Boiss) Elkan = Papaver decaisnei Hochst. et Steud. ex Boiss. ■☆

282550　Papaver dubium L. var. dielsianum Fedde = Papaver decaisnei Hochst. et Steud. ex Boiss. ■☆

282551　Papaver dubium L. var. glaucum Doum. = Papaver mairei Batt. ■☆

282552　Papaver dubium L. var. hirtellum Maire = Papaver mairei Batt. ■☆

282553　Papaver dubium L. var. laevigatum (M. Bieb.) Kadereit = Papaver dubium L. ■☆

282554　Papaver dubium L. var. lamottei (Boreau) Cariot = Papaver dubium L. ■☆

282555　Papaver dubium L. var. maroccanum Ball = Papaver dubium L. ■☆

282556　Papaver dubium L. var. modestum (Jord.) Fedde = Papaver dubium L. ■☆

282557　Papaver dubium L. var. obtusifolium (Desf.) Elkan = Papaver dubium L. ■☆

282558　Papaver dubium L. var. schweinfurthii (Fedde) Maire = Papaver dubium L. ■☆

282559　Papaver dubium L. var. subadpressiusculo-setosum (Fedde) Maire = Papaver dubium L. ■☆

282560　Papaver dubium L. var. subbipinnatifidum (Kuntze) Fedde = Papaver dubium L. ■☆

282561　Papaver dubium L. var. subintegrum (Kuntze) Fedde = Papaver dubium L. ■☆

282562　Papaver dubium L. var. tenue(Ball)Maire = Papaver dubium L. ■☆

282563　Papaver fauriei(Fedde) Fedde ex Miyabe et Tatew. ;法氏罂粟(日本罂粟);Faurie Poppy ■☆

282564　Papaver gariepinum Burch. = Papaver aculeatum Thunb. ■☆

282565　Papaver gaubae Cullen et Rech. f. = Papaver decaisnei Hochst. et Steud. ex Boiss. ■☆

282566　Papaver glaucum Boiss. et Hausskn. ex Boiss. ;粉叶罂粟（粉绿罂粟）; Tulip Poppy ■☆

282567　Papaver gorodkovii Tolm. et Petrovsky;高氏罂粟■☆

282568　Papaver hispidum Lam. = Papaver hybridum L. ■

282569　Papaver hookeri Baker = Papaver rhoeas L. ■

282570　Papaver hortense Hussenot = Papaver somniferum L. ■

282571　Papaver hultenii Knaben = Papaver lapponicum(Tolm.)Nordh. ■☆

282572　Papaver hultenii Knaben var. salmonicolor Hultén = Papaver lapponicum(Tolm.)Nordh. ■☆

282573　Papaver humile Fedde;矮小罂粟■☆

282574　Papaver hybridum L. ;杂种罂粟;Bristly Poppy, Hybrid Poppy, Rough Poppy, Round Pricklyhead Poppy, Round Prickly-headed Poppy, Round-headed Poppy ■

282575　Papaver hybridum L. var. grandifiorum Boiss. = Papaver pavoninum Schrenk ■

282576　Papaver hybridum L. var. microcarpum N. Busch = Papaver pavoninum Schrenk ■

282577　Papaver hybridum L. var. pinnatifidum Rouy et Foucaud = Papaver hybridum L. ■

282578　Papaver hybridum L. var. tenuifolium L. Chevall. = Papaver hybridum L. ■

282579　Papaver intermedium DC. ;间型罂粟■☆

282580　Papaver involucratum Popov;总苞罂粟■☆

282581　Papaver jugoricum(Tolm.)Stank. ;尤戈尔罂粟;Jugor Poppy ■☆

282582　Papaver keelei Porsild = Papaver macounii Greene subsp. discolor(Hultén)Rändel ex D. F. Murray ■☆

282583　Papaver kurdistanicum Fedde = Papaver macrostomum Boiss. et Huet ex Boiss. ■☆

282584　Papaver lacerum Popov;撕裂罂粟 ■☆

282585　Papaver laevigatum M. Bieb. ;光滑罂粟;Laevigate Poppy ■☆

282586　Papaver laevigatum M. Bieb. = Papaver dubium L. ■☆

282587　Papaver lamottei Boreau = Papaver dubium L. ■☆

282588　Papaver lapponicum(Tolm.) Nordh. ;拉普兰罂粟 ■☆

282589　Papaver lapponicum (Tolm.) Nordh. subsp. labradoricum (Fedde) Knaben = Papaver radicatum Rottb. ex DC. ■☆

282590　Papaver lapponicum (Tolm.) Nordh. subsp. occidentale (Lundstr.) Knaben = Papaver radicatum Rottb. ex DC. ■☆

282591　Papaver lapponicum (Tolm.) Nordh. subsp. porsildii Knaben = Papaver radicatum Rottb. ex DC. ■☆

282592　Papaver lasiothrix Fedde;多毛罂粟 ■☆

282593　Papaver ledebourianum Lundstr. ;莱德罂粟 ■☆

282594　Papaver leiocarpum Turcz. ;光果罂粟 ■☆

282595　Papaver lemmonii Greene = Papaver californicum A. Gray ■☆

282596　Papaver lisae N. Busch;丽萨罂粟 ■☆

282597　Papaver litwinowii Fedde ex Bornm. ;利特氏罂粟 (托里罂粟);Litwinov Poppy ■

282598　Papaver litwinowii Fedde ex Bornm. Engler = Papaver dubium L. ■☆

282599　Papaver macounii Greene;马氏罂粟 ■☆

282600　Papaver macounii Greene subsp. discolor(Hultén)Rändel ex D. F. Murray;异色马氏罂粟 ■☆

282601　Papaver macounii Greene var. discolor Hultén = Papaver macounii Greene subsp. discolor(Hultén)Rändel ex D. F. Murray ■☆

282602　Papaver macrostomum Boiss. et Huet = Papaver macrostomum Boiss. et Huet ex Boiss. ■☆

282603　Papaver macrostomum Boiss. et Huet ex Boiss. ;大药罂粟 ■☆

282604　Papaver mairei Batt. ;迈雷罂粟 ■☆

282605　Papaver mairei Batt. var. hirtellum Maire = Papaver mairei Batt. ■☆

282606　Papaver malviflorum Doum. = Papaver dubium L. ■☆

282607　Papaver malviflorum Doum. var. patulivillum Maire et Sam. = Papaver dubium L. ■☆

282608　Papaver mcconellii Hultén;加拿大罂粟 ■☆

282609　Papaver mexicanum(L.) E. H. L. Krause = Argemone mexicana L. ■

282610　Papaver microcarpum DC. ;小果罂粟 ■☆

282611　Papaver microcarpum DC. subsp. alaskanum (Hultén) Tolm. = Papaver radicatum Rottb. ex DC. subsp. alaskanum (Hultén) J. P. Anderson ■☆

282612　Papaver miyabeanum(Miyabe et Tatew.)Tatew. ;宫部氏罂粟 ■☆

282613　Papaver modestum Jord. = Papaver dubium L. ■☆

282614　Papaver monanthum Trautv. ;单花罂粟 ■☆

282615　Papaver nigroflavum D. Löve = Papaver radicatum Rottb. ex DC. ■☆

282616　Papaver nivale Tolm. ;雪白罂粟 ■☆

282617　Papaver nudicaule L. ;野罂粟(冰岛罂粟,橙黄罂粟,红橙罂粟,华罂粟,橘黄罂粟,裂叶野罂粟,毛罂粟,山大烟,山米壳,山罂粟,小罂粟,岩罂粟,野大烟);Iceland Poppy, Nakestem Poppy, Nudicaulous Poppy ■

282618　Papaver nudicaule L. f. seticarpum(P. Y. Fu)H. Chuang;毛果黑水罂粟;Hairfruit Poppy ■

282619　Papaver nudicaule L. subsp. album var. psilocarpum Fedde = Papaver nudicaule L. var. aquilegioides Fedde f. amurense (N. Busch) H. Chuang ■

282620　Papaver nudicaule L. subsp. americanum Rändel ex D. F. Murray;美洲野罂粟 ■☆

282621　Papaver nudicaule L. subsp. amurense Busch = Papaver nudicaule L. var. aquilegioides Fedde f. amurense (N. Busch) H. Chuang ■

282622　Papaver nudicaule L. subsp. amurense N. Busch;黑水罂粟(阿穆尔罂粟,黑水野罂粟,山大烟,野大烟花);Amur Poppy ■

282623　Papaver nudicaule L. subsp. amurense N. Busch = Papaver nudicaule L. var. aquilegioides Fedde ■

282624　Papaver nudicaule L. subsp. amurense N. Busch var. seticarpum (P. Y. Fu) H. Chuang = Papaver nudicaule L. f. seticarpum(P. Y. Fu) H. Chuang ■

282625　Papaver nudicaule L. subsp. aurantiacum(DC.) Fedde = Papaver nudicaule L. ■

282626　Papaver nudicaule L. subsp. rubroaurantiacum(Fisch. ex DC.) Fedde var. chinense(Regel) Fedde = Papaver nudicaule L. ■

282627　Papaver nudicaule L. subsp. rubroaurantiacum (Fisch. ex DC.) Fedde var. corydalifolium Fedde = Papaver nudicaule L. ■

282628　Papaver nudicaule L. subsp. rubroaurantiacum (Fisch. ex DC.) Fedde var. isopyroides Fedde = Papaver nudicaule L. ■

282629　Papaver nudicaule L. subsp. rubroaurantiacum (Fisch. ex DC.) Fedde var. miniatum Fedde = Papaver nudicaule L. var. aquilegioides Fedde ■

282630　Papaver nudicaule L. subsp. rubroaurantiacum (Fisch. ex DC.) Fedde var. subcoroydalifolium Fedde = Papaver nudicaule L. ■

282631　Papaver nudicaule L. subsp. rubroaurantiacum (Fisch. ex DC.) Fedde = Papaver nudicaule L. ■

282632　Papaver nudicaule L. var. amurense (N. Busch) H. Chuang = Papaver nudicaule L. subsp. amurense N. Busch ■

282633　Papaver nudicaule L. var. aquilegioides Fedde;光果野罂粟;Smoothfruit Poppy ■

282634　Papaver nudicaule L. var. aquilegioides Fedde f. amurense (N. Busch) H. Chuang = Papaver nudicaule L. subsp. amurense N. Busch ■

282635　Papaver nudicaule L. var. chinense(Regel) Fedde;山罂粟 ■

282636　Papaver nudicaule L. var. chinense (Regel) Fedde = Papaver nudicaule L. ■

282637　Papaver nudicaule L. var. corydalifolium Fedde = Papaver nudicaule L. ■

282638　Papaver nudicaule L. var. glabricarpum P. Y. Fu = Papaver nudicaule L. var. aquilegioides Fedde ■

282639　Papaver nudicaule L. var. isopyroides Fedde = Papaver nudicaule L. ■

282640　Papaver nudicaule L. var. labradoricum Fedde = Papaver radicatum Rottb. ex DC. ■☆

282641　Papaver nudicaule L. var. saxatile Kitag. = Papaver nudicaule L. ■

282642　Papaver nudicaule L. var. saxatulae Kitag. ;岩罂粟 ■

282643　Papaver nudicaule L. var. subcorydalifolium Fedde = Papaver nudicaule L. ■

282644　Papaver obtusifolium Desf. = Papaver dubium L. ■☆

282645　Papaver ocellatum Woronow = Papaver pavoninum Schrenk ■

282646　Papaver oreophilum Rupr. ;喜山罂粟 ■☆

282647　Papaver orientale L. ;鬼罂粟(东方罂粟,近东罂粟);Ghost Poppy,Oriental Poppy ■

282648　Papaver orientale L. ‘ Allegro Viva ’ = Papaver orientale L. ‘ Allegro’ ■☆

282649　Papaver orientale L. ‘ Allegro’;东方罂粟 ■☆

282650　Papaver orientale L. ‘ Beauty of Livermere’;红丽东方罂粟 ■☆

282651　Papaver orientale L. ‘ Indian Chief’;印第安酋长东方罂粟 ■☆

282652　Papaver orientale L. ‘May Queen’；五月皇后东方罂粟■☆

282653　Papaver orientale L. ‘Mrs Perry’；佩丽夫人东方罂粟■☆

282654　Papaver orientale L. ‘Perry's White’；佩丽白东方罂粟■☆

282655　Papaver paeoniaeflorum Hort.；牡丹罂粟；Paeony-flowered Poppy ■☆

282656　Papaver paniculatum D. Don = Meconopsis napaulensis DC. ■

282657　Papaver paniculatum D. Don = Meconopsis paniculata(D. Don) Prain ■

282658　Papaver paucifoliatum(Trautv.) Fedde；少叶罂粟■☆

282659　Papaver pavoninum Schrenk；黑环罂粟；Peacock Poppy ■

282660　Papaver pavoninum Schrenk f. album X. J. Ge；白花黑环罂粟；Whiteflower Peacock Poppy ■

282661　Papaver pavoninum Schrenk var. incornutum Fedde = Papaver pavoninum Schrenk ■

282662　Papaver persicum Lindl.；波斯罂粟■☆

282663　Papaver pinnatifidum Moris.；羽裂罂粟■☆

282664　Papaver piptostigma Bien. ex Fedde = Papaver macrostomum Boiss. et Huet ex Boiss. ■☆

282665　Papaver polare (Tolm.) Perfil. = Papaver cornwallisensis D. Löve ■☆

282666　Papaver pseudocanescens Popov；阿尔泰黄罂粟(阿尔泰罂粟)；Altai Poppy ■

282667　Papaver pseudocanescens Popov = Papaver canescens Tolm. ■

282668　Papaver pseudo-orientale(Fedde) Medw.；假东方罂粟；Oriental Poppy ■☆

282669　Papaver pseudo-orientale(Fedde)Medw. = Papaver orientale L. ■

282670　Papaver pseudo-orientale Medw. = Papaver pseudo-orientale (Fedde) Medw. ■☆

282671　Papaver pseudoradicatum Kitag. = Papaver radicatum Rottb. ex DC. var. pseudoradicatum(Kitag.) Kitag. ■

282672　Papaver pseudostubendorfii Popov；假斯图罂粟■☆

282673　Papaver pulvinatum Tolm.；叶枕罂粟■☆

282674　Papaver pygmaeum Rydb.；比利牛斯罂粟；Pyrenees Poppy ■☆

282675　Papaver radicatum Rottb. = Papaver radicatum Rottb. ex DC. ■☆

282676　Papaver radicatum Rottb. ex DC.；直根罂粟；Arctic Poppy, Yellow Poppy ■☆

282677　Papaver radicatum Rottb. ex DC. subsp. alaskanum (Hultén) J. P. Anderson；阿拉斯加罂粟■☆

282678　Papaver radicatum Rottb. ex DC. subsp. labradoricum (Fedde) Fedde = Papaver radicatum Rottb. ex DC. ■☆

282679　Papaver radicatum Rottb. ex DC. subsp. lapponicum Tolm. = Papaver lapponicum(Tolm.) Nordh. ■☆

282680　Papaver radicatum Rottb. ex DC. subsp. occidentale Lundstr. = Papaver radicatum Rottb. ex DC. ■☆

282681　Papaver radicatum Rottb. ex DC. subsp. porsildii (Knaben) Á. Löve = Papaver radicatum Rottb. ex DC. ■☆

282682　Papaver radicatum Rottb. ex DC. var. labradoricum (Fedde) J. Rousseau et Raymond = Papaver radicatum Rottb. ex DC. ■☆

282683　Papaver radicatum Rottb. ex DC. var. pseudoradicatum(Kitag.) Kitag.；长白山罂粟(白山罂粟,高山罂粟,假根罂粟,山大烟,山罂粟)；Changbaishan Poppy, Montane Poppy ■

282684　Papaver radicatum Rottb. ex DC. var. pygmaeum (Rydb.) S. L. Welsh = Papaver pygmaeum Rydb. ■☆

282685　Papaver rhoeas L.；虞美人(百媚娇,蝴蝶满园春,锦被花,孔冠花,丽春花,赛牡丹,莴苣莲,舞草,虞美人草,虞美人花)；Bledewort, Blind Eyes, Blind Man, Bull's Eyes, Butterfly Ladies, Butterfly Lady, Canker, Canker Rose, Chasbol, Chesbol, Chesbow, Cock Roise, Cockens, Cock-rose, Cock's Comb, Cock's Head, Collinhood, Common Poppy, Cootamundra Wattle, Cop Rose, Copper Rose, Copperoze, Corn Poppy, Corn Rose, Cornflower, Cup-rose, Curt Rose, Devil's Tongue, Earache, Eyebright, Field Poppy, Fireflout, Flanders Poppy, Flander's Poppy, Gollywog, Gye, Headache, Headacher, Headwarke, Headwork, Hogweed, Joan's Silver Pin, Knapbottle, Lightning, Lightning-flower, Old Woman's Petticoats, Paradise Lily, Pepper Box, Pepper-box, Pig Rose, Poison Poppy, Pope, Poppet, Popple, Poppy, Red Corn Rose, Red Cornflower, Red Dolly, Red Huntsman, Red Mailkes, Red Nap, Red Petti Coats, Red Poppy, Red Rags, Red Soldiers, Redcup Redcap, Redweed, Shirley Poppy, Shirley's Poppy, Sleepyhead, Smooth Round-beaded Poppy, Soldiers, Summer Poppy, Thunder Flower, Thunder-ball, Thunderbolts, Thunder-cup, Wart-flower, Wild Maws, Yedwark ■

282686　Papaver rhoeas L. var. agrivagum(Jord.)Beck = Papaver rhoeas L. ■

282687　Papaver rhoeas L. var. albiflorum Kuntze = Papaver rhoeas L. ■

282688　Papaver rhoeas L. var. alleizettei Maire = Papaver rhoeas L. ■

282689　Papaver rhoeas L. var. chanceliae Maire = Papaver rhoeas L. ■

282690　Papaver rhoeas L. var. dubium(L.) Schmalh. = Papaver dubium L. ■☆

282691　Papaver rhoeas L. var. hookeri(Baker)Fedde = Papaver rhoeas L. ■

282692　Papaver rhoeas L. var. leucanthemum Fedde = Papaver rhoeas L. ■

282693　Papaver rhoeas L. var. roubiaei (Vig.)Salis = Papaver rhoeas L. ■

282694　Papaver rhoeas L. var. strigosum Boenn. = Papaver rhoeas L. ■

282695　Papaver rhoeas L. var. subintegrum Lange = Papaver rhoeas L. ■

282696　Papaver rhoeas L. var. trichocarpum Pamp. = Papaver rhoeas L. ■

282697　Papaver rhoeas L. var. trifidum(Kuntze)Fedde = Papaver rhoeas L. ■

282698　Papaver rhoeas L. var. umbilicato-substipitatum Fedde = Papaver rhoeas L. ■

282699　Papaver rhoeas L. var. urophyllum Fedde = Papaver rhoeas L. ■

282700　Papaver roopianum(Bordz.) Sosn. = Papaver roopianum Bordz. ex Sosn. ■☆

282701　Papaver roopianum Bordz. ex Sosn.；卢罂粟■☆

282702　Papaver roubiaei Vig. = Papaver rhoeas L. ■

282703　Papaver rubroaurantiacum Fisch. ex DC. = Papaver nudicaule L. ■

282704　Papaver rubroaurantiacum Fisch. ex Steud. = Papaver rubroauriantiacum(Fisch. ex DC.) Steud. ■☆

282705　Papaver rubroauriantiacum(Fisch. ex DC.)Steud.；红黄罂粟■☆

282706　Papaver rupifragum Boiss. et Reut.；石隙罂粟；Spanish Poppy ■☆

282707　Papaver rupifragum Boiss. et Reut. subsp. atlanticum (Ball) Maire = Papaver rhoeas L. ■

282708　Papaver rupifragum Boiss. et Reut. var. atlanticum Ball = Papaver rhoeas L. ■

282709　Papaver rupifragum Boiss. et Reut. var. maroccanum Font Quer et Pau = Papaver rupifragum Boiss. et Reut. ■☆

282710　Papaver scammanianum D. Löve = Papaver macounii Greene subsp. discolor(Hultén) Rändel ex D. F. Murray ■☆

282711　Papaver schweinfurthii Fedde = Papaver dubium L. ■☆

282712　Papaver sendtneri A. Kern. ex Hayek = Papaver alpinum L. ■☆

282713　Papaver setigerum DC.；刚毛罂粟(具刺罂粟)；Common Poppy, Opium Poppy ■☆

282714　Papaver setigerum DC. = Papaver somniferum L. subsp. setigerum(DC.) Arcang. ■☆

282715　Papaver setigerum DC. var. valdesetosum Maire = Papaver somniferum L. subsp. setigerum(DC.) Arcang. ■☆

282716　Papaver simplicifolium G. Don = Meconopsis simplicifolia (G. Don) Walp. ■

282717　Papaver somniferum L.；罂粟(阿芙蓉,大烟,大烟花,米壳花,

米囊,米囊花,米囊拟,囊子,乌烟,象谷,鸦片,鸦片花,莺粟,罂子壳,罂子粟,罂子果,罂子花,御米,御米花);Carnation Poppy, Chesboul,Chesbow,Chesbowl,Common Poppy,Garden Poppy,Gear, Maw Seed,Mawseed,Oil Poppy,Opium,Opium Poppy,Scag,Smack, Somniferous Poppy,White Poppy ■

282718　Papaver somniferum L. 'Peony Flowered';芍药花罂粟■☆

282719　Papaver somniferum L. 'Pink Beauty';绯丽罂粟■☆

282720　Papaver somniferum L. 'White Cloud';白云罂粟■☆

282721　Papaver somniferum L. subsp. setigerum (DC.) Arcang. = Papaver setigerum DC. ■☆

282722　Papaver somniferum L. var. album DC. ;白罂粟(白花罂粟); White-race Poppy ■

282723　Papaver somniferum L. var. album DC. = Papaver somniferum L. ■

282724　Papaver somniferum L. var. atroviolaceum Maire = Papaver somniferum L. ■

282725　Papaver somniferum L. var. coccineum Maire = Papaver somniferum L. ■

282726　Papaver somniferum L. var. hortense (Hussenot) Rouy et Foucaud = Papaver somniferum L. ■

282727　Papaver somniferum L. var. nigrum DC. ;荷兰罂粟■☆

282728　Papaver somniferum L. var. nigrum DC. = Papaver somniferum L. ■

282729　Papaver somniferum L. var. poeoilospermum Alef. = Papaver somniferum L. ■

282730　Papaver somniferum L. var. setigerum (DC.) Webb = Papaver somniferum L. subsp. setigerum(DC.) Arcang. ■☆

282731　Papaver somniferum L. var. valdesetosum Maire = Papaver somniferum L. ■

282732　Papaver stewartianum Jafri et Qaiser;斯图尔特罂粟■☆

282733　Papaver strigosum(Boenn.) Schur = Papaver rhoeas L. ■

282734　Papaver strigosum Schur;密棘罂粟;Strigose Poppy ■☆

282735　Papaver strigosum Schur = Papaver rhoeas L. ■

282736　Papaver stubendorfii Tolm. ;斯图罂粟■☆

282737　Papaver suaveolens Lapeyr. ;甜罂粟;Pyrenean Poppy ■☆

282738　Papaver subadpressiusculo-setosum Fedde = Papaver dubium L. ■☆

282739　Papaver talyschense Grossh. ;塔里森罂粟■☆

282740　Papaver tenellum Tolm. ;小罂粟;Tender Poppy ■

282741　Papaver tenellum Tolm. = Papaver nudicaule L. ■

282742　Papaver tenue Ball = Papaver dubium L. ■☆

282743　Papaver tianschanicum Popov;天山罂粟;Tianshan Poppy ■☆

282744　Papaver tianschanicum Popov = Papaver canescens Tolm. ■

282745　Papaver tolmatchevii N. Semenova;陶尔罂粟■☆

282746　Papaver tubuliferum Fedde. = Papaver macrostomum Boiss. et Huet ex Boiss. ■☆

282747　Papaver turbinatum DC. = Papaver dubium L. ■☆

282748　Papaver walpolei A. E. Porsild;瓦尔罂粟■☆

282749　Papaver walpolei A. E. Porsild var. sulphureo-maculata Hultén = Papaver walpolei A. E. Porsild ■☆

282750　Papaveraceae Adans. = Papaveraceae Juss. (保留科名)●■

282751　Papaveraceae Juss. (1789)(保留科名);罂粟科;Poppy Family ●■

282752　Papaveraceae Voigt = Papaveraceae Juss. (保留科名)●■

282753　Papaya Adans. = Carica L. ●

282754　Papaya Mill. = Carica L. ●

282755　Papaya Tourn. ex L. = Carica L. ●

282756　Papaya carica Gaertn. = Carica papaya L. ●

282757　Papayaceae Blume = Caricaceae Dumort. (保留科名)●

282758　Papeda Hassk. = Citrus L. ●

282759　Papeda ichangensis C. J. Tseng = Citrus ichangensis Swingle ●

282760　Papeda rumphii Hassk. = Citrus hystrix DC. ●

282761　Paphia Seem. (1869);帕福斯杜鹃属(南树萝卜属)●☆

282762　Paphia Seem. = Agapetes D. Don ex G. Don ●

282763　Paphia vitiensis Seem. ;帕福斯杜鹃●☆

282764　Paphinia Lindl. (1843);帕福斯兰属(芭菲兰属)■☆

282765　Paphinia cristata Lindl. ;帕福斯兰■☆

282766　Paphinia grandiflora Barb. Rodr. ;大花帕福斯兰■☆

282767　Paphinia rugosa Rchb. f. ;粗糙帕福斯兰■☆

282768　Paphiopedilum Pfitzer(1886)(保留属名);兜兰属(兜舌兰属,拖鞋兰属,仙履兰属);Iowii Orchid, Lady's Slipper, Lady's Slipper Orchid, Paphiopedilum, Pocktorchid, Sliperorchids, Slipper Orchid ■

282769　Paphiopedilum acmodontum Schoser;锐齿兜兰■☆

282770　Paphiopedilum aestivum Z. J. Liu et J. Yong Zhang;夏花兜兰■

282771　Paphiopedilum aestivum Z. J. Liu et J. Yong Zhang = Paphiopedilum purpuratum(Lindl.)Stein ■

282772　Paphiopedilum affine De Wild. = Paphiopedilum gratrixianum Rolfe ■

282773　Paphiopedilum amabile Hallier f. ;可爱兜兰;Lovable Paphiopedilum ■☆

282774　Paphiopedilum angustatum Z. J. Liu et S. C. Chen = Paphiopedilum malipoense S. C. Chen et Z. H. Tsi var. angustatum (Z. J. Liu et S. C. Chen)Z. J. Liu et S. C. Chen ■

282775　Paphiopedilum appletonianum (Gower) Rolfe;卷萼兜兰;Rolledsepal Paphiopedilum,Rollsepal Pocktorchid ■

282776　Paphiopedilum appletonianum (Gower) Rolfe var. hainanense (Fowlie) Braem = Paphiopedilum appletonianum(Gower) Rolfe ■

282777　Paphiopedilum areeanum O. Gruss;根茎兜兰■

282778　Paphiopedilum argus(Rchb. f.) Stein;阿顾斯氏兜兰(疣边兜舌兰);Argus Paphiopedilum ■☆

282779　Paphiopedilum armeniacum S. C. Chen et F. Y. Liu;杏黄兜兰;Apricotyellow Paphiopedilum, Armenia Pocktorchid ■☆

282780　Paphiopedilum armeniacum S. C. Chen et F. Y. Liu f. markii(O. Gruss) Braem = Paphiopedilum armeniacum S. C. Chen et F. Y. Liu ■

282781　Paphiopedilum armeniacum S. C. Chen et F. Y. Liu var. markfun Fowlie = Paphiopedilum armeniacum S. C. Chen et F. Y. Liu ■

282782　Paphiopedilum armeniacum S. C. Chen et F. Y. Liu var. markii G. Gruss = Paphiopedilum armeniacum S. C. Chen et F. Y. Liu ■

282783　Paphiopedilum armeniacum S. C. Chen et F. Y. Liu var. parviflorum Z. J. Liu et J. Yong Zhang = Paphiopedilum armeniacum S. C. Chen et F. Y. Liu ■

282784　Paphiopedilum armeniacum S. C. Chen et F. Y. Liu var. parviflorum Z. J. Liu et J. Yong Zhang;金豆兜兰;Smallflower Armenia Pocktorchid ■

282785　Paphiopedilum armeniacum S. C. Chen et F. Y. Liu var. undulatum Z. J. Liu et J. Yong Zhang = Paphiopedilum armeniacum S. C. Chen et F. Y. Liu ■

282786　Paphiopedilum armeniacum S. C. Chen et F. Y. Liu var. undulatum Z. J. Liu et J. Yong Zhang;浅黄兜兰;Paleyellow Armenia Pocktorchid ■

282787　Paphiopedilum barbatum(Lindl.) Pfitzer;髯毛兜兰(粗毛兜舌兰);Bearded Paphiopedilum ■☆

282788　Paphiopedilum barbigerum Ts. Tang et F. T. Wang;小叶兜兰(兜兰,龙头兰,硬毛兜兰);Littleleaf Paphiopedilum, Littleleaf Pocktorchid ■

282789　Paphiopedilum barbigerum Ts. Tang et F. T. Wang var. lockianum Aver. = Paphiopedilum barbigerum Ts. Tang et F. T. Wang ■

282790　Paphiopedilum bellatulum (Rchb. f.) Pfitzer = Paphiopedilum bellatulum(Rchb. f.)Stein ■

282791　Paphiopedilum bellatulum(Rchb. f.)Stein;巨瓣兜兰（小唇兜舌兰，雅洁兜兰）;Giantpetal Pocktorchid,Prety Paphiopedilum ■

282792　Paphiopedilum boxallii (Rchb. f.) Pfitzer = Paphiopedilum villosum Lindl. var. boxallii(Rchb. f.)Pfitzer ■

282793　Paphiopedilum brevilabium Z. J. Liu et J. Yong Zhang;短唇兜兰;Shortlip Paphiopedilum ■

282794　Paphiopedilum brevilabium Z. J. Liu et J. Yong Zhang = Paphiopedilum wardii Summerh. ■

282795　Paphiopedilum bullenianum(Rchb. f.)Pfitzer;布林氏兜兰;Bullen Paphiopedilum ■☆

282796　Paphiopedilum burmanicum J. Yong Zhang et Z. J. Liu;缅甸兜兰■☆

282797　Paphiopedilum callosum(Rchb. f.)Pfitzer;褐疣瓣兜兰（草叶兜兰，硬皮兜兰,硬皮兜舌兰）;Brownwarty Petal Paphiopedilum ■☆

282798　Paphiopedilum callosum (Rchb. f.) Stein = Paphiopedilum callosum(Rchb. f.)Pfitzer ■☆

282799　Paphiopedilum chamberlainianum (O'Brien) Pfitzer;卡姆氏兜兰;Chamberlain Paphiopedilum ■☆

282800　Paphiopedilum chamberlainianum (O'Brien) Pfitzer var. flavescens Pfitzer;黄皱翼兜舌兰■☆

282801　Paphiopedilum chaoi S. H. Hu = Paphiopedilum henryanum Braem ■

282802　Paphiopedilum charlesworthii(Rolfe)Pfitzer;红旗兜兰（查尔斯兜兰,卡里斯氏兜兰）;Charlesworth Paphiopedilum ■

282803　Paphiopedilum chiwuanum Ts. Tang et F. T. Wang = Paphiopedilum hirsutissimum(Lindl. ex Hook.)Stein ■

282804　Paphiopedilum ciliolare(Rchb. f.)Stein;缘毛兜兰（缘毛兜舌兰）;Ciliolate Paphiopedilum ■☆

282805　Paphiopedilum concobellatulum Hort = Paphiopedilum wenshanense Z. J. Liu et J. Yong Zhang ■

282806　Paphiopedilum concolor(Bateman)Pfitzer;同色兜兰（斑点兜舌兰，狗舌草，黄花兜兰，老虎刻，狮子尾）;Concolor Paphiopedilum,Samecolor Pocktorchid ■

282807　Paphiopedilum concolor(Bateman)Pfitzer var. dahuaense Z. J. Liu et J. Yong Zhang = Paphiopedilum concolor(Bateman)Pfitzer ■

282808　Paphiopedilum concolor(Bateman)Pfitzer var. immaculatum Z. J. Liu et J. Yong Zhang = Paphiopedilum concolor(Bateman)Pfitzer ■

282809　Paphiopedilum curtisii(Rchb. f.) Pfitzer;库特斯兜兰;Curtis Paphiopedilum ■☆

282810　Paphiopedilum dayanum(Rchb. f.) Pfitzer;达氏兜兰（小疣兜舌兰）;Day Paphiopedilum ■☆

282811　Paphiopedilum delenatii Guillaumin;阔翼兜兰（德氏兜兰,阔翼兜舌兰,越南美人）■

282812　Paphiopedilum delicatum Z. J. Liu et J. Yong Zhang = Paphiopedilum helenae Aver. ■

282813　Paphiopedilum densissimum Z. J. Liu et S. C. Chen = Paphiopedilum villosum Lindl. var. densissimum (Z. J. Liu et S. C. Chen)Z. J. Liu et S. C. Chen ■

282814　Paphiopedilum dianthum Ts. Tang et F. T. Wang;长瓣兜兰（长瓣光兰，斗省草，红兜兰，双花兜兰）;Longpetal Paphiopedilum,Longpetal Pocktorchid ■

282815　Paphiopedilum dollii Lückel = Paphiopedilum henryanum Braem ■

282816　Paphiopedilum druryi(Bedd.)Stein.;德鲁里兜兰■☆

282817　Paphiopedilum emersonii Koop. et P. J. Cribb;白花兜兰;White Paphiopedilum,White Pocktorchid ■

282818　Paphiopedilum emersonii Koop. et P. J. Cribb f. album O. Gruss et Petchl. = Paphiopedilum emersonii Koop. et P. J. Cribb ■

282819　Paphiopedilum esquirolei Schltr. = Paphiopedilum hirsutissimum (Lindl. ex Hook.)Stein ■

282820　Paphiopedilum esquirolei Schltr. var. chiwuanum(Ts. Tang et F. T. Wang)Braem et Chiron = Paphiopedilum hirsutissimum(Lindl. ex Hook.)Stein ■

282821　Paphiopedilum exul(O'Brien)Pfitzer;狭叶兜舌兰■☆

282822　Paphiopedilum fairieanum(Lindl.)Stein.;弯瓣兜兰（翘翼兜舌兰,曲瓣兜兰）;Bent Petal Paphiopedilum ■☆

282823　Paphiopedilum glanzeanum O. Gruss et F. Roeth = Paphiopedilum micranthum Ts. Tang et F. T. Wang ■

282824　Paphiopedilum glaucophyllum J. J. Sm. ;粉绿叶兜兰（灰叶兜舌兰）;Glaucousleaf Paphiopedilum,Slipper Orchid ■☆

282825　Paphiopedilum glaucophyllum J. J. Sm. var. moquetteanum J. J. Sm. ;奇花兜兰■☆

282826　Paphiopedilum globulosum Z. J. Liu et S. C. Chen = Paphiopedilum micranthum Ts. Tang et F. T. Wang ■

282827　Paphiopedilum godefroyae(Godefroy)Pfitzer;紫点兜兰（短茎兜舌兰）;Purpledot Paphiopedilum ■☆

282828　Paphiopedilum godefroyae (Godefroy) Stein = Paphiopedilum bellatulum(Rchb. f.)Stein ■

282829　Paphiopedilum gratrixianum Rolfe;格力兜兰■

282830　Paphiopedilum hainanense Fowlie;海南兜兰;Slipper Orchid ■

282831　Paphiopedilum hainanense Fowlie = Paphiopedilum appletonianum (Gower)Rolfe ■

282832　Paphiopedilum hangianum Perner et O. Gruss;绿叶兜兰（心启兜兰）■

282833　Paphiopedilum hangianum Perner et O. Gruss f. album O. Gruss et Petchl. = Paphiopedilum hangianum Perner et O. Gruss ■

282834　Paphiopedilum haynaldianum (Rchb. f.) Pfitzer;哈娜迪兜兰（细瓣兜兰,细瓣兜舌兰）;Haynald Paphiopedilum ■☆

282835　Paphiopedilum helenae Aver. ;巧花兜兰■

282836　Paphiopedilum henryanum Braem; 亨 利 兜 兰; Henry Paphiopedilum,Henry Pocktorchid ■

282837　Paphiopedilum henryanum Braem var. christae Braem;无斑兜兰;Christ Paphiopedilum ■

282838　Paphiopedilum hiepii Aver. = Paphiopedilum malipoense S. C. Chen et Z. H. Tsi var. hiepii(Aver.)P. J. Cribb ■

282839　Paphiopedilum hirsutissimum(Lindl. ex Hook.)Stein;带叶兜兰（多毛兜舌兰）;Beltleaf Pocktorchid, Hirsute Paphiopedilum, Lady Slipper,Slipper Orchid ■

282840　Paphiopedilum hirsutissimum (Lindl. ex Hook.) Stein var. chiwanum (Ts. Tang et F. T. Wang) Cribb = Paphiopedilum hirsutissimum(Lindl. ex Hook.)Stein ■

282841　Paphiopedilum hirsutissimum (Lindl. ex Hook.) Stein var. esquirolei (Schltr.) Karas. et Saito = Paphiopedilum hirsutissimum (Lindl. ex Hook.) Stein ■

282842　Paphiopedilum hookerae (Rchb. f. ex Hook. f.) Stein subsp. appletonianum(Gower)M. W. Wood = Paphiopedilum appletonianum (Gower)Rolfe ■

282843　Paphiopedilum hookerae(Rchb. f.)Pfitzer;胡克氏兜兰;Hooker Paphiopedilum ■☆

282844　Paphiopedilum hybridum Hort. ;兜舌兰■☆

282845　Paphiopedilum insigne(Lindl.)Pfitzer;波瓣兜兰（波缘兜舌兰,美花兜兰，美丽兜兰，闹雀花）;Beautyflower Paphiopedilum, Wavepetal Pocktorchid ■

282846 Paphiopedilum insigne(Lindl.) Pfitzer var. barbigerum(Ts. Tang et F. T. Wang) Braem = Paphiopedilum barbigerum Ts. Tang et F. T. Wang ■

282847 Paphiopedilum insigne(Lindl.) Pfitzer var. chantinii Pfitzer;美丽兜舌兰■☆

282848 Paphiopedilum insigne(Lindl.) Pfitzer var. maulii Pfitzer;长春兜舌兰■☆

282849 Paphiopedilum insigne(Lindl.) Pfitzer var. maximum Pfitzer;大波缘兜舌兰■☆

282850 Paphiopedilum insigne(Lindl.) Pfitzer var. sanderae Pfitzer;绿唇兜舌兰■☆

282851 Paphiopedilum insigne(Wall. ex Lindl.) Pfitzer var. barbigerum (Ts. Tang et F. T. Wang) Braem = Paphiopedilum barbigerum Ts. Tang et F. T. Wang ■

282852 Paphiopedilum jackii H. S. Hua = Paphiopedilum malipoense S. C. Chen et Z. H. Tsi var. jackii(H. S. Hua) Aver. ■

282853 Paphiopedilum jackii H. S. Hua = Paphiopedilum malipoense S. C. Chen et Z. H. Tsi ■

282854 Paphiopedilum jackii H. S. Hua var. hiepii (Aver.) Koop. = Paphiopedilum malipoense S. C. Chen et Z. H. Tsi var. hiepii(Aver.) P. J. Cribb ■

282855 Paphiopedilum javanicum(Reinw.) Pfitzer;爪哇兜兰(爪哇兜舌兰); Java Paphiopedilum ■☆

282856 Paphiopedilum lawrenceanum(Rchb. f.) Pfitzer;劳伦斯兜兰; Lawrence Paphiopedilum ■☆

282857 Paphiopedilum lowii(Lindl.) Pfitzer;劳威氏兜兰(飞凤兜舌兰); Lowi Paphiopedilum ■☆

282858 Paphiopedilum malipoense S. C. Chen et Z. H. Tsi;麻栗坡兜兰; Malipo Paphiopedilum, Malipo Pocktorchid ■

282859 Paphiopedilum malipoense S. C. Chen et Z. H. Tsi var. angustatum(Z. J. Liu et S. C. Chen) Z. J. Liu et S. C. Chen;窄瓣兜兰; Narrowpetal Paphiopedilum, Narrowpetal Pocktorchid ■

282860 Paphiopedilum malipoense S. C. Chen et Z. H. Tsi var. hiepii (Aver.) P. J. Cribb;钩唇兜兰■

282861 Paphiopedilum malipoense S. C. Chen et Z. H. Tsi var. jackii(H. S. Hu) Aver. = Paphiopedilum malipoense S. C. Chen et Z. H. Tsi ■

282862 Paphiopedilum malipoense S. C. Chen et Z. H. Tsi var. jackii (H. S. Hua) Aver. ;浅斑兜兰■

282863 Paphiopedilum maranthum var. extendatum Fowlie = Paphiopedilum micranthum Ts. Tang et F. T. Wang ■

282864 Paphiopedilum markianum Fowlie = Paphiopedilum tigrinum Koop. et N. Haseg. ■

282865 Paphiopedilum mastersianum(Rchb. f.) Pfitzer;马氏兜兰■

282866 Paphiopedilum mastersianum Pfitzer = Paphiopedilum mastersianum(Rchb. f.) Pfitzer ■

282867 Paphiopedilum micranthum Ts. Tang et F. T. Wang;硬叶兜兰(斑叶兰, 花叶子, 小花兜兰); Hardleaf Pocktorchid, Smallflower Paphiopedilum ■

282868 Paphiopedilum micranthum Ts. Tang et F. T. Wang f. alboflavum (Braem) Braem = Paphiopedilum micranthum Ts. Tang et F. T. Wang ■

282869 Paphiopedilum micranthum Ts. Tang et F. T. Wang f. glanzeanum(O. Gruss et Roeth) O. Gruss et Roeth = Paphiopedilum micranthum Ts. Tang et F. T. Wang ■

282870 Paphiopedilum micranthum Ts. Tang et F. T. Wang subsp. eburneum Fowlie = Paphiopedilum micranthum Ts. Tang et F. T. Wang ■

282871 Paphiopedilum micranthum Ts. Tang et F. T. Wang var. alboflavum Braem = Paphiopedilum micranthum Ts. Tang et F. T. Wang ■

282872 Paphiopedilum micranthum Ts. Tang et F. T. Wang var. eburneum Fowlie = Paphiopedilum micranthum Ts. Tang et F. T. Wang ■

282873 Paphiopedilum micranthum Ts. Tang et F. T. Wang var. extendatum Fowlie = Paphiopedilum micranthum Ts. Tang et F. T. Wang ■

282874 Paphiopedilum micranthum Ts. Tang et F. T. Wang var. glanzeanum O. Gruss et Roeth = Paphiopedilum micranthum Ts. Tang et F. T. Wang ■

282875 Paphiopedilum micranthum Ts. Tang et F. T. Wang var. marginatum Fowlie = Paphiopedilum micranthum Ts. Tang et F. T. Wang ■

282876 Paphiopedilum micranthum Ts. Tang et F. T. Wang var. oblatum Z. J. Liu et J. Yong Zhang = Paphiopedilum micranthum Ts. Tang et F. T. Wang ■

282877 Paphiopedilum microchilum Z. J. Liu et S. C. Chen;玲珑兜兰; Smalllip Paphiopedilum ■

282878 Paphiopedilum microchilum Z. J. Liu et S. C. Chen = Paphiopedilum wardii Summerh. ■

282879 Paphiopedilum multifolium Z. J. Liu et J. Yong Zhang;多叶兜兰;Manyleaf Paphiopedilum ■

282880 Paphiopedilum multifolium Z. J. Liu et J. Yong Zhang = Paphiopedilum wardii Summerh. ■

282881 Paphiopedilum nigritum (Rchb. f.) Pfitzer;淡黑色兜兰; Blackish Paphiopedilum ■☆

282882 Paphiopedilum niveum(Rchb. f.) Pfitzer;雪白兜兰(白花兜兰, 白雪兜舌兰); Snow-white Paphiopedilum ■☆

282883 Paphiopedilum parishii(Rchb. f.) Pfitzer;飘带兜兰(巴掌草, 斑叶兰, 兜兰, 花叶鹿含草, 花叶子, 螺旋兜舌兰, 千灵丹); Parish Paphiopedilum, Ribbon Pocktorchid ■

282884 Paphiopedilum parishii(Rchb. f.) Stein var. dianthum(Ts. Tang et F. T. Wang) Karasawa et Saito = Paphiopedilum dianthum Ts. Tang et F. T. Wang ■

282885 Paphiopedilum parishii(Rchb. f.) Stein var. dianthum(Ts. Tang et F. T. Wang) Karas. et Saito = Paphiopedilum dianthum Ts. Tang et F. T. Wang ■

282886 Paphiopedilum philippinense (Rchb. f.) Pfitzer;菲律宾兜兰 (菲岛兜舌兰); Philippine Paphiopedilum ■☆

282887 Paphiopedilum praestans(Rchb. f.) Pfitzer;特殊兜兰(长袋兜舌兰); Exellent Paphiopedilum ■☆

282888 Paphiopedilum primulinum M. W. Wood et P. Taylor;黄花兜舌兰(黄花仔)■☆

282889 Paphiopedilum purpuratum(Lindl.) Stein;紫纹兜兰(拖鞋兰, 香港兜兰, 紫斑兜兰); Hongkong Paphiopedilum, Purplelines Pocktorchid ■

282890 Paphiopedilum purpuratum(Lindl.) Stein var. hainanense F. Y. Liu et Perner = Paphiopedilum purpuratum(Lindl.) Stein ■

282891 Paphiopedilum rhizomatosum S. C. Chen et Z. J. Liu = Paphiopedilum areeanum O. Gruss ■

282892 Paphiopedilum roebbelenii Pfitzer;亲王兜舌兰■☆

282893 Paphiopedilum rothschildianum(Rchb. f.) Pfitzer;罗斯德氏兜兰;Rothschild Paphiopedilum ■☆

282894 Paphiopedilum saccopetalum H. S. Hu = Paphiopedilum hirsutissimum(Lindl. ex Hook.) Stein ■

282895 Paphiopedilum sanderianum(Rchb. f.) Pfitzer;桑氏兜兰■☆

282896 Paphiopedilum singchii Z. J. Liu et J. Yong Zhang = Paphiopedilum hangianum Perner et O. Gruss ■

282897 Paphiopedilum sinicum (Hance ex Rchb. f.) Stein = Paphiopedilum purpuratum(Lindl.)Stein ■

282898 Paphiopedilum spicerianum(Rchb. f.) Pfitzer;白旗兜兰(鸡冠兜舌兰,巨腹萼兜兰);Handsome Paphiopedilum ■

282899 Paphiopedilum stonei(Hook. f.) Pfitzer;粉妆兜舌兰(窄瓣兜兰);Narrow Petal Paphiopedilum ■☆

282900 Paphiopedilum sukhakulii Schoser et Senghas;苏氏兜兰(麻翼兜兰,细梗兜舌兰);Sukhaku Paphiopedilum ■☆

282901 Paphiopedilum superbiens(Rchb. f.) Pfitzer;华丽兜兰(麻六甲兜舌兰);Pride Paphiopedilum ■☆

282902 Paphiopedilum tigrinum Koop. et N. Haseg. ;虎斑兜兰;Tiger Paphiopedilum, Tiger Pocktorchid ■

282903 Paphiopedilum tigrinum Koop. et N. Haseg. = Paphiopedilum markianum Fowlie ■

282904 Paphiopedilum tonsum(Rchb. f.) Pfitzer;光瓣兜兰(无毛兜舌兰);Glabrouspetal Paphiopedilum ■☆

282905 Paphiopedilum tonsum (Rchb. f.) Pfitzer var. cupreum Pfitzer;铜红兜舌兰■☆

282906 Paphiopedilum tonsum (Rchb. f.) Pfitzer var. curtissifolium Pfitzer;短叶兜舌兰■☆

282907 Paphiopedilum tranlienianum O. Gruss et Perner;天伦兜兰(陈莲)■

282908 Paphiopedilum tranlienianum O. Gruss et Perner var. alboviride O. Gruss = Paphiopedilum tranlienianum O. Gruss et Perner ■

282909 Paphiopedilum tranlienianum O. Gruss et Perner var. saxosum X. M. Xu = Paphiopedilum tranlienianum O. Gruss et Perner ■

282910 Paphiopedilum tridentatum S. C. Chen et Z. J. Liu;齿瓣兜兰;Tridentate Paphiopedilum ■

282911 Paphiopedilum venustum(Wall. ex Sims)Pfitzer;秀丽兜兰(美丽兜兰,惜阴兜舌兰);Charming Paphiopedilum, Spiffy Pocktorchid ■

282912 Paphiopedilum venustum (Wall. ex Sims) Pfitzer var. pardinum (Rchb. f.)Pfitzer;豹纹兜舌兰■☆

282913 Paphiopedilum victoriae-mariae(Hook. f.)Rolfe;维多利亚女王兜兰;Victoria-maria Paphiopedilum ■☆

282914 Paphiopedilum villosum(Lindl.) Stein;紫毛兜兰(斑点美丽兜兰,绒毛兜舌兰,越南兜兰);Purplehair Pocktorchid, Vilose Paphiopedilum ■

282915 Paphiopedilum villosum (Lindl.) Stein f. affine (De Wild.) O. Gruss et Roellke = Paphiopedilum gratrixianum Rolfe ■

282916 Paphiopedilum villosum Lindl. f. annamense (Rolfe) Braem = Paphiopedilum villosum Lindl. var. annamense Rolfe ■

282917 Paphiopedilum villosum Lindl. var. affine (De Wild.) Braem = Paphiopedilum gratrixianum Rolfe ■

282918 Paphiopedilum villosum Lindl. var. annamense Rolfe;白边兜兰(美丽兜兰,狭叶紫毛兜兰);Narrowleaf Villose Paphiopedilum ■

282919 Paphiopedilum villosum Lindl. var. boxallii(Rchb. f.) Pfitzer;包氏兜兰■

282920 Paphiopedilum villosum Lindl. var. densissimum(Z. J. Liu et S. C. Chen)Z. J. Liu et S. C. Chen;密毛兜兰;Densehair Paphiopedilum ■

282921 Paphiopedilum villosum Lindl. var. gratrixianum (Rolfe) Braem = Paphiopedilum gratrixianum Rolfe ■

282922 Paphiopedilum wardii Summerh. ; 彩 云 兜 兰; Ward Paphiopedilum, Ward Pocktorchid ■

282923 Paphiopedilum wenshanense Z. J. Liu et J. Yong Zhang;文山兜兰;■☆

282924 Paphiopedilum wenshanense Z. J. Liu et J. Yong Zhang f. album O. Gruss et Petchl. = Paphiopedilum concolor(Bateman)Pfitzer ■

282925 Paphiopedilum xichouense Z. J. Liu et S. C. Chen = Paphiopedilum delenatii Guillaumin ■

282926 Papilionaceae Giseke(1792)(保留科名);蝶形花科;Bean Family, Papilionaceous Plants ●■

282927 Papilionaceae Giseke(保留科名) = Fabaceae Lindl. (保留科名)●■

282928 Papilionaceae Giseke(保留科名) = Leguminosae Juss. (保留科名)●■

282929 Papilionanthe Schltr. (1915);凤蝶兰属;Papiliioorchis ■

282930 Papilionanthe Schltr. = Vanda Jones ex R. Br. ■

282931 Papilionanthe biswasiana(Ghose et Mukerjee)Garay;白花凤蝶兰;White Papiliioorchis ■

282932 Papilionanthe flavescens (Schltr.) Garay = Holcoglossum flavescens(Schltr.)Z. H. Tsi ■

282933 Papilionanthe taiwaniana(S. S. Ying)Ormerod;台湾凤蝶兰(台湾万代兰)■

282934 Papilionanthe teres (Roxb.) Schltr. ;凤蝶兰(棒叶万代兰,棒叶万带兰,棒叶玉兰,尖叶万带兰);Papiliioorchis ■

282935 Papilionanthe uniflora (Lindl.) Garay;单花凤蝶兰;Oneflower Papiliioorchis ■

282936 Papilionanthe vendarum(Rchb. f.) Garay;万代凤蝶兰■

282937 Papilionatae Taub. = Leguminosae Juss. (保留科名)●■

282938 Papilionopsis Steenis = Desmodium Desv. (保留属名)●■

282939 Papiliopsis E. Morren = Oncidium Sw. (保留属名)■☆

282940 Papiliopsis E. Morren ex Cogn. et Marchal = Oncidium Sw. (保留属名)■☆

282941 Papillaria Dulac = Scheuchzeria L. ■

282942 Papillaria patustris Dualac = Scheuchzeria palustris L. ■

282943 Papillilabium Dockrill(1967);乳唇兰属■☆

282944 Papillilabium beckleri(Benth.) Dockrill;乳唇兰■☆

282945 Papiria Thunb. = Gethyllis L. ■☆

282946 Papiria ciliaris Thunb. = Gethyllis ciliaris(Thunb.) Thunb. ■☆

282947 Papiria spiralis Thunb. = Gethyllis spiralis(Thunb.) Thunb. ■☆

282948 Papiria villosa Thunb. = Gethyllis villosa(Thunb.) Thunb. ■☆

282949 Papistylus Kellermann, Rye et K. R. Thiele = Cryptandra Sm. ●☆

282950 Papistylus Kellermann, Rye et K. R. Thiele(1942);澳洲缩苞木属●☆

282951 Pappagrostis Roshev. (1934);东北亚冠毛草属■☆

282952 Pappagrostis Roshev. = Stephanachne Keng ■

282953 Pappagrostis pappophorea(Hack.)Roshev. ;东北亚冠毛草■☆

282954 Pappagrostis pappophorea (Hack.) Roshev. = Stephanachne pappophorea(Hack.) Keng ■

282955 Pappea Eckl. et Zeyh. (1835);非洲冠毛无患子属●☆

282956 Pappea Sond. = Choritaenia Benth. ●☆

282957 Pappea Sond. et Harv. = Choritaenia Benth. ●☆

282958 Pappea capensis Eckl. et Zeyh. ;非洲冠毛无患子;Wild Plum ●☆

282959 Pappea capensis Eckl. et Zeyh. var. radlkoferi (Schweinf.) Schinz = Pappea capensis Eckl. et Zeyh. ●☆

282960 Pappea capensis Sond. et Harv. = Choritaenia capensis(Sond. et Harv.) Benth. ●☆

282961 Pappea fulva Conrath = Pappea capensis Eckl. et Zeyh. ●☆

282962 Pappea radlkoferi Penz. ex Schweinf. var. angolensis Schltr. = Pappea capensis Eckl. et Zeyh. ●☆

282963 Pappea radlkoferi Schweinf. = Pappea capensis Eckl. et Zeyh. ●☆

282964 Pappea ugandensis Baker f. = Pappea capensis Eckl. et Zeyh. ●☆

282965 Papperitzia Rchb. f. (1852);帕普兰属■☆

282966 Papperitzia leiboldii(Rchb. f.)Rchb. f. ;帕普兰■☆

282967　Pappobolus S. F. Blake(1916);脱冠菊属■●☆

282968　Pappobolus hypargyreus(S. F. Blake)Panero;下银脱冠菊●☆

282969　Pappobolus lehmannii(Hieron.)Panero;莱氏脱冠菊●☆

282970　Pappobolus nigrescens(Heiser)Panero;黑脱冠菊●☆

282971　Pappobolus schillingii Panero;希林脱冠菊●☆

282972　Pappochroma Raf. (1837);垫菀属■☆

282973　Pappochroma Raf. = Erigeron L. ■●

282974　Pappophonum persicum(Boiss.)Steud. ;垫菀■☆

282975　Pappophoraceae(Kunth)Herter = Gramineae Juss. (保留科名)■●

282976　Pappophoraceae(Kunth)Herter = Poaceae Barnhart(保留科名)■●

282977　Pappophoraceae Herter = Gramineae Juss. (保留科名)■●

282978　Pappophoraceae Herter = Poaceae Barnhart(保留科名)■●

282979　Pappophorum Schreb. (1791);冠芒草属;Pappus Grass, Pappus-grass ■☆

282980　Pappophorum abyssinicum Hochst. = Enneapogon cenchroides (Licht. ex Roem. et Schult.)C. E. Hubb. ■☆

282981　Pappophorum alopecuroideum Vahl;冠芒草■☆

282982　Pappophorum arabicum Hochst. ex Steud. = Enneapogon brachystachyus(Jaub. et Spach)Stapf ■

282983　Pappophorum aucheri Jaub. et Spach = Enneapogon persicus Boiss. ■☆

282984　Pappophorum benguellense(Rendle)K. Schum. = Enneapogon scaber Lehm. ■☆

282985　Pappophorum boreale Griseb. = Enneapogon borealis(Griseb.)Honda ■

282986　Pappophorum boreale Griseb. = Enneapogon desvauxii P. Beauv. ■

282987　Pappophorum brachystachyum Jaub. et Spach = Enneapogon brachystachyus(Jaub. et Spach)Stapf ■

282988　Pappophorum brachystachyum Jaub. et Spach = Enneapogon desvauxii P. Beauv. ■

282989　Pappophorum cenchroides Licht. ex Roem. et Schult. = Enneapogon cenchroides(Licht. ex Roem. et Schult.)C. E. Hubb. ■☆

282990　Pappophorum cenchroides Roem. et Schult. = Enneapogon cenchroides(Roem. et Schult.)C. E. Hubb. ■☆

282991　Pappophorum elegans Nees ex Steud. = Enneapogon persicus Boiss. ■☆

282992　Pappophorum elegans Nees ex Steud. = Enneapogon schimperanus(Hochst. ex A. Rich.)Renvoize ☆

282993　Pappophorum fasciculatum Chiov. = Enneapogon desvauxii P. Beauv. ■

282994　Pappophorum filifolium Pilg. = Enneapogon scoparius Stapf ■☆

282995　Pappophorum glumosum Hochst. = Enneapogon persicus Boiss. ■☆

282996　Pappophorum laxum Chiov. = Enneapogon scaber Lehm. ■☆

282997　Pappophorum molle(Lehm.)Kunth = Enneapogon cenchroides (Licht. ex Roem. et Schult.)C. E. Hubb. ■☆

282998　Pappophorum persicum(Boiss.)Steud. = Enneapogon persicus Boiss. ■☆

282999　Pappophorum pumilio Trin. = Boissiera squarrosa(Banks et Sol.)Nevski ■☆

283000　Pappophorum pumilio Trin. = Enneapogon borealis(Griseb.)Honda ■

283001　Pappophorum pusillum(Rendle)K. Schum. = Enneapogon desvauxii P. Beauv. ■

283002　Pappophorum robustum Hook. f. = Enneapogon cenchroides (Roem. et Schult.)C. E. Hubb. ■☆

283003　Pappophorum scabrum(Lehm.)Kunth = Enneapogon scaber Lehm. ■☆

283004　Pappophorum schimperanum Hochst. ex A. Rich. = Enneapogon persicus Boiss. ■☆

283005　Pappophorum schimperanum Hochst. ex A. Rich. = Enneapogon schimperanus(Hochst. ex A. Rich.)Renvoize ■☆

283006　Pappophorum senegalense Steud. = Enneapogon desvauxii P. Beauv. ■

283007　Pappophorum sinaicum Trin. = Boissiera squarrosa(Banks et Sol.)Nevski ■☆

283008　Pappophorum squarrosum Banks et Sol. = Boissiera squarrosa (Banks et Sol.)Nevski ■☆

283009　Pappophorum turcomanicum Trautv. = Enneapogon persicus Boiss. ■☆

283010　Pappophorum vincentinum J. A. Schmidt = Enneapogon desvauxii P. Beauv. ■

283011　Pappostipa(Speg.)Romasch. ,P. M. Peterson et Soreng = Stipa L. ■

283012　Pappostipa(Speg.)Romasch. , P. M. Peterson et Soreng (2008);冠针茅属■☆

283013　Pappostyles Pierre = Cremaspora Benth. ●☆

283014　Pappostylum Pierre = Cremaspora Benth. ●☆

283015　Pappostylum Pierre = Pappothrix(A. Gray)Rydb. ●■☆

283016　Pappothrix(A. Gray)Rydb. = Perityle Benth. ●■☆

283017　Pappothrix Rydb. = Perityle Benth. ●■☆

283018　Pappothrix cinerea(A. Gray)Rydb. = Perityle cinerea(A. Gray) A. M. Powell ■☆

283019　Pappothrix quinqueflora(Steyerm.)Everly = Perityle quinqueflora(Steyerm.)Shinners ■☆

283020　Pappothrix rupestris(A. Gray)Rydb. = Perityle rupestris(A. Gray)Shinners ●■☆

283021　Papuacalia Veldkamp(1991);粉蟹甲属■☆

283022　Papuacedrus H. L. Li = Libocedrus Endl. ●☆

283023　Papuaea Schltr. (1919);巴布亚兰属■☆

283024　Papuaea reticulata Schltr. ;巴布亚兰■☆

283025　Papuaithia Diels = Haplostichanthus F. Muell. ●☆

283026　Papualthia Diels(1912);巴布亚木属●☆

283027　Papualthia auriculata Diels;小耳巴布亚木●☆

283028　Papualthia bracteata Diels;具苞巴布亚木●☆

283029　Papualthia grandifolia Diels;大叶巴布亚木●☆

283030　Papualthia lanceolata Merr. ;披针叶巴布亚木●☆

283031　Papualthia longipes Quisumb. ;长梗巴布亚木●☆

283032　Papualthia micrantha Diels;小花巴布亚木●☆

283033　Papualthia mollis Diels;软巴布亚木●☆

283034　Papualthia reticulata Merr. ;网状巴布亚木●☆

283035　Papualthia tenuipes Merr. ;细梗巴布亚木●☆

283036　Papuanthes Danser(1931);巴布亚寄生属●☆

283037　Papuanthes albertisii(Tiegh.)Danser;巴布亚寄生●☆

283038　Papuasicyos Duyfjes(2003);巴布亚葫芦属●☆

283039　Papuastelma Bullock(1965);巴布亚萝藦属●☆

283040　Papuastelma secamonoides(Schltr.)Bullock;巴布亚萝藦●☆

283041　Papuechites Markgr. (1927);巴布亚夹竹桃属●☆

283042　Papuechites aambe Markgr. ;巴布亚夹竹桃●☆

283043　Papularia Forssk. = Trianthema L. ●

283044　Papularia crystallina Forssk. = Trianthema crystallina(Forssk.)Vahl ■☆

283045　Papulipetalum(Schltr.)M. A. Clem. et D. L. Jones = Bulbophyllum Thouars(保留属名)■

283046　Papulipetalum(Schltr.)M. A. Clem. et D. L. Jones(2002);疣

瓣兰属■☆

283047　Papuodendron C. T. White = Hibiscus L. (保留属名)●■

283048　Papuodendron C. T. White(1946);五苞萼木槿属●☆

283049　Papuodendron lepidotum C. T. White;五苞萼木槿●☆

283050　Papuzilla Ridl. = Lepidium L. ■

283051　Papyraceae Burnett = Cyperaceae Juss. (保留科名)■

283052　Papyria Raf. = Papyrius Lam. ●

283053　Papyrius Lam. = Broussonetia L'Hér. ex Vent. (保留属名)●

283054　Papyrius Lam. ex Kuntze = Broussonetia L'Hér. ex Vent. (保留属名)●

283055　Papyrius papyrifera(L.) Kuntze = Broussonetia papyrifera(L.) L'Hér. ex Vent. ●

283056　Papyrus Willd. = Cyperus L. ■

283057　Papyrus madagascariensis Willd. = Cyperus madagascariensis (Willd.) Roem. et Schult. ■☆

283058　Paquerina Cass. = Brachycome Cass. ●■☆

283059　Parabaena Miers(1851);连蕊藤属;Parabaena ■

283060　Parabaena sagittata Miers;连蕊藤;Arrowlike Parabaena, Sagittate Parabaena ■

283061　Parabambusa Widjaja(1997);拟竹属●☆

283062　Parabarium Pierre = Ecdysanthera Hook. et Arn. ●

283063　Parabarium Pierre = Parabarium Pierre ex Spire ●

283064　Parabarium Pierre = Urceola Roxb. (保留属名)●

283065　Parabarium Pierre ex Spire(1906);杜仲藤属;Parabarium ●

283066　Parabarium burmanicum Lý = Urceola tournieri (Pierre) D. J. Middleton ●

283067　Parabarium chunianum Tsiang;红杜仲藤(红喉崩);Chun Parabarium,Red Parabarium ●

283068　Parabarium chunianum Tsiang = Urceola quintaretii (Pierre) D. J. Middleton ●

283069　Parabarium hainanense Tsiang;海南杜仲藤;Hainan Parabarium ●

283070　Parabarium hainanense Tsiang = Urceola quintaretii (Pierre) D. J. Middleton ●

283071　Parabarium handelianum Tsiang = Urceola quintaretii (Pierre) D. J. Middleton ●

283072　Parabarium huaitingii Chun et Tsiang = Urceola huaitingii(Chun et Tsiang) D. J. Middleton ●

283073　Parabarium linearicarpum(Pierre) Pichon = Urceola linearicarpa (Pierre) D. J. Middleton ●

283074　Parabarium linocarpum Pierre = Urceola linearicarpa (Pierre) D. J. Middleton ●

283075　Parabarium micranthum (A. DC.) Pierre = Urceola micrantha (Wall. ex G. Don) D. J. Middleton ●

283076　Parabarium micranthum (Wall. ex G. Don) Pierre = Urceola micrantha(Wall. ex G. Don) D. J. Middleton ●

283077　Parabarium multiflorum (King et Gamble) Lý = Urceola micrantha(Wall. ex G. Don) D. J. Middleton ●

283078　Parabarium napeense (Quint.) Pierre = Urceola napeensis (Quintaret) D. J. Middleton ●

283079　Parabarium quintaretii (Pierre) Pierre = Urceola quintaretii (Pierre) D. J. Middleton ●

283080　Parabarium spireanum Pierre = Urceola micrantha(Wall. ex G. Don) D. J. Middleton ●

283081　Parabarium tournieri(Pierre) Pierre = Urceola tournieri (Pierre) D. J. Middleton ●

283082　Parabarium tournieri(Pierre) Pierre ex Spire = Urceola tournieri (Pierre) D. J. Middleton ●

283083　Parabarium utile (Hayata et Kawak.) Lý = Urceola micrantha (Wall. ex G. Don) D. J. Middleton ●

283084　Parabarium utile(Hayata et Kawak.) Lý var. kerrii Lý = Urceola micrantha(Wall. ex G. Don) D. J. Middleton ●

283085　Parabarleria Baill. = Barleria L. ●■

283086　Parabarleria boivinii(Lindau)Baill. = Barleria volkensii Lindau ●☆

283087　Parabeaumontia(Baill.) Pichon = Vallaris Burm. f. ●

283088　Parabeaumontia Pichon = Vallaris Burm. f. ●

283089　Parabeaumontia indecora (Baill.) Pichon = Vallaris indecora (Baill.)Tsiang et P. T. Li

283090　Parabeaumontia indecora Pichon = Vallaris indecora (Baill.) Tsiang et P. T. Li

283091　Parabenzoin Nakai = Lindera Thunb. (保留属名)●

283092　Parabenzoin Nakai(1925);裂果山胡椒属(假山胡椒属)●

283093　Parabenzoin praecox(Siebold et Zucc.) Nakai = Lindera praecox (Siebold et Zucc.) Blume ●

283094　Parabenzoin praecox (Siebold et Zucc.) Nakai var. pubescens Honda = Lindera praecox (Siebold et Zucc.) Blume var. pubescens (Honda) Kitam. ●☆

283095　Parabenzoin trilobum(Siebold et Zucc.) Nakai = Lindera triloba (Siebold et Zucc.) Blume ●☆

283096　Paraberlinia Pellegr. (1943);赛鞋木豆属●☆

283097　Paraberlinia Pellegr. = Julbernardia Pellegr. ●☆

283098　Paraberlinia bifoliolata Pellegr. ;二叶赛鞋木豆●☆

283099　Paraberlinia bifoliolata Pellegr. = Julbernardia pellegriniana Troupin ●☆

283100　Parabesleria Oerst. (1858);拟贝思乐苣苔属●■☆

283101　Parabesleria Oerst. = Besleria L. ●■☆

283102　Parabesleria costaricensis Oerst. ;拟贝思乐苣苔●■☆

283103　Parabesleria triflora Oerst. ;三花拟贝思乐苣苔●■☆

283104　Parabignonia Bureau = Parabignonia Bureau ex K. Schum. ●☆

283105　Parabignonia Bureau ex K. Schum. (1894);肖紫葳属●☆

283106　Parabignonia maximiliani Bureau;肖紫葳●☆

283107　Paraboea(C. B. Clarke)Ridl. (1905);蛛毛苣苔属(宽萼苣苔属);Paraboea ■

283108　Paraboea Ridl. = Paraboea(C. B. Clarke)Ridl. ■

283109　Paraboea auriculata Y. M. Shui et W. H. Chen;耳叶蛛毛苣苔;Auriculate Paraboea ■

283110　Paraboea barbatipes K. Y. Pan;髯丝蛛毛苣苔(滇桂蛛毛苣苔);Barbate Paraboea ■

283111　Paraboea barbatipes K. Y. Pan = Paraboea martinii(H. Lév. et Vaniot) B. L. Burtt ■

283112　Paraboea changjiangensis F. W. Xing et Z. X. Li;昌江蛛毛苣苔;Changjiang Paraboea ■

283113　Paraboea clavisepala D. Fang et D. H. Qin;棒萼蛛毛苣苔;Clavisepal Paraboea ■

283114　Paraboea crassifolia(Hemsl.) B. L. Burtt;厚叶蛛毛苣苔(厚叶牛耳草,厚叶旋蒴苣苔,石白菜,石斑,岩白菜);Thickleaf Boea, Thickleaf Paraboea ■

283115　Paraboea dictyoneura(Hance)B. L. Burtt;网脉蛛毛苣苔(大还魂,吊气还魂,山枇杷,石火草,石面枇杷,网脉旋蒴苣苔);Netted Boea,Netvein Boea,Netvein Paraboea ■

283116　Paraboea filipes(Hance)B. L. Burtt;丝梗蛛毛苣苔;Silkystipe Paraboea ■

283117　Paraboea glutinosa(Hand. -Mazz.) K. Y. Pan;白花蛛毛苣苔;White Paraboea ■

283118　Paraboea hainanensis (Chun) B. L. Burtt;海南蛛毛苣苔;

Hainan Paraboea ■

283119　Paraboea hekouensis(Oliv.) B. L. Burtt;河口蛛毛苣苔;Hekou Paraboea ■

283120　Paraboea martinii(H. Lév. et Vaniot) B. L. Burtt;马丁氏蛛毛苣苔(白花蛛毛苣苔,火艾);Martin's Paraboea ■

283121　Paraboea neorophylla(Collett et Hemsl.) B. L. Burtt;云南蛛毛苣苔;Yunnan Paraboea ■

283122　Paraboea nutans D. Fang et D. H. Qin;垂花蛛毛苣苔;Nutant Paraboea ■

283123　Paraboea paramartinii Z. R. Xu et B. L. Burtt;思茅蛛毛苣苔;Paramartin Paraboea ■

283124　Paraboea peltifolia D. Fang et L. Zeng;盾叶蛛毛苣苔(钝叶蛛毛苣苔,牛耳菜);Peltateleaf Paraboea ■

283125　Paraboea rufescens(Franch.) B. L. Burtt;锈色蛛毛苣苔(回生草,山白菜,锈毛旋蒴苣苔,岩枇杷,岩莴苣,蛛毛苣苔);Rust Paraboea,Rustyhair Boea,Rustyhairy Boea ■

283126　Paraboea rufescens(Franch.) B. L. Burtt var. umbellata(Drake) K. Y. Pan;伞花蛛毛苣苔;Umbellate Rust Paraboea ■

283127　Paraboea sinensis(Oliv.) B. L. Burtt;蛛毛苣苔(宽萼蛛毛苣苔,六百斤);Paraboea ●

283128　Paraboea sinensis(Oliv.) B. L. Burtt f. macra(Stapf) C. Y. Wu;大宽萼苣苔;Bigcalyx Paraboea ●

283129　Paraboea sinensis(Oliv.) B. L. Burtt f. macra(Stapf) C. Y. Wu = Paraboea sinensis(Oliv.) B. L. Burtt ●

283130　Paraboea sinensis(Oliv.) B. L. Burtt f. macra(Stapf) C. Y. Wu ex H. W. Li = Paraboea sinensis(Oliv.) B. L. Burtt ●

283131　Paraboea sinensis(Oliv.) B. L. Burtt f. macrophylla(Stapf) C. Y. Wu;大叶宽萼苣苔;Bigleaf Paraboea ●

283132　Paraboea sinensis(Oliv.) B. L. Burtt f. macrophylla(Stapf) C. Y. Wu ex H. W. Li = Paraboea sinensis(Oliv.) B. L. Burtt ●

283133　Paraboea sinensis(Oliv.) B. L. Burtt f. macrophylla(Stapf) C. Y. Wu = Paraboea sinensis(Oliv.) B. L. Burtt ●

283134　Paraboea swinhoii(Hance) B. L. Burtt;锥序蛛毛苣苔(火艾,旋葖木,牙硝,锥序旋蒴苣苔);Panicle Paraboea,Swinhoe Boea ●

283135　Paraboea thirionii (H. Lév.) B. L. Burtt;小花蛛毛苣苔;Smallflower Paraboea ■

283136　Paraboea tomentosa Barnett = Paraboea rufescens (Franch.) B. L. Burtt ■

283137　Paraboea tribracteata D. Fang et W. Y. Rao;三苞蛛毛苣苔(地枇杷);Three-bracted Paraboea ■

283138　Paraboea umbellata (Drake) B. L. Burtt = Paraboea rufescens (Franch.) B. L. Burtt var. umbellata(Drake) K. Y. Pan ■

283139　Paraboea velutina(W. T. Wang et C. Z. Gao) B. L. Burtt;密叶蛛毛苣苔;Denseleaf Paraboea ■

283140　Parabotrys J. C. Muell. = Parartabotrys Miq. ●☆

283141　Parabotrys J. C. Muell. = Xylopia L. (保留属名)●

283142　Parabouchetia Baill. (1887);拟布谢茄属☆

283143　Parabouchetia brasilensis Baill. ex Wettst. ;拟布谢茄☆

283144　Paracalanthe Kudo = Calanthe R. Br. (保留属名)■

283145　Paracalanthe Kudo = Ghiesbreghtia A. Rich. et Galeotti ■

283146　Paracalanthe Kudo(1930);假虾脊兰属■

283147　Paracalanthe gracilis (Lindl.) Kudo = Cephalantheropsis obcordata(Lindl.) Ormerod ■

283148　Paracalanthe gracilis Kudo = Cephalantheropsis gracilis(Lindl.) S. Y. Hu ■

283149　Paracalanthe lamellata (Hayata) Kudo = Calanthe tricarinata Lindl. ■

283150　Paracalanthe megalopha (Franch.) Miyabe et Kudo = Calanthe tricarinata Lindl. ■

283151　Paracalanthe reflexa(Maxim.) Kudo = Calanthe reflexa(Kuntze) Maxim. ■

283152　Paracalanthe reflexa (Maxim.) Kudo var. puberula (Lindl.) Kudo = Calanthe puberulla Lindl. ■

283153　Paracalanthe tricarinata (Lindl.) Kudo = Calanthe tricarinata Lindl. ■

283154　Paracalanthe venusta(Schltr.) Kudo = Cephalantheropsis gracilis (Lindl.)S. Y. Hu ■

283155　Paracalanthe venusta (Schltr.) Kudo = Cephalantheropsis obcordata(Lindl.) Ormerod ■

283156　Paracaleana Blaxell = Caleana R. Br. ■☆

283157　Paracaleana Blaxell(1972);假卡丽娜兰属■☆

283158　Paracaleana minor(R. Br.)Blaxell;假卡丽娜兰●☆

283159　Paracalia Cuatrec. (1960);藤蟹甲属●☆

283160　Paracalia jungioides (Hook. et Arn.) Cuatrec. ;藤蟹甲●☆

283161　Paracalyx Ali(1968);异萼豆属(副萼豆属)■☆

283162　Paracalyx microphyllus(Chiov.) Ali;小叶异萼豆■☆

283163　Paracalyx nogalensis(Chiov.) Ali;诺加尔异萼豆■☆

283164　Paracalyx scariosus(Roxb.) Ali;印度异萼豆■☆

283165　Paracalyx somalorum(Vierh.) Ali;热非异萼豆■☆

283166　Paracarpaea(K. Schum.) Pichon = Arrabidaea DC. ●☆

283167　Paracarpaea Pichon = Arrabidaea DC. ●☆

283168　Paracaryopsis(Riedl) R. R. Mill(1991);类并核果属■☆

283169　Paracaryopsis R. R. Mill = Paracaryopsis(Riedl) R. R. Mill ■☆

283170　Paracaryopsis lambertiana(C. B. Clarke) R. R. Mill;类并核果■☆

283171　Paracaryum(A. DC.)Boiss. (1849);并核果属(并核草属)●■☆

283172　Paracaryum Boiss. = Paracaryum(A. DC.) Boiss. ●■☆

283173　Paracaryum brachytubum Diels = Hackelia brachytuba(Diels) I. M. Johnst. ■

283174　Paracaryum bungei(Boiss.) Brand;邦奇并核果■☆

283175　Paracaryum erythraeum Schweinf. ex Brand;浅红并核果●☆

283176　Paracaryum glochidiatum (A. DC.) Benth. = Hackelia uncinata (Royle ex Benth.) C. E. C. Fisch. ■

283177　Paracaryum glochidiatum (A. DC.) Benth. et Hook. f. = Hackelia uncinata(Royle ex Benth.) C. E. C. Fisch. ■

283178　Paracaryum glochidiatum Benth. et Hook. f. = Hackelia uncinata (Benth.) C. E. C. Fisch. ■

283179　Paracaryum gracile Czerniak. ;纤细并核果■☆

283180　Paracaryum himalayense(Klotzsch) C. B. Clarke;喜马拉雅并核果■☆

283181　Paracaryum himalayense (Klotzsch) C. B. Clarke = Mattiastrum himalayense(Klotzsch) Brand ■

283182　Paracaryum incanum(Ledeb.) Boiss. ;灰毛并核果■☆

283183　Paracaryum intermedium(Fresen.) Lipsky;间型并核果■☆

283184　Paracaryum intermedium (Fresen.) Lipsky = Cynoglossum intermedium Fresen. ■☆

283185　Paracaryum karataviense Pavlov ex Popov;卡拉塔夫并核果■☆

283186　Paracaryum laxiflorum Trautv. ;疏花并核果■☆

283187　Paracaryum longiflorum Boiss. = Lindelofia longiflora Gurke ■☆

283188　Paracaryum longipes Boiss. ;长梗并核果●☆

283189　Paracaryum micranthum (DC.) Boiss. = Paracaryum intermedium(Fresen.) Lipsky ■☆

283190　Paracaryum strictum(C. Koch) Boiss. ;劲直并核果■☆

283191　Paracaryum turcomanicum Bornm. et Sint. ;土库曼并核果■☆

283192　Paracasearia Boerl. = Drypetes Vahl ●

283193　Paracautleya R. M. Sm. (1977)；肖距药姜属■☆

283194　Paracautleya bhatii R. M. Sm. ；肖距药姜■☆

283195　Paracelastrus Miq. = Microtropis Wall. ex Meisn. (保留属名)●

283196　Paracelastrus bivalvis(Jack) Miq. = Microtropis bivalvis (Jack) Wall. ●

283197　Paracelsea Zoll. (1844) = Exacum L. ●■

283198　Paracelsea Zoll. (1857) = Acalypha L. ●■

283199　Paracelsia Hassk. = Exacum L. ●■

283200　Paracelsia Hassk. = Paracelsea Zoll. (1844)●■

283201　Paracelsia Mart. ex Tul. = Mollinedia Ruiz et Pav. ●☆

283202　Paracephaelis Baill. (1879)；肖头九节属●☆

283203　Paracephaelis cinerea(A. Rich.) De Block；灰色头九节●☆

283204　Paracephaelis saxatilis(Scott-Elliot) De Block；岩地头九节●☆

283205　Paracephaelis tiliacea Baill. ；肖头九节●☆

283206　Paracephaelis trichantha(Baker) De Block；毛花头九节●☆

283207　Parachampionella Bremek. (1944)；兰嵌马蓝属(兰嵌马兰属)；Parachampionella ●■★

283208　Parachampionella Bremek. = Strobilanthes Blume ●■

283209　Parachampionella flexicaulis (Hayata) C. F. Hsieh et T. C. Huang；曲茎兰嵌马蓝(曲茎马蓝)●■

283210　Parachampionella flexicaulis(Hayata) C. F. Hsieh et T. C. Huang = Strobilanthes flexicaulis Hayata ●■

283211　Parachampionella rankanensis (Hayata) Bremek. ；兰嵌马蓝；Bankan Parachampionella ■●

283212　Parachampionella tashiroi (Hayata) Bremek. = Strobilanthes tashiroi Hayata ■☆

283213　Parachimarrhis Ducke(1922)；拟流茜属☆

283214　Parachimarrhis breviloba Ducke；拟流茜☆

283215　Parachionolaena M. O. Dillon et Sagást. (1992)；类衣鼠麹木属●☆

283216　Parachionolaena M. O. Dillon et Sagást. = Chionolaena DC. ●☆

283217　Paracladopus M. Kato(2006)；拟飞瀑草属■☆

283218　Paraclarisia Ducke = Sorocea A. St. -Hil. ●☆

283219　Paracleisthus Gagnep. = Cleistanthus Hook. f. ex Planch. ●

283220　Paracleisthus subgracilis Gagnep. = Cleistanthus sumatranus (Miq.) Müll. Arg. ●

283221　Paracleisthus tonkinensis (Jabl.) Gagnep. = Cleistanthus tonkinensis Jabl. ●

283222　Paracoffea(Miq.) J. -F. Leroy = Psilanthus Hook. f. (保留属名)●☆

283223　Paracoffea J. -F. Leroy = Psilanthus Hook. f. (保留属名)●☆

283224　Paracoffea melanocarpa (Welw. ex Hiern) J. -F. Leroy = Psilanthus melanocarpus(Welw. ex Hiern)J. -F. Leroy ●☆

283225　Paracolea Baill. = Phylloctenium Baill. ●☆

283226　Paracolea boivini Baill. = Phylloctenium bernieri Baill. ●☆

283227　Paracolea grevei Baill. = Phylloctenium bernieri Baill. ●☆

283228　Paracolpodium(Tzvelev) Tzvelev = Colpodium Trin. ■

283229　Paracolpodium(Tzvelev)Tzvelev(1965)；假鞘柄茅属(假拟沿沟草属)■

283230　Paracolpodium altaicum (Trin.) Tzvelev = Colpodium altaicum Trin. ■

283231　Paracolpodium altaicum (Trin.) Tzvelev subsp. leucolepis (Nevski)Tzvelev = Colpodium leucolepis Nevski ■

283232　Paracolpodium leucolepis (Nevski) Tzvelev = Colpodium leucolepis Nevski ■

283233　Paracolpodium leucolepis (Nevski) Tzvelev = Paracolpodium altaicum(Trin.)Tzvelev subsp. leucolepis(Nevski)Tzvelev ■

283234　Paracolpodium tibeticum (Bor) E. B. Alexeev = Colpodium tibeticum Bor ■

283235　Paracolpodium wallichii (Stapf) E. B. Alexeev = Colpodium wallichii(Stapf)Bor ■

283236　Paraconringia Lemee = Erysimum L. ■●

283237　Paraconringia Lemee = Palaeconringia E. H. L. Krause ■●

283238　Paracorokia M. Kral = Corokia A. Cunn. ●☆

283239　Paracorokia M. Kral(1966)；假宿萼果属●☆

283240　Paracorokia carpodetoides(F. Muell.) M. Král；假宿萼果●☆

283241　Paracorynanthe Capuron(1978)；肖宾树属●☆

283242　Paracorynanthe antankarana Capuron ex J. -F. Leroy；肖宾树●☆

283243　Paracorynanthe uropetala Capuron；尾瓣肖宾树●☆

283244　Paracostus C. D. Specht(2006)；假闭鞘姜属■☆

283245　Paracostus englerianus(K. Schum.) Engl. ；假闭鞘姜■☆

283246　Paracroton Gagnep. ex Pax et K. Hoffm. = Cleistanthus Hook. f. ex Planch. ●

283247　Paracroton Gagnep. ex Pax et K. Hoffm. = Paracleisthus Gagnep. ●

283248　Paracroton Miq. = Fahrenheitia Rchb. f. et Zoll. ex Müll. Arg. ●☆

283249　Paracryphia Baker f. (1921)；盔瓣花属；Paracryphia ●☆

283250　Paracryphia alticola(Schltr.) Steenis；高原盔瓣花●☆

283251　Paracryphia suaveolens Baker f. ；盔瓣花；Fragrant Paracryphia, Suaveolent Paracryphia ●☆

283252　Paracryphiaceae Airy Shaw(1965)；盔瓣花科(八蕊树科)●☆

283253　Paractaenium Benth. et Hook. f. = Paractaenum P. Beauv. ■☆

283254　Paractaenum P. Beauv. (1812)；澳洲弯穗草属■☆

283255　Paractaenum P. Beauv. = Panicum L. ■

283256　Paractaenum novae-hollandiae P. Beauv. ；澳洲弯穗草■☆

283257　Paracyclea Kudo et Yamam. (1932)；肖轮环藤属●☆

283258　Paracyclea Kudo et Yamam. = Cissampelos L. ●

283259　Paracyclea densiflora Yamam. = Cyclea gracillima Diels ●

283260　Paracyclea gracillima(Diels)Yamam. = Cyclea gracillima Diels ●

283261　Paracyclea insularis (Makino) Kudo et Yamam. = Cyclea insularis(Makino)Hatus. ●■

283262　Paracyclea ochiaiana (Yamam.) Kudo et Yamam. = Cyclea ochiaiana(Yamam.)S. F. Huang et T. C. Huang ●

283263　Paracyclea sutchuenensis (Gagnep.) Yamam. = Cyclea sutchuenensis Gagnep. ●■

283264　Paracyclea sutchuenensis(Gagnep.)Yamam. var. sessilis(Y. C. Wu)Yamam. = Cyclea sutchuenensis Gagnep. ●■

283265　Paracyclea wattii(Diels)Yamam. = Cyclea wattii Diels ●

283266　Paracynoglossum Popov = Cynoglossum L. ■

283267　Paracynoglossum Popov(1953)；假琉璃草属(假倒提壶属)■☆

283268　Paracynoglossum afrocaeruleum R. R. Mill = Cynoglossum coeruleum A. DC. ■☆

283269　Paracynoglossum asperrimum(Nakai)Popov；粗糙假琉璃草■☆

283270　Paracynoglossum denticulatum(DC.) Popov；小齿假琉璃草■☆

283271　Paracynoglossum imeretinum(Kusn.) Popov；假琉璃草(假倒提壶)■☆

283272　Paracynoglossum lanceolatum (Forssk.) R. R. Mill = Cynoglossum lanceolatum Forssk. ■

283273　Paradaniella Willis = Paradaniellia Rolfe ●☆

283274　Paradaniellia Rolfe = Daniellia Benn. ●☆

283275　Paradaniellia Rolfe(1912)；假丹尼苏木属●☆

283276　Paradaniellia oliveri Rolfe = Daniellia oliveri (Rolfe) Hutch. et Dalziel ●☆

283277　Paradarisia Ducke = Sorocea A. St. -Hil. ●☆

283278　Paradenocline Müll. Arg. = Adenocline Turcz. ■☆

283279　Paradenocline procumbens（L.）Müll. Arg. = Leidesia procumbens（L.）Prain ■☆

283280　Paraderris（Miq.）R. Geesink = Derris Lour.（保留属名）●

283281　Paraderris（Miq.）R. Geesink（1984）；拟鱼藤属●

283282　Paraderris canarensis（Dalzell）Adema；兰屿鱼藤（矩叶鱼藤）；Lanyu Jewelvine, Oblong Fishvine, Oblongate Fishvine ●

283283　Paraderris elliptica（Wall.）Adema = Derris elliptica（Wall.）Benth. ●

283284　Paraderris glauca（Merr. et Chun）T. C. Chen et Pedley；粉叶鱼藤（蟾蜍藤）；Glaucous Leaves Jewelvine, Glaucous-leaved Jewelvine, Greyblue Fishvine, Greyblue Jewelvine ●

283285　Paraderris hainanensis（Hayata）Adema；海南鱼藤；Hainan Fishvine, Hainan Jewelvine ●

283286　Paraderris hancei（Hemsl.）T. C. Chen et Pedley；粤东鱼藤（韩氏鱼藤，肇庆鱼藤）；Hance Fishvine, Hance Jewelvine ●

283287　Paraderris malaccensis（Benth.）Adema = Derris malaccensis（Benth.）Prain ●

283288　Paradigma Miers = Cordia L.（保留属名）●

283289　Paradina Pierre ex Pit.（1922）；类帽柱木属●

283290　Paradina Pierre ex Pit. = Mitragyna Korth.（保留属名）●

283291　Paradina krewanhensis Pierre ex Pit.；类帽柱木●☆

283292　Paradisanthus Rchb. f.（1852）；肖双花木属●☆

283293　Paradisanthus bahiensis Rchb. f.；肖双花木●☆

283294　Paradisanthus micranthus Schltr.；小花肖双花木●☆

283295　Paradisea Mazzuc.（1811）（保留属名）；藏百合属（藏鹭莺兰属，乐园百合属）；St. Bruno's Lily, St. -Bruno's-Lily ■☆

283296　Paradisea bulbulifera Lingelsh. = Notholirion bulbuliferum（Lingelsh.）Stearn ■

283297　Paradisea bulbulifera Lingelsh. ex H. Limpr. = Notholirion bulbuliferum（Lingelsh. ex H. Limpr.）Stearn ■

283298　Paradisea hemeroanthericoides Mazzuc.；藏百合■☆

283299　Paradisea liliastrum Bertol. = Paradises liliastrurn（L.）Bertol. ■

283300　Paradisea liliastrurn（L.）Bertol.；天堂百合（乐园百合，瘤藏百合）；Lily of Paradise, Saint Bruno's Lily, Spiderwort, St. Bruno's Lily ■☆

283301　Paradisea minor C. H. Wright = Diuranthera minor C. H. Wright ex Hemsl. ■

283302　Paradisia Benol. = Paradisea Mazzuc.（保留属名）■☆

283303　Paradolichandra Hassl. = Parabignonia Bureau ex K. Schum. ●☆

283304　Paradombeya Stapf（1902）；平当树属；Paradombeya ●

283305　Paradombeya burmarica Stapf；缅泰平当树；Burma Paradombeya ●☆

283306　Paradombeya rehderiana Hu = Paradombeya sinensis Dunn ●◇

283307　Paradombeya sinensis Dunn；平当树；China Paradombeya, Chinese Paradombeya ●◇

283308　Paradombeya szechuenica Hu = Paradombeya sinensis Dunn ●◇

283309　Paradrymonia Hanst.（1854）；假林苣苔属（假锥莫尼亚属）■●☆

283310　Paradrymonia Hanst. = Episcia Mart. ■☆

283311　Paradrymonia glabra（Benth.）Hanst.；假林苣苔（假锥莫尼亚）■☆

283312　Paradrypetes Kuhlm.（1935）；假核果木属●☆

283313　Paradrypetes ilicifolia Kuhlm.；假核果木●☆

283314　Paraeremostachys Adylov, Kamelin et Makhm. = Eremostachys Bunge ■

283315　Paraeremostachys desertorum（Regel）Adylov, Kamelin et Makhm. = Eremostachys desrtorum Regel ■

283316　Paraeremostachys phlomoides（Bunge）Adylov, Kamelin et Makhm. = Eremostachys phlomoides Bunge ■

283317　Parafaujasia C. Jeffrey（1992）；拟留菊属●☆

283318　Parafaujasia fontinalis（Cordem.）C. Jeffrey；拟留菊●☆

283319　Parafestuca E. B. Alexeev（1985）；异羊茅属■☆

283320　Parafestuca albida（Lowe）E. B. Alexeev；异羊茅■☆

283321　Paragelonium Léandri = Aristogeitonia Prain ☆

283322　Paragenipa Baill.（1879）；肖格尼木属●☆

283323　Paragenipa lancifolia（Baker）Tirveng. et Robbr.；肖格尼木●☆

283324　Parageum Nakai et H. Hara = Geum L. ■

283325　Parageum Nakai et H. Hara = Parageum Nakai et H. Hara ex H. Hara ■☆

283326　Parageum Nakai et H. Hara ex H. Hara = Geum L. ■

283327　Parageum Nakai et H. Hara ex H. Hara（1935）；假路边青属■☆

283328　Parageum calthifolium（Menzies ex Sm.）Nakai et H. Hara = Geum calthifolium Menzies ex Sm. ■☆

283329　Parageum calthifolium Nakai et H. Hara ex H. Hara = Geum calthifolium Menzies ex Sm. ■☆

283330　Parageum macranthum（Kearney）H. Hara；小花假路边青■☆

283331　Paraglycine F. J. Herm.（1962）；异大豆属■☆

283332　Paraglycine F. J. Herm. = Ophrestia H. M. L. Forbes ●■

283333　Paraglycine digitata（Harms）F. J. Herm. = Ophrestia digitata（Harms）Verdc. ■☆

283334　Paraglycine hedysaroides（Willd.）F. J. Herm. = Ophrestia hedysaroides（Willd.）Verdc. ■☆

283335　Paraglycine madagascarensis F. J. Herm. = Ophrestia madagascariensis（F. J. Herm.）Verdc. ●☆

283336　Paraglycine pinnata（Merr.）Herm. = Ophrestia pinnata（Merr.）H. M. L. Forbes ●■

283337　Paraglycine radicosa（A. Rich.）F. J. Herm. = Ophrestia radicosa（A. Rich.）Verdc. ■☆

283338　Paraglycine radicosa（A. Rich.）F. J. Herm. var. rufescens（Hauman）F. J. Herm. = Ophrestia radicosa（A. Rich.）Verdc. ■☆

283339　Paraglycine unicostata F. J. Herm. = Ophrestia unicostata（F. J. Herm.）Verdc. ■☆

283340　Paraglycine unifoliolata（Baker f.）F. J. Herm. = Ophrestia unifoliolata（Baker f.）Verdc. ■☆

283341　Paraglycine upembae（Hauman）F. J. Herm. = Ophrestia upembae（Hauman）Verdc. ■☆

283342　Paragnathis Spreng. = Diplomeris D. Don ■

283343　Paragnathis pulchella（D. Don）Spreng. = Diplomeris pulchella D. Don ■

283344　Paragoldfussia Bremek.（1944）；假金足草属■☆

283345　Paragoldfussia Bremek. = Strobilanthes Blume ■■

283346　Paragoldfussia barisanensis Bremek.；假金足草■☆

283347　Paragonia Bureau ex K. Schum. = Paragonia Bureau ■☆

283348　Paragonia Bureau（1872）；亚马孙紫葳属■☆

283349　Paragonia pyramidata（Rich.）Bureau；亚马孙紫葳■☆

283350　Paragonis J. R. Wheeler et N. G. Marchant = Agonis（DC.）Sweet（保留属名）●☆

283351　Paragonis J. R. Wheeler et N. G. Marchant（2007）；拟圆冠木属●☆

283352　Paragophyton K. Schum. = Spermacoce L. ●■

283353　Paragophyton spermacocinum K. Schum. = Spermacoce spermacocina（K. Schum.）Bridson et Puff ■☆

283354　Paragrewia Gagnep. ex R. S. Rao = Leptonychia Turcz. ●☆

283355　Paraguelba Scop. = Poraqueiba Aubl. ●☆

283356　Paragulubia Burret = Gulubia Becc. ●☆

283357　Paragulubia Burret（1936）；假古鲁比棕属（异单生槟榔属）●☆

283358　Paragulubia macrospadix Burret；假古鲁比棕●☆

283359 Paragutzlaffia H. P. Tsui = Strobilanthes Blume ●■

283360 Paragutzlaffia H. P. Tsui(1990);南一笼鸡属;Paragutzlaffia ●■★

283361 Paragutzlaffia henryi(Hemsl.)H. P. Tsui;南一笼鸡(丽江一笼鸡,腺序一笼鸡,异蕊一笼鸡);Henry Acoop of cock, Henry Gutzlaffia, Henry Paragutzlaffia, Lijiang Gutzlaffia ●■

283362 Paragutzlaffia lyi(H. Lév.)H. P. Tsui;异蕊南一笼鸡(异蕊一笼鸡);Ly's Paragutzlaffia ■

283363 Paragynoxys(Cuatrec.)Cuatrec.(1955);拟绒安菊属●☆

283364 Paragynoxys angosturae(Cuatrec.)Cuatrec.;拟绒安菊●☆

283365 Paragynoxys magnifolia Cuatrec.;大叶拟绒安菊●☆

283366 Parahancornia Ducke(1922);胶竹桃属●■☆

283367 Parahancornia amapa Ducke;胶竹桃●■☆

283368 Parahancornia fasciculata(Poir.)Benoist;簇生胶竹桃●☆

283369 Parahebe W. R. B. Oliv.(1944)(保留属名);拟长阶花属●☆

283370 Parahebe cataractae(G. Forst.)W. R. B. Oliv.;拟长阶花●☆

283371 Parahebe lyallii(Hook. f.)W. R. B. Oliv.;匍匐拟长阶花●☆

283372 Parahebe perfoliata(R. Br.)B. G. Briggs et Ehrend.;穿叶拟长阶花;Digger Speedwell, Digger's Speedwell ●☆

283373 Parahopea Heim = Shorea Roxb. ex C. F. Gaertn. ●

283374 Parahyparrhenia A. Camus(1950);假苞茅属(异雄草属)■☆

283375 Parahyparrhenia annua(Hack.)Clayton;一年假苞茅■☆

283376 Parahyparrhenia jaergeriana A. Camus = Parahyparrhenia annua(Hack.)Clayton ■☆

283377 Parahyparrhenia perennis Clayton;假苞茅■☆

283378 Paraia Rohwer, H. G. Richt. et van der Werff(1991);亚马孙樟属●☆

283379 Paraia bracteata Rohwer, H. G. Richter et van der Werff;亚马孙樟●☆

283380 Paraisometrum W. T. Wang(1998);弥勒苣苔属;Paraisometrum ■★

283381 Paraisometrum mileense W. T. Wang;弥勒苣苔■

283382 Paraixeris Nakai = Crepidiastrum Nakai ●■

283383 Paraixeris Nakai = Indoixeris Kitam. ■

283384 Paraixeris Nakai = Ixeris(Cass.)Cass. ■

283385 Paraixeris Nakai(1920);黄瓜菜属;Paraixeris ■

283386 Paraixeris chelidoniifolia(Makino)Nakai;少花黄瓜菜(岩苦荬菜)■

283387 Paraixeris chelidoniifolia(Makino)Nakai = Crepidiastrum chelidoniifolium(Makino)J. H. Pak et Kawano ☆

283388 Paraixeris denticulata(Houtt.)Nakai;黄瓜菜(败酱草,齿缘叶类苦荬菜,齿缘叶裂苦荬菜,黄花菜,黄花叶下红,剪刀草,苦菜,苦碟子,苦丁菜,苦荬菜,墓囤头,墓头回,牛舌菜,盘儿草,秋苦荬菜,山壳篮,山林水火草,稀须菜,细叶苦荬菜,野苦荬菜,一点红);Denticulate Ixeris ■

283389 Paraixeris denticulata(Houtt.)Nakai = Crepidiastrum denticulatum(Houtt.)J. H. Pak et Kawano ■☆

283390 Paraixeris denticulata(Houtt.)Nakai f. pallescens Momiy. et Tuyama = Crepidiastrum denticulatum(Houtt.)J. H. Pak et Kawano f. pallescens(Momiy. et Tuyama)Yonek. ■☆

283391 Paraixeris denticulata(Houtt.)Nakai f. pinnatipartita(Makino)Kitam. = Paraixeris pinnatipartita(Makino)Tzvelov ■

283392 Paraixeris denticulata(Houtt.)Nakai f. pinnatipartita(Makino)Nakai = Crepidiastrum denticulatum(Houtt.)J. H. Pak et Kawano f. pinnatipartitum(Makino)Sennikov ■

283393 Paraixeris denticulata(Houtt.)Nakai subsp. pubescens(Stebbins)C. Shih;柔毛黄瓜菜(毛苦荬菜)■

283394 Paraixeris denticulata(Houtt.)Nakai var. pinnatipartita(Makino)Barkalov = Crepidiastrum denticulatum(Houtt.)J. H. Pak et Kawano f. pinnatipartitum(Makino)■

283395 Paraixeris denticulata(Makino)Nakai f. pinnatipartita(Makino)Nakai = Paraixeris pinnatipartita(Makino)Tzvelov ■

283396 Paraixeris denticulata(Makino)Nakai subsp. pubescens(Stebbins)C. Shih = Paraixeris denticulata(Houtt.)Nakai subsp. pubescens(Stebbins)C. Shih ■

283397 Paraixeris humifusa(Dunn)C. Shih;心叶黄瓜菜(蔓生苦荬菜,平卧苦荬菜);Humifuse Paraixeris ■

283398 Paraixeris pinnatipartita(Makino)Tzvelev = Crepidiastrum denticulatum(Houtt.)J. H. Pak et Kawano f. pinnatipartitum(Makino)Sennikov ■

283399 Paraixeris saxatilis(A. I. Baranov)Tzvelev;岩黄瓜菜(岩阴苦荬菜)■

283400 Paraixeris serotina(Maxim.)Tzvelev;尖裂黄瓜菜(抱茎苦荬菜,猴尾草,苦碟子,秋抱茎苦荬菜,晚抱茎苦荬菜)■

283401 Paraixeris sonchifolia(Bunge)Tzvelev = Crepidiastrum sonchifolium(Bunge)J. H. Pak et Kawano ■☆

283402 Paraixeris sonchifolia(Maxim.)Tzvelev = Ixeridium sonchifolium(Maxim.)C. Shih ■

283403 Paraixeris yoshinoi(Makino)Nakai = Crepidiastrum yoshinoi(Makino)J. H. Pak et Kawano ■☆

283404 Parajaeschkea Burkill = Gentianella Moench(保留属名)■

283405 Parajaeschkea Burkill(1911);假口药属■☆

283406 Parajaeschkea smithii Burkill;假口药花■☆

283407 Parajubaea Burret(1930);脊果椰属(并朱北椰属,帕拉久巴椰子属,异杰椰子属)●☆

283408 Parajubaea cocoides Burret;脊果椰(帕拉久巴椰子);Cocumbe Palm ●☆

283409 Parajusticia Benoist = Justicia L. ●■

283410 Parajusticia Benoist(1936);假鸭嘴花属●☆

283411 Parajusticia peteloti Benoist;假鸭嘴花●☆

283412 Parakaempferia A. S. Rao et D. M. Verma(1971);肖山奈属■☆

283413 Parakaempferia synantha A. S. Rao et D. M. Verma;肖山奈■☆

283414 Parakibara Philipson(1985);拟盖裂桂属●☆

283415 Parakibara clavigera Philipson;拟盖裂桂●☆

283416 Parakmena Hu et W. C. Cheng = Magnolia L. ●

283417 Parakmeria Hu et W. C. Cheng(1951);拟单性木兰属;Parakmeria ●★

283418 Parakmeria kachirachirai(Kaneh. et Yamam.)Y. W. Law;恒春拟单性木兰(恒春厚朴,台湾木兰,乌心石舅);Hengchun Parakmeria, Kachirachia Magnolia, Taiwan Magnolia ●

283419 Parakmeria lotungensis(Chun et C. H. Tsoong)Y. W. Law;乐东拟单性木兰(乐东木兰);Ledong Magnolia, Ledong Parakmeria, Lotung Parakmeria ●

283420 Parakmeria lotungensis(Chun et C. H. Tsoong)Y. W. Law var. xiangxiensis C. L. Pang et L. H. Yan = Parakmeria lotungensis(Chun et C. H. Tsoong)Y. W. Law ●

283421 Parakmeria nitida(W. W. Sm.)Y. W. Law;光叶拟单性木兰(光叶木兰,光叶玉兰,亮叶木兰);Brightleaf Parakmeria, Glossy Magnolia, Japanese Cucumber Tree, Shining Parakmeria, Shiny-leaf Magnolia, Shiny-leaf Parakmeria, Shiny-leaved Magnolia ●

283422 Parakmeria omeiensis W. C. Cheng;峨眉拟单性木兰(峨嵋拟克林丽木,黄木兰);Emei Parakmeria, Omei Mountain Parakmeria, Omei Parakmeria ●◇

283423 Parakmeria yunnanensis Hu;云南拟单性木兰(黑心绿豆,云南拟克林丽木);Yunnan Parakmeria ●◇

283424　Paraknoxia Bremek. (1952);肖红芽大戟属■☆

283425　Paraknoxia parviflora (Stapf et Verdc.) Bremek. = Paraknoxia parviflora(Stapf ex Verdc.) Verdc. ex Bremek. ■☆

283426　Paraknoxia parviflora(Stapf ex Verdc.) Verdc. ex Bremek. ;肖红芽大戟■☆

283427　Paraknoxia ruziziensis Bremek. = Paraknoxia parviflora(Stapf ex Verdc.) Verdc. ex Bremek. ■☆

283428　Parakohleria Wiehler = Pearcea Regel ■☆

283429　Parakohleria Wiehler(1978);肖树苣苔属●☆

283430　Parakohleria parviflora(Rusby) Wiehler;小花肖树苣苔●☆

283431　Parakohleria purpurea (Poepp.) Wiehler;紫肖树苣苔●☆

283432　Paralabatia Pierre = Pouteria Aubl. ●

283433　Paralagarosolen Y. G. Wei(2004);方鼎苣苔属;Paralagarosolen ■★

283434　Paralagarosolen fangianum Y. G. Wei;方鼎苣苔; Fang Paralagarosolen ■

283435　Paralamium Dunn(1913);假野芝麻属;Falsedeadnettle ■

283436　Paralamium gracile Dunn;假野芝麻(假芝麻);Falsedeadnettle ■

283437　Paralbizzia Kosterm. (1954);胀荚合欢属;Paralbizzia ●

283438　Paralbizzia Kosterm. = Archidendron F. Muell. ●

283439　Paralbizzia Kosterm. = Cylindrokelupha Kosterm. ●

283440　Paralbizzia Kosterm. = Pithecellobium Mart. (保留属名)●

283441　Paralbizzia Kosterm. = Zygia P. Browne (废弃属名)●

283442　Paralbizzia croizatiana(F. P. Metcalf) P. C. Huang;滇桂胀荚合欢(桂合欢)●

283443　Paralbizzia robinsonii (Gagnep.) Kosterm. = Archidendron robinsonii(Gagnep.) I. C. Nielsen ●

283444　Paralbizzia robinsonii (Gagnep.) Kosterm. = Cylindrokelupha robinsonii(Gagnep.) Kosterm. ●

283445　Paralbizzia turgida(Merr.) Kosterm. ;胀荚合欢(鼎湖合欢, 两广合欢)●

283446　Paralbizzia turgida (Merr.) Kosterm. = Archidendron turgidum (Merr.) I. C. Nielsen ●

283447　Paralbizzia turgida (Merr.) Kosterm. = Cylindrokelupha turgida (Merr.) T. L. Wu ●

283448　Paralea Aubl. = Diospyros L. ●

283449　Paralepistemon Lejoly et Lisowski(1986);假鳞蕊藤属●☆

283450　Paralepistemon curtoi(Rendle) Lejoly et Lisowski;库尔假鳞蕊藤●☆

283451　Paralepistemon shirensis(Oliv.) Lejoly et Lisowski;假鳞蕊藤●☆

283452　Paraleucothoe(Nakai) Honda = Leucothoe D. Don ●

283453　Paraleucothoe(Nakai) Honda(1949);假木藜芦属●☆

283454　Paraleucothoe keiskei(Miq.) Honda;假木藜芦●☆

283455　Paralia Desv. = Diospyros L. ●

283456　Paralia Desv. ex Ham. = Diospyros L. ●

283457　Paraligusticopsis V. N. Tikhom. (1973);假藁本属■

283458　Paraligusticum V. N. Tikhom. = Ligusticum L. ■

283459　Paraligusticum discolor (Ledeb.) V. N. Tikhom. = Ligusticum discolor Ledeb. ■

283460　Paralinospadix Burret = Calyptrocalyx Blume ●

283461　Paralinospadix Burret(1935);海蓝肉穗棕属(异林椰子属)●☆

283462　Paralinospadix arfakiana(Becc.) Burret;假手杖棕●☆

283463　Paralinospadix leptostachys(Burret) Burret;细穗海蓝肉穗棕●☆

283464　Paralinospadix pachystachys(Burret) Burret;粗穗海蓝肉穗棕●☆

283465　Parallosa Alef. = Coppoleria Todaro ■

283466　Parallosa Alef. = Vicia L. ■

283467　Paralophia P. J. Cribb et Hermans(2005);假缠绕草属●■☆

283468　Paralophia epiphytica (P. J. Cribb, Du Puy et Bosser) P. J. Cribb;马岛假缠绕草●☆

283469　Paralophia palmicola(H. Perrier) P. J. Cribb;假缠绕草●☆

283470　Paralstonia Baill. = Alyxia Banks ex R. Br. (保留属名)●

283471　Paralychnophora MacLeish = Eremanthus Less. ●☆

283472　Paralysis Hill = Primula L. ■

283473　Paralyxia Baill. = Aspidosperma Mart. et Zucc. (保留属名)●☆

283474　Paramachaerium Ducke(1925);假军刀豆属(美洲豚豆属)■☆

283475　Paramachaerium schomburgkii(Benth.) Ducke;假军刀豆■☆

283476　Paramacrolobium J. Léonard(1954);赛大裂豆属●☆

283477　Paramacrolobium coeruleum(Taub.) J. Léonard;赛大裂豆●☆

283478　Paramammea J. -F. Leroy = Mammea L. ●

283479　Paramammea J. -F. Leroy(1977);假黄果木属●☆

283480　Paramammea megaphylla J. -F. Leroy;假黄果木●☆

283481　Paramanglietia Hu et W. C. Cheng = Manglietia Blume ●

283482　Paramanglietia aromatica (Dandy) Hu et W. C. Cheng = Manglietia aromatica Dandy ●◇

283483　Paramanglietia microcarpa Hung T. Chang = Manglietia fordiana (Hemsl.) Oliv. ●

283484　Paramansoa Baill. = Arrabidaea DC. ●☆

283485　Paramapania Uittien(1935);假擂鼓芳属■☆

283486　Paramapania radians(C. B. Clarke) Uittien;假擂鼓芳■☆

283487　Paramelhania Arènes(1949);肖梅蓝属●■☆

283488　Paramelhania decaryana Arènes;肖梅蓝■☆

283489　Parameria Benth. (1876);长节珠属(节荚藤属);Parameria ●

283490　Parameria barbata (Blume) K. Schum. = Parameria laevigata (Juss.) Moldenke ●

283491　Parameria esquirolii H. Lév. = Sindechites henryi Oliv. ●

283492　Parameria laevigata(Juss.) Moldenke;长节珠(赫当杜, 节荚藤, 金丝杜仲, 金丝藤仲, 银丝杜仲);Laevigate Parameria ●

283493　Parameria vulneraria Radlk. ;疗伤长节珠●☆

283494　Parameriopsis Pichon = Parameria Benth. ●

283495　Parameriopsis Pichon(1948);类长节珠属●☆

283496　Parameriopsis polyneura(Hook. f.) Pichon;类长节珠●☆

283497　Paramesus C. Presl = Trifolium L. ■

283498　Paramesus strictus(L.) Soják = Trifolium striatum L. ■☆

283499　Paramichelia Hu = Michelia L. ●

283500　Paramichelia Hu(1940);合果含笑属(合果木属, 假含笑属); Paramichelia ●

283501　Paramichelia baillonii(Pierre) Hu;合果含笑(合果木, 黑心树, 假含笑, 山白兰, 山桂花, 山缅桂, 山桃树);Baillon Paramichelia, Paramichelia ●◇

283502　Paramichelia baillonii(Pierre) Hu = Michelia baillonii(Pierre) Finet et Gagnep. ●◇

283503　Paramicropholis Aubrév. et Pellegr. = Micropholis (Griseb.) Pierre ●☆

283504　Paramicrorhynchus Kirp. (1964); 假小喙菊属; Paramicrorhynchus ■

283505　Paramicrorhynchus Kirp. = Launaea Cass. ■

283506　Paramicrorhynchus procumbens(Roxb.) Kirp. ;假小喙菊(栓果菊);Procumbent Paramicrorhynchus ■

283507　Paramiflos Cuatrec. = Espeletia Mutis ex Humb. et Bonpl. ●☆

283508　Paramignya Wight(1831);单叶藤橘属;Paramignya, Vinelime, Vine-lime ●

283509　Paramignya confertifolia Swingle;单叶藤橘(狗屎橘, 藤橘, 野橘); Denseleaf Paramignya, Dense-leaved Vine-lime, Simpleleaf Vinelime ●

283510　Paramignya hainanensis Swingle;海南单叶藤橘;Hainan

Paramignya ●

283511 Paramignya lobata Burkill;裂单叶藤橘●☆

283512 Paramignya rectispina Craib;直刺单叶藤橘(直刺藤橘);Straightspine Vinelime,Straightspiny Paramignya,Straight-spiny Vine-lime ●

283513 Paramitranthes Burret = Siphoneugena O. Berg ●☆

283514 Paramitranthes Burret(1941);假帽花木属●☆

283515 Paramitranthes kiaerskoviana Burret;假帽花木●☆

283516 Paramoltkia Greuter(1981);假弯果紫草属●☆

283517 Paramoltkia doerfleri(Wettst.)Greuter et Burdet;假弯果紫草●☆

283518 Paramomum S. Q. Tong = Amomum Roxb.(保留属名)■

283519 Paramomum petaloideum S. Q. Tong = Amomum petaloideum(S. Q. Tong)T. L. Wu ■

283520 Paramongaia Velarde(1948);秘鲁石蒜属■☆

283521 Paramongaia weberbaueri Velarde;秘鲁石蒜●☆

283522 Paramyrciaria Kausel(1967);假香桃木属●☆

283523 Paramyrciaria delicatula(A. DC.)Kausel;假香桃木●☆

283524 Paramyristica W. J. de Wilde = Myristica Gronov.(保留属名)●

283525 Paramyristica W. J. de Wilde(1994);假肉豆蔻属●☆

283526 Paramyristica sepicana(Foreman)W. J. de Wilde;假肉豆蔻●☆

283527 Paranecepsia Radcl. -Sm.(1976);假阿夫大戟属☆

283528 Paranecepsia alchorneifolia Radcl. -Sm.;假阿夫大戟☆

283529 Paranephelium Miq.(1861);假韶子属;Falserambutan,Paranephelium ●

283530 Paranephelium chinense Merr. = Amesiodendron chinense(Merr.)Hu ●

283531 Paranephelium hainanenses H. S. Lo;海南假韶子;Hainan Falserambutan,Hainan Paranephelium ●◇

283532 Paranephelium hystrix W. W. Sm.;云南假韶子;Yunnan Falserambutan,Yunnan Paranephelium ●◇

283533 Paranephelium macrophyllum King;大叶假韶子●☆

283534 Paranephelium xestophyllum Miq.;光叶假韶子;Smoothleaf Falserambutan ●☆

283535 Paranephelius Poepp.(1843);莲安菊属■☆

283536 Paranephelius Poepp. = Liabum Adans. ■●☆

283537 Paranephelius Poepp. et Endl. = Liabum Adans. ■●☆

283538 Paranephelius Poepp. et Endl. = Paranephelius Poepp. ■☆

283539 Paranephelius uniflorus Poepp. et Endl.;莲安菊■●☆

283540 Paraneurachne S. T. Blake(1972);假脉颖草属■☆

283541 Paraneurachne muelleri(Hack.)S. T. Blake;假脉颖草■☆

283542 Paranneslea Gagnep. = Anneslea Wall.(保留属名)●

283543 Paranomus Salisb.(1807);草地山龙眼属●☆

283544 Paranomus abrotanifolius Salisb. ex Knight;美叶草地山龙眼●☆

283545 Paranomus adiantifolius Salisb. ex Knight;铁线蕨叶草地山龙眼●☆

283546 Paranomus bolusii(Gand.)Levyns;博卢斯草地山龙眼●☆

283547 Paranomus bracteolaris Salisb. ex Knight;苞片草地山龙眼●☆

283548 Paranomus candicans(Thunb.)Kuntze;纯白草地山龙眼●☆

283549 Paranomus capitatus(R. Br.)Kuntze;头状草地山龙眼●☆

283550 Paranomus centaureoides Levyns;矢车菊草地山龙眼●☆

283551 Paranomus concavus(Lam.)Kuntze = Diastella thymelaeoides(P. J. Bergius)Rourke ●☆

283552 Paranomus crithmifolius Salisb. ex Knight = Paranomus spicatus(P. J. Bergius)Kuntze ●☆

283553 Paranomus diversifolius E. Phillips = Paranomus longicaulis Salisb. ex Knight ●☆

283554 Paranomus dregei(H. Buek ex Meisn.)Kuntze;德雷草地山龙眼●☆

283555 Paranomus esterhuyseniae Levyns;埃斯特草地山龙眼●☆

283556 Paranomus flabellifer Salisb. ex Knight = Paranomus spathulatus(Thunb.)Kuntze ●☆

283557 Paranomus lagopus(Thunb.)Salisb.;兔足草地山龙眼●☆

283558 Paranomus longicaulis Salisb. ex Knight;长茎草地山龙眼●☆

283559 Paranomus medius(R. Br.)Kuntze = Paranomus bracteolaris Salisb. ex Knight ●☆

283560 Paranomus micranthus(Schltr.)Compton = Paranomus abrotanifolius Salisb. ex Knight ●☆

283561 Paranomus reflexus(E. Phillips et Hutch.)Fourc.;反折草地山龙眼●☆

283562 Paranomus roodebergensis(Compton)Levyns;鲁德伯格草地山龙眼●☆

283563 Paranomus sceptrum-gustavianus(Sparrm.)Hyl.;王杖草地山龙眼●☆

283564 Paranomus spathulatus(Thunb.)Kuntze;匙形草地山龙眼●☆

283565 Paranomus spathulatus N. E. Br. = Paranomus adiantifolius Salisb. ex Knight ●☆

283566 Paranomus spicatus(P. J. Bergius)Kuntze;长穗草地山龙眼●☆

283567 Paranomus tomentosus(E. Phillips et Hutch.)N. E. Br.;绒毛草地山龙眼●☆

283568 Parantennaria Beauverd(1911);离冠蝶须属■☆

283569 Parantennaria uniceps(F. Muell.)Beauverd;离冠蝶须■☆

283570 Paranthe O. F. Cook = Chamaedorea Willd.(保留属名)●☆

283571 Parapachygone Forman = Pachygone Miers ●

283572 Parapachygone Forman(2007);拟粉绿藤属●☆

283573 Parapachygone longifolia(F. M. Bailey)Forman;拟粉绿藤●☆

283574 Parapactis W. Zimm. = Epipactis Zinn(保留属名)■

283575 Parapanax Miq. = Schefflera J. R. Forst. et G. Forst.(保留属名)●

283576 Parapanax Miq. = Trevesia Vis. ●

283577 Parapantadenia Capuron = Pantadenia Gagnep. ☆

283578 Parapantadenia chauvetiae Leandri et Capuron = Pantadenia chauvetiae(Leandri et Capuron)G. L. Webster ☆

283579 Parapentace Gagnep. = Burretiodendron Rehder ●

283580 Parapentace Gagnep. = Excentrodendron Hung T. Chang et R. H. Miao ●

283581 Parapentace tokinensis(A. Chev.)Gagnep. = Excentrodendron tonkinense(A. Chev.)Hung T. Chang et R. H. Miao ●

283582 Parapentapanax Hutch.(1967);假羽叶参属●

283583 Parapentapanax Hutch. = Pentapanax Seem. ●

283584 Parapentapanax racemosus(Seem.)Hutch.;假羽叶参●

283585 Parapentapanax racemosus(Seem.)Hutch. = Pentapanax racemosus Seem. ●

283586 Parapentapanax subcordatus(G. Don)Hutch. = Pentapanax subcordatus(Wall.)Seem. ●

283587 Parapentapanax subcordatus(Seem.)Hutch. = Pentapanax subcordatus(Wall.)Seem. ●

283588 Parapentas Bremek.(1952);肖五星花属■☆

283589 Parapentas battiscombei Verdc.;巴蒂肖五星花■☆

283590 Parapentas gabonica Bremek. = Parapentas setigera(Hiern)Verdc. ■☆

283591 Parapentas parviflora Bremek. = Parapentas silvatica(K. Schum.)Bremek. ■☆

283592 Parapentas procurrens(K. Schum.)Verdc. = Parapentas silvatica(K. Schum.)Bremek. ■☆

283593　Parapentas setigera(Hiern)Verdc. ;刚毛肖五星花■☆

283594　Parapentas silvatica(K. Schum.)Bremek. ;肖五星花■☆

283595　Parapentas silvatica(K. Schum.)Bremek. subsp. latifolia Verdc. ;宽叶肖五星花■☆

283596　Parapetalifera J. C. Wendl. (废弃属名)= Agathosma Willd. (保留属名)●◇☆

283597　Parapetalifera J. C. Wendl. (废弃属名)= Barosma Willd. (保留属名)●◇☆

283598　Paraphalaenopsis A. D. Hawkes(1964);拟蝶兰属(拟蝴蝶兰属)■☆

283599　Paraphalaenopsis denevei(J. J. Sm.)Hawkes;棒叶拟蝶兰(棒叶蝴蝶兰)■☆

283600　Paraphalaenopsis labukensis Shim, A. L. Lamb et C. L. Chan;拟蝶兰(拟蝴蝶兰)■☆

283601　Paraphalaenopsis laycockii(M. R. Hend.)Hawkes;香棒叶拟蝶兰(香棒叶蝴蝶兰)■☆

283602　Paraphalaenopsis serpentilingua(J. J. Sm.)Hawkes;铅笔拟蝶兰(铅笔蝴蝶兰)■☆

283603　Paraphlomis(Prain)Prain(1908);假糙苏属;Bethlehemsage, Paraphlomis ●■

283604　Paraphlomis Prain = Paraphlomis(Prain)Prain ●■

283605　Paraphlomis albida Hand. -Mazz. ;白毛假糙苏;Whitehairy Bethlehemsage, Whitehairy Paraphlomis ■

283606　Paraphlomis albida Hand. -Mazz. var. brevides Hand. -Mazz. ;短齿白毛假糙苏;Shorttooth Whitehairy Bethlehemsage ■

283607　Paraphlomis albiflora(Hemsl.)Hand. -Mazz. ;白花假糙苏(四轮麻);Whiteflower Paraphlomis ■

283608　Paraphlomis albiflora(Hemsl.)Hand. -Mazz. var. biflora(Y. Z. Sun)C. Y. Wu et H. W. Li;理阳参(二白花假糙苏,二花白花假糙苏);Bi-whiteflower Paraphlomis ■

283609　Paraphlomis albotomentosa C. Y. Wu;绒毛假糙苏(野芝麻);Floss Bethlehemsage, Floss Paraphlomis ■

283610　Paraphlomis biflora Y. Z. Sun = Paraphlomis albiflora(Hemsl.)Hand. -Mazz. var. biflora(Y. Z. Sun)C. Y. Wu et H. W. Li ■

283611　Paraphlomis brevidens Merr. = Paraphlomis pagantha Doan ■

283612　Paraphlomis brevifolia C. Y. Wu et H. W. Li;短叶假糙苏;Shortleaf Bethlehemsage, Shortleaf Paraphlomis ■

283613　Paraphlomis foliata(Dunn)C. Y. Wu et H. W. Li;曲茎假糙苏;Foliate Bethlehemsage, Foliate Paraphlomis ■

283614　Paraphlomis foliata(Dunn)C. Y. Wu et H. W. Li subsp. montigena X. H. Guo et S. B. Zhou;山地假糙苏■

283615　Paraphlomis formosana(Hayata)T. H. Hsieh et T. C. Huang;台湾假糙苏■

283616　Paraphlomis gracilis(Hemsl.)Kudo;纤细假糙苏(短柄舞子草,鸡脚草,细叶假糙苏,野木姜花);Slender Bethlehemsage, Slender Paraphlomis ■

283617　Paraphlomis gracilis(Hemsl.)Kudo = Paraphlomis formosana(Hayata)T. H. Hsieh et T. C. Huang ■

283618　Paraphlomis gracilis Kudo var. lutienensis(Y. Z. Sun)C. Y. Wu;罗甸假糙苏;Luodian Bethlehemsage, Luodian Paraphlomis ■

283619　Paraphlomis hirsuta Hand. -Mazz. = Paraphlomis albiflora(Hemsl.)Hand. -Mazz. ■

283620　Paraphlomis hirsutissima C. Y. Wu;多硬毛假糙苏;Hardhairy Bethlehemsage, Hirsute Paraphlomis ■

283621　Paraphlomis hispida C. Y. Wu;刚毛假糙苏;Hispid Bethlehemsage, Hispid Paraphlomis ■

283622　Paraphlomis intermedia C. Y. Wu et H. W. Li;中间假糙苏;Intermediate Bethlehemsage, Intermediate Paraphlomis ■

283623　Paraphlomis javanica(Blume)Prain;假糙苏(舞子草,皱纹糙苏,皱叶假糙苏,皱叶重楼);Java Bethlehemsage, Java Paraphlomis, Rugose Jerusalemsage ■

283624　Paraphlomis javanica(Blume)Prain var. angustifolia(C. Y. Wu)C. Y. Wu et H. W. Li;狭叶假糙苏(鬼灯笼树);Narrowleaf Bethlehemsage, Narrowleaf Paraphlomis ■

283625　Paraphlomis javanica(Blume)Prain var. coronata(Vaniot)C. Y. Wu et H. W. Li;小叶假糙苏(壶瓶花,金槐,玫檀花,荏子香,十二槐花);Small Java Paraphlomis ■

283626　Paraphlomis javanica(Blume)Prain var. henryi(Yamam.)C. Y. Wu et H. W. Li = Paraphlomis javanica(Blume)Prain ■

283627　Paraphlomis javanica(Blume)Prain var. henryi(Yamam.)C. Y. Wu et H. W. Li;短齿假糙苏(狭叶假糙苏)■

283628　Paraphlomis javanica(Blume)Prain var. henryi Yamam. = Paraphlomis javanica(Blume)Prain ■

283629　Paraphlomis kwangtungensis C. Y. Wu et H. W. Li;八角花;Eightangle Bethlehemsage, Eightangle Paraphlomis ■

283630　Paraphlomis lanceolata Hand. -Mazz. ;长叶假糙苏(壶瓶花,金槐,玫檀花,荏子香,十二槐花);Longleaf Bethlehemsage, Longleaf Paraphlomis ■

283631　Paraphlomis lanceolata Hand. -Mazz. var. sessilifolia Hand. -Mazz. ;无柄长叶假糙苏■

283632　Paraphlomis lanceolata Hand. -Mazz. var. subrosea Hand. -Mazz. ;红花长叶假糙苏■

283633　Paraphlomis lancidentata Y. Z. Sun;云和假糙苏;Lancedentate Paraphlomis, Lanceolatetooth Bethlehemsage ■

283634　Paraphlomis lancidentata Y. Z. Sun = Paraphlomis lanceolata Hand. -Mazz. ■

283635　Paraphlomis lutienensis Y. Z. Sun = Paraphlomis gracilis Kudo var. lutienensis(Y. Z. Sun)C. Y. Wu ■

283636　Paraphlomis membranacea C. Y. Wu et H. W. Li;薄萼假糙苏;Membranaceous Bethlehemsage, Membranaceous Paraphlomis ■

283637　Paraphlomis pagantha Doan;奇异假糙苏(乡间假糙苏);Frostbloom Bethlehemsage, Wongdful Paraphlomis ■

283638　Paraphlomis parviflora C. Y. Wu et H. W. Li;小花假糙苏;Littleflower Paraphlomis, Smallflower Bethlehemsage ■

283639　Paraphlomis parviflora C. Y. Wu et H. W. Li = Paraphlomis formosana(Hayata)T. H. Hsieh et T. C. Huang ■

283640　Paraphlomis patentisetulosa C. Y. Wu ex H. W. Li;展毛假糙苏;Patentsetulose Bethlehemsage, Patentsetulose Paraphlomis ■

283641　Paraphlomis paucisetosa C. Y. Wu ex H. W. Li;少刺毛假糙苏;Fewsetose Bethlehemsage, Fewsetose Paraphlomis ■

283642　Paraphlomis reflexa C. Y. Wu et H. W. Li;折齿假糙苏(野芝麻);Reflex Bethlehemsage, Reflex Paraphlomis ■

283643　Paraphlomis rugosa(Benth.)Prain;舞子草■

283644　Paraphlomis rugosa(Benth.)Prain = Paraphlomis javanica(Blume)Prain ■

283645　Paraphlomis rugosa(Benth.)Prain var. angustifolia C. Y. Wu = Paraphlomis javanica(Blume)Prain var. angustifolia(C. Y. Wu)C. Y. Wu et H. W. Li ■

283646　Paraphlomis rugosa(Benth.)Prain var. coronata(Vaniot)C. Y. Wu et H. W. Li = Paraphlomis javanica(Blume)Prain var. coronata(Vaniot)C. Y. Wu et H. W. Li ■

283647　Paraphlomis rugosa(Benth.)Prain var. henryi Yamam. = Paraphlomis javanica(Blume)Prain var. henryi(Yamam.)C. Y. Wu et H. W. Li ■

283648　Paraphlomis rugosa Prain = Paraphlomis rugosa(Benth.)Prain ■

283649　Paraphlomis seticalyx C. Y. Wu ex H. W. Li;刺萼假糙苏;Spinecalyx Bethlehemsage,Spinecalyx Paraphlomis ■

283650　Paraphlomis setulosa C. Y. Wu et H. W. Li;小刺毛假糙苏;Setulose Bethlehemsage,Setulose Paraphlomis ■

283651　Paraphlomis shunchangensis Z. Y. Li et M. S. Li;顺昌假糙苏;Shunchang Paraphlomis ■

283652　Paraphlomis subcoriacea C. Y. Wu ex H. W. Li;近革叶假糙苏;Similarleather Bethlehemsage,Subcoriaceous Paraphlomis ■

283653　Paraphlomis tomentoso-capitata Yamam.;绒头假糙苏(绒萼舞子草);Flockyhead Bethlehemsage,Tomentose Capitate Paraphlomis ●

283654　Parapholis C. E. Hubb.(1946);假牛鞭草属;Parapholia ■

283655　Parapholis filiformis(Roth)C. E. Hubb.;线假牛鞭草■☆

283656　Parapholis incurva(L.)C. E. Hubb.;假牛鞭草;Curved Hardgrass,Curved Sicklegrass,Incurved Parapholia,Sickle-grass ■☆

283657　Parapholis marginata Runemark;具边假牛鞭草■☆

283658　Parapholis pycnantha(Druce)C. E. Hubb.;密花假牛鞭草■☆

283659　Parapholis strigosa(Dumort.)C. E. Hubb.;糙毛假牛鞭草;Sea Hard-grass,Strigose Sicklegrass ■☆

283660　Paraphyadanthe Mildbr. = Caloncoba Gilg ●☆

283661　Paraphyadanthe coriacea Mildbr. = Oncoba flagelliflora(Mildbr.)Hul ●☆

283662　Paraphyadanthe flagelliflora Mildbr. = Oncoba flagelliflora(Mildbr.)Hul ●☆

283663　Paraphyadanthe flagelliflora Mildbr. var. hydrophila? = Oncoba flagelliflora(Mildbr.)Hul ●☆

283664　Paraphyadanthe lophocarpa(Oliv.)Gilg = Oncoba lophocarpa Oliv. ●☆

283665　Paraphyadanthe lophocarpa Gilg = Oncoba lophocarpa Oliv. ●☆

283666　Paraphyadanthe suffruticosa Milne-Redh. = Oncoba suffruticosa(Milne-Redh.)Hul et Breteler ●☆

283667　Paraphysis(DC.)Dostál = Amberboa(Pers.)Less. ■

283668　Parapiptadenia Brenan(1963);肖落腺豆属(赛落腺豆属)●☆

283669　Parapiptadenia rigida(Benth.)Brenan;肖落腺豆;Angico Gum ●☆

283670　Parapiqueria R. M. King et H. Rob.(1980);假皮氏菊属(拟皮格菊属)■☆

283671　Parapiqueria cavalcantei R. M. King et H. Rob.;假皮氏菊■☆

283672　Parapodium E. Mey.(1838);假足萝藦属●☆

283673　Parapodium costatum E. Mey.;单脉假足萝藦●☆

283674　Parapodium crispum N. E. Br.;皱波假足萝藦●☆

283675　Parapodium simile N. E. Br.;相似假足萝藦●☆

283676　Parapolydora H. Rob.(2005);锥束斑鸠菊属●■

283677　Parapolydora H. Rob. = Polydora Fenzl ■☆

283678　Parapolydora H. Rob. = Vernonia Schreb.(保留属名)●■

283679　Parapottsia Miq. = Pottsia Hook. et Arn. ●

283680　Paraprenanthes C. C. Chang ex C. Shih(1988);假福王草属;False Rattlesnakeroot,Paraprenanthes ■

283681　Paraprenanthes auriculiforma C. Shih;圆耳假福王草;Auriculateleaf Paraprenanthes,Roundear False Rattlesnakeroot ■

283682　Paraprenanthes glandulosissima(C. C. Chang)C. Shih;密毛假福王草;Gland,Paraprenanthes ■

283683　Paraprenanthes gracilipes C. Shih;长柄假福王草;Longstalk Paraprenanthes ■

283684　Paraprenanthes hastata C. Shih;三角叶假福王草(戟叶假福王草);Hastateleaf Paraprenanthes ■

283685　Paraprenanthes heptanhta C. Shih et D. J. Liou;雷山假福王草;Leishan Paraprenanthes,Sevenhead False Rattlesnakeroot ■

283686　Paraprenanthes longiloba Y. Ling et C. Shih;狭裂假福王草;Longlobed Paraprenanthes,Narrowlobed False Rattlesnakeroot ■

283687　Paraprenanthes luchunensis C. Shih;禄春假福王草(绿春假福王草);Luchun Paraprenanthes ■

283688　Paraprenanthes multiformis C. Shih;三裂假福王草;Muchform False Rattlesnakeroot,Threelobed Paraprenanthes ■

283689　Paraprenanthes pilipes(Migo)C. Shih;节毛假福王草(毛枝假福王草,毛轴山苦荬);Hairshoot False Rattlesnakeroot,Hairybranched Paraprenanthes ■

283690　Paraprenanthes pilipes(Migo)C. Shih = Lactuca sororia Miq. var. pilipes(Migo)Kitam. ■

283691　Paraprenanthes pilipes Migo = Paraprenanthes pilipes(Migo)C. Shih ■

283692　Paraprenanthes polypodifolia(Franch.)C. C. Chang ex C. Shih;蕨叶假福王草(毛枝乳苣,水龙骨叶苣,水龙骨叶乳苣);Fernleaf Mulgedium, Hairshooy Milklettuce, Hairybranch Mulgedium, Mulgedium,Wallfernleaf Milklettuce ■

283693　Paraprenanthes prenanthoides(Hemsl.)C. Shih;异叶假福王草(重庆苣);Differentleaf False Rattlesnakeroot, Rattesnakerooot-like Paraprenanthes ■

283694　Paraprenanthes sagittiformis C. Shih;箭耳假福王草;Arrowlear False Rattlesnakeroot,Sagittateleaf Paraprenanthes ■

283695　Paraprenanthes sororia(Miq.)C. Shih;假福王草(堆荬苣,山苦荬,异叶荬苣);Common Paraprenanthes,Diversifolious Lettuce,False Rattlesnakeroot ■

283696　Paraprenanthes sororia(Miq.)C. Shih = Lactuca sororia Miq. ■

283697　Paraprenanthes sylvicola C. Shih;林生假福王草;Forest False Rattlesnakeroot,Woodland Paraprenanthes ■

283698　Paraprenanthes thirionni(H. Lév.)C. Shih = Paraprenanthes glandulosissima(C. C. Chang)C. Shih ■

283699　Paraprenanthes thirionni(H. Lév.)C. Shih = Paraprenanthes sororia(Miq.)C. Shih ■

283700　Paraprenanthes yunnanensis(Franch.)C. Shih;云南假福王草;Yunnan False Rattlesnakeroot,Yunnan Paraprenanthes ■

283701　Paraprotium Cuatrec.(1952);类马蹄果属●☆

283702　Paraprotium Cuatrec. = Protium Burm. f.(保留属名)●

283703　Paraprotium vestitum Cuatrec.;类马蹄果●☆

283704　Parapteroceras Aver.(1990);虾尾兰属;Shrimptail-orchis ■

283705　Parapteroceras Aver. = Saccolabium Blume(保留属名)●■

283706　Parapteroceras elobe(Seidenf.)Aver.;虾尾兰;Shrimptail-orchis ■

283707　Parapteropyrum A. J. Li(1981);翅果蓼属;Parapteropyrum,Wingfruit-knotweed ●■★

283708　Parapteropyrum tibeticum A. J. Li;翅果蓼;Tibet Parapteropyrum,Wingfruit-knotweed ●■◇

283709　Parapyrenaria Hung T. Chang = Pyrenaria Blume ●

283710　Parapyrenaria Hung T. Chang(1963);多瓣核果茶属;Paradrupetea,Parapyrenaria ●★

283711　Parapyrenaria hainanensis Hung T. Chang = Parapyrenaria multisepala(Merr. et Chun)Hung T. Chang ●◇

283712　Parapyrenaria hainanensis Hung T. Chang = Pyrenaria jonquieriana Pierre ex Lanessan subsp. multisepala(Merr. et Chun)S. X. Yang ●

283713　Parapyrenaria multisepala(Merr. et Chun)Hung T. Chang;多瓣核果茶(拟核果茶);Hainan Paradrupetea,Many-sepal Parapyrenaria,Multisepal Paradrupetea,Paradrupetea ●◇

283714　Parapyrenaria multisepala(Merr. et Chun)Hung T. Chang = Pyrenaria jonquieriana Pierre ex Lanessan subsp. multisepala(Merr.

et Chun）S. X. Yang ●

283715 Parapyrola Miq.（1867）；假鹿蹄草属●☆

283716 Parapyrola Miq. = Epigaea L. ●☆

283717 Parapyrola asiatica（Maxim.）Kitam. = Epigaea asiatica Maxim. ●☆

283718 Parapyrola trichocarpa Miq.；假鹿蹄草●☆

283719 Paraqueiba Scop. = Poraqueiba Aubl. ●☆

283720 Paraquilegia J. R. Drumm. et Hutch.（1920）；拟楼斗菜属（假楼斗菜属）；Paraquilegia ■

283721 Paraquilegia anemonoides（Willd.）Engl. ex Ulbr.；乳突拟楼斗菜（大花拟楼斗菜,宿萼假楼斗菜,宿根假楼斗菜）；Anemonelike Paraquilegia, Bigflower Paraquilegia ■

283722 Paraquilegia anemonoides（Willd.）Ulbr. = Paraquilegia anemonoides（Willd.）Engl. ex Ulbr. ■

283723 Paraquilegia caespitosa（Boiss. et Hohen.）Drumm. et Hutch.；密丛拟楼斗菜■

283724 Paraquilegia grandiflora（Fisch. ex DC.）J. R. Drumm. et Hutch. = Paraquilegia anemonoides（Willd.）Engl. ex Ulbr. ■

283725 Paraquilegia grandiflora（Fisch.）Drumm. et Hutch. = Paraquilegia anemonoides（Willd.）Engl. ex Ulbr. ■

283726 Paraquilegia microphylla（Royle）Drumm. et Hutch.；拟楼斗菜（假楼斗菜,小叶假楼斗菜,叶矛对钩）；Littleleaf Paraquilegia ■

283727 Paraquilegia uniflora（Aitch. et Hemsl.）Drumm. et Hutch. = Isopyrum anemonoides Kar. et Kir. ■

283728 Pararchidendron I. C. Nielsen（1984）；假颌垂豆属（白粉牛蹄豆属）■☆

283729 Pararchidendron pruinosum（Benth.）I. C. Nielsen；假颌垂豆；Snow Wood ■☆

283730 Parardisia M. P. Nayar et G. S. Giri = Ardisia Sw.（保留属名）●■

283731 Parardisia M. P. Nayar et G. S. Giri（1988）；假紫金牛属●☆

283732 Parardisia involucrata（Kurz）M. P. Nayar et G. S. Giri；假紫金牛●☆

283733 Pararistolochia Hutch. et Dalziel（1927）；假马兜铃属（拟马兜铃属）●☆

283734 Pararistolochia ceropegioides（S. Moore）Hutch. et Dalziel；吊灯花假马兜铃●☆

283735 Pararistolochia congolana Hauman = Pararistolochia promissa（Mast.）Keay ●☆

283736 Pararistolochia flos-avis（A. Chev.）Hutch. et Dalziel = Pararistolochia macrocarpa（Duch.）Poncy ●☆

283737 Pararistolochia goldieana（Hook. f.）Hutch. et Dalziel；戈尔德假马兜铃●☆

283738 Pararistolochia ju-ju（S. Moore）Hutch. et Dalziel = Pararistolochia mannii（Hook. f.）Keay ●☆

283739 Pararistolochia leonensis（Mast.）Hutch. et Dalziel；莱昂假马兜铃●☆

283740 Pararistolochia macrocarpa（Duch.）Poncy；大果假马兜铃●☆

283741 Pararistolochia macrocarpa（Duch.）Poncy subsp. soyauxiana（Oliv.）Poncy；索亚假马兜铃●☆

283742 Pararistolochia mannii（Hook. f.）Keay；曼氏假马兜铃●☆

283743 Pararistolochia preussii（Engl.）Hutch. et Dalziel；普罗伊斯假马兜铃●☆

283744 Pararistolochia promissa（Mast.）Keay；长假马兜铃●☆

283745 Pararistolochia schweinfurthii（Engl.）Hutch. et Dalziel = Pararistolochia triactina（Hook. f.）Hutch. et Dalziel ●☆

283746 Pararistolochia soyauxiana（Oliv.）Hutch. et Dalziel = Pararistolochia macrocarpa（Duch.）Poncy subsp. soyauxiana（Oliv.）Poncy ●☆

283747 Pararistolochia staudtii（Engl.）Hutch. et Dalziel = Pararistolochia macrocarpa（Duch.）Poncy ●☆

283748 Pararistolochia talbotii（S. Moore）Keay = Pararistolochia promissa（Mast.）Keay ●☆

283749 Pararistolochia tenuicauda（S. Moore）Keay = Pararistolochia promissa（Mast.）Keay ●☆

283750 Pararistolochia triactina（Hook. f.）Hutch. et Dalziel；三线假马兜铃●☆

283751 Pararistolochia tribrachiata（S. Moore）Hutch. et Dalziel = Pararistolochia macrocarpa（Duch.）Poncy ●☆

283752 Pararistolochia zenkeri（Engl.）Hutch. et Dalziel；岑克尔假马兜铃●☆

283753 Parartabotrys Miq.（1861）；假鹰爪花属●☆

283754 Parartabotrys Miq. = Xylopia L.（保留属名）●

283755 Parartabotrys sumatranus Miq.；假鹰爪花●☆

283756 Parartocarpus Baill.（1875）；臭桑属（拟波罗蜜属）●☆

283757 Parartocarpus venenosus（Zoll. et Moritzi）Becc.；臭桑●☆

283758 Parartocarpus venenosus Becc. = Parartocarpua venenosus（Zoll. et Moritzi）Becc. ●☆

283759 Pararuellia Bremek. = Pararuellia Bremek. et Nann. -Bremek. ■

283760 Pararuellia Bremek. et Nann. -Bremek.（1948）；莲楠草属（地皮消属,莲南草属）；False Manyroot, Pararuellia ■

283761 Pararuellia alata H. P. Tsui；节翅地皮消；Winged Pararuellia ■

283762 Pararuellia cavaleriei（H. Lév.）E. Hossain；罗甸地皮消；Luodian Pararuellia ■

283763 Pararuellia delavayana（Baill.）E. Hossain；地皮消（刀口药,灯台草,地皮胶,红头翁,莲楠草,芦莉草,蛆药,喜栋小苞爵床,岩威灵仙,一扫光）；Delavay False Manyroot, Delavay Pararuellia ■

283764 Pararuellia delavayana（Baill.）E. Hossain = Pararuellia cavaleriei（H. Lév.）E. Hossain ■

283765 Pararuellia drymophila（Diels）C. Y. Wu et H. S. Lo = Pararuellia delavayana（Baill.）E. Hossain ■

283766 Pararuellia flagelliformis（Roxb.）Bremek. et Nann. -Bremek.；穗鞭地皮消；Whipformed False Manyroot, Whipformed Pararuellia ■

283767 Pararuellia hainanensis C. Y. Wu et H. S. Lo；海南莲楠草（海南地皮消）；Hainan False Manyroot, Hainan Pararuellia ■

283768 Parasamanea Kosterm.（1954）；假雨树属●☆

283769 Parasamanea Kosterm. = Albizia Durazz. ●

283770 Parasamanea landakensis（Kosterm.）Kosterm.；假雨树●☆

283771 Parasarcochilus Dockrill = Pteroceras Hasselt ex Hassk. ■

283772 Parasarcochilus Dockrill = Sarcochilus R. Br. ■☆

283773 Parasassafras D. G. Long（1984）；拟檫木属（假檫木属,密花檫属）●

283774 Parasassafras confertiflorum（Meisn.）D. G. Long；拟檫木（假檫木）●

283775 Parascheelea Dugand = Orbignya Mart. ex Endl.（保留属名）●☆

283776 Parascheelea Dugand（1940）；假希乐棕属●☆

283777 Parascheelea anchistropetala Dugand；假希乐棕●☆

283778 Parascopolia Baill.（废弃属名）= Lycianthes（Dunal）Hassl.（保留属名）●■

283779 Paraselinum H. Wolff（1921）；肖亮蛇床属■☆

283780 Paraselinum weberbaueri H. Wolff；肖亮蛇床■☆

283781 Parasenecio W. W. Sm. et J. Small（1922）；蟹甲草属（假千里光属）；Cacalia, Indian Plantain ■

283782 Parasenecio × abukumensis H. Koyama；阿武隈蟹甲草■☆

283783 Parasenecio × cuneatus（Honda）H. Koyama；楔形蟹甲草■☆

283784 Parasenecio × koidzumianus（Kitam.）H. Koyama；小泉氏蟹甲

草■☆

283785　Parasenecio × shiroumensis（Shizuo Ito et H. Koyama）H. Koyama;白马岳蟹甲草■☆

283786　Parasenecio adenostyloides（Franch. et Sav. ex Maxim.）H. Koyama;腺柱蟹甲草■

283787　Parasenecio ainsliiflorus（Franch.）Y. L. Chen;兔儿风蟹甲草（八角香,白花蟹甲草,兔儿风花蟹甲草,小八里麻,蜘蛛草）;Ainsliifolious Cacalia ■

283788　Parasenecio amagiensis（Kitam.）H. Koyama;天城山蟹甲草●☆

283789　Parasenecio ambiguus（Y. Ling）Y. L. Chen;两似蟹甲草（登云鞋）;Ambiguous Cacalia,Doubtful Cacalia ■

283790　Parasenecio ambiguus（Y. Ling）Y. L. Chen var. wangianus（Y. Ling）Y. L. Chen;作宾两似蟹甲草;Wang Ambiguous Cacalia ■

283791　Parasenecio auriculatus（DC.）H. Koyama = Parasenecio auriculatus（DC.）J. R. Grant ■

283792　Parasenecio auriculatus（DC.）J. R. Grant;耳叶蟹甲草（耳叶兔儿伞）;Auriculate Cacalia,Eared Indian Plantain ■

283793　Parasenecio auriculatus（DC.）J. R. Grant var. bulbifer（Koidz.）H. Koyama;珠芽耳叶蟹甲草（珠芽蟹甲草）■☆

283794　Parasenecio auriculatus（DC.）J. R. Grant var. kamtschaticus（Maxim.）H. Koyama;勘察加蟹甲草（库页岛蟹甲草）■

283795　Parasenecio begoniifolius（Franch.）Y. L. Chen;秋海棠叶蟹甲草;Begonialeaf Cacalia ■

283796　Parasenecio bulbiferoides（Hand. -Mazz.）Y. L. Chen;珠芽蟹甲草（大老虎草,拟球蟹甲草）■

283797　Parasenecio chenopodiformis（DC.）Y. L. Chen;藜叶蟹甲草■☆

283798　Parasenecio chokaiensis（Kudo）Kadota;鸟海山蟹甲草■☆

283799　Parasenecio chola（W. W. Sm.）R. C. Srivast. et C. Jeffrey;藏南蟹甲草（藜叶千里光）;Xizang Cacalia ■

283800　Parasenecio chola（W. W. Sm.）Y. L. Chen = Parasenecio chola（W. W. Sm.）R. C. Srivast. et C. Jeffrey ■

283801　Parasenecio cyclotus（Bureau et Franch.）Y. L. Chen;轮叶蟹甲草（轮耳蟹甲草）;Roundauricle Cacalia ■

283802　Parasenecio dasythyrsus（Hand. -Mazz.）Y. L. Chen;山西蟹甲草;Shanxi Cacalia ■

283803　Parasenecio decomposita Gray;多裂蟹甲草■☆

283804　Parasenecio delphiniifolius（Siebold et Zucc.）H. Koyama;翠雀花叶蟹甲草■☆

283805　Parasenecio delphiniifolius（Siebold et Zucc.）H. Koyama var. brevilobus（Sugim. et Sugino）Yonek.;短裂蟹甲草■☆

283806　Parasenecio delphiniphyllus（H. Lév.）Y. L. Chen;翠雀叶蟹甲草（兔儿伞,雨伞菜）;Larkspur-leaved Cacalia ■

283807　Parasenecio deltophyllus（Maxim.）Y. L. Chen;三角叶蟹甲草;Deltoidleaf Cacalia ■

283808　Parasenecio farfarifolius（Siebold et Zucc.）H. Koyama;吴风草状蟹甲草■☆

283809　Parasenecio farfarifolius（Siebold et Zucc.）H. Koyama subsp. petasitoides H. Lév. = Parasenecio petasitoides（H. Lév.）Y. L. Chen ■

283810　Parasenecio farfarifolius（Siebold et Zucc.）H. Koyama var. acerinus（Makino）H. Koyama;槭蟹甲草■☆

283811　Parasenecio farfarifolius（Siebold et Zucc.）H. Koyama var. bulbifer（Maxim.）H. Koyama;球根蟹甲草■☆

283812　Parasenecio firmus（Kom.）Y. L. Chen;大叶蟹甲草（大叶兔儿伞）;Largeleaf Cacalia ■

283813　Parasenecio floribundus A. Gray;多花蟹甲草■☆

283814　Parasenecio forrestii W. W. Sm. et J. Small;蟹甲草;Forrest Cacalia ■

283815　Parasenecio gansuensis Y. L. Chen;甘肃蟹甲草;Gansu Cacalia,Kansu Cacalia ■

283816　Parasenecio hastatus（L.）H. Koyama;山尖子（戟叶兔儿伞,山尖菜）;Hastate Cacalia ■

283817　Parasenecio hastatus（L.）H. Koyama subsp. glaber Ledeb. = Parasenecio hastatus（L.）H. Koyama var. glaber（Ledeb.）Y. L. Chen ■

283818　Parasenecio hastatus（L.）H. Koyama subsp. lancifolius Franch. = Parasenecio lancifolius（Franch.）Y. L. Chen ■

283819　Parasenecio hastatus（L.）H. Koyama subsp. orientalis（Kitam.）H. Koyama var. hayachinensis（Kitam.）H. Koyama = Parasenecio hayachinensis（Kitam.）Kadota ■☆

283820　Parasenecio hastatus（L.）H. Koyama subsp. orientalis（Kitam.）H. Koyama;东方山尖子■☆

283821　Parasenecio hastatus（L.）H. Koyama subsp. orientalis（Kitam.）H. Koyama var. nantaicus（Komatsu）H. Koyama;南体山山尖子（南体山蟹甲草）■☆

283822　Parasenecio hastatus（L.）H. Koyama subsp. orientalis（Kitam.）H. Koyama var. ramosus（Maxim.）H. Koyama;分枝东方山尖子■☆

283823　Parasenecio hastatus（L.）H. Koyama subsp. tanakae（Franch. et Sav.）H. Koyama var. chokaiensis（Kudo）H. Koyama = Parasenecio chokaiensis（Kudo）Kadota ■☆

283824　Parasenecio hastatus（L.）H. Koyama subsp. tanakae（Franch. et Sav.）H. Koyama;田中氏山尖子■☆

283825　Parasenecio hastatus（L.）H. Koyama var. glaber（Ledeb.）Y. L. Chen;无毛山尖子;Glabrous Hastate Cacalia ■

283826　Parasenecio hastiformis Y. L. Chen;戟状蟹甲草;Hastate Cacalia ■

283827　Parasenecio hayachinensis（Kitam.）Kadota;早池峰山蟹甲草■☆

283828　Parasenecio hwangshanicus（Y. Ling）C. I. Peng et S. W. Ching;黄山蟹甲草;Huangshan Cacalia ■

283829　Parasenecio hwangshanicus（Y. Ling）Y. L. Chen = Parasenecio hwangshanicus（Y. Ling）C. I. Peng et S. W. Ching ■

283830　Parasenecio ianthophyllus（Franch.）Y. L. Chen;紫背蟹甲草;Purpleback Cacalia ■

283831　Parasenecio jiulongensis Y. L. Chen;九龙蟹甲草;Jiulong Cacalia ■

283832　Parasenecio kangxianensis（Z. Y. Zhang et Y. H. Gou）Y. L. Chen;康县蟹甲草;Kangxian Cacalia ■

283833　Parasenecio kiusianus（Makino）H. Koyama;九州蟹甲草■☆

283834　Parasenecio koidzumianus（Kitam.）H. Koyama;小泉蟹甲草■☆

283835　Parasenecio komarovianus（Poljakov）Y. L. Chen;星叶蟹甲草（星叶兔儿伞）;Komarov Cacalia ■

283836　Parasenecio koualapensis（Franch.）Y. L. Chen;瓜拉坡蟹甲草;Gualapo Cacalia ■

283837　Parasenecio lancifolius（Franch.）Y. L. Chen;披针叶蟹甲草（披针叶山尖子,线叶山尖子）■

283838　Parasenecio latipes（Franch.）Y. L. Chen;阔柄蟹甲草（阔叶蟹甲草）;Broadpetiole Cacalia ■

283839　Parasenecio leucocephalus（Franch.）Y. L. Chen;白头蟹甲草（白毛千里光,泡桐七）;White Groundsel,Whitehead Cacalia ■

283840　Parasenecio lijiangensis（Hand. -Mazz.）Y. L. Chen;丽江蟹甲草;Lijiang Cacalia ■

283841　Parasenecio longispicus（Hand. -Mazz.）Y. L. Chen;长穗蟹甲草;Longspike Cacalia ■

283842　Parasenecio maowenensis Y. L. Chen;茂汶蟹甲草;Maowen Cacalia ■

283843　Parasenecio matsudai（Kitam.）Y. L. Chen;天目山蟹甲草;

Tianmushan Cacalia ■

283844 Parasenecio maximowiczianus(Nakai et F. Maek. ex H. Hara) H. Koyama;马氏蟹甲草■☆

283845 Parasenecio maximowiczianus(Nakai et F. Maek. ex H. Hara) H. Koyama var. alatus(F. Maek.)H. Koyama;具翅马氏蟹甲草■☆

283846 Parasenecio monanthus(Diels)C. I. Peng et S. W. Chung;山地蟹甲草(玉山蟹甲草);Yushan Cacalia ■

283847 Parasenecio morrisonensis Y. L. Chen;玉山蟹甲草;Morrison Cacalia, Yushan Cacalia ■

283848 Parasenecio nikomontanus(Matsum.)H. Koyama;日光山蟹甲草■☆

283849 Parasenecio nipponicus(Miq.)H. Koyama;本州蟹甲草■☆

283850 Parasenecio nokoensis(Masam. et Suzuki)C. I. Peng et S. W. Chung;能高蟹甲草(高雄蟹甲草)■

283851 Parasenecio nokoensis(Masam. et Suzuki)Y. L. Chen = Parasenecio nokoensis(Masam. et Suzuki)C. I. Peng et S. W. Chung ■

283852 Parasenecio ogamontanus Kadota;男鹿山蟹甲草■☆

283853 Parasenecio otopteryx(Hand. -Mazz.)Y. L. Chen;耳翼蟹甲草;Earedwing Cacalia ■

283854 Parasenecio palmatisectus(Jeffrey)Y. L. Chen;掌裂蟹甲草(虎草);Palmatisect Cacalia ■

283855 Parasenecio palmatisectus(Jeffrey)Y. L. Chen var. moupingensis(Franch.)Y. L. Chen;腺毛掌裂蟹甲草;Mouping Palmatisect Cacalia ■

283856 Parasenecio peltifolius(Makino)H. Koyama;盾叶蟹甲草■☆

283857 Parasenecio petasitoides(H. Lév.)Y. L. Chen;蜂斗菜状蟹甲草(蝙蝠草);Butterbur-like Cacalia ■

283858 Parasenecio phyllolepis(Franch.)Y. L. Chen;苞鳞蟹甲草;Leaf-scaled Cacalia ■

283859 Parasenecio pilgerianus(Diels)Y. L. Chen;太白山蟹甲草;Taibaishan Cacalia ■

283860 Parasenecio praetermissus(Poljakov)Y. L. Chen;长白蟹甲草(大叶兔儿伞);Changbaishan Cacalia ■

283861 Parasenecio profundorum(Dunn)Y. L. Chen;深山蟹甲草(泡桐七);Profund Cacalia ■

283862 Parasenecio quinquelobus(Wall. ex DC.)Y. L. Chen;五裂蟹甲草;Fivelobe Cacalia ■

283863 Parasenecio quinquelobus(Wall. ex DC.)Y. L. Chen var. sinuatus(Koyama)Y. L. Chen;深裂五裂蟹甲草;Sinuate Fivelobe Cacalia ■

283864 Parasenecio roborowskii(Maxim.)Y. L. Chen;蛛毛蟹甲草;Roborowski Cacalia ■

283865 Parasenecio rockianus(Hand. -Mazz.)Y. L. Chen;玉龙蟹甲草;Rock Cacalia ■

283866 Parasenecio rubescens(S. Moore)Y. L. Chen;矢镞叶蟹甲草(蝙蝠草,牛芳草);Rubescent Cacalia ■

283867 Parasenecio rufipilis(Franch.)Y. L. Chen;红毛蟹甲草;Redhair Cacalia ■

283868 Parasenecio shikokianus(Makino)H. Koyama;四国蟹甲草■☆

283869 Parasenecio sinicus(Y. Ling)Y. L. Chen;中华蟹甲草;China Cacalia ■

283870 Parasenecio souliei(Franch.)Y. L. Chen;川西蟹甲草;Soulie Cacalia ■

283871 Parasenecio subglaber(C. C. Chang)Y. L. Chen;无毛蟹甲草(蟹甲菊);Subglabrous Cacalia ■

283872 Parasenecio taliensis(Franch.)Y. L. Chen;大理蟹甲草;Dali Cacalia ■

283873 Parasenecio tanguticus(Maxim.)Hand. -Mazz.;羽裂蟹甲草■

283874 Parasenecio tebakoensis(Makino)H. Koyama;手笻山蟹甲草■☆

283875 Parasenecio tenianus(Hand. -Mazz.)Y. L. Chen;盐丰蟹甲草;Yanfeng Cacalia ■

283876 Parasenecio tongchuanensis Y. L. Chen;东川假千里光;Dongchuan Cacalia ■

283877 Parasenecio tongchuanensis Y. L. Chen = Parasenecio delphiniphyllus(H. Lév.)Y. L. Chen ■

283878 Parasenecio tripteris(Hand. -Mazz.)Y. L. Chen;昆明蟹甲草;Kunming Cacalia ■

283879 Parasenecio tsinlingensis(Hand. -Mazz.)Y. L. Chen;秦岭蟹甲草;Qinling Cacalia ■

283880 Parasenecio vespertilo(Franch.)Y. L. Chen;川鄂蟹甲草(蝙蝠蟹甲草)■

283881 Parasenecio xinjiashanensis(Z. Ying Zhang et Y. H. Guo)Y. L. Chen;辛家山蟹甲草;Xinjiashan Cacalia ■

283882 Parasenecio yakusimensis(Masam.)H. Koyama;屋久岛蟹甲草■☆

283883 Parasenecio yatabei(Matsum. et Koidz.)H. Koyama;谷田蟹甲草■☆

283884 Parasenecio yatabei(Matsum. et Koidz.)H. Koyama var. occidentalis(F. Maek. ex Kitam.)H. Koyama;西方谷田蟹甲草■☆

283885 Paraserianthes I. C. Nielsen(1984);异合欢属(南洋楹属);Paraserianthes ●

283886 Paraserianthes falcataria(L.)I. C. Nielsen = Falcataria moluccana(Miq.)Barneby et J. W. Grimes ●

283887 Paraserianthes lophantha(Willd.)I. C. Nielsen;簇花异合欢(二穗合欢,箭羽楹,羽叶合欢,羽状合欢);Brush Wattle, Brushwattle, Cape Leeuwin Wattle, Plume Albizia, Plume Albizzia, Plume-albizia ●☆

283888 Paraserianthes moluccana(Miq.)Barneby et Grimes;摩鹿加异合欢●☆

283889 Paraserianthes pullenii(Verdc.)I. C. Nielsen;异合欢;Pullen Paraserianthes ●☆

283890 Paraserianthes toona(Bailey)I. C. Nielsen;红材异合欢;Cedar Acacia, Red Siris ●☆

283891 Parashorea Kurz(1870);柳安属(赛罗双属,赛罗香属,望天树属);Lauan, Parashorea, White Seraya ●

283892 Parashorea aptera Slooten;无翼赛罗双●☆

283893 Parashorea chinensis H. Wang = Shorea chinensis(Wang Hsie)H. Zhu ●◇

283894 Parashorea chinensis H. Wang var. guangxiensis Lin Chi = Shorea chinensis Merr. ●◇

283895 Parashorea chinensis H. Wang var. kwangsiensis Lin Chi = Parashorea chinensis H. Wang ●◇

283896 Parashorea chinensis H. Wang var. kwangsiensis Lin Chi = Shorea chinensis(Wang Hsie)H. Zhu ●◇

283897 Parashorea densiflora Slooten et Symington;丛花塞罗双●☆

283898 Parashorea globosa Symington;球状塞罗双木●☆

283899 Parashorea lucida Kurz;光亮塞罗双●☆

283900 Parashorea malaanonan(Blanco)Merr.;赛娑罗双(马拉塞罗双);Bagtikan, Bagtikan White Lauan, Bagtikann, White Lauan ●☆

283901 Parashorea parvifolia Wyatt-Sm. ex P. S. Ashton;小叶塞罗双●☆

283902 Parashorea plicata Brandis;白柳安;Bagtikan, White Seraya ●☆

283903 Parashorea smythiesii Wyatt-Sm. ex P. S. Ashton;斯氏塞罗双●☆

283904 Parashorea stellata Kurz;柳安(星芒塞罗双木);Stellate Parashorea, Thingadu ●☆

283905 Parashorea tomentella(Symington)Meijer;小茸毛塞罗双●☆

283906　Parasia Post et Kuntze = Belmontia E. Mey. (保留属名)■

283907　Parasia Post et Kuntze = Parrasia Raf. (废弃属名)■

283908　Parasia Raf. = Belmontia E. Mey. (保留属名)■

283909　Parasia Raf. = Sebaea Sol. ex R. Br. ■

283910　Parasia cordata(L. f.)Raf. = Sebaea exacoides(L.)Schinz ■☆

283911　Parasia debilis(Welw.)Hiern = Sebaea debilis(Welw.)Schinz ■☆

283912　Parasia gracilis(Welw.) Hiern = Sebaea gracilis (Welw.) Paiva et I. Nogueira ■☆

283913　Parasia grandis (E. Mey.) Hiern = Sebaea grandis (E. Mey.) Steud. ■☆

283914　Parasia grandis(E. Mey.) Hiern var. major S. Moore = Sebaea grandis(E. Mey.)Steud. ■☆

283915　Parasia platyptera (Baker) Hiern = Sebaea platyptera (Baker) Boutique ■☆

283916　Parasia primuliflora (Welw.) Hiern = Sebaea primuliflora (Welw.)Sileshi ■☆

283917　Parasia thomasii S. Moore = Sebaea thomasii(S. Moore)Schinz ■☆

283918　Parasicyos Dieterle(1975);假刺瓜藤属■☆

283919　Parasicyos dieterleae Lira et R. Torres;假刺瓜藤(刺瓜藤)■☆

283920　Parasicyos maculatus Dieterle;斑点假刺瓜藤(斑点刺瓜藤)■☆

283921　Parasilaus Leute(1972);肖亮叶芹属■

283922　Parasilaus afghanicus(Gilli)Leute;肖亮叶芹■☆

283923　Parasilaus asiaticus(Korovin)Pimenov;亚洲肖亮叶芹■☆

283924　Parasirobilanthes Bremek. = Strobilanthes Blume ●■

283925　Parasitaxaceae A. V. Bobrov et Melikyan = Podocarpaceae Endl. (保留科名)●

283926　Parasitaxaceae Melikian et A. V. Bobrov = Podocarpaceae Endl. (保留科名)●

283927　Parasitaxus de Laub. (1972);寄生罗汉松属●☆

283928　Parasitaxus ustus(Vieill.)de Laub. ;寄生罗汉松●☆

283929　Parasitipomaea Hayata = Ipomoea L. (保留属名)●■

283930　Parasitipomaea formosana Hayata = Ipomoea indica (Burm.) Merr. ■

283931　Paraskevia W. Sauer et G. Sauer = Nonea Medik. ■

283932　Paraskevia W. Sauer et G. Sauer(1980);肖狼紫草属■☆

283933　Paraskevia cesalina(Fenzl et Friedrich)W. Sauer et G. Sauer;肖狼紫草■☆

283934　Parasopubia H. -P. Hofm. et Eb. Fisch. (2004);肖短冠草属■☆

283935　Parasopubia bonatii H. -P. Hofm. et Eb. Fisch. ;肖短冠草■☆

283936　Parasopubia delphiniifolia(L.)H. -P. Hofm. et Eb. Fisch. ;翠雀肖短冠草■☆

283937　Paraspalathus C. Presl = Aspalathus L. ●☆

283938　Paraspalathus aemula (E. Mey.) C. Presl = Aspalathus aemula E. Mey. ●☆

283939　Paraspalathus araneosa(L.)C. Presl = Aspalathus araneosa L. ●☆

283940　Paraspalathus argentea (L.) C. Presl = Aspalathus caledonensis R. Dahlgren ●☆

283941　Paraspalathus ascendens (E. Mey.) C. Presl = Aspalathus quinquefolia L. subsp. virgata(Thunb.). R. Dahlgren ●☆

283942　Paraspalathus callosa(L.)C. Presl = Aspalathus callosa L. ●☆

283943　Paraspalathus cancellata C. Presl = Aspalathus araneosa L. ●☆

283944　Paraspalathus capitata(L.)C. Presl = Aspalathus capitata L. ●☆

283945　Paraspalathus carnosa (P. J. Bergius) C. Presl = Aspalathus carnosa P. J. Bergius ●☆

283946　Paraspalathus cephalotes (Thunb.). C. Presl = Aspalathus cephalotes Thunb. ●☆

283947　Paraspalathus cinerea (Thunb.) C. Presl = Aspalathus cytisoides

283948　Paraspalathus crocea C. Presl = Aspalathus abietina Thunb. ●☆

283949　Paraspalathus cytisoides(Lam.)C. Presl = Aspalathus cytisoides Lam. ●☆

283950　Paraspalathus elongata (Eckl. et Zeyh.) C. Presl = Aspalathus quinquefolia L. subsp. virgata(Thunb.)R. Dahlgren ●☆

283951　Paraspalathus ericifolia(L.)C. Presl = Aspalathus ericifolia L. ●☆

283952　Paraspalathus erythrodes(Eckl. et Zeyh.) C. Presl = Aspalathus erythrodes Eckl. et Zeyh. ●☆

283953　Paraspalathus filicaulis (Eckl. et Zeyh.) C. Presl = Aspalathus filicaulis Eckl. et Zeyh. ●☆

283954　Paraspalathus galeata (E. Mey.) C. Presl = Aspalathus galeata E. Mey. ●☆

283955　Paraspalathus globosa (Andréws) C. Presl = Aspalathus globosa Andréws ●☆

283956　Paraspalathus heterophylla (L. f.) C. Presl = Aspalathus heterophylla L. f. ●☆

283957　Paraspalathus humifusa C. Presl = Aspalathus lotoides Thunb. ●☆

283958　Paraspalathus intermedia(Eckl. et Zeyh.) C. Presl = Aspalathus intermedia Eckl. et Zeyh. ●☆

283959　Paraspalathus jacobaea (E. Mey.) C. Presl = Aspalathus quinquefolia L. ●☆

283960　Paraspalathus lotoides (Thunb.) C. Presl = Aspalathus lotoides Thunb. ●☆

283961　Paraspalathus melanoides(Eckl. et Zeyh.)C. Presl = Aspalathus nigra L. ●☆

283962　Paraspalathus meyeriana(Eckl. et Zeyh.) C. Presl = Aspalathus ciliaris L. ●☆

283963　Paraspalathus multiflora (Thunb.) C. Presl = Aspalathus vermiculata Lam. ●☆

283964　Paraspalathus nigra(L.)C. Presl = Aspalathus nigra L. ●☆

283965　Paraspalathus nigrescens(E. Mey.)C. Presl = Aspalathus nigra L. ●☆

283966　Paraspalathus plukenetiana (Eckl. et Zeyh.) C. Presl = Aspalathus rugosa Thunb. ●☆

283967　Paraspalathus procumbens(E. Mey.)C. Presl = Aspalathus lotoides Thunb. ●☆

283968　Paraspalathus psoraleoides C. Presl = Aspalathus psoraleoides (C. Presl)Benth. ●☆

283969　Paraspalathus purpurascens(E. Mey.)C. Presl = Aspalathus ternata (Thunb.)Druce ●☆

283970　Paraspalathus sericea(P. J. Bergius)C. Presl = Aspalathus sericea P. J. Bergius ●☆

283971　Paraspalathus stellaris (Eckl. et Zeyh.) C. Presl = Aspalathus aspalathoides(L.)R. Dahlgren ●☆

283972　Paraspalathus stenophylla(Eckl. et Zeyh.)C. Presl = Aspalathus stenophylla Eckl. et Zeyh. ●☆

283973　Paraspalathus villosa (Thunb.) C. Presl = Aspalathus villosa Thunb. ●☆

283974　Paraspalathus virgata (Thunb.) C. Presl = Aspalathus quinquefolia L. subsp. virgata(Thunb.)R. Dahlgren ●☆

283975　Parasponia Miq. (1851);拟山黄麻属●☆

283976　Parasponia aspera Blume;粗糙拟山黄麻●☆

283977　Parasponia parviflora Miq. ;小花拟山黄麻●☆

283978　Parasponia rigida Merr. et L. M. Perry;坚挺拟山黄麻●☆

283979　Parastemon A. DC. (1842);异雄蔷薇属●☆

283980　Parastemon urophyllus A. DC. ;尾叶异雄蔷薇●☆

283981　Parastranthus G. Don = Lobelia L. ●■

283982　Parastranthus variifolius（Sims）G. Don = Monopsis variifolia（Sims）Urb. ■☆

283983　Parastrephia Nutt.（1841）;绒柏菀属■☆

283984　Parastrephia ericoides Nutt. ;绒柏菀■☆

283985　Parastrephia lepidophylla（Wedd.）Cabrera;鳞叶绒柏菀■☆

283986　Parastrephia lucida（Meyen）Cabrera;亮绒柏菀■☆

283987　Parastriga Mildbr.（1930）;肖独脚金属■☆

283988　Parastriga alectroides Mildbr. ;肖独脚金■☆

283989　Parastrobilanthes Bremek.（1944）;假马蓝属●☆

283990　Parastrobilanthes Bremek. = Strobilanthes Blume ●■

283991　Parastrobilanthes parabolica（Nees）Bremek. ;假马蓝●☆

283992　Parastyrax W. W. Sm.（1920）;茉莉果属（假野茉莉属,拟野茉莉属）;Jasminefruit,Parastyrax ●

283993　Parastyrax lacei（W. W. Sm.）W. W. Sm. ;茉莉果（拟野茉莉）;Common Jasminefruit,Common Parastyrax ●◇

283994　Parastyrax macrophyllus C. Y. Wu et K. W. Feng;大叶茉莉果（大咖啡,大叶拟野茉莉）;Largeleaf Jasminefruit, Largeleaf Parastyrax, Macrophyllous Parastyrax ●

283995　Parasympagis Bremek.（1944）;假合页草属■☆

283996　Parasympagis Bremek. = Strobilanthes Blume ●■

283997　Parasympagis kerrii Bremek. ;假合页草■☆

283998　Parasyringa W. W. Sm.（1919）;裂果女贞属●

283999　Parasyringa W. W. Sm. = Ligustrum L. ●

284000　Parasyringa sempervirens（Franch.）W. W. Sm. = Ligustrum sempervirens（Franch.）Lingelsh. ●

284001　Parasyringa sempervirens W. W. Sm. = Ligustrum sempervirens（Franch.）Lingelsh. ●

284002　Parasystasia Baill. = Asystasia Blume ●■

284003　Parasystasia kelleri Lindau = Asystasia guttata（Forssk.）Brummitt ●☆

284004　Parasystasia somalensis（Franch.）Baill. = Asystasia guttata（Forssk.）Brummitt ●☆

284005　Paratecoma Kuhlm.（1931）;赛黄钟花属●☆

284006　Paratecoma peroba（Record）Kuhlm. ;赛黄钟花（多脉白樫木）;Peroba ●☆

284007　Paratephrosia Domin = Tephrosia Pers.（保留属名）●■

284008　Paratephrosia Domin（1912）;假灰毛豆属●☆

284009　Paratephrosia lanata（Benth.）Domin;假灰毛豆●☆

284010　Paratheria Griseb.（1866）;水沼异颖草属■☆

284011　Paratheria glaberrima C. E. Hubb. ;光水沼异颖草■☆

284012　Paratheria prostrata Griseb. ;水沼异颖草■☆

284013　Parathesis（A. DC.）Hook. f.（1876）;芽冠紫金牛属●☆

284014　Parathesis Hook. f. = Parathesis（A. DC.）Hook. f. ●☆

284015　Parathesis acuminata Lundell;渐尖芽冠紫金牛●☆

284016　Parathesis acutissima Cuatrec. ;尖芽冠紫金牛●☆

284017　Parathesis angustifolia Lundell;窄叶芽冠紫金牛●☆

284018　Parathesis bicolor Lundell;二色芽冠紫金牛●☆

284019　Parathesis brevipes Lundell;短梗芽冠紫金牛●☆

284020　Parathesis calophylla Donn. Sm. ;美叶芽冠紫金牛●☆

284021　Parathesis chrysophylla Lundell;金叶芽冠紫金牛●☆

284022　Parathesis crassipes Lundell;粗梗芽冠紫金牛●☆

284023　Parathesis cubana（A. DC.）Molinet et M. Gómez;古巴芽冠紫金牛●☆

284024　Parathesis elliptica Lundell;椭圆芽冠紫金牛●☆

284025　Parathesis ferruginea Lundell;锈芽冠紫金牛●☆

284026　Parathesis fusca（Oerst.）Mez;褐芽冠紫金牛●☆

284027　Parathesis glaberrima Lundell;光滑芽冠紫金牛●☆

284028　Parathesis gracilis Lundell;细芽冠紫金牛●☆

284029　Parathesis lanceolata Brandegee;披针叶芽冠紫金牛●☆

284030　Parathesis latifolia Lundell;宽叶芽冠紫金牛●☆

284031　Parathesis laxa Lundell;松散芽冠紫金牛●☆

284032　Parathesis macrophylla Mez;大叶芽冠紫金牛●☆

284033　Parathesis membranacea Lundell;膜叶芽冠紫金牛●☆

284034　Parathesis mexicana Lundell;墨西哥芽冠紫金牛●☆

284035　Parathesis microcalyx Donn. Sm. ;小萼芽冠紫金牛●☆

284036　Parathesis montana Lundell;山地芽冠紫金牛●☆

284037　Parathesis multiflora Lundell;多花芽冠紫金牛●☆

284038　Parathesis nigropunctata Lundell;黑斑芽冠紫金牛●☆

284039　Parathesis obovalifolia Lundell;倒卵叶芽冠紫金牛●☆

284040　Parathesis obtusa Lundell;钝叶芽冠紫金牛●☆

284041　Parathesis oxyphylla Lundell;尖叶芽冠紫金牛●☆

284042　Parathesis pallida Lundell;苍白芽冠紫金牛●☆

284043　Parathesis platyphylla Lundell;阔叶芽冠紫金牛●☆

284044　Paratriaina Bremek.（1956）;拟三尖茜属☆

284045　Paratriaina xerophila Bremek. ;拟三尖茜☆

284046　Paratrophis Blume = Streblus Lour. ●

284047　Paratrophis caudata Merr. = Streblus macrophyllus Blume ●

284048　Paratropia（Blume）DC. = Schefflera J. R. Forst. et G. Forst.（保留属名）●

284049　Paratropia DC. = Heptapleurum Gaertn. ●■

284050　Paratropia DC. = Schefflera J. R. Forst. et G. Forst.（保留属名）●

284051　Paratropia cantoniensis Hook. et Arn. = Schefflera heptaphylla（L.）Frodin ●

284052　Paratropia contoniensis Hook. et Arn. = Schefflera octophylla（Lour.）Harms ●

284053　Paratropia cumingiana C. Presl = Polyscias cumingiana Fern. -Vill. ●

284054　Paratropia elata Hook. f. = Schefflera abyssinica（Hochst. ex A. Rich.）Harms ●☆

284055　Paratropia mannii Hook. f. = Schefflera mannii（Hook. f.）Harms ●☆

284056　Paratropia pubigera Brongn. et Planch. = Schefflera elliptica（Blume）Harms ●

284057　Paratropia venulosa Wight et Arn. = Schefflera venulosa（Wight et Arn.）Harms ●

284058　Paravallaris Pierre = Kibatalia G. Don ●

284059　Paravallaris Pierre ex Hua = Kibatalia G. Don ●

284060　Paravallaris macrophylla Pierre ex Hua = Kibatalia macrophylla（Pierre ex Hua）Woodson ●

284061　Paravallaris yunnanensis Tsiang et P. T. Li = Kibatalia macrophylla（Pierre ex Hua）Woodson ●

284062　Paravinia Hassk. = Praravinia Korth. ●☆

284063　Paravitex H. R. Fletcher（1937）;肖牡荆属●☆

284064　Paravitex siamica H. R. Fletcher;肖牡荆●☆

284065　Pardanthopsis（Hance）Lenz = Iris L. ●

284066　Pardanthopsis（Hance）Lenz（1972）;肖射干属■☆

284067　Pardanthopsis dichotoma（Pall.）Lenz = Iris dichotoma Pall. ■

284068　Pardanthopsis dichotoma（Pallas）Lenz;肖射干■☆

284069　Pardanthus Ker Gawl. = Belamcanda Adans.（保留属名）●■

284070　Pardanthus chinensis Ker Gawl. = Belamcanda chinensis（L.）DC. ■

284071　Pardanthus dichotomus Ledeb. = Iris decora Wall. ■

284072　Pardinia Herb. = Hydrotaenla Lindl. + Tigridia Juss. ■

284073　Pardisium Burm. f. = Gerbera L.（保留属名）■

284074　Pardoglossum Barbier et Mathez = Solenanthus Ledeb. ■

284075 Pardoglossum Barbier et Mathez(1973);豹舌草属■☆

284076 Pardoglossum atlanticum(Pit.)Barbier et Mathez;豹舌草■☆

284077 Pardoglossum atlanticum(Pit.)Barbier et Mathez = Cynoglossum pitardianum Greuter et Burdet ■☆

284078 Pardoglossum cheirifolium(L.)Barbier et Mathez = Cynoglossum cheirifolium L.■☆

284079 Pardoglossum cheirifolium (L.) Barbier et Mathez subsp. heterocarpum(Kunze)Mathez = Cynoglossum cheirifolium L. subsp. heterocarpum(Kunze)Maire ■☆

284080 Pardoglossum cheirifolium (L.) Barbier et Mathez var. arundanum (Coss.) Mathez = Cynoglossum cheirifolium L. subsp. heterocarpum(Kunze)Maire ■☆

284081 Pardoglossum lanatum (L.) Barbier et Mathez = Cynoglossum mathezii Greuter et Burdet ■☆

284082 Pardoglossum tubiflorum (Murb.) Barbier et Mathez = Cynoglossum tubiflorum(Murb.)Greuter et Burdet ■☆

284083 Pardoglossum tubiflorum(Murbeck)Barbier et Mathez;管花豹舌草■☆

284084 Pardoglossum watieri (Batt. et Maire) Barbier et Mathez = Cynoglossum watieri(Batt. et Maire)Braun-Blanq. et Maire ■☆

284085 Parduyna Salisb. = Kreysigia Rchb. ■☆

284086 Parduyna Salisb. = Schelhammera R. Br. (保留属名)■☆

284087 Parechites Miq. = Trachelospermum Lem. ●

284088 Parechites bowringii Hance = Gymnanthera oblonga(Burm. f.)P. S. Green ●

284089 Parectenium P. Beauv. ex Stapf = Paractaenum P. Beauv. ■☆

284090 Parectenium Stapf = Parectenium P. Beauv. ex Stapf ■☆

284091 Pareira Lour. ex Gomes = Vitis L. ●

284092 Parenterolobium Kosterm. = Albizia Durazz. ●

284093 Parentucellia Viv. (1824);帕伦列当属■☆

284094 Parentucellia flaviflora(Boiss.)Nevski;黄花帕伦列当■☆

284095 Parentucellia floribunda Viv. ;繁花帕伦列当■☆

284096 Parentucellia latifolia (L.) Caruel;宽花帕伦列当;Broadleaf Glandweed ■☆

284097 Parentucellia latifolia(L.)Caruel var. flaviflora(Boiss.)Dandy ex F. W. Andréws = Parentucellia latifolia(L.)Caruel ■☆

284098 Parentucellia viscosa (L.) Caruel;帕伦列当;Ellow Bartsia, Tweeny-legs, Twiny-legs, Viscid Bartsia, Yellow Bartsia, Yellow Glandweed ■☆

284099 Parepigynum Tsiang et P. T. Li(1973);富宁藤属;Funingvine, Parepigynum ●■★

284100 Parepigynum funingense Tsiang et P. T. Li;富宁藤;Funing Funingvine, Funing Parepigynum ●■

284101 Pareugenia Turrill = Syzygium R. Br. ex Gaertn. (保留属名)●

284102 Parexuris Nakai et Maek. = Sciaphila Blume ●

284103 Parfonsia Scop. = Cuphea Adans. ex P. Browne ●■

284104 Parfonsia Scop. = Parsonsia P. Browne(废弃属名)●

284105 Parhabenaria Gagnep. (1932);东南亚兰属■☆

284106 Parhabenaria cochinchinensis Gagnep. ;东南亚兰■☆

284107 Pariana Aubl. (1775);巴厘箫属■☆

284108 Pariana angustifolia Spreng. ;狭叶巴厘箫■☆

284109 Pariana argentea Hollowell et Davidse;银色巴厘箫■☆

284110 Pariana bicolor Tutin;二色巴厘箫■☆

284111 Pariana glauca Nees;灰巴厘箫■☆

284112 Pariana gracilis Döll;细巴厘箫■☆

284113 Pariana longiflora Tutin;长花巴厘箫■☆

284114 Pariana multiflora R. P. Oliveira, Longhi-Wagner et Hollowell;多花巴厘箫■☆

284115 Pariana nivea Huber ex Tutin;雪白巴厘箫■☆

284116 Pariana obtusa Swallen;钝巴厘箫■☆

284117 Pariana ovalifolia Swallen;卵叶巴厘箫■☆

284118 Pariana pallida Swallen;苍白巴厘箫■☆

284119 Pariana parviflora Trin. ;小花巴厘箫■☆

284120 Pariana velutina Swallen;黏巴厘箫■☆

284121 Parianaceae(Hack.)Nakai = Gramineae Juss. (保留科名)■●

284122 Parianaceae(Hack.)Nakai = Parianaceae Nakai ■●

284123 Parianaceae(Hack.)Nakai = Poaceae Barnhart(保留科名)■●

284124 Parianaceae Nakai = Gramineae Juss. (保留科名)■●

284125 Parianaceae Nakai = Poaceae Barnhart(保留科名)■●

284126 Parianaceae Nakai;巴厘箫科■

284127 Pariatica Post et Kuntze = Pariaticu Adans. ●

284128 Pariaticu Adans. = Nyctanthes L. ●

284129 Paridaceae Dumort. = Melanthiaceae Batsch ex Borkh. (保留科名)■

284130 Paridaceae Dumort. = Trilliaceae Chevall. (保留科名)■

284131 Parietaria L. (1753);墙草属;Pellitory, Pellitory-of-the-wall, Wallgrass ■

284132 Parietaria abyssinica A. Rich. = Pouzolzia guineensis Benth. ●☆

284133 Parietaria alsinifolia Delile;繁缕叶墙草■☆

284134 Parietaria chersonensis Dörfl. ;赫尔松墙草■☆

284135 Parietaria cochinchinensis Lour. = Pouzolzia zeylanica (L.)Benn. et R. Br. var. microphylla(Wedd.)W. T. Wang ■

284136 Parietaria coreana Nakai = Parietaria micrantha Ledeb. ■

284137 Parietaria cretica L. ;克里特墙草■☆

284138 Parietaria debilis G. Forst. = Parietaria lusitanica L. ■

284139 Parietaria debilis G. Forst. = Parietaria micrantha Ledeb. ■

284140 Parietaria debilis G. Forst. f. laxiflora (Engl.) Letouzey = Parietaria laxiflora Engl. ■☆

284141 Parietaria debilis G. Forst. var. micrantha (Ledeb.) Wedd. = Parietaria micrantha Ledeb. ■

284142 Parietaria diffusa Mert. et W. D. J. Koch = Parietaria judaica L. ■☆

284143 Parietaria diffusa Mert. et W. D. J. Koch var. fallax Gren. et Godr. = Parietaria judaica L. ■☆

284144 Parietaria elliptica K. Koch;椭圆墙草■☆

284145 Parietaria erecta Merr. et Koch;直立墙草;Erect Wallgrass ■☆

284146 Parietaria filamentosa Webb et Berthel. ;丝状墙草■☆

284147 Parietaria floridana Nutt. ;南美墙草■☆

284148 Parietaria hespera Hinton;北美墙草■☆

284149 Parietaria hespera Hinton var. californica Hinton;加州墙草■☆

284150 Parietaria indica L. = Pouzolzia zeylanica(L.)Benn. et R. Br. var. microphylla(Wedd.)W. T. Wang ■

284151 Parietaria indica L. = Pouzolzia zeylanica(L.)Benn. et R. Br. ■

284152 Parietaria judaica L. ;广布墙草(伸展墙草);Billy Beattie, Hammerwort, Lichwort, Parritory, Peletir, Pellitory of The Wall, Pellitory-of-the-wall, Peniterry, Sneezewort, Spreading Pellitory, Wall Pellitory, Wall Sage, Wallwort ■☆

284153 Parietaria judaica L. var. brevipetiolata Boiss. = Parietaria judaica L. ■☆

284154 Parietaria laxiflora Engl. ;少花墙草■☆

284155 Parietaria lusitanica L. = Parietaria micrantha Ledeb. ■

284156 Parietaria lusitanica L. subsp. chersonensis (Láng) Chrtek = Parietaria lusitanica L. ■

284157 Parietaria lusitanica L. subsp. chersonensis (Láng) Chrtek var. micrantha(Ledeb.)Chrtek = Parietaria micrantha Ledeb. ■

284158 Parietaria lusitanica L. var. micrantha (Ledeb.) Chrtek =

Parietaria micrantha Ledeb. ■

284159　Parietaria mauritanica Durieu;毛里塔尼亚墙草■☆

284160　Parietaria micrantha Ledeb.;墙草（白石薯,白猪仔菜,干菜子,软骨石薯,石薯,田薯,细叶贯菜子,小花墙草,指甲薯）;Smallflower Pellitory,Smallflower Wallgrass ■

284161　Parietaria micrantha Ledeb. = Parietaria lusitanica L. ■

284162　Parietaria micrantha Ledeb. var. coreana（Nakai）H. Hara;朝鲜墙草■☆

284163　Parietaria microphylla L. = Pilea microphylla（L.）Liebm. ■

284164　Parietaria nummularia Small = Parietaria floridana Nutt. ■☆

284165　Parietaria obtusa Rydb. ex Small = Parietaria pensylvanica Muhl. ex Willd. ■☆

284166　Parietaria occidentalis Rydb. = Parietaria pensylvanica Muhl. ex Willd. ■☆

284167　Parietaria officinalis L.;药用墙草;Medicinal Wallgrass,Parietary,Peilitory,Pright Pellitory,Wall Pellitory ■☆

284168　Parietaria officinalis L. subsp. judaica（L.）Bég. = Parietaria judaica L. ■☆

284169　Parietaria officinalis L. var. brevipetiolata（Boiss.）Briq. = Parietaria judaica L. ■☆

284170　Parietaria officinalis L. var. diffusa（Mert. et Koch）Wedd. = Parietaria judaica L. ■☆

284171　Parietaria officinalis L. var. fallax（Gren. et Godr.）Briq. = Parietaria judaica L. ■☆

284172　Parietaria officinalis L. var. judaica（L.）Hochr. = Parietaria judaica L. ■☆

284173　Parietaria officinalis L. var. ramiflora Asch. et Graebn. = Parietaria judaica L. ■☆

284174　Parietaria pensylvanica Muhl. ex Willd.;宾州墙草;Pellitory,Pennsylvanian Pellitory ■☆

284175　Parietaria pensylvanica Muhl. ex Willd. var. obtusa（Rydb. ex Small）Shinners = Parietaria pensylvanica Muhl. ex Willd. ■☆

284176　Parietaria ramiflora Moench = Parietaria judaica L. ■☆

284177　Parietaria ruwenzoriensis Cortesi;鲁文佐里墙草■☆

284178　Parietaria ruwenzoriensis Cortesi subsp. keniensis Gebauer;肯尼亚墙草■☆

284179　Parietaria scandens Engl. = Parietaria ruwenzoriensis Cortesi ■☆

284180　Parietaria sonneratii Poir. = Phenax sonneratii（Poir.）Wedd. ■☆

284181　Parietaria urticifolia L. f. = Pilea urticifolia（L. f.）Blume ■☆

284182　Parietaria zeylanica L. = Pouzolzia zeylanica（L.）Benn. et R. Br. ■

284183　Parietariaceae Bercht. et J. Presl = Urticaceae Juss.（保留科名）●■

284184　Parilax Raf. = Parillax Raf. ●

284185　Parilax Raf. = Smilax L. ●

284186　Parilia Dennst. = Elaeodendron J. Jacq. ●☆

284187　Parilium Gaertn. = Nyctanthes L. ●

284188　Parilium arbortristis Gaertn. = Nyctanthes arbor-tristis L. ●

284189　Parillax Raf. = Smilax L. ●

284190　Pariltaria Burm. f. = Parietaria L. ●

284191　Parinari Aubl.（1775）;姜饼木属（姜饼树属）;Parinarium ●☆

284192　Parinari albida Craib;白姜饼木●☆

284193　Parinari aubrevillei Pellegr. = Maranthes aubrevillei（Pellegr.）Prance ●☆

284194　Parinari bangweolensis R. E. Fr. = Magnistipula butayei De Wild. subsp. bangweolensis（R. E. Fr.）F. White ●☆

284195　Parinari baoulensis A. Chev. = Maranthes polyandra（Benth.）Prance ●☆

284196　Parinari benna Scott-Elliot = Bafodeya benna（Scott-Elliot）Prance ●☆

284197　Parinari bequaertii De Wild. = Maranthes floribunda（Baker）F. White ●☆

284198　Parinari bequaertii De Wild. var. longistaminea Hauman = Maranthes floribunda（Baker）F. White ●☆

284199　Parinari campestris Aubl.;平地姜饼木（田野姜饼树）●☆

284200　Parinari capensis Harv.;好望角姜饼木;Sand Apple ●☆

284201　Parinari capensis Harv. f. obtusifolia Cavaco = Parinari capensis Harv. ●☆

284202　Parinari capensis Harv. subsp. latifolia（Oliv.）R. A. Graham = Parinari capensis Harv. ●☆

284203　Parinari capensis Harv. var. latifolia Oliv. = Parinari capensis Harv. ●☆

284204　Parinari chapelieri Baill. = Parinari curatellifolia Planch. ex Benth. ●☆

284205　Parinari chrysophylla Oliv.;黄叶姜饼木●☆

284206　Parinari chrysophylla Oliv. = Maranthes chrysophylla（Oliv.）Prance ●☆

284207　Parinari congensis Didr.;刚果姜饼木●☆

284208　Parinari congoensis Engl. = Parinari congolana T. Durand et H. Durand ●☆

284209　Parinari congolana T. Durand et H. Durand;康戈尔姜饼木●☆

284210　Parinari corymbosa（Blume）Miq.;伞花姜饼木●☆

284211　Parinari curatellifolia Planch. ex Benth.;安吉利姜饼木;Mbura,Mobola,Mobola Plum ●☆

284212　Parinari curatellifolia Planch. ex Benth. subsp. mobola（Oliv.）R. A. Graham = Parinari curatellifolia Planch. ex Benth. ●☆

284213　Parinari curatellifolia Planch. ex Benth. var. fruticulosa R. E. Fr. = Parinari capensis Harv. ●☆

284214　Parinari elliottii Engl. = Parinari excelsa Sabine ●☆

284215　Parinari excelsa Sabine;大姜饼木;Grey Plum, Guinea Plum, Rough-skinned Plum ●☆

284216　Parinari excelsa Sabine subsp. holstii（Engl.）R. A. Graham = Parinari excelsa Sabine ●☆

284217　Parinari floribunda Baker;多花大姜饼木●☆

284218　Parinari floribunda Baker = Maranthes floribunda（Baker）F. White ●☆

284219　Parinari gabunensis Engl. = Maranthes gabunensis（Engl.）Prance ●☆

284220　Parinari gardineri Hemsl. = Parinari curatellifolia Planch. ex Benth. ●☆

284221　Parinari gilletii De Wild. = Maranthes glabra（Oliv.）Prance ●☆

284222　Parinari glabra（Oliv.）Prance var. gilletii（De Wild.）Hauman = Maranthes glabra（Oliv.）Prance ●☆

284223　Parinari glabra Oliv.;光姜饼木●☆

284224　Parinari glabra Oliv. = Maranthes glabra（Oliv.）Prance ●☆

284225　Parinari goetzeniana Engl.;戈茨姜饼木●☆

284226　Parinari goetzeniana Engl. = Maranthes goetzeniana（Engl.）Prance ●☆

284227　Parinari griffithiana Benth.;格氏姜饼木●☆

284228　Parinari guyanensis Fritsch;圭亚那姜饼木●☆

284229　Parinari holstii Engl.;霍氏姜饼木●☆

284230　Parinari holstii Engl. = Parinari excelsa Sabine ●☆

284231　Parinari holstii Engl. var. longifolia Engl. ex De Wild. = Parinari excelsa Sabine ●☆

284232　Parinari hypochrysea Mildbr. ex Letouzey et F. White;里金姜饼木●☆

284233 Parinari indica Bedd. ;印度姜饼木●☆

284234 Parinari ingangensis Pellegr. = Magnistipula tessmannii(Engl.) Prance ●☆

284235 Parinari iodocalyx Mildbr. = Maranthes chrysophylla (Oliv.) Prance subsp. coriacea F. White ●☆

284236 Parinari kerstingii Engl. ;克氏姜饼木●☆

284237 Parinari kerstingii Engl. = Maranthes kerstingii(Engl.)Prance ●☆

284238 Parinari klaineana Pierre = Maranthes gabunensis(Engl.)Prance ●☆

284239 Parinari klainei Aubrév. = Maranthes glabra(Oliv.) Prance ●☆

284240 Parinari latifolia(Oliv.) Exell = Parinari capensis Harv. ●☆

284241 Parinari laurina A. Gray;月桂姜饼木●☆

284242 Parinari laxiflora Ducke;疏花姜饼木●☆

284243 Parinari liberica Engl. ex Mildbr. ;利比里亚姜饼木●☆

284244 Parinari macrophylla Sabine;大叶姜饼木;Ginger-bread Plum, Gingerbread Plum-tree ●☆

284245 Parinari macrophylla Sabine = Neocarya macrophylla (Sabine) Prance ●☆

284246 Parinari mildbraedii Engl. = Parinari excelsa Sabine ●☆

284247 Parinari minus Baill. ex Aubrév. = Parinari congolana T. Durand et H. Durand ●☆

284248 Parinari mobola Oliv. ;沙地姜饼木;Cork Tree, Hissing Tree, Sand Apple ●☆

284249 Parinari mobola Oliv. = Parinari curatellifolia Planch. ex Benth. ●☆

284250 Parinari montana Aubl. ;山地姜饼木●☆

284251 Parinari montana Engl. = Maranthes glabra(Oliv.) Prance ●☆

284252 Parinari multiflora Miq. ;多花姜饼木●☆

284253 Parinari nalaensis De Wild. = Parinari excelsa Sabine ●☆

284254 Parinari nana Baill. ex A. Chev. ;矮姜饼木●☆

284255 Parinari nitida Hook. f. ;光亮姜饼木●☆

284256 Parinari occidentalis Prance;西方姜饼木●☆

284257 Parinari parvifolia Sandwith;小叶姜饼木(小叶姜饼树)●☆

284258 Parinari poggei Engl. ;波格姜饼木●☆

284259 Parinari polyandra Benth. ;多蕊姜饼木●☆

284260 Parinari polyandra Benth. = Maranthes polyandra (Benth.) Prance ●☆

284261 Parinari polyandra Benth. subsp. floribunda (Baker) R. A. Graham = Maranthes floribunda(Baker) F. White ●☆

284262 Parinari polyandra Benth. var. argentea Aubrév. = Maranthes polyandra(Benth.) Prance ●☆

284263 Parinari polyandra Benth. var. cinerea Engl. = Maranthes polyandra(Benth.) Prance ●☆

284264 Parinari polyandra Benth. var. villosa Aubrév. = Maranthes polyandra(Benth.) Prance ●☆

284265 Parinari polyneura Miq. ;多脉姜饼木●☆

284266 Parinari polystachya Poepp. ex Fritsch;多穗姜饼木●☆

284267 Parinari pumila Mildbr. = Parinari capensis Harv. ●☆

284268 Parinari riparia R. E. Fr. = Parinari excelsa Sabine ●☆

284269 Parinari robusta Oliv. ;山生姜饼木(粗壮姜饼木)●☆

284270 Parinari robusta Oliv. = Maranthes robusta(Oliv.) Prance ●☆

284271 Parinari robusta Oliv. var. klainei Aubrév. ex Pellegr. = Maranthes glabra(Oliv.) Prance ●☆

284272 Parinari rodolphi Huber;姜饼木(布达姜饼木)●☆

284273 Parinari sargosii Pellegr. ;萨氏姜饼木●☆

284274 Parinari sargosii Pellegr. = Magnistipula butayei De Wild. subsp. sargosii(Pellegr.) F. White ●☆

284275 Parinari senegalensis Perr. ex DC. = Neocarya macrophylla (Sabine) Prance ●☆

284276 Parinari subcordata Oliv. = Parinari congensis Didr. ●☆

284277 Parinari tenuifolia A. Chev. ;细叶姜饼木●☆

284278 Parinari tenuifolia A. Chev. = Parinari excelsa Sabine ●☆

284279 Parinari tessmannii Engl. = Magnistipula tessmannii (Engl.) Prance ●☆

284280 Parinari tibatensis Engl. = Maranthes glabra(Oliv.) Prance ●☆

284281 Parinari tisserantii Aubrév. et Pellegr. = Magnistipula butayei De Wild. subsp. tisserantii(Aubrév. et Pellegr.) F. White ●☆

284282 Parinari vassonii A. Chev. = Maranthes glabra(Oliv.)Prance ●☆

284283 Parinari verdickii De Wild. = Parinari excelsa Sabine ●☆

284284 Parinari versicolor Engl. = Magnistipula zenkeri Engl. ●☆

284285 Parinari whytei Engl. = Parinari excelsa Sabine ●☆

284286 Parinarium Comm. ex Juss. = Parinari Aubl. ●☆

284287 Parinarium Juss. = Parinari Aubl. ●☆

284288 Paripon Voigt(1826);帕利棕属●☆

284289 Paris L. (1753);重楼属(七叶一枝花属);Herb Paris, Love-apple, Paris ■

284290 Paris aprica H. Lév. = Paris polyphylla Sm. var. yunnanensis (Franch.) Hand. -Mazz. ■

284291 Paris arisanensis Hayata = Paris polyphylla Sm. var. stenophylla Franch. ■

284292 Paris atrata H. Lév. = Paris polyphylla Sm. var. yunnanensis (Franch.) Hand. -Mazz. ■

284293 Paris axialis H. Li;五指莲重楼(大叶重楼,大重楼,九道箍,铁灯台,五指莲,小重楼);Axial Paris ■

284294 Paris axialis H. Li var. rubra H. H. Zhou, K. Y. Wu et R. Tao;红果五指莲;Redfruited Axial Paris ■

284295 Paris axialis H. Li var. rubra H. H. Zhou, K. Y. Wu et R. Tao = Paris axilis H. Li ■

284296 Paris bashanensis F. T. Wang et Ts. Tang;巴山重楼(独龙钻山,露水珠);Bashan Paris, Pashan Paris ■

284297 Paris biondii Pamp. = Paris polyphylla Sm. ■

284298 Paris birmhnica(Takht.) H. Li et Noltie = Paris polyphylla Sm. var. yunnanensis(Franch.) Hand. -Mazz. ■

284299 Paris bockiana Diels = Paris polyphylla Sm. var. stenophylla Franch. ■

284300 Paris brachysepala Pamp. = Paris polyphylla Sm. var. chinensis (Franch.) H. Hara ■

284301 Paris brevipetala Y. K. Yang = Paris polyphylla Sm. var. chinensis(Franch.) H. Hara ■

284302 Paris cavaleriei H. Lév. et Vaniot = Paris polyphylla Sm. var. chinensis(Franch.) H. Hara ■

284303 Paris cavaleriei H. Lév. et Vaniot = Paris polyphylla Sm. var. yunnanensis(Franch.) Hand. -Mazz. ■

284304 Paris chinensis Franch. = Paris polyphylla Sm. var. chinensis (Franch.)H. Hara ■

284305 Paris chinensis Franch. = Paris polyphylla Sm. var. yunnanensis (Franch.) Hand. -Mazz. ■

284306 Paris christii H. Lév. = Paris polyphylla Sm. var. yunnanensis (Franch.) Hand. -Mazz. ■

284307 Paris cronquistii(Takht.) H. Li et Noltie;凌云重楼;Cronquist Paris ■

284308 Paris cronquistii(Takht.) H. Li et Noltie var. xichouensis H. Li;西畴重楼;Xichou Paris ■

284309 Paris dahurica Fisch. ex Turcz. = Paris verticillata M. Bieb. ■

284310 Paris daliensis H. Li et V. G. Soukup;大理重楼;Dali Paris ■

284311 Paris debeauxii H. Lév. = Paris polyphylla Sm. ■

284312 Paris delavayi Franch. ;金钱重楼;Delavay Paris ■

284313 Paris delavayi Franch. var. ovalifolia H. Li;卵叶重楼;Delavay Paris ■

284314 Paris delavayi Franch. var. ovalifolia H. Li = Paris fargesii Franch. var. petiolata（Baker ex C. H. Wright）F. T. Wang et Ts. Tang ■

284315 Paris delavayi Franch. var. petiolata（Baker ex C. H. Wright）H. Li = Paris fargesii Franch. var. petiolata（Baker ex C. H. Wright）F. T. Wang et Ts. Tang ■

284316 Paris dulongensis H. Li et Kurita;独龙重楼;Dulong Paris ■

284317 Paris dunniana H. Lév. ;海南重楼（七叶一枝花）;Dunn Paris ■

284318 Paris dunniana H. Lév. var. oligophylla F. T. Wang et Ts. Tang;少叶重楼■

284319 Paris fargesii Franch. ;球药隔重楼（白菜果,独角莲,法氏王子,红铁灯台,金线重楼,九重楼,七叶一枝花,三台消,铁灯台,五叶重楼,一枝花,重楼）;Farges Paris ■

284320 Paris fargesii Franch. var. brevipetalata（T. C. Huang et K. C. Yang）T. C. Huang et K. C. Yang = Paris polyphylla Sm. ■

284321 Paris fargesii Franch. var. brevipetalata（T. C. Huang et K. C. Yang）T. C. Huang et K. C. Yang = Paris fargesii Franch. ■

284322 Paris fargesii Franch. var. brevipetalata（T. C. Huang et K. C. Yang）T. C. Huang et K. C. Yang;短瓣球药隔七叶一枝花■

284323 Paris fargesii Franch. var. latipetala H. Li et V. G. Soukup;宽瓣球药隔重楼;Broad-petal Farges Paris ■

284324 Paris fargesii Franch. var. latipetala H. Li et V. G. Soukup = Paris fargesii Franch. ■

284325 Paris fargesii Franch. var. petiolata（Baker ex C. H. Wright）F. T. Wang et Ts. Tang;具柄重楼（九重楼,具柄王孙,五子莲,重楼）;Petiolate Paris ■

284326 Paris fauchtiana H. Lév. = Paris polyphylla Sm. var. chinensis（Franch.）H. Hara ■

284327 Paris formosana Hayata = Paris polyphylla Sm. var. chinensis（Franch.）H. Hara ■

284328 Paris formosana Hayata = Paris polyphylla Sm. ■

284329 Paris forrestii（Takht.）H. Li;长柱重楼;Forrest Paris ■

284330 Paris fracnhetiana H. Lév. = Paris polyphylla Sm. var. yunnanensis（Franch.）Hand. -Mazz. ■

284331 Paris gigas H. Lév. et Vaniot = Paris polyphylla Sm. var. yunnanensis（Franch.）Hand. -Mazz. ■

284332 Paris hainanensis Merr. = Paris dunniana H. Lév. ■

284333 Paris hamifer H. Lév. = Paris polyphylla Sm. var. stenophylla Franch. ■

284334 Paris henryi Diels = Paris delavayi Franch. ■

284335 Paris hexaphylla Cham. = Paris verticillata M. Bieb. ■

284336 Paris hexaphylla Cham. f. purpurea Miyabe et Tatew. = Paris verticillata M. Bieb. ■

284337 Paris hexaphylla Cham. var. manshurica（Kom.）Vorosch. = Paris verticillata M. Bieb. ■

284338 Paris hookeri H. Lév. = Paris fargesii Franch. ■

284339 Paris incompleta M. Bieb. ;不全重楼■☆

284340 Paris japonica（Franch. et Sav.）Franch. ;独脚莲（日本重楼）;Japan Paris ■

284341 Paris kwangtungensis R. H. Miao = Paris polyphylla Sm. var. kwangtungensis（R. H. Miao）S. C. Cheng et S. Yun Liang ■

284342 Paris lanceolata Hayata;高山七叶一枝花;Lanceolate Paris ■

284343 Paris lancifolia Hayata;柳叶重楼（玉山七叶莲）;Narrowleaf Paris ■

284344 Paris lancifolia Hayata = Paris polyphylla Sm. var. stenophylla Franch. ■

284345 Paris longistigmata H. Li = Paris forrestii（Takht.）H. Li ■

284346 Paris luquanensis H. Li;禄劝重楼（禄劝花叶重楼）;Luquan Paris ■

284347 Paris mairei H. Lév. ;毛脉重楼（九道箍,毛脉蚤休,毛叶重楼,毛重楼,重楼）;Maire Paris,Pubescent Paris ■

284348 Paris manshurica Kom. ;东北重楼■☆

284349 Paris manshurica Kom. = Paris verticillata M. Bieb. ■

284350 Paris marchandii H. Lév. = Paris polyphylla Sm. var. alba H. Li et R. J. Mitch. ■

284351 Paris marmorata Stearn;花叶重楼;Marbled Paris,Mottled Paris,Violetleaf Paris ■

284352 Paris mercieri H. Lév. = Paris polyphylla Sm. var. yunnanensis（Franch.）Hand. -Mazz. ■

284353 Paris obovata Ledeb. = Paris verticillata M. Bieb. ■

284354 Paris petiolata Baker ex C. H. Wright = Paris fargesii Franch. var. petiolata（Baker ex C. H. Wright）F. T. Wang et Ts. Tang ■

284355 Paris petiolata Baker ex C. H. Wright var. membranacea C. H. Wright = Paris fargesii Franch. ■

284356 Paris pinfaensis H. Lév. = Paris polyphylla Sm. var. yunnanensis（Franch.）Hand. -Mazz. ■

284357 Paris polyandra S. F. Wang;多蕊重楼;Manythrum Paris ■

284358 Paris polyphylla Sm. ;七叶一枝花（草甘遂,草河车,灯台七,独角莲,独叶一枝花,红独角莲,金环,九道箍,平伐重楼,天鹅蛋,一枝花,蚤休,重楼,重楼草,重楼金线,重台）;Leafy Paris, Manyleaf Paris,One Flower With Sevenleaves,Whorl Leaved Lily ■

284359 Paris polyphylla Sm. subsp. fargesii（Franch.）H. Hara = Paris fargesii Franch. ■

284360 Paris polyphylla Sm. subsp. marmorata（Steam）H. Hara = Paris marmorata Stearn ■

284361 Paris polyphylla Sm. var. alba H. Li et R. J. Mitch. ;白花重楼;Whiteflower Manyleaf Paris ■

284362 Paris polyphylla Sm. var. apetala Hand. -Mazz. ;缺瓣重楼;Apetalous Manyleaf Paris,Apetalous Paris ■

284363 Paris polyphylla Sm. var. appendiculata H. Hara = Paris thibetica Franch. ■

284364 Paris polyphylla Sm. var. brachystemon Franch. = Paris polyphylla Sm. var. stenophylla Franch. ■

284365 Paris polyphylla Sm. var. brachystemon Franch. = Paris polyphylla Sm. ■

284366 Paris polyphylla Sm. var. brevipetala Y. K. Yang;短瓣七叶一枝花;Shortpetal Paris ■

284367 Paris polyphylla Sm. var. chinensis（Franch.）H. Hara;华重楼（鳌休,白甘遂,白河车,草甘遂,草河车,蚩休,灯台七,独角莲,独脚莲,独立一枝花,独叶一枝花,多叶重楼,孩儿陶伞,海螺七,海南重楼,红重楼,金盘托荔枝,金盘托珠,金丝两重楼,金线重楼,九道箍,九重楼,螺丝七,螺陀三七,七层塔,七叶莲,七叶一盏灯,七叶一枝花,七叶遮花,七枝莲,七子莲,三层草,蛇药子,双层楼,双台,双喜草,台湾重楼,铁灯台,铁灯盏,一把伞,一枝箭,芋头三七,鸳鸯虫,蚤休,枝花头,中华王孙,重楼,重楼金线,重楼一枝箭,重台,重台草,紫河车）;China Paris ■

284368 Paris polyphylla Sm. var. kwangtungensis（R. H. Miao）S. C. Cheng et S. Yun Liang;广东重楼;Guangdong Paris ■

284369 Paris polyphylla Sm. var. latifolia F. T. Wang et C. C. Chang;宽叶重楼;Broadleaf Paris ■

284370 Paris polyphylla Sm. var. minora S. F. Wang;小重楼;Small Manyleaf Paris ■

284371 Paris polyphylla Sm. var. nana H. Li;矮重楼;Dwarf Paris ■

284372 Paris polyphylla Sm. var. platypetala Franch. = Paris polyphylla Sm. var. yunnanensis(Franch.) Hand. -Mazz. ■

284373 Paris polyphylla Sm. var. pseudothibetica f. macrosepala H. Li = Paris polyphylla Sm. var. pseudothibetica H. Li ■

284374 Paris polyphylla Sm. var. pseudothibetica H. Li;长药隔重楼(长药重楼,拟长药隔重楼)■

284375 Paris polyphylla Sm. var. pseudothibetica H. Li f. macrosepala H. Li = Paris polyphylla Sm. var. pseudothibetica H. Li ■

284376 Paris polyphylla Sm. var. pseudothibetica H. Li f. microsepala H. Li;大萼重楼■

284377 Paris polyphylla Sm. var. pubescens Hand. -Mazz. = Paris mairei H. Lév. ■

284378 Paris polyphylla Sm. var. stenophylla Franch. ;狭叶重楼(白重楼,半截烂,虫蒌,独角莲,高山七叶一枝花,海螺七,金酒壶,金线重楼,金子莲,九道箍,九重楼,烂屁股,六子含花,七叶一枝花,三台消,三重天,铁灯台,铜灯台,狭叶七叶一枝花,狭叶蚤休,小叶子重楼,小重楼,一盏灯);Narrowleaf Paris ■

284379 Paris polyphylla Sm. var. taitungensis(S. S. Ying) S. S. Ying;台东七叶一枝花(台东七叶莲)■

284380 Paris polyphylla Sm. var. thibetica (Franch.) H. Hara = Paris thibetica Franch. ■

284381 Paris polyphylla Sm. var. wallichii H. Hara;喜山重楼;Wallich Paris ■

284382 Paris polyphylla Sm. var. yunnanensis(Franch.) Hand. -Mazz. ;云南重楼(草河车,大重楼,滇重楼,独角莲,公鸡子,九道箍,宽瓣蚤休,宽瓣重楼,阔瓣蚤休,两把伞,麻婆婆,七叶一枝花,山重楼独足莲,土三七,王孙,一把伞,重楼,重楼一枝箭,重台);Yunnan Manyleaf Paris ■

284383 Paris polyphylla Sm. var. yunnanensis(Franch.) Hand. -Mazz. f. velutina H. Li et Noltie = Paris polyphylla Sm. var. yunnanensis (Franch.) Hand. -Mazz. ■

284384 Paris pubescens(Hand. -Mazz.)F. T. Wang et Ts. Tang;毛重楼■

284385 Paris pubescens(Hand. -Mazz.) F. T. Wang et Ts. Tang = Paris mairei H. Lév. ■

284386 Paris pubescens(Hand. -Mazz.)Takht. = Paris mairei H. Lév. ■

284387 Paris quadrifolia L. ;四叶重楼(轮叶王孙);Devil-in-a-bush, Fourleaf Paris,Four-leaved Grass,Four-leaved Truelove,Herb Paris,Herb Truelove,Herb True-love,Herb-Paris,Leopard's Bane,Love Troth,Love-troth,One-berry,True Lovers' Knot,True-love ■

284388 Paris quadrifolia L. = Paris verticillata M. Bieb. ■

284389 Paris quadrifolia L. var. angustiovata D. Z. Ma et H. L. Liu = Paris quadrifolia L. ■

284390 Paris quadrifolia L. var. angustiovata D. Z. Ma et H. L. Liu;宁夏四叶重楼;Ningxia Fourleaf Paris ■

284391 Paris quadrifolia L. var. angustiovata D. Z. Ma et H. L. Liu = Paris quadrifolia L. ■

284392 Paris quadrifolia L. var. dahurica(Fisch. ex Turcz.) Franch. = Paris verticillata M. Bieb. ■

284393 Paris quadrifolia L. var. dahurica (Fisch.) Franch. = Paris verticillata M. Bieb. ■

284394 Paris quadrifolia L. var. hexaphylla (Cham.) Fedtsch. = Paris verticillata M. Bieb. ■

284395 Paris quadrifolia L. var. obovata(Ledeb.) Regel et Tiling = Paris verticillata M. Bieb. ■

284396 Paris quadrifolia L. var. setchuanensis Franch. = Paris bashanensis F. T. Wang et Ts. Tang ■

284397 Paris rugosa H. Li et Kurita;皱叶重楼;Rugose Paris ■

284398 Paris setchuenensis(Franch.) Barkalov = Paris bashanensis F. T. Wang et Ts. Tang ■

284399 Paris taitungensis S. S. Ying = Paris polyphylla Sm. var. taitungensis(S. S. Ying) S. S. Ying ■

284400 Paris taitungensis S. S. Ying = Paris polyphylla Sm. ■

284401 Paris tetraphylla A. Gray;日本重楼(白功草,百节藕,长孙,海孙,旱莲,黄昏,黄孙,蔓延,牡蒙,四叶王孙);Japanese Paris ■☆

284402 Paris thibetica Franch. ;黑籽重楼(长药隔重楼,滇王孙,短梗重楼,金沙,九龙台,铁灯台,重楼); Shortpedicel Paris, Shortpedicle Paris,Tibet Paris,Xizang Paris ■

284403 Paris thibetica Franch. var. apetala Hand. -Mazz. ;无瓣重楼■

284404 Paris undulatis H. Li et V. G. Soukup;卷瓣重楼;Undulate Paris ■

284405 Paris vaniotii H. Lév. ;平伐重楼;Vaniot Paris ■

284406 Paris verticillata M. Bieb. ;北重楼(露水一颗珠,轮叶王孙,七叶一枝花,上天梯,王孙);Verticillate Paris ■

284407 Paris verticillata M. Bieb. f. purpurea (Miyabe et Tatew.) Honda = Paris verticillata M. Bieb. ■

284408 Paris verticillata M. Bieb. subsp. manshurica (Kom.) Kitag. = Paris verticillata M. Bieb. ■

284409 Paris verticillata M. Bieb. var. manshurica (Kom.) H. Hara = Paris verticillata M. Bieb. ■

284410 Paris verticillata M. Bieb. var. obovata (Ledeb.) H. Hara = Paris verticillata M. Bieb. ■

284411 Paris verticillata M. Bieb. var. setchuennensis (Franch.) Hand. -Mazz. = Paris verticillata M. Bieb. ■

284412 Paris vietnamensis (Takht.) H. Li;南重楼 (重楼); Vietnam Paris ■

284413 Paris violacea H. Lév. = Paris mairei H. Lév. ■

284414 Paris violacea H. Lév. = Paris polyphylla Sm. subsp. marmorata (Steam) H. Hara ■

284415 Paris wenxianensis Z. X. Peng et R. N. Zhao;文县重楼;Wenxian Paris ■

284416 Paris yunnanensis Franch. = Paris polyphylla Sm. var. yunnanensis(Franch.) Hand. -Mazz. ■

284417 Parisetta Augier = Paris L. ■

284418 Parishella A. Gray(1882);帕里桔梗属■☆

284419 Parishella californica A. Gray;帕里桔梗■☆

284420 Parishia Hook. f. (1860);帕里漆属●☆

284421 Parishia insignis Hook. f. ;帕里漆;Dhup,Red Dhup ●☆

284422 Parita Scop. = Pariti Adans. ●■

284423 Parita Scop. = Thespesia Sol. ex Corrêa(保留属名)●

284424 Parita populnea(L.)Scop. = Thespesia populnea(L.)Sol. ex Corrêa ●

284425 Pariti Adans. = Bupariti Duhamel(废弃属名)●

284426 Pariti Adans. = Hibiscus L. (保留属名)●■

284427 Pariti Adans. =Thespesia Sol. ex Corrêa(保留属名)●

284428 Pariti boninense(Nakai) Nakai = Hibiscus tiliaceus L. ●

284429 Pariti macrophyllum (Roxb. ex Hornem. = Hibiscus macrophyllus Roxb. ex Hornem. ●

284430 Pariti tiliaceum(L.) A. Juss. = Hibiscus tiliaceus L. ●

284431 Pariti tiliaceum (L.) A. Juss. var. heterophyllum (Nakai) Nakai = Hibiscus tiliaceus L. ●

284432 Paritium A. Juss. = Pariti Adans. ●■

284433 Paritium boninensis? = Hibiscus tiliaceus L. ●

284434 Paritium elatum Don = Hibiscus elatus Sw. ●

284435 Paritium gangeticum G. Don = Thespesia lampas (Cav.) Dalzell et A. Gibson ●

284436　Paritium glabrum Matsum. ex Hatt. = Hibiscus glaber（Matsum. ex Hatt.）Matsum. ex Nakai ●☆

284437　Paritium hamabo? = Hibiscus hamabo Siebold et Zucc. ●

284438　Paritium sterculiifolius Guillaumin et Perr. = Hibiscus sterculiifolius（Guillaumin et Perr.）Steud. ●☆

284439　Paritium tiliaceum（L.）A. St. -Hil. = Hibiscus tiliaceus L. ●

284440　Paritium tiliaceum A. St. -Hil. = Hibiscus glaber（Matsum. ex Hatt.）Matsum. ex Nakai ●☆

284441　Paritium tiliaceum Wight et Arn. = Hibiscus tiliaceus L. ●

284442　Paritium tiliifolium（Salisb.）Nakai = Hibiscus tiliaceus L. ●

284443　Parivoa Aubl. = Eperua Aubl. ●☆

284444　Parkia R. Br.（1826）；球花豆属（白球花属，爪哇合欢属）；Locust Bean，Nitta Tree，Nittatree，Nitta-tree ●

284445　Parkia africana R. Br.；非洲球花豆（非洲伯克林）；Africa Nittatree，African Nitta Tree ●

284446　Parkia africana R. Br. = Parkia biglobosa（Jacq.）R. Br. ex G. Don ●☆

284447　Parkia agboensis A. Chev. = Parkia bicolor A. Chev. ●☆

284448　Parkia bicolor A. Chev.；二色球花豆●☆

284449　Parkia bicolor A. Chev. var. agboensis（A. Chev.）Hagos? et de Wit = Parkia bicolor A. Chev. ●☆

284450　Parkia biglandulosa Wight et Arn.；双腺球花豆●☆

284451　Parkia biglobosa（Jacq.）R. Br. ex G. Don；双球球花豆●☆

284452　Parkia biglobosa（Roxb.）Benth. = Parkia timoriana（A. DC.）Merr. ●

284453　Parkia bussei Harms = Parkia filicoides Welw. ex Oliv. ●☆

284454　Parkia clappertoniana Keay = Parkia biglobosa（Jacq.）R. Br. ex G. Don ●☆

284455　Parkia filicoidea Welw. ex Oliv. = Parkia filicoides Welw. ex Oliv. ●☆

284456　Parkia filicoides Welw. ex Oliv.；蕨状球花豆（蕨叶派克木）；African Locust，African Locust Bean，Fern-leaf Nitta Tree ●☆

284457　Parkia filicoides Welw. ex Oliv. var. hildebrandtii（Harms）Chiov. = Parkia filicoides Welw. ex Oliv. ●☆

284458　Parkia fraterna Vatke = Xylia fraterna（Vatke）Baill. ●☆

284459　Parkia hildebrandtii Harms = Parkia filicoidea Welw. ex Oliv. ●☆

284460　Parkia hoffmannii Vatke = Xylia hoffmannii（Vatke）Drake ●☆

284461　Parkia javanica（Lam.）Merr. = Parkia timoriana（A. DC.）Merr. ●

284462　Parkia javanica（Lam.）Merr. et Anett = Parkia roxburghii G. Don ●

284463　Parkia klainei Pierre ex De Wild. = Parkia bicolor A. Chev. ●☆

284464　Parkia leiophylla Kurz；大叶球花豆（白球花，黄球花，爪哇派克豆）；Bigleaf Nittatree，Smooth-leaved Nitta-tree ●

284465　Parkia madagascariensis R. Vig.；马岛球花豆●☆

284466　Parkia multijuga Benth.；多列球花豆●☆

284467　Parkia oliveri J. F. Macbr. = Parkia biglobosa（Jacq.）R. Br. ex G. Don ●☆

284468　Parkia oppositifolia Spruce ex Benth.；对生叶球花豆●☆

284469　Parkia pendula Benth. ex Walp.；悬垂球花豆●☆

284470　Parkia roxburghii G. Don = Parkia timoriana（A. DC.）Merr. ●

284471　Parkia singularis Miq.；单生球花豆●☆

284472　Parkia speciosa Hassk.；美丽球花豆；Petai ●☆

284473　Parkia timoriana（DC.）Merr.；球花豆（白球花，大叶巴克豆，二球球花豆，派克木，双球豆，爪哇合欢，爪哇派克木，爪哇球花豆）；African Locust，Java Parkia，Kupong，Roxburgh Nitta-tree，Timor Nittatree，Two-ball Nitta Tree，Twoglobular Nittatree，West Indian Locust-bean Tree ●

284474　Parkia uniglobosa G. Don = Parkia biglobosa（Jacq.）R. Br. ex G. Don ●☆

284475　Parkia urticaria?；荨麻球花豆；African Nitta Tree ●☆

284476　Parkia zenkeri Harms = Parkia bicolor A. Chev. ●☆

284477　Parkinsonia L.（1753）；扁轴木属（巴荆木属，巴克豆属，扁叶轴木属，扁轴豆属）；Jerusalemthorn，Jerusalem Thorn，Palo Verde ●

284478　Parkinsonia Plum. ex L. = Parkinsonia L. ●

284479　Parkinsonia aculeata L.；扁轴木（巴金生豆，巴克豆，扁叶轴木）；Jerusalem Thorn，Jerusalemthorn，Mexican Palo Verde，Mexican Paloverde，Palo Verde，Retama ●

284480　Parkinsonia africana Sond.；非洲扁轴木●☆

284481　Parkinsonia florida S. Watson；多花扁轴木；Blue Palo Verde，Palo Verde ●☆

284482　Parkinsonia microphylla Torr.；小叶扁轴木；Foothill Palo Verde，Littleleaf Palo Verde ●☆

284483　Parkinsonia praecox（Ruiz et Pav.）Hawkins；散花扁轴木；Palo Brea ●☆

284484　Parkinsonia scioana（Chiov.）Brenan；赛欧扁轴木●☆

284485　Parlatorea Barb. Rodr. = Sanderella Kuntze ■☆

284486　Parlatoria Boiss.（1842）；帕拉托芥属■☆

284487　Parlatoria brachycarpa Boiss.；短果帕拉托芥■☆

284488　Parlatoria clavata Boiss.；棒状帕拉托芥■☆

284489　Parmena Greene = Rubus L. ●■

284490　Parmentiera DC.（1838）；蜡烛果属（蜡烛树属，桐花树属）；Parmentiera ●

284491　Parmentiera Raf. = Battata Hill ●■

284492　Parmentiera Raf. = Solanum L. ●■

284493　Parmentiera aculeata（Kunth）L. O. Williams；刺叶蜡烛果；Cat，Cow Okra，Cuachilote ●☆

284494　Parmentiera aculeata（Kunth）Seem. = Parmentiera aculeata（Kunth）L. O. Williams ●☆

284495　Parmentiera alata（Kunth）Miers = Crescentia alata Kunth ●

284496　Parmentiera alata Miers = Crescentia alata Kunth ●

284497　Parmentiera cereifera Seem.；蜡烛果（黑脚梗，黑榄，黑枝，红朔，蜡烛木，蜡烛树，浪柴，水菱，桐花树）；Candle Tree，Candle-tree，Ceriferous Parmentiera ●

284498　Parmentiera edulis DC.；黄瓜蜡烛果；Cuachilote，Grass-of-parnassus，Guajilote ●☆

284499　Parmentiera edulis DC. = Parmentiera aculeata（Kunth）Seem. ●☆

284500　Parnassia L.（1753）；梅花草属；Grass of Parnassus，Parnassia ■

284501　Parnassia affinis Hook. f. et Thomson = Parnassia pusilla Wall. ex Arn. ■

284502　Parnassia affinis Hook. f. et Thomson var. aucta（Diels）Nekr. = Parnassia mysorensis K. Heyne ex Wight et Arn. var. aucta Diels ■

284503　Parnassia alpicola Makino；姬梅花草■☆

284504　Parnassia americana Muhl. = Parnassia glauca Raf. ■☆

284505　Parnassia amoena Diels；南川梅花草；Nanchuan Parnassia ■

284506　Parnassia angustipetala T. C. Ku；窄瓣梅花草；Narrowpetal Parnassia ■

284507　Parnassia aphylla T. C. Ku = Parnassia scaposa Mattf. ■

284508　Parnassia appendiculata Batalin = Parnassia brevistyla（Brieger）Hand. -Mazz. ■

284509　Parnassia asarifolia Vent.；肾叶梅花草；Grass-of-parnassus，Kidney-leaf Grass-of-parnassus ■

284510　Parnassia bifolia Nekr.；双叶梅花草（二叶梅花草）■

284511　Parnassia brevistyla（Brieger）Hand. -Mazz.；短柱梅花草；Shorstyle Parnassia ■

284512　Parnassia cacuminum Hand. -Mazz.；高山梅花草；Alpine Parnassia ■

284513　Parnassia cacuminum Hand. -Mazz. f. yushuensis T. C. Ku；玉树梅花草；Yushu Parnassia ■

284514　Parnassia cacuminum Hand. -Mazz. f. yushuensis T. C. Ku = Parnassia cacuminum Hand. -Mazz. ■

284515　Parnassia caroliniana Michx.；卡罗林梅花草；Carolina Parnassia ■☆

284516　Parnassia chengkouensis T. C. Ku；城口梅花草；Chengkou Parnassia ■

284517　Parnassia chinensis Franch.；中国梅花草；China Parnassia, Chinese Parnassia ■

284518　Parnassia chinensis Franch. var. sechuanensis Z. P. Jien；四川梅花草；Sichuan Parnassia ■

284519　Parnassia cooperi W. E. Evans；指裂梅花草■

284520　Parnassia cordata(Drude) Z. P. Jien ex T. C. Ku；心叶梅花草；Heartleaf Parnassia ■

284521　Parnassia crassifolia Franch.；鸡心梅花草(鸡心草，水莲花)；Chiken Heart Grass, Thickleaf Parnassia ■

284522　Parnassia davidii Franch.；大卫梅花草(半边蝶)；David Parnassia ■

284523　Parnassia davidii Franch. var. arenicola Z. P. Jien；喜沙梅花草■

284524　Parnassia degeensis T. C. Ku；德格梅花草；Dege Parnassia ■

284525　Parnassia delavayi Franch.；突隔梅花草(白侧耳，苍耳七，肺小草，肺心草，金耳吊环，马蹄草，芒药苍耳七，梅花草，肾形草)；Delavay Parnassia ■

284526　Parnassia delavayi Franch. var. brevistyla Brieger = Parnassia brevistyla(Brieger) Hand. -Mazz. ■

284527　Parnassia delavayi Franch. var. brevistyla Brieger ex Limpr. = Parnassia brevistyla(Brieger) Hand. -Mazz. ■

284528　Parnassia deqenensis T. C. Ku；德钦梅花草；Deqin Parnassia ■

284529　Parnassia dilatata Hand. -Mazz.；宽叶梅花草；Broadleaf Parnassia ■

284530　Parnassia epunctulata J. T. Pan；无斑梅花草；Spotless Parnassia ■

284531　Parnassia esquirolii H. Lév.；龙场梅花草；Esquirol Parnassia ■

284532　Parnassia faberi Oliv.；峨眉梅花草(娥眉梅花草)；Faber Parnassia ■

284533　Parnassia faberi Oliv. f. abbreviata Engl.；短茎峨眉梅花草；Stemless Emei Parnassia ■

284534　Parnassia faberi Oliv. f. abbreviata Engl. = Parnassia faberi Oliv. ■

284535　Parnassia faberi Oliv. f. ramosa Engl. = Parnassia faberi Oliv. ■

284536　Parnassia farreri W. E. Evans；长爪梅花草(贡山梅花草)；Farrer Parnassia ■

284537　Parnassia filchneri Ulbr.；藏北梅花草；Filchner Parnassia, Xizang Parnassia ■

284538　Parnassia fimbriata Banks；落基山梅花草；Fringed Grass-of-parnassus, Rocky Mountain Parnassia ■☆

284539　Parnassia foliosa Hook. f. et Thomson；白耳菜(白侧耳，白须菜，白须草，苍耳七，光板，海里茶，叫天鸡，金钱灯塔草)；Foliosous Parnassia ■

284540　Parnassia foliosa Hook. f. et Thomson subsp. japonica(Nakai) Kitam. et Murata = Parnassia foliosa Hook. f. et Thomson var. japonica(Nakai) Ohwi ■☆

284541　Parnassia foliosa Hook. f. et Thomson subsp. nummularia(Maxim.) Kitam. et Murata = Parnassia foliosa Hook. f. et Thomson ■

284542　Parnassia foliosa Hook. f. et Thomson var. japonica(Nakai) Ohwi；日本白耳菜■☆

284543　Parnassia foliosa Hook. f. et Thomson var. nummularia(Maxim.) T. Ito = Parnassia foliosa Hook. f. et Thomson ■

284544　Parnassia gansuensis T. C. Ku；甘肃梅花草；Gansu Parnassia ■

284545　Parnassia glauca Raf.；美洲梅花草；American Grass-of-parnassus, Fen Grass-of-parnassus, Grass-of-parnassus ■☆

284546　Parnassia grandifolia DC.；美洲大叶梅花草；Grass-of-parnassus ■☆

284547　Parnassia guilinensis T. C. Ku；桂林梅花草；Guilin Parnassia ■

284548　Parnassia humilis T. C. Ku；矮小梅花草；Dwarf Parnassia ■

284549　Parnassia kangdingensis T. C. Ku；康定梅花草；Kangding Parnassia ■

284550　Parnassia kotzebuei Cham. et Schltdl.；考氏梅花草■☆

284551　Parnassia labiata Z. P. Jien；宝兴梅花草；Baoxing Parnassia ■

284552　Parnassia lanceolata T. C. Ku；披针瓣梅花草；Lanceolate Parnassia ■

284553　Parnassia lanceolata T. C. Ku var. oblongipetala T. C. Ku；长圆瓣梅花草；Oblongpetal Parnassia ■

284554　Parnassia laxmanni Pall. var. viridiflora(Batalin) Diels = Parnassia viridiflora Batalin ■

284555　Parnassia laxmannii Pall. = Parnassia laxmannii Pall. ex Schult. ■

284556　Parnassia laxmannii Pall. ex Schult.；新疆梅花草(单叶梅花草)；Xinjiang Parnassia ■

284557　Parnassia laxmannii Pall. var. viridiflora(Batalin) Diels = Parnassia viridiflora Batalin ■

284558　Parnassia leptophylla Hand. -Mazz.；细裂梅花草；Thinleaf Parnassia ■

284559　Parnassia lijiangensis T. C. Ku；丽江梅花草；Lijiang Parnassia ■

284560　Parnassia lijiangensis T. C. Ku = Parnassia mysorensis K. Heyne ex Wight et Arn. ■

284561　Parnassia longipetala Hand. -Mazz.；长瓣梅花草；Longpetal Parnassia ■

284562　Parnassia longipetala Hand. -Mazz. var. alba H. Chuang；白花长瓣梅花草；Whiteflower Longpetal Parnassia ■

284563　Parnassia longipetala Hand. -Mazz. var. brevipetala Z. P. Jien ex T. C. Ku；短瓣梅花草；Shortpetal Parnassia ■

284564　Parnassia longipetala Hand. -Mazz. var. striata H. Chuang；斑纹长瓣梅花草；Striate Longpetal Parnassia ■

284565　Parnassia longipetaloides J. T. Pan；似长瓣梅花草；Longpetal-like Parnassia ■

284566　Parnassia longipetaloides J. T. Pan = Parnassia yulongshanensis T. C. Ku ■

284567　Parnassia longshengensis T. C. Ku；龙胜梅花草；Longsheng Parnassia ■

284568　Parnassia longshengensis T. C. Ku = Parnassia dilatata Hand. -Mazz. ■

284569　Parnassia lutea Batalin；黄花梅花草(黄瓣梅花草)；Yellowpetal Parnassia ■

284570　Parnassia mairei H. Lév. = Parnassia delavayi Franch. ■

284571　Parnassia monochorifolia Franch.；大叶梅花草；Bigleaf Parnassia ■

284572　Parnassia mucronata Siebold et Zucc. = Parnassia palustris L. ■

284573　Parnassia multiseta(Ledeb.) Fernald = Parnassia palustris L. var. multiseta Ledeb. ■

284574　Parnassia multiseta(Ledeb.) Fernald = Parnassia palustris L. var. tenuis Wahlenb. ■☆

284575　Parnassia mysorensis K. Heyne ex Wight et Arn.；凹瓣梅花草(小苍耳七)；Concavepetal Parnassia ■

284576　Parnassia mysorensis K. Heyne ex Wight et Arn. = Parnassia chinensis Franch. ■

284577　Parnassia mysorensis K. Heyne ex Wight et Arn. var. aucta Diels；锐尖凹瓣梅花草；Sharp Concavepetal Parnassia ■

284578　Parnassia nana Griff. = Parnassia delavayi Franch. ■

284579　Parnassia noemiae Franch.；棒状梅花草；Clavate Parnassia ■

284580　Parnassia nubicola Wall. = Parnassia nubicola Wall. ex Royle ■

284581　Parnassia nubicola Wall. ex Royle；云梅花草；Pertaining-cloudy Parnassia ■

284582　Parnassia nubicola Wall. ex Royle subsp. occidentalis Schönb. -Tem.；西方梅花草■☆

284583　Parnassia nubicola Wall. ex Royle var. cordata Drude = Parnassia cordata(Drude)Z. P. Jien ex T. C. Ku ■

284584　Parnassia nubicola Wall. ex Royle var. nana T. C. Ku；矮云梅花草；Dwarf Parnassia ■

284585　Parnassia nummularia Maxim. = Parnassia foliosa Hook. f. et Thomson ■

284586　Parnassia obovata Hand. -Mazz.；倒卵叶梅花草；Obovate Parnassia ■

284587　Parnassia omeiensis T. C. Ku；金顶梅花草；Emei Parnassia ■

284588　Parnassia oreophila Hance；细叉梅花草（山地梅花草，四川苍耳七，铁棍子）；Mountain-loving Parnassia ■

284589　Parnassia ornata Wall. = Parnassia wightiana Wall. ex Wight et Arn. ■

284590　Parnassia ornata Wall. ex Arn. = Parnassia wightiana Wall. ex Wight et Arn. ■

284591　Parnassia palustris L.；梅花草（苍耳七）；Autumn's Goodbye, Farewell-summer, Grass of Parnassus, Grass-of-parnassus, Marsh Grass-of-parnassus, Summer's Goodbye, Swamp Grass-of-parnassus, White Buttercup, White Liverwort, Wideword Parnassia, Wide-world Parnassia ■

284592　Parnassia palustris L. f. nana T. C. Ku = Parnassia palustris L. ■

284593　Parnassia palustris L. f. rhodanthera H. Ohba et Umezu；红药梅花草■☆

284594　Parnassia palustris L. subsp. neogaea (Fernald) Hultén = Parnassia palustris L. var. tenuis Wahlenb. ■☆

284595　Parnassia palustris L. var. izuinsularis H. Ohba；伊豆岛梅花草■☆

284596　Parnassia palustris L. var. multiseta Ledeb.；多枝梅花草；Manybranch Wideword Parnassia ■

284597　Parnassia palustris L. var. multiseta Ledeb. = Parnassia palustris L. ■

284598　Parnassia palustris L. var. multiseta Ledeb. f. minima Masam. = Anaphalis sinica Hance var. yakusimensis(Masam.)Yahara ■☆

284599　Parnassia palustris L. var. neogaea Fernald = Parnassia palustris L. var. tenuis Wahlenb. ■☆

284600　Parnassia palustris L. var. parviflora(DC.)B. Boivin = Parnassia parviflora DC. ■☆

284601　Parnassia palustris L. var. tenuis Wahlenb.；细梅花草；Marsh Grass-of-parnassus, Swamp Grass-of-parnassus ■☆

284602　Parnassia palustris L. var. yakusimensis (Masam.) H. Ohba；屋久岛梅花草■☆

284603　Parnassia parviflora DC.；小花梅花草；Small-flowered Grass-of-parnassus, Small-flowered Parnassia ■☆

284604　Parnassia paxmanii Pall.；帕氏梅花草；Paxman Parnassia ■☆

284605　Parnassia perciliata Diels；厚叶梅花草（水年七）；Thickleaf Parnassia ■

284606　Parnassia petimenginii H. Lév.；贵阳梅花草；Guiyang Parnassia ■

284607　Parnassia pusilla Wall. ex Arn.；类三脉梅花草（小梅花草）；Small Parnassia ■

284608　Parnassia pusilla Wall. ex Arn. = Parnassia chinensis Franch. ■

284609　Parnassia qinghaiensis J. T. Pan；青海梅花草；Qinghai Parnassia ■

284610　Parnassia rhombipetala B. L. Chai；叙永梅花草；Rhombipetal Parnassia ■

284611　Parnassia rumicifolia Brieger = Parnassia viridiflora Batalin ■

284612　Parnassia rumicifolia Brieger ex Limpr. = Parnassia viridiflora Batalin ■

284613　Parnassia scaposa Mattf.；白花梅花草；Whiteflower Parnassia ■

284614　Parnassia schmidtii Zenker = Parnassia delavayi Franch. ■

284615　Parnassia setchuenensis Franch. = Parnassia oreophila Hance ■

284616　Parnassia simaoensis Y. Y. Qian；思茅梅花草；Simao Parnassia ■

284617　Parnassia simaoensis Y. Y. Qian = Parnassia wightiana Wall. ex Wight et Arn. ■

284618　Parnassia souliei Franch. ex Nekr. = Parnassia brevistyla (Brieger) Hand. -Mazz. ■

284619　Parnassia subacaulis Kar. et Kir. = Parnassia laxmannii Pall. ex Schult. ■

284620　Parnassia submysorensis J. T. Pan；近凹瓣梅花草；Subconcavepetal Parnassia ■

284621　Parnassia subscaposa C. Y. Wu ex T. C. Ku；倒卵瓣梅花草；Subscapes Parnassia ■

284622　Parnassia tenella Hook. f. et Thomson；青铜钱；Slender Parnassia, Tender Parnassia ■

284623　Parnassia tibetana Z. P. Jien ex T. C. Ku；西藏梅花草；Tibet Parnassia, Xizang Parnassia ■

284624　Parnassia trinervis Drude；三脉梅花草；Threenerve Parnassia ■

284625　Parnassia trinervis Drude var. viridiflora (Batalin) Hand. -Mazz. = Parnassia viridiflora Batalin ■

284626　Parnassia venusta Z. P. Jien；娇媚梅花草；Handsome Parnassia ■

284627　Parnassia viridiflora Batalin；绿花梅花草（绿花苍耳七，绿梅花草，绿叶梅花草）；Greenflower Parnassia ■

284628　Parnassia wightiana Wall. ex Wight et Arn.；鸡肫梅花草（白侧耳，白折耳，苍耳七，地核桃，肥猪草，黄梅花草，鸡眼草，鸡眼梅花草，鸡肫草，金钱七，荞麦叶，水侧耳，铜钱草）；Chiken Eye Parnassia, Wight Parnassia ■

284629　Parnassia wightiana Wall. ex Wight et Arn. var. brachyloba Franch. = Parnassia delavayi Franch. ■

284630　Parnassia wightiana Wall. ex Wight et Arn. var. flavida Franch. = Parnassia delavayi Franch. ■

284631　Parnassia wightiana Wall. ex Wight et Arn. var. microblephara Franch. = Parnassia delavayi Franch. ■

284632　Parnassia wightiana Wall. ex Wight et Arn. var. ornata(Wall. ex Arn.)Drude = Parnassia wightiana Wall. ex Wight et Arn. ■

284633　Parnassia wightiana Wall. ex Wight et Arn. var. ornata Dmde = Parnassia wightiana Wall. ex Wight et Arn. ■

284634　Parnassia xinganensis C. Z. Gao et G. Z. Li；兴安梅花草；Xing'an Parnassia ■

284635　Parnassia yakusimensis Masam. = Parnassia palustris L. var. yakusimensis(Masam.)H. Ohba ■☆

284636　Parnassia yanyuanensis T. C. Ku；盐源梅花草；Yanyuan Parnassia ■

284637　Parnassia yiliangensis T. C. Ku；彝良梅花草；Yiliang Parnassia ■

284638　Parnassia yiliangensis T. C. Ku = Parnassia monochorifolia Franch. ■

284639　Parnassia yui Z. P. Jien；俞氏梅花草；Yu Parnassia ■

284640　Parnassia yulongshanensis T. C. Ku；玉龙山梅花草；Yulongshan Parnassia ■

284641 Parnassia yunnanensis Franch.;云南梅花草;Yunnan Parnassia ■

284642 Parnassia yunnanensis Franch. var. angustipetala Z. P. Jien = Parnassia angustipetala T. C. Ku ■

284643 Parnassia yunnanensis Franch. var. longistipitata Z. P. Jien;长柄云南梅花草;Longstalk Yunnan Parnassia ■

284644 Parnassiaceae Gray = Parnassiaceae Martinov(保留科名)■

284645 Parnassiaceae Martinov(1820)(保留科名);梅花草科■

284646 Parniena Greene = Rubus L. ●■

284647 Parochetus Buch. -Ham. ex D. Don(1825);紫雀花属(金雀花属,蓝雀花属);Shamrock Pea,Shamrockpea,Shamrock-pea ■

284648 Parochetus africanus Polhill = Parochetus communis Buch. -Ham. ex D. Don subsp. africanus(Polhill)Chaudhary et Sanjappa ■☆

284649 Parochetus communis Buch. -Ham. = Parochetus communis Buch. -Ham. ex D. Don ■

284650 Parochetus communis Buch. -Ham. ex D. Don;紫雀花(金雀花,蓝雀花,三瓣草,散白草,生血草,一点血,一颗血);Blue Oxalis,Blue Shamrock,Common Shamrockpea,Shamrock Pea,Shamrockpea,Shamrock-pea ■

284651 Parochetus communis Buch. -Ham. ex D. Don subsp. africanus(Polhill)Chaudhary et Sanjappa;非洲紫雀花■☆

284652 Parochetus communis Buch. -Ham. ex D. Don var. grossecrenatus Cufod. = Parochetus communis Buch. -Ham. ex D. Don subsp. africanus(Polhill)Chaudhary et Sanjappa ■☆

284653 Parochetus major Don = Parochetus communis Buch. -Ham. ex D. Don ■

284654 Parochetus oxalidifolia Royle = Parochetus communis Buch. -Ham. ex D. Don ■

284655 Parodia Speg. (1923)(保留属名);锦绣玉属(宝玉属);Parodia ●

284656 Parodia aureicentra Backeb.;逆鉾球(逆鉾丸)■☆

284657 Parodia aureispina Backeb.;锦绣玉;Goldenspine Parodia ■

284658 Parodia ayopayana Cárdenas;玻利维亚锦绣玉■☆

284659 Parodia brasiliensis Speg.;梦绣玉;Brasil Parodia ■☆

284660 Parodia cardenasii F. Ritter;茶绣玉■☆

284661 Parodia carminata Backeb.;凤绣玉■☆

284662 Parodia catamarensis Backeb.;罗绣玉;Catamar Parodia ■☆

284663 Parodia chrysacanthion(K. Schum.)Backeb.;锦翁玉(锦绣玉);Golden Powder Puff ■☆

284664 Parodia commutans F. Ritter;群神球(群神丸)■☆

284665 Parodia concinna(Monv.)N. P. Taylor;美装玉(河内球,河内丸);Neat Parodia, Pretty Ball Cactus, Pretty Ballcactus, Pretty Parodia ■☆

284666 Parodia erinacea(Haw.)N. P. Taylor;地久球(针鼠玉)■☆

284667 Parodia erythrantha(Speg.)Y. Ito;彩绣玉;Redflower Parodia ■☆

284668 Parodia faustiana Backeb.;黄粧玉;Fast Parodia ■☆

284669 Parodia formosa F. Ritter;丽茶绣玉■☆

284670 Parodia graessneri (K. Schum. ex Rümpler) F. H. Brandt = Brasilicactus graessneri(K. Schum. ex Rümpler)Backeb. ex Schallert ■

284671 Parodia graessneri (K. Schum.) F. H. Brandt = Brasilicactus graessneri(K. Schum. ex Rümpler)Backeb. ex Schallert ■

284672 Parodia haselbergii (Haage ex Rümpler) F. H. Brandt = Brasilicactus haselbergii(Rümpler)Backeb. ex Schaffnit ■

284673 Parodia leninghausii(Haage)F. H. Brandt;软毛仙人球(黄翁,金晃,金晃球,金晃丸);Golden Ball Cactus ■☆

284674 Parodia maassii (Heese) Berger et F. M. Knuth;魔神球(魔神丸);Maass Parodia ■

284675 Parodia maassii(Heese) Berger et F. M. Knuth var. albescens F. Ritter;白刺魔神球(白刺魔神丸);White Maass Parodia ■

284676 Parodia maassii A. Berger var. nigrispina Y. Ito;黑神球(黑神丸)■☆

284677 Parodia magnifica(F. Ritter)F. H. Brandt;英冠玉■☆

284678 Parodia mammulosa(Lem.)N. P. Taylor;铭月(鬼云球,小粒玉);Tom Thumb Cactus ■☆

284679 Parodia maxima F. Ritter;武神球(武神丸)■☆

284680 Parodia microsperma Speg.;宝玉■

284681 Parodia microthele Backeb.;晓粧玉■☆

284682 Parodia multicostata F. Ritter et Jelin.;万绣玉■☆

284683 Parodia mutabilis Backeb.;丽绣玉■☆

284684 Parodia nivosa(Frič)Backeb.;银粧玉(银玉)■☆

284685 Parodia ocampoi Cardenas;圣天球(圣天丸)■☆

284686 Parodia ottonis (Lehm.) N. P. Taylor = Notocactus ottonis (Lehm.)A. Berger ■

284687 Parodia paraguaensis Speg.;黄绣玉■☆

284688 Parodia penicillata Fechser et Steeg;毛发锦绣玉■☆

284689 Parodia procera F. Ritter;高耸锦绣玉■☆

284690 Parodia rubriflora Backeb.;华粧玉■☆

284691 Parodia rutilans(Däniker et Krainz)N. P. Taylor;桃鬼球■☆

284692 Parodia saint-pieana Backeb.;多花锦绣玉■☆

284693 Parodia sanguiniflora Backeb.;绯绣玉;Red Flower Parodia ■

284694 Parodia sanguiniflora Backeb. = Parodia microsperma Speg. ■

284695 Parodia schuetziana Jajó;雪绣玉■☆

284696 Parodia schwebsiana(Werderm.)Backeb.;红绣玉■☆

284697 Parodia scopa(Spreng.)N. P. Taylor = Cactus scopa Spreng. ■☆

284698 Parodia scopa (Spreng.) N. P. Taylor = Notocactus scopa (Spreng.)Backeb. ■

284699 Parodia scopaoides Backeb.;绮绣玉■☆

284700 Parodia setifera Backeb.;明绣玉■☆

284701 Parodia stuemeri(Werderm.)Backeb.;橙绣玉■☆

284702 Parodia subterranea F. Ritter;黑云龙■☆

284703 Parodia taratensis Cárdenas;丽神球玉■☆

284704 Parodia tuberculata Cárdenas;小瘤锦绣玉■☆

284705 Parodia uebelmanniana F. Ritter;乌伯锦绣玉■☆

284706 Parodia uhligiana Backeb.;胧绣玉■☆

284707 Parodianthus Tronc. (1941);帕罗迪草属●☆

284708 Parodianthus ilicifolius(Moldenke)Tronc.;帕罗迪草●☆

284709 Parodiella Reeder et C. Reeder = Lorenzochloa Reeder et C. Reeder ■☆

284710 Parodiella Reeder et C. Reeder = Ortachne Nees ex Steud. ■☆

284711 Parodiochloa A. M. Molina = Koeleria Pers. ■

284712 Parodiochloa A. M. Molina = Raimundochloa A. M. Molina ■

284713 Parodiochloa C. E. Hubb. = Poa L. ■

284714 Parodiodendron Hunz. (1969);帕罗迪大戟属●☆

284715 Parodiodendron marginivillosum(Speg.)Hunz.;帕罗迪大戟●☆

284716 Parodiodoxa O. E. Schulz(1929);雪芥属■☆

284717 Parodiodoxa chionophila O. E. Schulz;雪芥■☆

284718 Parodiolyra Soderstr. et Zuloaga(1989);类栽利箣属■☆

284719 Parodiolyra colombiensis Davidse et Zuloaga;哥伦比亚类栽利箣■☆

284720 Parodiolyra micrantha(Kunth)Davidse et Zuloaga;小花类栽利箣■☆

284721 Parodiophyllochloa Zuloaga et Morrone(2008);帕罗迪禾属■☆

284722 Parolinia Engl. = Pachylepis Brongn. ●☆

284723 Parolinia Engl. = Widdringtonia Endl. ●☆

284724 Parolinia Webb(1840);加那利芥属■☆

284725 Parolinia filifolia G. Kunkel;线叶加那利芥■☆

284726　Parolinia glabriuscula Montelongo et Bramwell;光加那利芥■☆

284727　Parolinia intermedia Svent. et Bramwell;间型加那利芥■☆

284728　Parolinia ornata Webb;加那利芥■☆

284729　Parolinia platypetala G. Kunkel;宽翅加那利芥■☆

284730　Parolinia schizogynoides Svent. ;裂柱加那利芥■☆

284731　Paronichiaceae Juss. ;指甲草科■

284732　Paronichiaceae Juss. = Caryophyllaceae Juss. (保留科名)■●

284733　Paronychia Hill = Erophila DC. (保留属名)■☆

284734　Paronychia L. = Paronychia Mill.

284735　Paronychia Mill. （1754）;指甲草属（甲疽草属）; Nailwort, Whitlowwort,Whitlow-wort■

284736　Paronychia ahartii Ertter. ;阿哈特指甲草;Ahart's Nailwort ■☆

284737　Paronychia americana（Nutt. ）Fenzl ex Walp. ;美洲指甲草; American Nailwort ■☆

284738　Paronychia americana subsp. pauciflora（Small）Chaudhri = Paronychia americana（Nutt. ）Fenzl ex Walp. ■☆

284739　Paronychia arabica（L. ）DC. ;阿拉伯指甲草■☆

284740　Paronychia arabica（L. ）DC. subsp. annua（Delile）Maire et Weiller = Paronychia arabica（L. ）DC. subsp. longiseta Batt. ■☆

284741　Paronychia arabica（L. ）DC. subsp. aurasiaca（Coss. ）Batt. ;欧亚指甲草■☆

284742　Paronychia arabica（L. ）DC. subsp. breviseta（Asch. et Schweinf. ）Chaudhri;短毛阿拉伯指甲草■☆

284743　Paronychia arabica（L. ）DC. subsp. cossoniana（Batt. ）Batt. ;科森指甲草■☆

284744　Paronychia arabica（L. ）DC. subsp. desertorum（Boiss. ）Batt. = Paronychia arabica（L. ）DC. ■☆

284745　Paronychia arabica（L. ）DC. subsp. lenticulata Maire et Weiller = Paronychia arabica（L. ）DC. ■☆

284746　Paronychia arabica（L. ）DC. subsp. longiseta Batt. ;长毛阿拉伯指甲草■☆

284747　Paronychia arabica（L. ）DC. subsp. tibestica Quézel;提贝斯提指甲草■☆

284748　Paronychia arabica（L. ）DC. subsp. velata（Maire）Maire et Weiller = Paronychia velata（Maire）Chaudhri ■☆

284749　Paronychia arabica（L. ）DC. var. breviseta Asch. et Schweinf. = Paronychia arabica（L. ）DC. subsp. breviseta（Asch. et Schweinf. ）Chaudhri ■☆

284750　Paronychia arabica（L. ）DC. var. cossoniana Batt. = Paronychia arabica（L. ）DC. subsp. cossoniana（Batt. ）Batt. ■☆

284751　Paronychia arabica（L. ）DC. var. desertorum（Boiss. ）Durand et Barratte = Paronychia arabica（L. ）DC. ■☆

284752　Paronychia arabica（L. ）DC. var. elongata Chaudhri = Paronychia arabica（L. ）DC. ■☆

284753　Paronychia arabica（L. ）DC. var. fezzanica Chaudhri = Paronychia arabica（L. ）DC. ■☆

284754　Paronychia arabica（L. ）DC. var. inarmata Chaudhri = Paronychia arabica（L. ）DC. ■☆

284755　Paronychia arabica（L. ）DC. var. longiaristata Chaudhri = Paronychia arabica（L. ）DC. ■☆

284756　Paronychia arabica（L. ）DC. var. longifolia Borzí et Mattei = Paronychia arabica（L. ）DC. ■☆

284757　Paronychia arabica（L. ）DC. var. longiseta（Bertol. ）Batt. = Paronychia arabica（L. ）DC. subsp. longiseta Batt. ■☆

284758　Paronychia arabica（L. ）DC. var. macranthera Maire = Paronychia arabica（L. ）DC. ■☆

284759　Paronychia arabica（L. ）DC. var. marmarica Maire et Weiller =

284760　Paronychia arabica（L. ）DC. var. subvelata（Litard. et Maire）Maire et Weiller = Paronychia arabica（L. ）DC. ■☆

284761　Paronychia arabica（L. ）DC. var. tibestica Quézel;西藏指甲草■☆

284762　Paronychia arabica（L. ）DC. var. tripolitana E. A. Durand et Barratte = Paronychia arabica（L. ）DC. ■☆

284763　Paronychia aretioides DC. ;点地梅指甲草■☆

284764　Paronychia argentea Lam. ;银白指甲草■☆

284765　Paronychia argentea Lam. var. angustifolia Chaudhri = Paronychia argentea Lam. ■☆

284766　Paronychia argentea Lam. var. mauritanica（Willd. ）DC. = Paronychia argentea Lam. ■☆

284767　Paronychia argentea Lam. var. rotundata（DC. ）Chaudhri = Paronychia argentea Lam. ■☆

284768　Paronychia argentea Lam. var. scabra Sauvage = Paronychia argentea Lam. ■☆

284769　Paronychia argentea Lam. var. scariosissima Post = Paronychia argentea Lam. ■☆

284770　Paronychia argentea Lam. var. subvelata Litard. et Maire = Paronychia argentea Lam. ■☆

284771　Paronychia argentea Lam. var. suffruticosa Maire et Wilczek = Paronychia argentea Lam. ■☆

284772　Paronychia argentea Lam. var. velata Maire = Paronychia velata（Maire）Chaudhri ■☆

284773　Paronychia argentea Lam. var. velutina Ball = Paronychia argentea Lam. ■☆

284774　Paronychia argyrocoma（Michx. ）Nutt. ;北美银指甲草; Silverling,Silvery Nailwort ■☆

284775　Paronychia argyrocoma（Michx. ）Nutt. subsp. albimontana（Fernald）Maguire = Paronychia argentea Lam. ■☆

284776　Paronychia argyrocoma（Michx. ）Nutt. var. albimontana Fernald = Paronychia argentea Lam. ■☆

284777　Paronychia aurasiaca Coss. = Paronychia arabica（L. ）DC. subsp. aurasiaca（Coss. ）Batt. ■☆

284778　Paronychia baldwinii（Torr. et A. Gray）Fenzl ex Walp. ;鲍尔温指甲草;Baldwin's Nailwort ■☆

284779　Paronychia baldwinii（Torr. et A. Gray）Fenzl ex Walp. subsp. riparia（Chapm. ）Chaudhri;河岸指甲草■☆

284780　Paronychia baldwinii（Torr. et A. Gray）Fenzl ex Walp. var. ciliata Chaudhri = Paronychia baldwinii（Torr. et A. Gray）Fenzl ex Walp. ■☆

284781　Paronychia brasiliana DC. var. pubescens Chaudhri;短柔毛指甲草■☆

284782　Paronychia canadensis（L. ）A. W. Wood;加拿大指甲草; Forked Chickweed, Forked-chickweed, Smooth Forked Nail-wort, Smooth-forked Nailwort,Tall Forked Chickweed ■☆

284783　Paronychia canadensis（L. ）A. W. Wood var. pumila A. W. Wood = Paronychia fastigiata（Raf. ）Fernald var. pumila（A. W. Wood）Fernald ■☆

284784　Paronychia canariensis（L. f. ）Juss. ;加那利指甲草■☆

284785　Paronychia canariensis（L. f. ）Juss. var. congesta Bornm. = Paronychia canariensis（L. f. ）Juss. ■☆

284786　Paronychia canariensis（L. f. ）Juss. var. expansa Pit. = Paronychia canariensis（L. f. ）Juss. ■☆

284787　Paronychia canariensis（L. f. ）Juss. var. orthoclada Christ = Paronychia canariensis（L. f. ）Juss. ■☆

284788　Paronychia capitata（L. ）Lam. ;头状指甲草■☆

284789　Paronychia capitata(L.) Lam. subsp. atlantica (Ball) Chaudhri；大西洋指甲草■☆

284790　Paronychia capitata (L.) Lam. subsp. canariensis (Chaudhri) Sunding；加那利头状指甲草■☆

284791　Paronychia capitata(L.) Lam. subsp. chlorothyrsa(Murb.) Maire et Weiller = Paronychia chlorothyrsa Murb. ■☆

284792　Paronychia capitata (L.) Lam. subsp. nivea (DC.) Maire et Weiller = Paronychia capitata(L.) Lam. ■☆

284793　Paronychia capitata(L.) Lam. subsp. rifea Sennen et Mauricio = Paronychia chlorothyrsa Murb. ■☆

284794　Paronychia capitata(L.) Lam. var. acuminata Batt. = Paronychia capitata(L.) Lam. ■☆

284795　Paronychia capitata (L.) Lam. var. antiatlantica Maire = Paronychia capitata(L.) Lam. ■☆

284796　Paronychia capitata (L.) Lam. var. atlantica Ball = Paronychia capitata(L.) Lam. subsp. atlantica(Ball) Chaudhri ■☆

284797　Paronychia capitata (L.) Lam. var. bracteosa Batt. = Paronychia capitata(L.) Lam. ■☆

284798　Paronychia capitata (L.) Lam. var. chlorothyrsa Batt. = Paronychia capitata(L.) Lam. ■☆

284799　Paronychia capitata (L.) Lam. var. dichotoma Batt. = Paronychia capitata(L.) Lam. ■☆

284800　Paronychia capitata(L.) Lam. var. haggariensis (Diels) Maire = Paronychia haggariensis Diels ■☆

284801　Paronychia capitata (L.) Lam. var. libyca Borzí et Mattei = Paronychia capitata(L.) Lam. ■☆

284802　Paronychia capitata (L.) Lam. var. longistyla Emb. et Maire = Paronychia capitata(L.) Lam. ■☆

284803　Paronychia capitata (L.) Lam. var. obtusata Batt. = Paronychia capitata(L.) Lam. ■☆

284804　Paronychia capitata (L.) Lam. var. querioides (Ball) Maire et Weiller = Paronychia capitata(L.) Lam. ■☆

284805　Paronychia capitata (L.) Lam. var. rifea Sennen = Paronychia capitata(L.) Lam. ■☆

284806　Paronychia capitata (L.) Lam. var. tarhunensis (Pamp.) Chaudhri = Paronychia capitata(L.) Lam. ■☆

284807　Paronychia capitata Rchb. = Paronychia capitata(L.) Lam. ■☆

284808　Paronychia cephalotes(M. Bieb.) Besser；拟头状指甲草■☆

284809　Paronychia chartacea Fernald；纸指甲草；Paper Nailwort，Papery Whitlow-wort ■☆

284810　Paronychia chartacea Fernald subsp. minima L. C. Anderson = Paronychia chartacea Fernald var. minima (L. C. Anderson) R. L. Hartm. ■☆

284811　Paronychia chartacea Fernald var. minima(L. C. Anderson) R. L. Hartm. ；小纸指甲草■☆

284812　Paronychia chilensis DC. ；智利指甲草■☆

284813　Paronychia chlorothyrsa Murb. ；绿穗指甲草■☆

284814　Paronychia chlorothyrsa Murb. subsp. canariensis Chaudhri = Paronychia capitata(L.) Lam. subsp. canariensis(Chaudhri) Sunding ■☆

284815　Paronychia chlorothyrsa Murb. var. antiatlantica Maire = Paronychia chlorothyrsa Murb. ■☆

284816　Paronychia chlorothyrsa Murb. var. bracteosa Batt. ；多苞片指甲草■☆

284817　Paronychia chlorothyrsa Murb. var. coarctata Chaudhri；密集指甲草■☆

284818　Paronychia chlorothyrsa Murb. var. dichotoma Batt. = Paronychia chlorothyrsa Murb. ■☆

284819　Paronychia chlorothyrsa Murb. var. erythraea (Fiori) Chaudhri；浅红指甲草■☆

284820　Paronychia chlorothyrsa Murb. var. haggariensis(Diels) Maire = Paronychia haggariensis Diels ■☆

284821　Paronychia chlorothyrsa Murb. var. laxa Beauverd = Paronychia chlorothyrsa Murb. ■☆

284822　Paronychia chlorothyrsa Murb. var. pauciflora Sauvage = Paronychia chlorothyrsa Murb. ■☆

284823　Paronychia chlorothyrsa Murb. var. querioides (Ball) Batt. = Paronychia chlorothyrsa Murb. ■☆

284824　Paronychia chlorothyrsa Murb. var. tarhunensis Pamp. = Paronychia capitata(L.) Lam. ■☆

284825　Paronychia chorizanthoides Small = Paronychia lindheimeri Engelm. ex A. Gray ■☆

284826　Paronychia congesta Correll；格兰德指甲草；Rio Grande Nailwort ■☆

284827　Paronychia cossoniana(Batt.) Batt. = Paronychia arabica (L.) DC. subsp. cossoniana(Batt.) Batt. ■☆

284828　Paronychia cymosa(L.)DC. = Chaetonychia cymosa(L.)Sweet ■☆

284829　Paronychia decandra (Forssk.) Rohweder et Urmi-König = Gymnocarpos decandrus Forssk. ●☆

284830　Paronychia depressa(Torr. et A. Gray)Nutt. ex A. Nelson；凹指甲草；Spreading Nailwort ■☆

284831　Paronychia depressa(Torr. et A. Gray)Nutt. ex A. Nelson var. brevicuspis(A. Nelson)Chaudhri = Paronychia depressa(Torr. et A. Gray)Nutt. ex A. Nelson ■☆

284832　Paronychia depressa(Torr. et A. Gray)Nutt. ex A. Nelson var. diffusa(A. Nelson)Chaudhri = Paronychia depressa (Torr. et A. Gray)Nutt. ex A. Nelson ■☆

284833　Paronychia desertorum Boiss. = Paronychia arabica(L.)DC. ■☆

284834　Paronychia dichotoma (Michx.) A. Nelson = Paronychia canadensis(L.) A. W. Wood ■☆

284835　Paronychia diffusa A. Nelson；铺散指甲草■☆

284836　Paronychia drummondii Torr. et A. Gray；德拉蒙德指甲草；Drummond's Nailwort ■☆

284837　Paronychia drummondii Torr. et A. Gray subsp. parviflora Chaudhri = Paronychia drummondii Torr. et A. Gray ■☆

284838　Paronychia echinata (Poir.) Lam. = Paronychia echinulata Chater ■☆

284839　Paronychia echinata Lam. var. minutiflora H. Lindb. = Paronychia echinulata Chater ■☆

284840　Paronychia echinulata Chater；欧亚刺指甲草；Eurasian Nailwort ■☆

284841　Paronychia echinulata Chater var. minutiflora (H. Lindb.) Chaudhri = Paronychia echinulata Chater ■☆

284842　Paronychia erecta (Chapm.) Shinners；直立指甲草；Squareflower ■☆

284843　Paronychia erecta (Chapm.) Shinners var. corymbosa (Small) Chaudhri = Paronychia erecta(Chapm.) Shinners ■☆

284844　Paronychia erythraea Fiori = Paronychia chlorothyrsa Murb. var. erythraea(Fiori)Chaudhri ■☆

284845　Paronychia fastigiata (Raf.) Fernald；帚状指甲草；Forked Chickweed，Forked-chickweed，Hairy Forked Nailwort，Hairy Forked Nail-wort，Low Forked Chickweed ■☆

284846　Paronychia fastigiata (Raf.) Fernald var. nuttallii (Small) Fernald；纳托尔指甲草■☆

284847　Paronychia fastigiata(Raf.)Fernald var. paleacea Fernald；糠秕指甲草；Hairy Forked Nail-wort，Low Forked Chickweed ■☆

284848　Paronychia fastigiata (Raf.) Fernald var. paleacea Fernald = Paronychia fastigiata(Raf.) Fernald ■☆

284849　Paronychia fastigiata(Raf.) Fernald var. pumila (A. W. Wood) Fernald;矮小指甲草■☆

284850　Paronychia fastigiata (Raf.) Fernald var. typica Fernald = Paronychia fastigiata(Raf.) Fernald ■☆

284851　Paronychia franciscana Eastw. ;旧金山指甲草;San Francisco Nailwort ■☆

284852　Paronychia haggariensis Diels;哈加里指甲草■☆

284853　Paronychia haggariensis Diels var. latifolia Chaudhri;宽叶哈加里指甲草■☆

284854　Paronychia herniarioides (Michx.) Nutt. ;沿海平原指甲草;Coastal-plain Nailwort ■☆

284855　Paronychia jamesii Torr. et A. Gray;詹姆斯指甲草(詹姆士指甲草);James Nailwort,James' Nailwort,Nailwort ■☆

284856　Paronychia jamesii Torr. et A. Gray var. depressa Torr. et A. Gray = Paronychia depressa(Torr. et A. Gray)Nutt. ex A. Nelson ■☆

284857　Paronychia jamesii Torr. et A. Gray var. hirsuta Chaudhri = Paronychia jamesii Torr. et A. Gray ■☆

284858　Paronychia jamesii Torr. et A. Gray var. parviflora Chaudhri = Paronychia jamesii Torr. et A. Gray ■☆

284859　Paronychia jamesii Torr. et A. Gray var. praelongifolia Correll = Paronychia jamesii Torr. et A. Gray ■☆

284860　Paronychia jonesii M. C. Johnst. ;琼斯指甲草;Jones' Nailwort ■☆

284861　Paronychia kapela(Hacq.) A. Kern. ;卡佩拉指甲草■☆

284862　Paronychia kapela(Hacq.) A. Kern. var. africana Borzí et Mattei = Paronychia kapela(Hacq.) A. Kern. ■☆

284863　Paronychia kapela(Hacq.) A. Kern. var. hirta Emb. et Maire = Paronychia maroccana Chaudhri ■☆

284864　Paronychia kapela(Hacq.) A. Kern. var. libyos Borzí et Mattei = Paronychia kapela(Hacq.) A. Kern. ■☆

284865　Paronychia kapela(Hacq.) A. Kern. var. obtusata (Batt.) Maire = Paronychia kapela(Hacq.) A. Kern. ■☆

284866　Paronychia kurdica Boiss. ;库尔得指甲草■☆

284867　Paronychia lanuginosa Poir. = Gossypianthus lanuginosus (Poir.) Moq. ■☆

284868　Paronychia lenticulata Forssk. = Paronychia arabica (L.) DC. subsp. breviseta(Asch. et Schweinf.) Chaudhri ■☆

284869　Paronychia lindheimeri Engelm. ex A. Gray; 林德指甲草;Forked Nailwort ■☆

284870　Paronychia lindheimeri Engelm. ex A. Gray var. longibracteata Chaudhri = Paronychia lindheimeri Engelm. ex A. Gray ■☆

284871　Paronychia linearifolia DC. = Polycarpaea linearifolia (DC.) DC. ■☆

284872　Paronychia longiseta (Bertol.) Webb et Berthel. = Paronychia arabica(L.) DC. ■☆

284873　Paronychia lundellorum B. L. Turner = Paronychia setacea Torr. et A. Gray ■☆

284874　Paronychia maccartii Correll;马氏指甲草;McCart's Nailwort ■☆

284875　Paronychia macrosepala Boiss. var. querioides Ball = Paronychia chlorothyrsa Murb. ■☆

284876　Paronychia maroccana Chaudhri;摩洛哥指甲草■☆

284877　Paronychia montana (Small) Pax et K. Hoffm. = Paronychia fastigiata(Raf.) Fernald var. pumila(A. W. Wood)Fernald ■☆

284878　Paronychia monticola Cory;山地指甲草;Livermore Nailwort ■☆

284879　Paronychia nivea DC. = Paronychia capitata(L.) Lam. ■☆

284880　Paronychia nivea DC. var. libyca Borzí et Mattei = Paronychia capitata(L.) Lam. ■☆

284881　Paronychia nudata Correll = Paronychia monticola Cory ■☆

284882　Paronychia parksii Cory = Paronychia virginica Spreng. ■☆

284883　Paronychia patula Shinners;松林指甲草;Pineland Nailwort ■☆

284884　Paronychia polygonifolia(Vill.) DC. ;节叶指甲草■☆

284885　Paronychia polygonifolia (Vill.) DC. var. elegantior Maire = Paronychia polygonifolia(Vill.) DC. ■☆

284886　Paronychia polygonifolia(Vill.) DC. var. serratifolia Chaudhri = Paronychia polygonifolia(Vill.) DC. ■☆

284887　Paronychia polygonifolia (Vill.) DC. var. velucencis Boiss. = Paronychia polygonifolia(Vill.) DC. ■☆

284888　Paronychia pulvinata A. Gray;落基山指甲草;Rocky Mountain Nailwort ■☆

284889　Paronychia pulvinata A. Gray var. longiaristata Chaudhri = Paronychia pulvinata A. Gray ■☆

284890　Paronychia pumila (A. W. Wood) Core = Paronychia fastigiata (Raf.) Fernald var. pumila(A. W. Wood)Fernald ■☆

284891　Paronychia riparia Chapm. = Paronychia baldwinii(Torr. et A. Gray)Fenzl ex Walp. ■☆

284892　Paronychia rugelii (Chapm.) Shuttlew. ex Chapm. ;卢格指甲草;Rugel's Nailwort ■☆

284893　Paronychia rugelii (Chapm.) Shuttlew. ex Chapm. var. interior (Small) Chaudhri = Paronychia rugelii (Chapm.) Shuttlew. ex Chapm. ■☆

284894　Paronychia sclerocephala Decne. = Sclerocephalus arabicus Boiss. ■☆

284895　Paronychia scoparia Small = Paronychia virginica Spreng. ■☆

284896　Paronychia sedifolia R. Br. = Paronychia chlorothyrsa Murb. var. coarctata Chaudhri ■☆

284897　Paronychia sessiliflora Nutt. ;无梗花指甲草;Creeping Nailwort ■☆

284898　Paronychia setacea Torr. et A. Gray;刚毛指甲草;Bristle Nailwort ■☆

284899　Paronychia setacea Torr. et A. Gray var. longibracteata Chaudhri = Paronychia setacea Torr. et A. Gray ■☆

284900　Paronychia sinaica Fresen. ;叙利亚指甲草■☆

284901　Paronychia somaliensis Baker;索马里指甲草■☆

284902　Paronychia suffruticosa(L.) DC. ;亚灌木指甲草●☆

284903　Paronychia velata(Maire)Chaudhri;覆盖指甲草■☆

284904　Paronychia velata (Maire) Chaudhri var. subvelata (Litard. et Maire)Dobignard = Paronychia velata(Maire)Chaudhri ■☆

284905　Paronychia virginica Spreng. ;黄指甲草;Appalachian Nailwort,Forked Chickweed,Yellow Nailwort ■☆

284906　Paronychia virginica Spreng. var. parksii (Cory) Chaudhri = Paronychia virginica Spreng. ■☆

284907　Paronychia virginica Spreng. var. scoparia (Small) Cory = Paronychia virginica Spreng. ■☆

284908　Paronychia wilkinsonii S. Watson;威尔金森指甲草;Wilkinson's Nailwort ■☆

284909　Paronychiaceae A. St. -Hil. = Caryophyllaceae Juss. (保留科名)■●

284910　Paronychiaceae Juss. = Caryophyllaceae Juss. (保留科名)■●

284911　Parophiorrhiza C. B. Clarke = Mitreola L. ■

284912　Parophiorrhiza C. B. Clarke ex Hook. f. = Mitreola L. ■

284913　Parophiorrhiza khasiana C. B. Clarke = Mitreola pedicellata Benth. ■

284914　Parophiorrhiza khasiana C. B. Clarke ex Benth. = Mitreola pedicellata Benth. ■

284915　Paropsia Noronha ex Thouars(1805);基腺西番莲属●☆

284916　Paropsia Thouars = Paropsia Noronha ex Thouars ●☆

284917　Paropsia argutidens Sleumer = Paropsia brazzaeana Baill. ●☆

284918　Paropsia bequaertii De Wild. = Paropsia grewioides Welw. ex Mast. ●☆

284919　Paropsia braunii Gilg；布劳恩基腺西番莲●☆

284920　Paropsia brazzaeana Baill. ；布拉扎西番莲●☆

284921　Paropsia decandra（Baill.）Warb. = Paropsiopsis decandra（Baill.）Sleumer ●☆

284922　Paropsia dewevrei De Wild. et T. Durand var. condensata De Wild. = Paropsia grewioides Welw. ex Mast. ●☆

284923　Paropsia gabonica Breteler；加蓬基腺西番莲●☆

284924　Paropsia grewioides Welw. ex Mast.；扁担杆西番莲●☆

284925　Paropsia grewioides Welw. ex Mast. var. orientalis Sleumer；东方扁担杆西番莲●☆

284926　Paropsia guineensis Oliv.；几内亚基腺西番莲●☆

284927　Paropsia pritzelii Gilg = Paropsia guineensis Oliv. ●☆

284928　Paropsia reticulata Engl. = Paropsia brazzaeana Baill. ●☆

284929　Paropsia schliebeniana Sleumer = Paropsia braunii Gilg ●☆

284930　Paropsiaceae Dumort. = Passifloraceae Juss. ex Roussel（保留科名）●■

284931　Paropsiopsis Engl.（1891）；二列花属●☆

284932　Paropsiopsis africana Engl. = Paropsiopsis decandra（Baill.）Sleumer ●☆

284933　Paropsiopsis bipindensis Gilg；比平迪二列花●☆

284934　Paropsiopsis decandra（Baill.）Sleumer；二列花●☆

284935　Paropsiopsis ferruginea Exell；锈色二列花●☆

284936　Paropsiopsis jollyana Gilg；若利二列花●☆

284937　Paropsiopsis leucantha Gilg = Paropsiopsis decandra（Baill.）Sleumer ●☆

284938　Paropsiopsis pulchra Gilg；美丽二列花●☆

284939　Paropsiopsis zenkeri Gilg = Paropsiopsis decandra（Baill.）Sleumer ●☆

284940　Paropyrum Ulbr. = Isopyrum L.（保留属名）■

284941　Paropyrum anemonoides（Kar. et Kir.）Ulbr. = Isopyrum anemonoides Kar. et Kir. ■

284942　Parosela Cav. = Dalea L.（保留属名）●■☆

284943　Parosela bicolor Rydb. = Dalea bicolor Humb. et Bonpl. ■☆

284944　Parosela capitata Rose = Dalea capitata S. Watson ■☆

284945　Parosela enneandra（Nutt.）Britton = Dalea enneandra Nutt. ■☆

284946　Parosella Cav. ex DC. = Dalea L.（保留属名）●■☆

284947　Paroxygraphis W. W. Sm.（1913）；拟鸦跖属（山鸦跖草属）■☆

284948　Paroxygraphis sikkimensis W. W. Sm.；拟鸦跖花■☆

284949　Parquetina Baill.（1889）；帕尔凯萝藦属■☆

284950　Parquetina gabonica Baill. = Periploca nigrescens Afzel. ●☆

284951　Parquetina nigrescens（Afzel.）Bullock = Periploca nigrescens Afzel. ●☆

284952　Parqui Adans. = Cestrum L. ●

284953　Parrasia Greene = Nerisyrenia Greene ■☆

284954　Parrasia Raf.（废弃属名）= Belmontia E. Mey.（保留属名）■

284955　Parrasia Raf.（废弃属名）= Sebaea Sol. ex R. Br. ■

284956　Parria Steud. = Parrya R. Br. ●■

284957　Parrotia C. A. Mey.（1831）；银缕梅属（帕罗特木属，帕罗梯木属）；Parrotia, Shaniodendron, Witch Hazel ●

284958　Parrotia Walp. = Barrotia Gaudich. ●■

284959　Parrotia Walp. = Pandanus Parkinson ex Du Roi ●■

284960　Parrotia jacquemontiana Decne. = Parrotiopsis jacquemontiana（Decne.）Rehder ●☆

284961　Parrotia persica（DC.）C. A. Mey.；中亚银缕梅（波斯银缕梅，帕罗特木，帕罗梯木）；Iron Tree, Ironwood, Persian Ironwood, Persian Parrotia, Persian Witch Hazel, Persian Witch-hazel ●☆

284962　Parrotia persica C. A. Mey. ' Pendula'；垂枝银缕梅（垂枝帕罗特木）●☆

284963　Parrotia persica C. A. Mey. = Parrotia persica（DC.）C. A. Mey. ●☆

284964　Parrotia subaequalis（Hung T. Chang）R. M. Hao et H. T. Wei；银缕梅；Shaniodendron, Small-leaf Witchhazel ●◇

284965　Parrotiaceae Horan. = Hamamelidaceae R. Br.（保留科名）●

284966　Parrotiopsis（Nied.）C. K. Schneid.（1905）；白缕梅属（白苞缕梅属,异帕罗特木属）；Himalayan Hazel ●☆

284967　Parrotiopsis C. K. Schneid. = Parrotiopsis（Nied.）C. K. Schneid. ●☆

284968　Parrotiopsis jacquemontiana（Decne.）Rehder；白缕梅（白苞缕梅,异帕罗特木）；Himalayan Hazel ●☆

284969　Parrya R. Br.（1823）；条果芥属（巴料草属）；Parrya ●■

284970　Parrya ajanensis N. Busch = Parrya nudicaulis（L.）Regel ■

284971　Parrya albida Popov ex A. I. Baranov；白条芥■☆

284972　Parrya arabidiflora（DC.）G. Nicholson = Parrya nudicaulis（L.）Regel ■

284973　Parrya asperrima（B. Fedtsch.）Popov；粗糙条果芥■☆

284974　Parrya beketovii Krasn.；天山条果芥；Tianshan Parrya ■

284975　Parrya bellidifolia P. A. Dang. = Leiospora bellidifolia（P. A. Dang.）Botsch. et Pachom. ■

284976　Parrya chitralensis Jafri = Parrya pinnatifida Kar. et Kir. ■

284977　Parrya chitralensis Rech. f. = Parrya minjanensis Rech. f. ■

284978　Parrya chitralensis Rech. f. = Parrya pinnatifida Kar. et Kir. ■

284979　Parrya ciliaris Bureau et Franch. = Solms-Laubachia ciliaris（Bureau et Franch.）Botsch. ■

284980　Parrya ciliaris Bureau et Franch. = Solms-Laubachia pulcherrima Muschl. ■

284981　Parrya eriocalyx Regel et Schmidt = Leiospora eriocalyx（Regel et Schmidt）Dvorák ■

284982　Parrya eurycarpa Maxim. = Solms-Laubachia eurycarpa（Maxim.）Botsch. ■

284983　Parrya exscapa C. A. Mey. = Leiospora exscapa（C. A. Mey.）Dvorák ■

284984　Parrya exscapa Ledeb. = Leiospora exscapa（C. A. Mey.）Dvorák ■

284985　Parrya finchiana Dunn = Solms-Laubachia platycarpa（Hook. f. et Thomson）Botsch. ■

284986　Parrya finchiana Dunn = Solms-Laubachia pulcherrima Muschl. ■

284987　Parrya flabellata Regel = Desideria flabellata（Regel）Al-Shehbaz ■

284988　Parrya flabellata Regel et Herder = Desideria flabellata（Regel）Al-Shehbaz ■

284989　Parrya forrestii W. W. Sm. = Erysimum forrestii（W. W. Sm.）Polatschek ■

284990　Parrya fruticulosa Regel et Schmalh.；灌丛条果芥（灌木条果芥）；Bushwood Parrya ●

284991　Parrya golenkinii Lipsch.；高氏条果芥■☆

284992　Parrya integerrima G. Don = Parrya nudicaulis（L.）Regel ■

284993　Parrya karatavica Lipsch. et Pavlov；卡拉塔夫条果芥■☆

284994　Parrya lancifolia Popov；柳叶条果芥；Lance-leaved Parrya ■

284995　Parrya lanuginosa Hook. f. et Thomson = Christolea lanuginosa（Hook. f. et Thomson）Ovcz. ■

284996　Parrya lanuginosa Hook. f. et Thomson = Eurycarpus lanuginosus（Hook. f. et Thomson）Botsch. ■

284997　Parrya linearifolia W. W. Sm. = Solms-Laubachia linearifolia（W. W. Sm.）O. E. Schulz ■

284998   Parrya linnaeana Ledeb. = Parrya nudicaulis(L.) Regel ■

284999   Parrya macrocarpa R. Br. = Parrya nudicaulis(L.)Regel ■

285000   Parrya michaelis A. N. Vassiljeva = Parrya beketovii Krasn. ■

285001   Parrya microcarpa Ledeb. ;小果条果芥■☆

285002   Parrya minjanensis Rech. f. = Parrya chitralensis Jafri ■

285003   Parrya minjanensis Rech. f. = Parrya pinnatifida Kar. et Kir. ■

285004   Parrya nudicaulis(L.) Regel;裸茎条果芥（大果巴料草）; Nakedstem Parrya ■

285005   Parrya pamirica Botsch. et Vved. = Leiospora pamirica(Botsch. et Vved.) Botsch. et Pachom. ■

285006   Parrya pinnatifida Kar. et Kir. ;羽裂条果芥;Pinnatifid Parrya ■

285007   Parrya pinnatifida Kar. et Kir. var. glabra N. Busch;无毛条果芥;Glaubrous Pinnatifid Parrya ■

285008   Parrya pinnatifida Kar. et Kir. var. hirsuta N. Busch;有毛条果芥（毛叶条果芥）;Hirsute Pinnatifid Parrya ■

285009   Parrya pinnatifida Kar. et Kir. var. kizylarti Korsh. = Parrya pinnatifida Kar. et Kir. ■

285010   Parrya platycarpa Hook. f. et Thomson = Solms-Laubachia orbiculau Y. C. Lan et T. Y. Cheo ■

285011   Parrya platycarpa Hook. f. et Thomson = Solms-Laubachia platycarpa(Hook. f. et Thomson) Botsch. ■

285012   Parrya prolifera Maxim. = Christolea prolifera(Maxim.)Jafri ■

285013   Parrya prolifera Maxim. = Desideria prolifera (Maxim.) Al-Shehbaz ■

285014   Parrya pulvinata Popov;垫状条果芥;Cushion-shaped Parrya ■

285015   Parrya pumila Kurz = Christolea pumila(Kurz)Jafri ■

285016   Parrya pumila Kurz = Desideria pumila(Kurz) Al-Shehbaz ■

285017   Parrya ramosissima Franch. = Christolea crassifolia Cambess. ■

285018   Parrya runcinata(Regel et Schmalh.)N. Busch;倒齿条果芥■☆

285019   Parrya scapigera(Adams) G. Don = Parrya nudicaulis(L.)Regel ■

285020   Parrya schugnana Lipsch. ;舒格南条果芥■☆

285021   Parrya siliquosa Krasn. ;多荚条果芥■☆

285022   Parrya stenocarpa Kar. et Kir. ;狭果条果芥■☆

285023   Parrya stenocarpa Kar. et Kir. = Parrya pinnatifida Kar. et Kir. ■

285024   Parrya stenocarpa Kar. et Kir. subsp. gilgitica Jafri = Parrya minjanensis Rech. f. ■

285025   Parrya stenocarpa Kar. et Kir. var. minjanensis(Rech. f.)Kitam. = Parrya minjanensis Rech. f. ■

285026   Parrya stenocarpa Kar. et Kir. var. minjanensis? = Parrya chitralensis Jafri ■

285027   Parrya stenocarpa Kar. et Kir. var. pinnatisecta O. E. Schulz = Parrya chitralensis Jafri ■

285028   Parrya stenocarpa Kar. et Kir. var. pinnatisecta O. E. Schulz = Parrya minjanensis Rech. f. ■

285029   Parrya stenophylla Popov;狭叶条果芥■☆

285030   Parrya subsiliquosa Popov;长角条果芥（近长角条果芥）; Silique Parrya ■

285031   Parrya surculosa N. Busch;木质条果芥●☆

285032   Parrya turkestanica(Korsh.)N. Busch;土耳其斯坦条果芥■

285033   Parrya villosa Maxim. = Christolea villosa(Maxim.)Jafri ■

285034   Parrya villosa Maxim. = Phaeonychium villosum (Maxim.) Al-Shehbaz ■

285035   Parrya villosa Maxim. var. albiflora O. E. Schulz = Phaeonychium villosum(Maxim.) Al-Shehbaz ■

285036   Parrya xerophyta W. W. Sm. = Solms-Laubachia xerophyta(W. W. Sm.) Comber ■

285037   Parrycactus Doweld = Echinocactus Link et Otto ●

285038   Parrycactus Doweld(2000);佩雷掌属●☆

285039   Parryella Torr. et A. Gray ex A. Gray = Parryella Torr. et A. Gray ■☆

285040   Parryella Torr. et A. Gray(1868);丝叶豆属■☆

285041   Parryella filifolia Torr. et A. Gray;丝叶豆■☆

285042   Parryodes Jafri = Arabis L. ●■

285043   Parryodes Jafri(1957);腋花芥属（腋花荠属,珠峰荠属）; Parryodes ■

285044   Parryodes axilliflora Jafri;腋花芥（腋花南芥）;Axillaryflower Parryodes ■

285045   Parryodes axilliflora Jafri = Arabis axilliflora(Jafri)H. Hara ■

285046   Parryopsis Botsch. (1955);类条果芥属（假条果芥属）■

285047   Parryopsis Botsch. = Phaeonychium O. E. Schulz ■

285048   Parryopsis villosa(Maxim.)Botsch. ;类条果芥 ■

285049   Parryopsis villosa (Maxim.) Botsch. = Phaeonychium villosum (Maxim.) Al-Shehbaz ■

285050   Parsana Parsa et Maleki = Laportea Gaudich. (保留属名)●■

285051   Parsonsia P. Browne ex Adans. = Cuphea Adans. ex P. Browne ●■

285052   Parsonsia P. Browne ex Adans. = Parsonsia R. Br. (保留属名)●

285053   Parsonsia P. Browne(废弃属名) = Cuphea Adans. ex P. Browne ●■

285054   Parsonsia P. Browne(废弃属名) = Parsonsia R. Br. (保留属名)●

285055   Parsonsia R. Br. (1810)(保留属名);同心结属（爬森藤属）; Parsonsia ●

285056   Parsonsia alboflavescens(Dennst.)Mabb. ;海南同心结（爬森藤,同心结）; Hainan Parsonsia, Helicoid-stamenal Parsonsia, How Parsonsia,Smooth Parsonsia ●

285057   Parsonsia apiculata (Bakh. f.) D. J. Middleton = Grisseea apiculata Bakh. f. ●☆

285058   Parsonsia barbata Blume = Parameria laevigata(Juss.)Moldenke ●

285059   Parsonsia edulis(G. Benn.)Guillaumin;可食同心结●☆

285060   Parsonsia goniostemon Hand. -Mazz. ;广西同心结;Guangxi Parsonsia ●

285061   Parsonsia helicandra Hook. et Arn. = Parsonsia alboflavescens (Dennst.)Mabb. ●

285062   Parsonsia hookeriana Standl. = Cuphea hookeriana Walp. ●

285063   Parsonsia howii Tsiang = Parsonsia alboflavescens (Dennst.) Mabb. ●

285064   Parsonsia laevigata (Moon) Alston = Parsonsia alboflavescens (Dennst.)Mabb. ●

285065   Parsonsia micropetala Standl. = Cuphea micropetala Kunth ●

285066   Parsonsia spiralis Wall. ex G. Don = Parsonsia alboflavescens (Dennst.)Mabb. ●

285067   Partheniaceae Link = Asteraceae Bercht. et J. Presl(保留科名)●■

285068   Partheniaceae Link = Compositae Giseke(保留科名)●■

285069   Partheniastrum Fabr. = Parthenium L. ■●

285070   Parthenice A. Gray(1853);金胶菊属■☆

285071   Parthenice mollis A. Gray;金胶菊■☆

285072   Parthenium L. (1753);银胶菊属;Feverfew, Parthenium ■●

285073   Parthenium alpinum(Nutt.)Torr. et A. Gray;高山银胶菊■☆

285074   Parthenium alpinum(Nutt.)Torr. et A. Gray var. ligulatum M. E. Jones = Parthenium integrifolium L. ■☆

285075   Parthenium alpinum (Nutt.) Torr. et A. Gray var. tetraneuris (Barneby)Rollins = Parthenium alpinum(Nutt.)Torr. et A. Gray ■☆

285076   Parthenium argentatum A. Gray;灰白银胶菊（阿根廷银胶菊）; Argentatic Parthenium, Guayule, Guayule Parthenium, Guayule Rubber ■

285077   Parthenium auriculatum Britton = Parthenium integrifolium L. ■☆

285078 Parthenium confertum A. Gray；琴叶银胶菊；Lyreleaf Parthenium ■☆

285079 Parthenium confertum A. Gray var. divaricatum Rollins = Parthenium confertum A. Gray ■☆

285080 Parthenium confertum A. Gray var. lyratum（A. Gray）Rollins = Parthenium confertum A. Gray ■☆

285081 Parthenium confertum A. Gray var. microcephalum Rollins = Parthenium confertum A. Gray ■☆

285082 Parthenium fruticosum Less.；灌木状银胶菊■☆

285083 Parthenium hispidum Raf.；美洲银胶菊；American Feverfew ■☆

285084 Parthenium hispidum Raf. = Parthenium integrifolium L. ■☆

285085 Parthenium hispidum Raf. var. auriculatum（Britton）Rollins = Parthenium integrifolium L. ■☆

285086 Parthenium hysterophorus L.；银胶菊（后生银胶菊，假芹，解热银胶菊，野益母艾）；Common Parthenium, Parthenium, Santa Maria, Santa Maria Feverfew, Whim Broomweed, Whitehead, Wormwood ■

285087 Parthenium incanum Kunth；灰银胶菊；Mariola ■☆

285088 Parthenium integrifolium L.；全缘叶银胶菊（全叶银胶菊）；American Feverfew, Eastern Feverfew, Eastern Parthenium, Wild Quinine ■☆

285089 Parthenium integrifolium L. var. auriculatum（Britton）Cornelius ex Cronquist = Parthenium integrifolium L. ■☆

285090 Parthenium integrifolium L. var. henryanum Mears = Parthenium integrifolium L. ■☆

285091 Parthenium integrifolium L. var. hispidum（Raf.）Mears = Parthenium integrifolium L. ■☆

285092 Parthenium integrifolium L. var. mabryanum Mears = Parthenium integrifolium L. ■☆

285093 Parthenium matricaria Gesner ex Rupr. = Pyrethrum parthenium（L.）Sm. ■

285094 Parthenium radfordii Mears = Parthenium integrifolium L. ■☆

285095 Parthenium tetraneuris Barneby = Parthenium alpinum（Nutt.）Torr. et A. Gray ■☆

285096 Parthenium tomentosum DC.；绒毛银胶菊■☆

285097 Parthenocissus Planch.（1887）(保留属名)；地锦属（爬山虎属）；Creeper, Virginia Creeper, Woodbine ●

285098 Parthenocissus amamiana Hatus. = Parthenocissus heterophylla（Blume）Merr. ●

285099 Parthenocissus assamica Craib var. pilosissima Gagnep. = Cissus aristata Blume ●

285100 Parthenocissus austro-orientalis F. P. Metcalf；东南爬山虎●

285101 Parthenocissus austro-orientalis F. P. Metcalf = Yua austro-orientalis（F. P. Metcalf）C. L. Li ●

285102 Parthenocissus chinensis C. L. Li；小叶地锦；Chinese Creeper ●

285103 Parthenocissus cuspidifera（Miq.）Planch. var. pubifolia C. L. Li；毛脉地锦；Hairy-vein Creeper ●

285104 Parthenocissus cuspidifera（Miq.）Planch. var. pubifolia C. L. Li = Parthenocissus semicordata（Wall.）Planch. ●

285105 Parthenocissus dalzielii Gagnep.；异叶地锦（白花藤子，草叶藤，上树蛇，异叶爬山虎）；Diversileaf Creeper ●

285106 Parthenocissus engelmannii Koehne et Graebn. = Parthenocissus quinquefolia（L.）Planch. ●

285107 Parthenocissus engelmannii Koehne et Graebn. f. engelmannii（Koehne et Graebn.）Rehder = Parthenocissus quinquefolia（L.）Planch. ●

285108 Parthenocissus feddei（H. Lév.）C. L. Li；长柄地锦；Long-stalk Creeper ●

285109 Parthenocissus feddei（H. Lév.）C. L. Li var. pubescens C. L. Li；锈毛长柄地锦；Pubescent Long-stalk Creeper ●

285110 Parthenocissus henryana（Hemsl.）Diels et Gilg；花叶地锦（彩叶爬山虎，川鄂爬山虎，飞天香，红叶爬山虎，花叶爬山虎，猪蹄甲子）；Chinese Virginia Creeper, Henry Creeper, Silvervein Creeper, Silver-veined Creeper, Silvery Creeper ●

285111 Parthenocissus henryana（Hemsl.）Diels et Gilg var. glaucescens Diels et Gilg = Parthenocissus thomsonii（M. A. Lawson）P. Singh et B. V. Shetty ●

285112 Parthenocissus henryana（Hemsl.）Diels et Gilg var. glaucescens Diels et Gilg = Yua thomsonii（M. A. Lawson）C. L. Li var. glaucescens（Diels et Gilg）C. L. Li ●

285113 Parthenocissus henryana（Hemsl.）Diels et Gilg var. glaucescens Diels et Gilg = Yua chinensis C. L. Li ●

285114 Parthenocissus henryana（Hemsl.）Diels et Gilg var. hirsuta Diels et Gilg；毛脉花叶地锦；Hirsute Silvervein Creeper ●

285115 Parthenocissus henryana（Hemsl.）Diels et Gilg var. typica Diels et Gilg = Parthenocissus henryana（Hemsl.）Diels et Gilg ●

285116 Parthenocissus henryana（Hemsl.）Graebn. ex Diels et Gilg var. glaucescens Diels et Gilg = Yua thomsonii（M. A. Lawson）C. L. Li var. glaucescens（Diels et Gilg）C. L. Li ●

285117 Parthenocissus heterophylla（Blume）Merr.；异叶爬山虎（大叶爬山虎，地锦，吊岩风，红葡萄藤，爬山虎，日本爬山虎，三叉虎，三角风，三皮风，上木三叉虎，上木蛇，上树蜈蚣，上竹龙，小叶红藤，异叶地锦）；Diversifolious Creeper ●

285118 Parthenocissus heterophylla（Blume）Merr. = Parthenocissus dalzielii Gagnep. ●

285119 Parthenocissus heterophylla（Blume）Merr. = Parthenocissus feddei（H. Lév.）C. L. Li ●

285120 Parthenocissus himalayana（Royle）Planch. = Parthenocissus semicordata（Wall.）Planch. ●

285121 Parthenocissus himalayana Planch. = Parthenocissus semicordata（Wall. ex Roxb.）Planch. var. roylei（King ex Parker）Nazim. et Qaiser ●☆

285122 Parthenocissus himalayana Planch. var. rubifolia（H. Lév. et Vaniot）Gagnep. = Parthenocissus semicordata（Wall.）Planch. var. rubifolia（H. Lév. et Vaniot）C. L. Li ●

285123 Parthenocissus himalayana Planch. var. rubrifolia（H. Lév. et Vaniot）Gagnep. = Parthenocissus semicordata（Wall.）Planch. ●

285124 Parthenocissus himalayana Planch. var. vestita Hand.-Mazz. = Parthenocissus semicordata（Wall.）Planch. ●

285125 Parthenocissus hirsuta（Pursh）Graebn. = Parthenocissus quinquefolia（L.）Planch. ●

285126 Parthenocissus hirsuta（Pursh）Graebn. = Parthenocissus vitacea Hitchc. var. dubia Rehder ●☆

285127 Parthenocissus inserta（A. Kern.）Fritsch.；丛林爬山虎；Common Virginia Creeper, False Virginia Creeper, Thicket Creeper, Western Creeper ●☆

285128 Parthenocissus inserta（A. Kern.）Fritsch. = Parthenocissus vitacea（Knerr）Hitchc. ●☆

285129 Parthenocissus inserta（A. Kern.）Fritsch. f. dubia Rehder = Parthenocissus vitacea（Knerr）Hitchc. ●☆

285130 Parthenocissus laetevirens Rehder；绿叶地锦（大绿藤，亮绿爬山虎，绿爬山虎，绿叶爬山虎，青龙藤，青叶爬山虎）；Shinygreen Creeper, Shiny-green Creeper ●

285131 Parthenocissus landuk（Hassk.）Gagnep. = Parthenocissus dalzielii Gagnep. ●

285132　Parthenocissus landuk（Hassk.）Gagnep. = Parthenocissus heterophylla（Blume）Merr. ●

285133　Parthenocissus landuk Gagnep. = Parthenocissus dalzielii Gagnep. ●

285134　Parthenocissus multiflora Pamp. = Parthenocissus henryana（Hemsl.）Diels et Gilg ●

285135　Parthenocissus multiflora Pamp. = Parthenocissus heterophylla（Blume）Merr. ●

285136　Parthenocissus quinquefolia（L.）Planch.；五叶地锦（五叶爬山虎）；American Ivy，Five-leaved Ivy，True Virginia Creeper，Virginia Creeper，Woodbine ●

285137　Parthenocissus quinquefolia（L.）Planch. f. engelmannii（Koehne et Graebn.）Rehder = Parthenocissus quinquefolia（L.）Planch. ●

285138　Parthenocissus quinquefolia（L.）Planch. f. hirsuta（Pursh）Fernald = Parthenocissus quinquefolia（L.）Planch. ●

285139　Parthenocissus quinquefolia（L.）Planch. var. hirsuta（Pursh）Planch. = Parthenocissus quinquefolia（L.）Planch. ●

285140　Parthenocissus quinquefolia（L.）Planch. var. saint-paulii（Koehne et Graebn. ex Graebn.）Rehder = Parthenocissus quinquefolia（L.）Planch. ●

285141　Parthenocissus quinquefolia（L.）Planch. var. vitacea（Knerr）L. H. Bailey = Parthenocissus vitacea（Knerr）Hitchc. ●☆

285142　Parthenocissus rubifolia H. Lév. et Vaniot = Parthenocissus semicordata（Wall.）Planch. var. rubifolia（H. Lév. et Vaniot）C. L. Li ●

285143　Parthenocissus saint-paulii Koehne et Graebn. ex Graebn. = Parthenocissus quinquefolia（L.）Planch. ●

285144　Parthenocissus semicordata（Wall. ex Roxb.）Planch.；三叶地锦（大血藤，绿葡萄藤，爬山虎，三角风，三皮枫藤，三窝草，三叶爬墙虎，三叶爬山虎，三爪金龙，三爪龙，喜马拉雅地锦，喜马拉雅爬山虎，小红藤，岩山甲）；Himalayan Creeper，Himalayan Virginia Creeper，Himalayas Creeper，Three-leaf Creeper ●

285145　Parthenocissus semicordata（Wall. ex Roxb.）Planch. var. roylei（King ex Parker）Nazim. et Qaiser；罗伊尔地锦●☆

285146　Parthenocissus semicordata（Wall.）Planch. = Parthenocissus semicordata（Wall. ex Roxb.）Planch. ●

285147　Parthenocissus semicordata（Wall.）Planch. var. roylei（King ex Parker）Nazim. et Qaiser；罗氏三叶地锦●☆

285148　Parthenocissus semicordata（Wall.）Planch. var. rubifolia（H. Lév. et Vaniot）C. L. Li；红三叶地锦（红三叶爬山虎）●

285149　Parthenocissus semicordata（Wall.）Planch. var. rubrifolia（H. Lév. et Vaniot）C. L. Li = Parthenocissus semicordata（Wall.）Planch. ●

285150　Parthenocissus sinensis Diels et Gilg = Vitis piasezkii Maxim. ●

285151　Parthenocissus suberosa Hand. -Mazz.；栓翅地锦（栓翅爬山虎）；Corky Creeper，Suberose Creeper ●

285152　Parthenocissus subferruginea Merr. et Chun；琼南地锦；South Hainan Creeper，Subrusty Creeper ●

285153　Parthenocissus subferruginea Merr. et Chun = Cissus repanda Vahl var. subferruginea（Merr. et Chun）C. L. Li ●

285154　Parthenocissus thomsonii（M. A. Lawson）P. Singh et B. V. Shetty = Yua thomsonii（M. A. Lawson）C. L. Li ●

285155　Parthenocissus thomsonii（M. A. Lawson）Planch. = Yua thomsonii（M. A. Lawson）C. L. Li ●

285156　Parthenocissus thunbergii（Siebold et Zucc.）Nakai = Parthenocissus tricuspidata（Siebold et Zucc.）Planch. ●

285157　Parthenocissus tricuspidata（Siebold et Zucc.）Planch.；地锦（蝙蝠藤，长春藤，常春藤，大风藤，地嗓，多脚草，飞天蜈蚣，风藤，枫藤，腹水藤，过风藤，红葛，红葡萄藤，假葡萄藤，爬龙藤，爬墙虎，爬山虎，爬岩虎，日光子，三角枫藤，三叶地锦，三叶茄，山葡萄，石壁藤，土鼓藤，野枫藤，野葡萄，走游藤）；Boston Ivy，Japan Creeper，Japanese Creeper，Japanese Ivy，Virginia Creeper ●

285158　Parthenocissus tricuspidata（Siebold et Zucc.）Planch. 'Lowii'；洛氏爬山虎●☆

285159　Parthenocissus tricuspidata（Siebold et Zucc.）Planch. 'Veitchii'；维奇爬山虎；Boston Ivy，Japanese Creeper ●☆

285160　Parthenocissus tricuspidata（Siebold et Zucc.）Planch. var. ferruginea W. T. Wang = Parthenocissus suberosus Hand. -Mazz. ●

285161　Parthenocissus tricuspidata（Siebold et Zucc.）Planch. var. ferruginea W. T. Wang；锈毛爬山虎；Rustyhair Creeper ●

285162　Parthenocissus tricuspidata（Siebold et Zucc.）Planch. var. veitchii（Graebn.）? = Parthenocissus tricuspidatus（Siebold et Zucc.）Planch. 'Veitchii'●☆

285163　Parthenocissus veitchii Graebn. = Parthenocissus tricuspidata（Siebold et Zucc.）Planch. 'Veitchii'●☆

285164　Parthenocissus vitacea（Knerr）Hitchc.；葡萄叶爬山虎；Crape-leaf Creeper，Grape Woodbine，Woodbine ●☆

285165　Parthenocissus vitacea（Knerr）Hitchc. f. dubia（Rehder）Fernald = Parthenocissus vitacea（Knerr）Hitchc. ●☆

285166　Parthenocissus vitacea（Knerr）Hitchc. var. dubia Rehder；多毛爬山虎；Manyhair Creeper ●☆

285167　Parthenocissus vitacea（Knerr）Hitchc. var. lanicianis（Planch. ex DC.）Rehder；小叶爬山虎；Little-leaf Creeper ●☆

285168　Parthenocissus vitacea（Knerr）Hitchc. var. macrophylla Rehder；大叶爬山虎；Big-leaf Creeper ●☆

285169　Parthenocissus vitacea Hitchc. = Parthenocissus vitacea（Knerr）Hitchc. ●☆

285170　Parthenocissus vitacea Hitchc. var. dubia Rehder = Parthenocissus vitacea（Knerr）Hitchc. var. dubia Rehder ●☆

285171　Parthenocissus vitacea Hitchc. var. macrophyllus Rehder = Parthenocissus vitacea（Knerr）Hitchc. var. macrophylla Rehder ●☆

285172　Parthenocissus vitaceus Hitchc. var. lanicianis（Planch. ex DC.）Rehder = Parthenocissus vitacea（Knerr）Hitchc. var. lanicianis（Planch. ex DC.）Rehder ●☆

285173　Parthenopsis Kellogg = Venegasia DC. ●☆

285174　Parthenopsis Kellogg（1873）；类银胶菊属●☆

285175　Parthenopsis maritimus Kellogg；类银胶菊●☆

285176　Parthenostachys Fourr. = Ornithogalum L. ■

285177　Parthenoxylon Blume = Cinnamomum Schaeff.（保留属名）●

285178　Parthenoxylon porrectum（Roxb.）Blume = Cinnamomum parthenoxylum（Jack）Meisn. ●

285179　Parvatia Decne.（1837）；牛藤果属●☆

285180　Parvatia Decne. = Stauntonia DC. ●

285181　Parvatia brunoniana（Wall. ex Hemsl.）Decne. = Stauntonia brunoniana Wall. ●

285182　Parvatia brunoniana（Wall.）Decne. = Stauntonia brunoniana Wall. ●

285183　Parvatia brunoniana（Wall.）Decne. subsp. elliptica（Hemsl.）H. N. Qin = Stauntonia elliptica Hemsl. ●

285184　Parvatia chinensis Franch. = Sinofranchetia chinensis（Franch.）Hemsl. ●

285185　Parvatia decora Dunn = Stauntonia decora（Dunn）C. Y. Wu ●

285186　Parvatia elliptica（Hemsl.）Gagnep. = Stauntonia elliptica Hemsl. ●

285187　Parviopuntia Soulaire = Opuntia Mill. ●

285188　Parviopuntia Soulaire et Marn. -Lap. = Opuntia Mill. ●

285189　Parvisedum R. T. Clausen = Sedella Britton et Rose ■☆

285190　Parvotrisetum Chrtek = Trisetaria Forssk. ■☆

285191　Paryphantha Schauer = Thryptomene Endl. (保留属名) ●☆

285192　Paryphanthe Benth. = Paryphantha Schauer ●☆

285193　Paryphosphaera H. Karst. = Parkia R. Br. ●

285194　Pasaccardoa Kuntze(1891) ; 肋毛菊属■●☆

285195　Pasaccardoa baumii O. Hoffm. ; 鲍姆肋毛菊■☆

285196　Pasaccardoa dicomoides De Wild. et Muschl. = Pasaccardoa grantii(Benth. ex Oliv.) Kuntze ■☆

285197　Pasaccardoa grantii(Benth. ex Oliv.) Kuntze; 格氏肋毛菊■☆

285198　Pasaccardoa jeffreyi Wild; 杰氏肋毛菊■☆

285199　Pasaccardoa jeffreyi Wild subsp. kasaiensis Lisowski; 开赛肋毛菊■☆

285200　Pasaccardoa jeffreyi Wild subsp. procumbens Lisowski = Pasaccardoa procumbens(Lisowski) G. V. Pope ■☆

285201　Pasaccardoa kassneri De Wild. et Muschl. = Pasaccardoa grantii(Benth. ex Oliv.) Kuntze ■☆

285202　Pasaccardoa procumbens(Lisowski) G. V. Pope; 平铺肋毛菊■☆

285203　Pasania(Miq.) Oerst. (1867) ; 肖柯属(柯树属) ; Tan Oak, Tan-oak ●

285204　Pasania(Miq.) Oerst. = Lithocarpus Blume ●

285205　Pasania Oerst. = Lithocarpus Blume ●

285206　Pasania Oerst. = Pasania(Miq.) Oerst. ●

285207　Pasania amygdalifolia(V. Naray. ex Forbes et Hemsl.) Schottky = Lithocarpus amygdalifolius(V. Naray. ex Forbes et Hemsl.) Hayata ●

285208　Pasania amygdalifolia (V. Naray.) Schottky = Lithocarpus amygdalifolius(V. Naray. ex Forbes et Hemsl.) Hayata ●

285209　Pasania areca Hickel et A. Camus = Lithocarpus areca(Hickel et A. Camus) A. Camus ●◇

285210　Pasania attenuata(V. Naray.) Schottky = Lithocarpus attenuatus (V. Naray.) Rehder ●

285211　Pasania bacgiangensis Hickel et A. Camus = Lithocarpus bacgiangensis(Hickel et A. Camus) A. Camus ●

285212　Pasania bacgiangensis Hickel et A. Camus = Lithocarpus mekongensis(A. Camus) C. C. Huang et Y. T. Chang ●

285213　Pasania balansae (Drake) Hickel et A. Camus = Lithocarpus balansae(Drake) A. Camus ●

285214　Pasania baviensis (Drake) Schottky = Lithocarpus truncatus (King) Rehder var. baviensis(Drake) A. Camus ●

285215　Pasania blaoensis(A. Camus) Hu = Lithocarpus hancei(Benth.) Rehder ●

285216　Pasania bonnetii Hickel et A. Camus = Lithocarpus bonnetii (Hickel et A. Camus) A. Camus ●

285217　Pasania brevicaudata (V. Naray.) Schottky = Lithocarpus brevicaudatus(V. Naray.) Hayata ●

285218　Pasania brevicaudata (V. Naray.) Schottky var. arisanensis (Hayata) S. S. Ying = Lithocarpus hancei(Benth.) Rehder ●

285219　Pasania brunnea (Rehder) Chun = Lithocarpus taitoensis (Hayata) Hayata ●

285220　Pasania calathiformis (V. Naray.) Hickel et A. Camus = Castanopsis calathiformis(V. Naray.) Rehder et P. Wilson ●

285221　Pasania carolinae (V. Naray.) Schottky = Lithocarpus carolinae (V. Naray.) Rehder ●

285222　Pasania castanopsifolia (Hayata) Hayata = Lithocarpus lepidocarpus(Hayata) Hayata ●

285223　Pasania cathayana (Seem.) Schottky = Lithocarpus truncatus (King) Rehder ●

285224　Pasania caudatilimba(Merr.) Chun = Lithocarpus caudatilimbus (Merr.) A. Camus ●

285225　Pasania cerebrina Hickel et A. Camus = Castanopsis cerebrina (Hickel et A. Camus) Barnett ●

285226　Pasania cheliensis Hu = Lithocarpus fohaiensis(Hu) A. Camus ●

285227　Pasania chiaratuangensis(J. C. Liao) J. C. Liao; 加拉段柯(大武柯) ●

285228　Pasania chiaratuangensis (J. C. Liao) J. C. Liao = Lithocarpus harlandii(Hance) Rehder ●

285229　Pasania chiwui Hu = Cyclobalanopsis angustinii (V. Naray.) Schottky ●

285230　Pasania chiwui Hu = Cyclobalanopsis augustinii(Skan) Schottky ●

285231　Pasania cleistocarpa (Seem.) Schottky = Lithocarpus cleistocarpus(Seemen) Rehder et E. H. Wilson ●

285232　Pasania confertifolia Hu = Lithocarpus hancei(Benth.) Rehder ●

285233　Pasania cornea(Lour.) J. C. Liao; 后大埔石栎(后大埔柯, 烟斗石栎, 朱仔) ; Houdapu Tanoak, Scarlet Oak ●

285234　Pasania cornea (Lour.) Oerst. = Lithocarpus corneus (Lour.) Rehder ●

285235　Pasania cornea (Lour.) Oerst. = Pasania cornea (Lour.) J. C. Liao ●

285236　Pasania cuspidata (Thunb.) Oerst. = Castanopsis cuspidata (Thunb. ex A. Murray) Schottky ●☆

285237　Pasania cyrtocarpa (Drake) Schottky = Lithocarpus cyrtocarpus (Drake) A. Camus ●

285238　Pasania dealbata(DC.) Oerst. = Lithocarpus dealbatus(Hook. f. et Thomson ex DC.) Rehder ●

285239　Pasania dealbata (Hook. f. et Thomson ex Miq.) Oerst. = Lithocarpus dealbatus(Hook. f. et Thomson ex Miq.) Rehder ●

285240　Pasania dodonaeifolia (Hayata) Hayata = Lithocarpus dodonaeifolius(Hayata) Hayata ●

285241　Pasania dodonaeifolia Hayata = Lithocarpus dodonaeifolius (Hayata) Hayata ●

285242　Pasania echinocupula Hu = Lithocarpus echinotholus(Hu) Chun et C. C. Huang ex Y. C. Hsu et H. Wei Jen ●

285243　Pasania echinophora Hickel et A. Camus = Lithocarpus echinophorus(Hickel et A. Camus) A. Camus ●◇

285244　Pasania echinothola Hu = Lithocarpus echinotholus(Hu) Chun et C. C. Huang ex Y. C. Hsu et H. Wei Jen ●

285245　Pasania edulis (Makino) Makino = Lithocarpus edulis (Makino) Nakai ●☆

285246　Pasania elaeagnifolia (Seemen) Schottky = Lithocarpus elaeagnifolius(Seemen) Chun ●

285247　Pasania elata Hickel et A. Camus = Lithocarpus lycoperdon(V. Naray.) A. Camus ●

285248　Pasania elizabethiae (Tutcher) Schottky = Lithocarpus elizabethae(Tutcher) Rehder ●

285249　Pasania eyrei (Champ. ex Benth.) Oerst. = Castanopsis eyrei (Champ. ex Benth.) Tutcher ●

285250　Pasania eyrei (Champ.) Oerst. = Castanopsis eyrei (Champ.) Tutcher ●

285251　Pasania fangii Hu et W. C. Cheng = Lithocarpus fangii(Hu et W. C. Cheng) C. C. Huang et Y. T. Chang ●

285252　Pasania farinulenta(Hance) Hickel et A. Camus = Lithocarpus farinulentus(Hance) A. Camus ●

285253　Pasania fenestrata (Roxb.) Oerst. = Lithocarpus fenestratus (Roxb.) Rehder ●

285254　Pasania fissa (Champ. ex Benth.) Oerst. = Castanopsis fissa (Champ. ex Benth.) Rehder et E. H. Wilson ●

285255　Pasania fissa (Champ. ex Benth.) Oerst. var. tunkinensis (Drake) Hickel et A. Camus = Castanopsis fissa(Champ. ex Benth.) Rehder et E. H. Wilson ●

285256　Pasania fohaiensis Hu = Lithocarpus fohaiensis(Hu) A. Camus ●

285257　Pasania formosana (V. Naray. ex Forbes et Hemsl.) Schottky = Lithocarpus formosanus(V. Naray. ex Forbes et Hemsl.) Hayata ●

285258　Pasania formosana (V. Naray.) Schottky = Lithocarpus formosanus(V. Naray. ex Forbes et Hemsl.) Hayata ●

285259　Pasania garrettiana (Craib) Hickel et A. Camus = Lithocarpus garrettianus(Craib) A. Camus ●

285260　Pasania glabra (Thunb. ex A. Murray) Oerst. = Lithocarpus glaber(Thunb.) Nakai ●

285261　Pasania glabra (Thunb. ex A. Murray) Oerst. f. microphylla Hayashi = Lithocarpus glaber(Thunb.) Nakai ●

285262　Pasania glabra(Thunb.) Oerst. = Lithocarpus glaber (Thunb.) Nakai ●

285263　Pasania hancei (Benth.) Schottky = Lithocarpus hancei (Benth.) Rehder ●

285264　Pasania hancei(Benth.)Schottky var. arisanensis(Hayata) J. C. Liao;阿里山三斗石栎(阿里山三斗柯)●

285265　Pasania hancei(Benth.)Schottky var. arisanensis(Hayata) J. C. Liao = Lithocarpus hancei(Benth.) Rehder ●

285266　Pasania hancei(Benth.) Schottky var. ternaticupula(Hayata) J. C. Liao = Lithocarpus hancei(Benth.) Rehder ●

285267　Pasania hancei(Benth.) Schottky var. ternaticupula(Hayata) J. C. Liao;三斗石栎●

285268　Pasania harlandii (Hance ex Walp.) Oerst. = Lithocarpus harlandii(Hance) Rehder ●

285269　Pasania harlandii (Hance) Oerst. = Lithocarpus harlandii (Hance) Rehder ●

285270　Pasania hemishphaerica (Drake) Hickel et A. Camus = Lithocarpus corneus(Lour.) Rehder var. zonatus C. C. Huang et Y. T. Chang ●

285271　Pasania hemisphaerica (Drake) Hickel et A. Camus = Lithocarpus corneus(Lour.)Rehder var. zonatus C. C. Huang et Y. T. Chang ●

285272　Pasania henryi(Seem.) Schottky = Lithocarpus henryi (Seem.) Rehder et E. H. Wilson ●

285273　Pasania henryi(Seemen) Schottky = Lithocarpus henryi (Seem.) Rehder et E. H. Wilson ●

285274　Pasania hui (A. Camus) Hu = Lithocarpus variolosus (Franch.) Chun ●

285275　Pasania hypoglauca Hu = Lithocarpus hypoglaucus (Hu) C. C. Huang ex Y. C. Hsu et H. Wei Jen ●

285276　Pasania hypophaea (Hayata) H. L. Li = Cyclobalanopsis hypophaea(Hayata) Kudo ●

285277　Pasania impressineva (Hayata) Schottky = Lithocarpus brevicaudatus(V. Naray.) Hayata ●

285278　Pasania irwinii (Hance) Oerst. = Lithocarpus irwinii (Hance) Rehder ●◇

285279　Pasania ischnostachya Hu = Castanopsis fargesii Franch. ●

285280　Pasania iteaphylla (Hance) Schottky = Lithocarpus iteaphyllus (Hance) Rehder ●

285281　Pasania kawakamii (Hayata) Schottky = Lithocarpus kawakamii (Hayata) Hayata ●

285282　Pasania kawakamii(Hayata)Schottky var. chiaratuangensis J. C.

285282（续）Liao = Lithocarpus harlandii(Hance) Rehder ●

285283　Pasania kodaihoensis (Hayata) H. L. Li = Pasania cornea (Lour.) J. C. Liao ●

285284　Pasania konishii (Hayata) Schottky = Lithocarpus konishii (Hayata) Hayata ●

285285　Pasania krempfii Hickel et A. Camus = Lithocarpus lycoperdon (V. Naray.) A. Camus ●

285286　Pasania laotica Hickel et A. Camus = Lithocarpus laoticus (Hickel et A. Camus) A. Camus ●

285287　Pasania lepidocarpa (Hayata) Schott = Lithocarpus lepidocarpus (Hayata) Hayata ●

285288　Pasania lepidocarpa (Hayata) Schottky = Lithocarpus lepidocarpus(Hayata) Hayata ●

285289　Pasania leucostachya Hu = Lithocarpus variolosus (Franch.) Chun ●

285290　Pasania lithocarpaea Oerst. = Lithocarpus pasania C. C. Huang et Y. T. Chang ●

285291　Pasania litseifolia (Hance) Schottky = Lithocarpus litseifolius (Hance)Chun ●

285292　Pasania longicaudata (Hayata) Hayata = Castanopsis carlesii (Hemsl.) Hayata ●

285293　Pasania longinux Hu = Lithocarpus areca (Hickel et A. Camus) A. Camus ●◇

285294　Pasania longipedicellata Hickel et A. Camus = Lithocarpus longipedicellatus(Hickel et A. Camus) A. Camus ●

285295　Pasania lycoperdon (V. Naray.) Schottky = Lithocarpus lycoperdon(V. Naray.) A. Camus ●

285296　Pasania lysistachya Hu = Lithocarpus litseifolius(Hance) Chun ●

285297　Pasania magneinii Hickel et A. Camus = Lithocarpus magneinii (Hickel et A. Camus) A. Camus ●

285298　Pasania mairei Schottky = Lithocarpus hypoglaucus (Hu) C. C. Huang ex Y. C. Hsu et H. Wei Jen ●

285299　Pasania mairei Schottky = Lithocarpus mairei(Schottky)Rehder ●

285300　Pasania microsperma (A. Camus) Hu = Lithocarpus microspermus A. Camus ●

285301　Pasania mucronata Hickel et A. Camus = Lithocarpus litseifolius (Hance)Chun ●

285302　Pasania naiadarum (Hance) Schottky = Lithocarpus naiadarum (Hance)Chun ●

285303　Pasania nakaii(Hayata) Nakai = Lithocarpus taitoensis (Hayata) Hayata ●

285304　Pasania nantoensis (Hayata) Schottky = Lithocarpus nantoensis (Hayata) Hayata ●

285305　Pasania nitidinux Hu = Lithocarpus nitidinux (Hu) Chun ex C. C. Huang et Y. T. Chang ●

285306　Pasania pachyphylla(Kurz) Schottky = Lithocarpus pachyphyllus (Kurz) Rehder ●

285307　Pasania paniculata (Hand. -Mazz.) Chun = Lithocarpus paniculatus Hand. -Mazz. ●

285308　Pasania randaiensis (Hayata) Schottky = Castanopsis uraiana (Hayata) Kaneh. et Hatus. ●

285309　Pasania reinwardtii Hickel et A. Camus = Lithocarpus pseudoreinwardtii (Drake) A. Camus ●

285310　Pasania rhododendrophylla Hu = Lithocarpus hancei (Benth.) Rehder ●

285311　Pasania rhombocarpa (Hayata) Hayata = Lithocarpus taitoensis (Hayata) Hayata ●

285312　Pasania rhombocarpa (Hayata) Hayata = Pasania synbalanos (Hance) Schottky ●

285313　Pasania rosthornii Schottky = Lithocarpus rosthornii (Schottky) Barnett ●

285314　Pasania rostornii Schottky = Lithocarpus rosthornii (Schottky) Barnett ●

285315　Pasania sdvicolarum(Hance) Schottky = Lithocarpus silvicolarum (Hance) Chun ●

285316　Pasania shinsuiensis (Hayata et Kaneh.) Nakai = Lithocarpus shinsuiensis Hayata et Kaneh. ●

285317　Pasania sieboldiana (Blume) Nakai = Lithocarpus glaber (Thunb.) Nakai ●

285318　Pasania silvicolarum(Hance) Schottky = Lithocarpus silvicolarum (Hance) Chun ●

285319　Pasania skaniana (Dunn) Schottky = Lithocarpus skanianus (Dunn) Rehder ●

285320　Pasania sphaerocarpa Hickel et A. Camus = Lithocarpus sphaerocarpus(Hickel et A. Camus) A. Camus ●

285321　Pasania sphaerocarpus Hickel et A. Camus = Lithocarpus sphaerocarpus(Hickel et A. Camus) A. Camus ●

285322　Pasania spicatga var. brevipetiolata (DC.) Hu = Lithocarpus grandifolius(D. Don) S. N. Biswas ●

285323　Pasania subreticulata (Hayata) Hayata = Lithocarpus hancei (Benth.) Rehder ●

285324　Pasania suishaensis (Kaneh. et Yamam.) Nakai = Lithocarpus taitoensis(Hayata) Hayata ●

285325　Pasania synbalanos (Hance) Schottky = Lithocarpus litseifolius (Hance) Chun ●

285326　Pasania synbalanos (Hance) Schottky = Lithocarpus synbalanos (Hance) Chun ●

285327　Pasania taitoensis (Hayata) J. C. Liao = Lithocarpus taitoensis (Hayata) Hayata ●

285328　Pasania taitoensis Schottky = Lithocarpus taitoensis (Hayata) Hayata ●

285329　Pasania tephrocarpa (Drake) Hickel et A. Camus = Lithocarpus tephrocarpus(Drake) A. Camus ●

285330　Pasania ternaticupula (Hayata) Schottky = Lithocarpus hancei (Benth.) Rehder ●

285331　Pasania ternaticupula (Hayata) Schottky = Pasania hancei (Benth.) Schottky var. ternaticupula(Hayata) J. C. Liao ●

285332　Pasania ternaticupula(Hayata) Schottky f. matsudae (Hayata) J. C. Liao = Pasania hancei (Benth.) Schottky var. ternaticupula (Hayata) J. C. Liao ●

285333　Pasania ternaticupula (Hayata) Schottky var. arisanensis (Hayata) J. C. Liao = Lithocarpus hancei(Benth.) Rehder ●

285334　Pasania ternaticupula (Hayata) Schottky var. arisanensis (Hayata) J. C. Liao = Pasania hancei (Benth.) Schottky var. arisanensis(Hayata) J. C. Liao ●

285335　Pasania ternaticupula(Hayata) Schottky var. matsudai (Hayata) J. C. Liao = Lithocarpus hancei(Benth.) Rehder ●

285336　Pasania ternaticupula (Hayata) Schottky var. subreticulata (Hayata) J. C. Liao = Lithocarpus hancei(Benth.) Rehder ●

285337　Pasania thalassica (Hance) Oerst. = Lithocarpus glaber (Thunb.) Nakai ●

285338　Pasania thomsonii (Miq.) Hickel et A. Camus = Lithocarpus thomsonii(Miq.) Rehder ●

285339　Pasania tienchuanensis Hu = Lithocarpus megalophyllus Rehder

et E. H. Wilson ●

285340　Pasania tomentosinux Hu = Lithocarpus bacgiangensis(Hickel et A. Camus) A. Camus ●

285341　Pasania tomentosinux Hu = Lithocarpus mekongensis (A. Camus) C. C. Huang et Y. T. Chang ●

285342　Pasania trachycarpa Hickel et A. Camus = Lithocarpus trachycarpus(Hickel et A. Camus) A. Camus ●

285343　Pasania triquetra Hickel et A. Camus = Lithocarpus triqueter (Hickel et A. Camus) A. Camus ●

285344　Pasania truncata (King ex Hook. f.) Schottky = Lithocarpus truncatus(King) Rehder ●

285345　Pasania truncata(King) Schottky = Lithocarpus truncatus(King) Rehder ●

285346　Pasania tubulosa Hickel et A. Camus = Lithocarpus tubulosus (Hickel et A. Camus) A. Camus ●

285347　Pasania uraiana (Hayata) Schottky = Castanopsis uraiana (Hayata) Kaneh. et Hatus. ●

285348　Pasania uraiana Schottky = Castanopsis uraiana(Hayata) Kaneh. et Hatus. ●

285349　Pasania uvariifolia (Hance) Schottky = Lithocarpus uvariifolius (Hance) Rehder ●

285350　Pasania variolosa (Franch.) Schottky = Lithocarpus variolosus (Franch.) Chun ●

285351　Pasania wenshanensis Hu = Lithocarpus litseifolius (Hance) Chun ●

285352　Pasania wilsonii (Seem.) Schottky = Lithocarpus cleistocarpus (Seemen) Rehder et E. H. Wilson ●

285353　Pasania xylocarpa (Kurz) Hickel et A. Camus = Lithocarpus xylocarpus(Kurz) Markgr. ●

285354　Pasania yenshanensis Hu = Lithocarpus dealbatus (Hook. f. et Thomson ex DC.) Rehder ●

285355　Pasania yui Hu = Lithocarpus trachycarpus (Hickel et A. Camus) A. Camus ●

285356　Pasania yungjenensis Hu = Lithocarpus hypoglaucus (Hu) C. C. Huang ex Y. C. Hsu et H. Wei Jen ●

285357　Pasaniopsis Kudo = Castanopsis(D. Don) Spach(保留属名) ●

285358　Pascalia Ortega = Wedelia Jacq. (保留属名) ■●

285359　Pascalia Ortega(1797) ;微冠菊属■

285360　Pascalia glauca Ortega;微冠菊;Beach Creeping Oxeye ■

285361　Pascalia glauca Ortega = Wedelia glauca (Ortega) Hoffm. ex Hicken ■

285362　Pascalium Cass. = Psacalium Cass. ■☆

285363　Paschalococos J. Dransf. (1991) ;复活节岛椰子属●☆

285364　Paschalococos disperta J. Dransf. ;复活节岛椰子●☆

285365　Paschanthus Burch. = Adenia Forssk.●

285366　Paschanthus repandus Burch. = Adenia repanda(Burch.)Engl. ●☆

285367　Paschira G. Kuntze = Pachira Aubl. ●

285368　Pascopyrum Á. Löve = Elymus L. ■

285369　Pascopyrum smithii(Rydb.) Á. Löve = Elymus smithii (Rydb.) Gould ■

285370　Pasina Adans. = Horminum L. ■☆

285371　Pasithea D. Don(1832) ;参差蕊属■☆

285372　Pasithea caerulea(Ruiz et Pav.) D. Don;参差蕊■☆

285373　Pasovia H. Karst. = Phthirusa Mart. ●☆

285374　Paspalanthium Desv. = Paspalum L. ■

285375　Paspalidium Stapf = Setaria P. Beauv. (保留属名)■

285376　Paspalidium Stapf(1920) ;类雀稗属;Paspalidium ■

285377　Paspalidium ankarense A. Camus;安卡拉类雀稗■☆

285378　Paspalidium desertorum(A. Rich.) Stapf;阿拉伯类雀稗■☆

285379　Paspalidium distans (Trin.) Hughes; 广布类雀稗; Spreading Panicgrass ■☆

285380　Paspalidium flavidum (Retz.) A. Camus; 类雀稗; Yellowish Paspalidium ■

285381　Paspalidium geminatum(Forssk.)Stapf;双类雀稗(双生类雀稗); Binate Paspalidium,Egyptian Paspalidium,Egyptian Water Grass ■

285382　Paspalidium mucronatum (Roth ex Roem. et Schult.) Ohwi = Paspalidium punctatum(Burm. f.) A. Camus ■

285383　Paspalidium obtusifolium(Delile) N. D. Simpson;钝叶类雀稗■☆

285384　Paspalidium obtusifolium(Delile) N. D. Simpson var. acutifolium (Coss. et Durieu) Maire = Paspalidium obtusifolium (Delile) N. D. Simpson ■☆

285385　Paspalidium obtusifolium (Delile) Simpson var. acutifolium (Coss. et Durieu) Maire = Paspalidium obtusifolium (Delile) N. D. Simpson ■☆

285386　Paspalidium platyrhachis C. E. Hubb. = Paspalidium obtusifolium (Delile)N. D. Simpson ■☆

285387　Paspalidium punctatum(Burm. f.) A. Camus;尖头类雀稗(类雀稗);Sharplemma Paspalidium ■

285388　Paspalidium punctatum(Burm. f.) A. Camus = Setaria punctata (Burm. f.) Veldkamp ■

285389　Paspalidium tuyamae Ohwi = Paspalidium distans (Trin.) Hughes ■☆

285390　Paspalum L. (1759); 雀稗属 (水草属); Dallisgrass, Fingergrass,Jointgrass,Paspalum ■

285391　Paspalum africanum Poir. = Paspalum conjugatum P. J. Bergius ■

285392　Paspalum akoense Hayata = Paspalum scrobiculatum L. var. bispicatum Hack. ■

285393　Paspalum almum Chase;梳状雀稗;Comb's Crowngrass ■☆

285394　Paspalum ambiguum Lam. et DC. = Digitaria ischaemum (Schreb.) Schreb. ex Muhl. ■

285395　Paspalum annulatum Fluegge = Eriochloa procera (Retz.) C. E. Hubb. ■

285396　Paspalum aquaticum Masam. et Syozi = Paspalidium punctatum (Burm. f.) A. Camus ■

285397　Paspalum auriculatum J. Presl et C. Presl = Paspalum scrobiculatum L. ■

285398　Paspalum barbatum Schumach. = Paspalum scrobiculatum L. ■

285399　Paspalum bicorne Lam. = Digitaria bicornis (Lam.) Roem. et Schult. ■

285400　Paspalum bifidum (Bertol.) Nash; 二 裂 雀 稗; Pitchfork Paspalum ■☆

285401　Paspalum biforme (Willd.) Kunth = Digitaria bicornis (Lam.) Roem. et Schult. ■

285402　Paspalum brevifolium Flugge = Digitaria longiflora(Retz.)Pers. ■

285403　Paspalum bushii Nash = Paspalum setaceum Michx. var. stramineum(Nash) D. J. Banks ■☆

285404　Paspalum bushii Nash = Paspalum setaceum Michx. ■☆

285405　Paspalum cartilagineum J. Presl et C. Presl = Paspalum scrobiculatum L. ■

285406　Paspalum chinense Nees = Digitaria violascens Link ■

285407　Paspalum ciliatifolium Michx. = Paspalum setaceum Michx. ■☆

285408　Paspalum ciliatifolium Michx. var. muhlenbergii (Nash) Fernald = Paspalum setaceum Michx. ■☆

285409　Paspalum ciliatifolium Michx. var. muhlenbergii(Nash)Fernald =

285410　Paspalum ciliatifolium Michx. var. stramineum (Nash) Fernald = Paspalum setaceum Michx. var. stramineum(Nash) D. J. Banks ■☆

285411　Paspalum ciliatifolium Michx. var. stramineum (Nash) Fernald = Paspalum setaceum Michx. ■☆

285412　Paspalum ciliatifolium Trin. = Paspalum conjugatum P. J. Bergius ■

285413　Paspalum commersonii Lam. = Paspalum scrobiculatum L. var. bispicatum Hack. ■

285414　Paspalum commersonii Lam. = Paspalum scrobiculatum L. ■

285415　Paspalum compressum (Sw.) Raspail = Axonopus compressus (Sw.) P. Beauv. ■

285416　Paspalum concinnum (Schrad. ex Steud.) Steud. = Digitaria stricta Roth ex Roem. et Schult. ■

285417　Paspalum conjugatum P. J. Bergius;两耳草(叉仔草,结合雀稗, 双穗雀稗); Hilograss, Sour Dallisgrass, Sour Grass, Sour Paspalum ■

285418　Paspalum conspersum Schrad. ex Schult. ;散生雀稗;Scattered Paspalum ■☆

285419　Paspalum convexum Willd. ex Döll; 美 洲 雀 稗; American Crowngrass ■☆

285420　Paspalum deightonii(C. E. Hubb.)Clayton;戴顿雀稗■☆

285421　Paspalum delavayi Henrard; 云 南 雀 稗 (滇 雀 稗); Delavay Dallisgrass,Delavay Paspalum ■

285422　Paspalum digitaria C. H. Müll. ;指状雀稗■☆

285423　Paspalum dilatatum Poir. ;毛花雀稗;Dallis Grass, Dallisgrass, Hairflower Dallisgrass,Water Paspalum ■

285424　Paspalum dissectum(L.)L. ;泥滩雀稗;Mudbank Paspalum ■☆

285425　Paspalum dissectum Thunb. = Paspalum thunbergii Kunth ex Steud. ■

285426　Paspalum distichum L. ; 双 穗 雀 稗; Dualspike Dallisgrass, Dualspike Paspalum,Knotgrass,Thompsongrass,Water Finger-grass ■

285427　Paspalum distichum L. subsp. paspalodes (Michx.) Thell. = Paspalum distichum L. ■

285428　Paspalum distichum L. subsp. vaginatum (Sw.) Maire = Paspalum vaginatum Sw. ■

285429　Paspalum distichum L. var. vaginatum(Sw.)Griseb. = Paspalum vaginatum Sw. ■

285430　Paspalum distichum Rendle = Paspalum paspalodes (Michx.) Scribn. ■

285431　Paspalum effusum(L.)Raspail = Milium effusum L. ■

285432　Paspalum exile Kippist = Digitaria exilis(Kippist)Stapf ■☆

285433　Paspalum fasciculatum Willd. ex Fluegge;墨西哥雀稗;Mexican Crowngrass ■☆

285434　Paspalum filiculme Nees ex Miq. = Digitaria longiflora (Retz.) Pers. ■

285435　Paspalum fimbriatum Kunth;裂颖雀稗;Fimbriate Dallisgrass, Fimbriate Paspalum,Panama Crowngrass ■

285436　Paspalum fissifolium Raddi = Axonopus fissifolius (Raddi) Kuhlm. ■

285437　Paspalum floridanum Michx. ;佛罗里达雀稗;Florida Paspalum ■☆

285438　Paspalum fluitans (Elliott) Kunth = Paspalum repens P. J. Bergius ■☆

285439　Paspalum foliosum Kunth = Paspalum vaginatum Sw. ■

285440　Paspalum formosanum Honda = Paspalum hirsutum Retz. ■

285441　Paspalum fuscescens J. Presl = Digitaria fuscescens (J. Presl) Henrard ■☆

285442 Paspalum fuscum J. Presl = Digitaria violascens Link ■

285443 Paspalum glumaceum Clayton；颖状雀稗■☆

285444 Paspalum granulare Trin. ex Spreng. = Digitaria abludens
（Roem. et Schult.）Veldkamp ■

285445 Paspalum guadaloupense Steud. = Axonopus compressus（Sw.）
P. Beauv. ■

285446 Paspalum heteranthum Hook. f. = Digitaria heterantha（Hook.
f.）Merr. ■

285447 Paspalum hirsutum Poir. = Paspalum conjugatum P. J. Bergius ■

285448 Paspalum hirsutum Retz.；台湾雀稗；Taiwan Dallisgrass, Taiwan
Paspalum ■

285449 Paspalum humifusum（Pers.）Poir. = Digitaria ischaemum
（Schreb.）Schreb. ex Muhl. ■

285450 Paspalum hydrophilum Henrard；喜水雀稗；Water Paspalum ■☆

285451 Paspalum intermedium Munro ex Morong；间型雀稗；
Intermediate Paspalum ■☆

285452 Paspalum jardinii Steud. = Paspalum scrobiculatum L. ■

285453 Paspalum jubatum Griseb. = Digitaria jubata（Griseb.）Henrard ■

285454 Paspalum kisantuense Vanderyst = Axonopus compressus（Sw.）
P. Beauv. ■

285455 Paspalum kora Willd. = Paspalum scrobiculatum L. ■

285456 Paspalum laeve Michx.；平滑雀稗；Field Paspalum ■☆

285457 Paspalum lamprocaryon K. Schum. = Paspalum scrobiculatum
L. var. lanceolatum de Koning et Sosef ■☆

285458 Paspalum ledermannii Mez = Paspalum scrobiculatum L. ■

285459 Paspalum longiflorum Retz. = Digitaria longiflora（Retz.）Pers. ■

285460 Paspalum longifolium Roxb.；长叶雀稗；Longleaf Dallisgrass,
Longleaf Paspalum, Long-leaved Paspalum ■

285461 Paspalum macrophyllum Kunth；大叶雀稗；Bigleaf Paspalum ■☆

285462 Paspalum malacophyllum Trin.；棱稃雀稗；Ribbed Paspalum,
Softleaf Dallisgrass, Softleaf Paspalum ■

285463 Paspalum minus E. Fourn. ex Hemsl.；小雀稗■☆

285464 Paspalum minutiflorum Steud. = Digitaria violascens Link ■

285465 Paspalum molle J. Presl et C. Presl = Digitaria mollicoma
（Kunth）Henrard ■

285466 Paspalum mollicomum Kunth = Digitaria mollicoma（Kunth）
Henrard ■

285467 Paspalum muhlenbergii Nash = Paspalum setaceum Michx. var.
muhlenbergii（Nash）D. J. Banks ■☆

285468 Paspalum nicorae Parodi；布伦斯维克雀稗；Brunswickgrass ■☆

285469 Paspalum notatum Fluegge；百喜草（金冕草，山雀稗）；Bahia
Grass, Bahiagrass, Bahia-grass, Note Dallisgrass, Note Paspalum ■

285470 Paspalum notatum Fluegge var. latiflorum Döll；大花百喜草；
Bahiagrass ■☆

285471 Paspalum orbiculare G. Forst.；圆果雀稗；Ditch Dallisgrass,
Ditch Paspalum ■

285472 Paspalum orbiculare G. Forst. = Paspalum scrobiculatum L. var.
orbiculare（G. Forst.）Hack. ■

285473 Paspalum orbiculare G. Forst. = Paspalum scrobiculatum L. ■

285474 Paspalum ovatum Nees ex Trin. = Paspalum dilatatum Poir. ■

285475 Paspalum palustre Vanderyst；沼泽雀稗■☆

285476 Paspalum paniculatum L.；多穗雀稗（开穗雀稗）；Paniculate
Dallisgrass, Paniculate Paspalum ■

285477 Paspalum paspalodes（Michx.）Scribn. = Paspalum distichum L. ■

285478 Paspalum pedicellare Trin. ex Hook. f. = Digitaria abludens
（Roem. et Schult.）Veldkamp ■

285479 Paspalum platycaulon Poir. = Axonopus compressus（Sw.）P.
Beauv. ■

285480 Paspalum platyculmus Thouars ex Nees = Axonopus compressus
（Sw.）P. Beauv. ■

285481 Paspalum plicatulum Michx.；皱稃雀稗（棕籽雀稗）；
Brownseed Paspalum, Wrinklelemma Dallisgrass ■

285482 Paspalum plicatum Pers. = Paspalum plicatulum Michx. ■

285483 Paspalum polyphyllum Nees ex Trin.；多叶雀稗■☆

285484 Paspalum polystachyum R. Br. = Paspalum scrobiculatum L. ■

285485 Paspalum preslii Kunth = Digitaria longiflora（Retz.）Pers. ■

285486 Paspalum pseudodurva Nees = Digitaria longiflora（Retz.）Pers. ■

285487 Paspalum pseudodurva Nees var. minus Nees = Digitaria
longiflora（Retz.）Pers. ■

285488 Paspalum pseudosetaria Steud. = Digitaria stricta Roth ex
Roem. et Schult. ■

285489 Paspalum pubescenes J. Presl = Digitaria longiflora（Retz.）Pers. ■

285490 Paspalum pubescens Muhl. ex Willd. = Paspalum setaceum
Michx. var. muhlenbergii（Nash）D. J. Banks ■☆

285491 Paspalum pubiflorum Rupr. ex E. Fourn.；墨西哥毛花雀稗；
Paspalum ■☆

285492 Paspalum pulchellum（R. Br.）Raspail；丽雀稗；Grand
Paspalum ■☆

285493 Paspalum punctatum（Burm. f.）Stapf ex Ridl. = Paspalidium
punctatum（Burm. f.）A. Camus ■

285494 Paspalum quadrifarium Lam.；草丛雀稗；Tussock Paspalum ■☆

285495 Paspalum racemosum Lam.；秘鲁雀稗；Peruvian Paspalum ■☆

285496 Paspalum repens P. J. Bergius；水雀稗；Horsetail Paspalum,
Water Paspalum ■☆

285497 Paspalum rounkiserii Mez = Axonopus compressus（Sw.）P.
Beauv. ■

285498 Paspalum royleanum Nees ex Hook. f. = Digitaria stricta Roth ex
Roem. et Schult. ■

285499 Paspalum royleanum Nees ex Thwaites = Digitaria stricta Roth ex
Roem. et Schult. ■

285500 Paspalum sanguinale（L.）Lam. = Digitaria sanguinalis（L.）Scop. ■

285501 Paspalum sanguinale（L.）Lam. var. ciliare（Retz.）Hook. f. =
Digitaria ciliaris（Retz.）Koeler ■

285502 Paspalum sanguinale（L.）Lam. var. cruciatum（Nees ex Steud.）
Hook. f. = Digitaria cruciata（Nees ex Steud.）A. Camus ■

285503 Paspalum sanguinale（L.）Lam. var. debile Hook. f. = Digitaria
radicosa（J. Presl et C. Presl）Miq. ■

285504 Paspalum sanguinale（L.）Lam. var. pabulare（Aitch.）Hook. f.
= Digitaria nodosa Parl. ■☆

285505 Paspalum sanguinale（L.）Lam. var. pruriens（Trin.）Hook. f. =
Digitaria setigera Roth ex Roem. et Schult. ■

285506 Paspalum sanguinale（L.）Lam. var. rottleri Hook. f. = Digitaria
ciliaris（Retz.）Koeler ■

285507 Paspalum scrobiculatum L.；鸭［也 ＋ 母］草（长叶雀稗，鸭
草，圆果雀稗，皱稃雀稗）；Duckess Grass, Indian Paspalum, Kodo
Millet, Kodomillet, Rice Grass, Rice Grass Mau'u-Laiki, Ricegrass ■

285508 Paspalum scrobiculatum L. var. bispicatum Hack.；囡雀稗（南
雀稗，雀稗，台岛雀稗，台湾雀稗）；Commerson Dallisgrass,
Commerson Paspalum ■

285509 Paspalum scrobiculatum L. var. bispicatum Hack. = Paspalum
scrobiculatum L. ■

285510 Paspalum scrobiculatum L. var. commersonii（Lam.）Stapf =
Paspalum commersonii Lam. ■

285511 Paspalum scrobiculatum L. var. commersonii（Lam.）Stapf =

Paspalum scrobiculatum L. var. bispicatum Hack. ■

285512　Paspalum scrobiculatum L. var. commersonii (Lam.) Stapf = Paspalum scrobiculatum L. ■

285513　Paspalum scrobiculatum L. var. deightonii C. E. Hubb. = Paspalum deightonii(C. E. Hubb.) Clayton ■☆

285514　Paspalum scrobiculatum L. var. jardinii (Steud.) Franch. = Paspalum scrobiculatum L. ■

285515　Paspalum scrobiculatum L. var. lanceolatum de Koning et Sosef; 剑叶鸭嗼草■☆

285516　Paspalum scrobiculatum L. var. longifolium (Roxb.) Domin = Paspalum longifolium Roxb. ■

285517　Paspalum scrobiculatum L. var. orbiculare (G. Forst.) Hack. = Paspalum orbiculare G. Forst. ■

285518　Paspalum scrobiculatum L. var. polystachyum (R. Br.) Stapf = Paspalum scrobiculatum L. ■

285519　Paspalum scrobiculatum L. var. thunbergii (Kunth ex Steud.) Makino = Paspalum thunbergii Kunth ex Steud. ■

285520　Paspalum scrobiculatum L. var. velutinum Hack. ;毛鸭姆草■☆

285521　Paspalum scrobiculatum L. var. velutinum Hack. = Paspalum scrobiculatum L. ■

285522　Paspalum setaceum Michx. ; 丝叶雀稗; Hairy Bead Grass, Paspalum, Thin Paspalum ■☆

285523　Paspalum setaceum Michx. var. calvescens Fernald = Paspalum setaceum Michx. var. muhlenbergii (Nash) D. J. Banks ■☆

285524　Paspalum setaceum Michx. var. muhlenbergii (Nash) D. J. Banks; 米伦伯格雀稗; Hairy Bead Grass, Hairy Lens Grass, Hurrahgrass, Muhlenberg's Hairy Bead Grass, Thin Paspalum ■☆

285525　Paspalum setaceum Michx. var. stramineum(Nash) D. J. Banks; 淡黄丝叶雀稗; Downy Lens Grass, Hairy Bead Grass, Straw-colored Hairy Bead Grass, Yellow Sand Paspalum ■☆

285526　Paspalum squamatum Steud. = Paspalum vaginatum Sw. ■

285527　Paspalum stramineum Nash = Paspalum setaceum Michx. var. stramineum(Nash) D. J. Banks ■☆

285528　Paspalum ternatum (Hochst. ex A. Rich.) Hook. f. = Digitaria ternata(Hochst. ex A. Rich.) Stapf ex Dyer ■

285529　Paspalum thouarsianum Flüggé = Digitaria thouarsiana (Flüggé) A. Camus ■☆

285530　Paspalum thunbergii Kunth ex Steud. ; 雀稗（鱼眼草，猪儿草）;Japan Dallisgrass,Japanese Paspalum ■

285531　Paspalum thunbergii Kunth ex Steud. var. minor Makino = Paspalum orbiculare G. Forst. ■

285532　Paspalum thunbergii Kunth ex Steud. var. minus Makino = Paspalum scrobiculatum L. var. orbiculare(G. Forst.) Hack. ■

285533　Paspalum urvillei Steud. ; 丝毛雀稗（吴氏雀稗）; Silkhair Dallisgrass, Silkhair Paspalum, Vasey Grass, Vasey-grass, Vasey's Grass ■

285534　Paspalum vaginatum Sw. ; 海雀稗; Sea Dallisgrass, Sea Paspalum,Seashore Paspalum ■

285535　Paspalum villosum Thunb. = Eriochloa villosa(Thunb.) Kunth ■

285536　Paspalum virgatum L. ; 粗秆雀稗; Thickculm Dallisgrass, Thickculm Paspalum ■

285537　Paspalum wombaliense Vanderyst = Axonopus flexuosus(Peter) C. E. Hubb. ■☆

285538　Paspalum xizangense B. S. Sun et H. Sun = Axonopus fissifolius (Raddi) Kuhlm. ■

285539　Paspalus Flüggé = Paspalum L. ■

285540　Passacardoa Wild = Pasaccardoa Kuntze ■●☆

285541　Passaea Adans. = Ononis L. ■■

285542　Passaea Baill. = Bernardia L. ■

285543　Passaea Baill. = Polyscias J. R. Forst. et G. Forst. ■

285544　Passalia Sol. ex R. Br. = Rinorea Aubl. (保留属名)■

285545　Passaveria Mart. et Eichler = Ecclinusa Mart. ●☆

285546　Passaveria Mart. et Eichler ex Miq. = Ecclinusa Mart. ●☆

285547　Passerina L. (1753);麻雀木属;Sparrow-wort ●☆

285548　Passerina annua Wikstr. = Thymelaea passerina (L.) Coss. et Germ. ■

285549　Passerina annua Wikstr. var. algeriensis Chabert = Thymelaea gussonei Boreau ■☆

285550　Passerina anthylloides L. f. = Gnidia anthylloides(L. f.)Gilg ●☆

285551　Passerina burchellii Thoday;伯切尔麻雀木●☆

285552　Passerina canescens Schousb. = Thymelaea lanuginosa (Lam.) Ceballos et Vicioso ●☆

285553　Passerina capitata L. = Lachnaea capitata(L.) Crantz ●☆

285554　Passerina chamaedaphne Bunge = Wikstroemia chamaedaphne Meisn. ●

285555　Passerina ciliata L. = Struthiola ciliata(L.) Lam. ●☆

285556　Passerina comosa(Meisn.) C. H. Wright;簇毛麻雀木●☆

285557　Passerina corymbosa Eckl. ex C. H. Wright;伞序麻雀木●☆

285558　Passerina dodecandra L. = Struthiola dodecandra(L.) Druce ●☆

285559　Passerina drakensbergensis Hilliard et B. L. Burtt;德拉肯斯麻雀木●☆

285560　Passerina ericoides L. ;石南状麻雀木●☆

285561　Passerina eriocephala Thunb. = Lachnaea pedicellata Beyers ●☆

285562　Passerina esterhuyseniae Bredenk. et A. E. van Wyk;埃斯特麻雀木●☆

285563　Passerina falcifolia(Meisn.) C. H. Wright;镰叶麻雀木●☆

285564　Passerina filiformis L. ;丝状麻雀木●☆

285565　Passerina filiformis L. subsp. glutinosa(Thoday) Bredenk. et A. E. van Wyk;黏性麻雀木●☆

285566　Passerina filiformis L. var. comosa Meisn. = Passerina comosa (Meisn.) C. H. Wright ●☆

285567　Passerina filiformis L. var. falcifolia Meisn. = Passerina falcifolia (Meisn.) C. H. Wright ●☆

285568　Passerina filiformis L. var. glutinosa Thoday = Passerina filiformis L. subsp. glutinosa(Thoday) Bredenk. et A. E. van Wyk ●☆

285569　Passerina filiformis L. var. squarrosa Meisn. = Passerina rubra C. H. Wright ●☆

285570　Passerina filiformis L. var. vulgaris Meisn. = Passerina corymbosa Eckl. ex C. H. Wright ●☆

285571　Passerina galpinii C. H. Wright;盖尔麻雀木●☆

285572　Passerina ganpi Siebold et Zucc. = Diplomorpha ganpi(Siebold et Zucc.) Nakai ●

285573　Passerina globosa Lam. = Raspalia globosa(Lam.)Pillans ●☆

285574　Passerina grandiflora L. f. = Lachnaea grandiflora(L. f.)Baill. ●☆

285575　Passerina hirsuta L. = Thymelaea hirsuta(L.) Endl. ■☆

285576　Passerina japonica Siebold et Zucc. = Wikstroemia trichotoma (Thunb.) Makino ●

285577　Passerina japonica Thunb. = Diplomorpha trichotoma(Thunb.) Nakai ●

285578　Passerina laniflora C. H. Wright = Lachnaea laniflora (C. H. Wright)Bond ●☆

285579　Passerina laxa L. f. = Gnidia laxa(L. f.)Gilg ●☆

285580　Passerina lessertii Wikstr. = Stellera lessertii (Wikstr.) C. A. Mey. ●☆

285581　Passerina linearifolia Wikstr. = Gnidia linearifolia (Wikstr.) B. Peterson ●☆

285582　Passerina montana Thoday;山地麻雀木●☆

285583　Passerina montivaga Bredenk. et A. E. van Wyk;漫山麻雀木●☆

285584　Passerina nervosa Thunb. = Lachnaea nervosa(Thunb.)Meisn. ●☆

285585　Passerina nitida Vahl = Thymelaea argentata (Lam.)Pau ●☆

285586　Passerina obtusifolia Thoday;钝叶麻雀木●☆

285587　Passerina paleacea Wikstr.;膜片麻雀木●☆

285588　Passerina paludosa Thoday;沼泽麻雀木●☆

285589　Passerina pendula Eckl. et Zeyh. ex Thoday;下垂麻雀木●☆

285590　Passerina pentandra Thunb. = Lonchostoma pentandrum(Thunb.) Druce ●☆

285591　Passerina quadrifaria Bredenk. et A. E. van Wyk;四出麻雀木●☆

285592　Passerina racemosa Wikstr. = Stelleropsis altaica (Thieb.-Bern.)Pobed. ■

285593　Passerina rigida Wikstr.;坚硬麻雀木●☆

285594　Passerina rigida Wikstr. var. comosa Meisn. = Passerina pendula Eckl. et Zeyh. ex Thoday ●☆

285595　Passerina rigida Wikstr. var. tetragona Meisn. = Passerina truncata(Meisn.)Bredenk. et A. E. van Wyk ●☆

285596　Passerina rigida Wikstr. var. truncata Meisn. = Passerina truncata(Meisn.)Bredenk. et A. E. van Wyk ●☆

285597　Passerina rubra C. H. Wright;红麻雀木●☆

285598　Passerina spicata L. f. = Gnidia spicata(L. f.)Gilg ●☆

285599　Passerina striata Lam. = Lachnaea striata (Lam.)Meisn. ●☆

285600　Passerina truncata(Meisn.)Bredenk. et A. E. van Wyk;平截麻雀木●☆

285601　Passerina truncata(Meisn.)Bredenk. et A. E. van Wyk subsp. monticola Bredenk. et A. E. van Wyk;山地平截麻雀木●☆

285602　Passerina uniflora L. = Lachnaea uniflora(L.)Crantz ●☆

285603　Passerina vesiculosa Fisch. et C. A. Mey. = Diarthron vesiculosum C. A. Mey. ■

285604　Passerina virgata Desf. = Thymelaea virgata(Desf.)Endl. ■☆

285605　Passerina virgata Desf. var. broussonneti Ball = Thymelaea virgata(Desf.)Endl. subsp. broussonetii(Ball)Kit Tan ■☆

285606　Passerina vulgaris(Meisn.)Thoday = Passerina corymbosa Eckl. ex C. H. Wright ●☆

285607　Passiflora L. (1753);西番莲属;Granadilla, Passion Flower, Passion Fruit, Passionflower, Passion-flower ●■

285608　Passiflora acerifolia Cham. et Schltdl. = Passiflora adenopoda DC. ●

285609　Passiflora acuminata DC. = Passiflora laurifolia L. ●■

285610　Passiflora adenopoda DC. ;腺柄西番莲●

285611　Passiflora alata Aiton;翅茎西番莲(具刺西番莲);Wingstem Passionflower ■☆

285612　Passiflora alato-caerulea Lindl. ;蓝翅秋海棠;Passion Flower, Passion Vine, Skybluewing Passionflower ■

285613　Passiflora altebilobata Hemsl. ;月叶西番莲(蝴蝶暗消,苦胆七,藤子暗消,燕子尾,羊蹄暗消,月叶秋海棠);Bifid Passionflower ■

285614　Passiflora amethystina J. C. Mikan;紫水晶西番莲■☆

285615　Passiflora antioquiensis H. Karst. ;长管西番莲(哥伦比亚西番莲);Antioquien Passionflower, Banana Passion Fruit, Banana-passion Fruit ●☆

285616　Passiflora arborea Spreng. ;哥伦比亚西番莲●☆

285617　Passiflora assamica Chakr. = Passiflora perpera Mast. ■

285618　Passiflora assamica Chakr. = Passiflora wilsonii Hemsl. ■

285619　Passiflora bicornis Mill. ; 双球西番莲; Wingleaf Passion Flower, Wingleaf Passionflower ☆

285620　Passiflora biflora Lam. ;双花西番莲(二花西番莲);Twoflower Passionflower, Two-flowered Passion Vine ■☆

285621　Passiflora bryonioides Kunth;芭秧西番莲■☆

285622　Passiflora burmanica Chakr. = Passiflora jugorum W. W. Sm. ■

285623　Passiflora caerulea L. ;西番莲(秋海棠,时计草,时计果,西蕃莲,西洋鞠,洋酸茄花,玉蕊花,玉藥花,转心莲,转枝莲,子午花);Blue Crown Passion Flower, Blue Crown Passion-flower, Blue Passion Flower, Blue Passionflower, Blue Passion-vine, Bluecrown Passionflower, Blue-crown Passionflower, Christ-and-the-apostles, Common Passion Flower, Common Passionflower, Crown of Thorns, Eastern Star, Good Friday Flower, Gramophone Horns, Granadilla, Grenadille, Maracock, Passion Flower, Passionflower, Star of Bethlehem, Story-of-tile-cross, Twelve Disciples ●

285624　Passiflora caerulea L. 'Constance Elliot';白冠西番莲●

285625　Passiflora caerulea Lour. ex DC. = Passiflora caerulea L. ●

285626　Passiflora capsularis L. ;蒴果西番莲■☆

285627　Passiflora celata G. Cusset = Passiflora wilsonii Hemsl. ■

285628　Passiflora chinensis Mast. = Passiflora caerulea L. ●

285629　Passiflora chinensis Sweet = Passiflora moluccana Reinw. ex Blume var. teysmanniana(Miq.)W. J. de Wilde ■

285630　Passiflora cincinnata Mast. ;卷西番莲;Crato Passion Fruit ●☆

285631　Passiflora cinnabarina Lindl. ;朱砂西番莲;Red Passionflower, Vermilion Passionflower ■☆

285632　Passiflora coccinea Aubl. ;猩红西番莲(绯红西番莲);Red Granadilla, Red Passionflower, Scarlet Passionflower ■☆

285633　Passiflora cochinchinensis Spreng. = Passiflora moluccana Reinw. ex Blume var. teysmanniana(Miq.)W. J. de Wilde ■

285634　Passiflora coriacea Juss. ;革质西番莲;Bat Leaf Passion Flower ■☆

285635　Passiflora cupiformis Mast. ;杯叶西番莲(半边风,半截叶,杯叶秋海棠,叉痔草,对叉疗药,飞蛾草,蝴蝶暗消,金剪刀,马蹄暗消,双飞蝴蝶,四方台,燕尾草,羊蹄暗消,羊蹄草);Cupleaf Passionflower, Cup-leaved Passionflower ●■

285636　Passiflora eberhardtii Gagnep. ; 心 叶 西番莲; Eberhardt's Passionflower ●

285637　Passiflora edulis Sims;紫果西番莲(百香果,鸡蛋果,鸡蛋花,时计,时计果,土罗汉果,西番果,西番莲,洋石榴,紫果秋海棠);Edible Granadilla, Gramophone Horns, Granadilla, Grenadille, Passion Fruit, Passionflower, Passionfruit, Purple Granadilla, Sweet Cup ■

285638　Passiflora edulis Sims var. flavicarpa?;黄果西番莲;Yellow Passion Fruit ■☆

285639　Passiflora foetida L. ;龙珠果(大种毛葫芦,风雨花,假苦瓜,假苦果,龙吞珠,龙须果,龙眼果,龙珠草,龙爪球,毛蛉儿,肉果,山木鳖,天仙果,西番莲,香瓜子,香花果,野仙桃);Fetid Passionflower, Love-in-a-hedge, Passion Flower, Purple Granadilla, Runningpop, Stinking Passion Flower, Tagua Passionflower, Wild Passionflower, Wild Water Lemon ■

285640　Passiflora foetida L. f. glabra A. Fern. et R. Fern. ;光滑西番莲■☆

285641　Passiflora foetida L. var. hispida (DC. ex Triana et Planch.) Killip;毛西番莲;Weed Passionflower ■

285642　Passiflora foetida L. var. hispida (DC. ex Triana et Planch.) Killip = Passiflora foetida L. ■

285643　Passiflora foetida L. var. lanuginosa Killip;多毛西番莲;Scarletfruit Passionflower ■☆

285644　Passiflora franchetiana Hemsl. = Passiflora cupiformis Mast. ●■

285645　Passiflora gracilis Jacq. ex Link;纤细西番莲(细柱秋海棠,细柱西番莲);Crincle Passionflower, Crincled Passionflower, Crincled

Passion-flower, Slender Passion-flower ■☆

285646　Passiflora hahnii Mast. 哈氏西番莲■☆

285647　Passiflora hainanensis Hance = Passiflora cochinchinensis Spreng. ■

285648　Passiflora hainanensis Hance = Passiflora moluccana Reinw. ex Blume var. teysmanniana(Miq.)W. J. de Wilde ■

285649　Passiflora henryi Hemsl. ;圆叶西番莲(锅铲叶,老鼠铃,龙珠果,闹蛆叶,螃蟹眼睛草, 燕子尾, 圆叶秋海棠）; Henry Passionflower ■

285650　Passiflora hirsuta L. = Passiflora foetida L. ■

285651　Passiflora hispida DC. ex Triana et Planch. = Passiflora foetida L. ■

285652　Passiflora horsfieldii Blume = Passiflora cochinchinensis Spreng. ■

285653　Passiflora horsfieldii Blume var. elbertiana Hallier f. = Passiflora cochinchinensis Spreng. ■

285654　Passiflora incarnata L. ; 粉色西番莲;Apricot Vine, Mammey Apple, Mattop, May Apple, May-apple, Maypop, Maypops, Passion Flower, Passionflower, Purple Passion Flower, Wild Passionflower, Wild Passion-flower ■☆

285655　Passiflora jamesonii L. H. Bailey;詹氏西番莲;Passion Flower ☆

285656　Passiflora jianfengensis S. M. Hwang et Q. Huang;尖峰西番莲; Jianfeng Passionflower ●

285657　Passiflora jugorum W. W. Sm. ;山峰西番莲(石山南星, 燕子尾）; Yoke Passionflower ●■

285658　Passiflora kwangsiensis H. L. Li = Passiflora cupiformis Mast. ●■

285659　Passiflora kwangtungensis Merr. ;广东西番莲(广东秋海棠,散痧草）;Guangdong Passionflower ●■

285660　Passiflora laurifolia L. ;樟叶西番莲(水柠檬,樟叶秋海棠）; Bell-apple, Bull Apple, Cinnamonleaf Passionflower, Golden Bellapple, Jamaica Honeysuckle, Jamaica Honey-suckle, Sweet Cup, Water Lemon, Water-lemon, Water-lemon Jamaica Honey Suckle, Yellow Granadilla ●■

285661　Passiflora ledongensis S. M. Huang et Q. Huang;乐东西番莲; Ledong Passionflower ●■

285662　Passiflora ligularis A. Juss. ; 甜果西番莲; Sweet Granadilla, Sweet Passionflower ■☆

285663　Passiflora ligulifolia Mast. = Passiflora cochinchinensis Spreng. ■

285664　Passiflora ligulifolia Mast. = Passiflora moluccana Reinw. ex Blume var. teysmanniana(Miq.)W. J. de Wilde ■

285665　Passiflora loureiroi G. Don = Passiflora caerulea L. ●

285666　Passiflora lutea L. ; 深黄西番莲;Passion Flower, Yellow Passionflower ■☆

285667　Passiflora macrocarpa Mast. = Passiflora quadrangularis L. ●

285668　Passiflora maliformis L. ;多形西番莲;Apple-fruited Granadilla, Sweet Calabash, Sweet Cup ■☆

285669　Passiflora manicata Pers. = Passiflora manicata(Juss.)Pers. ●☆

285670　Passiflora minima Blanco = Passiflora edulis Sims ■

285671　Passiflora minima L. ;小西番莲■☆

285672　Passiflora mollissima(Kunth)L. H. Bailey;毛叶西番莲(毛鸡蛋果,绒毛西番莲）;Banana Passion Flower, Banana Passion Fruit, Banana Passionflower, Banana Passionfruit, Banana Poka, Banana-passion Fruit, Softleaf Passionflower, Softleaf Passion-flower, Sweet Calabash ●☆

285673　Passiflora moluccana Reinw. ex Blume;马六甲蛇王藤(马来蛇王藤）■

285674　Passiflora moluccana Reinw. ex Blume var. glaberrima (Gagnep.)W. J. de Wilde = Passiflora tonkinensis W. J. de Wilde ■

285675　Passiflora moluccana Reinw. ex Blume var. glaberrima

285676　Passiflora moluccana Reinw. ex Blume var. teysmanniana(Miq.)W. J. de Wilde = Passiflora cochinchinensis Spreng. ■

285677　Passiflora moluccana Reinw. ex Blume var. teysmanniana(Miq.)W. J. de Wilde;蛇王藤(海南西番莲,黄豆树,两眼蛇,山水瓜,蛇王,蛇眼藤,双目灵,治蛇灵）;Cochinchina Passionflower ■

285678　Passiflora monicata(Juss.)Pers. ;红花西番莲(紫心西番莲）; Red Passionflower, Red Passion-flower, Red Passion-vine ●☆

285679　Passiflora morifolia Mast. ;桑叶西番莲;Woodland Passionflower ●☆

285680　Passiflora nitida Kunth;光亮西番莲;Bel-apple ●☆

285681　Passiflora octandra Gagnep. = Passiflora siamica Craib ●

285682　Passiflora octandra Gagnep. var. attopensis Gagnep. = Passiflora siamica Craib ●

285683　Passiflora octandra Gagnep. var. cochinchinensis Gagnep. = Passiflora siamica Craib ●

285684　Passiflora octandra Gagnep. var. glaberrima Gagnep. = Passiflora tonkinensis W. J. de Wilde ■

285685　Passiflora pallida Lour. = Passiflora cochinchinensis Spreng. ■

285686　Passiflora papilio H. L. Li;蝴蝶藤(半边草,蝴蝶草,羊角断）; Butterfly Passionflower ■

285687　Passiflora passiflora Mast. = Passiflora cupiformis Mast. ●■

285688　Passiflora pennagiana Wall. = Adenia penangiana(Wall. ex G. Don)J. J. de Wilde ●

285689　Passiflora perpera Mast. ;半边风■

285690　Passiflora perpera Mast. = Passiflora wilsonii Hemsl. ■

285691　Passiflora philippinensis Elmer = Passiflora cochinchinensis Spreng. ■

285692　Passiflora pinnastipula Cav. ;智利西番莲●☆

285693　Passiflora popenovii Killip;波佩诺夫西番莲●☆

285694　Passiflora quadrangularis L. ;大西番莲(大果计时草,大果秋海棠,大果西番莲,大秋海棠,大西番果,大转心莲,黄鸡蛋果,日本瓜）;Barbadine, Common Granadilla, Giant Granadilla, Granadilla, Granadilla Vine, Maracuja, Square-stalked Passion Fruit, Square-stalked PassionflowerMaracuya ●

285695　Passiflora racemosa Brot. ;总状花西番莲(总序西番莲）; Princes Chalote's Passion-flower, Racemose Passionflower, Red Passion Flower, Red Passionflower ●☆

285696　Passiflora raddeana DC. = Passiflora coccinea Aubl. ■☆

285697　Passiflora rhombiformis S. Y. Bao;菱叶秋海棠(菱叶西番莲）; Rhombicleaf Passionflower ■

285698　Passiflora rhombiformis S. Y. Bao = Passiflora wilsonii Hemsl. ■

285699　Passiflora riparia Mart. ex Mast. ;河岸西番莲■☆

285700　Passiflora seguinii H. Lév. et Vaniot = Passiflora cupiformis Mast. ●■

285701　Passiflora serratifolia L. ;齿叶西番莲■☆

285702　Passiflora siamica Craib;长叶西番莲(八蕊西番莲,毛蛇王藤）;Siam Passionflower ●

285703　Passiflora suberosa L. ;细柱西番莲(南美西番莲,三角叶西番莲）;Corkystem Passionflower, Ink Vine, Meloncillo ■

285704　Passiflora subpeltata Ortega;白西番莲(近盾西番莲）;White Passionflower ■☆

285705　Passiflora tetrandra Banks et Sol. ex DC. ;四雄蕊西番莲■☆

285706　Passiflora tiliifolia L. ;椴叶西番莲■☆

285707　Passiflora tinifolia Juss. = Passiflora laurifolia L. ●■

285708　Passiflora tonkinensis W. J. de Wilde;东京蛇王藤(长叶蛇王藤）■

285709　Passiflora trifasciata Lem. ;三纹西番莲(三裂西番莲）;Three-banded Passionflower ■☆

285710   Passiflora tripartita(Juss.)Poir.;三裂西番莲;Banana Poka ■☆

285711   Passiflora tripartita (Juss.) Poir. var. mollissima (Kunth) Holm-Nielsen et P. Jorg. = Passiflora mollissima(Kunth) L. H. Bailey ●☆

285712   Passiflora tripartita Breiter var. mollissima(Kunth) Holm-Nielsen et P. Jorg. = Passiflora mollissima(Kunth) L. H. Bailey ●☆

285713   Passiflora tuberosa Jacq.;块状西番莲;Tuberous Passionflower ■☆

285714   Passiflora violacea Vell.;紫花西番莲(蓝冠西番莲,紫西番莲);Violet Passionflower ●☆

285715   Passiflora violacea Vell. = Passiflora amethystina J. C. Mikan ■☆

285716   Passiflora viridiflora Cav. = Passiflora suberosa L. ■

285717   Passiflora vitifolia Kunth;葡萄叶西番莲;Crimson passion Flower,Perfumed Passionflower,Vine Leaf Passion Flower ●☆

285718   Passiflora wangii Hu = Passiflora siamica Craib ●

285719   Passiflora wilsonii Hemsl.;镰叶西番莲(半节观音,半节叶,半截,半截叶,锅铲叶,金边莲,镰叶秋海棠);Sickleleaf Passionflower ■

285720   Passiflora xishuangbannaensis Krosnick;版纳西番莲■

285721   Passiflora yunnanensis Franch. = Passiflora cupiformis Mast. ●■

285722   Passifloraceae Juss. = Passifloraceae Juss. ex Roussel(保留科名)●■

285723   Passifloraceae Juss. ex DC. = Passifloraceae Juss. ex Roussel(保留科名)●■

285724   Passifloraceae Juss. ex Kunth = Passifloraceae Juss. ex Roussel(保留科名)●■

285725   Passifloraceae Juss. ex Roussel(1806)(保留科名);西番莲科;Passionflower Family ●■

285726   Passoura Aubl. = Rinorea Aubl. (保留属名)●

285727   Passovia H. Karst. = Loranthus Jacq. (保留属名)●

285728   Passovia H. Karst. ex Klotzsch = Loranthus Jacq. (保留属名)●

285729   Passowia H. Karst. = Passovia H. Karst. ex Klotzsch ●

285730   Passowia H. Karst. = Phthirusa Mart. ●☆

285731   Pastinaca L. (1753);欧洲防风属(欧防风属);Parsnip ■

285732   Pastinaca armena Fisch. et C. A. Mey.;亚美尼亚欧防风■☆

285733   Pastinaca atropurpurea Steud. ex A. Rich. = Erythroselinum atropurpureum(Steud. ex A. Rich.) Chiov. ■☆

285734   Pastinaca aurantiaca(Albov) Kolak.;橘色欧防风■☆

285735   Pastinaca capensis Sond. = Pastinaca sativa L. ■

285736   Pastinaca dasycarpa Regel et Schmalh. = Semenovia dasycarpa (Regel et Schmalh.) Korovin ex Pimenov et V. N. Tikhom. ■

285737   Pastinaca lanata (Michx.) Koso-Pol. = Heracleum sphondylium L. subsp. montanum(Schleich. ex Gaudin) Briq. ■

285738   Pastinaca opaca Bernh. ex Hornem.;毛欧防风;Opaque Parsnip,Shaded Parsnip ■☆

285739   Pastinaca pimpinellifolia M. Bieb.;茴芹叶欧防风■☆

285740   Pastinaca pratensis Martensen;草甸欧防风;Mea Parsnip dow ■☆

285741   Pastinaca sativa L.;欧防风(欧独活,欧洲防风);Bird's Tongue,Cow Cakes,Cow Flop,Cowflop,Garden Parsnip,Heel Trot,Hockweed,Kager,Kaiyer,Midden Mylies,Panes,Parsnip,Pasment,Pasmet,Passment,Skiwet,Wild Parsnip ■

285742   Pastinaca sativa L. hortensis?;田园欧防风;Parsnip ■☆

285743   Pastinaca sativa L. subsp. sylvestris? = Pastinaca sylvestris Mill. ■☆

285744   Pastinaca sylvestris Mill.;林欧防风;Wild Garden Parsnip,Wild Garden-parsnip,Wild Parsnip,Woods Parsnip ■☆

285745   Pastinaca umbrosa Stev. ex DC.;耐荫欧防风■☆

285746   Pastinacaceae Martinov = Apiaceae Lindl. (保留科名)●■

285747   Pastinacaceae Martinov = Umbelliferae Juss. (保留科名)●■

285748   Pastinacopsis Golosk. (1950);冰防风属(水防风属);Pastinacopsis ■

285749   Pastinacopsis glacialis Golosk.;冰防风;Glacial Pastinacopsis ■

285750   Pastoraea Tod. = Pastorea Tod. ex Bertol. ■

285751   Pastorea Tod. ex Bertol. = Ionopsidium Rchb. ■☆

285752   Patabea Aubl. = Ixora L. ●

285753   Patagnana Steud. = Petagnana J. F. Gmel. ●■

285754   Patagnana Steud. = Smithia Aiton(保留属名)●■

285755   Patagonia T. Durand et Jacks. = Adesmia DC. (保留属名)■☆

285756   Patagonica Boehm. = Patagonula L. ●☆

285757   Patagonium E. Mey. = Aeschynomene L. ●■

285758   Patagonium Schrank(废弃属名) = Adesmia DC. (保留属名)■☆

285759   Patagonula L. (1753);帕塔厚壳属●☆

285760   Patagonula americana L.;南美洲帕塔厚壳●☆

285761   Patagua Poepp. ex Baill. = Orites R. Br. ●☆

285762   Patagua Poepp. ex Rchb. = Villaresia Ruiz et Pav. ●☆

285763   Patamoguton Honck. = Potamogeton L. ■

285764   Patascoya Urb. = Freziera Willd. (保留属名)●☆

285765   Patellaria J. T. Williams = Beta L. ■

285766   Patellaria J. T. Williams A. J. Scott et Ford-Lloyd = Patellifolia A. J. Scott, Ford-Lloyd et J. T. Williams ■

285767   Patellaria J. T. Williams et Ford-Lloyd = Beta L. ■

285768   Patellaria J. T. Williams et Ford-Lloyd ex J. T. Williams, A. J. Scott et Ford-Lloyd = Patellifolia A. J. Scott, Ford-Lloyd et J. T. Williams ■

285769   Patellaria J. T. Williams, A. J. Scott et Ford-Lloyd = Beta L. ■

285770   Patellaria cordata J. T. Williams, A. J. Scott et Ford-Lloyd = Patellifolia patellaris(Moq.) A. J. Scott,Ford-Lloyd et J. T. Williams ■☆

285771   Patellaria procumbens(C. Sm.) J. T. Williams et A. J. Scott et Ford-Lloyd = Patellifolia procumbens (C. Sm.) A. J. Scott et Ford-Lloyd et J. T. Williams ■☆

285772   Patellaria webbiana(Moq.) J. T. Williams et A. J. Scott et Ford-Lloyd = Patellifolia webbiana(Moq.) A. J. Scott et al. ■☆

285773   Patellifolia A. J. Scott et Ford-Lloyd = Beta L. ■

285774   Patellifolia A. J. Scott, Ford-Lloyd et J. T. Williams = Beta L. ■

285775   Patellifolia A. J. Scott, Ford-Lloyd et J. T. Williams = Patellaria J. T. Williams ■

285776   Patellifolia patellaris (Moq.) A. J. Scott, Ford-Lloyd et J. T. Williams = Beta patellaris Moq. ■☆

285777   Patellifolia procumbens(C. Sm.) A. J. Scott, Ford-Lloyd et J. T. Williams = Beta procumbens C. Sm. ■☆

285778   Patellifolia webbiana(Moq.) A. J. Scott et al. = Beta webbiana Moq. ■☆

285779   Patellifolia webbiana (Moq.) A. J. Scott et al. = Patellaria webbiana(Moq.) J. T. Williams et A. J. Scott et Ford-Lloyd ■☆

285780   Patellocalamus W. T. Lin = Ampelocalamus S. L. Chen, T. H. Wen et G. Y. Sheng ●★

285781   Patellocalamus W. T. Lin = Dendrocalamus Nees ●

285782   Patellocalamus gongshanensis T. P. Yi = Ampelocalamus mianningensis(Q. Li et X. Jiang) D. Z. Li et Stapleton ●

285783   Patellocalamus mianningensis (Q. Li et X. Jiang) T. P. Yi. = Ampelocalamus mianningensis(Q. Li et X. Jiang) D. Z. Li et Stapleton ●

285784   Patellocalamus patellaris (Gamble) W. T. Lin = Dendrocalamus patellaris Gamble ●

285785   Patersonia Poir. = Patersonia R. Br. (保留属名)■☆

285786   Patersonia Poir. = Pattersonia J. F. Gmel. ■●

285787   Patersonia Poir. = Ruellia L. ■●

285788   Patersonia R. Br. (1807)(保留属名);澳洲鸢尾属;Australian Iris ■☆

285789   Patersonia flaccida Endl. ;柔软澳洲鸢尾■☆

285790   Patersonia glabrata Ker Gawl. ;无毛澳洲鸢尾■☆

285791   Patersonia glauca R. Br. ;灰绿澳洲鸢尾■☆

285792   Patersonia graminea Benth. ;禾叶澳洲鸢尾■☆

285793   Patersonia longifolia R. Br. ;长叶澳洲鸢尾■☆

285794   Patersonia media R. Br. ;间形澳洲鸢尾■☆

285795   Patersonia nana Endl. ;矮小澳洲鸢尾■☆

285796   Patersonia occidentalis R. Br. ;西方澳洲鸢尾■☆

285797   Patersonia sericea R. Br. ;绢毛澳洲鸢尾;Silky Purple Flag ■☆

285798   Patersonia sylvestris Endl. ;林地澳洲鸢尾■☆

285799   Patersonia tenuispatha Endl. ;细苞澳洲鸢尾■☆

285800   Patersonia umbrosa Endl. ;澳洲鸢尾■☆

285801   Patherannia Poir. = Patersonia Poir. ■☆

285802   Patientia Raf. = Rumex L. ■●

285803   Patima Aubl. = Sabicea Aubl. ●☆

285804   Patinoa Cuatrec. (1953);毛籽轮枝木棉属●☆

285805   Patinoa almirajo Cuatrec. ;毛籽轮枝木棉●☆

285806   Patinoa sphaerocarpa Cuatrec. ;球果毛籽轮枝木棉●☆

285807   Patis Ohwi = Stipa L. ■

285808   Patisna Jack ex Burkill = Urophyllum Jack ex Wall. ●

285809   Patmaceae Schultz Sch. = Rafflesiaceae Dumort. (保留科名)■

285810   Patonia Wight = Xylopia L. (保留属名)●

285811   Patosia Buchenau = Oxychloe Phil. ■☆

285812   Patosia Buchenau(1890);帕图斯灯芯草属■☆

285813   Patosia clandestina Buchenau. ;肖锐尖灯芯草■☆

285814   Patrinia Juss. (1807)(保留属名);败酱属;Patrinia ■

285815   Patrinia Raf. = Sophora L. ●■

285816   Patrinia Raf. = Vexibia Raf. ●■

285817   Patrinia angustifolia Hemsl. = Patrinia heterophylla Bunge subsp. angustifolia(Hemsl. ex Forbes et Hemsl.)H. J. Wang ■

285818   Patrinia angustifolia Hemsl. = Patrinia heterophylla Bunge ■

285819   Patrinia angustifolia Hemsl. ex Forbes et Hemsl. ;窄叶败酱(白升麻,大升麻,九层叶,苦菜,盲菜,狭叶败酱,蜘蛛香);Narrowleaf Patrinia ■

285820   Patrinia angustifolia Hemsl. ex Forbes et Hemsl. = Patrinia heterophylla Bunge ■

285821   Patrinia diandra Kitag. ;双蕊败酱■☆

285822   Patrinia dielsii Graebn. = Patrinia villosa(Thunb.)Juss. ■

285823   Patrinia dielsii Graebn. var. erosa Graebn. = Patrinia villosa (Thunb.)Juss. ■

285824   Patrinia dielsii Graebn. var. palustris Pamp. = Patrinia villosa (Thunb.)Juss. ■

285825   Patrinia dielsii Graebn. var. shensiensis Graebn. = Patrinia villosa (Thunb.)Juss. ■

285826   Patrinia formosana Kitam. ;台湾败酱(大样苦斋);Taiwan Patrinia ■

285827   Patrinia formosana Kitam. = Patrinia monandra C. B. Clarke var. formosana(Kitam.)H. J. Wang ■

285828   Patrinia formosana Kitam. = Patrinia monandra C. B. Clarke ■

285829   Patrinia gibbifera Nakai = Patrinia triloba(Miq.)Miq. ■☆

285830   Patrinia gibbosa Maxim. ;圆叶败酱(圆叶金铃花)■☆

285831   Patrinia glabrifolia Yamam. et Sasaki;光叶败酱(秃败酱);Smoothleaf Patrinia ■

285832   Patrinia graveolens Hance = Patrinia heterophylla Bunge ■

285833   Patrinia graveolens Hance = Patrinia villosa(Thunb.)Juss. ■

285834   Patrinia heterophylla Bunge;异叶败酱(摆子草,糙叶败酱,臭脚跟,鼓头灰,虎牙草,箭头风,脚汗草,木头回,墓回头,墓头灰,墓回头,墓头苗,盆棵,气布待棵,铜班道,追风箭);Diversifolious Patrinia ■

285835   Patrinia heterophylla Bunge subsp. angustifolia(Hemsl. ex Forbes et Hemsl.)H. J. Wang = Patrinia angustifolia Hemsl. ex Forbes et Hemsl. ■

285836   Patrinia heterophylla subsp. angustifolia (Hemsl.) H. J. Wang = Patrinia heterophylla Bunge ■

285837   Patrinia hispida Bunge = Patrinia scabiosifolia Fisch. ex Trevir. ■

285838   Patrinia hybrida Makino;杂种败酱■☆

285839   Patrinia intermedia(Hornem.)Roem. et Schult. ;中间败酱(多花败酱,黄花败酱,墓回头,中败酱);Intermediate Patrinia ■

285840   Patrinia jatamansi D. Don = Nardostachys jatamansi(D. Don)DC. ■

285841   Patrinia monandra C. B. Clarke;少蕊败酱(白升麻,单蕊败酱,单药败酱,黄凤仙,介头草,蚧头草,山芥花,土花蓝);Monostamen Patrinia ,Onestamened Patrinia ■

285842   Patrinia monandra C. B. Clarke var. formosana (Kitam.) H. J. Wang = Patrinia formosana Kitam. ■

285843   Patrinia monandra C. B. Clarke var. formosana (Kitam.) H. J. Wang = Patrinia monandra C. B. Clarke ■

285844   Patrinia monandra C. B. Clarke var. sinensis Batalin = Patrinia monandra C. B. Clarke ■

285845   Patrinia nudiuscula Fisch. = Patrinia intermedia (Hornem.) Roem. et Schult. ■

285846   Patrinia ovata Bunge = Patrinia villosa(Thunb.)Juss. ■

285847   Patrinia palmata Maxim. = Patrinia triloba(Miq.)Miq. ■☆

285848   Patrinia parviflora Siebold et Zucc. = Patrinia scabiosifolia Fisch. ex Trevir. ■

285849   Patrinia punctiflora P. S. Hsu et H. J. Wang;斑花败酱(箭头风,马竹霄,无心草,细样苦斋);Bicolor Patrinia, Spottedflower Patrinia ■

285850   Patrinia punctiflora P. S. Hsu et H. J. Wang = Patrinia monandra C. B. Clarke ■

285851   Patrinia punctiflora P. S. Hsu et H. J. Wang var. robusta P. S. Hsu et H. J. Wang = Patrinia monandra C. B. Clarke ■

285852   Patrinia punctiflora P. S. Hsu et H. J. Wang var. robusta P. S. Hsu et H. J. Wang;大斑花败酱(大败酱,萌菜);Bigflower Bicolor Patrinia ■

285853   Patrinia rupestris(Pall.)Dufr. = Patrinia rupestris(Pall.)Juss. ■

285854   Patrinia rupestris(Pall.)Juss. ;岩败酱;Cliiff Patrinia ■

285855   Patrinia rupestris (Pall.) Juss. subsp. scabra (Bunge) H. J. Wang;糙叶岩败酱(糙叶败酱,蒙古败酱,墓回头,山败酱);Scabrous Patrinia ■

285856   Patrinia rupestris Juss. = Patrinia intermedia(Hornem.) Roem. et Schult. ■

285857   Patrinia rupestris Pall. subsp. scabra (Bunge) H. J. Wang = Patrinia scabra Bunge ■

285858   Patrinia scabiosifolia Fisch. ex Link = Patrinia scabiosifolia Fisch. ex Trevir. ■

285859   Patrinia scabiosifolia Fisch. ex Trevir. ;败酱(败酱草,臭艾,臭根子,豆豉菜,豆豉草,豆渣菜,豆渣草,黄花败酱,黄花苦菜,黄花龙牙,黄花龙牙草,黄花龙芽,黄花香,黄屈花,鸡肠风,假苦菜,将军草,苦芙,苦菜,苦苣,苦麻菜,苦斋,苦斋菜,苦斋草,苦斋公,苦藏,苦猪菜,流注,龙牙败酱,鹿肠,鹿场,鹿酱,鹿首,麻鸡婆,马草,山芝麻,酸益,土龙草,野黄花,野苦菜,野芹,野青菜,泽败);Dahurian Patrinia ,Yellow Patrinia ■

285860   Patrinia scabiosifolia Fisch. ex Trevir. f. crassa (Masam. et Satomi)Kitam. ex T. Yamaz. ;粗败酱■☆

285861   Patrinia scabiosifolia Fisch. ex Trevir. f. glabra Kom. = Patrinia scabiosifolia Fisch. ex Trevir. ■

285862　Patrinia scabiosifolia Fisch. ex Trevir. f. hispida（Bunge）Kom. = Patrinia scabiosifolia Fisch. ex Trevir. ■

285863　Patrinia scabiosifolia Fisch. ex Trevir. var. crassa Masam. et Satomi = Patrinia scabiosifolia Fisch. ex Trevir. f. crassa（Masam. et Satomi）Kitam. ex T. Yamaz. ■☆

285864　Patrinia scabiosifolia Fisch. ex Trevir. var. hispida（Bunge）Franch. = Patrinia scabiosifolia Fisch. ex Trevir. ■

285865　Patrinia scabiosifolia Fisch. ex Trevir. var. nantcianensis Pamp. = Patrinia scabiosifolia Fisch. ex Trevir. ■

285866　Patrinia scabiosifolia Link = Patrinia scabiosifolia Fisch. ex Trevir. ■

285867　Patrinia scabra Bunge = Patrinia rupestris（Pall.）Juss. subsp. scabra（Bunge）H. J. Wang ■

285868　Patrinia serratulifolia（Trev.）Fisch. ex DC. = Patrinia scabiosifolia Fisch. ex Trevir. ■

285869　Patrinia serratulifolia Fisch. ex DC. = Patrinia scabiosifolia Fisch. ex Trevir. ■

285870　Patrinia sibirica（L.）Juss.；西伯利亚败酱；Siberia Patrinia，Siberian Patrinia ■

285871　Patrinia sinensis（H. Lév.）Koidz. = Patrinia villosa（Thunb.）Juss. ■

285872　Patrinia speciosa Hand. -Mazz.；秀苞败酱；Showy Patrinia ■

285873　Patrinia takeuchiana Makino = Patrinia triloba（Miq.）Miq. var. takeuchiana（Makino）Ohwi ■☆

285874　Patrinia trifoliata L. Jin et R. N. Zhao；三叶败酱；Threeleaf Patrinia ■

285875　Patrinia triloba（Miq.）Miq.；三裂败酱■☆

285876　Patrinia triloba（Miq.）Miq. var. gibbosa? = Patrinia triloba（Miq.）Miq. ■☆

285877　Patrinia triloba（Miq.）Miq. var. palmata（Maxim.）H. Hara；掌裂败酱■☆

285878　Patrinia triloba（Miq.）Miq. var. takeuchiana（Makino）Ohwi；竹内败酱■☆

285879　Patrinia villosa（Thunb.）Dufr. = Patrinia villosa（Thunb.）Juss. ■

285880　Patrinia villosa（Thunb.）Juss.；白花败酱（白花败酱草，白苦爹，败酱，败酱草，斑刀箭，斑竹甑，秤杆升麻，大升麻，豆豉草，豆渣菜，豆渣草，孩儿菊，黄花败酱草，蒻，苦菜，苦苣，苦蘵菜，苦斋，苦斋菜，苦斋草，苦斋公，苦蘵，苦猪菜，龙牙败酱，鹿肠，鹿酱，鹿首，马草，毛败酱，萌菜，攀倒峻，攀倒甑，酸益，土升麻，烟脂麻，胭脂麻，野苦菜，野苦斋，泽败，獐大耳）；White Patrinia，Whiteflower Patrinia ■

285881　Patrinia villosa（Thunb.）Juss. = Patrinia monandra C. B. Clarke var. formosana（Kitam.）H. J. Wang ■

285882　Patrinia villosa（Thunb.）Juss. subsp. punctifolia H. J. Wang；斑叶白花败酱（斑叶败酱）；Tuberless Whiteflower Patrinia ■

285883　Patrinia villosa（Thunb.）Juss. var. ambigua Pamp. = Patrinia villosa（Thunb.）Juss. ■

285884　Patrinia villosa（Thunb.）Juss. var. japonica H. Lév. = Patrinia villosa（Thunb.）Juss. ■

285885　Patrinia villosa（Thunb.）Juss. var. sinensis H. Lév. = Patrinia villosa（Thunb.）Juss. ■

285886　Patrisa Rich.（废弃属名）= Ryania Vahl（保留属名）●☆

285887　Patrisia Rohr ex Steud. = Dichapetalum Thouars ●

285888　Patrisiaceae Mart. = Flacourtiaceae Rich. ex DC.（保留科名）●

285889　Patrocles Salisb. = Narcissus L. ■

285890　Patrocles Salisb. = Schisanthes Haw. ■

285891　Patropyrum Á. Löve = Aegilops L.（保留属名）■

285892　Patropyrum tauschii（Coss.）Á. Löve = Aegilops tauschii Coss. ■

285893　Patsjotti Adans. = Strumpfia Jacq. ☆

285894　Pattalias S. Watson（1889）；木钉萝藦属●☆

285895　Pattalias angustifolius（Torr.）S. Watson；窄叶木钉萝藦●☆

285896　Pattalias palmeri S. Watson；木钉萝藦●☆

285897　Pattara Adans. = Embelia Burm. f.（保留属名）●■

285898　Pattara kilimandscharica（Gilg）Hiern = Embelia schimperi Vatke ●☆

285899　Pattara pellucida Hiern = Embelia guineensis Baker ●☆

285900　Pattara welwitschii Hiern = Embelia welwitschii（Hiern）K. Schum. ●☆

285901　Pattersonia J. F. Gmel. = Ruellia L. ■●

285902　Pattonia Wight = Grammatophyllum Blume ■☆

285903　Patulix Raf. = Clerodendrum L. ●■

285904　Patya Neck. = Verbena L. ■●

285905　Patzkea G. H. Loos = Anthoxanthum L. ■

285906　Patzkea G. H. Loos（2010）；帕茨克茅属■☆

285907　Paua Caball. = Andryala L. ■☆

285908　Paua Gand. = Torilis Adans. ■

285909　Paua maroccana Caball. = Andryala maroccana Pau ■☆

285910　Pauciflori Rydb. = Almutaster Á. Löve et D. Löve ■☆

285911　Pauella Ramam. et Sebastine = Theriophonum Blume ■☆

285912　Pauia Deb et R. M. Dutta（1965）；印东北茄属☆

285913　Pauia belladonna Deb et R. M. Dutta；印东北茄☆

285914　Pauladolfia Börner = Acetosella（Meisn.）Fourr. ●■

285915　Pauladolphia Börner = Acetosella（Meisn.）Fourr. ●■

285916　Pauldopia Steenis（1969）；翅叶木属；Pauldopia ●

285917　Pauldopia ghorta（Buch. -Ham. ex G. Don）Steenis；翅叶木（金丝岩柏，细口袋花，紫豇豆）；Pauldopia，Winged Pauldopia ●

285918　Pauletia Cav. = Bauhinia L. ●

285919　Pauletia bowkeri（Harv.）A. Schmitz = Bauhinia bowkeri Harv. ●☆

285920　Pauletia taitensis（Taub.）A. Schmitz = Bauhinia taitensis Taub. ●☆

285921　Pauletia tomentosa（L.）A. Schmitz = Bauhinia tomentosa L. ●

285922　Paulia Korovin = Paulita Soják ■☆

285923　Paulinia T. Durand = Roulinia Decne. ●■

285924　Paulinia T. Durand = Rouliniella Vail ●■

285925　Paulita Soják（1982）；中亚山草属■☆

285926　Paulita alpina（Schischk.）Soják；高山中亚山草■☆

285927　Paulita ovczinnikovii（Korovin）Soják；中亚山草■☆

285928　Paullinia L.（1753）；香无患子属（南美可可属，泡林藤属）；Pauilinia ●☆

285929　Paullinia asiatica L. = Toddalia asiatica（L.）Lam. ●

285930　Paullinia cupana Kunth；巴西香无患子（巴西可可，亚马孙香无患子）；Brazilian Cocoa，Guarana，Guarana Paullinia，Guarani ●☆

285931　Paullinia japonica Thunb. = Ampelopsis japonica（Thunb.）Makino ●

285932　Paullinia maritima Vell.；海边香无患子（海边泡林藤）●☆

285933　Paullinia pinnata L.；香无患子（泡林藤）●☆

285934　Paullinia thalictrifolia Juss.；唐松草叶泡林藤●☆

285935　Paullinia yoco R. E. Schult. et Killip；秘鲁香无患子（秘鲁可可）；Yoco ●☆

285936　Paulliniaceae Durande = Sapindaceae Juss.（保留科名）●■

285937　Paulomagnusia Kuntze = Micranthus（Pers.）Eckl.（保留属名）■☆

285938　Paulowilhelmia Hochst.（1844）；肖单口爵床属■☆

285939　Paulo-Wilhelmia Hochst. = Eremomastax Lindau. ☆

285940　Paulowilhelmia Hochst. = Eremomastax Lindau. ■☆

285941　Paulo-Wilhelmia Hochst. = Mollugo L. ■

285942　Paulo-wilhelmia Hochst. = Paulowilhelmia Hochst. ■☆

285943　Paulowilhelmia elata Lindau;高肖单口爵床■☆

285944　Paulowilhelmia glabra Lindau = Eremomastax speciosa (Hochst.) Cufod.■☆

285945　Paulowilhelmia nobilis C. B. Clarke;肖单口爵床■☆

285946　Paulowilhelmia polysperma Benth. = Eremomastax speciosa (Hochst.) Cufod.■☆

285947　Paulowilhelmia pubescens Lindau = Eremomastax speciosa (Hochst.) Cufod.■☆

285948　Paulowilhelmia sclerochiton (S. Moore) Lindau = Eremomastax speciosa(Hochst.) Cufod.■☆

285949　Paulowilhelmia speciosa Hochst. = Eremomastax speciosa(Hochst.) Cufod.■☆

285950　Paulowilhelmia togoensis Lindau = Eremomastax speciosa(Hochst.) Cufod.■☆

285951　Paulownia Siebold et Zucc. (1836);泡桐属;Empress Tree, Paulownia ●

285952　Paulownia × henanensis C. Y. Zhang et Y. H. Zhao;圆冠泡桐; Henan Paulownia ●

285953　Paulownia australis T. Gong;南方泡桐;S. China Pecteilis,South China Paulownia ●

285954　Paulownia australis T. Gong = Paulownia fargesii Franch. ●

285955　Paulownia australis T. Gong = Paulownia taiwaniana T. W. Hu et H. J. Chang ●

285956　Paulownia catalpifolia T. Gong ex D. Y. Hong;楸叶泡桐(山东泡桐,无籽泡桐,小叶泡桐);Catalpa Leaf Paulownia, Catalpaleaf Paulownia,Catalpa-leaved Paulownia ●

285957　Paulownia duclouxii Dode = Paulownia fortunei(Seem.)Hemsl. ●

285958　Paulownia elongata S. Y. Hu;兰考泡桐(长叶泡桐);Elongate Paulownia,Empress Tree, Lankao Pecteilis ●

285959　Paulownia fargesii Franch. ;川泡桐;Farges Paulownia,Sichuan Paulownia,Sichuan Pecteilis ●

285960　Paulownia fortunei(Seem.)Hemsl. ;白花泡桐(阿根廷泡桐,白花桐,大赣树,大果泡桐,笛螺木,饭桐子,华桐,火筒木,南投梧桐,泡桐,沙桐彭,水桐,水桐树,通心条,桐木,桐木树,梧桐,紫花树);Fortune's Paulownia, Foxglove Pecteilis, Powton, White-flowered Paulownia ●

285961　Paulownia fortunei(Seem.)Hemsl. var. tsinlingensis P. Y. Pai = Paulownia tomentosa(Thunb.)Steud. var. tsinlingensis(P. Y. Pai)T. Gong ●

285962　Paulownia glabrata Rehder = Paulownia tomentosa (Thunb.)Steud. var. tsinlingensis(P. Y. Pai)T. Gong ●

285963　Paulownia grandifolia Wettst. = Paulownia tomentosa (Thunb.)Steud. ●

285964　Paulownia imperialis Siebold et Zucc. = Paulownia tomentosa (Thunb.)Steud. ●

285965　Paulownia imperialis Siebold et Zucc. var. lanata Dode = Paulownia tomentosa(Thunb.)Steud. ●

285966　Paulownia kawakamii Ito;白桐(川泡桐,华东泡桐,黄毛泡桐,空桐树,水桐木,台湾泡桐,梧桐,淮东泡桐);Kawakami Paulownia, Sapphire Dragon Tree,Taiwan Paulownia,Taiwan Pecteilis ●

285967　Paulownia kawakamii Ito = Paulownia taiwaniana T. W. Hu et H. J. Chang ●

285968　Paulownia lilacina Sprague = Paulownia tomentosa (Thunb.) Steud. ●

285969　Paulownia longifolia Hand. -Mazz. = Paulownia fortunei (Seem.) Hemsl. ●

285970　Paulownia meridionalis Dode = Paulownia fortunei (Seem.) Hemsl. ●

285971　Paulownia mikado T. Ito = Paulownia fortunei(Seem.) Hemsl. ●

285972　Paulownia recurva Rehder = Paulownia tomentosa (Thunb.) Steud. ●

285973　Paulownia rehderiana Hand. -Mazz. = Paulownia kawakamii Ito ●

285974　Paulownia shensiensis P. Y. Pai = Paulownia tomentosa (Thunb.)Steud. var. tsinlingensis(P. Y. Pai)T. Gong ●

285975　Paulownia shexiensis P. Y. Pai;陕西泡桐;Shaanxi Paulownia ●

285976　Paulownia shexiensis P. Y. Pai = Paulownia tomentosa(Thunb.) Steud. var. tsinlingensis(P. Y. Pai)T. Gong ●

285977　Paulownia silvestrii Pamp. et Bonati = Shiuyinghua silvestrii (Pamp. et Bonati) Paclt ●☆

285978　Paulownia taiwaniana T. W. Hu et H. J. Chang;台湾泡桐(南方泡桐,台岛泡桐);Taiwan Pecteilis ●

285979　Paulownia thyrsoides Rehder = Paulownia kawakamii Ito ●

285980　Paulownia tomentosa (Thunb. ex Murray) Steud. = Paulownia tomentosa(Thunb.)Steud. ●

285981　Paulownia tomentosa(Thunb.)Steud. ;毛泡桐(艾椅,白桐,岗桐,花桐,空桐木,泡桐,日本泡桐,绒叶泡桐,荣桐,荣,水桐,桐,桐木,桐木树,锈毛泡桐,紫花树,紫花桐);Empress Tree, Empress-tree, Foxglove Tree, Foxglove-tree, Hairy Paulownia, Paulownia, Princess Tree,Princesstree,Princess-tree,Royal Paulownia ●

285982　Paulownia tomentosa (Thunb.) Steud. var. japonica Elwes = Paulownia tomentosa(Thunb.)Steud. ●

285983　Paulownia tomentosa (Thunb.) Steud. var. lanata (Dole) C. K. Schneid. ;小花泡桐●

285984　Paulownia tomentosa (Thunb.) Steud. var. lanata (Dole) C. K. Schneid. = Paulownia tomentosa(Thunb.)Steud. ●

285985　Paulownia tomentosa (Thunb.) Steud. var. tsinlingensis (P. Y. Pai) T. Gong;光泡桐(光桐);Glabrous Paulownia, Kill Wood, Qinling Pecteilis,Tsinling Paulownia ●

285986　Paulownia viscosa Hand. -Mazz. = Paulownia kawakamii Ito ●

285987　Paulowniaceae Nakai = Bignoniaceae Juss. (保留科名)●■

285988　Paulowniaceae Nakai(1949);泡桐科●

285989　Paulseniella Briq. = Elsholtzia Willd. ●■

285990　Pauridia Harv. (1838);三雄仙茅属■☆

285991　Pauridia hypoxidioides Harv. = Pauridia minuta (L. f.) T. Durand et Schinz ■☆

285992　Pauridia longituba M. F. Thomps. ;长管三雄仙茅■☆

285993　Pauridia minuta(L. f.) T. Durand et Schinz;微小三雄仙茅■☆

285994　Pauridiantha Hook. f. (1873);小花茜属●☆

285995　Pauridiantha afzelii(Hiern)Bremek. ;阿芙泽尔小花茜●☆

285996　Pauridiantha bequaertii (De Wild.)Bremek. ;贝卡尔小花茜●☆

285997　Pauridiantha bridelioides Verdc. ;土密树小花茜●☆

285998　Pauridiantha butaguensis (De Wild.) Bremek. = Pauridiantha paucinervis(Hiern)Bremek. subsp. butaguensis(De Wild.)Verdc. ●☆

285999　Pauridiantha butaguensis (De Wild.) Bremek. var. exsertostylosa? = Pauridiantha paucinervis (Hiern) Bremek. subsp. butaguensis(De Wild.) Verdc. ●☆

286000　Pauridiantha callicarpoides(Hiern)Bremek. ;美果小花茜●☆

286001　Pauridiantha camposii(G. Taylor) Exell;坎波斯小花茜●☆

286002　Pauridiantha claessensii Bremek. ;克莱森斯小花茜●☆

286003　Pauridiantha dewevrei(De Wild. et T. Durand) Bremek. ;德韦小花茜●☆

286004　Pauridiantha divaricata(K. Schum.)Bremek. ;叉开小花茜●☆

286005　Pauridiantha floribunda(K. Schum. et K. Krause)Bremek;繁花

小花茜●☆

286006 Pauridiantha hirtella(Benth.)Bremek.；多毛小花茜●☆

286007 Pauridiantha holstii (K. Schum.) Bremek. = Pauridiantha paucinervis(Hiern)Bremek. subsp. holstii(K. Schum.)Verdc.●☆

286008 Pauridiantha insculpta(Hutch. et Dalziel)Bremek.；雕刻小花茜●☆

286009 Pauridiantha insularis(Hiern)Bremek.；海岛小花茜●☆

286010 Pauridiantha kizuensis Bremek.；木津小花茜●☆

286011 Pauridiantha letestuana(N. Hallé)Ntore et Dessein；莱泰斯图小花茜●☆

286012 Pauridiantha liebrechtsiana (De Wild. et T. Durand) Ntore et Dessein；利布小花茜●☆

286013 Pauridiantha mayumbensis(R. D. Good)Bremek.；马永巴小花茜●☆

286014 Pauridiantha micrantha(Hiern)Bremek.；小花茜●☆

286015 Pauridiantha microphylla Good；小叶小花茜●☆

286016 Pauridiantha multiflora K. Schum.；非洲多花小花茜●☆

286017 Pauridiantha paucinervis(Hiern)Bremek.；少脉小花茜●☆

286018 Pauridiantha paucinervis (Hiern) Bremek. subsp. butaguensis (De Wild.)Verdc.；布塔古小花茜●☆

286019 Pauridiantha paucinervis (Hiern) Bremek. subsp. holstii (K. Schum.)Verdc.；霍尔小花茜●☆

286020 Pauridiantha pierlotii N. Hallé；皮氏小花茜●☆

286021 Pauridiantha pleiantha Ntore et Dessein；多花小花茜●☆

286022 Pauridiantha pyramidata(K. Krause)Bremek.；塔形小花茜●☆

286023 Pauridiantha rubens(Benth.)Bremek.；变淡红小花茜●☆

286024 Pauridiantha schnellii N. Hallé；施内尔小花茜●☆

286025 Pauridiantha siderophila N. Hallé；喜铁小花茜●☆

286026 Pauridiantha stipulosa(Hutch. et Dalziel)Hepper = Poecilocalyx stipulosa(Hutch. et Dalziel)N. Hallé■☆

286027 Pauridiantha sylvicola(Hutch. et Dalziel)Bremek.；林生小花茜●☆

286028 Pauridiantha symplocoides(S. Moore)Bremek.；山矾小花茜●☆

286029 Pauridiantha venusta N. Hallé；雅致小花茜●☆

286030 Pauridiantha verticillata(De Wild. et T. Durand)N. Hallé；轮生小花茜●☆

286031 Pauridiantha viridiflora(Hiern)Hepper；绿花小花茜●☆

286032 Pauridiantha ziamaeana (Jacq. -Fél.) Hepper = Stelechantha ziamaeana(Jacq. -Fél.)N. Hallé●☆

286033 Paurolepis S. Moore = Gutenbergia Sch. Bip. ex Walp.■☆

286034 Paurolepis S. Moore(1917)；线叶瘦片菊属■☆

286035 Paurolepis angusta S. Moore = Paurolepis filifolia(R. E. Fr.)Wild et G. V. Pope■☆

286036 Paurolepis filifolia(R. E. Fr.)Wild et G. V. Pope；线叶瘦片菊■☆

286037 Paurotis O. F. Cook = Acoelorrhaphe H. Wendl.●☆

286038 Paurotis O. F. Cook. (1902)；丛立刺棕属（丛立刺棕榈属、丛立刺椰子属）●☆

286039 Paurotis arborescens O. F. Cook；丛立刺棕●☆

286040 Paurotis wrightii Britton = Acoelorrhaphe wrightii Becc.●☆

286041 Pausandra Radlk. (1870)；贫雄大戟属☆

286042 Pausandra densiflora Lanj.；密花贫雄大戟☆

286043 Pausandra macropetala Ducke；大瓣贫雄大戟☆

286044 Pausandra macrostachya Ducke；大穗贫雄大戟☆

286045 Pausia Raf. (1836) = Thymelaea Mill. (保留属名)●■

286046 Pausia Raf. (1838) = Cartrema Raf. ●

286047 Pausia Raf. (1838) = Osmanthus Lour. ●

286048 Pausinystalia Pierre ex Beille(1906)；止睡茜属●☆

286049 Pausinystalia angolensis Wernham = Pausinystalia macroceras

(K. Schum.) Pierre ex Beille ●☆

286050 Pausinystalia bequaertii De Wild. = Pausinystalia macroceras (K. Schum.) Pierre ex Beille ●☆

286051 Pausinystalia brachythyrsum(K. Schum.)W. Brandt；短序止睡茜●☆

286052 Pausinystalia gilgii W. Brandt = Pausinystalia talbotii Wernham ●☆

286053 Pausinystalia ituriense De Wild. = Pausinystalia lane-poolei (Hutch.) Hutch. ex Lane-Poole subsp. ituriense(De Wild.)Stoffelen et Robbr. ●☆

286054 Pausinystalia johimbe(K. Schum.)Pierre ex Beille；育亨宾止睡茜；Yohimbe Bark ●☆

286055 Pausinystalia johimbe (Schum.) Beille = Pausinystalia johimbe (K. Schum.) Pierre ex Beille ●☆

286056 Pausinystalia lane-poolei(Hutch.)Hutch. ex Lane-Poole；兰普止睡茜●☆

286057 Pausinystalia lane-poolei(Hutch.)Hutch. ex Lane-Poole subsp. ituriense(De Wild.)Stoffelen et Robbr. ；伊图里止睡茜●☆

286058 Pausinystalia lane-poolei Hutch. ex Lane-Poole = Pausinystalia lane-poolei(Hutch.)Hutch. ex Lane-Poole ●☆

286059 Pausinystalia macroceras(K. Schum.)Pierre ex Beille；大角止睡茜●☆

286060 Pausinystalia macroceras (K. Schum.) Pierre ex Beille var. bequaertii (De Wild.) N. Hallé = Pausinystalia macroceras (K. Schum.)Pierre ex Beille ●☆

286061 Pausinystalia macroceras (K. Schum.) Pierre ex Beille. f. brachythyrsum(K. Schum.)N. Hallé = Pausinystalia brachythyrsum (K. Schum.)W. Brandt ●☆

286062 Pausinystalia mayumbensis R. D. Good = Corynanthe mayumbensis(R. D. Good)Raym. -Hamet ex N. Hallé ●☆

286063 Pausinystalia pachyceras (K. Schum.) De Wild. = Corynanthe pachyceras K. Schum. ●☆

286064 Pausinystalia reticulata Hutch. = Pausinystalia lane-poolei (Hutch.)Hutch. ex Lane-Poole ●☆

286065 Pausinystalia sankeyi Hutch. et Dalziel；桑基止睡茜●☆

286066 Pausinystalia sankeyi Hutch. et Dalziel = Pausinystalia talbotii Wernham ●☆

286067 Pausinystalia talbotii Wernham；长果止睡茜●☆

286068 Pausinystalia trillesii Pierre ex Dupouy et Beille = Pausinystalia johimbe(K. Schum.)Pierre ex Beille ●☆

286069 Pausinystalia zenkeri W. Brandt = Pausinystalia johimbe (K. Schum.)Pierre ex Beille ●☆

286070 Pautsauvia Juss. = Alangium Lam. (保留属名)●

286071 Pautsauvia Juss. = Stylidium Lour. (废弃属名)●

286072 Pavate Adans. = Pavetta L. ●

286073 Pavetta L. (1753)；大沙叶属（茜木属）；Pavetta ●

286074 Pavetta abyssinica Fresen.；阿比西尼亚大沙叶●☆

286075 Pavetta abyssinica Fresen. subsp. viridiflora Bridson；绿花阿比西尼亚大沙叶●☆

286076 Pavetta abyssinica Fresen. var. cinerascens A. Rich.；灰色大沙叶●☆

286077 Pavetta abyssinica Fresen. var. dolichosiphon (Bremek.) Bridson；长管大沙叶●☆

286078 Pavetta abyssinica Fresen. var. usambarica (Bremek.) Bridson；乌桑巴拉大沙叶●☆

286079 Pavetta acrochlora Bremek. = Pavetta bagshawei S. Moore var. leucosphaera(Bremek.)Bridson ●☆

286080 Pavetta adelensis Delile = Pavetta gardeniifolia A. Rich. ●☆

286081　Pavetta aethiopica Bremek.；埃塞俄比亚大沙叶●☆

286082　Pavetta akeassii J. B. Hall；阿克斯大沙叶●☆

286083　Pavetta albanensis Bremek. = Pavetta capensis（Houtt.）Bremek.●☆

286084　Pavetta albertina S. Moore = Pavetta subcana Hiern var. longiflora(Vatke)Bridson●☆

286085　Pavetta albicaulis S. Moore = Pavetta decumbens K. Schum. et K. Krause●☆

286086　Pavetta alexandrae Bremek. = Pavetta lanceolata Eckl.●☆

286087　Pavetta amaniensis Bremek.；阿马尼大沙叶●☆

286088　Pavetta amaniensis Bremek. var. trichocephala（Bremek.）Bridson；毛头大沙叶●☆

286089　Pavetta andongensis Hiern；安东大沙叶●☆

286090　Pavetta angolensis Hiern；安哥拉大沙叶●☆

286091　Pavetta annobonensis Bremek.；安诺本大沙叶●☆

286092　Pavetta apiculata Hutch. et Dalziel = Pavetta mollis Afzel. ex Hiern●☆

286093　Pavetta appendiculata De Wild. = Pavetta gardeniifolia A. Rich. var. appendiculata(De Wild.)Bridson●☆

286094　Pavetta arenicola K. Schum.；沙生大沙叶●☆

286095　Pavetta arenosa Lour.；大沙叶；Sand Pavetta●

286096　Pavetta arenosa Lour. f. glabrituba Chun et F. C. How；光萼大沙叶(光叶大沙叶)；Glabrous-calyx Pavetta●

286097　Pavetta assimilis Sond. = Pavetta gardeniifolia A. Rich.●☆

286098　Pavetta assimilis Sond. var. brevituba-glabra Bremek. = Pavetta gardeniifolia A. Rich.●☆

286099　Pavetta assimilis Sond. var. glabra Bremek. = Pavetta gardeniifolia A. Rich.●☆

286100　Pavetta assimilis Sond. var. puberula Bremek. = Pavetta gardeniifolia A. Rich. var. subtomentosa K. Schum.●☆

286101　Pavetta assimilis Sond. var. pubescens Bremek. = Pavetta gardeniifolia A. Rich. var. subtomentosa K. Schum.●☆

286102　Pavetta assimilis Sond. var. scabrida Bremek. = Pavetta gardeniifolia A. Rich. var. subtomentosa K. Schum.●☆

286103　Pavetta assimilis Sond. var. tomentella Bremek. = Pavetta gardeniifolia A. Rich. var. subtomentosa K. Schum.●☆

286104　Pavetta australiensis Bremek.；澳洲大沙叶；Australian Pavetta●☆

286105　Pavetta baconia Hiern = Pavetta corymbosa(DC.)F. N. Williams●☆

286106　Pavetta baconia Hiern var. congolana De Wild. et T. Durand = Pavetta nitidula Hiern●☆

286107　Pavetta baconia Hiern var. hispida Scott-Elliot = Pavetta lasioclada(K. Krause)Mildbr. ex Bremek.●☆

286108　Pavetta baconia Hiern var. oblongifolia? = Pavetta oblongifolia（Hiern）Bremek.●☆

286109　Pavetta baconia Hiern var. puberulosa De Wild. = Pavetta intermedia Bremek.●☆

286110　Pavetta bagshawei S. Moore；巴格大沙叶●☆

286111　Pavetta bagshawei S. Moore var. leucosphaera（Bremek.）Bridson；白球大沙叶●☆

286112　Pavetta bangweensis Bremek.；邦韦大沙叶●☆

286113　Pavetta barbertonensis Bremek.；巴伯顿大沙叶●☆

286114　Pavetta barteri Dawe = Pavetta crassipes K. Schum.●☆

286115　Pavetta batesiana Bremek.；贝茨大沙叶●☆

286116　Pavetta bechuanensis S. Moore = Pavetta eylesii S. Moore●☆

286117　Pavetta bequaertii De Wild.；贝卡尔大沙叶●☆

286118　Pavetta bidentata Hiern；双齿大沙叶●☆

286119　Pavetta bidentata Hiern var. sessilifolia S. D. Manning；无梗大沙叶●☆

286120　Pavetta bilineata Bremek.；双线大沙叶●☆

286121　Pavetta boonei De Wild. = Pavetta bagshawei S. Moore●☆

286122　Pavetta bowkeri Harv.；鲍氏大沙叶●☆

286123　Pavetta bowkeri Harv. var. glabra Bremek. = Pavetta natalensis Sond.●☆

286124　Pavetta bowkeri Harv. var. pubescens Bremek. = Pavetta bowkeri Harv.●☆

286125　Pavetta brachycalyx Hiern；短萼大沙叶●☆

286126　Pavetta brachycoryne K. Schum. = Rutidea glabra Hiern●☆

286127　Pavetta brachysiphon Bremek.；短管大沙叶●☆

286128　Pavetta breyeri Bremek. = Pavetta gracilifolia Bremek.●☆

286129　Pavetta breyeri Bremek. var. glabra? = Pavetta gracilifolia Bremek.●☆

286130　Pavetta breyeri Bremek. var. pubescens? = Pavetta gracilifolia Bremek.●☆

286131　Pavetta bruceana Bremek.；布鲁斯大沙叶●☆

286132　Pavetta bruneelii De Wild.；布吕内尔大沙叶●☆

286133　Pavetta buchneri K. Schum.；布赫纳大沙叶●☆

286134　Pavetta burttii Bremek.；伯特大沙叶●☆

286135　Pavetta butaguensis De Wild. = Pavetta ruwenzoriensis S. Moore●☆

286136　Pavetta buzica S. Moore = Pavetta gracillima S. Moore●☆

286137　Pavetta caffra L. f. = Pavetta capensis(Houtt.)Bremek.●☆

286138　Pavetta caffra L. f. var. pubescens Sond. = Pavetta capensis（Houtt.）Bremek. subsp. komghensis(Bremek.)Kok●☆

286139　Pavetta calothyrsa Bremek. = Pavetta nitidula Hiern●☆

286140　Pavetta camerounensis S. D. Manning；喀麦隆大沙叶●☆

286141　Pavetta candelabra Bremek.；烛台大沙叶●☆

286142　Pavetta canescens DC.；灰白大沙叶●☆

286143　Pavetta capensis(Houtt.)Bremek.；好望角大沙叶●☆

286144　Pavetta capensis（Houtt.）Bremek. subsp. komghensis（Bremek.）Kok；科姆大沙叶●☆

286145　Pavetta capillipes Bremek. = Pavetta sepium K. Schum. var. merkeri（K. Krause）Bridson●☆

286146　Pavetta cataractarum S. Moore；瀑布群高原大沙叶●☆

286147　Pavetta cataractarum S. Moore var. hirtiflora Bremek. = Pavetta cataractarum S. Moore●☆

286148　Pavetta catophylla K. Schum.；垂叶大沙叶●☆

286149　Pavetta catophylla K. Schum. var. glabra Bremek. = Pavetta catophylla K. Schum.●☆

286150　Pavetta catophylla K. Schum. var. pubescens Bremek. = Pavetta catophylla K. Schum.●☆

286151　Pavetta cecilae N. E. Br. = Pavetta radicans Hiern●☆

286152　Pavetta cellulosa Bremek.；纤维大沙叶●☆

286153　Pavetta cephalotes Bremek. = Pavetta refractifolia K. Schum.●☆

286154　Pavetta chapmanii Bridson；查普曼大沙叶●☆

286155　Pavetta chinensis Roem. et Schult. = Ixora chinensis Lam.●

286156　Pavetta chionantha K. Schum. et K. Krause = Pavetta owariensis P. Beauv. var. glaucescens(Hiern)S. D. Manning●☆

286157　Pavetta cinerascens（A. Rich.）Chiov. = Pavetta abyssinica Fresen. var. cinerascens A. Rich.●☆

286158　Pavetta cinerascens(A. Rich.)Chiov. var. glabrescens Chiov. = Pavetta villosa Vahl●☆

286159　Pavetta cinereifolia Berhaut；灰叶大沙叶●☆

286160　Pavetta claessensii De Wild.；克莱森斯大沙叶●☆

286161　Pavetta coelophlebia Bremek.；陷脉大沙叶●☆

286162　Pavetta comostyla S. Moore；簇柱大沙叶●☆

286163   Pavetta comostyla S. Moore subsp. nyassica（Bremek.）Bridson；
尼亚萨大沙叶●☆

286164   Pavetta comostyla S. Moore var. inyangensis（Bremek.）Bridson；
伊尼扬加大沙叶●☆

286165   Pavetta conflatiflora S. Moore = Pavetta cataractarum S. Moore ●☆

286166   Pavetta congensis Bremek. = Pavetta nitidula Hiern ●☆

286167   Pavetta constipulata Bremek.；联叶大沙叶●☆

286168   Pavetta constipulata Bremek. var. geoscopa？ = Pavetta constipulata
Bremek. ●☆

286169   Pavetta cooperi Harv. et Sond.；库珀大沙叶●☆

286170   Pavetta corethrogyne Bremek. = Pavetta olivaceo-nigra K.
Schum. ●☆

286171   Pavetta coriacea Bremek. = Pavetta nitidula Hiern ●☆

286172   Pavetta cornelia（Cham. et Schltdl.）Rchb. ex DC. = Keetia
cornelia（Cham. et Schltdl.）Bridson ●☆

286173   Pavetta corradiana Cufod. = Pavetta microphylla Chiov. ●☆

286174   Pavetta corymbosa（DC.）F. N. Williams；伞序大沙叶●☆

286175   Pavetta corymbosa（DC.）F. N. Williams var. glabra Bremek. =
Pavetta corymbosa（DC.）F. N. Williams ●☆

286176   Pavetta corymbosa（DC.）F. N. Williams var. neglecta Bremek.；
忽视大沙叶●☆

286177   Pavetta corynostylis K. Schum. = Rutidea fuscescens Hiern ●☆

286178   Pavetta crassipes K. Schum.；粗柄大沙叶●☆

286179   Pavetta crassipes K. Schum. var. major De Wild. = Pavetta
crassipes K. Schum. ●☆

286180   Pavetta crebrifolia Hiern；密叶大沙叶●☆

286181   Pavetta crebrifolia Hiern var. involucrata K. Schum. = Pavetta
stenosepala K. Schum. ●☆

286182   Pavetta crebrifolia Hiern var. pubescens Bridson；短柔毛大沙叶●☆

286183   Pavetta dalei Bremek. = Pavetta hymenophylla Bremek. ●☆

286184   Pavetta decumbens K. Schum. et K. Krause；外倾大沙叶●☆

286185   Pavetta deistelii K. Schum. = Pavetta bidentata Hiern ●☆

286186   Pavetta delagoensis Bremek. = Pavetta gracilifolia Bremek. ●☆

286187   Pavetta delicatifolia Bridson；美丽大沙叶●☆

286188   Pavetta denudata Bremek. = Pavetta oliveriana Hiern var.
denudata（Bremek.）Bridson ●☆

286189   Pavetta disarticulata Galpin = Pavetta edentula Sond. ●☆

286190   Pavetta dissimilis Bremek. = Pavetta zeyheri Sond. ●☆

286191   Pavetta divaricata Bremek. = Pavetta gracilifolia Bremek. ●☆

286192   Pavetta diversicalyx Bridson；异萼大沙叶●☆

286193   Pavetta diversipunctata Bridson；异斑大沙叶●☆

286194   Pavetta dolichantha Bremek.；长花大沙叶●☆

286195   Pavetta dolichosepala Hiern；长萼大沙叶●☆

286196   Pavetta dolichosiphon Bremek. = Pavetta abyssinica Fresen. var.
dolichosiphon（Bremek.）Bridson ●☆

286197   Pavetta durbanensis Bremek. = Pavetta capensis（Houtt.）
Bremek. subsp. komghensis（Bremek.）Kok ●☆

286198   Pavetta edentula Sond.；无齿大沙叶●☆

286199   Pavetta eketensis Bremek. = Pavetta owariensis P. Beauv. ●☆

286200   Pavetta ellenbeckii K. Schum. = Pavetta abyssinica Fresen. ●☆

286201   Pavetta elliottii K. Schum. et K. Krause；埃利大沙叶●☆

286202   Pavetta elliottii K. Schum. et K. Krause var. trichocalyx
（Bremek.）Bridson；毛萼大沙叶●☆

286203   Pavetta elliptica Hochst. = Feretia apodanthera Delile ●☆

286204   Pavetta eritreensis Bremek. = Pavetta abyssinica Fresen. var.
dolichosiphon（Bremek.）Bridson ●☆

286205   Pavetta erlangeri Bremek.；厄兰格大沙叶●☆

286206   Pavetta esqurollii H. Lév. = Clerodendrum bungei Steud. ●

286207   Pavetta exellii Bremek. = Pavetta hookeriana Hiern ●☆

286208   Pavetta eylesii S. Moore；艾尔斯大沙叶●☆

286209   Pavetta fascifolia Bremek.；簇叶大沙叶●☆

286210   Pavetta fastigiata R. D. Good = Ixora fastigiata（R. D. Good）
Bremek. ●☆

286211   Pavetta filistipulata Bremek.；线托大沙叶●☆

286212   Pavetta flammea K. Schum. = Pavetta canescens DC. ●☆

286213   Pavetta flaviflora（K. Schum.）Hutch. et Dalziel = Pavetta
brachycalyx Hiern ●☆

286214   Pavetta fossorum Bremek. = Pavetta gardeniifolia A. Rich. ●☆

286215   Pavetta friesiorum K. Krause = Tarenna pavettoides（Harv.）Sim
subsp. friesiorum（K. Krause）Bridson ●☆

286216   Pavetta fulgens Miq. = Ixora fulgens Roxb. ●

286217   Pavetta funebris Bremek. = Tarenna funebris（Bremek.）N.
Hallé ●☆

286218   Pavetta gabonica Bremek.；加蓬大沙叶●☆

286219   Pavetta galpinii Bremek.；盖尔大沙叶●☆

286220   Pavetta gardeniifolia A. Rich.；索马里大沙叶●☆

286221   Pavetta gardeniifolia A. Rich. var. angustata？ = Pavetta
hochstetteri Bremek. ●☆

286222   Pavetta gardeniifolia A. Rich. var. appendiculata（De Wild.）
Bridson；附属物索马里大沙叶●☆

286223   Pavetta gardeniifolia A. Rich. var. breviflora Vatke = Pavetta
gardeniifolia A. Rich. ●☆

286224   Pavetta gardeniifolia A. Rich. var. laxiflora K. Schum. = Pavetta
gardeniifolia A. Rich. ●☆

286225   Pavetta gardeniifolia A. Rich. var. longiflora Vatke = Pavetta
subcana Hiern var. longiflora（Vatke）Bridson ●☆

286226   Pavetta gardeniifolia A. Rich. var. subtomentosa K. Schum.；亚
光索马里大沙叶●☆

286227   Pavetta gardeniifolia Hochst. ex A. Rich. = Pavetta gardeniifolia
A. Rich. ●☆

286228   Pavetta garuensis Bridson；加鲁大沙叶●☆

286229   Pavetta genipifolia Hiern = Pavetta leonensis Keay ●☆

286230   Pavetta genipifolia Schumach.；格尼木大沙叶●☆

286231   Pavetta gerrardii Harv. = Pavetta bowkeri Harv. ●☆

286232   Pavetta gerstneri Bremek.；格斯大沙叶●☆

286233   Pavetta gillilandii Bremek. = Pavetta villosa Vahl ●☆

286234   Pavetta gillmanii Bremek. = Pavetta bagshawei S. Moore ●☆

286235   Pavetta glaucescens Hiern = Pavetta owariensis P. Beauv. var.
glaucescens（Hiern）S. D. Manning ●☆

286236   Pavetta globularis Bremek.；小球大沙叶●☆

286237   Pavetta gloveri Bremek. = Pavetta villosa Vahl ●☆

286238   Pavetta gossweileri Bremek.；戈斯大沙叶●☆

286239   Pavetta gracilifolia Bremek.；细叶大沙叶●☆

286240   Pavetta gracilifolia Bremek. var. glabra？ = Pavetta gracilifolia
Bremek. ●☆

286241   Pavetta gracilifolia Bremek. var. pubescens？ = Pavetta
gracilifolia Bremek. ●☆

286242   Pavetta gracilipes Hiern；细梗大沙叶●☆

286243   Pavetta gracilis A. Rich. ex DC. = Tarenna richardii Verdc. ●☆

286244   Pavetta gracilis Klotzsch = Pavetta klotzschiana K. Schum. ●☆

286245   Pavetta gracillima S. Moore；细长大沙叶●☆

286246   Pavetta grandiflora K. Schum. et K. Krause = Pavetta rigida
Hiern ●☆

286247   Pavetta graveolens S. Moore = Coptosperma graveolens（S.

Moore) Degreef ●☆

286248　Pavetta greenwayi Bremek. ;格林韦大沙叶●☆

286249　Pavetta grossissima S. D. Manning;粗大沙叶●☆

286250　Pavetta grumosa S. Moore;聚粒大沙叶●☆

286251　Pavetta haegarthii Bremek. = Pavetta cooperi Harv. et Sond. ●☆

286252　Pavetta handenina Bremek. = Pavetta refractifolia K. Schum. ●☆

286253　Pavetta harborii S. Moore;哈伯大沙叶●☆

286254　Pavetta heidelbergensis Bremek. = Pavetta gardeniifolia A. Rich. var. subtomentosa K. Schum. ●☆

286255　Pavetta herbacea Bremek. ;草色大沙叶●☆

286256　Pavetta hierniana Bremek. = Pavetta owariensis P. Beauv. ●☆

286257　Pavetta hirtiflora Bremek. = Pavetta trichardtensis Bremek. ●☆

286258　Pavetta hispida Hiern;粗毛大沙叶●☆

286259　Pavetta hochstetteri Bremek. ;霍赫大沙叶●☆

286260　Pavetta hochstetteri Bremek. var. glaberrima？ = Pavetta gardeniifolia A. Rich. ●☆

286261　Pavetta hochstetteri Bremek. var. graciliflora？ = Pavetta gardeniifolia A. Rich. ●☆

286262　Pavetta hochstetteri Bremek. var. mollirama？ = Pavetta gardeniifolia A. Rich. ●☆

286263　Pavetta holstii K. Schum. ex Engl. ;霍尔大沙叶●☆

286264　Pavetta hongkongensis Bremek. ;香港大沙叶(巴佛他树,大沙叶,大叶满天星,港大沙叶,广东大沙叶,满天星,茜木,青风木,山铁尺);Hongkong Pavetta ●

286265　Pavetta hookeri Oudem. = Ixora hookeri(Oudem.) Bremek. ●☆

286266　Pavetta hookeriana Hiern;胡克大沙叶●☆

286267　Pavetta hookeriana Hiern var. pubinervata S. D. Manning;毛脉胡克大沙叶●☆

286268　Pavetta humbertii Bremek. = Pavetta oliveriana Hiern ●☆

286269　Pavetta hygrophytica Bremek. = Pavetta owariensis P. Beauv. var. glaucescens(Hiern) S. D. Manning ●☆

286270　Pavetta hygrophytica Bremek. var. corymbosa？ = Pavetta corymbosa(DC.) F. N. Williams ●☆

286271　Pavetta hymenophylla Bremek. ;膜质大沙叶●☆

286272　Pavetta inandensis Bremek. ;伊南德大沙叶●☆

286273　Pavetta incana Klotzsch;灰毛大沙叶●☆

286274　Pavetta inconspicua Dinter ex Bremek. = Pavetta lasiopeplus K. Schum. ●☆

286275　Pavetta indica L. ;茜木(印度大沙叶);Indian Pavetta ●

286276　Pavetta indica L. = Pavetta tomentosa Roxb. ex Sm. ●

286277　Pavetta indica L. var. polantha Hook. = Pavetta polyantha R. Br. ex Bremek. ●

286278　Pavetta indica L. var. tomentosa (Roxb. ex Sm.) Hook. f. = Pavetta tomentosa Roxb. ex Sm. ●

286279　Pavetta indigotica Bridson;蓝色大沙叶●☆

286280　Pavetta insignis Bremek. = Pavetta molundensis K. Krause ●☆

286281　Pavetta insignis Bremek. var. glabra？ = Pavetta molundensis K. Krause ●☆

286282　Pavetta insignis Bremek. var. puberula？ = Pavetta molundensis K. Krause ●☆

286283　Pavetta intermedia Bremek. ;间型大沙叶●☆

286284　Pavetta involucrata Engl. = Pavetta stenosepala K. Schum. ●☆

286285　Pavetta inyangensis Bremek. = Pavetta comostyla S. Moore var. inyangensis(Bremek.) Bridson ●☆

286286　Pavetta iringensis Bremek. = Pavetta kyimbilensis Bremek. var. iringensis(Bremek.) Bridson ●☆

286287　Pavetta ituriensis Bremek. = Pavetta ruwenzoriensis S. Moore ●☆

286288　Pavetta javanica Blume = Ixora javanica(Blume) DC. ●☆

286289　Pavetta johnstonii Bremek. ;约翰斯顿大沙叶●☆

286290　Pavetta johnstonii Bremek. subsp. breviloba Bridson;浅裂约翰斯顿大沙叶●☆

286291　Pavetta junodii(Schinz) K. Schum. = Tarenna junodii(Schinz) Bremek. ●☆

286292　Pavetta kabarensis Bremek. = Pavetta subcana Hiern var. longiflora(Vatke) Bridson ●☆

286293　Pavetta kasaica Bremek. ;开赛大沙叶●☆

286294　Pavetta kenyensis Bremek. = Pavetta abyssinica Fresen. ●☆

286295　Pavetta kerenensis Becc. ex Martelli = Pavetta subcana Hiern var. longiflora(Vatke) Bridson ●☆

286296　Pavetta kiloensis De Wild. = Pavetta ruwenzoriensis S. Moore ●☆

286297　Pavetta kirschsteiniana K. Krause = Pavetta ruwenzoriensis S. Moore ●☆

286298　Pavetta kisarawensis Bremek. = Pavetta stenosepala K. Schum. subsp. kisarawensis(Bremek.) Bridson ●☆

286299　Pavetta kiwuensis K. Krause = Pavetta oliveriana Hiern ●☆

286300　Pavetta klotzschiana K. Schum. ;克洛彻大沙叶●☆

286301　Pavetta klotzschiana K. Schum. var. incana(Klotzsch) K. Schum. = Pavetta incana Klotzsch ●☆

286302　Pavetta komghensis Bremek. = Pavetta capensis (Houtt.) Bremek. subsp. komghensis(Bremek.) Kok ●☆

286303　Pavetta kotschyana Cufod. = Pavetta subcana Hiern var. longiflora(Vatke) Bridson ●☆

286304　Pavetta krauseana K. Krause = Pavetta gardeniifolia A. Rich. ●☆

286305　Pavetta kribiensis S. D. Manning;凯里卜大沙叶●☆

286306　Pavetta kroneana Miq. = Ixora chinensis Lam. ●

286307　Pavetta kupensis S. D. Manning;库普大沙叶●☆

286308　Pavetta kyimbilensis Bremek. ;基穆比拉大沙叶●☆

286309　Pavetta kyimbilensis Bremek. var. iringensis(Bremek.) Bridson;伊林加大沙叶●☆

286310　Pavetta laevis Benth. = Psychotria calva Hiern ●☆

286311　Pavetta lanceisepala Bremek. = Pavetta sphaerobotrys K. Schum. ●☆

286312　Pavetta lanceolata Eckl. ;蜡瓣大沙叶;Forest Bride's-bush ●☆

286313　Pavetta lasiobractea K. Schum. ;毛苞大沙叶●☆

286314　Pavetta lasioclada(K. Krause) Mildbr. ex Bremek. ;毛枝大沙叶●☆

286315　Pavetta lasiopeplus K. Schum. ;毛被大沙叶●☆

286316　Pavetta lasiorhachis K. Schum. et K. Krause = Tarenna lasiorhachis(K. Schum. et K. Krause) Bremek. ●☆

286317　Pavetta lateriflora G. Don = Memecylon lateriflorum (G. Don) Bremek. ●☆

286318　Pavetta laurentii De Wild. ;洛朗大沙叶●☆

286319　Pavetta laxa S. D. Manning;松散大沙叶●☆

286320　Pavetta lebrunii Bremek. = Pavetta oliveriana Hiern ●☆

286321　Pavetta ledermannii K. Krause = Pavetta lasioclada(K. Krause) Mildbr. ex Bremek. ●☆

286322　Pavetta leonensis Keay;莱昂大沙叶●☆

286323　Pavetta lescrauwaetii De Wild. ;莱斯大沙叶●☆

286324　Pavetta leucosphaera Bremek. = Pavetta bagshawei S. Moore var. leucosphaera(Bremek.) Bridson ●☆

286325　Pavetta ligustriodora Chiov. = Coptosperma graveolens (S. Moore) Degreef ●☆

286326　Pavetta lindina Bremek. ;林迪大沙叶●☆

286327　Pavetta linearifolia Bremek. ;线叶大沙叶●☆

286328　Pavetta loandensis(S. Moore) Bremek. ;罗安达大沙叶●☆

286329　Pavetta lomamiensis Bremek. ;洛马米大沙叶●☆

286330　Pavetta longibrachiata Bremek. ;长双展枝大沙叶●☆

286331　Pavetta longipedunculata R. D. Good = Ixora aneimenodesma K. Schum. subsp. kizuensis De Block ●☆

286332　Pavetta longistipulata Bremek. = Pavetta bidentata Hiern ●☆

286333　Pavetta longistyla S. D. Manning;长柱大沙叶●☆

286334　Pavetta lulandoensis Bridson;卢兰多大沙叶●☆

286335　Pavetta lutambensis Bremek. ;卢塔波大沙叶●☆

286336　Pavetta luteola Stapf = Tarenna luteola(Stapf)Bremek. ●☆

286337　Pavetta lynesii Bridson;莱恩斯大沙叶●☆

286338　Pavetta macrosepala Hiern;大萼大沙叶●☆

286339　Pavetta macrosepala Hiern var. glabra Bremek. = Pavetta macrosepala Hiern ●☆

286340　Pavetta macrosepala Hiern var. puberula K. Schum. ;短毛大萼大沙叶●☆

286341　Pavetta macrostemon K. Schum. = Pavetta rigida Hiern ●☆

286342　Pavetta macrothyrsa K. Krause = Pavetta nitidula Hiern ●☆

286343　Pavetta maitlandii Bremek. = Pavetta abyssinica Fresen. ●☆

286344　Pavetta manamoca Bremek. = Pavetta sphaerobotrys K. Schum. subsp. tanaica(Bremek.)Bridson ●☆

286345　Pavetta mangallana K. Schum. et K. Krause = Pavetta stenosepala K. Schum. ●☆

286346　Pavetta mannii Hiern = Pavetta neurocarpa Benth. ●☆

286347　Pavetta mannioides Hutch. et Dalziel = Pavetta neurocarpa Benth. ●☆

286348　Pavetta manyanguensis Bridson;马尼扬加大沙叶●☆

286349　Pavetta marlothii Bremek. = Pavetta harborii S. Moore ●☆

286350　Pavetta mayumbensis R. D. Good;马永巴大沙叶●☆

286351　Pavetta megistocalyx K. Krause = Pavetta genipifolia Schumach. ●☆

286352　Pavetta melanophylla K. Schum. = Nichallea soyauxii(Hiern)Bridson ●☆

286353　Pavetta membranifolia K. Krause;膜叶大沙叶●☆

286354　Pavetta merkeri K. Krause = Pavetta sepium K. Schum. var. merkeri(K. Krause)Bridson ●☆

286355　Pavetta micheliana J. -G. Adam;米歇尔大沙叶●☆

286356　Pavetta micrantha Bremek. ;小花大沙叶●☆

286357　Pavetta microlancea K. Schum. = Pavetta zeyheri Sond. subsp. microlancea(K. Schum.)Herman ●☆

286358　Pavetta microphylla Chiov. ;小叶大沙叶●☆

286359　Pavetta micropunctata Bridson;小斑大沙叶●☆

286360　Pavetta microthamnus K. Schum. ;小枝大沙叶●☆

286361　Pavetta middelburgensis Bremek. = Pavetta zeyheri Sond. subsp. middelburgensis(Bremek.)Herman ●☆

286362　Pavetta mildbraedii K. Krause;米尔德大沙叶●☆

286363　Pavetta modesta(Hiern)S. E. Dawson;适度大沙叶●☆

286364　Pavetta mollis Afzel. ex Hiern;柔软大沙叶●☆

286365　Pavetta mollissima Hutch. et Dalziel;极软大沙叶●☆

286366　Pavetta molundensis K. Krause;莫伦德大沙叶●☆

286367　Pavetta monticola Hiern;山生大沙叶●☆

286368　Pavetta muelleri Bridson;米勒大沙叶●☆

286369　Pavetta mufindiensis Bridson;穆芬迪大沙叶●☆

286370　Pavetta muiriana S. D. Manning;缪里大沙叶●☆

286371　Pavetta murleensis Cufod. = Pavetta subcana Hiern var. longiflora(Vatke)Bridson ●☆

286372　Pavetta murleensis Cufod. var. glabrescens? = Pavetta subcana Hiern var. longiflora(Vatke)Bridson ●☆

286373　Pavetta nana K. Schum. ;矮小大沙叶●☆

286374　Pavetta natalensis Sond. ;纳塔尔大沙叶●☆

286375　Pavetta nbumbulensis Bremek. = Pavetta capensis(Houtt.)Bremek. ●☆

286376　Pavetta neurocarpa Benth. ;脉果大沙叶●☆

286377　Pavetta neurophylla S. Moore = Coptosperma neurophyllum(S. Moore)Degreef ●☆

286378　Pavetta niansae K. Krause = Pavetta ternifolia(Oliv.)Hiern ●☆

286379　Pavetta nigrescens Hutch. et Dalziel = Tarenna hutchinsonii Bremek. ●☆

286380　Pavetta nigritana Bremek. = Pavetta owariensis P. Beauv. ●☆

286381　Pavetta nitida(Schumach. et Thonn.)Hutch. et Dalziel = Pavetta corymbosa(DC.)F. N. Williams ●☆

286382　Pavetta nitidissima Bridson;极亮大沙叶●☆

286383　Pavetta nitidula Hiern;稍亮大沙叶●☆

286384　Pavetta nyassica Bremek. = Pavetta comostyla S. Moore subsp. nyassica(Bremek.)Bridson ●☆

286385　Pavetta obanica Bremek. ;奥班大沙叶●☆

286386　Pavetta oblongifolia(Hiern)Bremek. ;矩圆叶大沙叶●☆

286387　Pavetta obovata E. Mey. ex Sond. = Pavetta revoluta Hochst. ●☆

286388　Pavetta odorata Willd. ;芳香大沙叶(孔雀花);Fragrabt Pavetta ●

286389　Pavetta olivaceo-nigra K. Schum. ;榄黑大沙叶(橄榄绿-黑大沙叶)●☆

286390　Pavetta olivieriana Hiern;奥氏大沙叶●☆

286391　Pavetta olivieriana Hiern var. denudata(Bremek.)Bridson;裸露奥氏大沙叶●☆

286392　Pavetta olivieriana Hiern var. glabrata K. Schum. = Pavetta erlangeri Bremek. ●☆

286393　Pavetta olivifolia Chiov. = Coptosperma graveolens(S. Moore)Degreef ●☆

286394　Pavetta ombrophila Bremek. ;喜雨大沙叶●☆

286395　Pavetta opaca Bremek. = Pavetta lanceolata Eckl. ●☆

286396　Pavetta orthanthera Bremek. ;直药大沙叶●☆

286397　Pavetta ovaliloba Bremek. = Pavetta gardeniifolia A. Rich. var. subtomentosa K. Schum. ●☆

286398　Pavetta owariensis P. Beauv. ;尾张大沙叶●☆

286399　Pavetta owariensis P. Beauv. var. glaucescens(Hiern)S. D. Manning;灰绿大沙叶●☆

286400　Pavetta owariensis P. Beauv. var. opaca S. D. Manning;暗色大沙叶●☆

286401　Pavetta pallida Bremek. = Pavetta capensis(Houtt.)Bremek. ●☆

286402　Pavetta pammalaka Bremek. = Pavetta venenata Hutch. et E. A. Bruce ●☆

286403　Pavetta parviflora Afzel. = Psydrax parviflora(Afzel.)Bridson ●☆

286404　Pavetta paupercula K. Schum. ;贫乏大沙叶●☆

286405　Pavetta permodesta Wernham = Pavetta bidentata Hiern ●☆

286406　Pavetta petraea Bremek. = Pavetta gardeniifolia A. Rich. ●☆

286407　Pavetta phillipsiae S. Moore = Pavetta villosa Vahl ●☆

286408　Pavetta pierlotii Bridson;皮氏大沙叶●☆

286409　Pavetta platycalyx Bremek. ;宽萼大沙叶●☆

286410　Pavetta platyphylla Chiov. = Pavetta crebrifolia Hiern var. pubescens Bridson ●☆

286411　Pavetta pleiantha Bremek. ;繁花大沙叶●☆

286412　Pavetta pleiantha Bremek. var. glabrifolia? = Pavetta gardeniifolia A. Rich. var. subtomentosa K. Schum. ●☆

286413　Pavetta pleiantha Bremek. var. velutina? = Pavetta gardeniifolia A. Rich. var. subtomentosa K. Schum. ●☆

286414　Pavetta plumosa Hutch. et Dalziel;羽状大沙叶●☆

286415   Pavetta pocsii Bridson;波奇大沙叶●☆

286416   Pavetta polyantha Bremek. = Pavetta polyantha R. Br. ex Bremek. ●

286417   Pavetta polyantha R. Br. = Pavetta polyantha R. Br. ex Bremek. ●

286418   Pavetta polyantha R. Br. ex Bremek.;多花大沙叶;Manyflower Pavetta,Multiflorous Pavetta ●

286419   Pavetta pseudo-albicaulis Bridson;假白茎大沙叶●☆

286420   Pavetta pseudozeyheri Bremek. = Pavetta zeyheri Sond. ●☆

286421   Pavetta puberula Hiern;微毛大沙叶●☆

286422   Pavetta pubiflora Bremek. = Pavetta subcana Hiern ●☆

286423   Pavetta pumila N. E. Br.;弱小大沙叶●☆

286424   Pavetta punctata K. Krause = Tarenna punctata (K. Krause) Bremek. ●☆

286425   Pavetta pygmaea Bremek.;低矮大沙叶●☆

286426   Pavetta radicans Hiern;具根大沙叶●☆

286427   Pavetta rattrayi Bremek. = Pavetta inandensis Bremek. ●☆

286428   Pavetta redheadii Bremek.;雷德黑德大沙叶●☆

286429   Pavetta reflexa R. Br. = Pavetta hochstetteri Bremek. ●☆

286430   Pavetta refractifolia K. Schum.;折叶大沙叶●☆

286431   Pavetta revoluta Hochst.;外卷大沙叶●☆

286432   Pavetta rhodesiaca Bremek. = Pavetta gardeniifolia A. Rich. var. subtomentosa K. Schum. ●☆

286433   Pavetta rhombifolia Bremek. = Pavetta corymbosa (DC.) F. N. Williams ●☆

286434   Pavetta richardsiae Bridson;理查兹大沙叶●☆

286435   Pavetta rigida Hiern;硬大沙叶●☆

286436   Pavetta robusta Bremek.;粗壮大沙叶●☆

286437   Pavetta rogersii Bremek. = Pavetta bowkeri Harv. ●☆

286438   Pavetta roseostellata Bridson;粉红星大沙叶●☆

286439   Pavetta rotundifolia B. Boivin = Tarenna grevei(Drake)Homolle ●☆

286440   Pavetta ruahaensis Bridson;鲁阿哈大沙叶●☆

286441   Pavetta rudolphina Cufod. = Pavetta subcana Hiern var. longiflora(Vatke) Bridson ●☆

286442   Pavetta rudolphina Cufod. var. robusta? = Pavetta subcana Hiern var. longiflora(Vatke) Bridson ●☆

286443   Pavetta rufipila Bremek. = Pavetta sansibarica K. Schum. var. rufipila(Bremek.) Bridson ●☆

286444   Pavetta ruwenzoriensis S. Moore;鲁文佐里大沙叶●☆

286445   Pavetta rwandensis Bridson;卢旺达大沙叶●☆

286446   Pavetta saligna S. Moore = Coptosperma nigrescens Hook. f. ●☆

286447   Pavetta sanguinolenta R. D. Good = Pavetta canescens DC. ●☆

286448   Pavetta sansibarica K. Schum.;桑给巴尔大沙叶●☆

286449   Pavetta sansibarica K. Schum. subsp. trichosphaera (Bremek.) Bridson;毛球桑给巴尔大沙叶●☆

286450   Pavetta sansibarica K. Schum. var. rufipila (Bremek.) Bridson;红毛大沙叶●☆

286451   Pavetta saxicola K. Krause = Pavetta gardeniifolia A. Rich. ●☆

286452   Pavetta scabrifolia Bremek.;糙叶大沙叶;Rough-leaved Pavetta,Scabrousleaf Pavetta ●

286453   Pavetta scaettae Bremek. = Pavetta bagshawei S. Moore ●☆

286454   Pavetta scandens Bremek. = Pavetta sepium K. Schum. ●☆

286455   Pavetta schliebenii Mildbr. ex Bremek.;施利本大沙叶●☆

286456   Pavetta schubotziana K. Krause = Pavetta ruwenzoriensis S. Moore ●☆

286457   Pavetta schumanniana F. Hoffm. ex K. Schum.;舒曼大沙叶●☆

286458   Pavetta schweinfurthii Bremek.;施韦大沙叶●☆

286459   Pavetta schweinfurthii Bremek. var. oblongifolia(Hiern) Aubrév.

= Pavetta oblongifolia(Hiern) Bremek. ●☆

286460   Pavetta schweinfurthii Bremek. var. pubescens?;短柔毛施韦大沙叶●☆

286461   Pavetta sennii(Chiov.) Bridson = Pavetta uniflora Bremek. ●☆

286462   Pavetta sennii Chiov.;森恩大沙叶●☆

286463   Pavetta sepium K. Schum.;篱笆大沙叶●☆

286464   Pavetta sepium K. Schum. var. glabra Bremek.;光滑大沙叶●☆

286465   Pavetta sepium K. Schum. var. massaica Bridson;马萨大沙叶●☆

286466   Pavetta sepium K. Schum. var. merkeri(K. Krause) Bridson;梅克尔大沙叶●☆

286467   Pavetta sepium K. Schum. var. pubescens Bremek. = Pavetta sepium K. Schum. ●☆

286468   Pavetta seretii De Wild.;赛雷大沙叶●☆

286469   Pavetta shimbensis Bremek. = Pavetta sansibarica K. Schum. subsp. trichosphaera(Bremek.) Bridson ●☆

286470   Pavetta silvae K. Schum. = Pavetta revoluta Hochst. ●☆

286471   Pavetta silvicola Bremek. = Pavetta abyssinica Fresen. ●☆

286472   Pavetta sinica Miq. = Pavetta arenosa Lour. ●

286473   Pavetta smeathmannii DC. = Keetia multiflora (Schumach. et Thonn.) Bridson ●☆

286474   Pavetta smythei Hutch. et Dalziel = Pavetta owariensis P. Beauv. ●☆

286475   Pavetta somaliensis Bremek. = Pavetta gardeniifolia A. Rich. ●☆

286476   Pavetta spaniotricha Bremek. = Pavetta subumbellata Bremek. ●☆

286477   Pavetta sparsipila Bremek.;疏毛大沙叶●☆

286478   Pavetta spathulata Bremek.;匙形大沙叶●☆

286479   Pavetta sphaerobotrys K. Schum.;球穗大沙叶●☆

286480   Pavetta sphaerobotrys K. Schum. subsp. lanceisepala(Bremek.) Bridson;剑萼球穗大沙叶●☆

286481   Pavetta sphaerobotrys K. Schum. subsp. tanaica (Bremek.) Bridson;塔纳大沙叶●☆

286482   Pavetta squarrosa K. Krause = Pavetta sepium K. Schum. var. glabra Bremek. ●☆

286483   Pavetta staudtii Hutch. et Dalziel;施陶大沙叶●☆

286484   Pavetta stemonogyne Mildbr. ex Bremek.;窄蕊大沙叶●☆

286485   Pavetta stenosepala K. Schum.;窄萼大沙叶●☆

286486   Pavetta stenosepala K. Schum. subsp. kisarawensis (Bremek.) Bridson;基萨拉韦大沙叶●☆

286487   Pavetta stephanantha Bremek. = Pavetta decumbens K. Schum. et K. Krause ●☆

286488   Pavetta stipulopallium K. Schum.;托叶大沙叶●☆

286489   Pavetta striatula Hutch. et Dalziel = Tarenna nitidula (Benth.) Hiern ●☆

286490   Pavetta stuhlmannii Bremek. = Pavetta bagshawei S. Moore ●☆

286491   Pavetta subcana Hiern;亚灰白大沙叶●☆

286492   Pavetta subcana Hiern var. longiflora(Vatke)Bridson;长叶灰白大沙叶●☆

286493   Pavetta subglabra Schumach.;近光大沙叶●☆

286494   Pavetta subumbellata Bremek.;小伞大沙叶●☆

286495   Pavetta subumbellata Bremek. var. subcoriacea Bridson;革质小伞大沙叶●☆

286496   Pavetta suffruticosa K. Schum. = Pavetta tetramera (Hiern) Bremek. ●☆

286497   Pavetta suluensis Bremek. = Pavetta natalensis Sond. ●☆

286498   Pavetta supra-axillaris Hemsl. = Coptosperma supra-axillare (Hemsl.) Degreef ●☆

286499   Pavetta suurbergensis Bremek. = Pavetta capensis (Houtt.) Bremek. ●☆

286500　Pavetta swatouica Bremek. ;汕头大沙叶;Shantou Pavetta ●

286501　Pavetta swynnertonii S. Moore = Tarenna pavettoides(Harv.) Sim subsp. affinis(K. Schum.) Bridson ●☆

286502　Pavetta syringoides Webb = Rutidea parviflora DC. ●☆

286503　Pavetta talbotii Wernham;塔尔博特大沙叶●☆

286504　Pavetta tanaica Bremek. = Pavetta sphaerobotrys K. Schum. subsp. tanaica(Bremek.) Bridson ●☆

286505　Pavetta tarennoides S. Moore;乌口树大沙叶●☆

286506　Pavetta tendagurensis Bremek. ;滕达古尔大沙叶●☆

286507　Pavetta tendagurensis Bremek. var. glabrescens Bridson = Pavetta tendagurensis Bremek. ●☆

286508　Pavetta tenuifolia Benth. = Psychotria leptophylla Hiern ●☆

286509　Pavetta tenuissima S. D. Manning;极细大沙叶●☆

286510　Pavetta termitaria Bremek. var. glabra? = Pavetta gardeniifolia A. Rich. ●☆

286511　Pavetta termitaria Bremek. var. pubescens? = Pavetta gardeniifolia A. Rich. var. subtomentosa K. Schum. ●☆

286512　Pavetta ternifolia(Oliv.) Hiern;三出大沙叶●☆

286513　Pavetta testui Bremek. ;泰斯蒂大沙叶●☆

286514　Pavetta tetramera(Hiern) Bremek. ;四数大沙叶●☆

286515　Pavetta thorbeckii K. Krause;托尔贝克大沙叶●☆

286516　Pavetta thyrsiflora Thunb. ex DC. = Pavetta capensis (Houtt.) Bremek. ●☆

286517　Pavetta tisserantii Bremek. = Pavetta lasioclada (K. Krause) Mildbr. ex Bremek. ●☆

286518　Pavetta tomentella Bremek. = Pavetta gardeniifolia A. Rich. var. subtomentosa K. Schum. ●☆

286519　Pavetta tomentosa A. Rich. = Pavetta canescens DC. ●☆

286520　Pavetta tomentosa Roxb. ex Sm. ;绒毛大沙叶; Tomentose Pavetta ●

286521　Pavetta tomentosa Roxb. ex Sm. var. roxburghii? = Pavetta tomentosa Roxb. ex Sm. ●

286522　Pavetta trichantha Baker = Paracephaelis trichantha(Baker) De Block ●☆

286523　Pavetta trichardtensis Bremek. ;特里哈特大沙叶●☆

286524　Pavetta trichocalyx Bremek. = Pavetta elliottii K. Schum. et K. Krause var. trichocalyx(Bremek.) Bridson ●☆

286525　Pavetta trichocephala Bremek. = Pavetta amaniensis Bremek. var. trichocephala(Bremek.) Bridson ●☆

286526　Pavetta trichosphaera Bremek. = Pavetta sansibarica K. Schum. subsp. trichosphaera(Bremek.) Bridson ●☆

286527　Pavetta trichotropis Bremek. = Pavetta abyssinica Fresen. ●☆

286528　Pavetta tristis Bremek. = Pavetta lanceolata Eckl. ●☆

286529　Pavetta troupinii Bridson;特鲁皮尼大沙叶●☆

286530　Pavetta ugandensis Bremek. = Pavetta bagshawei S. Moore ●☆

286531　Pavetta undulata Lehm. = Pavetta revoluta Hochst. ●☆

286532　Pavetta unguiculata Bremek. = Pavetta subcana Hiern var. longiflora(Vatke) Bridson ●☆

286533　Pavetta uniflora Bremek. ;单花大沙叶●☆

286534　Pavetta urbis-reginae Bremek. = Pavetta capensis (Houtt.) Bremek. ●☆

286535　Pavetta urophylla Bremek. ;尾叶大沙叶●☆

286536　Pavetta urophylla Bremek. subsp. bosii S. D. Manning;博斯大沙叶●☆

286537　Pavetta urundensis Bremek. ;乌隆迪大沙叶●☆

286538　Pavetta usambarica Bremek. = Pavetta abyssinica Fresen. var. usambarica(Bremek.) Bridson ●☆

286539　Pavetta utilis Hua = Pavetta crassipes K. Schum. ●☆

286540　Pavetta uwembae Gilli = Pavetta subumbellata Bremek. ●☆

286541　Pavetta vanderijstii Bremek. ;范德大沙叶●☆

286542　Pavetta vanwykiana Bridson ;万维大沙叶●☆

286543　Pavetta venenata Hutch. et E. A. Bruce;毒大沙叶●☆

286544　Pavetta venusta Bremek. = Pavetta bidentata Hiern ●☆

286545　Pavetta viburnoides A. Chev. ex Hutch. et Dalziel = Pavetta lasioclada(K. Krause) Mildbr. ex Bremek. ●☆

286546　Pavetta villosa Vahl;长柔毛大沙叶●☆

286547　Pavetta viridiflora R. D. Good = Pavetta microthamnus K. Schum. ●☆

286548　Pavetta viridiloba K. Krause;绿裂片大沙叶●☆

286549　Pavetta virungensis Bremek. = Pavetta urundensis Bremek. ●☆

286550　Pavetta wargalensis Bremek. = Pavetta villosa Vahl ●☆

286551　Pavetta warneckei K. Schum. et K. Krause = Pavetta subglabra Schumach. ●☆

286552　Pavetta whiteana Bridson;怀特大沙叶●☆

286553　Pavetta wildemannii Bremek. ;怀尔德曼大沙叶●☆

286554　Pavetta woodii Bremek. = Pavetta gracilifolia Bremek. ●☆

286555　Pavetta yalaensis Bremek. = Pavetta ternifolia(Oliv.) Hiern ●☆

286556　Pavetta yambatensis Bremek. ;扬巴塔大沙叶●☆

286557　Pavetta zeyheri Sond. ;泽赫大沙叶●☆

286558　Pavetta zeyheri Sond. subsp. microlancea(K. Schum.) Herman;小剑泽赫大沙叶●☆

286559　Pavetta zeyheri Sond. subsp. middelburgensis (Bremek.) Herman;米德尔堡大沙叶●☆

286560　Pavetta zeyheri Sond. var. brevituba Bremek. = Pavetta zeyheri Sond. ●☆

286561　Pavetta zeyheri Sond. var. pubescens Bremek. = Pavetta zeyheri Sond. ●☆

286562　Pavetta zeyheri Sond. var. sonderi Bremek. = Pavetta zeyheri Sond. ●☆

286563　Pavetta zimmermanniana Valeton;齐默尔曼大沙叶●☆

286564　Pavetta zombana K. Schum. = Rutidea fuscescens Hiern ●☆

286565　Pavetta zoutpansbergensis Bremek. = Pavetta trichardtensis Bremek. ●☆

286566　Pavia Boerh. ex Mill. = Aesculus L. ●

286567　Pavia Mill. = Aesculus L. ●

286568　Pavia indica Wall. ex Cambess. = Aesculus indica (Wall. ex Cambess.) Hook. f. ●☆

286569　Paviaceae Horan. = Hippocastanaceae A. Rich. (保留科名)●

286570　Paviaceae Horan. = Sapindaceae Juss. (保留科名)●■

286571　Paviana Raf. = Aesculus L. ●

286572　Paviana Raf. = Pavia Mill. ●

286573　Pavieasia Pierre(1895);檀栗属(棱果木属);Pavieasia ●

286574　Pavieasia anamensis Pierre;越南檀栗;Vietnam Pavieasia ●☆

286575　Pavieasia anamensis Pierre = Pavieasia yunnanensis H. S. Lo ●☆

286576　Pavieasia kwangxiensis H. S. Lo;广西檀栗;Guangxi Pavieasia, Kwangsi Pavieasia ●

286577　Pavieasia yunnanensis H. S. Lo;云南檀栗;Yunnan Pavieasia ●

286578　Pavinda Thunb. = Audouinia Brongn. ●☆

286579　Pavinda Thunb. ex Bartl. = Audouinia Brongn. ●☆

286580　Pavonia Cav. (1786)(保留属名);巴氏锦葵属(巴氏槿属,粉葵属,老虎花属,帕翁葵属,帕沃木属);Pavonia ●■☆

286581　Pavonia Dombey ex Lam. = Cordia L. (保留属名)●

286582　Pavonia Dombey ex Lam. = Pavonia Cav. (保留属名)●■☆

286583　Pavonia Ruiz = Laurelia Juss. (保留属名)●☆

286584　Pavonia Ruiz et Pav. = Laurelia Juss. (保留属名)●☆

286585　Pavonia × gledhillii Cheek；巴氏锦葵（巴氏槿，帕沃木）；Gledhill Pavonia ●☆

286586　Pavonia × gledhillii Cheek 'Kermesina'；暗红巴氏锦葵（暗红巴氏槿，暗红帕沃木）●☆

286587　Pavonia × gledhillii Cheek 'Rosea'；玫瑰红巴氏锦葵（玫瑰红帕沃木）●☆

286588　Pavonia arabica Hochst. et Steud. = Pavonia arabica Hochst. et Steud. ex Boiss. ●☆

286589　Pavonia arabica Hochst. et Steud. ex Boiss.；阿拉伯巴氏锦葵●☆

286590　Pavonia arabica Hochst. et Steud. ex Boiss. var. flavovelutina Ulbr. = Pavonia arabica Hochst. et Steud. ex Boiss. ●☆

286591　Pavonia arabica Hochst. et Steud. ex Boiss. var. glanduligera Gürke = Pavonia arabica Hochst. et Steud. ex Boiss. ●☆

286592　Pavonia arabica Hochst. et Steud. ex Boiss. var. procumbens A. Terracc. = Pavonia arabica Hochst. et Steud. ex Boiss. ●☆

286593　Pavonia baumii Gürke = Pavonia senegalensis (Cav.) Leistner ●☆

286594　Pavonia bequaertii De Wild. = Pavonia schimperiana Hochst. ex A. Rich. ●☆

286595　Pavonia blepharicarpa N. A. Brummitt et Vollesen；毛果巴氏锦葵●☆

286596　Pavonia brevibracteolata Hauman；短苞巴氏锦葵●☆

286597　Pavonia burchellii (DC.) R. A. Dyer；绒毛巴氏锦葵●☆

286598　Pavonia burchellii (DC.) R. A. Dyer var. glandulosa (Ulbr.) Heine = Pavonia burchellii (DC.) R. A. Dyer ☆

286599　Pavonia burchellii (DC.) R. A. Dyer var. schweinfurthii (Ulbr.) Heine；施韦绒毛巴氏锦葵●☆

286600　Pavonia burchellii (DC.) R. A. Dyer var. tomentosa (Ulbr.) Heine = Pavonia burchellii (DC.) R. A. Dyer ●☆

286601　Pavonia ceratocarpa Dalzell ex Mast. = Pavonia grewioides Hochst. ex Boiss. ●☆

286602　Pavonia cernua (Span.) Walp. = Pavonia procumbens (Wall. ex Wight et Arn.) Walp. ●☆

286603　Pavonia clathrata Mast.；格子巴氏锦葵●☆

286604　Pavonia columella Cav.；小圆柱巴氏锦葵●☆

286605　Pavonia commutata Conrath；变异巴氏锦葵●☆

286606　Pavonia corymbosa (Sw.) Willd.；密花巴氏锦葵（密花巴氏槿，密花帕沃木）；Tight-flower Swampmallow ●☆

286607　Pavonia coxii Tadul. et Jacob = Pavonia glechomifolia (A. Rich.) Garcke ●☆

286608　Pavonia cristata Schinz ex Gürke；冠状巴氏锦葵●☆

286609　Pavonia ctenophora Ulbr. = Pavonia procumbens (Wall. ex Wight et Arn.) Walp. ●☆

286610　Pavonia dentata Burtt Davy；尖齿巴氏锦葵●☆

286611　Pavonia digitata Hochst. ex Chiov. = Pavonia zeylanica (L.) Cav. ●☆

286612　Pavonia discolor Ulbr. = Pavonia leptocalyx (Sond.) Ulbr. ●☆

286613　Pavonia dregei Garcke；德雷巴氏锦葵●☆

286614　Pavonia elegans Garcke；雅致巴氏锦葵●☆

286615　Pavonia ellenbeckii Gürke；埃伦巴氏锦葵●☆

286616　Pavonia erlangeri Ulbr. = Pavonia arabica Hochst. et Steud. ex Boiss. ●☆

286617　Pavonia erythraeae Chiov. = Pavonia arabica Hochst. et Steud. ex Boiss. ●☆

286618　Pavonia flavescens Mattei = Pavonia serrata Franch. ●☆

286619　Pavonia flavoferruginea (Forssk.) Hepper et J. R. I. Wood；锈黄巴氏锦葵●☆

286620　Pavonia franchetiana Schinz ex Gürke = Pavonia arabica Hochst.

et Steud. ex Boiss. ●☆

286621　Pavonia friisii Thulin et Vollesen；弗里斯巴氏锦葵●☆

286622　Pavonia fruticulosa Ulbr. = Pavonia leptocalyx (Sond.) Ulbr. ●☆

286623　Pavonia gallaensis Ulbr.；加拉巴氏锦葵●☆

286624　Pavonia galpiniana Schinz = Pavonia columella Cav. ●☆

286625　Pavonia glandulosa Franch. = Pavonia arabica Hochst. et Steud. ex Boiss. ●☆

286626　Pavonia glechomifolia (A. Rich.) Garcke；活血丹叶巴氏锦葵●☆

286627　Pavonia glechomifolia (A. Rich.) Garcke = Pavonia flavoferruginea (Forssk.) Hepper et J. R. I. Wood ●☆

286628　Pavonia glechomifolia (A. Rich.) Garcke var. glabrescens Ulbr. = Pavonia flavoferruginea (Forssk.) Hepper et J. R. I. Wood ●☆

286629　Pavonia glechomifolia (A. Rich.) Garcke var. tomentosa Ulbr. = Pavonia procumbens (Wall. ex Wight et Arn.) Walp. ●☆

286630　Pavonia gossweileri Exell；戈斯巴氏锦葵●☆

286631　Pavonia grewioides Hochst. ex Boiss.；扁担杆巴氏锦葵●☆

286632　Pavonia grewioides Hochst. ex Boiss. = Pavonia propinqua Garcke ●☆

286633　Pavonia haematophthalmos Chiov. = Pavonia flavoferruginea (Forssk.) Hepper et J. R. I. Wood ●☆

286634　Pavonia hastata Cav.；红斑巴氏锦葵（巴氏槿，粉葵，红斑巴氏槿，红斑帕沃木）；Spearleaf Swamp Mallow, Spearleaf Swampmallow ●☆

286635　Pavonia hildebrandtii Gürke ex Ulbr. = Pavonia serrata Franch. ●☆

286636　Pavonia hirsuta Guillaumin et Perr. = Pavonia senegalensis (Cav.) Leistner ●☆

286637　Pavonia insignis Fenzl ex Webb = Pavonia senegalensis (Cav.) Leistner ●☆

286638　Pavonia intermedia A. St. -Hil.；中间巴氏锦葵（中间巴氏槿，中间帕沃木）●☆

286639　Pavonia kilimandscharica Gürke；基利巴氏锦葵●☆

286640　Pavonia kotschyi Hochst. ex Webb；科奇巴氏锦葵●☆

286641　Pavonia kotschyi Hochst. ex Webb var. glabrescens Ulbr.；渐光巴氏锦葵●☆

286642　Pavonia kotschyi Hochst. ex Webb var. glutinosa Mattei = Pavonia matteiana Thulin ●☆

286643　Pavonia kotschyi Hochst. ex Webb var. mollissima Mattei；绢毛巴氏锦葵●☆

286644　Pavonia kraussiana Hochst. = Pavonia burchellii (DC.) R. A. Dyer ●☆

286645　Pavonia kraussiana Hochst. var. genuina Ulbr. = Pavonia burchellii (DC.) R. A. Dyer ●☆

286646　Pavonia kraussiana Hochst. var. glandulosa Ulbr. = Pavonia burchellii (DC.) R. A. Dyer ●☆

286647　Pavonia kraussiana Hochst. var. schweinfurthii Ulbr. = Pavonia burchellii (DC.) R. A. Dyer var. schweinfurthii (Ulbr.) Heine ●☆

286648　Pavonia kraussiana Hochst. var. tomentosa Ulbr. = Pavonia burchellii (DC.) R. A. Dyer ●☆

286649　Pavonia lasiopetala Scheele；岩巴氏槿（岩帕沃木）；Rock Rosemallow ●☆

286650　Pavonia lavae Engl. = Pavonia schweinfurthii Ulbr. ●☆

286651　Pavonia leptocalyx (Sond.) Ulbr.；细萼巴氏锦葵●☆

286652　Pavonia leptoclada Ulbr. = Abutilon mauritianum (Jacq.) Medik. ■☆

286653　Pavonia longipilosa Thulin；长毛巴氏锦葵●☆

286654　Pavonia macrophylla E. Mey. ex Harv. = Pavonia burchellii (DC.) R. A. Dyer ●☆

286655　Pavonia marginata Thulin；具边巴氏锦葵●☆

286656 Pavonia matteiana Thulin;马特巴氏锦葵●☆

286657 Pavonia melhanioides Thulin;梅蓝巴氏锦葵●☆

286658 Pavonia meyeri Mast. = Pavonia columella Cav. ●☆

286659 Pavonia microphylla E. Mey. ex Harv. = Pavonia dregei Garcke ●☆

286660 Pavonia minimifolia Chiov. = Pavonia pirottae (A. Terracc.) Chiov. ●☆

286661 Pavonia mollis E. Mey. ex Harv. = Pavonia columella Cav. ●☆

286662 Pavonia mollissima(Garcke) Ulbr.;柔软巴氏锦葵●☆

286663 Pavonia multiflora A. St. -Hil.;多花巴氏锦葵(多花巴氏槿,多花帕沃木); Brazilian candles, Mexican Turk's Cap, Sleepy Hibiscus ●☆

286664 Pavonia neumannii Ulbr. = Pavonia urens Cav. ●☆

286665 Pavonia nigrescens Thulin;变黑巴氏锦葵●☆

286666 Pavonia odorata Willd.;香巴氏锦葵(孔雀花,帕翁葵,香帕翁葵)●☆

286667 Pavonia odorata Willd. var. molissima Garcke = Pavonia mollissima(Garcke) Ulbr. ●☆

286668 Pavonia paolii Mattei = Pavonia zeylanica(L.) Cav. ●☆

286669 Pavonia patens (Andréws) Chiov. var. tomentosa (Ulbr.) Cufod. = Pavonia procumbens(Wall. ex Wight et Arn.) Walp. ●☆

286670 Pavonia paucibracteata Thulin;疏苞巴氏锦葵●☆

286671 Pavonia pirottae(A. Terracc.) Chiov.;皮罗塔巴氏锦葵●☆

286672 Pavonia praemorsa(L. f.) Cav.;黄巴氏锦葵(黄巴氏槿,黄帕沃木); Yellow Mallow ●☆

286673 Pavonia praemorsa(L. f.) Willd. = Pavonia praemorsa (L. f.) Cav. ●☆

286674 Pavonia procumbens(Wall. ex Wight et Arn.)Walp.;匍匐锦葵●☆

286675 Pavonia procumbens (Wight) Walp. = Pavonia procumbens (Wall. ex Wight et Arn.) Walp. ●☆

286676 Pavonia propinqua Garcke = Pavonia grewioides Hochst. ex Boiss. ●☆

286677 Pavonia pseudo-arabica Mattei = Pavonia arabica Hochst. et Steud. ex Boiss. ●☆

286678 Pavonia rehmannii Szyszyl.;拉赫曼锦葵●☆

286679 Pavonia repanda(Roxb. ex Sm.)Spreng. = Urena repanda Roxb. ■

286680 Pavonia rogersii N. E. Br.;罗杰斯锦葵●☆

286681 Pavonia romborua Wall. = Pavonia odorata Willd. ●☆

286682 Pavonia rosea Schltdl.;粉红巴氏锦葵(玫瑰帕翁葵)●☆

286683 Pavonia rosea Wall. = Pavonia odorata Willd. ●☆

286684 Pavonia rotundifolia Thulin et Vollesen;圆叶巴氏锦葵●☆

286685 Pavonia rufescens Mattei = Pavonia ellenbeckii Gürke ●☆

286686 Pavonia rulingioides Ulbr. = Pavonia zeylanica(L.)Cav. ●☆

286687 Pavonia rutshuruensis De Wild. = Pavonia schimperiana Hochst. ex A. Rich. ●☆

286688 Pavonia ruwenzoriensis De Wild. = Pavonia urens Cav. ●☆

286689 Pavonia schiedenaa Steud.;希巴氏锦葵(希特帕翁葵)●☆

286690 Pavonia schimperiana Hochst. ex A. Rich.;欣珀巴氏锦葵●☆

286691 Pavonia schimperiana Hochst. ex A. Rich. var. glabrescens Ulbr. = Pavonia schimperiana Hochst. ex A. Rich. ●☆

286692 Pavonia schimperiana Hochst. ex A. Rich. var. hirsuta Hochst. ex Ulbr. = Pavonia urens Cav. ●☆

286693 Pavonia schimperiana Hochst. ex A. Rich. var. tomentosa Hochst. ex Ulbr. = Pavonia urens Cav. ●☆

286694 Pavonia schumanniana Gürke = Pavonia clathrata Mast. ●☆

286695 Pavonia schumanniana Gürke var. transvaalensis Ulbr. = Pavonia transvaalensis(Ulbr.)A. Meeuse ●☆

286696 Pavonia schweinfurthii Ulbr.;施韦巴氏锦葵●☆

286697 Pavonia semperflorens Garcke;常花巴氏锦葵(常花巴氏槿,常花帕沃木)●☆

286698 Pavonia senegalensis(Cav.)Leistner;塞内加尔巴氏锦葵●☆

286699 Pavonia sennii Chiov.;森恩巴氏锦葵●☆

286700 Pavonia serrata Franch.;具齿巴氏锦葵●☆

286701 Pavonia somalensis Franch.;索马里巴氏锦葵●☆

286702 Pavonia somalensis Mattei = Pavonia procumbens (Wall. ex Wight et Arn.) Walp. ●☆

286703 Pavonia spinifex(L.)Cav.;南美巴氏锦葵(南美巴氏槿,南美帕沃木); Gingerbush ●☆

286704 Pavonia stefanini Ulbr. = Pavonia procumbens(Wall. ex Wight et Arn.) Walp. ●☆

286705 Pavonia steudneri Ulbr.;斯托德巴氏锦葵●☆

286706 Pavonia stolzii Ulbr. = Pavonia urens Cav. ●☆

286707 Pavonia transvaalensis(Ulbr.)A. Meeuse;德兰士瓦巴氏锦葵●☆

286708 Pavonia triloba Guillaumin et Perr.;三裂巴氏锦葵●☆

286709 Pavonia ukambanica Ulbr. = Pavonia procumbens (Wall. ex Wight et Arn.) Walp. ●☆

286710 Pavonia urens Cav.;蛰毛巴氏锦葵●☆

286711 Pavonia urens Cav. var. glabrescens (Ulbr.) Brenan = Pavonia schimperiana Hochst. ex A. Rich. ●☆

286712 Pavonia urens Cav. var. hirsuta (Hochst. ex Ulbr.) Brenan = Pavonia urens Cav. ●☆

286713 Pavonia urens Cav. var. obtusiloba(Hiern)Brenan;钝裂巴氏锦葵●☆

286714 Pavonia urens Cav. var. schimperiana (Hochst. ex A. Rich.) Brenan = Pavonia schimperiana Hochst. ex A. Rich. ●☆

286715 Pavonia urens Cav. var. tomentosa(Hochst. ex Ulbr.) Brenan = Pavonia urens Cav. ●☆

286716 Pavonia urens Cav. var. variabilis (De Wild.) Hauman;易变巴氏锦葵☆

286717 Pavonia variabilis De Wild. = Pavonia urens Cav. ●☆

286718 Pavonia vespertilionacea Hochr. = Pavonia rehmannii Szyszyl. ●☆

286719 Pavonia zawadae Ulbr. = Pavonia senegalensis(Cav.)Leistner ●☆

286720 Pavonia zeylanica(L.)Cav.;斯里兰卡巴氏锦葵●☆

286721 Pavonia zeylanica (L.) Cav. subsp. afro-arabica Cufod. = Pavonia zeylanica(L.) Cav. ●☆

286722 Pavonia zeylanica (L.) Cav. var. microphylla Ulbr. = Pavonia zeylanica(L.) Cav. ●☆

286723 Pavonia zeylanica (L.) Cav. var. subquinqueloba Ulbr. = Pavonia zeylanica(L.) Cav. ●☆

286724 Pavonia zeylanica Cav. = Pavonia zeylanica(L.) Cav. ●☆

286725 Pawia Kuntze = Aesculus L. ●

286726 Pawia Kuntze = Pavia Mill. ●

286727 Paxia Gilg = Rourea Aubl. (保留属名)●

286728 Paxia Herter = Paxiuscula Herter ●☆

286729 Paxia O. Nilsson = Montia L. ■☆

286730 Paxia O. Nilsson = Neopaxia O. Nilsson ■☆

286731 Paxia calophylla Gilg ex G. Schellenb. = Rourea calophylla(Gilg ex G. Schellenb.) Jongkind ●☆

286732 Paxia calophylloides G. Schellenb. = Rourea calophylloides(G. Schellenb.) Jongkind ●☆

286733 Paxia cinnabarina G. Schellenb. = Rourea myriantha Baill. ●☆

286734 Paxia dewevrei De Wild. et T. Durand = Rourea thomsonii (Baker)Jongkind ●☆

286735 Paxia lancea G. Schellenb. = Rourea myriantha Baill. ●☆

286736 Paxia liberosepala(Baker f.) G. Schellenb. = Rourea myriantha

Baill. ●☆

286737　Paxia myriantha(Baill.) Pierre = Rourea myriantha Baill. ●☆

286738　Paxia scandens Gilg = Rourea myriantha Baill. ●☆

286739　Paxia zenkeri G. Schellenb. = Rourea myriantha Baill. ●☆

286740　Paxiactes Raf. = Heracleum L. ■

286741　Paxiactes Raf. = Tordyliopsis DC. ■

286742　Paxiodendron Engl. = Xymalos Baill. ●☆

286743　Paxiodendron ulugurense Engl. = Xymalos monospora(Harv.) Baill. ●☆

286744　Paxiodendron usambarense Engl. = Xymalos monospora(Harv.) Baill. ●☆

286745　Paxiodendron usambarense Engl. var. serratifolia? = Xymalos monospora(Harv.) Baill. ●☆

286746　Paxistima Raf. (1838);崖翠木属(厚柱头木属);Paxistima ●☆

286747　Paxistima macrophylla Farr. ;大叶崖翠木（大叶厚柱头木）;Largeleaf Paxistima ●☆

286748　Paxistima myrsinites(Pursh)Raf. ;铁仔崖翠木（厚柱头木,铁仔厚柱头木）;Myrsin-like Paxistima ●☆

286749　Paxistima myrtifolia (Nutt.) L. C. Wheeler;崖翠木（厚柱头木）;Falsebox,Oregon Boxleaf,Oregon Boxwood ●☆

286750　Paxiuscula Herter = Argythamnia P. Browne ●☆

286751　Paxiuscula Herter = Ditaxis Vahl ex A. Juss. ●☆

286752　Paxtonia Lindl. = Spathoglottis Blume ■

286753　Payanelia C. B. Clarke = Pajanelia DC. ●☆

286754　Payena A. DC. (1844);东南亚山榄属(巴椰榄属,巴因榄属,巴因那木属);Payena ●☆

286755　Payena leerii(Teijsm. et Binn.)Kurz;利氏东南亚山榄（利尔巴因榄)●☆

286756　Payena lucida A. DC. ;光亮东南亚山榄（光泽巴因榄）●☆

286757　Payena sericea H. J. Lam;绢毛东南亚山榄●☆

286758　Payera Baill. (1878)(保留属名);佩耶茜属●☆

286759　Payera bakeriana(Homolle)R. Buchner et Puff;贝克佩耶茜●☆

286760　Payera conspicua Baill. ;显著佩耶茜●☆

286761　Payera coriacea(Humbert)R. Buchner et Puff;革质佩耶茜●☆

286762　Payera decaryi(Homolle)R. Buchner et Puff;德卡里佩耶茜●☆

286763　Payera glabrifolia ex R. Buchner et Puff;光花佩耶茜●☆

286764　Payera madagascariensis(Cavaco)R. Buchner et Puff;马岛佩耶茜●☆

286765　Payera mandrarensis(Homolle ex Cavaco)R. Buchner et Puff;曼德拉佩耶茜●☆

286766　Payera marojejyensis R. Buchner et Puff;马罗佩耶茜●☆

286767　Payeria Baill. (废弃属名) = Payera Baill. (保留属名)●☆

286768　Payeria Baill. (废弃属名) = Quivisia Comm. ex Juss. ●

286769　Payeria Baill. (废弃属名) = Turraea L. ●

286770　Paypayrola Aubl. (1775);管蕊堇属■☆

286771　Paypayrola brasiliensis Steud. ;巴西管蕊堇■☆

286772　Paypayrola grandiflora Tul. ;大花管蕊堇■☆

286773　Paypayrola guianensis Aubl. ;圭亚那管蕊堇■☆

286774　Paypayrola longifolia Tul. ;长叶管蕊堇■☆

286775　Payrola Juss. = Paypayrola Aubl. ■☆

286776　Paysonia O'Kane et Al-Shehbaz = Vesicaria Tourn. ex Adans. ■☆

286777　Paysonia O'Kane et Al-Shehbaz(2002);佩森草属■☆

286778　Pearcea Regel(1867);皮尔斯苣苔属■☆

286779　Pearcea hypocyrtiflora(Hook.)Regel;皮尔斯苣苔■☆

286780　Pearsonia Dümmer(1912);皮尔逊豆属●☆

286781　Pearsonia aristata(Schinz)Dümmer;具芒皮尔逊豆●☆

286782　Pearsonia atherstonei Dümmer = Pearsonia sessilifolia (Harv.) Dümmer subsp. marginata(Schinz)Polhill ●☆

286783　Pearsonia bracteata(Benth.)Polhill;具苞皮尔逊豆●☆

286784　Pearsonia cajanifolia(Harv.)Polhill;木豆叶皮尔逊豆●☆

286785　Pearsonia cajanifolia(Harv.)Polhill subsp. cryptantha(Baker)Polhill;隐花皮尔逊豆●☆

286786　Pearsonia filifolia (Bolus) Dümmer = Pearsonia sessilifolia (Harv.)Dümmer subsp. filifolia(Bolus)Polhill ●☆

286787　Pearsonia flava(Baker f.)Polhill;黄皮尔逊豆●☆

286788　Pearsonia flava(Baker f.)Polhill subsp. mitwabaensis(Timp.)Polhill;米图瓦巴皮尔逊豆●☆

286789　Pearsonia grandifolia(Bolus)Polhill;大叶皮尔逊豆●☆

286790　Pearsonia grandifolia (Bolus) Polhill subsp. latibracteolata (Dümmer)Polhill;宽苞大叶皮尔逊豆●☆

286791　Pearsonia haygarthii(N. E. Br.)Dümmer = Pearsonia sessilifolia (Harv.)Dümmer subsp. filifolia(Bolus)Polhill ●☆

286792　Pearsonia hirsuta Germish. ;粗毛皮尔逊豆●☆

286793　Pearsonia madagascariensis(R. Vig.)Polhill;马岛皮尔逊豆●☆

286794　Pearsonia marginata (Schinz) Dümmer = Pearsonia sessilifolia (Harv.)Dümmer subsp. marginata(Schinz)Polhill ●☆

286795　Pearsonia metallifera Wild;光泽皮尔逊豆●☆

286796　Pearsonia mucronata Burtt Davy ex Baker f. = Pearsonia sessilifolia(Harv.)Dümmer ●☆

286797　Pearsonia obovata(Schinz)Polhill;卵形皮尔逊豆●☆

286798　Pearsonia podalyriifolia Dümmer = Pearsonia sessilifolia(Harv.)Dümmer subsp. marginata(Schinz)Polhill ●☆

286799　Pearsonia propinqua Dümmer = Pearsonia sessilifolia (Harv.)Dümmer subsp. marginata(Schinz)Polhill ●☆

286800　Pearsonia sessilifolia(Harv.)Dümmer;无梗皮尔逊豆●☆

286801　Pearsonia sessilifolia (Harv.) Dümmer subsp. filifolia (Bolus)Polhill;线叶无梗皮尔逊豆●☆

286802　Pearsonia sessilifolia(Harv.)Dümmer subsp. marginata(Schinz)Polhill;具边无梗皮尔逊豆●☆

286803　Pearsonia sessilifolia(Harv.)Dümmer subsp. swaziensis(Bolus)Polhill;斯威士皮尔逊豆●☆

286804　Pearsonia swaziensis (Bolus) Dümmer = Pearsonia sessilifolia (Harv.)Dümmer subsp. swaziensis(Bolus)Polhill ●☆

286805　Pearsonia uniflora(Kensit)Polhill;单花皮尔逊豆●☆

286806　Peautia Comm. ex Pfeiff. = Hydrangea L. ●

286807　Peccana Raf. = Euphorbia L. ●■

286808　Pechea Lapeyr. = Crypsis Aiton(保留属名)■

286809　Pechea Pour. = Crypsis Aiton(保留属名)■

286810　Pechea Pourr. ex Kunth = Crypsis Aiton(保留属名)■

286811　Pechea Steud. = Cybianthus Mart. (保留属名)●☆

286812　Pechea Steud. = Peckia Vell. (废弃属名)●☆

286813　Pecheya Scop. = Coussarea Aubl. ●☆

286814　Pechuelia Kuntze = Selago L. ●☆

286815　Pechuel-loeschea O. Hoffm. (1888);歧尾菊属●☆

286816　Pechuel-loeschea leubnitziae(Kuntze)O. Hoffm. ;歧尾菊■☆

286817　Pechuel-loeschea leubnitziae O. Hoffm. = Pechuel-loeschea leubnitziae(Kuntze)O. Hoffm. ■☆

286818　Peckelia Hutch. = Cajanus Adans. (保留属名)●

286819　Peckelia Hutch. = Peekelia Harms ●

286820　Peckeya Raf. = Coussarea Aubl. ●☆

286821　Peckeya Raf. = Pecheya Scop. ●☆

286822　Peckia Vell. (废弃属名) = Cybianthus Mart. (保留属名)●☆

286823　Peckoltia E. Fourn. (1885);佩克萝藦属☆

286824　Peckoltia pedalis E. Fourn. ;佩克萝藦☆

286825　Pectangis Thouars = Angraecum Bory ■

286826　Pectanisia Raf. = Reseda L. ■

286827　Pectantia Raf. = Mitella L. ■

286828　Pectantia Raf. = Pectiantia Raf. ■

286829　Pecteilis Raf.（1837）;白蝶兰属（白蝶花属）;Pecteilis ■

286830　Pecteilis bassacensis（Gagnep.）Ts. Tang et F. T. Wang = Pecteilis henryi Schltr. ■

286831　Pecteilis gigantea（Sm.）Raf. = Pecteilis susannae（L.）Raf. ■

286832　Pecteilis henryi Schltr.;滇南白蝶兰（亨氏白蝶兰,小白蝶兰,云南白蝶兰）;Henry Pecteilis ■

286833　Pecteilis lacei（Rolfe ex Downie）Ts. Tang et F. T. Wang = Pecteilis henryi Schltr. ■

286834　Pecteilis latilabris（Lindl.）Mitra = Platanthera latilabris Lindl. ■

286835　Pecteilis radiata（Thunb.）Raf.;狭叶白蝶兰（辐射状白蝶花,狭叶白蝶花）;Narrowleaf Pecteilis,Radiant Pecteilis ■

286836　Pecteilis susannae（L.）Raf.;龙头兰（白蝶花,白蝶兰,白花参,鹅毛白蝶花,和气草,鸡卵参,双肾参,双肾兰,土兰,土玉竹,兔儿草,兔耳草）;Common Pecteilis,Susann Pecteilis ■

286837　Pecteilis susannae（L.）Raf. subsp. henryi（Schltr.）Soó = Pecteilis henryi Schltr. ■

286838　Pecteilis susannae（L.）Raf. subsp. henryi Soó = Pecteilis henryi Schltr. ■

286839　Pecten Lam. = Scandix L. ■

286840　Pectianthia Rydb. = Pectiantia Raf. ■

286841　Pectiantia Raf. = Mitella L. ■

286842　Pectiantiaceae Raf. = Saxifragaceae Juss.（保留科名）●■

286843　Pectidium Less. = Pectis L. ■☆

286844　Pectidopsis DC. = Helioreos Raf. ■☆

286845　Pectidopsis DC. = Pectidium Less. ■☆

286846　Pectidopsis DC. = Pectis L. ■☆

286847　Pectinaria（Bernh.）Hack = Eremochloa Büse ■

286848　Pectinaria Bernh.（废弃属名）= Pectinaria Haw.（保留属名）■☆

286849　Pectinaria Bernh.（废弃属名）= Scandix L. ■

286850　Pectinaria Cordem. = Angraecum Bory ■

286851　Pectinaria Hack. = Eremochloa Büse ■

286852　Pectinaria Haw.（1819）（保留属名）;梳状萝藦属■☆

286853　Pectinaria arcuata N. E. Br. = Ophionella arcuata（N. E. Br.）Bruyns ■☆

286854　Pectinaria articulata（Aiton）Haw.;关节梳状萝藦■☆

286855　Pectinaria articulata（Aiton）Haw. subsp. asperiflora（N. E. Br.）Bruyns;糙叶梳状萝藦■☆

286856　Pectinaria articulata（Aiton）Haw. subsp. borealis Bruyns;北方梳状萝藦■☆

286857　Pectinaria articulata（Aiton）Haw. subsp. namaquensis（N. E. Br.）Bruyns;纳马夸梳状萝藦■☆

286858　Pectinaria articulata（Aiton）Haw. var. namaquensis N. E. Br. = Pectinaria articulata（Aiton）Haw. subsp. namaquensis（N. E. Br.）Bruyns ■☆

286859　Pectinaria asperiflora N. E. Br. = Pectinaria articulata（Aiton）Haw. subsp. asperiflora（N. E. Br.）Bruyns ■☆

286860　Pectinaria breviloba R. A. Dyer = Stapeliopsis breviloba（R. A. Dyer）Bruyns ■☆

286861　Pectinaria exasperata Bruyns = Stapeliopsis exasperata（Bruyns）Bruyns ■☆

286862　Pectinaria flavescens Plowes;浅黄梳状萝藦■☆

286863　Pectinaria longipes（N. E. Br.）Bruyns;长梗梳状萝藦■☆

286864　Pectinaria longipes（N. E. Br.）Bruyns subsp. villetii（C. A. Lückh.）Bruyns;维莱特梳状萝藦■☆

286865　Pectinaria mammillaris（L.）Sweet = Quaqua mammillaris（L.）Bruyns ■☆

286866　Pectinaria maughanii（R. A. Dyer）Bruyns;莫恩梳状萝藦■☆

286867　Pectinaria pillansii N. E. Br. = Stapeliopsis pillansii（N. E. Br.）Bruyns ■☆

286868　Pectinaria saxatilis N. E. Br. = Stapeliopsis saxatilis（N. E. Br.）Bruyns ■☆

286869　Pectinaria stayneri M. B. Bayer = Stapeliopsis stayneri（M. B. Bayer）Bruyns ■☆

286870　Pectinaria thouarsii Cordem. = Angraecum pectinatum Thouars ■☆

286871　Pectinaria tulipiflora C. A. Lückh. = Stapeliopsis saxatilis（N. E. Br.）Bruyns ■☆

286872　Pectinastrum Cass. = Centaurea L.（保留属名）●■

286873　Pectinea Gaertn.（废弃属名）= Erythrospermum Thouars（保留属名）●

286874　Pectinea Gaertn.（废弃属名）= Erythrospermurn Lam. ●

286875　Pectinella J. M. Black = Amphibolis C. Agardh ■☆

286876　Pectis L.（1759）;梳齿菊属（梳菊属）■☆

286877　Pectis angustifolia Torr.;狭叶梳菊;Crownseed Pectis,Lemonweed ■☆

286878　Pectis angustifolia Torr. var. fastigiata（A. Gray）D. J. Keil;得州梳齿菊;Texas Chinchweed ■☆

286879　Pectis angustifolia Torr. var. subaristata A. Gray = Pectis angustifolia Torr. ■☆

286880　Pectis angustifolia Torr. var. tenella（DC.）D. J. Keil;墨西哥梳齿菊;Low Pectis,Mexican Chinchweed ■☆

286881　Pectis bonplandiana Kunth;邦兰梳齿菊■☆

286882　Pectis cylindrica（Fernald）Rydb.;三线梳齿菊;Sonoran Chinchweed,Three-rayed Chinchweed ■☆

286883　Pectis discoidea（Spreng.）Hornem. = Leysera leyseroides（Desf.）Maire ●☆

286884　Pectis fastigiata A. Gray = Pectis angustifolia Torr. var. fastigiata（A. Gray）D. J. Keil ■☆

286885　Pectis filipes Harv. et A. Gray;线梗梳齿菊;Threadstalk Chinchweed ■☆

286886　Pectis filipes Harv. et A. Gray var. subnuda Fernald;近裸线梗梳齿菊■☆

286887　Pectis floridana D. J. Keil;佛罗里达梳齿菊■☆

286888　Pectis glaucescens（Cass.）D. J. Keil;沙丘梳齿菊;Sand Dune Chinchweed ■☆

286889　Pectis humifusa Sw.;扩散梳齿菊■☆

286890　Pectis imberbis A. Gray;无毛梳齿菊;Beardless Chinchweed,Tall Chinchweed ■☆

286891　Pectis leptocephala（Cass.）Urb. = Pectis glaucescens（Cass.）D. J. Keil ■☆

286892　Pectis lessingii Fernald = Pectis glaucescens（Cass.）D. J. Keil ■☆

286893　Pectis linearifolia Urb.;线叶梳齿菊;Florida Chinchweed ■☆

286894　Pectis linifolia L.;亚麻叶梳齿菊;Romero-macho ■☆

286895　Pectis linifolia L. var. marginalis Fernald = Pectis linifolia L. ■☆

286896　Pectis longipes A. Gray;长梗梳齿菊;Longstalk Chinchweed ■☆

286897　Pectis papposa Harv. et A. Gray;梳齿菊;Chinchweed,Common Chinchweed ■☆

286898　Pectis pinnata Lam. = Schkuhria pinnata（Lam.）Kuntze ex Thell. ■☆

286899　Pectis prostrata Cav.;平卧梳齿菊;Spreading Chinchweed ■☆

286900　Pectis prostrata Cav. var. cylindrica Fernald = Pectis cylindrica

（Fernald）Rydb. ■☆

286901　Pectis punctata Jacq. = Pectis linifolia L. ■☆

286902　Pectis rusbyi Greene ex A. Gray；鲁斯比梳齿菊；Rusby's Chinchweed ■☆

286903　Pectis sessiliflora Sch. Bip.；无柄花梳菊■☆

286904　Pectis tenella DC. = Pectis angustifolia Torr. var. tenella（DC.）D. J. Keil ■☆

286905　Pectocarya DC. = Pectocarya DC. ex Meisn. ●☆

286906　Pectocarya DC. ex Meisn.（1840）；沟果紫草属●☆

286907　Pectocarya boliviana I. M. Johnst.；玻利维亚沟果紫草●☆

286908　Pectocarya chilensis DC.；智利沟果紫草●☆

286909　Pectocarya dimorpha（I. M. Johnst.）I. M. Johnst.；二形沟果紫草●☆

286910　Pectocarya gracilis I. M. Johnst.；细沟果紫草●☆

286911　Pectocarya heterocarpa（I. M. Johnst.）I. M. Johnst.；异果沟果紫草●☆

286912　Pectocarya lateriflora DC.；侧花沟果紫草●☆

286913　Pectocarya linearis DC.；线形沟果紫草●☆

286914　Pectocarya platycarpa Munz et I. M. Johnst.；宽果沟果紫草●☆

286915　Pectophyllum Rchb. = Azorella Lam. ■☆

286916　Pectophyllum Rchb. = Pectophytum Kunth ●☆

286917　Pectophytum Kunth = Azorella Lam. ●☆

286918　Pedaliaceae R. Br.（1810）（保留科名）；胡麻科；Pedalium Family ●■

286919　Pedaliodiscus Ihlenf.（1968）；东非胡麻属■☆

286920　Pedaliodiscus Ihlenf. = Pedalium L. ●☆

286921　Pedaliodiscus macrocarpus Ihlenf.；东非胡麻■☆

286922　Pedaliophyton Engl. = Pterodiscus Hook. ■☆

286923　Pedaliophyton busseanum Engl. = Pterodiscus angustifolius Engl. ■☆

286924　Pedalium Adans. = Atraphaxis L. ●

286925　Pedalium D. Royen ex L.（1759）；印度胡麻属；Pedalium ■☆

286926　Pedalium L. = Pedalium D. Royen ex L. ■☆

286927　Pedalium busseanum（Engl.）Stapf = Pterodiscus angustifolius Engl. ■☆

286928　Pedalium intermedium（Engl.）Engl. = Pterodiscus intermedius Engl. ■☆

286929　Pedalium longiflorum（Royen）Decne. = Rogeria longiflora（Royen）J. Gay ex DC. ☆

286930　Pedalium murex D. Royen ex L.；印度胡麻；Indian Pedalium ●☆

286931　Pedalium murex L. = Pedalium murex D. Royen ex L. ●☆

286932　Pedalium ruspolii（Engl.）Engl. = Pterodiscus ruspolii Engl. ●☆

286933　Pedalium ruspolii Engl. var. aureus Chiov. = Pterodiscus ruspolii Engl. ■☆

286934　Pedastis Raf. = Cayratia Juss.（保留属名）●

286935　Peddiea Harv. = Peddiea Harv. ex Hook. ●☆

286936　Peddiea Harv. ex Hook.（1840）；佩迪木属●☆

286937　Peddiea africana Harv.；非洲佩迪木●☆

286938　Peddiea africana Harv. var. schliebenii Domke = Peddiea fischeri Engl. ●☆

286939　Peddiea arborescens A. Robyns = Peddiea fischeri Engl. ●☆

286940　Peddiea fischeri Engl.；菲舍尔佩迪木●☆

286941　Peddiea involucrata Baker；总苞佩迪木●☆

286942　Peddiea kaniamensis A. Robyns = Peddiea fischeri Engl. ●☆

286943　Peddiea kivuensis A. Robyns = Peddiea rapaneoides Gilg ex Engl. ●☆

286944　Peddiea lanceolata Domke；披针形佩迪木●☆

286945　Peddiea montana Domke；山地佩迪木●☆

286946　Peddiea orophila A. Robyns = Peddiea rapaneoides Gilg ex Engl. ●☆

286947　Peddiea parviflora Hook. f. = Peddiea africana Harv. ●☆

286948　Peddiea polyantha Gilg；多花佩迪木●☆

286949　Peddiea puberula Domke；微毛佩迪木●☆

286950　Peddiea rapaneoides Gilg ex Engl.；密花树佩迪木●☆

286951　Peddiea subcordata Domke；近心形佩迪木●☆

286952　Peddiea thomensis Exell = Peddiea africana Harv. ●☆

286953　Peddiea thulinii Temu；图林佩迪木●☆

286954　Peddiea volkensii Gilg = Peddiea africana Harv. ●☆

286955　Pederia Noronha = Paederia L.（保留属名）●■

286956　Pederlea Raf. = Acnistus Schott ●☆

286957　Pederota Scop. = Paederota L. ■☆

286958　Pederota Scop. = Veronica L. ■

286959　Pedersenia Holub（1998）；彼苋属■☆

286960　Pedersenia argentata（Mart.）Holub；彼苋■☆

286961　Pedicellaria Schrank（废弃属名）= Cleome L. ●■

286962　Pedicellaria Schrank（废弃属名）= Gynandropsis DC.（保留属名）■

286963　Pedicellaria gynandra（L.）Chiov. = Cleome gynandra L. ■

286964　Pedicellaria pentaphylla（L.）Schrank = Cleome gynandra L. ■

286965　Pedicellarum M. Hotta（1976）；虱子南星属■☆

286966　Pedicellarum paiei Hotte；虱子南星■☆

286967　Pedicellia Lour.（废弃属名）= Mischocarpus Blume（保留属名）●

286968　Pediculariaceae Juss.；马先蒿科■

286969　Pediculariaceae Juss. = Scrophulariaceae Juss.（保留科名）●■

286970　Pedicularidaceae Juss. = Orobanchaceae Vent.（保留科名）●■

286971　Pediculariopsis Á. Löve et D. Löve = Pedicularis L. ■

286972　Pedicularis L.（1753）；马先蒿属；Lousewort, Wood Betony, Woodbetony ■

286973　Pedicularis abrotanifolia M. Bieb. ex Steven；蒿叶马先蒿（苦艾叶马先蒿）；Wormwoodleaf Woodbetony ■

286974　Pedicularis achilleifolia Stephan ex Willd.；耆草叶马先蒿；Yarrowleaf Woodbetony ■

286975　Pedicularis adamsii Hultén；阿达马先蒿■☆

286976　Pedicularis adunca M. Bieb. ex Steven；钩状马先蒿■☆

286977　Pedicularis adunca M. Bieb. ex Steven subsp. sachalinensis（Miyabe et T. Miyake）Ivanina；库叶马先蒿■☆

286978　Pedicularis adunca M. Bieb. ex Steven var. sachalinensis（Miyabe et T. Miyake）Ohwi = Pedicularis adunca M. Bieb. ex Steven subsp. sachalinensis（Miyabe et T. Miyake）Ivanina ■☆

286979　Pedicularis aequibarbis Hand. -Mazz. = Pedicularis dunniana Bonati ■

286980　Pedicularis alaschanica Maxim.；阿拉善马先蒿；Alashan Woodbetony ■

286981　Pedicularis alaschanica Maxim. subsp. tibetica（Maxim.）P. C. Tsoong；西藏阿拉善马先蒿；Xizang Alashan Woodbetony ■

286982　Pedicularis alaschanica Maxim. var. tibetica Maxim. = Pedicularis alaschanica Maxim. subsp. tibetica（Maxim.）P. C. Tsoong ■

286983　Pedicularis alaschanica Maxim. var. typica Prain = Pedicularis alaschanica Maxim. ■

286984　Pedicularis alatauica Stadlm.；阿拉套马先蒿■☆

286985　Pedicularis albertii Regel；阿尔伯特马先蒿■☆

286986　Pedicularis aloensis Hand. -Mazz.；阿洛马先蒿；Aloe Woodbetony ■

286987　Pedicularis alopecuros Franch. = Pedicularis alopecuros Franch. ex Maxim. ■

286988　Pedicularis alopecuros Franch. ex Maxim.；狐尾马先蒿；

Alopecurus Woodbetony ■

286989　Pedicularis alopecuros Franch. ex Maxim. var. lasiandra P. C. Tsoong;毛药狐尾马先蒿;Hairy Alopecurus Woodbetony ■

286990　Pedicularis altaica Steph. = Pedicularis mariae Regel ■

286991　Pedicularis altaica Steph. ex Steven;阿尔泰马先蒿;Altai Woodbetony ■

286992　Pedicularis altifrontalis P. C. Tsoong;高额马先蒿;Highfront Woodbetony ■

286993　Pedicularis amoena Adams = Pedicularis verticillata L. ■

286994　Pedicularis amoena Adams ex Steven = Pedicularis anthemifolia Fisch. ■

286995　Pedicularis amoena Adams ex Steven var. elatior Regel = Pedicularis anthemifolia Fisch. subsp. elatior(Regel) P. C. Tsoong ■

286996　Pedicularis amoena Adams var. elatior Regel = Pedicularis anthemifolia Fisch. subsp. elatior(Regel) P. C. Tsoong ■

286997　Pedicularis amoeniflora Vved.;秀花马先蒿■☆

286998　Pedicularis amplitaba H. L. Li;半管马先蒿;Richtube Woodbetony ■

286999　Pedicularis anas Maxim.;鸭首马先蒿;Duck Woodbetony, Duckhead Woodbetony ■

287000　Pedicularis anas Maxim. var. tibetica Bonati;西藏鸭首马先蒿（鹅首马先蒿）;Xizang Duck Woodbetony ■

287001　Pedicularis anas Maxim. var. typica H. L. Li = Pedicularis anas Maxim. ■

287002　Pedicularis anas Maxim. var. xanthantha(H. L. Li) P. C. Tsoong;黄花鸭首马先蒿;Yelowerflower Duck Woodbetony ■

287003　Pedicularis angularis P. C. Tsoong;角盔马先蒿;Anglehelmet Woodbetony, Angular Woodbetony ■

287004　Pedicularis angustiflora Limpr. = Pedicularis oederi Vahl var. angustiflora(Limpr.) P. C. Tsoong ■

287005　Pedicularis angustiflora Limpr. = Pedicularis orthocoryne H. L. Li ■

287006　Pedicularis angustilabris H. L. Li;狭唇马先蒿;Narrowlip Woodbetony ■

287007　Pedicularis angustiloba P. C. Tsoong;狭裂马先蒿;Narrowlabiate Woodbetony, Narrowlobe Woodbetony ■

287008　Pedicularis anomala P. C. Tsoong et H. P. Yang;奇异马先蒿;Wongderful Woodbetony ■

287009　Pedicularis anthemifolia Fisch.;春黄菊叶马先蒿;Camomileleaf Woodbetony ■

287010　Pedicularis anthemifolia Fisch. subsp. elatior (Regel) P. C. Tsoong;高升春黄菊叶马先蒿;Tall Camomileleaf Woodbetony ■

287011　Pedicularis aphyllocaulis Hand. -Mazz. = Pedicularis praeruptorum Bonati ■

287012　Pedicularis apodochila Maxim.;无柄马先蒿■☆

287013　Pedicularis apodochila Maxim. f. albiflora Ohba;白花无柄马先蒿■☆

287014　Pedicularis aquilina Bonati;鹰嘴马先蒿（鸭嘴马先蒿）;Duckbill Woodbetony, Eaglebeak Woodbetony ■

287015　Pedicularis arguteserrata Vved.;细齿马先蒿■☆

287016　Pedicularis armata Maxim.;刺齿马先蒿;Spinetooth Woodbetony ■

287017　Pedicularis armata Maxim. var. trimaculata X. F. Lu;三点刺齿马先蒿

287018　Pedicularis artselaeri Maxim.;埃氏马先蒿(短茎马先蒿,蚂蚁窝);Shortstem Woodbetony ■

287019　Pedicularis artselaeri Maxim. var. wutaiensis Hurus.;五台埃氏马先蒿;Wutai Shortstem Woodbetony ■

287020　Pedicularis aschistorrhyncha Marquand et Airy Shaw;全喙马先蒿;Wholebeak Woodbetony, Wholebill Woodbetony ■

287021　Pedicularis atripurpurea Nordm. ;暗紫马先蒿■☆

287022　Pedicularis atroviridis P. C. Tsoong;深绿马先蒿（黑绿马先蒿）;Darkgreen Woodbetony ■

287023　Pedicularis atuntsiensis Bonati;阿墩子马先蒿;Adunzi Woodbetony, Atuntsi Woodbetony ■

287024　Pedicularis aurata (Bonati) H. L. Li;金黄马先蒿;Golden Woodbetony, Goldenyellow Woodbetony ■

287025　Pedicularis auriculata Sm. = Pedicularis lanceolata Michx. ■☆

287026　Pedicularis axillaris Franch. ex Maxim.;腋花马先蒿;Axilflower Woodbetony ■

287027　Pedicularis axillaris Franch. ex Maxim. subsp. balsouriana (Bonati) P. C. Tsoong;巴氏腋花马先蒿（绿腋花马先蒿）;Balsour Axilflower Woodbetony ■

287028　Pedicularis axillaris Franch. ex Maxim. var. balsouriana(Bonati) H. L. Li = Pedicularis axillaris Franch. ex Maxim. subsp. balsouriana (Bonati) P. C. Tsoong ■

287029　Pedicularis axillaris Franch. ex Maxim. var. typica H. L. Li = Pedicularis axillaris Franch. ex Maxim. ■

287030　Pedicularis balfouriana Bonati = Pedicularis axillaris Franch. ex Maxim. subsp. balsouriana(Bonati) P. C. Tsoong ■

287031　Pedicularis bambusetorum Hand. -Mazz. = Pedicularis aurata (Bonati) H. L. Li ■

287032　Pedicularis bartschioides Hand. -Mazz. = Xizangia bartschioides (Hand. -Mazz.) D. Y. Hong ■

287033　Pedicularis batangensis Bureau et Franch.;巴塘马先蒿;Batang Woodbetony ■

287034　Pedicularis bella Hook. f.;美丽马先蒿;Beautiful Woodbetony, Showy Woodbetony ■

287035　Pedicularis bella Hook. f. subsp. holophylla (C. Marquand et Airy Shaw) P. C. Tsoong;全叶美丽马先蒿;Entire Showy Woodbetony ■

287036　Pedicularis bella Hook. f. subsp. holophylla (C. Marquand et Airy Shaw) P. C. Tsoong var. cristifrons P. C. Tsoong;冠额马先蒿■

287037　Pedicularis bella Hook. f. subsp. holophylla (C. Marquand et Airy Shaw) P. C. Tsoong f. rosea(C. Marquand et Airy Shaw) P. C. Tsoong;绯色马先蒿■

287038　Pedicularis bella Hook. f. var. holophylla C. Marquand et Airy Shaw = Pedicularis bella Hook. f. subsp. holophylla (C. Marquand et Airy Shaw) P. C. Tsoong ■

287039　Pedicularis bella Hook. f. var. holophylla f. rosea C. Marquand et Airy Shaw = Pedicularis bella Hook. f. subsp. holophylla (C. Marquand et Airy Shaw) P. C. Tsoong f. rosea(C. Marquand et Airy Shaw) P. C. Tsoong ■

287040　Pedicularis bella Hook. f. var. typica H. L. Li = Pedicularis bella Hook. f. ■

287041　Pedicularis bicolor Diels;二色马先蒿;Bicolor Woodbetony, Twocolor Woodbetony ■

287042　Pedicularis bicornuta Klotzsch;二角马先蒿;Bicorn Woodbetony ■☆

287043　Pedicularis bidentata Maxim.;二齿马先蒿;Bidentate Woodbetony, Twoteeth Woodbetony ■

287044　Pedicularis bietii Franch.;皮氏马先蒿;Biet Woodbetony ■

287045　Pedicularis bifida(Buch. -Ham. ex D. Don) Pennell;二裂马先蒿;Twolobed Woodbetony ■

287046　Pedicularis binaria Maxim.;双生马先蒿;Geminate Woodbetony, Twin Woodbetony ■

287047　Pedicularis biondiana Diels = Pedicularis rhinanthoides Schrenk ex Fisch. subsp. labellata(Jacq.) Pennell ■

287048　Pedicularis biondiana Diels = Pedicularis rhinanthoides Schrenk ex Fisch. subsp. labellata(Jacq.) P. C. Tsoong ■

287049　Pedicularis birostris Bureau et Franch. = Pedicularis cranolopha Maxim. var. longicornuta Prain ■

287050　Pedicularis bomiensis H. P. Yang；波密马先蒿；Bomi Woodbetony ■

287051　Pedicularis bonatiana H. L. Li = Pedicularis verticillata L. subsp. tangutica(Bonati) P. C. Tsoong ■

287052　Pedicularis borodowskii Palib. = Pedicularis curvituba Maxim. subsp. provoti(Franch.) P. C. Tsoong ■

287053　Pedicularis borodowskii Palib. = Pedicularis curvituba Maxim. ■

287054　Pedicularis brachycrania H. L. Li；短盔马先蒿；Shorthelmet Woodbetony ■

287055　Pedicularis brachystachys Bunge；短穗马先蒿■☆

287056　Pedicularis bracteosa Pall. ex Bunge；多苞片马先蒿；Bracted Lousewort ■☆

287057　Pedicularis branchiophylla Pennell = Pedicularis oederi Vahl subsp. branchiophylla(Pennell) P. C. Tsoong ■

287058　Pedicularis breviflora Regel；短花马先蒿；Shortflower Woodbetony ■

287059　Pedicularis brevifolia Don = Pedicularis confertiflora Prain ■

287060　Pedicularis brevilabris Franch.；短唇马先蒿；Shortlip Woodbetony ■

287061　Pedicularis brunoniana Wall. = Pedicularis gracilis Wall. subsp. stricta P. C. Tsoong ■

287062　Pedicularis calosantha H. L. Li = Pedicularis verticillata L. ■

287063　Pedicularis canadensis L.；加拿大马先蒿；Canada Pedicularis, Canadian Lousewort, Common Lousewort, Forest Lousewort, Head Betony , Lousewort, Wood Betony , Wood-betony ■☆

287064　Pedicularis canadensis L. f. flava Farw. = Pedicularis canadensis L. ■☆

287065　Pedicularis canadensis L. f. praeclara A. H. Moore = Pedicularis canadensis L. ■☆

287066　Pedicularis canadensis L. var. dobbsii Fernald = Pedicularis canadensis L. ■☆

287067　Pedicularis capitata Adams；头状马先蒿■☆

287068　Pedicularis caucasica M. Bieb.；高加索马先蒿■☆

287069　Pedicularis cephalantha Franch. ex Maxim.；头花马先蒿；Headflower Woodbetony ■

287070　Pedicularis cephalantha Franch. ex Maxim. var. szetchuanica Bonati；四川头花马先蒿；Sichuan Headflower Woodbetony ■

287071　Pedicularis cephalantha Franch. ex Maxim. var. typica H. L. Li = Pedicularis cephalantha Franch. ex Maxim. ■

287072　Pedicularis cernua Bonati；俯垂马先蒿(虫莲,垂头马先蒿)；Nodding Woodbetony ■

287073　Pedicularis cernua Bonati subsp. latifolia (H. L. Li) P. C. Tsoong；宽叶俯垂马先蒿；Broadleaf Nodding Woodbetony ■

287074　Pedicularis cernua Bonati var. latifolia H. L. Li = Pedicularis cernua Bonati subsp. latifolia(H. L. Li) P. C. Tsoong ■

287075　Pedicularis cernua var. typica H. L. Li = Pedicularis cernua Bonati ■

287076　Pedicularis chamissonis Steven；沙米逊马先蒿■☆

287077　Pedicularis chamissonis Steven f. albiflora Tatew. ex H. Hara；白花沙米逊马先蒿■☆

287078　Pedicularis chamissonis Steven var. hokkaidoensis T. Shimizu；北

海道沙米逊马先蒿■☆

287079　Pedicularis chamissonis Steven var. japonica(Miq.) Maxim.；日本沙米逊马先蒿■☆

287080　Pedicularis chamissonis Steven var. japonica (Miq.) Maxim. f. rostrata T. Yamaz. = Pedicularis chamissonis Steven var. longirostrata T. Yamaz. ■☆

287081　Pedicularis chamissonis Steven var. longirostrata T. Yamaz.；长喙沙米逊马先蒿■☆

287082　Pedicularis chamissonis Steven var. rebunensis T. Yamaz.；礼文马先蒿■☆

287083　Pedicularis chamissonis Steven var. rebunensis T. Yamaz. f. alba T. Yamaz.；白花礼文马先蒿■☆

287084　Pedicularis cheilanthifolia Schrenk；碎米蕨叶马先蒿；Lipfernleaf Woodbetony ■

287085　Pedicularis cheilanthifolia Schrenk = Pedicularis plicata Maxim. subsp. apiculata P. C. Tsoong ■

287086　Pedicularis cheilanthifolia Schrenk subsp. svenhedinii(Paulsen) P. C. Tsoong；斯文氏碎米蕨叶马先蒿(文氏马先蒿)；Svenhedin Lipfernleaf Woodbetony ■

287087　Pedicularis cheilanthifolia Schrenk var. typica Prain = Pedicularis cheilanthifolia Schrenk ■

287088　Pedicularis chengxianensis Zh. G. Ma et Zh. Zh. Ma；成县马先蒿；Chengxian Woodbetony ■

287089　Pedicularis chenocephala Diels；鹅首马先蒿；Goosehead Woodbetony ■

287090　Pedicularis chinensis Maxim.；中国马先蒿；China Woodbetony, Chinese Woodbetony ■

287091　Pedicularis chinensis Maxim. = Pedicularis longiflora Rudolph var. tubiformis(G. Klotz) P. C. Tsoong ■

287092　Pedicularis chinensis Maxim. = Pedicularis longiflora Rudolph ■

287093　Pedicularis chinensis Maxim. f. erubescens P. C. Tsoong；浅红中国马先蒿；Palered Chinese Woodbetony ■

287094　Pedicularis chingii Bonati；秦氏马先蒿；Ching Woodbetony ■

287095　Pedicularis chumbica Prain；春丕马先蒿；Chumb Woodbetony ■

287096　Pedicularis cinerascens Franch.；灰色马先蒿；Grey Woodbetony ■

287097　Pedicularis clarkei Hook. f.；克氏马先蒿；C. B. Clarke Woodbetony ■

287098　Pedicularis collettii Prain var. nigra Bonati = Pedicularis nigra (Bonati) Vaniot ex Bonati ■

287099　Pedicularis comosa L.；丛毛马先蒿■☆

287100　Pedicularis compacta Stephan ex Willd.；密花马先蒿■☆

287101　Pedicularis comptoniifolia Franch.；四叶马先蒿(干黑马先蒿,康泊东叶马先蒿)；Black Woodbetony ■

287102　Pedicularis condensata M. Bieb.；密集马先蒿■☆

287103　Pedicularis confertiflora Prain；聚花马先蒿(红蒿枝)；Denseflower Woodbetony ■

287104　Pedicularis confertiflora Prain subsp. parvifolia (Hand. -Mazz.) P. C. Tsoong；小叶聚花马先蒿；Littleleaf Woodbetony, Smallleaf Woodbetony ■

287105　Pedicularis confluens P. C. Tsoong；连齿马先蒿；Confluent Woodbetony ■

287106　Pedicularis conifera Maxim. ex Hemsl.；结球马先蒿；Conebearing Woodbetony , Conlike Woodbetony ■

287107　Pedicularis connata H. L. Li；连叶马先蒿；Connate Woodbetony ■

287108　Pedicularis coppeyi Bonati = Pedicularis przewalskii Maxim. subsp. microphyton (Bureau et Franch.) P. C. Tsoong var. purpurea (Bonati) P. C. Tsoong ■

287109　Pedicularis corydaloides Hand. -Mazz.；拟紫堇马先蒿（野萝卜，紫堇状马先蒿）；Corydalis-like Woodbetony ■

287110　Pedicularis corymbifera H. P. Yang；伞房马先蒿；Corymb Woodbetony ■

287111　Pedicularis corymbosa Prain = Pedicularis lunglingensis Bonati ■

287112　Pedicularis cranolopha Maxim.；凸额马先蒿；Crestedgalea Woodbetony ■

287113　Pedicularis cranolopha Maxim. var. garnieri（Bonati）Bonati = Pedicularis cranolopha Maxim. var. garnieri（Bonati）P. C. Tsoong ■

287114　Pedicularis cranolopha Maxim. var. garnieri（Bonati）P. C. Tsoong；格氏凸额马先蒿；Garnier Crestedgalea Woodbetony ■

287115　Pedicularis cranolopha Maxim. var. garnieri Bonati = Pedicularis tricolor Hand. -Mazz. ■

287116　Pedicularis cranolopha Maxim. var. longicornuta Prain；长角凸额马先蒿（长角马先蒿）；Longcorn Crestedgalea Woodbetony ■

287117　Pedicularis cranolopha Maxim. var. typia Prain = Pedicularis cranolopha Maxim. ■

287118　Pedicularis craspedotricha Maxim.；缘毛马先蒿；Borderhair Woodbetony, Edgehair Woodbetony ■

287119　Pedicularis crassicaulis Vaniot ex Bonati = Pedicularis resupinata L. subsp. crassicaulis（Vaniot ex Bonati）P. C. Tsoong ■

287120　Pedicularis crassirostris Bunge；粗喙马先蒿■☆

287121　Pedicularis crenata Maxim.；波齿马先蒿（圆齿马先蒿）；Crenate Woodbetony ■

287122　Pedicularis crenata Maxim. subsp. crenatiformis（Bonati）P. C. Tsoong；全裂波齿马先蒿（裂叶波齿马先蒿，全萼圆齿马先蒿）；Dissected Crenate Woodbetony ■

287123　Pedicularis crenata Maxim. var. crenatiformis Bonati = Pedicularis crenata Maxim. subsp. crenatiformis（Bonati）P. C. Tsoong ■

287124　Pedicularis crenata Maxim. var. typica H. L. Li = Pedicularis crenata Maxim. ■

287125　Pedicularis crenularis H. L. Li；细波齿马先蒿（细圆齿马先蒿）；Crenulate Woodbetony, Smallcrenate Woodbetony ■

287126　Pedicularis cristata Maxim. = Pedicularis cristatella Pennell et H. L. Li ■

287127　Pedicularis cristatella Pennell et H. L. Li；具冠马先蒿；Cristate Woodbetony, Smallcristate Woodbetony ■

287128　Pedicularis croizatiana H. L. Li；克洛氏马先蒿（凹唇马先蒿，凹额马先蒿）；Concavelip Woodbetony ■

287129　Pedicularis cryptantha Marquand et Airy Shaw；隐花马先蒿；Hiddenflower Woodbetony ■

287130　Pedicularis cryptantha Marquand et Airy Shaw subsp. erecta P. C. Tsoong；直立隐花马先蒿；Erect Hiddenflower Woodbetony ■

287131　Pedicularis cupuliformis H. L. Li = Pedicularis thamnophila（Hand. -Mazz.）H. L. Li subsp. cupuliformis（H. L. Li）P. C. Tsoong ■

287132　Pedicularis curvituba Maxim.；弯管马先蒿；Curvedtube Woodbetony ■

287133　Pedicularis curvituba Maxim. subsp. provoti（Franch.）P. C. Tsoong；洛氏弯管马先蒿；Provot Curvedtube Woodbetony ■

287134　Pedicularis curvituba Maxim. subsp. provoti（Franch.）P. C. Tsoong = Pedicularis curvituba Maxim. ■

287135　Pedicularis cyathophylla Franch.；斗叶马先蒿；Cupleaf Woodbetony ■

287136　Pedicularis cyathophylloides W. Limpr.；拟斗叶马先蒿；Cupleaflike Woodbetony ■

287137　Pedicularis cyclorhyncha H. L. Li；环喙马先蒿；Cyclebill Woodbetony ■

287138　Pedicularis cymbalaria Bonati；舟形马先蒿（舟花马先蒿）；Boatshap Woodbetony ■

287139　Pedicularis daghestanica Bonati；达赫斯坦马先蒿■☆

287140　Pedicularis daltonii Prain；道氏马先蒿；Dalton Woodbetony ■

287141　Pedicularis daochengensis H. P. Yang；稻城马先蒿；Daocheng Woodbetony ■

287142　Pedicularis dasyantha Hadac；极地毛花马先蒿■☆

287143　Pedicularis dasystachys Schrenk；毛穗马先蒿（厚穗马先蒿）；Hairspike Woodbetony, Hairyspike Woodbetony ■

287144　Pedicularis daucifolia Bonati；胡萝卜叶马先蒿；Carrotleaf Woodbetony ■

287145　Pedicularis davidii Franch.；大卫氏马先蒿（粗野马先蒿，大卫马先蒿，邓氏马先蒿，黑参，黑参马先蒿，黑阳参，黑洋参，煤参，美观马先蒿，扭盔马先蒿，太白参，太白阳参，太白洋参）；David Woodbetony ■

287146　Pedicularis davidii Franch. var. flaccida Diels ex Bonati = Pedicularis dissecta（Bonati）Pennell et H. L. Li ■

287147　Pedicularis davidii Franch. var. pentodon P. C. Tsoong；五齿大卫氏马先蒿；Fivetooth David Woodbetony ■

287148　Pedicularis davidii Franch. var. platyodon P. C. Tsoong；宽齿大卫氏马先蒿；Broadtooth David Woodbetony ■

287149　Pedicularis debilis Franch. ex Maxim.；弱小马先蒿（细马先蒿）；Weak Woodbetony ■

287150　Pedicularis debilis Franch. ex Maxim. subsp. debilior P. C. Tsoong；极弱马先蒿（极弱弱小马先蒿，极细马先蒿）；Veryweak Woodbetony ■

287151　Pedicularis decora Franch.；美观马先蒿；Smallcalyx Woodbetony ■

287152　Pedicularis decorissima Diels；极丽马先蒿；Longcorollatube Woodbetony ■

287153　Pedicularis delavayi Franch. ex Maxim. = Pedicularis siphonantha D. Don var. delavayi（Franch.）P. C. Tsoong ■

287154　Pedicularis delavayi Franch. ex Maxim. = Pedicularis siphonantha D. Don var. delavayi（Franch. ex Maxim.）P. C. Tsoong ■

287155　Pedicularis deltoidea Franch. ex Maxim.；三角叶马先蒿；Triangularleaf Woodbetony, Trianguleleaf Woodbetony ■

287156　Pedicularis densispica Franch. ex Maxim.；密穗马先蒿；Densespike Woodbetony ■

287157　Pedicularis densispica Franch. ex Maxim. subsp. schneideri（Bonati）P. C. Tsoong；许氏密穗马先蒿（长苞宽穗马先蒿）；Schneider Densespike Woodbetony ■

287158　Pedicularis densispica Franch. ex Maxim. subsp. viridescens P. C. Tsoong；绿盔密穗马先蒿■

287159　Pedicularis densispica Franch. ex Maxim. var. schneideri Bonati = Pedicularis densispica Franch. ex Maxim. subsp. schneideri（Bonati）P. C. Tsoong ■

287160　Pedicularis densispica Franch. ex Maxim. var. typia H. L. Li = Pedicularis densispica Franch. ex Maxim. ■

287161　Pedicularis denudata Hook. f.；秃裸马先蒿■

287162　Pedicularis deqinensis H. P. Yang；德钦马先蒿；Deqin Woodbetony ■

287163　Pedicularis dichotoma Bonati；二歧马先蒿（大马蒿，怀阳草，两歧马先蒿）；Dichotomous Woodbetony ■

287164　Pedicularis dichotoma Bonati var. wardiana Bonati = Pedicularis dichotoma Bonati ■

287165　Pedicularis dichrocephala Hand. -Mazz.；重头马先蒿；Doublehead Woodbetony, Twohead Woodbetony ■

287166　Pedicularis dielsiana Bonati；第氏马先蒿；Diels Woodbetony ■

287167　Pedicularis dielsiana H. Limpr. = Pedicularis tibetica Franch.

287168　Pedicularis diffusa Prain；铺散马先蒿；Diffusive Woodbetony，Sprowled Woodbetony ■

287169　Pedicularis diffusa Prain subsp. elatior P. C. Tsoong；高升铺散马先蒿；Tall Diffusive Woodbetony ■

287170　Pedicularis dissecta(Bonati) Pennell et H. L. Li；全裂马先蒿（黑阳参，太白参，太白丽参，太白土高丽参）；Dissected Woodbetony ■

287171　Pedicularis dissectifolia H. L. Li；细裂叶马先蒿；Dissected Leaf Woodbetony，Dissectleaf Woodbetony ■

287172　Pedicularis dolichantha Bonati；修花马先蒿；Longflower Woodbetony ■

287173　Pedicularis dolichocymba Hand. -Mazz. ；长舟马先蒿；Longstyle Woodbetony ■

287174　Pedicularis dolichoglossa H. L. Li；长舌马先蒿；Longtongue Woodbetony ■

287175　Pedicularis dolichorrhiza Schrenk；长根马先蒿；Longroot Woodbetony ■

287176　Pedicularis dolichosiphon(Hand. -Mazz.) H. L. Li；长管马先蒿；Longtube Woodbetony ■

287177　Pedicularis dolichostachya H. L. Li；长穗马先蒿；Longspike Woodbetony ■

287178　Pedicularis dubia B. Fedtsch. ；可疑马先蒿■☆

287179　Pedicularis duclouxii Bonati；杜氏马先蒿（毛缘马先蒿）；Ducloux Woodbetony ■

287180　Pedicularis dulongensis H. P. Yang；独龙马先蒿；Dulong Woodbetony ■

287181　Pedicularis dunniana Bonati；邓氏马先蒿（大白洋参，等髯马先蒿，褐毛马先蒿）；Brownhair Woodbetony ■

287182　Pedicularis elata Willd. 高升马先蒿；Tall Woodbetony ■

287183　Pedicularis elegans Ten. = Pedicularis pseudomelampyriflora Bonati ■

287184　Pedicularis elliottii P. C. Tsoong；爱氏马先蒿；Elliott Woodbetony ■

287185　Pedicularis elwesii Hook. f. ；哀氏马先蒿（包唇马先蒿，裹盔马先蒿）；Elwes Woodbetony ■

287186　Pedicularis elwesii Hook. f. subsp. major（ H. L. Li）P. C. Tsoong；高大哀氏马先蒿；Big Elwes Woodbetony ■

287187　Pedicularis elwesii Hook. f. subsp. minor（ H. L. Li）P. C. Tsoong；矮小哀氏马先蒿；Small Elwes Woodbetony ■

287188　Pedicularis elwesii Hook. f. var. major H. L. Li = Pedicularis elwesii Hook. f. subsp. major(H. L. Li)P. C. Tsoong ■

287189　Pedicularis elwesii Hook. f. var. typica H. L. Li = Pedicularis elwesii Hook. f. ■

287190　Pedicularis erecta Gilib. = Pedicularis palustris L. ■

287191　Pedicularis eriophora Turcz. ；绵毛马先蒿■☆

287192　Pedicularis euphrasioides Stephan ex Willd. = Pedicularis labradorica Wirsing ■

287193　Pedicularis euphrasioides Stephan ex Willd. var. labradorica （ Houtt.) Willd. = Pedicularis labradorica Wirsing ■

287194　Pedicularis euphrasioides Steven ex Willd. ；小米马先蒿■

287195　Pedicularis euphrasioides Steven ex Willd. = Pedicularis labradorica Wirsing ■

287196　Pedicularis exaltata Besser ex Bunge；高马先蒿■☆

287197　Pedicularis excelsa Hook. f. ；卓越马先蒿；Exellent Woodbetony ■

287198　Pedicularis fargesii Franch. ；法氏马先蒿（华中马先蒿）；Farges Woodbetony ■

287199　Pedicularis fastigiata Franch. ；帚状马先蒿；Fastigiate Woodbetony ■

287200　Pedicularis fedschenkoi Bonati = Pedicularis physocalyx Bunge ■

287201　Pedicularis fengii H. L. Li；国楣马先蒿（旷地马先蒿）；Feng's Woodbetony ■

287202　Pedicularis fetisowii Regel；费氏马先蒿；Fetissow Woodbetony ■

287203　Pedicularis filicifolia Hemsl. ；羊齿叶马先蒿；Fernleaf Woodbetony ■

287204　Pedicularis filicula Franch. ；拟蕨马先蒿（小蕨马先蒿）；Fern-like Woodbetony ■

287205　Pedicularis filicula Franch. var. saganaica Hand. -Mazz. ；木里拟蕨马先蒿；Muli Fern-like Woodbetony ■

287206　Pedicularis filiculiformis P. C. Tsoong；假拟蕨马先蒿；False Fern-like Woodbetony，False Fern-like ■

287207　Pedicularis fissa Turcz. ；半裂马先蒿■☆

287208　Pedicularis flaccida Prain；软弱马先蒿；Flaccid Woodbetony，Sort Woodbetony ■

287209　Pedicularis flava Pall. ；黄花马先蒿；Yellow Woodbetony，Yellowflower Woodbetony ■

287210　Pedicularis flava Pall. = Pedicularis physocalyx Bunge ■

287211　Pedicularis flava Pall. var. altaica Bunge = Pedicularis physocalyx Bunge ■

287212　Pedicularis flava Pall. var. conica Bunge = Pedicularis physocalyx Bunge ■

287213　Pedicularis fletcheri P. C. Tsoong；阜莱氏马先蒿；Fletcher Woodbetony ■

287214　Pedicularis fletcheriana P. C. Tsoong = Pedicularis fletcheri P. C. Tsoong ■

287215　Pedicularis flexuosa Hook. f. ；曲茎马先蒿■

287216　Pedicularis floribunda Franch. ；多花马先蒿；Manyflower Woodbetony ■

287217　Pedicularis foliosa L. ；多叶马先蒿；Leafy Lousewort ■☆

287218　Pedicularis forrestiana Bonati；福氏马先蒿（兜唇马先蒿）；Forrest Woodbetony ■

287219　Pedicularis forrestiana Bonati subsp. flabellifera P. C. Tsoong；扇苞福氏马先蒿（扇苞兜唇马先蒿）；Fan Forrest Woodbetony ■

287220　Pedicularis fortunei Hemsl. ；山萝卜马先蒿■

287221　Pedicularis fragarioides P. C. Tsoong；草莓状马先蒿；Strawberrylike Woodbetony ■

287222　Pedicularis franchetiana Maxim. ；佛氏马先蒿；Franchet Woodbetony ■

287223　Pedicularis furbishiae S. Watson；法尔马先蒿；Furbish's Pedicularis ■☆

287224　Pedicularis furfuracea Wall. ex Prain；糠粃马先蒿；Furfuraceous Woodbetony ■

287225　Pedicularis furfuracea Wall. ex Prain var. integrifolia Hook. f. = Pedicularis pantlingii Prain ■

287226　Pedicularis futtereri Diels ex Futterer = Pedicularis kansuensis Maxim. ■

287227　Pedicularis gagnepainiana Bonati；戛氏马先蒿；Gagnepain Woodbetony ■

287228　Pedicularis galeata Bonati；显盔马先蒿；Galeate Woodbetony ■

287229　Pedicularis galeobdolon Diels = Pedicularis resupinata L. subsp. galeobdolon(Diels) P. C. Tsoong ■

287230　Pedicularis ganpinensis Vaniot ex Bonati；平坝马先蒿；Ganping Woodbetony ■

287231　Pedicularis garckeana Prain ex Maxim. ;戛克氏马先蒿(戛氏马先蒿);Garck Woodbetony ■

287232　Pedicularis garnieri Bonati = Pedicularis cranolopha Maxim. var. garnieri(Bonati)P. C. Tsoong ■

287233　Pedicularis garnieri Bonati = Pedicularis croizatiana H. L. Li ■

287234　Pedicularis garnieri Bonati = Pedicularis tricolor Hand. -Mazz. ■

287235　Pedicularis geosiphon Harry Sm. et P. C. Tsoong;地管马先蒿;Geotube Woodbetony ■

287236　Pedicularis giraldiana Bonati;奇氏马先蒿;Girald Woodbetony ■

287237　Pedicularis glabrescens H. L. Li; 退毛马先蒿;Glabrous Woodbetony ■

287238　Pedicularis globifera Hook. f. ; 球花马先蒿; Ballflower Woodbetony ■

287239　Pedicularis gloriosa Bisset et S. Moore;光荣马先蒿■

287240　Pedicularis gloriosa Bisset et S. Moore f. albiflora Takeda;白花光荣马先蒿■☆

287241　Pedicularis gloriosa Bisset et S. Moore var. iwatensis (Ohwi) Ohwi = Pedicularis iwatensis Ohwi ■☆

287242　Pedicularis goiantha Bureau et Franch. = Pedicularis kansuensis Maxim. ■

287243　Pedicularis gongshanensis H. P. Yang;贡山马先蒿;Gongshan Woodbetony ■

287244　Pedicularis goniantha Bureau et Franch. = Pedicularis kansuensis Maxim. ■

287245　Pedicularis gracilicaulis H. L. Li;细瘦马先蒿(细茎马先蒿);Gracilestem Woodbetony,Thin Woodbetony ■

287246　Pedicularis gracilis Wall. ; 纤细马先蒿;Fine Woodbetony, Slender Woodbetony ■

287247　Pedicularis gracilis Wall. subsp. macrocarpa P. C. Tsoong;大果纤细马先蒿;Bigfruit Slender Woodbetony ■

287248　Pedicularis gracilis Wall. subsp. sinensis P. C. Tsoong;中国纤细马先蒿(中华纤细马先蒿);Chinese Slender Woodbetony ■

287249　Pedicularis gracilis Wall. subsp. stricta P. C. Tsoong;坚挺纤细马先蒿;Strict Slender Woodbetony ■

287250　Pedicularis gracilis Wall. var. macrocarpa Prain = Pedicularis gracilis Wall. subsp. macrocarpa P. C. Tsoong ■

287251　Pedicularis gracilis Wall. var. sinensis H. L. Li = Pedicularis gracilis Wall. subsp. sinensis P. C. Tsoong ■

287252　Pedicularis gracilis Wall. var. typica H. L. Li = Pedicularis gracilis Wall. ■

287253　Pedicularis gracilis Wall. var. typica H. L. Li f. stricta Prain = Pedicularis gracilis Wall. subsp. stricta P. C. Tsoong ■

287254　Pedicularis gracilis Wall. var. yunnanensis Bonati ex H. L. Li;云南纤细马先蒿;Yunnan Strict Slender Woodbetony ■

287255　Pedicularis gracilituba H. L. Li; 细管马先蒿; Thintube Woodbetony ■

287256　Pedicularis gracilituba H. L. Li subsp. setosa P. C. Tsoong;刺毛细管马先蒿;Setose Thintube Woodbetony ■

287257　Pedicularis gracilituba H. L. Li var. setosa H. L. Li = Pedicularis gracilituba H. L. Li subsp. setosa P. C. Tsoong ■

287258　Pedicularis gracilituba H. L. Li var. typica H. L. Li = Pedicularis gracilituba H. L. Li ■

287259　Pedicularis grandiflora Fisch. ;野苏子(大花马先蒿,小花野苏子);Bigflower Woodbetony,Largeflower Woodbetony ■

287260　Pedicularis grandis Popov;大马先蒿■☆

287261　Pedicularis gruina Franch. ex Maxim. ;鹤首马先蒿;Crane Woodbetony,Cranehead Woodbetony ■

287262　Pedicularis gruina Franch. ex Maxim. subsp. pilosa(Bonati)P. C. Tsoong;多毛鹤首马先蒿;Manyhair Cranehead Woodbetony ■

287263　Pedicularis gruina Franch. ex Maxim. subsp. polyphylla(Franch.)P. C. Tsoong;多叶鹤首马先蒿;Manyleaf Cranehead Woodbetony ■

287264　Pedicularis gruina Franch. ex Maxim. var. cinerascens Franch. ex H. L. Li = Pedicularis gruina Franch. ex Maxim. subsp. pilosa(Bonati)P. C. Tsoong ■

287265　Pedicularis gruina Franch. ex Maxim. var. laxiflora Franch. = Pedicularis gruina Franch. ex Maxim. ■

287266　Pedicularis gruina Franch. ex Maxim. var. pilosa Bonati = Pedicularis gruina Franch. ex Maxim. subsp. pilosa(Bonati)P. C. Tsoong ■

287267　Pedicularis gruina Franch. ex Maxim. var. polyphylla(Franch. ex Maxim.)H. L. Li = Pedicularis gruina Franch. ex Maxim. subsp. polyphylla(Franch.)P. C. Tsoong ■

287268　Pedicularis gruina Franch. ex Maxim. var. polyphylla H. L. Li = Pedicularis gruina Franch. ex Maxim. subsp. polyphylla(Franch.)P. C. Tsoong ■

287269　Pedicularis gruina Franch. ex Maxim. var. typica H. L. Li = Pedicularis gruina Franch. ex Maxim. ■

287270　Pedicularis gyirongensis H. P. Yang;吉隆马先蒿;Jilong Woodbetony ■

287271　Pedicularis gyrorhyncha Franch. ex Maxim. ; 旋喙马先蒿;Revolvebill Woodbetony,Roundbeak Woodbetony ■

287272　Pedicularis habachanensis Bonati;哈巴山马先蒿;Habashan Woodbetony ■

287273　Pedicularis habachanensis Bonati subsp. multipinnata P. C. Tsoong;多羽片哈巴山马先蒿■

287274　Pedicularis handel-mazzettii Bonati = Pedicularis confertiflora Prain ■

287275　Pedicularis hemsleyana Prain;昂斯莱马先蒿(汉姆氏马先蒿);Hemsle Woodbetony ■

287276　Pedicularis henryi Maxim. ;亨氏马先蒿(凤尾参,互叶凤尾参,江南凤尾参,江南马先蒿,羊肚参,羊肝参,追风箭);Henry Woodbetony ■

287277　Pedicularis heterophylla Bonati = Pedicularis axillaris Franch. ex Maxim. ■

287278　Pedicularis heydei Prain = Pedicularis pheulpinii Bonati subsp. chilienensis P. C. Tsoong ■

287279　Pedicularis hirsuta L. ;硬毛马先蒿■☆

287280　Pedicularis hirtella Franch. ex Hemsl. ; 粗毛马先蒿;Hirsute Woodbetony ■

287281　Pedicularis holocalyx Hand. -Mazz. ; 全萼马先蒿;Holocalyx Woodbetony ■

287282　Pedicularis honanensis P. C. Tsoong;河南马先蒿;Henan Woodbetony,Honan Woodbetony ■

287283　Pedicularis hulteniana H. L. Li = Pedicularis anthemifolia Fisch. ■

287284　Pedicularis humilis Bonati;矮马先蒿;Dwarf Woodbetony ■

287285　Pedicularis humilis M. Bieb. = Pedicularis labradorica Wirsing ■

287286　Pedicularis hutteniana H. L. Li = Pedicularis anthemifolia Fisch. ■

287287　Pedicularis hypophylla T. Yamaz. ;里白马先蒿■

287288　Pedicularis ikomae Sasaki;生驹氏马先蒿(高山马先蒿,马先蒿草,南湖大山蒿草);Ikoma Woodbetony ■

287289　Pedicularis inaequilobata P. C. Tsoong; 不等裂马先蒿;Unequallobed Woodbetony ■

287290　Pedicularis incarnata L. ;肉色马先蒿■☆

287291　Pedicularis infirma H. L. Li；孱弱马先蒿；Frail Woodbetony, Weak Woodbetony ■

287292　Pedicularis inflexirostris F. S. Yang,D. Y. Hong et X. Q. Wang；弯喙马先蒿■

287293　Pedicularis ingens Maxim.；硕大马先蒿；Giant Woodbetony ■

287294　Pedicularis insignis Bonati；显著马先蒿；Insignis Woodbetony, Marked Woodbetony ■

287295　Pedicularis integerrima Pannell et H. L. Li = Pedicularis integrifolia Hook. f. subsp. integerrima（Pannell et H. L. Li）P. C. Tsoong ■

287296　Pedicularis integrifolia Hook. f.；全叶马先蒿（全缘叶马先蒿）；Entireleaf Woodbetony ■

287297　Pedicularis integrifolia Hook.f. subsp. integerrima（Pannell et H. L. Li）P. C. Tsoong；全缘全叶马先蒿（全缘马先蒿，全缘叶马先蒿）■

287298　Pedicularis interrupta Stephan ex Willd.；间断马先蒿■☆

287299　Pedicularis iwatensis Ohwi；岩手山马先蒿■☆

287300　Pedicularis japonica Miq.；日本马先蒿■☆

287301　Pedicularis kangtingensis P. C. Tsoong；康定马先蒿；Kangding Woodbetony, Kangting Woodbetony ■

287302　Pedicularis kansuensis Maxim.；甘肃马先蒿；Gansu Woodbetony, Kansu Woodbetony ■

287303　Pedicularis kansuensis Maxim. f. albiflora H. L. Li；白花甘肃马先蒿；Whiteflower Gansu Woodbetony ■

287304　Pedicularis kansuensis Maxim. subsp. kokonorica P. C. Tsoong；青甘马先蒿（青海甘肃马先蒿）；Qinghai-Gansu Woodbetony ■

287305　Pedicularis kansuensis Maxim. subsp. villosa P. C. Tsoong；厚毛甘肃马先蒿；Villose Gansu Woodbetony ■

287306　Pedicularis kansuensis Maxim. subsp. yargongensis（Bonati）P. C. Tsoong；雅江马先蒿（大果甘肃马先蒿，雅江甘肃马先蒿）；Yajiang Gansu Woodbetony ■

287307　Pedicularis karatavica Pavlov；卡拉塔夫马先蒿■☆

287308　Pedicularis kariensis Bonati；卡里马先蒿（维西马先蒿）；Kari Woodbetony ■

287309　Pedicularis karoi Freyn = Pedicularis palustris L. subsp. karoi（Freyn）P. C. Tsoong ■

287310　Pedicularis kaufmannii Pinzger；卡氏马先蒿■☆

287311　Pedicularis kawaguchii T. Yamaz.；喀瓦谷地马先蒿■

287312　Pedicularis keiskei Franch. et Sav.；伊藤氏马先蒿☆

287313　Pedicularis kialensis Franch.；甲拉马先蒿；Jiala Woodbetony ■

287314　Pedicularis kiangsiensis P. C. Tsoong et S. H. Cheng；江西马先蒿；Jiangxi Woodbetony, Kiangsi Woodbetony ■

287315　Pedicularis koidzumiana Tatew. et Ohwi；小泉氏马先蒿■☆

287316　Pedicularis kongboensis P. C. Tsoong；宫布马先蒿；Gongbu Woodbetony, Kongbo Woodbetony ■

287317　Pedicularis kongboensis P. C. Tsoong var. obtusata P. C. Tsoong；钝裂宫布马先蒿■

287318　Pedicularis korolkovii Regel；卡罗马先蒿■☆

287319　Pedicularis koueytchensis Bonati；滇东马先蒿；E. Yunnan Woodbetony,East Yunnan Woodbetony ■

287320　Pedicularis krylovii Bonati；克雷马先蒿■☆

287321　Pedicularis kuznetzovii Kom.；库氏马先蒿■☆

287322　Pedicularis labellata Jacq. = Pedicularis rhinanthoides Schrenk ex Fisch. subsp. labellata（Jacq.）P. C. Tsoong ■

287323　Pedicularis labellata Jacq. = Pedicularis rhinanthoides Schrenk ex Fisch. subsp. labellata（Jacq.）Pennell ■

287324　Pedicularis labordei Vaniot ex Bonati；拉氏马先蒿（长喙马先蒿,凤尾草,西南马先蒿）；Labord Woodbetony ■

287325　Pedicularis labradorica Houtt. = Pedicularis labradorica Wirsing ■

287326　Pedicularis labradorica Wirsing；拉不拉多马先蒿（北马先蒿）；Labrador Woodbetony ■

287327　Pedicularis labradorica Wirsing var. simplex Hultén = Pedicularis labradorica Wirsing ■

287328　Pedicularis lacerata Bonati = Pedicularis axillaris Franch. ex Maxim. ■

287329　Pedicularis lachnoglossa Hook. f.；绒舌马先蒿；Hairytongue Woodbetony, Hairytongued Woodbetony ■

287330　Pedicularis lachnoglossa Hook. f. var. macrantha Bonati = Pedicularis lachnoglossa Hook. f. ■

287331　Pedicularis laeta Steven ex Claus. = Pedicularis dasystachys Schrenk ■

287332　Pedicularis lamarum Limpr. = Pedicularis rex C. B. Clarke ex Maxim. subsp. lipskyana（Bonati）P. C. Tsoong ■

287333　Pedicularis lamioides Hand.-Mazz.；元宝草马先蒿；Deadnettlelike Woodbetony ■

287334　Pedicularis lanata Cham. et Schltdl.；绵花马先蒿■☆

287335　Pedicularis lanceolata Michx.；沼泽马先蒿；Fen Betony, Lapland Pedicularis,Swamp Betony,Swamp Lousewort,Swamp Wood Betony,Swamp-lousewort ■☆

287336　Pedicularis langsdorfii Pisch. ex Stev.；朗斯马先蒿■☆

287337　Pedicularis lanpingensis H. B. Yang；兰坪马先蒿；Lanping Woodbetony ■

287338　Pedicularis lapponica L.；拉普兰马先蒿；Lapland Pedicularis ■☆

287339　Pedicularis lasiantha H. L. Li = Pedicularis decora Franch. ■

287340　Pedicularis lasiophrys Maxim.；毛额马先蒿；Lanoschin Woodbetony, Woollychin Woodbetony ■

287341　Pedicularis lasiophrys Maxim. var. sinica Maxim.；毛背毛额马先蒿；Chinese Woollychin Woodbetony ■

287342　Pedicularis lasiophrys Maxim. var. typica H. L. Li = Pedicularis lasiophrys Maxim. ■

287343　Pedicularis lasiostachys Bunge；西方毛穗马先蒿；Hairspike Woodbetony ■☆

287344　Pedicularis latibracteata Yamaz.；宽苞马先蒿■

287345　Pedicularis latirostris P. C. Tsoong；宽喙马先蒿；Broadbeak Woodbetony, Broadbill Woodbetony ■

287346　Pedicularis latituba Bonati；粗管马先蒿；Broadtube Woodbetony ■

287347　Pedicularis laxiflora Franch.；疏花马先蒿；Laxflower Woodbetony, Scatterflower Woodbetony ■

287348　Pedicularis laxiflora P. C. Tsoong = Pedicularis membranacea H. L. Li ■

287349　Pedicularis laxispica H. L. Li；疏穗马先蒿；Laxspike Woodbetony, Scatterspike Woodbetony ■

287350　Pedicularis lecomtei Bonati；勒公氏马先蒿（鹤广马先蒿）；Lecomte Woodbetony ■

287351　Pedicularis legendrei Bonati；勒氏马先蒿；Legendre Woodbetony ■

287352　Pedicularis leptorhiza Rupr.；细根马先蒿■

287353　Pedicularis leptorhiza Rupr. = Pedicularis ludweigii Regel ■

287354　Pedicularis leptosiphon H. L. Li；纤管马先蒿；Finetube Woodbetony, Thintube Woodbetony ■

287355　Pedicularis liana Pennell ex H. L. Li = Pedicularis debilis Franch. ex Maxim. subsp. debilior P. C. Tsoong ■

287356　Pedicularis likiangensis Franch. ex Maxim.；丽江马先蒿；Lijiang Woodbetony, Likiang Woodbetony ■

287357　Pedicularis likiangensis Franch. subsp. pulchera P. C. Tsoong；美

丽丽江马先蒿;Beautiful Lijiang Woodbetony ■

287358　Pedicularis limprichtiana Fedde = Pedicularis tibetica Franch. ■

287359　Pedicularis limprichtiana Hand. -Mazz.;林氏马先蒿(会理马先蒿);Limpricht Woodbetony ■

287360　Pedicularis limprichtii Fedde = Pedicularis tibetica Franch. ■

287361　Pedicularis lineata Franch.;条纹马先蒿;Lineate Woodbetony ■

287362　Pedicularis lineata Franch. ex Maxim. var. dissecta Bonati = Pedicularis likiangensis Franch. ex Maxim. ■

287363　Pedicularis lineata Franch. var. dissecta Bonati = Pedicularis likiangensis Franch. ex Maxim. ■

287364　Pedicularis lingelsheimiana H. Limpr.;凌氏马先蒿;Lingelsheim Woodbetony ■

287365　Pedicularis lipskyana Bonati = Pedicularis rex C. B. Clarke ex Maxim. subsp. lipskyana(Bonati)P. C. Tsoong ■

287366　Pedicularis longicalyx H. P. Yang;长萼马先蒿;Longcalyx Woodbetony ■

287367　Pedicularis longicaulis Franch.;长茎马先蒿(长茎凤尾参,对叶凤尾参,凤尾参);Longstem Woodbetony ■

287368　Pedicularis longiflora Rudolph;长花马先蒿(露茹色博);Longlower Woodbetony ■

287369　Pedicularis longiflora Rudolph = Pedicularis chinensis Maxim. ■

287370　Pedicularis longiflora Rudolph = Pedicularis longiflora Rudolph var. tubiformis(G. Klotz)P. C. Tsoong ■

287371　Pedicularis longiflora Rudolph subsp. tubiformis (G. Klotz) Pennell = Pedicularis longiflora Rudolph var. tubiformis(G. Klotz)P. C. Tsoong ■

287372　Pedicularis longiflora Rudolph var. tubiformis (G. Klotz) P. C. Tsoong;管状长花马先蒿(斑唇马先蒿,长花马先蒿,长筒马先蒿);Punctatelip Woodbetony ■

287373　Pedicularis longiflora Rudolph var. yinshanensis Z. Y. Chu et Y. Z. Zhao;阴山长花马先蒿(阴山马先蒿);Yinshan Woodbetony ■

287374　Pedicularis longipes Maxim.;长梗马先蒿;Longstalk Woodbetony ■

287375　Pedicularis longipetiolata Franch.;长柄马先蒿;Longpetiole Woodbetony ■

287376　Pedicularis longistipitata P. C. Tsoong;长把马先蒿;Longstipe Woodbetony ■

287377　Pedicularis lophocentra Hand. -Mazz. = Pedicularis mussotii Franch. var. lophocentra(Hand. -Mazz.)H. L. Li ■

287378　Pedicularis lophotricha H. L. Li;盔须马先蒿;Cresthair Woodbetony,Helmetbeard Woodbetony ■

287379　Pedicularis lopingensis Hand. -Mazz. = Pedicularis rex C. B. Clarke ex Maxim. ■

287380　Pedicularis lucifuga Bonati = Pedicularis macrosiphon Franch. ■

287381　Pedicularis ludovicii H. Limpr. = Pedicularis tibetica Franch. ■

287382　Pedicularis ludweigii Regel;小根马先蒿(卢氏马先蒿);Littleroot Woodbetony,Ludweig Woodbetony,Rootlet Woodbetony ■

287383　Pedicularis lunglingensis Bonati;龙陵马先蒿;Longling Woodbetony,Lungling Woodbetony ■

287384　Pedicularis lusitanica Hoffmanns. et Link = Pedicularis sylvatica L. ■☆

287385　Pedicularis luteola H. L. Li = Pedicularis plicata Maxim. subsp. luteola(H. L. Li)P. C. Tsoong ■

287386　Pedicularis lutescens Franch.;浅黄马先蒿(淡黄马先蒿);Lightyellow Woodbetony,Yellowish Woodbetony ■

287387　Pedicularis lutescens Franch. subsp. brevifolia (Bonati) P. C. Tsoong;短叶浅黄马先蒿(短叶淡黄马先蒿);Shortleaf Yellowish Woodbetony ■

287388　Pedicularis lutescens Franch. subsp. longipetiolata(H. L. Li)P. C. Tsoong;长柄浅黄马先蒿(长柄淡黄马先蒿);Longpetiole Yellowish Woodbetony ■

287389　Pedicularis lutescens Franch. subsp. ramosa (Bonati) P. C. Tsoong;多枝浅黄马先蒿(多枝淡黄马先蒿);Manybranch Yellowish Woodbetony ■

287390　Pedicularis lutescens Franch. subsp. tongchuanensis(Bonati)P. C. Tsoong;东川淡黄马先蒿;Dongchuan Yellowish Woodbetony ■

287391　Pedicularis lutescens Franch. var. brevifolia Bonati = Pedicularis lutescens Franch. subsp. brevifolia(Bonati)P. C. Tsoong ■

287392　Pedicularis lutescens Franch. var. longipetiolata H. L. Li = Pedicularis lutescens Franch. subsp. longipetiolata(H. L. Li)P. C. Tsoong ■

287393　Pedicularis lutescens Franch. var. ramosa Bonati = Pedicularis lutescens Franch. subsp. ramosa(Bonati)P. C. Tsoong ■

287394　Pedicularis lutescens Franch. var. tongtchuanensis Bonati = Pedicularis lutescens Franch. subsp. tongchuanensis (Bonati) P. C. Tsoong ■

287395　Pedicularis lutescens Franch. var. typica H. L. Li = Pedicularis lutescens Franch. ■

287396　Pedicularis lyrata Prain = Pedicularis polyodonta H. L. Li ■

287397　Pedicularis lyrata Prain ex Maxim.;琴盔马先蒿;Lyrate Woodbetony ■

287398　Pedicularis lyrata Prain ex Maxim. var. cordifolia Franch. = Pedicularis polyodonta H. L. Li ■

287399　Pedicularis lyrata Prain var. cordifolia Franch. = Pedicularis polyodonta H. L. Li ■

287400　Pedicularis macilenta Franch. ex Hemsl.;瘠瘦马先蒿;Lean Woodbetony,Thin Woodbetony ■

287401　Pedicularis macrantha (Bonati) H. Lév. = Pedicularis lachnoglossa Hook. f. ■

287402　Pedicularis macrocalyx Bonati = Pedicularis dolichocymba Hand. -Mazz. ■

287403　Pedicularis macrochila Vved. = Pedicularis anthemifolia Fisch. subsp. elatior(Regel)P. C. Tsoong ■

287404　Pedicularis macrorhyncha H. L. Li;长喙马先蒿;Bigbill Woodbetony,Longbeak Woodbetony ■

287405　Pedicularis macrosiphon Franch.;大管马先蒿(长虫莲,长管马先蒿);Bigtube Woodbetony ■

287406　Pedicularis macrosiphon Franch. = Pedicularis muscicola Maxim. ■

287407　Pedicularis macrosiphon Franch. var. tribuloides(Bonati)H. L. Li = Pedicularis macrosiphon Franch. ■

287408　Pedicularis magnini Bonati = Pedicularis densispica Franch. ex Maxim. ■

287409　Pedicularis mahoangensis Bonati = Pedicularis rex C. B. Clarke ex Maxim. ■

287410　Pedicularis mairei Bonati;梅氏马先蒿(东川马先蒿);Maire Woodbetony ■

287411　Pedicularis mandshurica Maxim.;鸡冠马先蒿(东北马先蒿,鸡冠子花);Manchurian Woodbetony ■

287412　Pedicularis margaritae Bonati = Pedicularis gruina Franch. ex Maxim. subsp. pilosa(Bonati)P. C. Tsoong ■

287413　Pedicularis mariae Regel;玛丽马先蒿;Maria Woodbetony ■

287414　Pedicularis maximowiczii Krasn.;马克西姆马先蒿■

287415　Pedicularis maxonii Bonati;马克逊马先蒿(沙坝马先蒿);

Maxon Woodbetony ■

287416 Pedicularis mayana Hand. -Mazz. ;迈亚马先蒿(马牙马先蒿); May Woodbetony ■

287417 Pedicularis megalantha Don;硕花马先蒿;Largeflower Woodbetony ■

287418 Pedicularis megalantha Don var. typica Prain = Pedicularis megalantha Don ■

287419 Pedicularis megalochila L. ;大唇马先蒿;Biglip Woodbetony, Largelip Woodbetony ■

287420 Pedicularis megalochila L. var. ligulata P. C. Tsoong;舌状大唇马先蒿;Ligulate Largelip Woodbetony ■

287421 Pedicularis melampyriflora Franch. ex Hemsl. ;山萝花马先蒿;Cowwheatflower Woodbetony ■

287422 Pedicularis membranacea H. L. Li;膜叶马先蒿;Filmleaf Woodbetony, Membraceousleaf Woodbetony ■

287423 Pedicularis menziesii Benth. = Pedicularis verticillata L. ■

287424 Pedicularis merrilliana H. L. Li;迈氏马先蒿;Merrill Woodbetony ■

287425 Pedicularis metaszetschuanica P. C. Tsoong;后生四川马先蒿;After Szechuan Woodbetony, Later Sichuan Woodbetony ■

287426 Pedicularis meteororhyncha H. L. Li;翘喙马先蒿;Warpbill Woodbetony, Wingedbeak Woodbetony ■

287427 Pedicularis micrantha H. L. Li;小花马先蒿;Littleflower Woodbetony, Smallflower Woodbetony ■

287428 Pedicularis microcalyx Hook. f. ;小萼马先蒿;Littlecalyx Woodbetony, Smallcalyx Woodbetony ■

287429 Pedicularis microchila Franch. ;小唇马先蒿;Littlelip Woodbetony ■

287430 Pedicularis microphyton Bureau et Franch. = Pedicularis przewalskii Maxim. subsp. microphyton (Bureau et Franch.) P. C. Tsoong ■

287431 Pedicularis microphyton Bureau et Franch. var. purpurea Bonati = Pedicularis przewalskii Maxim. subsp. microphyton (Bureau et Franch.) P. C. Tsoong var. purpurea(Bonati)P. C. Tsoong ■

287432 Pedicularis minima P. C. Tsoong et S. H. Cheng;细小马先蒿;Mini Woodbetony, Minute Woodbetony ■

287433 Pedicularis minutilabris P. C. Tsoong;微唇马先蒿;Minilip Woodbetony, Minutelip Woodbetony ■

287434 Pedicularis mollis Wall. ;柔毛马先蒿;Hairy Woodbetony, Velvet Woodbetony ■

287435 Pedicularis monbeigiana Bonati;蒙氏马先蒿(澜沧马先蒿);Monbeig Woodbetony ■

287436 Pedicularis moupinensis Franch. ;穆坪马先蒿(宝兴马先蒿);Baoxing Woodbetony, Paohsing Woodbetony ■

287437 Pedicularis muliensis Hand. -Mazz. = Pedicularis duclouxii Bonati ■

287438 Pedicularis muscicola Maxim. ;藓生马先蒿(蓟参马先蒿,土人参);Muscicolous Woodbetony ■

287439 Pedicularis muscoides H. L. Li;藓状马先蒿;Musciform Woodbetony ■

287440 Pedicularis muscoides H. L. Li var. rosea H. L. Li;玫瑰色藓状马先蒿;Rose Musciform Woodbetony ■

287441 Pedicularis muscoides H. L. Li var. typica Li = Pedicularis muscoides H. L. Li ■

287442 Pedicularis mussotii Franch. ;耳喙马先蒿(谬氏马先蒿);Mussot Woodbetony ■

287443 Pedicularis mussotii Franch. var. lophocentra(Hand. -Mazz.) H.

L. Li;刺冠耳喙马先蒿(刺冠谬氏马先蒿)■

287444 Pedicularis mussotii Franch. var. mutata Bonati;变萼耳喙马先蒿(变形谬氏马先蒿)■

287445 Pedicularis mussotii Franch. var. typica H. L. Li = Pedicularis mussotii Franch. ■

287446 Pedicularis mychophila C. Marquand et Airy Shaw;菌生马先蒿;Fungiliving Woodbetony, Fungivorous Woodbetony ■

287447 Pedicularis myriophylla Pall. ;万叶马先蒿;Manyleaf Woodbetony ■

287448 Pedicularis myriophylla Pall. var. purpurea Bunge;紫色万叶马先蒿;Purple Manyleaf Woodbetony ■

287449 Pedicularis myriophylla Pall. var. tatarinowii(Maxim.) Hurus. = Pedicularis tatarinowii Maxim. ■

287450 Pedicularis myriophylla Pall. var. typica H. L. Li = Pedicularis myriophylla Pall. ■

287451 Pedicularis nanchuanensis P. C. Tsoong;南川马先蒿;Nanchuan Woodbetony ■

287452 Pedicularis nanfutashanensis T. Yamaz. ;南湖大山蒿草■

287453 Pedicularis nanfutashanensis T. Yamaz. = Pedicularis ikomai Sasaki ■

287454 Pedicularis nasturtiifolia Franch. ;薜菜叶马先蒿(乌龙毛);Marshcressleaf Woodbetony ■

287455 Pedicularis nasuta M. Bieb. ex Stev. ;鼻马先蒿■☆

287456 Pedicularis neoladtuba P. C. Tsoong;新粗管马先蒿;Newbroadtube Woodbetony ■

287457 Pedicularis nigra (Bonati) Vaniot ex Bonati;黑马先蒿(鸡脚参);Black Woodbetony ■

287458 Pedicularis nigrescens Nakai;翅茎草;Wingstem Woodbetony ■

287459 Pedicularis nipponica Makino;本州马先蒿■☆

287460 Pedicularis nipponica Makino f. alba M. Mizush. et Yokouchi;白花本州马先蒿■☆

287461 Pedicularis nordmanniana Bunge;诺尔马先蒿■☆

287462 Pedicularis numidica Pomel;努米底亚马先蒿■☆

287463 Pedicularis nyalamensis H. P. Yang;聂拉木马先蒿;Sanglamu Woodbetony ■

287464 Pedicularis nyingchiensis H. P. Yang et Tateishi;林芝马先蒿;Linzhi Woodbetony ■

287465 Pedicularis obscura Bonati;暗昧马先蒿;Obscure Woodbetony ■

287466 Pedicularis odontochila Diels;齿唇马先蒿;Toothlip Woodbetony ■

287467 Pedicularis odontophora Prain;具齿马先蒿;Toothed Woodbetony ■

287468 Pedicularis oederi Vahl;欧氏马先蒿(广布马先蒿,环极马先蒿,绵毛马先蒿);Oeder Woodbetony ■

287469 Pedicularis oederi Vahl subsp. branchyophylla (Pennell) P. C. Tsoong;鳃叶欧氏马先蒿■

287470 Pedicularis oederi Vahl subsp. heteroglossa (Prain) Pennell = Pedicularis oederi Vahl var. heteroglossa Prain ■

287471 Pedicularis oederi Vahl subsp. multipinna (H. L. Li) P. C. Tsoong;多羽片欧氏马先蒿;Manypinnate Woodbetony ■

287472 Pedicularis oederi Vahl var. angustiflora(Limpr.) P. C. Tsoong;狭花欧氏马先蒿■

287473 Pedicularis oederi Vahl var. bracteosa Bonati = Pedicularis orthocoryne H. L. Li ■

287474 Pedicularis oederi Vahl var. heteroglossa Prain;异盔欧氏马先蒿(异盔马先蒿)■

287475 Pedicularis oederi Vahl var. heteroglossa Prain = Pedicularis oederi Vahl var. sinensis(Maxim.)Hurus. ■

287476 Pedicularis oederi Vahl var. multipinna H. L. Li = Pedicularis oederi Vahl subsp. multipinna(H. L. Li)P. C. Tsoong ■

287477 Pedicularis oederi Vahl var. rubra(Maxim.)P. C. Tsoong;红色欧氏马先蒿■

287478 Pedicularis oederi Vahl var. sinensis(Maxim.)Hurus.;中国欧氏马先蒿（华马先蒿，穗花马先蒿，条参马先蒿）; China Woodbetony,Chinese Oeder Woodbetony ■

287479 Pedicularis oederi Vahl var. typica Prain = Pedicularis oederi Vahl ■

287480 Pedicularis oederi Vahl var. yezoana (Nakai) Hurus. = Pedicularis oederi Vahl var. heteroglossa Prain ■

287481 Pedicularis olgae Regel;奥尔嘎马先蒿■☆

287482 Pedicularis oligantha Franch. ex Maxim.;少花马先蒿; Fewflower Woodbetony ■

287483 Pedicularis oliveriana Prain;奥氏马先蒿（马先蒿，扭盔马先蒿,欧氏马先蒿,茸背马先蒿,西藏马先蒿）;Oliver Woodbetony ■

287484 Pedicularis oliveriana Prain subsp. lasiantha P. C. Tsoong = Pedicularis oliveriana Prain ■

287485 Pedicularis omiiana Bonati;峨眉马先蒿; Emei Woodbetony, Omei Woodbetony ■

287486 Pedicularis omiiana Bonati subsp. diffusa(Bonati)P. C. Tsoong;铺散峨眉马先蒿;Diffuse Emei Woodbetony ■

287487 Pedicularis omiiana Bonati var. diffusa Bonati = Pedicularis omiiana Bonati subsp. diffusa(Bonati)P. C. Tsoong ■

287488 Pedicularis omiiana Bonati var. typica H. L. Li = Pedicularis omiiana Bonati ■

287489 Pedicularis orthocoryne H. L. Li;直盔马先蒿; Straithelmet Woodbetony,Strictgalea Woodbetony ■

287490 Pedicularis oxycarpa Franch. ex Maxim.;尖果马先蒿; Acutefruit Woodbetony,Sharpfuit Woodbetony ■

287491 Pedicularis paiana H. L. Li;白氏马先蒿;Pai Woodbetony ■

287492 Pedicularis pallasii Vved.;帕拉西马先蒿■☆

287493 Pedicularis pallida Nutt. = Pedicularis lanceolata Michx. ■☆

287494 Pedicularis palustris L.;沼生马先蒿（沼地马先蒿,沼泽马先蒿）;Cock's Comb, Cow's Wort, Dead Man, Dead Man's Bellows, European Pedicularis, Honey Cap, Honey-cap, Honeysuckle, Lousewort,Marsh Loosewort, Marsh Red Rattle, Marsh Woodbetony, Moss-crop, Moss-flower, Rattle-grass, Rattleweed, Red Rattle, Red Rattle-grass,Suckles,Swamp Pedicularis,Wild Honeysuckle ■

287495 Pedicularis palustris L. subsp. karoi(Freyn)P. C. Tsoong;卡氏沼生马先蒿(沼地马先蒿);Karo's Marsh Woodbetony ■

287496 Pedicularis panjutinii E. Busch;帕恩马先蒿■☆

287497 Pedicularis pantlingii Prain;潘氏马先蒿（卵叶马先蒿）; Pantling Woodbetony ■

287498 Pedicularis pantlingii Prain subsp. brachycarpa P. C. Tsoong = Pedicularis pantlingii Prain subsp. brachycarpa P. C. Tsoong ex C. Y. Wu et Hong Wang ■

287499 Pedicularis pantlingii Prain subsp. brachycarpa P. C. Tsoong ex C. Y. Wu et Hong Wang;短果潘氏马先蒿(短果马先蒿);Shortfruit Pantling Woodbetony ■

287500 Pedicularis pantlingii Prain subsp. chimiliensis (Bonati) P. C. Tsoong;缅甸潘氏马先蒿（北缅卵叶马先蒿）; Burma Pantling Woodbetony ■

287501 Pedicularis pantlingii Prain var. chimiliensis Bonati = Pedicularis pantlingii Prain subsp. chimiliensis(Bonati)P. C. Tsoong ■

287502 Pedicularis pantlingii Prain var. typica H. L. Li = Pedicularis pantlingii Prain ■

287503 Pedicularis parvifolia Hand. -Mazz. = Pedicularis confertiflora Prain subsp. parvifolia(Hand. -Mazz.)P. C. Tsoong ■

287504 Pedicularis paxiana H. Limpr.;派氏马先蒿;Pax Woodbetony ■

287505 Pedicularis pectinata Wall.;箆形马先蒿■☆

287506 Pedicularis pectinatiformis Bonati;拟箆齿马先蒿; Pectinate Woodbetony ■

287507 Pedicularis peduncularis Popov;梗花马先蒿■☆

287508 Pedicularis pennellii Hultén;彭内尔马先蒿■☆

287509 Pedicularis pentagona H. L. Li;五角马先蒿; Fiveangle Woodbetony ■

287510 Pedicularis petelotii P. C. Tsoong;裴氏马先蒿（北越马先蒿）; Petelot Woodbetony ■

287511 Pedicularis petitmenginii Bonati;伯氏马先蒿; Petitmengin Woodbetony ■

287512 Pedicularis petitmenginii Bonati var. dissecta Bonati = Pedicularis dissecta(Bonati)Pennell et H. L. Li ■

287513 Pedicularis phacelifolia Franch.;菊叶马先蒿（草茱萸马先蒿,法且利亚叶马先蒿）;Phacelialeaf Woodbetony ■

287514 Pedicularis pheulpinii Bonati;费尔氏马先蒿（费尔马先蒿）; Pheulpin Woodbetony ■

287515 Pedicularis pheulpinii Bonati = Pedicularis maxonii Bonati ■

287516 Pedicularis pheulpinii Bonati subsp. chilienensis P. C. Tsoong;祁连费尔氏马先蒿■

287517 Pedicularis physocalyx Bunge;臌萼马先蒿; Inflatedcalyx Woodbetony,Swellingcalyx Woodbetony ■

287518 Pedicularis pilostachya Maxim.;绵穗马先蒿; Pilosestachys Woodbetony ■

287519 Pedicularis pinetorum Hand. -Mazz.;松林马先蒿; Pineforest Woodbetony ■

287520 Pedicularis platyrhyncha Schrenk;中亚宽喙马先蒿■☆

287521 Pedicularis plicata Maxim.;皱褶马先蒿;Plicate Woodbetony ■

287522 Pedicularis plicata Maxim. subsp. apiculata P. C. Tsoong;凸尖皱褶马先蒿;Apiculate Plicate Woodbetony ■

287523 Pedicularis plicata Maxim. subsp. luteola (H. L. Li) P. C. Tsoong;浅黄皱褶马先蒿;Yellowish Plicate Woodbetony ■

287524 Pedicularis plicata Maxim. var. apiculata P. C. Tsoong = Pedicularis plicata Maxim. subsp. apiculata P. C. Tsoong ■

287525 Pedicularis plicata Maxim. var. giraldiana (Diels) Limpr. = Pedicularis giraldiana Bonati ■

287526 Pedicularis plicata Maxim. var. typica H. L. Li = Pedicularis plicata Maxim. ■

287527 Pedicularis polygaloides Hook. f.;远志状马先蒿;Milkwortlike Woodbetony ■

287528 Pedicularis polyodonta H. L. Li;多齿马先蒿;Manytooth Woodbetony,Manytoothed Woodbetony ■

287529 Pedicularis polyphylla Franch. ex Maxim. = Pedicularis gruina Franch. ex Maxim. subsp. polyphylla(Franch.)P. C. Tsoong ■

287530 Pedicularis polyphylla Franch. ex Maxim. var. pilosa Bonati = Pedicularis gruina Franch. ex Maxim. subsp. pilosa (Bonati) P. C. Tsoong ■

287531 Pedicularis polyphylloides Bonati = Pedicularis gruina Franch. ex Maxim. subsp. pilosa(Bonati)P. C. Tsoong ■

287532 Pedicularis pontica Boiss.;蓬特马先蒿■☆

287533 Pedicularis popovii Vved.;波波夫马先蒿■☆

287534 Pedicularis porphyrantha H. L. Li = Pedicularis stenocorys Franch. ■

287535 Pedicularis potaninii Maxim. ;波氏马先蒿;Potanin Woodbetony ■

287536 Pedicularis praealta Bonati = Pedicularis smithiana Bonati ■

287537 Pedicularis praeclara Franch. = Pedicularis nipponica Makino ■☆

287538 Pedicularis praeruptorum Bonati;悬岩马先蒿;Precipice Woodbetony,Steep Woodbetony ■

287539 Pedicularis prainiana Maxim. ;帕兰氏马先蒿;Prain Woodbetony ■

287540 Pedicularis princeps Bureau et Franch. ;高超马先蒿;Glabrouscalyx Woodbetony ■

287541 Pedicularis proboscidea Steven;鼻喙马先蒿;Snout Woodbetony ■

287542 Pedicularis provoti Franch. = Pedicularis curvituba Maxim. ■

287543 Pedicularis provotii Franch. = Pedicularis curvituba Maxim. subsp. provoti(Franch.) P. C. Tsoong ■

287544 Pedicularis przewalski Maxim. ;普氏马先蒿(青海马先蒿);Przewalsk Woodbetony ■

287545 Pedicularis przewalskii Maxim. subsp. australis(H. L. Li) P. C. Tsoong;南方普氏马先蒿(南方青海马先蒿)■

287546 Pedicularis przewalskii Maxim. subsp. hirsuta(H. L. Li) P. C. Tsoong;粗毛普氏马先蒿(粗毛青海马先蒿);Hirsute Przewalsk Woodbetony ■

287547 Pedicularis przewalskii Maxim. subsp. hirsuta H. L. Li = Pedicularis przewalskii Maxim. subsp. hirsuta(H. L. Li)P. C. Tsoong ■

287548 Pedicularis przewalskii Maxim. subsp. microphyton (Bureau et Franch.) P. C. Tsoong;矮小普氏马先蒿(矮小马先蒿,矮小青海马先蒿, 短小青海马先蒿, 青海马先蒿); Dwarf Przewalsk Woodbetony ■

287549 Pedicularis przewalskii Maxim. subsp. microphyton (Bureau et Franch.) P. C. Tsoong var. purpurea(Bonati)P. C. Tsoong;紫色普氏马先蒿■

287550 Pedicularis przewalskii Maxim. var. australis H. L. Li = Pedicularis przewalskii Maxim. subsp. australis (H. L. Li) P. C. Tsoong ■

287551 Pedicularis przewalskii Maxim. var. cristata (H. L. Li) P. C. Tsoong;有冠普氏马先蒿■

287552 Pedicularis przewalskii Maxim. var. hirsuta H. L. Li = Pedicularis przewalskii Maxim. subsp. hirsuta(H. L. Li)P. C. Tsoong ■

287553 Pedicularis przewalskii Maxim. var. microphyton (Bureau et Franch.) P. C. Tsoong = Pedicularis przewalskii Maxim. subsp. microphyton(Bureau et Franch.)P. C. Tsoong ■

287554 Pedicularis przewalskii Maxim. var. purpurea(Bonati) H. L. Li = Pedicularis przewalskii Maxim. subsp. microphyton (Bureau et Franch.)P. C. Tsoong var. purpurea(Bonati)P. C. Tsoong ■

287555 Pedicularis przewalskii Maxim. var. typica H. L. Li = Pedicularis przewalskii Maxim. ■

287556 Pedicularis pseudocephalantha Bonati;假头花马先蒿(假头状马先蒿);False Headflower Woodbetony ■

287557 Pedicularis pseudocurvituba P. C. Tsoong;假弯管马先蒿;False Curvetube Woodbetony ■

287558 Pedicularis pseudoingens Bonati;假硕大马先蒿;False Giant Woodbetony ■

287559 Pedicularis pseudokaroi Bonati = Pedicularis palustris L. subsp. karoi(Freyn) P. C. Tsoong ■

287560 Pedicularis pseudomelampyriflora Bonati;假山萝花马先蒿;False Cowwheat Woodbetony ■

287561 Pedicularis pseudomuscicola Bonati;假藓生马先蒿;False Muscicolous Woodbetony ■

287562 Pedicularis pseudosteiningeri Bonati;假司氏马先蒿(大萼马先蒿);False Steininger Woodbetony ■

287563 Pedicularis pseudostenocorys Bonati = Pedicularis stenocorys Franch. ■

287564 Pedicularis pseudoversicolor Hand. -Mazz. ;假多色马先蒿;False Versicolor Woodbetony ■

287565 Pedicularis pteridifolia Bonati;蕨叶马先蒿;Frond Woodbetony, Pteridophytaleaf Woodbetony ■

287566 Pedicularis pubescens(Bunge)Y. Y. Pai = Pedicularis sceptrum-carolinum L. subsp. pubescens(Bunge)P. C. Tsoong ■

287567 Pedicularis pubiflora Vved. ;中亚毛花马先蒿■☆

287568 Pedicularis pulchra Paulsen;美花马先蒿■☆

287569 Pedicularis pycnantha Boiss. ;中亚密花马先蒿■☆

287570 Pedicularis pygmaea Maxim. ;侏儒马先蒿;Dwarf Woodbetony, Pygmy Woodbetony ■

287571 Pedicularis pygmaea Maxim. subsp. deqinensis Hong Wang;德钦侏儒马先蒿;Deqin Dwarf Woodbetony ■

287572 Pedicularis quxiangensis H. P. Yang;曲乡马先蒿;Quxiang Woodbetony ■

287573 Pedicularis ramalana Britten = Pedicularis rhodotricha Maxim. ■

287574 Pedicularis ramosissima Bonati;多枝马先蒿;Ramified Woodbetony,Ramose Woodbetony ■

287575 Pedicularis recurva Maxim. ;反曲马先蒿;Recurve Woodbetony,Recurved Woodbetony ■

287576 Pedicularis recurva Maxim. = Pedicularis kangtingensis P. C. Tsoong ■

287577 Pedicularis refracta(Maxim.)Maxim. ;反折马先蒿■☆

287578 Pedicularis refracta(Maxim.)Maxim. f. albiflora T. Watan. ex T. Yamaz. ;白花反折马先蒿■☆

287579 Pedicularis refracta (Maxim.) Maxim. var. transmorrisonensis (Hayata)Hurus. = Pedicularis transmorrisonensis Hayata ■

287580 Pedicularis refracta Maxim. = Pedicularis ganpinensis Vaniot ex Bonati ■

287581 Pedicularis refracta Maxim. var. transmorrisonensis Hurus. = Pedicularis transmorrisonensis Hayata ■

287582 Pedicularis remotiloba Hand. -Mazz. ;疏裂马先蒿■

287583 Pedicularis reptans P. C. Tsoong;爬行马先蒿;Creeping Woodbetony,Reptant Woodbetony ■

287584 Pedicularis resupinata L.;返顾马先蒿(虎麻,烂石草,练石草,马尿泡,马矢蒿,马屎蒿,马屎烧,马先蒿,马新蒿);Resupinate Woodbetony ■

287585 Pedicularis resupinata L. f. ramosa Kom. = Pedicularis resupinata L. var. ramosa(Kom.)Nakai ■

287586 Pedicularis resupinata L. subsp. crassicaulis(Vaniot ex Bonati) P. C. Tsoong;粗茎返顾马先蒿(白沙药);Thickstem Resupinate Woodbetony ■

287587 Pedicularis resupinata L. subsp. galeobdolon (Diels) P. C. Tsoong;鼬臭返顾马先蒿■

287588 Pedicularis resupinata L. subsp. lasiophylla P. C. Tsoong;光叶返顾马先蒿;Hairleaf Resupinate Woodbetony ■

287589 Pedicularis resupinata L. subsp. oppositifolia(Miq.) T. Yamaz. ;对叶返顾马先蒿■☆

287590 Pedicularis resupinata L. subsp. oppositifolia (Miq.) T. Yamaz. f. albiflora(Honda)H. Hara;白花对叶返顾马先蒿■☆

287591 Pedicularis resupinata L. subsp. oppositifolia (Miq.) T. Yamaz. f. alborosea Hiyama;粉对叶返顾马先蒿■☆

287592 Pedicularis resupinata L. subsp. oppositifolia(Miq.) T. Yamaz. var. microphylla Honda;小对叶返顾马先蒿■☆

287593　Pedicularis resupinata L. subsp. oppositifolia(Miq.) Yamaz. var. mikawana?;三河国返顾马先蒿■☆

287594　Pedicularis resupinata L. subsp. teucriifolia(M. Bieb. ex Steven) T. Yamaz.;香科叶马先蒿■☆

287595　Pedicularis resupinata L. subsp. teucriifolia(M. Bieb. ex Steven) T. Yamaz. var. caespitosa Koidz. ;丛生返顾马先蒿■☆

287596　Pedicularis resupinata L. var. albiflora Y. Z. Zhao;白花返顾马先蒿; Whiteflower Resupinate Woodbetony ■

287597　Pedicularis resupinata L. var. crassicaulis (Vaniot ex Bonati) H. Limpr. = Pedicularis resupinata L. subsp. crassicaulis(Vaniot ex Bonati) P. C. Tsoong ■

287598　Pedicularis resupinata L. var. galeobdolon Limpr. = Pedicularis resupinata L. subsp. galeobdolon(Diels) P. C. Tsoong ■

287599　Pedicularis resupinata L. var. oppositifolia Miq. = Pedicularis resupinata L. subsp. oppositifolia(Miq.)T. Yamaz. ■☆

287600　Pedicularis resupinata L. var. pubescens(Kom.) Nakai;毛返顾马先蒿;Pubescent Resupinate Woodbetony ■

287601　Pedicularis resupinata L. var. ramosa(Kom.) Nakai;多枝返顾马先蒿■

287602　Pedicularis resupinata L. var. teucriifolia(M. Bieb. ex Steven) Maxim. = Pedicularis resupinata L. subsp. teucriifolia (M. Bieb. ex Steven) T. Yamaz. ■☆

287603　Pedicularis resupinata L. var. typica H. L. Li = Pedicularis resupinata L. ■

287604　Pedicularis retingensis P. C. Tsoong;雷 丁 马 先 蒿; Reting Woodbetony ■

287605　Pedicularis rex C. B. Clarke ex Maxim. ;大王马先蒿(凤尾参,蒿枝龙胆草,还阳参,还阳草,四方盒子草,土茵陈,五凤朝阳草,羊肝狼头草);King Woodbetony, Rex Woodbetony ■

287606　Pedicularis rex C. B. Clarke ex Maxim. subsp. lipskyana(Bonati) P. C. Tsoong;立氏大王马先蒿;Lipsky Rex Woodbetony ■

287607　Pedicularis rex C. B. Clarke ex Maxim. subsp. parva(Bonati) P. C. Tsoong;矮小大王马先蒿(矮小狼头草);Dwarf Rex Woodbetony ■

287608　Pedicularis rex C. B. Clarke ex Maxim. subsp. pseudocyathus (Vaniot ex Bonati) P. C. Tsoong;假斗大王马先蒿■

287609　Pedicularis rex C. B. Clarke ex Maxim. subsp. zayuensis H. P. Yang;察隅大王马先蒿;Chayu Rex Woodbetony ■

287610　Pedicularis rex C. B. Clarke ex Maxim. var. lopingensis Hand. -Mazz. = Pedicularis rex C. B. Clarke ex Maxim. ■

287611　Pedicularis rex C. B. Clarke ex Maxim. var. parva Bonati = Pedicularis rex C. B. Clarke ex Maxim. subsp. parva (Bonati) P. C. Tsoong ■

287612　Pedicularis rex C. B. Clarke ex Maxim. var. pseudocyathus Vaniot ex Bonati = Pedicularis rex C. B. Clarke ex Maxim. subsp. zayuensis H. P. Yang ■

287613　Pedicularis rex C. B. Clarke ex Maxim. var. pseudocyathus Vaniot ex Bonati = Pedicularis rex C. B. Clarke ex Maxim. subsp. pseudocyathus(Vaniot ex Bonati) P. C. Tsoong ■

287614　Pedicularis rex C. B. Clarke ex Maxim. var. purpurea Bonati = Pedicularis rex C. B. Clarke ex Maxim. subsp. lipskyana(Bonati) P. C. Tsoong ■

287615　Pedicularis rex C. B. Clarke ex Maxim. var. rockii(Bonati) H. L. Li;洛氏大王马先蒿;Rock Rex Woodbetony ■

287616　Pedicularis rex C. B. Clarke ex Maxim. var. thamnophila Hand. -Mazz. = Pedicularis thamnophila(Hand. -Mazz.) H. L. Li ■

287617　Pedicularis rhinanthoides Schrenk ex Fisch. ;拟鼻花马先蒿(锉叶马先蒿,大拟鼻花马先蒿,象鼻马先蒿);Large Rhinanthus-like Woodbetony ■

287618　Pedicularis rhinanthoides Schrenk ex Fisch. subsp. labellata (Jacq.) P. C. Tsoong = Pedicularis rhinanthoides Schrenk ex Fisch. subsp. labellata(Jacq.) Pennell ■

287619　Pedicularis rhinanthoides Schrenk ex Fisch. subsp. labellata (Jacq.) Pennell;大唇拟鼻花马先蒿(大唇马先蒿,大拟鼻花马先蒿);Biglip Large Rhinanthus-like Woodbetony ■

287620　Pedicularis rhinanthoides Schrenk ex Fisch. subsp. tibetica (Bonati) P. C. Tsoong;西藏拟鼻花马先蒿(长毛象鼻马先蒿);Xizang Large Rhinanthus-like Woodbetony ■

287621　Pedicularis rhinanthoides Schrenk ex Fisch. var. labellata (Jacq.) Prain = Pedicularis rhinanthoides Schrenk ex Fisch. subsp. labellata(Jacq.) P. C. Tsoong ■

287622　Pedicularis rhinanthoides Schrenk ex Fisch. var. labellata (Jacq.) Prain = Pedicularis rhinanthoides Schrenk ex Fisch. subsp. labellata(Jacq.) Pennell ■

287623　Pedicularis rhinanthoides Schrenk ex Fisch. var. labellata Prain = Pedicularis rhinanthoides Schrenk ex Fisch. subsp. labellata (Jacq.) P. C. Tsoong ■

287624　Pedicularis rhinanthoides Schrenk ex Fisch. var. tibetica Bonati = Pedicularis rhinanthoides Schrenk ex Fisch. subsp. tibetica (Bonati) P. C. Tsoong ■

287625　Pedicularis rhinanthoides Schrenk ex Fisch. var. typica Prain = Pedicularis rhinanthoides Schrenk ex Fisch. ■

287626　Pedicularis rhinanthoides var. tibetica Bonati = Pedicularis rhinanthoides Schrenk ex Fisch. subsp. tibetica(Bonati) P. C. Tsoong ■

287627　Pedicularis rhizomatosa P. C. Tsoong;根茎马先蒿;Rhizome Woodbetony ■

287628　Pedicularis rhodotricha Maxim. ; 红 毛 马 先 蒿; Redhair Woodbetony ■

287629　Pedicularis rhynchodonta Bureau et Franch. ;喙齿马先蒿;Beaktooth Woodbetony, Billtooth Woodbetony ■

287630　Pedicularis rhynchodonta Bureau et Franch. f. maxima Bonati = Pedicularis rhynchodonta Bureau et Franch. ■

287631　Pedicularis rhynchodonta Bureau et Franch. f. typica H. L. Li = Pedicularis rhynchodonta Bureau et Franch. ■

287632　Pedicularis rhynchotricha P. C. Tsoong;喙毛马先蒿;Beakhair Woodbetony, Billhair Woodbetony ■

287633　Pedicularis rigida Franch. ex Maxim. ; 坚挺马先蒿; Hard Woodbetony, Rigid Woodbetony ■

287634　Pedicularis rigidiformis Bonati;拟坚挺马先蒿;Hardshape Woodbetony, Rigidform Woodbetony ■

287635　Pedicularis rizhaoensis H. P. Yang; 日 照 马 先 蒿; Rizhao Woodbetony ■

287636　Pedicularis roborowskii Maxim. ;劳氏马先蒿(聚齿马先蒿);Roborowsk Woodbetony ■

287637　Pedicularis robusta Hook. f. ;壮健马先蒿;Robust Woodbetony, Strong Woodbetony ■

287638　Pedicularis rockii Bonati = Pedicularis rex C. B. Clarke ex Maxim. var. rockii(Bonati) H. L. Li ■

287639　Pedicularis rotundifolia C. E. C. Fisch. ;圆叶马先蒿(团叶马先蒿);Roundleaf Woodbetony ■

287640　Pedicularis roylei Maxim. ;罗氏马先蒿(草甸马先蒿,肉根马先蒿);Royle Woodbetony ■

287641　Pedicularis roylei Maxim. subsp. megalantha P. C. Tsoong;大花罗氏马先蒿(大花马先蒿);Bigflower Royle Woodbetony ■

287642　Pedicularis roylei Maxim. subsp. shawii (P. C. Tsoong) P. C.

Tsoong;萧氏马先蒿(萧氏罗氏马先蒿);Shaw's Royle Woodbetony ■

287643　Pedicularis roylei Maxim. var. cinerascens Marquand =
Pedicularis roylei Maxim. subsp. shawii(P. C. Tsoong)P. C. Tsoong ■

287644　Pedicularis roylei Maxim. var. cinerascens Marquand et Shaw ex
Marquand = Pedicularis roylei Maxim. subsp. shawii(P. C. Tsoong)
P. C. Tsoong ■

287645　Pedicularis rubens Stephan ex Willd.;红色马先蒿(山马先
蒿);Red Woodbetony ■

287646　Pedicularis rubens Stephan ex Willd. var. japonica? =
Pedicularis apodochila Maxim. ■☆

287647　Pedicularis rubinskii Kom.;卢比马先蒿■☆

287648　Pedicularis rudis Maxim.;粗野马先蒿;Insertedstyle
Woodbetony ■

287649　Pedicularis ruoergaiensis H. P. Yang;若尔盖马先蒿;Ruoergai
Woodbetony ■

287650　Pedicularis rupicola Franch.;岩居马先蒿;Rockliving
Woodbetony ■

287651　Pedicularis rupicola Franch. subsp. zambalensis(Bonati)P. C.
Tsoong;川西岩居马先蒿;W. Sichuan Rockliving Woodbetony ■

287652　Pedicularis rupicola Franch. var. zambalensis Bonati = Pedicularis
rupicola Franch. subsp. zambalensis(Bonati)P. C. Tsoong ■

287653　Pedicularis sabaensis Bonati = Pedicularis maxonii Bonati ■

287654　Pedicularis salicifolia Bonati;柳叶马先蒿;Willowleaf
Woodbetony ■

287655　Pedicularis salviaeflora Franch. ex Hemsl.;丹参花马先蒿;
Sageflower Woodbetony ■

287656　Pedicularis salviaeflora Franch. ex Hemsl. var. leiocarpa H. P.
Yang;滑果马先蒿(滑果丹参马先蒿);Smoothfruit Sageflower
Woodbetony ■

287657　Pedicularis sceptrum-carolinum L.;旌节马先蒿(黄芪马先蒿,黄旗
马先蒿);Charles' Sceptre, Moor King, Rood-mark Woodbetony, Sceptre
Woodbetony ■

287658　Pedicularis sceptrum-carolinum L. f. pubescens(Bunge)Kitag.
= Pedicularis sceptrum-carolinum L. subsp. pubescens(Bunge)P. C.
Tsoong ■

287659　Pedicularis sceptrum-carolinum L. subsp. pubescens(Bunge)P.
C. Tsoong;有毛旌节马先蒿(毛旌节马先蒿)■

287660　Pedicularis sceptrum-carolinum L. subsp. pubescens P. C.
Tsoong = Pedicularis sceptrum-carolinum L. subsp. pubescens
(Bunge)P. C. Tsoong ■

287661　Pedicularis sceptrum-carolinum L. var. glabra Bunge =
Pedicularis sceptrum-carolinum L. ■

287662　Pedicularis sceptrum-carolinum L. var. pubescens Bunge =
Pedicularis sceptrum-carolinum L. subsp. pubescens(Bunge)P. C.
Tsoong ■

287663　Pedicularis sceptrum-carolinum L. var. typica H. L. Li =
Pedicularis sceptrum-carolinum L. ■

287664　Pedicularis schistostegia Vved. f. rubriflora Ohba ex T. Yamaz.;
红花马先蒿■☆

287665　Pedicularis schizorhyncha Prain;裂喙马先蒿;Lobedbeak
Woodbetony ■

287666　Pedicularis schugnana B. Fedtsch.;舒格南马先蒿■☆

287667　Pedicularis scolopax Maxim.;鹬形马先蒿;Kingfisher
Woodbetony, Kingfishershape Woodbetony ■

287668　Pedicularis semenovii Regel;赛氏马先蒿;Semenov Woodbetony ■

287669　Pedicularis semitorta Maxim.;半扭卷马先蒿;Halftwisting
Woodbetony ■

287670　Pedicularis shansiensis P. C. Tsoong;山西马先蒿;Shansi
Woodbetony,Shanxi Woodbetony ■

287671　Pedicularis shawii P. C. Tsoong = Pedicularis roylei Maxim.
subsp. shawii(P. C. Tsoong)P. C. Tsoong ■

287672　Pedicularis sherriffii P. C. Tsoong;昌都马先蒿(休氏马先蒿);
Sherriff Woodbetony ■

287673　Pedicularis sibirica Vved.;西伯利亚马先蒿■☆

287674　Pedicularis sibthorpii Boiss.;西伯马先蒿■☆

287675　Pedicularis sigmoidea Franch.;之喙马先蒿(弯嘴马先蒿,之
形喙马先蒿);Sigmoid Woodbetony ■

287676　Pedicularis sikangensis H. L. Li = Pedicularis verticillata L. ■

287677　Pedicularis sima Maxim.;矽镁马先蒿;Sima Woodbetony ■

287678　Pedicularis siphonantha D. Don;管花马先蒿;Tubeflower
Woodbetony ■

287679　Pedicularis siphonantha D. Don var. delavayi(Franch. ex
Maxim.)P. C. Tsoong;台氏管花马先蒿;Delavay ■

287680　Pedicularis siphonantha D. Don var. delavayi(Franch.)P. C.
Tsoong = Pedicularis siphonantha D. Don var. delavayi(Franch. ex
Maxim.)P. C. Tsoong ■

287681　Pedicularis siphonantha D. Don var. typica Prain = Pedicularis
siphonantha D. Don ■

287682　Pedicularis smithiana Bonati;史氏马先蒿(钩嘴马先蒿);
Smith Woodbetony ■

287683　Pedicularis sobulatidens P. C. Tsoong;斜齿马先蒿;
Obliquetooth Woodbetony ■

287684　Pedicularis soongarica Schrenk;准噶尔马先蒿;Dzungar
Woodbetony,Songar Woodbetony ■

287685　Pedicularis sorbifolia P. C. Tsoong;花揪叶马先蒿;Mountainash
Woodbetony, Mountainashleaf Woodbetony ■

287686　Pedicularis souliei Franch.;苏氏马先蒿;Soulie Woodbetony ■

287687　Pedicularis sparsissima P. C. Tsoong = Pedicularis lineata
Franch. ■

287688　Pedicularis sphaerantha P. C. Tsoong;团花马先蒿;
Sphaeroidalflower Woodbetony,Sphaeroidflower Woodbetony ■

287689　Pedicularis spicata Pall.;穗花马先蒿;Spicate Woodbetony ■

287690　Pedicularis spicata Pall. subsp. bracteata P. C. Tsoong;显苞穗
花马先蒿;Bracteate Spicate Woodbetony ■

287691　Pedicularis spicata Pall. subsp. stenocarpa P. C. Tsoong;狭果穗
花马先蒿;Narrowfruit Spicate Woodbetony ■

287692　Pedicularis spicata Pall. var. australis Bonati = Pedicularis
holocalyx Hand. -Mazz. ■

287693　Pedicularis spicata Pall. var. sesinowii Bonati = Pedicularis
spicata Pall. ■

287694　Pedicularis stadlmanniana Bonati;施氏马先蒿(洱海马先蒿);
Sdadlman Woodbetony ■

287695　Pedicularis stapfii Bonati = Pedicularis labordei Vaniot ex Bonati ■

287696　Pedicularis steiningeri Bonati;司氏马先蒿;Steininger
Woodbetony ■

287697　Pedicularis stenantha Franch. = Pedicularis oederi Vahl var.
angustiflora(Limpr.)P. C. Tsoong ■

287698　Pedicularis stenocorys Franch.;狭盔马先蒿;Narrowgalea
Woodbetony ■

287699　Pedicularis stenocorys Franch. subsp. melanotricha P. C. Tsoong;
黑毛狭盔马先蒿;Blackhair Narrowgalea Woodbetony ■

287700　Pedicularis stenocorys Franch. var. pseudostenocorys(Bonati)P.
C. Tsoong = Pedicularis stenocorys Franch. ■

287701　Pedicularis stenotheca P. C. Tsoong;狭室马先蒿;Narrowcase

Woodbetony, Narrowcelony Woodbetony ■

287702   Pedicularis stevenii Bunge = Pedicularis verticillata L. ■

287703   Pedicularis stewardii H. L. Li;斯氏马先蒿（司氏马先蒿）；Steward Woodbetony ■

287704   Pedicularis streptorhyncha P. C. Tsoong;扭喙马先蒿；Twistedbeak Woodbetony, Twistybill Woodbetony ■

287705   Pedicularis striata Pall.;红纹马先蒿（细叶马先蒿，野芝麻）；Redstriate Woodbetony ■

287706   Pedicularis striata Pall. subsp. arachnoidea（Franch.）P. C. Tsoong;蛛丝红纹马先蒿■

287707   Pedicularis striata Pall. var. arachnoidea Franch. = Pedicularis striata Pall. subsp. arachnoidea（Franch.）P. C. Tsoong ■

287708   Pedicularis striata Pall. var. poliocalyx Diels = Pedicularis striata Pall. subsp. arachnoidea（Franch.）P. C. Tsoong ■

287709   Pedicularis striata Pall. var. typica H. L. Li = Pedicularis striata Pall. ■

287710   Pedicularis stricta Wall. = Pedicularis gracilis Wall. subsp. stricta P. C. Tsoong ■

287711   Pedicularis strobilacea Franch. ex Forbes et Hemsl. var. riparia Bonati = Pedicularis pseudocephalantha Bonati ■

287712   Pedicularis strobilacea Franch. ex Hemsl.;球状马先蒿（球序马先蒿）；Globular Woodbetony ■

287713   Pedicularis strobilacea Franch. ex Hemsl. var. riparia Bonati = Pedicularis pseudocephalantha Bonati ■

287714   Pedicularis stylosa H. P. Yang;长柱马先蒿；Longstyle Woodbetony ■

287715   Pedicularis subacaulis Bonati = Pedicularis confertiflora Prain subsp. parvifolia（Hand. -Mazz.）P. C. Tsoong ■

287716   Pedicularis subrostrata C. A. Mey.;喙状马先蒿■☆

287717   Pedicularis subulatidens P. C. Tsoong;针齿马先蒿；Needletooth Woodbetony, Subulatetooth Woodbetony ■

287718   Pedicularis sudetica Willd.;苏底山马先蒿;Sudetic Pedieularis ■☆

287719   Pedicularis superba Franch. ex Maxim.;华丽马先蒿（莲座参）；Magnificent Woodbetony ■

287720   Pedicularis svenhedinii Paulsen = Pedicularis cheilanthifolia Schrenk subsp. svenhedinii（Paulsen）P. C. Tsoong ■

287721   Pedicularis sylvatica L.;林地马先蒿；Cock's Comb, Dwarf Red Rattle, Forest Woodbetony, Honeysookies, Honeysuckle, Lousewort, Painted Cup, Peppermint, Rattle-basket, Rattlepod, Shackle-box, Shaeklle Box ■☆

287722   Pedicularis sylvatica L. subsp. lusitanica（Hoffmanns. et Link）Cout.;葡萄牙马先蒿■☆

287723   Pedicularis szetschuanica Maxim.;四川马先蒿；Sichuan Woodbetony, Szechuan Woodbetony ■

287724   Pedicularis szetschuanica Maxim. = Pedicularis lineata Franch. ■

287725   Pedicularis szetschuanica Maxim. subsp. anastomosans P. C. Tsoong;网脉四川马先蒿；Reticulate Sichuan Woodbetony ■

287726   Pedicularis szetschuanica Maxim. subsp. angustifolia（Bonati）P. C. Tsoong = Pedicularis szetschuanica Maxim. subsp. anastomosans P. C. Tsoong ■

287727   Pedicularis szetschuanica Maxim. subsp. latifolia P. C. Tsoong;宽叶四川马先蒿；Broadleaf Sichuan Woodbetony ■

287728   Pedicularis szetschuanica Maxim. subsp. ovatifolia P. C. Tsoong = Pedicularis szetschuanica Maxim. subsp. latifolia P. C. Tsoong ■

287729   Pedicularis szetschuanica Maxim. subsp. typica var. angulata P. C. Tsoong = Pedicularis angularis P. C. Tsoong ■

287730   Pedicularis szetschuanica Maxim. subsp. typica var. dentigera P.

C. Tsoong = Pedicularis metaszetschuanica P. C. Tsoong ■

287731   Pedicularis szetschuanica Maxim. var. angustifolia Bonati = Pedicularis szetschuanica Maxim. ■

287732   Pedicularis szetschuanica Maxim. var. elata Bonati = Pedicularis holocalyx Hand. -Mazz. ■

287733   Pedicularis szetschuanica Maxim. var. longispica Bonati = Pedicularis verticillata L. ■

287734   Pedicularis szetschuanica Maxim. var. longispica Bonati = Pedicularis kansuensis Maxim. ■

287735   Pedicularis szetschuanica Maxim. var. longispicata Bonati ex H. Limpr. = Pedicularis kansuensis Maxim. ■

287736   Pedicularis szetschuanica Maxim. var. ovatifolia H. L. Li = Pedicularis triangubridens P. C. Tsoong ■

287737   Pedicularis tachanensis Bonati;大山马先蒿；Dashan Woodbetony, Tashan Woodbetony ■

287738   Pedicularis tahaiensis Bonati;大海马先蒿;Dahai Woodbetony, Tahai Woodbetony ■

287739   Pedicularis takpoensis P. C. Tsoong;塔布马先蒿（红茎马先蒿）；Redstem Woodbetony, Tabu Woodbetony ■

287740   Pedicularis talassica Vved.;塔拉斯马先蒿■☆

287741   Pedicularis taliensis Bonati;大理马先蒿；Dali Woodbetony, Tali Woodbetony ■

287742   Pedicularis tangutica Bonati = Pedicularis verticillata L. subsp. tangutica（Bonati）P. C. Tsoong ■

287743   Pedicularis tantalorhyncha Franch. ex Bonati;颤喙马先蒿；Tatntalobeak Woodbetony, Tremblingbill Woodbetony ■

287744   Pedicularis tapaoensis P. C. Tsoong;大炮马先蒿；Dapao Woodbetony, Tapao Woodbetony ■

287745   Pedicularis tatarinowii Maxim.;塔氏马先蒿（华北马先蒿）；Tatarinow Woodbetony ■

287746   Pedicularis tatsiensis A. I. Baranov et Franch.;打箭马先蒿；Kangding Woodbetony, Tatsien Woodbetony ■

287747   Pedicularis tayloriana P. C. Tsoong;泰氏马先蒿；Taylor Woodbetony ■

287748   Pedicularis taylorii P. C. Tsoong = Pedicularis tayloriana P. C. Tsoong ■

287749   Pedicularis temotiloba Hand. -Mazz.;寡裂马先蒿（疏裂马先蒿）；Looselobe Woodbetony ■

287750   Pedicularis tenacifolia P. C. Tsoong;宿叶马先蒿；Persistentleaf Woodbetony, Tenacular Woodbetony ■

287751   Pedicularis tenera H. L. Li;细茎马先蒿；Slenderstem Woodbetony, Thinstem Woodbetony ■

287752   Pedicularis tenuicalyx P. C. Tsoong = Pedicularis violascens Schrenk ■

287753   Pedicularis tenuicaulis Prain;纤茎马先蒿；Finestem Woodbetony, Tenuousstem Woodbetony ■

287754   Pedicularis tenuisecta Franch. ex Maxim.;纤裂马先蒿（凤尾参，细裂马先蒿）；Finesplit Woodbetony, Tenuidivided Woodbetony ■

287755   Pedicularis tenuisecta Franch. ex Maxim. f. albiflora Bonati;白花纤裂马先蒿；Whiteflower Tenuidivided Woodbetony ■

287756   Pedicularis tenuituba Pennell et H. L. Li;狭管马先蒿（窄管马先蒿）；Narrowtube Woodbetony ■

287757   Pedicularis ternata Maxim.;三叶马先蒿；Threeleaf Woodbetony ■

287758   Pedicularis thamnophila（Hand. -Mazz.）H. L. Li;灌丛马先蒿；Shrubby Woodbetony, Thicked Woodbetony ■

287759   Pedicularis thamnophila（Hand. -Mazz.）H. L. Li subsp. cupuliformis（H. L. Li）P. C. Tsoong;杯状灌丛马先蒿；Cup Thicked

Woodbetony ■

287760 Pedicularis tianschanica Rupr. ;天山马先蒿■

287761 Pedicularis tibetica Franch. ；西藏马先蒿；Tibet Woodbetony,
Xizang Woodbetony ■

287762 Pedicularis tomentosa H. L. Li；绒毛马先蒿；Tomentose
Woodbetony ■

287763 Pedicularis tongolensis Franch. ；东俄洛马先蒿；Tongol
Woodbetony ■

287764 Pedicularis tongtchouanensis Bonati = Pedicularis nigra(Bonati)
Vaniot ex Bonati ■

287765 Pedicularis torta Maxim. ；扭旋马先蒿；Torsional Woodbetony,
Twisted Woodbetony ■

287766 Pedicularis transmorrisonensis Hayata；台湾马先蒿；Taiwan
Woodbetony ■

287767 Pedicularis triangubridens P. C. Tsoong；三角齿马先蒿；
Triangularteeth Woodbetony,Trianguletooth Woodbetony ■

287768 Pedicularis triangubridens P. C. Tsoong subsp. chrysosplenioides
P. C. Tsoong；猫眼草三角齿马先蒿■

287769 Pedicularis triangubridens P. C. Tsoong var. angustiloba P. C.
Tsoong；狭裂三角齿马先蒿■

287770 Pedicularis tribuloides Bonati = Pedicularis macrosiphon Franch. ■

287771 Pedicularis trichocymba H. L. Li；毛舟马先蒿；Hairboad
Woodbetony,Hairycyme Woodbetony ■

287772 Pedicularis trichoglossa Hook. f. ；毛盔马先蒿；Hairygalea
Woodbetony ■

287773 Pedicularis trichomata H. L. Li；须毛马先蒿；Beard
Woodbetony,Hairy Woodbetony ■

287774 Pedicularis tricolor Hand. -Mazz. ；三色马先蒿；Tricolor
Woodbetony ■

287775 Pedicularis tricolor Hand. -Mazz. var. aequiretusa P. C. Tsoong；
等凹三色马先蒿■

287776 Pedicularis trigonophylla Hand. -Mazz. = Pedicularis maxonii
Bonati ■

287777 Pedicularis trimaculata X. F. Lu；三斑刺齿马先蒿■

287778 Pedicularis tristiformis Bonati = Pedicularis dolichocymba
Hand. -Mazz. ■

287779 Pedicularis tristis L. ；阴郁马先蒿；Cloudy Woodbetony,
Dullcolour Woodbetony ■

287780 Pedicularis tristis L. var. macrantha Maxim. = Pedicularis paiana
H. L. Li

287781 Pedicularis truchetii Bonati = Pedicularis lutescens Franch. ■

287782 Pedicularis tsaii H. L. Li；蔡氏马先蒿（碧罗马先蒿）；Tsai
Woodbetony ■

287783 Pedicularis tsangchanensis Franch. ；苍山马先蒿；Cangshan
Woodbetony,Tsangshan Woodbetony ■

287784 Pedicularis tsarungensis H. L. Li；察瓦龙马先蒿（察郎马先蒿）；
Chalang Woodbetony,Chawalong Woodbetony,Tsarung Woodbetony ■

287785 Pedicularis tsekouensis Bonati；茨口马先蒿；Cikou
Woodbetony,Tsekou Woodbetony ■

287786 Pedicularis tsiangii H. L. Li；蒋氏马先蒿；Jiang Woodbetony,
Tsiang Woodbetony ■

287787 Pedicularis tubiformis Klotzsch = Pedicularis longiflora Rudolph
var. tubiformis(G. Klotz)P. C. Tsoong ■

287788 Pedicularis uliginosa Bunge；水泽马先蒿；Marshy Woodbetony,
Wetland Woodbetony ■

287789 Pedicularis umbelliformis H. L. Li；伞花马先蒿；Umbellike
Woodbetony ■

287790 Pedicularis uralensia Vved. ；乌拉尔马先蒿■☆

287791 Pedicularis urceolata P. C. Tsoong；坛萼马先蒿；Jarcalyx
Woodbetony,Urceolate Woodbetony ■

287792 Pedicularis vagans Hemsl. ；蔓生马先蒿；Creeping
Woodbetony,Prostrate Woodbetony ■

287793 Pedicularis variegata H. L. Li；变色马先蒿；Colory
Woodbetony,Variegated Woodbetony ■

287794 Pedicularis venusta(Bunge)Schangin ex Bunge；秀丽马先蒿
（黑水马先蒿，美丽马先蒿）；Beautiful Woodbetony,Showy
Woodbetony ■

287795 Pedicularis verae Vved. ；维拉马先蒿■

287796 Pedicularis verbenifolia Franch. ；马鞭草叶马先蒿；
Verbenifolius Woodbetony ■

287797 Pedicularis versicolor Wahlenb. = Pedicularis oederi Vahl ■

287798 Pedicularis versicolor Wahlenb. var. europaea Maxim. =
Pedicularis oederi Vahl ■

287799 Pedicularis versicolor Wahlenb. var. rubra Maxim. = Pedicularis
oederi Vahl var. rubra(Maxim.)P. C. Tsoong ■

287800 Pedicularis versicolor Wahlenb. var. sinensis Maxim. =
Pedicularis oederi Vahl var. sinensis(Maxim.)Hurus. ■

287801 Pedicularis versicolor Wahlenb. var. yezoana？ = Pedicularis
oederi Vahl var. heteroglossa Prain ■

287802 Pedicularis verticillata L. ；轮叶马先蒿（轮生马先蒿，马先蒿，
玉山蒿草）；Whorled Pedicularis,Whorledleaf Woodbetony ■

287803 Pedicularis verticillata L. = Pedicularis anthemifolia Fisch. ■

287804 Pedicularis verticillata L. = Pedicularis szetschuanica Maxim. ■

287805 Pedicularis verticillata L. f. albiflora Makino；白花轮叶马先蒿■☆

287806 Pedicularis verticillata L. subsp. latisecta(Hultén)P. C. Tsoong；
宽裂轮叶马先蒿；Broadlobed Whorledleaf Woodbetony ■

287807 Pedicularis verticillata L. subsp. tangutica(Bonati)P. C. Tsoong；
唐古特轮叶马先蒿；Tangut Whorledleaf Woodbetony ■

287808 Pedicularis verticillata L. var. chinensis Maxim. = Pedicularis
kansuensis Maxim. ■

287809 Pedicularis verticillata L. var. latisecta Hultén = Pedicularis
verticillata L. subsp. latisecta(Hultén)P. C. Tsoong ■

287810 Pedicularis verticillata L. var. refracta？ = Pedicularis refracta
(Maxim.)Maxim. ■☆

287811 Pedicularis verticillata L. var. typica L. = Pedicularis verticillata L. ■

287812 Pedicularis vetonicifolia Franch. ；地黄叶马先蒿；Speedwellleaf
Woodbetony ■

287813 Pedicularis vialii Franch. ex Hemsl. ；维氏马先蒿（举喙马先
蒿,象头马先蒿）；Vial Woodbetony ■

287814 Pedicularis villosa Ledeb. ；长柔毛马先蒿■☆

287815 Pedicularis villosula Franch. ex Forbes et Hemsl. = Pedicularis
confertiflora Prain ■

287816 Pedicularis villosula Franch. ex Forbes et Hemsl. var. parvifolia
Hand. -Mazz. = Pedicularis confertiflora Prain subsp. parvifolia
(Hand. -Mazz.)P. C. Tsoong ■

287817 Pedicularis villosula Franch. ex Forbes et Hemsl. var. typica H.
L. Li = Pedicularis confertiflora Prain ■

287818 Pedicularis villosula Franch. var. parvifolia (Hand. -Mazz.)
Hand. -Mazz. = Pedicularis confertiflora Prain subsp. parvifolia
(Hand. -Mazz.)P. C. Tsoong ■

287819 Pedicularis violascens Schrenk；堇色马先蒿；Violet Woodbetony ■

287820 Pedicularis virginica Poir. = Pedicularis lanceolata Michx. ■☆

287821 Pedicularis vlassoviana Stev. ；乌拉索夫马先蒿■☆

287822 Pedicularis waldheimii Bonati；瓦尔马先蒿■☆

287823   Pedicularis wallichii Bunge;沃利克马先蒿（瓦氏马先蒿）；Wallich Woodbetony ■

287824   Pedicularis wallichioides Yamaz. = Pedicularis wallichii Bunge ■

287825   Pedicularis wangii H. L. Li = Pedicularis duclouxii Bonati ■

287826   Pedicularis wardii Bonati;华氏马先蒿（镰盔马先蒿）；Ward Woodbetony ■

287827   Pedicularis weixiensis H. P. Yang;维西马先蒿；Weixi Woodbetony ■

287828   Pedicularis wilhelmsiana Fisch. ex M. Bieb.;威廉马先蒿■☆

287829   Pedicularis wilidenovii Vved.;威利马先蒿■☆

287830   Pedicularis wilsonii Bonati;魏氏马先蒿;Wilson Woodbetony ■

287831   Pedicularis xanthantha H. L. Li = Pedicularis anas Maxim. var. xanthantha(H. L. Li) P. C. Tsoong ■

287832   Pedicularis xiangchengensis H. P. Yang;乡城马先蒿;Xiangcheng Woodbetony ■

287833   Pedicularis xiqingshanensis H. Y. Feng et J. Zh. Sun;西倾山马先蒿;Xiqingshan Woodbetony ■

287834   Pedicularis yanyuanensis H. P. Yang;盐源马先蒿;Yanyuan Woodbetony ■

287835   Pedicularis yargongensis Bonati = Pedicularis kansuensis Maxim. subsp. yargongensis(Bonati) P. C. Tsoong ■

287836   Pedicularis yargongensis Bonati var. longibracteata Bonati = Pedicularis kansuensis Maxim. subsp. yargongensis (Bonati) P. C. Tsoong ■

287837   Pedicularis yezoensis Maxim.;北海道马先蒿■☆

287838   Pedicularis yezoensis Maxim. var. pubescens H. Hara;毛北海道马先蒿■☆

287839   Pedicularis yui H. L. Li;季川马先蒿（贡山马先蒿）；Yu Woodbetony ■

287840   Pedicularis yui H. L. Li var. ciliata P. C. Tsoong;缘毛季川马先蒿（缘毛贡山马先蒿）；Ciliate Yu Woodbetony ■

287841   Pedicularis yunnanensis Franch. ex Maxim.;云南马先蒿；Yunnan Woodbetony ■

287842   Pedicularis zayuensis H. P. Yang;察隅马先蒿；Chayu Woodbetony ■

287843   Pedicularis zeravschanica Regel;泽拉夫尚马先蒿■☆

287844   Pedicularis zhongdianensis H. P. Yang;中甸马先蒿;Zhongdian Woodbetony ■

287845   Pedilanthus Neck. = Pedilanthus Neck. ex Poit. (保留属名)●

287846   Pedilanthus Neck. ex Poit. (1812)(保留属名)；红雀珊瑚属（白雀珊瑚属,银龙属,银雀珊瑚属)；Pedilanthus, Red Bird, Slipperplant ●

287847   Pedilanthus Poit. = Pedilanthus Neck. ex Poit. (保留属名)●

287848   Pedilanthus grandifolius?;大叶红雀珊瑚■☆

287849   Pedilanthus lycioides Baker = Euphorbia pedilanthoides Denis ●☆

287850   Pedilanthus macrocarpus Benth.;大果红雀珊瑚; Gallito, Largefruit Pedilanthus, Slipper Plant ●☆

287851   Pedilanthus pavonis Boiss.;墨西哥红雀珊瑚●☆

287852   Pedilanthus smallii Millsp. = Pedilanthus tithymaloides(L.) A. Poit. subsp. smallii(Millsp.) Dressler ●

287853   Pedilanthus tithymaloides(L.) A. Poit.;红雀珊瑚（百足草,大银龙,狗药,红雀掌,见肿消,扭曲草,青竹标,珊瑚枝,拖鞋花,洋珊瑚,玉带根,止血草)；Bird Cactus, Common Pedilanthus, Devil's Backbone, Devil's-backbone, Japanese Poinsettia, Red Bird, Redbird Flower, Ribbon Cactus, Slipper Flower ●

287854   Pedilanthus tithymaloides(L.) A. Poit. ' Nanus';密叶红雀珊瑚（对生龙凤木,怪银龙)；Dwarf Pedilanthus ●☆

287855   Pedilanthus tithymaloides(L.) A. Poit. ' Variegatus';花叶红雀珊瑚（斑叶红雀珊瑚,变叶银龙)；Redbird Flower, Ribbon Cactus, Slipper Flower, Variegated Pedilanthus ●

287856   Pedilanthus tithymaloides(L.) A. Poit. subsp. smallii(Millsp.) Dressler;卷叶珊瑚●

287857   Pedilanthus tithymaloides (L.) A. Poit. var. variegatus Hort. = Pedilanthus tithymaloides(L.) A. Poit. ' Variegatus'●

287858   Pedilea Lindl. = Dienia Lindl. ■

287859   Pedilea Lindl. = Malaxis Sol. ex Sw. ■

287860   Pedilochilus Schltr. (1905);足唇兰属■☆

287861   Pedilochilus brachiatus Schltr.;足唇兰■☆

287862   Pedilochilus flavus Schltr.;黄足唇兰■☆

287863   Pedilochilus grandifolius P. Royen;大叶足唇兰■☆

287864   Pedilochilus major J. J. Sm.;大足唇兰■☆

287865   Pedilochilus montanus Ridl.;山地足唇兰■☆

287866   Pedilochilus obovatus J. J. Sm.;倒卵足唇兰■☆

287867   Pedilochilus sulphureus J. J. Sm.;硫色足唇兰■☆

287868   Pedilonia C. Presl = Wachendorfia Burm. ■☆

287869   Pedilonum Blume = Dendrobium Sw. (保留属名)■

287870   Pedilonum goldschmidtianum (Kraenzl.) Rauschert = Dendrobium goldschmidtianum Kraenzl. ■

287871   Pedilonum longicalcaratum (Hayata) Rauschert = Dendrobium chameleon Ames ■

287872   Pedilonum miyakei (Schltr.) Rauschert = Dendrobium goldschmidtianum Kraenzl. ■

287873   Pedina Steven = Astragalus L. ●■

287874   Pedinogyne Brand = Trigonotis Steven ■

287875   Pedinogyne Brand(1925);藏紫草属■

287876   Pedinogyne tibetica (C. B. Clarke) Brand = Trigonotis tibetica (C. B. Clarke) I. M. Johnst. ■

287877   Pedinogyne tibetica Brand = Trigonotis tibetica(C. B. Clarke)I. M. Johnst. ■

287878   Pedinopetalum Urb. et H. Wolff(1929);平瓣芹属■☆

287879   Pedinopetalum domingense Urb. et H. Wolff;平瓣芹■☆

287880   Pediocactus Britton et Rose ex Britton et Brown = Pediocactus Britton et Rose ●☆

287881   Pediocactus Britton et Rose (1913);月华玉属（月华球属)；Foot Cactus, Hedgehog Cactus, Plains Cactus ●☆

287882   Pediocactus bradyi L. D. Benson;布氏月华玉；Brady's Hedgehog Cactus, Marble Canyon Cactus ●☆

287883   Pediocactus bradyi L. D. Benson subsp. despainii(S. L. Welsh et Goodrich) Hochstätter = Pediocactus despainii S. L. Welsh et Goodrich ●☆

287884   Pediocactus bradyi L. D. Benson subsp. winkleri (K. D. Heil) Hochstätter = Pediocactus winkleri K. D. Heil ●☆

287885   Pediocactus bradyi L. D. Benson subsp. winkleri;温氏月华玉；Winkler's Cactus ●☆

287886   Pediocactus bradyi L. D. Benson var. despainii(S. L. Welsh et Goodrich) Hochstätter = Pediocactus despainii S. L. Welsh et Goodrich ●☆

287887   Pediocactus bradyi L. D. Benson var. knowltonii(L. D. Benson) Backeb. = Pediocactus knowltonii L. D. Benson ●☆

287888   Pediocactus bradyi L. D. Benson var. winkleri (K. D. Heil) Hochstätter = Pediocactus winkleri K. D. Heil ●☆

287889   Pediocactus cloverae (K. D. Heil et J. M. Porter) Halda = Sclerocactus cloverae K. D. Heil et J. M. Porter ●☆

287890   Pediocactus cloverae(K. D. Heil et J. M. Porter) Halda subsp.

brackii K. D. Heil et J. M. Porter = Sclerocactus cloverae K. D. Heil et J. M. Porter ●☆

287891 Pediocactus despainii S. L. Welsh et Goodrich；德氏月华玉；Despain Footcactus，Despain's Pincushion Cactus，San Rafael Cactus ●☆

287892 Pediocactus glaucus（K. Schum.）Arp = Sclerocactus glaucus（K. Schum.）L. D. Benson ●☆

287893 Pediocactus knowltonii L. D. Benson；银河玉；Knowlton's Cactus，Knowlton's Minute Cactus ●☆

287894 Pediocactus mesae-verdae（Boissev. et C. Davidson）Arp = Sclerocactus mesae-verdae（Boissev. et C. Davidson）L. D. Benson ●☆

287895 Pediocactus nigrispinus（Hochstätter）Hochstätter；雪球玉；Snowball Cactus ●☆

287896 Pediocactus nigrispinus（Hochstätter）Hochstätter subsp. beastonii（Hochstätter）Hochstätter = Pediocactus nigrispinus（Hochstätter）Hochstätter ●☆

287897 Pediocactus nigrispinus（Hochstätter）Hochstätter subsp. puebloensis Hochstätter = Pediocactus nigrispinus（Hochstätter）Hochstätter ●☆

287898 Pediocactus nigrispinus（Hochstätter）Hochstätter var. beastonii Hochstätter = Pediocactus nigrispinus（Hochstätter）Hochstätter ●☆

287899 Pediocactus nyensis（Hochstätter）Halda = Sclerocactus nyensis Hochstatter ●☆

287900 Pediocactus papyracanthus（Engelm.）L. D. Benson = Sclerocactus papyracanthus（Engelm.）N. P. Taylor ●☆

287901 Pediocactus paradinei（B. W. Benson）Halda = Pediocactus paradinei B. W. Benson ●☆

287902 Pediocactus paradinei B. W. Benson；雏鹭球（帕氏月华玉）；Houserock Cactus，Kaibab Pincushion Cactus，Park Hedgehog Cactus ●☆

287903 Pediocactus parviflorus（Clover et Jotter）Halda = Sclerocactus parviflorus Clover et Jotter ●☆

287904 Pediocactus peeblesianus（Croizat）L. D. Benson；飞鸟球；Navajo Pincushion Cactus，Peeble's Navajo Cactus ●☆

287905 Pediocactus peeblesianus（Croizat）L. D. Benson subsp. fickeiseniae（Backeb. ex Hochstätter）Lüthy；斑鸠球；Fickeisen Pincushion Cactus，Fickeisen's Navajo Cactus ●☆

287906 Pediocactus polyancistrus（Engelm. et J. M. Bigelow）Arp = Sclerocactus polyancistrus（Engelm. et J. M. Bigelow）Britton et Rose ●☆

287907 Pediocactus pubispinus（Engelm.）D. Woodruff et L. D. Benson subsp. sileri（L. D. Benson）Halda = Sclerocactus sileri（L. D. Benson）K. D. Heil et J. M. Porter ●☆

287908 Pediocactus sileri（Engelm. ex J. M. Coult.）L. D. Benson；天狼；Gypsum Cactus，Siler Pincushion Cactus，Siler's Pincushion Cactus ●☆

287909 Pediocactus simpsonii（Engelm.）Britton et Rose；月华玉；Mountain Cactus，Simpson's Footcactus，Simpson's Hedegehog Cactus ●☆

287910 Pediocactus simpsonii（Engelm.）Britton et Rose subsp. bensonii（Engelm.）Hochstätter = Pediocactus simpsonii（Engelm.）Britton et Rose ●☆

287911 Pediocactus simpsonii（Engelm.）Britton et Rose subsp. bradyi（L. D. Benson）Halda = Pediocactus bradyi L. D. Benson ●☆

287912 Pediocactus simpsonii（Engelm.）Britton et Rose subsp. idahoensis Hochstätter = Pediocactus simpsonii（Engelm.）Britton et Rose ●☆

287913 Pediocactus simpsonii（Engelm.）Britton et Rose subsp. robustior（J. M. Coult.）Hochstätter = Pediocactus simpsonii（Engelm.）Britton et Rose ●☆

287914 Pediocactus simpsonii（Engelm.）Britton et Rose var. hermannii（T. Marshall）T. Marshall = Pediocactus simpsonii（Engelm.）Britton

et Rose ●☆

287915 Pediocactus simpsonii（Engelm.）Britton et Rose var. indraianus Hochstätter = Pediocactus simpsonii（Engelm.）Britton et Rose ●☆

287916 Pediocactus simpsonii（Engelm.）Britton et Rose var. knowltonii（L. D. Benson）Halda = Pediocactus knowltonii L. D. Benson ●☆

287917 Pediocactus simpsonii（Engelm.）Britton et Rose var. robustior（Engelm.）L. D. Benson = Pediocactus simpsonii（Engelm.）Britton et Rose ●☆

287918 Pediocactus simpsonii（Englem.）Britton et Rose var. nigrispinus Hochstätter = Pediocactus nigrispinus（Hochstätter）Hochstätter ●☆

287919 Pediocactus simpsonii Britton et Rose = Pediocactus simpsonii（Engelm.）Britton et Rose ●☆

287920 Pediocactus spinosior（Engelm.）Halda = Sclerocactus spinosior（Engelm.）D. Woodruff et L. D. Benson ●☆

287921 Pediocactus spinosior（Engelm.）Halda var. schlesseri（K. D. Heil et S. L. Welsh）Halda = Sclerocactus blainei S. L. Welsh et K. H. Thorne ●☆

287922 Pediocactus wetlandicus（Hochstätter）Halda = Sclerocactus wetlandicus Hochstätter ●☆

287923 Pediocactus whipplei（Engelm. et J. M. Bigelow）Arp = Sclerocactus whipplei（Engelm. et J. M. Bigelow）Britton et Rose ●☆

287924 Pediocactus whipplei（Engelm. et J. M. Bigelow）Arp subsp. busekii（Hochstätter）Halda = Sclerocactus sileri（L. D. Benson）K. D. Heil et J. M. Porter ●☆

287925 Pediocactus whipplei（Engelm. et J. M. Bigelow）Arp var. reevesii Castetter，P. Pierce et K. H. Sehwer. = Sclerocactus cloverae K. D. Heil et J. M. Porter ●☆

287926 Pediocactus winkleri K. D. Heil；温克尔月华玉；Winkler's Footcactus，Winkler's Pincushion Cactus ●☆

287927 Pediocactus wrightiae（L. D. Benson）Arp = Sclerocactus wrightiae L. D. Benson ●☆

287928 Pediomellum Rydb. = Orbexilum Raf. ■☆

287929 Pediomelum Rydb.（1919）；鹿角豆属■☆

287930 Pediomelum argophyllum（Pursh）J. W. Grimes；银叶鹿角豆；Silverleaf Scurf Pea，Silvery Psoralea，Silvery Scurf-pea ■☆

287931 Pediomelum canescens（Michx.）Rydb.；鹿角豆■☆

287932 Pediomelum esculentum（Pursh）Rydb.；可食鹿角豆；Breadnut，Breadroot，Breadroot Scurf-pea，Indian Breadroot，Pomme-de-prairie，Prairie Turnip，Prairie-turnip，Shaggy Prairie-turnip，Wild Potato ■☆

287933 Pedistylis Wiens = Loranthus Jacq.（保留属名）●
287934   Pedistylis Wiens（1979）；足柱寄生属●☆

287935 Pedistylis galpinii（Schinz ex Sprague）Wiens；足柱寄生●☆

287936 Pedochelus Wight = Podochilus Blume ■

287937 Pedrosia Lowe = Lotus L. ■

287938 Pedrosia tenella Lowe = Lotus tenellus（Lowe）Sandral，A. Santos et D. D. Sokoloff ■☆

287939 Peekelia Harms = Cajanus Adans.（保留属名）●

287940 Peekeliodendron Sleumer = Merrilliodendron Kaneh. ●☆

287941 Peekeliopanax Harms = Gastonia Comm. ex Lam. ●☆

287942 Peekeliopanax Harms = Polyscias J. R. Forst. et G. Forst. ●

287943 Peersia L. Bolus = Rhinephyllum N. E. Br. ●☆

287944 Peersia L. Bolus（1927）；肖鼻叶草属●☆

287945 Peersia frithii（L. Bolus）L. Bolus；弗里思肖鼻叶草●☆

287946 Peersia macradenia（L. Bolus）L. Bolus；大腺肖鼻叶草●☆

287947 Peersia vanheerdei（L. Bolus）H. E. K. Hartmann；黑尔德肖鼻叶草●☆

287948　Pegaeophyton Hayek et Hand. -Mazz. (1922)；单花荠属（单花芥属,葶荠属,无茎荠属）；Monoflorcress, Pegaeophyton ■

287949　Pegaeophyton angustiseptatum Al-Shehbaz, T. Y. Cheo, L. L. Lu et G. Yang；窄隔单花荠；Narroeseptate Pegaeophyton ■

287950　Pegaeophyton bhutanicum H. Hara；不丹单花荠■☆

287951　Pegaeophyton bhutanicum H. Hara = Pycnoplinthopsis bhuatanica Jafri ■

287952　Pegaeophyton garhwalense H. J. Chowdhery et Sur. Singh = Pegaeophyton minutum H. Hara ■

287953　Pegaeophyton minutum H. Hara；小单花荠；Mitute Monoflorcress, Mitute Pegaeophyton ■

287954　Pegaeophyton nepalense Al-Shehbaz, Kats. Arai et H. Ohba；尼泊尔单花荠；Nepal Pegaeophyton ■

287955　Pegaeophyton scapiflorum(Hook. f. et Thomson) C. Marquand et Airy Shaw var. pilosicalyx R. L. Guo et T. Y. Cheo = Pegaeophyton scapiflorum (Hook. f. et Thomson) C. Marquand et Airy Shaw ■

287956　Pegaeophyton scapiflorum(Hook. f. et Thomson) C. Marquand et Airy Shaw var. robusmm (O. E. Schulz) R. L. Guo et T. Y. Cheo = Pegaeophyton scapiflorum (Hook. f. et Thomson) C. Marquand et Airy Shaw var. robustum(O. E. Schulz) R. L. Guo et T. Y. Cheo ■

287957　Pegaeophyton scapiflorum (Hook. f. et Thomson) C. Marquand et Airy Shaw；单花荠（单花芥,高山辣根菜,高山无茎芥,无茎芥,无茎芥,喜马拉雅辣根菜）；Acaulescent Pegaeophyton, Monoflorcress ■

287958　Pegaeophyton scapiflorum (Hook. f. et Thomson) C. Marquand et Airy Shaw var. pilosicalyx R. L. Guo et T. Y. Cheo；毛萼单花荠（毛萼无茎芥）；Haircalyx Monoflorcress, Pilosecalyx Acaulescent Pegaeophyton ■

287959　Pegaeophyton scapiflorum(Hook. f. et Thomson) C. Marquand et Airy Shaw var. robustum (O. E. Schulz) R. L. Guo et T. Y. Cheo；粗壮单花荠（粗壮无茎荠）；Robust Acaulescent Pegaeophyton, Robust Monoflorcress ■

287960　Pegaeophyton scapiflorum (Hook. f. et Thomson) O. E. Schulz = Pegaeophyton scapiflorum (Hook. f. et Thomson) C. Marquand et Airy Shaw ■

287961　Pegaeophyton sinense (Hemsl.) Hayek et Hand. -Mazz. = Pegaeophyton scapiflorum(Hook. f. et Thomson) C. Marquand et Airy Shaw var. robustum(O. E. Schulz) R. L. Guo et T. Y. Cheo ■

287962　Pegaeophyton sinense (Hemsl.) Hayek et Hand. -Mazz. var. robustum O. E. Schulz = Pegaeophyton scapiflorum (Hook. f. et Thomson) C. Marquand et Airy Shaw var. robustum(O. E. Schulz)R. L. Guo et T. Y. Cheo ■

287963　Pegaeophyton sinense (Hemsl.) Hayek et Hand. -Mazz. var. stenophyllum O. E. Schulz = Solms-Laubachia pulcherrima Muschl. ■

287964　Pegaeophyton sinense Hayek et Hand. -Mazz. = Pegaeophyton scapiflorum(Hook. f. et Thomson) C. Marquand et Airy Shaw ■

287965　Pegaeophyton sinense Hayek et Hand. -Mazz. var. robustum O. E. Schulz = Pegaeophyton scapiflorum (Hook. f. et Thomson) C. Marquand et Airy Shaw var. robustum(O. E. Schulz) R. L. Guo et T. Y. Cheo ■

287966　Pegaeophyton sinense Hayek et Hand. -Mazz. var. stenophyllum O. E. Schulz = Pegaeophyton scapiflorum (Hook. f. et Thomson) C. Marquand et Airy Shaw ■

287967　Pegamea Vitman = Pagamea Aubl. ●☆

287968　Peganaceae(Engl.)Takht. = Zygophyllaceae R. Br. (保留科名)●■

287969　Peganaceae Tiegh. = Nitrariaceae Bercht et J. Presl ●

287970　Peganaceae Tiegh. = Peganaceae Tiegh. ex Takht. ●■

287971　Peganaceae Tiegh. = Pellicieraceae P. Beanvis. ●☆

287972　Peganaceae Tiegh. = Zygophyllaceae R. Br. (保留科名)●■

287973　Peganaceae Tiegh. ex Takht. (1987)；骆驼蓬科●■

287974　Peganon St. -Lag. = Peganum L. ●■

287975　Peganum L. (1753)；骆驼蓬属；Peganum ●■

287976　Peganum dauricum L. = Haplophyllum dauricum(L.)G. Don ●■

287977　Peganum dauricum Pall. = Peganum harmala L. ●■

287978　Peganum harmala L. ；骆驼蓬（臭菜,臭草,臭古朵,臭牡丹,苦苦菜,骆驼蒿,沙蓬豆豆）；African Rue, Common Peganum, Harmal, Harmal Peganum, Syrian Rue, Wild Rue ●■

287979　Peganum harmala L. var. garamantum Maire = Peganum harmala L. ●■

287980　Peganum harmala L. var. multisecta Maxim. = Peganum multisectum(Maxim.) Bobrov ●■

287981　Peganum harmala L. var. rotschildianum (Buxb.) Maire = Peganum harmala L. ●■

287982　Peganum multisectum(Maxim.) Bobrov；多裂骆驼蓬（匍根骆驼蓬,骆驼蓬）；Multifid peganum ●■

287983　Peganum nigellastrum Bunge；骆驼蒿（骆驼蓬,匍根骆驼蓬,细叶骆驼蓬,小骆驼蓬）；Little peganum ■

287984　Peganum retusum Forssk. = Nitraria retusa(Forssk.) Asch. ●

287985　Pegesia Steud. = Pagesia Raf. ■☆

287986　Pegia Colebr. (1827)；藤漆属（脉果漆属）；Pegia ●

287987　Pegia bijuga Hand. -Mazz. = Pegia sarmentosa (Lecomte) Hand. -Mazz. ●

287988　Pegia nitida Colebr. ；藤漆（泌脂藤）；Shining Pegia ●

287989　Pegia sarmentosa(Lecomte) Hand. -Mazz. ；刺黄藤（刺果藤,大飞天蜈蚣,肥力漆,红根叶,利黄藤,麦果漆,脉果漆,泌脂果,泌脂藤,退黄藤,追风藤）；Sarmentose Pegia ●

287990　Peglera Bolus = Nectaropetalum Engl. ●☆

287991　Peglera capensis Bolus = Nectaropetalum capense (Bolus) Stapf et Boodle ●☆

287992　Pegolettia Cass. (1825)；叉尾菊属■●☆

287993　Pegolettia acuminata DC. = Pegolettia retrofracta(Thunb.)Kies ■☆

287994　Pegolettia arenicola Dinter ex Range = Amellus flosculosus DC. ■☆

287995　Pegolettia baccharidifolia Less. ；种棉木叶叉尾菊■☆

287996　Pegolettia dentata Bolus = Anisothrix kuntzei O. Hoffm. ●☆

287997　Pegolettia dubiefiana Quézel = Pegolettia senegalensis Cass. ■☆

287998　Pegolettia gariepina Anderb. ；加里普叉尾菊■☆

287999　Pegolettia lanceolata Harv. ；剑叶叉尾菊■☆

288000　Pegolettia oxyodonta DC. ；圆齿叉尾菊■☆

288001　Pegolettia pinnatilobata(Klatt)O. Hoffm. ex Dinter；羽裂叉尾菊■☆

288002　Pegolettia plumosa M. D. Hend. ；羽状叉尾菊■☆

288003　Pegolettia polygalifolia Less. = Pegolettia retrofracta (Thunb.) Kies ■☆

288004　Pegolettia retrofracta(Thunb.)Kies；反曲叉尾菊■☆

288005　Pegolettia senegalensis Cass. ；塞内加尔叉尾菊■☆

288006　Pegolettia senegalensis Cass. f. pygmaea J. A. Schmidt = Pegolettia senegalensis Cass. ■☆

288007　Pegolettia tenella DC. = Vernonia galpinii Klatt ■☆

288008　Pegolettia tenuifolia Bolus；细叶叉尾菊■☆

288009　Pegolletia Less. = Pegolettia Cass. ■●☆

288010　Pehria Sprague(1923)；佩尔菜属●☆

288011　Pehria compacta Sprague；佩尔菜●☆

288012　Peiranisia Raf. = Cassia L. (保留属名)●■

288013　Peiresda Zucc. = Pereskia Mill. ●

288014　Peireskia Post et Kuntze = Hippocratea L. ●☆

288015　Peireskia Post et Kuntze = Pereskia Mill. ●

288016　Peireskia Steud. = Pereskia Mill. ●

288017　Peireskiopsis Vaupel = Pereskiopsis Britton et Rose ●☆

288018　Peirrea F. Heim = Hopea Roxb. (保留属名) ●

288019　Peixotoa A. Juss. (1833) ; 佩肖木属 ●☆

288020　Peixotoa adenopoda C. E. Anderson; 腺梗佩肖木 ●☆

288021　Peixotoa floribunda C. E. Anderson; 多花佩肖木 ●☆

288022　Peixotoa glabra A. Juss. ; 无毛佩肖木 ●☆

288023　Peixotoa leptoclada A. Juss. ; 细枝佩肖木 ●☆

288024　Peixotoa macrophylla Griseb. ; 大叶佩肖木 ●☆

288025　Peixotoa megalantha C. E. Anderson; 大花佩肖木 ●☆

288026　Peixotoa microphylla Turcz. ; 小叶佩肖木 ●☆

288027　Peixotoa octoflora C. E. Anderson; 八花佩肖木 ●☆

288028　Peixotoa parviflora A. Juss. ; 小花佩肖木 ●☆

288029　Peixotoa reticulata Griseb. ; 网状佩肖木 ●☆

288030　Peixotoa tomentosa A. Juss. ; 毛佩肖木 ●☆

288031　Pekea Aubl. = Caryocar F. Allam. ex L. ●☆

288032　Pekia Steud. = Pekea Aubl. ●☆

288033　Pelaë Adans. (废弃属名) = Xanthophyllum Roxb. (保留属名) ●

288034　Pelagatia O. E. Schulz = Weberbauera Gilg et Muschl. ■☆

288035　Pelagodendron Seem. (1866) ; 海茜草属 ●☆

288036　Pelagodendron Seem. = Randia L. ●

288037　Pelagodendron vitiense Seem. ; 海茜草 ●☆

288038　Pelagodoxa Becc. (1917) ; 海荣椰子属 (凡哈椰属, 帕拉哥椰树、全叶椰属, 培拉桐属, 银叶凤尾椰属) ●☆

288039　Pelagodoxa henryana Becc. ; 海荣椰子 (凡哈椰) ●☆

288040　Pelaphia Banks et Sol. = Coprosma J. R. Forst. et G. Forst. ●☆

288041　Pelaphia Banks et Sol. ex A. Cunn. = Coprosma J. R. Forst. et G. Forst. ●☆

288042　Pelaphoides Banks et Sol. ex Cheesem. = Pelaphia Banks et Sol. ex A. Cunn. ●☆

288043　Pelargonion St. -Lag. = Pelargonium L'Hér. ex Aiton ●■

288044　Pelargonium L'Hér. = Pelargonium L'Hér. ex Aiton ●■

288045　Pelargonium L'Hér. ex Aiton (1789) ; 天竺葵属 ; Cranes-bill, Garden Geranium, Geranium, Geranium of Florists, Geranium of House-plants, Geranium Oil, Mawah, Pelargonium, Stork's Bill, Storkbill, Stork's-bill ●■

288046　Pelargonium abrotanifolium (L. f.) Jacq. ; 南方木天竺葵; Southernwood Geranium ●☆

288047　Pelargonium abrotanifolium Jacq. = Pelargonium abrotanifolium (L. f.) Jacq. ●☆

288048　Pelargonium abyssinicum R. Br. = Pelargonium multibracteatum Hochst. ex A. Rich. ■☆

288049　Pelargonium acaule(Thunb.) DC. ; 无茎天竺葵 ■☆

288050　Pelargonium acerifolium L'Hér. ; 槭叶天竺葵; Maple-leaved Geranium ■☆

288051　Pelargonium acerifolium L'Hér. = Pelargonium cucullatum(L.) L'Hér. subsp. strigifolium Volschenk ■☆

288052　Pelargonium acetosum(L.) L'Hér. ; 酸洋葵 ■☆

288053　Pelargonium acetosum Sol. = Pelargonium acetosum(L.) L'Hér. ■☆

288054　Pelargonium aciculatum E. M. Marais; 针形天竺葵 ■☆

288055　Pelargonium aconitophyllum (Eckl. et Zeyh.) Steud. = Pelargonium luridum(Andréws) Sweet ■☆

288056　Pelargonium aconitophyllum (Eckl. et Zeyh.) Steud. var. angustisectum R. Knuth = Pelargonium luridum(Andréws) Sweet ■☆

288057　Pelargonium aconitophyllum (Eckl. et Zeyh.) Steud. var. latisectum R. Knuth = Pelargonium luridum(Andréws) Sweet ■☆

288058　Pelargonium aconitophyllum(Eckl. et Zeyh.) Steud. var. medium R. Knuth = Pelargonium luridum(Andréws) Sweet ■☆

288059　Pelargonium acuminatum (Thunb.) DC. = Pelargonium laevigatum(L. f.) Willd. ■☆

288060　Pelargonium aestivale E. M. Marais; 夏天竺葵 ■☆

288061　Pelargonium affine (Poir.) G. Don = Pelargonium longiflorum Jacq. ■☆

288062　Pelargonium album J. J. A. van der Walt; 白天竺葵 ■☆

288063　Pelargonium alchemillifolium Salisb. = Pelargonium alchemilloides (L.) L'Hér. ■☆

288064　Pelargonium alchemilloides(L.) L'Hér. ; 羽衣草天竺葵 ■☆

288065　Pelargonium alchemilloides (L.) L'Hér. subsp. multibracteatum (Hochst. ex A. Rich.) Kokwaro = Pelargonium multibracteatum Hochst. ex A. Rich. ■☆

288066　Pelargonium alpinum Eckl. et Zeyh. ; 高山天竺葵 ■☆

288067　Pelargonium alternans J. C. Wendl. ; 互生天竺葵 ■☆

288068　Pelargonium althaeoides(L.) L'Hér. ; 药葵天竺葵 ■☆

288069　Pelargonium amabile Dinter = Pelargonium sibthorpiifolium Harv. ■☆

288070　Pelargonium amatymbicum (Eckl. et Zeyh.) Harv. = Pelargonium schizopetalum Sweet ■☆

288071　Pelargonium anceps L'Hér. ex Aiton = Pelargonium grossularioides (L.) L'Hér. ex Aiton ■☆

288072　Pelargonium andrewsii (Sweet) G. Don = Pelargonium longifolium (Burm. f.) Jacq. ■☆

288073　Pelargonium anemonifolium Jacq. = Pelargonium myrrhifolium (L.) L'Hér. ■☆

288074　Pelargonium anethifolium(Eckl. et Zeyh.)Steud. ; 莳萝天竺葵 ■☆

288075　Pelargonium angulosum (Mill.) L'Hér. = Pelargonium cucullatum(L.) L'Hér. ■☆

288076　Pelargonium angulosum Szyszyl. = Pelargonium luridum(Andréws) Sweet ■☆

288077　Pelargonium angustifolium(Thunb.)DC. = Pelargonium longiflorum Jacq. ■☆

288078　Pelargonium angustipetalum E. M. Marais; 窄瓣天竺葵 ■☆

288079　Pelargonium angustissimum E. Mey. = Pelargonium coronopifolium Jacq. ■☆

288080　Pelargonium antidysentericum (Eckl. et Zeyh.) Kostel. subsp. inerme Scheltema; 无刺天竺葵 ■☆

288081　Pelargonium antidysentericum (Eckl. et Zeyh.) Kostel. subsp. zonale Scheltema; 环带天竺葵 ■☆

288082　Pelargonium apetalum P. Taylor; 无瓣天竺葵 ■☆

288083　Pelargonium aphanoides (Thunb.) DC. = Pelargonium alchemilloides(L.) L'Hér. ■☆

288084　Pelargonium apiifolium (Andréws) Loudon = Pelargonium petroselinifolium G. Don ■☆

288085　Pelargonium appendiculatum(L. f.) Willd. ; 附属物天竺葵 ■☆

288086　Pelargonium arenarium(Burm. f.) DC. ; 沙生天竺葵 ■☆

288087　Pelargonium arenicola Steud. = Pelargonium pulverulentum Colvill ex Sweet ■☆

288088　Pelargonium aridicola E. M. Marais; 旱生天竺葵 ■☆

288089　Pelargonium aridum R. A. Dyer; 耐旱天竺葵 ■☆

288090　Pelargonium aristatum(Sweet) G. Don; 具芒天竺葵 ■☆

288091　Pelargonium artemisiifolium DC. = Pelargonium divisifolium Vorster ■☆

288092　Pelargonium articulatum(Cav.) Willd. ; 关节天竺葵 ■☆

288093　Pelargonium astragalifolium (Cav.) Jacq. = Pelargonium pinnatum(L.) L'Hér. ■☆

288094　Pelargonium astragalifolium (Cav.) Jacq. var. foliolosum (DC.) Harv. = Pelargonium viciifolium DC. ■☆

288095　Pelargonium astragalifolium (Cav.) Jacq. var. minor Harv. = Pelargonium viciifolium DC. ■☆

288096　Pelargonium atrum L'Hér. = Pelargonium auritum(L.) Willd. ■☆

288097　Pelargonium attenuatum Harv. ;渐狭天竺葵■☆

288098　Pelargonium auriculatum Willd. = Pelargonium longifolium (Burm. f.) Jacq. ■☆

288099　Pelargonium auritum(L.) Willd. ;耳状天竺葵■☆

288100　Pelargonium auritum (L.) Willd. subsp. carneum (Harv.) J. J. A. van der Walt = Pelargonium auritum (L.) Willd. var. carneum (Harv.) E. M. Marais ■☆

288101　Pelargonium auritum (L.) Willd. var. carneum (Harv.) E. M. Marais;肉色耳状天竺葵■☆

288102　Pelargonium balsameum Jacq. = Pelargonium scabrum (L.) L'Hér. ■☆

288103　Pelargonium barbatum Jacq. = Pelargonium proliferum (Burm. f.) Steud. ■☆

288104　Pelargonium barklyi Scott-Elliot;巴克利天竺葵■☆

288105　Pelargonium bechuanicum Burtt Davy = Pelargonium dolomiticum R. Knuth ■

288106　Pelargonium bechuanicum Burtt Davy var. latisectum? = Pelargonium dolomiticum R. Knuth ■

288107　Pelargonium benguellense(Welw. ex Oliv.) Engl. = Pelargonium luridum(Andréws) Sweet ■☆

288108　Pelargonium betulinum(L.) L'Hér. ;桦天竺葵■☆

288109　Pelargonium bicolor(Jacq.) L'Hér. ;二色天竺葵■☆

288110　Pelargonium bifolium(Burm. f.) Willd. ;二叶天竺葵■☆

288111　Pelargonium bijugum (Eckl. et Zeyh.) Steud. = Pelargonium chelidonium(Houtt.) DC. ■☆

288112　Pelargonium bipinnatifidum(Eckl. et Zeyh.) Steud. = Pelargonium longifolium(Burm. f.) Jacq. ■☆

288113　Pelargonium boranense Friis et M. G. Gilbert;博兰天竺葵■☆

288114　Pelargonium bowkeri Harv. ;鲍克天竺葵■☆

288115　Pelargonium brevipetalum N. E. Br. ;短瓣天竺葵■☆

288116　Pelargonium brevirostre R. A. Dyer;短喙天竺葵■☆

288117　Pelargonium bubonifolium(Andréws) Pers. ;金币花天竺葵■☆

288118　Pelargonium bullatum Jacq. = Pelargonium myrrhifolium (L.) L'Hér. ■☆

288119　Pelargonium burchellii R. Knuth = Pelargonium dichondrifolium DC. ■☆

288120　Pelargonium burgerianum J. J. A. van der Walt;伯格天竺葵■☆

288121　Pelargonium burmannianum Steud. = Pelargonium carnosum (L.) L'Hér. ■☆

288122　Pelargonium burtoniae L. Bolus;伯顿天竺葵■☆

288123　Pelargonium buysii Hellbr. ;伯伊斯天竺葵■☆

288124　Pelargonium caespitosum Turcz. ;丛生天竺葵■☆

288125　Pelargonium caespitosum Turcz. subsp. concavum Hugo;凹天竺葵■☆

288126　Pelargonium caffrum(Eckl. et Zeyh.) Harv. ;开菲尔天竺葵■☆

288127　Pelargonium caledonicum L. Bolus;卡利登天竺葵■☆

288128　Pelargonium callosum E. Mey. = Pelargonium anethifolium (Eckl. et Zeyh.) Steud. ■☆

288129　Pelargonium campestre(Eckl. et Zeyh.)Steud. ;田野天竺葵■☆

288130　Pelargonium canariense Willd. = Pelargonium candicans Spreng. ■☆

288131　Pelargonium candicans Spreng. ;纯白天竺葵■☆

288132　Pelargonium capillare(Cav.)Willd. ;细毛天竺葵■☆

288133　Pelargonium capitatum(L.) Aiton;头状天竺葵(豆花牻牛儿苗,花头洋葵, 洋香葵); Rose Scented Geranium, Rose-scented Geranium,Rose-scented Pelargonium ■

288134　Pelargonium capitatum (L.) L'Hér. = Pelargonium capitatum (L.) Aiton ■

288135　Pelargonium capitatum Sol. = Pelargonium capitatum(L.) Aiton ■

288136　Pelargonium capituliforme R. Knuth;拟头状天竺葵■☆

288137　Pelargonium cardiophyllum Harv. = Pelargonium setulosum Turcz. ■☆

288138　Pelargonium carinatum J. C. Wendl. = Pelargonium rapaceum (L.) L'Hér. ■☆

288139　Pelargonium carneum Jacq. ;肉色天竺葵■☆

288140　Pelargonium carnosum(L.)L'Hér. ;佛肚天竺葵(佛肚洋葵)■☆

288141　Pelargonium carnosum Sol. = Pelargonium carnosum(L.)L'Hér. ■☆

288142　Pelargonium caroli-henrici B. Nord. ;卡罗尔天竺葵■☆

288143　Pelargonium caucalidifolium Schltr. = Pelargonium crassipes Harv. ■☆

288144　Pelargonium caucalifolium Jacq. subsp. convolvulifolium(Schltr. ex R. Knuth)J. J. A. van der Walt;旋花叶天竺葵■☆

288145　Pelargonium cavanillesii Knuth = Pelargonium heterophyllum Jacq. ■☆

288146　Pelargonium centauroides DC. = Pelargonium incrassatum(Andréws) Sims ■☆

288147　Pelargonium ceratophyllum L'Hér. ;角叶天竺葵■☆

288148　Pelargonium chamaedryfolium Jacq. ;石蚕叶天竺葵■☆

288149　Pelargonium chelidonium(Houtt.) DC. ;白屈菜天竺葵■☆

288150　Pelargonium ciliatum (Cav.) Pers. = Pelargonium proliferum (Burm. f.) Steud. ■☆

288151　Pelargonium ciliatum Jacq. = Pelargonium longifolium (Burm. f.) Jacq. ■☆

288152　Pelargonium ciliatum L'Hér. = Pelargonium heterophyllum Jacq. ■☆

288153　Pelargonium cinctum Baker = Pelargonium tabulare (Burm. f.) L'Hér. ■☆

288154　Pelargonium clavatum L'Hér. ex DC. ;棍棒天竺葵■☆

288155　Pelargonium columbinum Jacq. ;哥伦比亚天竺葵■☆

288156　Pelargonium concavifolium Pers. = Pelargonium radicatum Vent. ■☆

288157　Pelargonium condensatum Pers. = Pelargonium incrassatum (Andréws) Sims ■☆

288158　Pelargonium confertum E. M. Marais;密集天竺葵■☆

288159　Pelargonium confusum DC. ;混乱天竺葵■☆

288160　Pelargonium congestum (Sweet) G. Don = Pelargonium bubonifolium(Andréws) Pers. ■☆

288161　Pelargonium connivens E. M. Marais;靠合天竺葵■☆

288162　Pelargonium conspicuum (Sweet) G. Don = Pelargonium grenvillei(Andréws) Harv. ■☆

288163　Pelargonium convolvulifolium Schltr. ex R. Knuth = Pelargonium caucalifolium Jacq. subsp. convolvulifolium(Schltr. ex R. Knuth)J. J. A. van der Walt ■☆

288164　Pelargonium corallinum (Eckl. et Zeyh.) Steud. = Pelargonium alternans J. C. Wendl. ■☆

288165　Pelargonium cordatum L'Hér. ;心形天竺葵■☆

288166　Pelargonium cordatum L'Hér. = Pelargonium cordifolium(Cav.) Curtis ■☆

288167　Pelargonium cordatum L'Hér. var. lanatum (Thunb.) Harv. = Pelargonium cordifolium(Cav.) Curtis ■☆

288168　Pelargonium cordatum L'Hér. var. rubrocinctum (Link) Harv. =

Pelargonium cordifolium（Cav.）Curtis ■☆

288169　Pelargonium cordifolium（Cav.）Curtis；心叶天竺葵■☆

288170　Pelargonium coronillifolium（Andréws）Pers. = Pelargonium pinnatum（L.）L'Hér. ■☆

288171　Pelargonium coronopifolium Jacq.；鸟足叶天竺葵■☆

288172　Pelargonium cortusifolium Jacq. = Pelargonium tabulare（Burm. f.）L'Hér. ☆

288173　Pelargonium corydaliflorum（Sweet）DC. = Pelargonium rapaceum（L.）L'Hér. ■☆

288174　Pelargonium cradockense（Kuntze）R. Knuth = Pelargonium dichondrifolium DC. ■☆

288175　Pelargonium crassicaule L'Hér.；粗茎天竺葵■☆

288176　Pelargonium crassipes Harv.；粗梗天竺葵■☆

288177　Pelargonium crataegifolium（Thunb.）Eckl. et Zeyh. = Pelargonium scabrum（L.）L'Hér. ■☆

288178　Pelargonium crinitum Harv. = Pelargonium radiatum（Andréws）Pers. ■☆

288179　Pelargonium crispum（P. J. Bergius）L'Hér.；皱波天竺葵（皱叶天竺葵）；Crinkly-leaf Geranium, Finger Bowl Geranium, Lemon Geranium, Lemon-scented Geranium ■☆

288180　Pelargonium crispum（P. J. Bergius）L'Hér. ' Variegatum'；斑纹皱叶天竺葵■☆

288181　Pelargonium crispum L'Hér. ' Variegatum' = Pelargonium crispum（P. J. Bergius）L'Hér. ' Variegatum' ☆

288182　Pelargonium crispum L'Hér. = Pelargonium crispum（P. J. Bergius）L'Hér. ■☆

288183　Pelargonium crithmifolium Sm.；海茴香叶天竺葵；Samphire Leafed Geranium ■☆

288184　Pelargonium cucullatum（L.）L'Hér.；兜状天竺葵■☆

288185　Pelargonium cucullatum（L.）L'Hér. = Pelargonium cucullatum（L.）L'Hér. subsp. tabulare Volschenk ■☆

288186　Pelargonium cucullatum（L.）L'Hér. subsp. strigifolium Volschenk；直叶兜状天竺葵■☆

288187　Pelargonium cucullatum（L.）L'Hér. subsp. tabulare Volschenk；扁平兜状天竺葵■☆

288188　Pelargonium curviandrum E. M. Marais；弯蕊天竺葵■☆

288189　Pelargonium damarense R. Knuth = Pelargonium otaviense R. Knuth ■☆

288190　Pelargonium dasyphyllum E. Mey. ex R. Knuth；毛叶天竺葵■☆

288191　Pelargonium daucoides Jacq. = Pelargonium triste（L.）L'Hér. ■☆

288192　Pelargonium debticulatum（Poir.）J. Jacq.；细齿天竺葵；Pine Geranium ■☆

288193　Pelargonium denticulatum Jacq. = Pelargonium debticulatum（Poir.）J. Jacq. ■☆

288194　Pelargonium depressum Jacq. = Pelargonium longiflorum Jacq. ■☆

288195　Pelargonium desertorum Vorster；荒漠天竺葵■☆

288196　Pelargonium dichondrifolium DC.；马蹄金叶天竺葵■☆

288197　Pelargonium dioicum Aiton = Pelargonium auritum（L.）Willd. ■☆

288198　Pelargonium dipetalum L'Hér.；双瓣天竺葵■☆

288199　Pelargonium dispar N. E. Br.；异型天竺葵■

288200　Pelargonium dissectum（Eckl. et Zeyh.）Harv. = Pelargonium aridum R. A. Dyer ■☆

288201　Pelargonium distans L'Hér. ex DC.；远离天竺葵■

288202　Pelargonium divaricatum（Thunb.）DC. = Pelargonium fruticosum（Cav.）Willd. ■☆

288203　Pelargonium diversifolium J. C. Wendl. = Pelargonium laevigatum（L. f.）Willd. subsp. diversifolium（J. C. Wendl.）Schonken ■☆

288204　Pelargonium divisifolium Vorster；全裂天竺葵■☆

288205　Pelargonium dolomiticum R. Knuth = Pelargonium domesticum L. H. Bailey ■

288206　Pelargonium domesticum L. H. Bailey；家天竺葵（大花天竺葵，洋蝴蝶）；Common Washington Geranium, Fancy Geranium, Fancy Washington Geranium, Frecnch Geranium, Garden Storkbill, Lady Washington Geranium, Lady Washington Pelargonium, Regal Geranium, Regal Pelargonium, Show Geranium, Show Washington Geranium ■

288207　Pelargonium eberlanzii Dinter = Pelargonium sibthorpiifolium Harv. ■☆

288208　Pelargonium eberlanzii R. Knuth = Pelargonium carnosum（L.）L'Hér. ■☆

288209　Pelargonium echinatum Curtis；刺天竺葵（有刺天竺葵）；Castus Geranium, Spine Pelargonium, Sweetheart Geranium ■☆

288210　Pelargonium echinatum Curtis ' Albiflorum'；白花刺天竺葵■

288211　Pelargonium ecklonii Turcz. = Pelargonium myrrhifolium（L.）L'Hér. var. intermedium Harv. ■☆

288212　Pelargonium elatum（Sweet）DC. = Pelargonium tricolor Curtis ■☆

288213　Pelargonium elegans（Andréws）Willd.；雅致天竺葵■☆

288214　Pelargonium elongatum（Cav.）Salisb. = Pelargonium tabulare（Burm. f.）L'Hér. ■☆

288215　Pelargonium emarginatum Moench = Pelargonium longicaule Jacq. ■☆

288216　Pelargonium endlicherianum Fenzl；爱氏天竺葵■☆

288217　Pelargonium englerianum R. Knuth；恩格勒天竺葵■

288218　Pelargonium ensatum（Thunb.）DC. = Pelargonium auritum（L.）Willd. var. carneum（Harv.）E. M. Marais ■☆

288219　Pelargonium erectum R. Knuth = Pelargonium glutinosum（Jacq.）L'Hér. ■☆

288220　Pelargonium erlangerianum Engl. ex R. Knuth；厄兰格天竺葵■

288221　Pelargonium erythrophyllum（Eckl. et Zeyh.）Steud. = Pelargonium dipetalum L'Hér. ■☆

288222　Pelargonium fasciculaceum E. M. Marais；簇生天竺葵■

288223　Pelargonium fergusoniae L. Bolus；费格森天竺葵■☆

288224　Pelargonium ferulaceum（Burm. f.）Willd. = Pelargonium carnosum（L.）L'Hér. ■☆

288225　Pelargonium ferulaceum（Burm. f.）Willd. var. polycephalum E. Mey. ex Harv. = Pelargonium polycephalum（E. Mey. ex Harv.）R. Knuth ■☆

288226　Pelargonium ficaria Willd. = Pelargonium chelidonium（Houtt.）DC. ■☆

288227　Pelargonium filicaule R. Knuth = Pelargonium grossularioides（L.）L'Hér. ex Aiton ■☆

288228　Pelargonium filipendulifolium（Sims）Sweet = Pelargonium triste（L.）L'Hér. ■☆

288229　Pelargonium fischeri Engl. = Pelargonium quinquelobatum Hochst. ex A. Rich. ■☆

288230　Pelargonium fissifolium（Andréws）Pers.；半裂叶天竺葵■☆

288231　Pelargonium fissum Baker = Pelargonium alchemilloides（L.）L'Hér. ■☆

288232　Pelargonium flabellifolium Harv. = Pelargonium luridum（Andréws）Sweet ■☆

288233　Pelargonium flabellifolium Harv. var. benguellense Welw. ex Oliv. = Pelargonium luridum（Andréws）Sweet ■☆

288234　Pelargonium flavum（Burm. f.）L'Hér. = Pelargonium triste（L.）L'Hér. ☆

288235　Pelargonium floribundum（Andréws）Aiton ＝ Pelargonium fissifolium（Andréws）Pers. ■☆

288236　Pelargonium foliolosum DC. ＝ Pelargonium viciifolium DC. ■☆

288237　Pelargonium fourcadei R. Knuth；富尔卡德天竺葵■

288238　Pelargonium fragile（Andréws）Willd. ＝ Pelargonium trifidum Jacq. ■☆

288239　Pelargonium fragrans Willd. ；芳香天竺葵（马蹄纹天竺葵）；Nutmeg Geranium，Nutmeg-scented Geranium ■☆

288240　Pelargonium fruticosum（Cav.）Willd. ；灌丛天竺葵■☆

288241　Pelargonium fulgidum（L.）L'Hér. ；光亮天竺葵■☆

288242　Pelargonium fumarioides L'Hér. ex Harv. ＝ Pelargonium minimum（Cav.）Willd. ■☆

288243　Pelargonium fuscatum Jacq. ＝ Pelargonium tabulare（Burm. f.）L'Hér. ■☆

288244　Pelargonium gallense Chiov. ＝ Pelargonium whytei Baker ■☆

288245　Pelargonium galpinii Schltr. ex R. Knuth ＝ Pelargonium hypoleucum Turcz. ■☆

288246　Pelargonium geifolium E. Mey. ＝ Pelargonium lobatum（Burm. f.）L'Hér. ■☆

288247　Pelargonium georgense R. Knuth ＝ Pelargonium betulinum（L.）L'Hér. ■☆

288248　Pelargonium gibbosum（L.）L'Hér. ；有结天竺葵；Knotted Geranium，Knotted Storksbill ■☆

288249　Pelargonium gibbosum Aiton ＝ Pelargonium gibbosum（L.）L'Hér. ■☆

288250　Pelargonium gilgianum Schltr. ex R. Knuth；吉尔格天竺葵■☆

288251　Pelargonium githagineum E. M. Marais；浅绿红天竺葵■☆

288252　Pelargonium glabriphyllum E. M. Marais；光叶天竺葵■☆

288253　Pelargonium glaucum（Burm. f.）L'Hér. ；矛叶天竺葵；Spear-leaved Geranium ■☆

288254　Pelargonium glaucum（Burm. f.）L'Hér. ＝ Pelargonium lanceolatum（Cav.）Kern ■☆

288255　Pelargonium glechomoides A. Rich. ；活血丹天竺葵■☆

288256　Pelargonium glutinosum（Jacq.）L'Hér. ；胶质天竺葵；Pheasant's Foot Geranium ■☆

288257　Pelargonium glutinosum Aiton ＝ Pelargonium glutinosum（Jacq.）L'Hér. ■☆

288258　Pelargonium goetzeanum Engl. ＝ Pelargonium whytei Baker ■☆

288259　Pelargonium gracile（Eckl. et Zeyh.）Steud. ；纤细天竺葵■☆

288260　Pelargonium gracilipes R. Knuth；细梗天竺葵■☆

288261　Pelargonium gracillimum Fourc. ；细长天竺葵■☆

288262　Pelargonium gramineum Bolus ＝ Pelargonium coronopifolium Jacq. ■☆

288263　Pelargonium grandicalcaratum R. Knuth；大距天竺葵■☆

288264　Pelargonium grandiflorum（Andréws）Willd. ；大花天竺葵；Big White Pelargonium，Regal Pelargonium，Variable Bigwhite Pelargonium ■

288265　Pelargonium grandiflorum（Andréws）Willd. 'Fruehling'；早春大花天竺葵■☆

288266　Pelargonium grandiflorum（Andréws）Willd. 'Markgaertners'；喜悦大花天竺葵☆

288267　Pelargonium graniticum R. Knuth ＝ Pelargonium sibthorpiifolium Harv. ■☆

288268　Pelargonium graveolens（Thunb.）L'Hér. ；香叶天竺葵（玫红天竺葵，摸摸香，蔷薇天竺葵，香艾，香草，香洋葵，香叶）；Rose Geranium，Rose Pelargonium，Rose-scented Geranium，Samll Storkbill，Sweet Scented Geranium，Sweet-scented Geranium ■

288269　Pelargonium graveolens L'Hér. ＝ Pelargonium graveolens（Thunb.）L'Hér. ■

288270　Pelargonium grenvillei（Andréws）Harv. ；格伦维尔天竺葵■☆

288271　Pelargonium greytonense J. J. A. van der Walt；格雷敦天竺葵■☆

288272　Pelargonium griseum R. Knuth；灰天竺葵■☆

288273　Pelargonium grossularioides（L.）L'Hér. ex Aiton；茶藨子天竺葵；Coconut Geranium，Gooseberry Geranium ■☆

288274　Pelargonium grossularioides Aiton ＝ Pelargonium grossularioides（L.）L'Hér. ex Aiton ■☆

288275　Pelargonium hantamianum R. Knuth；汉塔姆天竺葵■☆

288276　Pelargonium hararense Engl. ex R. Knuth；哈拉雷天竺葵■☆

288277　Pelargonium harveyanum R. Knuth ＝ Pelargonium hypoleucum Turcz. ■☆

288278　Pelargonium heckmannianum Engl. ＝ Pelargonium luridum（Andréws）Sweet ■☆

288279　Pelargonium hederifolium Salisb. ＝ Pelargonium peltatum（L.）Aiton ■

288280　Pelargonium hemicyclicum Hutch. et C. A. Sm. ；半环天竺葵■☆

288281　Pelargonium heracleifolium Lodd. ＝ Pelargonium lobatum（Burm. f.）L'Hér. ■☆

288282　Pelargonium héritieri（Sweet）G. Don ＝ Pelargonium dipetalum L'Hér. ■☆

288283　Pelargonium heritieri Jacq. ＝ Pelargonium alchemilloides（L.）L'Hér. ■☆

288284　Pelargonium hermanniifolium（P. J. Bergius）Jacq. ；密钟木叶天竺葵■☆

288285　Pelargonium heterolobum DC. ＝ Pelargonium violiflorum（Sweet）DC. ■☆

288286　Pelargonium heterophyllum（Andréws）Loudon ＝ Pelargonium longifolium（Burm. f.）Jacq. ■☆

288287　Pelargonium heterophyllum Jacq. ；互叶天竺葵■☆

288288　Pelargonium hippocrepis L'Hér. ex DC. ＝ Pelargonium tabulare（Burm. f.）L'Hér. ■☆

288289　Pelargonium hirsutum（Burm. f.）Aiton ＝ Pelargonium auritum（L.）Willd. ■☆

288290　Pelargonium hirsutum（Burm. f.）Aiton var. carneum Harv. ＝ Pelargonium auritum（L.）Willd. var. carneum（Harv.）E. M. Marais ■☆

288291　Pelargonium hirsutum（Burm. f.）Aiton var. melananthum（Jacq.）Harv. ＝ Pelargonium auritum（L.）Willd. ■☆

288292　Pelargonium hirsutum Loudon ＝ Pelargonium undulatum（Andréws）Pers. ■☆

288293　Pelargonium hirtum（Burm. f.）Jacq. ；多毛天竺葵■☆

288294　Pelargonium hirtum Willd. ＝ Pelargonium heterophyllum Jacq. ■☆

288295　Pelargonium hispidum（L. f.）Willd. ；粗毛天竺葵■☆

288296　Pelargonium hollandii F. M. Leight. ＝ Pelargonium pulverulentum Colvill ex Sweet ■☆

288297　Pelargonium hortorum L. H. Bailey；天竺葵（木海棠，石腊红，洋葵，月月红）；Bedding Geranium，Common Fish Geranium，Common Geranium，Fish Geranium，Fish Storkbill，Garden Geranium，Horseshoe Geranium，House Geranium，Zonal Geranium ■

288298　Pelargonium hospitans Dinter ＝ Pelargonium ceratophyllum L'Hér. ■☆

288299　Pelargonium hurifolium Sweet ＝ Pelargonium luridum（Andréws）Sweet ■☆

288300　Pelargonium hybridum（L.）L'Hér. ；软枝洋葵；Scarlet Geranium ■☆

288301　Pelargonium hybridum Sol. ＝ Pelargonium hybridum（L.）L'Hér. ■☆

288302　Pelargonium hypoleucum Turcz. ；里白天竺葵■☆

288303　Pelargonium hystrix Harv. ;毛刺天竺葵■☆

288304　Pelargonium inaequilobum Mast. = Pelargonium quinquelobatum Hochst. ex A. Rich. ■☆

288305　Pelargonium incarnatum(L'Hér.)Moench;拟肉色天竺葵■☆

288306　Pelargonium incisum (Andréws) Willd. = Pelargonium abrotanifolium(L. f.)Jacq. ●☆

288307　Pelargonium incrassatum(Andréws)Sims;粗天竺葵■☆

288308　Pelargonium inodorum Willd. ;无味天竺葵;Scentless Geranium ■☆

288309　Pelargonium inquinans(L.) Aiton = Pelargonium inquinans(L.) L'Hér. ■☆

288310　Pelargonium inquinans (L.) L'Hér. ; 小花天竺葵; Polluting Geranium ,Scarlet Geranium ■☆

288311　Pelargonium intermedium R. Knuth = Pelargonium graveolens L'Hér. ■

288312　Pelargonium jacobii R. A. Dyer = Pelargonium klinghardtense R. Knuth ■☆

288313　Pelargonium jarmilae Halda;娅米拉天竺葵■☆

288314　Pelargonium juttae Dinter = Pelargonium dolomiticum R. Knuth ■

288315　Pelargonium karooicum Compton et P. E. Barnes;卡鲁天竺葵■☆

288316　Pelargonium karrooense R. Knuth = Pelargonium quercifolium (L. f.)L'Hér. ■☆

288317　Pelargonium klinghardtense R. Knuth;克林天竺葵■☆

288318　Pelargonium lacerum Jacq. = Pelargonium myrrhifolium (L.) L'Hér. ■☆

288319　Pelargonium laciniatum (Andréws) Pers. = Pelargonium proliferum (Burm. f.)Steud. ■☆

288320　Pelargonium laciniatum R. Knuth;撕裂天竺葵■☆

288321　Pelargonium laevigatum(L. f.)Willd. ;光滑天竺葵■☆

288322　Pelargonium laevigatum(L. f.)Willd. subsp. diversifolium(J. C. Wendl.)Schonken;异叶光滑天竺葵■☆

288323　Pelargonium laevigatum(L. f.)Willd. subsp. oxyphyllum(DC.)Schonken;尖叶光滑天竺葵■☆

288324　Pelargonium lanceolatum(Cav.)Kern;披针形天竺葵■☆

288325　Pelargonium lancifolium(Eckl. et Zeyh.)Steud. = Pelargonium longifolium(Burm. f.)Jacq. ■☆

288326　Pelargonium lateripes L'Hér. ;侧梗天竺葵■☆

288327　Pelargonium lateripes L'Hér. = Pelargonium peltatum (L.) L'Hér. ■

288328　Pelargonium lavaterifolium Steud. = Pelargonium inquinans (L.)L'Hér. ■☆

288329　Pelargonium laxum(Sweet)G. Don;松散天竺葵■☆

288330　Pelargonium leeanum (Sweet) G. Don = Pelargonium proliferum (Burm. f.)Steud. ■☆

288331　Pelargonium leipoldtii R. Knuth;莱波尔德天竺葵■☆

288332　Pelargonium leptopodium Bolus = Pelargonium capillare(Cav.)Willd. ■☆

288333　Pelargonium leptum L. Bolus;细天竺葵■☆

288334　Pelargonium lessertiifolium (Eckl. et Zeyh.) Steud. = Pelargonium pinnatum(L.)L'Hér. ■☆

288335　Pelargonium leucophyllum Turcz. ;白叶天竺葵■☆

288336　Pelargonium lineare (Andréws) Pers. = Pelargonium longiflorum Jacq. ■☆

288337　Pelargonium lobatum(Burm. f.)L'Hér. ;尖裂天竺葵■☆

288338　Pelargonium longicaule Jacq. ;长茎天竺葵■☆

288339　Pelargonium longicaule Jacq. var. angustipetalum C. Boucher;窄瓣长茎天竺葵■☆

288340　Pelargonium longiflorum Jacq. ;长花天竺葵■☆

288341　Pelargonium longiflorum Jacq. var. depressum(Jacq.) Loudon = Pelargonium longiflorum Jacq. ■☆

288342　Pelargonium longifolium(Burm. f.)Jacq. ;长叶天竺葵■☆

288343　Pelargonium longifolium(Burm. f.)Jacq. var. ciliatum(L'Hér.)Harv. = Pelargonium heterophyllum Jacq. ■☆

288344　Pelargonium longifolium (Burm. f.) Jacq. var. longiflorum (Jacq.)Harv. = Pelargonium longiflorum Jacq. ■☆

288345　Pelargonium longifolium(Burm. f.)Jacq. var. nivea(Sweet)R. Knuth = Pelargonium violiflorum(Sweet)DC. ■☆

288346　Pelargonium longifolium (Burm. f.) Jacq. var. parnassioides (DC.)Knuth = Pelargonium proliferum(Burm. f.)Steud. ■☆

288347　Pelargonium longifolium(Burm. f.)Jacq. var. virgineum(Pers.)Harv. = Pelargonium undulatum(Andréws)Pers. ■☆

288348　Pelargonium longiscapum Schltr. ex R. Knuth = Pelargonium luridum(Andréws)Sweet ■☆

288349　Pelargonium luridum(Andréws)Sweet;灰黄天竺葵■☆

288350　Pelargonium luteolum N. E. Br. ;淡黄天竺葵■☆

288351　Pelargonium luteum(Andréws)G. Don;黄天竺葵■☆

288352　Pelargonium macowanii Bolus = Pelargonium scabrum (L.) L'Hér. ■☆

288353　Pelargonium madagascariense Baker;马岛天竺葵■☆

288354　Pelargonium mahernium L'Hér. ex DC. = Pelargonium capillare (Cav.)Willd. ■☆

288355　Pelargonium malacoides R. Knuth = Pelargonium nanum L'Hér. ■☆

288356　Pelargonium malvifolium Jacq. = Pelargonium alchemilloides (L.)L'Hér. ■☆

288357　Pelargonium mammulosum J. C. Wendl. = Pelargonium carnosum (L.)L'Hér. ■☆

288358　Pelargonium marlothii R. Knuth = Pelargonium alpinum Eckl. et Zeyh. ■☆

288359　Pelargonium maximiliani Schltr. = Pelargonium carneum Jacq. ■☆

288360　Pelargonium melananthon Jacq. = Pelargonium auritum (L.) Willd. ■☆

288361　Pelargonium meyeri Harv. = Pelargonium chelidonium (Houtt.) DC. ■☆

288362　Pelargonium micropetalum E. Mey. = Pelargonium grossularioides (L.)L'Hér. ex Aiton ■☆

288363　Pelargonium microphyllum (Eckl. et Zeyh.) Steud. = Pelargonium alternans J. C. Wendl. ■☆

288364　Pelargonium middletonianum R. Knuth = Pelargonium dichondrifolium DC. ■☆

288365　Pelargonium migiurtinorum (Chiov.) Chiov. = Pelargonium somalense Franch. ■☆

288366　Pelargonium millefoliatum Sweet = Pelargonium triste (L.) L'Hér. ■☆

288367　Pelargonium minimum(Cav.)Willd. ;微小天竺葵■☆

288368　Pelargonium mirabile Dinter = Pelargonium crassicaule L'Hér. ■☆

288369　Pelargonium mollicomum Fourc. ;软天竺葵■☆

288370　Pelargonium moniliforme Harv. ;串珠状天竺葵■☆

288371　Pelargonium mossambicense Engl. ;莫桑比克天竺葵■☆

288372　Pelargonium multibracteatum Hochst. ex A. Rich. ;多苞天竺葵■☆

288373　Pelargonium multicaule Jacq. ;多茎天竺葵■☆

288374　Pelargonium multicaule Jacq. subsp. subherbaceum(R. Knuth)J. J. A. van der Walt;亚草本多茎天竺葵■☆

288375　Pelargonium multifidum Harv. = Pelargonium plurisectum Salter ■☆

288376　Pelargonium multiradiatum J. C. Wendl. ;多射线天竺葵■☆

288377　Pelargonium mutans Vorster;易变天竺葵■☆

288378　Pelargonium myrrhifolium（L.）L'Hér.；没药叶天竺葵■☆

288379　Pelargonium myrrhifolium（L.）L'Hér. var. betonicum（Burm. f.） Harv. = Pelargonium myrrhifolium（L.）L'Hér. ■☆

288380　Pelargonium myrrhifolium（L.）L'Hér. var. intermedium Harv.； 间型没药叶天竺葵■☆

288381　Pelargonium myrrhifolium（L.）L'Hér. var. longicaule（Jacq.） Harv. = Pelargonium longicaule Jacq. ■☆

288382　Pelargonium namaquense R. Knuth = Pelargonium bubonifolium （Andréws）Pers. ■☆

288383　Pelargonium nanum L'Hér.；矮小天竺葵■☆

288384　Pelargonium nelsonii Burtt Davy；纳尔逊天竺葵■☆

288385　Pelargonium nephrophyllum E. M. Marais；肾叶天竺葵■☆

288386　Pelargonium nervifolium Jacq.；脉叶天竺葵■☆

288387　Pelargonium nivenii Harv. = Pelargonium dipetalum L'Hér. ■☆

288388　Pelargonium niveum（Sweet）Loudon = Pelargonium violiflorum （Sweet）DC. ■☆

288389　Pelargonium nummulifolium Salisb.；铜钱叶天竺葵■☆

288390　Pelargonium nutans DC. = Pelargonium rapaceum（L.）L'Hér. ■☆

288391　Pelargonium oblongatum Harv.；矩圆天竺葵■☆

288392　Pelargonium ocellatum J. J. A. van der Walt；单眼天竺葵■☆

288393　Pelargonium ochroleucum Harv.；白绿天竺葵■☆

288394　Pelargonium odoratissimum（L.）L'Hér.；极香天竺葵■☆

288395　Pelargonium odoratissimum（L.）Sol. ex Aiton = Pelargonium odoratissimum Willd. ■

288396　Pelargonium odoratissimum Willd.；香天竺葵（豆蔻天竺葵,苹果香天竺葵）；Apple Geranium, Apple Pelargonium, Lemon-scented Geranium, Nutmeg Geranium, Nutmeg Pelargonium ■

288397　Pelargonium oenothera（L. f.）Jacq.；月见草天竺葵■☆

288398　Pelargonium oppositifolium Schltr.；对叶天竺葵■☆

288399　Pelargonium oreophilum Schltr.；喜山天竺葵■☆

288400　Pelargonium ornithopifolium（Eckl. et Zeyh.）Steud. = Pelargonium pinnatum（L.）L'Hér. ■☆

288401　Pelargonium otaviense R. Knuth；奥塔维天竺葵■☆

288402　Pelargonium ovale（Burm. f.）L'Hér.；卵形天竺葵■☆

288403　Pelargonium ovale（Burm. f.）L'Hér. subsp. hyalinum Hugo；透明卵形天竺葵■☆

288404　Pelargonium ovale（Burm. f.）L'Hér. subsp. ovatum Harv. = Pelargonium elegans（Andréws）Willd. ■☆

288405　Pelargonium ovale（Burm. f.）L'Hér. subsp. veronicifolium （Eckl. et Zeyh.）Hugo；婆婆纳叶天竺葵■☆

288406　Pelargonium ovalifolium（Sweet）DC. = Pelargonium auritum （L.）Willd. var. carneum（Harv.）E. M. Marais ■☆

288407　Pelargonium ovatifolium Steud. = Pelargonium undulatum （Andréws）Pers. ■☆

288408　Pelargonium ovatostipulatum R. Knuth = Pelargonium stipulaceum （L. f.）Willd. subsp. ovatostipulatum（R. Knuth）Vorster ■☆

288409　Pelargonium oxalidifolium（Andréws）Pers. = Pelargonium heterophyllum Jacq. ■☆

288410　Pelargonium oxaloides（Burm. f.）Willd.；酢浆草天竺葵■☆

288411　Pelargonium oxyphyllum DC. = Pelargonium laevigatum（L. f.） Willd. subsp. oxyphyllum（DC.）Schonken ■☆

288412　Pelargonium panduraeforme Eckl. et Zeyh. = Pelargonium quercifolium（L. f.）L'Hér. ■☆

288413　Pelargonium panduriforme Eckl. et Zeyh.；琴形天竺葵■☆

288414　Pelargonium paniculatum Jacq.；圆锥天竺葵■☆

288415　Pelargonium papaverifolium（Eckl. et Zeyh.）Steud. = Pelargonium triste（L.）L'Hér. ■☆

288416　Pelargonium papilionaceum（L.）L'Hér.；蝶形天竺葵■☆

288417　Pelargonium paradoxum Dinter = Pelargonium klinghardtense R. Knuth ■☆

288418　Pelargonium parnassioides DC. = Pelargonium proliferum （Burm. f.）Steud. ■☆

288419　Pelargonium parviflorum J. C. Wendl. = Pelargonium carnosum （L.）L'Hér. ■☆

288420　Pelargonium parvipetalum E. M. Marais；小瓣天竺葵■☆

288421　Pelargonium parvirostre R. A. Dyer；小喙天竺葵■☆

288422　Pelargonium parvulum DC. = Pelargonium nanum L'Hér. ■☆

288423　Pelargonium patentissimum J. C. Wendl. = Pelargonium tabulare （Burm. f.）L'Hér. ■☆

288424　Pelargonium patersonii R. Knuth；帕特森天竺葵■☆

288425　Pelargonium patulum Jacq.；张开天竺葵■☆

288426　Pelargonium patulum Jacq. var. grandiflorum N. van Wyk；大花张开天竺葵■☆

288427　Pelargonium patulum Jacq. var. tenuilobum（Eckl. et Zeyh.） Harv.；细裂张开天竺葵■☆

288428　Pelargonium pedicellatum Sweet = Pelargonium pulverulentum Colvill ex Sweet ■☆

288429　Pelargonium peltatum（L.）Aiton = Pelargonium peltatum（L.） L'Hér. ■

288430　Pelargonium peltatum（L.）L'Hér.；盾叶天竺葵（盾状天竺葵, 蔓性天竺葵）；Hanging Geranium, Ivy Geranium, Ivyleaf Geranium, Ivy-leaved Geranium, Ivy-leaved Pelargonium, Oakleaf Garden Geranium, Peltateleaf Storkbill, Trailing Pelargonium ■

288431　Pelargonium penniforme Pers. = Pelargonium proliferum（Burm. f.）Steud. ■☆

288432　Pelargonium petroselinifolium G. Don；欧芹叶天竺葵■☆

288433　Pelargonium phellandrium E. Mey. = Pelargonium senecioides L'Hér. ■☆

288434　Pelargonium pillansii Salter；皮朗斯天竺葵■☆

288435　Pelargonium pilosellifolium（Eckl. et Zeyh.）Steud.；山柳菊叶天竺葵■☆

288436　Pelargonium pilosum（Andréws）Pers. = Pelargonium petroselini-folium G. Don ■☆

288437　Pelargonium pilosum（Cav.）Steud. = Pelargonium heterophyllum Jacq. ■☆

288438　Pelargonium pingue（Thunb.）DC.；肥厚天竺葵■☆

288439　Pelargonium pinnatum（L.）L'Hér.；羽状天竺葵■☆

288440　Pelargonium plicatum（Thunb.）DC.；折叠天竺葵■☆

288441　Pelargonium plurisectum Salter；多裂天竺葵■☆

288442　Pelargonium polycephalum（E. Mey. ex Harv.）R. Knuth；多头天竺葵■☆

288443　Pelargonium polymorphum E. Mey. = Pelargonium luridum （Andréws）Sweet ■☆

288444　Pelargonium populifolium Eckl. et Zeyh. = Pelargonium ribifolium Jacq. ■☆

288445　Pelargonium praemorsum（Andréws）F. Dietr.；啮蚀天竺葵■☆

288446　Pelargonium praemorsum（Andréws）F. Dietr. subsp. speciosum Scheltema；美丽啮蚀天竺葵■☆

288447　Pelargonium procumbens（Andréws）Pers. = Pelargonium nanum L'Hér. ■☆

288448　Pelargonium proliferum（Burm. f.）Steud.；多育天竺葵■☆

288449　Pelargonium pseudofumarioides R. Knuth；紫堇天竺葵■☆

288450　Pelargonium pseudoglutinosum R. Knuth；假黏性天竺葵■☆

288451　Pelargonium pseudoglutinosum R. Knuth var. scabridum（R.

Knuth) R. Knuth = Pelargonium pseudoglutinosum R. Knuth ■☆

288452　Pelargonium pubipetalum E. M. Marais;毛瓣天竺葵■☆

288453　Pelargonium pulchellum Salisb. = Pelargonium reflexipetalum E. M. Marais ■☆

288454　Pelargonium pulchellum Sims;美丽天竺葵■☆

288455　Pelargonium pulcherrimum F. M. Leight. = Pelargonium radiatum(Andréws)Pers. ■☆

288456　Pelargonium pulverulentum Colvill ex Sweet;粉粒天竺葵■☆

288457　Pelargonium punctatum(Andréws)Willd. ;斑点天竺葵■☆

288458　Pelargonium purpurascens Pers. = Pelargonium proliferum (Burm. f.)Steud. ■☆

288459　Pelargonium quercifolium(L. f.)L'Hér. ;栎叶天竺葵;Almond Geranium, Almond-scented Geranium, Oakleaf Geranium, Oak-leaf Geranium, Oak-leaved Geranium, Scarlet-flowering Rose Geranium, Scented Oak, Village-oak Geranium ■☆

288460　Pelargonium quercifolium Aiton = Pelargonium quercifolium(L. f.)L'Hér. ☆

288461　Pelargonium quinatum Sims = Pelargonium praemorsum (Andréws)F. Dietr. ■☆

288462　Pelargonium quinquelobatum Hochst. ex A. Rich. ;五裂片天竺葵■☆

288463　Pelargonium quinquelobatum Hochst. ex A. Rich. var. migiurtinorum Chiov. = Pelargonium somalense Franch. ■☆

288464　Pelargonium quinquelobatum Hochst. ex A. Rich. var. somalense (Franch.)Verdc. = Pelargonium somalense Franch. ■☆

288465　Pelargonium radens H. E. Moore;毛 茛 天 竺 葵; Crowfoot Geranium ■☆

288466　Pelargonium radiatum(Andréws)Pers. = Pelargonium radicatum Vent. ■☆

288467　Pelargonium radicatum Vent. ;辐射天竺葵■☆

288468　Pelargonium radula(Cav.)L'Hér. ;菊 叶 天 竺 葵(菊花天竺葵);Daisyleaf Storkbill ■

288469　Pelargonium radula(Cav.)L'Hér. = Pelargonium radens H. E. Moore ■☆

288470　Pelargonium radulifolium(Eckl. et Zeyh.)Steud. ;刮刀叶天竺葵■☆

288471　Pelargonium ramosissimum(Cav.)Willd. ;多枝天竺葵■☆

288472　Pelargonium ranunculophyllum(Eckl. et Zeyh.)Baker;毛茛叶天竺葵■☆

288473　Pelargonium rapaceum(L.)L'Hér. ;芜菁天竺葵■☆

288474　Pelargonium rapaceum (L.) L'Hér. var. corydaliflorum (Sweet) DC. = Pelargonium rapaceum(L.)L'Hér. ☆

288475　Pelargonium rapaceum(L.)L'Hér. var. selinum(Andréws)Pers. = Pelargonium rapaceum(L.)L'Hér. ■☆

288476　Pelargonium recurvatum(Sweet)G. Don = Pelargonium aristatum (Sweet)G. Don ■☆

288477　Pelargonium redactum Vorster;回复天竺葵■☆

288478　Pelargonium reflexipetalum E. M. Marais;折瓣天竺葵■☆

288479　Pelargonium reflexum(Andréws)Pers. ;反折天竺葵■☆

288480　Pelargonium rehmannii Szyszyl. = Pelargonium luridum(Andréws) Sweet ■☆

288481　Pelargonium reliquifolium N. E. Br. = Pelargonium dichondrifolium DC. ■☆

288482　Pelargonium reniforme Curtis;肾形天竺葵■☆

288483　Pelargonium reniforme Curtis subsp. velutinum(Eckl. et Zeyh.) Dreyer;绒毛肾形天竺葵■☆

288484　Pelargonium reticulatum (Sweet) DC. = Pelargonium auritum

(L.)Willd. var. carneum(Harv.)E. M. Marais ■☆

288485　Pelargonium revolutum (Andréws) Pers. = Pelargonium chelidonium(Houtt.)DC. ■☆

288486　Pelargonium ribifolium Jacq. ;红叶天竺葵■☆

288487　Pelargonium riversdalense R. Knuth;里弗斯代尔天竺葵■☆

288488　Pelargonium roessingense Dinter = Pelargonium otaviense R. Knuth ■☆

288489　Pelargonium rogersii S. Moore = Pelargonium candicans Spreng. ■☆

288490　Pelargonium roseum(Andréws)Aiton = Pelargonium incrassatum (Andréws)Sims ■☆

288491　Pelargonium roseum(Andréws)R. Br. = Pelargonium graveolens (Thunb.)L'Hér. ■

288492　Pelargonium roseum Willd. = Pelargonium roseum(Andréws)R. Br. ■

288493　Pelargonium rubiginosum E. M. Marais;锈红天竺葵■☆

288494　Pelargonium rubrocinctum Link = Pelargonium cordifolium (Cav.)Curtis ■☆

288495　Pelargonium rubromaculatum Pers. = Pelargonium pulchellum Sims ■☆

288496　Pelargonium rumicifolium (Sweet) Loudon = Pelargonium longiflorum Jacq. ■☆

288497　Pelargonium rungvense R. Knuth;伦圭天竺葵■☆

288498　Pelargonium rutifolium Baker = Pelargonium rapaceum (L.) L'Hér. ■☆

288499　Pelargonium salmoneum R. A. Dyer;鲑色天竺葵■☆

288500　Pelargonium saniculifolium Willd. = Pelargonium tabulare (Burm. f.)L'Hér. ■☆

288501　Pelargonium scabroide R. Knuth;拟粗糙天竺葵■☆

288502　Pelargonium scabrum(L.)L'Hér. var. balsameum(Jacq.)Harv. = Pelargonium scabrum(L.)L'Hér. ■☆

288503　Pelargonium scabrum Aiton;粗糙天竺葵;Apricot Geranium, Strawberry Geranium ■☆

288504　Pelargonium schizopetalum Sweet;裂瓣天竺葵■☆

288505　Pelargonium schlechteri R. Knuth = Pelargonium luridum (Andréws)Sweet ■☆

288506　Pelargonium schonlandii R. Knuth = Pelargonium ribifolium Jacq. ■☆

288507　Pelargonium scutatum Sweet = Pelargonium peltatum(L.)L'Hér. ■

288508　Pelargonium selinum (Andréws) Pers. = Pelargonium rapaceum (L.)L'Hér. ■☆

288509　Pelargonium semitrilobum Jacq. ;半三裂天竺葵■☆

288510　Pelargonium senecioides L'Hér. ;千里光天竺葵■☆

288511　Pelargonium sericeum E. Mey. = Pelargonium sericifolium J. J. A. van der Walt ■☆

288512　Pelargonium sericifolium J. J. A. van der Walt;绢毛叶天竺葵■☆

288513　Pelargonium serratum(Thunb.)DC. ;齿叶天竺葵■☆

288514　Pelargonium setosum(Sweet)DC. ;刚毛天竺葵■☆

288515　Pelargonium setulosum Turcz. ;小刚毛天竺葵■☆

288516　Pelargonium sibthorpiifolium Harv. ;鸡玄参叶天竺葵■☆

288517　Pelargonium sidifolium (Thunb.) R. Knuth = Pelargonium sidoides DC. ■☆

288518　Pelargonium sidoides DC. ;黄花稔天竺葵■☆

288519　Pelargonium sisonifolium Baker = Pelargonium carnosum (L.) L'Hér. ■☆

288520　Pelargonium somalense Franch. ;索马里天竺葵■☆

288521　Pelargonium spathulatum (Andréws) Pers. = Pelargonium longiflorum Jacq. ■☆

288522　Pelargonium spathulatum (Andréws) Pers. var. affine (Poir.)

Loudon = Pelargonium longiflorum Jacq. ■☆

288523　Pelargonium spathulatum（Andréws）Pers. var. curviflorum（Andréws）Knuth = Pelargonium longiflorum Jacq. ■☆

288524　Pelargonium sphondyliifolium Salisb. = Pelargonium lobatum（Burm. f.）L'Hér. ■☆

288525　Pelargonium spinosum Willd. ；具刺天竺葵■☆

288526　Pelargonium squamulosum R. Knuth = Pelargonium radicatum Vent. ■☆

288527　Pelargonium squarrosum Dinter = Pelargonium grandicalcaratum R. Knuth ■☆

288528　Pelargonium staticifolium Steud. = Pelargonium coronopifolium Jacq. ■☆

288529　Pelargonium stipulaceum（L. f.）Willd. ；托叶天竺葵■☆

288530　Pelargonium stipulaceum（L. f.）Willd. subsp. ovatostipulatum（R. Knuth）Vorster；卵托叶天竺葵■☆

288531　Pelargonium strigosum（Eckl. et Zeyh.）Steud. = Pelargonium auritum（L.）Willd. var. carneum（Harv.）E. M. Marais ■☆

288532　Pelargonium subherbaceum R. Knuth = Pelargonium multicaule Jacq. subsp. subherbaceum（R. Knuth）J. J. A. van der Walt ■☆

288533　Pelargonium sublignosum R. Knuth；近木质天竺葵■☆

288534　Pelargonium suburbanum Clifford ex C. Boucher；城边天竺葵■☆

288535　Pelargonium suburbanum Clifford ex C. Boucher subsp. bipinnatifidum（Harv.）C. Boucher；羽裂天竺葵■☆

288536　Pelargonium sulphureum R. Knuth；硫色天竺葵■☆

288537　Pelargonium synnotii（Sweet）G. Don = Pelargonium myrrhifolium（L.）L'Hér. ■☆

288538　Pelargonium tabulare（Burm. f.）L'Hér. ；扁平天竺葵■☆

288539　Pelargonium tectum（Thunb.）DC. ；澳非天竺葵■☆

288540　Pelargonium tenellum（Andréws）G. Don；柔软天竺葵■☆

288541　Pelargonium tenuicaule R. Knuth；细茎天竺葵■☆

288542　Pelargonium tenuifolium L'Hér. = Pelargonium hirtum（Burm. f.）Jacq. ■☆

288543　Pelargonium ternatum（L. f.）Jacq. ；三出天竺葵■☆

288544　Pelargonium ternifolium Vorster；三出叶天竺葵■☆

288545　Pelargonium testaceum Baker = Pelargonium pulverulentum Colvill ex Sweet ■☆

288546　Pelargonium tetragonum（L. f.）L'Hér. ；四角天竺葵；Squarestalk Geranium ■☆

288547　Pelargonium tetragonum Aiton = Pelargonium tetragonum（L. f.）L'Hér. ■☆

288548　Pelargonium theianthum（Eckl. et Zeyh.）Steud. = Pelargonium ochroleucum Harv. ■☆

288549　Pelargonium theianthum Steud. = Pelargonium ochroleucum Harv. ■☆

288550　Pelargonium tomentosum Jacq. ；绒毛天竺葵（薄荷天竺葵，密毛洋葵，茸毛天竺葵）；Herb-scented Geranium, Peppermint Geranium, Peppermint Scented Geranium, Peppermint-scented Geranium ■☆

288551　Pelargonium tongaense Vorster；通加天竺葵☆

288552　Pelargonium torulosum E. M. Marais；结节天竺葵■☆

288553　Pelargonium tragacanthoides Burch. ；羊角刺天竺葵■☆

288554　Pelargonium transvaalense R. Knuth；德兰士瓦天竺葵■☆

288555　Pelargonium triandrum E. M. Marais；三蕊天竺葵■☆

288556　Pelargonium triangulare（Eckl. et Zeyh.）Steud. = Pelargonium multicaule Jacq. ■☆

288557　Pelargonium trichophorum Hutch. ；毛梗天竺葵■☆

288558　Pelargonium tricolor Curtis；三色天竺葵■☆

288559　Pelargonium tricuspidatum L'Hér. ；三尖天竺葵■☆

288560　Pelargonium trifidum Jacq. ；三裂天竺葵■☆

288561　Pelargonium trifoliatum Harv. = Pelargonium ternifolium Vorster ■☆

288562　Pelargonium trifoliatum Steud. = Pelargonium trifoliolatum（Eckl. et Zeyh.）E. M. Marais ■☆

288563　Pelargonium trifoliolatum（Eckl. et Zeyh.）E. M. Marais；三小叶天竺葵☆

288564　Pelargonium trilobum（Thunb.）DC. = Pelargonium chelidonium（Houtt.）DC. ■☆

288565　Pelargonium tripalmatum E. M. Marais；三掌裂天竺葵■☆

288566　Pelargonium tripartitum Willd. = Pelargonium trifidum Jacq. ■☆

288567　Pelargonium triphyllum Jacq. ；三叶天竺葵■☆

288568　Pelargonium triste（L.）L'Hér. ；暗淡天竺葵■☆

288569　Pelargonium triste（L.）L'Hér. var. filipendulifolium Sims = Pelargonium triste（L.）L'Hér. ■☆

288570　Pelargonium tysonii Szyszyl. = Pelargonium proliferum（Burm. f.）Steud. ■☆

288571　Pelargonium undulatum（Andréws）Pers. ；波状天竺葵■☆

288572　Pelargonium unduliflorum（Sweet）G. Don = Pelargonium auritum（L.）Willd. ■☆

288573　Pelargonium uniondalense R. Knuth = Pelargonium pseudoglutinosum R. Knuth ■☆

288574　Pelargonium uniondalense R. Knuth var. scabridum？= Pelargonium pseudoglutinosum R. Knuth ■☆

288575　Pelargonium urbanum（Eckl. et Zeyh.）Steud. var. bipinnatifidum Harv. = Pelargonium suburbanum Clifford ex C. Boucher subsp. bipinnatifidum（Harv.）C. Boucher ■☆

288576　Pelargonium urbanum（Eckl. et Zeyh.）Steud. var. pinnatifidum Harv. = Pelargonium suburbanum Clifford ex C. Boucher ■☆

288577　Pelargonium urenifolium Steud. = Pelargonium inquinans（L.）L'Hér. ■☆

288578　Pelargonium usambarense Engl. = Pelargonium multibracteatum Hochst. ex A. Rich. ■☆

288579　Pelargonium variifolium Steud. = Pelargonium violiflorum（Sweet）DC. ■☆

288580　Pelargonium veronicifolium（Eckl. et Zeyh.）Steud. = Pelargonium ovale（Burm. f.）L'Hér. subsp. veronicifolium（Eckl. et Zeyh.）Hugo ■☆

288581　Pelargonium viciifolium DC. ；蚕豆叶天竺葵■☆

288582　Pelargonium vinaceum E. M. Marais；葡萄酒色天竺葵■☆

288583　Pelargonium violareum Jacq. = Pelargonium tricolor Curtis ■☆

288584　Pelargonium violiflorum（Sweet）DC. ；堇花天竺葵■☆

288585　Pelargonium virgatum R. Knuth = Pelargonium otaviense R. Knuth ■☆

288586　Pelargonium virgineum Pers. = Pelargonium undulatum（Andréws）Pers. ■☆

288587　Pelargonium viscosissimum Sweet；极黏天竺葵■☆

288588　Pelargonium vitifolium（L.）L'Hér. ；葡萄叶天竺葵；Grapeleaf Geranium, Grape-leaved Geranium ■☆

288589　Pelargonium vitifolium Aiton = Pelargonium vitifolium（L.）L'Hér. ■☆

288590　Pelargonium whytei Baker；怀特天竺葵■☆

288591　Pelargonium woodii R. Knuth；伍得天竺葵■☆

288592　Pelargonium xerophyton Schltr. ex R. Knuth；沙地天竺葵；Desert Geranium ☆

288593　Pelargonium zeyheri Harv. = Pelargonium luridum（Andréws）Sweet ■☆

288594　Pelargonium zonale(L.) Aiton;马蹄纹天竺葵(马蹄天竺葵,洋葵);Horseshoe Geranium, Horseshoe Pelargonium, Horse-shoe Pelargonium, Zonal Geranium, Zonal Pelargonium ■

288595　Pelargonium zonale(L.)L'Hér. = Pelargonium zonale(L.) Aiton ■

288596　Pelargonium zonale L'Hér. ex Sol. = Pelargonium zonale(L.) Aiton ■

288597　Pelargonium zonale L'Hér. ex Sol. = Peltophorum pterocarpum(DC.) Backer ex K. Heyne ●

288598　Pelatantheria Ridl.(1896);钻柱兰属;Pelatantheria ■

288599　Pelatantheria bicuspidata(Rolfe ex Downie) Ts. Tang et F. T. Wang;尾丝钻柱兰;Twintail Pelatantheria ■

288600　Pelatantheria bicuspidata Ts. Tang et F. T. Wang = Pelatantheria bicuspidata(Rolfe ex Downie) Ts. Tang et F. T. Wang ■

288601　Pelatantheria cristata(Ridl.) Ridl.;鸡冠钻柱兰 ■☆

288602　Pelatantheria ctenoglossum Ridl.;锯尾钻柱兰;Narrowtongue Pelatantheria ■

288603　Pelatantheria insectifera(Rchb. f.) Ridl. = Pelatantheria rivesii(Guillaumin) Ts. Tang et F. T. Wang ■

288604　Pelatantheria rivesii(Guillaumin) Ts. Tang et F. T. Wang;钻柱兰;Pelatantheria, Rives Pelatantheria ■

288605　Pelatantheria scolopendrifolia(Makino) Aver.;蜈蚣兰(白脚蜈蚣,百脚蜈蚣,柏子兰,齿牙半枝莲,飞天蜈蚣,狗牙半枝,瓜子菜,金百脚,石蜈蚣,蜈蚣草,岩路,岩洛);Centipede Closedspurorchis, Scolopendrifolious Cleisostoma ■

288606　Pelea A. Gray = Melicope J. R. Forst. et G. Forst. ●

288607　Pelea madagascarica Baill. = Vepris madagascarica(Baill.) H. Perrier ●☆

288608　Pelecinus Mill. = Biserrula L. ■☆

288609　Pelecostemon Léonard(1958);斧蕊爵床属 ☆

288610　Pelecostemon trianae Léonard;斧蕊爵床 ☆

288611　Pelecynthis E. Mey. = Rafnia Thunb. ■☆

288612　Pelecynthis corymbosa E. Mey. = Rafnia capensis(L.) Schinz ■☆

288613　Pelecynthis gibba E. Mey. = Rafnia capensis(L.) Schinz subsp. dichotoma(Eckl. et Zeyh.) G. J. Campb. et B. -E. van Wyk ■☆

288614　Pelecynthis rhomboida E. Mey. = Rafnia capensis(L.) Schinz subsp. ovata(P. J. Bergius) G. J. Campb. et B. -E. van Wyk ■☆

288615　Pelecyphora C. Ehrenb.(1843);斧突球属;Hatchet Cactus ●

288616　Pelecyphora aselliformis C. Ehrenb.;精巧球(精巧丸,坡勒仙人掌,仙人斧);Hatchet Cactus, Peyotillo ●

288617　Pelecyphora pseudopectinata Backeb.;精巧殿;Falsepectinate Hatchet Cactus ●☆

288618　Pelecyphora strobiliformis(Werderm.) Frič et Schelle ex Kreuz.;松果斧突球;Pinecone Cactus ●☆

288619　Pelexia Lindl. = Pelexia Poit. ex Lindl.(保留属名)■

288620　Pelexia Poit. ex Lindl.(1826)(保留属名);肥根兰属(异兰属);Pelexia ■

288621　Pelexia Poit. ex Rich. = Pelexia Poit. ex Lindl.(保留属名)■

288622　Pelexia Rich. = Pelexia Poit. ex Lindl.(保留属名)■

288623　Pelexia adnata(Sw.) Spreng.;贴生肥根兰;Hachuela ■☆

288624　Pelexia calcarata(Sw.) Cogn. = Eltroplectris calcarata(Sw.) Garay et H. R. Sweet ■☆

288625　Pelexia cranichoides Griseb. = Cyclopogon cranichoides(Griseb.) Schltr. ■☆

288626　Pelexia falcata(Thunb.) Spreng. = Cephalanthera falcata(Thunb. ex A. Murray) Blume ■

288627　Pelexia japonica Spreng. = Cephalanthera falcata(Thunb. ex A. Murray) Blume ■

288628　Pelexia obliqua(J. J. Sm.) Garay;偏斜肥根兰 ■

288629　Pelexia setacea Lindl. var. glabra Cogn. = Eltroplectris calcarata(Sw.) Garay et H. R. Sweet ■☆

288630　Pelianthus E. Mey. ex Moq.(1849);铅花苋属(斧花苋属)●☆

288631　Pelianthus E. Mey. ex Moq. = Hermbstaedtia Rchb. ■●☆

288632　Pelianthus celosioides E. Mey. ex Moq.;铅花苋 ●☆

288633　Pelianthus celosioides E. Mey. ex Moq. = Hermbstaedtia caffra(Meisn.) Moq. ■☆

288634　Pelidnia Barnhart = Utricularia L. ■

288635　Peliosanthaceae Salisb.;球子草科 ■

288636　Peliosanthaceae Salisb. = Convallariaceae L. ■

288637　Peliosanthaceae Salisb. = Ruscaceae M. Roem.(保留科名)●

288638　Peliosanthes Andréws(1808);球子草属;Peliosanthes ■

288639　Peliosanthes arisanensis Hayata;阿里山球子草 ■

288640　Peliosanthes arisanensis Hayata = Peliosanthes macrostegia Hance ■

288641　Peliosanthes arisanensis Hayata = Peliosanthes teta Andréws var. humilis(Andréws) M. J. Lai ■

288642　Peliosanthes delavayi Franch. = Peliosanthes macrostegia Hance ■

288643　Peliosanthes humilis Andréws = Peliosanthes teta Andréws var. humilis(Andréws) M. J. Lai ■

288644　Peliosanthes kaoi Ohwi;高氏球子草(台东球子草);Kao Peliosanthes ■

288645　Peliosanthes macrophylla Wall. ex Baker;大叶球子草;Largeleaf Peliosanthes ■

288646　Peliosanthes macrostegia Hance;大盖球子草(矮球子草,扁担七,大叶球子草,蓼叶伸筋,入地蜈蚣,蜘蛛草);Bigcavor Peliosanthes, Smallflower Peliosanthes ■

288647　Peliosanthes mairei H. Lév. = Maianthemum atropurpureum(Franch.) LaFrankie ■

288648　Peliosanthes minor Yamam. = Peliosanthes teta Andréws ■

288649　Peliosanthes ophiopogonoides F. T. Wang et Ts. Tang;长苞球子草(毛标七);Longbract Peliosanthes ■

288650　Peliosanthes pachystachya W. H. Chen et Y. M. Shui;粗穗球子草;Bigstachys Peliosanthes ■

288651　Peliosanthes sinica F. T. Wang et Ts. Tang;匍匐球子草(老鼠竹);Chinese Peliosanthes, Creeping Peliosanthes ■

288652　Peliosanthes stenophylla Merr. = Ophiopogon stenophyllus(Merr.) Rodr. ■

288653　Peliosanthes tashiroi Hayata;台北球子草 ■

288654　Peliosanthes tashiroi Hayata = Peliosanthes macrostegia Hance ■

288655　Peliosanthes tashiroi Hayata = Peliosanthes teta Andréws var. humilis(Andréws) M. J. Lai ■

288656　Peliosanthes teta Andréws;簇花球子草(阿里山球子草,过界蜈蚣,过岭蜈蚣,球子草,上山蜈蚣,小球子草);Clustered Peliosanthes, Clusterflower Peliosanthes ■

288657　Peliosanthes teta Andréws subsp. humilis Andréws = Peliosanthes teta Andréws var. humilis(Andréws) M. J. Lai ■

288658　Peliosanthes teta Andréws var. humilis(Andréws) M. J. Lai;矮球子草(台北球子草);Low Peliosanthes ■

288659　Peliosanthes teta Andréws var. kaoi(Ohwi) S. S. Ying = Peliosanthes kaoi Ohwi ■

288660　Peliosanthes tonkinensis F. T. Wang et Ts. Tang = Peliosanthes teta Andréws ■

288661　Peliosanthes torulosa Y. Wan = Peliosanthes teta Andréws ■

288662　Peliosanthes yunnanensis F. T. Wang et Ts. Tang;云南球子草;Yunnan Peliosanthes ■

288663　Peliostomum Benth. = Peliostomum E. Mey. ex Benth. ■●☆

288664　Peliostomum E. Mey. ex Benth. (1836);铅口玄参属■●☆

288665　Peliostomum calycinum N. E. Br. ;萼状铅口玄参■☆

288666　Peliostomum leucorrhizum E. Mey. ex Benth. ;白根铅口玄参■☆

288667　Peliostomum leucorrhizum E. Mey. ex Benth. var. junceum Hiern = Aptosimum junceum(Hiern)Philcox ■☆

288668　Peliostomum leucorrhizum E. Mey. ex Benth. var. linearifolium F. E. Weber = Aptosimum lugardiae (N. E. Br. ex Hemsl. et V. Naray.)E. Phillips ■☆

288669　Peliostomum linearifolium Schinz ex Kuntze = Aptosimum lugardiae(N. E. Br. ex Hemsl. et V. Naray.)E. Phillips ■☆

288670　Peliostomum lugardiae N. E. Br. ex Hemsl. et V. Naray. = Aptosimum lugardiae(N. E. Br. ex Hemsl. et V. Naray.)E. Phillips ■☆

288671　Peliostomum marlothii Engl. = Aptosimum marlothii (Engl.) Hiern ■☆

288672　Peliostomum oppositifolium Engl. = Jamesbrittenia fruticosa (Benth.) Hilliard ■☆

288673　Peliostomum origanoides E. Mey. ex Benth. ;牛至铅口玄参■☆

288674　Peliostomum scoparium E. Mey. ex Benth. = Anticharis scoparia (E. Mey. ex Benth.)Hiern ex Benth. et Hook. f. ■☆

288675　Peliostomum virgatum E. Mey. ex Benth. ;条纹铅口玄参■☆

288676　Peliostomum viscosum E. Mey. ex Benth. ;黏铅口玄参■☆

288677　Peliotes Harv. et Sond. = Peliotis E. Mey. ●☆

288678　Peliotis E. Mey. = Lonchostoma Wikstr. (保留属名)●☆

288679　Peliotus E. Mey. = Lonchostoma Wikstr. (保留属名)●☆

288680　Peliotus E. Mey. = Peliotis E. Mey. ●☆

288681　Pella Gaertn. = Ficus L. ●

288682　Pellacalyx Korth. (1836);山红树属;Pellacalyx,Wildmangrove ●

288683　Pellacalyx axillaris Korth. ;腋花山红树;Axilflower Pellacalyx, Axilflower Wildmangrove ●☆

288684　Pellacalyx yunnanensis Hu;山红树;Yunnan Pellacalyx,Yunnan Wildmangrove ●◇

288685　Pellea André = Pellionia Gaudich. (保留属名)●●

288686　Pellegrinia Sleumer(1935);毛丝莓属●☆

288687　Pellegrinia buxifolia(Gardn. et Fielding) Sleumer;毛丝莓●☆

288688　Pellegrinia colombiana(A. C. Sm.)Cuatrec. ;哥伦比亚毛丝莓●☆

288689　Pellegrinia grandiflora(Ruiz et Pav.)Sleumer;大花毛丝莓●☆

288690　Pellegriniodendron J. Léonard(1955);热非二叶豆属●☆

288691　Pellegriniodendron diphyllum(Harms)J. Léonard;热非二叶豆■☆

288692　Pelleteria Poir. = Pelletiera A. St. -Hil. ■☆

288693　Pelletiera A. St. -Hil. (1822);三瓣花属■☆

288694　Pelletiera serpyllifolia (Poir.) Webb et Berthel. ;百里香叶三瓣花■☆

288695　Pelletiera wildpretii Valdés;三瓣花■☆

288696　Pelliciera Planch. et Triana ex Benth. = Pelliciera Planch. et Triana ●☆

288697　Pelliciera Planch. et Triana ex Benth. et Hook. f. = Pelliciera Planch. et Triana ●☆

288698　Pelliciera Planch. et Triana(1862)(' Pelliceria ');肖红树属(假红树属,中美洲红树属)●☆

288699　Pelliciera rhizophorae Planch. et Triana;肖红树(假红树,中美洲红树);Rhizophor Pelliciera ●☆

288700　Pellicieraceae (Triana et Planch.) L. Beauvis. ex Bullock. (1959);肖红树科(假红树科,中美洲红树科)●☆

288701　Pellicieraceae Bullock. = Pellicieraceae (Triana et Planch.) L. Beauvis. ex Bullock. ●☆

288702　Pellicieraceae L. Beauvis. = Pellicieraceae (Triana et Planch.) L. Beauvis. ex Bullock. ●☆

288703　Pellicieraceae L. Beauvis. = Tetrameristaceae Hutch. ●☆

288704　Pellinia Molina = Eucryphia Cav. ●☆

288705　Pellionia Gaudich. (1830)(保留属名);赤车属(赤车使者属);Pellionia, Redcarweed ●■

288706　Pellionia Gaudich. = Elatostema J. R. Forst. et G. Forst. (保留属名)●■

288707　Pellionia acaulis Hook. f. = Pellionia latifolia(Blume) Boerl. ■

288708　Pellionia acutidentata W. T. Wang;尖齿赤车;Acutidentate Pellionia,Sharptooth Redcarweed ■

288709　Pellionia annamica Gagnep. = Pellionia repens(Lour.) Merr. ■

288710　Pellionia arisanensis Hayata = Pellionia radicans (Siebold et Zucc.) Wedd. ■

288711　Pellionia arisanensis Hayata var. pygmaea Yamam. = Pellionia radicans(Siebold et Zucc.) Wedd. ■

288712　Pellionia balansae Gagnep. = Pellionia latifolia(Blume) Boerl. ■

288713　Pellionia bodinieri H. Lév. = Elatostema oblongifolium S. H. Fu ex W. T. Wang ■

288714　Pellionia brachyceras W. T. Wang;短角赤车;Shortangle Pellionia,Shortangle Redcarweed ■

288715　Pellionia brevifolia Benth. ;短叶赤车(猴接骨草,小叶赤车); Shortleaf Redcarweed,Smallleaf Pellionia ■

288716　Pellionia caulialata S. Y. Liu;翅茎赤车;Webstem Pellionia, Webstem Redcarweed ■

288717　Pellionia cephaloidea W. T. Wang;头序赤车;Head Pellionia, Head Redcarweed ●■

288718　Pellionia cephaloidea W. T. Wang = Pellionia scabra Benth. ●■

288719　Pellionia chikushiensis Yamam. = Pellionia radicans(Siebold et Zucc.)Wedd. ■

288720　Pellionia crispulihirtella W. T. Wang = Pellionia veronicoides Gagnep. ■

288721　Pellionia daveauana N. E. Br. = Pellionia repens(Lour.) Merr. ■

288722　Pellionia daveauana N. E. Br. var. viridis N. E. Br. = Pellionia repens(Lour.)Merr. ■

288723　Pellionia esquirolii H. Lév. = Elatostema parvum(Blume)Miq. ■

288724　Pellionia funingensis W. T. Wang;富宁赤车;Funing Pellionia, Funing Redcarweed ■

288725　Pellionia funingensis W. T. Wang = Pellionia grijsii Hance ■

288726　Pellionia griffithiana Wedd. = Pellionia heteroloba Wedd. ■

288727　Pellionia grijsii Hance;华南赤车;S. China Redcarweed,South China Pellionia ■

288728　Pellionia helferiana Wedd. = Pellionia latifolia(Blume) Boerl. ■

288729　Pellionia heteroloba Wedd. ;异被赤车(六耳零,楼梯草,日本赤车); Heterolobe Pellionia, Heterolobe Redcarweed, Japanese Pellionia ■

288730　Pellionia heteroloba Wedd. var. minor W. T. Wang;小异被赤车;Small Heterolobe Pellionia ■

288731　Pellionia heteroloba Wedd. var. minor W. T. Wang = Pellionia heteroloba Wedd. ■

288732　Pellionia heyneana Wedd. ;全缘赤车;Entireleaf Redcarweed, Entire Pellionia ●■

288733　Pellionia incisoserrata (H. Schroet.) W. T. Wang;羽脉赤车;Pinnateleaf Pellionia, Pinnateleaf Redcarweed ■

288734　Pellionia japonica Hatus. = Pellionia keitaoensis Yamam. ■

288735　Pellionia javanica(Wedd.) Wedd. = Pellionia latifolia(Blume) Boerl. ■

288736　Pellionia javanica(Wedd.) Wedd. var. acaulis Ridl. = Pellionia latifolia（Blume）Boerl. ■

288737　Pellionia javanica（Wedd.）Wedd. var. minor Ridl. = Pellionia latifolia（Blume）Boerl. ■

288738　Pellionia keitaoensis Yamam. = Pellionia heteroloba Wedd. ■

288739　Pellionia keitaoensis Yamam. f. yosiei（H. Hara）Hatus. = Pellionia yosiei H. Hara ■☆

288740　Pellionia latifolia（Blume）Boerl.；长柄赤车（半边风，绿赤车）；Longstalk Redcarweed，P. C. Tsoong Pellionia ■

288741　Pellionia leiocarpa W. T. Wang；光果赤车；Nakedfruit Pellionia，Nakedfruit Redcarweed ■

288742　Pellionia longgangensis W. T. Wang；弄岗赤车；Nonggang Pellionia ■

288743　Pellionia longgangensis W. T. Wang = Pellionia paucidentata（H. Schroet.）S. S. Chien ■

288744　Pellionia longipedunculata W. T. Wang；长梗赤车；Longpeduncle Pellionia，Longpeduncle Redcarweed ■

288745　Pellionia macrophylla W. T. Wang；大叶赤车；Bigleaf Pellionia，Bigleaf Redcarweed ■

288746　Pellionia mairei H. Lév. = Elatostema monandrum（D. Don）Hara ■

288747　Pellionia menglianensis Y. Y. Qian；孟连赤车；Menglian Pellionia，Menglian Redcarweed ■

288748　Pellionia menglianensis Y. Y. Qian = Pellionia heteroloba Wedd. ■

288749　Pellionia minima Makino；小赤车（卜罗草，山椒草，塌地草）；Small Pellionia，Small Redcarweed ■

288750　Pellionia minima Makino = Pellionia brevifolia Benth. ■

288751　Pellionia myrtillus H. Lév. = Elatostema myrtillus（H. Lév.）Hand. -Mazz. ■

288752　Pellionia paucidentata（H. Schroet.）S. S. Chien；滇南赤车；Fewdentate Pellionia，S. Yunnan Redcarweed ■

288753　Pellionia paucidentata（H. Schroet.）S. S. Chien var. hainanensis S. S. Chien et S. H. Wu = Pellionia paucidentata（H. Schroet.）S. S. Chien ■

288754　Pellionia paucidentata（H. Schroet.）S. S. Chien var. hainanensis S. S. Chien et S. H. Wu；海南赤车（无毛滇南赤车）；Hainan Pellionia ■

288755　Pellionia paucidentata var. hainanica S. S. Chien et S. H. Wu = Pellionia paucidentata（H. Schroet.）S. S. Chien ■

288756　Pellionia pauciflora W. T. Wang；少花赤车■

288757　Pellionia pauciflora W. T. Wang = Pellionia radicans（Siebold et Zucc.）Wedd. ■

288758　Pellionia pierrei Gagnep. = Pellionia latifolia（Blume）Boerl. ■

288759　Pellionia procropioides Gagnep. = Procris crenata C. B. Rob. ●■

288760　Pellionia pulchra N. E. Br.；美赤车；Beautiful Pellionia ■☆

288761　Pellionia pulchra N. E. Br. = Pellionia repens（Lour.）Merr. ■

288762　Pellionia radicans（Siebold et Zucc.）Wedd.；赤车（阿里山赤车使者，拔血红，半边山，赤车使者，吊血丹，风湿草，凤阳草，坑兰，冷坑清，毛骨草，少花赤车，天门草，乌梗子，小阿里山赤车使者，小铁木，岩下青，阴蒙藤）；Redcarweed，Rooted Pellionia ■

288763　Pellionia radicans（Siebold et Zucc.）Wedd. f. grandis Gagnep.；长茎赤车；Longstem Rooted Pellionia ■

288764　Pellionia radicans（Siebold et Zucc.）Wedd. f. grandis Gagnep. = Pellionia radicans（Siebold et Zucc.）Wedd. ■

288765　Pellionia radicans（Siebold et Zucc.）Wedd. var. grandis（Gagnep.）W. T. Wang = Pellionia radicans（Siebold et Zucc.）Wedd. f. grandis Gagnep. ■

288766　Pellionia radicans（Siebold et Zucc.）Wedd. var. grandis（Gagnep.）W. T. Wang = Pellionia radicans（Siebold et Zucc.）

Wedd. ■

288767　Pellionia repens（Lour.）Merr.；吐烟花（达沃赤车，喷烟花，披翼凤，匍匐赤车，吐烟草）；Creeping Pellionia，Smaoke Redcarweed，Trailing Watermelon Begonia，Watermelon Begonia ■

288768　Pellionia retrohispida W. T. Wang；曲毛赤车；Bendehair Redcarweed，Pellionia ■

288769　Pellionia scabra Benth.；蔓赤车（半边山，糙叶赤车使者，粗糙赤车使者，粗糙楼梯草，鸡骨香，接骨草，坑兰，坑冷，楼梯草，毛赤车，南投赤车使者，青麻，入脸麻，石解骨，水靛青，水田草，香蕉草，岩苋菜，羊眼草）；Rough Pellionia，Vine Redcarweed ●■

288770　Pellionia scabra Benth. subvar. pedunculata Yamam. = Pellionia scabra Benth. ●■

288771　Pellionia subundulata W. T. Wang；波缘赤车；Subundulate Pellionia，Waveedge Redcarweed ■

288772　Pellionia subundulata W. T. Wang = Pellionia paucidentata（H. Schroet.）S. S. Chien ■

288773　Pellionia subundulata W. T. Wang var. angustifolia W. T. Wang；狭叶赤车；Narrowleaf Subundulate Pellionia，Narrowleaf Waveedge Redcarweed ■

288774　Pellionia subundulata W. T. Wang var. angustifolia W. T. Wang = Pellionia longipedunculata W. T. Wang ■

288775　Pellionia trichosantha Gagnep. = Elatostema pycnodontum W. T. Wang ■

288776　Pellionia trilobulata Hayata；裂叶赤车使者■

288777　Pellionia trilobulata Hayata = Elatostema obtusum Wedd. var. trilobulatum（Hayata）W. T. Wang ■

288778　Pellionia trilobulata Hayata = Elatostema trilobulatum（Hayata）T. Yamaz. ■

288779　Pellionia tsoongii（Merr.）Merr. = Pellionia latifolia（Blume）Boerl. ■

288780　Pellionia veronicoides Gagnep.；硬毛赤车；Hardhair Pellionia，Hardhair Redcarweed ■

288781　Pellionia viridis C. H. Wright；绿赤车；Green Pellionia，Green Redcarweed ●■

288782　Pellionia viridis C. H. Wright var. basiinaequalis C. H. Wright；斜基绿赤车●

288783　Pellionia yosiei H. Hara；义江赤车■☆

288784　Pellionia yunnanensis（H. Schroet.）W. T. Wang；云南赤车；Yunnan Pellionia，Yunnan Redcarweed ●■

288785　Pellocalyx Post et Kuntze = Pellacalyx Korth. ●

288786　Pelma Finet = Bulbophyllum Thouars（保留属名）■

288787　Pelonastes Hook. f. = Myriophyllum L. ■

288788　Peloria Adans. = Linaria Mill. ■

288789　Pelotris Raf. = Muscari Mill. ■☆

288790　Pelozia Rose = Lopezia Cav. ■☆

288791　Pelta Dulac = Zannichellia L. ■

288792　Peltactila Raf. = Daucus L. ■

288793　Peltaea（C. Presl）Standl.（1916）（保留属名）；盾锦葵属●☆

288794　Peltaea（Gllrke）Standl. = Peltaea（C. Presl）Standl.（保留属名）●☆

288795　Peltaea Standl. = Peltaea（C. Presl）Standl.（保留属名）●☆

288796　Peltaea acutifolia（Gürke）Krapov. et Cristobal；尖叶盾锦葵●☆

288797　Peltaea ovata Standl.；卵形盾锦葵●☆

288798　Peltaea sessiliflora Standl.；无梗盾锦葵●☆

288799　Peltaea trinervis（C. Presl）Krapov. et Cristóbal；三脉盾锦葵●☆

288800　Peltandra Raf.（1819）（保留属名）；盾蕊南星属（盾蕊芋属）；Arrow Arum，Arrow-arum ■☆

288801　Peltandra Wight = Meineckia Baill. ■☆

288802　Peltandra Wight = Phyllanthus L. ●■

288803　Peltandra glauca (Elliott) Feay ex A. W. Wood = Peltandra sagittifolia(Michx.) Morong ■☆

288804　Peltandra luteospadix Fernald = Peltandra virginica(L.)Schott ■☆

288805　Peltandra sagittifolia (Michx.) Morong；白盾蕊南星；Spoonflower, White Arrow Arum ■☆

288806　Peltandra tharpii F. A. Barkley = Peltandra virginica(L.)Schott ■☆

288807　Peltandra virginica(L.)Kunth = Peltandra virginica(L.)Schott ■☆

288808　Peltandra virginica (L.) Schott；盾蕊南星；Arrow Arum, Green Arrow-arum, Tuckahoe ■☆

288809　Peltandra virginica (L.) Schott et Endl. f. hastifolia S. F. Blake = Peltandra virginica(L.)Schott ■☆

288810　Peltanthera Benth. (1876)(保留属名)；盾药草属●☆

288811　Peltanthera Roth(废弃属名) = Peltanthera Benth. (保留属名)●☆

288812　Peltanthera Roth(废弃属名) = Vallaris Burm. f. ●

288813　Peltanthera floribunda Benth. ；盾药草●☆

288814　Peltanthera solanacea Roth = Vallaris solanacea(Roth)Kuntze ●

288815　Peltantheria Post et Kuntze = Pelatantheria Ridl. ■

288816　Peltaria Burm. ex DC. = Wiborgia Thunb. (保留属名)■☆

288817　Peltaria DC. = Wiborgia Thunb. (保留属名)■☆

288818　Peltaria Jacq. (1762)；盾形草属；Shieldwort ■☆

288819　Peltaria aspera Grauer = Clypeola aspera(Grauer)Turrill ■☆

288820　Peltaria aucheri Boiss. ；奥氏盾形草■☆

288821　Peltaria capensis Burm. f. = Wiborgia fusca Thunb. ■☆

288822　Peltaria capensis Thunb. = Heliophila suborbicularis Al-Shehbaz et Mummenhoff ■☆

288823　Peltaria emarginata(Boiss.)Hausskn. ；微缺盾形草■☆

288824　Peltaria turkmena Lipsky；土库曼盾形草■☆

288825　Peltaria voronovii N. Busch；沃氏盾形草■☆

288826　Peltariopsis(Boiss.)N. Busch = Cochlearia L. ■

288827　Peltariopsis(Boiss.)N. Busch(1927)；假盾草属■

288828　Peltariopsis N. Busch = Peltariopsis(Boiss.)N. Busch ■

288829　Peltariopsis grossheimi N. Busch；格罗假盾草■☆

288830　Peltastes Woodson(1932)；盾竹桃属●☆

288831　Peltastes colombianus Woodson；哥伦盾竹桃●☆

288832　Pelticalyx Griff. (1854)；盾萼番荔枝属●☆

288833　Pelticalyx Griff. = ? Desmos Lour. ●

288834　Pelticalyx argentea Griff. ；盾萼番荔枝●☆

288835　Peltidium Zollik. = Chondrilla L. ■

288836　Peltiera Du Puy et Labat(1997)；盾豆木属●☆

288837　Peltiera alaotrensis Du Puy et Labat；盾豆木●☆

288838　Peltiera nitida Du Puy et Labat；光亮盾豆木●☆

288839　Peltimela Raf. (废弃属名) = Glossostigma Wight et Arn. (保留属名)■☆

288840　Peltiphyllum(Engl.)Engl. (1891)；盾叶属；Saxifrage ■☆

288841　Peltiphyllum(Engl.)Engl. = Darmera Voss ■☆

288842　Peltiphyllum Engl. = Darmera Voss ■☆

288843　Peltiphyllum Engl. = Peltiphyllum(Engl.)Engl. ■☆

288844　Peltiphyllum peltatum Engl. ；盾叶■☆

288845　Peltiphyllum peltatum Engl. = Darmera peltata (Torr. ex Benth.)Voss ■☆

288846　Peltispermum Moq. = Anthochlamys Fenzl ■☆

288847　Peltoboykinia(Engl.)Hara(1937)；涧边草属；Shoregrass ■

288848　Peltoboykinia tellimoides(Maxim.)H. Hara；涧边草(八幡草)；Japonese Boykinia, Shoregrass ■

288849　Peltoboykinia tellimoides (Maxim.) H. Hara var. watanabei

(Yatabe)H. Hara = Peltoboykinia watanabei(Yatabe)H. Hara ■☆

288850　Peltoboykinia watanabei(Yatabe)H. Hara；渡边涧边草■☆

288851　Peltobractea Rusby = Peltaea(C. Presl)Standl. (保留属名)●☆

288852　Peltobryon Klotzsch = Piper L. ●■

288853　Peltobryon Klotzsch ex Miq. = Piper L. ●■

288854　Peltocalathus Tamura = Ranunculus L. ■

288855　Peltocalathos Tamura(1992)；南非毛茛属■☆

288856　Peltocalathos baurii(MacOwan)Tamura；南非毛茛■☆

288857　Peltocalyx Post et Kuntze = ? Desmos Lour. ●

288858　Peltocalyx Post et Kuntze = Pelticalyx Griff. ●☆

288859　Peltodon Pohl(1827)；盾齿花属●■☆

288860　Peltodon radicans Pohl；生根盾齿花●☆

288861　Peltodon rugosus Tolm. ；粗糙盾齿花■☆

288862　Peltodon tomentosus Pohl；毛盾齿花■☆

288863　Peltogyne Vogel(1837)(保留属名)；紫心苏木属；Amaranth, Purpleheart ●☆

288864　Peltogyne catingae Ducke；卡提紫心苏木●☆

288865　Peltogyne confertiflora Benth. ；密花紫心苏木●☆

288866　Peltogyne discolor Vogel；异色紫心苏木●☆

288867　Peltogyne lecointei Ducke；紫心苏木●☆

288868　Peltogyne maranhensis Ducke；巴西紫心苏木●☆

288869　Peltogyne mexicana Martinez；墨西哥紫心苏木●☆

288870　Peltogyne paniculata Benth. ；圆锥紫心苏木；Purple Heart ●☆

288871　Peltogyne paradoxa Ducke；奇异紫心苏木●☆

288872　Peltogyne pubescens Benth. ；短柔毛紫心苏木(毛紫心苏木)●☆

288873　Peltogyne purpurea Pittier；巴拿马紫心苏木●☆

288874　Peltogyne venosa Benth. ；具脉紫心苏木●☆

288875　Peltomesa Raf. = Struthanthus Mart. (保留属名)●☆

288876　Peltophora Benth. et Hook. f. = Manisuris L. (废弃属名)■

288877　Peltophora Benth. et Hook. f. = Peltophorus Desv. ■☆

288878　Peltophoropsis Chiov. (1915)；类盾柱木属●☆

288879　Peltophoropsis Chiov. = Parkinsonia L. ●

288880　Peltophoropsis scioana Chiov. ；类盾柱木●☆

288881　Peltophoropsis scioana Chiov. = Parkinsonia scioana (Chiov.) Brenan ●☆

288882　Peltophorum(Vogel)Benth. (1840)(保留属名)；盾柱木属(双翼豆属,双翼苏木属)；Brasiletto, Peltophorum, Peltostyle ■

288883　Peltophorum Walp. = Peltophorum(Vogel)Benth. (保留属名)■

288884　Peltophorum africanum Sond. ；非洲盾柱木(非洲盾柱树,非洲双翼豆)；African Black Wattle, African Blackwood, African Peltophorum, Black Wood, Rhodesian Wattle ●☆

288885　Peltophorum dasyrhachis Baker = Peltophorum dasyrhachis Kurz ex Baker ●☆

288886　Peltophorum dasyrhachis Kurz ex Baker；粗轴盾柱木(粗轴双翼苏木)●☆

288887　Peltophorum dasyrrhachis (Miq.) Kurz var. tonkinensis (Pierre) K. Larsen et S. S. Larsen = Peltophorum tonkinense(Pierre)Gagnep. ■

288888　Peltophorum dasyrrhachis Kurz ex Baker var. tonkinensis (Pierre)K. Larsen et S. S. Larsen = Peltophorum tonkinense(Pierre) Gagnep. ■

288889　Peltophorum dubium(Spreng.)Taub. ；南美盾柱木(卷瓣双翼豆,有疑双翼苏木)；Horsebush ●☆

288890　Peltophorum dubium Taub. = Peltophorum dubium (Spreng.) Taub. ●☆

288891　Peltophorum ferrugineum (Decne.) Benth. = Peltophorum pterocarpum(DC.)Backer ex K. Heyne ■

288892　Peltophorum ferrugineum Benth. = Peltophorum pterocarpum

（DC.）Backer ex K. Heyne ■

288893　Peltophorum inerme（Benth.）Llanos = Peltophorum pterocarpum（DC.）Backer ex K. Heyne ■

288894　Peltophorum inerme（Roxb.）Náves = Peltophorum pterocarpum（DC.）Backer ex K. Heyne ■

288895　Peltophorum inerme（Roxb.）Náves ex Fern.-Vill. = Peltophorum pterocarpum（DC.）Baker ex K. Heyne ■

288896　Peltophorum massaiense Taub. = Bussea massaiensis（Taub.）Harms ●☆

288897　Peltophorum pterocarpum（DC.）Backer ex K. Heyne；盾柱木（盾柱豆，双翅果，双翼豆）；Braziletto Wood, Copperpod, Peltophorum, Rusty Shield-bearer, Soga, Wingfruit Peltophorum, Wingfruit Peltostyle, Wing-fruited Pterocarpous Peltophorum, Yellow Flamboyant, Yellow Flame, Yellow Flame Tree, Yellow Flame-tree, Yellow Gold Mahur, Yellow Poinciana, Yellow-flame ■

288898　Peltophorum pterocarpum（DC.）K. Heyne = Peltophorum pterocarpum（DC.）Backer ex K. Heyne ■

288899　Peltophorum roxburghii（G. Don）Degener = Peltophorum pterocarpum（DC.）Backer ex K. Heyne ■

288900　Peltophorum roxburghii G. Don = Peltophorum pterocarpum（DC.）Backer ex K. Heyne ■

288901　Peltophorum tonkinense（Pierre）Gagnep.；东京盾柱木（双翼豆,田螺掩,银珠,油楠）；Tonkin Peltophorum, Tonkin Peltostyle ■

288902　Peltophorum vogelianum Benth.；沃格盾柱木（沃格双翼苏木）●☆

288903　Peltophorus Desv. = Manisuris L.（废弃属名）■

288904　Peltophorus sulcatus Stapf = Heteropholis sulcata（Stapf）C. E. Hubb. ■☆

288905　Peltophyllum Gardner = Triuris Miers ■☆

288906　Peltophyllum Gardner（1843）；盾叶霉草属■☆

288907　Peltophyllum luteum Gardn.；盾叶霉草■☆

288908　Peltopsis Raf. = Potamogeton L. ■

288909　Peltopus（Schltr.）Szlach. et Marg.（2002）；盾足兰属■☆

288910　Peltopus（Schltr.）Szlach. et Marg. = Bulbophyllum Thouars（保留属名）■

288911　Peltospermum Benth. = Sacosperma G. Taylor ●☆

288912　Peltospermum DC. = Aspidosperma Mart. et Zucc.（保留属名）●☆

288913　Peltospermum Post et Kuntze = Anthochlamys Fenzl ■☆

288914　Peltospermum Post et Kuntze = Peltispermum Moq. ■☆

288915　Peltospermum paniculatum Benth. = Sacosperma paniculatum（Benth.）G. Taylor ●☆

288916　Peltostegia Turcz.（废弃属名）= Peltaea（C. Presl）Standl.（保留属名）●☆

288917　Peltostegia Turcz. = Kosteletzkya C. Presl（保留属名）■●☆

288918　Peltostigma Walp.（1846）；盾柱芸香属●☆

288919　Peltostigma parviflorum Q. Jiménez et Gereau；小花盾柱芸香●☆

288920　Peltostigma pentaphyllum Donn. Sm.；五叶盾柱芸香●☆

288921　Peltostigma ptelioides Walp.；盾柱芸香●☆

288922　Pelucha S. Watson（1889）；毛黄菊属●☆

288923　Pelucha trifida S. Watson；毛黄菊●☆

288924　Pembertonia P. S. Short（2004）；宽鳞鹅河菊属■☆

288925　Pembertonia latisquamata（F. Muell.）P. S. Short；宽鳞鹅河菊■☆

288926　Pemphis J. R. Forst. et G. Forst.（1775）；水芫花属；Pemphis ●

288927　Pemphis acidula J. R. Forst. et G. Forst.；水芫花（海纸钱鲁）；Common Pemphis, Pemphis, Reef Pemphis ●

288928　Penaea L.（1753）；管萼木属●☆

288929　Penaea acuta Thunb. = Brachysiphon acutus（Thunb.）A. Juss. ●☆

288930　Penaea acutifolia A. Juss.；尖叶管萼木●☆

288931　Penaea candolleana Stephens = Stylapterus candolleanus（Stephens）R. Dahlgren ●☆

288932　Penaea cneorum Meerb. subsp. gigantea R. Dahlgren；巨大管萼木●☆

288933　Penaea cneorum Meerb. subsp. lanceolata R. Dahlgren；披针形管萼木●☆

288934　Penaea cneorum Meerb. subsp. ovata（Eckl. et Zeyh. ex A. DC.）R. Dahlgren；卵形管萼木●☆

288935　Penaea cneorum Meerb. subsp. ruscifolia R. Dahlgren；假叶树管萼木●☆

288936　Penaea dahlgrenii Rourke；达尔管萼木●☆

288937　Penaea dubia Stephens = Stylapterus dubius（Stephens）R. Dahlgren ●☆

288938　Penaea ericoides（A. Juss.）Endl. = Stylapterus ericoides A. Juss. ●☆

288939　Penaea formosa Thunb. = Glischrocolla formosa（Thunb.）R. Dahlgren ●☆

288940　Penaea fruticulosa L. f. = Stylapterus fruticulosus（L. f.）A. Juss. ●☆

288941　Penaea fucata L. = Brachysiphon fucatus（L.）Gilg ●☆

288942　Penaea imbricata Graham = Brachysiphon fucatus（L.）Gilg ●☆

288943　Penaea lateriflora L. f. = Endonema lateriflora（L. f.）Gilg ●☆

288944　Penaea mucronata L.；短尖管萼木●☆

288945　Penaea ovata Eckl. et Zeyh. ex A. DC. = Penaea cneorum Meerb. subsp. ovata（Eckl. et Zeyh. ex A. DC.）R. Dahlgren ●☆

288946　Penaea sarcocolla L. = Saltera sarcocolla（L.）Bullock ●☆

288947　Penaea tetragona P. J. Bergius = Saltera sarcocolla（L.）Bullock ●☆

288948　Penaeaceae Guill. = Penaeaceae Sweet ex Guill.（保留科名）●☆

288949　Penaeaceae Sweet ex Guill.（1828）（保留科名）；管萼木科（管萼科）●☆

288950　Penanthes Vell. = Prenanthes L. ■

288951　Penar-Valli Adans. = Zanonia L. ●■

288952　Penarvallia Post et Kuntze = Penar-Valli Adans. ●■

288953　Penarvallia Post et Kuntze = Zanonia L. ●■

288954　Pendlphylis Thouars = Bulbophyllum Thouars（保留属名）■

288955　Penducella Luer et Thoerle = Lepanthes Sw. ■☆

288956　Penducella Luer et Thoerle（2010）；短枝鳞花兰属■☆

288957　Pendulina Willk. = Diplotaxis DC. ■

288958　Pendulina fontanesii Willk. = Diplotaxis harra（Forssk.）Boiss. ■☆

288959　Pendulina intricata Willk. = Diplotaxis harra（Forssk.）Boiss. ■☆

288960　Penelopeia Urb.（1921）；佩纳葫芦属■☆

288961　Penelopeia suburceolata Urb.；佩纳葫芦■☆

288962　Penianthus Miers（1867）；半花藤属●☆

288963　Penianthus camerounensis Dekker；喀麦隆半花藤●☆

288964　Penianthus fruticosus Hutch. et Dalziel = Penianthus longifolius Miers ●☆

288965　Penianthus gossweileri Exell = Penianthus longifolius Miers ●☆

288966　Penianthus klaineanus Pierre = Penianthus longifolius Miers ●☆

288967　Penianthus longifolius Miers；半花藤●☆

288968　Penianthus patulinervis Hutch. et Dalziel；展脉半花藤●☆

288969　Penianthus zenkeri（Engl.）Diels；岑克尔半花藤●☆

288970　Penicillanthemum Vieill. = Hugonia L. ●☆

288971　Penicillaria Willd. = Pennisetum Rich. ■

288972　Penicillaria deflexa Andersson ex A. Br. et Bouché = Pennisetum glaucum（L.）R. Br. ■

288973　Penicillaria fallax Fig. et De Not. = Pennisetum violaceum（Lam.）Rich. ■☆

288974　Penicillaria nigritana Schltdl. = Pennisetum glaucum(L.)R. Br. ■

288975　Penicillaria sieberiana Schltr. = Pennisetum sieberianum (Schltr.) Stapf et C. E. Hubb. ■☆

288976　Penicillaria spicata Willd. ;穗状半花藤●☆

288977　Penicillaria stenostachya Klotzsch ex A. Braun et C. D. Bouché = Pennisetum sieberianum(Schltr.) Stapf et C. E. Hubb. ■☆

288978　Penicillaria typhoides Schltdl. = Pennisetum americanum (L.) Leeke ■

288979　Penicillaria typhoides Schltdl. = Pennisetum glaucum(L.)R. Br. ■

288980　Penicillaria vulpina A. Braun et C. D. Bouché = Pennisetum vulpinum(A. Braun et C. D. Bouché) Stapf et C. E. Hubb. ■☆

288981　Peniculifera Ridl. = Trigonopleura Hook. f. ●☆

288982　Peniculus Swallen = Mesosetum Steud. ■☆

288983　Peniocereus(A. Berger) Britton et Rose (1909) ;块根柱属(鹿角掌属,丝柱属)■

288984　Peniocereus Britton et Rose = Peniocereus(A. Berger) Britton et Rose ■

288985　Peniocereus diguetii (F. A. C. Weber) Backeb. = Peniocereus striatus(Brandegee) Buxb. ■☆

288986　Peniocereus greggii(Engelm.) Britton et Rose;块根仙人鞭(鹿角掌) ; Arizona Queen-of-the-night, Desert Night-blooming Cereus, Night Blooming Cereus,Reina De La Noche ■

288987　Peniocereus greggii (Engelm.) Britton et Rose = Cereus greggii Engelm. ■

288988　Peniocereus greggii (Engelm.) Britton et Rose var. transmontanus(Engelm.) Backeb. ;夜花块根柱;Junco Espinoso, Night Blooming Cereus,Reina De La Noche ●☆

288989　Peniocereus striatus(Brandegee) Buxb. ;食用块根柱(具条纹丝柱) ;Cardoncillo,Dahlia-rooted Cereus,Jacamatraca ●☆

288990　Peniocereus viperinus(F. A. C. Weber) Buxb. ;白眉塔●☆

288991　Peniophyllum Pennell = Oenothera L. ●■

288992　Penkimia Phukan et Odyuo(2006) ;心启兰属■

288993　Penkimia nagalandensis Phukan et Odyuo;心启兰■

288994　Pennantia J. R. Forst. et G. Forst. (1775) ;澳茱萸属(盆南梯属)●☆

288995　Pennantia corymbosa J. R. Forst. et G. Forst. ;澳茱萸;Maori Fire ●☆

288996　Pennantia odorata Raoul;芳香澳茱萸●☆

288997　Pennantiaceae J. Agardh = Icacinaceae Miers(保留科名)●■

288998　Pennantiaceae J. Agardh(1858) ;澳茱萸科●☆

288999　Pennellia Nieuwl. (1918) ;彭内尔芥属■☆

289000　Pennellia boliviensis(Muschl.) Al-Shehbaz;玻利维亚彭内尔芥■☆

289001　Pennellia brachycarpa Beilstein et Al-Shehbaz;短果彭内尔芥■☆

289002　Pennellia gracilis (Wedd.) O. E. Schulz;细彭内尔芥■☆

289003　Pennellia longifolia(Benth.) Rollins;长叶彭内尔芥■☆

289004　Pennellia micrantha Nieuwl. ;小花彭内尔芥■☆

289005　Pennellianthus Crosswh. (1970) ;彭内尔婆婆纳属●☆

289006　Pennellianthus frutescens (Lamb.) Crosswh. = Penstemon frutescens Lamb. ●☆

289007　Pennilabium J. J. Sm. (1914) ;巾唇兰属;Pennilabium ■

289008　Pennilabium proboscideum A. S. Rao et J. Joseph;巾唇兰;Trunk Pennilabium ■

289009　Pennilabium yunnanensis S. C. Chen et Y. B. Luo;云南巾唇兰;Yunnan Pennilabium ■

289010　Pennisetum Pers. = Pennisetum Rich. ■

289011　Pennisetum Rich. (1805) ;狼尾草属;Fountain Grass,

Pennisetum, Wolftailgrass ■

289012　Pennisetum adoense Steud. = Pennisetum thunbergii Kunth ■☆

289013　Pennisetum advena Wipff et Veldkamp;泉边狼尾草;Fountain Grass ■☆

289014　Pennisetum albicauda Stapf et C. E. Hubb. = Pennisetum glaucum(L.) R. Br. ■

289015　Pennisetum alopecuroides(L.) Spreng. ;狼尾草(打死还阳,大狗尾草,董蓈,狗尾巴草,狗尾草,狗仔尾,狗仔尾草,狗子尾,光明草,黑狗尾草,狼茅,狼尾,稂,老鼠根,戾草,孟,芮草,守田,宿田翁,童粱,小芒草) ;China Wolftailgrass, Chinese Fountaingrass, Chinese Pennisetum, Chinese Wolftailgrass, Fountain Grass, Swamp Foxtail Grass ■

289016　Pennisetum alopecuroides(L.) Spreng. f. erythrochaetum(Ohwi) Ohwi = Pennisetum alopecuroides(L.) Spreng. ■

289017　Pennisetum alopecuroides (L.) Spreng. f. purpurascens (Thunb.) Ohwi = Pennisetum alopecuroides(L.) Spreng. ■

289018　Pennisetum alopecuroides (L.) Spreng. f. viridescens (Miq.) Ohwi;变绿狼尾草■

289019　Pennisetum alopecuroides (L.) Spreng. f. viridescens (Miq.) Ohwi = Pennisetum alopecuroides(L.) Spreng. ■

289020　Pennisetum alopecuroides (L.) Spreng. subsp. sordidum (Koidz.) T. Koyama = Pennisetum sordidum Koidz. ■☆

289021　Pennisetum alopecuroides (L.) Spreng. var. albiflorum Y. N. Lee;白花狼尾草■☆

289022　Pennisetum alopecuroides(L.) Spreng. var. erythrochaetum Ohwi = Pennisetum alopecuroides(L.) Spreng. ■

289023　Pennisetum alopecuroides (L.) Spreng. var. erythrochaetum Ohwi = Pennisetum alopecuroides (L.) Spreng. f. erythrochaetum (Ohwi) Ohwi ■

289024　Pennisetum alopecuroides(L.) Spreng. var. viridescens(Miq.) Ohwi = Pennisetum alopecuroides(L.) Spreng. f. viridescens(Miq.) Ohwi ■

289025　Pennisetum alopecuroides Spreng. = Pennisetum setosum(Sw.) Rich. ■

289026　Pennisetum alopecuroides Steud. = Pennisetum thunbergii Kunth ■☆

289027　Pennisetum alopecuros Nees ex Steud. = Pennisetum hohenackeri Hochst. ex Steud. ■☆

289028　Pennisetum alopecuros Steud. = Pennisetum hohenackeri Hochst. ex Steud. ■☆

289029　Pennisetum alopecuros Steud. var. occidentale Pilg. = Pennisetum hohenackeri Hochst. ex Steud. ■☆

289030　Pennisetum americanum(L.) K. Schum. = Pennisetum glaucum (L.) R. Br. ■

289031　Pennisetum americanum(L.) Leeke = Pennisetum glaucum(L.) R. Br. ■

289032　Pennisetum americanum (L.) Leeke subsp. monodii (Maire) Brunken = Pennisetum violaceum(Lam.) Rich. ■☆

289033　Pennisetum americanum (L.) Leeke subsp. stenostachyum (A. Br. et Bouché) Brunken = Pennisetum sieberianum(Schltr.) Stapf et C. E. Hubb. ■☆

289034　Pennisetum americanum(L.) Leeke subsp. typhoideum Maire et Weiller = Pennisetum glaucum(L.) R. Br. ■

289035　Pennisetum americanum(L.) Leeke var. typhoideum Maire et Weiller = Pennisetum glaucum(L.) R. Br. ■

289036　Pennisetum amoenum Hochst. ex A. Rich. = Pennisetum pedicellatum Trin. ■☆

289037　Pennisetum ancylochaete Stapf et C. E. Hubb. = Pennisetum glaucum(L.) R. Br. ■

289038 Pennisetum angolense Rendle = Pennisetum macrourum Trin. ■☆

289039 Pennisetum angolense Rendle var. laxispicatum? = Pennisetum macrourum Trin. ■☆

289040 Pennisetum arvense Pilg. = Pennisetum ramosum (Hochst.) Schweinf. ■☆

289041 Pennisetum asperifolium (Desf.) Kunth = Pennisetum setaceum (Forssk.)Chiov. subsp. asperifolium(Desf.)Maire ■☆

289042 Pennisetum asperum Schult. f. = Pennisetum macrourum Trin. ■☆

289043 Pennisetum atrichum Stapf et C. E. Hubb. = Pennisetum polystachion (L.) Schult. subsp. atrichum (Stapf et C. E. Hubb.) Brunken ■☆

289044 Pennisetum aureum A. Rich. = Setaria spacelata (Schumach.) Stapf et C. E. Hubb. ex M. B. Moss var. aurea(Hochst. ex A. Braun) Clayton ■☆

289045 Pennisetum baojiense W. X. Tong = Pennisetum longissimum S. L. Chen et Y. X. Jin ■

289046 Pennisetum barteri Stapf et C. E. Hubb. = Pennisetum violaceum (Lam.) Rich. ■☆

289047 Pennisetum blepharideum Gilli = Pennisetum purpureum Schum. ■

289048 Pennisetum brachystachyum Hack. = Pennisetum mezianum Leeke ■☆

289049 Pennisetum brevifolium Steud. = Cenchrus prieurii(Kunth)Maire ■☆

289050 Pennisetum catabasis Stapf et C. E. Hubb. = Pennisetum hohenackeri Hochst. ex Steud. ■☆

289051 Pennisetum cenchroides Rich. = Cenchrus ciliaris L. ■

289052 Pennisetum cenchroides Rich. var. echinoides Hook. f. = Cenchrus pennisetiformis Hochst. et Steud. ex Steud. ■☆

289053 Pennisetum centrasiaticum Tzvelev = Pennisetum flaccidum Griseb. ■

289054 Pennisetum centrasiaticum Tzvelev var. lanpinense S. L. Chen et Y. X. Jin = Pennisetum flaccidum Griseb. ■

289055 Pennisetum centrasiaticum Tzvelev var. lanpinense S. L. Chen et Y. X. Jin;兰坪狼尾草;Laping Pennisetum ■

289056 Pennisetum centrasiaticum Tzvelev var. qinghaiensis Y. H. Wu = Pennisetum flaccidum Griseb. ■

289057 Pennisetum chandestinum Hochst. ex Chiov. ;西非狼尾草(铺地狗尾草） ;West African Pennisetum ■☆

289058 Pennisetum chevalieri Stapf et C. E. Hubb. ;舍瓦利耶狼尾草■☆

289059 Pennisetum chilense Hack. ;智利狼尾草■☆

289060 Pennisetum chinense Steud. = Pennisetum alopecuroides (L.) Spreng. ■

289061 Pennisetum chudeaui Maire et Trab. = Pennisetum violaceum (Lam.)Rich. ■☆

289062 Pennisetum chudeaui Maire et Trab. subsp. monodii Maire = Pennisetum violaceum(Lam.)Rich. ■☆

289063 Pennisetum ciliare(L.)Link = Cenchrus ciliaris L. ■

289064 Pennisetum ciliare (L.) Link var. pallens Leeke = Cenchrus ciliaris L. ■

289065 Pennisetum ciliatum Parl. = Pennisetum polystachion (L.) Schult. ■

289066 Pennisetum cinereum Stapf et C. E. Hubb. = Pennisetum glaucum(L.)R. Br. ■

289067 Pennisetum clandestinum Hochst. ex Chiov. ;铺地狼尾草; Creeping Pennisetum, Creeping Wolftailgrass, Kikuyu Grass, Kikuyugrass, Kikuyu-grass ■

289068 Pennisetum cognatum Steud. = Pennisetum violaceum (Lam.) Rich. ■☆

289069 Pennisetum complanatum Hemsl. ;尼加拉瓜狼尾草; Nicaraguan Fountaingrass ■

289070 Pennisetum compressum R. Br. = Pennisetum alopecuroides (L.) Spreng. ■

289071 Pennisetum crus-galli(L.) Baumg. = Echinochloa crus-galli(L.) P. Beauv. ■

289072 Pennisetum cupreum Hitchc. ex L. H. Bailey = Pennisetum setaceum(Forssk.) Chiov. ■☆

289073 Pennisetum dalzielii Stapf et C. E. Hubb. = Pennisetum sieberianum(Schltr.) Stapf et C. E. Hubb. ■☆

289074 Pennisetum darfuricum Stapf et C. E. Hubb. = Pennisetum violaceum(Lam.) Rich. ■☆

289075 Pennisetum davyi Stapf et C. E. Hubb. = Pennisetum macrourum Trin. ■☆

289076 Pennisetum densiflorum (Fig. et De Not.)T. Durand et Schinz = Pennisetum pedicellatum Trin. ■☆

289077 Pennisetum dichotomum(Forssk.) Delile = Pennisetum divisum (Forssk. ex J. F. Gmel.) Henrard ■☆

289078 Pennisetum dichotomum(Forssk.) Delile var. scabrum Maire et Trab. = Pennisetum divisum(Forssk. ex J. F. Gmel.) Henrard ■☆

289079 Pennisetum dichotomum (Forssk.) Delile var. subplumosum Hack. = Pennisetum divisum(Forssk. ex J. F. Gmel.) Henrard ■☆

289080 Pennisetum dichotomum Delile = Pennisetum divisum(Forssk. ex J. F. Gmel.) Henrard ■☆

289081 Pennisetum dillonii Steud. = Pennisetum pedicellatum Trin. ■☆

289082 Pennisetum dioicum A. Rich. = Pennisetum petiolare (Hochst.) Chiov. ■☆

289083 Pennisetum dispiculatum L. C. Chia;双穗狼尾草;Twospike Pennisetum ■

289084 Pennisetum dispiculatum L. C. Chia = Pennisetum alopecuroides (L.) Spreng. ■

289085 Pennisetum divisum(Forssk. ex J. F. Gmel.)Henrard;全裂狼尾草■☆

289086 Pennisetum dowsonii Stapf et C. E. Hubb. = Pennisetum riparium Hochst. ex A. Rich. ■☆

289087 Pennisetum echinurus (K. Schum.) Stapf et C. E. Hubb. = Pennisetum glaucum(L.)R. Br. ■

289088 Pennisetum elatum Hochst. ex Steud. = Pennisetum divisum (Forssk. ex J. F. Gmel.) Henrard ■☆

289089 Pennisetum elatum Steud. = Pennisetum divisum(Forssk. ex J. F. Gmel.) Henrard ■☆

289090 Pennisetum elegans Nees ex Steud. = Pennisetum polystachion (L.) Schult. ■

289091 Pennisetum erythraeum Chiov. = Pennisetum setaceum (Forssk.)Chiov. ■☆

289092 Pennisetum exile Stapf et C. E. Hubb. = Pennisetum macrourum Trin. ■☆

289093 Pennisetum fallax (Fig. et De Not.) Stapf et C. E. Hubb. = Pennisetum violaceum(Lam.)Rich. ■☆

289094 Pennisetum fasciculatum Trin. = Pennisetum orientale Rich. ■☆

289095 Pennisetum felicianum Asong. ;费利奇狼尾草■☆

289096 Pennisetum flaccidum Griseb. ;白草(白草根, 白花草, 倒生草, 中亚狼尾草) ;Flaccid Pennisetum, Whitegrass ■

289097 Pennisetum flaccidum Griseb. ex Roshev. = Pennisetum centrasiaticum Tzvelev ■

289098 Pennisetum flaccidum Griseb. var. interruptum Griseb. = Pennisetum flaccidum Griseb. ■

289099 Pennisetum flavicomum Leeke = Pennisetum purpureum

Schumach. ■

289100 Pennisetum flexispica K. Schum. = Pennisetum purpureum Schumach. ■

289101 Pennisetum franchetianum Stapf et C. E. Hubb. = Pennisetum macrourum Trin. ■☆

289102 Pennisetum gabonense Franch. = Pennisetum polystachion (L.) Schult. ■

289103 Pennisetum gambiense Stapf et C. E. Hubb. = Pennisetum glaucum (L.)R. Br. ■

289104 Pennisetum geniculatum (Thunb.) Leeke = Pennisetum thunbergii Kunth ■☆

289105 Pennisetum germanicum (Mill.) Baumg. = Setaria italica (L.) P. Beauv. ■

289106 Pennisetum germanicum(Mill.) Baumgarten = Setaria italica (L.) P. Beauv. ■

289107 Pennisetum gibbosum Stapf et C. E. Hubb. = Pennisetum glaucum (L.)R. Br. ■

289108 Pennisetum giganteum A. Rich. = Pennisetum macrourum Trin. ■☆

289109 Pennisetum giganteum A. Rich. var. minor Leeke = Pennisetum macrourum Trin. ■☆

289110 Pennisetum giganteum A. Rich. var. trinervium Pilg. = Pennisetum macrourum Trin. ■☆

289111 Pennisetum glabrum Steud. = Pennisetum thunbergii Kunth ■☆

289112 Pennisetum glaucifolium Hochst. ex A. Rich. ;灰绿叶狼尾草■☆

289113 Pennisetum glaucifolium Hochst. ex A. Rich. var. glaberrima Chiov. = Pennisetum glaucifolium Hochst. ex A. Rich. ■☆

289114 Pennisetum glaucifolium Hochst. ex A. Rich. var. procera Chiov. = Pennisetum glaucifolium Hochst. ex A. Rich. ■☆

289115 Pennisetum glaucocladum Stapf et C. E. Hubb. ;灰绿枝狼尾草■☆

289116 Pennisetum glaucum(L.)R. Br. ;御谷(蜡烛稗,美国狼尾草,唐人稗,香蒲狼尾草,豫谷,珍珠粟);African Millet, America Wolftailgrass, American Pennisetum, Bajri, Bulrush Millet, Cattail Millet, Cat-tail Millet, Cattaimillet, Gero, Indian Millet, Ornamental Millet, Pearl Millet, Pearlmillet, Spiked Millet, Yellow Foxtail ■

289117 Pennisetum glaucum (L.) R. Br. = Pennisetum americanum (L.)Leeke ■

289118 Pennisetum gossweileri Stapf et C. E. Hubb. = Pennisetum purpureum Schumach. ■

289119 Pennisetum gracile Benth. = Pennisetum polystachion (L.) Schult. ■

289120 Pennisetum gracilescens Hochst. ;纤细狼尾草■☆

289121 Pennisetum grandiflorum Stapf et C. E. Hubb. = Pennisetum riparium Hochst. ex A. Rich. ■☆

289122 Pennisetum haareri Stapf et C. E. Hubb. = Pennisetum macrourum Trin. ■☆

289123 Pennisetum hainanensis H. R. Zhao et A. T. Liu = Pennisetum purpureum Schum. ■

289124 Pennisetum hohenackeri Hochst. ex Steud. ;霍恩狼尾草;Moya Grass ■☆

289125 Pennisetum hordeoides(Lam.)Steud. ;大麦狼尾草■☆

289126 Pennisetum humile Hochst. ex A. Rich. ;微小狼尾草■☆

289127 Pennisetum humile Hochst. ex A. Rich. var. nanum Engl. = Pennisetum humile Hochst. ex A. Rich. ■☆

289128 Pennisetum implicatum Steud. = Pennisetum pedicellatum Trin. ■☆

289129 Pennisetum inclusum Pilg. = Pennisetum clandestinum Hochst. ex Chiov. ■

289130 Pennisetum incomptum Nees ex Steud. ;劣狼尾草; Bad

Pennisetum ■☆

289131 Pennisetum intertextum Schltdl. = Pennisetum pedicellatum Trin. ■☆

289132 Pennisetum italicum(L.)R. Br. = Setaria italica(L.)P. Beauv. ■

289133 Pennisetum italicum L. = Setaria italica(L.) P. Beauv. ■

289134 Pennisetum jacquesii Mimeur = Pennisetum monostigma Pilg. ■☆

289135 Pennisetum japonicum Trin. ;日本狼尾草;Japanese Pennisetum ■

289136 Pennisetum japonicum Trin. = Pennisetum alopecuroides (L.) Spreng. ■

289137 Pennisetum kamerunense Mez = Pennisetum monostigma Pilg. ■☆

289138 Pennisetum kirkii Stapf = Pennisetum unisetum(Nees)Benth. ■☆

289139 Pennisetum kisantuense Vanderyst = Pennisetum macrourum Trin. ■☆

289140 Pennisetum lachnorhachis Peter = Pennisetum purpureum Schumach. ■

289141 Pennisetum lanatum G. Klotz;西藏狼尾草;Xizang Pennisetum, Xizang Wolftailgrass ■

289142 Pennisetum lanuginosum Hochst. = Pennisetum pedicellatum Trin. ■☆

289143 Pennisetum latifolium Spreng. ; 宽 叶 狼 尾 草; Uruguayan Fountaingrass ■☆

289144 Pennisetum laxior(Clayton) Clayton;疏散狼尾草■☆

289145 Pennisetum laxior(Clayton) Zon = Pennisetum laxior (Clayton) Clayton ■☆

289146 Pennisetum laxum Hochst. ex Leeke = Pennisetum glaucifolium Hochst. ex A. Rich. ■☆

289147 Pennisetum ledermannii Mez;莱德狼尾草■☆

289148 Pennisetum leekei Mez var. leucostachys Peter = Pennisetum thunbergii Kunth ■☆

289149 Pennisetum leonis Stapf et C. E. Hubb. = Pennisetum glaucum (L.) R. Br. ■

289150 Pennisetum longissimum S. L. Chen et Y. X. Jin;长序狼尾草 (宝鸡狼尾草);Baoji Pennisetum, Baoji Wolftailgrass, Longspike Pennisetum, Longspike Wolftailgrass ■

289151 Pennisetum longissimum S. L. Chen et Y. X. Jin = Pennisetum shaanxiense S. L. Chen et Y. X. Jin ■

289152 Pennisetum longissimum S. L. Chen et Y. X. Jin var. axiglabrum B. S. Sun et X. Yang = Pennisetum flaccidum Griseb. ■

289153 Pennisetum longissimum S. L. Chen et Y. X. Jin var. axiglabrum B. S. Sun et X. Yang;光轴狼尾草;Smoothaxis Pennisetum ■

289154 Pennisetum longissimum S. L. Chen et Y. X. Jin var. intermedium S. L. Chen et Y. X. Jin = Pennisetum shaanxiense S. L. Chen et Y. X. Jin ■

289155 Pennisetum longissimum S. L. Chen et Y. X. Jin var. intermedium S. L. Chen et Y. X. Jin;中型狼尾草;Intermediate Pennisetum ■

289156 Pennisetum longistylum Hochst. = Pennisetum villosum R. Br. ex Fresen. ■☆

289157 Pennisetum macropogon Stapf et C. E. Hubb. = Pennisetum macrourum Trin. ■☆

289158 Pennisetum macrostachyum Benth. = Pennisetum purpureum Schumach. ■

289159 Pennisetum macrourum Trin. ; 非 洲 大 狼 尾 草; African Feathergrass ■☆

289160 Pennisetum maiwa Stapf et C. E. Hubb. = Pennisetum glaucum (L.)R. Br. ■

289161 Pennisetum malacochaete Stapf et C. E. Hubb. = Pennisetum glaucum(L.)R. Br. ■

289162　Pennisetum massaicum Stapf;马萨狼尾草;Fountain Grass ■☆

289163　Pennisetum merkeri Leeke = Pennisetum sphacelatum(Nees)T. Durand et Schinz ■☆

289164　Pennisetum merkeri Trab. =Pennisetum purpureum Schumach. ■

289165　Pennisetum mezianum Leeke;梅茨狼尾草■☆

289166　Pennisetum mildbraedii Mez;米尔德狼尾草■☆

289167　Pennisetum molle Hitchc. = Pennisetum violaceum (Lam.) Rich. ■☆

289168　Pennisetum mollissimum Hochst. = Pennisetum violaceum (Lam.)Rich. ■☆

289169　Pennisetum mongolicum Franch. = Pennisetum centrasiaticum Tzvelev ■

289170　Pennisetum mongolicum Franch. = Pennisetum flaccidum Griseb. ■

289171　Pennisetum mongolicum Franch. ex Roshev. = Pennisetum flaccidum Griseb. ■

289172　Pennisetum monostigma Pilg. ;单柱头狼尾草■☆

289173　Pennisetum myurus Parl. = Pennisetum polystachion (L.) Schult. ■

289174　Pennisetum natalense Stapf;纳塔尔狼尾草■☆

289175　Pennisetum nepalense Spreng. = Pennisetum lanatum G. Klotz ■

289176　Pennisetum nervosum Trin. ; 弯 穗 狼 尾 草; Bentspike Fountaingrass ■☆

289177　Pennisetum nigritarum (Schltdl.) T. Durand et Schinz = Pennisetum glaucum(L.)R. Br. ■

289178　Pennisetum niloticum Stapf et C. E. Hubb. = Pennisetum sieberianum(Schltr.)Stapf et C. E. Hubb. ■☆

289179　Pennisetum nitens (Andersson) Hack. = Pennisetum purpureum Schum. ■

289180　Pennisetum nodiflorum Franch. = Pennisetum divisum (Forssk. ex J. F. Gmel.) Henrard ■☆

289181　Pennisetum notarisii T. Durand et Schinz = Pennisetum pedicellatum Trin. subsp. unispiculum Brunken ■☆

289182　Pennisetum nubicum(Hochst.) K. Schum. ex Engl. ;云雾狼尾草■☆

289183　Pennisetum ochrops Stapf et C. E. Hubb. = Pennisetum violaceum(Lam.)Rich. ■☆

289184　Pennisetum orientale (Willd.) A. Rich. subsp. parisii Trab. = Pennisetum setaceum(Forssk.)Chiov. ■☆

289185　Pennisetum orientale Rich. ;东 方 狼尾草; Oriental Fountain Grass , Oriental Pennisetum , Oriental Wolftailgrass ■☆

289186　Pennisetum orientale Rich. = Pennisetum setaceum (Forssk.) Chiov. subsp. orientale(Rich.)Maire ■☆

289187　Pennisetum orientale Rich. var. triflorum(Nees)Stapf ex Hook. f. ;三花东方狼尾草■☆

289188　Pennisetum orthochaete Stapf et C. E. Hubb. ;直毛狼尾草■☆

289189　Pennisetum ovale Rupr. ex Steud. = Pennisetum ramosum (Hochst.) Schweinf. ■☆

289190　Pennisetum oxyphyllum Peter = Cenchrus ciliaris L. ■

289191　Pennisetum pallescens Leeke = Pennisetum purpureum Schumach. ■

289192　Pennisetum pappianum Chiov. =Pennisetum yemense Deflers ■☆

289193　Pennisetum parviflorum Trin. = Pennisetum hordeoides(Lam.)Steud. ■☆

289194　Pennisetum paucisetum Peter =Pennisetum thunbergii Kunth ■☆

289195　Pennisetum pedicellatum Trin. ;梗花狼尾草;Kyasuma Grass ■☆

289196　Pennisetum pedicellatum Trin. subsp. unispiculum Brunken;单刺狼尾草;Kyasuma Grass ■☆

289197　Pennisetum pedicellatum Trin. var. pubirhachis Berhaut = Pennisetum pedicellatum Trin. ■☆

289198　Pennisetum pennisetiforme(Hochst. et Steud. ex Steud.) Wipff;羽毛狼尾草■☆

289199　Pennisetum pentastachyum Hochst. ex A. Rich. = Pennisetum squamulatum Fresen. ■☆

289200　Pennisetum perspeciosum Stapf et C. E. Hubb. ;美丽狼尾草■☆

289201　Pennisetum petiolare(Hochst.) Chiov. ;柄叶狼尾草;Petioled Fountaingrass ■☆

289202　Pennisetum petiolare Chiov. = Pennisetum petiolare(Hochst.)Chiov. ■☆

289203　Pennisetum pirottae Chiov. ;皮罗特狼尾草■☆

289204　Pennisetum polycladum Chiov. = Cenchrus ciliaris L. ■

289205　Pennisetum polystachion (L.) Schult. ; 牧 地 狼尾草; Bristl Pennisetum,Bristl Wolftailgrass,Mission Grass ■

289206　Pennisetum polystachion (L.) Schult. subsp. atrichum (Stapf et C. E. Hubb.)Brunken;无毛狼尾草■☆

289207　Pennisetum polystachion (L.) Schult. subsp. setosum (Sw.) Brunken = Pennisetum polystachion(L.)Schult. ■☆

289208　Pennisetum polystachyon (L.) Schult. subsp. setosum (Sw.) Brunken;刚毛牧地狼尾草;Mission Grass ■

289209　Pennisetum prieurii Kunth =Cenchrus prieurii(Kunth)Maire ■☆

289210　Pennisetum procerum(Stapf)Clayton;高大狼尾草■☆

289211　Pennisetum proximum Leeke = Pennisetum squamulatum Fresen. ■☆

289212　Pennisetum pruinosum Leeke = Pennisetum purpureum Schumach. ■

289213　Pennisetum pumilum Hack. ;矮狼尾草■☆

289214　Pennisetum purpurascens (Thunb.) Kuntze = Pennisetum alopecuroides(L.)Spreng. ■

289215　Pennisetum purpurascens (Thunb.) Kuntze f. chinense (Nees) Leek = Pennisetum alopecuroides(L.)Spreng. ■

289216　Pennisetum purpurascens Kunth = Pennisetum polystachion(L.) Schult. ■

289217　Pennisetum purpurascens Kunth = Pennisetum setosum (Sw.) Rich. ■

289218　Pennisetum purpureum Schumach. ;象草(紫狼尾草);Elephant Grass, Elephantgrass, Elephant-grass, Napier, Napier Grass, Napiergrass,Napier-grass,Napier's Fodder ■

289219　Pennisetum purpureum Schumach. subsp. benthamii (Steud.) Maire et Weiller = Pennisetum purpureum Schumach. ■

289220　Pennisetum purpureum Schumach. subsp. flexispica (K. Schum.)Maire et Weiller = Pennisetum purpureum Schumach. ■

289221　Pennisetum pycnostachyum (Steud.) Stapf et C. E. Hubb. = Pennisetum glaucum(L.)R. Br. ■

289222　Pennisetum qianningense S. L. Zhong;乾宁狼尾草;Qianning Pennisetum,Qianning Wolftailgrass ■

289223　Pennisetum quanxinense H. R. Zhao et A. T. Liu = Pennisetum longissimum S. L. Chen et Y. X. Jin var. intermedium S. L. Chen et Y. X. Jin ■

289224　Pennisetum quartinianum A. Rich. = Pennisetum macrourum Trin. ■☆

289225　Pennisetum ramosissimum Steud. = Pennisetum violaceum (Lam.)Rich. ■☆

289226　Pennisetum ramosum(Hochst.)Schweinf. ;分枝狼尾草■☆

289227 Pennisetum rangei Mez = Cenchrus ciliaris L. ■

289228 Pennisetum respiciens A. Rich. = Setaria verticillata (L.) P. Beauv. ■

289229 Pennisetum reversum Hack. = Pennisetum polystachion (L.) Schult. ■

289230 Pennisetum riparioides Hochst. ex A. Rich. = Pennisetum macrourum Trin. ■☆

289231 Pennisetum riparium Hochst. ex A. Rich. ;溪畔狼尾草■☆

289232 Pennisetum robustum Stapf et C. E. Hubb. ;粗壮狼尾草■☆

289233 Pennisetum rogeri Stapf et C. E. Hubb. = Pennisetum violaceum (Lam.) Rich. ■☆

289234 Pennisetum ruppellii Steud. ;鲁氏狼尾草; African Fountain-grass, Fountain-grass ■☆

289235 Pennisetum ruppellii Steud. = Pennisetum setaceum (Forssk.) Chiov. ■☆

289236 Pennisetum sagittifolium A. Rich. = Setaria sagittifolia (A. Rich.) Walp. ■☆

289237 Pennisetum salifex Stapf et C. E. Hubb. = Pennisetum riparium Hochst. ex A. Rich. ■☆

289238 Pennisetum sampsonii Stapf et C. E. Hubb. = Pennisetum sieberianum (Schltr.) Stapf et C. E. Hubb. ■☆

289239 Pennisetum scaettae Robyns = Pennisetum macrourum Trin. ■☆

289240 Pennisetum schimperi Hochst. ex A. Rich. = Pennisetum sphacelatum (Nees) T. Durand et Schinz ☆

289241 Pennisetum schimperi Hochst. ex A. Rich. var. glabrum (Steud.) T. Durand et Schinz = Pennisetum thunbergii Kunth ■☆

289242 Pennisetum schimperi Steud. = Pennisetum sphacelatum (Nees) T. Durand et Schinz ■☆

289243 Pennisetum schliebenii Pilg. = Pennisetum trisetum Leeke ■☆

289244 Pennisetum schweinfurthii Pilg. ;施韦狼尾草■☆

289245 Pennisetum sclerocladum Stapf et C. E. Hubb. = Pennisetum sieberianum (Schltr.) Stapf et C. E. Hubb. ■☆

289246 Pennisetum scoparium Chiov. = Pennisetum setaceum (Forssk.) Chiov. ■☆

289247 Pennisetum secundiflorum (Fig. et De Not.) T. Durand et Schinz = Pennisetum pedicellatum Trin. ■☆

289248 Pennisetum setaceum (Forssk.) Chiov. ;羽绒狼尾草 (牧地狼尾草) ; African Fountain Grass, African Fountain-grass, Annual Fountain Grass, Crimson Fountain Grass, Crimson Fountaingrass, Dwarf Red Fountain Grass, Fountain Grass, Fountain-grass, Green Fountain Grass, Purple Fountain Grass ■☆

289249 Pennisetum setaceum (Forssk.) Chiov. subsp. asperifolium (Desf.) Maire ;糙叶狼尾草■☆

289250 Pennisetum setaceum (Forssk.) Chiov. subsp. cupreum? ;铜色狼尾草 ; Fountain Grass ■☆

289251 Pennisetum setaceum (Forssk.) Chiov. subsp. orientale (Rich.) Maire ;东方羽绒狼尾草■☆

289252 Pennisetum setaceum (Forssk.) Chiov. var. orientale (Rich.) Maire = Pennisetum setaceum (Forssk.) Chiov. ■☆

289253 Pennisetum setaceum (Forssk.) Chiov. var. parisii (Trab.) Maire = Pennisetum setaceum (Forssk.) Chiov. ■☆

289254 Pennisetum setigerum (Vahl) Wipff ;刚毛狼尾草■☆

289255 Pennisetum setosum (Sw.) Rich. = Pennisetum polystachion (L.) Schult. ■

289256 Pennisetum setosum (Sw.) Rich. = Pennisetum polystachyon (L.) Schult. subsp. setosum (Sw.) Brunken ■

289257 Pennisetum shaanxiense S. L. Chen et Y. X. Jin ;陕西狼尾草 ; Shaanxi Pennisetum , Shaanxi Wolftailgrass ■

289258 Pennisetum sichuanense S. L. Chen et Y. X. Jin ;四川狼尾草 ; Sichuan Pennisetum , Sichuan Wolftailgrass ■

289259 Pennisetum sichuanense S. L. Chen et Y. X. Jin var. eduidistans B. S. Sun et X. Yang = Pennisetum flaccidum Griseb. ■

289260 Pennisetum sichuanense S. L. Chen et Y. X. Jin var. eduidistans B. S. Sun et X. Yang ;等距狼尾草■

289261 Pennisetum sieberianum (Schltr.) Stapf et C. E. Hubb. ;西氏狼尾草■☆

289262 Pennisetum siguiriense Mimeur = Pennisetum hordeoides (Lam.) Steud. ■☆

289263 Pennisetum sinense Mez = Pennisetum centrasiaticum Tzvelev ■

289264 Pennisetum sinense Mez = Pennisetum flaccidum Griseb. ■

289265 Pennisetum snowdenii C. E. Hubb. = Pennisetum thunbergii Kunth ■☆

289266 Pennisetum somalense (Clayton) Wipff ;索马里狼尾草■☆

289267 Pennisetum sordidum Koidz. ;污浊狼尾草■☆

289268 Pennisetum sphacelatum (Nees) T. Durand et Schinz ;球头狼尾草■☆

289269 Pennisetum sphacelatum (Nees) T. Durand et Schinz var. tenuifolium (Hack.) Stapf = Pennisetum sphacelatum (Nees) T. Durand et Schinz ■☆

289270 Pennisetum spicatum (L.) Körn. = Pennisetum glaucum (L.) R. Br. ■

289271 Pennisetum spicatum (L.) Körn. var. echinurus K. Schum. = Pennisetum glaucum (L.) R. Br. ■

289272 Pennisetum spicatum (L.) Körn. var. typhoideum T. Durand et Schinz = Pennisetum glaucum (L.) R. Br. ■

289273 Pennisetum squamulatum Fresen. ;鳞狼尾草■☆

289274 Pennisetum stapfianum F. Bolus = Pennisetum mezianum Leeke ■☆

289275 Pennisetum stenorrhachis Stapf et C. E. Hubb. = Pennisetum macrourum Trin. ■☆

289276 Pennisetum stenostachyum (Klotzsch ex A. Braun et C. D. Bouché) Stapf et C. E. Hubb. = Pennisetum sieberianum (Schltr.) Stapf et C. E. Hubb. ■☆

289277 Pennisetum stenostachyum Peter = Pennisetum polystachion (L.) Schult. ■

289278 Pennisetum stolzii Mez = Pennisetum macrourum Trin. ■☆

289279 Pennisetum stramineum Peter ;草黄狼尾草■☆

289280 Pennisetum subangustum (Schumach.) Stapf et C. E. Hubb. = Pennisetum polystachion (L.) Schult. ■

289281 Pennisetum subeglume Trab. = Pennisetum violaceum (Lam.) Rich. ■☆

289282 Pennisetum tenue Mez = Pennisetum macrourum Trin. ■☆

289283 Pennisetum tenuifolium Hack. = Pennisetum sphacelatum (Nees) T. Durand et Schinz ■☆

289284 Pennisetum tenuispiculatum Steud. = Pennisetum polystachion (L.) Schult. ■

289285 Pennisetum tetrastachyum K. Schum. ;四穗狼尾草■☆

289286 Pennisetum thulinii S. M. Phillips ;图林狼尾草■☆

289287 Pennisetum thunbergii Kunth ;通贝里狼尾草■☆

289288 Pennisetum togoense Mez = Pennisetum macrourum Trin. ■☆

289289 Pennisetum trachyphyllum Pilg. ;粗叶狼尾草■☆

289290 Pennisetum triflorum Nees ex Steud. = Pennisetum orientale Rich. ■☆

289291 Pennisetum trisetum Leeke ;三刚毛狼尾草■☆

289292 Pennisetum typhoides (Burm. f.) Stapf et C. E. Hubb. = Pennisetum glaucum (L.) R. Br. ■

289293　Pennisetum typhoideum L. = Pennisetum americanum（L.）Leeke ■

289294　Pennisetum typhoideum L. = Pennisetum glaucum（L.）R. Br. ■

289295　Pennisetum typhoideum Rich. = Pennisetum glaucum（L.）R. Br. ■

289296　Pennisetum uliginosum Hack. ;沼泽狼尾草■☆

289297　Pennisetum unisetum(Nees)Benth. ;单刚毛狼尾草■☆

289298　Pennisetum vahlii Kunth = Cenchrus setigerus Vahl ■

289299　Pennisetum validum Mez = Pennisetum macrourum Trin. ■☆

289300　Pennisetum verticillatum（L.）R. Br. = Setaria verticillata（L.）P. Beauv. ■

289301　Pennisetum verticillatum（L.）R. Br. ex Roem. et Schult. = Setaria verticillata（L.）P. Beauv. ■

289302　Pennisetum verticillatum R. Br. = Setaria verticillata（L.）P. Beauv. ■

289303　Pennisetum villosum R. Br. ex Fresen. ;长柔毛狼尾草(长毛狼尾草，银狐）; Abyssinian Feather-top, Feathertop, Feather-top, Villose Pennisetum ■☆

289304　Pennisetum violaceum(Lam.)Rich. ;堇色狼尾草■☆

289305　Pennisetum violaceum（Lam.）Rich. var. chudeaui（Maire et Trab.）Maire = Pennisetum violaceum(Lam.)Rich. ■☆

289306　Pennisetum violaceum（Lam.）Rich. var. monodianum（Maire）Maire = Pennisetum violaceum(Lam.)Rich. ■☆

289307　Pennisetum violaceum（Lam.）Rich. var. monodii（Maire）Maire = Pennisetum violaceum(Lam.)Rich. ■☆

289308　Pennisetum viride(L.)R. Br. = Setaria viridis(L.)P. Beauv. ■

289309　Pennisetum vulpinum（A. Braun et C. D. Bouché）Stapf et C. E. Hubb. ;狐色狼尾草■☆

289310　Pennisetum yemense Deflers ;也门狼尾草■☆

289311　Penplexis Wall. = Drypetes Vahl ●

289312　Penstemon Mitch. = Chelone L. ■☆

289313　Penstemon Raf. = Penstemon Schmidel ●■

289314　Penstemon Schmidel(1763);钓钟柳属(吊钟柳属，五蕊花属，钟铃花属); Beard Tongue, Beardtongue, Penstemon ●■

289315　Penstemon × hybridus Hort. ;杂种钓钟柳;Beardtongue ●☆

289316　Penstemon acuminatus Douglas;渐尖钓钟柳; Sharp-leaved Beardtongue ●☆

289317　Penstemon alluviorum Pennell = Penstemon digitalis Nutt. ex Sims ■☆

289318　Penstemon alpinus Torr. ;高山钓钟柳■☆

289319　Penstemon ambiguus Torr. ;沙地钓钟柳;Bush Penstemon, Sand Penstemon ■☆

289320　Penstemon angustifolius Nutt. ex Pursh;狭叶钓钟柳;Narrowleaf Penstemon ■☆

289321　Penstemon arkansanus Pennell;阿肯色钓钟柳;Arkansas Beard-tongue ■☆

289322　Penstemon arkansanus Pennell var. pubescens Pennell = Penstemon pallidus Small ●☆

289323　Penstemon baccharifolius Hook. ;灌木钓钟柳;Baccharis-leaf Penstemon,Rock Penstemon,Shrubby Penstemon ●☆

289324　Penstemon barbatus Nutt. ;髯毛钓钟柳(草本象牙红，髯舌花，五蕊花）;Bearded Penstemon,Beardtongue,Scarlet Bugler ■☆

289325　Penstemon bradburii Pursh = Penstemon grandiflorus Nutt. ■☆

289326　Penstemon brevisepalus Pennell = Penstemon pallidus Small ●☆

289327　Penstemon campanulatus（Cav.）Willd. ;钟铃花; Common Penstemon ●☆

289328　Penstemon canescens Britton;灰毛钟铃花;Beardtongue Penstemon,Beard-tongue Penstemon,Gray Beardtongue ●☆

289329　Penstemon cardinalis Wooton et Standl. ;绯红钓钟柳;Cardinal Penstemon ●☆

289330　Penstemon cardwellii Howell;蔓钓钟柳●☆

289331　Penstemon centranthifolius Benth. ;刺叶钓钟柳;Scarlet Bugler ●☆

289332　Penstemon cobaea Nutt. ;电灯花钓钟柳;Cobaea Beardtongue, Cobaea Beard-tongue,Cobaea Penstemon ■☆

289333　Penstemon cordifolius Benth. ;心叶钓钟柳;Heartleaf Penstemon,Vine Penstemon ■☆

289334　Penstemon corymbosus Benth. ;伞房钓钟柳;Red Shrubby Pemtemon,Thymeleaf Penstemon ■☆

289335　Penstemon davidsonii Greene;革叶五蕊花●☆

289336　Penstemon diffusus Douglas;铺散钓钟柳;Bushy Penstemon ■☆

289337　Penstemon digitalis Nutt. ex Sims;毛地黄钓钟柳;False Foxglove, Foxglove Beardtongue, Foxglove Beard-tongue, Mississippi Penstemon, Smooth Beard-tongue, Tall Beard-tongue, Tall White Beard-tongue,White Beardtongue ■☆

289338　Penstemon dolius M. E. Jones;琼斯钓钟柳;Jones' Penstemon ■☆

289339　Penstemon eatonii A. Gray;伊顿钓钟柳;Firecracker Penstemon ■☆

289340　Penstemon fendleri Torr. et A. Gray;芬氏钓钟柳;Fendler Penstemon ■☆

289341　Penstemon frutescens Lamb. ;灌木状钓钟柳●☆

289342　Penstemon frutescens Lamb. f. albiflorus Hayashi;白花灌木状钓钟柳●☆

289343　Penstemon fruticosus（Pursh）Greene;灌丛钓钟柳;Lowbush Penstemon ●☆

289344　Penstemon glaber Pursh;光滑钓钟柳;Blue Penstemon ■☆

289345　Penstemon gloxinioides?; 苣苔钓钟柳; Border Penstemon, Garden Penstemon ●☆

289346　Penstemon gracilis Nutt. ;纤细钓钟柳;Lilac Penstemon, Slender Beard-tongue ■☆

289347　Penstemon gracilis Nutt. subsp. wisconsinensis（Pennell）Pennell;威斯康星钓钟柳;Wisconsin Beard-tongue, Wisconsin Penstemon ■☆

289348　Penstemon gracilis Nutt. var. wisconsinensis（Pennell）Fassett = Penstemon gracilis Nutt. subsp. wisconsinensis(Pennell)Pennell ■☆

289349　Penstemon grandiflorus Nutt. ;大花钓钟柳;Large Beard-tongue, Large-flowered Beardtongue, Large-flowered Beard-tongue, Shell-leaf Penstemon,Wild Foxglove ■☆

289350　Penstemon hartwegii Benth. ;哈氏钓钟柳;Hartwegs Penstemon ■●☆

289351　Penstemon havardii A. Gray; 大弯钓钟柳; Big Bend Beardtongue ■☆

289352　Penstemon heterophyllus Lindl. ;异叶钓钟柳;Beard Tongue, Chaparral Penstemon,Foothill Penstemon ●☆

289353　Penstemon hirsutus（L.）Willd. ;粗毛钓钟柳;Eastern Penstemon, Hairy Beardtongue, Hairy Beard-tongue, Northeastern Beard-tongue ●☆

289354　Penstemon hirsutus（L.）Willd. var. minimus R. W. Benn. = Penstemon hirsutus(L.)Willd. ●☆

289355　Penstemon hirsutus（L.）Willd. var. pygmaeus R. W. Benn. = Penstemon hirsutus(L.)Willd. ●☆

289356　Penstemon hirsutus Willd. ;多毛钓钟柳;Hairy Beardtongue ●☆

289357　Penstemon isophyllus B. L. Rob. ;同叶钓钟柳●☆

289358　Penstemon laevigatus Aiton; 平滑钓钟柳; Glabrous Beardtongue,Smooth Beardtongue,Smooth Penstemon ●☆

289359　Penstemon laevigatus Aiton f. digitalis（Nutt. ex Sims）A. Gray = Penstemon digitalis Nutt. ex Sims ■☆

289360　Penstemon laevigatus Aiton subsp. digitalis (Nutt. ex Sims) R. W. Benn. = Penstemon digitalis Nutt. ex Sims ■☆

289361　Penstemon laevigatus Aiton var. angulatus R. W. Benn. = Penstemon digitalis Nutt. ex Sims ■☆

289362　Penstemon lemmonii A. Gray;莱蒙钓钟柳;Lemmon Penstemon ■☆

289363　Penstemon menziesii Hook.;门氏钓钟柳（圆叶五蕊花）; Menzies Penstemon, Mountain Pride ●☆

289364　Penstemon newberryi A. Gray;山钓钟柳;Mountain Pride ●☆

289365　Penstemon ovatus Douglas;卵叶钓钟柳●☆

289366　Penstemon pallidus Small;东部白钓钟柳;Eastern White Beard-tongue, Pale Beard-tongue ●☆

289367　Penstemon palmeri A. Gray;帕氏钓钟柳; Balloon Flower, Palmer Penstemon ☆

289368　Penstemon parryi A. Gray;帕雷钓钟柳;Parry's Penstemon ■☆

289369　Penstemon procerus Douglas ex Graham;长枝钓钟柳■☆

289370　Penstemon pseudospectabilis M. E. Jones;荒漠钓钟柳;Canyon Penstemon, Desert Beard Tongue, False Spectacular Penstemon ☆

289371　Penstemon pulchellus Greene;钓钟柳（钟铃花）■☆

289372　Penstemon richardsonii Douglas;理氏钓钟柳●☆

289373　Penstemon rupicola Howell;岩生五蕊花;Cliff Penstemon ●☆

289374　Penstemon rydbergii A. Nelson;吕德贝里钓钟柳;Rydberg's Penstemon ■☆

289375　Penstemon scouleri Douglas;斯库勒钓钟柳（紫花灌木钓钟柳）;Scouler Penstemon ●☆

289376　Penstemon serrulatus Menzies;细齿钓钟柳;Cascade Penstemon ●☆

289377　Penstemon spectabilis Thurb. ex Torr. et A. Gray;华丽钓钟柳;Royal Beard Tongue, Showy Penstemon ☆

289378　Penstemon strictus Benth.;落基山钓钟柳;Porch Penstemon, Rocky Mountain Penstemon ☆

289379　Penstemon superbus A. Nelson;优秀钓钟柳;Superb Beardtongue, Superb Penstemon ☆

289380　Penstemon thurberi Torr.;瑟伯钓钟柳;Desert Surprise, Thurber's Penstemon, Thurber's Beardtongue ☆

289381　Penstemon tubaeflorus Nutt.;管花钓钟柳;Beard-tongue, Tube Beard-tongue, Tube Penstemon, White Wand Beard-tongue ●☆

289382　Penstemon utahensis Eastw.;犹他荷钓钟柳;Utah Firecracker ■☆

289383　Penstemon watsonii A. Gray;瓦氏钓钟柳●☆

289384　Penstemon wisconsinensis Pennell = Penstemon gracilis Nutt. subsp. wisconsinensis (Pennell) Pennell ■☆

289385　Penstemon wrightii Hook.;赖氏钓钟柳;Wright's Penstemon ■☆

289386　Penstemum Raf. = Penstemon Schmidel ●■

289387　Pentabothra Hook. f. (1883);五孔萝藦属 ☆

289388　Pentabothra nana Hook. f.;五孔萝藦 ☆

289389　Pentabrachion Müll. Arg. (1864);五枝木属●☆

289390　Pentabrachion Müll. Arg. = Microdesmis Hook. f. ex Hook.

289391　Pentabrachion reticulatum Müll. Arg.;五枝木●

289392　Pentabrachium Müll. Arg. = Microdesmis Hook. f. ex Hook. ●

289393　Pentabrachium Müll. Arg. = Pentabrachion Müll. Arg. ●☆

289394　Pentacaellum Franch. et Sav. = Myoporum Banks et Sol. ex G. Forst. ●

289395　Pentacaellum Franch. et Sav. = Pentacoelium Siebold et Zucc. ●

289396　Pentacaena Bartl. = Cardionema DC. ■☆

289397　Pentacalia Cass. (1827);五蟹甲属●☆

289398　Pentacalia Cass. = Senecio L. ●■

289399　Pentacalia arborea (Kunth) H. Rob. et Cuatrec.;树五蟹甲●☆

289400　Pentacalia nitida (Kunth) Cuatrec.;光亮五蟹甲●☆

289401　Pentacalia peruviana (Pers.) Cuatrec.;秘鲁五蟹甲●☆

289402　Pentacarpaea Hiern = Pentanisia Harv. ■☆

289403　Pentacarpaea arenaria Hiern = Pentanisia arenaria (Hiern) Verdc. ■☆

289404　Pentacarpus T. Post et Kuntze = Pentacarpaea Hiern ■☆

289405　Pentacarpus T. Post et Kuntze = Pentanisia Harv. ■☆

289406　Pentacarya DC. ex Meisn. = Heliotropium L. ●■

289407　Pentace Hassk. (1858);五室椴属（硬椴属）;Pentace, Thitka ●☆

289408　Pentace burmanica Kurz;缅甸五室椴（缅甸硬椴）; Burmamahogany, Thitka ●☆

289409　Pentace esquirolii H. Lév. = Burretiodendron esquirolii (H. Lév.) Rehder ●◇

289410　Pentace polyantha Hassk.;五室椴（多花硬椴）●☆

289411　Pentace tonkinense A. Chev. = Excentrodendron tonkinense (A. Chev.) Hung T. Chang et R. H. Miao ●

289412　Pentace triptera Mast.;三翅五室椴（三翅硬椴）●☆

289413　Pentaceras Hook. f. (1862)（保留属名）;五角芸香属 ●☆

289414　Pentaceras Roem. et Schult. = Pentaceros G. Mey.（废弃属名）●☆

289415　Pentaceras australis Hook. f.;五角芸香●■

289416　Pentaceros G. Mey.（废弃属名）= Byttneria Loefl.（保留属名）●

289417　Pentaceros G. Mey.（废弃属名）= Pentaceras Hook. f.（保留属名）●☆

289418　Pentachaeta Nutt. (1840);毛冠菀属■☆

289419　Pentachaeta Nutt. = Chaetopappa DC. ■☆

289420　Pentachaeta alsinoides Greene;小毛冠菀;Tiny Pentachaeta ■☆

289421　Pentachaeta aurea Nutt.;毛冠菀;Golden-rayed Pentachaeta ■☆

289422　Pentachaeta bellidiflora Greene;白线毛冠菀;White-rayed Pentachaeta ■☆

289423　Pentachaeta exilis (A. Gray) A. Gray;柔弱毛冠菀;Meager Pentachaeta ■☆

289424　Pentachaeta exilis (A. Gray) A. Gray subsp. aeolica Van Horn et Ornduff;圣贝尼托毛冠菀;San Benito Pentachaeta ■☆

289425　Pentachaeta fragilis Brandegee;纤细毛冠菀;Fragile Pentachaeta ■☆

289426　Pentachaeta lyonii A. Gray;莱昂毛冠菀;Lyon's Pentachaeta ■☆

289427　Pentachlaena H. Perrier (1920);五被花属●☆

289428　Pentachlaena betamponensis Lowry et al.;五被花●☆

289429　Pentachlaena latifolia H. Perrier;宽叶五被花●☆

289430　Pentachlaena orientalis Capuron;东方五被花●☆

289431　Pentachondra R. Br. (1810);水螅石南属（核果尖苞木属,潘塔琼德木属,五软木属）;Pentachondra ●☆

289432　Pentachondra involucrata R. Br.;水螅石南●☆

289433　Pentaclathra Endl. = Polyclathra Bertol. ■☆

289434　Pentaclethra Benth. (1840);五柳豆属（五楹木属,五山柳苏木属）;Pentaclethra ●☆

289435　Pentaclethra eetveldeana De Wild. et T. Durand;五柳豆（五楹木,五山柳苏木）●☆

289436　Pentaclethra gigantea A. Chev. = Albizia gigantea A. Chev. ●☆

289437　Pentaclethra griffoniana Baill. = Newtonia griffoniana (Baill.) Baker f. ●☆

289438　Pentaclethra macroloba Kuntze;大裂五柳豆（爆荚豆,大裂五楹木,大裂五山柳苏木,山柳）;Pracaxi Fat ●☆

289439　Pentaclethra macrophylla Benth.;大叶五柳豆（大叶五楹木,大叶五山柳苏木,油豆树）;Congo Acacia, Gabon Yellow-wood, Oil Bean, Owala Oil Tree, Owala-oil Tree, Pracaxi Fat ●☆

289440　Pentacme A. DC. (1868);五齿香属;Pentacme ●☆

289441　Pentacme A. DC. = Shorea Roxb. ex C. F. Gaertn. ●

289442　Pentacme suavis A. DC.;五齿香●☆

289443　Pentacnida Post et Kuntze = Pentocnide Raf. ●■

289444　Pentacnida Post et Kuntze = Pouzolzia Gaudich. ●■

289445　Pentacocca Turcz. = Ochthocosmus Benth. ●☆

289446　Pentacocca leonenis Turcz. = Phyllocosmus africanus(Hook. f.) Klotzsch ●☆

289447　Pentacoelium Siebold et Zucc. (1846);类苦槛蓝属●

289448　Pentacoelium Siebold et Zucc. = Myoporum Banks et Sol. ex G. Forst. ●

289449　Pentacoelium bontioides Siebold et Zucc. = Myoporum bontioides(Siebold et Zucc.) A. Gray ●

289450　Pentacoilanthus Rappa et Camarrone(1954);南非日中花属■●☆

289451　Pentacoilanthus aitonis(Jacq.)Rappa et Camarrone = Mesembryanthemum aitonis Jacq. ■☆

289452　Pentacoilanthus crassicaulis (Haw.) Rappa et Camarrone = Sceletium crassicaule(Haw.)L. Bolus ●☆

289453　Pentacoilanthus crystallinus(L.)Rappa et Camarrone = Mesembryanthemum crystallinum L. ■

289454　Pentacoilanthus expansus(L.)Rappa et Camarrone = Sceletium expansum(L.)L. Bolus ●☆

289455　Pentacoilanthus splendens (L.) Rappa et Camarrone = Phyllobolus splendens(L.)Gerbaulet ●☆

289456　Pentacoilanthus tortuosus(L.)Rappa et Camarrone = Sceletium tortuosum(L.)N. E. Br. ●☆

289457　Pentacraspedon Steud. = Amphipogon R. Br. ■☆

289458　Pentacrophys A. Gray = Acleisanthes A. Gray ■●☆

289459　Pentacrostigma K. Afzel. = Ipomoea L. (保留属名)●■

289460　Pentacrostigma nyctanthum K. Afzel. = Ipomoea longituba Hallier f. ●■☆

289461　Pentacrypta Lehm. = Arracacia Bancr. ■☆

289462　Pentactina Nakai(1917);朝鲜岩蔷薇属■☆

289463　Pentactina rupicola Nakai;朝鲜岩蔷薇■☆

289464　Pentacyphus Schltr. (1906);五曲萝藦属■☆

289465　Pentacyphus boliviensis Schltr. ;五曲萝藦■☆

289466　Pentadactylon C. F. Gaertn. = Persoonia Sm. (保留属名)●■

289467　Pentadathra Endl. = Polyclathra Bertol. ■☆

289468　Pentadenia(Planch.)Hanst. (1854);五腺苣苔属●☆

289469　Pentadenia(Planch.)Hanst. = Columnea L. ●■☆

289470　Pentadenia Hanst. = Columnea L. ●■☆

289471　Pentadenia Hanst. = Pentadenia(Planch.)Hanst. ●☆

289472　Pentadenia angustata Wiehler;窄五腺苣苔●☆

289473　Pentadenia colombiana Wiehler;哥伦比亚五腺苣苔●☆

289474　Pentadenia strigosa Hanst. ;五腺苣苔●☆

289475　Pentadesma Sabine(1824);猪油果属(奶油树属,奶油藤黄属);Butter Tree,Lardfruit ●

289476　Pentadesma butyracea Sabine;猪油果 (奶油藤黄);Black Mango, Butter Tree, Candle Tree, Kanga Butter, Lamy Butter, Lardfruit,Sierra Leone Butter,Tallow Tree,Tallow-tree ●

289477　Pentadesma devredii Spirlet;德夫雷油果●☆

289478　Pentadesma excelliana Staner = Pentadesma grandifolia Baker f. ●☆

289479　Pentadesma gabonensis Pierre ex A. Chev. = Pentadesma butyracea Sabine ●

289480　Pentadesma grandifolia Baker f. ;大叶猪油果●☆

289481　Pentadesma lebrunii Staner;勒布伦猪油果●☆

289482　Pentadesma leptonema Pierre = Pentadesma butyracea Sabine ●

289483　Pentadesma leucantha A. Chev. = Pentadesma butyracea Sabine ●

289484　Pentadesma maritima Pierre;滨海猪油果●☆

289485　Pentadesma nigritana Baker f. = Pentadesma butyracea Sabine ●

289486　Pentadesma parviflora Exell = Mammea africana Sabine ●☆

289487　Pentadesma reyndersii Spirlet;润氏奶油藤黄●☆

289488　Pentadesma rutshuruensis Spirlet = Pentadesma lebrunii Staner ●☆

289489　Pentadesmos Spruce ex Planch. et Triana = Moronobea Aubl. ●☆

289490　Pentadiplandra(Baill.) Kuntze = Dizygotheca N. E. Br. ●☆

289491　Pentadiplandra(Baill.)Post et Kuntze = Dizygotheca N. E. Br. ●☆

289492　Pentadiplandra Baill. (1886);瘤药花属(瘤药树属)●☆

289493　Pentadiplandra brazzeana Baill. ;瘤药花(布拉瘤药树,瘤药树)●☆

289494　Pentadiplandra brazzeana Baill. var. valida Pellegr. ex Villiers;刚直瘤药花●☆

289495　Pentadiplandra gossweileri Exell = Pentadiplandra brazzeana Baill. ●☆

289496　Pentadiplandraceae Hutch. et Dalzell = Capparaceae Juss. (保留科名)●■

289497　Pentadiplandraceae Hutch. et Dalzell(1928);瘤药花科(瘤药树科)●☆

289498　Pentadynamis R. Br. = Crotalaria L. ●■

289499　Pentaglossum Forssk. = Lythrum L. ●■

289500　Pentaglottis Tausch(1829);五舌草属;Green Alkanet ■☆

289501　Pentaglottis Wall. = Melhania Forssk. ●■

289502　Pentaglottis sempervirens (L.) L. H. Bailey = Pentaglottis sempervirens(L.)Tausch ■☆

289503　Pentaglottis sempervirens(L.)Tausch;五舌草;Alkanet,Biddy's Eyes, Dyer's Bugloss, Evergreen Alkanet, Evergreen Bugloss, Green Alkanet,Pheasant's Eye,Pheasant's Eyes ■☆

289504　Pentaglottis sempervirens Tausch = Pentaglottis sempervirens (L.)Tausch ■☆

289505　Pentagonanthus Bullock(1962);五棱花属■☆

289506　Pentagonanthus caeruleus(E. A. Bruce)Bullock;蓝五棱花■☆

289507　Pentagonanthus grandiflorus(N. E. Br.)Bullock;大花五棱花■☆

289508　Pentagonanthus grandiflorus (N. E. Br.) Bullock subsp. glabrescens(Bullock)Bullock;光蓝五棱花■☆

289509　Pentagonanthus sudanicus A. Chev. ex Hutch. et Dalziel;五棱花■☆

289510　Pentagonaster Klotzsch = Kunzea Rchb. (保留属名)●☆

289511　Pentagonia Benth. (1845)(保留属名);五棱茜属■☆

289512　Pentagonia Fabr. = Nicandra Adans. (保留属名)■

289513　Pentagonia Heist. ex Fabr. (废弃属名) = Nicandra Adans. (保留属名)■

289514　Pentagonia Heist. ex Fabr. (废弃属名) = Pentagonia Benth. (保留属名)■☆

289515　Pentagonia Kuntze = Specularia Heist. ex A. DC. ●■☆

289516　Pentagonia Möhring ex Kuntze = Legousia T. Durand ●■☆

289517　Pentagonia Möhring ex Kuntze = Specularia Heist. ex A. DC. ●■☆

289518　Pentagonia alba Dwyer;白五棱茜■☆

289519　Pentagonia angustifolia C. M. Taylor;窄叶五棱茜■☆

289520　Pentagonia gigantophylla Standl. ex Steyerm. ;大叶五棱茜■☆

289521　Pentagonia parvifolia Steyerm. ;小叶五棱茜■☆

289522　Pentagonium Schauer = Philibertia Kunth ■

289523　Pentagonocarpus P. Micheli ex Parl. = Kosteletzkya C. Presl(保留属名)■●☆

289524　Pentagonocarpus Parl. = Kosteletzkya C. Presl(保留属名)■●☆

289525　Pentake Raf. = Cuscuta L. ■☆

289526　Pentalepis F. Muell. (1863);五鳞菊属●☆

289527　Pentalepis F. Muell. = Chrysogonum L. ●■☆

289528　Pentalepis trichodesmoides F. Muell. ;五鳞菊●☆

289529　Pentalinon Voigt = Prestonia R. Br. (保留属名)●☆

289530　Pentaloba Lour. = Rinorea Aubl.（保留属名）●

289531　Pentaloba sessilis Lour. = Rinorea sessilis（Lour.）Kuntze ●

289532　Pentaloncha Hook. f.（1873）；五矛茜属☆

289533　Pentaloncha humilis Hook. f.；五矛茜●☆

289534　Pentaloncha rubriflora R. D. Good；红花五矛茜●☆

289535　Pentaloncha stipulosa（Hutch. et Dalziel）Bremek. = Poecilocalyx stipulosa（Hutch. et Dalziel）N. Hallé ■☆

289536　Pentalophus A. DC. = Lithospermum L. ■

289537　Pentamera Willis = Pentanura Blume ●

289538　Pentamerea Baill. = Pentameria Klotzsch ex Baill. ●

289539　Pentamerea Klotzsch ex Baill. = Bridelia Willd.（保留属名）●

289540　Pentameria Klotzsch ex Baill. = Bridelia Willd.（保留属名）●

289541　Pentameria melanthesoides Baill. = Bridelia cathartica G. Bertol. f. melanthesoides（Baill.）Radcl.-Sm. ●☆

289542　Pentameris E. Mey. = Pavonia Cav.（保留属名）●■☆

289543　Pentameris P. Beauv.（1812）；五部芒属■☆

289544　Pentameris P. Beauv. = Danthonia DC.（保留属名）■

289545　Pentameris airoides Nees；银须草五部芒■☆

289546　Pentameris distichophylla（Lehm.）Nees；二列叶五部芒■☆

289547　Pentameris dregeana Stapf = Pentameris distichophylla（Lehm.）Nees ■☆

289548　Pentameris elegans（Nees）Steud. = Pentaschistis elegans（Nees）Stapf ■☆

289549　Pentameris glacialis N. P. Barker；冰雪五部芒■☆

289550　Pentameris hirtiglumis N. P. Barker；颖五部芒■☆

289551　Pentameris longiglumis（Nees）Stapf；长颖五部芒■☆

289552　Pentameris macrocalycina（Steud.）Schweick.；大萼五部芒■☆

289553　Pentameris obtusifolia（Hochst.）Schweick. = Pseudopentameris obtusifolia（Hochst.）N. P. Barker ■☆

289554　Pentameris oreophila N. P. Barker；喜山五部芒■☆

289555　Pentameris speciosa（Lehm. ex Nees）Steud. = Pentameris macrocalycina（Steud.）Schweick. ■☆

289556　Pentameris squarrosa Stapf = Pseudopentameris obtusifolia（Hochst.）N. P. Barker ■☆

289557　Pentameris swartbergensis N. P. Barker；斯瓦特五部芒■☆

289558　Pentameris thuarii P. Beauv.；南非五部芒■☆

289559　Pentameris uniflora N. P. Barker；单花五部芒■☆

289560　Pentamerista Maguire（1972）；五数花属●☆

289561　Pentamerista neotropica Maguire；五数花●☆

289562　Pentamorpha Scheidw. = Erythrochiton Nees et Mart. ●☆

289563　Pentanema Cass.（1818）；苇谷草属（苇谷草属）；Pentanema ■●

289564　Pentanema cernuum（Dalzell et A. Gibson）Y. Ling；垂头苇谷草；Drooping Pentanema ■

289565　Pentanema cernuum（Dalzell）Y. Ling = Pentanema cernuum（Dalzell et A. Gibson）Y. Ling ■

289566　Pentanema divaricatum Cass.；两歧苇谷草；Divaricate Pentanema ■☆

289567　Pentanema glanduligerum（Krasch.）Gorschk.；腺点苇谷草■☆

289568　Pentanema indicum（L.）Y. Ling；苇谷草（糙叶地丁，草金沙，草金杉，七头风，松香草，野罗伞，野杉根，止血草）；India Pentanema ■

289569　Pentanema indicum（L.）Y. Ling var. hypoleucum（Hand.-Mazz.）Y. Ling；白背苇谷草（糙叶地丁，草金杉，登亚严，七头风，松香草，野杉根，止血草）；Whiteback India Pentanema ■

289570　Pentanema ligneum Mesfin；木质苇谷草●☆

289571　Pentanema parietarioides（Nevski）Gorschk.；墙草状苇谷草■☆

289572　Pentanema propinquum（Nevski）Gorschk.；邻近苇谷草■☆

289573　Pentanema rabiatum Boiss. = Pentanema vestitum（Wall. ex DC.）Y. Ling ■

289574　Pentanema rupicola（Krasch.）Gorschk.；岩生苇谷草■☆

289575　Pentanema vestitum（Wall. ex DC.）Y. Ling；毛苇谷草；Hairy Pentanema ■

289576　Pentanisia Harv.（1842）；五异茜属■☆

289577　Pentanisia angustifolia（Hochst.）Hochst.；窄叶五异茜■☆

289578　Pentanisia annua K. Schum. = Neopentanisia annua（K. Schum.）Verdc. ■☆

289579　Pentanisia arenaria（Hiern）Verdc.；沙地五异茜■☆

289580　Pentanisia caerulea Hiern = Otiophora caerulea（Hiern）Bullock ■☆

289581　Pentanisia calcicola Verdc.；钙生五异茜■☆

289582　Pentanisia confertifolia（Baker）Verdc.；密叶五异茜■☆

289583　Pentanisia confertifolia（Baker）Verdc. f. glabrifolia Verdc. = Pentanisia confertifolia（Baker）Verdc. ■☆

289584　Pentanisia crassifolia K. Krause = Pentanisia schweinfurthii Hiern ■☆

289585　Pentanisia cymosa Klotzsch = Pentas lanceolata（Forssk.）K. Schum. subsp. cymosa（Klotzsch）Verdc. ●☆

289586　Pentanisia foetida Verdc.；臭五异茜■☆

289587　Pentanisia glaucescens Harv. = Pentanisia prunelloides（Klotzsch ex Eckl. et Zeyh.）Walp. ■☆

289588　Pentanisia longepedunculata Verdc.；长梗五异茜■☆

289589　Pentanisia longisepala K. Krause = Pentanisia prunelloides（Klotzsch ex Eckl. et Zeyh.）Walp. subsp. latifolia（Hochst.）Verdc. ■☆

289590　Pentanisia longituba（Franch.）Oliv.；长管五异茜■☆

289591　Pentanisia microphylla（Franch.）Chiov.；小叶五异茜■☆

289592　Pentanisia monogyna Engl.；单蕊五异茜■☆

289593　Pentanisia monticola（K. Krause）Verdc.；山地五异茜■☆

289594　Pentanisia ouranogyne S. Moore；穹蕊五异茜■☆

289595　Pentanisia ouranogyne S. Moore var. glabrifolia Cufod. = Pentanisia ouranogyne S. Moore ■☆

289596　Pentanisia parviflora Stapf ex Verdc. = Paraknoxia parviflora（Stapf ex Verdc.）Verdc. ex Bremek. ■☆

289597　Pentanisia pentagyne K. Schum. = Pentanisia arenaria（Hiern）Verdc. ■☆

289598　Pentanisia pentasiana Mattei = Pentanisia ouranogyne S. Moore ■☆

289599　Pentanisia procumbens R. D. Good；平铺五异茜■☆

289600　Pentanisia prunelloides（Klotzsch ex Eckl. et Zeyh.）Walp.；夏枯草五异茜■☆

289601　Pentanisia prunelloides（Klotzsch ex Eckl. et Zeyh.）Walp. subsp. latifolia（Hochst.）Verdc.；宽叶五异茜■☆

289602　Pentanisia prunelloides（Klotzsch ex Eckl. et Zeyh.）Walp. var. latifolia（Hochst.）Walp. = Pentanisia prunelloides（Klotzsch ex Eckl. et Zeyh.）Walp. subsp. latifolia（Hochst.）Verdc. ■☆

289603　Pentanisia renifolia Verdc.；肾叶五异茜■☆

289604　Pentanisia rhodesiana S. Moore = Pentanisia schweinfurthii Hiern ■☆

289605　Pentanisia schweinfurthii Hiern；施韦五异茜■☆

289606　Pentanisia schweinfurthii Hiern var. pubescens Verdc. = Pentanisia schweinfurthii Hiern ■☆

289607　Pentanisia sericocarpa S. Moore = Pentanisia schweinfurthii Hiern ■☆

289608　Pentanisia spicata S. Moore = Otiophora scabra Zucc. ■☆

289609　Pentanisia suffruticosa Klotzsch = Pentas lanceolata（Forssk.）K. Schum. ●■

289610　Pentanisia sykesii Hutch. subsp. otomerioides Verdc.；非洲耳茜状五异茜■☆

289611　Pentanisia variabilis Harv. = Pentanisia prunelloides（Klotzsch ex Eckl. et Zeyh.）Walp. ■☆

289612　Pentanisia variabilis Harv. var. glaucescens Cruse ex Sond. = Pentanisia angustifolia（Hochst.）Hochst. ■☆

289613　Pentanisia variabilis Harv. var. intermedia Sond. = Pentanisia prunelloides（Klotzsch ex Eckl. et Zeyh.）Walp. ■☆

289614　Pentanisia variabilis Harv. var. latifolia（Hochst.）Sond. = Pentanisia prunelloides（Klotzsch ex Eckl. et Zeyh.）Walp. subsp. latifolia（Hochst.）Verdc. ■☆

289615　Pentanisia veronicoides（Baker）K. Schum. ;婆婆纳叶五异茜■☆

289616　Pentanisia zanzibarica Klotzsch = Pentas zanzibarica（Klotzsch）Vatke ■☆

289617　Pentanome DC. = Zanthoxylum L. ●

289618　Pentanome Moc. et Sessé ex DC. = Zanthoxylum L. ●

289619　Pentanopsis Rendle（1898）;拟五异茜属（类五星花属）■☆

289620　Pentanopsis fragrans Rendle;香拟五异茜（类五星花）■☆

289621　Pentanopsis gracilicaulis（Verdc.）Thulin et B. Bremer;拟五异茜■☆

289622　Pentanthus Hook. et Arn. = Paracalia Cuatrec. ●☆

289623　Pentanthus Less. = Nassauvia Comm. ex Juss. ●☆

289624　Pentanthus Raf. = Jacquemontia Choisy ☆

289625　Pentanura Blume = Stelmatocrypton Baill. ●

289626　Pentanura khasiana Kurz = Stelmatocrypton khasianum（Kurz）Baill. ●

289627　Pentapanax Seem.（1864）;羽叶参属（五叶参属,羽叶五加属）;Pentapanax ●

289628　Pentapanax caesius（Hand. -Mazz.）C. B. Shang;圆叶羽叶参（圆叶楤木）;Round-leaf Pentapanax, Round-leaved Pentapanax ●

289629　Pentapanax castanopseicola Hayata;台湾羽叶参（台湾五叶参）;Taiwan Pentapanax ●

289630　Pentapanax forrestii W. W. Sm. = Pentapanax fragrans（D. Don）T. D. Ha var. forrestii（W. W. Sm.）C. B. Shang ●

289631　Pentapanax forrestii W. W. Sm. = Pentapanax leschenaultii（Wight et Arn.）Seem. var. forrestii（W. W. Sm.）H. L. Li ●

289632　Pentapanax fragrans（D. Don）T. D. Ha;羽叶参（光叶五叶参,石夹枫,五叶参,岩五加,岩五叶,羽叶五加）;Common Pentapanax ●

289633　Pentapanax fragrans（D. Don）T. D. Ha var. forrestii（W. W. Sm.）C. B. Shang;全缘羽叶参（全缘五叶参）;Forrest Common Pentapanax ●

289634　Pentapanax fragrans（D. Don）T. D. Ha var. longipedunculatus（N. S. Bui）T. D. Ha = Pentapanax longipedunculatus N. S. Bui ●

289635　Pentapanax glabrifoliolatus C. B. Shang;光叶羽叶参（光羽叶参）;Glabrous Pentapanax, Glabrous-leaved Pentapanax ●

289636　Pentapanax henryi Harms;锈毛五叶参（大麻漆,马肠子,马肠子树,锈毛羽叶参,圆锥五叶参）;Henry Pentapanax, Rusthair Pentapanax ●

289637　Pentapanax henryi Harms var. fangii G. Hoo;小果锈毛五叶参（小果五叶参）;Fang Henry Pentapanax ●

289638　Pentapanax henryi Harms var. fangii G. Hoo = Pentapanax henryi Harms ●

289639　Pentapanax henryi Harms var. hwangshanensis W. C. Cheng;黄山锈毛五叶参（黄山五叶参）;Huangshan Pentapanax ●

289640　Pentapanax henryi Harms var. larium（Hand. -Mazz.）Hand. -Mazz. = Pentapanax tomentellus（Franch.）C. B. Shang ●

289641　Pentapanax henryi Harms var. larius Hand. -Mazz. = Pentapanax henryi Harms ●

289642　Pentapanax henryi Harms var. tomentosus G. Hoo;毛叶锈毛五叶参（绒毛五叶参）;Tomentose Henry Pentapanax ●

289643　Pentapanax henryi Harms var. tomentosus G. Hoo = Pentapanax henryi Harms ●

289644　Pentapanax henryi Harms var. wangshanensis W. C. Cheng = Pentapanax henryi Harms ●

289645　Pentapanax hypoglaucus（C. J. Qi et T. R. Cao）C. B. Shang et X. P. Li;粉背羽叶参●

289646　Pentapanax lanceolatus（Wight et Arn.）Seem. ;披针叶五叶参（披针五叶参）;Lanceleaf Pentapanax, Lance-leaved Pentapanax, Lanceolateleaf Pentapanax ●

289647　Pentapanax lanceolatus G. Hoo = Pentapanax henryi Harms ●

289648　Pentapanax larius Hand. -Mazz. = Pentapanax henryi Harms ●

289649　Pentapanax larius Hand. -Mazz. = Pentapanax tomentellus（Franch.）C. B. Shang ●

289650　Pentapanax leschenaultii（DC.）Seem. = Pentapanax fragrans（D. Don）T. D. Ha ●

289651　Pentapanax leschenaultii（DC.）Seem. var. simplex K. M. Feng = Pentapanax fragrans（D. Don）T. D. Ha ●

289652　Pentapanax leschenaultii（DC.）Seem. var. villosus Y. R. Li = Pentapanax fragrans（D. Don）T. D. Ha ●

289653　Pentapanax leschenaultii（Wight et Arn.）Seem. = Pentapanax fragrans（D. Don）T. D. Ha ●

289654　Pentapanax leschenaultii（Wight et Arn.）Seem. var. forrestii（W. W. Sm.）H. L. Li = Pentapanax fragrans（D. Don）T. D. Ha var. forrestii（W. W. Sm.）C. B. Shang ●

289655　Pentapanax leschenaultii（Wight et Arn.）Seem. var. simplex K. M. Feng;单序五叶参;Simllex Common Pentapanax ●

289656　Pentapanax leschenaultii（Wight et Arn.）Seem. var. villosus Y. R. Li;毛背五叶参;Villose Common Pentapanax ●

289657　Pentapanax longipedunculatus N. S. Bui;长梗羽叶参;Longpedicel Pentapanax, Long-peduncled Pentapanax ●

289658　Pentapanax longipes（Merr.）C. B. Shang et C. F. Ji;独龙羽叶参;Dulong Pentapanax ●

289659　Pentapanax parasiticus（D. Don）Seem. ;寄生羽叶参（寄生五叶参）;Parasite Pentapanax, Parasitic Pentapanax ●

289660　Pentapanax parasiticus（D. Don）Seem. var. khasianus C. B. Clarke;毛梗寄生羽叶参（锈毛寄生五叶参）;Khasia Parasite Pentapanax ●

289661　Pentapanax parasiticus（D. Don）Seem. var. khasianus C. B. Clarke = Pentapanax parasiticus（D. Don）Seem. ●

289662　Pentapanax plumosus（H. L. Li）C. B. Shang;糙叶羽叶参（糙叶五叶参,糙羽叶参,羽叶楤木）;Feathery Pentapanax, Plumed Pentapanax ●

289663　Pentapanax racemosus Seem. ;总序羽叶参（总序五叶参）;Racemose Pentapanax ●

289664　Pentapanax subcordatus（Wall.）Seem. ;心叶五叶参;Heartleaf Pentapanax, Heart-leaved Pentapanax ●

289665　Pentapanax tomentellus（Franch.）C. B. Shang;绒毛羽叶参（马肠子树,绒毛五叶参）;Tomentose Pentapanax ●

289666　Pentapanax tomentellus（Franch.）C. B. Shang var. distinctum C. B. Shang;离柱马肠子树;Distingct Tomentose Pentapanax ●

289667　Pentapanax tomentellus（Franch.）C. B. Shang var. tomentosus（G. Hoo）Y. F. Deng = Pentapanax henryi Harms ●

289668　Pentapanax trifoliatus K. M. Feng = Pentapanax longipes（Merr.）C. B. Shang et C. F. Ji ●

289669　Pentapanax truncicola Hand. -Mazz. = Pentapanax fragrans（D. Don）T. D. Ha ●

289670　Pentapanax truncicola Hand. -Mazz. = Pentapanax leschenaultii (Wight et Arn.) Seem. var. forrestii (W. W. Sm.) H. L. Li ●

289671　Pentapanax verticillatus Dunn;轮伞羽叶参(轮伞五叶参); Verticillate Pentapanax ●

289672　Pentapanax wilsonii (Harms) C. B. Shang;西南羽叶参(川西楤木,西南楤木,西南五叶参);E. H. Wilson Pentapanax, Wilson Pentapanax ●

289673　Pentapanax wilsonii (Harms) C. B. Shang var. plumosus (H. L. Li) Y. F. Deng = Pentapanax plumosus (H. L. Li) C. B. Shang ●

289674　Pentapanax yunnanensis Franch. ;云南羽叶参(云南五叶参); Yunnan Pentapanax ●

289675　Pentapeltis (Endl.) Bunge = Xanthosia Rudge ■☆

289676　Pentapeltis Bunge = Xanthosia Rudge ■☆

289677　Pentapera Klotzsch = Erica L. ●☆

289678　Pentapera sicula (Guss.) Klotzsch = Erica sicula Guss. ●☆

289679　Pentaperygium interdictum Hand. -Mazz. = Agapetes interdicta (Hand. -Mazz.) Sleumer ●

289680　Pentaperygium interdictum Hand. -Mazz. var. stenolobum W. E. Evans = Agapetes interdicta (Hand. -Mazz.) Sleumer var. stenoloba (W. E. Evans) Sleumer ●

289681　Pentapetaceae Bercht. et J. Presl = Malvaceae Juss. (保留科名)●■

289682　Pentapetes L. (1753);午时花属;Middayflower, Pentapetes ■●

289683　Pentapetes acerifolia L. = Pterospermum acerifolium (L.) Willd. ●

289684　Pentapetes phoenicea L. ;午时花(川蜀葵,金钱花,毗尸沙,日中金钱,夜落金钱,子午花);Flor Impia, Purplered Middayflower, Purplered Pentapetes ●

289685　Pentaphalangium Warb. = Garcinia L. ●

289686　Pentaphiltrum Rchb. f. = Herschelia T. E. Bowdich ■

289687　Pentaphiltrum Rchb. f. = Physalis L. ■

289688　Pentaphitrum Rchb. = Physalis L. ■

289689　Pentaphorus D. Don = Gochnatia Kunth ●

289690　Pentaphorus D. Don(1830);腺菊木属●☆

289691　Pentaphorus fascicularis D. Don;腺菊木●☆

289692　Pentaphragma Wall. = Pentaphragma Wall. ex G. Don ■

289693　Pentaphragma Wall. ex G. Don(1834);五膜草属(五隔草属); Pentaphragma ■

289694　Pentaphragma Zucc. ex Rchb. = Araujia Brot. ●☆

289695　Pentaphragma corniculatum Chun et F. Chun = Pentaphragma spicatum Merr. ■

289696　Pentaphragma sinense Hemsl. et E. H. Wilson;五膜草(五隔草);China Pentaphragma, Chinese Pentaphragma ■

289697　Pentaphragma spicatum Merr. ;直序五膜草(直序五隔草); Spiked Pentaphragma ■

289698　Pentaphragmataceae J. Agardh(1858)(保留科名);五膜草科■

289699　Pentaphylacaceae Engl. (1897) (保留科名);五列木科; Pentaphylax Family ●

289700　Pentaphylax Gardner et Champ. (1849);五列木属; Pentaphylax ●

289701　Pentaphylax arborea Ridl. = Pentaphylax euryoides Gardner et Champ. ●

289702　Pentaphylax euryoides Gardner et Champ. ;五列木(毛生木); Common Pentaphylax ●

289703　Pentaphylax malayana Ridl. = Pentaphylax euryoides Gardner et Champ. ●

289704　Pentaphylax racemosa Merr. et Chun = Pentaphylax euryoides Gardner et Champ. ●

289705　Pentaphylax spicata Merr. = Pentaphylax euryoides Gardner et Champ. ●

289706　Pentaphylloides Duhamel = Potentilla L. ■●

289707　Pentaphylloides Duhamel(1755);金露梅属●

289708　Pentaphylloides dryanthoides (Juz.) Soják;帕米尔金露梅●

289709　Pentaphylloides floribunda (Pursh) Á. Löve;繁花金露梅; Shrubby Cinquefoil, Shrubby Five-fingers ●☆

289710　Pentaphylloides floribunda (Pursh) Á. Löve = Dasiphora fruticosa (L.) Rydb. ●

289711　Pentaphylloides floribunda (Pursh) Á. Löve = Pentaphylloides fruticosa (L.) O. Schwarz ●

289712　Pentaphylloides floribunda (Pursh) Á. Löve = Potentilla fruticosa L. ●

289713　Pentaphylloides fruticosa (L.) O. Schwarz;金露梅(扁麻,棍儿茶,金腊梅,金蜡梅,金老梅,老鸹爪,木本翻白草,木本委陵菜,药王茶,银老梅);Bush Cinquefoil, Golden Hardhack, Shrubby Cinquefoil, Shrubby Five-fingers, Shrubby Potentilla ●

289714　Pentaphylloides fruticosa (L.) O. Schwarz = Dasiphora fruticosa (L.) Rydb. ●

289715　Pentaphylloides fruticosa (L.) O. Schwarz = Potentilla fruticosa L. ●

289716　Pentaphylloides fruticosa (L.) O. Schwarz var. albicans (Rehder et E. H. Wilson) Y. L. Han = Potentilla fruticosa L. var. albicans Rehder et E. H. Wilson ●

289717　Pentaphylloides fruticosa (L.) O. Schwarz var. albicans (Rehder et E. H. Wilson) Y. L. Han;白毛金露梅(白叶金露梅); White-hair Bush Cinquefoil ●

289718　Pentaphylloides parvifolia (Fisch. ex Lehm.) Soják;小叶金露梅(小叶金老梅);Small-leaf Cinquefoil ●

289719　Pentaphylloides parvifolia (Fisch. ex Lehm.) Soják = Potentilla parvifolia Fisch. ex Lehm. ●

289720　Pentaphylloides phyllocalyx (Juz.) Soják;紫萼金露梅; Leafycalyx Cinquefoil ●

289721　Pentaphyllon Pers. = Lupinaster Fabr. ■☆

289722　Pentaphyllon Pers. = Trifolium L. ■

289723　Pentaphyllon lupinaster Pers. = Trifolium lupinaster L. ■

289724　Pentaphyllum Gaertn. = Potentilla L. ■●

289725　Pentaphyllum Hill = Potentilla L. ■●

289726　Pentaplaris L. O. Williams et Standl. (1952);五数木属●☆

289727　Pentaplaris doroteae L. O. Williams et Standl. ;五数木●☆

289728　Pentaple Rchb. = Cerastium L. ■

289729　Pentapleura Hand. -Mazz. (1913);五肋草属●☆

289730　Pentapleura subulifera Hand. -Mazz. ;五肋草●☆

289731　Pentapogon R. Br. (1810);四裂五芒草属■☆

289732　Pentapogon billardierei R. Br. ;四裂五芒草■☆

289733　Pentaptelion Turcz. = Leucopogon R. Br. (保留属名)●☆

289734　Pentaptera Roxb. = Terminalia L. (保留属名)●

289735　Pentaptera Roxb. ex DC. = Terminalia L. (保留属名)●

289736　Pentaptera arjuna Roxb. ex DC. = Terminalia arjuna (Roxb. ex DC.) Wight et Arn. ●☆

289737　Pentaptera roxburghii Tul. = Combretum roxburghii Spreng. ●

289738　Pentaptera saja Buch. -Ham. = Terminalia myriocarpa Van Heurck et Müll. Arg. ●◇

289739　Pentapteris Haller = Myriophyllum L. ■

289740　Pentapteris Haller = Pentapterophyllon Hill ■

289741　Pentapterophyllon Hill = Myriophyllum L. ■

289742　Pentapterophyllum Fabr. = Pentapterophyllon Hill ■

289743　Pentapterygium Klotzsch = Agapetes D. Don ex G. Don ●

289744　Pentapterygium Klotzsch(1851);五翅莓属;Pentapterygium ●

289745　Pentapterygium flavum Hook. f. = Agapetes flava (Hook. f.) Sleumer ●

289746　Pentapterygium interdictum Hand. -Mazz. = Agapetes interdicta (Hand. -Mazz.) Sleumer ●

289747　Pentapterygium interdictum Hand. -Mazz. var. stenolobum W. E. Evans = Agapetes interdicta (Hand. -Mazz.) Sleumer ●

289748　Pentapterygium listeri King ex C. B. Clarke = Agapetes listeri (King ex C. B. Clarke) Sleumer ●

289749　Pentapterygium rugosum (Hook. f. et Thomson ex Hook.) Hook. f. = Agapetes incurvata (Griff.) Sleumer ●

289750　Pentapterygium serpens (Wight) Klotzsch;五翅莓(垂枝树萝卜,匍匐树萝卜);Creeping Pentapterygium, Drooping Branches Agapetes,Fivewing Agapetes,Serpent Agapetes ●

289751　Pentapterygium serpens (Wight) Klotzsch = Agapetes serpens (Wight) Sleumer ●

289752　Pentapterygium sikkimense W. W. Sm. = Agapetes smithiana Sleumer ●☆

289753　Pentaptilon E. Pritz. (1904);五翼草海桐属■☆

289754　Pentaptilon careyi (F. Muell.) E. Pritz. ;五翼草海桐■☆

289755　Pentaptychaceae Dulac = Plumbaginaceae Juss. (保留科名)●■

289756　Pentapyxis Hook. f. = Leycesteria Wall. ●

289757　Pentapyxis stipulata (Hook. f. et Thomson) Hook. f. ex C. B. Clarke = Leycesteria stipulata (Hook. f. et Thomson) Fritsch ●

289758　Pentaraphia Lindl. = Pentarhaphia Lindl. ■☆

289759　Pentarhaphia Lindl. (1827);五针苣苔属;Pentarhaphia ■☆

289760　Pentarhaphia Lindl. = Gesneria L. ●☆

289761　Pentarhaphia longiflora Lindl. ;长叶五针苣苔■☆

289762　Pentarhopalopilia (Engl.) Hiepko(1987);五杖毛属●☆

289763　Pentarhopalopilia marquesii (Engl.) Hiepko;五杖毛●☆

289764　Pentarhopalopilia umbellulata (Baill.) Hiepko;伞形五杖毛●☆

289765　Pentaria (DC.) M. Roem. = Passiflora L. ●■

289766　Pentaria (DC.) M. Roem. = Peremis Raf. ●■

289767　Pentaria M. Roem. = Passiflora L. ●■

289768　Pentaria M. Roem. = Peremis Raf. ●■

289769　Pentarrhaphis Kunth(1816);灌丛垂穗草属■☆

289770　Pentarrhaphis annua Swallen;灌丛垂穗草■☆

289771　Pentarrhaphis polymorpha Griffiths;多型灌丛垂穗草■☆

289772　Pentarrhaphis scabra Kunth;一年灌丛垂穗草●☆

289773　Pentarrhinum E. Mey. (1838);五鼻萝藦属●☆

289774　Pentarrhinum abyssinicum Decne. ;阿比西尼亚五鼻萝藦●☆

289775　Pentarrhinum abyssinicum Decne. subsp. angolense (N. E. Br.) Liede et Nicholas;安哥拉五鼻萝藦●☆

289776　Pentarrhinum abyssinicum Decne. var. angolense N. E. Br. = Pentarrhinum abyssinicum Decne. subsp. angolense (N. E. Br.) Liede et Nicholas ●☆

289777　Pentarrhinum balense (Liede) Liede;巴兰五鼻萝藦■☆

289778　Pentarrhinum coriaceum Schltr. ;革质五鼻萝藦●☆

289779　Pentarrhinum fasciculatum K. Schum. = Pentatropis nivalis (J. F. Gmel.) D. V. Field et J. R. I. Wood ■☆

289780　Pentarrhinum gonoloboides (Schltr.) Liede;美洲萝藦状五鼻萝藦●☆

289781　Pentarrhinum insipidum E. Mey. ;无味五鼻萝藦●☆

289782　Pentarrhinum iringense Markgr. ;伊林加五鼻萝藦●☆

289783　Pentarrhinum somaliense (N. E. Br.) Liede;索马里五鼻萝藦●☆

289784　Pentas Benth. (1844);五星花属;Pentas, Star Clusters, Star-cluster,Star-clusters ●■

289785　Pentas ainsworthii Scott-Elliot = Pentas lanceolata (Forssk.) K. Schum. ●■

289786　Pentas angustifolia (A. Rich. ex DC.) Verdc. ;窄叶五星花●☆

289787　Pentas arvensis Hiern;田野五星花■☆

289788　Pentas arvensis Hiern var. violacea Hiern ex Scott-Elliot = Pentas cleistostoma K. Schum. ●☆

289789　Pentas burmanica Kurz;缅甸五星花;Burma Mahogany ●☆

289790　Pentas bussei K. Krause;布瑟五星花●☆

289791　Pentas bussei K. Krause. f. brevituba Verdc. = Pentas bussei K. Krause ●☆

289792　Pentas bussei K. Krause. f. glabra Verdc. = Pentas bussei K. Krause ●☆

289793　Pentas bussei K. Krause. f. minor Verdc. = Pentas bussei K. Krause ●☆

289794　Pentas carnea Benth. = Pentas lanceolata (Forssk.) K. Schum. subsp. cymosa (Klotzsch) Verdc. ●☆

289795　Pentas carnea Benth. = Pentas lanceolata (Forssk.) K. Schum. ●■

289796　Pentas carnea Benth. var. welwitschii Scott-Elliot = Pentas angustifolia (A. Rich. ex DC.) Verdc. ●☆

289797　Pentas cleistostoma K. Schum. ;闭口五星花●☆

289798　Pentas cleistostoma K. Schum. var. poggeana ?;波格五星花●☆

289799　Pentas coccinea Stapf = Pentas bussei K. Krause ●☆

289800　Pentas coerulea Chiov. = Pentas lanceolata (Forssk.) K. Schum. var. leucaster (K. Krause) Verdc. ●☆

289801　Pentas concinna K. Schum. ;整洁五星花●☆

289802　Pentas confertifolia Baker = Pentanisia confertifolia (Baker) Verdc. ■☆

289803　Pentas decaryana Homolle ex Verdc. ;德卡里五星花●☆

289804　Pentas decora S. Moore;装饰五星花●☆

289805　Pentas decora S. Moore var. lasiocarpa Verdc. ;毛果五星花●☆

289806　Pentas decora S. Moore var. triangularis (De Wild.) Verdc. ;三角五星花●☆

289807　Pentas decora S. Moore. f. opposita Verdc. = Pentas decora S. Moore ●☆

289808　Pentas dewevrei De Wild. et T. Durand = Otomeria volubilis (K. Schum.) Verdc. ■☆

289809　Pentas elata K. Schum. ;高五星花●☆

289810　Pentas elatior (A. Rich. ex DC.) Walp. = Otomeria elatior (A. Rich. ex DC.) Verdc. ■☆

289811　Pentas extensa K. Krause = Pentas zanzibarica (Klotzsch) Vatke var. intermedia Verdc. ■☆

289812　Pentas fililoba K. Krause = Virectaria major (K. Schum.) Verdc. ●☆

289813　Pentas flammea Chiov. = Pentas bussei K. Krause ●☆

289814　Pentas geophila Verdc. = Pentas lindenioides (S. Moore) Verdc. ●☆

289815　Pentas glabrescens Baker;渐光五星花●☆

289816　Pentas glabrescens Baker subsp. brevituba Verdc. ;短管渐光五星花●☆

289817　Pentas globifera Hutch. = Pentas decora S. Moore var. triangularis (De Wild.) Verdc. ●☆

289818　Pentas graniticola E. A. Bruce;花岗岩五星花●☆

289819　Pentas herbacea (Hiern) K. Schum. ;草本五星花■☆

289820　Pentas hindsioides K. Schum. ;海因兹茜五星花●☆

289821　Pentas hindsioides K. Schum. var. glabrescens Verdc. ;渐光海

因兹茜五星花●☆

289822 Pentas hindsioides K. Schum. var. williamsii Verdc.；威廉斯五星花●☆

289823 Pentas hirtiflora Baker；马岛毛花五星花●☆

289824 Pentas homblei De Wild. = Pentas decora S. Moore ●☆

289825 Pentas involucrata Baker = Spermacoce dibrachiata Oliv. ■☆

289826 Pentas ionolaena K. Schum.；堇被五星花●☆

289827 Pentas ionolaena K. Schum. subsp. schumanniana（K. Krause）Verdc. = Pentas schumanniana K. Krause ●☆

289828 Pentas klotzschii Vatke = Pentas lanceolata（Forssk.）K. Schum. ●■

289829 Pentas lanceolata（Forssk.）Deflers = Pentas lanceolata（Forssk.）K. Schum. ●■

289830 Pentas lanceolata（Forssk.）Deflers var. quartiniana（A. Rich.）Verdc. = Pentas lanceolata（Forssk.）K. Schum. var. quartiniana（A. Rich.）Verdc. ■☆

289831 Pentas lanceolata（Forssk.）K. Schum.；五星花（繁星花）；Egyptian Star, Egyptian Starcluster, Egyptian Star-cluster, Lanceolate Pentas, Pentas, Star Cluster, Star Clusters, Star-cluster ●■

289832 Pentas lanceolata（Forssk.）K. Schum. 'New Look Pink'；粉红五星花●■☆

289833 Pentas lanceolata（Forssk.）K. Schum. 'New Look Red'；红色五星花●■☆

289834 Pentas lanceolata（Forssk.）K. Schum. f. velutina Verdc.；短绒毛五星花●☆

289835 Pentas lanceolata（Forssk.）K. Schum. f. velutina Verdc. = Pentas lanceolata（Forssk.）K. Schum. ●■

289836 Pentas lanceolata（Forssk.）K. Schum. subsp. cymosa（Klotzsch）Verdc.；聚伞五星花●☆

289837 Pentas lanceolata（Forssk.）K. Schum. var. alba Verdc.；白五星花●☆

289838 Pentas lanceolata（Forssk.）K. Schum. var. angustifolia Verdc.；狭叶五星花●☆

289839 Pentas lanceolata（Forssk.）K. Schum. var. coccinea（Stapf）Verdc.；绯红繁星花●■☆

289840 Pentas lanceolata（Forssk.）K. Schum. var. leucaster（K. Krause）Verdc.；白星五星花●☆

289841 Pentas lanceolata（Forssk.）K. Schum. var. membranacea Verdc. = Pentas lanceolata（Forssk.）K. Schum. ●■

289842 Pentas lanceolata（Forssk.）K. Schum. var. nemorosa（Chiov.）Verdc.；森林五星花●☆

289843 Pentas lanceolata（Forssk.）K. Schum. var. quartiniana（A. Rich.）Verdc.；夸尔廷五星花●☆

289844 Pentas lanceolata（Forssk.）K. Schum. var. stenostygma（Chiov.）Cufod.；窄柱头五星花●☆

289845 Pentas lanceolata（Forssk.）K. Schum. var. usambarica Verdc.；乌桑巴拉五星花●☆

289846 Pentas lanceolata K. Schum. = Pentas lanceolata（Forssk.）K. Schum. ●■

289847 Pentas ledermannii K. Krause；莱德曼五星花●☆

289848 Pentas leucaster K. Krause = Pentas lanceolata（Forssk.）K. Schum. var. leucaster（K. Krause）Verdc. ●☆

289849 Pentas liebrechtsiana De Wild.；利布五星花●☆

289850 Pentas lindenioides（S. Moore）Verdc.；林登五星花●☆

289851 Pentas longiflora Oliv.；长花五星花●☆

289852 Pentas longiflora Oliv. f. glabrescens Verdc. = Pentas longiflora Oliv. ●☆

289853 Pentas longiflora Oliv. var. nyassana Scott-Elliot = Pentas longiflora Oliv. ●☆

289854 Pentas longituba K. Schum. ex Engl.；长管五星花●☆

289855 Pentas longituba K. Schum. ex Engl. var. magnifica（Bullock）Bullock et Taylor = Pentas longituba K. Schum. ex Engl. ●☆

289856 Pentas magnifica Bullock = Pentas longituba K. Schum. ex Engl. ●☆

289857 Pentas mechowiana K. Schum. = Pentas purpurea Oliv. subsp. mechowiana（K. Schum.）Verdc. ●☆

289858 Pentas micrantha Baker；小花五星花●☆

289859 Pentas micrantha Baker subsp. wyliei（N. E. Br.）Verdc.；怀利五星花■☆

289860 Pentas modesta Baker = Kohautia coccinea Royle ■☆

289861 Pentas mombassana Oliv. = Pentas parvifolia Hiern ●☆

289862 Pentas monbassana Hiern ex Oliv. = Pentas parvifolia Hiern ●☆

289863 Pentas mussaendoides Baker；玉叶金花五星花●☆

289864 Pentas nervosa Hepper；多脉五星花●☆

289865 Pentas nobilis S. Moore；名贵五星花●☆

289866 Pentas nobilis S. Moore var. grandifolia Verdc. = Pentas nobilis S. Moore ●☆

289867 Pentas occidentalis（Hook. f.）Benth. et Hook. ex Hiern = Pentas schimperiana（A. Rich.）Vatke subsp. occidentalis（Hook. f.）Verdc. ●☆

289868 Pentas parviflora Benth. = Sacosperma parviflorum（Benth.）G. Taylor ●☆

289869 Pentas parvifolia Hiern；小叶五星花●☆

289870 Pentas parvifolia Hiern var. nemorosa Chiov. = Pentas lanceolata（Forssk.）K. Schum. ●■

289871 Pentas parvifolia Hiern. f. intermedia Verdc.；间型五星花●☆

289872 Pentas parvifolia Hiern. f. spicata Verdc.；穗状五星花●☆

289873 Pentas pauciflora Baker；少花五星花■☆

289874 Pentas pseudomagnifica M. Taylor；假华丽五星花●☆

289875 Pentas pubiflora S. Moore；非洲毛花五星花●☆

289876 Pentas pubiflora S. Moore subsp. bamendensis Verdc. = Pentas ledermannii K. Krause ●☆

289877 Pentas pubiflora S. Moore var. longistyla？= Pentas pubiflora S. Moore ●☆

289878 Pentas purpurea Oliv.；紫五星花●☆

289879 Pentas purpurea Oliv. subsp. mechowiana（K. Schum.）Verdc.；梅休五星花●☆

289880 Pentas purpurea Oliv. var. buchananii Scott-Elliot = Pentas purpurea Oliv. ●☆

289881 Pentas purseglovei Verdc.；帕斯格洛夫五星花●☆

289882 Pentas quadrangularis Rendle = Conostomium quadrangulare（Rendle）Cufod. ■☆

289883 Pentas quartiniana（A. Rich.）Oliv. = Pentas lanceolata（Forssk.）K. Schum. var. quartiniana（A. Rich.）Verdc. ●☆

289884 Pentas schimperi Engl. = Pentas schimperiana（A. Rich.）Vatke ●☆

289885 Pentas schimperiana（A. Rich.）Vatke；欣珀五星花●☆

289886 Pentas schimperiana（A. Rich.）Vatke subsp. occidentalis（Hook. f.）Verdc.；西部五星花●☆

289887 Pentas schumanniana K. Krause；舒曼五星花●☆

289888 Pentas schweinfurthii Scott-Elliot = Pentas lanceolata（Forssk.）K. Schum. ●■

289889 Pentas speciosa Baker = Otomeria elatior（A. Rich. ex DC.）Verdc. ■☆

289890 Pentas stolzii K. Schum. et K. Krause = Pentas purpurea Oliv. ●☆

289891 Pentas suswaensis Verdc. ;苏苏瓦五星花●☆

289892 Pentas tenuis Verdc. ;细五星花●☆

289893 Pentas thomsonii Scott-Elliot = Pentas schimperiana（A. Rich.）Vatke ●☆

289894 Pentas tibestica Quézel;提贝斯提五星花■☆

289895 Pentas tibestica Quézel var. parviflora ? = Pentas tibestica Quézel ■☆

289896 Pentas transvaalensis Baer = Pentas angustifolia（A. Rich. ex DC.）Verdc. ●☆

289897 Pentas triangularis De Wild. = Pentas decora S. Moore var. triangularis（De Wild.）Verdc. ●☆

289898 Pentas ulugurica（Verdc.）Hepper;乌卢古尔五星花■☆

289899 Pentas verruculosa Chiov. = Pentas lanceolata（Forssk.）K. Schum. var. quartiniana（A. Rich.）Verdc. ●☆

289900 Pentas verticillata Scott-Elliot = Pentas decora S. Moore ●☆

289901 Pentas verticillata Scott-Elliot var. pubescens S. Moore = Pentas decora S. Moore var. triangularis（De Wild.）Verdc. ●☆

289902 Pentas volubilis K. Schum. = Otomeria volubilis（K. Schum.）Verdc. ■☆

289903 Pentas warburgiana K. Schum. ;沃伯格五星花●☆

289904 Pentas woodii Scott-Elliot = Pentas angustifolia（A. Rich. ex DC.）Verdc. ●☆

289905 Pentas wyliei N. E. Br. = Pentas micrantha Baker subsp. wyliei（N. E. Br.）Verdc. ■☆

289906 Pentas zanzibarica（Klotzsch）Vatke;桑给巴尔五星花■☆

289907 Pentas zanzibarica（Klotzsch）Vatke var. intermedia Verdc. ;间型桑给巴尔五星花■☆

289908 Pentas zanzibarica（Klotzsch）Vatke var. latifolia Verdc. ;宽叶桑给巴尔五星花■☆

289909 Pentas zanzibarica（Klotzsch）Vatke var. membranacea Verdc. = Pentas micrantha Baker subsp. wyliei（N. E. Br.）Verdc. ■☆

289910 Pentas zanzibarica（Klotzsch）Vatke var. pembensis Verdc. = Pentas micrantha Baker subsp. wyliei（N. E. Br.）Verdc. ■☆

289911 Pentas zanzibarica（Klotzsch）Vatke var. rubra Verdc. ;红桑给巴尔五星花■☆

289912 Pentas zanzibarica（Klotzsch）Vatke var. tenuifolia Verdc. ;细叶桑给巴尔五星花■☆

289913 Pentasachme G. Don = Pentasacme Wall. ex Wight et Arn. ■

289914 Pentasachme Wall. ex Wight = Pentasacme Wall. ex Wight et Arn. ■

289915 Pentasachme brachyantha Hand. -Mazz. = Cynanchum stauntonii（Decne.）Schltr. ex H. Lév. ●■

289916 Pentasachme caudatum sensu Merr. = Pentasacme caudatum Wall. ex Wight ■

289917 Pentasachme caudatum Wall. ex Wight = Pentasacme caudatum Wall. ex Wight ■

289918 Pentasachme championii Benth. = Pentasacme caudatum Wall. ex Wight ■

289919 Pentasachme esquirolii H. Lév. = Heterostemma esquirolii（H. Lév.）Tsiang ●

289920 Pentasachme glaucescens Decne. = Cynanchum glaucescens（Decne.）Hand. -Mazz. ●■

289921 Pentasachme stauntonii Decne. = Cynanchum stauntonii（Decne.）Schltr. ex H. Lév. ●■

289922 Pentasacme G. Don = Pentasacme Wall. ex Wight et Arn. ■

289923 Pentasacme Wall. ex Wight = Pentasacme Wall. ex Wight et Arn. ■

Arn. ■

289924 Pentasacme Wall. ex Wight et Arn.（1834）;石萝藦属（凤尾草属）;Pentasacme ■

289925 Pentasacme caudatum Wall. ex Wight;石萝藦（凤尾草,假了刁竹,满草,南石萝藦,水杨柳,五来）;Caudate Pentasacme ■

289926 Pentasacme championii Benth. = Pentasacme caudatum Wall. ex Wight ■

289927 Pentasacme esquirolii H. Lév. = Heterostemma esquirolii（H. Lév.）Tsiang ●

289928 Pentasacme stauntonii Decne. = Cynanchum stauntonii（Decne.）Schltr. ex H. Lév. ●■

289929 Pentaschistis（Nees）Spach(1841);南非禾属（五裂草属）■☆

289930 Pentaschistis Stapf = Pentaschistis（Nees）Spach ■☆

289931 Pentaschistis acinosa Stapf;葡萄南非禾■☆

289932 Pentaschistis airoides（Nees）Stapf;银须草南非禾■☆

289933 Pentaschistis alticola H. P. Linder;高原南非禾■☆

289934 Pentaschistis ampla（Nees）McClean;大南非禾■☆

289935 Pentaschistis andringitrensis A. Camus;安德林吉特拉山南非禾■☆

289936 Pentaschistis angulata（Nees）Adamson = Pentaschistis tortuosa（Trin.）Stapf ■☆

289937 Pentaschistis angustifolia（Nees）Stapf;窄叶南非禾■☆

289938 Pentaschistis angustifolia（Nees）Stapf = Pentaschistis pallida（Thunb.）H. P. Linder ■☆

289939 Pentaschistis angustifolia（Nees）Stapf var. albescens Stapf = Pentaschistis pallida（Thunb.）H. P. Linder ■☆

289940 Pentaschistis angustifolia（Nees）Stapf var. cirrhulosa ? = Pentaschistis cirrhulosa（Nees）H. P. Linder ■☆

289941 Pentaschistis angustifolia（Nees）Stapf var. micrathera ? = Pentaschistis glandulosa（Schrad.）H. P. Linder ■☆

289942 Pentaschistis angustifolia Stapf = Pentaschistis angustifolia（Nees）Stapf ■☆

289943 Pentaschistis argentea Stapf;银白南非禾■☆

289944 Pentaschistis aristidoides（Thunb.）Stapf;三芒草南非禾■☆

289945 Pentaschistis aristifolia Schweick. ;芒叶南非禾■☆

289946 Pentaschistis aspera（Thunb.）Stapf;粗糙南非禾■☆

289947 Pentaschistis aurea（Steud.）McClean;黄南非禾■☆

289948 Pentaschistis aurea（Steud.）McClean subsp. pilosogluma（McClean）H. P. Linder;毛颖黄南非禾■☆

289949 Pentaschistis bachmannii McClean = Pentaschistis ecklonii（Nees）McClean ■☆

289950 Pentaschistis barbata（Nees）H. P. Linder;髯毛南非禾■☆

289951 Pentaschistis barbata（Nees）H. P. Linder subsp. orientalis H. P. Linder;东部髯毛南非禾■☆

289952 Pentaschistis basutorum Stapf;巴苏托南非禾■☆

289953 Pentaschistis borussica（K. Schum.）Pilg. ;博鲁斯南非禾■☆

289954 Pentaschistis borussica（K. Schum.）Pilg. var. minor Ballard et C. E. Hubb. = Pentaschistis pictigluma（Steud.）Pilg. var. minor（Ballard et C. E. Hubb.）S. M. Phillips ■☆

289955 Pentaschistis brachyathera Stapf = Pentaschistis tomentella Stapf ■☆

289956 Pentaschistis burchellii Stapf = Pentaschistis cirrhulosa（Nees）H. P. Linder ■☆

289957 Pentaschistis calcicola H. P. Linder;钙生南非禾■☆

289958 Pentaschistis calcicola H. P. Linder var. hirsuta H. P. Linder;粗毛钙生南非禾■☆

289959 Pentaschistis capensis（Nees）Stapf;好望角南非禾■☆

289960 Pentaschistis capillaris（Thunb.）McClean;发状南非禾■☆

289961　Pentaschistis caulescens H. P. Linder;具茎南非禾■☆

289962　Pentaschistis chrysurus（K. Schum.）Peter;金黄南非禾■☆

289963　Pentaschistis cirrhulosa（Nees）H. P. Linder;流苏南非禾■☆

289964　Pentaschistis clavata Galley;棍棒南非禾■☆

289965　Pentaschistis colorata（Steud.）Stapf;着色南非禾■☆

289966　Pentaschistis colorata（Steud.）Stapf var. polytricha Stapf ＝ Pentaschistis colorata（Steud.）Stapf ■☆

289967　Pentaschistis curvifolia（Schrad.）Stapf;折叶南非禾☆

289968　Pentaschistis densifolia（Nees）Stapf;密叶南非禾■☆

289969　Pentaschistis densifolia（Nees）Stapf var. intricata Stapf ＝ Pentaschistis densifolia（Nees）Stapf ■☆

289970　Pentaschistis dolichochaeta S. M. Phillips;长刚毛南非禾■☆

289971　Pentaschistis ecklonii（Nees）McClean;埃氏南非禾■☆

289972　Pentaschistis effusa Peter ＝ Pentaschistis borussica（K. Schum.）Pilg.■☆

289973　Pentaschistis elegans（Nees）Stapf;雅致南非禾☆

289974　Pentaschistis eriostoma（Nees）Stapf;毛口南非禾■☆

289975　Pentaschistis euadenia Stapf ＝ Pentaschistis patula（Nees）Stapf ■☆

289976　Pentaschistis exserta H. P. Linder;伸出南非禾■☆

289977　Pentaschistis fibrosa Stapf ＝ Pentaschistis tysonii Stapf ■☆

289978　Pentaschistis filiformis（Nees）Stapf ＝ Pentaschistis pallida（Thunb.）H. P. Linder ■☆

289979　Pentaschistis galpinii（Stapf）McClean;盖尔南非禾☆

289980　Pentaschistis glandulosa（Schrad.）H. P. Linder;腺点南非禾■☆

289981　Pentaschistis gracilis S. M. Phillips ＝ Pentaschistis pictigluma（Steud.）Pilg. var. gracilis（S. M. Phillips）S. M. Phillips ■☆

289982　Pentaschistis heptamera（Nees）Stapf;七数南非禾■☆

289983　Pentaschistis heterochaeta Stapf ＝ Pentaschistis pallida（Thunb.）H. P. Linder ■☆

289984　Pentaschistis hirsuta（Nees）Stapf ＝ Pentaschistis rupestris（Nees）Stapf ■☆

289985　Pentaschistis horrida Galley;多刺南非禾☆

289986　Pentaschistis humbertii A. Camus;亨伯特南非禾■☆

289987　Pentaschistis imatongensis C. E. Hubb. ＝ Pentaschistis pictigluma（Steud.）Pilg. var. mannii（Stapf ex C. E. Hubb.）S. M. Phillips ■☆

289988　Pentaschistis imperfecta Stapf ＝ Pentaschistis pallida（Thunb.）H. P. Linder ■☆

289989　Pentaschistis involuta（Steud.）Adamson ＝ Pentaschistis pallescens（Schrad.）Stapf ■☆

289990　Pentaschistis juncifolia Stapf;灯心草叶南非禾■☆

289991　Pentaschistis leucopogon Stapf ＝ Pentaschistis barbata（Nees）H. P. Linder ■☆

289992　Pentaschistis lima（Nees）Stapf;侧生南非禾■☆

289993　Pentaschistis longipes Stapf;长梗南非禾■☆

289994　Pentaschistis malouinensis（Steud.）Clayton;马鲁因南非禾■☆

289995　Pentaschistis mannii Stapf ex C. E. Hubb. ＝ Pentaschistis pictigluma（Steud.）Pilg. var. mannii（Stapf ex C. E. Hubb.）S. M. Phillips ■☆

289996　Pentaschistis meruensis C. E. Hubb. ＝ Pentaschistis borussica（K. Schum.）Pilg.■☆

289997　Pentaschistis microphylla（Nees）McClean;小叶南非禾■☆

289998　Pentaschistis minor（Ballard et C. E. Hubb.）Ballard et C. E. Hubb. ＝ Pentaschistis pictigluma（Steud.）Pilg. var. minor（Ballard et C. E. Hubb.）S. M. Phillips ☆

289999　Pentaschistis montana H. P. Linder;山地南非禾■☆

290000　Pentaschistis natalensis Stapf;纳塔尔南非禾■☆

290001　Pentaschistis nutans（Nees）Stapf ＝ Pentaschistis tortuosa（Trin.）Stapf ■☆

290002　Pentaschistis oreodoxa Schweick.;山景南非禾■☆

290003　Pentaschistis pallescens（Schrad.）Stapf;变苍白南非禾■☆

290004　Pentaschistis pallida（Thunb.）H. P. Linder;苍白南非禾☆■☆

290005　Pentaschistis papillosa（Steud.）H. P. Linder;乳头南非禾■☆

290006　Pentaschistis patula（Nees）Stapf;张开南非禾■☆

290007　Pentaschistis patula（Nees）Stapf var. acuta Stapf ＝ Pentaschistis patula（Nees）Stapf ■☆

290008　Pentaschistis patula（Nees）Stapf var. glabrata Stapf ＝ Pentaschistis airoides（Nees）Stapf ■☆

290009　Pentaschistis patuliflora Rendle ＝ Pentaschistis cirrhulosa（Nees）H. P. Linder ■☆

290010　Pentaschistis perrieri A. Camus ＝Pentaschistis natalensis Stapf ■☆

290011　Pentaschistis pictigluma（Steud.）Pilg.;色颖南非禾■☆

290012　Pentaschistis pictigluma（Steud.）Pilg. var. gracilis（S. M. Phillips）S. M. Phillips;纤细色颖南非禾■☆

290013　Pentaschistis pictigluma（Steud.）Pilg. var. mannii（Stapf ex C. E. Hubb.）S. M. Phillips;曼氏色颖南非禾■☆

290014　Pentaschistis pictigluma（Steud.）Pilg. var. minor（Ballard et C. E. Hubb.）S. M. Phillips;小色颖南非禾■☆

290015　Pentaschistis pilosogluma McClean ＝Pentaschistis aurea（Steud.）McClean subsp. pilosogluma（McClean）H. P. Linder ■☆

290016　Pentaschistis praecox H. P. Linder;早南非禾☆

290017　Pentaschistis pseudopallescens H. P. Linder;假苍白南非禾■☆

290018　Pentaschistis pungens H. P. Linder;刚毛南非禾■☆

290019　Pentaschistis pusilla（Nees）H. P. Linder;微小南非禾☆

290020　Pentaschistis pyrophila H. P. Linder;喜炎南非禾■☆

290021　Pentaschistis reflexa H. P. Linder;反折南非禾■☆

290022　Pentaschistis rigidissima Pilg. ex H. P. Linder;坚挺南非禾☆

290023　Pentaschistis rosea H. P. Linder;粉红南非禾■☆

290024　Pentaschistis rosea H. P. Linder subsp. purpurascens;紫南非禾■☆

290025　Pentaschistis rupestris（Nees）Stapf;岩生南非禾■☆

290026　Pentaschistis ruwenzoriensis C. E. Hubb. ＝ Pentaschistis borussica（K. Schum.）Pilg.■☆

290027　Pentaschistis scandens H. P. Linder;攀缘南非禾■☆

290028　Pentaschistis setifolia（Thunb.）McClean;毛叶南非禾■☆

290029　Pentaschistis silvatica Adamson ＝ Pentaschistis pallescens（Schrad.）Stapf ■☆

290030　Pentaschistis steudelii（Nees）McClean ＝ Pentaschistis malouinensis（Steud.）Clayton ■☆

290031　Pentaschistis subulifolia Stapf ＝ Pentaschistis papillosa（Steud.）H. P. Linder ■☆

290032　Pentaschistis thunbergii（Kunth）Stapf ＝ Pentaschistis triseta（Thunb.）Stapf ■☆

290033　Pentaschistis thunbergii（Kunth）Stapf var. brevifolia Stapf ＝ Pentaschistis pallida（Thunb.）H. P. Linder ■☆

290034　Pentaschistis thunbergii（Kunth）Stapf var. bulbothrix Stapf ＝ Pentaschistis pallida（Thunb.）H. P. Linder ■☆

290035　Pentaschistis thunbergii（Kunth）Stapf var. ebarbata Stapf ＝ Pentaschistis pallida（Thunb.）H. P. Linder ■☆

290036　Pentaschistis tomentella Stapf;绒毛南非禾■☆

290037　Pentaschistis tortuosa（Trin.）Stapf;扭曲南非禾■☆

290038　Pentaschistis trifida Galley;三裂南非禾■☆

290039　Pentaschistis triseta（Thunb.）Stapf;三刚毛南非禾■☆

290040　Pentaschistis trisetoides（Hochst. ex Steud.）Pilg.;拟三刚毛

南非禾■☆

290041　Pentaschistis tysonii Stapf;泰森南非禾■☆

290042　Pentaschistis velutina H. P. Linder;短绒毛南非禾■☆

290043　Pentaschistis veneta H. P. Linder;海色南非禾■☆

290044　Pentaschistis viscidula（Nees）Stapf;黏南非禾■☆

290045　Pentaschistis welwitschii Rendle ＝ Piptophyllum welwitschii（Rendle）C. E. Hubb.■☆

290046　Pentaschistis zeyheri Stapf ＝ Pentaschistis papillosa（Steud.）H. P. Linder ■☆

290047　Pentascyphus Radlk.（1879）;五皿木属●☆

290048　Pentascyphus thyrsiflorus Radlk.;五皿木●☆

290049　Pentaspadon Hook. f.（1860）;五裂漆属●☆

290050　Pentaspadon motleyi Hook. f.;五裂漆●☆

290051　Pentaspadon velutinus Hook. f.;毛五裂漆●☆

290052　Pentaspatella Gleason ＝ Sauvagesia L.●

290053　Pentastachya Hochst. ex Steud. ＝ Pennisetum Rich.■

290054　Pentastachya Steud. ＝ Pennisetum Rich.■

290055　Pentastelma Tsiang et P. T. Li(1974);白水藤属;Pentastelma ●★

290056　Pentastelma auritum Tsiang et P. T. Li;白水藤;Longeared Pentastelma,Long-eared Pentastelma ●◇

290057　Pentastemon Batsch ＝ Penstemon Schmidel ●■

290058　Pentastemon L'Hér. ＝ Penstemon Schmidel ●■

290059　Pentastemona Steenis(1982);五出百部属■☆

290060　Pentastemona egregia（Schott）Steenis;五出百部■☆

290061　Pentastemona sumatrana Steenis;苏门答腊五出百部;Sumatra Pentastemona ■☆

290062　Pentastemonaceae Duyfjes ＝ Stemonaceae Caruel(保留科名)■

290063　Pentastemonaceae Duyfjes(1992);五出百部科(鳞百部科)■☆

290064　Pentastemonodiscus Rech. f.（1965）;五蕊线球草属■☆

290065　Pentastemonodiscus monochlamydeus Rech. f.;五蕊线球草■☆

290066　Pentastemum Steud. ＝ Penstemon Schmidel ●■

290067　Pentastemum Steud. ＝ Pentastemon Batsch ●■

290068　Pentasticha Turcz. ＝ Fuirena Rottb.■

290069　Pentastigma Maxim. ex Kom. ＝ Clematoclethra（Franch.）Maxim.●★

290070　Pentastira Ridl. ＝ Dichapetalum Thouars ●

290071　Pentataentum Tamamsch. ＝ Stenotaenia Boiss.■☆

290072　Pentataphrus Schltdl. ＝ Astroloma R. Br.●☆

290073　Pentataxis D. Don ＝ Helichrysum Mill.(保留属名)●■

290074　Pentatherum Nábelek ＝ Agrostis L.(保留属名)■

290075　Pentatherum dshungaricum Tzvelev ＝ Agrostis dshungarica（Tzvelev）Tzvelev ■

290076　Pentatherum hissaricum Nevski ＝ Polypogon hissaricus（Roshev.）Bor ■

290077　Pentatherum pilosulum（Trin.）Tzvelev ＝ Agrostis pilosula Trin.■

290078　Pentathymelaea Lecomte ＝ Daphne L.●

290079　Pentathymelaea Lecomte(1916);五出瑞香属●

290080　Pentathymelaea thibetensis Lecomte;五出瑞香●

290081　Pentathymelaea thibetensis Lecomte ＝ Daphne holoserica（Diels）Hamaya var. thibetensis（Lecomte）Hamaya ●

290082　Pentatrichia Klatt(1895);齿叶鼠麹木属●■☆

290083　Pentatrichia alata S. Moore;具翅齿叶鼠麹木■☆

290084　Pentatrichia avasmontana Merxm.;埃文齿叶鼠麹木■☆

290085　Pentatrichia confertifolia Merxm. ＝ Pentatrichia avasmontana Merxm.■☆

290086　Pentatrichia petrosa Klatt;岩生齿叶鼠麹木■☆

290087　Pentatrichia rehmii（Merxm.）Merxm.;雷姆齿叶鼠麹木■☆

290088　Pentatropis R. Br. ＝ Pentatropis R. Br. ex Wight et Arn.■☆

290089　Pentatropis R. Br. ex Wight et Arn.（1834）;朱砂莲属■☆

290090　Pentatropis atropurpurea（F. Muell.）Benth.;紫朱砂莲■☆

290091　Pentatropis bentii（N. E. Br.）Liede;本特朱砂莲■☆

290092　Pentatropis capensis（L. f.）Bullock;好望角朱砂莲■☆

290093　Pentatropis cynanchoides R. Br. ＝ Pentatropis nivalis（J. F. Gmel.）D. V. Field et J. R. I. Wood ■☆

290094　Pentatropis cynanchoides R. Br. ＝ Pentatropis spiralis（Forssk.）Decne.■☆

290095　Pentatropis cynanchoides R. Br. var. longepetiolata Engl. ＝ Pentatropis nivalis（J. F. Gmel.）D. V. Field et J. R. I. Wood ■☆

290096　Pentatropis cynanchoides R. Br. var. senegalensis（Decne.）N. E. Br. ＝ Pentatropis nivalis（J. F. Gmel.）D. V. Field et J. R. I. Wood ■☆

290097　Pentatropis fasciculatus（K. Schum.）N. E. Br. ＝ Pentatropis nivalis（J. F. Gmel.）D. V. Field et J. R. I. Wood ■☆

290098　Pentatropis hoyoides K. Schum. ＝ Pentatropis nivalis（J. F. Gmel.）D. V. Field et J. R. I. Wood ■☆

290099　Pentatropis linearis Decne.;线性朱砂莲■☆

290100　Pentatropis madagascariensis Decne. ＝ Pentatropis nivalis（J. F. Gmel.）D. V. Field et J. R. I. Wood subsp. madagascariensis（Decne.）Liede et Meve ■☆

290101　Pentatropis madagascariensis Decne. ＝ Pentatropis nivalis（J. F. Gmel.）D. V. Field et J. R. I. Wood ■☆

290102　Pentatropis microphylla（Roth）Wall. ＝ Pentatropis capensis（L. f.）Bullock ■☆

290103　Pentatropis nivalis（J. F. Gmel.）D. V. Field et J. R. I. Wood;雪白朱砂莲■☆

290104　Pentatropis nivalis（J. F. Gmel.）D. V. Field et J. R. I. Wood subsp. madagascariensis（Decne.）Liede et Meve;马达加斯加雪白朱砂莲■☆

290105　Pentatropis officinalis Hemsl. ＝ Cynanchum officinale（Hemsl.）Tsiang et H. D. Zhang ●■

290106　Pentatropis rigida Chiov. ＝ Pentatropis nivalis（J. F. Gmel.）D. V. Field et J. R. I. Wood ■☆

290107　Pentatropis senegalensis Decne. ＝ Pentatropis nivalis（J. F. Gmel.）D. V. Field et J. R. I. Wood ■☆

290108　Pentatropis spiralis（Forssk.）Decne.;螺旋朱砂莲■☆

290109　Penteca Raf. ＝ Croton L.●

290110　Pentelesia Raf. ＝ Arrabidaea DC.●☆

290111　Pentena Raf. ＝ Scabiosa L.●■

290112　Penthea（D. Don）Spach ＝ Barnadesia Mutis ex L. f.●☆

290113　Penthea Lindl.（1835）;哀兰属(潘西亚兰属);Penthea ■☆

290114　Penthea Lindl. ＝ Disa P. J. Bergius ■☆

290115　Penthea Spach ＝ Barnadesia Mutis ex L. f.●☆

290116　Penthea atricapilla Harv. ex Lindl. ＝ Disa atricapilla（Harv. ex Lindl.）Bolus ■☆

290117　Penthea filicornis（L. f.）Lindl. ＝ Disa filicornis（L. f.）Thunb.■☆

290118　Penthea filicornis（L.）Lindl.;哀兰(潘西亚兰);Common Penthea ■☆

290119　Penthea melaleuca（Thunb.）Lindl. ＝ Disa bivalvata（L. f.）T. Durand et Schinz ■☆

290120　Penthea minor Sond. ＝ Disa minor（Sond.）Rchb. f.■☆

290121　Penthea obtusa Lindl. ＝ Disa richardiana Lehm. ex Bolus ■☆

290122　Penthea patens（L. f.）Lindl. ＝ Disa tenuifolia Sw.■☆

290123　Penthea pumilio Lindl. = Schwartzkopffia pumilio（Lindl.）Schltr. ■☆

290124　Penthea reflexa Lindl. = Disa filicornis（L. f.）Thunb. ■☆

290125　Penthea triloba Sond. = Disa oligantha Rchb. f. ■☆

290126　Pentheriella O. Hoffm. et Muschl.（1910－1911）；小扯根菜属■☆

290127　Pentheriella O. Hoffm. et Muschl. = Heteromma Benth. ■☆

290128　Pentheriella krookii O. Hoffm. et Muschl. = Heteromma krookii（O. Hoffm. et Muschl.）Hilliard et B. L. Burtt ■☆

290129　Penthoraceae Rydb. ex Britton（1901）（保留科名）；扯根菜科（扯根草科）■

290130　Penthoraceae Rydb. ex Britton（保留科名）= Saxifragaceae Juss.（保留科名）●■

290131　Penthoraceae Tiegh. = Penthoraceae Rydb. ex Britton（保留科名）■

290132　Penthorum Gronov. ex L. = Penthorum L. ■

290133　Penthorum L.（1753）；扯根菜属；Penthorum ■

290134　Penthorum chinense Pursh；扯根菜（干黄草，赶黄草，红柳信，山黄鳝，水苋菜，水杨柳，水泽兰，水滓蓝）；China Penthorum，Chinese Penthorum ■

290135　Penthorum humile Regel et Maack = Penthorum chinense Pursh ■

290136　Penthorum humile Regel et Maack ex Regel = Penthorum chinense Pursh ■

290137　Penthorum intermedium Turcz. = Penthorum chinense Pursh ■

290138　Penthorum sedoides L.；景天扯根菜（北美扯根菜）；Ditch Stonecrop，Penthorum ■☆

290139　Penthorum sedoides L. = Penthorum chinense Pursh ■

290140　Penthorum sedoides L. subsp. chinense（Pursh）S. Y. Li et Adair = Penthorum chinense Pursh ■

290141　Penthorum sedoides L. var. chinense（Pursh）Maxim. = Penthorum chinense Pursh ■

290142　Penthysa Raf. = Echiopsis Rchb. ■☆

290143　Penthysa Raf. = Lobostemon Lehm. ■☆

290144　Pentiphragma Hook. = Pentaphragma Wall. ex G. Don ■

290145　Pentisea（Lindl.）Szlach. = Caladenia R. Br. ■☆

290146　Pentisea Lindl. = Caladenia R. Br. ■☆

290147　Pentitdis Zipp. ex Blume = Opilia Roxb. ●

290148　Pentlandia Herb. = Urceolina Rchb.（保留属名）■☆

290149　Pentochna Tiegh. = Ochna L. ●

290150　Pentocnide Raf. = Pouzolzia Gaudich. ●■

290151　Pentodon Ehrenb. ex Boiss. = Kochia Roth ●■

290152　Pentodon Hochst.（1844）；五齿茜属■☆

290153　Pentodon abyssinicus Hochst. = Pentodon pentandrus（Schumach. et Thonn.）Vatke ■☆

290154　Pentodon decumbens Hochst. = Pentodon pentandrus（Schumach. et Thonn.）Vatke var. minor Bremek. ■☆

290155　Pentodon laurentioides Chiov.；五齿茜■☆

290156　Pentodon pentandrus（Schumach. et Thonn.）Vatke；五蕊五齿茜■☆

290157　Pentodon pentandrus（Schumach. et Thonn.）Vatke var. minor Bremek.；小五蕊五齿茜■☆

290158　Pentopetia Decne.（1844）；坡梯草属（盆托坡梯草属）；Pentopetia ■☆

290159　Pentopetia alba Jum. et H. Perrier = Pentopetia elastica Jum. et H. Perrier ■☆

290160　Pentopetia albicans（Jum. et H. Perrier）Klack.；苍白坡梯草■☆

290161　Pentopetia androsaemifolia Decne.；盆托坡梯草■☆

290162　Pentopetia androsaemifolia Decne. subsp. lanceolata Costantin et Gallaud = Pentopetia androsaemifolia Decne. ■☆

290163　Pentopetia androsaemifolia Decne. subsp. multiflora Boivin ex Costantin et Gallaud = Pentopetia androsaemifolia Decne. ■☆

290164　Pentopetia androsaemifolia Decne. subsp. ovalifolia Costantin et Gallaud = Pentopetia androsaemifolia Decne. ■☆

290165　Pentopetia androsaemifolia Decne. subsp. pilosa Costantin et Gallaud = Pentopetia androsaemifolia Decne. ■☆

290166　Pentopetia androsaemifolia Decne. var. cordifolia Costantin et Gallaud = Pentopetia androsaemifolia Decne. ■☆

290167　Pentopetia androsaemifolia Decne. var. cowanii Costantin et Gallaud = Pentopetia androsaemifolia Decne. ■☆

290168　Pentopetia androsaemifolia Decne. var. scabra Costantin et Gallaud = Pentopetia androsaemifolia Decne. ■☆

290169　Pentopetia boinensis Jum. et H. Perrier = Pentopetia androsaemifolia Decne. ■☆

290170　Pentopetia boivinii Costantin et Gallaud；博伊文坡梯草■☆

290171　Pentopetia bosseri Klack.；博瑟坡梯草■☆

290172　Pentopetia calycina Klack.；萼状坡梯草■☆

290173　Pentopetia cotoneaster Decne.；枸子状坡梯草■☆

290174　Pentopetia cotoneaster Decne. subsp. acustelma（Costantin et Gallaud）Costantin et Gallaud = Pentopetia cotoneaster Decne. ■☆

290175　Pentopetia cotoneaster Decne. subsp. glabra Costantin et Gallaud = Pentopetia androsaemifolia Decne. ■☆

290176　Pentopetia cotoneaster Decne. subsp. pentopetiopsis（Costantin et Gallaud）Costantin et Gallaud = Pentopetia ovalifolia（Costantin et Gallaud）Klack. ■☆

290177　Pentopetia cotoneaster Decne. subsp. thouarsii Costantin et Gallaud = Pentopetia boivinii Costantin et Gallaud ■☆

290178　Pentopetia cotoneaster Decne. var. acustelma Costantin et Gallaud = Pentopetia cotoneaster Decne. ■☆

290179　Pentopetia cotoneaster Decne. var. pentopetiopsis Costantin et Gallaud = Pentopetia ovalifolia（Costantin et Gallaud）Klack. ■☆

290180　Pentopetia dasynema Choux；毛坡梯草■☆

290181　Pentopetia dolichopodia Klack.；长足坡梯草■☆

290182　Pentopetia ecoronata Klack.；无冠坡梯草■☆

290183　Pentopetia elastica Jum. et H. Perrier；弹性坡梯草■☆

290184　Pentopetia glaberrima Choux；无毛坡梯草■☆

290185　Pentopetia gracilis Decne. = Pentopetia androsaemifolia Decne. ■☆

290186　Pentopetia graminifolia Costantin et Gallaud = Ischnolepis graminifolia（Costantin et Gallaud）Klack. ■☆

290187　Pentopetia grevei（Baill.）Venter；格雷弗坡梯草■☆

290188　Pentopetia longipetala Klack.；长瓣坡梯草■☆

290189　Pentopetia lutea Klack. et Civeyrel；黄坡梯草■☆

290190　Pentopetia mollis Jum. et H. Perrier；柔软坡梯草■☆

290191　Pentopetia natalensis Schltr. = Ischnolepis natalensis（Schltr.）Venter ■☆

290192　Pentopetia ovalifolia（Costantin et Gallaud）Klack.；卵叶坡梯草■☆

290193　Pentopetia pinnata Costantin et Gallaud；羽状坡梯草■☆

290194　Pentopetia reticulata Jum. et H. Perrier；网状坡梯草■☆

290195　Pentopetia urceolata Klack.；坛状坡梯草■☆

290196　Pentopetiopsis Costantin et Gallaud = Pentopetia Decne. ■☆

290197　Pentopetiopsis Costantin et Gallaud（1906）；拟坡梯草属；Pentopetia ■☆

290198　Pentopetiopsis ovalifolia Costantin et Gallaud；拟坡梯草■☆

290199　Pentopetiopsis ovalifolia Costantin et Gallaud = Pentopetia ovalifolia（Costantin et Gallaud）Klack. ■☆

290200 　Pentossaea Judd(1989)；五数野牡丹属●☆

290201 　Pentossaea angustifolia（DC.）Judd；窄叶五数野牡丹●☆

290202 　Pentossaea brachystachya（DC.）Judd；短穗五数野牡丹●☆

290203 　Pentostemon Raf. = Ossaea DC. ●☆

290204 　Pentostemon Raf. = Penstemon Schmidel ●■

290205 　Pentrias Benth. et Hook. f. = Amaranthus L. ■

290206 　Pentrias Benth. et Hook. f. = Pentrius Raf. ■

290207 　Pentrius Raf. = Amaranthus L. ■

290208 　Pentropis Raf. = Campanula L. ■●

290209 　Pentsteira Griff. = Torenia L. ■

290210 　Pentstemon Aiton = Penstemon Schmidel ●■

290211 　Pentstemon Mitch. = Penstemon Schmidel ●■

290212 　Pentstemonacanthus Nees(1847)；五刺蕊爵床属☆

290213 　Pentstemonacanthus modestus Nees；五刺蕊爵床☆

290214 　Pentstemonopsis Rydb. = Chionophila Benth. ■☆

290215 　Pentstemum Steud. = Penstemon Schmidel ●■

290216 　Pentsteria Griff. = Pentsteira Griff. ■

290217 　Pentsteria Griff. = Torenia L. ■

290218 　Pentstira Post et Kuntze = Pentsteria Griff. ■

290219 　Penttatherum Nábelek = Agrostis L.（保留属名）■

290220 　Pentulops Raf. = Maxillaria Ruiz et Pav. ■☆

290221 　Pentulops Raf. = Xylobium Lindl. ■☆

290222 　Pentzia Thunb.（1800）；杯子菊属；African Sheepbush, Sheepbush ■●☆

290223 　Pentzia acutiloba（DC.）Hutch. = Myxopappus acutilobus（DC.）Källersjö ■☆

290224 　Pentzia albida（DC.）Hutch. = Foveolina dichotoma（DC.）Källersjö ■☆

290225 　Pentzia annua DC. = Foveolina dichotoma（DC.）Källersjö ■☆

290226 　Pentzia argentea Hutch.；银白杯子菊■☆

290227 　Pentzia athanasioides S. Moore = Phymaspermum athanasioides（S. Moore）Källersjö ■☆

290228 　Pentzia bolusii Hutch.；博卢斯杯子菊■☆

290229 　Pentzia burchellii（DC.）Fenzl = Pentzia punctata Harv. ■☆

290230 　Pentzia calcarea Kies；石灰杯子菊■☆

290231 　Pentzia calva S. Moore；光秃杯子菊■☆

290232 　Pentzia caudiculata Thell. = Pentzia pinnatisecta Hutch. ■☆

290233 　Pentzia cinerascens DC. = Pentzia sphaerocephala DC. ■☆

290234 　Pentzia cooperi Harv.；库珀杯子菊■☆

290235 　Pentzia dentata（L.）Kuntze；尖齿杯子菊■☆

290236 　Pentzia dichotoma DC. = Foveolina dichotoma（DC.）Källersjö ■☆

290237 　Pentzia eenii S. Moore = Rennera eenii（S. Moore）Källersjö ■☆

290238 　Pentzia elegans DC.；雅致杯子菊■☆

290239 　Pentzia flabelliformis Willd. = Pentzia dentata（L.）Kuntze ■☆

290240 　Pentzia galpinii Hutch. = Myxopappus hereroensis（O. Hoffm.）Källersjö ■☆

290241 　Pentzia globifera（Thunb.）Hutch. = Oncosiphon piluliferum（L. f.）Källersjö ■☆

290242 　Pentzia globifera Licht. ex Less. = Pentzia globosa Less. ■☆

290243 　Pentzia globosa Less.；球形杯子菊■☆

290244 　Pentzia grandiflora（Thunb.）Hutch. = Oncosiphon grandiflorum（Thunb.）Källersjö ■☆

290245 　Pentzia grisea Muschl. ex Dinter = Pentzia sphaerocephala DC. ■☆

290246 　Pentzia hereroensis O. Hoffm. = Myxopappus hereroensis（O. Hoffm.）Källersjö ■☆

290247 　Pentzia hesperidum Maire et Wilczek；金星杯子菊■☆

290248 　Pentzia incana（Thunb.）Kuntze；非洲杯子菊；African Sheepbush ■☆

290249 　Pentzia integrifolia Muschl. ex Dinter = Pentzia monocephala S. Moore ■☆

290250 　Pentzia intermedia Hutch. = Oncosiphon intermedium（Hutch.）Källersjö ■☆

290251 　Pentzia lanata Hutch.；绵毛杯子菊■☆

290252 　Pentzia laxa Bremek. et Oberm. = Rennera laxa（Bremek. et Oberm.）Källersjö ■☆

290253 　Pentzia macrocephala Dinter ex Merxm. et Eberle = Pentzia spinescens Less. ■☆

290254 　Pentzia membranacea Hutch. = Foveolina albidiformis（Thell.）Källersjö ■☆

290255 　Pentzia microcephala Dinter ex Range = Pentzia calva S. Moore ■☆

290256 　Pentzia monocephala S. Moore；单头杯子菊■☆

290257 　Pentzia monodiana Maire；莫诺杯子菊■☆

290258 　Pentzia nana Burch.；矮小杯子菊■☆

290259 　Pentzia peduncularis B. Nord.；梗花杯子菊■☆

290260 　Pentzia pilulifera（L. f.）Fourc. = Oncosiphon piluliferum（L. f.）Källersjö ■☆

290261 　Pentzia pinnatifida Oliv. = Phymaspermum pinnatifidum（Oliv.）Källersjö ●☆

290262 　Pentzia pinnatisecta Hutch.；羽裂杯子菊■☆

290263 　Pentzia punctata Harv.；斑点杯子菊☆

290264 　Pentzia quinquefida（Thunb.）Less.；五裂杯子菊■☆

290265 　Pentzia quinquefida（Thunb.）Less. var. nana（Burch.）Harv. = Pentzia nana Burch. ■☆

290266 　Pentzia sabulosa（Wolley-Dod）Hutch. = Oncosiphon sabulosum（Wolley-Dod）Källersjö ■☆

290267 　Pentzia schinziana（Thell.）Merxm. et Eberle = Foveolina schinziana（Thell.）Källersjö ■☆

290268 　Pentzia schistostephioides M. Taylor = Inulanthera nuda Källersjö ■☆

290269 　Pentzia somalensis E. A. Bruce ex Thulin；索马里杯子菊■☆

290270 　Pentzia sphaerocephala DC.；球头杯子菊■☆

290271 　Pentzia spinescens Less.；细刺杯子菊■☆

290272 　Pentzia stenocephala Thell. = Phymaspermum acerosum（DC.）Källersjö ■☆

290273 　Pentzia suffruticosa（L.）Hutch. ex Merxm. = Oncosiphon schlechteri（Bolus ex Schltr.）Källersjö ■☆

290274 　Pentzia suffruticosa（L.）Hutch. ex Merxm. = Oncosiphon suffruticosum（L.）Källersjö ■☆

290275 　Pentzia tanacetifolia（L.）Hutch. = Oncosiphon schlechteri（Bolus ex Schltr.）Källersjö ■☆

290276 　Pentzia tanacetifolia（L.）Hutch. = Oncosiphon suffruticosum（L.）Källersjö ■☆

290277 　Pentzia tomentosa B. Nord.；绒毛杯子菊■☆

290278 　Pentzia tortuosa（DC.）Fenzl ex Harv.；扭曲杯子菊■☆

290279 　Pentzia tysonii Thell. = Phymaspermum woodii（Thell.）Källersjö ■☆

290280 　Pentzia virgata Less. = Pentzia incana（Thunb.）Kuntze ■☆

290281 　Pentzia viridis Kies；绿杯子菊■☆

290282 　Pentzia woodii Thell. = Phymaspermum woodii（Thell.）Källersjö ■☆

290283 　Peperidia Kostel. = Piper L. ●■

290284 　Peperidia Rchb. = Chloranthus Sw. ■●

290285　Peperidium Lindl. = Renealmia L. f. (保留属名)■☆

290286　Peperomia Ruiz et Pav. (1794);草胡椒属(豆瓣绿属,椒草属);Paperelder,Peperomia,Piperomia,Radiator Plant,Rock Balsam ■

290287　Peperomia abyssinica Miq.;阿比西尼亚草胡椒■☆

290288　Peperomia abyssinica Miq. var. stuhlmannii (C. DC.) Düll = Peperomia abyssinica Miq. ■☆

290289　Peperomia alata Ruiz et Pav.;翅草胡椒■☆

290290　Peperomia amplexicaulis (Sw.) A. Dietr.;杰氏草胡椒;Jackie's Saddle ■☆

290291　Peperomia amplexicaulis A. Dietr. = Peperomia amplexicaulis (Sw.) A. Dietr. ■☆

290292　Peperomia ankaranensis G. Mathieu;安卡兰草胡椒■☆

290293　Peperomia annobonensis Mildbr. = Peperomia vulcanica Baker et C. H. Wright ■☆

290294　Peperomia arabica Decne. ex Miq. = Peperomia blanda (Jacq.) Kunth ■

290295　Peperomia arabica Decne. ex Miq. var. floribunda Miq. = Peperomia blanda (Jacq.) Kunth ■

290296　Peperomia arabica Decne. var. parvifolia C. DC. = Peperomia blanda (Jacq.) Kunth ■

290297　Peperomia arabica Hottes var. floribunda Miq. = Peperomia blanda (Jacq.) Kunth ■

290298　Peperomia argyrea Hottes;阿氏草胡椒;Rugby Football Plant,Silver Peperomia ■☆

290299　Peperomia argyreia E. Morren;西瓜皮豆瓣绿(善氏豆瓣绿,西瓜皮椒草);Rugby Football Plant,Watermelon Peperomia,Watermelon Peperomia,Watermelon Plant ■☆

290300　Peperomia argyroneura Hort.;银线豆瓣绿;Silver-thread Peperomia ■☆

290301　Peperomia bachmannii C. DC. = Peperomia retusa (L. f.) A. Dietr. var. bachmannii (C. DC.) Düll ■☆

290302　Peperomia bequaertii De Wild. = Peperomia blanda (Jacq.) Kunth ■

290303　Peperomia bernieriana Miq.;伯尼尔草胡椒■☆

290304　Peperomia bicolor Sodiro;双色椒草■☆

290305　Peperomia blanda (Jacq.) Kunth;石蝉草(豆瓣绿,豆瓣七,红豆瓣,红茎椒草,胡椒草,火伤草,火烧叶,散血丹,散血胆,石瓜子,石马菜);Dindygul Peperomia,Stonecicada Paperelder ■

290306　Peperomia blanda (Jacq.) Kunth var. floribunda (Miq.) H. Huber = Peperomia blanda (Jacq.) Kunth ■

290307　Peperomia blanda (Jacq.) Kunth var. leptostachya (Hook. et Arn.) Düll = Peperomia blanda (Jacq.) Kunth ■

290308　Peperomia boninsimensis Makino;小笠原椒草■☆

290309　Peperomia botteri C. DC.;鲍氏草胡椒■☆

290310　Peperomia brachytricha Baker;短毛草胡椒■☆

290311　Peperomia brachytrichoides Engl. = Peperomia blanda (Jacq.) Kunth ■

290312　Peperomia brevipes C. DC.;短叶柄草胡椒■☆

290313　Peperomia bueana C. DC. = Peperomia retusa (L. f.) A. Dietr. ■☆

290314　Peperomia butaguensis De Wild. = Peperomia fernandopoiana C. DC. var. butaguensis (De Wild.) Düll ■☆

290315　Peperomia caperata Ruiz et Pav. ex Yunck.;皱叶椒草;Emerald Ripple,Little Fantasy,Rat-tail Plant ■☆

290316　Peperomia cavalieriei C. DC.;硬毛草胡椒(指甲草);Cavalerie Paperelder ■

290317　Peperomia clusiifolia Hook.;琴叶椒草;Baby Rubber-plant,Baby Rubber Plant,Clusia-leaved Peperomia ■☆

290318　Peperomia clusiifolia Hook. 'Jewelry';三色椒草■☆

290319　Peperomia clusiifolia Hook. 'Variegata';五彩琴叶椒草;Baby Ruber Plant ■☆

290320　Peperomia commersonii Baill.;科梅逊椒草■☆

290321　Peperomia costata G. Mathieu;单脉草胡椒■☆

290322　Peperomia crassifolia Baker;毛叶椒草■☆

290323　Peperomia cubensis C. DC.;软茎椒草■☆

290324　Peperomia cumulicola Small = Peperomia humilis (Vahl) A. Dietr. ■☆

290325　Peperomia dindygulensis Miq. = Peperomia blanda (Jacq.) Kunth ■

290326　Peperomia dryadum C. DC.;森林草胡椒■☆

290327　Peperomia dubia Balle;可疑草胡椒■☆

290328　Peperomia duclouxii C. DC.;短穗草胡椒;Ducloux Paperelder ■

290329　Peperomia duclouxii C. DC. = Peperomia heyneana Miq. ■

290330　Peperomia dusenii C. DC.;杜森草胡椒■☆

290331　Peperomia eburnea Lindl.;象牙色椒草■☆

290332　Peperomia elliptica (Lam.) A. Dietr.;椭圆草胡椒■☆

290333　Peperomia elongata Kunth;伸长草胡椒■☆

290334　Peperomia emarginata De Wild. = Peperomia abyssinica Miq. ■☆

290335　Peperomia erythrocaulis G. Mathieu;红茎草胡椒■☆

290336　Peperomia esquirolii H. Lév. = Peperomia blanda (Jacq.) Kunth ■

290337　Peperomia estaminea C. DC.;无雄草胡椒■☆

290338　Peperomia exiguum Blume = Peperomia pellucida (L.) Kunth ■

290339　Peperomia fauriei C. DC. = Peperomia blanda (Jacq.) Kunth ■

290340　Peperomia fernandopoiana C. DC.;费尔南草胡椒■☆

290341　Peperomia fernandopoiana C. DC. var. butaguensis (De Wild.) Düll;布塔古草胡椒■☆

290342　Peperomia fernandopoiana C. DC. var. subopacifolia ? = Peperomia molleri C. DC. ■☆

290343　Peperomia floridana Small = Peperomia obtusifolia (L.) A. Dietr. ■☆

290344　Peperomia foliiflora Ruiz et Pav.;叶花草胡椒■☆

290345　Peperomia formosana C. DC. = Peperomia blanda (Jacq.) Kunth ■

290346　Peperomia forsythii C. DC. = Peperomia trichophylla Baker ■☆

290347　Peperomia freireifolia A. Rich. = Peperomia pellucida (L.) Kunth ■

290348　Peperomia galioides Kunth;狭叶椒草■☆

290349　Peperomia glabella (Sw.) A. Dietr.;无毛草胡椒(玲珑椒草);Wax Privet ■☆

290350　Peperomia glabella Griseb. = Peperomia glabella (Sw.) A. Dietr. ■☆

290351　Peperomia glabrilimba C. DC.;光边草胡椒■☆

290352　Peperomia goetzeana Engl.;格兹草胡椒■☆

290353　Peperomia gracilipetiolata De Wild. = Peperomia retusa (L. f.) A. Dietr. ■☆

290354　Peperomia griseoargentea Yunck.;银叶椒草;Ivy Peperomia,Ivy-leaf Peperomia,Silver-leaf Peperomia ■☆

290355　Peperomia harmandii Merr. = Peperomia blanda (Jacq.) Kunth ■

290356　Peperomia hederifolia ? = Peperomia griseoargentea Yunck. ■☆

290357　Peperomia heyneana Miq.;蒙自草胡椒(狗骨头,海尼豆瓣绿,散血丹);Mengzi Paperelder ■

290358　Peperomia hildebrandtii Vatke ex C. DC.;希尔德草胡椒■☆

290359　Peperomia holstii C. DC. = Peperomia molleri C. DC. ■☆

290360　Peperomia humbertii G. Mathieu;亨伯特草胡椒■☆

290361　Peperomia humilis (Vahl) A. Dietr.;小草胡椒■☆

290362 Peperomia hygrophila Engl. = Peperomia vulcanica Baker et C. H. Wright ■☆

290363 Peperomia imerinae C. DC. = Peperomia molleri C. DC. ■☆

290364 Peperomia incana A. Dietr. ;灰白椒草■☆

290365 Peperomia japonica Makino;日本草胡椒(椒草);Japan Paperelder,Japanese Peperomia ■

290366 Peperomia japonica Makino = Peperomia blanda (Jacq.) Kunth ■

290367 Peperomia kamerunana C. DC. ;喀麦隆草胡椒■☆

290368 Peperomia knoblecheriana Schott = Peperomia pellucida (L.) Kunth ■

290369 Peperomia kotoensis Yamam. = Peperomia rubrivenosa C. DC. ■

290370 Peperomia kyimbilana C. DC. = Peperomia blanda (Jacq.) Kunth ■

290371 Peperomia laeteviridis Engl. ;鲜绿草胡椒■☆

290372 Peperomia laticaulis C. DC. = Peperomia blanda (Jacq.) Kunth ■

290373 Peperomia leptostachya Hook. et Arn. = Peperomia blanda (Jacq.) Kunth ■

290374 Peperomia leptostachya Hook. et Arn. f. cambodiana A. DC. = Peperomia blanda (Jacq.) Kunth ■

290375 Peperomia leptostachya Hook. et Arn. var. cambodiana (A. DC.) Merr. = Peperomia blanda (Jacq.) Kunth ■

290376 Peperomia leptostachya Hook. et Arn. var. cambodiana (C. DC.) Merr. ;柬埔寨草胡椒(东亚细穗草胡椒);Cambodia Paperelder ■

290377 Peperomia leptostachya Hook. et Arn. var. cambodiana (C. DC.) Merr. = Peperomia blanda (Jacq.) Kunth ■

290378 Peperomia lyallii C. DC. ;莱尔草胡椒■☆

290379 Peperomia macrostachya (Vahl) A. Dietr. ;大穗草胡椒■☆

290380 Peperomia maculosa (L.) Hook. ;斑叶豆瓣绿;Radiator Plant, Spotleaf Peperomia ■☆

290381 Peperomia magilensis Baker = Peperomia molleri C. DC. ■☆

290382 Peperomia magnoliifolia (Jacq.) A. Dietr. ;木兰叶草胡椒(翡翠椒草);Desert Privet ■☆

290383 Peperomia magnoliifolia (Jacq.) A. Dietr. = Peperomia obtusifolia A. Dietr. ■☆

290384 Peperomia mannii Hook. f. = Peperomia retusa (L. f.) A. Dietr. ■☆

290385 Peperomia marmorata Hook. f. ;云纹椒草;Silver Heart ■☆

290386 Peperomia metallica L. Linden et Rodigas;铜红椒草■☆

290387 Peperomia mocquerysii C. DC. ;莫克里斯草胡椒■☆

290388 Peperomia molleri C. DC. ;摩尔草胡椒■☆

290389 Peperomia nakaharai Hayata;山草椒(阿里山草胡椒,山椒草);Nakahara Paperelder ■

290390 Peperomia nana C. DC. = Peperomia pellucida (L.) Kunth ■

290391 Peperomia nicolliae G. Mathieu;尼科尔草胡椒■☆

290392 Peperomia nigropunctata Miq. ;黑点草胡椒■☆

290393 Peperomia nossibeana C. DC. ;诺西波草胡椒■☆

290394 Peperomia nummularifolia (Sw.) Kunth = Peperomia rotundifolia (L.) Kunth ■☆

290395 Peperomia obtusifolia (L.) A. Dietr. ;卵叶豆瓣绿(豆瓣绿,钝头椒草,钝叶豆瓣绿,钝叶椒草,圆叶椒草);American Rubber Plant,Baby Rubber Plant,Obtuseleaf Peperomia,Pepper Face ■☆

290396 Peperomia obtusifolia A. Dietr. 'Green and Gold';金边钝叶椒草(花叶豆瓣绿,绿金豆瓣绿)■☆

290397 Peperomia okinawensis T. Yamaz. ;冲绳草胡椒■☆

290398 Peperomia ornata Yunck. ;丽叶椒草;Adorned Peperomia, Ornate Peperomia ■☆

290399 Peperomia pacifica Nakai;太平洋豆瓣绿■☆

290400 Peperomia pedunculata C. DC. ;梗花草胡椒■☆

290401 Peperomia pellucida (L.) Kunth;草胡椒(透明草);Shiny Paperelder,Shiny Peperomia ■

290402 Peperomia peltata C. DC. ;盾状草胡椒■☆

290403 Peperomia petigera C. DC. ;茄状草胡椒■☆

290404 Peperomia preussii C. DC. = Peperomia fernandopoiana C. DC. ■☆

290405 Peperomia reflexa (L. f.) A. Dietr. ;小椒草■☆

290406 Peperomia reflexa (L. f.) A. Dietr. = Peperomia tetraphylla (G. Forst.) Hook. et Arn. ■

290407 Peperomia reflexa (L. f.) A. Dietr. f. sinensis A. DC. = Peperomia tetraphylla (G. Forst.) Hook. et Arn. ■

290408 Peperomia reflexa (L. f.) A. Dietr. f. sinensis C. DC. = Peperomia tetraphylla (G. Forst.) Hook. et Arn. var. sinensis (C. DC.) P. S. Chen et P. C. Zhu ■

290409 Peperomia reflexa (L. f.) A. Dietr. f. tenuipes C. DC. = Peperomia tetraphylla (G. Forst.) Hook. et Arn. ■

290410 Peperomia reflexa (L. f.) A. Dietr. var. capense (Miq.) C. DC. = Peperomia tetraphylla (G. Forst.) Hook. et Arn. ■

290411 Peperomia reflexa (L. f.) A. Dietr. var. parvifolia sensu H. Lév. = Peperomia tetraphylla (G. Forst.) Hook. et Arn. ■

290412 Peperomia resedaeflora Lindl. et André;薄叶椒草■☆

290413 Peperomia retusa (L. f.) A. Dietr. ;微凹草胡椒■☆

290414 Peperomia retusa (L. f.) A. Dietr. var. bachmannii (C. DC.) Düll;巴氏草胡椒■☆

290415 Peperomia retusa (L. f.) A. Dietr. var. ciliolata C. DC. = Peperomia retusa (L. f.) A. Dietr. ■☆

290416 Peperomia retusa (L. f.) A. Dietr. var. mannii (Hook. f.) Düll. = Peperomia retusa (L. f.) A. Dietr. ■☆

290417 Peperomia richardsonii G. Mathieu;理查草胡椒■☆

290418 Peperomia rotundifolia (L.) Kunth;圆叶草胡椒(圆叶豆瓣绿);Baby's Tears ■☆

290419 Peperomia rotundilimba C. DC. ;圆边草胡椒■☆

290420 Peperomia rubella Hook. ;红茎椒草■☆

290421 Peperomia rubrivenosa C. DC. ;兰屿椒草(红脉草胡椒)■

290422 Peperomia rungwensis Engl. = Peperomia goetzeana Engl. ■☆

290423 Peperomia ruwenzoriensis Rendle = Peperomia fernandopoiana C. DC. var. butaguensis (De Wild.) Düll ☆

290424 Peperomia sandersii C. DC. = Peperomia argyreia E. Morren ■☆

290425 Peperomia sandersii DC. var. argyreia L. H. Bailey = Peperomia argyreia E. Morren ■☆

290426 Peperomia scandens Ruiz et Pav. ;蔓椒草;Cupid Peperomia ■☆

290427 Peperomia schmidtii C. DC. ;施密特垂椒草■☆

290428 Peperomia serpens C. DC. ;垂椒草■☆

290429 Peperomia serpens C. DC. 'Variegata';斑叶垂椒草■☆

290430 Peperomia silvicola C. DC. ;森林垂椒草■☆

290431 Peperomia spathulifolia Small = Peperomia magnoliifolia (Jacq.) A. Dietr. ■☆

290432 Peperomia staudtii Engl. = Peperomia fernandopoiana C. DC. ■☆

290433 Peperomia stolzii C. DC. = Peperomia molleri C. DC. ■☆

290434 Peperomia stuhlmannii C. DC. = Peperomia abyssinica Miq. ■☆

290435 Peperomia subdichotoma De Wild. = Peperomia retusa (L. f.) A. Dietr. ■☆

290436 Peperomia sui T. T. Lin et S. Y. Lu = Peperomia blanda (Jacq.) Kunth ■

290437 Peperomia tanalensis Baker;塔纳尔椒草■☆

290438 Peperomia tenuispica C. DC. = Peperomia retusa (L. f.) A.

Dietr. ■☆

290439　Peperomia tetraphylla（G. Forst.）Hook. et Arn.；豆瓣绿（豆瓣菜，豆瓣草，豆瓣打不死，豆瓣鹿衔草，豆瓣如意，豆瓣如意草，瓜子鹿衔，瓜子细辛，客阶，如意草，三年草，石瓜子，石山开花，石山瓦浆，四瓣金钗，四块瓦，小椒草，岩豆瓣，岩花，岩筋草，一柱香，指甲草）；Fourleaf Paperelder，Fourleaf Peperomia ■

290440　Peperomia tetraphylla（G. Forst.）Hook. et Arn. var. sinense（A. DC.）P. S. Chen et P. C. Zhu = Peperomia tetraphylla（G. Forst.）Hook. et Arn. ■

290441　Peperomia tetraphylla（G. Forst.）Hook. et Arn. var. sinensis（C. DC.）P. S. Chen et P. C. Zhu = Peperomia tetraphylla（G. Forst.）Hook. et Arn. ■

290442　Peperomia tetraphylla（G. Forst.）Hook. et Arn. var. sinensis（C. DC.）P. S. Chen et P. C. Zhu；毛叶豆瓣绿；Hairyleaf Paperelder ■

290443　Peperomia tetraphylla var. sinensis（C. DC.）P. S. Chen et P. C. Zhu = Peperomia tetraphylla（G. Forst.）Hook. et Arn. ■

290444　Peperomia thollonii C. DC.；托伦草胡椒■☆

290445　Peperomia thomeana C. DC.；汤姆草胡椒■☆

290446　Peperomia triadophylla Peter = Peperomia pellucida（L.）Kunth ■

290447　Peperomia trichophylla Baker；毛叶草胡椒■☆

290448　Peperomia trichopoda C. DC. = Peperomia trichophylla Baker ■☆

290449　Peperomia trifolia A. Dietr.；三叶草胡椒■☆

290450　Peperomia ukingensis Engl. = Peperomia retusa（L. f.）A. Dietr. ■☆

290451　Peperomia ulugurensis Engl. = Peperomia retusa（L. f.）A. Dietr. ■☆

290452　Peperomia ulugurensis Engl. var. acutifolia Balle = Peperomia retusa（L. f.）A. Dietr. ■☆

290453　Peperomia ulugurensis Engl. var. diversifolia Balle = Peperomia retusa（L. f.）A. Dietr. ■☆

290454　Peperomia ulugurensis Engl. var. ukingensis（Engl.）Balle = Peperomia retusa（L. f.）A. Dietr. ■☆

290455　Peperomia usambarensis Engl. = Peperomia retusa（L. f.）A. Dietr. ■☆

290456　Peperomia vaccinifolia C. DC. = Peperomia thomeana C. DC. ■☆

290457　Peperomia velutina Linden；厄瓜多尔草胡椒■☆

290458　Peperomia verschaffeltii Lem.；斑马椒草■

290459　Peperomia verticillata A. Dietr.；轮叶椒草■☆

290460　Peperomia vogelii Miq. = Peperomia pellucida（L.）Kunth ■

290461　Peperomia vulcanica Baker et C. H. Wright；火山草胡椒■☆

290462　Peperomiaceae（Miq.）Wettst. = Peperomiaceae Wettst. ■●

290463　Peperomiaceae A. C. Sm. = Peperomiaceae Wettst. ■

290464　Peperomiaceae A. C. Sm. = Piperaceae Giseke（保留科名）●■

290465　Peperomiaceae Wettst.；草胡椒科（三瓣绿科）■●

290466　Peperomiaceae Wettst. = Piperaceae Giseke（保留科名）●■

290467　Pepinia Brongn.（1870）；异翠凤草属■☆

290468　Pepinia Brongn. ex André. = Pepinia Brongn. ■☆

290469　Pepinia Brongn. ex André. = Pitcairnia L'Hér.（保留属名）■☆

290470　Pepinia occidentalis（L. B. Sm.）G. S. Varad. et Gilmartin；西方异翠凤草■☆

290471　Peplidium Delile（1996）；小莩荠艾属；Peplidium ■

290472　Peplidium humifusum Delile；小莩荠艾■☆

290473　Peplidium humifusum Delile = Peplidium maritimum（L. f.）Asch. ■☆

290474　Peplidium maritimum（L. f.）Asch.；沼泽小莩荠艾■☆

290475　Peplidium muelleri Benth.；米勒小莩荠艾■☆

290476　Peplis L.（1753）；莩艾属；Peplis，Water Purslane，Water-purslane ■

290477　Peplis L. = Lythrum L. ●■

290478　Peplis alternifolia M. Bieb.；莩艾；Alternateleaf Peplis ■

290479　Peplis boraei（Guépin）Jord. = Lythrum borysthenicum（Schrank）Litv. ■☆

290480　Peplis diandra Nutt. ex DC.；二蕊莩艾；Water Hedge，Water Purslane，Water-purslane ■☆

290481　Peplis diandra Nutt. f. aquatica Koehne = Peplis diandra Nutt. ex DC. ■☆

290482　Peplis diandra Nutt. f. terrestris Koehne = Peplis diandra Nutt. ex DC. ■☆

290483　Peplis fradinii Pomel = Lythrum portula（L.）D. A. Webb ■☆

290484　Peplis hispidula Durieu = Lythrum borysthenicum（Schrank）Litv. ■☆

290485　Peplis hyrcanica Sosn.；希尔康莩艾■☆

290486　Peplis indica Willd. = Rotala indica（Willd.）Koehne ■

290487　Peplis longidentata J. Gay = Lythrum portula（L.）D. A. Webb ■☆

290488　Peplis numulariifolia Jord. = Lythrum borysthenicum（Schrank）Litv. ■☆

290489　Peplis portula L.；水莩艾；Pepilles，Water Purslane ■☆

290490　Peplis portula L. = Lythrum portula（L.）D. A. Webb ■☆

290491　Peplis portula L. var. fradinii（Pomel）Pau et Font Quer = Lythrum portula（L.）D. A. Webb ■☆

290492　Peplis portula L. var. longidentata Gay = Lythrum portula（L.）D. A. Webb ■☆

290493　Peplonia Decne.（1844）；袍萝藦属■☆

290494　Peplonia amazonica Benth.；亚马逊袍萝藦■☆

290495　Peplonia nitida Decne.；亮袍萝藦■☆

290496　Pepo Mill. = Cucurbita L. ■

290497　Pepo foetidissima（Kunth）Britton = Cucurbita foetidissima Kunth ■☆

290498　Peponia Naudin = Peponium Engl. ■☆

290499　Peponia bojeri Cogn. = Peponium vogelii（Hook. f.）Engl. ■☆

290500　Peponia bracteata Cogn. = Peponium vogelii（Hook. f.）Engl. ■☆

290501　Peponia bracteata Cogn. var. hirsuta ? = Peponium vogelii（Hook. f.）Engl. ■☆

290502　Peponia caledonica（Sond.）Cogn. = Peponium caledonicum（Sond.）Engl. ■☆

290503　Peponia chirindensis Baker f. = Peponium chirindense（Baker f.）Cogn. ■☆

290504　Peponia cienkowskii（Schweinf.）Hook. f. = Peponium cienkowskii（Schweinf.）Engl. ■☆

290505　Peponia cucullata Bojer ex Hook. f. = Peponium vogelii（Hook. f.）Engl. ■☆

290506　Peponia dissecta Cogn. = Peponium vogelii（Hook. f.）Engl. ■☆

290507　Peponia grandiflora Cogn. = Peponium vogelii（Hook. f.）Engl. ■☆

290508　Peponia kilimandscharica Cogn. = Peponium vogelii（Hook. f.）Engl. ■☆

290509　Peponia kilimandscharica Cogn. var. holstii ? = Peponium vogelii（Hook. f.）Engl. ■☆

290510　Peponia lagenarioides Hook. f. = Peponium lagenarioides（Hook. f.）Cogn. ■☆

290511　Peponia laurentii De Wild. = Peponium vogelii（Hook. f.）Engl. ■☆

290512　Peponia leucantha Gilg = Peponium leucanthum（Gilg）Cogn. ■☆

290513　Peponia macroura Gilg ＝ Peponium vogelii（Hook. f.）Engl. ■☆

290514　Peponia parviflora Cogn. var. trilobata ？ ＝ Coccinia trilobata（Cogn.）C. Jeffrey ■☆

290515　Peponia rufotomentosa Gilg ＝ Peponium vogelii（Hook. f.）Engl. ■☆

290516　Peponia trilobata（Cogn.）Engl. ＝ Coccinia trilobata（Cogn.）C. Jeffrey ■☆

290517　Peponia umbellata Cogn. ＝ Momordica calantha Gilg ■☆

290518　Peponia urticoides Gilg ＝ Peponium urticoides（Gilg）Cogn. ■☆

290519　Peponia usambarensis Engl. ＝ Peponium vogelii（Hook. f.）Engl. ■☆

290520　Peponia vogelii Hook. f. ＝ Peponium vogelii（Hook. f.）Engl. ■☆

290521　Peponia vogelii Hook. f. var. cucullata（Bojer ex Hook. f.）Naudin ＝ Peponium vogelii（Hook. f.）Engl. ■☆

290522　Peponidium（Baill.）Arènes（1960）;小瓠果属●■☆

290523　Peponidium Baill. ex Arènes ＝ Peponidium（Baill.）Arènes ●■☆

290524　Peponidium calcaratum Homolle ex Arènes;距小瓠果■☆

290525　Peponidium crassifolium Lantz,Klack. et Razafim. ;厚叶小瓠果■☆

290526　Peponidium cuspidatum Arènes;骤尖小瓠果■☆

290527　Peponidium flavum Homolle ex Arènes;黄小瓠果■☆

290528　Peponidium horridum（Baill.）Arènes;多刺小瓠果■☆

290529　Peponidium humbertii Homolle ex Arènes;亨伯特小瓠果■☆

290530　Peponidium lanceolatifolium Cavaco;披针叶小瓠果■☆

290531　Peponidium madagascariense Cavaco;马岛小瓠果■☆

290532　Peponidium montanum Arènes;山地小瓠果■☆

290533　Peponidium occidentale Homolle ex Arènes;西方小瓠果■☆

290534　Peponidium orientale Cavaco;东方小瓠果■☆

290535　Peponidium pallens Baill. ex Arènes;稍白小瓠果■☆

290536　Peponidium pallidum Arènes;苍白小瓠果■☆

290537　Peponidium parvifolium Arènes;小叶小瓠果■☆

290538　Peponidium perrieri Arènes;佩里耶小瓠果■☆

290539　Peponidium pervilleanoides Arènes;拟佩尔小瓠果■☆

290540　Peponidium pervilleanum（Baill.）Homolle ex Arènes;佩尔小瓠果■☆

290541　Peponidium tsaratananense Arènes;察拉塔纳纳小瓠果■☆

290542　Peponidium velutinum Arènes;短绒毛小瓠果■☆

290543　Peponiella Kuntze ＝ Peponium Engl. ■☆

290544　Peponium Engl.（1897）;瓠果属■☆

290545　Peponium adpressipilosum A. Zimm. ＝ Peponium vogelii（Hook. f.）Engl. ■☆

290546　Peponium betsiliense Rabenant. ;贝齐里瓠果■☆

290547　Peponium boivinii（Cogn.）Engl. ;博伊文瓠果■☆

290548　Peponium bracteatum（Cogn.）Cogn. ＝ Peponium vogelii（Hook. f.）Engl. ■☆

290549　Peponium caledonicum（Sond.）Engl. ;卡利登瓠果■☆

290550　Peponium chirindense（Baker f.）Cogn. ;奇林达瓠果■☆

290551　Peponium cienkowskii（Schweinf.）Engl. ;西恩瓠果■☆

290552　Peponium grandidieri Rabenant. ;格朗瓠果■☆

290553　Peponium hirtellum Rabenant. ;多毛瓠果■☆

290554　Peponium humbertii Rabenant. ;亨伯特瓠果■☆

290555　Peponium laceratum Rabenant. ;撕裂瓠果■☆

290556　Peponium lagenarioides（Hook. f.）Cogn. ;葫芦状瓠果■☆

290557　Peponium leucanthum（Gilg）Cogn. ;白花瓠果■☆

290558　Peponium mackenii（Naudin）Engl. ;马肯瓠果■☆

290559　Peponium pageanum C. Jeffrey ;纸瓠果■☆

290560　Peponium perrieri Rabenant. ;佩里耶瓠果■☆

290561　Peponium racemosum Rabenant. ;总花瓠果■☆

290562　Peponium rectipilosum A. Zimm. ＝ Peponium vogelii（Hook. f.）Engl. ■☆

290563　Peponium seyrigii Rabenant. ;塞里格瓠果■☆

290564　Peponium urticoides（Gilg）Cogn. ;荨麻瓠果■☆

290565　Peponium vogelii（Hook. f.）Engl. ;瓠果■☆

290566　Peponopsis Naudin（1859）;拟瓠果属■☆

290567　Peponopsis adhaerens Naudin;拟瓠果■☆

290568　Pera Mutis（1784）;袋戟属●☆

290569　Pera benensis Rusby;本袋戟■☆

290570　Pera bumeliifolia Baill. ;山榄叶袋戟;Jiqi ■☆

290571　Pera microcarpa Urb. ;小果袋戟■☆

290572　Peracarpa Hook. f. et Thomson（1858）;袋果草属（肉荚草属,肉荚果属,山桔梗属）;Peracarpa ■

290573　Peracarpa carnosa（Wall.）Hook. f. et Thomson;袋果草（肉荚草,山桔梗）;Peracarpa ■

290574　Peracarpa carnosa（Wall.）Hook. f. et Thomson ＝ Peracarpa circaeoides（F. Schmidt）Feer ■

290575　Peracarpa carnosa（Wall.）Hook. f. et Thomson f. macrantha H. Hara ＝ Peracarpa carnosa（Wall.）Hook. f. et Thomson ■

290576　Peracarpa carnosa（Wall.）Hook. f. et Thomson var. circaeoides（F. Schmidt ex Miq.）Makino ex H. Hara ＝ Peracarpa carnosa（Wall.）Hook. f. et Thomson ■

290577　Peracarpa carnosa（Wall.）Hook. f. et Thomson var. formosana H. Hara ＝ Peracarpa carnosa（Wall.）Hook. f. et Thomson ■

290578　Peracarpa carnosa（Wall.）Hook. f. et Thomson var. kiusiana H. Hara ＝ Peracarpa carnosa（Wall.）Hook. f. et Thomson ■

290579　Peracarpa carnosa（Wall.）Hook. f. et Thomson var. pumila H. Hara ＝ Peracarpa carnosa（Wall.）Hook. f. et Thomson ■

290580　Peracarpa circaeoides（F. Schmidt ex Miq.）Feer ＝ Peracarpa carnosa（Wall.）Hook. f. et Thomson ■

290581　Peracarpa circaeoides（F. Schmidt）Feer ＝ Peracarpa carnosa（Wall.）Hook. f. et Thomson ■

290582　Peracarpa luzonica Rolfe ＝ Peracarpa carnosa（Wall.）Hook. f. et Thomson ■

290583　Peraceae Benth. ex Klotzsch ＝ Euphorbiaceae Juss.（保留科名）●■

290584　Peraceae Benth. ex Klotzsch;袋戟科●☆

290585　Peraceae Klotzsch ＝ Euphorbiaceae Juss.（保留科名）●■

290586　Peraceae Klotzsch ＝ Peraceae Benth. ex Klotzsch ●☆

290587　Perakanthus Robyns ex Ridl. ＝ Perakanthus Robyns ●☆

290588　Perakanthus Robyns ex Ridl. et Robyns ＝ Perakanthus Robyns ●☆

290589　Perakanthus Robyns（1925）;佩拉花属●☆

290590　Perakanthus pauciflorus Robyns ex Ridl. ;佩拉花☆

290591　Peraltea Kunth ＝ Brongniartia Kunth ●☆

290592　Perama Aubl.（1775）;佩茜属●☆

290593　Perama hirsuta Aubl. ;佩茜●☆

290594　Peramibus Raf. ＝ Rudbeckia L. ■

290595　Peramium R. A. Salisbury ex MacMill. ＝ Epipactis Ség.（废弃属名）■

290596　Peramium Salisb. ＝ Goodyera R. Br. ■

290597　Peramium Salisb. ex Britton et Brown ＝ Goodyera R. Br. ■

290598　Peramium Salisb. ex Coult. ＝ Goodyera R. Br. ■

290599　Peramium alboreticulatum（Hayata）Makino ＝ Goodyera hachijoensis Yatabe ■

290600　Peramium arisanense（Hayata）Makino ＝ Goodyera arisanensis Hayata ■

290601　Peramium bilamellatum（Hayata）Makino ＝ Goodyera robusta

Hook. f. ■

290602　Peramium cyrtoglossum（Hayata）Makino = Goodyera fumata Thwaites ■

290603　Peramium decipiens（Hook.）Piper = Goodyera oblongifolia Raf. ■☆

290604　Peramium formosanum（Rolfe）Makino = Goodyera fumata Thwaites ■

290605　Peramium hachijoense（Yatabe）Makino = Goodyera hachijoensis Yatabe ■

290606　Peramium longibracteatum（Hayata）Makino = Goodyera rubicunda（Rchb. f.）J. J. Sm. ■

290607　Peramium longicolumnum（Hayata）Makino = Goodyera rubicunda（Rchb. f.）J. J. Sm. ■

290608　Peramium longirostratum（Hayata）Makino = Goodyera viridiflora（Blume）Blume ■

290609　Peramium macranthum（Maxim. ex Regel）Makino = Goodyera biflora（Lindl.）Hook. f. ■

290610　Peramium matsumuranum（Schltr.）Makino = Goodyera hachijoensis Yatabe ■

290611　Peramium maximowiczianum（Makino）Makino = Goodyera henryi Rolfe ■

290612　Peramium morrisonicola（Hayata）Makino = Goodyera velutina Maxim. ■

290613　Peramium nantoense（Hayata）Makino = Goodyera repens（L.）R. Br. ■

290614　Peramium ogatae（Yamam.）Makino = Goodyera viridiflora（Blume）Blume ■

290615　Peramium ophioides（Fernald）Rydb. = Goodyera repens（L.）R. Br. ■

290616　Peramium pachyglossum（Hayata）Makino = Goodyera foliosa（Lindl.）Benth. ex C. B. Clarke ■

290617　Peramium pendulum（Maxim.）Makino = Goodyera pendula Maxim. ■

290618　Peramium procerum（Ker Gawl.）Makino = Goodyera procera（Ker Gawl.）Hook. ■

290619　Peramium pubescens（Willd.）MacMill. = Goodyera pubescens（Willd.）R. Br. ■☆

290620　Peramium repens（L.）Salisb. = Goodyera repens（L.）R. Br. ■

290621　Peramium repens（L.）Salisb. var. ophioides（Fernald）A. Heller = Goodyera repens（L.）R. Br. ■

290622　Peramium schlechtendalianum（Rchb. f.）Makino = Goodyera schlechtendaliana Rchb. f. ■

290623　Peramium tesselatum（Lodd.）A. Heller = Goodyera tesselata Lodd. ■☆

290624　Peramium velutinum（Maxim. ex Regel）Makino = Goodyera velutina Maxim. ■

290625　Perantha Craib = Oreocharis Benth.（保留属名）■

290626　Perantha aurantiaca（Franch.）Pellegr. = Oreocharis aurantiaca Franch. ■

290627　Perantha cordatula Craib = Oreocharis cordatula（Craib）Pellegr. ■

290628　Perantha forrestii Craib = Oreocharis aurantiaca Franch. ■

290629　Perantha minor Craib = Oreocharis minor（Craib）Pellegr. ■

290630　Perapentacoilanthus Rappa et Camarrone = Mesembryanthemum L.（保留属名）■●

290631　Perapentacoilanthus acuminatus（Haw.）Rappa et Camarrone = Phyllobolus splendens（L.）Gerbaulet ●☆

290632　Perapentacoilanthus aitonis（Jacq.）Rappa et Camarrone = Mesembryanthemum aitonis Jacq. ■☆

290633　Perapentacoilanthus brevifolius（L. Bolus）Rappa et Camarrone = Phyllobolus splendens（L.）Gerbaulet ●☆

290634　Perapentacoilanthus crystallinus（L.）Rappa et Camarrone = Mesembryanthemum crystallinum L. ■

290635　Perapentacoilanthus delus（L. Bolus）Rappa et Camarrone = Phyllobolus delus（L. Bolus）Gerbaulet ■☆

290636　Perapentacoilanthus fastigiatus（Thunb.）Rappa et Camarrone = Mesembryanthemum fastigiatum Thunb. ■☆

290637　Perapentacoilanthus granulicaulis（Haw.）Rappa et Camarrone = Psilocaulon granulicaule（Haw.）Schwantes ■☆

290638　Perapentacoilanthus grossus（Aiton）Rappa et Camarrone = Phyllobolus grossus（Aiton）Gerbaulet ■☆

290639　Perapentacoilanthus longispinulus（Haw.）Rappa et Camarrone = Phyllobolus grossus（Aiton）Gerbaulet ■☆

290640　Perapentacoilanthus scintillans（Dinter）Rappa et Camarrone = Phyllobolus oculatus（N. E. Br.）Gerbaulet ●☆

290641　Perapentacoilanthus spinuliferus（Haw.）Rappa et Camarrone = Phyllobolus spinuliferus（Haw.）Gerbaulet ●☆

290642　Perapentacoilanthus splendens（L.）Rappa et Camarrone = Phyllobolus splendens（L.）Gerbaulet ●☆

290643　Perapentacoilanthus sulcatus（Haw.）Rappa et Camarrone = Phyllobolus splendens（L.）Gerbaulet ●☆

290644　Perapentacoilanthus vanrensburgii（L. Bolus）Rappa et Camarrone = Prenia vanrensburgii L. Bolus ■☆

290645　Perapentacoilanthus viridiflorus（Aiton）Rappa et Camarrone = Phyllobolus viridiflorus（Aiton）Gerbaulet ●☆

290646　Peraphora Miers = Cyclea Arn. ex Wight ●■

290647　Peraphyllum Nutt.（1840）;囊叶蔷薇属●☆

290648　Peraphyllum Nutt. = Amelanchier Medik. ●

290649　Peraphyllum Nutt. ex Torr. et A. Gray = Amelanchier Medik. ●

290650　Peraphyllum ramosissimum Nutt. ;囊叶蔷薇●☆

290651　Peratanthe Urb.（1921）;对面花属■☆

290652　Peratanthe cubensis Urb. ;古巴对面花■☆

290653　Peratanthe ekmanii Urb. ;对面花■☆

290654　Peraxilla Tiegh.（1894）;腋寄生属●☆

290655　Peraxilla colensoi（Hook. f.）Tiegh. ;腋寄生●☆

290656　Percepier Dill. ex Moench = Alchemilla L. ■

290657　Percepier Moench = Alchemilla L. ■

290658　Percepier Moench = Aphanes L. ●■☆

290659　Perdicesca Prov. = Mitchella L. ■

290660　Perdicesea E. A. Delamare, Renauld et Cardot = Perdicesca Prov. ■

290661　Perdiciaceae Link = Asteraceae Bercht. et J. Presl（保留科名）●■

290662　Perdiciaceae Link = Compositae Giseke（保留科名）●■

290663　Perdicium L.（1760）;白丁草属■☆

290664　Perdicium L. = Gerbera L.（保留属名）+ Acanthaceae Juss.（保留科名）●■

290665　Perdicium abyssinicum（Sch. Bip.）Hiern = Gerbera viridifolia（DC.）Sch. Bip. ■☆

290666　Perdicium anandria（L.）R. Br. = Leibnitzia anandria（L.）Turcz. ■

290667　Perdicium capense L. ;好望角白丁草■☆

290668　Perdicium leiocarpum DC. ;光果白丁草■☆

290669　Perdicium nervosum Thunb. = Haplocarpha nervosa（Thunb.）Beauverd ■☆

290670　Perdicium piloselloides（L.）Hiern ＝ Gerbera piloselloides（L.）Cass. ■

290671　Perdicium semiflosculare L. ＝ Perdicium capense L. ■☆

290672　Perdicium taraxaci Vahl ＝ Perdicium capense L. ■☆

290673　Perdicium tomentosum Thunb. ＝ Gerbera anandria（L.）Sch. Bip. ■

290674　Perdicium tomentosum Thunb. ＝ Gerbera curvisquama Hand. -Mazz. ■

290675　Perdicium tomentosum Thunb. ＝ Leibnitzia anandria（L.）Turcz. ■

290676　Perdicium triflorum Buch. -Ham. ex D. Don ＝ Ainsliaea latifolia（D. Don）Sch. Bip. ■

290677　Perebea Aubl.（1775）;黄乳桑属（热美桑属）●☆

290678　Perebea acanthogyne Ducke;刺蕊黄乳桑●☆

290679　Perebea angustifolia（Poepp. et Endl.）C. C. Berg;窄叶黄乳桑●☆

290680　Perebea australis（Hemsl.）J. F. Macbr.;澳洲黄乳桑●☆

290681　Perebea calophylla Benth. et Hook. f. ;美叶黄乳桑●☆

290682　Perebea glabrifolia（Ducke）C. C. Berg;光叶黄乳桑●☆

290683　Perebea laurifolia Tul. ;桂叶黄乳桑●☆

290684　Perebea mollis（Poepp. et Endl.）J. E. Huber;软黄乳桑●☆

290685　Perebea rubra（Trécul）C. C. Berg;红黄乳桑●☆

290686　Peregrina W. R. Anderson(1985);奇异金虎尾属●☆

290687　Peregrina linearifolia（A. St. -Hil.）W. R. Anderson;奇异金虎尾●☆

290688　Pereilema J. Presl ＝ Pereilema J. Presl et C. Presl ■☆

290689　Pereilema J. Presl et C. Presl(1830);半面穗属■☆

290690　Pereilema crinitum J. Presl et C. Presl;半面穗■☆

290691　Pereiria Lindl. ＝ Coscinium Colebr. ●☆

290692　Perella（Tiegh.）Tiegh. ＝ Peraxilla Tiegh. ●☆

290693　Perella Tiegh. ＝ Peraxilla Tiegh. ●☆

290694　Peremis Raf. ＝ Passiflora L. ●■

290695　Perenideboles Ram. Goyena(1911);尼加拉瓜爵床属■☆

290696　Perenideboles ciliatum Ram. Goyena;尼加拉瓜爵床■☆

290697　Perepusa Steud. ＝ Prepusa Mart. ■☆

290698　Perescia Lem. ＝ Pereskia Mill. ●

290699　Pereskia Mill.（1754）;木麒麟属（虎刺属，叶仙人掌属）;Blade Apple, Leaf Cactus, Pereskia ●

290700　Pereskia Vell. ＝ Hippocratea L. ●☆

290701　Pereskia aculeata Mill. ;木麒麟（虎刺,具刺木麒麟,仙人藤,叶仙人掌,针叶仙人掌）;Barbados Gooseberry, Barbados Shrub, Blade Apple Cactus, Leafy Cactus, Lemon Vine, Tsunya, West Indian Gooseberry ●

290702　Pereskia aculeata Mill. 'Godseffiana';紫背叶木麒麟（锦叶木麒麟,紫背叶仙人掌）●

290703　Pereskia aculeata Mill. 'Rubescens';花叶木麒麟（花叶仙人掌）●

290704　Pereskia aureiflora F. Ritter;黄花木麒麟;Fachno ●☆

290705　Pereskia bleo（Kunth）DC. ;樱麒麟●☆

290706　Pereskia bleo（Kunth）DC. ＝ Pereskia grandifolia Haw. ■

290707　Pereskia grandiflora Haw. ;大花木麒麟●☆

290708　Pereskia grandifolia Haw. ;大叶木麒麟（大叶叶仙人掌,叶仙人棒）;Rose Cactus ■

290709　Pereskia nemorosa Rojas Acosta;森林木麒麟;Amapola ●☆

290710　Pereskia pereskia（L.）H. Karst. ＝ Pereskia aculeata Mill. ●

290711　Pereskia portulacifolia（L.）DC. ;马齿苋叶木麒麟;Camelia Roja ●☆

290712　Pereskia rotundifolia DC. ;圆叶叶仙人掌■☆

290713　Pereskia sacharosa Griseb. ;蔷薇木麒麟（蔷薇仙人棒,仙人树）;Sacharosa ■

290714　Pereskia stenantha F. Ritter;狭花木麒麟;Espina De Santo Antonio ●☆

290715　Pereskia subulata Muehlenpf. ＝ Austrocylindropuntia subulata（Muehlenpf.）Backeb. ■☆

290716　Pereskia weberiana K. Schum. ;韦伯木麒麟;Cervetano ●☆

290717　Pereskiopsis Britton et Rose(1907);麒麟掌属（拟叶仙人掌属）●☆

290718　Pereskiopsis kellermanii Rose;凯勒曼麒麟掌;Cola Lagarto ●☆

290719　Pereskiopsis porteri Britton et Rose;波特麒麟掌;Xoconoxtle ●☆

290720　Pereskiopsis rotundifolia Britton et Rose;圆叶麒麟掌;Chapistle ●☆

290721　Pereskiopsis spathulata Britton et Rose;匙形麒麟掌●☆

290722　Pereskiopsis velutina Rose;短毛麒麟掌（短毛麒麟）●☆

290723　Pereuphora Hoffmanns. ＝ Serratula L.

290724　Perezia Lag.（1811）;莲座钝柱菊属（墨西哥菊属）;Perezia ■☆

290725　Perezia adnata A. Gray;贴生莲座钝柱菊（贴生墨西哥菊）;Adnate Perezia ■☆

290726　Perezia microcephala（DC.）Gray;小头莲座钝柱菊（小头墨西哥菊）■☆

290727　Perezia multiflora Less. ;多花莲座钝柱菊（多花墨西哥菊）■☆

290728　Perezia oxylepis A. Gray;尖鳞莲座钝柱菊（墨西哥菊）■☆

290729　Perezinpsis J. M. Coult. ＝ Onoseris Willd. ●■☆

290730　Pereziopsis J. M. Coult.（1895）;类莲座钝柱菊属■☆

290731　Pereziopsis donnell-smithii J. M. Coult. ;类莲座钝柱菊■☆

290732　Perfoliata Burm. ex Kuntze ＝ Hermas L. ■☆

290733　Perfoliata Dod. ex Fourr. ＝ Bupleurum L. ●■

290734　Perfoliata Fourr. ＝ Bupleurum L. ●■

290735　Perfoliata Kuntze ＝ Hermas L. ■☆

290736　Perfolisa Raf. ＝ Bupleurum L. ●■

290737　Perfonon Raf. ＝ Rhamnus L. ●

290738　Pergamena Finet ＝ Dactylostalix Rchb. f. ■☆

290739　Pergularia L.（1767）;非洲夜来香属（夜来香属,紫荆萝藦属）;Pergularia ■☆

290740　Pergularia adenophylla Schltr. et K. Schum. ;腺叶非洲夜来香■☆

290741　Pergularia africana N. E. Br. ＝ Telosma africana（N. E. Br.）N. E. Br. ●☆

290742　Pergularia barbata（Klotzsch）N. E. Br. ex Brenan ＝ Pergularia daemia（Forssk.）Chiov. subsp. barbata（Klotzsch）Goyder ■☆

290743　Pergularia daemia（Forssk.）Blatt. et McCann ＝ Pergularia daemia（Forssk.）Chiov. ■☆

290744　Pergularia daemia（Forssk.）Chiov.;地美夜来香（地美紫荆萝藦,夜来香）■☆

290745　Pergularia daemia（Forssk.）Chiov. subsp. barbata（Klotzsch）Goyder;髯毛地美夜来香■☆

290746　Pergularia daemia（Forssk.）Chiov. subsp. garipensis（E. Mey.）Goyder;加里普夜来香■☆

290747　Pergularia daemia（Forssk.）Chiov. var. leiocarpa（K. Schum.）H. Huber ＝ Pergularia daemia（Forssk.）Chiov. subsp. garipensis（E. Mey.）Goyder ■☆

290748　Pergularia daemia（Forssk.）Chiov. var. macrantha Chiov. ＝ Pergularia daemia（Forssk.）Chiov. ■☆

290749　Pergularia divaricata Lour. ＝ Strophanthus divaricatus（Lour.）Hook. et Arn. ●

290750　Pergularia edulis Thunb. ＝ Fockea edulis（Thunb.）K. Schum. ●☆

290751　Pergularia extensa（Jacq.）N. E. Br. ＝ Pergularia daemia

（Forssk.）Chiov.■☆

290752　Pergularia extensa N. E. Br. ;广生夜来香(广生紫荆萝藦)■☆

290753　Pergularia filipes Schltr. = Telosma procumbens（Blanco）Merr. ●

290754　Pergularia garipensis（E. Mey.）N. E. Br. = Pergularia daemia（Forssk.）Chiov. subsp. garipensis（E. Mey.）Goyder■☆

290755　Pergularia glabra（Forssk.）Chiov. ;光滑非洲夜来香■☆

290756　Pergularia japonica Thunb. = Metaplexis japonica（Thunb.）Makino ●■

290757　Pergularia minor Andr. = Telosma cordata（Burm. f.）Merr. ●

290758　Pergularia minor Andréws = Telosma cordata（Burm. f.）Merr. ●

290759　Pergularia odoratissima（Lour.）Sm. = Telosma cordata（Burm. f.）Merr. ●

290760　Pergularia odoratissima Sm. ;番瑞香■☆

290761　Pergularia odoratissima Sm. = Telosma cordata（Burm. f.）Merr. ●

290762　Pergularia pallida（Roxb.）Wight et Arn. = Telosma cordata（Burm. f.）Merr. ●

290763　Pergularia pallida（Roxb.）Wight et Arn. = Telosma pallida（Roxb.）Craib ●

290764　Pergularia pallida Wight et Arn. = Telosma pallida（Roxb.）Craib ●

290765　Pergularia procumbens Blanco = Telosma procumbens（Blanco）Merr. ●

290766　Pergularia sanguinolenta Britten = Telosma africana（N. E. Br.）N. E. Br. ●☆

290767　Pergularia sanguinolenta Lindl. = Cryptolepis sanguinolenta（Lindl.）Schltr. ●☆

290768　Pergularia sinensis Lour. = Cryptolepis sinensis（Lour.）Merr. ●

290769　Pergularia tacazzeana Chiov. = Telosma africana（N. E. Br.）N. E. Br. ●☆

290770　Pergularia tomentosa L. ;毛非洲夜来香●☆

290771　Pergularia tomentosa L. var. virescens Maire = Pergularia tomentosa L. ●☆

290772　Periandra Cambess. = Thylacospermum Fenzl ■

290773　Periandra Mart. = Periandra Mart. ex Benth. ■☆

290774　Periandra Mart. ex Benth. (1837);甜甘豆属■☆

290775　Periandra acutifolia Benth. ;尖叶甜甘豆■☆

290776　Periandra caespitosa Cambess. = Thylacospermum caespitosum（Cambess.）Schischk. ■

290777　Periandra densiflora Benth. ;密花甜甘豆■☆

290778　Periandra mucronata Benth. ;钝尖甜甘豆■☆

290779　Periandra parviflora Micheli;小花甜甘豆■☆

290780　Perianthomega Bureau ex Baill. (1888);绕花紫葳属●☆

290781　Perianthomega vellozii Bureau;绕花紫葳●☆

290782　Perianthopodus Silva Manso = Cayaponia Silva Manso（保留属名）●☆

290783　Perianthostelma Baill. = Cynanchum L. ●■

290784　Periaria Fabr. = Aegilops L. (保留属名)■

290785　Periarius Kuntze = Pipturus Wedd. ●

290786　Periarrabidaea A. Samp. (1936);肖阿拉树属●☆

290787　Periarrabidaea duckei A. Samp. ;肖阿拉树●☆

290788　Periarrabidaea truncata A. Samp. ;平截肖阿拉树●☆

290789　Peribaea Lindl. = Hyacinthus L. ■☆

290790　Peribaea Lindl. = Periboea Kunth ■☆

290791　Periballanthus Franch. et Sav. = Polygonatum Mill. ■

290792　Periballanthus involucratus Franch. et Sav. = Polygonatum

involucratum（Franch. et Sav.）Maxim. ■

290793　Periballia Trin. (1820);地中海发草属■☆

290794　Periballia Trin. = Deschampsia P. Beauv. ■

290795　Periballia hispanica Trin. ;地中海发草■☆

290796　Periballia laevis（Brot.）Asch. et Graebn. = Molineriella laevis（Brot.）Rouy ■☆

290797　Periballia minuta（L.）Asch. et Graebn. ;小地中海发草■☆

290798　Periballia minuta（L.）Asch. et Graebn. = Molineriella minuta（L.）Rouy ■☆

290799　Periballia minuta（L.）Asch. et Graebn. subsp. australis Paunero = Molineriella minuta（L.）Rouy subsp. australis（Paunero）Rivas Mart. ■☆

290800　Periballia minuta（L.）Asch. et Graebn. var. baetica（Willk.）Font Quer = Molineriella minuta（L.）Rouy subsp. australis（Paunero）Rivas Mart. ■☆

290801　Periballia minuta（L.）Asch. et Graebn. var. genuina Maire et Weiller = Molineriella minuta（L.）Rouy ■☆

290802　Periballia minuta（L.）Asch. et Graebn. var. lanata Maire = Molineriella minuta（L.）Rouy subsp. australis（Paunero）Rivas Mart. ■☆

290803　Periballia minuta（L.）Asch. et Graebn. var. sabulicola Braun-Blanq. et Maire = Molineriella minuta（L.）Rouy subsp. australis（Paunero）Rivas Mart. ■☆

290804　Periblema DC. = Boutonia DC. ●☆

290805　Periblema cuspidata（DC.）DC. = Boutonia cuspidata DC. ●☆

290806　Periblepharis Tiegh. = Luxemburgia A. St. -Hil. ●☆

290807　Periboea Kunth = Hyacinthus L. ■☆

290808　Periboea Kunth(1843);弱小风信子属■☆

290809　Periboea brevifolia Kunth;弱小风信子■☆

290810　Periboea corymbosa（L.）Kunth = Lachenalia corymbosa（L.）J. C. Manning et Goldblatt ■☆

290811　Periboea oliveri U. Müll. -Doblies et D. Müll. -Doblies = Polyxena paucifolia（W. F. Barker）A. M. Van der Merwe et J. C. Manning ■☆

290812　Periboea paucifolia（W. F. Barker）U. Müll. -Doblies et D. Müll. -Doblies = Polyxena paucifolia（W. F. Barker）A. M. Van der Merwe et J. C. Manning ■☆

290813　Pericalia Cass. = Roldana La Llave ●■☆

290814　Pericallis D. Don(1834);瓜叶菊属(细圆菊属);Cineraria ●■

290815　Pericallis Webb et Berthel. = Senecio L. ■●

290816　Pericallis appendiculata（L. f.）B. Nord. ;附属物瓜叶菊■☆

290817　Pericallis aurita（L'Hér.）B. Nord. ;金黄瓜叶菊■☆

290818　Pericallis cruenta（L'Hér.）Bolle;血红瓜叶菊■☆

290819　Pericallis echinata（L. f.）B. Nord. ;刺瓜叶菊■☆

290820　Pericallis hansenii（G. Kunkel）Sunding;汉森瓜叶菊■☆

290821　Pericallis hybrida B. Nord. ;瓜叶菊(杂种千里光);Cineraria, Common Cineraria, Common Ragwort, Florists Cineraria, Florists' Cineraria, Roundtoothleaf Droundsel ■

290822　Pericallis hybrida B. Nord. 'Brilliant';布里兰特瓜叶菊■☆

290823　Pericallis hybrida B. Nord. 'Royalty';皇族瓜叶菊■☆

290824　Pericallis hybrida B. Nord. 'Spring Glory';春辉瓜叶菊■☆

290825　Pericallis hybrida B. Nord. 'Star Wars';星战瓜叶菊■☆

290826　Pericallis lanata（L'Hér.）B. Nord. ;绵毛瓜叶菊■☆

290827　Pericallis lanata（L'Hér.）B. Nord. var. cyanophtalma（Hook.）A. Hansen et Sunding = Pericallis lanata（L'Hér.）B. Nord. ■☆

290828　Pericallis multiflora（L'Hér.）B. Nord. ;多花瓜叶菊■☆

290829　Pericallis papyracea（DC.）B. Nord. ;纸质瓜叶菊■☆

290830 Pericallis steetzii（Bolle）B. Nord. ;斯蒂兹瓜叶菊■☆

290831 Pericallis tussilaginis（L'Hér. ）D. Don;款冬瓜叶菊■☆

290832 Pericallis webbii（Sch. Bip. ）Bolle;韦布瓜叶菊■☆

290833 Pericalymma（Endl. ）Endl. （1840）;纱罩木属●☆

290834 Pericalymma Endl. = Leptospermum J. R. Forst. et G. Forst. （保留属名）●☆

290835 Pericalymma Endl. = Pericalymma（Endl. ）Endl. ●☆

290836 Pericalymma dlipticum（Endl. ）Schauer;纱罩木●☆

290837 Pericalmna Meisn. = Pericalymma（Endl. ）Endl. ●☆

290838 Pericalypta Benoist（1962）;双花爵床属（周盖爵床属）☆

290839 Pericalypta biflora Benoist;双花爵床（周盖爵床）☆

290840 Pericampylus Miers（1851）（保留属名）;细圆藤属（蓬莱藤属）;Pericampylus ●

290841 Pericampylus formosanus Diels;台湾细圆藤（蓬莱藤,三肋蓬莱藤,台湾青藤）;Taiwan Pericampylus ●

290842 Pericampylus formosanus Diels = Pericampylus glaucus（Lam. ）Merr. ●

290843 Pericampylus glaucus（Lam. ）Merr. ;细圆藤（东线藤,蛤仔藤,广藤,黑风散,土藤,小广藤）;Greyblue Pericampylus, Greyblue Pericampylus, Pericampylus ●

290844 Pericampylus incannus（Colebr. ）Hook. f. et Thomson = Pericampylus glaucus（Lam. ）Merr. ●

290845 Pericampylus omeiensis W. Y. Lien;峨眉细圆藤;Emei Pericampylus ●

290846 Pericampylus omeiensis W. Y. Lien = Pericampylus glaucus（Lam. ）Merr. ●

290847 Pericampylus trinervatus Yamam. ;三脉细圆藤（三脉青藤）;Three-nerved Pericampylus ●

290848 Pericampylus trinervatus Yamam. = Pericampylus formosana Diels ●

290849 Pericampylus trinervatus Yamam. = Pericampylus glaucus（Lam. ）Merr. ●

290850 Pericaulon Raf. = Baptisia Vent. ■☆

290851 Perichasma Miers = Stephania Lour. ●■

290852 Perichasma Miers（1866）;肖千金藤属●☆

290853 Perichasma laetificata Miers;肖千金藤●☆

290854 Perichasma laetificata Miers var. obovata Kundu et Guha Bakshi;倒卵肖千金藤●☆

290855 Perichasma miersii Kundu et Guha Bakshi;非洲肖千金藤●☆

290856 Perichlaena Baill. （1888）;周被紫葳属●☆

290857 Perichlaena richardii Baill. ;周被紫葳●☆

290858 Pericla Raf. = Cleome L. ●■

290859 Periclesia A. C. Smith = Ceratostema Juss. ●☆

290860 Pericllstia Benth. = Paypayrola Aubl. ■☆

290861 Periclyma Raf. = Periclymenum Mill. ●■

290862 Periclymenum Mill. = Lonicera L. ●■

290863 Periclymenum sempervirens（L. ）Mill. = Lonicera sempervirens L. ●■

290864 Pericodia Raf. = Passiflora L. ●■

290865 Pericome A. Gray（1853）;环毛菊属■●☆

290866 Pericome caudata A. Gray;尾叶环毛菊;Tail-leaf Pericome, Taper Leaf ■☆

290867 Pericome caudata A. Gray var. glandulosa（Goodman）H. D. Harr. = Pericome caudata A. Gray ■☆

290868 Pericome glandulosa Goodman = Pericome caudata A. Gray ■☆

290869 Pericopsis Thwaites（1864）;美木豆属●☆

290870 Pericopsis angolensis（Baker）Meeuwen;安哥拉美木豆●●☆

290871 Pericopsis angolensis（Baker）Meeuwen f. brasseuriana（De Wild. ）Brummitt;布拉瑟尔美木豆●☆

290872 Pericopsis angolensis（Baker）Meeuwen f. intermedia Yakovlev;间型安哥拉美木豆●☆

290873 Pericopsis angolensis（Baker）Meeuwen subsp. laxiflora（Benth. ）Yakovlev = Pericopsis laxiflora（Benth. ）Meeuwen ●☆

290874 Pericopsis angolensis（Baker）Meeuwen subsp. subtomentosa（De Wild. ）Yakovlev = Pericopsis angolensis（Baker）Meeuwen ●☆

290875 Pericopsis angolensis（Baker）Meeuwen var. subtomentosa（De Wild. ）Meeuwen = Pericopsis angolensis（Baker）Meeuwen ●☆

290876 Pericopsis elata（Harms）Meeuwen;高美木豆（非洲红豆树,西非红豆树）;Afrormosia, Asamela, Kokrodua, West Africa Afrormosia ●

290877 Pericopsis laxiflora（Benth. ex Baker）Meeuwen;疏花美木豆;False Dalbergia ●☆

290878 Pericopsis laxiflora（Benth. ）Meeuwen = Pericopsis laxiflora（Benth. ex Baker）Meeuwen ●☆

290879 Pericopsis mooniana Thwaites;斯里兰卡美木豆●☆

290880 Pericopsis schliebenii（Harms）Meeuwen = Pericopsis angolensis（Baker）Meeuwen f. brasseuriana（De Wild. ）Brummitt ●☆

290881 Perictenia Miers = Odontadenia Benth. ●☆

290882 Pericycla Blume = Licuala Thunb. ●

290883 Perideraea Webb = Chamaemelum Mill. ■

290884 Perideraea Webb = Ormenis（Cass. ）Cass. ●■☆

290885 Perideraea fuscata Webb = Chamaemelum fuscatum（Brot. ）Vasc. ■☆

290886 Perideridia Rchb. （1837）;项圈草属■☆

290887 Perideridia americana（Nutt. ex DC. ）Rchb. ;美洲项圈草■☆

290888 Perideridia gairdneri（Hook. et Arn. ）Mathias;项圈草■☆

290889 Perideridia neurophylla（Maxim. ）T. I. Chuang et Constance = Pterygopleurum neurophyllum（Maxim. ）Kitag. ■

290890 Peridesia A. C. Sm. = Ceratostema Juss. ●☆

290891 Peridictyon Seberg, Fred. et Baden = Elymus L. ■

290892 Peridictyon Seberg, Fred. et Baden = Festucopsis（C. E. Hubb. ）Melderis ■

290893 Peridictyon Seberg, Fred. et Baden（1991）;网禾属■☆

290894 Peridictyon sanctum（Janka）Seberg, Fred. et Baden;网禾■☆

290895 Peridiscaceae Kuhlm. （1950）（保留科名）;围盘树科（巴西肉盘科,围花盘树科,周位花盘科）●☆

290896 Peridiscus Benth. （1862）;围盘树属;Peridiscus ●☆

290897 Peridiscus lucidus Benth. ;围盘树;Lucens Peridiscus, Lucid Peridiscus ●☆

290898 Peridium Schott = Pera Mutis ●☆

290899 Perieilema Benth. et Hook. f. = Pereilema J. Presl et C. Presl ■☆

290900 Periestes Baill. = Hypoestes Sol. ex R. Br. ●■

290901 Periestes baronii Baill. = Hypoestes diclipteroides Nees ●☆

290902 Perieteris Raf. = Nicotiana L. ●■

290903 Perigaria Span. = Gustavia L. （保留属名）●☆

290904 Perigaria Span. = Planchonia Blume ●☆

290905 Periglossum Decne. （1844）;舌萝藦属■☆

290906 Periglossum angustifolium Decne. ;窄叶舌萝藦■☆

290907 Periglossum kassnerianum Schltr. = Periglossum mackenii Harv. ■☆

290908 Periglossum mackenii Harv. ;马克舌萝藦■☆

290909 Periglossum macrum（E. Mey. ）Decne. = Sisyranthus macer（E. Mey. ）Schltr. ■☆

290910 Periglossum mossambicense Schltr. = Periglossum mackenii Harv. ■☆

290911 Perihema Raf. = Haemanthus L. ■

290912 Perihemia B. D. Jacks. = Perihema Raf. ∎

290913 Perihemia Raf. = Haemanthus L. ∎

290914 Perijea (Tul.) A. Juss. = Zanthoxylum L. ●

290915 Perijea (Tul.) Tul. = Fagara L. (保留属名) ●

290916 Perijea Tul. = Fagara L. (保留属名) ●

290917 Perilepta Bremek. (1944); 耳叶爵床属 (耳叶马蓝属); Perilepta ●∎

290918 Perilepta Bremek. = Strobilanthes Blume ●∎

290919 Perilepta auriculata (Nees) Bremek.; 耳叶爵床 (耳叶马兰, 耳叶马蓝); Auriculate Perilepta, Earleaf Conehead ∎

290920 Perilepta auriculata (Wall.) Bremek. = Strobilanthes auriculata (Wall.) Nees ∎

290921 Perilepta dyeriana (Mast.) Bremek.; 红背耳叶马蓝 (红背草, 红背马蓝, 花山蓝, 缅甸马蓝, 疏花金足草, 紫背爵床); Burma Conehead, Persian Shield, Redback Conehead ●∎

290922 Perilepta dyeriana (Mast.) Bremek. = Strobilanthes dyeriana Mast. ●∎

290923 Perilepta edgeworthiana (Nees) Bremek.; 墨江耳叶马蓝 (墨江耳叶马兰) ∎

290924 Perilepta ferruginea (D. Fang et H. S. Lo) C. Y. Wu et C. C. Hu; 锈背耳叶马蓝; Ferrugineous Peridiscus ∎

290925 Perilepta longgangensis (D. Fang et H. S. Lo) C. Y. Wu et C. C. Hu; 弄岗耳叶马蓝; Longgang Perilepta ∎

290926 Perilepta longzhouensis (H. S. Lo et D. Fang) C. Y. Wu et C. C. Hu; 龙州耳叶马蓝; Longzhou Perilepta ∎

290927 Perilepta refracta (D. Fang, Y. G. Wei et J. Murata) C. Y. Wu et C. C. Hu; 折苞耳叶马蓝 (折苞马蓝); Refracte Conehead ●

290928 Perilepta retusa (D. Fang) C. Y. Wu et C. C. Hu; 凹苞耳叶马蓝 (凹苞马蓝); Retuse Conehead ∎

290929 Perilepta siamensis (C. B. Clarke) Bremek.; 泰国耳叶马蓝 ∎

290930 Perilimnastes Ridl. = Anerincleistus Korth. ●☆

290931 Perilimnastes Ridl. = Oritrephes Ridl. ●☆

290932 Perilla L. (1764); 紫苏属 ∎

290933 Perilla albiflora Odash. = Perilla frutescens (L.) Britton var. purpurascens (Hayata) H. W. Li ∎

290934 Perilla arguta Benth. = Perilla frutescens (L.) Britton var. crispa (Benth.) W. Deane ex Bailey ∎

290935 Perilla avium Dunn = Perilla frutescens (L.) Britton ∎

290936 Perilla cavaleriei H. Lév. = Perilla frutescens (L.) Britton var. purpurascens (Hayata) H. W. Li ∎

290937 Perilla citriodora (Makino) Nakai = Perilla frutescens (L.) Britton ∎

290938 Perilla crispa Tanaka = Perilla frutescens (L.) Britton var. crispa (Benth.) W. Deane ex Bailey ∎

290939 Perilla elata D. Don = Elsholtzia blanda (Benth.) Benth. ∎

290940 Perilla frutescens (L.) Britton; 紫苏 (白苏, 白紫苏, 薄荷, 赤苏, 臭苏, 大紫苏, 桂荏, 黑苏, 红勾苏, 红苏, 鸡苏, 家苏, 假紫苏, 聋耳麻, 南苏, 青苏, 犬屎薄, 犬屎苏, 荏, 荏子, 山紫苏, 水升麻, 苏麻, 苏子, 香苏, 香菱, 野藿麻, 野苏, 野苏麻, 玉苏, 孜珠); Beefsteak, Beefsteak Plant, Beefsteakplant, Beefsteak-plant, Common Perilla, Perilla, Perilla Mint, Perilla-mint, Purple Mint, Shiso, Yegoma Oil ∎

290941 Perilla frutescens (L.) Britton = Perilla ocimoides L. ∎

290942 Perilla frutescens (L.) Britton f. citriodora (Makino) Makino = Perilla citriodora (Makino) Nakai ∎

290943 Perilla frutescens (L.) Britton var. acuta (Thunb.) Kudo = Perilla frutescens (L.) Britton var. crispa (Benth.) W. Deane ex

Bailey ∎

290944 Perilla frutescens (L.) Britton var. acuta (Thunb.) Kudo = Perilla frutescens (L.) Britton var. purpurascens (Hayata) H. W. Li ∎

290945 Perilla frutescens (L.) Britton var. arguta (Benth.) Hand.-Mazz. = Perilla frutescens (L.) Britton var. crispa (Benth.) W. Deane ex Bailey ∎

290946 Perilla frutescens (L.) Britton var. auriculata-dentata C. Y. Wu et S. J. Hsuan ex H. W. Li = Elsholtzia hunanensis Hand.-Mazz. ∎

290947 Perilla frutescens (L.) Britton var. auriculata-dentata C. Y. Wu et S. J. Hsuan ex H. W. Li; 耳齿紫苏; Auriculate-dentate Perilla ∎

290948 Perilla frutescens (L.) Britton var. auriculato-dentata C. Y. Wu et S. J. Hsuan ex H. W. Li = Elsholtzia hunanensis Hand.-Mazz. ∎

290949 Perilla frutescens (L.) Britton var. citriodora (Makino) Ohwi = Perilla frutescens (L.) Britton ∎

290950 Perilla frutescens (L.) Britton var. citriodora (Makino) Ohwi = Perilla citriodora (Makino) Nakai ∎

290951 Perilla frutescens (L.) Britton var. crispa (Benth.) H. W. Li = Perilla frutescens (L.) Britton var. crispa (Benth.) W. Deane ex Bailey ∎

290952 Perilla frutescens (L.) Britton var. crispa (Benth.) W. Deane ex Bailey; 回回苏 (波缘青紫苏, 波缘紫苏, 彩色紫苏, 鸡冠紫苏, 青紫苏, 苏, 皱叶白苏, 皱紫苏, 紫苏); Curled Basil, Japanese Basil, Lettnee-leaf Basil, Lettuce-leaf Basil, Perilla, Uygur Perilla ∎

290953 Perilla frutescens (L.) Britton var. crispa (Benth.) W. Deane ex Bailey 'Discolor'; 彩色回回苏 ∎☆

290954 Perilla frutescens (L.) Britton var. crispa (Benth.) W. Deane ex Bailey 'Viridi-crispa'; 绿冠回回苏 ∎☆

290955 Perilla frutescens (L.) Britton var. crispa (Benth.) W. Deane ex Bailey f. purpurea (Makino) Makino; 紫回回苏 (日本紫苏) ∎☆

290956 Perilla frutescens (L.) Britton var. crispa (Benth.) W. Deane ex Bailey f. rosea (G. Nicholson) Kudo; 粉回回苏 ∎☆

290957 Perilla frutescens (L.) Britton var. crispa (Benth.) W. Deane ex Bailey f. viridis (Makino) Makino; 绿回回苏 ∎☆

290958 Perilla frutescens (L.) Britton var. crispa (Thunb.) Hand.-Mazz. = Perilla frutescens (L.) Britton var. crispa (Benth.) H. W. Li ∎

290959 Perilla frutescens (L.) Britton var. crispa (Thunb.) Hand.-Mazz. = Perilla frutescens (L.) Britton var. crispa (Benth.) W. Deane ex Bailey ∎

290960 Perilla frutescens (L.) Britton var. crispa W. Deane = Perilla frutescens (L.) Britton var. crispa (Benth.) W. Deane ex Bailey ∎

290961 Perilla frutescens (L.) Britton var. crispa W. Deane ex Bailey = Perilla frutescens (L.) Britton var. crispa (Benth.) H. W. Li ∎

290962 Perilla frutescens (L.) Britton var. crispa W. Deane ex Bailey f. crispa Makino = Perilla frutescens (L.) Britton var. crispa (Benth.) H. W. Li ∎

290963 Perilla frutescens (L.) Britton var. hirtella (Nakai) Makino = Perilla frutescens (L.) Britton ∎

290964 Perilla frutescens (L.) Britton var. hirtella (Nakai) Makino = Perilla hirtella Nakai ∎☆

290965 Perilla frutescens (L.) Britton var. japonica (Hassk.) H. Hara = Perilla frutescens (L.) Britton ∎

290966 Perilla frutescens (L.) Britton var. nankinensis (Lour.) Britton = Perilla frutescens (L.) Britton var. crispa (Benth.) W. Deane ex Bailey ∎

290967 Perilla frutescens (L.) Britton var. nankinensis Britton = Perilla frutescens (L.) Britton var. crispa (Benth.) W. Deane ex

Bailey ■

290968　Perilla frutescens（L.）Britton var. purpurascens（Hayata）H. W. Li；野生紫苏(白丝草,臭草,蛤树,红香师菜,尖紫苏,青叶紫苏,苏菅,苏麻,蚊草,香丝菜,野香丝,野猪疏,野紫苏,紫禾草,紫苏)；Purple Perilla ■

290969　Perilla frutescens（L.）Britton var. typica Makino ＝ Perilla frutescens（L.）Britton ■

290970　Perilla fruticosa D. Don ＝ Elsholtzia fruticosa（D. Don）Rehder ●

290971　Perilla heteromorpha Carrière ＝ Perilla frutescens（L.）Britton var. purpurascens（Hayata）H. W. Li ■

290972　Perilla hirtella Nakai；多毛紫苏■☆

290973　Perilla lanceolata Benth. ＝ Mosla scabra（Thunb.）C. Y. Wu et H. W. Li ■

290974　Perilla leptostachya D. Don ＝ Elsholtzia stachyodes（Link）C. Y. Wu ■

290975　Perilla macrostachys Benth. ＝ Perilla frutescens（L.）Britton ■

290976　Perilla nankinensis（Lour.）Decne. ＝ Perilla frutescens（L.）Britton var. crispa（Benth.）W. Deane ex Bailey ■

290977　Perilla nankinensis Decne. ＝ Perilla frutescens（L.）Britton var. crispa（Benth.）W. Deane ex Bailey ■

290978　Perilla ocimoides L. ＝ Perilla frutescens（L.）Britton ■

290979　Perilla ocimoides L. var. crispa Benth. ＝ Perilla frutescens（L.）Britton var. crispa（Benth.）W. Deane ex Bailey ■

290980　Perilla ocimoides L. var. purpurascens Hayata ＝ Perilla frutescens（L.）Britton var. purpurascens（Hayata）H. W. Li ■

290981　Perilla ocymoides L. ＝ Perilla frutescens（L.）Britton var. purpurascens（Hayata）H. W. Li ■

290982　Perilla ocymoides L. f. citriodora ？ ＝ Perilla frutescens（L.）Britton ■

290983　Perilla ocymoides L. var. purpurascens Hayata ＝ Perilla frutescens（L.）Britton ■

290984　Perilla ocymoides L. var. purpurascens Hayata ＝ Perilla frutescens（L.）Britton var. purpurascens（Hayata）H. W. Li ■

290985　Perilla polystachya D. Don ＝ Elsholtzia ciliata（Thunb. ex Murray）Hyl. ■

290986　Perilla schimadae Kudo ＝ Perilla frutescens（L.）Britton var. purpurascens（Hayata）H. W. Li ■

290987　Perilla urticifolia Salisb. ＝ Perilla frutescens（L.）Britton ■

290988　Perillula Maxim.（1875）；小紫苏属■☆

290989　Perillula reptans Maxim. ；小紫苏■☆

290990　Periloba Raf. ＝ Nolana L. ex L. f. ■☆

290991　Perilomia Kunth ＝ Scutellaria L. ●■

290992　Perima Raf. ＝ Entada Adans.（保留属名）●

290993　Perimenium Steud. ＝ Perymenium Schrad. ■●☆

290994　Perinerion Baill. ＝ Baissea A. DC. ●☆

290995　Perinerion welwitschii Baill. ＝ Baissea welwitschii（Baill.）Stapf ex Hiern ●☆

290996　Perinka Raf. ＝ Perinkara Adans. ●

290997　Perinkara Adans. ＝ Elaeocarpus L. ●

290998　Periomphale Baill.（1888）；新喀海桐属●☆

290999　Periomphale Baill. ＝ Wittsteinia F. Muell. ●☆

291000　Periomphale balansae Baill. ；新喀海桐●☆

291001　Peripea Steud. ＝ Buchnera L. ●

291002　Peripea Steud. ＝ Piripea Aubl. ■

291003　Peripentadenia L. S. Sm.（1957）；环腺木属●☆

291004　Peripentadenia mearsii（C. T. White）L. S. Sm. ；环腺木●☆

291005　Peripeplus Pierre（1898）；套茜属●☆

291006　Peripeplus bracteosus（Hiern）E. M. Petit；多苞片套茜●☆

291007　Peripeplus klaineanus Pierre ＝ Peripeplus bracteosus（Hiern）E. M. Petit ●☆

291008　Peripetasma Ridl. ＝ Dioscorea L.（保留属名）■

291009　Periphaa Raf. ＝ Convolvulus L. ■●

291010　Periphanes Salisb. ＝ Hessea Herb.（保留属名）■☆

291011　Periphanes brachyscypha（Baker）F. M. Leight. ＝ Hessea breviflora Herb. ■☆

291012　Periphanes cinnamomea（L'Hér.）F. M. Leight. ＝ Hessea cinnamomea（L'Hér.）T. Durand et Schinz ■☆

291013　Periphanes dregeana（Kunth）F. M. Leight. ＝ Hessea breviflora Herb. ■☆

291014　Periphanes gemmata（Ker Gawl.）F. M. Leight. ＝ Strumaria gemmata Ker Gawl. ■☆

291015　Periphanes karooica（W. F. Barker）F. M. Leight. ＝ Strumaria karooica（W. F. Barker）Snijman ■☆

291016　Periphanes leipoldtii（L. Bolus）F. M. Leight. ＝ Strumaria leipoldtii（L. Bolus）Snijman ■☆

291017　Periphanes spiralis F. M. Leight. ＝ Strumaria pygmaea Snijman ■☆

291018　Periphanes stellaris（Jacq.）Salisb. ＝ Hessea stellaris（Jacq.）Herb. ■☆

291019　Periphanes strumosa（Aiton）F. M. Leight. ＝ Strumaria tenella（L. f.）Snijman ■☆

291020　Periphanes unguiculata（W. F. Barker）F. M. Leight. ＝ Strumaria unguiculata（W. F. Barker）Snijman ■☆

291021　Periphanes zeyheri（Baker）F. M. Leight. ＝ Hessea breviflora Herb. ■☆

291022　Periphas Raf. ＝ Convolvulus L. ■●

291023　Periphragmos Ruiz et Pav. ＝ Cantua Juss. ex Lam. ●☆

291024　Periphyllium Gand. ＝ Paronychia Mill. ■

291025　Peripleads Wall. ＝ Drypetes Vahl ●

291026　Peripleura（N. T. Burb.）G. L. Nesom(1994)；单头层菀属■☆

291027　Peripleura Clifford et Ludlow ＝ Vittadinia A. Rich. ■

291028　Peripleura bicolor（N. T. Burb.）G. L. Nesom；二色单头层菀■☆

291029　Peripleura diffusa（N. T. Burb.）G. L. Nesom；铺散单头层菀■☆

291030　Peripleura obovata（N. T. Burb.）G. L. Nesom；倒卵单头层菀■☆

291031　Periploca L.（1753）；杠柳属；Silk Vine，Silkvine，Silk-vine ●

291032　Periploca Tourn. ex L. ＝ Periploca L. ●

291033　Periploca africana L. ＝ Cynanchum africanum（L.）Hoffmanns. ●☆

291034　Periploca afzelii G. Don ＝ Periploca nigrescens Afzel. ●☆

291035　Periploca alboflavescens Dennst. ＝ Parsonsia alboflavescens（Dennst.）Mabb. ●

291036　Periploca angustifolia Labill. ；窄叶杠柳●☆

291037　Periploca aphylla Decne. ；无叶杠柳●☆

291038　Periploca aphylla Decne. subsp. laxiflora（Bornm.）Browicz；疏花无叶杠柳●☆

291039　Periploca aphylla Decne. var. laxiflora Bornm. ＝ Periploca aphylla Decne. subsp. laxiflora（Bornm.）Browicz ●☆

291040　Periploca arborea Dennst. ＝ Wrightia arborea（Dennst.）Mabb. ●

291041　Periploca astacus H. Lév. ＝ Trachelospermum axillare Hook. f. ●

291042　Periploca audiacea Raeusch. ＝ Periploca angustifolia Labill. ●☆

291043　Periploca batesii Wernham ＝ Cynanchum polyanthum K. Schum. ■☆

291044　Periploca brevicoronata Goyder et Boulos ＝ Periploca somaliense Browicz ●☆

291045　Periploca calophylla（Baill.）Roberty ＝ Omphalogonus calophyllus Baill. ●

291046　Periploca calophylla（Wight）Falc.；青蛇藤（管人香，黑骨头，黑乌骨，黑乌骚，鸡骨头，宽叶凤仙藤，铁夹藤，乌骨鸡，乌骚风）；Greensnake Vine，Prettyleaf Silkvine，Pretty-leaved Silkvine ●

291047　Periploca calophylla（Wight）Falc. subsp. floribunda（Tsiang）Browicz = Periploca floribunda Tsiang ●

291048　Periploca calophylla（Wight）Falc. subsp. forrestii（Schltr.）Browicz = Periploca forrestii Schltr. ●

291049　Periploca calophylla（Wight）Falc. var. forrestii（Schltr.）Browicz = Periploca forrestii Schltr. ●

291050　Periploca calophylla（Wight）Falc. var. mucronata P. T. Li；凸尖叶青蛇藤；Mucronate Prettyleaf Silkvine ●

291051　Periploca chevalieri Browicz；舍瓦利耶杠柳●☆

291052　Periploca chinensis Spreng. = Cryptolepis sinensis（Lour.）Merr. ●

291053　Periploca chrysantha D. S. Yao, X. C. Chen et J. W. Ren；黄花杠柳；Yellowflower Silkvine ●

291054　Periploca cochinchinensis Lour. = Calotropis gigantea（L.）Dryand. ex W. T. Aiton ●

291055　Periploca divaricata（Lour.）Spreng. = Strophanthus divaricatus（Lour.）Hook. et Arn. ●

291056　Periploca divaricata Spreng. = Strophanthus divaricatus（Lour.）Hook. et Arn. ●

291057　Periploca ephedriformis（Deflers）Schweinf. ex Deflers = Periploca visciformis（Vatke）K. Schum. ●☆

291058　Periploca esculenta L. f. = Oxystelma esculentum（L. f.）R. Br. ■

291059　Periploca fasciculata Viv. ex Coss. = Periploca angustifolia Labill. ●☆

291060　Periploca floribunda Tsiang；多花青蛇藤；Flowery Silkvine，Manyflower Silkvine，Multiflorous Silkvine ●

291061　Periploca forrestii Schltr.；飞仙藤（达风藤，滇杠柳，黑骨藤，黑骨头，黑龙骨，柳叶过山龙，柳叶夹，奶浆藤，牛尾蕨，青风藤，青色丹，青蛇胆，青香藤，清香藤，山筋线，山杨柳，铁骨头，铁散沙，西南杠柳，小黑牛，小青蛇）；Forrest Silkvine ●

291062　Periploca gabonica（Baill.）A. Chev. = Periploca nigrescens Afzel. ●☆

291063　Periploca gracea L.；希腊杠柳（长藤杠柳）；Grecian Silk Vine，Grecian Silk-vine，Greece Silkvine，Silk Vine，Silkvine，Silk-vine ●☆

291064　Periploca indica L. = Hemidesmus indicus（L.）R. Br. ●☆

291065　Periploca khasiana Benth. = Stelmatocrypton khasianum（Kurz）Baill. ●

291066　Periploca laevigata Aiton f. angustifolia（Labill.）Ross = Periploca angustifolia Labill. ●☆

291067　Periploca laevigata Aiton subsp. chevalieri（Browicz）G. Kunkel = Periploca chevalieri Browicz ●☆

291068　Periploca laevigata Aiton var. angustifolia（Labill.）Fiori = Periploca angustifolia Labill. ●☆

291069　Periploca latifolia K. Schum. = Mondia whitei（Hook. f.）Skeels ●☆

291070　Periploca laxiflora K. Schum. = Mondia whitei（Hook. f.）Skeels ●☆

291071　Periploca linearifolia Quart.-Dill. et A. Rich. ex A. Rich.；线叶杠柳●☆

291072　Periploca linearis Hochst. = Periploca linearifolia Quart.-Dill. et A. Rich. ex A. Rich. ●☆

291073　Periploca nigrescens Afzel.；黑杠柳●☆

291074　Periploca nigricans Schltr. = Periploca nigrescens Afzel. ●☆

291075　Periploca ovata Poir. = Pleurostelma cernuum（Decne.）Bullock ●☆

291076　Periploca pallida Salisb. = Cynanchum africanum（L.）Hoffmanns. ●☆

291077　Periploca petersiana Vatke = Dregea macrantha Klotzsch ●☆

291078　Periploca preussi K. Schum. = Periploca nigrescens Afzel. ●☆

291079　Periploca refractifolia Gilli；折叶杠柳●☆

291080　Periploca rigida Viv. = Periploca angustifolia Labill. ●☆

291081　Periploca secamone L. = Secamone alpini Schult. ●☆

291082　Periploca sepium Bunge；杠柳（北五加皮，臭槐，臭加皮，臭五加，狗奶子，红柳，立柳，山五加皮，桃不桃柳不柳，狭叶萝藦，香加皮，香五加皮，羊角梢，羊角桃，羊角条，羊角叶，羊奶藤，羊奶条，羊奶子，羊桃梢，阴柳，钻墙柳）；China Silkvine，Chinese Silk Vine，Chinese Silkvine ●

291083　Periploca sinensis（Lour.）Steud. = Cryptolepis sinensis（Lour.）Merr. ●

291084　Periploca sinensis Steud. = Cryptolepis sinensis（Lour.）Merr. ●

291085　Periploca somaliense Browicz；索马里杠柳●☆

291086　Periploca sylvestris Retz. = Gymnema sylvestre（Retz.）Schult. ●

291087　Periploca sylvestris Retz. = Marsdenia sylvestris（Retz.）P. I. Forst. ●

291088　Periploca sylvestris Willd. = Gymnema sylvestre（Retz.）Schult. ●

291089　Periploca tsiangii D. Fang et H. Z. Ling；大花杠柳（小花杠柳）；Tsiang's Silkvine ●

291090　Periploca venosa Hochst. ex Decne. = Tacazzea venosa Decne. ●☆

291091　Periploca visciformis（Vatke）K. Schum.；槲寄生状杠柳●☆

291092　Periploca visciformis（Vatke）K. Schum. var. glabra Browicz = Periploca visciformis（Vatke）K. Schum. ●☆

291093　Periploca wildemanii A. Chev. = Periploca nigrescens Afzel. ●☆

291094　Periplocaceae Schltr.（1905）（保留科名）；杠柳科■●

291095　Periplocaceae Schltr.（保留科名）= Apocynaceae Juss.（保留科名）●■

291096　Periplocaceae Schltr.（保留科名）= Asclepiadaceae Borkh.（保留科名）●■

291097　Periptera DC.（1824）；环翅锦葵属■●☆

291098　Periptera punicea DC.；环翅锦葵■●☆

291099　Peripterygia（Baill.）Loes.（1906）；环翼卫矛属●☆

291100　Peripterygia Loes. = Peripterygia（Baill.）Loes. ●☆

291101　Peripterygia marginata Loes.；环翼卫矛●☆

291102　Peripterygiaceae F. N. Williams = Cardiopteridaceae Blume（保留科名）●■

291103　Peripterygiaceae G. King = Cardiopteridaceae Blume（保留科名）●■

291104　Peripterygium Hassk. = Cardiopteris Wall. ex Royle ●■

291105　Peripterygium platycarpum（Gagnep.）Sleumer = Cardiopteris platycarpa Gagnep. ●■

291106　Peripterygium quinquelobum Hassk. = Cardiopteris quinqueloba Hassk. ■

291107　Perispermum O. Deg. = Bonamia Thouars（保留属名）●☆

291108　Perissandrn Gagnep. = Vatica L. ●

291109　Perissocarpa Steyerm. et Maguire = Elvasia DC. ●☆

291110　Perissocarpa Steyerm. et Maguire（1984）；奇果金莲木属●☆

291111　Perissocarpa umbellifera Steyerm. et Maguire；奇果金莲木 ●☆

291112　Perissocoeleum Mathias et Constance（1952）；奇腹草属■●☆

291113　Perissocoeleum crinoideum（Mathias et Constance）Mathias et

Constance;奇腹草■☆

291114 Perissolobus N. E. Br. = Machairophyllum Schwantes ■☆

291115 Perissolobus bijliae N. E. Br. = Machairophyllum bijliae（N. E. Br.）L. Bolus ■☆

291116 Perissus Miers = Asthotheca Miers ex Planch. et Triana ●☆

291117 Perissus Miers = Clusia L. ●☆

291118 Peristeira Hook. f. = Pentsteira Griff. ■

291119 Peristeira Hook. f. = Torenia L. ■

291120 Peristera（DC.）Eckl. et Zeyh. = Corthumia Rchb. ●■

291121 Peristera（DC.）Eckl. et Zeyh. = Pelargonium L'Hér. ex Aiton ●■

291122 Peristera Eckl. et Zeyh. = Corthumia Rchb. ●■

291123 Peristera Eckl. et Zeyh. = Pelargonium L'Hér. ex Aiton ●■

291124 Peristera Endl. = Peristeria Hook. ■☆

291125 Peristeranthus T. E. Hunt(1954);鸽花兰属■☆

291126 Peristeranthus hillii（F. Muell.）T. E. Hunt;鸽花兰■☆

291127 Peristeria Hook.（1831）;鸽兰属;Dove Flower, Dove Orchid, Holy Ghost Orchid, Holy Gosta Orchid ■☆

291128 Peristeria cerina Lindl.;蜡色鸽兰;Waxcolor Dove Orchid ■☆

291129 Peristeria elata Hook.;高鸽兰;Dove Flower, Dove Orchid, Holy Ghost Orchid, Tall Dove Orchid ■☆

291130 Peristeria pendula Hook.;下垂鸽兰;Pendulous Dove Orchid ■☆

291131 Peristethium Tiegh. = Loranthus Jacq.（保留属名）●

291132 Peristethium Tiegh. = Struthanthus Mart.（保留属名）●☆

291133 Peristima Raf. = Heliotropium L. ●■

291134 Peristrophe Nees(1832);观音草属（九头狮子草属，山蓝属）;Peristrophe ■

291135 Peristrophe aculeata（C. B. Clarke）Brummitt = Dicliptera aculeata C. B. Clarke ■☆

291136 Peristrophe angolensis（S. Moore）K. Balkwill = Dicliptera angolensis S. Moore ■☆

291137 Peristrophe angustifolia Nees = Peristrophe hyssopifolia（Burm. f.）Bremek. ■☆

291138 Peristrophe baphica（Spreng.）Bremek.;观音草（白牛膝，长花九头狮子草，大青，大叶辣椒草，对叶接骨草，高山辣椒，红蓝，红丝绒草，红丝线，红线草，黄丁苦草，九头狮子草，蓝茶，绿骨大青，青红线，青丝线，染色九头狮子草，山蓝，四川草，围包草，项开口，野靛青，指甲花）;Colored Peristrophe, Dyeing Peristrophe, Roxburgh Peristrophe ■

291139 Peristrophe bicalyculata（Retz.）Nees;双萼观音草;Twocalyx Peristrophe ■

291140 Peristrophe bicalyculata（Retz.）Nees = Dicliptera paniculata（Forssk.）I. Darbysh. ■☆

291141 Peristrophe bicalyculata（Retz.）Nees = Peristrophe paniculata（Forssk.）Brummitt ■☆

291142 Peristrophe bicalyculata Retz. = Dicliptera paniculata（Forssk.）I. Darbysh. ■☆

291143 Peristrophe bivalvis（L.）Merr. = Peristrophe baphica（Spreng.）Bremek. ■

291144 Peristrophe caulopsila E. Mey. ex Nees = Peristrophe cernua Hook. ex Nees ■☆

291145 Peristrophe cernua Hook. ex Nees;纳塔尔九头狮子草■☆

291146 Peristrophe chinensis Nees = Peristrophe japonica（Thunb.）Bremek. ■

291147 Peristrophe cliffordii K. Balkwill;克利福德九头狮子草■☆

291148 Peristrophe cumingiana Nees;灌状观音草（野山蓝）;Cuming Peristrophe, Fruticose Peristrophe ■

291149 Peristrophe cumingiana Nees = Hypoestes cumingiana Benth. et Hook. f. ■

291150 Peristrophe doriae A. Terracc. = Dicliptera paniculata（Forssk.）I. Darbysh. ■☆

291151 Peristrophe fera C. B. Clarke;野山蓝（大叶观音草）;Wild Peristrophe ■

291152 Peristrophe fera C. B. Clarke var. intermedia C. B. Clarke;大叶观音草;Intermediate Peristrophe ■

291153 Peristrophe floribunda（Hemsl.）C. Y. Wu et H. S. Lo;海南山蓝;Hainan Peristrophe ■

291154 Peristrophe grandibracteata Lindau;大苞九头狮子草■☆

291155 Peristrophe guangxiensis H. S. Lo et D. Fang;广西山蓝;Guangxi Peristrophe ■

291156 Peristrophe hensii（Lindau）C. B. Clarke = Dicliptera hensii Lindau ■☆

291157 Peristrophe hereroensis（Schinz）K. Balkwill = Dicliptera hereroensis Schinz ■☆

291158 Peristrophe hyssopifolia（Burm. f.）Bremek.;神香草叶观音草（爪哇观音草）;Marble Leaf ■☆

291159 Peristrophe hyssopifolia（Burm. f.）Bremek.'Aureovariegata';黄斑爪哇观音草■☆

291160 Peristrophe jalappifolia Nees ex C. B. Clarke = Peristrophe fera C. B. Clarke var. intermedia C. B. Clarke ■

291161 Peristrophe japonica（Thunb.）Bremek.;九头狮子草（菜豆青，川白牛膝，肺痨草，红丝线草，化痰青，尖惊药，接骨草，金钗草，九节篱，咳风尘，辣叶青药，六角英，绿豆青，日本狗肝菜，日本狮子草，三面青，蛇舌草，四季青，天青菜，铁焊椒，铁脚万年青，土红蓝，土细辛，万年青，王灵仁，项开口，野青仔，晕病药，竹节青）;Japan Dogliverweed, Japan Peristrophe, Japanese Dicliptera, Japanese Peristrophe ■

291162 Peristrophe japonica（Thunb.）Bremek. var. subrotunda（Matsuda）Murata et Terao;亚圆九头狮子草■☆

291163 Peristrophe kotschyana Nees = Dicliptera paniculata（Forssk.）I. Darbysh. ■☆

291164 Peristrophe krebsii C. Presl = Peristrophe cernua Hook. ex Nees ■☆

291165 Peristrophe lanceolaria（Roxb.）Nees;五指山蓝■

291166 Peristrophe lanceolata（Lindau）Dandy;披针形九头狮子草■☆

291167 Peristrophe luteoviridis C. B. Clarke = Peristrophe lanceolata（Lindau）Dandy ■☆

291168 Peristrophe montana Nees;岩观音草;Montain Peristrophe ■

291169 Peristrophe namibiensis K. Balkwill;纳米比亚九头狮子草■☆

291170 Peristrophe natalensis T. Anderson = Peristrophe cernua Hook. ex Nees ■☆

291171 Peristrophe oblonga Nees = Peristrophe cernua Hook. ex Nees ■☆

291172 Peristrophe paniculata（Forssk.）Brummitt;圆锥九头狮子草■☆

291173 Peristrophe pilosa Turrill;疏毛九头狮子草■☆

291174 Peristrophe pumila（Lindau）Gilli = Dicliptera pumila（Lindl.）Dandy ■☆

291175 Peristrophe pumila（Lindau）Lindau = Dicliptera pumila（Lindl.）Dandy ■☆

291176 Peristrophe purpurea（L.）Hochr. = Hypoestes purpurea（L.）R. Br. ●■

291177 Peristrophe roxburghiana（Schult.）Bremek.;长花九头狮子草（红丝绒，山蓝）■

291178 Peristrophe roxburghiana（Schult.）Bremek. = Peristrophe baphica（Spreng.）Bremek. ■

291179 Peristrophe serpenticola K. Balkwill et Campb.-Young = Dicliptera serpenticola（K. Balkwill et Campb.-Young）I. Darbysh. ■☆

291180　Peristrophe strigosa C. Y. Wu et H. S. Lo;糙叶山蓝;Roughleaf Peristrophe ■

291181　Peristrophe tianmuensis H. S. Lo;天目山蓝;Tianmushan Peristrophe ■

291182　Peristrophe tinctoria（Roxb.）Nees ＝ Peristrophe baphica（Spreng.）Bremek. ■

291183　Peristrophe tinctoria Span. ex Nees ＝ Peristrophe baphica（Spreng.）Bremek. ■

291184　Peristrophe transvaalensis（C. B. Clarke）K. Balkwill ＝ Dicliptera transvaalensis C. B. Clarke ■☆

291185　Peristrophe tridentata（E. Mey.）Baill. ;三齿九头狮子草■☆

291186　Peristrophe usta C. B. Clarke ＝ Dicliptera pumila（Lindl.）Dandy ■☆

291187　Peristrophe yunnanensis W. W. Sm. ;滇观音草;Yunnan Peristrophe ■

291188　Peristylus Blume（1825）（保留属名）;阔蕊兰属;Peristylus, Perotis ■

291189　Peristylus affinis（D. Don）Seidenf. ;小花阔蕊兰（鸡肾草,卵子草,小龙盘参）;Sampson Peristylus, Sampson Perotis, Smallflower Peristylus, Smallflower Perotis ■

291190　Peristylus alaschanicus（Maxim.）N. Pearce et P. J. Cribb ＝ Herminium alaschanicum Maxim. ■

291191　Peristylus albidulus Chiov. ＝ Habenaria montolivaea Kraenzl. ex Engl. ■☆

291192　Peristylus alpinus（Hand.-Mazz.）K. Y. Lang ＝ Bhutanthera alpina（Hand.-Mazz.）Renz ■

291193　Peristylus arachnoideus A. Rich. ＝ Holothrix arachnoidea（A. Rich.）Rchb. f. ■☆

291194　Peristylus brachylobos Summerh. ＝ Habenaria brachylobos（Summerh.）Summerh. ■☆

291195　Peristylus bracteatus Lindl. ＝ Dactylorhiza viridis（L.）R. M. Bateman, Pridgeon et M. W. Chase ■

291196　Peristylus bulleyi（Rolfe）K. Y. Lang;条叶阔蕊兰（条叶角盘兰）;Beltleaf Peristylus, Beltleaf Perotis ■

291197　Peristylus calcaratus（Rolfe）S. Y. Hu;长须阔蕊兰（猫须兰）;Longbeard Peristylus, Longbeard Perotis ■

291198　Peristylus chloranthus Lindl. ＝ Peristylus lacertifer（Lindl.）J. J. Sm. ■

291199　Peristylus coeloceras Finet;凸孔阔蕊兰（凸孔角盘兰）;Emptyhorn Peristylus, Emptyhorn Perotis ■

291200　Peristylus constrictus（Lindl.）Lindl. ;大花阔蕊兰;Bigflower Peristylus, Bigflower Perotis ■

291201　Peristylus cordatus Lindl. ＝ Gennaria diphylla（Link）Parl. ■☆

291202　Peristylus densus（Lindl.）Santapau et Kapadia;狭穗阔蕊兰（倒杆章,鬼箭玉凤花,土天麻,狭穗露兰,狭穗鹭兰,狭穗舌唇兰,狭穗玉凤兰）;Narroespike Peristylus, Narroespike Perotis, Narroespike Platanthera ■

291203　Peristylus ecalcaratus Finet ＝ Herminium ecalcaratum（Finet）Schltr. ■

291204　Peristylus ecalcaratus Finet ＝ Herminium latifolium Gagnep. ■

291205　Peristylus ecalcaratus Tang et F. T. Wang ＝ Herminium tangianum（S. Y. Hu）K. Y. Lang ■

291206　Peristylus elisabethae（Duthie）R. K. Gupta;西藏阔蕊兰;Tibet Peristylus, Xizang Perotis ■

291207　Peristylus fallax Lindl. ;盘腺阔蕊兰;False Peristylus, False Perotis ■

291208　Peristylus fallax Lindl. var. dwarikae Deva et H. B. Naithani ＝ Peristylus fallax Lindl. ■

291209　Peristylus filiformis Kraenzl. ＝ Cynorkis papillosa（Ridl.）Summerh. ■☆

291210　Peristylus flagellier（Makino）Ohwi var. acutifolia（Hayata）Hatus. ＝ Peristylus formosanus（Schltr.）T. P. Lin ■

291211　Peristylus flagellifer（Makino）Ohwi;鞭须阔蕊兰（鞭毛玉凤兰）;Flagellate Peristylus, Flagellate Perotis ■

291212　Peristylus flagellifer（Makino）Ohwi ＝ Peristylus densus（Lindl.）Santapau et Kapadia ■

291213　Peristylus flagellifer（Makino）Ohwi ex K. Y. Lang ＝ Habenaria flagellifera Makino ■

291214　Peristylus flagellifer（Makino）Ohwi var. acutifolius（Hayata）Hatus. ＝ Peristylus formosanus（Schltr.）T. P. Lin ■

291215　Peristylus forceps Finet;一掌参;Forcipate Herminium, Palm Perotis ■

291216　Peristylus formosanus（Schltr.）T. P. Lin;台湾阔蕊兰（触须兰,台湾鹭草）;Taiwan Peristylus, Taiwan Perotis ■

291217　Peristylus formosanus（Schltr.）T. P. Lin ＝ Habenaria lacertifera（Lindl.）Benth. ■

291218　Peristylus forrestii（Schltr.）K. Y. Lang;条唇阔蕊兰;Beltlip Perotis, Forrest Peristylus ■

291219　Peristylus garrettii（Rolfe ex Downie）J. J. Wood et Ormerod ＝ Peristylus tentaculatus（Lindl.）J. J. Sm. ■

291220　Peristylus glaberrima（Ridl.）Rolfe ＝ Benthamia glaberrima（Ridl.）H. Perrier ●☆

291221　Peristylus goodyeroides（D. Don）Lindl. ;阔蕊兰（白缘边玉凤兰,斑叶玉凤兰,绿花阔蕊兰,南投阔蕊兰,南投玉凤兰,山砂姜,珍珠草）;Common Perotis, Greenflower Peristylus ■

291222　Peristylus goodyeroides（D. Don）Lindl. var. affinis（D. Don）Cooke ＝ Peristylus affinis（D. Don）Seidenf. ■

291223　Peristylus goodyeroides（D. Don）Lindl. var. affinis（King et Pantl.）Cooke ＝ Peristylus affinis（D. Don）Seidenf. ■

291224　Peristylus gracilis Blume;纤细阔蕊兰■☆

291225　Peristylus gracillimus（Hook. f.）Kraenzl. ＝ Peristylus mannii（Rchb. f.）Mukerjee ■

291226　Peristylus gracillimus（Hook. f.）Kraenzl. f. lankongensis Finet ＝ Peristylus bulleyi（Rolfe）K. Y. Lang ■

291227　Peristylus gracillimus Kraenzl. f. lankongensis Finet ＝ Peristylus bulleyi（Rolfe）K. Y. Lang ■

291228　Peristylus gramineus（Thouars）S. Moore ＝ Cynorkis graminea（Thouars）Schltr. ●☆

291229　Peristylus hatusimanus T. Hashim. ;初岛阔蕊兰■☆

291230　Peristylus hispidulus Rendle ＝ Brachycorythis pubescens Harv. ■☆

291231　Peristylus humidicola K. Y. Lang et D. S. Deng ＝ Frigidorchis humidicola（K. Y. Lang et D. S. Deng）Z. J. Liu et S. C. Chen ■

291232　Peristylus iyoensis Ohwi ＝ Habenaria iyoensis Ohwi ■

291233　Peristylus jinchuanicus K. Y. Lang;金川阔蕊兰;Jinchuan Peristylus, Jinchuan Perotis ■

291234　Peristylus lacertifer（Lindl.）J. J. Sm. ;撕唇阔蕊兰（鹅毛玉凤兰,裂唇阔蕊兰,青花阔蕊兰,撕唇玉凤花,细花玉凤兰）;Laceratedlip Habenaria, Splitlip Peristylus, Splitlip Perotis ■

291235　Peristylus lacertifer（Lindl.）J. J. Sm. ＝ Habenaria lacertifera（Lindl.）Benth. ■

291236　Peristylus lacertifer（Lindl.）J. J. Sm. var. formosanum（Makino et Hayata）S. S. Ying ＝ Peristylus formosanus（Schltr.）T. P. Lin ■

291237　Peristylus lacertifer（Lindl.）J. J. Sm. var. formosanus（Schltr.）S. S. Ying ＝ Peristylus formosanus（Schltr.）T. P. Lin ■

291238　Peristylus lacertifer（Lindl.）J. J. Sm. var. taipoensis（S. Y. Hu et Barretto) S. C. Chen,S. W. Gale et P. J. Cribb;短裂阔蕊兰■

291239　Peristylus lefebureanus A. Rich. = Habenaria lefebureana（A. Rich.）T. Durand et Schinz ■☆

291240　Peristylus longiracemus（Fukuy.）K. Y. Lang;长穗阔蕊兰（长穗玉凤花,总状花玉凤兰）;Longspike Peristylus,Longspike Perotis ■

291241　Peristylus longiracemus（Fukuy.）K. Y. Lang = Habenaria longiracema Fukuy. ■

291242　Peristylus longiracemus（Fukuy.）K. Y. Lang = Habenaria lucida Wall. ex Lindl. ■

291243　Peristylus macropetalus Finet = Benthamia madagascariensis（Rolfe）Schltr. ●☆

291244　Peristylus madagascariensis（Rolfe）Rolfe = Benthamia madagascariensis（Rolfe）Schltr. ●☆

291245　Peristylus mannii（Rchb. f.）Mukerjee;纤茎阔蕊兰;Weakstem Peristylus,Weakstem Perotis ■

291246　Peristylus monanthus Finet = Amitostigma monanthum（Finet）Schltr. ■

291247　Peristylus monophyllus（Collett et Hemsl.）Kraenzl. = Orchis monophylla（Collett et Hemsl.）Rolfe ■

291248　Peristylus monophyllus（Collett et Hemsl.）Kraenzl. = Ponerorchis monophylla（Collett et Hemsl.）Soó ■

291249　Peristylus natalensis（Rchb. f.）Rolfe = Habenaria petitiana（A. Rich.）T. Durand et Schinz ■☆

291250　Peristylus neglectus（King et Pantl.）Kraenzl. = Peristylus densus（Lindl.）Santapau et Kapadia ■

291251　Peristylus nematocaulon（Hook. f.）Banerji et P. Pradhan;小巧阔蕊兰■

291252　Peristylus neotineoides（Ames et Schltr.）K. Y. Lang;川西阔蕊兰;W. Sichuan Peristylus,W. Sichuan Perotis ■

291253　Peristylus orchidis（Lindl.）Kraenzl. = Gymnadenia orchidis Lindl. ■

291254　Peristylus parishii Rchb. f. ;滇桂阔蕊兰;Yun-Gui Peristylus,Yun-Gui Perotis ■

291255　Peristylus petitianus A. Rich. = Habenaria petitiana（A. Rich.）T. Durand et Schinz ■☆

291256　Peristylus preussii（Kraenzl.）Rolfe = Habenaria microceras Hook. f. ■☆

291257　Peristylus pricei Hayata = Peristylus calcaratus（Rolfe）S. Y. Hu ■

291258　Peristylus purpureus S. Moore = Cynorkis purpurea（Thouars）Kraenzl. ■☆

291259　Peristylus sampsonii Hance = Peristylus affinis（D. Don）Seidenf. ■

291260　Peristylus secundiflorus（Hook. f.）Kraenzl. = Neottianthe secundiflora（Hook. f.）Schltr. ■

291261　Peristylus secundiflorus Kraenzl. = Neottianthe secundiflora（Hook. f.）Schltr. ■

291262　Peristylus snowdenii Rolfe = Habenaria petitiana（A. Rich.）T. Durand et Schinz ■☆

291263　Peristylus sphaerocentron Ts. Tang et F. T. Wang = Peristylus goodyeroides（D. Don）Lindl. ■

291264　Peristylus spiralis（Thouars）S. Moore = Benthamia perularioides Schltr. ●☆

291265　Peristylus spiranthes（Schauroth）S. Y. Hu = Peristylus lacertifer（Lindl.）J. J. Sm. ■

291266　Peristylus spiranthes（Schauroth）S. Y. Hu var. taipoensis S. Y. Hu et Barretto = Peristylus lacertifer（Lindl.）J. J. Sm. ■

291267　Peristylus squamatus Hochst. ex A. Rich. = Holothrix squamata（Hochst. ex A. Rich.）Rchb. f. ■☆

291268　Peristylus stenostachys（Lindl.）Santapau et Kapadia = Peristylus densus（Lindl.）Santapau et Kapadia ■

291269　Peristylus stenostachyus（Lindl. ex Benth.）Kraenzl. = Peristylus densus（Lindl.）Santapau et Kapadia ■

291270　Peristylus steudneri（Rchb. f.）Rolfe;斯氏阔蕊兰;Steudner Peristylus ■☆

291271　Peristylus steudneri（Rchb. f.）Rolfe = Habenaria petitiana（A. Rich.）T. Durand et Schinz ■☆

291272　Peristylus tangianus S. Y. Hu = Herminium latifolium Gagnep. ■

291273　Peristylus tangianus S. Y. Hu = Herminium tangianum（S. Y. Hu）K. Y. Lang ■

291274　Peristylus tentaculatus（Lindl.）J. J. Sm. ;触须阔蕊兰（触须玉凤花,鸡蛋参,鸡卵草,卷须玉凤花）;Palp Peristylus,Palp Perotis ■

291275　Peristylus tetralobus Finet = Amitostigma tetralobum（Finet）Schltr. ■

291276　Peristylus tetralobus Finet f. basifoliatus Finet = Amitostigma basifoliatum（Finet）Schltr. ■

291277　Peristylus tetralobus Finet f. parceflorus Finet = Amitostigma parceflorum（Finet）Schltr. ■

291278　Peristylus tetralobus Finet var. typicus Finet = Amitostigma tetralobum（Finet）Schltr. ■

291279　Peristylus tridentatus Hook. f. = Holothrix tridentata（Hook. f.）Rchb. f. ■☆

291280　Peristylus ugandensis Rolfe = Habenaria petitiana（A. Rich.）T. Durand et Schinz ■☆

291281　Peristylus unifolius Hochst. = Holothrix unifolia（Rchb. f.）Rchb. f. ■☆

291282　Peristylus viridis（L.）Lindl. = Coeloglossum viride（L.）Hartm. ■

291283　Peristylus viridis（L.）Lindl. = Dactylorhiza viridis（L.）R. M. Bateman,Pridgeon et M. W. Chase ■

291284　Peristylus volkensianus（Kraenzl.）Rolfe = Habenaria petitiana（A. Rich.）T. Durand et Schinz ■☆

291285　Peristylus xanthochlorus Blatter et McCann = Peristylus densus（Lindl.）Santapau et Kapadia ■

291286　Peritassa Miers(1872);佩里木属●☆

291287　Peritassa compta Miers;粗茎佩里木●☆

291288　Perithrix Pierre = Batesanthus N. E. Br. ●☆

291289　Perithrix glabra Pierre = Batesanthus purpureus N. E. Br. ●☆

291290　Peritoma DC. = Cleome L. ●■

291291　Peritomia G. Don = Perilomia Kunth ●■

291292　Peritris Raf. = Arnica L. ●■☆

291293　Perittium Vogel = Melanoxylon Schotr ●☆

291294　Perittostema I. M. Johnst.（1954）;奇冠紫草属■☆

291295　Perittostema pinetorum（I. M. Johnst.）I. M. Johnst. ;奇冠紫草■☆

291296　Perityle Benth.（1844）;岩雏菊属●■☆

291297　Perityle aglossa A. Gray;峭壁岩雏菊;Bluff Rock Daisy,Rayless Rock Daisy ●■☆

291298　Perityle ajoensis Todsen;阿霍岩雏菊;Ajo Rock Daisy ●■☆

291299　Perityle ambrosiifolia Greene ex A. M. Powell et Yarborough;缎带岩雏菊;Lace-leaved Rock Daisy ●■☆

291300　Perityle angustifolia（A. Gray）Shinners;窄叶岩雏菊;Narrow-leaved Rock Daisy ■☆

291301　Perityle bisetosa（Torr. ex A. Gray）Shinners;双刚毛岩雏菊;

Two-bristle Rock Daisy ■☆

291302　Perityle cernua（Greene）Shinners；奥尔甘岩雏菊；Organ Mountain Rock Daisy ■☆

291303　Perityle ciliata（L. H. Dewey）Rydb.；睫毛岩雏菊■☆

291304　Perityle cinerea（A. Gray）A. M. Powell；灰岩雏菊；Gray Rock Daisy ■☆

291305　Perityle cochisensis（W. E. Niles）A. M. Powell；支那岩雏菊；Cochise Rock Daisy ●■☆

291306　Perityle congesta（M. E. Jones）Shinners；峡谷岩雏菊；Grand Canyon Rock Daisy，Kaibab Rock Daisy ●■☆

291307　Perityle coronopifolia A. Gray；鸟足岩雏菊；Crow-foot Rock Daisy ●☆

291308　Perityle dissecta（Torr.）A. Gray；细裂岩雏菊；Dissected Rock Daisy，Slim-lobe Rock Daisy ●■☆

291309　Perityle emoryi Torr.；埃默里岩雏菊；Emory's Rock Daisy ■☆

291310　Perityle fosteri A. M. Powell；福斯特岩雏菊；Foster's Rock Daisy ■☆

291311　Perityle gilensis（M. E. Jones）J. F. Macbr.；基拉岩雏菊；Gila Rock Daisy ●■☆

291312　Perityle gilensis（M. E. Jones）J. F. Macbr. var. longilobus（W. E. Niles）A. M. Powell et Yarbor. = Perityle gilensis（M. E. Jones）J. F. Macbr. var. salensis A. M. Powell ■☆

291313　Perityle gilensis（M. E. Jones）J. F. Macbr. var. salensis A. M. Powell；索尔特岩雏菊；Salt River Rock Daisy ■☆

291314　Perityle gracilis（M. E. Jones）Rydb.；纤细岩雏菊；Slender Rock Daisy，Three-lobed Rock Daisy ●■☆

291315　Perityle halimifolia（A. Gray）Shinners = Perityle lindheimeri（A. Gray）Shinners var. halimifolia（A. Gray）A. M. Powell ●■☆

291316　Perityle huecoensis A. M. Powell；休考岩雏菊；Hueco Rock Daisy ●■☆

291317　Perityle intricata（Brandegee）Shinners；沙漠岩雏菊；Desert Rock Daisy ●■☆

291318　Perityle inyoensis（Ferris）A. M. Powell；音约岩雏菊；Inyo Rock Daisy ●■☆

291319　Perityle lemmonii（A. Gray）J. F. Macbr.；莱蒙岩雏菊；Lemmon's Rock Daisy ●■☆

291320　Perityle lindheimeri（A. Gray）Shinners；林氏岩雏菊；Lindheimer's Rock Daisy ●■☆

291321　Perityle lindheimeri（A. Gray）Shinners var. halimifolia（A. Gray）A. M. Powell；滨藜叶岩雏菊●■☆

291322　Perityle megalocephala（S. Watson）J. F. Macbr.；大头岩雏菊；Nevada Rock Daisy ●■☆

291323　Perityle megalocephala（S. Watson）J. F. Macbr. var. intricata（Brandegee）A. M. Powell = Perityle intricata（Brandegee）Shinners ●■☆

291324　Perityle megalocephala（S. Watson）J. F. Macbr. var. oligophylla A. M. Powell；小叶大头岩雏菊；Small-leaved Rock Daisy ●■☆

291325　Perityle microglossa Benth.；小舌岩雏菊；Short-ray Rock Daisy ■☆

291326　Perityle microglossa Benth. var. effusa A. Gray = Perityle microglossa Benth. ■☆

291327　Perityle parryi A. Gray；帕里岩雏菊；Parry's Rock Daisy ●■☆

291328　Perityle quinqueflora（Steyerm.）Shinners；五花岩雏菊；Five-flower Rock Daisy ■☆

291329　Perityle rotundata（Rydb.）Shinners = Perityle lindheimeri（A. Gray）Shinners ●■☆

291330　Perityle rupestris（A. Gray）Shinners；多叶岩雏菊；Leafy Rock Daisy，Yellow leafy Rock Daisy ●■☆

291331　Perityle rupestris（A. Gray）Shinners var. albiflora A. M. Powell；白花多叶岩雏菊；White leafy Rock Daisy ■☆

291332　Perityle saxicola（Eastw.）Shinners；岩雏菊；Roosevelt Dam Rock Daisy ●■☆

291333　Perityle specuicola S. L. Welsh et Neese；园艺岩雏菊；Alcove Rock Daisy，Hanging-garden Rock Daisy ●■☆

291334　Perityle stansburii（A. Gray）J. F. Macbr.；斯氏岩雏菊；Stansbury's Rock Daisy ●■☆

291335　Perityle staurophylla（Barneby）Shinners；新墨西哥岩雏菊；New Mexico Rock Daisy ●■☆

291336　Perityle tenella（M. E. Jones）J. F. Macbr.；美国南部岩雏菊；Dixie Daisy，Springdale Rock Daisy ●■☆

291337　Perityle vaseyi J. M. Coult.；瓦齐岩雏菊；Vasey's Rock Daisy ●■☆

291338　Perityle villosa（S. F. Blake）Shinners；长柔毛岩雏菊；Hanaupah Rock Daisy ●■☆

291339　Perityle vitreomontana Warnock；石英岩雏菊；Glass Mountain Rock Daisy ■☆

291340　Perityle warnockii A. M. Powell；瓦氏岩雏菊；Warnock's Rock Daisy ●■☆

291341　Perizoma（Miers）Lindl. = Salpichroa Miers ●☆

291342　Perizoma Miers ex Lindl. = Salpichroa Miers ●☆

291343　Perizomanthus Pursh = Dicentra Bernh.（保留属名）■

291344　Perlaria Fabr. = Aegilops L.（保留属名）■

291345　Perlaria Heist. ex Fabr. = Aegilops L.（保留属名）■

291346　Perlarius Kuntze = Botrymorus Miq. ●

291347　Perlarius Kuntze = Pipturus Wedd. ●

291348　Perlarius Rumph. = Pipturus Wedd. ●

291349　Perlebia DC. = Heptaptera Margot et Reut. ■☆

291350　Perlebia Mart. = Bauhinia L. ●

291351　Perlebia exellii（Torre et Hillc.）A. Schmitz = Bauhinia exellii Torre et Hillc. ●☆

291352　Perlebia galpinii（N. E. Br.）A. Schmitz = Bauhinia galpinii N. E. Br. ●☆

291353　Perlebia macrantha（Oliv.）A. Schmitz = Bauhinia petersiana Bolle subsp. macrantha（Oliv.）Brummitt et J. H. Ross ●☆

291354　Perlebia macrantha（Oliv.）A. Schmitz subsp. serpae（Ficalho et Hiern）A. Schmitz = Bauhinia petersiana Bolle subsp. macrantha（Oliv.）Brummitt et J. H. Ross ●☆

291355　Perlebia mendoncae（Torre et Hillc.）A. Schmitz = Bauhinia mendoncae Torre et Hillc. ●☆

291356　Perlebia natalensis（Oliv. ex Hook.）A. Schmitz = Bauhinia natalensis Oliv. ex Hook. ●☆

291357　Perlebia petersiana（Bolle）A. Schmitz = Bauhinia petersiana Bolle ●☆

291358　Perlebia purpurea（L.）A. Schmitz = Bauhinia purpurea L. ●

291359　Perlebia urbaniana（Schinz）A. Schmitz = Bauhinia urbaniana Schinz ●☆

291360　Perlebia variegata（L.）A. Schmitz = Bauhinia variegata L. ●

291361　Perlebia variegata（L.）A. Schmitz var. alboflava（de Wit）A. Schmitz = Bauhinia variegata L. var. candida（Roxb.）Voigt ●

291362　Permia Raf. = Entada Adans.（保留属名）●

291363　Pernettya Gaudich.（1825）（'Pernettia'）（保留属名）；南鹃属；Arrayan，Pernettya ●☆

291364　Pernettya Gaudich.（保留属名）= Gaultheria L. ●

291365　Pernettya ciliaris Don.；缘毛南鹃 ●☆

291366　Pernettya macrostigma Colenso；大柱头南鹃（大柱头白珠）；

Big-stigma Gaulteria ●☆

291367 Pernettya mucronata Gaudich. ex G. Don;锐尖南鹃;Broadleaf Pernettya,Prickly Heath,Sharp-pointed Pernettya ●☆

291368 Pernettya mucronata Gaudich. ex G. Don 'Rosea';粉果锐尖南鹃;Pink-frtd Pernettya ●☆

291369 Pernettya mucronata Gaudich. ex G. Don = Gaultheria mucronata J. Rémy ●☆

291370 Pernettya nana Colenso;匍状南鹃(匐状南鹃);Mat-forming Pernettya ●☆

291371 Pernettya prostrata Sleumer;平卧南鹃;Prostrate Pernettya ●☆

291372 Pernettya pumila Hook. ;矮小南鹃;Dwarf Pernettya ●☆

291373 Pernettyopsis King et Gamble = Diplycosia Blume ●☆

291374 Pernettyopsis King et Gamble(1906);类南鹃属●☆

291375 Pernettyopsis breviflora Ridl. ;短花类南鹃●☆

291376 Pernetya Scop. (废弃属名) = Canarina L. (保留属名)■☆

291377 Pernetya Scop. (废弃属名) = Pernettya Gaudich. (保留属名)●☆

291378 Peroa Pers. = Leucopogon R. Br. (保留属名)●☆

291379 Perobachne C. Presl = Themeda Forssk. ■

291380 Perobachne J. Presl et C. Presl = Anthistiria L. f. ■

291381 Perocarpa Feer = Peracarpa Hook. f. et Thomson ■

291382 Peroillaea Decne. = Toxocarpus Wight et Arn. ●

291383 Perojoa Cav. (废弃属名) = Leucopogon R. Br. (保留属名)●☆

291384 Peronema Jack(1822);东南亚马鞭属●☆

291385 Peronema canescens Jack;东南亚马鞭●☆

291386 Peronema heterophyllum Miq. = Peronema canescens Jack ●☆

291387 Peronia Delar. = Thalia L. ■☆

291388 Peronia Delar. ex DC. = Thalia L. ■☆

291389 Peronia R. Br. = Sarcosperma Hook. f. ●

291390 Peroniaceae Dostal = Sarcospermataceae H. J. Lam(保留科名)●

291391 Peronocactus Doweld = Notocactus (K. Schum.) A. Berger et Backeb. ■

291392 Peronocactus scopa (Spreng.) Doweld = Notocactus scopa (Spreng.) Backeb. ■

291393 Perostema Raeusch. = Nectandra Rol. ex Rottb. (保留属名)●☆

291394 Perostis P. Beauv. = Perotis Aiton ■

291395 Perostis laxifolia P. Beauv. = Perotis latifolia Aiton ■

291396 Perotis Aiton(1789);茅根属(茅根草属);Perotis ■

291397 Perotis acanthoneuron Cope;刺脉茅根■☆

291398 Perotis burmanica Gand. = Perotis hordeiformis Nees ■

291399 Perotis chinensis Gand. = Perotis hordeiformis Nees ex Hook. et Arn. ■

291400 Perotis glabrata Steud. = Perotis hordeiformis Nees ex Hook. et Arn. ■

291401 Perotis hildebrandtii Mez;希尔德茅根■☆

291402 Perotis holstii Mez = Perotis hildebrandtii Mez ■☆

291403 Perotis hordeiformis Nees;麦穗茅根(大麦状茅根,麦茅根);Barleylike Perotis ■

291404 Perotis hordeiformis Nees ex Hook. et Arn. = Perotis hordeiformis Nees ■

291405 Perotis indica (L.) Kuntze;茅根;India Perotis,Indian Perotis ■

291406 Perotis latifolia Aiton = Perotis indica (L.) Kuntze ■

291407 Perotis leptopus Pilg. ;细梗茅根■☆

291408 Perotis longiflora Nees = Perotis rara R. Br. ■

291409 Perotis macrantha Honda = Perotis rara R. Br. ■

291410 Perotis patens Gand. ;铺展茅根■☆

291411 Perotis patens Gand. var. parvispicula Robyns et Tournay = Perotis patens Gand. ■☆

291412 Perotis patula Nees = Perotis rara R. Br. ■

291413 Perotis phleoides Hack. = Mosdenia leptostachys (Ficalho et Hiern) Clayton ■☆

291414 Perotis pilosa Cope;疏毛茅根■☆

291415 Perotis polystachya Willd. = Pogonatherum paniceum (Lam.) Hack. ■

291416 Perotis rara R. Br. ;大花茅根(大穗茅根);Largeflower Perotis ■

291417 Perotis scabra Willd. ex Trin. ;粗糙茅根■☆

291418 Perotis scabra Willd. ex Trin. var. parvispicula (Robyns et Tournay) Cufod. = Perotis patens Gand. ■☆

291419 Perotis somalensis Chiov. ;索马里茅根■☆

291420 Perotis transvaalensis Gand. = Perotis patens Gand. ■☆

291421 Perotis vaginata Hack. ;具鞘茅根■☆

291422 Perotriche Cass. = Stoebe L. ●■☆

291423 Perotriche microphylla Sch. Bip. = Stoebe schultzii Levyns ●☆

291424 Perotriche tortilis Cass. = Stoebe capitata P. J. Bergius ●☆

291425 Perottetia Post et Kuntze = Desmodium Desv. (保留属名)●■

291426 Perottetia Post et Kuntze = Perrottetia DC. ●

291427 Perovskia Kar. (1841);分药花属;Perovskia,Russian Sage ●

291428 Perovskia abrotanoides Kar. ;分药花;Caspian Perovskia ●

291429 Perovskia angustifolia Kudr. ;狭叶分药花;Narrowleaf Perovskia ●

291430 Perovskia atriplicifolia Benth. ;滨藜叶分药花(滨藜分药花);Russian Sage,Saltbushleaf Perovskia,Silver Sage ●☆

291431 Perovskia kudrjaschevii Gorschk. et Pjataeva;库德分药花●☆

291432 Perovskia linczevskli Kudr. ;林氏分药花●☆

291433 Perovskia pamirica C. Y. Yang et B. Wang;帕米尔分药花;Pamir Perovskia ●

291434 Perovskia pamirica C. Y. Yang et B. Wang = Perovskia atriplicifolia Benth. ●

291435 Perovskia virgata Kudr. ;条纹分药花●☆

291436 Perowskia Benth. = Perovskia Kar. ●

291437 Perowskia scrophulariifolia Bunge = Nepeta fordii Hemsl. ■

291438 Perpensum Burm. f. = Gunnera L. ■☆

291439 Perplexia Iljin = Jurinea Cass. ●■

291440 Perplexia Iljin(1962);肖苓菊属■☆

291441 Perplexia microcephala (Boiss.) Iljin;肖苓菊☆

291442 Perralderia Coss. (1859);直壁菊属●☆

291443 Perralderia coronopifolia Coss. ;鸟足叶直壁菊■☆

291444 Perralderia coronopifolia Coss. subsp. eu-coronopifolia Maire = Perralderia coronopifolia Coss. ■☆

291445 Perralderia coronopifolia Coss. subsp. purpurascens (Coss. ex Batt.) Maire;紫鸟足叶直壁菊■☆

291446 Perralderia coronopifolia Coss. subvar. longibracteata Sauvage = Perralderia coronopifolia Coss. subsp. purpurascens (Coss. ex Batt.) Maire ■☆

291447 Perralderia coronopifolia Coss. var. dessignyana (Hochr.) Maire = Perralderia coronopifolia Coss. subsp. purpurascens (Coss. ex Batt.) Maire ■☆

291448 Perralderia coronopifolia Coss. var. typica Maire = Perralderia coronopifolia Coss. ■☆

291449 Perralderia dessignyana Hochr. = Perralderia coronopifolia Coss. ■☆

291450 Perralderia dessignyana Hochr. = Perralderia coronopifolia Coss. subsp. purpurascens (Coss. ex Batt.) Maire ■☆

291451 Perralderia garamantum Asch. ;噶拉门特直壁菊●☆

291452 Perralderia paui Font Quer;波氏直壁菊■☆

291453 Perralderia purpurascens Coss. ex Batt. = Perralderia coronopifolia Coss. subsp. purpurascens (Coss. ex Batt.) Maire ■☆

291454 Perralderiopsis Rauschert = Iphiona Cass. (保留属名) ●■☆

291455 Perreymondia Barntoud = Schizopetalon Sims ■☆

291456 Perriera Courchet(1905);佩氏木属●☆

291457 Perriera madagascariensis Courchet;佩氏木●☆

291458 Perriera orientalis Capuron;东方佩氏木●☆

291459 Perrieranthus Hochr. = Perrierophytum Hochr. ●☆

291460 Perrierastrum Guillaumin = Plectranthus L'Hér. (保留属名)●■

291461 Perrierastrum Guillaumin(1931);佩氏草属●☆

291462 Perrierastrum oreophilum Guillaumin;佩氏草●☆

291463 Perrierastrum oreophilum Guillaumin = Plectranthus bipinnatus A. J. Paton ■☆

291464 Perrierbambus A. Camus(1924);泊尔竹属●☆

291465 Perrierbambus madagascariensis A. Camus;马岛泊尔竹●☆

291466 Perrierbambus tsarasaotrensis A. Camus;泊尔竹●☆

291467 Perrieriella Schltr. (1925);佩里耶兰属■☆

291468 Perrieriella Schltr. = Oeonia Lindl. (保留属名)■☆

291469 Perrieriella madagascariensis Schltr. = Oeonia madagascariensis (Schltr.) Bosser ■☆

291470 Perrierodendron Cavaco(1951);佩氏苞杯花属●☆

291471 Perrierodendron boinense (H. Perrier) Cavaco;佩氏苞杯花●☆

291472 Perrierodendron capuronii J. -F. Leroy, Lowry, Haev. , Labat et G. E. Schatz;凯普伦佩氏苞杯花●☆

291473 Perrierodendron occidentale J. -F. Leroy, Lowry, Haev. , Labat et G. E. Schatz;西方佩氏苞杯花●☆

291474 Perrierodendron quartzitorum J. -F. Leroy, Lowry, Haev. , Labat et G. E. Schatz;阔茨佩氏苞杯花●☆

291475 Perrierophytum Hochr. (1916);佩氏锦葵属●☆

291476 Perrierophytum rubrum Hochr. ;红佩氏锦葵●☆

291477 Perrierophytum viridiflorum Hochr. ;绿花佩氏锦葵●☆

291478 Perrierosedum (A. Berger) H. Ohba(1978);马岛佩氏景天属●☆

291479 Perrierosedum madagascariense (H. Perrier) H. Ohba;马岛佩氏景天■☆

291480 Perrottetia DC. = Desmodium Desv. (保留属名)●■

291481 Perrottetia Kunth (1824);核子木属 (佩罗特木属);Nucleonwood, Perrottetia ●

291482 Perrottetia arisanensis Hayata;台湾核子木 (佩罗特木);Taiwan Nucleonwood, Taiwan Perrottetia ●

291483 Perrottetia japonica Makino;日本核子木;Japan Nucleonwood, Japanese Nucleonwood ●

291484 Perrottetia macrocarpa C. Yu Chang;大果核子木;Bigfruit Nucleonwood, Bigfruit Perrottetia, Big-fruited Perrottetia ●

291485 Perrottetia quindiuensis Kunth;金迪奥核子木;Quindiu Nucleonwood, Quindiu Perrottetia ●

291486 Perrottetia racemosa (Oliv.) Loes. ;核子木;Common Nucleonwood, Racemose Perrottetia ●

291487 Perrottetia racemosa (Oliv.) Loes. var. arisanensis Hayata = Perrottetia arisanensis Hayata ●

291488 Perrottetia sandwicensis A. Gray;茶叶核子木●☆

291489 Perrottetia sympodislis Hu;合轴核子木;Sympodial Nucleonwood, Sympodial Perrottetia ●

291490 Perryodendron T. G. Hartley = Melicope J. R. Forst. et G. Forst. ●

291491 Persea Mill. (1754)(保留属名);鳄梨属(樟梨属);Avocado, Persea, Red Bay ●

291492 Persea Plum. ex L. = Persea Mill. (保留属名)●

291493 Persea acuminatissima (Hayata) Kosterm. = Cinnamomum philippinense (Merr.) C. E. Chang ●

291494 Persea americana Mill. ;鳄梨(黄油梨,酪梨,油梨,樟梨);Aguacate, Alligator Pear, Alligator-pear, American Avocado, American Avocado-tree, American Persea, Avocado, Avocado Pear, Avocado-pear, Butter-fruit, Midshipman's Butter, Palta, Subal Tern's Butter, Subaltern's Butter ●

291495 Persea americana Mill. var. drymifolia (Schltdl. et Cham.) Blake;墨西哥酪梨;Mexican Avocado ●

291496 Persea arisanensis (Hayata) Kosterm. = Machilus thunbergii Siebold et Zucc. ●

291497 Persea austroguizhouensis (S. K. Lee and F. N. Wei) Kosterm. = Machilus austroguizhouensis S. K. Lee and F. N. Wei ●◇

291498 Persea azorica Seub. = Laurus azorica (Seub.) Franco ●☆

291499 Persea baviensis (Lecomte) Kosterm. = Caryodaphnopsis baviensis (Lecomte) Airy Shaw ●

291500 Persea blumei Kosterm. ;布氏鳄梨●☆

291501 Persea bombycina (King ex Hook. f.) Kosterm. = Machilus bombycina King ex Hook. f. ●

291502 Persea bombycina (King ex Hook. f.) Kosterm. = Machilus gamblei King ex Hook. f. ●

291503 Persea bonii (Lecomte) Kosterm. = Machilus bonii Lecomte ●◇

291504 Persea bonii (Lecomte) Kosterm. = Machilus dumicola (W. W. Sm.) H. W. Li ●

291505 Persea borbonia (L.) Spreng. ;红鳄梨(波旁鳄梨,红湾鳄梨木);Red Bay, Red Bay Persea, Redbay, Red-bay Persea, Redbay Swampbay, Shorebay ●☆

291506 Persea borbonia (L.) Spreng. var. humilis (Nash) L. E. Kopp = Persea humilis Nash ●☆

291507 Persea borbonia (L.) Spreng. var. pubescens (Pursh) Little = Persea palustris (Raf.) Sarg. ●☆

291508 Persea bournei (Hemsl.) Kosterm. = Phoebe bournei (Hemsl.) Yen C. Yang ●

291509 Persea bracteata (Lecomte) Kosterm. = Machilus yunnanensis Lecomte ●

291510 Persea breviflora (Benth.) Pax = Machilus breviflora (Benth.) Hemsl. ●

291511 Persea camphora (L.) Spreng. = Cinnamomum camphora (L.) T. Nees et C. H. Eberm. ●

291512 Persea camphora Spreng. = Cinnamomum camphora (L.) T. Nees et C. H. Eberm. ●

291513 Persea cassia (L.) Spreng. = Cinnamomum cassia C. Presl ●

291514 Persea cavaleriei (H. Lév.) Kosterm. = Nothaphoebe cavaleriei (H. Lév.) Yen C. Yang ●◇

291515 Persea chayuensis (S. K. Lee) Kosterm. = Machilus chayuensis S. K. Lee ●

291516 Persea chekiangensis (S. K. Lee) Kosterm. = Machilus chienkweiensis S. K. Lee ●

291517 Persea chienkweiensis (S. K. Lee) Kosterm. = Machilus chienkweiensi S. K. Lee ●

291518 Persea chinensis (Benth.) Pax = Machilus chinensis (Champ. ex Benth.) Hemsl. ●

291519 Persea chinensis (Champ. ex Benth.) Pax = Machilus chinensis (Champ. ex Benth.) Hemsl. ●

291520 Persea chrysotricha (H. W. Li) Kosterm. = Machilus chrysotricha H. W. Li ●

291521 Persea chuanchienensis (S. K. Lee) Kosterm. = Machilus chuanchienensis S. K. Lee ●

291522 Persea cicatricosa (S. K. Lee) Kosterm. = Machilus cicatricosa S. K. Lee ●

291523　Persea cubeba Lour. = Litsea cubeba（Lour.）Pers. ●

291524　Persea decursinervis（Chun）Kosterm. = Machilus decursinervis Chun ●

291525　Persea dinganensis（S. K. Lee et F. N. Wei）Kosterm. = Machilus dinganensis S. K. Lee et F. N. Wei ●

291526　Persea dumicola（W. W. Sm.）Airy Shaw = Machilus dumicola（W. W. Sm.）H. W. Li ●

291527　Persea dumicola（W. W. Sm.）Kosterm. = Machilus dumicola（W. W. Sm.）H. W. Li ●

291528　Persea duthiei（King ex Hook. f.）Kosterm. = Machilus duthiei King ex Hook. f. ●

291529　Persea fasciculata（H. W. Li）Kosterm. = Machilus fasciculata H. W. Li ●

291530　Persea foonchewii（S. K. Lee）Kosterm. = Machilus foonchewii S. K. Lee ●

291531　Persea fukienensis（Hung T. Chang ex S. K. Lee et al.）Kosterm. = Machilus fukienensis Hung T. Chang ●

291532　Persea glaucescens（Nees）D. G. Long = Machilus glaucescens（Nees）H. W. Li ●

291533　Persea glaucifolia（S. K. Lee et F. N. Wei）Kosterm. = Machilus glaucifolia S. K. Lee et F. N. Wei ●◇

291534　Persea gongshanensis（H. W. Li）Kosterm. = Machilus gongshanensis H. W. Li ●

291535　Persea gracillima（Chun）Kosterm. = Machilus gracillima Chun ●

291536　Persea grandibracteata（S. K. Lee et F. N. Wei）Kosterm. = Machilus grandibracteata S. K. Lee et F. N. Wei ●

291537　Persea gratissima C. F. Gaertn. = Persea americana Mill. ●

291538　Persea grijsii（Hance）Kosterm. = Machilus grijsii Hance ●

291539　Persea henryi（Airy Shaw）Kosterm. = Caryodaphnopsis henryi Airy Shaw ●

291540　Persea humilis Nash；矮鳄梨；Persea，Silk Bay ●☆

291541　Persea ichangensis（Rehder et E. H. Wilson）Kosterm. = Machilus ichangensis Rehder et E. H. Wilson ●

291542　Persea inaequialis A. C. Sm. = Caryodaphnopsis inaequialis（A. C. Sm.）van der Werff et H. G. Richt. ●☆

291543　Persea indica indica（L.）Spreng.；印度鳄梨（印度月桂）；Madeira Bay，Madeira Bay Persea，Madeira Bay-persea ●☆

291544　Persea indica Spreng. = Persea indica indica（L.）Spreng. ●☆

291545　Persea japonica（Siebold et Zucc.）Kosterm. = Machilus japonica Siebold et Zucc. ex Meisn. ●

291546　Persea japonica（Siebold et Zucc.）Siebold = Machilus japonica Siebold et Zucc. ex Meisn. ●

291547　Persea japonica Siebold ex Siebold et Zucc. = Machilus japonica Siebold et Zucc. ex Meisn. ●

291548　Persea kadooriei Kosterm. = Machilus wangchiana Chun ●

291549　Persea kobu（Maxim.）Kosterm. = Machilus kobu Maxim. ●☆

291550　Persea konishii（Hayata）Kosterm. = Machilus konishii Hayata ●

291551　Persea konishii（Hayata）Kosterm. = Nothaphoebe konishii（Hayata）Hayata ●

291552　Persea kurzii（King ex Hook. f.）Kosterm. = Machilus kurzii King ex Hook. f. ●

291553　Persea kusanoi（Hayata）H. L. Li = Machilus japonica Siebold et Zucc. ex Meisn. var. kusanoi（Hayata）J. C. Liao ●

291554　Persea kusanoi（Hayata）H. L. Li = Machilus kusanoi Hayata ●

291555　Persea kusanoi（Hayata）Kosterm. = Machilus japonica Siebold et Zucc. ex Meisn. var. kusanoi（Hayata）J. C. Liao ●

291556　Persea kusanoi Hayata = Machilus japonica Siebold et Zucc. ex Meisn. var. kusanoi（Hayata）J. C. Liao ●

291557　Persea kwangtungensis（Yen C. Yang）Kosterm. = Machilus kwangtungensis Yen C. Yang ●

291558　Persea lenticellata（S. K. Lee et F. N. Wei）Kosterm. = Machilus lenticellata S. K. Lee et F. N. Wei ●

291559　Persea leptophylla（Hand. -Mazz.）Kosterm. = Machilus leptophylla Hand. -Mazz. ●

291560　Persea levinei（Merr.）Kosterm. = Machilus phoenicis Dunn ●

291561　Persea liangkwangensis（Chun）Kosterm. = Machilus robusta W. W. Sm. ●

291562　Persea lichuanensis（W. C. Cheng ex S. K. Lee et al.）Kosterm. = Machilus lichuanensis W. C. Cheng ex S. K. Lee et al. ●

291563　Persea lingue（Miers ex Bertero）Nees；舌状鳄梨；Lingue ●☆

291564　Persea litseifolia（S. K. Lee）Kosterm. = Machilus litseifolia S. K. Lee ●

291565　Persea littoralis Small = Persea borbonia（L.）Spreng. ●☆

291566　Persea lohuiensis（S. K. Lee）Kosterm. = Machilus lohuiensis S. K. Lee ●

291567　Persea longipedicellata（Lecomte）Kosterm. = Machilus longipedicellata Lecomte ●

291568　Persea longipedicellata（Lecomte）Kosterm. = Machilus yunnanensis Lecomte ●

291569　Persea longipedunculata（S. K. Lee et F. N. Wei）Kosterm. = Machilus chienkweiensi S. K. Lee ●

291570　Persea major（Meisn.）L. E. Kopp；大鳄梨 ●☆

291571　Persea melanophylla（H. W. Li）Kosterm. = Machilus melanophylla H. W. Li ●

291572　Persea microcarpa（Hemsl.）Kosterm. = Machilus microcarpa Hemsl. ●

291573　Persea minkweiensis（S. K. Lee）Kosterm. = Machilus minkweiensis S. K. Lee ●

291574　Persea minutiloba（S. K. Lee）Kosterm. = Machilus minutiloba S. K. Lee ●◇

291575　Persea monticola（S. K. Lee）Kosterm. = Machilus monticola S. K. Lee ●◇

291576　Persea multinervia（H. Liu）Kosterm. = Machilus multinervia H. Liu ●

291577　Persea nakao（S. K. Lee）Kosterm. = Machilus nakao S. K. Lee ●

291578　Persea nanchuanensis（N. Chao ex S. K. Lee et al.）Kosterm. = Machilus nanchnanensis N. Chao ●

291579　Persea nanmu Oliv. = Machilus nanmu（Oliv.）Hemsl. ●◇

291580　Persea nanmu Oliv. = Phoebe nanmu（Oliv.）Gamble ●◇

291581　Persea obovatifolia（Hayata）Kosterm. = Machilus obovatifolia（Hayata）Kaneh. et Sasaki ●

291582　Persea obscurinervis（S. K. Lee）Kosterm. = Machilus obscurinervia S. K. Lee ●

291583　Persea oculodracontis（Chun）Kosterm. = Machilus oculodracontis Chun ●

291584　Persea odoratissima（Nees）Kosterm.；极香鳄梨 ●☆

291585　Persea oreophila（Hance）Kosterm. = Machilus oreophila Hance ●

291586　Persea ovatiloba（S. K. Lee）Kosterm. = Machilus ovatiloba S. K. Lee ●

291587　Persea palustris（Raf.）Sarg.；沼泽鳄梨；Swamp Red Bay，Swamp Redbay ●☆

291588　Persea pauhoi（Kaneh.）Kosterm. = Machilus pauhoi Kaneh. ●

291589　Persea pedicellata Kosterm. = Machilus longipes Hung T. Chang ●

291590　Persea petiolaris（Meisn.）Debb. = Alseodaphne petiolaris

（Meisn.）Hook. f. ●

291591　Persea philippinensis（Merr.）Elmer ＝Cinnamomum philippinense（Merr.）C. E. Chang ●

291592　Persea phoenicis（Dunn）Kosterm. ＝Machilus phoenicis Dunn ●

291593　Persea pingii（W. C. Cheng ex Yen C. Yang）Kosterm. ＝Machilus nanmu（Oliv.）Hemsl. ●◇

291594　Persea pingii（W. C. Cheng ex Yen C. Yang）Kosterm. ＝Machilus pingii W. C. Cheng ex Yen C. Yang ●

291595　Persea platycarpa（Chun）Kosterm. ＝Machilus platycarpa Chun ●

291596　Persea pomifera（Kosterm.）Kosterm. ＝Machilus pomifera（Kosterm.）S. K. Lee ●

291597　Persea pomifera Kosterm. ＝Machilus pomifera（Kosterm.）S. K. Lee ●

291598　Persea pseudokobu（Koidz.）Kosterm. ＝Machilus pseudokobu Koidz. ●☆

291599　Persea pseudolongifolia（Hayata）Kosterm. ＝Machilus japonica Siebold et Zucc. ex Meisn. ●

291600　Persea pseudolongifolia（Hayata）Kosterm. ＝Machilus zuihoensis Hayata ●

291601　Persea pubescens（Pursh）Sarg. ＝Persea palustris（Raf.）Sarg. ●☆

291602　Persea pyramidalis（H. W. Li）Kosterm. ＝Machilus pyramidalis H. W. Li ●

291603　Persea pyriformis Elmer ＝Caryodaphnopsis tonkinensis（Lecomte）Airy Shaw ●

291604　Persea rehderi（C. K. Allen）Kosterm. ＝Machilus rehderi C. K. Allen ●

291605　Persea robusta（W. W. Sm.）Kosterm. ＝Machilus robusta W. W. Sm. ●

291606　Persea rufipes（H. W. Li）Kosterm. ＝Machilus rufipes H. W. Li ●

291607　Persea salicina（Hance）Kosterm. ＝Machilus salicina Hance ●

291608　Persea salicoides（S. K. Lee）Kosterm. ＝Machilus salicoides S. K. Lee ●◇

291609　Persea schiedeana Nees；希德鳄梨；Coyo Avocado ●☆

291610　Persea shiwandashanica（Hung T. Chang）Kosterm. ＝Machilus shiwandashanica Hung T. Chang ●

291611　Persea shweliensis（W. W. Sm.）Kosterm. ＝Machilus shweliensis W. W. Sm. ●

291612　Persea sichourensis（H. W. Li）Kosterm. ＝Machilus sichourensis H. W. Li ●

291613　Persea sichuanensis（N. Chao ex S. K. Lee et al.）Kosterm. ＝Machilus sichuanensis N. Chao ex S. K. Lee ●◇

291614　Persea tenuipilis（H. W. Li）Kosterm. ＝Machilus tenuipilis H. W. Li ●

291615　Persea thunbergii（Siebold et Zucc.）Kosterm. ＝Machilus thunbergii Siebold et Zucc. ●

291616　Persea tonkinensis（Lecomte）Kosterm. ＝Caryodaphnopsis tonkinensis（Lecomte）Airy Shaw ●

291617　Persea velutina（Champ. ex Benth.）Kosterm. ＝Machilus velutina Champ. ex Benth. ●

291618　Persea velutinoides（S. K. Lee et F. N. Wei）Kosterm. ＝Machilus velutinoides S. K. Lee et F. N. Wei ●

291619　Persea verruculosa（H. W. Li）Kosterm. ＝Machilus verruculosa H. W. Li ●

291620　Persea villosa（Roxb.）Kosterm. ＝Machilus glaucescens

（Nees）H. W. Li ●

291621　Persea villosa（Roxb.）Kosterm. ＝Machilus villosa（Roxb.）Hook. f. ●

291622　Persea viridis（Hand.-Mazz.）Kosterm. ＝Machilus viridis Hand.-Mazz. ●

291623　Persea wangchiana（Chun）Kosterm. ＝Machilus wangchiana Chun ●

291624　Persea wenshanensis（H. W. Li）Kosterm. ＝Machilus wenshanensis H. W. Li ●

291625　Persea yunnanensis（Lecomte）Kosterm. ＝Machilus yunnanensis Lecomte ●

291626　Persea zuihoensis（Hayata）H. L. Li ＝Machilus zuihoensis Hayata ●

291627　Persea zuihoensis（Hayata）Kosterm. ＝Machilus zuihoensis Hayata ●

291628　Perseaceae Horan. ＝Lauraceae Juss.（保留科名）●■

291629　Persica Mill. ＝Amygdalus L. ●

291630　Persica Mill. ＝Prunus L. ●

291631　Persica davidiana Carrière ＝Amygdalus davidiana（Carrière）de Vos ex Henry ●

291632　Persica domestica Risso ＝Amygdalus davidiana（Carrière）de Vos ex Henry ●

291633　Persica ferganensis（Kostina et Rjabov）Kov. et Kostina ＝Amygdalus ferganensis（Kostina et Rjabov）Te T. Yu et A. M. Lu ●

291634　Persica ferganensis（Kostina et Rjabov）Kov. et Kostina ＝Prunus persica Siebold et Zucc. subsp. ferganensis Kostina et Rjabov ●

291635　Persica kansuensis（Rehder）Kov. et Kostina ＝Amygdalus kansuensis（Rehder）Skeels ●

291636　Persica kansuensis（Rehder）Kov. et Kostina ＝Prunus kansuensis Rehder ●

291637　Persica mira（Koehne）Kov. et Kostina ＝Amygdalus mira（Koehne）Kov. et Kostina ●

291638　Persica mira（Koehne）Kov. et Kostina ＝Prunus mira Koehne ●

291639　Persica platycarpa Decne. ＝Amygdalus persica L. var. compressa（Loudon）Te T. Yu et A. M. Lu ●

291640　Persica platycarpa Decne. ＝Prunus persica Siebold et Zucc. var. compressa（Loudon）Bean ●

291641　Persica potanini（Batalin）Kov. et Kostina ＝Amygdalus davidiana（Carrière）de Vos ex Henry var. potaninii（Batalin）Te T. Yu et A. M. Lu ●

291642　Persica potanini（Batalin）Kov. et Kostina ＝Prunus davidiana（Carrière）Franch. var. potaninii Rehder ●

291643　Persica simonii Decne. ＝Prunus simonii Carrière ●

291644　Persica tangutica Kov. et Kostina ＝Amygdalus tangutica（Batalin）Korsh. ●

291645　Persica tangutica Kov. et Kostina ＝Prunus tangutica（Batalin）Koehne ●

291646　Persica vulgaris Mill. ＝Amygdalus persica L. ●

291647　Persica vulgaris Mill. ＝Prunus persica（L.）Batsch ●

291648　Persica vulgaris Mill. var. compressa Loudon ＝Amygdalus persica L. var. compressa（Loudon）Te T. Yu et A. M. Lu ●

291649　Persicana Scop. ＝Persicaria（L.）Mill. ■

291650　Persicana Tourn. ex Scop. ＝Persicaria（L.）Mill. ■

291651　Persicaria（L.）Mill.（1754）；蓼属（蓼属，马蓼属）；Knotgrass, Knotweed, Persicaria, Smartweed ■

291652　Persicaria L. ＝Persicaria（L.）Mill. ■

291653　Persicaria Mill. ＝Persicaria（L.）Mill. ■

291654 Persicaria Neck. = Atraphaxis L. ●

291655 Persicaria acaulis Gross = Polygonum hookeri Meisn. ■

291656 Persicaria acaulis Hook. f. = Polygonum hookeri Meisn. ■

291657 Persicaria acuminata (Kunth) M. Gómez;渐尖蔍蓄■☆

291658 Persicaria aestiva Ohki = Persicaria sagittata (L.) H. Gross var. sibirica (Meisn.) Miyabe f. aestiva (Ohki) H. Hara ■☆

291659 Persicaria affinis (D. Don) Ronse Decr. 'Darjeeling Red';大吉岭红密穗蓼●■

291660 Persicaria affinis (D. Don) Ronse Decr. 'Donald Lowndes';唐纳德·朗兹密穗蓼●■

291661 Persicaria affinis (D. Don) Ronse Decr. = Polygonum affine D. Don ●■

291662 Persicaria alata (Buch.-Ham. ex D. Don) Nakai = Polygonum nepalense Meisn. ■

291663 Persicaria alpina (All.) H. Gross;高山蓼(草原蓼,兴安蓼);Alp Knotweed, Alpine Fleeceflower, Alpine Knotweed ■

291664 Persicaria alpina (All.) H. Gross = Polygonum alpinum All. ■

291665 Persicaria alpina (All.) H. Gross var. sinica (Dammer) H. Gross = Polygonum campanulatum Hook. f. var. fulvidum Hook. f. ■

291666 Persicaria amblyophylla H. Hara;钝叶蓼■☆

291667 Persicaria amphibia (L.) Delarbre = Polygonum amphibium L. ■

291668 Persicaria amphibia (L.) Gray = Polygonum amphibium L. ■

291669 Persicaria amphibia (L.) Gray subsp. amurensis (Korsh.) Soják = Persicaria amphibia (L.) Delarbre ■

291670 Persicaria amphibia (L.) Gray var. amurensis (Korsh.) H. Hara = Persicaria amphibia (L.) Delarbre ■

291671 Persicaria amphibia (L.) Gray var. emersa (Michx.) J. C. Hickman = Persicaria amphibia (L.) Gray ■

291672 Persicaria amphibia (L.) Gray var. emersa (Michx.) J. C. Hickman = Polygonum amphibium L. ■

291673 Persicaria amphibia (L.) Gray var. stipulacea (N. Coleman) H. Hara = Polygonum amphibium L. ■

291674 Persicaria amphibia (L.) Gray var. stipulacea (N. Coleman) H. Hara = Polygonum amphibium L. var. stipulaceum N. Coleman ■

291675 Persicaria amphibia (L.) Gray var. terrestris (Leyss.) Munshi et Javeid = Polygonum amphibium L. ■

291676 Persicaria amphibia Gray = Polygonum amphibium L. ■

291677 Persicaria amphibia Gray var. stipulacea (N. Coleman) H. Hara = Persicaria amphibia (L.) Gray ■

291678 Persicaria amphibia Gray var. stipulacea (N. Coleman) H. Hara = Polygonum amphibium L. ■

291679 Persicaria amplexicaulis (D. Don) Ronse Decr. 'Firetail';火尾两栖蓼■

291680 Persicaria amplexicaulis (D. Don) Ronse Decr. = Polygonum amplexicaule D. Don ■

291681 Persicaria amurensis (Korsh.) Nieuwl. = Polygonum amphibium L. ■

291682 Persicaria anguillana (Koidz.) Honda = Persicaria sagittata (L.) H. Gross var. sibirica (Meisn.) Miyabe ■

291683 Persicaria angustifolia (Pall.) Ronse Decraene = Polygonum angustifolium Pall. ■

291684 Persicaria arifolia (L.) Haraldson;戟叶蔍蓄(戟叶蓼);Halberd-leaf Tearthumb, Halberd-leaved Tear-thumb ■☆

291685 Persicaria arifolia (L.) Haraldson = Polygonum arifolium L. ■

291686 Persicaria arifolia (L.) Haraldson var. pubescens (A. Keller) Fernald = Persicaria arifolia (L.) Haraldson ■☆

291687 Persicaria arifolia (L.) Haraldson var. pubescens (A. Keller) Fernald = Polygonum arifolium L. ■

291688 Persicaria assamica (Meisn.) Soják = Polygonum assamicum Meisn. ■

291689 Persicaria attenuata (R. Br.) Soják subsp. africana K. L. Wilson = Persicaria madagascariensis (Meisn.) S. Ortiz et Paiva ■☆

291690 Persicaria attenuata (R. Br.) Soják subsp. pulchra (Blume) K. L. Wilson = Polygonum pulchrum Blume ■

291691 Persicaria auriculata S. K. Dixit et al. = Polygonum praetermissum Hook. f. ■

291692 Persicaria barbata (L.) H. Hara = Polygonum barbatum L. ■

291693 Persicaria barbata (L.) H. Hara subsp. gracilis (Danser) Soják = Polygonum barbatum L. var. gracile (Danser) Steward ■

291694 Persicaria barbata (L.) H. Hara var. gracilis (Danser) H. Hara = Polygonum barbatum L. var. gracile (Danser) Steward ■

291695 Persicaria biconvexa (Hayata) Nemoto = Polygonum biconvexum Hayata ■

291696 Persicaria bicornis (Raf.) Nieuwl.;双角蔍蓄;Pink Smartweed ■☆

291697 Persicaria bicornis (Raf.) Nieuwl. = Polygonum pensylvanicum L. ■☆

291698 Persicaria bistorta (L.) Samp. = Bistorta officinalis Delarbre ■☆

291699 Persicaria bistorta (L.) Samp. = Polygonum bistorta L. ■

291700 Persicaria bistorta Samp. 'Superba';极品拳参■☆

291701 Persicaria bistorta Samp. = Polygonum bistorta L. ■

291702 Persicaria bistortoides (Pursh) H. R. Hinds = Bistorta bistortoides (Pursh) Small ■☆

291703 Persicaria blumei (Meisn.) H. Gross = Persicaria longiseta (Bruijn) Kitag. ■

291704 Persicaria blumei (Meisn.) H. Gross = Polygonum longisetum Bruijn ■

291705 Persicaria breviochreata (Makino) Ohki;短鞘托叶蔍蓄■☆

291706 Persicaria buisanensis (Ohki) Sasaki = Polygonum longisetum Bruijn ■

291707 Persicaria bungeana (Turcz.) Nakai;布氏蔍蓄;Prickly Smartweed ■☆

291708 Persicaria bungeana (Turcz.) Nakai = Polygonum bungeanum Turcz. ■

291709 Persicaria bungeana (Turcz.) Nakai ex Mori = Polygonum bungeanum Turcz. ■

291710 Persicaria bungeana (Turcz.) Nakai ex T. Mori = Polygonum bungeanum Turcz. ■

291711 Persicaria caespitosa (Blume) Nakai var. longiseta (Bruijn) C. F. Reed = Persicaria longiseta (Bruijn) Kitag. ■

291712 Persicaria caespitosa (Blume) Nakai var. longiseta (Bruijn) C. F. Reed = Polygonum longisetum Bruijn ■

291713 Persicaria campanulata (Hook. f.) Ronse Decr. = Polygonum campanulatum Hook. f. ■

291714 Persicaria capitata (Buch.-Ham. ex D. Don) H. Gross;头状蔍蓄;Pink-head Knotweed ■☆

291715 Persicaria capitata (Buch.-Ham. ex D. Don) H. Gross = Polygonum capitatum Buch.-Ham. ex D. Don ■

291716 Persicaria careyi (Olney) Greene;卡雷蔍蓄;Carey's Smartweed ■☆

291717 Persicaria careyi (Olney) Greene = Polygonum careyi Olney ■☆

291718 Persicaria cespitosa (Blume) Nakai = Polygonum cespitosum Blume ■☆

291719 Persicaria chinensis (L.) H. Gross = Polygonum chinense L. ■

291720 Persicaria chinensis (L.) H. Gross var. ovalifolia (Hemsl.) H.

Hara = Polygonum chinense L. var. ovalifolium Meisn. ■

291721　Persicaria chinensis（L.）H. Gross var. siamensis H. Lév. = Polygonum chinense L. ■

291722　Persicaria chinensis（L.）H. Gross var. thunbergiana（Meisn.）Okuyama = Persicaria chinensis（L.）H. Gross ■

291723　Persicaria coccinea（Muhl. ex Willd.）Greene = Persicaria amphibia（L.）Gray ■

291724　Persicaria coccinea（Muhl. ex Willd.）Greene = Polygonum amphibium L. ■

291725　Persicaria coccinea（Muhl.）Greene = Polygonum amphibium L. var. emersum Michx. ■☆

291726　Persicaria cochinchinensis（Lour.）Kitag. = Persicaria orientalis（L.）Spach ■

291727　Persicaria cochinchinensis（Lour.）Kitag. = Polygonum orientale L. ■

291728　Persicaria conspicua（Nakai）Nakai ex Ohki = Persicaria macrantha（Meisn.）Haraldson subsp. conspicua（Nakai）Yonek. ■

291729　Persicaria conspicua（Nakai）Nakai ex Ohki f. albiflora Hiyama;白花樱蓼■☆

291730　Persicaria conspicua（Nakai）Nakai ex T. Mori = Polygonum japonicum Meisn. var. conspicuum Nakai ■

291731　Persicaria criopolitana（Hance）Migo = Polygonum criopolitanum Hance ■

291732　Persicaria debilis（Meisn.）H. Gross ex W. T. Lee;柔软蓼■☆

291733　Persicaria decipiens（R. Br.）K. L. Wilson;迷惑蓼蓄■☆

291734　Persicaria densiflora（Meisn.）Moldenke = Persicaria glabra（Willd.）M. Gómez ■

291735　Persicaria densiflora（Meisn.）Moldenke = Polygonum densiflorum Meisn. ■☆

291736　Persicaria densiflora（Meisn.）Moldenke = Polygonum glabrum Willd. ■

291737　Persicaria dichotoma（Blume）Masam. = Polygonum dichotomum Blume ■

291738　Persicaria dissitiflora（Hemsl.）H. Gross ex T. Mori = Polygonum dissitiflorum Hemsl. ■

291739　Persicaria divaricata（L.）H. Gross = Aconogonon divaricatum（L.）Nakai ■

291740　Persicaria divaricata（L.）H. Gross = Polygonum divaricatum L. ■

291741　Persicaria dolichopoda（Ohki）Sasaki = Polygonum persicaria L. ■

291742　Persicaria duclouxii（H. Lév. et Vaniot）H. Gross = Polygonum campanulatum Hook. f. var. fulvidum Hook. f. ■

291743　Persicaria duclouxii（H. Lév. et Vaniot）H. Gross var. hypoleuca H. Lév. = Polygonum campanulatum Hook. f. var. fulvidum Hook. f. ■

291744　Persicaria erectominor（Makino）Nakai;直立小蓼蓄■☆

291745　Persicaria erectominor（Makino）Nakai f. viridiflora（Nakai）I. Ito;绿花蓼蓄■☆

291746　Persicaria erectominor（Makino）Nakai var. koreensis（Nakai）I. Ito;朝鲜蓼蓄(科雷蓼蓄)■☆

291747　Persicaria erectominor（Makino）Nakai var. roseoviridis（Kitag.）I. Ito;粉绿蓼蓄■☆

291748　Persicaria erectominor（Makino）Nakai var. trigonocarpa（Makino）H. Hara;三角蓼蓄■☆

291749　Persicaria excurrens（Steward）Koidz. = Persicaria viscofera（Makino）H. Gross var. robusta（Makino）Hiyama ■

291750　Persicaria excurrens（Steward）Koidz. = Polygonum viscoferum

Makino ■

291751　Persicaria extremiorientalis（Vorosch.）Tzvelev = Polygonum persicaria L. ■

291752　Persicaria fauriei（H. Lév. et Vaniot）Nakai ex T. Mori = Polygonum dissitiflorum Hemsl. ■

291753　Persicaria filiformis（Thunb.）Nakai = Antenoron filiforme（Thunb.）Rob. et Vautier ■

291754　Persicaria filiformis（Thunb.）Nakai = Persicaria filiformis（Thunb.）Nakai ex W. T. Lee ■

291755　Persicaria filiformis（Thunb.）Nakai ex W. T. Lee;线蓼■

291756　Persicaria filiformis（Thunb.）Nakai ex W. T. Lee = Antenoron filiforme（Thunb.）Rob. et Vautier ■

291757　Persicaria filiformis（Thunb.）Nakai ex W. T. Lee f. albiflora（Hiyama）Yonek.;白花线蓼■☆

291758　Persicaria filiformis（Thunb.）Nakai ex W. T. Lee var. neofiliformis（Nakai）T. B. Lee = Persicaria neofiliformis（Nakai）Ohki ■

291759　Persicaria flaccida（Meisn.）H. Gross = Polygonum pubescens Blume ■

291760　Persicaria flaccida（Meisn.）Nakai ex Sasaki = Polygonum pubescens Blume ■

291761　Persicaria foliosa（H. Lindb.）Kitag. = Polygonum foliosum H. Lindb. ■

291762　Persicaria foliosa（H. Lindb.）Kitag. var. nikaii（Makino）H. Hara = Polygonum foliosum H. Lindb. var. nikaii（Makino）Kitam. ■☆

291763　Persicaria foliosa（H. Lindb.）Kitag. var. paludicola（Makino）H. Hara = Polygonum foliosum H. Lindb. var. paludicola（Makino）Kitam. ■

291764　Persicaria fusiformis（Greene）Greene = Persicaria maculosa Gray ■☆

291765　Persicaria gentiliana H. Lév. = Polygonum longisetum Bruijn ■

291766　Persicaria glabra（Willd.）M. Gómez;光蓼(红辣蓼,竹节蓼子草);Smooth Knotweed,Smooth Smartweed ■

291767　Persicaria glabra（Willd.）M. Gómez = Polygonum glabrum Willd. ■

291768　Persicaria glacialis（Meisn.）H. Hara = Polygonum glaciale（Meisn.）Hook. f. ■

291769　Persicaria glandulo-pilosa（De Wild.）Soják;腺毛蓼蓄■☆

291770　Persicaria glomerata（Dammer）S. Ortiz et Paiva;团集蓼蓄■☆

291771　Persicaria hartwrightii（A. Gray）Greene = Persicaria amphibia（L.）Gray ■

291772　Persicaria hartwrightii（A. Gray）Greene = Polygonum amphibium L. ■

291773　Persicaria hastatoauriculata（Makino ex Nakai）H. Gross ex Nakai = Polygonum praetermissum Hook. f. ■

291774　Persicaria hastatoauriculata（Makino ex Nakai）Nakai = Persicaria praetermissa（Hook. f.）H. Hara ■

291775　Persicaria hastatosagittata（Makino）Nakai;长箭叶蓼(长叶犁避,戟箭叶蓼,箭叶蓼);Long Arrowleaf Knotweed ■

291776　Persicaria hastatosagittata（Makino）Nakai ex T. Mori = Polygonum hastatosagittatum Makino ■

291777　Persicaria hirsuta（Walter）Small;毛蓼蓄;Hairy Smartweed ■☆

291778　Persicaria hookeri（Meisn.）Ronse Decraene = Polygonum hookeri Meisn. ■

291779　Persicaria humilis（Meisn.）H. Hara = Polygonum humile Meisn. ■

291780　Persicaria hydropiper（L.）Opiz = Polygonum hydropiper L. ■

291781 Persicaria hydropiper (L.) Spach = Polygonum hydropiper L. ■

291782 Persicaria hydropiper (L.) Spach f. purpurascens (Makino) Nemoto = Polygonum hydropiper L. f. purpurascens Makino ■

291783 Persicaria hydropiper (L.) Spach subsp. flaccida (Meisn.) Munshi et Javeid = Polygonum pubescens Blume ■

291784 Persicaria hydropiper (L.) Spach var. diffusa Kitag. = Polygonum hydropiper L. ■

291785 Persicaria hydropiper (L.) Spach var. fastigiata (Makino) Araki;帚状水蓼■☆

291786 Persicaria hydropiper (L.) Spach var. fastigiata (Makino) Araki f. angustissima (Makino) Araki;窄帚状水蓼■☆

291787 Persicaria hydropiper (L.) Spach var. filiformis Araki;线状水蓼■☆

291788 Persicaria hydropiper (L.) Spach var. latifolia Araki = Persicaria hydropiper (L.) Spach f. purpurascens (Makino) Nemoto ■

291789 Persicaria hydropiper (L.) Spach var. maximowiczii (Regel) Nemoto;马氏蓼■☆

291790 Persicaria hydropiper (L.) Spach var. maximowiczii (Regel) Nemoto f. viridis (Makino) Araki;绿花马氏蓼■☆

291791 Persicaria hydropiper (L.) Spach var. vulgaris (Meisn.) Ohki = Polygonum hydropiper L. ■

291792 Persicaria hydropiperoides (Michx.) Small; 拟水蓼; False Water Pepper, False Water-pepper, Mild Water pepper, Mild Water-pepper, Swamp Smartweed, Wild Water Pepper ■☆

291793 Persicaria hydropiperoides (Michx.) Small = Polygonum hydropiperoides Michx. ■☆

291794 Persicaria hydropiperoides (Michx.) Small var. breviciliata (Fernald) C. F. Reed = Persicaria hydropiperoides (Michx.) Small ■☆

291795 Persicaria hydropiperoides (Michx.) Small var. euronotora (Fernald) C. F. Reed = Persicaria hydropiperoides (Michx.) Small ■☆

291796 Persicaria hydropiperoides (Michx.) Small var. opelousana (Riddell) J. S. Wilson = Persicaria hydropiperoides (Michx.) Small ■☆

291797 Persicaria hystricula (J. Schust.) Soják;豪猪萹蓄■☆

291798 Persicaria incarnata (Elliott) Small = Polygonum lapathifolium L. ■

291799 Persicaria japonica (Meisn.) H. Gross ex Nakai = Polygonum japonicum Meisn. ■

291800 Persicaria japonica (Meisn.) Nakai ex Ohki = Polygonum japonicum Meisn. ■

291801 Persicaria jucunda (Meisn.) Migo = Polygonum jucundum Meisn. ■

291802 Persicaria kawagoeana (Makino) Nakai = Persicaria tenella (Blume) H. Hara var. kawagoeana (Makino) H. Hara ■

291803 Persicaria kawagoeana (Makino) Nakai = Polygonum kawagoeanum Makino ■

291804 Persicaria kawagoeana Makino var. densiflora H. Hara et I. Ito = Persicaria tenella (Blume) H. Hara ■

291805 Persicaria koreensis (Nakai) Nakai = Persicaria erectominor (Makino) Nakai var. koreensis (Nakai) I. Ito ■☆

291806 Persicaria kuekenthalii H. Lév. = Polygonum viscosum Buch.-Ham. ex D. Don ■

291807 Persicaria lanata (Roxb.) Tzvelev = Polygonum lapathifolium L. var. lanatum (Roxb.) Stewart ■

291808 Persicaria lanigera (R. Br.) Soják;绵毛萹蓄■☆

291809 Persicaria lapathifolia (L.) Delarbre = Polygonum lapathifolium L. ■

291810 Persicaria lapathifolia (L.) Gray = Polygonum lapathifolium L. ■

291811 Persicaria lapathifolia (L.) Gray subsp. africana (Meisn.) Soják = Persicaria senegalensis (Meisn.) Soják. f. albotomentosa (R. A. Graham) K. L. Wilson ■☆

291812 Persicaria lapathifolia (L.) Gray subsp. lanata (Roxb.) Soják = Persicaria lapathifolia (L.) Gray var. lanata (Roxb.) H. Hara ■

291813 Persicaria lapathifolia (L.) Gray subsp. lanata (Roxb.) Soják = Polygonum lapathifolium L. var. lanatum (Roxb.) Steward ■

291814 Persicaria lapathifolia (L.) Gray subsp. nodosa (Pers.) Á. Löve = Persicaria lapathifolia (L.) Delarbre ■

291815 Persicaria lapathifolia (L.) Gray subsp. pallida (With.) S. Ekman et Knutsson var. incana (Roth) S. Ekman et Knutsson = Polygonum lapathifolium L. var. salicifolium Sibth. ■

291816 Persicaria lapathifolia (L.) Gray var. incana (K. Koch) H. Hara = Persicaria lapathifolia (L.) Gray var. tomentosa (Schrank) H. Gross ■

291817 Persicaria lapathifolia (L.) Gray var. lanata (Roxb.) H. Hara = Polygonum lapathifolium L. var. lanatum (Roxb.) Stewart ■

291818 Persicaria lapathifolia (L.) Gray var. tomentosa (Schrank) H. Gross;毛酸模叶蓼■

291819 Persicaria laxmannii (Lepech.) H. Gross = Polygonum ochreatum L. ■

291820 Persicaria limbata (Meisn.) H. Hara;具边萹蓄■☆

291821 Persicaria limicola (Sam.) Yonek. et H. Ohashi = Polygonum limicola Sam. ■

291822 Persicaria longiseta (Bruijn) Kitag.;长毛蓼■

291823 Persicaria longiseta (Bruijn) Kitag. = Polygonum blumei Meisn. ■

291824 Persicaria longiseta (Bruijn) Kitag. = Polygonum longisetum Bruijn ■

291825 Persicaria longiseta (Bruijn) Kitag. f. albiflora (Honda) Masam.;白花长毛蓼■☆

291826 Persicaria longiseta (Bruijn) Moldenke = Polygonum longisetum Bruijn ■

291827 Persicaria longiseta (Bruyn) Kitag. f. brevifolia (Makino) Hiyama;短叶长毛蓼■☆

291828 Persicaria longiseta (Bruyn) Kitag. f. divaricatoramosa Hiyama;叉长毛蓼■☆

291829 Persicaria longiseta (Bruyn) Moldenke = Polygonum cespitosum Blume var. longisetum (Bruyn) Stewart ■☆

291830 Persicaria longistyla (Small) Small = Persicaria bicornis (Raf.) Nieuwl. ■☆

291831 Persicaria longistyla (Small) Small = Polygonum pensylvanicum L. ■☆

291832 Persicaria maackiana (Regel) Nakai = Polygonum maackianum Regel ■

291833 Persicaria maackiana (Regel) Nakai ex T. Mori = Polygonum maackianum Regel ■

291834 Persicaria maackiana (Regel) Nakai f. albiflora H. Hara;白花马氏蓼■☆

291835 Persicaria macrantha (Meisn.) Haraldson;大花蓼(长花蓼); Bigflower Knotweed ■☆

291836 Persicaria macrantha (Meisn.) Haraldson subsp. conspicua (Nakai) Yonek. = Polygonum conspicuum (Nakai) Nakai ■

291837 Persicaria maculata (Raf.) Á. Löve et D. Löve = Polygonum persicaria L. ■

291838 Persicaria maculata (Raf.) Gray = Persicaria lapathifolia (L.)

Gray ■

291839 Persicaria maculata (Raf.) Gray = Polygonum persicaria L. ■

291840 Persicaria maculosa Gray;斑点蓄蓄;Redleg, Redshank, Spotted Lady's Thumb ■☆

291841 Persicaria maculosa Gray = Polygonum persicaria L. ■

291842 Persicaria maculosa Gray subsp. hirticaulis (Danser) S. Ekman et Knutsson = Polygonum persicaria L. ■

291843 Persicaria madagascariensis (Meisn.) S. Ortiz et Paiva;马岛蓄蓄■☆

291844 Persicaria makinoi (Nakai) Nakai = Persicaria viscofera (Makino) H. Gross var. robusta (Makino) Hiyama ■

291845 Persicaria makinoi Nakai = Polygonum viscoferum Makino ■

291846 Persicaria makinoi Nakai f. laevis (Kitag.) Kitag. = Persicaria viscofera (Makino) H. Gross var. robusta (Makino) Hiyama f. laevis (Kitag.) Hiyama ■☆

291847 Persicaria makinoi Nakai var. laevis (Kitag.) Honda = Persicaria viscofera (Makino) H. Gross var. robusta (Makino) Hiyama f. laevis (Kitag.) Hiyama ■☆

291848 Persicaria manshuricola Kitag. = Polygonum longisetum Bruijn ■

291849 Persicaria meisneriana (Cham. et Schltdl.) M. Gómez = Polygonum meisnerianum Cham. et Schltdl. ■☆

291850 Persicaria mesochora Greene = Polygonum amphibium L. var. stipulaceum N. Coleman ■

291851 Persicaria microcephala (D. Don) H. Gross = Polygonum microcephalum D. Don ■

291852 Persicaria microcephala (D. Don) H. Gross var. wallichii (Meisn.) H. Hara = Polygonum wallichii Meisn. ■

291853 Persicaria minor (Huds.) Opiz;蒙古小蓄蓄■

291854 Persicaria minor (Huds.) Opiz = Polygonum minus Huds. ■

291855 Persicaria minuta (Hayata) Nakai = Polygonum filicaule Wall. ex Meisn. ■

291856 Persicaria mississippiensis (Stanford) Small = Persicaria pensylvanica (L.) M. Gómez ■☆

291857 Persicaria mississippiensis (Stanford) Small = Polygonum pensylvanicum L. ■☆

291858 Persicaria mitis Delarbre = Polygonum persicaria L. ■

291859 Persicaria mitis Gilib. var. amblyophylla (H. Hara) Hiyama = Persicaria amblyophylla H. Hara ■☆

291860 Persicaria mitis Gilib. var. hirticaulis (Danser) H. Hara et I. Ito = Polygonum persicaria L. ■

291861 Persicaria mollis (D. Don) H. Gross = Polygonum molle D. Don ●■

291862 Persicaria morrisonensis (Hayata) Nakai = Polygonum runcinatum Buch.-Ham. ex D. Don ■

291863 Persicaria muhlenbergia (S. Watson) Small = Persicaria amphibia (L.) Gray ■

291864 Persicaria muhlenbergia (S. Watson) Small = Polygonum amphibium L. ■

291865 Persicaria muhlenbergii (Meisn.) Small = Polygonum amphibium L. ■

291866 Persicaria muhlenbergii (S. Watson) Small = Polygonum amphibium L. var. emersum Michx. ■☆

291867 Persicaria muricata (Meisn.) Nemoto = Polygonum muricatum Meisn. ■

291868 Persicaria musashinoensis Hiyama;武藏野蓄蓄■☆

291869 Persicaria nebraskensis Greene = Polygonum amphibium L. var. stipulaceum N. Coleman ■

291870 Persicaria neofiliformis (Nakai) Ohki = Antenoron filiforme

(Thunb.) Rob. et Vautier var. neofiliforme (Nakai) A. J. Li ■

291871 Persicaria nepalensis (Meisn.) H. Gross = Polygonum nepalense Meisn. ■

291872 Persicaria nepalensis (Meisn.) H. Gross f. adenothrix (Nakai) Hiyama;腺毛尼泊尔蓼■

291873 Persicaria nepalensis (Meisn.) H. Gross f. adenothrix (Nakai) Hiyama = Polygonum nepalense Meisn. ■

291874 Persicaria nipponensis (Makino) H. Gross ex Nakai = Polygonum muricatum Meisn. ■

291875 Persicaria nipponensis (Makino) H. Gross ex Nakai f. albiflora (Makino) Nemoto = Polygonum nipponense Makino f. albiflorum Makino ■☆

291876 Persicaria nipponensis (Makino) Nakai = Persicaria muricata (Meisn.) Nemoto ■

291877 Persicaria nodosa (Pers.) Opiz = Persicaria lapathifolia (L.) Delarbre ■

291878 Persicaria nodosa (Pers.) Opiz = Polygonum lapathifolium L. ■

291879 Persicaria obtusifolia (Täckh. et Boulos) Greuter et Burdet;钝叶蓄蓄■☆

291880 Persicaria odorata (Lour.) Soják = Polygonum odoratum Lour. ■☆

291881 Persicaria omerostroma (Ohki) Sasaki = Polygonum barbatum L. ■

291882 Persicaria opaca (Sam.) Koidz. = Polygonum persicaria L. var. opacum (Sam.) A. J. Li ■

291883 Persicaria opelousana (Riddell ex Small) Small = Persicaria hydropiperoides (Michx.) Small ■☆

291884 Persicaria opelousana (Riddell ex Small) Small = Polygonum hydropiperoides Michx. ■☆

291885 Persicaria oreophila (Makino) Hiyama;喜山蓄蓄■☆

291886 Persicaria orientale (L.) Spach = Polygonum orientale L. ■

291887 Persicaria palmata (Dunn) Yonek. et H. Ohashi = Polygonum palmatum Dunn ■

291888 Persicaria paludicola (Makino) Nakai = Polygonum foliosum H. Lindb. var. paludicola (Makino) Kitam. ■

291889 Persicaria paludicola Small = Persicaria hydropiperoides (Michx.) Small ■☆

291890 Persicaria paludicola Small = Polygonum hydropiperoides Michx. ■☆

291891 Persicaria peduncularis (Wall. ex Meisn.) Nemoto = Polygonum dichotomum Blume ■

291892 Persicaria pensylvanica (L.) M. Gómez = Polygonum pensylvanicum L. ■☆

291893 Persicaria pensylvanica (L.) M. Gómez var. dura (Stanford) C. F. Reed = Persicaria pensylvanica (L.) M. Gómez ■☆

291894 Persicaria pensylvanica (L.) M. Gómez var. dura (Stanford) C. F. Reed = Polygonum pensylvanicum L. ■☆

291895 Persicaria pensylvanica (L.) M. Gómez var. laevigata (Fernald) M. C. Ferguson;光滑蓄蓄■☆

291896 Persicaria perfoliata (L.) H. Gross = Polygonum perfoliatum L. ■

291897 Persicaria persicaria (L.) Small = Polygonum persicaria L. ■

291898 Persicaria persicarioides (Kunth) Small = Persicaria hydropiperoides (Michx.) Small ■☆

291899 Persicaria pilosa (Roxb. ex Meisn.) Kitag. = Polygonum orientale L. ■

291900 Persicaria pilosa (Roxb.) Kitag. = Persicaria orientalis (L.) Spach ■

291901 Persicaria pinetorum (Hemsl.) H. Gross = Polygonum

pinetorum Hemsl. ■

291902　Persicaria poiretii（Meisn.）K. L. Wilson;普瓦雷萹蓄■☆

291903　Persicaria polystachya（Wall. ex Meisn.）H. Gross = Polygonum polystachyum Wall. ex Meisn. ●■

291904　Persicaria polystachya（Wall.）H. Gross = Persicaria wallichii Greuter et Burdet ■☆

291905　Persicaria polystachya（Wall.）H. Gross = Polygonum polystachyum Wall. ex Meisn. ●■

291906　Persicaria portoricensis（Bertero ex Small）Small = Persicaria glabra（Willd.）M. Gómez ■

291907　Persicaria portoricensis（Bertero ex Small）Small = Polygonum glabrum Willd. ■

291908　Persicaria posumbu（Buch. -Ham. ex D. Don）H. Gross = Polygonum posumbu Buch. -Ham. ex D. Don ■

291909　Persicaria posumbu（Buch. -Ham. ex D. Don）H. Gross var. laxiflora（Meisn.）H. Hara = Polygonum posumbu Buch. -Ham. ex D. Don ■

291910　Persicaria posumbu（Buch. -Ham. ex D. Don）H. Gross var. laxiflora（Meisn.）H. Hara = Persicaria posumbu（Buch. -Ham. ex D. Don）H. Gross ■

291911　Persicaria posumbu（Buch. -Ham. ex D. Don）H. Gross var. stenophylla（Makino）Yonek. et H. Ohashi = Polygonum posumbu Buch. -Ham. ex D. Don var. stenophyllum（Makino）Murata ■☆

291912　Persicaria praetermissa（Hook. f.）H. Hara = Polygonum praetermissum Hook. f. ■

291913　Persicaria pratincola Greene = Polygonum amphibium L. var. emersum Michx. ■☆

291914　Persicaria pubescens（Blume）H. Hara = Polygonum pubescens Blume ■

291915　Persicaria pubescens（Blume）H. Hara f. macrantha M. Mizush. ;大花毛萹蓄■☆

291916　Persicaria pubescens（Blume）H. Hara var. acuminata（Franch. et Sav.）H. Hara;渐尖毛萹蓄■☆

291917　Persicaria pulchra（Blume）Soják = Persicaria attenuata（R. Br.）Soják subsp. pulchra（Blume）K. L. Wilson ■

291918　Persicaria pulchra（Blume）Soják = Polygonum pulchrum Blume ■

291919　Persicaria punctata（Elliott）Small = Polygonum punctatum Elliott ■☆

291920　Persicaria punctata（Elliott）Small var. eciliata Small = Polygonum punctatum Elliott ■☆

291921　Persicaria punctata（Elliott）Small var. leptostachya（Meisn.）Small = Polygonum punctatum Elliott var. confertiflorum（Meisn.）Fassett ■☆

291922　Persicaria punctata（Elliott）Small var. robustior（Small）Small = Persicaria robustior（Small）E. P. Bicknell ■☆

291923　Persicaria quarrei（De Wild.）Soják = Persicaria setosula（A. Rich.）K. L. Wilson ■☆

291924　Persicaria robustior（Small）E. P. Bicknell;胖萹蓄;Stout Smartweed ■☆

291925　Persicaria roseoviridis Kitag. = Persicaria erectominor（Makino）Nakai var. roseoviridis（Kitag.）I. Ito ■☆

291926　Persicaria roseoviridis Kitag. = Polygonum longisetum Bruijn ■

291927　Persicaria roseoviridis Kitag. var. menshuricola（Kitag.）C. F. Fang = Polygonum longisetum Bruijn ■

291928　Persicaria ruderalis（Salisb.）C. F. Reed = Polygonum persicaria L. ■

291929　Persicaria ruderalis（Salisb.）C. F. Reed var. vulgaris（Webb et Moq.）C. F. Reed = Polygonum persicaria L. ■

291930　Persicaria rudis（Meisn.）H. Gross = Polygonum molle D. Don var. rude（Meisn.）A. J. Li ■

291931　Persicaria runcinata（Buch. -Ham. ex D. Don）H. Gross = Polygonum runcinatum Buch. -Ham. ex D. Don ■

291932　Persicaria sagittata（L.）H. Gross;箭头蓼（箭叶蓼）;Arrow-leaf Tearthumb, Arrow-leaved Tear-thumb, Arrow-vine ■

291933　Persicaria sagittata（L.）H. Gross = Polygonum sagittatum L. ■

291934　Persicaria sagittata（L.）H. Gross ex Nakai = Polygonum sagittatum L. ■

291935　Persicaria sagittata（L.）H. Gross ex Nakai var. sieboldii（Meisn.）Nakai = Polygonum sagittatum L. ■

291936　Persicaria sagittata（L.）H. Gross var. aestiva（Ohki）Masam. = Persicaria sagittata（L.）H. Gross var. sibirica（Meisn.）Miyabe f. aestiva（Ohki）H. Hara ■☆

291937　Persicaria sagittata（L.）H. Gross var. sibirica（Meisn.）Miyabe;西伯利亚箭头蓼■

291938　Persicaria sagittata（L.）H. Gross var. sibirica（Meisn.）Miyabe f. aestiva（Ohki）H. Hara;夏西伯利亚箭头蓼■☆

291939　Persicaria sagittata（L.）H. Gross var. sibirica（Meisn.）Miyabe f. pilosa（H. Hara）H. Hara;疏毛西伯利亚箭头蓼■☆

291940　Persicaria sagittata（L.）H. Gross var. sibirica（Meisn.）Miyabe f. tomentosa（H. Hara）H. Hara;绒毛西伯利亚箭头蓼■☆

291941　Persicaria sagittata（L.）H. Gross var. sibirica（Meisn.）Miyabe f. viridialba（Honda）H. Hara;白花西伯利亚箭头蓼■☆

291942　Persicaria sagittata（L.）H. Gross var. sieboldii（Meisn.）Nakai = Polugonum sibiricum Meisn. ■

291943　Persicaria sagittata（L.）Haraldson = Polygonum sagittatum L. ■

291944　Persicaria sagittifolia H. Gross = Polygonum darrisii H. Lév. ■

291945　Persicaria salicifolia（Brouss. ex Willd.）Assenov = Persicaria decipiens（R. Br.）K. L. Wilson ■☆

291946　Persicaria sambesicum（J. Schust.）Soják = Persicaria senegalensis（Meisn.）Soják ■☆

291947　Persicaria scabra（Moench）Moldenke = Persicaria lapathifolia（L.）Gray var. tomentosa（Schrank）H. Gross ■

291948　Persicaria schinzii J. Schust. = Polygonum hydropiper L. ■

291949　Persicaria senegalensis（Meisn.）Soják;塞内加尔蓼■☆

291950　Persicaria senegalensis（Meisn.）Soják. f. albotomentosa（R. A. Graham）K. L. Wilson;白毛塞内加尔蓼■☆

291951　Persicaria senticosa（Meisn.）H. Gross = Persicaria senticosa（Meisn.）H. Gross ex Nakai ■

291952　Persicaria senticosa（Meisn.）H. Gross = Polygonum senticosum（Meisn.）Franch. et Sav. ■

291953　Persicaria senticosa（Meisn.）H. Gross ex Nakai = Polygonum senticosum（Meisn.）Franch. et Sav. ■

291954　Persicaria senticosa（Meisn.）H. Gross ex Nakai f. albiflora（Honda）Honda;白花刺蓼■☆

291955　Persicaria senticosa（Meisn.）H. Gross ex Nakai var. inermis Sugim. ;无刺蓼■☆

291956　Persicaria senticosa（Meisn.）H. Gross ex Nakai var. sagittifolia Yonek. et H. Ohashi = Polygonum darrisii H. Lév. ■

291957　Persicaria serrulata（Lag.）Webb et Moq. = Persicaria decipiens（R. Br.）K. L. Wilson ■☆

291958　Persicaria setacea（Baldwin ex Elliott）Small;沼地萹蓄;Bog Smartweed, Water Ppepper ■☆

291959　Persicaria setacea（Baldwin）Small = Persicaria setacea

（Baldwin ex Elliott）Small ■☆

291960　Persicaria setosula（A. Rich.）K. L. Wilson;刚毛蓄蓄■☆

291961　Persicaria sibirica（Laxm.）H. Gross = Polygonum sibiricum Laxm. ■

291962　Persicaria sieboldii（Meisn.）Ohki = Persicaria sagittata（L.）H. Gross var. sibirica（Meisn.）Miyabe ■

291963　Persicaria sieboldii（Meisn.）Ohki = Polygonum sagittatum L. ■

291964　Persicaria sieboldii（Meisn.）Ohki var. aestiva（Ohki）Okuyama = Persicaria sagittata（L.）H. Gross var. sibirica（Meisn.）Miyabe f. aestiva（Ohki）H. Hara ■☆

291965　Persicaria sieboldii（Meisn.）Ohki var. brevifolia Kitag. = Polygonum sagittatum L. ■

291966　Persicaria sieboldii（Meisn.）Ohwi = Polygonum sieboldii Meisn. ■

291967　Persicaria sinica Migo = Polygonum thunbergii Siebold et Zucc. ■

291968　Persicaria sphaerocephala（Wall. ex Meisn.）H. Gross = Polygonum microcephalum D. Don var. sphaerocephalum（Wall. ex Meisn.）Murata ■

291969　Persicaria sterilis（Nakai）Nakai et Ohki = Polygonum japonicum Meisn. var. conspicuum Nakai ■

291970　Persicaria strigosa（R. Br.）Nakai = Polygonum strigosum R. Br. ■

291971　Persicaria sungareense Kitag. = Polygonum longisetum Bruijn var. rotundatum A. J. Li ■

291972　Persicaria tanganyikae（J. Schust.）Soják = Persicaria senegalensis（Meisn.）Soják ■☆

291973　Persicaria taquetii（H. Lév.）Koidz. = Polygonum taquetii H. Lév. ■

291974　Persicaria taquetii（H. Lév.）Koidz. var. minutula（Makino）Honda = Persicaria taquetii（H. Lév.）Koidz. ■

291975　Persicaria tenella（Blume）H. Hara = Polygonum tenellum Blume ■

291976　Persicaria tenella（Blume）H. Hara var. kawagoeana（Makino）H. Hara = Polygonum micranthum Boiss. ex Meisn. ■

291977　Persicaria tenella（Blume）H. Hara var. kwangoeana（Makino）H. Hara = Persicaria tenella（Blume）H. Hara var. kawagoeana（Makino）H. Hara ■

291978　Persicaria tenella（Blume）H. Hara var. kwangoeana（Makino）H. Hara = Polygonum tenellum Blume var. micranthum（Boiss. ex Meisn.）C. Y. Wu ■

291979　Persicaria thunbergii（Siebold et Zucc.）H. Gross = Polygonum thunbergii Siebold et Zucc. ■

291980　Persicaria thunbergii（Siebold et Zucc.）H. Gross var. oreophila（Makino）Murai = Persicaria oreophila（Makino）Hiyama ■☆

291981　Persicaria thunbergii（Siebold et Zucc.）H. Gross var. stolonifera（F. Schmidt）Nakai ex H. Hara = Polygonum thunbergii Siebold et Zucc. var. stoloniferum（F. Schmidt）Makino ■

291982　Persicaria thunbergii（Siebold et Zucc.）H. Gross var. stolonifera（F. Schmidt）H. Gross ex Nakai = Polygonum thunbergii Siebold et Zucc. ■

291983　Persicaria tinctoria（Aiton）Spach = Polygonum tinctorium Aiton ■

291984　Persicaria tomentosa（Schrank）E. P. Bicknell = Persicaria lapathifolia（L.）Gray var. tomentosa（Schrank）H. Gross ■

291985　Persicaria tomentosa（Schrank）E. P. Bicknell = Polygonum lapathifolium L. ■

291986　Persicaria tomentosa（Willd.）E. P. Bicknell = Persicaria

attenuata（R. Br.）Soják subsp. pulchra（Blume）K. L. Wilson ■

291987　Persicaria trigonocarpa（Makino）Nakai = Persicaria erectominor（Makino）Nakai var. trigonocarpa（Makino）H. Hara ■☆

291988　Persicaria umbellata（Houtt.）Nakai = Persicaria chinensis（L.）H. Gross ■

291989　Persicaria ussuriensis（Regel）Nakai ex T. Mori = Polygonum hastatosagittatum Makino ■

291990　Persicaria vacciniifolia（Wall. ex Meisn.）Ronse Decr. = Polygonum vaccinifolium Wall. ex Meisn. ■

291991　Persicaria vaniotiana H. Lév. = Polygonum lapathifolium L. ■

291992　Persicaria vernalis Nakai = Polygonum hydropiper L. ■

291993　Persicaria virginiana（L.）Gaertn. = Polygonum virginianum L. ■☆

291994　Persicaria viscofera（Makino）H. Gross = Polygonum viscoferum Makino ■

291995　Persicaria viscofera（Makino）H. Gross ex Nakai = Persicaria viscofera（Makino）H. Gross ■

291996　Persicaria viscofera（Makino）H. Gross ex Nakai = Polygonum viscoferum Makino ■

291997　Persicaria viscofera（Makino）H. Gross f. viridescens（Nakai）Hiyama;绿黏蓼■☆

291998　Persicaria viscofera（Makino）H. Gross var. robusta（Makino）Hiyama f. laevis（Kitag.）Hiyama;平滑粗壮黏蓼■☆

291999　Persicaria viscofera（Makino）H. Gross var. robusta（Makino）Hiyama = Polygonum viscoferum Makino ■

292000　Persicaria viscosa（Buch. -Ham. ex D. Don）H. Gross ex Nakai = Polygonum viscosum Buch. -Ham. ex D. Don ■

292001　Persicaria viscosa（Buch. -Ham. ex D. Don）H. Gross ex T. Mori = Polygonum viscosum Buch. -Ham. ex D. Don ■

292002　Persicaria viscosa（Buch. -Ham.）H. Gross ex Nakai = Polygonum viscosum Buch. -Ham. ex D. Don ■

292003　Persicaria vivipara（L.）Ronse Decr. = Bistorta vivipara（L.）Delarbre ■

292004　Persicaria vivipara（L.）Ronse Decr. = Polygonum viviparum L. ■

292005　Persicaria vulgaris Webb et Moq. = Persicaria maculosa Gray ■☆

292006　Persicaria vulgaris Webb et Moq. = Polygonum persicaria L. ■

292007　Persicaria wallichii Greuter et Burdet;沃利克蓼; Himalayan Knotweed ■☆

292008　Persicaria wallichii Greuter et Burdet = Polygonum polystachyum Wall. ex Meisn. ●■

292009　Persicaria wellensii（De Wild.）Soják = Polygonum wellensii De Wild. ■☆

292010　Persicaria weyrichii（F. Schmidt）H. Gross = Aconogonon weyrichii（F. Schmidt）H. Hara ■☆

292011　Persicaria yokusaiana（Makino）Nakai = Persicaria posumbu（Buch. -Ham. ex D. Don）H. Gross ■

292012　Persicaria yokusaiana（Makino）Nakai = Polygonum posumbu Buch. -Ham. ex D. Don ■

292013　Persicaria yokusaiana（Makino）Nakai f. laxiflora（Meisn.）Hiyama = Persicaria posumbu（Buch. -Ham. ex D. Don）H. Gross ■

292014　Persicaria yokusaiana（Makino）Nakai var. stenophylla（Makino）Honda = Persicaria posumbu（Buch. -Ham. ex D. Don）H. Gross var. stenophylla（Makino）Yonek. et H. Ohashi ■☆

292015　Persicariaceae Adans. ex Post et Kuntze = Polygonaceae Juss.（保留科名）●■

292016　Persicariaceae Martinov = Persicariaceae Adans. ex Post et

Kuntze ●■

292017　Persimon Raf. = Diospyros L. ●

292018　Personaceae Dulac = Scrophulariaceae Juss.（保留科名）●■

292019　Personaria Lam. = Gorteria L. ■☆

292020　Personatae Vent. = Scrophulariaceae Juss.（保留科名）●■

292021　Personia Raf. = Marshallia Schreb.（保留属名）■☆

292022　Personia Raf. = Persoonia Sm.（保留属名）●☆

292023　Personula Raf. = Utricularia L. ■

292024　Persoonia Michx. = Marshallia Schreb.（保留属名）■☆

292025　Persoonia Sm.（1798）（保留属名）；佩松木属（皮索尼亚属，匹索尼亚属）；Snodgollion, Snot-goblin, Snottygobble, Persoonia, Geebung ●☆

292026　Persoonia Willd. = Carapa Aubl. ●☆

292027　Persoonia chamaepitys A. Cunn.；匍匐佩松木（匍匐皮索尼亚）●☆

292028　Persoonia lanceolata Andréws；披针叶佩松木（披针叶皮索尼亚）；Geebung, Lance-leaf Gee-bung ●☆

292029　Persoonia levis（Cav.）Domin；平滑佩松木（平滑皮索尼亚）；Broad-leaf Gee-bung ●☆

292030　Persoonia linearis Andréws；狭叶佩松木（狭叶皮索尼亚）；Narrow-leaf Geebung ●☆

292031　Persoonia mollis R. Br.；毛叶佩松木（毛皮索尼亚，毛叶皮索尼亚）；Geebung ●☆

292032　Persoonia nutans R. Br.；垂花佩松木（垂花皮索尼亚）●☆

292033　Persoonia pinifolia R. Br.；松叶佩松木（松叶皮索尼亚）；Pineleaf Geebung, Pine-leaf Geebung ●☆

292034　Persoonia toru A. Cunn.；毛佩松木（毛利匹索尼亚）；Toru Persoonia ●☆

292035　Perspicillum Fabr. = Biscutella L.（保留属名）■☆

292036　Perspicillum Heist. ex Fabr. = Biscutella L.（保留属名）■☆

292037　Pertusadina Ridsdale（1979）；槽裂木属；Pertusadina ●

292038　Pertusadina hainanensis（F. C. How）Ridsdale；海南槽裂木（槽裂木，骨罗树，海南水团花）；Hainan Adina, Hainan Pertusadina ●

292039　Pertusadina metcalfii（Merr. ex H. L. Li）Y. F. Deng et C. M. Hu；广西槽裂木 ●

292040　Pertya Sch. Bip.（1862）；帚菊属（高野帚属，帚菊木属）；Pertybush, Perty-bush ●■

292041　Pertya × koribana（Nakai）Makino et Nemoto；郡场帚菊 ●☆

292042　Pertya × suzukii Kitam.；铃木帚菊 ●☆

292043　Pertya angustifolia Y. C. Tseng；狭叶帚菊；Narrowleaf Pertybush ●

292044　Pertya berberidoides（Hand. -Mazz.）Y. C. Tseng；异叶帚菊（小檗状帚菊）；Diverseleaf Pertybush ●

292045　Pertya bodinferi Vaniot；昆明帚菊；Bodinifer Pertybush ●

292046　Pertya bodinferi Vaniot var. berberidoides Hand. -Mazz. = Pertya berberidoides（Hand. -Mazz.）Y. C. Tseng ●

292047　Pertya cordifolia Mattf.；心叶帚菊；Heart Pertybush ●

292048　Pertya cordifolia Mattf. var. pubescens Y. Ling = Pertya pubescens Y. Ling ●

292049　Pertya corymbosa Y. C. Tseng；疏花帚菊；Laxflower Pertybush ●

292050　Pertya desmocephala Diels；聚头帚菊 ■

292051　Pertya dioica（Bunge）T. G. Gao = Myripnois dioica Bunge ●

292052　Pertya discolor Rehder；两色帚菊（二色帚菊，异色帚菊）；Bicolor Pertybush ●

292053　Pertya discolor Rehder var. calvescens L.；同色帚菊 ●

292054　Pertya esquirolii H. Lév. = Ainsliaea elegans Hemsl. ■

292055　Pertya glabrescens Sch. Bip.；渐光帚菊（长花帚菊）；Glabrous Pertybush ●

292056　Pertya glabrescens Sch. Bip. var. viridis Nakai = Pertya glabrescens Sch. Bip. ●

292057　Pertya henanensis Y. C. Tseng；瓜叶帚菊；Melonleaf Pertybush ■

292058　Pertya hybrida Makino；杂种帚菊；Hybrid Pertybush ●☆

292059　Pertya koribana（Nakai）Makino et Nemoto；日本帚菊 ●☆

292060　Pertya mattfeldii Bornm.；阿富汗帚菊 ●☆

292061　Pertya monocephala W. W. Sm.；单头帚菊；One-head Pertybush ●

292062　Pertya ovata Maxim. = Pertya scandens（Thunb.）Sch. Bip. ●

292063　Pertya phylicoides Jeffrey；针叶帚菊（小叶帚菊）；Littleleaf Pertybush, Little-leaved Perty-bush ●

292064　Pertya phylicoides Jeffrey var. monocephala W. W. Sm. = Pertya monocephala W. W. Sm. ●

292065　Pertya pubescens Y. Ling；腺叶帚菊（慧香）；Pubescent Pertybush ●

292066　Pertya pungens Y. C. Tseng；尖苞帚菊；Spinybract Pertybush ●

292067　Pertya rigidula（Miq.）Makino；轮叶大托菊（鬼督邮）●☆

292068　Pertya rigidula Makino = Pertya rigidula（Miq.）Makino ●☆

292069　Pertya robusta（Maxim.）Makino；粗壮帚菊 ●☆

292070　Pertya robusta（Maxim.）Makino var. kiushiana Kitam.；九州粗壮帚菊 ●☆

292071　Pertya sacndens（Thunb.）Sch. Bip. var. schultziana Franch. = Pertya glabrescens Sch. Bip. ●

292072　Pertya scandens（Thunb.）Sch. Bip.；长花帚菊 ●

292073　Pertya scandens（Thunb.）Sch. Bip. var. shimozawai（Masam.）Kitam. = Pertya shimozawai Masam. ●

292074　Pertya scandens Sch. Bip. = Pertya glabrescens Sch. Bip. ●

292075　Pertya scandens Sch. Bip. var. shimozawai（Masam.）Kitam. = Pertya shimozawai Masam. ●

292076　Pertya scandens Sch. Bip. var. viridis Nakai = Pertya scandens（Thunb.）Sch. Bip. ●

292077　Pertya shimozawai Masam.；台湾帚菊（半高野帚）；Taiwan Pertybush ●

292078　Pertya sinensis Oliv.；华帚菊；China Pertybush, Chinese Perty-bush ●

292079　Pertya suzukii Kitam. = Pertya × suzukii Kitam. ●☆

292080　Pertya triloba（Makino）Makino；三裂帚菊 ●☆

292081　Pertya triloba Makino var. koribana（Nakai）Makino = Pertya koribana（Nakai）Makino et Nemoto ●☆

292082　Pertya tsoongiana Y. Ling；巫山帚菊；Wushan Pertybush ●

292083　Pertya uniflora（Maxim.）Mattf.；单花帚菊；Monoflower Pertybush ●

292084　Pertya yakushimensis H. Koyama et Nagam.；屋久岛帚菊 ●☆

292085　Perula Raf. = Ficus L. ●

292086　Perula Raf. = Urostigma Gasp. ●

292087　Perula Schreb. = Pera Mutis ●☆

292088　Perularia Lindl. = Platanthera Rich.（保留属名）■

292089　Perularia Lindl. = Tulotis Raf. ■

292090　Perularia fuscescens（L.）Lindl. = Platanthera souliei Kraenzl. ■

292091　Perularia fuscescens（L.）Lindl. = Tulotis fuscescens Raf. ■

292092　Perularia scutellata（Nutt.）Small = Platanthera flava（L.）Lindl. ■☆

292093　Perularia shensiana（Kraenzl.）Schltr. = Platanthera ussuriensis（Regel et Maack）Maxim. ■

292094　Perularia shensiana（L.）Schltr. = Tulotis ussuriensis（Regel et Maack）H. Hara ■

292095　Perularia souliei（Kraenzl.）Schltr. = Platanthera souliei Kraenzl. ■

292096　Perularia souliei（Kraenzl.）Schltr. = Tulotis fuscescens Raf. ■

292097　Perularia ussuriensis （Maxim.） Schltr. = Platanthera ussuriensis（Regel et Maack）Maxim. ■

292098　Perularia ussuriensis （Maxim.） Schltr. = Tulotis ussuriensis（Regel et Maack）H. Hara ■

292099　Perularia ussuriensis （Regel） Schltr. = Platanthera ussuriensis（Regel et Maack）Maxim. ■

292100　Perularia whangshanensis S. S. Chien = Calanthe triplicata（Willemet）Ames ■

292101　Perularia whangshanensis S. S. Chien = Platanthera tipuloides（L. f.）Lindl. ■

292102　Perulifera A. Camus = Pseudechinolaena Stapf ■

292103　Peruvocereus Akers = Haageocereus Backeb. ●☆

292104　Pervillaea Decne.（1844）;肖弓果藤属●☆

292105　Pervillaea Decne. = Toxocarpus Wight et Arn. ●

292106　Pervillaea decaryi（Choux）Klack. ;德氏肖弓果藤●☆

292107　Pervillaea phillipsonii Klack. ;菲舍尔肖弓果藤●☆

292108　Pervillaea tomentosa Decne. ;毛肖弓果藤●☆

292109　Pervillaea venenata（Baill.）Klack. ;肖弓果藤●☆

292110　Pervinca Mill. = Vinca L. ■

292111　Pervinca rosea（L.）Moench = Catharanthus roseus（L.）G. Don ■

292112　Pervinca rosea（L.）Moench = vinca rosea L. ■

292113　Pervinea Mill. = Vinca L. ■

292114　Perxo Raf. = Basilicum Moench ■

292115　Perxo Raf. = Moschosma Rchb. ■

292116　Perymeniopsis H. Rob.（1978）;落芒菊属●☆

292117　Perymeniopsis H. Rob. = Perymenium Schrad. ■●☆

292118　Perymeniopsis Sch. Bip. ex Klatt = Gymnolomia Kunth ■☆

292119　Perymeniopsis ovalifolia（A. Gray）H. Rob. ;落芒菊■☆

292120　Perymenium Schrad.（1830）;月菊属■●☆

292121　Perymenium acuminatum（La Llave）S. F. Blake;渐尖月菊■☆

292122　Perymenium album S. Watson;白月菊■☆

292123　Perymenium angustifolium Brandegee;窄叶月菊■☆

292124　Perymenium asperifolium Sch. Bip. ;糙叶月菊■☆

292125　Perymenium discolor Sch. Bip;月菊■☆

292126　Perymenium gracile Hemsl. ;细月菊■☆

292127　Perymenium grande Hemsl. ;大月菊■☆

292128　Perymenium jelskii（Hieron.）S. F. Blake;耶尔斯克月菊■☆

292129　Perymenium lancifolium S. F. Blake;披针叶月菊■☆

292130　Perymenium lasiolepis S. F. Blake;毛梗月菊■☆

292131　Perymenium leptopodum S. F. Blake;细梗月菊■☆

292132　Perymenium macrocephalum Greenm. ;大头月菊■☆

292133　Perymenium microcephalum Sch. Bip. ex Klatt;小头月菊■☆

292134　Perymenium microphyllum B. L. Rob. et Greenm. ;小叶月菊■☆

292135　Perymenium ovalifolium（A. Gray）B. L. Turner;卵叶月菊■☆

292136　Perytis Raf.（废弃属名）= Cyclobalanopsis Oerst.（保留属名）●

292137　Pescatorea Rchb. f.（1869）= Pescatoria Rchb. f. ■☆

292138　Pescatoria Rchb. f.（1852）;鲨口兰属（帕卡兰属,佩斯卡托兰属）■☆

292139　Pescatoria lamellosa Rchb. f. ;鲨口兰■☆

292140　Pescatoria violacea（Lindl.）Dressler;堇色鲨口兰■☆

292141　Pescatoria whitei（Rolfe）Dressler;瓦特鲨口兰■☆

292142　Peschiera A. DC.（1844）;山马茶属（白希木属）●☆

292143　Peschiera A. DC. = Tabernaemontana L. ●

292144　Peschiera laeta Miers;可爱山马茶（可爱白希木）●☆

292145　Peschiera lundii Miers;伦迪山马茶（龙氏白希木）●☆

292146　Peschkovia（Tzvelev）Tzvelev = Lychnis L.（废弃属名）■

292147　Peschkovia（Tzvelev）Tzvelev = Silene L.（保留属名）■

292148　Peschkovia（Tzvelev）Tzvelev（2006）;岩石竹属■☆

292149　Pesomeria Lindl. = Cyanorchis Thouars ■

292150　Pesomeria Lindl. = Cyanorkis Thouars ■

292151　Pesomeria Lindl. = Phaius Lour. ■

292152　Pessularia Salisb. = Anthericum L. ■☆

292153　Pestallozzia Willis = Gynostemma Blume ■

292154　Pestallozzia Willis = Pestalozzia Zoll. et Moritzi ■

292155　Pestalozza Moritzi = Gynostemma Blume ■

292156　Pestalozza Moritzi = Pestalozzia Zoll. et Moritzi ■

292157　Pestalozzia Zoll. et Moritzi = Gynostemma Blume ■

292158　Pestalozzia laxa（Wall.）Thwaites = Gynostemma laxum（Wall.）Cogn. ■

292159　Pestalozzia pedata（Blume）Zoll. et Moritzi = Gynostemma pentaphyllum（Thunb.）Makino ■

292160　Petagna Endl. = Petagnia Raf. ●■

292161　Petagna Endl. = Solanum L. ●■

292162　Petagnaea Caruel = Petagnia Guss. ■☆

292163　Petagnaea Caruel（1889）;佩塔草属■☆

292164　Petagnaea gussonei（Spreng.）Rauschert;佩塔草■☆

292165　Petagnana J. F. Gmel. = Smithia Aiton（保留属名）●■

292166　Petagnia Guss.（1827）;佩塔芹属■☆

292167　Petagnia Guss. = Petagnaea Caruel ■☆

292168　Petagnia Raf. = Solanum L. ●■

292169　Petagnia saniculifolia Guss. ;佩塔芹☆

292170　Petagniana Raf. = Petagnana J. F. Gmel. ●■

292171　Petagniana Raf. = Smithia Aiton（保留属名）●■

292172　Petagomoa Bremek. = Psychotria L.（保留属名）●

292173　Petalacte D. Don（1826）;具托鼠麴木属●☆

292174　Petalacte canescens DC. = Langebergia canescens（DC.）Anderb. ●☆

292175　Petalacte coronata（L.）D. Don;具托鼠麴木●☆

292176　Petalacte vlokii Hilliard = Anderbergia vlokii（Hilliard）B. Nord. ●☆

292177　Petalactella N. E. Br. = Ifloga Cass. ■☆

292178　Petalactella woodii N. E. Br. = Trichogyne decumbens（Thunb.）Less. ■☆

292179　Petaladenium Ducke（1938）;巴西耳壶豆属■☆

292180　Petaladenium urceoliferum Ducke;巴西耳壶豆■☆

292181　Petalandra F. Muell. = Euphorbia L. ●■

292182　Petalandra F. Muell. ex Boiss. = Euphorbia L. ●■

292183　Petalandra Hassk. = Hopea Roxb.（保留属名）●

292184　Petalanisia Raf. = Hypericum L. ■●

292185　Petalanthera Nees = Ocotea Aubl. ●☆

292186　Petalanthera Nutt. = Cevallia Lag. ●☆

292187　Petalanthera Raf. = Justicia L. ●■

292188　Petalepis Raf. = Lepuropetalon Elliott ■☆

292189　Petalidium Nees（1832）;扁爵床属■☆

292190　Petalidium angustitubum P. G. Mey. ;窄管扁爵床■☆

292191　Petalidium aromaticum Oberm. ;芳香扁爵床■☆

292192　Petalidium barlerioides（Roth）Nees;假杜鹃扁爵床■☆

292193　Petalidium bracteatum Oberm. ;具苞扁爵床■☆

292194　Petalidium canescens（Engl.）C. B. Clarke;灰扁爵床■☆

292195　Petalidium coccineum S. Moore;绯红扁爵床■☆

292196　Petalidium crispum A. Meeuse ex P. G. Mey. ;皱波扁爵床■☆

292197　Petalidium currorii（Lindau）Benth. ex S. Moore;库洛里扁爵

床■☆

292198　Petalidium cymbiforme Schinz;船状扁爵床■☆

292199　Petalidium damarense S. Moore = Petalidium variabile（Engl.）C. B. Clarke ■☆

292200　Petalidium eenii S. Moore = Petalidium canescens（Engl.）C. B. Clarke ■☆

292201　Petalidium elatum Benoist;高扁爵床■☆

292202　Petalidium englerianum（Schinz）C. B. Clarke;恩氏扁爵床■☆

292203　Petalidium eurychlamys Mildbr. = Petalidium englerianum（Schinz）C. B. Clarke ■☆

292204　Petalidium giessii P. G. Mey.;吉斯扁爵床■☆

292205　Petalidium glandulosum S. Moore;具腺扁爵床■☆

292206　Petalidium glutinosum（Engl.）C. B. Clarke = Petalidium variabile（Engl.）C. B. Clarke ■☆

292207　Petalidium gossweileri S. Moore;戈斯扁爵床■☆

292208　Petalidium halimoides（Nees）S. Moore;哈利木扁爵床■☆

292209　Petalidium huillense C. B. Clarke;威拉扁爵床■☆

292210　Petalidium incanum（Engl.）Mildbr. = Petalidium variabile（Engl.）C. B. Clarke ■☆

292211　Petalidium lanatum（Engl.）C. B. Clarke;绵毛扁爵床■☆

292212　Petalidium latifolium（Schinz）C. B. Clarke = Petalidium englerianum（Schinz）C. B. Clarke ■☆

292213　Petalidium linifolium T. Anderson;亚麻叶扁爵床■☆

292214　Petalidium loranthifolium S. Moore = Petalidium halimoides（Nees）S. Moore ■☆

292215　Petalidium lucens Oberm.;光亮扁爵床■☆

292216　Petalidium luteoalbum A. Meeuse;黄白扁爵床■☆

292217　Petalidium microtrichum Benoist;小毛扁爵床■☆

292218　Petalidium oblongifolium C. B. Clarke;矩圆叶扁爵床■☆

292219　Petalidium otaviense Dinter;奥塔维扁爵床■☆

292220　Petalidium ovatum（Schinz）C. B. Clarke = Petalidium englerianum（Schinz）C. B. Clarke ■☆

292221　Petalidium parvifolium C. B. Clarke ex Schinz;小叶扁爵床■☆

292222　Petalidium parvifolium C. B. Clarke ex Schinz var. angustifolia Schinz = Petalidium linifolium T. Anderson ■☆

292223　Petalidium physaloides S. Moore;酸浆扁爵床■☆

292224　Petalidium pilosi-bracteolatum Merxm. et Hainz;疏毛苞扁爵床■☆

292225　Petalidium ramulosum Schinz;多枝扁爵床■☆

292226　Petalidium rautanenii Schinz;劳塔宁扁爵床■☆

292227　Petalidium rossmannianum P. G. Mey.;罗斯曼扁爵床■☆

292228　Petalidium rubescens Oberm. = Petalidium coccineum S. Moore ■☆

292229　Petalidium rupestre S. Moore;岩生扁爵床■☆

292230　Petalidium setosum C. B. Clarke ex Schinz;刚毛扁爵床■☆

292231　Petalidium spiniferum C. B. Clarke;刺扁爵床■☆

292232　Petalidium spiniferum C. B. Clarke var. obtusa ?;钝扁爵床■☆

292233　Petalidium subcrispum P. G. Mey.;拟皱波扁爵床■☆

292234　Petalidium tomentosum S. Moore;绒毛扁爵床■☆

292235　Petalidium variabile（Engl.）C. B. Clarke;易变扁爵床■☆

292236　Petalidium variabile（Engl.）C. B. Clarke var. spectabile Mildbr.;壮观扁爵床■☆

292237　Petalidium welwitschii S. Moore;韦尔扁爵床■☆

292238　Petalidium wilmaniae Oberm. = Petalidium parvifolium C. B. Clarke ex Schinz ■☆

292239　Petalinia Becc. = Ochanostachys Mast. ●☆

292240　Petalocaryum Pierre ex A. Chev.（1917）;扁果铁青树属●☆

292241　Petalocentrum Schltr. = Sigmatostalix Rchb. f. ■☆

292242　Petalochilus R. S. Rogers = Caladenia R. Br. ■☆

292243　Petalodactylis Arènes = Cassipourea Aubl. ●☆

292244　Petalodiscus Baill. = Savia Willd. ●☆

292245　Petalodiscus Pax = Wielandia Baill. ●☆

292246　Petalodiscus Pax（1890）;肖维兰德大戟属●☆

292247　Petalodiscus fadenii（Radcl.-Sm.）Radcl.-Sm.;肖维兰德大戟●☆

292248　Petalodon Luer = Masdevallia Ruiz et Pav. ■☆

292249　Petalodon Luer（2006）;扁齿兰属■☆

292250　Petalogyne F. Muell. = Petalostylis R. Br. ●☆

292251　Petalolepis Cass. = Helichrysum Mill.（保留属名）●■

292252　Petalolepis Less. = Petalacte D. Don ●☆

292253　Petalolophus K. Schum.（1905）;瘤瓣花属●☆

292254　Petalolophus megalopus K. Schum.;瘤瓣花●☆

292255　Petaloma Raf. ex Boiss. = Euphorbia L. ●■

292256　Petaloma Roxb. = Lumnitzera Willd. ●

292257　Petaloma Sw. = Mouriri Aubl. ●☆

292258　Petaloma alba Blanco = Lumnitzera racemosa Willd. ●

292259　Petaloma albiflora Zipp. ex Span. = Lumnitzera racemosa Willd. ●

292260　Petaloma alternifolia Roxb. = Lumnitzera racemosa Willd. ●

292261　Petaloma coccinea（Gaudich.）Blanco = Lumnitzera littorea（Jacq.）Voigt ●◇

292262　Petalonema Gilg = Gravesia Naudin ●☆

292263　Petalonema Gilg = Neopetalonema Brenan ●☆

292264　Petalonema Peter = Impatiens L. ■

292265　Petalonema Schltr. = Quisumbingia Merr. ●☆

292266　Petalonema fissibracteatum Peter = Impatiens briartii De Wild. et T. Durand ■☆

292267　Petalonema glanduligerum Pellegr. = Dicellandra glanduligera（Pellegr.）Jacq.-Fél. ●☆

292268　Petalonema pulchrum Gilg = Gravesia pulchra（Gilg）Wickens ■●☆

292269　Petalonyx A. Gray（1854）;扁爪刺莲花属●☆

292270　Petalonyx linearis Greene;线形扁爪刺莲花●☆

292271　Petalonyx nitidus S. Watson;亮扁爪刺莲花●☆

292272　Petalonyx parryi A. Gray;扁爪刺莲花●☆

292273　Petalopogon Reiss. = Phylica L. ●☆

292274　Petalopogon Reiss. ex Endl. et Fenzl = Phylica L. ●☆

292275　Petalosteira Raf. = Tiarella L. ■

292276　Petalostelma E. Fourn. = Cynanchum L. ●■

292277　Petalostelma E. Fourn. = Glossonema Decne. ■☆

292278　Petalostelma E. Fourn. = Metastelma R. Br. ●☆

292279　Petalostemon Michx.（1803）（'Petalostemum'）（保留属名）;瓣蕊豆属;Prairie Clover ●☆

292280　Petalostemon Michx.（保留属名）= Dalea L.（保留属名）●■☆

292281　Petalostemon candidum（Michx. ex Willd.）Michx;瓣蕊豆●☆

292282　Petalostemon candidum（Michx. ex Willd.）Michx. = Dalea candida Michx. ex Willd. ■☆

292283　Petalostemon candidum（Michx. ex Willd.）Michx. var. oligophyllus（Torr.）F. J. Herm. = Dalea candida Michx. ex Willd. var. oligophylla（Torr.）Shinners ■☆

292284　Petalostemon gracile Nutt. var. oligophyllum Torr. = Dalea candida Michx. ex Willd. var. oligophylla（Torr.）Shinners ■☆

292285　Petalostemon molle Rydb. = Dalea purpurea Vent. ■☆

292286　Petalostemon multiflorum Nutt. = Dalea multiflora（Nutt.）Shinners ■☆

292287　Petalostemon occidentale（A. Heller）Fernald = Dalea candida Michx. ex Willd. var. oligophylla（Torr.）Shinners ■☆

292288　Petalostemon oligophyllum（Torr.）Rydb. = Dalea candida

Michx. ex Willd. var. oligophylla（Torr.）Shinners ■☆

292289 Petalostemon purpureum（Vent.）Rydb. = Dalea purpurea Vent. ■☆

292290 Petalostemon purpureum（Vent.）Rydb. f. pubescens（A. Gray）Fassett = Dalea purpurea Vent. ■☆

292291 Petalostemon purpureum（Vent.）Rydb. var. molle（Rydb.）B. Boivin = Dalea purpurea Vent. ■☆

292292 Petalostemon villosum Nutt. = Dalea villosa（Nutt.）Spreng. ■☆

292293 Petalostemum Michx. = Petalostemon Michx.（保留属名）■☆

292294 Petalostigma F. Muell.（1857）；瓣柱戟属；Quinine Bush, Quinine Tree ●☆

292295 Petalostigma pachyphyllum Airy Shaw；厚叶瓣柱戟●☆

292296 Petalostigma pubescens Domin；短毛瓣柱戟；Quinine Bush, Quinine Tree ●☆

292297 Petalostigma quadriloculare F. Muell.；四室瓣柱戟●☆

292298 Petalostigma sericea ?；绢毛瓣柱戟●☆

292299 Petalostima Raf. = Wahlenbergia Schrad. ex Roth（保留属名）■●

292300 Petalostyle Benth. = Petalostylis R. Br. ●☆

292301 Petalostyles Benth. = Petalostylis R. Br. ●☆

292302 Petalostylis Lindl. = Leianthus Griseb. ●☆

292303 Petalostylis Lindl. = Petasostylis Griseb. ●☆

292304 Petalostylis R. Br.（1849）；瓣柱豆属●☆

292305 Petalostylis labicheoides R. Br.；瓣柱豆☆

292306 Petalotoma DC. = Carallia Roxb.（保留属名）●

292307 Petalotoma brachiata（Lour.）DC. = Carallia brachiata（Lour.）Merr. ●

292308 Petaloxis Raf. = Dichorisandra J. C. Mikan（保留属名）■☆

292309 Petalvitemon Raf. = Petalostemon Michx.（保留属名）■☆

292310 Petamenes Salisb. = Gladiolus L. ■

292311 Petamenes Salisb. ex J. W. Loudon = Gladiolus L. ■

292312 Petamenes Salisb. ex N. E. Br. = Gladiolus L. ■

292313 Petamenes abbreviatus（Andréws）N. E. Br. = Gladiolus abbreviatus Andréws ■☆

292314 Petamenes aethiopica（L.）Allan = Chasmanthe aethiopica（L.）N. E. Br. ■☆

292315 Petamenes aethiopica（L.）E. Phillips = Chasmanthe aethiopica（L.）N. E. Br. ■☆

292316 Petamenes bicolor（Gasp. ex Ten.）E. Phillips = Chasmanthe bicolor（Gasp. ex Ten.）N. E. Br. ■☆

292317 Petamenes buckerveldii（L. Bolus）N. E. Br. = Gladiolus buckerveldii（L. Bolus）Goldblatt ■☆

292318 Petamenes caffra（Ker Gawl. ex Baker）E. Phillips = Tritoniopsis caffra（Ker Gawl. ex Baker）Goldblatt ■☆

292319 Petamenes cunonia（L.）E. Phillips = Gladiolus cunonius（L.）Gaertn. ■☆

292320 Petamenes duftii（Schinz）E. Phillips = Gladiolus saccatus（Klatt）Goldblatt et M. P. de Vos ■☆

292321 Petamenes floribunda（Salisb.）E. Phillips = Chasmanthe floribunda（Salisb.）N. E. Br. ■☆

292322 Petamenes fucata（Herb.）E. Phillips = Crocosmia fucata（Herb.）M. P. de Vos ■☆

292323 Petamenes gracilis（N. E. Br.）E. Phillips = Gladiolus saccatus（Klatt）Goldblatt et M. P. de Vos ■☆

292324 Petamenes guthriei（L. Bolus）N. E. Br. = Gladiolus overbergensis Goldblatt et M. P. de Vos ■☆

292325 Petamenes huillensis（Welw. ex Baker）N. E. Br. = Gladiolus huillensis（Welw. ex Baker）Goldblatt ■☆

292326 Petamenes intermedia（Baker）E. Phillips = Tritoniopsis intermedia（Baker）Goldblatt ■☆

292327 Petamenes latifolius N. E. Br. = Gladiolus abyssinicus（Brongn. ex Lem.）Goldblatt et M. P. de Vos ■☆

292328 Petamenes magnificus（Harms）R. C. Foster = Gladiolus magnificus（Harms）Goldblatt ■☆

292329 Petamenes peglerae（N. E. Br.）E. Phillips = Chasmanthe aethiopica（L.）N. E. Br. ■☆

292330 Petamenes pilosus（Klatt）Goldblatt = Gladiolus bonaspei Goldblatt et M. P. de Vos ■☆

292331 Petamenes saccatus（Klatt）E. Phillips = Gladiolus saccatus（Klatt）Goldblatt et M. P. de Vos ■☆

292332 Petamenes schweinfurthii（Baker）N. E. Br. = Gladiolus schweinfurthii（Baker）Goldblatt et M. P. de Vos ■☆

292333 Petamenes spectabilis（Schinz）E. Phillips = Gladiolus magnificus（Harms）Goldblatt ■☆

292334 Petamenes splendens（Sweet）E. Phillips = Gladiolus splendens（Sweet）Herb. ■☆

292335 Petamenes steingroeveri（Pax）E. Phillips = Gladiolus saccatus（Klatt）Goldblatt et M. P. de Vos ■☆

292336 Petamenes vaginifer Milne-Redh. = Gladiolus huillensis（Welw. ex Baker）Goldblatt ■☆

292337 Petamenes vittigera（Salisb.）E. Phillips = Chasmanthe aethiopica（L.）N. E. Br. ■☆

292338 Petamenes zambesiacus（Baker）N. E. Br. = Gladiolus magnificus（Harms）Goldblatt ■☆

292339 Petasachme Wall. ex Wight = Pentasacme Wall. ex Wight et Arn. ■

292340 Petasioides Vitman = Petesioides Jacq. ●☆

292341 Petasioides Vitman = Wallenia Sw.（保留属名）●☆

292342 Petasites Mill.（1754）；蜂斗菜属（蜂斗叶属，款冬属）；Butter Bur, Butterbur, Butter-bur, Coltsfoot, Petasites, Winter Heliotrope ■

292343 Petasites alaskanus Rydb. = Petasites frigidus（L.）Fr. ■☆

292344 Petasites albus（L.）Gaertn.；白花蜂斗菜■☆

292345 Petasites albus A. Gray = Petasites japonicus（Siebold et Zucc.）Maxim. ■

292346 Petasites amplus Kitam. = Petasites japonicus（Siebold et Zucc.）Maxim. subsp. giganteus（G. Nicholson）Kitam. ■☆

292347 Petasites arcticus A. E. Porsild = Petasites frigidus（L.）Fr. var. palmatus（Aiton）Cronquist ■☆

292348 Petasites arcticus Porsild = Petasites frigidus（L.）Fr. ■☆

292349 Petasites corymbosus（R. Br.）Rydb. = Petasites frigidus（L.）Fr. ■☆

292350 Petasites dentatus Blank. = Petasites frigidus（L.）Fr. var. sagittatus（Banks ex Pursh）Chern. ■☆

292351 Petasites fominii Bords.；福明蜂斗菜■☆

292352 Petasites formosanus Kitam.；台湾蜂斗菜（山菊，台湾款冬）；Taiwan Butterbur ■

292353 Petasites fragrans（Vill.）C. Presl；香蜂斗菜（香蜂斗叶）；Guernsey Coltsfoot, Ploughman's Spikenard, Rhubarb-weed, Sweet Colsfoot, Sweet Coltsfoot, Winter Heliotriope, Winter-heliotriope ■☆

292354 Petasites frigidus（L.）Fr.；冷地蜂斗菜；Arctic Butterbur, Arctic Sweet Coltsfoot, Arctic Sweet-colt's-foot, Northern Sweet-colt's-foot ■☆

292355 Petasites frigidus（L.）Fr. subsp. arcticus（Porsild）Cody = Petasites frigidus（L.）Fr. ■☆

292356 Petasites frigidus（L.）Fr. subsp. palmatus（Aiton）Cody =

Petasites frigidus（L.）Fr. ■☆

292357　Petasites frigidus（L.）Fr. var. hyperboreoides Hultén = Petasites frigidus（L.）Fr. ■☆

292358　Petasites frigidus（L.）Fr. var. nivalis（Greene）Cronquist = Petasites frigidus（L.）Fr. ■☆

292359　Petasites frigidus（L.）Fr. var. palmatus（Aiton）Cronquist;掌叶冷蜂斗菜;Western Sweet Coltsfoot ■☆

292360　Petasites frigidus（L.）Fr. var. sagittatus（Banks ex Pursh）Chern.;箭叶冷蜂斗菜;Arrowhead Sweet Coltsfoot, Arrowhead Sweet-colt's-foot, Arrowleaf Sweet Coltsfoot, Arrow-leaf Sweet-colt's-foot ■☆

292361　Petasites frigidus（L.）Fr. var. sagittatus（Banks ex Pursh）Chern. = Petasites sagittatus（Banks ex Pursh）A. Gray ■☆

292362　Petasites frigidus（L.）Fr. var. vitifolius（Greene）Chern.;葡萄叶冷蜂斗菜■☆

292363　Petasites frigidus Fr. = Petasites frigidus（L.）Fr. ■☆

292364　Petasites frigidus Fr. subsp. arcticus（A. E. Porsild）Cody = Petasites frigidus（L.）Fr. var. palmatus（Aiton）Cronquist ■☆

292365　Petasites georgicus Manden.;乔治蜂斗菜■☆

292366　Petasites giganteus G. Nicholson = Petasites japonicus（Siebold et Zucc.）Maxim. subsp. giganteus（G. Nicholson）Kitam. ■☆

292367　Petasites glacialis（Ledeb.）Polunin;冰雪蜂斗菜■☆

292368　Petasites gmelinii（Turcz. ex DC.）Polunin;格氏蜂斗菜■☆

292369　Petasites gracilis Britton = Petasites frigidus（L.）Fr. ■☆

292370　Petasites himailacus Kitam. = Petasites tricholobus Franch. ■

292371　Petasites hookerianus（Nutt.）Rydb. = Petasites frigidus（L.）Fr. ■☆

292372　Petasites hybridus（L.）Gaertn.,Mey. et Scherb.;杂种蜂斗菜（欧蜂斗菜,杂种蜂斗叶,紫蜂斗菜）;Batter Dock, Batter-dock, Blatterdock, Bog Horn, Bog Rhubarb, Butter Dock, Butterbur, Butterbur Coltsfoot, Butterdock, Cap Docken, Cap-docken, Cleat, Cleat-leaves,Close Sciences,Clote,Coses Sciences, Devil's Rhubarb, Dummies, Early Mushroom, Eidin Docken, Elden, Elden Eldin, Eldin, Eldin-docken, Ell Docken, Ell-docken, Flapper Dock, Flatterbaw, Flea Dock, Gallon, Gaund, Gypsy's Rhibarb, Gypsy's Rhubarb,Kadle Dock, Kettle Dock, Kettle-case, Lagwort, Pestilence Wort,Pestilence-wort,Pestilentwort, Plague Flower, Poison Rhubarb, Purple Butterbur, Snake's Food, Snake's Rhubarb, Son-before-the-father, Sweet-scented Coltsfoot, Turkey Rhubarb, Umbrella Plant, Umbrella-leaves,Wild Rhubarb ■☆

292373　Petasites hyperboreus Rydb. = Petasites frigidus（L.）Fr. ■☆

292374　Petasites japonicus（Siebold et Zucc.）Maxim.;蜂斗菜(八角亭,白蜂斗菜,白蜂斗叶,白花蜂斗菜,毒解草,蜂斗叶,黑南瓜,葫芦包叶,葫芦叶,蔛,南瓜七,南瓜三七,蛇头草,水斗叶,水钟流头,网丝皮,野饭瓜,野金瓜头,野南瓜,钻冻);Creamy Butterbur, Giant Butterbur, Japan Butterbur, Japanese Butterbur, Japanese Coltsfoot,Japanese Sweet Coltsfoot,White Butterbur ■

292375　Petasites japonicus（Siebold et Zucc.）Maxim. f. glabrus Sugim.;光蜂斗菜(宽大蜂斗菜)■☆

292376　Petasites japonicus（Siebold et Zucc.）Maxim. f. purpurascens Makino;紫蜂斗菜■☆

292377　Petasites japonicus（Siebold et Zucc.）Maxim. subsp. giganteus（G. Nicholson）Kitam.;巨蜂斗菜■☆

292378　Petasites japonicus（Siebold et Zucc.）Maxim. var. giganteus G. Nicholson = Petasites japonicus（Siebold et Zucc.）Maxim. subsp. giganteus（G. Nicholson）Kitam. ■☆

292379　Petasites laevigatus（Willd.）Rchb.;平滑蜂斗菜■☆

292380　Petasites liukiuensis Kitam. = Petasites japonicus（Siebold et Zucc.）Maxim. ■

292381　Petasites mairei H. Lév. = Petasites tricholobus Franch. ■

292382　Petasites nivalis Greene = Petasites frigidus（L.）Fr. ■☆

292383　Petasites nivalis Greene subsp. hyperboreus（Rydb.）J. Toman = Petasites frigidus（L.）Fr. ■☆

292384　Petasites nivalis Greene subsp. vitifolius（Greene）J. Toman = Petasites frigidus（L.）Fr. var. vitifolius（Greene）Chern. ■☆

292385　Petasites niveus（Vill.）Baumg. = Petasites paradoxus（Retz.）Baumg. ■☆

292386　Petasites officinalis Moench;药用蜂斗菜;Purple Butterbur ■☆

292387　Petasites palmatus（Aiton）A. Gray = Petasites frigidus（L.）Fr. var. palmatus（Aiton）Cronquist ■☆

292388　Petasites palmatus（Aiton）A. Gray = Petasites frigidus（L.）Fr. ■☆

292389　Petasites palmatus（Aiton）A. Gray = Petasites tatewakianus Kitam. ■

292390　Petasites palmatus（Aiton）A. Gray subsp. speciosus（Nutt.）Toman = Petasites frigidus（L.）Fr. var. palmatus（Aiton）Cronquist ■☆

292391　Petasites paradoxus（Retz.）Baumg.;奇异蜂斗菜■☆

292392　Petasites petelotii（Merr.）Kitam. = Petasites tricholobus Franch. ■

292393　Petasites pyrenaicus（L.）G. López = Petasites fragrans（Vill.）C. Presl ■☆

292394　Petasites rubellus（J. F. Gmel.）J. Toman;长白蜂斗菜(长白蜂斗叶)■

292395　Petasites sagittatus（Banks ex Pursh）A. Gray = Petasites frigidus（L.）Fr. var. sagittatus（Banks ex Pursh）Chern. ■☆

292396　Petasites saxatilis（Turcz.）Kom. = Petasites rubellus（J. F. Gmel.）J. Toman ■

292397　Petasites saxatilis Kom. = Petasites rubellus（J. F. Gmel.）J. Toman ■

292398　Petasites saxatilis Turcz. ex A. DC. = Petasites rubellus（J. F. Gmel.）J. Toman ■

292399　Petasites speciosus（Nutt.）Piper = Petasites frigidus（L.）Fr. var. palmatus（Aiton）Cronquist ■☆

292400　Petasites speciosus（Nutt.）Piper = Petasites frigidus（L.）Fr. ■☆

292401　Petasites spurius（Retz.）Rchb. f.;假蜂斗叶菜(假蜂斗菜,毛裂蜂斗菜)■☆

292402　Petasites spurius Miq. = Petasites japonicus（Siebold et Zucc.）Maxim. ■

292403　Petasites tatewakianus Kitam.;掌叶蜂斗菜;Japanese Coltsfoot, Palmate Butterbur,Palmate Sweet-coltsfoot,Palmleaf Butterbur,Sweet Coltsfoot ■

292404　Petasites tricholobus Franch.;毛裂蜂斗菜(冬花,蜂斗菜,旱荷叶,葫芦叶);Hairylobed Butterbur ■

292405　Petasites tricholobus Franch. = Petasites formosanus Kitam. ■

292406　Petasites trigonophyllus Greene = Petasites vitifolius Greene ■☆

292407　Petasites vanioti H. Lév. = Petasites tricholobus Franch. ■

292408　Petasites versipilus Hand.-Mazz.;盐源蜂斗菜;Yanyuan Butterbur ■

292409　Petasites vitifolius Greene;葡萄叶蜂斗菜;Sweet-colt's-foot ■☆

292410　Petasites vitifolius Greene = Petasites frigidus（L.）Fr. var. vitifolius（Greene）Chern. ■☆

292411　Petasites warrenii H. St. John = Petasites frigidus（L.）Fr. var. vitifolius（Greene）Chern. ■☆

292412　Petasites warrenii H. St. John　= Petasites vitifolius Greene ■☆

292413　Petasitis Mill. = Petasites Mill. ■

292414　Petasostylis Griseb. = Leianthus Griseb. ●☆

292415　Petastoma Miers = Arrabidaea DC. ●☆

292416　Petasula Noronha = Schfflera J. R. Forst. et G. Forst. + Trevesia Vis. ●

292417　Petchia Livera(1926);佩奇木属(皮氏木属)●☆

292418　Petchia africana Leeuwenb.;非洲佩奇木(非洲皮氏木)●☆

292419　Petchia ceylanica（Wight）Livera;锡兰佩奇木●☆

292420　Petchia cryptophlebia（Baker）Leeuwenb.;隐脉佩奇木●☆

292421　Petchia erythrocarpa（Vatke）Leeuwenb.;红果佩奇木●☆

292422　Petchia humbertii（Markgr.）Leeuwenb.;亨伯特佩奇木●☆

292423　Petchia madagascariensis（A. DC.）Leeuwenb.;马岛佩奇木●☆

292424　Petchia montana（Pichon）Leeuwenb.;山地佩奇木●☆

292425　Petchia plectaneiifolia（Pichon）Leeuwenb.;织叶佩奇木●☆

292426　Petelotia Gagnep. = Petelotiella Gagnep. ■

292427　Petelotiella Gagnep.（1929）;越南麻属■☆

292428　Petelotiella tonkinensis Gagnep.;越南麻■☆

292429　Petelotoma DC. = Carallia Roxb.（保留属名）●

292430　Petenaea Lundell(1962);危地马拉椴属●☆

292431　Petenaea cordata Lundell;危地马拉椴●☆

292432　Peteniodendron Lundell = Pouteria Aubl. ●

292433　Peteravenia R. M. King et H. Rob.（1971）;光瓣亮泽兰属■●☆

292434　Peteria A. Gray(1852);彼得豆属■☆

292435　Peteria Raf. = Rondeletia L. ●

292436　Peteria glandulosa（A. Gray ex S. Watson）Rydb.;腺点彼得豆■☆

292437　Peteria glandulosa Rydb. = Peteria glandulosa（A. Gray ex S. Watson）Rydb. ■☆

292438　Peteria scoparia A. Gray;彼得豆■☆

292439　Petermannia F. Muell.（1860）（保留属名）;刺藤属（花须藤属）■●☆

292440　Petermannia Klotzsch = Begonia L. ●■

292441　Petermannia Rchb.（废弃属名）= Cycloloma Moq. ■☆

292442　Petermannia Rchb.（废弃属名）= Petermannia F. Muell.（保留属名）●■☆

292443　Petermannia cirrosa F. Muell.;刺藤●■☆

292444　Petermanniaceae Hutch.（1934）（保留科名）;刺藤科●■☆

292445　Petermanniaceae Hutch.（保留科名）= Petiveriaceae C. Agardh ■☆

292446　Petermanniaceae Hutch.（保留科名）= Philesiaceae Dumort.（保留科名）●■☆

292447　Peterodendron Sleumer(1936);彼得木属●☆

292448　Peterodendron ovatum（Sleumer）Sleumer;彼得木●☆

292449　Petersia Klotzsch = Capparis L. ●

292450　Petersia Welw. ex Benth. = Petersianthus Merr. ●☆

292451　Petersia Welw. ex Benth. et Hook. f.（1865）;攀木属●☆

292452　Petersia Welw. ex Benth. et Hook. f. = Combretodendron A. Chev. ex Exell ●☆

292453　Petersia Welw. ex Benth. et Hook. f. = Petersianthus Merr. ●☆

292454　Petersia africana Welw. ex Benth. et Hook. f.;非洲攀木●☆

292455　Petersia africana Welw. ex Benth. et Hook. f. = Petersianthus macrocarpus（P. Beauv.）Liben ●☆

292456　Petersia klainei A. Chev. = Petersianthus macrocarpus（P. Beauv.）Liben ●☆

292457　Petersia minor Nied. = Petersianthus macrocarpus（P. Beauv.）Liben ●☆

292458　Petersia rosea Klotzsch;粉红攀木●☆

292459　Petersia rosea Klotzsch = Capparis erythrocarpos Isert var. rosea（Klotzsch）DeWolf ●☆

292460　Petersia viridiflora（A. Chev.）A. Chev. = Petersianthus macrocarpus（P. Beauv.）Liben ●☆

292461　Petersia viridiflora A. Chev.;绿花攀木●☆

292462　Petersianthus Merr.（1916）;彼得斯玉蕊属●☆

292463　Petersianthus Merr. = Combretodendron A. Chev. ex Exell ●☆

292464　Petersianthus Merr. = Petersia Welw. ex Benth. et Hook. f. ●☆

292465　Petersianthus africanus（Welw. ex Benth. et Hook. f.）Merr. = Petersianthus macrocarpus（P. Beauv.）Liben ●☆

292466　Petersianthus macrocarpus（P. Beauv.）Liben;彼得斯玉蕊;Essia ●☆

292467　Petesia P. Browne = Rondeletia L. ●

292468　Petesiodes Kuntze = Petesioi Jacq. ●☆

292469　Petesiodes Kuntze = Wallenia Sw.（保留属名）●☆

292470　Petesioi Jacq. = Wallenia Sw.（保留属名）●☆

292471　Petesioides Jacq. = Wallenia Sw.（保留属名）●☆

292472　Petesioides Jacq. ex Kuntze = Wallenia Sw.（保留属名）●☆

292473　Petilium Ludw. = Fritillaria L. ■

292474　Petilium eduardi（A. Regel）Vved. = Fritillaria eduardi Regel ■☆

292475　Petilium imperiale J. St. -Hil. = Fritillaria imperialis L. ■☆

292476　Petiniotia J. Léonard = Sterigmostemum M. Bieb. ■

292477　Petiniotia J. Léonard(1980);紫色芥属■☆

292478　Petinotia purpurascens（Boiss.）Léonard;紫色芥■☆

292479　Petitia J. Gay = Xatardia Meisn. et Zeyh. ■☆

292480　Petitia Jacq.（1832）;珀蒂草属●☆

292481　Petitia Neck. = Hibiscus L.（保留属名）●■

292482　Petitia domingensis Jacq.;珀蒂草;Fiddlewood ●☆

292483　Petitiocodon Robbr.（1988）;珀蒂茜属●☆

292484　Petitiocodon parviflorum（Keay）Robbr.;珀蒂茜●☆

292485　Petitmenginia Bonati(1911);钟山草属（毛冠四蕊草属）;Petitmenginia,Zhongshangrass ■

292486　Petitmenginia comosa Bonati;滇毛冠四蕊草■

292487　Petitmenginia matsumurae T. Yamaz.;钟山草;Nanjing Petitmenginia,Nanking Petitmenginia,Zhongshangrass ■

292488　Petiveria L.（1753）;毛头独子属（叉果商陆属,蒜臭母鸡草属）■●☆

292489　Petiveria Plum. ex L. = Petiveria L. ■●☆

292490　Petiveria alliacea L.;毛头独子（叉果商陆,蒜臭母鸡草）;Guinea Hen Weed,Guinea-hen-weed,Gully-root ●☆

292491　Petiveria sanguinea H. Walter;红毛头独子（血红蒜臭母鸡草）■☆

292492　Petiveriaceae C. Agardh = Phytolaccaceae R. Br.（保留科名）●■

292493　Petiveriaceae C. Agardh(1825);毛头独子科（蒜臭母鸡草科）■☆

292494　Petkovia Stef. = Campanula L. ■●

292495　Petlomelia Nieuwl. = Fraxinus L. ●

292496　Petopentia Bullock = Tacazzea Decne. ●☆

292497　Petopentia Bullock(1954);肖塔卡萝藦属●☆

292498　Petopentia natalensis（Schltr.）Bullock = Ischnolepis natalensis（Schltr.）Venter ■☆

292499　Petracanthus Nees = Gymnostachyum Nees ■

292500　Petradoria Greene(1895);岩黄花属;Rock Goldenrod ●☆

292501　Petradoria discoidea（H. M. Hall）L. C. Anderson = Cuniculotinus gramineus（H. M. Hall）Urbatsch, R. P. Roberts et Neubig ●☆

292502　Petradoria graminea Wooton et Standl. = Petradoria pumila（Nutt.）Greene var. graminea（Wooton et Standl.）S. L. Welsh ■☆

292503　Petradoria pumila（Nutt.）Greene;小岩黄花;Rock Goldenrod ■☆

292504　Petradoria pumila（Nutt.）Greene subsp. graminea（Wooton et Standl.）L. C. Anderson ＝ Petradoria pumila（Nutt.）Greene var. graminea（Wooton et Standl.）S. L. Welsh ■☆

292505　Petradoria pumila（Nutt.）Greene var. graminea（Wooton et Standl.）S. L. Welsh；禾叶小岩黄花；Grass-leaved Rock Goldenrod ■☆

292506　Petradosia T. Durand et Jacks. ＝ Petradoria Greene ■☆

292507　Petraea B. Juss. ex Juss. ＝ Petrea L. ●

292508　Petraea Juss. ＝ Petrea L. ●

292509　Petraeomyrtus Craven ＝ Melaleuca L.（保留属名）●

292510　Petraeomyrtus Craven（1999）；澳千层属●☆

292511　Petraeovitex Oliv.（1883）；比得牡荆属●☆

292512　Petraeovitex multifora（Sm.）Merr.；比得牡荆●☆

292513　Petramnia Raf. ＝ Pentarhaphia Lindl. ■☆

292514　Petrantha DC. ＝ Riencourtia Cass. ■☆

292515　Petrantha DC. ＝ Tetrantha Poit. ■☆

292516　Petranthe Salisb. ＝ Scilla L. ■

292517　Petrea L.（1753）；蓝花藤属（紫霞藤属）；Blueflowervine, Petrea, Purple Wreath, Queen's Wreath ●

292518　Petrea volubilis L.；蓝花藤（紫霞藤）；Purple Wreath, Purple Wreath Petrea, Purple Wreath Vine, Queen's Wreath, Sandpaper Vine, Star-flower Ver Vine, Starflower Vine, Twisting Blueflowervine, Twisting Petrea ●

292519　Petreaceae J. Agardh ＝ Verbenaceae J. St. -Hil.（保留科名）●■

292520　Petreaceae J. Agardh；蓝花藤科●

292521　Petriella Zotov ＝ Ehrharta Thunb.（保留属名）■☆

292522　Petrina J. B. Phipps ＝ Danthoniopsis Stapf ■☆

292523　Petrina lignosa（C. E. Hubb.）J. B. Phipps ＝ Danthoniopsis lignosa C. E. Hubb. ■☆

292524　Petrina parva J. B. Phipps ＝ Danthoniopsis parva（J. B. Phipps）Clayton ■☆

292525　Petrina pruinosa（C. E. Hubb.）J. B. Phipps ＝ Danthoniopsis pruinosa C. E. Hubb. ■☆

292526　Petrobium Bong. ＝ Lithobium Bong. ☆

292527　Petrobium R. Br.（1817）（保留属名）；岩菊木属；Roek Beauty ●☆

292528　Petrobium arboreum R. Br.；岩菊木●☆

292529　Petrocallis R. Br. ＝ Petrocallis W. T. Aiton ■☆

292530　Petrocallis W. T. Aiton（1812）；岩丽芥属（岩美草属）；Rock Beauty ■☆

292531　Petrocallis pyrenaica（L.）R. Br.；岩丽芥■☆

292532　Petrocallis pyrenaica（L.）R. Br. ＝ Draba pyrenaica L. ■☆

292533　Petrocarvi Tausch ＝ Athamanta L. ■☆

292534　Petrocarvi Tausch ＝ Killinga Adans.（废弃属名）■

292535　Petrocarya Schreb. ＝ Parinari Aubl. ●☆

292536　Petrocodon Hance（1883）；石山苣苔属（石钟花属）；Petrocodon ■★

292537　Petrocodon dealbatus Hance；石山苣苔（石钟花）；Common Petrocodon ■

292538　Petrocodon dealbatus Hance var. denticulatus（W. T. Wang）W. T. Wang；齿缘石山苣苔■

292539　Petrocodon denticulatus W. T. Wang ＝ Petrocodon dealbatus Hance var. denticulatus（W. T. Wang）W. T. Wang ■

292540　Petrocodon ferrugineus Y. G. Wei；锈色石山苣苔■

292541　Petrocodon longistylus Kraenzl. ＝ Petrocodon dealbatus Hance ■

292542　Petrocoma Rupr. ＝ Silene L.（保留属名）■

292543　Petrocoptis A. Braun ＝ Petrocoptis A. Braun ex Endl. ■☆

292544　Petrocoptis A. Braun ex Endl.（1842）；岩黄连属（岩剪秋箩属）；Rocky Lychnis ■☆

292545　Petrocoptis lagascae Willk.；岩黄连（岩剪秋箩）；Rocky Lychnis ■☆

292546　Petrocoptis pyrenaica（Berger）A. Br.；南欧岩黄连（南欧岩剪秋箩）；South European Rocky Lychnis ■☆

292547　Petrocosmea Oliv.（1887）；石蝴蝶属；Petrocosmea, Stonebutterfly ■

292548　Petrocosmea barbata Craib；髯毛石蝴蝶（昆明石蝴蝶）；Barbate Petrocosmea, Barbate Stonebutterfly ■

292549　Petrocosmea begoniifolia C. Y. Wu ex H. W. Li；秋海棠叶石蝴蝶；Begonialeaf Petrocosmea, Begonialeaf Stonebutterfly ■

292550　Petrocosmea cavaleriei H. Lév.；贵州石蝴蝶；Guizhou Petrocosmea, Guizhou Stonebutterfly ■

292551　Petrocosmea cavaleriei H. Lév. ＝ Petrocosmea martinii（H. Lév.）H. Lév. ■

292552　Petrocosmea coerulea C. Y. Wu ex W. T. Wang；蓝石蝴蝶（巴瑞石蝴蝶）；Blue Petrocosmea, Blue Stonebutterfly, Parry Petrocosmea ■

292553　Petrocosmea confluens W. T. Wang；江药石蝴蝶；Coanther Petrocosmea, Coanther Stonebutterfly ■

292554　Petrocosmea duclouxii Craib；石蝴蝶；Common Petrocosmea, Common Stonebutterfly ■

292555　Petrocosmea flaccida Craib；萎软石蝴蝶；Soft Petrocosmea, Soft Stonebutterfly ■

292556　Petrocosmea forrestii Craib；大理石蝴蝶；Dali Petrocosmea, Dali Stonebutterfly ■

292557　Petrocosmea grandiflora Hemsl.；大花石蝴蝶；Largeflower Petrocosmea, Largeflower Stonebutterfly ■

292558　Petrocosmea grandifolia W. T. Wang；大叶石蝴蝶；Bigleaf Petrocosmea, Bigleaf Stonebutterfly ■

292559　Petrocosmea henryi Craib ＝ Petrocosmea minor Hemsl. ■

292560　Petrocosmea iodioides Hemsl.；蒙自石蝴蝶；Mengzi Petrocosmea, Mengzi Stonebutterfly ■

292561　Petrocosmea ionantha（H. Wendl.）Rodigas ＝ Saintpaulia ionantha H. Wendl. ■☆

292562　Petrocosmea kerii Craib；滇泰石蝴蝶（凯尔石蝴蝶）；Kerr Petrocosmea, Kerr Stonebutterfly ■

292563　Petrocosmea kerii Craib var. crinita W. T. Wang；绵毛石蝴蝶；Hairy Kerr Petrocosmea ■

292564　Petrocosmea latisepala W. T. Wang ＝ Petrocosmea oblata Craib var. latisepala（W. T. Wang）W. T. Wang ■

292565　Petrocosmea longipedicellata W. T. Wang；长梗石蝴蝶；Longstalk Petrocosmea, Longstalk Stonebutterfly ■

292566　Petrocosmea mairei H. Lév.；东川石蝴蝶；E. Sichuan Petrocosmea, E. Sichuan Stonebutterfly ■

292567　Petrocosmea mairei H. Lév. var. intraglabra W. T. Wang；会东石蝴蝶；Huidong Petrocosmea ■

292568　Petrocosmea martinii（H. Lév.）H. Lév.；滇黔石蝴蝶；Martin Petrocosmea, Martin Stonebutterfly ■

292569　Petrocosmea martinii（H. Lév.）H. Lév. var. leiandra W. T. Wang；光蕊滇黔石蝴蝶；Smoothsepal Martin Petrocosmea ■

292570　Petrocosmea menglianensis H. W. Li；孟连石蝴蝶；Menglian Petrocosmea, Menglian Stonebutterfly ■

292571　Petrocosmea minor Hemsl.；小石蝴蝶；Mini Petrocosmea, Mini Stonebutterfly ■

292572　Petrocosmea nervosa Craib；显脉石蝴蝶；Nerved Petrocosmea, Nerved Stonebutterfly ■

292573　Petrocosmea oblata Craib；扁圆石蝴蝶；Flatrund Petrocosmea, Flatrund Stonebutterfly ■

292574 Petrocosmea oblata Craib var. latisepala（W. T. Wang）W. T. Wang；宽萼石蝴蝶；Broadsepale Petrocosmea ■

292575 Petrocosmea parryorum C. E. C. Fisch. = Petrocosmea coerulea C. Y. Wu ex W. T. Wang ■

292576 Petrocosmea peltata Merr. et Chun = Metapetrocosmea peltata W. T. Wang ■☆

292577 Petrocosmea qinlingensis W. T. Wang；秦岭石蝴蝶；Qinling Petrocosmea，Qinling Stonebutterfly ■

292578 Petrocosmea rosettifolia C. Y. Wu ex H. W. Li；莲座石蝴蝶；Rosetterfly Petrocosmea，Rosetterfly Stonebutterfly ■

292579 Petrocosmea sericea C. Y. Wu ex H. W. Li；丝毛石蝴蝶；Silkyhair Petrocosmea，Silkyhair Stonebutterfly ■

292580 Petrocosmea sichuanensis Chun ex W. T. Wang；四川石蝴蝶；Sichuan Petrocosmea，Sichuan Stonebutterfly ■

292581 Petrocosmea sinensis Oliv.；中华石蝴蝶（石头花，石花）；China Stonebutterfly，Chinese Petrocosmea ■

292582 Petrocosmea sinensis Oliv. = Petrocosmea qinlingensis W. T. Wang ■

292583 Petrocosmea wardii W. W. Sm. = Petrocosmea kerii Craib ■

292584 Petrodavisia Holub = Centaurea L.（保留属名）●■

292585 Petrodavisia Holub = Ptosimopappus Boiss. ●■

292586 Petrodora Fourr. = Veronica L. ■

292587 Petrodoxa J. Anthony = Beccarinda Kuntze ■

292588 Petrodoxa argentea Anthony = Beccarinda argentea（Anthony）B. L. Burtt ■

292589 Petroedmondia Tamamsch.（1987）；爱石芹属■☆

292590 Petroedmondia syriaca（Boiss.）Tamamsch.；爱石芹■☆

292591 Petrogenia I. M. Johnst.（1941）；石旋花属●☆

292592 Petrogenia repens I. M. Johnst.；石旋花●☆

292593 Petrogenla I. M. Johnst. = Bonamia Thouars（保留属名）●☆

292594 Petrogeton alpinum Eckl. et Zeyh. = Crassula umbellata Thunb. ■☆

292595 Petrogeton nemorosum Eckl. et Zeyh. = Crassula nemorosa（Eckl. et Zeyh.）Endl. ex Walp. ●☆

292596 Petrogeton nivale Eckl. et Zeyh. = Crassula nemorosa（Eckl. et Zeyh.）Endl. ex Walp. ●☆

292597 Petrogeton patens Eckl. et Zeyh. = Crassula dentata Thunb. ■☆

292598 Petrogeton typicum Eckl. et Zeyh. = Crassula dentata Thunb. ■☆

292599 Petrogeton typicum Eckl. et Zeyh. var. minus ？ = Crassula dentata Thunb. ■☆

292600 Petrogeton umbella（Jacq.）Eckl. et Zeyh. = Crassula umbella Jacq. ●☆

292601 Petrogrton Eckl. et Zeyh. = Crassula L. ●■☆

292602 Petrollinia Chiov. = Inula L. ●■

292603 Petrollinia heteromalla（Vatke）Chiov. = Inula mannii（Hook. f.）Oliv. et Hiern ■☆

292604 Petromarula Nieuwl. et Lunell = Lobelia L. ●■

292605 Petromarula Vent. ex R. Hedw.（1806）；石苣草属■☆

292606 Petromarula pinnata（L.）A. DC.；石苣草■☆

292607 Petromecon Greene = Eschscholzia Cham. ■

292608 Petronia Barb. Rodr. = Batemannia Lindl. ■☆

292609 Petronia Jungh. = Pteronia L.（保留属名）●☆

292610 Petronymphe H. E. Moore（1951）；石葱属■☆

292611 Petronymphe decora H. E. Moore；石葱■☆

292612 Petrophila R. Br. = Petrophile R. Br. ex Knight ●☆

292613 Petrophila biloba R. Br. = Petrophile biloba R. Br. ●☆

292614 Petrophila linearis R. Br. = Petrophile linearis R. Br. ●☆

292615 Petrophila media R. Br. = Petrophile media R. Br. ●☆

292616 Petrophila pulchella（Schrad. et J. C. Wendl.）R. Br. = Petrophile pulchella（Schrad. et J. C. Wendl.）R. Br. ●☆

292617 Petrophila sessilis Sieber ex Schult. = Petrophile sessilis Sieber ex Schult. ●☆

292618 Petrophila shuttleworthiana Meisn. = Petrophile shuttleworthiana Meisn. ●☆

292619 Petrophile Knight = Petrophile R. Br. ex Knight ●☆

292620 Petrophile R. Br. = Petrophile R. Br. ex Knight ●☆

292621 Petrophile R. Br. ex Knight（1809）；石龙眼属（彼得费拉属）；Combbush，Cone Sticks ●☆

292622 Petrophile biloba R. Br.；二列叶石龙眼（二列叶彼得费拉）；Twinlobe Combbush ●☆

292623 Petrophile linearis R. Br.；厚叶石龙眼（厚叶彼得费拉）；Pixie Mops ●☆

292624 Petrophile media R. Br.；圆叶石龙眼（圆叶彼得费拉）；Roundleaf Combbush ●☆

292625 Petrophile pulchella（Schrad. et J. C. Wendl.）R. Br.；美丽石龙眼；Cone Sticks ●☆

292626 Petrophile sessilis Sieber ex Schult.；刺叶石龙眼（刺叶彼得费拉）●☆

292627 Petrophile shuttleworthiana Meisn.；梳状叶石龙眼（梳状叶彼得费拉）；Shuttleworth Combbush ●☆

292628 Petrophiloides Bowerb. ex Reid et Chandler = Platycarya Siebold et Zucc. ●

292629 Petrophiloides Reid et Chandler = Platycarya Siebold et Zucc. ●

292630 Petrophiloides Reid et Chandler（1933）；假石龙眼属●

292631 Petrophiloides strobilacea（Siebold et Zucc.）Reid et Chandler var. kawakamii（Hayata）Kaneh. = Platycarya strobilacea Siebold et Zucc. ●

292632 Petrophyes Webb et Berthel. = Monanthes Haw. ■☆

292633 Petrophyes agriostachys Webb et Berthel. = Monanthes agriostaphys（Webb et Berthel.）Christ ■☆

292634 Petrophyes brachycaulon Webb et Berthel. = Monanthes brachycaulon（Webb et Berthel.）Lowe ■☆

292635 Petrophyes microbotrys Bolle et Webb = Monanthes laxiflora Bolle var. microbotrys（Bolle et Webb）Burch. ■☆

292636 Petrophyes minimum Bolle = Monanthes minima（Bolle）Christ ■☆

292637 Petrophyes murale Bolle = Monanthes muralis（Bolle）Christ ■☆

292638 Petrophyes palens Christ = Monanthes pallens（H. Christ）H. Christ ■☆

292639 Petrophyes polyphyllum（Haw.）Webb et Berthel. = Monanthes polyphylla Haw. ■☆

292640 Petrophyes purpurescens Bolle et Webb = Monanthes purpurascens（Bolle et Webb）H. Christ ■☆

292641 Petrophyton Rydb. = Petrophytum（Nutt. ex Torr. et A. Gray）Rydb. ●☆

292642 Petrophyton caespitosum（Nutt.）Rydb. = Petrophytum caespitosum（Nutt.）Rydb. ●☆

292643 Petrophyton cinerascens Rydb. = Petrophytum cinerascens Rydb. ●☆

292644 Petrophyton hendersonii（Canby）Rydb. = Petrophytum hendersonii（Canby）Rydb. ●☆

292645 Petrophytum（Nutt. ex Torr. et A. Gray）Rydb.（1900）（'Petrophyton'）；岩绣线菊属；Rock Mat，Rock-mat ●☆

292646 Petrophytum（Nutt.）Rydb. = Petrophytum（Nutt. ex Torr. et A. Gray）Rydb. ●☆

292647 Petrophytum（Torr. et A. Gray）Rydb. = Petrophytum（Nutt. ex

Torr. et A. Gray）Rydb. ●☆

292648　Petrophytum caespitosum（Nutt.）Rydb.；岩绣线菊●☆

292649　Petrophytum cinerascens Rydb.；灰叶岩绣线菊●☆

292650　Petrophytum hendersonii（Canby）Rydb.；三脉岩绣线菊●☆

292651　Petroravenia Al-Shehbaz(1994)；石拉芥属■☆

292652　Petroravenia eseptata Al-Shehbaz；石拉芥■☆

292653　Petrorchis D. L. Jones et M. A. Clem.（2002）；双角石兰属■☆

292654　Petrorchis D. L. Jones et M. A. Clem. = Pterostylis R. Br.（保留属名）■☆

292655　Petrorhagia（Ser. ex DC.）Link = Petrorhagia（Ser.）Link ■

292656　Petrorhagia（Ser.）Link（1831）；膜萼花属（裙花属，洋石竹属）；Coat Flower,Filmcalyx,Pink,Tunic Flower,Tunicflower,Tunic-flower ■

292657　Petrorhagia Link = Petrorhagia（Ser.）Link ■

292658　Petrorhagia alpina（Hablitz）P. W. Ball et Heywood；直立膜萼花；Erect Filmcalyx,Erect Tunicflower ■

292659　Petrorhagia cyrenaica（E. A. Durand et Barratte）P. W. Ball；昔兰尼膜萼花■☆

292660　Petrorhagia dubia（Raf.）G. López et Romo；可疑膜萼花；Childing Pink,Hairypink ■☆

292661　Petrorhagia dubia（Raf.）G. López et Romo = Dianthus dubius Raf.■☆

292662　Petrorhagia illyrica（Ard.）P. W. Ball et Heywood；伊里利亚膜萼花■☆

292663　Petrorhagia illyrica（Ard.）P. W. Ball et Heywood subsp. angustifolia（Poir.）P. W. Ball et Heywood；窄叶伊里利亚膜萼花■☆

292664　Petrorhagia nanteulii（Burnat）P. W. Ball et Heywood；南氏膜萼花；Childing Pink,Proliferous Pink ■☆

292665　Petrorhagia prolifera（L.）P. W. Ball et Heywood；多育膜萼花；Chiiding Sweet William,Childing Pink,Little-leaved Tunic-flower,Proliferous Pink ■☆

292666　Petrorhagia rhiphaea（Pau et Font Quer）P. W. Ball et Heywood；山地膜萼花■☆

292667　Petrorhagia rupestris Brullo et Furnari；岩生膜萼花■☆

292668　Petrorhagia saxifraga（L.）Link；膜萼花（裙花,外套花,洋石竹）；Coat Flower,Filmcalyx,Saxifrage Pink,Saxifrage Tunicflower,Tunic Flower,Tunic-flower ■

292669　Petrorhagia saxifraga（L.）Link ‘Rosette’；莲座裙花■☆

292670　Petrorhagia velutina（Guss.）P. W. Ball et Heywood = Petrorhagia dubia（Raf.）G. López et Romo ■☆

292671　Petrosavia Becc.（1871）；无叶莲属（樱井草属）；Leaflesslotus,Petrosavia ■

292672　Petrosavia sakurai（Makino）Dandy = Petrosavia sakurai（Makino）J. J. Sm. ex Steenis ■

292673　Petrosavia sakurai（Makino）J. J. Sm. ex Steenis；疏花无叶莲（无叶莲,樱井草）；Laxflower Leaflesslotus,Laxflower Petrosavia ■

292674　Petrosavia sinii（K. Krause）Gagnep.；无叶莲；Leaflesslotus,Sin Petrosavia ■

292675　Petrosavia sinii（K. Krause）Gagnep. = Petrosavia sakurai（Makino）J. J. Sm. ex Steenis ■

292676　Petrosaviaceae Hutch.（1934）（保留科名）；无叶莲科（樱井草科）■

292677　Petrosaviaceae Hutch.（保留科名）= Melanthiaceae Batsch ex Borkh.（保留科名）■

292678　Petrosciadium Edgew. = Eriocycla Lindl. ■

292679　Petrocladium Edgew. = Pimpinella L. ■

292680　Petrosedum Grulich = Sedum L. ●■

292681　Petrosedum amplexicaule（DC.）Velayos = Sedum amplexicaule DC.■☆

292682　Petrosedum forsterianum（Sm.）Grulich = Sedum forsterianum Sm.■☆

292683　Petrosedum sediforme（Jacq.）Grulich = Sedum sediforme（Jacq.）Pau ■☆

292684　Petrosedum tenuiflorum（Sm.）Grulich = Sedum amplexicaule DC. subsp. tenuifolium（Sm.）Greuter ■☆

292685　Petroselinum A. W. Hill(1756)；欧芹属；Parsley ■

292686　Petroselinum crispum（Mill.）A. W. Hill ‘Tuberosum’ = Petroselinum crispum（Mill.）Nyman ex A. W. Hill ‘Tuberosum’■☆

292687　Petroselinum crispum（Mill.）A. W. Hill = Petroselinum crispum（Mill.）Nyman ex A. W. Hill ■

292688　Petroselinum crispum（Mill.）A. W. Hill var. filicinum ? = Petroselinum crispum（Mill.）Nyman ex A. W. Hill var. filicinum ? ■☆

292689　Petroselinum crispum（Mill.）A. W. Hill var. latifolium ? = Petroselinum crispum（Mill.）Nyman ex A. W. Hill var. latifolium ? ■☆

292690　Petroselinum crispum（Mill.）A. W. Hill var. neapolitanum Danert = Petroselinum crispum（Mill.）Nyman ex A. W. Hill var. neapolitanum Danert ■☆

292691　Petroselinum crispum（Mill.）Fuss subsp. tuberosum（Bernh. ex Rchb.）Soó = Petroselinum crispum（Mill.）A. W. Hill ‘Tuberosum’■☆

292692　Petroselinum crispum（Mill.）Fuss var. angustifolium（Hayne）Reduron = Petroselinum crispum（Mill.）Nyman ex A. W. Hill var. angustifolium（Hayne）Reduron ■☆

292693　Petroselinum crispum（Mill.）Nyman = Petroselinum crispum（Mill.）Nyman ex A. W. Hill ■

292694　Petroselinum crispum（Mill.）Nyman ex A. W. Hill；欧芹（庭园欧芹,皱叶欧芹,皱叶石蛇床）；Common Garden Parsley,Curled Parsley,Devil's Oatmeal,Garden Parsley,Parsley,Percely,Persele,Persll,Wrinkleleaf Parsley ■

292695　Petroselinum crispum（Mill.）Nyman ex A. W. Hill ‘Tuberosum’；块根欧芹；Hamburg Parsley,Turnip-rooted Parsley ■☆

292696　Petroselinum crispum（Mill.）Nyman ex A. W. Hill var. angustifolium（Hayne）Reduron；窄叶欧芹■☆

292697　Petroselinum crispum（Mill.）Nyman ex A. W. Hill var. filicinum ?；蕨叶欧芹；Fern-leaved Parsley ■☆

292698　Petroselinum crispum（Mill.）Nyman ex A. W. Hill var. latifolium ?；阔叶欧芹；Common Garden Parsley ■☆

292699　Petroselinum crispum（Mill.）Nyman ex A. W. Hill var. latifolium ? = Petroselinum crispum（Mill.）Nyman ex A. W. Hill ■

292700　Petroselinum crispum（Mill.）Nyman ex A. W. Hill var. neapolitanum Danert；意大利欧芹；Italian Parsley ■☆

292701　Petroselinum crispum（Mill.）Nyman ex A. W. Hill var. radicosum（Alef.）Danert；放射欧芹；Turnip Garden Parsley,Turnip-rooted Parsley ■☆

292702　Petroselinum crispum（Mill.）Nyman var. giganteum（Pau）Maire = Petroselinum crispum（Mill.）A. W. Hill ■

292703　Petroselinum crispum（Mill.）Nyman var. gracillimum Maire = Petroselinum crispum（Mill.）Nyman ex A. W. Hill ■

292704　Petroselinum crispum（Mill.）Nyman var. latifolium ? = Petroselinum crispum（Mill.）Nyman ex A. W. Hill ■

292705　Petroselinum filiforme A. Rich.；镰状欧芹；Hamburg Parsley,Turnip-rooted Parsley ■☆

292706　Petroselinum hortense Hoffm. = Petroselinum crispum（Mill.）Nyman ex A. W. Hill ■

292707　Petroselinum hortense Hoffm. var. crispum（Mill.）L. H. Bailey
＝Petroselinum crispum（Mill.）Nyman ex A. W. Hill ■

292708　Petroselinum hortense Hoffm. var. crispum L. H. Bailey ＝
Petroselinum crispum（Mill.）Nyman ex A. W. Hill ■

292709　Petroselinum humile Meisn. ＝ Sonderina humilis（Meisn.）H.
Wolff ■☆

292710　Petroselinum sativum Hoffm. ＝ Petroselinum crispum（Mill.）
Nyman ex A. W. Hill ■

292711　Petroselinum segetum W. D. J. Koch；玉米地欧芹；Corn
Carraway, Corn Honewort, Corn Parsley ■☆

292712　Petroselinum segetum W. D. J. Koch ＝ Carum segetum Benth. et
Hook. f. ■☆

292713　Petrosilene Fourr. ＝ Silene L.（保留属名）■

292714　Petrosimonia Bunge（1862）；叉毛蓬属；Petrosimonia ■

292715　Petrosimonia brachiata（Pall.）Bunge；短叉毛蓬■☆

292716　Petrosimonia brachyphylla（Bunge）Iljin；短叶叉毛蓬■☆

292717　Petrosimonia crassifolia（Pall.）Bunge；厚叶叉毛蓬■☆

292718　Petrosimonia crassifolia（Pall.）Bunge ＝ Petrosimonia
oppositifolia（Pall.）Litv. ■

292719　Petrosimonia crassifolia（Pall.）Bunge var. glaucenscens Bunge
＝Petrosimonia glaucescens（Bunge）Iljin ■

292720　Petrosimonia glauca（Pall.）Bunge；灰绿叉毛蓬■☆

292721　Petrosimonia glaucescens（Bunge）Iljin；变灰绿叉毛蓬（灰蓝
叉毛蓬）；Glaucescent Petrosimonia ■

292722　Petrosimonia hirsutissima（Bunge）Iljin；多毛叉毛蓬■☆

292723　Petrosimonia litwinowii Korsh.；平卧叉毛蓬；Litwinov
Petrosimonia ■

292724　Petrosimonia monandra（Pall.）Bunge；单蕊叉毛蓬■☆

292725　Petrosimonia oppositifolia（Pall.）Litv.；对生叶叉毛蓬（短苞
叉毛蓬）；Shortbract Petrosimonia ■

292726　Petrosimonia sibirica（Pall.）Bunge；叉毛蓬；Siberia
Petrosimonia ■

292727　Petrosimonia squarrosa（Schrenk）Bunge；粗糙叉毛蓬；
Scabrous Petrosimonia ■

292728　Petrosimonia triandra（Pall.）Simonk.；三蕊叉毛蓬■☆

292729　Petrostylis Pritz. ＝ Pterostylis R. Br.（保留属名）■☆

292730　Petrotheca Steud. ＝ Tetratheca Sm. ●☆

292731　Petrusia Baill. ＝ Tetraena Maxim. ●★

292732　Petrusia Baill. ＝ Zygophyllum L. ●■

292733　Petrusia madagascariensis Baill. ＝ Tetraena madagascariensis
（Baill.）Beier et Thulin ■☆

292734　Pettera Rchb.（废弃属名）＝ Arenaria L. ■

292735　Pettera Rchb.（废弃属名）＝ Petteria C. Presl（保留属名）●☆

292736　Petteria C. Presl（1845）（保留属名）；巴尔干豆属；Dalmatian
Laburnum ●☆

292737　Petteria ramentacea（Sieber）C. Presl；巴尔干豆（毒豆）；
Dalmatian Laburnum ●☆

292738　Petteria ramentacea C. Presl ＝ Petteria ramentacea（Sieber）C.
Presl ■☆

292739　Pettospermum Roxb. ＝ Pittosporum Banks ex Gaertn.（保留属
名）●

292740　Petunga DC. ＝ Hypobathrum Blume ●☆

292741　Petunia Juss.（1803）（保留属名）；碧冬茄属（矮牵牛属）；
Petunia ■

292742　Petunia atkinsiana D. Don ex Loudon；普通碧冬茄；Common
Petunia, Garden Petunia ■☆

292743　Petunia axillaris（Lam.）Britton, Sterns et Poggenb. ＝ Petunia

nyctuginiflora A. Juss. ■☆

292744　Petunia hybrida（Hook. f.）Vilm. ＝ Petunia hybrida Vilm. ■

292745　Petunia hybrida Vilm.；碧冬茄（矮牵牛，彩花茄，灵芝牡丹，
杂种碧冬茄,杂种撞羽朝颜）；Common Garden Petunia, Common
Petunia, Petunia, Water Docken ■

292746　Petunia hybrida Vilm. ＝ Petunia atkinsiana D. Don ex Loudon ■☆

292747　Petunia hybrida Vilm. var. pendula ?；层叠碧冬茄；Balcony
Petunia, Cascade Petunia ■☆

292748　Petunia inflata R. E. Fr.；大花矮牵牛■☆

292749　Petunia integrifolia（Hook.）Schinz et Thell.；全缘叶碧冬茄；
Entireleaf Petunia, Violetflower Petunia ■☆

292750　Petunia minima Reiche ＝ Combera minima（Reiche）Sandwith ■☆

292751　Petunia nyctuginiflora A. Juss.；腋生碧冬茄（大白花矮牵牛，
腋花牵牛）；Axillary Petunia, Large White Petunia, Night-flowering
Petunia, White Petunia, White-moon Petunia ■☆

292752　Petunia occidentalis R. E. Fr.；西方碧冬茄■☆

292753　Petunia parviflora Juss.；小花碧冬茄■☆

292754　Petunia violacea Lindl.；紫花矮牵牛（矮牵牛）；Violet
Petunia, Violet-flowered Petunia ■☆

292755　Petunia violacea Lindl. var. hybrida Hook. f. ＝ Petunia hybrida
（Hook. f.）Vilm. ■

292756　Peuce Rich. ＝ Picea A. Dietr. ＋ Abies Mill. ＋ Cedrus Trew ＋
Larix Mill. ●

292757　Peucedanon Raf. ＝ Peucedanum L. ■

292758　Peucedanum L.（1753）；前胡属（石防风属）；Hogfennel, Hog's
Fennel, Hog's-Fennel ■

292759　Peucedanum abbreviatum E. Mey. ex Meisn. ＝ Glia prolifera
（Burm. f.）B. L. Burtt ■☆

292760　Peucedanum abbreviatum E. Mey. ex Sond. ＝ Peucedanum
camdebooense B. L. Burtt ■☆

292761　Peucedanum aberdarense H. Wolff ＝ Peucedanum friesiorum H.
Wolff var. bipinnatum C. C. Towns. ■☆

292762　Peucedanum aberdaricum Chiov. ＝ Peucedanum linderi C.
Norman ■☆

292763　Peucedanum abyssinicum Vatke；阿比西尼亚前胡■☆

292764　Peucedanum acaule R. H. Shan et M. L. Sheh；会泽前胡；Huize
Hogfennel ■

292765　Peucedanum adae Woronow；阿达前胡■☆

292766　Peucedanum alsaticum L.；亚尔萨斯前胡■☆

292767　Peucedanum altum Hiern ＝ Peucedanum petitianum A. Rich. ■☆

292768　Peucedanum ampliatum K. T. Fu；天竺山前胡；Tianzhushan
Hogfennel ■

292769　Peucedanum anethum Baill. ＝ Anethum graveolens L. ■

292770　Peucedanum angelicoides H. Wolff ex Kretschmer；芷叶前胡；
Angelicalike Hogfennel ■

292771　Peucedanum angolense（Welw. ex Ficalho）Cannon ＝ Lefebvrea
grantii（Hiern）S. Droop ■☆

292772　Peucedanum angustisectum（Engl.）C. Norman；细毛前胡■☆

292773　Peucedanum araliaceum（Hochst.）Benth. et Hook. f. ＝
Steganotaenia araliacea Hochst. ■☆

292774　Peucedanum arenarium Waldst. et Kit.；沙生前胡（洋前胡）■☆

292775　Peucedanum articulatum C. C. Towns.；关节前胡■☆

292776　Peucedanum atlanticum Pomel ＝ Imperatoria hispanica Boiss. ■☆

292777　Peucedanum atropurpureum（Steud. ex A. Rich.）Hiern ＝
Erythroselinum atropurpureum（Steud. ex A. Rich.）Chiov. ■☆

292778　Peucedanum aucheri Boiss.；奥切尔前胡■☆

292779　Peucedanum austriacum Koch；奥地利前胡；Austrian Hogfennel ■☆

292780　Peucedanum baicalense（I. Redowsky ex Willd.）W. D. J. Koch;兴安前胡（草原石防风,兴安石防风）;Baical Hogfennel, Xing'an Hogfennel ■

292781　Peucedanum baicalense（I. Redowsky）C. Koch = Peucedanum baicalense（I. Redowsky ex Willd.）W. D. J. Koch ■

292782　Peucedanum baicalense（I. Redowsky）K. Koch = Peucedanum baicalense（I. Redowsky ex Willd.）W. D. J. Koch ■

292783　Peucedanum benguelense（Engl.）Eyles = Lefebvrea grantii（Hiern）S. Droop ■☆

292784　Peucedanum bequaertii C. Norman = Physotrichia muriculata（Hiern）S. Droop et C. C. Towns. ■☆

292785　Peucedanum bojerianum Baker = Tana bojeriana（Baker）B. -E. van Wyk ■☆

292786　Peucedanum boninense（Tuyama）Tuyama = Angelica japonica A. Gray var. boninensis（Tuyama）T. Yamaz. ■☆

292787　Peucedanum borysthenicum Klokov;第聂伯前胡■☆

292788　Peucedanum brachystylum（Hiern）Drude = Lefebvrea brachystyla Hiern ■☆

292789　Peucedanum buchananii Baker = Lefebvrea grantii（Hiern）S. Droop ■☆

292790　Peucedanum bupleuriforme H. Wolff = Seseli mairei H. Wolff ■

292791　Peucedanum bupleuroides H. Wolff = Seseli mairei H. Wolff ■

292792　Peucedanum caespitosum H. Wolff;北京前胡;Beijing Hogfennel ■

292793　Peucedanum caffrum（Meisn.）E. Phillips;开菲尔前胡■☆

292794　Peucedanum calcareum Albov;石灰前胡■☆

292795　Peucedanum camdebooense B. L. Burtt;卡姆德布前胡■☆

292796　Peucedanum camerunense Jacq. -Fél.;喀麦隆前胡■☆

292797　Peucedanum canaliculatum Verdc. = Peucedanum harmsianum H. Wolff subsp. australis C. C. Towns. ■☆

292798　Peucedanum caneroonsum M. Hiroe = Lefebvrea grantii（Hiern）S. Droop ■☆

292799　Peucedanum canescens Ledeb. = Ferula canescens（Ledeb.）Ledeb. ■

292800　Peucedanum capense（Eckl. et Zeyh.）D. Dietr. = Peucedanum polyactinum B. L. Burtt ■☆

292801　Peucedanum capense（Thunb.）Sond.;好望角前胡●■☆

292802　Peucedanum capense（Thunb.）Sond. var. lanceolatum Sond.;披针形好望角前胡●■☆

292803　Peucedanum capense（Thunb.）Sond. var. latifolium Sond. = Peucedanum capense（Thunb.）Sond. ●■☆

292804　Peucedanum capensoides Sales et Hedge;拟好望角前胡■☆

292805　Peucedanum capillaceum Thunb.;纤毛前胡■☆

292806　Peucedanum capillaceum Thunb. var. rigidum Sond.;硬毛前胡■☆

292807　Peucedanum cartilaginomarginata Makino ex Nakagawa = Angelica cartilaginomarginata（Makino ex Y. Yabe）Nakai ■

292808　Peucedanum cartilaginomarginatum Makino ex Y. Yabe = Angelica cartilaginomarginata（Makino ex Y. Yabe）Nakai ■

292809　Peucedanum caspicum（M. Bieb.）Link = Ferula caspica M. Bieb. ■

292810　Peucedanum caucasicum K. Koch;高加索前胡■☆

292811　Peucedanum cavaleriei H. Wolff = Ligusticum brachylobum Franch. ■

292812　Peucedanum cervaria（L.）Guss. = Peucedanum cervaria（L.）Lapeyr. ■☆

292813　Peucedanum cervaria（L.）Lapeyr.;鹿芹前胡;Much-good ■☆

292814　Peucedanum cervariifolium C. A. Mey.;鹿芹叶前胡■☆

292815　Peucedanum chinense M. Hiroe;林地前胡;Woodland Hogfennel ■

292816　Peucedanum claessensii C. Norman;克莱森斯前胡☆

292817　Peucedanum clematidifolium（C. Norman）M. Hiroe = Pseudocarum eminii（Engl.）H. Wolff ■☆

292818　Peucedanum collinum（Eckl. et Zeyh.）D. Dietr. = Peucedanum striatum（Thunb.）Sond. ■☆

292819　Peucedanum condensatum（L.）Koso-Pol. = Libanotis condensata Crantz ■

292820　Peucedanum connatum E. Mey. ex Sond. = Peucedanum caffrum（Meisn.）E. Phillips ■☆

292821　Peucedanum coreanum Nakai;朝鲜前胡■☆

292822　Peucedanum crucifolium（Kom.）H. Boissieu = Angelica cartilaginomarginata（Makino）ex Y. Yabe Nakai var. matsumurae（H. Boissieu）Kitag. ■

292823　Peucedanum cynorhiza Sond. = Peucedanum typicum（Eckl. et Zeyh.）B. L. Burtt ■☆

292824　Peucedanum dasycarpum Regel et Schmalh.;毛果前胡■

292825　Peucedanum decumbens Maxim.;倾卧前胡;Decumbent Hogfennel ■

292826　Peucedanum decursivum（Miq.）Maxim. = Angelica decursiva（Miq.）Franch. et Sav. ■

292827　Peucedanum decursivum（Miq.）Maxim. var. albiflora Maxim. = Angelica decursiva（Miq.）Franch. et Sav. var. albiflora（Maxim.）Nakai ■

292828　Peucedanum dehoideum Makino ex Y. Yabe = Peucedanum terebinthaceum（Fisch. ex Trevir.）Fisch. ex Turcz. var. deltoideum（Makino ex Y. Yabe）Makino ■

292829　Peucedanum delavayi Franch.;滇西前胡（丽江前胡）;W. Yunnan Hogfennel ■

292830　Peucedanum deltoideum Makino ex Y. Yabe;三角叶前胡■

292831　Peucedanum deltoideum Makino ex Y. Yabe = Peucedanum terebinthaceum（Fisch. ex Trevir.）Fisch. ex Turcz. var. deltoideum（Makino ex Y. Yabe）Makino ■

292832　Peucedanum deltoideum Makino ex Y. Yabe = Peucedanum terebinthaceum（Fisch. ex Trevir.）Fisch. ex Turcz. ■

292833　Peucedanum dielsianum Fedde ex H. Wolff;竹节前胡（川防风,竹节防风）;Diels Hogfennel ■

292834　Peucedanum dinteri H. Wolff = Peucedanum upingtoniae（Schinz）Drude ■☆

292835　Peucedanum dissectum（C. H. Wright）Dawe = Peucedanum kerstenii Engl. ■☆

292836　Peucedanum dissectum Ledeb. = Ferula dissecta（Ledeb.）Ledeb. ■

292837　Peucedanum dissolutum（Diels）H. Wolff;南川前胡;S. Sichuan Hogfennel ■

292838　Peucedanum diversifolium H. Wolff = Peucedanum chinense M. Hiroe ■

292839　Peucedanum doctoris C. Norman = Peucedanum scottianum Engl. ■☆

292840　Peucedanum dregeanum D. Dietr.;德雷前胡■☆

292841　Peucedanum ecklonianum Sond. = Peucedanum dregeanum D. Dietr. ■☆

292842　Peucedanum ecklonis Walp. = Peucedanum capillaceum Thunb. var. rigidum Sond. ■☆

292843　Peucedanum elegans Kom.;刺尖前胡（刺尖石防风,雅致前

胡）；Elegant Hogfennel ■

292844 Peucedanum elgonense H. Wolff；埃尔贡前胡■☆

292845 Peucedanum elliotii Engl. = Afroligusticum elliotii（Engl.）C. Norman ■☆

292846 Peucedanum elongatum E. Mey. ex Meisn. = Peucedanum ferulaceum（Thunb.）Eckl. et Zeyh. ■☆

292847 Peucedanum eminii Engl. = Pseudocarum eminii（Engl.）H. Wolff ■☆

292848 Peucedanum englerianum H. Wolff；恩格勒前胡■☆

292849 Peucedanum eylesii C. Norman；艾尔斯前胡■☆

292850 Peucedanum falcaria Turcz.；镰叶前胡；Sikleleaf Hogfennel ■

292851 Peucedanum farinosum Hook.；被粉前胡；Biscuit-root ■☆

292852 Peucedanum ferulaceum（Thunb.）Eckl. et Zeyh.；阿魏前胡■☆

292853 Peucedanum ferulaceum（Thunb.）Eckl. et Zeyh. var. stadense（Eckl. et Zeyh.）Sond.；施塔德前胡■☆

292854 Peucedanum ferulifolium Gilli；杖叶前胡■☆

292855 Peucedanum feruloides Steud. = Ferula ferulaeoides（Steud.）Korovin ■

292856 Peucedanum filicinum H. Wolff = Conioselinum chinense（L.）Britton，Sterns et Poggenb. ■

292857 Peucedanum formosanum Hayata；台湾前胡；Taiwan Hogfennel ■

292858 Peucedanum franchetii C. Y. Wu et F. T. Pu；异叶前胡；Differentleaf Hogfennel ■

292859 Peucedanum fraxinifolium Oliv. = Steganotaenia araliacea Hochst. ■☆

292860 Peucedanum friesiorum H. Wolff；弗里斯前胡■☆

292861 Peucedanum friesiorum H. Wolff var. bipinnatum C. C. Towns.；羽状弗里斯前胡■☆

292862 Peucedanum galbanum（L.）Drude；热非前胡■☆

292863 Peucedanum giraldii Diels. = Tongoloa dunnii（H. Boissieu）H. Wolff ■

292864 Peucedanum glaucum Hook. f. et Thomson ex C. B. Clarke = Peucedanum natalense（Sond.）Engl. ■☆

292865 Peucedanum gossweileri C. Norman；戈斯前胡■☆

292866 Peucedanum gracile Ledeb. = Ferula gracilis（Ledeb.）Ledeb. ■

292867 Peucedanum grande C. B. Clarke；大前胡■☆

292868 Peucedanum grandifolioides H. Wolff = Angelica decursiva（Miq.）Franch. et Sav. ■

292869 Peucedanum grantii Hiern = Lefebvrea grantii（Hiern）S. Droop ■☆

292870 Peucedanum graveolens（L.）Benth. et Hook. f. = Anethum graveolens L. ■

292871 Peucedanum graveolens（L.）Hiern = Anethum graveolens L. ■

292872 Peucedanum graveolens Benth. et Hook. f. = Anethum graveolens L. ■

292873 Peucedanum guangxiense R. H. Shan et M. L. Sheh；广西前胡（土防风）；Guangxi Hogfennel ■

292874 Peucedanum gummiferum（L.）Wijnands；产胶前胡■☆

292875 Peucedanum harmsianum H. Wolff；哈姆斯前胡■☆

292876 Peucedanum harmsianum H. Wolff subsp. australis C. C. Towns.；南方哈姆斯前胡■☆

292877 Peucedanum harry-smithii Fedde ex H. Wolff；华北前胡（毛白花前胡）；N. China Hogfennel ■

292878 Peucedanum harry-smithii Fedde ex H. Wolff var. grande（K. T. Fu）R. H. Shan et M. L. Sheh；广序北前胡■

292879 Peucedanum harry-smithii Fedde ex H. Wolff var. subglabrum（R. H. Shan et M. L. Sheh）R. H. Shan et M. L. Sheh；少毛华北前胡■

292880 Peucedanum henryi H. Wolff；鄂西前胡；Henry Hogfennel ■

292881 Peucedanum heracleoides Baker = Physotrichia muriculata（Hiern）S. Droop et C. C. Towns. ■☆

292882 Peucedanum herbidum M. Hiroe = Peucedanum harmsianum H. Wolff ■☆

292883 Peucedanum heterophyllum Franch. = Peucedanum franchetii C. Y. Wu et F. T. Pu ■

292884 Peucedanum hirsutiusculum（Ma）V. M. Vinogr. = Peucedanum harry-smithii Fedde ex H. Wolff ■

292885 Peucedanum hirsutiusculum（Ma）V. M. Vinogr. var. subglabrum R. H. Shan et M. L. Sheh = Peucedanum harry-smithii Fedde ex H. Wolff var. subglabrum（R. H. Shan et M. L. Sheh）R. H. Shan et M. L. Sheh ■

292886 Peucedanum hispanicum（Boiss.）Endl. = Imperatoria hispanica Boiss. ■☆

292887 Peucedanum hispanicum（Boiss.）Endl. var. atlanticum（Pomel）Maire = Imperatoria hispanica Boiss. ■☆

292888 Peucedanum hissaricum Korovin；希萨尔前胡■☆

292889 Peucedanum hockii C. Norman = Steganotaenia hockii（C. Norman）C. Norman ■☆

292890 Peucedanum hypoleucum（Meisn.）Drude = Peucedanum tenuifolium Thunb. ■☆

292891 Peucedanum jaeschkeanum（Vatke）Baill. = Ferula jaeschkeana Vatke ■

292892 Peucedanum japonicum Thunb.；滨海前胡（防葵，房苑，牡丹发疯，日本前胡）；Japan Hogfennel，Japanese Hogfennel ■

292893 Peucedanum japonicum Thunb. f. purpurascens（Makino）Honda；紫滨海前胡■☆

292894 Peucedanum japonicum Thunb. var. australe M. Hotta et A. Seo；南方滨海前胡■☆

292895 Peucedanum japonicum Thunb. var. latifolium M. Hotta et Shiuchi；宽叶滨海前胡■☆

292896 Peucedanum karataviense Regel et Schmalh. = Ferula karataviensis（Regel et Schmalh.）Korovin ex Pavlov ■

292897 Peucedanum kerstenii Engl.；克斯滕前胡■☆

292898 Peucedanum khamiesbergense B. L. Burtt；卡米前胡■☆

292899 Peucedanum kingaense Engl. = Physotrichia muriculata（Hiern）S. Droop et C. C. Towns. ■☆

292900 Peucedanum kingdon-wardii（H. Wolff）Korovin = Ferula kingdon-wardii H. Wolff ■

292901 Peucedanum kirungae（Engl.）M. Hiroe = Heracleum abyssinicum（Boiss.）C. Norman ■☆

292902 Peucedanum kotschyi Boiss. = Peucedanum aucheri Boiss. ■☆

292903 Peucedanum kupense I. Darbysh. et Cheek；库普前胡■☆

292904 Peucedanum laevigatum（Aiton）H. Wolff = Peucedanum capense（Thunb.）Sond. var. lanceolatum Sond. ●■☆

292905 Peucedanum lancifoliolum（Mattf.）M. Hiroe = Peucedanum upingtoniae（Schinz）Drude ■☆

292906 Peucedanum lasiocarpum Boiss. = Platytaenia lasiocarpa（Boiss.）Rech. f. et Riedl ■☆

292907 Peucedanum lateriflorum（Eckl. et Zeyh.）Sond. = Annesorhiza lateriflora（Eckl. et Zeyh.）B. -E. van Wyk ■☆

292908 Peucedanum latifolium（M. Bieb.）DC.；宽叶前胡；Broadleaf Hogfennel ■☆

292909 Peucedanum ledebourielloides K. T. Fu；利氏前胡（华山前胡）■

292910 Peucedanum lefebvria Drude = Lefebvrea abyssinica A. Rich. ■☆

292911 Peucedanum lefebvrioides（Engl.）M. Hiroe = Erythroselinum atropurpureum（Steud. ex A. Rich.）Chiov. ■☆

292912　Peucedanum lhasense C. B. Clarke ex H. Wolff;拉萨前胡■

292913　Peucedanum linderi C. Norman;林德前胡■☆

292914　Peucedanum longipedicellatum（Engl.）Drude ＝ Lefebvrea longipedicellata Engl. ■☆

292915　Peucedanum longshengense R. H. Shan et M. L. Sheh;南岭前胡;Longsheng Hogfennel ■

292916　Peucedanum lubimenkoanum Kotov;柳比前胡■☆

292917　Peucedanum lundense Cannon;隆德前胡■☆

292918　Peucedanum luxurians Tamamsch.;茂盛前胡■☆

292919　Peucedanum lynesii Norman;莱恩斯前胡■☆

292920　Peucedanum macilentum Franch.;细裂前胡（瘦弱前胡）;Thinsplit Hogfennel ■

292921　Peucedanum malcolmii Hemsl. et Pearson ＝ Heracleum millefolium Diels ■

292922　Peucedanum mashanense R. H. Shan et M. L. Sheh;马山前胡（马山防风）;Mashan Hogfennel ■

292923　Peucedanum mattirolii Chiov.;马蒂奥里前胡■☆

292924　Peucedanum medicum Dunn;华中前胡（光头独活,土前胡,岩棕）;Central China Hogfennel ■

292925　Peucedanum medicum Dunn var. gracile Dunn ex R. H. Shan et M. L. Sheh;岩前胡（光头前胡）■

292926　Peucedanum megaphyllum（Diels）H. Boissieu ＝ Angelica megaphylla Diels ■

292927　Peucedanum meisnerianum（A. Br.）H. Scholz ＝ Peucedanum caffrum（Meisn.）E. Phillips ■☆

292928　Peucedanum melanotilingia（H. Boissieu）H. Boissieu ＝ Angelica decursiva（Miq.）Franch. et Sav. ■

292929　Peucedanum microphyllum Sm.;小叶前胡■☆

292930　Peucedanum mildbraedii H. Wolff ＝ Peucedanum kerstenii Engl. ■☆

292931　Peucedanum millefolium Sond.;粟草叶前胡■☆

292932　Peucedanum miqueliana H. Wolff ＝ Ostericum sieboldii（Miq.）Nakai ■

292933　Peucedanum miquelianum（Maxim.）H. Wolff ＝ Ostericum sieboldii（Miq.）Nakai ■

292934　Peucedanum montanum（Eckl. et Zeyh.）Druce ＝ Peucedanum dregeanum D. Dietr. ■☆

292935　Peucedanum montanum（Sond.）Drude ＝ Peucedanum tenuifolium Thunb. ■☆

292936　Peucedanum monticola C. Norman ＝ Peucedanum scottianum Engl. ■☆

292937　Peucedanum morisonii Besser ex Schult.;准噶尔前胡（莫氏前胡,胃寒草,细叶前胡）;Dzungar Hogfennel ■

292938　Peucedanum morrisonicola（Hayata）M. Hiroe ＝ Angelica morrisonicola Hayata ■

292939　Peucedanum morrisonicola（Hayata）M. Hiroe var. nanhutashanense（S. L. Liu, C. Y. Chao et T. I. Chuang）Q. X. Liu ＝ Angelica morrisonicola Hayata var. nanhutashanensis S. L. Liu, C. Y. Chao et T. I. Chuang ■

292940　Peucedanum multiradiatum Drude ＝ Peucedanum polyactinum B. L. Burtt ■☆

292941　Peucedanum multivittatum Cufod. ＝ Peucedanum harmsianum H. Wolff ■☆

292942　Peucedanum multivittatum Maxim.;白山前胡（白山防风）■☆

292943　Peucedanum multivittatum Maxim. f. dissectum Makino;深裂白山前胡■☆

292944　Peucedanum multivittatum Maxim. f. linearilobum（Tatew.）Ohwi;条裂白山前胡■☆

292945　Peucedanum multivittatum Maxim. var. linearilobum Tatew. ＝ Peucedanum multivittatum Maxim. f. linearilobum（Tatew.）Ohwi ■☆

292946　Peucedanum munbyi Boiss.;芒比前胡■☆

292947　Peucedanum muriculatum Hiern ＝ Physotrichia muriculata（Hiern）S. Droop et C. C. Towns. ■☆

292948　Peucedanum muriculatum Hiern var. goetzeanum Engl. ＝ Physotrichia muriculata（Hiern）S. Droop et C. C. Towns. ■☆

292949　Peucedanum naegeleanum（H. Wolff）M. Hiroe ＝ Lefebvrea longipedicellata Engl. ■☆

292950　Peucedanum nanum R. H. Shan et M. L. Sheh;矮前胡;Dwarf Hogfennel ■

292951　Peucedanum natalense（Sond.）Engl.;纳塔尔前胡■☆

292952　Peucedanum nigeriae（H. Wolff）M. Hiroe ＝ Lefebvrea grantii（Hiern）S. Droop ■☆

292953　Peucedanum normanii M. Hiroe ＝ Afroligusticum elliotii（Engl.）C. Norman ■☆

292954　Peucedanum nucliusculum Koso-Pol. ＝ Phlojodicarpus sibiricus（Stephan ex Spreng.）Koso-Pol. ■

292955　Peucedanum nyassicum H. Wolff;尼亚萨前胡■☆

292956　Peucedanum oblongisectum C. C. Towns.;矩圆前胡■☆

292957　Peucedanum officinale L.;药用前胡（欧前胡,欧洲前胡,药砜）;Bastwort, Brimstonewort, Cammock, Common Hog's-fennel, Dog Fennel, Harestrong, Harstrong, Hoarstrange, Hoarstrong, Hog's Fennel, Horestrange, Horestrong, Horse-strong, Marsh Hog's Fennel, Rastewort, Rostewort, Sea Hog's Fennel, Sea Sulphurwort, Sow Fennel, Spreusidany, Sulphur Root, Sulphurweed, Sulphurwort ■☆

292958　Peucedanum officinale L. subsp. vogelianum（Emb. et Maire）Maire ＝ Peucedanum vogelianum Emb. et Maire ■☆

292959　Peucedanum olifantianum（Koso-Pol.）M. Hiroe ＝ Peucedanum typicum（Eckl. et Zeyh.）B. L. Burtt ■☆

292960　Peucedanum olivaceum Diels ＝ Ferula olivacea（Diels）H. Wolff ex Hand.-Mazz. ■

292961　Peucedanum oreogelinum（L.）Moench;山生前胡;Mountain Parsley ■☆

292962　Peucedanum ostruthium（L.）C. Koch ＝ Imperatoria ostruthium L. ■☆

292963　Peucedanum ovinum Boiss. ＝ Ferula ovina（Boiss.）Boiss. ■

292964　Peucedanum paishanense Nakai ＝ Peucedanum terebinthaceum（Fisch. ex Trevir.）Fisch. ex Turcz. ■

292965　Peucedanum palimboides Boiss.;额尔齐斯芹前胡■☆

292966　Peucedanum palustre（L.）Moench;沼地前胡（欧洲亮蛇床,沼生亮蛇床）;Brimstonewort, Dog Fennel, Hogfennel, Hog's Fennel, Marsh Hog's Fennel, Marsh Parsley, Milk Parsley, Milkweed ■☆

292967　Peucedanum pastinaca（L.）Benth. et Hook. f. ＝ Pastinaca sativa L. ■

292968　Peucedanum paucifolium Ledeb.;寡叶前胡■☆

292969　Peucedanum pauciradiatum Tamamsch.;稀射线前胡■☆

292970　Peucedanum pearsonii Adamson;皮尔逊前胡■☆

292971　Peucedanum petitianum A. Rich.;佩蒂蒂前胡■☆

292972　Peucedanum petitianum A. Rich. var. kilimandscharicum Engl. ＝ Peucedanum winkleri H. Wolff ■☆

292973　Peucedanum piliferum Hand.-Mazz.;乳头前胡;Pipple Hogfennel ■

292974　Peucedanum platycarpum E. Mey. ex Sond.;阔果前胡■☆

292975　Peucedanum podolicum Eichw.;柄前胡■☆

292976　Peucedanum polyactinum B. L. Burtt；多射线前胡■☆

292977　Peucedanum polyanthum（Korovin）Korovin；多花前胡■☆

292978　Peucedanum polyphyllum Ledeb. = Peucedanum baicalense（I. Redowsky ex Willd.）W. D. J. Koch ■

292979　Peucedanum porphyroscias（Miq.）Makino = Angelica decursiva（Miq.）Franch. et Sav. ■

292980　Peucedanum porphyroscias（Miq.）Makino var. albiflorum（Maxim.）Makino = Angelica decursiva（Miq.）Franch. et Sav. f. albiflora（Maxim.）Nakai ■

292981　Peucedanum porphyroscias Makino = Angelica decursiva（Miq.）Franch. et Sav. ■

292982　Peucedanum praeruptorum Dunn；前胡（艾鹰爪根，白花前胡，长前胡，官前胡，光前胡，鸡脚前胡，山独活，小前胡，棕色前胡）；Hogfennel ■

292983　Peucedanum praeruptorum Dunn subsp. hirsutiusculum Ma；白毛花前胡；Hirsute Hogfennel ■

292984　Peucedanum praeruptorum Dunn subsp. hirsutiusculum Ma = Peucedanum harry-smithii Fedde ex H. Wolff ■

292985　Peucedanum praeruptorum Dunn var. grande K. T. Fu = Peucedanum harry-smithii Fedde ex H. Wolff var. grande（K. T. Fu）R. H. Shan et M. L. Sheh ■

292986　Peucedanum pricei N. D. Simpson；蒙古前胡；Mongol Hogfennel ■

292987　Peucedanum puberulum Turcz.；小毛前胡■☆

292988　Peucedanum pubescens Hand.-Mazz.；毛前胡；Hair Hogfennel ■

292989　Peucedanum pulchrum H. Wolff = Peucedanum turgeniifolium H. Wolff ■

292990　Peucedanum pungens E. Mey. ex Sond.；刚毛前胡■☆

292991　Peucedanum quarrei（C. Norman）M. Hiroe；卡雷前胡■☆

292992　Peucedanum renardii Regel et Schmalh.；雷氏前胡■☆

292993　Peucedanum reptans Diels = Ligusticum reptans（Diels）H. Wolff ■

292994　Peucedanum rhodesicum Cannon；罗得西亚前胡■☆

292995　Peucedanum rigidum Bunge = Ferula bungeana Kitag. ■

292996　Peucedanum rigidum E. Mey. = Peucedanum capense（Thunb.）Sond. ●■☆

292997　Peucedanum rigidum Eckl. et Zeyh. = Peucedanum capillaceum Thunb. var. rigidum Sond. ■☆

292998　Peucedanum rivae（Engl.）M. Hiroe = Heracleum abyssinicum（Boiss.）C. Norman ■☆

292999　Peucedanum rubricaule R. H. Shan et M. L. Sheh；红前胡；Red Hogfennel ■

293000　Peucedanum runssoricum Engl.；伦索前胡■☆

293001　Peucedanum ruspolii Engl. = Erythroselinum atropurpureum（Steud. ex A. Rich.）Chiov. ■☆

293002　Peucedanum ruthenicum M. Bieb.；俄国前胡（俄罗斯前胡）；Russia Hogfennel ■☆

293003　Peucedanum salinum Pall. ex Spreng.；盐地前胡■☆

293004　Peucedanum saposhnikoviodes K. T. Fu；华山前胡■

293005　Peucedanum sativum（L.）B. D. Jacks. = Pastinaca sativa L. ■

293006　Peucedanum saxicola（Makino ex Y. Yabe）Makino = Angelica saxicola Makino ex Y. Yabe ■☆

293007　Peucedanum schottii Besser ex DC.；修化前胡■☆

293008　Peucedanum scottianum Engl.；司科特前胡■☆

293009　Peucedanum serratum（H. Wolff）C. Norman = Peucedanum scottianum Engl. ■☆

293010　Peucedanum sieberianum Sond. = Peucedanum strictum（Spreng.）B. L. Burtt ■☆

293011　Peucedanum sieboldii Miq. = Ostericum sieboldii（Miq.）Nakai ■

293012　Peucedanum sikkimense C. B. Clarke = Arcuatopterus sikkimensis（C. B. Clarke）Pimenov et Ostroumova ■

293013　Peucedanum silaifolium Hiern = Peucedanum abyssinicum Vatke ■☆

293014　Peucedanum silaus L. = Silaum silaus（L.）Schinz et Thell. ■

293015　Peucedanum sonderi（M. Hiroe）B. L. Burtt = Peucedanum tenuifolium Thunb. ■☆

293016　Peucedanum songoricum G. Don = Peucedanum morisonii Besser ex Schult. ■

293017　Peucedanum songpanense R. H. Shan et F. T. Pu；松潘前胡；Songpan Hogfennel ■

293018　Peucedanum sowa（Roxb.）Kurz. = Anethum graveolens L. ■

293019　Peucedanum stadense Eckl. et Zeyh. = Peucedanum ferulaceum（Thunb.）Eckl. et Zeyh. var. stadense（Eckl. et Zeyh.）Sond. ■☆

293020　Peucedanum stenocarpum Boiss. et Reut. ex Boiss.；核果前胡■☆

293021　Peucedanum stenospermum C. C. Towns.；狭籽前胡■☆

293022　Peucedanum stepposum C. C. Huang；草原前胡（草原石防风）；Grassland Hogfennel ■

293023　Peucedanum stolzii（Engl. et H. Wolff）M. Hiroe = Heracleum abyssinicum（Boiss.）C. Norman ■☆

293024　Peucedanum striatum（Thunb.）Sond.；条纹前胡■☆

293025　Peucedanum strictum（Spreng.）B. L. Burtt；劲直前胡■☆

293026　Peucedanum stuhlmannii（Engl.）Drude = Lefebvrea abyssinica A. Rich. ■☆

293027　Peucedanum sulcatum Sond.；纵沟前胡■☆

293028　Peucedanum taquetii H. Wolff = Angelica polymorpha Maxim. ■

293029　Peucedanum taquetii H. Wolff = Ostericum grosseserratum（Maxim.）Kitag. ■

293030　Peucedanum tauricum M. Bieb.；克里木前胡；Klimu Hogfennel ■☆

293031　Peucedanum tenue C. C. Towns.；细前胡■☆

293032　Peucedanum tenuifolium Thunb.；细叶前胡■☆

293033　Peucedanum terebinthaceum（Fisch. ex Trevir.）Fisch. ex Turcz.；石防风（防风，前胡，山胡芹，山芹菜，山葵，山香菜，珊瑚菜，小芹菜，小叶芹幌子，岩防风，硬苗前胡）；Resin Hogfennel ■

293034　Peucedanum terebinthaceum（Fisch. ex Trevir.）Fisch. ex Turcz. f. aciculare（T. Inoue）H. Ohba；针状石防风■☆

293035　Peucedanum terebinthaceum（Fisch. ex Trevir.）Fisch. ex Turcz. subsp. formosanum Kitag. = Peucedanum formosanum Hayata ■

293036　Peucedanum terebinthaceum（Fisch. ex Trevir.）Fisch. ex Turcz. var. aciculare T. Inoue = Peucedanum terebinthaceum（Fisch. ex Trevir.）Fisch. ex Turcz. f. aciculare（T. Inoue）H. Ohba ■☆

293037　Peucedanum terebinthaceum（Fisch. ex Trevir.）Fisch. ex Turcz. var. deltoideum（Makino ex Y. Yabe）Makino = Peucedanum terebinthaceum（Fisch. ex Trevir.）Fisch. ex Turcz. ■

293038　Peucedanum terebinthaceum（Fisch. ex Trevir.）Fisch. ex Turcz. var. deltoideum（Makino ex Y. Yabe）Makino；宽叶石防风（风芹）；Broadleaf Resin Hogfennel ■

293039　Peucedanum terebinthaceum（Fisch. ex Trevir.）Fisch. ex Turcz. var. flagellare Nakai = Peucedanum terebinthaceum（Fisch. ex Trevir.）Fisch. ex Turcz. ■

293040　Peucedanum terebinthaceum（Fisch. ex Trevir.）Fisch. ex Turcz. var. paishanense（Nakai）C. C. Huang = Peucedanum terebinthaceum（Fisch. ex Trevir.）Fisch. ex Turcz. ■

293041　Peucedanum terebinthaceum（Fisch. ex Trevir.）Fisch. ex Turcz. var. paishanense（Nakai）C. C. Huang；白山石防风■

293042　Peucedanum terebinthaceum subsp. formosanum（Hayata）

Kitag. = Peucedanum formosanum Hayata ■

293043　Peucedanum terebinthaceum var. paishanense（Nakai）Y. Huei Huang = Peucedanum terebinthaceum（Fisch. ex Trevir.）Fisch. ex Turcz. ■

293044　Peucedanum thodei Arnold；索德前胡■☆

293045　Peucedanum thomsonii C. B. Clarke = Ferula ovina（Boiss.）Boiss. ■

293046　Peucedanum torilifolium H. Boissieu；窃衣叶前胡（窃衣前胡）；Hedgeparsleyleaf Hogfennel ■

293047　Peucedanum transiliense Herder = Talassia transiliensis（Herder）Korovin ■

293048　Peucedanum transiliense Regel et Herder = Talassia transiliensis（Regel et Herder）Korovin ■

293049　Peucedanum trinioides H. Wolff = Peucedanum caespitosum H. Wolff ■

293050　Peucedanum trisectum C. C. Towns.；三裂前胡■☆

293051　Peucedanum triternatum Eckl. et Zeyh.；三出前胡■☆

293052　Peucedanum turcomanicum Schischk.；土库曼前胡■☆

293053　Peucedanum turgeniifolium H. Wolff；长前胡（川西前胡）；Beautiful Hogfennel，Turgenleaf Hogfennel ■

293054　Peucedanum typicum（Eckl. et Zeyh.）B. L. Burtt；标准前胡■☆

293055　Peucedanum ugandium M. Hiroe = Peucedanum scottianum Engl. ■☆

293056　Peucedanum uhligii H. Wolff = Oenanthe palustris（Chiov.）C. Norman ■☆

293057　Peucedanum uliginosum（Eckl. et Zeyh.）D. Dietr. = Peucedanum tenuifolium Thunb. ■☆

293058　Peucedanum upingtoniae（Schinz）Drude；阿平顿前胡■☆

293059　Peucedanum vaginatum Ledeb.；具鞘前胡■☆

293060　Peucedanum vaginoides Sales et Hedge；拟具鞘前胡■☆

293061　Peucedanum veitchii H. Boissieu；华西前胡；Veitch Hogfennel ■

293062　Peucedanum venosum Burtt Davy = Peucedanum upingtoniae（Schinz）Drude ■☆

293063　Peucedanum vervaria Cusson；鹿前胡■☆

293064　Peucedanum violaceum R. H. Shan et M. L. Sheh；紫茎前胡；Purplestem Hogfennel ■

293065　Peucedanum virgatum Cham. et Schltdl. = Peucedanum capense（Thunb.）Sond. var. lanceolatum Sond. ●■☆

293066　Peucedanum vogelianum Emb. et Maire；沃格尔前胡■☆

293067　Peucedanum volkensii Engl.；福尔前胡■☆

293068　Peucedanum wallichianum DC. = Selinum candollei DC. ■

293069　Peucedanum wallichianum DC. = Selinum wallichianum（DC.）Raizada et H. O. Saxena ■

293070　Peucedanum wawrae（H. Wolff）H. Y. Su = Peucedanum wawrae（H. Wolff）H. Y. Su ex M. L. Sheh ■

293071　Peucedanum wawrae（H. Wolff）H. Y. Su ex M. L. Sheh；泰山前胡（防风，狗头前胡，前胡，山东邪蒿）；Shandong Seseli，Taishan Hogfennel ■

293072　Peucedanum welwitschii（Engl.）M. Hiroe = Lefebvrea grantii（Hiern）S. Droop ■☆

293073　Peucedanum whytei M. Hiroe = Lefebvrea grantii（Hiern）S. Droop ■☆

293074　Peucedanum wildemanianum C. Norman；怀尔德曼前胡■☆

293075　Peucedanum wilmsianum H. Wolff；维尔姆斯前胡■☆

293076　Peucedanum winkleri H. Wolff；温克勒前胡■☆

293077　Peucedanum wrightii M. Hiroe = Peucedanum kerstenii Engl. ■☆

293078　Peucedanum wulongense R. H. Shan et M. L. Sheh；武隆前胡；

Wulong Hogfennel ■

293079　Peucedanum yakushimense（Masam. et Ohwi）T. Yamaz. = Angelica longiradiata（Maxim.）Kitag. var. yakushimensis（Masam. et Ohwi）Kitag. ■☆

293080　Peucedanum yunnanense H. Wolff；云南前胡（滇前胡）；Yunnan Hogfennel ■

293081　Peucedanum zedelmeyerianum Mand.；泽德前胡■☆

293082　Peucedanum zenkeri H. Wolff = Lefebvrea grantii（Hiern）S. Droop ■☆

293083　Peucedanum zeyheri Sond. = Peucedanum typicum（Eckl. et Zeyh.）B. L. Burtt ■☆

293084　Peucedanum zeyheri Steud. = Peucedanum capillaceum Thunb. var. rigidum Sond. ■☆

293085　Peuceluma Baill. = Lucuma Molina ●

293086　Peuceluma Baill. = Pouteria Aubl. ●

293087　Peucephyllum A. Gray（1859）；枞叶菊属（矮松菊属）●☆

293088　Peucephyllum schottii A. Gray；枞叶菊（矮松菊）■☆

293089　Peudanum Dingl. = Peucedanum L. ●

293090　Peumus Molina = Boldu Adans.（废弃属名）+ Cryptocarya R. Br.（保留属名）●

293091　Peumus Molina（1782）（保留属名）；波多茶属（杯轴花科，比乌木属）●☆

293092　Peumus boldus Molina；波多茶（比乌木，波耳多树，博路都树，博路多树）；Boldo ●☆

293093　Peurousea Steud. = Peyrousea DC.（保留属名）●☆

293094　Peutalis Raf. = Persicaria（L.）Mill. ■

293095　Peutalis Raf. = Polygonum L.（保留属名）■●

293096　Peuteron Raf. = Capparis L. ●

293097　Peuteron Raf. = Pleuteron Raf. ●

293098　Pevraea Comm. ex Juss. = Combretum Loefl.（保留属名）●

293099　Pevraea Comm. ex Juss. = Poivrea Comm. ex Thouars ●

293100　Peyritschia E. Fourn. = Trisetum Pers. ■

293101　Peyritschia E. Fourn. ex Benth. et Hook. f. = Deschampsia P. Beauv. ■

293102　Peyrousea DC.（1836）= Osmites L.（废弃属名）●☆

293103　Peyrousea DC.（1838）（保留属名）；佩罗菊属●☆

293104　Peyrousea DC.（保留属名）= Schistostephium Less. ●☆

293105　Peyrousea argentea Compton = Schistostephium umbellatum（L. f.）Bremer et Humphries ●☆

293106　Peyrousea calycina DC.；佩罗菊●☆

293107　Peyrousea calycina DC. = Schistostephium umbellatum（L. f.）Bremer et Humphries ●☆

293108　Peyrousea oxylepis DC.；尖鳞佩罗菊●☆

293109　Peyrousea oxylepis DC. = Schistostephium umbellatum（L. f.）Bremer et Humphries ●☆

293110　Peyrousea umbellata（L. f.）Fourc. = Schistostephium umbellatum（L. f.）Bremer et Humphries ●☆

293111　Peyrousia Poir.（废弃属名）= Lapeyrousia Pourr. ■☆

293112　Peyrousia Poir.（废弃属名）= Peyrousea DC.（保留属名）●☆

293113　Peyrusa Rich. ex Dunal = Hornemannia Vahl ●☆

293114　Peyrusa Rich. ex Dunal = Symphysia C. Presl ●☆

293115　Peyssonelia Boiv. ex Webb et Berthel. = Cytisus Desf.（保留属名）●

293116　Pezisicarpus Vernet（1904）；秃果夹竹桃属●☆

293117　Pezisicarpus montana Vernet；秃果夹竹桃●☆

293118　Pfaffia Mart.（1825）；无柱苋属（巴西人参属，巴西苋属，普法苋属）■☆

293119　Pfaffia iresinoides Spreng.；血红无柱苋■☆

293120　Pfaffia paniculata（Mart.）Kuntze；无柱苋（巴西人参,巴西苋,普菲西）■☆

293121　Pfaffia paniculata（Mart.）Kuntze = Hebanthe paniculata Mart.■☆

293122　Pfeiffera Salm-Dyck = Lepismium Pfeiff.●☆

293123　Pfeiffera Salm-Dyck（1845）；普氏仙人掌属●☆

293124　Pfeiffera ianthothele Web.；麦乳头麒麟●☆

293125　Pfeifferago Kuntze = Codia J. R. Forst. et G. Forst.●☆

293126　Pfeifferia Buchinger = Cuscuta L.■

293127　Pfitzeria Senghas = Comparettia Poepp. et Endl.■☆

293128　Pfitzeria Senghas（1998）；普菲兰属■☆

293129　Pfosseria Speta = Scilla L.■

293130　Pfosseria Speta（1998）；乳突风信子属■☆

293131　Pfosseria bithynica（Boiss.）Speta；乳突风信子■☆

293132　Pfundia Opiz = Ranunculus L.■

293133　Pfundia Opiz ex Nevski = Hericinia Fourr.■

293134　Phaca L. = Astragalus L.●■

293135　Phaca alpina（L.）Rydb. = Astragalus frigidus（L.）A. Gray■

293136　Phaca bisulcata Hook. = Astragalus bisulcatus（Hook.）A. Gray■☆

293137　Phaca boetica L. = Erophaca baetica（L.）Boiss.●☆

293138　Phaca brachycarpa Turcz. = Astragalus zacharensis Bunge■

293139　Phaca frigida L. = Astragalus frigidus（L.）A. Gray■

293140　Phaca hoffmeisteri Klotzsch = Astragalus hoffmeisteri（Klotzsch）Ali■

293141　Phaca lanata Pall. = Oxytropis lanata（Pall.）DC.■

293142　Phaca lapponica Wahlenb. = Oxytropis lapponica（Wahlenb.）J. Gay■

293143　Phaca membranacea Fisch. = Astragalus membranaceus（Fisch. ex Link）Bunge■

293144　Phaca microphylla Pall. = Oxytropis microphylla（Pall.）DC.■

293145　Phaca montana Wahlenb. = Oxytropis lapponica（Wahlenb.）J. Gay■

293146　Phaca muricata Pall. = Oxytropis muricata（Pall.）DC.■

293147　Phaca myriophylla Pall. = Oxytropis myriophylla（Pall.）DC.■

293148　Phaca neglecta Torr. et A. Gray = Astragalus neglectus（Torr. et A. Gray）E. Sheld.■☆

293149　Phaca salsula Pall. = Sphaerophysa salsula（Pall.）DC.●■

293150　Phaca vogelii Webb = Astragalus vogelii（Webb）Bornm.■☆

293151　Phacelia Juss.（1789）；钟穗花属（伐塞利阿花属,法塞利亚花属,芹叶草属,束花属）；Phacelia,Scorpionweed■☆

293152　Phacelia artemisioides Griseb.；蒿状钟穗花■☆

293153　Phacelia bicknellii Small = Phacelia purshii Buckley■☆

293154　Phacelia bipinnatifida Frank ex A. DC.；紫钟穗花；Fern-leaved Phacelia,Loose-flowered Phacelia,Phacelia,Purple Phacelia■☆

293155　Phacelia bipinnatifida Michx. = Phacelia bipinnatifida Frank ex A. DC.■☆

293156　Phacelia boykinii（A. Gray）Small = Phacelia purshii Buckley■☆

293157　Phacelia californica Cham.；加州钟穗花（加州束花）●☆

293158　Phacelia campanularia A. Gray；钟穗花（加州蓝铃花,加洲蓝钟）；California Bluebell,Desert Bell,Desert Blue Bells,Phacelia■☆

293159　Phacelia cinerea Eastw. ex J. F. Macbr. = Phacelia distans Benth.■☆

293160　Phacelia congesta Hook.；密集钟穗花■☆

293161　Phacelia crenulata Torr. ex S. Watson；圆耻钟穗花；Scalloped Leaf Phacelia■☆

293162　Phacelia distans Benth.；普通钟穗花；Common Phacelia,Distant Phacelia,Wild Heliotrope■☆

293163　Phacelia distans Benth. var. australis Brand = Phacelia distans Benth.■☆

293164　Phacelia dubia Trel. ex Trel.,Branner et Coville；可疑钟穗花；Small-flowered Phacelia■☆

293165　Phacelia fimbriata Michx.；流苏钟穗花；Fringed Phacelia■☆

293166　Phacelia gilioides Brand；毛钟穗花；Hairy Phacelia■☆

293167　Phacelia glandulosa Nutt.；腺点钟穗花；Waterleaf Scorpion-weed■☆

293168　Phacelia hirsuta Nutt.；拟毛钟穗花；Phacelia■☆

293169　Phacelia integrifolia Torr.；全缘叶钟穗花；Gypsum Phacelia,Scalloped Phacelia,Texan Phacelia■☆

293170　Phacelia linearis Holz.；线叶钟穗花；Threadleaf Phacelia■☆

293171　Phacelia minor（Harv.）Thell.；小钟穗花；Californian Bluebell■☆

293172　Phacelia popei Torr. et Gray；波普钟穗花；Pope's Phacelia■☆

293173　Phacelia purshii Buckley；普氏钟穗花；Miami Mist,Miami-mist,Scorpionweed,Scorpion-weed■☆

293174　Phacelia ranunculacea（Nutt.）Constance；毛茛钟穗花；Phacelia■☆

293175　Phacelia sericea A. Gray；绢毛钟穗花；Silky Phacelia■☆

293176　Phacelia tanacetifolia Benth.；芹叶钟穗花；Lacy Phacelia,Phacelia,Tansy Phacelia■☆

293177　Phacelia tanacetifolia Benth. et Lindl. = Phacelia tanacetifolia Benth.■☆

293178　Phacelia viscida（Benth. ex Lindl.）Torr.；黏钟穗花■☆

293179　Phacelia whitlavia A. Gray；惠氏钟穗花；California Blue Bell,Californian Bluebell,Whitlavia■☆

293180　Phacellanthus Klotzsch ex Kunth = Picramnia Sw.（保留属名）●☆

293181　Phacellanthus Siebold et Zucc.（1846）；黄筒花属；Phacellanthus■

293182　Phacellanthus Siebold et Zucc. = Gahnia J. R. Forst. et G. Forst.■

293183　Phacellanthus Steud. = Phacellanthus Steud. ex Zoll. et Moritzi■

293184　Phacellanthus Steud. ex Zoll. et Moritzi = Gahnia J. R. Forst. et G. Forst.■

293185　Phacellanthus continentalis Kom. = Phacellanthus tubiflorus Siebold et Zucc.■

293186　Phacellanthus tubiflorus Siebold et Zucc.；黄筒花；Phacellanthus,Tubiflorous Phacellanthus■

293187　Phacellaria Benth.（1880）；重寄生属（法色草属,鳞叶寄生木属）；Parasite,Phacellaria●■

293188　Phacellaria Steud. = Chloris Sw.●■

293189　Phacellaria Willd. ex Steud. = Chloris Sw.●■

293190　Phacellaria caulescens Collett et Hemsl.；粗序重寄生（侧序隐茎木）；Compress Parasite,Phacellaria●

293191　Phacellaria compressa Benth.；扁穗重寄生（扁序重寄生）；Compress Parasite,Flattened Phacellaria●■

293192　Phacellaria fargesii Lecomte；重寄生（法色草）；Farges Parasite,Farges Phacellaria●■

293193　Phacellaria ferruginea W. W. Sm. = Phacellaria compressa Benth.●■

293194　Phacellaria glomerata D. D. Tao；聚果重寄生■

293195　Phacellaria rigidula Benth.；微挺重寄生（硬序重寄生）；Hard Parasite●■

293196　Phacellaria tonkinensis Lecomte；长序重寄生（北越重寄生）；Tonkin Parasite,Tonkin Phacellaria●■

293197　Phacellaria wattii Hook. f. = Phacellaria compressa Benth.●■

293198　Phacellothrix F. Muell. (1878);束毛菊属●☆

293199　Phacellothrix cladochaeta (F. Muell.) F. Muell.;束毛菊●☆

293200　Phacelophrynium K. Schum. (1902);束柊叶属■☆

293201　Phacelophrynium cylindricum Merr.;圆柱束柊叶■☆

293202　Phacelophrynium laxum Clausager et Borchs.;松散束柊叶■☆

293203　Phacelophrynium longispicum K. Schum.;长穗束柊叶■☆

293204　Phacelophrynium maximum K. Schum.;大束柊叶■☆

293205　Phacelophrynium whitei Ridl.;瓦特束柊叶■☆

293206　Phacelura Benth. = Phacelurus Griseb. ■

293207　Phacelurus Griseb. (1846);束尾草属;Phacelurus ■

293208　Phacelurus angustifolius (Debeaux) Nakai = Phacelurus latifolius (Steud.) Ohwi ■

293209　Phacelurus caespitosus C. E. Hubb. = Loxodera caespitosa (C. E. Hubb.) Simon ■☆

293210　Phacelurus congoensis (Hack.) Zon = Phacelurus gabonensis (Steud.) Clayton ■☆

293211　Phacelurus digitatus (Sibth. et Sm.) Griseb.;掌状束尾草(希腊束尾草);Digitate Phacelurus ■

293212　Phacelurus franksae (J. M. Wood) Clayton;弗兰克斯束尾草■☆

293213　Phacelurus gabonensis (Steud.) Clayton;加蓬束尾草■☆

293214　Phacelurus huillensis (Rendle) Clayton;威拉束尾草■☆

293215　Phacelurus latifolius (Steud.) Ohwi;束尾草;Broadleaf Phacelurus ■

293216　Phacelurus latifolius (Steud.) Ohwi var. angustifolius (Debeaux) Keng;狭叶束尾草(立秋,芦秋,鸟秋);Narrowleaf Phacelurus ■

293217　Phacelurus latifolius (Steud.) Ohwi var. angustifolius (Debeaux) Kitag. = Phacelurus latifolius (Steud.) Ohwi ■

293218　Phacelurus latifolius (Steud.) Ohwi var. monostachys Keng ex S. L. Chen = Phacelurus latifolius (Steud.) Ohwi ■

293219　Phacelurus latifolius (Steud.) Ohwi var. monostachyus Keng;单穗束尾草;Singlespike Phacelurus ■

293220　Phacelurus latifolius (Steud.) Ohwi var. monostachyus Keng = Phacelurus latifolius (Steud.) Ohwi ■

293221　Phacelurus latifolius (Steud.) Ohwi var. trichophyllus (S. L. Zhong) B. S. Sun et Z. H. Hu = Phacelurus trichophyllus S. L. Zhong ■

293222　Phacelurus schliebenii (Pilg.) Clayton;施利本束尾草■☆

293223　Phacelurus speciosus (Steud.) C. E. Hubb.;美丽束尾草■☆

293224　Phacelurus speciosus (Steud.) C. E. Hubb. var. afghanicus Melderis = Phacelurus speciosus (Steud.) C. E. Hubb. ■☆

293225　Phacelurus trichophyllus S. L. Zhong;毛叶束尾草;Hairleaf Phacelurus ■

293226　Phacelurus trichophyllus S. L. Zhong = Phacelurus latifolius (Steud.) Ohwi var. trichophyllus (S. L. Zhong) B. S. Sun et Z. H. Hu ■

293227　Phacelurus zea (C. B. Clarke) Clayton;黍束尾草■

293228　Phacocapnos Bernh. = Corydalis DC. (保留属名)■

293229　Phacolobus Post et Kuntze = Anthyllis L. ■☆

293230　Phacolobus Post et Kuntze = Fakeloba Raf. ■☆

293231　Phacomene Rydb. = Astragalus L. ●■

293232　Phacopsis Rydb. = Astragalus L. ●■

293233　Phacopsis Rydb. = Phacomene Rydb. ●■

293234　Phacopsis Rydb. = Rydbergiella Fedde et Syd. ex Rydb. ●■

293235　Phacosperma Haw. = Calandrinia Kunth(保留属名)●■☆

293236　Phadrosanthus Neck. = Oncidium Sw. (保留属名)■☆

293237　Phadrosanthus Neck. ex Raf. = Epidendrum L. (保留属名)■☆

293238　Phaeanthus Hook. f. et Thomson(1855);亮花木属;Phaeanthus ●

293239　Phaeanthus Post et Kuntze = Moraea Mill. (保留属名)■

293240　Phaeanthus Post et Kuntze = Phaianthes Raf. ■

293241　Phaeanthus saccopetaloides W. T. Wang;囊瓣亮花木;Saccatepetal Phaeanthus ●

293242　Phaeanthus saccopetaloides W. T. Wang = Desmos saccopetaloides (W. T. Wang) P. T. Li ●

293243　Phaeanthus yunnanensis Hu = Desmos yunnanense (Hu) P. T. Li ●

293244　Phaecasium Cass. (1826);华美参属■

293245　Phaecasium Cass. = Crepis L. ■

293246　Phaecasium lampsanoides Cass.;肖还阳参■☆

293247　Phaecasium pulchrum (L.) Rchb. = Crepis pulchra L. ■☆

293248　Phaecasium pulchrum (L.) Rchb. = Phaecasium lampsanoides Cass. ■☆

293249　Phaedra Klotzsch = Phaedra Klotzsch ex Endl. ●

293250　Phaedra Klotzsch ex Endl. = Bernardia L. ●

293251　Phaedra Klotzsch ex Endl. = Polyscias J. R. Forst. et G. Forst. ●

293252　Phaedranassa Herb. (1845);后喜花属;Queen Lily ■☆

293253　Phaedranassa carmioli Baker;佳美后喜花■☆

293254　Phaedranassa viridiflora Baker;绿花后喜花■☆

293255　Phaedranthus Miers(1863)(保留属名);肖红钟藤属●☆

293256　Phaedranthus Miers(保留属名) = Distictis Mart. ex Meisn. ●☆

293257　Phaedranthus buccinatorius (DC.) Miers = Distictis buccinatoria (DC.) A. H. Gentry ●☆

293258　Phaedranthus cinerascens Miers;灰肖红钟藤●☆

293259　Phaedranthus exsertus Miers;肖红钟藤●☆

293260　Phaedrosanthus Post et Kuntze = Cattleya Lindl. ■

293261　Phaedrosanthus Post et Kuntze = Epidendrum L. (保留属名)■☆

293262　Phaelypaea P. Br. = Stemodia L. (保留属名)■☆

293263　Phaenanthoecium C. E. Hubb. (1936);显颖草属■☆

293264　Phaenanthoecium koestlinii (Hochst. ex A. Rich.) C. E. Hubb.;显颖草■☆

293265　Phaeneilema G. Brückn. = Murdannia Royle(保留属名)■

293266　Phaeneilema malabarica (L.) V. Naray. ex Biswas = Murdannia nudiflora (L.) Brenan ■

293267　Phaenicanthus Thwaites = Premna L. (保留属名)●■

293268　Phaenicaulis Greene = Phoenicaulis Nutt. ■☆

293269　Phaeniopsis Cass. = Scariola F. W. Schmidt ■●

293270　Phaenix Hill = Phoenix L. ●

293271　Phaenixopus Cass. = Lactuca L. ■

293272　Phaenixopus Cass. = Scariola F. W. Schmidt ■●

293273　Phaenocodon Salisb. = Lapageria Ruiz et Pav. ●☆

293274　Phaenocoma D. Don(1826);紫花帚鼠麹属(粉红苞菊属);Cape Everlasting ●☆

293275　Phaenocoma prolifera (L.) D. Don;紫花帚鼠麹(粉红苞菊);Cape Everlasting ●☆

293276　Phaenocoma prolifera D. Don = Phaenocoma prolifera (L.) D. Don ●☆

293277　Phaenohoffmannia Kuntze = Pearsonia Dümmer ●☆

293278　Phaenohoffmannia Kuntze = Pleiospora Harv. ●☆

293279　Phaenohoffmannia cajanifolia (Harv.) Kuntze subsp. cryptantha (Baker) J. B. Gillett = Pearsonia cajanifolia (Harv.) Polhill subsp. cryptantha (Baker) Polhill ●☆

293280　Phaenohoffmannia grandifolia (Bolus) J. B. Gillett = Pearsonia grandifolia (Bolus) Polhill ●☆

293281　Phaenohoffmannia latibracteolata (Dümmer) J. B. Gillett = Pearsonia grandifolia (Bolus) Polhill subsp. latibracteolata

（Dümmer）Polhill ●☆

293282　Phaenohoffmannia obovata（Schinz）J. B. Gillett = Pearsonia obovata（Schinz）Polhill ●☆

293283　Phaenomeria Steud. = Nicolaia Horan.（保留属名）■☆

293284　Phaenomeria Steud. = Phaeomeria Lindl. ex K. Schum. ■☆

293285　Phaenopoda Cass. = Podotheca Cass.（保留属名）■☆

293286　Phaenopoda Cassini = Podosperma Labill.（废弃属名）■☆

293287　Phaenopus DC. = Lactuca L. ■

293288　Phaenopus DC. = Phaenixopus Cass. ■

293289　Phaenopus DC. = Scariola F. W. Schmidt ■●

293290　Phaenopus orientalis Boiss. = Scariola orientalis（Boiss.）Soják ■

293291　Phaenopyrum M. Roem. = Crataegus L. ●

293292　Phaenopyrum Schrad. ex Nees = Lagenocarpus Nees ■☆

293293　Phaenosperma Munro ex Benth.（1881）;显籽草属(褐子草属,显子草属);Phaenosperma ■

293294　Phaenosperma Munro ex Benth. et Hook. f. = Phaenosperma Munro ex Benth. ■

293295　Phaenosperma globosum Munro = Phaenosperma globosum Munro ex Benth. ■

293296　Phaenosperma globosum Munro ex Benth.;显籽草(乌珠茅,显子草,岩高粱);Globose Phaenosperma ■

293297　Phaenostoma Steud. = Chaenostoma Benth.（保留属名）■☆

293298　Phaenostoma Steud. = Sutera Roth ■●☆

293299　Phaeocarpus Mart. = Magonia A. St. -Hil. ●☆

293300　Phaeocephalum Ehrh. = Rhynchospora Vahl(保留属名)■

293301　Phaeocephalum Ehrh. = Schoenus L. ■

293302　Phaeocephalum Ehrh. ex House = Rhynchospora Vahl(保留属名)■

293303　Phaeocephalum Ehrh. ex House = Schoenus L. ■

293304　Phaeocephalum House = Rhynchospora Vahl(保留属名)■

293305　Phaeocephalum album（L.）House = Rhynchospora alba（L.）Vahl ■

293306　Phaeocephalum album（L.）House var. macrum（C. B. Clarke ex Britton）Farw. = Rhynchospora macra（C. B. Clarke ex Britton）Small ■☆

293307　Phaeocephalum baldwinii（A. Gray）House = Rhynchospora baldwinii A. Gray ■☆

293308　Phaeocephalum brachychaetum（C. Wright）House = Rhynchospora brachychaeta C. Wright ■☆

293309　Phaeocephalum caducum（Elliott）House = Rhynchospora caduca Elliott ■☆

293310　Phaeocephalum capillaceum（Torr.）Farw. = Rhynchospora capillacea Torr. ■☆

293311　Phaeocephalum chapmanii（M. A. Curtis）House = Rhynchospora chapmanii M. A. Curtis ■☆

293312　Phaeocephalum ciliatum（Michx.）House = Rhynchospora ciliaris（Michx.）C. Mohr ■☆

293313　Phaeocephalum compressum（J. Carey ex Chapm.）House = Rhynchospora compressa J. Carey ex Chapm. ■☆

293314　Phaeocephalum curtissii（Britton）House = Rhynchospora curtissii Britton ■☆

293315　Phaeocephalum cymosum（Elliott）House = Rhynchospora recognita（Gale）Král ■☆

293316　Phaeocephalum decurrens（Chapm.）House = Rhynchospora decurrens Chapm. ■☆

293317　Phaeocephalum distans（Michx.）House = Rhynchospora fascicularis（Michx.）Vahl ■☆

293318　Phaeocephalum dodecandrum（Baldwin ex A. Gray）House = Rhynchospora megalocarpa A. Gray ■☆

293319　Phaeocephalum earlei（Britton）House = Rhynchospora harveyi W. Boott ■☆

293320　Phaeocephalum fasciculare（Michx.）House = Rhynchospora fascicularis（Michx.）Vahl ■☆

293321　Phaeocephalum filifolium（A. Gray）House = Rhynchospora filifolia A. Gray ■☆

293322　Phaeocephalum fuscum（L.）House = Rhynchospora fusca（L.）W. T. Aiton ■☆

293323　Phaeocephalum glomeratum（L.）House = Rhynchospora glomerata（L.）Vahl ■☆

293324　Phaeocephalum glomeratum（L.）House var. minus（Britton）Farw. = Rhynchospora capitellata（Michx.）Vahl ■☆

293325　Phaeocephalum gracilentum（A. Gray）House = Rhynchospora gracilenta A. Gray ■☆

293326　Phaeocephalum grayi（Kunth）House = Rhynchospora grayi Kunth ■☆

293327　Phaeocephalum intermedium（Chapm.）House = Rhynchospora pineticola C. B. Clarke ■☆

293328　Phaeocephalum microcarpum（Baldwin ex A. Gray）House = Rhynchospora microcarpa Baldwin ex A. Gray ■☆

293329　Phaeocephalum miliaceum（Lam.）House = Rhynchospora miliacea（Lam.）A. Gray ■☆

293330　Phaeocephalum pallidum（M. A. Curtis）House = Rhynchospora pallida M. A. Curtis ■☆

293331　Phaeocephalum patulum（A. Gray）House = Rhynchospora microcarpa Baldwin ex A. Gray ■☆

293332　Phaeocephalum perplexum（Britton）House = Rhynchospora perplexa Britton ■☆

293333　Phaeocephalum plankii（Britton）House = Rhynchospora harveyi W. Boott ■☆

293334　Phaeocephalum plumosa（Elliott）House = Rhynchospora plumosa Elliott ■☆

293335　Phaeocephalum proliferum（Small）House = Rhynchospora mixta Britton ■☆

293336　Phaeocephalum punctatum（Elliott）House = Rhynchospora punctata Elliott ■☆

293337　Phaeocephalum pusillum（Chapm. ex M. A. Curtis）House = Rhynchospora pusilla Chapm. ex M. A. Curtis ■☆

293338　Phaeocephalum rariflorum（Michx.）House = Rhynchospora rariflora（Michx.）Elliott ■☆

293339　Phaeocephalum schoenoides（A. W. Wood）House = Rhynchospora elliottii A. Dietr. ■☆

293340　Phaeocephalum stipitatum（Chapm.）House = Rhynchospora odorata C. Wright ex Griseb. ■☆

293341　Phaeocephalum torreyanum（A. Gray）House = Rhynchospora torreyana A. Gray ■☆

293342　Phaeocephalum tracyi（Britton）House = Rhynchospora tracyi Britton ■☆

293343　Phaeocephalus S. Moore = Hymenolepis Cass. ●☆

293344　Phaeocephalus gnidioides S. Moore = Hymenolepis gnidioides（S. Moore）Källersjö ●☆

293345　Phaeocles Salisb. = Caruelia Parl. ■

293346　Phaeocles Salisb. = Ornithogalum L. ■

293347　Phaeocordylis Griff. = Rhopalocnemis Jungh. ■

293348　Phaeocordylis areolata Griff. = Rhopalocnemis phalloides Jungh. ■

293349 Phaeolorum Ehth. = Carex L. ■

293350 Phaeolorum Ehth. = Nicolaia Horan.（保留属名）■☆

293351 Phaeomeria（Ridl.）K. Schum. = Etlingera Roxb. ■

293352 Phaeomeria Lindl. = Nicolaia Horan.（保留属名）■☆

293353 Phaeomeria Lindl. ex K. Schum. = Nicolaia Horan.（保留属名）■☆

293354 Phaeomeria magnifica（Roscoeoe）K. Schum. = Nicolaia elatior（Jack.）Horan. ■

293355 Phaeomeria speciosa（Blume）Koord. = Etlingera elatior（Jack）R. M. Sm. ■

293356 Phaeomeria speciosa（Blume）Koord. = Nicolaia elatior（Jack.）Horan. ■

293357 Phaeomeria speciosa（Blume）Merr. = Alpinia zerumbet（Pers.）B. L. Burtt et R. M. Sm. ■

293358 Phaeoneuron Gilg = Ochthocharis Blume ●☆

293359 Phaeoneuron dicellandroides Gilg = Ochthocharis dicellandroides（Gilg）Hansen et Wickens ●☆

293360 Phaeoneuron gracile（A. Chev.）Hutch. et Dalziel = Ochthocharis dicellandroides（Gilg）Hansen et Wickens ●☆

293361 Phaeoneuron libericum（Gilg）Engl. = Ochthocharis setosa（Hook. f.）Hansen et Wickens ●☆

293362 Phaeoneuron moloneyi Stapf = Ochthocharis dicellandroides（Gilg）Hansen et Wickens ●☆

293363 Phaeoneuron schweinfurthii Stapf = Ochthocharis dicellandroides（Gilg）Hansen et Wickens ●☆

293364 Phaeoneuron setosum（Hook. f.）Stapf = Ochthocharis setosa（Hook. f.）Hansen et Wickens ●☆

293365 Phaeonychium O. E. Schulz（1927）；藏芥属；Phaeonychium, Xizanggrass ■

293366 Phaeonychium albiflorum（T. Anderson）Jafri；白花藏芥■

293367 Phaeonychium fengii Al-Shehbaz；冯氏藏芥■

293368 Phaeonychium jafrii Al-Shehbaz；杰氏藏芥■

293369 Phaeonychium kashgaricum（Botsch.）Al-Shehbaz；喀什藏芥■

293370 Phaeonychium parryoides（Kurz ex Hook. f. et T. Anderson）O. E. Schulz；藏芥；Parrya-like Phaeonychium, Xizanggrass ■

293371 Phaeonychium villosum（Maxim.）Al-Shehbaz；柔毛藏芥■

293372 Phaeonychium villosum（Maxim.）Al-Shehbaz var. albiflora O. E. Schulz = Phaeonychium villosum（Maxim.）Al-Shehbaz ■

293373 Phaeopappus（DC.）Boiss. = Centaurea L.（保留属名）●■

293374 Phaeopappus（DC.）Boiss. = Psephellus Cass. ●■☆

293375 Phaeopappus Boiss. = Centaurea L.（保留属名）●■

293376 Phaeopappus Boiss. = Psephellus Cass. ●■☆

293377 Phaeophleps Post et Kuntze = Phaiophleps Raf. ■☆

293378 Phaeopsis Nutt. ex Benth. = Stenogyne Benth.（保留属名）■☆

293379 Phaeoptilon Engl. = Phaeoptilum Radlk. ●☆

293380 Phaeoptilum Radlk.（1883）；褐羽花属●☆

293381 Phaeoptilum heimerlii Engl. = Phaeoptilum spinosum Radlk. ●☆

293382 Phaeoptilum spinosum Radlk.；褐羽花●☆

293383 Phaeorneria Lindl. ex K. Schum. = Etlingera Roxb. ■

293384 Phaeosperma Post et Kuntze = Phaiosperma Raf. ■☆

293385 Phaeosperma Post et Kuntze = Polytaenia DC. ■☆

293386 Phaeosphaerion Hassk. = Commelina L. ■

293387 Phaeosphaeriona B. D. Jacks. = Phaeosphaerion Hassk. ■

293388 Phaeosphaeriona Hassk. = Commelina L. ■

293389 Phaeospheriona Willis = Phaeosphaerion Hassk. ■

293390 Phaeostemma E. Fourn.（1885）；暗冠萝藦属●☆

293391 Phaeostemma grandifolia Rusby；大叶暗冠萝藦●☆

293392 Phaeostigma Muldashev = Ajania Poljakov ●■

293393 Phaeostigma Muldashev（1981）；栎叶菊属■●

293394 Phaeostigma quercifolium（W. W. Sm.）Muldashev = Ajania quercifolia（W. W. Sm.）Y. Ling et C. Shih ■

293395 Phaeostigma quercifolium（W. W. Sm.）Muldashev = Tanacetum quercifolium W. W. Sm. ●

293396 Phaeostigma salicifolium（Mattf.）Muldashev = Ajania salicifolia（Mattf. ex Rehder et Kobuski）Poljakov ■

293397 Phaeostigma salicifolium（Mattf.）Muldashev = Tanacetum salicifolium Mattf. ■

293398 Phaeostigma variifolium（C. C. Chang）Muldashev = Ajania variifolia（C. C. Chang）Tzvelev ■

293399 Phaeostigma variifolium（C. C. Chang）Muldashev = Chrysanthemum variifolium C. C. Chang ■

293400 Phaeostigma variifolium（C. C. Chang）Muldashev var. ramosum（C. C. Chang）Muldashev = Chrysanthemum variifolium C. C. Chang var. ramosum C. C. Chang ■

293401 Phaeostigma variifolium（C. C. Chang）Muldashev var. ramosum（C. C. Chang）Muldashev = Ajania ramosa（C. C. Chang）C. Shih ■

293402 Phaeostoma Spach = Clarkia Pursh ■

293403 Phaestoma Spach = Clarkia Pursh ■

293404 Phaethusa Gaertn. = Verbesina L.（保留属名）●■☆

293405 Phaethusia Raf. = Phaethusa Gaertn. ●■☆

293406 Phaetusa Schreb. = Phaethusa Gaertn. ●■☆

293407 Phaeus Post et Kuntze = Phaius Lour. ■

293408 Phagnalon Cass.（1819）；绵毛菊属（棉毛草属，棉毛菊属）；Cottondaisy, Phagnalon ●■

293409 Phagnalon abyssinicum Sch. Bip. ex A. Rich.；阿比西尼亚绵毛菊■☆

293410 Phagnalon acuminatum Boiss.；渐尖绵毛菊■☆

293411 Phagnalon androssovii B. Fedtsch.；阿氏绵毛菊■☆

293412 Phagnalon androssovii B. Fedtsch. = Phagnalon schweinfurthii Sch. Bip. ex Schweinf. ■☆

293413 Phagnalon atlanticum Ball = Phagnalon bicolor Ball ■☆

293414 Phagnalon atlanticum Ball var. rehamnarum Maire = Phagnalon bicolor Ball ■☆

293415 Phagnalon barbeyanum Asch. et Graebn.；巴比绵毛菊■☆

293416 Phagnalon bennettii DC. = Phagnalon hansenii Qaiser et Lack ■☆

293417 Phagnalon bicolor Ball；二色绵毛菊■☆

293418 Phagnalon calycinum（Cav.）DC.；萼状绵毛菊■☆

293419 Phagnalon calycinum（Cav.）DC. subsp. ballsianum Maire；鲍尔斯绵毛菊■☆

293420 Phagnalon calycinum（Cav.）DC. subsp. caroli-paui（Font Quer）Emb. et Maire；波氏绵毛菊■☆

293421 Phagnalon calycinum（Cav.）DC. subsp. spathulatum（H. Lindb.）Maire；匙形萼状绵毛菊■☆

293422 Phagnalon caroli-paui Font Quer = Phagnalon calycinum（Cav.）DC. subsp. caroli-paui（Font Quer）Emb. et Maire ■☆

293423 Phagnalon darvazicum Krasch.；达地绵毛菊■☆

293424 Phagnalon denticulatum C. B. Clarke = Phagnalon niveum Edgew. ●■

293425 Phagnalon denticulatum Decne. ex C. B. Clarke = Phagnalon niveum Edgew. ●■

293426 Phagnalon embergeri Humbert et Maire = Aliella embergeri（Humbert et Maire）Qaiser et Lack ■☆

293427 Phagnalon embergeri Humbert et Maire var. lepineyi（Emb.）Maire = Aliella embergeri（Humbert et Maire）Qaiser et Lack ■☆

293428 Phagnalon garamantum Maire；噶拉门特绵毛菊■☆

293429 Phagnalon glabrifolium Rech. f. ;阿富汗绵毛菊■☆

293430 Phagnalon glabrifolium Rech. f. = Phagnalon darvazicum Krasch.■☆

293431 Phagnalon graecum Boiss. et Heldr. ;希腊绵毛菊■☆

293432 Phagnalon hansenii Qaiser et Lack;汉森绵毛菊■☆

293433 Phagnalon helichrysoides (Ball) Coss. = Aliella ballii (Klatt) Greuter ■☆

293434 Phagnalon helichrysoides (Ball) Coss. var. nitidum Emb. = Aliella ballii (Klatt) Greuter subsp. nitida (Emb.) Greuter ■☆

293435 Phagnalon hypoleucum Sch. Bip. ex Oliv. et Hiern = Phagnalon abyssinicum Sch. Bip. ex A. Rich.■☆

293436 Phagnalon latifolium Maire;宽叶绵毛菊■☆

293437 Phagnalon lavranosii Qaiser et Lack;拉夫拉诺斯绵毛菊■☆

293438 Phagnalon lepidotum Pomel = Phagnalon saxatile (L.) Cass.■☆

293439 Phagnalon lepineyi Emb. = Aliella embergeri (Humbert et Maire) Qaiser et Lack ■☆

293440 Phagnalon luridum Webb = Phagnalon melanoleucum Webb ■☆

293441 Phagnalon melanoleucum Webb;黑白绵毛菊■☆

293442 Phagnalon melanoleucum Webb var. luridum (Webb) A. Chev. = Phagnalon melanoleucum Webb ■☆

293443 Phagnalon metlesicsii Pignatti;梅特绵毛菊■☆

293444 Phagnalon nitidum Fresen. ;光亮绵毛菊■☆

293445 Phagnalon niveum Edgew. ; 绵毛菊(棉毛菊); Snowwhite Cottondaisy , Snowwhite Phagnalon ●■

293446 Phagnalon phagnaloides (Sch. Bip. ex A. Rich.) Cufod. ;普通绵毛菊■☆

293447 Phagnalon platyphyllum Maire = Aliella platyphylla (Maire) Qaiser et Lack ■☆

293448 Phagnalon purpurascens Sch. Bip. = Phagnalon saxatile (L.) Cass. subsp. purpurascens (Sch. Bip.) Batt.■☆

293449 Phagnalon pycnophyllon Rech. f. ;密叶绵毛菊■☆

293450 Phagnalon quartinianum A. Rich. ;夸尔廷绵毛菊■☆

293451 Phagnalon rupestre (L.) DC. ;岩生绵毛菊■☆

293452 Phagnalon rupestre (L.) DC. subsp. graecum (Boiss.) Batt. = Phagnalon graecum Boiss. et Heldr.■☆

293453 Phagnalon rupestre (L.) DC. subsp. illyricum (H. Lindb.) Ginzb. = Phagnalon tenorei (Spreng.) C. Presl ■☆

293454 Phagnalon rupestre (L.) DC. var. annoticum (Burnat) Briq. = Phagnalon tenorei (Spreng.) C. Presl ■☆

293455 Phagnalon rupestre (L.) DC. var. calycinum (Cav.) Ball = Phagnalon calycinum (Cav.) DC.■☆

293456 Phagnalon rupestre (L.) DC. var. graecum (Boiss.) Fiori = Phagnalon graecum Boiss. et Heldr.■☆

293457 Phagnalon rupestre (L.) DC. var. linnaei (Rouy) Maire = Phagnalon tenorei (Spreng.) C. Presl ■☆

293458 Phagnalon rupestre (L.) DC. var. tenorei (Spreng.) Fiori = Phagnalon tenorei (Spreng.) C. Presl ■☆

293459 Phagnalon saxatile (L.) Cass. ;岩地绵毛菊■☆

293460 Phagnalon saxatile (L.) Cass. subsp. purpurascens (Sch. Bip.) Batt. ;浅紫岩地绵毛菊■☆

293461 Phagnalon saxatile (L.) Cass. subsp. spathulata H. Lindb. = Phagnalon saxatile (L.) Cass. subsp. purpurascens (Sch. Bip.) Batt. ■☆

293462 Phagnalon saxatile (L.) Cass. var. denudatum (Welw.) Mariz = Phagnalon saxatile (L.) Cass.■☆

293463 Phagnalon saxatile (L.) Cass. var. intermedium (Lag.) DC. = Phagnalon saxatile (L.) Cass. ■☆

293464 Phagnalon saxatile (L.) Cass. var. lepidotum (Pomel) Batt. = Phagnalon saxatile (L.) Cass. ■☆

293465 Phagnalon saxatile (L.) Cass. var. perez-mendezi Pau = Phagnalon saxatile (L.) Cass. ■☆

293466 Phagnalon saxatile (L.) Cass. var. saxatile = Phagnalon saxatile (L.) Cass. ■☆

293467 Phagnalon scalarum Schweinf. ex Schwartz = Phagnalon stenolepis Chiov. ■☆

293468 Phagnalon scalarum Schweinf. ex Schwartz subsp. glabrum Miré et Quézel = Phagnalon stenolepis Chiov. ■☆

293469 Phagnalon scalarum Schweinf. ex Schwartz var. meridionale (Quézel) Wickens = Phagnalon stenolepis Chiov. ■☆

293470 Phagnalon schweinfurthii Sch. Bip. ex Schweinf. ;施韦绵毛菊■☆

293471 Phagnalon scoparium Sch. Bip. ex Oliv. et Hiern = Phagnalon phagnaloides (Sch. Bip. ex A. Rich.) Cufod. ■☆

293472 Phagnalon sordidum (L.) Rchb. ;污浊绵毛菊■☆

293473 Phagnalon stenolepis Chiov. ;窄鳞绵毛菊■☆

293474 Phagnalon tenorei (Spreng.) C. Presl;泰诺雷绵毛菊■☆

293475 Phagnalon tibesticum Chevassut et Quézel = Phagnalon stenolepis Chiov. ■☆

293476 Phagnalon tibesticum Chevassut et Quézel subsp. meridionale Quézel = Phagnalon stenolepis Chiov. ■☆

293477 Phagnalon umbelliforme DC. ;伞花绵毛菊■☆

293478 Phaianthes Raf. = Moraea Mill. (保留属名)■

293479 Phaianthes lurida (Ker Gawl.) Raf. = Moraea lurida Ker Gawl. ■☆

293480 Phainantha Gleason(1948);辉花野牡丹属●☆

293481 Phainantha laxiflora (Triana) Gleason;辉花野牡丹●☆

293482 Phaiophleps Raf. = Olsynium Raf. ■☆

293483 Phaiosperma Raf. = Polytaenia DC. ■☆

293484 Phaius Lour. (1790);鹤顶兰属;Phaius ■

293485 Phaius actinomorphus (Fukuy.) T. P. Lin = Calanthe actinomorpha Fukuy. ■

293486 Phaius albus Lindl. = Thunia alba (Lindl.) Rchb. f. ■

293487 Phaius albus Rchb. f. ;白花鹤顶兰■☆

293488 Phaius amboinensis (Zipp.) Blume;阿姆波鹤顶兰;Amboi Phaius ■☆

293489 Phaius calanthoides Ames = Cephalantheropsis calanthoides (Ames) Tang S. Liu et H. J. Su ■

293490 Phaius calanthoides Ames = Cephalantheropsis halconensis (Ames) S. S. Ying ■

293491 Phaius columnaris C. Z. Tang et S. J. Cheng;仙笔鹤顶兰; Columnar Phaius ■

293492 Phaius crinitus (Gagnep.) Seidenf. = Phaius mishmensis (Lindl. et Paxton) Rchb. f. ■

293493 Phaius delavayi (Finet) P. J. Cribb et Perner;少花鹤顶兰■

293494 Phaius flavus (Blume) Lindl. ;黄花鹤顶兰(斑叶鹤顶兰,斑叶兰,红花鹤顶兰,虎牙庄,黄鹤顶兰,黄鹤兰,九子莲,洒金鹤顶兰,小花鹤顶兰);Woodford Phaius, Yellow Phaius, Yellowflower Phaius ■

293495 Phaius flavus (Blume) Lindl. f. punctatus (Ohwi) Hatus. ;斑点黄花鹤顶兰■☆

293496 Phaius francoisii (Schltr.) Summerh. = Gastrorchis francoisii Schltr. ■☆

293497 Phaius gibbosulus H. Perrier = Imerinaea madagascarica Schltr. ■☆

293498 Phaius gracilis (Lindl.) S. S. Ying = Cephalantheropsis gracilis (Lindl.) S. Y. Hu ■

293499 Phaius gracilis (Lindl.) S. S. Ying = Cephalantheropsis obcordata（Lindl.）Ormerod ■

293500　Phaius gracilis（Lindl.）S. S. Ying var. calanthoides（Ames）S. S. Ying ＝ Cephalantheropsis halconensis（Ames）S. S. Ying ■

293501　Phaius gracilis Hayata ＝ Phaius mishmensis（Lindl. et Paxton）Rchb. f. ■

293502　Phaius gracilis Hayata var. calanthoides（Ames）S. S. Ying ＝ Cephalantheropsis calanthoides（Ames）Tang S. Liu et H. J. Su ■

293503　Phaius grandifolius Lour. ＝ Phaius tankervilleae（Banks ex L'Hér.）Blume ■

293504　Phaius grandifolius Lour. var. superbus Van Houtte ＝ Phaius tankervilleae（Banks ex L'Hér.）Blume ■

293505　Phaius guizhouensis G. Z. Li；贵州鹤顶兰；Guizhou Phaius ■

293506　Phaius guizhouensis G. Z. Li ＝ Phaius columnaris C. Z. Tang et S. J. Cheng ■

293507　Phaius hainanensis C. Z. Tang et S. J. Cheng；海南鹤顶兰；Hainan Phaius ■

293508　Phaius hainanensis C. Z. Tang et S. J. Cheng 'Red Lip'；白花红唇鹤顶兰■☆

293509　Phaius halconensis Ames ＝ Cephalantheropsis halconensis（Ames）S. S. Ying ■

293510　Phaius humblotii Rchb. f. ＝ Gastrorchis humblotii（Rchb. f.）Schltr. ■☆

293511　Phaius longicruris Z. H. Tsi ＝ Phaius takeoi（Hayata）H. J. Su ■

293512　Phaius longipes（Hook. f.）Holttum ＝ Cephalantheropsis gracilis（Lindl.）S. Y. Hu ■

293513　Phaius longipes（Hook. f.）Holttum ＝ Cephalantheropsis longipes（Hook. f.）Ormerod ■

293514　Phaius longipes（Hook. f.）Holttum var. calanthoides（Ames）T. P. Lin ＝ Cephalantheropsis halconensis（Ames）S. S. Ying ■

293515　Phaius longipes（Hook. f.）Holttum var. calanthoides（Ames）T. P. Lin ＝ Cephalantheropsis calanthoides（Ames）Tang S. Liu et H. J. Su ■

293516　Phaius maculatus Lindl. ＝ Phaius flavus（Blume）Lindl. ■

293517　Phaius maculatus Lindl. var. minor（Blume）Franch. et Sav. ＝ Phaius flavus（Blume）Lindl. ■

293518　Phaius magniflorus Z. H. Tsi et S. C. Chen ＝ Phaius wallichii Lindl. ■

293519　Phaius mannii Rchb. f.；曼氏鹤顶兰■☆

293520　Phaius marshalliana（Rchb. f.）N. E. Br. ＝ Thunia alba（Lindl.）Rchb. f. ■

293521　Phaius mindorensis Ames ＝ Cephalantheropsis longipes（Hook. f.）Ormerod ■

293522　Phaius minor Blume ＝ Phaius flavus（Blume）Lindl. ■

293523　Phaius mishmensis（Lindl. et Paxton）Rchb. f.；紫花鹤顶兰（细茎鹤顶兰,细距鹤顶兰）；Purple Phaius ■

293524　Phaius occidentalis Schltr.；西方鹤顶兰■☆

293525　Phaius pauciflorus（Blume）Blume；疏花鹤顶兰；Fewflower Phaius ■☆

293526　Phaius pulchellus Kraenzl.；美丽鹤顶兰■☆

293527　Phaius pulcher（Humbert et H. Perrier）Summerh. ＝ Gastrorchis pulchra Humbert et H. Perrier ■☆

293528　Phaius ramosii（Ames）Ames ＝ Cephalantheropsis obcordata（Lindl.）Ormerod ■

293529　Phaius roseus Rolfe ＝ Phaius mishmensis（Lindl. et Paxton）Rchb. f. ■

293530　Phaius simulans Rolfe ＝ Gastrorchis simulans（Rolfe）Schltr. ■☆

293531　Phaius sinensis Rolfe ＝ Phaius tankervilleae（Banks ex L'Hér.）Blume ■

293532　Phaius somai Hayata ＝ Phaius flavus（Blume）Lindl. ■

293533　Phaius steppicola Hand.-Mazz. ＝ Eulophia spectabilis（Dennst.）Suresh ■

293534　Phaius takeoi（Hayata）H. J. Su；长茎鹤顶兰（粗茎鹤顶兰）；Longstem Phaius ■

293535　Phaius tancarvilleae（Banks ex L'Hér.）Blume var. superbus（Van Houtte）S. Y. Hu ＝ Phaius tankervilleae（Banks ex L'Hér.）Blume ■

293536　Phaius tancarvilleae（L'Hér.）Blume f. veronicae S. Y. Hu et Barretto ＝ Phaius flavus（Blume）Lindl. ■

293537　Phaius tankervilleae（Banks ex L'Hér.）Blume；鹤顶兰（大白芨,大白及,拐子药,红鹤顶兰,红鹤兰）；Common Phaius, Nun's-hood Orchid, Phaius ■

293538　Phaius tankervilleae（Banks ex L'Hér.）Blume f. veronicae S. Y. Hu et Barretto ＝ Phaius flavus（Blume）Lindl. ■

293539　Phaius tankervilleae（Banks ex L'Hér.）Blume var. superbus（Van Houtte）S. Y. Hu ＝ Phaius tankervilleae（Banks ex L'Hér.）Blume ■

293540　Phaius tankervilliae（Banks ex L'Hér.）Blume ＝ Phaius tankervilleae（Banks ex L'Hér.）Blume ■

293541　Phaius tankervilliae（L'Hér.）Blume ＝ Phaius tankervilleae（Banks ex L'Hér.）Blume ■

293542　Phaius tuberculatus Blume ＝ Gastrorchis tuberculosa（Thouars）Schltr. ■☆

293543　Phaius tuberculosus（Thouars）Blume ＝ Gastrorchis tuberculosa（Thouars）Schltr. ■☆

293544　Phaius undulatomarginatus Hayata ＝ Phaius flavus（Blume）Lindl. ■

293545　Phaius wallichii Lindl.；大花鹤顶兰；Largeflower Phaius ■

293546　Phaius warpuri Weathers ＝ Gastrorchis tuberculosa（Thouars）Schltr. ■☆

293547　Phaius wenshanensis F. Y. Liu；文山鹤顶兰；Wenshan Phaius ■

293548　Phaius woodfordii（Hook.）Merr.；斑叶鹤顶兰■

293549　Phaius woodfordii（Hook.）Merr. ＝ Phaius flavus（Blume）Lindl. ■

293550　Phajus Lindl. ＝ Phaius Lour. ■

293551　Phakellanthus Steud. ＝ Phacellanthus Siebold et Zucc. ■

293552　Phalachroloma Cass. ＝ Erigeron L. ■●

293553　Phalacrachena Iljin（1937）；秃菊属■☆

293554　Phalacrachena calva（Ledeb.）Iljin；卡尔瓦秃菊■☆

293555　Phalacrachena inuloides（Fisch. ex Schmalh.）Iljin；秃菊■☆

293556　Phalacraea DC.（1836）；秃冠菊属■☆

293557　Phalacraea DC. ＝ Piqueria Cav. ●■☆

293558　Phalacraea latifolia DC.；秃冠菊■☆

293559　Phalacrocarpum（DC.）Willk.（1864）；秃果菊属■☆

293560　Phalacrocarpum Willk. ＝ Phalacrocarpum（DC.）Willk. ■☆

293561　Phalacrocarpum oppositifolium Willk.；秃果菊■☆

293562　Phalacrocarpus（Boiss.）Tiegh. ＝ Cephalaria Schrad.（保留属名）■

293563　Phalacroderis DC. ＝ Crepis L. ■

293564　Phalacroderis DC. ＝ Rodigia Spreng. ■

293565　Phalacroderis DC. ＝ Wibelia P. Gaertn., B. Mey. et Scherb. ■

293566　Phalacrodiscus Less. ＝ Chrysanthemum L.（保留属名）■●

293567　Phalacrodiscus Less. ＝ Leucanthemum Mill. ■●

293568　Phalacroglossum Sch. Bip. ＝ Chrysanthemum L.（保留属名）■●

293569　Phalacroloma Cass. ＝ Erigeron L. ■●

293570　Phalacroloma annuum（L.）Dumort. = Erigeron annuus（L.）Pers. ■

293571　Phalacroloma strigosum（Muhl. ex Willd.）Tzvelev = Erigeron strigosum Muhl. ex Willd. ■☆

293572　Phalacromesus Cass. = Tessaria Ruiz et Pav. ●☆

293573　Phalacros Wenzig = Crataegus L. ●

293574　Phalacroseris A. Gray（1868）;秃头苣属■☆

293575　Phalacroseris bolanderi A. Gray;秃头苣;Bolander Dandelion ■☆

293576　Phalacroseris bolanderi A. Gray var. coronata H. M. Hall = Phalacroseris bolanderi A. Gray ■☆

293577　Phalaenopsis Blume（1825）; 蝴蝶兰属（蝶兰属）; Butterflyorchis,Moth Orchid,Moth-Orchid,Phalaenopsis ■

293578　Phalaenopsis amabilis（L.）Blume;马岛蝴蝶兰（白蝴蝶兰, 报穗蝴蝶兰）■☆

293579　Phalaenopsis amabilis（L.）Blume var. aphrodite（Rchb. f.）Ames = Phalaenopsis aphrodite Rchb. f. ■

293580　Phalaenopsis amabilis Blume var. aphrodite（Rchb. f.）Ames = Phalaenopsis aphrodite Rchb. f. ■

293581　Phalaenopsis amboinensis J. J. Sm.;虎斑蝴蝶兰（安汶蝴蝶兰）■☆

293582　Phalaenopsis aphrodite Rchb. f.;蝴蝶兰（阿芙若蝴蝶兰,白蝴蝶兰,报岁蝴蝶兰,蝶兰,可爱蝴蝶兰,台湾蝴蝶兰）;Aphrodita Moth Orchid,Butterflyorchis,Moth Orchid,Phalaenopsis ■

293583　Phalaenopsis aphrodite Rchb. f. subsp. formosana Christenson; 台湾蝴蝶兰 ■

293584　Phalaenopsis bellina（Rchb. f.）Christenson;美丽蝴蝶兰■☆

293585　Phalaenopsis braceana（Hook. f.）Christenson（红河蝶兰,尖囊兰）;Common Kingidium ■

293586　Phalaenopsis braceana（Hook. f.）Christenson = Kingidium braceanum（Hook. f.）Seidenf. ■

293587　Phalaenopsis buyssoniana Rchb. f. = Doritis pulcherrima Lindl. ■

293588　Phalaenopsis celebensis H. R. Sweet;苏拉威西蝴蝶兰■☆

293589　Phalaenopsis chibae T. Yukawa;千叶蝴蝶兰■☆

293590　Phalaenopsis chuxiongensis F. Y. Liu;楚雄蝶兰;Chuxiong Moth Orchid ■

293591　Phalaenopsis chuxiongensis F. Y. Liu = Phalaenopsis hainanensis Ts. Tang et F. T. Wang ■

293592　Phalaenopsis chuxiongensis F. Y. Liu = Phalaenopsis wilsonii Rolfe ■

293593　Phalaenopsis cochlearis Holttum;贝壳蝴蝶兰■☆

293594　Phalaenopsis corningiana Rchb. f.;皱叶蝴蝶兰■☆

293595　Phalaenopsis cornucervi（Breda）Blume et Rchb. f.;角距蝴蝶兰（鹿角蝴蝶兰,羊角蝴蝶兰）;Tawnycolour Horn Moth Orchid ■☆

293596　Phalaenopsis decumbens（Griff.）Holttum;偃花蝴蝶兰■☆

293597　Phalaenopsis decumbens（Griff.）Holttum var. lobbii（Rchb. f.）H. R. Sweet;洛布蝴蝶兰（罗氏蝴蝶兰）■☆

293598　Phalaenopsis deliciosa Rchb. f.;大尖囊蝴蝶兰（大尖囊兰,俯茎胼胝兰,树葱,小蛾兰）;Large Kingidium ■☆

293599　Phalaenopsis deliciosa Rchb. f. = Kingidium deliciosum（Rchb. f.）H. R. Sweet ■

293600　Phalaenopsis equestris（Schauer）Rchb. f.;小兰屿蝴蝶兰（粉红蝴蝶兰,红花蝴蝶兰,兰屿小蝴蝶兰,桃红蝴蝶兰）;Redflower Phalaenopsis,Small Lanyu Butterflyorchis ■

293601　Phalaenopsis esmeralda Rchb. f. = Doritis pulcherrima Lindl. ■

293602　Phalaenopsis fasciate Rchb. f.;缟蝴蝶兰（横纹蝴蝶兰）■☆

293603　Phalaenopsis fimbriata（Lindl.）J. J. Sm.;裂缘蝴蝶兰（流苏蝴蝶兰）■☆

293604　Phalaenopsis formosana Miwa = Phalaenopsis aphrodite Rchb. f. ■

293605　Phalaenopsis fuscata Rchb. f.;淡褐蝴蝶兰（褐斑蝴蝶兰,香蝴蝶兰）;Brownish Phalaenopsis ■☆

293606　Phalaenopsis gigantea J. J. Sm.;象耳蝴蝶兰■☆

293607　Phalaenopsis hainanensis Ts. Tang et F. T. Wang;海南蝴蝶兰（海南蝶兰）;Hainan Butterflyorchis ■

293608　Phalaenopsis hieroglyphica（Rchb. f.）H. R. Sweet;豹纹蝴蝶兰（象形文字蝴蝶兰）■☆

293609　Phalaenopsis honghenensis F. Y. Liu;红河蝴蝶兰;Honghe Butterflyorchis ■

293610　Phalaenopsis hongkenensis F. Y. Liu = Kingidium braceanum（Hook. f.）Seidenf. ■

293611　Phalaenopsis hybrida Hort.;杂交蝴蝶兰（蝴蝶万朵兰）■☆

293612　Phalaenopsis intermedia Lindl.;小舌蝴蝶兰■☆

293613　Phalaenopsis javanica J. J. Sm.;爪哇蝴蝶兰■☆

293614　Phalaenopsis lindenii Loker;细花蝴蝶兰（林登蝴蝶兰）■☆

293615　Phalaenopsis lobbii（Rchb. f.）H. R. Sweet = Phalaenopsis decumbens（Griff.）Holttum var. lobbii（Rchb. f.）H. R. Sweet ■☆

293616　Phalaenopsis lowii Rchb. f.;洛威氏蝴蝶兰（劳氏蝴蝶兰）;Lowi Phalaenopsis ■☆

293617　Phalaenopsis lueddemanniana Rchb. f.;露德蝴蝶兰（短梗蝴蝶兰,菲律宾蝴蝶兰,路易德氏蝴蝶兰）;Lueddemann Phalaenopsis ■☆

293618　Phalaenopsis maculata Rchb. f.;斑点蝴蝶兰（斑花蝴蝶兰）■☆

293619　Phalaenopsis malipoensis Z. J. Liu et S. C. Chen;麻栗坡蝴蝶兰■

293620　Phalaenopsis mannii Rchb. f.;版纳蝴蝶兰（卷瓣蝴蝶兰,曼尼氏蝴蝶兰,曼氏蝴蝶兰）;Mann Butterflyorchis,Mann Phalaenopsis ■

293621　Phalaenopsis mariae Burb.;玛利蝴蝶兰（玛莉亚蝴蝶兰,绒瓣蝴蝶兰）;Maria Phalaenopsis ■☆

293622　Phalaenopsis micholitzii Rolfe;米库氏蝴蝶兰■☆

293623　Phalaenopsis minor F. Y. Liu;小蝶兰;Small Butterflyorchis ■

293624　Phalaenopsis minor F. Y. Liu = Phalaenopsis wilsonii Rolfe ■

293625　Phalaenopsis pallens（Lindl.）Rchb. f.;乳黄蝴蝶兰（淡白蝴蝶兰）■☆

293626　Phalaenopsis parishii Rchb. f.;侏儒蝴蝶兰（柏氏蝴蝶兰）■☆

293627　Phalaenopsis parishii Rchb. f. var. lobbii Rchb. f. = Phalaenopsis lobbii（Rchb. f.）H. R. Sweet ■

293628　Phalaenopsis philippinensis Golamco ex Fowlie et C. Z. Tang;菲律宾蝴蝶兰■☆

293629　Phalaenopsis pulcherrima（Lindl.）J. J. Sm. = Doritis pulcherrima Lindl. ■

293630　Phalaenopsis pulchra（Rchb. f.）H. R. Sweet;美花蝴蝶兰（美丽蝴蝶兰）■☆

293631　Phalaenopsis riteiwanensis Masam. = Phalaenopsis equestris（Schauer）Rchb. f. ■

293632　Phalaenopsis sanderiana Rchb. f.;火焰蝴蝶兰（桑德利亚蝴蝶兰）■☆

293633　Phalaenopsis schilleriana Rchb. f.;希莱氏蝴蝶兰（长梗蝴蝶兰,花叶蝶兰,花叶蝴蝶兰,西蕾丽蝴蝶兰,希氏蝴蝶兰）;Moth Orchid,Schiller Phalaenopsis ■☆

293634　Phalaenopsis speciosa Rchb. f.;紫红蝴蝶兰（美丽蝴蝶兰）;Beautiful Phalaenopsis ■☆

293635　Phalaenopsis stobariana Rchb. f.;滇西蝴蝶兰;Stobar Butterflyorchis ■

293636　Phalaenopsis stuartiana Rchb. f.;斯图阿氏蝴蝶兰（史塔基蝴蝶兰,小叶蝴蝶兰）;Stuart Phalaenopsis ■☆

293637　Phalaenopsis sumatrana Korth. et Rchb. f.;苏门答腊蝴蝶兰（南洋蝴蝶兰）;Sumatra Phalaenopsis ■☆

293638 Phalaenopsis sumatrana Korth. et Rchb. f. var. gersenii（Teijsm. et Binn. ）Rchb. f. ；网纹蝴蝶兰■

293639 Phalaenopsis taenialis（Lindl. ）Christenson et Pradhan；小尖囊兰（扁根蝴蝶兰）；Small Kingidium ■

293640 Phalaenopsis taenialis（Lindl. ）Christenson et Pradhan = Kingidium taeniale（Lindl. ）P. F. Hunt ■

293641 Phalaenopsis tetraspsis Rchb. f. ；盾花蝴蝶兰■☆

293642 Phalaenopsis venosa P. S. Shim et Fowlie；红脉蝴蝶兰■☆

293643 Phalaenopsis violacea Teijsm. et Binn. ；菫花蝴蝶兰（大叶蝴蝶兰，紫纹蝴蝶兰）；Violetflower Phalaenopsis ■☆

293644 Phalaenopsis viridis J. J. Sm. ；绿花蝴蝶兰■☆

293645 Phalaenopsis wightii Rchb. f. = Phalaenopsis decumbens（Griff. ）Holttum ■☆

293646 Phalaenopsis wightii Rchb. f. = Phalaenopsis deliciosa Rchb. f. ■

293647 Phalaenopsis wightii Rchb. f. var. stobartiana（Rchb. f. ）Burb. = Phalaenopsis stobariana Rchb. f. ■

293648 Phalaenopsis wilsonii Rolfe；华西蝴蝶兰（楚雄蝶兰，蝶兰，金环草，捆仙绳，西南蝴蝶兰，小蝶兰）；E. H. Wilson Butterflyorchis ■

293649 Phalaenopsis zebrina Teijsm. et Binn. ；斑纹蝴蝶兰■☆

293650 Phalanganthus Schrank = Anthericum L. ■☆

293651 Phalangion St. -Lag. = Anthericum L. ■☆

293652 Phalangion St. -Lag. = Phalangium Mill. ■☆

293653 Phalangites Bubani = Anthericum L. ■☆

293654 Phalangites Bubani = Phalangion St. -Lag. ■☆

293655 Phalangium Adans. = Anthericum L. ■☆

293656 Phalangium Adans. = Urginea Steinh. ■☆

293657 Phalangium Boehm. （1760）；双列百合属■☆

293658 Phalangium Burm. f. = Melasphaerula Ker Gawl. ■☆

293659 Phalangium Kuntze = Bulbine Wolf（保留属名）■☆

293660 Phalangium Mill. = Anthericum L. ■☆

293661 Phalangium Möhring ex Kuntze = Bulbine Wolf（保留属名）■☆

293662 Phalangium abyssinicum Kunth = Chlorophytum tetraphyllum（L. f. ）Baker ■☆

293663 Phalangium algeriense Boiss. et Reut. = Anthericum maurum Rothm. ■☆

293664 Phalangium altissimum（Mill. ）Kuntze = Bulbine asphodeloides（L. ）Spreng. ■☆

293665 Phalangium angustifolium（Hochst. ex A. Rich. ）Schweinf. = Anthericum angustifolium Hochst. ex A. Rich. ■☆

293666 Phalangium asphodeloides Kuntze = Bulbine asphodeloides（L. ）Spreng. ■☆

293667 Phalangium baeticum Boiss. = Anthericum baeticum（Boiss. ）Boiss. ■☆

293668 Phalangium baeticum Boiss. var. rhiphaeum Pau et Font Quer = Anthericum baeticum（Boiss. ）Boiss. ■☆

293669 Phalangium bipedunculatum（Jacq. ）Poir. = Chlorophytum triflorum（Aiton）Kunth ■☆

293670 Phalangium bulbosum（R. Br. ）Kuntze；双列百合■☆

293671 Phalangium canaliculatum（Aiton）Poir. = Trachyandra ciliata（L. f. ）Kunth ■☆

293672 Phalangium capillare Poir. = Bulbinella triquetra（L. f. ）Kunth ■☆

293673 Phalangium comosum（Thunb. ）Poir. = Chlorophytum comosum（Thunb. ）Jacques ■☆

293674 Phalangium croceum Michx. = Schoenolirion croceum（Michx. ）A. W. Wood ■☆

293675 Phalangium elatum（Aiton）Redouté = Chlorophytum capense（L. ）Voss ■

293676 Phalangium fastigiatum Poir. = Chlorophytum capense（L. ）Voss ■

293677 Phalangium flavescens Kunth = Echeandia flavescens（Schult. et Schult. f. ）Cruden ■☆

293678 Phalangium fontqueri Sennen et Mauricio = Anthericum baeticum（Boiss. ）Boiss. ■☆

293679 Phalangium humile（Hochst. ex A. Rich. ）Schweinf. et Asch. = Anthericum angustifolium Hochst. ex A. Rich. ■☆

293680 Phalangium indicum（Schult. et Schult. f. ）Kunth = Chlorophytum tuberosum（Roxb. ）Baker ■☆

293681 Phalangium liliago（L. ）Schreb. var. algeriense（Boiss. et Reut. ）Batt. et Trab. = Anthericum baeticum（Boiss. ）Boiss. ■☆

293682 Phalangium longifolium（Jacq. ）Poir. = Trachyandra ciliata（L. f. ）Kunth ■☆

293683 Phalangium nepalensis Lindl. = Chlorophytum nepalense（Lindl. ）Baker ■

293684 Phalangium ornithogaloides（Hochst. ex A. Rich. ）Schweinf. et Asch. = Chlorophytum tuberosum（Roxb. ）Baker ■☆

293685 Phalangium parviflorum Wight = Chlorophytum laxum R. Br. ■

293686 Phalangium pomeridianum（DC. ）Sweet = Chlorogalum pomeridianum Kunth ■☆

293687 Phalangium quamash Pursh = Camassia quamash（Pursh）Greene ■☆

293688 Phalangium revolutum（L. ）Poir. = Trachyandra revoluta（L. ）Kunth ■☆

293689 Phalangium scilloides Poir. = Lilaea scilloides（Poir. ）Hauman ■☆

293690 Phalangium squameum（L. f. ）Poir. = Trachyandra hispida（L. ）Kunth ■☆

293691 Phalangium triflorum（Aiton）Pers. = Chlorophytum triflorum（Aiton）Kunth ■☆

293692 Phalangium tuberosum（Roxb. ）Kunth = Chlorophytum tuberosum（Roxb. ）Baker ■☆

293693 Phalangium undulatum（Jacq. ）Poir. = Chlorophytum undulatum（Jacq. ）Oberm. ■☆

293694 Phalangium viviparum Hort. = Chlorophytum comosum（Thunb. ）Jacques ■☆

293695 Phalarella Boiss. = Phleum L. ■

293696 Phalariaceae Meisn. ；䕡草科■

293697 Phalariaceae Meisn. = Thymelaea Mill. （保留属名）●■

293698 Phalaridaceae Burnett = Gramineae Juss. （保留科名）■●

293699 Phalaridaceae Burnett = Poaceae Barnhart（保留科名）■●

293700 Phalaridaceae Link = Gramineae Juss. （保留科名）■●

293701 Phalaridaceae Link = Poaceae Barnhart（保留科名）■●

293702 Phalaridantha St. -Lag. = Phalaris L. ■

293703 Phalaridium Nees = Dissanthelium Trin. ■☆

293704 Phalaridium Nees et Meyen = Dissanthelium Trin. ■☆

293705 Phalaris L. （1753）；䕡草属（草芦属，鹬草属）；Canary Grass, Canarygrass, Canary-Grass, Ribbon Grass ■☆

293706 Phalaris angusta Nees ex Trin. ；狭䕡草；Narrow Canarygrass ■☆

293707 Phalaris appendiculata Schult. ；附属物䕡草■☆

293708 Phalaris aquatica L. ；水生䕡草；Bulbous Canarygrass, Bulbous Canary-grass, Harding Grass, Toowoomba Canary Grass ■☆

293709 Phalaris arundinacea L. ；䕡草（草芦，马羊草，五色草，鹬，䕡，鹬草）；Adder's Grass, Gardener's Garters, Joan's Ribbon, Lady's Garters, Lady's Grass, Lady's Laces, Lady's Ribands, Painted Grass, Reed Canary Grass, Reed Canarygrass, Reed Canary-grass, Reed Phalaris, Reed-grass, Ribbon Grass, Ribbon-grass, Sparked Grass,

Spires ■

293710　Phalaris arundinacea L.'Picta';丝带草(彩叶虉草,花叶虉草,银边草);Gardener's Garter,Painted Canarygrass,Ribbon Grass,Ribbon-grass,Various-leaved Canary – Grass ■

293711　Phalaris arundinacea L. f. variegata (Parn.) Druce = Phalaris arundinacea L. ■

293712　Phalaris arundinacea L. f. variegata (Parn.) Druce = Phalaris arundinacea L. var. variegata Parn. ■☆

293713　Phalaris arundinacea L. subsp. oehleri Pilg. = Phalaris caesia Nees ■☆

293714　Phalaris arundinacea L. var. japonica (Steud.) Hack.;日本虉草■☆

293715　Phalaris arundinacea L. var. leioclada Maire = Phalaris caesia Nees ■☆

293716　Phalaris arundinacea L. var. naturalised ? = Phalaris arundinacea L. ■

293717　Phalaris arundinacea L. var. picta L. = Phalaris arundinacea L.'Picta' ■

293718　Phalaris arundinacea L. var. picta L. = Phalaris arundinacea L. ■

293719　Phalaris arundinacea L. var. thyrsoidea Willk. = Phalaris caesia Nees ■☆

293720　Phalaris arundinacea L. var. variegata Parn.;花叶虉草;Gardener's Garters,Ribbon-grass ■☆

293721　Phalaris avicularis Salisb. = Phalaris canariensis L. ■

293722　Phalaris brachystachys Link;短穗杜草;Confused Canary-grass,Shortspike Canary Grass,Shortspike Canarygrass ■☆

293723　Phalaris bulbosa L.;球虉草■☆

293724　Phalaris bulbosa L. = Phalaris aquatica L. ■☆

293725　Phalaris bulbosa L. var. alata (Trab.) Maire et Weiller = Phalaris aquatica L. ■☆

293726　Phalaris bulbosa L. var. clausonis (Maire et Trab.) Maire = Phalaris aquatica L. ■☆

293727　Phalaris bulbosa L. var. hirtiglumis (Trab.) Maire et Weiller = Phalaris elongata Braun-Blanq. ■☆

293728　Phalaris caesia Nees;蔡斯虉草■☆

293729　Phalaris canariensis L.;洋虉草(加拿大丽鹬草,加那利杜草,加那列杜草,金丝雀虉草);Annual Canarygrass,Canary Grass,Canary Phalaris,Canarygrass,Canary-grass,Common Canary Grass,Grass Corn,Petty Panick ■

293730　Phalaris canariensis L. subsp. brachystachys (Link) Posp. = Phalaris brachystachys Link ■☆

293731　Phalaris caroliniana Walter;卡罗林虉草;Canary Grass,May Grass ■☆

293732　Phalaris coerulescens Desf.;浅蓝虉草■☆

293733　Phalaris daviesii S. T. Blake;戴维斯虉草■☆

293734　Phalaris dentata L. f. = Prionanthium dentatum (L. f.) Henrard ■☆

293735　Phalaris elongata Braun-Blanq.;长虉草■☆

293736　Phalaris gracilis Parl. = Phalaris minor Retz. ■

293737　Phalaris hematites Duval-Jouve et Paris = Phalaris minor Retz. ■

293738　Phalaris hematites Duval-Jouve et Paris var. granulosa Sennen et Mauricio = Phalaris minor Retz. ■

293739　Phalaris hirtiglumis (Trab.) Baldini = Phalaris elongata Braun-Blanq. ■☆

293740　Phalaris hispida Thunb. = Arthraxon hispidus (Thunb.) Makino ■

293741　Phalaris maderensis (Menezes) Menezes;梅德虉草■☆

293742　Phalaris mauritii Sennen = Phalaris minor Retz. ■

293743　Phalaris minor Retz.;小虉草(细虉草,小籽虉草);Lesser Canary Grass,Lesser Canary-grass,Littleseed Canarygrass,Littleseeded Canary Grass ■

293744　Phalaris minor Retz. var. gracilis (Parl.) Pamp. = Phalaris minor Retz. ■

293745　Phalaris minor Retz. var. hematites (Duval-Jouve et Paris) Trab. = Phalaris minor Retz. ■

293746　Phalaris minor Retz. var. integra Trab. = Phalaris minor Retz. ■

293747　Phalaris nodosa L. = Phalaris aquatica L. ■☆

293748　Phalaris nodosa Murray = Phalaris aquatica L. ■☆

293749　Phalaris oryzoides L. = Leersia oryzoides (L.) Sw. ■

293750　Phalaris paradoxa L.;奇异虉草(变形虉草,奇异杜草,珍杜草);Awned Canary-grass,Hood Canarygrass ■

293751　Phalaris paradoxa L. f. nana Chiov. = Phalaris appendiculata Schult. ■☆

293752　Phalaris paradoxa L. var. appendiculata (Schult.) Chiov. = Phalaris appendiculata Schult. ■☆

293753　Phalaris paradoxa L. var. intacta Coss. et Durieu = Phalaris paradoxa L. ■

293754　Phalaris paradoxa L. var. intermedia Coss. et Durieu = Phalaris paradoxa L. ■

293755　Phalaris paradoxa L. var. praemorsa (Lam.) Coss. et Durieu;啮蚀虉草■☆

293756　Phalaris paradoxa L. var. praemorsa (Lam.) Coss. et Durieu = Phalaris paradoxa L. ■

293757　Phalaris paradoxa L. var. praemorsa Coss. et Durieu = Phalaris paradoxa L. ■

293758　Phalaris phleoides L. = Phleum phleoides (L.) H. Karst. ■

293759　Phalaris pseudoparadoxa Fig. et De Not. = Phalaris appendiculata Schult. ■☆

293760　Phalaris pubescens Lam. = Rostraria salzmannii (Boiss. et Reut.) Holub ■☆

293761　Phalaris semiverticillata Forssk. = Agrostis viridis Gouan ■

293762　Phalaris semiverticillata Forssk. = Polypogon viridis (Gouan) Breistr. ■

293763　Phalaris setacea Forssk. = Pennisetum setaceum (Forssk.) Chiov. ■☆

293764　Phalaris stenoptera Hack.;窄翅块茎虉草;Cut-scale Canary Grass,Cut-scale Canary-grass,Harding's Grass,Harding's-grass,Toowoomba Canary Grass ■☆

293765　Phalaris truncata Guss.;平截虉草■☆

293766　Phalaris truncata Guss. var. angustata Trab. = Phalaris truncata Guss. ■☆

293767　Phalaris truncata Guss. var. viliglumis Trab. = Phalaris truncata Guss. ■☆

293768　Phalaris tuberosa L.;块茎虉草;Harding Grass,Toowoomba Canary Grass ■☆

293769　Phalaris tuberosa L. = Phalaris aquatica L. ■☆

293770　Phalaris tuberosa L. var. clausonis Maire et Trab. = Phalaris aquatica L. ■☆

293771　Phalaris tuberosa L. var. hirtiglumis Trab. = Phalaris elongata Braun-Blanq. ■☆

293772　Phalaris tuberosa L. var. stenoptera ? = Phalaris stenoptera Hack. ■☆

293773　Phalaris urundinacea ?;乌隆迪虉草;Reed-grass ■☆

293774　Phalaris vaginiflora Forssk. = Crypsis schoenoides (L.) Lam. ■

293775　Phalaris vaginiflora Forssk. = Crypsis vaginiflora (Forssk.)

Opiz ■☆

293776　Phalaris velutina Forssk. = Digitaria velutina（Forssk.）P. Beauv. ■☆

293777　Phalaris verticillata Forssk. = Polypogon viridis（Gouan）Breistr. ■

293778　Phalaris zizanioides L. = Chrysopogon zizanioides（L.）Roberty ■

293779　Phalaris zizanioides L. = Vetiveria zizanioides（L.）Nash ■

293780　Phalaroides Wolf = Phalaris L. + Phleum L. ■

293781　Phalaroides Wolf = Phalaris L. ■

293782　Phalaroides Wolf(1781);拟虉草属■☆

293783　Phalaroides arundinacea（L.）Rauschert = Phalaris arundinacea L. ■

293784　Phalaroides arundinacea（L.）Rauschert subsp. caesia（Nees）Tzvelev = Phalaris caesia Nees ■☆

293785　Phalaroides arundinacea（L.）Rauschert subsp. japonica（Steud.）Tzvelev = Phalaris arundinacea L. var. japonica（Steud.）Hack. ■☆

293786　Phalaroides arundinacea（L.）Rauschert subsp. oehleri（Pilg.）Valdés et H. Scholz = Phalaroides caesia（Nees）Holub ■☆

293787　Phalaroides arundinacea（L.）Rauschert var. picta（L.）Tzvelev = Phalaris arundinacea L. ■

293788　Phalaroides caesia（Nees）Holub = Phalaris caesia Nees ■☆

293789　Phalaroides japonica（Steud.）De Moor = Phalaris arundinacea L. var. japonica（Steud.）Hack. ■☆

293790　Phaleria Jack(1822);白斑瑞香属●☆

293791　Phaleria acuminata Gilg;渐尖白斑瑞香●☆

293792　Phaleria angustifolia A. C. Sm.;窄叶白斑瑞香●☆

293793　Phaleria axillaris Elmer;腋生白斑瑞香●☆

293794　Phaleria biflora（C. T. White）Herber;双花白斑瑞香●☆

293795　Phaleria lanceolata Gilg;披针叶白斑瑞香●☆

293796　Phaleria laurifolia Hook. f. ;桂叶白斑瑞香●☆

293797　Phaleria longifolia Boerl.;长叶白斑瑞香●☆

293798　Phaleria longituba P. F. Stevens;长管白斑瑞香●☆

293799　Phaleria montana Gilg;山地白斑瑞香●☆

293800　Phaleria parvifolia Backer;小叶白斑瑞香●☆

293801　Phaleria platyphylla Merr.;宽叶白斑瑞香●☆

293802　Phaleriaceae Meisn. = Thymelaea Mill.（保留属名）●■

293803　Phalerocarpus G. Don = Gaultheria L. ●

293804　Phallaria Schumach. et Thonn. = Psydrax Gaertn. ●☆

293805　Phallaria lucida（Harv.）Hochst. = Psydrax obovata（Klotzsch ex Eckl. et Zeyh.）Bridson ●☆

293806　Phallaria schimperi Hochst. = Psydrax schimperiana（A. Rich.）Bridson ●☆

293807　Phallaria spinosa Schumach. et Thonn. = Vangueriella spinosa（Schumach. et Thonn.）Verdc. ●☆

293808　Phallerocarpus G. Don = Gaultheria L. ●

293809　Phallerocarpus G. Don = Phalerocarpus G. Don ●

293810　Phalocallis Herb. = Cypella Herb. ■☆

293811　Phalodallls T. Durand et Jacks. = Cypella Herb. ■☆

293812　Phalodallls T. Durand et Jacks. = Phalocallis Herb. ■☆

293813　Phaloe Dumort. = Sagina L. ■

293814　Phalolepis Cass. = Centaurea L.（保留属名）●■

293815　Phalona Dumort. = Cynosurus L. ■

293816　Phalona Dumort. = Falona Adans. ■

293817　Phanera Lour. = Bauhinia L. ●

293818　Phanera burkeana Benth. = Tylosema esculentum（Burch.）A. Schreib. ●☆

293819　Phanera championii Benth. = Bauhinia championii（Benth.）Benth. ●

293820　Phanera corymbosa（Roxb. ex DC.）Benth. = Bauhinia corymbosa Roxb. ex DC. ●

293821　Phanera corymbosa Benth. = Bauhinia corymbosa Roxb. ●

293822　Phanera glauca Wall. ex Benth. = Bauhinia glauca（Wall. ex Benth.）Benth. ●

293823　Phanera macrostachya Benth. = Bauhinia wallichii J. F. Macbr. ●

293824　Phanera nervosa Wall. ex Benth. = Bauhinia nervosa（Wall. ex Benth.）Baker ●

293825　Phanera pyrrhoclada（Drake）de Wit = Bauhinia pyrrhoclada Drake ●

293826　Phanera retusa（Roxb.）Benth. = Bauhinia retusa Roxb. ●☆

293827　Phanera tenuiflora（Watt ex C. B. Clarke）de Wit = Bauhinia glauca（Wall. ex Benth.）Benth. subsp. tenuiflora（Watt ex C. B. Clarke）K. Larsen et S. S. Larsen ●

293828　Phanera variegata（L.）Benth. = Bauhinia variegata L. ●

293829　Phanerandra Stschegl. = Leucopogon R. Br.（保留属名）●☆

293830　Phaneranthera DC. ex Meisn. = Nonea Medik. ■

293831　Phanerocalyx S. Moore = Heisteria Jacq.（保留属名）●☆

293832　Phanerodiscus Cavaco(1954);显盘树属●☆

293833　Phanerodiscus capuronii Malécot,G. E. Schatz et Bosser;凯普伦显盘树●☆

293834　Phanerodiscus diospyroidea Capuron;柿状显盘树●☆

293835　Phanerodiscus perrieri Cavaco;佩里耶显盘树●☆

293836　Phaneroglossa B. Nord.(1978);腋毛千里光属●☆

293837　Phaneroglossa bolusii（Oliv.）B. Nord. ;腋毛千里光●☆

293838　Phanerogonocarpus Cavaco = Tambourissa Sonn. ●☆

293839　Phanerogonocarpus Cavaco(1958);节果树属●☆

293840　Phanerogonocarpus capuronii Cavaco;节果树●☆

293841　Phanerostylis（A. Gray）R. M. King et H. Rob.(1972);显柱菊属■☆

293842　Phanerostylis（A. Gray）R. M. King et H. Rob. = Brickellia Elliott(保留属名)■●

293843　Phanerostylis coahuilensis（A. Gray）R. M. King et H. Rob. ;显柱菊■☆

293844　Phanerotaenia St. John = Polytaenia DC. ■☆

293845　Phania DC.(1836);背腺菊属■●☆

293846　Phania arbutifolia DC. ;背腺菊■●☆

293847　Phania multicaulis DC. ;多茎背腺菊■☆

293848　Phania trinervia Moc. ex DC. ;三脉背腺菊■☆

293849　Phaniasia Blume ex Miq. = Gymnadenia R. Br. ■

293850　Phaniasia Blume ex Miq. = Habenaria Willd. ■

293851　Phanopyrum（Raf.）Nash = Panicum L. ■

293852　Phanopyrum Nash = Panicum L. ■

293853　Phanrangia Tardieu = Mangifera L. ●

293854　Phantis Adans. = Atalantia Corrêa(保留属名)●

293855　Pharaceae（Stapf）Herter = Gramineae Juss.（保留科名）■●

293856　Pharaceae（Stapf）Herter = Poaceae Barnhart(保留科名)■●

293857　Pharaceae Herter = Gramineae Juss.（保留科名）■●

293858　Pharaceae Herter = Poaceae Barnhart(保留科名)■●

293859　Pharbitis Choisy = Diatremis Raf.（废弃属名）●■

293860　Pharbitis Choisy = Ipomoea L.（保留属名）●■

293861　Pharbitis Choisy(1833)（保留属名）;牵牛属（紫牵牛属）;Morning Glory,Pharbitis ■

293862　Pharbitis acuminata（Vahl）Choisy = Ipomoea indica（Burm.）Merr. ■

293863 Pharbitis acuminata (Vahl) Choisy var. congesta (R. Br.) Choisy = Ipomoea indica (Burm.) Merr. ■
293864 Pharbitis barbigera (Sweet) G. Don = Ipomoea nil (L.) Roth ■
293865 Pharbitis catharica (Poir.) Choisy = Ipomoea indica (Burm.) Merr. ■
293866 Pharbitis congesta (R. Br.) Hara = Ipomoea indica (Burm.) Merr. ■
293867 Pharbitis fragrans Bojer = Ipomoea rubens Choisy ■☆
293868 Pharbitis hederacea (Jacq.) Choisy = Ipomoea nil (L.) Roth ■
293869 Pharbitis hederacea (L.) Choisy = Ipomoea hederacea (L.) Jacq. ■
293870 Pharbitis hederacea Franch. = Ipomoea nil (L.) Roth ■
293871 Pharbitis hispida Choisy = Ipomoea purpurea (L.) Roth ■
293872 Pharbitis indica (Burm.) Hagiw. = Ipomoea indica (Burm.) Merr. ■
293873 Pharbitis insularia Choisy = Ipomoea indica (Burm.) Merr. ■
293874 Pharbitis learii (Paxton) Lindl. = Ipomoea indica (Burm.) Merr. ■
293875 Pharbitis learii (Paxton) Lindl. = Ipomoea nil (L.) Roth ■
293876 Pharbitis nil (L.) Choisy = Ipomoea nil (L.) Roth ■
293877 Pharbitis nil (L.) Choisy var. japonica (Hallier f.) H. Hara = Ipomoea nil (L.) Roth ■
293878 Pharbitis preauxii Webb;普雷牵牛■☆
293879 Pharbitis purpurea (L.) Voigt = Ipomoea purpurea (L.) Roth ■
293880 Pharbitis rubrocoerulea Choisy = Ipomoea tricolor Cav. ■☆
293881 Pharbitis triloba (Thunb.) Miq. = Ipomoea nil (L.) Roth ■
293882 Pharetranthus F. W. Klatt = Petrobium R. Br. (保留属名)●☆
293883 Pharetrella Salisb. = Cyanella L. ■☆
293884 Pharia Steud. = Phania DC. ■●☆
293885 Pharium Herb. = Bessera Schult. f. (保留属名)■☆
293886 Pharmacaceae Dulac = Ranunculaceae Juss. (保留科名)●■
293887 Pharmaceum Kuntze = Astronia Blume ●
293888 Pharmacosycea Miq. = Ficus L. ●
293889 Pharmacum Kuntze = Astronia Blume ●
293890 Pharmacum Kuntze = Astronia Noronha ●
293891 Pharmacum Kuntze = Murraya J. König ex L. (保留属名)●
293892 Pharmacum Rumph. ex Kuntze = Astronia Noronha ●
293893 Pharmacum Rumph. ex Kuntze = Murraya J. König ex L. (保留属名)●
293894 Pharnaceaceae Dulac = Molluginaceae Bartl. (保留科名)■
293895 Pharnaceaceae Dulac = Ranunculaceae Juss. (保留科名)●■
293896 Pharnaceaceae Martinov = Aizoaceae Martinov(保留科名)●■
293897 Pharnaceae Martinov = Aizoaceae Martinov(保留科名)●■
293898 Pharnaceum L. (1753);线叶粟草属■●☆
293899 Pharnaceum albens L. f.;微白线叶粟草■☆
293900 Pharnaceum alpinum Adamson;高山线叶粟草■☆
293901 Pharnaceum aurantium (DC.) Druce;橙色线叶粟草■☆
293902 Pharnaceum brevicaule (DC.) Bartl.;短茎线叶粟草■☆
293903 Pharnaceum cerviana L. = Mollugo cerviana (L.) Ser. ■
293904 Pharnaceum ciliare Adamson;缘毛线叶粟草■☆
293905 Pharnaceum confertum (DC.) Eckl. et Zeyh.;密集线叶粟草■☆
293906 Pharnaceum confertum (DC.) Eckl. et Zeyh. var. brachyphyllum Adamson;短叶线叶粟草■☆
293907 Pharnaceum croceum E. Mey. ex Fenzl;镉黄线叶粟草■☆
293908 Pharnaceum depressum L. = Polycarpon prostratum (Forssk.) Asch. et Schweinf. ■
293909 Pharnaceum dichotomum L. f.;二歧线叶粟草■☆
293910 Pharnaceum distichum Thunb. = Pharnaceum thunbergii Adamson ■☆
293911 Pharnaceum elongatum (DC.) Adamson;伸长线叶粟草■☆
293912 Pharnaceum exiguum Adamson;小线叶粟草■☆
293913 Pharnaceum fluviale Eckl. et Zeyh.;河边线叶粟草■☆
293914 Pharnaceum gracile Fenzl;纤细线叶粟草■☆
293915 Pharnaceum hirtum Spreng. = Glinus lotoides L. ■
293916 Pharnaceum incanum L.;灰毛线叶粟草■☆
293917 Pharnaceum lanatum Bartl.;绵毛线叶粟草■☆
293918 Pharnaceum lichtensteinianum Schult. = Adenogramma lichtensteiniana (Schult.) Druce ■☆
293919 Pharnaceum lineare L. f.;细线叶粟草■☆
293920 Pharnaceum longearistatum Dinter = Coelanthum grandiflorum E. Mey. ex Fenzl ■☆
293921 Pharnaceum maritimum Walter = Sesuvium maritimum (Walter) Britton, Stearns et Poggenb. ■☆
293922 Pharnaceum merxmuelleri Friedrich = Pharnaceum brevicaule (DC.) Bartl. ■☆
293923 Pharnaceum microphyllum L. f.;小叶线叶粟草■☆
293924 Pharnaceum microphyllum L. f. var. albens Adamson;白小叶线叶粟草■☆
293925 Pharnaceum mollugo L. = Glinus oppositifolius (L.) Aug. DC. ■
293926 Pharnaceum mucronatum Thunb. = Psammotropha mucronata (Thunb.) Fenzl ●☆
293927 Pharnaceum rigidum Bartl. = Adenogramma rigida (Bartl.) Sond. ■☆
293928 Pharnaceum rubens Adamson;淡红线叶粟草■☆
293929 Pharnaceum salsoloides Burch. = Hypertelis salsoloides (Burch.) Adamson ■☆
293930 Pharnaceum salsoloides Burch. var. mossamedense Welw. ex Hiern = Hypertelis salsoloides (Burch.) Adamson var. mossamedensis (Welw. ex Hiern) Gonc. ■☆
293931 Pharnaceum scleranthoides Sond. = Suessenguthiella scleranthoides (Sond.) Friedrich ■☆
293932 Pharnaceum serpyllifolium L. f.;百里香线叶粟草■☆
293933 Pharnaceum spathulatum Spreng. = Mollugo nudicaulis Lam. ■
293934 Pharnaceum subtile E. Mey.;细小线叶粟草■☆
293935 Pharnaceum suffruticosum Pall. = Flueggea suffruticosa (Pall.) Baill. ●
293936 Pharnaceum thunbergii Adamson;通贝里线叶粟草■☆
293937 Pharnaceum trigonum Eckl. et Zeyh.;三角线叶粟草■☆
293938 Pharnaceum verrucosum Eckl. et Zeyh. = Hypertelis salsoloides (Burch.) Adamson ■☆
293939 Pharnaceum viride Adamson;绿线叶粟草■☆
293940 Pharnoceum cerviana L. = Mollugo cerviana Ser. ■
293941 Pharochilum D. L. Jones et M. A. Clem. (2002);织唇兰属■☆
293942 Pharochilum D. L. Jones et M. A. Clem. = Pterostylis R. Br. (保留属名)■☆
293943 Pharseophora Miers = Memora Miers ●☆
293944 Pharus P. Browne(1756);被禾属(法若竿属)■☆
293945 Pharus angustifolius Döll;窄叶被禾■☆
293946 Pharus aristatus Retz. = Hygroryza aristata (Retz.) Nees ex Wight et Arn. ■
293947 Pharus brasiliensis Raddi;巴西被禾■☆
293948 Pharus ciliatus Retz.;睫毛被禾■☆
293949 Pharus glaber Kunth;光被禾■☆
293950 Pharus lancifolius Ham.;剑叶被禾■☆

293951 Pharus latifolius L.；宽叶被禾■☆

293952 Pharus longifolius Swallen；长叶被禾■☆

293953 Pharus micranthus Schrad. ex Nees；小花被禾■☆

293954 Pharus ovalifolius Ham.；卵叶被禾■☆

293955 Pharus parvifolius Nash；小叶被禾■☆

293956 Phasellus Medik.（废弃属名）= Phaseolus L. ■

293957 Phasellus Medik.（废弃属名）= Strophostyles Elliott（保留属名）■☆

293958 Phaseolaceae Mart. = Fabaceae Lindl.（保留科名）●■

293959 Phaseolaceae Mart. = Leguminosae Juss.（保留科名）●■

293960 Phaseolaceae Ponce de León et Alvares = Fabaceae Lindl.（保留科名）●■

293961 Phaseolaceae Ponce de León et Alvares = Leguminosae Juss.（保留科名）●■

293962 Phaseolaceae Schnizl. = Fabaceae Lindl.（保留科名）●■

293963 Phaseolaceae Schnizl. = Leguminosae Juss.（保留科名）●■

293964 Phaseolodes Kuntze = Millettia Wight et Arn.（保留属名）●■

293965 Phaseoloides Duhamel（废弃属名）= Wisteria Nutt.（保留属名）●

293966 Phaseolus L.（1753）；菜豆属；Bean ■

293967 Phaseolus abyssinicus Savi = Vigna radiata（L.）R. Wilczek var. sublobata（Roxb.）Verdc. ■☆

293968 Phaseolus aconitifolius Jacq. = Vigna aconitifolia（Jacq.）Maréchal ■

293969 Phaseolus acutifolius A. Gray；尖叶菜豆；Tepary Bean, Texas Bean ■☆

293970 Phaseolus acutifolius A. Gray var. latifolius F. L. Freeman；宽尖叶菜豆；Tepary Bean ■☆

293971 Phaseolus adenanthus G. Mey. = Vigna adenantha（G. Mey.）Maréchal, Mascherpa et Stainier ■

293972 Phaseolus amboensis Schinz = Bolusia amboensis（Schinz）Harms ■☆

293973 Phaseolus angularis（Willd.）W. Wight = Vigna angularis（Willd.）Ohwi et H. Ohashi ■

293974 Phaseolus angularis（Willd.）W. Wight f. nipponensis（Ohwi）Kitam. = Vigna angularis（Willd.）Ohwi et H. Ohashi var. nipponensis（Ohwi）Ohwi et H. Ohashi ■

293975 Phaseolus angularis（Willd.）W. Wight var. nipponensis（Ohwi）Ohwi = Vigna angularis（Willd.）Ohwi et H. Ohashi var. nipponensis（Ohwi）Ohwi et H. Ohashi ■

293976 Phaseolus atropurpureus DC. = Macroptilium atropurpureum（DC.）Urb. ■

293977 Phaseolus atropurpureus Moc. et Sessé ex DC.；赛鸟豆■

293978 Phaseolus atropurpureus Sessé et Moc. ex DC. = Macroptilium atropurpureum（Sessé et Moc. ex DC.）Urb. ■

293979 Phaseolus aureus Roxb. = Vigna radiata（L.）R. Wilczek ■

293980 Phaseolus calcaratus Roxb. = Vigna umbellata（Thunb.）Ohwi et H. Ohashi ■

293981 Phaseolus calcaratus Roxb. var. gracilis Prain = Vigna gracilicaulis（Ohwi）Ohwi et Ohashi ■

293982 Phaseolus capensis Burm. f. = Rhynchosia capensis（Burm. f.）Schinz ■☆

293983 Phaseolus caracalla L. = Vigna caracalla（L.）Verdc. ■☆

293984 Phaseolus cibellii Chiov. = Vatovaea pseudolablab（Harms）J. B. Gillett ■☆

293985 Phaseolus coccineus L.；荷包豆（多花菜豆,红花菜豆,看豆,看花豆,龙爪豆）；Flowering Bran, French Bean, Pouch Bean,

Runner Bean, Scarlet Runner, Scarlet Runner Bean, Snail Vine ■

293986 Phaseolus coccineus L. f. albus（Alef.）L. H. Bailey；白荷包豆■☆

293987 Phaseolus coccineus L. f. albus L. H. Bailey = Phaseolus coccineus L. f. albus（Alef.）L. H. Bailey ■☆

293988 Phaseolus cylindricus L. = Vigna unguiculata（L.）Walp. subsp. cylindrica（L.）Verdc. ■

293989 Phaseolus demissus Kitag.；下垂菜豆；Hanging Down Bean ■

293990 Phaseolus dinteri Harms = Decorsea dinteri（Harms）Verdc. ■☆

293991 Phaseolus flavescens Piper；浅黄菜豆■☆

293992 Phaseolus fuscus Wall. = Dunbaria fusca（Wall.）Kurz ■

293993 Phaseolus gracilicaulis Ohwi = Vigna gracilicaulis（Ohwi）Ohwi et Ohashi ■

293994 Phaseolus gracilicaulis Ohwi = Vigna minima（Roxb.）Ohwi et H. Ohashi ■

293995 Phaseolus grahamianus Wight et Arn. = Wajira grahamiana（Wight et Arn.）Thulin et Lavin ■☆

293996 Phaseolus grandidieri Baill. = Decorsea grandidieri（Baill.）R. Vig. ex M. Pelt. ■☆

293997 Phaseolus grandis Wall. ex Benth. = Dysolobium grande（Benth.）Prain ●

293998 Phaseolus helvulus L. = Strophostyles helvula（L.）Elliott ■☆

293999 Phaseolus heterophyllus Hayata；异叶菜豆；Diverseleaf Bean ■

294000 Phaseolus heterophyllus Hayata = Vigna minima（Roxb.）Ohwi et H. Ohashi ■

294001 Phaseolus heterophyllus Willd. = Vigna minima（Roxb.）Ohwi et H. Ohashi var. dimorphophylla T. L. Wu ■

294002 Phaseolus juruanus Harms = Vigna juruana（Harms）Verdc. ■☆

294003 Phaseolus kirkii Baker = Vigna kirkii（Baker）J. B. Gillett ■☆

294004 Phaseolus lathyroides L. = Macroptilium lathyroides（L.）Urb. ■

294005 Phaseolus leiospermus Torr. et A. Gray = Strophostyles leiosperma（Torr. et A. Gray）Piper ■☆

294006 Phaseolus limensis Macfad.；大棉豆（菜豆,利马豆）；Lima Bean ■☆

294007 Phaseolus limensis Macfad. = Phaseolus lunatus L. var. limenanus（L. H. Bailey）Burkart ■☆

294008 Phaseolus limensis Macfad. = Phaseolus lunatus L. ■

294009 Phaseolus longifolius Benth. = Vigna longifolia（Benth.）Verdc. ■☆

294010 Phaseolus lunatus L.；棉豆（菜豆,大白芸豆,观音藤,皇帝豆,金甲豆,香豆,雪豆,植豆）；Burma Bean, Butter Bean, Cape Bean, Carolina Bean, Civet Bean, Cotton Bean, Duffin Bean, Hibbert Bean, Java Bean, Lima Bean, Madagascar Bean, Paigya Bean, Paigyabean, Pole Bean, Rangoon Bean, Scimitar-podded Kidney Bean, Sieva Bean, Sievabean, Sieve Bean, Sugar Bean, White Bean ■

294011 Phaseolus lunatus L. var. limenanus（L. H. Bailey）Burkart = Phaseolus lunatus L. ■

294012 Phaseolus lunatus L. var. limenanus（L. H. Bailey）Burkart = Phaseolus limensis Macfad. ■☆

294013 Phaseolus lunatus L. var. macrocarpus Benth. = Phaseolus limensis Macfad. ■☆

294014 Phaseolus luteolus（Jacq.）Gagnep. = Vigna luteola（Jacq.）Benth. ■

294015 Phaseolus macrorhynchus Harms = Wajira grahamiana（Wight et Arn.）Thulin et Lavin ■☆

294016 Phaseolus marinus Burm. = Vigna marina（Burm.）Merr. ■

294017 Phaseolus massaiensis Taub.；马萨菜豆■☆

294018 Phaseolus max L. = Glycine max（L.）Merr. ■

294019　Phaseolus minimus Roxb. ; 山绿豆（贼小豆）；Little Phaseolus, Wild Green Bean ■☆

294020　Phaseolus minimus Roxb. = Vigna minima（Roxb.）Ohwi et H. Ohashi ■

294021　Phaseolus minimus Roxb. f. heterophyllus（Hayata）Hosok. = Vigna minima（Roxb.）Ohwi et H. Ohashi var. dimorphophylla T. L. Wu ■

294022　Phaseolus minimus Roxb. f. linearis Hosok. = Vigna minima（Roxb.）Ohwi et H. Ohashi ■

294023　Phaseolus minimus Roxb. f. linearis Hosok. = Vigna minima（Roxb.）Ohwi et H. Ohashi f. linelais（Hosok.）T. C. Huang et H. Ohashi ■

294024　Phaseolus minimus Roxb. f. rotundifolius（Hayata,）Hosok. = Vigna minima（Roxb.）Ohwi et H. Ohashi ■

294025　Phaseolus minimus Roxb. f. rotundifolius Hosok. = Vigna minima（Roxb.）Ohwi et H. Ohashi var. minor（Matsum.）Tateishi ■

294026　Phaseolus minimus Roxb. f. typicus Hosok. = Vigna minima（Roxb.）Ohwi et H. Ohashi ■

294027　Phaseolus multiflorus Lam. = Phaseolus coccineus L. ■

294028　Phaseolus multiflorus Willd. = Phaseolus coccineus L. ■

294029　Phaseolus mungo L. = Vigna mungo（L.）Hepper ■

294030　Phaseolus nanus L. ；矮菜豆（倭小豆）■☆

294031　Phaseolus nipponensis Ohwi；日本菜豆；Japanese Bean ■☆

294032　Phaseolus nipponensis Ohwi = Vigna angularis（Willd.）Ohwi et H. Ohashi var. nipponensis（Ohwi）Ohwi et H. Ohashi ■

294033　Phaseolus pauciflorus Benth. = Strophostyles leiosperma（Torr. et A. Gray）Piper ■☆

294034　Phaseolus polystachios（L.）Britton, Sterns et Poggenb. ；多穗菜豆；Wild Bean ■☆

294035　Phaseolus psoraleoides Wight et Arn. = Macroptilium lathyroides（L.）Urb. ■

294036　Phaseolus pubescens Blume = Vigna umbellata（Thunb.）Ohwi et H. Ohashi ■

294037　Phaseolus quadriflorus Hochst. ex A. Rich. = Vigna vexillata（L.）A. Rich. ■

294038　Phaseolus radiatus L. = Vigna radiata（L.）R. Wilczek ■

294039　Phaseolus radiatus L. var. aureus Prain；日本赤小豆（赤小豆，红豆仔，米豆）■☆

294040　Phaseolus radiatus L. var. flexuosus Matsum. ；蔓小豆 ■

294041　Phaseolus radiatus L. var. typicus Prain = Vigna radiata（L.）R. Wilczek ■

294042　Phaseolus reflexopilosus（Hayata）Ohwi = Vigna reflexopilosa Hayata ■

294043　Phaseolus riukiuensis Ohwi = Vigna minima（Roxb.）Ohwi et H. Ohashi var. minor（Matsum.）Tateishi ■

294044　Phaseolus riukiuensis Ohwi = Vigna riukiuensis（Ohwi）Ohwi et H. Ohashi ■

294045　Phaseolus rotundifolius Hayata = Vigna minima（Roxb.）Ohwi et H. Ohashi ■

294046　Phaseolus schimperi Taub. = Wajira grahamiana（Wight et Arn.）Thulin et Lavin ■☆

294047　Phaseolus schlechteri Harms = Decorsea schlechteri（Harms）Verdc. ■☆

294048　Phaseolus semierectus L. = Macroptilium lathyroides（L.）Urb. ■

294049　Phaseolus stenocarpus Harms = Wajira grahamiana（Wight et Arn.）Thulin et Lavin ■☆

294050　Phaseolus sublobatus Roxb. = Vigna radiata（L.）R. Wilczek

var. sublobata（Roxb.）Verdc. ■

294051　Phaseolus trichocarpus C. Wright = Vigna longifolia（Benth.）Verdc. ■☆

294052　Phaseolus trilobatus（L.）Schreb. = Vigna trilobata（L.）Verdc. ■

294053　Phaseolus trilobus Aiton = Vigna trilobata（L.）Verdc. ■

294054　Phaseolus trinervius Wight et Arn. = Vigna radiata（L.）R. Wilczek ■

294055　Phaseolus vexillatus L. = Vigna vexillata（L.）A. Rich. ■

294056　Phaseolus vulgaris L. ；菜豆（白豆，白饭豆，白云豆，豆角，二生豆，粉豆，高脚龙牙豆，花云豆，家雀豆，莲豆，六月鲜，龙骨豆，龙爪豆，三生豆，四季豆，唐豆，唐豇，隐元豆，云藊豆，云豆，芸扁豆，芸豆敏豆）；Arber Bean, Bean, Borlotti Bean, Borlotti Beans, Bush Bean, Bushbean, Canellini Bean, Cape Pea, Common Bean, Dwarf Bean, Faselles, Feasil, Field Bean, Flageolet, Flageolet Bean, French Bean, Frijol, Frijoles, Garden Bean, Haricot Bean, Kidney Bean, Long Peasen, Navy Bean, Nunas Bean, Pea-bean, Pinto Bean, Pinto Beans, Small Kidney Bean, Snap Bean, String Bean, Wax Bean ■

294057　Phaseolus vulgaris L. humilis Alef. ；龙牙豆（矮菜豆）；Bush Kidney Bean ■☆

294058　Phaseolus vulgaris L. var. nanus（L.）Asch. ；矮小菜豆 ■☆

294059　Phaseolus yunnanensis F. T. Wang et Ts. Tang；滇绿豆；Yunnan Bean ■

294060　Phaulanthus Ridl. = Anerincleistus Korth. ●☆

294061　Phaulopsis Lindau = Phaulopsis Willd.（保留属名）■

294062　Phaulopsis Willd.（1800）（'Phaylopsis'）（保留属名）；肾苞草属；Phaulopsis ■

294063　Phaulopsis aequivoca Manktelow；等肾苞草 ■☆

294064　Phaulopsis angolana S. Moore；安哥拉肾苞草 ■☆

294065　Phaulopsis barteri T. Anderson；巴特肾苞草 ■☆

294066　Phaulopsis barteri T. Anderson var. pauciglandula Manktelow；寡腺肾苞草 ■☆

294067　Phaulopsis ciliata（Willd.）Hepper；睫毛肾苞草 ■☆

294068　Phaulopsis dorsiflora（Retz.）Santapau = Phaulopsis oppositifolius（J. C. Wendl.）Lindau ■

294069　Phaulopsis dorsiflora（Retz.）Santapau = Ruellia dorsiflora Retz. ■

294070　Phaulopsis falcisepala C. B. Clarke = Phaulopsis ciliata（Willd.）Hepper ■☆

294071　Phaulopsis grandiflora Manktelow；大花肾苞草 ■☆

294072　Phaulopsis imbricata（Forssk.）Sweet = Phaulopsis oppositifolius（J. C. Wendl.）Lindau ■

294073　Phaulopsis imbricata（Forssk.）Sweet subsp. pallidifolia Manktelow；苍白叶肾苞草 ■☆

294074　Phaulopsis imbricata（Forssk.）Sweet subsp. poggei（Lindau）Manktelow；波格肾苞草 ■☆

294075　Phaulopsis imbricata（Forssk.）Sweet var. inaequalis（Hochst. ex Pic. Serm.）Cufod. = Phaulopsis imbricata（Forssk.）Sweet ■

294076　Phaulopsis imbricata Sweet = Phaulopsis oppositifolius（J. C. Wendl.）Lindau ■

294077　Phaulopsis inaequalis Hochst. ex Pic. Serm. = Phaulopsis imbricata（Forssk.）Sweet ■

294078　Phaulopsis johnstonii C. B. Clarke；约翰斯顿肾苞草 ■☆

294079　Phaulopsis lankesterioides（Lindau）Lindau；兰克爵床肾苞草 ■☆

294080　Phaulopsis latiloba Manktelow；宽裂肾苞草 ■☆

294081　Phaulopsis lindaviana De Wild. ；林达维肾苞草 ■☆

294082　Phaulopsis longifolia Thomson = Phaulopsis imbricata

（Forssk.）Sweet ■

294083　Phaulopsis micrantha（Benth.）C. B. Clarke；小花肾苞草■☆

294084　Phaulopsis oppositifolius（J. C. Wendl.）Lindau；肾苞草；Dorsiflorous Phaulopsis ■

294085　Phaulopsis parviflora Willd. = Phaulopsis imbricata（Forssk.）Sweet ■

294086　Phaulopsis parviflora Willd. = Phaulopsis oppositifolius（J. C. Wendl.）Lindau ■

294087　Phaulopsis poggei（Lindau）Lindau = Phaulopsis imbricata（Forssk.）Sweet subsp. poggei（Lindau）Manktelow ☆

294088　Phaulopsis pulchella Manktelow；美丽肾苞草■☆

294089　Phaulopsis semiconica P. G. Mey.；半锥肾苞草■☆

294090　Phaulopsis silvestris（Lindau）Lindau = Phaulopsis angolana S. Moore ■☆

294091　Phaulopsis talbotii S. Moore；塔尔博特肾苞草■☆

294092　Phaulothamnus A. Gray（1885）；蛇眼果属；Snake-eyes ●☆

294093　Phaulothamnus spinescens A. Gray；蛇眼果；Snake-eyes ●☆

294094　Phaylopsis Willd. = Phaulopsis Willd.（保留属名）■

294095　Phaylopsis betonica S. Moore = Phaulopsis johnstonii C. B. Clarke ■☆

294096　Phaylopsis falcisepala C. B. Clarke = Phaulopsis ciliata（Willd.）Hepper ■☆

294097　Phaylopsis glandulosa（Lindau）C. B. Clarke = Phaulopsis lankesterioides（Lindau）Lindau ☆

294098　Phaylopsis glutinosa Loudon = Phaulopsis talbotii S. Moore ■☆

294099　Phaylopsis lankesterioides C. B. Clarke var. longituba Benoist = Phaulopsis semiconica P. G. Mey. ☆

294100　Phaylopsis longifolia Sims = Phaulopsis barteri T. Anderson ☆

294101　Phaylopsis microphylla T. Anderson ex C. B. Clarke = Phaulopsis angolana S. Moore ■☆

294102　Phaylopsis obliqua T. Anderson ex S. Moore = Phaulopsis micrantha（Benth.）C. B. Clarke ■☆

294103　Phaylopsis poggei（Lindau）C. B. Clarke = Phaulopsis imbricata（Forssk.）Sweet subsp. poggei（Lindau）Manktelow ■☆

294104　Phaylopsis talbotii S. Moore = Phaulopsis imbricata（Forssk.）Sweet ■

294105　Phebalium Vent.（1805）；假桃金娘属；Phebalium ●☆

294106　Phebalium nudum Hook.；丛生假桃金娘；Mairehau ●☆

294107　Phebalium squamulosum Vent.；艳丽假桃金娘；Scaly Phebalium ●☆

294108　Pheboantha Rchb. = Ajuga L. ■●

294109　Pheboantha Rchb. = Phleboanthe Tausch ■●

294110　Phebolitis DC. = ? Mimusops L. ●☆

294111　Phebolitis DC. = Phlebolithis Gaertn. ●☆

294112　Phedimus Raf.（1817）；费菜属■

294113　Phedimus Raf. = Sedum L. ■

294114　Phedimus aizoon（L.）’t Hart；费菜（八仙草,长生景天,大马菜,大三七,豆瓣还阳,豆包还阳,多花景天三七,广三七,还阳草,活血丹,见血散,金不换,景天三七,九莲花,六月还阳,六月淋,破血丹,墙头三七,乳毛土三七,三百棒,生三七,石菜兰,收丹皮,四季还阳,田三七,土三七,吐血草,细叶费菜,小种三七,蝎子草,血山草）；Aizoon Stonecrop ■

294115　Phedimus aizoon（L.）’t Hart var. floribundum（Nakai）H. Ohba；繁花费菜■☆

294116　Phedimus aizoon（L.）’t Hart var. latifolium（Maxim.）H. Ohba, K. T. Fu et B. M. Barthol. = Phedimus aizoon（L.）’t Hart ■

294117　Phedimus aizoon（L.）’t Hart var. latifolium（Maxim.）H.

Ohba, K. T. Fu et B. M. Barthol.；宽叶费菜（宽叶景天,宽叶土三七）；Broadleaf Stonecrop, Stonecrop ■

294118　Phedimus aizoon（L.）’t Hart var. scabrum（Maxim.）H. Ohba, K. T. Fu et B. M. Barthol.；乳毛费菜■

294119　Phedimus aizoon（L.）’t Hart var. yamatutae（Kitag.）H. Ohba, K. T. Fu et B. M. Barthol.；狭叶费菜（狭叶土三七）■

294120　Phedimus floriferum（Praeger）’t Hart；多花费菜（多花景天）；Manyflower Stonecrop ■

294121　Phedimus hsinganicum（Y. C. Chu ex S. H. Fu et Y. H. Huang）H. Ohba, K. T. Fu et B. M. Barthol.；兴安费菜（兴安景天）；Xing'an Stonecrop ■

294122　Phedimus hybridum（L.）’t Hart；杂交费菜（千里光,杂交景天,杂景天,杂种景天）；Evergreen Stonecrop, Hybrid Stonecrop ■

294123　Phedimus kamtschaticum（Fisch. et C. A. Mey.）’t Hart；堪察加费菜（白三七,北景天,倒山黑豆,费菜,横根费菜,胡椒七,黄菜子,回生草,金不换,堪察加景天,马三七,七叶草,石板菜,血草,晏海豆片,养心草）；Kamschatka Sedum, Kamtschatka Stonecrop, Orange Stonecrop ■

294124　Phedimus middendorffianum（Maxim.）’t Hart；吉林费菜（狗景天,狗千里光,吉林景天,岩景天,岩千里光）；Jilin Stonecrop, Middendorff Stonecrop, Stonecrop ■

294125　Phedimus odontophyllum（Fröd.）’t Hart；齿叶费菜（齿叶景天,打不死,红胡豆七,红三七,六月还阳,石风丹,天黄七,天簧七）；Toothedleaf Stonecrop, Toothleaf Stonecrop ■

294126　Phedimus selskianum（Regel et Maack）’t Hart；灰毛费菜（灰毛景天,毛景天）；Selsk Stonecrop ■

294127　Phedimus sikokianus（Maxim.）’t Hart；四国景天■☆

294128　Phedimus stevenianum（Rouy et Camus）’t Hart；史梯景天；Steven Stonecrop ■

294129　Phegopyrum Peterm. = Fagopyrum Mill.（保留属名）●■

294130　Phegos St. -Lag. = Fagus L. ●

294131　Pheidochloa S. T. Blake（1944）；俭约草属■☆

294132　Pheidochloa gracilis S. T. Blake；俭约草■☆

294133　Pheidonocarpa L. E. Skog = Gesneria L. ●☆

294134　Pheidonocarpa L. E. Skog（1976）；俭果苣苔属●☆

294135　Pheidonocarpa corymbosa（Sw.）L. E. Skog；俭果苣苔●☆

294136　Pheladenia D. L. Jones et M. A. Clem.（2001）；澳洲裂缘兰属■☆

294137　Pheladenia D. L. Jones et M. A. Clem. = Caladenia R. Br. ■☆

294138　Phelandrium Neck. = Oenanthe L. ■

294139　Phelandrium Neck. = Phellandrium L. ■

294140　Pheliandra Werderm. = Solanum L. ●■

294141　Phelima Noronha = Horsfieldia Willd. ●

294142　Phelipaca Fourr. = Phelipaea Desf. ■☆

294143　Phelipaea Desf. = Orobanche L. ■

294144　Phelipaea Desf. = Phelypaea L. ■☆

294145　Phelipaea Post et Kuntze = Aeginetia L. ■

294146　Phelipaea Post et Kuntze = Cytinus L.（保留属名）■☆

294147　Phelipaea Post et Kuntze = Phaelypaea P. Br. ■☆

294148　Phelipaea Post et Kuntze = Phelypaea Boehm. ■☆

294149　Phelipaea Post et Kuntze = Phelypaea L. ■☆

294150　Phelipaea Post et Kuntze = Phelypea Thunb. ■☆

294151　Phelipaea Post et Kuntze = Stemodia L.（保留属名）■☆

294152　Phelipaea Tourn. ex Desf. = Phelypaea L. ■☆

294153　Phelipaea aegyptiaca（Pers.）Walp. = Orobanche aegyptiaca Pers. ■

294154　Phelipaea aegyptiaca（Pers.）Walp. var. tricholoba（Reut.）Beck = Orobanche hirtiflora（Reut.）Burkill ■☆

294155　Phelipaea ambigua Bunge ＝ Cistanche salsa（C. A. Mey.）Beck ■

294156　Phelipaea caerulea（Vill.）C. A. Mey. ＝ Orobanche purpurea Jacq. ■☆

294157　Phelipaea caesia（Rchb.）Reut. ＝ Orobanche caesia Rchb. ■

294158　Phelipaea coelestis（Boiss. et Reut.）Reut. ＝ Orobanche coelestis Boiss. et Reut. ex Reut. ■

294159　Phelipaea coelestis Reut. ＝ Orobanche clarkei Hook. f. ■

294160　Phelipaea coelestis Reut. ＝ Orobanche coelestis（Reut.）Boiss. et Reut. ex Beck ■

294161　Phelipaea heldreichii Reut. ＝ Orobanche coelestis Boiss. et Reut. ex Reut. ■

294162　Phelipaea hirtiflora Reut. ＝ Orobanche hirtiflora（Reut.）Burkill ■☆

294163　Phelipaea indica（Buch. -Ham. ex Roxb.）G. Don ＝ Orobanche aegyptiaca Pers. ■

294164　Phelipaea indica（Buch. -Ham.）G. Don ＝ Orobanche aegyptiaca Pers. ■

294165　Phelipaea indica Spreng. ex Steud. ＝ Aeginetia indica Roxb. ■

294166　Phelipaea languinosa C. A. Mey. ＝ Orobanche caesia Rchb. ■

294167　Phelipaea lanuginosa C. A. Mey. ＝ Orobanche lanuginosa（C. A. Mey.）Greuter et Burdet ■

294168　Phelipaea lutea Desf. ＝ Cistanche phelypaea（L.）Cout. ■☆

294169　Phelipaea mutelii（F. W. Schultz）Reut. ＝ Orobanche ramosa L. var. brevispicata（Ledeb.）Graham ■☆

294170　Phelipaea mutelii Reut. var. spissa Beck ＝ Kopsia mutelii（F. W. Schultz）Bég. var. spissa（Beck）Bég. et Vacc. ■☆

294171　Phelipaea oxyloba Reut. ＝ Orobanche orientalis Beck ■☆

294172　Phelipaea oxyloba Reut. ＝ Orobanche oxyloba（Reut.）Beck ■☆

294173　Phelipaea pallens Bunge ＝ Orobanche uralensis Beck ■

294174　Phelipaea pallens Bunge ex Ledeb. ＝ Orobanche uralensis Beck ■

294175　Phelipaea ramosa（L.）C. A. Mey. ＝ Orobanche ramosa L. ■☆

294176　Phelipaea ramosa（L.）C. A. Mey. var. brevispicata Ledeb. ＝ Orobanche ramosa L. var. brevispicata（Ledeb.）Graham ■☆

294177　Phelipaea salsa C. A. Mey. ＝ Cistanche salsa（C. A. Mey.）Beck ■

294178　Phelipaea senegalensis Reut. ＝ Cistanche phelypaea（L.）Cout. ■☆

294179　Phelipaea syspirensis C. Koch ＝ Orobanche clarkei Hook. f. ■

294180　Phelipaea syspirensis C. Koch ＝ Orobanche coelestis Boiss. et Reut. ex Reut. ■

294181　Phelipaea tricholoba Reut. ＝ Orobanche hirtiflora（Reut.）Tzvelev ■☆

294182　Phelipaea tricholoba Reut. var. simplex Reut. ＝ Orobanche clarkei Hook. f. ■

294183　Phelipaea tricholoba Reut. var. simplex Reut. ＝ Orobanche coelestis Boiss. et Reut. ex Reut. ■

294184　Phelipaea tubulosa Schenk ＝ Cistanche tubulosa（Schernk）Wight ■

294185　Phelipanche Pomel ＝ Orobanche L. ■

294186　Phelipanche aegyptiaca（Pers.）Pomel ＝ Orobanche aegyptiaca Pers. ■

294187　Phelipanche arenaria（Borkh.）Pomel ＝ Orobanche arenaria Borkh. ■☆

294188　Phelipanche brassicae（Novopokr.）Soják ＝ Orobanche brassicae Novopokr. ■

294189　Phelipanche caesia（Rchb.）Soják ＝ Orobanche lanuginosa（C. A. Mey.）Greuter et Burdet ■

294190　Phelipanche cernua（Loefl.）Pomel ＝ Orobanche cernua Loefl. ■

294191　Phelipanche coelesti Reut. ＝ Orobanche coelestis Boiss. et Reut. ex Reut. ■

294192　Phelipanche coelestis（Reut.）Soják ＝ Orobanche coelestis（Reut.）Boiss. et Reut. ex Beck ■

294193　Phelipanche coelestis（Reut.）Soják ＝ Orobanche coelestis Boiss. et Reut. ex Reut. ■

294194　Phelipanche floribunda Pomel ＝ Orobanche ramosa L. subsp. mutelii（F. W. Schultz）Cout. ■☆

294195　Phelipanche kelleri（Novopokr.）Soják ＝ Orobanche kelleri Novopokr. ■

294196　Phelipanche lavandulacea（Rchb.）Pomel ＝ Orobanche lavandulacea Rchb. ■☆

294197　Phelipanche mutelii（F. W. Schultz）Reut. ＝ Orobanche ramosa L. subsp. mutelii（F. W. Schultz）Cout. ■☆

294198　Phelipanche nana（Reut.）Soják ＝ Orobanche ramosa L. subsp. nana（Reut.）Cout. ■☆

294199　Phelipanche pallens（Bunge）Soják ＝ Orobanche uralensis Beck ■

294200　Phelipanche pallens Bunge ＝ Orobanche uralensis Beck ■

294201　Phelipanche pulchra Pomel ＝ Orobanche ramosa L. subsp. nana（Reut.）Cout. ■☆

294202　Phelipanche purpurea（Jacq.）Soják ＝ Orobanche purpurea Jacq. ■☆

294203　Phelipanche ramosa（L.）Pomel ＝ Orobanche ramosa L. ■☆

294204　Phelipanche schultzii（Mutel）Pomel ＝ Orobanche schultzii Mutel ■☆

294205　Phelipanche schweinfurthii（Beck）Soják ＝ Orobanche schweinfurthii Beck ■☆

294206　Phelipanche tenuiflora Pomel ＝ Orobanche ramosa L. subsp. mutelii（F. W. Schultz）Cout. ■☆

294207　Phelipanche trichocalyx（Webb et Berthel.）Soják ＝ Orobanche trichocalyx（Webb et al.）Beck ■☆

294208　Phelipanche tunetana（Beck）Soják ＝ Orobanche tunetana Beck ■☆

294209　Phelipanche uralensis（Beck.）De Moor ＝ Orobanche uralensis Beck ■

294210　Phelipea Pers. ＝ Phelipaea Desf. ■☆

294211　Phellandrium L. ＝ Oenanthe L. ■

294212　Phellandrium stoloniferum Roxb. ＝ Oenanthe javanica（Blume）DC. ■

294213　Phellandryum Gilib. ＝ Phellandrium L. ■

294214　Phellinaceae（Loes.）Takht. ＝ Aquifoliaceae Bercht. et J. Presl（保留科名）●

294215　Phellinaceae（Loes.）Takht. ＝ Phellinaceae Takht. ●☆

294216　Phellinaceae Takht.（1967）；石冬青科（新冬青科）●☆

294217　Phellinaceae Takht. ＝ Aquifoliaceae Bercht. et J. Presl（保留科名）●

294218　Phelline Labill.（1824）；石冬青属（软冬青属，新冬青属）●☆

294219　Phelline comosa Labill.；石冬青●☆

294220　Phelline floribunda Baill.；繁花石冬青●☆

294221　Phelline macrophylla Baill.；大叶石冬青●☆

294222　Phelline microcarpa Baill.；大果石冬青●☆

294223　Phelline robusta Baill.；粗壮石冬青●☆

294224　Phellocalyx Bridson（1980）；软萼茜属☆

294225　Phellocalyx vollesenii Bridson；软萼茜☆

294226　Phellocarpus Benth.（1837）；栓果豆属●☆

294227　Phellocarpus Benth. = Pterocarpus Jacq.（保留属名）●

294228　Phellocarpus Benth. = Pterocarpus L. ●

294229　Phellocarpus amazonum C. Mart. ex Benth. ;栓果豆●☆

294230　Phellodendron Rupr.（1857）;黄柏属（黄檗属，黄蘗属）;Cork Tree,Corktree,Cork-tree ●

294231　Phellodendron amurense Rupr. ;黄檗（檗木,关黄柏,红椿树,黄柏,黄波罗,黄菠萝树,黄伯栗,黄檗木,黄蘗,蘗木,暖木,山茱萸树,檀桓,铁麸盐树,元柏）;Amur Cork Tree, Amur Corktree, Amur Cork-tree,Yellow-bark ●

294232　Phellodendron amurense Rupr. f. molle（Nakai）Y. C. Chu = Phellodendron molle Nakai ●

294233　Phellodendron amurense Rupr. var. japonicum（Maxim.）Ohwi;日本黄柏（大叶黄柏）;Japanese Corktree ●☆

294234　Phellodendron amurense Rupr. var. lavallei（Dode）Sprague;厚皮黄檗(拉瓦黄柏);Lavalle Corktree ●☆

294235　Phellodendron amurense Rupr. var. sachalinense F. Schmidt = Phellodendron sachalinensis（F. Schmidt）Sarg. ●

294236　Phellodendron amurense Rupr. var. wilsonii（Hayata et Kaneh.）C. E. Chang;台湾黄蘗●

294237　Phellodendron amurense Rupr. var. wilsonii C. E. Chang = Phellodendron chinense C. K. Schneid. var. glabriusculum C. K. Schneid. ●

294238　Phellodendron chinense C. K. Schneid. ;川黄檗（檗木,黄柏,黄皮树,灰皮树,小黄连树）;China Corktree, Chinese Cork Tree, Chinese Corktree,Chinese Cork-tree,Chuan Corktree ●

294239　Phellodendron chinense C. K. Schneid. var. falcatum C. C. Huang = Phellodendron chinense C. K. Schneid. var. glabriusculum C. K. Schneid. ●

294240　Phellodendron chinense C. K. Schneid. var. glabriusculum C. K. Schneid. ;秃叶黄檗（檗木,峨眉黄檗,峨眉黄皮树,光叶黄皮树,黄柏,黄檗,黄檗皮,黄皮,黄皮树,镰刀叶黄皮树,镰叶黄皮树,台湾黄柏,台湾黄檗,台湾黄蘗,秃叶黄皮树,威氏黄柏,辛氏黄檗,元柏,云南黄皮树）;Glabrousleaf China Corktree, Glabrousleaf Chinese Corktree ●

294241　Phellodendron chinense C. K. Schneid. var. omeiense C. C. Huang = Phellodendron chinense C. K. Schneid. var. glabriusculum C. K. Schneid. ●

294242　Phellodendron chinense C. K. Schneid. var. yunnanense C. C. Huang = Phellodendron chinense C. K. Schneid. var. glabriusculum C. K. Schneid. ●

294243　Phellodendron fargesii Dode = Phellodendron chinense C. K. Schneid. ●

294244　Phellodendron insulare Nakai;海岛黄柏●☆

294245　Phellodendron japonicum Maxim. = Phellodendron amurense Rupr. var. japonicum（Maxim.）Ohwi ●☆

294246　Phellodendron japonicum Maxim. = Phellodendron amurense Rupr. ●

294247　Phellodendron lavallei Dode = Phellodendron amurense Rupr. var. lavallei（Dode）Sprague ●☆

294248　Phellodendron macrophyllum Dode = Evodia sutchuenensis Dode ●

294249　Phellodendron molle Nakai;毛叶黄檗（毛黄檗）;Hairy Corktree ●

294250　Phellodendron sachalinense（F. Schmidt）Sarg. = Phellodendron amurense Rupr. ●

294251　Phellodendron sachalinensis（F. Schmidt）Sarg. ;库页岛黄柏（川黄柏,库页黄檗）;Sachalin Corktree,Sakhalin Corktree ●

294252　Phellodendron sachalinensis Sarg. = Phellodendron sachalinensis（F. Schmidt）Sarg. ●

294253　Phellodendron sinense Chun = Phellodendron chinense C. K. Schneid. var. glabriusculum C. K. Schneid. ●

294254　Phellodendron sinense Dode = Phellodendron chinense C. K. Schneid. ●

294255　Phellodendron sinense Dode = Phellodendron chinense C. K. Schneid. var. glabriusculum C. K. Schneid. ●

294256　Phellodendron wilsonii Hayata et Kaneh. ;威氏黄柏(台湾黄柏,台湾黄蘗,魏氏黄柏）;E. H. Wilson Corktree,Taiwan Cork-tree ●

294257　Phellodendron wilsonii Hayata et Kaneh. = Phellodendron amurense Rupr. var. wilsonii（Hayata et Kaneh.）C. E. Chang ●

294258　Phellodendron wilsonii Hayata et Kaneh. = Phellodendron chinense C. K. Schneid. var. glabriusculum C. K. Schneid. ●

294259　Phelloderma Miers = Castelia Cav.（废弃属名）■☆

294260　Phelloderma Miers = Priva Adans. ■☆

294261　Phellolophium Baker(1886);软木花属☆

294262　Phellolophium decaryi Sales et Hedge;软木花☆

294263　Phellolophium madagascariense Baker;马岛软木花☆

294264　Phellopterus（Nutt. ex Torr. et A. Gray）J. M. Coult. et Rose = Cymopterus Raf. ■☆

294265　Phellopterus（Nutt. ex Torr. et A. Gray）J. M. Coult. et Rose（1900);美国聚散翼属■☆

294266　Phellopterus（Torr. et A. Gray）J. M. Coult. et Rose = Phellopterus（Nutt. ex Torr. et A. Gray）J. M. Coult. et Rose ■☆

294267　Phellopterus Benth. = Glehnia F. Schmidt ex Miq. ■

294268　Phellopterus Nutt. ex Torr. et A. Gray = Cymopterus Raf. ■☆

294269　Phellopterus Nutt. ex Torr. et A. Gray = Phellopterus（Nutt. ex Torr. et A. Gray）J. M. Coult. et Rose ■☆

294270　Phellopterus littoralis（F. Schmidt ex Miq.）Benth. = Glehnia littoralis F. Schmidt ex Miq. ■

294271　Phellopterus littoralis Benth. = Glehnia littoralis F. Schmidt ex Miq. ■

294272　Phellosperma Britton et Rose = Mammillaria Haw.（保留属名）●

294273　Phellosperma Britton et Rose(1923);栓籽掌属■☆

294274　Phellosperma longiflora（Britton et Rose）Buxb. ;长花栓籽掌■☆

294275　Phellosperma tetrancistra（Engelm.）Britton et Rose;栓籽掌■☆

294276　Phelpsiella Maguire(1958);费尔偏穗草属■☆

294277　Phelpsiella ptericaulis Maguire;费尔偏穗草■☆

294278　Phelypaea Boehm. = Aeginetia L. ■

294279　Phelypaea D. Don = Phelipaea Desf. ■☆

294280　Phelypaea L.（1758）;矢车菊列当属■☆

294281　Phelypaea Thunb. = Cytinus L.（保留属名）■☆

294282　Phelypaea Thunb. = Haematolepis C. Presl ■☆

294283　Phelypaea aegyptiaca（Pers.）Walp. = Orobanche aegyptiaca Pers. ■

294284　Phelypaea arenaria（Borkh.）Walp. = Orobanche arenaria Borkh. ■☆

294285　Phelypaea arenaria（Borkh.）Walp. var. atlantica（Pomel）Batt. = Orobanche arenaria Borkh. ■☆

294286　Phelypaea brunneri Webb = Cistanche phelypaea（L.）Cout. ■☆

294287　Phelypaea caerulea（Vill.）C. A. Mey. = Orobanche purpurea Jacq. ■☆

294288　Phelypaea capensis G. Don = Harveya squamosa（Thunb.）Steud. ■☆

294289　Phelypaea cernua Pomel = Orobanche lavandulacea Rchb. ■☆

294290　Phelypaea compacta Viv. = Cistanche compacta（Viv.）Bég. et

Vacc. ■☆

294291 Phelypaea floribunda Pomel ＝ Orobanche ramosa L. subsp. mutelii (F. W. Schultz) Cout. ■☆

294292 Phelypaea lavandulacea F. W. Schultz ＝ Orobanche lavandulacea Rchb. ■☆

294293 Phelypaea lavandulacea F. W. Schultz subsp. frasii (Walp.) Batt. ＝ Orobanche lavandulacea Rchb. ■☆

294294 Phelypaea lutea Desf. ＝ Cistanche phelypaea (L.) Cout. ■☆

294295 Phelypaea lutea Parry ＝ Orobanche fasciculata Nutt. ■☆

294296 Phelypaea mauritanica Coss. et Durieu ＝ Cistanche mauritanica (Coss. et Durieu) Beck ■☆

294297 Phelypaea mutelii (F. W. Schultz) Reut. ＝ Orobanche ramosa L. subsp. mutelii (F. W. Schultz) Cout. ■☆

294298 Phelypaea mutelii (F. W. Schultz) Reut. subsp. nana (Reut.) Batt. ＝ Orobanche ramosa L. subsp. nana (Reut.) Cout. ■☆

294299 Phelypaea mutelii (F. W. Schultz) Reut. subsp. pulchra (Pomel) Batt. ;美丽矢车菊列当■☆

294300 Phelypaea mutelii (F. W. Schultz) Reut. var. angustifolia？＝ Orobanche ramosa L. subsp. mutelii (F. W. Schultz) Cout. ■☆

294301 Phelypaea mutelii (F. W. Schultz) Reut. var. floribunda (Pomel) Batt. ＝ Orobanche ramosa L. subsp. mutelii (F. W. Schultz) Cout. ■☆

294302 Phelypaea mutelii (F. W. Schultz) Reut. var. nana (Reut.) Ball ＝ Orobanche ramosa L. subsp. mutelii (F. W. Schultz) Cout. ■☆

294303 Phelypaea mutelii (F. W. Schultz) Reut. var. spissa Beck ＝ Orobanche ramosa L. subsp. mutelii (F. W. Schultz) Cout. ■☆

294304 Phelypaea mutelii (F. W. Schultz) Reut. var. tenuiflora (Pomel) Batt. ＝ Orobanche ramosa L. subsp. mutelii (F. W. Schultz) Cout. ■☆

294305 Phelypaea pulchra Pomel ＝ Orobanche ramosa L. ■☆

294306 Phelypaea ramosa (L.) C. A. Mey. ＝ Orobanche ramosa L. ■☆

294307 Phelypaea schultzii (Mutel) Walp. ＝ Orobanche schultzii Mutel ■☆

294308 Phelypaea stricta Bertol. ＝ Orobanche schultzii Mutel ■☆

294309 Phelypaea tenuiflora Pomel ＝ Orobanche ramosa L. subsp. mutelii (F. W. Schultz) Cout. ■☆

294310 Phelypaea tinctoria Willd. ＝ Cistanche phelypaea (L.) Cout. ■☆

294311 Phelypaea tublosa Schrenk ＝ Cistanche tubulosa (Schrenk) Hook. f. ■

294312 Phelypaea violacea Desf. ＝ Cistanche violacea (Desf.) Hoffmanns. et Link ■☆

294313 Phelypaeaceae Horan. ＝ Orobanchaceae Vent. (保留科名)●■

294314 Phelypea Adans. ＝ Aeginetia L. ■

294315 Phelypea Adans. ＝ Phelypaea Boehm. ■☆

294316 Phelypea Thunb. ＝ Cytinus L. (保留属名)■☆

294317 Phelypea sanguinea Thunb. ＝ Cytinus sanguineus (Thunb.) Fourc. ■☆

294318 Phemeranthus Raf. (1814);焰花苋属;Fameflower,Flameflower ■

294319 Phemeranthus Raf. ＝ Talinum Adans. (保留属名)■●

294320 Phemeranthus aurantiacus (Engelm.) Kiger;黄焰花苋■☆

294321 Phemeranthus brevicaulis (S. Watson) Kiger;短茎焰花苋■☆

294322 Phemeranthus brevifolius (Torr.) Hershk. ;短叶焰花苋■☆

294323 Phemeranthus calcaricus (S. Ware) Kiger;石灰焰花苋■☆

294324 Phemeranthus calycinus (Engelm.) Kiger ＝ Talinum calycinum Engelm. ■☆

294325 Phemeranthus confertiflorus (Greene) Hershk. ＝ Phemeranthus parviflorus (Nutt.) Kiger ■☆

294326 Phemeranthus humilis (Greene) Kiger;矮焰花苋■☆

294327 Phemeranthus longipes (Wooton et Standl.) Kiger;长梗焰花苋■☆

294328 Phemeranthus marginatus (Greene) Kiger;北美焰花苋■☆

294329 Phemeranthus mengesii (W. Wolf) Kiger;焰花苋■☆

294330 Phemeranthus parviflorus (Nutt.) Kiger;小花焰花苋;Prairie Fame-flower,Sunbright ■☆

294331 Phemeranthus rugospermus (Holz.) Kiger;糙籽焰花苋;Prairie Fame-flower, Rough-seeded Fameflower, Rough-seeded Fame-flower, Sand Fame-flower ■☆

294332 Phemeranthus sediformis (Poelln.) Kiger;景天焰花苋■☆

294333 Phemeranthus spinescens (Torr.) Hershk. ;刺焰花苋■☆

294334 Phemeranthus teretifolius (Pursh) Raf. ;岩地焰花苋;Rock-portulaca ■☆

294335 Phemeranthus thompsonii (N. D. Atwood et S. L. Welsh) Kiger; 汤普森焰花苋■☆

294336 Phemeranthus validulus (Greene) Kiger;强壮焰花苋■☆

294337 Phenakospermum Endl. (1833);圭亚那红籽莲属■☆

294338 Phenakospermum guyanense Endl. ;圭亚那红籽莲■☆

294339 Phenakosperum Endl. ＝ Phenakospermum Endl. ■☆

294340 Phenax Wedd. (1854);无被麻属●☆

294341 Phenax hirtus (Sw.) Wedd. ;毛无被麻●☆

294342 Phenax laevigatus Wedd. ;平滑无被麻●☆

294343 Phenax laxiflorus Wedd. ;疏花无被麻●☆

294344 Phenax sonneratii (Poir.) Wedd. ;亚洲无被麻; Asian Ghostweed ■☆

294345 Phenianthus Raf. ＝ Lonicera L. ●■

294346 Phenopus Hook. f. ＝ Lactuca L. ■

294347 Phenopus Hook. f. ＝ Phaenopus DC. ■

294348 Phenotrichis Steud. ＝ Pherotrichis Decne. ●☆

294349 Pherelobus Phillips ＝ Dorotheanthus Schwantes ■☆

294350 Pherelobus Phillips ＝ Pherolobus N. E. Br. ■☆

294351 Pherolobus N. E. Br. ＝ Dorotheanthus Schwantes ■☆

294352 Pherolobus maughanii N. E. Br. ＝ Dorotheanthus maughanii (N. E. Br.) Ihlenf. et Struck ■☆

294353 Pherolobus maughanii N. E. Br. var. stayneri L. Bolus ＝ Dorotheanthus maughanii (N. E. Br.) Ihlenf. et Struck ■☆

294354 Pherosphaera Hook. f. ＝ Microstrobos J. Garden et L. A. S. Johnson ●☆

294355 Pherosphaera W. Archer ＝ Diselma Hook. f. ＋ Microcachrys Hook. f. ●☆

294356 Pherosphaera W. Archer ＝ Microcachrys Hook. f. ●☆

294357 Pherosphaera W. Archer ＝ Microstrobos J. Garden et L. A. S. Johnson ●☆

294358 Pherosphaera W. Archer bis ＝ Microcachrys Hook. f. ●☆

294359 Pherosphaeraceae Nakai ＝ Podocarpaceae Endl. (保留科名)●

294360 Pherotrichis Decne. (1838);多毛萝藦属●☆

294361 Pherotrichis leptogenia B. L. Rob. ;多毛萝藦●☆

294362 Phialacanthus Benth. (1876);宽刺爵床属☆

294363 Phialacanthus griffithii Benth. ;宽刺爵床■☆

294364 Phialacanthus major C. B. Clarke;大宽刺爵床■☆

294365 Phialacanthus minor C. B. Clarke;小宽刺爵床■☆

294366 Phialanthus Griseb. (1861);皿花茜属☆

294367 Phialanthus ellipticus Urb. ;椭圆皿花茜☆

294368 Phialanthus glaberrimus Borhidi;光皿花茜☆

294369 Phialanthus grandifolius Alain;大叶皿花茜☆

294370 Phialanthus macrocalyx Borhidi;大萼皿花茜☆

294371 Phialanthus oblongatus Urb. ;矩圆皿花茜☆

294372 Phialanthus parvifolius Urb. ;小叶皿花茜☆

294373　Phialanthus rigidus Griseb. ;硬皿花茜☆

294374　Phialis Spreng. = Eriophyllum Lag. ●■☆

294375　Phialis Spreng. = Trichophyllum Nutt. ■☆

294376　Phialocarpus Defiers = Kedrostis Medik. ■☆

294377　Phialocarpus Deflers = Corallocarpus Welw. ex Benth. et Hook. f. ■☆

294378　Phialodiscus Radlk. = Blighia K. König ●☆

294379　Phialodiscus laurentii (De Wild.) Radlk. = Blighia welwitschii (Hiern) Radlk. ●☆

294380　Phialodiscus mortehanii De Wild. = Blighia welwitschii (Hiern) Radlk. ●☆

294381　Phialodiscus plurijugatus Radlk. = Blighia unijugata Baker ●☆

294382　Phialodiscus unijugatus (Baker) Radlk. = Blighia unijugata Baker ●☆

294383　Phialodiscus verschuerenii De Wild. = Blighia unijugata Baker ●☆

294384　Phialodiscus welwitschii Hiern = Blighia welwitschii (Hiern) Radlk. ●☆

294385　Phialodiscus zambesiacus (Baker) Radlk. = Blighia unijugata Baker ●☆

294386　Phiambolia Klak(2003);皮姆番杏属■☆

294387　Phiambolia franciscii (L. Bolus) Klak;弗氏皮姆番杏■☆

294388　Phiambolia hallii (L. Bolus) Klak;霍尔皮姆番杏■☆

294389　Phiambolia incumbens (L. Bolus) Klak;斜倚皮姆番杏■☆

294390　Phiambolia persistens (L. Bolus) Klak;宿存皮姆番杏■☆

294391　Phiambolia stayneri (L. Bolus ex Toelken et Jessop) Klak;斯泰纳皮姆番杏●☆

294392　Phiambolia unca (L. Bolus) Klak;钩状皮姆番杏■☆

294393　Phidiasia Urb. = Odontonema Nees(保留属名)●■☆

294394　Philacanthus B. D. Jacks. = Phialacanthus Benth. ☆

294395　Philacra Dwyer(1944);平顶金莲木属●☆

294396　Philacra auriculata Dwyer;耳状平顶金莲木●☆

294397　Philacra duidae (Gleason) Dwyer;平顶金莲木●☆

294398　Philacra longifolia (Gleason) Dwyer;长叶平顶金莲木●☆

294399　Philacra steyermarkii Maguire;斯氏平顶金莲木●☆

294400　Philactis Schrad. (1833);单芒菊属●☆

294401　Philactis longipes A. Gray;长梗单芒菊●☆

294402　Philactis zinnioides Schrad. ;单芒菊●☆

294403　Philadelphaceae D. Don = Hydrangeaceae Dumort. (保留科名)●■

294404　Philadelphaceae D. Don;山梅花科; Mock-orange Family, Philadelphus Family ●

294405　Philadelphaceae Martinov = Hydrangeaceae Dumort. (保留科名)●■

294406　Philadelphaceae Martinov = Philadelphaceae D. Don ●

294407　Philadelphus L. (1753);山梅花属;Mock Orange, Mockorange, Mock-orange, Philadelphus, Syringa ●

294408　Philadelphus × virginalis Rehder;洁白山梅花;Mockorange, Virginal ●☆

294409　Philadelphus argyrocalyx Wooton;银萼山梅花●☆

294410　Philadelphus brachybotrys (Koehne) Koehne = Philadelphus brachybotrys Koehne ex Vilm. et Bois ●

294411　Philadelphus brachybotrys (Koehne) Koehne var. laxiflorus S. Y. Hu = Philadelphus zhejiangensis (W. C. Cheng) S. M. Hwang ●

294412　Philadelphus brachybotrys (Koehne) Koehne var. purpurascens Koehne = Philadelphus purpurascens (Koehne) Rehder ●

294413　Philadelphus brachybotrys Koehne ex Vilm. et Bois;短序山梅花（宝仙,短序太平花）;Shortinflorescenced Beijing Mockorange,

Shortraceme Mockorange ●

294414　Philadelphus brachybotrys Koehne ex Vilm. et Bois var. laxiflorus (W. C. Cheng) S. Y. Hu = Philadelphus zhejiangensis (W. C. Cheng) S. M. Hwang ●

294415　Philadelphus brachybotrys Koehne ex Vilm. et Bois var. purpurascens Koehne = Philadelphus purpurascens (Koehne) Rehder ●

294416　Philadelphus brachybotrys Koehne ex Vilm. et Bois var. purpurascens Koehne = Philadelphus delavayi L. Henry var. melanocalyx Lemoine ex L. Henry ●

294417　Philadelphus brachybotrys Koehne ex Vilm. et Bois var. venustus (Koehne) S. Y. Hu;美丽短序山梅花（美丽山梅花）;Beautiful Purplecup Mockorange ●

294418　Philadelphus californicus Benth. ;加州山梅花●☆

294419　Philadelphus calvescens (Rehder) S. M. Hwang;丽江山梅花（光叶山梅花）;Lijiang Mockorange ●

294420　Philadelphus calvescens (Rehder) S. M. Hwang var. compositus S. M. Hwang = Philadelphus calvescens (Rehder) S. M. Hwang ●

294421　Philadelphus calvescens (Rehder) S. M. Hwang var. compositus S. M. Hwang;复序山梅花●

294422　Philadelphus caucasicus Koehne; 高加索山梅花; Caucasia Mockorange,Caucasian Mock-orange,Caucasus Mock Orange ●☆

294423　Philadelphus caudatus S. M. Hwang;尾萼山梅花; Caudate Mockorange ●

294424　Philadelphus chianshanensis F. T. Wang et H. L. Li = Philadelphus tsianschanensis F. T. Wang et H. L. Li ●

294425　Philadelphus cordifolius Lange;心叶山梅花;Heart-leaved Mock Orange ●☆

294426　Philadelphus coronarius L. ;欧洲山梅花（山梅花,西洋山梅花）; Common Mock Orange, European Mock-orange, Falscher Jasmin, Fragrant Mockorange, Heartleaf Mock Orange, Heart-leaved Mock-orange, Mock Orange, Mock-orange, Orange Blossom, Orange Flower Tree,Roman Jasmine,Roman Jessamine,Sweet Mock Orange, Sweet Mockorange,Sweet Mock-orange,Syringa ●

294427　Philadelphus coronarius L. 'Aureus';金叶欧洲山梅花（黄叶西洋山梅花）; Golden-leaf Sweet Mockorange, Golden-leaved Mockorange ●☆

294428　Philadelphus coronarius L. 'Bowle's Variety';鲍尔斯欧洲山梅花●☆

294429　Philadelphus coronarius L. 'Deutziiflorus';重瓣欧洲山梅花●☆

294430　Philadelphus coronarius L. 'Multiflorus Plenus' = Philadelphus coronarius L. 'Deutziiflorus'●☆

294431　Philadelphus coronarius L. 'Variegatus';白边欧洲山梅花（斑叶西洋山梅花）●☆

294432　Philadelphus coronarius L. = Philadelphus pekinensis Rupr. ●

294433　Philadelphus coronarius L. nothovar. kiotensis (Murata) Kitam. et Murata = Philadelphus satsumi Siebold ex Lindl. et Paxton f. nikoensis (Rehder) Ohwi ●☆

294434　Philadelphus coronarius L. var. chinensis H. Lév. = Philadelphus sericanthus Koehne ●

294435　Philadelphus coronarius L. var. lancifolius (Uyeki) Kitam. et Murata = Philadelphus satsumi Siebold ex Lindl. et Paxton f. shikokianus (Nakai) H. Ohba et Akiyama ●☆

294436　Philadelphus coronarius L. var. mandshuricus Maxim. = Philadelphus schrenkii Rupr. var. mandshuricus (Maxim.) Kitag. ●

294437　Philadelphus coronarius L. var. parviflorus (Dippel) Kitam. et Murata = Philadelphus satsumi Siebold ex Lindl. et Paxton f.

nikoensis（Rehder）Ohwi ●☆

294438 Philadelphus coronarius L. var. pekinensis（Rupr.）Maxim. = Philadelphus pekinensis Rupr. ●

294439 Philadelphus coronarius L. var. pekinensis Maxim. = Philadelphus pekinensis Rupr. ●

294440 Philadelphus coronarius L. var. satsumi（Siebold ex Lindl. et Paxton）Maxim. = Philadelphus satsumi Siebold ex Lindl. et Paxton ●☆

294441 Philadelphus coronarius L. var. tenuifolius（Rupr. ex Maxim.）Maxim. = Philadelphus tenuifolius Rupr. ex Maxim. ●

294442 Philadelphus coronarius L. var. tenuifolius Maxim. = Philadelphus tenuifolius Rupr. ex Maxim. ●

294443 Philadelphus coronarius L. var. tomentosus（Wall. ex G. Don）Hook. f. et Thomson = Philadelphus tomentosus Wall. ex G. Don ●

294444 Philadelphus coronarius L. var. tomentosus Hook. f. et Thomson = Philadelphus incanus Koehne ●

294445 Philadelphus coulteri S. Watson；库尔特山梅花；Coulter Mock Orange，Coulter's Mock-orange ●☆

294446 Philadelphus coulteri S. Watson = Philadelphus purpureomaculatus Lemoine ●☆

294447 Philadelphus cymosus Rehder；聚伞山梅花●☆

294448 Philadelphus dasycalyx（Rehder）S. Y. Hu；毛萼山梅花（老密杆）；Hairy-calyx Mockorange ●

294449 Philadelphus delavayi（Rehder）S. Y. Hu f. cruciflorus S. Y. Hu = Philadelphus delavayi L. Henry var. melanocalyx Lemoine ex L. Henry ●

294450 Philadelphus delavayi（Rehder）S. Y. Hu f. melanocalyx（Lemoine ex L. Henry）Rehder = Philadelphus delavayi L. Henry var. melanocalyx Lemoine ex L. Henry ●

294451 Philadelphus delavayi L. Henry；云南山梅花（西南山梅花）；Delavay Mockorange ●

294452 Philadelphus delavayi L. Henry = Philadelphus calvescens（Rehder）S. M. Hwang ●

294453 Philadelphus delavayi L. Henry var. calvescens Rehder = Philadelphus calvescens（Rehder）S. M. Hwang ●

294454 Philadelphus delavayi L. Henry var. cruciflorus S. Y. Hu；十字山梅花●

294455 Philadelphus delavayi L. Henry var. melanocalyx Lemoine ex L. Henry f. cruciflorus S. Y. Hu = Philadelphus delavayi L. Henry var. melanocalyx Lemoine ex L. Henry ●

294456 Philadelphus delavayi L. Henry var. melanocalyx Lemoine ex L. Henry f. melanocalyx（Lemoine）Rehder = Philadelphus delavayi L. Henry var. melanocalyx Lemoine ex L. Henry ●

294457 Philadelphus delavayi L. Henry var. melanocalyx Lemoine ex L. Henry；黑萼山梅花（紫萼山梅花）；Black-calyx Mockorange，Purplecup Mockorange，Purple-cupped Mockorange ●

294458 Philadelphus delavayi L. Henry var. trichocladus Hand. -Mazz.；毛枝山梅花；Hairy-branch Mockorange ●

294459 Philadelphus falconeri Hort.；法氏山梅花；Falconer's Mock Orange，Falconer's Mock-orange ●☆

294460 Philadelphus floridus Beadle；美国山梅花；Beadle Mock Orange ●☆

294461 Philadelphus gloriosus Beadle = Philadelphus inodorus L. ●☆

294462 Philadelphus gordonianus Lindl.；戈登山梅花；Arrow-wood，Gordon Mock Orange，Gordon Mockorange，Gordon's Mock-orange ●☆

294463 Philadelphus grandiflorus Willd.；大花山梅花；Big Scentless Mock Orange，Big Scentless Mock-orange，Bigflower Mockorange ●☆

294464 Philadelphus grandiflorus Willd. = Philadelphus inodorus L. ●☆

294465 Philadelphus henryi Koehne；滇南山梅花（亨利山梅花，卷毛

山梅花，毛叶木通，山梅花）；Henry Mockorange ●

294466 Philadelphus henryi Koehne var. cinereus Hand. -Mazz.；灰毛山梅花（灰叶山梅花）；Grey-leaf Henry Mockorange ●

294467 Philadelphus henryi Koehne var. lissocalyx Hand. -Mazz. = Philadelphus calvescens（Rehder）S. M. Hwang ●

294468 Philadelphus hirsutus Nutt.；硬毛山梅花；Hirsute Mockorange，Mock Orange，Mockorange ●☆

294469 Philadelphus hupehensis（Koehne）S. Y. Hu = Philadelphus henryi Koehne var. cinereus Hand. -Mazz. ●

294470 Philadelphus hupehensis（Koehne）S. Y. Hu = Philadelphus sericanthus Koehne ●

294471 Philadelphus incanus Koehne；山梅花（白毛山梅花，鸡骨头，密密材，兴隆茶）；Chinese Mock-orange，Grey Mockorange，Grey Mock-orange，Mockorange ●

294472 Philadelphus incanus Koehne = Philadelphus laxiflorus Rehder ●

294473 Philadelphus incanus Koehne = Philadelphus reevesianus S. Y. Hu ●

294474 Philadelphus incanus Koehne var. baileyi Rehder；鸡公山山梅花（短轴山梅花）；Bailey Grey Mockorange ●

294475 Philadelphus incanus Koehne var. mitsai（S. Y. Hu）S. M. Hwang；米柴山梅花；Mitsai Mockorange ●

294476 Philadelphus incanus Koehne var. sargentianus f. kulingensis Koehne = Philadelphus sericanthus Koehne var. kulingensis（Koehne）Hand. -Mazz. ●

294477 Philadelphus incanus Koehne var. sargentianus Koehne f. hupehensis Koehne = Philadelphus sericanthus Koehne ●

294478 Philadelphus inodoratus L.；无香山梅花；Inodorate Mockorange ●☆

294479 Philadelphus inodorus L.；无味山梅花；Appalachian Mockorange，Mock Orange，Mockorange，Scentless Mock Orange，Scentless Mock-orange ●☆

294480 Philadelphus inodorus L. var. carolinus S. Y. Hu = Philadelphus inodorus L. ●☆

294481 Philadelphus inodorus L. var. grandiflorus（Willd.）A. Gray = Philadelphus inodorus L. ●☆

294482 Philadelphus inodorus L. var. laxus（Schrad.）S. Y. Hu = Philadelphus inodorus L. ●☆

294483 Philadelphus inodorus L. var. strigosus Beadle = Philadelphus inodorus L. ●☆

294484 Philadelphus kansuensis（Rehder）S. Y. Hu；甘肃山梅花；Gansu Mockorange ●

294485 Philadelphus karwinskianus Koehne；卡尔温斯基山梅花；Evergreen Mock Orange，Mock Orange ●☆

294486 Philadelphus kunmingensis S. M. Hwang；昆明山梅花；Kunming Mockorange ●

294487 Philadelphus kunmingensis S. M. Hwang var. parvifolius S. M. Hwang；小叶山梅花；Small-leaf Kunming Mockorange ●

294488 Philadelphus latifolius Schrad. ex DC.；宽叶山梅花；Hoary Mock Orange，Hoary Mock-orange ●☆

294489 Philadelphus laxiflorus Rehder；疏花山梅花（光盘山梅花）；Lax-flowered Mockorange，Loose-flower Mockorange ●

294490 Philadelphus laxus Schrad. = Philadelphus inodorus L. ●☆

294491 Philadelphus lemoinei Lemoine；香雪山梅花（雷蒙山梅花）；Fragrant Mockorange，Lemoine Mockorange ●☆

294492 Philadelphus lemoinei Lemoine 'Enchantment'；重瓣香雪山梅花；Double-flowered Fragrant Mockorange ●☆

294493 Philadelphus lewisii Pursh；路易斯山梅花；Indian Arrowwood，Lewis Mock Orange，Lewis Mockorange，Lewis Syringa，Lewis'

Syringa，Mock Orange，Mock-orange，Western Mockorange●

294494 Philadelphus lewisii Pursh var. gordonianus (Lindl.) Jeps.；毛叶路易斯山梅花●☆

294495 Philadelphus lushuiensis T. C. Ku et S. M. Hwang；泸水山梅花；Lushui Mockorange●

294496 Philadelphus magdalenae Koehne = Philadelphus sericanthus Koehne●

294497 Philadelphus magdalenae Koehne = Philadelphus subcanus Koehne var. magdalenae (Koehne) S. Y. Hu●

294498 Philadelphus mandshuricus (Maxim.) Nakai = Philadelphus schrenkii Rupr. var. mandshuricus (Maxim.) Kitag.●

294499 Philadelphus maximus Rehder；大山梅花；Mexican Mock Orange●☆

294500 Philadelphus mexicanus Schechter；墨西哥山梅花；Evergreen Mockorange，Mexican Mock Orange，Mexican Mock-orange●☆

294501 Philadelphus microphyllus A. Gray；细叶山梅花（小细叶山梅花，小叶山梅花）；Little-leaved Mock Orange，Little-leaved Mock-orange，Small-leaved Mockorange●☆

294502 Philadelphus millis var. erythrocalyx H. Lév. ex Rehder = Philadelphus henryi Koehne●

294503 Philadelphus mitsai S. Y. Hu = Philadelphus incanus Koehne var. mitsai (S. Y. Hu) S. M. Hwang●

294504 Philadelphus nepalensis Koehne = Philadelphus henryi Koehne●

294505 Philadelphus nepalensis Koehne = Philadelphus tomentosus Wall. ex G. Don●

294506 Philadelphus nepalensis Rehder = Philadelphus calvescens (Rehder) S. M. Hwang●

294507 Philadelphus pallidus Hayek；西洋山梅花●☆

294508 Philadelphus paniculatus Rehder = Philadelphus subcanus Koehne●

294509 Philadelphus pekinensis Rupr.；京山梅花（白花结，丰瑞花，太平花，太平圣瑞花，银盘盘花）；Beijing Mockorange，Peking Mock Orange，Peking Mockorange，Peking Mock-orange●

294510 Philadelphus pekinensis Rupr. f. lanceolatus S. Y. Hu；披针叶山梅花（长叶太平花）；Lanceleaf Mockorange●

294511 Philadelphus pekinensis Rupr. f. lanceolatus S. Y. Hu = Philadelphus pekinensis Rupr.●

294512 Philadelphus pekinensis Rupr. var. brachybotrys Koehne = Philadelphus brachybotrys Koehne ex Vilm. et Bois●

294513 Philadelphus pekinensis Rupr. var. dasycalyx Rehder = Philadelphus dasycalyx (Rehder) S. Y. Hu●

294514 Philadelphus pekinensis Rupr. var. kansuensis Rehder = Philadelphus kansuensis (Rehder) S. Y. Hu●

294515 Philadelphus pekinensis Rupr. var. laxiflorus (W. C. Cheng) S. Y. Hu = Philadelphus zhejiangensis (W. C. Cheng) S. M. Hwang●

294516 Philadelphus pekinensis Rupr. var. laxiflorus W. C. Cheng = Philadelphus zhejiangensis (W. C. Cheng) S. M. Hwang●

294517 Philadelphus pubescens Loisel.；毛叶山梅花；Hoary Mock Orange，Mock Orange，Pubescent Mockorange●☆

294518 Philadelphus pubescens Loisel. = Philadelphus latifolius Schrad. ex DC.●☆

294519 Philadelphus purpurascens (Koehne) Rehder；紫萼山梅花●

294520 Philadelphus purpurascens (Koehne) Rehder = Philadelphus delavayi L. Henry var. melanocalyx Lemoine ex L. Henry●

294521 Philadelphus purpurascens (Koehne) Rehder var. szechuanensis (W. P. Fang) S. M. Hwang；四川山梅花；Sichuan Mockorange●

294522 Philadelphus purpurascens (Koehne) Rehder var. venustus (Koehne) S. Y. Hu = Philadelphus brachybotrys Koehne ex Vilm. et Bois var. venustus (Koehne) S. Y. Hu●

294523 Philadelphus purpureomaculatus Lemoine；紫斑山梅花●☆

294524 Philadelphus reevesianus S. Y. Hu；毛药山梅花；Hairystrum Mockorange●

294525 Philadelphus rosiflorus Hort.；粉色山梅花；Rosy Mockorange●☆

294526 Philadelphus rubricaulis Carrière = Philadelphus pekinensis Rupr.●

294527 Philadelphus satsumai Siebold ex Lindl. et Paxton f. nikoensis (Rehder) Ohwi = Philadelphus satsumi Siebold ex Lindl. et Paxton f. nikoensis (Rehder) Ohwi●☆

294528 Philadelphus satsumanus Miq. = Philadelphus satsumi Siebold ex Lindl. et Paxton●☆

294529 Philadelphus satsumanus Miq. f. nikoensis Ohwi = Philadelphus satsumai Siebold ex Lindl. et Paxton f. nikoensis (Rehder) Ohwi●☆

294530 Philadelphus satsumi Siebold ex Lindl. et Paxton；萨摩山梅花（山梅花）；Satsuma Mock Orange，Satsuma Mock-orange，Satsuman Mockorange●☆

294531 Philadelphus satsumi Siebold ex Lindl. et Paxton f. nikoensis (Rehder) Ohwi；日光山梅花●☆

294532 Philadelphus satsumi Siebold ex Lindl. et Paxton f. shikokianus (Nakai) H. Ohba et Akiyama；四国山梅花；Sikoku Mockorange●☆

294533 Philadelphus satsumi Siebold ex Lindl. et Paxton nothovar. kiotensis Murata = Philadelphus satsumi Siebold ex Lindl. et Paxton f. nikoensis (Rehder) Ohwi●☆

294534 Philadelphus satsumi Siebold ex Lindl. et Paxton var. lancifolius (Uyeki) Murata = Philadelphus satsumi Siebold ex Lindl. et Paxton f. shikokianus (Nakai) H. Ohba et Akiyama●☆

294535 Philadelphus satsumi Siebold ex Lindl. et Paxton var. parviflorus (Dippel) Kitam. et Murata = Philadelphus satsumi Siebold ex Lindl. et Paxton f. nikoensis (Rehder) Ohwi●☆

294536 Philadelphus schrenkii Rupr.；东北山梅花（辽东山梅花，石氏山梅花）；Schrenk Mock Orange，Schrenk Mockorange，Schrenk Mock-orange●

294537 Philadelphus schrenkii Rupr. = Philadelphus tenuifolius Rupr. ex Maxim.●

294538 Philadelphus schrenkii Rupr. var. jackii Koehne；河北山梅花（疏毛山梅花）；Hebei Mockorange●

294539 Philadelphus schrenkii Rupr. var. mandshuricus (Maxim.) Kitag.；毛盘山梅花；Hairydisk Mockorange●

294540 Philadelphus sericanthus Koehne；绢毛山梅花（白花杆，大常山，鸡骨头，建德山梅花，绿花山梅花，毛萼山梅花，山梅花，探花，土常山，小吉通）；Sericeous Mockorange，Silk Mockorange●

294541 Philadelphus sericanthus Koehne var. bockii Koehne = Philadelphus sericanthus Koehne●

294542 Philadelphus sericanthus Koehne var. kulingensis (Koehne) Hand.-Mazz.；牯岭山梅花；Guling Mockorange●

294543 Philadelphus sericanthus Koehne var. leiocalyx Migo；光萼山梅花；Smooth-calyx Silk Mockorange●

294544 Philadelphus sericanthus Koehne var. rehderianus Koehne；大叶山梅花；Rehder Silk Mockorange●

294545 Philadelphus sericanthus Koehne var. rehderianus Koehne = Philadelphus sericanthus Koehne var. kulingensis (Koehne) Hand.-Mazz.●

294546 Philadelphus sericanthus Koehne var. rehderianus Koehne = Philadelphus subcanus Koehne●

294547 Philadelphus sericanthus Koehne var. rosthornii Koehne =

　　　　Philadelphus sericanthus Koehne ●

294548　Philadelphus shikokianus Nakai ＝Philadelphus satsumi Siebold ex Lindl. et Paxton f. shikokianus（Nakai）H. Ohba et Akiyama ●☆

294549　Philadelphus subcanus Koehne;毛柱山梅花（大叶山梅花,河南山梅花）;Hairystyle Mockorange, Hairy-styled Mockorange ●

294550　Philadelphus subcanus Koehne var. dubius Koehne;密毛山梅花（川黔山梅花）;Doubtful Hairystyle Mockorange ●

294551　Philadelphus subcanus Koehne var. magdalenae（Koehne）S. Y. Hu;城口山梅花（矮生毛柱山梅花,川山梅花,川西山梅花,马格达莱纳山梅花,全缘叶山梅花）;Chuanxi Hairystyle Mockorange ●

294552　Philadelphus subcanus Koehne var. wilsonii（Koehne）Rehder ＝Philadelphus subcanus Koehne ●

294553　Philadelphus subcanus Koehne var. wilsonii Rehder;威尔逊山梅花;Wilson Hairystyle Mockorange ●

294554　Philadelphus subcanus Koehne var. wilsonii Rehder ＝Philadelphus subcanus Koehne ●

294555　Philadelphus szechuanensis W. P. Fang ＝Philadelphus purpurascens（Koehne）Rehder var. szechuanensis（W. P. Fang）S. M. Hwang ●

294556　Philadelphus tenuifolius Rupr. ex Maxim.;薄叶山梅花（堇叶山梅花,细叶山梅花）;Mock Orange, Thinleaf Mock Orange, Thinleaf Mockorange, Thin-leaved Mockorange, Violetleaf Mockorange ●

294557　Philadelphus tenuifolius Rupr. ex Maxim. var. latipetalus S. Y. Hu;宽瓣山梅花;Broad-sepal Mockorange ●

294558　Philadelphus tetragonus S. M. Hwang;四棱山梅花;Fourangular Mockorange ●

294559　Philadelphus tomentella var. glabrescens C. C. Yan ＝Pileostegia viburnoides Hook. f. et Thomson var. glabrescens（C. C. Yang）S. M. Hwang ●

294560　Philadelphus tomentosus Wall. ex G. Don;绒毛山梅花（毛叶山梅花,尼泊尔山梅花）;Fuzzy Mock Orange, Hairy-leaved Mockorange, Mockorange, Tomentose Mockorange, Woolly-leaf Mock Orange, Woolly-leaf Mock-orange ●

294561　Philadelphus tsianschanensis F. T. Wang et H. L. Li;千山山梅花;Qianshan Mockorange ●

294562　Philadelphus venustus Koehne ＝Philadelphus purpurascens（Koehne）Rehder var. venustus（Koehne）S. Y. Hu ●

294563　Philadelphus verrucosus Schrad. ＝Philadelphus verrucosus Schrad. ex DC. ●☆

294564　Philadelphus verrucosus Schrad. ex DC.;瘤山梅花;Warty Mock Orange, Warty Mock-orange ●☆

294565　Philadelphus virginalis Rehder;雪白山梅花（洁白山梅花）;Pure White Mockorange, Virginal Mockorange ●☆

294566　Philadelphus virginalis Rehder 'Avalanche';艾维兰奇雪白山梅花●☆

294567　Philadelphus virginalis Rehder 'Beauclerk';比乌科勒克雪白山梅花●☆

294568　Philadelphus virginalis Rehder 'Belle Etoile';美女星雪白山梅花●☆

294569　Philadelphus virginalis Rehder 'Boule d'Argent';博德银雪白山梅花●☆

294570　Philadelphus virginalis Rehder 'Bouquet Blanc';香布兰克雪白山梅花●☆

294571　Philadelphus virginalis Rehder 'Buckley's Quill';布克利纤管雪白山梅花●☆

294572　Philadelphus virginalis Rehder 'Dame Blanche';布兰奇女爵士雪白山梅花●☆

294573　Philadelphus virginalis Rehder 'Dwarf Minnesota Snowflake';矮明尼苏达雪白山梅花;Dwarf Minnesota Pure White Mockorange ●☆

294574　Philadelphus virginalis Rehder 'Etoile Rose';星玫瑰雪白山梅花●☆

294575　Philadelphus virginalis Rehder 'Fimbriatus';小睫毛雪白山梅花●☆

294576　Philadelphus virginalis Rehder 'Glacier';矮生雪白山梅花（冰川雪白山梅花）;Dwarf Pure White Mockorange ●☆

294577　Philadelphus virginalis Rehder 'Innocence';清白雪白山梅花●☆

294578　Philadelphus virginalis Rehder 'Manteau d'Hermine';曼特德赫明雪白山梅花●☆

294579　Philadelphus virginalis Rehder 'Miniature Snowflake';小雪花莲雪白山梅花●☆

294580　Philadelphus virginalis Rehder 'Minnesota Snowflake';明尼苏达雪白山梅花（明尼苏达雪花莲雪白山梅花）;Minnesota Pure White Mockorange ●☆

294581　Philadelphus virginalis Rehder 'Natchez';单瓣雪白山梅花（纳齐兹雪白山梅花）;Single-flowered Pure White Mockorange ●☆

294582　Philadelphus virginalis Rehder 'Purity';纯净雪白山梅花●☆

294583　Philadelphus virginalis Rehder 'Romeo';罗密欧雪白山梅花●☆

294584　Philadelphus virginalis Rehder 'Rosace';蔷薇花饰雪白山梅花●☆

294585　Philadelphus virginalis Rehder 'Schneestrum';斯奇纳斯图雪白山梅花●☆

294586　Philadelphus virginalis Rehder 'Sybille';斯比勒雪白山梅花●☆

294587　Philadelphus virginalis Rehder 'Thelma';塞尔玛雪白山梅花●☆

294588　Philadelphus virginalis Rehder 'Virginal';重瓣雪白山梅花（纯洁雪白山梅花）;Double-flowered Pure White Mockorange ●☆

294589　Philadelphus wilsonii Koehne ＝Philadelphus subcanus Koehne ●

294590　Philadelphus zeyheri Schrad. ex DC.;泽氏山梅花;Zeyher Mock Orange, Zeyher Mock-orange ●☆

294591　Philadelphus zhejiangensis（W. C. Cheng）S. M. Hwang;浙江山梅花;Zhejiang Mockorange ●

294592　Philaginopsis Walp. ＝Filaginopsis Torr. et A. Gray ■

294593　Philaginopsis Walp. ＝Filago L.（保留属名）■

294594　Philagonia Blume ＝Evodia J. R. Forst. et G. Forst. ●

294595　Philagonia fraxinifolia（D. Don）Hook. ＝Evodia fraxinifolia（D. Don）Hook. f. ●

294596　Philammos（Steven）Steven ＝Astragalus L. ●■

294597　Philammos Steven ＝Astragalus L. ●■

294598　Philastrea Pierre ＝Munronia Wight ●

294599　Philbornea Hallier f.（1912）;岛麻属●☆

294600　Philbornea magnifolia Hallier f.;大叶岛麻●☆

294601　Philbornea palawanica Hallier f.;岛麻●☆

294602　Philcoxia P. Taylor et V. C. Souza（2000）;巴西参属■☆

294603　Philcoxia goiasensis P. Taylor;巴西参■☆

294604　Philenoptera Fenzl ＝Lonchocarpus Kunth（保留属名）●■☆

294605　Philenoptera Fenzl ex A. Rich. ＝Lonchocarpus Kunth（保留属名）●■☆

294606　Philenoptera Fenzl ex A. Rich. ＝Philenoptera Hochst. ex A. Rich. ●■☆

294607　Philenoptera Hochst. ex A. Rich.（1847）;肖矛果豆属●■☆

294608　Philenoptera bussei（Harms）Schrire;布瑟肖矛果豆●☆

294609　Philenoptera cyanescens（Schumach. et Thonn.）Roberty;浅蓝肖矛果豆●☆

294610　Philenoptera eriocalyx（Harms）Schrire;毛萼肖矛果豆●☆

294611　Philenoptera kanurii（Brenan et J. B. Gillett）Schrire;卡努里肖

矛果豆●☆

294612　Philenoptera katangensis（De Wild.）Schrire;加丹加肖矛果豆●☆

294613　Philenoptera kotschyana Fenzl = Philenoptera laxiflora（Guillaumin et Perr.）Roberty ●☆

294614　Philenoptera laxiflora（Guillaumin et Perr.）Roberty;疏花肖矛果豆●☆

294615　Philenoptera madagascariensis（Vatke）Schrire;马岛肖矛果豆●☆

294616　Philenoptera nelsii（Schinz）Schrire;内尔斯肖矛果豆●☆

294617　Philenoptera pallescens（Welw. ex Baker）Schrire;变苍白肖矛果豆●☆

294618　Philenoptera schimperi Hochst. ex A. Rich. = Philenoptera laxiflora（Guillaumin et Perr.）Roberty ●☆

294619　Philenoptera sutherlandii（Harv.）Schrire;萨瑟兰肖矛果豆●☆

294620　Philenoptera violacea（Klotzsch）Schrire;堇色肖矛果豆●☆

294621　Phileozera Buckley = Actinea Juss. ■

294622　Phileozera Buckley = Actinella Pers. ■

294623　Philesia Comm. ex Juss.（1789）;智利花属（金钟木属）●☆

294624　Philesia magellanica J. F. Gmel.;智利花（金钟木）●☆

294625　Philesiaceae Dumort.（1829）（保留科名）;智利花科（垂花科，金钟木科,喜爱花科）●■☆

294626　Philetaeria Liebm. = Fouquieria Kunth ●☆

294627　Philexia Raf. = Lythrum L. ●■

294628　Philgamia Baill.（1894）;马岛金虎尾属●☆

294629　Philgamia Baill. ex Dubard,Dop et Arènes = Philgamia Baill. ●☆

294630　Philgamia Baill. ,Dubard et Dop = Philgamia Baill. ●☆

294631　Philgamia brachystemon Arènes;短冠马岛金虎尾●☆

294632　Philgamia denticulata Arènes;小齿马岛金虎尾●☆

294633　Philgamia glabrifolia Arènes;光叶马岛金虎尾●☆

294634　Philibertella Vail = Funastrum E. Fourn. ■

294635　Philibertella Vail = Sarcostemma R. Br. ■

294636　Philibertia Kunth = Sarcostemma R. Br. ■

294637　Philibertia fiebrigii（Schltr.）Liede = Stelmatocodon fiebrigii Schltr. ■☆

294638　Philippia Klotzsch = Erica L. ●☆

294639　Philippia Klotzsch（1834）;肖石南属●☆

294640　Philippia absinthoides（Thunb.）E. G. H. Oliv. = Erica tristis Bartl. ●☆

294641　Philippia abyssinica Pic. Serm. et Heiniger = Erica trimera（Engl.）Beentje subsp. abyssinica（Pic. Serm. et Heiniger）Dorr ●☆

294642　Philippia adenophylla Baker = Erica perhispida Dorr et E. G. H. Oliv. ●☆

294643　Philippia alticola E. G. H. Oliv. = Erica altiphila E. G. H. Oliv. ●☆

294644　Philippia aristata Benth. = Erica bojeri Dorr et E. G. H. Oliv. ●☆

294645　Philippia benguelensis（Welw. ex Engl.）Britten = Erica benguelensis（Welw. ex Engl.）E. G. H. Oliv. ●☆

294646　Philippia benguelensis（Welw. ex Engl.）Britten var. albescens R. Ross = Erica benguelensis（Welw. ex Engl.）E. G. H. Oliv. var. albescens（R. Ross）E. G. H. Oliv. ●☆

294647　Philippia benguelensis（Welw. ex Engl.）Britten var. intermedia Weim. = Erica hexandra（S. Moore）E. G. H. Oliv. ●☆

294648　Philippia betsileana H. Perrier = Erica betsileana（H. Perrier）Dorr et E. G. H. Oliv. ●☆

294649　Philippia capitata Baker = Erica armandiana Dorr et E. G. H. Oliv. ●☆

294650　Philippia cauliflora Hochr. = Erica goudotiana（Klotzsch）Dorr et E. G. H. Oliv. ●☆

294651　Philippia cauliflora Hochr. subsp. gigas H. Perrier = Erica goudotiana（Klotzsch）Dorr et E. G. H. Oliv. ●☆

294652　Philippia cauliflora Hochr. subsp. tenuis H. Perrier = Erica goudotiana（Klotzsch）Dorr et E. G. H. Oliv. ●☆

294653　Philippia chamissonis Klotzsch = Erica tristis Bartl. ●☆

294654　Philippia ciliata Benth. = Erica boutonii Dorr et E. G. H. Oliv. ●☆

294655　Philippia ciliata Benth. subsp. cinerea H. Perrier = Erica boutonii Dorr et E. G. H. Oliv. ●☆

294656　Philippia congoensis S. Moore = Erica benguelensis（Welw. ex Engl.）E. G. H. Oliv. ●☆

294657　Philippia cryptoclada Baker = Erica cryptoclada（Baker）Dorr et E. G. H. Oliv. ●☆

294658　Philippia cryptoclada Baker var. hybrida H. Perrier = Erica cryptoclada（Baker）Dorr et E. G. H. Oliv. ●☆

294659　Philippia danguyana H. Perrier = Erica danguyana（H. Perrier）Dorr et E. G. H. Oliv. ●☆

294660　Philippia densa Benth. = Erica densata Dorr et E. G. H. Oliv. ●☆

294661　Philippia elgonensis Mildbr. = Erica trimera（Engl.）Beentje subsp. elgonensis（Mildbr.）Beentje ●☆

294662　Philippia elsieana E. G. H. Oliv. = Erica elsieana（E. G. H. Oliv.）E. G. H. Oliv. ●☆

294663　Philippia esterhuyseniae E. G. H. Oliv. = Erica esteriana E. G. H. Oliv. ●☆

294664　Philippia esterhuyseniae E. G. H. Oliv. subsp. swartbergensis E. G. H. Oliv. = Erica esteriana E. G. H. Oliv. subsp. swartbergensis（E. G. H. Oliv.）E. G. H. Oliv. ●☆

294665　Philippia evansii N. E. Br. = Erica evansii（N. E. Br.）E. G. H. Oliv. ●☆

294666　Philippia excelsa Alm et T. C. E. Fr. = Erica rossii Dorr ●☆

294667　Philippia floribunda Benth. = Erica baroniana Dorr et E. G. H. Oliv. ●☆

294668　Philippia floribunda Benth. subsp. glandulosa H. Perrier = Erica lyallii Dorr et E. G. H. Oliv. ●☆

294669　Philippia floribunda Benth. subsp. goudotiana（Klotzsch）H. Perrier = Erica goudotiana（Klotzsch）Dorr et E. G. H. Oliv. ●☆

294670　Philippia floribunda Benth. subsp. heterophylla（H. Perrier）H. Perrier = Erica sylvainiana Dorr et E. G. H. Oliv. ●☆

294671　Philippia floribunda Benth. subsp. macrantha H. Perrier = Erica baroniana Dorr et E. G. H. Oliv. ●☆

294672　Philippia floribunda Benth. subsp. orientalis H. Perrier = Erica goudotiana（Klotzsch）Dorr et E. G. H. Oliv. ●☆

294673　Philippia floribunda Benth. subsp. parviflora（Benth.）H. Perrier = Erica lyallii Dorr et E. G. H. Oliv. ●☆

294674　Philippia floribunda Benth. subsp. pilulifera（H. Perrier）H. Perrier = Erica wangfatiana Dorr et E. G. H. Oliv. ●☆

294675　Philippia friesii Weim. = Erica simii（S. Moore）E. G. H. Oliv. ●☆

294676　Philippia gracilis（Benth.）H. Perrier = Erica rakotozafyana Dorr et E. G. H. Oliv. ●☆

294677　Philippia heterophylla H. Perrier = Erica sylvainiana Dorr et E. G. H. Oliv. ●☆

294678　Philippia hexagona T. C. E. Fr. = Erica trimera（Engl.）Beentje subsp. keniensis（S. Moore）Beentje ●☆

294679　Philippia hexandra S. Moore = Erica hexandra（S. Moore）E. G. H. Oliv. ●☆

294680　Philippia hispida Baker = Erica perhispida Dorr et E. G. H. Oliv. ●☆

294681　Philippia holstii Engl. = Erica benguelensis（Welw. ex Engl.）

E. G. H. Oliv. ●☆

294682 Philippia holstii Engl. var. glanduligera ? = Erica benguelensis (Welw. ex Engl.) E. G. H. Oliv. ●☆

294683 Philippia humbertii H. Perrier = Erica humbertii (H. Perrier) Dorr et E. G. H. Oliv. ●☆

294684 Philippia humbertii Staner = Erica trimera (Engl.) Beentje ●☆

294685 Philippia ibityensis H. Perrier = Erica ibityensis (H. Perrier) Dorr et E. G. H. Oliv. ●☆

294686 Philippia imerinensis H. Perrier = Erica imerinensis (H. Perrier) Dorr et E. G. H. Oliv. ●☆

294687 Philippia irrorata E. G. H. Oliv. = Erica madida E. G. H. Oliv. ●☆

294688 Philippia isaloensis H. Perrier = Erica isaloensis (H. Perrier) Dorr et E. G. H. Oliv. ●☆

294689 Philippia jaegeri Engl. = Erica trimera (Engl.) Beentje subsp. jaegeri (Engl.) Beentje ●☆

294690 Philippia johnstonii Schweinf. ex Engl. = Erica johnstonii (Schweinf. ex Engl.) Dorr ●☆

294691 Philippia jumellei H. Perrier = Erica jumellei (H. Perrier) Dorr et E. G. H. Oliv. ●☆

294692 Philippia keniensis S. Moore = Erica trimera (Engl.) Beentje subsp. keniensis (S. Moore) Beentje ●☆

294693 Philippia keniensis S. Moore subsp. abyssinica (Pic. Serm. et Heiniger) R. Ross = Erica trimera (Engl.) Beentje subsp. abyssinica (Pic. Serm. et Heiniger) Dorr ●☆

294694 Philippia keniensis S. Moore subsp. elgonensis (Mildbr.) Ross = Erica trimera (Engl.) Beentje subsp. elgonensis (Mildbr.) Beentje ●☆

294695 Philippia keniensis S. Moore subsp. meruensis R. Ross = Erica trimera (Engl.) Beentje subsp. meruensis (R. Ross) Dorr ●☆

294696 Philippia kundelungensis S. Moore = Erica benguelensis (Welw. ex Engl.) E. G. H. Oliv. ●☆

294697 Philippia latifolia H. Perrier = Erica perrieri Dorr et E. G. H. Oliv. ●☆

294698 Philippia lebrunii Staner = Erica trimera (Engl.) Beentje ●☆

294699 Philippia lecomtei H. Perrier = Erica lecomtei (H. Perrier) Dorr et E. G. H. Oliv. ●☆

294700 Philippia leucoclada Baker = Erica leucoclada (Baker) Dorr et E. G. H. Oliv. ●☆

294701 Philippia macrocalyx Baker = Erica macrocalyx (Baker) Dorr et E. G. H. Oliv. ●☆

294702 Philippia madagascariensis H. Perrier = Erica madagascariensis (H. Perrier) Dorr et E. G. H. Oliv. ●☆

294703 Philippia mafiensis Engl. = Erica mafiensis (Engl.) Dorr ●☆

294704 Philippia mannii (Hook. f.) Alm et T. C. E. Fr. = Erica mannii (Hook. f.) Beentje ●☆

294705 Philippia mannii (Hook. f.) Alm et T. C. E. Fr. subsp. pallidiflora (Engl.) Ross = Erica mannii (Hook. f.) Beentje subsp. pallidiflora (Engl.) E. G. H. Oliv. ●☆

294706 Philippia mannii (Hook. f.) Alm et T. C. E. Fr. subsp. usambarensis (Alm et T. C. E. Fr.) R. Ross = Erica mannii (Hook. f.) Beentje subsp. usambarensis (Alm et T. C. E. Fr.) Beentje ●☆

294707 Philippia milanjiensis Britten et Rendle = Erica benguelensis (Welw. ex Engl.) E. G. H. Oliv. ●☆

294708 Philippia minutifolia Baker = Erica minutifolia (Baker) Dorr et F. Oliv. ●☆

294709 Philippia multiglandulosa (Klotzsch) Alm et T. C. E. Fr. ；多腺肖石南●☆

294710 Philippia myriadenia Baker = Erica myriadenia (Baker) Dorr et E. G. H. Oliv. ●☆

294711 Philippia neohumbertii Staner = Erica trimera (Engl.) Beentje ●☆

294712 Philippia norlindhii Weim. = Erica hexandra (S. Moore) E. G. H. Oliv. ●☆

294713 Philippia nyassana Alm et T. C. E. Fr. = Erica nyassana (Alm et T. C. E. Fr.) E. G. H. Oliv. ●☆

294714 Philippia oophylla Baker = Erica minutifolia (Baker) Dorr et F. Oliv. ●☆

294715 Philippia pallida L. Guthrie = Erica peltata Andréws ●☆

294716 Philippia pallidiflora Engl. = Erica mannii (Hook. f.) Beentje subsp. pallidiflora (Engl.) E. G. H. Oliv. ●☆

294717 Philippia pallidiflora Engl. subsp. usambarensis (Alm et T. C. E. Fr.) R. Ross = Erica mannii (Hook. f.) Beentje subsp. usambarensis (Alm et T. C. E. Fr.) Beentje ●☆

294718 Philippia parkeri Baker = Erica parkeri (Baker) Dorr et E. G. H. Oliv. ●☆

294719 Philippia parviflora Benth. = Erica lyallii Dorr et E. G. H. Oliv. ●☆

294720 Philippia petrophila E. G. H. Oliv. = Erica petricola E. G. H. Oliv. ●☆

294721 Philippia pilosa Baker = Erica madagascariensis (H. Perrier) Dorr et E. G. H. Oliv. ●☆

294722 Philippia pilulifera H. Perrier = Erica wangfatiana Dorr et E. G. H. Oliv. ●☆

294723 Philippia quadratiflora H. Perrier = Erica quadratiflora (H. Perrier) Dorr et E. G. H. Oliv. ●☆

294724 Philippia senescens Baker = Erica cryptoclada (Baker) Dorr et E. G. H. Oliv. ●☆

294725 Philippia simii S. Moore = Erica simii (S. Moore) E. G. H. Oliv. ●☆

294726 Philippia spinifera H. Perrier = Erica spinifera (H. Perrier) Dorr et E. G. H. Oliv. ●☆

294727 Philippia stuhlmannii Engl. = Erica benguelensis (Welw. ex Engl.) E. G. H. Oliv. ●☆

294728 Philippia tenuifolia Benth. = Erica goudotiana (Klotzsch) Dorr et E. G. H. Oliv. ●☆

294729 Philippia tenuissima Klotzsch = Erica rakotozafyana Dorr et E. G. H. Oliv. ●☆

294730 Philippia thomensis Henriq. = Erica thomensis (Henriq.) Dorr et E. G. H. Oliv. ●☆

294731 Philippia trichoclada Baker = Erica hebeclada Dorr et E. G. H. Oliv. ●☆

294732 Philippia trichoclada Baker subsp. latisepala H. Perrier = Erica hebeclada Dorr et E. G. H. Oliv. ●☆

294733 Philippia trichoclada Baker var. albescens H. Perrier = Erica hebeclada Dorr et E. G. H. Oliv. ●☆

294734 Philippia trichoclada Baker var. subalbida H. Perrier = Erica hebeclada Dorr et E. G. H. Oliv. ●☆

294735 Philippia trimera Engl. = Erica trimera (Engl.) Beentje ●☆

294736 Philippia trimera Engl. subsp. abyssinica (Pic. Serm. et Heiniger) Hedberg = Erica trimera (Engl.) Beentje subsp. abyssinica (Pic. Serm. et Heiniger) Dorr ●☆

294737 Philippia trimera Engl. subsp. elgonensis (Mildbr.) Hedberg = Erica trimera (Engl.) Beentje subsp. elgonensis (Mildbr.) Beentje ●☆

294738 Philippia trimera Engl. subsp. jaegeri (Engl.) Hedberg = Erica trimera (Engl.) Beentje subsp. jaegeri (Engl.) Beentje ●☆

294739 Philippia trimera Engl. subsp. keniensis (S. Moore) Hedberg =

Erica trimera（Engl.）Beentje subsp. keniensis（S. Moore）Beentje ●☆

294740　Philippia trimera Engl. subsp. kilimanjarica Hedberg = Erica trimera（Engl.）Beentje subsp. kilimanjarica（Hedberg）Beentje ●☆

294741　Philippia tristis Bolus = Erica caespitosa Hilliard et B. L. Burtt ●☆

294742　Philippia uhehensis Engl. = Erica mannii（Hook. f.）Beentje subsp. pallidiflora（Engl.）E. G. H. Oliv. ●☆

294743　Philippia usambarensis Alm et T. C. E. Fr. = Erica mannii（Hook. f.）Beentje subsp. usambarensis（Alm et T. C. E. Fr.）Beentje ●☆

294744　Philippia viguieri H. Perrier = Erica viguieri（H. Perrier）Dorr et E. G. H. Oliv. ●☆

294745　Philippiamra Kuntze = Silvaea Phil. ●☆

294746　Philippicereus Backeb. = Eulychnia Phil. ●☆

294747　Philippiella Speg.（1897）;异株指甲木属●☆

294748　Philippiella patagonica Speg. ;异株指甲木●☆

294749　Philippimalva Kuntze = Gaya Kunth ■●☆

294750　Philippimalva Kuntze = Tetraptera Phil. ■●☆

294751　Philippinaea Schltr. et Ames = Orchipedum Breda ■☆

294752　Philippodendraceae A. Juss. = Malvaceae Juss.（保留科名）●■

294753　Philippodendraceae Endl. = Malvaceae Juss.（保留科名）●■

294754　Philippodendron Endl. = Philippodendrum Poit. ●☆

294755　Philippodendron Endl. = Plagianthus J. R. Forst. et G. Forst. ●☆

294756　Philippodendrum Poit. = Plagianthus J. R. Forst. et G. Forst. ●☆

294757　Phillipsia Rolfe = Dyschoriste Nees ■●

294758　Phillipsia Rolfe = Satanocrater Schweinf. ■☆

294759　Phillipsia fruticulosa Rolfe = Dyschoriste hildebrandtii（S. Moore）Lindau ■☆

294760　Phillyraea Moench = Phillyrea L. ●☆

294761　Phillyrea L.（1753）;欧女贞属（非丽属,假女贞属,总序桂属）;Mock Privet, Phillyrea ●☆

294762　Phillyrea angustifolia L. ;狭叶假女贞（狭叶欧女贞,狭叶总序桂）;Narrow-leaved Phillyrea ●☆

294763　Phillyrea angustifolia L. subsp. berengueri Sennen = Phillyrea latifolia L. ●☆

294764　Phillyrea angustifolia L. subsp. fontqueri Sennen et Mauricio = Phillyrea angustifolia L. ●☆

294765　Phillyrea angustifolia L. subsp. jorroi Sennen = Phillyrea angustifolia L. ●☆

294766　Phillyrea angustifolia L. subsp. latifolia（L.）Maire = Phillyrea latifolia L. ●☆

294767　Phillyrea angustifolia L. subsp. mauritii Sennen = Phillyrea latifolia L. ●☆

294768　Phillyrea angustifolia L. subsp. media（L.）Rouy = Phillyrea latifolia L. ●☆

294769　Phillyrea angustifolia L. subsp. orientalis Sébastian = Phillyrea angustifolia L. ●☆

294770　Phillyrea angustifolia L. var. brachiata Aiton = Phillyrea angustifolia L. ●☆

294771　Phillyrea cordifolia Sennen = Phillyrea latifolia L. ●☆

294772　Phillyrea latifolia L. ;假女贞（阔叶欧女贞,总序桂）;Mock Privet, Tree Phillyrea, Tree Phillyroa ●☆

294773　Phillyrea media L. ;药用欧女贞（非利女贞）●☆

294774　Phillyrea media L. = Phillyrea latifolia L. ●☆

294775　Phillyrea media L. var. grandifolia Alleiz. = Phillyrea latifolia L. ●☆

294776　Phillyrea medianifolia Sennen = Phillyrea latifolia L. ●☆

294777　Phillyrea medianifolia Sennen subsp. caballeroi Sennen et Mauricio = Phillyrea latifolia L. ●☆

294778　Phillyrea medianifolia Sennen subsp. frondosa ? = Phillyrea latifolia L. ●☆

294779　Phillyrea ramiflora Roxb. ex C. B. Clarke = Chionanthus ramiflorus Roxb. ●

294780　Phillyrophyllum O. Hoffm. = Philyrophyllum O. Hoffm. ■☆

294781　Philocrena Bong. = Tristicha Thouars ■☆

294782　Philocrenaceae Bong. = Podostemaceae Rich. ex Kunth（保留科名）■

294783　Philodendraceae Vines = Araceae Juss.（保留科名）■●

294784　Philodendron Schott ex Endl. = Philodendron Schott（保留属名）■●

294785　Philodendron Schott（1829）（'Philodendrum'）（保留属名）;喜林芋属（蔓绿绒属）;Philodendron ■●

294786　Philodendron 'Imperial Green';绿帝王■☆

294787　Philodendron 'Lemon lime';金锄蔓绿绒■☆

294788　Philodendron 'Pig Skin';猪皮蔓绿绒■☆

294789　Philodendron 'Pluto';神锯喜林芋■☆

294790　Philodendron 'Red Duchess';紫黑叶蔓绿绒■☆

294791　Philodendron 'Wend-imbe';箭叶蔓绿绒■☆

294792　Philodendron andreanum Devansaye;金叶喜林芋（绒叶蔓绿绒,喜林芋）;Goldenleaf Philodendron ●■

294793　Philodendron angustialatum Engl. ;长叶蔓绿绒●■☆

294794　Philodendron angustisectum Engl. ;细裂蔓绿绒●■☆

294795　Philodendron asperatum C. Koch;粗糙喜林芋;Rugged Philodendron ●■

294796　Philodendron bipinnatifidum Schott;羽叶喜林芋（复羽裂喜林芋,琴叶蔓绿绒,羽叶喜树蕉）;Cutleaf Philodendron, Pinnate-leaf Philodendron, Tree Philodendron ●■☆

294797　Philodendron cannifolium Engl. ;立叶喜林芋●■☆

294798　Philodendron cordatum Kunth;心叶喜树蕉（心叶蔓绿绒）;Heart-leaf Philodendron, Philodendron ●■☆

294799　Philodendron crassinervium Lindl. ;粗肋蔓绿绒●■☆

294800　Philodendron cruentum Poepp. et Engl. ;紫背蔓绿绒●■☆

294801　Philodendron cymbispathum Engl. ;舟苞喜林芋●■☆

294802　Philodendron domesticum Bunting;锦叶蔓绿绒;Elephant's Ear, Spade-leaf Philodendron ●■☆

294803　Philodendron eichleri Engl. ;小羽裂蔓绿绒●■☆

294804　Philodendron elegans K. Krause;深裂喜林芋;Elegant Philodendron ●■☆

294805　Philodendron erubescens C. Koch et Augustin;红苞喜林芋（红柄蔓绿绒,红柄喜林芋,红叶树藤,羽叶蔓绿绒）;Blushing Philodendron, Redbract Philodendron ●■

294806　Philodendron evansii Hort. ;浅裂蔓绿绒●■☆

294807　Philodendron florida Graf;鱼叶蔓绿绒●■☆

294808　Philodendron fragrans Bunting;玉簪叶蔓绿绒●■☆

294809　Philodendron giganteum Schott;大蔓绿绒●■☆

294810　Philodendron gloriosum André;缎叶喜林芋（扇叶蔓绿绒,心叶喜林芋）;Philodendron, Satin Philodendron ●■

294811　Philodendron grandifolium（Jacq.）Schott;大叶蔓绿绒●■☆

294812　Philodendron grazielae Bunting;团扇蔓绿绒■☆

294813　Philodendron ilsemannii Hort. ;爱丽喜林芋;Ilsemann Philodendron ●■☆

294814　Philodendron imbe Schott;喜林芋（蔓绿绒,喜树蕉）;Weak Philodendron ●■☆

294815　Philodendron imbe Schott 'Silver Metal';银色蔓绿绒■☆

294816　Philodendron karstenianum Schott;墨西哥喜林芋●■☆

294817　Philodendron laciniatum（Vell.）Engl. = Philodendron pedatum

Kunth ●■☆

294818 Philodendron ligulatum Schott;舌叶喜树蕉●■☆

294819 Philodendron linnaei Kunth;林奈喜林芋■☆

294820 Philodendron longilaminatum Schott;长叶喜林芋;Long-leaf Philodendron ●■☆

294821 Philodendron mamei André;玛美蔓绿绒■☆

294822 Philodendron martianum Engl.;战神喜林芋(黑金喜林芋,立叶蔓绿绒,泡泡蔓绿绒);Martian Philodendron ●■☆

294823 Philodendron melanochrysum Linden et André;铜绿蔓绿绒●■☆

294824 Philodendron meliononi Brongn. ex Regel;明脉喜林芋■☆

294825 Philodendron micans K. Koch;小叶喜林芋●■☆

294826 Philodendron ornatum Schott;美饰蔓绿绒●■☆

294827 Philodendron oxycardium Schott;圆叶蔓绿绒(心叶蔓绿绒)●■☆

294828 Philodendron oxycardium Schott 'Lime';小圆叶蔓绿绒(小黄心叶蔓绿绒)■☆

294829 Philodendron panduraeforme (Kunth) Kunth;琴叶蔓绿绒(琴叶树滕,琴叶喜林芋)●■☆

294830 Philodendron pedatum Kunth;趾叶喜林芋(掌裂蔓绿绒,掌叶喜林芋,趾叶喜林芋);Slashed Philodendron ●■☆

294831 Philodendron pertusum Kunth et Bouché = Monstera deliciosa Liebm. ■

294832 Philodendron pinnatifisum (Jacq.) Kunth;羽中裂喜林芋(羽蔓绿绒裂);Pinnatifid Philodendron ●■☆

294833 Philodendron pinnatilobum Engl. 'Fern Leaf';蕨叶喜林芋●■☆

294834 Philodendron rubrum Hort.;红叶蔓绿绒●■☆

294835 Philodendron sagittifolium Liebm.;箭叶喜林芋(箭叶蔓绿绒);Arrowleaf Philodendron ●■

294836 Philodendron sanguineum Regel;血红箭叶蔓绿绒●■☆

294837 Philodendron scandens C. Koch et Sello f. micans (Koch) Bunting;小叶蔓绿绒(心叶绿萝)●■☆

294838 Philodendron scandens K. Koch et Sello;攀缘喜林芋(攀缘蔓绿绒);Climbing Philodendron,Heart Leaf,Heart-leaved Philodendron,Sweetheart Plant ●■☆

294839 Philodendron selloum K. Koch;羽裂喜林芋(春羽,羽裂蔓绿绒);Cut-leaf Philodendron,Lacy Tree Philodendron,Split Leaf Philodendron ●■☆

294840 Philodendron simsii C. Koch = Philodendron simsii Hort. ex K. Koch ●■☆

294841 Philodendron simsii Hort. ex K. Koch;西美喜林芋;Sims Philodendron ●■☆

294842 Philodendron sodiroi Hort.;银叶喜林芋(银叶蔓绿绒);Silver-leaf Philodendron ●■☆

294843 Philodendron speciosum Schott;美雅喜林芋●■☆

294844 Philodendron squamiferum Poepp. et Engl.;绵毛喜林芋(锦毛喜林芋,绵毛蔓绿绒);Bristle Philodendron ●■☆

294845 Philodendron talamancae Engl.;长三角蔓绿绒●■☆

294846 Philodendron trifoliatum ? = Syngonium auritum (L.) Schott ■☆

294847 Philodendron tripartitum (Jacq.) Schott;三裂喜林芋(三裂蔓绿绒,三裂树滕);Trileaf Philodendron,Tripartite Philodendron ●■

294848 Philodendron tuxmog Hort.;尖叶蔓绿绒●■☆

294849 Philodendron variifolium Schott;变叶蔓绿绒●■☆

294850 Philodendron verrucosum L. Mathieu ex Schott;刺柄喜林芋(刺柄蔓绿绒,疣叶喜林芋);Velvet Leaf Philodendron ●■☆

294851 Philodendrum Schott = Plagianthus J. R. Forst. et G. Forst. ●☆

294852 Philodice Mart. (1834);无鞘谷精草属■☆

294853 Philodice hoffmannseggii Mart.;无鞘谷精草■☆

294854 Philoglossa DC. (1836);匍匐黑药菊属■☆

294855 Philoglossa peruviana DC.;匍匐黑药菊■☆

294856 Philoglossa pterocarpa Sandwith;翅果匍匐黑药菊■☆

294857 Philoglossa purpureodisca H. Rob.;紫盘匍匐黑药菊■☆

294858 Philogyne Salisb. = Narcissus L. ■

294859 Philomeda Noronha ex Thouars = Ouratea Aubl. (保留属名)●

294860 Philomidoschema Vved. = Stachys L. ●■

294861 Philonoma DC. ex Meisn. = Macromeria D. Don ■☆

294862 Philonotion Schott = Schismatoglottis Zoll. et Moritzi ■

294863 Philopodium Hort. = Muehlenbeckia Meisn. (保留属名)●☆

294864 Philostemon Raf. = Rhus L. ●

294865 Philostemum Steud. = Philostemon Raf. ●

294866 Philostizus Cass. = Centaurea L. (保留属名)●■

294867 Philotheca Rudge (1816);佳囊芸香属;Wax Flower ●☆

294868 Philotheca myoporoides (DC.) Bayly;长叶佳囊芸香;Long Leaf Wax Flower,Wax Flower ●☆

294869 Philotheca verrucosa (A. Rich.) Paul G. Wilson;多疣佳囊芸香;Fairy Wax Flower ●☆

294870 Philotria Raf. = Elodea Michx. ■☆

294871 Philotria angustifolia (Muhl.) Britton ex Rydb. = Elodea nuttallii (Planch.) H. St. John ■☆

294872 Philotria canadensis (Michx.) Britton = Elodea canadensis Michx. ■☆

294873 Philotria linearis Rydb. = Elodea canadensis Michx. ■☆

294874 Philotria minor (Engelm. ex Casp.) Small = Elodea nuttallii (Planch.) H. St. John ■☆

294875 Philotria nuttallii (Planch.) Rydb. = Elodea nuttallii (Planch.) H. St. John ■☆

294876 Philotria occidentalis (Pursh) House = Elodea nuttallii (Planch.) H. St. John ■☆

294877 Philoxerus R. Br. (1810);安旱苋属(安旱草属);Philoxerus ■

294878 Philoxerus R. Br. = Gomphrena L. ●■

294879 Philoxerus vermicularis (L.) R. Br. ex Sm. = Blutaparon vermiculare (L.) Mears ■☆

294880 Philoxerus vermicularis (L.) Sm. = Blutaparon vermiculare (L.) Mears ■☆

294881 Philoxerus wrightii Hook. f. = Philoxerus wrightii Hook. f. ex Maxim. ■

294882 Philoxerus wrightii Hook. f. ex Maxim.;安旱苋(安旱草,赖特布氏兰);Wright Philoxerus ■

294883 Philoxerus wrightii Hook. f. ex Maxim. = Blutaparon wrightii (Hook. f. ex Maxim.) Mears ■

294884 Philyca Boehm. = Phylica L. ●☆

294885 Philyca L. = Phylica L. ●☆

294886 Philydraceae Link (1821) (保留科名);田葱科;Fildchive Family,Philydrum Family ■

294887 Philydrella Caruel = Cerochilus Lindl. ■

294888 Philydrella Caruel = Hetaeria Blume(保留属名)■

294889 Philydrella Caruel(1878);小田葱属■☆

294890 Philydrella pygmaea (R. Br.) Caruel;小田葱■☆

294891 Philydrum Banks = Philydrum Banks ex Gaertn. ■

294892 Philydrum Banks ex Gaertn. (1788);田葱属;Fildchive,Philydrum ■

294893 Philydrum cavaleriei H. Lév. = Philydrum lanuginosum Banks et Sol. ex Gaertn. ■

294894 Philydrum cavaleriei H. Lév. = Utricularia bifida L. ■

294895 Philydrum lanuginosum Banks et Sol. ex Gaertn.;田葱(白根子草,扁合草,扇合草,水葱,水芦荟,中葱);Common Philydrum,

Fildchive ■

294896　Philydrum lanuginosum Banks ex Sol. = Philydrum lanuginosum Banks et Sol. ex Gaertn. ■

294897　Philydrum lanuginosum Gaertn. = Philydrum lanuginosum Banks et Sol. ex Gaertn. ■

294898　Philyra Klotzsch(1841);巴西菲利大戟属●☆

294899　Philyra brasiliensis Klotzsch;巴西菲利大戟●☆

294900　Philyrea Blume = Phillyrea L. ●☆

294901　Philyrophyllum O. Hoffm. (1890);金绒草属■☆

294902　Philyrophyllum alatum Burtt Davy = Pentatrichia alata S. Moore ■☆

294903　Philyrophyllum brandbergense Herman;金绒草■☆

294904　Philyrophyllum schinzii O. Hoffm. ;欣兹金绒草■☆

294905　Phinaea Benth. (1876);飞尼亚苣苔属;Phinaea ■☆

294906　Phinaea multiflora C. V. Morton;多花飞尼亚苣苔;Many-flowered Phinaea ■☆

294907　Phippsia (Trin.) R. Br. (1823);松鼠尾草属;Icegrass, Icegrass ■☆

294908　Phippsia R. Br. = Phippsia (Trin.) R. Br. ■☆

294909　Phippsia algida (Sol.) R. Br. ;松鼠尾草■☆

294910　Phippsia concinna (Th. Fr.) Lindeb. ;整洁松鼠尾草■☆

294911　Phippsia himalaica Hook. f. = Catabrosella himalaica (Hook. f.) Tzvelev ■☆

294912　Phisalis Nocca = Physalis L. ■☆

294913　Phitopis Hook. f. (1871);秘鲁茜属●☆

294914　Phitopis multiflora Hook. f. ;秘鲁茜☆

294915　Phitosia Kamari et Greuter = Crepis L. ■

294916　Phitosia Kamari et Greuter(2000);光株还阳参属■☆

294917　Phitosia crocifolia (Boiss. et Heldr.) Kamari et Greuter;光株还阳参■☆

294918　Phleaceae Link = Gramineae Juss. (保留科名)●■

294919　Phleaceae Link = Poaceae Barnhart(保留科名)●■

294920　Phlebanthe Post et Kuntze = Ajuga L. ■●

294921　Phlebanthe Post et Kuntze = Phleboanthe Tausch ■●

294922　Phlebanthe Rchb. = Phlebanthe Post et Kuntze ■●

294923　Phlebanthia Rchb. = Minuartia L. ■

294924　Phlebidia Lindl. = Disa P. J. Bergius ■☆

294925　Phlebiophragmus O. E. Schulz = Mostacillastrum O. E. Schulz ■☆

294926　Phlebiophragmus O. E. Schulz(1924);篱脉芥属(脉障芥属)■☆

294927　Phlebiophragmus macrorrhizus O. E. Schulz;篱脉芥■☆

294928　Phleboanthe Tausch = Ajuga L. ■●

294929　Phlebocalymna Griff. ex Miers = Gonocaryum Miq. ●

294930　Phlebocalymna Griff. ex Miers = Sphenostemon Baill. ●☆

294931　Phlebocalymna calleryana Baill. = Gonocaryum calleryanum (Baill.) Becc. ●

294932　Phlebocarya R. Br. (1810);棱果血草属■☆

294933　Phlebocarya ciliata R. Br. ;棱果血草■☆

294934　Phlebochilus (Benth.) Szlach. (2001);棱唇兰属■☆

294935　Phlebochilus (Benth.) Szlach. = Caladenia R. Br. ■☆

294936　Phlebochiton Wall. = Pegia Colebr. ●

294937　Phlebochiton extensum Wall. = Pegia nitida Colebr. ●

294938　Phlebochiton sarmentosum Lecomte = Pegia sarmentosa (Lecomte) Hand. -Mazz. ●

294939　Phlebochiton sinense Diels = Pegia sarmentosa (Lecomte) Hand. -Mazz. ●

294940　Phlebolithis Gaertn. (1788);印度山榄属●☆

294941　Phlebolithis Gaertn. = ? Mimusops L. ●☆

294942　Phlebolithis indica Gaertn. ;印度山榄●☆

294943　Phlebolobium O. E. Schulz(1933);脉裂芥属■☆

294944　Phlebolobium maclovianum (d' Urv.) O. E. Schulz;脉裂芥■☆

294945　Phlebophyllum Nees = Strobilanthes Blume ●■

294946　Phlebophyllum apricum (Hance) Benth. = Gutzlaffia aprica Hance ■

294947　Phlebosporium Hassk. = Phlebophyllum Nees ●■

294948　Phlebosporium Jungh. = Campylotropis Bunge ●

294949　Phlebosporum Jungh. = Lespedeza Michx. ●■

294950　Phlebotaenia Griseb. (1860);带脉远志属●■☆

294951　Phlebotaenia Griseb. = Polygala L. ●■

294952　Phlebotaenia cuneata Griseb. ;带脉远志●■☆

294953　Phledinium Spach = Delphinium L. ■

294954　Phlegmatospermum O. E. Schulz(1933);黏籽芥属■☆

294955　Phlegmatospermum villosulum (F. Muell. et Tate) O. E. Schulz;黏籽芥■☆

294956　Phleobanthe Ledeb. = Ajuga L. ■●

294957　Phleobanthe Ledeb. = Phleboanthe Tausch ■●

294958　Phleoides Ehrh. = Phalaris L. ■

294959　Phleoides Ehrh. = Phleum L. ■

294960　Phleum L. (1753);梯牧草属(大粟草属,大粟米草属);Cat's Tail Grass, Cat's-tail, Herd's-Grass, Timothy ■

294961　Phleum alopecuroides Piller et Mitterp. = Crypsis alopecuroides (Piller et Mitterp.) Schrad. ■☆

294962　Phleum alpinum L. ;高山梯牧草;Alpine Cat's Tail, Alpine Cat's-tail, Alpine Timothy ■

294963　Phleum alpinum L. subsp. trabutii Litard. et Maire = Phleum pratense L. subsp. trabutii (Litard. et Maire) Kerguélen ■☆

294964　Phleum arenarium Hook. f. = Phleum himalaicum Mez ■☆

294965　Phleum arenarium Hook. f. var. thomsonii Griseb. = Phleum himalaicum Mez ■☆

294966　Phleum arenarium L. ;沙生梯牧草(沙梯牧草);Sand Cat's Tail, Sand Cat's-tail, Sand Timothy ■☆

294967　Phleum asperum Jacq. = Phleum paniculatum Huds. ■

294968　Phleum bertolonii DC. ;伯氏梯牧草;Bertolon Timothy, Cat's Tail, Cat's Tails, Smaller Cat's-tail ■☆

294969　Phleum bertolonii DC. = Phleum pratense L. subsp. serotinum (Jord.) Berher ■☆

294970　Phleum bertolonii DC. subsp. trabutii (Litard. et Maire) Kerguélen = Phleum pratense L. subsp. trabutii (Litard. et Maire) Kerguélen ■☆

294971　Phleum boehmeri Wibel = Phleum phleoides (L.) Simonk. ■

294972　Phleum boissieri Bornm. ;布瓦西耶梯牧草■☆

294973　Phleum cochinchinense Lour. = Heteropholis cochinchinensis (Lour.) Clayton ■

294974　Phleum cochinchinense Lour. = Mnesithea laevis (Retz.) Kunth ■

294975　Phleum commutatum Gaudin = Phleum alpinum L. ■

294976　Phleum echinatum Host;刺梯牧草■☆

294977　Phleum gerardii All. = Colobachne gerardii (All.) Link ■☆

294978　Phleum glomeruliflorum Steud. = Elytrophorus spicatus (Willd.) A. Camus ■

294979　Phleum graecum Boiss. et Heldr. ;格氏梯牧草■☆

294980　Phleum himalaicum Mez;喜马拉雅梯牧草■☆

294981　Phleum indicum Houtt. ;印度梯牧草(印度三穗草);Java Grass ■☆

294982　Phleum indicum Houtt. = Ischaemum indicum (Houtt.) Merr. ■

294983　Phleum indicum Houtt. = Polytrias indica (Houtt.) Veldkamp ■

294984   Phleum japonicum Franch. et Sav. = Phleum paniculatum Huds. ■

294985   Phleum michelii All. ;米氏梯牧草■☆

294986   Phleum montanum K. Koch;山梯牧草■☆

294987   Phleum nodosum L. = Phleum bertolonii DC. ■☆

294988   Phleum paniculatum Huds. ;鬼蜡烛（假看麦娘，蜡烛草）；
British Timothy ,Ghost candle ■

294989   Phleum phleoides（L.）H. Karst. ;假梯牧草；False Timothy ,
Pointed Cat's Tail ,Purple-stem Cat's-tail ■

294990   Phleum phleoides（L.）Simonk. = Phleum phleoides（L.）H.
Karst. ■

294991   Phleum phleoides（L.）Simonk. var. blepharodes（Asch. et
Graebn.）Halácsy = Phleum phleoides（L.）Simonk. ■

294992   Phleum pratense L. ;梯牧草（猫尾草，丝草）；Cat's-tail-grass,
Common Timothy, Herd's-grass, Herd's Grass, Timothy, Timothy
Grass ,Timothygrass ,Timothy-grass ■

294993   Phleum pratense L. f. viviperum（Gray）Louis-Marie = Phleum
pratense L. ■

294994   Phleum pratense L. subsp. bertolonii（DC.）Bornm. = Phleum
bertolonii DC. ■☆

294995   Phleum pratense L. subsp. bertolonii（DC.）Bornm. = Phleum
pratense L. subsp. serotinum（Jord.）Berher ■☆

294996   Phleum pratense L. subsp. serotinum（Jord.）Berher;晚梯牧草■☆

294997   Phleum pratense L. subsp. trabutii（Litard. et Maire）
Kerguélen;特拉布特梯牧草■☆

294998   Phleum pratense L. subsp. vulgare（Celak.）Asch. et Graebn.
= Phleum pratense L. ■

294999   Phleum pratense L. var. nodosum（L.）Huds. = Phleum
pratense L. ■

295000   Phleum schoenoides L. = Crypsis schoenoides（L.）Lam. ■

295001   Phleum subulatum（Savi）Asch. et Graebn. ;意大利梯牧草；
Italian Timothy ■☆

295002   Phleum subulatum（Savi）Asch. et Graebn. subsp. ciliatum
（Boiss.）Humphries;睫毛意大利梯牧草■☆

295003   Phleum subulatum（Savi）Asch. et Graebn. var. ciliatum Boiss.
= Phleum subulatum（Savi）Asch. et Graebn. subsp. ciliatum
（Boiss.）Humphries ■☆

295004   Phleum tenue（Host）Schrad. = Phleum subulatum（Savi）
Asch. et Graebn. ■☆

295005   Phleum tenue Schrad. ;细梯牧草■☆

295006   Phloeodicarpus Bess. = Phlojodicarpus Turcz. ex Ledeb. ■

295007   Phloeophila Hoehne et Schltr. = Pleurothallis R. Br. ■☆

295008   Phloga Noronha ex Benth. ex Hook. f.（1883）;簇叶椰属（簇叶
桐属,夫落哥桐属）●☆

295009   Phloga Noronha ex Benth. ex Hook. f. = Dypsis Noronha ex
Mart. ●☆

295010   Phloga Noronha ex Hook. f. = Phloga Noronha ex Benth. ex
Hook. f. ●☆

295011   Phloga Noronha ex Thouars = Phloga Noronha ex Benth. ex
Hook. f. ●☆

295012   Phloga gracilis（Jum.）H. Perrier = Dypsis oreophila Beentje ●☆

295013   Phloga nodifera（Mart.）Salomon = Dypsis nodifera Mart. ●☆

295014   Phloga polystachya Becc. = Dypsis pinnatifrons Mart. ●☆

295015   Phloga polystachya Becc. var. stenophylla Becc. = Dypsis
nodifera Mart. ●☆

295016   Phloga sambiranensis Jum. = Dypsis ambanjae Beentje ●☆

295017   Phloga scottiana Becc. = Dypsis scottiana（Becc.）Beentje et
J. Dransf. ●☆

295018   Phlogacanthus Nees（1832）;火焰花属（焰爵床属）；
Phlogacanthus ,Falmeflower ●■

295019   Phlogacanthus abbreviatus（Craib）Benoist;缩序火焰花●

295020   Phlogacanthus asperulus Nees = Phlogacanthus vitellinus
（Roxb.）T. Anderson ●■

295021   Phlogacanthus colaniae Benoist et Benoist;广西火焰花■

295022   Phlogacanthus curviflorus（Wall.）Nees;火焰花（华木张，黄
张，乔木张，弯花焰爵床，焰爵床）；Curvedflower Phlogacanthus,
Falmeflower ●

295023   Phlogacanthus paniculatus（T. Anderson）Imlay = Cystacanthus
paniculatus T. Anderson ●

295024   Phlogacanthus pubinervius T. Anderson;毛脉火焰花；
Hairynerve Falmeflower ,Hairynerve Phlogacanthus ●

295025   Phlogacanthus pyramidalis Benoist;金塔火焰花（火焰花）■

295026   Phlogacanthus thyrsiflorus Nees = Phlogacanthus thyrsiformis
（Hardw.）Mabb. ●☆

295027   Phlogacanthus thyrsiformis（Hardw.）Mabb. ;印度火焰花●☆

295028   Phlogacanthus vitellinus（Roxb.）T. Anderson;糙叶火焰花
（淡黄焰爵床，地蓝，黄尿草）；Roughleaf Falmeflower, Roughleaf
Phlogacanthus ●■

295029   Phlogella Baill.（1894）;小簇叶椰属（拟夫落哥桐属）●

295030   Phlogella Baill. = Chrysalidocarpus H. Wendl. ●

295031   Phlogella Baill. = Dypsis Noronha ex Mart. ●☆

295032   Phlogella humblotiana Baill. ;小簇叶椰●☆

295033   Phloiodicarpus Rchb. = Phlojodicarpus Turcz. ex Ledeb. ■

295034   Phlojodicarpus Turcz. ex Besser = Phlojodicarpus Turcz. ex
Ledeb. ■

295035   Phlojodicarpus Turcz. ex Ledeb.（1844）;胀果芹属（燥芹属）；
Swellenfruit Celery ■

295036   Phlojodicarpus abolinii Korovin = Libanotis abolinii（Korovin）
Korovin ■

295037   Phlojodicarpus sibiricus（Fisch. ex Spreng.）Koso-Pol. =
Phlojodicarpus sibiricus（Stephan ex Spreng.）Koso-Pol. ■

295038   Phlojodicarpus sibiricus（Stephan ex Spreng.）Koso-Pol. ;胀果
芹（膨果芹，燥芹）；Swellenfruit Celery ■

295039   Phlojodicarpus sibiricus（Stephan ex Spreng.）Koso-Pol. subsp.
villosus（Turcz. ex Fisch. et C. A. Mey.）Vorosch. = Phlojodicarpus
villosus（Turcz. ex Fisch. et C. A. Mey.）Turcz. ex Ledeb. ■

295040   Phlojodicarpus sibiricus（Stephan ex Spreng.）Koso-Pol. var.
villosus（Turcz. ex Fisch. et C. A. Mey.）Y. C. Chu =
Phlojodicarpus villosus（Turcz. ex Fisch. et C. A. Mey.）Turcz. ex
Ledeb. ■

295041   Phlojodicarpus villosus（Turcz. ex Fisch. et C. A. Mey.）Turcz.
ex Ledeb. ;柔毛胀果芹（毛序燥芹）；Velvety Swellenfruit celery ■

295042   Phlomidopsis Link = Phlomis L. ●■

295043   Phlomidopsis tuberosa（L.）Link = Phlomis tuberosa L. ■

295044   Phlomidopsis tuberosa Link = Phlomis tuberosa L. ■

295045   Phlomidoschema（Benth.）Vved.（1941）;中亚糙苏属；
Littleflower Betony ■☆

295046   Phlomidoschema（Benth.）Vved. = Stachys L. ●■

295047   Phlomidoschema Vved. = Phlomidoschema（Benth.）Vved. ■☆

295048   Phlomidoschema parviflorum（Benth.）Vved. ;中亚糙苏（小花
水苏）；Littleflower Betony ■☆

295049   Phlomis L.（1753）;糙苏属；Jerusalem Sage, Jerusalemsage,
Phlomis ,Sage ●■

295050   Phlomis africana P. Beauv. = Leonotis nepetifolia（L.）R. Br.
var. africana（P. Beauv.）J. K. Morton ■☆

295051　Phlomis agraria Bunge;耕地糙苏;Field Jerusalemsage ■

295052　Phlomis alaica Knorring;阿拉糙苏■☆

295053　Phlomis albiflora Hemsl. = Paraphlomis albiflora（Hemsl.）Hand.-Mazz. ■

295054　Phlomis alpina Pall.;高山糙苏;Alpine Jerusalemsage ■

295055　Phlomis ambigua Hand.-Mazz.;沧江糙苏;Cangjiang Jerusalemsage

295056　Phlomis angrenica Knorring;安哥拉糙苏■☆

295057　Phlomis antiatlantica M. Peltier;安蒂糙苏■☆

295058　Phlomis armeniaca Benth.;亚美尼亚糙苏■☆

295059　Phlomis aspera Willd. = Leucas aspera（Willd.）Link ■

295060　Phlomis atropurpurea Dunn;深紫糙苏;Darkpurple Jerusalemsage ■

295061　Phlomis atropurpurea Dunn f. palidior C. Y. Wu;浅色紫糙苏■

295062　Phlomis atropurpurea Dunn f. pilosa C. Y. Wu;疏毛紫糙苏■

295063　Phlomis betonicifolia Regel;药水苏糙苏■☆

295064　Phlomis betonicoides Diels;假秦艽（白玄参,白洋参,白元参,甘草,土甘草,雪山甘草）;Betonicalike Jerusalemsage ■

295065　Phlomis betonicoides Diels f. alba C. Y. Wu;白花假秦艽;Whiteflower Betonicalike Jerusalemsage ■

295066　Phlomis betonicoides Diels f. alba C. Y. Wu = Phlomis betonicoides Diels ■

295067　Phlomis bicolor（Viv.）Benth. = Phlomis floccosa D. Don ■☆

295068　Phlomis biloba Desf.;二裂糙苏■☆

295069　Phlomis bovei Noë;博韦糙苏■☆

295070　Phlomis bovei Noë subsp. maroccana Maire;摩洛哥糙苏■☆

295071　Phlomis brachystegia Bunge;短盖糙苏■☆

295072　Phlomis bracteosa Royle = Phlomis bracteosa Royle ex Benth. ■

295073　Phlomis bracteosa Royle = Phlomis likiangensis C. Y. Wu ■

295074　Phlomis bracteosa Royle ex Benth.;有苞糙苏（多苞糙苏,香苏）;Bracteate Jerusalemsage ■☆

295075　Phlomis brevidentata H. W. Li;短齿糙苏;Shortooth Jerusalemsage ■

295076　Phlomis breviflora Benth.;短花糙苏;Shortflower Jerusalemsage ■☆

295077　Phlomis bucharica Regel;布哈尔糙苏■☆

295078　Phlomis caballeroi Pau = Phlomis purpurea L. subsp. caballeroi（Pau）Rivas Mart. ■☆

295079　Phlomis caballeroi Pau var. submontana Pau et Font Quer = Phlomis purpurea L. subsp. caballeroi（Pau）Rivas Mart. ■☆

295080　Phlomis cancellata Bunge;格纹糙苏■☆

295081　Phlomis canescens Regel;灰糙苏■☆

295082　Phlomis capensis Thunb. = Leucas capensis（Benth.）Engl. ●☆

295083　Phlomis cashmeriana Royle ex Benth.;克什米尔糙苏（喀什米尔糙苏）;Cashmer Jerusalemsage,Kashmir Sage ●☆

295084　Phlomis caucasica Rchb. f.;高加索糙苏■☆

295085　Phlomis cephalotes Roth = Leucas cephalotes Spreng. ●

295086　Phlomis chinensis Retz. = Leucas chinensis（Retz.）R. Br. ■

295087　Phlomis chinghoensis C. Y. Wu;清河糙苏;Chingho Jerusalemsage,Qinghe Jerusalemsage ■

295088　Phlomis chrysophylla Boiss.;黄叶糙苏●☆

295089　Phlomis ciliata Heyne ex Wall. = Leucas ciliata Benth. ■

295090　Phlomis composita Pau;复合糙苏■☆

295091　Phlomis congesta C. Y. Wu;乾精菜（野苏麻）;Conferted Jerusalemsage,Crowded Jerusalemsage ■

295092　Phlomis crinita Cav. = Phlomis biloba Desf. ■☆

295093　Phlomis crinita Cav. subsp. mauritanica（Munby）Murb. = Phlomis biloba Desf. ■☆

295094　Phlomis cuneata C. Y. Wu;楔叶糙苏;Cunealleaf Jerusalemsage,Cuneateleaf Jerusalemsage ■

295095　Phlomis cyclodon Knorring;环齿糙苏■☆

295096　Phlomis dentosa Franch.;尖齿糙苏（毛尖糙苏）;Dentate Jerusalemsage ■

295097　Phlomis dentosa Franch. var. glabrescens Danguy;渐光尖齿糙苏（渐光糙苏）■

295098　Phlomis drobovii Popov;德罗糙苏■☆

295099　Phlomis ferganensis Popov;费尔干糙苏■☆

295100　Phlomis fimbriata C. Y. Wu;裂唇糙苏;Fimbriate Jerusalemsage ■

295101　Phlomis floccosa D. Don;丛毛糙苏■☆

295102　Phlomis forrestii Diels;苍山糙苏;Cangshan Jerusalemsage,Forrest Jerusalemsage ■

295103　Phlomis forrestii Diels var. taronensis C. Y. Wu;独龙糙苏;Dulong Jerusalemsage ■

295104　Phlomis forrestii Diels var. taronensis C. Y. Wu = Phlomis forrestii Diels ■

295105　Phlomis franchetiana Diels;大理糙苏;Dali Jerusalemsage ■

295106　Phlomis franchetiana Diels var. aristata C. Y. Wu;芒尖大理糙苏（芒尖糙苏）;Aristate Dali Jerusalemsage ■

295107　Phlomis franchetiana Diels var. aristata C. Y. Wu = Phlomis franchetiana Diels ■

295108　Phlomis franchetiana Diels var. leptophylla C. Y. Wu;薄叶大理糙苏（薄叶糙苏）;Thinleaf Dali Jerusalemsage ■

295109　Phlomis franchetiana Diels var. leptophylla C. Y. Wu = Phlomis franchetiana Diels ■

295110　Phlomis franchetiana Hand.-Mazz. var. hirticalyx Hand.-Mazz. = Phlomis tatsienensis Bureau et Franch. var. hirticalyx（Hand.-Mazz.）C. Y. Wu ■

295111　Phlomis fruticosa L.;橙花糙苏（耶路撒冷糙苏）;Common Jerusalem Sage, Jerusalem Sage, Jerusalem-sage, Orangeflower Jerusalemsage,Shrubby Jerusalem Sage ●■

295112　Phlomis glabrata Vahl = Leucas glabrata（Vahl）Sm. ■●☆

295113　Phlomis gracilis Hemsl. = Paraphlomis formosana（Hayata）T. H. Hsieh et T. C. Huang ■

295114　Phlomis gracilis Hemsl. = Paraphlomis gracilis（Hemsl.）Kudo ■

295115　Phlomis herba-venti L.;地中海糙苏■☆

295116　Phlomis hybrida Zelen.;杂种糙苏■☆

295117　Phlomis hypoleuca Vved.;里白糙苏■☆

295118　Phlomis inaequalisepala C. Y. Wu;斜萼糙苏;Slantingsepal Jerusalemsage,Unequalsepal Jerusalemsage ■

295119　Phlomis italica L.;意大利糙苏●☆

295120　Phlomis italica Rivas Mart. = Phlomis antiatlantica M. Peltier ■☆

295121　Phlomis italica Rivas Mart. subsp. antiatlantica（M. Peltier）Rivas Mart. = Phlomis antiatlantica M. Peltier ■☆

295122　Phlomis javanica（Blume）Prain = Paraphlomis javanica（Blume）Prain ■

295123　Phlomis jeholensis Nakai et Kitag.;口外糙苏（热河糙苏）;Jehol Jerusalemsage ■

295124　Phlomis kansuensis C. Y. Wu;甘肃糙苏;Gansu Jerusalemsage,Kansu Jerusalemsage ■

295125　Phlomis kawaguchii Murata = Phlomis younghusbandii Mukerjee ■

295126　Phlomis knorringiana Popov;克诺氏糙苏■☆

295127　Phlomis kopetdagensis Knorring;科佩特糙苏■☆

295128　Phlomis koraiensis Nakai;长白糙苏;Changbai Jerusalemsage,Korean Jerusalemsage ■

295129　Phlomis lanata Willd.;小糙苏;Dwarf Jerusalem Sage ■☆

295130　Phlomis lenkoranica Knorring;连科兰糙苏■☆

295131　Phlomis leonitis L. = Leonotis ocymifolia（Burm. f.）Iwarsson ●■☆

295132 Phlomis likiangensis C. Y. Wu;丽江糙苏(多苞糙苏,豨莶草,香苏,野苏子);Lijiang Jerusalemsage,Likiang Jerusalemsage ■

295133 Phlomis linearifolia Zakirov;线叶糙苏 ■☆

295134 Phlomis linifolia Roth = Leucas lavandulifolia Sm. ■

295135 Phlomis longicalyx C. Y. Wu;长萼糙苏;Longcalyx Jerusalemsage ■

295136 Phlomis longifolia Boiss. et Blanche;长叶糙苏 ●☆

295137 Phlomis lychnitis L.;灯芯糙苏 ■☆

295138 Phlomis lycia D. Don;灰毛糙苏 ●■☆

295139 Phlomis majkopensis (Novopokr.) Grossh. 马伊科普糙苏 ■☆

295140 Phlomis marrubioides Regel = Stachyopsis marrubioides (Regel) Ikonn. -Gal. ■

295141 Phlomis mauritanica Munby = Phlomis biloba Desf. ■☆

295142 Phlomis maximowiczii Regel;大叶糙苏(大丁黄,丁黄草,山苏子,苏木帐子,野苏子);Largeleaf Jerusalemsage ■

295143 Phlomis medicinalis Diels;萝卜秦艽;Medicinal Jerusalemsage ■

295144 Phlomis megalantha Diels;大花糙苏(老鼠刺);Largeflower Jerusalemsage ■

295145 Phlomis megalantha Diels var. pauciflora C. Y. Wu;少花糙苏(少花大花糙苏);Fewflower Jerusalemsage ■

295146 Phlomis melanantha Diels;黑花糙苏;Blackflower Jerusalemsage ■

295147 Phlomis melanantha Diels var. angustifolia C. Y. Wu;狭叶黑花糙苏(少花黑花糙苏);Narrow Blackflower Jerusalemsage ■

295148 Phlomis melanantha Diels var. angustifolia C. Y. Wu = Phlomis melanantha Diels ■

295149 Phlomis milingensis C. Y. Wu et H. W. Li;米林糙苏(螃蟹甲);Milin Jerusalemsage,Miling Jerusalemsage ■

295150 Phlomis mongolica Turcz.;串铃草(毛尖茶,蒙古糙苏,野洋芋);Mongol Jerusalemsage,Mongolian Jerusalemsage ■

295151 Phlomis mongolica Turcz. = Phlomis umbrosa Turcz. var. ovalifolia C. Y. Wu ■

295152 Phlomis mongolica Turcz. var. macracephala C. Y. Wu;大头串铃草;Bighead Mongol Jerusalemsage ■

295153 Phlomis muliensis C. Y. Wu;木里糙苏;Muli Jerusalemsage ■

295154 Phlomis nana C. Y. Wu;侏儒糙苏;Dwarf Jerusalemsage ■

295155 Phlomis nepetifolia L. = Leonotis nepetifolia (L.) R. Br. ■☆

295156 Phlomis nyslamensis H. W. Li;聂拉木糙苏;Nielamu Jerusalemsage ■

295157 Phlomis oblongata Schrenk ex Fisch. et C. A. Mey. = Stachyopsis oblongata (Schrenk ex Fisch. et C. A. Mey.) Popov et Vved. ■

295158 Phlomis oblongata Schrenk var. canescens Regel = Stachyopsis marrubioides (Regel) Ikonn. -Gal. ■

295159 Phlomis oblongata Schrenk. = Stachyopsis oblongata (Schrenk) Popov et Vved. ■

295160 Phlomis oblongata Schrenk. var. canescens Regel = Stachyopsis marrubioides (Regel) Ikonn. -Gal. ■

295161 Phlomis ocymifolia Burm. f. = Leonotis ocymifolia (Burm. f.) Iwarsson ●■☆

295162 Phlomis olgae Regel;奥氏糙苏 ■☆

295163 Phlomis oreophila Kar. et Kir.;山地糙苏;Mountain Jerusalemsage,Mountainous Jerusalemsage ■

295164 Phlomis oreophila Kar. et Kir. var. evillosa C. Y. Wu;无毛山地糙苏;Hairless Mountain Jerusalemsage ■

295165 Phlomis ornata C. Y. Wu;美观糙苏;Fair Jerusalemsage,Showy Jerusalemsage ■

295166 Phlomis ornata C. Y. Wu var. minor C. Y. Wu;小花美观糙苏;Smallflower Showy Jerusalemsage ■

295167 Phlomis ostrovskiana Regel;奥斯特罗夫斯基糙苏 ■☆

295168 Phlomis pallida Schumach. et Thonn. = Leonotis nepetifolia (L.) R. Br. var. africana (P. Beauv.) J. K. Morton ■☆

295169 Phlomis paohsingensis C. Y. Wu;宝兴糙苏;Baoxing Jerusalemsage,Paohsing Jerusalemsage ■

295170 Phlomis pararotata Y. Z. Sun ex C. H. Hu;假轮状糙苏;Pararotary Jerusalemsage,Pararotate Jerusalemsage ■

295171 Phlomis pedunculata Y. Z. Sun ex C. H. Hu;具梗糙苏;Peduncled Jerusalemsage,Pedunculate Jerusalemsage ■

295172 Phlomis pratensis Kar. et Kir.;草原糙苏;Meadow Jerusalemsage ■

295173 Phlomis pseudopungens Knorring;假锐尖糙苏 ■☆

295174 Phlomis puberula Krylov et Serg.;微毛糙苏 ■☆

295175 Phlomis pungens Willd.;锐尖糙苏 ■☆

295176 Phlomis pungens Willd. = Phlomis herba-venti L. ■☆

295177 Phlomis purpurea L.;紫糙苏 ■☆

295178 Phlomis purpurea L. subsp. almeriensis (Pau) Losa et Rivas Goday;阿梅糙苏 ■☆

295179 Phlomis purpurea L. subsp. caballeroi (Pau) Rivas Mart.;卡瓦列罗糙苏 ■☆

295180 Phlomis pygmaea C. Y. Wu;矮糙苏;Dwarf Jerusalemsage ■

295181 Phlomis regelii Popov;雷格尔糙苏 ■☆

295182 Phlomis rotata Benth. ex Hook. f. = Lamiophlomis rotata (Benth. ex Hook. f.) Kudo ■

295183 Phlomis rugosa Benth. = Paraphlomis javanica (Blume) Prain ■

295184 Phlomis ruptilis C. Y. Wu;裂萼糙苏;Ruptile Jerusalemsage ■

295185 Phlomis russeliana Lag. ex Benth.;心叶糙苏(西亚糙苏);Turkish Sage ●☆

295186 Phlomis sagittata Regel = Metastachydium sagittatum (Regel) C. Y. Wu et H. W. Li ■

295187 Phlomis salicifolia Regel;柳叶糙苏 ■☆

295188 Phlomis samia L.;黏糙苏 ■☆

295189 Phlomis samia L. subsp. bovei (Noë) Maire = Phlomis bovei Noë ■☆

295190 Phlomis samia L. subsp. maroccana Maire = Phlomis bovei Noë subsp. maroccana Maire ■☆

295191 Phlomis setifera Bureau et Franch.;刺毛糙苏;Setose Jerusalemsage,Spinatehair Jerusalemsage ■

295192 Phlomis setigera Falc. ex Benth.;刚毛糙苏;Setose Jerusalemsage ■☆

295193 Phlomis setigera Falc. ex Benth. var. filiformis ? = Phlomis breviflora Benth. ■☆

295194 Phlomis simplex Royle ex Benth. = Phlomis bracteosa Royle ex Benth. ■

295195 Phlomis souliei H. Lév. = Phlomis tatsienensis Bureau et Franch. ■

295196 Phlomis speciosa Hand. -Mazz. = Phlomis ornata C. Y. Wu ■

295197 Phlomis spectabilis Falc. ex Benth.;美丽糙苏;Showy Jerusalemsage ■☆

295198 Phlomis spinidens Nevski;刺齿糙苏 ■☆

295199 Phlomis stenocalyx Diels = Phlomis umbrosa Turcz. var. stenocalyx (Diels) C. Y. Wu ■

295200 Phlomis stewartii Hook. f.;斯太沃特糙苏;Stewart Jerusalemsage ■☆

295201 Phlomis strigosa C. Y. Wu;糙毛糙苏(粗毛糙苏);Strigose Jerusalemsage ■

295202 Phlomis szechuanensis C. Y. Wu;柴续断;Firewood Teasel,Sichuan Jerusalemsage,Szechuan Jerusalemsage ■

295203 Phlomis tatsienensis Bureau et Franch.;康定糙苏;Kangding

Jerusalemsage ■

295204 Phlomis tatsienensis Bureau et Franch. var. hirticalyx（Hand.-Mazz.）C. Y. Wu；毛萼康定糙苏■

295205 Phlomis taurica Hartwiss.；克里木糙苏■☆

295206 Phlomis tenuis Knorring；细糙苏■☆

295207 Phlomis thapsoides Bunge；毒胡萝卜糙苏■☆

295208 Phlomis tibetica C. Marquand et Airy Shaw；西藏糙苏；Tibet Jerusalemsage，Xizang Jerusalemsage ■

295209 Phlomis tibetica C. Marquand et Airy Shaw var. wardii C. Marquand et Airy Shaw；毛盔西藏糙苏；Ward Xizang Jerusalemsage ■

295210 Phlomis tomeotosa Regel；毛糙苏■☆

295211 Phlomis tschimganica Vved.；契穆干糙苏■☆

295212 Phlomis tuberosa L.；块根糙苏（块茎糙苏，野山药）；Tuber Jerusalemsage，Tuberous Jerusalem Sage，Tuberousroot Jerusalemsage ■

295213 Phlomis tuberosa L. = Phlomis mongolica Turcz. ■

295214 Phlomis tuberosa Moench = Phlomis tuberosa L. ■

295215 Phlomis umbrosa Turcz.；糙苏（白薮，白苕，常山，大叶糙苏，山苏子，山芝麻，豨莶草，小兰花烟，小蓝花烟，续断）；Jerusalemsage ■

295216 Phlomis umbrosa Turcz. var. australia Hemsl.；南方糙苏（白升麻，糙苏，大黑理肺散，灯笼大秦艽，牛王肺筋草，山甘草，土玄参，豨莶草）■

295217 Phlomis umbrosa Turcz. var. latibracteata Y. Z. Sun；宽苞糙苏；Broadbract Jerusalemsage ■

295218 Phlomis umbrosa Turcz. var. ovalifolia C. Y. Wu；卵叶糙苏（印川糙苏）；Ovateleaf Jerusalemsage ■

295219 Phlomis umbrosa Turcz. var. stenocalyx（Diels）C. Y. Wu；狭萼糙苏；Narrowcalyx Jerusalemsage ■

295220 Phlomis umbrosa Turcz. var. subaustralia K. T. Fu et J. Q. Fu；拟南方糙苏；Ovateleaf-like Jerusalemsage ■

295221 Phlomis umbrosa Turcz. var. sylvatica K. T. Fu et J. Q. Fu；森林糙苏；Forest Jerusalemsage ■

295222 Phlomis uniceps C. Y. Wu；单头糙苏（单花糙苏）；Singlehead Jerusalemsage ■

295223 Phlomis urodonta Popov；尾齿糙苏■☆

295224 Phlomis urticifolia Vahl = Leucas urticifolia（Vahl）R. Br. ■☆

295225 Phlomis vavilovii Popov；瓦维糙苏■☆

295226 Phlomis wangii Hu et H. T. Tsai = Phlomis medicinalis Diels ■

295227 Phlomis younghusbandii Mukerjee；青藏糙苏（藏糙苏，块根糙苏，露木，螃蟹甲，西藏糙苏）；Crab，Crust，Younghusband Jerusalemsage ■

295228 Phlomis zanaidae Knorring；扎氏糙苏■☆

295229 Phlomis zeylanica Jacq. = Leucas lavandulifolia Sm. ■

295230 Phlomis zeylanica L. = Leucas zeylanica（L.）R. Br. ■

295231 Phlomitis（Rchb.）Spenn. = Phlomoides Moench ●■

295232 Phlomitis Rchb. ex T. Nees = Phlomis L. ●■

295233 Phlomoides Moench = Phlomis L. ●■

295234 Phlomoides maximowiczii（Regel）Kamelin et Makhm. = Phlomis maximowiczii Regel ■

295235 Phlomoides tuberosa（L.）Moench = Phlomis tuberosa L. ■

295236 Phlomostachys Beer = Puya Molina ■☆

295237 Phlomostachys C. Koch = Pitcairnia L'Hér.（保留属名）■☆

295238 Phlomostachys K. Koch = Pitcairnia L'Hér.（保留属名）■☆

295239 Phlox L.（1753）；天蓝绣球属（福禄考属）；Phlox ■

295240 Phlox × procumbens Lehm.；平卧福禄考■☆

295241 Phlox × procumbens Lehm.'Millstream'；水流平卧福禄考■☆

295242 Phlox × procumbens Lehm.'Variegata'；彩叶平卧福禄考■☆

295243 Phlox acuminata Pursh = Phlox paniculata L. ■

295244 Phlox adsurgens Torr. ex A. Gray；斜升福禄考；Perwinkle Phlox ■☆

295245 Phlox adsurgens Torr. ex A. Gray'Wagon Wheel'；马车轮斜升福禄考■☆

295246 Phlox amoena Sims；可爱福禄考；Amoena Phlox ■☆

295247 Phlox amoena Sims'Variegata' = Phlox × procumbens Lehm.'Variegata'■☆

295248 Phlox amplifolia Britton；宽叶福禄考；Broadleaf Phlox ■☆

295249 Phlox argillacea Clute et Ferriss = Phlox pilosa L. ■☆

295250 Phlox bifida Beck subsp. arkansana Marshall = Phlox bifida L. C. Beck ■☆

295251 Phlox bifida Beck var. glandifera Wherry = Phlox bifida L. C. Beck ■☆

295252 Phlox bifida L. C. Beck；二裂福禄考；Cleft Phlox，Sand Phlox，Ten-point Phlox ■☆

295253 Phlox bifida L. C. Beck var. stellaria（A. Gray）Wherry = Phlox bifida L. C. Beck ■☆

295254 Phlox caespitosa Nutt.；丛生福禄考；Pink Phlox，Tufted Phlox ■☆

295255 Phlox carolina L.；薄叶天蓝绣球（卡罗莱纳福禄考，卡罗利纳天蓝绣球）；Carolina Phlox ■☆

295256 Phlox carolina L. var. angusta（Wherry）Steyerm. = Phlox carolina L. ■☆

295257 Phlox divaricata L.；蓝紫福禄考；Blue Phlox，Divaricate Phlox，Forest Phlox，Sweet William Phlox，Wild Blue Phlox，Wild Sweet William，Wild Sweet-william，Willow-blossom，Woodland Phlox ■☆

295258 Phlox divaricata L. subsp. laphamii（A. W. Wood）Wherry；狭裂福禄考；Blue Phlox，Forest Phlox，Wild Blue Phlox，Wild Sweet-william，Woodland Phlox ■☆

295259 Phlox divaricata L. subsp. laphamii（A. W. Wood）Wherry'Chatahoochee'；红心狭裂福禄考■☆

295260 Phlox divaricata L. var. laphamii A. W. Wood = Phlox divaricata L. ■☆

295261 Phlox divaricata L. var. laphamii A. W. Wood = Phlox divaricata L. subsp. laphamii（A. W. Wood）Wherry ■☆

295262 Phlox douglasii Hook.；道氏福禄考■☆

295263 Phlox douglasii Hook.'Boothman's Variety'；布恩曼道氏福禄考■☆

295264 Phlox douglasii Hook.'Crackerjack'；克拉克尔杰克道氏福禄考■☆

295265 Phlox douglasii Hook.'May Snow'；五月雪道氏福禄考■☆

295266 Phlox douglasii Hook.'Red Admiral'；红将军道氏福禄考■☆

295267 Phlox drummondii Hook.；小天蓝绣球（草夹竹桃，福禄考，金山海棠，小洋花，雁来红）；Annual Phlox，Drummond Phlox，Drummond's Phlox，Pride-of-texas，Texan Pride ■

295268 Phlox drummondii Hook.'African Sunset'；非洲落日福禄考■☆

295269 Phlox drummondii Hook.'Carnival'；狂欢福禄考■☆

295270 Phlox drummondii Hook.'Chanal'；钱纳福禄考■☆

295271 Phlox drummondii Hook.'Petticoat'；佩蒂克德福禄考■☆

295272 Phlox drummondii Hook.'Stellaris'；星瓣福禄考；Star Phlox ■☆

295273 Phlox drummondii Hook.'Sternenzauber'；闪耀福禄考■☆

295274 Phlox drummondii Hook. var. stellaris Voss = Phlox drummondii Hook.'Stellaris'■☆

295275 Phlox glaberrima L.；光秃福禄考；Appalachian Smooth Phlox，Marsh Phlox，Smooth Phlox ■☆

295276 Phlox glaberrima L. subsp. interior（Wherry）Wherry；间型福禄考；Smooth Phlox ■☆

295277 Phlox glaberrima L. var. interior Wherry = Phlox glaberrima L.

subsp. interior（Wherry）Wherry ■☆

295278 Phlox glaberrima L. var. interior Wherry = Phlox glaberrima L. ■☆

295279 Phlox heterophylla P. Beauv. ex Brand；异叶福禄考■☆

295280 Phlox hoodii Richardson；落基山福禄考■☆

295281 Phlox longifolia Nutt.；长叶福禄考；Long-leaved Phlox ■☆

295282 Phlox maculata L.；斑茎福禄考；Meadow Phlox,Spotted Phlox,Summer Perennial Phlox,Sweet William Phlox,Wild Sweet William,Wild Sweet-william ■☆

295283 Phlox maculata L. 'Alpha'；阿尔法斑茎福禄考■☆

295284 Phlox maculata L. 'Omega'；欧米加斑茎福禄考■☆

295285 Phlox maculata L. var. pyramidalis（Sm.）Wherry = Phlox maculata L. ■☆

295286 Phlox nivalis Lodd.；蔓生福禄考；Trailing Phlox ■☆

295287 Phlox ovata L.；卵叶福禄考（山天蓝绣球）；Mountain Phlox ■☆

295288 Phlox paniculata L.；天蓝绣球（草夹竹桃，圆锥福禄考，锥花福禄考）；Fall Phlox,Garden Phlox,Perennial Phlox,Phlox,Summer Perennial Phlox,Summer Phlox ■

295289 Phlox paniculata L. 'Aida'；爱达天蓝绣球■☆

295290 Phlox paniculata L. 'Amethyst'；紫晶天蓝绣球■☆

295291 Phlox paniculata L. 'Balmoral'；圆帽天蓝绣球■☆

295292 Phlox paniculata L. 'Brigadier'；组长天蓝绣球■☆

295293 Phlox paniculata L. 'Bright Eyes'；明目天蓝绣球■☆

295294 Phlox paniculata L. 'Eva Cullum'；夏娃天蓝绣球■☆

295295 Phlox paniculata L. 'Evetide'；黄昏天蓝绣球■☆

295296 Phlox paniculata L. 'Fujiyama'；富士山天蓝绣球■☆

295297 Phlox paniculata L. 'Graf Zeppelin'；乔帕林伯爵天蓝绣球■☆

295298 Phlox paniculata L. 'Hampton Court'；汉普顿天蓝绣球■☆

295299 Phlox paniculata L. 'Harlequin'；五彩天蓝绣球■☆

295300 Phlox paniculata L. 'Le Mahdi'；酋长天蓝绣球■☆

295301 Phlox paniculata L. 'Mia Ruys'；米亚·吕斯天蓝绣球■☆

295302 Phlox paniculata L. 'Mother of Pearl'；珍珠母天蓝绣球■☆

295303 Phlox paniculata L. 'Norah Leigh'；若拉利天蓝绣球■☆

295304 Phlox paniculata L. 'Prince of Orange'；橙王天蓝绣球■☆

295305 Phlox paniculata L. 'Russian Violet'；俄罗斯紫罗兰天蓝绣球■☆

295306 Phlox paniculata L. 'Sandringham'；桑德灵厄姆天蓝绣球■☆

295307 Phlox paniculata L. 'Sir John Falstaff'；约翰福斯塔夫天蓝绣球■☆

295308 Phlox paniculata L. 'White Admiral'；白蛱蝶天蓝绣球■☆

295309 Phlox paniculata L. 'Windsor'；温莎公爵天蓝绣球■☆

295310 Phlox pilosa L.；长毛福禄考■☆

295311 Phlox pilosa L. f. albiflora（MacMill.）Standl. = Phlox pilosa L. ■☆

295312 Phlox pilosa L. subsp. fulgida（Wherry）Wherry；光亮长毛福禄考；Downy Phlox,Prairie Phlox ■☆

295313 Phlox pilosa L. var. amplexicaulis（Raf.）Wherry = Phlox pilosa L. ■☆

295314 Phlox pilosa L. var. fulgida Wherry = Phlox pilosa L. subsp. fulgida（Wherry）Wherry ■☆

295315 Phlox pilosa L. var. fulgida Wherry = Phlox pilosa L. ■☆

295316 Phlox pilosa L. var. ozarkana Wherry = Phlox pilosa L. ■☆

295317 Phlox pilosa L. var. virens（Michx.）Wherry = Phlox pilosa L. ■☆

295318 Phlox sibirica L.；西伯利亚蓝绣球；Siberia Phlox ■☆

295319 Phlox siebmanni Benth. = Phlox paniculata L. ■

295320 Phlox stellaria A. Gray；星形福禄考■☆

295321 Phlox stolonifera Sims；匍枝福禄考；Creeping Phlox ■☆

295322 Phlox stolonifera Sims 'Ariane'；阿里匍枝福禄考■☆

295323 Phlox stolonifera Sims 'Blue Ridge'；蓝岭匍枝福禄考■☆

295324 Phlox subulata L.；钻叶天蓝绣球（丛生福禄考,钻叶福禄考）；Ground Phlox,Ground Pink,Mos Phlox,Moss Phlox,Moss Pink,Mosspink,Moss-pink,Phlox,Thrift

295325 Phlox subulata L. 'Marjorie'；马乔里钻叶福禄考■☆

295326 Phlox suffruticosa Vent.；亚灌木状福禄考；Early Perennial Phlox ■☆

295327 Phlox tenuifolia E. Nelson；细叶福禄考；Desert PhloxSanta Catalina Mountain Phlox ■☆

295328 Phlox villosissima（A. Gray）Small = Phlox pilosa L. ■☆

295329 Phlyarodoxa S. Moore = Ligustrum L. ●

295330 Phlyarodoxa leucantha S. Moore = Ligustrum leucanthum（S. Moore）P. S. Green ●

295331 Phlyctidocarpa Cannon et Theobald（1967）；泡果芹属■☆

295332 Phlyctidocarpa flava Cannon et W. L. Theob.；泡果芹■☆

295333 Phoberos Lour. = Scolopia Schreb.（保留属名）●

295334 Phoberos chinensis Lour. = Scolopia chinensis（Lour.）Clos ●

295335 Phoberos cochinchinensis Lour. = Scolopia chinensis（Lour.）Clos ●

295336 Phoberos ecklonii（Nees）C. Presl = Scolopia zeyheri（Nees）Harv. ●☆

295337 Phoberos mundii（Eckl. et Zeyh.）C. Presl = Scolopia mundtii（Eckl. et Zeyh.）Warb. ●☆

295338 Phoberos saevus Hance = Scolopia saeva（Hance）Hance ●

295339 Phoberos zeyherii（Nees）Arn. = Scolopia zeyheri（Nees）Harv. ●☆

295340 Phocea Seem. = Macaranga Thouars ●

295341 Phoebanthus S. F. Blake（1916）；向日菊属；False Sunflower ■☆

295342 Phoebanthus grandiflorus（Torr. et A. Gray）S. F. Blake；大花向日菊；Florida False Sunflower ■☆

295343 Phoebanthus tenuifolius（Torr. et A. Gray）S. F. Blake；细叶向日菊；Pineland False Sunflower ■☆

295344 Phoebe Nees（1836）；楠木属（楠属,雅楠属）；Nanmu,Phoebe ●

295345 Phoebe acuminata Merr. = Phoebe bournei（Hemsl.）Yen C. Yang ●

295346 Phoebe angustifolia Meisn.；沼楠；Marsh Nanmu,Narrowleaf Phoebe,Narrow-leaved Phoebe ●

295347 Phoebe angustifolia Meisn. var. annamensis H. Liu = Phoebe angustifolia Meisn. ●

295348 Phoebe blepharopus Hand.-Mazz. = Phoebe bournei（Hemsl.）Yen C. Yang ●

295349 Phoebe bournei（Hemsl.）Yen C. Yang；闽楠（毛丝桢楠,楠木,兴安楠木,竹叶楠）；Bourne Phoebe,Fujian Nanmu,Nanmu Phoebe ●

295350 Phoebe brachythyrsa H. W. Li；短序楠；Short-inflorescenced Phoebe,Shortthyrse Nanmu,Short-thyrse Shortraceme ●

295351 Phoebe calcarea S. K. Lee et F. N. Wei；石山楠；Liny Nan,Liny Nanmu ●

295352 Phoebe chekiangensis C. B. Shang；浙江楠；Zhejiang Nanmu,Zhejiang Phoebe ●◇

295353 Phoebe chinensis Chun；山楠；Chinese Phoebe,Wild Nanmu ●

295354 Phoebe crassipedicella S. K. Lee et F. N. Wei；粗柄楠；Thick-pedicel Nanmu,Thick-pedicel Phoebe,Thick-pediceled Phoebe ●

295355 Phoebe crassipedicellata S. K. Lee et F. N. Wei；密梗楠；Thick-pedicel Nanmu,Thick-pedicel Phoebe ●

295356 Phoebe cuneata Blume = Phoebe tavoyana（Meisn.）Hook. f. ●

295357 Phoebe cuneata Blume var. glabra H. Liu = Phoebe yaiensis S. K. Lee ●

295358　Phoebe cuneata Blume var. poilanei H. Liu ＝ Phoebe tavoyana（Meisn.）Hook. f. ●

295359　Phoebe dunniana（H. Lév.）Kosterm. ＝ Nothaphoebe cavaleriei（H. Lév.）Yen C. Yang ●◇

295360　Phoebe faberi（Hemsl.）Chun；竹叶楠（小樟木）；Bambooleaf Nanmu，Faber Phoebe ●

295361　Phoebe formosana（Matsum. et Hayata）Hayata；台楠（火炭楠，石楠，台湾雅楠）；Taiwan Nanmu，Taiwan Phoebe ●

295362　Phoebe forrestii W. W. Sm.；长毛楠（红楠木）；Forrest Phoebe，Longhair Nanmu ●

295363　Phoebe glaucescens（Nees）Nees ＝ Machilus glaucescens（Nees）H. W. Li ●

295364　Phoebe glaucifolia S. K. Lee et F. N. Wei；白背楠；Yunnan Nanmu，Yunnan Phoebe ●

295365　Phoebe glaucophylla H. W. Li.；粉叶楠；Glaucous leaf Nanmu，Glaucous-leaf Phoebe，Glaucous-leaved Phoebe ●

295366　Phoebe goalparensis Hutch.；陈萨姆楠●☆

295367　Phoebe hainanensis Merr.；茶槁楠（长柄楠，长叶楠）；Hainan Nanmu，Hainan Phoebe ●

295368　Phoebe henryi（Hemsl.）Merr. ＝ Phoebe tavoyana（Meisn.）Hook. f. ●

295369　Phoebe hui W. C. Cheng ex Yen C. Yang；细叶楠；Hu's Nanmu，Hu's Phoebe ●

295370　Phoebe hunanensis Hand.-Mazz.；湘楠（湖南楠）；Hunan Nanmu，Hunan Phoebe ●

295371　Phoebe hungmaoensis Lecomte；红毛山楠（红丹，毛丹）；Hongmao Mountain Phoebe，Hongmaoshan Nanmu，Hongmaoshan Phoebe ●

295372　Phoebe kwangsiensis H. Liu；桂楠；Guangxi Nanmu，Guangxi Phoebe，Kwangsi Phoebe ●◇

295373　Phoebe lanceolata（Wall. ex Nees）Nees；披针叶楠；Lanceolate Nanmu，Lanceolate Phoebe，Lanceolate-leaf Phoebe ●

295374　Phoebe latifolia Champ. ＝ Cinnamomum parthenoxylum（Jack）Meisn. ●

295375　Phoebe latifolia Champ. ＝ Cinnamomum porrectum（Roxb.）Kosterm. ●

295376　Phoebe latifolia Champ. ex Benth. ＝ Cinnamomum parthenoxylum（Jack）Meisn. ●

295377　Phoebe legendrei Lecomte；雅砻江楠（雅砻山楠）；Legendre Nanmu，Legendre Phoebe ●

295378　Phoebe lichuanensis S. K. Lee；利川楠；Lichuan Nanmu，Lichuan Phoebe ●

295379　Phoebe macrocarpa C. Y. Wu；大果楠；Bigfruit Nanmu，Bigfruit Phoebe，Large Phoebe ●◇

295380　Phoebe macrophylla（Blume）Blume ＝ Phoebe chinensis Chun ●

295381　Phoebe macrophylla Gamble ＝ Phoebe chinensis Chun ●

295382　Phoebe megacalyx H. W. Li；大萼楠；Bigcalyx Phoebe，Largecalyx Nanmu，Large-calyx Phoebe，Large-calyxed Phoebe ●◇

295383　Phoebe microphylla H. W. Li；小叶楠（薄叶楠）；Littleleaf Phoebe，Smallleaf Nanmu，Small-leaved Phoebe ●

295384　Phoebe minutiflora H. W. Li；小花楠（薄叶楠）；Filmleaf Nanmu，Littleflower Phoebe，Miniflower Nanmu，Small-flowered Phoebe ●

295385　Phoebe motuonan S. K. Lee et F. N. Wei；墨脱楠（红丹，毛丹）；Medog Phoebe，Motuo Nanmu，Motuo Phoebe ●

295386　Phoebe nanmu（Oliv.）Gamble；楠木（滇楠，柟，楠木树，楠树，润楠，雅楠，桢楠）；Coffee Tree，Coffee Wood，Nanmu，Nanmu Phoebe，Nanmu Wood，Yunnan Phoebe ●◇

295387　Phoebe nanmu（Oliv.）Gamble ＝ Machilus nanmu（Oliv.）Hemsl. ●◇

295388　Phoebe neurantha（Hemsl.）Gamble；白楠；Nervate-flowered Phoebe，Nervedflower Phoebe，White Nan，White Nanmu ●

295389　Phoebe neurantha（Hemsl.）Gamble var. brevifolia H. W. Li；短叶白楠（短叶楠）；Short-leaved Nervedflower Phoebe ●

295390　Phoebe neurantha（Hemsl.）Gamble var. cavaleriei H. Liu；兴义白楠（兴义楠）；Cavaleri Nervedflower Phoebe，Xingyi Nan，Xingyi Nanmu，Xingyi Phoebe ●

295391　Phoebe neurantha（Hemsl.）Gamble var. omeiensis Yen C. Yang ＝ Phoebe sheareri（Hemsl.）Gamble var. omeiensis（Yen C. Yang）N. Chao ●

295392　Phoebe neuranthoides S. K. Lee et F. N. Wei；光枝楠；Glabroustwig Nanmu，Glabroustwig Phoebe，Glabrous-twigged Phoebe ●

295393　Phoebe nigrifolia S. K. Lee et F. N. Wei；黑叶楠；Blackleaf Phoebe，Darkleaf Nanmu，Dark-leaf Phoebe，Dark-leaved Phoebe ●

295394　Phoebe omeiensis R. H. Miao；峨眉楠；Emei Nanmu，Emei Phoebe ●

295395　Phoebe omeiensis R. H. Miao ＝ Phoebe faberi（Hemsl.）Chun ●

295396　Phoebe pallida Nees；灰楠●☆

295397　Phoebe pandurata S. K. Lee et F. N. Wei；琴叶楠；Fiddle-shaped Leaf Nan，Violinleaf Nanmu，Zitherform Leaf Phoebe ●

295398　Phoebe pandurata S. K. Lee et F. N. Wei ＝ Phoebe neuranthoides S. K. Lee et F. N. Wei ●

295399　Phoebe poilanei Kosterm. ＝ Phoebe macrocarpa C. Y. Wu ●◇

295400　Phoebe porphyria Mez；坡非楠●☆

295401　Phoebe puwenensis W. C. Cheng；普文楠（黄心楠，细三合）；Puwen Nanmu，Puwen Phoebe ●

295402　Phoebe rufescens H. W. Li；红梗楠；Reddish-pediseled Phoebe，Redpedisel Nanmu，Red-stalk Phoebe ●

295403　Phoebe sheareri（Hemsl.）Gamble；紫楠（大叶紫楠，黄心楠，金丝楠，金心楠，楠木，枇杷木，赛金楠，山枇杷，小叶嫩蒲紫，野枇杷，紫金楠）；Purple Nan，Purple Nanmu，Shearer Phoebe ●

295404　Phoebe sheareri（Hemsl.）Gamble ＝ Phoebe hungmaoensis Lecomte ●

295405　Phoebe sheareri（Hemsl.）Gamble var. formosana（Hayata）Nakai ＝ Phoebe formosana（Matsum. et Hayata）Hayata ●

295406　Phoebe sheareri（Hemsl.）Gamble var. formosana（Matsum. et Hayata）Nakai ＝ Phoebe formosana（Matsum. et Hayata）Hayata ●

295407　Phoebe sheareri（Hemsl.）Gamble var. longepaniculata H. Liu ＝ Phoebe puwenensis W. C. Cheng ●

295408　Phoebe sheareri（Hemsl.）Gamble var. omeiensis（Yen C. Yang）N. Chao；峨眉紫楠（峨眉楠，桢楠树）；Emei Nanmu，Emei Phoebe，Omei Phoebe ●

295409　Phoebe sheareri（Hemsl.）Gamble var. stenophylla Nakai ＝ Phoebe formosana（Matsum. et Hayata）Hayata ●

295410　Phoebe tavoyana（Meisn.）Hook. f.；乌心楠（白椰槁，尖尾槁）；Bare-buded Phoebe，Blackheart Nanmu，Nakedbud Phoebe，Tavoy Phoebe ●

295411　Phoebe tenuirhachis R. H. Miao；纤轴楠●

295412　Phoebe tenuirhachis R. H. Miao ＝ Nothaphoebe cavaleriei（H. Lév.）Yen C. Yang ●◇

295413　Phoebe villosa（Roxb.）Wight ＝ Machilus glaucescens（Nees）H. W. Li ●

295414　Phoebe villosa Wight ＝ Persea villosa（Roxb.）Kosterm. ●

295415　Phoebe yaiensis S. K. Lee；崖楠；Yaxian Nanmu，Yaxian Phoebe ●

295416　Phoebe yunnanensis H. W. Li；景东楠；Jingdong Nanmu，Jingdong Phoebe ●

295417　Phoebe zhennan S. K. Lee et F. N. Wei；桢楠（楠材，楠木，雅楠，樟木）；Nanmu，Zhennan ●◇

295418　Phoenicaceae Burnett = Arecaceae Bercht. et J. Presl（保留科名）●

295419　Phoenicaceae Burnett = Palmae Juss.（保留科名）●

295420　Phoenicaceae Schultz Sch. = Arecaceae Bercht. et J. Presl（保留科名）●

295421　Phoenicaceae Schultz Sch. = Palmae Juss.（保留科名）●

295422　Phoenicanthemum（Blume）Blume = Helixanthera Lour. ●

295423　Phoenicanthemum（Blume）Rchb. = Helixanthera Lour. ●

295424　Phoenicanthemum（Blume）Rchb. = Loranthus Jacq.（保留属名）●

295425　Phoenicanthemum Blume = Helixanthera Lour. ●

295426　Phoenicanthus Alston（1931）；凤凰花属●☆

295427　Phoenicanthus Post et Kuntze = Phaenicanthus Thwaites ●■

295428　Phoenicanthus Post et Kuntze = Premna L.（保留属名）●■

295429　Phoenicanthus coriacea（Thwaites）H. Huber；凤凰花●☆

295430　Phoenicanthus obliquus（Hook. f. et Thomson）Alston；斜凤凰花●☆

295431　Phoenicaulis Nutt.（1838）；紫茎草属■☆

295432　Phoenicaulis Nutt. = Cheiranthus L. ●■

295433　Phoenicaulis Nutt. ex Torr. et A. Gray = Cheiranthus L. ●■

295434　Phoenicaulis cheiranthoides Nutt. ；紫茎草■☆

295435　Phoenicimon Ridl. = Glycosmis Corrêa（保留属名）●

295436　Phoenicocissus Mart. ex Meisn. = Lundia DC.（保留属名）●☆

295437　Phoenicophorium H. Wendl.（1865）；紫红棕属（凤凰刺椰属，凤凰椰属，凤尾椰属）；Phoenicophorium ●☆

295438　Phoenicophorium borsigianum（K. Koch）H. Wendl. = Phoenicophorium borsigianum（K. Koch）Stuntz ●☆

295439　Phoenicophorium borsigianum（K. Koch）Stuntz；紫红棕；Borsig Phoenicophorium，Stevensonia，Stevevson Palm，Thief Palm ●☆

295440　Phoenicophorium borsigianum Stuntz = Phoenicophorium borsigianum（K. Koch）Stuntz ●☆

295441　Phoenicopus Spach = Lactuca L. ■

295442　Phoenicopus Spach = Phaenixopus Cass. ■

295443　Phoenicopus intricatus Pomel = Lactuca viminea（L.）J. Presl et C. Presl subsp. chondrilliflora（Boreau）Bonnier ■☆

295444　Phoenicoseris（Skottsb.）Skottsb. = Dendroseris D. Don ●☆

295445　Phoenicosperma Miq. = Berkheya Ehrh.（保留属名）●■☆

295446　Phoenicosperma Miq. = Stobaea Thunb. ●■☆

295447　Phoenicospermum B. D. Jacks. = Phoenicosperma Miq. ●■☆

295448　Phoenicospermum Miq. = Berkheya Ehrh.（保留属名）●■☆

295449　Phoenicospermum Miq. = Stobaea Thunb. ●■☆

295450　Phoenix Haller = Chrysopogon Trin.（保留属名）■

295451　Phoenix L.（1753）；刺葵属（海枣属，枣椰属，枣椰子属，战捷木属，针葵属）；Date，Date Palm，Datepalm，Date-palm ●

295452　Phoenix abyssinica Drude = Phoenix reclinata Jacq. ●☆

295453　Phoenix acaulis Roxb. ；无茎刺葵（刺葵，无柄刺葵）；Acaulous Date，Dwarf Date Palm，Sessile Date，Stemless Date，Stemless Datepalm ●

295454　Phoenix atlantica A. Chev. ；大西洋刺葵；Atlantic Date Palm，False Date-palm ●☆

295455　Phoenix atlantica A. Chev. var. maroccana ？ = Phoenix dactylifera L. ●

295456　Phoenix atlantidis A. Chev. ；亚特兰大刺葵●☆

295457　Phoenix baoulensis A. Chev. = Phoenix reclinata Jacq. ●☆

295458　Phoenix caespitosa Chiov. ；丛生刺葵●☆

295459　Phoenix canariensis Chabaud；加那利刺葵（槟榔竹，长叶刺葵，加拿里椰子，加拿列海枣，加那利海枣，加那列海枣，洋槟榔竹，针葵）；Canary Date，Canary Date-palm，Canary Island Date，Canary Island Date Palm，Canary Island Date-palm，Canary Palm ●☆

295460　Phoenix comorensis Becc. = Phoenix reclinata Jacq. ●☆

295461　Phoenix dactylifera L. ；椰枣（波斯果，波斯枣，番枣，凤尾蕉，海枣，海棕，海棕木，金果，窟莽树，苦鲁麻，苦鲁麻枣，千年枣，屈莽树，万年枣，万岁枣，无漏果，无漏子，仙枣，香枣，伊拉克蜜枣，伊拉克枣，枣椰子，战捷木，中东海枣，紫京）；Common Date Palm，Dacts Tree，Date，Date Palm，Datepalm，Date-palm，Edible Date ●

295462　Phoenix djalonensis A. Chev. = Phoenix reclinata Jacq. ●☆

295463　Phoenix dybowskii A. Chev. ex A. Chev. = Phoenix reclinata Jacq. ●☆

295464　Phoenix formosana（Becc.）Masam. = Phoenix hanceana Naudin ●

295465　Phoenix hanceana Naudin；刺葵（桃椰，糠椰，糠椰，台湾海枣，小针葵）；Formosan Date，Formosan Date Palm，Hance Date，Hance Datepalm ●

295466　Phoenix hanceana Naudin var. formosana Becc. = Phoenix hanceana Naudin ●

295467　Phoenix humilis（L.）Cav. = Chamaerops humilis L. ●☆

295468　Phoenix humilis Royle；矮刺葵；Minor Date Palm ●☆

295469　Phoenix humilis Royle = Phoenix loureirii Kunth ●☆

295470　Phoenix Humilis Royle var. hanceana（Naudin）Becc. = Phoenix hanceana Naudin ●

295471　Phoenix humilis Royle var. hanceana Naudin = Phoenix hanceana Naudin ●

295472　Phoenix humilis Royle var. loureiri（Kunth）Becc. = Phoenix loureirii Kunth ●☆

295473　Phoenix humilis Royle var. loureirii Becc. = Phoenix roebelenii O'Brien ●

295474　Phoenix jubae Webb ex H. Christ；枣椰子；Canarian Date Palm ●☆

295475　Phoenix loureirii Kunth；软叶刺葵（刺葵，罗比亲王海枣）；Dwarf Date，Hance Date，Pygmy Date Palm ●☆

295476　Phoenix ouseleyana Griff. = Phoenix loureirii Kunth ●☆

295477　Phoenix paludosa Roxb. ；沼泽刺葵（大刺葵）；Malayan Date，Mangrove Date Palm ●☆

295478　Phoenix pusilla Gaertn. ；微小刺葵；Slender Date Palm ●☆

295479　Phoenix reclinata Jacq. ；非洲刺葵（非洲海枣，塞内加尔刺葵）；African Wild Date，Dwarf Date，Dwarf Date Palm，Reclining Date Palm，Senegal Date，Senegal Date Palm，Senegal Date-palm，Wild Date Palm ●☆

295480　Phoenix reclinata Jacq. var. comorensis（Becc.）Jum. et H. Perrier = Phoenix reclinata Jacq. ●☆

295481　Phoenix reclinata Jacq. var. madagascariensis Becc. = Phoenix reclinata Jacq. ●☆

295482　Phoenix reclinata Jacq. var. somalensis Chiov. = Phoenix reclinata Jacq. ●☆

295483　Phoenix robusta Hook. f. ；健刺葵●☆

295484　Phoenix roebelenii O'Brien；江边刺葵（罗比亲王海枣，美丽针葵，亲王海枣，软叶刺葵，软叶枣椰，软叶针葵）；Dwarf Date Palm，Miniature Date，Miniature Date Palm，Miniature Date-palm，Pygmy Date Palm，Pygmy Datepalm，Pygmy Date-palm，Roebelen Date，Roebelen's Palm，Softleaf Datepalm ●

295485　Phoenix rupicola T. Anderson；岩海枣（岩枣椰）；Cliff Date Palm，Indian Date Palm，Pigmy Date Palm ■☆

295486　Phoenix spinosa Schumach. et Thonn. = Phoenix reclinata Jacq. ●☆

295487 Phoenix sylvestris (L.) Roxb.;林刺葵(橙枣椰,野生刺葵,银海枣,枣椰);Date Sugar Palm,India Date,India Date-palm,Indian Wild Date,Silver Date Palm,Sugar Date,Sugar Date-palm,Wild Date Palm,Wood Datepalm ●

295488 Phoenix sylvestris Roxb. = Phoenix sylvestris (L.) Roxb. ●

295489 Phoenix theophrastii Greuter;克里特枣椰;Cretan Date Palm,Cretan Palm ●☆

295490 Phoenix zeylanica Trimen;锡兰刺葵;Ceylon Date Palm ●☆

295491 Phoenixopus Rchb. = Lactuca L. ■

295492 Phoenixopus Rchb. = Phaenixopus Cass. ■

295493 Phoenixopus Rchb. = Scariola F. W. Schmidt ■●

295494 Phoenocoma G. Don = Phaenocoma D. Don ●☆

295495 Phoenopus Nyman = Lactuca L. ■

295496 Phoenopus Nyman = Phaenopus DC. ■

295497 Pholacilia Griseb. = Trichilia P. Browne(保留属名)●

295498 Pholidandra Neck. = Raputia Aubl. ●☆

295499 Pholidia R. Br. = Eremophila R. Br. ●☆

295500 Pholidiopsis F. Muell. = Pholidia R. Br. ●☆

295501 Pholidocarpus Blume(1830);角鳞果棕属(金钱棕属,鳞果桐属,球棕属);Pholidocarpus ●☆

295502 Pholidocarpus macrocarpus Becc.;大果角鳞果棕;Bigfruit Pholidocarpus ●☆

295503 Pholidophyllum Vis. (1847);鳞叶凤梨属■☆

295504 Pholidophyllum Vis. = Cryptanthus Otto et A. Dietr. (保留属名)■☆

295505 Pholidophyllum zonatum Vis. ;鳞叶凤梨■☆

295506 Pholidostachys H. Wendl. ex Benth. et Hook. f. (1883);鳞穗棕属(红柄椰属,丽椰属,丽棕属)●☆

295507 Pholidostachys H. Wendl. ex Hook. f. = Pholidostachys H. Wendl. ex Benth. et Hook. f. ●☆

295508 Pholidostachys pulchra H. Wendl. ;鳞穗棕●☆

295509 Pholidota Lindl. = Pholidota Lindl. ex Hook. ■

295510 Pholidota Lindl. ex Hook. (1825);石仙桃属(石山桃属);Pholidota,Rattlesnake Orchid ■

295511 Pholidota articulata Lindl. ;节茎石仙桃(石蚌接骨丹,石丰腿,石楞腿);Articulate Pholidota,Nodose Pholidota ■

295512 Pholidota articulata Lindl. var. grifithii (Hook. f.) King et Pantl. = Pholidota articulata Lindl. ■

295513 Pholidota articulata Lindl. var. obovata (Hook. f.) Ts. Tang et F. T. Wang = Pholidota articulata Lindl. ■

295514 Pholidota bracteata (D. Don) Seidenf. = Pholidota imbricata Hook. ■

295515 Pholidota bracteata (D. Don) Seidenf. = Pholidota pallida Lindl. ■

295516 Pholidota cantonensis Rolfe;细叶石仙桃(对叶草,广东石仙桃,果上叶,双叶岩珠,乌来石山桃,小石仙桃,岩豆,岩珠);Slenderleaf Pholidota ■

295517 Pholidota carnea (Blume) Lindl. ;肉色石仙桃;Flesh-colour Pholidota ■☆

295518 Pholidota chinensis Lindl. ;石仙桃(川甲草,大吊兰,浮石斛,果上叶,马榴根,麦斛,千年矮,肉色石仙桃,上树蛤蟆,石芭蕉,石槟榔,石穿盘,石橄榄,石莲,石山桃,石上莲,石上仙桃,石黄肉,细颈葫芦,小扣子兰,圆柱石仙桃);China Pholidota,Chinese Pholidota ■

295519 Pholidota chinensis Lindl. var. cylindracea Ts. Tang et F. T. Wang = Pholidota chinensis Lindl. ■

295520 Pholidota convallariae (Rchb. f.) Hook. f.;凹唇石仙桃;Concavelip Pholidota ■

295521 Pholidota grifithii Hook. f. = Pholidota articulata Lindl. ■

295522 Pholidota henryi Kraenzl. = Pholidota imbricata Hook. ■

295523 Pholidota imbricata Hook.;宿苞石仙桃;Bract Pholidota ■

295524 Pholidota imbricata Hook. var. henryi (Kraenzl.) Ts. Tang et F. T. Wang = Pholidota imbricata Hook. ■

295525 Pholidota khasiana Rchb. f. = Pholidota articulata Lindl. ■

295526 Pholidota kouytcheensis Gagnep. = Pholidota yunnanensis Rolfe ■

295527 Pholidota leveilleana Schltr. ;单叶石仙桃;Monoleaf Pholidota ■

295528 Pholidota longipes S. C. Chen et Z. H. Tsi; 长足石仙桃;Longfoot Pholidota ■

295529 Pholidota lugardii Rolfe = Pholidota articulata Lindl. ■

295530 Pholidota missionariorum Gagnep. ;尖叶石仙桃;Tineleaf Pholidota ■

295531 Pholidota obovata Hook. f. = Pholidota articulata Lindl. ■

295532 Pholidota pallida Lindl. ; 粗脉石仙桃(柱茎石仙桃);Thickvein Pholidota ■

295533 Pholidota pallida Lindl. = Pholidota bracteata (D. Don) Seidenf. ■

295534 Pholidota protracta Hook. f. ;尾尖石仙桃;Tail Pholidota ■

295535 Pholidota roseans Schltr. ;贵州石仙桃;Guizhou Pholidota ■

295536 Pholidota rupestris Hand. -Mazz. ;岩生石仙桃;Saxicolous Pholidota ■

295537 Pholidota rupestris Hand. -Mazz. = Pholidota missionariorum Gagnep. ■

295538 Pholidota schlechteri Gagnep. = Pholidota pallida Lindl. ■

295539 Pholidota suaveolens Lindl. = Coelogyne suaveolens (Lindl.) Hook. f. ■

295540 Pholidota tixieri Guillaumin = Pholidota pallida Lindl. ■

295541 Pholidota uraiensis Hayata;乌来石山桃■

295542 Pholidota uraiensis Hayata = Pholidota cantonensis Rolfe ■

295543 Pholidota wenshanica S. C. Chen et Z. H. Tsi;文山石仙桃;Wenshan Pholidota ■

295544 Pholidota wenshanica S. C. Chen et Z. H. Tsi = Pholidota leveilleana Schltr. ■

295545 Pholidota yunnanensis Rolfe;云南石仙桃(滇石仙桃,果上叶,六棱锥,乱角莲,乱脚莲,石草果,石海椒,石灵芝,石枣子,雅雀还阳,岩火炮);Yunnan Pholidota ■

295546 Pholidota yunnanensis Schltr. = Pholidota bracteata (D. Don) Seidenf. ■

295547 Pholidota yunnanensis Schltr. = Pholidota pallida Lindl. ■

295548 Pholidota yunpeensis Hu = Pholidota bracteata (D. Don) Seidenf. ■

295549 Pholidota yunpeensis Hu = Pholidota pallida Lindl. ■

295550 Pholisma Nutt. ex Hook. (1844);鳞叶多室花属■☆

295551 Pholisma arenarium Nutt. ;鳞叶多室花■☆

295552 Pholisma paniculatum B. C. Templeton;圆锥鳞叶多室花■☆

295553 Pholistoma Lilja(1839);鳞口麻属■☆

295554 Pholistoma auritum (Lindl.) Lilja;鳞口麻■☆

295555 Pholiurus Trin. (1820);鳞尾草属;Sickle-Grass ■☆

295556 Pholiurus glabriglumis Nevski = Henrardia glabriglumis (Nevski) C. E. Hubb. ■☆

295557 Pholiurus incurvatus (L.) Hitchc. = Parapholis incurva (L.) C. E. Hubb. ■

295558 Pholiurus incurvus (L.) Schinz = Parapholis incurva (L.) C. E. Hubb. ■

295559 Pholiurus incurvus (L.) Schinz et Thell. = Parapholis incurva (L.) C. E. Hubb. ■

295560 Pholiurus incurvus (L.) Schinz et Thell. subsp. filiformis (Roth) A. Camus = Parapholis filiformis (Roth) C. E. Hubb. ■☆

295561 Pholiurus pannonicus (Host) Trin. ;帕地鳞尾草■☆

295562 Pholiurus persicus (Boiss.) A. Camus = Henrardia persica (Boiss.) C. E. Hubb. ■☆

295563 Pholomphis Raf. = Miconia Ruiz et Pav. (保留属名) ●☆

295564 Phoniphora Neck. = Phoenix L. ●

295565 Phonus Hill = Carthamus L. ■

295566 Phonus Hill(1762);黄刺菊属●☆

295567 Phonus arborescens (L.) G. López = Carthamus arborescens L. ■☆

295568 Phonus fruticosus (Maire) G. López = Carthamus fruticosus Maire ■☆

295569 Phonus lanatus Hill;黄刺菊●☆

295570 Phonus mareoticus (Delile) G. López = Carthamus mareoticus Delile ■☆

295571 Phonus rhiphaeus (Font Quer et Pau) G. López = Carthamus rhiphaeus Font Quer et Pau ■☆

295572 Phoradendraceae H. Karst.;美洲桑寄生科●☆

295573 Phoradendraceae H. Karst. = Santalaceae R. Br. (保留科名)●■

295574 Phoradendraceae H. Karst. = Viscaceae Miq. ●

295575 Phoradendron Nutt. (1848);美洲桑寄生属(栗寄生属,美洲寄生属);American Mistletoe,Flores De Palo,Wood Flowers ●☆

295576 Phoradendron brachystachyum Nutt.;短穗美洲寄生(短穗栗寄生)●☆

295577 Phoradendron californicum Nutt.;加州美洲桑寄生;Desert Mistletoe ●☆

295578 Phoradendron flavescens (Pursh) Nutt. ex A. Gray;黄美洲桑寄生(美洲槲寄生);American Mistletoe ●☆

295579 Phoradendron flavescens Nutt. ex Engelm. = Phoradendron leucarpum (Raf.) Reveal et M. C. Johnst. ●☆

295580 Phoradendron juniperinum Engelm. ex A. Gray;刺柏美洲桑寄生;Juniper Mistletoe,Mistletoe ●☆

295581 Phoradendron leucarpum (Raf.) Reveal et M. C. Johnst.;白果美洲桑寄生;American Mistletoe,Eastern Mistletoe,Mistletoe ●☆

295582 Phoradendron schumannii Trel.;舒曼美洲寄生●☆

295583 Phoradendron serotinum (Raf.) M. C. Johnst.;美洲桑寄生(美桑寄生);American Christmas-mistletoe, American Mistletoe, Mistletoe,Oak Mistletoe ●☆

295584 Phoradendron serotinum (Raf.) M. C. Johnst. = Phoradendron leucarpum (Raf.) Reveal et M. C. Johnst. ●☆

295585 Phoradendron tomentosum (DC.) Engelm. ex Gray;绒毛美洲寄生●☆

295586 Phoradendron villosum Nutt.;长毛美洲桑寄生(柔毛美洲寄生);Mistletoe ●☆

295587 Phoringopsis D. L. Jones et M. A. Clem. (2002);澳洲节唇兰属■☆

295588 Phoringopsis D. L. Jones et M. A. Clem. = Arthrochilus F. Muell. ■☆

295589 Phormangis Schltr. = Ancistrorhynchus Finet ■☆

295590 Phormiaceae A. E. Murray;惠灵麻科(麻兰科,山菅兰科,新西兰麻科)●■

295591 Phormiaceae J. Agardh = Agavaceae Dumort. (保留科名)●■

295592 Phormiaceae J. Agardh = Hemerocallidaceae R. Br. ■

295593 Phormiaceae J. Agardh = Phrymaceae Schauer(保留科名)●

295594 Phormium J. R. Forst. et G. Forst. (1775);惠灵麻属(麻兰属,新西兰麻属);Fiber Lily,Flax Lily,New Zealand Flax ■☆

295595 Phormium aloides L. f. = Lachenalia aloides (L. f.) Engl. ■☆

295596 Phormium bulbiferum Cirillo = Lachenalia bulbifera (Cirillo) Engl. ■☆

295597 Phormium colensoi Hook. f. = Phormium cookianum Le Jol. ■☆

295598 Phormium cookianum Le Jol.;山麻兰;Lesser New Zealand Flax,Mountain Flax,New Zealand Flax ■☆

295599 Phormium cookianum Le Jol. 'Variegatum';斑叶山麻兰;

Mountain Flax ■☆

295600 Phormium tenax J. R. Forst. et G. Forst.;新西兰麻(麻兰);Bush Flax, Flax Lily, Maori Flax, New Zealand Fiber Lily, New Zealand Flax,New Zealand Hemp,Tough Flax-lily ■☆

295601 Phormium tenax J. R. Forst. et G. Forst. 'Atropurpureum';紫叶新西兰麻■☆

295602 Phormium tenax J. R. Forst. et G. Forst. 'Aureum';黄纹新西兰麻■☆

295603 Phormium tenax J. R. Forst. et G. Forst. 'Rubrum';红叶新西兰麻■☆

295604 Phormium tenax J. R. Forst. et G. Forst. 'Variegatum';花叶新西兰麻■☆

295605 Phornothamnus Baker = Gravesia Naudin ●☆

295606 Phornothamnus thymoides Baker = Gravesia thymoides (Baker) H. Perrier ●☆

295607 Phorodendrum Post et Kuntze = Phoradendron Nutt. ●☆

295608 Phosanthus Raf. = Isertia Schreb. ●☆

295609 Photinia Lindl. (1820);石楠属(扇骨木属,石斑木属);Chokeberry,Photinia,Stranvaesia ●☆

295610 Photinia Lindl. = Aronia Medik. (保留属名)●☆

295611 Photinia M. Roem. = Photinia Lindl. ●

295612 Photinia × fraseri Dress;弗雷泽石楠(福塞石楠,杂种石楠);Fraser Photinia,Fraser's Photinia ●☆

295613 Photinia × fraseri Dress 'Birmingham';伯明翰杂种石楠●☆

295614 Photinia × fraseri Dress 'Red Robin';红罗宾弗雷泽石楠(红知更鸟杂种石楠)●☆

295615 Photinia × fraseri Dress 'Robusta';健壮弗雷泽石楠●☆

295616 Photinia amphidoxa (C. K. Schneid.) Rehder et E. H. Wilson = Stranvaesia amphidoxa C. K. Schneid. ●

295617 Photinia amphidoxa (C. K. Schneid.) Rehder et E. H. Wilson var. amphileia Hand. -Mazz. = Stranvaesia amphidoxa C. K. Schneid. var. amphileia (Hand. -Mazz.) Te T. Yu ●

295618 Photinia amphidoxa (C. K. Schneid.) Rehder et E. H. Wilson var. kwangsiensis F. P. Metcalf = Stranvaesia amphidoxa C. K. Schneid. ●

295619 Photinia amphidoxa (C. K. Schneid.) Rehder et E. H. Wilson var. kwangsiensis Metcalf;广西红果树●

295620 Photinia amphidoxa (C. K. Schneid.) Rehder et E. H. Wilson var. stylosa Cardot = Stranvaesia amphidoxa C. K. Schneid. ●

295621 Photinia anlungensis Te T. Yu;安龙石楠;Anlong Photinia, Anlung Photinia ●

295622 Photinia arbutifolia Lindl. = Heteromeles arbutifolia M. Roem. ●☆

295623 Photinia ardisiifolia Hayata = Photinia serratifolia (Desf.) Kalkman var. ardisiifolia (Hayata) H. Ohashi ●

295624 Photinia ardissifolia Hayata = Photinia serratifolia (Desf.) Kalkman var. ardisiifolia (Hayata) H. Ohashi ●

295625 Photinia arguta Lindl.;锐齿石楠;Sharptooth Photinia, Sharp-toothed Photinia ●

295626 Photinia arguta Lindl. var. hookeri (Decne.) Vidal;云南锐齿石楠(毛果锐齿石楠);Hooker. Sharptooth Photinia ●

295627 Photinia arguta Lindl. var. salicifolia (Decne.) Vidal;柳叶锐齿石楠;Willow-leaf Sharptooth Photinia ●

295628 Photinia austroguizhouensis Y. K. Li = Photinia chingiana Hand. -Mazz. ●

295629 Photinia bartletti Merr. = Sorbus corymbifera (Miq.) T. H. Nguyên et Yakovlev ●

295630 Photinia beauverdiana C. K. Schneid.;中华石楠(波氏石楠,假

思桃,牛筋木);Beauverd Photinia,China Photinia ●

295631 Photinia beauverdiana C. K. Schneid. var. brevifolia Cardot;短叶中华石楠;Shortleaf Beauverd Photinia ●

295632 Photinia beauverdiana C. K. Schneid. var. lofauensis Metcalf = Photinia schneideriana Rehder et E. H. Wilson ●

295633 Photinia beauverdiana C. K. Schneid. var. lohfauensis F. P. Metcalf = Photinia schneideriana Rehder et E. H. Wilson ●

295634 Photinia beauverdiana C. K. Schneid. var. notabilis (C. K. Schneid.) Rehder et E. H. Wilson = Photinia beauverdiana C. K. Schneid. ●

295635 Photinia beauverdiana C. K. Schneid. var. notabilis (C. K. Schneid.) Rehder et E. H. Wilson;厚叶中华石楠(华石楠,台湾老叶儿树);Thickleaf Beauverd Photinia ●

295636 Photinia beckii C. K. Schneid.;椭圆叶石楠;Beck Photinia ●

295637 Photinia benthamiana Hance;闽粤石楠(边沁石斑木,石斑木);Bentham Photinia ●

295638 Photinia benthamiana Hance var. obovata H. L. Li;倒卵叶闽粤石楠;Obovateleaf Photinia ●

295639 Photinia benthamiana Hance var. salicifolia Cardot;柳叶闽粤石楠;Willowleaf Bentham Photinia ●

295640 Photinia berberidifolia Rehder et E. H. Wilson;小檗叶石楠;Barberryleaf Photinia,Barberry-leaved Photinia ●

295641 Photinia bergerae C. K. Schneid.;湖北石楠;Berger Photinia ●

295642 Photinia blinii (H. Lév.) Rehder;短叶石楠;Blin Photinia,Shortleaf Photinia ●

295643 Photinia bodinieri H. Lév.;贵州石楠;Bodinier Photinia,Guizhou Photinia ●

295644 Photinia bodinieri H. Lév. var. longifolia Cardot;长叶贵州石楠;Longleaf Bodinier Photinia,Longleaf Photinia ●

295645 Photinia brevipetiolata Cardot;短柄石楠●

295646 Photinia brevipetiolata Cardot = Photinia calleryana (Decne.) Cardot ●

295647 Photinia buisanensis Hayata = Eriobotrya deflexa (Hemsl.) Nakai f. buisanensis (Hayata) Nakai ●

295648 Photinia buisanensis Hayata = Eriobotrya deflexa (Hemsl.) Nakai ●

295649 Photinia calleryana (Decne.) Cardot;城口石楠●

295650 Photinia calleryana (Decne.) Cardot = Photinia benthamiana Hance ●

295651 Photinia callosa Chun ex K. C. Kuan;厚齿石楠;Callous Photinia ●

295652 Photinia cardotii F. P. Metcalf = Photinia villosa (Thunb.) DC. var. sinica Rehder et E. H. Wilson ●

295653 Photinia cardotii Metcalf = Photinia villosa (Thunb.) DC. var. sinica Rehder et E. H. Wilson ●

295654 Photinia cavaleriei H. Lév. = Photinia beauverdiana C. K. Schneid. ●

295655 Photinia cavaleriei H. Lév. = Photinia crassifolia H. Lév. ●

295656 Photinia chihsiniana K. C. Kuan;临桂石楠(钟氏石楠);Lingui Photinia ●

295657 Photinia chingiana Hand. -Mazz.;宜山石楠(黔南石楠);Ching Photinia,S. Guizhou Photinia,South Guizhou Photinia,South-Guizhou Photinia ●

295658 Photinia chingiana Hand. -Mazz. var. lipingensis (Y. K. Li et X. M. Wang) L. T. Lu et C. L. Li;黎平石楠;Liping Photinia ●

295659 Photinia chingshuiensis (T. Shimizu) Tang S. Liu et H. J. Su;清水石楠●

295660 Photinia consimilis Hand. -Mazz. = Photinia prunifolia (Hook. et Arn.) Lindl. ●

295661 Photinia crassifolia H. Lév.;厚叶石楠(玉枇杷);Thickleaf Photinia,Thick-leaved Photinia ●

295662 Photinia crassifolia H. Lév. var. denticulata Cardot = Photinia crassifolia H. Lév. ●

295663 Photinia crenatoserrata Hance = Pyracantha fortuneana (Maxim.) H. L. Li ●

295664 Photinia dabieshanensis M. B. Deng et G. Yao;大别山石楠;Dabieshan Photinia ●

295665 Photinia dabieshanensis M. B. Deng et G. Yao = Photinia schneideriana Rehder et E. H. Wilson ●

295666 Photinia daphniphylloides Hayata = Photinia serratifolia (Desf.) Kalkman ●

295667 Photinia daphniphylloides Hayata = Photinia serratifolia (Desf.) Kalkman var. daphniphylloides (Hayata) L. T. Lu ●

295668 Photinia davidiana (Decne.) Cardot = Stranvaesia davidiana Decne. ●

295669 Photinia davidiana (Decne.) Cardot var. formosana (Cardot) H. Ohashi et Iketani;台湾红果树●

295670 Photinia davidiana Cardot = Photinia davidiana (Decne.) Cardot ●

295671 Photinia davidiana Cardot = Stranvaesia davidiana Decne. ●

295672 Photinia davidiana Decne. = Photinia davidiana (Decne.) Cardot ●

295673 Photinia davidiana Decne. = Stranvaesia davidiana Decne. ●

295674 Photinia davidiana Decne. var. salicifolia (Hutch.) Rehder;玉山假沙梨●

295675 Photinia davidsoniae Rehder et E. H. Wilson;椤木石楠(椤木,梅子树,山官木,水红树花,凿树);Davidson Photinia ●

295676 Photinia davidsoniae Rehder et E. H. Wilson = Photinia bodinieri H. Lév. ●

295677 Photinia davidsoniae Rehder et E. H. Wilson var. ambigua Cardot;毛瓣椤木石楠(毛瓣石楠)●

295678 Photinia davidsoniae Rehder et E. H. Wilson var. pungens Cardot;锐尖椤木石楠(锐尖石楠)●

295679 Photinia delexa Hemsl. = Eriobotrya deflexa (Hemsl.) Nakai ●

295680 Photinia esquirolii (H. Lév.) Rehder;黔南石楠●

295681 Photinia euphlebia Merr. et Chun = Photinia impressivena Hayata ●

295682 Photinia fauriei Cardot = Photinia schneideriana Rehder et E. H. Wilson ●

295683 Photinia fautiei Cardot = Photinia beauverdiana C. K. Schneid. var. notabilis (C. K. Schneid.) Rehder et E. H. Wilson ●

295684 Photinia flavidiflora W. W. Sm. = Photinia integrifolia Lindl. var. flavidiflora (W. W. Sm.) Vidal ●

295685 Photinia floribunda (Lindl.) K. R. Robertson et J. B. Phipps = Aronia prunifolia (Marshall) Rehder ●☆

295686 Photinia fokienensis (Franch.) Franch. ex Cardot;福建石楠;Fujian Photinia,Fukien Photinia ●

295687 Photinia fortuneana Maxim. = Pyracantha fortuneana (Maxim.) H. L. Li ●

295688 Photinia franchetiana Diels;密花石楠;Franchet Photinia ●

295689 Photinia franchetiana Diels = Photinia glomerata Rehder et E. H. Wilson ●

295690 Photinia fraseri Dress;红顶石楠;Red Tip Photinia ●☆

295691 Photinia glabra (Thunb.) Maxim.;光叶石楠(醋林子,光凿

树,红檬子,山官木,扇骨木,石斑木);Japan Photinia, Japanese Photinia ●

295692 Photinia glabra (Thunb.) Maxim. 'Rubens';红色光叶石楠●☆

295693 Photinia glabra (Thunb.) Maxim. var. chinensis Maxim. = Photinia serratifolia (Desf.) Kalkman ●

295694 Photinia glabra (Thunb.) Maxim. var. chinensis Maxim. = Photinia serrulata Lindl. ●

295695 Photinia glabra (Thunb.) Maxim. var. fokienensis Franch. = Photinia fokienensis (Franch.) Franch. ex Cardot ●

295696 Photinia glabra Hemsl. var. fokienensis Finet et Franch. = Photinia fokienensis (Franch.) Franch. ex Cardot ●

295697 Photinia glomerata Rehder et E. H. Wilson;球花石楠; Glomerate Photinia ●

295698 Photinia glomerata Rehder et E. H. Wilson var. cuneata Te T. Yu = Photinia glomerata Rehder et E. H. Wilson ●

295699 Photinia glomerata Rehder et E. H. Wilson var. microphylla Te T. Yu = Photinia glomerata Rehder et E. H. Wilson ●

295700 Photinia griffithii Decne. = Photinia glomerata Rehder et E. H. Wilson ●

295701 Photinia hirsuta Hand. -Mazz.;褐毛石楠;Hirsute Photinia ●

295702 Photinia hirsuta Hand. -Mazz. var. lobalata Te T. Yu;裂叶褐毛石楠;Lobedleaf Photinia ●

295703 Photinia hookeri (Decne.) Merr. = Photinia arguta Lindl. var. hookeri (Decne.) Vidal ●

295704 Photinia impressivena Hayata;陷脉石楠(凹叶假沙梨,青凿木);Impressednerve Photinia, Impresed-nerved Photinia ●

295705 Photinia impressivena Hayata var. urceolocarpa (Vidal) Vidal;毛序陷脉石楠;Pitchershaped-fruit Photinia ●

295706 Photinia integrifolia Lindl.;全缘石楠(蓝靛树);Entire Photinia ●

295707 Photinia integrifolia Lindl. var. flavidiflora (W. W. Sm.) Vidal;黄花全缘石楠(黄花石楠);Yellowflower Photinia ●

295708 Photinia integrifolia Lindl. var. notoniana (Wight et Arn.) Vidal;长柄全缘石楠;Longstalk Photinia ●

295709 Photinia integrifolia Lindl. var. notoniana (Wight et Arn.) Vidal = Photinia integrifolia Lindl. ●

295710 Photinia integrifolia Lindl. var. yunnanensis Te T. Yu;云南全缘石楠;Yunnan Yellowflower Photinia ●

295711 Photinia integrifolia Lindl. var. yunnanensis Te T. Yu = Photinia integrifolia Lindl. ●

295712 Photinia komarovii (H. Lév. et Vaniot) L. T. Lu et C. L. Li;垂丝石楠;Komarov's Photinia ●

295713 Photinia kudoi Masam.;工东石楠;Gongdong Photinia, Kudo Photinia ●

295714 Photinia kudoi Masam. = Photinia beauverdiana C. K. Schneid. var. notabilis (C. K. Schneid.) Rehder et E. H. Wilson ●

295715 Photinia kwangsiensis H. L. Li;广西石楠;Guangxi Photinia, Kwangsi Photinia ●

295716 Photinia lactiflora Pall.;乳花石楠;Milk-flower Photinia ●☆

295717 Photinia lancifolia Rehder et E. H. Wilson = Photinia arguta Lindl. var. salicifolia (Decne.) Vidal ●

295718 Photinia lancilimbum J. E. Vidal = Photinia arguta Lindl. var. salicifolia (Decne.) Vidal ●

295719 Photinia lancilimbum J. E. Vidal var. urceolocarpa J. E. Vidal = Photinia impressivena Hayata var. urceolocarpa (Vidal) Vidal ●

295720 Photinia lanuginosa Te T. Yu;绵毛石楠;Lanose Photinia, Shortwoolled Photinia, Short-woolled Photinia, Woolly Photinia ●

295721 Photinia lasiogyna (Franch.) C. K. Schneid.;倒卵叶石楠; Hairypistil Photinia, Hairy-pistiled Photinia, Obovateleaf Photinia ●

295722 Photinia lasiogyna (Franch.) C. K. Schneid. var. glabrescens L. T. Lu et C. L. Li;脱毛石楠;Glabrescent Hairypistil Photinia ●

295723 Photinia lasiopetala Hayata = Photinia serratifolia (Desf.) Kalkman var. lasiopetala (Hayata) H. Ohashi ●

295724 Photinia latouchei Franch. = Photinia fokienensis (Franch.) Franch. ex Cardot ●

295725 Photinia lindleyana Wight et Arn.;印度石楠●

295726 Photinia lipingensis Y. K. Li et M. Z. Yang = Photinia chingiana Hand. -Mazz. var. lipingensis (Y. K. Li et X. M. Wang) L. T. Lu et C. L. Li ●

295727 Photinia lochengensis Te T. Yu;罗城石楠;Locheng Photinia, Luocheng Photinia ●

295728 Photinia loriformis W. W. Sm.;带叶石楠(红牛筋,黄牛筋,牛筋条);Strapleaf Photinia, Strap-leaved Photinia ●

295729 Photinia lucida (Decne.) C. K. Schneid.;台湾石楠(虾尾); Taiwan Photinia ●

295730 Photinia mairei H. Lév. = Photinia lasiogyna (Franch.) C. K. Schneid. ●

295731 Photinia megaphylla Te T. Yu et L. T. Lu;大叶石楠;Bigleaf Photinia, Largeleaf Photinia, Large-leaved Photinia ●

295732 Photinia melanocarpa (Michx.) K. R. Robertson et J. B. Phipps;黑果石楠(黑果腺肋花椒,黑果腺肋花楸,黑苦味果,黑涩果);Black Chokeberry, Chokeberry, Hybrid Chokeberry ●☆

295733 Photinia melanocarpa (Michx.) K. R. Robertson et J. B. Phipps = Aronia melanocarpa (Michx.) Elliott ●☆

295734 Photinia mollis Hook. f. = Photinia arguta Lindl. var. hookeri (Decne.) Vidal ●

295735 Photinia mollis Hook. f. var. angustifolia Fisch. = Photinia arguta Lindl. var. salicifolia (Decne.) Vidal ●

295736 Photinia niitakayamensis Hayata = Photinia davidiana Decne. var. salicifolia (Hutch.) Rehder ●

295737 Photinia niitakayamensis Hayata = Stranvaesia davidiana Decne. ●

295738 Photinia notabilis C. K. Schneid. = Photinia beauverdiana C. K. Schneid. ●

295739 Photinia notabilis C. K. Schneid. = Photinia beauverdiana C. K. Schneid. var. notabilis (C. K. Schneid.) Rehder et E. H. Wilson ●

295740 Photinia notoniana Wight et Arn. = Photinia integrifolia Lindl. var. notoniana (Wight et Arn.) Vidal ●

295741 Photinia notoniana Wight et Arn. = Photinia integrifolia Lindl. ●

295742 Photinia nussia (D. Don) Kalkman;努斯石楠(印缅红果树,印缅石楠)●☆

295743 Photinia obliqua Stapf;斜脉石楠;Oblique Photinia ●

295744 Photinia parviflora Cardot = Photinia schneideriana Rehder et E. H. Wilson var. parviflora (Cardot) L. T. Lu et C. L. Li ●

295745 Photinia parvifolia (E. Pritz.) C. K. Schneid. var. kankaoensis (Hatus.) Te T. Yu et K. C. Kuan = Photinia villosa (Thunb.) DC. var. sinica Rehder et E. H. Wilson ●

295746 Photinia parvifolia (E. Pritz.) C. K. Schneid. var. tenuipes (P. S. Hsu et L. C. Li) P. L. Chiu = Photinia komarovii (H. Lév. et Vaniot) L. T. Lu et C. L. Li ●

295747 Photinia parvifolia (Pritz.) C. K. Schneid.;小叶石楠(棒梨子,棒头果,港口老叶儿,牛筋木,牛李子,山红子,小花石楠); Littleleaf Photinia ●

295748 Photinia parvifolia (Pritz.) C. K. Schneid. var. kankoensis

（Hatus.）Te T. Yu et T. C. Kuan = Photinia parvifolia（Pritz.）C. K. Schneid. ●

295749 Photinia parvifolia（Pritz.）C. K. Schneid. var. kankoensis（Hatus.）Te T. Yu et T. C. Kuan;台湾小叶石楠;Taiwan Photinia ●

295750 Photinia parvifolia（Pritz.）C. K. Schneid. var. subparvifolia（Y. K. Li et X. M. Wang）L. T. Lu et C. L. Li;假小叶石楠;False Littleleaf Photinia, False Little-leaved Photinia ●

295751 Photinia parvifolia C. K. Schneid. var. chingshiensis（T. Shimizu）S. S. Ying = Photinia chingshuiensis（T. Shimizu）Tang S. Liu et H. J. Su ●

295752 Photinia pilosicalyx Te T. Yu;毛果石楠;Hairycalyx Photinia, Hairy-fruited Photinia ●

295753 Photinia podocarpifolia Te T. Yu;罗汉松叶石楠;Podocarpifolious Photinia ●

295754 Photinia prionophylla（Franch.）C. K. Schneid.;刺叶石楠;Spinyleaf Photinia, Spiny-leaved Photinia ●

295755 Photinia prionophylla（Franch.）C. K. Schneid. var. nudifolia Hand.-Mazz.;无毛刺叶石楠;Nakedleaf Photinia ●

295756 Photinia prunifolia（Franch.）C. K. Schneid. var. denticulata Te T. Yu = Photinia prunifolia（Hook. et Arn.）Lindl. var. denticulata Te T. Yu ●

295757 Photinia prunifolia（Hook. et Arn.）Lindl.;桃叶石楠（石斑木）;Peachleaf Photinia, Peach-leaved Photinia ●

295758 Photinia prunifolia（Hook. et Arn.）Lindl. var. denticulata Te T. Yu;水花石楠（齿叶桃叶石楠,山杠木,重齿桃叶石楠）;Miuntely Toothed Photinia ●

295759 Photinia pustulata Lindl. = Photinia serratifolia（Desf.）Kalkman ●

295760 Photinia pyrifolia（Lam.）K. R. Robertson et J. B. Phipps;梨叶石楠;Red Chokeberry ● ☆

295761 Photinia raupingensis K. C. Kuan;饶平石楠（细叶石斑木）;Raoping Photinia ●

295762 Photinia rosifoliolata H. Lév. = Cotoneaster glaucophylla Franch. ●

295763 Photinia rubrolutea H. Lév. = Malus sieboldii（Regel）Rehder ●

295764 Photinia salicifolia（Decne.）C. K. Schneid. = Photinia arguta Lindl. var. salicifolia（Decne.）Vidal ●

295765 Photinia salicifolia（Decne.）Schneid. = Photinia arguta Lindl. var. salicifolia（Decne.）Vidal ●

295766 Photinia salicifolia C. Presl;柳叶石楠;Christmas Berry ● ☆

295767 Photinia sambuciflora W. W. Sm.;瑞丽石楠;Elderflower Photinia, Ruili Photinia ●

295768 Photinia sambuciflora W. W. Sm. = Photinia integrifolia Lindl. var. notoniana（Wight et Arn.）Vidal ●

295769 Photinia sambuciflora W. W. Sm. = Photinia integrifolia Lindl. ●

295770 Photinia scandens Stapf = Photinia integrifolia Lindl. ●

295771 Photinia schneideriana Rehder et E. H. Wilson;绒毛石楠;Floss Photinia, Schneider Photinia ●

295772 Photinia schneideriana Rehder et E. H. Wilson var. parviflora（Cardot）L. T. Lu et C. L. Li;小花石楠（小花绒毛石楠）;Littleflower Photinia, Small-flowered Photinia ●

295773 Photinia serratifolia（Desf.）Kalkman;石楠（笔树,扁骨木,风药,鬼目,红树,花莲港石楠,将军梨,宽叶石楠,栾茶,千年红,山官木,扇骨木,石纲,石南,石楠柴,石岩树,石眼树,水红树,太鲁阁石楠,油蜡树,凿角,凿木）;Chinese Hawthorn, Chinese Photinia, Hualian Photinia, Oriental Photinia, Photinia, Taiwanese Photinia ●

295774 Photinia serratifolia（Desf.）Kalkman f. daphniphylloides（Hayata）L. T. Lu = Photinia serratifolia（Desf.）Kalkman ●

295775 Photinia serratifolia（Desf.）Kalkman var. ardisiifolia（Hayata）H. Ohashi;台东石楠(卵叶石楠,树杞叶石楠,窄叶石楠,紫金牛叶石楠）;Narrowleaf Chinese Photinia, Taidong Photinia ●

295776 Photinia serratifolia（Desf.）Kalkman var. daphniphylloides（Hayata）L. T. Lu = Photinia serratifolia（Desf.）Kalkman ●

295777 Photinia serratifolia（Desf.）Kalkman var. daphniphylloides（Hayata）L. T. Lu;宽叶石楠;Broad-leaved Chinese Photinia ●

295778 Photinia serratifolia（Desf.）Kalkman var. lasiopetala（Hayata）H. Ohashi;毛瓣石楠;Hairypetal Photinia ●

295779 Photinia serrulata Lindl. = Photinia serratifolia（Desf.）Kalkman ●

295780 Photinia serrulata Lindl. f. ardisiifolia（Hayata）H. L. Li = Photinia serratifolia（Desf.）Kalkman var. ardisiifolia（Hayata）H. Ohashi ●

295781 Photinia serrulata Lindl. f. daphniphylloides（Hayata）H. L. Li = Photinia serratifolia（Desf.）Kalkman var. daphniphylloides（Hayata）L. T. Lu ●

295782 Photinia serrulata Lindl. f. daphniphylloides（Hayata）H. L. Li = Photinia serrulata Lindl. var. daphniphylloides（Hayata）K. C. Kuan ●

295783 Photinia serrulata Lindl. f. lasiopetala（Hayata）T. Shimizu = Photinia serratifolia（Desf.）Kalkman var. lasiopetala（Hayata）H. Ohashi ●

295784 Photinia serrulata Lindl. var. aculeata Lawr. = Photinia serratifolia（Desf.）Kalkman ●

295785 Photinia serrulata Lindl. var. aculeata Lawr. = Photinia serrulata Lindl. ●

295786 Photinia serrulata Lindl. var. ardisiifolia（Hayata）K. C. Kuan = Photinia serratifolia（Desf.）Kalkman var. ardisiifolia（Hayata）H. Ohashi ●

295787 Photinia serrulata Lindl. var. congestiflora Cardot = Photinia glomerata Rehder et E. H. Wilson ●

295788 Photinia serrulata Lindl. var. daphniphylloides（Hayata）K. C. Kuan = Photinia serratifolia（Desf.）Kalkman ●

295789 Photinia serrulata Lindl. var. daphniphylloides（Hayata）K. C. Kuan = Photinia serratifolia（Desf.）Kalkman var. daphniphylloides（Hayata）L. T. Lu ●

295790 Photinia serrulata Lindl. var. lasiopetala（Hayata）K. C. Kuan = Photinia serratifolia（Desf.）Kalkman var. lasiopetala（Hayata）H. Ohashi ●

295791 Photinia serrulata Lindl. var. prunifolia Hook. et Arn. = Photinia prunifolia（Hook. et Arn.）Lindl. ●

295792 Photinia simplex Y. K. Li et X. M. Wang;单齿石楠;Simpletooth Photinia ●

295793 Photinia simplex Y. K. Li et X. M. Wang = Photinia chingiana Hand.-Mazz. ●

295794 Photinia stenophylla Hand.-Mazz.;窄叶石楠;Narrowleaf Photinia, Narrow-leaved Photinia, Stenophtllous Photinia ●

295795 Photinia subparvifolia Y. K. Li et X. M. Wang = Photinia parvifolia（Pritz.）C. K. Schneid. var. subparvifolia（Y. K. Li et X. M. Wang）L. T. Lu et C. L. Li ●

295796 Photinia subumbellata Rehder et E. H. Wilson = Photinia parvifolia（Pritz.）C. K. Schneid. ●

295797 Photinia subumbellata Rehder et E. H. Wilson var. villosa Cardot = Photinia villosa（Thunb.）DC. var. sinica Rehder et E. H. Wilson ●

295798 Photinia taiwanensis Hayata = Photinia lucida（Decne.）C. K. Schneid. ●

295799 Photinia tsaii Rehder;福贡石楠;Tsai Photinia ●

295800　Photinia tushanensis Te T. Yu；独山石楠；Dushan Photinia，Tushan Photinia ●

295801　Photinia undulata（Decne.）Cardot = Stranvaesia davidiana Decne. var. undulata（Decne.）Rehder et E. H. Wilson ●

295802　Photinia undulata（Decne.）Cardot var. formosana Cardot = Stranvaesia davidiana Decne. ●

295803　Photinia undulata Cardot var. formosana Cardot = Photinia davidiana Decne. var. salicifolia（Hutch.）Rehder ●

295804　Photinia undulata Cardot var. formosana Cardot = Photinia niitakayamensis Hayata ●

295805　Photinia variabilis Hemsl. = Photinia lucida（Decne.）C. K. Schneid. ●

295806　Photinia variabilis Hemsl. = Photinia villosa（Thunb.）DC. ●

295807　Photinia variabilis Hemsl. ex Forbes et Hemsl. = Photinia villosa（Thunb.）DC. ●

295808　Photinia villosa（Thunb.）DC.；毛叶石楠（邓向观，活鸡丁，鸡丁子，吉铃子，糯米珠，清水石楠，柔毛石楠，细毛扇骨木）；Oriental Photinia ●

295809　Photinia villosa（Thunb.）DC. = Pourthiaea villosa（Thunb.）Decne. ●

295810　Photinia villosa（Thunb.）DC. var. formosana Hance = Photinia lucida（Decne.）C. K. Schneid. ●

295811　Photinia villosa（Thunb.）DC. var. glabricalycina L. T. Lu et C. L. Li；光萼石楠；Glabrous-calyx Photinia ●

295812　Photinia villosa（Thunb.）DC. var. parvifolia（E. Pritz.）P. S. Hsu et L. C. Li = Photinia parvifolia（Pritz.）C. K. Schneid. ●

295813　Photinia villosa（Thunb.）DC. var. sinica Rehder et E. H. Wilson；庐山石楠（无毛石楠）；Chinese Oriental Photinia ●

295814　Photinia villosa（Thunb.）DC. var. sinica Rehder et E. H. Wilson = Photinia beauverdiana C. K. Schneid. var. notabilis（C. K. Schneid.）Rehder et E. H. Wilson ●

295815　Photinia villosa（Thunb.）DC. var. tenuipes P. S. Hsu et L. C. Li；垂丝毛叶石楠●

295816　Photinia villosa（Thunb.）DC. var. tenuipes P. S. Hsu et L. C. Li = Photinia komarovii（H. Lév. et Vaniot）L. T. Lu et C. L. Li ●

295817　Photinia wrightiana Maxim. ；赖氏石楠●☆

295818　Photinia wuyishanensis Z. X. Yu；武夷山石楠；Wuyishan Photinia ●

295819　Photinia wuyishanensis Z. X. Yu = Photinia komarovii（H. Lév. et Vaniot）L. T. Lu et C. L. Li ●

295820　Photinia zhejiangensis P. L. Chiu；浙江石楠；Zhejiang Photinia ●

295821　Photinia zhijiangensis T. C. Ku = Photinia schneideriana Rehder et E. H. Wilson ●

295822　Phoxanthus Benth. = Ophiocaryon R. H. Schomb. ex Endl. ●☆

295823　Phragmanthera Tiegh.（1895）；裂花桑寄生属●☆

295824　Phragmanthera Tiegh. = Tapinanthus（Blume）Rchb.（保留属名）●☆

295825　Phragmanthera albizziae（De Wild.）Balle = Phragmanthera usuiensis（Oliv.）M. G. Gilbert ●☆

295826　Phragmanthera batangae（Engl.）Balle；巴坦加裂花桑寄生●☆

295827　Phragmanthera baumii（Engl. et Gilg）Polhill et Wiens；鲍姆裂花桑寄生●☆

295828　Phragmanthera brieyi（De Wild.）Polhill et Wiens；布里裂花桑寄生●☆

295829　Phragmanthera capitata（Spreng.）Balle；头状裂花桑寄生●☆

295830　Phragmanthera cinerea（Engl.）Balle；灰色裂花桑寄生●☆

295831　Phragmanthera cinerea（Engl.）Tiegh. ex Durand et B. D. Jacks. = Phragmanthera cinerea（Engl.）Balle ●☆

295832　Phragmanthera cistoides（Welw. ex Engl.）Tiegh. = Phragmanthera glaucocarpa（Peyr.）Balle ●☆

295833　Phragmanthera cornetii（Dewèvre）Polhill et Wiens；科尔内裂花桑寄生●☆

295834　Phragmanthera crassicaulis（Engl.）Balle；粗茎裂花桑寄生●☆

295835　Phragmanthera dombeyae（K. Krause et Dinter）Polhill et Wiens；东拜亚裂花桑寄生●☆

295836　Phragmanthera dschallensis（Engl.）M. G. Gilbert；查伦裂花桑寄生●☆

295837　Phragmanthera edouardii（Balle）Polhill et Wiens；爱德华裂花桑寄生●☆

295838　Phragmanthera eminii（Engl.）Polhill et Wiens；埃明裂花桑寄生●☆

295839　Phragmanthera engleri（Hiern）Polhill et Wiens；恩格勒裂花桑寄生●☆

295840　Phragmanthera erythraea（Sprague）M. G. Gilbert；浅红裂花桑寄生●☆

295841　Phragmanthera exellii Balle ex Polhill et Wiens；埃克塞尔裂花桑寄生●☆

295842　Phragmanthera fulva Tiegh. = Phragmanthera dombeyae（K. Krause et Dinter）Polhill et Wiens ●☆

295843　Phragmanthera glaucocarpa（Peyr.）Balle；灰绿裂花桑寄生●☆

295844　Phragmanthera guerichii（Engl.）Balle；盖里克裂花桑寄生●☆

295845　Phragmanthera incana（Schumach. et Thonn.）Balle = Phragmanthera capitata（Spreng.）Balle ●☆

295846　Phragmanthera irebuensis（De Wild.）Balle = Phragmanthera batangae（Engl.）Balle ●☆

295847　Phragmanthera kamerunensis（Engl.）Balle；喀麦隆裂花桑寄生●☆

295848　Phragmanthera lapathifolia（Engl. et K. Krause）Balle = Phragmanthera capitata（Spreng.）Balle ●☆

295849　Phragmanthera leonensis（Sprague）Balle；莱昂裂花桑寄生●☆

295850　Phragmanthera longiflora（Balle）Polhill et Wiens；长裂花桑寄生●☆

295851　Phragmanthera luteovittata（Engl. et K. Krause）Polhill et Wiens；黄线裂花桑寄生●☆

295852　Phragmanthera macrosolen（Steud. ex A. Rich.）M. G. Gilbert；大管裂花桑寄生●☆

295853　Phragmanthera marginata（Danser）Balle = Phragmanthera crassicaulis（Engl.）Balle ●☆

295854　Phragmanthera nigritana（Hook. f. ex Benth.）Balle；尼格里塔裂花桑寄生●☆

295855　Phragmanthera nigritana（Hook. f. ex Benth.）Balle var. leonensis（Sprague）Balle = Phragmanthera leonensis（Sprague）Balle ●☆

295856　Phragmanthera polycrypta（Didr.）Balle；安哥拉裂花桑寄生●☆

295857　Phragmanthera polycrypta（Didr.）Balle subsp. raynaliana Balle = Phragmanthera raynaliana（Balle）Polhill et Wiens ●☆

295858　Phragmanthera polycrypta（Didr.）Balle subsp. subglabrifolia Balle ex Polhill et Wiens；光叶安哥拉裂花桑寄生●☆

295859　Phragmanthera proteicola（Engl.）Polhill et Wiens；海神木裂花桑寄生●☆

295860　Phragmanthera raynaliana（Balle）Polhill et Wiens；雷纳尔裂花桑寄生●☆

295861　Phragmanthera redingii（De Wild.）Balle = Phragmanthera capitata（Spreng.）Balle ●☆

295862　Phragmanthera regularis（Steud. ex Sprague）M. G. Gilbert；对称裂花桑寄生●☆

295863　Phragmanthera rufescens（DC.）Balle；焦黄裂花桑寄生●☆

295864　Phragmanthera rufescens（DC.）Balle subsp. longiflora Balle ＝ Phragmanthera longiflora（Balle）Polhill et Wiens ●☆

295865　Phragmanthera rufescens（DC.）Balle subsp. usuiensis（Oliv.）Balle ＝ Phragmanthera usuiensis（Oliv.）M. G. Gilbert ●☆

295866　Phragmanthera sarertaensis（Hutch. et E. A. Bruce）M. G. Gilbert；萨雷塔裂花桑寄生●☆

295867　Phragmanthera seretii（De Wild.）Balle；赛雷裂花桑寄生●☆

295868　Phragmanthera sterculiae（Hiern）Polhill et Wiens；斯泰裂花桑寄生●☆

295869　Phragmanthera talbotiorum（Sprague）Balle；塔尔博特裂花桑寄生●☆

295870　Phragmanthera usuiensis（Oliv.）M. G. Gilbert；乌苏裂花桑寄生●☆

295871　Phragmanthera usuiensis（Oliv.）M. G. Gilbert subsp. sigensis（Engl.）Polhill et Wiens；锡格裂花桑寄生●☆

295872　Phragmanthera vignei Balle；维涅裂花桑寄生●☆

295873　Phragmanthera zygiarum（Hiern）Polhill et Wiens；齐格裂花桑寄生●☆

295874　Phragmipedilum Rolfe ＝ Phragmipedium Rolfe（保留属名）■☆

295875　Phragmipedium Rolfe（1896）（保留属名）；拖鞋兰属（长翼兰属，马褂兰属，南美拖鞋兰属）；Lady Slipper, Mandarin Orchid ■☆

295876　Phragmipedium boissierianum（Rchb. f.）Rolfe；布瓦西耶拖鞋兰（包氏马褂兰）；Boissier Mandarin Orchid ■☆

295877　Phragmipedium caricinum（Lindl. et Paxton）Rolfe；苔草状拖鞋兰■☆

295878　Phragmipedium caudatum（Lindl.）Rolfe；尾尖拖鞋兰（长翼兰，尾尖马褂兰，尾状马褂兰）；Caudate Mandarin Orchid, Phragmipedium Orchid ■☆

295879　Phragmipedium longifolium（Warsz. et Rchb. f.）Rolfe；长叶拖鞋兰（长叶马褂兰）；Longleaf Mandarin Orchid ■☆

295880　Phragmipedium schlimii（Rchb. f.）Rolfe；长瓣拖鞋兰■☆

295881　Phragmites Adans.（1763）；芦苇属；Common Reed, Reed ■

295882　Phragmites Trin. ＝ Gynerium Willd. ex P. Beauv. ■☆

295883　Phragmites Trin. ＝ Phragmites Adans. ■

295884　Phragmites australis（Cav.）Steud. ＝ Phragmites australis（Cav.）Trin. ex Steud. ■

295885　Phragmites australis（Cav.）Steud. subsp. altissima（Benth.）Clayton ＝ Phragmites australis（Cav.）Trin. ex Steud. var. altissima（Benth.）D. Rivera et M. A. Carreras ■

295886　Phragmites australis（Cav.）Steud. subsp. chrysantha（Mabille）Soják ＝ Phragmites australis（Cav.）Steud. subsp. altissima（Benth.）Clayton ■

295887　Phragmites australis（Cav.）Steud. subsp. humilis（De Not.）Kerguélen ＝ Phragmites australis（Cav.）Trin. ex Steud. ■

295888　Phragmites australis（Cav.）Trin. ex Steud.；芦苇（长瓣芦苇，大芦柴，葭，葭华，蒹，芦，芦草，芦蓬茸，芦箬，芦笋，芦苇子，芦竹，芦，南方芦苇，日本芦苇，水萌强，顺江龙，为鲁居，苇，苇子，苇子草）；Common Reed, Common Reed Grass, Danube Grass, Giant Reed, Longpetal Reed, Nal, Norfolk Reed, Phragmites, Pole Reed, Pole-reed, Reed, Reed-grass, Rix, Southern Reed, Spear ■

295889　Phragmites australis（Cav.）Trin. ex Steud. f. pilifer（Ohwi）H. Hara；纤毛芦苇■☆

295890　Phragmites australis（Cav.）Trin. ex Steud. var. altissima（Benth.）D. Rivera et M. A. Carreras；高芦苇■

295891　Phragmites australis（Cav.）Trin. ex Steud. var. berlandieri（E. Fourn.）C. F. Reed ＝ Phragmites australis（Cav.）Trin. ex Steud. ■

295892　Phragmites australis（Cav.）Trin. ex Steud. var. stenophylla（Boiss.）Bor ＝ Phragmites australis（Cav.）Trin. ex Steud. ■

295893　Phragmites breviglumis Pomel ＝ Phragmites australis（Cav.）Steud. ■

295894　Phragmites cinctus（Hook. f.）B. S. Sun ＝ Phragmites karka（Retz.）Trin. ex Steud. ■

295895　Phragmites communis（L.）Trin. ＝ Phragmites australis（Cav.）Trin. ex Steud. ■

295896　Phragmites communis（L.）Trin. var. berlandieri（E. Fourn.）Fernald；伯氏芦苇；Berlandier Reed ■

295897　Phragmites communis（L.）Trin. var. longivalvis Miq. ＝ Phragmites communis（L.）Trin. ■

295898　Phragmites communis Trin. ＝ Phragmites australis（Cav.）Steud. subsp. altissima（Benth.）Clayton ■

295899　Phragmites communis Trin. ＝ Phragmites australis（Cav.）Trin. ex Steud. ■

295900　Phragmites communis Trin. subsp. clarianus Sennen ＝ Phragmites australis（Cav.）Steud. subsp. altissima（Benth.）Clayton ■

295901　Phragmites communis Trin. subsp. maximus（Forssk.）Clayton ＝ Phragmites australis（Cav.）Steud. subsp. altissima（Benth.）Clayton ■

295902　Phragmites communis Trin. var. altissimus（Benth.）Trab. ＝ Phragmites australis（Cav.）Steud. subsp. altissima（Benth.）Clayton ■

295903　Phragmites communis Trin. var. berlandieri（E. Fourn.）Fernald ＝ Phragmites australis（Cav.）Trin. ex Steud. ■

295904　Phragmites communis Trin. var. isiaca（Delile）Coss. et Durieu ＝ Phragmites australis（Cav.）Steud. subsp. altissima（Benth.）Clayton ■

295905　Phragmites communis Trin. var. pungens L. Chevall. ＝ Phragmites australis（Cav.）Steud. ■

295906　Phragmites communis Trin. var. stenophylla Boiss. ＝ Phragmites australis（Cav.）Trin. ex Steud. ■

295907　Phragmites hirsuta Kitag.；毛芦苇；Hairy Reed ■

295908　Phragmites isiaca（Delile）Kunth ＝ Phragmites australis（Cav.）Steud. subsp. altissima（Benth.）Clayton ■

295909　Phragmites isiaca Rchb.；女神芦苇■☆

295910　Phragmites jahandiezii Sennen et Mauricio ＝ Phragmites australis（Cav.）Steud. subsp. altissima（Benth.）Clayton ■

295911　Phragmites japonica Steud.；日本苇■

295912　Phragmites japonica Steud. var. prostrata（Makino）L. Liou；爬苇■

295913　Phragmites japonica Steud. var. prostrata（Makino）L. Liou ＝ Phragmites japonica Steud. ■

295914　Phragmites jeholensis Honda；河北芦苇（热河芦苇）；Hebei Reed, Hobei Reed ■

295915　Phragmites karka（Retz.）Steud. ＝ Phragmites karka（Retz.）Trin. ex Steud. ■

295916　Phragmites karka（Retz.）Trin. ex Steud.；卡开芦（大芦，过江龙，过江芦荻，卡开卡芦，开卡芦，芦竹芦苇，水芦荻，水竹）；Karka Reed ■

295917　Phragmites karka（Retz.）Trin. ex Steud. ＝ Phragmites vallatoria（Pluk. ex L.）Veldkamp ■

295918　Phragmites karka (Retz.) Trin. ex Steud. var. cinctus Hook. f. ; 丝毛芦■

295919　Phragmites longivalvis Steud. ;长瓣芦苇■☆

295920　Phragmites macer Munro ex S. Moore = Hakonechloa macra (Munro ex S. Moore) Makino ex Honda ■☆

295921　Phragmites mauritiana Kunth;毛里求斯芦苇■☆

295922　Phragmites maxima (Forssk.) Blatt. et McCann = Phragmites australis (Cav.) Trin. ex Steud. ■

295923　Phragmites maxima (Forssk.) Blatt. et McCann = Phragmites karka (Retz.) Trin. ex Steud. ■

295924　Phragmites nepalensis Nees ex Steud. = Phragmites karka (Retz.) Trin. ex Steud. ■

295925　Phragmites phragmites (L.) H. Karst. = Phragmites australis (Cav.) Trin. ex Steud. ■

295926　Phragmites prostrata Makino = Phragmites japonica Steud. ■

295927　Phragmites pungens Hack. = Phragmites mauritiana Kunth ■☆

295928　Phragmites roxburghii (Kunth) Steud. = Phragmites karka (Retz.) Trin. ex Steud. ■

295929　Phragmites roxburghii Kunth = Phragmites karka (Retz.) Trin. ex Steud. ■

295930　Phragmites serotina Kom. = Phragmites japonica Steud. ■

295931　Phragmites vallatoria (L.) Veldkamp = Phragmites karka (Retz.) Trin. ex Steud. ■

295932　Phragmites vallatoria (Pluk. ex L.) Veldkamp = Phragmites communis (L.) Trin. ■

295933　Phragmites vulgaris (Lam.) Crép. = Phragmites australis (Cav.) Steud. ■

295934　Phragmites vulgaris (Lam.) Crép. var. humilis (De Not.) Parl. = Phragmites australis (Cav.) Steud. ■

295935　Phragmites vulgaris (Lam.) Crép. var. isaicus (Delile) Dur. et Barratte = Phragmites australis (Cav.) Steud. subsp. altissima (Benth.) Clayton ■

295936　Phragmites vulgaris Crép. = Phragmites australis (Cav.) Steud. subsp. altissima (Benth.) Clayton ■

295937　Phragmites xenochloa Trin. ex Steud. ;澳非芦苇■☆

295938　Phragmites zollingeri Steud. = Neyraudia reynaudiana (Kunth) Keng ex Hitchc. ■

295939　Phragmocarpidium Krapov. (1969);篱果锦葵属●☆

295940　Phragmocarpidium heringeri Krapov. ;篱果锦葵●☆

295941　Phragmocassia Britton et Rose = Cassia L. (保留属名)●■

295942　Phragmocassia Britton et Rose = Senna Mill. ●■

295943　Phragmopedilum (Pfitzer) Pfitzer = Phragmipedium Rolfe (保留属名)■☆

295944　Phragmopedilum Pfitzer = Phragmipedium Rolfe(保留属名)■☆

295945　Phragmopedilum Rolfe = Phragmipedium Rolfe(保留属名)■☆

295946　Phragmorchis L. O. Williams(1938);篱笆兰属■☆

295947　Phragmorchis teretifolia L. O. Williams;篱笆兰■☆

295948　Phragmotheca Cuatrec. (1946);篱囊木棉属●☆

295949　Phragmotheca amazonica (W. S. Alverson) Fern. Alonso;亚马逊篱囊木棉●☆

295950　Phragmotheca leucoflora D. R. Simpson;白花篱囊木棉●☆

295951　Phragmotheca rubriflora Fern. Alonso;红花篱囊木棉●☆

295952　Phreatia Lindl. (1830);馥兰属(芙乐兰属);Phreatia ■

295953　Phreatia caulescens Ames;垂茎馥兰（垂茎芙乐兰）; Nutantstem Phreatia ■☆

295954　Phreatia densiflora (Blume) Lindl. ;密花馥兰;Denseflower Phreatia ■☆

295955　Phreatia elegans (Blume) Lindl. = Phreatia formosana Rolfe ■

295956　Phreatia elegans Lindl. = Phreatia formosana Rolfe ■

295957　Phreatia evrardii Gagnep. = Phreatia formosana Rolfe ■

295958　Phreatia formosana Rolfe;套叶馥兰（宝岛芙乐兰,馥兰）; Common Phreatia,Elegant Phreatia,Evrard Phreatia ■

295959　Phreatia kotoensularis Fukuy. = Phreatia formosana Rolfe ■

295960　Phreatia morii Hayata;大馥兰(大芙乐兰);Big Phreatia ■

295961　Phreatia secunda (Blume) Lindl. ;偏馥兰;Secund Phreatia ■☆

295962　Phreatia taiwaniana Fukuy. ;台湾馥兰(台湾芙乐兰);Taiwan Phreatia ■

295963　Phreatia uniflora Wight = Conchidium pusillum Griff. ■

295964　Phrenanthes Wigg. = Prenanthes L. ■

295965　Phrissocarpus Miers = Tabernaemontana L. ●

295966　Phrodus Miers(1849);智利小叶茄属■☆

295967　Phrodus microphyllus (Miers) Miers. ;智利小叶茄■☆

295968　Phryganocydia Mart. ex Baill. (1872);品红紫葳属●☆

295969　Phryganocydia Mart. ex DC. = Macfadyena A. DC. ●

295970　Phryganocydia corymbosa (Vent.) Schum. ;品红紫葳●☆

295971　Phryganthus Baker = Phyganthus Poepp. et Endl. ■

295972　Phryganthus Baker = Tecophilaea Bertero ex Colla ■☆

295973　Phrygia (Pers.) Gray = Centaurea L. (保留属名)●■

295974　Phrygia Gray = Centaurea L. (保留属名)●■

295975　Phrygilanthus Eichler = Notanthera (DC.) G. Don ●☆

295976　Phrygiobureaua Kuntze = Phryganocydia Mart. ex Baill. ●☆

295977　Phryma Forssk. = Priva Adans. ■☆

295978　Phryma L. (1753);透骨草属(毒蛆草属,蝎毒草属);Lopseed ■

295979　Phryma asiatica (H. Hara) O. Deg. et I. Deg. = Phryma leptostachya L. subsp. asiatica (H. Hara) Kitam. ■

295980　Phryma asiatica (H. Hara) Prob. = Phryma leptostachya L. subsp. asiatica (H. Hara) Kitam. ■

295981　Phryma dehiscens L. f. = Chascanum cuneifolium (L. f.) E. Mey. ●☆

295982　Phryma esquirolii H. Lév. = Phryma leptostachya L. subsp. asiatica (H. Hara) Kitam. ■

295983　Phryma humilis Koidz. = Phryma leptostachya L. subsp. asiatica (H. Hara) Kitam. ■

295984　Phryma leptostachya L. ;北美透骨草(毒蛆草,老婆子针线,麻荆芥,透骨草,小青,药蛆,一马光,一抹光,一扫光,粘人裙); American Lop-seed,Lopseed,Smallbetony Lopseed ■

295985　Phryma leptostachya L. f. melanostachya (Kitag.) Kitag. = Phryma leptostachya L. subsp. asiatica (H. Hara) Kitam. ■

295986　Phryma leptostachya L. subsp. asiatica (H. Hara) Kitam. ;透骨草(长圆叶透骨草,大一扫光,倒刺草,倒扣草,毒蛆草,毒蝇草,剪草,接生草,老婆子针线,前草,山箭草,神砂一把抓,仙人一把遮,小肥猪,小青,小蛆药,亚洲透骨草,药曲草,野倒钩草,一马光,一抹光,一扫光,蝇毒草,粘人裙);Asian Lopseed,Lopseed, Oblongleaf Lopseed ■

295987　Phryma leptostachya L. subsp. asiatica (H. Hara) Kitam. = Phryma leptostachya L. var. oblongifolia (Koidz.) Honda ■

295988　Phryma leptostachya L. subsp. asiatica (H. Hara) Kitam. f. nana (Koidz.) Akasawa;小透骨草■☆

295989　Phryma leptostachya L. subsp. asiatica (H. Hara) Kitam. f. oblongifolia (Koidz.) Ohwi = Phryma leptostachya L. var. oblongifolia (Koidz.) Honda ■

295990　Phryma leptostachya L. subsp. asiatica (H. Hara) Kitam. var. nana (Koidz.) H. Hara = Phryma leptostachya L. subsp. asiatica (H. Hara) Kitam. f. nana (Koidz.) Akasawa ■☆

295991　Phryma leptostachya L. var. asiatica H. Hara ＝ Phryma leptostachya L. ■

295992　Phryma leptostachya L. var. asiatica H. Hara ＝ Phryma leptostachya L. subsp. asiatica（H. Hara）Kitam. ■

295993　Phryma leptostachya L. var. asiatica H. Hara ＝ Phryma leptostachya L. var. oblongifolia（Koidz.）Honda ■

295994　Phryma leptostachya L. var. confertifolia Fernald ＝ Phryma leptostachya L. ■

295995　Phryma leptostachya L. var. humilis（Koidz.）H. Hara ＝ Phryma leptostachya L. subsp. asiatica（H. Hara）Kitam. ■

295996　Phryma leptostachya L. var. melanostachya Kitag. ;黑子透骨草■

295997　Phryma leptostachya L. var. melanostachya Kitag. ＝ Phryma leptostachya L. subsp. asiatica（H. Hara）Kitam. ■

295998　Phryma leptostachya L. var. nana（Koidz.）H. Hara ＝ Phryma leptostachya L. subsp. asiatica（H. Hara）Kitam. ■

295999　Phryma leptostachya L. var. oblongifolia（Koidz.）Honda;长圆叶透骨草■

296000　Phryma leptostachya L. var. oblongifolia（Koidz.）Honda ＝ Phryma leptostachya L. subsp. asiatica（H. Hara）Kitam. ■

296001　Phryma media Raf. ＝ Phryma leptostachya L. ■

296002　Phryma nana Koidz. ＝ Phryma leptostachya L. subsp. asiatica（H. Hara）Kitam. ■

296003　Phryma oblongifolia Koidz. ＝ Phryma leptostachya L. subsp. asiatica（H. Hara）Kitam. ■

296004　Phryma parvifora Raf. ＝ Phryma leptostachya L. ■

296005　Phryma pubescens Raf. ＝ Phryma leptostachya L. ■

296006　Phrymaceae Schauer（1847）（保留科名）;透骨草科;Lopseed Family ■

296007　Phrymaceae Schauer（保留科名）＝ Verbenaceae J. St. -Hil.（保留科名）●■

296008　Phrymataceae Schauer ＝ Phrymaceae Schauer（保留科名）■

296009　Phryna（Boiss.）Pax et K. Hoffm. ＝ Phrynella Pax et K. Hoffm. ■☆

296010　Phryne Bubani ＝ Sisymbrium L. ■

296011　Phryne Bubani（1901）;蟾芥属■☆

296012　Phryne huetii（Boiss.）O. E. Schulz;休氏蟾芥■☆

296013　Phryne pinnatifida（Lam.）O. E. Schulz ＝ Murbeckiella boryi（Boiss.）Rothm. ■☆

296014　Phrynella Pax et K. Hoffm.（1934）;棱石竹属■☆

296015　Phrynella ortegioides（Fisch. et C. A. Mey.）Pax et K. Hoffm. ;棱石竹■☆

296016　Phrynium Loefl ＝ Heteranthera Ruiz et Pav.（保留属名）■☆

296017　Phrynium Loefl. ex Kuntze ＝ Heteranthera Ruiz et Pav.（保留属名）■☆

296018　Phrynium Willd.（1797）（保留属名）;柊叶属;Phrynium ■

296019　Phrynium adenocarpum（K. Schum.）Baker ＝ Megaphrynium macrostachyum（Benth.）Milne-Redh. ■☆

296020　Phrynium baccatum（K. Schum.）Baker ＝ Sarcophrynium schweinfurthianum（Kuntze）Milne-Redh. ■☆

296021　Phrynium benthamii Baker ＝ Megaphrynium macrostachyum（Benth.）Milne-Redh. ■☆

296022　Phrynium bisubulatum（K. Schum.）Baker ＝ Sarcophrynium bisubulatum（K. Schum.）K. Schum. ■☆

296023　Phrynium brachystachyum（Benth.）Körn. ＝ Sarcophrynium brachystachyum（Benth.）K. Schum. ■☆

296024　Phrynium capitatum Willd. ＝ Phrynium rheedei Suresh et Nicolson ■

296025　Phrynium cerasiferum A. Chev. ＝ Sarcophrynium prionogonium（K. Schum.）K. Schum. ■☆

296026　Phrynium confertum（Benth.）K. Schum. ＝ Ataenidia conferta（Benth.）Milne-Redh. ■☆

296027　Phrynium coriscense Baker ＝ Halopegia azurea（K. Schum.）K. Schum. ■☆

296028　Phrynium crista-galli A. Chev. ＝ Ataenidia conferta（Benth.）Milne-Redh. ■☆

296029　Phrynium daniellii Bennet ＝ Thaumatococcus daniellii（Bennet）Benth. ■☆

296030　Phrynium dispermum Gagnep. ＝ Phrynium oliganthum Merr. ■

296031　Phrynium filipes Benth. ＝ Marantochloa filipes（Benth.）Hutch. ■☆

296032　Phrynium flexuosum Benth. ＝ Marantochloa cuspidata（Roscoe）Milne-Redh. ■☆

296033　Phrynium hainanense T. L. Wu et S. J. Chen;海南柊叶;Hainan Phrynium ■

296034　Phrynium hensii Baker ＝ Marantochloa mannii（Benth.）Milne-Redh. ■☆

296035　Phrynium holostachyum Baker ＝ Marantochloa monophylla（K. Schum.）D'Orey ■☆

296036　Phrynium inaequilaterum Baker ＝ Marantochloa congensis（K. Schum.）J. Léonard et Mullend. var. pubescens（Loes.）J. Léonard et Mullend. ■☆

296037　Phrynium leiogonium（K. Schum.）Baker ＝ Sarcophrynium prionogonium（K. Schum.）K. Schum. ■☆

296038　Phrynium macrophyllum（K. Schum.）Baker ＝ Megaphrynium macrostachyum（Benth.）Milne-Redh. ■☆

296039　Phrynium macrostachyum Benth. ＝ Megaphrynium macrostachyum（Benth.）Milne-Redh. ■☆

296040　Phrynium mannii（Benth.）K. Schum. ＝ Marantochloa mannii（Benth.）Milne-Redh. ■☆

296041　Phrynium molle A. Chev. ＝ Sarcophrynium brachystachyum（Benth.）K. Schum. ■☆

296042　Phrynium monophyllum（K. Schum.）Baker ＝ Marantochloa monophylla（K. Schum.）D'Orey ■☆

296043　Phrynium oliganthum Merr. ;少花柊叶（两籽柊叶）;Fewflower Lopseed,Twoseed Phrynium ■

296044　Phrynium ovatum（L.）Druce ＝ Phrynium rheedei Suresh et Nicolson ■

296045　Phrynium oxycarpum（K. Schum.）Baker ＝ Megaphrynium macrostachyum（Benth.）Milne-Redh. ■☆

296046　Phrynium parviflorum Roxb. ;小花柊叶■☆

296047　Phrynium parviflorum Roxb. ＝ Phrynium placentarium（Lour.）Merr. ■

296048　Phrynium picturata Lindl. ＝ Calathea picturata K. Koch et Linden ■☆

296049　Phrynium placentarium（Lour.）Merr. ;尖苞柊叶（冬叶,小花柊叶,柊叶,棕叶）;Sharpbract Phrynium ■

296050　Phrynium prionogonium（K. Schum.）Baker ＝ Sarcophrynium prionogonium（K. Schum.）K. Schum. ■☆

296051　Phrynium ramosissimum Benth. ＝ Marantochloa ramosissima（Benth.）Hutch. ■☆

296052　Phrynium rheedei Suresh et Nicolson;柊叶（冬叶,棕粑叶,棕叶）;Capitate Phrynium ■

296053　Phrynium sinicum Miq. ＝ Phrynium placentarium（Lour.）Merr. ■

296054　Phrynium sulphureum Baker ＝ Marantochloa sulphurea（Baker）

Koechlin ■☆

296055　Phrynium textile Ridl. = Ataenidia conferta (Benth.) Milne-Redh. ■☆

296056　Phrynium tonkinense Gagnep.；云南柊叶；Tonkin Phrynium, Yunnan Phrynium ■

296057　Phrynium variegatum N. E. Br. = Maranta arundinacea L. ■

296058　Phrynium velutinum Baker = Megaphrynium velutinum (Baker) Koechlin ■☆

296059　Phrynium villosum Benth. = Sarcophrynium villosum (Benth.) K. Schum. ■☆

296060　Phtheirospermum Bunge = Phtheirospermum Bunge ex Fisch. et C. A. Mey. ■

296061　Phtheirospermum Bunge ex Fisch. et C. A. Mey. (1835)；松蒿属；Phtheirospermum ■

296062　Phtheirospermum auratum Bonati = Pedicularis aurata (Bonati) H. L. Li ■

296063　Phtheirospermum auratum Bonati = Phtheirospermum japonicum (Thunb.) Kanitz ■

296064　Phtheirospermum chinense Bunge = Phtheirospermum japonicum (Thunb.) Kanitz ■

296065　Phtheirospermum chinense Bunge ex Fisch. et C. A. Mey. = Phtheirospermum japonicum (Thunb.) Kanitz ■

296066　Phtheirospermum esquirolii Bonati ex Petitm.；贵州松蒿■

296067　Phtheirospermum glandulosum Benth. et Hook. f. = Pseudobartsia yunnanensis D. Y. Hong ■

296068　Phtheirospermum japonicum (Thunb.) Kanitz；松蒿(糯蒿,日本松蒿,土茵陈,小盐灶菜,小盐灶草)；Japanese Phtheirospermum ■

296069　Phtheirospermum japonicum (Thunb.) Kanitz f. albiflorum (Honda) H. Hara；白花松蒿■☆

296070　Phtheirospermum muliense C. Y. Wu et D. D. Tao；木里松蒿；Muli Phtheirospermum ■

296071　Phtheirospermum parishii Hook. f.；黑籽松蒿；Parish Phtheirospermum ■

296072　Phtheirospermum tenuisectum Bureau et Franch.；细裂叶松蒿(草柏枝,黄花,黄花松蒿,松叶蒿,松叶接骨草,蜈蚣草)；Finelydivided Phtheirospermum ■

296073　Phtheirotheca Maxim. ex Regel = Caulophyllum Michx. ●

296074　Phthirusa Mart. (1830)；热美桑寄生属●☆

296075　Phthirusa Mart. = Hemitria Raf. ●☆

296076　Phthirusa caribaea (Krug et Urb.) Britton et E. Wilson；热美桑寄生；West Indian Mistletoe ●☆

296077　Phtirium Raf. = Delphinium L. ■

296078　Phu Ludw. = Valeriana L. ●■

296079　Phu Rupp. = Valeriana L. ●■

296080　Phucagrostis Cavolini(废弃属名) = Cymodocea K. D. König(保留属名)■

296081　Phucagrostis Cavolini(废弃属名) = Zostera L. + Cymodocea K. D. König(保留属名)■

296082　Phucagrostis Willd. = Cymodocea K. D. König(保留属名)■

296083　Phucagrostis major Cavolini = Cymodocea nodosa (Ucria) Asch. ■☆

296084　Phuodendron (Graebn.) Dalla Torre et Harms = Valeriana L. ●■

296085　Phuodendron Graebn. = Valeriana L. ●■

296086　Phuopsis (Griseb.) Benth. et Hook. f. (1873)；长柱草属(球序茜属)■

296087　Phuopsis (Griseb.) Hook. f. = Phuopsis (Griseb.) Benth. et Hook. f. ■

296088　Phuopsis Benth. et Hook. f. = Phuopsis (Griseb.) Benth. et Hook. f. ■

296089　Phuopsis Griseb. = Phuopsis (Griseb.) Hook. f. ■

296090　Phuopsis stylosa (Trin.) B. D. Jacks. = Phuopsis stylosa (Trin.) Hook. f. ■

296091　Phuopsis stylosa (Trin.) Hook. f.；长柱草(长柱花,球序茜)■

296092　Phuphanochloa Sungkaew et Teerawat. (2008)；泰禾属■☆

296093　Phusicarpos Poir. = Hovea R. Br. ex W. T. Aiton ●■☆

296094　Phycagrostis Post et Kuntze = Phucagrostis Cavolini (废弃属名)■

296095　Phycagrostis Post et Kuntze = Zostera L. + Cymodocea K. D. König(保留属名)■

296096　Phycella Lindl. (1825)；肖朱顶红属■☆

296097　Phycella Lindl. = Hippeastrum Herb. (保留属名)■

296098　Phycella angustifolia Phil.；窄叶肖朱顶红■☆

296099　Phycella australis Ravenna；南方肖朱顶红■☆

296100　Phycella bicolor Herb.；二色肖朱顶红■☆

296101　Phycella biflora Lindl.；双花肖朱顶红■☆

296102　Phycella graciliflora Herb.；细花肖朱顶红■☆

296103　Phycella obtusifolia Pritz.；钝叶肖朱顶红■☆

296104　Phycoschoenus (Asch.) Nakai = Cymodocea K. D. König(保留属名)■

296105　Phyfiaurea Lour. = Codiaeum A. Juss. (保留属名)●

296106　Phyganthus Poepp. et Endl. = Tecophilaea Bertero ex Colla ■☆

296107　Phygelius E. Mey. = Phygelius E. Mey. ex Benth. ■●☆

296108　Phygelius E. Mey. ex Benth. (1836)；南非吊金钟属(避日花属,费格利木属,南非金钟属)；Cape Figwort, Fuchsia ■●☆

296109　Phygelius × rectus Coombes；粉红花南非吊金钟；Cape Fuchsia ●☆

296110　Phygelius aequalis Harv. ex Hiern；黄花南非吊金钟(亮叶费格利木,同型避日花)；Cape Fuchsia ●☆

296111　Phygelius aequalis Harv. ex Hiern 'Yellow Trumpet'；黄喇叭亮叶费格利木(黄喇叭同型避日花)●☆

296112　Phygelius capensis E. Mey. ex Benth.；南非吊金钟(好望角费格利木)；Cape Figwort, Cape Fuchsia, River Bells ●☆

296113　Phygelius capensis E. Mey. ex Benth. 'Coccineus'；绯红避日花；Cape Figwort ●☆

296114　Phyla Lour. (1790)；过江藤属(鸭舌癀属)；Phyla ■

296115　Phyla canescens (Kunth) Greene；灰白过江藤(灰二郎箭)●☆

296116　Phyla chinensis Lour. = Phyla nodiflora (L.) Greene ■

296117　Phyla filiformis (Schrad.) Meikle；丝状过江藤■☆

296118　Phyla lanceolata (Michx.) Greene；剑叶过江藤；Lance-leaf Fog-fruit, Northern Fog Fruit, Northern Frog Fruit ■☆

296119　Phyla lanceolata (Michx.) Greene var. recognita (Fernald et Griscom) Soper = Phyla lanceolata (Michx.) Greene ■☆

296120　Phyla nodiflora (L.) E. L. Richards = Phyla nodiflora (L.) Greene ■

296121　Phyla nodiflora (L.) Greene；过江藤(大二郎箭,番梨草,番梨仔草,凤梨草,过江龙,苦舌草,雷公锤草,蓬莱草,水黄芹,水马齿苋,铜锤草,旺梨草,虾子草,鸭脚板,鸭舌癀,鸭嘴黄,野千年锤)；Aztec Sweet Herb, Common Fog Fruit, Common Frog Fruit, Creeping Lip Plant, Knottedflower Phyla, Lippia, Turkey Tangle, Turkey Tangle Fogfruit ■

296122　Phyla nodiflora (L.) Greene var. canescens (Kunth) Moldenke = Phyla canescens (Kunth) Greene ●☆

296123　Phyla nodiflora (L.) Greene var. longifolia Moldenke；长叶二郎箭■☆

296124　Phyla nodiflora（L.）Greene var. rosea（D. Don）Moldenke；粉红过江藤（粉红二郎箭）■●☆

296125　Phyla nodiflora（L.）Greene var. sarmentosa ?；大二郎箭■●☆

296126　Phyla nodiflora（L.）Greene var. sericea（Kuntze）Moldenke；绢毛二郎箭■●☆

296127　Phyla reptans（Kunth）Greene；匍匐过江藤■●☆

296128　Phylacanthus Benth. = Angelonia Bonpl. ■●☆

296129　Phylacium A. W. Benn.（1840）；苞护豆属（长柄荚属）；Phylacium ■

296130　Phylacium majus Collett et Hemsl.；苞护豆；Common Phylacium，Phylacium ■

296131　Phylactis Schrad. = Philactis Schrad. ●☆

296132　Phylanthera Noronha = Hypobathrum Blume ●☆

296133　Phylanthus Murr. = Phyllanthus L. ●■

296134　Phylax Noronha = Polygala L. ●■

296135　Phylepidum Raf. = Phyllepidum Raf. ■☆

296136　Phylepidum Raf. = Polygonella Michx. ■☆

296137　Phylesiaceae Dumort. = Philesiaceae Dumort.（保留科名）●■☆

296138　Phylica L.（1753）；菲利木属●☆

296139　Phylica abietina Eckl. et Zeyh.；冷杉菲利木●☆

296140　Phylica acmaephylla Eckl. et Zeyh.；叶菲利木●☆

296141　Phylica aemula Schltr.；匹敌菲利木●☆

296142　Phylica aemula Schltr. var. multibracteolata Pillans；多苞片菲利木●☆

296143　Phylica affinis Sond.；近缘菲利木●☆

296144　Phylica agathosmoides Pillans；香芸木菲利木●☆

296145　Phylica alba Pillans；白菲利木●☆

296146　Phylica alpina Eckl. et Zeyh.；高山菲利木●☆

296147　Phylica alticola Pillans；高原菲利木●☆

296148　Phylica ambigua Sond.；可疑菲利木●☆

296149　Phylica amoena Pillans；秀丽菲利木●☆

296150　Phylica anomala Pillans；异常菲利木●☆

296151　Phylica apiculata Sond.；细尖菲利木●☆

296152　Phylica atrata Licht. ex Roem. et Schult.；黑菲利木●☆

296153　Phylica atrata Licht. ex Roem. et Schult. var. litoralis（Eckl. et Zeyh.）Sond. = Phylica litoralis（Eckl. et Zeyh.）D. Dietr. ●☆

296154　Phylica axillaris Lam.；腋生菲利木●☆

296155　Phylica axillaris Lam. var. cooperi Pillans；库珀腋生菲利木●☆

296156　Phylica axillaris Lam. var. densifolia Pillans；密叶腋生菲利木●☆

296157　Phylica axillaris Lam. var. gracilis Pillans；纤细腋生菲利木●☆

296158　Phylica axillaris Lam. var. hirsuta Sond.；毛腋生菲利木●☆

296159　Phylica axillaris Lam. var. lutescens（Eckl. et Zeyh.）Pillans；浅黄腋生菲利木●☆

296160　Phylica axillaris Lam. var. maritima Pillans；滨海腋生菲利木●☆

296161　Phylica axillaris Lam. var. microphylla（Eckl. et Zeyh.）Pillans；小叶腋生菲利木●☆

296162　Phylica axillaris Lam. var. parvifolia Sond. = Phylica pinea Thunb. ●☆

296163　Phylica axillaris Lam. var. pedicellaris Sond. = Phylica pinea Thunb. ●☆

296164　Phylica axillaris Lam. var. pulchra Pillans；美丽腋生菲利木●☆

296165　Phylica barbata Pillans；髯毛菲利木●☆

296166　Phylica barnardii Pillans；巴纳德菲利木●☆

296167　Phylica bicolor L. = Phylica strigosa P. J. Bergius ●☆

296168　Phylica bolusii Pillans；博卢斯菲利木●☆

296169　Phylica brachycephala Sond.；短头菲利木●☆

296170　Phylica brevifolia Eckl. et Zeyh.；短叶菲利木●☆

296171　Phylica burchellii Pillans；伯切尔菲利木●☆

296172　Phylica buxifolia L.；黄杨叶菲利木；Box Hard-leaf ●☆

296173　Phylica calcarata Pillans；距菲利木●☆

296174　Phylica callosa L. f.；硬皮菲利木●☆

296175　Phylica capitata Thunb. = Phylica pubescens Aiton ●☆

296176　Phylica capitata Thunb. var. angustifolia Sond. = Phylica pubescens Aiton var. angustifolia（Sond.）Pillans ●☆

296177　Phylica capitata Thunb. var. brachycephala Sond. = Phylica dodii N. E. Br. ●☆

296178　Phylica cephalantha Sond.；头花菲利木●☆

296179　Phylica chionocephala Schltr.；雪头菲利木●☆

296180　Phylica chionophila Schltr.；喜雪菲利木●☆

296181　Phylica comosa Sond.；簇毛菲利木●☆

296182　Phylica comptonii Pillans；康普顿菲利木●☆

296183　Phylica confusa Pillans；混乱菲利木●☆

296184　Phylica constricta Pillans；缢缩菲利木●☆

296185　Phylica cordata L. var. laevis Schltdl. = Phylica laevis（Schltdl.）Steud. ●☆

296186　Phylica costata Pillans；单脉菲利木●☆

296187　Phylica cryptandroides Sond.；隐蕊菲利木●☆

296188　Phylica curvifolia Pillans；折叶菲利木●☆

296189　Phylica cuspidata Eckl. et Zeyh.；骤尖菲利木●☆

296190　Phylica cuspidata Eckl. et Zeyh. var. minor Pillans；小骤尖菲利木●☆

296191　Phylica cylindrica J. C. Wendl.；柱形菲利木●☆

296192　Phylica cylindrica Sond. var. glabrata ? = Phylica cylindrica J. C. Wendl. ●☆

296193　Phylica debilis Eckl. et Zeyh.；弱小菲利木●☆

296194　Phylica debilis Eckl. et Zeyh. var. fourcadei Pillans；富尔卡德菲利木●☆

296195　Phylica diffusa Pillans；松散菲利木●☆

296196　Phylica dioica L.；异株菲利木●☆

296197　Phylica diosmoides Sond.；逸香木菲利木●☆

296198　Phylica disticha Eckl. et Zeyh.；二列菲利木●☆

296199　Phylica disticha Eckl. et Zeyh. var. cuneata Pillans；楔形菲利木●☆

296200　Phylica dodii N. E. Br.；多德菲利木●☆

296201　Phylica elimensis Pillans；埃利姆菲利木●☆

296202　Phylica emirnensis（Tul.）Pillans；埃米菲利木●☆

296203　Phylica emirnensis（Tul.）Pillans var. nyassae Pillans ex Verdc. = Phylica emirnensis（Tul.）Pillans ●☆

296204　Phylica ericoides L.；石南状菲利木●☆

296205　Phylica ericoides L. var. montana Pillans；山地石南状菲利木●☆

296206　Phylica ericoides L. var. muirii Pillans；缪里菲利木●☆

296207　Phylica ericoides L. var. pauciflora Pillans；少花石南状菲利木●☆

296208　Phylica ericoides L. var. zeyheri Pillans；泽耶尔菲利木●☆

296209　Phylica eriophoros P. J. Bergius = Phylica imberbis P. J. Bergius var. eriophoros（P. J. Bergius）Pillans ●☆

296210　Phylica eriophoros P. J. Bergius var. imberbis（P. J. Bergius）Sond. = Phylica imberbis P. J. Bergius ●☆

296211　Phylica excelsa J. C. Wendl.；高大菲利木●☆

296212　Phylica excelsa J. C. Wendl. var. brevifolia Sond. = Phylica excelsa J. C. Wendl. var. papillosa（J. C. Wendl.）Sond. ●☆

296213　Phylica excelsa J. C. Wendl. var. laxa Sond. = Phylica excelsa J. C. Wendl. var. papillosa（J. C. Wendl.）Sond. ●☆

296214　Phylica excelsa J. C. Wendl. var. papillosa（J. C. Wendl.）

Sond.；乳头高大菲利木●☆

296215　Phylica excelsa J. C. Wendl. var. stricta Sond. = Phylica excelsa J. C. Wendl. var. papillosa（J. C. Wendl.）Sond.●☆

296216　Phylica floccosa Pillans；丛毛菲利木●☆

296217　Phylica floribunda Pillans；繁花菲利木●☆

296218　Phylica fourcadei Pillans = Phylica debilis Eckl. et Zeyh. var. fourcadei Pillans●☆

296219　Phylica fruticosa Schltr.；灌丛菲利木●☆

296220　Phylica fulva Eckl. et Zeyh. = Phylica excelsa J. C. Wendl. var. papillosa（J. C. Wendl.）Sond.●☆

296221　Phylica galpinii Pillans；盖尔菲利木●☆

296222　Phylica glabrata Thunb.；光滑菲利木●☆

296223　Phylica gnidioides Eckl. et Zeyh.；格尼瑞香菲利木●☆

296224　Phylica gracilis（Eckl. et Zeyh.）D. Dietr.；纤细菲利木●☆

296225　Phylica greyii Pillans；格雷菲利木●☆

296226　Phylica guthriei Pillans；格斯里菲利木●☆

296227　Phylica harveyi（Arn.）Pillans；哈维菲利木●☆

296228　Phylica hirta Pillans；多毛菲利木●☆

296229　Phylica humilis Sond.；低矮菲利木●☆

296230　Phylica imberbis P. J. Bergius；无须菲利木●☆

296231　Phylica imberbis P. J. Bergius var. eriophoros（P. J. Bergius）Pillans；毛菲利木●☆

296232　Phylica imberbis P. J. Bergius var. secunda Sond.；单侧菲利木●☆

296233　Phylica incurvata Pillans；内折菲利木●☆

296234　Phylica insignis Pillans；显著菲利木●☆

296235　Phylica intrusa Pillans；外来菲利木●☆

296236　Phylica karroica Pillans；卡罗菲利木●☆

296237　Phylica keetii Pillans；克特菲利木●☆

296238　Phylica keetii Pillans var. mollis？；柔软菲利木●☆

296239　Phylica lachneaeoides Pillans；毛瑞香菲利木●☆

296240　Phylica laevifolia Pillans；光叶菲利木●☆

296241　Phylica laevigata Pillans；稍平滑菲利木●☆

296242　Phylica laevis（Schltdl.）Steud.；平滑菲利木●☆

296243　Phylica lanata Pillans；绵毛菲利木●☆

296244　Phylica lasiantha Pillans；毛花菲利木●☆

296245　Phylica lasiocarpa Sond.；毛果菲利木●☆

296246　Phylica leipoldtii Pillans；莱波尔德菲利木●☆

296247　Phylica levynsiae Pillans；勒温斯菲利木●☆

296248　Phylica linifolia Pillans；亚麻叶菲利木●☆

296249　Phylica litoralis（Eckl. et Zeyh.）D. Dietr.；滨海菲利木●☆

296250　Phylica longimontana Pillans；山地长菲利木●☆

296251　Phylica lucens Pillans；光亮菲利木●☆

296252　Phylica lucida Pillans；明亮菲利木●☆

296253　Phylica lutescens（Eckl. et Zeyh.）D. Dietr. = Phylica axillaris Lam. var. lutescens（Eckl. et Zeyh.）Pillans●☆

296254　Phylica mairei Pillans；迈雷菲利木●☆

296255　Phylica marlothii Pillans；马洛斯菲利木●☆

296256　Phylica marlothii Pillans var. crassa？；粗利木●☆

296257　Phylica maximiliani Schltr.；马克西米亚诺菲利木●☆

296258　Phylica meyeri Sond.；迈尔菲利木●☆

296259　Phylica microphylla（Eckl. et Zeyh.）D. Dietr. = Phylica axillaris Lam. var. microphylla（Eckl. et Zeyh.）Pillans●☆

296260　Phylica minutiflora Schltdl.；微花菲利木●☆

296261　Phylica montana Sond.；山地菲利木●☆

296262　Phylica mundii Pillans；蒙德菲利木●☆

296263　Phylica natalensis Pillans；纳塔尔菲利木●☆

296264　Phylica nervosa Pillans；多脉菲利木●☆

296265　Phylica nigrita Sond.；尼格里塔菲利木●☆

296266　Phylica nigromontana Pillans；黑色山地菲利木●☆

296267　Phylica nodosa Pillans；多节菲利木●☆

296268　Phylica obtusifolia Pillans；钝叶菲利木●☆

296269　Phylica odorata Schltr.；芳香菲利木●☆

296270　Phylica oleifolia Vent.；木犀榄叶菲利木●☆

296271　Phylica oleoides DC. = Phylica oleifolia Vent.●☆

296272　Phylica paniculata Willd.；圆锥菲利木●☆

296273　Phylica papillosa J. C. Wendl. = Phylica excelsa J. C. Wendl. var. papillosa（J. C. Wendl.）Sond.●☆

296274　Phylica parviflora P. J. Bergius；小花菲利木●☆

296275　Phylica parvula Pillans；较小菲利木●☆

296276　Phylica pauciflora Pillans；少花菲利木●☆

296277　Phylica pearsonii Pillans；皮尔逊菲利木●☆

296278　Phylica pedicellata DC. = Phylica villosa Thunb. var. pedicellata（DC.）Sond.●☆

296279　Phylica pinea Thunb.；松菲利木●☆

296280　Phylica pinifolia L. f. = Pseudobaeckea africana（Burm. f.）Pillans●☆

296281　Phylica piquetbergensis Pillans；皮克特菲利木●☆

296282　Phylica plumigera Pillans；羽状菲利木●☆

296283　Phylica plumosa L.；密毛菲利木；Flannel Flower●☆

296284　Phylica plumosa L. var. horizontalis（Vent.）Sond.；平展菲利木●☆

296285　Phylica plumosa L. var. squarrosa（Vent.）Sond.；粗鳞菲利木●☆

296286　Phylica propinqua Sond.；邻近菲利木●☆

296287　Phylica pubescens Aiton；短柔毛菲利木●☆

296288　Phylica pubescens Aiton var. angustifolia（Sond.）Pillans；窄叶菲利木●☆

296289　Phylica pubescens Aiton var. orientalis Pillans；东方菲利木●☆

296290　Phylica pulchella Schltr.；美丽菲利木●☆

296291　Phylica purpurea Sond.；紫菲利木●☆

296292　Phylica purpurea Sond. var. floccosa Pillans；丛卷毛菲利木●☆

296293　Phylica purpurea Sond. var. reclinata（J. C. Wendl.）Sond. = Phylica pinea Thunb.●☆

296294　Phylica pustulata E. Phillips；泡状菲利木●☆

296295　Phylica radiata L. = Staavia radiata（L.）Dahl●☆

296296　Phylica reclinata Bernh. ex Krauss = Phylica curvifolia Pillans●☆

296297　Phylica reclinata J. C. Wendl. = Phylica pinea Thunb.●☆

296298　Phylica recurvifolia Eckl. et Zeyh.；曲叶菲利木●☆

296299　Phylica reflexa Lam. = Phylica dioica L.●☆

296300　Phylica retorta Pillans；反折菲利木●☆

296301　Phylica retrorsa E. Mey. ex Sond.；倒向菲利木●☆

296302　Phylica reversa Pillans；倒转菲利木●☆

296303　Phylica rigida Eckl. et Zeyh.；硬菲利木●☆

296304　Phylica rigidifolia Sond.；硬叶菲利木●☆

296305　Phylica rogersii Pillans；罗杰斯菲利木●☆

296306　Phylica rubra Willd. ex Roem. et Schult.；红菲利木●☆

296307　Phylica salteri Pillans；索尔特菲利木●☆

296308　Phylica schlechteri Pillans；施莱菲利木●☆

296309　Phylica selaginoides Sond.；石松菲利木●☆

296310　Phylica sericea Pillans；绢毛菲利木●☆

296311　Phylica simii Pillans；西姆菲利木●☆

296312　Phylica spicata L. f.；穗花菲利木●☆

296313　Phylica spicata L. f. var. piquetbergensis Pillans；皮克特穗花菲利木●☆

296314 Phylica squarrosa Vent. = Phylica plumosa L. var. squarrosa (Vent.) Sond. ●☆

296315 Phylica stenantha Pillans;窄花菲利木●☆

296316 Phylica stenopetala Schltr. ;窄瓣菲利木●☆

296317 Phylica stenopetala Schltr. var. sieberi Pillans;西伯尔菲利木●☆

296318 Phylica stipularis L. = Trichocephalus stipularis (L.) Brongn. ●☆

296319 Phylica stokoei Pillans;斯托克菲利木●☆

296320 Phylica strigosa P. J. Bergius;糙伏毛菲利木●☆

296321 Phylica strigosa P. J. Bergius var. australis Pillans;南方糙伏毛菲利木●☆

296322 Phylica strigosa P. J. Bergius var. dregei Pillans;德雷糙伏毛菲利木●☆

296323 Phylica strigosa P. J. Bergius var. elongata Pillans;伸长菲利木●☆

296324 Phylica strigosa P. J. Bergius var. macowanii Pillans;麦克欧文菲利木●☆

296325 Phylica strigulosa Sond. ;硬毛菲利木●☆

296326 Phylica subulifolia Pillans;钻叶菲利木●☆

296327 Phylica thodei E. Phillips;索德菲利木●☆

296328 Phylica thunbergiana Sond. ;通贝里菲利木●☆

296329 Phylica tortuosa E. Mey. ex Harv. et Sond. ;扭曲菲利木●☆

296330 Phylica trachyphylla (Eckl. et Zeyh.) D. Dietr. ;糙叶菲利木●☆

296331 Phylica trichotoma Thunb. = Staavia capitella (Thunb.) Sond. ●☆

296332 Phylica tropica Baker = Phylica emirnensis (Tul.) Pillans ●☆

296333 Phylica tuberculata Pillans;多疣菲利木●☆

296334 Phylica tubulosa Schltr. ;管状菲利木●☆

296335 Phylica tysonii Pillans;泰森菲利木●☆

296336 Phylica tysonii Pillans var. brevifolia ?;短叶泰森菲利木●☆

296337 Phylica variabilis Pillans;易变菲利木●☆

296338 Phylica velutina Sond. ;短绒毛菲利木●☆

296339 Phylica villosa Thunb. ;长柔毛菲利木●☆

296340 Phylica villosa Thunb. var. pedicellata (DC.) Sond. ;梗花长柔毛菲利木●☆

296341 Phylica villosa Thunb. var. squarrosa Sond. = Phylica axillaris Lam. var. microphylla (Eckl. et Zeyh.) Pillans ●☆

296342 Phylica virgata (Eckl. et Zeyh.) D. Dietr. ;条纹菲利木●☆

296343 Phylica vulgaris Pillans;普通菲利木●☆

296344 Phylica vulgaris Pillans var. major ?;大菲利木●☆

296345 Phylica willdenowiana Eckl. et Zeyh. ;威尔菲利木●☆

296346 Phylica wittebergensis Pillans;维特伯格菲利木●☆

296347 Phylicaceae J. Agardh = Rhamnaceae Juss. (保留科名)●

296348 Phylicaceae J. Agardh;菲利木科●☆

296349 Phylidraceae Lindl. = Philydraceae Link(保留科名)■

296350 Phylidrum Willd. = Philydrum Banks ex Gaertn. ■

296351 Phylirastrum (Pierre) Pierre = Caloncoba Gilg ●☆

296352 Phyllacantha Hook. f. = Phyllacanthus Hook. f. ■☆

296353 Phyllacanthus Hook. f. (1871);刺叶茜属■☆

296354 Phyllacanthus grisebachianus Hook. f. ;刺叶茜■☆

296355 Phyllachne J. R. Forst. et G. Forst. (1775);球垫花柱草属(叶壳草属)■☆

296356 Phyllachne rubra Cheeseman;红球垫花柱草■☆

296357 Phyllachne uliginosa J. R. Forst. et G. Forst. ;球垫花柱草■☆

296358 Phyllactinia Benth. = Pasaccardoa Kuntze ■●☆

296359 Phyllactinia grantii Benth. ex Oliv. = Pasaccardoa grantii (Benth. ex Oliv.) Kuntze ■☆

296360 Phyllactis Pers. (1805);叶线草属■☆

296361 Phyllactis Pers. = Valeriana L. ●■

296362 Phyllactis spathulata Pers. ;叶线草■☆

296363 Phyllactis tenuifolia Pers. ;细叶叶线草■☆

296364 Phyllactts Steud. = Philactis Schrad. ●☆

296365 Phyllagathis Blume(1831);锦香草属(金锦香属);Metalleaf ●■

296366 Phyllagathis anisophylla Diels;毛柄锦香草;Hairy-stalk Metalleaf, Unequalleaf Metalleaf, Unequal-leaved Metalleaf ●

296367 Phyllagathis anisophylla Diels = Phyllagathis oligotricha Merr. ●

296368 Phyllagathis asarifolia C. Chen;细辛锦香草;Wildginger Metalleaf, Wildgingerleaf Metalleaf ■

296369 Phyllagathis calisaurea C. Chen;金盏锦香草;Golden Metalleaf, Goldencup Metalleaf ●

296370 Phyllagathis calisaurea C. Chen = Phyllagathis ovalifolia H. L. Li ●■

296371 Phyllagathis cavalerii (H. Lév. et Vaniot) Guillaumin;锦香草(猫耳朵草,铺地毡,熊巴耳,熊巴掌);Cavalerie Metalleaf, Metalleaf ■

296372 Phyllagathis cavalerii (H. Lév. et Vaniot) Guillaumin var. tankahkeei (Merr.) C. Y. Wu = Phyllagathis cavalerii (H. Lév. et Vaniot) Guillaumin ■

296373 Phyllagathis cavalerii (H. Lév. et Vaniot) Guillaumin var. tankahkeei (Merr.) C. Y. Wu ex C. Chen;短毛锦香草(豆角消,短毛熊巴掌,虎耳,猫耳朵,熊巴耳,叶下红,猪婆耳);Shorthair Metalleaf, Shorthairy Metalleaf ■

296374 Phyllagathis cavalerii (H. Lév. et Vaniot) Guillaumin var. wilsoniana Guillaumin = Phyllagathis cavalerii (H. Lév. et Vaniot) Guillaumin ■

296375 Phyllagathis cavalerii (H. Lév. et Vaniot) Guillaumin var. wilsoniana Guillaumin;长柄锦香草(长柄熊巴掌);Long Stalk Metalleaf, Long-stalked Metalleaf ●

296376 Phyllagathis chinensis Dunn = Sarcopyramis nepalensis Wall. ■

296377 Phyllagathis cymigera C. Chen;聚伞锦香草;Cyme Metalleaf, Cymose Metalleaf ●

296378 Phyllagathis dajiakensis Ohwi;达贾锦香草●☆

296379 Phyllagathis deltoda C. Chen;三角齿锦香草;Triangletooth Metalleaf, Triangular Metalleaf ●

296380 Phyllagathis elattandra Diels;广东锦香草(碑边救生,大崩沙,红敷地发,石发,石莲);Guangdong Metalleaf, Rocklotus Metalleaf ■

296381 Phyllagathis erecta (S. Y. Hu) C. Y. Wu ex C. Chen;直立锦香草(直立无距花);Erect Metalleaf ●■

296382 Phyllagathis erythrotricha Merr. et Chun = Scorpiothyrsus erythrotrichus (Merr. et Chun) H. L. Li ●

296383 Phyllagathis fengii C. Hansen;刚毛锦香草;Hispid Metalleaf ●

296384 Phyllagathis fordii (Hance) C. Chen = Bredia fordii (Hance) Diels ●

296385 Phyllagathis fordii (Hance) C. Chen var. micrantha C. Chen;小花叶底红;Small-flower Ford Metalleaf ●

296386 Phyllagathis fordii (Hance) C. Chen var. micrantha C. Chen. = Bredia fordii (Hance) Diels ●

296387 Phyllagathis gracilis (Hand. -Mazz.) C. Chen;细梗锦香草;Slender Metalleaf, Thinstipe Metalleaf ●■

296388 Phyllagathis hainanensis (Merr. et Chun) C. Chen;海南锦香草;Hainan Metalleaf ●

296389 Phyllagathis hispida (S. Y. Hu) C. Y. Wu = Phyllagathis fengii C. Hansen ●

296390 Phyllagathis hispida (S. Y. Hu) C. Y. Wu ex C. Chen = Phyllagathis fengii C. Hansen ●

296391 Phyllagathis hispidissima (C. Chen) C. Chen;密毛锦香草(密毛野海棠);Densehair Bredia, Dense-haired Metalleaf, Dense-hairs Metalleaf, Densehispid Metalleaf ●

296392 Phyllagathis latisepala C. Chen;宽萼锦香草;Broadcalyx

Metalleaf，Broad-sepal Metalleaf ●■

296393　Phyllagathis longearistata C. Chen；长芒锦香草；Long-aristate Metalleaf，Longawn Metalleaf，Longbeard Metalleaf ●

296394　Phyllagathis longipes H. L. Li = Phyllagathis cavaleriei（H. Lév. et Vaniot）Guillaumin ■

296395　Phyllagathis longipes H. L. Li = Phyllagathis cavaleriei（H. Lév. et Vaniot）Guillaumin var. wilsoniana Guillaumin ■

296396　Phyllagathis longiradiosa（C. Chen）C. Chen；大叶熊巴掌（大叶野海棠）；Bigleaf Bredia，Big-leaved Bredia ●■

296397　Phyllagathis longiradiosa（C. Chen）C. Chen var. pulchella C. Chen；丽萼熊巴掌；Beautiful Bigleaf Bredia ●■

296398　Phyllagathis melastomatoides（Merr. et Chun）W. C. Ko；毛锦香草；Hair Metalleaf，Melastoma-like Metalleaf ●

296399　Phyllagathis melastomatoides（Merr. et Chun）W. C. Ko var. brevipes W. C. Ko；短柄锦香草（短柄毛锦香草）；Short-stalk Hairy Metalleaf，Short-stalk Melastoma-like Metalleaf ●

296400　Phyllagathis nudipes C. Chen；秃柄锦香草；Baldstipe Metalleaf，Bare-stalked Metalleaf，Naked Stalk Metalleaf，Smooth-stalk Metalleaf ●

296401　Phyllagathis nudipes C. Chen = Phyllagathis oligotricha Merr. ●

296402　Phyllagathis oligotricha Merr. = Phyllagathis anisophylla Diels ●

296403　Phyllagathis oligotricha Merr. ex Merr. et Chun = Phyllagathis anisophylla Diels ●

296404　Phyllagathis oligotricha Merr. ex Merr. et Chun = Phyllagathis oligotricha Merr. ●

296405　Phyllagathis ovalifolia H. L. Li；卵叶锦香草；Ovate Metalleaf，Ovate-leaf Metalleaf，Ovate-leaved Metalleaf ●■

296406　Phyllagathis ovalifolia H. L. Li var. pauciflora R. H. Miao；少花卵叶锦香草；Fewflower Ovate-leaf Metalleaf ■

296407　Phyllagathis ovalifolia H. L. Li var. pauciflora R. H. Miao = Phyllagathis ovalifolia H. L. Li ●■

296408　Phyllagathis plagiopetala C. Chen；偏斜锦香草（水角风）；Oblique Metalleaf，Oblique Petal Metalleaf ●

296409　Phyllagathis pluriumbellata R. H. Miao；伞花锦香草●

296410　Phyllagathis pluriumbellata R. H. Miao = Phyllagathis cymigera C. Chen ●

296411　Phyllagathis scorpiothyrsoides C. Chen；斑叶锦香草；Spotleaf Metalleaf，Spotted Metalleaf ■

296412　Phyllagathis setotheca H. L. Li；刺蕊锦香草；Setoseconnective Metalleaf，Setose-connective Metalleaf，Spiny Metalleaf，Spotted Metalleaf ●

296413　Phyllagathis setotheca H. L. Li var. setotuba C. Chen；毛萼锦香草；Hairy Tube Metalleaf，Hairy-sepal Metalleaf ●

296414　Phyllagathis setotheca H. L. Li var. setotuba C. Chen = Phyllagathis setotheca H. L. Li ●

296415　Phyllagathis stenophylla（Merr. et Chun）H. L. Li；窄叶锦香草；Narrow Metalleaf，Narrow-leaf Metalleaf，Stenophyllous Metalleaf ●

296416　Phyllagathis tankahkeei Merr. = Phyllagathis cavaleriei（H. Lév. et Vaniot）Guillaumin var. tankahkeei（Merr.）C. Y. Wu ex C. Chen ■

296417　Phyllagathis tankahkeei Merr. = Phyllagathis cavaleriei（H. Lév. et Vaniot）Guillaumin ■

296418　Phyllagathis tentaculifera C. Hansen；须花锦香草●

296419　Phyllagathis tenuicaulis C. Chen = Plagiopetalum tenuicaule（C. Chen）C. Hansen ■

296420　Phyllagathis ternata C. Chen；三瓣锦香草；Ternate Metalleaf，Threepetal Metalleaf，Tripetaled Metalleaf ●■

296421　Phyllagathis tetrandra Diels；四蕊熊巴掌；Four Stamens Metalleaf，Fourstamen Metalleaf ■

296422　Phyllagathis velutina（Diels）C. Chen；腺毛锦香草；Glandhair Metalleaf，Velety Metalleaf，Velvety Metalleaf ●■

296423　Phyllagathis wenshanensis S. Y. Hu；猫耳朵；Catear Metalleaf，Wenshan Metalleaf ■

296424　Phyllagathis wenshanensis S. Y. Hu = Phyllagathis cavaleriei（H. Lév. et Vaniot）Guillaumin ■

296425　Phyllagathis xanthosticta Merr. et Chun = Scorpiothyrsus xanthostictus（Merr. et Chun）H. L. Li ●◇

296426　Phyllagathis xinyiensis Z. J. Feng；信宜锦香草；Xinyi Metalleaf ●

296427　Phyllagathis xinyiensis Z. J. Feng = Phyllagathis hispidissima（C. Chen）C. Chen ●

296428　Phyllamphora Lour. = Nepenthes L. ●■

296429　Phyllamphora mirabilis Lour. = Nepenthes mirabilis（Lour.）Druce ●■

296430　Phyllangium Dunlop = Mitrasacme Labill. ■

296431　Phyllangium Dunlop（1996）；澳姬苗属■☆

296432　Phyllanoa Croizat（1943）；哥伦比亚大戟属☆

296433　Phyllanoa colombiana Croizat；哥伦比亚大戟 ☆

296434　Phyllanthaceae J. Agardh = Euphorbiaceae Juss.（保留科名）●■

296435　Phyllanthaceae Martinov = Euphorbiaceae Juss.（保留科名）●■

296436　Phyllanthaceae Martinov（1820）；叶下珠科（叶萝藦科）●■

296437　Phyllanthera Blume（1827）；叶药萝藦属●☆

296438　Phyllanthera bifida Blume；叶药萝藦●☆

296439　Phyllanthera multinervosa（P. I. Forst.）Venter；多脉叶药萝藦●☆

296440　Phyllantherum Raf. = Trillium L. ■

296441　Phyllanthidea Didr. = Andrachne L. ●☆

296442　Phyllanthodendron Hemsl.（1898）；珠子木属（叶珠木属，余甘树属）；Pearlwood，Phyllanthodendron ●

296443　Phyllanthodendron Hemsl. = Phyllanthus L. ●■

296444　Phyllanthodendron album Craib et Hosseus = Phyllanthodendron roseum Craib et Hutch. ●

296445　Phyllanthodendron anthopotamicum（Hand.-Mazz.）Croizat；珠子木（花池叶下珠，花溪珠子木，叶下花，叶珠木，鱼骨树）；Common Phyllanthodendron，Pearlwood ●

296446　Phyllanthodendron anthopotamicum Hand.-Mazz. = Phyllanthodendron anthopotamicum（Hand.-Mazz.）Croizat ●

296447　Phyllanthodendron breynioides P. T. Li；龙州珠子木（黑面神叶珠子木）；Longzhou Pearlwood，Longzhou Phyllanthodendron ●

296448　Phyllanthodendron caudatifolium P. T. Li；尾叶珠子木；Caudate-leaf Pearlwood，Caudate-leaf Phyllanthodendron ●

296449　Phyllanthodendron cavaleriei H. Lév. = Phyllanthodendron dunnianum H. Lév. ●

296450　Phyllanthodendron dunnianum（H. Lév.）Hand.-Mazz. = Phyllanthodendron dunnianum H. Lév. ●

296451　Phyllanthodendron dunnianum H. Lév.；枝翅珠子木（枝翅叶下珠）；Dunn's Pearlwood，Dunn's Phyllanthodendron ●

296452　Phyllanthodendron dunnianum H. Lév. var. hypoglaucum H. Lév. = Phyllanthodendron dunnianum H. Lév. ●

296453　Phyllanthodendron hypoglaucum H. Lév.；粉背叶下珠子木；Hypoglaucous Pearlwood，Hypoglaucous Phyllanthodendron ●

296454　Phyllanthodendron lativenium Croizat；宽脉珠子木；Thick-nerve Pearlwood，Thick-nerve Phyllanthodendron ●

296455　Phyllanthodendron mirabile Hemsl.；泰国珠子木●☆

296456　Phyllanthodendron moi（P. T. Li）P. T. Li；弄岗珠子木（弄岗叶下珠）；Longgang Leaf-flower，Longgang Pearlwood ●

296457　Phyllanthodendron orbicularifolium P. T. Li；圆叶珠子木；Round-leaf Pearlwood，Round-leaf Phyllanthodendron ●

296458　Phyllanthodendron petraeum P. T. Li；岩生珠子木；Rockliving Pearlwood，Rockliving Phyllanthodendron ●

296459　Phyllanthodendron roseum （Craib et Hutch.）Beille = Phyllanthodendron roseum Craib et Hutch. ●

296460　Phyllanthodendron roseum Craib et Hutch.；玫花珠子木；Pink Phyllanthodendron，Rose Pearlwood，Rose Phyllanthodendron ●

296461　Phyllanthodendron roseum Craib et Hutch. var. glabrum Craib ex Hosseus = Phyllanthodendron roseum Craib et Hutch. ●

296462　Phyllanthodendron yunnanense Croizat；云南珠子木（滇珠子木）；Yunnan Pearlwood，Yunnan Phyllanthodendron ●

296463　Phyllanthopsis （Scheele）Voronts. et Petra Hoffm. （2008）；拟叶下珠属●■☆

296464　Phyllanthos St. -Lag. = Phyllanthus L. ●■

296465　Phyllanthus L. （1753）；叶下珠属（油甘属，油柑属）；Leaf Flower，Leaf-flower，Underleaf Pearl ●■

296466　Phyllanthus acidus （L.）Skeels；酸果叶下珠（酸叶下珠，西印度醋栗）；Indian Gooseberry，Otaheite Gooseberry，Otaheite Gooseberry，Otaheite Gooseberryleaf Flower，Otaheite-gooseberry Leal Flower，Star Gooseberry，Tahitian Gooseberry Tree ●☆

296467　Phyllanthus acidus Skeels = Phyllanthus acidus （L.）Skeels ●☆

296468　Phyllanthus acuminatus Vahl；渐尖叶下珠●☆

296469　Phyllanthus alpestris Beille；高山叶下珠●☆

296470　Phyllanthus amapondensis Sim = Margaritaria discoidea （Baill.）G. L. Webster var. fagifolia （Pax）Radcl. -Sm. ●☆

296471　Phyllanthus amarus K. Schum. et Thonn.；苦味叶下珠（苦叶叶下珠，小返魂）；Bitter Leaf-flower，Carry Me Seed ●■

296472　Phyllanthus andersonii Müll. Arg. = Glochidion assamicum （Müll. Arg.）Hook. f. ●

296473　Phyllanthus angolensis Müll. Arg.；安哥拉叶下珠●☆

296474　Phyllanthus angustatus Hutch. = Phyllanthus friesii Hutch. ●☆

296475　Phyllanthus angustifolius Sw.；狭叶叶下珠；Foliage Flower ●☆

296476　Phyllanthus annamensis Beille；崖县叶下珠；Yaxian Leaf-flower ●

296477　Phyllanthus annamensis Beille = Phyllanthus cochinchinensis （Lour.）Spreng. ●

296478　Phyllanthus anthopotamicum Hand. -Mazz. = Phyllanthodendron anthopotamicum （Hand. -Mazz.）Croizat ●

296479　Phyllanthus arborescens （Blume）Müll. Arg. = Glochidion arborescens Blume ●

296480　Phyllanthus arenarius Beille；沙地叶下珠；Sand Leaf-flower，Sand Underleaf Pearl ●■

296481　Phyllanthus arenarius Beille var. yunnanensis T. L. Chin；云南沙地叶下珠；Yunnan Underleaf Sand pearl ●■

296482　Phyllanthus argyi H. Lév. = Flueggea suffruticosa （Pall.）Baill. ●

296483　Phyllanthus arnottianus （Müll. Arg.）Müll. Arg. = Glochidion hirsutum （Roxb.）Voigt ●

296484　Phyllanthus arvensis Müll. Arg.；田野叶下珠●☆

296485　Phyllanthus aspericaulis Pax = Phyllanthus rotundifolius Klein ex Willd. ●☆

296486　Phyllanthus asperulatus Hutch.；糙叶下珠●☆

296487　Phyllanthus asperus Brunel et J. P. Roux；粗糙叶下珠●☆

296488　Phyllanthus assamicus Müll. Arg. = Glochidion assamicum （Müll. Arg.）Hook. f. ●

296489　Phyllanthus asteranthos Croizat = Phyllanthus pulcher Wall. ex Müll. Arg. ●

296490　Phyllanthus bacciformis L. = Sauropus bacciformis （L.）Airy Shaw ■

296491　Phyllanthus beillei Hutch.；贝勒叶下珠●☆

296492　Phyllanthus benguelensis Müll. Arg.；本格拉叶下珠●☆

296493　Phyllanthus bequaertii Robyns et Lawalrée；贝卡尔叶下珠●☆

296494　Phyllanthus bernierianus Baill. ex Müll. Arg.；伯尼尔叶下珠●☆

296495　Phyllanthus bernierianus Baill. ex Müll. Arg. var. glaber Radcl. -Sm.；光滑叶下珠●☆

296496　Phyllanthus bicolor Müll. Arg. = Glochidion triandrum （Blanco）C. B. Rob. ●

296497　Phyllanthus bodinieri （H. Lév.）Rehder；贵州叶下珠；Bodinier Leaf-flower，Guizhou Leafflower，Guizhou Underleaf Pearl ●

296498　Phyllanthus boehmii Pax；贝姆叶下珠●☆

296499　Phyllanthus boehmii Pax var. humilis Radcl. -Sm.；低矮叶下珠●☆

296500　Phyllanthus boninsimae Nakai = Phyllanthus deblis Klein ex Willd. ●☆

296501　Phyllanthus borenensis M. G. Gilbert；北方叶下珠●☆

296502　Phyllanthus brandegei Millsp.；波氏叶下珠●☆

296503　Phyllanthus brasiliensis Müll. Arg.；巴西叶下珠●☆

296504　Phyllanthus braunii Pax；布劳恩叶下珠●☆

296505　Phyllanthus brazzae Brunel = Phyllanthus dinklagei Pax ●☆

296506　Phyllanthus burchellii Müll. Arg. = Phyllanthus parvulus Sond. var. garipensis （E. Mey. ex Drège）Radcl. -Sm. ●☆

296507　Phyllanthus buxifolius （Blume）Müll. Arg.；黄杨叶下珠●☆

296508　Phyllanthus caesiifolius Petra Hoffm. et Cheek；淡蓝叶下珠●☆

296509　Phyllanthus caespitosus Brenan；丛生叶下珠●☆

296510　Phyllanthus camerunensis Brunel；喀麦隆叶下珠●☆

296511　Phyllanthus cantoniensis Hornem. = Phyllanthus urinaria L. ●■

296512　Phyllanthus cantqniensis Schweigg. = Phyllanthus urinaria L. ●■

296513　Phyllanthus capillariformis Vatke et Pax = Meineckia phyllanthoides Baill. subsp. capillariformis （Vatke et Pax）G. L. Webster ■☆

296514　Phyllanthus capillaris Schumach. et Thonn. = Phyllanthus nummulariifolius Poir. var. capillaris （Schumach. et Thonn.）Radcl. -Sm. ●☆

296515　Phyllanthus caroliniensis Walter；卡罗来纳叶下珠；Leaf-flower ■☆

296516　Phyllanthus cedrelifolius I. Verd. = Phyllanthus polyanthus Pax ●☆

296517　Phyllanthus chekiangensis Croizat et Metcalf；浙江叶下珠；Zhejiang Leafflower，Zhejiang Leaf-flower，Zhejiang Underleaf Pearl ●◇

296518　Phyllanthus chevalieri Beille；舍瓦利耶叶下珠●☆

296519　Phyllanthus cinerascens Hook. et Arn. = Phyllanthus cochinchinensis （Lour.）Spreng. ●

296520　Phyllanthus cinereoviridis Pax；灰绿叶下珠●☆

296521　Phyllanthus clarkei Hook. f.；滇藏叶下珠（刺果叶下珠，思茅叶下珠，云南叶下珠）；C. B. Clarke Underleaf Pearl，C. B. Clarke's Leafflower，Clarke Leaf-flower，Forrest Leaf-flower，Spiny-fruited Leaf-flower ●

296522　Phyllanthus cochinchinensis （Lour.）Spreng.；越南叶下珠（苍蝇草，狗脚迹，铁扫把，乌蝇叶，乌蝇翼，崖县叶下珠）；Cochinchina Leafflower，Cochin-China Leaf-flower，Viet Nam Leaf-flower，Vietnam Underleaf Pearl ●

296523　Phyllanthus cochinchinensis Müll. Arg. = Phyllanthus cochinchinensis （Lour.）Spreng. ●

296524　Phyllanthus compressicaulis （Kurz ex Teijsm. et Binn.）Müll. Arg. = Glochidion philippicum （Cav.）C. B. Rob. ●

296525　Phyllanthus concinnus Ridl. = Phyllanthus gracilipes （Miq.）Müll. Arg. ●

296526　Phyllanthus confusus Brenan；混乱叶下珠●☆

296527　Phyllanthus corcovadensis Müll. Arg. = Phyllanthus tenellus Roxb. ●

296528　Phyllanthus crassinervius Radcl. -Sm. ；粗脉叶下珠●☆

296529　Phyllanthus daltonii Müll. Arg. = Glochidion daltonii （Müll. Arg.）Kurz ●

296530　Phyllanthus debilis Klein ex Willd. ；锐叶小还魂；Niruri ●☆

296531　Phyllanthus dekindtii Hutch. ；德金叶下珠●☆

296532　Phyllanthus delagoensis Hutch. ；迪拉果叶下珠●☆

296533　Phyllanthus delpyanus Hutch. = Phyllanthus polyanthus Pax ●☆

296534　Phyllanthus dewildeorum M. G. Gilbert；德维尔德叶下珠●☆

296535　Phyllanthus diandrus Pax；双蕊叶下珠●☆

296536　Phyllanthus dictyophlebs Radcl. -Sm. ；网状叶下珠●☆

296537　Phyllanthus dinklagei Pax；丁克叶下珠●☆

296538　Phyllanthus dinteri Pax；丁特叶下珠●☆

296539　Phyllanthus discofractus Croizat = Phyllanthus gracilipes （Miq.）Müll. Arg. ●

296540　Phyllanthus discoides （Baill.）Müll. Arg. ；圆盘叶下珠●☆

296541　Phyllanthus discoides （Baill.）Müll. Arg. = Margaritaria discoidea （Baill.）G. L. Webster ●☆

296542　Phyllanthus discoides Müll. Arg. = Phyllanthus discoides （Baill.）Müll. Arg. ●☆

296543　Phyllanthus distichus （L.）Müll. Arg. ；二列叶下珠；Otaheite Gooseberry ●☆

296544　Phyllanthus distichus （L.）Müll. Arg. = Phyllanthus acidus （L.）Skeels ●☆

296545　Phyllanthus diversifolius Miq. = Glochidion rubrum Blume ●

296546　Phyllanthus dongfangensis P. T. Li；后生叶下珠；Oriental Leafflower，Oriental Underleaf Pearl ●

296547　Phyllanthus dongfangensis P. T. Li = Cleistanthus dongfangensis （P. T. Li）H. S. Kiu ●

296548　Phyllanthus dregeanus Scheele = Andrachne ovalis （E. Mey. ex Sond.）Müll. Arg. ●☆

296549　Phyllanthus dunnianus （H. Lév.）Hand. -Mazz. = Phyllanthodendron dunnianum H. Lév. ●

296550　Phyllanthus dusenii Hutch. ；杜森叶下珠●☆

296551　Phyllanthus echinocarpus T. L. Chin = Phyllanthus forestii W. W. Sm. ●

296552　Phyllanthus eliae （Brunel et J. P. Roux）Brunel ex Govaerts et Radcl. -Sm. ；埃利亚叶下珠●☆

296553　Phyllanthus embergeri Haicour et Rossignol；拟叶下珠●

296554　Phyllanthus emblica L. ；余甘子（庵罗果，庵摩簕，庵罗迦果，庵摩勒，庵摩罗迦果，庵摩落迦果，滇橄榄，橄榄，黑面长，喉甘子，米含，牛甘子，实瓶，望果，油甘，油甘子，油柑，油柑树，油柑子，余甘，余柑子，鱼木果）；Ambal，Amioki，Emblic，Emblic Leaf Flower，Emblic Leafflower，Emblic Leaf-flower，Emblic Underleaf Pearl，Myrobalan，Woody Leaf Flower ●

296555　Phyllanthus emblica L. = Emblica officinalis Gaertn. ●☆

296556　Phyllanthus engleri Pax；恩格勒叶下珠（毒叶下珠）●☆

296557　Phyllanthus eriocarpus （Champ. ex Benth.）Müll. Arg. = Glochidion eriocarpum Champ. ex Benth. ●

296558　Phyllanthus eylesii S. Moore = Phyllanthus leucanthus Pax ●☆

296559　Phyllanthus fagifolius （Miq.）Müll. Arg. = Glochidion sphaerogynum （Müll. Arg.）Kurz ●

296560　Phyllanthus fangchengensis P. T. Li；尖叶下珠；Fanchen Leafflower，Fanchen Leaf-flower，Fanchen Underleaf Pearl ●

296561　Phyllanthus fasciculatus Müll. Arg. = Phyllanthus cochinchinensis （Lour.）Spreng. ●

296562　Phyllanthus fimbricalyx P. T. Li；穗萼叶下珠（流萼叶下珠）；Fimbricalyx Leafflower，Underleaf Pearl ●

296563　Phyllanthus fischeri Pax；菲舍尔叶下珠●☆

296564　Phyllanthus flacourtioides Hutch. = Margaritaria discoidea （Baill.）G. L. Webster var. nitida （Pax）Radcl. -Sm. ●☆

296565　Phyllanthus flexuosus （Siebold et Zucc.）Müll. Arg. ；落萼叶下珠（河杆巴，红五眼，红鱼眼，漂浮叶下珠，曲折叶下珠，曲枝叶下珠，弯曲叶下珠，细鱼眼）；Flexuosus Leptopus，Sinuate Leaf-flower，Underleaf Pearl ●

296566　Phyllanthus floribundus Müll. Arg. = Phyllanthus muellerianus （Kuntze）Exell ●☆

296567　Phyllanthus flueggeiformis Müll. Arg. = Phyllanthus glaucus Wall. ex Müll. Arg. ●

296568　Phyllanthus fluggeoides Müll. Arg. = Flueggea suffruticosa （Pall.）Baill. ●

296569　Phyllanthus fluitans Müll. Arg. ；南美叶下珠●☆

296570　Phyllanthus fluminis-athi Radcl. -Sm. ；河流叶下珠●☆

296571　Phyllanthus forestii W. W. Sm. ；刺果叶下珠（云南叶下珠）；Forrest Underleaf Pearl ●

296572　Phyllanthus forestii W. W. Sm. = Phyllanthus clarkei Hook. f. ●

296573　Phyllanthus franchetianus H. Lév. ；云贵叶下珠（成凤叶下珠，雷波叶下珠）；Franchet Leaf-flower，Franchet Leptopus，Franchet Underleaf Pearl ●

296574　Phyllanthus fraternus G. L. Webster；海湾叶下珠；Gulf Leaf-flower ●☆

296575　Phyllanthus fraternus G. L. Webster subsp. togoensis Brunel et J. P. Roux = Phyllanthus fraternus G. L. Webster ●☆

296576　Phyllanthus friesii Hutch. ；弗里斯叶下珠●☆

296577　Phyllanthus gagnioevae Brunel et J. P. Roux；甘吉叶下珠●☆

296578　Phyllanthus garipensis E. Mey. ex Drège = Phyllanthus parvulus Sond. var. garipensis （E. Mey. ex Drège）Radcl. -Sm. ●☆

296579　Phyllanthus gasstroemii Müll. Arg. ；格司叶下珠●☆

296580　Phyllanthus geniculatostemon Brunel = Phyllanthus goniostemon Radcl. -Sm. ●☆

296581　Phyllanthus genistoides Sond. = Phyllanthus incurvus Thunb. ●☆

296582　Phyllanthus gilletii De Wild. = Phyllanthus macranthus Pax var. gilletii （De Wild.）Brunel ex Radcl. -Sm. ●☆

296583　Phyllanthus gillettianus Brunel ex Radcl. -Sm. ；吉莱特叶下珠●☆

296584　Phyllanthus glabrocapsulus Metcalf = Phyllanthus leptoclados F. P. Metcalf ●

296585　Phyllanthus glaucescens Kunth；灰色叶下珠●☆

296586　Phyllanthus glaucophyllus Sond. ；灰叶下珠●☆

296587　Phyllanthus glaucophyllus Sond. var. major Müll. Arg. = Phyllanthus glaucophyllus Sond. ●☆

296588　Phyllanthus glaucophyllus Sond. var. suborbicularis Hutch. = Phyllanthus glaucophyllus Sond. ●☆

296589　Phyllanthus glaucus Wall. ex Müll. Arg. ；青灰叶下珠；Greyblue Leafflower，Grey-blue Leaf-flower，Greyblue Underleaf Pearl ●

296590　Phyllanthus goniocladus Merr. et Chun = Sauropus bacciformis （L.）Airy Shaw ■

296591　Phyllanthus goniostemon Radcl. -Sm. ；膝冠叶下珠●☆

296592　Phyllanthus gossweileri Hutch. ；戈斯叶下珠●☆

296593　Phyllanthus gracilipes （Miq.）Müll. Arg. ；毛果叶下珠；Hairfruit Underleaf Pearl，Hairy-fruited Leaf-flower ●

296594　Phyllanthus grahamii Hutch. et M. B. Moss = Phyllanthus beillei Hutch. ●☆

296595　Phyllanthus graminicola Hutch. ex S. Moore；草莺叶下珠●☆

296596　Phyllanthus guangdongensis P. T. Li;隐脉叶下珠;Guangdong Leaf-flower, Obscurevein Leafflower, Obscurevein Underleaf Pearl ●

296597　Phyllanthus guenzii Müll. Arg. = Phyllanthus maderaspatensis L. ●☆

296598　Phyllanthus guineensis Pax = Phyllanthus ovalifolius Forssk. ●☆

296599　Phyllanthus hainanensis Merr. ;海南叶下珠(海南油柑); Hainan Leafflower, Hainan Leaf-flower, Hainan Underleaf Pearl ●

296600　Phyllanthus hamiltonianus Müll. Arg. = Phyllanthus sikkimensis Müll. Arg. ●☆

296601　Phyllanthus harrisii Radcl. -Sm. ;哈里斯叶下珠●☆

296602　Phyllanthus heterophyllus E. Mey. ex Müll. Arg. ;互生叶下珠●☆

296603　Phyllanthus hildebrandtii Pax;希尔德叶下珠●☆

296604　Phyllanthus hirsutus (Roxb.) Müll. Arg. = Glochidion hirsutum (Roxb.) Voigt ●

296605　Phyllanthus holostylus Milne-Redh. ;全柱叶下珠●☆

296606　Phyllanthus hongkongensis (Müll. Arg.) Müll. Arg. = Glochidion zeylanicum (Gaertn.) A. Juss. ●

296607　Phyllanthus hookeri Müll. Arg. ;疣果叶下珠●

296608　Phyllanthus hookeri Müll. Arg. = Phyllanthus lepidocarpus Siebold et Zucc. ●

296609　Phyllanthus hookeri Müll. Arg. = Phyllanthus tsarongensis W. W. Sm. ●

296610　Phyllanthus hutchinsonianus S. Moore;哈钦森叶下珠●☆

296611　Phyllanthus hypoleucus (Miq.) Müll. Arg. = Glochidion lutescens Blume ●

296612　Phyllanthus incurvus Thunb. ;弯曲叶下珠●☆

296613　Phyllanthus indicus (Dalzell) Müll. Arg. = Margaritaria indica (Dalzell) Airy Shaw ●

296614　Phyllanthus inflatus Hutch. ;膨胀叶下珠●☆

296615　Phyllanthus irriguus Radcl. -Sm. ;贮水叶下珠●☆

296616　Phyllanthus jaegeri Brunel et J. P. Roux;耶格叶下珠●☆

296617　Phyllanthus japonicus (Baill.) Müll. Arg. = Phyllanthus flexuosus (Siebold et Zucc.) Müll. Arg. ●

296618　Phyllanthus juniperinus Wall. = Phyllanthus parvifolius Buch. -Ham. ex D. Don ●

296619　Phyllanthus kaessneri Hutch. ;卡斯纳叶下珠●☆

296620　Phyllanthus kaessneri Hutch. var. polycytotrichum Radcl. -Sm. ;多囊毛叶下珠●☆

296621　Phyllanthus kerstingii Brunel;克斯廷叶下珠●☆

296622　Phyllanthus khasicus Müll. Arg. = Glochidion khasicum (Müll. Arg.) Hook. f. ●

296623　Phyllanthus kiangsiensis Croizat et F. P. Metcalf = Phyllanthus chekiangensis Croizat et Metcalf ●◇

296624　Phyllanthus kirkianus Müll. Arg. = Phyllanthus pinnatus (Wight) G. L. Webster ●☆

296625　Phyllanthus klainei Hutch. = Phyllanthus polyanthus Pax ●☆

296626　Phyllanthus kurzianus Müll. Arg. = Glochidion philippicum (Cav.) C. B. Rob. ●

296627　Phyllanthus lalambensis Schweinf. = Phyllanthus ovalifolius Forssk. ●☆

296628　Phyllanthus lanceolarius (Roxb.) Müll. Arg. = Glochidion lanceolarium (Roxb.) Voigt ●

296629　Phyllanthus lebrunii Robyns et Lawalrée;勒布伦叶下珠●☆

296630　Phyllanthus leiboensis T. L. Chin = Phyllanthus franchetianus H. Lév. ●

296631　Phyllanthus leonardianus Lisowski, Malaisse et Symoens;莱奥叶下珠●☆

296632　Phyllanthus lepidocarpus Siebold et Zucc. ;察瓦龙叶下珠(鲤下子叶下珠);Hooker Leaf-flower ●

296633　Phyllanthus leptoclados F. P. Metcalf;细枝叶下珠;Slender-twigged Leaf-flower, Thin-branch Leafflower, Thin-branch Underleaf Pearl ●

296634　Phyllanthus leptoclados F. P. Metcalf = Phyllanthus chekiangensis Croizat et Metcalf ●◇

296635　Phyllanthus leptoclados F. P. Metcalf var. pubescens P. T. Li et D. Y. Liu = Phyllanthus chekiangensis Croizat et Metcalf ●◇

296636　Phyllanthus leucanthus Pax;白花叶下珠●☆

296637　Phyllanthus leucocalyx (Müll. Arg.) Hutch. ;白萼叶下珠; White-calyx Leafflower, White-calyx Underleaf Pearl ●☆

296638　Phyllanthus leucocalyx Hutch. = Phyllanthus leucocalyx (Müll. Arg.) Hutch. ●☆

296639　Phyllanthus leucochlamys Radcl. -Sm. ;白被叶下珠●☆

296640　Phyllanthus leucopyrus (Willd.) König ex Roxb. = Flueggea leucopyrus Willd. ●

296641　Phyllanthus liukiuensis Matsum. ex Hayata;琉球叶下珠●☆

296642　Phyllanthus loandensis Welw. ex Müll. Arg. ;罗安达叶下珠●☆

296643　Phyllanthus lucens Poir. = Breynia fruticosa (L.) Hook. f. ●

296644　Phyllanthus lutescens (Blume) Müll. Arg. = Glochidion lutescens Blume ●

296645　Phyllanthus macranthus Pax;大花叶下珠●☆

296646　Phyllanthus macranthus Pax var. gilletii (De Wild.) Brunel ex Radcl. -Sm. ;吉勒特叶下珠●☆

296647　Phyllanthus maderaspatensis L. ;麻德拉斯叶下珠;Maderaspat Leaf-flower ●☆

296648　Phyllanthus magnificens Brunel et J. P. Roux;华丽叶下珠●☆

296649　Phyllanthus mairei H. Lév. = Phyllanthus emblica L. ●

296650　Phyllanthus mannianus Müll. Arg. ;曼氏叶下珠●☆

296651　Phyllanthus martinii Radcl. -Sm. ;马丁叶下珠●☆

296652　Phyllanthus matsumurae Hayata ex Fabe = Phyllanthus ussuriensis Rupr. et Maxim. ●■

296653　Phyllanthus mendesii Brunel ex Radcl. -Sm. ;门代斯叶下珠●☆

296654　Phyllanthus mendoncae Brunel ex Radcl. -Sm. ;门东萨叶下珠●☆

296655　Phyllanthus merripaensis Brunel = Phyllanthus leucanthus Pax ●☆

296656　Phyllanthus meruensis Pax = Phyllanthus sepialis Müll. Arg. ●☆

296657　Phyllanthus meyerianus Müll. Arg. ;迈尔叶下珠●☆

296658　Phyllanthus micrandrus Müll. Arg. ;小蕊叶下珠●☆

296659　Phyllanthus microcarpus (Benth.) Müll. Arg. = Phyllanthus reticulatus Poir. ●

296660　Phyllanthus microdendron Welw. ex Müll. Arg. ;小树叶下珠●☆

296661　Phyllanthus microphyllinus Müll. Arg. ;小叶叶下珠●☆

296662　Phyllanthus moi P. T. Li = Phyllanthodendron moi (P. T. Li) P. T. Li ●

296663　Phyllanthus mooneyi M. G. Gilbert;穆尼叶下珠●☆

296664　Phyllanthus mozambicensis Gand. = Phyllanthus parvulus Sond. ●☆

296665　Phyllanthus muellerianus (Kuntze) Exell;热非叶下珠●☆

296666　Phyllanthus multicaulis Müll. Arg. = Phyllanthus incurvus Thunb. ●☆

296667　Phyllanthus multiflorus Poir. ;多花油柑(白仔)●

296668　Phyllanthus multiflorus Poir. = Phyllanthus oligospermus Hayata ●

296669　Phyllanthus multiflorus Poir. = Phyllanthus reticulatus Poir. ●

296670　Phyllanthus multiflorus Willd. = Phyllanthus reticulatus Poir. ●

296671　Phyllanthus myrtaceus Sond. ;香桃木叶下珠●☆

296672　Phyllanthus myrtifolius (Wight) Müll. Arg. ;瘤腺叶下珠(锡兰

桃金娘，锡兰叶下珠）；Myrtle-leaved Leafflower，Myrtle-leaved Leaf-flower ●

296673   Phyllanthus myrtifolius Moon = Phyllanthus myrtifolius（Wight）Müll. Arg. ●

296674   Phyllanthus myrtilloides Chiov. ；黑果越橘叶下珠●☆

296675   Phyllanthus nanellus P. T. Li；单花叶下珠（单花水油甘）；Monoflower Leafflower，Monoflower Underleaf Pearl，Solitary Leafflower ●

296676   Phyllanthus nepalensis Müll. Arg. = Glochidion velutinum Wight ●

296677   Phyllanthus nigericus Brenan；尼日利亚叶下珠●☆

296678   Phyllanthus niruri Klotzsch = Phyllanthus amarus K. Schum. et Thonn. ●■

296679   Phyllanthus niruri L. ；珠子草（霸贝菜，榛屎，小返魂，叶下珠）；Necklace Leaf-flower ●■

296680   Phyllanthus niruri L. = Phyllanthus fraternus G. L. Webster ●☆

296681   Phyllanthus niruri L. subsp. amarus（Schum.）Léandri = Phyllanthus amarus K. Schum. et Thonn. ●■

296682   Phyllanthus niruri L. var. amarus（Schumach. et Thonn.）Léandri = Phyllanthus amarus K. Schum. et Thonn. ●■

296683   Phyllanthus niruroides Müll. Arg. ；普通叶下珠●☆

296684   Phyllanthus niruroides Müll. Arg. subsp. pierlotii Brunel；皮氏叶下珠●☆

296685   Phyllanthus nivosus W. Bull = Breynia disticha J. R. Forst. et G. Forst. f. nivosa（W. Bull）Croizat ex Radcl. -Sm. ●☆

296686   Phyllanthus nummulariifolius Poir. ；铜钱叶下珠●☆

296687   Phyllanthus nummulariifolius Poir. var. capillaris（Schumach. et Thonn.）Radcl. -Sm. ；发状铜钱叶下珠●☆

296688   Phyllanthus nyassae Pax et K. Hoffm. = Phyllanthus beillei Hutch. ●☆

296689   Phyllanthus nyikae Radcl. -Sm. ；尼卡叶下珠●☆

296690   Phyllanthus oblongiglans M. G. Gilbert；矩圆叶下珠●☆

296691   Phyllanthus obovatus（Siebold et Zucc.）Müll. Arg. = Glochidion obovatum Siebold et Zucc. ●

296692   Phyllanthus odontadenius Müll. Arg. = Phyllanthus gagnioevae Brunel et J. P. Roux ●☆

296693   Phyllanthus odontadenius Müll. Arg. subsp. gagnioevae Brunel et J. P. Roux = Phyllanthus gagnioevae Brunel et J. P. Roux ●☆

296694   Phyllanthus odontadenius Müll. Arg. var. braunii（Pax）Hutch. = Phyllanthus braunii Pax ●☆

296695   Phyllanthus oligospermus Hayata；少子叶下珠（白仔，多花油柑，烂头钵，龙眼睛，山兵豆，小果叶下珠，新竹叶下珠，新竹油柑，新竹油树）；Hayata Leaf-flower，Hsinchu Leafflower，Xinzhu Leafflower ●

296696   Phyllanthus omahakensis Dinter et Pax；奥马哈克叶下珠●☆

296697   Phyllanthus orbiculatus Rich. ；圆叶下珠●☆

296698   Phyllanthus ovalifolius Forssk. ；椭圆叶下珠●☆

296699   Phyllanthus ovalis E. Mey. ex Sond. = Andrachne ovalis（E. Mey. ex Sond.）Müll. Arg. ●☆

296700   Phyllanthus oxycoccifolius Hutch. ；红莓苔子叶下珠●☆

296701   Phyllanthus parvifolius Buch. -Ham. ex D. Don；水油甘；Small-leaf Leafflower，Small-leaf Underleaf Pearl，Small-leaved Leaf-flower ●

296702   Phyllanthus parvulus Sond. ；较小叶下珠●☆

296703   Phyllanthus parvulus Sond. var. garipensis（E. Mey. ex Drège）Radcl. -Sm. ；加里普叶下珠●☆

296704   Phyllanthus parvus Hutch. ；小叶下珠●☆

296705   Phyllanthus patens Roxb. = Breynia retusa（Dennst.）Alston ●

296706   Phyllanthus paxianus Dinter = Phyllanthus maderaspatensis L. ●☆

296707   Phyllanthus paxii Hutch. ；帕克斯叶下珠●☆

296708   Phyllanthus pentandrus Roxb. ex Thwaites = Phyllanthus reticulatus Poir. ●

296709   Phyllanthus pentandrus Schumach. et Thonn. ；五蕊叶下珠●☆

296710   Phyllanthus petraeus A. Chev. ex Beille；岩生叶下珠●☆

296711   Phyllanthus philippinensis（Benth.）Müll. Arg. = Glochidion philippicum（Cav.）C. B. Rob. ●

296712   Phyllanthus physocarpus Müll. Arg. ；囊果叶下珠●☆

296713   Phyllanthus pinnatus（Wight）G. L. Webster；羽状叶下珠●☆

296714   Phyllanthus piscatorum Kunth；鱼叶下珠●☆

296715   Phyllanthus polyanthus Pax；多花叶下珠●☆

296716   Phyllanthus polygonoides Nutt. ex Spreng. ；蓼状叶下珠；Buck Brush ●☆

296717   Phyllanthus pomaceus Moon = Breynia retusa（Dennst.）Alston ●

296718   Phyllanthus profusus N. E. Br. ；横卧叶下珠●☆

296719   Phyllanthus prostratus Welw. ex Müll. Arg. ；平卧叶下珠●☆

296720   Phyllanthus pseudoniruri Müll. Arg. ；拟假叶下珠●☆

296721   Phyllanthus puberus Müll. Arg. = Glochidion puberum（L.）Hutch. ●

296722   Phyllanthus puberus Müll. Arg. var. fortunei（Hance）Müll. Arg. = Glochidion puberum（L.）Hutch. ●

296723   Phyllanthus puberus Müll. Arg. var. sinicus（Hook. et Arn.）Müll. Arg. = Glochidion puberum（L.）Hutch. ●

296724   Phyllanthus pulcher Müll. Arg. = Phyllanthus pulcher Wall. ex Müll. Arg. ●

296725   Phyllanthus pulcher Wall. ex Müll. Arg. ；云桂叶下珠（美丽叶下珠）；Beautiful Leafflower，Beautiful Leaf-flower，Beautiful Underleaf Pearl，Tropical Leaf-flower ●

296726   Phyllanthus purpureus Müll. Arg. ；紫叶下珠●☆

296727   Phyllanthus pynaertii De Wild. ；皮那叶下珠●☆

296728   Phyllanthus quadrangularis Willd. = Sauropus quadrangularis（Willd.）Müll. Arg. ●

296729   Phyllanthus quercinus Müll. Arg. = Glochidion philippicum（Cav.）C. B. Rob. ●

296730   Phyllanthus ramiflorus（J. R. Forst. et G. Forst.）Müll. Arg. = Glochidion ramiflorum J. R. Forst. et G. Forst. ●

296731   Phyllanthus ramiflorus Pers. = Flueggea suffruticosa（Pall.）Baill. ●

296732   Phyllanthus raynalii Brunel et J. P. Roux；雷纳尔叶下珠●☆

296733   Phyllanthus reoperianus Müll. Arg. = Phyllanthus cochinchinensis（Lour.）Spreng. ●

296734   Phyllanthus reticulatus Poir. ；小果叶下珠（白仔，多花油柑，飞檫木，烂头砵，烂头钵，龙眼睛，山兵豆，通城虎，网状叶下珠）；Netted-veined Leafflower，Netted-veined Leaf-flower，Reticulate Leafflower，Reticulate Leaf-flower，Reticulate Underleaf Pearl ●

296735   Phyllanthus reticulatus Poir. var. glaber（Thwaites）Müll. Arg. ；无毛小果叶下珠（光叶小果叶下珠，红鱼眼，无毛龙眼睛）；Glabrous Reticulate Leafflower，Glabrous Reticulate Underleaf Pearl ●

296736   Phyllanthus reticulatus Poir. var. glaber Müll. Arg. = Phyllanthus reticulatus Poir. var. glaber（Thwaites）Müll. Arg. ●

296737   Phyllanthus retinervis Hutch. ；网脉叶下珠●☆

296738   Phyllanthus retusus Dennst. = Breynia retusa（Dennst.）Alston ●

296739   Phyllanthus rhamnoides Roxb. = Sauropus quadrangularis（Willd.）Müll. Arg. ●

296740   Phyllanthus rhamnoides Willd. = Breynia vitis-idaea（Burm. f.）C. E. C. Fisch. ●

296741   Phyllanthus rhizomatosus Radcl. -Sm. ；根茎叶下珠●☆

296742　Phyllanthus ringoetii De Wild. ;林戈叶下珠●☆

296743　Phyllanthus rivae Pax = Phyllanthus leucanthus Pax ●☆

296744　Phyllanthus roeperianus Müll. Arg. = Phyllanthus cochinchinensis（Lour.）Spreng. ●

296745　Phyllanthus roeperianus Müll. Arg. var. parvifolius（Buch. -Ham. ex D. Don）Hand. -Mazz. = Phyllanthus parvifolius Buch. -Ham. ex D. Don ●

296746　Phyllanthus rogersii Hutch. = Phyllanthus graminicola Hutch. ex S. Moore ●☆

296747　Phyllanthus roseus （Craib et Hutch.） Beille = Phyllanthodendron roseum Craib et Hutch. ●

296748　Phyllanthus rotundifolius Klein ex Willd. ;圆叶叶下珠●☆

296749　Phyllanthus rotundifolius Klein ex Willd. var. leucocalyx Müll. Arg. ;白萼圆叶叶下珠●☆

296750　Phyllanthus rouxii Brunel ;鲁叶下珠●☆

296751　Phyllanthus ruber（Lour.）Spreng. ;红叶下珠（山杨桃,鹧鸪鸣）; Red Leaf-flower,Red-leaf Leafflower,Red-leaf Underleaf Pearl ●

296752　Phyllanthus sacleuxii Radcl. -Sm. ;萨克勒叶下珠●☆

296753　Phyllanthus scabrellus Webb = Phyllanthus amarus K. Schum. et Thonn. ●■

296754　Phyllanthus schliebenii Mansf. ex Radcl. -Sm. ;施利本叶下珠●☆

296755　Phyllanthus sellowianus Müll. Arg. ;塞洛叶下珠●☆

296756　Phyllanthus sepialis Müll. Arg. ;篱边叶下珠●☆

296757　Phyllanthus serpentinicola Radcl. -Sm. ;蛇纹岩叶下珠●☆

296758　Phyllanthus sikkimensis Müll. Arg. ;锡金叶下珠●☆

296759　Phyllanthus silheticus Müll. Arg. = Glochidion arborescens Blume ●

296760　Phyllanthus simplex Retz. = Phyllanthus virgatus J. Forst. ●

296761　Phyllanthus simplex Retz. var. chinensis Müll. Arg. = Phyllanthus ussuriensis Rupr. et Maxim. ●■

296762　Phyllanthus simplex Retz. var. nussuriensis （Rupr. et Maxim.）Müll. Arg. = Phyllanthus ussuriensis Rupr. et Maxim. ●■

296763　Phyllanthus simplex Retz. var. tonkinensis Baill. = Phyllanthus clarkei Hook. f. ●

296764　Phyllanthus simplex Retz. var. virgatus（J. Forst.）Müll. Arg. = Phyllanthus virgatus J. Forst. ●

296765　Phyllanthus sinensis Müll. Arg. = Phyllanthus reticulatus Poir. ●

296766　Phyllanthus sinicus（Baill.）Müll. Arg. = Margaritaria indica （Dalzell）Airy Shaw ●

296767　Phyllanthus somalensis Hutch. ;索马里叶下珠●☆

296768　Phyllanthus sootepensis Craib;云泰叶下珠（美丽叶下珠）; Soótep Leaf-flower ●

296769　Phyllanthus sphaerogynus Müll. Arg. = Glochidion sphaerogynum （Müll. Arg.） Kurz ●

296770　Phyllanthus spinosus Chiov. ;具刺叶下珠●☆

296771　Phyllanthus stolzianus Pax et K. Hoffm. = Phyllanthus beillei Hutch. ●☆

296772　Phyllanthus stuhlmannii Pax = Phyllanthus nummulariifolius Poir. ●☆

296773　Phyllanthus sublanatus Schumach. et Thonn. ;绵毛叶下珠●☆

296774　Phyllanthus sublanatus Schumach. et Thonn. subsp. eliae Brunel et J. P. Roux = Phyllanthus eliae（Brunel et J. P. Roux）Brunel ex Govaerts et Radcl. -Sm. ☆

296775　Phyllanthus subpulchellus Croizat = Phyllanthus sootepensis Craib ●

296776　Phyllanthus suffrutescens Pax;灌木叶下珠●☆

296777　Phyllanthus swartzii Kostel. = Phyllanthus amarus K. Schum. et Thonn. ●■

296778　Phyllanthus taitensis Hutch. = Phyllanthus sacleuxii Radcl. -Sm. ●☆

296779　Phyllanthus takaoensis Hayata;高雄叶下珠（高雄油柑）; Gaoxiong Leafflower,Kao-hsiung Leafflower ●

296780　Phyllanthus takaoensis Hayata = Phyllanthus oligospermus Hayata ●

296781　Phyllanthus takaoensis Hayata = Phyllanthus reticulatus Poir. ●

296782　Phyllanthus tanzaniensis Brunel = Phyllanthus nummulariifolius Poir. ●☆

296783　Phyllanthus taxodiifolius Beille;落羽松叶下珠（滇橄榄）●

296784　Phyllanthus taylorianus Brunel ex Radcl. -Sm. ;泰勒叶下珠●☆

296785　Phyllanthus tenellus Roxb. ;五蕊油柑（小叶下珠）;Mascarene Island Leaf-flower ●

296786　Phyllanthus tenellus Roxb. var. exiguus Müll. Arg. = Phyllanthus parvulus Sond. var. garipensis（E. Mey. ex Drège）Radcl. -Sm. ●☆

296787　Phyllanthus tenellus Roxb. var. garipensis（E. Mey. ex Drège）Müll. Arg. = Phyllanthus parvulus Sond. var. garipensis（E. Mey. ex Drège）Radcl. -Sm. ☆

296788　Phyllanthus tener Radcl. -Sm. ;极细叶下珠●☆

296789　Phyllanthus tenuis Radcl. -Sm. ;细叶下珠●☆

296790　Phyllanthus tessmannii Hutch. ;泰斯曼叶下珠●☆

296791　Phyllanthus thomsonii Müll. Arg. = Glochidion thomsonii （Müll. Arg.）Hook. f. ●

296792　Phyllanthus thonningii Schumach. et Thonn. = Phyllanthus maderaspatensis L. ●☆

296793　Phyllanthus thulinii Radcl. -Sm. ;图林叶下珠●☆

296794　Phyllanthus triandrus （Blanco） Müll. Arg. = Glochidion triandrum（Blanco）C. B. Rob. ●

296795　Phyllanthus trichotepalus Brenan;毛叶下珠●☆

296796　Phyllanthus trinervius Wall. = Sauropus trinervius （Wall.） Hook. f. et Thomson ex Müll. Arg. ●

296797　Phyllanthus tsarongensis W. W. Sm. ;西南叶下珠●

296798　Phyllanthus tsiangii P. T. Li = Phyllanthus ruber（Lour.）Spreng. ●

296799　Phyllanthus turbinatus Sims = Breynia fruticosa（L.）Hook. f. ●

296800　Phyllanthus ukagurensis Radcl. -Sm. ;乌卡古鲁叶下珠●☆

296801　Phyllanthus urinaria L. ;叶下珠（蓖萁草,刺果叶下珠,地槐菜,飞阳草,故芏,关门草,胡羞羞,鲫鱼草,假油柑,假油树,老鸦珠,落地油柑,日开夜闭,山皂角,十字珍珠草,五时合,小利柑,杨梅珠草,叶后珠,叶下珍珠,夜关门,夜合草,夜合珍珠,阴阳草,油柑草,鱼鳞草,珍珠草,真珠草,珠仔草）;Chamber Bitter, Common Leaf-flower,Underleaf Pearl,Wrinkle-fruited Leafflower ●■

296802　Phyllanthus urinaria L. subsp. nudicarpus Rossign. et Haicour; 裸果叶下珠（光果叶下珠）●☆

296803　Phyllanthus ussuriensis Rupr. et Maxim. ;蜜柑草（飞蛇子,蜜甘草,乌苏里蜜柑草,乌苏里叶下珠,夜关门）;Matsumura Leafflower,Matsumura Underleaf Pearl,Ussuri Leaf-flower ●■

296804　Phyllanthus vanderysti Hutch. et De Wild. ;范德叶下珠●☆

296805　Phyllanthus velutinus Müll. Arg. = Glochidion velutinum Wight ●

296806　Phyllanthus verdickii De Wild. ;韦氏叶下珠●☆

296807　Phyllanthus verrucosus Thunb. = Flueggea verrucosa（Thunb.）G. L. Webster ●☆

296808　Phyllanthus villosus Poir. = Glochidion puberum（L.）Hutch. ●

296809　Phyllanthus virgatus J. Forst. ;黄珠子草（单叶下珠,地珍珠,假芋,日开夜闭,细叶油柑,细叶油树,鱼骨草,珍珠草）;Simple Leafflower Simple,Underleaf Pearl ●

296810 Phyllanthus virgatus J. Forst. var. chinensis （Müll. Arg.） Webster = Phyllanthus ussuriensis Rupr. et Maxim. ●■

296811 Phyllanthus virgulatus Müll. Arg. ；条纹叶下珠●☆

296812 Phyllanthus virosus Roxb. ex Willd. = Flueggea virosa （Roxb. ex Willd.） Voigt ●

296813 Phyllanthus volkensii Engl. ；福尔叶下珠●☆

296814 Phyllanthus welwitschianus Müll. Arg. ；韦尔叶下珠●☆

296815 Phyllanthus welwitschianus Müll. Arg. var. beillei （Hutch.） Radcl. -Sm. = Phyllanthus beillei Hutch. ●☆

296816 Phyllanthus wilfordii Croizat et F. P. Metcalf = Phyllanthus ussuriensis Rupr. et Maxim. ●■

296817 Phyllanthus wittei Robyns et Lawalrée；维特叶下珠●☆

296818 Phyllanthus wrightii （Benth.） Müll. Arg. = Glochidion wrightii Benth. ●

296819 Phyllanthus xiphephorus Brunel ex Radcl. -Sm. ；剑梗叶下珠●☆

296820 Phyllanthus xylorrhizus Thulin；木根叶下珠■☆

296821 Phyllanthus zambicus Radcl. -Sm. ；赞比亚叶下珠■☆

296822 Phyllanthus zeylanicus Müll. Arg. = Glochidion zeylanicum （Gaertn.） A. Juss. ●

296823 Phyllanthus zornioides Radcl. -Sm. ；丁癸草叶下珠●☆

296824 Phyllapophysis Mansf. = Catanthera F. Muell. ●☆

296825 Phyllarthron DC. = Phyllarthron DC. ex Meisn. ●☆

296826 Phyllarthron DC. ex Meisn. （1839）；叶节木属（菲拉尔木属）●☆

296827 Phyllarthron antongiliense Capuron；安通吉尔叶节木●☆

296828 Phyllarthron articulatum （Desf. ex Poir.） K. Schum. ；关节叶节木●☆

296829 Phyllarthron bernierianum Seem. ；伯尼尔叶节木●☆

296830 Phyllarthron bilabiatum A. H. Gentry；双唇叶节木●☆

296831 Phyllarthron bojeranum A. DC. ；博耶尔叶节木●☆

296832 Phyllarthron cauliflorum Capuron；茎花叶节木●☆

296833 Phyllarthron comorense DC. ；叶节木（菲拉尔木）●☆

296834 Phyllarthron humblotianum H. Perrier；洪布叶节木●☆

296835 Phyllarthron ilicifolium （Pers.） H. Perrier；冬青叶节木●☆

296836 Phyllarthron laxinervium H. Perrier；疏脉叶节木●☆

296837 Phyllarthron madagascariense K. Schum. ；马岛叶节木●☆

296838 Phyllarthron megaphyllum Capuron；大叶叶节木●☆

296839 Phyllarthron megapterum H. Perrier；大翅叶节木●☆

296840 Phyllarthron multiflorum H. Perrier；多花叶节木●☆

296841 Phyllarthron noronhianum A. DC. = Phyllarthron articulatum （Desf. ex Poir.） K. Schum. ●☆

296842 Phyllarthron poivreanum A. DC. = Phyllarthron articulatum （Desf. ex Poir.） K. Schum. ●☆

296843 Phyllarthron schatzii A. H. Gentry；沙茨叶节木●☆

296844 Phyllarthron suarezense H. Perrier；苏亚雷斯叶节木●☆

296845 Phyllarthron subumbellatum H. Perrier；小伞叶节木●☆

296846 Phyllarthron thouarsianum A. DC. = Filicium thouarsianum （A. DC.） Capuron ●☆

296847 Phyllarthus Neck. = Opuntia Mill. ●

296848 Phyllarthus Neck. ex M. Gomez = Opuntia Mill. ●

296849 Phyllaurea Lour. （废弃属名） = Codiaeum A. Juss. （保留属名）●

296850 Phyllepidum Raf. = Polygonella Michx. ■☆

296851 Phyllera Endl. = Philyra Klotzsch ●☆

296852 Phyllimena Blume ex DC. = Enydra Lour. ■

296853 Phyllirea Adans. = Phillyrea L. ●☆

296854 Phyllirea Duhamel = Phillyrea L. ●☆

296855 Phyllirea Tourn. ex Adans. = Phillyrea L. ●☆

296856 Phyllis L. （1753）；叶茜属●☆

296857 Phyllis galopina （Thunb.） Cruse = Galopina circaeoides Thunb. ●☆

296858 Phyllis nobla L. ；叶茜●☆

296859 Phyllis nobla L. var. subviscosa Kuntze = Phyllis viscosa H. Christ ●☆

296860 Phyllis viscosa H. Christ；黏叶茜●☆

296861 Phyllobaea Benth. = Phylloboea Benth. ■☆

296862 Phylloboea Benth. （1876）；叶苣苔属■☆

296863 Phylloboea C. B. Clarke = Phylloboea Benth. ■☆

296864 Phylloboea amplexicaulis （Parish ex C. B. Clarke） Benth. ；叶苣苔■☆

296865 Phylloboea glandulosa B. L. Burtt；腺点叶苣苔■☆

296866 Phylloboea henryi Duthie ex Bedd. = Paraboea rufescens （Franch.） B. L. Burtt ■

296867 Phylloboea sinensis Oliv. = Paraboea sinensis （Oliv.） B. L. Burtt ●

296868 Phyllobolus N. E. Br. （1925）；凤卵草属●☆

296869 Phyllobolus abbreviatus （L. Bolus） Gerbaulet；缩短凤卵草●☆

296870 Phyllobolus amabilis Gerbaulet et Struck；秀丽凤卵草●☆

296871 Phyllobolus canaliculatus （Haw.） Bittrich；具沟凤卵草●☆

296872 Phyllobolus caudatus （L. Bolus） Gerbaulet；尾状凤卵草●☆

296873 Phyllobolus chrysophthalmus Gerbaulet et Struck；金眼凤卵草●☆

296874 Phyllobolus congestus （L. Bolus） Gerbaulet；密集凤卵草●☆

296875 Phyllobolus deciduus （L. Bolus） Gerbaulet；脱落凤卵草●☆

296876 Phyllobolus decurvatus （L. Bolus） Gerbaulet；下延凤卵草●☆

296877 Phyllobolus delus （L. Bolus） Gerbaulet；明显凤卵草■☆

296878 Phyllobolus digitatus （Aiton） Gerbaulet = Dactylopsis digitata （Aiton） N. E. Br. ■☆

296879 Phyllobolus digitatus （Aiton） Gerbaulet subsp. littlewoodii （L. Bolus） Gerbaulet = Dactylopsis digitata （Aiton） Gerbaulet subsp. littlewoodii （L. Bolus） Klak ■☆

296880 Phyllobolus grossus （Aiton） Gerbaulet；宽萼凤卵草■☆

296881 Phyllobolus herbertii （N. E. Br.） Gerbaulet；赫伯特凤卵草●☆

296882 Phyllobolus latipetalus （L. Bolus） Gerbaulet；阔瓣凤卵草●☆

296883 Phyllobolus lesliei N. E. Br. = Phyllobolus resurgens （Kensit） Schwantes ●☆

296884 Phyllobolus lignescens （L. Bolus） Gerbaulet；木质凤卵草●☆

296885 Phyllobolus melanospermus （Dinter et Schwantes） Gerbaulet；黑籽凤卵草●☆

296886 Phyllobolus nitidus （Haw.） Gerbaulet；光亮凤卵草●☆

296887 Phyllobolus noctiflorus （L.） Bittrich = Aridaria noctiflora （L.） Schwantes ●☆

296888 Phyllobolus oculatus （N. E. Br.） Gerbaulet；小眼凤卵草●☆

296889 Phyllobolus pallens （Aiton） Bittrich = Prenia pallens （Aiton） N. E. Br. ■☆

296890 Phyllobolus pearsonii N. E. Br. = Phyllobolus resurgens （Kensit） Schwantes ●☆

296891 Phyllobolus prasinus （L. Bolus） Gerbaulet；草绿凤卵草●☆

296892 Phyllobolus publicalyx N. E. Br. = Phyllobolus resurgens （Kensit） Schwantes ●☆

296893 Phyllobolus pumilus （L. Bolus） Gerbaulet；矮小凤卵草●☆

296894 Phyllobolus quartziticus （L. Bolus） Gerbaulet；阔茨凤卵草●☆

296895 Phyllobolus rabiei （L. Bolus） Gerbaulet；拉比耶凤卵草●☆

296896 Phyllobolus resurgens （Kensit） Schwantes；重生凤卵草●☆

296897 Phyllobolus roseus （L. Bolus） Gerbaulet；粉红凤卵草●☆

296898 Phyllobolus saturatus （L. Bolus） Gerbaulet；富色凤卵草●☆

296899　Phyllobolus sinuosus（L. Bolus）Gerbaulet；深波凤卵草●☆

296900　Phyllobolus spinuliferus（Haw.）Gerbaulet；小刺凤卵草●■☆

296901　Phyllobolus splendens（L.）Gerbaulet；亮凤卵草●☆

296902　Phyllobolus splendens（L.）Gerbaulet subsp. pentagonus（L. Bolus）Gerbaulet；五角小刺凤卵草●■☆

296903　Phyllobolus suffruticosus（L. Bolus）Gerbaulet；亚灌木凤卵草●☆

296904　Phyllobolus tenuiflorus（Jacq.）Gerbaulet；细花凤卵草●☆

296905　Phyllobolus tortuosus（L.）Bittrich ＝ Sceletium tortuosum（L.）N. E. Br. ●☆

296906　Phyllobolus trichotomus（Thunb.）Gerbaulet；毛片凤卵草●☆

296907　Phyllobolus viridiflorus（Aiton）Gerbaulet；绿花凤卵草●☆

296908　Phyllobotrium Willis ＝ Phyllobotryon Müll. Arg. ●☆

296909　Phyllobotryon Müll. Arg.（1864）；叶序大风子属●☆

296910　Phyllobotryon bracteatum（Lecomte）Hul；具苞叶序大风子●☆

296911　Phyllobotryon lebrunii Staner；勒布伦叶序大风子●☆

296912　Phyllobotryon paradoxum（Baill.）Hul；奇异叶序大风子●☆

296913　Phyllobotryon spathulatum Müll. Arg.；匙形叶序大风子●☆

296914　Phyllobotrys（Spach）Fourr. ＝ Genista L. ●

296915　Phyllobotrys Fourr. ＝ Genista L. ●

296916　Phyllobotryum Müll. Arg. ＝ Phyllobotryon Müll. Arg. ●☆

296917　Phyllobotryum basiflorum Gilg ＝ Phyllobotryon spathulatum Müll. Arg. ●☆

296918　Phyllobotryum soyauxianum Baill. ＝ Phyllobotryon spathulatum Müll. Arg. ●☆

296919　Phyllobotryum zenkeri Gilg ＝ Phyllobotryon spathulatum Müll. Arg. ●☆

296920　Phyllobryon Miq. ＝ Peperomia Ruiz et Pav. ■

296921　Phyllocactus Link ＝ Epiphyllum Haw. ●

296922　Phyllocactus oxypetalus（DC.）Link ＝ Epiphyllum oxypetalum（DC.）Haw. ■

296923　Phyllocactus oxypetalus（DC.）Link ex Walp. ＝ Epiphyllum oxypetalum（DC.）Haw. ■

296924　Phyllocactus phyllanthus（L.）Link ＝ Epiphyllum phyllanthus（L.）Haw. ■☆

296925　Phyllocalymma Benth. ＝ Angianthus J. C. Wendl.（保留属名）■●☆

296926　Phyllocalyx A. Rich.（1847）；叶萼豆属●☆

296927　Phyllocalyx A. Rich. ＝ Crotalaria L. ●■

296928　Phyllocalyx O. Berg ＝ Eugenia L. ●

296929　Phyllocalyx quartinianus A. Rich.；夸尔廷叶萼豆●☆

296930　Phyllocara Gusul. ＝ Anchusa L. ■

296931　Phyllocara aucheri（A. DC.）Gusul. ＝ Anchusa aucheri A. DC. ■☆

296932　Phyllocarpa Nutt. ex Moq. ＝ Atriplex L. ■●

296933　Phyllocarpus Riedel ex Endl.（1842）；叶果豆属（叶荚豆属）●☆

296934　Phyllocarpus Riedel ex Tul. ＝ Phyllocarpus Riedel ex Endl. ●☆

296935　Phyllocarpus pterocarpus（A. DC.）Endl. ex B. D. Jacks.；叶果豆●☆

296936　Phyllocasia Rchb. ＝ Xanthosoma Schott ■

296937　Phyllocephalium Miq. ＝ Phyllocephalum Blume ■☆

296938　Phyllocephalum Blume ＝ Centratherum Cass. ■☆

296939　Phyllocephalum Blume（1826）；叶苞瘦片菊属■☆

296940　Phyllocephalum frutescens Blume；叶苞瘦片菊■☆

296941　Phyllocephalum indicum（Less.）K. Kirkman；印度叶苞瘦片菊■☆

296942　Phyllocephalum microcephalum（Dalzell）H. Rob.；小头叶苞瘦片菊■☆

296943　Phyllocereus Miq. ＝ Epiphyllum Haw. ●

296944　Phyllocharis Diels ＝ Ruthiella Steenis ■☆

296945　Phyllochilium Cabrera ＝ Chiliophyllum Phil.（保留属名）●☆

296946　Phyllochlamys Bureau ＝ Streblus Lour. ●

296947　Phyllochlamys Bureau（1873）；酒饼树属（叶珠木属）●

296948　Phyllochlamys spinosa Bureau；酒饼树●☆

296949　Phyllochlamys taxoides（Heyne）Koord. ＝ Streblus taxoides（K. Heyne）Kurz ●

296950　Phyllochlamys taxoides（Roth）Koord. ＝ Streblus taxoides（K. Heyne）Kurz ●

296951　Phyllochlamys tridentata Gagnep.；三齿酒饼树●☆

296952　Phyllocladaceae Bessey ＝ Podocarpaceae Endl.（保留科名）●

296953　Phyllocladaceae Bessey（1907）；叶枝杉科（伪叶竹柏科）●☆

296954　Phyllocladus Mirb. ＝ Phyllocladus Rich. ex Mirb.（保留属名）●☆

296955　Phyllocladus Rich. ＝ Phyllocladus Rich. ex Mirb.（保留属名）●☆

296956　Phyllocladus Rich. ex Mirb.（1825）（保留属名）；叶枝杉属（叶状枝杉属）；Celery Pine，Celery-pine ●☆

296957　Phyllocladus aspleniifolius（Labill.）Hook. f.；铁线蕨叶叶枝杉；Celery-top Pine ●☆

296958　Phyllocladus trichomanoides；毛叶枝杉●☆

296959　Phylloclinium Baill.（1890）；斜叶大风子属●☆

296960　Phylloclinium bracteatum Lecomte ＝ Phyllobotryon bracteatum（Lecomte）Hul ●☆

296961　Phylloclinium bracteatum Lecomte var. coriaceum ？ ＝ Phyllobotryon bracteatum（Lecomte）Hul ●☆

296962　Phylloclinium brevipetiolatum R. Germ. ＝ Phyllobotryon paradoxum（Baill.）Hul ●☆

296963　Phylloclinium paradoxum Baill.；斜叶大风子●☆

296964　Phylloclinium paradoxum Baill. ＝ Phyllobotryon paradoxum（Baill.）Hul ●☆

296965　Phyllocomos Mast. ＝ Anthochortus Nees ■☆

296966　Phyllocomos insignis Mast. ＝ Anthochortus insignis（Mast.）H. P. Linder ■☆

296967　Phyllocoryne Hook. f. ＝ Scybalium Schott et Endl. ■☆

296968　Phyllocosmua Klotzsch ＝ Ochthocosmus Benth. ●☆

296969　Phyllocosmus Klotzsch（1857）；叶饰木属●☆

296970　Phyllocosmus africanus（Hook. f.）Klotzsch；非洲叶饰木●☆

296971　Phyllocosmus calothyrsus Mildbr.；美序叶饰木●☆

296972　Phyllocosmus candidus Engl. et Gilg ＝ Phyllocosmus lemaireanus（De Wild. et T. Durand）T. Durand et H. Durand ●☆

296973　Phyllocosmus congolensis（De Wild. et T. Durand）T. Durand et H. Durand；刚果叶饰木●☆

296974　Phyllocosmus dewevrei Engl. ＝ Phyllocosmus africanus（Hook. f.）Klotzsch ●☆

296975　Phyllocosmus lemaireanus（De Wild. et T. Durand）T. Durand et H. Durand；勒迈尔叶饰木●☆

296976　Phyllocosmus senensis Engl. ＝ Phyllocosmus lemaireanus（De Wild. et T. Durand）T. Durand et H. Durand ●☆

296977　Phyllocosmus sessiliflorus Oliv.；无花梗叶饰木●☆

296978　Phyllocrater Wernham（1914）；叶杯茜属☆

296979　Phyllocrater gibbsiae Wernham；叶杯茜☆

296980　Phylloctenium Baill.（1887）；篦叶紫葳属●☆

296981　Phylloctenium bernieri Baill.；篦叶紫葳●☆

296982　Phylloctenium decaryanum H. Perrier；马岛篦叶紫葳●☆

296983　Phyllocyclus Kurz ＝ Canscora Lam. ■

296984　Phyllocytisus（W. D. J. Koch）Fourn. ＝ Cytisus Desf.（保留属名）●

296985　Phyllocytisus（W. D. J. Koch）Fourn. ＝ Cytisus L.（废弃属名）●

296986　Phyllocytisus Fourn. ＝ Cytisus Desf.（保留属名）●

296987　Phyllocytisus Fourn. = Cytisus L. (废弃属名) ●

296988　Phyllodes L. = Phrynium Willd. (保留属名) ■

296989　Phyllodes Lour. (废弃属名) = Phrynium Willd. (保留属名) ■

296990　Phyllodes adenocarpum K. Schum. = Megaphrynium macrostachyum (Benth.) Milne-Redh. ■☆

296991　Phyllodes baccatum K. Schum. = Sarcophrynium schweinfurthianum (Kuntze) Milne-Redh. ■☆

296992　Phyllodes bisubulatum K. Schum. = Sarcophrynium bisubulatum (K. Schum.) K. Schum. ■☆

296993　Phyllodes leiogonium K. Schum. = Sarcophrynium prionogonium (K. Schum.) K. Schum. ■☆

296994　Phyllodes macrophyllum K. Schum. = Megaphrynium macrostachyum (Benth.) Milne-Redh. ■☆

296995　Phyllodes macrostachyum Benth. = Megaphrynium macrostachyum (Benth.) Milne-Redh. ■☆

296996　Phyllodes monophyllum K. Schum. = Marantochloa monophylla (K. Schum.) D'Orey ■☆

296997　Phyllodes oxycarpum K. Schum. = Megaphrynium macrostachyum (Benth.) Milne-Redh. ■☆

296998　Phyllodes placentaria Lour. = Phrynium placentarium (Lour.) Merr. ■

296999　Phyllodes placentarium Lour. = Phrynium placentarium (Lour.) Merr. ■

297000　Phyllodes prionogonium K. Schum. = Sarcophrynium prionogonium (K. Schum.) K. Schum. ■☆

297001　Phyllodesmis Tiegh. = Taxillus Tiegh. ●

297002　Phyllodesmis caloreas (Diels) Danser = Taxillus caloreas (Diels) Danser ●

297003　Phyllodesmis coriacea Tiegh. = Taxillus delavayi (Tiegh.) Danser ●

297004　Phyllodesmis delavayi Tiegh. = Taxillus delavayi (Tiegh.) Danser ●

297005　Phyllodesmis kaempferi (DC.) Tiegh. = Taxillus kaempferi (DC.) Danser ●

297006　Phyllodesmis paucifolia Tiegh. = Taxillus delavayi (Tiegh.) Danser ●

297007　Phyllodium Desv. (1813);排钱树属(排钱草属);Phyllodium ●

297008　Phyllodium elegans (Lour.) Desv.;毛排钱树(叠钱草,连里尾树,鳞狸鳞,毛排钱草,排钱草,麒麟片,雅致山蚂蝗);Elegant Phyllodium ●

297009　Phyllodium grande (Kurz.) Schindl. = Phyllodium kurzianum (Kuntze) H. Ohashi ●

297010　Phyllodium kurzianum (Kuntze) H. Ohashi;云南排钱树(长柱排钱树,云南排钱草);Kurz Phyllodium,Longstyle Phyllodium ●

297011　Phyllodium kurzii (Craib) Chun = Phyllodium kurzianum (Kuntze) H. Ohashi ●

297012　Phyllodium longipes (Craib) Schindl.;长叶排钱树(大排钱树);Longleaf Phyllodium,Long-leaved Phyllodium,Longstalk Phyllodium ●

297013　Phyllodium pulchellum (L.) Desv.;排钱树(阿婆钱,串钱草,叠钱草,虎尾金钱,尖叶阿婆钱,金钱豹,金钱草,拉里兰子,笠碗子树,猎狸尾草,龙鳞草,排钱草,牌钱树,钱串草,钱串木,钱串子,钱排草,钱排木,双金钱,双排钱,四季春,铜钱树,午时合,午时灵,燕子尾,圆苞小槐花,掌牛奴,纸钱剑,猪狸尾草,竺碗子树);Beautiful Phyllodium,Beech-leaved Flemingia,Round-bracted Tricklover,Spiffy Phyllodium ●

297014　Phyllodoce Link = Acacia Mill. (保留属名) ●■

297015　Phyllodoce Salisb. (1806);松毛翠属(栂樱属,母樱属);Blue Heath,Mountain Heath,Mountainheath,Needlejade ●

297016　Phyllodoce aleutica (Spreng.) A. Heller;阿留申松毛翠;Aleutian Phyllodoce ●☆

297017　Phyllodoce alpina Koidz.;信浓松毛翠;Alpine Phyllodoce ●☆

297018　Phyllodoce breweri (A. Gray) A. Heller;紫红松毛翠;Brewer's Mountain-heather,Red Heather ●☆

297019　Phyllodoce caerulea (L.) Bab.;松毛翠(栂樱,天蓝松毛翠);Blue Heath,Blue Mountain Heath,Blue Mountainheath,Blue Needlejade,Lapp Heather,Menziesia,Mountain Heath,Sky-blue Phyllodoce ●◇

297020　Phyllodoce caerulea (L.) Bab. f. takedana (Tatew.) Ohwi = Phyllodoce caerulea (L.) Bab. var. takedana Tatew. ●☆

297021　Phyllodoce caerulea (L.) Bab. f. yesoensis Nakai;北海道松毛翠;Blue Heath,Yezo Needlejade ●☆

297022　Phyllodoce caerulea (L.) Bab. var. takedana Tatew.;武田松毛翠 ●☆

297023　Phyllodoce deflexa Ching ex H. P. Yang;反折松毛翠;Deflexed Empetrum,Deflexed Needlejade ●

297024　Phyllodoce empetriformis (Sm.) D. Don;岩高兰状松毛翠(岩高兰状母樱);Like Empetrum Phyllodoce,Pink Mountain-heather,Red Mountain-heather ●☆

297025　Phyllodoce glandulifera ?;腺体松毛翠;Yellow Mountain-heather ●☆

297026　Phyllodoce glanduliflora (Hook.) Coville;腺花松毛翠(乳黄松毛翠);Glandular Flowers Phyllodoce ●☆

297027　Phyllodoce intermedia (Hook.) Rydb.;细梗松毛翠(间型松毛翠) ●☆

297028　Phyllodoce intermedia (Hook.) Rydb. 'Drummondii';德拉蒙德细梗松毛翠(德拉蒙德间型松毛翠,朱蒙迪细梗松毛翠) ●☆

297029　Phyllodoce intermedia (Hook.) Rydb. 'Fred Stoker';弗瑞德·斯托克间型松毛翠 ●☆

297030　Phyllodoce nipponica Makino;日本松毛翠(栂樱,母樱,松毛翠);Japanese Phyllodoce ●☆

297031　Phyllodoce nipponica Makino subsp. tsugifolia (Nakai) Toyok.;长叶日本松毛翠 ●☆

297032　Phyllodoce nipponica Makino subsp. tsugifolia Ohwi = Phyllodoce nipponica Makino subsp. tsugifolia (Nakai) Toyok. ●☆

297033　Phyllodoce nipponica Makino var. gracilis Nakai;纤细日本松毛翠 ●☆

297034　Phyllodoce nipponica Makino var. oblongo-ovata (Tatew.) Toyok. ex Ohwi = Phyllodoce nipponica Makino subsp. tsugifolia (Nakai) Toyok. ●☆

297035　Phyllodoce nipponica Makino var. tsugifolia (Nakai) Ohwi = Phyllodoce nipponica Makino subsp. tsugifolia (Nakai) Toyok. ●☆

297036　Phyllodoce taxifolia Salisb. = Phyllodoce caerulea (L.) Bab. ●◇

297037　Phyllodoce tsugifolia Nakai = Phyllodoce nipponica Makino subsp. tsugifolia (Nakai) Toyok. ●☆

297038　Phyllodolon Salisb. = Allium L. ■

297039　Phyllogeiton (Weberb.) Herzog = Berchemia Neck. ex DC. (保留属名) ●

297040　Phyllogeiton (Weberb.) Herzog (1903);象牙木属 ●☆

297041　Phyllogeiton Herzog = Berchemia Neck. ex DC. (保留属名) ●

297042　Phyllogeiton Herzog = Phyllogeiton (Weberb.) Herzog ●☆

297043　Phyllogeiton discolor (Klotzsch) Herzog = Berchemia discolor (Klotzsch) Hemsl. ●☆

297044　Phyllogeiton zeyheri (Sond.) Suess.;象牙木;Dina Red

Ivorywood ●☆

297045　Phyllogeiton zeyheri （Sond.） Suess. = Berchemia zeyheri （Sond.） Grubov ●☆

297046　Phyllogeiton zeyheri （Sond.） Suess. = Rhamnus zeyheri Sond. ●☆

297047　Phylloglottis Salisb. = Eriospermum Jacq. ex Willd. ■☆

297048　Phyllogonum Coville = Gilmania Coville ■☆

297049　Phyllogonum luteolum Coville = Gilmania luteola （Coville） Coville ■☆

297050　Phyllolepidum Trinajsti ć = Alyssum L. ●●

297051　Phyllolepidum Trinajsti ć（1990）;鳞叶荠属■☆

297052　Phyllolobium Fisch. = Astragalus L. ●■

297053　Phyllolobium Fisch. ex Spreng. = Astragalus L. ●■

297054　Phyllolobium chinense Fisch. = Astragalus complanatus R. Br. ex Bunge ■

297055　Phylloma Ker Gawl. = Lomatophyllum Willd. ■☆

297056　Phyllomatia （Wight et Arn.） Benth. = Rhynchosia Lour. （保留属名）●■

297057　Phyllomatia Benth. = Rhynchosia Lour. （保留属名）●■

297058　Phyllomelia Griseb. （1866）;楝叶茜属●☆

297059　Phyllomelia coronata Griseb. ;楝叶茜●☆

297060　Phyllomeria Griseb. = Phyllomelia Griseb. ●☆

297061　Phyllomphax Schltr. = Brachycorythis Lindl. ■

297062　Phyllomphax affinis （D. Don） Schltr. = Peristylus affinis （D. Don） Seidenf. ■

297063　Phyllomphax championii （Lindl.） Schltr. = Brachycorythis galeandra （Rchb. f.） Summerh. ■

297064　Phyllomphax galeandra （Rchb. f.） Hand. -Mazz. = Brachycorythis galeandra （Rchb. f.） Summerh. ■

297065　Phyllomphax galeandra （Rchb. f.） Schltr. = Brachycorythis galeandra （Rchb. f.） Summerh. ■

297066　Phyllomphax henryi Schltr. = Brachycorythis henryi （Schltr.） Summerh. ■

297067　Phyllomphax truncatolabellata （Hayata） Schltr. = Brachycorythis galeandra （Rchb. f.） Summerh. ■

297068　Phyllonoma Schult. = Phyllonoma Willd. ex Schult. ●☆

297069　Phyllonoma Willd. ex Schult. （1820）;叶茶藨属（假茶藨属,叶顶花属）●☆

297070　Phyllonoma integerrima Britton = Phyllonoma ruscifolia Willd. ex Schult. ●☆

297071　Phyllonoma laticuspis （Turcz.） Engl. ;宽尖叶茶藨●☆

297072　Phyllonoma ruscifolia Willd. ex Schult. ;叶茶藨●☆

297073　Phyllonomaceae Rusby = Dulongiaceae J. Agardh ●☆

297074　Phyllonomaceae Rusby = Grossulariaceae DC. （保留科名）●

297075　Phyllonomaceae Rusby = Phyllonomataceae Small ●☆

297076　Phyllonomaceae Small = Grossulariaceae DC. （保留科名）●

297077　Phyllonomaceae Small（1905）;叶茶藨科（假茶藨科）●☆

297078　Phyllonomataceae Small = Dulongiaceae J. Agardh ●☆

297079　Phyllonomataceae Small = Phyllonomaceae Small ●☆

297080　Phyllopappus Walp. = Microseris D. Don ■☆

297081　Phyllopentas （Verdc.） K ？ rehed et B. Bremer（2007）;叶星花属●■☆

297082　Phyllophiorhiza Kuntze = Ophiorrhiziphyllon Kurz ■

297083　Phyllophyton Kudo = Marmorites Benth. ■

297084　Phyllophyton Kudo = Nepeta L. ■●

297085　Phyllophyton Kudo（1929）;肖扭连钱属（扭连钱属）■★

297086　Phyllophyton complanatum （Dunn） Kudo = Marmoritis complanata （Dunn） A. L. Budantzev ■

297087　Phyllophyton decolorans （Hemsl.） Kudo = Marmoritis decolorans （Hemsl.） H. W. Li ■

297088　Phyllophyton nivale （Jacq. ex Benth.） C. Y. Wu = Marmoritis nivalis （Jacq. ex Benth.） Hedge ■

297089　Phyllophyton pharicum （Prain） Kudo = Marmoritis pharica （Prain） A. L. Budantzev ■

297090　Phyllophyton tibeticum （Jacq. ex Benth.） C. Y. Wu = Marmoritis rotundifolia Benth. ■

297091　Phyllopodium Benth. （1836）;叶梗玄参属■☆

297092　Phyllopodium alpinum N. E. Br. ;高山叶梗玄参■☆

297093　Phyllopodium anomalum Hilliard;异常叶梗玄参■☆

297094　Phyllopodium augei Hiern = Manulea augei （Hiern） Hilliard ■☆

297095　Phyllopodium baurii Hiern = Selago baurii （Hiern） Hilliard ■☆

297096　Phyllopodium bracteatum Benth. ;具苞叶梗玄参■☆

297097　Phyllopodium caespitosum Hilliard;丛生叶梗玄参■☆

297098　Phyllopodium calvum Hiern = Melanospermum transvaalense （Hiern） Hilliard ■☆

297099　Phyllopodium capillare （L. f.） Hilliard;纤毛叶梗玄参■☆

297100　Phyllopodium capitatum （L. f.） Benth. = Phyllopodium cephalophorum （Thunb.） Hilliard ■☆

297101　Phyllopodium cephalophorum （Thunb.） Hilliard;头状叶梗玄参■☆

297102　Phyllopodium collinum （Hiern） Hilliard;山丘叶梗玄参■☆

297103　Phyllopodium cordatum （Thunb.） Hilliard;心形叶梗玄参■☆

297104　Phyllopodium cuneifolium （L. f.） Benth. ;楔形叶梗玄参■☆

297105　Phyllopodium diffusum Benth. ;松散叶梗玄参■☆

297106　Phyllopodium dolomiticum Hilliard;多罗米蒂叶梗玄参■☆

297107　Phyllopodium elegans （Choisy） Hilliard;雅致叶梗玄参■☆

297108　Phyllopodium glutinosum Schltr. = Trieenea glutinosa （Schltr.） Hilliard ■☆

297109　Phyllopodium heterophyllum （L. f.） Benth. ;互叶梗玄参■☆

297110　Phyllopodium hispidulum （Thell.） Hilliard;细毛叶梗玄参■☆

297111　Phyllopodium krebsianum Benth. = Glekia krebsiana （Benth.） Hilliard ●☆

297112　Phyllopodium linearifolium Bolus = Phyllopodium elegans （Choisy） Hilliard ■☆

297113　Phyllopodium lupuliforme （Thell.） Hilliard;狼叶梗玄参■☆

297114　Phyllopodium maxii （Hiern） Hilliard;马克斯叶梗玄参■☆

297115　Phyllopodium micranthum （Schltr.） Hilliard;大花叶梗玄参■☆

297116　Phyllopodium mimetes Hilliard;相似叶梗玄参■☆

297117　Phyllopodium minimum Hiern = Zaluzianskya minima （Hiern） Hilliard ■☆

297118　Phyllopodium multifolium Hiern;多叶梗玄参■☆

297119　Phyllopodium namaense （Thell.） Hilliard;纳马叶梗玄参●☆

297120　Phyllopodium phyllopodioides （Schltr.） Hilliard;热非叶梗玄参■☆

297121　Phyllopodium pubiflorum Hilliard;短毛花叶梗玄参■☆

297122　Phyllopodium pumilum Benth. ;矮叶梗玄参■☆

297123　Phyllopodium rangei Engl. = Phyllopodium namaense （Thell.） Hilliard ●☆

297124　Phyllopodium rudolphii Hiern = Pseudoselago humilis （Rolfe） Hilliard ●☆

297125　Phyllopodium rupestre Hiern = Melanospermum rupestre （Hiern） Hilliard ■☆

297126　Phyllopodium rustii （Rolfe） Hilliard;鲁斯特梗玄参●☆

297127　Phyllopodium schlechteri Hiern = Trieenea schlechteri （Hiern） Hilliard ■☆

297128 Phyllopodium sordidum Hiern = Manulea augei（Hiern）Hilliard ■☆

297129 Phyllopodium tweedense Hilliard；特威迪叶梗玄参■☆

297130 Phyllopodium viscidissimum Hilliard；极黏叶梗玄参■☆

297131 Phyllopus DC. = Henriettea DC. ●☆

297132 Phyllorachis Trimen（1879）；叶梗禾属（刺状假叶柄草属）■☆

297133 Phyllorachis sagittata Trimen；叶梗禾■☆

297134 Phyllorchis Thouars = Bulbophyllum Thouars（保留属名）■

297135 Phyllorchis Thouars = Phyllorkis Thouars（废弃属名）■

297136 Phyllorchis Thouars ex Kuntze = Bulbophyllum Thouars（保留属名）■

297137 Phyllorchis Thouars ex Kuntze = Phyllorkis Thouars（废弃属名）■

297138 Phyllorchis baronii（Ridl.）Kuntze = Bulbophyllum baronii Ridl. ■☆

297139 Phyllorchis erecta（Thouars）Kuntze = Bulbophyllum erectum Thouars ■☆

297140 Phyllorchis helenae Kuntze = Bulbophyllum helenae（Kuntze）J. J. Sm. ■

297141 Phyllorchis hildebrandtii（Rchb. f.）Kuntze = Bulbophyllum hildebrandtii Rchb. f. ■☆

297142 Phyllorchis josephii Kuntze = Bulbophyllum josephii（Kuntze）Summerh. ■☆

297143 Phyllorchis longiflora（Thouars）Kuntze = Bulbophyllum longiflorum Thouars ■☆

297144 Phyllorchis minuta（Thouars）Kuntze = Bulbophyllum minutum Thouars ■☆

297145 Phyllorchis multiflora（Ridl.）Kuntze = Bulbophyllum multiflorum Ridl. ■☆

297146 Phyllorchis nutans（Thouars）Kuntze = Bulbophyllum nutans Thouars ■☆

297147 Phyllorchis occlusa（Ridl.）Kuntze = Bulbophyllum occlusum Ridl. ■☆

297148 Phyllorchis stenobulbon（Parl. et Rchb. f.）Kuntze = Bulbophyllum stenobulbon Parl. et Rchb. f. ■

297149 Phyllorchis thompsonii（Ridl.）Kuntze = Bulbophyllum thompsonii Ridl. ■☆

297150 Phyllorchis variegata（Thouars）Kuntze = Bulbophyllum variegatum Thouars ■☆

297151 Phyllorhachis Trimen = Phyllorachis Trimen ■☆

297152 Phyllorkis Thouars（废弃属名）= Bulbophyllum Thouars（保留属名）■

297153 Phyllorkis affinis（Lindl.）Kuntze = Bulbophyllum affine Lindl. ■

297154 Phyllorkis andersonii（Hook. f.）Kuntze = Bulbophyllum andersonii（Hook. f.）J. J. Sm. ■

297155 Phyllorkis bicolor（Lindl.）Kuntze = Bulbophyllum bicolor（Lindl.）Hook. f. ■

297156 Phyllorkis chinensis（Lindl.）Kuntze = Bulbophyllum chinense（Lindl.）Rchb. f. ■

297157 Phyllorkis cylindracea（Lindl.）Kuntze = Bulbophyllum cylindraceum Lindl. ■

297158 Phyllorkis gymnopus（Hook. f.）Kuntze = Bulbophyllum gymnopus Hook. f. ■

297159 Phyllorkis helenae Kuntze = Bulbophyllum helenae（Kuntze）J. J. Sm. ■

297160 Phyllorkis hirta（Sm.）Kuntze = Bulbophyllum hirtum（Sm.）

297161 Phyllorkis monantha Kuntze = Bulbophyllum pteroglossum Schltr. ■

297162 Phyllorkis psittacoglossa（Rchb. f.）Kuntze = Bulbophyllum psittacoglossum Rchb. f. ■

297163 Phyllorkis reptans（Lindl.）Kuntze = Bulbophyllum reptans（Lindl.）Lindl. ■

297164 Phyllorkis retusiuscula（Rchb. f.）Kuntze = Bulbophyllum retusiusculum Rchb. f. ■

297165 Phyllorkis rolfei Kuntze = Bulbophyllum rolfei（Kuntze）Seidenf. ■

297166 Phyllorkis secunda（Hook. f.）Kuntze = Bulbophyllum secundum Hook. f. ■

297167 Phyllorkis stenobulbon（E. C. Parish et Rchb. f.）Kuntze = Bulbophyllum stenobulbon Parl. et Rchb. f. ■

297168 Phyllorkis taeniophylla（E. C. Parish et Rchb. f.）Kuntze = Bulbophyllum taeniophyllum Parl. et Rchb. f. ■

297169 Phyllorkis tristis（Rchb. f.）Kuntze = Bulbophyllum triste Rchb. f. ■

297170 Phyllorkis umbellata（Lindl.）Kuntze = Bulbophyllum umbellatum Lindl. ■

297171 Phyllorkis wallichii（Lindl.）Kuntze = Bulbophyllum wallichii Rchb. f. ■

297172 Phylloschoenus Fourr. = Juncus L. ■

297173 Phylloscirpus C. B. Clarke（1908）；藨叶莎属■☆

297174 Phylloscirpus Döll ex Börner = Scirpus L.（保留属名）■

297175 Phylloscirpus andesinus C. B. Clarke；藨叶莎■☆

297176 Phylloscirpus boliviensis（Barros）Dhooge et Goetgh.；玻利维亚藨叶莎■☆

297177 Phyllosma L. Bolus = Phyllosma L. Bolus ex Schltr. ●☆

297178 Phyllosma L. Bolus ex Schltr.（1897）；烈味芸香属●☆

297179 Phyllosma barosmoides（Dümmer）I. Williams；非洲烈味芸香●☆

297180 Phyllosma capensis L. Bolus；烈味芸香●☆

297181 Phyllospadix Hook.（1838）；虾海藻属；Shrimpalga, Sud Grass, Surfgrass Surf-grass ■

297182 Phyllospadix iwatensis Makino；红纤维虾海藻；Redfibre Shrimpalga ■

297183 Phyllospadix japonicus Makino；黑纤维虾海藻（虾海藻）；Blachfibre Shrimpalga, Japanese Surfgrass ■

297184 Phyllospadix scouleri Asch. = Phyllospadix scouleri Hook. ■☆

297185 Phyllospadix scouleri Hook.；斯库虾海藻（虾海藻）；Scouler's Surf-grass, Scoulfer Surfgrass ■☆

297186 Phyllospadix serrulatus Rupr. ex Asch.；齿虾海藻■☆

297187 Phyllospadix torreyi S. Watson；托里虾海藻；Torrey Surfgrass, Torrey's Surf-grass ■☆

297188 Phyllostachys Siebold et Zucc.（1843）（保留属名）；刚竹属（淡竹属,苦竹属,毛竹属,孟宗竹属）；Bamboo, Black Bamboo, Firmbamboo, Hairy Bamboo, Hairy-bamboo, Phyllostachys, Surf-grass ●

297189 Phyllostachys Torr. = Carex L. ■

297190 Phyllostachys Torr. ex Steud. = Carex L. ■

297191 Phyllostachys acuta C. D. Chu et C. S. Chao；尖头青竹；Acutesprout Bamboo, Acute-sproutted Bamboo, Casp Bamboo ●

297192 Phyllostachys altiligulata G. G. Tang et Y. L. Hsu；高孟哺鸡竹●

297193 Phyllostachys altiligulata G. G. Tang et Y. L. Hsu = Phyllostachys viridi-glaucescens（Carrière）Rivière et C. Rivière ●

297194 Phyllostachys angusta McClure；黄苦竹；Bamboo, Chinese Edible Bamboo, Chinese Fish-rod Bamboo, Narrow Sheath Bamboo,

Narrowsheath Bamboo，Narrow-sheathed Bamboo，Stone Bamboo ●

297195　Phyllostachys arcana McClure；石绿竹；Arcane Bamboo ●

297196　Phyllostachys arcana McClure‘Luteosulcata’；黄槽石绿竹；Yellowgroove Arcane Bamboo ●

297197　Phyllostachys arcana McClure f. luteosulcata C. D. Chu et C. S. Chao ＝ Phyllostachys arcana McClure‘Luteosulcata’●

297198　Phyllostachys aristata W. T. Lin；刺芒刚竹●

297199　Phyllostachys aristata W. T. Lin ＝ Phyllostachys rubromarginata McClure ●

297200　Phyllostachys assamica Gamble ex Brandis；阿萨姆刚竹；Assam Hairy Bamboo ●

297201　Phyllostachys assamica Gamble ex Brandis ＝ Phyllostachys mannii Gamble ●

297202　Phyllostachys atrovaginata C. S. Chao et H. Y. Chou；乌芽竹；Blackbud Bamboo，Black-budded Bamboo，Black-sheath Bamboo ●

297203　Phyllostachys aurea Carrière ex Rivière et C. Rivière；人面竹（八面竹，布袋竹，短舌刚竹，佛肚竹，鼓槌竹，虎山竹，姜竹，罗汉竹，台湾人面竹，吴竹，五三竹，仙人杖）；Fishpole Bamboo，Fish-poly Bamboo，Fourtain Bamboo，Golden Bamboo，Golden Fishpole Bamboo，Monk's Belly Bamboo，Running Bamboo，Yellow Bamboo ●

297204　Phyllostachys aureosulcata McClure；黄槽竹；Golden-groove Bamboo，Groove Bamboo，Yellow Groove Bamboo，Yellow-groove Bamboo ●

297205　Phyllostachys aureosulcata McClure‘Aureocaulis’；黄杆金竹（黄竿金竹，黄竿京竹）●

297206　Phyllostachys aureosulcata McClure‘Pekinensis’；京竹●

297207　Phyllostachys aureosulcata McClure‘Spectabilis’；金镶玉竹；Greengroove Bamboo，Green-grooved Bamboo ●

297208　Phyllostachys aureosulcata McClure f. alata T. H. Wen ＝ Phyllostachys aureosulcata McClure‘Pekinensis’●

297209　Phyllostachys aureosulcata McClure f. aureocaulis Z. P. Wang et N. X. Ma ＝ Phyllostachys aureosulcata McClure‘Aureocaulis’●

297210　Phyllostachys aureosulcata McClure f. pekinensis（J. L. Lu）T. H. Wen ＝ Phyllostachys aureosulcata McClure‘Pekinensis’●

297211　Phyllostachys aureosulcata McClure f. pekinensis J. L. Lu ＝ Phyllostachys aureosulcata McClure‘Pekinensis’●

297212　Phyllostachys aureosulcata McClure f. spectabilis C. D. Chu et C. S. Chao ＝ Phyllostachys aureosulcata McClure‘Spectabilis’●

297213　Phyllostachys aurita J. L. Lu；毛环水竹（流苏唐竹）；Ariculate Bamboo，Hairring Fishscale Bamboo，Long-eared Bamboo ●

297214　Phyllostachys aurita J. L. Lu ＝ Phyllostachys rubromarginata McClure ●

297215　Phyllostachys bambusoides Siebold et Zucc.；桂竹（碍角竹，斑竹，刚竹，钢铁头竹，光竹，鬼角竹，黑壳竹，箭竹，苦竹，日本苦竹，台竹，真竹）；Castillon Bamboo，Embroidered Sheath Bamboo，Giant Timber Bamboo，Giant-timbered Bamboo，Golden Bamboo，Hard Giant-timbered Bamboo，Hardy Giant Timber Bamboo，Hardy Timber Bamboo，Japanese Timber Bamboo，Madake Bamboo，Timber Bamboo ●

297216　Phyllostachys bambusoides Siebold et Zucc.‘Allgold’ ＝ Phyllostachys bambusoides Siebold et Zucc.‘Castilloni-inversa’●

297217　Phyllostachys bambusoides Siebold et Zucc.‘Allgold’ ＝ Phyllostachys sulphurea（Carrière）Rivière et C. Rivière ●

297218　Phyllostachys bambusoides Siebold et Zucc.‘Castillon’；砾竹；Castillon Bamboo，Marliac's Castillon Bamboo，Variegated Castillon Bamboo ●

297219　Phyllostachys bambusoides Siebold et Zucc.‘Castilloni-inversa’ ＝ Phyllostachys bambusoides Siebold et Zucc.‘Castillon’●

297220　Phyllostachys bambusoides Siebold et Zucc.‘Holochrysa’；全金桂竹；All Gold Bamboo ●☆

297221　Phyllostachys bambusoides Siebold et Zucc.‘Lacrimadeae’；斑竹；Spot Giant Timber Bamboo ●

297222　Phyllostachys bambusoides Siebold et Zucc.‘Marliacea’；皱竹；Marliac's Bamboo，Wrinkled Bamboo ●

297223　Phyllostachys bambusoides Siebold et Zucc.‘Slender Crookstem’；纤细竹；Slender Crookstem Japanese Bamboo ●

297224　Phyllostachys bambusoides Siebold et Zucc.‘White Crookstem’；白桂竹●

297225　Phyllostachys bambusoides Siebold et Zucc. ＝ Phyllostachys reticulata（Rupr.）K. Koch ●

297226　Phyllostachys bambusoides Siebold et Zucc. f. castillonis（Mitford）Muroi ＝ Phyllostachys bambusoides Siebold et Zucc.‘Castilloni-inversa’●

297227　Phyllostachys bambusoides Siebold et Zucc. f. castillonis（Mitford）Muroi ＝ Phyllostachys bambusoides Siebold et Zucc.‘Castillon’●

297228　Phyllostachys bambusoides Siebold et Zucc. f. lacrimadeae P. C. Keng et T. H. Wen ＝ Phyllostachys bambusoides Siebold et Zucc.‘Lacrimadeae’●

297229　Phyllostachys bambusoides Siebold et Zucc. f. mixta Z. P. Wang et N. X. Ma；黄槽斑竹；Mixed Giant Timber Bamboo ●

297230　Phyllostachys bambusoides Siebold et Zucc. f. shouzhu T. P. Yi；寿竹；Shouzhu BambooMi ●

297231　Phyllostachys bambusoides Siebold et Zucc. f. tamake Makino ex Tsuboi ＝ Phyllostachys bambusoides Siebold et Zucc. f. lacrimadeae P. C. King et T. H. Wen ●

297232　Phyllostachys bambusoides Siebold et Zucc. f. tamake Makino ex Tsuboi ＝ Phyllostachys bambusoides Siebold et Zucc.‘Lacrimadeae’●

297233　Phyllostachys bambusoides Siebold et Zucc. f. zitchiku Makino；实竹（木竹，实心竹，印材竹）；Zitchiku Bamboo ●

297234　Phyllostachys bambusoides Siebold et Zucc. f. zitchiku Makino ＝ Phyllostachys bambusoides Siebold et Zucc. ●

297235　Phyllostachys bambusoides Siebold et Zucc. f. zitchiku Makino ＝ Phyllostachys heteroclada Oliv. f. solida（S. L. Chen）Z. P. Wang et Z. H. Yu ●

297236　Phyllostachys bambusoides Siebold et Zucc. var. aurea（Carrière ex Rivière et C. Rivière）Makino ＝ Phyllostachys aurea Carrière ex Rivière et C. Rivière ●

297237　Phyllostachys bambusoides Siebold et Zucc. var. aurea（Siebold）Makino ＝ Phyllostachys aurea Carrière ex Rivière et C. Rivière ●

297238　Phyllostachys bambusoides Siebold et Zucc. var. castilloni-holochrysa（Pfitzer）J. Houz. ＝ Phyllostachys sulphurea（Carrière）Rivière et C. Rivière ●

297239　Phyllostachys bambusoides Siebold et Zucc. var. castillonis（Lat. -Marl.）J. Houz.；黄金间碧玉竹（金明竹，青叶竹，水竹）；Inverted Castillon Bamboo ●

297240　Phyllostachys bambusoides Siebold et Zucc. var. mariliacea（Mitford）Makino ＝ Phyllostachys bambusoides Siebold et Zucc.‘Marliacea’●

297241　Phyllostachys bambusoides Siebold et Zucc. var. sulphurea Makino ex Tsuboi ＝ Phyllostachys sulphurea（Carrière）Rivière et C. Rivière ●

297242　Phyllostachys bapida T. P. Yi ＝ Phyllostachys flexuosa（Carrière）Rivière et C. Rivière ●

297243 Phyllostachys bawa E. G. Camus ＝Phyllostachys mannii Gamble ●

297244 Phyllostachys bissetii McClure；蓉城竹（百夹竹）；Bisset's Bamboo，Rongcheng Bamboo ●

297245 Phyllostachys bissetii McClure var. denigrata T. P. Yi et H. R. Qi；黑蓉城竹；Blak Rongcheng Bamboo ●

297246 Phyllostachys bissettii McClure 'Dwarf'；矮蓉城竹；Bamboo ●

297247 Phyllostachys breviligula W. T. Lin et Z. M. Wu ＝Phyllostachys aurea Carrière ex Rivière et C. Rivière ●

297248 Phyllostachys cantoniensis W. T. Lin ＝Phyllostachys nidularia Munro ●

297249 Phyllostachys carnea G. H. Ye et Z. P. Wang；湖南刚竹●

297250 Phyllostachys carnea G. H. Ye et Z. P. Wang ＝Phyllostachys heteroclada Oliv. ●

297251 Phyllostachys castillomi Mitford var. holochrysa Pfitzer ＝Phyllostachys sulphurea（Carrière）Rivière et C. Rivière ●

297252 Phyllostachys cerate McClure ＝Phyllostachys heteroclada Oliv. ●

297253 Phyllostachys chlorina T. H. Wen；黄鞍竹●

297254 Phyllostachys chlorina T. H. Wen ＝Phyllostachys sulphurea（Carrière）Rivière et C. Rivière var. viridis R. A. Young ●

297255 Phyllostachys chlorina T. H. Wen ＝Phyllostachys sulphurea（Carrière）Rivière et C. Rivière 'Viridis' ●

297256 Phyllostachys circumpilis C. Y. Yao et S. Y. Chen；毛壳花哺鸡竹；Circumpilose Bamboo，Hairsheath Broodhen Bamboo ●

297257 Phyllostachys concava Z. H. Yu et Z. P. Wang ＝Phyllostachys rubicunda T. H. Wen ●

297258 Phyllostachys congesta McClure ＝Phyllostachys atrovaginata C. S. Chao et H. Y. Chou ●

297259 Phyllostachys congesta Rendle ＝Phyllostachys heteroclada Oliv. ●

297260 Phyllostachys decora McClure ＝Phyllostachys mannii Gamble ●

297261 Phyllostachys dubia Keng；疑竹；Doubt Bamboo ●☆

297262 Phyllostachys dubia Keng ＝Phyllostachys heteroclada Oliv. ●

297263 Phyllostachys dulcis McClure；白哺鸡竹（甜竹）；Chinese Edible Bamboo，Sweetshoot Bamboo，Sweet-shooted Bamboo，White Broodhen Bamboo ●

297264 Phyllostachys edulis（Carrière）J. Houz. ＝Phyllostachys heterocycla（Carrière）Matsum. 'Pubescens' ●

297265 Phyllostachys edulis（Carrière）J. Houz. f. huamozhu（T. H. Wen）C. S. Chao et Renvoize ＝Phyllostachys heterocycla（Carièrre）Matsum. 'Tao Kiang' ●

297266 Phyllostachys edulis Rivière et C. Rivière；孟宗竹（江南竹，毛竹，茅茹竹，貌儿竹，貌头竹，南竹）；Edible Bamboo，Maso Bamboo，Tortoise Shell Bamboo ●

297267 Phyllostachys edulis Rivière et C. Rivière f. luteosulcata（T. H. Wen）C. S. Chao et Renvoize ＝Phyllostachys heterocycla（Carrière）Matsum. 'Luteosulcata' ●

297268 Phyllostachys edulis Rivière et C. Rivière f. viridisulcata（T. H. Wen）C. S. Chao et Renvoize ＝Phyllostachys heterocycla（Carrière）Matsum. 'Viridisulcata' ●

297269 Phyllostachys edulis Rivière et C. Rivière var. heterocycla（Carrière）J. Houz. ＝Phyllostachys heteroclada Oliv. ●

297270 Phyllostachys elegans McClure；甜笋竹；Graceful Bamboo，Sweet Bamboo ●

297271 Phyllostachys erecta T. H. Wen；乌壳鳗竹；Erect Bamboo ●

297272 Phyllostachys erecta T. H. Wen ＝Phyllostachys robustiramea S. Y. Chen et C. Y. Yao ●

297273 Phyllostachys faberi Rendle；费氏竹；Faber Bamboo ●

297274 Phyllostachys faberi Rendle ＝Phyllostachys sulphurea（Carrière）Rivière et C. Rivière var. viridis R. A. Young ●

297275 Phyllostachys faberi Rendle ＝Phyllostachys sulphurea（Carrière）Rivière et C. Rivière 'Viridis' ●

297276 Phyllostachys fastuosa（Mitford）Pfitzer ＝Semiarundinaria fastuosa（Mitford）Makino ex Nakai ●

297277 Phyllostachys fauriei Hack. ＝Phyllostachys nigra（Lodd. ex Lindl.）Munro var. henonis（Bean）Stapf ex Rendle ●

297278 Phyllostachys filifera McClure；丝竹；Thread Bamboo ●

297279 Phyllostachys filifera McClure ＝Phyllostachys nigra（Lodd. ex Lindl.）Munro ●

297280 Phyllostachys fimbriligula T. H. Wen；角竹；Fimbriate-ligular Bamboo，Fimbriligulate Bamboo，Horn Bamboo ●

297281 Phyllostachys flexuosa（Carrière）Rivière et C. Rivière；曲秆竹（曲竿竹，甜竹）；Drooping Timber Bamboo，Flexuose Bamboo，Sinuate Bamboo，Zigzag Bamboo，Zig-zag Bamboo ●

297282 Phyllostachys formofana Hayata；台湾竹（人面竹，台湾刚竹，台湾人面竹）；Formosan Bamboo，Taiwan Hair Bamboo ●

297283 Phyllostachys formofana Hayata ＝Phyllostachys aurea Carrière ex Rivière et C. Rivière ●

297284 Phyllostachys glabrata S. Y. Chen et C. Y. Yao；花哺鸡竹；Colony Broodhen Bamboo，Glabrate Bamboo，Glabrous Bamboo ●

297285 Phyllostachys glauca McClure；淡竹（粉绿竹，黄竿，黄金竹，黄皮竹，黄竹，金竹，筠竹）；Glaucous Bamboo，Palegreen Bamboo ●

297286 Phyllostachys glauca McClure 'Yunzhu'；筠竹；Yunzhu ●

297287 Phyllostachys glauca McClure f. yunzhu J. L. Lu ＝Phyllostachys glauca McClure 'Yunzhu' ●

297288 Phyllostachys glauca McClure var. variabilis J. L. Lu；变竹；Changeable Bamboo，Variabile Glaucous Bamboo ●

297289 Phyllostachys guizhouensis C. S. Chao et J. Q. Zhang；贵州刚竹；Guizhou Bamboo ●

297290 Phyllostachys helva T. H. Wen；红鸡竹●

297291 Phyllostachys helva T. H. Wen ＝Phyllostachys mannii Gamble ●

297292 Phyllostachys henonis Bean ＝Phyllostachys nigra（Lodd. ex Lindl.）Munro var. henonis（Bean）Stapf ex Rendle ●

297293 Phyllostachys henonis Mitford ＝Phyllostachys nigra（Lodd. ex Lindl.）Munro var. henonis（Bean）Stapf ex Rendle ●

297294 Phyllostachys henryi Rendle；亨利刚竹；Henry Bamboo ●

297295 Phyllostachys henryi Rendle ＝Phyllostachys nigra（Lodd. ex Lindl.）Munro var. henonis（Bean）Stapf ex Rendle ●

297296 Phyllostachys heteroclada Oliv.；水竹（湖南刚竹，角竹）；Fishscale Bamboo，Horn Bamboo，Water Bamboo ●

297297 Phyllostachys heteroclada Oliv. f. decurtata（S. L. Chen）T. H. Wen ＝Phyllostachys heteroclada Oliv. f. solida（S. L. Chen）Z. P. Wang et Z. H. Yu ●

297298 Phyllostachys heteroclada Oliv. f. purpurata（McClure）T. H. Wen；黎子竹；Purple Fishscale Bamboo ●

297299 Phyllostachys heteroclada Oliv. f. solida（S. L. Chen）Z. P. Wang et Z. H. Yu ＝Phyllostachys heteroclada Oliv. f. decurtata（S. L. Chen）T. H. Wen ●

297300 Phyllostachys heteroclada Oliv. f. solida（S. L. Chen）Z. P. Wang et Z. H. Yu；实心竹（木竹）；Solid Fishscale Bamboo ●

297301 Phyllostachys heterocycla（Carrière）Matsum.；龟甲竹；Heterocycle Bamboo，Tortoiseshell Bamboo，Turtleback Bamboo ●

297302 Phyllostachys heterocycla（Carrière）Matsum. 'Gracilis'；金丝毛竹●

297303 Phyllostachys heterocycla（Carrière）Matsum. 'Luteosulcata'；黄槽毛竹●

297304　Phyllostachys heterocycla（Carrière）Matsum.'Obliguinoda'；强竹●

297305　Phyllostachys heterocycla（Carrière）Matsum.'Obtusangula'；梅花毛竹●

297306　Phyllostachys heterocycla（Carrière）Matsum.'Pubescens'；毛竹（刮肠篦,江南竹,狸头竹,猫儿竹,猫头竹,茅茹竹,茅竹,孟宗竹,南竹）；Giant Hairysheath Edible Bamboo,Mose Bamboo,Moso Bamboo ●

297307　Phyllostachys heterocycla（Carrière）Matsum.'Tao Kiang'；花毛竹●

297308　Phyllostachys heterocycla（Carrière）Matsum.'Tetrangula'；方秆毛竹（方竿毛竹）●

297309　Phyllostachys heterocycla（Carrière）Matsum.'Tubaeformis'；圣音毛竹●

297310　Phyllostachys heterocycla（Carrière）Matsum.'Ventricosa'；佛肚毛竹●

297311　Phyllostachys heterocycla（Carrière）Matsum.'Viridisulcata'；绿槽毛竹（金鞭毛竹）●

297312　Phyllostachys heterocycla（Carrière）Matsum. f. huamaozhu（T. H. Wen）T. H. Wen ＝ Phyllostachys heterocycla（Carrière）Matsum.'Tao Kiang'●

297313　Phyllostachys heterocycla（Carrière）Matsum. f. luteosulcata（T. H. Wen）T. H. Wen ＝ Phyllostachys heterocycla（Carrière）Matsum.'Luteosulcata'●

297314　Phyllostachys heterocycla（Carrière）Matsum. f. nabeshimana（Muroi）Muroi ＝ Phyllostachys heterocycla（Carrière）Matsum.'Tao Kiang'●

297315　Phyllostachys heterocycla（Carrière）Matsum. f. pubescens（Mazel ex J. Houz.）Muroi ＝Phyllostachys heterocycla（Carrière）Matsum. ●

297316　Phyllostachys heterocycla（Carrière）Matsum. f. pubescens（Mazel ex J. Houz.）Muroi ＝ Phyllostachys heterocycla（Carrière）Matsum.'Pubescens'●

297317　Phyllostachys heterocycla（Carrière）Matsum. f. viridisulcata（T. H. Wen）T. H. Wen ＝ Phyllostachys heterocycla（Carrière）Matsum.'Viridisulcata'●

297318　Phyllostachys heterocycla（Carrière）Matsum. var. pubescens（Mazel ex J. Houz.）Ohwi ＝Phyllostachys heterocycla（Carrière）Matsum. ●

297319　Phyllostachys heterocycla（Carrière）Matsum. var. pubescens（Mazel ex J. Houz.）Ohwi ＝ Phyllostachys heterocycla（Carrière）Matsum.'Pubescens'●

297320　Phyllostachys heterocycla（Carrière）Matsum. var. pubescens（Mazel ex J. Houz.）Ohwi f. ventricosa Z. P. Wang et N. X. Ma ＝Phyllostachys heterocycla（Carrière）Matsum.'Ventricosa'●

297321　Phyllostachys heterocycla（Carrière）Matsum. var. pubescens（Mazel ex J. Houz.）Ohwi ＝Phyllostachys edulis Rivière et C. Rivière ●

297322　Phyllostachys heterocycla（Carrière）Matsum. var. pubescens（Mazel）Ohwi ＝ Phyllostachys heterocycla（Carrière）Matsum.'Pubescens'●

297323　Phyllostachys heterocycla（Carrière）Matsum. var. pubescnes f. obliguinoda Z. P. Wanget N. X. Ma ＝ Phyllostachys heterocycla（Carrière）Matsum.'Obliguinoda'●

297324　Phyllostachys heterocycla（Carrière）Mitford ＝ Phyllostachys edulis Rivière et C. Rivière ●

297325　Phyllostachys hispida S. C. Li, S. H. Wu et S. Y. Chen；毛壳竹；Hispid Bamboo ●

297326　Phyllostachys hispida S. C. Li, S. H. Wu et S. Y. Chen ＝ Phyllostachys varioauriculata S. C. Li et S. H. Wu ●

297327　Phyllostachys incarnata T. H. Wen；红壳雷竹；Flesh-coloured Bamboo,Incarnate Bamboo,Redsheath Bamboo ●

297328　Phyllostachys iridescens C. Y. Yao et S. Y. Chen；红哺竹（红哺鸡竹,红壳竹）；Monal Bamboo, Red Broodhen Bamboo, Redsheath Bamboo ●

297329　Phyllostachys iridescens C. Y. Yao et S. Y. Chen f. striata T. H. Wen；康岭红竹；Kangling Bamboo ●

297330　Phyllostachys kumasaca（Zoll. ex Steud.）Munro ＝ Shibataea kumasasa（Zoll. ex Steud.）Makino ex Nakai ●

297331　Phyllostachys kumasasa（Zoll. ex Steud.）Munro ＝ Shibataea kumasasa（Zoll. ex Steud.）Makino ex Nakai ●

297332　Phyllostachys kumasaca（Zoll.）Munro ＝ Shibataea kumasasa（Zoll. ex Steud.）Makino ex Nakai ●

297333　Phyllostachys kwangsiensis W. Y. Hsiung, Q. H. Dai et J. K. Liu；假毛竹；Guangxi Bamboo, Kwangsi Bamboo ●

297334　Phyllostachys lithophila Hayata；轿杠竹（石生竹,石竹,石竹子）；Lithophilous Bamboo, Rock-loving Bamboo, Thill Bamboo ●

297335　Phyllostachys lithophila Hayata ＝ Phyllostachys reticulata（Rupr.）K. Koch ●

297336　Phyllostachys lofushanensis Z. P. Wang et al.；大节刚竹（罗浮刚竹）；Largenode Bamboo, Lofushan Bamboo ●

297337　Phyllostachys makinoi Hayata；台湾桂竹（桂竹,桂竹仔,笙笋,笙竹,仙人杖）；Makino Bamboo, Taiwan Bamboo ●

297338　Phyllostachys mannii Gamble；美竹（吴竹）；Beautiful Bamboo, Mann's Bamboo, Spiffy Bamboo ●

297339　Phyllostachys marmorea（Mill.）Asch. et Graebn. ＝ Chimonobambusa marmorea（Mitford）Makino ex Nakai ●◇

297340　Phyllostachys marmorea（Mitford）Asch. et Graebn. ＝ Chimonobambusa marmorea（Mitford）Makino ex Nakai ●◇

297341　Phyllostachys mau-diae Dunn ＝Pseudosasa hindsii（Munro）S. L. Chen et G. Y. Sheng ex T. G. Liang ●

297342　Phyllostachys maudie Dunn；广东竹●

297343　Phyllostachys megastachya Steud. ＝ Phyllostachys reticulata（Rupr.）K. Koch ●

297344　Phyllostachys meyeri McClure；毛环竹（梅氏竹,浙江淡竹）；Meyer Bamboo, Meyer's Bamboo ●

297345　Phyllostachys meyeri McClure f. sphaeroidea T. H. Wen；美姑扁竹●

297346　Phyllostachys meyeri McClure f. sphaeroidea T. H. Wen ＝ Phyllostachys sulphurea（Carrière）Rivière et C. Rivière 'Viridis'●

297347　Phyllostachys mitis Poir. var. sulphurea（Carrière）Carrière ＝ Phyllostachys sulphurea（Carrière）Rivière et C. Rivière ●

297348　Phyllostachys mitis Riv. et C. Riv. ＝ Phyllostachys sulphurea（Carrière）Rivière et C. Rivière 'Viridis'●

297349　Phyllostachys mitis Riv. et C. Riv. var. sulphurea（Carrière）J. Houz. ＝ Phyllostachys sulphurea（Carrière）Rivière et C. Rivière ●

297350　Phyllostachys montana Rendle ＝ Phyllostachys nigra（Lodd. ex Lindl.）Munro var. henonis（Bean）Stapf ex Rendle ●

297351　Phyllostachys montana Renedle ＝ Phyllostachys nigra（Lodd. ex Lindl.）Munro var. henonis（Bean）Stapf ex Rendle ●

297352　Phyllostachys nana Rendle；矮竹；Dwarf Bamboo ●

297353　Phyllostachys nana Rendle ＝Phyllostachys nigra（Lodd. ex Lindl.）Munro ●

297354　Phyllostachys neveinii Hance；尼氏毛竹；Nevin Bamboo ●

297355　Phyllostachys neveinii Hance ＝ Phyllostachys nigra（Lodd. ex Lindl.）Munro var. henonis（Bean）Stapf ex Rendle ●

297356　Phyllostachys nevinii Hance ＝ Phyllostachys nigra（Lodd. ex Lindl.）Munro var. henonis（Bean）Stapf ex Rendle ●

297357　Phyllostachys nevinii Hance var. hupehensis Rendle ＝ Phyllostachys nigra (Lodd. ex Lindl.) Munro var. henonis (Bean) Stapf ex Rendle ●

297358　Phyllostachys nevinii var. hupehensis Rendle = Phyllostachys nigra (Lodd. ex Lindl.) Munro var. henonis (Bean) Stapf ex Rendle ●

297359　Phyllostachys nidularia Munro；篔竹（春花小竹，花竹，水竹）；Flower Bamboo ●

297360　Phyllostachys nidularia Munro 'Smoothsheath'；光箨篔竹；Smooth-sheath Flower Bamboo ●

297361　Phyllostachys nidularia Munro f. farcta H. R. Zhao et A. T. Liu；实肚竹；Stuffed Flower Bamboo ●

297362　Phyllostachys nidularia Munro f. glabrovagina (McClure) T. H. Wen = Phyllostachys nidularia Munro 'Smoothsheath' ●

297363　Phyllostachys nidularia Munro f. mirabilis T. P. Yi et C. Q. Shen；丝秆黄槽百夹竹●

297364　Phyllostachys nidularia Munro f. subeurea T. P. Yi et C. G. Chen；金黄百夹竹●

297365　Phyllostachys nidularia Munro f. vexillaria T. H. Wen；蝶竹；Vexillary Flower Bamboo ●

297366　Phyllostachys nigella T. H. Wen；富阳乌哺鸡竹（富阳乌脯鸡竹）；Fuyang Bamboo，Fuyang Broodhen Bamboo ●

297367　Phyllostachys nigra (Lodd. ex Lindl.) Munro；紫竹（虫笋，黑竹，箭，篆，水竹子，乌竹）；Black Bamboo，Calcutta Bamboo，Munro Black Bamboo，Purple Bamboo，Wangee，Whangee ●

297368　Phyllostachys nigra (Lodd. ex Lindl.) Munro 'Flavescens'；浅黄紫竹●☆

297369　Phyllostachys nigra (Lodd. ex Lindl.) Munro 'Henon' = Phyllostachys nigra (Lodd. ex Lindl.) Munro var. henonis (Bean) Stapf ex Rendle ●

297370　Phyllostachys nigra (Lodd. ex Lindl.) Munro 'Pendula'；垂紫竹●☆

297371　Phyllostachys nigra (Lodd. ex Lindl.) Munro f. henonis (Mitford) Muroi = Phyllostachys nigra (Lodd. ex Lindl.) Munro var. henonis (Bean) Stapf ex Rendle ●

297372　Phyllostachys nigra (Lodd. ex Lindl.) Munro f. henonis Mnroi ex Sugim. = Phyllostachys nigra (Lodd. ex Lindl.) Munro var. henonis (Bean) Stapf ex Rendle ●

297373　Phyllostachys nigra (Lodd. ex Lindl.) Munro var. henonis (Bean) Stapf ex Rendle；毛金竹（白夹竹茹，斑真青秆竹茹，瘪竹，虫笋，淡竹，淡竹皮茹，淡竹茹，钓鱼竹，钓鱼竹茹，杜圆竹茹，甘竹，甘竹茹，光苦竹茹，恒生骨，姜竹茹，金竹，金竹花，荆竹茹，罗汉竹茹，麻巴，平竹茹，青竹茹，如金竹茹，水竹茹，退秧竹，仙人杖，竹二皮，竹二青，竹皮，竹茹）；Goldenhair Bamboo，Henon Bamboo ●

297374　Phyllostachys nigra (Lodd. ex Lindl.) Munro var. puberula (Miq.) Fiori = Phyllostachys nigra (Lodd. ex Lindl.) Munro var. henonis (Bean) Stapf ex Rendle ●

297375　Phyllostachys nigripes Hayata；黑梗竹（鞭竹，乌竹，乌竹仔）；Taiwan Black Bamboo ●

297376　Phyllostachys nigripes Hayata = Phyllostachys nigra (Lodd. ex Lindl.) Munro ●

297377　Phyllostachys nigrivagina T. H. Wen = Phyllostachys bambusoides Siebold et Zucc. ●

297378　Phyllostachys nigrivagina T. H. Wen = Phyllostachys viridi-glaucescens (Carrière) Rivière et C. Rivière ●

297379　Phyllostachys nuda McClure；灰竹（净竹，裸箨竹，石竹）；Grey Bamboo，Nude Phyllostachys ●

297380　Phyllostachys nuda McClure 'Localis'；紫浦头灰竹●

297381　Phyllostachys nuda McClure f. localis Z. P. Wang et Z. H. Yu = Phyllostachys nuda McClure 'Localis' ●

297382　Phyllostachys nuda McClure f. lucida T. H. Wen；光秆石竹●

297383　Phyllostachys nuda McClure f. lucida T. H. Wen = Phyllostachys nuda McClure ●

297384　Phyllostachys nuda McClure f. lucida T. H. Wen = Phyllostachys tianmuensis Z. P. Wang et N. X. Ma ●

297385　Phyllostachys parvifolia C. D. Chu et H. Y. Chou；安吉金竹；Littleleaf Bamboo，Small-leaf Bamboo，Small-leaved Bamboo ●

297386　Phyllostachys parvifolia C. D. Chu et H. Y. Chou f. lignosa T. H. Wen；实壁竹●

297387　Phyllostachys parvifolia C. D. Chu et H. Y. Chou f. lignosa T. H. Wen = Phyllostachys heteroclada Oliv. f. solida (S. L. Chen) Z. P. Wang et Z. H. Yu ●

297388　Phyllostachys pinyanensis T. H. Wen；水桂竹；Pinyuan Bamboo ●

297389　Phyllostachys pinyanensis T. H. Wen = Phyllostachys bambusoides Siebold et Zucc. ●

297390　Phyllostachys pinyanensis T. H. Wen = Phyllostachys reticulata (Rupr.) K. Koch ●

297391　Phyllostachys platyglossa C. P. Wang et Z. H. Yu；灰水竹；Broad-ligular Bamboo，Martar Bamboo，Platyglossey Bamboo ●

297392　Phyllostachys praecox C. D. Chu et C. S. Chao 'Notata'；黄条早竹（黄条早竹）●

297393　Phyllostachys praecox C. D. Chu et C. S. Chao 'Praecox' = Phyllostachys praecox C. D. Chu et C. S. Chao ●

297394　Phyllostachys praecox C. D. Chu et C. S. Chao 'Prevernalis'；雷竹（花秆早竹）●

297395　Phyllostachys praecox C. D. Chu et C. S. Chao = Phyllostachys violascens (Carrière) Rivière et C. Rivière ●

297396　Phyllostachys praecox C. D. Chu et C. S. Chao f. notata S. Y. Chen et C. Y. Yao = Phyllostachys praecox C. D. Chu et C. S. Chao 'Notata' ●

297397　Phyllostachys praecox C. D. Chu et C. S. Chao f. prevernalis S. Y. Chen et C. Y. Yao = Phyllostachys praecox C. D. Chu et C. S. Chao 'Prevernalis' ●

297398　Phyllostachys primotina T. H. Wen；遂昌雷竹（遂昌早竹）；Suichang Bamboo ●

297399　Phyllostachys primotina T. H. Wen = Phyllostachys incarnata T. H. Wen ●

297400　Phyllostachys prominens W. Y. Xiong；高节竹；Prominent Bamboo，Tallnode Bamboo ●

297401　Phyllostachys propinqua McClure；早园竹（沙竹）；Early Garden Bamboo，Propinquity Bamboo ●

297402　Phyllostachys propinqua McClure f. lanuginosa T. H. Wen；望江哺鸡竹●

297403　Phyllostachys propinqua McClure f. lanuginosa T. H. Wen = Phyllostachys propinqua McClure ●

297404　Phyllostachys puberula (Miq.) Munro = Phyllostachys nigra (Lodd. ex Lindl.) Munro var. henonis (Bean) Stapf ex Rendle ●

297405　Phyllostachys puberula (Miq.) Munro var. nigra (Lodd. ex Lindl.) Makino = Phyllostachys nigra (Lodd. ex Lindl.) Munro ●

297406　Phyllostachys puberula (Miq.) Munro var. nigra (Lodd.) J. J. Houz. = Phyllostachys nigra (Lodd. ex Lindl.) Munro ●

297407　Phyllostachys puberula Makino = Phyllostachys nigra (Lodd. ex Lindl.) Munro ●

297408　Phyllostachys pubescens Mazel ex J. Houz. 'Tao King' = Phyllostachys heterocycla (Carrière) Matsum. 'Tao Kiang' ●

297409　Phyllostachys pubescens Mazel ex J. Houz. = Phyllostachys edulis Rivière et C. Rivière ●

297410　Phyllostachys pubescens Mazel ex J. Houz. = Phyllostachys heterocycla（Carrière）Matsum. ●

297411　Phyllostachys pubescens Mazel ex J. Houz. = Phyllostachys heterocycla（Carrière）Matsum. 'Pubescens' ●

297412　Phyllostachys pubescens Mazel ex J. Houz. f. gracilis W. Y. Hsiung = Phyllostachys heterocycla（Carrière）Matsum. 'Gracilis' ●

297413　Phyllostachys pubescens Mazel ex J. Houz. f. huamaozhu T. H. Wen = Phyllostachys heterocycla（Carrière）Matsum. 'Tao Kiang' ●

297414　Phyllostachys pubescens Mazel ex J. Houz. f. juamozhu T. H. Wen = Phyllostachys heterocycla（Carrière）Matsum. 'Tao Kiang' ●

297415　Phyllostachys pubescens Mazel ex J. Houz. f. lutea T. H. Wen = Phyllostachys heterocycla（Carrière）Matsum. 'Pubescens' ●

297416　Phyllostachys pubescens Mazel ex J. Houz. f. luteosulcata T. H. Wen = Phyllostachys heterocycla（Carrière）Matsum. 'Luteosulcata' ●

297417　Phyllostachys pubescens Mazel ex J. Houz. f. obtusangula S. Y. Wang = Phyllostachys heterocycla（Carrière）Matsum. 'Obtusangula' ●

297418　Phyllostachys pubescens Mazel ex J. Houz. f. quadrangulata S. Y. Wang = Phyllostachys heterocycla（Carrière）Matsum. 'Tetrangula' ●

297419　Phyllostachys pubescens Mazel ex J. Houz. f. tetrangulata S. Y. Wang = Phyllostachys heterocycla（Carrière）Matsum. 'Tetrangula' ●

297420　Phyllostachys pubescens Mazel ex J. Houz. f. tubaeformis S. Y. Wang = Phyllostachys heterocycla（Carrière）Matsum. 'Tubaeformis' ●

297421　Phyllostachys pubescens Mazel ex J. Houz. f. viridosulcata T. H. Wen = Phyllostachys heterocycla（Carrière）Matsum. 'Viridisulcata' ●

297422　Phyllostachys purpurata McClure 'Solidstem' = Phyllostachys heteroclada Oliv. f. solida（S. L. Chen）Z. P. Wang et Z. H. Yu ●

297423　Phyllostachys purpurata McClure 'Straighstem' = Phyllostachys heterocycla（Carrière）Matsum. ●

297424　Phyllostachys purpurata McClure = Phyllostachys heteroclada Oliv. f. purpurata（McClure）T. H. Wen ●

297425　Phyllostachys purpurata McClure = Phyllostachys heteroclada Oliv. ●

297426　Phyllostachys purpurata McClure f. decurtata S. L. Chen;盘珠竹●

297427　Phyllostachys purpurata McClure f. decurtata S. L. Chen = Phyllostachys heteroclada Oliv. f. solida（S. L. Chen）Z. P. Wang et Z. H. Yu ●

297428　Phyllostachys purpurata McClure f. solida S. L. Chen = Phyllostachys heteroclada Oliv. f. decurtata（S. L. Chen）T. H. Wen ●

297429　Phyllostachys purpurata McClure f. solida S. L. Chen = Phyllostachys heteroclada Oliv. f. solida（S. L. Chen）Z. P. Wang et Z. H. Yu ●

297430　Phyllostachys purpureomaculata W. T. Lin et Z. J. Feng = Phyllostachys heteroclada Oliv. ●

297431　Phyllostachys purpureomaoulata W. T. Lin et Z. J. Feng;小斑刚竹●

297432　Phyllostachys quadrangularis（Franceschi）Rendle = Chimonobambusa quadrangularis（Franch.）Makino ex Nakai ●

297433　Phyllostachys quilioi Rivière et C. Rivière = Phyllostachys reticulata（Rupr.）K. Koch ●

297434　Phyllostachys quilioi Rivière et C. Rivière var. castillonis-holochrysa Regel et J. Houz. = Phyllostachys sulphurea（Carrière）

Rivière et C. Rivière ●

297435　Phyllostachys reticulata（Rupr.）K. Koch;网状竹（刚竹,钢铁头竹,光竹,鬼角竹,桂竹,台竹）;Net-veined Bamboo, Reticulate Bamboo ●

297436　Phyllostachys reticulata（Rupr.）K. Koch var. aurea（Carrière ex Rivière et C. Rivière）Makino = Phyllostachys aurea Carrière ex Rivière et C. Rivière ●

297437　Phyllostachys reticulata（Rupr.）K. Koch var. sulphurea（Carrière）Makino = Phyllostachys sulphurea（Carrière）Rivière et C. Rivière ●

297438　Phyllostachys reticulata C. Koch var. holochrysa（Pfitzer）Nakai = Phyllostachys sulphurea（Carrière）Rivière et C. Rivière ●

297439　Phyllostachys reticulata C. Koch var. sulphurea（Carrière）Makino = Phyllostachys sulphurea（Carrière）Rivière et C. Rivière ●

297440　Phyllostachys reticulata K. Koch = Phyllostachys reticulata（Rupr.）K. Koch ●

297441　Phyllostachys reticulata Y. Chen = Phyllostachys bambusoides Siebold et Zucc. ●

297442　Phyllostachys retusa T. H. Wen;华东水竹●

297443　Phyllostachys retusa T. H. Wen = Phyllostachys rubicunda T. H. Wen ●

297444　Phyllostachys rigida X. Jiang et Q. Li = Phyllostachys aurita J. L. Lu ●

297445　Phyllostachys rigida X. Jiang et Q. Li = Phyllostachys veitchiana Rendle ●

297446　Phyllostachys rivalis H. R. Zhao et A. T. Liu;河竹;River Bamboo ●

297447　Phyllostachys robustiramea S. Y. Chen et C. Y. Yao;芽竹;Bud Bamboo, Robuste-branch Bamboo, Thick-ramel Bamboo ●

297448　Phyllostachys rubicunda T. H. Wen;红后竹(安吉水胖竹,华东水竹);Blush-red Bamboo, Red Bamboo ●

297449　Phyllostachys rubromarginata McClure;红边竹（观音竹）; Bamboo, Reddish Bamboo, Reddishmargin Bamboo, Reddish-margined Bamboo, Red-margin Phyllostachys ●

297450　Phyllostachys rubromarginata McClure f. castigata T. H. Wen = Phyllostachys shuchengensis S. C. Li et S. H. Wu ●

297451　Phyllostachys rubromarginata McClure f. castigata T. H. Wen = Phyllostachys rubromarginata McClure ●

297452　Phyllostachys rubromarginata Z. P. Wang et al. = Phyllostachys aurita J. L. Lu ●

297453　Phyllostachys rutila T. H. Wen;衢县红壳竹;Quxian Bamboo, Reddish-orange Bamboo ●

297454　Phyllostachys sapida T. P. Yi;彭县刚竹●

297455　Phyllostachys sapida T. P. Yi = Phyllostachys propinqua McClure ●

297456　Phyllostachys shuchengensis S. C. Li et S. H. Wu;舒城刚竹●

297457　Phyllostachys shuchengensis S. C. Li et S. H. Wu = Phyllostachys rubromarginata McClure ●

297458　Phyllostachys spectabilis C. D. Chu et C. S. Chao = Phyllostachys aureosulcata McClure 'Spectabilis' ●

297459　Phyllostachys stauntoni Munro = Phyllostachys nigra（Lodd. ex Lindl.）Munro var. henonis（Bean）Stapf ex Rendle ●

297460　Phyllostachys stimulosa H. R. Zhao et A. T. Liu;漫竹;Setose Bamboo, Stimulose Bamboo ●

297461　Phyllostachys stimulosa H. R. Zhao et A. T. Liu f. unifoliata T. H. Wen;水后竹;One-leaved Bamboo ●

297462　Phyllostachys subulata W. T. Lin et Z. M. Wu;金竹仔●

297463   *Phyllostachys subulata* W. T. Lin et Z. M. Wu = *Phyllostachys nidularia* Munro ●

297464   *Phyllostachys sulphurea* (Carrière) Rivière et C. Rivière;金竹(黄竿,黄金竹,黄竹);Golden Bamboo,Sulphur Bamboo ●

297465   *Phyllostachys sulphurea* (Carrière) Rivière et C. Rivière 'Houzeau';绿皮黄筋竹●

297466   *Phyllostachys sulphurea* (Carrière) Rivière et C. Rivière 'Robert Young';黄皮绿筋竹(黄皮刚竹,黄皮胖竹)●

297467   *Phyllostachys sulphurea* (Carrière) Rivière et C. Rivière 'Sulphrea' = *Phyllostachys sulphurea* (Carrière) Rivière et C. Rivière ●

297468   *Phyllostachys sulphurea* (Carrière) Rivière et C. Rivière 'Sulphrea';黄金竹●

297469   *Phyllostachys sulphurea* (Carrière) Rivière et C. Rivière 'Viridis';刚竹(胖竹);Green Sulphur Bamboo ●

297470   *Phyllostachys sulphurea* (Carrière) Rivière et C. Rivière f. robert-young (McClure) T. P. Yi = *Phyllostachys sulphurea* (Carrière) Rivière et C. Rivière 'Robert Young' ●

297471   *Phyllostachys sulphurea* (Carrière) Rivière et C. Rivière var. viridis R. A. Young f. houzeauana (C. D. Chu et C. S. Chao) C. S. Chao et S. A. Renv. = *Phyllostachys sulphurea* (Carrière) Rivière et C. Rivière 'Houzeau' ●

297472   *Phyllostachys sulphurea* (Carrière) Rivière et C. Rivière var. viridis R. A. Young f. robertii C. S. Chao et S. A. Renv. = *Phyllostachys sulphurea* (Carrière) Rivière et C. Rivière 'Robert Young' ●

297473   *Phyllostachys sulphurea* (Carrière) Rivière et C. Rivière var. viridis R. A. Young = *Phyllostachys sulphurea* (Carrière) Rivière et C. Rivière 'Viridis' ●

297474   *Phyllostachys sulphurea* (Carrière) Rivière et C. Rivière var. viridis R. A. Young f. robert-young (McClure) T. P. Yi = *Phyllostachys sulphurea* (Carrière) Rivière et C. Rivière 'Robert Young' ●

297475   *Phyllostachys tianmuensis* Z. P. Wang et N. X. Ma;天目早竹(天目旱竹);Tianmu Bamboo,Tianmushan Bamboo ●

297476   *Phyllostachys varioauriculata* S. C. Li et S. H. Wu;乌竹(光毛竹); Black Bamboo, Variable-auricled Bamboo, Varioauriculate Bamboo ●

297477   *Phyllostachys veitchiana* Rendle;硬头青竹;Haedhead Bamboo, Veitch's Bamboo ●

297478   *Phyllostachys verrucosa* G. H. Ye et Z. P. Wang;长沙刚竹; Changsha Bamboo,Changsha Firmbamboo,Warted Bamboo ●

297479   *Phyllostachys villosa* T. H. Wen;黄腊竹;Villose Bamboo ●

297480   *Phyllostachys villosa* T. H. Wen = *Phyllostachys sulphurea* (Carrière) Rivière et C. Rivière var. viridis R. A. Young ●

297481   *Phyllostachys villosa* T. H. Wen = *Phyllostachys sulphurea* (Carrière) Rivière et C. Rivière 'Viridis' ●

297482   *Phyllostachys villosa* T. H. Wen = *Phyllostachys sulphurea* (Carrière) Rivière et C. Rivière 'Sulphrea' ●

297483   *Phyllostachys violascens* (Carrière) Rivière et C. Rivière;早竹; Early Bamboo,Early Spring Shoot Bamboo,Pre-vernal Bamboo ●

297484   *Phyllostachys virella* T. H. Wen;东阳青皮竹;Dongyang Bamboo,Dongyang Greenbark Bamboo ●

297485   *Phyllostachys viridi-glaucescens* (Carrière) Rivière et C. Rivière;粉绿竹;Greenwax Golden Bamboo,Grey-blue Bamboo ●

297486   *Phyllostachys viridis* (R. A. Young) McClure;胖竹(刚竹,仙人杖);Firmb Bamboo, Green Bamboo, Green-sulphur Bamboo ●

297487   *Phyllostachys viridis* (R. A. Young) McClure 'Houzeau' = *Phyllostachys sulphurea* (Carrière) Rivière et C. Rivière 'Houzeau' ●

297488   *Phyllostachys viridis* (R. A. Young) McClure 'Robert Young' = *Phyllostachys sulphurea* (Carrière) Rivière et C. Rivière 'Robert Young' ●

297489   *Phyllostachys viridis* (R. A. Young) McClure = *Phyllostachys sulphurea* (Carrière) Rivière et C. Rivière 'Viridis' ●

297490   *Phyllostachys viridis* (R. A. Young) McClure = *Phyllostachys sulphurea* (Carrière) Rivière et C. Rivière var. viridis R. A. Young ●

297491   *Phyllostachys viridis* (R. A. Young) McClure f. aurata T. H. Wen = *Phyllostachys sulphurea* (Carrière) Rivière et C. Rivière 'Robert Young' ●

297492   *Phyllostachys viridis* (R. A. Young) McClure f. houzeanana C. D. Chu et C. S. Chao = *Phyllostachys sulphurea* (Carrière) Rivière et C. Rivière 'Houzeau' ●

297493   *Phyllostachys viridis* (R. A. Young) McClure f. laqueata T. H. Wen = *Phyllostachys meyeri* McClure ●

297494   *Phyllostachys viridis* (R. A. Young) McClure f. surata T. H. Wen;黄皮刚竹●

297495   *Phyllostachys viridis* (R. A. Young) McClure f. viridisulcata P. X. Zhang;绿槽刚竹●

297496   *Phyllostachys viridis* (R. A. Young) McClure f. vitata T. H. Wen = *Phyllostachys propinqua* McClure ●

297497   *Phyllostachys viridis* (R. A. Young) McClure f. youngii C. D. Chu et C. S. Chao = *Phyllostachys sulphurea* (Carrière) Rivière et C. Rivière 'Robert Young' ●

297498   *Phyllostachys viridis* (Young) McClure f. laqueata T. H. Wen = *Phyllostachys meyeri* McClure ●

297499   *Phyllostachys vivax* McClure;乌哺鸡竹(乌脯鸡竹);Bamboo, Black Broodhen Bamboo, Smooth Sheath Bamboo, Smoothsheath Bamboo, Vivax Bamboo ●

297500   *Phyllostachys vivax* McClure 'Aureoculis';黄秆乌哺鸡竹(黄竿乌脯鸡竹,黄竿乌哺鸡竹)●

297501   *Phyllostachys vivax* McClure 'Huanwenzhu';黄纹竹●

297502   *Phyllostachys vivax* McClure 'Vivax' = *Phyllostachys vivax* McClure ●

297503   *Phyllostachys vivax* McClure f. aureocaulis N. X. Ma = *Phyllostachys vivax* McClure 'Aureoculis' ●

297504   *Phyllostachys vivax* McClure f. huanwenzhu J. L. Lu = *Phyllostachys vivax* McClure 'Huanwenzhu' ●

297505   *Phyllostachys vivax* McClure f. vittata T. H. Wen;褐条乌哺鸡竹●

297506   *Phyllostachys yunhoensis* S. Y. Chen et C. Y. Yao;云和乌哺鸡竹(云和哺鸡竹,云和乌脯鸡竹);Yunhe Bamboo,Yunhe Broodhen Bamboo,Yunho Bamboo ●

297507   *Phyllostegia* Benth. (1830);叶覆草属■●☆

297508   *Phyllostegia adenophora* H. St. John;腺梗叶覆草■☆

297509   *Phyllostegia alba* H. St. John;白叶叶覆草■☆

297510   *Phyllostegia axillaris* H. St. John;腋生叶覆草■☆

297511   *Phyllostegia brevicalycis* H. St. John;短萼叶覆草■☆

297512   *Phyllostegia brevidens* A. Gray;短齿叶覆草■☆

297513   *Phyllostegia capitata* H. St. John;头状叶覆草■☆

297514   *Phyllostegia elliptica* H. St. John;椭圆叶覆草■☆

297515   *Phyllostegia glabra* Benth.;光叶覆草■☆

297516   *Phyllostegia leptostachys* Benth.;细穗叶覆草■☆

297517   *Phyllostegia macrophylla* Benth.;大叶叶覆草■☆

297518   *Phyllostegia micrantha* H. St. John;小花叶覆草■☆

297519   *Phyllostegia microphylla* Benth.;小叶叶覆草■☆

297520　Phyllostegia mollis Benth. ;柔软叶覆草■☆

297521　Phyllostegia montana H. St. John;山地叶覆草■☆

297522　Phyllostegia orientalis H. St. John;东方叶覆草■☆

297523　Phyllostegia ovata H. St. John;卵形叶覆草■☆

297524　Phyllostegia polyantha H. St. John;多花叶覆草■☆

297525　Phyllostegia truncata A. Gray;平截叶覆草■☆

297526　Phyllostelidium Beauverd ＝ Baccharis L. (保留属名)●■☆

297527　Phyllostema Neck. ＝ Quassia L. ●☆

297528　Phyllostema Neck. ＝ Simaba Aubl. ●☆

297529　Phyllostemonodaphne Kosterm. (1936);叶蕊楠属●☆

297530　Phyllostemonodaphne geminiflora (Mez) Kosterm. ;叶蕊楠●☆

297531　Phyllostephanus Tiegh. ＝ Aetanthus (Eichler) Engl. ●☆

297532　Phyllostylon Capan. ＝ Phyllostylon Capan. ex Benth. et Hook. f. ●☆

297533　Phyllostylon Capan. ex Benth. ＝ Phyllostylon Capan. ex Benth. et Hook. f. ●☆

297534　Phyllostylon Capan. ex Benth. et Hook. f. (1880);叶柱榆属●☆

297535　Phyllostylon brasiliensis Capan. ＝ Phyllostylon brasiliensis Capan. ex Benth. ●☆

297536　Phyllostylon brasiliensis Capan. ex Benth. ;巴西叶柱榆木; Baitoa, Bsitoa, San Domingo Box, San Domingo Boxwood ●☆

297537　Phyllota (DC.) Benth. (1837);耳叶豆属●☆

297538　Phyllota Benth. ＝ Phyllota (DC.) Benth. ●☆

297539　Phyllota DC. ＝ Phyllota (DC.) Benth. ●☆

297540　Phyllota aspera Benth. ;耳叶豆●☆

297541　Phyllotaenium André ＝ Xanthosoma Schott ■

297542　Phyllotephrum Gand. ＝ Clinopodium L. ■●

297543　Phyllotheca Nutt. ex Moq. ＝ Atriplex L. ■●

297544　Phyllotrichum Thorel ex Lecomte(1911);毛叶无患子属●☆

297545　Phyllotrichum mekongense Lecomte;毛叶无患子●☆

297546　Phylloxylon Baill. (1861);叶木豆属●☆

297547　Phylloxylon arenicola Du Puy, Labat et Schrire;砂地叶木豆●☆

297548　Phylloxylon cloiselii Drake ＝ Phylloxylon xylophylloides (Baker) Du Puy, Labat et Schrire ●☆

297549　Phylloxylon decipiens Baill. ;迷惑叶木豆●☆

297550　Phylloxylon ensifolium Baill. ＝ Phylloxylon xylophylloides (Baker) Du Puy, Labat et Schrire ●☆

297551　Phylloxylon perrieri Drake;佩里耶叶木豆●☆

297552　Phylloxylon perrieri Drake subsp. albiflorum R. Vig. ＝ Phylloxylon xiphoclada (Baker) Du Puy, Labat et Schrire ●☆

297553　Phylloxylon phillipsonii Du Puy, Labat et Schrire;菲利叶木豆●☆

297554　Phylloxylon spinosa Du Puy, Labat et Schrire;具刺叶木豆●☆

297555　Phylloxylon xiphoclada (Baker) Du Puy, Labat et Schrire;剑枝叶木豆●☆

297556　Phylloxylon xylophylloides (Baker) Du Puy, Labat et Schrire;木叶叶木豆●☆

297557　Phyllmena Blume ex Miq. ＝ Enydra Lour. ■

297558　Phyllmena Blume ex Miq. ＝ Phyllimena Blume ex DC. ■

297559　Phyllyrea G. Don ＝ Phillyrea L. ●☆

297560　Phyllyrea paniculata Roxb. ＝ Ligustrum lucidum W. T. Aiton ●

297561　Phyllyrea robusta Roxb. ＝ Ligustrum robustum (Roxb.) Blume ●

297562　Phylocarpos Raf. ＝ Physocarpus (Cambess.) Raf. (保留属名)●

297563　Phylohydrax Puff(1986);叶水茜属☆

297564　Phylohydrax carnosa (Hochst.) Puff;叶水茜■☆

297565　Phylohydrax madagascariensis (Willd. ex Roem. et Schult.) Puff;马岛叶水茜■☆

297566　Phyloma Gmel. ＝ Cymbaria L. ■

297567　Phymaspermum Less. (1832);瘤子菊属●☆

297568　Phymaspermum acerosum (DC.) Källersjö;针状瘤子菊■☆

297569　Phymaspermum aciculare (E. Mey. ex Harv.) Benth. et Hook. ex B. D. Jacks. ;针形瘤子菊●☆

297570　Phymaspermum argenteum Brusse;银白瘤子菊●☆

297571　Phymaspermum athanasioides (S. Moore) Källersjö;永菊状瘤子菊■☆

297572　Phymaspermum bolusii (Hutch.) Källersjö;博卢斯瘤子菊●☆

297573　Phymaspermum equisetoides Thell. ;木贼瘤子菊●☆

297574　Phymaspermum erubescens (Hutch.) Källersjö;变红瘤子菊●☆

297575　Phymaspermum junceum Less. ;灯心草瘤子菊●☆

297576　Phymaspermum leptophyllum (DC.) Benth. et Hook. ex B. D. Jacks. ;细叶瘤子菊●☆

297577　Phymaspermum montanum (Hutch.) Källersjö;山地瘤子菊■☆

297578　Phymaspermum parvifolium (DC.) Benth. et Hook. ex B. D. Jacks. ;小叶瘤子菊●☆

297579　Phymaspermum peglerae (Hutch.) Källersjö;佩格拉瘤子菊■☆

297580　Phymaspermum pinnatifidum (Oliv.) Källersjö;羽裂瘤子菊●☆

297581　Phymaspermum pinnatifidum (Oliv.) Källersjö ＝ Athanasia pinnatifida (Oliv.) Hilliard ●☆

297582　Phymaspermum pinnatifidum (Oliv.) Källersjö ＝ Pentzia pinnatifida Oliv. ●☆

297583　Phymaspermum pubescens Kuntze;毛瘤子菊●☆

297584　Phymaspermum scoparium (DC.) Källersjö;帚状瘤子菊■☆

297585　Phymaspermum villosum (Hilliard) Källersjö;长柔毛瘤子菊■☆

297586　Phymaspermum woodii (Thell.) Källersjö;伍得瘤子菊■☆

297587　Phymatanthus Lindl. Sweet ＝ Pelargonium L' Hér. ex Aiton ●■

297588　Phymatanthus Sweet ＝ Pelargonium L' Hér. ex Aiton ●■

297589　Phymatanthus elatus Sweet ＝ Pelargonium tricolor Curtis ●☆

297590　Phymatarum M. Hotta(1965);瘤南星属■☆

297591　Phymatarum borneense M. Hotta;瘤南星■☆

297592　Phymatarum montanum M. Hotta;山地瘤南星■☆

297593　Phymatidiopsis Szlach. (2006);拟瘤兰属■☆

297594　Phymatidiopsis Szlach. ＝ Phymatidium Lindl. ■☆

297595　Phymatidium Lindl. (1833);小瘤兰属■☆

297596　Phymatidium delicatulum Lindl. ;瘤兰■☆

297597　Phymatidium falcifolium Lindl. ;镰叶瘤兰■☆

297598　Phymatidium microphyllum (Barb. Rodr.) Toscano;小叶瘤兰■☆

297599　Phymatis E. Mey. ＝ Carum L. ■

297600　Phymatocarpus F. Muell. (1862);瘤果桃金娘属●☆

297601　Phymatocarpus porphyrocephalus F. Muell. ;瘤果桃金娘●

297602　Phymatochilum Christenson(2005);瘤唇兰属■☆

297603　Phymosia Desv. ＝ Phymosia Desv. ex Ham. ●☆

297604　Phymosia Desv. ex Ham. (1825);菲莫斯木属●☆

297605　Phymosia Ham. ＝ Phymosia Desv. ex Ham. ●☆

297606　Phymosia umbellata (Cav.) Kearney;深红菲莫斯木●☆

297607　Phyodina Raf. ＝ Callisia Loefl. ■☆

297608　Phyodina cordifolia (Sw.) Rohweder ＝ Callisia cordifolia (Sw.) E. S. Anderson et Woodson ■☆

297609　Phyodina micrantha (Torr.) D. R. Hunt ＝ Callisia micrantha (Torr.) D. R. Hunt ■☆

297610　Phyrrheima Hassk. ＝ Pyrrheima Hassk. ■☆

297611　Phyrrheima Hassk. ＝ Siderasis Raf. ■☆

297612　Physa Noronha ex Thouars ＝ Glinus L. ■

297613　Physa Thouars ＝ Glinus L. ■

297614　Physacanthus Benth. (1876);泡刺爵床属■☆

297615　Physacanthus batanganus (J. Braun et K. Schum.) Lindau;泡刺爵床■☆

297616　Physacanthus cylindricus C. B. Clarke ＝ Physacanthus batanganus（J. Braun et K. Schum.）Lindau ■☆

297617　Physacanthus inflatus C. B. Clarke ＝ Physacanthus batanganus（J. Braun et K. Schum.）Lindau ■☆

297618　Physacanthus lucernarius N. Hallé ＝ Physacanthus batanganus（J. Braun et K. Schum.）Lindau ■☆

297619　Physacanthus nematosiphon（Lindau）Rendle et Britten；非洲泡刺爵床■☆

297620　Physacanthus talbotii S. Moore；塔尔泡刺爵床■☆

297621　Physalastrum Monteiro ＝ Krapovickasia Fryxell ■☆

297622　Physalastrum Monteiro ＝ Sida L. ●■

297623　Physaliastrum Makino ＝ Leucophysalis Rydb. ■☆

297624　Physaliastrum Makino（1914）；散血丹属（白姑娘属，刺酸浆属，地海椒属）；Blooddisperser, Physaliastrum ■

297625　Physaliastrum chamaesarachoides（Makino）Makino；广西地海椒（林氏灯笼草）；Guangxi Archiphysalis, Kwangsi Archiphysalis, Lin's Physaliastrum ●■

297626　Physaliastrum echinatum（Yatabe）Makino；具刺散血丹（日本散血丹）■

297627　Physaliastrum heterophyllum（Hemsl.）Migo；江南散血丹（刺酸浆，龙须参）；Diversifolious Blooddisperser, Diversifolious Physaliastrum ■

297628　Physaliastrum japoncum（Franch. et Sav.）Honda；日本散血丹（山茄子）；Japan Blooddisperser, Japanese Physaliastrum ■

297629　Physaliastrum japoncum（Franch. et Sav.）Honda var. occultibaccum X. H. Guo et S. B. Zhou；隐果散血丹■

297630　Physaliastrum japonicum（Franch. et Sav.）Honda ＝ Physaliastrum echinatum（Yatabe）Makino ■

297631　Physaliastrum kimurae Franch. et Sav. ＝ Physaliastrum japonicum（Franch. et Sav.）Honda ■

297632　Physaliastrum kweichouense Kuang et A. M. Lu；散血丹；Blooddisperser, Guizhou Physaliastrum, Kweichou Physaliastrum ■

297633　Physaliastrum savatieri（Makino）Makino ＝ Physaliastrum japonicum（Franch. et Sav.）Honda ■

297634　Physaliastrum sinense（Hemsl.）D'Arcy et Zhi Y. Zhang ＝ Archiphysalis sinensis（Hemsl.）Kuang ●

297635　Physaliastrum sinicum Kuang et A. M. Lu；华北散血丹（地海椒，山茄子）；China Blooddisperser, Chinese Physaliastrum ■

297636　Physaliastrum yunnanense Kuang et A. M. Lu；云南散血丹；Yunnan Blooddisperser, Yunnan Physaliastrum ■

297637　Physalidium Fenzl ＝ Graellsia Boiss. ■☆

297638　Physalidium Fenzl（1866）；小泡芥属■☆

297639　Physalidium graellsiifolium Lipsky；小泡芥■☆

297640　Physalis L.（1753）酸浆属（灯笼草属）；Cape Gooseberry, Chinese Lantern, Chinese Lanterns, Chinese-lantern, Ground Cherry, Groundcherry, Ground-cherry, Husk Tomato, Husk-tomato, Japanese-lantern, Physalis, Winter Cherry ■

297641　Physalis acutifolia（Miers）Sandwith；狭叶酸浆；Ground Cherry, Sharpleaf Ground Cherry ■☆

297642　Physalis aequata J. Jacq. ex Nees；平酸浆；Husk Tomato ■☆

297643　Physalis alkekengi L.；酸浆（醋浆，灯笼草，豆姑娘，姑娘菜，姑娘花，挂金灯，寒浆，红姑娘，苦耽，苦蔵，苦蘵，洛神珠，欧亚酸浆，皮瓣草，酸蒋，天泡果，王母珠，蔵）；Alkakeng, Alkekengi, Alkekengy, Bladder Cherry, Bladderherb, Bladder-herb, Bladderwort, China Lantern, Chinese Lantern, Chinese Lantern Plant, Chinese Lantern-plant, Chinese Lanterns, Chinese-lantern, Granny's Bonnet, Granny's Bonnets, Groundcherry, Japanese Lanterns, Japanese-lantern, Physalis, Red Nightshade, Strawberry Ground Cherry, Strawberry Groundcherry, Strawberry Ground-cherry, Strawberry Tomato, Strawberry-tomato, Winter Cherry, Wintercherry ■

297644　Physalis alkekengi L. var. angthoxantha H. Lév. ＝ Physalis alkekengi L. ■

297645　Physalis alkekengi L. var. franchetii（Mast.）Makino；挂金灯（包铃子，醋浆，打拍草，打扑草，灯笼草，灯笼儿，灯笼果，灯笼花，端浆实，姑娘菜，姑娘花，挂金灯酸浆，鬼灯笼，寒浆，荷朴，红姑娘，红娘子，花姑娘，浆水罐，金灯，金灯草，金灯笼，锦灯笼，锦灯笼草，九古牛，苦耽，苦蔵，苦蘵，勒马回，铃儿草，泡泡草，泡子草，皮弁草，扑扑子草，朴朴子草，山瑚柳，珊瑚架，水辣子，酸浆，酸浆草，酸浆实，天灯笼，天灯笼草，天泡，天泡草，天泡草铃儿，天泡果，野胡椒，叶下灯，蔵）；Alkekengi, Bladder Cherry, Bladderherb, Chinese Lantern Plant, Chinese-lantern, Frachet Groundcherry, Japanese-lantern, Physalis, Strawberry Ground Cherry, Strawberry Ground-cherry, Strawberry-tomato, Winter Cherry ■

297646　Physalis alkekengi L. var. franchetii（Mast.）Makino ＝ Physalis alkekengi L. ■

297647　Physalis alkekengi L. var. glabripes（Pojark.）Grubov ＝ Physalis alkekengi L. var. franchetii（Mast.）Makino ■

297648　Physalis alkekengi L. var. orientalis Pamp. ＝ Physalis alkekengi L. ■

297649　Physalis ambigua（A. Gray）Britton ＝ Physalis heterophylla Nees ■☆

297650　Physalis angulata L.；苦蘵（灯笼草，灯笼泡，灯笼酸浆，黄蔴，黄姑娘，苦耽，绿灯，炮仔草，天泡草，响铃草，小苦耽，小酸浆，蘵）；Cow Pops, Cowpops, Cutleaf Ground Cherry, Cutleaf Groundcherry, Ground-cherry, Tomatillo ■

297651　Physalis angulata L. ＝ Physalis minima L. ■

297652　Physalis angulata L. var. lanceifolia（Nees）Waterf.；剑叶苦蘵■☆

297653　Physalis angulata L. var. pendula（Rydb.）Waterf.；下垂苦蘵■☆

297654　Physalis angulata L. var. philadelphica（Lam.）A. Gray ＝ Physalis philadelphica Lam. ■

297655　Physalis angulata L. var. villosa Bonati；毛苦蘵；Villous Groundcherry ■

297656　Physalis angulata L. var. villosa Bonati ＝ Physalis minima L. ■

297657　Physalis bodinieri H. Lév. ＝ Physalis angulata L. ■

297658　Physalis bunyardii Makino；秘鲁酸浆；Cape Gooseberry, Gooseberry Tomato, Ground Cherry, Jack-in-a-lantern, Peruvian Cherry, Poha, Strawberry Tomato, Wild Gooseberry ■☆

297659　Physalis bunyardii Makino ＝ Physalis alkekengi L. var. franchetii（Mast.）Makino ■

297660　Physalis cavaleriei H. Lév. ＝ Physalis philadelphica Lam. ■

297661　Physalis chamaesarachoides Makino ＝ Physaliastrum chamaesarachoides（Makino）Makino ●■

297662　Physalis ciliata Siebold et Zucc. ＝ Physalis alkekengi L. ■

297663　Physalis ciliata Siebold et Zucc. ＝ Physalis pubescens L. ■☆

297664　Physalis cinerascens（Dunal）Hitchc.；浅灰酸浆；Ground Cherry ■☆

297665　Physalis cordata Mill.；棱萼酸浆（心叶酸浆）；Ground Cherry ■

297666　Physalis divaricata D. Don；南亚酸浆■☆

297667　Physalis edulis Sim ＝ Physalis peruviana L. ■

297668　Physalis esquirolii H. Lév. et Vaniot ＝ Physalis angulata L. ■

297669　Physalis fendleri A. Gray；芬氏酸浆；Ground Cherry ■☆

297670　Physalis franchetii L. var. bunyardii Makino ＝ Physalis alkekengi L. var. franchetii（Mast.）Makino ■

297671　Physalis franchetii Mast. ＝ Physalis alkekengi L. var. franchetii

（Mast.）Makino ■

297672 Physalis franchetii Mast. = Physalis alkekengi L. ■

297673 Physalis franchetii Mast. var. bunyardii Makino = Physalis alkekengi L. var. franchetii（Mast.）Makino ■

297674 Physalis glabripes Pojark. = Physalis alkekengi L. var. franchetii（Mast.）Makino ■☆

297675 Physalis grandiflora Hook. = Leucophysalis grandiflora（Hook.）Rydb. ●☆

297676 Physalis grisea（Waterf.）Martinez；灰酸浆；Downy Ground-cherry，Husk Ground-cherry，Strawberry Tomato ■☆

297677 Physalis hederifolia A. Gray；常春藤叶酸浆；Ivy-leaved Ground Cherry ■☆

297678 Physalis heterophylla Nees；异叶酸浆；Bladder Cherry，Clammy Ground Cherry，Clammy Ground-cherry，Ground Cherry，Lanternweed ■☆

297679 Physalis heterophylla Nees var. ambigua（A. Gray）Rydb. = Physalis heterophylla Nees ■☆

297680 Physalis heterophylla Nees var. clavipes Fernald = Physalis heterophylla Nees ■☆

297681 Physalis heterophylla Nees var. nyctaginea（Dunal）Rydb. = Physalis heterophylla Nees ■☆

297682 Physalis heterophylla Nees var. villosa Waterf. = Physalis heterophylla Nees ■☆

297683 Physalis hispida（Waterf.）Cronquist；草原酸浆；Plains-sandhill Ground-cherry，Prairie Ground-cherry ■☆

297684 Physalis indica Lam. var. microcarpa Nees = Physalis divaricata D. Don ■☆

297685 Physalis intermedia Rydb. = Physalis virginiana Mill. ■☆

297686 Physalis ixocarpa Brot. ex Hornem.；黏果酸浆；Ground-cherry，Jamberry，Mexican Ground Cherry，Tomathlo，Tomatillo ■☆

297687 Physalis ixocarpa Brot. ex Hornem. = Physalis philadelphica Lam. ■

297688 Physalis ixocarpa Hornem. = Physalis philadelphica Lam. ■

297689 Physalis kansuensis Pojark. = Physalis alkekengi L. ■

297690 Physalis lagascae Roem. et Schult. = Physalis minima L. ■

297691 Physalis lanceolata Michx.；披针叶酸浆；Prairie Ground Cherry ■☆

297692 Physalis linii Y. C. Liu et C. H. Ou = Physaliastrum chamaesarachoides（Makino）Makino ●■

297693 Physalis lobata Torr.；浅裂酸浆；Purple Ground-chen-y ■☆

297694 Physalis longifolia Nutt.；长叶酸浆；Common Ground Cherry，Long-leaved Ground-cherry，Smooth Long-leaved Ground-cherry，Tall Ground-cherry ■☆

297695 Physalis longifolia Nutt. var. hispida（Waterf.）Steyerm. = Physalis hispida（Waterf.）Cronquist ■☆

297696 Physalis longifolia Nutt. var. hispida（Waterf.）Steyerm. = Physalis pumila Nutt. ■☆

297697 Physalis longifolia Nutt. var. subglabrata（Mack. et Bush）Cronquist；光长叶酸浆；Long-leaved Ground-cherry，Smooth Ground Cherry，Smooth Ground-cherry，Tall Ground-cherry ■☆

297698 Physalis lunatus ?；新月酸浆；Scimitar-podded Kidney Bean ■☆

297699 Physalis macrophysa Rydb.；大果酸浆（紫果酸浆，紫桃子）；Bigfruit Groundcherry，Largefruit Groundcherry ■

297700 Physalis macrophysa Rydb. = Physalis longifolia Nutt. var. subglabrata（Mack. et Bush）Cronquist ■☆

297701 Physalis macrophysa Rydb. = Physalis longifolia Nutt. ■☆

297702 Physalis micrantha Link = Physalis lagascae Roem. et Schult. ■

297703 Physalis minima L.；小酸浆（卜子草，打卜草，打额泡，灯笼草，灯笼果，灯笼泡，挂金灯，挂金泡，黄灯笼，黄姑娘，沙灯笼，沙

灯笼草，水灯笼，天泡草，天泡果，天泡子，王不留行，王母珠）；Ground Cherry，Little Groundcherry，Sun Berry，Sunberry，Wild Gooseberry ■

297704 Physalis minima L. = Physalis angulata L. ■

297705 Physalis minima L. = Physalis pubescens L. ■☆

297706 Physalis missouriensis Mack. et Bush；密苏里酸浆■☆

297707 Physalis monticola C. Mohr = Physalis virginiana Mill. ■☆

297708 Physalis nyctaginea Dunal = Physalis heterophylla Nees ■☆

297709 Physalis parviflora R. Br. = Physalis minima L. ■

297710 Physalis pendula Rydb. = Physalis angulata L. ■

297711 Physalis peruviana L.；灯笼果（爆卜草，打卜草，打额泡，打头泡，灯笼草，灯笼泡，鬼灯笼，荷卜草，苦灯笼草，秘鲁苦蘵，炮掌果，泡泡草，沙灯笼，水灯笼草，响铃子，小果酸浆，心不干）；Cape Gooseberry，Cape-gooseberry，Ground-cherry，Lantern Fruit，Peru Groundcherry，Peruvian Ground Cherry，Peruvian Groundcherry，Peruvian Ground-cherry，Physalis，Strawberry-tomato，Winter Cherry ■

297712 Physalis philadelphica Lam.；毛酸浆（爆竹草，打额泡，灯笼草，灯笼泡草，费城酸浆，鬼灯笼，黄蘵，黄姑娘，苦蘵，绿灯，墨西哥灯笼果，劈拍草，朴朴草，天泡草，天泡子，响铃草，响泡子，小苦耽，小酸浆，洋姑娘，野绿灯，蘵，蘵草）；Husk-tomato，Jamberry，Mexican Groundcherry，Mexican Ground-cherry，Strawberry Tomato，Tomatillo ■

297713 Physalis philadelphica Lam. var. immaculata Waterf.；无斑墨西哥灯笼果；Mexican Groundcherry ■☆

297714 Physalis philadelphica Lam. var. immaculata Waterf. = Physalis philadelphica Lam. ■

297715 Physalis praetermissa Pojark. = Physalis alkekengi L. var. franchetii（Mast.）Makino ■

297716 Physalis pruinosa Bailey；食用酸浆；Dwarf Cape-gooseberry，Husk Tomato，Strawberry Tomato，Strawberry-tomato ■☆

297717 Physalis pruinosa Bailey = Physalis pubescens L. ■☆

297718 Physalis pruinosa L. sensu Fassett = Physalis grisea（Waterf.）Martinez ■☆

297719 Physalis pruinosa L. sensu Swink et Wilh. = Physalis grisea（Waterf.）Martinez ■☆

297720 Physalis pubescens L.；欧美毛酸浆；Alkekengi，Annual Ground Cherry，Barnados Gooseberry，Cape Gooseberry，Downy Ground Cherry，Downy Ground-cherry，Dwarf Cape Gooseberry，Hair Groundcherry，Hairy Ground Cherry，Husk Ground-cherry，Husk Tomato，Husk-tomato，Pop Vine，Pops，Pop-vine，Pubescent Groundcherry，Strawberry Tomato，Winter Cherry ■☆

297721 Physalis pubescens L. = Physalis angulata L. ■

297722 Physalis pubescens L. = Physalis philadelphica Lam. ■

297723 Physalis pubescens L. var. glabra（Michx.）Waterf. = Physalis pubescens L. ■☆

297724 Physalis pubescens L. var. grisea Waterf. = Physalis grisea（Waterf.）Martinez ■☆

297725 Physalis pubescens L. var. integrifolia（Dunal）Waterf. = Physalis pubescens L. ■☆

297726 Physalis pubescens L. var. missouriensis（Mack. et Bush）Waterf. = Physalis missouriensis Mack. et Bush ■☆

297727 Physalis pumila Nutt.；草地酸浆；Prairie Ground Cherry ■☆

297728 Physalis pumila Nutt. subsp. hispida（Waterf.）W. F. Hinton = Physalis hispida（Waterf.）Cronquist ■☆

297729 Physalis pumila Nutt. var. sonorae Torr. = Physalis longifolia Nutt. ■☆

297730 Physalis rigida Pollard et C. R. Ball = Physalis longifolia Nutt. ■☆

297731 Physalis sinensis (Hemsl.) Averett = Archiphysalis sinensis (Hemsl.) Kuang ●

297732 Physalis sinensis (Hemsl.) Averett = Physaliastrum sinense (Hemsl.) D'Arcy et Zhi Y. Zhang ●

297733 Physalis sinuata Rydb. = Physalis heterophylla Nees ■☆

297734 Physalis somnifera L. = Withania somnifera (L.) Dunal ●■

297735 Physalis stramonifera Wall. = Anisodus luridus Link et Otto ■

297736 Physalis subglabrata Mack. et Bush = Physalis longifolia Nutt. var. subglabrata (Mack. et Bush) Cronquist ■☆

297737 Physalis subglabrata Mack. et Bush = Physalis longifolia Nutt. ■☆

297738 Physalis szechuensis Pojark. = Physalis alkekengi L. var. franchetii (Mast.) Makino ■

297739 Physalis turbinata Medik. = Physalis pubescens L. ■☆

297740 Physalis virginiana Mill.；弗州酸浆；Lance-leaved Ground-cherry, Obedient Plant, Virginia Ground Cherry, Virginia Ground-cherry ■☆

297741 Physalis virginiana Mill. var. ambigua A. Gray = Physalis heterophylla Nees ■☆

297742 Physalis virginiana Mill. var. hispida Waterf. = Physalis hispida (Waterf.) Cronquist ■☆

297743 Physalis virginiana Mill. var. sonorae (Torr.) Waterf. = Physalis longifolia Nutt. ■☆

297744 Physalis virginiana Mill. var. subglabrata (Mack. et Bush) Waterf. = Physalis longifolia Nutt. var. subglabrata (Mack. et Bush) Cronquist ■☆

297745 Physalis viscosa L.；黏酸浆；Large Bladder Ground Cherry ■☆

297746 Physalis viscosa L. subsp. mollis (Nutt.) Waterf. var. cinerascens (Dunal) Waterf. = Physalis cinerascens (Dunal) Hitchc. ■☆

297747 Physalobium Steud. = Kennedia Vent. ●☆

297748 Physalobium Steud. = Physolobium Benth. ●☆

297749 Physalodes Boehm(废弃属名) = Nicandra Adans. (保留属名)■

297750 Physalodes Boehm. ex Kuntze = Nicandra Adans. (保留属名)■

297751 Physalodes physalodes (L.) Britton = Nicandra physaloides (L.) Gaertn. ■

297752 Physaloides Moench = Alicabon Raf. ●■

297753 Physaloides Moench = Withania Pauquy(保留属名)●■

297754 Physaloides somnifera (L.) Moench = Withania somnifera (L.) Dunal ●■

297755 Physandra Botsch. = Salsola L. ●■

297756 Physanthemum Klotzsch = Courbonia Brongn. ●☆

297757 Physanthemum Klotzsch = Maerua Forssk. ●☆

297758 Physanthemum glaucum Klotzsch = Maerua edulis (Gilg et Gilg-Ben.) DeWolf ●☆

297759 Physanthera Bert. ex Steud. = Rodriguezia Ruiz et Pav. ■☆

297760 Physanthillis Boiss. = Anthyllis L. ■☆

297761 Physanthyllis Boiss. = Anthyllis L. ■☆

297762 Physanthyllis Boiss. = Tripodion Medik. ■☆

297763 Physanthyllis tetraphylla (L.) Boiss. = Tripodion tetraphyllum (L.) Fourr. ■☆

297764 Physaria (Nutt. ex Torr. et A. Gray) A. Gray = Vesicaria Tourn. ex Adans. ■☆

297765 Physaria (Nutt. ex Torr. et A. Gray) A. Gray(1848)；胀荚荠属(洋球果荠属)■

297766 Physaria (Nutt.) A. Gray = Physaria (Nutt. ex Torr. et A. Gray) A. Gray ■

297767 Physaria (Nutt.) A. Gray = Vesicaria Tourn. ex Adans. ■☆

297768 Physaria A. Gray = Physaria (Nutt. ex Torr. et A. Gray) A. Gray ■

297769 Physaria A. Gray = Vesicaria Tourn. ex Adans. ■☆

297770 Physaria Rchb. = Coulterina Kuntze ■

297771 Physaria Rchb. = Physaria (Nutt. ex Torr. et A. Gray) A. Gray ■

297772 Physaria filiformis (Rollins) O'Kane et Al-Shehbaz；密苏里胀荚荠；Missouri Bladderpod ■☆

297773 Physarus Steud. = Physurus Rich. ex Lindl. ■

297774 Physcium Post et Kuntze = Physkium Lour. ■

297775 Physcium Post et Kuntze = Vallisneria L. ■

297776 Physedra Hook. f. = Coccinia Wight et Arn. ■

297777 Physedra barteri (Hook. f.) Cogn. = Coccinia barteri (Hook. f.) Keay ■☆

297778 Physedra bequaertii De Wild. = Bambekea racemosa Cogn. ■☆

297779 Physedra chaetocarpa Harms et Gilg = Raphidiocystis chrysocoma (Schumach.) C. Jeffrey ■☆

297780 Physedra djalonis A. Chev. = Ruthalicia eglandulosa (Hook. f.) C. Jeffrey ■☆

297781 Physedra eglandulosa (Hook. f.) Hutch. et Dalziel = Ruthalicia eglandulosa (Hook. f.) C. Jeffrey ■☆

297782 Physedra elegans Harms et Gilg = Coccinia barteri (Hook. f.) Keay ☆

297783 Physedra gracilis A. Chev. = Coccinia grandis (L.) Voigt ■

297784 Physedra heterophylla A. Chev. = Ruthalicia eglandulosa (Hook. f.) C. Jeffrey ■☆

297785 Physedra heterophylla Hook. f. = Coccinia barteri (Hook. f.) Keay ■☆

297786 Physedra heterophylla Hook. f. var. hookeri Hiern = Coccinia barteri (Hook. f.) Keay ■☆

297787 Physedra ivorensis A. Chev. = Ruthalicia eglandulosa (Hook. f.) C. Jeffrey ■☆

297788 Physedra longipes Hook. f. = Ruthalicia longipes (Hook. f.) C. Jeffrey ■☆

297789 Physedra macrantha Gilg = Ruthalicia eglandulosa (Hook. f.) C. Jeffrey ■☆

297790 Physedra sylvatica A. Chev. = Ruthalicia eglandulosa (Hook. f.) C. Jeffrey ■☆

297791 Physena Noronha ex Thouars(1806)；独子果属(非生木属)●■☆

297792 Physena madagascariensis Steud. = Physena madagascariensis Thouars ex Tul. ●☆

297793 Physena madagascariensis Thouars ex Tul.；马岛独子果(马岛非生木)●☆

297794 Physena madagascariensis Thouars ex Tul. var. longifolia Scott-Elliot = Physena madagascariensis Thouars ex Tul. ●☆

297795 Physena sessiliflora Tul.；无柄独子果(独子果,无柄花非生木)●☆

297796 Physenaceae Takht. (1985)；独子果科(非桐科)●☆

297797 Physenaceae Takht. = Passifloraceae Juss. ex Roussel(保留科名)●■

297798 Physeterostemon R. Goldenb. et Amorim(2006)；孔蕊野牡丹属●■☆

297799 Physeterostemon fiaschii R. Goldenb. et Amorim；孔蕊野牡丹●☆

297800 Physetobasis Hassk. = Holarrhena R. Br. ●

297801 Physianthus Mart. = Araujia Brot. ●☆

297802 Physicarpos DC. = Hovea R. Br. ex W. T. Aiton ●■☆

297803 Physicarpos DC. = Phusicarpos Poir. ●■☆

297804 Physichilus Nees = Hygrophila R. Br. ●■

297805 Physichilus barbata Nees = Hygrophila barbata (Nees) T.

Anderson ●☆

297806 Physichilus senegalensis Nees = Hygrophila senegalensis (Nees) T. Anderson ■☆

297807 Physidium Schrad. = Angelonia Bonpl. ■●☆

297808 Physiglochis Neck. = Carex L. ■

297809 Physinga Lindl. = Epidendrum L.（保留属名）■☆

297810 Physiphora Sol. ex DC. = Rinorea Aubl.（保留属名）●

297811 Physiphora Sol. ex R. Br. = Rinorea Aubl.（保留属名）●

297812 Physkium Lour. = Vallisneria L. ■

297813 Physkium natans Lour. = Vallisneria natans (Lour.) H. Hara ■

297814 Physocalycium Vest = Bryophyllum Salisb. ■

297815 Physocalymma Pohl(1827)；胀被千屈菜属●☆

297816 Physocalymma scaberrima Pohl；胀被千屈菜●☆

297817 Physocalymna DC. = Physocalymma Pohl ●☆

297818 Physocalyx Pohl(1827)；胀萼列当属(胀萼玄参属)●☆

297819 Physocalyx aurantiacus Pohl；胀萼列当●☆

297820 Physocardamum Hedge(1968)；土耳其碎米荠属■☆

297821 Physocardamum davisii Hedge；土耳其碎米荠■☆

297822 Physocarpa (Cambess.) Raf. = Physocarpus (Cambess.) Raf.（保留属名）●

297823 Physocarpa Raf. = Physocarpus (Cambess.) Raf.（保留属名）●

297824 Physocarpon Neck. = Lychnis L.（废弃属名）■

297825 Physocarpon Neck. = Melandrium Röhl. ■

297826 Physocarpon Neck. ex Raf. = Lychnis L.（废弃属名）■

297827 Physocarpon Neck. ex Raf. = Melandrium Röhl. ■

297828 Physocarpum (DC.) Bercht. et J. Presl = Sumnera Nieuwl. ■

297829 Physocarpum (DC.) Bercht. et J. Presl = Thalictrum L. ■

297830 Physocarpum Bercht. et J. Presl = Sumnera Nieuwl. ■

297831 Physocarpum Bercht. et J. Presl = Thalictrum L. ■

297832 Physocarpus (Cambess.) Maxim. = Physocarpus (Cambess.) Raf.（保留属名）●

297833 Physocarpus (Cambess.) Raf. (1838)（'Physocarpa'）（保留属名）；风箱果属(鳔鱼梅属,托盘幌属)；Bellowsfruit,Ninebark ●

297834 Physocarpus Maxim. = Physocarpus (Cambess.) Raf.（保留属名）●

297835 Physocarpus Post et Kuntze = Hovea R. Br. ex W. T. Aiton ●■☆

297836 Physocarpus Post et Kuntze = Phusicarpos Poir. ●■☆

297837 Physocarpus amurensis (Maxim.) Maxim.；风箱果(阿穆尔风箱果,托盘幌)；Amur Bellowsfruit, Amur Ninebark, Manchurian Ninebark ●

297838 Physocarpus capitatus Kuntze；球花风箱果；Western Ninebark ●☆

297839 Physocarpus intermedius (Rydb.) C. K. Schneid. = Physocarpus opulifolius (L.) Maxim. var. intermedius B. L. Rob. ●☆

297840 Physocarpus malvaceus (Greene) Kuntze；锦葵状风箱果(锦葵风箱果)；Mallow Bellowsfruit ●☆

297841 Physocarpus monogynus Kuntze；高山风箱果；Mountain Ninebark ●☆

297842 Physocarpus opulifolius (L.) Maxim.；无毛风箱果(荚蒾叶风箱果,美国风箱果,美洲风箱果,山荣树叶风箱果)；America Bellowsfruit, Common Ninebark, Eastern Ninebark, Ninebark ●☆

297843 Physocarpus opulifolius (L.) Maxim. 'Dart's Gold'；达特金荚蒾叶风箱果(金翅美国风箱果)；Dart's Gold Ninebark ●☆

297844 Physocarpus opulifolius (L.) Maxim. 'Diabolo'；空竹荚蒾叶风箱果；Purple-leaved Ninebark ●☆

297845 Physocarpus opulifolius (L.) Maxim. 'Luteus'；深黄荚蒾叶风箱果●☆

297846 Physocarpus opulifolius (L.) Maxim. 'Nanus'；矮生荚蒾叶风箱果；Dwarf Ninebark ●☆

297847 Physocarpus opulifolius (L.) Maxim. 'Nugget'；金叶无毛风箱果；Gold-leaved Ninebark ●☆

297848 Physocarpus opulifolius (L.) Maxim. f. atropurpureus Geerinck；暗紫荚蒾叶风箱果●☆

297849 Physocarpus opulifolius (L.) Maxim. var. intermedius (Rydb.) B. L. Rob.；间型风箱果；Atlantic Ninebark, Ninebark ●☆

297850 Physocarpus opulifolius (L.) Maxim. var. intermedius B. L. Rob.；居间荚蒾叶风箱果；Ninebark ●☆

297851 Physocarpus ribesifolia Kom.；茶藨叶风箱果●☆

297852 Physocarpus stellatus (Rydb. ex Small) Rehder；阿拉巴马风箱果；Alabama Bellowsfruit, Alabama Ninebark ●☆

297853 Physocaulis (DC.) Tausch = Myrrhoides Heist. ex Fabr. ■☆

297854 Physocaulis (DC.) Tausch(1834)；膨茎草属■☆

297855 Physocaulis Tausch = Physocaulis (DC.) Tausch ■☆

297856 Physocaulis nodosus (L.) Tausch；膨茎草■☆

297857 Physocaulis nodosus Koch = Myrrhoides nodosa (L.) Cannon ■☆

297858 Physocaulos Fiori et Paol. = Physocaulis (DC.) Tausch ■☆

297859 Physocaulus Koch = Physocaulis (DC.) Tausch ■☆

297860 Physoceras Schltr. (1925)；膨距兰属■☆

297861 Physoceras bellum Schltr.；雅致膨距兰■☆

297862 Physoceras lageniferum H. Perrier；长颈瓶膨距兰■☆

297863 Physoceras perrieri Schltr.；佩里耶膨距兰■☆

297864 Physoceras rotundifolium H. Perrier；圆叶膨距兰■☆

297865 Physoceras violaceum Schltr.；堇色膨距兰■☆

297866 Physochellus Nutt. ex Benth. = Orthocarpus Nutt. ■☆

297867 Physochilus Post et Kuntze = Hygrophila R. Br. ●■

297868 Physochilus Post et Kuntze = Physichilus Nees ●■

297869 Physochlaena C. Koch = Physochlaina G. Don ■

297870 Physochlaena K. Koch = Physochlaina G. Don ■

297871 Physochlaena Miers = Physochlaina G. Don ■

297872 Physochlaena Miers. = Physochlaina G. Don ■

297873 Physochlaena dahurica Miers = Physochlaina physaloides (L.) G. Don ■

297874 Physo-chlaena physaloides (L.) Miers = Physochlaina physaloides (L.) G. Don ■

297875 Physochlaina G. Don (1838)；泡囊草属（华山参属）；Bubbleweed, Physochlaina ■

297876 Physochlaina capitata A. M. Lu；伊犁泡囊草；Ili Bubbleweed, Yili Bubbleweed, Yili Physochlaina ■

297877 Physochlaina grandiflora Hook. = Physochlaina praealta (Decne.) Miers ■

297878 Physochlaina infundibularis Kuang；漏斗泡囊草(白毛参,大红参,大紫参,二月旺,华参,华山参,秦参,热参)；Funnel Bubbleweed, Funneled Physochlaina ■

297879 Physochlaina macrocalyx Pascher；长萼泡囊草；Bigcalyx Bubbleweed, Largecalyx Bubbleweed, Largecalyx Physochlaina ■

297880 Physochlaina macrophylla Bonati；大叶泡囊草；Largeleaf Bubbleweed, Largeleaf Physochlaina ■

297881 Physochlaina orientalis Don；东方泡囊草(东泡囊草)；Oriental Bubbleweed, Oriental Physochlaina ■☆

297882 Physochlaina physaloides (L.) G. Don；泡囊草(大头狼毒,汤乌普)；Bubbleweed, Common Physochlaina ■

297883 Physochlaina praealta (Decne.) Miers；西藏泡囊草；Xizang Bubbleweed, Xizang Physochlaina ■

297884 Physochlaina pseudophysaloides Pascher = Physochlaina physaloides (L.) G. Don ■

297885 Physochlaina semenowii Regel;塞氏泡囊草■☆

297886 Physochlaina urceolata Kuang et A. M. Lu;坛萼泡囊草;Jarcalyx Bubbleweed,Jarcalyx Physochlaina■

297887 Physochlaina urceolata Kuang et A. M. Lu = Physochlaina praealta（Decne.）Miers■

297888 Physoclada（DC.）Lindl = Cordia L.（保留属名）●

297889 Physoclaina Boiss. = Physochlaina G. Don■

297890 Physocodon Turcz. = Melochia L.（保留属名）●■

297891 Physodeira Hanst. = Episcia Mart.■☆

297892 Physodia Salisb. = Urginea Steinh.■☆

297893 Physodium C. Presl = Melochia L.（保留属名）●■

297894 Physogeton Jaub. et Spach = Halanthium K. Koch■☆

297895 Physoglochin Post et Kuntze = Carex L.■

297896 Physoglochin Post et Kuntze = Physiglochis Neck.■

297897 Physogyne Garay = Pseudogoodyera Schltr.■☆

297898 Physogyne Garay = Schiedeella Schltr.■☆

297899 Physokentia Becc.（1934）;菱籽椰属（瓦奴亚椰属,胞堪蒂桐属）●☆

297900 Physokentia tete（Becc.）Becc.;菱籽椰●☆

297901 Physolepidion Schrenk = Cardaria Desv.■

297902 Physolepidion Schrenk = Lepidium L.■

297903 Physolepidion repens Schrenk = Cardaria chalepense（L.）Hand.-Mazz.■

297904 Physolepidion repens Schrenk = Cardaria draba（L.）Desv. subsp. chalepensis（L.）O. E. Schulz■

297905 Physolepidium Endl. = Physolepidion Schrenk■

297906 Physoleucas（Benth.）Jaub. et Spach = Leucas Burm. ex R. Br.●■

297907 Physoleucas Jaub. et Spach = Leucas Burm. ex R. Br.●■

297908 Physoleucas acrodonta Jaub. et Spach = Leucas inflata Benth.■☆

297909 Physoleucas arabica Jaub. et Spach = Leucas inflata Benth.■☆

297910 Physoleucas inflata（Benth.）Jaub. et Spach ex Briq. = Leucas inflata Benth.■☆

297911 Physoleucas pachystachya Jaub. et Spach = Leucas inflata Benth.■☆

297912 Physoleucas schimperi（C. Presl）Jaub. et Spach = Leucas inflata Benth.■☆

297913 Physolobium Benth. = Kennedia Vent.●☆

297914 Physolophium Turcz. = Angelica L.■

297915 Physolophium Turcz. = Coelopleurum Ledeb.■

297916 Physolophium saxatile（Turcz. ex Ledeb.）Turcz. = Coelopleurum saxatile（Turcz. ex Ledeb.）Drude■

297917 Physolychnis（Benth.）Rupr. = Gastrolychnis（Fenzl）Rchb.■

297918 Physolychnis（Benth.）Rupr. = Silene L.（保留属名）■

297919 Physolychnis Rupr. = Lychnis L.（废弃属名）■

297920 Physolychnis altaica（Pers.）Rupr. = Silene altaica Pers.■

297921 Physolychnis gonosperma Rupr. = Silene gonosperma（Rupr.）Bocquet■

297922 Physondra Raf. = Phaca L.●■

297923 Physophora Link = Physospermum Cusson ex Juss.■☆

297924 Physophora Post et Kuntze = Physiphora Sol. ex R. Br.●

297925 Physophora Post et Kuntze = Rinorea Aubl.（保留属名）●

297926 Physoplexis（Endl.）Schur = Synotoma（G. Don）R. Schulz■☆

297927 Physoplexis（Endl.）Schur(1853);喙檐花属■☆

297928 Physoplexis Schur = Physoplexis（Endl.）Schur■☆

297929 Physoplexis comosa（L.）Schur;喙檐花■☆

297930 Physoplexis comosa Schur = Physoplexis comosa（L.）Schur■☆

297931 Physopodium Desv. = Combretum Loefl.（保留属名）●

297932 Physopsis Turcz.（1849）;风箱草属●☆

297933 Physopsis spicata Turcz.;风箱草●☆

297934 Physoptychis Boiss.（1867）;泡褶芥属（泡折芥属）■☆

297935 Physoptychis gnaphalodes（DC.）Boiss.;泡褶芥■☆

297936 Physopyrum Popov(1935);泡子蓼属☆

297937 Physopyrum teretifolium Popov;泡子蓼☆

297938 Physorhynchus Hook.（1851）;膀胱喙芥属■☆

297939 Physorhynchus brahuicus Hook.;膀胱喙芥■☆

297940 Physorhyncus Hook. f. et Andersson = Physorhynchus Hook.■☆

297941 Physosiphon Lindl. = Pleurothallis R. Br.■☆

297942 Physospermopsis H. Wolff（1925）;滇芎属;Dianxiong, Physospermopsis■★

297943 Physospermopsis alepidioides（H. Wolff et Hand.-Mazz.）R. H. Shan;全叶滇芎（单叶邪蒿）;Entire Dianxiong, Entireleaf Physospermopsis■

297944 Physospermopsis bhutanensis Farille et S. B. Malla = Physospermopsis kingdon-wardii（H. Wolff）C. Norman■

297945 Physospermopsis cruciata H. Wolff = Meeboldia yunnanensis（H. Wolff）Constance et F. T. Pu■

297946 Physospermopsis cuneata H. Wolff;楔叶滇芎（滇芎）;Cuneateleaf Physospermopsis,Wedgeleaf Dianxiong■

297947 Physospermopsis delavayi（Franch.）H. Wolff;前胡叶滇芎（滇芎,金钱参,拟囊果芹）;Delavay Physospermopsis, Dianxiong■

297948 Physospermopsis dielsii Pimenov et Kljuykov = Physospermopsis shaniana C. Y. Wu et F. T. Pu■

297949 Physospermopsis farillei P. K. Mukh. et Constance = Physospermopsis obtusiuscula（Wall. ex DC.）C. Norman■

297950 Physospermopsis forrestii（Diels）C. Norman = Physospermopsis shaniana C. Y. Wu et F. T. Pu■

297951 Physospermopsis fuscopurpurea（Hand.-Mazz.）Pimenov et Kljuykov = Pleurospermum heterosciadium H. Wolff■

297952 Physospermopsis hirsutula（C. B. Clarke）Farille = Physospermopsis obtusiuscula（Wall. ex DC.）C. Norman■

297953 Physospermopsis kingdon-wardii（H. Wolff）C. Norman;小滇芎■

297954 Physospermopsis lalabhduriana Farille et S. B. Malla = Pleurospermum wilsonii H. Boissieu■

297955 Physospermopsis lalabhduriana Farille et S. B. Malla = Pleurospermum crassicaule H. Wolff■

297956 Physospermopsis muktinathensis Farille et S. B. Malla = Physospermopsis rubrinervis（Franch.）C. Norman■

297957 Physospermopsis muliensis R. H. Shan et S. L. Liou;木里滇芎;Muli Dianxiong, Muli Physospermopsis■

297958 Physospermopsis nana（Franch.）Pimenov et Kljuykov = Pleurospermum nanum Franch.■

297959 Physospermopsis obtusiuscula（C. B. Clarke）C. Norman;波棱滇芎;Slightlyobtuse Physospermopsis,Wavyrib Dianxiong■

297960 Physospermopsis obtusiuscula（Wall. ex DC.）C. Norman = Physospermopsis obtusiuscula（C. B. Clarke）C. Norman■

297961 Physospermopsis purpurascens（Franch.）Pimenov et Kljuykov = Pleurospermum nanum Franch.■

297962 Physospermopsis rubrinervis（Franch.）C. Norman;紫脉滇芎（紫脉拟囊果芹）;Redvein Dianxiong, Redvein Physospermopsis■

297963 Physospermopsis shaniana C. Y. Wu et F. T. Pu;丽江滇芎（长苞粗子芹,丽江拟囊果芹,怒江滇芎）;Forrest Dianxiong, Forrest Physospermopsis, Forrest Trachydium■

297964 Physospermopsis wolffiana Fedde ex H. Wolff;高原滇芎;Wolff

Physospermopsis ■

297965　Physospermopsis wolffiana Fedde ex H. Wolff = Tongoloa stewardii H. Wolff ■

297966　Physospermum Cusson ex Juss. = Physospermum Cusson ■☆

297967　Physospermum Cusson（1787）；囊果草属；Bladder Seed, Bladderseed ■☆

297968　Physospermum Lag. = Pleurospermum Hoffm. ■

297969　Physospermum actifolium C. Presl = Physospermum verticillatum（Waldst. et Kit.）Vis. ■☆

297970　Physospermum cornubiense DC.；囊果草；Bladder Seed, Bladderseed ■☆

297971　Physospermum verticillatum（Waldst. et Kit.）Vis.；轮状囊果草 ■☆

297972　Physospermum verticillatum Vis. = Physospermum verticillatum（Waldst. et Kit.）Vis. ■☆

297973　Physostegia Benth.（1829）；假龙头花属（囊萼花属）；False Dragonhead, Lion's-Heart, Obedient Plant ■☆

297974　Physostegia angustifolia Fernald；窄叶假龙头花；False Dragonhead ■☆

297975　Physostegia denticulata（Aiton）Britton = Physostegia virginiana（L.）Benth. ■☆

297976　Physostegia formosior Lunell = Physostegia virginiana（L.）Benth. ■☆

297977　Physostegia granulosa Fassett = Physostegia virginiana（L.）Benth. ■☆

297978　Physostegia intermedia（Nutt.）Engelm. et A. Gray；全叶假龙头花；False Dragonhead ■☆

297979　Physostegia intermedia Engelm. et A. Gray = Physostegia intermedia（Nutt.）Engelm. et A. Gray ■☆

297980　Physostegia latidens House；宽齿假龙头花 ■☆

297981　Physostegia parviflora Nutt. ex DC.；小花假龙头花；False Dragonhead, Small-flowered Dragonhead ■☆

297982　Physostegia speciosa（Sweet）Sweet = Physostegia virginiana（L.）Benth. ■☆

297983　Physostegia speciosa（Sweet）Sweet var. glabriflora Fassett = Physostegia virginiana（L.）Benth. ■☆

297984　Physostegia virginiana（L.）Benth.；假龙头花（囊萼花,芝麻花）；Dragonhead, False Dragonhead, False Dragon-head, Lion's Heart, Obedience, Obedient Plant, Virginia False Dragonhead, Virginia False-dragonhead, Virginia Lionsheart ■☆

297985　Physostegia virginiana（L.）Benth. 'Summer Snow'；夏雪假龙头花 ■☆

297986　Physostegia virginiana（L.）Benth. 'Variegata'；银边假龙头花 ■☆

297987　Physostegia virginiana（L.）Benth. 'Vivid'；矮生假龙头花（艳色假龙头花）；Dwarf Obedient Plant, Obedient Plant ■☆

297988　Physostegia virginiana（L.）Benth. f. candida Benke = Physostegia virginiana（L.）Benth. ■☆

297989　Physostegia virginiana（L.）Benth. var. elongata B. Boivin = Physostegia virginiana（L.）Benth. ■☆

297990　Physostegia virginiana（L.）Benth. var. formosior（Lunell）B. Boivin = Physostegia virginiana（L.）Benth. ■☆

297991　Physostegia virginiana（L.）Benth. var. granulosa（Fassett）Fernald = Physostegia virginiana（L.）Benth. ■☆

297992　Physostegia virginiana（L.）Benth. var. speciosa（Sweet）A. Gray 'Variegata' = Physostegia virginiana（L.）Benth. 'Variegata' ■☆

297993　Physostegia virginiana（L.）Benth. var. speciosa（Sweet）A. Gray = Physostegia virginiana（L.）Benth. ■☆

297994　Physostelma Wight（1834）；泡冠萝藦属 ●☆

297995　Physostelma wallichii Wight；泡冠萝藦 ●☆

297996　Physostemon Mart.（1824）；囊蕊白花菜属；Calabar Bean ■☆

297997　Physostemon Mart. = Cleome L. ●■

297998　Physostemon Mart. et Zucc. = Physostemon Mart. ■☆

297999　Physostemon guianense Briq.；圭亚那囊蕊白花菜 ■☆

298000　Physostemon lanceolatum Mart.；披针叶囊蕊白花菜 ■☆

298001　Physostemon melanospermum（S. Watson）Pax et K. Hoffm.；黑籽囊蕊白花菜 ■☆

298002　Physostemon rotundifolium Mart. et Zucc.；圆叶囊蕊白花菜 ■☆

298003　Physostemon tenuifolium Mart.；细叶囊蕊白花菜 ■☆

298004　Physostigma Balf.（1861）；毒扁豆属（毒毛扁豆属,加拉拔儿豆属,加刺拔儿豆属）；Calabar Bean, Calabarbean ■

298005　Physostigma coriaceum Merxm.；革质毒扁豆 ■☆

298006　Physostigma cylindrospermum（Welw. ex Baker）Holmes；圆柱毒扁豆 ■☆

298007　Physostigma cylindrospermum Holmes = Physostigma cylindrospermum（Welw. ex Baker）Holmes ■☆

298008　Physostigma laxius Merxm.；疏松毒扁豆 ■☆

298009　Physostigma venenosum Balf.；毒扁豆（加拉拔儿豆,泡豆）；Calabar Bean, Chop Nut, Culnbar Bean, Deadly Calabar Bean, Deadly Calabarbean, Ordeal Bean ■

298010　Physothallis Garay = Pleurothallis R. Br. ■☆

298011　Physotheca Raf. = Physocarpus（Cambess.）Raf.（保留属名）●

298012　Physotrichia Hiern（1873）；毛囊草属 ■☆

298013　Physotrichia arenaria Engl. = Diplolophium zambesianum Hiern ■☆

298014　Physotrichia atropurpurea（C. Norman）Cannon；暗紫毛囊草 ■☆

298015　Physotrichia buchananii Benth. ex Oliv. = Diplolophium buchananii（Benth. ex Oliv.）C. Norman ■☆

298016　Physotrichia diplolophioides H. Wolff = Diplolophium diplolophioides（H. Wolff）Jacq.-Fél. ■☆

298017　Physotrichia gorungosensis Engl. = Diplolophium buchananii（Benth. ex Oliv.）C. Norman subsp. swynnertonii（Baker f.）Cannon ■☆

298018　Physotrichia helenae Buscal. et Muschl. = Diplolophium zambesianum Hiern ■☆

298019　Physotrichia heracleoides H. Wolff；独活毛囊草 ■☆

298020　Physotrichia kassneri H. Wolff = Pimpinella kassneri（H. Wolff）Cannon ■☆

298021　Physotrichia longiradiatum H. Wolff = Aframmi longiradiatum（H. Wolff）Cannon ■☆

298022　Physotrichia muriculata（Hiern）S. Droop et C. C. Towns.；粗糙毛囊草 ■☆

298023　Physotrichia swynnertonii Baker f.；斯温纳顿毛囊草 ■☆

298024　Physotrichia verdickii C. Norman；韦尔迪毛囊草 ■☆

298025　Physotrichia welwitschii Hiern；韦尔毛囊草 ■☆

298026　Physurus L. = Erythrodes Blume ■

298027　Physurus Rich. = Erythrodes Blume ■

298028　Physurus Rich. = Physurus Rich. ex Lindl. ■

298029　Physurus Rich. ex Lindl. = Erythrodes Blume ■

298030　Physurus blumei Lindl. = Erythrodes blumei（Lindl.）Schltr. ■

298031　Physurus bracteatus Blume = Herpysma longicaulis Lindl. ■

298032　Physurus chinensis Rolfe = Erythrodes blumei（Lindl.）Schltr. ■

298033　Physurus commelynifolius Rchb. f. = Platythelys querceticola（Lindl.）Garay ■☆

298034　Physurus henryi（Schltr.）K. Schum. et Lauterb. = Erythrodes blumei（Lindl.）Schltr. ■

298035　Physurus herpysmoides King et Pantl. = Erythrodes hirsuta（Griff.）Ormerod ■

298036　Physurus hirsutus（Griff.）Lindl. = Erythrodes hirsuta（Griff.）Ormerod ■

298037　Physurus jamaicensis Fawc. et Rendle = Platythelys querceticola（Lindl.）Garay ■☆

298038　Physurus querceticola Lindl. = Platythelys querceticola（Lindl.）Garay ■☆

298039　Physurus sagreanus A. Rich. = Platythelys querceticola（Lindl.）Garay ■☆

298040　Physurus viridiflorus（Blume）Lindl. = Goodyera viridiflora（Blume）Blume ■

298041　Phytarrhiza Vis. = Tillandsia L. ■☆

298042　Phytelephaceae Perleb = Arecaceae Bercht. et J. Presl（保留科名）●

298043　Phytelephaceae Perleb = Palmae Juss.（保留科名）●

298044　Phytelephantaceae Brongn. ex Martinet = Arecaceae Bercht. et J. Presl（保留科名）●

298045　Phytelephantaceae Brongn. ex Martinet = Palmae Juss.（保留科名）●

298046　Phytelephantaceae Martinet = Palmae Juss.（保留科名）●

298047　Phytelephantaceae Martinet = Phytelephantaceae Brongn. ex Martinet ●

298048　Phytelephantaceae Martinet ex Perleb = Arecaceae Bercht. et J. Presl（保留科名）●

298049　Phytelephantaceae Martinet ex Perleb = Palmae Juss.（保留科名）●

298050　Phytelephas Ruiz et Pav.（1798）；象牙棕属（石棕桐属，象牙椰属，象牙椰子属）；Ivary Nut Palm, Ivary Palm, Ivory-Nut Palm, Large-fruit Ivary Palm, Negro's Head Palm, Vagetable Ivary, Vagetable Ivary Palm ●☆

298051　Phytelephas macrocarpa Ruiz et Pav.；象牙棕（大果石棕，象牙椰子）；Common Ivory Palm, Corozo Nut Palm, Ivary Nut Palm, Ivory Nut, Ivory Palm, Ivory-nut Palm, Negro's Head Palm, Tabua Palm, Tagua Nut, Tugua Palm, Vegetable Ivory, Vegetable Ivory Palm ●☆

298052　Phytelephasiaceae Brongn. ex Chadef. et Emberg. = Arecaceae Bercht. et J. Presl（保留科名）●

298053　Phytelephasiaceae Brongn. ex Chadef. et Emberg. = Palmae Juss.（保留科名）●

298054　Phyteuma L.（1753）；牧根草属；Horned Rampion, Mixed-Flower, Rampion ■☆

298055　Phyteuma Lour. = Sambucus L. ●■

298056　Phyteuma betonicifolium St. -Lag.；蓝牧根草；Blue Spiked Rampion ■☆

298057　Phyteuma canescens Waldst. et Kit.；灰白牧根草■☆

298058　Phyteuma comosum L.；束毛牧根草■☆

298059　Phyteuma comosum L. = Physoplexia comosa（L.）Schur ■☆

298060　Phyteuma haleri All.；哈氏牧根草■☆

298061　Phyteuma hemisphaericum L.；半球牧根草■☆

298062　Phyteuma japonicum Miq. = Asyneuma japonicum（Miq.）Briq. ●

298063　Phyteuma limonifolium Sibth. et Sm.；补血草叶牧根草■☆

298064　Phyteuma nigrum Schmidt.；黑牧根草；Black Mixed-flower ■☆

298065　Phyteuma orbiculare L.；圆叶牧根草；Ball-headed Mixed-flower, Pride of Sussex, Rampion, Round-headed Rampion ■☆

298066　Phyteuma ovatum Honck.；卵形牧根草；Dark Rampion ■☆

298067　Phyteuma rigidum Willd. = Asyneuma rigidum（Willd.）Grossh. ■☆

298068　Phyteuma scheuchzeri All.；球序牧根草；Horned Rampion, Oxford Rampion ■☆

298069　Phyteuma spicatum L.；穗花牧根草；Spiked Mixed-flower, Spiked Rampion ■☆

298070　Phyteuma tenerum Rich. Schulz = Phyteuma orbiculare L. ■☆

298071　Phyteuma tenuifolium A. DC.；细叶牧根草■☆

298072　Phyteuma tetramerum Schur.；四数牧根草■☆

298073　Phyteumoides Smeathman ex DC. = Virecta Sm. ■☆

298074　Phyteumoides Smeathman ex DC. = Virectaria Bremek. ■☆

298075　Phyteumopsis Juss. ex Poir.（1816）；类牧根草属■☆

298076　Phyteumopsis Juss. ex Poir. = Marshallia Schreb.（保留属名）■☆

298077　Phyteumopsis angustifolia Poir.；狭叶类牧根草■☆

298078　Phyteumopsis lanceolata Poir.；剑叶类牧根草■☆

298079　Phyteumopsis latifolia Poir.；宽叶类牧根草■☆

298080　Phytholacca Brot. = Phytolacca L. ●■

298081　Phytocrenaceae Arn. ex R. Br. = Icacinaceae Miers（保留科名）●■

298082　Phytocrene Wall.（1831）（保留属名）；泉茱萸属；Phytocrenum ●☆

298083　Phytocrene dasycarpa Miq.；毛果泉茱萸●☆

298084　Phytocrene macrocarpa Griff.；大果泉茱萸●☆

298085　Phytocrene macrophylla Blume；大叶泉茱萸●☆

298086　Phytocrene ovalifolia Koord.；卵叶泉茱萸●☆

298087　Phytogyne Salisb. ex Haw. = Narcissus L. ■

298088　Phytolacca L.（1753）；商陆属；Pock Berry, Pokeberry, Pokeweed ●■

298089　Phytolacca Tourn. ex L. = Phytolacca L. ●■

298090　Phytolacca abyssinica Hoffm. = Phytolacca dodecandra L'Hér. ■☆

298091　Phytolacca abyssinica Hoffm. var. apiculata Engl. = Phytolacca dodecandra L'Hér. ■☆

298092　Phytolacca acinosa Hook. f. = Phytolacca latbenia（Moq.）Walter ■

298093　Phytolacca acinosa Roxb.；商陆（白昌，白菖，白母鸡，白章陆，抱母鸡，昌陆，长不老，常蓼，春牛头，大苋菜，当陆，倒水莲，地萝卜，杜大黄，莪羊菜，鹅羊菜，肥猪菜，狗头三七，红苋菜，花商陆，见肿消，金鸡姆，金七娘，蓟，鹿神，马尾，牛大黄，牛萝卜，牛舌根，菩柳，山包谷，山鹿脯，山萝卜，商六，湿萝卜，湿苋菜，食用商陆，水萝卜，台湾商陆，蕿，天萝卜，甜鹿脯，土当归，土冬瓜，土母鸡，忘母牛，文章柳，乌鸡婆，下山虎，苋菜蓝，苋陆，野萝卜，夜呼，张国老，张果老，章柳，章陆，蓸柳，樟柳，猪母甲，猪姆耳，蓬蒢，抓肿消）；Edible Pokeweed, India Pockberry, India Pokeweed, Indian Pockberry, Indian Poke, Indian Pokeweed ■

298094　Phytolacca americana L.；垂序商陆（白鸡腿，白癞鸡婆，花商陆，美国商陆，美商陆，美洲商陆，十蕊商陆，洋商陆，野胭脂）；American Nightshade, American Pokeberry, American Spinach, Bear's Grape, Cancer Jalap, Cancer-root, Chongras, Cocum Coakum, Common Pockberry, Common Pokeberry, Crimsonberry, Crowberry, Drooopraceme Pokeweed, Dyer's Grape, Dyer's Grapes, Garget, Garget-plant, Inkberry, Kermes Bush, Pigeon Berry, Pigeonberry, Pigeon-berry, Pocan, Pockberry, Poke, Poke Weed, Pokeberry, Pokeroot, Poke-root, Pokeweed, Red Ink Berry, Red Ink Plant, Red Inkberry, Redink Plant, Red-ink Plant, Red-ink-plant, Redweed, Scoke, Skoke, Virginian Poke, Virginian Pokeweed ■

298095　Phytolacca americana L. var. mexicana ? = Phytolacca octandra L. ■☆

298096　Phytolacca americana L. var. rigida（Small）Caulkins et R. E. Wyatt = Phytolacca rigida Small ■☆

298097　Phytolacca arborea Moq. = Phytolacca dioica L. ●☆

298098　Phytolacca asiatica L. = Leea asiatica（L.）Ridsdale ●

298099　Phytolacca bogotensis Kunth；南方商陆；Southern Pokeweed ■☆

298100　Phytolacca clavipera W. W. Sm. = Phytolacca polyandra Batalin ■

298101　Phytolacca cyclopetala H. Walter；圆瓣商陆■☆

298102　Phytolacca decandra L. = Phytolacca americana L. ■

298103　Phytolacca dioica L.；阿根廷商陆（树商陆）；Bella Sombra Tree, Bella Umbra, Ombu, Ombu Tree, Ombutree Pockberry, Phytolacca ●☆

298104　Phytolacca dodecandra L' Hér.；十二雄蕊商陆；Endod ■☆

298105　Phytolacca dodecandra L' Hér. var. brevipedicellata H. Walter = Phytolacca dodecandra L'Hér. ■☆

298106　Phytolacca dodecandra L'Hér. var. apiculata（Engl.）Baker et C. H. Wright = Phytolacca dodecandra L'Hér. ■☆

298107　Phytolacca dodecandra L'Hér. var. brevipedicellata H. Walter = Phytolacca dodecandra L'Hér. ■☆

298108　Phytolacca esculenta Van Houtte = Phytolacca acinosa Roxb. ■

298109　Phytolacca heptandra Retz.；七蕊商陆；Umbra Tree, Wild Sweet Potato ■☆

298110　Phytolacca heterotepala H. Walter；墨西哥商陆；Mexican Pokeweed ■☆

298111　Phytolacca hunanensis Hand. -Mazz. = Phytolacca japonica Makino ■

298112　Phytolacca icosandra L.；热带商陆；Tropical Pokeweed ■☆

298113　Phytolacca insularis Nakai；岛生商陆■☆

298114　Phytolacca japonica Makino；日本商陆（红色倒水莲，圆果商陆）；Japan Pokeweed，Japanese Pockberry ■☆

298115　Phytolacca nutans H. Walter；俯垂商陆■☆

298116　Phytolacca octandra L.；红商陆（八蕊商陆）；Inkweed, Red Inkplant ■☆

298117　Phytolacca pekinensis Hance = Phytolacca acinosa Roxb. ■

298118　Phytolacca polyandra Batalin；多雄蕊商陆（多蕊商陆，多药商陆）；Manystamen Pokeweed ■

298119　Phytolacca rigida Small；硬商陆；Inkberry, Pigeonberry, Poke, Pokeberry, Pokeweed ■☆

298120　Phytolacca rigida Small = Phytolacca americana L. var. rigida（Small）Caulkins et R. E. Wyatt ■☆

298121　Phytolacca rivinoides Kunth et Bouché；沟商陆■☆

298122　Phytolacca stricta Hoffm. = Phytolacca heptandra Retz. ■☆

298123　Phytolacca zhejiangensis W. T. Fan = Phytolacca japonica Makino ■

298124　Phytolaccaceae R. Br.（1818）（保留科名）；商陆科；Pokeweed Family ●■

298125　Phytosalpinx Lunell = Lycopus L. ■

298126　Phytoxis Molina（废弃属名）= Sphacele Benth.（保留属名）●■☆

298127　Phytoxys Spreng. = Sphacele Benth.（保留属名）●■☆

298128　Piaggiaea Chiov. = Wrightia R. Br. ●

298129　Piaggiaea boranensis Chiov. = Wrightia demartiniana Chiov. ●☆

298130　Piaggiaea demartiniana（Chiov.）Chiov. = Wrightia demartiniana Chiov. ●☆

298131　Piaradena Raf. = Salvia L. ●■

298132　Piaranthus R. Br.（1810）；脂花萝藦属■☆

298133　Piaranthus aridus（Masson）G. Don = Quaqua arida（Masson）Bruyns ■☆

298134　Piaranthus atrosanguineus（N. E. Br.）Bruyns；暗血红脂花萝藦■☆

298135　Piaranthus barrydalensis Meve = Piaranthus geminatus

（Masson）N. E. Br. ■☆

298136　Piaranthus comptus N. E. Br.；装饰脂花萝藦■☆

298137　Piaranthus comptus N. E. Br. var. ciliatus ? = Piaranthus comptus N. E. Br. ■☆

298138　Piaranthus cornutus N. E. Br.；角状脂花萝藦■☆

298139　Piaranthus cornutus N. E. Br. var. grandis ? = Piaranthus cornutus N. E. Br. ■☆

298140　Piaranthus cornutus N. E. Br. var. ruschii（Nel）Bruyns；鲁施角状脂花萝藦■☆

298141　Piaranthus decipiens（N. E. Br.）Bruyns；迷惑脂花萝藦■☆

298142　Piaranthus decorus（Masson）N. E. Br. = Piaranthus geminatus（Masson）N. E. Br. subsp. decorus（Masson）Bruyns ■☆

298143　Piaranthus decorus（Masson）N. E. Br. subsp. cornutus（N. E. Br.）Meve = Piaranthus cornutus N. E. Br. ■☆

298144　Piaranthus disparilis N. E. Br. = Piaranthus geminatus（Masson）N. E. Br. ■☆

298145　Piaranthus disparilis N. E. Br. var. immaculata C. A. Lückh. = Piaranthus geminatus（Masson）N. E. Br. ■☆

298146　Piaranthus foetidus N. E. Br. = Piaranthus geminatus（Masson）N. E. Br. ■☆

298147　Piaranthus foetidus N. E. Br. var. diversus ? = Piaranthus geminatus（Masson）N. E. Br. ■☆

298148　Piaranthus foetidus N. E. Br. var. multipunctatus ? = Piaranthus geminatus（Masson）N. E. Br. ■☆

298149　Piaranthus foetidus N. E. Br. var. pallidus ? = Piaranthus geminatus（Masson）N. E. Br. ■☆

298150　Piaranthus foetidus N. E. Br. var. purpureus ? = Piaranthus geminatus（Masson）N. E. Br. ■☆

298151　Piaranthus framesii Pillans = Piaranthus punctatus（Masson）Schult. var. framesii（Pillans）Bruyns ■☆

298152　Piaranthus geminatus（Masson）N. E. Br.；双生脂花萝藦■☆

298153　Piaranthus geminatus（Masson）N. E. Br. subsp. decorus（Masson）Bruyns；装饰双生脂花萝藦■☆

298154　Piaranthus geminatus（Masson）N. E. Br. var. foetidus（N. E. Br.）Meve = Piaranthus geminatus（Masson）N. E. Br. ■☆

298155　Piaranthus globosus A. C. White et B. Sloane = Piaranthus geminatus（Masson）N. E. Br. ■☆

298156　Piaranthus grivanus N. E. Br. = Piaranthus decipiens（N. E. Br.）Bruyns ■☆

298157　Piaranthus incarnatus（L. f.）G. Don = Quaqua incarnata（L. f.）Bruyns ■☆

298158　Piaranthus incarnatus（L. f.）G. Don var. albus G. Don = Quaqua incarnata（L. f.）Bruyns ■☆

298159　Piaranthus mammillaris（L.）G. Don = Quaqua mammillaris（L.）Bruyns ■☆

298160　Piaranthus mennellii C. A. Lückh. = Piaranthus cornutus N. E. Br. ■☆

298161　Piaranthus nebrownii Dinter = Piaranthus cornutus N. E. Br. ■☆

298162　Piaranthus pallidus C. A. Lückh. = Piaranthus cornutus N. E. Br. ■☆

298163　Piaranthus parviflorus（Masson）Sweet = Quaqua parviflora（Masson）Bruyns ■☆

298164　Piaranthus parvulus N. E. Br.；较小脂花萝藦■☆

298165　Piaranthus pillansii N. E. Br. = Piaranthus geminatus（Masson）N. E. Br. ■☆

298166　Piaranthus pillansii N. E. Br. var. inconstans ? = Piaranthus geminatus（Masson）N. E. Br. ■☆

298167　Piaranthus pulcher N. E. Br. = Piaranthus cornutus N. E. Br. ■☆

298168　Piaranthus pulcher N. E. Br. var. nebrownii（Dinter）A. C. White et B. Sloane = Piaranthus cornutus N. E. Br. ■☆

298169　Piaranthus pullus（Aiton）Haw. = Quaqua mammillaris（L.）Bruyns ■☆

298170　Piaranthus punctatus（Masson）R. Br. ;斑点脂花萝藦■☆

298171　Piaranthus punctatus（Masson）Schult. var. framesii（Pillans）Bruyns;弗雷斯脂花萝藦■☆

298172　Piaranthus ramosus（Masson）Sweet = Quaqua ramosa（Masson）Bruyns ■☆

298173　Piaranthus ruschii Nel = Piaranthus cornutus N. E. Br. var. ruschii（Nel）Bruyns ■☆

298174　Piaranthus serrulatus（Jacq.）N. E. Br. = Piaranthus geminatus（Masson）N. E. Br. subsp. decorus（Masson）Bruyns ■☆

298175　Piaranthus streyianus Nel = Orbea maculata（N. E. Br.）L. C. Leach subsp. rangeana（Dinter et A. Berger）Bruyns ■☆

298176　Piarimula Raf. = Lippia L. ●■☆

298177　Piarimula Raf. = Phyla Lour. ■

298178　Piarophyla Raf. = Bergenia Moench（保留属名）■

298179　Piaropus Raf.（废弃属名）= Eichhornia Kunth（保留属名）■

298180　Piaropus crassipes（Mart.）Raf. = Eichhornia crassipes（Mart.）Solms ■

298181　Picardaea Urb.（1903）;皮卡尔茜属●☆

298182　Picardaea cubensis（Griseb.）Britton ex Urb. ;皮卡尔茜●☆

298183　Picardenia Steud. = Actinea Juss. ■

298184　Picardenia Steud. = Actinella Pers. ■

298185　Picardenia Steud. = Picradenia Hook. ☆

298186　Piccia Neck. = Moronobea Aubl. ●☆

298187　Picconia A. DC.（1844）;皮康木犀属●☆

298188　Picconia excelsa（Aiton）DC. ;皮康木犀●☆

298189　Picea A. Dietr.（1824）;云杉属;Spruce, Fir ●

298190　Picea D. Don ex Loudon = Abies Mill. ●

298191　Picea abies（L.）H. Karst. ;欧洲云杉（挪威云杉）;Baltic White Wood, Baltic Whitewood, Burgundy Pitch, Christmas Spruce, Christmas Tree, Common Spruce, Cone-dealapple, Deal Apple, Deal-apple, Europe Spruce, European Spruce, Fir Apple, Fir-apple, Jura Turpentine, Norway Spruce, Norway Spruce Fir, Norway Spruce-fir, Pine Apple, Pine-apple, Spruce, Spruce Beer, Spruce Fir, White Deal, Whitewood ●

298192　Picea abies（L.）H. Karst. 'Acrocona';早果欧洲云杉;Early-coning Spruce ●☆

298193　Picea abies（L.）H. Karst. 'Argentea';银叶欧洲云杉●☆

298194　Picea abies（L.）H. Karst. 'Clanbrassiliana';科兰布雷斯欧洲云杉（克兰巴西挪威云杉）●☆

298195　Picea abies（L.）H. Karst. 'Cranstonii';科兰斯通欧洲云杉;Cranston Spruce ●☆

298196　Picea abies（L.）H. Karst. 'Cupressina';柏树欧洲云杉●☆

298197　Picea abies（L.）H. Karst. 'Echiniformis';刺叶欧洲云杉;Hedgehog Norway Spruce ●☆

298198　Picea abies（L.）H. Karst. 'Gregoryana';格雷角利欧洲云杉（格雷戈里挪威云杉）●☆

298199　Picea abies（L.）H. Karst. 'Humilis';矮生欧洲云杉●☆

298200　Picea abies（L.）H. Karst. 'Inversa';倒垂挪威云杉●☆

298201　Picea abies（L.）H. Karst. 'Little Gem';小宝石欧洲云杉（小宝石挪威云杉）●☆

298202　Picea abies（L.）H. Karst. 'Maxwellii';麦克斯维尔欧洲云杉;Maxwell Spruce, Maxwell's Norway Spruce ●☆

298203　Picea abies（L.）H. Karst. 'Nidiformis';鸟巢欧洲云杉（鸟巢挪威云杉）;Birdsnest Spruce ●☆

298204　Picea abies（L.）H. Karst. 'Ohlendorffii';奥伦道夫挪威云杉●☆

298205　Picea abies（L.）H. Karst. 'Pachyphylla';厚叶欧洲云杉●☆

298206　Picea abies（L.）H. Karst. 'Pendula';垂枝欧洲云杉●☆

298207　Picea abies（L.）H. Karst. 'Pricumbens';平卧欧洲云杉●☆

298208　Picea abies（L.）H. Karst. 'Pumila';密矮欧洲云杉;Pumila Norway Spruce ●☆

298209　Picea abies（L.）H. Karst. 'Pyramidalis Gracilis';细塔欧洲云杉●☆

298210　Picea abies（L.）H. Karst. 'Reflexa';下弯欧洲云杉（反曲挪威云杉）●☆

298211　Picea abies（L.）H. Karst. 'Repens';匍匐欧洲云杉●☆

298212　Picea abies（L.）H. Karst. 'Tabuliformis';平台欧洲云杉●☆

298213　Picea abies（L.）H. Karst. subsp. obovata（Ledeb.）Hultén = Picea obovata Ledeb. ●

298214　Picea abies（L.）H. Karst. var. chlorocarpa（Purk.）Th. Fr. ;绿果挪威云杉;Greencone Norway Spruce ●☆

298215　Picea abies（L.）H. Karst. var. erythrocarpa（Purk.）Rehder;紫果挪威云杉;Purplecone Norway Spruce ●☆

298216　Picea abies（L.）H. Karst. var. nigra（Loudon）Th. Fr. ;黑挪威云杉;Black Norway Spruce ●☆

298217　Picea abies（L.）H. Karst. var. obovata（Ledeb.）Lindquist = Picea obovata Ledeb. ●

298218　Picea ajanensis Fisch. ex Carrière = Picea ajanensis（Lindau et Gordon）Fisch. ex Carrière ●

298219　Picea ajanensis Fisch. ex Carrière = Picea jezoensis（Siebold et Zucc.）Carrière ●

298220　Picea ajanensis Fisch. ex Carrière = Picea jezoensis（Siebold et Zucc.）Carrière var. microsperma（Lindl.）W. C. Cheng et L. K. Fu ●

298221　Picea ajanensis Fisch. ex Trautv. et C. A. Mey. = Picea jezoensis（Siebold et Zucc.）Carrière var. microsperma（Lindl.）W. C. Cheng et L. K. Fu ●

298222　Picea alba（Aiton）Link = Picea glauca（Moench）Voss ●

298223　Picea alba（Aiton）Link var. albertiana（S. Br.）Beissn. = Picea glauca（Moench）Voss ●

298224　Picea albertiana S. Br. = Picea glauca（Moench）Voss ●

298225　Picea alcockiana Carrière;阿尔考云杉;Alcock Spruce, Alcock's Spruce ●☆

298226　Picea alcoquiana（Veitch ex Lindl.）Carrière;二色云杉;Alcock Spruce, Alcock's Spruce ●☆

298227　Picea alcoquiana（Veitch ex Lindl.）Carrière f. chlorocarpa（Hayashi）Yonek. ;绿果二色云杉●☆

298228　Picea alcoquiana（Veitch ex Lindl.）Carrière var. acicularis（Shiras. et Koyama）Fitschen = Picea koyamae Hayashi var. acicularis（Shiras. et Koyama）T. Shimizu ●☆

298229　Picea alcoquiana（Veitch ex Lindl.）Carrière var. acicularis（Shiras. et Koyama）Fitschen = Picea shirasawae Hayashi ●☆

298230　Picea alcoquiana（Veitch ex Lindl.）Carrière var. reflexa（Shiras. et Koyama）Fitschen;反卷二色云杉●☆

298231　Picea ascendens Patschke = Picea brachytyla（Franch.）E. Pritz. ●◇

298232　Picea asperata Mast. ;云杉（茂县云杉）;China Spruce, Chinese Spruce, Dragon Spruce ●

298233　Picea asperata Mast. var. aurantiaca（Mast.）Boom;白皮云杉（黄枝云杉）;Whitebark Spruce, Yellow-twig Spruce ●

298234　Picea asperata Mast. var. heterolepis（Mast.）W. C. Cheng =

Picea asperata Mast. ●

298235　Picea asperata Mast. var. heterolepis（Rehder et E. H. Wilson）L. K. Fu et Nan Li = Picea asperata Mast. var. heterolepis（Rehder et E. H. Wilson）Rehder ●

298236　Picea asperata Mast. var. heterolepis（Rehder et E. H. Wilson）Rehder;裂鳞云杉(茂县云杉)●

298237　Picea asperata Mast. var. notabilis Rehder et E. H. Wilson = Picea asperata Mast. var. heterolepis（Rehder et E. H. Wilson）Rehder ●

298238　Picea asperata Mast. var. notabilis Rehder et E. H. Wilson = Picea asperata Mast. var. heterolepis（Rehder et E. H. Wilson）L. K. Fu et Nan Li ●

298239　Picea asperata Mast. var. notabilis Rehder et E. H. Wilson = Picea aurantiaca Mast. ●

298240　Picea asperata Mast. var. ponderosa Rehder et E. H. Wilson = Picea asperata Mast. ●

298241　Picea asperata Mast. var. ponderosa Rehder et E. H. Wilson = Picea aurantiaca Mast. ●

298242　Picea asperata Mast. var. retroflexa（Mast.）Boom = Picea retroflexa Mast. ●

298243　Picea asperata Mast. var. retroflexa（Mast.）W. C. Cheng = Picea asperata Mast. ●

298244　Picea asperata Mast. var. retroflexa（Mast.）W. C. Cheng = Picea retroflexa Mast. ●

298245　Picea aurantiaca Mast. = Picea asperata Mast. var. aurantiaca（Mast.）Boom ●

298246　Picea australis Small = Picea rubens Sarg. ●☆

298247　Picea balfouriana f. bicolor S. Chen = Picea likiangensis（Franch.）E. Pritz. var. rubescens Rehder et E. H. Wilson ●

298248　Picea balfouriana Rehder et E. H. Wilson;川西云杉(水平杉,西康云杉);Balfour Spruce,Xikang Spruce ●

298249　Picea balfouriana Rehder et E. H. Wilson = Picea likiangensis（Franch.）E. Pritz. var. rubescens Rehder et E. H. Wilson ●

298250　Picea balfouriana Rehder et E. H. Wilson f. bicolor S. Chen = Picea balfouriana Rehder et E. H. Wilson ●

298251　Picea balfouriana Rehder et E. H. Wilson f. bicolor S. Chen = Picea likiangensis（Franch.）E. Pritz. ●

298252　Picea balfouriana Rehder et E. H. Wilson f. bicolor S. Chen = Picea likiangensis（Franch.）E. Pritz. var. rubescens Rehder et E. H. Wilson ●

298253　Picea balfouriana Rehder et E. H. Wilson var. hirtella（Rehder et E. H. Wilson）W. C. Cheng = Picea likiangensis（Franch.）E. Pritz. var. hirtella（Rehder et E. H. Wilson）W. C. Cheng ●

298254　Picea bicolor（Maxim.）Mayr = Picea alcoquiana（Veitch ex Lindl.）Carrière ●☆

298255　Picea bicolor（Maxim.）Mayr f. chlorocarpa Hayashi = Picea alcocquiana（Veitch ex Lindl.）Carrière f. chlorocarpa（Hayashi）Yonek. ●☆

298256　Picea bicolor（Maxim.）Mayr var. acicularis ? = Picea shirasawae Hayashi ●☆

298257　Picea bicolor（Maxim.）Mayr var. reflexa Shiras. et Koyama = Picea alcoquiana（Veitch ex Lindl.）Carrière var. reflexa（Shiras. et Koyama）Fitschen ●☆

298258　Picea brachytyla（Franch.）E. Pritz.;麦吊云杉(垂枝云杉,巅枞树,麦吊杉);Sargent Spruce,Sargent's Spruce ●◇

298259　Picea brachytyla（Franch.）Pritz. = Picea sargentiana Rehder et E. H. Wilson ●◇

298260　Picea brachytyla（Franch.）Pritz. var. complanata（Mast.）W. C. Cheng ex Rehder;油麦吊云杉(米条云杉,油麦吊杉);Graybark Sargent Spruce,Graybark Spruce ●

298261　Picea brachytyla（Franch.）Pritz. var. latisquamea Stapf = Picea brachytyla（Franch.）E. Pritz. ●◇

298262　Picea brachytyla（Franch.）Pritz. var. pachyclada（Patschke）Silba = Picea brachytyla（Franch.）E. Pritz. ●◇

298263　Picea brachytyla（Franch.）Pritz. var. rhombisquamea Stapf = Picea brachytyla（Franch.）E. Pritz. ●◇

298264　Picea brevifolia Peck = Picea mariana（Mill.）Britton, Sterns et Poggenb. ●☆

298265　Picea breweriana S. Watson;布鲁尔氏云杉(北美垂枝云杉);Brewer Pine,Brewer Spruce,Brewer's Spruce,Brewer's Weeping Spruce,Weeping Spruce ●☆

298266　Picea canadensis（Mill.）Britton,Sterns et Poggenb. = Picea glauca（Moench）Voss ●

298267　Picea canadensis（Mill.）Britton, Sterns et Poggenb. var. glauca（Moench）Sudw. = Picea glauca（Moench）Voss ●

298268　Picea canadensis Britton, Sterns et Poggenb. = Picea glauca（Moench）Voss ●

298269　Picea columbiana Lemmon = Picea engelmannii Parry ex Engelm. ●☆

298270　Picea complanata Mast. = Picea brachytyla（Franch.）Pritz. var. complanata（Mast.）W. C. Cheng ex Rehder ●

298271　Picea concolor Gordon et Glend. = Abies concolor（Gordon et Glend.）Lindl. ex Hildebrandt ●☆

298272　Picea crassifolia Kom.;青海云杉(杆树);Thick-leaf Spruce,Thick-leaved Spruce ●

298273　Picea engelmannii Parry ex Engelm.;恩氏云杉(北美山地云杉,恩格曼氏云杉);Columbian Spruce, Engelmann Spruce, Engelmann's Spruce,Mountain Spruce,Silver Spruce ●☆

298274　Picea engelmannii Parry ex Engelm. var. glabra Goodman = Picea engelmannii Parry ex Engelm. ●☆

298275　Picea excelsa（Lam.）Link = Picea abies（L.）H. Karst. ●

298276　Picea excelsa（Lam.）Link var. obovata（Ledeb.）Blytt = Picea obovata Ledeb. ●

298277　Picea excelsa Link = Picea abies（L.）H. Karst. ●

298278　Picea excelsa Link var. altaica Tepl. = Picea obovata Ledeb. ●

298279　Picea excelsa Link var. obovata（Ledeb.）Koch = Picea obovata Ledeb. ●

298280　Picea falcata（Raf.）Suringar = Picea sitchensis（Bong.）Carrière ●☆

298281　Picea farreri C. N. Page et Rushforth;缅甸云杉●

298282　Picea fennica Hort.;芬兰云杉;Finland Spruce ●☆

298283　Picea fortunei A. Murray bis = Keteleeria fortunei（A. Murray）Carrière ●

298284　Picea fortunei Murray = Keteleeria fortunei（A. Murray）Carrière ●

298285　Picea gemmata Rehder et E. H. Wilson = Picea asperata Mast. ●

298286　Picea glauca（Moench）Voss;北美白云杉(白云杉,苍白云杉,灰绿云杉,加拿大云杉,美洲灰云杉,银白云杉);Black Hills Spruce,Canada Spruce,Canadian Spruce,Cat Spruce,Dwarf Alberta Spruce,Pasture Spruce,Porsild Spruce,Skunk Spruce,Western White Spruce,White Spruce ●

298287　Picea glauca（Moench）Voss 'Alberta Blue';阿尔伯特蓝白云杉●☆

298288　Picea glauca（Moench）Voss 'Conica';圆锥白云杉;Dwarf Alberta White Spruce ●☆

298289　Picea glauca（Moench）Voss 'Densata';稠密白云杉●☆

298290　Picea glauca（Moench）Voss 'Echiniformis';刺叶白云杉●☆

298291　Picea glauca（Moench）Voss 'Nana';矮小白云杉●☆

298292　Picea glauca（Moench）Voss 'Rainbow's End';虹端白云杉; Rainbow's End Spruce ●☆

298293　Picea glauca（Moench）Voss subsp. engelmannii（Parry ex Engelm.）T. M. C. Taylor ＝ Picea engelmannii Parry ex Engelm. ●☆

298294　Picea glauca（Moench）Voss var. albertiana（S. Br.）Sarg. ＝ Picea glauca（Moench）Voss ●

298295　Picea glauca（Moench）Voss var. densata Bailey ＝ Picea glauca（Moench）Voss ●

298296　Picea glauca（Moench）Voss var. densata L. H. Bailey ＝ Picea glauca（Moench）Voss ●

298297　Picea glauca（Moench）Voss var. porsildii Raup ＝ Picea glauca（Moench）Voss ●

298298　Picea glehnii（F. Schmidt）Mast.;库页云杉（格林云杉,萨哈林云杉）;Glehn's Spruce, Saghalien Fir, Saghalin Spruce, Sakhalin Spruce, Sakhaline Spruce ●☆

298299　Picea glehnii（F. Schmidt）Mast. f. chlorocarpa Miyabe et Kudo;绿色库页云杉（绿色萨哈林云杉）●

298300　Picea heterolepis Rehder et E. H. Wilson ＝ Picea asperata Mast. var. heterolepis（Rehder et E. H. Wilson）Rehder ●

298301　Picea hirtella Rehder et E. H. Wilson ＝ Picea balfouriana Rehder et E. H. Wilson var. hirtella（Rehder et E. H. Wilson）W. C. Cheng ●

298302　Picea hirtella Rehder et E. H. Wilson ＝ Picea likiangensis（Franch.）E. Pritz. var. hirtella（Rehder et E. H. Wilson）W. C. Cheng ●

298303　Picea hondoensis Mayr;本州鱼鳞松（本州云杉）;Hondo Spruce ●☆

298304　Picea hookeriana（A. Murray bis）Bertrand ＝ Tsuga mertensiana（Bong.）Carrière ●☆

298305　Picea intercedens Nakai ＝ Picea koraiensis Nakai ●

298306　Picea intercedens Nakai var. glabra Uyeki ＝ Picea koraiensis Nakai ●

298307　Picea jezoensis（Siebold et Zucc.）Carrière;鱼鳞云杉（北海道云杉,旱谷鱼鳞云杉,卵果鱼鳞云杉,日本鱼鳞云杉,虾夷松,虾夷鱼鳞云杉,虾夷云杉,兴安鱼鳞云杉,鱼鳞松）;Eggfruit Spruce, Hondo Spruce, Honshu Spruce, Jezo Spruce, Little Seed Spruce, Littleseed Spruce, Yeddo Spruce, Yeso Spruce, Yezo Spruce ●

298308　Picea jezoensis（Siebold et Zucc.）Carrière subsp. hondoensis（Mayr）P. Schmidt ＝ Picea hondoensis Mayr ●☆

298309　Picea jezoensis（Siebold et Zucc.）Carrière var. ajanensis（Fisch. ex Trautv. et C. A. Mey.）W. C. Cheng et L. K. Fu ＝ Picea ajanensis（Lindau et Gordon）Fisch. ex Carrière ●

298310　Picea jezoensis（Siebold et Zucc.）Carrière var. ajanensis（Fisch. ex Trautv. et C. A. Mey.）W. C. Cheng et L. K. Fu ＝ Picea jezoensis（Siebold et Zucc.）Carrière ●

298311　Picea jezoensis（Siebold et Zucc.）Carrière var. ajanensis（Fisch. ex Trautv. et C. A. Mey.）W. C. Cheng et L. K. Fu ＝ Picea jezoensis（Siebold et Zucc.）Carrière var. microsperma（Lindl.）W. C. Cheng et L. K. Fu ●

298312　Picea jezoensis（Siebold et Zucc.）Carrière var. hondoensis（Mayr）Rehder ＝ Picea hondoensis Mayr ●☆

298313　Picea jezoensis（Siebold et Zucc.）Carrière var. hondoensis（Mayr）Rehder f. ozeensis Hayashi;尾瀬鱼鳞云杉●☆

298314　Picea jezoensis（Siebold et Zucc.）Carrière var. komarovii（V. N. Vassil.）W. C. Cheng et L. K. Fu ＝ Picea ajanensis（Lindau et Gordon）Fisch. ex Carrière ●

298315　Picea jezoensis（Siebold et Zucc.）Carrière var. komarovii（V. N. Vassil.）W. C. Cheng et L. K. Fu ＝ Picea jezoensis（Siebold et Zucc.）Carrière ●

298316　Picea jezoensis（Siebold et Zucc.）Carrière var. komarovii（V. N. Vassil.）W. C. Cheng et L. K. Fu;长白鱼鳞云杉（长白鱼鳞松,鱼鳞松,鱼鳞云杉）;Komarov Spruce ●

298317　Picea jezoensis（Siebold et Zucc.）Carrière var. microsperma（Lindl.）W. C. Cheng et L. K. Fu ＝ Picea ajanensis（Lindau et Gordon）Fisch. ex Carrière ●

298318　Picea jezoensis（Siebold et Zucc.）Carrière var. microsperma（Lindl.）W. C. Cheng et L. K. Fu;小籽鱼鳞云杉（鱼鳞云杉）●

298319　Picea kamtchchatkensis Lacass.;堪察加云杉;Kamtschatka Spruce ●

298320　Picea kamtchchatkensis Lacass. ＝ Picea jezoensis（Siebold et Zucc.）Carrière var. microsperma（Lindl.）W. C. Cheng et L. K. Fu ●

298321　Picea khutrow（Royle ex Turra）Carrière ＝ Picea smithiana（Wall.）Boiss. ●◇

298322　Picea komarovii V. N. Vassil. ＝ Picea jezoensis（Siebold et Zucc.）Carrière var. komarovii（V. N. Vassil.）W. C. Cheng et L. K. Fu ●

298323　Picea koraiensis Nakai;红皮云杉（八岳云杉,高丽云杉,红皮臭,虎尾松,灰叶云杉,沙树,针松）;Hondo Spruce, Korean Spruce, Koyama Spruce, Koyama's Spruce ●

298324　Picea koraiensis Nakai var. intercedens（Nakai）Y. L. Chou ＝ Picea koraiensis Nakai ●

298325　Picea koraiensis Nakai var. nenjiangensis S. Q. Nie et X. Y. Yuan;嫩江云杉;Nenjiang Spruce ●

298326　Picea koyamae Hayashi var. acicularis（Shiras. et Koyama）T. Shimizu ＝ Picea shirasawae Hayashi ●☆

298327　Picea koyamae Shiras var. koraiensis（Nakai）Liou et Q. L. Wang ＝ Picea koraiensis Nakai ●

298328　Picea koyamae Shiras. ＝ Picea koraiensis Nakai ●

298329　Picea koyamae Shiras. var. acicularis（Shiras. et Koyama）T. Shimizu;尖灰叶云杉●☆

298330　Picea koyamae Shiras. var. koraiensis（Nakai）Liou et Z. Wang ＝ Picea koraiensis Nakai ●

298331　Picea likiangensis（Franch.）E. Pritz.;丽江云杉;Balfour's Spruce, Lijiang Spruce, Likiang Spruce, Southern Likiang Spruce ●

298332　Picea likiangensis（Franch.）E. Pritz. f. bicolor（S. Chen）L. K. Fu;二色川西云杉●

298333　Picea likiangensis（Franch.）E. Pritz. var. balfouriana（Rehder et E. H. Wilson）Hillier ex Slavin ＝ Picea balfouriana Rehder et E. H. Wilson ●

298334　Picea likiangensis（Franch.）E. Pritz. var. balfouriana（Rehder et E. H. Wilson）Slavin ＝ Picea likiangensis（Franch.）E. Pritz. var. rubescens Rehder et E. H. Wilson ●

298335　Picea likiangensis（Franch.）E. Pritz. var. hirtella（Rehder et E. H. Wilson）W. C. Cheng ex F. H. Chen ＝ Picea likiangensis（Franch.）E. Pritz. var. hirtella（Rehder et E. H. Wilson）W. C. Cheng ●

298336　Picea likiangensis（Franch.）E. Pritz. var. hirtella（Rehder et E. H. Wilson）W. C. Cheng;黄果云杉;Hairy Xikang Spruce, Yellowcone Spruce ●

298337　Picea likiangensis（Franch.）E. Pritz. var. linzhiensis f. bicolro W. C. Cheng et L. K. Fu ＝ Picea brachytyla（Franch.）Pritz. var. complanata（Mast.）W. C. Cheng ex Rehder ●

298338　Picea likiangensis（Franch.）E. Pritz. var. linzhiensis W. C. Cheng et L. K. Fu ＝ Picea linzhiensis（W. C. Cheng et L. K. Fu）

W. C. Cheng et L. K. Fu ●

298339　Picea likiangensis（Franch.）E. Pritz. var. linzhiensis W. C. Cheng et L. K. Fu f. bicolor W. C. Cheng et L. K. Fu = Picea brachytyla（Franch.）Pritz. var. complanata（Mast.）W. C. Cheng ex Rehder ●

298340　Picea likiangensis（Franch.）E. Pritz. var. montigena（Mast.）W. C. Cheng = Picea likiangensis（Franch.）E. Pritz. var. montigena（Mast.）W. C. Cheng ex F. H. Chen ●◇

298341　Picea likiangensis（Franch.）E. Pritz. var. montigena（Mast.）W. C. Cheng ex F. H. Chen = Picea montigena Mast. ●◇

298342　Picea likiangensis（Franch.）E. Pritz. var. purpurea（Mast.）Dallim. et A. B. Jacks. = Picea purpurea Mast. ●

298343　Picea likiangensis（Franch.）E. Pritz. var. rubescens Rehder et E. H. Wilson = Picea balfouriana Rehder et E. H. Wilson ●

298344　Picea likiangensis var. balfouriana（Rehder et E. H. Wilson）Slavin = Picea likiangensis（Franch.）E. Pritz. var. rubescens Rehder et E. H. Wilson ●

298345　Picea linzhiensis（W. C. Cheng et L. K. Fu）W. C. Cheng et L. K. Fu；林芝云杉；Linzhi Spruce ●

298346　Picea lowiana Gordon = Abies lowiana A. Murray ●☆

298347　Picea manchurica Nakai = Picea koraiensis Nakai ●

298348　Picea manshurica Nakai = Picea jezoensis（Siebold et Zucc.）Carrière var. microsperma（Lindl.）W. C. Cheng et L. K. Fu ●

298349　Picea mariana（Mill.）Britton, Sterns et Poggenb.；黑云杉（沼泽云杉）；American Black Spruce, Black Spruce, Bog Black Spruce, Bog Spruce, Canadian Black Spruce, Morinda Spruce, Swamp Spruce ●☆

298350　Picea mariana（Mill.）Britton, Sterns et Poggenb. 'Doumetii'；杜美特黑云杉（多米特黑云杉）●☆

298351　Picea mariana（Mill.）Britton, Sterns et Poggenb. 'Nana'；矮小黑云杉（矮生黑云杉）；Blue Nest Spruce, Dwarf Black Spruce ●☆

298352　Picea mariana（Mill.）Britton, Sterns et Poggenb. var. brevifolia（Peck）Rehder = Picea mariana（Mill.）Britton, Sterns et Poggenb. ●☆

298353　Picea mastersii Mayr = Picea wilsonii Mast. ●

298354　Picea maximowiczii Regel ex Carrière；马氏云杉（灌木云杉, 马克西莫氏云杉）；Japonese Bush Spruce, Maximowicz Spruce ●☆

298355　Picea maximowiczii Regel ex Carrière var. senanensis Hayashi = Picea maximowiczii Regel ex Carrière ●☆

298356　Picea menziesii（Douglas ex D. Don）Carrière = Picea sitchensis（Bong.）Carrière ●☆

298357　Picea mexicana Mart.；墨西哥云杉；Mexica Spruce ●☆

298358　Picea meyeri Rehder et E. H. Wilson；白杆（白儿松, 白杆云杉, 刺儿松, 钝叶杉, 红杆云杉, 利儿松, 罗汉松, 毛枝云杉）；Meyer Spruce ●

298359　Picea meyeri Rehder et E. H. Wilson f. pyramidalis（H. W. Jen et C. G. Bai）L. K. Fu et Nan Li = Picea meyeri Rehder et E. H. Wilson ●

298360　Picea meyeri Rehder et E. H. Wilson var. mongolica H. Q. Wu；蒙古云杉；Mongolian Spruce ●

298361　Picea meyeri Rehder et E. H. Wilson var. mongolica H. Q. Wu = Picea meyeri Rehder et E. H. Wilson ●

298362　Picea meyeri Rehder et E. H. Wilson var. pyramidalis H. Wei Jen et C. G. Bai = Picea meyeri Rehder et E. H. Wilson ●

298363　Picea meyeri Rehder et E. H. Wilson var. pyramidalis H. Wei Jen et C. G. Bai；塔形白杆 ●

298364　Picea microsperma（Lindl.）Carrière = Picea jezoensis（Siebold et Zucc.）Carrière var. microsperma（Lindl.）W. C. Cheng et L. K. Fu ●

298365　Picea microsperma Carrière = Picea ajanensis（Lindau et Gordon）Fisch. ex Carrière ●

298366　Picea microsperma Carrière = Picea jezoensis（Siebold et Zucc.）Carrière var. microsperma（Lindl.）W. C. Cheng et L. K. Fu ●

298367　Picea mongolica（H. Q. Wu）W. D. Xu = Picea meyeri Rehder et E. H. Wilson ●

298368　Picea montigena Mast.；康定云杉；Candalabra Spruce, Kangding Spruce ●◇

298369　Picea montigena Mast. = Picea likiangensis（Franch.）E. Pritz. var. montigena（Mast.）W. C. Cheng ex F. H. Chen ●◇

298370　Picea montigena Mast. = Picea likiangensis（Franch.）E. Pritz. ●

298371　Picea morinda（Loudon）Link = Picea smithiana（Wall.）Boiss. ●◇

298372　Picea morinda Link = Picea smithiana（Wall.）Boiss. ●◇

298373　Picea morinda Link subsp. tianschanica（Rupr.）Berezin = Picea schrenkiana Fisch. et C. A. Mey. ●

298374　Picea morindoides Rehder = Picea spinulosa（Griff.）Henry ●

298375　Picea morrisonicola Hayata；台湾云杉（白松柏, 松萝杜, 玉山云杉）；Formosan Spruce, Morrison Spruce, Mount Morrison Spruce, Taiwan Spruce ●

298376　Picea neoveitchii Mast.；大果云杉（大果青杆, 青杆, 青杆杉）；Bigcone Spruce, Big-coned Spruce, New Veitch Spruce ●◇

298377　Picea neoveitchii Mast. = Picea wilsonii Mast. ●

298378　Picea nigra（Aiton）Link = Picea mariana（Mill.）Britton, Sterns et Poggenb. ●☆

298379　Picea nigra（Aiton）Link var. rubra（Du Roi）Engelm. = Picea rubens Sarg. ●☆

298380　Picea notabilis（Rehder et E. H. Wilson）Lacass. = Picea asperata Mast. var. heterolepis（Rehder et E. H. Wilson）Rehder ●

298381　Picea notabilis（Rehder et E. H. Wilson）Lacass. = Picea asperata Mast. ●

298382　Picea obovata Ledeb.；西伯利亚云杉（沙松, 新疆云杉）；Siberian Spruce, Xinjiang Spruce ●

298383　Picea obovata Ledeb. = Picea koraiensis Nakai ●

298384　Picea obovata Ledeb. var. schrenkiana（Fisch. et C. A. Mey.）Carrière = Picea schrenkiana Fisch. et C. A. Mey. ●

298385　Picea obovata Ledeb. var. schrenkiana Carrière = Picea meyeri Rehder et E. H. Wilson ●

298386　Picea obovata Ledeb. var. schrenkiana Carrière = Picea wilsonii Mast. ●

298387　Picea omorika（Pancic）F. Bolle；塞尔维亚云杉；Dwarf Serbian Spruce, Serbian Spruce ●☆

298388　Picea omorika（Pancic）F. Bolle 'Gnom'；格诺姆云杉 ●☆

298389　Picea omorika（Pancic）F. Bolle 'Nana'；矮塞尔维亚云杉；Dwarf Servian Spruce ●☆

298390　Picea omorika（Pancic）F. Bolle 'Pendula'；垂枝塞尔维亚云杉；Weeping Serbian Spruce ●☆

298391　Picea orientalis Carrière；东方云杉；Caucasian Spruce, Eastern Spruce, Oriental Spruce ●☆

298392　Picea orientalis Carrière 'Aurea'；黄叶东方云杉 ●☆

298393　Picea orientalis Carrière 'Nana'；矮东方云杉；Dwarf Norway Spruce ●☆

298394　Picea orientalis Carrière 'Skylands'；金叶东方云杉 ●☆

298395　Picea pachyclada Patschke = Picea brachytyla（Franch.）E. Pritz. ◇

298396　Picea parryana Sarg. = Picea pungens Engelm. ●☆

298397　Picea pichta（Lodd.）Loudon = Abies sibirica Ledeb. ●

298398　Picea polita（Siebold et Zucc.）Carrière = Picea torano（Siebold ex K. Koch）Koehne ●

298399　Picea polita（Siebold et Zucc.）Carrière f. rubriflora Kobayasi = Picea torano（Siebold ex K. Koch）Koehne f. rubriflora（Kobayasi）Yonek. ●☆

298400　Picea ponderosa（Rehder et E. H. Wilson）Lacass. = Picea asperata Mast. ●

298401　Picea pungens Engelm. ；北美云杉（尖云杉, 锐尖北美云杉, 硬尖云杉）；Blue Colorado Spruce, Blue Spruce, Colorado Blue Spruce, Colorado Spruce, Silver Spruce ●☆

298402　Picea pungens Engelm. = Picea pungens Sarg. ●☆

298403　Picea pungens Sarg. 'Blue Ice'；蓝冰锐尖北美云杉 ●☆

298404　Picea pungens Sarg. 'Compacta'；矮北美云杉；Dwarf Spruce ●☆

298405　Picea pungens Sarg. 'Fat Albert'；胖艾伯特北美云杉；Fat Albert Spruce ●☆

298406　Picea pungens Sarg. 'Glauca'；灰叶锐尖北美云杉；Blue Colorado Spruce, Blue Spruce, Colorado Spruce ●☆

298407　Picea pungens Sarg. 'Globosa'；球形锐尖北美云杉 ●☆

298408　Picea pungens Sarg. 'Hoopsi'；胡普斯锐尖北美云杉（霍普斯硬尖云杉）●☆

298409　Picea pungens Sarg. 'Koster'；科斯特锐尖北美云杉（科斯特硬尖云杉）●☆

298410　Picea pungens Sarg. 'Mission Blue'；蓝北美云杉；Mission Blue Spruce ●☆

298411　Picea pungens Sarg. 'Moerhermii'；莫尔赫密锐尖北美云杉 ●☆

298412　Picea pungens Sarg. 'Montgomery'；蒙哥马利硬尖云杉 ●☆

298413　Picea pungens Sarg. 'Royal Blue'；品蓝锐尖北美云杉 ●☆

298414　Picea pungens Sarg. = Picea pungens Engelm. ●☆

298415　Picea pungens Sarg. var. glauca ? = Picea pungens Sarg. 'Glauca' ●☆

298416　Picea pungsanensis Uyeki = Picea koraiensis Nakai ●

298417　Picea purpurea Mast. ；紫果云杉（红松, 紫果杉）；Purple-cone Spruce, Purple-coned Spruce ●

298418　Picea purpurea Mast. var. balfouriana（Rehder et E. H. Wilson）Silba = Picea balfouriana Rehder et E. H. Wilson ●

298419　Picea purpurea Mast. var. balfouriana（Rehder et E. H. Wilson）Silba = Picea likiangensis（Franch.）E. Pritz. var. rubescens Rehder et E. H. Wilson ●

298420　Picea purpurea Mast. var. hirtella（Rehder et E. H. Wilson）Silba = Picea likiangensis（Franch.）E. Pritz. var. hirtella（Rehder et E. H. Wilson）W. C. Cheng ●

298421　Picea retroflexa Mast. ；鳞皮云杉（箭炉云杉, 密毛杉）；Scalybark Spruce, Scaly-barked Spruce ●

298422　Picea retroflexa Mast. = Picea asperata Mast. ●

298423　Picea rubens Sarg. ；红云杉（美国红果云杉）；American Red Spruce, Eastern Spruce, He-balsam, Red Spruce, Yellow Spruce ●☆

298424　Picea rubra（Du Roi）Link = Picea rubens Sarg. ●☆

298425　Picea sargentiana Rehder et E. H. Wilson = Picea brachytyla（Franch.）E. Pritz. ◇

298426　Picea schrenkiana Fisch. et C. A. Mey. ；雪岭云杉（天山云杉, 雪岭杉）；Schrenk Spruce, Schrenk's Spruce, Tien Shad Spruce, Tien Shanzspruce ●

298427　Picea schrenkiana Fisch. et C. A. Mey. subsp. tianschanica（Rupr.）Bykov = Picea schrenkiana Fisch. et C. A. Mey. ●

298428　Picea schrenkiana Fisch. et C. A. Mey. var. tianschanica（Rupr.）W. C. Cheng et S. H. Fu = Picea schrenkiana Fisch. et C. A. Mey. ●

298429　Picea schrenkiana Fisch. et C. A. Mey. var. tianschanica（Rupr.）W. C. Cheng et S. H. Fu；天山云杉（红果雪岭杉）；Tianshan Mountain Spruce, Tianshan Schrenk Spruce ●

298430　Picea shirasawae Hayashi；白泽云杉 ●☆

298431　Picea shirasawae Hayashi = Picea koyamae Hayashi var. acicularis（Shiras. et Koyama）T. Shimizu ●☆

298432　Picea sikangensis W. C. Cheng = Picea balfouriana Rehder et E. H. Wilson ●

298433　Picea sikangensis W. C. Cheng = Picea likiangensis（Franch.）E. Pritz. var. balfouriana（Rehder et E. H. Wilson）Hillier ex Slavin ●

298434　Picea sikangensis W. C. Cheng = Picea likiangensis（Franch.）E. Pritz. var. rubescens Rehder et E. H. Wilson ●

298435　Picea sitchensis（Bong.）Carrière；西特喀云杉（北美云杉, 西加云杉, 锡加云杉）；Alaska Spruce, Coast Spruce, Lowland Spruce, Menzies Spruce, Silver Spruce, Sitka Spruce, Tideland Spruce, Western Spruce, Yellow Spruce ●☆

298436　Picea smithiana（Wall.）Boiss. ；长叶云杉（长叶杉, 喜马拉雅云杉, 喜马云杉）；Himalaya Spruce, Himalayan Spruce, India Spruce, Indian Spruce, Morinda Spruce, West Himalayan Spruce ●◇

298437　Picea spinulosa（Griff.）Henry；喜马拉雅云杉（西藏云杉, 小刺云杉）；East Himalayan Spruce, Sikkim Spruce, Tibet Spruce ●

298438　Picea spinulosa（Griff.）Henry var. yatungensis Silba = Picea spinulosa（Griff.）Henry ●

298439　Picea spinulosa Griff. var. yatungensis Silba = Picea spinulosa（Griff.）Henry ●

298440　Picea tianschanica Rupr. = Picea schrenkiana Fisch. et C. A. Mey. ●

298441　Picea tonaiensis Nakai = Picea koraiensis Nakai ●

298442　Picea torano（Siebold ex K. Koch）Koehne；日本云杉（虎尾云杉）；Japan Spruce, Japanese Spruce, Tigertail Spruce, Tiger-tail Spruce ●

298443　Picea torano（Siebold ex K. Koch）Koehne = Picea polita（Siebold et Zucc.）Carrière ●

298444　Picea torano（Siebold ex K. Koch）Koehne f. rubriflora（Kobayasi）Yonek. ；红花日本云杉 ●☆

298445　Picea vulgaris Link var. altaica Tepl. = Picea obovata Ledeb. ●

298446　Picea wallichiana A. B. Jacks. ；尼泊尔云杉 ●

298447　Picea watsoniana Mast. = Picea wilsonii Mast. ●

298448　Picea webbiana（Wall.）Loudon = Abies spectabilis（D. Don）Spach ●

298449　Picea wilsonii Mast. ；青杆（白杆云杉, 刺儿松, 方形杉, 方叶松, 黑儿松, 黑杆松, 红毛杉, 华北云杉, 杆树松, 青杆云杉, 细叶松, 细叶云杉）；Wilson Spruce, Wilson's Spruce ●

298450　Picea wilsonii Mast. var. shanxiensis Silba = Picea wilsonii Mast. ●

298451　Picea wilsonii Mast. var. watsoniana（Mast.）Silba = Picea wilsonii Mast. ●

298452　Picea yunnanensis E. H. Wilson = Picea likiangensis（Franch.）E. Pritz. ●

298453　Picea yunnanensis Lacass. = Picea likiangensis（Franch.）E. Pritz. ●

298454　Piceaceae Gorozh. = Pinaceae Spreng. ex F. Rudolphi（保留科名）●

298455　Pichinia S. Y. Wong et P. C. Boyce（2010）；皮基尼南星属 ■☆

298456　Pichisermollia H. C. Monteiro = Areca L. ●

298457　Pichleria Stapf et Wettst. = Zosima Hoffm. ■

298458　Pichonia Pierre = Lucuma Molina ●

298459　Pichonia Pierre(1890);皮雄榄属●☆

298460　Pichonia balansana Pierre;皮雄榄●☆

298461　Pichonia elliptica Pierre;椭圆皮雄榄●☆

298462　Pichonia occidentalis (H. J. Lam) Aubrév.;西方皮雄榄●☆

298463　Pichonia sessiliflora (C. T. White) Aubrév.;无梗花皮雄榄●☆

298464　Pickeringia Nutt. (1834) = Ardisia Sw. (保留属名)●■

298465　Pickeringia Nutt. (1840)(保留属名);加州山豆属●☆

298466　Pickeringia Nutt. ex Torr. et A. Gray = Pickeringia Nutt. (保留属名)●☆

298467　Pickeringia montana Nutt. ex Torr. et A. Gray;加州山豆;Chaparral Pea●☆

298468　Picnoeomon Wallr. ex DC. = Cephalaria Schrad. (保留属名)■

298469　Picnomon Adans. (1763);密苞蓟属■☆

298470　Picnomon Adans. = Cirsium Mill. ■

298471　Picnomon acama (L.) Cass.;密苞蓟■☆

298472　Picotia Roem. et Schult. = Omphalodes Mill. ■☆

298473　Picradenia Hook. = Hymenoxys Cass. ■☆

298474　Picradenia helenioides Rydb. = Hymenoxys helenioides (Rydb.) Cockerell ■☆

298475　Picradenia lemmonii Greene = Hymenoxys lemmonii (Greene) Cockerell ■☆

298476　Picradenia richardsonii Hook. = Hymenoxys richardsonii (Hook.) Cockerell ■☆

298477　Picradeniopsis Rydb. = Bahia Lag. ■☆

298478　Picradeniopsis Rydb. = Picradeniopsis Rydb. ex Britton ■☆

298479　Picradeniopsis Rydb. ex Britton(1901);拟尖膜菊属■☆

298480　Picradeniopsis oppositifolia (Nutt.) Rydb.;对叶拟尖膜菊■☆

298481　Picradeniopsis woodhousei (A. Gray) Rydb.;拟尖膜菊■☆

298482　Picraena Lindl. = Aeschrion Vell. ●☆

298483　Picraena Lindl. = Picrasma Blume ●

298484　Picraena Lindl. = Picrita Sehumacher ●☆

298485　Picraena Steven = Astragalus L. ●■

298486　Picralima Pierre(1897);赤非夹竹桃属●☆

298487　Picralima elliotii (Stapf) Stapf = Hunteria umbellata (K. Schum.) Hallier f. ●☆

298488　Picralima gracilis A. Chev. = Hunteria umbellata (K. Schum.) Hallier f. ●☆

298489　Picralima klaineana Pierre = Picralima nitida (Stapf) T. Durand et H. Durand ●☆

298490　Picralima laurifolia A. Chev. = Hunteria simii (Stapf) H. Huber ●☆

298491　Picralima macrocarpa A. Chev. = Picralima nitida (Stapf) T. Durand et H. Durand ●☆

298492　Picralima nitida (Stapf) T. Durand et H. Durand;赤非夹竹桃●☆

298493　Picralima umbellata (K. Schum.) Stapf = Hunteria umbellata (K. Schum.) Hallier f. ●☆

298494　Picramnia Sw. (1788)(保留属名);美洲苦木属;Bitterbush, Macary Bitter ●☆

298495　Picramnia antidesma Sw.;美洲苦木;Macary Bitter ●☆

298496　Picramnia ciliata Mart.;缘毛美洲苦木●☆

298497　Picramnia grandifolia Engl.;大叶美洲苦木●☆

298498　Picramnia longifolia Standl.;长叶美洲苦木●☆

298499　Picramnia macrocarpa Urb. et Ekman;大果美洲苦木●☆

298500　Picramnia pentandra Sw.;五蕊美洲苦木(佛罗里达苦木);Florida Bitterbush ●☆

298501　Picramnia polyantha Planch.;多花美洲苦木●☆

298502　Picramniaceae Fernando et Quinn(1995);美洲苦木科(夷苦木科)●☆

298503　Picramnlaceae (Engl.) Fernando et Quinn = Picramniaceae Fernando et Quinn ●☆

298504　Picranena Endl. = Aeschrion Vell. ●☆

298505　Picranena Endl. = Picraena Lindl. ●

298506　Picrasma Blume(1825);苦木属(苦树属);Quassia, Quassia Wood, Quassiawood, Quassia-wood ●

298507　Picrasma ailanthoides (Bunge) Planch. = Picrasma quassioides (D. Don) A. W. Benn. ●

298508　Picrasma andamanica Kurz ex A. W. Benn. = Picrasma javanica Blume ●

298509　Picrasma chinensis P. Y. Chen;中国苦木(中国苦树);China Quassia, Chinese Quassia Wood, Chinese Quassiawood ●

298510　Picrasma excelsa (Sw.) Planch.;牙买加苦木(高大苦树,牙买加苦树);Bitter Ash, Jamaica Quassia Wood, Jamaica Quassiawood, Quassia, Quassia Wood ●☆

298511　Picrasma excelsa Planch. = Picrasma excelsa (Sw.) Planch. ●☆

298512　Picrasma japonica A. Gray = Picrasma quassioides (D. Don) A. W. Benn. ●

298513　Picrasma javanica Blume;长绿苦木(长绿苦树,爪哇苦木,爪哇苦树);Javan Quassia Wood, Javan Quassiawood ●

298514　Picrasma javanica Blume = Picrasma chinensis P. Y. Chen ●

298515　Picrasma nepalensis A. W. Benn. = Picrasma javanica Blume ●

298516　Picrasma quassioides (D. Don) A. W. Benn.;苦木(臭辣子,胆木,赶狗木,狗胆木,寒苦树,花楸树,黄楝树,苦胆木,苦胆树,苦楝树,苦楝子,苦皮树,苦皮子,苦树,苦檀木,青鱼胆,山核桃树,山苦楝,山熊胆,土樗仔,熊胆树,崖漆树,鱼胆树);India Quassia Wood, India Quassiawood, Indian Quassia Wood, Indian Quassiawood, Quassia ●

298517　Picrasma quassioides (D. Don) A. W. Benn. f. dasycarpa Kitag.;毛果苦木(毛果苦树)●☆

298518　Picrasma quassioides (D. Don) A. W. Benn. f. glabrescens (Pamp.) Kitag. = Picrasma quassioides (D. Don) A. W. Benn. ●

298519　Picrasma quassioides (D. Don) A. W. Benn. var. glabrescens Pamp.;光序苦木(光序苦楝,光序苦树);Glabrousinflorescence Indian Quassiawood ●

298520　Picrasma quassioides (D. Don) A. W. Benn. var. glabrescens Pamp. = Picrasma quassioides (D. Don) A. W. Benn. ●

298521　Picrella Baill. = Helietta Tul. ●☆

298522　Picreus Juss. = Pycreus P. Beauv. ■

298523　Picria Benth. et Hook. f. = Coutoubea Aubl. ■☆

298524　Picria Benth. et Hook. f. = Picrium Schreb. ■☆

298525　Picria Lour. (1790);苦玄参属;Bitterfigwort, Picria ■

298526　Picria Lour. = Curanga Juss. ■☆

298527　Picria felterrae Lour.;苦玄参(地胆草,苦草,苦胆草,苦味草,蛇总管,四环素草,鱼胆草);Common Bitterfigwort, Common Picria ■

298528　Picria surinamensis Spreng.;苏里南苦玄参■☆

298529　Picricarya Dennst. = Olea L. ●

298530　Picridaceae Martinov = Asteraceae Bercht. et J. Presl(保留科名)●■

298531　Picridaceae Martinov = Compositae Giseke(保留科名)●■

298532　Picridium Desf. = Reichardia Roth ■☆

298533　Picridium discolor Pomel = Reichardia tingitana (L.) Roth subsp. discolor (Pomel) Batt. ■☆

298534　Picridium intermedium Sch. Bip. = Reichardia intermedia (Sch. Bip.) Cout. ■☆

298535　Picridium intermedium Sch. Bip. var. humile Caball. = Reichardia intermedia (Sch. Bip.) Cout. ■☆

298536 Picridium ligulatum Vent. = Reichardia ligulata (Vent.) G. Kunkel et Sunding ■☆

298537 Picridium orientale (L.) Desf. = Reichardia tingitana (L.) Roth ■☆

298538 Picridium rupestre Pomel = Reichardia picroides (L.) Roth ■☆

298539 Picridium saharae Pomel = Reichardia tingitana (L.) Roth ■☆

298540 Picridium tingitanum (L.) DC. = Reichardia tingitana (L.) Roth ■☆

298541 Picridium tingitanum (L.) DC. subsp. discolor (Pomel) Batt. = Reichardia tingitana (L.) Roth subsp. discolor (Pomel) Batt. ■☆

298542 Picridium tingitanum (L.) DC. var. maritimum Ball = Reichardia tingitana (L.) Roth subsp. discolor (Pomel) Batt. ■☆

298543 Picridium tingitanum (L.) DC. var. maroccanum Ball = Reichardia tingitana (L.) Roth subsp. discolor (Pomel) Batt. ■☆

298544 Picridium tingitanum (L.) DC. var. saharae (Pomel) Batt. = Reichardia tingitana (L.) Roth subsp. discolor (Pomel) Batt. ■☆

298545 Picridium vulgare Desf. = Reichardia picroides (L.) Roth ■☆

298546 Picridium vulgare Desf. subsp. intermedium (Sch. Bip.) Batt. = Reichardia intermedia (Sch. Bip.) Cout. ■☆

298547 Picridium vulgare Desf. subsp. maritimum (Ball) Batt. = Reichardia tingitana (L.) Roth ■☆

298548 Picridium vulgare Desf. var. serioides Maire = Reichardia tingitana (L.) Roth ■☆

298549 Picrina Rchb. ex Steud. = Picris L. ■

298550 Picris L. (1753) ;毛连菜属(毛莲菜属);Oxtongue, Ox-tongue ■

298551 Picris abyssinica Sch. Bip. ;阿比西尼亚毛连菜■☆

298552 Picris aculeata Vahl = Helminthotheca aculeata (Vahl) Lack ■☆

298553 Picris aculeata Vahl subsp. latisquamea (H. Lindb.) Sauvage = Helminthotheca aculeata (Vahl) Lack ■☆

298554 Picris aculeata Vahl subsp. maroccana Sauvage = Helminthotheca aculeata (Vahl) Lack subsp. maroccana (Sauvage) Greuter ■☆

298555 Picris albida Ball;苍白毛连菜■☆

298556 Picris albida Ball var. chevallieri (Batt.) Maire = Picris albida Ball ■☆

298557 Picris albida Ball var. concolor Maire = Picris albida Ball ■☆

298558 Picris altissima Delile = Picris rhagadioloides (L.) Desf. ■☆

298559 Picris aspera Gilib. = Picris hieracioides L. ■

298560 Picris asplenioides L. ;铁线蕨叶状毛连菜■☆

298561 Picris asplenioides L. subsp. saharae (Coss. et Kralik) Dobignard;佐原铁线蕨叶状毛连菜■☆

298562 Picris balansae (Coss. et Durieu) Maire = Helminthotheca balansae (Coss. et Durieu) Lack ■☆

298563 Picris canescens (Steven) Vassiliev;灰色毛连菜■

298564 Picris comosa (Boiss.) B. D. Jacks. = Helminthotheca comosa (Boiss.) Lack ■☆

298565 Picris comosa (Boiss.) B. D. Jacks. var. racemosa (Pomel) Maire = Helminthotheca comosa (Boiss.) Lack ■☆

298566 Picris comosa (Boiss.) B. D. Jacks. var. rubiginosa (Pomel) Maire = Helminthotheca comosa (Boiss.) Lack ■☆

298567 Picris coronopifolia (Desf.) DC. = Picris asplenioides L. ■☆

298568 Picris coronopifolia (Desf.) DC. subsp. albida (Ball) Maire = Picris albida Ball ■☆

298569 Picris coronopifolia (Desf.) DC. subsp. saharae (Coss.) Maire = Picris asplenioides L. subsp. saharae (Coss. et Kralik) Dobignard ■☆

298570 Picris coronopifolia (Desf.) DC. var. aviorum Maire = Picris asplenioides L. subsp. saharae (Coss. et Kralik) Dobignard ■☆

298571 Picris coronopifolia (Desf.) DC. var. chevallieri (Batt.) Maire = Picris asplenioides L. ■☆

298572 Picris coronopifolia (Desf.) DC. var. citrina Maire = Picris asplenioides L. ■☆

298573 Picris coronopifolia (Desf.) DC. var. getula (Pomel) Maire = Picris asplenioides L. ■☆

298574 Picris coronopifolia (Desf.) DC. var. macrorrhyncha Maire et Wilczek = Picris asplenioides L. ■☆

298575 Picris coronopifolia (Desf.) DC. var. transiens Maire = Picris asplenioides L. ■☆

298576 Picris cupuligera (Durieu) Walp. ;杯状毛连菜■☆

298577 Picris cyanocarpa Boiss. ;蓝果毛连菜■☆

298578 Picris cyrenaica (Pamp.) Lack;昔兰尼毛连菜■☆

298579 Picris dahurica Fisch. ex Hornem. = Picris davurica Fisch. ex Hornem. ■

298580 Picris davurica Fisch. ex Hornem. ;兴安毛连菜(毛连菜,枪刀菜,黏叶子草);Dahurian Oxtongue ■

298581 Picris davurica Fisch. ex Hornem. = Picris japonica Thunb. ■

298582 Picris davurica Fisch. ex Hornem. var. koreana (Kitam.) Kitag. = Picris hieracioides L. subsp. japonica (Thunb.) Krylov var. koreana (Kitam.) Kitam. ■☆

298583 Picris divaricata Vaniot;滇苦菜(叉枝毛连菜,尖刀苦马菜,剪刀菜,蒲公英,野莴苣菜);Divaricate Oxtongue ■

298584 Picris echinoides L. ;刺缘毛连菜(蓝蓟毛连菜);Bristly Oxtongue, Bristly Ox-tongue, Lang De Beef, Langley Beef, Oxtongue, Ox-tongue ■☆

298585 Picris echioides L. = Helminthotheca echioides (L.) Holub ■☆

298586 Picris echioides L. var. pratensis (DC.) Maire = Helminthotheca echioides (L.) Holub ■☆

298587 Picris echioides L. var. tuberculata (Moench) Fiori = Helminthotheca echioides (L.) Holub ■☆

298588 Picris helminthioides (Ball) Greuter = Leontodon hispanicus Poir. subsp. helminthioides (Coss. et Durieu) Maire ■☆

298589 Picris hieracioides L. ;毛连菜(毛柴胡,毛耳大黄,毛莲菜,枪刀菜,枪刀菜花,我立神花);Cat's Ear, Great Hawkweed, Hawkweed Oxtongue, Hawkweed Ox-tongue, Yellow Succory ■

298590 Picris hieracioides L. subsp. fuscipilosa Hand. -Mazz. ;单毛毛连菜(褐毛毛连菜,褐毛毛连菜,毛柴胡,毛牛耳大黄,牛踏鼻,羊下巴);Brownhair Oxtongue ■

298591 Picris hieracioides L. subsp. japonica (Thunb.) Hand. -Mazz. = Picris japonica Thunb. ■

298592 Picris hieracioides L. subsp. japonica (Thunb.) Krylov = Picris japonica Thunb. ■

298593 Picris hieracioides L. subsp. japonica (Thunb.) Krylov f. laevicaulis M. Mizush. ex T. Shimizu;光茎日本毛连菜■☆

298594 Picris hieracioides L. subsp. japonica (Thunb.) Krylov f. maritima Sugim. ;海滨光茎日本毛连菜■☆

298595 Picris hieracioides L. subsp. japonica (Thunb.) Krylov var. akaishiensis Kitam. ;明石毛连菜■☆

298596 Picris hieracioides L. subsp. japonica (Thunb.) Krylov var. jessoensis (Tatew.) Kitam. ;北海道连菜■☆

298597 Picris hieracioides L. subsp. japonica (Thunb.) Krylov var. koreana (Kitam.) Kitam. = Picris japonica Thunb. var. koreana (Kitam.) Kitag. ■

298598 Picris hieracioides L. subsp. japonica (Thunb.) Krylov var. koreana Kitam. = Picris japonica Thunb. var. koreana (Kitam.) Kitag. ■

298599　Picris hieracioides L. subsp. japonica（Thunb.）Krylov var. litoralis Kitam. = Picris hieracioides L. subsp. japonica（Thunb.）Krylov f. maritima Sugim. ■☆

298600　Picris hieracioides L. subsp. japonica（Thunb.）Krylov var. mayebarae（Kitam.）Ohwi;前原毛连菜■☆

298601　Picris hieracioides L. subsp. jessoensis（Tatew.）Kitam. = Picris hieracioides L. subsp. japonica（Thunb.）Krylov var. jessoensis（Tatew.）Kitam. ■☆

298602　Picris hieracioides L. subsp. kamtschatica（Ledeb.）Hultén;勘察加毛连菜■☆

298603　Picris hieracioides L. subsp. kamtschatica（Ledeb.）Hultén = Picris hieracioides L. ■

298604　Picris hieracioides L. subsp. koreana（Thunb.）Krylov = Picris japonica Thunb. var. koreana（Kitam.）Kitag. ■

298605　Picris hieracioides L. subsp. mayebarae Kitam. = Picris hieracioides L. subsp. japonica（Thunb.）Krylov var. mayebarae（Kitam.）Ohwi ■☆

298606　Picris hieracioides L. subsp. morrisonensis（Hayata）Kitam.;阿里山毛连菜（二歧毛连菜,台湾毛连菜,羽状毛连菜,玉山毛连菜）;Alishan Oxtongue ■

298607　Picris hieracioides L. subsp. ohwiana（Kitam.）Kitam.;黄毛毛连菜（高山毛连菜,莲座毛连菜,木质毛连菜）;Ohwi Alishan Oxtongue ■

298608　Picris hieracioides L. subsp. tsekouensis Kitam. = Picris hieracioides L. ■

298609　Picris hieracioides L. var. alpina Koidz. = Picris hieracioides L. ■

298610　Picris hieracioides L. var. alpina Koidz. = Picris hieracioides L. subsp. kamtschatica（Ledeb.）Hultén ■☆

298611　Picris hieracioides L. var. glabrescens（Regel）Ohwi = Picris hieracioides L. subsp. japonica（Thunb.）Krylov ■

298612　Picris hieracioides L. var. indica DC. = Picris hieracioides L. subsp. kamtschatica（Ledeb.）Hultén ■☆

298613　Picris hieracioides L. var. japonica（Thunb.）Regel ex Herder = Picris hieracioides L. subsp. japonica（Thunb.）Krylov ■

298614　Picris hieracioides L. var. japonica Regel et Herder = Picris japonica Thunb. ■

298615　Picris hispanica（Willd.）P. D. Sell = Leontodon hispanicus Poir. ■☆

298616　Picris humilis DC.;低矮毛连菜■☆

298617　Picris integrifolia Desf. = Picris rhagadioloides（L.）Desf. ■☆

298618　Picris japonica Thunb.;日本毛连菜（毛柴胡,毛连菜,毛牛耳大黄,牛踏鼻,枪刀菜,羊下巴）;Japan Oxtongue, Japanese Oxtongue ■

298619　Picris japonica Thunb. = Picris hieracioides L. subsp. japonica（Thunb.）Krylov ■

298620　Picris japonica Thunb. var. akaishiensis（Kitam.）Ohwi = Picris hieracioides L. subsp. japonica（Thunb.）Krylov var. akaishiensis Kitam. ■☆

298621　Picris japonica Thunb. var. alpina（Koidz.）Ohwi = Picris hieracioides L. subsp. kamtschatica（Ledeb.）Hultén ■☆

298622　Picris japonica Thunb. var. davurica（Fisch. ex Hornem.）Kitag. = Picris davurica Fisch. ex Hornem. ■

298623　Picris japonica Thunb. var. davurica（Fisch. ex Hornem.）Kitag. = Picris japonica Thunb. ■

298624　Picris japonica Thunb. var. jessoensis（Tatew.）Ohwi = Picris hieracioides L. subsp. japonica（Thunb.）Krylov var. jessoensis（Tatew.）Kitam. ■☆

298625　Picris japonica Thunb. var. koreana（Kitam.）Kitag.;朝鲜毛连菜（朝鲜兴安毛连菜,绿苞毛连菜,沿海毛连菜）;Korean Oxtongue ■

298626　Picris japonica Thunb. var. mayebarae（Kitam.）Ohwi = Picris hieracioides L. subsp. japonica（Thunb.）Krylov var. mayebarae（Kitam.）Ohwi ■☆

298627　Picris junnanensis V. N. Vassil.;云南毛连菜（丽江毛连菜,硬毛毛连菜）■☆

298628　Picris kamtschatica Ledeb. = Picris hieracioides L. subsp. kamtschatica（Ledeb.）Hultén ■☆

298629　Picris kotschyi Boiss.;科奇毛连菜■☆

298630　Picris mairei H. Lév. = Picris japonica Thunb. ■

298631　Picris morrisonensis Hayata = Picris hieracioides L. subsp. morrisonensis（Hayata）Kitam. ■

298632　Picris ohwiana Kitam. = Picris hieracioides L. subsp. ohwiana（Kitam.）Kitam. ■

298633　Picris pauciflora Willd.;少花毛连菜;Fewflower Oxtongue, Smallflower Oxtongue ■☆

298634　Picris pitardiana Gand. = Picris cupuligera（Durieu）Walp. ■☆

298635　Picris rhagadioloides（L.）Desf.;北美毛连菜■☆

298636　Picris rigida Ledeb.;硬毛毛连菜■☆

298637　Picris rubra Lam. = Crepis rubra L. ■☆

298638　Picris saharae（Coss. et Kralik）Batt. = Picris asplenioides L. subsp. saharae（Coss. et Kralik）Dobignard ■☆

298639　Picris saharae（Coss. et Kralik）Batt. var. oranensis Hochr. = Picris asplenioides L. subsp. saharae（Coss. et Kralik）Dobignard ■☆

298640　Picris similis V. N. Vassil.;新疆毛连菜（相似毛连菜）;Xinjiang Oxtongue ■

298641　Picris sinuata（Lam.）Lack;深波毛连菜■☆

298642　Picris sprengeriana（L.）Poir.;苦毛连菜;Bitterweed ■☆

298643　Picris strigosa M. Bieb.;密棘毛连菜■☆

298644　Picris sulphurea Delile;硫色毛连菜■☆

298645　Picris tenuis Caball. = Picris cupuligera（Durieu）Walp. ■☆

298646　Picris willkommii（Sch. Bip.）Nyman;维尔考姆毛连菜■☆

298647　Picris xylopoda Lack;木梗毛连菜■☆

298648　Picrita Sehumacher = Aeschrion Vell. ●☆

298649　Picrium Schreb. = Coutoubea Aubl. ●☆

298650　Picriza Raf. = Gentiana L. ●☆

298651　Picrocardia Radlk. = Soulamea Lam. ●☆

298652　Picrococcus Nutt. = Vaccinium L. ●

298653　Picrodendraceae Small ex Britton et Millsp. = Euphorbiaceae Juss.（保留科名）●■

298654　Picrodendraceae Small ex Britton et Millsp. = Picrodendraceae Small（保留科名）●☆

298655　Picrodendraceae Small(1917)（保留科名）;脱皮树科（三叶脱皮树科）●☆

298656　Picrodendraceae Small（保留科名）= Euphorbiaceae Juss.（保留科名）●■

298657　Picrodendron Griseb.(1860)（保留属名）;脱皮树属●☆

298658　Picrodendron Planch.（废弃属名）= Picrodendron Griseb.（保留属名）●☆

298659　Picrodendron arboreum（Mill.）Planch.;脱皮树●☆

298660　Picrodendron baccatum（L.）Krug et Urb.;浆果脱皮树●☆

298661　Picroderma Thorel ex Gagnep. = Trichilia P. Browne（保留属名）●

298662　Picrolemma Hook. f.(1862);苦籽木属●☆

298663　Picrolemma sprucei Hook. f.;苦籽木●☆

298664　Picrophloeus Blume　= Fagraea Thunb. ●

298665　Picrophyta F. Muell. = Goodenia Sm. ●■☆

298666　Picrorhiza Royle = Neopicrorhiza D. Y. Hong ■

298667　Picrorhiza Royle = Picrorhiza Royle ex Benth. ■

298668　Picrorhiza Royle ex Benth. (1835);胡黄莲属;Picrorhiza ■

298669　Picrorhiza Royle ex Benth. = Neopicrorhiza D. Y. Hong ■

298670　Picrorhiza kurroa Royle ex Benth.;库洛胡黄连(胡黄连,胡连);Kurro Picrorhiza,Kutki ■

298671　Picrorhiza scrophulariiflora Pennell = Neopicrorhiza scrophulariiflora (Pennell) D. Y. Hong ■

298672　Picrorhiza scrophulariiflora Pennell = Picrorhiza kurroa Royle ex Benth. ■

298673　Picrorhiza Wittst. = Picrorhiza Royle ex Benth. ■

298674　Picrosia D. Don(1830);糙毛苣属■☆

298675　Picrosia longifolia D. Don;糙毛苣■☆

298676　Picrothamnus Nutt. (1841);苦味蒿属;Budsage ●☆

298677　Picrothamnus Nutt. = Artemisia L. ●■

298678　Picrothamnus desertorum Nutt.;苦味蒿●☆

298679　Picroxylon Warb. = Eurycoma Jack ●☆

298680　Pictetia DC. (1825);佛堤豆属■☆

298681　Pictetia angustifolia Griseb.;窄叶佛堤豆■☆

298682　Pictetia cubensis Bisse;古巴佛堤豆■☆

298683　Pictetia desvauxii DC.;佛堤豆■☆

298684　Pictetia microphylla Benth.;小叶佛堤豆■☆

298685　Pictetia pubescens Hochst. = Ormocarpum pubescens (Hochst.) Cufod. ●☆

298686　Pictetia sessilifolia C. Wright ex Greenm.;无柄佛堤豆■☆

298687　Piddingtonia A. DC. = Pratia Gaudich. ■

298688　Piddingtonia montana Miq. = Pratia montana (Reinw. ex Blume) Hassk. ■

298689　Piddingtonia nummularia A. DC. = Lobelia angulata G. Forst. ■☆

298690　Piddingtonia nummularia A. DC. = Pratia nummularia (Lam.) A. Br. et Asch. ■

298691　Piddingtonia patens Miq. = Pratia montana (Reinw. ex Blume) Hassk. ■

298692　Pierardia Post et Kuntze = Ethulia L. f. ■

298693　Pierardia Post et Kuntze = Pirarda Adans. ■

298694　Pierardia Raf. = Dendrobium Sw. (保留属名)■

298695　Pierardia Roxb. = Baccaurea Lour. ●

298696　Pierardia Roxb. ex Jack = Baccaurea Lour. ●

298697　Pierardia barteri Baill. = Maesobotrya barteri (Baill.) Hutch. ●☆

298698　Pierardia griffoniana Baill. = Maesobotrya griffoniana (Baill.) Hutch. ●☆

298699　Pierardia motleyana Müll. Arg. = Baccaurea motleyana (Müll. Arg.) Müll. Arg. ●

298700　Pierardia sapida Roxb. = Baccaurea ramiflora Lour. ●

298701　Piercea Mill. = Rivina L. ●

298702　Pieridia Rchb. = Pieris D. Don ●

298703　Pieris D. Don (1834);马醉木属(梫木属,鮸木属);Andromeda,Pieris ●

298704　Pieris annamensis Dop = Lyonia ovalifolia (Wall.) Drude var. rubrovenia (Merr.) Judd ●

298705　Pieris annamensis Dop = Lyonia rubrovenia (Merr.) Chun ●

298706　Pieris bodinieri H. Lév. = Pieris formosa (Wall.) D. Don ●

298707　Pieris bracteata W. W. Sm. = Vaccinium mandarinorum Diels ●

298708　Pieris buxifolia H. Lév. = Vaccinium triflorum Rehder ●

298709　Pieris buxifolia H. Lév. et Vaniot = Vaccinium triflorum Rehder ●

298710　Pieris cavaleriei H. Lév. et Vaniot = Leucothoe griffithiana C. B. Clarke ●

298711　Pieris compta W. W. Sm. et Jeffrey = Lyonia compta (W. W. Sm. et Jeffrey) Hand. -Mazz. ●

298712　Pieris divaricata H. Lév. = Vaccinium bracteatum Thunb. ●

298713　Pieris doyonensis Hand. -Mazz. = Lyonia doyonensis (Hand. -Mazz.) Hand. -Mazz. ●

298714　Pieris duclouxii H. Lév. = Vaccinium duclouxii (H. Lév.) Hand. -Mazz. ●

298715　Pieris elliptica (Siebold et Zucc.) Nakai = Lyonia ovalifolia (Wall.) Drude subsp. neziki (Nakai et H. Hara) H. Hara ●☆

298716　Pieris elliptica (Siebold et Zucc.) Nakai = Lyonia ovalifolia (Wall.) Drude var. elliptica (Siebold et Zucc.) Hand. -Mazz. ●

298717　Pieris esquirolii H. Lév. et Vaniot = Vaccinium laetum Diels ●

298718　Pieris esquirolii H. Lév. et Vaniot = Vaccinium mandarinorum Diels ●

298719　Pieris esquirolii H. Lév. et Vaniot var. discolor H. Lév. et Vaniot = Vaccinium mandarinorum Diels ●

298720　Pieris esquirolii H. Lév. et Vaniot var. leucocalyx H. Lév. = Vaccinium pubicalyx Franch. var. leucocalyx (H. Lév.) Rehder ●

298721　Pieris floribunda (Pursh ex Sims) Benth. et Hook.;多花马醉木(美国马醉木);Arctic Heather, Fetter Bush, Fetterbush, Fetterbush, Free-flowering Andromeda, Mountain Andromeda, Mountain Fetterbush,Mountain Fetter-bush,Mountain Pieris ●

298722　Pieris formosa (Wall.) D. Don;美丽马醉木(长苞美丽马醉木,炮仗花,细梅树,兴山马醉木,珍珠花);Beautiful Pieris, Himalayan Andromeda,Himalayan Pieris,Himalayas Pieris ●

298723　Pieris formosa (Wall.) D. Don f. longiracemosa W. P. Fang = Pieris formosa (Wall.) D. Don ●

298724　Pieris formosa (Wall.) D. Don var. forrestii (R. L. Harrow) Airy Shaw = Pieris formosa (Wall.) D. Don ●

298725　Pieris formosa (Wall.) D. Don var. forrestii (R. L. Harrow) Airy Shaw = Pieris forrestii R. L. Harrow ex W. W. Sm. ●

298726　Pieris formosana Komatsu = Lyonia ovalifolia (Wall.) Drude subsp. neziki (Nakai et H. Hara) H. Hara ●☆

298727　Pieris formosana Komatsu = Lyonia ovalifolia (Wall.) Drude var. elliptica (Siebold et Zucc.) Hand. -Mazz. ●

298728　Pieris forrestii R. L. Harrow = Pieris formosa (Wall.) D. Don ●

298729　Pieris forrestii R. L. Harrow ex W. W. Sm.;白萼马醉木(福氏马醉木);Himalayan Pieris,White Sepals Pieris ●

298730　Pieris fortunatii H. Lév. = Gaultheria leucocarpa Blume var. yunnanensis (Franch.) T. Z. Hsu et R. C. Fang ●

298731　Pieris gagnepainiana H. Lév. = Vaccinium fragile Franch. ●

298732　Pieris henryi H. Lév. = Lyonia ovalifolia (Wall.) Drude var. hebecarpa (Franch. ex Forbes et Hemsl.) Chun ●

298733　Pieris huana W. P. Fang = Pieris formosa (Wall.) D. Don ●

298734　Pieris japonica (Thunb.) D. Don ex G. Don;马醉木(梫木,日本马醉木,台湾马醉木,台湾梫木);Formosa Pieris,Japan Pieris, Japanese Andromeda, Japanese Pieris, Lily-of-the-valley Bush, Lily-of-the-valley Tree,Polished Pieris,Taiwan Pieris ●

298735　Pieris japonica (Thunb.) D. Don ex G. Don 'Bent Chandler';钱德勒伯特马醉木;Bent Chandler Pieris ●☆

298736　Pieris japonica (Thunb.) D. Don ex G. Don 'Christmas Cheer';欢度圣诞节马醉木●☆

298737　Pieris japonica (Thunb.) D. Don ex G. Don 'Daisen';大山马醉木●☆

298738　Pieris japonica (Thunb.) D. Don ex G. Don 'Dorothy

Wyckoff'；多萝西・威克弗马醉木●☆

298739　Pieris japonica（Thunb.）D. Don ex G. Don 'Karenoma'；卡罗努玛马醉木●☆

298740　Pieris japonica（Thunb.）D. Don ex G. Don 'Little Heath'；小奚斯马醉木●☆

298741　Pieris japonica（Thunb.）D. Don ex G. Don 'Mountain Fire'；山火马醉木●☆

298742　Pieris japonica（Thunb.）D. Don ex G. Don 'Purity'；纯白马醉木●☆

298743　Pieris japonica（Thunb.）D. Don ex G. Don 'Robinswood'；罗宾斯伍德马醉木●☆

298744　Pieris japonica（Thunb.）D. Don ex G. Don 'Scarlett O'Hara'；斯卡雷特奥哈娜马醉木●☆

298745　Pieris japonica（Thunb.）D. Don ex G. Don 'Valley Rose'；山玫瑰马醉木●☆

298746　Pieris japonica（Thunb.）D. Don ex G. Don 'Valley Valentine'；瓦伦丁山谷马醉木●☆

298747　Pieris japonica（Thunb.）D. Don ex G. Don 'Variegata'；斑叶马醉木(银边马醉木)●☆

298748　Pieris japonica（Thunb.）D. Don ex G. Don 'Whitecaps'；白浪马醉木●☆

298749　Pieris japonica（Thunb.）D. Don ex G. Don f. elegantissima（Carrière）Nakai；白边日本马醉木●☆

298750　Pieris japonica（Thunb.）D. Don ex G. Don f. monostachya（Nakai）H. Hara；单穗日本马醉木；Onestachys Japan Pieris ●☆

298751　Pieris japonica（Thunb.）D. Don ex G. Don f. pygmea Yatabe；小花日本马醉木；Smallflower Japan Pieris ●☆

298752　Pieris japonica（Thunb.）D. Don ex G. Don subsp. formosa（Wall.）Kitam. = Pieris formosa（Wall.）D. Don ●

298753　Pieris japonica（Thunb.）D. Don ex G. Don subsp. koidzumiana（Ohwi）Hatus. = Pieris koidzumiana Ohwi ●☆

298754　Pieris japonica（Thunb.）D. Don ex G. Don var. amamiana ?；奄美马醉木；Lily of the Valley Shrub ☆

298755　Pieris japonica（Thunb.）D. Don ex G. Don var. koidzumiana（Ohwi）Masam. = Pieris koidzumiana Ohwi ●☆

298756　Pieris japonica（Thunb.）D. Don ex G. Don var. taiwanensis（Hayata）Kitam. = Pieris japonica（Thunb.）D. Don ex G. Don ●

298757　Pieris japonica（Thunb.）D. Don ex G. Don var. yakushimensis T. Yamaz.；屋久岛马醉木●☆

298758　Pieris koidzumiana Ohwi；琉球马醉木；Liukiu Japan Pieris ●☆

298759　Pieris koidzumiana Ohwi = Pieris japonica（Thunb.）D. Don ex G. Don var. koidzumiana（Ohwi）Masam. ●☆

298760　Pieris kouyangensis H. Lév. = Lyonia ovalifolia（Wall.）Drude var. lanceolata（Wall.）Hand. -Mazz. ●

298761　Pieris lanceolata（Wall.）D, Don = Lyonia ovalifolia（Wall.）Drude var. lanceolata（Wall.）Hand. -Mazz. ●

298762　Pieris longicornu H. Lév. et Vaniot = Vaccinium mandarinorum Diels ●

298763　Pieris longicornuta H. Lév. et Vaniot = Vaccinium mandarinorum Diels ●

298764　Pieris lucida H. Lév. = Vaccinium bracteatum Thunb. ●

298765　Pieris macrocalyx J. Anthony = Lyonia macrocalyx（Anthony）Airy Shaw ●

298766　Pieris mairei H. Lév. = Lyonia ovalifolia（Wall.）Drude var. hebecarpa（Franch. ex Forbes et Hemsl.）Chun ●

298767　Pieris mairei H. Lév. var. parvifolia H. Lév. = Lyonia ovalifolia（Wall.）Drude var. hebecarpa（Franch. ex Forbes et Hemsl.）Chun ●

298768　Pieris martini H. Lév. = Vaccinium dunalianum Wight var. urophyllum Rehder et E. H. Wilson ●

298769　Pieris nana（Maxim.）Makino = Arcterica nana（Maxim.）Makino ●☆

298770　Pieris nitida Benth. et Hook. f.；亮叶马醉木；Horse Wicky，Hurrah-bush，Tetterbush ●☆

298771　Pieris obliquinervis Merr. et Chun = Lyonia ovalifolia（Wall.）Drude var. lanceolata（Wall.）Hand. -Mazz. ●

298772　Pieris obliquinervis Merr. et Chun = Lyonia rubrovenia（Merr.）Chun ●

298773　Pieris oligodonta H. Lév. = Maesa japonica（Thunb.）Moritzi ex Zoll. ●

298774　Pieris ovalifolia（Wall.）D. Don = Lyonia ovalifolia（Wall.）Drude ●

298775　Pieris ovalifolia（Wall.）D. Don var. hebecarpa Franch. ex Forbes et Hemsl. = Lyonia ovalifolia（Wall.）Drude var. hebecarpa（Franch. ex Forbes et Hemsl.）Chun ●

298776　Pieris ovalifolia（Wall.）D. Don var. lanceolata（Wall.）C. B. Clarke = Lyonia ovalifolia（Wall.）Drude var. lanceolata（Wall.）Hand. -Mazz. ●

298777　Pieris ovalifolia（Wall.）D. Don var. pubescens Franch. = Lyonia villosa（Hook. f. ex C. B. Clarke）Hand. -Mazz. ●

298778　Pieris ovalifolia（Wall.）D. Don var. tomentosa W. P. Fang = Lyonia ovalifolia（Wall.）Drude var. tomentosa（W. P. Fang）C. Y. Wu ●

298779　Pieris ovalifolia（Wall.）Drude var. elliptica（Siebold et Zucc.）Rehder et E. H. Wilson = Lyonia ovalifolia（Wall.）Drude var. elliptica（Siebold et Zucc.）Hand. -Mazz. ●

298780　Pieris ovalifolia D. Don = Lyonia ovalifolia（Wall.）Drude ●

298781　Pieris ovalifolia D. Don var. denticulata H. Lév. = Vaccinium bracteatum Thunb. ●

298782　Pieris ovalifolia D. Don var. elliptica（Siebold et Zucc.）Rehder et E. H. Wilson = Lyonia ovalifolia（Wall.）Drude var. elliptica（Siebold et Zucc.）Hand. -Mazz. ●

298783　Pieris ovalifolia D. Don var. elliptica（Siebold et Zucc.）Rehder et E. H. Wilson = Lyonia ovalifolia（Wall.）Drude subsp. neziki（Nakai et H. Hara）H. Hara ●☆

298784　Pieris ovalifolia D. Don var. hebecarpa Franch. ex Forbes et Hemsl. = Lyonia ovalifolia（Wall.）Drude var. hebecarpa（Franch. ex Forbes et Hemsl.）Chun ●

298785　Pieris ovalifolia D. Don var. lanceolata（Wall.）C. B. Clarke = Lyonia ovalifolia（Wall.）Drude var. lanceolata（Wall.）Hand. -Mazz. ●

298786　Pieris ovalifolia D. Don var. pubescens Franch. = Lyonia villosa（Hook. f. ex C. B. Clarke）Hand. -Mazz. var. pubescens（Franch.）Judd ●

298787　Pieris ovalifolia D. Don var. tomenttosa Fang = Lyonia ovalifolia（Wall.）Drude var. lanceolata（Wall.）Hand. -Mazz. ●

298788　Pieris phillyreifolia（Hook.）DC.；长圆叶马醉木；Like Phillyrea Leavea Pieris ●☆

298789　Pieris pilosa Komatsu = Lyonia ovalifolia（Wall.）Drude subsp. neziki（Nakai et H. Hara）H. Hara ●☆

298790　Pieris pilosa Komatsu = Lyonia ovalifolia（Wall.）Drude var. elliptica（Siebold et Zucc.）Hand. -Mazz. ●

298791　Pieris pilosa Komatsu = Rhododendron albrechtii Maxim. ●☆

298792　Pieris polita W. W. Sm. et Jeffrey = Pieris japonica（Thunb.）D. Don ex G. Don ●

298793　Pieris popowii Palib. = Pieris japonica（Thunb.）D. Don ex G. Don ●

298794　Pieris repens H. Lév. = Vaccinium fragile Franch. ●

298795　Pieris rubrovenia Merr. = Lyonia ovalifolia（Wall.）Drude var. rubrovenia（Merr.）Judd ●

298796　Pieris rubrovenia Merr. = Lyonia rubrovenia（Merr.）Chun ●

298797　Pieris swinhoei Hemsl.；长萼马醉木；Longcalyx Pieris, Longcalyxed Pieris ●

298798　Pieris taiwanensis Hayata；台湾马醉木●

298799　Pieris taiwanensis Hayata = Pieris japonica（Thunb.）D. Don ex G. Don ●

298800　Pieris ulbrichii H. Lév. = Lyonia ovalifolia（Wall.）Drude var. lanceolata（Wall.）Hand. -Mazz. ●

298801　Pieris vaccinium H. Lév. = Gaultheria leucocarpa Blume var. yunnanensis（Franch.）T. Z. Hsu et R. C. Fang ●

298802　Pieris villosa Wall. ex C. B. Clarke = Lyonia villosa（Hook. f. ex C. B. Clarke）Hand. -Mazz. ●

298803　Pieris villosa Wall. ex C. B. Clarke var. pubescens（Franch.）Rehder et E. H. Wilson = Lyonia villosa（Hook. f. ex C. B. Clarke）Hand. -Mazz. var. pubescens（Franch.）Judd ●

298804　Pieris villosa Wall. ex C. B. Clarke var. pubescens（Franch.）Rehder et E. H. Wilson = Lyonia villosa（Hook. f. ex C. B. Clarke）Hand. -Mazz. ●

298805　Pierotia Blume = Ixonanthes Jack ●

298806　Pierranthus Bonati(1912)；皮埃拉婆婆纳属（皮埃拉玄参属）■☆

298807　Pierranthus capitatus Bonati；皮埃拉婆婆纳■☆

298808　Pierrea F. Heim(1891)（保留属名）；肖坡垒属●☆

298809　Pierrea F. Heim（保留属名）= Hopea Roxb.（保留属名）●

298810　Pierrea Hance（废弃属名）= Homalium Jacq. ●

298811　Pierrea Hance（废弃属名）= Pierrea F. Heim（保留属名）●☆

298812　Pierrea pachycarpa Heim；肖坡垒●☆

298813　Pierreanthus Willis = Pierranthus Bonati ■☆

298814　Pierrebraunia Esteves = Cipocereus F. Ritter ●☆

298815　Pierrebraunia Esteves = Floribunda F. Ritter ●☆

298816　Pierrebraunia Esteves(1997)；皮玻掌属●☆

298817　Pierredmondia Tamamsch. = Petroedmondia Tamamsch. ■☆

298818　Pierreocarpus Ridl. ex Symington = Hopea Roxb.（保留属名）●

298819　Pierreodendron A. Chev. = Letestua Lecomte ●☆

298820　Pierreodendron A. Chev. = Quassia L. ●☆

298821　Pierreodendron Engl.(1907)；皮埃尔木属●☆

298822　Pierreodendron Engl. = Quassia L. ●☆

298823　Pierreodendron africanum（Hook. f.）Little；非洲皮埃尔木●☆

298824　Pierreodendron durissimum A. Chev. = Letestua durissima（A. Chev.）Lecomte ●☆

298825　Pierreodendron grandifolium Engl. = Pierreodendron africanum（Hook. f.）Little ●☆

298826　Pierreodendron kerstingii（Engl.）Little；非洲别尔苦木●☆

298827　Pierrina Engl.(1909)；球果革瓣花属●☆

298828　Pierrina zenkeri Engl.；球果革瓣花●☆

298829　Pietrosia Nyar. = Andryala L. ■☆

298830　Pigafetta（Blume）Becc.(1877)（'Pigafettia'）（保留属名）；马来刺椰属（比加飞椰子属，比加飞棕属，金刺椰属，马来西亚葵属，皮非塔藤属）；Pigafettia ●☆

298831　Pigafetta（Blume）Mart. ex Becc. = Pigafetta（Blume）Becc.（保留属名）●☆

298832　Pigafetta Adans.（废弃属名）= Eranthemum L. ●■

298833　Pigafetta Adans.（废弃属名）= Pigafetta（Blume）Becc.（保留属名）●☆

298834　Pigafetta Becc. = Pigafetta（Blume）Becc.（保留属名）●☆

298835　Pigafetta Benth. et Hook. f. = Pigafetta（Blume）Becc.（保留属名）●☆

298836　Pigafetta filaris（Giseke）Becc.；马来刺椰（比加飞棕）●☆

298837　Pigafetta filaris Becc. = Pigafetta filaris（Giseke）Becc. ●☆

298838　Pigafettaea Post et Kuntze = Eranthemum L. ●■

298839　Pigafettaea Post et Kuntze = Pigafetta（Blume）Becc.（保留属名）●☆

298840　Pigea DC. = Hybanthus Jacq.（保留属名）●■

298841　Pigeum Laness. = Lauro-Cerasus Duhamel ●

298842　Pigeum Laness. = Pygeum Gaertn. ●

298843　Pikria G. Don = Curanga Juss. ■☆

298844　Pikria G. Don = Picria Lour. ■

298845　Pilanthus Poit. ex Endl. = Centrosema（DC.）Benth.（保留属名）●■☆

298846　Pilasia Raf. = Urginea Steinh. ■☆

298847　Pilderia Klotzsch = Begonia L. ●■

298848　Pilea Lindl.(1821)（保留属名）；冷水花属（冷水麻属）；Artillery Plant, Clearweed, Coldwaterflower, Stingless Nettle ■

298849　Pilea alongensis Gagnep. = Pilea boniana Gagnep. ■

298850　Pilea amamiana Ohwi = Pilea aquarum Dunn subsp. brevicornuta（Hayata）C. J. Chen ■

298851　Pilea amplistipulata C. J. Chen；大托叶冷水花（镜面草）；Bigstipular Coldwaterflower, Big-stipulete Clearweed ●

298852　Pilea angolensis（Hiern）Rendle；安哥拉冷水花■☆

298853　Pilea angolensis（Hiern）Rendle subsp. christiaensenii（Lambinon）Friis；克里冷水花■☆

298854　Pilea angulata（Blume）Blume；圆瓣冷水花（长柄冷水麻，湖北冷水花，棱枝冷水花，圆瓣冷水麻）；Angulate Clearweed, Roundpetal Coldwaterflower ■

298855　Pilea angulata（Blume）Blume = Pilea notata C. H. Wright ■

298856　Pilea angulata（Blume）Blume subsp. latiuscula C. J. Chen；华中冷水花；Central China Clearweed, Central China Coldwaterflower ■

298857　Pilea angulata（Blume）Blume subsp. petiolaris（Siebold et Zucc.）C. J. Chen；长柄冷水花（长柄冷水麻）；Longstalk Clearweed, Longstalk Roundpetal Coldwaterflower ■

298858　Pilea anisophylla Wedd.；异叶冷水花（高雄冷水麻）；Differleaf Coldwaterflower, Unequalleaf Clearweed ■

298859　Pilea anisophylla Wedd. var. khasiana Hook. f. = Pilea insolens Wedd. ■

298860　Pilea anisophylla Wedd. var. khasiana Hook. f. = Pilea khasiana（Hook. f.）C. J. Chen ■

298861　Pilea anisophylla Wedd. var. robusta Hook. f. = Pilea anisophylla Wedd. ■

298862　Pilea anivoranensis Léandri = Pilea rivularis Wedd. ■☆

298863　Pilea antisophylla Wedd.；对叶冷水花■☆

298864　Pilea approximata C. B. Clarke；顶叶冷水花；Adjoin Clearweed, Adjoin Coldwaterflower ■

298865　Pilea approximata C. B. Clarke var. incisoserrata C. J. Chen；锐裂齿顶叶冷水花（锐裂齿冷水花）；Sharplobed Adjoin Clearweed ■

298866　Pilea aquarum Dunn；湿生冷水花；Wet Clearweed, Wet Coldwaterflower ■

298867　Pilea aquarum Dunn = Pilea symmeria Wedd. ■

298868　Pilea aquarum Dunn subsp. acutidentata C. J. Chen；锐齿湿生冷水花；Sharptooth Wet Clearweed ■

298869　Pilea aquarum Dunn subsp. brevicornuta（Hayata）C. J. Chen；

短角湿生冷水花(短角冷水麻,巨叶短角冷水麻,疏花短角冷水麻,四轮炸);Shortcorn Wet Clearweed ■

298870　Pilea auricularis C. J. Chen;耳基冷水花;Earbase Clearweed, Earbase Coldwaterflower ■

298871　Pilea bambuseti Engl. ;邦布塞特冷水花■☆

298872　Pilea bambuseti Engl. subsp. aethiopica Friis;埃塞俄比亚冷水花■☆

298873　Pilea bambusifolia C. J. Chen;竹叶冷水花;Bambooleaf Clearweed,Bambooleaf Coldwaterflower ■

298874　Pilea basicordata W. T. Wang ex C. J. Chen;基心叶冷水花(登赫赫,红鲜草,接骨风,心叶冷水花);Baseheartleaf Coldwaterflower,Basicordateleaf Clearweed ●■

298875　Pilea baviensis Gagnep. = Pilea boniana Gagnep. ■

298876　Pilea blind H. Lév. = Pilea plataniflora C. H. Wright ■

298877　Pilea boniana Gagnep.;五萼冷水花;Fivecalyx Coldwaterflower,Fivesepal Clearweed ■

298878　Pilea bracteosa Wedd. ;多苞冷水花;Bract-bearing Clearweed, Manybract Coldwaterflower ■

298879　Pilea bracteosa Wedd. = Pilea oxyodon Wedd. ■

298880　Pilea bracteosa Wedd. var. oxyodon (Wedd.) Hara = Pilea oxyodon Wedd. ■

298881　Pilea bracteosa Wedd. var. striolata Hand. -Mazz. = Pilea bracteosa Wedd. ■

298882　Pilea brevicornuta Hayata = Pilea aquarum Dunn subsp. brevicornuta (Hayata) C. J. Chen ■

298883　Pilea brevicornuta Hayata f. laxiflora Yamam. = Pilea aquarum Dunn subsp. brevicornuta (Hayata) C. J. Chen ■

298884　Pilea brevicornuta Hayata f. magnifolia Yamam. = Pilea aquarum Dunn subsp. brevicornuta (Hayata) C. J. Chen ■

298885　Pilea cadierei Gagnep. et Guillaumin;花叶冷水花(白雪草,冷水花);Aluminium Plant, Artillery Plant, Friendship Plant, Spotleaf Clearweed,Spotleaf Coldwaterflower ●■

298886　Pilea cadierei Gagnep. et Guillaumin 'Compacta';密叶冷水花■

298887　Pilea cadierei Gagnep. et Guillaumin 'Minima';姬冷水花■

298888　Pilea cavaleriei H. Lév. ;波缘冷水花(打不死,肥奴奴草,肥猪菜,痹积草,冷冻草,肉质冷水花,石花菜,石西洋菜,石苋菜,石油菜,小石芥,岩鸡心草,岩油菜);Waveedge Clearweed, Waveedge Coldwaterflower ■

298889　Pilea cavaleriei H. Lév. subsp. crenata C. J. Chen;圆齿石油菜;Roundtooth Clearweed,Roundtooth Coldwaterflower ■

298890　Pilea cavaleriei H. Lév. subsp. valida C. J. Chen;石油菜(石凉草);Valid Roundtooth Clearweed ■

298891　Pilea cavaleriei H. Lév. subsp. valida C. J. Chen = Pilea cavaleriei H. Lév. ■

298892　Pilea ceratomera Wedd. = Pilea rivularis Wedd. ■☆

298893　Pilea ceratomera Wedd. var. mildbraedii Engl. = Pilea rivularis Wedd. ■☆

298894　Pilea chartacea C. J. Chen;纸质冷水花;Paperlike Clearweed, Paperlike Coldwaterflower ■

298895　Pilea chevalieri Schnell = Pilea sublucens Wedd. ■☆

298896　Pilea christiaensenii Lambinon = Pilea angolensis (Hiern) Rendle subsp. christiaensenii (Lambinon) Friis ■☆

298897　Pilea cordifolia Hook. f. ;弯叶冷水花(歪叶冷水花);Cordateleaf Clearweed,Wryleaf Coldwaterflower ■

298898　Pilea cordifolia Hook. f. = Pilea bracteosa Wedd. ■

298899　Pilea cordistipulata C. J. Chen;心托冷水花;Heartstipular Clearweed,Heartstipular Coldwaterflower ■

298900　Pilea crassifolia Hance = Pilea sinocrassifolia C. J. Chen ■

298901　Pilea crateraforma F. P. Metcalf = Pilea swinglei Merr. ■

298902　Pilea crateriforma Metcalf = Pilea swinglei Merr. ■

298903　Pilea cuneatifolia Yamam. = Pilea aquarum Dunn subsp. brevicornuta (Hayata) C. J. Chen ■

298904　Pilea cuneatifolia Yamam. = Pilea melastomoides (Poir.) Wedd. ●■

298905　Pilea depressa Blume;凹陷冷水花;Depressed Clearweed ■☆

298906　Pilea dielsiana Hand. -Mazz. = Pilea plataniflora C. H. Wright ■

298907　Pilea distachys Yamam. = Pilea rotundinucula Hayata ■

298908　Pilea divaricata Hauman = Pilea angolensis (Hiern) Rendle subsp. christiaensenii (Lambinon) Friis ■☆

298909　Pilea dolichocarpa C. J. Chen;光疣冷水花(瘤果冷水花);Tumorhull Clearweed,Tumorhull Coldwaterflower ●■

298910　Pilea elatostematifolia Hauman = Pilea johnstonii Oliv. subsp. kiwuensis (Engl.) Friis ■☆

298911　Pilea elegantissima C. J. Chen;石林冷水花;Shilin Clearweed, Shilin Coldwaterflower ■

298912　Pilea elliptifolia B. L. Shih et Yuen P. Yang = Pilea notata C. H. Wright ■

298913　Pilea elliptilimba C. J. Chen;椭圆叶冷水花(椭圆冷水花,椭圆叶冷水麻);Ellipseleaf Clearweed, Ellipseleaf Coldwaterflower ■

298914　Pilea engleri Rendle = Pilea usambarensis Engl. var. engleri (Rendle) Friis ■☆

298915　Pilea engleri Rendle. f. contracta Hauman = Pilea usambarensis Engl. var. engleri (Rendle) Friis ■☆

298916　Pilea fasciata Franch. = Pilea sinofasciata C. J. Chen ■

298917　Pilea fontana (Lunell) Rydb. ;沼泽冷水花;Bog Clearweed, Clearweed,Lesser Clearweed ■☆

298918　Pilea forgetii N. E. Br. ;银纹草■☆

298919　Pilea funkikensis Hayata;奋起湖冷水花(奋起湖冷水麻,奋起湖水麻);Fenqihu Clearweed, Fenqihu Coldwaterflower ■

298920　Pilea funkikensis Hayata var. rotundinucula (Hayata) S. S. Ying = Pilea rotundinucula Hayata ■

298921　Pilea funkikensis Hayata var. somae (Hayata) S. S. Ying = Pilea somai Hayata ■

298922　Pilea gansuensis C. J. Chen et Z. X. Peng;陇南冷水花;Gansu Clearweed,Gansu Coldwaterflower ■

298923　Pilea glaberrima (Blume) Blume;点乳冷水花(小齿冷水花);Perfect Clearweed,Perfect Coldwaterflower ●■

298924　Pilea goetzei Engl. ;格兹冷水花■☆

298925　Pilea goglado Blume = Pilea glaberrima (Blume) Blume ●■

298926　Pilea gracilis Hand. -Mazz. ;纤细冷水花;Gracile Clearweed, Thin Coldwaterflower ■

298927　Pilea gracillis Hand. -Mazz. = Pilea verucosa Hand. -Mazz. ■

298928　Pilea grandifolia Blume = Pilea grandis Wedd. ■☆

298929　Pilea grandis Wedd. ;蛤蟆草(虾蟆草);Aluminum Plant ■☆

298930　Pilea hamaoi Makino = Pilea pumila (L.) A. Gray var. hamaoi (Makino) C. J. Chen ■

298931　Pilea hamaoi Makino = Pilea pumila (L.) A. Gray var. obtusifolia C. J. Chen ■

298932　Pilea hamaoi Makino = Pilea pumila (L.) A. Gray ■

298933　Pilea henryana C. H. Wright = Pilea swinglei Merr. ■

298934　Pilea hexagona C. J. Chen;六棱茎冷水花;Hexagon Coldwaterflower,Six-angular Clearweed ●■

298935　Pilea hilliana Hand. -Mazz. ;翠茎冷水花(托叶冷水花);Greenstem Coldwaterflower, Stipule Clearweed ■

298936   Pilea holstii Engl. ;霍尔冷水花■☆

298937   Pilea hookeriana Wedd. ;中印冷水花(须弥冷水花);Hooker. Clearweed ■

298938   Pilea hookeriana Wedd. = Pilea martinii (H. Lév.) Hand. - Mazz. ■

298939   Pilea howelliana Hand. -Mazz. ;泡果冷水花(腾冲冷水花); Howell Coldwaterflower,Tengchong Clearweed ■

298940   Pilea howelliana Hand. -Mazz. var. denticulata C. J. Chen;细齿泡果冷水花;Thintooth Clearweed,Thintooth Howell Coldwaterflower ■

298941   Pilea hugelii Blume = Pilea wightii Wedd. ■

298942   Pilea insolens Wedd. ; 盾基冷水花; Excellent Clearweed, Excellent Coldwaterflower ■

298943   Pilea involucrata (Sims) C. H. Wright et Dewar;巴拿马冷水花(绿蛤蟆,毛叶冷水花);Friendship Plant,Pan-american Friendship Plant,Panamiga ■☆

298944   Pilea involucrata Urb. = Pilea involucrata (Sims) C. H. Wright et Dewar ■☆

298945   Pilea japonica (Maxim.) Hand. -Mazz. ;山冷水花(红水草,华东冷水花,日本冷水花,日本冷水麻,山美豆,苔水草);Clearweed Wild, Wild Coldwaterflower ■

298946   Pilea japonica (Maxim.) Hand. -Mazz. f. major Hatus. ;大山冷水花■☆

298947   Pilea japonica (Maxim.) Hand. -Mazz. f. pilosa Hiyama;毛山冷水花■☆

298948   Pilea javanica Wedd. = Pellionia latifolia (Blume) Boerl. ■

298949   Pilea johnstonii Oliv. ;约翰斯顿冷水花■☆

298950   Pilea johnstonii Oliv. subsp. kiwuensis (Engl.) Friis;热非冷水花■☆

298951   Pilea johnstonii Oliv. subsp. rwandensis Friis;卢旺达冷水花■☆

298952   Pilea kankaoensis Hayata = Pilea plataniflora C. H. Wright ■

298953   Pilea khasiana (Hook. f.) C. J. Chen;具柄冷水花; Khas Coldwaterflower,Khas Clearweed ■

298954   Pilea khasiana (Hook. f.) C. J. Chen = Pilea insolens Wedd. ■

298955   Pilea kiotensis Ohwi;京都冷水花■☆

298956   Pilea kiwuensis Engl. = Pilea johnstonii Oliv. subsp. kiwuensis (Engl.) Friis ■☆

298957   Pilea langsonensis Gagnep. = Pilea plataniflora C. H. Wright ■

298958   Pilea ledermannii H. Winkl. ;莱德冷水花■☆

298959   Pilea linearifolia C. J. Chen;条叶冷水花;Linearleaf Clearweed, Linearleaf Coldwaterflower ■

298960   Pilea lomatogramma Hand. -Mazz. ;隆脉冷水花(肥猪草,急尖冷水花,鼠舌草);Nervose Clearweed, Veinrise Coldwaterflower ■

298961   Pilea longicaulis Hand. -Mazz. ;长茎冷水花(长柄冷水花,接骨风);Long-stem Clearweed, Longstem Coldwaterflower ●■

298962   Pilea longicaulis Hand. -Mazz. var. erosa C. J. Chen;啮齿冷水花(啮齿叶冷水花,啮蚀冷水花)●

298963   Pilea longicaulis Hand. -Mazz. var. flaviflora C. J. Chen;黄花冷水花;Yellow-flower Long-stem Clearweed, Yellow-flower Long-stem Coldwaterflower ●

298964   Pilea longipedunculata S. S. Chien et C. J. Chen;鱼眼果冷水花;Fisheye Coldwaterflower,Longpeduncle Clearweed ■

298965   Pilea macrocarpa C. J. Chen;大果冷水花;Bigfruit Clearweed, Bigfruit Coldwaterflower ■

298966   Pilea manniana Wedd. ;曼氏冷水花■☆

298967   Pilea martinii (H. Lév.) Hand. -Mazz. ;大叶冷水花(大水边麻,到老嫩,异被冷水花);Bigleaf Coldwaterflower, Martin Clearweed ■

298968   Pilea matsudae Yamam. ;细尾冷水花(细尾冷水麻);Littletail Clearweed, Littletail Coldwaterflower ■

298969   Pilea media C. J. Chen;中间型冷水花; Middle Clearweed, Middle Coldwaterflower ■

298970   Pilea medongensis C. J. Chen;墨脱冷水花;Motuo Clearweed, Motuo Coldwaterflower ■

298971   Pilea melastomoides (Poir.) Wedd. ;长序冷水花(大冷水麻,冷清花,三脉冷水花);Longspike Coldwaterflower, Three-vein Clearweed ●■

298972   Pilea menghaiensis C. J. Chen;勐海冷水花; Menghai Clearweed, Menghai Coldwaterflower ■

298973   Pilea microcardia Hand. -Mazz. ; 广西冷水花; Guangxi Clearweed,Guangxi Coldwaterflower ■

298974   Pilea microphylla (L.) Liebm. ;小叶冷水花(玻璃草,透明草,小叶冷水麻); Artillery Clearweed, Artillery Plant, Artillery Weed, Clearweed, Gunpowder Plant, Littleleaf Clearweed, Littleleaf Coldwaterflower,Pistol Plant,Rockweed ■

298975   Pilea mildbraedii (Engl.) Engl. = Pilea rivularis Wedd. ■☆

298976   Pilea minor Yamam. = Pilea aquarum Dunn subsp. brevicornuta (Hayata) C. J. Chen ■

298977   Pilea minutepilosa Hayata = Pilea plataniflora C. H. Wright ■

298978   Pilea miyakei Yamam. ;台湾冷水花(三宅氏冷水花)■

298979   Pilea mongolica Wedd. = Pilea pumila (L.) A. Gray ■

298980   Pilea mongolica Wedd. var. yakushimensis Hatus. ;屋久岛冷水花■☆

298981   Pilea monilifera Hand. -Mazz. ;念珠冷水花(项链冷水花); Beads Coldwaterflower, Necklace Clearweed ■

298982   Pilea morseana Hand. -Mazz. = Pilea boniana Gagnep. ■

298983   Pilea multicellularis C. J. Chen;串珠毛冷水花;Multicellutate Clearweed, Multicellutate Coldwaterflower ■

298984   Pilea muscosa Lindl. = Pilea microphylla (L.) Liebm. ■

298985   Pilea myriantha (Dunn) C. J. Chen = Aboriella myriantha (Dunn) Bennet ■

298986   Pilea nanchuanensis C. J. Chen;南川冷水花(半边座,倒老嫩,水珠麻); Nanchuan Clearweed, Nanchuan Coldwaterflower ■

298987   Pilea nanchuanensis C. J. Chen = Pilea verucosa Hand. -Mazz. ■

298988   Pilea nokozanensis Yamam. = Pilea angulata (Blume) Blume subsp. petiolaris (Siebold et Zucc.) C. J. Chen ■

298989   Pilea notata C. H. Wright;冷水花(白山羊,长柄冷水花,冷水花草,山羊血,水麻,水麻叶,土甘草,紫色冷草);Coldwaterflower ■

298990   Pilea notata C. H. Wright = Pilea pseudonotata C. J. Chen ●

298991   Pilea nummulariifolia (Sw.) Wedd. ;铜钱叶冷水花(泡叶冷水花);Creeping Charley,Creeping Charlie ■☆

298992   Pilea obesa Wedd. = Pilea umbrosa Blume var. obesa Wedd. ■

298993   Pilea obliqua Hook. f. = Pilea bracteosa Wedd. ■

298994   Pilea opaca (Lunell) Rydb. = Pilea fontana (Lunell) Rydb. ■☆

298995   Pilea ovatinucula Hayata = Pilea melastomoides (Poir.) Wedd. ●■

298996   Pilea oxyodon Wedd. ;雅致冷水花;Clearweed Elegant, Elegant Coldwaterflower ■

298997   Pilea oxyodon Wedd. = Pilea lomatogramma Hand. -Mazz. ■

298998   Pilea paniculigera C. J. Chen;滇东南冷水花; E. China Clearweed, E. China Coldwaterflower ■

298999   Pilea pauciflora C. J. Chen;少花冷水花;Fewflower Clearweed, Fewflower Coldwaterflower ■

299000   Pilea pellionioides C. J. Chen;赤车冷水花;Chiche Clearweed, Redcarweed Coldwaterflower ●■

299001   Pilea peltata Hance;盾叶冷水花(背花疮,盾状冷水花,石苋

菜);Peltate Clearweed,Peltate Coldwaterflower ■

299002 Pilea peltata Hance var. ovatifolia C. J. Chen;卵盾叶冷水花(卵形盾叶冷水花,卵叶盾叶冷水花);Ovateleaf Peltate Clearweed,Ovateleaf Peltate Coldwaterflower ■

299003 Pilea penninervis C. J. Chen;钝齿冷水花;Obtusetooth Clearweed,Obtusetooth Coldwaterflower ■

299004 Pilea pentasepala Hand. -Mazz. = Pilea boniana Gagnep. ■

299005 Pilea peperomioides Diels;镜面草(翠屏草,跌打散,金钱草,一点金);Lens Coldwaterflower,Pepermioides Clearweed,Roundleaf Pilea ■

299006 Pilea peploides (Gaudich.) Hook. et Arn.;矮冷水花(矮冷水麻,葶艾冷水花,苦水花,冷水花,苔水花,圆叶豆瓣草,坐镇草);Dwarf Clearweed Dwarf,Dwarf Coldwaterflower,Spatulate Pilea ■

299007 Pilea peploides (Gaudich.) Hook. et Arn. var. cavaleriei (H. Lév.) H. Lév. = Pilea cavaleriei H. Lév. ■

299008 Pilea peploides (Gaudich.) Hook. et Arn. var. cavaleriei H. Lév. = Pilea cavaleriei H. Lév. subsp. crenata C. J. Chen ■

299009 Pilea peploides (Gaudich.) Hook. et Arn. var. cavaleriei H. Lév. = Pilea sinocrassifolia C. J. Chen ■

299010 Pilea peploides (Gaudich.) Hook. et Arn. var. major Wedd.;齿叶矮冷水花(矮冷水麻,葶艾冷水花,齿叶矮冷水麻,地油子,虎牙草,蚯蚓草,水下拉,水苋菜,苔水花,透明草);Toothleaf Coldwaterflower,Toothleaf Dwarf Clearweed ■

299011 Pilea peploides (Gaudich.) Hook. et Arn. var. major Wedd. = Pilea peploides (Gaudich.) Hook. et Arn. ■

299012 Pilea peploides (Gaudich.) Hook. et Arn. var. minutissima Hsu = Pilea swinglei Merr. ■

299013 Pilea petelotii Gagnep. = Pilea plataniflora C. H. Wright ■

299014 Pilea petiolaris (Siebold et Zucc.) Blume = Pilea angulata (Blume) Blume subsp. petiolaris (Siebold et Zucc.) C. J. Chen ■

299015 Pilea petiolaris (Siebold et Zucc.) Blume = Pilea angulata (Blume) Blume ■

299016 Pilea petiolaris (Siebold et Zucc.) Blume subsp. pseudopetiolaris (Hatus.) Kitam. = Pilea notata C. H. Wright ■

299017 Pilea plataniflora C. H. Wright;石筋草(拔毒草,草本三股筋,大包药,到老嫩,狗骨节,过金桥,恒春冷水麻,六月冷,全缘冷水花,软枝三股筋,三线草,蛇踝节,石稔草,石头花,歪叶冷水麻,西南冷水花,西南冷水麻,血桐子,洋肚参);Planetreeflower Clearweed,Stonemuscle Coldwaterflower ■

299018 Pilea procumbens Peter = Laportea ovalifolia (Schumach. et Thonn.) Chew ●☆

299019 Pilea producta Blume = Pilea umbrosa Blume ■

299020 Pilea producta Diels = Pilea sinofasciata C. J. Chen ■

299021 Pilea pseudonotata C. J. Chen;拟冷水花(假冷水花);False Clearweed,False Coldwaterflower ●

299022 Pilea pseudopetiolaris Hatus. = Pilea notata C. H. Wright ■

299023 Pilea pterocaulis (S. S. Chien) C. J. Chen = Pilea subcoriacea (Hand. -Mazz.) C. J. Chen ■

299024 Pilea pumila (L.) A. Gray;透茎冷水花(肥肉草,亮杆芹,蒙古冷水花,水麻叶,水荨麻,透茎冷水麻,野麻,直苎麻);Canadian Clearweed, Clearweed, Richweed, Stingless Nettle, Throughstem Clearweed,Throughstem Coldwaterflower ■

299025 Pilea pumila (L.) A. Gray var. deamii (Lunell) Fernald = Pilea pumila (L.) A. Gray ■

299026 Pilea pumila (L.) A. Gray var. hamaoi (Makino) C. J. Chen;荫地冷水花(钩状冷水花);Hooked Clearweed ■

299027 Pilea pumila (L.) A. Gray var. obtusifolia C. J. Chen;钝尖冷水花;Obtuseleaf Throughstem Clearweed ■

299028 Pilea pumila Liebm. = Pilea pumila (L.) A. Gray ■

299029 Pilea purpurella C. J. Chen;紫背冷水花;Purpleback Clearweed,Purpleback Coldwaterflower ■

299030 Pilea purpurella C. J. Chen = Pilea verucosa Hand. -Mazz. ■

299031 Pilea racemiformis C. J. Chen;总状序冷水花;Raceme Clearweed,Raceme Coldwaterflower ■

299032 Pilea racemosa (Royle) Tuyama;亚高山冷水花;Racemose Clearweed,Subalpine Coldwaterflower ■

299033 Pilea raceptacularis C. J. Chen;序托冷水花(地水麻);Raceptacle Clearweed,Raceptacle Coldwaterflower ■

299034 Pilea radicans (Sw.) Wedd. = Pilea wightii Wedd. ■

299035 Pilea rivularis Wedd.;溪边冷水花■☆

299036 Pilea rostellata C. J. Chen;短喙冷水花;Shortbeak Clearweed,Shortbeak Coldwaterflower ■

299037 Pilea rotundinucula Hayata;圆果冷水花(微齿冷水麻,圆果冷清草,圆果冷水麻);Roundfruit Clearweed, Roundfruit Coldwaterflower ■

299038 Pilea rubriflora C. H. Wright;红花冷水花;Red Coldwaterflower,Red-flower Clearweed ●■

299039 Pilea salwinensis (Hand. -Mazz.) C. J. Chen;怒江冷水花(九节风);Nujiang Clearweed,Nujiang Coldwaterflower ●■

299040 Pilea scripta (Buch. -Ham. ex D. Don) Wedd.;细齿冷水花(九节风);Thintooth Clearweed,Thintooth Coldwaterflower ■

299041 Pilea scripta Wedd.;雕饰冷水花■☆

299042 Pilea secunda S. S. Chien = Pilea anisophylla Wedd. ■

299043 Pilea semisessilis Hand. -Mazz.;镰叶冷水花;Falcateleaf Clearweed,Sickleleaf Coldwaterflower ■

299044 Pilea sinocrassifolia C. J. Chen;厚叶冷水花(石芫茜);Thickleaf Clearweed,Thickleaf Coldwaterflower ■

299045 Pilea sinocrassifolia C. J. Chen var. glaberrima Wedd. = Pilea subedentata C. J. Chen ■

299046 Pilea sinofasciata C. J. Chen;粗齿冷水花(阿伯秀,扁花冷水花,大茴香,宫麻,扇花冷水花,水甘草,水麻,水麻叶,紫绿草,走马胎);Roughtooth Clearweed,Roughtooth Coldwaterflower ■

299047 Pilea smilacifolia Wedd. = Pilea glaberrima (Blume) Blume ●■

299048 Pilea somai Hayata;细叶冷水花(细叶冷水麻);Thinleaf Clearweed,Thinleaf Coldwaterflower ■

299049 Pilea spinulosa C. J. Chen;刺果冷水花(小刺果冷水花);Spinefruit Coldwaterflower,Spiny-fruit Clearweed ●■

299050 Pilea spruceana Wedd. 'Norfolk';思鲁冷水花■

299051 Pilea squamosa C. J. Chen;鳞片冷水花;Squama Clearweed,Squama Coldwaterflower ■

299052 Pilea squamosa C. J. Chen var. sparsa C. J. Chen;少鳞冷水花;Fewscale Clearweed ■

299053 Pilea squamosa C. J. Chen var. sparsa C. J. Chen = Pilea squamosa C. J. Chen ■

299054 Pilea stipulosa (Miq.) Miq. = Pilea angulata (Blume) Blume ■

299055 Pilea stipulosa Miq. = Pilea hilliana Hand. -Mazz. ■

299056 Pilea strangulata Franch. = Pilea angulata (Blume) Blume subsp. petiolaris (Siebold et Zucc.) C. J. Chen ■

299057 Pilea subalpina Hand. -Mazz. = Pilea racemosa (Royle) Tuyama ■

299058 Pilea subcoriacea (Hand. -Mazz.) C. J. Chen;翅茎冷水花(小赤麻,亚革质冷水花);Webstem Clearweed, Webstem Coldwaterflower ■

299059 Pilea subedentata C. J. Chen;小齿冷水花;Smalltoth Clearweed,Smalltoth Coldwaterflower ■

299060　Pilea sublucens Wedd. ;光亮冷水花■☆

299061　Pilea swinglei Merr. ;三角形冷水花(玻璃草,三角叶冷水花,散血丹,史维冷水花,铁丝草); Triangle Clearweed, Triangle Coldwaterflower ■

299062　Pilea symmeria Wedd. ;喙萼冷水花(大果冷水花,九节风,蛇毛草,水麻); Beakcalyx Coldwaterflower, Snakehair Clearweed ■

299063　Pilea symmeria Wedd. var. pterocaulis S. S. Chien = Pilea subcoriacea (Hand. -Mazz.) C. J. Chen ■

299064　Pilea symmeria Wedd. var. salwinensis Hand. -Mazz. = Pilea salwinensis (Hand. -Mazz.) C. J. Chen ●■

299065　Pilea symmeria Wedd. var. subcoriacea Hand. -Mazz. = Pilea verucosa Hand. -Mazz.

299066　Pilea symmeria Wedd. var. subcoriacea Hand. -Mazz. f. stenobasis Hand. -Mazz. = Pilea verucosa Hand. -Mazz. ■

299067　Pilea taitoensis Hayata = Pilea plataniflora C. H. Wright ■

299068　Pilea ternifolia Wedd. ;羽脉冷水花; Trifoliate Clearweed, Trifoliate Coldwaterflower ■

299069　Pilea tetraphylla (Steud.) Blume;四叶冷水花■☆

299070　Pilea trianthemoides (Sw.) Lindl. ;大炮冷水花;Artillery Plant ■☆

299071　Pilea trinervia (Roxb.) Wight = Pilea melastomoides (Poir.) Wedd. ●■

299072　Pilea trinervia Wight = Pilea scripta (Buch. -Ham. ex D. Don) Wedd. ■

299073　Pilea trinervia Wight = Pilea subcoriacea (Hand. -Mazz.) C. J. Chen ■

299074　Pilea tsiangiana F. P. Metcalf;海南冷水花; Hainan Coldwaterflower, Tsiang Clearweed ●■

299075　Pilea umbrosa Blume;荫生冷水花(荫地冷水花); Shade Clearweed, Shady Coldwaterflower ■

299076　Pilea umbrosa Blume var. obesa Wedd. ;少毛冷水花(少毛荫生冷水花); Fewhair Shade Clearweed ■

299077　Pilea umbrosa Blume var. obesa Wedd. = Pilea medongensis C. J. Chen ■

299078　Pilea umbrosa Wedd. = Pilea umbrosa Blume ■

299079　Pilea unciformis C. J. Chen;鹰嘴萼冷水花; Eaglebeak Clearweed, Eaglebeak Coldwaterflower ■

299080　Pilea urticifolia (L. f.) Blume;荨麻叶冷水花■☆

299081　Pilea usambarensis Engl. ;乌桑巴拉冷水花■☆

299082　Pilea usambarensis Engl. var. engleri (Rendle) Friis;恩格勒冷水花■☆

299083　Pilea usambarensis Engl. var. veronicifolia (Engl.) Friis;婆婆纳叶冷水花■☆

299084　Pilea velutinipes Hand. -Mazz. = Pilea aquarum Dunn ■

299085　Pilea veronicifolia Engl. = Pilea usambarensis Engl. var. veronicifolia (Engl.) Friis ■☆

299086　Pilea verucosa Hand. -Mazz. ;疣果冷水花(瘤果冷水花,土甘草); Verucose Clearweed, Wartpull Coldwaterflower ■

299087　Pilea verucosa Hand. -Mazz. subsp. fujianensis C. J. Chen;闽北冷水花;Fujian Clearweed, Fujian Coldwaterflower ■

299088　Pilea verucosa Hand. -Mazz. subsp. subtriplinervia C. J. Chen;离基脉冷水花■

299089　Pilea villicaulis Hand. -Mazz. ;毛茎冷水花; Hairstem Clearweed, Hairstem Coldwaterflower ■

299090　Pilea villicaulis Hand. -Mazz. var. subglabra C. J. Chen;秃茎冷水花(秃净冷水花); Glabrous Hairystem Clearweed ■

299091　Pilea villicaulis Hand. -Mazz. var. subglabra C. J. Chen = Pilea villicaulis Hand. -Mazz. ■

299092　Pilea viridissima Makino = Pilea pumila (L.) A. Gray ■

299093　Pilea wattersii Hance;中华冷水花(中华冷水麻); Watters Clearweed, Watters Coldwaterflower ■

299094　Pilea wattersii Hance = Boehmeria blinii H. Lév. var. podocarpa W. T. Wang ●

299095　Pilea wattersii Hance = Boehmeria zollingeriana Wedd. var. podocarpa (W. T. Wang) W. T. Wang et C. J. Chen ●

299096　Pilea wattersii Hance = Pilea miyakei Yamam. ●

299097　Pilea wightii Wedd. ;生根冷水花; Wight Clearweed, Wight Coldwaterflower ■

299098　Pilea wightii Wedd. = Pilea chartacea C. J. Chen ■

299099　Pilea wightii Wedd. = Pilea umbrosa Blume var. obesa Wedd. ■

299100　Pilea wightii Wedd. var. royle Hook. f. = Pilea racemosa (Royle) Tuyama ■

299101　Pilea worsdellii N. E. Br. = Pilea rivularis Wedd. ■☆

299102　Pileanthus Labill. (1806);帽花属●☆

299103　Pileanthus limacis Labill. ;帽花●☆

299104　Pileocalyx Gasp. (废弃属名) = Cucurbita L. ■

299105　Pileocalyx Gasp. (废弃属名) = Piliocalyx Brongn. et Gris(保留属名)■☆

299106　Pileocalyx Post et Kuntze = Piliocalyx Brongn. et Gris(保留属名)☆

299107　Pileocalyx elegans Gasp. ;帽萼■☆

299108　Pileostegia Hook. f. et Thomson = Schizophragma Siebold et Zucc. ●

299109　Pileostegia Hook. f. et Thomson (1857);冠盖藤属(青棉花属); Pileostegia ●

299110　Pileostegia Turcz. = Ilex L. ●

299111　Pileostegia obtusifolia (Hu) Hu = Decumaria sinensis Oliv. ●

299112　Pileostegia tomentella Hand. -Mazz. ;星毛冠盖藤(山枇杷,星毛青棉花); Tomentose Pileostegia ●

299113　Pileostegia tomentella Hand. -Mazz. var. glabrescens C. C. Yang = Pileostegia viburnoides Hook. f. et Thomson var. glabrescens (C. C. Yang) S. M. Hwang ●

299114　Pileostegia urceolata Hayata = Pileostegia viburnoides Hook. f. et Thomson ●

299115　Pileostegia viburnoides Hook. f. et Thomson;冠盖藤(阿里山青棉花,大青叶,大藤,旱禾树,红棉花藤,红棉毛藤,猴大绳,猴头藤,青棉花,青棉花藤,纸加藤,竹麻,竹马); Common Pileostegia, Eg Climbing Hydrangea ●

299116　Pileostegia viburnoides Hook. f. et Thomson var. glabrescens (C. C. Yang) S. M. Hwang;柔毛冠盖藤●

299117　Pileostegia viburnoides Hook. f. et Thomson var. parviflora Oliv. ex Maxim. = Pileostegia viburnoides Hook. f. et Thomson ●

299118　Pileostigma B. D. Jacks. = Piliostigma Hochst. (保留属名)■☆

299119　Pileostigma Hochst. = Piliostigma Hochst. (保留属名)■☆

299120　Piletocarpus Hassk. = Aneilema R. Br. ■☆

299121　Piletocarpus Hassk. = Dictyospermum Wight ■

299122　Piletocarpus protensus Hassk. = Dictyospermum scaberrimum (Blume) J. K. Morton ex H. Hara ■

299123　Piletocarpus protensus Hassk. = Rhopalephora scaberrima (Blume) Faden ■

299124　Piletocarpus protensus Hassk. var. intermedius ? = Dictyospermum scaberrimum (Blume) J. K. Morton ex H. Hara ■

299125　Pileus Ramirez = Jacaratia A. DC. ●☆

299126　Pilgerina Z. S. Rogers, Nickrent et Malécot(2008);马岛檀香属●☆

299127　Pilgerochloa Eig = Ventenata Koeler(保留属名)■☆

299128　Pilgerochloa Eig. (1929);皮尔禾属■☆

299129　Pilgerodendraceae A. V. Bobrov et Melikyan ＝ Cupressaceae Gray(保留科名)●

299130　Pilgerodendraceae A. V. Bobrov et Melikyan;南智利柏科●☆

299131　Pilgerodendron Florin(1930);南智利柏属(智利南部柏属,智南柏属)●☆

299132　Pilgerodendron uviferum (D. Don) Florin;智利南柏(智利南部柏);Alerce,Chilean Cedar,Patagonian Pilgerodendron ●☆

299133　Pilgerodendron uviferum (D. Don) Florin ＝ Juniperus uvifera D. Don ●☆

299134　Pilgerodendron uviferum Florin ＝ Juniperus uvifera D. Don ●☆

299135　Pilgerodendron uviferum Florin ＝ Pilgerodendron uviferum (D. Don) Florin ●☆

299136　Pilicordia (DC.) Lindl ＝ Cordia L. (保留属名)●

299137　Pilidiostigma Burret(1941);帽柱桃金娘属●☆

299138　Pilidiostigma cuneatum Burret;帽柱桃金娘●☆

299139　Pilidiostigma glabrum Burret;光帽柱桃金娘●☆

299140　Pilidiostigma parviflorum Burret;小花帽柱桃金娘●☆

299141　Pilinophyton Klotzsch ＝ Croton L. ●

299142　Pilinophytum Klotzseh ＝ Croton L. ●

299143　Piliocalyx Brongn. et Gris(1865)(保留属名);帽萼葫芦属(帽萼属)■☆

299144　Piliocalyx baudouinii Brongn. et Gris;帽萼葫芦■☆

299145　Piliocalyx laurifolius Brongn. et Gris;桂叶帽萼葫芦■☆

299146　Piliocalyx robustus Brong. et Gris;粗壮帽萼葫芦■☆

299147　Piliosanthes Hassk. ＝ Peliosanthes Andréws ■

299148　Piliostigma Hochst. (1846)(保留属名);帽柱豆属(毛拉豆属,毛柱豆属)■☆

299149　Piliostigma Hochst. ＝ Bauhinia L. ●

299150　Piliostigma malabaricum (Roxb.) Benth. ;马拉巴毛柱豆■☆

299151　Piliostigma racemosa (Lam.) Benth. ＝ Bauhinia racemosa Lam. ●

299152　Piliostigma reticulatum (DC.) Hochst. ;毛柱豆■☆

299153　Piliostigma thonningii (Schumach.) Milne-Redh. ;托尼毛柱豆(托尼毛拉豆)■☆

299154　Piliostigma thonningii Schum. ＝ Piliostigma thonningii (Schumach.) Milne-Redh. ■☆

299155　Pilitis Lindl. ＝ Richea R. Br. (保留属名)●☆

299156　Pillansia L. Bolus(1914);皮朗斯鸢尾属■☆

299157　Pillansia templemannii (Baker) L. Bolus;皮朗斯鸢尾☆

299158　Pillera Endl. ＝ Mucuna Adans. (保留属名)●■

299159　Piloblephis Raf. (1838);肖香草属●☆

299160　Piloblephis Raf. ＝ Satureja L. ●■

299161　Piloblephis ericoides Raf. ;肖香草●☆

299162　Piloblephis rigida (W. Bartram ex Benth.) Raf. ;硬肖香草●☆

299163　Pilocanthus B. W. Benson et Backeb. ＝ Pediocactus Britton et Rose ●☆

299164　Pilocanthus paradinei (B. W. Benson) B. W. Benson et Backeb. ＝ Pediocactus paradinei B. W. Benson ●☆

299165　Pilocarpaceae J. Agardh ＝ Rutaceae Juss. (保留科名)●■

299166　Pilocarpaceae J. Agardh;毛果芸香科●☆

299167　Pilocarpus Vahl(1796);毛果芸香属;Jaborandi,Pilocarpus ●☆

299168　Pilocarpus atropurpureus K. Koch;深紫毛果芸香●☆

299169　Pilocarpus jaborandii Holmes;毛果芸香●☆

299170　Pilocarpus microphyllus Stapf;小叶毛果芸香●☆

299171　Pilocarpus pennatifolius Lem. ;羽叶毛果芸香●☆

299172　Pilocarpus pinnatifolius Lem. et Hassl. ;巴拉圭毛果芸香(巴西彼罗卡巴,翼叶毛果芸香);Spike Pilocarpus ●☆

299173　Pilocarpus racemosus Vahl;总花毛果芸香●☆

299174　Pilocarpus trachylophus Holmes;粗冠毛果芸香●☆

299175　Pilocereus K. Schum. ＝ Pilosocereus Byles et G. D. Rowley ●☆

299176　Pilocereus Lem. ＝ Cephalocereus Pfeiff. ●

299177　Pilocereus giganteus (Engelm.) Rümpler ＝ Carnegiea gigantea (Engelm.) Britton et Rose ●☆

299178　Pilocereus robinii Lem. ＝ Pilosocereus robinii (Lem.) Byles et G. D. Rowley ●☆

299179　Pilocereus schottii (Engelm.) Lem. ＝ Pachycereus schottii (Engelm.) D. R. Hunt ●☆

299180　Pilocereus senilis Lem. ＝ Cephalocereus senilis (Haw.) Pfeiff. ●

299181　Pilocereus thurberi (Engelm.) Rümpler ＝ Stenocereus thurberi (Engelm.) Buxb. ●☆

299182　Pilocopiapoa F. Ritter ＝ Copiapoa Britton et Rose ●

299183　Pilocosta Almeda et Whiffin(1981);毛肋野牡丹属●☆

299184　Pilocosta nana (Standl.) Almeda et Whiffin;矮毛肋野牡丹●☆

299185　Pilocosta oerstedii (Triana) Almeda et Whiffin;毛肋野牡丹●☆

299186　Pilogyne Eckl. ex Schrad. ＝ Zehneria Endl. ■

299187　Pilogyne Gagnep. ＝ Myrsine L. ●

299188　Pilogyne Schrad. ＝ Zehneria Endl. ■

299189　Pilogyne affinis Schrad. ＝ Zehneria scabra (L. f.) Sond. ●☆

299190　Pilogyne lucida Naudin ＝ Zehneria maysorensis (Wight et Arn.) Arn. ■

299191　Pilogyne peneyana Naudin ＝ Zehneria peneyana (Naudin) Asch. et Schweinf. ■☆

299192　Pilogyne suavis Schrad. ＝ Zehneria scabra (L. f.) Sond. ●☆

299193　Pilogyne tenuiflora Schrad. ＝ Zehneria scabra (L. f.) Sond. ●☆

299194　Pilogyne velutina Schrad. ＝ Zehneria scabra (L. f.) Sond. ●☆

299195　Piloisa B. D. Jacks. ＝ Piloisia Raf. ●

299196　Piloisia Raf. ＝ Cordia L. (保留属名)●

299197　Pilophora Jacq. ＝ Manicaria Gaertn. ●☆

299198　Pilophyllum Schltr. (1914);毛叶兰属■☆

299199　Pilophyllum Schltr. ＝ Chrysoglossum Blume ■

299200　Pilophyllum villosum Schltr. ;毛叶兰●☆

299201　Pilopleura Schischk. (1951);毛棱芹属■

299202　Pilopleura goloskokovii (Korovin) Pimenov ＝ Platytaenia goloskokovii Korovin ■☆

299203　Pilopleura kozo-poljanskii Schischk. ;毛棱芹■

299204　Pilopleura kozo-poljanskii Schischk. ＝ Peucedanum dasycarpum Regel et Schmalh. ■

299205　Pilopleura tordyloides (Korovin) Pimenov ＝ Zosima tordyloides Korovin ■

299206　Pilopsis Y. Ito ＝ Arthrocereus A. Berger(保留属名)●☆

299207　Pilopsis Y. Ito ＝ Echinopsis Zucc. ●

299208　Pilopus Raf. (废弃属名) ＝ Phyla Lour. ■

299209　Pilorea Raf. (废弃属名) ＝ Edraianthus A. DC. (保留属名)■☆

299210　Pilorea Raf. (废弃属名) ＝ Wahlenbergia Schrad. ex Roth(保留属名)■●

299211　Pilosanthus Stead. ＝ Liatris Gaertn. ex Schreb. (保留属名)■☆

299212　Pilosanthus Stead. ＝ Psilosanthus Neck. ■☆

299213　Pilosella F. W. Schultz et Sch. Bip. ＝ Hieracium L. ■

299214　Pilosella Hill ＝ Hieracium L. ■

299215　Pilosella Hill(1756);匍茎山柳菊属;Mouse-ear Hawkweed ■☆

299216　Pilosella Kostel. ＝ Arabidopsis Heynh. (保留属名)■

299217　Pilosella Kostel. ex Rydb. ＝ Arabidopsis Heynh. (保留属名)■

299218　Pilosella Vaill. ＝ Pilosella Hill ■☆

299219　Pilosella aurantiaca （L.） F. W. Schultz et Sch. Bip. ＝ Hieracium aurantiacum L. ■

299220　Pilosella caespitosa （Dumort.） P. D. Sell et C. West；丛生匍茎山柳菊；Yellow Fox-and-cubs ■☆

299221　Pilosella caespitosa （Dumort.） P. D. Sell et C. West ＝ Hieracium caespitosum Dumort. ■☆

299222　Pilosella flagellaris （Willd.） P. D. Sell et C. West ＝Hieracium flagellare Willd. ■☆

299223　Pilosella flagellaris （Willd.） P. D. Sell et C. West ＝ Lippia citriodora （Paláu） Kunth ●☆

299224　Pilosella floribunda （Wimm. et Grab.） Fr.；繁花匍茎山柳菊■☆

299225　Pilosella officinarum F. W. Schultz et Sch. Bip. ＝ Hieracium pilosella L. ■☆

299226　Pilosella peleteriana F. W. Schultz et Sch. Bip.；佩氏匍茎山柳菊；Shaggy Mouse-ear Hawkweed ■☆

299227　Pilosella piloselloides （Vill.） Soják ＝ Hieracium piloselloides Vill. ■☆

299228　Pilosella praealta F. W. Schultz et Sch. Bip. ＝ Hieracium praealatum Vill. ■☆

299229　Pilosella spathulata F. W. Schultz et Sch. Bip. ＝ Hieracium traillii Greene ■☆

299230　Piloselloides （Less.） C. Jeffrey ex Cufod. （1967）；兔耳一枝箭属■

299231　Piloselloides （Less.） C. Jeffrey ex Cufod. ＝ Gerbera L. （保留属名）■

299232　Piloselloides cordata （Thunb.） C. Jeffrey ＝ Gerbera cordata （Thunb.） Less. ■☆

299233　Piloselloides hirsuta （Forssk.） C. Jeffrey ＝ Gerbera piloselloides （L.） Cass. ■

299234　Piloselloides hirsuta （Forssk.） C. Jeffrey ex Cufod.；兔耳一枝箭■

299235　Piloselloides hirsuta （Forssk.） C. Jeffrey ex Cufod. ＝ Gerbera piloselloides （L.） Cass. ■

299236　Pilosia Tausch ＝ Picris L. ■

299237　Pilosia Tausch ＝ Pieris D. Don ●

299238　Pilosocereus Byles et G. D. Rowley（1957）；疏毛刺柱属（毛刺柱属，毛柱属）；Tree Cactus ●☆

299239　Pilosocereus Byles et Rowl. ＝ Cephalocereus Pfeiff. ●

299240　Pilosocereus alensis （F. A. C. Weber） Byles et G. D. Rowley；白天龙●☆

299241　Pilosocereus arenicola （Werderm.） Byles et G. D. Rowley；沙生疏毛刺柱●☆

299242　Pilosocereus arrabidae （Lem.） Byles et G. D. Rowley；英贵龙●☆

299243　Pilosocereus aurisetus （Werderm.） Byles et G. D. Rowley；金色疏毛刺柱●☆

299244　Pilosocereus backebergii Byles et G. D. Rowley；白毛龙●☆

299245　Pilosocereus calcisaxicolus ?；石生疏毛刺柱●☆

299246　Pilosocereus catingicola （Gurke） Byles et G. D. Rowley；星云阁●☆

299247　Pilosocereus chrysacanthus （F. A. C. Weber） Byles et G. D. Rowley；金刺疏毛刺柱（金凤龙）●☆

299248　Pilosocereus chrysostele （Vaupel） Byles et G. D. Rowley；黄金龙●☆

299249　Pilosocereus collinsii （Britton et Rose） Byles et G. D. Rowley；光琳龙●☆

299250　Pilosocereus cometes （Scheidw.） Byles et G. D. Rowley；彗星柱●☆

299251　Pilosocereus deeringii （Small） Byles et G. D. Rowley ＝ Pilosocereus robinii （Lem.） Byles et G. D. Rowley ●☆

299252　Pilosocereus fulvilanatus （Buining et Brederoo） F. Ritter；黄锦毛柱●☆

299253　Pilosocereus glaucescens （Labour.） Byles et G. D. Rowley；灰斑疏毛刺柱（苍白毛柱）；Blue Torch Cactus ●☆

299254　Pilosocereus glaucochrous （Werderm.） Blyles et G. D. Rowley；巴西青毛柱●☆

299255　Pilosocereus gounellei （F. A. C. Weber） Byles et G. D. Rowley；豪壮龙；Pilosocereus，Xique-Xique ●☆

299256　Pilosocereus keyensis （Britton et Rose） Byles et G. D. Rowley ＝ Pilosocereus robinii （Lem.） Byles et G. D. Rowley ●☆

299257　Pilosocereus leucocephalus （Poselg.） Byles et G. D. Rowley ＝ Pilosocereus palmeri （Rose） Byles et G. D. Rowley ●☆

299258　Pilosocereus magnificus （Buining et Brederoo） F. Ritter；华丽疏毛刺柱●☆

299259　Pilosocereus maxonii （Rose） Byles et G. D. Rowley；翁狮子●☆

299260　Pilosocereus moritzianus Byles et G. D. Rowley；白升龙●☆

299261　Pilosocereus nobilis （Haw.） Byles et G. D. Rowley；红笔●☆

299262　Pilosocereus palmeri （Rose） Byles et G. D. Rowley；春衣（疏长毛柱）；Woolly Torch ●☆

299263　Pilosocereus pentaedrophorus （Labour.） Byles et G. D. Rowley；五角毛柱●☆

299264　Pilosocereus polygonus；多节疏毛刺柱；Bahama Dildo，Robin Tree Cactus ●☆

299265　Pilosocereus purpusii （Britton et Rose） Byles et G. D. Rowley；细枝疏毛刺柱（黄焰龙）；Viejos ●☆

299266　Pilosocereus robinii （Lem.） Byles et G. D. Rowley；老翁；Key Tree Cactus ●☆

299267　Pilosocereus robinii （Lem.） Byles et G. D. Rowley var. deeringii （Small） Kartesz et Gandhi ＝ Pilosocereus robinii （Lem.） Byles et G. D. Rowley ●☆

299268　Pilosocereus royenii （L.） Byles et G. D. Rowley ＝ Pilosocereus robinii （Lem.） Byles et G. D. Rowley ●☆

299269　Pilosocereus sartorianus （Rose） Byles et G. D. Rowley；猿取阁●☆

299270　Pilosocereus tehuacanus Byles et G. D. Rowley；大蛾阁；Caxacubri ●☆

299271　Pilosperma Planch. et Triana（1860）；毛籽藤黄属●☆

299272　Pilosperma caudatum Planch. et Triana；毛籽藤黄●☆

299273　Pilostachys B. D. Jacks. ＝ Pilostaxis Raf. ●■

299274　Pilostachys Raf. ＝ Polygala L. ●■

299275　Pilostaxis Raf. ＝ Polygala L. ●■

299276　Pilostemon Iljin ＝ Jurinea Cass. ●■

299277　Pilostemon Iljin（1961）；毛蕊菊属；Pilostemon ■

299278　Pilostemon filifolia （C. Winkl.） Iljin；毛蕊菊；Filifolious Pilostemon ■

299279　Pilostemon karateginii （Lipsky） Iljin；短冠毛蕊菊■

299280　Pilostigma Costantin ＝ Costantina Bullock ■

299281　Pilostigma Costantin ＝ Lygisma Hook. f. ■

299282　Pilostigma Tiegh. ＝ Amyema Tiegh. ●☆

299283　Pilostigma inflexum Costantin ＝ Lygisma inflexum （Costantin） Kerr ■

299284　Pilostigma racemosa ? ＝ Bauhinia racemosa Vahl ●☆

299285　Pilostyles Guill. （1834）；毛柱大花草属（豆生花属）■☆

299286　Pilostyles aethiopica Welw. ＝ Berlinianche aethiopica （Welw.） Vattimo ●☆

299287　Pilostyles haussknechtii Boiss.；毛柱大花草■☆

299288　Pilostyles holtzii Engl. ＝ Berlinianche holtzii （Engl.） Vattimo ■☆

299289　Pilotheca T. L. Mitch. = Philotheca Rudge ●☆

299290　Pilothecium (Kiaersk.) Kausel = Myrtus L. ●

299291　Pilotrichum Hook. f. et T. Anderson = Ptilotrichum C. A. Mey. ●■

299292　Pilouratea Tiegh. = Ouratea Aubl. (保留属名) ●

299293　Pilumna Lindl. = Trichopilia Lindl. ■☆

299294　Pimecaria Raf. = Ximenia L. ●

299295　Pimela Lour. = Canarium L. ●

299296　Pimela alba Lour. = Canarium album (Lour.) Raeusch. ●

299297　Pimela nigra Lour. = Canarium pimela K. D. Koenig ●

299298　Pimela stricta Blume = Canarium strictum Roxb. ●

299299　Pimelaea Kuntze = Pimelea Banks ex Gaertn. (保留属名) ●☆

299300　Pimelandra A. DC. = Ardisia Sw. (保留属名) ●■

299301　Pimelea Banks et Sol. = Pimelea Banks ex Gaertn. (保留属名) ●☆

299302　Pimelea Banks et Sol. ex Gaertn. = Pimelea Banks ex Gaertn. (保留属名) ●☆

299303　Pimelea Banks ex Gaertn. (1788) (保留属名); 稻花木属 (稻花属); Rice Flower, Rice-flower ●☆

299304　Pimelea Banks ex Sol. = Pimelea Banks ex Gaertn. (保留属名) ●☆

299305　Pimelea alpina F. Muell. ex Meisn.; 高山稻花木; Alpine Rice Flower ●☆

299306　Pimelea axiflora F. Muell.; 健壮稻花木; Bootlace Plant, Tough Rice Flower ●☆

299307　Pimelea ferruginea Labill.; 玫瑰红稻花木 (稻花); Pink Rice Flower, Rosy Rice Flower ●☆

299308　Pimelea ligustrina Labill.; 高稻花木; Tall Rice Flower ●☆

299309　Pimelea linifolia Sm.; 亚麻叶稻花木; Rice Flower, Slender Rice Flower, Slender Rice-flower ●☆

299310　Pimelea microcephala R. Br.; 小头稻花木 (小头稻花) ●☆

299311　Pimelea nivea Labill.; 雪白稻花木; White Cotton Bush ●☆

299312　Pimelea physodes Hook.; 切花稻花木; Qualulo Bell, Qualup Bells ●☆

299313　Pimelea prostrata Willd.; 新西兰稻花木 (平卧稻花); New Zealand Daphne, Strathmore Weed ●☆

299314　Pimelea rosea R. Br.; 粉花稻花木 ●☆

299315　Pimelea simplex F. Muell.; 单枝稻花木 (单枝稻花) ●☆

299316　Pimelea trichostachya Lindl.; 毛穗稻花木 (毛穗稻花) ●☆

299317　Pimeledendrum Hassk. = Pimelodendron Hassk. ●

299318　Pimeleodendron Maell. Arg. = Pimelodendron Hassk. ●

299319　Pimelodendron Hassk. (1856); 油载木属 ●

299320　Pimelodendron amboinicum Hassk.; 油载木 ●

299321　Pimenta Lindl. (1821); 众香树属 (多香果属, 香椒属); Pimento ●☆

299322　Pimenta acris (Sw.) Kostel. = Pimenta racemosa (Mill.) J. W. Moore ●☆

299323　Pimenta acris Kostel.; 辛众香树; Bay Rum Tree, Bayberry, Black Cinnamon ●☆

299324　Pimenta acuminata Bello; 渐尖众香树 ●☆

299325　Pimenta dioica (L.) Merr.; 众香树 (披门他树, 牙买加胡椒, 药用众香树, 玉桂); Allspice, Allspice Pimenta, Jamaica Allspice, Jamaica Pepper, Jamaica Pimento, Pimenta, Pimento ●☆

299326　Pimenta officinalis Lindl. = Eugenia pimenta DC. ●☆

299327　Pimenta officinalis Lindl. = Pimenta dioica (L.) Merr. ●☆

299328　Pimenta racemosa (Mill.) J. W. Moore; 香叶众香树 (多香果, 香叶多香果); Bay Rum Tree, Bayrum, Bay-rum, Bay-rum Tree, Bayrum-tree, West Indian Bay Tree ●☆

299329　Pimenta racemosa J. W. Moore = Pimenta racemosa (Mill.) J. W. Moore ●☆

299330　Pimentelea Willis = Pimentelia Wedd. ●☆

299331　Pimentelia Wedd. (1849); 皮门茜属 ●☆

299332　Pimentelia glomerata Wedd.; 皮门茜 ☆

299333　Pimentella Walp. = Pimentelia Wedd. ●☆

299334　Pimentella Wedd. = Pimentelia Wedd. ●☆

299335　Pimentus Raf. = Eugenia L. ●

299336　Pimentus Raf. = Melaleuca L. (保留属名) ●

299337　Pimentus Raf. = Pimenta Lindl. ●☆

299338　Pimia Seem. (1862); 皮姆梧桐属 ●☆

299339　Pimia rhamnoides Seem.; 皮姆梧桐 ●☆

299340　Pimphaele St. -Lag. = Pimpinella L. ■

299341　Pimpinele St. -Lag. = Pimpinella L. ■

299342　Pimpinella L. (1753); 茴芹属; Burnet Saxifrage, Burnet-saxifrage, Pimpinella ■

299343　Pimpinella Ség. = Poterium L. ■☆

299344　Pimpinella achilleifolia (DC.) C. B. Clarke = Meeboldia achilleifolia (DC.) P. K. Mukh. et Constance ■

299345　Pimpinella achilleifolia (Wall.) C. B. Clarke; 耆叶茴芹; Yarrowleaf Pimpinella ■

299346　Pimpinella acuminata (Edgew.) C. B. Clarke; 尖叶茴芹; Sharpleaf Pimpinella ■

299347　Pimpinella acutidentata C. Norman; 尖齿茴芹 ■☆

299348　Pimpinella affinis Ledeb.; 近缘茴芹 ■☆

299349　Pimpinella africana M. Hiroe = Pimpinella ledermannii H. Wolff subsp. engleriana (H. Wolff) C. C. Towns. ■☆

299350　Pimpinella albescens Franch. = Eriocycla albescens (Franch.) H. Wolff ■

299351　Pimpinella anisum L.; 茴芹 (洋茴芹, 洋茴香); Aneys, Anise, Anise Burnet Saxifrage, Aniseed, Anisette, Anny, Annyle, Pastis, Pimpinella, Sweet Alice ■

299352　Pimpinella anthriscoides Boiss.; 峨参茴芹 ■☆

299353　Pimpinella arguta Diels; 锐叶茴芹 (尖齿茴芹); Sharptooth Pimpinella ■

299354　Pimpinella armena Schischk.; 亚美尼亚茴芹 ■☆

299355　Pimpinella aromatica M. Bieb.; 芳香茴芹 ■☆

299356　Pimpinella asianensis M. Hiroe = Pimpinella kingdon-wardii H. Wolff ■

299357　Pimpinella astilbifolia Hayata; 落新妇茴芹; Astilbileaf Pimpinella ■

299358　Pimpinella astilbifolia Hayata = Pimpinella niitakayamensis Hayata ■

299359　Pimpinella atropurpurea C. Y. Wu ex R. H. Shan et F. T. Pu; 深紫茴芹; Darkviolet Pimpinella ■

299360　Pimpinella aurea DC.; 金黄茴芹 ■☆

299361　Pimpinella battandieri Chabert; 巴坦茴芹 ■☆

299362　Pimpinella betsileensis Sales et Hedge; 贝齐尔茴芹 ■☆

299363　Pimpinella bisecta Baker = Pimpinella ebracteata Baker ■☆

299364　Pimpinella bisinuata H. Wolff; 重波茴芹; Doublewave Pimpinella ■

299365　Pimpinella brachycarpa (Kom.) Nakai; 短果茴芹 (大叶芹, 短果羊角芹, 假茴芹); Shortfruit Pimpinella ■

299366　Pimpinella brachystyla Hand. -Mazz.; 短柱茴芹; Shortstyle Pimpinella ■

299367　Pimpinella bubonoides Brot. = Pimpinella villosa Schousb. ■☆

299368　Pimpinella buchananii H. Wolff; 布坎南茴芹 ■☆

299369　Pimpinella buchananii H. Wolff subsp. septentrionalis C. C. Towns.; 北方茴芹 ■☆

299370　Pimpinella buchananii H. Wolff var. longistyla C. C. Towns. ;长柱布坎南茴芹■☆

299371　Pimpinella buchananii H. Wolff var. triradiata C. Norman = Pimpinella buchananii H. Wolff ■☆

299372　Pimpinella caffra (Eckl. et Zeyh.) D. Dietr. ;开菲尔茴芹■☆

299373　Pimpinella calycina Maxim. ;具萼茴芹;Calycinate Pimpinella ■

299374　Pimpinella calycina Maxim. = Spuriopimpinella calycina (Maxim.) Kitag. ■

299375　Pimpinella calycina Maxim. var. brachycarpa Kom. = Pimpinella brachycarpa (Kom.) Nakai ■

299376　Pimpinella candolleana Wight et Arn. ;杏叶茴芹(白花草,白花箭,大寒药,单膻臭,地胡椒,九月白花草,犁头尖,马蹄防风,马蹄叶,满身串,清当归,三足蝉,骚羊古,癀疬股,山当归,山茴香,蛇倒退,天蓬草,土当归,兔耳防风,小菊花,小羊膻,杏叶防风,羊膻臭,阳山臭,蜘蛛香);Apricotleaf Pimpinella ■

299377　Pimpinella capillifolia Regel et Schmalh. = Aphanopleura capillifolia (Regel et Schmalh.) Lipsky ■

299378　Pimpinella cartilaginomarginata (Makino ex Y. Yabe) H. Wolff = Angelica cartilaginomarginata (Makino ex Y. Yabe) Nakai ■

299379　Pimpinella cartilaginomarginata (Makino) H. Wolff = Angelica cartilaginomarginata (Makino ex Y. Yabe) Nakai ■

299380　Pimpinella caudata (Franch.) H. Wolff;尾尖茴芹(尾叶茴芹);Tailleaf Pimpinella ■

299381　Pimpinella chateriana Cannon et Farille = Pimpinella atropurpurea C. Y. Wu ex R. H. Shan et F. T. Pu ■

299382　Pimpinella chungdienensis C. Y. Wu;中甸茴芹;Zhongdian Pimpinella ■

299383　Pimpinella clarkeana Watt ex Banerji = Pternopetalum vulgare (Dunn) Hand. -Mazz. ■

299384　Pimpinella cnidioides H. Pearson ex H. Wolff;蛇床茴芹;Cnidiumleaf Pimpinella ■

299385　Pimpinella confusa Woronow;混乱茴芹■☆

299386　Pimpinella coriacea (Franch.) H. Boissieu;革叶茴芹(羊膻臭);Coriaceousleaf Pimpinella ■

299387　Pimpinella crinita Boiss. = Psammogeton canescens (DC.) Vatke ■☆

299388　Pimpinella crispulifolia H. Boissieu;皱叶茴芹;Wrinkleleaf Pimpinella ■

299389　Pimpinella cyclophylla Chiov. = Pimpinella heywoodii Dawit ■☆

299390　Pimpinella daghestanica Schischk. ;达赫斯坦茴芹■☆

299391　Pimpinella decursiva (Miq.) H. Wolff = Angelica decursiva (Miq.) Franch. et Sav. ■

299392　Pimpinella dichotoma L. = Stoibrax dichotomum (L.) Raf. ■☆

299393　Pimpinella dichotoma L. var. vegeta Pau et Font Quer = Stoibrax dichotomum (L.) Raf. ■☆

299394　Pimpinella dissecta Retz. ;深裂茴芹■☆

299395　Pimpinella diversifolia DC. ;异叶茴芹(八月白,白花菜,白花雷公根,白花仔,百路通六月寒,大叶半边莲,冬青草,鹅脚板,虎羊丁,茴芹,金锁匙,空心草,苦爹菜,苦爷菜,犁头草,六月寒,三脚蛤蟆,三叶茴芹,三叶茴香,骚羊股,山当归,蛇咬草,铁铲头,香草,羊膻草,羊膻七);Diversileaf Pimpinella ■

299396　Pimpinella diversifolia DC. var. angustipetala R. H. Shan et F. T. Pu;尖瓣异叶茴芹;Sharppetal Pimpinella ■

299397　Pimpinella diversifolia DC. var. divisa C. B. Clarke = Pimpinella diversifolia DC. ■

299398　Pimpinella diversifolia DC. var. sermentifera Goel et U. C. Bhattach. = Pimpinella diversifolia DC. var. stolonifera Hand. -Mazz. ■

299399　Pimpinella diversifolia DC. var. simplicifolia Kuntze = Pimpinella diversifolia DC. ■

299400　Pimpinella diversifolia DC. var. stolonifera Hand. -Mazz. ;走茎异叶茴芹(匍枝鹅脚板);Stolonifer Pimpinella ■

299401　Pimpinella duclouxii H. Boissieu;东川茴芹;Ducloux Pimpinella ■

299402　Pimpinella duclouxii H. Boissieu = Pimpinella flaccida C. B. Clarke ■

299403　Pimpinella dunnii H. Boissieu = Tongoloa dunnii (H. Boissieu) H. Wolff ■

299404　Pimpinella ebracteata Baker;无苞茴芹■☆

299405　Pimpinella edosmioides H. Boissieu. = Cyclorhiza peucedanifolia (Franch.) Constance ■

299406　Pimpinella elata (H. Wolff) M. Hiroe = Tongoloa elata H. Wolff ■

299407　Pimpinella engleriana Fedde ex H. Wolff = Pimpinella kingdon-wardii H. Wolff ■

299408　Pimpinella engleriana H. Wolff = Pimpinella ledermannii H. Wolff subsp. engleriana (H. Wolff) C. C. Towns. ■☆

299409　Pimpinella erlangeri Engl. ;厄兰格茴芹■☆

299410　Pimpinella erythraeae Armari;浅红茴芹■☆

299411　Pimpinella fargesii H. Boissieu;城口茴芹;Chengkou Pimpinella ■

299412　Pimpinella fargesii H. Boissieu var. alba H. Boissieu = Pimpinella fargesii H. Boissieu ■

299413　Pimpinella favifolia C. Norman = Pimpinella buchananii H. Wolff ■☆

299414　Pimpinella feddei W. C. Wu et C. Y. Wu;腾冲茴芹;Fedde Pimpinella ■

299415　Pimpinella feddei W. C. Wu et C. Y. Wu = Pimpinella kingdon-wardii H. Wolff ■

299416　Pimpinella filicina (Franch.) Diels = Pternopetalum filicinum (Franch.) Hand. -Mazz. ■

299417　Pimpinella filipedicellata S. L. Liou;细柄茴芹;Thinpedicel Pimpinella ■

299418　Pimpinella flaccida C. B. Clarke;细软茴芹(柔软茴芹);Tender Pimpinella ■

299419　Pimpinella fortunatii H. Boissieu = Tongoloa silaifolia (H. Boissieu) H. Wolff ■

299420　Pimpinella friesiorum H. Wolff = Pimpinella oreophila Hook. f. ■☆

299421　Pimpinella gossweileri H. Wolff = Pimpinella huillensis Engl. ■☆

299422　Pimpinella grisea H. Wolff;灰叶茴芹;Greyleaf Pimpinella ■

299423　Pimpinella grossheimii Schischk. ;格罗氏茴芹■☆

299424　Pimpinella gymnosciadium Hiern = Pimpinella pimpinelloides (Hochst.) H. Wolff ■☆

299425　Pimpinella hazariensis H. Wolff. = Pimpinella acuminata (Edgew.) C. B. Clarke ■

299426　Pimpinella heliosciadea H. Boissieu;沼生茴芹;Marshy Pimpinella ■

299427　Pimpinella henryi Diels;川鄂茴芹;Henry Pimpinella ■

299428　Pimpinella heywoodii Dawit;海伍得茴芹■☆

299429　Pimpinella hiernii M. Hiroe = Angoseseli mossamedensis (Welw. ex Hiern) C. Norman ■☆

299430　Pimpinella hirtella A. Rich. ;多毛茴芹■☆

299431　Pimpinella homblei C. Norman;洪布勒茴芹■☆

299432　Pimpinella hookeri C. B. Clarke = Acronema hookeri (C. B. Clarke) H. Wolff ■

299433　Pimpinella hookeri C. B. Clarke var. gramimfolia W. W. Sm. = Acronema graminifolium (H. Wolff) S. L. Liou et R. H. Shan ■

299434　Pimpinella huillensis Engl. ;粗壮茴芹■☆

299435　Pimpinella huillensis Engl. var. elatior Hiern = Pimpinella huillensis Engl. ■☆

299436　Pimpinella huillensis Engl. var. welwitschii (Engl.) Engl. = Pimpinella huillensis Engl. ■☆

299437　Pimpinella humbertii Sales et Hedge;亨伯特茴芹■☆

299438　Pimpinella hydrophila H. Wolff;喜水茴芹■☆

299439　Pimpinella idae Takht.;伊达茴芹■☆

299440　Pimpinella involucrata (Roxb.) Wight et Arn. = Trachyspermum roxburghianum (DC.) H. Wolff ■

299441　Pimpinella involucrata Hiern ex Engl. = Angoseseli mossamedensis (Welw. ex Hiern) C. Norman ■☆

299442　Pimpinella kashmirica Stewart ex Dunn = Aegopodium alpestre Ledeb. ■

299443　Pimpinella kassneri (H. Wolff) Cannon;卡斯纳茴芹■☆

299444　Pimpinella keniensis C. Norman;肯尼亚茴芹■☆

299445　Pimpinella kilimandscharica Engl. = Pimpinella oreophila Hook. f. var. kilimandscharica (Engl.) C. C. Towns. ■☆

299446　Pimpinella kingdon-wardii H. Wolff;德钦茴芹;Deqin Pimpinella ■

299447　Pimpinella komarovii (Kitag.) R. H. Shan et F. T. Pu;辽冀茴芹;Komalov Pimpinella ■

299448　Pimpinella koreana (Y. Yabe) Nakai;朝鲜茴芹;Korea Pimpinella ■

299449　Pimpinella koreana (Y. Yabe) Nakai = Spuriopimpinella koreana (Y. Yabe) Kitag. ■

299450　Pimpinella korshinskyi Schischk.;考尔茴芹■☆

299451　Pimpinella kraussiana Meisn. = Pimpinella caffra (Eckl. et Zeyh.) D. Dietr. ■☆

299452　Pimpinella krookii H. Wolff;克鲁科茴芹■☆

299453　Pimpinella kuramensis Kitam. = Platytaenia kuramensis (Kitam.) Nasir ■☆

299454　Pimpinella kyimbilaensis H. Wolff;基穆比拉茴芹■☆

299455　Pimpinella ledermannii H. Wolff;莱德茴芹■☆

299456　Pimpinella ledermannii H. Wolff subsp. engleriana (H. Wolff) C. C. Towns.;恩格勒茴芹■☆

299457　Pimpinella leptophylla Pers. = Apium leptophyllum (Pers.) F. Muell. ex Benth. ■

299458　Pimpinella leptophylla Pers. = Cyclospermum leptophyllum (Pers.) Sprague ex Britton et P. Wilson ■

299459　Pimpinella liiana M. Hiroe;景东茴芹;Jingdong Pimpinella ■

299460　Pimpinella lindblomii H. Wolff;林德布卢姆茴芹■☆

299461　Pimpinella lineariloba Cannon;线裂片茴芹■☆

299462　Pimpinella lithophila Schischk.;喜石茴芹■☆

299463　Pimpinella litvinovii Schischk.;里特茴芹■☆

299464　Pimpinella loloenis H. Boissieu = Tongoloa loloensis (H. Boissieu) H. Wolff ■

299465　Pimpinella lutea Desf.;黄茴芹■☆

299466　Pimpinella magna L.;大茴芹;Burnet Saxifrage, Greater Pimpinella ■☆

299467　Pimpinella major (L.) Huds.;欧洲大茴芹(大茴芹);Greater Burnet Saxifrage, Greater Burnet-saxifrage, Greater Pimpinella, Hollowstem Burnet Saxifrage ■☆

299468　Pimpinella major Huds. = Pimpinella major (L.) Huds. ■☆

299469　Pimpinella markgrafiana Fedde ex H. Wolff;黑水茴芹■

299470　Pimpinella markgrafiana Fedde ex H. Wolff = Pimpinella purpurea (Franch.) H. Boissieu ■

299471　Pimpinella mechowii (Engl.) H. Wolff = Pimpinella huillensis Engl. ■☆

299472　Pimpinella monoica Dalzell = Pimpinella weishanensis R. H. Shan et F. T. Pu ■

299473　Pimpinella mossamedensis (Welw. ex Hiern) M. Hiroe = Angoseseli mossamedensis (Welw. ex Hiern) C. Norman ■☆

299474　Pimpinella muscicola Hand.-Mazz. = Acronema muscicola (Hand.-Mazz.) Hand.-Mazz. ■

299475　Pimpinella nakaiana Kitag. = Pimpinella brachystyla Hand.-Mazz. ■

299476　Pimpinella nandensis C. Norman = Heracleum abyssinicum (Boiss.) C. Norman ■☆

299477　Pimpinella neglecta C. Norman;忽视茴芹■☆

299478　Pimpinella neumannii Engl. ex H. Wolff;纽曼茴芹■☆

299479　Pimpinella nigra Mill.;黑茴芹■☆

299480　Pimpinella niitakayamensis Hayata;台湾茴芹(玉山茴芹);Taiwan Pimpinella ■

299481　Pimpinella nikoensis Y. Yabe ex Makino et Nemoto;日光茴芹■☆

299482　Pimpinella nikoensis Y. Yabe ex Makino et Nemoto = Spuriopimpinella koreana (Y. Yabe) Kitag. ■

299483　Pimpinella nikoensis Y. Yabe ex Makino et Nemoto var. koreana Y. Yabe = Pimpinella koreana (Y. Yabe) Nakai ■

299484　Pimpinella nudicaulis Trautv.;裸茎茴芹■☆

299485　Pimpinella nyassica C. Norman = Pimpinella stadensis (Eckl. et Zeyh.) D. Dietr. ■☆

299486　Pimpinella nyingchiensis Z. H. Pan et K. Yao;林芝茴芹;Linzhi Pimpinella ■

299487　Pimpinella oreophila Hook. f.;喜山茴芹■☆

299488　Pimpinella oreophila Hook. f. var. kilimandscharica (Engl.) C. C. Towns.;基利茴芹■☆

299489　Pimpinella paludosa C. C. Towns.;沼泽茴芹■☆

299490　Pimpinella peregrina L.;洋茴芹■☆

299491　Pimpinella perrieri Sales et Hedge;佩里耶茴芹■☆

299492　Pimpinella petrosa Dawit;岩生茴芹■☆

299493　Pimpinella peucedanifolia Fisch.;前胡叶茴芹■☆

299494　Pimpinella peucedanifolia H. Boissieu = Tongoloa silaifolia (H. Boissieu) H. Wolff ■

299495　Pimpinella physotrichioides C. Norman;囊毛茴芹■☆

299496　Pimpinella pimpinellisimulacrum (Farille et S. B. Malla) Farille;喜马拉雅茴芹■☆

299497　Pimpinella pimpinelloides (H. Boissieu) M. Hiroe = Melanosciadium pimpinelloideum H. Boissieu ■

299498　Pimpinella pimpinelloides (Hochst.) H. Wolff;热非茴芹■☆

299499　Pimpinella platyphylla Hiern = Pimpinella huillensis Engl. ■☆

299500　Pimpinella pseudocaffra C. Norman;假开菲尔茴芹■☆

299501　Pimpinella pseudocandolleana H. Wolff = Pimpinella yunnanensis (Franch.) H. Wolff ■

299502　Pimpinella puberula (DC.) H. Boissieu;微毛茴芹;Pubescent Pimpinella ■

299503　Pimpinella purpurea (Franch.) H. Boissieu;紫瓣茴芹;Violetpetal Pimpinella ■

299504　Pimpinella pusilla (Pic. Serm.) M. Hiroe = Pimpinella pimpinelloides (Hochst.) H. Wolff ■☆

299505　Pimpinella radiatum W. W. Sm. = Acronema radiatum (W. W. Sm.) H. Wolff ■

299506　Pimpinella ramosa Schischk.;分枝茴芹■☆

299507　Pimpinella ranunculifolia Boiss.;毛茛叶茴芹■☆

299508　Pimpinella reenensis Rech. f. = Pimpinella stadensis (Eckl. et

Zeyh.）D. Dietr. ■☆

299509　Pimpinella refracta H. Wolff；下曲茴芹（澜沧茴芹）；Drooping Pimpinella ■

299510　Pimpinella renifolia H. Wolff；肾叶茴芹；Kidneyleaf Pimpinella ■

299511　Pimpinella rhodantha Boiss.；粉花茴芹■☆

299512　Pimpinella rhomboidea Diels；菱叶茴芹（鄂西独活）；Rhombicleaf Pimpinella ■

299513　Pimpinella rhomboidea Diels var. tenuiloba R. H. Shan et F. T. Pu；小菱叶茴芹；Smallrhombicleaf Pimpinella ■

299514　Pimpinella richardsiae C. C. Towns.；理查兹茴芹■☆

299515　Pimpinella rigidistyla C. C. Towns.；挺柱茴芹■☆

299516　Pimpinella rigidiuscula C. C. Towns.；稍坚挺茴芹■☆

299517　Pimpinella rivae Engl.；沟茴芹■☆

299518　Pimpinella robusta C. Norman ＝ Pimpinella huillensis Engl. ■☆

299519　Pimpinella robynsii C. Norman；罗宾斯茴芹■☆

299520　Pimpinella rockii H. Wolff；丽江茴芹；Rock Pimpinella ■

299521　Pimpinella rosthornii Diels ＝ Pternopetalum rosthornii（Diels）Hand. -Mazz.

299522　Pimpinella rubescens（Franch.）H. Wolff ex Hand. -Mazz.；少花茴芹；Fewflower Pimpinella ■

299523　Pimpinella saxifraga L.；虎耳草茴芹（普通茴芹，锐齿茴芹）；Bennet, Breakstone, Burnet Saxifrage, Burnet-saxifrage, Lesser Burnet, Old Man's Plaything, Saxifrage Pimpinella, Self-heal, Solidstem Burnet Saxifrage, Solid-stem Burnet-saxifrage ■☆

299524　Pimpinella saxifraga L. subsp. nigra（Mill.）Tzvelev；黑虎耳草茴芹；Solidstem Burnet Saxifrage ■☆

299525　Pimpinella saxifraga L. var. dissectifolia C. B. Clarke ＝ Vicatia wolffiana（H. Wolff ex Fedde）C. Norman ■☆

299526　Pimpinella scaberula（Franch.）H. Boissieu. ＝ Trachyspermum scaberulum（Franch.）H. Wolff ex Hand. -Mazz. ■

299527　Pimpinella scaberula（Franch.）H. Boissieu. var. ambrosiifolia（Franch.）H. Wolff. ＝ Trachyspermum scaberulum（Franch.）H. Wolff ex Hand. -Mazz. var. ambrosiifolium（Franch.）R. H. Shan ■

299528　Pimpinella scaberula（Franch.）H. Wolff ＝ Trachyspermum scaberulum（Franch.）H. Wolff ex Hand. -Mazz. ■

299529　Pimpinella scaberula H. Boissieu ＝ Trachyspermum scaberulum（Franch.）H. Wolff ex Hand. -Mazz. ■

299530　Pimpinella scaberula H. Boissieu var. ambrosiifolia（Franch.）H. Wolff ＝ Trachyspermum scaberulum（Franch.）H. Wolff ex Hand. -Mazz. var. ambrosiifolium（Franch.）R. H. Shan ■

299531　Pimpinella schimperi Dawit；欣珀茴芹■☆

299532　Pimpinella schlechteri H. Wolff；施莱茴芹■☆

299533　Pimpinella schweinfurthii Asch.；施韦茴芹■☆

299534　Pimpinella serra Franch. et Sav.；锯边茴芹；Serrate Pimpinella ■

299535　Pimpinella silaifolia H. Boissieu ＝ Tongoloa silaifolia（H. Boissieu）H. Wolff ■

299536　Pimpinella simensis（J. Gay ex A. Rich.）Benth. ＝ Oreoschimperella verrucosa（J. Gay ex A. Rich.）Rauschert ■☆

299537　Pimpinella simensis Benth. et Hook. f. ＝ Oreoschimperella verrucosa（J. Gay ex A. Rich.）Rauschert ■☆

299538　Pimpinella sinica Hance ＝ Pimpinella diversifolia DC. ■

299539　Pimpinella smithii H. Wolff；直立茴芹；Erect Pimpinella ■

299540　Pimpinella stadensis（Eckl. et Zeyh.）D. Dietr. 施塔德茴芹■☆

299541　Pimpinella stewardii（H. Wolff）M. Hiroe ＝ Tongoloa stewardii H. Wolff ■

299542　Pimpinella stewartii（Dunn）Nasir；斯氏茴芹■☆

299543　Pimpinella stocksii Boiss. ＝ Psammogeton stocksii（Boiss.）Nasir ■☆

299544　Pimpinella stolzii H. Wolff ＝ Pimpinella buchananii H. Wolff ■☆

299545　Pimpinella stricta H. Wolff ＝ Pimpinella smithii H. Wolff ■

299546　Pimpinella sutchuensis H. Boissieu ＝ Pimpinella henryi Diels

299547　Pimpinella sylvatica Hand. -Mazz.；木里茴芹（林间茴芹，戀理茴芹）；Woodland Pimpinella ■

299548　Pimpinella taeniophylla H. Boissieu ＝ Tongoloa taeniophylla（H. Boissieu）H. Wolff ■

299549　Pimpinella tagawai M. Hiroe；田代氏茴芹■

299550　Pimpinella tanakae（Franch. et Sav.）Diels ＝ Pternopetalum tanakae（Franch. et Sav.）Hand. -Mazz. ■

299551　Pimpinella taurica（Ledeb.）Steud.；克里木茴芹；Crimean Bumet-saxifrage, Crimean Burnet Saxifrage ■☆

299552　Pimpinella tenera（Wall.）Benth. et Hook. f. ex C. B. Clarke ＝ Acronema tenerum（Wall.）Edgew. ■

299553　Pimpinella tenuicaulis Baker；细茎茴芹■☆

299554　Pimpinella tenuissima C. Norman ＝ Pimpinella erythraeae Armari ■☆

299555　Pimpinella thellungiana H. Wolff；羊红膻（东北茴芹，六月寒，缺刻叶茴芹，羊洪膻）；Thellung. Pimpinella ■

299556　Pimpinella thellungiana H. Wolff var. tenuisecta Y. C. Chu；细裂东北茴芹■

299557　Pimpinella thellungiana H. Wolff var. tenuisecta Y. C. Chu ＝ Pimpinella cnidioides H. Pearson ex H. Wolff ■

299558　Pimpinella thellungiana H. Wolff var. tenuisecta Y. C. Chu ＝ Pimpinella thellungiana H. Wolff ■

299559　Pimpinella thyrsiflora H. Wolff；锥序茴芹；Paniculate Pimpinella ■

299560　Pimpinella thyrsiflora H. Wolff ＝ Pimpinella kingdon-wardii H. Wolff ■

299561　Pimpinella tibetanica H. Wolff；藏芹茴（藏芹）；Himalayas Pimpinella

299562　Pimpinella tilia M. Hiroe ＝ Tongoloa gracilis H. Wolff ■

299563　Pimpinella titanophila Woronow；白垩茴芹■☆

299564　Pimpinella tomiophylla（Woronow）Stank.；百里香叶茴芹■☆

299565　Pimpinella tonkinensis Cherm.；瘤果茴芹；Tumorfruit Pimpinella ■

299566　Pimpinella tragium Vill. subsp. lithophila（Schischk.）Tutin ＝ Pimpinella lithophila Schischk. ■☆

299567　Pimpinella transvaalensis H. Wolff；德兰士瓦茴芹■☆

299568　Pimpinella trichomanifolia（Franch.）Diels ＝ Pternopetalum trichomanifolium（Franch.）Hand. -Mazz. ■

299569　Pimpinella trifurcata H. Wolff ＝ Pimpinella buchananii H. Wolff subsp. septentrionalis C. C. Towns. ■☆

299570　Pimpinella tripartita Aitch. et Hemsl. ＝ Platytaenia kuramensis（Kitam.）Nasir ■☆

299571　Pimpinella triternata Diels；三出叶茴芹■

299572　Pimpinella tsusimensis（Y. Yabe）M. Hiroe et Constance ＝ Tilingia tsusimensis（Y. Yabe）Kitag. ■☆

299573　Pimpinella turcomanica Schischk.；土库曼茴芹■☆

299574　Pimpinella urbaniana Fedde ex H. Wolff；乌蒙茴芹；Wumeng Pimpinella ■

299575　Pimpinella valleculosa K. T. Fu；谷生茴芹；Valley Pimpinella ■

299576　Pimpinella villosa Schousb.；长柔毛茴芹■☆

299577　Pimpinella volkensii Engl. ＝ Pimpinella hirtella A. Rich. ■☆

299578　Pimpinella weishanensis R. H. Shan et F. T. Pu；巍山茴芹；Weishan Pimpinella ■

299579　Pimpinella weishanensis R. H. Shan et F. T. Pu ＝ Pimpinella kingdon-wardii H. Wolff ■

299580 Pimpinella welwitschii Engl. = Pimpinella huillensis Engl. ■☆

299581 Pimpinella welwitschii Engl. var. buchneri ? = Pimpinella huillensis Engl. ■☆

299582 Pimpinella welwitschii Engl. var. mechowii ? = Pimpinella huillensis Engl. ■☆

299583 Pimpinella wolffiana Fedde ex H. Wolff;思茅茴芹;Simao Pimpinella ■

299584 Pimpinella wolffiana Fedde ex H. Wolff = Pimpinella rockii H. Wolff ■

299585 Pimpinella xizangense R. H. Shan et F. T. Pu;西藏茴芹(多花茴芹);Xizang Pimpinella ■

299586 Pimpinella yunnanensis (Franch.) H. Wolff;云南茴芹(滇茴芹);Yunnan Pimpinella ■

299587 Pimpinella zernyi Gilli = Pimpinella buchananii H. Wolff ■☆

299588 Pimpinellaceae Bercht. et J. Presl = Apiaceae Lindl.(保留科名)●■

299589 Pimpinellaceae Bercht. et J. Presl = Umbelliferae Juss.(保留科名)■●

299590 Pinacantha Gilli(1959);板花草属☆

299591 Pinacantha porandica Gilli;板花草☆

299592 Pinaceae Adans. = Pinaceae Spreng. ex F. Rudolphi(保留科名)●

299593 Pinaceae Lindl. = Pinaceae Spreng. ex F. Rudolphi(保留科名)●

299594 Pinaceae Spreng. ex F. Rudolphi(1830)(保留科名);松科;Pine Family ●

299595 Pinacopodium Exell et Mendonça(1951);扁梗古柯属●☆

299596 Pinacopodium congolense (S. Moore) Exell et Mendonça;扁梗古柯●☆

299597 Pinacopodium gabonense (Cavaco et Normand) Normand et Cavaco;加蓬古柯●☆

299598 Pinalia Lindl.(1826);苹兰属■

299599 Pinalia Lindl. = Eria Lindl.(保留属名)■

299600 Pinalia acervata (Lindl.) Kuntze;钝叶苹兰(钝叶毛兰);Congested Eria, Congregate Hairorchis ■

299601 Pinalia acervata (Lindl.) Kuntze = Eria acervata Lindl. ■

299602 Pinalia albidotomentosa (Blume) Kuntze = Dendrolirium lasiopetalum (Willd.) S. C. Chen et J. J. Wood ■

299603 Pinalia amica (Rchb. f.) Kuntze;粗茎苹兰(粗茎毛兰,黄绿花毛兰,小脚筒兰,易湿毛兰);Confused Eria, Thickstem Hairorchis, Yellowish-greenflower Eria ■

299604 Pinalia amica (Rchb. f.) Kuntze = Eria amica Rchb. f. ■

299605 Pinalia andersonii (Hook. f.) Kuntze = Pinalia amica (Rchb. f.) Kuntze ■

299606 Pinalia bambusifolia (Lindl.) Kuntze = Callostylis bambusifolia (Lindl.) S. C. Chen et J. J. Wood ■

299607 Pinalia barbata (Lindl.) Kuntze = Eriodes barbata (Lindl.) Rolfe ■

299608 Pinalia bipunctata (Lindl.) Kuntze;双点苹兰(多节毛兰,双点毛兰);Pairspot Eria, Pairspot Hairorchis ■

299609 Pinalia calamifolia (Hook. f.) Kuntze = Eria pannea Lindl. ■

299610 Pinalia calamifolia (Hook. f.) Kuntze = Mycaranthes pannea (Lindl.) S. C. Chen et J. J. Wood ■

299611 Pinalia conferta (S. C. Chen et Z. H. Tsi) S. C. Chen et J. J. Wood;密苞苹兰(密苞毛兰);Densebract Eria, Densebract Hairorchis ■

299612 Pinalia confusa (Hook. f.) Kuntze = Eria amica Rchb. f. ■

299613 Pinalia confusa (Hook. f.) Kuntze = Pinalia amica (Rchb. f.) Kuntze ■

299614 Pinalia copelandii (Leav.) W. Suarez et Cootes;台湾苹兰■

299615 Pinalia dasyphylla (E. C. Parish et Rchb. f.) Kuntze = Trichotosia dasyphylla (Parl. et Rchb. f.) Kraenzl. ■

299616 Pinalia donnaiensis (Gagnep.) S. C. Chen et J. J. Wood;中越苹兰■

299617 Pinalia excavata (Lindl.) Kuntze;反苞苹兰(反苞毛兰);Concave Eria, Concave Hairorchis ■

299618 Pinalia fragrans (Rchb. f.) Kuntze = Eria javanica (Sw.) Blume ■

299619 Pinalia graminifolia (Lindl.) Kuntze;禾叶苹兰(禾叶毛兰,禾叶墨斛,禾颐苹兰);Grassleaf Eria, Grassleaf Hairorchis ■

299620 Pinalia longlingensis (S. C. Chen) S. C. Chen et J. J. Wood;龙陵苹兰(龙陵毛兰);Longling Eria, Longling Hairorchis ■

299621 Pinalia marginata (Rolfe) Kuntze = Cylindrolobus marginatus (Rolfe) S. C. Chen et J. J. Wood ■

299622 Pinalia microphylla (Blume) Kuntze = Trichotosia microphylla Blume ■

299623 Pinalia muscicola (Lindl.) Kuntze = Conchidium muscicola (Lindl.) Rauschert ■

299624 Pinalia obvia (W. W. Sm.) S. C. Chen et J. J. Wood;长苞苹兰(长苞毛兰);Longbract Eria, Longbract Hairorchis ■

299625 Pinalia ovata (Lindl.) W. Suarez et Cootes;大脚筒(大脚筒兰,卵苞毛兰);Largefoottube Eria, Largefoottube Hairorchis, Ovatebract Eria ■

299626 Pinalia pachyphylla (Aver.) S. C. Chen et J. J. Wood;厚叶苹兰(厚叶毛兰);Thickleaf Eria, Thickleaf Hairorchis ■

299627 Pinalia paniculata (Lindl.) Kuntze = Mycaranthes floribunda (D. Don) S. C. Chen et J. J. Wood ■

299628 Pinalia pannea (Lindl.) Kuntze = Eria pannea Lindl. ■

299629 Pinalia pannea (Lindl.) Kuntze = Mycaranthes pannea (Lindl.) S. C. Chen et J. J. Wood ■

299630 Pinalia pubescens (Hook.) Kuntze = Dendrolirium lasiopetalum (Willd.) S. C. Chen et J. J. Wood ■

299631 Pinalia pulvinata (Lindl.) Kuntze = Trichotosia pulvinata (Lindl.) Kraenzl. ■

299632 Pinalia pusilla (Griff.) Kuntze = Conchidium pusillum Griff. ■

299633 Pinalia quinquelamellosa (Ts. Tang et F. T. Wang) S. C. Chen et J. J. Wood;五脊苹兰(五脊毛兰);Fiverib Eria, Fiverib Hairorchis ■

299634 Pinalia retroflexa (Lindl.) Kuntze = Pinalia ovata (Lindl.) W. Suarez et Cootes ■

299635 Pinalia rosea (Lindl.) Kuntze = Cryptochilus roseus (Lindl.) S. C. Chen et J. J. Wood ■

299636 Pinalia rosea (Lindl.) Kuntze = Eria rosea Lindl. ■

299637 Pinalia sinica (Lindl.) Kuntze = Eria sinica (Lindl.) Lindl. ■

299638 Pinalia sinica (Lindley) Kuntze = Conchidium pusillum Griff. ■

299639 Pinalia spicata (D. Don) S. C. Chen et J. J. Wood;密花苹兰(密花毛兰,穗花毛兰);Denseflower Hairorchis, Spike Eria ■

299640 Pinalia stellata (Lindl.) Kuntze = Eria javanica (Sw.) Blume ■

299641 Pinalia stricta (Lindl.) Kuntze;鹅白苹兰(鹅白毛兰);Goosewhite Eria, Goosewhite Hairorchis ■

299642 Pinalia striolata (Rchb. f.) Kuntze = Eria javanica (Sw.) Blume ■

299643 Pinalia szetschuanica (Schltr.) S. C. Chen et J. J. Wood;马齿苹兰(马齿毛兰);Sichuan Eria, Sichuan Hairorchis, Szechwan Eria ■

299644 Pinalia tomentosa（J. König）Kuntze = Dendrolirium tomentosum（J. König）S. C. Chen et J. J. Wood ■

299645 Pinalia tomentosa（K. D. König）Kuntze = Eria tomentosa（K. D. König）Hook. f. ■

299646 Pinalia ustulata（E. C. Parish et Rchb. f.）Kuntze = Porpax ustulata（Parl. et Rchb. f.）Rolfe ■

299647 Pinalia vittata（Lindl.）Kuntze = Eria vittata Lindl. ■

299648 Pinalia yunnanensis（S. C. Chen et Z. H. Tsi）S. C. Chen et J. J. Wood;滇南苹兰(滇南毛兰);S. Yunnan Eria,S. Yunnan Hairorchis ■

299649 Pinanga Blume（1838）;山槟榔属(类槟榔属);Pinang Palm,Pinanga,Pinangapalm,Pinanga-palm ●

299650 Pinanga acaulis Ridl.;无茎山槟榔;Stemless Pinanga ●☆

299651 Pinanga adangensis Ridl.;阿当山槟榔;Adang Pinanga ●☆

299652 Pinanga baviensis Becc.;山槟榔;Bavi Pinanga,Wild Pinangapalm ●☆

299653 Pinanga baviensis Becc. = Pinanga discolor Burret ●

299654 Pinanga beccariana Furtado;白卡山槟榔;Beccar Pinanga ●☆

299655 Pinanga caesia Blume;红冠山槟榔●☆

299656 Pinanga canina Becc.;犬山山槟榔;Dog Pinanga ●☆

299657 Pinanga chinensis Becc.;华山槟榔(华山竹);China Pinangapalm,Chinese Pinangapalm,Chinese Pinanga-palm ●

299658 Pinanga coronata Blume;冠山槟榔●☆

299659 Pinanga discolor Burret;变色山槟榔(假山葵,山槟榔,异色山槟榔);Changecolor Pinangapalm,Diversecolor Pinangapalm,Diversecolor Pinanga-palm,Particoloured Pinanga ●

299660 Pinanga disticha Blume;二列山槟榔●☆

299661 Pinanga fruticans Ridl.;灌木状山槟榔;Shrubby Pinanga ●☆

299662 Pinanga gracilis（Roxb.）Blume;纤细山槟榔;Fine Pinanga-palm,Slender Pinanga,Slender Pinanga-palm,Thin Pinangapalm ●☆

299663 Pinanga hexasticha（Kurz）Scheff.;六列山槟榔;Hexastichous Pinanga-palm,Sixlow Pinangapalm ●

299664 Pinanga kuhlii Blume;库里山槟榔;Ivory Cane Palm,Kuhl Palm ●☆

299665 Pinanga limosa Ridl.;沼泽山槟榔;Marshy Pinanga ●☆

299666 Pinanga macroclada Burret;长枝山竹;Long-branch Pinanga,Long-branched Pinanga-palm,Longtwig Pinangapalm ●

299667 Pinanga maculata Porte;斑点山槟榔●☆

299668 Pinanga malaiana（Mart.）Scheff.;马来山槟榔;Malay Pinanga ●☆

299669 Pinanga paradoxa Scheff.;奇怪山槟榔;Paradox Pinanga ●☆

299670 Pinanga patula Blume;开展山槟榔;Spreading Pinanga ●☆

299671 Pinanga pectinata Becc.;篦形山槟榔;Pectinate Pinanga ●☆

299672 Pinanga perakensis Becc.;波拉克山槟榔;Perak Pinanga ●☆

299673 Pinanga polymorpha Becc.;多型山槟榔;Polymorphous Pinanga ●☆

299674 Pinanga scortechinii Becc.;斯考氏山槟榔(斯考里山槟榔);Scortechin Pinanga ●☆

299675 Pinanga simplicifrons（Miq.）Becc.;单叶山槟榔;Simpleleaf Pinanga ●☆

299676 Pinanga sinii Burret;燕尾山槟榔(瑶山山槟榔);Sin Pinangapalm,Sin Pinanga-palm,Swallowtail Pinangapalm ●

299677 Pinanga subintegra Ridl.;近全缘山槟榔;Subentire Pinanga ●☆

299678 Pinanga subruminata Becc.;近嚼烂状山槟榔;Subruminate Pinanga ●☆

299679 Pinanga tashiroi Hayata;兰屿山槟榔(山槟榔);Lanyu Pinanga,Lanyu Pinangapalm,Lanyu Pinanga-palm ●

299680 Pinanga viridis Burret;绿色山槟榔(绿山槟榔);Green Pinanga,Green Pinangapalm,Green Pinanga-palm ●

299681 Pinanga wrayi Furtado;沃瑞氏山槟榔;Wray Pinanga ●☆

299682 Pinarda Vell. = Micranthemum Michx.（保留属名）■☆

299683 Pinardia Cass. = Chrysanthemum L.（保留属名）■●

299684 Pinardia Neck. = Aster L. ●■

299685 Pinaria（DC.）Rchb. = Matthiola W. T. Aiton（保留属名）■●

299686 Pinaropappus Less.（1832）;岩莴苣属(污毛菊属);Rocklettuce ■☆

299687 Pinaropappus parvus S. F. Blake;小岩莴苣;Small rocklettuce ■☆

299688 Pinaropappus roseus（Less.）Less.;白岩莴苣(红毛污毛菊,红色潘纳菊);White Dandelion,White Rocklettuce ■☆

299689 Pinaropappus roseus（Less.）Less. var. foliosus Shinners = Pinaropappus roseus（Less.）Less. ■☆

299690 Pinarophyllon Brandegee(1914);劣叶茜属 ☆

299691 Pinarophyllon flavum Brandegee;劣叶茜 ☆

299692 Pinasgelon Raf. = Cnidium Cusson ex Juss. ■

299693 Pinasgelon monnieri（L.）Raf. = Cnidium monnieri（L.）Cusson ■

299694 Pincecnitia Hort. ex Lem. = Pincenectitia Hort. ex Lem. ●■☆

299695 Pincecnitia Lem. = Pincenectitia Hort. ex Lem. ●■☆

299696 Pincenectia Hort. ex Lem. = Pincenectitia Hort. ex Lem. ●■☆

299697 Pincenectia Lem. = Pincenectitia Hort. ex Lem. ●■☆

299698 Pincenectitia Hort. ex Lem. = Nolina Michx. ●☆

299699 Pincenictitia Baker = Pincenectitia Hort. ex Lem. ●■☆

299700 Pincinectia Hort. ex Lem. = Pincenectitia Hort. ex Lem. ●■☆

299701 Pinckneya Michx.（1803）;黄疟树属(宾克莱木属,黄疟属);Fever Tree,Pinckneya ●☆

299702 Pinckneya braceata（Bartram）Raf.;猩猩木;Fever Tree,Georgia Bark Tree,Poinsettia Tree ●☆

299703 Pinckneya pubens Michx.;黄疟树(宾克莱木);Fever Tree,Georgia Bark,Pinckneya ●☆

299704 Pinckneya pubens Michx. = Pinckneya braceata（Bartram）Raf. ●☆

299705 Pinda P. K. Mukh. et Constance(1986);印度草属■☆

299706 Pinda concanensis（Dalzell）P. K. Mukh. et Constance;印度草■☆

299707 Pindarea Barb. Rodr. = Attalea Kunth ●☆

299708 Pinea Opiz = Pinus L. ●

299709 Pinea Wolf = Pinus L. ●

299710 Pineda Ruiz et Pav.（1794）;安第斯大风子属●☆

299711 Pineda incana Ruiz et Pav.;安第斯大风子●☆

299712 Pinelea Willis = Pinelia Lindl. ■☆

299713 Pinelia Lindl.（1853）;皮内尔兰属■☆

299714 Pinelia alticola Garay et Dunst.;高原皮内尔兰■☆

299715 Pinelia hypolepta Lindl.;皮内尔兰■☆

299716 Pinelianthe Rauschert = Pinelia Lindl. ■☆

299717 Pinellia Ten.（1839）（保留属名）;半夏属;Halfummer,Pinellia ■

299718 Pinellia browniana Dunn = Pinellia cordeta N. E. Br. ■

299719 Pinellia cordeta N. E. Br.;心叶半夏(斑叶滴水珠,滴水珠,独角莲,独龙珠,独叶一枝花,山半夏,蛇珠,石半夏,石里开,石蜘蛛,水半夏,水滴珠,天灵芋,岩芋,岩珠,野慈姑,一滴珠,一粒珠,制蛇子);Cordare Halfummer,Cordare Pinellia ■

299720 Pinellia integrifolia N. E. Br.;石蜘蛛(白铃子,一面锣);Stonespider ■

299721 Pinellia pedatisecta Schott;掌叶半夏(半夏,半夏子,大三步跳,滇半夏,独败家子,独角莲,独脚莲,狗爪半夏,虎掌,虎掌南星,绿芋子,麻芋果,麻芋子,南星,鸟足叶半夏,田南星);Palmata-leaf Pinellia,Tigerpalm ■

299722 Pinellia peltata C. P'ei;盾叶半夏(白滴水珠,白岩芋);Peltateleaf Halfummer,Pinellia ■

299723 Pinellia polyphylla S. L. Hu;大半夏;Leafy Pinellia ■

299724　Pinellia ternata（Thunb.）Breitenb.；半夏（白傍儿子，半子，地巴豆，地茨菇，地慈姑，地雷公，地文，地星，地鹞鸪，地珠半夏，法半夏，泛石子，戈制夏，狗芋头，和姑，尖叶半夏，姜半，扣子莲，老瓜蒜，老鸹头，老鸹眼，老和尚扣，老和尚头，老黄咀，老黄嘴，老鸦头，老鸦眼，老鸦芋头，老捏嘴豆子，裂刀菜，麻草子，麻玉果，麻芋果，麻芋子，清半夏，球半夏，三不掉，三步魂，三步跳，三角草，三棱草，三片叶，三兴草，三叶半夏，三叶老，三叶头草，生半夏，示姑，守田，水玉，宋半夏，痰宫劈历，天老星，天落星，田里心，土半夏，无心菜，仙半夏，小天老星，小天南星，蝎子草，燕子尾，羊眼半夏，洋梨头，药狗丹，药狗蛋，野半夏，野芋头，制半夏，雉毛奴邑，雉毛邑，珠半夏，捉嘴豆子）；Crowdipper，Halfummer，Ternate Pinellia ■

299725　Pinellia ternata（Thunb.）Breitenb. f. angustata（Schott）Makino；线叶半夏（狭叶半夏）；Linearifolious Pinellia ■☆

299726　Pinellia ternata（Thunb.）Breitenb. f. atropurpurea（Makino）Ohwi；紫苞半夏■☆

299727　Pinellia ternata（Thunb.）Breitenb. f. subcuspidata Honda；亚尖半夏■☆

299728　Pinellia tripartita（Blume）Schott；三裂半夏；Trilobe Pinellia ■

299729　Pinellia tripartita（Blume）Schott f. atropurpurea（Makino）Ohwi；暗紫三裂半夏■☆

299730　Pinellia tuberifera Ten. = Pinellia ternata（Thunb.）Breitenb. ■

299731　Pinellia tuberifera Ten. var. pedatisecta（Schott）Engl. = Pinellia pedatisecta Schott ■

299732　Pinellia wawrae Engl. = Pinellia pedatisecta Schott ■

299733　Pinellia yaoluopingensis X. H. Guo et X. L. Liu；鹞落坪半夏；Yaoluo Pinellia ■

299734　Pinga Widjaja（1997）；新几内亚禾属■☆

299735　Pingraea Cass. = Baccharis L.（保留属名）●■☆

299736　Pinguicula L.（1753）；捕虫堇属（捕虫堇菜属）；Bog Violet，Butter Wort，Butterwort ■

299737　Pinguicula alpina L.；高山捕虫堇（捕虫草，捕虫堇）；Alpine Butterwort，Mountain Butterwort ■☆

299738　Pinguicula antarctica Vahl；南极长距捕虫堇■☆

299739　Pinguicula caerulea Walter；堇色捕虫堇；Violet Butterwort ■☆

299740　Pinguicula caudata Schltdl.；长距捕虫堇■☆

299741　Pinguicula colimensis McVaugh et Mickel；墨西哥捕虫堇■☆

299742　Pinguicula elatior Michx.；高捕虫堇（较高捕虫堇）■☆

299743　Pinguicula glandulosa Trautv. et C. A. Mey.；腺点捕虫堇■☆

299744　Pinguicula grandiflora Lam.；大花捕虫堇；Great Butterwort，Irish Butterwort，Large-flowered Butterwort ■☆

299745　Pinguicula gypsicola Brandegee；石生捕虫堇■☆

299746　Pinguicula heterophylla Benth.；异叶捕虫堇■☆

299747　Pinguicula hirtiflora Ten.；毛花捕虫堇■☆

299748　Pinguicula ionantha Godfrey；美国捕虫堇■☆

299749　Pinguicula kondoi Casper；近藤氏捕虫堇■☆

299750　Pinguicula longifolia Ramond ex DC.；长叶捕虫堇■☆

299751　Pinguicula lusitanica L.；西部捕虫堇；Pale Butterwort，Western Butterwort ■☆

299752　Pinguicula lutea Walter；黄花捕虫堇；Yellow Butterwort ■☆

299753　Pinguicula macroceras Pall. ex Link = Pinguicula vulgaris L. var. macroceras（Pall. ex Link）Herder ■☆

299754　Pinguicula ramosa Miyoshi ex Yatabe；庚申草■☆

299755　Pinguicula ramosa Miyoshi f. albiflora Komiya et C. Shibata；白花庚申草■☆

299756　Pinguicula variegata Turcz.；斑点捕虫堇■☆

299757　Pinguicula villosa L.；北捕虫堇；North Butterwort ■

299758　Pinguicula villosa L. var. ramosa？ = Pinguicula ramosa Miyoshi ex Yatabe ■☆

299759　Pinguicula vulgaris L.；捕虫堇（紫花捕虫堇）；Beanweed，Bog Violet，Butter Plant，Butter Root，Butter Wort，Butterwort，Clowns，Common Butter Wort，Common Butterwort，Earning Grass，Eccle Grass，Eccle-grass，Ekkel-girse，Flycatcher，Luss-ny-ollee，Red Rot，Rot-grass，Sheep-root，St. Patrick's Spit，St. Patrick's Staff，Steepgrass，Steepweed，Steepwort，Thickening Grass，Violet Butterwort，White Rot，White Sincle，Whiteroot，Yirnin-girse，Yorkshire Sanicle ■☆

299760　Pinguicula vulgaris L. = Pinguicula alpina L. ■

299761　Pinguicula vulgaris L. subsp. macroceras（Pall. ex Link）Calder et Taylor = Pinguicula vulgaris L. var. macroceras（Pall. ex Link）Herder ■☆

299762　Pinguicula vulgaris L. var. americana A. Gray = Pinguicula vulgaris L. ■☆

299763　Pinguicula vulgaris L. var. floribunda S. Watan. et A. Takeda；繁花捕虫堇■☆

299764　Pinguicula vulgaris L. var. macroceras（Pall. ex Link）Herder；大角捕虫堇■☆

299765　Pinguicula vulgaris L. var. macroceras（Pall. ex Link）Herder f. albiflora Komiya；白花大角捕虫堇■☆

299766　Pinguiculaceae Dumort.；捕虫堇科■

299767　Pinguiculaceae Dumort. = Lentibulariaceae Rich.（保留科名）■

299768　Pinguin Adans. = Bromelia L. ■☆

299769　Pinillosa Ossa ex DC. = Pinillosia Ossa ■☆

299770　Pinillosia Ossa（1836）；佳乐菊属■☆

299771　Pinillosia berteri Urb.；佳乐菊■☆

299772　Pinknea Pers. = Pinckneya Michx. ●☆

299773　Pinkneya Raf. = Pinknea Pers. ●☆

299774　Pinochia M. E. Endress et B. F. Hansen = Echites P. Browne ●☆

299775　Pinochia M. E. Endress et B. F. Hansen（2007）；美洲蛇木属●☆

299776　Pinosia Urb.（1930）；古巴石竹属■☆

299777　Pinosia glandulosa Alain；古巴石竹■☆

299778　Pintoa Gay（1846）；平托蒺藜属●☆

299779　Pintoa chilensis Gay；平托蒺藜●☆

299780　Pinus L.（1753）；松属；Gopher Wood，Pine ●

299781　Pinus × densi-thunbergii Uyeki；通贝里松●☆

299782　Pinus abies L. = Picea abies（L.）H. Karst. ●

299783　Pinus abies L. f. schrenkiana（Fisch. et C. A. Mey.）Voss = Picea schrenkiana Fisch. et C. A. Mey. ●

299784　Pinus alba Aiton = Picea glauca（Moench）Voss ●

299785　Pinus alba L.；银白松（白松）；White Pine ●☆

299786　Pinus albicaulis Engelm.；美国白皮松（高山白皮松）；Alpine Whitebark Pine，Creeping Pine，Scrub Pine，White Stem Pine，Whitebark Pine，White-bark Pine ●☆

299787　Pinus albicaulis Engelm. 'Nana'；矮生美国白皮松●☆

299788　Pinus albicaulis Engelm. 'Noble's Darf'；诺贝尔矮美国白皮松●☆

299789　Pinus amamiana Koidz.；奄美岛松；Amami Pine ●☆

299790　Pinus amamiana Koidz. = Pinus armandii Franch. ●

299791　Pinus anhweiensis W. C. Cheng et Y. W. Law = Pinus dabeshanensis W. C. Cheng et Y. W. Law ●

299792　Pinus apacheca Lemmon = Pinus engelmannii Carrière ●☆

299793　Pinus argyi Lemée et H. Lév. = Pinus massoniana Lamb. ●

299794　Pinus argyi Lemée et H. Lév. var. longevagians H. Lév. = Pinus massoniana Lamb. ●

299795   Pinus aristata Engelm.；刺果松（刚果松，芒松，硬毛松）；Balfour Pine, Bristcone Pine, Bristlecone Pine, Bristle-cone Pine, Colorado Bristlecone Pine, Eastern Brisdecone Pine, Foxtail Pine, Hickory Pine, Rocky Mountain Bristlecone Pine ●☆

299796   Pinus aristata Engelm. var. longaeva（D. K. Bailey）Little = Pinus longaeva D. K. Bailey ●☆

299797   Pinus arizonica Engelm.；亚利桑那松（亚马逊松）；Arizona Pine, Arizona Ponderosa Pine ●☆

299798   Pinus arizonica Engelm. = Pinus ponderosa Douglas ex C. Lawson et C. Lawson arizonica（Engelm.）Shaw ●☆

299799   Pinus armandii Franch.；华山松（白皮刺毛松，白杉，白松，枞，果松，牛松，青松，松柏，榀枞树，五须松，五叶松）；Armand Pine, Armand's Pine, China Armand Pine, China White Pine, Chinese White Pine, David Pine, David's Pine, Huashan Pine, Jack Pine ●

299800   Pinus armandii Franch. var. amamiana（Koidz.）Hatus. = Pinus armandii Franch. ●

299801   Pinus armandii Franch. var. dabeshanensis（W. C. Cheng et Y. W. Law）Silba = Pinus fenzeliana Hand. -Mazz. var. dabeshanensis（W. C. Cheng et Y. W. Law）L. K. Fu et Nan Li ◇

299802   Pinus armandii Franch. var. mastersiana（Hayata）Hayata；台湾果松（台湾华山松）；Masters Pine ●

299803   Pinus armandii Franch. var. mastersiana（Hayata）Hayata = Pinus armandii Franch. ●

299804   Pinus aspleniifolius ?；铁线蕨叶松；Celery-top Pine ●☆

299805   Pinus atlantica Endl. = Cedrus atlantica（Endl.）Manetti ex Carrière ●

299806   Pinus attenuata Lemmon；瘤果松（莱伯克松，窄果松）；Knobcone Pine ●☆

299807   Pinus australis F. Michx.；大王松；Australian Pine ●

299808   Pinus australis F. Michx. = Pinus palustris Mill. ●

299809   Pinus ayacahuite C. Ehrenb. ex Schltdl.；墨西哥松（短翅墨西哥白松，墨西哥白松）；Mexican Pine, Mexican White Pine, Mexico Pine ●☆

299810   Pinus ayacahuite C. Ehrenb. var. brachyptera Shaw = Pinus strobiformis Engelm. ●☆

299811   Pinus ayacahuite C. Ehrenb. var. reflexa（Engelm.）Voss = Pinus strobiformis Engelm. ●☆

299812   Pinus ayacahuite C. Ehrenb. var. strobiformis（Engelm.）Lemmon = Pinus strobiformis Engelm. ●☆

299813   Pinus balfouriana A. Murray；狐尾松（巴尔弗氏松，巴耳弗氏松）；Balfour Pine, Foxtail Pine ●☆

299814   Pinus balfouriana A. Murray subsp. austrina R. J. Mastrog. et J. D. Mastrog. = Pinus balfouriana A. Murray ●☆

299815   Pinus balfouriana A. Murray var. austrina（R. J. Mastrog. et J. D. Mastrog.）Silba = Pinus balfouriana A. Murray ●☆

299816   Pinus balfouriana Grev. et Balf. var. aristata（Engelm.）Engelm. = Pinus aristata Engelm. ●☆

299817   Pinus balsamea L. = Abies balsamea（L.）Mill. ●☆

299818   Pinus banksiana Lamb.；北美短叶松（班克松，邦克松，短叶松，杰克松，美国短叶松，盘古斯松）；Banks Pine, Banksian Pine, Gray Pine, Jack Pine, Jackpine, Joseph Bank Pine, Princess Pine, Scrub Pine ●

299819   Pinus bhutanica Grierson, D. G. Long et C. N. Page；不丹松 ●

299820   Pinus bifida（Siebold et Zucc.）Antoine = Abies firma Siebold et Zucc. ●

299821   Pinus bolanderi Douglas ex L. Parl. = Pinus contorta Douglas ex Loudon ●☆

299822   Pinus brachyptera Engelm.；落基山黄松；Rock Pine, Rocky Mountain Yellow Pine ●☆

299823   Pinus brachyptera Engelm. = Pinus ponderosa Douglas ex C. Lawson et C. Lawson var. scopulorum Engelm. ●☆

299824   Pinus bracteata D. Don = Abies bracteata（D. Don）Poit. ●

299825   Pinus brevispica Hayata = Pinus taiwanensis Hayata ●

299826   Pinus brunoniana Wall. = Tsuga dumosa（D. Don）Eichler ●

299827   Pinus brutia Ten.；土耳其松（地中海松，南意松）；Calabrian Pine, Erect Cone Pine, Erect-cone Aleppe Pine, Erect-cone Aleppo Pine, Turkey Pine, Turkish Pine ●☆

299828   Pinus brutia Ten. var. eldarica（Medw.）Silba；阿富汗松；Afghan Pine, Mondell Pine ●☆

299829   Pinus bungeana Zucc. = Pinus bungeana Zucc. ex Endl. ●

299830   Pinus bungeana Zucc. ex Endl.；白皮松（白骨松，白果松，白里松，白松，虎皮松，蟠龙松，三针松，蛇皮松）；Bunge Pine, Chinese Whitebark Pine, Lace Pine, Lacebark Pine, Lace-bark Pine ●

299831   Pinus californiarum D. K. Bailey = Pinus monophylla Torr. et Frém. ●

299832   Pinus canadensis L. = Tsuga canadensis（L.）Carrière ●☆

299833   Pinus canaliculata Miq. = Pinus massoniana Lamb. ●

299834   Pinus canariensis C. Sm. = Pinus canariensis C. Sm. ex DC. ●☆

299835   Pinus canariensis C. Sm. ex DC.；加那利松（加拿利刺松，加那列松，卡内里松）；Canary Island Pine, Canary Pine, Caribbean Pine, Tea Wood ●☆

299836   Pinus caribaea Morelet；加勒比松（鞭子松，古巴松，加籇比松，加籇比油松）；Bahama Pitchpine, Bahamas Pitch Pine, Caribbean Pine, Caribbean Pitch Pine, Cuban Pine, Slash Pine ●

299837   Pinus caribaea Morelet var. bahamensis（Griseb.）Barratt et Golfari；巴哈马加勒比松（巴哈马加籇比松）；Bahama Pine ●☆

299838   Pinus caribaea Morelet var. hondurensis（Sénécl.）Barratt et Golfari；洪都拉斯加勒比松（洪都拉斯加籇比松）；Honduras Pine ●

299839   Pinus cavaleriei Lemée et H. Lév. = Pinus massoniana Lamb. ●

299840   Pinus cedrus L. = Cedrus libani A. Rich. ●☆

299841   Pinus cedrus L. var. atlantica Parl. = Cedrus atlantica（Endl.）Carrière ●

299842   Pinus cembra L.；瑞士石松（瑞士松，瑞士五针松，瑞士岩松，生松）；Alpine Pine, Arolla Pine, Cembran Pine, Cembrian Pine, Nut Pine, Siberian Cedar, Swiss Stone Pine, Swiss Stone-pine ●☆

299843   Pinus cembra L. ‘Chlorocarpa’；绿果瑞士五针松 ●☆

299844   Pinus cembra L. ‘Kairamo’；凯拉莫瑞士五针松 ●☆

299845   Pinus cembra L. ‘Pendula’；垂枝瑞士五针松 ●☆

299846   Pinus cembra L. subsp. sibirica（Du Tour）Krylov = Pinus sibirica（Loudon）Mayr ●

299847   Pinus cembra L. var. pumila Pall. = Pinus pumila（Pall.）Regel ●

299848   Pinus cembra L. var. pygmaea Loudon = Pinus pumila（Pall.）Regel ●

299849   Pinus cembra L. var. sibirica（Du Tour）G. Don = Pinus sibirica（Loudon）Mayr ●

299850   Pinus cembra L. var. sibirica Loudon = Pinus sibirica（Loudon）Mayr ●

299851   Pinus cembroides Zucc.；墨西哥果松（墨西哥矮松，墨西哥黑松，墨西哥岩松）；Arizona Pine, Mexican Nut Pine, Mexican Pine, Mexican Pinyon, Mexican Pinyon Pine, Mexican Stone Pine, Mexico Nut Pine, Mexico Pine, Nut Pine, Pinyon, Stoneseed Pinyon ●☆

299852   Pinus cembroides Zucc. var. bicolor Little = Pinus cembroides Zucc. ●☆

299853   Pinus cembroides Zucc. var. edulis（Engelm.）Voss = Pinus edulis Engelm. ●☆

299854　Pinus cembroides Zucc. var. monophylla（Torr. et Frém.）Voss；墨西哥单针松；Nut Pine，Singleleaf Pinyon ●☆

299855　Pinus cembroides Zucc. var. monophylla Voss ＝ Pinus cembroides Zucc. var. monophylla（Torr. et Frém.）Voss ●☆

299856　Pinus cembroides Zucc. var. monophylla Voss ＝ Pinus monophylla Torr. et Frém. ●

299857　Pinus cembroides Zucc. var. parryana Voss；墨西哥四针松（四针松）；Parry Pinyon ●☆

299858　Pinus cembroides Zucc. var. parryana Voss ＝ Pinus quadrifolia Parl. ex Sudw. ●☆

299859　Pinus cembroides Zucc. var. remota Little ＝ Pinus cembroides Zucc. ●☆

299860　Pinus chiapensis（Martínez）Andresen ＝ Pinus strobus L. ●

299861　Pinus chihuahuana Engelm.；济华华松；Chihuahua Pine ●☆

299862　Pinus chihuahuana Engelm. ＝ Pinus leiophylla Schltdl. et Champ. var. chihuahuana（Engelm.）Shaw ●☆

299863　Pinus chylla Lodd. ＝ Picea wallichiana A. B. Jacks. ●☆

299864　Pinus clausa（Chapm. ex Engelm.）Sarg.；美国沙松（美国沙地松，沙地松，沙松，砂地松）；Sand Pine，Scub Pine，Spruce Pine ●☆

299865　Pinus clausa（Chapm. ex Engelm.）Sarg. var. immuginata D. B. Ward ＝ Pinus clausa（Chapm. ex Engelm.）Sarg. ●☆

299866　Pinus clausa（Chapm.）Vasey ＝ Pinus clausa（Chapm. ex Engelm.）Sarg. ●☆

299867　Pinus clusiana Clemente ＝ Pinus nigra J. F. Arnold ●

299868　Pinus clusiana Clemente subsp. mauretanica（Maire et Peyerimh.）O. Schwarz ＝ Pinus nigra J. F. Arnold subsp. mauretanica（Maire et Peyerimh.）Heywood ●☆

299869　Pinus contorta Douglas ex Loudon；扭叶松（扭松，小干松，旋叶松）；Beach Pine，Coast Pine，Lodgepole Pine，Screw Pine，Shore Pine，Shorepine，Tamarack Pine ●☆

299870　Pinus contorta Douglas ex Loudon 'Spaan's Dwarf'；斯潘矮种旋叶松●☆

299871　Pinus contorta Douglas ex Loudon subsp. bolanderi（Parl.）Critchf. ＝ Pinus contorta Douglas ex Loudon ●☆

299872　Pinus contorta Douglas ex Loudon subsp. latifolia（Engelm.）Critchf. ＝ Pinus contorta Douglas ex Loudon var. latifolia Engelm. ex S. Watson ●☆

299873　Pinus contorta Douglas ex Loudon var. bolanderi Lemmon ＝ Pinus contorta Douglas ex Loudon ●☆

299874　Pinus contorta Douglas ex Loudon var. latifolia Engelm. ex S. Watson；宽叶扭叶松（宽叶旋叶松）；Lodgepole Pine，Rocky Mountain Lodgepole Pine ●☆

299875　Pinus cooperi C. E. Blanco；墨西哥山松；Mexican Mountain Pine ●☆

299876　Pinus coulteri D. Don；大果松（歌德松）；Bigcone Pine，Bigcone Pine，Coulter Pine，Coulter's Pine，Pitch Pine ●☆

299877　Pinus crassicorticea Y. C. Zhong et Critchf.；拉雅松●

299878　Pinus crassicorticea Y. C. Zhong et Critchf. ＝ Pinus massoniana Lamb. ●

299879　Pinus culminicola Andresen et Beaman；高寒松；Potosi Pinyon ●☆

299880　Pinus dabeshanensis W. C. Cheng et Y. W. Law ＝ Pinus armandii Franch. ●

299881　Pinus dabeshanensis W. C. Cheng et Y. W. Law ＝ Pinus fenzeliana Hand.-Mazz. var. dabeshanensis（W. C. Cheng et Y. W. Law）L. K. Fu et Nan Li ●◇

299882　Pinus dahurica Fisch. ex Turcz. ＝ Larix gmelinii（Rupr.）Rupr. ●

299883　Pinus dammara Lamb. ＝ Agathis dammara（Lamb.）Rich. et A. Rich. ●

299884　Pinus deflexa Torr. ＝ Pinus jeffreyi Murray ●☆

299885　Pinus densa（Little et Dorman）de Laub. et Silba ＝ Pinus elliottii Engelm. var. densa Littlew. et Dorman ●

299886　Pinus densata Mast.；高山松（西康赤松，西康油松）；Alpine Pine ●

299887　Pinus densata Mast. var. pygmaea J. R. Xue ＝ Pinus yunnanensis Franch. var. pygmaea（J. R. Xue）J. R. Xue ●

299888　Pinus densata Mast. var. pygmaea J. R. Xue ex C. Y. Cheng, W. C. Cheng et L. K. Fu ＝ Pinus yunnanensis Franch. var. pygmaea（J. R. Xue）J. R. Xue ●

299889　Pinus densiflora f. ussuriensis（Liou et Q. L. Wang）Kitag. ＝ Pinus densiflora Siebold et Zucc. var. ussuriensis Liou et Z. Wang ●

299890　Pinus densiflora Siebold et Zucc.；赤松（短叶赤松，灰果赤松，辽东赤松，日本赤松）；Japan Pine，Japanese Pine，Japanese Red Pine，Red Pine ●

299891　Pinus densiflora Siebold et Zucc. 'Alice Verkade'；阿里斯·维克德赤松●☆

299892　Pinus densiflora Siebold et Zucc. 'Aurea'；金叶赤松；Golden Japanese Red Pine ●

299893　Pinus densiflora Siebold et Zucc. 'Globosa'；球冠赤松；Globosa Japanese Pine ●

299894　Pinus densiflora Siebold et Zucc. 'Oculus-draconis'；龙眼松；Dragon-eye Pine，Dragon's-eye Pine ●☆

299895　Pinus densiflora Siebold et Zucc. 'Pendula'；垂枝赤松；Weeping Japanese Red Pine ●

299896　Pinus densiflora Siebold et Zucc. 'Tagyosho' ＝ Pinus densiflora Siebold et Zucc. 'Umbraculifera' ●

299897　Pinus densiflora Siebold et Zucc. 'Umbraculifera'；千头赤松（美松，伞冠赤松，伞形赤松）；Umbellare Japanese Pine ●

299898　Pinus densiflora Siebold et Zucc. ＝ Pinus taiwanensis Hayata ●

299899　Pinus densiflora Siebold et Zucc. f. globosa（Mayr）Beissn. ＝ Pinus densiflora Siebold et Zucc. 'Globosa' ●

299900　Pinus densiflora Siebold et Zucc. f. liaotungensis（Liou et Z. Wang）Kitag. ＝ Pinus densiflora Siebold et Zucc. ●

299901　Pinus densiflora Siebold et Zucc. f. subtrifoliata Hurus. ＝ Pinus densiflora Siebold et Zucc. ●

299902　Pinus densiflora Siebold et Zucc. f. sylvestriformis Taken. ＝ Pinus sylvestris L. var. sylvestriformis（Taken.）W. C. Cheng et C. D. Chu ●◇

299903　Pinus densiflora Siebold et Zucc. f. sylvestriformis Taken. ＝ Pinus sylvestriformis（Taken.）Z. Wang ex W. C. Cheng ●◇

299904　Pinus densiflora Siebold et Zucc. f. umbraculifera（Mayr）Sugim. ＝ Pinus densiflora Siebold et Zucc. 'Umbraculifera' ●

299905　Pinus densiflora Siebold et Zucc. f. ussuriensis（Liou et Z. Wang）Kitag. ＝ Pinus densiflora Siebold et Zucc. var. ussuriensis Liou et Z. Wang ●

299906　Pinus densiflora Siebold et Zucc. var. brevifolia Liou et Z. Wang ＝ Pinus densiflora Siebold et Zucc. ●

299907　Pinus densiflora Siebold et Zucc. var. funebris（Kom.）Liou et Z. Wang ＝ Pinus densiflora Siebold et Zucc. ●

299908　Pinus densiflora Siebold et Zucc. var. liaotungensis Liou et Z. Wang ＝ Pinus densiflora Siebold et Zucc. ●

299909　Pinus densiflora Siebold et Zucc. var. sylvestriformis（Taken.）Q. L. Wang ＝ Pinus sylvestris L. var. sylvestriformis（Taken.）W.

C. Cheng et C. D. Chu ●◇

299910   Pinus densiflora Siebold et Zucc. var. sylvestriformis（Taken.） Q. L. Wang = Pinus sylvestriformis（Taken.）Z. Wang ex W. C. Cheng ●◇

299911   Pinus densiflora Siebold et Zucc. var. tabulaeformis（Carrière）Fortune ex Mast. = Pinus tabuliformis Carrière ●

299912   Pinus densiflora Siebold et Zucc. var. tabuliformis（Carrière）Mast. = Pinus tabuliformis Carrière ●

299913   Pinus densiflora Siebold et Zucc. var. umbraculifera Mayr = Pinus densiflora Siebold et Zucc. 'Umbraculifera' ●

299914   Pinus densiflora Siebold et Zucc. var. ussuriensis Liou et Z. Wang;兴凯赤松（兴凯湖松）●

299915   Pinus densiflora Siebold et Zucc. var. zhangwuensis S. J. Zhang et al.;彰武赤松●

299916   Pinus deodara Roxb. = Cedrus deodara（Roxb.）G. Don ●

299917   Pinus dicksonii Carrière = Pinus wallichiana A. B. Jacks. ●

299918   Pinus discolor D. K. Bailey et Hawksw. = Pinus cembroides Zucc. ●☆

299919   Pinus divaricata（Aiton）Sudw. = Pinus banksiana Lamb. ●

299920   Pinus douglasiana Mart.;道格拉斯松;Douglas Pine ●☆

299921   Pinus douglasiana Mart. var. maximinoi（H. E. Moore）Silba = Pinus maximinoi H. E. Moore ●☆

299922   Pinus douglasii Sabine ex D. Don = Pseudotsuga menziesii （Mirb.）Franco ●

299923   Pinus dumosa D. Don = Tsuga dumosa（D. Don）Eichler ●

299924   Pinus durangensis Mart.;杜兰戈松（杜兰果松）;Durango Pine ●☆

299925   Pinus echinata Mill.;萌芽松（短叶松,芒刺松,毛松,猥毛松）;Shortleaf Pine,Short-leaf Pine,Shortleaved Pine,Shortneedle Pine,Shortstraw Pine,Southern Yellow Pine,Yellow Pine ●

299926   Pinus edulis Engelm.;食松（薄皮果松,短叶果松）;Arizona Nut Pine,Colorado Pine,Colorado Pinyon,Colorado Pinyon Pine,Eating Pine,Nut Pine,Pinyon,Pinyon Pine,Rocky Mountain Pinyon,Twoleaf Nut Pine,Twoleaf Pine,Two-leaf Pinion,Twoleaf Pinyon,Two-needle Pinyon ●☆

299927   Pinus eldarica Medw.;爱大利松●☆

299928   Pinus elliottii Engelm.;湿地松（美松）;American Pitch,Caribbean Pine,Marsh Pine,Slash Pine,Southern Pine,Swamp Pine,Yellow Slash Pine ●

299929   Pinus elliottii Engelm. = Pinus caribaea Morelet ●

299930   Pinus elliottii Engelm. var. densa Littlew. et Dorman;南湿地松;S. Florida Marsh Pine,South Florida Slash Pine ●

299931   Pinus engelmannii Carrière;大叶松（恩氏松）;Apache Pine,Arizona Longleaf Pine,Engelmann Pine,Engelmann's Pine,Largeleaf Pine ●☆

299932   Pinus excelsa Lam. = Picea abies（L.）H. Karst. ●

299933   Pinus excelsa Lam. var. ehinensis Patschke = Pinus armandii Franch. ●

299934   Pinus excelsa Wall. = Pinus wallichiana A. B. Jacks. ●

299935   Pinus excelsa Wall. ex D. Don var. chinensis Patschke = Pinus armandii Franch. ●

299936   Pinus excelsa Wall. ex Lamb. = Pinus wallichiana A. B. Jacks. ●

299937   Pinus faberi（Mast.）Voss = Abies faberi（Mast.）Craib ●

299938   Pinus fenzeliana Hand. -Mazz.;海南五针松（海南松,海南五须松,葵花松,油松,粤松）;Fenzel Pine ●

299939   Pinus fenzeliana Hand. -Mazz. var. dabeshanensis（W. C. Cheng et Y. W. Law）L. K. Fu et Nan Li;大别山五针松（安徽五针松,大别山松,大别五针松,软木松）;Dabieshan Pine, Dapieh Mountain

Pine ●◇

299940   Pinus finlaysoniana Wall. = Pinus latteri Mason ●

299941   Pinus firma（Siebold et Zucc.）Antoine = Abies firma Siebold et Zucc. ●

299942   Pinus flexilis E. James;柔松（美国果松）;Flexible Pine,Limber Pine,Rocky Mountain White Pine,Soft Pine,Vanderwolf Pine,White Pine ●☆

299943   Pinus flexilis E. James 'Pendula';垂枝柔松;Weeping Limber Pine ●☆

299944   Pinus flexilis E. James var. reflexa Engelm. = Pinus strobiformis Engelm. ●☆

299945   Pinus fnebbris Kom.;比顺特松●☆

299946   Pinus fnebbris Kom. = Pinus densiflora Siebold et Zucc. ●

299947   Pinus formosana Hayata = Pinus morrisonicola Hayata ●

299948   Pinus fortunei（Murray）Parl. = Keteleeria fortunei（A. Murray）Carrière ●

299949   Pinus fraseri Pursh = Abies fraseri（Pursh）Poir. ●☆

299950   Pinus funebris Kom. = Pinus densiflora Siebold et Zucc. ●

299951   Pinus funebris Kom. = Pinus sylvestris L. var. sylvestriformis （Taken.）W. C. Cheng et C. D. Chu ●◇

299952   Pinus gerardiana Wall. = Pinus gerardiana Wall. ex Lamb. ●

299953   Pinus gerardiana Wall. ex Lamb.;喜马拉雅白皮松（西藏白皮松）;Chilgoza Pine,Gerard Pine,Gerard's Pine ●

299954   Pinus glabra Walter;光松;Bottom White Pine,Cedar Pine,Spruce Pine,Walter Pine,Walter's Pine ●☆

299955   Pinus glauca Moench = Picea glauca（Moench）Voss ●

299956   Pinus grandis Douglas ex D. Don = Abies grandis（Douglas ex D. Don）Lindl. ●

299957   Pinus greggii Engelm. ex Parl.;硬枝展松（格雷基松）;Gregg Pine ●☆

299958   Pinus griffithiana（Carrière）Voss = Larix griffithiana（Lindl. et Gordon）Carrière ●

299959   Pinus griffithiana（Lindl. et Gordon）Voss = Larix griffithiana （Lindl. et Gordon）Carrière ●

299960   Pinus griffithii（Hook. f.）Parl. = Larix griffithiana（Lindl. et Gordon）Carrière ●

299961   Pinus griffithii McClell. = Pinus wallichiana A. B. Jacks. ●

299962   Pinus griffithii Parl. = Larix griffithiana（Lindl. et Gordon）Carrière ●

299963   Pinus hakkodoensis Makino = Pinus pumila（Pall.）Regel ●

299964   Pinus halepensis Mill.;地中海松（阿勒颇松,布鲁地中海松,地中海白松,哈列布松）;Aleppo Pine,Cyprus Pine,Jerusalem Pine ●☆

299965   Pinus halepensis Mill. var. brutia（Ten.）= Pinus brutia Ten. ●☆

299966   Pinus hamata（Steven）Sosn.;钩形松●☆

299967   Pinus hartwegii Lindl.;灰叶山松;Hartweg Pine ●☆

299968   Pinus heldreichii H. Christ;巴尔干松（白皮巴尔干松,波斯尼亚松,赫德赖克松,灰皮巴尔干松）;Balkan Pine,Bosnian Pine,Heldreich Pine ●☆

299969   Pinus heldreichii H. Christ 'Compact Gem';小宝石巴尔干松（袖珍波斯尼亚松）●☆

299970   Pinus heldreichii H. Christ 'Smidtii';斯密德特巴尔干松●☆

299971   Pinus heldreichii H. Christ subsp. leucodermis（Antoine）E. Murray = Pinus leucodermis Antoine ●☆

299972   Pinus henryi Mast.;巴山松（短叶马尾松）;Bashan Pine,Henry Pine ●

299973   Pinus henryi Mast. = Pinus tabuliformis Carrière var. henryi （Mast.）C. T. Kuan ●

299974　Pinus herrerai Martinez；哈利科斯松；Jalisco Pine ●☆

299975　Pinus heterophylla（Elliott）Sudw. = Pinus elliottii Engelm. ●

299976　Pinus himekomatsu Miyabe et Kudo = Pinus parviflora Siebold et Zucc. ●

299977　Pinus hingganensis H. J. Zhang = Pinus sibirica（Loudon）Mayr ●

299978　Pinus holfordiana A. B. Jacks.；霍氏松；Holford Pine ●☆

299979　Pinus holophylla（Maxim.）Parl. = Abies holophylla Maxim. ●

299980　Pinus hookeriana（A. Murray bis）McNab = Tsuga mertensiana（Bong.）Carrière ●☆

299981　Pinus hwangshanensis W. Y. Hsia = Pinus taiwanensis Hayata ●

299982　Pinus hwangshanensis W. Y. Hsia var. wulingensis S. C. Li；杉松；Wuling Pine ●

299983　Pinus ikedae Y. Yamam. = Pinus latteri Mason ●

299984　Pinus inops Aiton var. clausa Chapm. ex Engelm. = Pinus clausa（Chapm. ex Engelm.）Sarg. ●☆

299985　Pinus insignis Douglas ex Loudon = Pinus radiata D. Don ●

299986　Pinus insularis Endl.；岛松（海岛陶松）；Benquet Pine, Island Torrey Pine, Khasi Pine, Khasya Pine, Luson Pine, Philippine Pine ●☆

299987　Pinus insularis Endl. = Pinus yunnanensis Franch. ●

299988　Pinus insularis Endl. var. khasyana（Griffiths）Silba = Pinus kesiya Royle ex Gordon ●

299989　Pinus insularis Endl. var. langbianensis（A. Chev.）Silba = Pinus kesiya Royle ex Gordon ●

299990　Pinus insularis Endl. var. tenuifolia（W. C. Cheng et Y. W. Law）Silba = Pinus yunnanensis Franch. var. tenuifolia W. C. Cheng et Y. W. Law ●

299991　Pinus insularis Endl. var. yunnanensis（Franch.）Silba = Pinus yunnanensis Franch. ●

299992　Pinus jeffreyi Murray；黑材松（杰弗里松, 美国蓝叶松, 约弗松, 约弗亚松）；Black Pine, Bull Pine, Jeffrey Pine, Jeffrey's Pine, Westen Yellow Pine ●☆

299993　Pinus jeffreyi Murray var. deflexa（Torr.）Lemmon = Pinus jeffreyi Murray ●☆

299994　Pinus jezoensis f. microsperma（Mast.）Voss = Picea jezoensis（Siebold et Zucc.）Carrière var. microsperma（Lindl.）W. C. Cheng et L. K. Fu ●

299995　Pinus juarezensis Lanner = Pinus quadrifolia Parl. ex Sudw. ●☆

299996　Pinus kaempferi Lamb. = Larix kaempferi（Lamb.）Sarg. ●

299997　Pinus kaempferi Lamb. = Pseudolarix amabilis（J. Nelson）Rehder ●◇

299998　Pinus kaempferi Lamb. = Pseudolarix kaempferi（Lindl.）Gordon ●◇

299999　Pinus kesiya Royle ex Gordon；思茅松（卡西松, 卡西亚松）；Khasi Pine, Khasya Pine, Northern Burma Pine, Simao Pine ●

300000　Pinus kesiya Royle ex Gordon var. langbianensis（A. Chev.）Gaussen ex Bui = Pinus kesiya Royle ex Gordon ●

300001　Pinus kesiya Royle ex Gordon var. langbianensis（A. Chev.）Gaussen = Pinus kesiya Royle ex Gordon ●

300002　Pinus khasya Royle = Pinus kesiya Royle ex Gordon ●

300003　Pinus khutrow Royle = Picea smithiana（Wall.）Boiss. ●◇

300004　Pinus khutrow Royle ex Turra = Picea smithiana（Wall.）Boiss. ●◇

300005　Pinus komarovii H. Lév. = Pinus armandii Franch. ●

300006　Pinus koraiensis Siebold et Zucc.；红松（朝鲜松, 果松, 海松, 韩松, 红果松, 前红松, 新罗松）；Bosian Pine, Corean Pine, Korea Pine, Korean Pine ●◇

300007　Pinus krempfii Lecomte；越南松；Krempf Pine, Vietnam Pine ●☆

300008　Pinus kwangshanensis K. C. Hsia = Pinus taiwanensis Hayata ●

300009　Pinus kwangtungensis Chun ex Tsiang；华南五针松（广东松）；Guangdong Pine, Kwangtung Pine ●◇

300010　Pinus kwangtungensis Chun ex Tsiang var. varifolia Nan Li et Y. C. Zhong；变叶华南五针松 ●

300011　Pinus lamberitiana Douglas ex Taylor et Philipson；糖松；Giant Pine, Giantic Pine, Lambert Pine, Purpleconed Sugar Pine, Shade Pine, Sugar Pine ●☆

300012　Pinus lambertiana Douglas = Pinus lamberitiana Douglas ex Taylor et Philipson ●☆

300013　Pinus lanceolata Lamb. = Cunninghamia lanceolata（Lamb.）Hook. ●

300014　Pinus langbianensis A. Chev. = Pinus kesiya Royle ex Gordon var. langbianesis（A. Chev.）Gaussen ●

300015　Pinus langbianensis A. Chev. = Pinus kesiya Royle ex Gordon ●

300016　Pinus laricina Du Roi = Larix laricina（Du Roi）K. Koch ●☆

300017　Pinus laricio Poir. = Pinus nigra Aiton var. maritima（Aiton）Melville ●☆

300018　Pinus laricio Poir. var. poiretiana Antoine = Pinus nigra Aiton var. poiretiana（Antoine）C. K. Schneid. ●

300019　Pinus laricio Poir = Pinus nigra Ation var. maritima（Aiton）Melville ●☆

300020　Pinus larix Douglas = Larix occidentalis Nutt. ●☆

300021　Pinus larix L. = Larix decidua Mill. ●

300022　Pinus larix L. = Larix europaea DC. ●

300023　Pinus larix L. var. russica Endl. = Larix sibirica（Münchh.）Ledeb. ●

300024　Pinus larix Poir. var. europaea Pall. = Larix sibirica（Münchh.）Ledeb. ●

300025　Pinus larix Siev. = Larix dahurica Lawson ●

300026　Pinus larix Siev. = Larix gmelinii（Rupr.）Rupr. ●

300027　Pinus larix Thunb. = Larix kaempferi（Lamb.）Carrière ●

300028　Pinus lasiocarpa Hook. = Abies lasiocarpa（Hook.）Nutt. ●☆

300029　Pinus latifolia Sarg. = Pinus engelmannii Carrière ●☆

300030　Pinus latteri Mason；南亚松（南洋二针松, 越南松）；Latter Pine, S. Asia Pine ●

300031　Pinus lawsonii Roezl ex Gordon et Glend.；劳森松（拉威逊松）；Lawson Pine ●☆

300032　Pinus ledebourii（Rupr.）Endl. = Larix sibirica（Münchh.）Ledeb. ●

300033　Pinus leiophylla Schiede et Deppe；光叶松（平滑叶松）；Chihuahua Pine, Ocote Chino, Pino Chino, Smoothleaf Pine, Smoothleaf Pine, Yellow Pine ●☆

300034　Pinus leiophylla Schltdl. et Champ. = Pinus leiophylla Schiede et Deppe ●☆

300035　Pinus leiophylla Schltdl. et Champ. var. chihuahuana（Engelm.）Shaw；奇瓦瓦松；Chihuahua Pine, Pino Prieto, Pino Real ●☆

300036　Pinus leucodermis Antoine；欧洲白皮松（白皮巴尔干松）；Bosnian Pine, Europe Pine, Heldreich Pine ●☆

300037　Pinus leucodermis Antoine = Pinus heldreichii H. Christ ●☆

300038　Pinus leucosperma Maxim. = Pinus tabulaeformis Carrière ●

300039　Pinus levis Lemee et H. Lév. = Pinus armandii Franch. ●

300040　Pinus longaeva D. K. Bailey；长寿松（毛松）；Ancient Pine, Bristle-cone Pine, Great Basin Bristlecone Pine, Intermountain Bristlecone Pine ●☆

300041　Pinus longbianensis A. Chev.；郎边松；Longbian Pine ●☆

300042　Pinus longifolia Roxb. ex Lamb. = Pinus roxburghii Sarg. ●

300043   Pinus longifolia Salisb. = Pinus palustris Mill. ●

300044   Pinus luchuensis Mayr;琉球松;Liuqiu Pine,Luchu Pine ●

300045   Pinus luchuensis Mayr subsp. hwangshanensis (W. Y. Hsia) D. Z. Li = Pinus taiwanensis Hayata ●

300046   Pinus luchuensis Mayr subsp. taiwanensis (Hayata) D. Z. Li = Pinus taiwanensis Hayata ●

300047   Pinus luchuensis Mayr var. hwangshanensis (W. Y. Hsia) Y. C. Wu = Pinus taiwanensis Hayata ●

300048   Pinus lumholtzii B. L. Rob. et Fernald;垂枝松（卢氏松）; Lumboltz Pine,Mexican Pine,Pino Barda Pine ● ☆

300049   Pinus macrophylla Engelm. = Pinus engelmannii Carrière ● ☆

300050   Pinus mandschurica Rupr. = Pinus koraiensis Siebold et Zucc. ● ◇

300051   Pinus mandshurica Rupr. = Pinus koraiensis Siebold et Zucc. ● ◇

300052   Pinus mariana Du Roi = Picea mariana (Mill.) Britton, Sterns et Poggenb. ● ☆

300053   Pinus maritima Lam. = Pinus pinaster Aiton ● ☆

300054   Pinus massoniana Lamb.;马尾松（白松,枫树,枞松,黑松,厚皮松,青松,山松,松,松柏,松萝,松树,台湾赤松,台湾五叶松,铁甲松）;Chinese Red Pine,Horsetail Pine,Masson Pine ●

300055   Pinus massoniana Lamb. = Pinus thunbergii Parl. ●

300056   Pinus massoniana Lamb. var. hainanensis C. Y. Cheng, W. C. Cheng et L. K. Fu;雅加松;Hainan Masson Pine ● ◇

300057   Pinus massoniana Lamb. var. henryi (Mast.) C. L. Wu = Pinus henryi Mast. ●

300058   Pinus massoniana Lamb. var. huanglinsong Hort.;黄鳞松; Huanglin Masson Pine ●

300059   Pinus massoniana Lamb. var. lingnanensis Hort.;岭南马尾松; Lingnan Masson Pine ●

300060   Pinus massoniana Lamb. var. shaxianensis D. X. Zhou;沙县黄松（沙黄松）;Shaxian Masson Pine ●

300061   Pinus massoniana Lamb. var. wulingensis C. J. Qi et Q. Z. Lin = Pinus henryi Mast. ●

300062   Pinus massoniana Lamb. var. wulingensis C. J. Qi et Q. Z. Lin = Pinus tabuliformis Carrière var. henryi (Mast.) C. T. Kuan ●

300063   Pinus mastersiana Hayata = Pinus armandii Franch. var. mastersiana (Hayata) Hayata ●

300064   Pinus mastersiana Hayata = Pinus armandii Franch. ●

300065   Pinus maximinoi H. E. Moore;狭叶松;Thin-leaf Pine ● ☆

300066   Pinus menziesii Douglas ex D. Don = Picea sitchensis (Bong.) Carrière ● ☆

300067   Pinus merkusiana Cooling et Gaussen = Pinus latteri Mason ●

300068   Pinus merkusii Jungh. et de Vriese;苏门答腊松（吕宋松,南洋松）;Merkus Pine,Sumatra Pine,Tenasserium Pine ● ☆

300069   Pinus merkusii Jungh. et de Vriese = Pinus latteri Mason ●

300070   Pinus merkusii Jungh. et de Vriese subsp. latteri (Mason) D. Z. Li = Pinus latteri Mason ●

300071   Pinus merkusii Jungh. et de Vriese var. latteri (Mason) Silba = Pinus latteri Mason ●

300072   Pinus merkusii Jungh. et de Vriese var. tonkinensis (A. Chev.) A. Chev. ex Gaussen = Pinus latteri Mason ●

300073   Pinus merkusii Jungh. var. latteri (Mason) Silba = Pinus latteri Mason ●

300074   Pinus merkusii Jungh. var. tonkinensis (A. Chev.) Gaussen ex Bui = Pinus latteri Mason ●

300075   Pinus mertensiana Bong. = Tsuga mertensiana (Bong.) Carrière ● ☆

300076   Pinus mesogeensis Fieschi et Gaussen = Pinus pinaster Aiton ● ☆

300077   Pinus michoacana Martinez;米却肯松;Michoaca Pine, Michoacan Pine ● ☆

300078   Pinus mitis Michx. = Pinus echinata Mill. ●

300079   Pinus monophylla Torr. et Frém.;单针松（单叶松）;Desert Nut Pine,Nevada Nut Pine,Oneleaf Pine,One-leaf Pine,Pinon,Singleleaf Pine,Single-leaf Pine,Singleleaf Pinyon ●

300080   Pinus montana Mill. = Pinus mugo Turra ● ☆

300081   Pinus montezumae Lamb.;山松（蒙地兹松,孟特松）; Montezuma Pine,Rough-brached Mexican Pine ● ☆

300082   Pinus monticola Douglas ex D. Don;加州山松（加利福尼亚州山松,美国山地白松,山白松,西部白松）;California Wild Pine, Californian Mountain Pine,Finger-cone Pine,Idaho White Pine, Mountain White Pine,Silver Pine,Western White Pine ● ☆

300083   Pinus morrisonicola Hayata;台湾五针松（马尾松,山松柏,台湾白松,台湾松,台湾五须松,台湾五叶松）;Morrison Pine,Taiwan Short-leaf Pine,Taiwan White Pine ●

300084   Pinus morrisonicola Hayata = Pinus kwangtungensis Chun ex Tsiang ● ◇

300085   Pinus mughus Jacq. = Pinus mugo Turra ● ☆

300086   Pinus mugo Turra;欧洲山松（山松,中欧山松）;Dwarf Mountain Pine,Dwarf Mountain-pine,Dwarf Pine,Dwarf Wild Pine, Knee-pine,Mountain Pine,Mugho Pine,Mugo Pine,Scrub Mugo Pine, Swiss Mountain Pine,Swiss Mountain-pine,Swiss Wild Pine ● ☆

300087   Pinus mugo Turra 'Gnom';袖珍欧洲山松（袖珍中欧山松）● ☆

300088   Pinus mugo Turra 'Green Candles';绿蜡烛欧洲山松 ● ☆

300089   Pinus mugo Turra 'Honeycomb';蜂窝欧洲山松 ● ☆

300090   Pinus mugo Turra 'Mops';拖把欧洲山松（拖把中欧山松）; Mops Mugo Pine ● ☆

300091   Pinus mugo Turra 'Paul's Dwarf';矮鲍尔欧洲山松 ● ☆

300092   Pinus mugo Turra 'Slowmound';斯鲁姆德欧洲山松 ● ☆

300093   Pinus mugo Turra 'Tannenbaum';塔尼班欧洲山松 ● ☆

300094   Pinus mugo Turra 'Teeny';极小欧洲山松 ● ☆

300095   Pinus mugo Turra subsp. pumilio (Haenke) E. Murray;矮欧洲山松;Dwarf Swiss Mountain Pine,Little Mugo Pine ● ☆

300096   Pinus mukdensis Uyeki = Pinus tabuliformis Carrière var. mukdensis Uyeki ex Nakai ●

300097   Pinus mukdensis Uyeki ex Nakai = Pinus tabuliformis Carrière var. mukdensis Uyeki ex Nakai ●

300098   Pinus muricata D. Don;加州沼松（粗糙松,加利福尼亚州沼松,加州二叶松,尖瘤松）;Bishop Pine,California Swamp Pine, Obispo Pine,Prickle-cone Pine,Santa Cruz Island Pine ● ☆

300099   Pinus muricata D. Don var. borealis Axelrod = Pinus muricata D. Don ● ☆

300100   Pinus muricata D. Don var. cedrosensis J. T. Howell = Pinus muricata D. Don ● ☆

300101   Pinus muricata D. Don var. stantonii Axelrod = Pinus muricata D. Don ● ☆

300102   Pinus murrayana Balf. = Pinus contorta Douglas ex Loudon var. latifolia Engelm. ex S. Watson ● ☆

300103   Pinus nelsonii Shaw;连叶松;Mexican Pine,Nelson Pine,Nelson Pinyon ● ☆

300104   Pinus nepalensis Chambray = Picea wallichiana A. B. Jacks. ●

300105   Pinus nepalensis J. Forbes = Pinus massoniana Lamb. ●

300106   Pinus nigra Aiton 'Hornibrookiana';赫氏欧洲黑松 ● ☆

300107   Pinus nigra Aiton = Picea mariana (Mill.) Britton, Sterns et Poggenb. ● ☆

300108   Pinus nigra Aiton = Pinus nigra J. F. Arnold ●

300109   Pinus nigra Aiton subsp. calabrica (Loudon) E. Murray = Pinus nigra Aiton var. maritima（Aiton）Melville ●☆

300110　Pinus nigra Aiton var. austriaca Asch. et Graebn. = Pinus nigra Aiton ●☆

300111　Pinus nigra Aiton var. caramanica（Loudon）Rehder;克里米亚黑松;Caraman Pine,Crimean Pine ●☆

300112　Pinus nigra Aiton var. cedennensis ?;比利牛斯黑松;Pyrenean Pine ●☆

300113　Pinus nigra Aiton var. dalmatica（Vis.）Businský;达尔马提黑松;Dalmatian Pine ●☆

300114　Pinus nigra Aiton var. maritima（Aiton）Melville;科西嘉黑松;Corsican Pine ●☆

300115　Pinus nigra Aiton var. maritima（Aiton）Melville = Pinus laricio Poir. ●☆

300116　Pinus nigra Aiton var. poiretiana（Antoine）C. K. Schneid.;南欧黑松;Poiret. Pine ●

300117　Pinus nigra J. F. Arnold;欧洲黑松(奥地利黑松,奥地利松,黑皮松);Austria Pine,Austrian corsican pine,Austrian Pine,Black Austrian Pine,Black Pine,Calabrian Black Pine,Corsican Pine,European Black Pine ●

300118　Pinus nigra J. F. Arnold subsp. laricio ? = Pinus nigra Aiton var. maritima（Aiton）Melville ●☆

300119　Pinus nigra J. F. Arnold subsp. mauretanica（Maire et Peyerimh.）Heywood;马里坦黑松●☆

300120　Pinus nigra J. F. Arnold subsp. salzmannii（Dunal）Franco;萨尔欧洲黑松;Corsican Pine ●☆

300121　Pinus nigra J. F. Arnold var. corsicana（Loudon）Hyl. = Pinus nigra J. F. Arnold subsp. salzmannii（Dunal）Franco ●☆

300122　Pinus nigra J. F. Arnold var. mauritanica（Maire et Peyerimh.）Farjon = Pinus nigra J. F. Arnold subsp. mauretanica（Maire et Peyerimh.）Heywood ●☆

300123　Pinus nigra J. F. Arnold var. mauritanica Maire et Peyerimh. = Pinus nigra J. F. Arnold subsp. mauretanica（Maire et Peyerimh.）Heywood ●☆

300124　Pinus oaxacana Mirov;中美高地松(中美洲高地松);Honduras Pine,Oaxaca Pine ●☆

300125　Pinus occidentalis Sw.;古巴松(海地松);Antilles Pine,Jalisco Pine,Sierra Juniper ●☆

300126　Pinus oocarpa Schiede ex Schltdl.;卵果松;Nicaragua Pine,Nicaraguan Pine,Ovalcone Pine,Oval-coned Pine ●

300127　Pinus pallasiana Lamb.;克里木松●☆

300128　Pinus palustris Mill.;长叶松(大王松,沼泽松);American Pitch Pine,Broom Pine,Chir Pine,Fat Pine,Florida Longleaf Pine,Florida Pine,Georgia Pine,Longleaf Pine,Long-leaf Pine,Longleaf Yellow Pine,Longleaved Pine,Longstraw Pine,Pitch Pine,Rosemary Pine,Southern Pine,Southern Pitch-pine,Southern Yellow Pine,Swamp Pine,True Pitch Pine,Türpentine Pine ●

300129　Pinus parryana Engelm. = Pinus quadrifolia Parl. ex Sudw. ●☆

300130　Pinus parviflora Siebold et Zucc.;日本五针松(日本五钗松,日本五须松,日本五叶松,五钗松,五针松);Japan White Pine,Japanese White Pine,Smallflowered Japan Pine ●

300131　Pinus parviflora Siebold et Zucc. 'Adcock's Dwarf';阿迪科克矮型日本五叶松●☆

300132　Pinus parviflora Siebold et Zucc. = Pinus morrisonicola Hayata ●

300133　Pinus parviflora Siebold et Zucc. subsp. pentaphylla（Mayr）Businsky = Pinus parviflora Siebold et Zucc. var. pentaphylla（Mayr）A. Henry ●☆

300134　Pinus parviflora Siebold et Zucc. var. fenzeliana（Hand. -Mazz.）

C. L. Wu = Pinus fenzeliana Hand. -Mazz. ●

300135　Pinus parviflora Siebold et Zucc. var. laevis H. Hara = Pinus parviflora Siebold et Zucc. var. pentaphylla（Mayr）A. Henry f. laevis（H. Hara）Sugim. ●☆

300136　Pinus parviflora Siebold et Zucc. var. morrisonicola（Hayata）Y. C. Wu = Pinus morrisonicola Hayata ●

300137　Pinus parviflora Siebold et Zucc. var. pentaphylla（Mayr）A. Henry;姬小松(日本五针松);Japanese White Pine ●☆

300138　Pinus parviflora Siebold et Zucc. var. pentaphylla（Mayr）A. Henry f. laevis（H. Hara）Sugim.;平滑姬小松●☆

300139　Pinus pattoniana（A. Murray bis）Parl. = Tsuga mertensiana（Bong.）Carrière ●☆

300140　Pinus patula Schltdl. et Cham.;展叶松(垂枝松);Mexican Pine,Mexican Weeping Pine,Mexican Yellow Pine,Mexico Weeping Pine,Patula Pine,Spreading Pine,Spreading-leaf Pine,Spreading-leafed Pine,Spreading-leaved Pine,Weeping Pine ●☆

300141　Pinus pentaphylla Mayr = Pinus parviflora Siebold et Zucc. var. pentaphylla（Mayr）A. Henry ●☆

300142　Pinus pentaphylla Mayr f. laevis（H. Hara）Kusaka = Pinus parviflora Siebold et Zucc. var. pentaphylla（Mayr）A. Henry f. laevis（H. Hara）Sugim. ●☆

300143　Pinus pentaphylla Mayr var. himekomatsu（Miyabe et Kudo）Makino = Pinus parviflora Siebold et Zucc. ●

300144　Pinus peuce Griseb.;扫帚松(巴尔干松,白斯松);Balkan Pine,Bela Movra,Macedonia Pine,Macedonian Pine ●☆

300145　Pinus pichta Lodd. = Abies sibirica Ledeb. ●

300146　Pinus pinaster Aiton;海岸松(法国海岸松,海滨松,南欧海松,松果松);Bournemouth Pine,Cluster Pine,French Turpentine,Maritime Pine,Pinaster,Sea-pine,Star Pine ●☆

300147　Pinus pinaster Aiton subsp. atlantica Villar = Pinus pinaster Aiton ●☆

300148　Pinus pinceana Gordon;长枝松(垂枝松);Pince Pine,Pince Pinyon Pine,Weeping Pinyon ●☆

300149　Pinus pindrow Royle = Abies spectabilis（D. Don）Spach ●

300150　Pinus pindrow Royle = Abies webbiana（Wall.）Lindl. ●

300151　Pinus pinea L.;意大利伞松(果松,笠松,石松,意大利果松,意大利松,意大利五针松);Italian Stone Pine,Italy Stone Pine,Pignon Pignoli,Roman Pine,Stone Pine,Umbrella Pine,Western Yellow Pine ●

300152　Pinus ponderosa Douglas = Pinus ponderosa Douglas ex C. Lawson et C. Lawson ●

300153　Pinus ponderosa Douglas ex C. Lawson et C. Lawson;西黄松(黄松,美国长三叶松,美国黄松,西部黄松,西南松);Arizona Pine,Blackjack Pine,Bull Pine,Heavy-wooded Pine,Pitch Pine,Ponderosa Pine,Western Yellow Pine,Yellow Pine ●

300154　Pinus ponderosa Douglas ex C. Lawson et C. Lawson arizonica（Engelm.）Shaw = Pinus arizonica Engelm. ●☆

300155　Pinus ponderosa Douglas ex C. Lawson et C. Lawson subsp. jeffreyi（Balf.）E. Murray = Pinus jeffreyi Murray ●☆

300156　Pinus ponderosa Douglas ex C. Lawson et C. Lawson var. jeffreyi Balf. ex Vasey = Pinus jeffreyi Murray ●☆

300157　Pinus ponderosa Douglas ex C. Lawson et C. Lawson var. scopulorum Engelm.;山地西黄松;Rocky Mountain Ponderosa Pine ●☆

300158　Pinus praetermissa Styles et McVaugh;旱生松●☆

300159　Pinus pringlei Shaw;曲枝松(普林松);Pringle Pine,Pringle's Pine ●☆

300160　Pinus prokoraiensis Y. T. Zhao, J. M. Lu et A. G. Gu = Pinus koraiensis Siebold et Zucc. ●◇

300161　Pinus promienus Mast. = Pinus densata Mast. ●

300162　Pinus pseudostrobus Lindl.；拟北美乔松（假北美乔松）；False Weymouth Pine,Pseudostrobus Pine,Smooth-barked Mexican Pine ●☆

300163　Pinus pumila（Pall.）Regel；偃松（矮松,爬地松,爬松,千叠松）；Dwarf Siberian Pine, Dwarf Stone Pine, Japan Stone Pine, Japanese Pine,Japanese Stone Pine ●

300164　Pinus pumila（Pall.）Regel 'Globe'；球型偃松●☆

300165　Pinus pumila（Pall.）Regel var. kubinaga Ishii et Kusaka = Pinus pumila（Pall.）Regel ●

300166　Pinus pumila（Pall.）Regel var. yezoalpina Ishii et Kusaka = Pinus pumila（Pall.）Regel ●

300167　Pinus pumilio Haenke；矮松●☆

300168　Pinus pungens Lamb.；辛松（刺针松）；Hickory Pine, Mountain Pine, Prickle Pine, Table Mountain Pine ●☆

300169　Pinus pygmaea Fisch. ex Spach = Pinus pumila（Pall.）Regel ●

300170　Pinus pyrenaica Lapeyr. subsp. mauretanica（Maire et Peyerimh.）O. Schwartz = Pinus nigra J. F. Arnold subsp. mauretanica（Maire et Peyerimh.）Heywood ●☆

300171　Pinus quadrifolia Parl. ex Sudw.；四针松；Californian Pinyon Pine, Four-needle Pyron, Nut Pine, Parry Pine, Parry Pinyon ●☆

300172　Pinus quinquefolia David = Pinus armandii Franch. ●

300173　Pinus radiata D. Don；辐射松（放射松,蒙达利松）；Insignis Pine, Monterrey Pine, Radiata Pine ●

300174　Pinus radiata D. Don var. binata（Engelm.）Brewer et S. Watson = Pinus muricata D. Don ●☆

300175　Pinus reflexa（Engelm.）Engelm.；撒古松●☆

300176　Pinus reflexa（Engelm.）Engelm. = Pinus strobiformis Engelm. ●☆

300177　Pinus remorata H. Mason；留恋松；Papershell Pinyon ●☆

300178　Pinus remorata H. Mason = Pinus muricata D. Don ●☆

300179　Pinus remota（Little）D. K. Bailey et Hawksw. = Pinus cembroides Zucc. ●☆

300180　Pinus resinosa Aiton；多脂松（美加红松,树脂松）；Canadian Pine, Canadian Red Pine, Norway Pine, Red Pine ●☆

300181　Pinus rigida Mill.；刚松（北美油松,刚叶松,美国短三叶松,萌芽松,硬叶松,直立松）；Hard Pine, Northern Pine, Northern Pitch Pine, Pitch Pine, Sap Pine, Southern Yellow Pine ●

300182　Pinus rigida Mill. subsp. serotina（Michx.）R. T. Clausen = Pinus serotina Michx. ●

300183　Pinus rigida Mill. var. serotina（Michx.）Hoopes = Pinus serotina Michx. ●

300184　Pinus rigida Mill. var. serotina（Michx.）Loudon ex Hoopes = Pinus rigida Mill. ●

300185　Pinus rigida Mill. var. serotina（Michx.）Loudon ex Hoopes = Pinus serotina Michx. ●

300186　Pinus roxburghii Sarg.；喜马拉雅长叶松（长叶松,罗氏松,西藏长叶松,西藏红豆杉,喜马拉雅松）；Asia Longleaved Pine, Asiatic Longleaved Pine, Chir Pine, Emodi Pine, Himalayan Chir Pine, Himalayan Longleaf Pine, Himalayan Long-leaf Pine, Himalayan Pine, Imodi Pine, Indian Chir Pine, Longleaf India Pine, Longleaf Indian Pine, Longleaved Indian Pine, Roxburgh Pine ●

300187　Pinus rudis Endl.；粗糙松●☆

300188　Pinus sabiniana Douglas = Pinus sabiniana Douglas ex D. Don ●☆

300189　Pinus sabiniana Douglas ex D. Don；加州大籽松（加利福尼亚州大松,沙滨松）；Bull Pine, Digger Pine, Gray Pine, Gray-leaf Pine, Grey Pine, Greyleaf Pine, Nut Pine, Sabine Pine, Sabine's Pine ●☆

300190　Pinus sacra（Franch.）Voss = Keteleeria davidiana（Bertrand）Beissn. ●

300191　Pinus schrenkiana（Fisch. et C. A. Mey.）Antoine = Picea schrenkiana Fisch. et C. A. Mey. ●

300192　Pinus scipioniformis Mast. = Pinus armandii Franch. ●

300193　Pinus scopifera Miq. = Pinus densiflora Siebold et Zucc. ●

300194　Pinus scopulorum（Engelm.）Lemmon = Pinus ponderosa Douglas ex C. Lawson et C. Lawson var. scopulorum Engelm. ●☆

300195　Pinus serotina Michx.；晚松；Black Pine, Marsh Pine, Pocosin Pine, Pond Pine ●

300196　Pinus serotina Michx. = Pinus rigida Mill. ●

300197　Pinus sibirica（Loudon）Mayr；西伯利亚红松（西伯利亚松,新疆五针松,兴安松）；Siberia Pine, Siberia Stone Pine, Siberian Cedar, Siberian Pine, Siberian Stone Pine, Siberian White Pine, Siberian Yellow Pine, Xing' an Pine ●

300198　Pinus sibirica（Loudon）Mayr var. hingganensis（H. J. Zhang）Silba = Pinus sibirica（Loudon）Mayr ●

300199　Pinus sibirica var. hingganensis（H. J. Zhang）Silba = Pinus sibirica（Loudon）Mayr ●

300200　Pinus sinensis D. Don = Pinus massoniana Lamb. ●

300201　Pinus sinensis D. Don var. densata（Mast.）Shaw = Pinus densata Mast. ●

300202　Pinus sinensis D. Don var. yunnanensis（Franch.）Shaw = Pinus yunnanensis Franch. ●

300203　Pinus sinensis Lamb. var. yunnanensis（Franch.）Shaw = Pinus yunnanensis Franch. ●

300204　Pinus sinensis Mayr = Pinus tabuliformis Carrière ●

300205　Pinus sitchensis Bong. = Picea sitchensis（Bong.）Carrière ●☆

300206　Pinus smithiana Wall. = Picea smithiana（Wall.）Boiss. ●◇

300207　Pinus spectabilis D. Don = Abies spectabilis（D. Don）Spach ●

300208　Pinus squamata X. W. Li；五针白皮松（巧家五针松）；Squamate Pine ●

300209　Pinus stankewiczi（Sukaczev）Fomin；司坦氏松；Stankewicz Pine ●☆

300210　Pinus strobiformis Engelm.；类球果松（类球松）；Border White Pine, Mexican White Pine, Southwestern White Pine ●☆

300211　Pinus strobus L.；北美乔松（白松,美国白松,美国五针松,美国五针松,美洲五针松）；Canadian Yellow Pine, Eastern White Pine, New England Pine, Northern White Pine, Pumpkin Pine, Weymouth Pine, White Pine, Yellow Pine ●

300212　Pinus strobus L. 'Fastigiata'；高大北美乔松●☆

300213　Pinus strobus L. 'Horsford'；豪斯福德北美乔松●☆

300214　Pinus strobus L. 'Nana' = Pinus strobus L. 'Radiata'●☆

300215　Pinus strobus L. 'Ontario'；安大略乔松；Ontario White Pine ●☆

300216　Pinus strobus L. 'Pendula'；垂枝北美乔松；Weeping White Pine ●☆

300217　Pinus strobus L. 'Prostrata'；平卧北美乔松●☆

300218　Pinus strobus L. 'Radiata'；辐射北美乔松（矮生北美乔松,辐枝美国白松）；Dwarf Eastern White Pine, Radiata White Pine ●☆

300219　Pinus strobus L. var. chiapensis Martínez；墨西哥白松；Mexican White Pine ●☆

300220　Pinus strobus L. var. chiapensis Martínez = Pinus strobus L. ●

300221　Pinus succinifera ?；琥珀松；Amber Pine ●☆

300222　Pinus sylvestriformis（Taken.）Z. Wang ex W. C. Cheng；长白松（长白赤松,长果赤松,灰果赤松,美人松）；Changbai Scotch Pine, Changpai Scotch Pine ●◇

300223　Pinus sylvestris L.；欧洲赤松（长白松,拉普兰欧洲松,美人

松,欧洲松）；Archangel Fir,Baltic Pine,Baltic Redwood,Baltic Red-
wood,Baltic Yellow Deal,Bay Lambs,Bel Bor,Berk Apple,Berk-
apple,Bur,Chats,Chatts,Cuckoo,Danzig Fir,Danzig Pine,Deal
Apple,Deal Wood,Deal-apple,Dealies,Dealseys,Deal-tree,Dell,
Delseed,Durkens,Eyeglasses,Fir Apple,Fir Bob,Fir Deal Tree,Fir
Deal-tree,Fir-apple,Fir-ball,Fir-bob,Fir-dale Tree,Fir-top,Giant
Mugo Pine,Key-balls,Lapland Pine,Mountain Pine,Norway Fir,
Norway Pine,Oyster,Pie Apple,Pin Apple,Pine Apple,Pine-apple,
Pur Apple,Pur-apple,Red Deal,Redwood,Riga Fir,Riga Pine,
Rotkiefer,Scotch Fir,Scotch Pine,Scots Fir,Scots Pine,Sheep,Tory-
top,Vippe,Wild Pine,Yellow Deal ●

300224　Pinus sylvestris L. ' Argentea';银色欧洲赤松●☆

300225　Pinus sylvestris L. 'Aurea';金叶欧洲赤松●☆

300226　Pinus sylvestris L. 'Beuvronensis';伯夫龙欧洲赤松●☆

300227　Pinus sylvestris L. 'Doone Walley';山形欧洲赤松●☆

300228　Pinus sylvestris L. ' Edwin Hillier ' = Pinus sylvestris L.
'Argentea' ●☆

300229　Pinus sylvestris L. 'Fastigiatea';高大欧洲赤松（狭冠欧洲
松）●☆

300230　Pinus sylvestris L. 'Glauca Nana';矮蓝欧洲赤松;Dwarf Blue
Scotch Pine ●☆

300231　Pinus sylvestris L. 'Gold Goin';金币欧洲赤松●☆

300232　Pinus sylvestris L. ' Hillside Creeper';坡地匍匐欧洲赤松;
Hillside Creeper Pine ●☆

300233　Pinus sylvestris L. 'Moseri';莫里斯欧洲赤松●☆

300234　Pinus sylvestris L. 'Nana' = Pinus sylvestris L. 'Watereri'
●☆

300235　Pinus sylvestris L. 'Saxatilis';岩生欧洲赤松●☆

300236　Pinus sylvestris L. 'Troopsii';曲普斯欧洲赤松●☆

300237　Pinus sylvestris L. 'Watereri';沃特雷利欧洲赤松（沃特尔欧
洲赤松）;Dwarf Scotch Pine,Walter's Pine ●☆

300238　Pinus sylvestris L. var. divaricata Aiton = Pinus banksiana
Lamb. ●

300239　Pinus sylvestris L. var. manguiensis S. Y. Li et Adair = Pinus
sylvestris L. var. mongolica Litv. ●

300240　Pinus sylvestris L. var. mongolica Litv.;樟子松（海拉尔松,蒙
古松）;Mongol Scotch Pine,Mongolian Scotch Pine ●

300241　Pinus sylvestris L. var. sylvestriformis (Taken.) W. C. Cheng et
C. D. Chu = Pinus sylvestriformis (Taken.) Z. Wang ex W. C.
Cheng ●◇

300242　Pinus szmaoensis W. C. Cheng et Y. W. Law = Pinus kesiya
Royle ex Gordon ●

300243　Pinus tabulaeformis Carriè = Pinus tabulaeformis Carrière ●

300244　Pinus tabulaeformis Carrière f. purpurea Liou et Z. Wang =
Pinus tabulaeformis Carrière ●

300245　Pinus tabulaeformis Carrière var. bracreata Taken. = Pinus
tabuliformis Carrière ●

300246　Pinus tabulaeformis Carrière var. densata (Mast.) Rehder =
Pinus densata Mast. ●

300247　Pinus tabulaeformis Carrière;油松（东北黑松,短叶马尾松,短
叶松,红皮松,红皮油松,巨果油松,紫翅油松）;China Pine,
Chinese Pine,Chinese Red Pine,Northern Chinese Pine,Oil Pine,
Tabularformed Pine,Tabular-formed Pine ●

300248　Pinus tabulaeformis Carrière f. densa Q. Q. Li et H. Y. Ye;密枝
油松;Dense China Pine ●

300249　Pinus tabulaeformis Carrière var. henryi (Mast.) C. T. Kuan =
Pinus henryi Mast. ●

300250　Pinus tabuliformis Carrière var. kwangtungensis W. C. Cheng =
Pinus kwangtungensis Chun ex Tsiang ●◇

300251　Pinus tabuliformis Carrière var. mukdensis Uyeki ex Nakai;黑皮
油松（东北黑皮油松,辽东黑皮油松）;Blackbark China Pine,
Blackbark Chinese Pine ●

300252　Pinus tabuliformis Carrière var. pygmaea (J. R. Xue) Silba =
Pinus yunnanensis Franch. var. pygmaea (J. R. Xue) J. R. Xue ●

300253　Pinus tabuliformis Carrière var. rubescens Uyeki ex Nakai =
Pinus tabulaeformis Carrière ●

300254　Pinus tabuliformis Carrière var. tokunakai (Nakai) Taken. =
Pinus tabuliformis Carrière ●

300255　Pinus tabuliformis Carrière var. umbraculifera (Liou et Z.
Wang) Q. L. Wang;扫帚油松;Broomshape China Pine,Broomshape
Chinese Pine,Broomshaped Chinese Pine ●

300256　Pinus tabuliformis Carrière var. yunnanensis (Franch.) Dallim.
= Pinus yunnanensis Franch. ●

300257　Pinus tabuliformis Carrière f. jeholensis Liou et Z. Wang = Pinus
tabuliformis Carrière ●

300258　Pinus tabuliformis Carrière f. purpurea Liou et Z. Wang = Pinus
tabuliformis Carrière ●

300259　Pinus tabuliformis Carrière var. bracteata Taken. = Pinus
tabuliformis Carrière ●

300260　Pinus tabuliformis Carrière var. densata (Mast.) Rehder =
Pinus densata Mast. ●

300261　Pinus tabuliformis Carrière var. pygmaea (J. R. Xue) Silba =
Pinus yunnanensis Franch. var. pygmaea (J. R. Xue) J. R. Xue ●

300262　Pinus tabuliformis Carrière var. tokunagai (Nakai) Taken. =
Pinus tabuliformis Carrière ●

300263　Pinus tabuliformis Carrière var. yunnanensis (Franch.) Dallim.
et A. B. Jacks. = Pinus yunnanensis Franch. ●

300264　Pinus taeda L.;火炬松（德达松,台大松）;Bastard Pine,Bull
Pine,Frankincense Pine,Loblolly Pine,Monterey Pine,Old Field
Pine,Rosemary Pine,Shortleaf Pine,Short-leaf Pine,Southern Pine,
Torch Pine ●

300265　Pinus taeda L. 'Dixie';南部火炬松;Dwarf Loblolly Pine ●

300266　Pinus taeda L. 'Nana';矮火炬松;Dwarf Loblolly Pine ●

300267　Pinus taeda L. var. heterophylla Elliott = Pinus elliottii Engelm. ●

300268　Pinus taihangshanensis Hu et C. Y. Yao = Pinus tabuliformis
Carrière ●

300269　Pinus taiwanensis Hayata;黄山松（长穗松,松柏,台湾二叶
松,台湾二针松,台湾松）;Huangshan Mountain Pine,Huangshan
Pine,Taiwan Pine,Taiwan Red Pine ●

300270　Pinus taiwanensis Hayata var. damingshanensis W. C. Cheng et
L. K. Fu;大明松;Daming Pine,Damingshan Pine ●

300271　Pinus taiwanensis Hayata var. damingshanensis W. C. Cheng et
L. K. Fu = Pinus taiwanensis Hayata ●

300272　Pinus takahasii Nakai;兴凯湖松（黑河赤松,兴凯赤松,兴凯
松）;Takahas Pine ●◇

300273　Pinus takahasii Nakai = Pinus densiflora Siebold et Zucc. var.
ussuriensis Liou et Z. Wang ●

300274　Pinus takahasii Nakai = Pinus sylvestris L. var. mongolica Litv. ●

300275　Pinus taxifolia Lamb. = Pseudotsuga menziesii (Mirb.) Franco ●

300276　Pinus tenuifolia Benth.;薄叶松●☆

300277　Pinus teocote Schltdl. et Champ.;卷叶松;Aztec Pine,Mexican
Common Pine,Twisted-leaf Pine ●☆

300278　Pinus termifolia Benth.;细叶松;Thinleaf Pine ●☆

300279　Pinus thunbergiana Franco ' Thunderhead';雷暴松;

Thunderhead Pine ●☆

300280 Pinus thunbergiana Franco ＝ Pinus thunbergii Parl. ●

300281 Pinus thunbergii Parl.；黑松(白芽松,黑皮刺松树,马尾松,日本黑松,松树)；Black Pine,Japan Black Pine,Japanese Black Pine,Thunberg Pine ●

300282 Pinus thunbergii Parl. 'Kotobuki'；考陶布克黑松●☆

300283 Pinus thunbergii Parl. 'Majestic Beauty'；大美人黑松●☆

300284 Pinus thunbergii Parl. 'Tsukasa'；朱卡萨黑松●☆

300285 Pinus thunbergii Parl. f. multicaulis Uyeki；多茎黑松●☆

300286 Pinus tokunagai Nakai ＝ Pinus tabuliformis Carrière ●

300287 Pinus tonkinensis A. Chev. ＝ Pinus latteri Mason ●

300288 Pinus torreyana Parry ex Carrière；陶松(吐丽松)；Mainland Torrey. Pine,Soleded Pine,Torrey Pine ●☆

300289 Pinus tropicalis Morelet；热带松；Cuba Pine,Tropical Pine ●

300290 Pinus tuberculata D. Don；球锥松；Knob-cone Pine ●☆

300291 Pinus tuberculata Gordon ＝ Pinus attenuata Lemmon ●☆

300292 Pinus uncinata Ramond ex DC. ＝ Pinus pumilio Haenke ●☆

300293 Pinus uncinata Steud. ＝ Pinus sylvestris L. ●

300294 Pinus ussuriensis (Liou et Z. Wang) W. C. Cheng et Y. W. Law ＝ Pinus takahasii Nakai ●◇

300295 Pinus uyematsui Hayata ＝ Pinus morrisonicola Hayata ●

300296 Pinus virginiana Mill.；维州松(北美二针松,北美洲松,杰塞松,维吉尼亚松,矮松)；Jersey Pine,Scrub Pine,Virginia Pine ●

300297 Pinus wallichiana A. B. Jacks.；乔松；Bhutan Pine,Blue Pine,Griffith Pine,Himalayan Blue Pine,Himalayan Pine,Himalayas Pine,Lofty Pine,Western Himalayan Pine ●

300298 Pinus wangii Hu et W. C. Cheng；毛枝五针松(滇南松,软木松,云南五针松)；Wang Pine ●◇

300299 Pinus wangii Hu et W. C. Cheng var. kwangtungensis (Chun et Tsiang) Silba ＝ Pinus kwangtungensis Chun ex Tsiang ●◇

300300 Pinus wangii Hu et W. C. Cheng var. kwangtungensis (Chun ex Tsiang) W. C. Cheng et Y. W. Law ＝ Pinus kwangtungensis Chun ex Tsiang ●◇

300301 Pinus washoensis H. Mason et Stockw.；华树松；Sierra Nevada Pine,Washoe Pine ●

300302 Pinus webbiana Wall. ex Lamb. ＝ Abies spectabilis (D. Don) Spach ●

300303 Pinus wilsonii Shaw ＝ Pinus densata Mast. ●

300304 Pinus yamazutai Uyeki ＝ Pinus sylvestris L. var. mongolica Litv. ●

300305 Pinus yunnanensis Franch.；云南松(长毛松,飞松,青松)；Burma Pine,Yunnan Pine ●

300306 Pinus yunnanensis Franch. var. pygmaea (J. R. Xue) J. R. Xue；地盘松；Pygmy Yunnan Pine ●

300307 Pinus yunnanensis Franch. var. pygmaea J. R. Xue ＝ Pinus yunnanensis Franch. var. pygmaea (J. R. Xue) J. R. Xue ●

300308 Pinus yunnanensis Franch. var. tenuifolia W. C. Cheng et Y. W. Law；细叶云南松；Thinleaf Yunnan Pine ●

300309 Pinzona Mart. et Zucc. (1832)；高攀五桠果属■☆

300310 Pinzona coriacea Mart. et Zucc.；高攀五桠果■☆

300311 Pioctonon Raf. ＝ Heliotropium L. ●■

300312 Piofontia Cuatrec. ＝ Diplostephium Kunth ●☆

300313 Pionandra Miers ＝ Cyphomandra Mart. ex Sendtn. ●■

300314 Pionocarpus S. F. Blake ＝ Iostephane Benth. ●☆

300315 Piora J. Kost. (1966)；香菀木属●☆

300316 Piora ericoides J. Kost.；香菀木●☆

300317 Pioriza Raf. ＝ Gentiana L. ■

300318 Pioriza Raf. ＝ Picriza Raf. ■

300319 Piotes Sol. ex Britton ＝ Augea Thunb. (保留属名)■☆

300320 Pipaceae Dulac ＝ Aristolochiaceae Juss. (保留科名)■●

300321 Pipalia Swkes ＝ Litsea Lam. (保留属名)●

300322 Piparea Aubl. ＝ Casearia Jacq. ●

300323 Piper L. (1753)；胡椒属；Pepper ●■

300324 Piper acquemontianum (Kunth) DC.；阿克胡椒●☆

300325 Piper acutifolium Ruiz et Pav.；尖叶胡椒■☆

300326 Piper acutifolium Ruiz et Pav. var. subverbascifolium C. DC.；毛蕊花叶胡椒■☆

300327 Piper aduncum L.；钩状胡椒■☆

300328 Piper aduncum Vell. ＝ Piper aduncum L. ■☆

300329 Piper albispicum C. DC. ＝ Piper sarmentosum Roxb. ■

300330 Piper amalgo L.；印度胡椒(西印度胡椒)；Indian Pepper ●■☆

300331 Piper amplexicaule Sw. ＝ Peperomia amplexicaulis (Sw.) A. Dietr. ■☆

300332 Piper andreanum C. DC.；安岛胡椒■☆

300333 Piper angustifolium Vahl；狭叶胡椒；Matico,Matico Pepper ●☆

300334 Piper arbelaezii Tel. et Yunck.；哥伦比亚胡椒●☆

300335 Piper arborescens Roxb.；兰屿胡椒(兰屿风藤)；Lanyu Pepper ■

300336 Piper arborescens Roxb. var. angustilimbum Quisumb. ＝ Piper arborescens Roxb. ■

300337 Piper arboreum Aubl.；乔木胡椒●☆

300338 Piper arboricola C. DC.；小叶崖爬香(薄叶风藤,风藤,小叶爬崖香)；Littleleaf Pepper,Small-leaved Pepper ●■

300339 Piper arboricola C. DC. ＝ Piper kadsura (Choisy) Ohwi ●■

300340 Piper arieianum C. DC.；艾氏胡椒●☆

300341 Piper attenuatum Buch. -Ham. ex Wall.；卵叶胡椒(狭胡椒)；Ovateleaf Pepper,Ovateleaved Pepper ■

300342 Piper aurantiacum Wall. ex C. DC. ＝ Piper wallichii (Miq.) Hand. -Mazz. ■

300343 Piper aurantiacum Wall. ex C. DC. var. hupeense C. DC. ＝ Piper wallichii (Miq.) Hand. -Mazz. ■

300344 Piper auritum Kunth；耳状胡椒；Vera Cruz Pepper ■☆

300345 Piper auritum Kunth var. amplifolium C. DC.；大叶耳状胡椒(大叶胡椒)；Vera Cruz Pepper ■☆

300346 Piper austrosinense Y. C. Tseng；华南胡椒(弯穗蒟)；S. China Pepper,South China Pepper ●

300347 Piper bambusifolium Y. C. Tseng；竹叶胡椒(山胡椒,山药)；Bambooleaf Pepper,Bamboo-leaved Pepper ●

300348 Piper banksii Miq.；班克胡椒；Banks' Pepper ●☆

300349 Piper bavinum C. DC.；腺脉蒟；Glandularnerve Pepper,Glandular-nerved Pepper ■

300350 Piper bavinum C. DC. ＝ Piper thomsonii (C. DC.) Hook. f. ■

300351 Piper begoniifolia (Blume ex Schult. et Schult. f.) C. DC. ＝ Zippelia begoniifolia Blume ex Schult. et Schult. f. ■

300352 Piper begoniifolium (Blume) Quisumb. ＝ Zippelia begoniifolia Blume ex Schult. et Schult. f. ■

300353 Piper bequaertii De Wild. ＝ Piper capense L. f. ●☆

300354 Piper betle L.；蒌叶(槟榔蒟,槟榔蒌,大荜芨,大荜菝,大芦子,扶留藤,枸酱,蒟酱,蒟酱叶,蒟青,蒟叶,蒟子,荖,荖草,荖藤,荖叶,娄子,蒌蒟,蒌叶胡椒,芦子,芦子女,橙叶,青蒟,青荖,青蒌,土荜菝,土荜芨,香荖,香蒌)；Betel,Betel Pepper ●

300355 Piper betle L. var. psilocarpum C. DC. ＝ Piper nudibaccatum Y. C. Tseng ■

300356 Piper blandum Jacq. ＝ Peperomia blanda (Jacq.) Kunth ■

300357 Piper boehmeriifolium (Miq.) C. DC.；苎叶蒟(大肠风,大麻疙瘩,芦子兰,芦子藤,荨麻叶胡椒,荨麻叶蒟,十八症,野胡椒,

叶子兰,苎叶蒌,苎叶山蒟);Falsenettleleaf Pepper ●

300358 Piper boehmeriifolium（Miq.）C. DC. var. glabricaule（C. DC.）M. G. Gilbert et N. H. Xia;光茎胡椒;Glabrous-stem Pepper ●

300359 Piper boehmeriifolium（Miq.）C. DC. var. tonkinense C. DC.;光轴苎叶蒟(大肠风,光轴苎蒟,十八症,歪叶子兰,小麻疙瘩,萱叶蒌);Tonkin Falsenettleleaf Pepper,Tonkin Pepper ●

300360 Piper boehmeriifolium（Miq.）C. DC. var. tonkinense C. DC. = Piper boehmeriifolium（Miq.）C. DC. ●

300361 Piper bonii C. DC.;复毛胡椒;Bon Pepper ■

300362 Piper bonii C. DC. var. macrophyllum Y. C. Tseng;大叶复毛胡椒;Bigleaf Bon Pepper ■

300363 Piper brachyrhachis C. H. Wright = Piper capense L. f. var. brachyrhachis（C. H. Wright）Verdc. ●☆

300364 Piper braehystachyum Wall. ex Hook. f. = Piper mullesua D. Don ●

300365 Piper bredemeyeri Jacq.;布雷德胡椒■☆

300366 Piper brevieaule A. DC. = Piper sarmentosum Roxb. ■

300367 Piper caninum Blume;犬胡椒●☆

300368 Piper capense L. f.;好望角胡椒●☆

300369 Piper capense L. f. var. brachyrhachis（C. H. Wright）Verdc.;短柱胡椒●☆

300370 Piper cathayanum M. G. Gilbert et N. H. Xia;华山蒌;China Pepper,Chinese Pepper ■

300371 Piper chaba Hunter = Piper retrofractum Vahl ■

300372 Piper chaudocanum C. DC.;勐海胡椒;Menghai Pepper ■

300373 Piper chinense Miq.;中华胡椒;China Pepper,Chinese Pepper ●

300374 Piper clusii（Miq.）C. DC.;克氏胡椒;African Cubebs,West African Black Pepper ●☆

300375 Piper clusii（Miq.）C. DC. = Piper guineense Schumach. et Thonn. ●☆

300376 Piper clusii C. DC. = Piper clusii（Miq.）C. DC. ●☆

300377 Piper cubeba L. f.;荜澄茄(毕茄,澄茄,婆澄茄,毗陵茄子);Cubeb,Cubeb Pepper,Cubebs,Java Pepper,Tailed Pepper ●

300378 Piper curtipedunculum A. DC. = Piper pedicellatum C. DC. ■

300379 Piper curtipedunculum C. DC.;细苞胡椒;Slenderbract Pepper,Slender-bracted Pepper ●

300380 Piper curtipedunculum C. DC. = Piper pedicellatum C. DC. ■

300381 Piper damiaoshanense Y. C. Tseng;大苗山胡椒;Damiao Mountain Pepper,Damiaoshan Pepper ■

300382 Piper dilatatum Rich. ex Kunth;膨大胡椒;Elderbush,Rock-bush ■☆

300383 Piper dolichastachyum M. G. Gilbert et N. H. Xia;长穗胡椒■

300384 Piper elongatum Vahl;长胡椒●☆

300385 Piper emeiense Y. C. Tseng;峨眉胡椒;Emei Pepper,Omei Pepper ●

300386 Piper emeiense Y. C. Tseng = Piper wallichii（Miq.）Hand.-Mazz. ■

300387 Piper emirnense Baker = Piper capense L. f. ●☆

300388 Piper exiguicaule Yunck. = Peperomia pellucida（L.）Kunth ●

300389 Piper famechonii C. DC. = Piper guineense Schumach. et Thonn. ●☆

300390 Piper ferriei C. DC.;费氏胡椒●☆

300391 Piper flagelliforme Yamam. = Piper hainanense Hemsl. ●

300392 Piper flaviflorum C. DC.;黄花胡椒;Yellowflower Pepper,Yellow-flowered Pepper ■

300393 Piper freireifolia Hochst. = Peperomia pellucida（L.）Kunth ●

300394 Piper futokadsura Siebold = Piper kadsura（Choisy）Ohwi ●■

300395 Piper glabellum Sw. = Peperomia glabella（Sw.）A. Dietr. ■☆

300396 Piper glabricaule C. DC. = Piper boehmeriifolium（Miq.）C. DC. var. glabricaule（C. DC.）M. G. Gilbert et N. H. Xia ●

300397 Piper guianeense K. Schum. et Thonn.;几内亚胡椒;Ashanti Pepper,Benin Pepper,Bush Pepper,Guinea Cubeb,Guinea Cubebs,Guinea Pepper,West African Black Pepper ●☆

300398 Piper guianiense K. Schum. et Thonn.;圭亚那胡椒■☆

300399 Piper guigual Buch.-Ham. ex D. Don = Piper mullesua D. Don ●

300400 Piper guigual D. Don = Piper mullesua D. Don ●

300401 Piper guineense Schumach. et Thonn.;阿善堤胡椒;Ashanti Pepper,Benin Pepper,Guinea Cubebs ●☆

300402 Piper guineense Schumach. et Thonn. var. clusii（Miq.）Engl. = Piper guineense Schumach. et Thonn. ●☆

300403 Piper guineense Schumach. et Thonn. var. congolense De Wild. ex C. DC. = Piper guineense Schumach. et Thonn. ●☆

300404 Piper guineense Schumach. et Thonn. var. gilletii C. DC. = Piper guineense Schumach. et Thonn. ●☆

300405 Piper guineense Schumach. et Thonn. var. thomeanum C. DC. = Piper guineense Schumach. et Thonn. ●☆

300406 Piper gymnostachyum A. DC. = Piper sarmentosum Roxb. ■

300407 Piper hainanense Hemsl.;海南蒟(海南胡椒,山胡椒,上树胡椒);Hainan Pepper ●

300408 Piper hancei Maxim.;山蒟(穿壁风,二十四症,风气药,过节风,海风藤,汉斯胡椒,酒饼藤,蓝藤,毛蒟,南藤,爬岩香,山蒌,上树风,石蒟,石南藤);Hance Pepper,Wild Pepper ●■

300409 Piper harmandii C. DC.;哈氏胡椒●

300410 Piper henryci C. DC. = Piper wallichii（Miq.）Hand.-Mazz. ■

300411 Piper henryi sensu Metcalf = Piper wallichii（Miq.）Hand.-Mazz. ■

300412 Piper hispidum Hayata = Piper kadsura（Choisy）Ohwi ●■

300413 Piper hispidum Hayata = Piper sintenense Hatus. ■

300414 Piper hochiense Y. C. Tseng;河池胡椒;Hechi Pepper ■

300415 Piper hongkongense C. DC.;毛蒟(绒毛胡椒,野芦子);Hairy Pepper,Pubescent Pepper ■

300416 Piper hookeri Miq.;胡克胡椒;Hooker. Pepper ●☆

300417 Piper humile Vahl = Peperomia humilis（Vahl）A. Dietr. ■☆

300418 Piper ichangense C. DC. = Piper wallichii（Miq.）Hand.-Mazz. ■

300419 Piper infossibaccatum A. Huang;嵌果胡椒●

300420 Piper infossum Y. C. Tseng;沉果胡椒;Buried Pepper ●

300421 Piper infossum Y. C. Tseng var. nudum Y. C. Tseng;裸果沉果胡椒(落叶沉果胡椒);Naked Buried Pepper ●

300422 Piper interruptum Opiz;疏果胡椒（多脉风藤）;Interrupted Pepper,Loosefruit Pepper ●

300423 Piper interruptum Opiz var. multinervum C. DC.;多脉风藤;Manynerved Loosefruit Pepper ■

300424 Piper interruptum Opiz var. multinervum C. DC. = Piper interruptum Opiz ■

300425 Piper jaborandii Vell.;耶仆兰胡椒（巴西胡椒）;Jaborand Pepper ●☆

300426 Piper jacquemontianum（Kunth）DC.;杰克胡椒☆

300427 Piper kadsura（Choisy）Ohwi;风藤(巴岩香,大风藤,海风藤,苈藤,爬崖香,石楠藤,细叶青风藤,细叶青蒌藤,野杜衡,真风藤);Japanese Pepper,Kadsura Pepper,Pepper ●■

300428 Piper kadsura（Choisy）Ohwi f. macrophyllum（Nakai）M. Mizush.;大叶风藤●☆

300429 Piper kadsura（Choisy）Ohwi var. boninense M. Mizush. =

Piper kadsura（Choisy）Ohwi ●■

300430 Piper kadsura（Choisy）Ohwi var. macrophyllum（Nakai）Nemoto = Piper kadsura（Choisy）Ohwi f. macrophyllum（Nakai）M. Mizush. ●☆

300431 Piper kadzura（Choisy）Ohwi var. postelsiana（Maxim.）M. Hiroe = Piper postelsianum Maxim. ●

300432 Piper kawakamii Hayata；恒春胡椒（川上氏爬崖香，恒春风藤）；Hengchun Pepper，Kawakami Pepper ■

300433 Piper kotoense Yamam. = Piper arborescens Roxb. ■

300434 Piper kwashoense Hayata；绿岛胡椒（绿岛风藤）；Kwasho Pepper ■

300435 Piper laetispicum C. DC.；大叶蒟（山胡椒，小肠风，野胡椒）；Big-leaved Pepper，Largeleaf Pepper ■

300436 Piper lampong L.；印尼大叶胡椒■☆

300437 Piper lappacerum C. DC. = Zippelia begoniifolia Blume ex Schult. et Schult. f. ■

300438 Piper lappaceum（Bennett）C. DC. = Zippelia begoniifolia Blume ex Schult. et Schult. f. ■

300439 Piper latispicum C. DC. = Piper laetispicum C. DC. ●

300440 Piper laurentii De Wild. = Piper guineense Schumach. et Thonn. ●☆

300441 Piper leonense C. DC. = Piper guineense Schumach. et Thonn. ●☆

300442 Piper lingshuiense Y. C. Tseng；陵水胡椒；Lingshui Pepper ●

300443 Piper lolot A. DC. = Piper sarmentosum Roxb. ■

300444 Piper lolot C. DC.；洛洛胡椒■☆

300445 Piper lolot C. DC. = Piper sarmentosum Roxb. ■

300446 Piper longum L.；荜茇（毕拔，荜拔，荜拔梨，荜菝子，荜勃，荜茇，勃梨，蛤蒌，椹圣，鼠尾）；Biba Pepper，Indian Long Pepper，Jaborandi Pepper，Long Pepper ■

300447 Piper lowong Blume；洛旺胡椒●☆

300448 Piper maclurei Merr. = Piper laetispicum C. DC. ●

300449 Piper macropodum C. DC.；粗梗胡椒；Macropodous Pepper，Thickstem Pepper ■

300450 Piper madidum Y. C. Tseng；西藏胡椒；Tibet Pepper，Xizang Pepper ●

300451 Piper madidum Y. C. Tseng = Piper rhytidocarpum Hook. f. ■●

300452 Piper magnoliifolium Jacq. = Peperomia magnoliifolia（Jacq.）A. Dietr. ■☆

300453 Piper margiatum Jacq.；边缘胡椒●☆

300454 Piper martinii C. DC.；毛山蒟；Hair Wild Pepper，Martin Pepper ■

300455 Piper martinii C. DC. = Piper wallichii（Miq.）Hand.-Mazz. ■

300456 Piper matthewii Dunn = Piper hancei Maxim. ●■

300457 Piper mekongense A. DC. = Piper polysyphonum C. DC. ●

300458 Piper methysticum G. Forst.；卡瓦胡椒（麻醉椒，醉椒）；Kava，Kava Pepper，Kawa Pepper，Yanggona，Yangona ●☆

300459 Piper mischocarpum Y. C. Tseng；柄果胡椒；Long-pedicelled Pepper，Stipefruit Pepper ■

300460 Piper mullesua D. Don；短蒟（钮子跌打，细芦子藤）；Globular Pepper ●

300461 Piper mutabile C. DC.；变叶胡椒；Variablele Pepper，Variableleaf Pepper ■

300462 Piper nepalense Miq.；尼泊尔胡椒；Nepal Pepper ●

300463 Piper nepalense Miq. = Piper suipigua Buch.-Ham. ex D. Don ●

300464 Piper nigrum L.；胡椒（白川，白古月，白胡椒，浮椒，古月，黑川，黑胡，黑胡椒，糊椒，昧履支，木椒，玉椒，章）；Black Pepper，Madagascar Pepper，Pepper，Pepper Plant，Pepper Vine，White Pepper ●■

300465 Piper nigrum L. var. macrostachyum A. DC. = Piper rhytidocarpum Hook. f. ■●

300466 Piper novae-hollandiae Miq.；新希腊胡椒☆

300467 Piper nudibaccatum Y. C. Tseng；裸果胡椒；Bare-fruited Pepper，Nakefruit Pepper ■

300468 Piper nummularifolium Sw. = Peperomia rotundifolia（L.）Kunth ■☆

300469 Piper obliquum Ruiz et Pav.；叙胡椒●☆

300470 Piper obtusifolium L. = Peperomia obtusifolia（L.）A. Dietr. ■☆

300471 Piper odoratum C. DC.；芳香胡椒●☆

300472 Piper officinarum（Miq.）C. DC. = Piper retrofractum Vahl ■

300473 Piper ornatum N. E. Br.；观赏胡椒（彩脉风藤，彩脉胡椒，苏拉威西胡椒）；Celebes Pepper ●

300474 Piper paepuloides Roxb.；芦子藤（短序柄胡椒）●☆

300475 Piper pedicellatum C. DC.；角果胡椒；Pedicelled Pepper，Pedicled Pepper ■

300476 Piper pellucidum L. = Peperomia pellucida（L.）Kunth ■

300477 Piper peltatum L.；盾状胡椒●☆

300478 Piper petiolatum Hook. f.；具柄胡椒●☆

300479 Piper philippinum Miq.；台东胡椒（菲律宾胡椒）；Philippine Pepper ■

300480 Piper pierrei C. DC. = Piper sarmentosum Roxb. ■

300481 Piper pingbienense Y. C. Tseng；屏边胡椒；Pingbian Pepper ■

300482 Piper piscatorum Trel. et Yunck.；毒鱼胡椒●☆

300483 Piper pleiocarpum C. C. Chang ex Y. C. Tseng；线梗胡椒；Linearstalk Pepper，Maltifruited Pepper，Many-fruit Pepper ●

300484 Piper polysyphonum C. DC.；樟叶胡椒；Camphorleaf Pepper，Camphortreeleaf Pepper ●

300485 Piper ponesheense C. DC.；肉轴胡椒；Poneshe Pepper，Succulentstem Pepper ■

300486 Piper postelsianum Maxim. = Piper umbellatum L. ●

300487 Piper postelsianum Maxim. = Pothomorphe subpeltata（Willd.）Miq. ●

300488 Piper psitorhache C. DC.；墨西哥胡椒；Mexican Pepper ●☆

300489 Piper puberulilimbum C. DC.；毛叶胡椒（毛胡椒，玉溪天仙藤）；Hairleaf Pepper，Hairy-leaved Pepper ■

300490 Piper puberulum（Benth.）Maxim. = Piper hongkongense C. DC. ■

300491 Piper pubicatulum C. DC.；岩椒（岩参）；Hairy Pepper，Rocky Pepper ■

300492 Piper punctulivenum C. DC. = Piper thomsonii（C. DC.）Hook. f. ■

300493 Piper punctulivenum C. DC. var. parvifolium C. DC. = Piper thomsonii（C. DC.）Hook. f. ■

300494 Piper reflexum L. f. = Peperomia tetraphylla（G. Forst.）Hook. et Arn. ■

300495 Piper retrofractum Vahl；假荜茇（荜澄茄，大荜茇，胡椒藤，假荜拔，洽巴胡椒，爪哇长果胡椒，爪哇长胡椒）；False Biba Pepper，Long Pepper，Retrofracted Pepper ■

300496 Piper retusum L. f. = Peperomia retusa（L. f.）A. Dietr. ■☆

300497 Piper rhytidocarpum Hook. f.；皱果胡椒■●

300498 Piper ribesioides；茶蘸胡椒●☆

300499 Piper rohrii C. DC.；罗尔胡椒●☆

300500 Piper rotundifolium L. = Peperomia rotundifolia（L.）Kunth ■☆

300501 Piper rubrum C. DC.；红果胡椒；Redfruit Pepper，Red-fruited Pepper ■

300502 Piper sacleuxii C. DC. = Piper capense L. f. ●☆

300503 Piper saigonense A. DC. = Piper sarmentosum Roxb. ■

300504 Piper sarmentosum Roxb. ；假蒟（巴岩香，毕拔，毕拔子，毕博敬，荜拔，臭蒌，大柄蒌，封口好，蛤蒟，蛤荖，蛤蒌，假姜，假老，假蒌，假蒌，马蹄蒌，酿苦瓜，山蒌，酸苦瓜，猪拔菜，钻骨风）；Ranner Pepper, Sarmentose Pepper ■

300505 Piper semiimmersum C. DC. ；缘毛胡椒；Ciliate Pepper ■

300506 Piper sempervirens (Trel.) Lundell；常绿胡椒●☆

300507 Piper senporeiense Yamam. ；斜叶蒟；Oblique-leaved Pepper, Slantleaf Pepper ●

300508 Piper sidifolium Link et Otto；黄花稔胡椒●☆

300509 Piper sinense (Champ. ex Benth.) C. DC. = Piper cathayanum M. G. Gilbert et N. H. Xia ■

300510 Piper sintenense Hatus. ；小叶爬崖香（薄叶风藤）■

300511 Piper sintenense Hatus. = Piper arboricola C. DC. ●■

300512 Piper spirei C. DC. ；滇南胡椒；Spire Pepper ●

300513 Piper spirei C. DC. = Piper boehmeriifolium (Miq.) C. DC. ●

300514 Piper spirei C. DC. var. pilosius C. DC. = Piper boehmeriifolium (Miq.) C. DC. ●

300515 Piper stipitiforme C. C. Chang ex Y. C. Tseng；短柄胡椒；Short-stalked Pepper, Shortstipe Pepper ■

300516 Piper subcordata Hayata = Piper kawakamii Hayata ■

300517 Piper subglaucescens C. DC. = Piper kadsura (Choisy) Ohwi ●■

300518 Piper submultinerve C. DC. ；多脉胡椒；Multinerved Pepper, Veing Pepper ■

300519 Piper submultinerve C. DC. var. nandanicum Y. C. Tseng；狭叶多脉胡椒；Narrowleaf Veing Pepper ■

300520 Piper subpeltatum Willd. = Piper umbellatum L. ●

300521 Piper subpeltatum Willd. = Pothomorphe subpeltata (Willd.) Miq. ●

300522 Piper suipigua Buch. -Ham. ；滇西胡椒■

300523 Piper suipigua Buch. -Ham. ex D. Don = Piper suipigua Buch. -Ham. ■

300524 Piper sylvaticum Roxb. ；长柄胡椒（银叶胡椒）；Longstalk Pepper ■

300525 Piper sylvaticum Roxb. = Piper suipigua Buch. -Ham. ex D. Don ■

300526 Piper szemaoense C. DC. ；思茅胡椒；Simao Pepper ■

300527 Piper szemaoense C. DC. = Piper macropodum C. DC. ●

300528 Piper taiwanense T. T. Lin et S. Y. Lu；台湾蒟藤（台湾胡椒）；Taiwan Pepper ■

300529 Piper terminaliflorum Y. C. Tseng；顶花胡椒；Terminal-flower Pepper, Topflower Pepper, Top-flowered Pepper ●

300530 Piper terminaliflorum Y. C. Tseng = Piper boehmeriifolium (Miq.) C. DC. ●

300531 Piper tetraphyllum G. Forst. = Peperomia tetraphylla (G. Forst.) Hook. et Arn. ■

300532 Piper thomsonii (C. DC.) Hook. f. ；球穗胡椒；Ballspike Pepper ■

300533 Piper thomsonii (C. DC.) Hook. f. var. microphyllum Y. C. Tseng；小叶球穗胡椒；Littleleaf Pepper, Smallleaf Pepper ■

300534 Piper trichopodum C. DC. = Piper capense L. f. ●☆

300535 Piper trichostachyon C. DC. ；毛穗胡椒；Hairy-spike Pepper ●☆

300536 Piper tricolor Y. C. Tseng；三色胡椒；Threecolored Pepper, Tricolour Pepper ■

300537 Piper tsangyuanense P. S. Chen et P. C. Zhu；粗穗胡椒（狗芦子）；Congyuan Pepper, Thickspike Pepper ■

300538 Piper tsengianum M. G. Gilbert et N. H. Xia；瑞丽胡椒●

300539 Piper tuberculatum Jacq. ；瘤状胡椒●☆

300540 Piper umbellatum L. ；台湾胡椒（大胡椒，伞花胡椒）；Taiwan Pepper ●

300541 Piper umbellatum L. var. glabrum C. DC. = Piper umbellatum L. ●

300542 Piper umbellatum Willd. var. subpeltatum (Willd.) C. DC. = Piper umbellatum L. ●

300543 Piper villiramulum C. DC. ；毛枝胡椒●☆

300544 Piper volkensii C. DC. = Piper capense L. f. ●☆

300545 Piper volkensii C. DC. f. crassiusculum Peter = Piper capense L. f. ●☆

300546 Piper volkensii C. DC. f. eucordatum Peter = Piper capense L. f. ●☆

300547 Piper volkensii C. DC. f. ovatum Peter = Piper capense L. f. ●☆

300548 Piper wallichii (Miq.) Hand. -Mazz. ；石南藤（巴岩香，丁父，丁公寄，丁公藤，风藤，湖北胡椒，蓝藤，毛蒟，毛蒌，南藤，爬崖香，爬崖香藤，爬岩香，绒毛胡椒，石蒌，搜山虎，瓦氏胡椒，小毛蒟）；Wallich Pepper ■

300549 Piper wallichii (Miq.) Hand. -Mazz. var. hupeense (C. DC.) Hand. -Mazz. = Piper wallichii (Miq.) Hand. -Mazz. ■

300550 Piper wangii M. G. Gilbert et N. H. Xia；景洪胡椒■

300551 Piper wichmannii C. DC. ；新几内亚胡椒●☆

300552 Piper yinkiangense Y. C. Tseng；盈江胡椒；Yingjiang Pepper, Yingkiang Pepper ■

300553 Piper yui M. G. Gilbert et N. H. Xia；椭圆叶胡椒■

300554 Piper yunnanense Y. C. Tseng；蒟子（大麻疙瘩）；Yunnan Pepper ●

300555 Piper zippelia C. DC. = Zippelia begoniifolia Blume ex Schult. et Schult. f. ■

300556 Piperaceae C. Agardh = Piperaceae Giseke(保留科名)●■

300557 Piperaceae Giseke(1792)(保留科名)；胡椒科；Pepper Family ●■

300558 Piperanthera C. DC. = Peperomia Ruiz et Pav. ●

300559 Piperella (C. Presl ex Rchb.) Spach = Thymus L. ●

300560 Piperella C. Presl = Micromeria Benth. (保留属名)■●

300561 Piperi St. -Lag. = Piper L. ●■

300562 Piperia Rydb. (1901)；派珀兰属■☆

300563 Piperia Rydb. = Platanthera Rich. (保留属名)■

300564 Piperia candida Rand. Morgan et Ackerman；白派珀兰■☆

300565 Piperia colemanii Rand. Morgan et Glic. ；科尔曼派珀兰■☆

300566 Piperia cooperi (S. Watson) Rydb. ；库珀派珀兰■☆

300567 Piperia elegans (Lindl.) Rydb. ；雅致派珀兰■☆

300568 Piperia elegans (Lindl.) Rydb. var. elata (Jeps.) Luer = Piperia elongata Rydb. ■☆

300569 Piperia elongata Rydb. ；伸长派珀兰■☆

300570 Piperia elongata Rydb. subsp. michaelii (Greene) Ackerman = Piperia michaelii (Greene) Rydb. ■☆

300571 Piperia lancifolia Rydb. = Piperia cooperi (S. Watson) Rydb. ■☆

300572 Piperia leptopetala Rydb. ；细瓣派珀兰■☆

300573 Piperia maritima Rydb. = Piperia elegans (Lindl.) Rydb. ■☆

300574 Piperia michaelii (Greene) Rydb. ；迈克尔派珀兰■☆

300575 Piperia multiflora Rydb. = Piperia elegans (Lindl.) Rydb. ■☆

300576 Piperia transversa Suksd. ；横派珀兰■☆

300577 Piperia yadonii Rand. Morgan et Ackerman；亚当派珀兰■☆

300578 Piperiphorum Neck. = Piper L. ●■

300579 Piperodendron Fabr. = Schinus L. ●

300580 Piperodendron Heist. ex Fabr. = Schinus L. ●

300581 Piperomia Pritz. = Peperomia Ruiz et Pav. ■

300582 Piperomia argyreia E. Morren = Peperomia argyreia E. Morren ■☆

300583　Piperomia argyroneura Hort. ＝Peperomia argyroneura Hort. ■☆

300584　Piperomia cavaleriei C. DC. ＝Peperomia cavaleriei C. DC. ■

300585　Piperomia dindygulensis Miq. ＝Peperomia blanda（Jacq.）Kunth ■

300586　Piperomia duclouxii C. DC. ＝Peperomia duclouxii C. DC. ■

300587　Piperomia duclouxii C. DC. ＝Peperomia heyneana Miq. ■

300588　Piperomia heyneana Miq. ＝Peperomia heyneana Miq. ■

300589　Piperomia japonica Makino ＝Peperomia japonica Makino ■

300590　Piperomia leptostachya Hook. et Arn. ＝Peperomia blanda（Jacq.）Kunth ■

300591　Piperomia leptostachya Hook. et Arn. var. cambodiana（C. DC.）Merr. ＝Peperomia leptostachya Hook. et Arn. var. cambodiana（C. DC.）Merr. ■

300592　Piperomia leptostachya Hook. et Arn. var. cambodiana（C. DC.）Merr. ＝Peperomia blanda（Jacq.）Kunth ■

300593　Piperomia maculosa（L.）Hook. ＝Peperomia maculosa（L.）Hook. ■☆

300594　Piperomia magnoliifolia A. Dietr. ＝Peperomia magnoliifolia（Jacq.）A. Dietr. ■☆

300595　Piperomia nakaharai Hayata ＝Peperomia nakaharai Hayata ■

300596　Piperomia obtusifolia A. Dietr. ＝Peperomia obtusifolia A. Dietr. ■☆

300597　Piperomia pellucida（L.）Kunth ＝Peperomia pellucida（L.）Kunth ■

300598　Piperomia reflexa（L. f.）A. Dietr. ＝Peperomia reflexa（L. f.）A. Dietr. ■☆

300599　Piperomia reflexa（L. f.）A. Dietr. f. sinensis C. DC. ＝Peperomia tetraphylla（G. Forst.）Hook. et Arn. var. sinensis（C. DC.）P. S. Chen et P. C. Zhu ■

300600　Piperomia tetraphylla（G. Forst.）Hook. et Arn. ＝Peperomia tetraphylla（G. Forst.）Hook. et Arn. ■

300601　Piperomia tetraphylla（G. Forst.）Hook. et Arn. var. sinensis（C. DC.）P. S. Chen et P. C. Zhu ＝Peperomia tetraphylla（G. Forst.）Hook. et Arn. var. sinensis（C. DC.）P. S. Chen et P. C. Zhu ■

300602　Piperomia tetraphylla（G. Forst.）Hook. et Arn. var. sinensis（C. DC.）P. S. Chen et P. C. Zhu ＝Peperomia tetraphylla（G. Forst.）Hook. et Arn. ■

300603　Piperonia Pritz. ＝Peperomia Ruiz et Pav. ■

300604　Pippenalia McVaugh（1972）；翠雀菊属■☆

300605　Pippenalia delphiniifolia（Rydb.）McVaugh；翠雀菊■☆

300606　Pipseva Raf. ＝Chimaphila Pursh ●■

300607　Piptadenia Benth.（1840）；落腺豆属（落腺蕊属）；Piptadenia ●☆

300608　Piptadenia Benth. ＝Parapiptadenia Brenan ●☆

300609　Piptadenia africana Hook. f. ；非洲落腺豆；African Piptadenia ●☆

300610　Piptadenia africana Hook. f. ＝Piptadeniastrum africanum（Hook. f.）Brenan ●☆

300611　Piptadenia amazonica Ducke；亚马逊落腺豆●☆

300612　Piptadenia aubrevillei Pellegr. ＝Newtonia aubrevillei（Pellegr.）Keay ●☆

300613　Piptadenia bequaertii De Wild. ＝Pseudoprosopis claessensii（De Wild.）G. C. C. Gilbert et Boutique ●☆

300614　Piptadenia boiviniana Baill. ＝Entada chrysostachys（Benth.）Drake ●☆

300615　Piptadenia buchananii Baker f. ；布氏落腺豆●☆

300616　Piptadenia buchananii Baker f. ＝Newtonia buchananii（Baker f.）G. C. C. Gilbert et Boutique ●☆

300617　Piptadenia chevalieri A. Chev. ＝Calpocalyx brevibracteatus Harms ●☆

300618　Piptadenia chevalieri Harms ＝Tetrapleura chevalieri（Harms）Baker f. ●☆

300619　Piptadenia claessensii De Wild. ＝Pseudoprosopis uncinata Evrard ●☆

300620　Piptadenia colubrina Vell. ＝Anadenanthera colubrina（Vell.）Brenan ●☆

300621　Piptadenia communis Benth. ；普通落腺豆；Common Piptadenia ●☆

300622　Piptadenia contorta Benth. ；巴西红心木●☆

300623　Piptadenia duparquetiana（Baill.）Pellegr. ＝Newtonia duparquetiana（Baill.）Keay ●☆

300624　Piptadenia elliotii Harms ＝Newtonia elliotii（Harms）Keay ●☆

300625　Piptadenia erlangeri Harms ＝Newtonia erlangeri（Harms）Brenan ●☆

300626　Piptadenia excelsa Lillo；大落腺豆（美丽红心木）●☆

300627　Piptadenia flabellata Baill. ＝Entada chrysostachys（Benth.）Drake ●☆

300628　Piptadenia gabunensis（Harms）Roberty ＝Cylicodiscus gabunensis Harms ●☆

300629　Piptadenia glandulifera Pellegr. ＝Newtonia glandulifera（Pellegr.）G. C. C. Gilbert et Boutique ●☆

300630　Piptadenia goetzei Harms ＝Elephantorrhiza goetzei（Harms）Harms ●☆

300631　Piptadenia gonoacantha J. F. Macbr. ；落腺豆●☆

300632　Piptadenia grandidieri Baill. ＝Entada chrysostachys（Benth.）Drake ●☆

300633　Piptadenia greveana Baill. ＝Entada chrysostachys（Benth.）Drake ●☆

300634　Piptadenia griffoniana（Baill.）Baker f. ＝Newtonia griffoniana（Baill.）Baker f. ●☆

300635　Piptadenia hildebrandtii Vatke ＝Newtonia hildebrandtii（Vatke）Torre ●☆

300636　Piptadenia kerstingii Harms ＝Aubrevillea kerstingii（Harms）Pellegr. ●☆

300637　Piptadenia klaineana Pierre ex A. Chev. ＝Newtonia griffoniana（Baill.）Baker f. ●☆

300638　Piptadenia leptoclada Baker ＝Gagnebina commersoniana（Baill.）R. Vig. ●☆

300639　Piptadenia leucocarpa Harms ＝Newtonia leucocarpa（Harms）Gilbert et Boutique ●☆

300640　Piptadenia macrocarpa Benth. ＝Anadenanthera macrocarpa（Benth.）Brenan ●☆

300641　Piptadenia mannii Oliv. ＝Entada mannii（Oliv.）Tisser. ●☆

300642　Piptadenia oudhensis Brandis ＝Indopiptadenia oudhensis（Brandis）Brenan ●☆

300643　Piptadenia patens Benth. ；广红心木●☆

300644　Piptadenia paucijuga Harms ＝Newtonia paucijuga（Harms）Brenan ●☆

300645　Piptadenia peregrina Benth. ＝Anadenanthera peregrina（L.）Speg. ●☆

300646　Piptadenia pervillei Vatke ＝Entada pervillei（Vatke）R. Vig. ●☆

300647　Piptadenia pittieri Harms；委内瑞拉落腺豆●☆

300648　Piptadenia rigida Benth. ；坚硬落腺豆（坚挺落腺豆）●☆

300649　Piptadenia schlechteri Harms ＝Adenopodia schlechteri（Harms）Brenan ■☆

300650　Piptadenia suaveolens Miq. ；香甜落腺豆●☆

300651　Piptadenia unijuga Pierre ex A. Chev. ＝Newtonia duparquetiana（Baill.）Keay ●☆

300652　Piptadenia winkleri Harms ＝Calpocalyx winkleri（Harms）

Harms ●☆

300653　Piptadenia zenkeri （Harms）Pellegr. = Newtonia griffoniana （Baill.）Baker f. ●☆

300654　Piptadeniastrum Brenan（1955）;落腺瘤豆属●☆

300655　Piptadeniastrum africanum （Hook. f.）Brenan;非洲落腺瘤豆; Agboin,Daboma,Dahoma,Ekhimi ●☆

300656　Piptadeniopsis Burkart（1944）;类落腺豆属(拟落腺豆属)●☆

300657　Piptadeniopsis lomentifera Burkart;类落腺豆●☆

300658　Piptandra Turcz. = Scholtzia Schauer ●☆

300659　Piptanthocereus （A. Berger）Riccob. = Cereus Mill. ●

300660　Piptanthus D. Don ex Sweet（1828）;黄花木属;Piptanthus ●

300661　Piptanthus Sweet = Piptanthus D. Don ex Sweet ●

300662　Piptanthus bicolor Craib = Piptanthus concolor R. L. Harrow ex Craib subsp. yunnanensis （Craib）Stapf ●

300663　Piptanthus bicolor Craib = Piptanthus nepalensis （Hook.）Sweet ●

300664　Piptanthus bombycinus Marquand = Piptanthus concolor R. L. Harrow ex Craib ●

300665　Piptanthus bombycinus Marquand = Piptanthus nepalensis （Hook.）Sweet ●

300666　Piptanthus chinens Przew. = Ammopiptanthus mongolicus （Maxim. ex Kom.）S. H. Cheng ●◇

300667　Piptanthus concolor R. L. Harrow ex Craib;黄花木(圆荚树); Common Piptanthus,Greenleaf Piptanthus,Green-leaved Piptanthus ●

300668　Piptanthus concolor R. L. Harrow ex Craib = Piptanthus nepalensis （Hook.）Sweet ●

300669　Piptanthus concolor R. L. Harrow ex Craib subsp. harrowii Stapf = Piptanthus nepalensis （Hook.）Sweet ●

300670　Piptanthus concolor R. L. Harrow ex Craib subsp. harrowii Stapf = Piptanthus concolor R. L. Harrow ex Craib ●

300671　Piptanthus concolor R. L. Harrow ex Craib subsp. yunnanensis （Craib）Stapf = Piptanthus concolor R. L. Harrow ex Craib ●

300672　Piptanthus concolor R. L. Harrow ex Craib subsp. yunnanensis （Craib）Stapf;云南黄花木;Yunnan Piptanthus ●

300673　Piptanthus concolor R. L. Harrow ex Craib subsp. yunnanensis Stapf = Piptanthus concolor R. L. Harrow ex Craib ●

300674　Piptanthus forrestii Craib;蒙自黄花木;Forrest Piptanthus ●

300675　Piptanthus forrestii Craib = Piptanthus concolor R. L. Harrow ex Craib ●

300676　Piptanthus forrestii Craib = Piptanthus nepalensis （Hook.）Sweet ●

300677　Piptanthus laburnifolius （D. Don）Stapf;金链叶黄花木(披针叶黄花木);Lanceleaf Piptanthus,Lance-leaved Piptanthus ●

300678　Piptanthus laburnifolius （D. Don）Stapf = Piptanthus nepalensis （Hook.）Sweet ●

300679　Piptanthus laburnifolius （D. Don）Stapf f. nepalensis Stapf = Piptanthus nepalensis （Hook.）Sweet ●

300680　Piptanthus laburnifolius （D. Don）Stapf f. sikkimensis Stapf = Piptanthus nepalensis （Hook.）Sweet ●

300681　Piptanthus leiocarpus Stapf = Piptanthus nepalensis （Hook.）Sweet ●

300682　Piptanthus leiocarpus Stapf = Piptanthus nepalensis Sweet f. leiocarpus （Stapf）S. Q. Wei ●

300683　Piptanthus leiocarpus Stapf f. sericopetalus P. C. Li = Piptanthus nepalensis Sweet f. sericopetalus （P. C. Li）S. Q. Wei ●

300684　Piptanthus leiocarpus Stapf var. sericopetalus P. C. Li = Piptanthus nepalensis （Hook.）Sweet ●

300685　Piptanthus leiocarpus Stapf var. sericopetalus P. C. Li = Piptanthus nepalensis Sweet f. sericopetalus （P. C. Li）S. Q. Wei ●

300686　Piptanthus mongolicus Maxim. = Ammopiptanthus mongolicus （Maxim. ex Kom.）S. H. Cheng ●◇

300687　Piptanthus mongolicus Maxim. ex Kom. = Ammopiptanthus mongolicus （Maxim. ex Kom.）S. H. Cheng ●◇

300688　Piptanthus nanus Popov;小黄花木●☆

300689　Piptanthus nanus Popov = Ammopiptanthus mongolicus （Maxim. ex Kom.）S. H. Cheng ●◇

300690　Piptanthus nanus Popov = Ammopiptanthus nanus （Popov）S. H. Cheng ●◇

300691　Piptanthus nepalensis （Hook.）D. Don;尼泊尔黄花木(金链叶黄花木）; Evergreen Laburnum, Lanceleaf Piptanthus, Nepal Piptanthus ●

300692　Piptanthus nepalensis （Hook.）D. Don ex Sweet = Piptanthus nepalensis （Hook.）D. Don ●

300693　Piptanthus nepalensis （Hook.）Sweet = Piptanthus nepalensis （Hook.）D. Don ●

300694　Piptanthus nepalensis （Hook.）Sweet f. leiocarpus （Stapf）S. Q. Wei = Piptanthus nepalensis （Hook.）Sweet ●

300695　Piptanthus nepalensis （Hook.）Sweet f. sericopetalus （P. C. Li）S. Q. Wei = Piptanthus nepalensis （Hook.）Sweet ●

300696　Piptanthus nepalensis Sweet = Piptanthus nepalensis （Hook.）D. Don ●

300697　Piptanthus nepalensis Sweet f. leiocarpus （Stapf）S. Q. Wei;光果黄花木;Smooth-fruit Piptanthus,Smooth-fruited Piptanthus ●

300698　Piptanthus nepalensis Sweet f. sericopetalus （P. C. Li）S. Q. Wei;毛瓣黄花木;Hairypetal ●

300699　Piptanthus tomentosus Franch. ;绒叶黄花木(毛叶黄花木,绒毛叶黄花木);Tomentose Piptanthus ●

300700　Piptatherum P. Beauv.（1812）;落芒草属■

300701　Piptatherum P. Beauv. = Oryzopsis Michx. ■

300702　Piptatherum aequiglume （Duthie ex Hook. f.）Roshev. ;等颖落芒草(同颖落芒草);Equalglume Ricegrass ■

300703　Piptatherum aequiglume （Duthie ex Hook. f.）Roshev. var. fasciculatum （Hack.）Freitag = Piptatherum aequiglume （Duthie ex Hook. f.）Roshev. ■

300704　Piptatherum aequiglume （Duthie ex Hook. f.）Roshev. var. parviflora （Z. L. Wu）S. M. Phillips et Z. L. Wu;下花落芒草;Smallflower Ricegrass ■

300705　Piptatherum aequiglume Duthie ex Hook. f. var. ligulatum （P. C. Kuo et Z. L. Wu）S. M. Phillips et Z. L. Wu;长舌落芒草;Liguulate Ricegrass ■

300706　Piptatherum aequiglumis （Hook. f.）Roshev. = Oryzopsis aequiglumis Duthie ex Hook. f. ■

300707　Piptatherum alpestre （Grig.）Roshev. ;高山落芒草■☆

300708　Piptatherum coerulescens （Desf.）P. Beauv. ;浅蓝落芒草■☆

300709　Piptatherum coerulescens P. Beauv. = Piptatherum coerulescens （Desf.）P. Beauv. ■☆

300710　Piptatherum elegans P. Beauv. ;雅致落芒草■☆

300711　Piptatherum fasciculatum （Hack.）Roshev. = Piptatherum aequiglume （Duthie ex Hook. f.）Roshev. ■

300712　Piptatherum fedtschenkoi Roshev. ;范氏落芒草■☆

300713　Piptatherum ferganense （Litv.）Roshev. ;费尔干落芒草■☆

300714　Piptatherum gracile Mez;小落芒草;Slender Ricegrass ■

300715　Piptatherum gracile Mez = Oryzopsis gracilis （Mez）Pilg. ■

300716　Piptatherum grandispiculum （P. C. Kuo et Z. L. Wu）S. M.

Phillips et Z. L. Wu；大穗落芒草；Bigspike Ricegrass，Largespike Ricegrass ■

300717　Piptatherum hilariae Pazij；少穗落芒草（矮落芒草，希拉利落芒草）；Dwarf Ricegrass，Fewspike Ricegrass，Lowly Ricegrass ■

300718　Piptatherum humile（Bor）S. Kumar et Raizada ＝ Piptatherum hilariae Pazij ■

300719　Piptatherum keniense（Pilg.）Roshev. ＝ Stipa keniensis（Pilg.）Freitag ■☆

300720　Piptatherum kokanica（Regel）Nevski ＝ Oryzopsis tianschanica Drobow et Vved. ■

300721　Piptatherum kokanica（Regel）Ovcz. et Czukav. ＝ Oryzopsis tianschanica Drobow et Vved. ■

300722　Piptatherum kuoi S. M. Phillips et Z. L. Wu；钝颖落芒草（钝头落芒草）；Obtuse Ricegrass ■

300723　Piptatherum laterale（Munro ex Regel）Roshev. ＝ Piptatherum laterale（Regel ex Regel）Munro ex Nevski ■

300724　Piptatherum laterale（Regel ex Regel）Munro ex Nevski；细弱落芒草（偏侧落芒草）；Lateral Ricegrass ■

300725　Piptatherum laterale（Regel）Munro ex Nevski ＝ Piptatherum laterale（Regel ex Regel）Munro ex Nevski ■

300726　Piptatherum laterale Munro ex Aitch. ＝ Oryzopsis lateralis（Regel）Stapf ex Hook. f. ■

300727　Piptatherum latifolium Roshev.；宽叶落芒草■☆

300728　Piptatherum miliaceum（L.）Coss.；粟落芒草（黍落芒草）；Smilo Grass，Smilograss，Smilo-grass ■☆

300729　Piptatherum miliaceum（L.）Coss. subsp. thomasii（Duby）Freitag；托马斯粟落芒草■☆

300730　Piptatherum multiflorum P. Beauv.；多花落芒草■☆

300731　Piptatherum munroi（Stapf ex Hook. f.）Mez ＝ Piptatherum munroi（Stapf）Mez ■

300732　Piptatherum munroi（Stapf ex Hook. f.）Mez var. parviflorum（Z. L. Wu）S. M. Phillips et Z. L. Wu ＝ Piptatherum munroi（Stapf）Mez var. parviflorum（Z. L. Wu）S. M. Phillips et Z. L. Wu ■

300733　Piptatherum munroi（Stapf）Mez；落芒草；Common Ricegrass ■

300734　Piptatherum munroi（Stapf）Mez ＝ Oryzopsis munroi Stapf ex Hook. f. ■

300735　Piptatherum munroi（Stapf）Mez var. parviflorum（Z. L. Wu）S. M. Phillips et Z. L. Wu；小花落芒草■

300736　Piptatherum obtusum（Stapf）Roshev. ＝ Piptatherum kuoi S. M. Phillips et Z. L. Wu ■

300737　Piptatherum pamiroalaicum（Grig.）Roshev.；帕米尔落芒草■☆

300738　Piptatherum paradoxum（L.）P. Beauv.；奇异落芒草■☆

300739　Piptatherum parviflorum Roshev. ＝ Oryzopsis chinensis Hitchc. ■

300740　Piptatherum platyanthum Nevski；宽花落芒草■☆

300741　Piptatherum purpurascens（Hack.）Roshev.；紫落芒草■☆

300742　Piptatherum racemosum（Sm.）Eaton ＝ Oryzopsis racemosa（Sm.）Ricker ex Hitchc. ■☆

300743　Piptatherum racemosum Ricker ex Hitchc. ＝ Oryzopsis racemosa（Sm.）Ricker ex Hitchc. ■☆

300744　Piptatherum sinense Mez ＝ Oryzopsis aequiglumis Duthie ex Hook. f. ■

300745　Piptatherum sinense Mez ＝ Piptatherum aequiglume（Duthie ex Hook. f.）Roshev. ■

300746　Piptatherum sogdianum（Grig.）Roshev.；索格落芒草■☆

300747　Piptatherum songaricum（Trin. et Rupr.）Roshev.；新疆落芒草；Dzungar Ricegrass，Xinjiang Ricegrass ■

300748　Piptatherum songaricum（Trin. et Rupr.）Roshev. ＝ Oryzopsis songarica（Trin. et Rupr.）B. Fedtsch. ■

300749　Piptatherum songaricum（Trin. et Rupr.）Roshev. subsp. tianschanicum（Drobow et Vved.）Tzvelev ＝ Piptatherum songaricum（Trin. et Rupr.）Roshev. ■

300750　Piptatherum songaricum（Trin. et Rupr.）Roshev. var. tianschanicum（Drobow et Vved.）Tzvelev ＝ Oryzopsis tianschanica Drobow et Vved. ■

300751　Piptatherum thomasii（Duby）Kunth ＝ Piptatherum miliaceum（L.）Coss. subsp. thomasii（Duby）Freitag ■☆

300752　Piptatherum tibeticum Roshev.；藏落芒草；Tibetic Ricegrass ■

300753　Piptatherum tibeticum Roshev. ＝ Oryzopsis tibetica（Roshev.）P. C. Kuo ■

300754　Piptatherum tibeticum Roshev. var. psilolepis（P. C. Kuo et Z. L. Wu）S. M. Phillips et Z. L. Wu；光稃落芒草；Nakedscale Ricegrass ■

300755　Piptatherum tremuloides Ovcz. et Czukav.；颤落芒草■☆

300756　Piptatherum vicarium（Grig.）Roshev.；中南亚落芒草■☆

300757　Piptocalyx Benth. ＝ Trimenia Seem.（保留属名）●☆

300758　Piptocalyx Oliv. ex Benth.（废弃属名）＝ Trimenia Seem.（保留属名）●☆

300759　Piptocalyx Torr. ＝ Greeneocharis Gürke et Harms ■☆

300760　Piptocarpha Hook. et Arn. ＝ Chuquiraga Juss. ●☆

300761　Piptocarpha Hook. et Arn. ＝ Dasyphyllum Kunth ●☆

300762　Piptocarpha R. Br.（1817）；落苞菊属（落枝菊属，南美菊属）●☆

300763　Piptocarpha chontalensis Baker；落苞菊（落枝菊，南美菊）●☆

300764　Piptocarpha leubnitziae Kuntze ＝ Pechuel-Loeschea leubnitziae（Kuntze）O. Hoffm. ■☆

300765　Piptocephalum Sch. Bip. ＝ Catananche L. ■☆

300766　Piptoceras Cass. ＝ Centaurea L.（保留属名）●■

300767　Piptochaetium J. Presl（1830）（保留属名）；落毛禾属（美洲落芒草属）■☆

300768　Piptochaetium bicolor（Vahl）E. Desv.；二色落毛禾■☆

300769　Piptochaetium setosum（Trin.）Arechav.；具毛落毛禾；Bristly Speargrass ■☆

300770　Piptochaetium stipoides（Trin. et Rupr.）Hack.；具梗落毛禾；Purple Speargrass ■☆

300771　Piptochlaena Post et Kuntze ＝ Piptolaena Harv. ●

300772　Piptochlaena Post et Kuntze ＝ Voacanga Thouars ●

300773　Piptochlamys C. A. Mey. ＝ Thymelaea Mill.（保留属名）●■

300774　Piptoclaina G. Don ＝ Heliotropium L. ●■

300775　Piptocoma Cass.（1817）；脱冠落苞菊属●☆

300776　Piptocoma Less. ＝ Lychnophora Mart. ●☆

300777　Piptocoma acuminata（Kunth）Pruski；渐尖脱冠落苞菊●☆

300778　Piptocoma discolor（Kunth）Pruski；异色脱冠落苞菊●☆

300779　Piptocoma macrophylla（Sch. Bip.）Pruski；大叶脱冠落苞菊●☆

300780　Piptocoma rufescens Cass.；脱冠落苞菊●☆

300781　Piptolaena Harv. ＝ Voacanga Thouars ●

300782　Piptolaena dregei（E. Mey.）A. DC. ＝ Voacanga thouarsii Roem. et Schult. ●☆

300783　Piptolepis Benth.（废弃属名）＝ Forestiera Poir.（保留属名）●☆

300784　Piptolepis Benth.（废弃属名）＝ Piptolepis Sch. Bip.（保留属名）●☆

300785　Piptolepis Sch. Bip.（1863）（保留属名）；密叶巴西菊属●☆

300786　Piptolepis buxoides Sch. Bip.；密叶巴西菊●☆

300787　Piptomeris Turcz. ＝ Jacksonia R. Br. ex Sm. ●☆

300788　Piptophyllum C. E. Hubb.（1957）；落叶草属■☆

300789　Piptophyllum welwitschii（Rendle）C. E. Hubb.；落叶草■☆

300790　Piptopogon Cass. = Hypochaeris L. ■

300791　Piptopogon Cass. = Seriola L. ■

300792　Piptopogon macrospermus C. A. Mey. ex Turcz. = Scorzonera albicaulis Bunge ■

300793　Piptoptera Bunge（1877）；落翅蓬属■☆

300794　Piptoptera turkestana Bunge；落翅蓬■☆

300795　Piptosaecos Turcz. = Dysoxylum Blume ●

300796　Piptoseras Cass. = Centaurea L.（保留属名）●■

300797　Piptospatha N. E. Br.（1879）；落苞南星属■☆

300798　Piptospatha acutifolia Engl. ；尖叶落苞南星■☆

300799　Piptospatha angustifolia Engl. ex Alderw. ；窄叶落苞南星■☆

300800　Piptospatha rigidifolia Engl. ；硬叶落苞南星■☆

300801　Piptostachya（C. E. Hubb.）J. B. Phipps = Zonotriche（C. E. Hubb.）J. B. Phipps ■☆

300802　Piptostachya inamoena（K. Schum.）J. B. Phipps = Zonotriche inamoena（K. Schum.）Clayton ■☆

300803　Piptostegia Hoffmanns. = Merremia Dennst. ex Endl.（保留属名）●■

300804　Piptostegia Hoffmanns. = Operculina Silva Manso（废弃属名）●■

300805　Piptostemma（D. Don）Spach = Panargyrum D. Don ●☆

300806　Piptostemma Spach = Nassauvia Comm. ex Juss. ●☆

300807　Piptostemma Turcz.（1851）；落冠菊属■☆

300808　Piptostemma Turcz. = Angianthus J. C. Wendl.（保留属名）■●☆

300809　Piptostemma carpesioides Turcz. ；落冠菊■☆

300810　Piptostemum Steud. = Nassauvia Comm. ex Juss. ●☆

300811　Piptostemum Steud. = Piptostemma Spach ●☆

300812　Piptostigma Oliv.（1865）；落柱木属●☆

300813　Piptostigma aubrevillei Ghesq. ex Aubrév. = Piptostigma fasciculatum（De Wild.）Boutique ●☆

300814　Piptostigma calophyllum Mildbr. et Diels；美叶落柱木●☆

300815　Piptostigma exellii R. E. Fr. ；宽瓣落柱木●☆

300816　Piptostigma fasciculatum（De Wild.）Boutique；菲氏落柱木●☆

300817　Piptostigma fugax A. Chev. ex Hutch. et Dalziel；早萎落柱木●☆

300818　Piptostigma giganteum Hutch. et Dalziel；巨落柱木●☆

300819　Piptostigma glabrescens Oliv. ；光落柱木●☆

300820　Piptostigma glabrescens Oliv. var. lanceolata Le Thomas；剑光落柱木●☆

300821　Piptostigma latipetalum（Exell）R. E. Fr. = Piptostigma exellii R. E. Fr. ●☆

300822　Piptostigma longepilosum Engl. var. subnudum Tisser. = Piptostigma mortehanii De Wild. ●☆

300823　Piptostigma mayumbense Exell；马永巴落柱木●☆

300824　Piptostigma mortehanii De Wild. ；长毛落柱木●☆

300825　Piptostigma mortehanii De Wild. var. pilosa Sillans = Piptostigma mortehanii De Wild. ●☆

300826　Piptostigma multinervium Engl. et Diels；多脉落柱木●☆

300827　Piptostigma oyemense Pellegr. ；奥也姆落柱木●☆

300828　Piptostigma pilosum Oliv. ；疏毛落柱木●☆

300829　Piptostigma preussii Engl. et Diels = Piptostigma glabrescens Oliv. ●☆

300830　Piptostylis Dalzell = Clausena Burm. f. ●

300831　Piptothrix A. Gray（1886）；落毛菊属■●☆

300832　Piptothrix palmeri A. Gray；落毛菊■●☆

300833　Pipturus Wedd.（1854）；落尾木属（落尾麻属）；Pipturus ●

300834　Pipturus albidus Gray；白落尾木●☆

300835　Pipturus arborescens（Link）C. B. Rob. ；落尾木（落尾麻）；Arborescent Pipturus, Pipturus ●

300836　Pipturus argenteus（G. Forst. ）Wedd. ；银毛落尾木；Native Mulberry ●☆

300837　Pipturus asper Wedd. = Pipturus arborescens（Link）C. B. Rob. ●

300838　Pipturus fauriei Yamam. = Pipturus arborescens（Link）C. B. Rob. ●

300839　Pipturus repandus Wedd. ；兰屿落尾麻●

300840　Piqueria Cav.（1795）；皮氏菊属（皮奎菊属，皮奎属）●■☆

300841　Piqueria peruviana（J. F. Gmel. ）Robison；秘鲁皮氏菊（秘鲁皮奎菊）●☆

300842　Piqueria trinervia Cav. ；三脉皮氏菊（皮奎菊，三脉皮奎菊）；Stevia, Tabardillo ■☆

300843　Piqueriella R. M. King et H. Rob.（1974）；小皮氏菊属■☆

300844　Piqueriella brasiliensis R. M. King et H. Rob. ；小皮氏菊■☆

300845　Piqueriopsis R. M. King（1965）；矮皮氏菊属■☆

300846　Piqueriopsis michoacana R. M. King；矮皮氏菊■☆

300847　Piquetia（Pierre）Hallier f. = Camellia L. ●

300848　Piquetia Hallier f. = Camellia L. ●

300849　Piquetia N. E. Br. = Erepsia N. E. Br. ●☆

300850　Piquetia N. E. Br. = Kensitia Fedde ●☆

300851　Piranhea Baill.（1866）；皮兰大戟属●☆

300852　Piranhea trifoliata Baill. ；三叶皮兰大戟●☆

300853　Pirarda Adans. = Ethulia L. f. ■

300854　Piratinera Aubl.（废弃属名）= Brosimum Sw.（保留属名）●☆

300855　Piratinera guianensis Aubl. = Brosimum guianense（Aubl. ）Huber ●☆

300856　Pirazzia Chiov. = Matthiola R. Br. ■●

300857　Pirazzia Chiov. = Matthiola W. T. Aiton（保留属名）■●

300858　Pirazzia elliptica（R. Br. ex DC. ）Chiov. = Diceratella elliptica（R. Br. ex DC. ）Jonsell ■☆

300859　Pircunia Bertero = Phytolacca L. ●■

300860　Pircunia Bertero ex Arn. = Phytolacca L. ●■

300861　Pircunia abyssinica（Hoffm. ）Moq. = Phytolacca dodecandra L'Hér. ■☆

300862　Pircunia stricta（Hoffm. ）Moq. = Phytolacca heptandra Retz. ■☆

300863　Pirea T. Durand = Nasturtium W. T. Aiton（保留属名）■

300864　Pirea T. Durand = Rorippa Scop. ■

300865　Pirea olgae T. Durand = Nasturtium microphyllum Boenn. ex Rchb. ■☆

300866　Pirenoa C. Koch = Pyrenia Clairv. ●

300867　Pirenoa C. Koch = Pyrus L. ●

300868　Piresia Swallen（1964）；皮雷禾属（派雷斯笮属）■☆

300869　Piresia goeldii Swallen；皮雷禾■☆

300870　Piresia leptophylla Soderstr. ；细叶皮雷禾■☆

300871　Piresia macrophylla Soderstr. ；大叶皮雷禾■☆

300872　Piresiella Judz. , Zuloaga et Morrone（1993）；小皮雷禾属■☆

300873　Piresiella strephioides（Griseb. ）Judz. , Zuloaga et Morrone；小皮雷禾■☆

300874　Piresodendron Aubrév. = Pouteria Aubl. ●

300875　Piresodendron Aubrév. ex Le Thomas et Leroy = Pouteria Aubl. ●

300876　Piriadacus Pichon = Cuspidaria DC.（保留属名）●☆

300877　Pirigara Aubl. = Gustavia L.（保留属名）●☆

300878　Pirigarda C. B. Clarke = Pirigara Aubl. ●☆

300879　Piringa Juss. = Gardenia Ellis（保留属名）●

300880　Piringa caquepiria Juss. = Gardenia thunbergia Thunb. ●☆

300881　Pirinia M. Král（1984）；多子莲豆草属■☆

300882　Pirinia koenigii Král；多子莲豆草■☆

300883 Piripea Aubl. = Buchnera L. ■

300884 Piriqueta Aubl.（1775）；腺叶时钟花属■●☆

300885 Piriqueta Aubl. = Erblichia Seem. ●☆

300886 Piriqueta capensis（Harv.）Urb.；腺叶时钟花●☆

300887 Piriquetaceae Martinov = Theaceae Mirb.（保留科名）●

300888 Piritanera R. H. Schomb. = Brosimum Sw.（保留属名）●☆

300889 Piritanera R. H. Schomb. = Piratinera Aubl.（废弃属名）●☆

300890 Pirocydonia H. K. A. Winkl. ex L. L. Daniel（1913）；梨楂梓属●☆

300891 Pirola Neck. = Pyrola L. ●■

300892 Pironneaua Benth. et Hook. f. = Pironneava Gaudich. ■☆

300893 Pironneauella Kuntze = Pironneava Gaudich. ■☆

300894 Pironneava Gaudich. = Hohenbergia Schult. et Schult. f. ■☆

300895 Pironneava Gaudich. ex Regel = Hohenbergia Schult. et Schult. f. ■☆

300896 Pirophorum Neck. = Pyrus L. ●

300897 Pirottantha Speg. = Plathymenia Benth.（保留属名）●★

300898 Pirronearia Benth. et Hook. f. = Pironneava Gaudich. ■☆

300899 Pirus Hall = Pyrus L. ●

300900 Pirus astateria Cardot = Sorbus astateria（Cardot）Hand. -Mazz. ●

300901 Pirus aucuparia（L.）Gaertn. var. randaiensis Hayata = Sorbus randaiensis（Hayata）Koidz. ●

300902 Pirus cavaleriei H. Lév. = Stranvaesia davidiana Decne. ●

300903 Pirus coronata Cardot = Sorbus coronata（Cardot）Te T. Yu et H. T. Tsai ●

300904 Pirus delavayi Franch. = Docynia delavayi（Franch.）C. K. Schneid. ●

300905 Pirus doumeri Boiss. = Malus doumeri（Bois）A. Chev. ●

300906 Pirus feddei H. Lév. = Stranvaesia amphidoxa C. K. Schneid. ●

300907 Pirus foliolosa Wall. var. subglabra Cardot = Sorbus poteriifolia Hand. -Mazz. ●

300908 Pirus formosana Kawak. et Koidz. = Malus doumeri（Bois）A. Chev. ●

300909 Pirus glabrescens Cardot = Sorbus oligodonta（Cardot）Hand. -Mazz. ●

300910 Pirus halliana Voss = Malus halliana Koehne ●

300911 Pirus hupehensis Pamp. = Malus hupehensis（Pamp.）Rehder ●

300912 Pirus hypoglauca Cardot = Sorbus rehderiana Koehne ●

300913 Pirus koehneana Cardot = Sorbus koehneana C. K. Schneid. ●

300914 Pirus koehnei Schneid. = Sorbus hemsleyi（C. K. Schneid.）Rehder ●

300915 Pirus laosensis Cardot = Malus doumeri（Bois）A. Chev. ●

300916 Pirus melliana Hand. -Mazz. = Malus melliana（Hand. -Mazz.）Rehder ●

300917 Pirus mesogea Cardot = Malus hupehensis（Pamp.）Rehder ●

300918 Pirus monbeigii Cardot = Sorbus monbeigii（Cardot）Te T. Yu ●

300919 Pirus obsoletidentata Cardot = Sorbus obsoletidentata（Cardot）Te T. Yu ●

300920 Pirus oligodonta Cardot = Sorbus oligodonta（Cardot）Hand. -Mazz. ●

300921 Pirus rehderiana Cardot = Sorbus rehderiana Koehne ●

300922 Pirus rufiofiia H. Lév. = Docynia indica（Wall.）Decne. ●

300923 Pirus subcrataegifolia H. Lév. = Malus sieboldii（Regel）Rehder ●

300924 Pirus taqueti H. Lév. = Amelanchier asiatica（Siebold et Zucc.）Endl. ex Walp. ●

300925 Pirus thibetica Cardot = Sorbus thibetica（Cardot）Hand. -Mazz. ●

300926 Pirus trilocularis Hayata = Sorbus randaiensis（Hayata）Koidz. ●

300927 Pirus vainior H. Lév. = Amelanchier asiatica（Siebold et Zucc.）Endl. ex Walp. ●

300928 Pirus wilsoniana Cardot = Sorbus wilsoniana C. K. Schneid. ●

300929 Pisaura Bonato = Lopezia Cav. ■☆

300930 Pisaura Bonato ex Endl. = Lopezia Cav. ■☆

300931 Piscaria Piper = Eremocarpus Benth. ■☆

300932 Piscidia L.（1759）（保留属名）；毒鱼豆属■☆

300933 Piscidia americana Sessé et Moc.；美洲毒鱼豆■☆

300934 Piscidia communis Harms；普通毒鱼豆■☆

300935 Piscidia erythrina L. = Piscidia piscipula Sarg. ■☆

300936 Piscidia ovalifolia Larranaga；卵叶毒鱼豆■☆

300937 Piscidia piscipula Sarg.；牙买加毒鱼豆（毒鱼豆，牙买加山茱萸）；Jamaica Dogwood ■☆

300938 Piscidia punicea Cav. = Sesbania punicea（Cav.）Benth. ●☆

300939 Piscipula Loefl. = Ichthyomethia P. Browne（废弃属名）■☆

300940 Piscipula Loefl. = Piscidia L.（保留属名）■☆

300941 Pisonia L.（1753）；腺果藤属（避霜花属，皮孙木属，腺果木属）；Pisonia ●

300942 Pisonia Plum. ex L. = Pisonia L. ●

300943 Pisonia Rottb. = Diospyros L. ●

300944 Pisonia aculeata L.；腺果藤（避霜花，齿托，刺藤，栖头果，腺果水冬瓜，猪勾搭）；Cockspur, Devil's Claw Pisonia, Devil's-claw, Devilsclaw Pisonia, Glandular Fruit Piso Tree, Glandular-fruited Pisotree, Old-hook, Pull-and-hold-back ●

300945 Pisonia aculeata L. f. inermis Kuntze = Pisonia zapallo Griseb. ●☆

300946 Pisonia alba Span. = Ceodes grandis（R. Br.）D. Q. Lu ●

300947 Pisonia alba Span. = Pisonia umbellifera（J. R. Forst. et G. Forst.）Seem. ●

300948 Pisonia brunoniana Endl. ' Variegata'；斑点大叶避霜花（斑叶大叶避霜花）●☆

300949 Pisonia brunoniana Endl. = Pisonia umbellifera（J. R. Forst. et G. Forst.）Seem. ●

300950 Pisonia buxifolia Rottb. = Diospyros ferrea（Willd.）Bakh. ●

300951 Pisonia capitata（S. Watson）Standl.；头状腺果藤●☆

300952 Pisonia discolor Spreng. = Guapira discolor（Spreng.）Little ●☆

300953 Pisonia excelsa Blume = Ceodes umbellifera J. R. Forst. et G. Forst. ●

300954 Pisonia excelsa Blume = Pisonia umbellifera（J. R. Forst. et G. Forst.）Seem. ●

300955 Pisonia grandis R. Br.；抗风桐（白避霜花，高大腺果藤，麻枫桐，无刺藤）；Birdlime Tree, Grand Devil's-claws, Lettuce Tree, Southern Sea Catchbird Tree ●

300956 Pisonia grandis R. Br. 'Alba'；白无刺藤；Moluccan Cabbage ●☆

300957 Pisonia grandis R. Br. = Ceodes grandis（R. Br.）D. Q. Lu ●

300958 Pisonia morindifolia R. Br. = Pisonia grandis R. Br. ●

300959 Pisonia morindifolia R. Br. ex Wight；巴戟天腺果藤（奶树）●

300960 Pisonia morindifolia Roxb. ex Wight = Pisonia grandis R. Br. ●

300961 Pisonia nishimurae Koidz. = Pisonia umbellifera（J. R. Forst. et G. Forst.）Seem. ●

300962 Pisonia obtusata Jacq. = Guapira obtusata（Jacq.）Little ●☆

300963 Pisonia rotundata Griseb.；圆叶腺果藤●☆

300964 Pisonia umbellifera（J. R. Forst. et G. Forst.）Seem. = Ceodes umbellifera J. R. Forst. et G. Forst. ●

300965 Pisonia zapallo Griseb.；阿根廷腺果藤●☆

300966 Pisoniaceae J. Agardh = Nyctaginaceae Juss.（保留科名）●■

300967 Pisoniaceae J. Agardh；腺果藤科（避霜花科）●

300968 Pisoniella（Heimerl）Standl.（1911）；小腺果藤属（小避霜花

属）●☆

300969　Pisoniella Standl. = Pisoniella（Heimerl）Standl. ●☆

300970　Pisoniella arborescens（Lag. et Rodrigues）Standl. ；小腺果藤（小避霜花）●☆

300971　Pisophaca Rydb. = Astragalus L. ●■

300972　Pisosperma Sond. = Kedrostis Medik. ■☆

300973　Pisosperma capense Sond. = Kedrostis capensis（Sond.）A. Meeuse ■☆

300974　Pistacia L.（1753）；黄连木属；Pistache，Pistache Tree，Pistachio ●

300975　Pistacia aethiopica Dale et Greenway = Pistacia aethiopica Kokwaro ●☆

300976　Pistacia aethiopica Kokwaro；埃塞俄比亚黄连木●☆

300977　Pistacia atlantica Desf. ；大西洋黄连木（阿特拉斯黄连木）；Atlantic Pistache，Atlas Mastic，Atlas Mastic Tree，Atlas Pistache，Bombay Mastic，Large Terebinth，Mount Atlas Mastic，Mount Atlas Mastic Tree，Mount Atlas Pistache，Mount Atlas Pistachio，Mt. Atlas Mastic Tree ●☆

300978　Pistacia atlantica Desf. subsp. cabulica（Stocks）Rech. f. ；卡布尔黄连木●☆

300979　Pistacia cabulica Stocks = Pistacia atlantica Desf. subsp. cabulica（Stocks）Rech. f. ●☆

300980　Pistacia chinensis Bunge；黄连木（茶树，黄儿茶，黄果树，黄华，黄鹂，黄鹂木，黄连，黄连茶，黄连树，黄连头，黄莲，黄练，黄楝树，黄腻芽树，黄银树，回味，鸡冠果，鸡冠木，楷，楷木，孔菜，孔木，苦楝子树，蓝香，烂心木，凉茶树，木黄连，木蓼树，山崖子树，胜铁力木，石连，田苗树，岩拐角，药木，药树）；China Pistachio，Chinese Pistache，Chinese Pistachio ●

300981　Pistacia chinensis Bunge f. latifoliolata Loes. = Pistacia chinensis Bunge ●

300982　Pistacia chinensis Bunge subsp. integerrima（J. L. Stewart ex Brand.）Rech. f. = Pistacia integerrima J. L. Stewart ex Brand. ●☆

300983　Pistacia chinensis Bunge subsp. integerrima（J. L. Stewart）Rech. f. = Pistacia integerrima J. L. Stewart ex Brand. ●☆

300984　Pistacia chinensis Bunge var. falcata（Mart.）Zohary = Pistacia falcata Becc. ex Martelli ●☆

300985　Pistacia coccinea Collett et Hemsl. = Pistacia weinmannifolia Poiss. ex Franch. ●

300986　Pistacia falcata Becc. ex Martelli；镰形黄连木●☆

300987　Pistacia formosa Matsum. = Pistacia chinensis Bunge ●

300988　Pistacia integerrima J. L. Stewart ex Brand. ；全缘黄连木●☆

300989　Pistacia integerrima J. L. Stewart ex Brandis = Pistacia chinensis Bunge subsp. integerrima（J. L. Stewart ex Brand.）Rech. f. ●☆

300990　Pistacia khinjuk Stocks；薰陆香（乳香黄连木，薰陆）●☆

300991　Pistacia khinjuk Stocks var. glaberrima Schweinf. ex Boiss. = Pistacia khinjuk Stocks var. glabra Schweinf. ex Engl. ●☆

300992　Pistacia khinjuk Stocks var. glabra Schweinf. ex Engl. ；光滑薰陆香●☆

300993　Pistacia khinjuk Stocks var. microphylla Boiss. ；小叶薰陆香●☆

300994　Pistacia lentiscus L. ；乳香黄连木（匹思答吉，乳香，香黄连木，熏陆香，洋乳香，洋乳香树，粘胶乳香树）；Evergreen Pistache，Lentisc，Lentisc Pistache，Lentisco，Lentiscus Pistachio，Lentisk，Lentisk Lentisco，Lentisk Pistache，Mastic，Mastic Tree，Mastic-tree，Pistachia Gall ●

300995　Pistacia lentiscus L. var. emarginata Engl. = Pistacia aethiopica Kokwaro ●☆

300996　Pistacia lentiscus L. var. falcatula Chiov. ；镰乳香黄连木●☆

300997　Pistacia lentiscus L. var. latifolius Coss. ；洋乳香黄连木●☆

300998　Pistacia mutica Fisch. et C. A. Mey. ；短截黄连木（钝乳香树）；Turk Terebinth Pistaehe，Turk Terebinth-pistache ●☆

300999　Pistacia mutica Fisch. et C. A. Mey. = Pistacia atlantica Desf. ●☆

301000　Pistacia mutica Fisch. et C. A. Mey. subsp. cabulica（Stocks）Engl. = Pistacia atlantica Desf. subsp. cabulica（Stocks）Rech. f. ●☆

301001　Pistacia philippinensis Merr. et Rolfe = Pistacia chinensis Bunge ●

301002　Pistacia terebinthus L. ；笃乳香树（巴西乳香，笃耨香，笃耨香黄连，笃乳香，马尾香，薰陆树）；Chian Turpentine，Chian Turpentine Tree，Cyprus Turpentine，Palestine Terebinth，Pistachio Galls，Terebinth，Terebinth Pistache，Terebinth Tree，Turpentine Tree ●

301003　Pistacia terebinthus L. var. palaestina ？；巴勒斯坦黄连木；Palestine Terebinth，Turpentine Tree ●☆ ！

301004　Pistacia texana Swingle；得克萨斯黄连木；American Pistachio，Lentisco，Texas Pistache，Wild Pistachio ●☆

301005　Pistacia vera L. ；阿月浑子（无名木，无名子）；Common Pistache，Festike Nut，Pistachio，Pistachio Nut，Real Mastic Tree ●

301006　Pistacia weinmannifolia J. Poiss. ex Franch. ；清香木（对节皮，昆明乌木，梅江叶，清香树，细叶楷木，香叶树，紫叶，紫油木）；Yunnan Pistache ●

301007　Pistaciaceae（Marchand）Caruel = Pistaciaceae Caruel ●

301008　Pistaciaceae Caruel = Anacardiaceae R. Br.（保留科名）●

301009　Pistaciaceae Caruel；黄连木科●

301010　Pistaciaceae Mart. ex Caruel = Anacardiaceae R. Br.（保留科名）●

301011　Pistaciaceae Mart. ex Caruel = Pistaciaceae Caruel ●

301012　Pistaciaceae Mart. ex Perleb = Anacardiaceae R. Br.（保留科名）●

301013　Pistaciaceae Mart. ex Perleb = Pistaciaceae Caruel ●

301014　Pistaciaceae Martinov = Anacardiaceae R. Br.（保留科名）●

301015　Pistaciaceae Martinov = Pistaciaceae Caruel ●

301016　Pistaciopsis Engl. = Haplocoelum Radlk. ●☆

301017　Pistaciopsis dekindtiana Engl. = Haplocoelum foliolosum（Hiern）Bullock ●☆

301018　Pistaciopsis gallaensis Engl. = Haplocoelum foliolosum（Hiern）Bullock subsp. strongylocarpum（Bullock）Verdc. ●☆

301019　Pistaciovitex Kuntze = Aglaia Lour.（保留属名）●

301020　Pistaciovitex Kuntze = Vitex L. ●

301021　Pistia L.（1753）；大藻属（大萍属）；Tropical Duchweed，Water Lettuce，Waterlettuce，Water-lettuce ■

301022　Pistia crispata Blume = Pistia stratiotes L. ■

301023　Pistia minor Blume = Pistia stratiotes L. ■

301024　Pistia spathulata Michx. = Pistia stratiotes L. ■

301025　Pistia stratiotes L. ；大藻（大浮萍，大连，大萍，大萍叶，大蒲藻，大蕊萍，大叶萍，番萍，肥猪草，浮萍，母猪莲，萍蓬草，水浮莲，水浮萍，水荷莲，水葫芦，水莲，天浮萍，猪乸莲）；Nile Cabbage，Water Lettuce，Waterlettuce，Water-lettuce ■

301026　Pistiaceae C. Agardh = Cytinus L. + Nepenthes L. + Pistia L. ■

301027　Pistiaceae Dumort. ；大漂科■

301028　Pistiaceae Dumort. = Araceae Juss.（保留科名）■●

301029　Pistiaceae Dumort. = Pittosporaceae R. Br.（保留科名）●

301030　Pistiaceae Rich. ex C. Agardh = Araceae Juss.（保留科名）■●

301031　Pistiaceae Rich. ex C. Agardh = Pistiaceae Dumort. ■

301032　Pistolochia Bernh.（废弃属名）= Corydalis DC.（保留属名）■●

301033　Pistolochia Raf. = Aristolochia L. ■●

301034　Pistolochia buschii（Nakai）Soják = Corydalis buschii Nakai ■

301035　Pistolochia decumbens（Thunb.）Holub = Corydalis

decumbens（Thunb.）Pers. ■

301036　Pistolochia glaucescens（Regel）Soják ＝ Corydalis glaucescens Regel ■

301037　Pistolochia kiautschouensis（Poelln.）Holub ＝ Corydalis kiautschouensis Poelln. ■

301038　Pistolochia ledebouriana（Kar. et Kir.）Soják ＝ Corydalis ledebouriana Kar. et Kir. ■

301039　Pistolochia pauciflora（Steph.）Soják ＝ Corydalis pauciflora（Stephan）Pers. ■

301040　Pistolochia repens（Mandl. et Muehld.）Soják ＝ Corydalis repens Mandl et Muehld. ■

301041　Pistolochia schanginii（Pall.）Soják ＝ Corydalis schanginii（Pall.）B. Fedtsch. ■

301042　Pistolochia sewerzovii（Regel）Soják ＝ Corydalis sewerzovii Regel ■

301043　Pistolochiaceae J. B. Mull. ＝ Aristolochiaceae Juss.（保留科名）■●

301044　Pistolochiaceae Link ＝ Aristolochiaceae Juss.（保留科名）■●

301045　Pistorinia DC.（1828）;基丝景天属●☆

301046　Pistorinia DC. ＝ Cotyledon L. ●■☆

301047　Pistorinia attenuata（H. Lindb.）Greuter;狭变基丝景天●☆

301048　Pistorinia attenuata（H. Lindb.）Greuter subsp. mairei（H. Lindb.）Greuter;迈雷基丝景天●☆

301049　Pistorinia brachyantha Coss. ;短花基丝景天●☆

301050　Pistorinia breviflora Boiss. ＝ Cotyledon breviflora（Boiss.）Maire ●☆

301051　Pistorinia breviflora Boiss. subsp. intermedia（Boiss. et Reut.）Greuter et Burdet;间型短花基丝景天●☆

301052　Pistorinia salzmannii Boiss. ＝ Pistorinia breviflora Boiss. subsp. intermedia（Boiss. et Reut.）Greuter et Burdet ●☆

301053　Pisum L.（1753）;豌豆属；Garden Pea,Pea ■

301054　Pisum abyssinicum A. Br. ＝ Pisum sativum L. var. abyssinicum（A. Br.）Alef. ■☆

301055　Pisum arvense L. ;饲料豌豆(番仔豆,留豆,田野豌豆,豌豆,野豌豆);Austrian Pea,Dun Pea,Field Pen,Grey Pea,Maple Pea,Mutter Pea,Partridge Pea,Partridge-pea,Peluskina ■

301056　Pisum arvense L. ＝ Pisum sativum L. var. arvense（L.）Poir. ■

301057　Pisum arvense L. ＝ Pisum sativum L. var. arvense（L.）Trautv. ■

301058　Pisum arvense L. ＝ Pisum sativum L. ■

301059　Pisum arvense L. var. abyssinicum（A. Br.）Alef. ＝ Pisum sativum L. var. abyssinicum（A. Br.）Alef. ■☆

301060　Pisum aucheri Jaub. et Spach. ;奥库豌豆■☆

301061　Pisum commune Govorov;欧豌豆■☆

301062　Pisum elatius M. Bieb. ＝ Pisum sativum L. subsp. elatius（M. Bieb.）Asch. et Graebn. ■☆

301063　Pisum formosum（Stev.）Boiss. ;丽豌豆■☆

301064　Pisum fulvum Sibth. et Sm. ;茶色豌豆;Tawny Pea ■☆

301065　Pisum humile Boiss. et Noë ＝ Pisum sativum L. subsp. humile（Holmboe）Greuter et al. ■☆

301066　Pisum maritimum L. ＝ Lathyrus japonicus Willd. var. maritimus（L.）Kartesz et Gandhi ■

301067　Pisum maritimum L. ＝ Lathyrus japonicus Willd. ■

301068　Pisum maritimum L. var. glabrum Ser. ＝ Lathyrus japonicus Willd. var. maritimus（L.）Kartesz et Gandhi ■

301069　Pisum saccharatum Rchb. ＝ Pisum sativum L. ■

301070　Pisum sativum L. ;豌豆(豍豆,毕豆,寒豆,寒豆儿,荷兰豆,胡豆,回鹘豆,兰豆,蔄累,麻累,麦豆,麦豌子,雪豆);Crown Pea,Dun Pea,Field Pea,Garden Pea,Partridge Pea,Pea,Pease,Sugar Pea ■

301071　Pisum sativum L. f. abyssinicum（A. Br.）Gams ＝ Pisum sativum L. var. abyssinicum（A. Br.）Alef. ■☆

301072　Pisum sativum L. subsp. abyssinicum（A. Br.）Govorov ＝ Pisum sativum L. var. abyssinicum（A. Br.）Alef. ■☆

301073　Pisum sativum L. subsp. arvense（L.）Asch. et Graebn. ＝ Pisum sativum L. subsp. elatius（M. Bieb.）Asch. et Graebn. ■☆

301074　Pisum sativum L. subsp. elatius（M. Bieb.）Asch. et Graebn. ;高豌豆■☆

301075　Pisum sativum L. subsp. hortense Asch. et Graebn. ＝ Pisum sativum L. ■

301076　Pisum sativum L. subsp. humile（Holmboe）Greuter et al. ;小豌豆■☆

301077　Pisum sativum L. var. abyssinicum（A. Br.）Alef. ;阿比西尼亚豌豆■☆

301078　Pisum sativum L. var. arvense（L.）Poir. ＝ Pisum arvense L. ■

301079　Pisum sativum L. var. arvense（L.）Trautv. ＝ Pisum arvense L. ■

301080　Pisum sativum L. var. macrocarpum Ser. ;大果豌豆;Mangetout,Snow Pea,Sugar Pea ■☆

301081　Pisum sativum L. var. medullare ?;大粒豌豆;Marrow Pea,Marrowfat ■☆

301082　Pisum sativum L. var. var. pumilio Meikle;矮豌豆■☆

301083　Pisum syriacum（A. Berger）E. Lehm. ＝ Pisum sativum L. subsp. humile（Holmboe）Greuter et al. ■☆

301084　Pisum transcaucasicum（Govorov）Stank. ;南高加索豌豆■☆

301085　Pisum umbellatum Mill. ＝ Pisum sativum L. ■

301086　Pitardella Tirveng.（2003）;皮他茜属●☆

301087　Pitardia Batt. ex Pit.（1918）;肖荆芥属■☆

301088　Pitardia Batt. ex Pit. ＝ Nepeta L. ■●

301089　Pitardia coerulescens Maire;兰肖荆芥■☆

301090　Pitardia gracilis Andr. ＝ Pitardia nepetoides Batt. ■☆

301091　Pitardia nepetoides Batt. ;肖荆芥■☆

301092　Pitavia Molina（1810）;皮氏草属(皮达维草属)■☆

301093　Pitavia Nutt. ex Torr. et A. Gray ＝ Cneoridium Hook. f. ●☆

301094　Pitavia dumosa Nutt. ex Torrey et A. Gray ＝ Cneoridium dumosum（Nutt. ex Torr. et A. Gray）B. D. Jacks. ●☆

301095　Pitavia punctata Molina;皮氏草(皮达维草)■☆

301096　Pitaviaster T. G. Hartley ＝ Euodia J. R. Forst. et G. Forst. ●

301097　Pitcairinia Regel ＝ Pitcairnia L' Hér.（保留属名）■☆

301098　Pitcairnia J. R. Forst. et G. Forst. ＝ Pennantia J. R. Forst. et G. Forst. ●☆

301099　Pitcairnia L' Hér.（1789）（保留属名）;翠凤草属(比氏凤梨属,短茎凤梨属,皮开儿属,皮开尼属,匹氏凤梨属,穗花凤梨属,穗花属,艳红凤梨属);Pitcairnia ■☆

301100　Pitcairnia albiflos Herb. ;白花翠凤草■☆

301101　Pitcairnia altensteinii Lem. ;阿氏翠凤草■☆

301102　Pitcairnia andreana Lindau;安氏翠凤草（黑白艳凤）;Andréan Pitcairnia ■☆

301103　Pitcairnia angustifolia（Sw.）Aiton;狭叶翠凤草■☆

301104　Pitcairnia aphelandraeflora Lem. ;爵床花翠凤草■☆

301105　Pitcairnia australis Hort. Par. ex K. Koch;澳大利亚翠凤草■☆

301106　Pitcairnia corallina Lind. et André;白背菠萝■☆

301107　Pitcairnia densiflora Brongn. ;密花翠凤草■☆

301108　Pitcairnia feliciana（A. Chev.）Harms et Mildbr. ;费利奇翠凤草■☆

301109　Pitcairnia flammea Lindl. ;焰花翠凤草■☆

301110　Pitcairnia heterophylla（Lindl.）Beer;异叶艳凤■☆

301111　Pitcairnia muscosa Mart. ex Schult. f. ;巴西翠凤草（艳红菠萝）;Brazilian Pitcairnia ■☆

301112　Pitcairnia punicea Scheidw. ;红亮凤梨 ■☆

301113　Pitcairnia recurvata (Scheidw.) K. Koch;反曲翠凤草 ■☆

301114　Pitcairnia xanthocalyx Mart. ;黄萼翠凤草 ■☆

301115　Pitcarnia J. F. Gmel. = Pitcairnia L' Hér. (保留属名) ■☆

301116　Pitcheria Nutt. = Rhynchosia Lour. (保留属名) ●■

301117　Pithecellobium Mart. (1837) ('Pithecollobium') (保留属名);牛蹄豆属（猴耳环属,金龟属,围涎树属）;Ape's Earring, Ape's-earring, Black Bead, Blackbead, Black-bead, Monkey Earrings Pea, Monkey's Ear-rings ●

301118　Pithecellobium angulatum Benth. = Archidendron clypearia (Jack) I. C. Nielsen ●

301119　Pithecellobium angulatum Benth. = Pithecellobium clypearia (Jack) Benth. ●

301120　Pithecellobium angulatum Benth. = Pithecellobium utile Chun et F. C. How ●

301121　Pithecellobium arboreum Urb. ;乔状围涎树 ●☆

301122　Pithecellobium attopeuense Pierre = Albizia attopeuense (Pierre) Nielsen ●

301123　Pithecellobium attopeuense Pierre = Albizia attopeuensis (Pierre) I. C. Nielsen var. lauii (Merr.) I. C. Nielsen ●

301124　Pithecellobium balansae Oliv. = Archidendron balansae (Oliv.) I. C. Nielsen ●

301125　Pithecellobium balansae Oliv. = Cylindrokelupha balansae (Oliv.) Kosterm. ●

301126　Pithecellobium bigeminum Mart. ;皂皮围涎树;Soapbark Tree ●☆

301127　Pithecellobium caribaeum Urb. = Albizia caribaea (Urb.) Britton et Rose ●☆

301128　Pithecellobium clypearia (Jack) Benth. ;围涎树（猴耳环,鸡心树,角冷,金龟树,尿糖松,尿桶公）;Ape's Earring, Common Apea-earring, Common Monkey Earrings Pea, Guaymochil, Haumachil, Madras Thorn, Manilla Tamarind, Monkey's Ear-rings ●

301129　Pithecellobium clypearia (Jack) Benth. = Archidendron clypearia (Jack) I. C. Nielsen ●

301130　Pithecellobium clypearia (Jack) Benth. var. acuminatum Gagnep. = Archidendron clypearia (Jack) I. C. Nielsen ●

301131　Pithecellobium clypearia (Jack) Benth. var. acuminatum Gagnep. = Pithecellobium clypearia (Jack) Benth. ●

301132　Pithecellobium corymbosum Gagnep. ;伞花围涎树 ●☆

301133　Pithecellobium dinklagei (Harms) Harms = Albizia dinklagei (Harms) Harms ●☆

301134　Pithecellobium dulce (Roxb.) Benth. ;甜肉围涎树（金龟树,牛蹄豆）;Blackbead, Gaumachil Apea-earring, Gaumachil Monkey Earrings Pea, Guayamochil, Madras Thorn, Madras-thorn, Manila Tamarind, Monkeypod ●

301135　Pithecellobium eriorhachis (Harms) Harms = Cathormion eriorhachis (Harms) Dandy ●☆

301136　Pithecellobium flexicaule (Benth.) J. M. Coult. ;弯茎猴耳环;Ebony Blavkbead, Ebony Monkey Earrings Pea, Taxas Ebony ●

301137　Pithecellobium glaberrimum (Schumach. et Thonn.) Aubrév. = Albizia glaberrima (Schumach. et Thonn.) Benth. ●☆

301138　Pithecellobium guadalupense Champ. ;瓜岛围涎树 ●☆

301139　Pithecellobium inaequale (Kunth) Benth. ;不等围涎树 ●☆

301140　Pithecellobium jupunba Urb. ;朱庞围涎树 ●☆

301141　Pithecellobium kerrii Gagnep. = Archidendron kerrii (Gagnep.) I. C. Nielsen ●

301142　Pithecellobium kerrii Gagnep. = Cylindrokelupha kerrii (Gagnep.) T. L. Wu ●

301143　Pithecellobium laoticum Gagnep. = Archidendron laoticum (Gagnep.) I. C. Nielsen ●

301144　Pithecellobium laoticum Gagnep. = Cylindrokelupha laoticum (Gagnep.) C. Chen et H. Sun ●

301145　Pithecellobium leptophyllum (Cav.) Daveau;狭叶围涎树 ●☆

301146　Pithecellobium lobatum Benth. ;爪哇围涎树（爪哇猴耳环）●☆

301147　Pithecellobium lucidum Benth. ;亮叶围涎树（番仔环,颌垂豆,黑汉豆,猴耳环,环沟树,火汤木,鸡三树,蛟龙木,雷公凿,雷公凿树,亮叶猴耳环,落地金钱,尿桶弓,尿桶公,婆劈树,三不正,三角果,山木香,水肿木,围涎树,乌鸡骨,羊角）;China Monkey Earrings Pea, Chinese Apes-earring ●

301148　Pithecellobium lucidum Benth. = Archidendron lucidum (Benth.) I. C. Nielsen ●

301149　Pithecellobium mathewsi Benth. ;秘鲁围涎树 ●☆

301150　Pithecellobium multifoliolatum H. Q. Wen;多叶猴耳环;Manyleaf Monkey Earrings Pea ●

301151　Pithecellobium obliquifoliolatum (De Wild.) J. Léonard = Cathormion obliquifoliolatum (De Wild.) G. C. C. Gilbert et Boutique ●☆

301152　Pithecellobium pedicellare (DC.) Benth. ;柄花围涎树 ●☆

301153　Pithecellobium pervilleanum Benth. = Albizia boivinii E. Fourn. ●☆

301154　Pithecellobium pruinosum Benth. ;白粉围涎树 ●☆

301155　Pithecellobium robinsonii Gagnep. = Archidendron robinsonii (Gagnep.) I. C. Nielsen ●

301156　Pithecellobium robinsonii Gagnep. = Cylindrokelupha robinsonii (Gagnep.) Kosterm. ●

301157　Pithecellobium saman (Jacq.) Benth. = Samanea saman (Jacq.) Merr. ●

301158　Pithecellobium stuhlmannii Taub. = Cathormion altissimum (Hook. f.) Hutch. et Dandy ●☆

301159　Pithecellobium trapezifolium (Vahl) Benth. ;梯叶围涎树 ●☆

301160　Pithecellobium turgidum Merr. = Archidendron turgidum (Merr.) I. C. Nielsen ●

301161　Pithecellobium turgidum Merr. = Cylindrokelupha turgida (Merr.) T. L. Wu ●

301162　Pithecellobium unguiscati (L.) Benth. ;猫爪猴耳环;Black Bead, Black Jessie, Bread-and-cheese, Cat's Claw, Catsclaw ●

301163　Pithecellobium unguiscati Benth. = Pithecellobium unguis-cati (L.) Benth. ●

301164　Pithecellobium utile Chun et F. C. How;薄叶围涎树（薄叶猴耳环）;Common Monkey Earrings Pea, Thinleaf Apea-earring, Thin-leaved Apea-earring ●

301165　Pithecellobium utile Chun et F. C. How = Archidendron utile (Chun et F. C. How) I. C. Nielsen ●

301166　Pithecellobium vinhatico Record;绿白围涎树 ●☆

301167　Pithecellobium zanzibaricum S. Moore = Acacia zanzibarica (S. Moore) Taub. ●☆

301168　Pithecoctenium Mart. ex DC. = Pithecoctenium Mart. ex Meisn. ●☆

301169　Pithecoctenium Mart. ex Meisn. (1840);猴梳藤属;Monkey Comb ●☆

301170　Pithecoctenium buccinatorium DC. = Distictis buccinatoria (DC.) A. H. Gentry ●☆

301171　Pithecoctenium crucigerum (L.) A. H. Gentry = Bignonia capreolata L. ●

301172　Pithecoctenium echinatum (Jacq.) Baill. = Pithecoctenium crucigerum（L.）A. H. Gentry ●

301173　Pithecoctenium echinatum（Jacq.）K. Schum. 猴梳藤●☆

301174　Pithecodendron Speg. = Acacia Mill.（保留属名）●■

301175　Pithecolobium Benth. = Pithecellobium Mart.（保留属名）●

301176　Pithecolobium Mart. = Pithecellobium Mart.（保留属名）●

301177　Pithecolobium altissimum（Hook. f.）Oliv. = Cathormion altissimum（Hook. f.）Hutch. et Dandy ●☆

301178　Pithecoseris Mart. = Pithecoseris Mart. ex DC. ■☆

301179　Pithecoseris Mart. ex DC.（1836）;猴菊属■☆

301180　Pithecoseris pacourinoides Mart. ex DC. ;猴菊■☆

301181　Pithecoxanium Corrêa de Mello = Clytostoma Miers ex Bureau ●

301182　Pithecurus Kunth = Schizachyrium Nees ■

301183　Pithecurus Willd. ex Kunth = Andropogon L.（保留属名）■

301184　Pithocarpa Lindl.（1839）;疏头鼠麴草属■☆

301185　Pithocarpa corymbulosa Lindl. ;疏头鼠麴草■☆

301186　Pithocarpa major Steetz;大疏头鼠麴草■☆

301187　Pithocarpa melanostigma P. Lewis et Summerh. ;黑药疏头鼠麴草■☆

301188　Pithocarpa pulchella Lindl. ;美丽疏头鼠麴草■☆

301189　Pithodes O. F. Cook = Coccothrinax Sarg. ●☆

301190　Pithosillum Cass. = Emilia（Cass.）Cass. ■

301191　Pithosillum Cass. = Senecio L. ■●

301192　Pithuranthos DC. = Pituranthos Viv. ■☆

301193　Pitraea Turcz.（1863）;皮特马鞭草属■☆

301194　Pitraea chilensis Turcz. ;皮特马鞭草■☆

301195　Pittiera Cogn. = Polyclathra Bertol. ■☆

301196　Pittiera Cogn. ex T. Durand et Pitt. = Polyclathra Bertol. ■☆

301197　Pittierella Schltr. = Cryptocentrum Benth. ■☆

301198　Pittierothamnus Steyerm. = Amphidasya Standl. ●☆

301199　Pittocaulon H. Rob. et Brettell（1973）;肉脂菊属●☆

301200　Pittocaulon praecox（Cav.）H. Rob. et Brettell;肉脂菊●☆

301201　Pittonia Mill. = Tournefortia L. ●■

301202　Pittoniotis Griseb. = Antirhea Comm. ex Juss. ●

301203　Pittosporaceae R. Br.（1814）（保留科名）;海桐花科（海桐科）;Pittosporum Family,Seatung Family,Tobira Family ●

301204　Pittosporoides Sol. ex Gaertn. = Pittosporum Banks ex Gaertn.（保留属名）●

301205　Pittosporopsis Craib（1911）;假海桐属;False Seatung,Pittosporopsis ●

301206　Pittosporopsis kerrii Craib;假海桐（杙果）;False Seatung,Kerr Pittosporopsis ●

301207　Pittosporum Banks ex Gaertn.（1788）（保留属名）;海桐花属（海桐属）;Brisbane,Cheesewood,Pittosporum,Queensland Laurel,Seatung ●

301208　Pittosporum Banks ex Sol. = Pittosporum Banks ex Gaertn.（保留属名）●

301209　Pittosporum abyssinicum Delile;阿比西尼亚海桐●☆

301210　Pittosporum abyssinicum Delile subsp. cardiocarpum（Cufod.）Cufod. = Pittosporum abyssinicum Delile ●☆

301211　Pittosporum abyssinicum Delile subsp. engleri（Cufod.）Cufod. = Pittosporum abyssinicum Delile ●☆

301212　Pittosporum abyssinicum Delile subsp. fulvo-tomentosum（Engl.）Cufod. = Pittosporum abyssinicum Delile ●☆

301213　Pittosporum abyssinicum Delile subsp. gilletii Cufod. = Pittosporum abyssinicum Delile ●☆

301214　Pittosporum abyssinicum Delile subsp. lanatum（Hutch. et E. A. Bruce）Cufod. = Pittosporum abyssinicum Delile ●☆

301215　Pittosporum abyssinicum Delile var. angolense Oliv. = Pittosporum viridiflorum Sims ●☆

301216　Pittosporum adaphnephylloides Hu et F. T. Wang = Pittosporum daphniphylloides Hayata var. adaphniphylloides（Hu et F. T. Wang）W. T. Wang ●

301217　Pittosporum adaphniphylloides Hu et F. T. Wang = Pittosporum daphniphylloides Hayata var. adaphniphylloides（Hu et F. T. Wang）W. T. Wang ●

301218　Pittosporum angustilimbum C. Y. Wu;窄叶海桐;Angustifolius Pittosporum,Narrow-leaf Pittosporum,Narrow-leaf Seatung ●

301219　Pittosporum antunesii Engl. = Pittosporum viridiflorum Sims ●☆

301220　Pittosporum baileyanum Gowda = Pittosporum balansae A. DC. var. angustifolium Gagnep. ●

301221　Pittosporum balansae A. DC. ;聚花海桐（山霸王,山辣椒）;Balansa Pittosporum,Balansa Seatung ●

301222　Pittosporum balansae A. DC. var. angustifolium Gagnep. ;窄叶聚花海桐（皱叶海桐花）;Narrowleaf Balansa Pittosporum,Narrowleaf Balansa Seatung ●

301223　Pittosporum balansae A. DC. var. chatterjeeanum（Gowda）Z. Y. Zhang et Turland;披针叶聚花海桐●

301224　Pittosporum beecheyi Tuyama = Pittosporum parvifolium Hayata var. beecheyi（Tuyama）H. Ohba ●☆

301225　Pittosporum bicolor Hook. ;二色海桐（狭叶海桐）;Banyalla,Cheese Wood,Cheesewood ●☆

301226　Pittosporum boninense Koidz. ;小笠原海桐●☆

301227　Pittosporum boninense Koidz. var. chichijimense（Nakai ex Tuyama）H. Ohba;父岛海桐●☆

301228　Pittosporum boninense Koidz. var. denudatum（Nakai）H. Ohba = Pittosporum boninense Koidz. var. lutchuense（Koidz.）H. Ohba ●☆

301229　Pittosporum boninense Koidz. var. lutchuense（Koidz.）H. Ohba;琉球海桐●☆

301230　Pittosporum brevicalyx（Oliv.）Gagnep. ;短萼海桐（拔毒散,大朵林,鸡骨头,木辣椒,山桂花,万里香,万年青,小朵林,小年药,小黏药）;Shortcalyx Pittosporum,Short-calyx Pittosporum,Shortcalyx Seatung ●

301231　Pittosporum brevicalyx（Oliv.）Gagnep. var. brevistamineum Gagnep. = Pittosporum brevicalyx（Oliv.）Gagnep. ●

301232　Pittosporum buxifolium K. M. Feng ex W. Q. Yin = Pittosporum kweichowense Gowda var. buxifolium（K. M. Feng ex W. Q. Yin）Z. Y. Zhang et Turland ●

301233　Pittosporum cacondense Exell et MendonAʒωça;卡孔达海桐●☆

301234　Pittosporum calcicola C. Y. Wu;灰岩海桐;Calcicole Pittosporum,Calcicolous Pittosporum,Calcicolous Seatung ●

301235　Pittosporum cardiocarpum Cufod. = Pittosporum abyssinicum Delile ●☆

301236　Pittosporum cavaleriei H. Lév. = Pittosporum glabratum Lindl. var. neriifolium Rehder et E. H. Wilson ●

301237　Pittosporum chatterjeeanum Gowda;岗房海桐;Chatterjee Pittosporum ●

301238　Pittosporum chatterjeeanum Gowda = Pittosporum balansae A. DC. var. chatterjeeanum（Gowda）Z. Y. Zhang et Turland ●

301239　Pittosporum chichijimense Nakai ex Tuyama = Pittosporum boninense Koidz. var. chichijimense（Nakai ex Tuyama）H. Ohba ●☆

301240　Pittosporum chinense Donn = Pittosporum tobira（Thunb.）W. T. Aiton ●

301241　Pittosporum colensoi Hook. f. ;卷瓣海桐●☆

301242　Pittosporum commutatum Putt. = Pittosporum viridiflorum Sims ●☆

301243　Pittosporum confertum Merr. et Chun ＝ Pittosporum balansae A. DC. ●

301244　Pittosporum coriaceum Aiton；革质海桐●☆

301245　Pittosporum cornifolium A. Cunn.；山茱萸叶海桐●☆

301246　Pittosporum crassifolium A. Cunn. 'Variegatum'；斑叶厚叶海桐(银边厚叶海桐)●☆

301247　Pittosporum crassifolium A. Cunn. ＝ Pittosporum crassifolium Banks et Sol. ex A. Cunn. ●☆

301248　Pittosporum crassifolium Banks et Sol. ex A. Cunn.；厚叶海桐；Karo, Parciment Bark, Pittosporum, Stiffleaf Cheesewood, Thick-leaved Pittosporum ●☆

301249　Pittosporum crispulum Gagnep.；皱叶海桐(黄木)；Cryspateleaf Pittosporum, Cryspateleaf Seatung, Cryspate-leaved Pittosporum ●

301250　Pittosporum crispulum sensu Gowda ＝ Pittosporum kunmingense Hung T. Chang et S. Z. Yan ●

301251　Pittosporum dallii Cheeseman；大丽海桐(南岛海桐)●☆

301252　Pittosporum dalzielii Hutch. ＝ Pittosporum viridiflorum Sims ●☆

301253　Pittosporum daphniphylloides Hayata；牛耳枫叶海桐(大叶海桐,楠叶海桐,爬崖花子,山青皮)；Bigleaf Pittosporum, Bigleaf Seatung, Big-leaved Pittosporum, Daphne Pittosporum ●

301254　Pittosporum daphniphylloides Hayata sensu Rehder et E. H. Wilson ＝ Pittosporum daphniphylloides Hayata ●

301255　Pittosporum daphniphylloides Hayata var. adaphniphylloides (Hu et F. T. Wang) W. T. Wang；大叶海桐(山枝茶,山枝仁)；Bigleaf Pittosporum, Bigleaf Seatung, Big-leaved Pittosporum ●

301256　Pittosporum densinervatum H. T. Chang et S. Z. Yan ＝ Pittosporum kweichowense Gowda ●

301257　Pittosporum densinervatum Hung T. Chang et S. Z. Yan；密脉海桐；Densenerve Seatung, Densevein Pittosporum, Dense-veined Pittosporum ●

301258　Pittosporum densinervatum Hung T. Chang et S. Z. Yan ＝ Pittosporum kweichowense Gowda ●

301259　Pittosporum elevaticostatum Hung T. Chang et S. Z. Yan；突肋海桐(突脉海桐)；Convex-nerved Pittosporum, Elevatednerve Pittosporum, Elevatednerve Seatung ●

301260　Pittosporum engleri J. Léonard ex Cufod. ＝ Pittosporum abyssinicum Delile ●☆

301261　Pittosporum eugenioides A. Cunn.；番樱桃状海桐(柠檬木)；Lemonwood, New Zealand Lemonwood, Tarata ●☆

301262　Pittosporum eugenioides A. Cunn. 'Variegata'；斑叶番樱桃状海桐(黄斑番樱桃状海桐)●☆

301263　Pittosporum feddeanum Pax ＝ Pittosporum viridiflorum Sims ●☆

301264　Pittosporum ferrugineum sensu Merr. ＝ Pittosporum balansae A. DC. ●

301265　Pittosporum flavum Hook. ＝ Hymenosporum flavum (Hook.) R. Br. ex F. Muell. ●☆

301266　Pittosporum floribundum Wight et Arn.；多花海桐(拔毒散,小年药,云南海桐)；Himalayan pittosporum, Manyflower Pittosporum, Manyflower Seatung ●

301267　Pittosporum floribundum Wight et Arn. ＝ Pittosporum napaulense (DC.) Rehder et E. H. Wilson ●

301268　Pittosporum floribundum Wight et Arn. ＝ Pittosporum viridiflorum Sims ●☆

301269　Pittosporum floribundum Wight et Arn. ex Royle ＝ Pittosporum napaulense (DC.) Rehder et E. H. Wilson ●

301270　Pittosporum floribundum Wight et Arn. sensu Hand.-Mazz. ＝ Pittosporum brevicalyx (Oliv.) Gagnep. ●

301271　Pittosporum formosanum Hayata ＝ Pittosporum pentandrum (Blanco) Merr. var. formosanum (Hayata) Z. Y. Zhang et Turland ●

301272　Pittosporum formosanum Hayata ＝ Pittosporum pentandrum (Blanco) Merr. var. hainanense (Gagnep.) H. L. Li ●

301273　Pittosporum formosanum Hayata ＝ Pittosporum pentandrum (Blanco) Merr. ●

301274　Pittosporum formosanum Hayata var. hainanense Gagnep. ＝ Pittosporum pentandrum (Blanco) Merr. var. hainanense (Gagnep.) H. L. Li ●

301275　Pittosporum formosanum Hayata var. hainanense Gagnep. ＝ Pittosporum pentandrum (Blanco) Merr. var. formosanum (Hayata) Z. Y. Zhang et Turland ●

301276　Pittosporum fortunei Turcz. ＝ Pittosporum glabratum Lindl. ●

301277　Pittosporum fulvipilosum Hung T. Chang et S. Z. Yan；褐毛海桐；Brownhair Seatung, Brown-haired Pittosporum, Fulvileaf Pittosporum ●

301278　Pittosporum fulvo-tomentosum Engl. ＝ Pittosporum abyssinicum Delile ●☆

301279　Pittosporum glabratum Lindl.；光叶海桐(芭豆,长果满天香,臭皮,大皮子药,大天王,广枝,见血飞,七星胆,山饭树,山海桐,山枝,山枝茶,土连翘,鸭脚板树,崖花,崖花子,野连翘,一朵云,榨木)；Glabrous Leaf Pittosporum, Glabrousleaf Pittosporum, Glabrousleaf Seatung, Glabrous-leaved Pittosporum ●

301280　Pittosporum glabratum Lindl. sensu E. H. Wilson ＝ Pittosporum illicioides Makino ●

301281　Pittosporum glabratum Lindl. sensu Rehder ＝ Pittosporum trigonocarpum H. Lév. ●

301282　Pittosporum glabratum Lindl. var. angustifolium Pritz. ＝ Pittosporum podocarpum Gagnep. ●

301283　Pittosporum glabratum Lindl. var. chinense Pamp. ＝ Pittosporum omeiense Hung T. Chang et S. Z. Yan ●

301284　Pittosporum glabratum Lindl. var. chinense Pamp. ＝ Pittosporum podocarpum Gagnep. ●

301285　Pittosporum glabratum Lindl. var. ciliicalyx Franch. ＝ Pittosporum podocarpum Gagnep. ●

301286　Pittosporum glabratum Lindl. var. neriifolium Rehder et E. H. Wilson；披针叶海桐(黑皮子,黄栀子,金刚摆,金刚口摆,山栀子,台湾圆果海桐,狭叶海桐,狭叶崖子花,斩蛇剑)；Oleanderleaf Pittosporum, Oleanderleaf Seatung ●

301287　Pittosporum glabratum Lindl. var. wenxianense (G. H. Wang et Y. S. Lian) Z. Y. Zhang et Turland；文县海桐；Wenxian Pittosporum, Wenxian Seatung ●

301288　Pittosporum goetzei Engl.；格兹海桐●☆

301289　Pittosporum hejiangense H. Y. Su ＝ Pittosporum podocarpum Gagnep. var. hejiangense (H. Y. Su) Z. Y. Zhang et Turland ●

301290　Pittosporum henryi Gowda；小柄果海桐；Henry Pittosporum, Henry Seatung ●

301291　Pittosporum heterophyllum Franch.；异叶海桐(臭椿皮,臭皮,椿根白皮,红杉树,鸡骨头,细杉树,野桂花)；Rock Pittosporum, Rock Seatung ●

301292　Pittosporum heterophyllum Franch. var. ledoides Hand.-Mazz.；带叶海桐(臭皮)●

301293　Pittosporum heterophyllum Franch. var. sessile Gowda；无柄异叶海桐●

301294　Pittosporum illicioides Makino；岩花海桐(白背风,吊灯笼,海金子,海桐树,接骨丹,两广海桐,满山香,莽草海桐,山海桐,山海桐花,山枝木,山枝条,山栀茶,山栀花,疏果海桐,台湾圆果海桐,五月上树风,崖花子,崖花海桐,崖花子,野黄栀,粘子柴,柞木)；Anisetreelike Pittosporum, Anisetree-like Pittosporum, Anisetree-like Seatung, Fewfruited Pittosporum ●

301295 Pittosporum illicioides Makino var. angustifolium T. C. Cuang = Pittosporum illicioides Makino var. angustifolium T. C. Huang ex S. Y. Lu ●

301296 Pittosporum illicioides Makino var. angustifolium T. C. Cuang ex S. Y. Lu = Pittosporum illicioides Makino ●

301297 Pittosporum illicioides Makino var. angustifolium T. C. Huang ex S. Y. Lu;细叶疏果海桐(细叶海桐)●

301298 Pittosporum illicioides Makino var. oligocarpum (Hayata) Kitam. = Pittosporum illicioides Makino ●

301299 Pittosporum illicioides Makino var. stenophyllum P. L. Chiu = Pittosporum illicioides Makino ●

301300 Pittosporum illicioides Makino var. stenophyllum P. L. Chiu ex Hung T. Chang et S. Z. Yan = Pittosporum illicioides Makino ●

301301 Pittosporum illicioides Makino var. stenophyllum P. L. Chiu ex Hung T. Chang et S. Z. Yan; 狭叶海金子; Narrowleaf Anisetree-like Pittosporum, Narrowleaf Anisetree-like Seatung ●

301302 Pittosporum johnstonianum Gowda; 滇西海桐 (贡山海桐); Johnston. Pittosporum, Johnston. Seatung, West Yunnan Pittosporum ●

301303 Pittosporum johnstonianum Gowda var. glomerulatum C. Y. Wu;密花海桐;Dense-flower Johnston. Pittosporum, Dense-flower Johnston. Seatung ●

301304 Pittosporum kapiriense Cufod. = Pittosporum viridiflorum Sims ● ☆

301305 Pittosporum kerrii Craib;羊脆木海桐(白箐檀木,羊脆骨,羊脆木,羊耳朵树,杨翠木);Kerr Pittosporum, Kerr Seatung ●

301306 Pittosporum kobuskianum Gowda = Pittosporum illicioides Makino ●

301307 Pittosporum kruegeri Engl. = Pittosporum viridiflorum Sims ● ☆

301308 Pittosporum kunmingense Hung T. Chang et S. Z. Yan;昆明海桐;Kunming Pittosporum, Kunming Seatung ●

301309 Pittosporum kwangxiense Hung T. Chang et S. Z. Yan;广西海桐(木辣蓼,山辣蓼);Guangxi Pittosporum, Guangxi Seatung, Kwangsi Seatung ●

301310 Pittosporum kweichowense Gowda;贵州海桐;Guizhou Pittosporum, Guizhou Seatung ●

301311 Pittosporum kweichowense Gowda var. buxifolium (K. M. Feng ex W. Q. Yin) Z. Y. Zhang et Turland;黄杨叶海桐;Box-leaf Pittosporum, Box-leaf Seatung, Box-leaved Pittosporum ●

301312 Pittosporum kweichowense Gowda var. podocarpifolium (C. Y. Wu) Z. Y. Zhang et Turland;罗汉松叶海桐;Podocarileaf Pittosporum, Podocarileaf Seatung ●

301313 Pittosporum lanatum Hutch. et E. A. Bruce = Pittosporum abyssinicum Delile ● ☆

301314 Pittosporum lanatum Hutch. et E. A. Bruce var. engleri (J. Léonard ex Cufod.) Cufod. = Pittosporum abyssinicum Delile ● ☆

301315 Pittosporum ledoides (Hand.-Mazz.) C. Y. Wu;杜香叶海桐●

301316 Pittosporum ledoides (Hand.-Mazz.) C. Y. Wu = Pittosporum heterophyllum Franch. var. ledoides Hand.-Mazz. ●

301317 Pittosporum lenticellatum Chun ex H. Peng et Y. F. Deng;卵果海桐●

301318 Pittosporum leptosepalum Gowda; 薄萼海桐; Thin-sepal Pittosporum, Thinsepal Seatung, Thin-sepaled Pittosporum ●

301319 Pittosporum lignilobum Hu et F. T. Wang = Pittosporum crispulum Gagnep. ●

301320 Pittosporum littorale Merr. ;海岸海桐 ● ☆

301321 Pittosporum littorale Merr. = Pittosporum moluccanum Miq. ●

301322 Pittosporum littorale Merr. sensu H. L. Li = Pittosporum viburnifolium Hayata ●

301323 Pittosporum longicarpum S. K. Wu ex W. C. Yin = Pittosporum kweichowense Gowda ●

301324 Pittosporum longicarpum S. K. Wu ex W. Q. Yin;长果海桐;Long-fruit Pittosporum, Long-fruit Seatung ●

301325 Pittosporum lutchuense Koidz. = Pittosporum boninense Koidz. var. lutchuense (Koidz.) H. Ohba ● ☆

301326 Pittosporum lutchuense Koidz. var. denudatum (Nakai) S. Kobay. = Pittosporum boninense Koidz. var. lutchuense (Koidz.) H. Ohba ● ☆

301327 Pittosporum lynesii Cufod. = Pittosporum viridiflorum Sims ● ☆

301328 Pittosporum makinoi Nakai = Pittosporum tobira (Thunb.) W. T. Aiton var. calvescens Ohwi ●

301329 Pittosporum makinoi Nakai = Pittosporum tobira (Thunb.) W. T. Aiton ●

301330 Pittosporum malosanum Baker = Pittosporum viridiflorum Sims ● ☆

301331 Pittosporum mannii Hook. f. = Pittosporum viridiflorum Sims ● ☆

301332 Pittosporum mannii Hook. f. subsp. ripicola (J. Léonard) Cufod. = Pittosporum viridiflorum Sims ● ☆

301333 Pittosporum melanospermum F. Muell. ;黑籽海桐●☆

301334 Pittosporum membranifolium S. C. Huang ex W. C. Yin = Pittosporum perryanum Gowda ●

301335 Pittosporum membrenifolium S. C. Huang = Pittosporum perryanum Gowda ●

301336 Pittosporum membrenifolium S. C. Huang ex C. Y. Wu;膜叶海桐;Membrana-leaf Pittosporum, Membrana-leaf Seatung ●

301337 Pittosporum membrenifolium S. C. Huang ex C. Y. Wu = Pittosporum perryanum Gowda ●

301338 Pittosporum merrillianum Gowda;滇越海桐;Merrill Pittosporum ●

301339 Pittosporum mildbraedii Engl. ;米尔德海桐●☆

301340 Pittosporum moluccanum Miq. ;兰屿海桐;Lanyu Pittosporum ●

301341 Pittosporum moluccanum sensu Bakker et Steenis = Pittosporum viburnifolium Hayata ●

301342 Pittosporum monanthum C. Y. Wu; 单花海桐; One-flower Pittosporum, One-flower Seatung ●

301343 Pittosporum monanthum C. Y. Wu = Pittosporum podocarpum Gagnep. ●

301344 Pittosporum napaulense (DC.) Rehder et E. H. Wilson;滇藏海桐(尼泊尔海桐);Nepal Pittosporum, Nepal Seatung ●

301345 Pittosporum napaulense (DC.) Rehder et E. H. Wilson var. rawalpindiense Gowda = Pittosporum napaulense (DC.) Rehder et E. H. Wilson ●

301346 Pittosporum neelgherrense Wight et Arn. ;尼基山海桐(尼山海桐花)●☆

301347 Pittosporum neelgherrense Wight et Arn. var. laxiflorum Franch. = Pittosporum brevicalyx (Oliv.) Gagnep. ●

301348 Pittosporum oblongifolium C. H. Wright = Rinorea oblongifolia (C. H. Wright) Marquand ex Chipp ● ☆

301349 Pittosporum oligocarpum Hayata = Pittosporum illicioides Makino ●

301350 Pittosporum oligophlebium Hung T. Chang et S. Z. Yan;贫脉海桐(黄脉海桐,台湾圆果海桐);Fewvein Seatung, Yellow-veined Pittosporum, Yellow-veined Seatung ●

301351 Pittosporum oligospermum Hayata = Pittosporum illicioides Makino ●

301352 Pittosporum omeiense Hung T. Chang et S. Z. Yan;峨眉海桐(刺海桐);Emei Pittosporum, Emei Seatung, Omei Mountain Pittosporum, Thorny Pittosporum ●

301353 Pittosporum ovoideum Gowda = Pittosporum pauciflorum Hook. et Arn. ●

301354 Pittosporum ovoideum Hung T. Chang et S. Z. Yan =

Pittosporum lenticellatum Chun ex H. Peng et Y. F. Deng ●

301355　Pittosporum paniculiferum Hung T. Chang et S. Z. Yan；圆锥海桐；Panicled Pittosporum, Panicute Pittosporum, Panicutate Seatung ●

301356　Pittosporum parvicapsulare Hung T. Chang et S. Z. Yan；小果海桐（五月上树风，崖子花）；Littlefruit Pittosporum, Littlefruit Seatung, Little-fruited Pittosporum ●

301357　Pittosporum parvifolium Hayata；东方小叶海桐●☆

301358　Pittosporum parvifolium Hayata var. beecheyi（Tuyama）H. Ohba；毕氏小叶海桐●☆

301359　Pittosporum parvilimbum Hung T. Chang et S. Z. Yan；小叶海桐；Littleleaf Pittosporum, Little-leaved Pittosporum, Smallflower Seatung ●

301360　Pittosporum pauciflorum Hook. et Arn.；少花海桐（卵果海桐，满山香，崖花子）；Fewflower Pittosporum, Fewflower Seatung, Ovoidfruit Pittosporum, Ovoidfruit Seatung, Ovoid-fruited Pittosporum, Pauciflorous Pittosporum ●

301361　Pittosporum pauciflorum Hook. et Arn. var. brevicalyx Oliv. = Pittosporum brevicalyx（Oliv.）Gagnep. ●

301362　Pittosporum pauciflorum Hook. et Arn. var. oblongum Hung T. Chang et S. Z. Yan；长果小叶海桐（长果海桐）；Longfruit Fewflower Pittosporum, Longfruit Fewflower Seatung ●

301363　Pittosporum pentandrum（Blanco）Merr.；台湾海桐（七里香）；Five Stamens Pittosporum, Mamalis, Philippine Pittosporum, Taiwan Pittosporum, Taiwan Seatung, Taiwanese Cheesewood ●

301364　Pittosporum pentandrum（Blanco）Merr. var. formosanum（Hayata）Z. Y. Zhang et Turland；台琼海桐（七里香，台湾海桐花）；Hainan Fewflower Pittosporum, Hainan Pittosporum, Hainan Seatung ●

301365　Pittosporum pentandrum（Blanco）Merr. var. hainanense（Gagnep.）H. L. Li = Pittosporum pentandrum（Blanco）Merr. var. formosanum（Hayata）Z. Y. Zhang et Turland ●

301366　Pittosporum perglabratum Hung T. Chang et S. Z. Yan；全秃海桐（土连翘，鸭脚板树）；Bald Seatung, Glabrous Pittosporum, Smooth Pittosporum ●

301367　Pittosporum perryanum Gowda；缝线海桐（吊灯笼，黄珠子，黄珠子海桐，珠木）；Sutured Pittosporum, Sutured Seatung ●

301368　Pittosporum perryanum Gowda var. linearifolium Hung T. Chang et S. Z. Yan；狭叶缝线海桐（窄叶缝线海桐）；Narrowleaf Sutured Pittosporum, Narrowleaf Sutured Seatung ●

301369　Pittosporum phillyraeoides DC.；狭叶海桐（菲力桂海桐）；Butterbush, Native Apricot, Weeping Pittosporum, Willow Pittosporum ●

301370　Pittosporum planilobum Hung T. Chang et S. Z. Yan；扁片海桐；Flatlobed Pittosporum, Flat-lobed Pittosporum, Flatlobed Seatung ●

301371　Pittosporum podocarpifolium C. Y. Wu = Pittosporum kweichowense Gowda var. podocarpifolium（C. Y. Wu）Z. Y. Zhang et Turland ●

301372　Pittosporum podocarpum Gagnep.；柄果海桐（寡鸡蛋树，广枳仁，山枝条）；Stalkedfruit Pittosporum, Stalkedfruit Seatung, Stalk-fruited Pittosporum ●

301373　Pittosporum podocarpum Gagnep. var. angustatum Gowda；线叶柄果海桐（狭叶柄果海桐，崖子花）；Narrowleaf Pittosporum, Narrowleaf Seatung ●

301374　Pittosporum podocarpum Gagnep. var. hejiangense（H. Y. Su）Z. Y. Zhang et Turland；合江海桐；Hejiang Pittosporum, Hejiang Seatung ●

301375　Pittosporum podocarpum Gagnep. var. molle W. D. Han；毛花柄果海桐；Hairyflower Narrowleaf Pittosporum ●

301376　Pittosporum polycarpum Hung T. Chang et S. Z. Yan；多果海桐；Manyfruit Seatung ●

301377　Pittosporum polycarpum Hung T. Chang et S. Z. Yan = Pittosporum paniculiferum Hung T. Chang et S. Z. Yan ●

301378　Pittosporum pulcherum Gagnep.；秀丽海桐；Beautiful Pittosporum, Beautiful Seatung ●

301379　Pittosporum qinlingense Y. Ren et X. Liu；秦岭海桐；Qinling Pittosporum ●

301380　Pittosporum quartinianum Cufod. = Pittosporum viridiflorum Sims ●☆

301381　Pittosporum ralphii Kirk；拉尔夫海桐（北岛海桐，拉斐海桐）●☆

301382　Pittosporum ralphii Kirk 'Variegatum'；斑叶拉尔夫海桐●☆

301383　Pittosporum reflexisepalum C. Y. Wu；折萼海桐；Reflexed-sepal Pittosporum, Reflexed-sepal Seatung ●

301384　Pittosporum rehderianum Gowda；厚圆果海桐（厚果海桐，秦岭海桐，铁棒锤）；Rehder Pittosporum, Rehder Seatung ●

301385　Pittosporum rehderianum Gowda var. ternstroemioides（C. Y. Wu）Z. Y. Zhang et Turland；厚皮香海桐；Ternstroemia-like Pittosporum, Ternstroemia-like Seatung ●

301386　Pittosporum resiniferum Hemsl.；树脂海桐●☆

301387　Pittosporum reticosum sensu Bakker et Steenis = Pittosporum kerrii Craib ●

301388　Pittosporum revolutum Dryand.；黄海桐；Rough-fruit Pittosporum, Yellow Pittosporum ●☆

301389　Pittosporum rhodesicum Cufod. = Pittosporum viridiflorum Sims ●☆

301390　Pittosporum rhombifolium A. Cunn. ex Hook.；菱叶海桐；Diamond-leaf Laurel, Holly Wood, Queensland Pittosporum ●☆

301391　Pittosporum ripicola J. Léonard = Pittosporum viridiflorum Sims ●☆

301392　Pittosporum ripicola J. Léonard subsp. katangense ? = Pittosporum viridiflorum Sims ●☆

301393　Pittosporum sahnianum Gowda = Pittosporum illicioides Makino ●

301394　Pittosporum saxicola Rehder et E. H. Wilson；石生海桐；Rock Pittosporum, Saxatile Pittosporum, Saxicolous Seatung ●

301395　Pittosporum sinense Desf. = Pittosporum viridiflorum Sims ●☆

301396　Pittosporum spathicalyx De Wild. = Pittosporum viridiflorum Sims ●☆

301397　Pittosporum subulisepalum Hu et F. T. Wang；尖萼海桐；Sharpsepal Pittosporum, Sharpsepal Seatung, Sharp-sepaled Pittosporum ●

301398　Pittosporum tenuifolium Gaertn.；细叶海桐（薄叶海桐）；Kohuhu, Pittosporum, Tawhiwhi Kohuhu ●☆

301399　Pittosporum tenuifolium Gaertn. 'Deborah'；德博拉细叶海桐●☆

301400　Pittosporum tenuifolium Gaertn. 'Elia Keightley'；凯特丽伊利亚细叶海桐●☆

301401　Pittosporum tenuifolium Gaertn. 'Irene Paterson'；帕特森艾琳细叶海桐●☆

301402　Pittosporum tenuifolium Gaertn. 'James Stirling'；斯特灵詹姆斯细叶海桐●☆

301403　Pittosporum tenuifolium Gaertn. 'Limelight'；白炽灯细叶海桐●☆

301404　Pittosporum tenuifolium Gaertn. 'Margaret Turnbull'；马格里特·特布尔薄叶海桐●☆

301405　Pittosporum tenuifolium Gaertn. 'Marjorie Channon'；查诺马乔里细叶海桐●☆

301406　Pittosporum tenuifolium Gaertn. 'Sunburst' = Pittosporum tenuifolium Gaertn. 'Elia Keightley' ●☆

301407　Pittosporum tenuifolium Gaertn. 'Tom Thunb'；大拇指汤姆细叶海桐（拇指汤姆薄叶海桐）●☆

301408　Pittosporum tenuifolium Gaertn. 'Variegatum';斑叶细叶海桐●☆

301409　Pittosporum tenuifolium Gaertn. 'Warnham Gold';沃汉姆金细叶海桐●☆

301410　Pittosporum tenuivalvatum Hung T. Chang et S. Z. Yan;薄片海桐;Filmy Seatung,Thinleaf Pittosporum,Thin-valvated Pittosporum ●

301411　Pittosporum ternstroemioides C. Y. Wu = Pittosporum rehderianum Gowda var. ternstroemioides（C. Y. Wu）Z. Y. Zhang et Turland ●

301412　Pittosporum tetraspermum sensu Gowda = Pittosporum tonkinense Gagnep. ●

301413　Pittosporum tobira（Thunb.）W. T. Aiton;海桐（臭榕子,垂青树,海桐花,七里香）;Japanese Cheesewood,Japanese Pittosporum,Mock Orange,Mock-orange,Tobira,Tobira Pittosporum,Tobira Seatung ●

301414　Pittosporum tobira（Thunb.）W. T. Aiton 'Variegatum';斑叶海桐●☆

301415　Pittosporum tobira（Thunb.）W. T. Aiton 'Wheeler's Dwarf';惠勒矮海桐●☆

301416　Pittosporum tobira（Thunb.）W. T. Aiton f. macrophyllum（Nakai）Sugim.;日本大叶海桐（大叶海桐花）●☆

301417　Pittosporum tobira（Thunb.）W. T. Aiton var. calvescens Ohwi;秃序海桐;Glabrescent Pittosporum,Glabrescent Seatung ●

301418　Pittosporum tobira（Thunb.）W. T. Aiton var. calvescens Ohwi = Pittosporum tobira（Thunb.）W. T. Aiton ●

301419　Pittosporum tobira（Thunb.）W. T. Aiton var. chinense S. Kobay. = Pittosporum tobira（Thunb.）W. T. Aiton ●

301420　Pittosporum tobira（Thunb.）W. T. Aiton var. fukienense Gowda = Pittosporum tobira（Thunb.）W. T. Aiton var. calvescens Ohwi ●

301421　Pittosporum tobira Aiton = Pittosporum tobira（Thunb.）W. T. Aiton ●

301422　Pittosporum tomentosum Engl. = Pittosporum abyssinicum Delile ●☆

301423　Pittosporum tonkinense Gagnep.;四子海桐;Tonkin Pittosporum,Tonkin Seatung ●

301424　Pittosporum trigonocarpum H. Lév.;棱果海桐（公栀子,鸡骨头,三角果海桐,瘦鱼蓼）;Triangular Pittosporum,Triangularfruit Seatung,Triangular-fruited Pittosporum ●

301425　Pittosporum trigonocarpum H. Lév. sensu Gowda = Pittosporum xylocarpum Hu et F. T. Wang ●

301426　Pittosporum truncatum E. Pritz. ex Diels;崖花子（菱叶海桐,七里香,山茶辣,山枝子,崖花树,崖花子树,崖子花,岩子花）;Truncate Pittosporum,Truncate Seatung ●

301427　Pittosporum truncatum E. Pritz. ex Diels var. tsai Gowda;云南崖花树;Yunnan Pittosporum ●

301428　Pittosporum truncatum E. Pritz. ex Diels var. tsai Gowda = Pittosporum heterophyllum Franch. ●

301429　Pittosporum truncatum Pritz. var. tsaii Gowda = Pittosporum heterophyllum Franch. ●

301430　Pittosporum tubiflorum Hung T. Chang et S. Z. Yan;管花海桐;Tube-flower Pittosporum,Tube-flowered Pittosporum,Tubiflorous Seatung ●

301431　Pittosporum umbellatum Gaertn.;伞形花海桐●☆

301432　Pittosporum undulatifolium Hung T. Chang et S. Z. Yan;波叶海桐;Undulateleaf Pittosporum,Undulateleaf Seatung,Undulate-leaved Pittosporum ●

301433　Pittosporum undulatum Vent.;波状海桐（波叶海桐,岛海桐,岛海桐花）;Australian Cheesewood,Cheesewood,Moch Orange,Mochorange,Orangeberry Pittosporum,Orange-berry Pittosporum,Seatung,Sweet Pittosporum,Victorian Box ●☆

301434　Pittosporum undulatum Vent. sensu Gowda = Pittosporum undulatifolium Hung T. Chang et S. Z. Yan ●

301435　Pittosporum ustulatum Cufod. = Pittosporum viridiflorum Sims ●☆

301436　Pittosporum venulosum F. Muell. ;细脉海桐●☆

301437　Pittosporum verticillatum Wall. = Pittosporum napaulense（DC.）Rehder et E. H. Wilson ●

301438　Pittosporum viburnifolium Hayata;荚蒾叶海桐;Viburnum-leaf Pittosporum,Viburnumleaf Seatung,Viburnum-leaved Pittosporum ●

301439　Pittosporum viburnifolium Hayata = Pittosporum moluccanum Miq. ●

301440　Pittosporum viridiflorum Sims;绿花海桐;Cape Cheesewood,Cape Pittosporum,Cheesewood ●☆

301441　Pittosporum viridiflorum Sims subsp. afrorientale Cufod. = Pittosporum viridiflorum Sims ●☆

301442　Pittosporum viridiflorum Sims subsp. angolense（Oliv.）Cufod. = Pittosporum viridiflorum Sims ●☆

301443　Pittosporum viridiflorum Sims subsp. dalzielii（Hutch.）Cufod. = Pittosporum viridiflorum Sims ●☆

301444　Pittosporum viridiflorum Sims subsp. feddeanum（Pax）Cufod. = Pittosporum viridiflorum Sims ●☆

301445　Pittosporum viridiflorum Sims subsp. malosanum（Baker）Cufod. = Pittosporum viridiflorum Sims ●☆

301446　Pittosporum viridiflorum Sims subsp. quartinianum（Cufod.）Cufod. = Pittosporum viridiflorum Sims ●☆

301447　Pittosporum viridiflorum Sims subsp. somalense Cufod. = Pittosporum viridiflorum Sims ●☆

301448　Pittosporum viridiflorum Sims var. afrorientale（Cufod.）Cufod. = Pittosporum viridiflorum Sims ●☆

301449　Pittosporum viridiflorum Sims var. commutatum（Putt.）Moeser = Pittosporum viridiflorum Sims ●☆

301450　Pittosporum viridiflorum Sims var. kruegeri（Engl.）Moeser ex Engl. = Pittosporum viridiflorum Sims ●☆

301451　Pittosporum viridiflorum Sims var. malosanum（Baker）Cufod. = Pittosporum viridiflorum Sims ●☆

301452　Pittosporum vosseleri Engl. = Pittosporum viridiflorum Sims ●☆

301453　Pittosporum wenxianense G. H. Wang et Y. S. Lian = Pittosporum glabratum Lindl. var. wenxianense（G. H. Wang et Y. S. Lian）Z. Y. Zhang et Turland ●

301454　Pittosporum xylocarpum Hu et F. T. Wang;木果海桐（大果海桐,广栀仁,山枝茶,山枝仁）;Woodfruit Pittosporum,Woodyfruit Seatung,Woody-fruited Pittosporum ●

301455　Pittosporum yunnanense Franch. = Osmanthus yunnanensis（Franch.）P. S. Green ◇

301456　Pittunia Miers = Petunia Juss.（保留属名）■

301457　Pitumba Aubl. = Casearia Jacq. ■

301458　Pituranthos Viv.（1824）;肖德弗草属■☆

301459　Pituranthos Viv. = Deverra DC. ■

301460　Pituranthos Viv. = Eriocycla Lindl. ■

301461　Pituranthos aphyllus（Cham. et Schltdl.）Benth. et Hook. f. ex Schinz = Deverra denudata（Viv.）Pfisterer et Podlech subsp. aphylla（Cham. et Schltdl.）Pfisterer et Podlech ■☆

301462　Pituranthos battandieri Maire = Deverra battandieri（Maire）Chrtek ■☆

301463　Pituranthos battandieri Maire subsp. abbreviatus ? = Deverra battandieri（Maire）Chrtek ■☆

301464　Pituranthos battandieri Maire subsp. leptactis ? = Deverra battandieri（Maire）Chrtek ■☆

301465　Pituranthos battandieri Maire var. tuszonii（Andr.）Maire ＝Deverra battandieri（Maire）Chrtek ■☆

301466　Pituranthos burchellii（DC.）Benth. et Hook. f. ex Schinz ＝Deverra burchellii（DC.）Eckl. et Zeyh. ■☆

301467　Pituranthos chloranthus（Coss. et Durieu）Benth. et Hook. ＝Deverra denudata（Viv.）Pfisterer et Podlech ■☆

301468　Pituranthos chloranthus（Coss. et Durieu）Benth. et Hook. subsp. cossonianus Maire ＝Deverra denudata（Viv.）Pfisterer et Podlech ■☆

301469　Pituranthos chloranthus（Coss. et Durieu）Benth. et Hook. subsp. intermedius（Chevall.）Maire ＝Deverra triradiata Boiss. subsp. intermedia（Chevall.）Pfisterer et Podlech ■☆

301470　Pituranthos chloranthus（Coss. et Durieu）Benth. et Hook. subsp. robustus Maire ＝Deverra battandieri（Maire）Chrtek ■☆

301471　Pituranthos chloranthus（Coss. et Durieu）Benth. et Hook. var. calvescens Maire et Weiller ＝Deverra triradiata Boiss. subsp. intermedia（Chevall.）Pfisterer et Podlech ■☆

301472　Pituranthos crassifolius Andr. ＝Deverra denudata（Viv.）Pfisterer et Podlech ■☆

301473　Pituranthos denudatus Viv. ＝Deverra denudata（Viv.）Pfisterer et Podlech ■☆

301474　Pituranthos denudatus Viv. subsp. battandieri（Maire）Jafri ＝Deverra battandieri（Maire）Chrtek ■☆

301475　Pituranthos nudus（Lindl.）Benth. ex C. B. Clarke ＝Eriocycla nuda Lindl. ■

301476　Pituranthos pelliotii H. Boissieu ＝Eriocycla pelliotii（H. Boissieu）H. Wolff ■

301477　Pituranthos reboudii（Coss. et Durieu）Maire ＝Deverra reboudii Coss. et Durieu ■☆

301478　Pituranthos rholfsianus（Asch.）Schinz ＝Deverra rholfsiana Asch. ■☆

301479　Pituranthos scoparius（Coss. et Durieu）Benth. et Hook. ex Schinz var. fallax（Batt.）Maire ＝Deverra scoparia Coss. et Durieu ■☆

301480　Pituranthos scoparius（Coss. et Durieu）Benth. et Hook. ex Schinz var. junceus（Ball）Maire ＝Deverra scoparia Coss. et Durieu ■☆

301481　Pituranthos scoparius（Coss. et Durieu）Benth. et Hook. ex Schinz var. rubellus Thell. ＝Deverra scoparia Coss. et Durieu ■☆

301482　Pituranthos scoparius（Coss. et Durieu）Benth. et Hook. ex Schinz ＝Deverra scoparia Coss. et Durieu ■☆

301483　Pituranthos stewartii Dunn ＝Pimpinella stewartii（Dunn）Nasir ■☆

301484　Pituranthos thomsonii C. B. Clarke ＝Eriocycla thomsonii（Clarke）H. Wolff ■☆

301485　Pituranthos tortuosus（Desf.）Maire ＝Deverra tortuosa（Desf.）DC. ■☆

301486　Pituranthos tortuosus（Desf.）Maire var. virgatus（Coss. et Durieu）Durand et Barratte ＝Deverra scoparia Coss. et Durieu ■☆

301487　Pituranthos tripolitanus Andr. ＝Deverra rholfsiana Asch. ■☆

301488　Pituranthos tuzsoni Andr. ＝Deverra battandieri（Maire）Chrtek ■☆

301489　Pituranthos virgatus（Coss. et Durieu）Hochr. ＝Deverra scoparia Coss. et Durieu ■☆

301490　Pituranthus albescens（Franch.）H. Boissieu ＝Eriocycla albescens（Franch.）H. Wolff ■

301491　Pituranthus nuda（Lindl.）Benth. ＝Eriocycla nuda Lindl. ■

301492　Pituranthus pelliotii H. Boissieu ＝Eriocycla pelliotii（H. Boissieu）H. Wolff ■

301493　Pituranthus provostii H. Boissieu ＝Eriocycla albescens（Franch.）H. Wolff ■

301494　Pitygentias Gilg ＝Gentianella Moench（保留属名）■

301495　Pitygentias Gilg ＝Selatium D. Don ex G. Don ■

301496　Pityopsis（Nutt.）Torr. et A. Gray ＝Pityopsis Nutt. ■☆

301497　Pityopsis Nutt.（1840）;禾叶金菀属;Grass-leaved Goldenasters ■☆

301498　Pityopsis Nutt. ＝Chrysopsis（Nutt.）Elliott（保留属名）■☆

301499　Pityopsis adenolepis（Fernald）Semple ＝Pityopsis aspera（A. Gray）Small var. adenolepis（Fernald）Semple et F. D. Bowers ■☆

301500　Pityopsis argentea（Pers.）Nutt. ＝Pityopsis graminifolia（Michx.）Nutt. var. latifolia（Fernald）Semple et F. D. Bowers ■☆

301501　Pityopsis aspera（A. Gray）Small;粗糙禾叶金菀 ■☆

301502　Pityopsis aspera（A. Gray）Small var. adenolepis（Fernald）Semple et F. D. Bowers;腺鳞粗糙禾叶金菀 ☆

301503　Pityopsis falcata（Pursh）Nutt. ;镰形禾叶金菀 ■☆

301504　Pityopsis flexuosa（Nash）Small;之字金菀 ■☆

301505　Pityopsis graminifolia（Michx.）Nutt. ;禾叶金菀（禾叶金菊）;Grassleaf Goldaster, Grass-leaved Goldenaster ■☆

301506　Pityopsis graminifolia（Michx.）Nutt. var. latifolia（Fernald）Semple et F. D. Bowers;宽禾叶金菀 ■☆

301507　Pityopsis graminifolia（Michx.）Nutt. var. microcephala（Small）Semple ＝Pityopsis graminifolia（Michx.）Nutt. var. tenuifolia（Torr.）Semple et F. D. Bowers ■☆

301508　Pityopsis graminifolia（Michx.）Nutt. var. tenuifolia（Torr.）Semple et F. D. Bowers;窄禾叶金菀 ■☆

301509　Pityopsis graminifolia（Michx.）Nutt. var. tracyi（Small）Semple;特拉西禾叶金菀 ■☆

301510　Pityopsis nervosa（Willd.）Dress ＝Pityopsis graminifolia（Michx.）Nutt. var. latifolia（Fernald）Semple et F. D. Bowers ■☆

301511　Pityopsis oligantha（Chapm. ex Torr. et A. Gray）Small;少花金菀;Few-headed Grass-leaved Goldenaster ■☆

301512　Pityopsis pinifolia（Elliott）Nutt. ;松叶金菀;Pine-leaved Goldenaster ■☆

301513　Pityopsis ruthii（Small）Small;鲁斯金菀;Ruth's Grass-leaved Goldenaster ■☆

301514　Pityopsis tracyi（Small）Small ＝Pityopsis graminifolia（Michx.）Nutt. var. tracyi（Small）Semple ■☆

301515　Pityopus Small（1914）;松林杜鹃属（毛晶兰属）●☆

301516　Pityopus oregonus Small;松林杜鹃 ●☆

301517　Pityothamnus Small ＝Asimina Adans. ●☆

301518　Pityothamnus angustifolius（A. Gray）Small ＝Asimina longifolia Král ●☆

301519　Pityothamnus incanus（W. Bartram）Small ＝Asimina incana（W. Bartram）Exell ●☆

301520　Pityothamnus obovatus（Willd.）Small ＝Asimina obovata（Willd.）Nash ●☆

301521　Pityothamnus pygmaeus（W. Bartram）Small ＝Asimina pygmaea（W. Bartram）Dunal ●☆

301522　Pityothamnus reticulatus（Chapm.）Small ＝Asimina reticulata Shuttlew. ex Chapm. ●☆

301523　Pityothamnus tetramerus（Small）Small ＝Asimina tetramera Small ●☆

301524　Pityphyllum Schltr.（1920）;松叶兰属 ■☆

301525　Pityphyllum amesianum Schltr. ;松叶兰 ■☆

301526　Pityranthe Thwaites ＝Diplodiscus Turcz. ●

301527　Pityranthe Thwaites（1858）;麸椴属 ●

301528　Pityranthe trichosperma（Merr.）Kubitzki ＝Diplodiscus trichospermus（Merr.）Y. Tang, M. G. Gilbert et Dorr ●◇

301529　Pityranthe trichosperma（Merr.）Kubitzki ＝Hainania

trichosperma Merr. ●◇

301530　Pityranthe verrucosa Thwaites;萩椴●☆

301531　Pityranthes Willis = Pituranthos Viv. ■☆

301532　Pityranthus H. Wolff = Pituranthos Viv. ■☆

301533　Pityranthus Mart. = Alternanthera Forssk. ■

301534　Pityrocarpa（Benth.）Britton et Rose ＝Piptadenia Benth. ●☆

301535　Pityrocarpa（Benth.）Britton, Rose et Brenan ＝Piptadenia Benth. ●☆

301536　Pityrocarpa Britton et Rose ＝Piptadenia Benth. ●☆

301537　Pityrodia R. Br.（1810）;叉毛灌属●☆

301538　Pityrophyllum Beer = Tillandsia L. ■☆

301539　Pityrosperma Siebold et Zucc. = Cimicifuga L. ●■

301540　Pityrosperma acerinum Siebold et Zucc. ＝Cimicifuga japonica（Thunb.）Spreng. ■

301541　Piuttia Mattei =Thalictrum L. ■

301542　Pivonneava Hook. f. = Pironneava Gaudich. ■☆

301543　Placea Miers ex Lindl. = Placea Miers ■☆

301544　Placea Miers（1841）;扁石蒜属■☆

301545　Placea grandiflora Lem.;大花扁石蒜■☆

301546　Placea lutea Phil.;黄扁石蒜■☆

301547　Placea ornata Miers;扁石蒜■☆

301548　Placocarpa Hook. f.（1873）;扁果茜属☆

301549　Placocarpa mexicana Hook. f.;扁果茜☆

301550　Placodiscus Radlk.（1878）;盾盘木属●☆

301551　Placodiscus amaniensis Radlk.;阿马尼盾盘木●☆

301552　Placodiscus angustifolius Radlk.;窄叶盾盘木●☆

301553　Placodiscus attenuatus J. B. Hall;渐狭盾盘木●☆

301554　Placodiscus bancoensis Aubrév. et Pellegr.;邦克盾盘木●☆

301555　Placodiscus boya Aubrév. et Pellegr.;博亚盾盘木●☆

301556　Placodiscus bracteosus J. B. Hall;多苞片盾盘木●☆

301557　Placodiscus caudatus Pierre ex Radlk.;尾状盾盘木●☆

301558　Placodiscus cuneatus Radlk. = Placodiscus angustifolius Radlk. ●☆

301559　Placodiscus glandulosus Radlk.;具腺盾盘木●☆

301560　Placodiscus leptostachys Radlk.;细穗盾盘木●☆

301561　Placodiscus letestui Pellegr. = Placodiscus opacus Radlk. ●☆

301562　Placodiscus oblongifolius J. B. Hall;矩圆叶盾盘木●☆

301563　Placodiscus opacus Radlk.;暗色盾盘木●☆

301564　Placodiscus paniculatus Hauman;圆锥盾盘木●☆

301565　Placodiscus pedicellatus F. G. Davies;梗花盾盘木●☆

301566　Placodiscus pseudostipularis Radlk.;假托叶盾盘木●☆

301567　Placodiscus pynaertii De Wild.;皮那盾盘木●☆

301568　Placodiscus riparius Keay;河岸盾盘木●☆

301569　Placodiscus splendidus Keay;闪光盾盘木●☆

301570　Placodiscus turbinatus Radlk.;陀螺形盾盘木●☆

301571　Placodium Benth. et Hook. f. = Plocama Aiton ●☆

301572　Placodium Hook. f. = Plocama Aiton ●☆

301573　Placolobium Miq.（1858）;盾荚豆属●☆

301574　Placolobium Miq. = Ormosia Jacks.（保留属名）●

301575　Placolobium sumatranum Miq.;盾荚豆●☆

301576　Placoma J. F. Gmel. = Plocama Aiton ●☆

301577　Placopoda Balf. f.（1882）;扁足茜属☆

301578　Placopoda virgata Balf. f.;扁足茜☆

301579　Placospermum C. T. White et W. D. Francis（1924）;盾籽龙眼属●☆

301580　Placospermum corLaceum C. T. White et W. D. Francis;盾籽龙眼●☆

301581　Placostigma Blume = Podochilus Blume ■

301582　Placseptalia Espinosa = Ochagavia Phil. ■☆

301583　Placus Lour.（废弃属名）= Blumea DC.（保留属名）■●

301584　Placus oxyodonta（DC.）Kuntze = Blumea oxyodonta DC. ■

301585　Placus procera Kuntze = Blumea repanda（Roxb.）Hand. -Mazz. ■

301586　Pladaroxylon（Endl.）Hook. f.（1870）;白树菊属●☆

301587　Pladaroxylon Hook. f. = Pladaroxylon（Endl.）Hook. f. ●☆

301588　Pladaroxylon Hook. f. = Senecio L. ■●

301589　Pladaroxylon leucadendron（DC.）Hook. f. = Pladaroxylon leucadendron（Willd.）Hook. f. ●☆

301590　Pladaroxylon leucadendron（G. Forst.）Hook. f. = Pladaroxylon leucadendron（Willd.）Hook. f. ●☆

301591　Pladaroxylon leucadendron（Willd.）Hook. f.;白树菊●☆

301592　Pladera Sol. = Canscora Lam. ■

301593　Pladera Sol. ex Roxb. = Canscora Lam. ■

301594　Pladera decussata Roxb. = Canscora alata（Roth ex Roem. et Schult.）Wall. ■☆

301595　Pladera decussata Roxb. = Canscora decussata（Roxb.）Roem. et Schult. f. ■☆

301596　Plaea Pers. = Pleea Michx. ■☆

301597　Plaesiantha Hook. f. = Pellacalyx Korth. ●

301598　Plaesianthera（C. B. Clarke）Livera ＝Brillantaisia P. Beauv. ●■☆

301599　Plaesianthera（C. B. Clarke）Livera ＝Hygrophila R. Br. ●■

301600　Plaesianthera Livera ＝Hygrophila R. Br. ●■

301601　Plagiacanthus Nees = Dianthera L. ■☆

301602　Plagiacanthus Nees = Justicia L. ●■

301603　Plagiantha Renvoize（1982）;斜花黍属■☆

301604　Plagiantha tenella Renvoize;斜花黍■☆

301605　Plagianthaceae J. Agardh ＝Malvaceae Juss.（保留科名）●■

301606　Plagianthera Rchb. f. et Zoll.（1856）;斜药大戟属●☆

301607　Plagianthera Rchb. f. et Zoll. = Mallotus Lour. ●

301608　Plagianthera affinis Baill.;近缘斜药大戟●☆

301609　Plagianthera oppositifolia Rchb. et Zoll.;斜药大戟●☆

301610　Plagianthus J. R. Forst. et G. Forst.（1775）;新西兰锦葵属;Ribbonwood,Twinebark ●☆

301611　Plagianthus betulinus A. Cunn. = Plagianthus divaricatus J. R. Forst. et G. Forst. ●☆

301612　Plagianthus divaricatus J. R. Forst. et G. Forst.;海滨新西兰锦葵;Makaka,Ribbonwood,Shore Ribbonwood ●☆

301613　Plagianthus regius（Poit.）Hochr.;新西兰锦葵;Ribbonwood ●☆

301614　Plagianthus regius Hochr. = Plagianthus regius（Poit.）Hochr. ●☆

301615　Plagiarthron P. A. Duvign. = Loxodera Launert ■☆

301616　Plagidia Raf. = Paronychia Mill. ■

301617　Plagielytrum Post et Kuntze = Plagiolytrum Nees ■

301618　Plagielytrum Post et Kuntze = Tripogon Roem. et Schult. ■

301619　Plagiobasis Schrenk（1845）;斜果菊属;Slantdaisy ■

301620　Plagiobasis centauroides Schrenk;斜果菊;Centaury-like Slantdaisy ■

301621　Plagiobasis dschungaricus Iljin = Plagiobasis centauroides Schrenk ■

301622　Plagiobasis sogdiana Bunge = Russowia sogdiana（Bunge）B. Fedtsch. ■

301623　Plagiobothrys Fisch. et C. A. Mey.（1836）;斜紫草属;Popcorn Flower,White Forget-me-not ■☆

301624　Plagiobothrys axillaris Benth.;腋生斜紫草■☆

301625　Plagiobothrys figuratus（Piper）I. M. Johnst. ex M. Peck;斜紫草;Fragrant Popcorn-flower,Scorpion-grass ■☆

301626　Plagiobothrys orientalis（L.）I. M. Johnst. ;东方斜紫草■☆

301627　Plagiobothrys scouleri（Hook. et Arn.）I. M. Johnst. ;斯考勒斜紫草; Meadow Plagiobothrys, Scouler's Popcorn-flower, White Forget-me-not ■☆

301628　Plagiobothrys scouleri（Hook. et Arn.）I. M. Johnst. var. penicillatus（Greene）Cronquist;束斜紫草; Meadow Plagiobothrys, Scouler's Popcorn-flower ■☆

301629　Plagiobothrys stipitatus（Greene）I. M. Johnst. ;具柄斜紫草■☆

301630　Plagiocarpus Benth.（1873）;偏果豆属(独花腋生豆属)■☆

301631　Plagiocarpus axillaris Benth. ;偏果豆■☆

301632　Plagioceltis Mildbr. ex Baehni ＝ Ampelocera Klotzsch ●☆

301633　Plagiocheilus Arn. ＝ Plagiocheilus Arn. ex DC. ■☆

301634　Plagiocheilus Arn. ex DC.（1838）;偏唇菊属■☆

301635　Plagiocheilus bogotesis（Kunth）Wedd. ;波哥达偏唇菊■☆

301636　Plagiochilus Lindl. ＝ Plagiocheilus Arn. ex DC. ■☆

301637　Plagiochloa Adamson et Sprague ＝ Tribolium Desv. ■☆

301638　Plagiochloa Adamson et Sprague(1941);肖三尖草属■☆

301639　Plagiochloa acutiflora（Nees）Adamson et Sprague ＝ Tribolium acutiflorum（Nees）Renvoize ■☆

301640　Plagiochloa alternans（Nees）Adamson et Sprague ＝ Tribolium uniolae（L. f.）Renvoize ■☆

301641　Plagiochloa brachystachya（Nees）Adamson et Sprague ＝ Tribolium brachystachyum（Nees）Renvoize ■☆

301642　Plagiochloa ciliaris（Stapf）Adamson et Sprague ＝ Tribolium ciliare（Stapf）Renvoize ■☆

301643　Plagiochloa oblitera（Hemsl.）Adamson et Sprague ＝ Tribolium obliterum（Hemsl.）Renvoize ■☆

301644　Plagiochloa uniolae（L. f.）Adamson et Sprague;肖三尖草■☆

301645　Plagiochloa uniolae（L. f.）Adamson et Sprague var. villosa（Stapf）Adamson ＝ Tribolium uniolae（L. f.）Renvoize ■☆

301646　Plagiocladus Jean F. Brunel ＝ Plagiocladus Jean F. Brunel ex Petra Hoffm. ●☆

301647　Plagiocladus Jean F. Brunel ex Petra Hoffm.（1987）;喀麦隆叶下珠属●☆

301648　Plagiocladus Jean F. Brunel ex Petra Hoffm. ＝ Phyllanthus L. ●■

301649　Plagiolirion Baker ＝ Urceolina Rchb.（保留属名）●☆

301650　Plagiolirion Baker(1883);肖耳壶石蒜属■☆

301651　Plagiolirion horsmanni Baker;肖耳壶石蒜■☆

301652　Plagioloba（C. A. Mey.）Rchb. ＝ Hesperis L. ■

301653　Plagioloba Rchb. ＝ Hesperis L. ■

301654　Plagiolobium Sweet ＝ Hovea R. Br. ex W. T. Aiton ●■☆

301655　Plagiolophus Greenm.（1904）;斜冠菊属■☆

301656　Plagiolophus millspaughii Greenm. ;斜冠菊■☆

301657　Plagiolytrum Nees ＝ Tripogon Roem. et Schult. ■

301658　Plagiolytrum filiformis Nees ＝ Tripogon filiformis Nees ex Steud. ■

301659　Plagiolytrum unidentatum Nees ＝ Tripogon filiformis Nees ex Steud. ■

301660　Plagion St. -Lag. ＝ Chrysanthemum L.（保留属名）■●

301661　Plagion St. -Lag. ＝ Plagius L'Hér. ex DC. ■☆

301662　Plagiopetalum Rehder ＝ Anerincleistus Korth. ●☆

301663　Plagiopetalum Rehder（1917）;偏瓣花属; Plagiopetalum, Slantpetal ●■

301664　Plagiopetalum blinii（H. Lév.）C. Y. Wu;刺柄偏瓣花; Blin Plagiopetalum ●

301665　Plagiopetalum blinii（H. Lév.）C. Y. Wu ＝ Plagiopetalum esquirolii（H. Lév.）Rehder ●

301666　Plagiopetalum blinii（H. Lév.）C. Y. Wu ex C. Chen ＝ Plagiopetalum esquirolii（H. Lév.）Rehder ●

301667　Plagiopetalum esquirolii（H. Lév.）Rehder;偏瓣花; Esquirol Plagiopetalum, Esquirol Slantpetal ●

301668　Plagiopetalum esquirolii（H. Lév.）Rehder var. quadrangulum（Rehder）C. Chen;四棱偏瓣花; Four-angled Serrate Plagiopetalum ●

301669　Plagiopetalum esquirolii（H. Lév.）Rehder var. septemnervium C. Chen;七脉偏瓣花; Sevenvein Slantpetal, Seven-veins Esquirol Plagiopetalum ●

301670　Plagiopetalum esquirolii（H. Lév.）Rehder var. septemnervium C. Chen ＝ Plagiopetalum esquirolii（H. Lév.）Rehder ●

301671　Plagiopetalum esquirolii（H. Lév.）Rehder var. serratum（Diels）C. Hansen ＝ Plagiopetalum esquirolii（H. Lév.）Rehder ●

301672　Plagiopetalum hainanense（Merr. et Chun）Merr. ex H. L. Li ＝ Phyllagathis hainanensis（Merr. et Chun）C. Chen ●

301673　Plagiopetalum henryi（Kraenzl.）S. Y. Hu ＝ Plagiopetalum esquirolii（H. Lév.）Rehder ●

301674　Plagiopetalum quadrangulum Rehder ＝ Plagiopetalum esquirolii（H. Lév.）Rehder var. quadrangulum（Rehder）C. Chen ●

301675　Plagiopetalum quadrangulum Rehder ＝ Plagiopetalum esquirolii（H. Lév.）Rehder ●

301676　Plagiopetalum quadrangulum Rehder ＝ Plagiopetalum serratum（Diels）Diels var. quadrangulum（Rehder）C. Chen ●

301677　Plagiopetalum serratum（Diels）Diels;光叶偏瓣花; Serrate Plagiopetalum, Serrate Slantpetal ●

301678　Plagiopetalum serratum（Diels）Diels ＝ Plagiopetalum esquirolii（H. Lév.）Rehder ●

301679　Plagiopetalum serratum（Diels）Diels var. quadrangulum（Rehder）C. Chen ＝ Plagiopetalum esquirolii（H. Lév.）Rehder ●

301680　Plagiopetalum serratum（Diels）Diels var. quadrangulum（Rehder）C. Chen ＝ Plagiopetalum esquirolii（H. Lév.）Rehder var. quadrangulum（Rehder）C. Chen ●

301681　Plagiopetalum serratum Diels ＝ Plagiopetalum esquirolii（H. Lév.）Rehder var. quadrangulum（Rehder）C. Chen ●

301682　Plagiopetalum tenuicaule（C. Chen）C. Hansen;细茎偏瓣花（柔茎锦香草,四棱偏瓣花）; Fine Stem Metalleaf, Softstem Metalleaf ■

301683　Plagiophyllum Schltdl. ＝ Centradenia G. Don ●■☆

301684　Plagiopoda（R. Br.）Spach ＝ Grevillea R. Br. ex Knight(保留属名)●

301685　Plagiopoda Spach ＝ Grevillea R. Br. ex Knight(保留属名)●

301686　Plagiopteraceae Airy Shaw ＝ Celastraceae R. Br.（保留科名)●

301687　Plagiopteraceae Airy Shaw（1965）;斜翼科（印桐科）; Plagiopteron Family, Slantwing Family ●

301688　Plagiopteron Griff.（1843）;斜翼属; Oblique-wing, Plagiopteron, Slantwing ●

301689　Plagiopteron chinense X. X. Chen ＝ Plagiopteron suaveolens Griff. ●◇

301690　Plagiopteron fragrans Griff. ＝ Plagiopteron suaveolens Griff. ●◇

301691　Plagiopteron suaveolens Griff. ;斜翼（华斜翼,扣丝,山钩藤）; China Slantwing, Chinese Oblique-wing, Fragrant Plagiopteron ●◇

301692　Plagiorhegma Maxim.（1859）;鲜黄连属■

301693　Plagiorhegma Maxim. ＝ Jeffersonia Barton ■☆

301694　Plagiorhegma dubia Maxim. ;鲜黄连(常黄连,朝鲜黄连,假黄连,假细辛,立田草,毛黄连,丝卷草,铁丝草,铁丝草,细辛幌子,洋虎耳草)■

301695　Plagiorrhiza（Pierre）Hallier. f. ＝ Mesua L. ●■

301696　Plagiorrhiza Hallier. f. ＝ Mesua L. ●■

301697 Plagioscyphus Radlk. (1878);斜杯木属●☆

301698 Plagioscyphus calciphilus Capuron;喜岩斜杯木●☆

301699 Plagioscyphus cauliflorus Radlk.;茎花斜杯木●☆

301700 Plagioscyphus danguyanus Capuron;当吉斜杯木●☆

301701 Plagioscyphus humbertii Capuron;亨伯特斜杯木●☆

301702 Plagioscyphus jumellei (Choux) Capuron;朱迈尔斜杯木●☆

301703 Plagioscyphus louvelii Danguy et Choux;卢韦尔斜杯木●☆

301704 Plagioscyphus meridionalis Capuron;南方斜杯木●☆

301705 Plagioscyphus nudicalyx Capuron;裸萼斜杯木●☆

301706 Plagioscyphus stelechanthus (Radlk.) Capuron;干花斜杯木●☆

301707 Plagioscyphus unijugatus Capuron;成双斜杯木●☆

301708 Plagiosetum Benth. (1877);斜毛草属■☆

301709 Plagiosetum refractum (F. Muell.) Benth.;斜毛草■☆

301710 Plagiosiphon Harms(1897);偏管豆属■☆

301711 Plagiosiphon discifer Harms;盘状偏管豆■☆

301712 Plagiosiphon emarginatus (Hutch. et Dalziel) J. Léonard;微缺偏管豆■☆

301713 Plagiosiphon gabonensis (A. Chev.) J. Léonard;加蓬偏管豆■☆

301714 Plagiosiphon longitubus (Harms) J. Léonard;长管偏管豆■☆

301715 Plagiosiphon multijugus (Harms) J. Léonard;多对偏管豆■☆

301716 Plagiospermum Oliv. = Prinsepia Royle ●

301717 Plagiospermum Oliv. = Sinoplagiospermum Rauschert ●

301718 Plagiospermum Pierre = Benzoin Hayne ●

301719 Plagiospermum Pierre = Styrax L. ●

301720 Plagiospermum sinense Oliv. = Prinsepia sinensis (Oliv.) Oliv. ex Bean ●

301721 Plagiostachys Ridl. (1899);偏穗姜属;Plagiostachys, Slantspike ■

301722 Plagiostachys austrosinensis T. L. Wu et S. J. Chen;偏穗姜(山姜);S. China Slantspike, South China Plagiostachys ■

301723 Plagiostachys elliptica S. Q. Tong et Y. M. Xia = Zingiber ellipticum (S. Q. Tong et Y. M. Xia) Q. G. Wu et T. L. Wu ■

301724 Plagiostemon Klotzsch = Simocheilus Klotzsch ●☆

301725 Plagiostemon bicolor (Klotzsch) Klotzsch = Erica inaequalis (N. E. Br.) E. G. H. Oliv. ●☆

301726 Plagiostemon puberulus Klotzsch = Erica inaequalis (N. E. Br.) E. G. H. Oliv. ●☆

301727 Plagiostigma C. Presl = Aspalathus L. ●☆

301728 Plagiostigma Zucc. = Ficus L. ●

301729 Plagiostigma pinea (Thunb.) C. Presl = Aspalathus pinea Thunb. ●☆

301730 Plagiostyles Pierre(1897);非洲斜柱大戟属●☆

301731 Plagiostyles africana (Müll. Arg.) Prain;非洲斜柱大戟●☆

301732 Plagiostyles africana Prain ex De Wild. = Plagiostyles africana (Müll. Arg.) Prain ●☆

301733 Plagiostyles klaineana Pierre = Plagiostyles africana (Müll. Arg.) Prain ●☆

301734 Plagiotaxis Wall. = Chukrasia A. Juss. ●

301735 Plagiotaxis Wall. ex Kuntze = Chukrasia A. Juss. ●

301736 Plagiotaxis velutina Wall. = Chukrasia tabularis A. Juss. var. velutina (Wall.) King ●

301737 Plagiotheca Chiov. (1935);肖叉序草属■☆

301738 Plagiotheca Chiov. = Isoglossa Oerst. (保留属名)■★

301739 Plagiotheca fallax Chiov.;肖叉序草■☆

301740 Plagistra Raf. = Aristolochia L. ■●

301741 Plagius L'Hér. ex DC. (1838);合肋菊属■☆

301742 Plagius L'Hér. ex DC. = Chrysanthemum L. (保留属名)■●

301743 Plagius flosculosus (L.) Alava et Heywood;多小花合肋菊■☆

301744 Plagius grandiflorus (Desf.) L'Hér. = Plagius grandis (L.) Alavi et Heywood ■☆

301745 Plagius grandis (L.) Alavi et Heywood;大合肋菊■☆

301746 Plagius maghrebinus Vogt et Greuter;合肋菊■☆

301747 Plaius calanthoides Ames = Cephalantheropsis calanthoides (Ames) Tang S. Liu et H. J. Su ■

301748 Plaius gracilis (Lindl.) S. Y. Hu var. calanthoides (Ames) T. P. Lin = Cephalantheropsis calanthoides (Ames) Tang S. Liu et H. J. Su ■

301749 Plaius longipes (Hook. f.) Holttum var. calanthoides (Ames) T. P. Lin = Cephalantheropsis calanthoides (Ames) Tang S. Liu et H. J. Su ■

301750 Plakothira J. Florence(1986);马克萨斯刺莲花属●■☆

301751 Plakothira frutescens J. Florence;灌木马克萨斯刺莲花●☆

301752 Plakothira parviflora J. Florence;小花马克萨斯刺莲花■☆

301753 Plakothira perlmanii J. Florence;马克萨斯刺莲花■☆

301754 Planaltoa Taub. (1895);多花修泽兰属●☆

301755 Planaltoa salviifolia Taub.;多花修泽兰●☆

301756 Planarium Desv. (1826);平豆属■☆

301757 Planarium Desv. = Chaetocalyx DC. ■☆

301758 Planarium latisiliquum Desv.;平豆■☆

301759 Planchonella Pierre = Pouteria Aubl. ●

301760 Planchonella Pierre(1890)(保留属名);山榄属(假水石梓属,树青属);Planchonella, Wildolive ●

301761 Planchonella Tiegh. = Godoya Ruiz et Pav. ●☆

301762 Planchonella africana (A. DC.) Baehni = Chrysophyllum africanum A. DC. ●☆

301763 Planchonella albida (G. Don) Baehni = Chrysophyllum albidum G. Don ●☆

301764 Planchonella annamensis Pierre = Pouteria annamensis (Pierre ex Dubard) Baehni ●

301765 Planchonella annamensis Pierre ex Dubard = Pouteria annamensis (Pierre ex Dubard) Baehni ●

301766 Planchonella aurata Pierre = Eberhardtia aurata (Pierre ex Dubard) Lecomte ●

301767 Planchonella aurata Pierre ex Dubard = Eberhardtia aurata (Pierre ex Dubard) Lecomte ●

301768 Planchonella australis (R. Br.) Pierre = Pouteria australis (R. Br.) Baehni ●☆

301769 Planchonella boninensis (Nakai) Masam. et Yanagih.;小笠原山榄●☆

301770 Planchonella clemensii (Lecomte) P. Royen;狭叶山榄;Clemens Planchonella, Clemens Wildolive ●

301771 Planchonella duclitan (Blanco) Bakh. f.;兰屿山榄;Lanyu Planchonella, Lanyu Pouteria ●

301772 Planchonella ferruginea (Hook. et Arn.) Pierre = Planchonella obovata (R. Br.) Pierre ●

301773 Planchonella grandifolia (Wall.) Pierre = Pouteria grandifolia (Wall.) Baehni ●

301774 Planchonella kerrii H. R. Fletcher = Pouteria grandifolia (Wall.) Baehni ●

301775 Planchonella laurifolia (A. Rich.) Pierre;月桂叶山榄●☆

301776 Planchonella nitida Dubard;光亮山榄(光假水石梓)●☆

301777 Planchonella obovata (R. Br.) Pierre;山榄(石松,树青);Obovate Planchonella, Obovate Wildolive, Pouteria ●

301778 Planchonella obovata (R. Br.) Pierre var. dubia (Koidz. ex H.

Hara）Hatus. ex T. Yamaz. ;可疑山榄●☆

301779　Planchonella pedunculata（Hemsl.）H. J. Lam. et Kerpel ＝ Sinosideroxylon pedunculatum（Hemsl.）H. Chuang ●

301780　Planchonella pohlmaniana（F. Muell.）Pierre ex Dubard;波曼山榄●☆

301781　Planchonella rostrata（Merr.）Lam. ＝ Xantolis boniana（Dubard）Royen var. rostrata（Merr.）Royen ●

301782　Planchonella stenosepala（Hu）Hu ＝ Xantolis stenosepala（Hu）Royle ●

301783　Planchonella subnuda（Baker）Baehni ＝ Chrysophyllum subnudum Baker ●☆

301784　Planchonella thyrsoidea C. T. White ex F. S. Walker;凯特山榄●☆

301785　Planchonella yunnanensis C. Y. Wu ＝ Sinosideroxylon wightianum（Hook. et Arn.）Aubrév. ●

301786　Planchonella yunnanensis C. Y. Wu ＝ Sinosideroxylon yunnanense（C. Y. Wu）H. Chuang ●

301787　Planchonia Blume(1851-1852);澳洲玉蕊属(普兰木属,普朗金刀木属)●☆

301788　Planchonia Dunal ＝ Salpichroa Miers ●☆

301789　Planchonia J. Gay ex Benth. et Hook. f. ＝ Polycarpaea Lam.（保留名）■●

301790　Planchonia australis ?;澳洲玉蕊(澳洲普兰木)●☆

301791　Planchonia careya（F. Muell.）R. Knuth;凯里澳洲山榄(凯里普兰木);Cocky Apple ●☆

301792　Planchonia papuana Knuth;巴布澳洲玉蕊(巴布亚金刀木)●☆

301793　Planchonia spectabilis Merr. ;美丽澳洲玉蕊(普朗金刀木)●☆

301794　Planchonia valida Blume;粗状澳洲玉蕊(粗状金刀木)●☆

301795　Plancia Neck. ＝ Leontodon L.（保留属名）■

301796　Planea Karis(1990);寡头帚鼠麴属●☆

301797　Planea P. O. Karis ＝ Costus L. ■

301798　Planea schlechteri（L. Bolus）P. O. Karis;寡头帚鼠麴■☆

301799　Planera Giseke ＝ Costus L. ■

301800　Planera Giseke ＝ Hellenia Retz. ■

301801　Planera J. F. Gmel.（1791）;水榆属(沼榆属);Planer Tree, Water Elm, Water-elm ●☆

301802　Planera P. O. Karis ＝ Costus L. ■

301803　Planera acuminata Lindl. ＝ Zelkova serrata（Thunb.）Makino ●

301804　Planera aquatica（Walter）J. F. Gmel. ;水榆;Planer Tree, Planertree, Water Elm, Water-elm ●☆

301805　Planera aquatica J. F. Gmel. ＝ Planera aquatics（Walter）J. F. Gmel. ●

301806　Planera davidii Hance ＝ Hemiptelea davidii（Hance）Planch. ●

301807　Planera japonica Miq. ＝ Zelkova serrata（Thunb.）Makino ●

301808　Planera japonica Miq. ＝ Zelkova sinica C. K. Schneid. ●

301809　Planera parvifolia（Jacq.）Sweet ＝ Ulmus parvifolia Jacq. ●

301810　Planera parvifolia Sweet ＝ Ulmus parvifolia Jacq. ●

301811　Planetanthemum（Endl.）Kuntze ＝ Pseuderanthemum Radlk. ●■

301812　Planichloa B. K. Simon ＝ Ectrosia R. Br. ■☆

301813　Planocarpa C. M. Weiller(1996);扁果石南属●☆

301814　Planocarpa nitida（Jarman）C. M. Weiller;光亮扁果石南●☆

301815　Planocarpa petiolaris（DC.）C. M. Weiller;细柄扁果石南●☆

301816　Planocarpa sulcata（Mihaich）C. M. Weiller;扁果石南●☆

301817　Planodes Greene ＝ Sibara Greene ■☆

301818　Planodes Greene(1912);平芥属■☆

301819　Planodes virginicum（L.）Greene;平芥■☆

301820　Planotia Munro ＝ Neurolepis Meisn. ●☆

301821　Plantaginaceae Juss.（1789）（保留科名）;车前科(车前草

科）;Plantago Family, Plantain Family ■

301822　Plantaginastrum Fabr. ＝ Alisma L. ■

301823　Plantaginastrum Heist. ex Fabr. ＝ Alisma L. ■

301824　Plantaginella Fourr. ＝ Plantago L. ■●

301825　Plantaginella Hill ＝ Limosella L. ■

301826　Plantaginorchis Szlach.（2004）;车前兰属■☆

301827　Plantaginorchis Szlach. ＝ Habenaria Willd. ■

301828　Plantaginorchis dentata（Sw.）Szlach. ＝ Habenaria dentata（Sw.）Schltr. ■

301829　Plantaginorchis finetiana（Schltr.）Szlach. ＝ Habenaria finetiana Schltr. ■

301830　Plantaginorchis longicalcarata（Hayata）Szlach. et Kras-Lap. ＝ Platanthera longicalcarata Hayata ■

301831　Plantaginorchis radiata（Thunb.）Szlach. ＝ Zeuxine strateumatica（L.）Schltr. ■

301832　Plantago L.（1753）;车前属(车前草属）;Plantain, Ribwort ■●

301833　Plantago accendens Raf. ＝ Plantago virginica L. ■

301834　Plantago afra L. ;法车前;Glandular Plantain ■☆

301835　Plantago afra L. var. obtusata（Svent.）A. Hansen et Sunding ＝ Plantago afra L. ■☆

301836　Plantago afra L. var. parviflora（Desf.）Lewalle;小花法车前■☆

301837　Plantago afra L. var. sicula（J. Presl et C. Presl）Guss. ＝ Plantago afra L. ■☆

301838　Plantago afra L. var. stricta（Schousb.）Verdc. ;条纹法车前■☆

301839　Plantago africana Verdc. ;非洲车前■☆

301840　Plantago akkensis Coss. et Murb. ;阿卡车前■☆

301841　Plantago albicans L. ;白色车前■☆

301842　Plantago albicans L. subsp. lanuginosa Chevall. ＝ Plantago albicans L. ■☆

301843　Plantago albicans L. var. angustifolia Guss. ＝ Plantago albicans L. ■☆

301844　Plantago albicans L. var. desertica Pamp. ＝ Plantago albicans L. ■☆

301845　Plantago albicans L. var. humilis Ball ＝ Plantago albicans L. ■☆

301846　Plantago albicans L. var. lanata Pamp. ＝ Plantago albicans L. ■☆

301847　Plantago albicans L. var. lanuginosa L. Chevall. ＝ Plantago albicans L. ■☆

301848　Plantago albicans L. var. latifolia Willk. ＝ Plantago albicans L. ■☆

301849　Plantago albicans L. var. macropoda Pamp. ＝ Plantago albicans L. ■☆

301850　Plantago albicans L. var. major Boiss. ＝ Plantago albicans L. ■☆

301851　Plantago albicans L. var. viridis Batt. ＝ Plantago albicans L. ■☆

301852　Plantago albicantiformis Sennen ＝ Plantago albicans L. ■☆

301853　Plantago alpina L. ;高山车前;Alpine Plantain, Alps Plantain ■☆

301854　Plantago altissima L. ;高大车前■☆

301855　Plantago amplexicaulia Cav. ;抱茎车前■☆

301856　Plantago amplexicaulis Cav. var. glabra Le Houér. ＝ Plantago amplexicaulis Cav. ■☆

301857　Plantago annua Ryding;一年车前■☆

301858　Plantago arachnoidea Schrenk;蛛毛车前(带叶车前,蛛丝毛车前）;Branched Plantain, Whorled Plantain ■☆

301859　Plantago arachnoidea Schrenk var. lorata J. Z. Liu ＝ Plantago arachnoidea Schrenk ■

301860　Plantago arborescens Poir. ;乔木状车前■

301861　Plantago arborescens Poir. subsp. maderensis（Decne.）Hansen et G. Kunkel;梅德车前■☆

301862　Plantago arborescens Poir. var. canescens Svent. ＝ Plantago arborescens Poir. ■

301863　Plantago arborescens Poir. var. mascaensis Svent. = Plantago arborescens Poir. ■

301864　Plantago arenaria Waldst. et Kit.；对叶车前（法车前，法车前草，欧车前，亚麻车前，亚麻籽车前，印车前）；Asiatic Plantain, Branched Plantain, Clammy Plantain, Flaxseed Plantain, Flax-seed Plantain, Flea Nit, Flea Nut, Fleaseed, Fleawort, Indian Plantain, Leafy-stemmed Plantain, Psyllium, Psyllium Seed, Sand Plantain, Spanish Psyllium, Whorled Plantain ■

301865　Plantago arenaria Waldst. et Kit. = Plantago indica L. ■☆

301866　Plantago arenaria Waldst. et Kit. = Plantago psyllium L. ■☆

301867　Plantago argentea Desf. = Plantago ovata Forssk. ■

301868　Plantago aristata Michx.；芒苞车前（车前子，具芒车前，线叶车前，小芒苞车前）；Bracted Plantain, Buckhorn, Buck-horn, Large-bracted Plantain ■

301869　Plantago aristata Michx. var. minuta T. K. Cheng et X. S. Wan = Plantago aristata Michx. ■

301870　Plantago aristata Michx. var. nuttallii（Rapin）E. Morris = Plantago aristata Michx. ■

301871　Plantago asiatica L.；车前（白贯草，蟾蜍草，车茶草，车串串，车轱辘菜，车轮菜，车轮草，车前草，打官司草，大车前，大粒车前，当道，地胆头，饭匙草，凤眼前，苤苜，蛤蟆草，黄蟆龟草，黄蟆叶，灰盆草，陵舄，驴耳朵草，马蹄草，马舄，牛耳朵草，牛耳朵棵，牛舌草，牛甜菜，牛舄，牛遗，蒲杓草，七星草，钱贯草，青茶草，胜舄，田菠菜，五根草，五斤草，虾蟆衣，鸭脚板，医马草，猪肚菜，猪耳草，猪耳朵，猪耳朵草）；Asia Plantain, Asiatic Plantain ■

301872　Plantago asiatica L. f. albostriata（Makino）H. Hara；白条车前■☆

301873　Plantago asiatica L. f. amamiana（Yamam.）H. Hara；奄美车前■☆

301874　Plantago asiatica L. f. folioscopa（T. Ito）Honda = Plantago asiatica L. ■

301875　Plantago asiatica L. f. paniculata（Makino）Hara = Plantago asiatica L. ■

301876　Plantago asiatica L. f. rosea Makino ex Nakai = Plantago asiatica L. ■

301877　Plantago asiatica L. f. variegata（Ikeno）H. Hara；斑点车前■☆

301878　Plantago asiatica L. subsp. densiflora（J. Z. Liu）Z. Yu Li；密花车前（长果车前）；Denseflower Plantain ■

301879　Plantago asiatica L. subsp. erosa（Wall.）Z. Y. Li；啮蚀车前（车轮菜，车前草，滇车前，蛤蟆草，蛤蟆叶，疏花车前，小车前）；Yunnan Plantain ■

301880　Plantago asiatica L. var. brevior Pilg. = Plantago asiatica L. ■

301881　Plantago asiatica L. var. densiuscula Pilg. = Plantago asiatica L. ■

301882　Plantago asiatica L. var. laxa Pilg. = Plantago asiatica L. ■

301883　Plantago asiatica L. var. lobulata Pilg. = Plantago asiatica L. ■

301884　Plantago asiatica L. var. sphaerocarpa Kitag.；球果车前■☆

301885　Plantago asiatica L. var. yakusimensis（Masam.）Ohwi；屋久岛车前■☆

301886　Plantago asphodeloides Svent.；阿福花车前■☆

301887　Plantago atrata Hoppe = Plantago montana Huds. ■☆

301888　Plantago australis Lam.；澳洲车前；Dwarf Plantain ■☆

301889　Plantago axachnoides Schrenk；蛛丝毛车前■

301890　Plantago bellardii All.；贝拉尔车前■☆

301891　Plantago brachyphylla Edgew. ex Decne.；短叶车前■☆

301892　Plantago brachyphylla Edgew. ex Decne. = Plantago himalaica Pilg. ■☆

301893　Plantago bungei Steud. = Plantago tenuiflora Waldst. et Kit. ■

301894　Plantago camtschatica Cham. ex Link；海滨车前（勘察加车前，绿豆菜，绿叶根）■

301895　Plantago camtschatica Cham. ex Link f. glabra（Makino et Honda）Ohwi；光海滨车前■☆

301896　Plantago canadensis Hort. = Plantago cordata Lam. ■☆

301897　Plantago canescens Adams；灰车前■☆

301898　Plantago carnosa Lam. = Plantago crassifolia Forssk. ■☆

301899　Plantago caroliniana Walter = Plantago virginica L. ■

301900　Plantago cavaleriei H. Lév.；尖萼车前（长柱车前，丽江车前）■

301901　Plantago centralis Pilg. = Plantago asiatica L. subsp. erosa（Wall.）Z. Y. Li ■

301902　Plantago centralis Pilg. = Plantago erosa Wall. ■

301903　Plantago chotticus Pomel = Plantago maritima L. ■

301904　Plantago ciliata Desf.；睫毛车前■☆

301905　Plantago commutata Guss.；变异车前■☆

301906　Plantago connivens Moench = Plantago virginica L. ■

301907　Plantago cordata Lam.；心形车前；Heartleaf Plantain, Heart-leaved Plantain, King-root ■☆

301908　Plantago coreana H. Lév.；朝鲜车前■☆

301909　Plantago cornuti Gouan；长柄车前（湿车前）；Heart-leaved Plantain, King-root ■

301910　Plantago coronopus L.；臭荠车前；Buckhorn Plantain, Buck-horn Plantain, Buck's Horn, Buck's Horn Plantain, Buck's-horn Plantain, Capuchin's Beard, Crowfoot, Crowfoot Plantain, Crowfoot Waybread, Crow's Foot Plantain, Crow's-foot Plantain, Grace of God, Hartshorn, Hartshorn Plantain, Herb Eve, Herb Ive, Herb Ivy, St. Bridget's Wort, Stag's Horn Plantain, Stag's-horn Plantain, Star of the Earth ■☆

301911　Plantago coronopus L. subsp. columnae（Gouan）Batt. = Plantago coronopus L. ■☆

301912　Plantago coronopus L. subsp. commutata（Guss.）Pilg. = Plantago commutata Guss. ■☆

301913　Plantago coronopus L. subsp. cupanii（Guss.）Nyman = Plantago cupanii Guss. ■☆

301914　Plantago coronopus L. subsp. macrorrhiza（Poir.）Arcang. = Plantago macrorrhiza Poir. ■☆

301915　Plantago coronopus L. subsp. purpurascens（Nyman）Pilg. = Plantago weldenii Rchb. subsp. purpurascens（Nyman）Greuter et Burdet ■☆

301916　Plantago coronopus L. subsp. rosulata Batt. = Plantago cupanii Guss. ■☆

301917　Plantago coronopus L. subsp. scleropus（Murb.）Le Houér. = Plantago coronopus L. ■☆

301918　Plantago coronopus L. var. aristata Pilg. = Plantago coronopus L. ■☆

301919　Plantago coronopus L. var. ceratophylla（Hoffmanns. et Link）Rapin = Plantago coronopus L. ■☆

301920　Plantago coronopus L. var. columnae（Gouan）Willd. = Plantago coronopus L. ■☆

301921　Plantago coronopus L. var. commutata（Guss.）Bég. = Plantago commutata Guss. ■☆

301922　Plantago coronopus L. var. crassipes Coss. et Daveau = Plantago coronopus L. ■☆

301923　Plantago coronopus L. var. cupanii（Guss.）Decne. = Plantago cupanii Guss. ■☆

301924　Plantago coronopus L. var. erecta Pilg. = Plantago coronopus L. ■☆

301925　Plantago coronopus L. var. filiformis（Koch）Pamp. = Plantago coronopus L. ■☆

301926　Plantago coronopus L. var. firma Pilg. = Plantago coronopus L. ■☆

301927　Plantago coronopus L. var. laciniata（Willk.）Pilg. = Plantago coronopus L. ■☆

301928　Plantago coronopus L. var. maritima Gren. et Godr. = Plantago coronopus L. ■☆

301929　Plantago coronopus L. var. maroccana Ball = Plantago coronopus L. ■☆

301930　Plantago coronopus L. var. pseudomacrorrhiza Cout. = Plantago coronopus L. ■☆

301931　Plantago coronopus L. var. rigida Pilg. = Plantago weldenii Rchb. ■☆

301932　Plantago coronopus L. var. rosulata（Batt.）Pilg. = Plantago coronopus L. ■☆

301933　Plantago coronopus L. var. scleropus Murb. = Plantago coronopus L. ■☆

301934　Plantago coronopus L. var. stricta Pilg. = Plantago coronopus L. ■☆

301935　Plantago coronopus L. var. subspinulosa Maire = Plantago coronopus L. ■☆

301936　Plantago coronopus L. var. tlemceniana Barnéoud = Plantago coronopus L. ■☆

301937　Plantago coronopus L. var. weldenii（Rchb.）Pott. -Alap. = Plantago weldenii Rchb. ■☆

301938　Plantago coronopus Sauss. = Plantago coronopus L. ■☆

301939　Plantago crassifolia Forssk.；厚叶车前■☆

301940　Plantago crassifolia Forssk. var. hirsuta（Thunb.）Bég.；粗毛车前☆

301941　Plantago crithmoides Desf. = Plantago macrorrhiza Poir. ●☆

301942　Plantago crypsoides Boiss.；隐花草车前■☆

301943　Plantago cupanii Guss.；库潘车前■☆

301944　Plantago cylindrica Forssk.；柱形车前■☆

301945　Plantago cynops L.；狗尾状车前（赛诺普车前）；Cynop Plantain ■

301946　Plantago cynops L. = Plantago afra L. ■☆

301947　Plantago cynops L. = Plantago sempervirens Crantz ●☆

301948　Plantago cyrenaica E. A. Durand et Barratte；昔兰尼车前■☆

301949　Plantago debilis Nees；柔软车前；Weak Plantain ■☆

301950　Plantago decumbens Forssk. = Plantago ovata Forssk. ■

301951　Plantago densiflora J. Z. Liu = Plantago asiatica L. subsp. densiflora（J. Z. Liu）Z. Yu Li ■

301952　Plantago depressa Willd.；平车前（白贯草，蟾蜍草，车茶草，车串串，车轱辘菜，车轮菜，车前，车前草，打官司草，当道，地胆头，饭匙草，凤眼前，苤苢，蛤蟆草，黄蟆龟草，黄蟆叶，灰盆草，陵舄，驴耳朵草，马蹄草，马舄，牛耳朵草，牛耳朵棵，牛舌草，牛甜菜，牛舄，牛遗，蒲勺草，七星草，钱贯草，青茶菜，胜舄，田菠菜，五根草，五斤草，虾蟆衣，小车前，小粒车前子，鸭脚板，医马草，猪肚菜，猪耳草，猪耳朵，猪耳朵草，主根车前）；Depressed Plantain ■

301953　Plantago depressa Willd. f. glaberrirna Kom. = Plantago depressa Willd. ■

301954　Plantago depressa Willd. f. minor Kom. = Plantago depressa Willd. ■

301955　Plantago depressa Willd. subsp. camtschatica（Link）Pilg. = Plantago camtschatica Cham. ex Link ■

301956　Plantago depressa Willd. subsp. turczainowii（Ganesch.）Tzvelev；毛平车前；Turczaninov Depressed Plantain ■

301957　Plantago depressa Willd. var. eudepressa Ganesch. = Plantago depressa Willd. ■

301958　Plantago depressa Willd. var. magnibracteata T. Tanaka et T. K. Zheng = Plantago depressa Willd. ■

301959　Plantago depressa Willd. var. montana Kitag. = Plantago depressa Willd. subsp. turczainowii（Ganesch.）Tzvelev ■

301960　Plantago depressa Willd. var. turczainowii（Ganesch.）Tzvelev = Plantago depressa Willd. subsp. turczainowii（Ganesch.）Tzvelev ■

301961　Plantago depressa Willd. var. turczaninowii Ganesch. = Plantago depressa Willd. subsp. turczaninowii（Ganesch.）Tzvelev ■

301962　Plantago divaricata Zucc. = Plantago afra L. var. parviflora（Desf.）Lewalle ■☆

301963　Plantago dregeana Decne. = Plantago major L. ■

301964　Plantago durandoi Pomel = Plantago afra L. ■☆

301965　Plantago eocoronopus Pilg. = Plantago maritima L. subsp. ciliata Pritz. ■

301966　Plantago eriocarpa Viv. = Plantago afra L. ■☆

301967　Plantago eripoda ?；毛梗车前；Saline Plantain ■☆

301968　Plantago erosa Wall. = Plantago asiatica L. subsp. erosa（Wall.）Z. Y. Li ■

301969　Plantago erosa Wall. var. fengdouensis Z. E. Chao et Yong Wang = Plantago fengdouensis（Z. E. Chao et Yong Wang）Yong Wang et Z. Y. Li ■

301970　Plantago exigua Murray；弱小车前■☆

301971　Plantago fastigiata Morris = Plantago ovata Forssk. ■

301972　Plantago fengdouensis（Z. E. Chao et Yong Wang）Yong Wang et Z. Y. Li；丰都车前■

301973　Plantago filiformis Decne. = Plantago aristata Michx. ■

301974　Plantago firma Kuntze ex Walp.；智利车前；Chilean Plantain ■☆

301975　Plantago fischeri Engl.；菲舍尔车前■☆

301976　Plantago fischeri Engl. f. supina Pilg. = Plantago fischeri Engl. ■☆

301977　Plantago formosana Tateishi et Masam.；台湾车前（钱贯草）■

301978　Plantago formosana Tateishi et Masam. = Plantago asiatica L. ■

301979　Plantago frankii Steud. = Plantago aristata Michx. ■

301980　Plantago gentianoides Sibth. et Sm.；龙胆状车前■

301981　Plantago gentianoides Sibth. et Sm. subsp. griffithii（Decne.）Rech. f.；革叶车前■

301982　Plantago gentianoides Sibth. et Sm. var. eugentianoides Pilg. = Plantago gentianoides Sibth. et Sm. ■

301983　Plantago gentianoides Sibth. et Sm. var. laxa Pilg. = Plantago gentianoides Sibth. et Sm. subsp. griffithii（Decne.）Rech. f. ■

301984　Plantago gentianoides Sibth. et Sm. var. laxa Pilg. = Plantago gentianoides Sibth. et Sm. ■

301985　Plantago gentianoides Sibth. et Sm. var. tatarica（Decne.）Pilg. = Plantago gentianoides Sibth. et Sm. subsp. griffithii（Decne.）Rech. f. ■

301986　Plantago gentianoides Sibth. et Sm. var. tatarica（Decne.）Pilg. = Plantago gentianoides Sibth. et Sm. ■

301987　Plantago gigas H. Lév. = Plantago major L. ■

301988　Plantago gigas H. Lév. var. cavaleriei（H. Lév.）H. Lév. = Plantago cavaleriei H. Lév. ■

301989　Plantago gnaphalioides Nutt. = Plantago patagonica Jacq. ■☆

301990　Plantago gnaphalioides Sibth. et Sm. var. aristata（Michx.）Hook. = Plantago aristata Michx. ■

301991　Plantago gracilis Poir. = Plantago serraria L. ■☆

301992　Plantago griffithii Decne. = Plantago gentianoides Sibth. et Sm. subsp. griffithii（Decne.）Rech. f. ■

301993　Plantago griffithii Decne. var. alpina Bornm. = Plantago gentianoides Sibth. et Sm. subsp. griffithii（Decne.）Rech. f. ■

301994　Plantago griffithii Decne. var. alpina Bornm. = Plantago gentianoides Sibth. et Sm. ■

301995 Plantago griffithii Decne. var. pamirica Fedtsch. = Plantago gentianoides Sibth. et Sm. ■

301996 Plantago griffithii Decne. var. pamirica Fedtsch. = Plantago gentianoides Sibth. et Sm. subsp. griffithii（Decne.）Rech. f. ■

301997 Plantago hakusanensis Koidz. ;白山车前■☆

301998 Plantago hakusanensis Koidz. f. glabra T. Yamaz. ;光白山车前■☆

301999 Plantago hakusanensis Koidz. f. viridescens（Takeda）H. Hara ;绿白山车前■☆

302000 Plantago helleri Small;海氏车前;Heller's Plantain ■☆

302001 Plantago heterophylla Nutt. ;异叶车前;Small Plantain ■☆

302002 Plantago himalaica Pilg. ;喜山车前;Himalayan Plantain ■☆

302003 Plantago himalaica Pilg. = Plantago gentianoides Sibth. et Sm. ■

302004 Plantago hirsuta Thunb. = Plantago crassifolia Forssk. var. hirsuta（Thunb.）Bég. ■☆

302005 Plantago holostea Desf. = Plantago bellardii All. ■☆

302006 Plantago hostifolia Nakai et Kitag. = Plantago asiatica L. ■

302007 Plantago huadianica S. H. Li et Y. Yang = Plantago depressa Willd. ■

302008 Plantago humilis Guss. subsp. atlantis（Emb. et Maire）Brullo et al. = Plantago subulata L. subsp. atlantis（Emb. et Maire）Greuter et Burdet ■☆

302009 Plantago indica L. ;印度车前;Bunched Plantain ■☆

302010 Plantago indica L. = Plantago arachnoidea Schrenk ■

302011 Plantago indica L. = Plantago arenaria Waldst. et Kit. ■

302012 Plantago intermedia Gilib. = Plantago major L. ■

302013 Plantago ispaghula Roxb. ex Fleming = Plantago ovata Forssk. ■

302014 Plantago japonica Franch. et Sav. ;日本车前（大车前）;Japan Plantain ■

302015 Plantago japonica Franch. et Sav. = Plantago major L. var. japonica（Franch. et Sav.）Miyabe ■

302016 Plantago japonica Franch. et Sav. f. fastigiata Honda = Plantago japonica Franch. et Sav. ■

302017 Plantago japonica Franch. et Sav. f. yezomaritima（Koidz.）Kitam. = Plantago japonica Franch. et Sav. ■

302018 Plantago japonica Franch. et Sav. var. yezomaritima（Koidz.）H. Hara = Plantago japonica Franch. et Sav. ■

302019 Plantago jehohlensis Koidz. ;毛车前（车轱辘菜）;Hairy Plantain ■

302020 Plantago jehohlensis Koidz. = Plantago major L. ■

302021 Plantago juncoides Lam. = Plantago maritima L. ■

302022 Plantago kentukensis Michx. = Plantago cordata Lam. ■☆

302023 Plantago kerstenii Asch. = Plantago palmata Hook. f. ■☆

302024 Plantago komarovii Pavlov ;翅柄车前;Komarov Plantain ■

302025 Plantago lachnantha Bunge ;毛花车前■☆

302026 Plantago lagocephala Bunge ;毛瓣车前;Hairypetal Plantain ■

302027 Plantago lagopoides Desf. = Plantago amplexicaulis Cav. ■☆

302028 Plantago lagopus L. ;兔足车前（哈氏车前）;Hare's Foot Plantain,Hare's-foot Plantain ■☆

302029 Plantago lagopus L. var. caulescens Christ = Plantago lagopus L. ■☆

302030 Plantago lagopus L. var. glabrata Maire = Plantago lagopus L. ■☆

302031 Plantago lagopus L. var. lusitanica（L.）Ball = Plantago lagopus L. ■☆

302032 Plantago lagopus L. var. minor Ten. = Plantago lagopus L. ■☆

302033 Plantago lanata Poir. = Plantago bellardii All. ■☆

302034 Plantago lanceolata L. ;长叶车前（长叶车前草,绿豆菜,欧前,欧洲车前,披针叶车前,婆婆丁花,窄叶车前）;Basket,Bennet,Bent,Black Bout,Black Boys,Black Gypsy,Black Jack,Black Man,Black Plantain,Blackle Tops,Blaekie Top,Bobbies,Bobbins,Buckhorn,Buckhorn Plantain,Buck-horn Plantain,Buckitorn,Carldod,Chimney Sweep,Chimney Sweeper,Christ's Heel,Clock,Cock Battler,Cock Fighters,Cock-battler,Cock-fighter,Cock-grass,Cocks,Cock's Head,Cocks-and-hens,Conkers,Conqueror-flower,Curldoddy,Devil's Head,Devils-and-angels,Dog's Rib,Donkey's Ear,Donkey's Ears,English Plantain,Fechter,Fightee Cocks,Fighting Cocks,Fire-grass,Fire-leaf,Fireweed,French-and-English Grass,French-and-English Soldiers,French-and-English Weed,Gypsy,Hardhead,Headman,Hen Plant,Jack Straw,Jack-straws,Johnsmas Flowers,Johnsmas Pairs,Kemps,Kempseed,Lady's Mantle,Lamb's Tails,Lamb's Tongue,Lance-leaved Plantain,Launceley,Long Plantain,Longleaf Plantain,Lords-and-ladies,Man-of-war,Narrow-leaf Plantain,Narrow-leaved Plantain,Nigger,Nigger's Head,Nigger's Heal,Oak of Mamre,Pash-leaf,Plantain,Pull Poker,Rat's Tail,Rat's Tails,Rib Grass,Rib Plantain,Ribgrass,Rib-grass,Ribwort,Ribwort Plantain,Ribworth,Ripple Grass,Rippling Grass,Rupple-grass,Snake Plantain,Soldiers,Soldier's Ribwort,Soldier's Tap Pie,Soldier's Tappie,Sweep's Brush,Sword-and-spear,Swords-and-spears,Tinker-tailor Grass,Violin Strings,Wild Sago,Windles ■

302035 Plantago lanceolata L. f. composita Farw. = Plantago lanceolata L. ■

302036 Plantago lanceolata L. subsp. altissima（L.）Nyman = Plantago altissima L. ■☆

302037 Plantago lanceolata L. subsp. communis（Schltr.）Jahand. et Maire = Plantago lanceolata L. ■

302038 Plantago lanceolata L. subsp. cyrenaica Maire et Weiller;昔兰尼长叶车前■☆

302039 Plantago lanceolata L. subsp. intermedia（Gilib.）Quézel et Santa = Plantago major L. subsp. intermedia（Gilib.）Lange ■

302040 Plantago lanceolata L. var. angustifolia Poir. = Plantago lanceolata L. ■

302041 Plantago lanceolata L. var. lanuginosa Bastard = Plantago lanceolata L. ■

302042 Plantago lanceolata L. var. mediterranea Pilg. = Plantago altissima L. ■☆

302043 Plantago lanceolata L. var. mediterranea Pilg. = Plantago lanceolata L. ■

302044 Plantago lanceolata L. var. sphaerostachya Mert. et W. D. J. Koch f. eriophora（Hoffmanns. et Link）Beck = Plantago lanceolata L. ■

302045 Plantago lanceolata L. var. sphaerostachya Mert. et W. D. J. Koch = Plantago lanceolata L. ■

302046 Plantago laxiflora Decne. ;疏花车前■☆

302047 Plantago leiocephala Wallr. ex K. Koch ;光头车前■☆

302048 Plantago leiopetala Lowe ;光瓣车前■☆

302049 Plantago leptostachys Ledeb. = Plantago depressa Willd. ■☆

302050 Plantago lessingii Fisch. et C. A. Mey. ;条叶车前（来森车前,细叶车前）;Lessing Plantain ■

302051 Plantago lessingii Fisch. et C. A. Mey. = Plantago minuta Pall. ■

302052 Plantago libyca Bég. et Vacc. ;利比亚车前■☆

302053 Plantago litoraria Fourc. = Plantago crassifolia Forssk. ■☆

302054 Plantago loeflingii L. ;廖氏车前■☆

302055 Plantago loeflingii L. subsp. notata（Lag.）O. Bolòs et Vigo = Plantago notata Lag. ■☆

302056 Plantago longissima Decne. ;长车前■☆

302057 Plantago longissima Decne. var. burkei Pilg. = Plantago longissima Decne. ■☆

302058   Plantago longissima Decne. var. densiuscula Pilg. = Plantago longissima Decne. ■☆

302059   Plantago lorata（J. Z. Liu）Shipunov = Plantago arachnoidea Schrenk ■

302060   Plantago ludoviciana Raf. = Plantago virginica L. ■

302061   Plantago lusitanica L. = Plantago lagopus L. ■☆

302062   Plantago macrocarpa Cham. et Schltdl. ;大果车前■☆

302063   Plantago macro-nipponica Yamam. = Plantago major L. ■

302064   Plantago macrorrhiza Poir. ;大根车前■☆

302065   Plantago macrorrhiza Poir. var. hirsuta Pilg. = Plantago macrorrhiza Poir. ■☆

302066   Plantago macrorrhiza Poir. var. tangerina Pau = Plantago macrorrhiza Poir. ■☆

302067   Plantago macrorrhiza Poir. var. tenuispica Faure et Maire = Plantago macrorrhiza Poir. ■☆

302068   Plantago major L. ;大车前（车轮菜,车前,车前草,莶,大车前草,大叶车前,大猪耳朵草,当道,地衣,苶苴,巨叶车前草,马舄,牛舌曹,牛遗,钱串子,钱贯草,胜舄,田贯草）;Aisatic Plantain, American Wayfaring, Bennet, Big Plantain, Bird's Meat, Birdseed, Bleeding Grass, Broad-leaf Plantain, Broad-leaved Plantain, Canary Broad-leaf, Canary Flower, Canary Food, Carrag, Christ's Heel, Cocks, Common Plantain, Cow Grass, Cuckoo-bread, Curldoddy, Cut-grass, Dooryard Plantain, English Plantain, Englishman's Foot, Fiddlestrings, Giantleaf Plantain, Great Plantain, Great Waybread, Greater Plantain, Hardhead, Healing Blade, Healing Leaf, Johnsmas Flowers, Johnsmas Pairs, Kemps, Lamb's Foot, Lamb's Tails, Larkseed, Plantain, Plantine, Pony's Tail, Pony's Tails, Poverty-grass, Rat's Tail, Rat's Tails, Rat-tail Plantain, Ratten Tails, Ratten-tail, Rib Grass, Ribwort, Ripple Grass, Rippleseed Plantain, Ripple-seed Plantain, Slanlus, Snakeweed, Traveller's Foot, Wabran-leaf, Wabret-leaf, Warba-leaves, Waveran-leaf, Wayberry, Wayborn, Waybread, Waybroad, Waybroadleaf, Waybrow, Wayburn-leaf, Wayforn, Wayfron, Wayside Bread, Weybred, White Man's Foot, White Man's Footprints, Wibrow-wobrow, Wild Sago, Wlbrow ■

302069   Plantago major L. 'Atropurpurea';紫叶大车前;Purple-leaved Plantain, Red Leaf Plantain ■☆

302070   Plantago major L. subsp. intermedia（Gilib.）Lange = Plantago major L. ■

302071   Plantago major L. subsp. pleiosperma Pilg. = Plantago major L. ■

302072   Plantago major L. var. asiatica（L.）Decne. = Plantago asiatica L. ■

302073   Plantago major L. var. asiatica（L.）Decne. f. paniculata Makino = Plantago asiatica L. ■

302074   Plantago major L. var. egastachya（Wimm.）Graebn. = Plantago major L. ■

302075   Plantago major L. var. folioscopa T. Ito = Plantago asiatica L. ■

302076   Plantago major L. var. gigas（H. Lév.）H. Lév. = Plantago major L. ■

302077   Plantago major L. var. japonica（Franch. et Sav.）Miyabe = Plantago japonica Franch. et Sav. ■

302078   Plantago major L. var. japonica（Franch. et Sav.）Miyabe f. polystachya（Makino）T. Yamaz. = Plantago japonica Franch. et Sav. ■

302079   Plantago major L. var. jehohlensis（Koidz.）S. H. Li = Plantago major L. ■

302080   Plantago major L. var. kimurae Yamam. = Plantago major L. ■

302081   Plantago major L. var. paludosa Bég. = Plantago major L. ■

302082   Plantago major L. var. pauciflora（Gilib.）Geguinot = Plantago major L. ■

302083   Plantago major L. var. phyllostachya Wallr. = Plantago major L. ■

302084   Plantago major L. var. salina Wirtg. = Plantago major L. ■

302085   Plantago major L. var. salsa（Pall.）Pilg. ;盐生大车前（盐生车前）■

302086   Plantago major L. var. sawadae Yamam. = Plantago major L. ■

302087   Plantago major L. var. scopulorum Fr. et Broberg？= Plantago major L. ■

302088   Plantago major L. var. sinuata（Lam.）Decne. ;深波大车前■

302089   Plantago major L. var. sinuata（Lam.）Decne. = Plantago major L. ■

302090   Plantago major L. var. variegata Hort. ;多变车前草■☆

302091   Plantago maritima L. ;沿海车前; Bog Horn, Buck's-horn Plantain, European Seaside Plantain, Rush-like Plantain, Sea Kemps, Sea Plantain, Seaside Plantain, Sheep's Herb ■

302092   Plantago maritima L. subsp. ciliata Pritz. ;盐生车前（蒙古车前）;Europe Plantain, European Seaside Plantain, Salt Plantain ■

302093   Plantago maritima L. subsp. salsa（Pall.）Rech. f. = Plantago maritima L. subsp. ciliata Pritz. ■

302094   Plantago maritima L. var. chottica（Pomel）Hochr. = Plantago maritima L. ■

302095   Plantago maritima L. var. communis Will. = Plantago maritima L. ■

302096   Plantago maritima L. var. integralis（DC.）Pilg. = Plantago maritima L. ■

302097   Plantago maritima L. var. salsa（Pall.）Pilg. = Plantago maritima L. subsp. ciliata Pritz. ■

302098   Plantago mauritanica Boiss. et Reut. ;毛里塔尼亚前草■☆

302099   Plantago mauritanica Boiss. et Reut. var. maroccana Batt. = Plantago mauritanica Boiss. et Reut. ■☆

302100   Plantago mauritii Sennen ;毛里特车前■☆

302101   Plantago maxima Juss. ex Jacq. ;巨车前;Giant Plantain ■

302102   Plantago maxonnoi Sennen = Plantago cupanii Guss. ■☆

302103   Plantago media L. ;北车前（草车前,药用车前,中车前,中间车前）; Ashy Poker, Boots-and-stockings, Chimney Sweep, Cotton Flower, Dwarf Plantain, Fire-leaf, Fireweed, Hoary Plantain, Honey Plantain, Honey-plantain, Lamb's Ear, Lamb's Tongue, Lamb's Tongue Plantain, Lamb's-tongue, Lamb's-tongue Plantain, Lords-and-ladies, Medicinal Plantain, Nigger's Head, Nigger's Heal, Oak of Mamre, Scent Bottle, Shoes-and-stockings, Sweep's Brush, Sweet Plantain ■

302104   Plantago media L. 'Ashy';灰色北车前;Ashy Poker ■☆

302105   Plantago media L. var. monnieri（Giraudias）Roug.？= Plantago media L. ■

302106   Plantago media L. var. urvilleana Rapin = Plantago media L. ■

302107   Plantago minuta Pall. ;小车前（草车前,条叶车前,细叶车前）;Little Plantain ■

302108   Plantago minuta Pall. subsp. lessingii（Fisch. et C. A. Mey.）Tzvelev = Plantago minuta Pall. ■

302109   Plantago missouriensis Steud. = Plantago virginica L. ■

302110   Plantago mohnikei Miq. = Plantago hakusanensis Koidz. ■☆

302111   Plantago mongolica Decne. = Plantago minuta Pall. ■

302112   Plantago montana Huds. ;山地车前;Mountain Plantain ■☆

302113   Plantago nivalis Boiss. ;雪白车前■☆

302114   Plantago notata Lag. ;斑纹车前■☆

302115   Plantago nuttallii Rapin = Plantago aristata Michx. ■

302116   Plantago oblonga E. Morris = Plantago patagonica Jacq. ■☆

302117 Plantago oligantha Phil. ;少花车前;Few-flowered Plantain ■☆

302118 Plantago orientalis Stapf = Plantago lanceolata L. ■

302119 Plantago ovata Forssk. ;圆苞车前(卵叶车前,印车前,棕色车前草); Desert Indian Wheat, Isphaghul Plantain, Isphaghul Seeds, Ovateleaf Plantain, Spogel Seed ■

302120 Plantago palmata Hook. f. ;掌裂车前■☆

302121 Plantago parviflora Desf. = Plantago afra L. var. parviflora (Desf.) Lewalle ■☆

302122 Plantago patagonica Jacq. ;巴塔哥尼亚车前; Patagonian Plantain, Pursh Plantain, Salt-and-pepper Plant, Woolly Plantain ■☆

302123 Plantago patagonica Jacq. var. aristata (Michx.) A. Gray = Plantago aristata Michx. ■

302124 Plantago patagonica Jacq. var. breviscapa (Shinners) Shinners = Plantago patagonica Jacq. ■☆

302125 Plantago patagonica Jacq. var. gnaphalioides (Nutt.) A. Gray = Plantago patagonica Jacq. ■☆

302126 Plantago patagonica Jacq. var. oblonga (E. Morris) Shinners = Plantago patagonica Jacq. ■☆

302127 Plantago patagonica Jacq. var. spinulosa (Decne.) A. Gray = Plantago patagonica Jacq. ■☆

302128 Plantago pauciflora Gilib. = Plantago major L. ■

302129 Plantago perssonii Pilg. ;苣叶车前;Persson Plantain ■

302130 Plantago phaeostoma Boiss. et Heldr. ;褐口车前■☆

302131 Plantago picta E. Morris = Plantago patagonica Jacq. ■☆

302132 Plantago polysperma Kar. et Kir. ;多籽车前;Manyseed Plantain ■

302133 Plantago psyllium L. = Plantago afra L. ☆

302134 Plantago psyllium L. = Plantago arenaria Waldst. et Kit. ■

302135 Plantago psyllium L. = Plantago indica L. ■☆

302136 Plantago psyllium L. subsp. stricta (Schousb.) Batt. = Plantago afra L. var. stricta (Schousb.) Verdc. ■☆

302137 Plantago psyllium L. var. durandoi (Pomel) Batt. = Plantago afra L. ■☆

302138 Plantago psyllium L. var. libyca (Bég. et Vacc.) Pamp. = Plantago afra L. ■☆

302139 Plantago psyllium L. var. monocephala Alb. = Plantago afra L. ■☆

302140 Plantago psyllium L. var. parviflora (Desf.) Batt. = Plantago afra L. var. parviflora (Desf.) Lewalle ■☆

302141 Plantago psyllium L. var. stricta (Schousb.) Maire = Plantago afra L. var. stricta (Schousb.) Verdc. ■☆

302142 Plantago pumila L. f. ;矮小车前■☆

302143 Plantago purpurascens Nutt. ex Rapin = Plantago virginica L. ■

302144 Plantago purshii Roem. et Schult. ;普氏车前;Pursh Plantain, Pursh Ribgrass, Woolly Indian Wheat ■☆

302145 Plantago purshii Roem. et Schult. = Plantago patagonica Jacq. ■☆

302146 Plantago purshii Roem. et Schult. var. aristata (Michx.) M. E. Jones = Plantago aristata Michx. ■

302147 Plantago purshii Roem. et Schult. var. breviscapa Shinners = Plantago patagonica Jacq. ■☆

302148 Plantago purshii Roem. et Schult. var. oblonga (E. Morris) Shinners = Plantago patagonica Jacq. ■☆

302149 Plantago purshii Roem. et Schult. var. picta (E. Morris) Pilg. = Plantago patagonica Jacq. ■☆

302150 Plantago purshii Roem. et Schult. var. spinulosa (Decne.) Shinners = Plantago patagonica Jacq. ■☆

302151 Plantago pusilla Bunge = Plantago tenuiflora Waldst. et Kit. ■

302152 Plantago pusilla Nutt. ; 微小车前; Slender Plantain, Small Plantain ■☆

302153 Plantago ramosa (Gilib.) Asch. = Plantago indica L. ■☆

302154 Plantago remota Lam. ;稀疏车前■☆

302155 Plantago remotiflora Stocks = Plantago stocksii Boiss. ex Decne. ■☆

302156 Plantago rhizoxylon Emb. ;木根车前■☆

302157 Plantago rhizoxylon Emb. var. litardierei Maire = Plantago rhizoxylon Emb. ■☆

302158 Plantago rhodosperma Decne. ;红籽车前;Red-seeded Plantain ■☆

302159 Plantago rugelii Decne. ; 黑子车前; American Plantain, Black-seeded Plantain, Broad-leaved Plantain, Pale Plantain, Red-stalked Plantain, Red-stemmed Plantain, Rugel's Plantain ■☆

302160 Plantago rugelii Decne. var. alterniflora Farw. = Plantago rugelii Decne. ■☆

302161 Plantago rugelii Decne. var. aspera Farw. = Plantago rugelii Decne. ■☆

302162 Plantago salesarensis Gand. = Plantago gentianoides Sibth. et Sm. subsp. griffithii (Decne.) Rech. f. ■

302163 Plantago salsa Pall. = Plantago maritima L. subsp. ciliata Pritz. ■

302164 Plantago sarcophylla Zohary = Plantago squarrosa Murray ■☆

302165 Plantago sawadai (Yamam.) Yamam. = Plantago major L. ■

302166 Plantago scabra Moench = Plantago arenaria Waldst. et Kit. ■

302167 Plantago schneideri Pilg. = Plantago cavaleriei H. Lév. ■

302168 Plantago schneideri Pilg. var. delicatior Pilg. = Plantago cavaleriei H. Lév. ■

302169 Plantago schwarzenbergiana Schur;施瓦车前■☆

302170 Plantago sempervirens Crantz;灌木车前;Shrubby Plantain ●☆

302171 Plantago serraria L. ;细齿车前■☆

302172 Plantago serraria L. var. africana Barnéoud = Plantago serraria L. ■☆

302173 Plantago serraria L. var. hispanica Decne. = Plantago serraria L. ■☆

302174 Plantago serraria L. var. laciniata (Willk.) Pau = Plantago serraria L. ■☆

302175 Plantago serraria L. var. microdon Pilg. = Plantago serraria L. ■☆

302176 Plantago sibirica Poir. = Plantago depressa Willd. ■

302177 Plantago sinuata Lam. = Plantago major L. ■

302178 Plantago spinulosa Decne. ;细刺车前■☆

302179 Plantago spinulosa Decne. = Plantago patagonica Jacq. ■☆

302180 Plantago squamosa Nutt. ex Decne. = Plantago aristata Michx. ■

302181 Plantago squarrosa Murray;粗糙车前■☆

302182 Plantago squarrosa Nutt. = Plantago aristata Michx. ■

302183 Plantago stepposa Kuprian. ;草车前■

302184 Plantago stepposa Kuprian. = Plantago media L. ■

302185 Plantago stepposa Kuprian. = Plantago minuta Pall. ■

302186 Plantago stepposa Lam. = Plantago major L. ■

302187 Plantago stocksii Boiss. ex Decne. ;斯托克斯车前■☆

302188 Plantago stricta Schousb. = Plantago afra L. var. stricta (Schousb.) Verdc. ■☆

302189 Plantago subspathulata Pilg. ;匙形车前■☆

302190 Plantago subulata L. ;钻形车前■☆

302191 Plantago subulata L. subsp. atlantis (Emb. et Maire) Greuter et Burdet;大西洋车前■☆

302192 Plantago subulata L. var. atlantis Emb. et Maire = Plantago subulata L. subsp. atlantis (Emb. et Maire) Greuter et Burdet ☆

302193 Plantago syrtica Viv. = Plantago notata Lag. ■☆

302194 Plantago tanalensis Baker;塔纳尔车前■☆

302195 Plantago tatarica Decne. = Plantago gentianoides Sibth. et Sm. subsp. griffithii (Decne.) Rech. f. ■

302196 Plantago tenuiflora Waldst. et Kit. ; 小花车前; Smallflower

Plantain ■

302197 Plantago tenuiflora Waldst. et Kit. f. pilosa Pilg. = Plantago tenuiflora Waldst. et Kit. ■

302198 Plantago tenuiflora Waldst. et Kit. f. pilosa subf. nana Pilg. = Plantago tenuiflora Waldst. et Kit. ■

302199 Plantago tibetica（Aiton）Willd.；西藏车前；Tibet Plantain, Xizang Plantain ■

302200 Plantago tibetica Hook. f. = Plantago depressa Willd. ■

302201 Plantago tibetica Hook. f. et Thomson = Plantago depressa Willd. ■

302202 Plantago tibetica Hook. f. et Thomson ex Hook. f. = Plantago cavaleriei H. Lév. ■

302203 Plantago tibetica Hook. f. et Thomson ex Hook. f. = Plantago depressa Willd. ■

302204 Plantago togashii Miyabe et Tatew. = Plantago major L. var. japonica（Franch. et Sav.）Miyabe ■

302205 Plantago trichophylla Nábělek = Plantago ovata Forssk. ■

302206 Plantago tunetana Murb.；图内特车前■☆

302207 Plantago villifera Franch. = Plantago camtschatica Cham. ex Link ■

302208 Plantago villifera Kitag. = Plantago major L. ■

302209 Plantago virginica L.；北美车前（毛车前，毛车前草）；Dwarf Plantain, Hoary Plantain, Pale-seed Plantain, Pale-seeded Plantain, Virginia Plantain ■

302210 Plantago virginica L. var. viridescens Fernald = Plantago virginica L. ■

302211 Plantago webbii Barnéoud；韦布车前■☆

302212 Plantago weldenii Rchb.；韦尔登车前■☆

302213 Plantago weldenii Rchb. subsp. purpurascens（Nyman）Greuter et Burdet；紫韦尔登车前■☆

302214 Plantago wrightiana Decne.；赖特车前；Wright's Plantain ■☆

302215 Plantago wyomingensis Gand. = Plantago patagonica Jacq. ■☆

302216 Plantago yakusimensis Masam. = Plantago asiatica L. var. yakusimensis（Masam.）Ohwi ■☆

302217 Plantago zeylanica L.；锡兰车前（白花丹）；Ceylon Plantain ■☆

302218 Plantia Herb. = Hexaglottis Vent. ■☆

302219 Plantinia Bubani = Phleum L. ■

302220 Plappertia Rchb. = Dichapetalum Thouars ●

302221 Plappertia Rchb. = Leucosia Thouars ●

302222 Plarodrigoa Looser = Cristaria Cav.（保留属名）■●☆

302223 Plaso Adans.（废弃属名）= Butea Roxb. ex Willd.（保留属名）●

302224 Plaso monosperma Knntze = Butea monosperma（Lam.）Taub. ●

302225 Plastobrassica（O. E. Schulz）Tzvelev = Erucastrum（DC.）C. Presl ■☆

302226 Plastolaena Pierre ex A. Chev. = Schumanniophyton Harms ●☆

302227 Plastolaena klaineana Pierre = Schumanniophyton magnificum（K. Schum.）Harms var. klaineanum（Pierre ex A. Chev.）N. Hallé ●☆

302228 Platanaceae Dumort. = Platanaceae T. Lestib.（保留科名）●

302229 Platanaceae T. Lestib.（1826）（保留科名）；悬铃木科（法国梧桐科，条悬铃木科，洋桐木科）；Plane Tree Family, Planetree Family, Plane-Tree Family ●

302230 Platanaceae T. Lestib. ex Dumort. = Platanaceae T. Lestib.（保留科名）●

302231 Platanaria Gray = Sparganium L. ■

302232 Platanocarpum（Endl.）Korth. = Nauclea L. ●

302233 Platanocarpum（Endl.）Korth. = Sarcocephalus Afzel. ex Sabine ●☆

302234 Platanocarpum Korth. = Nauclea L. ●

302235 Platanocarpum Korth. = Sarcocephalus Afzel. ex Sabine ●☆

302236 Platanocarpum africanum（Willd.）Hook. f. = Mitragyna inermis（Willd.）K. Schum. ●☆

302237 Platanocarpus Korth. = Platanocarpum Korth. ●☆

302238 Platanocephalus Crantz = Nauclea L. ●

302239 Platanocephalus Vaill. ex Crantz = Nauclea L. ●

302240 Platanos St.-Lag. = Platanus L. ●

302241 Platanthera Rich.（1817）（保留属名）；舌唇兰属（长距兰属，粉蝶兰属）；Bog Orchid, Butter Orchid, Butterfly Orchid, Butterfly-orchid, Fringed Orchid, Platanthera, Rain Orchid ■

302242 Platanthera × okubo-hachijoensis K. Inoue；八丈岛小舌唇兰■☆

302243 Platanthera × ophryo-tipuloides K. Inoue；拟眉兰舌唇兰■☆

302244 Platanthera acuminata Lindl. = Platanthera latilabris Lindl. ■

302245 Platanthera albida（L.）Lindl. var. straminea（Fernald）Luer = Pseudorchis albida（L.）Á. Löve et D. Löve subsp. straminea（Fernald）Á. Löve et D. Löve ■☆

302246 Platanthera algeriensis Batt. et Trab.；阿尔及利亚舌唇兰；Algerian Butterfly-orchid ■☆

302247 Platanthera altigena Schltr. = Platanthera roseotincta（W. W. Sm.）Ts. Tang et F. T. Wang ■

302248 Platanthera amabilis Koidz.；秀丽舌唇兰■☆

302249 Platanthera amamiana Ohwi = Platanthera mandarinorum Rchb. f. subsp. hachijoensis（Honda）Murata var. amamiana（Ohwi）K. Inoue ■☆

302250 Platanthera anboensis Masam. = Platanthera mandarinorum Rchb. f. subsp. hachijoensis（Honda）Murata var. amamiana（Ohwi）K. Inoue ■☆

302251 Platanthera andrewsii（M. White）Luer；安氏舌唇兰；Andréws' Bog Orchid ■☆

302252 Platanthera angolensis Schltr. = Brachycorythis angolensis（Schltr.）Schltr. ■☆

302253 Platanthera angustata（Blume）Lindl.；厚唇粉蝶兰■☆

302254 Platanthera angustata Lindl. = Platanthera mandarinorum Rchb. f. subsp. pachyglossa（Hayata）T. P. Lin et K. Inoue ■

302255 Platanthera angustifolia（Lindl.）Rchb. f. = Herminium lanceum（Thunb. ex Sw.）Vuijk ■

302256 Platanthera aquilonis Sheviak；北部舌唇兰；Northern Green Orchid ■☆

302257 Platanthera arachnoidea（A. Rich.）Engl. = Holothrix arachnoidea（A. Rich.）Rchb. f. ■☆

302258 Platanthera arcuata Lindl.；弧形舌唇兰■

302259 Platanthera bakeriana（King et Pantl.）Kraenzl.；滇藏舌唇兰；Baker Platanthera ■

302260 Platanthera bifolia（L.）Rich.；细距舌唇兰（二叶长距兰，双叶舌唇兰，苏舌兰）；Butterfly Orchid, Double-leaf Platanthera, Lesser Butterfly Orchid, Lesser Butterfly-orchid, Thinspur Platanthera ■

302261 Platanthera blephariglottis（Willd.）Lindl.；睫毛舌唇兰；White Fringed Orchis ■☆

302262 Platanthera boninensis Koidz.；小笠原舌唇兰■☆

302263 Platanthera borealis（Cham.）Rchb. f.；北美舌唇兰；North America Platanthera ■☆

302264 Platanthera bracteata Torr.；长距兰；Long-bracted Orchid ■☆

302265 Platanthera brevicalcarata Hayata；短距舌唇兰（短距粉蝶兰）；Shortspur Platanthera ■

302266 Platanthera brevicalcarata Hayata subsp. yakumontana

（Masam.）Masam.；屋久岛山地短距舌唇兰■☆

302267　Platanthera brevicalcarata Hayata var. yakumontana（Masam.）Masam. = Platanthera brevicalcarata Hayata subsp. yakumontana（Masam.）Masam.■☆

302268　Platanthera brevifolia（Greene）Kraenzl.；短叶舌唇兰■☆

302269　Platanthera buchananii Schltr. = Brachycorythis buchananii（Schltr.）Rolfe■☆

302270　Platanthera bulbinella（Rchb. f.）Schltr. = Schizochilus bulbinellus（Rchb. f.）Bolus☆

302271　Platanthera camtschatica（Cham.）Makino = Gymnadenia camtschatica（Cham.）Miyabe et Kudo■☆

302272　Platanthera championii Lindl. = Brachycorythis galeandra（Rchb. f.）Summerh.■

302273　Platanthera chapmanii（Small）Luer；查普曼舌唇兰■☆

302274　Platanthera chiloglossa（Ts. Tang et F. T. Wang）K. Y. Lang；察瓦龙舌唇兰；Chawalong Platanthera■

302275　Platanthera chingshuishania S. S. Ying；清水山舌唇兰（清水山粉蝶兰）；Qingshuishan Platanthera■

302276　Platanthera chingshuishania S. S. Ying = Platanthera stenoglossa Hayata■

302277　Platanthera chlorantha（Custer）Rchb. = Platanthera chlorantha Custer ex Rchb.■

302278　Platanthera chlorantha（Custer）Rchb. subsp. algeriensis（Batt. et Trab.）Emb. = Platanthera algeriensis Batt. et Trab.■☆

302279　Platanthera chlorantha（Custer）Rchb. var. longicalcarata Emb. = Platanthera chlorantha（Custer）Rchb.■

302280　Platanthera chlorantha Custer ex Rchb.；二叶舌唇兰（大叶长距兰，绿花长距兰，蛇儿参，土白芨，土白及）；Bileaf Platanthera, Butterfly Orchid, Greater Butterfly Orchid, Greater Butterfly-orchid, Night Violet, Twoleaf Platanthera, White Angel Orchid■

302281　Platanthera chlorantha Custer ex Rchb. var. orientalis Schltr. = Platanthera densa Freyn■

302282　Platanthera chloranthella Nakai = Platanthera densa Freyn■

302283　Platanthera chorisiana（Cham.）Rchb. f. = Pseudodiphryllum chorisianum（Cham.）Nevski■☆

302284　Platanthera chorisiana（Cham.）Rchb. f. var. elata Finet = Platanthera chorisiana（Cham.）Rchb. f.■☆

302285　Platanthera ciliaris（L.）Lindl.；缘毛舌唇兰；Yellow Fringed Orchid■☆

302286　Platanthera ciliaris（L.）Lindl. = Habenaria ciliaris（L.）R. Br.■☆

302287　Platanthera clavellata（Michx.）Luer；林地舌唇兰（小舌唇兰）；Club-spur Orchid, Green Woodland Orchid, Green Wood-orchis, Little Club-spear Orchid, Small Green Fringed Orchid, Small Green Wood Orchid, Small Green Wood-orchis, Woodland Orchid■☆

302288　Platanthera clavellata（Michx.）Luer = Habenaria clavellata（Michx.）Spreng.■☆

302289　Platanthera clavigera Lindl.；藏南舌唇兰（鸡肾草，鸡肾子，鸡腰子，肾经草，双仁，腰子草）；S. Xizang Platanthera■

302290　Platanthera comu-bovis Nevski；东北舌唇兰（长白舌唇兰）；NE. China Platanthera■

302291　Platanthera constricta（Lindl.）Wall. = Peristylus constrictus（Lindl.）Lindl.■

302292　Platanthera contigua Ts. Tang et F. T. Wang = Diphylax contigua（Ts. Tang et F. T. Wang）Ts. Tang, F. T. Wang et K. Y. Lang■

302293　Platanthera convallariifolia Fisch. ex Lindl. = Limnorchis convallariifolius（Fisch. ex Lindl.）Rydb.■☆

302294　Platanthera convallariifolia Lindl. = Platanthera hyperborea（L.）Lindl.■☆

302295　Platanthera cornu-bovis Nevski = Platanthera mandarinorum Rchb. f.■

302296　Platanthera cristata（Michx.）Lindl.；冠舌唇兰；Crested Yellow Orchis■☆

302297　Platanthera curvata K. Y. Lang；弓背舌唇兰；Bendback Platanthera■

302298　Platanthera curvata K. Y. Lang = Platanthera platantheroides（Ts. Tang et F. T. Wang）K. Y. Lang■

302299　Platanthera damingshanica K. Y. Lang et Han S. Guo；大明山舌唇兰；Damingshan Platanthera■

302300　Platanthera deflexilabella K. Y. Lang；反唇舌唇兰；Flexlip Platanthera■

302301　Platanthera delavayi Schltr. = Platanthera mandarinorum Rchb. f.■

302302　Platanthera densa（Lindl.）Soó = Platanthera clavigera Lindl.■

302303　Platanthera densa（Wall. ex Lindl.）Soó = Platanthera clavigera Lindl.■

302304　Platanthera densa Freyn；多叶舌唇兰■

302305　Platanthera densa Freyn subsp. orientalis（Schltr.）Efimov = Platanthera densa Freyn■

302306　Platanthera dentata（Sw.）Lindl. = Habenaria dentata（Sw.）Schltr.■

302307　Platanthera devolii（T. P. Lin et T. W. Hu）T. P. Lin et K. Inoue；长叶舌唇兰（长叶蜻蜓兰，台湾蜻蛉兰，台湾蜻蜓兰）；Taiwan Dragonflyoechis■

302308　Platanthera devolii（T. P. Lin et T. W. Hu）T. P. Lin et K. Inoue = Tulotis devolii T. P. Lin et T. W. Hu■

302309　Platanthera dielsiana Soó = Brachycorythis henryi（Schltr.）Summerh.■

302310　Platanthera dilatata（Pursh）Lindl. ex L. C. Beck = Limnorchis dilatatus（Pursh）Rydb.■☆

302311　Platanthera dilatata（Pursh）Lindl. ex L. C. Beck var. albiflora（Cham.）Ledeb. = Limnorchis dilatata（Pursh）Rydb. subsp. albiflora（Cham.）Á. Löve et W. Simon■☆

302312　Platanthera dilatata（Pursh）Lindl. ex L. C. Beck var. angustifolia Hook. = Platanthera dilatata（Pursh）Lindl. ex L. C. Beck■☆

302313　Platanthera dilatata（Pursh）Lindl. ex L. C. Beck var. angustifolia Hook. = Limnorchis dilatatus（Pursh）Rydb.■☆

302314　Platanthera diphylla Link = Gennaria diphylla（Link）Parl.■☆

302315　Platanthera ditmariana Kom. = Platanthera chorisiana（Cham.）Rchb. f.■☆

302316　Platanthera edgeworthii（Hook. f. ex Collett）R. K. Gupta = Habenaria edgeworthii Hook. f. ex Collett■☆

302317　Platanthera elachyantha Ts. Tang et F. T. Wang = Platanthera exelliana Soó■

302318　Platanthera elegans Lindl.；雅致舌唇兰；Elegant Platanthera, Rein Orchid■☆

302319　Platanthera elegans Lindl. = Piperia elegans（Lindl.）Rydb.■☆

302320　Platanthera engleriana（Kraenzl.）Rolfe = Brachycorythis tenuior Rchb. f.■☆

302321　Platanthera exelliana Soó；高原舌唇兰；Plateau Platanthera■

302322　Platanthera extremiorientalis Nevski = Platanthera metabifolia F. Maek.■

302323　Platanthera fallax（Lindl.）Schltr. = Peristylus fallax Lindl.■

302324 Platanthera fimbrista（Dryand.）Lindl.；流苏舌唇兰；Fimbriate Platanthera，Large Purple Fringed Orchid ■☆

302325 Platanthera finetiana Schltr.；对耳舌唇兰；Finet Platanthera ■

302326 Platanthera flava（L.）Lindl.；黄舌唇兰（黄玉凤兰）；Pale Green Orchid，Pale Green Orchis，Southern Rain-orchis，Tubercled Orchid，Tubercled Orchis ■☆

302327 Platanthera flava（L.）Lindl. = Tulotis fuscescens Raf. ■

302328 Platanthera flava（L.）Lindl. var. herbiola（R. Br.）Luer；浅绿舌唇兰；Pale Green Orchid，Tubercled Orchid ■☆

302329 Platanthera florentii Franch. et Sav.；福劳舌唇兰■☆

302330 Platanthera freynii Kraenzl. = Platanthera chlorantha Custer ex Rchb. ■

302331 Platanthera freynli Kraenzl.；弗雷舌唇兰■☆

302332 Platanthera friesii Schltr. = Brachycorythis friesii（Schltr.）Summerh. ■☆

302333 Platanthera fuscescens（L.）Kraenzl. = Platanthera souliei Kraenzl. ■

302334 Platanthera fuscescens（L.）Kraenzl. = Tulotis fuscescens Raf. ■

302335 Platanthera galeandra Rchb. f. = Brachycorythis galeandra（Rchb. f.）Summerh. ■

302336 Platanthera geniculata（D. Don）Lindl. = Habenaria dentata（Sw.）Schltr. ■

302337 Platanthera gerrardii（Rchb. f.）Schltr. = Schizochilus gerrardii（Rchb. f.）Bolus ■☆

302338 Platanthera gigantea Lindl. ex Wall. = Pecteilis gigantea（Sm.）Raf. ■

302339 Platanthera glaberrima（Ridl.）Kraenzl. = Benthamia glaberrima（Ridl.）H. Perrier ●☆

302340 Platanthera glossophora（W. W. Sm.）Schltr. = Platanthera hologlottis Maxim. ■

302341 Platanthera gracilis Lindl. = Platanthera stricta Lindl. ■☆

302342 Platanthera graminea（Thouars）Lindl. = Cynorkis graminea（Thouars）Schltr. ■☆

302343 Platanthera grandiflora（Bigelow）Lindl.；大花舌唇兰■☆

302344 Platanthera hachijoensis Honda = Platanthera mandarinorum Rchb. f. subsp. hachijoensis（Honda）Murata ■☆

302345 Platanthera handel-mazzettii K. Inoue；贡山舌唇兰；Gongshan Platanthera ■

302346 Platanthera helleborina（Hook. f.）Rolfe = Brachycorythis macrantha（Lindl.）Summerh. ■☆

302347 Platanthera hemlinioides Ts. Tang et F. T. Wang；高黎贡舌唇兰；Gaoligong Platanthera ■

302348 Platanthera henryi（Rolfe）Kraenzl. = Platanthera minor（Miq.）Rchb. f. ■

302349 Platanthera henryi（Rolfe）Rolfe；亨氏舌唇兰■

302350 Platanthera henryi（Rolfe）Rolfe = Platanthera minor（Miq.）Rchb. f. ■

302351 Platanthera herbiola（R. Br.）Lindl. = Tulotis fuscescens Raf. ■

302352 Platanthera herbiola（R. Br.）Lindl. var. japonica Finet = Tulotis ussuriensis（Regel et Maack）H. Hara ■

302353 Platanthera herbiola Lindl. var. japonica Finet = Platanthera ussuriensis（Regel et Maack）Maxim. ■

302354 Platanthera hispidula（Rendle）Gilg = Brachycorythis pubescens Harv. ■☆

302355 Platanthera hologlottis Maxim.；密花舌唇兰（沼兰）；Denseflower Platanthera ■

302356 Platanthera hologlottis Maxim. var. glossophora（W. W. Sm.）

K. Inoue = Platanthera hologlottis Maxim. ■

302357 Platanthera hondoensis（Ohwi）K. Inoue；本州舌唇兰■☆

302358 Platanthera hookeri（Torr. ex A. Gray）Lindl.；胡克舌唇兰；Hooker's Orchid，Pad-leaf ■☆

302359 Platanthera hookeri（Torr. ex A. Gray）Lindl. var. abbreviata（Fernald）W. J. Schrenk = Platanthera hookeri（Torr. ex A. Gray）Lindl. ■☆

302360 Platanthera huronensis（Nutt.）Lindl.；休伦湖舌唇兰；Huron Green Orchid，Tall Nothern Bog Orchid ■☆

302361 Platanthera hyperborea（L.）Lindl.；北方大舌唇兰；Northern Butterfly-orchid，Northern Green Orchid，Northern Green Orchis，Tall Northern Bog Orchid，Tall Northern Green Orchid ■☆

302362 Platanthera hyperborea（L.）Lindl. = Platanthera aquilonis Sheviak ■☆

302363 Platanthera hyperborea（L.）Lindl. var. huronensis（Nutt.）Luer = Platanthera huronensis（Nutt.）Lindl. ■☆

302364 Platanthera hyperborea（L.）Lindl. var. major Lange = Platanthera huronensis（Nutt.）Lindl. ■☆

302365 Platanthera hyperborea（L.）Lindl. var. minor Lange = Platanthera hyperborea（L.）Lindl. ■☆

302366 Platanthera hyperborea（L.）Lindl. var. purpurascens（Rydb.）Luer = Platanthera purpurascens（Rydb.）Sheviak et W. F. Jenn. ■☆

302367 Platanthera hyperborea（L.）Lindl. var. viridiflora（Cham.）Kitam. = Platanthera hyperborea（L.）Lindl. ■☆

302368 Platanthera hyperborea（L.）Lindl. var. viridiflora（Cham.）Kitam. = Platanthera stricta Lindl. ■☆

302369 Platanthera iantha Wight = Brachycorythis henryi（Schltr.）Summerh. ■

302370 Platanthera iinumae（Makino）Makino；饭沼舌唇兰■☆

302371 Platanthera inhambanensis Schltr. = Brachycorythis inhambanensis（Schltr.）Schltr. ■☆

302372 Platanthera integra（Nutt.）A. Gray = Platanthera integra（Nutt.）A. Gray ex L. C. Beck ■☆

302373 Platanthera integra（Nutt.）A. Gray ex L. C. Beck；全缘舌唇兰；Entire Platanthera ■☆

302374 Platanthera interrupta Maxim. = Platanthera minor（Miq.）Rchb. f. ■

302375 Platanthera iriomotensis Masam. = Platanthera stenoglossa Hayata subsp. iriomotensis（Masam.）K. Inoue ■☆

302376 Platanthera iriomotensis Masam. = Platanthera stenoglossa Hayata ■

302377 Platanthera japonica（Thunb. ex A. Murray）Lindl.；舌唇兰（长距兰，阔叶长距兰，龙爪参，骑马参，蛇儿参，水麦冬，猪寮参，走肾草）；Japan Platanthera，Japanese Platanthera ■

302378 Platanthera juncea（King et Pantl.）Kraenzl.；小巧舌唇兰；Skilful Platanthera ■

302379 Platanthera juncea（King et Pantl.）Kraenzl. = Peristylus nematocaulon（Hook. f.）Banerji et P. Pradhan ■

302380 Platanthera kwangsiensis K. Y. Lang；广西舌唇兰；Guangxi Platanthera ■

302381 Platanthera lacei Rolfe ex Downie = Pecteilis henryi Schltr. ■

302382 Platanthera lacera（Michx.）G. Don；撕裂舌唇兰；Green Fringed Orchid，Ragged Fringed Orchid，Ragged Orchid，Ragged Orchis ■☆

302383 Platanthera lalashaniana S. S. Ying；拉拉山舌唇兰（拉拉山粉蝶兰）；Lalashan Platanthera ■

302384 Platanthera lalashaniana S. S. Ying = Platanthera yangmeiensis

Rchb. f. ■

302439 Platanthera minor（Miq.）Rchb. f.；小舌唇兰（大一枝箭，高山粉蝶兰，观音竹，卵唇粉蝶兰，蛇蓼子，土洋参，小长距兰，鸭肾参，猪獠参）；Minor Platanthera，Little Platanthera ■

302440 Platanthera minor（Miq.）Rchb. f. var. mikurensis Hid. Takah.；御藏舌唇兰■☆

302441 Platanthera minutiflora Schltr.；小花舌唇兰；Smallflower Platanthera ■

302442 Platanthera minutiflora Schltr. = Platanthera handel-mazzettii K. Inoue ■

302443 Platanthera montana（F. W. Schmidt）Rchb. f.；山地舌唇兰；Montane Platanthera ■☆

302444 Platanthera multibracteata（W. W. Sm.）Schltr. = Platanthera minor（Miq.）Rchb. f. ■

302445 Platanthera nankotaizanensis（Masam.）Masam. = Coeloglossum viride（L.）Hartm. ■

302446 Platanthera nankotaizanensis（Masam.）Masam. = Dactylorhiza viridis（L.）R. M. Bateman，Pridgeon et M. W. Chase ■

302447 Platanthera natalensis（Rchb. f.）Schltr. = Habenaria petitiana（A. Rich.）T. Durand et Schinz ■☆

302448 Platanthera neglecta Schltr. = Platanthera mandarinorum Rchb. f. var. neglecta（Schltr.）F. Maek. ex K. Inoue ■

302449 Platanthera neglecta Schltr. = Platanthera mandarinorum Rchb. f. ■

302450 Platanthera nematocaulon（Hook. f.）Kraenzl. = Peristylus nematocaulon（Hook. f.）Banerji et P. Pradhan ■

302451 Platanthera nipponica Makino = Platanthera tipuloides（L. f.）Lindl. subsp. nipponica（Makino）Murata ■☆

302452 Platanthera nipponica Makino var. linearifolia（Ohwi）Masam. = Platanthera tipuloides（L. f.）Lindl. subsp. linearifolia（Ohwi）K. Inoue ☆

302453 Platanthera nivea（Nutt.）Luer；雪白舌唇兰；Snow-white Platanthera ■☆

302454 Platanthera obcordata Lindl. = Brachycorythis galeandra（Rchb. f.）Summerh. ■

302455 Platanthera obtusata（Banks ex Pursh）Lindl.；钝叶舌唇兰（钝叶玉凤花）；Blunt Bog Orchid，Blunt-leaved Orchid，Blunt-leaved Orchis，One-leaved Butterfly-orchid，Small Northern Bog Orchid，Small Northern Bog-orchis ■☆

302456 Platanthera obtusata（Banks ex Pursh）Lindl. subsp. oligantha（Turcz.）Hultén = Platanthera obtusata（Banks ex Pursh）Lindl. ■☆

302457 Platanthera obtusata（Pursh）Lindl. = Platanthera obtusata（Banks ex Pursh）Lindl. ☆

302458 Platanthera okuboi Makino；大久舌唇兰■☆

302459 Platanthera omeiensis（Rolfe）Schltr. = Platanthera japonica（Thunb. ex A. Murray）Lindl. ■

302460 Platanthera ophrydioides F. Schmidt = Platanthera mandarinorum Rchb. f. subsp. ophrydioides（F. Schmidt）K. Inoue ■

302461 Platanthera ophrydioides F. Schmidt f. australis（Ohwi）Makino = Platanthera mandarinorum Rchb. f. subsp. ophrydioides（F. Schmidt）K. Inoue var. monophylla（Honda）K. Inoue f. australis（Ohwi）K. Inoue ■☆

302462 Platanthera ophrydioides F. Schmidt subsp. takedae（Makino）Soó = Platanthera takedae Makino ■☆

302463 Platanthera ophrydioides F. Schmidt var. amabilis（Koidz.）Masam. = Platanthera amabilis Koidz. ■☆

302464 Platanthera ophrydioides F. Schmidt var. australis Ohwi =

302464续 Platanthera mandarinorum Rchb. f. subsp. ophrydioides（F. Schmidt）K. Inoue var. monophylla（Honda）K. Inoue f. australis（Ohwi）K. Inoue ■☆

302465 Platanthera ophrydioides F. Schmidt var. hachijoensis（Honda）F. Maek. = Platanthera mandarinorum Rchb. f. subsp. hachijoensis（Honda）Murata ■☆

302466 Platanthera ophrydioides F. Schmidt var. monophylla Honda = Platanthera mandarinorum Rchb. f. subsp. ophrydioides（F. Schmidt）K. Inoue var. monophylla（Honda）K. Inoue ■☆

302467 Platanthera ophrydioides F. Schmidt var. ophrydioides = Platanthera mandarinorum Rchb. f. subsp. ophrydioides（F. Schmidt）K. Inoue ■☆

302468 Platanthera ophrydioides F. Schmidt var. takedae（Makino）Ohwi = Platanthera takedae Makino ■☆

302469 Platanthera ophrydioides F. Schmidt var. uzenensis Ohwi = Platanthera takedae Makino subsp. uzenensis（Ohwi）K. Inoue ■☆

302470 Platanthera ophrydioidss F. Schmidt；眉兰舌唇兰■☆

302471 Platanthera opsimantha Ts. Tang et F. T. Wang = Diphylax uniformis（Ts. Tang et F. T. Wang）Ts. Tang，F. T. Wang et K. Y. Lang ■

302472 Platanthera orbiculata（Pursh）Lindl.；大圆叶舌唇兰（圆叶玉凤花）；Large Round-leaf Orchid，Large Round-leaved Orchid，Round Leaved Orchis ■☆

302473 Platanthera orbiculata（Pursh）Lindl. var. lehorsii（Fernald）Catling = Platanthera orbiculata（Pursh）Lindl. ■☆

302474 Platanthera orbiculata（Pursh）Lindl. var. macrophylla（Goldie）Luer = Platanthera macrophylla（Goldie）P. M. Br. ■☆

302475 Platanthera orchidis Lindl. = Gymnadenia orchidis Lindl. ■

302476 Platanthera orchidis Lindl. ex Wall. = Gymnadenia orchidis Lindl. ■

302477 Platanthera oreophila（W. W. Sm.）Schltr.；齿瓣舌唇兰；Toothpetal Platanthera ■

302478 Platanthera oreophila Schltr. = Platanthera oreophila（W. W. Sm.）Schltr. ■

302479 Platanthera pachyglossa Hayata = Platanthera mandarinorum Rchb. f. subsp. pachyglossa（Hayata）T. P. Lin et K. Inoue ■

302480 Platanthera pallida P. M. Br. = Platanthera cristata（Michx.）Lindl. ■☆

302481 Platanthera peichiatieniana S. S. Ying；北插天山舌唇兰（北插天山粉蝶兰）；Beichatian Platanthera ■

302482 Platanthera peramoena（A. Gray）A. Gray；紫舌唇兰；Purple Fringeless Orchid，Purple Fringeless Orchis ■☆

302483 Platanthera petitiana（A. Rich.）Engl. = Habenaria petitiana（A. Rich.）T. Durand et Schinz ■☆

302484 Platanthera platantheroides（Ts. Tang et F. T. Wang）K. Y. Lang = Platanthera curvata K. Y. Lang ■

302485 Platanthera platantheroides K. Y. Lang = Platanthera curvata K. Y. Lang ■

302486 Platanthera pleistophylla（Rchb. f.）Schltr. = Brachycorythis pleistophylla Rchb. f. ■

302487 Platanthera praeclara Sheviak et M. L. Bowles；西部草原舌唇兰；Western Prairie Fringed Orchid ■☆

302488 Platanthera praeustipetala Kraenzl. = Peristylus bulleyi（Rolfe）K. Y. Lang ■

302489 Platanthera preussii Kraenzl. = Habenaria microceras Hook. f. ■☆

302490 Platanthera pricei Hayata = Peristylus calcaratus（Rolfe）S. Y. Hu ■

302491　Platanthera psycodes（L.）Lindl.；小紫舌唇兰；Lesser Purple Fringed Orchid，Purple Fringed Orchid，Small Purple Fringed Orchid，Smaller Purple-fringed Orchis，White Fringed Orchid ■☆

302492　Platanthera pugionifera（W. W. Sm.）Schltr. = Platanthera souliei Kraenzl.

302493　Platanthera pugionifera（W. W. Sm.）Schltr. = Tulotis fuscescens Raf. ■

302494　Platanthera purpurascens（Rydb.）Sheviak et W. F. Jenn.；浅紫舌唇兰■☆

302495　Platanthera radiata（Thunb.）Lindl. = Pecteilis radiata（Thunb.）Raf. ■

302496　Platanthera radiata（Thunb.）Lindl. = Zeuxine strateumatica（L.）Schltr. ■

302497　Platanthera rhodostachys Schltr. = Brachycorythis rhodostachys（Schltr.）Summerh. ■☆

302498　Platanthera rhynchocarpa Thwaites = Habenaria stenopetala（Lindl.）Benth. ■

302499　Platanthera robusta Lindl. = Ponerorchis taiwanensis（Fukuy.）Ohwi ■

302500　Platanthera roseotincta（W. W. Sm.）Ts. Tang et F. T. Wang；棒距舌唇兰；Stickspur Platanthera ■

302501　Platanthera saccata（Greene）Hultén = Platanthera stricta Lindl. ■☆

302502　Platanthera sachalinensis F. Schmidt；库页舌唇兰（长苞粉蝶兰，高山粉蝶兰，高山舌唇兰）；Alp Platanthera ■

302503　Platanthera sachalinensis F. Schmidt var. hondoensis Ohwi = Platanthera hondoensis（Ohwi）K. Inoue ■☆

302504　Platanthera setchuanica Kraenzl. = Platanthera japonica（Thunb. ex A. Murray）Lindl. ■

302505　Platanthera shensiana（Kraenzl.）Ts. Tang et F. T. Wang = Platanthera ussuriensis（Regel et Maack）Maxim. ■

302506　Platanthera sigeyosii Masam.；卵唇粉蝶兰■

302507　Platanthera sigeyosii Masam. = Platanthera minor（Miq.）Rchb. f. ■

302508　Platanthera sigmoidea Maek. = Platanthera chlorantha Custer ex Rchb. ■

302509　Platanthera sikkimensis（Hook. f.）Kraenzl.；长瓣舌唇兰；Sikkim Platanthera ■

302510　Platanthera silagnsis Hand. -Mazz. = Platanthera leptocaulon（Hook. f.）Soó ■

302511　Platanthera sinica Ts. Tang et F. T. Wang；滇西舌唇兰；W. Yunnan Platanthera ■

302512　Platanthera sonoharae Masam.；园原舌唇兰■☆

302513　Platanthera souliei Kraenzl.；蜻蜓舌唇兰（浅褐舌唇兰，蜻蛉兰，蜻蜓兰，竹叶兰）；Asiatic Dragonflyoechis，Dragonflyoechis ■

302514　Platanthera souliei Kraenzl. = Tulotis fuscescens Raf. ■

302515　Platanthera sparsiflora（S. Watson）Schltr.；疏花舌唇兰■☆

302516　Platanthera sparsiflora（S. Watson）Schltr. var. ensifolia（Rydb.）Luer = Platanthera sparsiflora（S. Watson）Schltr. ■☆

302517　Platanthera stenantha（Hook. f.）Soó；条瓣舌唇兰；Beltpetal Platanthera ■

302518　Platanthera stenantha（Hook. f.）Soó subsp. omeiensis（Rolfe）Soó = Platanthera japonica（Thunb. ex A. Murray）Lindl. ■

302519　Platanthera stenoglossa Hayata；狭瓣舌唇兰（薄唇粉蝶兰，狭瓣粉蝶兰，狭唇粉蝶兰，狭叶舌唇兰）；Narrowpetal Platanthera ■

302520　Platanthera stenoglossa Hayata subsp. iriomotensis（Masam.）K. Inoue；西表舌唇兰■☆

302521　Platanthera stenoglossa Schltr. = Platanthera stenoglossa Hayata ■

302522　Platanthera stenophylla Ts. Tang et F. T. Wang；独龙江舌唇兰；Narrowleaf Platanthera ■

302523　Platanthera stenosepala Schltr.；狭瓣粉蝶兰■

302524　Platanthera stenosepala Schltr. = Platanthera stenoglossa Hayata ■

302525　Platanthera stenostachya Lindl. = Peristylus densus（Lindl.）Santapau et Kapadia ■

302526　Platanthera stenostachya Lindl. ex Benth. = Peristylus densus（Lindl.）Santapau et Kapadia ■

302527　Platanthera stricta Lindl.；直立舌唇兰；Strict Platanthera ■☆

302528　Platanthera subulifera（W. W. Sm.）Schltr. = Platanthera chlorantha Custer ex Rchb. ■

302529　Platanthera susannae（L.）Lindl. = Pecteilis susannae（L.）Raf. ■

302530　Platanthera susannae（L.）Lindl. = Ponerorchis taiwanensis（Fukuy.）Ohwi ■

302531　Platanthera taiwaniana（S. S. Ying）S. C. Chen；台湾舌唇兰；Taiwan Platanthera ■

302532　Platanthera taiwaniana S. S. Ying = Platanthera taiwaniana（S. S. Ying）S. C. Chen ■

302533　Platanthera takedae Makino；武田舌唇兰■☆

302534　Platanthera takedae Makino subsp. uzenensis（Ohwi）K. Inoue；羽前舌唇兰■☆

302535　Platanthera tenuior（Rchb. f.）Schltr. = Brachycorythis tenuior Rchb. f. ■☆

302536　Platanthera tipuloides（L. f.）Lindl.；筒距舌唇兰；Tubespur Platanthera ■

302537　Platanthera tipuloides（L. f.）Lindl. subsp. linearifolia（Ohwi）K. Inoue；线叶筒距舌唇兰■☆

302538　Platanthera tipuloides（L. f.）Lindl. subsp. nipponica（Makino）Murata；本州筒距舌唇兰■☆

302539　Platanthera tipuloides（L. f.）Lindl. var. linearifolia Ohwi = Platanthera tipuloides（L. f.）Lindl. subsp. linearifolia（Ohwi）K. Inoue ■☆

302540　Platanthera tipuloides（L. f.）Lindl. var. nipponica（Makino）Ohwi = Platanthera tipuloides（L. f.）Lindl. subsp. nipponica（Makino）Murata ■☆

302541　Platanthera tipuloides（L. f.）Lindl. var. sororia（Schltr.）Soó；堆积舌唇兰■☆

302542　Platanthera tipuloides（L. f.）Lindl. var. ussuriensis Regel et Maack. = Tulotis ussuriensis（Regel et Maack）H. Hara ■

302543　Platanthera tipuloides Lindl. var. ussuriensis Regel = Platanthera ussuriensis（Regel et Maack）Maxim. ■

302544　Platanthera transnokoensis Ohwi et Fukuy. = Platanthera sachalinensis F. Schmidt ■

302545　Platanthera tridentata（Hook. f.）Engl. = Holothrix tridentata（Hook. f.）Rchb. f. ■☆

302546　Platanthera truncatolabellata Hayata = Brachycorythis galeandra（Rchb. f.）Summerh. ■

302547　Platanthera uncata Rolfe = Cynorkis uncata（Rolfe）Kraenzl. ■☆

302548　Platanthera uniformis Ts. Tang et F. T. Wang = Diphylax uniformis（Ts. Tang et F. T. Wang）Ts. Tang, F. T. Wang et K. Y. Lang ■

302549　Platanthera ussuriensis（Regel et Maack）Maxim.；东亚舌唇兰（半层莲，半春莲，半脊莲，大叶黄龙缠树，乌苏里舌唇兰，小花蜻蜓兰，小手掌参，野苞芦）；Smallflower Dragonflyoechis，Wusuli Dragonflyoechis ■

302550 Platanthera ussuriensis（Regel et Maack）Maxim. = Tulotis ussuriensis（Regel et Maack）H. Hara ■

302551 Platanthera uzenensis（Ohwi）F. Maek. = Platanthera takedae Makino subsp. uzenensis（Ohwi）K. Inoue ■☆

302552 Platanthera virginea Bolus = Dracomonticola virginea（Bolus）H. P. Linder et Kurzweil ■☆

302553 Platanthera viridis（L.）Lindl. = Coeloglossum viride（L.）Hartm. ■

302554 Platanthera viridis（L.）Lindl. = Dactylorhiza viridis（L.）R. M. Bateman，Pridgeon et M. W. Chase ■

302555 Platanthera volkensiana Kraenzl. = Habenaria petitiana（A. Rich.）T. Durand et Schinz ■☆

302556 Platanthera winkeriana Schltr. = Platanthera mandarinorum Rchb. f. ■

302557 Platanthera yakumontana Masam. = Platanthera brevicalcarata Hayata subsp. yakumontana（Masam.）Masam. ■☆

302558 Platanthera yangmeiensis T. P. Lin；荫生舌唇兰（阴粉蝶兰，阴生舌唇兰）；Shady Platanthera ■

302559 Platanthera zeyheri（Sond.）Schltr. = Schizochilus zeyheri Sond. ■☆

302560 Platanthera zothecina（L. C. Higgins et S. L. Welsh）Kartesz et Gandhi；犹他荷舌唇兰■☆

302561 Platantheroides Szlach.（2004）；假舌唇兰属■☆

302562 Platantheroides Szlach. = Habenaria Willd. ■

302563 Platantheroides alata（Hook.）Szlach.；翅假舌唇兰■☆

302564 Platantheroides clavigera（Lindl.）Szlach. = Platanthera clavigera Lindl. ■

302565 Platantheroides densa（Wall. ex Lindl.）Szlach.；密集假舌唇兰■☆

302566 Platantheroides densa（Wall. ex Lindl.）Szlach. = Platanthera clavigera Lindl. ■

302567 Platantheroides eustachya（Rchb. f.）Szlach.；良穗假舌唇兰■☆

302568 Platantheroides floribunda（Lindl.）Szlach.；多花假舌唇兰■☆

302569 Platantheroides latilabris（Lindl.）Szlach. = Platanthera latilabris Lindl. ■

302570 Platantheroides linifolia（C. Presl）Szlach.；亚麻叶假舌唇兰■☆

302571 Platantheroides lucida（Wall. ex Lindl.）Szlach.；亮假舌唇兰■☆

302572 Platantheroides lucida（Wall. ex Lindl.）Szlach. = Habenaria lucida Wall. ex Lindl. ■

302573 Platantheroides obtusa（Lindl.）Szlach.；粗壮假舌唇兰■☆

302574 Platanus L.（1753）；悬铃木属（法国梧桐属）；Button Wood，Buttonwood，Plane，Plane Tree，Planetree，Plane-Tree，Platan，Sycamore ●

302575 Platanus acerifolia（Aiton）Willd.；悬铃木（二球悬铃木）●

302576 Platanus acerifolia（Aiton）Willd. = Platanus hispanica Münchh. ●

302577 Platanus acerifolia（Aiton）Willd. var. hispanica（Münchh.）？ = Platanus hispanica Münchh. ●

302578 Platanus cuneata Willd.；楔形悬铃木●☆

302579 Platanus cuneata Willd. = Platanus orientalis L. ●

302580 Platanus digitata Gordon；指裂悬铃木；Cyprus Plane Tree ●☆

302581 Platanus glabrata Fernald = Platanus occidentalis L. ●

302582 Platanus hispanica Münchh.；二球悬铃木（法松，槭叶悬铃木，悬铃木，英国梧桐）；English Planetree，European Planetree，Lacewood，Londen Plane，Londen Planetree，London Plane，London Plane Tree，Spanish Plane Tree ●

302583 Platanus hispanica Münchh.'Pyramidalis'；塔冠二球悬铃木●☆

302584 Platanus hispanica Münchh.'Suttneri'；萨腾纳英国梧桐●☆

302585 Platanus hybrida Brot. = Platanus acerifolia（Aiton）Willd. ●

302586 Platanus lindeniana Watson；林登悬铃木；Jalapa Plane Tree ●☆

302587 Platanus mexicana Moric.；墨西哥悬铃木；Mexican Plane，Mexican Plane Tree ●☆

302588 Platanus occidentalis L.；一球悬铃木（扣子树，美国梧桐，美国悬铃木，美桐，美洲桐木，美洲悬铃木，球悬铃木，水山毛榉树，悬条木）；America Planetree，American Plane，American Plane Tree，American Plane-tree，American Sycamore，Button Wood Buttonwood，Buttonball，Button-ball，Button-wood，Lacewood，Plane，Sycamore，Water Beech，Western Plane ●

302589 Platanus occidentalis L. f. attenuata Sarg. = Platanus occidentalis L. ●

302590 Platanus occidentalis L. var. glabrata（Fernald）Sarg. = Platanus occidentalis L. ●

302591 Platanus orieatalior Dode = Platanus orientalis L. ●

302592 Platanus orientalis L.；三球悬铃木（法国梧桐，净土树，祛汗树，条悬木，悬铃木）；Eastern Planetree，Oriental Plane，Oriental Plane Tree，Oriental Planetree ●

302593 Platanus orientalis L. var. acerifolia Aiton = Platanus acerifolia（Aiton）Willd. ●

302594 Platanus orientalis L. var. acerifolia Aiton = Platanus hispanica Münchh. ●

302595 Platanus racemosa Nutt.；加州悬铃木；Aliso，California Plane，California Plane Tree，California Sycamore，Western Sycamore ●

302596 Platanus wrightii S. Watson；亚利桑那悬铃木；Alamo，Arizona Plane，Arizona Planetree，Arizona Sycamore ●

302597 Platcalaria W. T. Steam = Anemopaegma Mart. ex Meisn.（保留属名）●☆

302598 Platcalaria W. T. Steam = Platolaria Raf.（废弃属名）●☆

302599 Platea Blume（1826）；肖榄属；Platea ●

302600 Platea hainanensis Howard = Platea latifolia Blume ●

302601 Platea latifolia Blume；阔叶肖榄（海南肖榄，木棍树，蒜头树）；Broadleaf Platea，Broad-leaved Platea ●

302602 Platea lobbiana Miers. = Gonocaryum lobbianum（Miers）Kurz ●

302603 Platea parviflora Dahl = Platea latifolia Blume ●

302604 Platea parvifolia Merr. et Chun；东方肖榄；Orient Platea，Small-leaf Platea，Small-leaved Platea ●

302605 Plateana Salisb. = Narcissus L. ■

302606 Plateilema（A. Gray）Cockerell（1904）；美洲阔苞菊属（阔封菊属）■☆

302607 Plateilema Cockercll = Plateilema（A. Gray）Cockerell ■☆

302608 Plateilema palmeri（A. Gray）Cockerell；美洲阔苞菊■☆

302609 Platenia H. Karst. = Syagrus Mart. ●

302610 Plathymenia Benth.（1840）（保留属名）；黄苏木属（黄木豆属）●☆

302611 Plathymenia foliolosa Benth.；小叶黄苏木●☆

302612 Plathymenia reticulata Benth.；网状黄苏木；Vinhalico ●☆

302613 Platolaria Raf.（废弃属名）= Anemopaegma Mart. ex Meisn.（保留属名）●☆

302614 Platonia Kunth = Neurolepis Meisn. ●☆

302615 Platonia Kunth = Planotia Munro ●☆

302616 Platonia Mart.（1832）（保留属名）；普拉顿藤黄属●☆

302617 Platonia Raf.（废弃属名）= Anemopaegma Mart. ex Meisn.（保留属名）●☆

302618 Platonia Raf.（废弃属名）= Helianthemum Mill. ●■

302619 Platonia Raf.（废弃属名）= Phyla Lour. ■

302620 Platonia Raf. (废弃属名) = Platonia Mart. (保留属名)●☆

302621 Platonia esculenta (Arruda) Rickett et Stafleu; 普拉顿藤黄; Bacury, Bakury ●☆

302622 Platonia insignis Mart.; 巴西普拉顿藤黄; Bacury, Bakuri Guiana Orange ●☆

302623 Platorheedia Rojas = ? Rheedia L. ●☆

302624 Platostoma P. Beauv. (1808); 平口花属■☆

302625 Platostoma africanum P. Beauv.; 非洲平口花☆

302626 Platostoma buettnerianum Briq. = Platostoma africanum P. Beauv. ■☆

302627 Platostoma coeruleum (R. E. Fr.) A. J. Paton; 浅蓝平口花■☆

302628 Platostoma denticulatum Robyns; 细齿平口花●☆

302629 Platostoma denticulatum Robyns var. minimum ? = Platostoma denticulatum Robyns ●☆

302630 Platostoma dilungense (Lisowski et Mielcarek) A. J. Paton; 迪龙平口花■☆

302631 Platostoma djalonense A. Chev. = Platostoma africanum P. Beauv. ■☆

302632 Platostoma fastigiatum A. J. Paton et Hedge; 帚状平口花●☆

302633 Platostoma flaccidum (A. Rich.) Benth. = Platostoma africanum P. Beauv. ■☆

302634 Platostoma gabonense A. J. Paton; 加蓬平口花●☆

302635 Platostoma glomerulatum A. J. Paton et Hedge; 团集平口花■☆

302636 Platostoma grandiflorum (Doan) A. J. Paton; 大花平口花■☆

302637 Platostoma hildebrandtii (Vatke) A. J. Paton; 希尔平口花●☆

302638 Platostoma laxiflorum A. J. Paton et Hedge; 疏花平口花●☆

302639 Platostoma lisowskianum (Bamps) A. J. Paton; 利索平口花■☆

302640 Platostoma madagascariense (Benth.) A. J. Paton et Hedge; 马岛平口花●☆

302641 Platostoma montanum (Robyns) A. J. Paton; 山地平口花●☆

302642 Platostoma rotundifolium (Briq.) A. J. Paton; 圆叶平口花●☆

302643 Platostoma strictum (Hiern) A. J. Paton; 刚直平口花●☆

302644 Platostoma tenellum (Benth.) A. J. Paton et Hedge; 柔软平口花●☆

302645 Platostoma thymifolium (Benth.) A. J. Paton et Hedge; 百里香叶平口花●☆

302646 Platunum A. Juss. = Holmskioldia Retz. ●

302647 Platunum rubrum Juss. = Holmskioldia sanguinea Retz. ●

302648 Platyadenia B. L. Burtt(1971); 宽腺苣苔属■☆

302649 Platyadenia descendens B. L. Burtt; 宽腺苣苔■☆

302650 Platyaechmea (Baker) L. B. Sm. et Kress = Aechmea Ruiz et Pav. (保留属名)■☆

302651 Platyaechmea (Baker) L. B. Sm. et W. J. Kress(1990); 宽矛光萼荷属■☆

302652 Platyaechmea fasciata (Lindl.) L. B. Sm. et Kress; 宽矛光萼荷; Urn Plant ■☆

302653 Platycalyx N. E. Br. (1905); 宽萼杜鹃属●☆

302654 Platycalyx N. E. Br. = Erica L. ●☆

302655 Platycalyx pumila N. E. Br. = Erica platycalyx E. G. H. Oliv. ●☆

302656 Platycapnos (DC.) Bernh. (1833); 头花烟堇属■☆

302657 Platycapnos Bernh. = Platycapnos (DC.) Bernh. ■☆

302658 Platycapnos saxicola Willk.; 岩地头花烟堇●☆

302659 Platycapnos spicatus (L.) Bernh.; 头花烟堇■☆

302660 Platycapnos spicatus (L.) Bernh. subsp. echeandiae (Pau) Heywood = Platycapnos spicatus (L.) Bernh. ●☆

302661 Platycapnos spicatus (L.) Bernh. var. albiflorus Lange = Platycapnos spicatus (L.) Bernh. ■☆

302662 Platycapnos spicatus (L.) Bernh. var. ochroleuca Lange = Platycapnos spicatus (L.) Bernh. ■☆

302663 Platycapnos spicatus (L.) Bernh. var. tenuilobus (Pomel) Batt. et Trab. = Platycapnos tenuilobus Pomel ■☆

302664 Platycapnos tenuilobus Pomel; 细裂头花烟堇■☆

302665 Platycarpha Less. (1831); 紫莲菊属■☆

302666 Platycarpha carlinoides Oliv. et Hiern; 刺苞菊状紫莲菊■☆

302667 Platycarpha glomerata (Thunb.) Less.; 紫莲菊■☆

302668 Platycarpha parvifolia S. Moore; 小叶紫莲菊■☆

302669 Platycarphella V. A. Funk et H. Rob. (1857); 小紫莲菊属■☆

302670 Platycarpidium F. Muell. = Platysace Bunge ■☆

302671 Platycarpum Bonpl. (1811); 宽果茜属☆

302672 Platycarpum Humb. et Bonpl. = Platycarpum Bonpl. ☆

302673 Platycarpum orinocense Humb. et Bonpl.; 宽果茜☆

302674 Platycarya Siebold et Zucc. (1843); 化香树属(化香属); Dyetree, Dye-tree, Platycarya ●

302675 Platycarya kwangtungensis Chun = Platycarya strobilacea Siebold et Zucc. ●

302676 Platycarya longipes C. Y. Wu; 圆果化香树(白皮树); Ballfruit Dyetree, Roundfruit Dyetree, Round-fruited Dye-tree ●

302677 Platycarya longipes C. Y. Wu = Platycarya strobilacea Siebold et Zucc. ●

302678 Platycarya longzhouensis S. Ye Liang et G. J. Liang; 龙州化香树; Longzhou Dye Tree ●

302679 Platycarya simplicifolia G. R. Long; 单叶化香树; One-leaf Dyetree ●

302680 Platycarya simplicifolia G. R. Long = Platycarya strobilacea Siebold et Zucc. ●

302681 Platycarya simplicifolia G. R. Long var. ternata G. R. Long; 三小叶化香树; Ternate One-leaf Dyetree ●

302682 Platycarya simplicifolia G. R. Long var. ternata G. R. Long = Platycarya strobilacea Siebold et Zucc. ●

302683 Platycarya sinensis Mottet = Platycarya strobilacea Siebold et Zucc. ●

302684 Platycarya strobilacea Siebold et Zucc.; 化香树(白皮树, 板栗树, 兜娄婆香, 兜炉, 放香树, 花果, 花果儿树, 花椰果, 花龙树, 花笼树, 花木香, 花香, 化龙树, 化皮树, 化树, 还香树, 换香树, 栲花, 栲花树, 栲蒲, 栲香, 麻柳树, 皮杆条, 山柳, 山柳树, 山麻柳, 台湾化香树, 鱼化树); Black Dye-tree, Dye Tree, Dyetree, Dye-tree ●

302685 Platycarya strobilacea Siebold et Zucc. var. kawakamii Hayata = Platycarya strobilacea Siebold et Zucc. ●

302686 Platycaryaceae Nakai = Juglandaceae DC. ex Perleb (保留科名)●

302687 Platycaryaceae Nakai = Platycaryaceae Nakai ex Doweld ●

302688 Platycaryaceae Nakai ex Doweld = Juglandaceae DC. ex Perleb (保留科名)●

302689 Platycaryaceae Nakai ex Doweld; 化香树科●

302690 Platycaulos H. P. Linder(1984); 扁茎帚灯草属■☆

302691 Platycaulos acutus Esterh.; 尖扁茎帚灯草■☆

302692 Platycaulos anceps (Mast.) H. P. Linder; 二棱扁茎帚灯草■☆

302693 Platycaulos callistachyus (Kunth) H. P. Linder; 美穗扁茎帚灯草☆

302694 Platycaulos cascadensis (Pillans) H. P. Linder; 喀斯喀特扁茎帚灯草■☆

302695 Platycaulos compressus (Rottb.) H. P. Linder; 扁茎帚灯草■☆

302696 Platycaulos depauperatus (Kunth) H. P. Linder; 萎缩扁茎帚灯草■☆

302697　Platycaulos major（Mast.）H. P. Linder；大扁茎帚灯草■☆

302698　Platycaulos subcompressus（Pillans）H. P. Linder；亚扁茎帚灯草■☆

302699　Platycelyphium Harms（1905）；蓝花宽荚豆属（宽荚豆属）■☆

302700　Platycelyphium cyananthum Harms ＝ Platycelyphium voense（Engl.）Wild ■☆

302701　Platycelyphium voense（Engl.）Wild；蓝花宽荚豆■☆

302702　Platycentrum Klotzsch ＝ Begonia L. ●■

302703　Platycentrum Naudin ＝ Leandra Raddi ●■☆

302704　Platychaeta Boiss. ＝ Pulicaria Gaertn. ■●

302705　Platychaete Bornm. ＝ Pulicaria Gaertn. ■●

302706　Platycheilis Cass. ＝ Holocheilus Cass. ■☆

302707　Platycheilus Cass. ＝ Trixis P. Browne ■●☆

302708　Platychilum Delaun. ＝ Hovea R. Br. ex W. T. Aiton ●■☆

302709　Platychilum Laun. ＝ Hovea R. Br. ex W. T. Aiton ●■☆

302710　Platychilus Post et Kuntze ＝ Platycheilus Cass. ■●☆

302711　Platychilus Post et Kuntze ＝ Trixis P. Browne ■●☆

302712　Platychorda B. G. Briggs et L. A. S. Johnson（1998）；三室帚灯草属■☆

302713　Platychorda applanata（Spreng.）B. G. Briggs et L. A. S. Johnson；宽三室帚灯草■☆

302714　Platychorda rivalis B. G. Briggs et L. A. S. Johnson；三室帚灯草■☆

302715　Platycladaceae A. V. Bobrov et Melikyan ＝ Cupressaceae Gray（保留科名）●

302716　Platycladaceae A. V. Bobrov et Melikyan；侧柏科●

302717　Platycladus Spach ＝ Thuja L. ●

302718　Platycladus Spach（1841）；侧柏属；Arborvitae, Chinese Arborvitae, Chinese Arbor-vitae ●

302719　Platycladus dolabrata（L. f.）Spach ＝ Thujopsis dolabrata（L. f.）Siebold et Zucc. ●

302720　Platycladus dolabrata（Thunb. ex L. f.）Spach ＝ Thujopsis dolabrata（Thunb. ex L. f.）Siebold et Zucc. ●

302721　Platycladus orientalis（L.）Franco；侧柏（柏,柏刺,柏树,扁柏,扁桧,东方崖柏,黄柏,榈,香柏,香柯树,香树,崖柏）；Biota, China Arborvitae, Chinese Arborvitae, Chinese Arbor-vitae, Chinese Cedar, Chinese Thuja, Dwarf Golden Arborvitae, Eastern Arbor-vitae, Golden Arborvitae, Oriental Arborvitae, Oriental Arbor-vitae, Oriental-cedar ●

302722　Platycladus orientalis（L.）Franco 'Aurea Nana'；矮金柏（矮黄千头柏,矮生金黄侧柏,洒金侧柏）；Dwarf Golden Arborvitae ●☆

302723　Platycladus orientalis（L.）Franco 'Aurea'；金黄侧柏；Golden Arborvitae ●☆

302724　Platycladus orientalis（L.）Franco 'Bakeri'；巴克侧柏；Baker Arborvitae ●☆

302725　Platycladus orientalis（L.）Franco 'Berkmanii'；贝克曼侧柏；Berkman Arborvitae ●☆

302726　Platycladus orientalis（L.）Franco 'Beverleyensis'；金塔柏（金塔侧柏,金枝侧柏）；Beverly Hills Arborvitae ●

302727　Platycladus orientalis（L.）Franco 'Blue Cone'；圆锥侧柏；Conical Arborvitae ●☆

302728　Platycladus orientalis（L.）Franco 'Bonita'；圆球侧柏（圆锥金柏）；Round Arborvitae ●☆

302729　Platycladus orientalis（L.）Franco 'Elegantissima'；极美侧柏●☆

302730　Platycladus orientalis（L.）Franco 'Excelsa'；矮塔柏●

302731　Platycladus orientalis（L.）Franco 'Filiformis'；垂丝柏●

302732　Platycladus orientalis（L.）Franco 'Rosedalis'；罗斯达侧柏●☆

302733　Platycladus orientalis（L.）Franco 'Semperaurescens'；金黄球柏（四季黄千头柏）；Golden Chinese Arborvitae ●

302734　Platycladus orientalis（L.）Franco 'Sieboldii'；千头柏（凤尾柏,扫帚柏,子孙柏）；Siebold Chinese Arborvitae ●

302735　Platycladus orientalis（L.）Franco 'Texana Glauca'；蓝绿柏●

302736　Platycladus orientalis（L.）Franco 'Zhaiguancebai'；窄冠侧柏；Narrowcrown Chinese Arborvitae ●

302737　Platycladus orientalis（L.）Franco ＝ Thuja orientalis L. ●

302738　Platycladus orientalis（L.）Franco f. pendula Q. Q. Liu et H. Y. Ye；垂枝侧柏；Pendulous Oriental Arborvitae ●

302739　Platycladus stricta Spach ＝ Platycladus orientalis（L.）Franco ●

302740　Platyclinis Benth.（1881）；平床兰属■☆

302741　Platyclinis Benth. ＝ Dendrochilum Blume ■

302742　Platyclinis cobbiana Hemsl. ＝ Dendrochilum cobbianum Rchb. f. ■☆

302743　Platyclinis filiforme Benth. ＝ Dendrochilum filiforme Lindl. ■☆

302744　Platyclinis formosana Schltr. ＝ Dendrochilum uncatum Rchb. f. ■

302745　Platyclinis glumaceum Benth. ＝ Dendrochilum glumaceum Lindl. ■☆

302746　Platyclinium T. Moore ＝ Begonia L. ●■

302747　Platycodon A. DC.（1830）；桔梗属（兰花参属）；Balloon Flower, Balloonflower, Balloon-Flower, Chinese Bellfower ■

302748　Platycodon Rchb. ＝ Daucus L. ■

302749　Platycodon autumnalis Decne. ＝ Platycodon grandiflorus（Jacq.）A. DC. ■

302750　Platycodon chinense Lindl. et Paxton ＝ Platycodon grandiflorus（Jacq.）A. DC. ■

302751　Platycodon expansus（Rudolph）Fed.；北桔梗■

302752　Platycodon glaucus（Thunb.）Nakai ＝ Platycodon grandiflorus（Jacq.）A. DC. ■

302753　Platycodon grandiflorus（Jacq.）A. DC.；桔梗（白桔梗,白药,包袱花,大药,道拉基,方图,房图,粉桔梗,符萜,符蒠,梗草,和尚花,和尚头花,鸡把腿,吉祥杵,苦菜根,苦蒉,苦梗,苦桔梗,喇叭花,利加,利如,利茹,铃铛花,卢如,卢茹,芦如,木便,南桔梗,荙苊,沙油菜,山铃铛花,四叶菜,土人参,玉桔梗）；Balloon Flower, Balloonflower, Balloon-flower, China Bellflower, Chinese Bellflower, Japanese Bellflower ■

302754　Platycodon grandiflorus（Jacq.）A. DC. 'Duplex'；二重桔梗■☆

302755　Platycodon grandiflorus（Jacq.）A. DC. f. albiflorus（Honda）H. Hara；白花桔梗■☆

302756　Platycodon grandiflorus（Jacq.）A. DC. var. glaucus Siebold et Zucc. ＝ Platycodon grandiflorus（Jacq.）A. DC. ■

302757　Platycodon mariesii Hort.；蓝桔梗■☆

302758　Platycodon sinensis Lem. ＝ Platycodon grandiflorus（Jacq.）A. DC. ■

302759　Platycoryne Rchb. f.（1855）；扁棒兰属■☆

302760　Platycoryne affinis Summerh. ；近缘扁棒兰■☆

302761　Platycoryne ambigua（Kraenzl.）Summerh. ；可疑扁棒兰■☆

302762　Platycoryne aurea（Kraenzl.）Rolfe ＝ Platycoryne paludosa（Lindl.）Rolfe ■☆

302763　Platycoryne brevirostris Summerh. ；短喙扁棒兰■☆

302764　Platycoryne buchananiana（Kraenzl.）Rolfe；布坎南扁棒兰■☆

302765　Platycoryne crocea（Schweinf. ex Rchb. f.）Rolfe；镉黄扁棒兰■☆

302766　Platycoryne crocea（Schweinf. ex Rchb. f.）Rolfe subsp. elegantula（Kraenzl.）Summerh. ；雅致扁棒兰■☆

302767　Platycoryne crocea（Schweinf. ex Rchb. f.）Rolfe subsp. montiselgon（Schltr.）Summerh. ；埃尔贡扁棒兰■☆

302768　Platycoryne crocea（Schweinf. ex Rchb. f.）Rolfe subsp.

ochrantha（Schltr.）Summerh.；苍白花扁棒兰■☆

302769　Platycoryne elegantula Kraenzl. = Platycoryne crocea（Schweinf. ex Rchb. f.）Rolfe subsp. elegantula（Kraenzl.）Summerh.■☆

302770　Platycoryne grandilfora Summerh. = Platycoryne latipetala Summerh. var. grandiflora（Summerh.）Geerinck■☆

302771　Platycoryne heterophylla Summerh.；互叶扁棒兰■☆

302772　Platycoryne kitondo（De Wild.）Summerh. = Platycoryne buchananiana（Kraenzl.）Rolfe■☆

302773　Platycoryne latipetala Summerh.；阔瓣扁棒兰■☆

302774　Platycoryne latipetala Summerh. var. grandiflora（Summerh.）Geerinck；大花扁棒兰■☆

302775　Platycoryne macroceras Summerh.；大角扁棒兰■☆

302776　Platycoryne mediocris Summerh.；中位扁棒兰■☆

302777　Platycoryne megalorrhyncha Summerh.；大喙扁棒兰■☆

302778　Platycoryne micrantha Summerh.；小花扁棒兰■☆

302779　Platycoryne montiselgon（Schltr.）Summerh. = Platycoryne crocea（Schweinf. ex Rchb. f.）Rolfe subsp. montiselgon（Schltr.）Summerh.■☆

302780　Platycoryne ochrantha（Schltr.）Summerh. = Platycoryne crocea（Schweinf. ex Rchb. f.）Rolfe subsp. ochrantha（Schltr.）Summerh.■☆

302781　Platycoryne paludosa（Lindl.）Rolfe；沼泽扁棒兰■☆

302782　Platycoryne pervillei Rchb. f.；佩尔扁棒兰■☆

302783　Platycoryne protearum（Rchb. f.）Rolfe；海神扁棒兰■☆

302784　Platycoryne protearum（Rchb. f.）Rolfe var. recurvirostrum G. Will.；曲喙扁棒兰■☆

302785　Platycoryne robynsiana Geerinck；罗宾斯扁棒兰■☆

302786　Platycoryne stuhlmannii Schltr.；斯图尔曼扁棒兰■☆

302787　Platycoryne tenuicaulis（Rendle）Rolfe = Platycoryne pervillei Rchb. f.☆

302788　Platycoryne trilobata Summerh.；三裂扁棒兰■☆

302789　Platycoryne ukingensis（Schltr.）Summerh. = Platycoryne protearum（Rchb. f.）Rolfe■☆

302790　Platycoryne wilfordii（Ridl.）Rolfe = Platycoryne paludosa（Lindl.）Rolfe■☆

302791　Platycorynoides Szlach.（2005）；拟扁棒兰属■☆

302792　Platycraspedum O. E. Schulz（1922）；宽框荠属（阔脉芥属）；Keeledsiliclecress, Platycraspedum■★

302793　Platycraspedum tibeticum O. E. Schulz；宽框荠；Tibet Keeledsiliclecress, Tibet Platycraspedum■

302794　Platycraspedum wuchengyii Al-Shehbaz；吴氏宽框荠■

302795　Platycrater Siebold et Zucc.（1838）；蛛网萼属（梅花甜茶属）；Platycrater, Cobwebcalyx●

302796　Platycrater arguta Siebold et Zucc.；蛛网萼（盾儿花，梅花甜茶）；Acute Cobwebcalyx, Acute Platycrater●

302797　Platycyamus Benth.（1862）；拟扁豆木属●☆

302798　Platycyamus regnellii Benth.；拟扁豆木；Angelim-rosa●☆

302799　Platycyparis A. V. Bobrov et Melikyan = Cupressus L.●

302800　Platycyparis A. V. Bobrov et Melikyan（2006）；中国柏属●★

302801　Platydadus Spach = Thuja L.●

302802　Platydaueon Rchb. = Daucus L.■

302803　Platydesma H. Mann（1866）；宽带芸香属●☆

302804　Platydesma auriculifolia Hillebr.；耳叶宽带芸香●☆

302805　Platydesma campanulata H. Mann；宽带芸香●☆

302806　Platydesma grandifolium（A. Gray）Skottsb.；大叶宽带芸香●☆

302807　Platydinis Benth. = Dendrochilum Blume■

302808　Platyelasma（Briq.）Kitag. = Elsholtzia Willd.●■

302809　Platyelasma Kitag. = Elsholtzia Willd.●■

302810　Platyelasma calycocarpum（Diels）Kitag. = Elsholtzia densa Benth.■

302811　Platyelasma densum（Benth.）Kitag. = Elsholtzia densa Benth.■

302812　Platyelasma eriostachyum（Benth.）Kitag. = Elsholtzia eriostachya Benth.■

302813　Platyelasma eriostachyum（Benth.）Kitag. var. pusillum（Benth.）Kitag. = Elsholtzia eriostachya Benth.■

302814　Platyelasma manshuricum Kitag. = Elsholtzia densa Benth.■

302815　Platyestes Salisb. = Lachenalia J. Jacq. ex Murray■☆

302816　Platyglottis L. O. Williams（1942）；宽舌兰属■☆

302817　Platyglottis coriacea L. O. Williams；宽舌兰■☆

302818　Platygonia Naudin = Trichosanthes L.●●

302819　Platygonia kaempferi Naudin = Trichosanthes cucumeroides（Ser.）Maxim. ex Franch. et Sav.■

302820　Platygyna P. Mercier（1830）；宽蕊大戟属☆

302821　Platygyna dentata Alain；齿宽蕊大戟☆

302822　Platygyna obovata Borhidi；倒卵宽蕊大戟☆

302823　Platygyna parvifolia Alain；小叶宽蕊大戟☆

302824　Platygyna triandra Borhidi；三蕊宽蕊大戟☆

302825　Platygyne Howard = Platygyna P. Mercier☆

302826　Platyhymenia Walp. = Plathymenia Benth.（保留属名）●★

302827　Platykeleba N. E. Br.（1895）；宽杯萝藦属■☆

302828　Platykeleba N. E. Br. = Cynanchum L.●■

302829　Platykeleba insignis N. E. Br.；宽杯萝藦■☆

302830　Platykeleba insignis N. E. Br. = Cynanchum insigne（N. E. Br.）Liede et Meve■☆

302831　Platylepis A. Rich.（1828）（保留属名）；阔鳞兰属■☆

302832　Platylepis Kunth = Ascolepis Nees ex Steud.（保留属名）■☆

302833　Platylepis angolensis（Rchb. f.）T. Durand et Schinz = Platylepis glandulosa（Lindl.）Rchb. f.■☆

302834　Platylepis australis Rolfe = Platylepis glandulosa（Lindl.）Rchb. f.■☆

302835　Platylepis bigibbosa H. Perrier；双囊阔鳞兰■☆

302836　Platylepis brasiliensis Kunth = Ascolepis brasiliensis（Kunth）Benth.■☆

302837　Platylepis capensis Kunth = Ascolepis capensis（Kunth）Ridl.■☆

302838　Platylepis densiflora Rolfe；密花阔鳞兰■☆

302839　Platylepis glandulosa（Lindl.）Rchb. f.；腺阔鳞兰■☆

302840　Platylepis goodyeroides A. Rich. = Polystachya anceps Ridl.■☆

302841　Platylepis humicola Schltr. = Goodyera humicola Schltr.■☆

302842　Platylepis margaritifera Schltr.；珍珠阔鳞兰■☆

302843　Platylepis nyassana Schltr. = Platylepis glandulosa（Lindl.）Rchb. f.■☆

302844　Platylepis occulta（Thouars）Rchb. f.；隐蔽阔鳞兰■☆

302845　Platylepis perrieri Schltr. = Goodyera perrieri Schltr.■☆

302846　Platylepis polyadenia Rchb. f.；多腺阔鳞兰■☆

302847　Platylepis talbotii Rendle = Platylepis glandulosa（Lindl.）Rchb. f.■☆

302848　Platylobium Sm.（1793）；扁豆木属（澳扁豆木属）；Flat Pra●■☆

302849　Platylobium formosum Sm.；美丽扁豆木；Handsome Flat Pra●■☆

302850　Platylobium obtusangulum Hook.；扁豆木；Common Flat Pra●☆

302851　Platylophus Cass.（废弃属名）= Centaurea L.（保留属名）●■

302852　Platylophus Cass.（废弃属名）= Platylophus D. Don（保留属名）●☆

302853　Platylophus D. Don（1830）（保留属名）；阔脊木属●☆

302854　Platylophus trifoliatus（L. f.）D. Don;阔脊木●☆

302855　Platyluma Baill. = Micropholis（Griseb.）Pierre ●☆

302856　Platymerium Bartl. ex DC. = Hypobathrum Blume ●☆

302857　Platymetra Noronha ex Salisb. = Tupistra Ker Gawl. ■

302858　Platymetraceae Salisb. = Aspidistraceae J. Agardh ■

302859　Platymetraceae Salisb. = Convallariaceae L. ■

302860　Platymiscium Vogel（1837）;阔变豆属;Macawood ●☆

302861　Platymiscium duckei Huber;阔变豆木●☆

302862　Platymiscium floribundum Vogel;多花阔变豆●☆

302863　Platymiscium pinnatum（Jacq.）Dugand;羽状阔变豆;Robie ●☆

302864　Platymiscium trinitatis Benth. ;特立阔变豆●☆

302865　Platymiscium ulei Harms;乌氏阔变豆●☆

302866　Platymitium Warb. = Dobera Juss. ●☆

302867　Platymitium loranthifolium Warb. = Dobera loranthifolia（Warb.）Harms ●☆

302868　Platymitra Boerl.（1899）;宽帽花属●☆

302869　Platymitra arborea（Blanco）Kessler;乔木宽帽花●☆

302870　Platymitra macrocarpa Boerl. ;宽帽花●☆

302871　Platymitrium Willis = Dobera Juss. ●☆

302872　Platymitrium Willis = Platymitium Warb. ●☆

302873　Platynema Schrad. = Mertensia Roth(保留属名)■

302874　Platynema Schrad. = Winkleria Rchb. ■

302875　Platynema Wight et Arn. = Tristellateia Thouars ●

302876　Platyopuntia（Eng.）Fri？ et Schelle ex Kreuz. = Opuntia Mill. ●

302877　Platyopuntia Kreuz. = Opuntia Mill. ●

302878　Platyosprion（Maxim.）Maxim. = Cladrastis Raf. ●

302879　Platyosprion Maxim. = Cladrastis Raf. ●

302880　Platyosprion platycarpum（Maxim.）Maxim. = Cladrastis platycarpa（Maxim.）Makino ●

302881　Platyosprion platycarpum Maxim. = Cladrastis platycarpa（Maxim.）Makino ●

302882　Platypetalum R. Br. = Braya Sternb. et Hoppe ■

302883　Platypholis Maxim.（1887）;小笠原列当属■☆

302884　Platypholis boninsimae Maxim. ;小笠原列当■☆

302885　Platypholis boninsimae Maxim. = Orobanche boninsimae（Maxim.）Tuyama ■☆

302886　Platypodanthera R. M. King et H. Rob.（1972）;宽药柄泽兰属■●☆

302887　Platypodanthera melissifolia（DC.）R. M. King et H. Rob.;宽药柄泽兰■☆

302888　Platypodium Vogel（1837）;宽柄豆属■☆

302889　Platypodium elegans Vogel;宽柄豆■☆

302890　Platyptelea J. Drumm. ex Harv. = Aphanopetalum Endl. ●☆

302891　Platypteris Kunth = Verbesina L.（保留属名）●■☆

302892　Platypterocarpus Dunkley et Brenan(1948);宽翅果卫矛属●☆

302893　Platypterocarpus tanganyikensis Dunkley et Brenan;宽翅果卫矛●☆

302894　Platypus Small et Nash = Eulophia R. Br.（保留属名）■

302895　Platypus altus（L.）Small = Eulophia alta（L.）Fawc. et Rendle ■☆

302896　Platypus papilliferus Small = Eulophia alta（L.）Fawc. et Rendle ■☆

302897　Platyraphe Miq. = Pimpinella L. ■

302898　Platyraphium Cass. = Lamyra Cass. ■☆

302899　Platyrhaphe Miq. = Pimpinella L. ■

302900　Platyrhaphe japonica Miq. = Pimpinella diversifolia DC. ■

302901　Platyrhiza Barb. Rodr.（1881）;扁根兰属■☆

302902　Platyrhiza quadricolor Barb. Rodr. ;扁根兰■☆

302903　Platyrhodon（Decne.）Hurst = Rosa L. ●

302904　Platyrhodon（Decne.）Hurst(1928);巴西刺梨属●

302905　Platyrhodon Hurst = Platyrhodon（Decne.）Hurst ●

302906　Platyrhodon Hurst = Rosa L. ●

302907　Platyrhodon microphyllum（Roxb.）Hurst;巴西刺梨●

302908　Platyruscus A. P. Khokhr. et V. N. Tikhom.（1993）;宽假叶树属●☆

302909　Platyruscus A. P. Khokhr. et V. N. Tikhom. = Ruscus L. ●

302910　Platysace Bunge(1845);宽盾草属■☆

302911　Platysace heterophylla（Benth.）C. Norman;互叶宽盾草■☆

302912　Platysace linearifolia（Cav.）C. Norman;线叶宽盾草;Carrot Tops ■☆

302913　Platysace tenuissima（Benth.）C. Norman;细宽盾草■☆

302914　Platyschkuhria（A. Gray）Rydb.（1906）;盆雏菊属;Basindaisy ■☆

302915　Platyschkuhria Rydb. = Platyschkuhria（A. Gray）Rydb. ■☆

302916　Platyschkuhria integrifolia（A. Gray）Rydb. ;盆雏菊■☆

302917　Platyschkuhria integrifolia（A. Gray）Rydb. var. desertorum（M. E. Jones）W. L. Ellison = Platyschkuhria integrifolia（A. Gray）Rydb. ■☆

302918　Platyschkuhria integrifolia（A. Gray）Rydb. var. oblongifolia（A. Gray）W. L. Ellison = Platyschkuhria integrifolia（A. Gray）Rydb. ■☆

302919　Platyschkuhria integrifolia Rydb. = Platyschkuhria integrifolia（A. Gray）Rydb. ■☆

302920　Platysema Benth. = Centrosema（DC.）Benth.（保留属名）●■☆

302921　Platysepalum Welw. ex Baker(1871);宽萼豆属■☆

302922　Platysepalum chevalieri Harms;舍瓦利耶宽萼豆■☆

302923　Platysepalum chevalieri Harms var. aureum Hauman = Platysepalum chevalieri Harms ■☆

302924　Platysepalum chrysophyllum Hauman;金叶宽萼豆■☆

302925　Platysepalum cuspidatum Taub. ;骤尖宽萼豆■☆

302926　Platysepalum ferrugineum Taub. ;锈色宽萼豆■☆

302927　Platysepalum hirsutum（Dunn）Hepper;多毛宽萼豆■☆

302928　Platysepalum hypoleucum Taub. ;白背宽萼豆■☆

302929　Platysepalum inopinatum Harms;意外宽萼豆■☆

302930　Platysepalum ledermannii Harms = Platysepalum violaceum Welw. ex Baker var. vanhouttei（De Wild.）Hauman ■☆

302931　Platysepalum poggei Taub. ;波格宽萼豆■☆

302932　Platysepalum polyanthum Harms = Platysepalum violaceum Welw. ex Baker var. vanhouttei（De Wild.）Hauman ■☆

302933　Platysepalum pulchrum Louis ex Hauman;美丽宽萼豆■☆

302934　Platysepalum scaberulum Harms;粗糙宽萼豆■☆

302935　Platysepalum tessmannii Harms = Platysepalum violaceum Welw. ex Baker var. vanhouttei（De Wild.）Hauman ■☆

302936　Platysepalum vanderystii De Wild. ;范德宽萼豆■☆

302937　Platysepalum vanhouttei De Wild. = Platysepalum violaceum Welw. ex Baker var. vanhouttei（De Wild.）Hauman ■☆

302938　Platysepalum violaceum Welw. ex Baker;堇色宽萼豆■☆

302939　Platysepalum violaceum Welw. ex Baker var. ebracteolatum P. Sousa;无苞宽萼豆■☆

302940　Platysepalum violaceum Welw. ex Baker var. vanhouttei（De Wild.）Hauman;瓦氏宽萼豆■☆

302941　Platysma Blume = Placostigma Blume ■

302942　Platysma Blume = Podochilus Blume ■

302943　Platysperma Rchb. = Daucus L. ■

302944　Platysperma Rchb. = Platyspermum Hoffm. ■

302945　Platyspermatiaceae Doweld = Alseuosmiaceae Airy Shaw ●☆

302946　Platyspermation Guillaumin(1950);扁子岛海桐属●☆

302947　Platyspermation crassifolium Guillaumin;扁子岛海桐●☆

302948　Platyspermum Hoffm. = Daucus L. ■

302949　Platyspermum Hook. = Idahoa A. Nelson et J. F. Macbr. ■☆

302950　Platyspermum aureum (Desf.) Pomel = Daucus aureus Desf. ■☆

302951　Platyspermum grandiflorum Pomel = Daucus carota L. subsp. maximus (Desf.) Ball. ■☆

302952　Platysraa Blume = Podochilus Blume ■

302953　Platystachys C. Koch = Tillandsia L. ■☆

302954　Platystachys K. Koch = Allardtia A. Dietr. ■☆

302955　Platystachys K. Koch = Tillandsia L. ■☆

302956　Platystele Schltr. (1910);阔柱兰属■☆

302957　Platystele jungermannioides (Schltr.) Garay;阔柱兰■☆

302958　Platystemma Wall. (1831);堇叶苣苔属(花叶苣苔属,叶花苣苔属);Platystemma ■

302959　Platystemma majus Wall. = Platystemma violoides Wall. ■

302960　Platystemma violoides Wall.;堇叶苣苔(叶花苣苔);Violet-like Platystemma ■

302961　Platystemon Benth. (1835);宽蕊罂粟属(平蕊罂粟属);Cream Cups,Creamcups,Platystemon ■☆

302962　Platystemon arizonicus Greene = Platystemon californicus Benth. ■☆

302963　Platystemon australis Greene = Platystemon californicus Benth. ■☆

302964　Platystemon californicus Benth.;宽蕊罂粟(平蕊罂粟,乳杯花);Californian Poppy,Cream Cup,Cream Cups,Creamcups ■☆

302965　Platystemon californicus Benth. var. ciliatus Dunkle = Platystemon californicus Benth. ■☆

302966　Platystemon californicus Benth. var. crinitus Greene = Platystemon californicus Benth. ■☆

302967　Platystemon californicus Benth. var. horridulus (Greene) Jeps. = Platystemon californicus Benth. ■☆

302968　Platystemon californicus Benth. var. nutans K. Brandegee = Platystemon californicus Benth. ■☆

302969　Platystemon californicus Benth. var. ornithopus (Greene) Munz = Platystemon californicus Benth. ■☆

302970　Platystemon leiocarpus Fisch. et E. Mey. = Platystemon californicus Benth. ■☆

302971　Platystemon linearis (Benth.) Curran = Hesperomecon linearis (Benth.) Greene ■☆

302972　Platystemon mohavensis Greene = Platystemon californicus Benth. ■☆

302973　Platystemonaceae A. C. Sm.;宽蕊罂粟科■

302974　Platystemonaceae A. C. Sm. = Papaveraceae Juss. (保留科名)●■

302975　Platystemonaceae Lilja = Papaveraceae Juss. (保留科名)●■

302976　Platystephium Gardner = Egletes Cass. ■☆

302977　Platystigma Benth. = Meconella Nutt. ■☆

302978　Platystigma R. Br. = Platea Blume ●

302979　Platystigma R. Br. ex Benth. = Platea Blume ●

302980　Platystigma R. Br. ex Benth. et Hook. f. = Platea Blume ●

302981　Platystigma lineare Benth. = Hesperomecon linearis (Benth.) Greene ■☆

302982　Platystoma Benth. et Hook. f. = Platostoma P. Beauv. ■☆

302983　Platystyliparis Marg. (2006);宽柱兰属■☆

302984　Platystyliparis Marg. = Malaxis Sol. ex Sw. ■

302985　Platystyliparis delicatula (Hook. f.) Marg. = Liparis delicatula Hook. f. ■

302986　Platystyliparis fissipetala (Finet) Marg. = Ypsilorchis fissipetala

(Finet) Z. J. Liu,S. C. Chen et L. J. Chen ■

302987　Platystyliparis perpusilla (Hook. f.) Marg. = Liparis perpusilla Hook. f. ■

302988　Platystyliparis platyrachis (Hook. f.) Marg. = Liparis platyrachis Hook. f. ■

302989　Platystyliparis resupinata (Ridl.) Marg. = Liparis resupinata Ridl. ■

302990　Platystylis (Blume) Lindl. = Liparis Rich. (保留属名)■

302991　Platystylis Lindl. = Liparis Rich. (保留属名)■

302992　Platystylis Sweet = Lathyrus L. ■

302993　Platytaenia Nevski et Vved. (1937);宽带芹属(阔带芹属)■☆

302994　Platytaenia Nevski et Vved. = Neoplatytaenia Geld. ■

302995　Platytaenia Nevski et Vved. = Semenovia Regel et Herder ■

302996　Platytaenia absinthifolia ?;苦叶宽带芹■☆

302997　Platytaenia bucharica Schischk. ;布赫宽带芹■☆

302998　Platytaenia dasycarpa (Regel et Schmalh.) Korovin = Semenovia dasycarpa (Regel et Schmalh.) Korovin ex Pimenov et V. N. Tikhom. ■

302999　Platytaenia depauperata Schischk. ;萎缩宽带芹■☆

303000　Platytaenia heterodonta Korovin;异齿宽带芹■☆

303001　Platytaenia komarovii (Manden.) Schischk. = Semenovia dasycarpa (Regel et Schmalh.) Korovin ex Pimenov et V. N. Tikhom. ■

303002　Platytaenia komarovii Schischk. ;科马罗夫宽带芹■☆

303003　Platytaenia kuramensis (Kitam.) Nasir;库拉姆宽带芹■☆

303004　Platytaenia lasiocarpa (Boiss.) Rech. f. et Riedl;毛果宽带芹■☆

303005　Platytaenia lasiocarpa (Boiss.) Rech. f. et Riedl subsp. radiates Rech. f. et Riedl;辐射宽带芹■☆

303006　Platytaenia lasiocarpa (Boiss.) Rech. f. et Riedl subsp. thomsonii (C. B. Clarke) Rech. f. et Riedl;托马森宽带芹■☆

303007　Platytaenia lasiocarpa (Boiss.) Rech. f. et Riedl subsp. thomsonii (C. B. Clarke) Rech. f. et Riedl var. glabrior (C. B. Clarke) Nasir;无毛宽带芹■☆

303008　Platytaenia multicaule Nasir;多茎宽带芹■☆

303009　Platytaenia olgae (Regel et Schmalh.) Korovin = Heracleum olgae Regel et Schmalh. ex Regel ■

303010　Platytaenia pamirica (Lipsky) Nevski et Vved. ;帕米尔宽带芹■☆

303011　Platytaenia pimpinelloides Nevski;茴芹状宽带芹■☆

303012　Platytaenia pimpinelloides Nevski = Semenovia pimpinelloides (Nevski) Manden. ■

303013　Platytaenia rubtzovii Schischk. ;鲁氏宽带芹■☆

303014　Platytaenia rubtzovii Schischk. = Semenovia rubtzovii (Schischk.) Manden. ■

303015　Platytaenia tripartita (Aitch. et Hemsl.) Rech. f. et Riedl = Platytaenia kuramensis (Kitam.) Nasir ■☆

303016　Platythea O. F. Cook = Chamaedorea Willd. (保留属名)●☆

303017　Platytheca Steetz(1845);阔囊孔药花属●☆

303018　Platytheca crassifolia Steetz;厚叶阔囊孔药花●☆

303019　Platythelys Garay(1977);阔喙兰属■☆

303020　Platythelys querceticola (Lindl.) Garay;阔喙兰;Jug Orchid ■☆

303021　Platythelys sagreana (A. Rich.) Garay = Platythelys querceticola (Lindl.) Garay ■☆

303022　Platythyra N. E. Br. (1925);平盾番杏属■☆

303023　Platythyra N. E. Br. = Aptenia N. E. Br. ●☆

303024　Platythyra barklyi (N. E. Br.) Schwantes = Mesembryanthemum barklyi N. E. Br. ■☆

303025　Platythyra haeckeliana (A. Berger) N. E. Br. = Aptenia haeckeliana（A. Berger）Bittrich ex Gerbaulet ■☆

303026 Platythyra pallens（Aiton）L. Bolus = Prenia pallens（Aiton）N. E. Br. ■☆

303027 Platythyra relaxata（Willd.）Schwantes = Prenia pallens（Aiton）N. E. Br. ■☆

303028 Platytinospora（Engl.）Diels(1910)；非洲青牛胆属●☆

303029 Platytinospora Diels = Platytinospora（Engl.）Diels ●☆

303030 Platytinospora buchholzii（Engl.）Diels；非洲青牛胆●☆

303031 Platytinospora buchholzii（Engl.）Diels var. macrophylla Diels；大叶非洲青牛胆●☆

303032 Platyzamia Zucc. = Dioon Lindl.（保留属名）●☆

303033 Platzchaeta Sch. Bip. = Platychaeta Boiss. ■●

303034 Plazaea Post et Kuntze = Plazia Ruiz et Pav. ●☆

303035 Plazeria Steud. = Plazerium Willd. ex Kunth ■

303036 Plazeria Steud. = Saccharum L. ■

303037 Plazerium Kunth = Eriochrysis P. Beauv. ■☆

303038 Plazerium Willd. ex Kunth = Saccharum L. ■

303039 Plazia Ruiz et Pav.（1794）；脂菊木属●☆

303040 Plazia brasiliensis Spreng.；巴西脂菊木●☆

303041 Plazia conferta Ruiz et Pav.；脂菊木●☆

303042 Pleconax Adans. = Silene L.（保留属名）■

303043 Pleconax Raf. = Silene L.（保留属名）■

303044 Plecospermum Trécul = Maclura Nutt.（保留属名）●

303045 Plecostachys Hilliard et B. L. Burtt(1981)；密头火绒草属■☆

303046 Plecostachys polifolia（Thunb.）Hilliard et B. L. Burtt；灰叶密头火绒草■☆

303047 Plecostachys serpyllifolia（P. J. Bergius）Hilliard et B. L. Burtt；百里香叶密头火绒草■☆

303048 Plecostigma Turcz. = Gagea Salisb. ■

303049 Plecostigma pauciflorum Turcz. = Gagea pauciflora Turcz. ■

303050 Plecostigma pauciflorum Turcz. ex Trautv. = Gagea pauciflora Turcz. ■

303051 Plectaneia Thouars(1806)；编织夹竹桃属●☆

303052 Plectaneia boivinii Jum. = Plectaneia thouarsii Roem. et Schult. ●☆

303053 Plectaneia breviloba Markgr. = Plectaneia thouarsii Roem. et Schult. ●☆

303054 Plectaneia elastica Jum. et H. Perrie rf. firingalavensis Jum. = Plectaneia thouarsii Roem. et Schult. ●☆

303055 Plectaneia elastica Jum. et H. Perrier = Plectaneia thouarsii Roem. et Schult. ●☆

303056 Plectaneia elastica Jum. et H. Perrier var. insularis Markgr. = Plectaneia thouarsii Roem. et Schult. ●☆

303057 Plectaneia elastica Jum. et H. Perrier var. inutilis（Jum. et H. Perrier）Pichon = Plectaneia thouarsii Roem. et Schult. ●☆

303058 Plectaneia firingalavensis（Jum.）Jum. = Plectaneia thouarsii Roem. et Schult. ●☆

303059 Plectaneia firingalavensis（Jum.）Jum. f. setulosa Markgr. = Plectaneia thouarsii Roem. et Schult. ●☆

303060 Plectaneia firingalavensis（Jum.）Jum. var. lanceolata（Pichon）Markgr. = Plectaneia thouarsii Roem. et Schult. ●☆

303061 Plectaneia hildebrandtii（Jum.）Jum. f. hirsuta Jum. et H. Perrier = Plectaneia thouarsii Roem. et Schult. ●☆

303062 Plectaneia hildebrandtii K. Schum. = Plectaneia thouarsii Roem. et Schult. ●☆

303063 Plectaneia inutilis Jum. et H. Perrier = Plectaneia thouarsii Roem. et Schult. ●☆

303064 Plectaneia inutilis Jum. et H. Perrier var. hirsuta Jum. =

Plectaneia thouarsii Roem. et Schult. ●☆

303065 Plectaneia isalensis Jum. = Plectaneia thouarsii Roem. et Schult. ●☆

303066 Plectaneia isalensis Jum. f. glomerata Jum. = Plectaneia thouarsii Roem. et Schult. ●☆

303067 Plectaneia lanceolata Pichon = Plectaneia thouarsii Roem. et Schult. ●☆

303068 Plectaneia longisepala Markgr.；长瓣编织夹竹桃●☆

303069 Plectaneia macrocarpa Jum. = Plectaneia thouarsii Roem. et Schult. ●☆

303070 Plectaneia microphylla Jum. et H. Perrier = Plectaneia thouarsii Roem. et Schult. ●☆

303071 Plectaneia pervillei K. Schum. = Plectaneia thouarsii Roem. et Schult. ●☆

303072 Plectaneia rhomboidalis Jum. et H. Perrier = Plectaneia thouarsii Roem. et Schult. ●☆

303073 Plectaneia stenophylla Jum.；窄叶编织夹竹桃●☆

303074 Plectaneia thouarsii Roem. et Schult.；小叶编织夹竹桃●☆

303075 Plectaneia thouarsii Roem. et Schult. var. macrocarpa（Jum.）Markgr. = Plectaneia thouarsii Roem. et Schult. ●☆

303076 Plectaneia volubilis Poir. = Plectaneia thouarsii Roem. et Schult. ●☆

303077 Plectanthera Mart. = Luxemburgia A. St. -Hil. ●☆

303078 Plectis O. F. Cook = Euterpe Mart.（保留属名）●☆

303079 Plectocephalus D. Don = Centaurea L.（保留属名）●■

303080 Plectocephalus D. Don(1830)；网苞菊属；Basketflower ■☆

303081 Plectocephalus abyssinicus Boiss. = Plectocephalus varians（A. Rich.）C. Jeffrey ex Cufod. ■☆

303082 Plectocephalus americanus（Nutt.）D. Don；美洲网苞菊（大矢车菊，美洲矢车菊）；American Basket Flower, American Basketflower, American Knapweed, Basket Flower American Starthistle, Basket-flower, Basketflower Centaurea, Powderpuff Thistle, Thornless Thistle ■☆

303083 Plectocephalus americanus（Nutt.）D. Don = Centaurea americana Nutt. ■☆

303084 Plectocephalus cyanoides Boiss. = Plectocephalus varians（A. Rich.）C. Jeffrey ex Cufod. ■☆

303085 Plectocephalus rothrockii（Greenm.）D. J. N. Hind；墨西哥网苞菊；Mexican Basketflower, Rothrock's Basketflower, Rothrock's Knapweed ■☆

303086 Plectocephalus varians（A. Rich.）C. Jeffrey ex Cufod.；网苞菊■☆

303087 Plectocomia Mart. et Blume(1830)；钩叶藤属（钩叶棕属,巨藤属,毛蕊桐属,毛藤属）；Giant Mountain Rattan, Hookleafvine, Plectocomia ●☆

303088 Plectocomia assamica Griff.；大钩叶藤；Assam Plectocomia, Big Hookleafvine ●

303089 Plectocomia elongata Mart. et Blume；长叶钩叶藤；Penang Rattan Palm ●☆

303090 Plectocomia griffithii Becc.；格瑞氏钩叶藤●

303091 Plectocomia himalayana Griff.；高地钩叶藤；Highland Hookleafvine, Himalaya Plectocomia, Himalayan Plectocomia ●◇

303092 Plectocomia kerrana Becc.；钩叶藤；Hookleafvine, Kerr Plectocomia ●

303093 Plectocomia kerrana Becc. = Plectocomia microstachys Burret ●◇

303094 Plectocomia microstachys Burret；小钩叶藤（钩叶藤）；Giant Mountain Rattan, Littlespike Plectocomia, Little-spiked Plectocomia,

Small Hookleafvine ●◇

303095 Plectocomia montana Hook. f. et Thomson = Plectocomia himalayana Griff. ●◇

303096 Plectocomiopsis Becc.(1893);拟钩叶藤属(编织藤属,假钩叶藤属,假毛蕊桐属,来茛藤属,类钩叶藤属,囊凸藤属,拟毛藤属);Plectocomiopsis ●☆

303097 Plectocomiopsis corneri Furtado;考奈拟钩叶藤;Corner Plectocomiopsis ●☆

303098 Plectocomiopsis dubius Becc.;可疑拟钩叶藤;Dubious Plectocomiopsis ●☆

303099 Plectocomiopsis geminiflorus (Griff.) Becc.;二花拟钩叶藤;Rotang Palm,Twoflower Plectocomiopsis ●☆

303100 Plectocomiopsis wrayi Becc.;小拟钩叶藤(沃瑞氏钩叶藤);Wray Plectocomiopsis ●☆

303101 Plectogyne Link. = Aspidistra Ker Gawl. ●■

303102 Plectogyne variegata Link = Aspidistra elatior Blume ●■

303103 Plectoma Raf. = Utricularia L. ■

303104 Plectoma montana Hook. f. et Thomson = Plectocomia himalayana Griff. ●◇

303105 Plectomirtha W. R. B. Oliv. = Pennantia J. R. Forst. et G. Forst. ●☆

303106 Plectopoma Hanst. = Achimenes Pers. (保留属名)■☆

303107 Plectopoma Hanst. = Gloxinia L'Hér. ☆

303108 Plectorrhiza Dockrill(1967);澳兰属■☆

303109 Plectorrhiza brevilabris (F. Muell.) Dockrill;澳兰■☆

303110 Plectorrhiza erecta (Fitzg.) Dockrill;直立澳兰■☆

303111 Plectorrhiza tridentata (Lindl.) Dockrill;三齿澳兰■☆

303112 Plectrachne Henrard(1929);圆丘草属;Spinifex ●☆

303113 Plectrachne mollis Lazarides;柔软圆丘草■☆

303114 Plectrachne rigidissima (Pilg.) C. E. Hubb.;坚挺圆丘草■☆

303115 Plectrachne schinzii Henrard;圆丘草■☆

303116 Plectranthastrum T. C. E. Fr. (1924);肖阿尔韦斯草属●☆

303117 Plectranthastrum T. C. E. Fr. = Alvesia Welw. (保留属名)●■☆

303118 Plectranthastrum clerodendroides T. C. E. Fr. = Alvesia clerodendroides (T. C. E. Fr.) B. Mathew ●☆

303119 Plectranthastrum cyclindricalyx B. Mathew;柱萼肖阿尔韦斯草●☆

303120 Plectranthastrum rosmarinifolium (Welw.) B. Mathew;肖阿尔韦斯草●☆

303121 Plectranthera Benth. et Hook. f. = Luxemburgia A. St. -Hil. ●☆

303122 Plectranthera Benth. et Hook. f. = Plectanthera Mart. ●☆

303123 Plectranthrastrum Willis = Alvesia Welw. (保留属名)●■☆

303124 Plectranthrastrum Willis = Plectranthastrum T. C. E. Fr. ●☆

303125 Plectranthus L'Hér. (1788)(保留属名);香茶属(香茶菜属,香茶树属,延命草属);Spurflower ●■

303126 Plectranthus acaulis Brummitt et Seyani;无茎香茶菜■☆

303127 Plectranthus adenanthus Diels = Isodon adenanthus (Diels) Kudo ■

303128 Plectranthus adenanthus Diels = Rabdosia adenantha (Diels) H. Hara ■

303129 Plectranthus adenoloma Hand. -Mazz. ;腺叶香茶菜;Common Rabdosia,Glandularleaf Rabdosia,Glandular-leaved Rabdosia ●

303130 Plectranthus adenoloma Hand. -Mazz. = Rabdosia adenoloma (Hand. -Mazz.) H. Hara ●

303131 Plectranthus adenophorus Gürke;腺梗香茶菜■☆

303132 Plectranthus aegyptiacus (Forssk.) C. Chr.;埃及香茶菜■●☆

303133 Plectranthus albidus Baker = Capitanopsis albida (Baker) Hedge ●☆

303134 Plectranthus alboviolaceus Gürke;浅堇色香茶菜■☆

303135 Plectranthus albus Gürke = Plectranthus laxiflorus Benth. ■☆

303136 Plectranthus allenii C. H. Wright = Orthosiphon allenii (C. H. Wright) Codd ●☆

303137 Plectranthus almamii A. Chev. = Plectranthus glandulosus Hook. f. ■☆

303138 Plectranthus ambiguus (Bolus) Codd;可疑香茶菜■☆

303139 Plectranthus amboinicus (Lour.) Spreng. ;墨西哥香茶(安汶香茶菜);Mexican Mint ■☆

303140 Plectranthus amethystoides Benth. = Isodon amethystoides (Benth.) H. Hara ■

303141 Plectranthus amethystoides Benth. = Rabdosia amethystoides (Benth.) H. Hara ■

303142 Plectranthus amplexicaulis Hedge;抱茎香茶菜■☆

303143 Plectranthus andongensis (Hiern) Baker;安东香茶菜■☆

303144 Plectranthus angulatus Hedge;三棱香茶菜■☆

303145 Plectranthus angustifolius Dunn = Isodon angustifolius (Dunn) Kudo ■

303146 Plectranthus angustifolius Dunn = Isodon nervosus (Hemsl.) Kudo ■

303147 Plectranthus angustifolius Dunn = Rabdosia angustifolia (Dunn) H. Hara ■

303148 Plectranthus angustifolius Dunn = Rabdosia nervosa (Hemsl.) C. Y. Wu et H. W. Li ■

303149 Plectranthus antongilicus Hedge;安通吉尔香茶菜■☆

303150 Plectranthus argentatus S. T. Blake;银香茶;Silver Plectranthus ●☆

303151 Plectranthus argentifolius Ryding;银叶香茶菜■☆

303152 Plectranthus aromaticus (Benth.) Roxb. = Plectranthus amboinicus (Lour.) Spreng. ■☆

303153 Plectranthus arthropodus Briq. = Plectranthus fruticosus L'Hér. ●☆

303154 Plectranthus assurgens (Baker) J. K. Morton;上升香茶菜■☆

303155 Plectranthus atroviolaceus Hedge;暗堇色香茶菜●☆

303156 Plectranthus auriculatus Robyns et Lebrun;耳形香茶菜■☆

303157 Plectranthus aurifer Dinter ex Launert = Plectranthus hereroensis Engl. ■☆

303158 Plectranthus australis R. Br. ;瑞典香茶;Swedish Ivy ●☆

303159 Plectranthus axillariflorus Honda = Isodon shikokianus (Makino) H. Hara var. intermedius (Kudo) Murata ■☆

303160 Plectranthus barbatus Andr. = Coleus forskohlii (Willd.) Briq. ■

303161 Plectranthus barbatus Andréws = Coleus forskohlii (Willd.) Briq. ■

303162 Plectranthus barbatus Andréws = Plectranthus pseudobarbatus J. K. Morton ■☆

303163 Plectranthus barbatus Andréws var. grandis (L. H. Cramer) Lukhoba et A. J. Paton;大毛香茶■☆

303164 Plectranthus baumii Gürke = Holostylon baumii (Gürke) G. Taylor ■☆

303165 Plectranthus behrii Compton = Plectranthus fruticosus L'Hér. ●☆

303166 Plectranthus bifidocalyx Dunn = Isodon macrocalyx (Dunn) Kudo ■

303167 Plectranthus biflorus Baker = Plectranthus tetragonus Gürke ■☆

303168 Plectranthus bipinnatus A. J. Paton;双羽香茶■☆

303169 Plectranthus bojeri (Benth.) Hedge;博耶尔香茶■☆

303170 Plectranthus bolusii T. Cooke = Orthosiphon suffrutescens (Thonn.) J. K. Morton ●☆

303171 Plectranthus bongensis Baker = Solenostemon latifolius (Hochst. ex Benth.) J. K. Morton ■☆

303172 Plectranthus bosseri Hedge = Plectranthus sylvestris Gürke ■☆

303173　Plectranthus brandisii Prain ＝ Isodon walkeri（Arn.）H. Hara ■

303174　Plectranthus brevicaulis（Baker）Hedge；短茎香茶■☆

303175　Plectranthus brevifolius Hand.-Mazz. ＝ Isodon brevifolius（Hand.-Mazz.）H. W. Li ●

303176　Plectranthus brevifolius Hand.-Mazz. ＝ Rabdosia brevifolia（Hand.-Mazz.）H. Hara ●

303177　Plectranthus brevimentum T. J. Edwards；短香茶■☆

303178　Plectranthus brevipes Baker；短梗香茶■☆

303179　Plectranthus buchananii Baker；布坎南香茶■☆

303180　Plectranthus buergeri Miq. ＝ Isodon japonicus（Burm. f.）H. Hara ■

303181　Plectranthus buergeri Miq. ＝ Rabdosia japonica（Burm. f.）H. Hara ■

303182　Plectranthus bullatus Robyns et Lebrun ＝ Isodon ramosissimus（Hook. f.）Codd ●☆

303183　Plectranthus bulleyanus Diels ＝ Isodon bulleyanus（Diels）Kudo ●

303184　Plectranthus bulleyanus Diels ＝ Rabdosia bulleyana（Diels）H. Hara var. foliosa C. Y. Wu ●

303185　Plectranthus bulleyanus Diels ＝ Rabdosia bulleyana（Diels）H. Hara ●

303186　Plectranthus burnati Briq. ＝ Plectranthus lanceolatus Bojer ex Benth. ●☆

303187　Plectranthus caespitosus Lukhoba et A. J. Paton；丛生香茶■☆

303188　Plectranthus calcaratus Hemsl. ＝ Ceratanthus calcaratus（Hemsl.）G. Taylor ■

303189　Plectranthus calcicola Hand.-Mazz. ＝ Isodon calcicola（Hand.-Mazz.）H. Hara ■

303190　Plectranthus calcicola Hand.-Mazz. ＝ Rabdosia calcicola（Hand.-Mazz.）H. Hara ■

303191　Plectranthus calcicola Hand.-Mazz. var. subcalva Hand.-Mazz. ＝ Rabdosia calcicola（Hand.-Mazz.）H. Hara var. subcalva（Hand.-Mazz.）C. Y. Wu et H. W. Li ■

303192　Plectranthus calcicola Hand.-Mazz. var. subcalvus Hand.-Mazz. ＝ Isodon calcicola（Hand.-Mazz.）H. Hara var. subcalva（Hand.-Mazz.）H. W. Li ■

303193　Plectranthus calcicola Hand.-Mazz. var. subcalvus Hand.-Mazz. ＝ Rabdosia calcicola（Hand.-Mazz.）H. Hara var. subcalva（Hand.-Mazz.）C. Y. Wu et H. W. Li ■

303194　Plectranthus calycinus Benth.；尊状香茶●■

303195　Plectranthus calycinus Benth. ＝ Rabdosiella calycina（Benth.）Codd ●■

303196　Plectranthus calycinus Benth. var. pachystachyus（Briq.）T. Cooke ＝ Rabdosiella calycina（Benth.）Codd ●■

303197　Plectranthus candelabriformis Launert；烛台香茶■☆

303198　Plectranthus canescens Benth.；灰白香茶■☆

303199　Plectranthus canescens Benth. var. membranacea Scott-Elliot ＝ Plectranthus membranaceus（Scott-Elliot）Hedge ■☆

303200　Plectranthus caninus Roth；印度厚叶香茶■☆

303201　Plectranthus capuronii Hedge；凯普伦香茶●☆

303202　Plectranthus cardiaphyllus Hemsl. ＝ Heterolamium debile（Hemsl.）C. Y. Wu var. cardiophyllum（Hemsl.）C. Y. Wu ■

303203　Plectranthus carnosifolius Hemsl. ＝ Coleus carnosifolius（Hemsl.）Dunn ■

303204　Plectranthus carnosus Sm. ＝ Anisochilus carnosus（L.）Wall. ●

303205　Plectranthus cataractarum B. J. Pollard；瀑布群高原香茶■☆

303206　Plectranthus caudatus S. Moore；尾状香茶■☆

303207　Plectranthus cavaleriei H. Lév. ＝ Isodon coetsus（Buch.-Ham. ex D. Don）H. Hara var. cavaleriei（H. Lév.）H. W. Li ●■

303208　Plectranthus cavaleriei H. Lév. ＝ Rabdosia coetsa（Buch.-Ham. ex D. Don）H. Hara var. cavaleriei（H. Lév.）C. Y. Wu et H. W. Li ●■

303209　Plectranthus charianthus Briq. ＝ Plectranthus fruticosus L'Hér. ●☆

303210　Plectranthus chenmui Y. Z. Sun ex C. H. Hu ＝ Isodon phyllopodus（Diels）Kudo ●■

303211　Plectranthus chenmui Y. Z. Sun ex C. H. Hu ＝ Rabdosia phyllopoda（Diels）H. Hara ●■

303212　Plectranthus chevalieri（Briq.）B. J. Pollard et A. J. Paton；舍瓦利耶香茶■☆

303213　Plectranthus chienii Y. Z. Sun ex C. H. Hu ＝ Isodon hispidus（Benth.）Murata ■

303214　Plectranthus chienii Y. Z. Sun ex C. H. Hu ＝ Rabdosia hispida（Benth.）H. Hara ■

303215　Plectranthus chimanimanensis S. Moore；奇马尼曼香茶■☆

303216　Plectranthus ciliatus E. Mey. ex Benth.；缘毛香茶■☆

303217　Plectranthus clementiae Hedge；克莱门特香茶●☆

303218　Plectranthus coerulescens（Gürke）R. H. Willemse ＝ Plectranthus barbatus Andréws ■☆

303219　Plectranthus coeruleus（Gürke）Agnew；青蓝香茶■☆

303220　Plectranthus coetsa Buch.-Ham. ex D. Don ＝ Isodon coetsus（Buch.-Ham. ex D. Don）Kudo ●■

303221　Plectranthus coetsa Buch.-Ham. ex D. Don ＝ Rabdosia coetsa（Buch.-Ham. ex D. Don）H. Hara ●■

303222　Plectranthus coetsa Buch.-Ham. ex D. Don var. cavaleriei（H. Lév.）McKean ＝ Rabdosia coetsa（Buch.-Ham. ex D. Don）H. Hara var. cavaleriei（H. Lév.）C. Y. Wu et H. W. Li ●■

303223　Plectranthus coetsa Buch.-Ham. ex D. Don var. cavaleriei（H. Lév.）McKean ＝ Isodon coetsus（Buch.-Ham. ex D. Don）H. Hara var. cavaleriei（H. Lév.）H. W. Li ●■

303224　Plectranthus colleoides；肥根兰香茶；Spurflower ■☆

303225　Plectranthus coloratus D. Don ＝ Geniosporum coloratum（D. Don）Kuntze ●

303226　Plectranthus coloratus E. Mey. ex Benth. ＝ Plectranthus ambiguus（Bolus）Codd ■☆

303227　Plectranthus comosus Sims；簇毛香茶■☆

303228　Plectranthus congestus R. Br.；密生香茶菜■☆

303229　Plectranthus conglomeratus（T. C. E. Fr.）Hutch. et Dandy；聚集香茶■☆

303230　Plectranthus cooperi T. Cooke ＝ Plectranthus dolichopodus Briq. ■☆

303231　Plectranthus copinii Cornu ＝ Solenostemon rotundifolius（Poir.）J. K. Morton ■☆

303232　Plectranthus cordifolius D. Don ＝ Plectranthus mollis Spreng. ■☆

303233　Plectranthus crassifolius Vahl ＝ Plectranthus aegyptiacus（Forssk.）C. Chr. ■●☆

303234　Plectranthus crassus N. E. Br.；粗香茶菜■☆

303235　Plectranthus crenatus Gürke；圆齿香茶菜■☆

303236　Plectranthus cuneatus（Baker f.）Ryding；楔形香茶菜■☆

303237　Plectranthus cyaneus Gürke；蓝色香茶菜■☆

303238　Plectranthus cylindraceus Hochst. ex Benth.；柱形香茶菜■☆

303239　Plectranthus cylindrostachys Robyns et Lebrun；柱穗香茶菜■☆

303240　Plectranthus cymosus Baker ＝ Plectranthus bojeri（Benth.）Hedge ■☆

303241　Plectranthus daitonensis Hayata ＝ Isodon amethystoides

（Benth.）H. Hara ■

303242 Plectranthus daitonensis Hayata = Rabdosia amethystoides （Benth.）H. Hara ■

303243 Plectranthus daviesii（E. A. Bruce）B. Mathew;戴维斯香茶■☆

303244 Plectranthus dawoensis Hand.-Mazz. = Isodon dawoensis （Hand.-Mazz.）H. Hara ●

303245 Plectranthus dawoensis Hand.-Mazz. = Rabdosia dawoensis （Hand.-Mazz.）H. Hara ●

303246 Plectranthus decaryi Hedge;德卡里香茶●☆

303247 Plectranthus decumbens Hook. f.;外倾香茶●☆

303248 Plectranthus decurrens（Gürke）J. K. Morton;下延香茶■☆

303249 Plectranthus defoliatus Hochst. ex Benth. = Isodictyophorus defoliatus（Hochst. ex Benth.）Agnew ■☆

303250 Plectranthus delicatissimus Hedge;姣美香茶■☆

303251 Plectranthus densiflorus T. Cooke = Plectranthus cylindraceus Hochst. ex Benth. ■☆

303252 Plectranthus densus N. E. Br.;密集香茶菜■☆

303253 Plectranthus denudatus A. Chev. ex Hutch. et Dalziel = Englerastrum nigericum Alston ■☆

303254 Plectranthus dichromophyllus Diels = Isodon rubescens （Hemsl.）H. Hara ●

303255 Plectranthus dichromophyllus Diels = Rabdosia rubescens （Hemsl.）H. Hara ●

303256 Plectranthus dinteri Briq. ;丁特香茶■☆

303257 Plectranthus discolor Dunn = Isodon parvifolius（Batalin）H. Hara ●

303258 Plectranthus discolor Dunn = Rabdosia parvifolia（Batalin）H. Hara ●

303259 Plectranthus dissectus Brenan;深裂香茶菜■☆

303260 Plectranthus dissitiflorus（Gürke）J. K. Morton;稀花香茶菜■☆

303261 Plectranthus dolichopodus Briq. ;长足香茶菜■☆

303262 Plectranthus dolomiticus Codd;多罗米蒂香茶■☆

303263 Plectranthus draconis Briq. = Plectranthus hadiensis（Forssk.）Schweinf. ex Spreng. var. woodii（Gürke）Codd ■☆

303264 Plectranthus dregei Codd = Plectranthus ambiguus（Bolus）Codd ■☆

303265 Plectranthus drogotschiensis Hand.-Mazz. = Isodon barbeyanus （H. Lév.）H. W. Li ■

303266 Plectranthus drogotschiensis Hand.-Mazz. = Rabdosia drogotschiensis（Hand.-Mazz.）H. Hara ●

303267 Plectranthus drosocarpus Hand.-Mazz. = Isodon macrocalyx （Dunn）Kudo ■

303268 Plectranthus drosocarpus Hand.-Mazz. = Rabdosia macrocalyx （Dunn）H. Hara ■

303269 Plectranthus dubius Spreng. = Anisochilus carnosus（L.）Wall. ●

303270 Plectranthus dubius Vahl ex Benth. = Isodon amethystoides （Benth.）H. Hara ■

303271 Plectranthus dubius Vahl ex Benth. = Rabdosia amethystoides （Benth.）H. Hara ■

303272 Plectranthus ecklonii Benth. ;尖叶香茶菜●☆

303273 Plectranthus edulis（Vatke）Agnew;可食香茶菜■☆

303274 Plectranthus effusus（Maxim.）Honda = Isodon effusus （Maxim.）H. Hara ■☆

303275 Plectranthus effusus（Maxim.）Honda = Rabdosia effusa （Maxim.）H. Hara ■☆

303276 Plectranthus elegans Britten;雅致香茶菜■☆

303277 Plectranthus elegantulus Briq. ;稍雅致香茶菜■☆

303278 Plectranthus ellenbeckii Gürke = Plectranthus longipes Baker ■☆

303279 Plectranthus ellipticus Hedge;椭圆香茶菜●☆

303280 Plectranthus eminii Gürke;埃明香茶菜●☆

303281 Plectranthus emirnensis（Baker）Hedge;埃米香茶菜■☆

303282 Plectranthus enanderianus Hand.-Mazz. = Isodon enanderianus （Hand.-Mazz.）H. W. Li ●

303283 Plectranthus enanderianus Hand.-Mazz. = Rabdosia enanderiana （Hand.-Mazz.）H. Hara ●

303284 Plectranthus epilithicus B. J. Pollard;热非香茶菜■☆

303285 Plectranthus equisetiformis（E. A. Bruce）Launert;木贼香茶菜■☆

303286 Plectranthus eriocalyx Dunn = Isodon eriocalyx（Dunn）Kudo ●■

303287 Plectranthus eriocalyx Dunn = Rabdosia eriocalyx（Dunn）H. Hara ●■

303288 Plectranthus erlangeri Gürke;厄兰格香茶菜■☆

303289 Plectranthus esculentus N. E. Br.;食用香茶菜;Hausa Potato ■☆

303290 Plectranthus esquirolii H. Lév. = Isodon longitubus（Miq.）Kudo ■

303291 Plectranthus esquirolii H. Lév. = Isodon lophanthoides（Buch.-Ham. ex D. Don）H. Hara ■

303292 Plectranthus esquirolii H. Lév. = Rabdosia longituba（Miq.）H. Hara ■

303293 Plectranthus excisoides Y. Z. Sun ex C. H. Hu = Isodon excisoides（Y. Z. Sun ex C. H. Hu）H. Hara ■

303294 Plectranthus excisoides Y. Z. Sun ex C. H. Hu = Rabdosia excisoides（Y. Z. Sun ex C. H. Hu）C. Y. Wu et H. W. Li ■

303295 Plectranthus excisus Maxim. = Isodon excisus（Maxim.）Kudo ■

303296 Plectranthus excisus Maxim. = Isodon henryi（Hemsl.）Kudo ■

303297 Plectranthus excisus Maxim. = Isodon macrophyllus（Migo）H. Hara ●■

303298 Plectranthus excisus Maxim. = Rabdosia excisa（Maxim.）H. Hara ■

303299 Plectranthus excisus Maxim. = Rabdosia henryi（Hemsl.）H. Hara ■

303300 Plectranthus excisus Maxim. = Rabdosia macrophylla（Migo）C. Y. Wu et H. W. Li ●■

303301 Plectranthus excisus Maxim. var. racemosus（Hemsl.）Dunn = Isodon racemosus（Hemsl.）H. W. Li ■

303302 Plectranthus excisus Maxim. var. racemosus（Hemsl.）Dunn = Rabdosia racemosa（Hemsl.）H. Hara ■

303303 Plectranthus excisus Maxim. var. shikokianus ? = Isodon shikokianus（Makino）H. Hara ■☆

303304 Plectranthus fangii Y. Z. Sun = Isodon longitubus（Miq.）Kudo ■

303305 Plectranthus fangii Y. Z. Sun = Isodon lophanthoides（Buch.-Ham. ex D. Don）H. Hara ■

303306 Plectranthus fangii Y. Z. Sun = Rabdosia longituba（Miq.）H. Hara ■

303307 Plectranthus fimbriatus（Lebrun et L. Touss.）Troupin et Ayob. ;流苏香茶菜■☆

303308 Plectranthus fischeri Gürke;菲舍尔香茶菜■☆

303309 Plectranthus flaccidus（Vatke）Gürke;柔软香茶菜■☆

303310 Plectranthus flavidus Hand.-Mazz. = Isodon flavidus（Hand.-Mazz.）H. Hara ■

303311 Plectranthus flavidus Hand.-Mazz. = Rabdosia flavida（Hand.-Mazz.）H. Hara ■

303312 Plectranthus floribundus N. E. Br. = Plectranthus esculentus N. E. Br. ■☆

303313 Plectranthus floribundus N. E. Br. var. longipes ? = Plectranthus esculentus N. E. Br. ■☆

303314　Plectranthus forrestii Diels ＝ Isodon forrestii（Diels）Kudo ■

303315　Plectranthus forrestii Diels ＝ Rabdosia forrestii（Diels）H. Hara ■

303316　Plectranthus forskalaei Willd. ＝ Coleus barbatus（Andréws）Benth. ■

303317　Plectranthus forskalaei Willd. ＝ Coleus forskohlii（Willd.）Briq. ■

303318　Plectranthus forsteri Benth.;福斯特香茶菜;Blue Spur Flower ■☆

303319　Plectranthus forsteri Benth. 'Marginatus';具边福斯特香茶菜; Spur Flower ■☆

303320　Plectranthus forsythii Hedge;福赛斯香茶菜●☆

303321　Plectranthus fragilis Baker;脆香茶菜■☆

303322　Plectranthus fragrans Lebrun et L. Touss. ;甜香茶菜■☆

303323　Plectranthus fraternus T. C. E. Fr. ＝ Plectranthus laxiflorus Benth. ■☆

303324　Plectranthus fruticosus L'Hér. ;灌木香茶菜;Cockspur Flower ●☆

303325　Plectranthus fruticosus Wight ex Hook. f. ＝ Plectranthus fruticosus L'Hér. ●☆

303326　Plectranthus galpinii Schltr. ＝ Plectranthus fruticosus L'Hér. ●☆

303327　Plectranthus garckeanus（Vatke）J. K. Morton;加尔凯香茶菜●☆

303328　Plectranthus gerardianus Benth. ＝ Isodon lophanthoides（Buch. -Ham. ex D. Don）H. Hara var. gerardiana（Benth.）H. Hara ■

303329　Plectranthus gerardianus Benth. ＝ Rabdosia lophanthoides（Buch. -Ham. ex D. Don）H. Hara var. graciliflora（Benth.）H. Hara ■

303330　Plectranthus gerardianus Benth. var. graciliflorus（Benth.）Hook. f. ＝ Isodon lophanthoides（Buch. -Ham. ex D. Don）H. Hara var. graciliflora（Benth.）H. Hara ■

303331　Plectranthus gerardianus Benth. var. graciliflorus（Benth.）Hook. f. ＝ Rabdosia lophanthoides（Buch. -Ham. ex D. Don）H. Hara var. graciliflora（Benth.）H. Hara ■

303332　Plectranthus gesneroides J. Sinclair ＝ Isodon gesneroides（J. Sinclair）H. Hara ■

303333　Plectranthus gesneroides J. Sinclair ＝ Rabdosia gesneroides（J. Sinclair）H. Hara ■

303334　Plectranthus gibbosus Hedge;浅囊香茶菜●☆

303335　Plectranthus gillettii J. K. Morton;吉莱特香茶菜●☆

303336　Plectranthus glandulosus Britten ＝ Solenostemon autrani（Briq.）J. K. Morton ■☆

303337　Plectranthus glandulosus Hook. f. ;具腺香茶菜■☆

303338　Plectranthus glaucocalyx Maxim. ＝ Isodon japonicus（Burm. f.）H. Hara var. glaucocalyx（Maxim.）H. W. Li ■

303339　Plectranthus glaucocalyx Maxim. ＝ Rabdosia japonica（Burm. f.）H. Hara var. glaucocalyx（Maxim.）H. Hara ■

303340　Plectranthus glaucocalyx Maxim. var. japonicus（Burm. f.）Maxim. ＝ Isodon japonicus（Burm. f.）H. Hara ■

303341　Plectranthus glaucocalyx Maxim. var. japonicus Maxim. ＝ Isodon japonicus（Burm. f.）H. Hara ■

303342　Plectranthus glaucocalyx Maxim. var. japonicus Maxim. ＝ Rabdosia japonica（Burm. f.）H. Hara ■

303343　Plectranthus globosus Ryding;球形香茶菜■☆

303344　Plectranthus glomeratus R. A. Dyer ＝ Plectranthus cylindraceus Hochst. ex Benth. ■☆

303345　Plectranthus goetzei Gürke;格兹香茶菜●☆

303346　Plectranthus graciliflorus Benth. ＝ Isodon lophanthoides（Buch. -Ham. ex D. Don）H. Hara var. graciliflora（Benth.）H.

Hara ■

303347　Plectranthus graciliflorus Benth. ＝ Rabdosia lophanthoides（Buch. -Ham. ex D. Don）H. Hara var. graciliflora（Benth.）H. Hara ■

303348　Plectranthus gracilis Suess. ;纤细香茶菜■☆

303349　Plectranthus gracillimus（T. C. E. Fr.）Hutch. et Dandy;细长香茶菜■☆

303350　Plectranthus grallatus Briq. ;高腿香茶菜■☆

303351　Plectranthus grandibracteatus Hedge;大苞香茶菜●☆

303352　Plectranthus grandicalyx（E. A. Bruce）J. K. Morton ＝ Coleus grandicalyx E. A. Bruce ■☆

303353　Plectranthus grandidentatus Gürke;大齿香茶菜■☆

303354　Plectranthus grandifolius Hand. -Mazz. ＝ Isodon grandifolius（Hand. -Mazz.）H. Hara ●

303355　Plectranthus grandifolius Hand. -Mazz. ＝ Rabdosia grandifolia（Hand. -Mazz.）H. Hara ●

303356　Plectranthus grandis（L. H. Cramer）R. H. Willemse ＝ Plectranthus barbatus Andréws var. grandis（L. H. Cramer）Lukhoba et A. J. Paton ■☆

303357　Plectranthus grosseserratus Dunn ＝ Isodon grosseserratus（Dunn）Kudo ■

303358　Plectranthus grosseserratus Dunn ＝ Rabdosia grosseserrata（Dunn）H. Hara ■

303359　Plectranthus gurkei Briq. ＝ Neohyptis paniculata（Baker）J. K. Morton ●☆

303360　Plectranthus hadiensis（Forssk.）Schweinf. ex Spreng. var. tomentosus（Benth.）Codd;绒毛香茶■☆

303361　Plectranthus hadiensis（Forssk.）Schweinf. ex Spreng. var. woodii（Gürke）Codd;伍得香茶■☆

303362　Plectranthus hadiensis（Forssk.）Spreng. ＝ Ocimum forsskaolii Benth. ■☆

303363　Plectranthus hallii J. K. Morton;霍尔香茶■☆

303364　Plectranthus hanceiformis H. Lév. ＝ Teucrium bidentatum Hemsl. ■

303365　Plectranthus hararensis Gürke ＝ Plectranthus hadiensis（Forssk.）Schweinf. ex Spreng. ■☆

303366　Plectranthus harrisii J. K. Morton;哈里斯香茶■☆

303367　Plectranthus henryi Hemsl. ＝ Isodon henryi（Hemsl.）Kudo ■

303368　Plectranthus henryi Hemsl. ＝ Rabdosia henryi（Hemsl.）H. Hara ■

303369　Plectranthus herbaceus（Hiern）Briq. ;草色香茶●☆

303370　Plectranthus herbaceus Briq. ＝ Coleus herbaceus（Briq.）G. Taylor ■☆

303371　Plectranthus herbaceus Schweinf. ＝ Plectranthus longipes Baker ■☆

303372　Plectranthus hereroensis Engl. ;赫雷罗香茶■☆

303373　Plectranthus hexaphyllus Baker;六叶香茶■☆

303374　Plectranthus hilliardiae Codd;希利亚德香茶■☆

303375　Plectranthus hilliardiae Codd subsp. australis Van Jaarsv. et A. E. van Wyk;南方希利亚德香茶■☆

303376　Plectranthus hirsutus Hedge;粗毛香茶■☆

303377　Plectranthus hirtellus Hand. -Mazz. ＝ Isodon hirtellus（Hand. -Mazz.）H. Hara ●

303378　Plectranthus hirtellus Hand. -Mazz. ＝ Rabdosia hirtella（Hand. -Mazz.）H. Hara ●

303379　Plectranthus hirtus Benth. ＝ Plectranthus madagascariensis（Pers.）Benth. ■☆

303380　Plectranthus hispidus Benth. ＝ Isodon hispidus（Benth.）

Murata ■

303381　Plectranthus hispidus Benth. = Rabdosia hispida（Benth.）H. Hara ■

303382　Plectranthus hjalmari（T. C. E. Fr.）Hutch. et Dandy；亚尔马香茶■☆

303383　Plectranthus hockii De Wild. = Holostylon katangense（De Wild.）Robyns et Lebrun ■☆

303384　Plectranthus holstii Gürke；霍尔斯特香茶■☆

303385　Plectranthus horridus（Hiern）Baker；多刺香茶●☆

303386　Plectranthus hoslundioides Baker = Isodon ramosissimus（Hook. f.）Codd ●☆

303387　Plectranthus hosseusii Muschl. = Isodon ternifolius（D. Don）Kudo ●■

303388　Plectranthus hosseusii Muschl. = Rabdosia ternifolia（D. Don）H. Hara ●■

303389　Plectranthus humbertii Hedge；亨伯特香茶■☆

303390　Plectranthus hurtellus Hand. -Mazz. = Isodon hirtellus（Hand. -Mazz.）H. Hara ●

303391　Plectranthus hurtellus Hand. -Mazz. = Rabdosia hirtella（Hand. -Mazz.）H. Hara ●

303392　Plectranthus hylophilus Gürke = Plectranthus laxiflorus Benth. ■☆

303393　Plectranthus igniarioides Ryding；拟火红香茶■☆

303394　Plectranthus igniarius（Schweinf.）Agnew；火红香茶■☆

303395　Plectranthus incanus Link = Plectranthus mollis Spreng. ■☆

303396　Plectranthus inconspicuus Miq. = Isodon inflexus（Thunb.）Kudo ■

303397　Plectranthus inconspicuus Miq. = Isodon trichocarpus（Maxim.）Kudo ■☆

303398　Plectranthus inconspicuus Miq. = Rabdosia inflexa（Thunb.）H. Hara ■

303399　Plectranthus inflexus（Thunb.）Vahl ex Benth. = Isodon inflexus（Thunb.）Kudo ■

303400　Plectranthus inflexus（Thunb.）Vahl ex Benth. f. macrophyllus（Maxim.）Kitag. = Isodon inflexus（Thunb.）Kudo ■

303401　Plectranthus inflexus（Thunb.）Vahl ex Benth. f. subglabrus？= Isodon inflexus（Thunb.）Kudo ■

303402　Plectranthus inflexus（Thunb.）Vahl ex Benth. f. vilior？= Isodon inflexus（Thunb.）Kudo ■

303403　Plectranthus inflexus（Thunb.）Vahl ex Benth. var. canescens Nakai = Isodon inflexus（Thunb.）Kudo ■

303404　Plectranthus inflexus（Thunb.）Vahl ex Benth. var. macrophyllus Maxim. = Isodon inflexus（Thunb.）Kudo ■

303405　Plectranthus inflexus（Thunb.）Vahl ex Benth. var. microphyllus？= Isodon inflexus（Thunb.）Kudo ■

303406　Plectranthus inflexus（Thunb.）Vahl ex Benth. var. verticillatus？= Isodon inflexus（Thunb.）Kudo ■

303407　Plectranthus inflexus Thunb. = Rabdosia inflexa（Thunb.）H. Hara ■

303408　Plectranthus insignis Hook. f. ；显著香茶■☆

303409　Plectranthus insolitus C. H. Wright；异常香茶■☆

303410　Plectranthus irroratus Forrest ex Diels = Isodon irroratus（Forrest ex Diels）Kudo ●

303411　Plectranthus irroratus Forrest ex Diels = Rabdosia irrorata（Forrest ex Diels）H. Hara ●

303412　Plectranthus japonicus（Burm. f.）Koidz. = Isodon japonicus（Burm. f.）H. Hara ■

303413　Plectranthus japonicus（Burm. f.）Koidz. var. glaucocalyx（Maxim.）Koidz. = Isodon japonicus（Burm. f.）H. Hara var. glaucocalyx（Maxim.）H. Hara ■

303414　Plectranthus japonicus（Burm. f.）Koidz. var. glaucocalyx（Maxim.）Koidz. = Rabdosia japonica（Burm. f.）H. Hara var. glaucocalyx（Maxim.）H. Hara ■

303415　Plectranthus japonicus Burm. f. = Rabdosia japonica（Burm. f.）H. Hara ■

303416　Plectranthus johnstonii Baker = Plectranthus laxiflorus Benth. ■☆

303417　Plectranthus kameba（Okuyama ex Ohwi）Ohwi var. hakusanensis（Kudo）Ohwi = Isodon umbrosus（Maxim.）H. Hara var. hakusanensis（Kudo）K. Asano ■☆

303418　Plectranthus kameba（Okuyama ex Ohwi）Ohwi var. latifolius（Okuyama）Ohwi = Isodon umbrosus（Maxim.）H. Hara var. latifolius Okuyama ■☆

303419　Plectranthus kameba（Okuyama）Ohwi = Isodon umbrosus（Maxim.）H. Hara var. leucanthus（Murai）K. Asano f. kameba（Okuyama ex Ohwi）K. Asano ■☆

303420　Plectranthus kameba（Okuyama）Ohwi var. excisinflexus（Nakai）Ohwi = Isodon umbrosus（Maxim.）H. Hara var. excisinflexus（Nakai）K. Asano ■☆

303421　Plectranthus kamerunensis Gürke；喀麦隆香茶■☆

303422　Plectranthus kapatensis（R. E. Fr.）J. K. Morton；卡帕特香茶■☆

303423　Plectranthus kassneri（T. C. E. Fr.）Hutch. et Dandy；卡斯纳香茶■☆

303424　Plectranthus katangensis De Wild. = Holostylon katangense（De Wild.）Robyns et Lebrun ■☆

303425　Plectranthus kilimandschari（Gürke）C. A. Maass；基利香茶■☆

303426　Plectranthus kivuensis（Lebrun et L. Touss.）R. H. Willemse；基伍香茶■☆

303427　Plectranthus kondowensis Baker = Plectranthus laxiflorus Benth. ■☆

303428　Plectranthus koualensis（A. Chev. ex Hutch. et Dalziel）B. J. Pollard et A. J. Paton；夸拉香茶■☆

303429　Plectranthus krookii Gürke ex Zahlbr. = Plectranthus grallatus Briq. ■☆

303430　Plectranthus kuntzeanus Domin = Plectranthus strigosus Benth. ■☆

303431　Plectranthus kuntzei Gürke = Plectranthus petiolaris E. Mey. ex Benth. ■☆

303432　Plectranthus labordei（Vaniot）Diels = Elsholtzia rugulosa Hemsl. ●■

303433　Plectranthus lactiflorus（Vatke）Agnew；乳白花香茶菜■☆

303434　Plectranthus lanceolatus Bojer ex Benth. ；披针形香茶菜●☆

303435　Plectranthus lanceus Nakai = Isodon shikokianus（Makino）H. Hara var. intermedius（Kudo）Murata ■☆

303436　Plectranthus lanuginosus（Hochst. ex Benth.）Agnew；索马里香茶菜■☆

303437　Plectranthus lasianthus（Gürke）Vollesen；毛花香茶菜■☆

303438　Plectranthus lasiocarpus Hayata = Isodon serrus（Maxim.）Kudo ■

303439　Plectranthus lasiocarpus Hayata = Rabdosia serra（Maxim.）H. Hara ■

303440　Plectranthus lastii Baker；拉斯特香茶菜■☆

303441　Plectranthus laurifolius Hedge；月桂叶香茶●☆

303442　Plectranthus lavanduloides Baker = Plectranthus lanceolatus Bojer ex Benth. ●☆

303443　Plectranthus laxiflorus Benth. ；疏花香茶菜■☆

303444　Plectranthus laxiflorus Benth. var. genuinus Briq. = Plectranthus laxiflorus Benth. ■☆

303445 Plectranthus laxiflorus Benth. var. stenodontus Briq. = Plectranthus laxiflorus Benth. ■☆

303446 Plectranthus leptobotrys Diels = Isodon coetsus（Buch. -Ham. ex D. Don）Kudo ●■

303447 Plectranthus leptobotrys Diels = Rabdosia coetsa（Buch. -Ham. ex D. Don）H. Hara ●■

303448 Plectranthus leucanthus Diels = Isodon phyllopodus（Diels）Kudo ●■

303449 Plectranthus leucanthus Diels = Rabdosia phyllopoda（Diels）H. Hara ●■

303450 Plectranthus leucophyllus Dunn = Isodon leucophyllus（Dunn）Kudo ●

303451 Plectranthus leucophyllus Dunn = Rabdosia leucophylla（Dunn）H. Hara ●

303452 Plectranthus leviculus N. E. Br. ;光滑香茶菜■☆

303453 Plectranthus lilacinus Gürke;紫丁香香茶■☆

303454 Plectranthus lindblomii T. C. E. Fr. ;林德布卢姆香茶菜■☆

303455 Plectranthus linearifolius（J. K. Morton）B. J. Pollard et A. J. Paton;线叶香茶菜■☆

303456 Plectranthus linearis Hedge;线状香茶菜■☆

303457 Plectranthus longiflorus Benth. ;长花香茶菜■☆

303458 Plectranthus longipes Baker;长柄香茶菜■☆

303459 Plectranthus longipetiolatus Hedge;长梗香茶菜●☆

303460 Plectranthus longitubus Miq. = Isodon longitubus（Miq.）Kudo ■

303461 Plectranthus longitubus Miq. = Rabdosia longituba（Miq.）H. Hara ■

303462 Plectranthus longitubus Miq. var. contractus Maxim. = Isodon longitubus（Miq.）Kudo ■

303463 Plectranthus longitubus Miq. var. contractus Maxim. = Rabdosia longituba（Miq.）H. Hara ■

303464 Plectranthus longitubus Miq. var. effusus Maxim. = Isodon effusus（Maxim.）H. Hara ■☆

303465 Plectranthus lophanthoides Buch. -Ham. ex D. Don var. graciliflora Benth. = Rabdosia lophanthoides（Buch. -Ham. ex D. Don）H. Hara var. graciliflora（Benth.）H. Hara ■

303466 Plectranthus loxothyrsus Hand. -Mazz. = Isodon loxothyrsus（Hand. -Mazz.）H. Hara ●

303467 Plectranthus loxothyrsus Hand. -Mazz. = Rabdosia loxothyrsa（Hand. -Mazz.）H. Hara ●

303468 Plectranthus lucidus（Benth.）Van Jaarsv. et T. J. Edwards;光亮香茶菜■☆

303469 Plectranthus luteus Gürke = Plectranthus melleri Baker ■☆

303470 Plectranthus macilentus Hedge;瘦香茶■☆

303471 Plectranthus macranthus Hook. f. = Siphocranion macranthum（Hook. f.）C. Y. Wu ■

303472 Plectranthus macreei Benth. = Isodon coetsus（Buch. -Ham. ex D. Don）H. Hara var. cavaleriei（H. Lév.）H. W. Li ●■

303473 Plectranthus macreei Benth. = Rabdosia coetsa（Buch. -Ham. ex D. Don）H. Hara var. cavaleriei（H. Lév.）C. Y. Wu et H. W. Li ●■

303474 Plectranthus macrocalyx Dunn = Isodon macrocalyx（Dunn）Kudo ■

303475 Plectranthus macrocalyx Dunn = Rabdosia macrocalyx（Dunn）H. Hara ■

303476 Plectranthus macrocalyx Dunn = Rabdosia macrophylla（Migo）C. Y. Wu et H. W. Li ●■

303477 Plectranthus macrophyllus Migo = Rabdosia macrophylla

（Migo）C. Y. Wu et H. W. Li ●■

303478 Plectranthus madagascariensis（Pers.）Benth. ;马达加斯加延命草;Mintleaf ■☆

303479 Plectranthus madagascariensis（Pers.）Benth. 'Variegated Mintleaf';白斑马达加斯加延命草■☆

303480 Plectranthus madagascariensis（Pers.）Benth. var. ramosior Benth. ;多枝马达加斯加香茶■☆

303481 Plectranthus madagascariensis Benth. = Plectranthus madagascariensis（Pers.）Benth. ■☆

303482 Plectranthus maddenii Benth. ex Hook. f. = Isodon coetsus（Buch. -Ham. ex D. Don）Kudo ●■

303483 Plectranthus mairei H. Lév. = Isodon coetsus（Buch. -Ham. ex D. Don）H. Hara var. cavaleriei（H. Lév.）H. W. Li ●■

303484 Plectranthus malawiensis B. Mathew;马拉维香茶■☆

303485 Plectranthus malinvaldii Briq. ;马林香茶■☆

303486 Plectranthus malvinus Van Jaarsv. et T. J. Edwards;锦葵香茶■☆

303487 Plectranthus mannii Baker = Plectranthus kamerunensis Gürke ■☆

303488 Plectranthus marmoritis Hance = Orthosiphon marmoritis（Hance）Dunn ■

303489 Plectranthus marrubioides Hochst. ex Benth. = Plectranthus cylindraceus Hochst. ex Benth. ■☆

303490 Plectranthus masukensis Baker;马苏克香茶●☆

303491 Plectranthus matabelensis Baker = Plectranthus hereroensis Engl. ■☆

303492 Plectranthus mauritianus Bojer = Plectranthus madagascariensis（Pers.）Benth. ■☆

303493 Plectranthus megathyrsus Diels = Isodon megathyrsus（Diels）H. W. Li ■

303494 Plectranthus megathyrsus Diels = Rabdosia megathyrsa（Diels）H. Hara ■

303495 Plectranthus melanocarpus Gürke = Plectranthus tetragonus Gürke ■☆

303496 Plectranthus melissoides Benth. = Isodon melissoides（Benth.）H. Hara ■

303497 Plectranthus melleri Baker;梅勒香茶■☆

303498 Plectranthus membranaceus（Scott-Elliot）Hedge;膜质香茶■☆

303499 Plectranthus menthoides Benth. = Isodon coetsus（Buch. -Ham. ex D. Don）Kudo ●■

303500 Plectranthus menthoides Benth. = Rabdosia coetsa（Buch. -Ham. ex D. Don）H. Hara ●■

303501 Plectranthus microphyllus Baker = Solenostemon porpeodon（Baker）J. K. Morton ■☆

303502 Plectranthus mildbraedii Perkins;米尔德香茶■☆

303503 Plectranthus minimus Gürke = Plectranthus tenuicaulis（Hook. f.）J. K. Morton ■☆

303504 Plectranthus minutiflorus Ryding;微花香茶■☆

303505 Plectranthus mirabilis（Briq.）Launert;奇异香茶■☆

303506 Plectranthus miserabilis Briq. ;贫弱香茶■☆

303507 Plectranthus mocquerysii Briq. ;莫克里斯香茶●☆

303508 Plectranthus modestus Baker;适度香茶■☆

303509 Plectranthus mollis Spreng. ;柔软香茶■☆

303510 Plectranthus monostachyus（P. Beauv.）B. J. Pollard;单穗延命草■☆

303511 Plectranthus monostachyus（P. Beauv.）B. J. Pollard subsp. latericola（A. Chev.）B. J. Pollard;侧生香茶■☆

303512 Plectranthus monostachyus（P. Beauv.）B. J. Pollard subsp. marrubiifolius（Brenan）B. J. Pollard;夏至草叶香茶■☆

303513　Plectranthus monticola Gürke;山地延命草■☆

303514　Plectranthus moschosmoides Baker ＝ Plectranthus cylindraceus Hochst. ex Benth. ■☆

303515　Plectranthus moslifolius H. Lév. ＝ Isodon eriocalyx（Dunn）Kudo ●■

303516　Plectranthus moslifolius H. Lév. ＝ Isodon nervosus（Hemsl.）Kudo ■

303517　Plectranthus moslifolius H. Lév. ＝ Rabdosia eriocalyx（Dunn）H. Hara ●■

303518　Plectranthus moslifolius H. Lév. ＝ Rabdosia nervosa（Hemsl.）C. Y. Wu et H. W. Li ■

303519　Plectranthus muliensis W. W. Sm. ＝ Isodon muliensis（W. W. Sm.）Kudo ●

303520　Plectranthus muliensis W. W. Sm. ＝ Rabdosia muliensis（W. W. Sm.）H. Hara ●

303521　Plectranthus mutabilis Codd;易变香茶■☆

303522　Plectranthus myrianthus Briq. ＝ Plectranthus hereroensis Engl. ■☆

303523　Plectranthus mzimvubensis Van Jaarsv. ;姆津香茶■☆

303524　Plectranthus nankinensis（Lour.）Spreng. ＝ Perilla frutescens（L.）Britton var. crispa（Benth.）W. Deane ex Bailey ■

303525　Plectranthus nankinensis Spreng. ＝ Perilla frutescens（L.）Britton var. crispa（Benth.）W. Deane ex Bailey ■

303526　Plectranthus natalensis Gürke ＝ Plectranthus ciliatus E. Mey. ex Benth. ■☆

303527　Plectranthus natalensis Gürke. f. glandulosa E. Phillips ＝ Plectranthus grallatus Briq. ■☆

303528　Plectranthus neglectus Dinter ＝ Aeollanthus neglectus（Dinter）Launert ■☆

303529　Plectranthus nervosus Hemsl. ＝ Isodon nervosus（Hemsl.）Kudo ■

303530　Plectranthus nervosus Hemsl. ＝ Rabdosia nervosa（Hemsl.）C. Y. Wu et H. W. Li ■

303531　Plectranthus neumannii Gürke ＝ Plectranthus laxiflorus Benth. ■☆

303532　Plectranthus neumannii Gürke ex Engl. ;纽曼香茶●☆

303533　Plectranthus nudipes Hemsl. ＝ Siphocranion nudipes（Hemsl.）Kudo ■

303534　Plectranthus nummularius Briq. ＝ Plectranthus verticillatus（L. f.）Druce ■☆

303535　Plectranthus nyasicus（Baker）M. Ashby;尼亚斯香茶■☆

303536　Plectranthus nyikensis Baker;尼卡香茶●☆

303537　Plectranthus oblanceolatus Hedge;倒披针形香茶●☆

303538　Plectranthus occidentalis B. J. Pollard;西方香茶■☆

303539　Plectranthus oertendahlii T. C. E. Fr. ;银脉延命草;Candle Plant,Prostrate Coleus,Swedish Ivy ■☆

303540　Plectranthus ombrophilus Hedge;喜雨香茶●☆

303541　Plectranthus orbicularis Gürke;圆形香茶■☆

303542　Plectranthus oreophilus Diels ＝ Skapanthus oreophilus（Diels）C. Y. Wu et H. W. Li ■

303543　Plectranthus oreophilus Diels var. elongatus Hand. -Mazz. ＝ Skapanthus oreophilus（Diels）C. Y. Wu et H. W. Li var. elonggatus（Hand. -Mazz.）C. Y. Wu et H. W. Li ■

303544　Plectranthus oresbius W. W. Sm. ＝ Isodon oresbius（W. W. Sm.）Kudo ●

303545　Plectranthus oresbius W. W. Sm. ＝ Rabdosia oresbia（W. W. Sm.）H. Hara ●

303546　Plectranthus ornatus Codd;装饰香茶■☆

303547　Plectranthus otaviensis Dinter ＝ Plectranthus hereroensis Engl. ■☆

303548　Plectranthus otostegioides（Gürke）Ryding;八盖香茶■☆

303549　Plectranthus ovatifolius Oliv. ＝ Ocimum lamiifolium Hochst. ex Benth. ■☆

303550　Plectranthus pachyphyllus Gürke ex T. Cooke;厚叶香茶●☆

303551　Plectranthus pachystachyus Briq. ＝ Rabdosiella calycina（Benth.）Codd ●■

303552　Plectranthus pachythyrsus Hand. -Mazz. ＝ Isodon leucophyllus（Dunn）Kudo ●

303553　Plectranthus pachythyrsus Hand. -Mazz. ＝ Rabdosia leucophylla（Dunn）H. Hara ●

303554　Plectranthus panganensis Gürke;庞岸香茶■☆

303555　Plectranthus paniculatus Baker ＝ Isodon ramosissimus（Hook. f.）Codd ●☆

303556　Plectranthus pantadenius Hand. -Mazz. ＝ Isodon pantadenius（Hand. -Mazz.）H. W. Li ■

303557　Plectranthus pantadenius Hand. -Mazz. ＝ Rabdosia pantadenia（Hand. -Mazz.）H. Hara ■

303558　Plectranthus papilionaceus Ranirison et Phillipson;蝶形香茶■☆

303559　Plectranthus parviflorus Gürke ＝ Plectranthus strigosus Benth. ■☆

303560　Plectranthus parviflorus R. Br. ＝ Basilicum polystachyon（L.）Moench ■

303561　Plectranthus parviflorus Willd. ;小花香茶;Little Spurflower☆

303562　Plectranthus parvifolia Batalin ＝ Isodon parvifolius（Batalin）H. Hara ●

303563　Plectranthus parvifolia Batalin ＝ Rabdosia parvifolia（Batalin）H. Hara ●

303564　Plectranthus parvifolius（Batalin）C. P'ei ＝ Isodon parvifolius（Batalin）H. Hara ●

303565　Plectranthus parvifolius（Batalin）C. P'ei ＝ Rabdosia parvifolia（Batalin）H. Hara ●

303566　Plectranthus patchouli C. B. Clarke ex Hook. f. ＝ Microtoena patchoulii（C. B. Clarke）C. Y. Wu et S. J. Hsuan ■

303567　Plectranthus pauciflorus Baker;少花香茶■☆

303568　Plectranthus peglerae T. Cooke ＝ Plectranthus fruticosus L'Hér. ●☆

303569　Plectranthus pekinensis Maxim. ＝ Isodon amethystoides（Benth.）H. Hara ■

303570　Plectranthus pekinensis Maxim. ＝ Rabdosia amethystoides（Benth.）H. Hara ■

303571　Plectranthus pendulus Gürke;下垂香茶■☆

303572　Plectranthus pentheri（Gürke）Van Jaarsv. et T. J. Edwards;彭泰尔香茶■☆

303573　Plectranthus perrieri Hedge;佩里耶香茶●☆

303574　Plectranthus persoonii（Benth.）Hedge;帕松香茶■☆

303575　Plectranthus petiolaris E. Mey. ex Benth. ;柄叶香茶■☆

303576　Plectranthus petrensis S. Moore;皮特拉香茶■☆

303577　Plectranthus pharicus Prain ＝ Isodon pharicus（Prain）Murata ●

303578　Plectranthus phyllopodus Diels ＝ Isodon phyllopodus（Diels）Kudo ●■

303579　Plectranthus phyllopodus Diels ＝ Rabdosia phyllopoda（Diels）H. Hara ●■

303580　Plectranthus phyllostachys Diels ＝ Isodon phyllostachys（Diels）Kudo ●

303581　Plectranthus phyllostachys Diels ＝ Rabdosia phyllostachys（Diels）H. Hara ●

303582　Plectranthus piliferus Chiov. ;纤毛香茶■☆

303583　Plectranthus pleiophyllus Diels ＝ Isodon pleiophyllus（Diels）Kudo ●

303584　Plectranthus pleiophyllus Diels ＝ Rabdosia pleiophylla（Diels）C. Y. Wu et H. W. Li ●

303585　Plectranthus polystachys Y. Z. Sun ex C. H. Hu ＝ Isodon coetsus（Buch. -Ham. ex D. Don）Kudo ●■

303586　Plectranthus polystachys Y. Z. Sun ex C. H. Hu ＝ Rabdosia coetsa（Buch. -Ham. ex D. Don）H. Hara ●■

303587　Plectranthus polystachys Y. Z. Sun ex C. H. Hu ＝ Rabdosia phyllostachys（Diels）H. Hara ●

303588　Plectranthus polystachyus（L.）Rchb. ＝ Basilicum polystachyon（L.）Moench ■

303589　Plectranthus porpeodon Baker ＝ Solenostemon porpeodon（Baker）J. K. Morton ■☆

303590　Plectranthus porphyranthus T. J. Edwards et N. R. Crouch;紫花香茶■☆

303591　Plectranthus praetermissus Codd;疏忽香茶■☆

303592　Plectranthus praetervisus Briq. ＝ Plectranthus grallatus Briq. ■☆

303593　Plectranthus prainianus（H. Lév.）Dunn ＝ Siphocranion macranthum（Hook. f.）C. Y. Wu ■

303594　Plectranthus pratensis Gürke;草原香茶■☆

303595　Plectranthus primulinus Baker;报春香茶■☆

303596　Plectranthus prostratus Gürke;平卧香茶■☆

303597　Plectranthus provicarii H. Lév. ＝ Isodon bulleyanus（Diels）Kudo ●

303598　Plectranthus provicarii H. Lév. ＝ Rabdosia bulleyana（Diels）H. Hara ●

303599　Plectranthus provicarii H. Lév. ＝ Rabdosia provicarii（H. Lév.）H. Hara ●■

303600　Plectranthus psammophilus Codd;喜沙香茶■☆

303601　Plectranthus pseudobarbatus J. K. Morton;髯毛香茶■☆

303602　Plectranthus pseudomarrubioides R. H. Willemse;假夏至草香茶■☆

303603　Plectranthus puberulentus J. K. Morton;微毛香茶■☆

303604　Plectranthus pubescens Baker;短柔毛香茶■☆

303605　Plectranthus punctatus（L. f.）L'Hér.;斑点香茶■☆

303606　Plectranthus punctatus（L. f.）L'Hér. subsp. lanatus J. K. Morton;绵毛斑点香茶■☆

303607　Plectranthus purpuratus Harv.;紫香茶■☆

303608　Plectranthus purpuratus Harv. subsp. montanus Van Jaarsv. et T. J. Edwards;山地紫香茶■☆

303609　Plectranthus pyramidatus Gürke ＝ Rabdosiella calycina（Benth.）Codd ●■

303610　Plectranthus quadridentatus Schweinf. ex Baker ＝ Plectranthus prostratus Gürke ■☆

303611　Plectranthus racemosus Hemsl. ＝ Isodon racemosus（Hemsl.）H. W. Li ■

303612　Plectranthus racemosus Hemsl. ＝ Rabdosia racemosa（Hemsl.）H. Hara ■

303613　Plectranthus ramosissimus Hook. f. ＝ Isodon ramosissimus（Hook. f.）Codd ●☆

303614　Plectranthus reflexus Van Jaarsv. et T. J. Edwards;反折香茶■☆

303615　Plectranthus rehmannii Gürke;拉赫曼香茶■☆

303616　Plectranthus rhomboideus Gürke;菱形香茶■☆

303617　Plectranthus ricinispermus Pamp. ＝ Isodon rubescens（Hemsl.）H. Hara ●

303618　Plectranthus ricinispermus Pamp. ＝ Rabdosia rubescens（Hemsl.）H. Hara ●

303619　Plectranthus rosthornii Diels ＝ Isodon flabelliformis（C. Y. Wu）H. Hara ■

303620　Plectranthus rosthornii Diels ＝ Isodon rosthornii（Diels）Kudo ■

303621　Plectranthus rosthornii Diels ＝ Rabdosia flabelliformis C. Y. Wu ■

303622　Plectranthus rosthornii Diels ＝ Rabdosia rosthornii（Diels）H. Hara ■

303623　Plectranthus rosulatus Hedge;莲座香茶■☆

303624　Plectranthus rotundifolius（Poir.）Spreng.;圆叶香茶;Hausa Potato ■☆

303625　Plectranthus rotundifolius（Poir.）Spreng. ＝ Solenostemon rotundifolius（Poir.）J. K. Morton ■☆

303626　Plectranthus rubescens Hemsl. ＝ Isodon rubescens（Hemsl.）H. Hara ●

303627　Plectranthus rubescens Hemsl. ＝ Rabdosia rubescens（Hemsl.）H. Hara ●

303628　Plectranthus rubicundus D. Don ＝ Orthosiphon rubicundus（D. Don）Benth. ■

303629　Plectranthus rubicundus D. Don ＝ Orthosiphon wulfenioides（Diels）Hand. -Mazz. ■

303630　Plectranthus rubropunctatus Codd;红斑香茶■☆

303631　Plectranthus rubroviolaceus Hedge;红堇色香茶●☆

303632　Plectranthus rugosiformis Hand. -Mazz. ＝ Isodon rugosiformis（Hand. -Mazz.）H. Hara ●

303633　Plectranthus rugosiformis Hand. -Mazz. ＝ Rabdosia rugosiformis（Hand. -Mazz.）H. Hara ●

303634　Plectranthus rugosus Wall. ＝ Isodon grandifolius（Hand. -Mazz.）H. Hara ●

303635　Plectranthus rugosus Wall. ＝ Isodon loxothyrsus（Hand. -Mazz.）H. Hara ●

303636　Plectranthus rugosus Wall. ＝ Isodon rugosus（Wall. ex Benth.）Codd ●

303637　Plectranthus rugosus Wall. ＝ Rabdosia grandifolia（Hand. -Mazz.）H. Hara ●

303638　Plectranthus rugosus Wall. ex Benth. ＝ Isodon rugosus（Wall. ex Benth.）Codd ●

303639　Plectranthus rupestris Vatke ex Baker;岩生香茶■☆

303640　Plectranthus rupicola Dinter ex Gooss. ＝ Aeollanthus buchnerianus Briq. ■☆

303641　Plectranthus rupicola Gürke;岩地香茶■☆

303642　Plectranthus rutenbergianus Vatke;鲁滕贝格香茶■☆

303643　Plectranthus saccatus Benth.;大花香茶●☆

303644　Plectranthus saccatus Benth. subsp. pondoensis Van Jaarsv. et Milstein;庞多香茶●☆

303645　Plectranthus saccatus Benth. var. longitubus Codd;长管香茶●☆

303646　Plectranthus sakarensis Gürke;萨卡尔香茶●☆

303647　Plectranthus salicarius Hand. -Mazz. ＝ Isodon nervosus（Hemsl.）Kudo ■

303648　Plectranthus salicarius Hand. -Mazz. ＝ Rabdosia nervosa（Hemsl.）C. Y. Wu et H. W. Li ■

303649　Plectranthus salubenii Brummitt et Seyani;萨卢本香茶■☆

303650　Plectranthus salviiflorus Chiov. ＝ Plectranthus garckeanus（Vatke）J. K. Morton ●☆

303651　Plectranthus sanguineus Britten;血红香茶●☆

303652　Plectranthus saxicola B. J. Pollard et A. J. Paton;岩栖香茶■☆

303653　Plectranthus scaposus Hedge;粗糙香茶●☆

303654　Plectranthus schimperi Vatke ＝ Isodon schimperi（Vatke）J. K. Morton ●☆

303655　Plectranthus schizophyllus Baker;裂叶香茶■☆

303656　Plectranthus schlechteri（T. C. E. Fr.）Hutch. et Dandy；施莱香茶■☆

303657　Plectranthus schlechteri Gürke = Englerastrum alstonianum Hutch. et Dandy ■☆

303658　Plectranthus scrophularioides Wall. ex Benth. = Isodon scoparius（C. Y. Wu et H. W. Li）H. Hara ●

303659　Plectranthus scrophularioides Wall. ex Benth. = Isodon scrophularioides（Wall. ex Benth.）Murata ■

303660　Plectranthus sculponeatus Vaniot = Isodon sculponeatus（Vaniot）Kudo ■

303661　Plectranthus sculponeatus Vaniot = Rabdosia sculponeata（Vaniot）H. Hara ■

303662　Plectranthus scutellarioides（L.）R. Br.；矩圆盾状香茶■☆

303663　Plectranthus scutellarioides（L.）R. Br. = Coleus scutellarioides（L.）Benth. ■

303664　Plectranthus secundiflorus（Baker）Hedge；侧花香茶●☆

303665　Plectranthus semayatensis Cufod. ；印非香茶■☆

303666　Plectranthus seretii（De Wild.）Vollesen；赛雷香茶■☆

303667　Plectranthus serrulatus（Robyns）Troupin et Ayob. ；细齿香茶■☆

303668　Plectranthus serrus Maxim. = Isodon serrus（Maxim.）Kudo ■

303669　Plectranthus serrus Maxim. = Rabdosia serra（Maxim.）H. Hara ■

303670　Plectranthus setschwanensis Hand. -Mazz. = Isodon setschwanensis（Hand. -Mazz.）H. Hara ●

303671　Plectranthus setschwanensis Hand. -Mazz. = Rabdosia setschwanensis（Hand. -Mazz.）H. Hara ●

303672　Plectranthus shikokianus（Makino）Makino = Isodon shikokianus（Makino）H. Hara ■☆

303673　Plectranthus shikokianus（Makino）Makino var. intermedius（Kudo）Ohwi = Isodon shikokianus（Makino）H. Hara var. intermedius（Kudo）Murata ■☆

303674　Plectranthus shikokianus（Makino）Makino var. occidentalis（Murata）Ohwi = Isodon shikokianus（Makino）H. Hara var. occidentalis Murata ■☆

303675　Plectranthus sinensis Miq. = Isodon amethystoides（Benth.）H. Hara ■

303676　Plectranthus sinensis Miq. = Rabdosia amethystoides（Benth.）H. Hara ■

303677　Plectranthus smithianus Hand. -Mazz. = Isodon smithianus（Hand. -Mazz.）H. Hara ●

303678　Plectranthus smithianus Hand. -Mazz. = Rabdosia smithiana（Hand. -Mazz.）H. Hara ●

303679　Plectranthus sphaerophyllus Baker；球叶香茶■☆

303680　Plectranthus spicatus E. Mey. ex Benth. ；长穗叶香茶■☆

303681　Plectranthus spiciformis R. A. Dyer = Plectranthus cylindraceus Hochst. ex Benth. ■☆

303682　Plectranthus stachyoides Oliv. ；穗状香茶■☆

303683　Plectranthus stenophyllus Baker；窄叶香茶■☆

303684　Plectranthus stenosiphon Baker；窄管香茶■☆

303685　Plectranthus stocksii Hook. f. = Isodon longitubus（Miq.）Kudo ■

303686　Plectranthus stocksii Hook. f. = Isodon lophanthoides（Buch. -Ham. ex D. Don）H. Hara ■

303687　Plectranthus stocksii Hook. f. = Rabdosia longituba（Miq.）H. Hara ■

303688　Plectranthus stolzii Gürke；斯托尔兹香茶■☆

303689　Plectranthus stracheyi Benth. = Isodon nervosus（Hemsl.）Kudo ■

303690　Plectranthus stracheyi Benth. = Rabdosia nervosa（Hemsl.）C. Y. Wu et H. W. Li ■

303691　Plectranthus stracheyi Benth. ex Hook. f. = Isodon walkeri（Arn.）H. Hara ■

303692　Plectranthus stracheyi Benth. ex Hook. f. = Rabdosia stracheyi（Benth. ex Hook. f.）H. Hara ■

303693　Plectranthus striatus Benth. = Isodon longitubus（Miq.）Kudo ■

303694　Plectranthus striatus Benth. = Isodon lophanthoides（Buch. -Ham. ex D. Don）H. Hara ■

303695　Plectranthus striatus Benth. = Rabdosia longituba（Miq.）H. Hara ■

303696　Plectranthus striatus Benth. var. gerardianus（Benth.）Hand. -Mazz. = Isodon lophanthoides（Buch. -Ham. ex D. Don）H. Hara var. graciliflora（Benth.）H. Hara ■

303697　Plectranthus striatus Benth. var. gerardianus（Benth.）Hand. -Mazz. = Rabdosia lophanthoides（Buch. -Ham. ex D. Don）H. Hara var. graciliflora（Benth.）H. Hara ■

303698　Plectranthus striatus Benth. var. graciliflorus（Benth.）Hand. -Mazz. = Isodon lophanthoides（Buch. -Ham. ex D. Don）H. Hara var. graciliflora（Benth.）H. Hara ■

303699　Plectranthus striatus Benth. var. graciliflorus（Benth.）Hand. -Mazz. = Rabdosia lophanthoides（Buch. -Ham. ex D. Don）H. Hara var. graciliflora（Benth.）H. Hara ■

303700　Plectranthus strigosus Benth. ；糙伏毛香茶■☆

303701　Plectranthus strigosus Benth. var. lucidus ? = Plectranthus lucidus（Benth.）Van Jaarsv. et T. J. Edwards ■☆

303702　Plectranthus strobiliferus Roxb. = Anisochilus carnosus（L.）Wall. ●

303703　Plectranthus stuhlmannii Gürke；斯图尔曼香茶■☆

303704　Plectranthus subacaulis Baker = Aeollanthus subacaulis（Baker）Hua et Briq. ■☆

303705　Plectranthus subspicatus Hochst. = Plectranthus spicatus E. Mey. ex Benth. ■☆

303706　Plectranthus subtenuis Berhaut = Englerastrum nigericum Alston ■☆

303707　Plectranthus succulentus R. A. Dyer et E. A. Bruce = Thorncroftia succulenta（R. A. Dyer et E. A. Bruce）Codd ■☆

303708　Plectranthus swynnertonii S. Moore；斯温纳顿香茶■☆

303709　Plectranthus sylvestris Gürke；林地香茶■☆

303710　Plectranthus tatei Hemsl. = Isodon lophanthoides（Buch. -Ham. ex D. Don）H. Hara var. gerardiana（Benth.）H. Hara ■

303711　Plectranthus tatei Hemsl. = Rabdosia lophanthoides（Buch. -Ham. ex D. Don）H. Hara var. gerardiana（Benth.）H. Hara ■

303712　Plectranthus tenuicaulis（Hook. f.）J. K. Morton；细茎香茶■☆

303713　Plectranthus tenuiflorus（Vatke）Agnew；细花香茶■☆

303714　Plectranthus tenuifolius W. W. Sm. = Isodon tenuifolius（W. W. Sm.）Kudo ●

303715　Plectranthus tenuifolius W. W. Sm. = Rabdosia tenuifolia（W. W. Sm.）H. Hara ●

303716　Plectranthus tenuis Hutch. et Dandy = Plectranthus gracillimus（T. C. E. Fr.）Hutch. et Dandy ■☆

303717　Plectranthus ternifolius D. Don = Isodon ternifolius（D. Don）Kudo ●■

303718　Plectranthus ternifolius D. Don = Rabdosia ternifolia（D. Don）H. Hara ●■

303719　Plectranthus tetensis（Baker）Agnew；太特香茶■☆

303720　Plectranthus tetensis（Baker）Agnew = Plectranthus lasianthus（Gürke）Vollesen ■☆

303721　Plectranthus tetragonus Gürke;四角香茶■☆

303722　Plectranthus thiothyrsus Hand. -Mazz. = Isodon leucophyllus（Dunn）Kudo ●

303723　Plectranthus thiothyrsus Hand. -Mazz. = Rabdosia leucophylla（Dunn）H. Hara ●

303724　Plectranthus thorncroftii S. Moore = Thorncroftia thorncroftii（S. Moore）Codd ■☆

303725　Plectranthus thunbergii Benth. = Plectranthus verticillatus（L. f.）Druce ■☆

303726　Plectranthus thyrsoideus（Baker）B. Mathew;中非香茶（聚伞香茶,中非延命草）■☆

303727　Plectranthus tomentosus Benth. = Plectranthus hadiensis（Forssk.）Schweinf. ex Spreng. var. tomentosus（Benth.）Codd ■☆

303728　Plectranthus transvaalensis Briq.;德兰士瓦香茶■☆

303729　Plectranthus transvaalensis Briq. = Plectranthus grallatus Briq. ■☆

303730　Plectranthus transvaalensis Briq. var. grandifolia T. Cooke = Plectranthus grallatus Briq. ■☆

303731　Plectranthus trichocarpus Maxim. = Isodon trichocarpus（Maxim.）Kudo ■☆

303732　Plectranthus triflorus Baker = Plectranthus laxiflorus Benth. ■☆

303733　Plectranthus trilobus Hedge;三裂香茶●☆

303734　Plectranthus tysonii Gürke = Solenostemon latifolius（Hochst. ex Benth.）J. K. Morton ■☆

303735　Plectranthus ugandensis Lye;乌干达香茶■☆

303736　Plectranthus ugandensis S. Moore = Plectranthus ugandensis Lye ■☆

303737　Plectranthus umbrosus（Maxim.）Makino = Isodon umbrosus（Maxim.）H. Hara ■☆

303738　Plectranthus umbrosus（Maxim.）Makino var. komaensis（Okuyama）Okuyama = Isodon umbrosus（Maxim.）H. Hara var. latifolius Okuyama ■☆

303739　Plectranthus umbrosus（Maxim.）Makino var. latifolius（Okuyama）Okuyama = Isodon umbrosus（Maxim.）H. Hara var. latifolius Okuyama ■☆

303740　Plectranthus umbrosus Hand. -Mazz. = Isodon setschwanensis（Hand. -Mazz.）H. Hara ●

303741　Plectranthus umbrosus Hand. -Mazz. = Rabdosia setschwanensis（Hand. -Mazz.）H. Hara ●

303742　Plectranthus unguentarius Codd;爪状香茶■☆

303743　Plectranthus urticoides Baker = Plectranthus laxiflorus Benth. ■☆

303744　Plectranthus usambarensis Gürke;乌桑巴拉香茶■☆

303745　Plectranthus vagatus（E. A. Bruce）Codd = Plectranthus tetensis（Baker）Agnew ■☆

303746　Plectranthus venteri Van Jaarsv. et Hankey;文特尔香茶■☆

303747　Plectranthus veronicifolius Hance = Isodon walkeri（Arn.）H. Hara ■

303748　Plectranthus verticillatus（L. f.）Druce;澳洲延命草（瑞士常青藤）;Whorled Plectranthus ■☆

303749　Plectranthus verticillatus Druce = Plectranthus verticillatus（L. f.）Druce ■☆

303750　Plectranthus vestitus Benth. ;包被香茶●☆

303751　Plectranthus vicinus T. C. E. Fr.;邻近香茶●☆

303752　Plectranthus villosus T. Cooke = Plectranthus cylindraceus Hochst. ex Benth. ■☆

303753　Plectranthus vinaceus Hedge;葡萄酒色香茶●☆

303754　Plectranthus violaceus Gürke = Plectranthus laxiflorus Benth. ■☆

303755　Plectranthus volkensianus Muschl. = Isodon lophanthoides（Buch. -Ham. ex D. Don）H. Hara ■

303756　Plectranthus volkensianus Muschl. = Rabdosia lophanthoides（Buch. -Ham. ex D. Don）H. Hara ■

303757　Plectranthus volkmannae Dinter = Aeollanthus buchnerianus Briq. ■☆

303758　Plectranthus walkeri Arn. = Isodon walkeri（Arn.）H. Hara ■

303759　Plectranthus wardii Marquand et Airy Shaw = Isodon wardii（C. Marquand et Airy Shaw）H. Hara ●

303760　Plectranthus wardii Marquand et Airy Shaw = Rabdosia wardii（C. Marquand et Airy Shaw）H. Hara ●

303761　Plectranthus websteri Hemsl. = Isodon websteri（Hemsl.）Kudo ■

303762　Plectranthus websteri Hemsl. = Rabdosia websteri（Hemsl.）H. Hara ■

303763　Plectranthus welwitschii（Briq.）Codd;韦尔香茶■☆

303764　Plectranthus welwitschii Vatke = Aeollanthus engleri Briq. ■☆

303765　Plectranthus wikstroemioides Hand. -Mazz. = Isodon wikstroemioides（Hand. -Mazz.）H. Hara ●

303766　Plectranthus wkikstroemioides Hand. -Mazz. = Rabdosia wikstroemioides（Hand. -Mazz.）H. Hara ●

303767　Plectranthus woodii Gürke = Plectranthus hadiensis（Forssk.）Schweinf. ex Spreng. var. woodii（Gürke）Codd ■☆

303768　Plectranthus wui Y. Z. Sun ex C. H. Hu = Isodon adenanthus（Diels）Kudo ■

303769　Plectranthus wui Y. Z. Sun ex C. H. Hu = Rabdosia adenantha（Diels）H. Hara ■

303770　Plectranthus xerophilus Codd;旱生香茶■☆

303771　Plectranthus yunnanensis Hand. -Mazz. = Isodon yunnanensis（Hand. -Mazz.）H. Hara ■

303772　Plectranthus yunnanensis Hand. -Mazz. = Rabdosia weisiensis C. Y. Wu ■

303773　Plectranthus zatarhendi（Forssk.）E. A. Bruce = Plectranthus aegyptiacus（Forssk.）C. Chr. ■●☆

303774　Plectranthus zebrarum Brummitt et Seyani;泽布拉香茶■☆

303775　Plectranthus zernyi Gilli;策尼香茶■☆

303776　Plectranthus zombensis Baker;宗巴香茶■☆

303777　Plectranthus zuluensis T. Cooke;祖卢香茶■☆

303778　Plectreca Raf. = Vernonia Schreb.（保留属名）●■

303779　Plectrelminthes Merr. = Plectrelminthus Raf. ■☆

303780　Plectrelminthus Raf.（1838）;蠕距兰属■☆

303781　Plectrelminthus caudatus（Lindl.）Summerh. ;蠕距兰■☆

303782　Plectrelminthus caudatus（Lindl.）Summerh. var. trilobatus Szlach. et Olszewski;三裂蠕距兰■☆

303783　Plectrelminthus spiculatus（Finet）Summerh. = Aerangis spiculata（Finet）Senghas ■☆

303784　Plectritis（Lindl.）DC.（1830）;距缬草属■☆

303785　Plectritis DC. = Plectritis（Lindl.）DC. ■☆

303786　Plectrocarpa Gillies = Plectrocarpa Gillies ex Hook. ●☆

303787　Plectrocarpa Gillies ex Hook. = Plectrocarpa Gillies ex Hook. et Arn. ●☆

303788　Plectrocarpa Gillies ex Hook. et Arn.（1833）;距果蒺藜属●☆

303789　Plectrocarpa tetracantha Gill. = Plectrocarpa tetracantha Gill. ex Hook. ●☆

303790　Plectrocarpa tetracantha Gill. ex Hook. ;距果蒺藜●☆

303791　Plectronema Raf. = Zephyranthes Herb.（保留属名）■

303792　Plectronema candida（Lindl.）Raf. = Zephyranthes candida（Lindl.）Herb. ■

303793　Plectronia Buching. ex Krauss = Olinia Thunb. ●☆

303794　Plectronia L. = Canthium Lam. + Olinia Thunb. ●☆

303795   Plectronia L. = Olinia Thunb. ● ☆

303796   Plectronia Lour. = Acanthopanax (Decne. et Planch.) Miq. ●

303797   Plectronia abbreviata K. Schum. = Pygmaeothamnus zeyheri (Sond.) Robyns ● ☆

303798   Plectronia acarophyta De Wild. = Psydrax acutiflora (Hiern) Bridson ● ☆

303799   Plectronia acuminata De Wild. = Keetia acuminata (De Wild.) Bridson ● ☆

303800   Plectronia acutiflora (Hiern) K. Schum. = Psydrax acutiflora (Hiern) Bridson ● ☆

303801   Plectronia amaniensis K. Krause = Rytigynia celastroides (Baill.) Verdc. ● ☆

303802   Plectronia angiensis De Wild. = Psydrax schimperiana (A. Rich.) Bridson ● ☆

303803   Plectronia angustiflora De Wild. = Keetia tenuiflora (Hiern) Bridson ● ☆

303804   Plectronia anomocarpum (DC.) K. Schum. = Psydrax horizontalis (Schumach.) Bridson ● ☆

303805   Plectronia apiculatifolia De Wild. = Keetia ripae (De Wild.) Bridson ● ☆

303806   Plectronia arnoldiana De Wild. et T. Durand = Psydrax arnoldiana (De Wild. et T. Durand) Bridson ● ☆

303807   Plectronia barteri (Hiern) De Wild. = Keetia venosa (Oliv.) Bridson ● ☆

303808   Plectronia bibracteata Baker = Pyrostria bibracteata (Baker) Cavaco ● ☆

303809   Plectronia bicolor De Wild. = Multidentia dichrophylla (Mildbr.) Bridson ■ ☆

303810   Plectronia bispathacea Mildbr. = Pyrostria bispathacea (Mildbr.) Bridson ● ☆

303811   Plectronia bogosensis Martelli = Pyrostria phyllanthoidea (Baill.) Bridson ● ☆

303812   Plectronia brevifolium (Engl.) Engl. ex De Wild. et T. Durand = Keetia gracilis (Hiern) Bridson ● ☆

303813   Plectronia brieyi De Wild. = Psydrax parviflora (Afzel.) Bridson ● ☆

303814   Plectronia buarica Mildbr. = Multidentia crassa (Hiern) Bridson et Verdc. var. ampla (S. Moore) Bridson et Verdc. ● ☆

303815   Plectronia buxifolia Baker = Pyrostria buxifolia Hochr. ● ☆

303816   Plectronia calycophila K. Schum. = Vangueriella laxiflora (K. Schum.) Verdc. ● ☆

303817   Plectronia campylacantha Mildbr. = Vangueriella campylacantha (Mildbr.) Verdc. ● ☆

303818   Plectronia caudatiflora (Hiern) K. Schum. = Psydrax horizontalis (Schumach.) Bridson ● ☆

303819   Plectronia chamaedendrum Kuntze = Pygmaeothamnus chamaedendrum (Kuntze) Robyns ● ☆

303820   Plectronia charadrophila K. Krause = Keetia gueinzii (Sond.) Bridson ● ☆

303821   Plectronia chinensis Lour. = Acanthopanax trifoliatus (L.) Merr. ●

303822   Plectronia chlorantha K. Schum. = Vangueriella chlorantha (K. Schum.) Verdc. ● ☆

303823   Plectronia ciliata (Klotzsch ex Eckl. et Zeyh.) Sond. = Canthium ciliatum (Klotzsch ex Eckl. et Zeyh.) Kuntze ● ☆

303824   Plectronia ciliata (Klotzsch ex Eckl. et Zeyh.) Sond. var. glabrata Sond. = Canthium ciliatum (Klotzsch ex Eckl. et Zeyh.) Kuntze ● ☆

303825   Plectronia ciliata (Klotzsch ex Eckl. et Zeyh.) Sond. var. mollis Sond. = Canthium ciliatum (Klotzsch ex Eckl. et Zeyh.) Kuntze ● ☆

303826   Plectronia ciliata (Klotzsch) D. Dietr. = Canthium ciliatum (Klotzsch ex Eckl. et Zeyh.) Kuntze ● ☆

303827   Plectronia citrifolia Eckl. et Zeyh. = Psychotria capensis (Eckl.) Vatke ● ☆

303828   Plectronia connata De Wild. et T. Durand = Psydrax subcordata (DC.) Bridson var. connata (De Wild. et T. Durand) Bridson ● ☆

303829   Plectronia cornelioides De Wild. = Keetia zanzibarica (Klotzsch) Bridson subsp. cornelioides (De Wild.) Bridson ● ☆

303830   Plectronia cuspido-stipulata K. Schum. ex Engl. = Keetia venosa (Oliv.) Bridson ● ☆

303831   Plectronia decidua K. Schum. = Vangueriella soyauxii (K. Schum.) Verdc. ● ☆

303832   Plectronia dewevrei De Wild. = Colletoecema dewevrei (De Wild.) E. M. Petit ● ☆

303833   Plectronia dichrophylla Mildbr. = Multidentia dichrophylla (Mildbr.) Bridson ■ ☆

303834   Plectronia diplodiscus K. Schum. = Canthium mombazense Baill. ● ☆

303835   Plectronia dubiosa De Wild. = Rytigynia dubiosa (De Wild.) Robyns ■ ☆

303836   Plectronia dundusanensis De Wild. = Keetia venosa (Oliv.) Bridson ● ☆

303837   Plectronia eickii K. Schum. et K. Krause = Rytigynia eickii (K. Schum. et K. Krause) Bullock ■ ☆

303838   Plectronia fililoba (K. Krause) Mildbr. = Psydrax gilletii (De Wild.) Bridson ● ☆

303839   Plectronia flaviflora K. Schum. et K. Krause = Keetia mannii (Hiern) Bridson ● ☆

303840   Plectronia foetida (Hiern) K. Schum. = Keetia foetida (Hiern) Bridson ● ☆

303841   Plectronia foliosa Burtt Davy = Pavetta eylesii S. Moore ● ☆

303842   Plectronia formicarum K. Krause = Psydrax subcordata (DC.) Bridson ● ☆

303843   Plectronia fragrantissima K. Schum. = Psydrax fragrantissima (K. Schum.) Bridson ● ☆

303844   Plectronia gentilii De Wild. = Keetia zanzibarica (Klotzsch) Bridson subsp. gentilii (De Wild.) Bridson ● ☆

303845   Plectronia gilfillanii N. E. Br. = Canthium gilfillanii (N. E. Br.) O. B. Mill. ● ☆

303846   Plectronia gilletii De Wild. = Psydrax gilletii (De Wild.) Bridson ● ☆

303847   Plectronia glabriflora (Hiern) K. Schum. = Psydrax subcordata (DC.) Bridson ● ☆

303848   Plectronia glauca (Hiern) K. Schum. = Canthium glaucum Hiern ● ☆

303849   Plectronia golungensis (Hiern) K. Schum. = Psydrax parviflora (Afzel.) Bridson ● ☆

303850   Plectronia graniticola Chiov. = Psydrax graniticola (Chiov.) Bridson ● ☆

303851   Plectronia gueinzii (Sond.) Sim = Keetia gueinzii (Sond.) Bridson ● ☆

303852   Plectronia guidottii Chiov. = Polysphaeria multiflora Hiern subsp. pubescens Verdc. ● ☆

303853   Plectronia heliotropiodora K. Schum. et K. Krause = Psydrax livida (Hiern) Bridson ● ☆

303854 Plectronia henriquesiana K. Schum. = Psydrax acutiflora (Hiern) Bridson ● ☆

303855 Plectronia hirsuta DC. = Keetia cornelia (Cham. et Schltdl.) Bridson ● ☆

303856 Plectronia hispida (Benth.) K. Schum. = Keetia hispida (Benth.) Bridson ● ☆

303857 Plectronia hispida (Benth.) K. Schum. var. glabrescens K. Krause = Keetia zanzibarica (Klotzsch) Bridson subsp. cornelioides (De Wild.) Bridson ● ☆

303858 Plectronia hispido-nervosa De Wild. = Keetia hispida (Benth.) Bridson ● ☆

303859 Plectronia huillensis (Hiern) K. Schum. = Psydrax livida (Hiern) Bridson ● ☆

303860 Plectronia junodii Burtt Davy = Psydrax livida (Hiern) Bridson ● ☆

303861 Plectronia kidaria K. Schum. et K. Krause = Rytigynia uhligii (K. Schum. et K. Krause) Verdc. ■ ☆

303862 Plectronia kraussioides (Hiern) K. Schum. = Psydrax kraussioides (Hiern) Bridson ● ☆

303863 Plectronia lactescens (Hiern) K. Schum. = Canthium lactescens Hiern ● ☆

303864 Plectronia lamprophylla K. Schum. = Psydrax micans (Bullock) Bridson ● ☆

303865 Plectronia lanciflora (Hiern) Eyles = Vangueriopsis lanciflora (Hiern) Robyns ● ☆

303866 Plectronia laurentii De Wild. = Psydrax subcordata (DC.) Bridson var. connata (De Wild. et T. Durand) Bridson ● ☆

303867 Plectronia laurentii De Wild. var. katangensis ? = Psydrax subcordata (DC.) Bridson ● ☆

303868 Plectronia laxiflora K. Schum. = Vangueriella laxiflora (K. Schum.) Verdc. ● ☆

303869 Plectronia leucantha K. Krause = Keetia leucantha (K. Krause) Bridson ● ☆

303870 Plectronia levinei Merr. = Fagerlindia scandens (Thunb.) Tirveng. ●

303871 Plectronia livida (Hiern) K. Schum. = Psydrax livida (Hiern) Bridson ● ☆

303872 Plectronia locuples K. Schum. = Psydrax locuples (K. Schum.) Bridson ● ☆

303873 Plectronia longistaminea K. Schum. et K. Krause = Psydrax kaessneri (S. Moore) Bridson ● ☆

303874 Plectronia lucida De Wild. et T. Durand = Psydrax splendens (K. Schum.) Bridson ● ☆

303875 Plectronia lucida K. Schum. et K. Krause = Psydrax horizontalis (Schumach.) Bridson ● ☆

303876 Plectronia macrocarpa K. Schum. = Vangueriella zenkeri Verdc. ● ☆

303877 Plectronia macrophylla K. Schum. = Keetia hispida (Benth.) Bridson ● ☆

303878 Plectronia macrostipulata De Wild. = Keetia molundensis (K. Krause) Bridson var. macrostipulata (De Wild.) Bridson ● ☆

303879 Plectronia malacocarpa K. Schum. et K. Krause = Psydrax kraussioides (Hiern) Bridson ● ☆

303880 Plectronia medusula (Hiern) K. Schum. = Keetia hispida (Benth.) Bridson ● ☆

303881 Plectronia microterantha K. Schum. et K. Krause = Canthium pseudoverticillatum S. Moore ● ☆

303882 Plectronia mildbraedii K. Krause = Keetia hispida (Benth.)

Bridson ● ☆

303883 Plectronia mortehani De Wild. = Keetia molundensis (K. Krause) Bridson var. macrostipulata (De Wild.) Bridson ● ☆

303884 Plectronia mundiana (Cham. et Schltdl.) Pappe = Canthium mundianum Cham. et Schltdl. ● ☆

303885 Plectronia mundii Sims = Canthium mundianum Cham. et Schltdl. ● ☆

303886 Plectronia neglecta (Hiern) K. Schum. ex Engl. = Rytigynia neglecta (Hiern) Robyns ■ ☆

303887 Plectronia nitens (Hiern) K. Schum. = Psydrax schimperiana (A. Rich.) Bridson ● ☆

303888 Plectronia obliquifolia De Wild. = Keetia hispida (Benth.) Bridson ● ☆

303889 Plectronia obovata (Klotzsch ex Eckl. et Zeyh.) Sim = Psydrax obovata (Klotzsch ex Eckl. et Zeyh.) Bridson ● ☆

303890 Plectronia obovata De Wild. = Keetia hispida (Benth.) Bridson ● ☆

303891 Plectronia oddonii De Wild. = Psydrax palma (K. Schum.) Bridson ● ☆

303892 Plectronia opima (S. Moore) Mildbr. = Multidentia crassa (Hiern) Bridson et Verdc. ● ☆

303893 Plectronia orbicularis K. Schum. = Rytigynia orbicularis (K. Schum.) Robyns ■ ☆

303894 Plectronia orthacantha Mildbr. = Vangueriella orthacantha (Mildbr.) Bridson et Verdc. ● ☆

303895 Plectronia ovata Burtt Davy = Plectroniella armata (K. Schum.) Robyns ● ☆

303896 Plectronia pallida K. Schum. = Canthium mombazense Baill. ● ☆

303897 Plectronia palma K. Schum. = Psydrax palma (K. Schum.) Bridson ● ☆

303898 Plectronia pauciflora Klotzsch ex Eckl. et Zeyh. = Canthium kuntzeanum Bridson ● ☆

303899 Plectronia platyphylla (Hiern) K. Schum. = Vangueriopsis lanciflora (Hiern) Robyns ● ☆

303900 Plectronia psychotrioides De Wild. = Canthium lactescens Hiern ● ☆

303901 Plectronia pulchra K. Schum. = Psydrax kraussioides (Hiern) Bridson ● ☆

303902 Plectronia pynaertii De Wild. = Keetia gueinzii (Sond.) Bridson ● ☆

303903 Plectronia randii (S. Moore) Eyles = Canthium lactescens Hiern ● ☆

303904 Plectronia reygaertii De Wild. = Keetia venosa (Oliv.) Bridson ● ☆

303905 Plectronia rhamnifolia Chiov. = Coffea rhamnifolia (Chiov.) Bridson ● ☆

303906 Plectronia rhamnoides (Hiern) K. Schum. = Vangueriella rhamnoides (Hiern) Verdc. ● ☆

303907 Plectronia richardii A. Chev. = Pyrostria angustifolia (A. Rich. ex DC.) Cavaco ● ☆

303908 Plectronia ripae De Wild. = Keetia ripae (De Wild.) Bridson ● ☆

303909 Plectronia rubrinervis K. Krause = Keetia hispida (Benth.) Bridson ● ☆

303910 Plectronia rutshuruensis De Wild. = Keetia tenuiflora (Hiern) Bridson ● ☆

303911 Plectronia schimperiana (A. Rich.) Vatke = Psydrax schimperiana (A. Rich.) Bridson ● ☆

303912 Plectronia sclerocarpa K. Schum. = Multidentia sclerocarpa (K. Schum.) Bridson ● ☆

303913 Plectronia sennii Chiov. = Pavetta uniflora Bremek. ● ☆

303914　Plectronia setiflora（Hiern）K. Schum. = Canthium setiflorum Hiern ●☆

303915　Plectronia sordida K. Schum. = Canthium sordidum（K. Schum.）Bullock ●☆

303916　Plectronia soyauxii K. Schum. = Vangueriella soyauxii（K. Schum.）Verdc. ●☆

303917　Plectronia spinosa（Schumach. et Thonn.）K. Schum. = Vangueriella spinosa（Schumach. et Thonn.）Verdc. ●☆

303918　Plectronia spinosa Klotzsch ex Eckl. et Zeyh. = Canthium spinosum（Klotzsch ex Eckl. et Zeyh.）Kuntze ●☆

303919　Plectronia stipulata De Wild. = Keetia venosa（Oliv.）Bridson ●☆

303920　Plectronia strychnoides K. Schum. = Psydrax horizontalis（Schumach.）Bridson ●☆

303921　Plectronia subcordata（DC.）K. Schum. = Psydrax subcordata（DC.）Bridson ●☆

303922　Plectronia subcordatifolia De Wild. = Keetia gueinzii（Sond.）Bridson ●☆

303923　Plectronia subopaca K. Schum. et K. Krause = Tricalysia ovalifolia Hiern ●☆

303924　Plectronia swynnertonii（S. Moore）Eyles = Canthium inerme（L. f.）Kuntze ●☆

303925　Plectronia sylvatica（Hiern）K. Schum. = Keetia venosa（Oliv.）Bridson ●☆

303926　Plectronia syringodora K. Schum. = Psydrax livida（Hiern）Bridson ●☆

303927　Plectronia tenuiflora（Hiern）K. Schum. = Keetia tenuiflora（Hiern）Bridson ●☆

303928　Plectronia umbrosa（Hiern）K. Schum. = Canthium lactescens Hiern ●☆

303929　Plectronia vanguerioides（Hiern）K. Schum. = Vangueriella vanguerioides（Hiern）Verdc. ●☆

303930　Plectronia vatkeana（Hiern）K. Schum. ex Engl. = Rytigynia neglecta（Hiern）Robyns var. vatkeana（Hiern）Verdc. ●☆

303931　Plectronia venosa Oliv. = Keetia venosa（Oliv.）Bridson ●☆

303932　Plectronia ventosa L. = Olinia ventosa（L.）Cufod. ●☆

303933　Plectronia virgata（Hiern）K. Schum. = Psydrax virgata（Hiern）Bridson ●☆

303934　Plectronia vulgaris K. Schum. = Psydrax parviflora（Afzel.）Bridson ●☆

303935　Plectronia welwitschii（Hiern）K. Schum. = Psydrax subcordata（DC.）Bridson ●☆

303936　Plectronia wildii Suess. = Psydrax livida（Hiern）Bridson ●☆

303937　Plectronia xanthotricha K. Schum. = Rytigynia xanthotricha（K. Schum.）Verdc. ●☆

303938　Plectronia zanzibarica（Klotzsch）Vatke = Keetia zanzibarica（Klotzsch）Bridson ●☆

303939　Plectroniaceae Hiern = Oliniaceae Harv. et Sond.（保留科名）●☆

303940　Plectroniella Robyns（1928）；小距茜属●☆

303941　Plectroniella armata（K. Schum.）Robyns；小距茜●☆

303942　Plectroniella capillaris Bremek.；发状小距茜●☆

303943　Plectrophora H. Focke = Jansenia Barb. Rodr. ■☆

303944　Plectrophora H. Focke（1848）；距兰属■☆

303945　Plectrophora alata（Rolfe）Garay；翅距兰■☆

303946　Plectrophora iridifolia H. Focke；距兰■☆

303947　Plectrornis Raf. = Delphinium L. ■

303948　Plectrornis Raf. ex Luneli = Delphinastrum（DC.）Spach ■

303949　Plectrornis Raf. ex Luneli = Delphinium L. ■

303950　Plectrotropis Schumach. et Thonn. = Vigna Savi（保留属名）■

303951　Plectrotropis angustifolia Schumach. et Thonn. = Vigna vexillata（L.）A. Rich. var. angustifolia（Schumach. et Thonn.）Baker ■☆

303952　Plectrotropis hirsuta Schumach. et Thonn. = Vigna vexillata（L.）A. Rich. ■

303953　Plectrurus Raf. = Tipularia Nutt. ■

303954　Plecturus Raf. = Tipularia Nutt. ■

303955　Pleea Michx.（1803）；北美普氏百合属；Rush-featherling ■☆

303956　Pleea tenuifolia Michx.；北美普氏百合■☆

303957　Plegerina B. D. Jacks. = Licania Aubl. + Couepia Aubl. ●☆

303958　Plegerina B. D. Jacks. = Pleragina Arruda ex Kost. ●☆

303959　Plegmatolemma Bremek. = Justicia L. ●■

303960　Plegorhiza Molina = Limonium Mill.（保留属名）●■

303961　Pleiacanthus（Hook. ex Nutt.）Rydb.（1917）；刺骨苣属；Thorny Skeletonweed ●☆

303962　Pleiacanthus（Hook. ex Nutt.）Rydb. = Lygodesmia D. Don ■☆

303963　Pleiacanthus（Hook. ex Nutt.）Rydb. = Pleiacanthus（Nutt.）Rydb. ●☆

303964　Pleiacanthus（Nutt.）Rydb. = Lygodesmia D. Don ■☆

303965　Pleiacanthus Rydb. = Pleiacanthus（Nutt.）Rydb. ●☆

303966　Pleiacanthus spinosus（Nutt.）Rydb.；刺骨苣；Thorny Skeletonweed ■☆

303967　Pleiadelphia Stapf = Elymandra Stapf ■☆

303968　Pleiadelphia gossweileri Stapf = Elymandra gossweileri（Stapf）Clayton ■☆

303969　Pleianthemum K. Schum. ex A. Chev. = Duboscia Bocquet ●☆

303970　Pleianthemum macrophyllum K. Schum. ex A. Chev. = Desplatsia chrysochlamys（Mildbr. et Burret）Mildbr. et Burret ●☆

303971　Pleiariana N. Chao et G. T. Gong = Salix L.（保留属名）●

303972　Pleiarina N. Chao et G. T. Gong = Salix L.（保留属名）●

303973　Pleiarina Raf. = Salix L.（保留属名）●

303974　Pleiarina balansae（Seemen）N. Chao et G. T. Gong = Salix balansae Seemen ●

303975　Pleiarina boseensis（N. Chao）N. Chao et G. T. Gong = Salix boseensis N. Chao ●

303976　Pleiarina cavaleriei（H. Lév.）N. Chao et G. T. Gong = Salix cavaleriei H. Lév. ●

303977　Pleiarina dictyoneura（Seemen）N. Chao et G. T. Gong = Salix rosthornii Seemen ex Diels ●

303978　Pleiarina dunnii（C. K. Schneid.）N. Chao et G. T. Gong = Salix dunnii C. K. Schneid. ●

303979　Pleiarina glandulosa（Seemen）N. Chao et G. T. Gong = Salix chaenomeloides Kimura ●

303980　Pleiarina humaensis（Y. L. Chou et R. C. Chou）N. Chao et G. T. Gong = Salix humaensis Y. L. Chou et R. C. Chou ●

303981　Pleiarina kusanoi（Hayata）N. Chao et G. T. Gong = Salix kusanoi（Hayata）C. K. Schneid. et Kimura ●

303982　Pleiarina mesinyi（Hance）N. Chao et G. T. Gong = Salix mesinyi Hance ●

303983　Pleiarina paraplesia（C. K. Schneid.）N. Chao et G. T. Gong = Salix paraplesia C. K. Schneid. ●

303984　Pleiarina pendtandra（L.）N. Chao et G. T. Gong = Salix pentandra L. ●

303985　Pleiarina pentandra（L.）N. Chao et G. T. Gong = Salix pentandra L. ●

303986　Pleiarina songarica（Andersson）N. Chao et G. T. Gong = Salix songarica Andersson ●

303987　Pleiarina tetrasperma（Roxb.）N. Chao et G. T. Gong ＝ Salix tetrasperma Roxb. ●

303988　Pleiarina tsoongii（W. C. Cheng）N. Chao et G. T. Gong ＝ Salix dunnii C. K. Schneid. var. tsoongii（W. C. Cheng）C. Y. Yu et S. D. Zhao ●

303989　Pleiarina warburgii（Seemen）N. Chao et G. T. Gong ＝ Salix warburgii Seemen ●

303990　Pleienia Raf. ＝ Sabatia Adans. ■☆

303991　Pleimeris Raf. ＝ Gardenia Ellis（保留属名）●

303992　Pleimeris Raf. ＝ Thunbergia Retz.（保留属名）●■

303993　Pleioblastus Nakai ＝ Arundinaria Michx. ●

303994　Pleioblastus Nakai（1925）；苦竹属（川竹属，大明竹属）；Bamboo，Bitter Bamboo，Bitterbamboo，Bitter-bamboo，Striped Bamboo ●

303995　Pleioblastus actinotrichus（Merr. et Chun）P. C. Keng；射毛苦竹；Star-hair Bamboo，Star-hair Bitterbamboo ●

303996　Pleioblastus actinotrichus（Merr. et Chun）P. C. Keng ＝ Ampelocalamus actinotrichus（Merr. et Chun）S. L. Chen，T. H. Wen et G. Y. Sheng ●

303997　Pleioblastus altiligulatus S. L. Chen et S. Y. Chen；高舌苦竹；Altiligulate Bitter Bamboo，Highligule Bitterbamboo，Taller-ligular Bitter-bamboo ●

303998　Pleioblastus altiligulatus S. L. Chen et S. Y. Chen var. spongiosus B. M. Yang ＝ Oligostachyum spongiosum（C. D. Chu et C. S. Chao）G. H. Ye et Z. P. Wang ●

303999　Pleioblastus altiligulatus S. L. Chen et S. Y. Chen var. spongiosus L. H. Liu；源陵苦竹；Yuanling Bitterbamboo ●

304000　Pleioblastus amarus（Keng）P. C. Keng；苦竹（光箨苦竹，苦笋，苦竹笋，伞柄竹，仙人杖）；Bitter Bamboo，Bitterbamboo，Bitter-bamboo，Common Chinese Bamboo ●

304001　Pleioblastus amarus（Keng）P. C. Keng f. huangshanensis C. L. Huang ＝ Pleioblastus amarus（Keng）P. C. Keng ●

304002　Pleioblastus amarus（Keng）P. C. Keng var. hangzhouensis S. L. Chen et S. Y. Chen；杭州苦竹；Hangzhou Bitter Bamboo，Hangzhou Bitterbamboo ●

304003　Pleioblastus amarus（Keng）P. C. Keng var. pendulifolius S. Y. Chen；垂枝苦竹；Drooptwig Bitterbamboo，Pendentbranch Bitter Bamboo ●

304004　Pleioblastus amarus（Keng）P. C. Keng var. subglabratus S. Y. Chen ＝ Pleioblastus hsienchuensis T. H. Wen var. subglabratus（S. Y. Chen）C. S. Chao et G. Y. Yang ●

304005　Pleioblastus amarus（Keng）P. C. Keng var. tubatus T. H. Wen；胖苦竹；Fat Bitterbamboo，Trumpet Bitter Bamboo ●

304006　Pleioblastus angustatus W. T. Lin；尖舌苦竹；Anguste Bitterbamboo ●

304007　Pleioblastus argenteostriatus（Regel）Nakai；白纹竹；Dwarf Variegated Bamboo，Whitestripe Bamboo ●☆

304008　Pleioblastus argenteostriatus（Regel）Nakai 'Argenteostriatus'；尖舌白纹竹（尖舌苦竹）●☆

304009　Pleioblastus argenteostriatus（Regel）Nakai 'Distichus'；无毛翠竹；Hairlrss Jadegreen Sasa ●

304010　Pleioblastus argenteostriatus（Regel）Nakai f. glaber（Makino）Murata；光白纹竹●☆

304011　Pleioblastus auricomus（Mitford）D. C. McClint.；金发苦竹；Compact Golden-striped Bamboo，Kanuro-Zasa ●☆

304012　Pleioblastus brevinodus W. T. Lin et Z. J. Feng；短节苦竹；Shortnode Bitterbamboo ●

304013　Pleioblastus brevinodus W. T. Lin et Z. J. Feng ＝ Pleioblastus amarus（Keng）P. C. Keng ●

304014　Pleioblastus chino（Franch. et Sav.）Makino；青苦竹；Chino Bitterbamboo ●

304015　Pleioblastus chino（Franch. et Sav.）Makino f. nebulosus（Makino）Muroi；星云青苦竹●☆

304016　Pleioblastus chino（Franch. et Sav.）Makino f. pumilis（Mitford）Sad. Suzuki；矮青苦竹●☆

304017　Pleioblastus chino（Franch. et Sav.）Makino var. gracilis（Makino）Nakai ＝ Pleioblastus chino（Franch. et Sav.）Makino ●

304018　Pleioblastus chino（Franch. et Sav.）Makino var. gracilis（Makino）Nakai 'Variegatus'；花叶青苦竹●☆

304019　Pleioblastus chino（Franch. et Sav.）Makino var. hisauchii Makino；狭叶青苦竹；Narrowleaf Green Bitterbamboo ●

304020　Pleioblastus chino（Franch. et Sav.）Makino var. vaginatus（Hack.）Sad. Suzuki ＝ Pleioblastus chino（Franch. et Sav.）Makino ●

304021　Pleioblastus chino（Franch. et Sav.）Makino var. viridis（Makino）Sad. Suzuki ＝ Pleioblastus argenteostriatus（Regel）Nakai f. glaber（Makino）Murata ●☆

304022　Pleioblastus communis（Makino）Nakai；山川竹；Mountain Bitter Bamboo ●

304023　Pleioblastus distichus（Mitford）Nakai ＝ Pleioblastus argenteostriatus（Regel）Nakai 'Distichus' ●

304024　Pleioblastus distichus（Mitford）Nakai ＝ Sasa pygmaea（Miq.）E. G. Camus var. disticha（Mitford）C. S. Chao et G. G. Tang ●

304025　Pleioblastus dolichanthus（Keng）P. C. Keng ＝ Sinobambusa tootsik（Siebold）Makino ex Nakai ●

304026　Pleioblastus fortunei（Van Houtte）Nakai 'Fortunei' ＝ Pleioblastus fortunei（Van Houtte）Nakai ●

304027　Pleioblastus fortunei（Van Houtte）Nakai 'Pygmaeus' ＝ Pleioblastus argenteostriatus（Regel）Nakai 'Distichus' ●

304028　Pleioblastus fortunei（Van Houtte）Nakai ＝ Sasa fortunei（Van Houtte）Fiori ●

304029　Pleioblastus globinodus C. H. Hu；球节苦竹；Ballnode Bitterbamboo，Globose-node Bitter Bamboo，Globose-noded Bitter-bamboo ●

304030　Pleioblastus globinodus C. H. Hu ＝ Oligostachyum gracilipes（McClure）G. H. Ye et Z. P. Wang ●

304031　Pleioblastus gozadakensis Nakai ＝ Pleioblastus linearis（Hack.）Nakai ●

304032　Pleioblastus gramineus（Bean）Nakai；大明竹（大妨竹，青叶竹，四季竹，四时竹，通丝竹）；Daming Bitterbamboo，Graminoid Bitter-bamboo，Grass-like Bamboo ●

304033　Pleioblastus hattorianus Koidz.；服部苦竹●☆

304034　Pleioblastus higoensis Makino；肥后苦竹●☆

304035　Pleioblastus hindsii（Munro）Nakai ＝ Pseudosasa hindsii（Munro）S. L. Chen et G. Y. Sheng ex T. G. Liang ●

304036　Pleioblastus hispidulus W. T. Lin ＝ Pseudosasa hindsii（Munro）S. L. Chen et G. Y. Sheng ex T. G. Liang ●

304037　Pleioblastus hsienchuensis T. H. Wen；仙居苦竹；Hsienchu Bitter Bamboo，Xianju Bitterbamboo ●

304038　Pleioblastus hsienchuensis T. H. Wen var. juxianensis（T. H. Wen et al.）S. L. Chen ex T. G. Liang et al. ＝ Pleioblastus juxianensis T. H. Wen，C. Y. Yao et S. Y. Chen ●

304039　Pleioblastus hsienchuensis T. H. Wen var. subglabratus（S. Y. Chen）C. S. Chao et G. Y. Yang；光箨苦竹；Subglabros Bitter

Bamboo, Velvetsheath Bitterbamboo ●

304040　Pleioblastus humilis (Mitford) Nakai;矮苦竹;Iow Bamboo ●☆

304041　Pleioblastus hupehensis J. L. Lu;风竹●

304042　Pleioblastus hupehensis J. L. Lu = Oligostachyum hupehense (J. L. Lu) Z. P. Wang et G. H. Ye ●

304043　Pleioblastus incarnatus S. L. Chen et G. Y. Sheng;绿苦竹; Fleshcoloured Bitter Bamboo, Flesh-coloured Bitter-bamboo, Green Bitterbamboo ●

304044　Pleioblastus intermedius S. Y. Chen;华丝竹;China Bitterbamboo, Chinese Bitterbamboo, Intermediate Bitter Bamboo, Intermediate Bitter-bamboo ●

304045　Pleioblastus juxianensis T. H. Wen,C. Y. Yao et S. Y. Chen;衢县苦竹;Juxian Bitter-bamboo, Quxian Bitter Bamboo, Quxian Bitterbamboo ●

304046　Pleioblastus kodzumae Makino f. higoensis (Makino) Sad. Suzuki = Pleioblastus higoensis Makino ●☆

304047　Pleioblastus kongosanensis Makino;金刚山苦竹●☆

304048　Pleioblastus kongosanensis Makino var. xystrophyllus (Koidz.) Sad. Suzuki;耙叶苦竹●☆

304049　Pleioblastus kunishii (Hayata) Ohki = Gelidocalamus kunishii (Hayata) P. C. Keng et T. H. Wen ●

304050　Pleioblastus kunishii (Hayata) Ohki et Nemoto = Gelidocalamus kunishii (Hayata) P. C. Keng et T. H. Wen ●

304051　Pleioblastus kwangsiensis W. Y. Hsiung et C. S. Chao;广西苦竹;Guangxi Bitter Bamboo, Guangxi Bitterbamboo, Kwangsi Bitterbamboo ●

304052　Pleioblastus kwangsiensis W. Y. Hsiung et C. S. Chao = Arundinaria oleosa (T. H. Wen) C. S. Chao et G. Y. Yang ●

304053　Pleioblastus kwangsiensis W. Y. Hsiung et C. S. Chao = Pleioblastus maculatus (McClure) C. D. Chu et C. S. Chao ●

304054　Pleioblastus linearis (Hack.) Nakai;琉球矢竹(仰叶竹); Linear Bitter-bamboo, Linear-leaf Bamboo, Linear-leaved Bamboo, Liuqiu Bitterbamboo ●

304055　Pleioblastus linearis (Hack.) Nakai f. albostriatus Muroi;白条琉球矢竹●☆

304056　Pleioblastus longifimbriatus S. Y. Chen;硬头苦竹;Hardhead Bitterbamboo, Long-fimbriate Bitter-bamboo, Longifimbriate Bitter Bamboo ●

304057　Pleioblastus longifimbriatus S. Y. Chen = Sinobambusa intermedia McClure ●

304058　Pleioblastus longinternodius B. M. Yang = Arundinaria oleosa (T. H. Wen) C. S. Chao et G. Y. Yang ●

304059　Pleioblastus longinternodius B. M. Yang = Pleioblastus oleosus T. H. Wen ●

304060　Pleioblastus longispiculatus B. M. Yang = Arundinaria oleosa (T. H. Wen) C. S. Chao et G. Y. Yang ●

304061　Pleioblastus longispiculatus B. M. Yang = Pleioblastus maculatus (McClure) C. D. Chu et C. S. Chao ●

304062　Pleioblastus longispiculatus B. M. Yang = Pleioblastus oleosus T. H. Wen ●

304063　Pleioblastus longqishanensis N. X. Zhao et Z. Y. Li;龙栖山苦竹;Longqishan Bitterbamboo ●

304064　Pleioblastus longqishanensis N. X. Zhao et Z. Yu Li = Pleioblastus amarus (Keng) P. C. Keng ●

304065　Pleioblastus maculatus (McClure) C. D. Chu et C. S. Chao;斑苦竹;Spotted Bitter Bamboo,Spotty Bitter-bamboo ●

304066　Pleioblastus maculatus (McClure) C. D. Chu et C. S. Chao =

Arundinaria oleosa (T. H. Wen) C. S. Chao et G. Y. Yang ●

304067　Pleioblastus maculatus (McClure) C. D. Chu et C. S. Chao var. longitubus Li et Wu = Pleioblastus maculatus (McClure) C. D. Chu et C. S. Chao ●

304068　Pleioblastus maculosoides T. H. Wen;丽水苦竹;Lishui Bitter Bamboo,Lishui Bitterbamboo,Lishui Bitter-bamboo ●

304069　Pleioblastus matsunoi Nakai ex Makino et Nemoto;松野苦竹●☆

304070　Pleioblastus maximowiczii Nakai = Pleioblastus chino (Franch. et Sav.) Makino ●

304071　Pleioblastus nagashima (Mitford) Nakai;长岛苦竹●☆

304072　Pleioblastus nagashima (Mitford) Nakai f. dokyoanus (Koidz.) Sad. Suzuki;道京苦竹●☆

304073　Pleioblastus nagashima (Mitford) Nakai var. koidzumii (Makino ex Koidz.) Sad. Suzuki;小泉氏苦竹■☆

304074　Pleioblastus naibunensis (Hayata) Kaneh. et Sasaki = Ampelocalamus naibunensis (Hayata) T. H. Wen ●

304075　Pleioblastus naibunensis (Hayata) Nakai = Drepanostachyum naibunense (Hayata) P. C. Keng ●

304076　Pleioblastus niitakayamensis (Hayata) Ohki = Yushania niitakayamensis (Hayata) P. C. Keng ●

304077　Pleioblastus niitakeayhamensis (Hayata) Ohki = Yushania niitakayamensis (Hayata) P. C. Keng ●

304078　Pleioblastus oedogonatus Z. P. Wang et G. H. Ye = Oligostachyum oedogonatum (Z. P. Wang et G. H. Ye) Q. F. Zheng et K. F. Huang ●

304079　Pleioblastus oiwakensis (Hayata) Ohki = Yushania niitakayamensis (Hayata) P. C. Keng ●

304080　Pleioblastus oleosus T. H. Wen = Arundinaria oleosa (T. H. Wen) C. S. Chao et G. Y. Yang ●

304081　Pleioblastus pandus (Keng) P. C. Keng = Pseudosasa hindsii (Munro) S. L. Chen et G. Y. Sheng ex T. G. Liang ●

304082　Pleioblastus pseudosasaoides Sad. Suzuki;拟茶秆竹●☆

304083　Pleioblastus pubescens Nakai;毛苦竹;Pubescent Bitterbamboo ●☆

304084　Pleioblastus pumilus (Mitford) Nakai;匍匐苦竹;Ground Bamboo,Low Bamboo ●☆

304085　Pleioblastus pygmaeus (Miq.) E. G. Camus = Sasa pygmaea (Miq.) E. G. Camus ●

304086　Pleioblastus pygmaeus (Miq.) Nakai;矮小苦竹;Dwarf Bamboo ●

304087　Pleioblastus pygmaeus (Miq.) Nakai 'Distichus' = Pleioblastus argenteostriatus (Regel) Nakai 'Distichus' ●

304088　Pleioblastus pygmaeus (Miq.) Nakai = Pleioblastus argenteostriatus (Regel) Nakai 'Distichus' ●

304089　Pleioblastus pygmaeus (Miq.) Nakai = Pleioblastus fortunei (Van Houtte) Nakai ●

304090　Pleioblastus pygmaeus (Miq.) Nakai = Sasa pygmaea (Miq.) E. G. Camus ●

304091　Pleioblastus pygmaeus (Miq.) Nakai var. distichus (Mitford) Nakai = Pleioblastus distichus (Mitford) Nakai ●

304092　Pleioblastus pygmaeus (Miq.) Nakai var. distichus (Mitford) Nakai = Sasa pygmaea (Miq.) E. G. Camus var. disticha (Mitford) C. S. Chao et G. G. Tang ●

304093　Pleioblastus pygmaeus (Miq.) Nakai var. distichus (Mitford.) Nakai = Pleioblastus argenteostriatus (Regel) Nakai 'Distichus' ●

304094　Pleioblastus rugatus T. H. Wen et S. Y. Chen;皱苦竹;Rugous Bitter-bamboo, Wrinkle Bitterbamboo, Wrinkled Bitter Bamboo ●

304095　Pleioblastus ruyuanensis W. T. Lin et Z. J. Feng = Oligostachyum scabriflorum (McClure) Z. P. Wang et G. H. Ye ●

304096　Pleioblastus sanmingensis S. L. Chen et G. Y. Sheng;三明苦竹; Sanming Bitter Bamboo, Sanming Bitterbamboo, Sanming Bitter-bamboo ●

304097　Pleioblastus shibuyanus Makino ex Nakai var. basihirsutus Sad. Suzuki;基毛苦竹●☆

304098　Pleioblastus simonii (Carrière) Nakai;川竹(空心苦竹,苦竹,女竹,皮竹,水苦竹); Arrow Bamboo, Medake Bamboo, Sichuan Bitterbamboo, Simon Bamboo, Simon Bitter Bamboo, Simon Bitter-bamboo, Simon Cane-bamboo ●

304099　Pleioblastus simonii (Carrière) Nakai 'Heterophyllus' = Pleioblastus simonii (Carrière) Nakai var. heterophyllus (Makino) Nakai ●☆

304100　Pleioblastus simonii (Carrière) Nakai f. variegatus (Hook. f.) Muroi;银条山竹●☆

304101　Pleioblastus simonii (Carrière) Nakai var. heterophyllus (Makino) Nakai;异叶川竹●☆

304102　Pleioblastus simonii (Carrière) Nakai var. variegata Makino = Pleioblastus simonii (Carrière) Nakai f. variegatus (Hook. f.) Muroi ●☆

304103　Pleioblastus solidus S. Y. Chen;实心苦竹;Solid Bitter Bamboo, Solid Bitterbamboo, Solid Bitter-bamboo ●

304104　Pleioblastus subrectangularis T. P. Yi et H. Long;方箨苦竹●

304105　Pleioblastus subrectangularis T. P. Yi et H. Long = Pleioblastus amarus (Keng) P. C. Keng ●

304106　Pleioblastus truncatus T. H. Wen;尖子竹;Truncate Bitterbamboo ●

304107　Pleioblastus usawae (Hayata) Ohki = Pseudosasa japonica (Siebold et Zucc. ex Steud.) Makino ex Nakai ●

304108　Pleioblastus usawai (Hayata) Ohki = Pseudosasa usawai (Hayata) Makino et Nemoto ●

304109　Pleioblastus variegatus (Siebold ex Miq.) Makino;小苦竹(稚子竹); Compact White-striped Bamboo, Dwarf Green-leaved Pleioblastus, Dwarf Plain-green Leaf Pleioblastus, Dwarf White-stripe Bamboo, Plain-green Leaf Pleioblastus, Variegated Bitter Bamboo ●

304110　Pleioblastus variegatus (Siebold ex Miq.) Makino 'Fortunei'; 佛氏小苦竹;Dwarf Whitestripe Bamboo, Dwarf White-stripe-leaved Bamboo, Fortunei Bamboo, White-striped-leaf Pleioblastus ●☆

304111　Pleioblastus variegatus (Siebold ex Miq.) Makino = Pleioblastus fortunei (Van Houtte) Nakai ●

304112　Pleioblastus variegatus (Siebold) Makino = Sasa fortunei (Van Houtte) Fiori ●

304113　Pleioblastus varius (Keng) P. C. Keng = Pleioblastus amarus (Keng) P. C. Keng ●

304114　Pleioblastus viridistriatus (Siebold ex André) Makino;绿条小苦竹●☆

304115　Pleioblastus viridistriatus (Siebold ex André) Makino f. chrysophyllus Makino;金叶小苦竹●☆

304116　Pleioblastus viridistriatus (Siebold ex André) Makino f. vagans (Gamble) Muroi;铺散小苦竹●☆

304117　Pleioblastus wuyishanensis Q. F. Zheng et K. F. Huang;武夷山苦竹;Wuyishan Bitter Bamboo, Wuyishan Bitterbamboo, Wuyishan Bitter-bamboo ●

304118　Pleioblastus yingdeensis W. T. Lin et Z. M. Wu = Pleioblastus amarus (Keng) P. C. Keng ●

304119　Pleioblastus yixingensis S. L. Chen et S. Y. Chen;宜兴苦竹;Yixing Bitter Bamboo, Yixing Bitterbamboo, Yixing Bitter-bamboo ●

304120　Pleiocardia Greene = Streptanthus Nutt. ■☆

304121　Pleiocardia Greene(1904);多心芥属■☆

304122　Pleiocardia breweri (A. Gray) Greene;多心芥■☆

304123　Pleiocardia gracilis Greene;细多心芥■☆

304124　Pleiocardia orbiculata Greene;圆多心芥■☆

304125　Pleiocarpa Benth. (1876);多果树属●☆

304126　Pleiocarpa bagshawei S. Moore = Pleiocarpa pycnantha (K. Schum.) Stapf ●☆

304127　Pleiocarpa bakueana A. Chev. = Pleiocarpa mutica Benth. ●☆

304128　Pleiocarpa bicarpellata Stapf;双小果多果树●☆

304129　Pleiocarpa breviloba (Hallier f.) Stapf = Pleiocarpa pycnantha (K. Schum.) Stapf ●☆

304130　Pleiocarpa brevistyla Omino;短柱多果树●☆

304131　Pleiocarpa camerunensis (K. Schum. ex Hallier f.) Brenan = Hunteria camerunensis K. Schum. ex Hallier f. ●☆

304132　Pleiocarpa flavescens Stapf = Pleiocarpa pycnantha (K. Schum.) Stapf ●☆

304133　Pleiocarpa hockii De Wild.;霍克多果树●☆

304134　Pleiocarpa micrantha Stapf = Pleiocarpa pycnantha (K. Schum.) Stapf ●☆

304135　Pleiocarpa microcarpa Stapf = Pleiocarpa pycnantha (K. Schum.) Stapf ●☆

304136　Pleiocarpa mutica Benth. ;多果树●☆

304137　Pleiocarpa picralimoides (Pichon) Omino;赤非夹竹桃多果树●☆

304138　Pleiocarpa pycnantha (K. Schum.) Stapf;密花多果树●☆

304139　Pleiocarpa pycnantha (K. Schum.) Stapf var. tubicina (Stapf) Pichon = Pleiocarpa pycnantha (K. Schum.) Stapf ●☆

304140　Pleiocarpa rostrata Benth. ;喙状多果树●☆

304141　Pleiocarpa salicifolia Stapf = Pleiocarpa mutica Benth. ●☆

304142　Pleiocarpa simii (Stapf) Hutch. et Dalziel = Hunteria simii (Stapf) H. Huber ●☆

304143　Pleiocarpa swynnertonii S. Moore = Pleiocarpa pycnantha (K. Schum.) Stapf ●☆

304144　Pleiocarpa talbotii Wernham = Pleiocarpa rostrata Benth. ●☆

304145　Pleiocarpa ternata A. Chev. = Pleiocarpa mutica Benth. ●☆

304146　Pleiocarpa tricarpellata Stapf = Pleiocarpa mutica Benth. ●☆

304147　Pleiocarpa tubicina Stapf = Pleiocarpa pycnantha (K. Schum.) Stapf ●☆

304148　Pleiocarpa welwitschii Stapf ex Hiern = Pleiocarpa pycnantha (K. Schum.) Stapf ●☆

304149　Pleiocarpidia K. Schum. (1897);繁果茜属●☆

304150　Pleiocarpidia capitata Bremek. ;头状繁果茜●☆

304151　Pleiocarpidia longipetala (Ridl.) Bremek. ;长瓣繁果茜●☆

304152　Pleiocarpidia magnifolia Bremek. ;大叶繁果茜●☆

304153　Pleiocarpidia pilosa (Ridl.) Bremek. ;柔毛繁果茜●☆

304154　Pleiocarpidia polyneura (Miq.) Bremek. ;多脉繁果茜●☆

304155　Pleioceras Baill. (1888);多角夹竹桃属●☆

304156　Pleioceras afzelii (K. Schum.) Stapf;阿芙泽尔多角夹竹桃●☆

304157　Pleioceras barteri Baill. ;巴尔多角夹竹桃●☆

304158　Pleioceras gilletii Stapf;吉勒特多角夹竹桃●☆

304159　Pleioceras glaberrima Wernham = Pleioceras gilletii Stapf ●☆

304160　Pleioceras oblonga Wernham = Pleioceras zenkeri Stapf ●☆

304161　Pleioceras orientale Vollesen;东方多角夹竹桃●☆

304162　Pleioceras stapfiana Wernham = Pleioceras gilletii Stapf ●☆

304163　Pleioceras talbotii Wernham = Pleioceras zenkeri Stapf ●☆

304164　Pleioceras whytei Stapf = Pleioceras afzelii (K. Schum.) Stapf ●☆

304165　Pleioceras zenkeri Stapf;岑克尔多角夹竹桃●☆

304166　Pleiochasia (Kamienski) Barnhart = Utricularia L. ■

304167　Pleiochasia Barnhart = Utricularia L. ■

304168　Pleiochiton Naudin ex A. Gray(1853);多被野牡丹属☆

304169　Pleiochiton crassifolium Naudin ex A. Gray;厚叶多被野牡丹☆

304170　Pleiochiton micranthum Cogn.;小花多被野牡丹☆

304171　Pleiochiton roseum Cogn.;粉红多被野牡丹☆

304172　Pleiococca F. Muell. = Acronychia J. R. Forst. et G. Forst.(保留属名)●

304173　Pleiocoryne Rauschert(1982);多棒茜属●☆

304174　Pleiocoryne fernandense(Hiern) Rauschert;费尔南多多棒茜●☆

304175　Pleiocoryne fernandense(Hiern) Rauschert var. pobeguinii(N. Hallé) J. -P. Lebrun et Stork;波别多棒茜☆

304176　Pleiocraterium Bremek.(1939);多杯茜属■☆

304177　Pleiocraterium plantaginifolium(Arn.) Bremek.;多杯茜■☆

304178　Pleiocraterium sumatranum Bremek.;苏门答腊多杯茜■☆

304179　Pleiodon Rchb. = Bouteloua Lag.(保留属名)■

304180　Pleiodon Rchb. = Polyodon Kunth ■

304181　Pleiogyne C. Koch = Cotula L. ■

304182　Pleiogyne K. Koch = Cotula L. ■

304183　Pleiogyne K. Koch(1843);肖山芫荽属■☆

304184　Pleiogyne anthemoides(L.) K. Koch = Cotula anthemoides L. ■

304185　Pleiogyne anthemoides K. Koch;肖山芫荽■☆

304186　Pleiogyne australis(Spreng.) K. Koch;南方肖山芫荽■☆

304187　Pleiogyne australis K. Koch = Pleiogyne australis(Spreng.) K. Koch ■☆

304188　Pleiogyne cardiospermum Edgew.;心籽肖山芫荽■☆

304189　Pleiogyne microcephala K. Koch;小头肖山芫荽■☆

304190　Pleiogyne multifida Sond.;多裂肖山芫荽■☆

304191　Pleiogyne nudicaulis K. Koch;裸茎肖山芫荽■☆

304192　Pleiogynium Engl.(1883);倍柱木属(伯德金李属)●☆

304193　Pleiogynium cerasiferum(F. Muell.) R. Parker = Pleiogynium timorense(DC.) Leenh.●☆

304194　Pleiogynium cerasiferum R. Parker = Pleiogynium timorense(DC.) Leenh.●☆

304195　Pleiogynium solandri(Benth.) Engl. = Pleiogynium timorense(DC.) Leenh.●☆

304196　Pleiogynium solandri Engl. = Pleiogynium timorense(DC.) Leenh.●☆

304197　Pleiogynium timorense(DC.) Leenh.;倍柱木(帝摩);Bordekin Plum, Burdekin Plum ●☆

304198　Pleiokirkia Capuron(1961);马岛苦木属●☆

304199　Pleiokirkia leandrii Capuron;马岛苦木●☆

304200　Pleioluma(Baill.) C. Baehni = Sersalisia R. Br.●

304201　Pleioluma Baill. = Pouteria Aubl.●

304202　Pleiomeris A. DC.(1841);管基紫金牛属●☆

304203　Pleiomeris canariensis(Willd.) A. DC.;管基紫金牛●☆

304204　Pleione D. Don(1825);独蒜兰属(一叶兰属);Pleione ■

304205　Pleione × christianii H. Perner;滇西独蒜兰■

304206　Pleione × taliensis P. J. Cribb et Butterf.;大理独蒜兰■

304207　Pleione alba H. Li et G. H. Feng = Pleione forrestii Schltr. var. alba(H. Li et G. H. Feng) P. J. Cribb ■

304208　Pleione alba H. Li et G. H. Feng = Pleione forrestii Schltr.■

304209　Pleione albiflora P. J. Cribb et C. Z. Tang;白花独蒜兰;White Pleione ■

304210　Pleione amoena Schltr. = Pleione bulbocodioides(Franch.) Rolfe ■

304211　Pleione amoena Schltr. = Pleione pleionoides(Kraenzl. ex Diels) Braem et H. Mohr ■

304212　Pleione aurita P. J. Cribb et Pfenning;艳花独蒜兰■

304213　Pleione aurita P. J. Cribb et Pfenning = Pleione chunii C. L. Tso ■

304214　Pleione autumnalis S. C. Chen et G. H. Zhu;长颈独蒜兰■

304215　Pleione barbarae Braem = Pleione grandiflora(Rolfe) Rolfe ■

304216　Pleione birmanica(Rchb. f.) B. S. Williams = Pleione praecox(Sm.) D. Don ■

304217　Pleione braemii Pinkep. = Pleione grandiflora(Rolfe) Rolfe ■

304218　Pleione bulbocodioides(Franch.) Rolfe;独蒜兰(冰球子,大独蒜兰,德氏独蒜兰,金灯花,毛慈姑,米兰状独蒜兰,台湾独蒜兰,台湾一叶兰);Common Pleione, Henry Pleione, Pleione, Taiwan Pleione ■

304219　Pleione bulbocodioides(Franch.) Rolfe = Pleione formosana Hayata ■

304220　Pleione bulbocodioides(Franch.) Rolfe f. nivea(Fukuy.) S. S. Ying = Pleione bulbocodioides(Franch.) Rolfe ■

304221　Pleione bulbocodioides(Franch.) Rolfe var. limprichtii(Schltr.) P. J. Cribb = Pleione limprichtii Schltr.■

304222　Pleione bulbocodioides(Franch.) Rolfe var. nivea(Fukuy.) Masam. = Pleione bulbocodioides(Franch.) Rolfe ■

304223　Pleione bulbocodioides(Franch.) Rolfe var. nivea(Fukuy.) S. S. Ying = Pleione formosana Hayata ■

304224　Pleione chinensis Kuntze = Coelogyne fimbriata Lindl.■

304225　Pleione chiwuana Ts. Tang et F. T. Wang;启无独蒜兰;Qiwu Pleione ■

304226　Pleione chiwuana Ts. Tang et F. T. Wang = Pleione yunnanensis(Rolfe) Rolfe ■

304227　Pleione chunii C. L. Tso;陈氏独蒜兰;Chun Pleione ■

304228　Pleione communis Gagnep. = Pleione bulbocodioides(Franch.) Rolfe ■

304229　Pleione communis Gagnep. var. subobtusum Gagnep. = Pleione bulbocodioides(Franch.) Rolfe ■

304230　Pleione concolor B. S. Williams = Pleione praecox(Sm.) D. Don ■

304231　Pleione confusa Cribb et C. Z. Tang;芳香独蒜兰;Fragrant Pleione ■

304232　Pleione corymbosa(Lindl.) Kuntze = Coelogyne corymbosa Lindl.■

304233　Pleione delavayi(Rolfe) Rolfe = Pleione bulbocodioides(Franch.) Rolfe ■

304234　Pleione diantha Schltr. = Pleione humilis(Sm.) D. Don ■

304235　Pleione diphylla Lindl. = Pleione maculata(Lindl.) Lindl. et Paxton ■

304236　Pleione diphylla Lindl. et Paxton = Pleione maculata(Lindl.) Lindl. et Paxton ■

304237　Pleione fargesii Gagnep. = Pleione bulbocodioides(Franch.) Rolfe ■

304238　Pleione fimbriata(Lindl.) Kuntze = Coelogyne fimbriata Lindl.■

304239　Pleione flaccida(Lindl.) Kuntze = Coelogyne flaccida Lindl.■

304240　Pleione flavida(Hook. f. ex Lindl.) Kuntze = Coelogyne prolifera Lindl.■

304241　Pleione formosana Hayata;台湾独蒜兰(石龙珠,台湾一叶兰)■

304242　Pleione formosana Hayata = Pleione bulbocodioides(Franch.) Rolfe ■

304243　Pleione formosana Hayata f. alba Torelli et Riccaboni = Pleione formosana Hayata ■

304244　Pleione formosana Hayata f. nivea Fukuy. = Pleione bulbocodioides(Franch.) Rolfe ■

304245　Pleione formosana Hayata f. nivea Fukuy. = Pleione formosana

Hayata ■

304246 Pleione formosana Hayata var. nivea（Fukuy.）Masam. ＝ Pleione bulbocodioides（Franch.）Rolfe ■

304247 Pleione formosana Hayata var. nivea（Fukuy.）Masam. ＝ Pleione formosana Hayata ■

304248 Pleione forrestii Schltr.；黄花独蒜兰（弗氏独蒜兰）；Forrest Pleione，Yellow Pleione ■

304249 Pleione forrestii Schltr. ＝ Pleione confusa Cribb et C. Z. Tang ■

304250 Pleione forrestii Schltr. f. alba（H. Li et G. H. Feng）Torelli et Riccaboni ＝ Pleione forrestii Schltr. var. alba（H. Li et G. H. Feng）P. J. Cribb ■

304251 Pleione forrestii Schltr. var. alba（H. Li et G. H. Feng）P. J. Cribb；白瓣独蒜兰 ■

304252 Pleione fuliginosa（Lodd. ex Hook.）Kuntze ＝ Coelogyne fimbriata Lindl. ■

304253 Pleione ganchuenensis Gagnep. ＝ Pleione bulbocodioides（Franch.）Rolfe ■

304254 Pleione gardneriana（Lindl.）Kuntze ＝ Neogyna gardneriana（Lindl.）Rchb. f. ■

304255 Pleione goweri（Rchb. f.）Kuntze ＝ Coelogyne punctulata Lindl. ■

304256 Pleione graminifolia（E. C. Parish et Rchb. f.）Kuntze ＝ Coelogyne viscosa Rchb. f. ■

304257 Pleione grandiflora（Rolfe）Rolfe；大花独蒜兰；Bigflower Pleione，Largeflower Pleione ■

304258 Pleione harberdii Braem ＝ Pleione grandiflora（Rolfe）Rolfe ■

304259 Pleione henryi（Rolfe）Schltr. ＝ Pleione bulbocodioides（Franch.）Rolfe ■

304260 Pleione henryi Rolfe ＝ Pleione bulbocodioides（Franch.）Rolfe ■

304261 Pleione hookeriana（Lindl.）B. S. Williams；毛唇独蒜兰（南独蒜兰）；Hairlip Pleione ■

304262 Pleione hookeriana（Lindl.）Rollisson f. nivea（Fukuy.）M. Hiroe ＝ Pleione formosana Hayata ■

304263 Pleione hookeriana（Lindl.）Rollisson var. brachyglossa（Rchb. f.）Rolfe ＝ Pholidota imbricata Hook. ■

304264 Pleione hookeriana（Lindl.）Rollisson var. sinensis G. Kleinh. ex Torelli et Riccaboni ＝ Pleione chunii C. L. Tso ■

304265 Pleione hubeiensis Torelli et Riccaboni ＝ Pleione pleionoides（Kraenzl. ex Diels）Braem et H. Mohr ■

304266 Pleione hui Schltr. ＝ Pleione formosana Hayata ■

304267 Pleione humilis（Sm.）D. Don；矮小独蒜兰（矮独蒜兰，矮生独蒜兰）；Dwarf Pleione ■

304268 Pleione humilis（Sm.）D. Don var. adnata Pfitzer ＝ Pleione humilis（Sm.）D. Don ■

304269 Pleione humilis（Sm.）D. Don var. pulchella E. W. Cooper ＝ Pleione humilis（Sm.）D. Don ■

304270 Pleione humilis（Sm.）D. Don var. purpurascens Pfitzer ＝ Pleione humilis（Sm.）D. Don ■

304271 Pleione kaatiae P. H. Peeters；卡氏独蒜兰 ■

304272 Pleione kohlsii Braem；春花独蒜兰；Spring Pleione ■

304273 Pleione lagenaria Lindl. et Paxton；酒瓶独蒜兰；Bottle Pleione ■☆

304274 Pleione laotica Kerr ＝ Pholidota imbricata Hook. ■

304275 Pleione laotica Kerr ＝ Pleione hookeriana（Lindl.）B. S. Williams ■

304276 Pleione limprichtii Schltr.；四川独蒜兰；Sichuan Pleione ■

304277 Pleione longipes（Lindl.）Kuntze ＝ Coelogyne longipes Lindl. ■

304278 Pleione maculata（Lindl.）Lindl. et Paxton；秋花独蒜兰（斑点独蒜兰）；Autumn Pleione，Spotted Pleione ■

304279 Pleione maculata（Lindl.）Lindl. et Paxton var. arthuriana（Rchb. f.）Rolfe ex Kraenzl. ＝ Pleione maculata（Lindl.）Lindl. et Paxton ■

304280 Pleione maculata（Lindl.）Lindl. et Paxton var. virginea Rchb. f. ＝ Pleione maculata（Lindl.）Lindl. et Paxton ■

304281 Pleione mairei Schltr. ＝ Pleione bulbocodioides（Franch.）Rolfe ■

304282 Pleione mandarinorum（Kraenzl.）Kraenzl. ＝ Ischnogyne mandarinorum（Kraenzl.）Schltr. ■

304283 Pleione microphylla S. C. Chen et Z. H. Tsi；小叶独蒜兰；Smallleaf Pleione ■

304284 Pleione milanii Braem ＝ Pleione chunii C. L. Tso ■

304285 Pleione moelleri Braem ＝ Pleione grandiflora（Rolfe）Rolfe ■

304286 Pleione mohrii Braem ＝ Pleione grandiflora（Rolfe）Rolfe ■

304287 Pleione nitida Kuntze ＝ Coelogyne punctulata Lindl. ■

304288 Pleione ochracea（Lindl.）Kuntze ＝ Coelogyne nitida（Wall. ex D. Don）Lindl. ■

304289 Pleione pinkepankii Braem et H. Mohr. ＝ Pleione grandiflora（Rolfe）Rolfe ■

304290 Pleione pleionoides（Kraenzl. ex Diels）Braem et H. Mohr；美丽独蒜兰；Spiffy Pleione ■

304291 Pleione praecox（Sm.）D. Don；疣鞘独蒜兰（秋花独蒜兰，早生独蒜兰）；Early Pleione，Reichenbach Pleione ■

304292 Pleione praecox（Sm.）D. Don var. birmanica（Rchb. f.）Grant ＝ Pleione praecox（Sm.）D. Don ■

304293 Pleione praecox（Sm.）D. Don var. reichenbachiana（T. Moore et Veitch）Torelli et Riccaboni ＝ Pleione praecox（Sm.）D. Don ■

304294 Pleione praecox（Sm.）D. Don var. wallichiana（Lindl.）E. W. Cooper ＝ Pleione praecox（Sm.）D. Don ■

304295 Pleione pricei Rolfe ＝ Pleione bulbocodioides（Franch.）Rolfe ■

304296 Pleione pricei Rolfe ＝ Pleione formosana Hayata ■

304297 Pleione prolifera（Lindl.）Kuntze ＝ Coelogyne prolifera Lindl. ■

304298 Pleione reichenbachiana（T. Moore et Veitch）Kuntze ＝ Pleione praecox（Sm.）D. Don ■

304299 Pleione reichenbachiana Kuntze ＝ Pleione praecox（Sm.）D. Don ■

304300 Pleione rhombilabia Hand.-Mazz. ＝ Pleione bulbocodioides（Franch.）Rolfe ■

304301 Pleione saxicola Ts. Tang et F. T. Wang ex S. C. Chen；岩生独蒜兰；Saxicolous Pleione ■

304302 Pleione scopulorum W. W. Sm.；二叶独蒜兰（岩石独蒜兰）；Rocky Pleione，Twoleaf Pleione ■

304303 Pleione smithii Schltr. ＝ Pleione bulbocodioides（Franch.）Rolfe ■

304304 Pleione speciosa Ames et Schltr. ＝ Pleione pleionoides（Kraenzl. ex Diels）Braem et H. Mohr ■

304305 Pleione suaveolens（Lindl.）Kuntze ＝ Coelogyne suaveolens（Lindl.）Hook. f. ■

304306 Pleione thuniana（Rchb. f.）Kuntze ＝ Panisea uniflora（Lindl.）Lindl. ■

304307 Pleione treutleri（Hook. f.）Kuntze ＝ Epigeneium treutleri（Hook. f.）Ormerod ■

304308 Pleione uniflora（Lindl.）Kuntze ＝ Panisea uniflora（Lindl.）Lindl. ■

304309 Pleione viscosa（Rchb. f.）Kuntze ＝ Coelogyne viscosa Rchb. f. ■

304310 Pleione votolinii Torelli et Riccaboni = Pleione pleionoides (Kraenzl. ex Diels) Braem et H. Mohr ■

304311 Pleione wallichiana (Lindl.) Lindl. = Pleione praecox (Sm.) D. Don ■

304312 Pleione wallichiana (Lindl.) Lindl. et Paxton = Pleione praecox (Sm.) D. Don ■

304313 Pleione yunnanensis (Rolfe) Rolfe；云南独蒜兰（冰球子，滇独蒜兰，独菇，独叶白芨，毛慈姑，糯白芨，小白芨）；Yunnan Pleione ■

304314 Pleione yunnanensis (Rolfe) Rolfe = Pleione bulbocodioides (Franch.) Rolfe ■

304315 Pleione yunnanensis (Rolfe) Rolfe var. chiwuana (Ts. Tang et F. T. Wang) G. Kleinh. ex Torelli et Riccaboni = Pleione yunnanensis (Rolfe) Rolfe ■

304316 Pleioneura (C. E. Hubb.) J. B. Phipps = Danthoniopsis Stapf ■☆

304317 Pleioneura Rech. f. (1951)；多脉石头花属■☆

304318 Pleioneura griffithiana (Boiss.) Rech. f.；多脉石头花■☆

304319 Pleioneura petiolata (J. B. Phipps) J. B. Phipps = Danthoniopsis petiolata (J. B. Phipps) Clayton ■☆

304320 Pleioneura ramosa (Stapf) J. B. Phipps = Danthoniopsis ramosa (Stapf) Clayton ■☆

304321 Pleioneura simulans (C. E. Hubb.) J. B. Phipps = Danthoniopsis simulans (C. E. Hubb.) Clayton ■☆

304322 Pleiophaca F. Muell. ex Baill. = Archidendron F. Muell. ●

304323 Pleiosepalum Hand. -Mazz. = Aruncus L. ●■

304324 Pleiosepalum Moss = Krauseola Pax et K. Hoffm. ■☆

304325 Pleiosepalum gombalanum Hand. -Mazz. = Aruncus gombalanus Hand. -Mazz. ■

304326 Pleiosepalum mosambicinum Moss = Krauseola mosambicina (Moss) Pax et K. Hoffm. ■☆

304327 Pleiosmilax Seem. = Smilax L. ●

304328 Pleiosorbus L. H. Zhou et C. Y. Wu = Sorbus L. ●

304329 Pleiosorbus L. H. Zhou et C. Y. Wu (2000)；多蕊石灰树属；Pleiosorbus ■★

304330 Pleiosorbus megacarpus L. H. Zhou et C. Y. Wu；多蕊石灰树；Pleiosorbus ●

304331 Pleiosorbus megacarpus L. H. Zhou et C. Y. Wu = Sorbus medogensis L. T. Lu et T. C. Ku ●

304332 Pleiospermium (Engl.) Swingle (1916)；多籽橘属（多子橘属）●☆

304333 Pleiospermium Swingle = Pleiospermium (Engl.) Swingle ●☆

304334 Pleiospermium alatum (Wight et Arn.) Swingle；翅多籽橘●☆

304335 Pleiospermium longisepalum Swingle；长花多籽橘●☆

304336 Pleiospilos N. E. Br. (1925)；对叶花属（凤卵玉属，凤卵属）；Loving-rock ■☆

304337 Pleiospilos archeri L. Bolus = Tanquana archeri (L. Bolus) H. E. K. Hartmann et Liede ■☆

304338 Pleiospilos barbarae Karrer = Pleiospilos bolusii (Hook. f.) N. E. Br. ■☆

304339 Pleiospilos beaufortensis L. Bolus = Pleiospilos bolusii (Hook. f.) N. E. Br. ■☆

304340 Pleiospilos bolusii (Hook. f.) N. E. Br.；对叶花（凤卵，凤卵草）；Bolus Loving-rock, Living Rock, Mimicry Plant ■☆

304341 Pleiospilos borealis L. Bolus = Pleiospilos compactus (Aiton) Schwantes subsp. canus (Haw.) H. E. K. Hartmann et Liede ■☆

304342 Pleiospilos canus (Haw.) L. Bolus = Pleiospilos compactus (Aiton) Schwantes subsp. canus (Haw.) H. E. K. Hartmann et Liede ■☆

304343 Pleiospilos clavatus L. Bolus = Tanquana archeri (L. Bolus) H. E. K. Hartmann et Liede ■☆

304344 Pleiospilos compactus (Aiton) Schwantes；贵凤卵；Mimicry Plant ■☆

304345 Pleiospilos compactus (Aiton) Schwantes subsp. canus (Haw.) H. E. K. Hartmann et Liede；白贵凤卵■☆

304346 Pleiospilos compactus (Aiton) Schwantes subsp. fergusoniae (L. Bolus) H. E. K. Hartmann et Liede = Pleiospilos fergusoniae L. Bolus ■☆

304347 Pleiospilos compactus (Aiton) Schwantes subsp. minor (L. Bolus) H. E. K. Hartmann et Liede；小贵凤卵■☆

304348 Pleiospilos compactus (Aiton) Schwantes subsp. sororius (N. E. Br.) H. E. K. Hartmann et Liede；堆积贵凤卵■☆

304349 Pleiospilos compactus Schwantes = Pleiospilos compactus (Aiton) Schwantes ■☆

304350 Pleiospilos dekenahi (N. E. Br.) Schwantes = Pleiospilos compactus (Aiton) Schwantes subsp. canus (Haw.) H. E. K. Hartmann et Liede ■☆

304351 Pleiospilos dekenahi Schwantes；如来■☆

304352 Pleiospilos dimidiatus L. Bolus = Pleiospilos compactus (Aiton) Schwantes subsp. sororius (N. E. Br.) H. E. K. Hartmann et Liede ■☆

304353 Pleiospilos fergusoniae L. Bolus；鸾城■☆

304354 Pleiospilos fergusoniae L. Bolus = Pleiospilos compactus (Aiton) Schwantes subsp. fergusoniae (L. Bolus) H. E. K. Hartmann et Liede ■☆

304355 Pleiospilos framesii L. Bolus = Pleiospilos compactus (Aiton) Schwantes subsp. canus (Haw.) H. E. K. Hartmann et Liede ■☆

304356 Pleiospilos grandiflorus L. Bolus = Pleiospilos compactus (Aiton) Schwantes subsp. canus (Haw.) H. E. K. Hartmann et Liede ■☆

304357 Pleiospilos hilmarii L. Bolus；希氏对叶花■☆

304358 Pleiospilos hilmarii L. Bolus = Tanquana hilmarii (L. Bolus) H. E. K. Hartmann et Liede ■☆

304359 Pleiospilos kingiae L. Bolus = Pleiospilos compactus (Aiton) Schwantes subsp. canus (Haw.) H. E. K. Hartmann et Liede ■☆

304360 Pleiospilos latifolius L.；大凤卵草■☆

304361 Pleiospilos latifolius L. Bolus = Pleiospilos compactus (Aiton) Schwantes subsp. canus (Haw.) H. E. K. Hartmann et Liede ■☆

304362 Pleiospilos latipetalus L. Bolus = Pleiospilos compactus (Aiton) Schwantes subsp. canus (Haw.) H. E. K. Hartmann et Liede ■☆

304363 Pleiospilos leipoldtii L. Bolus = Pleiospilos compactus (Aiton) Schwantes subsp. canus (Haw.) H. E. K. Hartmann et Liede ■☆

304364 Pleiospilos loganii L. Bolus = Tanquana archeri (L. Bolus) H. E. K. Hartmann et Liede ■☆

304365 Pleiospilos longibracteatus L. Bolus = Pleiospilos compactus (Aiton) Schwantes ■☆

304366 Pleiospilos longisepalus L. Bolus = Pleiospilos compactus (Aiton) Schwantes subsp. canus (Haw.) H. E. K. Hartmann et Liede ■☆

304367 Pleiospilos magnipunctatus (Haw.) Schwantes；斑点凤卵草（凤翼）■☆

304368 Pleiospilos magnipunctatus (Haw.) Schwantes = Pleiospilos compactus (Aiton) Schwantes subsp. canus (Haw.) H. E. K. Hartmann et Liede ■☆

304369 Pleiospilos magnipunctatus (Haw.) Schwantes var. inaequalis L. Bolus = Pleiospilos compactus (Aiton) Schwantes subsp. canus (Haw.) H. E. K. Hartmann et Liede ■☆

304370 Pleiospilos magnipunctatus (Haw.) Schwantes var. sesquiuncialis L. Bolus = Pleiospilos compactus (Aiton) Schwantes subsp. canus (Haw.) H. E. K. Hartmann et Liede ■☆

304371 Pleiospilos minor L. Bolus = Pleiospilos compactus (Aiton) Schwantes subsp. minor (L. Bolus) H. E. K. Hartmann et Liede ■☆

304372 Pleiospilos nelii Schwantes;大花风卵草(帝玉);Split Rock ■☆

304373 Pleiospilos nobilis (Haw.) Schwantes = Pleiospilos compactus (Aiton) Schwantes subsp. canus (Haw.) H. E. K. Hartmann et Liede ■☆

304374 Pleiospilos optatus (N. E. Br.) Schwantes = Pleiospilos compactus (Aiton) Schwantes ■☆

304375 Pleiospilos pedunculatus L. Bolus = Pleiospilos nelii Schwantes ■☆

304376 Pleiospilos peersii L. Bolus = Pleiospilos compactus (Aiton) Schwantes subsp. canus (Haw.) H. E. K. Hartmann et Liede ■☆

304377 Pleiospilos prismaticus (Schwantes) Schwantes = Tanquana prismatica (Schwantes) H. E. K. Hartmann et Liede ■☆

304378 Pleiospilos roodiae (N. E. Br.) Schwantes = Tanquana prismatica (Schwantes) H. E. K. Hartmann et Liede ■☆

304379 Pleiospilos rouxii L. Bolus = Pleiospilos compactus (Aiton) Schwantes subsp. canus (Haw.) H. E. K. Hartmann et Liede ■☆

304380 Pleiospilos sesquiuncialis (N. E. Br.) Schwantes = Pleiospilos compactus (Aiton) Schwantes subsp. canus (Haw.) H. E. K. Hartmann et Liede ■☆

304381 Pleiospilos simulans (Marloth) N. E. Br.;波疣风卵草(青鸾)■☆

304382 Pleiospilos sororius (N. E. Br.) Schwantes = Pleiospilos compactus (Aiton) Schwantes subsp. sororius (N. E. Br.) H. E. K. Hartmann et Liede ■☆

304383 Pleiospilos tricolor N. E. Br. = Pleiospilos nelii Schwantes ■☆

304384 Pleiospilos willowmorensis L. Bolus = Pleiospilos compactus (Aiton) Schwantes subsp. canus (Haw.) H. E. K. Hartmann et Liede ■☆

304385 Pleiospora Harv. = Pearsonia Dümmer ●☆

304386 Pleiospora Harv. = Phaenohoffmannia Kuntze ●☆

304387 Pleiospora bolusii Dümmer = Pearsonia cajanifolia (Harv.) Polhill subsp. cryptantha (Baker) Polhill ●☆

304388 Pleiospora buchananii Harms = Pearsonia cajanifolia (Harv.) Polhill subsp. cryptantha (Baker) Polhill ●☆

304389 Pleiospora cajanifolia Harv. = Pearsonia cajanifolia (Harv.) Polhill ●☆

304390 Pleiospora cryptantha (Baker) Baker = Pearsonia cajanifolia (Harv.) Polhill subsp. cryptantha (Baker) Polhill ●☆

304391 Pleiospora gracilior Dümmer = Pearsonia cajanifolia (Harv.) Polhill ●☆

304392 Pleiospora grandifolia (Bolus) Dümmer = Pearsonia grandifolia (Bolus) Polhill ●☆

304393 Pleiospora holosericea Schinz = Pearsonia cajanifolia (Harv.) Polhill subsp. cryptantha (Baker) Polhill ●☆

304394 Pleiospora latibracteolata Dümmer = Pearsonia grandifolia (Bolus) Polhill subsp. latibracteolata (Dümmer) Polhill ●☆

304395 Pleiospora macrophylla Dümmer = Pearsonia cajanifolia (Harv.) Polhill subsp. cryptantha (Baker) Polhill ●☆

304396 Pleiospora obovata Schinz = Pearsonia obovata (Schinz) Polhill ●☆

304397 Pleiospora paniculata Bolus ex Dümmer = Pearsonia cajanifolia (Harv.) Polhill subsp. cryptantha (Baker) Polhill ●☆

304398 Pleiostachya K. Schum. (1902);繁花竹芋属■☆

304399 Pleiostachya leiostachya (Donn. Sm.) Hammel;光穗繁花竹芋■☆

304400 Pleiostachya morlaei K. Schum. ;繁花竹芋■☆

304401 Pleiostachyopiper Trel. (1934);繁花胡椒属●☆

304402 Pleiostachyopiper Trel. = Piper L. ●■

304403 Pleiostachyopiper nudilimbum (C. DC.) Trel. ;繁花胡椒●☆

304404 Pleiostemon Sond. (1850);多蕊大戟属●☆

304405 Pleiostemon Sond. = Flueggea Willd. ●

304406 Pleiostemon verrucosum (Thunb.) Sond. = Flueggea verrucosa (Thunb.) G. L. Webster ●☆

304407 Pleiosyngyne Baum. -Bod. (1992);大叶假山毛榉属●☆

304408 Pleiosyngyne Baum. -Bod. = Nothofagus Blume;(保留属名)●☆

304409 Pleiotaenia J. M. Coult. et Rose = Polytaenia DC. ■☆

304410 Pleiotaenia nuttallii (DC.) J. M. Coult. et Rose = Polytaenia nuttallii DC. ■☆

304411 Pleiotaxis Steetz(1864);多肋菊属●■☆

304412 Pleiotaxis affinis O. Hoffm. ;近缘多肋菊■☆

304413 Pleiotaxis ambigua S. Moore;可疑多肋菊■☆

304414 Pleiotaxis amoena R. E. Fr. = Pleiotaxis eximia O. Hoffm. ■☆

304415 Pleiotaxis angolensis Rodr. Oubina et S. Ortiz;安哥拉多肋菊■☆

304416 Pleiotaxis angustirugosa C. Jeffrey;窄皱褶多肋菊■☆

304417 Pleiotaxis antunesii O. Hoffm. ;安图内思多肋菊■☆

304418 Pleiotaxis antunesii O. Hoffm. var. planifolia S. Moore = Pleiotaxis dewevrei O. Hoffm. ex T. Durand et De Wild. ■☆

304419 Pleiotaxis arenaria Milne-Redh. = Pleiotaxis fulva Hiern ■☆

304420 Pleiotaxis argentea M. Taylor = Pleiotaxis huillensis O. Hoffm. subsp. argentea (M. Taylor) S. Ortiz et Rodr. Oubina ■☆

304421 Pleiotaxis bampsiana Lisowski;邦氏多肋菊■☆

304422 Pleiotaxis baumii S. Moore = Pleiotaxis ambigua S. Moore ■☆

304423 Pleiotaxis buscalioni Chiov. = Ochrocephala imatongensis (Philipson) Dittrich ■☆

304424 Pleiotaxis chlorolepis C. Jeffrey;绿鳞多肋菊■☆

304425 Pleiotaxis clivicola S. Moore = Pleiotaxis subpaniculata Chiov. ☆

304426 Pleiotaxis davyi S. Moore = Pleiotaxis rogersii S. Moore ■☆

304427 Pleiotaxis decipiens C. Jeffrey;迷惑多肋菊■☆

304428 Pleiotaxis dewevrei O. Hoffm. ex T. Durand et De Wild. ;德韦多肋菊■☆

304429 Pleiotaxis duvigneaudii Lisowski;迪维尼奥多肋菊■☆

304430 Pleiotaxis eximia O. Hoffm. ;优异多肋菊■☆

304431 Pleiotaxis eximia O. Hoffm. subsp. kassneri (S. Moore) G. V. Pope;卡斯纳多肋菊☆

304432 Pleiotaxis fulva Hiern;黄褐多肋菊■☆

304433 Pleiotaxis gossweileri S. Moore = Pleiotaxis huillensis O. Hoffm. ■☆

304434 Pleiotaxis huillensis O. Hoffm. ;威拉多肋菊■☆

304435 Pleiotaxis huillensis O. Hoffm. subsp. argentea (M. Taylor) S. Ortiz et Rodr. Oubina;银白多肋菊☆

304436 Pleiotaxis huillensis O. Hoffm. subsp. axillaris S. Ortiz et Rodr. Oubina;腋生多肋菊☆

304437 Pleiotaxis huillensis O. Hoffm. var. macrocephala S. Ortiz et Rodr. Oubina;大头威拉多肋菊■☆

304438 Pleiotaxis jeffreyana Lisowski;杰弗里多肋菊■☆

304439 Pleiotaxis kassneri S. Moore = Pleiotaxis eximia O. Hoffm. subsp. kassneri (S. Moore) G. V. Pope ■☆

304440 Pleiotaxis kassneri S. Moore var. angustifolia ? = Pleiotaxis eximia O. Hoffm. subsp. kassneri (S. Moore) G. V. Pope ■☆

304441 Pleiotaxis latisquama S. Moore = Pleiotaxis pulcherrima Steetz ■☆

304442 Pleiotaxis lawalreeana Lisowski;拉瓦尔多肋菊■☆

304443 Pleiotaxis lejolyana Lisowski;勒若利多肋菊■☆

304444 Pleiotaxis linearifolia O. Hoffm. ;线叶多肋菊■☆

304445 Pleiotaxis macrophylla Muschl. ex S. Moore;大叶多肋菊■☆

304446 Pleiotaxis newtonii O. Hoffm. ;纽敦多肋菊■☆

304447 Pleiotaxis overlaetii Staner;奥弗莱特多肋菊☆

304448 Pleiotaxis oxylepis C. Jeffrey;尖鳞多肋菊■☆

304449 Pleiotaxis paucinervia C. Jeffrey;少脉多肋菊■☆

304450 Pleiotaxis perfoliata Lisowski;穿叶多肋菊■☆

304451 Pleiotaxis petitiana Lisowski;佩蒂蒂多肋菊■☆

304452 Pleiotaxis pulcherrima Steetz;美丽多肋菊■☆

304453 Pleiotaxis pulcherrima Steetz var. angustifolia O. Hoffm. = Pleiotaxis pulcherrima Steetz ■☆

304454 Pleiotaxis pulcherrima Steetz var. poggeana O. Hoffm. = Pleiotaxis pulcherrima Steetz ■☆

304455 Pleiotaxis racemosa O. Hoffm. ;总花多肋菊■☆

304456 Pleiotaxis robynsiana Lisowski;罗宾斯多肋菊■☆

304457 Pleiotaxis rogersii S. Moore;罗杰斯多肋菊■☆

304458 Pleiotaxis rugosa O. Hoffm. ;皱褶多肋菊■☆

304459 Pleiotaxis rugosa O. Hoffm. var. angustifolia ? = Pleiotaxis rugosa O. Hoffm. ■☆

304460 Pleiotaxis rugosa O. Hoffm. var. auriculata Welw. ex Hiern = Pleiotaxis rugosa O. Hoffm. ■☆

304461 Pleiotaxis sapinii S. Moore = Pleiotaxis pulcherrima Steetz ■☆

304462 Pleiotaxis sciaphila S. Moore = Pleiotaxis dewevrei O. Hoffm. ex T. Durand et De Wild. ■☆

304463 Pleiotaxis subpaniculata Chiov. ;圆锥多肋菊■☆

304464 Pleiotaxis subscaposa C. Jeffrey;亚莲茎多肋菊■☆

304465 Pleiotaxis upembensis Lisowski;乌彭贝多肋菊■☆

304466 Pleiotaxis vernonioides S. Moore = Pleiotaxis dewevrei O. Hoffm. ex T. Durand et De Wild. ■☆

304467 Pleiotaxis welwitschii S. Moore;韦尔多肋菊■☆

304468 Pleisolirion Raf. = Paradisea Mazzuc. (保留属名)■☆

304469 Pleistachyopiper Trel. = Piper L. ●■

304470 Plenckia Moc. et Sessé ex DC. = Choisya Kunth ●☆

304471 Plenckia Raf. (废弃属名) = Glinus L. ■

304472 Plenckia Raf. (废弃属名) = Plenckia Reissek(保留属名)●☆

304473 Plenckia Reissek(1861)(保留属名);普伦卫矛属●☆

304474 Plenckia integerrima Lundell;全缘普伦卫矛●☆

304475 Plenckia microcarpa Lundell;小果普伦卫矛●☆

304476 Plenckia populnea Reissek;普伦卫矛●☆

304477 Pleocarphus D. Don = Jungia L. f. (保留属名)■●☆

304478 Pleocarphus D. Don(1830);卷叶菊属●☆

304479 Pleocarphus revolutus D. Don;卷叶菊●☆

304480 Pleocarpus Walp. = Pleocarphus D. Don ●☆

304481 Pleocaulus Bremek. = Strobilanthes Blume ●■

304482 Pleodendron Tiegh. (1899);多瓣樟属●☆

304483 Pleodendron macranthum Tiegh. ;多瓣樟●☆

304484 Pleodiporochna Tiegh. = Ochna L. ●

304485 Pleodiporochna buettneri (Engl. et Gilg) Tiegh. = Ochna latisepala (Tiegh.) Bamps ●☆

304486 Pleogyne Miers ex Benth. = Pleogyne Miers ●☆

304487 Pleogyne Miers(1851);多心藤属●☆

304488 Pleogyne australis Benth. ;多心藤●☆

304489 Pleomele Salisb. (1796);剑叶木属(龙血树属)●

304490 Pleomele Salisb. = Dracaena Vand. ex L. ●■

304491 Pleomele angustifolia (Roxb.) N. E. Br. = Dracaena angustifolia Roxb. ●

304492 Pleomele angustifolia N. E. Br. = Dracaena angustifolia Roxb. ●

304493 Pleomele cambodiana (Gagnep.) Merr. et Chun = Dracaena cambodiana Pierre ex Gagnep. ●◇

304494 Pleomele cambodiana (Pierre ex Gagnep.) Merr. et Chun = Dracaena cambodiana Pierre ex Gagnep. ●◇

304495 Pleomele cochinchinensis (Lour.) Merr. = Dracaena cochinchinensis (Lour.) S. C. Chen ●◇

304496 Pleomele cochinchinensis Merr. ex Gagnep. = Dracaena cochinchinensis (Lour.) S. C. Chen ●◇

304497 Pleomele godseffiana N. E. Br. = Dracaena godseffiana Baker ●

304498 Pleomele goldieana N. E. Br. = Dracaena goldieana Sander ex Mast. ●

304499 Pleomele hookeriana (K. Koch) N. E. Br. = Dracaena aletriformis (Haw.) Bos ●☆

304500 Pleomele humilis (Baker) N. E. Br. = Dracaena aubryana Brongn. ex E. Morren ●☆

304501 Pleomele sanderiana N. E. Br. = Dracaena sanderiana Sander ●

304502 Pleomele thalioides (E. Morren) N. E. Br. = Dracaena aubryana Brongn. ex E. Morren ●☆

304503 Pleomele thalioides (E. Morren) N. E. Br. = Dracaena thalioides E. Morren ●☆

304504 Pleonanthus Ehrh. = Dianthus L. ■

304505 Pleonanthus Ehrh. = Kohlrauschia Kunth ■☆

304506 Pleonotoma Miers(1863);多节花属●☆

304507 Pleonotoma albiflora (Salzm. ex DC.) A. H. Gentry;白花多节花●☆

304508 Pleonotoma auriculatum K. Schum. ex Sprague;耳状多节花●☆

304509 Pleonotoma brittoni Rusby;布氏多节花●☆

304510 Pleonotoma diversifolium (Kunth) Bureau et K. Schum. ;异叶多节花●☆

304511 Pleonotoma flava Miers;黄多节花●☆

304512 Pleonotoma orientalis Sandwith;东方多节花●☆

304513 Pleopadium Raf. = Croton L. ●

304514 Pleopetalum Tiegh. = Ochna L. ●

304515 Pleopogon Nutt. = Lycurus Kunth ■☆

304516 Pleorothyrium Endl. = Ocotea Aubl. ●☆

304517 Pleorothyrium Endl. = Pleurothyrium Nees ex Lindl. ●☆

304518 Pleotheca Wall. = Spiradiclis Blume ●■

304519 Pleouratea Tiegh. = Ouratea Aubl. (保留属名)●

304520 Pleradenophora Esser = Sebastiania Spreng. ●

304521 Pleradenophora Esser(2001);长尖地阳桃属●☆

304522 Pleragina Arruda = Licania Aubl. + Couepia Aubl. ●☆

304523 Pleragina Arruda ex Kost. = Licania Aubl. + Couepia Aubl. ●☆

304524 Plerandra A. Gray = Schefflera J. R. Forst. et G. Forst. (保留属名)●

304525 Plerandra jatrophifolia Hance = Trevesia palmata (Roxb. ex Lindl.) Vis. ●

304526 Plerandropsis R. Vig. = Trevesia Vis. ●

304527 Pleroma D. Don = Tibouchina Aubl. ●■☆

304528 Plesiagopus Raf. = Ipomoea L. (保留属名)●■

304529 Plesiatropha Pierre = Plesiatropha Pierre ex Hutch. ■☆

304530 Plesiatropha Pierre ex Hutch. (1912);肖麻疯树属■☆

304531 Plesiatropha Pierre ex Hutch. = Mildbraedia Pax ■☆

304532 Plesiatropha carpinifolia (Pax) Breteler;鹅耳枥叶肖麻疯树■☆

304533 Plesiatropha carpinifolia (Pax) Breteler var. strigosa (A. R. Sm.) Breteler;糙伏毛肖麻疯树■☆

304534 Plesiatropha paniculata (Pax) Breteler;圆锥肖麻疯树■☆

304535 Plesiatropha paniculata (Pax) Breteler var. occidentalis (J. Léonard) Breteler;西方肖米尔大戟■☆

304536 Plesilia Raf. = Breweria R. Br. ●☆

304537 Plesilia Raf. = Stylisma Raf. ■☆

304538 Plesiopsora Raf. = Scabiosa L. ●■

304539 Plesisa Raf. = Utricularia L. ■

304540 Plesmonium Schott = Amorphophallus Blume ex Decne. (保留属名)■●

304541 Plesmonium Schott(1856);印度芋属■☆

304542 Plesmonium margaritiferum (Roxb.) Schott;印度芋■☆

304543 Plesmonium margaritiferum (Roxb.) Schott = Arum margaritiferum Roxb. ■☆

304544 Plethadenia Urb. (1912);群腺芸香属■☆

304545 Plethadenia cubensis Urb. ;古巴群腺芸香●☆

304546 Plethadenia granulata Urb. ;群腺芸香●☆

304547 Plethiandra Hook. f. (1867);群雄野牡丹属●☆

304548 Plethiandra acuminata Merr. ;渐尖群腺芸香●☆

304549 Plethiandra cuneata Stapf;楔形群腺芸香●☆

304550 Plethiandra hookeri Stapf;胡克群腺芸香●☆

304551 Plethiandra robusta (Cogn.) Nayar;粗壮群腺芸香●☆

304552 Plethiandra sessiliflora Merr. ;无梗群腺芸香●☆

304553 Plethiandra tomentosa G. Kadereit;绒毛群腺芸香●☆

304554 Plethiosphace (Benth.) Opiz = Salvia L. ●■

304555 Plethiosphace Opiz = Salvia L. ●■

304556 Plethostephia Miers = Cordia L. (保留属名)●

304557 Plethyrsis Raf. = Richardia L. ■

304558 Plettkea Mattf. (1934);坚果繁缕属■☆

304559 Plettkea cryptantha Mattf. ;阴花坚果繁缕■☆

304560 Plettkea macrophylla (Muschl.) Mattf. ;大叶坚果繁缕■☆

304561 Pleudia Raf. = Salvia L. ●■

304562 Pleurachne Schrad. = Ficinia Schrad. (保留属名)■☆

304563 Pleurachne sieberi Schrad. = Ficinia secunda (Vahl) Kunth ■☆

304564 Pleuradena Raf. = Euphorbia L. ●■

304565 Pleuradenia B. D. Jacks. = Pleuradena Raf. ●■

304566 Pleuradenia Raf. = Collinsonia L. ■☆

304567 Pleuralluma Plowes = Caralluma R. Br. ■

304568 Pleurandra Labill. = Hibbertia Andréws ●☆

304569 Pleurandra Raf. = Gaura L. ■

304570 Pleurandra Raf. = Pleurostemon Raf. ■

304571 Pleurandropsis Baill. = Asterolasia F. Muell. ●☆

304572 Pleurandros St. -Lag. = Hibbertia Andréws ●☆

304573 Pleurandros St. -Lag. = Pleurandra Labill. ●☆

304574 Pleuranthe Salisb. = Protea L. (保留属名)●☆

304575 Pleuranthe subulifolia Salisb. ex Knight = Protea subulifolia (Salisb. ex Knight) Rourke ●☆

304576 Pleuranthemum (Pichon) Pichon = Hunteria Roxb. ●

304577 Pleuranthemum ballayi (Hua) Pichon = Hunteria ballayi Hua ●☆

304578 Pleuranthium Benth. = Epidendrum L. (保留属名)■☆

304579 Pleuranthium (Rchb. f. Benth. = Epidendrum L. (保留属名)■☆

304580 Pleuranthodendron L. O. Williams(1961);侧花椴属●☆

304581 Pleuranthodendron mexicana (A. Gray) L. O. Williams;侧花椴●☆

304582 Pleuranthodes Weberb. (1896);腋花鼠李属●☆

304583 Pleuranthodes hillebrandtii Weberbauer;腋花鼠李●☆

304584 Pleuranthodes orbiculare Weberbauer;圆腋花鼠李●☆

304585 Pleuranthodium (K. Schum.) R. M. Sm. (1991);侧花姜属■☆

304586 Pleuranthodium (K. Schum.) R. M. Sm. = Alpinia Roxb. (保留属名)■

304587 Pleuranthodium branderhorstii (Valeton) R. M. Sm. ;侧花姜■☆

304588 Pleuranthodium floribundum (K. Schum.) R. M. Sm. ;繁花侧花姜■☆

304589 Pleuranthodium pterocarpum (K. Schum.) R. M. Sm. ;翅果侧花姜■☆

304590 Pleuranthus Rich. ex Pers. = Dulichium Pers. ■☆

304591 Pleuraphis Torr. (1824);侧芒禾属(海氏草属,黑拉禾属)■☆

304592 Pleuraphis Torr. = Hilaria Kunth ■☆

304593 Pleuraphis rigida Thurb. ;硬侧芒禾;Big Galleta ■☆

304594 Pleuraphis sericea Nutt. ex Benth. ;绢毛侧芒禾■☆

304595 Pleurastis Raf. = Lycoris Herb. ■

304596 Pleureia Raf. = Psychotria L. (保留属名)●

304597 Pleuremidis Raf. = Thunbergia Retz. (保留属名)●■

304598 Pleurendotria Raf. (废弃属名) = Lithophragma (Nutt.) Torr. et A. Gray(保留属名)●☆

304599 Pleurenodon Raf. = Hypericum L. ■●

304600 Pleuriarum Nakai = Arisaema Mart. ●■

304601 Pleuricospora A. Gray(1868);歪子杜鹃属(黄晶兰属)●☆

304602 Pleuricospora densa Small;密歪子杜鹃●☆

304603 Pleuricospora fimbriolata A. Gray;歪子杜鹃●☆

304604 Pleuricospora longipetala Howell;长瓣歪子杜鹃●☆

304605 Pleurima Raf. = Campanula L. ■●

304606 Pleurimaria B. D. Jacks. = Blackstonia Huds. ■☆

304607 Pleurimaria B. D. Jacks. = Plurimaria Raf. ■☆

304608 Pleuripetalum T. Durand = Anaxagorea A. St. -Hil. ●

304609 Pleuripetalum T. Durand = Eburopetalum Becc. ●

304610 Pleurisanthaceae Tiegh. = Icacinaceae Miers(保留科名)●■

304611 Pleurisanthes Baill. (1874);侧花茱萸属●☆

304612 Pleurisanthes brasiliensis Tiegh. ;侧花茱萸●☆

304613 Pleurisanthes flava Sandwith;黄侧花茱萸●☆

304614 Pleurisanthes parviflora (Ducke) R. A. Howard;小花侧花茱萸●☆

304615 Pleuroblepharis Baill. = Crossandra Salisb. ●

304616 Pleuroblepharis grandidieri Baill. = Crossandra grandidieri (Baill.) Benoist ●☆

304617 Pleurobotryum Barb. Rodr. = Pleurothallis R. Br. ■☆

304618 Pleurocalyptus Brongn. et Gris(1868);隐脉桃金娘属●☆

304619 Pleurocalyptus deplanchei Brongn. et Gris;隐脉桃金娘●☆

304620 Pleurocarpaea Benth. (1867);少花糙毛菊属■☆

304621 Pleurocarpaea denticulata Benth. ;少花糙毛菊■☆

304622 Pleurocarpus Klotzsch = Cinchona L. ■●

304623 Pleurocarpus Klotzsch = Rhyssocarpus Endl. ●☆

304624 Pleurochaenia Griseb. = Miconia Ruiz et Pav. (保留属名)●☆

304625 Pleurocitrus Tanaka = Citrus L. ●

304626 Pleurocoffea Baill. = Coffea L. ●

304627 Pleurocoronis R. M. King et H. Rob. (1966);侧冠菊属●☆

304628 Pleurocoronis pluriseta (A. Gray) R. M. King et H. Rob. ;北美侧冠菊;Bush Arrowleaf ●☆

304629 Pleurodesmia Arn. = Schumacheria Vahl ■☆

304630 Pleurodiscus Pierre ex A. Chev. = Laccodiscus Radlk. ●☆

304631 Pleurogyna Eschsch. ex Cham. et Schltdl. = Lomatogonium A. Braun ■

304632 Pleurogyna carinthiaca G. Don = Lomatogonium carinthiacum (Wulfen) Rchb. ■

304633 Pleurogyne Eschsch. ex Griseb. = Lomatogonium A. Braun ■

304634 Pleurogyne Eschsch. ex Griseb. = Swertia L. ■

304635 Pleurogyne Griseb. = Lomatogonium A. Braun ■

304636 Pleurogyne Griseb. = Pleurogyna Eschsch. ex Cham. et Schltdl. ■

304637 Pleurogyne bodinieri H. Lév. = Lomatogonium forrestii (Balf. f.) Fernald var. bonatianum (Burkill) T. N. Ho ■

304638 Pleurogyne brachyanthera C. B. Clarke = Lomatogonium

brachyantherum（C. B. Clarke）Fernald ■

304639　Pleurogyne carinata Edgew. = Lomatogonium carinthiacum（Wulfen）Rchb. ■

304640　Pleurogyne carinthiaca（Wulfen）Griseb. var. cordifolia Franch. = Lomatogonium carinthiacum（Wulfen）Rchb. ■

304641　Pleurogyne carinthiaca G. Don var. cordifolia Franch. = Lomatogonium cordifolium（Franch.）H. W. Li ex T. N. Ho ■

304642　Pleurogyne carinthiaca Griseb. = Lomatogonium carinthiacum（Wulfen）Rchb. ■

304643　Pleurogyne carinthiaca Griseb. var. cordifolia Franch. = Lomatogonium cordifolium（Franch.）H. W. Li ex T. N. Ho ■

304644　Pleurogyne diffusa Maxim. = Lomatogonium brachyantherum（C. B. Clarke）Fernald ■

304645　Pleurogyne diffusa Maxim. = Lomatogonium thomsonii（C. B. Clarke）Fernald ■

304646　Pleurogyne forrestii Balf. f. = Lomatogonium forrestii（Balf. f.）Fernald ■

304647　Pleurogyne lubahniana Vatke = Swertia rosulata（Baker）Klack. ■☆

304648　Pleurogyne macrantha Diels et Gilg = Lomatogonium macranthum（Diels et Gilg）Fernald ■

304649　Pleurogyne mairei H. Lév. = Swertia patens Burkill ■

304650　Pleurogyne mairei H. Lév. var. rubropunctata H. Lév. = Swertia patens Burkill ■

304651　Pleurogyne oreocharis Diels = Lomatogonium oreocharis（Diels）Marquand ■

304652　Pleurogyne patens H. Lév. = Lomatogonium forrestii（Balf. f.）Fernald var. bonatianum（Burkill）T. N. Ho ■

304653　Pleurogyne rotata Griseb. = Lomatogonium rotatum（L.）Fr. ex Nyman ■

304654　Pleurogyne rotata L. var. floribunda Franch. = Lomatogonium rotatum（L.）Fr. ex Nyman var. floribundum（Franch.）T. N. Ho ■

304655　Pleurogyne thomsonii C. B. Clarke = Lomatogonium brachyantherum（C. B. Clarke）Fernald ■

304656　Pleurogyne thomsonii C. B. Clarke = Lomatogonium thomsonii（C. B. Clarke）Fernald ■

304657　Pleurogyne vaniotii H. Lév. = Swertia patens Burkill ■

304658　Pleurogynella Ikonn. = Lomatogonium A. Braun ■

304659　Pleurogynella Ikonn. = Swertia L. ■

304660　Pleurogynella brachyanthera（C. B. Clarke）Ikonn. = Lomatogonium brachyantherum（C. B. Clarke）Fernald ■

304661　Pleurogynella thomsonii（C. B. Clarke）Ikonn. = Lomatogonium thomsonii（C. B. Clarke）Fernald ■

304662　Pleurolobus J. St. -Hil.（废弃属名）= Desmodium Desv.（保留属名）●■

304663　Pleuromenes Raf. = Prosopis L. ●

304664　Pleuropappus F. Muell.（1855）;齿鳞鼠麹草属■☆

304665　Pleuropappus F. Muell. = Angianthus J. C. Wendl.（保留属名）■●☆

304666　Pleuropappus phyllocalymnus F. Muell. ;齿鳞鼠麹草■☆

304667　Pleuropetalon Blume = Chariessa Miq. ●☆

304668　Pleuropetalum Benth. et Hook. f. = Pleuropetalon Blume ●☆

304669　Pleuropetalum Hook. f.（1846）;肋瓣苋属●☆

304670　Pleuropetalum darwinii Hook. f. ;肋瓣苋●☆

304671　Pleurophora D. Don（1837）;肋梗千屈菜属■☆

304672　Pleurophora annulosa Koehne;肋梗千屈菜■☆

304673　Pleurophora aspera Phil. ;粗糙肋梗千屈菜■☆

304674　Pleurophora polyandra Hook. et Arn. ;多蕊肋梗千屈菜■☆

304675　Pleurophragma Rydb.（1907）;肋隔芥属■☆

304676　Pleurophragma gracilipes Rydb. ;细梗肋隔芥■☆

304677　Pleurophragma integrifolium Rydb. ;全缘肋隔芥■☆

304678　Pleurophragma platypodum Rydb. ;平梗肋隔芥■☆

304679　Pleurophyllum Hook. f.（1844）;纵脉菀属■☆

304680　Pleurophyllum Mart. ex K. Schum. = Warscewiczia Klotzsch ■☆

304681　Pleurophyllum criniferum Hook. f. ;纵脉菀■☆

304682　Pleurophyllum splendens Mart. ex K. Schum. ;纤细纵脉菀■☆

304683　Pleuroplitis Trin. = Arthraxon P. Beauv. ■

304684　Pleuroplitis centrasiatica Griseb. = Arthraxon hispidus（Thunb.）Makino var. centrasiaticus（Griseb.）Tzvelev ■

304685　Pleuroplitis ciliata J. A. Schmidt = Arthraxon lancifolius（Trin.）Hochst. ■☆

304686　Pleuroplitis lancifolia（Trin.）Regel = Arthraxon lanceolatus（Roxb.）Hochst. ■

304687　Pleuroplitis lancifolia Regel = Arthraxon lancifolius（Trin.）Hochst. ■

304688　Pleuroplitis langsdorffii Trin. = Arthraxon hispidus（Thunb.）Makino ■

304689　Pleuroplitis langsdorffii Trin. var. centrasi-atica（Griseb.）Regel. = Arthraxon hispidus（Thunb.）Makino var. centrasiaticus（Griseb.）Tzvelev ■

304690　Pleuroplitis langsdorffii Trin. var. centrasiatica Regel = Arthraxon hispidus（Thunb.）Makino var. centrasiaticus（Griseb.）Tzvelev ■

304691　Pleuroplitis langsdorffii Trin. var. chinensis Regel. = Arthraxon hispidus（Thunb.）Makino ■

304692　Pleuroplitis microphylla（Trin.）Regel. = Arthraxon microphyllus（Trin.）Hochst. ■

304693　Pleuroplitis microphylla Regel = Arthraxon microphyllus（Trin.）Hochst. ■

304694　Pleuroplitis schimperi（Hochst. ex A. Rich.）Regel = Arthraxon lancifolius（Trin.）Hochst. ■☆

304695　Pleuropogon R. Br.（1823）;北极甜茅属;Semaphore Grass, Semaphore-grass ■☆

304696　Pleuropogon californicum Benth. ex Vasey;加州北极甜茅■☆

304697　Pleuropogon sabinei R. Br. ;北极甜茅■☆

304698　Pleuropsa Merr. = Pleurospa Raf.（废弃属名）●☆

304699　Pleuropterantha Franch.（1882）;肋翅苋属●☆

304700　Pleuropterantha revoilii Franch. ;肋翅苋●☆

304701　Pleuropterantha revoilii Franch. var. rhodoptera Chiov. = Pleuropterantha revoilii Franch. ●☆

304702　Pleuropterantha thulinii C. C. Towns. ;修林肋翅苋●☆

304703　Pleuropterantha undulatifolia Chiov. ;波叶肋翅苋●☆

304704　Pleuropteropyrum Gross = Aconogonon（Meisn.）Rchb. ☆

304705　Pleuropteropyrum Gross = Persicaria（L.）Mill. ■

304706　Pleuropteropyrum Gross（1913）;肋翼蓼属（虎杖属,神血宁属）■☆

304707　Pleuropteropyrum ajanense（Regel et Tiling）Nakai = Aconogonon ajanense（Regel et Tiling）H. Hara ■

304708　Pleuropteropyrum ajanense（Regel et Tiling）Nakai = Polygonum ajanense（Regel et Tiling）Grig. ■

304709　Pleuropteropyrum alpinum（All.）Kitag. = Polygonum alpinum All. ■

304710　Pleuropteropyrum alpinum（Maxim.）Koidz. = Aconogonon weyrichii（F. Schmidt）H. Hara var. alpinum（Maxim.）H. Hara ■☆

304711　Pleuropteropyrum alpinum （Maxim.） Koidz. var. chokaense Koidz. = Aconogonon weyrichii （F. Schmidt） H. Hara var. alpinum （Maxim.） H. Hara ■☆

304712　Pleuropteropyrum alpinum （Maxim.） Koidz. var. chokaense Koidz. = Polygonum weyrichii F. Schmidt var. alpinum Maxim. ■☆

304713　Pleuropteropyrum angustifolium （Pall.） Kitag. = Polygonum angustifolium Pall. ■

304714　Pleuropteropyrum bucharicum （Grig.） Nevski = Polygonum coriarium Grig. ■

304715　Pleuropteropyrum divaricatum （L.） Nakai = Polygonum divaricatum L. ■

304716　Pleuropteropyrum jeholense Kitag. = Polygonum alpinum All. ■

304717　Pleuropteropyrum laxmannii （Lepech.） Kitag. = Polygonum ochreatum L. ■

304718　Pleuropteropyrum limosum （Kom.） Kitag. = Polygonum limosum Kom. ■

304719　Pleuropteropyrum nakaii H. Hara = Aconogonon nakaii （H. Hara） H. Hara ■☆

304720　Pleuropteropyrum platyphyllum （S. X. Li et Y. L. Chang） Kitag. = Polygonum platyphyllum S. X. Li et Y. L. Chang ■

304721　Pleuropteropyrum polystachyum （Wall. ex Meisn.） Munshi et Javeid = Polygonum polystachyum Wall. ex Meisn. ●■

304722　Pleuropteropyrum polystachyum （Wall. ex Meisn.） Munshi et Javied = Persicaria wallichii Greuter et Burdet ■☆

304723　Pleuropteropyrum sibiricum （Laxm.） Kitag. = Polygonum sibiricum Laxm. ■

304724　Pleuropteropyrum tortuosum （D. Don） Munshi et Javeid = Polygonum tortuosum D. Don ■

304725　Pleuropteropyrum undulatum Murray = Polygonum alpinum All. ■

304726　Pleuropteropyrum weyrichii （F. Schmidt） H. Gross = Aconogonon weyrichii （F. Schmidt） H. Hara ■☆

304727　Pleuropteropyrum weyrichii （F. Schmidt） H. Gross var. alpicola Koidz. = Aconogonon weyrichii （F. Schmidt） H. Hara ■☆

304728　Pleuropteropyrum weyrichii （F. Schmidt） H. Gross var. alpinum （Maxim.） H. Gross = Aconogonon weyrichii （F. Schmidt） H. Hara var. alpinum （Maxim.） H. Hara ■☆

304729　Pleuropterus Turcz. = Fallopia Adans. ●■

304730　Pleuropterus Turcz. = Polygonum L. （保留属名）■●

304731　Pleuropterus ciliinervis Nakai = Fallopia ciliinervis （Nakai） K. Hammer ■

304732　Pleuropterus ciliinervis Nakai = Fallopia multiflora （Thunb.） Haraldson var. ciliinervis （Nakai） A. J. Li ■

304733　Pleuropterus cordatus （Thunb.） Turcz. = Fallopia multiflora （Thunb.） Haraldson ■

304734　Pleuropterus cordatus Turcz. = Fallopia multiflora （Thunb.） Haraldson ■

304735　Pleuropterus cuspidatus （Siebold et Zucc.） H. Gross = Fallopia japonica （Houtt.） Ronse Decr. ■

304736　Pleuropterus cuspidatus （Siebold et Zucc.） H. Gross = Reynoutria japonica Houtt. ■

304737　Pleuropterus cuspidatus H. Gross = Reynoutria japonica Houtt. ■

304738　Pleuropterus multiflorus （Thunb.） Nakai = Fallopia multiflora （Thunb.） Haraldson ■

304739　Pleuropterus multiflorus （Thunb.） Turcz. ex Nakai = Fallopia multiflora （Thunb.） Haraldson ■

304740　Pleuropterus sachalinensis （F. W. Schmidt ex Maxim.） H. Gross = Polygonum sachalinense F. Schmidt ex Maxim. ■☆

304741　Pleuropterus zuccarinii （Small） Small = Fallopia japonica （Houtt.） Ronse Decr. ■

304742　Pleuroraphis Post et Kuntze = Pleuraphis Torr. ■☆

304743　Pleuroridgea Tiegh. = Brackenridgea A. Gray ●☆

304744　Pleuroridgea alboserrata （Engl.） Tiegh. = Brackenridgea zanguebarica Oliv. ●☆

304745　Pleuroridgea bussei （Gilg） Tiegh. = Brackenridgea zanguebarica Oliv. ●☆

304746　Pleuroridgea ferruginea （Engl.） Tiegh. = Brackenridgea arenaria （De Wild. et T. Durand） N. Robson ●☆

304747　Pleuroridgea lastii Tiegh. = Brackenridgea zanguebarica Oliv. ●☆

304748　Pleuroridgea zanguebarica （Oliv.） Tiegh. = Brackenridgea zanguebarica Oliv. ●☆

304749　Pleurospa Raf. （废弃属名） = Montrichardia Crueg. （保留属名）■☆

304750　Pleurospermopsis C. Norman（1938）；簇苞芹属■

304751　Pleurospermopsis sikkimensis （C. B. Clarke） C. Norman；簇苞芹■

304752　Pleurospermum Hoffm. （1814）；棱子芹属；Pleurospermum，Ribseedcelery ■

304753　Pleurospermum affine H. Wolff = Pleurospermum hookeri C. B. Clarke var. thomsonii C. B. Clarke ■

304754　Pleurospermum albimarginatum H. Wolff；白边棱子芹（东俄洛棱子芹）；Whiteedge Pleurospermum，Whiteedge Ribseedcelery ■

304755　Pleurospermum album C. B. Clarke ex H. Wolff；白苞棱子芹（白棱子芹，亚东子芹）；White Pleurospermum，White Ribseedcelery ■☆

304756　Pleurospermum amabile Craib ex W. W. Sm.；美丽棱子芹；Beautiful Pleurospermum，Beautiful Ribseedcelery ■

304757　Pleurospermum angelicoides （Wall. ex DC.） Benth. ex C. B. Clarke；归叶棱子芹；Angelicalike Pleurospermum，Angelicalike Ribseedcelery ■

304758　Pleurospermum angelicoides （Wall. ex DC.） C. B. Clarke = Pleurospermum angelicoides （Wall. ex DC.） Benth. ex C. B. Clarke ■

304759　Pleurospermum angelicoides （Wall.） Benth. ex C. B. Clarke = Pleurospermum angelicoides （Wall. ex DC.） Benth. ex C. B. Clarke ■

304760　Pleurospermum apiolens C. B. Clarke；紫色棱子芹；Darkpurple Pleurospermum，Violet Ribseedcelery ■

304761　Pleurospermum apiolens C. B. Clarke var. nipaulense Farille et S. B. Malla = Pleurospermum apiolens C. B. Clarke ■

304762　Pleurospermum aromaticum W. W. Sm.；芳香棱子芹■☆

304763　Pleurospermum astrantioideum （H. Boissieu） K. T. Fu et Y. C. Ho；雅江棱子芹；Yajiang Pleurospermum，Yajiang Ribseedcelery ■

304764　Pleurospermum atropurpureum K. T. Fu et Y. C. Ho = Pleurospermum apiolens C. B. Clarke ■

304765　Pleurospermum austriacum （L.） Hoffm.；奥地利棱子芹（欧洲棱子芹）；Austrian Pleurospermum ■☆

304766　Pleurospermum austriacum （L.） Hoffm. subsp. uralense （Hoffm.） Sommier；乌拉尔棱子芹（棱子芹）；Ural Pleurospermum，Ural Ribseedcelery ■

304767　Pleurospermum austriacum Hoffm. subsp. uralense （Hoffm.） Sommier = Pleurospermum austriacum （L.） Hoffm. subsp. uralense （Hoffm.） Sommier ■

304768　Pleurospermum benthamii （Wall . ex DC.） C. B. Clarke；宝兴棱子芹（绵参，棉参）；David Pleurospermnu，David Ribseedcelery ■

304769　Pleurospermum bicolor （Franch.） C. Norman ex Z. H. Pan et M. F. Watson；二色棱子芹；Twocolored Ribseedcelery，Twocoloured Pleurospermum ■

304770　Pleurospermum calcareum H. Wolff;疣叶棱子芹(灰岩棱子芹,灰质棱子芹);Calcareous Pleurospermum ■

304771　Pleurospermum calophlebicum (H. Wolff) M. Hiroe = Ligusticum likiangense (H. Wolff) F. T. Pu et M. F. Watson ■

304772　Pleurospermum camtschaticum Hoffm.;棱子芹(走马芹);Kamchatka Pleurospermum, Kamchatka Ribseedcelery ■

304773　Pleurospermum camtschaticum Hoffm. = Pleurospermum austriacum (L.) Hoffm. subsp. uralense (Hoffm.) Sommier ■

304774　Pleurospermum camtschaticum Hoffm. = Pleurospermum uralense Hoffm. ■

304775　Pleurospermum candollei (DC.) C. B. Clarke;康多勒棱子芹 ■☆

304776　Pleurospermum capillaceum (H. Wolff) M. Hiroe = Ligusticum capillaceum H. Wolff ■

304777　Pleurospermum cavaleri M. Hiroe = Tongoloa stewardii H. Wolff ■

304778　Pleurospermum cicutarium Lindl. = Selinum wallichianum (DC.) Raizada et H. O. Saxena ■

304779　Pleurospermum claeareum H. Wolff;钙土棱子芹, Chalky Pleurospermum, Chalky Ribseedcelery ■☆

304780　Pleurospermum cnidiifoium H. Wolff = Pleurospermum crassicaule H. Wolff ■

304781　Pleurospermum cnidiifolium H. Wolff = Pleurospermum wilsonii H. Boissieu ■

304782　Pleurospermum crassicaule H. Wolff = Pleurospermum wilsonii H. Boissieu ■

304783　Pleurospermum cristatum H. Boissieu;鸡冠棱子芹; Cristate Pleurospermum, Cristate Ribseedcelery ■

304784　Pleurospermum cruciatum (H. Wolff) M. Hiroe = Meeboldia yunnanensis (H. Wolff) Constance et F. T. Pu ■

304785　Pleurospermum darwasicum B. Fedtsch.;戴尔棱了子芹 ■

304786　Pleurospermum davidii Franch. = Pleurospermum benthamii (Wall. ex DC.) C. B. Clarke ■

304787　Pleurospermum decurrens Franch.;翼叶棱子芹(多子芹,下延棱子芹,异叶棱子芹); Wingleaf Pleurospermum, Wingleaf Ribseedcelery ■

304788　Pleurospermum delavayi (Franch.) Hiroe = Physospermopsis cuneata H. Wolff ■

304789　Pleurospermum delavayi (Franch.) Hiroe = Physospermopsis delavayi (Franch.) H. Wolff ■

304790　Pleurospermum delavayi (Franch.) M. Hiroe. = Physospermopsis delavayi (Franch.) H. Wolff ■

304791　Pleurospermum discolor (Ledeb.) M. Hiroe = Ligusticum discolor Ledeb. ■

304792　Pleurospermum dochenense W. W. Sm. = Pleurospermum hookeri C. B. Clarke var. thomsonii C. B. Clarke ■

304793　Pleurospermum foetens Franch.;臭棱子芹(丽江棱子芹,萝卜参); Lijiang Pleurospermum, Lijiang Ribseedcelery, Likiang Pleurospermum ■

304794　Pleurospermum forrestii (Diels) M. Hiroe. = Physospermopsis shaniana C. Y. Wu et F. T. Pu ■

304795　Pleurospermum franchetianum Hemsl.;松潘棱子芹(黄芜,异伞棱子芹);Franchet Pleurospermum, Franchet Ribseedcelery ■

304796　Pleurospermum giraldii Diels;太白棱子芹(药茴芹,药茴香); Girald Pleurospermum, Girald Ribseedcelery ■

304797　Pleurospermum glaciale (Bonner) M. Hiroe. = Cortiella hookeri (C. B. Clarke) C. Norman ■

304798　Pleurospermum glaucescens H. Wolff;带粉棱子芹; Glaucous Pleurospermum ■

304799　Pleurospermum glaucescens H. Wolff. = Selinum cryptotaenium H. Boissieu ■

304800　Pleurospermum govanianum (DC.) Benth. ex C. B. Clarke = Pleurospermum stellatum (D. Don) C. B. Clarke ■

304801　Pleurospermum govanianum (DC.) C. B. Clarke = Pleurospermum stellatum (D. Don) C. B. Clarke ■

304802　Pleurospermum govanianum Benth. ex C. B. Clarke var. bicolor H. Wolff = Pleurospermum bicolor (Franch.) C. Norman ex Z. H. Pan et M. F. Watson ■

304803　Pleurospermum handelii H. Wolff;高山棱子芹 ■

304804　Pleurospermum handelii H. Wolff ex Hand.-Mazz.;细裂棱子芹;Handel Pleurospermum ■

304805　Pleurospermum hedinii Diels;垫状棱子芹; Hedin Pleurospermum, Hedin Ribseedcelery ■

304806　Pleurospermum heracleifolium Franch. ex H. Boissieu;芷叶棱子芹;Cowparsnipleaf Pleurospermum, Cowparsnipleaf Ribseedcelery ■

304807　Pleurospermum heterosciadium H. Wolff;异伞棱子芹(褐紫粗子芹); Abnormalumbrella Ribseedcelery, Darkpurple Trachydium, Diverseumbel Pleurospermum ■

304808　Pleurospermum hookeri C. B. Clarke;紫茎棱子芹(西藏棱子芹,喜马拉雅棱子芹,紫堇棱子芹); Hooker Pleurospermum, Hooker Ribseedcelery ■

304809　Pleurospermum hookeri C. B. Clarke var. thomsonii C. B. Clarke;西藏棱子芹(扇叶粗子芹); Fanleaf Trachydium, Thomson Pleurospermum, Thomson Ribseedcelery ■

304810　Pleurospermum kansuense H. Wolff = Pleurospermum pulszkyi Kanitz ■

304811　Pleurospermum kingdon-wardii (H. Wolff) M. Hiroe. = Physospermopsis kingdon-wardii (H. Wolff) C. Norman ■

304812　Pleurospermum lecomtianum H. Wolff = Pleurospermum crassicaule H. Wolff ■

304813　Pleurospermum lecomtianum H. Wolff = Pleurospermum wilsonii H. Boissieu ■

304814　Pleurospermum likiangense H. Wolff = Ligusticum likiangense (H. Wolff) F. T. Pu et M. F. Watson ■

304815　Pleurospermum likiangense H. Wolff = Pleurospermum hookeri C. B. Clarke var. thomsonii C. B. Clarke ■

304816　Pleurospermum limprichtii H. Wolff = Pleurospermum giraldii Diels ■

304817　Pleurospermum lindleyanum (Lipsky) F. Fedtsch.;天山棱子芹(矮薄苞芹,小膜苞芹,心草); Lindley Pleurospermum, Tianshan Ribseedcelery ■

304818　Pleurospermum linearilobum W. W. Sm.;线裂棱子芹; Linearlobed Pleurospermum, Linearlobed Ribseedcelery ■

304819　Pleurospermum longicarpum R. H. Shan et Z. H. Pan;长果棱子芹;Longfruit Pleurospermum ■

304820　Pleurospermum longicaule H. Wolff = Ligusticum thomsonii C. B. Clarke ■

304821　Pleurospermum longipetiolatum H. Wolff = Pleurospermum franchetianum Hemsl. ■

304822　Pleurospermum macrochlaenum K. T. Fu et Y. C. Ho;大苞棱子芹;Bigbract Pleurospermum, Bigbract Ribseedcelery ■

304823　Pleurospermum markgrafianum H. Wolff = Pleurospermum hookeri C. B. Clarke var. thomsonii C. B. Clarke ■

304824　Pleurospermum meoides Diels = Pleurospermum giraldii Diels ■

304825　Pleurospermum nanum (Rupr.) Benth. et Hook. =

Pleurospermum lindleyanum（Lipsky）F. Fedtsch. ■

304826　Pleurospermum nanum Franch. ;矮棱子芹（紫果粗子芹,紫叶粗子芹,紫棕棱子芹）;Dwarf Pleurospermum, Dwarf Ribseedcelery, Purple Trachydium ■

304827　Pleurospermum nubigenum H. Wolff;皱果棱子芹;Wrinklefruit Pleurospermum, Wrinklefruit Ribseedcelery ■

304828　Pleurospermum obtusiusculum （Wall. ex DC.）M. Hiroe = Physospermopsis obtusiuscula（Wall. ex DC.）C. Norman ■

304829　Pleurospermum pilgerianum Fedde ex H. Wolff = Pleurospermum franchetianum Hemsl. ■

304830　Pleurospermum pilosum C. B. Clarke ex H. Wolff;疏毛棱子芹;Pilose Pleurospermum, Scatterhair Ribseedcelery ■

304831　Pleurospermum prattii H. Wolff; 康定棱子芹; Pratt Pleurospermum, Pratt Ribseedcelery ■

304832　Pleurospermum prattii H. Wolff = Pleurospermum wrightianum H. Boissieu ■

304833　Pleurospermum pseudoangelica（H. Boissieu）H. Boissieu 当归状棱子芹■

304834　Pleurospermum pseudoinvolucratum H. Wolff = Pleurospermum hookeri C. B. Clarke var. thomsonii C. B. Clarke ■

304835　Pleurospermum pseudoyunnanense H. Wolff;茨开棱子芹;Cikai Pleurospermum ■

304836　Pleurospermum pseudoyunnanense H. Wolff = Pleurospermum yunnanense Franch. ■

304837　Pleurospermum pulchrum Aitch. et Hemsl. = Pleurospermum stylosum C. B. Clarke ■

304838　Pleurospermum pulszkyi Kanitz; 青藏棱子芹; Pulszky Pleurospermum, Pulszky Ribseedcelery ■

304839　Pleurospermum rivulorum（Diels）K. T. Fu et Y. C. Ho;心叶棱子芹（滇羌活,蛇头羌活）;Cordateleaf Pleurospermum, Heartleaf Ribseedcelery ■

304840　Pleurospermum rivulorum （Diels）M. Hiroe = Pleurospermum rivulorum（Diels）K. T. Fu et Y. C. Ho ■

304841　Pleurospermum rockii Fedde ex H. Wolff = Pleurospermum franchetianum Hemsl. ■

304842　Pleurospermum rotundatum（DC.）C. B. Clarke;圆叶棱子芹■

304843　Pleurospermum rubrinerve （Franch.） M. Hiroe. = Physospermopsis rubrinervis（Franch.）C. Norman ■

304844　Pleurospermum rupestre（Popov）K. T. Fu et Y. C. Ho;岩生棱子芹;Cliffliving Pleurospermum, Rocky Ribseedcelery ■

304845　Pleurospermum sikkimense C. B. Clarke = Pleurospermopsis sikkimensis（C. B. Clarke）C. Norman ■

304846　Pleurospermum simplex（Rupr.）Benth. et Hook. f. ex Drude;单茎棱子芹;Singlestem Pleurospermum, Singlestem Ribseedcelery ■

304847　Pleurospermum souliaei H. Wolff;川康棱子芹（索氏棱子芹）■

304848　Pleurospermum stellatum（D. Don）C. B. Clarke;尖头棱子芹■

304849　Pleurospermum stellatum （D. Don）C. B. Clarke var. lindleyanum（Klotzsch）C. B. Clarke = Pleurospermum lindleyanum（Lipsky）F. Fedtsch. ■

304850　Pleurospermum stylosum C. B. Clarke;新疆棱子芹■

304851　Pleurospermum szechenyii Kanitz; 青海棱子芹; Chinghai Pleurospermum, Qinghai Pleurospermum, Qinghai Ribseedcelery ■

304852　Pleurospermum tanacetifolium H. Wolff = Pleurospermum crassicaule H. Wolff ■

304853　Pleurospermum tanacetifolium H. Wolff = Pleurospermum wilsonii H. Boissieu ■

304854　Pleurospermum tibetanicum H. Wolff = Pleurospermum hookeri C. B. Clarke var. thomsonii C. B. Clarke ■

304855　Pleurospermum tsekuense R. H. Shan; 泽库棱子芹; Tseku Pleurospermum, Zeku Pleurospermum, Zeku Ribseedcelery ■

304856　Pleurospermum uralense Hoffm. = Pleurospermum austriacum （L.）Hoffm. subsp. uralense（Hoffm.）Sommier ■

304857　Pleurospermum uralense Hoffm. = Pleurospermum camtschaticum Hoffm. ■

304858　Pleurospermum wilsonii H. Boissieu;粗茎棱子芹;Thickstem Pleurospermum, Thickstem Ribseedcelery ■

304859　Pleurospermum wolffianum Fedde ex H. Wolff = Pleurospermum hookeri C. B. Clarke ■

304860　Pleurospermum wrightianum H. Boissieu; 瘤果棱子芹; Tumorfruit Ribseedcelery, Wright Pleurospermum ■

304861　Pleurospermum yulungense C. Y. Wu;玉龙棱子芹;Yulong Pleurospermum ■

304862　Pleurospermum yunnanense Franch. ;云南棱子芹（苍山棱子芹）;Yunnan Pleurospermum, Yunnan Ribseedcelery ■

304863　Pleurostachys Brongn.（1833）;侧穗莎属■☆

304864　Pleurostachys angustifolia Boeck.；窄叶侧穗莎■☆

304865　Pleurostachys boliviana Palla;玻利维亚侧穗莎■☆

304866　Pleurostachys elegans Kunth;雅致侧穗莎■☆

304867　Pleurostachys foliosa Kunth;多叶侧穗莎■☆

304868　Pleurostachys graminifolia Brogn.;禾叶侧穗莎■☆

304869　Pleurostachys grandifolia Boeck. ;大叶侧穗莎■☆

304870　Pleurostachys guianensis Uittien;圭亚那侧穗莎■☆

304871　Pleurostachys macrantha Kunth;大花侧穗莎■☆

304872　Pleurostachys montana Palla;山地侧穗莎■☆

304873　Pleurostachys paniculata Boeck. ;圆锥侧穗莎■☆

304874　Pleurostachys robusta Palla;粗壮侧穗莎■☆

304875　Pleurostachys tenuiflora Brongn. ;细花侧穗莎■☆

304876　Pleurostelma Baill.（1890）;侧冠萝藦属■☆

304877　Pleurostelma Schltr. = Schlechterella K. Schum. ■●☆

304878　Pleurostelma africanum Schltr. = Schlechterella africana（Schltr.）K. Schum. ☆

304879　Pleurostelma cernuum（Decne.）Bullock;侧冠萝藦■☆

304880　Pleurostelma grevei Baill. = Pleurostelma cernuum（Decne.）Bullock ■☆

304881　Pleurostelma schimperi（Vatke）Liede;欣珀侧冠萝藦■☆

304882　Pleurostemon Raf. = Gaura L. ■

304883　Pleurostena Raf. = Polygonum L.（保留属名）■●

304884　Pleurostigma Hochst. = Bouchea Cham.（保留属名）●☆

304885　Pleurostima Raf.（1837）;亚顶柱属■☆

304886　Pleurostima Raf. = Barbacenia Vand. ■☆

304887　Pleurostima brachycalyx（Goethart et Henrard）N. L. Menezes;短萼亚顶柱■☆

304888　Pleurostima brevifolia（Taub.）N. L. Menezes;短叶亚顶柱■☆

304889　Pleurostima monticola（L. B. Sm. et Ayensu）N. L. Menezes;山地亚顶柱■☆

304890　Pleurostima purpurea Raf. ;紫亚顶柱■☆

304891　Pleurostima stenophylla（Goeth. et Henrard）N. L. Menezes;窄叶亚顶柱■☆

304892　Pleurostylia Wight et Arn.（1834）;盾柱属（盾柱木属,盾柱卫矛属）;Pleurostylia, Shieldstyle ●

304893　Pleurostylia africana Loes. ;非洲盾柱●☆

304894　Pleurostylia capensis（Turcz.）Loes. ;好望角盾柱●☆

304895　Pleurostylia cochinchinensis Pierre = Pleurostylia opposita（Wall.）Alston ●

304896　Pleurostylia henryi Wight et Arn. = Pleurostylia opposita（Wall.）Alston ●

304897　Pleurostylia heynei Wight et Arn. var. acutifolia Suess. = Pleurostylia africana Loes. ●☆

304898　Pleurostylia opposita（Wall.）Alston;盾柱（盾柱卫矛,越南盾柱）;Cochin-China Pleurostylia, Oposite Pleurostylia, Oposite Shieldstyle, Vietnamse Pleurostylia ●

304899　Pleurostylia opposita（Wall.）Merr. et F. P. Metcalf = Pleurostylia opposita（Wall.）Alston ●

304900　Pleurostylia serrulata Loes. ;细齿盾柱●☆

304901　Pleurostylia wightii Wight et Arn. = Pleurostylia opposita（Wall.）Alston ●

304902　Pleurostylis Walp. = Pleurostylia Wight et Arn. ●

304903　Pleurotaenia Hohen. ex Benth. et Hook. f. = Peucedanum L. ■

304904　Pleurothallis R. Br.（1813）;肋枝兰属（腋花兰属）;Pleurothallis ■☆

304905　Pleurothallis amesiana L. O. Williams;阿米斯肋枝兰;Ames Pleurothallis ■☆

304906　Pleurothallis barberiana Rchb. f. ;巴比氏肋枝兰;Barber Pleurothallis ■☆

304907　Pleurothallis brighamii S. Watson;布瑞氏肋枝兰;Brigham Pleurothallis ■☆

304908　Pleurothallis ciliaris（Lindl.）L. O. Williams;缘毛肋枝兰;Ciliate Pleurothallis ■☆

304909　Pleurothallis compacta（Ames）Ames et C. Schweinf. ;密花肋枝兰;Denseflower Pleurothallis ■☆

304910　Pleurothallis corniculata（Sw.）Lindl. ;具角肋枝兰;Horned Pleurothallis ■☆

304911　Pleurothallis disticha（Lam.）A. Rich. = Oberonia disticha（Lam.）Schltr. ■☆

304912　Pleurothallis elegans（Kunth）Lindl. ;雅致肋枝兰;Elegant Pleurothallis ■☆

304913　Pleurothallis gelida Lindl. ;寒地肋枝兰■☆

304914　Pleurothallis grobyi Bateman;格若氏肋枝兰（古美腋花兰）;Groby Pleurothallis ■☆

304915　Pleurothallis hawkesii Fichkinger;哈克斯肋枝兰;Hawkes Pleurothallis ■☆

304916　Pleurothallis hirsuta Ames;硬毛肋枝兰;Hirsute Pleurothallis ■☆

304917　Pleurothallis inflata Rolfe;垂花肋枝兰;Droopingflower Pleurothallis ■☆

304918　Pleurothallis insignis Rolfe;美花肋枝兰;Beatyflower Pleurothallis ■☆

304919　Pleurothallis johnsonii Ames;约翰森肋枝兰;Johnson Pleurothallis ■☆

304920　Pleurothallis longissima Lindl. ;长序肋枝兰■☆

304921　Pleurothallis melanantha Rchb. f. = Lepanthopsis melanantha（Rchb. f.）Ames ■☆

304922　Pleurothallis ophiocephala Lindl. ;蛇头肋枝兰;Snake-head Pleurothallis ■☆

304923　Pleurothallis ospinae R. E. Schult. ;欧斯潘肋枝兰;Ospin Pleurothallis ■☆

304924　Pleurothallis pachyrachis A. Rich. = Bulbophyllum pachyrachis（A. Rich.）Griseb. ■☆

304925　Pleurothallis prolifera Herb. ex Lindl. ;肋枝兰;Bearing Progeny Pleurothallis ■☆

304926　Pleurothallis rubens Lindl. ; 红花肋枝兰; Redflower Pleurothallis ■☆

304927　Pleurothallis ruscifolia（Jacq.）R. Br. ;假叶树叶肋枝兰;Pleurothallis ■☆

304928　Pleurothallis scapha Rchb. f. ;舟形肋枝兰■☆

304929　Pleurothallis stenostachya Rchb. f. ;柔穗花序肋枝兰;Narrowspike Pleurothallis ■☆

304930　Pleurothallis venosa Rolfe;细脉肋枝兰■☆

304931　Pleurothallopsis Porto = Pleurothallopsis Porto et Brade ■☆

304932　Pleurothallopsis Porto et Brade = Octomeria R. Br. ■☆

304933　Pleurothallopsis Porto et Brade(1937);类肋枝兰属■☆

304934　Pleurothallopsis nemorosa（Barb. Rodr.）Porto et Brade;类肋枝兰■☆

304935　Pleurothyrium Nees = Ocotea Aubl. ●☆

304936　Pleurothyrium Nees ex Lindl. = Pleurothyrium Nees ●☆

304937　Pleurothyrium Nees(1836);蚁心樟属●☆

304938　Pleuteron Raf. = Breynia L.（废弃属名）●

304939　Pleuteron Raf. = Capparis L. ●

304940　Pleuteron Raf. = Linnaeobreynia Hutch. ●

304941　Plexaure Endl. = Phreatia Lindl. ■

304942　Plexipus Raf.（废弃属名）= Bouchea Cham.（保留属名）●☆

304943　Plexipus Raf.（废弃属名）= Chascanum E. Mey.（保留属名）●☆

304944　Plexipus adenostachyus（Schauer）R. Fern. = Chascanum adenostachyum（Schauer）Moldenke ●☆

304945　Plexipus angolensis（Moldenke）R. Fern. = Chascanum angolense Moldenke ●☆

304946　Plexipus angolensis（Moldenke）R. Fern. subsp. zambesiacus R. Fern. = Chascanum angolense Moldenke subsp. zambesiacum（R. Fern.）R. Fern. ●☆

304947　Plexipus arabicus（Moldenke）R. Fern. = Chascanum laetum Walp. ●☆

304948　Plexipus caespitosus（H. Pearson）R. Fern. = Chascanum caespitosum（H. Pearson）Moldenke ●☆

304949　Plexipus cernuus（L.）R. Fern. = Chascanum cernuum（L.）E. Mey. ●☆

304950　Plexipus cuneifolius（L. f.）Raf. = Chascanum cuneifolium（L. f.）E. Mey. ●☆

304951　Plexipus garipensis（E. Mey.）R. Fern. = Chascanum garipense E. Mey. ●☆

304952　Plexipus gillettii（Moldenke）R. Fern. = Chascanum gillettii Moldenke ●☆

304953　Plexipus hanningtonii（Oliv.）R. Fern. = Chascanum hanningtonii（Oliv.）Moldenke ●☆

304954　Plexipus hederaceus（Sond.）R. Fern. = Chascanum hederaceum（Sond.）Moldenke ●☆

304955　Plexipus hederaceus（Sond.）R. Fern. var. natalensis（H. Pearson）R. Fern. = Chascanum hederaceum（Sond.）Moldenke var. natalense（H. Pearson）●☆

304956　Plexipus hildebrandtii（Vatke）R. Fern. = Chascanum hildebrandtii（Vatke）J. B. Gillett ●☆

304957　Plexipus incisus（H. Pearson）R. Fern. = Chascanum incisum（H. Pearson）Moldenke ●☆

304958　Plexipus integrifolius（H. Pearson）R. Fern. = Chascanum integrifolium（H. Pearson）Moldenke ●☆

304959　Plexipus krookii（Gürke ex Zahlbr.）R. Fern. = Chascanum krookii（Gürke ex Zahlbr.）Moldenke ●☆

304960　Plexipus latifolius（Harv.）R. Fern. = Chascanum latifolium（Harv.）Moldenke ●☆

304961　Plexipus latifolius（Harv.）R. Fern. var. glabrescens（H.

Pearson) R. Fern. = Chascanum latifolium (Harv.) Moldenke var. glabrescens (H. Pearson) Moldenke ●☆

304962　Plexipus latifolius (Harv.) R. Fern. var. transvaalensis (Moldenke) R. Fern. = Chascanum latifolium (Harv.) Moldenke var. transvaalense Moldenke ●☆

304963　Plexipus marrubiifolius (Fenzl ex Walp.) R. Fern. = Chascanum marrubiifolium Fenzl ex Walp. ●☆

304964　Plexipus namaquanus (Bolus ex H. Pearson) R. Fern. = Chascanum namaquanum (Bolus ex H. Pearson) Moldenke ●☆

304965　Plexipus pinnatifidus (L. f.) R. Fern. = Chascanum pinnatifidum (L. f.) E. Mey. ●☆

304966　Plexipus pinnatifidus (L. f.) R. Fern. var. racemosus (Schinz ex Moldenke) R. Fern. = Chascanum pinnatifidum (L. f.) E. Mey. var. racemosum Schinz ex Moldenke ●☆

304967　Plexipus pumilus (E. Mey.) R. Fern. = Chascanum pumilum E. Mey. ●☆

304968　Plexipus rariflorus (A. Terracc.) R. Fern. = Chascanum rariflorum (A. Terracc.) Moldenke ●☆

304969　Plexipus schlechteri (Gürke) R. Fern. = Chascanum schlechteri (Gürke) Moldenke ●☆

304970　Plexipus schlechteri (Gürke) R. Fern. var. torrei (Moldenke) R. Fern. = Chascanum schlechteri (Gürke) Moldenke var. torrei Moldenke ●☆

304971　Plexipus sessilifolius (Vatke) R. Fern. = Chascanum sessilifolium (Vatke) Moldenke ●☆

304972　Plextnium Raf. = Androcymbium Willd. ■☆

304973　Plextstenn Raf. = Allium L. ■

304974　Pliarina Post et Kuntze = Pleiarina Raf. ●

304975　Pliarina Post et Kuntze = Salix L. (保留属名)●

304976　Plicangis Thouars = Angraecum Bory ■

304977　Plicosepalus Tiegh. (1894);扭萼寄生属●☆

304978　Plicosepalus acaciae (Zucc.) Wiens et Polhill;金合欢扭萼寄生●☆

304979　Plicosepalus acaciaedetinentis (Dinter) Danser = Plicosepalus kalachariensis (Schinz) Danser ●☆

304980　Plicosepalus amplexicaulis Wiens;抱茎扭萼寄生●☆

304981　Plicosepalus curviflorus (Benth. ex Oliv.) Tiegh.;弯花扭萼寄生●☆

304982　Plicosepalus faurotii (Franch.) Tiegh. = Plicosepalus curviflorus (Benth. ex Oliv.) Tiegh. ●☆

304983　Plicosepalus foliosus Wiens et Polhill;多叶扭萼寄生●☆

304984　Plicosepalus kalachariensis (Schinz) Danser;卡拉恰扭萼寄生●☆

304985　Plicosepalus meridianus (Danser) Wiens et Polhill;南方扭萼寄生●☆

304986　Plicosepalus nummulariifolius (Franch.) Wiens et Polhill;铜钱叶扭萼寄生●☆

304987　Plicosepalus ogadenensis M. G. Gilbert;欧加登扭萼寄生●☆

304988　Plicosepalus robustus Wiens et Polhill;粗壮扭萼寄生●☆

304989　Plicosepalus sagittifolius (Engl.) Danser;箭叶扭萼寄生●☆

304990　Plicosepalus somalensis Wiens et Polhill;索马里扭萼寄生●☆

304991　Plicosepalus undulatus (E. Mey. ex Harv.) Tiegh.;波状扭萼寄生●☆

304992　Plicouratea Tiegh. = Ouratea Aubl. (保留属名)●

304993　Plicula Raf. = Acnistus Schott ●☆

304994　Plienta Post et Kuntze = Pleienia Raf. ■☆

304995　Plienta Post et Kuntze = Sabatia Adans. ■☆

304996　Plinia Blanco = Kayea Wall. ●☆

304997　Plinia Blanco = Mesua L. ●

304998　Plinia L. (1837);普林木属●☆

304999　Plinia acutissima Urb.;尖普林木●☆

305000　Plinia cauliflora (Mart.) Kausel;茎花普林木●☆

305001　Plinia cordifolia (D. Legrand) Sobral;心叶普林木●☆

305002　Plinia nana Sobral;矮普林木●☆

305003　Plinthanthesis Steud. (1853);干花扁芒草属■☆

305004　Plinthanthesis Steud. = Danthonia DC. (保留属名)■

305005　Plinthanthesis tenuior Steud.;干花扁芒草■☆

305006　Plinthine Rchb. = Arenaria L. ■

305007　Plinthocroma Dulac = Rhododendron L. ●

305008　Plinthus Fenzl(1839);鳞叶番杏属●☆

305009　Plinthus arenarius Adamson;沙地鳞叶番杏■☆

305010　Plinthus cryptocarpus Fenzl;隐果鳞叶番杏■☆

305011　Plinthus karooicus I. Verd.;卡鲁鳞叶番杏■☆

305012　Plinthus karooicus I. Verd. var. alternifolia Adamson = Plinthus karooicus I. Verd. ■☆

305013　Plinthus laxifolius I. Verd. = Plinthus sericeus Pax ■☆

305014　Plinthus psammophilus Dinter = Plinthus rehmannii G. Schellenb. ■☆

305015　Plinthus rehmannii G. Schellenb.;拉赫曼鳞叶番杏■☆

305016　Plinthus sericeus Pax;绢毛鳞叶番杏■☆

305017　Pliocarpidia Post et Kuntze = Pleiocarpidia K. Schum. ●☆

305018　Pliodon Post et Kuntze = Botelua Lag. ■

305019　Pliodon Post et Kuntze = Pleiodon Rchb. ■

305020　Pliogyna Post et Kuntze = Pleogyne Miers ex Benth. ●☆

305021　Pliogynopsts Kuntze = Pleiogynium Engl. ●☆

305022　Pliophaca Post et Kuntze = Archidendron F. Muell. ●

305023　Pliophaca Post et Kuntze = Pleiophaca F. Muell. ex Baill. ●

305024　Ploca Lour. ex Gomes = Christia Moench ■●

305025　Ploca Lour. ex Gomes = Lourea Neck. ex Desv. ■●

305026　Plocaglottis Steud. = Plocoglottis Blume ■☆

305027　Plocama Aiton(1789);卷毛茜属●☆

305028　Plocama pendula Aiton;卷毛茜●☆

305029　Plocandra E. Mey. = Chironia L. ●■☆

305030　Plocandra albens E. Mey. = Chironia palustris Burch. ■☆

305031　Plocandra palustris (Burch.) Griseb. = Chironia palustris Burch. ■☆

305032　Plocandra purpurascens E. Mey. = Chironia purpurascens (E. Mey.) Benth. et Hook. f. ■☆

305033　Plocaniophyllon Brandegee = Deppea Cham. et Schltdl. ●☆

305034　Plocoglottis Blume(1825);环唇兰属■☆

305035　Plocoglottis acuminata Blume;尖环唇兰■☆

305036　Plocoglottis angulata J. J. Sm.;窄环唇兰■☆

305037　Plocoglottis atroviridis Schltr.;墨绿环唇兰■☆

305038　Plocoglottis fimbriata Teijsm. et Binn.;　环唇兰■☆

305039　Plocoglottis foetida Ridl.;烈味环唇兰■☆

305040　Plocoglottis glaucescens Schltr.;渐灰环唇兰■☆

305041　Plocoglottis hirta Ridl.;环硬毛唇兰■☆

305042　Plocoglottis lancifolia J. J. Sm.;披针叶环唇兰■☆

305043　Plocoglottis latifrons J. J. Sm.;宽花环唇兰■☆

305044　Plocoglottis maculata Schltr.;斑点环唇兰■☆

305045　Plocosperma Benth. (1876);戴毛子属(毛子树属)●☆

305046　Plocosperma buxifolium Benth.;戴毛子(戴毛子树)●☆

305047　Plocospermaceae Hutch. = Plocospermataceae Hutch. ●☆

305048　Plocospermataceae Hutch. (1973);戴毛子科(环生籽科,毛子树科)●☆

305049 Plocospermataceae Hutch. = Loganiaceae R. Br. ex Mart.（保留科名）●■

305050 Plocostemma Blume = Hoya R. Br. ●

305051 Plocostigma Benth. = Placostigma Blume ■

305052 Plocostigma Post et Kuntze = Plokiostigma Schuch. ■☆

305053 Plocostigma Post et Kuntze = Stackhousia Sm. ■☆

305054 Ploearium Post et Kuntze = Ploiarium Korth. ●☆

305055 Ploesslia Endl. = Boswellia Roxb. ex Colebr. ●☆

305056 Ploiarium Korth.（1842）;舟胶树属●☆

305057 Ploiarium elegans Korth.;舟胶树●☆

305058 Ploionixus Tiegh. ex Lecomte = Viscum L. ●

305059 Plokiostigma Schuch. = Stackhousia Sm. ■☆

305060 Plostaxis Raf. = Polygala L. ●■

305061 Ploteas J. F. Gmel. = Plotia Adans. ●

305062 Ploteas J. F. Gmel. = Salvadora Garcin ex L. ●

305063 Plothirium Raf. = Delphinium L. ■

305064 Plotia Adans. = Salvadora Garcin ex L. ●

305065 Plotia Neck. = Embelia Burm. f.（保留属名）●■

305066 Plotia Schreb. ex Steud. = Poa L. ●

305067 Plotia Steud. = Glyceria R. Br.（保留属名）■

305068 Plottzia Arn. = Paronychia Mill. ■

305069 Plowmania Hunz. et Subils(1986);普洛曼茄属●☆

305070 Plowmania nyctaginoides（Standl.）Hunz. et Subils;普洛曼茄●☆

305071 Plowmanianthus Faden et C. R. Hardy(2004);普洛曼花属●■☆

305072 Pluchea Cass.（1817）;阔苞菊属（燕茜属）;Fleabane,Pluchea ●■

305073 Pluchea aphanantha（Baker）Humbert = Pluchea rufescens（DC.）A. J. Scott ■☆

305074 Pluchea arabica（Boiss.）Qaiser et Lack;阿拉伯阔苞菊●■●☆

305075 Pluchea arguta Boiss.;亮阔苞菊●☆

305076 Pluchea arguta Boiss. subsp. glabra Qaiser;光滑阔苞菊■●☆

305077 Pluchea baccharis（Mill.）Pruski;粉红阔苞菊;Rosy Camphorweed ■☆

305078 Pluchea balsamifera（L.）Less. = Blumea balsamifera（L.）DC. ■

305079 Pluchea bequaertii Robyns;贝卡尔阔苞菊●☆

305080 Pluchea bojeri（DC.）Humbert;博耶尔阔苞菊■☆

305081 Pluchea bravae Bolle = Pluchea ovalis（Pers.）DC. ■☆

305082 Pluchea bulleyana Jeffrey = Anaphalis bulleyana（Jeffrey）C. C. Chang ■

305083 Pluchea camphorata（L.）DC.;樟脑味阔苞菊;Camphor Pluchea,Camphor Weed,Inland Marsh Fleabane,Marsh Fleabane,Plowman's-wort,Stinkweed ●■

305084 Pluchea camphorata DC. = Pluchea camphorata（L.）DC. ●■

305085 Pluchea carolinensis（Jacq.）G. Don;加罗里阔苞菊（美洲阔苞菊）;Carolina Pluchea,Cough Bush,Cure-for-all,Sepi,Sourbush,Wild Tobacco ●

305086 Pluchea caspia（Pall.）O. Hoffm. ex Paulsen = Karelinia caspia（Pall.）Less. ■

305087 Pluchea crenata Quézel = Conyza pyrrhopappa Sch. Bip. ex A. Rich. ■☆

305088 Pluchea dioscoridis（L.）DC.;薯蓣阔苞菊●☆

305089 Pluchea dioscoridis（L.）DC. var. glabra Oliv. et Hiern = Pluchea dioscoridis（L.）DC. ●☆

305090 Pluchea dioscoridis（L.）DC. var. pseudovalis Cufod. = Pluchea dioscoridis（L.）DC. ●☆

305091 Pluchea eggersii Urb. = Pluchea foetida（L.）DC. ■☆

305092 Pluchea eupatorioides Kurz;泽兰阔苞菊（长叶阔苞菊,香艾）;Longleaf Pluchea ●

305093 Pluchea foetida（L.）DC.;沼泽阔苞菊;Marsh Fleabane,Stinking Camphorweed,Stinking Fleabane ■☆

305094 Pluchea foetida（L.）DC. var. imbricata Kearney = Pluchea foetida（L.）DC. ■☆

305095 Pluchea frutescens Benth. = Pluchea arguta Boiss. ●☆

305096 Pluchea frutescens Benth. var. parvifolia Chiov. = Pluchea somaliensis（Thell.）Thulin ■☆

305097 Pluchea grevei（Baill.）Humbert;格雷弗阔苞菊●☆

305098 Pluchea heterophylla Vatke;互叶阔苞菊■☆

305099 Pluchea hirsuta Less. = Blumea clarkei Hook. f. ■

305100 Pluchea indica（L.）Less.;阔苞菊（格杂树,鲫鱼胆,栾犀,栾樨,五里香,烟茜,烟樨,燕茜,苑樨）;India Pluchea,Indian Camphorweed,Indian Pluchea ●■

305101 Pluchea integrifolia Mattf. = Pluchea bojeri（DC.）Humbert ●☆

305102 Pluchea kelleri（Thell.）Thulin;凯勒阔苞菊■☆

305103 Pluchea krausii（Sch. Bip. ex Walp.）Sch. Bip. ex Steetz = Pluchea dioscoridis（L.）DC. ●☆

305104 Pluchea lanceolata（DC.）C. B. Clarke;披针叶阔苞菊（披针阔苞菊）■☆

305105 Pluchea lanceolata（DC.）Oliv. et Hiern = Pluchea lanceolata（DC.）C. B. Clarke ■☆

305106 Pluchea leptophylla D. Y. Hong et F. H. Chen = Pluchea pteropoda Hemsl. ex Forbes et Hemsl. ●■

305107 Pluchea leubnitziae（Kuntze）N. E. Br. = Pechuel-Loeschea leubnitziae（Kuntze）O. Hoffm. ■☆

305108 Pluchea littoralis Thulin;滨海阔苞菊■☆

305109 Pluchea longifolia Nash;长叶阔苞菊;Long-leaf camphorweed ■☆

305110 Pluchea lucens Thulin;光亮阔苞菊■☆

305111 Pluchea lycioides（Hiern）Merxm.;枸杞状阔苞菊■☆

305112 Pluchea monocephala E. A. Bruce = Porphyrostemma monocephala（E. A. Bruce）Leins ■☆

305113 Pluchea nitens O. Hoffm. = Nicolasia nitens（O. Hoffm.）Eyles ■☆

305114 Pluchea nogalensis Chiov.;诺加尔阔苞菊■☆

305115 Pluchea odorata（L.）Cass.;芳香阔苞菊（香阔苞菊）;Shrubby Camphor-weed,Sweetscent ●■☆

305116 Pluchea odorata（L.）Cass. var. succulenta（Fernald）Cronquist;多汁芳香阔苞菊■☆

305117 Pluchea ovalis（Pers.）DC.;椭圆阔苞菊■☆

305118 Pluchea pectinata F. G. Davies et J.-P. Lebrun = Pluchea somaliensis（Thell.）Thulin ■☆

305119 Pluchea petiolata Cass.;叶柄阔苞菊■☆

305120 Pluchea pinnatidifida Hook. f.;羽裂阔苞菊■☆

305121 Pluchea pinnatifida Hook. f. = Pluchea arabica（Boiss.）Qaiser et Lack ■●☆

305122 Pluchea pteroclada Chiov. = Pluchea nogalensis Chiov. ■☆

305123 Pluchea pteropoda Hemsl. ex Forbes et Hemsl.;光梗阔苞菊（山兰苞菊,翼柄燕茜）;Wingedstalk Pluchea ●■

305124 Pluchea purpurascens（Sw.）DC. = Pluchea odorata（L.）Cass. ●■☆

305125 Pluchea purpurascens（Sw.）DC. var. succulenta Fernald = Pluchea odorata（L.）Cass. var. succulenta（Fernald）Cronquist ■☆

305126 Pluchea quitoc DC. = Pluchea sagittalis（Lam.）Cabrera ■

305127 Pluchea rosea R. K. Godfrey = Pluchea baccharis（Mill.）Pruski ■☆

305128 Pluchea rubicunda Schneid. = Vernonia bockiana Diels ●

305129 Pluchea rufescens（DC.）A. J. Scott;浅红阔苞菊■☆

305130　Pluchea sagittalis（Lam.）Cabrera；翼茎阔苞菊（箭形阔苞菊，矢状阔苞菊）；Wing-stem Camphorweed，Wingstem Pluchea ■

305131　Pluchea sarcophylla Chiov.；肉叶阔苞菊■☆

305132　Pluchea scabrida DC. = Conyza scabrida DC. ■☆

305133　Pluchea senegalensis Klatt = Pluchea odorata（L.）Cass. ●■☆

305134　Pluchea sericea（Nutt.）Coville；绢毛阔苞菊；Arrowweed，Arrow-weed ■☆

305135　Pluchea serra Franch. = Iphionopsis rotundifolia（Oliv. et Hiern）Anderb. ■☆

305136　Pluchea somaliensis（Thell.）Thulin；索马里阔苞菊■☆

305137　Pluchea sordida（Vatke）Oliv. et Hiern；暗色阔苞菊■☆

305138　Pluchea suaveolens（Vell.）Kuntze = Pluchea sagittalis（Lam.）Cabrera ■

305139　Pluchea subumbellata Klatt = Microglossa pyrifolia（Lam.）Kuntze ●

305140　Pluchea succulenta Mesfin；多汁阔苞菊■☆

305141　Pluchea symphytifolia（Mill.）Gillis；合生叶阔苞菊■☆

305142　Pluchea tenuifolia Small = Pluchea foetida（L.）DC. ■☆

305143　Pluchea tomentosa DC. = Pluchea ovalis（Pers.）DC. ■☆

305144　Pluchea yucatanensis G. L. Nesom；尤卡坦阔苞菊；Yucatan Camphorweed ■☆

305145　Pluchia Vell. = Diclidanthera Mart. ●■☆

305146　Pluechea Zoll. et Moritzi = Pluchea Cass. ●■

305147　Plukenetia L.（1753）；普拉克大戟属●☆

305148　Plukenetia africana Sond.；非洲普拉克大戟●☆

305149　Plukenetia ankaranensis L. J. Gillespie；安卡兰普拉克大戟●☆

305150　Plukenetia conophora Müll. Arg.；束梗普拉克大戟；Owusa Nut ●☆

305151　Plukenetia conophora Müll. Arg. = Tetracarpidium conophorum（Müll. Arg.）Hutch. et Dalziel ●☆

305152　Plukenetia decidua L. J. Gillespie；脱落普拉克大戟●☆

305153　Plukenetia hastata Müll. Arg. = Plukenetia africana Sond. ●☆

305154　Plukenetia madagascariensis Léandri；马岛普拉克大戟●☆

305155　Plukenetia procumbens Prain = Pterococcus procumbens（Prain）Pax et K. Hoffm. ●☆

305156　Plukenetia zenkeri Pax = Hamilcoa zenkeri（Pax）Prain ●☆

305157　Pluknelia Boehm. = Plukenetia L. ●☆

305158　Plumaria Fabr. = Eriophorum L. ■

305159　Plumaria Heist. ex Fabr. = Eriophorum L. ■

305160　Plumaria Opiz = Dianthus L. ■

305161　Plumatichilos Szlach.（2001）；羽兰属■☆

305162　Plumbagella Spach（1841）；鸡娃草属（鸡娃花属，类白花丹属，小蓝花丹属，小蓝雪花属，小蓝血花属）；Plumbagella ■

305163　Plumbagella micrantha（Ledeb.）Spach；鸡娃草（刺矾松，鸡娃花，蓝雪草，蓝雪花，小蓝花丹，小蓝雪草，小蓝雪花）；Littleflower Plumbagella ■

305164　Plumbagidium Spach = Plumbago L. ●■

305165　Plumbaginaceae Juss.（1789）（保留科名）；白花丹科（矾松科，蓝雪科）；Leadwort Family，Thrift Family ●■

305166　Plumbaginella Ledeb. = Plumbagella Spach ■

305167　Plumbago L.（1753）；白花丹属（蓝雪花属，蓝雪属，乌面马属）；Lead Wort，Leadwort，Plumbago ●■

305168　Plumbago Tourn. ex L. = Plumbago L. ●■

305169　Plumbago alba Pasq. = Plumbago auriculata Lam. f. alba（Pasq.）Z. X. Peng ●

305170　Plumbago amplexicaulis Oliv.；抱茎白花丹●☆

305171　Plumbago aphylla Bojer ex Boiss.；无叶白花丹●☆

305172　Plumbago auriculata Lam.；蓝花丹（花绣球，蓝花矾松，蓝茉莉,蓝雪,蓝雪丹,蓝雪花）；Blue Leadwort，Blueflower Leadwort，Cape Leadwort，Cape Plumbago，Leadwort，Leafless leadwort，Plumbago ●

305173　Plumbago auriculata Lam. ‘Alba’；白花蓝花丹●☆

305174　Plumbago auriculata Lam. ‘Royal Cape’；王国之角蓝花丹●☆

305175　Plumbago auriculata Lam. f. alba（Pasq.）Z. X. Peng；雪花丹（白耳雪花丹，白雪花丹）；Cape Plumbago，Whiteear Leadwort ●

305176　Plumbago capensis Thunb. = Plumbago auriculata Lam. ●

305177　Plumbago capensis Thunb. var. alba Cam = Plumbago auriculata Lam. f. alba（Pasq.）Z. X. Peng ●

305178　Plumbago ciliata Wilmot-Dear；缘毛蓝花丹●☆

305179　Plumbago coccinea Salisb. = Plumbago indica L. ●■

305180　Plumbago dawei Rolfe；道氏白花丹●☆

305181　Plumbago esquirolii H. Lév.；贵州白花丹●

305182　Plumbago esquirolii H. Lév. = Anisadenia pubescens Griff. ■

305183　Plumbago europaea L.；欧洲白花丹（欧白花丹）；Europe，European Leadwort ●☆

305184　Plumbago glandulicaulis Wilmot-Dear；腺茎白花丹●☆

305185　Plumbago indica L.；紫花丹（赤花藤，红花丹，谢三娘，印度乌面马，紫花藤，紫雪花）；Fire-plant，India Leadwort，Indian Leadwort，Rose Plumbago，Rosy-flowered Leadwort，Whorled Plantain ●■

305186　Plumbago larpentae Lindl. = Ceratostigma plumbaginoides Bunge ●■

305187　Plumbago micrantha Ledeb. = Plumbagella micrantha（Ledeb.）Spach ■

305188　Plumbago montis-elgonis Bull.；埃尔贡白花丹●☆

305189　Plumbago parvifolia Hemsl. = Plumbago aphylla Bojer ex Boiss. ●☆

305190　Plumbago pearsonii L. Bolus；皮尔逊白花丹●☆

305191　Plumbago rosea L. = Plumbago indica L. ●■

305192　Plumbago rosea L. = Plumbago zeylanica L. ●

305193　Plumbago rosea L. var. coccinea（Lour.）Hook. = Plumbago indica L. ●■

305194　Plumbago scandens L.；夏蓝雪花；Devil's Herb，Leadwort，Summer Snow ●☆

305195　Plumbago scandens L. = Plumbago zeylanica L. ●

305196　Plumbago spinosa K. S. Hao = Plumbagella micrantha（Ledeb.）Spach ■

305197　Plumbago stenophylla Wilmot-Dear；窄叶白花丹●☆

305198　Plumbago toxicaria Bertol.；毒白花丹●☆

305199　Plumbago tristis Aiton；暗淡白花丹●☆

305200　Plumbago viscosa Blanco = Plumbago zeylanica L. ●

305201　Plumbago zeyalanica L. var. glaucescens Boiss. = Plumbago zeylanica L. ●

305202　Plumbago zeylanica L.；白花丹（白花金丝岩陀，白花九股牛，白花藤，白花谢三娘，白花岩陀，白花皂药，白花仔，白雪丹，白雪花，白皂药，耳丁藤，隔布草，钩藤托，火灵丹，假茉莉，猛老虎，千槟榔，千里及，三素英，山坡苓，天槟榔，天山娘，乌面马，野苜莉，一见消，照药，照药根子，总管）；Ceylon Leadwort，Doctorbush，Whiteflower Leadwort，White-flowered Leadwort ●

305203　Plumbago zeylanica L. var. dawei（Rolfe）Mildbr. = Plumbago dawei Rolfe ●☆

305204　Plumbago zeylanica L. var. oxypetala Boiss.；尖瓣白花丹；Shappetal Leadwort ●

305205　Plumea Lunan = Guarea F. Allam.（保留属名）●☆

305206　Plumeria L.（1753）；鸡蛋花属（缅栀属，缅栀子属）；Frangipani，Plumeria，Temple Tree ●

305207　Plumeria Tourn. ex L. = Plumeria L. ●

305208　Plumeria acuminata Aiton ＝Plumeria rubra L. 'Acutifolia'●

305209　Plumeria acuminata Aiton ＝Plumeria rubra L.●

305210　Plumeria acutifolia Aiton ＝Plumeria rubra L.●

305211　Plumeria acutifolia Poir. ＝Plumeria rubra L. 'Acutifolia'●

305212　Plumeria acutifolia Poir. ＝Plumeria rubra L.●

305213　Plumeria alba L.；白鸡蛋花(白花缅栀,粉花鸡蛋花,西印度茉莉)；Frangipani, Temple Tree, White Frangipani, White Indian Jasmine ●☆

305214　Plumeria cubensis Urb. ；古巴鸡蛋花●☆

305215　Plumeria lancifolia Müll. Arg. ；柳叶鸡蛋花；Lanceoleaf Frangipani ●☆

305216　Plumeria leucantha G. Don；白花鸡蛋花●☆

305217　Plumeria leuconeura Urb. ；白脉鸡蛋花●☆

305218　Plumeria lutea A. Chev. ＝Plumeria rubra L.●

305219　Plumeria lutea Ruiz et Pav. ；黄花鸡蛋花●☆

305220　Plumeria multiflora Standl. ；多花鸡蛋花●☆

305221　Plumeria obtusa L.；钝叶鸡蛋花(白鸡蛋花)；Frangipani, Pagoda Tree, Singapore Plumeria, White Frangipani ●

305222　Plumeria obtusa L. 'Dwarf Singapore Pink'；新加坡矮粉白鸡蛋花●☆

305223　Plumeria obtusa L. 'Singapore White'；新加坡白鸡蛋花●☆

305224　Plumeria retusa Lam. ＝Tabernaemontana retusa (Lam.) Palacky ●☆

305225　Plumeria rubra L.；红鸡蛋花(蕃花,蕃仔花,鸡蛋花,尖叶印度素馨,缅栀,印度素馨)；Frangipani, Mexican Frangipani, Nosegay Frangipani, Pagoda Tree, Plumeria, Red Frangipani, Red Jasmine, Temple Flower, Temple Tree, Templetree, West Indian Jasmine ●

305226　Plumeria rubra L. 'Acutifolia'；鸡蛋花(大季花,蛋黄花,番花,番茉莉,黄鸡蛋花,尖叶红鸡蛋花,擂捶花,缅栀子,鸭脚木)；Frangipani Tree, Jasmine Tree, Mexican Frangipani, Pagoda Tree, Spanish Jasmine, Temple Tree ●

305227　Plumeria rubra L. 'Alba' ＝Plumeria alba L.●☆

305228　Plumeria rubra L. 'Bridal White'；白色新娘红鸡蛋花●☆

305229　Plumeria rubra L. 'Celandine'；白屈菜红鸡蛋花●☆

305230　Plumeria rubra L. 'Dark Red'；暗红红鸡蛋花●☆

305231　Plumeria rubra L. 'Rosy Dawn'；浅玫瑰红红鸡蛋花●☆

305232　Plumeria rubra L. 'Rubra' ＝Plumeria rubra L.●

305233　Plumeria rubra L. 'Starlight'；星光红鸡蛋花●☆

305234　Plumeria rubra L. f. acutifolia (Poir. ex Lam.) Woodson ＝Plumeria rubra L. 'Acutifolia'●

305235　Plumeria rubra L. f. acutifolia Woodson ＝Plumeria rubra L. 'Acutifolia'●

305236　Plumeria rubra L. var. acutifolia (Poir. ex Lam.) Bailey；缅栀●

305237　Plumeria rubra L. var. acutifolia (Poir. ex Lam.) L. H. Bailey ＝Plumeria rubra L. 'Acutifolia'●

305238　Plumeria rubra L. var. acutifolia (Poir.) L. H. Bailey ＝Plumeria rubra L.●

305239　Plumeria rubra L. var. acutifolia (Poir.) L. H. Bailey ＝Plumeria rubra L. 'Acutifolia'●

305240　Plumeria stenophylla Urb. ；狭叶鸡蛋花●☆

305241　Plumeria sucuuba Spruce ex Müll. Arg. ；巴西鸡蛋花●☆

305242　Plumeriaceae Horan. ；鸡蛋花科●

305243　Plumeriaceae Horan. ＝Apocynaceae Juss. (保留科名)●■

305244　Plumeriopsis Rusby et Woodson ＝Ahouai Mill. (废弃属名)●■

305245　Plumeriopsis Rusby et Woodson ＝Thevetia L. (保留属名)●■

305246　Plumiera Adans. ＝Plumeria L.●

305247　Plumiera L. ＝Plumeria L.●

305248　Plummera A. Gray ＝Hymenoxys Cass. ■☆

305249　Plummera A. Gray(1882)；普卢默菊属■☆

305250　Plummera ambigens S. F. Blake ＝Hymenoxys ambigens (S. F. Blake) Bierner ■☆

305251　Plummera floribunda A. Gray ＝Hymenoxys ambigens (S. F. Blake) Bierner var. floribunda (A. Gray) W. L. Wagner ■☆

305252　Plumosipappus Czerep. ＝Centaurea L. (保留属名)●■

305253　Plumosipappus Czerep. ＝Phaeopappus (DC.) Boiss. ●■

305254　Plumosipappus De Moor ＝Centaurea L. (保留属名)●■

305255　Pluridens Neck. ＝Bidens L. ■●

305256　Plurimaria Raf. ＝Blackstonia Huds. ■☆

305257　Plutarchia A. C. Sm. (1936)；烟花莓属●☆

305258　Plutarchia angulata A. C. Sm. ；窄烟花莓●☆

305259　Plutarchia dasyphylla A. C. Sm. ；毛叶烟花莓●☆

305260　Plutarchia minor A. C. Sm. ；小烟花莓●☆

305261　Plutarchia monantha A. C. Sm. ；山地烟花莓●☆

305262　Plutarchia rigida (Benth.) A. C. Sm. ；硬烟花莓●☆

305263　Plutonia Noronha ＝Phaleria Jack ●☆

305264　Plutonopuntia P. V. Heath ＝Opuntia Mill. ●

305265　Plutonopuntia P. V. Heath(1999)；暗掌属●☆

305266　Pneumaria Hill(废弃属名) ＝Mertensia Roth(保留属名)■

305267　Pneumonanthe Gled. ＝Gentiana L. ■

305268　Pneumonanthe andrewsii (Griseb.) W. A. Weber ＝Gentiana andrewsii Griseb. ■☆

305269　Pneumonanthe depressa D. Don ＝Gentiana depressa D. Don et C. E. C. Fisch. ■

305270　Pneumonanthe linearis (Froel.) Greene ＝Gentiana linearis Froel. ■☆

305271　Pneumonanthe linearis Greene ＝Gentiana linearis Froel. ■☆

305272　Pneumonanthe ornata Wall. ex G. Don ＝Gentiana ornata (Wall. ex G. Don) Griseb. ■

305273　Pneumonanthopsis Miq. ＝Voyria Aubl. ■☆

305274　Poa L. (1753)；早熟禾属；Blue Gress, Bluegrass, Blue-grass, Meadow Gras, Meadow-grass, Poa, Spear-grass ■

305275　Poa abbreviata R. Br. ；短早熟禾(矮莓系,短缩早熟禾)；Sharp-glume Bluegrass ■

305276　Poa abyssinica Jacq. ＝Eragrostis tef (Zuccagni) Trotter ■☆

305277　Poa acmocalyx Keng ex L. Liou ＝Poa lapponica Prokudin subsp. acmocalyx (Keng ex L. Liou) Olonova et G. Zhu ■

305278　Poa acroleuca Steud. ；白顶早熟禾；Whitetopped Bluegrass ■

305279　Poa acroleuca Steud. f. submoniliformis (Makino) T. Koyama ＝Poa acroleuca Steud. var. submoniliformis Makino ■☆

305280　Poa acroleuca Steud. var. ryukyuensis Koba et Tateoka；如昆早熟禾■

305281　Poa acroleuca Steud. var. spiciformis Honda ＝Poa hisauchii Honda ■

305282　Poa acroleuca Steud. var. submoniliformis Makino；亚串珠白顶早熟禾■☆

305283　Poa acuminata Ovcz. ＝Poa fragilis Ovcz. et Czukav. ■

305284　Poa acuminata Ovcz. ＝Poa versicolor Besser subsp. relaxa (Ovcz.) Tzvelev ■

305285　Poa acuticaulis Ovcz. et Czukav. ；尖茎早熟禾■☆

305286　Poa aegyptiaca Willd. ＝Eragrostis aegyptiaca (Willd.) Delile ■☆

305287　Poa afghanica Bor；阿富汗早熟禾；Afghan Bluegrass ■

305288　Poa airoides (Nees) Kunth ＝Catabrosa aquatica (L.) P. Beauv. ■

305289　Poa airoides Kunth ＝Catabrosa aquatica (L.) P. Beauv. ■

305290　Poa aitchisonii Boiss. ;艾松早熟禾■

305291　Poa albertii Regel;阿拉套早熟禾■

305292　Poa albertii Regel = Poa litwinowiana Ovcz. ■

305293　Poa albertii Regel subsp. arnoldii（Melderis）Olonova et G. Zhu;阿诺早熟禾■

305294　Poa albertii Regel subsp. kunlunensis（N. R. Cui）Olonova et G. Zhu;高寒早熟禾(印度早熟禾）;Indian Bluegrass ■

305295　Poa albertii Regel subsp. lahulensis（Bor）Olonova et G. Zhu;拉哈尔早熟禾(拉哈尔莓系）;Lahule Bluegrass ■

305296　Poa albertii Regel subsp. poophagorum（Bor）Olonova et G. Zhu;波伐早熟禾;Tibet-sikkim Bluegrass, Xizang-sikkim Bluegrass ■

305297　Poa albida Turcz. ex Trin. = Festuca sibirica Hack. ex Boiss. ■

305298　Poa albida Turcz. ex Trin. = Leucopoa albida（Turcz. ex Trin.）V. I. Krecz. et Bobrov ■

305299　Poa almasovii Golub;阿玛早熟禾■

305300　Poa alpigena（Blytt）Lindm. ;高原早熟禾;Plateau Bluegrass, Plateau-living Bluegrass ■

305301　Poa alpigena（Blytt）Lindm. subsp. staintonii Melderis = Poa pratensis L. subsp. staintonii（Melderis）Dickoré ■

305302　Poa alpigena Lindm. = Poa alpigena（Blytt）Lindm. ■

305303　Poa alpina L. ;高山早熟禾(高山莓系,高原早熟禾）;Alpine Bluegrass, Alpine Blue-grass, Alpine Meadow Grass, Alpine Meadow-grass ■

305304　Poa alpina L. f. vivipara（L.）B. Boivin = Poa bulbosa L. ■

305305　Poa alpina L. subsp. atlantica（Trab.）Romo;北非高山早熟禾■☆

305306　Poa alpina L. subsp. stenobotrya Maire;窄叶高山早熟禾■☆

305307　Poa alpina L. subsp. vivipara（L.）Scribn. et Merr. = Poa bulbosa L. ■

305308　Poa alpina L. subsp. vivipara（L.）Tzvelev = Poa bulbosa L. ■

305309　Poa alpina L. var. atlantica Trab. = Poa alpina L. subsp. atlantica（Trab.）Romo ■☆

305310　Poa alpina L. var. djurdjurae Trab. = Poa alpina L. subsp. atlantica（Trab.）Romo ■☆

305311　Poa alpina L. var. obtusata Litard. et Maire = Poa alpina L. subsp. atlantica（Trab.）Romo ■☆

305312　Poa alpina L. var. saposhnikovii Serg. = Poa smirnowii Roshev. subsp. mariae（Reverd.）Tzvelev ■

305313　Poa alpina L. var. vivipara L. = Poa bulbosa L. ■

305314　Poa alsodes A. Gray;北美早熟禾;Grove Meadow-grass ■☆

305315　Poa alta Hitchc. ;高株早熟禾(高株莓系）;High Bluegrass ■

305316　Poa altaica Trin. ;阿尔泰早熟禾;Altai Bluegrass ■

305317　Poa alternans（Nees）Steud. = Tribolium uniolae（L. f.）Renvoize ■☆

305318　Poa amabilis L. = Eragrostis amabilis（L.）Wight et Arn. ■

305319　Poa amabilis L. = Eragrostis tenella（L.）P. Beauv. ex Roem. et Schult. ■

305320　Poa amabilis L. = Eragrostis unioloides（Retz.）Nees ex Steud. ■

305321　Poa amboinica L. = Eragrostis ciliaris（L.）R. Br. ■

305322　Poa amoena Bor = Poa pseudamoena Bor ■

305323　Poa ampla Merr. ;大早熟禾(巨早熟禾）;Big Bluegrass ■

305324　Poa ampla Merr. = Poa secunda J. Presl subsp. juncifolia（Scribn.）Soreng ■

305325　Poa anceps（Gaudich.）Bor = Poa pratensis L. var. anceps Gaudich. ex Griseb. ■

305326　Poa anceps G. Forst. = Poa pratensis L. var. anceps Gaudich. ex Griseb. ■

305327　Poa angustata R. Br. = Puccinellia angustata（R. Br.）E. L. Rand et Redfield ■

305328　Poa angustifolia L. ;细叶早熟禾(窄叶早熟禾）;Narrow-leaved Bluegrass, Narrow-leaved Blue-grass, Narrow-leaved Meadow-grass, Thinleaf Bluegrass ■

305329　Poa angustifolia L. = Poa pratensis L. subsp. angustifolia（L.）Arcang. var. strigosa（Hoffm.）Gaudin ■

305330　Poa angustifolia L. subsp. laxuispicula D. F. Cui;疏穗细叶早熟禾■

305331　Poa angustiglumis Roshev. ;狭颖早熟禾(窄颖早熟禾）;Narrowglume Bluegrass ■

305332　Poa angustiglumis Roshev. = Poa pratensis L. ■

305333　Poa ankaratrensis A. Camus et H. Perrier;安卡拉特拉早熟禾■☆

305334　Poa annua L. ;早熟禾（发汗草）;Annual Bluegrass, Annual Blue-grass, Annual Meadow Grass, Annual Meadow-grass, Blue-grass, Low Spear-grass, Meadow-grass, Monkey Nut, Six Weeks' Grass ■

305335　Poa annua L. f. reptans（Hausskn.）T. Koyama = Poa annua L. var. reptans Hausskn. ■

305336　Poa annua L. f. reptans（Hausskn.）T. Koyama = Poa annua L. ■

305337　Poa annua L. subsp. exilis（Freyn）Murb. = Poa infirma Kunth ■

305338　Poa annua L. subsp. exilis（Tomm. ex Freyn）Asch. et Graebn. = Poa infirma Kunth ■

305339　Poa annua L. subsp. pilantha（Ronniger）H. Scholz;毛花早熟禾■☆

305340　Poa annua L. subsp. typica（Beck）Braun-Blanq. = Poa annua L. ■

305341　Poa annua L. var. aquatica Asch. = Poa annua L. ■

305342　Poa annua L. var. exilis（Freyn）Murb. = Poa infirma Kunth ■

305343　Poa annua L. var. exilis Tomm. ex Freyn = Poa infirma Kunth ■

305344　Poa annua L. var. maroccana（Nannf.）Litard. = Poa maroccana Nannf. ■☆

305345　Poa annua L. var. nepalensis Griseb. = Poa nepalensis Wall. ex Duthie ■

305346　Poa annua L. var. ovalis（Tineo）Trab. = Poa annua L. ■

305347　Poa annua L. var. pilantha Ronniger = Poa annua L. subsp. pilantha（Ronniger）H. Scholz ■☆

305348　Poa annua L. var. reptans Hausskn. ;匍生早熟禾(爬地早熟禾）;Reptant Annual Bluegrass ■

305349　Poa annua L. var. reptans Hausskn. = Poa annua L. ■

305350　Poa annua L. var. rivulorum（Maire et Trab.）Litard. et Maire = Poa rivulorum Maire et Trab. ■☆

305351　Poa annua L. var. sikkimensis Stapf = Poa sikkimensis（Stapf）Bor ■

305352　Poa annua L. var. supina（Schrad.）Link = Poa supina Schrad. ■

305353　Poa annua L. var. supina（Schrad.）Rchb. = Poa supina Schrad. ■

305354　Poa antennata Delile ex Poir. = Eragrostis aegyptiaca（Willd.）Delile ■☆

305355　Poa aquatica L. = Glyceria maxima（Hartm.）Holmb. ■

305356　Poa arachnifera Torr. ;得州早熟禾;Texas Blue Grass ■☆

305357　Poa araratica Trautv. ;阿洼早熟禾■

305358　Poa araratica Trautv. subsp. altior（Keng）Olonova et G. Zhu;高阿洼早熟禾■

305359　Poa araratica Trautv. subsp. ianthina（Keng ex Shan Chen）Olonova et G. Zhu;堇色早熟禾;Violet Bluegrass ■

305360　Poa araratica Trautv. subsp. oligophylla（Keng）Olonova et G. Zhu;贫叶早熟禾;Few-leaved Bluegrass ■

305361　Poa araratica Trautv. subsp. psilolepis（Keng）Olonova et G.

Zhu；光稃早熟禾；Smoothlemma Bluegrass ■

305362　Poa arctica R. Br.；极地早熟禾（北极早熟禾）；Arctic Bluegrass，Arctic Blue-grass ■

305363　Poa arctica R. Br. subsp. caespitans Nannf. = Poa tolmatchewii Roshev. ■

305364　Poa arctica R. Br. subsp. caespitans Simmons ex Nannf.；簇生极地早熟禾（极地早熟禾，托玛早熟禾）■

305365　Poa arctica R. Br. subsp. caespitans Simmons ex Nannf. = Poa tolmatchewii Roshev. ■

305366　Poa arctica R. Br. subsp. smirnowii（Roshev.）Malyschev = Poa smirnowii Roshev. ■

305367　Poa argunensis Roshev. = Poa versicolor Besser subsp. reverdattoi（Roshev.）Olonova et G. Zhu ■

305368　Poa argunensis Roshev. ex Kom. et Roshev.；额尔古纳早熟禾；Srgunen Bluegrass ■

305369　Poa arjinsanensis D. F. Cui；阿尔金山早熟禾；Arjinshan Bluegrass ■

305370　Poa arjinsanensis D. F. Cui = Poa araratica Trautv. subsp. oligophylla（Keng）Olonova et G. Zhu ■

305371　Poa arnoldii Melderis = Poa albertii Regel subsp. arnoldii（Melderis）Olonova et G. Zhu ■

305372　Poa articulata Ovcz.；节状早熟禾■☆

305373　Poa aspera Jacq. = Eragrostis aspera（Jacq.）Nees ■☆

305374　Poa asperifolia Bor；糙叶早熟禾■

305375　Poa atrovens Desf. = Eragrostis atrovirens（Desf.）Trin. ex Steud. ■

305376　Poa atrovirens Desf. = Eragrostis atrovirens（Desf.）Trin. ex Steud. ■

305377　Poa attenuata Boiss. = Poa araratica Trautv. ■

305378　Poa attenuata Boiss. subsp. argunensis（Roshev.）Tzvelev = Poa argunensis Roshev. ex Kom. et Roshev. ■

305379　Poa attenuata Boiss. subsp. botryoides Tzvelev = Poa versicolor Besser subsp. stepposa（Krylov）Tzvelev ■

305380　Poa attenuata Trin. ex Bunge；渐尖早熟禾（华灰早熟禾，渐狭早熟禾，葡系早熟禾，兴安莓系）；Chines Glaucous Bluegrass，Dahurian Bluegrass，Racemoce Bluegrass ■

305381　Poa attenuata Trin. ex Bunge subsp. argunensis（Roshev.）Tzvelev = Poa argunensis Roshev. ex Kom. et Roshev. ■

305382　Poa attenuata Trin. ex Bunge subsp. botryoides（Trin. ex Griseb.）Tzvelev = Poa attenuata Trin. ex Bunge ■

305383　Poa attenuata Trin. ex Bunge var. dahurica（Trin.）Griseb.；达呼里早熟禾■

305384　Poa attenuata Trin. ex Bunge var. stepposa Krylov = Poa stepposa（Krylov）Roshev. ■

305385　Poa attenuata Trin. var. altior Keng = Poa araratica Trautv. subsp. altior（Keng）Olonova et G. Zhu ■

305386　Poa attenuata Trin. var. stepposa Krylov = Poa versicolor Besser subsp. stepposa（Krylov）Tzvelev ■

305387　Poa attenuata Trin. var. versicolor（Besser）Regel = Poa versicolor Besser ■

305388　Poa aulacosperma Fresen. = Eragrostis papposa（Roem. et Schult.）Steud. ■☆

305389　Poa bactriana Roshev.；荒漠早熟禾（荒漠胎生早熟禾）■☆

305390　Poa bactriana Roshev. subsp. glabriflora（Roshev. ex Ovcz.）Tzvelev = Poa glabriflora Roshev. ex Kom. et Roshev. ■

305391　Poa bactriana Roshev. subsp. glabriflora（Roshev.）Tzvelev = Poa glabriflora Roshev. ex Kom. et Roshev. ■

305392　Poa bactriana Roshev. subsp. zaprjagajevii（Ovcz.）Tzvelev = Poa glabriflora Roshev. ex Kom. et Roshev.

305393　Poa bactriana Roshev. subsp. zaprjagajevii（Ovcz.）Tzvelev = Poa zaprjagajevii Ovcz. ■

305394　Poa badensis Haenke ex Willd.；巴顿早熟禾■

305395　Poa bakuensis Litv.；巴库早熟禾■☆

305396　Poa balfourii Parn. = Poa glauca Vahl ■

305397　Poa barguzinensis Popov = Poa paucispicula Scribn. et Merr. ■

305398　Poa bedeliensis Litv. = Poa lipskyi Roshev. ■

305399　Poa bergiana Kunth = Eragrostis bergiana（Kunth）Trin. ■☆

305400　Poa bidentata Stapf = Poa pratensis L. ■

305401　Poa bifaria Vahl = Eragrostiella bifaria（Vahl）Bor ■

305402　Poa biformis Kunth = Eragrostis atrovirens（Desf.）Trin. ex Steud. ■

305403　Poa binodis Keng；双节早熟禾；Binode Bluegrass ■

305404　Poa bipollicaris Hochst. = Poa annua L. ■

305405　Poa bomiensis C. Ling；波密早熟禾；Bomi Bluegrass ■

305406　Poa boreali-tibetica C. Ling；藏北早熟禾；N. Xizang Bluegrass ■

305407　Poa borealitibetica C. Ling = Poa albertii Regel subsp. lahulensis（Bor）Olonova et G. Zhu ■

305408　Poa botryoides（Trin. ex Griseb.）Kom. = Poa versicolor Besser subsp. stepposa（Krylov）Tzvelev ■

305409　Poa botryoides Trin. ex Besser；葡系早熟禾■

305410　Poa botryoides Trin. ex Besser = Poa attenuata Trin. ex Bunge ■

305411　Poa bracteosa Kom.；膜苞早熟禾■

305412　Poa breviligula（Keng）L. Liou；短舌早熟禾（西伯利亚早熟禾）；Shortligule Bluegrass ■

305413　Poa breviligula Keng ex L. Liou = Poa albertii Regel ■

305414　Poa brizoides L. f. = Eragrostis capensis（Thunb.）Trin. ■☆

305415　Poa brownii Kunth = Eragrostis brownii（Kunth）Nees ■

305416　Poa bryophila Trin.；苔地莓系；Moss-loving Bluegrass ■

305417　Poa bucharica Roshev.；布哈尔早熟禾（布查早熟禾）■

305418　Poa bucharica Roshev. subsp. karateginensis（Roshev. ex Ovcz.）Tzvelev；卡拉蒂早熟禾■

305419　Poa bulbosa L.；鳞茎早熟禾；Bulbous Bluegrass，Bulbous Meadow-grass ■

305420　Poa bulbosa L. subsp. crispa（Thuill.）Dumort. = Poa bulbosa L. var. vivipara Koeler ■

305421　Poa bulbosa L. subsp. nevskii（Roshev. ex Ovcz.）Tzvelev = Poa nevskii Roshev. ex Kom. et Roshev. ■

305422　Poa bulbosa L. subsp. vivipara（Koeler）Arcang.；胎生鳞茎早熟禾■

305423　Poa bulbosa L. var. brisaeformis Trab. = Poa bulbosa L. ■

305424　Poa bulbosa L. var. glabriflora Roshev. = Poa glabriflora Roshev. ex Kom. et Roshev. ■

305425　Poa bulbosa L. var. vivipara Koch = Poa bulbosa L. subsp. vivipara（Koeler）Arcang. ■

305426　Poa bulbosa L. var. vivipara Koch = Poa bulbosa L. ■

305427　Poa bulbosa L. var. vivipara Koeler = Poa bulbosa L. subsp. vivipara（Koeler）Arcang. ■

305428　Poa burmanica Bor；缅甸早熟禾■

305429　Poa cachectica Schumach. = Eragrostis cilianensis（All.）Vignolo ex Janch. ■

305430　Poa caesia Sm. = Poa glauca Vahl ■

305431　Poa calliopsis Litv. ex Ovcz.；美丽早熟禾■

305432　Poa cambessediana Kunth = Eragrostis gangetica（Roxb.）Steud. ■☆

305433 Poa caucasica Trin.;高加索早熟禾■☆
305434 Poa cenisia All.;菘山早熟禾;Songshan Bluegrass ■
305435 Poa cenisia Sarg. =Poa granitica Braun-Blanq. ■
305436 Poa chaixii Vill.;查氏早熟禾(扁鞘早熟禾);Broadleaf Bluegrass,Broad-leaved Meadow-grass,Chaix Bluegrass ■
305437 Poa chalarantha Keng;疏花早熟禾;Fewflower Bluegrass ■
305438 Poa chalarantha Keng ex L. Liou =Poa polycolea Stapf ■
305439 Poa chapelieri Kunth =Eragrostis chapelieri (Kunth) Nees ■☆
305440 Poa chapmaniana Scribn.;查普曼早熟禾;Chapman Bluegrass ■☆
305441 Poa chariis Schult. =Eragrostis atrovirens (Desf.) Trin. ex Steud. ■
305442 Poa chilensis Trin.;智利早熟禾;Chilean Bluegrass ■☆
305443 Poa chinensis L. =Leptochloa chinensis (L.) Nees ■
305444 Poa chumbiensis Noltie =Poa szechuensis Rendle ■
305445 Poa chushualana Rajeshwari,R. R. Rao et G. Arti =Poa tibetica Munro ex Stapf ■
305446 Poa cilianensis All. =Eragrostis cilianensis (All.) Vignolo ex Janch. ■
305447 Poa ciliaris L. =Eragrostis ciliaris (L.) R. Br. ■
305448 Poa ciliata Roxb. =Eragrostis ciliata (Roxb.) Nees ■
305449 Poa ciliatiflora Roshev.;睫毛花早熟禾■
305450 Poa ciliatiflora Roshev. =Poa tibetica Munro ex Stapf ■
305451 Poa compressa L.;加拿大早熟禾(加拿大莓系);Canada Bluegrass,Canada Blue-grass,Flat-stemmed Meadow-grass,Flattened Meadow-grass ■
305452 Poa contracta Ovcz. et Czukav. =Poa lipskyi Roshev. ■
305453 Poa convoluta Hornem. =Puccinellia convoluta (Hornem.) Fourr. ■
305454 Poa crassinervis Honda;粗脉早熟禾■☆
305455 Poa crassinervis Honda =Poa annua L. ■
305456 Poa crispa Thuill. =Poa bulbosa L. subsp. vivipara (Koeler) Arcang. ■
305457 Poa cristata (L.) L. =Koeleria cristata (L.) Pers. ■
305458 Poa cristata (L.) L. =Koeleria macrantha (Ledeb.) Schult. ■
305459 Poa crymophila Keng ex C. Ling;冷地早熟禾;Polar Bluegrass ■
305460 Poa crymophila Keng ex C. Ling =Poa araratica Trautv. ■
305461 Poa curvula Schrad. =Eragrostis curvula (Schrad.) Nees ■
305462 Poa curvula Schrad. =Koeleria cristata (L.) Pers. ■
305463 Poa cylindrica Roxb. =Eragrostis cylindrica (Roxb.) Nees ex Hook. et Arn. ■
305464 Poa cynosuroides Retz. =Desmostachya bipinnata (L.) Stapf ■
305465 Poa cyperoides Thunb. =Cladoraphis cyperoides (Thunb.) S. M. Phillips ●☆
305466 Poa cyrenaica E. A. Durand et Barratte =Libyella cyrenaica (E. A. Durand et Barratte) Pamp. ■☆
305467 Poa dahurica Trin. =Poa attenuata Trin. ex Bunge var. dahurica (Trin.) Griseb. ■
305468 Poa dahurica Trin. =Poa attenuata Trin. ex Bunge ■
305469 Poa debilior Hitchc.;细早熟禾■
305470 Poa debilior Hitchc. =Poa szechuensis Rendle var. debilior (Hitchc.) Soreng et G. Zhu ■
305471 Poa debilis Torr.;软早熟禾;Weak Meadow-grass ■☆
305472 Poa declinata Keng;垂枝早熟禾■
305473 Poa declinata Keng ex L. Liou =Poa szechuensis Rendle var. debilior (Hitchc.) Soreng et G. Zhu ■
305474 Poa densa Troitzky;密序早熟禾■
305475 Poa densissima Roshev. ex Ovcz.;小密早熟禾■
305476 Poa densissima Roshev. ex Ovcz. =Poa albertii Regel ■
305477 Poa densissima Roshev. ex Ovcz. =Poa glauciculmis Ovcz. ■☆
305478 Poa densissima Roshev. ex Ovcz. =Poa litwinowiana Ovcz. ■
305479 Poa desertorum Trin. =Poa bulbosa L. subsp. vivipara (Koeler) Arcang. ■
305480 Poa diandra Roxb. =Eragrostis japonica (Thunb.) Trin. ■
305481 Poa diaphora Trin. =Eremopoa altaica (Trin.) Roshev. ■
305482 Poa diaphora Trin. subsp. oxyglumis (Boiss.) Soreng et G. Zhu =Eremopoa oxyglumis (Boiss.) Roshev. ■
305483 Poa diarrhena Schult. =Eragrostis japonica (Thunb.) Trin. ■
305484 Poa digena Melderis;第吉那早熟禾■
305485 Poa digena Melderis =Poa stapfiana Bor ■
305486 Poa dimorphantha Murb.;二型花早熟禾■☆
305487 Poa distans Jacq. =Puccinellia distans (Jacq.) Parl. ■
305488 Poa distans L. =Puccinellia distans (L.) Parl. ■
305489 Poa divaricata Gouan =Sphenopus divaricatus (Gouan) Rchb. ■☆
305490 Poa diversifolia (Boiss. et Balansa) Hack. ex Boiss.;异叶早熟禾■
305491 Poa djurdjurae Trab. =Poa alpina L. subsp. atlantica (Trab.) Romo ■☆
305492 Poa dolichachyra Keng;长稃早熟禾;Longlemma Bluegrass ■
305493 Poa dolichachyra Keng ex P. C. Keng et G. Q. Song =Poa pratensis L. subsp. staintonii (Melderis) Dickoré ■
305494 Poa dolichachyra Keng var. longiflora S. L. Chen et D. Z. Ma;长花长稃早熟禾■
305495 Poa domingensis Pers. =Eragrostis domingensis (Pers.) Steud. ■☆
305496 Poa dschungarica Roshev. =Poa lipskyi Roshev. subsp. dschungarica (Roshev.) Tzvelev ■
305497 Poa dshilgensis Roshev. =Poa timoleontis Heldr. ex Boiss. var. dshilgensis (Roshev.) Tzvelev ■
305498 Poa dshilgensis Roshev. ex Kom.;季莨早熟禾■
305499 Poa dzongicola Noltie;雅江早熟禾■
305500 Poa eduardii Golub. =Poa platyantha Kom. ■
305501 Poa elanata Keng ex L. Liou;光盘早熟禾;Hairleass Bluegrass ■
305502 Poa elanata Keng ex Tzvelev =Poa hylobates Bor ■
305503 Poa eleanorae Bor;易乐早熟禾■
305504 Poa elegantula Kunth =Eragrostis atrovirens (Desf.) Trin. ex Steud. ■
305505 Poa elongata Willd. =Eragrostis elongata (Willd.) Jacq. ■
305506 Poa eminens J. Presl et C. Presl;类早熟禾;Large-flowered Meadow-grass ■
305507 Poa eragraostioides L.;画眉草状早熟禾■
305508 Poa eragrostioides L. Liou =Poa sikkimensis (Stapf) Bor ■
305509 Poa eragrostis L. =Eragrostis minor Host ■
305510 Poa exilis (Tomm. ex Freyn) Murb. =Poa infirma Kunth ■
305511 Poa exilis Murb. =Poa infirma Kunth ■
305512 Poa extremiorientalis Ohwi =Poa malacantha Kom. var. shinanoana (Ohwi) Ohwi f. vivipara (Ohwi) Ohwi ■☆
305513 Poa faberi Rendle;华东早熟禾(法氏早熟禾);Faber Bluegrass ■
305514 Poa faberi Rendle var. ligulata Rendle;尖舌华东早熟禾(尖舌早熟禾)■
305515 Poa faberi Rendle var. longifolia (Keng) Olonova et G. Zhu;毛颖华东早熟禾(毛颖早熟禾)■
305516 Poa falconeri Hook. f.;福克纳早熟禾■
305517 Poa fasciculata Torr. =Puccinellia fasciculata (Torr.) E. P. Bicknell ■☆

305518 Poa fascinata Keng;蛊早熟禾;Fascinate Bluegrass ■

305519 Poa fascinata Keng ex L. Liou = Poa faberi Rendle var. longifolia (Keng) Olonova et G. Zhu ■

305520 Poa fauriei Hack.;法氏早熟禾■☆

305521 Poa fedtschenkoi Roshev.;费氏早熟禾;Fedtschenko Bluegrass ■

305522 Poa ferruginea Thunb. = Eragrostis ferruginea (Thunb.) P. Beauv. ■

305523 Poa festucaeformis Host. = Puccinellia festuciformis (Host) Parl. ■

305524 Poa festuciformis Host = Puccinellia festuciformis (Host) Parl. ■

305525 Poa festucoides N. R. Cui = Poa albertii Regel ■

305526 Poa festucoides N. R. Cui = Poa parafestuca L. Liou ■

305527 Poa festucoides N. R. Cui subsp. kunlunensis N. R. Cui = Poa albertii Regel subsp. kunlunensis (N. R. Cui) Olonova et G. Zhu ■

305528 Poa festucoides N. R. Cui var. kunlunensis N. R. Cui;昆仑羊茅状早熟禾■

305529 Poa flabellata Hook. f. ;扇状早熟禾;Tussac-grass ■☆

305530 Poa flaccidula Boiss. et Reut.;柔弱早熟禾;Flaccid Bluegrass ■

305531 Poa flavida Keng;黄色早熟禾;Yellow Bluegrass ■

305532 Poa flavida Keng ex L. Liou = Poa alta Hitchc. ■

305533 Poa flavidula Kom. = Poa shumushuensis Ohwi ■

305534 Poa flexousa Wahlenb. = Poa granitica Braun-Blanq. ■

305535 Poa florida N. R. Cui;多花早熟禾■

305536 Poa florida N. R. Cui = Poa pratensis L. ■

305537 Poa formosae Ohwi = Poa khasiana Stapf ■

305538 Poa fragilis Ovcz. = Poa versicolor Besser subsp. relaxa (Ovcz.) Tzvelev ■

305539 Poa fragilis Ovcz. et Czukav.;脆早熟禾■

305540 Poa friesiorum Pilg. = Poa leptoclada Hochst. ex A. Rich. ■☆

305541 Poa gamblei Bor;甘波早熟禾■

305542 Poa gammieana Hook. f. ;莨密早熟禾■

305543 Poa ganeschini Roshev. ex Kom. et Roshev.;盖氏早熟禾;Ganeschin Bluegrass ■☆

305544 Poa ganeschinii Roshev.;嘎奈早熟禾■☆

305545 Poa gangetica Roxb. = Eragrostis gangetica (Roxb.) Steud. ■☆

305546 Poa gilgitica Dickoré = Poa polycolea Stapf ■

305547 Poa glabriflora (Roshev.) Roshev. ex Ovcz. = Poa glabriflora Roshev. ex Kom. et Roshev. ■

305548 Poa glabriflora Roshev. ex Kom. et Roshev.;光滑早熟禾■

305549 Poa glabriflora Roshev. ex Kom. et Roshev. = Poa zaprjagajevii Ovcz. ■

305550 Poa glacialis Stapf = Poa ruwenzoriensis Robyns et Tournay ■☆

305551 Poa glauca Vahl;灰早熟禾(格陵兰早熟禾,灰莓系);Balfour's Meadow Grass, Balfour's Meadow-grass, Glaucous Meadow-grass, Greenland Bluegrass ■

305552 Poa glauca Vahl subsp. altaica (Trin.) Olonova et G. Zhu = Poa altaica Trin. ■

305553 Poa glauca Vahl subsp. conferta (Blytt) Lindm. = Poa glauca Vahl var. conferta (Blytt) Nannf. ■☆

305554 Poa glauca Vahl subsp. kitadakensis (Ohwi) T. Koyama = Poa glauca Vahl var. kitadakensis (Ohwi) Ohwi ■☆

305555 Poa glauca Vahl subsp. litwinowiana (Ovcz.) Tzvelev = Poa litwinowiana Ovcz. ■

305556 Poa glauca Vahl subsp. reverdattoi (Roshev.) Tzvelev = Poa reverdattoi Roshev. ex Kom. et Roshev. ■

305557 Poa glauca Vahl var. conferta (Blytt) Nannf.;密集早熟禾■☆

305558 Poa glauca Vahl var. kitadakensis (Ohwi) Ohwi;信州北岳早熟禾■☆

305559 Poa glauciculmis Ovcz.;浅粉绿早熟禾■☆

305560 Poa glomerata Thunb. = Tribolium obtusifolium (Nees) Renvoize ■☆

305561 Poa glumaris Trin. = Poa eminens J. Presl et C. Presl ■

305562 Poa gorbunovii Ovcz. = Puccinellia subspicata V. I. Krecz. ex Ovcz. et Czukav. ■

305563 Poa gorbunovii Ovcz. ex Kom. et Ovcz. = Puccinellia subspicata V. I. Krecz. ex Ovcz. et Czukav. ■

305564 Poa gracilior Keng;茬弱早熟禾;Slender Bluegrass ■

305565 Poa gracilior Keng ex L. Liou = Poa himalayana Nees ex Steud. et Bor ■

305566 Poa gracillima Rendle = Poa szechuensis Rendle ■

305567 Poa grandis Hand. -Mazz.;阔叶早熟禾(大早熟禾,大莓系);Big Bluegrass ■

305568 Poa grandispica Keng ex L. Liou = Poa sphondylodes Trin. ex Bunge var. subtrivialis Ohwi ■

305569 Poa granitica Braun-Blanq.;曲折早熟禾(岩地早熟禾);Flexuose Bluegrass, Wavy Meadow-grass ■

305570 Poa grisea Korotky = Poa pratensis L. subsp. pruinosa (Korotky) Dickoré ■

305571 Poa grisea Korotky = Poa pruinosa Korotky ■

305572 Poa hakusanensis Hack.;白山早熟禾■☆

305573 Poa hayachinensis Koidz.;哈亚早熟禾■

305574 Poa hedbergii S. M. Phillips;赫德早熟禾■☆

305575 Poa hengshanica Keng;恒山早熟禾;Hengshan Bluegrass ■

305576 Poa hengshanica Keng ex L. Liou = Poa sichotensis Prob. ■

305577 Poa heterogama Hack. = Poa binata Nees ■☆

305578 Poa himalayana Nees ex Steud. = Poa himalayana Nees ex Steud. et Bor ■

305579 Poa himalayana Nees ex Steud. et Bor;须弥早熟禾(史蒂瓦早熟禾,喜马拉雅莓系,喜马拉雅早熟禾);Himalayan Bluegrass ■

305580 Poa hippuris Schumach. = Eragrostis aspera (Jacq.) Nees ■☆

305581 Poa hirta Thunb. = Arundinella hirta (Thunb.) Tanaka ■

305582 Poa hirtiglumis Hook. f.;颖毛早熟禾(毛花早熟禾)■

305583 Poa hirtiglumis Hook. f. var. nimuana (C. Ling) Soreng et G. Zhu;尼木早熟禾;Nimu Bluegrass ■

305584 Poa hisauchii Honda;久内早熟禾;Hisauch Bluegrass ■

305585 Poa hissarica Roshev. ex Kom. et Roshev.;希萨尔早熟禾■

305586 Poa hissarica Roshev. ex Ovcz. = Poa hissarica Roshev. ex Kom. et Roshev. ■

305587 Poa humilis (M. Bieb.) K. Koch = Catabrosella humilis (M. Bieb.) Tzvelev ■

305588 Poa humilis Ehrh.;铺散早熟禾;Spreading Meadow-grass ■☆

305589 Poa hybrida Gaudich.;杂早熟禾■

305590 Poa hylobates Bor;喜巴早熟禾■

305591 Poa hypnoides Lam. = Eragrostis hypnoides (Lam.) Britton, Sterns et Poggenb. ■☆

305592 Poa ianthina Keng ex Shan Chen = Poa araratica Trautv. subsp. ianthina (Keng ex Shan Chen) Olonova et G. Zhu ■

305593 Poa iberica Fisch. et C. A. Mey.;伊比利亚早熟禾■☆

305594 Poa imperialis Bor;苗壮早熟禾■

305595 Poa incerta Keng;疑早熟禾;Incertitude Bluegrass ■

305596 Poa incerta Keng ex L. Liou = Poa versicolor Besser subsp. orinosa (Keng) Olonova et G. Zhu ■

305597 Poa indattenuata Keng = Poa albertii Regel subsp. kunlunensis (N. R. Cui) Olonova et G. Zhu ■

305598　Poa indattenuata Keng ex P. C. Keng et G. Q. Song ＝ Poa albertii Regel subsp. kunlunensis（N. R. Cui）Olonova et G. Zhu ■

305599　Poa infirma Kunth；低矮早熟禾；Early Meadow-grass, Weak Bluegrass ■

305600　Poa insignis Litv. ＝ Poa sibirica Roshev. subsp. uralensis Tzvelev ■

305601　Poa insignis Litv. ex Roshev.；显稃早熟禾（大花西伯利亚早熟禾）■

305602　Poa interior Rydb.；内陆早熟禾；Inland Bluegrass ■☆

305603　Poa interrupta Lam. ＝ Eragrostis japonica（Thunb.）Trin. ■

305604　Poa ircutica Roshev.；伊尔库早熟禾；Inland Bluegrass ■

305605　Poa irrigata Lindm.；浸水早熟禾（湿地早熟禾）；Spreading Bluegrass ■

305606　Poa iwateana Ohwi；岩手山早熟禾■☆

305607　Poa japonica Thunb. ＝ Eragrostis japonica（Thunb.）Trin. ■

305608　Poa jaunsarensis Bor；江萨早熟禾■

305609　Poa jaunsarensis Bor ＝ Poa lhasaensis Bor ■

305610　Poa juldusicola Regel ＝ Poa albertii Regel ■

305611　Poa juncifolia Scribn. ＝ Poa secunda J. Presl subsp. juncifolia（Scribn.）Soreng ■

305612　Poa kanboensis Ohwi ＝ Poa urssulensis Trin. var. kanboensis（Ohwi）Olonova et G. Zhu ■

305613　Poa karatavica Bunge ＝ Leucopoa karatavica（Bunge）V. I. Krecz. et Bobrov ■

305614　Poa karateginensis Roshev. ex Ovcz. ＝ Poa bucharica Roshev. subsp. karateginensis（Roshev. ex Ovcz.）Tzvelev ■

305615　Poa kelungensis Ohwi ＝ Poa sphondylodes Trin. ex Bunge var. kelungensis（Ohwi）Ohwi ■

305616　Poa kelungensis Ohwi ＝ Poa sphondylodes Trin. ex Bunge ■

305617　Poa khasiana Stapf；喀斯早熟禾（喀西早熟禾，卡夏早熟禾，台湾早熟禾）；Khasian Bluegrass, Taiwan Bluegrass ■

305618　Poa kilimanjarica（Hedberg）Markgr. -Dann.；基利曼早熟禾■☆

305619　Poa koelzii Bor ＝ Poa albertii Regel subsp. kunlunensis（N. R. Cui）Olonova et G. Zhu ■

305620　Poa koelzii Bor ＝ Poa rangkulensis Ovcz. et Czukav. ■

305621　Poa koenigii Kunth ＝ Eragrostis japonica（Thunb.）Trin. ■

305622　Poa kolymensis Tzvelev；科雷马早熟禾（科利早熟禾）■

305623　Poa komarovii Roshev.；科马罗夫早熟禾（柯马氏早熟禾）；Komarov Bluegrass ■

305624　Poa komarovii Roshev. var. shinanoana（Ohwi）Ohwi ＝ Poa shinanoana Ohwi ■

305625　Poa komarovii Roshev. var. shinanoana（Ohwi）Ohwi f. vivipara（Ohwi）Ohwi ＝ Poa malacantha Kom. var. shinanoana（Ohwi）Ohwi f. vivipara（Ohwi）Ohwi ■☆

305626　Poa korshunensis Golosk. ＝ Poa nemoralis L. subsp. korshunensis（Golosk.）Tzvelev ■

305627　Poa korshunensis Golosk. ＝ Poa urssulensis Trin. var. korshunensis（Golosk.）Olonova et G. Zhu ■

305628　Poa krylovii Reverd. ＝ Poa urssulensis Trin. var. kanboensis（Ohwi）Olonova et G. Zhu ■

305629　Poa krylovii Reverdin；克瑞早熟禾■

305630　Poa kumgansani Ohwi；库姆早熟禾■☆

305631　Poa kungeica Golosk. ＝ Poa lipskyi Roshev. ■

305632　Poa ladakhensis H. Hartm. ＝ Puccinellia ladakhensis（H. Hartm.）Dickoré ■

305633　Poa lahulensis Bor ＝ Poa albertii Regel subsp. lahulensis（Bor）Olonova et G. Zhu ■

305634　Poa lanata Kom. ＝ Poa platyantha Kom. ■

305635　Poa lanata Kom. ＝ Poa triviliformis Kom. ■

305636　Poa lanata Scribn. et Merr.；绵毛早熟禾■

305637　Poa lanatiflora Roshev.；绵毛花早熟禾■☆

305638　Poa langtangensis Melderis；朗坦早熟禾■

305639　Poa lapponica Prokudin；拉普早熟禾■

305640　Poa lapponica Prokudin subsp. acmocalyx（Keng ex L. Liou）Olonova et G. Zhu；尖颖早熟禾■

305641　Poa lapponica Prokudin subsp. pilipes（Keng ex Shan Chen）Olonova et G. Zhu；毛轴早熟禾；Hairystalk Bluegrass ■

305642　Poa latifolia G. Forst. ＝ Centotheca lappacea（L.）Desv. ■

305643　Poa laudanensis Roshev. ex Kom. et Roshev.；劳丹早熟禾■

305644　Poa laudanensis Roshev. ex Kom. et Roshev. ＝ Poa hissarica Roshev. ex Kom. et Roshev. ■

305645　Poa laudanensis Roshev. ex Ovcz. ＝ Poa hissarica Roshev. ex Kom. et Roshev. ■

305646　Poa laxa Haenke；稀穗早熟禾；Mount Washington Bluegrass, Mount Washington Blue-grass ■

305647　Poa laxa Haenke var. tristis Griseb. ＝ Poa tristis Trin. ■

305648　Poa lepida A. Rich. ＝ Eragrostis lepida（A. Rich.）Hochst. ex Steud. ■☆

305649　Poa lepta Keng；柔软早熟禾；Thin Bluegrass ■

305650　Poa lepta Keng ex L. Liou ＝ Poa faberi Rendle var. longifolia（Keng）Olonova et G. Zhu ■

305651　Poa leptoclada Hochst. ex A. Rich.；细枝早熟禾■☆

305652　Poa leptocoma Hochst. ex A. Rich. subsp. paucispicula（Scribn. et Merr.）Tzvelev ＝ Poa paucispicula Scribn. et Merr. ■

305653　Poa levipes（Keng）L. Liou；光轴早熟禾；Smooth-stalked Bluegrass ■

305654　Poa levipes Keng ex L. Liou ＝ Poa pagophila Bor ■

305655　Poa lhasaensis Bor；拉萨早熟禾（江萨早熟禾）■

305656　Poa ligulata Boiss.；尖舌早熟禾■☆

305657　Poa ligulata Boiss. subsp. paui（Font Quer）Maire ＝ Poa ligulata Boiss. var. paui（Font Quer）Maire ■☆

305658　Poa ligulata Boiss. var. djurdjurae（Trab.）Maire ＝ Poa alpina L. subsp. atlantica（Trab.）Romo ■☆

305659　Poa ligulata Boiss. var. mauretanica Maire ＝ Poa ligulata Boiss. ■☆

305660　Poa ligulata Boiss. var. paui（Font Quer）Maire ＝ Poa ligulata Boiss. ■☆

305661　Poa limbata Link；花边莓系；Limbate Bluegrass ■

305662　Poa linearis Schum. ＝ Eragrostis domingensis（Pers.）Steud. ■☆

305663　Poa linearis Trin. ＝ Poa faberi Rendle ■

305664　Poa lipskyi Roshev.；疏穗早熟禾（天山早熟禾）；Tianshan Bluegrass ■

305665　Poa lipskyi Roshev. subsp. dschungarica（Roshev.）Tzvelev；准噶尔早熟禾（准噶尔疏穗早熟禾）■

305666　Poa lipskyi Roshev. subsp. dschungarica（Roshev.）Tzvelev ＝ Poa dschungarica Roshev. ■

305667　Poa lipskyi Roshev. var. contracta Tzvelev ＝ Poa lipskyi Roshev. ■

305668　Poa lithophila Keng；石生早熟禾；Rockliving Bluegrass ■

305669　Poa lithophila Keng ex L. Liou ＝ Poa polycolea Stapf ■

305670　Poa lithuanica Gorski ＝ Glyceria lithuanica（Gorski）Gorski ■

305671　Poa littoralis Gouan ＝ Aeluropus littoralis（Gouan）Parl. ■☆

305672　Poa litvinoviana Ovcz. ＝ Poa albertii Regel ■

305673　Poa litwinowiana Ovcz.；昆仑早熟禾（中亚早熟禾）■

305674　Poa longifolia A. Rich. ＝ Eragrostis longifolia Hochst. ex Steud. ■☆

305675　Poa longifolia Trin.；长叶早熟禾；Longleaf Bluegrass ■

305676   Poa longifolia Trin. subsp. meyeri (Trin. ex Roshev.) Tzvelev = Poa meyeri Trin. ex Roshev. ■

305677   Poa longiglumis Keng;长颖早熟禾■

305678   Poa longiglumis Keng = Poa plurinodis Keng ■

305679   Poa longiglumis Keng ex L. Liou = Poa sphondylodes Trin. ex Bunge var. erikssonii Melderis ■

305680   Poa lubrica Ovcz. ;黏早熟禾■☆

305681   Poa ludens R. R. Stewart = Poa mairei Hack. ■

305682   Poa macilenta A. Rich. = Eragrostis macilenta (A. Rich.) Steud. ■☆

305683   Poa macroanthera D. F. Cui;大药早熟禾■

305684   Poa macroanthera D. F. Cui subsp. meilitzyka D. F. Cui;美丽大药早熟禾;Meiliqike Bluegrass ■

305685   Poa macroanthera D. F. Cui subsp. meilitzyka D. F. Cui = Poa lipskyi Roshev. ■

305686   Poa macrocalyx Trautv. et C. A. Mey. ;大萼早熟禾(大萼莓系早熟禾);Largecalyx Bluegrass ■☆

305687   Poa macrocalyx Trautv. et C. A. Mey. var. fallax (Hack.) Ohwi;假大萼早熟禾■☆

305688   Poa macrocalyx Trautv. et C. A. Mey. var. sachalinensis Koidz. = Poa sachalinensis (Koidz.) Honda ■

305689   Poa macrocalyx Trautv. et C. A. Mey. var. scabriflora (Hack.) Ohwi = Poa macrocalyx Trautv. et C. A. Mey. ■☆

305690   Poa macrocalyx Trautv. et C. A. Mey. var. tatewakiana (Ohwi) Ohwi ex Tateoka = Poa tatewakiana Ohwi ■☆

305691   Poa macrocalyx Trautv. et C. A. Mey. var. tianschanica Regel = Poa pratensis L. subsp. pruinosa (Korotky) Dickoré ■

305692   Poa macrocalyx Trautv. et C. A. Mey. var. tianschanica Regel = Poa pruinosa Korotky ■

305693   Poa macrolepis Keng ex C. Ling;大颖早熟禾;Largeglume Bluegrass ■

305694   Poa macrolepis Keng ex C. Ling = Poa hirtiglumis Hook. f. var. nimuana (C. Ling) Soreng et G. Zhu ■

305695   Poa madecassa A. Camus;马德卡萨早熟禾■☆

305696   Poa maerkangica L. Liou;马尔康早熟禾■

305697   Poa maerkangica L. Liou = Poa polycolea Stapf ■

305698   Poa mairei Hack. ;毛稃早熟禾(东川早熟禾)■

305699   Poa major D. F. Cui;大序早熟禾(大早熟禾)■

305700   Poa major D. F. Cui = Poa nemoraliformis Roshev. ■

305701   Poa malabarica L. = Leptochloa fusca (L.) Kunth ■

305702   Poa malabarica L. = Ottochloa nodosa (Kunth) Dandy var. micrantha (Balansa ex A. Camus) S. L. Chen et S. M. Phillips ■

305703   Poa malaca Keng;纤弱早熟禾;Bluegrass ■

305704   Poa malaca Keng = Poa faberi Rendle var. longifolia (Keng) Olonova et G. Zhu ■

305705   Poa malacantha Kom. ;软稃早熟禾■

305706   Poa malacantha Kom. subsp. shinanoana (Ohwi) T. Koyama = Poa malacantha Kom. var. shinanoana (Ohwi) Ohwi ■

305707   Poa malacantha Kom. var. shinanoana (Ohwi) Ohwi = Poa shinanoana Ohwi ■

305708   Poa marginata Ovcz. ;具边早熟禾■☆

305709   Poa marginata Ovcz. = Poa litwinowiana Ovcz. ■

305710   Poa mariae Reverd. = Poa smirnowii Roshev. subsp. mariae (Reverd.) Tzvelev ■

305711   Poa mariae Reverdin;马利早熟禾;Maria Bluegrass ■

305712   Poa mariesii Rendle;东川早熟禾(马利斯早熟禾);Maries Bluegrass ■

305713   Poa mariesii Rendle = Poa nepalensis (G. C. Wall. ex Griseb.) Duthie ■

305714   Poa maritima Huds. = Puccinellia maritima (Huds.) Parl. ■

305715   Poa markgrafii H. Hartm. = Poa pratensis L. subsp. pruinosa (Korotky) Dickoré ■

305716   Poa markgrafii H. Hartm. = Poa pruinosa Korotky ■

305717   Poa maroccana Nannf. ;摩洛哥早熟禾■☆

305718   Poa masenderana Freyn et Sint. ;玛森早熟禾■

305719   Poa massauensis Fresen. = Aeluropus lagopoides (L.) Trin. ex Thwaites ■☆

305720   Poa matsumurae Hack. ;松村氏早熟禾■☆

305721   Poa maydelii Roshev. ;马伊早熟禾■☆

305722   Poa media Schur;中间早熟禾■

305723   Poa megalothyrsa Keng ex Tzvelev;大锥早熟禾;Largepanicle Bluegrass ■

305724   Poa megalothyrsa Keng ex Tzvelev = Poa asperifolia Bor ■

305725   Poa megastachya Koeler = Eragrostis cilianensis (All.) Link ex Vignolo ■

305726   Poa membranigluma D. F. Cui;膜颖早熟禾■

305727   Poa mexicana Hornem. = Eragrostis mexicana (Hornem.) Link ■☆

305728   Poa meyeri Trin. ex Roshev. ;玫珥早熟禾;Meyen. Bluegrass ■

305729   Poa micrandra Keng;小药早熟禾;Microanther Bluegrass ■

305730   Poa micrandra Keng = Poa nepalensis (G. C. Wall. ex Griseb.) Duthie ■

305731   Poa microstachya Desv. ;小穗早熟禾■☆

305732   Poa milioides Honda = Aniselytron treutleri (Kuntze) Soják ■

305733   Poa mongolica (Rendle) Keng et S. L. Chen;李枝早熟禾(蒙古早熟禾);Mongol Bluegrass, Mongolian Bluegrass ■

305734   Poa mongolica (Rendle) Keng ex Shan Chen = Poa alta Hitchc. ■

305735   Poa mucronata Poir. = Megastachya mucronata (Poir.) P. Beauv. ■☆

305736   Poa multiflora Forssk. = Eragrostis tremula Hochst. ex Steud. ■☆

305737   Poa mustangensis Rajbh. = Poa albertii Regel subsp. arnoldii (Melderis) Olonova et G. Zhu ■

305738   Poa naltchikensis Roshev. ;纳尔契克早熟禾■☆

305739   Poa nankoensis Ohwi;南湖大山早熟禾;Nanko Bluegrass ■

305740   Poa neglecta Steud. ;忽莓;Neglected Bluegrass ■

305741   Poa nemoraliformis Roshev. ;林早熟禾(假林地早熟禾,早熟禾)■

305742   Poa nemoraliformis Roshev. = Poa ussuriensis Roshev. ■

305743   Poa nemoralis L. ;林地早熟禾(林莓系);Forest Bluegrass, Wood Bluegrass, Wood Blue-grass, Wood Meadow Grass, Wood Meadow-grass,Woods Bluegrass ■

305744   Poa nemoralis L. subsp. interior (Rydb.) W. A. Weber;间型林地早熟禾■☆

305745   Poa nemoralis L. subsp. korshunensis (Golosk.) Tzvelev = Poa korshunensis Golosk. ■

305746   Poa nemoralis L. subsp. ochotensis (Trin.) Tzvelev = Poa versicolor Besser subsp. ochotensis (Trin.) Tzvelev ■

305747   Poa nemoralis L. subsp. parca N. R. Cui;疏穗林地早熟禾(疏穗早熟禾)■

305748   Poa nemoralis L. var. acuta Trab. = Poa nemoralis L. ■

305749   Poa nemoralis L. var. agrostoides Asch. et Graebn. = Poa nemoralis L. ■

305750   Poa nemoralis L. var. coaictata Gand. ;紧缩早熟禾■

305751   Poa nemoralis L. var. coarctata (Gaudin) Gaudin = Poa nemoralis L. ■

305752　Poa nemoralis L. var. firmula Gand. ；长穗早熟禾■

305753　Poa nemoralis L. var. miliacea（Vill.）Nyman ＝Poa nemoralis L. ■

305754　Poa nemoralis L. var. mongolica Rendle ＝Poa alta Hitchc. ■

305755　Poa nemoralis L. var. mongolica Rendle ＝Poa mongolica（Rendle）Keng et S. L. Chen ■

305756　Poa nemoralis L. var. montana Gaudin ＝Poa nemoralis L. var. miliacea（Vill.）Nyman ■

305757　Poa nemoralis L. var. parca N. R. Cui ＝Poa nemoralis L. subsp. parca N. R. Cui ■

305758　Poa nemoralis L. var. rigidula Mert. et Koch ＝Poa nemoralis L. ■

305759　Poa nemoralis L. var. stenophylla Keng；窄叶早熟禾；Narrowleaf Bluegrass ■

305760　Poa nemoralis L. var. tenella Rchb. ；细弱早熟禾；Tender Bluegrass ■

305761　Poa nemoralis L. var. uniflora Mart. et Koch；单花早熟禾■

305762　Poa nemoralis L. var. vulgaris Gaudin ＝Poa nemoralis L. ■

305763　Poa nemoralis L. var. wutaiensis Keng；五台早熟禾；Wutai Bluegrass ■

305764　Poa nepalensis（G. C. Wall. ex Griseb.）Duthie；尼泊尔早熟禾■

305765　Poa nepalensis（G. C. Wall. ex Griseb.）Duthie var. nipponica（Koidz.）Soreng et G. Zhu ＝Poa nipponica Koidz. ■

305766　Poa nepalensis Wall. ex Duthie ＝Poa nepalensis（G. C. Wall. ex Griseb.）Duthie ■

305767　Poa nephelophila Bor；那菲早熟禾■

305768　Poa nephelophila Bor ＝Poa nepalensis（G. C. Wall. ex Griseb.）Duthie ■

305769　Poa nervata Willd. ＝Glyceria striata（Lam.）Hitchc. ■☆

305770　Poa nervosa（Hook.）Vasey；显脉早熟禾；Prominentlynerveid Bluegrass ■

305771　Poa nevskii Roshev. ex Kom. et Roshev. ；涅氏早熟禾（尼氏早熟禾）■

305772　Poa nevskii Roshev. ex Ovcz. ＝Poa nevskii Roshev. ex Kom. et Roshev. ■

305773　Poa nigropurpurea C. Ling；紫黑早熟禾；Blackpurple Bluegrass ■

305774　Poa nigropurpurea C. Ling ＝Poa pagophila Bor ■

305775　Poa nimuana C. Ling ＝Poa hirtiglumis Hook. f. var. nimuana（C. Ling）Soreng et G. Zhu ■

305776　Poa nipponica Koidz. ；日本早熟禾；Japanese Bluegrass ■

305777　Poa nitens Weber ＝Anthoxanthum monticola（Bigelow）Veldkamp ■

305778　Poa nitens Weber ＝Anthoxanthum nitens（Weber）Y. Schouten et Veldkamp ■

305779　Poa nitida Lam. ＝Koeleria macrantha（Ledeb.）Schult. ■

305780　Poa nitidespiculata Bor；亮穗早熟禾（闪穗早熟禾）；Shining Spiculate Bluegrass ■

305781　Poa nivicola Kom. ＝Poa shumushuensis Ohwi ■

305782　Poa nivicola Roshev. ＝Poa paucispicula Scribn. et Merr. ■

305783　Poa nubigena Keng；云生早熟禾；Cloudy Bluegrass ■

305784　Poa nubigena Keng var. levipes Keng ＝Poa levipes（Keng）L. Liou ■

305785　Poa nudiflora Hack. ＝Puccinellia nudiflora（Hack.）Tzvelev ■

305786　Poa nutans Retz. ＝Eragrostis nutans（Retz.）Nees ex Steud. ■

305787　Poa ochotensis Trin. ＝Poa versicolor Besser subsp. ochotensis（Trin.）Tzvelev ■

305788　Poa ogamontana Mochizuki；男鹿山早熟禾■☆

305789　Poa oligantha Hochst. ex Steud. ＝Poa schimperana A. Rich. ■☆

305790　Poa oligophylla Keng ＝Poa araratica Trautv. subsp. oligophylla（Keng）Olonova et G. Zhu ■

305791　Poa omeiensis Rendle ＝Poa szechuensis Rendle ■

305792　Poa oreades Peter ＝Poa leptoclada Hochst. ex A. Rich. ■☆

305793　Poa orinosa Keng ＝Poa versicolor Besser subsp. orinosa（Keng）Olonova et G. Zhu ■

305794　Poa orinosa Keng ex L. Liou；山地早熟禾；Mountain Bluegrass ■

305795　Poa orinosa Keng ex L. Liou var. levipes Keng；长叶山地早熟禾（长叶早熟禾）■

305796　Poa orinosa Keng var. longifolia Keng ＝Poa faberi Rendle var. longifolia（Keng）Olonova et G. Zhu ■

305797　Poa ovczinnikovii Ikonn. ＝Poa lipskyi Roshev. ■

305798　Poa ovina A. Rich. ＝Eragrostis gangetica（Roxb.）Steud. ■☆

305799　Poa pachyantha Keng ex S. Chen；密花早熟禾；Thickflowered Bluegrass ■

305800　Poa pachyantha Keng ex Shan Chen ＝Poa pratensis L. subsp. pruinosa（Korotky）Dickoré ■

305801　Poa pachyantha Keng ex Shan Chen ＝Poa pruinosa Korotky ■

305802　Poa pagophila Bor；曲枝早熟禾■

305803　Poa paludigena Fernald et Wiegand；湿地早熟禾；Bog Bluegrass, Marsh Bluegrass, Patterson's Bluegrass ■☆

305804　Poa palustris L. ；沼泽早熟禾（河源早熟禾，湿地早熟禾，泽地早熟禾，沼生早熟禾）；Fowl Bluegrass, Fowl Blue-grass, Fowl Meadow Grass, Fowl Meadow-grass, Swamp Meadow-grass ■

305805　Poa palustris L. var. strictula（Steud.）Hack. ＝Poa sphondylodes Trin. ex Bunge ■

305806　Poa pamirica Roshev. ex Ovcz. ；帕米尔早熟禾■

305807　Poa pamirica Roshev. ex Ovcz. ＝Poa pratensis L. subsp. pruinosa（Korotky）Dickoré ■

305808　Poa pamirica Roshev. ex Ovcz. ＝Poa pruinosa Korotky ■

305809　Poa pamirica Roshev. ex Ovcz. ＝Poa tianschanica（Regel）Hack. ex O. Fedtsch. ■

305810　Poa panicea Retz. ＝Leptochloa panicea（Retz.）Ohwi ■

305811　Poa paniciformis A. Braun ＝Eragrostis paniciformis（A. Braun）Steud. ■☆

305812　Poa parafestuca L. Liou；羊茅状早熟禾■

305813　Poa parafestuca L. Liou ＝Poa albertii Regel ■

305814　Poa parvissima Chin C. Kuo ex D. F. Cui；小早熟禾■

305815　Poa patens Keng ex P. C. Keng；开展早熟禾；Spreading Bluegrass ■

305816　Poa patens Keng ex P. C. Keng ＝Poa mairei Hack. ■

305817　Poa paucifolia Keng ex S. Chen；少叶早熟禾；Fewleaf Bluegrass ■

305818　Poa paucifolia Keng ex Shan Chen ＝Poa faberi Rendle ■

305819　Poa paucispicula Halter ＝Poa shumushuensis Ohwi ■

305820　Poa paucispicula Scribn. et Merr. ；寡穗早熟禾■

305821　Poa paui Font Quer ＝Poa ligulata Boiss. var. paui（Font Quer）Maire ■☆

305822　Poa pectinacea Michx. ＝Eragrostis pectinacea（Michx.）Nees ■☆

305823　Poa penicillata Kom. ＝Poa platyantha Kom. ■

305824　Poa perennis Keng ex P. C. Keng；宿生早熟禾；Perennial Bluegrass ■

305825　Poa perlaxa Pilg. ＝Poa schimperana A. Rich. ■☆

305826　Poa persica Trin. ＝Eremopoa persica（Trin.）Roshev. ■

305827　Poa persica Trin. var. oxyglumis Boiss. ＝Eremopoa oxyglumis（Boiss.）Roshev. ■

305828　Poa persica Trin. var. oxyglumis Boiss. = Poa diaphora Trin. subsp. oxyglumis（Boiss.）Soreng et G. Zhu ■

305829　Poa persica Trin. var. songarica（Schrenk）Stapf = Poa diaphora Trin. ■

305830　Poa petraea Trin. ;岩生早熟禾■☆

305831　Poa petraea Trin. ex Kom. = Poa lanata Kom. ■

305832　Poa phariana Bor;帕里早熟禾■

305833　Poa phariana Bor = Poa calliopsis Litv. ex Ovcz. ■

305834　Poa phryganodes Trin. = Puccinellia phryganodes（Trin.）Scribn. et Merr. ■

305835　Poa phryganoides Trin. ;堪察加早熟禾;Kamtschatka Bluegrass ■

305836　Poa pilipes Keng ex Shan Chen = Poa lapponica Prokudin subsp. pilipes（Keng ex Shan Chen）Olonova et G. Zhu ■

305837　Poa pilosa L. = Eragrostis pilosa（L.）P. Beauv. ■

305838　Poa pinegensis Roshev. ;皮涅加早熟禾■☆

305839　Poa pitardiana H. Scholz;皮塔德早熟禾■☆

305840　Poa platyantha Kom. ;阔花早熟禾■

305841　Poa platyglumis（L. Liou）L. Liou;宽颖早熟禾（宽颖碱茅）■

305842　Poa platyglumis（L. Liou）L. Liou = Poa pseudamoena Bor ■

305843　Poa plumosa Retz. = Eragrostis amabilis（L.）Wight et Arn. ■

305844　Poa plumosa Retz. = Eragrostis tenella（L.）P. Beauv. ex Roem. et Schult. ■

305845　Poa plurifolia Keng = Poa sphondylodes Trin. ex Bunge var. erikssonii Melderis ■

305846　Poa plurinodis Keng;多节早熟禾（长颖早熟禾）;Longglume Bluegrass , Multi-noded Bluegrass ■

305847　Poa podolica Blocki;波多尔早熟禾■☆

305848　Poa poiphagorum Bor = Poa albertii Regel subsp. poiphagorum（Bor）Olonova et G. Zhu ■

305849　Poa poiphagorum Bor var. hunczilapensis Keng ex D. F. Cui;红旗拉甫早熟禾■

305850　Poa polozhiae Revjankina = Poa smirnowii Roshev. subsp. polozhiae（Revjankina）Olonova ■

305851　Poa polycolea Stapf;多鞘早熟禾;Many-sheath Bluegrass ■

305852　Poa polyneuron Bor;多脉早熟禾■

305853　Poa poophagorum Bor = Poa albertii Regel subsp. poophagorum（Bor）Olonova et G. Zhu ■

305854　Poa poophagorum Bor subsp. hunczilapensis Keng ex D. F. Cui = Poa albertii Regel ■

305855　Poa pratensis L. ;草地早熟禾（六月禾）;Common Meadow Grass , June Grass , Kentucky Blue Grass , Kentucky Bluegrass , Kentucky Blue-grass , Smooth Meadow-grass , Smooth-stalked Meadow-grass ■

305856　Poa pratensis L. subsp. alpigena（Blytt）Hitonen = Poa alpigena（Blytt）Lindm. ■

305857　Poa pratensis L. subsp. alpigena（Lindm.）Hitonen = Poa alpigena（Blytt）Lindm. ■

305858　Poa pratensis L. subsp. angustifolia（L.）Arcang. var. hatusimae（Ohwi）Ohwi;初岛早熟禾■☆

305859　Poa pratensis L. subsp. angustifolia（L.）Arcang. var. strigosa（Hoffm.）Gaudin = Poa strigosa Hoffm. ■☆

305860　Poa pratensis L. subsp. angustifolia（L.）Gaudin = Poa pratensis L. var. anceps Gaudich. ex Griseb. ■

305861　Poa pratensis L. subsp. angustifolia（L.）Lej. = Poa angustifolia L. ■

305862　Poa pratensis L. subsp. angustifolia（L.）Sm. = Poa angustifolia L. ■

305863　Poa pratensis L. subsp. angustiglumis（Roshev.）Tzvelev = Poa angustiglumis Roshev. ■

305864　Poa pratensis L. subsp. atlantis Maire;亚特兰大早熟禾■☆

305865　Poa pratensis L. subsp. irrigata（Lindm.）H. Lindb. = Poa humilis Ehrh. ■☆

305866　Poa pratensis L. subsp. irrigata（Lindm.）H. Lindb. = Poa irrigata Lindm. ■

305867　Poa pratensis L. subsp. pruinosa（Korotky）Dickoré = Poa pruinosa Korotky ■

305868　Poa pratensis L. subsp. sergievskajae（Prob.）Tzvelev;色草地早熟禾■

305869　Poa pratensis L. subsp. staintonii（Melderis）Dickoré;长秆草地早熟禾■

305870　Poa pratensis L. subsp. stenachyra（Keng ex P. C. Keng et G. Q. Song）Soreng et G. Zhu;窄颖草地早熟禾■

305871　Poa pratensis L. subsp. subulosa（Roshev.）Tzvelev = Poa subulosa（Roshev.）Turcz. ex Roshev. ■☆

305872　Poa pratensis L. var. alpigena Blytt = Poa alpigena（Blytt）Lindm. ■

305873　Poa pratensis L. var. alpigena Blytt = Poa pratensis L. subsp. alpigena（Blytt）Hitonen ■

305874　Poa pratensis L. var. anceps（Gaudin）Griseb. = Poa pratensis L. ■

305875　Poa pratensis L. var. anceps Gaudich. ex Griseb. ;扁秆早熟禾（扁早熟禾）;Flattened Culm Bluegrass ■

305876　Poa pratensis L. var. angustifolia（L.）Gaudin;窄叶草地早熟禾■☆

305877　Poa pratensis L. var. angustifolia（L.）Sm. = Poa angustifolia L. ■

305878　Poa pratensis L. var. angustifolia（L.）Sm. = Poa pratensis L. var. anceps Gaudich. ex Griseb. ■

305879　Poa pratensis L. var. contracta Keng = Poa alpigena（Blytt）Lindm. ■

305880　Poa pratensis L. var. domestica Laest. ;土著早熟禾■☆

305881　Poa pratensis L. var. iantha Laest. = Poa alpigena（Blytt）Lindm. ■

305882　Poa pratensis L. var. latifolia Rchb. ;宽叶草地早熟禾■☆

305883　Poa pratensis L. var. maritima Litv. ;滨海早熟禾;Sea Shore Bluegrass ■

305884　Poa pratensis L. var. subulosa（Turcz.）Roshev. = Poa subulosa（Roshev.）Turcz. ex Roshev. ■☆

305885　Poa pratensis L. var. vivipara（Malmgren）B. Boivin;胎生早熟禾■☆

305886　Poa procera Roxb. ;高莓系;Tall Bluegrass ■

305887　Poa prolifera Sw. = Eragrostis prolifera（Sw.）Steud. ■☆

305888　Poa prolixior Rendle;细长早熟禾;Slender Bluegrass ■

305889　Poa prolixior Rendle = Poa faberi Rendle ■

305890　Poa pruinosa Korotky;粉绿早熟禾■

305891　Poa pruinosa Korotky = Poa pratensis L. subsp. pruinosa（Korotky）Dickoré ■

305892　Poa pruinosa Korotky = Poa tianschanica（Regel）Hack. ex O. Fedtsch. ■

305893　Poa psammophila Schur = Poa bulbosa L. ■

305894　Poa pseudamoena Bor;拟早熟禾■

305895　Poa pseudoabbreviata Roshev. ;假缩短早熟禾■☆

305896　Poa pseudodisiecta Ovcz. = Poa lipskyi Roshev. ■

305897　Poa pseudonemoralis Skvortsov = Poa alta Hitchc. ■

305898　Poa pseudonemoralis Skvortsov = Poa pseudopalustris Keng ■

305899　Poa pseudonemoralis Skvortsov = Poa skvortzovii Prob. ■

305900　Poa pseudopalustris Keng = Poa pseudopalustris Keng ex L. Liou ■

305901　Poa pseudopalustris Keng ex L. Liou;假泽早熟禾; Pseudoswampy Bluegrass ■

305902　Poa pseudopalustris Keng ex Shan Chen = Poa alta Hitchc. ■

305903　Poa pseudopratensis Hook. f. = Poa ludens R. R. Stewart ■

305904　Poa pseudopratensis Hook. f. = Poa mairei Hack. ■

305905　Poa pseudotibetica Noltie = Poa tibetica Munro ex Stapf var. aristulata Stapf ■

305906　Poa pseudotremuloides Ovcz. et Czukav. ;假颤早熟禾■☆

305907　Poa psilolepis Keng = Poa araratica Trautv. subsp. psilolepis (Keng) Olonova et G. Zhu ■

305908　Poa psilophylla Hochst. = Poa simensis Hochst. ex A. Rich. ■☆

305909　Poa pubicalyx Keng ex L. Liou;毛颖早熟禾; Hairyglume Bluegrass ■

305910　Poa pubicalyx Keng ex L. Liou = Poa faberi Rendle var. longifolia (Keng) Olonova et G. Zhu ■

305911　Poa pumila K. Koch;矮早熟禾■

305912　Poa pungens M. Bieb. = Aeluropus pungens (M. Bieb.) C. Koch ■

305913　Poa qinghaiensis Soreng et G. Zhu;青海早熟禾■

305914　Poa quadripedalis Ehrh. ex Koeler = Poa remota Forselles ■

305915　Poa racemosa Thunb. = Eragrostis racemosa (Thunb.) Steud. ■☆

305916　Poa radula Franch. et Sav. ;匍根早熟禾■

305917　Poa raduliformis Prob. ;糙早熟禾■

305918　Poa rajbhandarii Noltie;喜马拉雅早熟禾■

305919　Poa rangkulensis Ovcz. et Czukav. ;雪地早熟禾■

305920　Poa rangkulensis Ovcz. et Czukav. = Poa albertii Regel subsp. kunlunensis (N. R. Cui) Olonova et G. Zhu ■

305921　Poa relaxa Ovcz. = Poa versicolor Besser subsp. relaxa (Ovcz.) Tzvelev ■

305922　Poa remota Forselles;疏序早熟禾(散早熟禾)■

305923　Poa remota Forselles subsp. raduliformis (Prob.) Vorosch. = Poa raduliformis Prob. ■

305924　Poa remotiflora (Hack.) Murb. = Poa infirma Kunth ■

305925　Poa reverdattoi Roshev. = Poa versicolor Besser subsp. reverdattoi (Roshev.) Olonova et G. Zhu ■

305926　Poa reverdattoi Roshev. ex Kom. et Roshev. ;瑞沃达早熟禾■

305927　Poa rhadina Bor;等颖早熟禾(西喜马拉雅莓系); West-himalayan Bluegrass ■

305928　Poa rhomboidea Roshev. ;圆穗早熟禾■

305929　Poa rigida L. = Catapodium rigidum (L.) C. E. Hubb. ex Dony ■☆

305930　Poa rigida L. = Desmazeria rigida (L.) Tutin ■☆

305931　Poa rigida L. = Scleropoa rigida (L.) Griseb. ■☆

305932　Poa rivulorum Maire et Trab. ;溪边早熟禾■☆

305933　Poa roemeri Bor;诺米早熟禾■

305934　Poa roemeri Bor = Poa albertii Regel subsp. kunlunensis (N. R. Cui) Olonova et G. Zhu ■

305935　Poa rohmooana Noltie = Poa szechuensis Rendle var. rossbergiana (K. S. Hao) Soreng et G. Zhu ■

305936　Poa rossbergiana K. S. Hao = Poa szechuensis Rendle var. rossbergiana (K. S. Hao) Soreng et G. Zhu ■

305937　Poa royleana Nees ex Steud. = Poa annua L. ■

305938　Poa rubens Lam. = Eragrostis unioloides (Retz.) Nees ex Steud. ■

305939　Poa ruwenzoriensis Robyns et Tournay;鲁文佐里早熟禾■☆

305940　Poa sabulosa (Roshev.) Turcz. ex Roshev. ;砾沙早熟禾■

305941　Poa sabulosa Turcz. ex Roshev. = Poa sabulosa (Roshev.) Turcz. ex Roshev. ■

305942　Poa sachalinensis (Koidz.) Honda;库页早熟禾(萨哈林早熟禾)■

305943　Poa sachalinensis (Koidz.) Honda var. yatsugatakensis (Honda) Ohwi = Poa yatsugatakensis Honda ■☆

305944　Poa sachalinensis (Koidz.) Honda var. yezoensis Ohwi;北海道早熟禾■☆

305945　Poa sajanensis Roshev. ;萨因早熟禾■☆

305946　Poa saltuensis Fernald et Wiegand;西方林地早熟禾; Forest Bluegrass, Forest Meadow-grass ■☆

305947　Poa sarmentosa Thunb. = Eragrostis sarmentosa (Thunb.) Trin. ■☆

305948　Poa saxicola R. Br. = Saxipoa saxicola (R. Br.) Soreng, L. J. Gillespie et S. W. L. Jacobs ■☆

305949　Poa scabriculmis N. R. Cui = Poa albertii Regel subsp. kunlunensis (N. R. Cui) Olonova et G. Zhu ■

305950　Poa scabriculmis N. R. Cui = Poa scabristemmed N. R. Cui ■

305951　Poa scabristemmed N. R. Cui;糙茎早熟禾■

305952　Poa schimperana A. Rich. ;欣珀早熟禾■☆

305953　Poa schimperana A. Rich. var. longigluma Chiov. = Poa leptoclada Hochst. ex A. Rich. ■☆

305954　Poa schimperana A. Rich. var. micrantha Chiov. = Poa leptoclada Hochst. ex A. Rich. ■☆

305955　Poa schischkinii Tzvelev;希斯肯早熟禾■

305956　Poa schliebenii Pilg. = Poa leptoclada Hochst. ex A. Rich. ■☆

305957　Poa schoenites Keng;蔺状早熟禾; Schoenus-like Bluegrass ■

305958　Poa schoenites Keng ex L. Liou = Poa versicolor Besser subsp. orinosa (Keng) Olonova et G. Zhu ■

305959　Poa scitula Bor = Poa glabriflora Roshev. ex Kom. et Roshev. ■

305960　Poa secunda J. Presl subsp. juncifolia (Scribn.) Soreng;巨早熟禾■

305961　Poa senegalensis Desv. = Eragrostis pilosa (L.) P. Beauv. ■

305962　Poa sergievskajae Prob. = Poa pratensis L. subsp. sergievskajae (Prob.) Tzvelev ■

305963　Poa serotina Ehrh. = Poa palustris L. ■

305964　Poa serotina Ehrh. ex Hoffm. var. botryoides Trin. ex Griseb. = Poa versicolor Besser subsp. stepposa (Krylov) Tzvelev ■

305965　Poa serotina Ehrh. var. botryoides Trin. ex Griseb. = Poa botryoides Trin. ex Besser ■

305966　Poa setacea Hoffm. ;刚毛早熟禾■☆

305967　Poa setulosa Bor;细刺早熟禾(尖早熟禾); Fine-bristle Bluegrass ■

305968　Poa shansiensis Hitchc. ;山西早熟禾; Shanxi Bluegrass ■

305969　Poa shansiensis Hitchc. = Poa tangii Hitchc. ■

305970　Poa shinanoana Ohwi;深山早熟禾; Shinano Bluegrass ■

305971　Poa shinanoana Ohwi f. vivipara Ohwi = Poa malacantha Kom. var. shinanoana (Ohwi) Ohwi f. vivipara (Ohwi) Ohwi ■☆

305972　Poa shumushuensis Ohwi;苏姆早熟禾■

305973　Poa sibirica Roshev. ;西伯利亚早熟禾; Siberia Bluegrass, Siberian Bluegrass ■

305974　Poa sibirica Roshev. subsp. insignis (Litv.) Olonova = Poa sibirica Roshev. subsp. uralensis Tzvelev ■

305975　Poa sibirica Roshev. subsp. uralensis Tzvelev = Poa insignis Litv. ex Roshev. ■

305976　Poa sibirica Roshev. var. insignis (Litv. ex Roshev.) Serg. =

Poa insignis Litv. ex Roshev. ■

305977　Poa sibirica Roshev. var. insignis（Litv.）Serg. = Poa sibirica Roshev. subsp. uralensis Tzvelev ■

305978　Poa sichotensis Prob.；西可早熟禾■

305979　Poa sicula Jacq. = Desmazeria sicula（Jacq.）Dumort. ■☆

305980　Poa sikkimensis（Stapf）Bor；锡金早熟禾；Sikkim Bluegrass ■

305981　Poa silvatica Vill. = Poa chaixii Vill. ■

305982　Poa silvicola Guss.；林粗茎早熟禾■☆

305983　Poa simensis Hochst. ex A. Rich.；锡米早熟禾■☆

305984　Poa sinaica Steud.；西奈早熟禾■

305985　Poa sinattenuata Keng；中华早熟禾；Chinese Attenuate Bluegrass ■

305986　Poa sinattenuata Keng = Poa albertii Regel ■

305987　Poa sinattenuata Keng var. breviligula Keng = Poa albertii Regel ■

305988　Poa sinattenuata Keng var. breviligula Keng = Poa breviligula（Keng）L. Liou ■

305989　Poa sinattenuata Keng var. vivipara（Rendle）Keng et S. L. Chen；胎生中华早熟禾；Viviparous Bluegrass ■

305990　Poa sinoglauca Ohwi；华灰早熟禾■

305991　Poa sinoglauca Ohwi = Poa araratica Trautv. subsp. ianthina（Keng ex Shan Chen）Olonova et G. Zhu ■

305992　Poa sinoglauca Ohwi = Poa attenuata Trin. ex Bunge ■

305993　Poa skvortzovii Prob.；斯哥佐早熟禾■

305994　Poa skvortzovii Prob. = Poa alta Hitchc. ■

305995　Poa smirnowii Roshev.；史米诺早熟禾■

305996　Poa smirnowii Roshev. subsp. mariae（Reverd.）Tzvelev；美丽史米诺早熟禾■

305997　Poa smirnowii Roshev. subsp. polozhiae（Revjankina）Olonova；朴咯早熟禾■

305998　Poa soczawai Roshev.；索氏早熟禾■☆

305999　Poa songarica（Schrenk）Boiss. = Poa diaphora Trin. ■

306000　Poa spectabilis Pursh = Eragrostis spectabilis（Pursh）Steud. ■☆

306001　Poa sphondylodes Trin. = Poa ochotensis Trin. ■

306002　Poa sphondylodes Trin. = Poa sphondylodes Trin. ex Bunge ■

306003　Poa sphondylodes Trin. ex Bunge；硬质早熟禾（龙须草）；Hard Bluegrass ■

306004　Poa sphondylodes Trin. ex Bunge var. erikssonii Melderis；多叶早熟禾（长颖早熟禾）；Multi-leaved Bluegrass ■

306005　Poa sphondylodes Trin. ex Bunge var. kelungensis（Ohwi）Ohwi；基隆早熟禾；Jilong, Kelong Bluegrass ■

306006　Poa sphondylodes Trin. ex Bunge var. kelungensis（Ohwi）Ohwi = Poa kelungensis Ohwi ■

306007　Poa sphondylodes Trin. ex Bunge var. kelungensis（Ohwi）Ohwi = Poa sphondylodes Trin. ex Bunge ■

306008　Poa sphondylodes Trin. ex Bunge var. macerrima Keng；瘦弱早熟禾■

306009　Poa sphondylodes Trin. ex Bunge var. strictula（Steud.）Koidz.；直立莓系；Straight Bluegrass ■

306010　Poa sphondylodes Trin. ex Bunge var. subtrivialis Ohwi；大穗早熟禾；Largespike Bluegrass ■

306011　Poa sphondylodes Trin. var. subtrivialis Ohwi = Poa sphondylodes Trin. ex Bunge var. subtrivialis Ohwi ■

306012　Poa spiciformis D. F. Cui；密穗早熟禾■

306013　Poa spiciformis D. F. Cui = Poa tibetica Munro ex Stapf ■

306014　Poa spontanea Bor；自生早熟禾■

306015　Poa spontanea Bor = Poa grandis Hand.-Mazz. ■

306016　Poa squamata Lam. = Eragrostis squamata（Lam.）Steud. ■☆

306017　Poa squarrosa Roem. et Schult. = Pogonarthria squarrosa（Roem. et Schult.）Pilg. ■☆

306018　Poa stapfiana Bor；斯塔夫早熟禾■

306019　Poa stenachyra Keng ex L. Liou；窄颖早熟禾；Narrow-glumme Bluegrass ■

306020　Poa stenachyra Keng ex P. C. Keng et G. Q. Song = Poa pratensis L. subsp. stenachyra（Keng ex P. C. Keng et G. Q. Song）Soreng et G. Zhu ■

306021　Poa stenostachya S. L. Lu et X. F. Lu = Poa tibetica Munro ex Stapf ■

306022　Poa stenostachya S. L. Lu et X. F. Lu var. koko-norica S. L. Lu et X. F. Lu = Poa tibetica Munro ex Stapf ■

306023　Poa stepposa（Krylov）Roshev. = Poa versicolor Besser subsp. stepposa（Krylov）Tzvelev ■

306024　Poa stepposa Krylov = Poa versicolor Besser subsp. stepposa（Krylov）Tzvelev ■

306025　Poa stereophylla Keng；硬叶早熟禾；Solidleaf Bluegrass ■

306026　Poa stereophylla Keng ex L. Liou = Poa versicolor Besser subsp. orinosa（Keng）Olonova et G. Zhu ■

306027　Poa sterilis M. Bieb.；贫育早熟禾(不实早熟禾)■

306028　Poa sterilis M. Bieb. var. versicolor Griseb. = Poa versicolor Besser ■

306029　Poa stewartiana Bor；史蒂瓦早熟禾■

306030　Poa stewartiana Bor = Poa himalayana Nees ex Steud. et Bor ■

306031　Poa strictula Steud. = Poa ochotensis Trin. ■

306032　Poa strictula Steud. = Poa sphondylodes Trin. ex Bunge ■

306033　Poa strigosa Hoffm.；直早熟禾■☆

306034　Poa subaphylla Honda；缺叶早熟禾（缺叶莓系）；Fewleaved Bluegrass ■

306035　Poa subaphylla Honda = Poa versicolor Besser subsp. ochotensis（Trin.）Tzvelev ■

306036　Poa subcaerulea Sm. = Poa humilis Ehrh. ■☆

306037　Poa subfastigiata Trin. = Poa subfastigiata Trin. ex Ledeb. ■

306038　Poa subfastigiata Trin. ex Ledeb.；散穗早熟禾；Spreadspike Bluegrass ■

306039　Poa subulata Desv. = Eragrostis domingensis（Pers.）Steud. ■☆

306040　Poa subulosa（Roshev.）Turcz. ex Roshev. = Poa sabulosa（Roshev.）Turcz. ex Roshev. ■☆

306041　Poa sudetica Haenke = Poa chaixii Vill. ■

306042　Poa sudetica Haenke var. ramota Fr. = Poa remota Forselles ■

306043　Poa sudetica Haenke var. remota（Forselles）Fr. = Poa remota Forselles ■

306044　Poa sunbisinii Soreng et G. Zhu；孙必兴早熟禾■

306045　Poa supina Schrad.；仰卧早熟禾■

306046　Poa supina Schrad. subsp. ustulata（S. E. Fröhner）Á. Löve et D. Löve = Poa supina Schrad. ■

306047　Poa suruana H. Hartm. = Poa bucharica Roshev. subsp. karateginensis（Roshev. ex Ovcz.）Tzvelev ■

306048　Poa sylvestris A. Gray；森林早熟禾；Sylvan Bluegrass, Woodland Bluegrass ■☆

306049　Poa sylvicola Guss. = Poa trivialis L. subsp. sylvicola（Guss.）H. Lindb. ■

306050　Poa szechuensis Rendle；四川早熟禾；Sichuan Bluegrass, Szechwan Bluegrass ■

306051　Poa szechuensis Rendle = Poa debilior Hitchc. ■

306052　Poa szechuensis Rendle var. debilior（Hitchc.）Soreng et G. Zhu；垂枝四川早熟禾（垂枝早熟禾）；Declinate Bluegrass ■

306053　Poa szechuensis Rendle var. rossbergiana（K. S. Hao）Soreng et

G. Zhu;罗氏早熟禾;Rossberg Bluegrass ■

306054　Poa taimyrensis Roshev.;太米尔早熟禾■☆

306055　Poa taimyrensis Roshev. = Poa paucispicula Scribn. et Merr. ■

306056　Poa taiwanicola Ohwi;台湾早熟禾(高山早熟禾,宜兰早熟禾);Taiwan Bluegrass ■

306057　Poa taiwanicola Ohwi = Poa glauca Vahl ■

306058　Poa takasagomontana Ohwi;高砂早熟禾■

306059　Poa takashimana Honda;朝鲜早熟禾;Korean Bluegrass ■

306060　Poa tanfiljiewii Roshev. ex Kom. et Roshev.;谭氏早熟禾■☆

306061　Poa tangii Hitchc.;唐氏早熟禾;Tang Bluegrass ■

306062　Poa tatewakiana Ohwi;馆肋早熟禾■☆

306063　Poa tef Zuccagni = Eragrostis tef(Zuccagni)Trotter ■☆

306064　Poa tenella L. = Eragrostis amabilis(L.)Wight et Arn. ■

306065　Poa tenella L. = Eragrostis tenella(L.)P. Beauv. ex Roem. et Schult. ■

306066　Poa tenellula Kunth = Eragrostis japonica(Thunb.)Trin. ■

306067　Poa tenuicula Ohwi;细秆早熟禾(细莓系);Tenuity Bluegrass ■

306068　Poa tenuifolia A. Rich. = Eragrostis tenuifolia(A. Rich.)Steud. ■☆

306069　Poa tetrantha Keng;四花早熟禾;Tetrafloret Bluegrass ■

306070　Poa tetrantha Keng ex L. Liou = Poa attenuata Trin. ex Bunge ■

306071　Poa tianschanica(Regel)Hack. ex O. Fedtsch.;天山早熟禾■

306072　Poa tianschanica(Regel)Hack. ex O. Fedtsch. = Poa pratensis L. subsp. pruinosa(Korotky)Dickoré ■

306073　Poa tianschanica(Regel)Hack. ex O. Fedtsch. = Poa pruinosa Korotky ■

306074　Poa tibetica Munro ex Stapf;西藏早熟禾;Tibet Bluegrass,Xizang Bluegrass ■

306075　Poa tibetica Munro ex Stapf var. aristulata Stapf;芒柱早熟禾■

306076　Poa tibetica Munro ex Stapf var. aristulata Stapf = Poa tibetica Munro ex Stapf ■

306077　Poa tibeticola Bor;藏南早熟禾(藏早熟禾);Xizang Bluegrass ■

306078　Poa tibeticola Bor = Poa szechuensis Rendle ■

306079　Poa timoleontis Heldr. ex Boiss.;厚鞘早熟禾■

306080　Poa timoleontis Heldr. ex Boiss. var. dshilgensis(Roshev.)Tzvelev = Poa dshilgensis Roshev. ex Kom. ■

306081　Poa tolmatchewii Roshev. = Poa arctica R. Br. subsp. caespitans Simmons ex Nannf. ■

306082　Poa transbaicalica Roshev.;外贝加早熟禾(外贝加尔湖莓系);Transbaical Bluegrass ■

306083　Poa transbaicalica Roshev. = Poa versicolor Besser subsp. stepposa(Krylov)Tzvelev ■

306084　Poa tremula Lam. = Eragrostis tremula Hochst. ex Steud. ■☆

306085　Poa tremula Stapf = Poa stapfiana Bor ■

306086　Poa tremula Stapf var. micranthera Stapf = Poa stapfiana Bor ■

306087　Poa tremuloides Litv.;颤早熟禾■☆

306088　Poa trenula Stapf = Poa stapfiana Bor ■

306089　Poa trichophylla Heldr. et Sart.;三叶早熟禾■

306090　Poa triflora Gilib. = Poa palustris L. ■

306091　Poa triglumis Keng ex L. Liou = Poa polycolea Stapf ■

306092　Poa triglumis P. C. Keng;三颖早熟禾;Triglume Bluegrass ■

306093　Poa trinii Scribn. et Merr. = Poa eminens J. Presl et C. Presl ■

306094　Poa tristis Trin.;暗穗早熟禾■

306095　Poa tristis Trin. = Poa altaica Trin. ■

306096　Poa tristis Trin. ex Regel = Poa altaica Trin. ■

306097　Poa trivialiformis Kom.;匍茎早熟禾■

306098　Poa trivialis L.;普通早熟禾(粗茎莓系,粗茎早熟禾);Rough Bluegrass, Rough Meadow Grass, Rough Meadowgrass, Rough Meadow-grass,Roughstalk Bluegrass, Rough-stalk Bluegrass, Rough-stalked Blue-grass,Rough-stalked Meadow-grass ■

306099　Poa trivialis L. subsp. sylvicola(Guss.)H. Lindb.;欧早熟禾■

306100　Poa trivialis L. subsp. sylvicola(Guss.)H. Lindb. = Poa sylvicola Guss. ■

306101　Poa trivialis L. var. glabra Döll = Poa trivialis L. ■

306102　Poa trivialis L. var. obtusata Maire = Poa trivialis L. ■

306103　Poa trivialis L. var. sylvicola(Guss.)Hack. = Poa trivialis L. subsp. sylvicola(Guss.)H. Lindb. ■

306104　Poa trivialis L. var. sylvicola(Guss.)Roshev. = Poa sylvicola Guss. ■

306105　Poa trivialis L. var. vulgaris Rchb. = Poa trivialis L. ■

306106　Poa tuberifera Faurie ex Hack.;块茎早熟禾■☆

306107　Poa tunicata Keng ex C. Ling;套鞘早熟禾;Tunicate Bluegrass ■

306108　Poa tunicata Keng ex C. Ling = Poa sikkimensis(Stapf)Bor ■

306109　Poa turfosa Litv.;泥炭早熟禾(泥炭莓系);Bog Bluegrass ■

306110　Poa turfosa Litv. = Poa pratensis L. ■

306111　Poa turgida Schumach. = Eragrostis turgida(Schumach.)De Wild. ■☆

306112　Poa unioloides Retz. = Eragrostis unioloides(Retz.)Nees ex Steud. ■

306113　Poa urjanchaica Roshev.;蒙莓系;Urjanchaic Bluegrass ■

306114　Poa urjanchaica Roshev. = Poa pratensis L. ■

306115　Poa ursina Velen. = Poa media Schur ■

306116　Poa urssulensis Trin.;乌尔苏早熟禾■

306117　Poa urssulensis Trin. var. kanboensis(Ohwi)Olonova et G. Zhu;坎博早熟禾■

306118　Poa urssulensis Trin. var. korshunensis(Golosk.)Olonova et G. Zhu;柯顺早熟禾(密穗早熟禾)■

306119　Poa ussuriensis Roshev.;乌苏里早熟禾(乌苏早熟禾);Ussuri Bluegrass ■

306120　Poa ussuriensis Roshev. f. angustifolia I. C. Chung = Poa ussuriensis Roshev. ■

306121　Poa ussuriensis Roshev. f. scabra I. C. Chung = Poa ussuriensis Roshev. ■

306122　Poa ustulata S. E. Fröhner = Poa supina Schrad. ■

306123　Poa vaginans Keng;长鞘早熟禾;Sheathing Bluegrass ■

306124　Poa vaginans Keng = Poa alta Hitchc. ■

306125　Poa vaginata Pamp.;具鞘早熟禾■☆

306126　Poa varia Keng = Poa versicolor Besser subsp. varia(Keng ex L. Liou)Olonova et G. Zhu ■

306127　Poa varia Keng ex L. Liou = Poa versicolor Besser subsp. varia(Keng ex L. Liou)Olonova et G. Zhu ■

306128　Poa variegata Haller f. = Poa supina Schrad. ■

306129　Poa vedenskyi Drobow;维登早熟禾■

306130　Poa veresczaginii Tzvelev;薇早熟禾■

306131　Poa versicolor Besser;变色早熟禾(杂花早熟禾,杂色早熟禾)■

306132　Poa versicolor Besser subsp. araratica(Trautv.)Tzvelev = Poa araratica Trautv. ■

306133　Poa versicolor Besser subsp. ochotensis(Trin.)Tzvelev;乌库早熟禾■

306134　Poa versicolor Besser subsp. ochotensis(Trin.)Tzvelev = Poa ochotensis Trin. ■

306135　Poa versicolor Besser subsp. orinosa(Keng)Olonova et G. Zhu;山地变色早熟禾(山地早熟禾)■

306136　Poa versicolor Besser subsp. relaxa(Ovcz.)Tzvelev;新疆早熟

禾■

306137 Poa versicolor Besser subsp. relaxa (Ovcz.) Tzvelev = Poa relaxa Ovcz. ■

306138 Poa versicolor Besser subsp. reverdattoi (Roshev.) Olonova et G. Zhu = Poa reverdattoi Roshev. ex Kom. et Roshev. ■

306139 Poa versicolor Besser subsp. stepposa (Krylov) Tzvelev;低山早熟禾(草原早熟禾);Steppe Bluegrass ■

306140 Poa versicolor Besser subsp. stepposa (Krylov) Tzvelev = Poa stepposa (Krylov) Roshev. ■

306141 Poa versicolor Besser subsp. varia (Keng ex L. Liou) Olonova et G. Zhu;多变早熟禾;Variable Bluegrass ■

306142 Poa verticillata Cav. = Eragrostis pilosa (L.) P. Beauv. ■

306143 Poa violacea Bellardi;紫早熟禾;Purple Bluegrass ■☆

306144 Poa virgata Poir. = Leptochloa panicea (Retz.) Ohwi ■

306145 Poa viridiflora Hochst. = Poa schimperana A. Rich. ■☆

306146 Poa viridula Palib.;绿早熟禾;Greenish Bluegrass ■

306147 Poa viridula Palib. = Poa pratensis L. ■

306148 Poa viscosa Retz. = Eragrostis viscosa (Retz.) Trin. ■☆

306149 Poa vivipara (L.) Willd. = Poa bulbosa L. ■

306150 Poa vrangelica Tzvelev;弗兰格尔早熟禾■

306151 Poa wardiana Bor;瓦氏早熟禾(瓦迪早熟禾);Ward Bluegrass ■

306152 Poa wolfii Scribn.;沃尔夫早熟禾;Meadow Bluegrass, Wolf's Bluegrass ■☆

306153 Poa woronowii Roshev. ex Komarov et Roshev.;沃氏早熟禾■☆

306154 Poa xingkaiensis Y. X. Ma;星早熟禾■

306155 Poa yakiangensis L. Liou = Poa dzongicola Noltie ■

306156 Poa yatsugatakensis Honda;八束早熟禾■☆

306157 Poa yatsugatakensis Honda var. shinanoana (Ohwi) T. Shimizu = Poa shinanoana Ohwi ■

306158 Poa yatsugatakensis Honda var. shinanoana (Ohwi) T. Shimizu f. vivipara (Ohwi) T. Shimizu = Poa shinanoana Ohwi ■

306159 Poa zaprjagajevii Ovcz.;塔吉早熟禾■

306160 Poa zaprjagajevii Ovcz. = Poa glabriflora Roshev. ex Kom. et Roshev. ■

306161 Poa zhongbaensis C. Ling;仲巴早熟禾;Zhongba Bluegrass ■

306162 Poa zhongbaensis C. Ling = Poa hirtiglumis Hook. f. var. nimuana (C. Ling) Soreng et G. Zhu ■

306163 Poa zhongdianensis L. Liou;中甸早熟禾■

306164 Poaceae Barnhart(1895)(保留科名);禾本科■●

306165 Poaceae Barnhart(保留科名) = Gramineae Juss.(保留科名)■●

306166 Poaceae Caruel = Gramineae Juss.(保留科名)■●

306167 Poaceae Caruel = Poaceae Barnhart(保留科名)■●

306168 Poacynum Baill.(1888);白麻属●

306169 Poacynum Baill. = Apocynum L. ●■

306170 Poacynum hendersonii (Hook. f.) Woodson = Apocynum pictum Schrenk ●

306171 Poacynum pictum (Schrenk) Baill. = Apocynum pictum Schrenk ●

306172 Poaephyllum Ridl.(1907);肖禾叶兰属■☆

306173 Poaephyllum fuscum Ridl.;褐肖禾叶兰■☆

306174 Poaephyllum parviflorum Rolfe;小花肖禾叶兰■☆

306175 Poaephyllum pauciflora Ridl.;少花肖禾叶兰■☆

306176 Poaephyllum tenuipes Rolfe;细梗肖禾叶兰■☆

306177 Poaephyllum trilobum J. J. Sm.;三裂肖禾叶兰■☆

306178 Poaephyllum uniflorum J. J. Wood;单花肖禾叶兰■☆

306179 Poagris Raf. = Poa L. ■

306180 Poagrostis Stapf(1899);澳非属(南极小草属)■☆

306181 Poagrostis pusilla (Nees) Stapf = Pentaschistis pusilla (Nees) H. P. Linder ■☆

306182 Poarchon Allemëo = Trimezia Salisb. ex Herb. ■☆

306183 Poarchon Mart. ex Seub. = Abolboda Humb. ■☆

306184 Poarion Rchb. = Koeleria Pers. ■

306185 Poarion Rchb. = Rostraria Trin. ■☆

306186 Poarium Desv. = Poarium Desv. ex Ham. ■☆

306187 Poarium Desv. = Stemodia L.(保留属名)■☆

306188 Poarium Desv. ex Ham.(1825);禾叶玄参属■☆

306189 Poarium Desv. ex Ham. = Stemodia L.(保留属名)■☆

306190 Poarium veronicoides Desv.;禾叶玄参■☆

306191 Poarium verticillatum (Mill.) Pennell;轮生禾叶玄参■☆

306192 Pobeguinea Jacq.-Fél. = Anadelphia Hack. ■☆

306193 Pobeguinea afzeliana (Rendle) Jacq.-Fél. = Anadelphia afzeliana (Rendle) Stapf ■☆

306194 Pobeguinea arrecta (Stapf) Jacq.-Fél. = Anadelphia afzeliana (Rendle) Stapf ■☆

306195 Pobeguinea chevalieri (Reznik) Jacq.-Fél. = Anadelphia chevalieri Reznik ■☆

306196 Pobeguinea gabonensis Koechlin = Anadelphia trispiculata Stapf ■☆

306197 Pobeguinea hamata (Stapf) Jacq.-Fél. = Anadelphia hamata Stapf ■☆

306198 Pobeguinea trichaeta (Reznik) Jacq.-Fél. = Anadelphia trichaeta (Reznik) Clayton ■☆

306199 Pobeguinea trispiculata (Stapf) Jacq.-Fél. = Anadelphia trispiculata Stapf ■☆

306200 Pobeguinsa (Stapf) Jacq.-Fél. = Anadelphia Hack. ■☆

306201 Pochota Ram. Goyena(废弃属名) = Bombacopsis Pittier(保留属名)●☆

306202 Pochota Ram. Goyena(废弃属名) = Pachira Aubl. ●

306203 Pochota glabra (Pasq.) Bullock = Pachira glabra Pasq. ●☆

306204 Pocilla (Dumort.) Fourr. = Cochlidiosperma (Rchb.) Rchb. ■

306205 Pocilla (Dumort.) Fourr. = Veronica L. ■

306206 Pocilla Fourr. = Veronica L. ■

306207 Pocillaria Ridl. = Rhyticaryum Becc. ●☆

306208 Pocockia Ser. = Melissitus Medik. ■☆

306209 Pocockia Ser. = Trigonella L. ■

306210 Pocockia Ser. ex DC. = Trigonella L. ■

306211 Pocockia cachemiriana (Cambess.) Boiss. = Trigonella cachemiriana Cambess. ■

306212 Pocockia liaoxiensis P. Y. Fu et Y. A. Chen = Melilotoides ruthenica (L.) Soják var. liaoxiensis (P. Y. Fu et Y. A. Chen) H. C. Fu et Y. Q. Jiang ■

306213 Pocockia ruthenica (L.) Boiss. = Medicago ruthenica (L.) Trautv. ■

306214 Pocockia ruthenica (L.) Boiss. = Trigonella ruthenica L. ■

306215 Pocockia ruthenica (L.) Boiss. var. inschanica (H. C. Fu et Y. Q. Jiang) H. C. Fu et Y. Q. Jiang = Melilotoides ruthenica (L.) Soják var. inschanica H. C. Fu et Y. Q. Jiang ■

306216 Pocockia ruthenica (L.) Boiss. var. liaoxiensis (P. Y. Fu et Y. A. Chen) H. C. Fu et Y. Q. Jiang = Melilotoides ruthenica (L.) Soják var. liaoxiensis (P. Y. Fu et Y. A. Chen) H. C. Fu et Y. Q. Jiang ■

306217 Pocockia ruthenica (L.) Boiss. var. oblongifolia (Fr.) H. C. Fu et Y. Q. Jiang = Melilotoides ruthenica (L.) Soják var. oblongifolia (Freyn) H. C. Fu et Y. Q. Jiang ■

306218 Pocophorum Neck. = Rhus L. ●

306219　Pocronostylis squarrosus（Vahl）Bertol. = Fimbristylis squarrosa Vahl ■

306220　Poculodiscus Danguy et Choux = Plagioscyphus Radlk. ●☆

306221　Podachaenium Benth. = Podachaenium Benth. ex Oerst. ●☆

306222　Podachaenium Benth. ex Oerst.（1852）；白花冠鳞菊属（坡达开菊属）●☆

306223　Podachaenium eminens（Lag.）Sch. Bip.；白花冠鳞菊（坡达开菊）；Daisy Tree ●☆

306224　Podachaenium eminens Baill. = Podachaenium eminens（Lag.）Sch. Bip. ●☆

306225　Podadenia Thwaites(1861)；足腺大戟属 ☆

306226　Podadenia sapida Thwaites；足腺大戟 ☆

306227　Podaechmea（Mez）L. B. Sm. et W. J. Kress = Aechmea Ruiz et Pav.（保留属名）■☆

306228　Podaechmea（Mez）L. B. Sm. et W. J. Kress = Ursulaea Read et Baensch ■☆

306229　Podagraria Hill = Aegopodium L. ■

306230　Podagrostis（Griseb.）Scribn. et Merr. = Agrostis L.（保留属名）■

306231　Podagrostis Scribn. et Merr. = Agrostis L.（保留属名）■

306232　Podaletra Raf. = Convolvulus L. ■●

306233　Podaliria Willd. = Podalyria Willd.（保留属名）●☆

306234　Podalyria Lam. = Podalyria Willd.（保留属名）●☆

306235　Podalyria Lam. ex Willd. = Podalyria Willd.（保留属名）●☆

306236　Podalyria Willd.（1799）（保留属名）；香豆木属（花槐属）；Sweet-pea Bush ●☆

306237　Podalyria amoena Eckl. et Zeyh.；秀丽香豆木●☆

306238　Podalyria angustifolia Eckl. et Zeyh. = Podalyria argentea Salisb. ●☆

306239　Podalyria anomala Lehm. = Podalyria sericea（Andréws）R. Br. ex W. T. Aiton ●☆

306240　Podalyria argentea Eckl. et Zeyh. = Podalyria biflora Lam. ●☆

306241　Podalyria argentea Salisb.；银白香豆木●☆

306242　Podalyria aurea（Aiton）Willd. = Calpurnia aurea（Aiton）Benth. ■☆

306243　Podalyria biflora Lam.；双花香豆木●☆

306244　Podalyria biflora Sims = Podalyria argentea Salisb. ●☆

306245　Podalyria burchellii DC.；伯切尔香豆木●☆

306246　Podalyria buxifolia（Retz.）Willd.；黄杨叶香豆木●☆

306247　Podalyria buxifolia Eckl. et Zeyh. = Podalyria myrtillifolia（Retz.）Willd. ●☆

306248　Podalyria buxifolia Lam. = Podalyria glauca（Thunb.）DC. ●☆

306249　Podalyria calyptrata（Retz.）Willd.；香豆木（帽状香豆木）；Sweet Pea Bush，Sweet-pea Bush ●☆

306250　Podalyria calyptrata（Retz.）Willd. var. lanceolata E. Mey. = Podalyria calyptrata（Retz.）Willd. ●☆

306251　Podalyria calyptrata Willd. = Podalyria calyptrata（Retz.）Willd. ●☆

306252　Podalyria canescens E. Mey.；灰色香豆木●☆

306253　Podalyria canescens Eckl. et Zeyh. = Podalyria sericea（Andréws）R. Br. ex W. T. Aiton ●☆

306254　Podalyria capensis（L.）Willd. = Virgilia oroboides（P. J. Bergius）T. M. Salter ●☆

306255　Podalyria chrysantha Adamson = Stirtonanthus chrysanthus（Adamson）B. -E. van Wyk et A. L. Schutte ●☆

306256　Podalyria cordata R. Br.；心形香豆木●☆

306257　Podalyria cuneifolia Eckl. et Zeyh. = Podalyria argentea Salisb. ●☆

306258　Podalyria cuneifolia Vent. = Podalyria myrtillifolia（Retz.）

306259　Podalyria genistoides（L.）Willd. = Cyclopia genistoides（L.）R. Br. ●☆

306260　Podalyria glauca（Thunb.）DC.；灰蓝香豆木●☆

306261　Podalyria haematoxylon Schumach. et Thonn. = Baphia nitida Lodd. ●☆

306262　Podalyria hamata E. Mey.；顶钩香豆木●☆

306263　Podalyria hirsuta（Aiton）Willd.；粗毛香豆木●☆

306264　Podalyria insignis Compton = Stirtonanthus insignis（Compton）B. -E. van Wyk et A. L. Schutte ●☆

306265　Podalyria intermedia Eckl. et Zeyh. = Podalyria canescens E. Mey. ●☆

306266　Podalyria lanceolata Benth. = Podalyria calyptrata（Retz.）Willd. ●☆

306267　Podalyria lancifolia Eckl. et Zeyh. = Podalyria burchellii DC. ●☆

306268　Podalyria leipoldtii L. Bolus；莱波尔德香豆木●☆

306269　Podalyria liparioides Eckl. et Zeyh. = Podalyria argentea Salisb. ●☆

306270　Podalyria meyeriana Eckl. et Zeyh. = Podalyria canescens E. Mey. ●☆

306271　Podalyria microphylla E. Mey.；小叶香豆木●☆

306272　Podalyria montana Hutch.；山地香豆木●☆

306273　Podalyria mundiana Eckl. et Zeyh. = Podalyria glauca（Thunb.）DC. ●☆

306274　Podalyria myrtillifolia（Retz.）Willd.；黑果越桔香豆木●☆

306275　Podalyria myrtillifolia Eckl. et Zeyh. = Podalyria calyptrata（Retz.）Willd. ●☆

306276　Podalyria nana（Popov）Popov = Ammopiptanthus nanus（Popov）S. H. Cheng ●◇

306277　Podalyria obcordata Lam. ex Poir. = Requienia obcordata（Lam. ex Poir.）DC. ■☆

306278　Podalyria oleifolia Salisb.；木犀榄叶香豆木●☆

306279　Podalyria orbicularis E. Mey.；圆形香豆木●☆

306280　Podalyria pearsonii E. Phillips；皮尔逊香豆木●☆

306281　Podalyria pedunculata Eckl. et Zeyh. = Podalyria argentea Salisb. ●☆

306282　Podalyria pulcherrima Schinz；艳丽香豆木●☆

306283　Podalyria racemulosa Eckl. et Zeyh.；小总花香豆木●☆

306284　Podalyria reticulata Harv.；网状香豆木●☆

306285　Podalyria sericea（Andréws）R. Br. ex W. T. Aiton；金毛香豆木(绢毛花槐)；Silky Sophora ●☆

306286　Podalyria sericea R. Br. = Podalyria sericea（Andréws）R. Br. ex W. T. Aiton ●☆

306287　Podalyria sparsiflora Eckl. et Zeyh. = Podalyria glauca（Thunb.）DC. ●☆

306288　Podalyria speciosa Eckl. et Zeyh.；美丽香豆木●☆

306289　Podalyria styracifolia Sims = Podalyria calyptrata（Retz.）Willd. ●☆

306290　Podalyria subbiflora Benth. = Podalyria argentea Salisb. ●☆

306291　Podalyria tayloriana L. Bolus = Stirtonanthus taylorianus（L. Bolus）B. -E. van Wyk et A. L. Schutte ●☆

306292　Podalyria thunbergiana Eckl. et Zeyh. = Podalyria canescens E. Mey. ●☆

306293　Podalyria velutina Burch. ex Benth.；短绒毛香豆木●☆

306294　Podandra Baill.（1890）；足蕊萝藦属 ■☆

306295　Podandra boliviana Baill.；足蕊萝藦 ■☆

306296　Podandria Rolfe = Habenaria Willd. ■

306297　Podandria macrandra（Lindl.）Rolfe = Habenaria macrandra

Lindl. ■☆

306298 Podandriella Szlach. = Podandria Rolfe ■

306299 Podandriella batesii （ la Croix ） Szlach. et Olszewski = Habenaria batesii la Croix ■☆

306300 Podandriella letouzeyana Szlach. et Olszewski = Habenaria letouzeyana （Szlach. et Olszewski） P. J. Cribb et Stévart ■☆

306301 Podandrogyne Ducke(1930)；足蕊南星属■☆

306302 Podandrogyne brachycarpa （ DC. ） Woodson；短果足蕊南星■☆

306303 Podandrogyne glabra Ducke；光足蕊南星■☆

306304 Podandrogyne gracilis （ Triana et Planch. ） Woodson；细足蕊南星■☆

306305 Podandrogyne macrophylla （ Turcz. ） Woodson；大叶足蕊南星■☆

306306 Podandrogyne pubescens Asplund；毛足蕊南星■☆

306307 Podangis Schltr. (1918)；足距兰属(裂距兰属)■☆

306308 Podangis dactyloceras （ Rchb. f. ） Schltr. ；足距兰■☆

306309 Podanisia Raf. = Polanisia Raf. ■

306310 Podanthera Wight ＝Epipogium J. G. Gmel. ex Borkh. ■

306311 Podanthera pallida Wight ＝Epipogium roseum （D. Don） Lindl. ■

306312 Podanthes Haw. （废弃属名） ＝ Orbea Haw. ■☆

306313 Podanthes Haw. （废弃属名）＝Podanthus Lag. （保留属名）●☆

306314 Podanthes geminata （ Masson ） G. Nicholson ＝ Piaranthus geminatus （Masson） N. E. Br. ■☆

306315 Podanthes incarnata （ L. f. ） Sweet ＝Quaqua incarnata （L. f. ） Bruyns ■☆

306316 Podanthes pulchra Haw. ＝ Orbea verrucosa （ Masson ） L. C. Leach ■☆

306317 Podanthum （ G. Don ） Boiss. ＝ Asyneuma Griseb. et Schenk ■

306318 Podanthum Boiss. ＝ Asyneuma Griseb. et Schenk ■

306319 Podanthum aurasiacum Batt. et Trab. ＝ Asyneuma rigidum （ Willd. ） Grossh. subsp. aurasiacum （ Batt. et Trab. ） Damboldt ■☆

306320 Podanthus Lag. (1816)（保留属名）；柄花菊属●☆

306321 Podanthus mitiqui Lindl. ；米氏柄花菊●☆

306322 Podanthus ovatifolius Lag. ；柄花菊(卵南美菊)●☆

306323 Podasaemium Rchb. ＝ Muhlenbergia Schreb. ■

306324 Podasaemium Rchb. ＝ Podosemum Desv. ■

306325 Podasaemum Rchb. ＝ Podasaemium Rchb. ■

306326 Podia Neck. ＝ Centaurea L. （保留属名）●■

306327 Podianthus Schnital. ＝ Trichopus Gaertn. ■☆

306328 Podionapus Dulac ＝ Deschampsia P. Beauv. ■

306329 Podiopetalum Hochst. ＝ Dalbergia L. f. （保留属名）●

306330 Podisonia Dumort. ex Steud. ＝ Posidonia K. D. König（保留属名）■

306331 Podistera S. Watson(1887)；星梗芹属■☆

306332 Podistera nevadensis S. Watson；星梗芹■☆

306333 Podlechiella Maassoumi et Kaz. Osaloo ＝ Astragalus L. ●■

306334 Podlechiella Maassoumi et Kaz. Osaloo ＝ Phaca L. ●■

306335 Podlechiella Maassoumi et Kaz. Osaloo(2003)；佛德角黄耆属■☆

306336 Podoaceae Baill. ex Franck ＝ Anacardiaceae R. Br. （保留科名）●

306337 Podoaceae Baill. ex Franck；九子母科（九子不离母科）；Podoa Family ●■

306338 Podocaelia （ Benth. ） A. Fern. et R. Fern. (1962)；空足野牡丹属■☆

306339 Podocaelia （ Benth. ） A. Fern. et R. Fern. ＝ Deamia Britton et Rose ■☆

306340 Podocaelia （ Benth. ） A. Fern. et R. Fern. ＝ Derosiphia Raf. ●■

306341 Podocaelia tubulosa （ Sm. ） A. Fern. et R. Fern. ＝ Osbeckia

tubulosa Sm. ●☆

306342 Podocallis Salisb. ＝ Massonia Thunb. ex Houtt. ■☆

306343 Podocalyx Klotzsch(1841)；柄萼大戟属☆

306344 Podocalyx loranthoides Klotzsch；柄萼大戟☆

306345 Podocarpaceae Endl. （ 1847 ）（ 保留科名 ）；罗汉松科； Longstalked Yew Family, Podocarp Family, Podocarpus Family, Yaccatree Family, Yellow-wood Family ●

306346 Podocarpia Benth. ＝ Hylodesmum H. Ohashi et R. R. Mill ●■

306347 Podocarpium （ Benth. ） Yen C. Yang et P. H. Huang ＝ Hylodesmum H. Ohashi et R. R. Mill ●■

306348 Podocarpium （ Benth. ） Yen. C. Yang et P. H. Huang ＝ Desmodium Desv. （保留属名）●■

306349 Podocarpium densum （ C. Chen et X. J. Cui ） P. H. Huang ＝ Hylodesmum densum （C. Chen et X. J. Cui） H. Ohashi et R. R. Mill ●

306350 Podocarpium densum （ C. Chen et X. J. Cui ） P. H. Huang ＝ Podocarpium fallax （ C. K. Schneid. ） C. Chen et X. J. Cui ■

306351 Podocarpium duclouxii （ Pamp. ） Yen C. Yang et P. H. Huang ＝ Hylodesmum longipes （ Franch. ） H. Ohashi et R. R. Mill ■

306352 Podocarpium duclouxii （ Pamp. ） Yen C. Yang et P. H. Hunag ＝ Desmodium duclouxii Pamp. ■

306353 Podocarpium fallax （ C. K. Schneid. ） C. Chen et X. J. Cui ＝ Desmodium fallax C. K. Schneid. ●■

306354 Podocarpium fallax （ C. K. Schneid. ） C. Chen et X. J. Cui ＝ Hylodesmum podocarpum （ DC. ） H. Ohashi et R. R. Mill subsp. fallax （ C. K. Schneid. ） H. Ohashi et R. R. Mill ■

306355 Podocarpium fallax （ C. K. Schneid. ） C. Chen et X. J. Cui var. densum C. Chen et X. J. Cui ＝ Hylodesmum densum （ C. Chen et X. J. Cui） H. Ohashi et R. R. Mill ■

306356 Podocarpium fallax （ C. K. Schneid. ） C. Chen et X. J. Cui var. mandshuricum （ Maxim. ） Nakai ＝ Podocarpium podocarpum （ DC. ） Yen C. Yang et P. H. Huang var. mandshuricum （ Maxim. ） P. H. Huang ●■

306357 Podocarpium lancangense Y. Y. Qian ＝ Hylodesmum lancangense （ Y. Y. Qian） X. Y. Zhu et H. Ohashi. ●

306358 Podocarpium laxum （ DC. ） Yen C. Yang et P. H. Huang ＝ Desmodium laxum DC. ●■

306359 Podocarpium laxum （ DC. ） Yen C. Yang et P. H. Huang ＝ Hylodesmum laxum （ DC. ） H. Ohashi et R. R. Mill ■

306360 Podocarpium laxum （ DC. ） Yen C. Yang et P. H. Huang var. laterale （ C. K. Schneid. ） Yen C. Yang et P. H. Huang ＝ Desmodium hainanensis Isely ■

306361 Podocarpium laxum （ DC. ） Yen C. Yang et P. H. Huang var. laterale （ C. K. Schneid. ） Yen C. Yang et P. H. Huang ＝ Desmodium laterale C. K. Schneid. ■

306362 Podocarpium laxum （ DC. ） Yen C. Yang et P. H. Huang var. laterale （ C. K. Schneid. ） Yen C. Yang et P. H. Huang ＝ Hylodesmum laterale （ C. K. Schneid. ） H. Ohashi et R. R. Mill ■

306363 Podocarpium leptopum （ A. Gray ex Benth. ） Yen C. Yang et P. H. Huang ＝ Desmodium leptopum A. Gray ex Benth. ●■

306364 Podocarpium leptopum （ A. Gray ex Benth. ） Yen C. Yang et P. H. Huang ＝ Desmodium laxum DC. subsp. leptopum （ A. Gray ex Benth. ） H. Ohashi ●■

306365 Podocarpium leptopum （ A. Gray ex Benth. ） Yen C. Yang et P. H. Huang ＝ Desmodium gardneri Benth. ●■

306366 Podocarpium leptopus （ A. Gray ex Benth. ） Yen C. Yang et P. H. Huang ＝ Hylodesmum leptopus （ A. Gray ex Benth. ） H. Ohashi et R. R. Mill ●

306367　Podocarpium mandschuricum（Maxim.）Czerep. = Hylodesmum podocarpum（DC.）H. Ohashi et R. R. Mill subsp. oxyphyllum（DC.）H. Ohashi et R. R. Mill ■

306368　Podocarpium mandshuricum（Maxim.）Schindl. = Podocarpium podocarpum（DC.）Yen C. Yang et P. H. Huang var. mandshuricum（Maxim.）P. H. Huang ●■

306369　Podocarpium menglaense C. Chen et X. J. Cui = Hylodesmum menglaense（H. Ohashi）H. Ohashi et R. R. Mill ■

306370　Podocarpium oldhami（Oliv.）Yen C. Yang et P. H. Huang = Hylodesmum oldhamii（Oliv.）H. Ohashi et R. R. Mill ●■

306371　Podocarpium oldhamii（Oliv.）Yen C. Yang et P. H. Huang = Desmodium oldhamii Oliv. ●■

306372　Podocarpium podocarpum（A. DC.）Yen C. Yang et P. H. Huang = Hylodesmum podocarpum（DC.）H. Ohashi et R. R. Mill ■

306373　Podocarpium podocarpum（DC.）Yen C. Yang et P. H. Huang = Desmodium podocarpum DC. ●■

306374　Podocarpium podocarpum（DC.）Yen C. Yang et P. H. Huang subsp. oxyphyllum var. mandshuricum Maxim. = Podocarpium podocarpum（DC.）Yen C. Yang et P. H. Huang var. mandshuricum（Maxim.）P. H. Huang ●■

306375　Podocarpium podocarpum（DC.）Yen C. Yang et P. H. Huang var. fallax（C. K. Schneid.）Yen C. Yang et P. H. Huang = Desmodium fallax C. K. Schneid. var. mandshuricum（Maxim.）Nakai ●■

306376　Podocarpium podocarpum（DC.）Yen C. Yang et P. H. Huang var. fallax（C. K. Schneid.）Yen C. Yang et P. H. Huang = Desmodium mandshuricum Nakai ●■

306377　Podocarpium podocarpum（DC.）Yen C. Yang et P. H. Huang var. fallax（C. K. Schneid.）Yen C. Yang et P. H. Huang = Desmodium podocarpum DC. subsp. fallax（C. K. Schneid.）H. Ohashi ●■

306378　Podocarpium podocarpum（DC.）Yen C. Yang et P. H. Huang var. fallax（C. K. Schneid.）Yen C. Yang et P. H. Huang = Hylodesmum podocarpum（DC.）H. Ohashi et R. R. Mill subsp. fallax（C. K. Schindl.）H. Ohashi et R. R. Mill ■

306379　Podocarpium podocarpum（DC.）Yen C. Yang et P. H. Huang var. japonicum（Matsum.）P. H. Huang = Desmodium oxyphyllum DC. var. japonicum Matsum. ●

306380　Podocarpium podocarpum（DC.）Yen C. Yang et P. H. Huang var. japonicum（Matsum.）P. H. Huang = Desmodium podocarpum DC. subsp. oxyphyllum（DC.）H. Ohashi var. japonicum（Miq.）Maxim. ■

306381　Podocarpium podocarpum（DC.）Yen C. Yang et P. H. Huang var. japonicum（Matsum.）P. H. Huang = Hylodesmum podocarpum（DC.）H. Ohashi et R. R. Mill subsp. oxyphyllum（DC.）H. Ohashi et R. R. Mill ■

306382　Podocarpium podocarpum（DC.）Yen C. Yang et P. H. Huang var. mandschuricum（Maxim.）P. H. Huang = Hylodesmum podocarpum（DC.）H. Ohashi et R. R. Mill subsp. oxyphyllum（DC.）H. Ohashi et R. R. Mill ■

306383　Podocarpium podocarpum（DC.）Yen C. Yang et P. H. Huang var. mandshuricum（Maxim.）P. H. Huang = Desmodium mandshuricum（Maxim.）C. K. Schneid. ●■

306384　Podocarpium podocarpum（DC.）Yen C. Yang et P. H. Huang var. oxyphyllum（DC.）Yen C. Yang et P. H. Huang = Desmodium oxyphyllum DC. ●■

306385　Podocarpium podocarpum（DC.）Yen C. Yang et P. H. Huang var. oxyphyllum（DC.）Yen C. Yang et P. H. Huang = Hylodesmum podocarpum（DC.）H. Ohashi et R. R. Mill subsp. oxyphyllum（DC.）H. Ohashi et R. R. Mill ■

306386　Podocarpium podocarpum（DC.）Yen C. Yang et P. H. Huang var. szechuenense（Craib）Yen C. Yang et P. H. Huang = Desmodium szechuenense C. K. Schneid. ●■

306387　Podocarpium podocarpum（DC.）Yen C. Yang et P. H. Huang var. szechuenense（Craib）Yen C. Yang et P. H. Huang = Hylodesmum podocarpum（DC.）H. Ohashi et R. R. Mill subsp. szechuenense（Craib）H. Ohashi et R. R. Mill ■

306388　Podocarpium racemosum（DC.）Yen C. Yang et P. H. Huang var. mandshuricum（Maxim.）Ohwi = Podocarpium podocarpum（DC.）Yen C. Yang et P. H. Huang var. mandshuricum（Maxim.）P. H. Huang ●■

306389　Podocarpium repandum（Vahl）Yen C. Yang et P. H. Huang = Desmodium repandum（Vahl）DC. ●

306390　Podocarpium repandum（Vahl）Yen C. Yang et P. H. Huang = Hylodesmum repandum（Vahl）H. Ohashi et R. R. Mill ●

306391　Podocarpium williamsii（H. Ohashi）Yen C. Yang et P. H. Huang = Desmodium williamsii H. Ohashi ●■

306392　Podocarpium williamsii（H. Ohashi）Yen C. Yang et P. H. Huang = Hylodesmum williamsii（H. Ohashi）H. Ohashi et R. R. Mill ■

306393　Podocarpus L'Hér. ex Pers. = Podocarpus Pers.（保留属名）●

306394　Podocarpus Labill. = Podocarpus Pers.（保留属名）●

306395　Podocarpus Pers.（1806）（保留属名）；罗汉松属（竹柏属）；African Yellow-wood, Longstalked Yew, Podocarpus, White Pine, Yaccatree, Yellow Wood, Yellowwood, Yellow-wood, Yew Pine ●

306396　Podocarpus Pers. = Phyllocladus Rich. ex Mirb.（保留属名）●☆

306397　Podocarpus acutifolius Kirk；刺叶罗汉松；Sharp-leaved Yellow-wood ●☆

306398　Podocarpus alata ?；翅罗汉松；Brown Pine ●☆

306399　Podocarpus alpinus R. Br. ex Hook. f.；高山罗汉松（塔斯马尼亚罗汉松）；Tasmanian Podocarp, Tasmanian Podocarpus ●☆

306400　Podocarpus amara Blume = Sundacarpus amara（Blume）C. N. Page ●☆

306401　Podocarpus andinus Posp. = Prumnopitys andina（Poepp. ex Endl.）de Laub. ●☆

306402　Podocarpus annamiensis N. E. Gray；海南罗汉松；Hainan Podocarpus, Hainan Yaccatree ●◇

306403　Podocarpus argotaenia Hance = Amentotaxus argotaenia（Hance）Pilg. ●◇

306404　Podocarpus argotaenia Hance = Amentotaxus formosana H. L. Li ●◇

306405　Podocarpus blumei Endl. = Nageia wallichiana（C. Presl）Kuntze ●

306406　Podocarpus bracteata Blume = Podocarpus neriifolius D. Don ●

306407　Podocarpus brevifolius（Stapf）Foxw.；小叶罗汉松；Shortleaf Podocarpus, Shortleaf Yaccatree, Short-leaved Podocarpus ●☆

306408　Podocarpus brevifolius（Stapf）Foxw. = Podocarpus wangii C. C. Chang ●

306409　Podocarpus capuronii de Laub.；凯普伦罗汉松 ●☆

306410　Podocarpus chilinus Rich.；智利罗汉松 ●☆

306411　Podocarpus chinensis Sweet = Podocarpus nakaii Hayata ●

306412　Podocarpus chinensis Sweet var. maki（Siebold et Zucc.）K. S. Hao = Podocarpus macrophyllus（Thunb.）D. Don var. maki（Siebold）Endl. ●

306413　Podocarpus chinensis Sweet var. maki（Siebold）K. S. Hao =

Podocarpus nakaii Hayata ●

306414　Podocarpus chinensis Wall. ex J. Forbes = Podocarpus macrophyllus (Thunb.) Sweet var. maki Endl. ●

306415　Podocarpus chinensis Wall. ex J. Forbes = Podocarpus macrophyllus (Thunb.) D. Don var. maki (Siebold) Endl. ●

306416　Podocarpus chinensis Wall. ex J. Forbes var. makii (Siebold) K. S. Hao = Podocarpus macrophyllus (Thunb.) D. Don var. maki (Siebold) Endl. ●

306417　Podocarpus chingianus S. Y. Hu = Podocarpus macrophyllus (Thunb.) D. Don var. chingii N. E. Gray ●

306418　Podocarpus coriaceus L. ;牙买加罗汉松;Yacca Podocarpus ●☆

306419　Podocarpus costalis C. Presl;兰屿罗汉松;Lanyu Nagi, Lanyu Podocarpus, Lanyu Yaccatree ●◇

306420　Podocarpus dacrydioides Rich. ;泪柏罗汉松(泪柏);Ahikatea, Huonpine Yaccatree, Kahikatea, New Zealand White Podocarpus, White Pine ●☆

306421　Podocarpus dacrydioides Rich. = Dacrycarpus dacrydioides (A. Rich.) de Laub. ●☆

306422　Podocarpus discolor Blume = Podocarpus neriifolius D. Don ●

306423　Podocarpus drouynianus F. Muell. ;鸸鹋罗汉松;Emu Berry, Wild Plum ●☆

306424　Podocarpus elatus R. Br. ;澳大利亚罗汉松(澳洲罗汉松,高大罗汉松);Australia Yaccatree, Brown Pine, Plum Pine, Plum-pine, She Pine, She-pine ●☆

306425　Podocarpus elongatus (Aiton) L'Hér. ex Pers. ;好望角罗汉松(长罗汉松,南非罗汉松,球冠罗汉松,伸展罗汉松,微叶罗汉松);Bastard River Yellowwood, Bastard Yellowwood, Cape of Good Hope Yacca, Cape of Good Podocarpus, Cape Yellowwood, Sfrican Yellowwood, South African Yellowwood ●☆

306426　Podocarpus falcatus A. Cunn. ex Parl. ;镰叶竹柏(镰形罗汉松,镰叶罗汉松);Bastaed Yellowwood, Qutenigua Yellowwood, Outeniqua Yellow-wood, Sickleleaf Yaccatree ●☆

306427　Podocarpus fasciculus de Laub. ;丛花百日青●☆

306428　Podocarpus ferrugineus G. Benn. ex D. Don = Prumnopitys ferruginea (G. Benn. ex D. Don) de Laub. ●☆

306429　Podocarpus fleuryi Hickel;长叶竹柏(佛劳利罗汉松,桐木树);Fleury Podocarpus, Fleury Yaccatree ●◇

306430　Podocarpus fleuryi Hickel = Nageia fleuryi (Hickel) de Laub. ●◇

306431　Podocarpus formosensis Dümmer;窄叶竹柏(恒春竹柏,台湾竹柏);Hengchun Nagi, Taiwan Podocarpus, Taiwan Yaccatree ●

306432　Podocarpus formosensis Dümmer = Nageia nagi (Thunb.) Kuntze ●

306433　Podocarpus forrestii Craib et W. W. Sm. ;大理罗汉松;Forrest Podocarpus, Forrest Yaccatree ●

306434　Podocarpus forrestii Craib et W. W. Sm. = Podocarpus macrophyllus (Thunb.) D. Don ●

306435　Podocarpus gracilior Pilg. ;细叶罗汉松(东非罗汉松,细长罗汉松,纤细罗汉松);African Fern Pine, East African Yellow-wood, Fern Pine, Musengerra Podocarpus, Musengerra Yaccatree, Podo, Yellow-wood ●☆

306436　Podocarpus grayae de Laub. ;垂叶罗汉松●☆

306437　Podocarpus guatemaleasis Steud. ;危地马拉罗汉松;British Honduras Yellow Wood, Guatemalan Yellow Wood, Yaccatree ●☆

306438　Podocarpus hallii Kirk;哈尔罗汉松(哈氏罗汉松,霍氏罗汉松);Hall's Totara, Totara ●☆

306439　Podocarpus henkelii Pilg. ;长叶罗汉松(肯氏罗汉松,纳塔尔罗汉松);Falcate Yellowwood, Henkel Yellowwood, Long-leafed Yellowwood ●

306440　Podocarpus humbertii de Laub. ;亨伯特罗汉松●☆

306441　Podocarpus imbricatus Blume = Dacrycarpus imbricatus (Blume) de Laub. ●◇

306442　Podocarpus insignis Hemsl. = Amentotaxus argotaenia (Hance) Pilg. ●◇

306443　Podocarpus japonicus J. Nelson = Nageia nagi (Thunb.) Kuntze ●

306444　Podocarpus japonicus Siebold ex Endl. = Podocarpus macrophyllus (Thunb.) D. Don var. maki (Siebold) Endl. ●

306445　Podocarpus javanicus Merr. = Podocarpus imbricatus Blume ●◇

306446　Podocarpus kawai Hayata = Podocarpus imbricatus Blume ●◇

306447　Podocarpus koshunensis (Kaneh.) Kaneh. = Nageia nagi (Thunb.) Kuntze ●

306448　Podocarpus koshunensis (Kaneh.) Kaneh. = Podocarpus formosensis Dümmer ●

306449　Podocarpus lamberti Klotzsch ex Endl. ;朗伯罗汉松(巴西南部罗汉松)●☆

306450　Podocarpus latifolius (Thunb.) R. Br. ;非洲罗汉松(阔叶罗汉松);Real Yellowwood, Real Yellow-wood, Upright Yellowwood, Upright Yellow-wood, Yellowwood ●☆

306451　Podocarpus latifolius Blume = Nageia wallichiana (C. Presl) Kuntze ●

306452　Podocarpus latifolius R. Br. = Nageia wallichiana (C. Presl) Kuntze ●

306453　Podocarpus lawrencei Hook. f. ;劳伦斯罗汉松;Mountain Plum Pine ●☆

306454　Podocarpus leptostachyus Blume = Podocarpus neriifolius D. Don ●

306455　Podocarpus macrophyllus (Thunb. ex A. Murray) D. Don = Podocarpus macrophyllus (Thunb.) D. Don ●

306456　Podocarpus macrophyllus (Thunb. ex A. Murray) Sweet ' Tetragona ' = Podocarpus macrophyllus (Thunb.) D. Don ' Tetragona ' ●☆

306457　Podocarpus macrophyllus (Thunb.) D. Don;罗汉松(百日青,大叶罗汉松,罗汉杉,土杉,土松);Broad-leaved Podocarpus, Buddhist Pine, Japanese Large-leaved Podocarp, Japanese Yew, Kusamaki, Large-leaf Podocarpus, Lohan Pine, Longleaf Podocarpus, Long-leaved Podocarpus, Shrubby Yew Podocarpus, Shrubby Yew Pine, Southern Yew, Yaccatree, Yew Pine, Yew Plum Pine, Yew Podocarpus, Yew-pine ●

306458　Podocarpus macrophyllus (Thunb.) D. Don ' Tetragona ';四角短叶罗汉松●☆

306459　Podocarpus macrophyllus (Thunb.) D. Don = Podocarpus annamiensis N. E. Gray ●◇

306460　Podocarpus macrophyllus (Thunb.) D. Don = Podocarpus forrestii Craib et W. W. Sm. ●

306461　Podocarpus macrophyllus (Thunb.) D. Don = Podocarpus nakaii Hayata ●

306462　Podocarpus macrophyllus (Thunb.) D. Don f. angustifolius (Blume) Pilg. = Podocarpus macrophyllus (Thunb.) D. Don var. angustifolius Blume ●

306463　Podocarpus macrophyllus (Thunb.) D. Don f. grandifolius Pilg. = Podocarpus neriifolius D. Don ●

306464　Podocarpus macrophyllus (Thunb.) D. Don subsp. makii Pilg. = Podocarpus nakaii Hayata ●

306465　Podocarpus macrophyllus (Thunb.) D. Don subsp. makii Pilg. = Podocarpus macrophyllus (Thunb.) D. Don ●

306466　Podocarpus macrophyllus (Thunb.) D. Don var. acuminatissima

Pritz. = Podocarpus neriifolius D. Don ●

306467　Podocarpus macrophyllus（Thunb.）D. Don var. angustifolius Blume；狭叶罗汉松；Narrowleaf Podocarpus, Yaccatree ●

306468　Podocarpus macrophyllus（Thunb.）D. Don var. chingii N. E. Gray；柱冠罗汉松；Columnar Podocarpus, Yaccatree ●

306469　Podocarpus macrophyllus（Thunb.）D. Don var. maki（Siebold）Endl.；小罗汉松（短叶罗汉松，短叶土杉，江南柏，江南侧柏，小叶罗汉松）；China Yaccatree, Chinese Podocarpus, Japanese Podocarp, Kusamaki, Maki Podocarpus, Podocarpus, Shrubby, Shrubby Yew Podocarpus, Yew Plum Pine ●

306470　Podocarpus macrophyllus（Thunb.）D. Don var. nakaii（Hayata）H. L. Li et H. Keng = Podocarpus nakaii Hayata ●

306471　Podocarpus macrophyllus（Thunb.）D. Don var. piliramulus Z. X. Chen et Z. Q. Li；毛枝罗汉松●

306472　Podocarpus macrophyllus（Thunb.）Sweet‘Tetragona’= Podocarpus macrophyllus（Thunb.）D. Don‘Tetragona’● ☆

306473　Podocarpus macrophyllus（Thunb.）Sweet = Podocarpus macrophyllus（Thunb.）D. Don ●

306474　Podocarpus macrophyllus（Thunb.）Sweet var. nakaii（Hayata）H. L. Li et H. Keng = Podocarpus nakaii Hayata ●

306475　Podocarpus macrophyllus D. Don = Podocarpus annamiensis N. E. Gray ● ◇

306476　Podocarpus macrophyllus D. Don = Podocarpus forrestii Craib et W. W. Sm. ●

306477　Podocarpus macrophyllus D. Don = Podocarpus nakaii Hayata ●

306478　Podocarpus macrophyllus D. Don = Podocarpus neriifolius D. Don ●

306479　Podocarpus macrophyllus D. Don f. angustifolius（Blume）Pilg. = Podocarpus macrophyllus（Thunb.）D. Don var. angustifolius Blume ●

306480　Podocarpus macrophyllus D. Don f. grandifolius Pilg. = Podocarpus neriifolius D. Don ●

306481　Podocarpus macrophyllus D. Don subsp. maki（Siebold et Zucc.）Pilg. = Podocarpus macrophyllus（Thunb.）D. Don var. maki（Siebold）Endl. ●

306482　Podocarpus macrophyllus D. Don subsp. maki Pilg. = Podocarpus nakaii Hayata ●

306483　Podocarpus macrophyllus D. Don var. acuminatissima Pritz. = Podocarpus neriifolius D. Don ●

306484　Podocarpus macrophyllus D. Don var. nakaii（Hayata）H. L. Li et H. Keng = Podocarpus nakaii Hayata ●

306485　Podocarpus madagascariensis Baker；马达加斯加罗汉松；Madagascar Yellowwood, Yaccatree ● ☆

306486　Podocarpus matudai Lundell；松田罗汉松● ☆

306487　Podocarpus maximus（de Laub.）Gaussen = Decussocarpus maximus de Laub. ● ☆

306488　Podocarpus maximus（de Laub.）Whitmore；大叶竹柏● ☆

306489　Podocarpus maximus（de Laub.）Whitmore = Decussocarpus maximus de Laub. ● ☆

306490　Podocarpus milanjianus Rendle；米兰加罗汉松（东非罗汉松，米加罗汉松，米兰加汉松）；East African Yellowwood ● ☆

306491　Podocarpus nageia R. Br. ex Endl. = Nageia nagi（Thunb.）Kuntze ●

306492　Podocarpus nageia R. Br. ex Mirb. = Podocarpus nagi（Thunb.）Zoll. et Moritzi ex Zoll. ●

306493　Podocarpus nagi（Thunb.）Pilg.；竹柏（宝芳，船家树，大果竹柏，黄杂树，罗汉柴，椤树，山杉，糖鸡子，铁甲树，乌心石，椰树，

猪肝树）；Broadleaf Pine, Broadleaf Podocarpus, Japenese Podocarpus, Mountain Podocarpus, Nagai Podocarpus, Nagi, Nagi Podocarpus, Nagi Yaccatree ●

306494　Podocarpus nagi（Thunb.）Pilg. = Nageia nagi（Thunb.）Kuntze ●

306495　Podocarpus nagi（Thunb.）Pilg. = Podocarpus nagi（Thunb.）Zoll. et Moritzi ex Zoll. ●

306496　Podocarpus nagi（Thunb.）Zoll. et Moritzi ex Makino = Nageia nagi（Thunb.）Kuntze ●

306497　Podocarpus nagi（Thunb.）Zoll. et Moritzi ex Makino = Podocarpus macrophyllus（Thunb.）D. Don‘Tetragona’● ☆

306498　Podocarpus nagi（Thunb.）Zoll. et Moritzi ex Zoll. = Podocarpus macrophyllus（Thunb.）D. Don‘Tetragona’● ☆

306499　Podocarpus nagi Makino = Podocarpus nagi（Thunb.）Zoll. et Moritzi ex Zoll. ●

306500　Podocarpus nagi Makino var. koshunensis Kaneh. = Nageia nagi（Thunb.）Kuntze ●

306501　Podocarpus nagi Makino var. koshunensis Kaneh. = Podocarpus formosensis Dümmer ●

306502　Podocarpus nakaii Hayata；台湾罗汉松（百日青，尖叶罗汉松，桃实，土杉）；Nakai Podocarpus, Taiwan Yaccatree ●

306503　Podocarpus nankoensis Hayata；南口罗汉松（南港竹柏，山杉）；Nanko Podocarpus, Nanko Yaccatree, Nankung Nagi ●

306504　Podocarpus nankoensis Hayata = Nageia nagi（Thunb.）Kuntze ●

306505　Podocarpus neglecta Blume = Podocarpus neriifolius D. Don ●

306506　Podocarpus neriifolius D. Don；百日青（白松，大叶竹柏松，脉叶罗汉松，桃柏松，桃实，璎珞柏，油松，竹柏松，竹叶罗汉松，竹叶松，紫柏）；Blackleaf Podocarp, Oleander Podocarpus, Oleanderleaf Yaccatree, Thitmin ●

306507　Podocarpus neriifolius D. Don var. brevifolius Stapf = Podocarpus brevifolius（Stapf）Foxw. ● ☆

306508　Podocarpus neriifolius D. Don var. brevifolius Stapf = Podocarpus wangii C. C. Chang ●

306509　Podocarpus nivalis Hook. f.；雪白罗汉松（高山罗汉松）；Alp Yaccatree, Alpine Totara ● ☆

306510　Podocarpus nubigenus Lindl.；云雾罗汉松；Chilean Totara, Manio, Manio Podocarpus, Nabilous Yaccatree ● ☆

306511　Podocarpus parlatorei Pilg.；弯叶罗汉松（阿根廷罗汉松）● ☆

306512　Podocarpus philippinensis Foxw.；菲律宾罗汉松；Philippine Podocarpus, Philippine Yaccatree ●

306513　Podocarpus polystachyus R. Br.；多穗罗汉松；Manyspike Podocarpus, Manyspike Yaccatree ● ☆

306514　Podocarpus polystachyus R. Br. = Podocarpus costalis C. Presl ● ◇

306515　Podocarpus purdieanus Hook.；普迪罗汉松；Purdie Podocarpus, Purdie Yaccatree ● ☆

306516　Podocarpus rumphii Blume；红皮罗汉松● ☆

306517　Podocarpus salignus D. Don；柳叶罗汉松；Chilean Yew-pine, Manio, Willow Podocarp, Willowleaf Podocarp, Willowleaf Podocarpus, Willowleaf Yaccatree ● ☆

306518　Podocarpus spicatus R. Br.；穗花罗汉松；Black Pine, Matai, Matai Podocarpus, Spicate Yaccatree ● ☆

306519　Podocarpus spinulosus（Sm.）R. Br. ex Mirb.；微刺罗汉松；Spinule Podocarpus, Yaccatree ● ☆

306520　Podocarpus sutchuanensis Franch. = Keteleeria davidiana（Bertrand）Beissn. ●

306521　Podocarpus totara G. Benn. ex D. Don = Podocarpus totarus D. Don ex Lamb. ● ☆

306522　Podocarpus totarus D. Don ex Lamb.；新西兰罗汉松；Totara，Totara Pine，Totara Podocarpus，Totara Yaccatree，Yew-pine ●☆

306523　Podocarpus totarus D. Don ex Lamb.‘Aureus’；金叶新西兰罗汉松●☆

306524　Podocarpus urbanii Pilg.；乌尔班罗汉松；Yacca ●☆

306525　Podocarpus usambarensis Pilg.；乌桑巴拉罗汉松●☆

306526　Podocarpus wallichianus C. Presl；肉托竹柏（大叶罗汉松，大叶竹柏）；Wallich Podocarpus，Wallich Yaccatree ●◇

306527　Podocarpus wallichianus C. Presl ＝ Nageia fleuryi（Hickel）de Laub. ●◇

306528　Podocarpus wallichianus C. Presl ＝ Nageia wallichiana（C. Presl）Kuntze ●◇

306529　Podocarpus wangii C. C. Chang；短叶罗汉松（小叶兰罗汉松，小叶竹松柏）●

306530　Podocentrum Borch. ex Meisn. ＝ Emex Campd.（保留属名）■☆

306531　Podochilopsis Guillaumin ＝ Adenoncos Blume ■☆

306532　Podochilopsis Guillaumin（1963）；类柄唇兰属■☆

306533　Podochilopsis dalatensis Guillaumin；类柄唇兰■☆

306534　Podochilus Blume（1825）；柄唇兰属；Podochilus ■

306535　Podochilus chinensis Schltr. ＝ Podochilus khasianus Hook. f. ■

306536　Podochilus cornuta（Blume）Schltr. ＝ Appendicula cornuta（Blume）Schltr. ■

306537　Podochilus formosana（Hayata）S. S. Ying ＝ Appendicula formosana Hayata ■

306538　Podochilus khasianus Hook. f.；柄唇兰；Khas Podochilus ■

306539　Podochilus kotoensis（Hayata）S. S. Ying ＝ Appendicula formosana Hayata ■

306540　Podochilus microphyllus Lindl.；小叶柄唇兰■☆

306541　Podochilus muricatus（Teijsm. et Binn.）Schltr.；密叶柄唇兰■☆

306542　Podochilus oxystophylloides Ormerod；云南柄唇兰■

306543　Podochilus taiwanianus S. S. Ying ＝ Appendicula formosana Hayata ■

306544　Podochrea Fourr. ＝ Astragalus L. ●■

306545　Podochrosia Baill. ＝ Rauvolfia L. ●

306546　Podococcus G. Mann et H. Wendl.（1864）；梗椰属（柄裂果属，凸花椰属）●☆

306547　Podococcus acaulis Hua ＝ Podococcus barteri G. Mann et H. Wendl. ●☆

306548　Podococcus barteri G. Mann et H. Wendl.；梗椰●☆

306549　Podocoma Cass.（1817）；层菀属■☆

306550　Podocoma R. Br. ＝ Ixiochlamys F. Muell. et Sond. ex Sond. ■●☆

306551　Podocoma cuneifolia R. Br.；楔叶层菀●☆

306552　Podocoma foliosa Malme；多叶层菀●☆

306553　Podocoma glandulosa Baker；腺层菀■☆

306554　Podocoma hirsuta Baker；毛层菀■☆

306555　Podocoma nana Ewart et Jean White；矮层菀●☆

306556　Podocybe K. Schum. ＝ Gleditsia L. ●

306557　Podocybe K. Schum. ＝ Pogocybe Pierre ●

306558　Podocytisus Boiss. et Heldr.（1849）；扫帚豆属■☆

306559　Podocytisus caramanicus Boiss. et Heldr.；扫帚豆■☆

306560　Podogyne Hoffmanns. ＝ Gynandropsis DC.（保留属名）■

306561　Podogynium Taub. ＝ Zenkerella Taub. ■☆

306562　Podolasia N. E. Br.（1882）；毛足南星属■☆

306563　Podolasia stipitata N. E. Br.；毛足南星■☆

306564　Podolepis Labill.（1806）（保留属名）；柄鳞菊属（纸苞金绒草属）■☆

306565　Podolepis acuminata R. Br.；尖柄鳞菊■☆

306566　Podolepis affinis Sond.；近缘柄鳞菊■☆

306567　Podolepis angustifolia Hort.；窄叶柄鳞菊■☆

306568　Podolepis auriculata DC.；小耳柄鳞菊■☆

306569　Podolepis canescens A. Cunn. ex DC.；灰柄鳞菊■☆

306570　Podolepis ferruginea DC.；锈色柄鳞菊■☆

306571　Podolepis filiformis Steetz；线形柄鳞菊■☆

306572　Podolepis gracilis Graham；细柄鳞菊■☆

306573　Podolepis laevigata Gand.；平滑柄鳞菊■☆

306574　Podolepis monticola R. J. F. Hend.；山地柄鳞菊■☆

306575　Podolepis robusta（Maiden et Betche）J. H. Willis；粗壮柄鳞菊■☆

306576　Podolepis rosea Steetz；粉红柄鳞菊■☆

306577　Podolobium R. Br. ＝ Oxylobium Andréws（保留属名）■☆

306578　Podolobium R. Br. ＝ Podolobium R. Br. ex W. T. Aiton ●☆

306579　Podolobium R. Br. ex W. T. Aiton（1811）；裂足豆属●☆

306580　Podolobium obovatum A. Gray；倒卵裂足豆●☆

306581　Podolobium trilobatum R. Br.；裂足豆●☆

306582　Podolobus Raf. ＝ Stanleya Nutt. ■☆

306583　Podolopus Steud. ＝ Podolobus Raf. ■☆

306584　Podolotus Benth. ＝ Astragalus L. ●■

306585　Podolotus Royle ＝ Lotus L. ■

306586　Podolotus Royle ex Benth. ＝ Astragalus L. ●■

306587　Podoluma Baill. ＝ Lucuma Molina ●

306588　Podoluma Baill. ＝ Pouteria Aubl. ●

306589　Podonephelium Baill.（1874）；足韶子属●☆

306590　Podonephelium concolor Radlk.；同色足韶子●☆

306591　Podonephelium deplanchei Baill.；足韶子●☆

306592　Podonephelium parvifolium Radlk.；小花足韶子●☆

306593　Podonephelium stipitatum Baill. ＝ Podonephelium deplanchei Baill. ●☆

306594　Podonix Raf. ＝ Tulipa L. ■

306595　Podonosma Boiss.（1849）；肖滇紫草属■☆

306596　Podonosma Boiss. ＝ Onosma L. ■

306597　Podonosma galalensis Boiss.；肖滇紫草■☆

306598　Podoon Baill. ＝ Dobinea Buch. -Ham. ex D. Don ●■

306599　Podoon delavayi Baill. ＝ Dobinea delavayi（Baill.）Baill. ■

306600　Podoonaceae Baill. ex Franch. ＝ Podoaceae Baill. ex Franck ●■

306601　Podopappus Hook. et Arn. ＝ Podocoma Cass. ■☆

306602　Podopetalum F. Muell. ＝ Ormosia Jacks.（保留属名）●

306603　Podopetalum Gandin ＝ Trochiscanthes W. D. J. Koch ■☆

306604　Podophania Baill. ＝ Hofmeisteria Walp. ■●☆

306605　Podophorus Phil.（1856）；雀麦禾属■☆

306606　Podophorus bromoides Phil.；雀麦禾■☆

306607　Podophyllaceae DC.（1817）（保留科名）；鬼臼科（桃儿七科）■

306608　Podophyllaceae DC.（保留科名）＝ Berberidaceae Juss.（保留科名）●■

306609　Podophyllum L.（1753）；鬼臼属（八角莲属，北美桃儿七属，足叶草属）；Mandrake，May Apple，May-Apple ■☆

306610　Podophyllum aurantiocaule Hand. -Mazz. ＝ Dysosma aurantiocaulis（Hand. -Mazz.）Hu ■

306611　Podophyllum aurantiocaule Hand. -Mazz. subsp. furfuraceum（S. Y. Bao）J. M. H. Shaw ＝ Dysosma aurantiocaulis（Hand. -Mazz.）Hu ■

306612　Podophyllum chengii S. S. Chien ＝ Dysosma chengii（S. S. Chien）M. Hiroe ■

306613　Podophyllum chengii S. S. Chien ＝ Dysosma pleiantha（Hance）Woodson ■

306614　Podophyllum delavayi Franch. ＝ Dysosma delavayi（Franch.）

Hu ■

306615 Podophyllum delavayi Franch. = Dysosma veitchii（Hemsl. et E. H. Wilson）S. H. Fu ex T. S. Ying ■

306616 Podophyllum delavayi Franch. var. longipetalum J. M. H. Shaw = Dysosma delavayi（Franch.）Hu ■

306617 Podophyllum difforme Hemsl. et E. H. Wilson = Dysosma diformis（Hemsl. et E. H. Wilson）T. H. Wang ex T. S. Ying ■

306618 Podophyllum diphyllum L. = Jeffersonia diphylla（L.）Pers. ■☆

306619 Podophyllum emodi Falc. ex Royle = Sinopodophyllum hexandrum（Royle）T. S. Ying ■

306620 Podophyllum emodi Falc. ex Royle var. chinensis Sprague = Sinopodophyllum hexandrum（Royle）T. S. Ying ■

306621 Podophyllum emodii Wall. ex Hook. f. et Thomson = Dysosma emodii（Wall. ex Hook. f. et Thomson）M. Hiroe ■

306622 Podophyllum emodii Wall. ex Hook. f. et Thomson = Sinopodophyllum hexandrum（Royle）T. S. Ying ■

306623 Podophyllum emodii Wall. ex Hook. f. et Thomson var. chinensis Sprague = Sinopodophyllum hexandrum（Royle）T. S. Ying ■

306624 Podophyllum emodii Wall. ex Hook. f. et Thomson var. hexandrum（Royle）Chatt. et Mukerjee = Podophyllum emodi Wall. ex Royle ■☆

306625 Podophyllum emodii Wall. ex Royle；喜马拉雅鬼臼■☆

306626 Podophyllum esquirolii H. Lév. = Dysosma versipellis（Hance）M. Cheng ex T. S. Ying ■

306627 Podophyllum guangxiense（Y. S. Wang）J. M. H. Shaw = Dysosma majoensis（Gagnep.）M. Hiroe ■

306628 Podophyllum hexandrum Royle = Podophyllum emodii Wall. ex Hook. f. et Thomson ■

306629 Podophyllum hexandrum Royle = Sinopodophyllum hexandrum（Royle）T. S. Ying ■

306630 Podophyllum hispidum K. S. Hao = Dysosma hispida（K. S. Hao）Chun ■

306631 Podophyllum hispidum K. S. Hao = Dysosma pleiantha（Hance）Woodson ■

306632 Podophyllum mairei Gagnep. = Dysosma aurantiocaulis（Hand.-Mazz.）Hu ■

306633 Podophyllum majoense Gagnep. = Dysosma aurantiocaulis（Hand.-Mazz.）Hu ■

306634 Podophyllum majoense Gagnep. = Dysosma majoensis（Gagnep.）T. S. Ying ■

306635 Podophyllum onzoi Hayata = Dysosma pleiantha（Hance）Woodson ■

306636 Podophyllum peltatum L.；盾叶鬼臼（八角莲，北美鬼臼，北美桃儿七，盾鬼臼，足叶草）；American Mandrake, Common May Apple, Devil's-apple, Duck's Foot, Ground Lemon, Hog Apple, Indian-apple, Mandrake, May Apple, Mayapple, May-apple, Raccoon-berry, Shield-leaved Duck's Foot, Wild Lemon, Wild Mandrake, Wild-mandrake ■☆

306637 Podophyllum peltatum L. f. aphyllum Plitt = Podophyllum peltatum L. ■☆

306638 Podophyllum peltatum L. f. biltmoreanum Steyerm. = Podophyllum peltatum L. ■☆

306639 Podophyllum peltatum L. f. deamii Raymond = Podophyllum peltatum L. ■☆

306640 Podophyllum peltatum L. f. polycarpum（Clute）Plitt = Podophyllum peltatum L. ■☆

306641 Podophyllum pleianthum Hance = Dysosma pleiantha（Hance）

Woodson ■

306642 Podophyllum pleianthum Hance var. album Masam. = Dysosma pleiantha（Hance）Woodson ■

306643 Podophyllum sikkimense Chatterjee et Mukerjee = Dysosma aurantiocaulis（Hand.-Mazz.）Hu ■

306644 Podophyllum sikkimense Chatterjee et Mukerjee = Sinopodophyllum hexandrum（Royle）T. S. Ying ■

306645 Podophyllum sikkimense Chatterjee et Mukerjee var. majus Chatterjee et Mukerjee = Dysosma aurantiocaulis（Hand.-Mazz.）Hu ■

306646 Podophyllum tonkinense Gagnep. = Dysosma diformis（Hemsl. et E. H. Wilson）T. H. Wang ex T. S. Ying ■

306647 Podophyllum triangulum Hand.-Mazz. = Dysosma diformis（Hemsl. et E. H. Wilson）T. H. Wang ex T. S. Ying ■

306648 Podophyllum veitchii Hemsl. et E. H. Wilson = Dysosma delavayi（Franch.）Hu ■

306649 Podophyllum veitchii Hemsl. et E. H. Wilson = Dysosma veitchii（Hemsl. et E. H. Wilson）S. H. Fu ex T. S. Ying ■

306650 Podophyllum versipelle Hance = Dysosma versipellis（Hance）M. Cheng ex T. S. Ying ■

306651 Podopogon Raf.（废弃属名）= Piptochaetium J. Presl（保留属名）■☆

306652 Podopogon Raf.（废弃属名）= Stipa L. ■

306653 Podopterus Bonpl.（1812）；刺蓼树属●☆

306654 Podopterus Humb. et Bonpl. = Podopterus Bonpl. ●☆

306655 Podopterus cordifolius Rose et Standl.；心叶刺蓼树●☆

306656 Podopterus mexicanus Humb. et Bonpl.；墨西哥刺蓼树●☆

306657 Podopterus paniculatus（Donn. Sm.）Roberty et Vautier；圆锥刺蓼树●☆

306658 Podoria Pers. = Boscia Lam. ex J. St.-Hil.（保留属名）■☆

306659 Podoria senegalensis Pers. = Boscia senegalensis（Pers.）Lam. ●☆

306660 Podortocarpus Lam. ex Pers. = Podoria Pers. ■☆

306661 Podorungia Baill.（1891）；足孩儿草属■☆

306662 Podorungia clandestina（Stapf）Benoist；足孩儿草■☆

306663 Podorungia decaryi Benoist = Podorungia humblotii Benoist ■☆

306664 Podorungia humblotii Benoist；洪布足孩儿草■☆

306665 Podorungia lantzei Baill.；兰兹足孩儿草■☆

306666 Podorungia serotina（Benoist）Benoist；迟花足孩儿草■☆

306667 Podosaemon Spreng. = Muhlenbergia Schreb.

306668 Podosaemon Spreng. = Podosemum Desv. ■

306669 Podosaemum Kunth = Podosaemon Sprang. ■

306670 Podosciadium A. Gray = Perideridia Rchb. ■☆

306671 Podosemum Desv. = Muhlenbergia Schreb. ■

306672 Podospadix Raf. = Anthurium Schott ■

306673 Podosperma Labill.（废弃属名）= Podotheca Cass.（保留属名）■☆

306674 Podospermaceae Dulac = Santalaceae R. Br.（保留科名）●■

306675 Podospermum DC.（保留属名）= Scorzonera L.

306676 Podospermum laciniatum（L.）DC. = Scorzonera laciniata L. ■☆

306677 Podospermum laciniatum DC. = Scorzonera laciniata L. ■☆

306678 Podospermum laciniatum DC. var. calcitrapiifolia？ = Scorzonera laciniata L. subsp. decumbens（Guss.）Greuter ■☆

306679 Podospermum laciniatum DC. var. intermedium（Guss.）Batt. = Scorzonera laciniata L. ■☆

306680 Podospermum laciniatum DC. var. octangulare（Willd.）Batt. = Scorzonera laciniata L. ■☆

306681 Podospermum laciniatum DC. var. songaricum Kar. et Kir. =

Scorzonera songorica（Kar. et Kir.）Lipsch. et Vassilcz. ■

306682　Podostachys Klotzsch ＝ Croton L. ●

306683　Podostaurus Jungh. ＝ Boenninghausenia Rchb. ex Meisn.（保留属名）●■

306684　Podostelma K. Schum.（1893）；足冠萝藦属☆

306685　Podostelma schimperi（Vatke）K. Schum. ＝ Pleurostelma schimperi（Vatke）Liede ■☆

306686　Podostelma schimperi K. Schum.；足冠萝藦☆

306687　Podostemaceae Rich. ex C. Agardh ＝ Podostemaceae Rich. ex Kunth（保留科名）■

306688　Podostemaceae Rich. ex Kunth（1816）（保留科名）；川苔草科；Riverweed Family ■

306689　Podostemma Greene ＝ Asclepias L. ■

306690　Podostemon Michx. ＝ Podostemum Michx. ■☆

306691　Podostemon ceratophyllum Michx. ＝ Podostemum ceratophyllum Michx. ■☆

306692　Podostemon griffithii Wall. ex Griff. ＝ Hydrobryum griffithii（Wall. ex Griff.）Tul. ■

306693　Podostemonaceae Rich. ex C. Agardh ＝ Podostemaceae Rich. ex Kunth（保留科名）■

306694　Podostemum Michx.（1803）；川苔草属；Riverweed ■☆

306695　Podostemum ceratophyllum Michx.；喙叶川苔草（川苔草）；Riverweed ■☆

306696　Podostemum griffithii Wall. ex Griff. ＝ Hydrobryum griffithii（Wall. ex Griff.）Tul. ■

306697　Podostemum thollonii Baill. ＝ Ledermanniella thollonii（Baill.）C. Cusset ■☆

306698　Podostigma Elliott ＝ Asclepias L. ■

306699　Podostima Raf. ＝ Breweria R. Br. ●☆

306700　Podostima Raf. ＝ Stylisma Raf. ●☆

306701　Podotheca Cass.（1822）（保留属名）；草苞鼠麴草属■☆

306702　Podotheca angustifolia（Labill.）Less.；狭叶草苞鼠麴草■☆

306703　Podotheca chrysantha（Steetz）Benth.；金花草苞鼠麴草■☆

306704　Podranea Sprague（1904）；非洲凌霄属（肖粉凌霄属）；Podranea ●

306705　Podranea brycei（N. E. Br.）Sprague；布氏非洲凌霄●☆

306706　Podranea ricasoliana（Tanfani）Sprague；非洲凌霄（肖粉凌霄）；Pink Trumpet Vine, Pink Trumpetvine, Ricason Podranea, Zimbabwe Creeper ●

306707　Podranea ricasoliana Sprague ＝ Podranea ricasoliana（Tanfani）Sprague ●

306708　Poecadenia Wittst. ＝ Poikadenia Elliott ●■

306709　Poecadenia Wittst. ＝ Psoralea L. ●■

306710　Poechia Endl. ＝ Psilotrichum Blume ●■

306711　Poechia Opiz ＝ Murraya J. König ex L.（保留属名）●

306712　Poechia Opiz ＝ Sicklera M. Roem. ●

306713　Poecilacanthus Post et Kuntze ＝ Poikilacanthus Lindau ●☆

306714　Poecilandra Tul.（1847）；小蕊金莲木属●☆

306715　Poecilandra retusa Tul.；小蕊金莲木●☆

306716　Poecilanthe Benth.（1860）；小花杂花豆属●☆

306717　Poecilanthe parviflora Benth.；小花杂花豆●☆

306718　Poecilla Post et Kuntze ＝ Jacaima Rendle ☆

306719　Poecilla Post et Kuntze ＝ Poicilla Griseb. ☆

306720　Poecilocalyx Bremek.（1940）；杂萼茜属■☆

306721　Poecilocalyx crystallinus N. Hallé；水晶杂萼茜■☆

306722　Poecilocalyx schumannii Bremek.；舒曼杂萼茜■☆

306723　Poecilocalyx setiflorus（R. D. Good）Bremek.；毛花杂萼茜■☆

306724　Poecilocalyx stipulosa（Hutch. et Dalziel）N. Hallé；托叶曼杂萼茜■☆

306725　Poecilocarpus Nevski ＝ Astragalus L. ●■

306726　Poecilochroma Miers ＝ Saracha Ruiz et Pav. ●☆

306727　Poecilocnemis Mart. ex Nees ＝ Geissomeria Lindl. ☆

306728　Poecilodermis Schott ＝ Brachychiton Schott et Endl. ●☆

306729　Poecilodermis Schott ＝ Sterculia L. ●

306730　Poecilodermis Schott et Endl. ＝ Brachychiton Schott et Endl. ●☆

306731　Poecilodermis Schott et Endl. ＝ Sterculia L. ●

306732　Poecilodermis populnea Schott et Endl. ＝ Brachychiton populneum（Schott et Endl.）R. Br. ●☆

306733　Poecilolepis Grau（1977）；葡菀属■☆

306734　Poecilolepis ficoidea（DC.）Grau；葡菀■☆

306735　Poecilolepis maritima（Bolus）Grau；沼泽葡菀■☆

306736　Poeciloneuron Bedd.（1865）；印度杂脉藤黄属（格脉树属）●

306737　Poeciloneuron indicum Bedd.；印度杂脉藤黄●☆

306738　Poeciloneuron pauciflorum Bedd.；疏花印度杂脉藤黄●☆

306739　Poecilospermum Post et Kuntze ＝ Poikilospermum Zipp. ex Miq. ●

306740　Poecilostachys Hack.（1884）；杂色穗草属■☆

306741　Poecilostachys alleizettei A. Camus ＝ Poecilostachys mainborondroensis A. Camus ■☆

306742　Poecilostachys ambositrensis A. Camus ＝ Poecilostachys bakeri（Schinz）C. E. Hubb. ■☆

306743　Poecilostachys bakeri（Schinz）C. E. Hubb.；贝克杂色穗草■☆

306744　Poecilostachys decaryana A. Camus ＝ Poecilostachys mainborondroensis A. Camus ■☆

306745　Poecilostachys flaccidula Stapf ex Rendle ＝ Poecilostachys oplismenoides（Hack.）Clayton ■☆

306746　Poecilostachys geminata Hack.；双杂色穗草■☆

306747　Poecilostachys hildebrandtii Hack.；希尔德杂色穗草■☆

306748　Poecilostachys humbertii A. Camus；亨伯特杂色穗草■☆

306749　Poecilostachys leandrii A. Camus ＝ Poecilostachys bakeri（Schinz）C. E. Hubb. ■☆

306750　Poecilostachys mainborondroensis A. Camus；梅恩杂色穗草■☆

306751　Poecilostachys manongarivensis A. Camus ＝ Poecilostachys geminata Hack. ■☆

306752　Poecilostachys marojejyensis A. Camus；马罗杂色穗草■☆

306753　Poecilostachys oplismenoides（Hack.）Clayton；求米草状杂色穗草■☆

306754　Poecilostachys perrieri A. Camus ＝ Poecilostachys bakeri（Schinz）C. E. Hubb. ■☆

306755　Poecilostachys viguieri A. Camus ＝ Poecilostachys bakeri（Schinz）C. E. Hubb. ■☆

306756　Poecilostemon Triana et Planch. ＝ Chrysochlamys Poepp. ●☆

306757　Poecilotriche Dulac ＝ Saussurea DC.（保留属名）●■

306758　Poeckia Benth. et Hook. f. ＝ Poechia Endl. ●■

306759　Poeckia Benth. et Hook. f. ＝ Psilotrichum Blume ●■

306760　Poederia Reuss ＝ Paederia L.（保留属名）●■

306761　Poederiopsis Rusby ＝ Manettia Mutis ex L.（保留属名）●■☆

306762　Poederiopsis Rusby ＝ Paederia L.（保留属名）●■

306763　Poelinitzia Uitewaal（1940）；红花松塔掌属（合片阿福花属）●■☆

306764　Poellnitzia rubriflora（L. Bolus）Uitewaal；红花松塔掌■☆

306765　Poellnitzia rubriflora（L. Bolus）Uitewaal ＝ Astroloba rubriflora（L. Bolus）G. F. Sm. et J. C. Manning ■☆

306766　Poellnitzia rubriflora（L. Bolus）Uitewaal var. jacobseniana（Poelln.）Uitewaal ＝ Astroloba rubriflora（L. Bolus）G. F. Sm. et J. C. Manning ■☆

306767　Poenosedum Holub　= Rhodiola L. ■

306768　Poeonia Crantz　= Paeonia L. ●■

306769　Poeppigia Bertero　= Rhaphithamnus Miers ●☆

306770　Poeppigia Bertero ex Férussac　= Rhaphithamnus Miers ●☆

306771　Poeppigia C. Presl(1830);珀高豆属■☆

306772　Poeppigia Kuntze ex Rchb.　= Tecophilaea Bertero ex Colla ■☆

306773　Poeppigia procera（Spreng.）C. Presl;珀高豆☆

306774　Poevrea Tul.　= Combretum Loefl.（保留属名）●

306775　Poevrea Tul.　= Pevraea Comm. ex Juss. ●

306776　Poga Pierre(1896);赤非红树属●☆

306777　Poga oleosa Pierre;赤非红树（油赤非红树）;Afo Nut, Inoi Nut, Inoy Nut ●☆

306778　Pogadelpha Raf.　= Sisyrinchium L. ■

306779　Pogalis Raf.　= Polygonum L.（保留属名）■●

306780　Pogenda Raf.　= Olea L. ●

306781　Poggea Gürke　= Poggea Gürke ex Warb. ●☆

306782　Poggea Gürke ex Warb.（1893）;波格木属●☆

306783　Poggea alata Gürke;具翅波格木●☆

306784　Poggea gossweileri Exell;戈斯波格木●☆

306785　Poggea kamerunensis Gilg　= Poggea alata Gürke ●☆

306786　Poggea klaineana Pierre ex Gilg　= Poggea alata Gürke ●☆

306787　Poggea longepedunculata Bamps;长梗波格木●☆

306788　Poggea ovata Sleumer　= Peterodendron ovatum（Sleumer）Sleumer ●☆

306789　Poggea stenura Gilg　= Poggea alata Gürke ●☆

306790　Poggendorffia H. Karst.　= Passiflora L. ●■

306791　Poggendorffia H. Karst.　= Tacsonia Juss. ●■

306792　Poggeophyton Pax　= Erythrococca Benth. ●☆

306793　Poggeophyton aculeatum Pax　= Erythrococca poggeophyton Prain ●☆

306794　Pogoblephis Raf.　= Gentianella Moench(保留属名)■

306795　Pogochilus Falc.　= Galeola Lour. ■

306796　Pogochloa S. Moore　= Gouinia E. Fourn. ex Benth. et Hook. f. ■☆

306797　Pogocybe Pierre　= Gleditsia L. ●

306798　Pogogyne Benth.（1834）;须柱草属■☆

306799　Pogogyne douglasii Benth.;须柱草■☆

306800　Pogogyne floribunda Jokerst;繁花须柱草■☆

306801　Pogogyne multiflora Benth.;多花须柱草■☆

306802　Pogogyne parviflora Benth.;小花须柱草■☆

306803　Pogoina B. Grant　= Pogonia Juss. ■

306804　Pogoina Griff. ex B. Grant　= Pogonia Juss. ■

306805　Pogomesia Raf.（废弃属名）= Tinantia Scheidw.（保留属名）■☆

306806　Pogonachne Bor(1949);总状须颖草属■☆

306807　Pogonachne raccmosa Bor;总状须颖草■☆

306808　Pogonanthera（G. Don）Spach　= Pogonanthera Blume ■☆

306809　Pogonanthera（G. Don）Spach　= Scaevola L.（保留属名）●■

306810　Pogonanthera Blume(1831);毛药野牡丹属■☆

306811　Pogonanthera H. W. Li et X. H. Guo　= Paraphlomis（Prain）Prain ●■

306812　Pogonanthera H. W. Li et X. H. Guo　= Sinopogonanthera H. W. Li ■

306813　Pogonanthera Spach　= Scaevola L.（保留属名）●■

306814　Pogonanthera cauropteris H. W. Li et X. H. Guo　= Sinopogonanthera cauropteris H. W. Li ■

306815　Pogonanthera intermedia（C. Y. Wu et H. W. Li）H. W. Li et X. H. Guo　= Sinopogonanthera intermedia（C. Y. Wu et H. W. Li）H. W. Li ■

306816　Pogonanthera pulverulenta Blume;毛药野牡丹■☆

306817　Pogonanthus Montrouz.　= Morinda L. ●■

306818　Pogonarthria Stapf ex Rendle　= Pogonarthria Stapf ■☆

306819　Pogonarthria Stapf(1900);镰穗草属■☆

306820　Pogonarthria brainii Stent　= Eragrostis brainii（Stent）Launert ■☆

306821　Pogonarthria falcata Rendle　= Pogonarthria squarrosa（Roem. et Schult.）Pilg. ■☆

306822　Pogonarthria fleckii（Hack.）Hack.;弗莱克镰穗草■☆

306823　Pogonarthria hackelii Chiov.　= Pogonarthria squarrosa（Roem. et Schult.）Pilg. ■☆

306824　Pogonarthria orthoclada Peter　= Pogonarthria squarrosa（Roem. et Schult.）Pilg. ■☆

306825　Pogonarthria refracta Launert;反折镰穗草■☆

306826　Pogonarthria squarrosa（Roem. et Schult.）Pilg.;粗鳞镰穗草■☆

306827　Pogonarthria tuberculata Pilg.　= Pogonarthria fleckii（Hack.）Hack. ■☆

306828　Pogonatherum P. Beauv.（1812）;金发草属（金丝茅属）;Goldenhairgrass, Pogonatherum ■

306829　Pogonatherum aureum（Bory）Roberty　= Eulalia aurea（Bory）Kunth ■

306830　Pogonatherum biaristatum S. L. Chen et G. Y. Sheng;二芒金发草;Twoawn Goldenhairgrass, Twoawn Pogonatherum ■

306831　Pogonatherum contortum Brongn.　= Pseudopogonatherum contortum（Brongn.）A. Camus ■

306832　Pogonatherum crinitum（Thunb.）Kunth;金丝草（笔毛草,笔尾草,笔须草,笔仔草,笔子草,鬼子草,猴毛草,胡毛草,黄毛草,金发草,金黄草,金丝茅,猫毛草,猫茅草,猫尾草,猫仔草,毛毛草,眉毛草,牛毛草,牛母草,牛尾草,墙头竹,水路草,竹蒿草,竹叶草）;Crinite Pogonatherum, Golden-hair Grass, Goldensilk Grass ■

306833　Pogonatherum huillense（Rendle）Roberty　= Homozeugos huillense（Rendle）Stapf ■☆

306834　Pogonatherum paniceum（Lam.）Hack.;金发草（笔茅草,笔仔草,猴毛草,胡毛草,黄毛草,吉祥草,金发竹,金黄草,金丝草,龙奶草,露水草,猫仔草,眉毛草,牛尾草,墙头草,墙头竹,蓑衣草,竹篙草,竹叶草）;Golden Pogonatherum, Goldenhairgrass ■

306835　Pogonatherum paniceum Alston　= Pogonatherum crinitum（Thunb.）Kunth ■

306836　Pogonatherum polystachyum（Willd.）Roem. et Schult.　= Pogonatherum paniceum（Lam.）Hack. ■

306837　Pogonatherum saccharoideum P. Beauv.　= Pogonatherum paniceum（Lam.）Hack. ■

306838　Pogonatherum saccharoideum P. Beauv. var. crinitum（Thunb.）F. N. Williams　= Pogonatherum crinitum（Thunb.）Kunth ■

306839　Pogonatherum saccharoideum P. Beauv. var. genuinum Hack.　= Pogonatherum paniceum（Lam.）Hack. ■

306840　Pogonatherum saccharoideum P. Beauv. var. monandrum（Roxb.）Hack.　= Pogonatherum crinitum（Thunb.）Kunth ■

306841　Pogonatherum villosum（Thunb.）Roberty　= Eulalia villosa（Thunb.）Nees ■☆

306842　Pogonatum Steud.　= Pogonatherum P. Beauv. ■

306843　Pogonella Salisb.　= Simethis Kunth(保留属名)■☆

306844　Pogonema Raf.　= Zephyranthes Herb.（保留属名）■

306845　Pogonetes Lindl.　= Scaevola L.（保留属名）●■

306846　Pogonia Andréws　= Myoporum Banks et Sol. ex G. Forst. ●

306847　Pogonia Juss.（1789）;朱兰属（须唇兰属）;Beard Flower, Beard-flower, Pogonia ■

306848　Pogonia abyssinica Chiov.　= Nervilia kotschyi（Rchb. f.）Schltr. ■☆

306849　Pogonia affinis Austin ex A. Gray ＝Isotria medeoloides (Pursh) Raf. ■☆

306850　Pogonia barklyana Rchb. f. ＝ Nervilia bicarinata (Blume) Schltr. ■☆

306851　Pogonia bicarinata Blume ＝Nervilia bicarinata (Blume) Schltr. ■☆

306852　Pogonia biflora Wight ＝Nervilia plicata (Andréws) Schltr. ■

306853　Pogonia bollei Rchb. f. ＝Nervilia simplex (Thouars) Schltr. ■☆

306854　Pogonia buchananii Rolfe ＝Nervilia shirensis (Rolfe) Schltr. ■☆

306855　Pogonia carinata (Roxb.) Lindl. ＝Nervilia aragoana Gaudich. ■

306856　Pogonia chariensis A. Chev. ＝Nervilia bicarinata (Blume) Schltr. ■☆

306857　Pogonia commersonii Blume ＝Nervilia bicarinata (Blume) Schltr. ■☆

306858　Pogonia crispata Blume ＝Nervilia crociformis (Zoll. et Moritzi) Seidenf. ■

306859　Pogonia crispata Blume ＝Nervilia simplex (Thouars) Schltr. ■☆

306860　Pogonia dallachyana F. Muell. ex Benth. ＝Nervilia plicata (Andréws) Schltr. ■

306861　Pogonia discolor (Blume) Blume ＝Nervilia plicata (Andréws) Schltr. ■

306862　Pogonia divaricata (L.) R. Br. ＝Cleistes divaricata (L.) Ames ■☆

306863　Pogonia djalonensis A. Chev. ＝Nervilia bicarinata (Blume) Schltr. ■☆

306864　Pogonia finetii A. Chev. ＝Nervilia simplex (Thouars) Schltr. ■☆

306865　Pogonia flabelliformis Lindl. ＝Nervilia aragoana Gaudich. ■

306866　Pogonia fordii Hance ＝Nervilia fordii (Hance) Schltr. ■

306867　Pogonia gammieana Hook. f. ＝Nervilia gammieana (Hook. f.) Pfitzer ■☆

306868　Pogonia ghindana Fiori ＝Nervilia bicarinata (Blume) Schltr. ■☆

306869　Pogonia gracilis Blume ＝Nervilia aragoana Gaudich. ■

306870　Pogonia japonica Rchb. f.；朱兰(降龙草,青蛇剑,蛇剑草,四月一枝花,斩龙剑,祖师箭)；Japan Pogonia,Japanese Pogonia ■

306871　Pogonia japonica Rchb. f. f. pallescens Tatew.；苍白朱兰■☆

306872　Pogonia japonica Rchb. f. var. minor Makino ＝Pogonia minor (Makino) Makino ■

306873　Pogonia kotschyi Rchb. f. ＝Nervilia kotschyi (Rchb. f.) Schltr. ■☆

306874　Pogonia kungii Ts. Tang et F. T. Wang ＝Pogonia japonica Rchb. f. ■

306875　Pogonia lanceolata Kraenzl. ＝Cremastra appendiculata (D. Don) Makino ■

306876　Pogonia lanceolata Kraenzl. ＝Cremastra appendiculata (D. Don) Makino var. variabilis (Blume) I. D. Lund ■

306877　Pogonia leguminosarum Moreau ＝Nervilia leguminosarum Jum. et H. Perrier ■☆

306878　Pogonia mackinnonii Duthie ＝Nervilia mackinnonii (Duthie) Schltr. ■

306879　Pogonia minor (Makino) Makino；小朱兰(小须唇兰)；Small Pogonia ■

306880　Pogonia minor (Makino) Makino f. pallescens (Nakai) Okuyama；苍白小朱兰■☆

306881　Pogonia nervilia Blume ＝Nervilia aragoana Gaudich. ■

306882　Pogonia ophioglossoides (L.) Juss. var. brachypogon Fernald ＝Pogonia ophioglossoides (L.) Ker Gawl. ■☆

306883　Pogonia ophioglossoides (L.) Ker Gawl.；美洲朱兰(瓶尔小草朱兰)；Adder's-mouth,American Pogonia,Rose Crested Orchid,Rose Pogonia,Snake Mouth,Snake-mouth,Snakemouth Orchid ■☆

306884　Pogonia ophioglossoides (L.) Ker Gawl. ＝Pogonia japonica Rchb. f. ■

306885　Pogonia ophioglossoides (L.) Ker Gawl. f. albiflora E. L. Rand et Redfield ＝Pogonia ophioglossoides (L.) Ker Gawl. ■☆

306886　Pogonia ophioglossoides (L.) Ker Gawl. var. brachypogon Fernald ＝Pogonia ophioglossoides (L.) Ker Gawl. ■☆

306887　Pogonia ophioglossoides (L.) Ker Gawl. var. japonica (Rchb. f.) Finet ＝Pogonia japonica Rchb. f. ■

306888　Pogonia parvula Schltr. ＝Pogonia japonica Rchb. f. ■

306889　Pogonia pleionoides Kraenzl. ＝Pleione pleionoides (Kraenzl. ex Diels) Braem et H. Mohr ■

306890　Pogonia pleionoides Kraenzl. ex Diels ＝Pleione pleionoides (Kraenzl. ex Diels) Braem et H. Mohr ■

306891　Pogonia plicata (Andréws) Lindl. ＝Nervilia plicata (Andréws) Schltr. ■

306892　Pogonia plicata Lindl. ＝Nervilia plicata (Andréws) Schltr. ■

306893　Pogonia prainiana King et Pantl. ＝Nervilia crociformis (Zoll. et Moritzi) Seidenf. ■

306894　Pogonia prainiana King et Pantl. ＝Nervilia simplex (Thouars) Schltr. ■☆

306895　Pogonia pudica Ames ＝Nervilia plicata (Andréws) Schltr. ■

306896　Pogonia pulchella Hook. f. ＝Nervilia plicata (Andréws) Schltr. ■

306897　Pogonia purpurata Rchb. f. et Sond. ＝Nervilia kotschyi (Rchb. f.) Schltr. var. purpurata (Rchb. f. et Sond.) Börge Pett. ■☆

306898　Pogonia purpurea Hayata ＝Nervilia plicata (Andréws) Schltr. var. purpurea (Hayata) S. S. Ying ■

306899　Pogonia purpurea Hayata ＝Nervilia plicata (Andréws) Schltr. ■

306900　Pogonia renschiana Rchb. f. ＝Nervilia renschiana (Rchb. f.) Schltr. ■☆

306901　Pogonia sakoae Moreau ＝Nervilia kotschyi (Rchb. f.) Schltr. ■☆

306902　Pogonia scottii Rchb. f. ＝Nervilia aragoana Gaudich. ■

306903　Pogonia shirensis Rolfe ＝Nervilia shirensis (Rolfe) Schltr. ■☆

306904　Pogonia similis Blume ＝Pogonia japonica Rchb. f. ■

306905　Pogonia simplex (Schltr.) Rchb. f. ＝Nervilia simplex (Thouars) Schltr. ■☆

306906　Pogonia taitoensis Hayata ＝Nervilia taitoensis (Hayata) Schltr. ■

306907　Pogonia thouarsii Blume ＝Nervilia simplex (Thouars) Schltr. ■☆

306908　Pogonia trianthophora (Sw.) Britton, Sterns et Poggenb. ＝Triphora trianthophora (Sw.) Rydb. ■☆

306909　Pogonia umbrosa Rchb. f. ＝Nervilia bicarinata (Blume) Schltr. ■☆

306910　Pogonia velutina E. C. Parish et Rchb. f. ＝Nervilia plicata (Andréws) Schltr. ■

306911　Pogonia verticillata (Muhl. ex Willd.) Nutt. ＝Isotria verticillata (Muhl. ex Willd.) Raf. ■☆

306912　Pogonia viridiflava Rchb. f. ＝Nervilia bicarinata (Blume) Schltr. ■☆

306913　Pogonia yunnanensis Finet；云南朱兰；Yunnan Pogonia ■

306914　Pogoniopsis Rchb. f. (1881)；拟朱兰属■☆

306915　Pogoniopsis nidus-avis Rchb. f.；拟朱兰■☆

306916　Pogonitis Rchb. ＝Anthyllis L. ■☆

306917　Pogonochloa C. E. Hubb. (1940)；热非须毛草属■☆

306918　Pogonochloa greenwayi C. E. Hubb.；热非须毛草■☆

306919　Pogonolepis Steetz ＝Angianthus J. C. Wendl. (保留属名)■●☆

306920　Pogonolepis Steetz(1845)；须鳞鼠麴草属■☆

306921　Pogonolepis stricta Steetz；须鳞鼠麴草■☆

306922　Pogonolobus F. Muell. ＝Coelospermum Blume ●

306923　Pogononeura Napper(1963);东非双花草属■☆

306924　Pogononeura biflora Napper;东非双花草■☆

306925　Pogonophora Miers ex Benth. (1854);非洲毛梗大戟属■☆

306926　Pogonophora africana Letouzey ＝ Pogonophora letouzeyi Feuillet ■☆

306927　Pogonophora letouzeyi Feuillet;非洲毛梗大戟■☆

306928　Pogonophyllum Didr. ＝ Micrandra Benth. (保留属名)●☆

306929　Pogonopsis C. Presl ＝ Pogonatherum P. Beauv. ■

306930　Pogonopsis J. Presl et C. Presl ＝ Pogonatherum P. Beauv. ■

306931　Pogonopus Klotzsch(1854);髯毛花属■☆

306932　Pogonopus febrifugus (Wedd.) Benth. et Hook. f. ex Hieron.; 解热髯毛花☆

306933　Pogonopus spectosus (Jacq.) Schum.;美洲髯毛花☆

306934　Pogonopus tubulosus (A. Rich. ex DC.) K. Schum.;管形髯毛花■☆

306935　Pogonorhynchus Crueg. ＝ Miconia Ruiz et Pav. (保留属名)●☆

306936　Pogonorrhinum Betsche ＝ Linaria Mill. ■

306937　Pogonorrhinum Betsche ＝ Nanorrhinum Betsche ●■☆

306938　Pogonorrhinum heterophyllum (Schousb.) Betsche ＝ Kickxia heterophylla (Schousb.) Dandy ■☆

306939　Pogonorrhinum somalense (Vatke) Betsche ＝ Nanorrhinum ramosissimum (Wall.) Betsche ●☆

306940　Pogonosperraum Hochst. ＝ Monechma Hochst. ■●☆

306941　Pogonostemon Hassk. ＝ Pogostemon Desf. ●■

306942　Pogonostigma Boiss. ＝ Tephrosia Pers. (保留属名)●■

306943　Pogonostigma arabicum Boiss. ＝ Tephrosia arabica (Boiss.) Martelli ●☆

306944　Pogonostigma nubicum Boiss. ＝ Tephrosia nubica (Boiss.) Baker ■☆

306945　Pogonostylis Bertol. ＝ Fimbristylis Vahl(保留属名)■

306946　Pogonotium J. Dransf. (1980);无鞭藤属(异苞藤属,鬃毛藤属)●☆

306947　Pogonotium divaricatum J. Dransf.;无鞭藤●☆

306948　Pogonotrophe Miq. ＝ Ficus L. ●

306949　Pogonotrophe foveolata Wall. ex Miq. ＝ Ficus sarmentosa Buch. - Ham. ex Sm. ●

306950　Pogonotrophe pubigera Wall. ＝ Ficus pubigera (Wall. ex Miq.) Miq. ●

306951　Pogonotrophe pubigera Wall. ex Miq. ＝ Ficus pubigera (Wall. ex Miq.) Miq. ●

306952　Pogonotrophe reticulata Miq. ＝ Ficus sarmentosa Buch. -Ham. ex Sm. ●

306953　Pogonura DC. ex Lindl. ＝ Perezia Lag. ■☆

306954　Pogopetalum Benth. ＝ Emmotum Desv. ex Ham. ●☆

306955　Pogospermum Brongn. ＝ Catopsis Griseb. ■☆

306956　Pogostemon Desf. (1815);刺蕊草属(广藿香属,水珍珠菜属);Pogostemon, Spinestemon ●■

306957　Pogostemon amarantoides Benth.;苋状刺蕊草■☆

306958　Pogostemon aquaticus (C. H. Wright) Press;水刺蕊草■☆

306959　Pogostemon atropurpureus Benth.;暗紫刺蕊草■☆

306960　Pogostemon auricularius (L.) Hassk.;水珍珠菜(耳叶刺蕊草,耳状水蜡烛,老鼠癀,毛蛇草,毛射草,毛水珍珠菜,毛水珍珠草,密花节节红,牛触臭,蛇尾草,水蜡烛);Auriculate Dysophylla, Auriculate Pogostemon, Auriculate Watercandle, Water Loosestrife ■

306961　Pogostemon benthaminanus Kuntze ＝ Dysophylla stellata (Lour.) Benth. ■

306962　Pogostemon brachystachys Benth.;短序刺蕊草(短穗花序刺蕊草)■☆

306963　Pogostemon brevicorollus Y. Z. Sun;短冠刺蕊草(马鹿菜,水大靛);Shortcorolla Pogostemon ●■

306964　Pogostemon cablin (Blanco) Benth.;广藿香(刺蕊草,刺香,到手香,方藿香,广霍香,藿香,枝香);Cabin Pogostemon, Cabin Potchouli, Patchouly, Potchouli ●■

306965　Pogostemon championii Prain;短穗刺蕊草;Shortspike Pogostemon, Shortspike Spinestemon ●

306966　Pogostemon chinensis C. Y. Wu et Y. C. Huang;长苞刺蕊草;China Spinestemon, Chinese Pogostemon ■

306967　Pogostemon cruciatum (Benth.) Kuntze ＝ Dysophylla cruciata Benth. ■

306968　Pogostemon cyprianii (Pavol.) Pamp. ＝ Elsholtzia cypriani (Pavol.) S. Chow ex Y. C. Hsu ■

306969　Pogostemon dielsianus Dunn;狭叶刺蕊草;Narrowleaf Pogostemon, Narrowleaf Spinestemon ●

306970　Pogostemon elsholtzioides Benth.;香薷状刺蕊草■☆

306971　Pogostemon esquirolii (H. Lév.) C. Y. Wu et Y. C. Huang;膜叶刺蕊草(鸡骨头菜,野蓝靛);Esquirol Pogostemon, Esquirol Spinestemon ●

306972　Pogostemon esquirolii (H. Lév.) C. Y. Wu et Y. C. Huang var. tsingpingensis C. Y. Wu et Y. C. Huang;金平膜叶刺蕊草(金平刺蕊草);Jinping Esquirol Pogostemon, Jinping Esquirol Spinestemon ●■

306973　Pogostemon falcatus (C. Y. Wu) C. Y. Wu et H. W. Li;镰叶水珍珠菜;Dysophylla, Falcate Pogostemon, Falcate Watercandle, Sickleleaf Spinestemon ■

306974　Pogostemon formosanus Oliv.;台湾刺蕊草(尖尾凤,节节红,台湾广藿香);Taiwan Pogostemon, Taiwan Spinestemon ■

306975　Pogostemon fraternus Miq. ＝ Pogostemon menthoides Blume ■

306976　Pogostemon fraternus Miq. var. nigrescens (Dunn) Kudo ＝ Pogostemon nigrescens Dunn ■

306977　Pogostemon gardneri Hook. f.;加尔得纳刺蕊草■☆

306978　Pogostemon glaber Benth.;刺蕊草(鸡挂骨草,鸡排骨草,野靛,走马胎);Glabrous Pogostemon, Glabrous Spinestemon ■

306979　Pogostemon glaber Benth. ＝ Pogostemon esquirolii (H. Lév.) C. Y. Wu et Y. C. Huang ●■

306980　Pogostemon griffithii Prain;长柱刺蕊草;Griffith Spinestemon, Longstyle Pogostemon ■

306981　Pogostemon griffithii Prain var. latifolius C. Y. Wu et Y. C. Huang;宽叶长柱刺蕊草(宽叶刺蕊草);Broadleaf Longstyle Pogostemon ■

306982　Pogostemon heyneanus Benth.;印度刺蕊草;Indian Plant ■☆

306983　Pogostemon heyneanus Benth. ＝ Pogostemon patchouly Pellet. ■☆

306984　Pogostemon hirsutus Benth.;硬毛刺蕊草■☆

306985　Pogostemon hispidocalyx C. Y. Wu et Y. C. Huang;刚毛萼刺蕊草;Hispidcalyx Pogostemon, Setacalyx Spinestemon ■

306986　Pogostemon japonicus Benth. et Hook. f. ＝ Comanthosphace japonica (Miq.) S. Moore ex Hook. f. ■

306987　Pogostemon javanicus Backer ex Adelb. ＝ Pogostemon cablin (Blanco) Benth. ●■

306988　Pogostemon menthoides Blume;小刺蕊草;Mint-like Pogostemon, Mint-like Spinestemon ■

306989　Pogostemon mollis Benth.;毛刺蕊草■☆

306990　Pogostemon nigrescens Dunn;黑刺蕊草(臭野芝麻,紫花一柱香);Black Pogostemon, Black Spinestemon ■

306991　Pogostemon paludosus Benth.;沼泽毛刺蕊草■☆

306992　Pogostemon paniculatus Benth.;圆锥花序刺蕊草■☆

306993　Pogostemon parviflorus Benth. ＝ Pogostemon championii Prain ●

306994 Pogostemon patchouly Pellet.；印度广藿香；Dilem, Heyne Patchouli, Patchouli, Patchouli Plant ■☆

306995 Pogostemon patchouly Pellet. = Pogostemon cablin (Blanco) Benth. ●■

306996 Pogostemon patchouly Pellet. var. suavis Hook. f. = Pogostemon cablin (Blanco) Benth. ●■

306997 Pogostemon plectranthoides Benth.；香茶菜状刺蕊草■☆

306998 Pogostemon purpurascens Benth.；紫刺蕊草■☆

306999 Pogostemon reflexus Benth.；反折刺蕊草■☆

307000 Pogostemon rogersii N. E. Br.；罗杰斯刺蕊草■☆

307001 Pogostemon rotundatus Benth.；圆刺蕊草■☆

307002 Pogostemon rupestris Benth.；岩石刺蕊草■☆

307003 Pogostemon septentrionalis C. Y. Wu et Y. C. Huang；北刺蕊草；Northern Pogostemon, Northern Spinestemon ●■

307004 Pogostemon speciosus Benth.；美丽刺蕊草■☆

307005 Pogostemon stellatus (Lour.) Kuntze = Dysophylla stellata (Lour.) Benth. ■

307006 Pogostemon strigosus Benth.；粗伏毛刺蕊草■☆

307007 Pogostemon tisserantii (Pellegr.) J. -P. Lebrun et Stork；蒂斯朗特刺蕊草■☆

307008 Pogostemon tuberculosus Benth.；小瘤刺蕊草■☆

307009 Pogostemon verticillatus Miq. = Dysophylla stellata (Lour.) Benth. ■

307010 Pogostemon vestitus Benth.；被毛刺蕊草■☆

307011 Pogostemon villosus Benth.；长柔毛刺蕊草■☆

307012 Pogostemon wightii Benth.；威特刺蕊草■☆

307013 Pogostemon xanthiifolius C. Y. Wu et Y. C. Huang；苍耳叶刺蕊草；Cockleburleaf Pogostemon, Cockleburleaf Spinestemon ■

307014 Pogostemon yatabeanus (Makino) Press = Dysophylla yatabeana Makino ■

307015 Pogostoma Schrad. = Capraria L. ■☆

307016 Pohlana Leandro = Pohlana Mart. et Nees ●☆

307017 Pohlana Leandro = Zanthoxylum L. ●

307018 Pohlana Mart. et Nees = Zanthoxylum L. ●

307019 Pohlidium Davidse, Soderstr. et R. P. Ellis(1986)；波尔禾属（拟叶柄草属）■☆

307020 Pohlidium petiolatum Davidse, Soderstr. et R. P. Ellis；波尔禾（拟叶柄草）■☆

307021 Pohliella Engl. (1926)；波尔苔草属■☆

307022 Pohliella flabellata G. Taylor = Saxicolella flabellata (G. Taylor) C. Cusset ■☆

307023 Pohliella laciniata Engl.；波尔苔草■☆

307024 Pohliella laciniata Engl. = Saxicolella laciniata (Engl.) C. Cusset ■☆

307025 Poicilla Griseb. = Jacaima Rendle ☆

307026 Poicillopsis Schltr. = Poicillopsis Schltr. ex Rendle ■☆

307027 Poicillopsis Schltr. et Rendle = Poicillopsis Schltr. ex Rendle ■☆

307028 Poicillopsis Schltr. ex Rendle(1936)；拟亚卡萝藦属■☆

307029 Poicillopsis acuminata Schltr.；尖拟亚卡萝藦■☆

307030 Poicillopsis mollis Schltr.；软拟亚卡萝藦■☆

307031 Poicillopsis oblongata (Griseb.) Schltr.；矩圆拟亚卡萝藦■☆

307032 Poicillopsis ovatifolia Schltr.；卵叶拟亚卡萝藦■☆

307033 Poidium Nees = Poa L. ■

307034 Poikadenia Elliott = Psoralea L. ●■

307035 Poikilacanthus Lindau(1895)；杂刺爵床属●☆

307036 Poikilacanthus flexuosus (Nees) Lindau；杂刺爵床●☆

307037 Poikilacanthus glandulosa (Nees) Ariza；多腺杂刺爵床●☆

307038 Poikilogyne Baker f. (1917)；杂蕊野牡丹属●☆

307039 Poikilogyne arfakensis Baker f.；杂蕊野牡丹●☆

307040 Poikilospermum Zipp. ex Miq. (1864)；锥头麻属；Awlnettle, Poikilospermum ●

307041 Poikilospermum acuminatum (Trécul) Merr.；锥头麻（锐叶疥意奇罗麻）●

307042 Poikilospermum lanceolatum (Trécul) Merr.；毛叶锥头麻；Hairleaf Awlnettle, Lanceolate Poikilospermum ●

307043 Poikilospermum naucleiflorum (Roxb. ex Lindl.) Chew；大序锥头麻●

307044 Poikilospermum sinense (C. H. Wright) Merr. = Poikilospermum suaveolens (Blume) Merr. ●

307045 Poikilospermum suaveolens (Blume) Merr.；香甜锥头麻（锥头麻）；Awlnettle, Fragrant Poikilospermum ●

307046 Poilanedora Gagnep. (1948)；五苞山柑属■☆

307047 Poilania Gagnep. = Epaltes Cass. ■

307048 Poilania laggeroides Gagnep. = Epaltes divaricata (L.) Cass. ■

307049 Poilaniella Gagnep. (1925)；脆刺木属（博兰木属）●

307050 Poilaniella fragilis Gagnep.；脆刺木●

307051 Poilannammia C. Hansen. (1988)；博伊野牡丹属●☆

307052 Poilannammia allomorphioidea C. Hansen；博伊野牡丹●☆

307053 Poincettia Klotzsch et Garcke = Euphorbia L. ●■

307054 Poincettia Klotzsch et Garcke = Poinsettia Graham ●■

307055 Poinciana L. (1753)；金凤花属；Poinciana ●

307056 Poinciana L. = Caesalpinia L. ●

307057 Poinciana Tourn. ex L. = Poinciana L. ●

307058 Poinciana adansonioides R. Vig. = Delonix floribunda (Baill.) Capuron ●☆

307059 Poinciana baccal Chiov. = Delonix baccal (Chiov.) Baker f. ●☆

307060 Poinciana boiviniana Baill. = Delonix boiviniana (Baill.) Capuron ●☆

307061 Poinciana brachycarpa R. Vig. = Delonix brachycarpa (R. Vig.) Capuron ●☆

307062 Poinciana coriaria Jacq. = Caesalpinia coriaria (Jacq.) Willd. ex Kunth ●

307063 Poinciana decaryi R. Vig. = Delonix decaryi (R. Vig.) Capuron ●☆

307064 Poinciana elata L. = Delonix elata (L.) Gamble ●☆

307065 Poinciana gilliesii Hook. = Caesalpinia gilliesii (Wall. ex Hook.) Benth. ●☆

307066 Poinciana gilliesii Wall. ex Hook. = Caesalpinia gilliesii (Wall. ex Hook.) D. Dietr. ●☆

307067 Poinciana leucantha R. Vig. = Delonix leucantha (R. Vig.) Du Puy, Phillipson et R. Rabev. ●☆

307068 Poinciana pulcherrima L. = Caesalpinia pulcherrima (L.) Sw. ●

307069 Poinciana regia Bojer = Delonix regia (Bojer ex Hook.) Raf. ●

307070 Poinciana regia Bojer ex Hook. = Delonix regia (Bojer ex Hook.) Raf. ●

307071 Poinciana roxburghii G. Don = Peltophorum pterocarpum (DC.) Backer ex K. Heyne ●

307072 Poinciana spinosa Molina = Caesalpinia spinosa (Molina) Kuntze ●☆

307073 Poinciana tomentosa R. Vig. = Delonix tomentosa (R. Vig.) Capuron ●☆

307074 Poincianella Britton et Rose = Caesalpinia L. ●

307075 Poinsettia Graham = Euphorbia L. ●■

307076 Poinsettia Graham(1836)；猩猩木属（一品红属）；Poinsettia ●■

307077 Poinsettia cyathophora (Murray) Bartl. = Euphorbia

cyathophora Murray ■

307078　Poinsettia cyathophora（Murray）Klotzsch et Garcke ＝ Euphorbia cyathophora Murray ■

307079　Poinsettia cyathophora（Murray）Klotzsch et Garcke var. graminifolia（Michx.）Mohlenbr. ＝ Euphorbia cyathophora Murray ■

307080　Poinsettia dentata（Michx.）Klotzsch et Garcke ＝ Euphorbia dentata Michx. ■

307081　Poinsettia dentata（Michx.）Klotzsch et Garcke var. cuphosperma（Engelm.）Mohlenbr. ＝ Euphorbia dentata Michx. ■

307082　Poinsettia geniculata（Ortega）Klotzsch et Garcke ＝ Euphorbia cyathophora Murray ■

307083　Poinsettia geniculata（Ortega）Klotzsch et Garcke ＝ Euphorbia heterophylla L. ■

307084　Poinsettia heterophylla（L.）Klotzsch et Garcke ＝ Euphorbia cyathophora Murray ■

307085　Poinsettia heterophylla（L.）Klotzsch et Garcke ＝ Euphorbia heterophylla L. ■

307086　Poinsettia pulcherrima（Willd. ex Klotzsch）Graham ＝ Euphorbia pulcherrima Willd. ●

307087　Poiretia Cav. ＝ Sprengelia Sm. ●☆

307088　Poiretia J. F. Gmel.（废弃属名）＝ Houstonia L. ■☆

307089　Poiretia J. F. Gmel.（废弃属名）＝ Poiretia Vent.（保留属名）●■☆

307090　Poiretia Sm. ＝ Hovea R. Br. ex W. T. Aiton ●■☆

307091　Poiretia Vent.（1807）（保留属名）；普瓦豆属●■☆

307092　Poiretia elegans Cl. Müll.；雅致普瓦豆■☆

307093　Poiretia latifolia Vogel；宽叶普瓦豆●☆

307094　Poiretia longipes Harms；长梗普瓦豆●☆

307095　Poissonella Pierre ＝ Lucuma Molina ●

307096　Poissonella Pierre ＝ Pouteria Aubl. ●

307097　Poissonia Baill. ＝ Coursetia DC. ●☆

307098　Poitaea DC. ＝ Poitea Vent. ●☆

307099　Poitea Vent.（1807）；加勒比普豆属●☆

307100　Poitea carinalis（Griseb.）Lavin；加勒比普豆●☆

307101　Poivrea Comm. ex DC. ＝ Combretum Loefl.（保留属名）●

307102　Poivrea Comm. ex Thouars ＝ Combretum Loefl.（保留属名）●

307103　Poivrea aculeata（Vent.）DC. ＝ Combretum aculeatum Vent. ●☆

307104　Poivrea bracteosa Hochst. ＝ Combretum bracteosum（Hochst.）Brandis ●☆

307105　Poivrea coccinea DC. ＝ Combretum coccineum（Sonn.）Lam. ●☆

307106　Poivrea comosa（G. Don）Walp. ＝ Combretum comosum G. Don ●☆

307107　Poivrea conferta Benth. ＝ Combretum confertum（Benth.）M. A. Lawson ●☆

307108　Poivrea constricta Benth. ＝ Combretum constrictum（Benth.）M. A. Lawson ●☆

307109　Poivrea glutinosa Klotzsch ＝ Combretum mossambicense（Klotzsch）Engl. ●☆

307110　Poivrea hartmanniana Schweinf. ＝ Combretum aculeatum Vent. ●☆

307111　Poivrea mossambicensis Klotzsch ＝ Combretum mossambicense（Klotzsch）Engl. ●☆

307112　Poivrea ovalis（G. Don）Walp. ＝ Combretum aculeatum Vent. ●☆

307113　Poivrea pilosa（Roxb.）Wight et Arn. ＝ Combretum pilosum Roxb. ●

307114　Poivrea roxburghii DC. ＝ Combretum roxburghii Spreng. ●

307115　Poivrea senensis Klotzsch ＝ Combretum mossambicense（Klotzsch）Engl. ●☆

307116　Poivrea sericea Walp. ＝ Getonia floribunda Roxb. ●

307117　Poivrea squamosa（Roxb. ex G. Don）Walp. ＝ Combretum

punctatum Blume subsp. squamosum（Roxb. ex G. Don）Exell ●

307118　Pojarkovia Askerova（1984）；白蟹甲属■☆

307119　Pojarkovia stenocephala（Boiss.）Askerova；白蟹甲■☆

307120　Pokornya Montrouz. ＝ Lumnitzera Willd. ●

307121　Pokornya ettingshausenii Montrouz. ＝ Lumnitzera racemosa Willd. ●

307122　Polakia Stapf ＝ Salvia L. ●■

307123　Polakiastrum Nakai ＝ Salvia L. ●■

307124　Polakiastrum Nakai（1917）；日本鼠尾草属■☆

307125　Polakiastrum longipes Nakai；日本鼠尾草■☆

307126　Polakiastrum longipes Nakai ＝ Salvia japonica Thunb. f. polakioides（Honda）T. Yamaz. ■☆

307127　Polakowskia Pittier ＝ Sechium P. Browne（保留属名）■

307128　Polakowskia Pittier（1910）；肖佛手瓜属■☆

307129　Polakowskia tacaco Pittier；肖佛手瓜；Tacaco ■☆

307130　Polameia Rchb. ＝ Potameia Thouars ●☆

307131　Polanina Raf. ＝ Polanisia Raf. ■

307132　Polanisia Raf.（1819）；臭矢菜属（黄花菜属）；Clammyweed ■

307133　Polanisia augustinensis Hochr. ＝ Cleome augustinensis（Hochr.）Briq. ■☆

307134　Polanisia bicolor（Pax）Pax ＝ Cleome oxyphylla Burch. ■☆

307135　Polanisia bororensis（Klotzsch）Gilg ＝ Cleome bororensis（Klotzsch）Oliv. ■☆

307136　Polanisia carnosa（Pax）Pax ＝ Cleome carnosa（Pax）Gilg et Gilg-Ben. ■☆

307137　Polanisia diandra（Burch.）T. Durand et Schinz ＝ Cleome angustifolia Forssk. var. diandra（Burch.）Kers ■☆

307138　Polanisia dianthera DC. ＝ Cleome angustifolia Forssk. var. diandra（Burch.）Kers ■☆

307139　Polanisia dianthera DC. var. delagoensis（Kuntze）Schinz et Junod ＝ Cleome angustifolia Forssk. subsp. petersiana（Klotzsch）Kers ■☆

307140　Polanisia didynama（Hochst. ex Oliv.）T. Durand et Schinz ＝ Cleome angustifolia Forssk. ■☆

307141　Polanisia dodecandra（L.）DC.；糙籽臭矢菜（十二蕊臭矢菜）；Clammyweed，Polanisia，Rough-seed Clammy-weed ■☆

307142　Polanisia dodecandra（L.）DC. subsp. dodecandra var. trachysperma（Torr. et A. Gray）H. H. Iltis ＝ Polanisia dodecandra（L.）DC. ■☆

307143　Polanisia dodecandra（L.）DC. subsp. trachysperma（Torr. et A. Gray）H. H. Iltis；大糙籽臭矢菜；Large Clammy-weed ■☆

307144　Polanisia dodecandra（L.）DC. var. trachysperma（Torr. et A. Gray）H. H. Iltis ＝ Polanisia dodecandra（L.）DC. subsp. trachysperma（Torr. et A. Gray）H. H. Iltis ■☆

307145　Polanisia graveolens Raf. ＝ Polanisia dodecandra（L.）DC. ■☆

307146　Polanisia hirta（Klotzsch）Pax ＝ Cleome hirta（Klotzsch）Oliv. ■☆

307147　Polanisia icosandra（L.）Wight et Arn. ＝ Cleome viscosa L. ■

307148　Polanisia icosandra Wight et Arn. f. deglabrata（Backer）Backer ＝ Arivela viscosa（L.）Raf. var. deglabrata（Backer）M. L. Zhang et G. C. Tucker ■

307149　Polanisia jamesii（Torr. et A. Gray）H. H. Iltis；詹姆斯臭矢菜；James' Clammy-weed，James' Cristatella ■☆

307150　Polanisia kalachariensis Schinz ＝ Cleome kalachariensis（Schinz）Gilg et Gilg-Ben. ■☆

307151　Polanisia kelleriana Schinz ＝ Cleome kelleriana（Schinz）Gilg et Gilg-Ben. ■☆

307152　Polanisia linearifolia Stephens ＝ Cleome semitetranda Sond. ■☆

307153　Polanisia luederitziana（Schinz）Schinz ＝ Cleome foliosa Hook. f. ■☆

307154　Polanisia lutea Sond. ＝ Cleome foliosa Hook. f. var. lutea（Sond.）Codd et Kers ■☆

307155　Polanisia maculata Sond. ＝ Cleome maculata（Sond.）Szyszyl. ■☆

307156　Polanisia maximiliani Wawra ＝ Cleome foliosa Hook. f. ■☆

307157　Polanisia orthocarpa Webb ＝ Cleome viscosa L. ■

307158　Polanisia oxyphylla（Burch.）DC. ＝ Cleome oxyphylla Burch. ■☆

307159　Polanisia paxii Schinz ＝ Cleome paxii（Schinz）Gilg et Gilg-Ben. ■☆

307160　Polanisia petersiana（Klotzsch）Pax ＝ Cleome angustifolia Forssk. subsp. petersiana（Klotzsch）Kers ■☆

307161　Polanisia seretii De Wild. ＝ Cleome polyanthera Schweinf. et Gilg ■☆

307162　Polanisia strigosa Bojer ＝ Cleome strigosa（Bojer）Oliv. ■☆

307163　Polanisia suffruticosa（Schinz）Pax ＝ Cleome suffruticosa Schinz ●☆

307164　Polanisia trachysperma Torr. et A. Gray ＝ Polanisia dodecandra（L.）DC. subsp. trachysperma（Torr. et A. Gray）H. H. Iltis ■☆

307165　Polanisia triphylla Conrath ＝ Cleome conrathii Burtt Davy ■☆

307166　Polanisia viscosa（L.）DC. ＝ Cleome viscosa L. ■

307167　Polanisia viscosa（L.）DC. var. deglabrata Backer ＝ Arivela viscosa（L.）Raf. var. deglabrata（Backer）M. L. Zhang et G. C. Tucker ■

307168　Polanysia Raf. ＝ Polanisia Raf. ■

307169　Polaskia Backeb.（1949）;雷神阁属（雷神角柱属）●☆

307170　Polaskia chichipe（Rol. -Goss.）Backeb. ;雷神阁（角鸾凤,雷神柱）;Chichibe,Chichipe,Chichituna ●☆

307171　Polathera Raf. ＝ Polatherus Raf. ■

307172　Polatherus Raf. ＝ Gaillardia Foug. ■

307173　Polemannia Bergius ex Schltdl. ＝ Dipcadi Medik. ■☆

307174　Polemannia Eckl. et Zeyh.（1837）(保留属名);波尔曼草属 ■☆

307175　Polemannia K. Bergius ex Schltdl.（废弃属名）＝ Polemannia Eckl. et Zeyh.（保留属名）■☆

307176　Polemannia grossulariifolia Eckl. et Zeyh. ;波尔曼草 ■☆

307177　Polemannia marlothii H. Wolff ＝ Polemanniopsis marlothii（H. Wolff）B. L. Burtt ■☆

307178　Polemannia montana Schltr. et H. Wolff;山地波尔曼草 ■☆

307179　Polemannia simplicior Hilliard et B. L. Burtt;简单波尔曼草 ■☆

307180　Polemannia verticillata Sond. ＝ Anginon verticillatum（Sond.）B. L. Burtt ■☆

307181　Polemanniopsis B. L. Burtt(1989);拟波尔曼草属 ■☆

307182　Polemanniopsis marlothii（H. Wolff）B. L. Burtt;拟波尔曼草 ■☆

307183　Polembrium Steud. ＝ Polembryum A. Juss. ●☆

307184　Polembryon Benth. et Hook. f. ＝ Polembryum A. Juss. ●☆

307185　Polembryum A. Juss. ＝ Esenbeckia Kunth ●☆

307186　Polemoniaceae Juss.（1789）(保留科名);花荵科;Jacob's-ladder Family,Phlox Family,Polemonium Family ●■

307187　Polemoniella A. Heller ＝ Polemonium L. ■

307188　Polemoniella A. Heller(1904);小花荵属 ■☆

307189　Polemoniella micrantha（Benth.）A. Heller;小花荵 ■☆

307190　Polemonium L.（1753）;花荵属;Greek Valerian, Jacob's Ladder,Jacob's-ladder,Moss Pink,Polemonium ■

307191　Polemonium acutiflorum Willd. ex Roem. et Schult. ＝ Polemonium caeruleum L. subsp. campanulatum Th. Fr. ■☆

307192　Polemonium acutiflorum Willd. ex Roem. et Schult. ＝ Polemonium caeruleum L. var. acutiflorum（Willd. ex Roem. et

Schult.）Ledeb. ■

307193　Polemonium acutiflorum Willd. ex Roem. et Schult. var. laxiflorum（Regel）Ohwi ＝ Polemonium caeruleum L. subsp. laxiflorum（Regel）Koji Ito ■☆

307194　Polemonium acutiflorum Willd. ex Roem. et Schult. var. nipponicum（Kitam.）Ohwi ＝ Polemonium caeruleum L. subsp. yezoense（Miyabe et Kudo）H. Hara var. nipponicum（Kitam.）Koji Ito ■☆

307195　Polemonium boreale Adams;北方花荵;Jacob's Ladder ■☆

307196　Polemonium boreale Adams subsp. hinganicum P. H. Huang et S. Y. Li;兴安花荵 ■

307197　Polemonium caeruleum L. ;花荵(电灯花,手参,新疆花荵,穴菜,鱼翅菜);Blue Jackets, Charity, Common Polemonium, Gilliflower, Greek Valerian, Greek-valerian Polemonium, Jacobo-ladder,Jacob's Ladder,Jacob's Walking Stick,Jacob's Walking-stick,Jacob's-ladder,Joseph's Walking Stick,Joseph's Walkingstick,Ladder To Heaven, Ladder-to-Heaven, Makebate, Polemonium, Poverty, Road-to-Heaven,Valerian,Western Polemonium ■

307198　Polemonium caeruleum L. subsp. campanulatum Th. Fr. ;风铃草状花荵 ■☆

307199　Polemonium caeruleum L. subsp. campanulatum Th. Fr. var. humile（Willd.）Herder ＝ Polemonium boreale Adams ■☆

307200　Polemonium caeruleum L. subsp. kiushianum（Kitam.）H. Hara ＝ Polemonium chinense（Brand）Brand ■

307201　Polemonium caeruleum L. subsp. laxiflorum（Regel）Koji Ito;疏花花荵 ■☆

307202　Polemonium caeruleum L. subsp. laxiflorum（Regel）Koji Ito f. albiflorum Tatew. ;白疏花花荵 ■☆

307203　Polemonium caeruleum L. subsp. laxiflorum（Regel）Koji Ito f. insulare Koji Ito;海岛疏花花荵 ■☆

307204　Polemonium caeruleum L. subsp. laxiflorum（Regel）Koji Ito var. paludosum（Koji Ito）T. Yamaz. ;沼泽疏花花荵 ■☆

307205　Polemonium caeruleum L. subsp. villosum（Rudolph ex Georgi）Brand ＝ Polemonium caeruleum L. var. acutiflorum（Willd. ex Roem. et Schult.）Ledeb. ■

307206　Polemonium caeruleum L. subsp. yezoense（Miyabe et Kudo）H. Hara var. laxiflorum（Regel）Miyabe et Kudo ＝ Polemonium caeruleum L. subsp. laxiflorum（Regel）Koji Ito ■☆

307207　Polemonium caeruleum L. subsp. yezoense（Miyabe et Kudo）H. Hara var. nipponicum（Kitam.）Koji Ito;本州花荵 ■☆

307208　Polemonium caeruleum L. subsp. yezoense（Miyabe et Kudo）H. Hara ＝ Polemonium yezoense（Miyabe et Kudo）Kitam. ■☆

307209　Polemonium caeruleum L. var. acutiflorum（Willd. ex Roem. et Schult.）Ledeb. ;尖裂花荵 ■

307210　Polemonium caeruleum L. var. chinense Brand ＝ Polemonium chinense（Brand）Brand ■

307211　Polemonium campanulatum（Th. Fr.）H. Lindb. ＝ Polemonium caeruleum L. subsp. campanulatum Th. Fr. ■☆

307212　Polemonium campanuloides L. f. ＝ Prismatocarpus campanuloides（L. f.）Sond. ●☆

307213　Polemonium carneum A. Gray;肉红花荵 ■☆

307214　Polemonium caucasicum N. Busch;高加索花荵 ■☆

307215　Polemonium chinense（Brand）Brand;中华花荵(电灯花,花荵,山菠菜,丝花花荵,小花荵);China Polemonium, Chinese Polemonium,Samall Polemonium ■

307216　Polemonium chinense（Brand）Brand ＝ Polemonium caeruleum L. subsp. kiushianum（Kitam.）H. Hara ■

307217　Polemonium chinense（Brand）Brand var. hirticaulum G. H. Liu et Ma；毛茎花葱■

307218　Polemonium chinense Brand ＝ Polemonium chinense（Brand）Brand ■

307219　Polemonium cuspidatum ?；骤尖花葱；Stiffpoint Jacob's-Ladder ■☆

307220　Polemonium foliosissimum A. Gray；多叶花葱■☆

307221　Polemonium hultenii H. Hara ＝ Polemonium caeruleum L. subsp. campanulatum Th. Fr. var. humile（Willd.）Herder ■☆

307222　Polemonium humile Willd. ex Roem. et Schult.；低花葱■☆

307223　Polemonium kiushianum Kitam. ＝ Polemonium caeruleum L. subsp. kiushianum（Kitam.）H. Hara ■

307224　Polemonium laxiflorum（Regel）Kitam. ＝ Polemonium caeruleum L. subsp. laxiflorum（Regel）Koji Ito ■☆

307225　Polemonium laxiflorum（Regel）Kitam. ＝ Polemonium caeruleum L. ■

307226　Polemonium laxiflorum（Regel）Kitam. ＝ Polemonium chinense（Brand）Brand ■

307227　Polemonium laxiflorum（Regel）Kitam. f. albiflorum H. Hara ＝ Polemonium caeruleum L. subsp. laxiflorum（Regel）Koji Ito f. albiflorum Tatew. ■☆

307228　Polemonium liniflorum V. N. Vassil. ＝ Polemonium chinense（Brand）Brand ■

307229　Polemonium majus Tolm.；大花葱■☆

307230　Polemonium nipponicum Kitam. ＝ Polemonium caeruleum L. subsp. yezoense（Miyabe et Kudo）H. Hara var. nipponicum（Kitam.）Koji Ito ■☆

307231　Polemonium obscurum Blanco ＝ Lepistemon obscurus（Blanco）Merr. ■

307232　Polemonium occidentale Greene；西方花葱；Western Jacob's-ladder，Western Polemonium ■☆

307233　Polemonium pacificum V. N. Vassil.；太平洋花葱■☆

307234　Polemonium parviflorum Tolm.；小花花葱■☆

307235　Polemonium pauciflorum S. Watson；贫花葱■☆

307236　Polemonium pseudopulchellum V. N. Vassil.；假美花葱■☆

307237　Polemonium pulchellum Bunge；美花葱（美丽花葱）■☆

307238　Polemonium racemosum Kitam. ＝ Polemonium caeruleum L. subsp. kiushianum（Kitam.）H. Hara ■

307239　Polemonium racemosum Kitam. ＝ Polemonium caeruleum L. var. acutiflorum（Willd. ex Roem. et Schult.）Ledeb. ■

307240　Polemonium racemosum Kitam. var. laxiflorum（Regel）Nakai ex W. T. Lee ＝ Polemonium caeruleum L. subsp. laxiflorum（Regel）Koji Ito ■☆

307241　Polemonium reptans L.；匍匐花葱；American Greek Valerian，Bluebell，Creeping Jacob's Ladder，Creeping Polemonium，False Jacob's Ladder，Greek Valerian，Greek-valerian，Jacob's Ladder，Native Jacob's Ladder，Spreading Jacob's-ladder，Sweatroot ■☆

307242　Polemonium roelloides L. f. ＝ Prismatocarpus pedunculatus（P. J. Bergius）A. DC. ●☆

307243　Polemonium sibiricum D. Don；西伯利亚花葱；Siberian Polemonium ■☆

307244　Polemonium sumushanense G. H. Liu et Ma；苏木山花葱；Sumushan Polemonium ■

307245　Polemonium vanbruntiae Britton；美洲花葱；American Jacob's Ladder，Jacob's Ladder ■☆

307246　Polemonium villosum Rudolph ex Georgi；柔毛花葱■

307247　Polemonium villosum Rudolph ex Georgi ＝ Polemonium caeruleum L. var. acutiflorum（Willd. ex Roem. et Schult.）Ledeb. ■

307248　Polemonium villosum Rudolph ex Georgi var. glabrum S. D. Zhao；光花葱；Glabrous Polemonium ■

307249　Polemonium villosum Rudolph ex Georgi var. glabrum S. D. Zhao ＝ Polemonium caeruleum L. var. acutiflorum（Willd. ex Roem. et Schult.）Ledeb. ■

307250　Polemonium yezoense（Miyabe et Kudo）Kitam.；北海道花葱■☆

307251　Polemonium yezoense（Miyabe et Kudo）Kitam. ＝ Polemonium caeruleum L. subsp. yezoense（Miyabe et Kudo）H. Hara ■☆

307252　Polevansia De Winter（1966）；挺秆草属■☆

307253　Polevansia rigida De Winter；挺秆草■☆

307254　Polgidon Raf. ＝ Chaerophyllum L. ■

307255　Polhillia C. H. Stirt.（1986）；南非银豆属●☆

307256　Polhillia brevicalyx（C. H. Stirt.）B. -E. van Wyk et A. L. Schutte；短萼南非银豆●☆

307257　Polhillia canescens C. H. Stirt.；短灰南非银豆●☆

307258　Polhillia connata（Harv.）C. H. Stirt.；合生南非银豆●☆

307259　Polhillia involucrata（Thunb.）B. -E. van Wyk et A. L. Schutte；总苞南非银豆●☆

307260　Polhillia obsoleta（Harv.）B. -E. van Wyk；不全南非银豆●☆

307261　Polhillia pallens C. H. Stirt.；苍白南非银豆●☆

307262　Polhillia waltersii（C. H. Stirt.）C. H. Stirt. ＝ Polhillia obsoleta（Harv.）B. -E. van Wyk ●☆

307263　Polia Lour.（废弃属名）＝ Polycarpaea Lam.（保留属名）■●

307264　Polia Ten. ＝ Cypella Herb. ■☆

307265　Polianthes L.（1753）；晚香玉属（晚红玉属）；Tuberose ■

307266　Polianthes geminiflora（Lex.）Rose ＝ Bravoa geminiflora Lex. ■☆

307267　Polianthes maculosa（Hook.）Shinners ＝ Manfreda maculosa Rose ■☆

307268　Polianthes runyonii Shinners ＝ Manfreda longiflora（Rose）Verh. -Will. ■☆

307269　Polianthes tuberosa L.；晚香玉（晚红玉，夜来香，玉簪花，月来香，月下香）；Omixochitl，Tuberose ■

307270　Polianthes tuberosa L. var. azucena ?；月下香■

307271　Polianthes variegata（Jacobi）Shinners ＝ Manfreda variegata（Jacobi）Rose ■☆

307272　Policarpaea Lam. ＝ Polycarpaea Lam.（保留属名）■●

307273　Policarpea Lam. ＝ Polycarpaea Lam.（保留属名）■●

307274　Polichia Schrank（废弃属名）＝ Lamiastrum Heist. ex Fabr. ■

307275　Polichia Schrank（废弃属名）＝ Pollichia Aiton（保留属名）●☆

307276　Poligala Neck. ＝ Polygala L. ●■

307277　Poligonum Neck. ＝ Polygonum L.（保留属名）■●

307278　Poliodendron Webb et Berthel. ＝ Teucrium L. ●■

307279　Poliomintha A. Gray（1870）；灰薄荷属●☆

307280　Poliomintha bicolor S. Watson；二色灰薄荷■☆

307281　Poliomintha glabrescens A. Gray；变光灰薄荷；Rosemary Mint ■☆

307282　Poliomintha incana（Torr.）A. Gray；甜灰薄荷；Desert Rosemary，Rosemary Mint，Sweet Sage ■☆

307283　Poliomintha incana A. Gray ＝ Poliomintha incana（Torr.）A. Gray ■☆

307284　Poliomintha longiflora A. Gray；长花灰薄荷；Mexican Oregano，Rosemary Mint ■☆

307285　Poliomintha maderensis Henrickson；墨西哥灰薄荷；Lavender Spice，Madrean Rosemary ■☆

307286　Poliomintha versicolor A. Gray；变色灰薄荷■☆

307287　Poliophyton O. E. Schulz ＝ Mancoa Wedd.（保留属名）■☆

307288　Poliophyton O. E. Schulz（1933）；灰毛芥属■☆

307289　Poliothyrsidaceae Doweld；山拐枣科●

307290　Poliothyrsis Oliv.（1889）；山拐枣属；Pearl Bloom Tree, Pearlbloomtree, Pearlbloom-tree, Wild Turnjujube ●★

307291　Poliothyrsis sinensis Oliv.；山拐枣；Chinese Pearl bloom Tree, Chinese Pearlbloomtree, Chinese Pearlbloom-tree, Poliothyrsis, Wild Turnjujube ●

307292　Poliothyrsis sinensis Oliv. var. subglabra S. S. Lai；南方山拐枣；Subglabrous Pearlbloomtree ●

307293　Polium Mill. = Teucrium L. ●■

307294　Polium Stokes = Polia Lour.（废弃属名）■●

307295　Polium Stokes = Polycarpaea Lam.（保留属名）■●

307296　Poljakanthema Kamelin = Tanacetum L. ■●

307297　Poljakanthema Kamelin(1993)；土耳其菊蒿属■☆

307298　Poljakanthema aphanassievii（Krasch.）Kamelin；土耳其菊蒿■☆

307299　Poljakovia Grubov et Filatova = Tanacetum L. ■●

307300　Poljakovia alashanensis（Y. Ling）Grubov et Filatova = Hippolytia kaschgarica（Krasch.）Poljakov ■

307301　Poljakovia falcatolobata（Krasch.）Grubov et Filatova = Cancrinia maximowiczii C. Winkl. ●

307302　Poljakovia kaschgarica（Krasch.）Grubov et Filatova = Hippolytia kaschgarica（Krasch.）Poljakov ■

307303　Pollalesta Kunth(1818)；波拉菊属●☆

307304　Pollalesta acuminata（Kunth）Pruski；渐尖波拉菊●☆

307305　Pollalesta argentea Aristeg.；银白波拉菊●☆

307306　Pollalesta colombiana Aristeg.；哥伦比亚波拉菊●☆

307307　Pollalesta discolor（Kunth）Aristeg.；异色波拉菊●☆

307308　Pollalesta ferruginea（Gleason）Aristeg.；锈色波拉菊●☆

307309　Pollalesta macrophylla（Sch. Bip.）Aristeg.；大叶波拉菊●☆

307310　Pollardia Withner et P. A. Harding = Epidendrum L.（保留属名）■☆

307311　Pollardia Withner et P. A. Harding(2004)；异柱瓣兰属■☆

307312　Pollia Thunb.（1781）；杜若属；Pollia ■

307313　Pollia aclisia Hassk. = Pollia hasskarlii R. S. Rao ■

307314　Pollia bambusifolia（H. Lév.）H. Lév. = Rhopalephora scaberrima（Blume）Faden ■

307315　Pollia bracteata K. Schum.；具苞杜若■☆

307316　Pollia cavaleriei（H. Lév. et Vaniot）H. Lév. = Murdannia hookeri（C. B. Clarke）Brückn. ■

307317　Pollia condensata C. B. Clarke；密集杜若■☆

307318　Pollia cyanocarpa K. Schum.；蓝果杜若■☆

307319　Pollia dielsii H. Lév. = Spatholirion longifolium（Gagnep.）Dunn ■

307320　Pollia elegans Hassk. = Pollia secundiflora（Blume）Bakh. f. ■

307321　Pollia gracilis C. B. Clarke；纤细杜若■☆

307322　Pollia hasskarlii R. S. Rao；大杜若（粗柄杜若,大剑叶木,黑珍珠,七喜草,水芭蕉,占点领,竹节兰）；Big Pollia ■

307323　Pollia indica Thunb. = Pollia secundiflora（Blume）Bakh. f. ■

307324　Pollia japonica Thunb.；杜若（楚蘅,地藕,杜蘅,杜莲,莲花姜,良姜,山竹壳菜,竹叶花,竹叶莲）；Japan Pollia, Japanese Pollia ■

307325　Pollia japonica Thunb. var. minor（Honda）E. Walker = Pollia miranda（H. Lév.）H. Hara ■

307326　Pollia japonica Thunb. var. minor（Honda）Hayata ex Masam. = Pollia miranda（H. Lév.）H. Hara ■

307327　Pollia japonica Thunb. var. minor Hayata ex Masam. = Pollia miranda（H. Lév.）H. Hara ■

307328　Pollia japonica Thunb. var. miranda（H. Lév.）Kitam. = Pollia miranda（H. Lév.）H. Hara ■

307329　Pollia macrobracteata D. Y. Hong；大苞杜若■

307330　Pollia mannii C. B. Clarke；曼氏杜若■☆

307331　Pollia minor（Hayata）Honda = Pollia miranda（H. Lév.）H. Hara ■

307332　Pollia minor Honda = Pollia miranda（H. Lév.）H. Hara ■

307333　Pollia miranda（H. Lév.）H. Hara；小杜若（川杜若,杜若,石竹叶,竹叶兰）；Omei Mountain Pollia, Sichuan Pollia ■

307334　Pollia omeiensis D. Y. Hong = Pollia miranda（H. Lév.）H. Hara ■

307335　Pollia pumila Hallier f. = Dictyospermum conspicuum（Blume）Hassk. ■

307336　Pollia sambiranensis H. Perrier；马岛杜若■☆

307337　Pollia secundiflora（Blume）Bakh. f.；长花枝杜若（丛林杜若,菲岛杜若）；Longflower Pollia ■

307338　Pollia secundiflora Blume = Pollia siamensis（Craib）Faden ex D. Y. Hong ■

307339　Pollia siamensis（Craib）Faden ex D. Y. Hong；长柄杜若；Longstalk Pollia ■

307340　Pollia sorzogonensis（E. Mey.）Endl. = Pollia secundiflora（Blume）Bakh. f. ■

307341　Pollia subumbellata C. B. Clarke；伞花杜若；Umbelflower Pollia ■

307342　Pollia thyrsiflora（Blume）Endl. ex Hassk.；密花杜若；Denseflower Pollia ■

307343　Pollia urnbellata H. Lév. = Pollia secundiflora（Blume）Bakh. f. ■

307344　Pollia zollingeri C. B. Clarke = Pollia miranda（H. Lév.）H. Hara ■

307345　Pollichia（Sol.）Aiton = Pollichia Aiton（保留属名）●☆

307346　Pollichia Aiton(1789)（保留属名）；指甲藤属●☆

307347　Pollichia Medik. = Trichodesma R. Br.（保留属名）●■

307348　Pollichia Schrank = Galeobdolon Adans. ■

307349　Pollichia Schrank = Lamium L. ■

307350　Pollichia Sol. = Pollichia Aiton（保留属名）●●☆

307351　Pollichia amplexicaulis（L.）Willd. = Lamium amplexicaule L. ■

307352　Pollichia amplexicaulis Willd. = Lamium amplexicaule L. ■

307353　Pollichia campestris Aiton；田野指甲藤●☆

307354　Pollinia Spreng.（废弃属名）= Chrysopogon Trin.（保留属名）■

307355　Pollinia Trin.（1833）；异味草属；Sugar Grass, Sugar-grass ■☆

307356　Pollinia Trin. = Microstegium Nees ■

307357　Pollinia argentea（Brongn.）Trin. = Eulalia trispicata（Schult.）Henrard ■

307358　Pollinia arisanensis Hayata = Microstegium nudum（Trin.）A. Camus ■

307359　Pollinia articulata Trin. = Pseudopogonatherum contortum（Brongn.）A. Camus ■

307360　Pollinia articulata Trin. subsp. fragilis var. setifolia Hack. = Pseudopogonatherum setifolium（Nees）A. Camus ■

307361　Pollinia bequaertii De Wild. = Microstegium vagans（Nees ex Steud.）A. Camus ■

307362　Pollinia birmanica Hook. f. = Eulalia speciosa（Debeaux）Kuntze ■

307363　Pollinia brevifolia（Sw.）Spreng. = Schizachyrium brevifolium（Sw.）Nees ex Büse ■

307364　Pollinia brevifolium（Sw.）Spreng. = Schizachyrium brevifolium（Sw.）Nees ex Büse ■

307365　Pollinia cantonensis Rendle = Microstegium vimineum（Trin.）A. Camus ■

307366　Pollinia ciliata Trin. = Microstegium ciliatum（Trin.）A. Camus ■

307367　Pollinia ciliata Trin. var. breviaristata Rendle ＝ Microstegium fasciculatum（L.）Henrard ■

307368　Pollinia ciliata Trin. var. formosana（Hack.）Honda ＝ Microstegium ciliatum（Trin.）A. Camus ■

307369　Pollinia ciliata Trin. var. seminuda Hack. ＝ Microstegium ciliatum（Trin.）A. Camus ■

307370　Pollinia collina Balansa ＝ Pseudopogonatherum contortum（Brongn.）A. Camus ■

307371　Pollinia cumingii Nees ＝ Eulalia leschenaultiana（Decne.）Ohwi ■

307372　Pollinia curningii Nees ＝ Eulalia leschenaultiana（Decne.）Ohwi ■

307373　Pollinia curningii Nees var. genuina Hack. ＝ Eulalia leschenaultiana（Decne.）Ohwi ■

307374　Pollinia delicatula Hook. f. ＝ Microstegium delicatulum（Hook. f.）A. Camus ■

307375　Pollinia delicatulum Hook. f. ＝ Microstegium delicatulum（Hook. f.）A. Camus ■

307376　Pollinia distachya（L.）Spreng. ＝ Andropogon distachyos L. ■☆

307377　Pollinia eriopoda Hance ＝ Eulaliopsis binata（Retz.）C. E. Hubb. ■

307378　Pollinia fauriei Hayata ＝ Microstegium fauriei（Hayata）Honda ■

307379　Pollinia formosana（Hack.）Hayata ＝ Microstegium ciliatum（Trin.）A. Camus ■

307380　Pollinia geniculata Hayata ＝ Microstegium fauriei（Hayata）Honda subsp. geniculatum（Hayata）T. Koyama ■

307381　Pollinia geniculatum Hayata ＝ Microstegium geniculatum（Hayata）Honda ■

307382　Pollinia glaberrimum Honda ＝ Microstegium glaberrimum（Honda）Koidz. ■

307383　Pollinia grata Hack. ＝ Microstegium fasciculatum（L.）Henrard ■

307384　Pollinia grata Hack. ＝ Microstegium vagans（Nees ex Steud.）A. Camus ■

307385　Pollinia hendersonii C. E. Hubb. ＝ Microstegium fauriei（Hayata）Honda subsp. geniculatum（Hayata）T. Koyama ■

307386　Pollinia homblei De Wild. ＝ Eulalia aurea（Bory）Kunth ■

307387　Pollinia huillensis Rendle ＝ Homozeugos huillense（Rendle）Stapf ■☆

307388　Pollinia imberbis Nees ex Steud. ＝ Microstegium nodosum（Kom.）Tzvelev ■

307389　Pollinia imberbis Nees ex Steud. ＝ Microstegium vimineum（Trin.）A. Camus ■

307390　Pollinia imberbis Nees ex Steud. var. willdenowiana（Nees ex Steud.）Hack. ＝ Microstegium vimineum（Trin.）A. Camus ■

307391　Pollinia japonica Miq. ＝ Microstegium japonicum（Miq.）Koidz. ■

307392　Pollinia lancea Nees ex Steud. ＝ Microstegium ciliatum（Trin.）A. Camus ■

307393　Pollinia laxa Nees ex Steud. ＝ Microstegium ciliatum（Trin.）A. Camus ■

307394　Pollinia mollis（Griseb.）Hack. ＝ Eulalia mollis（Griseb.）Kuntze ■

307395　Pollinia monandra（Roxb.）Spreng. ＝ Pogonatherum crinitum（Thunb.）Kunth ■

307396　Pollinia monantha Nees ex Steud. ＝ Microstegium ciliatum（Trin.）A. Camus ■

307397　Pollinia monantha Nees ex Steud. ＝ Microstegium fasciculatum（L.）Henrard ■

307398　Pollinia monantha Nees ex Steud. ＝ Microstegium monanthum（Nees ex Steud.）A. Camus ■

307399　Pollinia monantha Nees ex Steud. var. formosana Hack. ＝ Microstegium ciliatum（Trin.）A. Camus ■

307400　Pollinia nepalensis（Trin.）Benth. ex Duthie ＝ Miscanthus nepalensis（Trin.）Hack. ■

307401　Pollinia nuda Trin. ＝ Microstegium nudum（Trin.）A. Camus ■

307402　Pollinia pallens Hack. ＝ Eulalia pallens（Hack.）Kuntze ■

307403　Pollinia parceciliata Pilg. ＝ Microstegium vagans（Nees ex Steud.）A. Camus ■

307404　Pollinia phaeothris Hack. ＝ Eulalia phaeothrix（Hack.）Kuntze ■

307405　Pollinia phaeothrix Hack. ＝ Eulalia phaeothrix（Hack.）Kuntze ■

307406　Pollinia phaeothrix Hack. var. aurea A. Camus ＝ Eulalia speciosa（Debeaux）Kuntze ■

307407　Pollinia polyneura Pilg. ＝ Eulalia polyneura（Pilg.）Stapf ■☆

307408　Pollinia polystachya（Willd.）Spreng. ＝ Pogonatherum paniceum（Lam.）Hack. ■

307409　Pollinia praemorsa Nees ex Steud. ＝ Polytrias amaura（Büse）Kuntze ■

307410　Pollinia praemorsa Nees ex Steud. ＝ Polytrias indica（Houtt.）Veldkamp ■

307411　Pollinia quadrinervis Hack. ＝ Eulalia quadrinervis（Hack.）Kuntze ■

307412　Pollinia quadrinervis Hack. var. latifolia Rendle ＝ Eulalia siamensis Bor var. latifolia（Rendle）S. M. Phillips et S. L. Chen ■

307413　Pollinia quadrinervis Hack. var. wightii Hook. f. ＝ Eulalia wightii（Hook. f.）Bor ■

307414　Pollinia sericea Chiov. ＝ Eulalia villosa（Thunb.）Nees ■☆

307415　Pollinia setifolia Nees ＝ Pseudopogonatherum contortum（Brongn.）A. Camus ■

307416　Pollinia setifolia Nees ＝ Pseudopogonatherum koretrostachys（Trin.）Henrard ■

307417　Pollinia setifolia Nees ＝ Pseudopogonatherum setifolium（Nees）A. Camus ■

307418　Pollinia speciosa（Debeaux）Hack. ＝ Eulalia speciosa（Debeaux）Kuntze ■

307419　Pollinia tristachya（Steud.）Thwaites ＝ Eulalia trispicata（Schult.）Henrard ■

307420　Pollinia vagans Nees ex Steud. ＝ Microstegium fasciculatum（L.）Henrard ■

307421　Pollinia vagans Nees ex Steud. ＝ Microstegium vagans（Nees ex Steud.）A. Camus ■

307422　Pollinia velutina Rendle ＝ Eulalia speciosa（Debeaux）Kuntze ■

307423　Pollinia villosa（Thunb.）Spreng. ＝ Eulalia villosa（Thunb.）Nees ■☆

307424　Pollinia villosa Munro ＝ Eulalia quadrinervis（Hack.）Kuntze ■

307425　Pollinia villosa Munro var. chefuensis Franch. ＝ Eulalia quadrinervis（Hack.）Kuntze ■

307426　Pollinia viminea（Trin.）Merr. ＝ Microstegium vimineum（Trin.）A. Camus ■

307427　Pollinia wallichiana Nees ex Steud. ＝ Microstegium ciliatum（Trin.）A. Camus ■

307428　Pollinia willdenowiana（Nees ex Steud.）Benth. ＝ Microstegium vimineum（Trin.）A. Camus ■

307429　Pollinia willdenowiana（Nees）Benth. ＝ Microstegium vimineum（Trin.）A. Camus ■

307430　Pollinidium Haines ＝ Eulaliopsis Honda ■

307431　Pollinidium Stapf ex Haines ＝ Eulaliopsis Honda ■

307432　Pollinidium angustifolium（Trin.）Haines ＝ Eulaliopsis binata
（Retz.）C. E. Hubb. ■

307433　Pollinidium binatum（Retz.）C. E. Hubb. ＝ Eulaliopsis binata
（Retz.）C. E. Hubb. ■

307434　Polliniopsia Hayata ＝ Microstegium Nees ■

307435　Polliniopsis Hayata（1918）；相马荞竹属■

307436　Polliniopsis somae Hayata ＝ Microstegium somae（Hayata）
Ohwi ■

307437　Pollinirhiza Dulac ＝ Listera R. Br.（保留属名）■

307438　Pollinirhiza Dulac ＝ Neottia Guett.（保留属名）■

307439　Pollinirhiza cordata（L.）Mizush. ＝ Listera cordata（L.）R.
Br. ■☆

307440　Poloa DC. ＝ Pulicaria Gaertn. ■●

307441　Polpoda C. Presl（1829）；南非粟米草属（聚叶粟米草属）●☆

307442　Polpoda capensis C. Presl；南非粟米草●☆

307443　Polpoda stipulacea（Leight.）Adamson；非洲南非粟米草●☆

307444　Polpodaceae Nakai ＝ Aizoaceae Martinov（保留科名）●■

307445　Polpodaceae Nakai ＝ Molluginaceae Bartl.（保留科名）■

307446　Polpodaceae Nakai；南非粟米草科●

307447　Poltolobium C. Presl ＝ Andira Lam.（保留属名）●☆

307448　Polulago Mill. ＝ Caltha L. ■

307449　Polyacantha Gray ＝ Centaurea L.（保留属名）●■

307450　Polyacantha Hill ＝ Carduus L. ＋ Cirsium Mill. ■

307451　Polyacanthus C. Presl（废弃属名）＝ Gymnosporia（Wight et
Arn.）Benth. et Hook. f.（保留属名）●

307452　Polyacanthus angustifolius C. Presl ＝ Gymnosporia linearis（L.
f.）Loes. ●☆

307453　Polyacanthus stenophyllus（Eckl. et Zeyh.）C. Presl ＝
Gymnosporia linearis（L. f.）Loes. ●☆

307454　Polyachyrus Lag.（1811）；繁花钝柱菊属●■☆

307455　Polyachyrus annuus I. M. Johnst. ；一年繁花钝柱菊■☆

307456　Polyachyrus foliosus Phil. ；多叶繁花钝柱菊■☆

307457　Polyachyrus glabratus Phil. ；光繁花钝柱菊■☆

307458　Polyachyrus glandulosus Nutt. ；多腺繁花钝柱菊■☆

307459　Polyachyrus latifolius Phil. ；宽叶繁花钝柱菊■☆

307460　Polyachyrus multifidus D. Don；多裂繁花钝柱菊■☆

307461　Polyachyrus niveus Lag. ex DC. ；雪白繁花钝柱菊■☆

307462　Polyactidium DC. ＝ Polyactis Less. ■

307463　Polyactis Less. ＝ Erigeron L. ■●

307464　Polyactium（DC.）Eckl. et Zeyh. ＝ Pelargonium L'Hér. ex
Aiton ●■

307465　Polyactium Eckl. et Zeyh. ＝ Pelargonium L'Hér. ex Aiton ●■

307466　Polyactium aconitophyllum Eckl. et Zeyh. ＝ Pelargonium
luridum（Andréws）Sweet ■☆

307467　Polyactium amatymbicum Eckl. et Zeyh. ＝ Pelargonium
schizopetalum Sweet ■☆

307468　Polyactium anethifolium Eckl. et Zeyh. ＝ Pelargonium
anethifolium（Eckl. et Zeyh.）Steud. ■☆

307469　Polyactium arenarium Eckl. et Zeyh. ＝ Pelargonium
pulverulentum Colvill ex Sweet ■☆

307470　Polyactium caffrum Eckl. et Zeyh. ＝ Pelargonium caffrum
（Eckl. et Zeyh.）Harv. ■☆

307471　Polyactium coniophyllum Eckl. et Zeyh. ＝ Pelargonium triste
（L.）L'Hér. ■☆

307472　Polyactium hybridifolium Eckl. et Zeyh. ＝ Pelargonium
radulifolium（Eckl. et Zeyh.）Steud. ■☆

307473　Polyactium papaverifolium Eckl. et Zeyh. ＝ Pelargonium triste
（L.）L'Hér. ■☆

307474　Polyactium peucedanifolium Eckl. et Zeyh. ＝ Pelargonium
anethifolium（Eckl. et Zeyh.）Steud. ■☆

307475　Polyactium primuliforme Eckl. et Zeyh. ＝ Pelargonium
pulverulentum Colvill ex Sweet ■☆

307476　Polyactium radulifolium Eckl. et Zeyh. ＝ Pelargonium
radulifolium（Eckl. et Zeyh.）Steud. ■☆

307477　Polyadelphaceae Dulac ＝ Hypericaceae Juss.（保留科名）●■

307478　Polyadenia Nees ＝ Lindera Thunb.（保留属名）●

307479　Polyadoa Stapf ＝ Hunteria Roxb. ●

307480　Polyadoa camerunensis（K. Schum. ex Hallier f.）Brenan ＝
Hunteria camerunensis K. Schum. ex Hallier f. ●☆

307481　Polyadoa elliotii Stapf ＝ Hunteria umbellata（K. Schum.）
Hallier f. ●☆

307482　Polyadoa simii Stapf ＝ Hunteria simii（Stapf）H. Huber ●☆

307483　Polyadoa umbellata（K. Schum.）Stapf ＝ Hunteria umbellata
（K. Schum.）Hallier f. ●☆

307484　Polyalthia Blume（1830）；暗罗属（鸡爪树属）；Greenstar ●

307485　Polyalthia acuminata Oliv. ＝ Polyalthia suaveolens Engl. et Diels ●☆

307486　Polyalthia aubrevillei Ghesq. ex Aubrév. ；科特迪瓦暗罗●☆

307487　Polyalthia aubrevillei Ghesq. ex Aubrév. ＝ Polyalthia suaveolens
Engl. et Diels ●☆

307488　Polyalthia cerasoides（Roxb.）Bedd. ＝ Polyalthia cerasoides
（Roxb.）Benth. et Hook. f. ●

307489　Polyalthia cerasoides（Roxb.）Benth. et Hook. f. ；细基丸（暗
香,红英,黄肖,老人皮,老人皮树,山芭蕉,猪槟榔）；Cherrylike
Greenstar,Cherry-like Greenstar ●

307490　Polyalthia cerasoides（Roxb.）Benth. et Hook. f. ex Bedd. ＝
Polyalthia cerasoides（Roxb.）Benth. et Hook. f. ●

307491　Polyalthia cheliensis Hu；景洪暗罗；Cheli Greenstar, Jinghong
Greenstar ●

307492　Polyalthia cheliensis Hu ＝ Polyalthia simiarum（Ham. ex Hook.
f. et Thomson）Benth. ex Hook. f. et Thomson ●

307493　Polyalthia chinensis S. K. Wu et P. T. Li；西藏暗罗；China
Greenstar,Chinese Greenstar ●

307494　Polyalthia consanguinea Merr. ＝ Polyalthia obliqua Hook. f. et
Thomson ●

307495　Polyalthia crassipes Engl. ＝ Cleistopholis staudtii Engl. et Diels ●☆

307496　Polyalthia crassipetala Merr. ＝ Polyalthia cerasoides（Roxb.）
Benth. et Hook. f. ●

307497　Polyalthia florulenta C. Y. Wu et P. T. Li；小花暗罗；Flosculous
Greenstar,Littleflower Greenstar,Small-flowered Greenstar ●

307498　Polyalthia fragrans（Dalzell）Benth. et Hook. f. ex Hook. f. ＝
Polyalthia fragrans（Dalziel）Benth. et Hook. f. ex Hook. f. et
Thomson ●

307499　Polyalthia fragrans（Dalzell）Hook. f. et Thomson ＝ Polyalthia
fragrans（Dalziel）Benth. et Hook. f. ex Hook. f. et Thomson ●

307500　Polyalthia fragrans（Dalziel）Benth. et Hook. f. ex Hook. f. et
Thomson；云南暗罗（伞花暗罗）；Yunnan Greenstar ●

307501　Polyalthia heteropetala（Diels）Ghesq. ＝ Fenerivia heteropetala
Diels ●☆

307502　Polyalthia hypogaea King；地下暗罗●☆

307503　Polyalthia jenkinsii（Hook. f. et Thomson）Hook. f. et Thomson
＝ Polyalthia rumphii（Blume ex Hensch.）Merr. ●

307504　Polyalthia jucunda Finet et Gagnep. ＝ Polyalthia laui Merr. ●

307505　Polyalthia lancilimba C. Y. Wu ex P. T. Li；剑叶暗罗；Lanceleaf
Greenstar,Lance-leaved Greenstar,Swordleaf Greenstar ●

307506 Polyalthia laui Merr. ;海南暗罗（藤椿）；Hainan Greenstar ●

307507 Polyalthia litseifolia C. Y. Wu ex P. T. Li；木姜叶暗罗；Litsea-leaved Greenstar, Litseileaf Greenstar ●

307508 Polyalthia liukiuensis Hatus. ;琉球暗罗（台湾暗罗）；Liuqiu Greenstar, Taiwan Greenstar ●

307509 Polyalthia longifolia（Sonn.）Thwaites；长叶暗罗（长花暗罗）；India Green Star, India Green-star, Indian Willow, Longleaf Greenstar ●

307510 Polyalthia macropoda（Miq.）F. Muell. ;大柄暗罗●☆

307511 Polyalthia mayumbensis Exell ＝ Xylopia quintasii Pierre ex Engl. et Diels ●☆

307512 Polyalthia micrantha（Hassk.）Boerl. ;爪哇暗罗●☆

307513 Polyalthia mortehanii De Wild. ＝ Polyalthia suaveolens Engl. et Diels ●☆

307514 Polyalthia mossambicensis Vollesen；莫桑比克暗罗●☆

307515 Polyalthia nemoralis A. DC. ;陵水暗罗（黑根皮，黑皮根，落坎薯）；Lingshui Greenstar, Woodland Greenstar, Woods Greenstar ●

307516 Polyalthia nitidissima（Dunal）Benth. ;亮果暗罗●☆

307517 Polyalthia obliqua Hook. f. et Thomson；沙煲暗罗（滑桃，山蕉树，血春藤）；Consanguineous Greenstar, Sandpot Greenstar ●

307518 Polyalthia oligogyna Merr. et Chun ＝ Polyalthia nemoralis A. DC. ●

307519 Polyalthia oliveri Engl. ;奥里弗暗罗●☆

307520 Polyalthia petelotii Merr. ;云桂暗罗；Petelot Greenstar ●

307521 Polyalthia pingpienensis P. T. Li；多脉暗罗；Manynerved Greenstar, Pingbian Greenstar ●

307522 Polyalthia pingpienensis P. T. Li ＝ Polyalthia plagioneura Diels ●

307523 Polyalthia plagioneura Diels；斜脉暗罗（厚皮树，九层皮，九重皮）；Obliquenerved Greenstar, Oblique-nerved Greenstar ●

307524 Polyalthia rumphii（Blume ex Hensch.）Merr. ;香花暗罗（大花暗罗）；Rumph Greenstar ●

307525 Polyalthia sasakii Yamam. ＝ Goniothalamus amuyon（Blanco）Merr. ●◇

307526 Polyalthia simiarum（Ham. ex Hook. f. et Thomson）Benth. ex Hook. f. et Thomson subsp. cheliensis（Hu）Tien Ban ＝ Polyalthia simiarum（Ham. ex Hook. f. et Thomson）Benth. ex Hook. f. et Thomson ●

307527 Polyalthia simiarum（Ham. ex Hook. f. et Thomson）Benth. ex Hook. f. et Thomson；腺叶暗罗；Glandular Greenstar ●

307528 Polyalthia stuhlmannii（Engl.）Verdc. ;斯图尔曼暗罗●☆

307529 Polyalthia suaveolens Engl. et Diels；甜香暗罗●☆

307530 Polyalthia suaveolens Engl. et Diels var. gabonica Pellegr. ex Le Thomas；加蓬暗罗●☆

307531 Polyalthia suberosa（Roxb.）Thwaites；暗罗（鸡爪树，老人皮，眉尾木，山观音）；Suberous Greenstar ●

307532 Polyalthia tanganyikensis Vollesen；坦噶尼喀暗罗●☆

307533 Polyalthia verdcourtii Vollesen；韦尔德暗罗●☆

307534 Polyalthia verrucipes C. Y. Wu ex P. T. Li；疣叶暗罗（瘤叶暗罗）；Wartyleaf Greenstar, Warty-leaved Greenstar ●

307535 Polyalthia viridis Craib；毛脉暗罗；Green Greenstar, Hairvein Greenstar ●

307536 Polyandra Leal（1951）；多雄大戟属 ☆

307537 Polyandra bracteosa Leal；多雄大戟 ☆

307538 Polyandrococos Barb. Rodr.（1901）；多蕊椰属（多蕊果属）●☆

307539 Polyandrococos caudscens（Mart.）Barb. Rodr. ;多蕊椰；Buri Palm ●☆

307540 Polyanthemum Bubani ＝ Leucojum L. ■●

307541 Polyanthemum Medik. ＝ Armeria Willd.（保留属名）■☆

307542 Polyantherix Nees ＝ Elymus L. ■

307543 Polyantherix Nees ＝ Sitanion Raf. ■☆

307544 Polyanthes Hill ＝ Polyanthes Jacq. ■☆

307545 Polyanthes Jacq. ＝ Polyxena Kunth ■☆

307546 Polyanthes L. ＝ Polianthes L. ■

307547 Polyanthes pygmaea Jacq. ＝ Polyxena ensifolia（Thunb.）Schönland ■☆

307548 Polyanthina R. M. King et H. Rob.（1970）；多花尖泽兰属■☆

307549 Polyanthina nemorosa（Klatt）R. M. King et H. Rob. ;多花尖泽兰■☆

307550 Polyanthus C. H. Hu ＝ Pleioblastus Nakai ●

307551 Polyanthus C. H. Hu et Y. C. Hu ＝ Arundinaria Michx. ●

307552 Polyanthus longispiculatus（B. M. Yang）C. H. Hu ＝ Arundinaria oleosa（T. H. Wen）C. S. Chao et G. Y. Yang ●

307553 Polyanthus longispiculatus（B. M. Yang）C. H. Hu ＝ Pleioblastus oleosus T. H. Wen ●

307554 Polyarrhena Cass.（1828）；帚菀木属●☆

307555 Polyarrhena Cass. ＝ Felicia Cass.（保留属名）●■

307556 Polyarrhena imbricata（DC.）Grau；非洲帚菀木●☆

307557 Polyarrhena prostrata Grau；帚菀木●☆

307558 Polyarrhena prostrata Grau subsp. dentata ?；尖齿帚菀木●☆

307559 Polyarrhena reflexa（L.）Cass. ;反折帚菀木●☆

307560 Polyarrhena reflexa（L.）Cass. subsp. brachyphylla（Sond. ex Harv.）Grau；短叶帚菀木●☆

307561 Polyarrhena stricta Grau；刚直帚菀木●☆

307562 Polyasmaceae Blume ＝ Grossulariaceae DC.（保留科名）●

307563 Polyaster Hook. f.（1862）；多星芸香属●☆

307564 Polyaulax Backer ＝ Meiogyne Miq. ●

307565 Polyaulax Backer（1945）；多犁木属●☆

307566 Polybactrum Salisb. ＝ Pseudorchis Śeg. ■☆

307567 Polybaea Klotasch ex Benh. et Hook. f. ＝ Cavendishia Lindl.（保留属名）●☆

307568 Polybaea Klotasch ex Benh. et Hook. f. ＝ Polyboea Klotzsch ●

307569 Polyboea Klotzsch ＝ Cavendishia Lindl.（保留属名）●☆

307570 Polyboea Klotzsch ex Endl. ＝ Bernardia L. ●

307571 Polyboea Klotzsch ex Endl. ＝ Polyscias J. R. Forst. et G. Forst. ●

307572 Polycalymma F. Muell. et Sond.（1853）；顶序鼠麴草属■☆

307573 Polycalymma F. Muell. et Sond. ＝ Myriocephalus Benth. ■☆

307574 Polycalymma stuartii F. Muell. et Sond. ;顶序鼠麴草■☆

307575 Polycandia Steud. ＝ Polycardia Juss. ●☆

307576 Polycantha Hill ＝ Carduus L. ＋ Cirsium Mill. ■

307577 Polycantha Hill ＝ Polyacantha Hill ■

307578 Polycardia Juss.（1789）；多心卫矛属●☆

307579 Polycarena Benth.（1836）；多头玄参属■☆

307580 Polycarena aemulans Hilliard；匹敌多头玄参■☆

307581 Polycarena aethiopica（L.）Druce ＝ Sutera aethiopica（L.）Kuntze ■☆

307582 Polycarena alpina（N. E. Br.）Levyns ＝ Phyllopodium alpinum N. E. Br. ■☆

307583 Polycarena arenaria Hiern ＝ Polycarena pubescens Benth. ■☆

307584 Polycarena augei（Hiern）Levyns ＝ Manulea augei（Hiern）Hilliard ■☆

307585 Polycarena aurea Benth. ;黄多头玄参■☆

307586 Polycarena batteniana Hilliard；巴滕多头玄参■☆

307587 Polycarena baurii（Hiern）Levyns ＝ Selago baurii（Hiern）Hilliard ■☆

307588　Polycarena bracteata （Benth.） Levyns = Phyllopodium bracteatum Benth. ■☆

307589　Polycarena calva （Hiern） Levyns = Melanospermum transvaalense （Hiern） Hilliard ■☆

307590　Polycarena capensis （L.） Benth. ;好望角多头玄参■☆

307591　Polycarena capillaris （L. f.） Benth. = Phyllopodium capillare （L. f.） Hilliard ■☆

307592　Polycarena capitata （L. f.） Levyns = Phyllopodium heterophyllum （L. f.） Benth. ■☆

307593　Polycarena cephalophora （Thunb.） Levyns = Phyllopodium cephalophorum （Thunb.） Hilliard ■☆

307594　Polycarena collina Hiern = Phyllopodium collinum （Hiern） Hilliard ■☆

307595　Polycarena comptonii Hilliard;康普顿多头玄参■☆

307596　Polycarena cuneifolia （L. f.） Levyns = Phyllopodium cuneifolium （L. f.） Benth. ■☆

307597　Polycarena diffusa （Benth.） Levyns = Phyllopodium diffusum Benth. ■☆

307598　Polycarena dinteri Thell. = Zaluzianskya peduncularis （Benth.） Walp. ■☆

307599　Polycarena discolor Schinz = Melanospermum foliosum （Benth.） Hilliard ■☆

307600　Polycarena exigua Hilliard;小多头玄参■☆

307601　Polycarena filiformis Diels;线形多头玄参■☆

307602　Polycarena foliosa Benth. = Melanospermum foliosum （Benth.） Hilliard ■☆

307603　Polycarena formosa Hilliard;美丽多头玄参■☆

307604　Polycarena gilioides Benth. ;吉莉花多头玄参■☆

307605　Polycarena glaucescens Hiern = Polycarena pubescens Benth. ■☆

307606　Polycarena glutinosa （Schltr.） Levyns = Trieenea glutinosa （Schltr.） Hilliard ■☆

307607　Polycarena gracilipes N. E. Br. ex Hiern = Polycarena filiformis Diels ■☆

307608　Polycarena gracilis Hilliard;纤细多头玄参■☆

307609　Polycarena heterophylla （L. f.） Levyns = Phyllopodium heterophyllum （L. f.） Benth. ■☆

307610　Polycarena hispidula Thell. = Phyllopodium hispidulum （Thell.） Hilliard ■☆

307611　Polycarena intertexta Benth. = Cromidon decumbens （Thunb.） Hilliard ■☆

307612　Polycarena leipoldtii Hiern = Polycarena rariflora Benth. ■☆

307613　Polycarena lilacina Hilliard;紫丁香色多头玄参■☆

307614　Polycarena linearifolia （Bolus） Levyns = Phyllopodium elegans （Choisy） Hilliard ■☆

307615　Polycarena lupuliformis Thell. = Phyllopodium lupuliforme （Thell.） Hilliard ■☆

307616　Polycarena maxii Hiern = Phyllopodium maxii （Hiern） Hilliard ■☆

307617　Polycarena minima （Hiern） Levyns = Zaluzianskya minima （Hiern） Hilliard ■☆

307618　Polycarena multifolia （Hiern） Levyns = Phyllopodium multifolium Hiern ■☆

307619　Polycarena namaensis Thell. = Phyllopodium namaense （Thell.） Hilliard ●☆

307620　Polycarena nardouwensis Hilliard;纳尔多多头玄参■☆

307621　Polycarena parvula Schltr. = Phyllopodium capillare （L. f.） Hilliard ■☆

307622　Polycarena phyllanthoides （Lam.） DC. ;叶花多头玄参■☆

307623　Polycarena plantaginea （L. f.） Benth. = Cromidon plantaginis （L. f.） Hilliard ■☆

307624　Polycarena pubescens Benth. ;短柔毛多头玄参■☆

307625　Polycarena pumila （Benth.） Levyns = Phyllopodium pumilum Benth. ■☆

307626　Polycarena rangei （Engl.） Levyns = Phyllopodium namaense （Thell.） Hilliard ●☆

307627　Polycarena rariflora Benth. ;稀花多头玄参■☆

307628　Polycarena rariflora Benth. var. micrantha Schltr. = Phyllopodium micranthum （Schltr.） Hilliard ■☆

307629　Polycarena rupestris （Hiern） Levyns = Melanospermum rupestre （Hiern） Hilliard ■☆

307630　Polycarena schlechteri （Hiern） Levyns = Trieenea schlechteri （Hiern） Hilliard ■☆

307631　Polycarena selaginoides Schltr. ex Hiern = Phyllopodium phyllopodioides （Schltr.） Hilliard ■☆

307632　Polycarena silenoides Harv. ex Benth. ;蝇子草多头玄参■☆

307633　Polycarena sordida （Hiern） Levyns = Manulea augei （Hiern） Hilliard ■☆

307634　Polycarena subtilis Hilliard;细多头玄参■☆

307635　Polycarena tenella Hiern;柔弱多头玄参■☆

307636　Polycarena transvaalensis Hiern = Melanospermum transvaalense （Hiern） Hilliard ■☆

307637　Polycarpa L. = Polycarpon L. ■

307638　Polycarpa Linden ex Carrière = Idesia Maxim. （保留属名）●

307639　Polycarpa Loefl. = Polycarpon L. ■

307640　Polycarpa maximowiczii Linden ex Carrière = Idesia polycarpa Maxim. ●

307641　Polycarpa pusilla （Roxb. ex Wight et Arn.） Hiern = Polycarpon prostratum （Forssk.） Asch. et Schweinf. ■

307642　Polycarpaea Lam. （1792）（保留属名）;白鼓钉属（白鼓丁属）;Polycarpaea,Whitedrumnail ■●

307643　Polycarpaea akkensis Maire;阿卡白鼓钉■☆

307644　Polycarpaea aristata （Aiton） DC. ;具芒白鼓钉■☆

307645　Polycarpaea candida Webb et Berthel. = Polycarpaea nivea （Aiton） Webb ■☆

307646　Polycarpaea candida Webb et Berthel. var. diffusa Pit. = Polycarpaea nivea （Aiton） Webb ■☆

307647　Polycarpaea candida Webb et Berthel. var. pygmaea Pit. = Polycarpaea nivea （Aiton） Webb ■☆

307648　Polycarpaea candida Webb et Berthel. var. robusta Pit. = Polycarpaea nivea （Aiton） Webb ■☆

307649　Polycarpaea candida Webb et Berthel. var. webbiana Pit. = Polycarpaea nivea （Aiton） Webb ■☆

307650　Polycarpaea carnosa Buch;肉质白鼓钉■☆

307651　Polycarpaea carnosa Buch var. carnosa ? = Polycarpaea carnosa Buch ■☆

307652　Polycarpaea carnosa Buch var. diversifolia ? = Polycarpaea carnosa Buch ■☆

307653　Polycarpaea carnosa Buch var. spathulata Svent. = Polycarpaea carnosa Buch ■☆

307654　Polycarpaea confusa Maire = Polycarpon robbaireum Kuntze ■☆

307655　Polycarpaea confusa Maire subsp. garamantum Quézel = Polycarpaea robbairea （Kuntze） Greuter et Burdet subsp. garamantum （Quézel） Dobignard ■☆

307656　Polycarpaea corymbosa （L.） Lam. ;白鼓钉（白凤花,白鼓丁, 白花仔,百花草,广白头翁,过饥草,满天星草,声色草,辛苦草,

星色草）；Corymbe Polycarpaea，Oldman's Cap，Whitedrumnail ■

307657　Polycarpaea corymbosa（L.）Lam. var. contracta Balle；紧缩白鼓钉■☆

307658　Polycarpaea corymbosa（L.）Lam. var. effusa Oliv. = Polycarpaea eriantha Hochst. ex A. Rich. var. effusa（Oliv.）Turrill ■☆

307659　Polycarpaea corymbosa（L.）Lam. var. expansa Balle；扩展白鼓钉■☆

307660　Polycarpaea corymbosa（L.）Lam. var. parviflora Oliv. = Polycarpaea tenuifolia（Willd.）DC. ■☆

307661　Polycarpaea corymbosa（L.）Lam. var. pseudolinearifolia Berhaut；假线叶白鼓钉■☆

307662　Polycarpaea divaricata（Aiton）Poir.；叉开白鼓钉■☆

307663　Polycarpaea djalonis A. Chev. = Polycarpaea tenuifolia（Willd.）DC. ■☆

307664　Polycarpaea eriantha Hochst. ex A. Rich.；毛花白鼓钉■☆

307665　Polycarpaea eriantha Hochst. ex A. Rich. var. effusa（Oliv.）Turrill；开展白鼓钉■☆

307666　Polycarpaea filifolia H. Christ；丝叶白鼓钉■☆

307667　Polycarpaea fragilis Delile = Polycarpaea repens（Forssk.）Asch. et Schweinf. ■☆

307668　Polycarpaea gamopetala Berhaut；瓣白鼓钉■☆

307669　Polycarpaea garuensis J. -P. Lebrun；加鲁白鼓钉■☆

307670　Polycarpaea gaudichaudii Gagnep.；大花白鼓钉（贾氏白鼓钉）；Bigflower Whitedrumnail，Gaudichaud Polycarpaea ■

307671　Polycarpaea gayi Webb；盖伊白鼓钉■☆

307672　Polycarpaea gayi Webb var. halimoides ? = Polycarpaea gayi Webb ■☆

307673　Polycarpaea gayi Webb var. helichrysoides ? = Polycarpaea gayi Webb ■☆

307674　Polycarpaea gayi Webb var. lycioides ? = Polycarpaea gayi Webb ■☆

307675　Polycarpaea glabrifolia DC.；光叶白鼓钉■☆

307676　Polycarpaea gnaphaloides（Schousb.）Poir. = Polycarpaea nivea（Aiton）Webb ■☆

307677　Polycarpaea gomerensis Burch. = Polycarpaea filifolia H. Christ ■☆

307678　Polycarpaea grahamii Turrill；格雷厄姆白鼓钉■☆

307679　Polycarpaea grossartii Dinter = Polycarpaea eriantha Hochst. ex A. Rich. ■☆

307680　Polycarpaea holliensis A. Chev. = Polycarpaea tenuifolia（Willd.）DC. ■☆

307681　Polycarpaea inaequalifolia Engl. et Gilg；不等叶白鼓钉■☆

307682　Polycarpaea lancifolia Christ = Polycarpaea nivea（Aiton）Webb ■☆

307683　Polycarpaea latifolia Willd.；宽叶白鼓钉■☆

307684　Polycarpaea linearifolia（DC.）DC.；线叶白鼓钉■☆

307685　Polycarpaea memphitica Delile = Polycarpon prostratum（Forssk.）Asch. et Schweinf. ■

307686　Polycarpaea microphylla Cav. = Polycarpaea nivea（Aiton）Webb ■☆

307687　Polycarpaea mozambica Kunth et Bouché = Polycarpon prostratum（Forssk.）Asch. et Schweinf. ■

307688　Polycarpaea nebulosa Lakela = Polycarpaea corymbosa（L.）Lam. ■

307689　Polycarpaea nivea（Aiton）Webb；雪白白鼓钉■☆

307690　Polycarpaea platyphylla Pax = Polycarpaea glabrifolia DC. ■☆

307691　Polycarpaea pobeguinii Berhaut；波别白鼓钉■☆

307692　Polycarpaea poggei Pax；波格白鼓钉■☆

307693　Polycarpaea prostrata Decne. = Polycarpaea robbairea（Kuntze）Greuter et Burdet ■☆

307694　Polycarpaea prostrata Decne. var. brevipes Maire = Polycarpaea robbairea（Kuntze）Greuter et Burdet ■☆

307695　Polycarpaea prostrata Decne. var. minor Asch. et Schweinf. = Polycarpaea robbairea（Kuntze）Greuter et Burdet ■☆

307696　Polycarpaea pulvinata M. G. Gilbert；叶枕白鼓钉■☆

307697　Polycarpaea repens（Forssk.）Asch. et Schweinf.；匍匐白鼓钉■☆

307698　Polycarpaea repens Forssk. = Polycarpaea repens（Forssk.）Asch. et Schweinf. ■☆

307699　Polycarpaea rhodesica Suess. = Polycarpaea eriantha Hochst. ex A. Rich. var. effusa（Oliv.）Turrill ■☆

307700　Polycarpaea robbairea（Kuntze）Greuter et Burdet；罗巴尔白鼓钉■☆

307701　Polycarpaea robbairea（Kuntze）Greuter et Burdet subsp. garamantum（Quézel）Dobignard；噶拉门特白鼓钉■☆

307702　Polycarpaea robusta（Pit.）G. Kunkel；粗壮白鼓钉■☆

307703　Polycarpaea rupicola J. -P. Lebrun et Stork；岩生白鼓钉■☆

307704　Polycarpaea rupicola Pomel = Polycarpon polycarpoides（Biv.）Fiori subsp. herniarioides（Ball）Maire et Weiller ■☆

307705　Polycarpaea smithii Link；史密斯白鼓钉■☆

307706　Polycarpaea somalensis Engl.；索马里白鼓钉■☆

307707　Polycarpaea spicata Wight ex Arn.；长穗白鼓钉■☆

307708　Polycarpaea staticaeformis Webb = Polycarpaea spicata Wight ex Arn. ■☆

307709　Polycarpaea stellata（Willd.）DC.；星状白鼓钉■☆

307710　Polycarpaea tenerifae Lam. = Polycarpaea divaricata（Aiton）Poir. ■☆

307711　Polycarpaea teneriffae Lam. var. aristata Bornm. = Polycarpaea divaricata（Aiton）Poir. ■☆

307712　Polycarpaea teneriffae Lam. var. crassifolia Pit. = Polycarpaea divaricata（Aiton）Poir. ■☆

307713　Polycarpaea teneriffae Lam. var. intermedia Kuntze = Polycarpaea divaricata（Aiton）Poir. ■☆

307714　Polycarpaea teneriffae Lam. var. laxiflora Pit. = Polycarpaea divaricata（Aiton）Poir. ■☆

307715　Polycarpaea teneriffae Lam. var. linearifolia Bornm. = Polycarpaea divaricata（Aiton）Poir. ■☆

307716　Polycarpaea tenuifolia（Willd.）DC.；细叶白鼓钉■☆

307717　Polycarpaea tenuis H. Christ；细白鼓钉■☆

307718　Polycarpaea tenuistyla Turrill；细穗白鼓钉■☆

307719　Polycarpaeaceae Mart. = Caryophyllaceae Juss.（保留科名）■●

307720　Polycarpaeaceae Schur = Caryophyllaceae Juss.（保留科名）■●

307721　Polycarpea Pomel = Polycarpaea Lam.（保留属名）■●

307722　Polycarpia Webb et Berthel. = Polycarpaea Lam.（保留属名）■●

307723　Polycarpoea Lam. = Polycarpaea Lam.（保留属名）■●

307724　Polycarpon L.（1759）；多荚草属；Allseed，Four-leaved Allseed，Fruitfulgrass，Manyseed，Polycarpon ■

307725　Polycarpon L. = Polycarpon Loefl. ex L. ■

307726　Polycarpon Loefl. = Polycarpon L. ■

307727　Polycarpon Loefl. ex L. = Polycarpon L. ■

307728　Polycarpon alsinifolium（Biv.）DC. = Polycarpon tetraphyllum（L.）L. subsp. alsinifolium（Biv.）Ball ■☆

307729　Polycarpon bivonae J. Gay = Polycarpon polycarpoides（Biv.）Fiori ☆

307730　Polycarpon bivonae J. Gay var. rupicola（Pomel）Chabert = Polycarpon polycarpoides（Biv.）Fiori ■☆

307731　Polycarpon delileanum（Milne-Redh.）Monod ＝ Polycarpon robbaireum Kuntze ■☆

307732　Polycarpon depressum（L.）Rohrb. ＝ Polycarpon prostratum（Forssk.）Asch. et Schweinf. ■

307733　Polycarpon depressum Nutt.；加州多荚草；California Manyseed ■☆

307734　Polycarpon diphyllum Cav. ＝Polycarpon teraphyllum（L.）L. ■☆

307735　Polycarpon diphyllum Cav. ＝ Polycarpon tetraphyllum（L.）L. subsp. diphyllum（Cav.）O. Bolòs et Font Quer ■☆

307736　Polycarpon herniarioides Ball ＝ Polycarpon polycarpoides（Biv.）Fiori subsp. herniarioides（Ball）Maire et Weiller ■☆

307737　Polycarpon indicum（Retz.）Merr. ＝ Polycarpon prostratum（Forssk.）Asch. et Schweinf. ■

307738　Polycarpon loeflingiae Benth. et Hook. f. ＝ Polycarpon prostratum（Forssk.）Asch. et Schweinf. ■

307739　Polycarpon loeflingii Benth. ＝ Polycarpon prostratum（Forssk.）Asch. et Schweinf. ■

307740　Polycarpon loeflingii Wight et Arn. ex Benth. ＝ Polycarpon prostratum（Forssk.）Asch. et Schweinf. ■

307741　Polycarpon memphiticum（Delile）Fenzl ex Broun et Massey ＝ Polycarpon prostratum（Forssk.）Asch. et Schweinf. ■

307742　Polycarpon peploides DC. ＝ Polycarpon polycarpoides（Biv.）Fiori ■☆

307743　Polycarpon peploides DC. subsp. bivonae（J. Gay）Maire et Weiller ＝ Polycarpon polycarpoides（Biv.）Fiori ■☆

307744　Polycarpon peploides DC. var. rupicola（Pomel）Batt. ＝ Polycarpon polycarpoides（Biv.）Fiori subsp. herniarioides（Ball）Maire et Weiller ■☆

307745　Polycarpon polycarpoides（Biv.）Fiori；多果多荚草 ■☆

307746　Polycarpon polycarpoides（Biv.）Fiori subsp. bivonae（J. Gay）Maire et Weiller ＝ Polycarpon polycarpoides（Biv.）Fiori ■☆

307747　Polycarpon polycarpoides（Biv.）Fiori subsp. herniarioides（Ball）Maire et Weiller；治疝草状多荚草 ■☆

307748　Polycarpon prostratum（Forssk.）Asch. et Schweinf.；多荚草（印度多荚草）；Fruitfulgrass，India Fruitfulgrass，Indian Polycarpon，Prostrate Polycarpon ■

307749　Polycarpon prostratum（Forssk.）Asch. et Schweinf. var. littorale J. Raynal et A. Raynal；滨海多荚草 ■☆

307750　Polycarpon pusillum Roxb. ex Wight et Arn. ＝ Polycarpon prostratum（Forssk.）Asch. et Schweinf. ■

307751　Polycarpon robbaireum Kuntze；罗贝多荚草 ■☆

307752　Polycarpon rupicola（Pomel）Batt. ＝ Polycarpon polycarpoides（Biv.）Fiori subsp. herniarioides（Ball）Maire et Weiller ■☆

307753　Polycarpon sauvagei Mathez；索瓦热多荚草 ■☆

307754　Polycarpon succulentum Webb et Berthel. ＝ Polycarpon tetraphyllum（L.）L. ■☆

307755　Polycarpon teraphyllum（L.）L.；四叶多荚草；Fourleaf Manyseed，Four-leaved Allseed，Guernsey Chickweed ☆

307756　Polycarpon tetraphyllum（L.）L. subsp. alsinifolium（Biv.）Ball；繁缕叶多荚草；Fourleaf Manyseed ■☆

307757　Polycarpon tetraphyllum（L.）L. subsp. diphyllum（Cav.）O. Bolòs et Font Quer；二叶多荚草 ■☆

307758　Polycarpon tetraphyllum（L.）L. var. alsinoides Gren. et Godr. ＝Polycarpon tetraphyllum（L.）L. ■☆

307759　Polycarpon tetraphyllum（L.）L. var. rotundatum Batt. ＝ Polycarpon tetraphyllum（L.）L. ■☆

307760　Polycarpon tetraphyllum（L.）L. var. verticillatum Fenzl ＝ Polycarpon tetraphyllum（L.）L. ■☆

307761　Polycenia Choisy ＝ Hebenstretia L. ●☆

307762　Polycenia cordata（L.）E. Mey. ＝ Hebenstretia cordata L. ●☆

307763　Polycenia dregei Gand. ＝ Hebenstretia cordata L. ●☆

307764　Polycenia fenestrata E. Mey. ＝ Hebenstretia repens Jaroscz ●☆

307765　Polycenia fruticosa E. Mey. ＝ Hebenstretia dregei Rolfe ●☆

307766　Polycenia hebenstretioides Choisy ＝ Hebenstretia repens Jaroscz ●☆

307767　Polycenia lanceolata E. Mey. ＝ Hebenstretia lanceolata（E. Mey.）Rolfe ●☆

307768　Polycenia tenera Walp. ＝ Hebenstretia repens Jaroscz ●☆

307769　Polycephalium Engl.（1897）；多头茱萸属 ●☆

307770　Polycephalium capitatum（Baill.）Keay；头状多头茱萸 ●☆

307771　Polycephalium integrum De Wild. et T. Durand ＝ Polycephalium lobatum（Pierre）Pierre ex Engl. ●☆

307772　Polycephalium lobatum（Pierre）Pierre ex Engl.；尖裂多头茱萸 ●☆

307773　Polycephalium mildbraedii Engl. ＝ Polycephalium lobatum（Pierre）Pierre ex Engl. ●☆

307774　Polycephalium poggei Engl. ＝ Polycephalium lobatum（Pierre）Pierre ex Engl. ●☆

307775　Polycephalos Forssk. ＝ Sphaeranthus L. ■

307776　Polycephalos suaveolens Forssk. ＝ Sphaeranthus suaveolens（Forssk.）DC. ■☆

307777　Polyceratocarpus Engl. et Diels（1900）；多角果属 ●☆

307778　Polyceratocarpus angustifolius Paiva；窄叶多角果 ●☆

307779　Polyceratocarpus germainii Boutique；杰曼多角果 ●☆

307780　Polyceratocarpus gossweileri（Exell）Paiva；戈斯多角果 ●☆

307781　Polyceratocarpus laurifolius Paiva；月桂叶多角果 ●☆

307782　Polyceratocarpus microtrichus（Engl. et Diels）Ghesq. ex Pellegr.；小毛多角果 ●☆

307783　Polyceratocarpus parviflorus（Baker f.）Ghesq.；小花多角果 ●☆

307784　Polyceratocarpus pellegrinii Le Thomas；佩尔格兰多角果 ●☆

307785　Polyceratocarpus scheffleri Engl. et Diels；谢夫勒多角果 ●☆

307786　Polyceratocarpus vermoesenii Robyns et Ghesq. ＝ Polyceratocarpus gossweileri（Exell）Paiva ●☆

307787　Polyceratocarpus vermoesenii Robyns et Ghesq. var. letestui Pellegr. ＝ Polyceratocarpus pellegrinii Le Thomas ●☆

307788　Polychaetia Less.（1832）；刚毛菊属 ●☆

307789　Polychaetia Less. ＝ Nestlera Spreng. ■☆

307790　Polychaetia Tausch ex Less. ＝ Tolpis Adans. ●■☆

307791　Polychaetia acerosa DC. ＝ Relhania acerosa（DC.）K. Bremer ●☆

307792　Polychaetia brevifolia DC. ＝ Geigeria brevifolia（DC.）Harv. ●☆

307793　Polychaetia garnotii Less. ＝ Relhania garnotii（Less.）K. Bremer ●☆

307794　Polychaetia oppositifolia DC. ＝ Rosenia oppositifolia（DC.）K. Bremer ●☆

307795　Polychaetia passerinoides（L'Hér.）DC. ＝ Geigeria ornativa O. Hoffm. ■☆

307796　Polychaetia pectidea DC. ＝ Geigeria pectidea（DC.）Harv. ■☆

307797　Polychaetia tricephala DC. ＝ Relhania tricephala（DC.）K. Bremer ●☆

307798　Polychilos Breda ＝ Phalaenopsis Blume ■

307799　Polychilos Breda，Kuhl et Hasselt ＝ Phalaenopsis Blume ■

307800　Polychilos lobbii（Rchb. f.）Shim ＝ Phalaenopsis lobbii（Rchb. f.）H. R. Sweet ■

307801　Polychilos mannii（Rchb. f.）Shim ＝ Phalaenopsis mannii Rchb. f. ■

307802　Polychilos stobartianus（Rchb. f.）Shim ＝ Phalaenopsis

stobariana Rchb. f. ■

307803 Polychilos taenialis（Lindl.）Shim ＝ Phalaenopsis taenialis（Lindl.）Christenson et Pradhan ■

307804 Polychilos wilsonii（Rolfe）Shim ＝ Phalaenopsis wilsonii Rolfe ■

307805 Polychisma C. Muell. ＝ Pelargonium L'Hér. ex Aiton ●■

307806 Polychisma C. Muell. ＝ Polyschisma Turcz. ●■

307807 Polychlaena G. Don ＝ Melochia L.（保留属名）●■

307808 Polychlaena Garcke ＝ Hibiscus L.（保留属名）●■

307809 Polychnemum Zumagl. ＝ Polycnemum L. ■

307810 Polychroa Lour.（废弃属名）＝ Pellionia Gaudich.（保留属名）●■

307811 Polychroa repens Lour. ＝ Pellionia repens（Lour.）Merr. ■

307812 Polychroa scabra（Benth.）Hu ＝ Pellionia scabra Benth. ●■

307813 Polychroa tsoongii Merr. ＝ Pellionia tsoongii（Merr.）Merr. ■

307814 Polychrysum（Tzvelev）Kovalevsk.（1962）；密金蒿属■☆

307815 Polychrysum tadshikorum（Kudr.）Kovalevsk.；塔什克密金蒿■☆

307816 Polyclados Phil. ＝ Lepidophyllum Cass. ●☆

307817 Polyclathra Bertol.（1840）；多格瓜属■☆

307818 Polycline Oliv. ＝ Athroisma DC. ■●☆

307819 Polycline gracilis（Oliv.）Oliv. ＝ Athroisma gracile（Oliv.）Mattf. ●☆

307820 Polycline haareri Dandy ＝ Athroisma gracile（Oliv.）Mattf. subsp. psyllioides（Oliv.）T. Erikss. ●☆

307821 Polycline haareri Dandy var. javellensis Lanza ＝ Athroisma boranense Cufod. ●☆

307822 Polycline lobata（Klatt）Chiov. ＝ Athroisma lobatum（Klatt）Mattf. ●☆

307823 Polycline psyllioides Oliv. ＝ Athroisma gracile（Oliv.）Mattf. subsp. psyllioides（Oliv.）T. Erikss. ●☆

307824 Polycline stuhlmannii O. Hoffm. ＝ Athroisma stuhlmannii（O. Hoffm.）Mattf. ●☆

307825 Polyclita A. C. Sm.（1936）；大杯莓属●☆

307826 Polyclita turbinata（Kuntze）A. C. Sm.；大杯莓●☆

307827 Polyclonos Raf. ＝ Orobanche L. ■

307828 Polycnemaceae Menge ＝ Amaranthaceae Juss.（保留科名）●■

307829 Polycnemon F. Muell. ＝ Polycnemum L. ■

307830 Polycnemum L.（1753）；多节草属［Needleleaf，Polycnemum］■

307831 Polycnemum americanum Nutt. ＝ Corispermum americanum（Nutt.）Nutt. ■☆

307832 Polycnemum arvense L.；多节草；Soft Needleleaf ■☆

307833 Polycnemum crassifolium Pall. ＝ Petrosimonia oppositifolia（Pall.）Litv. ■

307834 Polycnemum erinaceum Pall. ＝ Nanophyton erinaceum（Pall.）Bunge ●■

307835 Polycnemum fontanesii Durieu et Moq.；丰塔纳多节草■☆

307836 Polycnemum fontanesii Durieu et Moq. subsp. maroccanum Murb. ＝ Polycnemum fontanesii Durieu et Moq. ■☆

307837 Polycnemum fontanesii Durieu et Moq. var. oxysepalum Maire ＝ Polycnemum fontanesii Durieu et Moq. ■☆

307838 Polycnemum glaucum Pall. ＝ Petrosimonia glaucescens（Bunge）Iljin ■

307839 Polycnemum heuffelii Láng；霍氏多节草■☆

307840 Polycnemum majus A. Braun ex Bogenh.；大多节草；Giant Needleleaf ■☆

307841 Polycnemum oppositifolium Pall. ＝ Petrosimonia oppositifolia（Pall.）Litv. ■

307842 Polycnemum perenne Litv.；多年生多节草■☆

307843 Polycnemum sibiricum Pall. ＝ Petrosimonia sibirica（Pall.）Bunge ■

307844 Polycnemum verrucosum Láng；多疣多节草■☆

307845 Polycodium Raf. ＝ Vaccinium L. ●

307846 Polycodium Raf. ex Greene ＝ Vaccinium L. ●

307847 Polycoelium A. DC. ＝ Myoporum Banks et Sol. ex G. Forst. ●

307848 Polycoelium A. DC. ＝ Pentacoelium Siebold et Zucc. ●

307849 Polycoelium bontioides（Siebold et Zucc.）A. DC. ＝ Myoporum bontioides（Siebold et Zucc.）A. Gray ●

307850 Polycoelium bontioides（Siebold et Zucc.）A. DC. ＝ Pentacoelium bontioides Siebold et Zucc. ●

307851 Polycoelium chinense A. DC. ＝ Myoporum bontioides（Siebold et Zucc.）A. Gray ●

307852 Polycoelium chinense A. DC. ＝ Pentacoelium bontioides Siebold et Zucc. ●

307853 Polycoryne Keay ＝ Pleiocoryne Rauschert ●☆

307854 Polycoryne fernandensis（Hiern）Keay ＝ Pleiocoryne fernandense（Hiern）Rauschert ●☆

307855 Polycoryne fernandensis（Hiern）Keay var. pobeguinii N. Hallé ＝ Pleiocoryne fernandense（Hiern）Rauschert var. pobeguinii（N. Hallé）J.-P. Lebrun et Stork ●☆

307856 Polyctenium Greene（1912）；多栉芥属■☆

307857 Polyctenium fremontii Greene；多栉芥■☆

307858 Polycycliska Ridl. ＝ Lerchea L.（保留属名）●■

307859 Polycycnis Rchb. f.（1855）；白鸟兰属■☆

307860 Polycycnis barbata Rchb. f.；长毛鹅花兰（长毛白鸟兰）■☆

307861 Polycycnopsis Szlach.（1845）；拟天鹅兰属■☆

307862 Polycycnopsis Szlach. ＝ Polycycnis Rchb. f. ■☆

307863 Polycyema Voigt ＝ Clausena Burm. f. ●

307864 Polycyrtus Schltdl. ＝ Ferula L. ■

307865 Polydendris Thouars ＝ Dendrobium Sw.（保留属名）■

307866 Polydendris Thouars ＝ Polystachya Hook.（保留属名）■

307867 Polydiclis（G. Don）Miers ＝ Nicotiana L. ●■

307868 Polydiclis Miers ＝ Nicotiana L. ●■

307869 Polydontia Blume ＝ Lauro-Cerasus Duhamel ●

307870 Polydontia Blume ＝ Pygeum Gaertn. ●

307871 Polydora Fenzl ＝ Vernonia Schreb.（保留属名）●■

307872 Polydora Fenzl（2004）；多毛瘦片菊属■☆

307873 Polydora angustifolia（Steetz）H. Rob. ＝ Vernonia rhodanthoidea Muschl. ■☆

307874 Polydora bainesii（Oliv. et Hiern）H. Rob. ＝ Vernonia bainesii Oliv. et Hiern ●☆

307875 Polydora chloropappa（Baker）H. Rob. ＝ Vernonia chloropappa Baker ●☆

307876 Polydora jelfiae（S. Moore）H. Rob. ＝ Vernonia jelfiae S. Moore ●☆

307877 Polydora serratuloides（DC.）H. Rob. ＝ Vernonia perrottetii Sch. Bip. ex Walp. ●☆

307878 Polydora stoechadifolia Fenzl ＝ Vernonia perrottetii Sch. Bip. ex Walp. ●☆

307879 Polydora sylvicola（G. V. Pope）H. Rob. ＝ Vernonia sylvicola G. V. Pope ●☆

307880 Polydragma Hook. f. ＝ Spathiostemon Blume ●☆

307881 Polyechma Hochst. ＝ Hygrophila R. Br. ●■

307882 Polyechma abyssinicum Hochst. ex Nees ＝ Hygrophila abyssinica（Hochst. ex Nees）T. Anderson ●☆

307883 Polyechma caeruleum Hochst. ＝ Hygrophila caerulea（Hochst.）T. Anderson ■☆

307884 Polyechma micranthum Nees ＝ Hygrophila micrantha（Nees）

T. Anderson ■☆

307885 Polyechma odorum Nees = Hygrophila odora（Nees）T. Anderson ■☆

307886 Polyembrium Schott ex Steud. = Polyembryum Schott ex Stand. ●☆

307887 Polyembryum Schotr ex Stand. = Esenbeckia Kunth ●☆

307888 Polyembryum Schott ex Steud. = Polembryum A. Juas. ●☆

307889 Polyembryum Schott ex Steud. = Esenbeckia Kunth ●☆

307890 Polyembryum Schott ex Steud. = Polembryum A. Juss. ●☆

307891 Polygala L.（1753）；远志属；Milkwort，Polygala ●■

307892 Polygala abyssinica R. Br. ex Fresen.；阿比西尼亚远志■☆

307893 Polygala abyssinica R. Br. ex Fresen. var. gerardiana（Hassk.）Chodat = Polygala abyssinica R. Br. ex Fresen.■☆

307894 Polygala acicularis Oliv.；针形远志■☆

307895 Polygala adamsonii Exell；亚当森远志■☆

307896 Polygala affinis DC. = Polygala scabra L.■☆

307897 Polygala africana Chodat；非洲远志■☆

307898 Polygala alba Nutt.；白远志；White Milkwort ■☆

307899 Polygala albida Schinz；浅白远志■☆

307900 Polygala albida Schinz subsp. stanleyana（Chodat）Paiva；斯坦利远志■☆

307901 Polygala albida Schinz var. angustifolia Exell = Polygala albida Schinz subsp. stanleyana（Chodat）Paiva ■☆

307902 Polygala alopecuroides L. = Muraltia alopecuroides（L.）DC. ●☆

307903 Polygala alpicola Rupr.；高山远志■☆

307904 Polygala amara L. = Polygala amarella Crantz ■☆

307905 Polygala amarella Crantz；奥地利远志(苦牛奶草,苦远志,约克夏远志)；Dwarf Milkwort，Kentish Milkwort，Yorkshire Milkwort ■☆

307906 Polygala amatymbica Eckl. et Zeyh.；热非远志●☆

307907 Polygala ambigua Nutt.；可疑远志■☆

307908 Polygala ambigua Nutt. = Polygala verticillata L. var. ambigua（Nutt.）A. W. Wood ■☆

307909 Polygala amoenissima Tamamsch.；秀丽远志■☆

307910 Polygala anatolica Boiss. et Heldr.；小亚细亚远志■☆

307911 Polygala andrachnoides Willd.；黑钩叶远志■☆

307912 Polygala andringitrensis Paiva；安德林吉特拉山远志■☆

307913 Polygala angolensis Chodat；安哥拉远志●☆

307914 Polygala ankaratrensis H. Perrier；安卡拉特拉远志■☆

307915 Polygala antunesii Gürke；安图内思远志■☆

307916 Polygala aphrodisiaca Gürke = Polygala conosperma Bojer ●☆

307917 Polygala apopetala Brandegee；离瓣远志●☆

307918 Polygala arcuata Chodat = Polygala rodrigueana Paiva ■☆

307919 Polygala arcuata Hayata；台湾远志(巨花远志,巨叶花远志)；Largeflower Greenstar，Taiwan Milkwort ●◇

307920 Polygala arenaria Willd.；沙地远志■☆

307921 Polygala arenicola Gürke；沙生远志■☆

307922 Polygala argentea Thulin；银白远志■☆

307923 Polygala arillata Buch. -Ham. ex D. Don；荷苞山桂花(白糯消,吊吊果,吊吊黄,桂花岩陀,荷包山桂花,花岩陀,黄花鸡骨,黄花远志,黄金卵,黄杨参,鸡肚子根,鸡肚子果,鸡根,金不换,老母鸡嘴,树参,小荷包,小荷苞,小鸡花,阳雀花)；Coinbag Milkwort，Yellowflower Milkwort，Yellow-flowered Milkwort ●

307924 Polygala arillata Buch. -Ham. ex D. Don f. kachinensis Mukerjee = Polygala globulifera Dunn var. longiracemosa S. K. Chen ●

307925 Polygala arillata Buch. -Ham. ex D. Don var. ovata Gagnep.；卵叶荷包山桂花；Ovateleaf Milkwort ●

307926 Polygala armata Chodat = Polygala leptophylla Burch. var. armata（Chodat）Paiva ■☆

307927 Polygala arvensis Willd.；无柄花瓜子金(白花远志,辰沙草,辰砂草,多年红,金牛草,坡白草,七寸金,细金草,细金牛,细金牛草,细牛草,细叶金不换,小花远志,小金不换,小金牛草,小兰青,紫背金牛,紫花地丁)；Miniflower Milkwort ■

307928 Polygala arvicola Bojer；旱沙远志■☆

307929 Polygala asbestina Burch.；阿斯别斯特■☆

307930 Polygala asbestina Burch. var. rigens（Burch.）Harv. = Polygala rigens Burch.■☆

307931 Polygala aschersoniana Chodat；阿舍森远志■☆

307932 Polygala asperifolia Chodat = Polygala stenopetala Klotzsch ●☆

307933 Polygala atacorensis Jacq. -Fél.；阿塔科远志■☆

307934 Polygala aurata Gagnep. = Polygala linarifolia Willd.■

307935 Polygala aurata Gagnep. var. macrostachya Gagnep. = Polygala linarifolia Willd.■

307936 Polygala aureocauda Dunn = Polygala fallax Hemsl. ●

307937 Polygala austriaca Crantz = Polygala amarella Crantz ■☆

307938 Polygala axillaris Poir. = Floscopa axillaris（Poir.）C. B. Clarke ■☆

307939 Polygala baetica Willk.；伯蒂卡远志■☆

307940 Polygala baetica Willk. var. pallescens Emb. et Maire = Polygala baetica Willk.■☆

307941 Polygala baetica Willk. var. sennenii（Pau）Maire = Polygala baetica Willk.■☆

307942 Polygala baetica Willk. var. stenoptera Humbert et Maire = Polygala baetica Willk.■☆

307943 Polygala baikiei Chodat subsp. pobeguinii（A. Chev. et Jacq. -Fél.）Paiva；波别远志■☆

307944 Polygala bakeriana Chodat；贝克远志■☆

307945 Polygala balansae Coss.；巴兰萨远志■☆

307946 Polygala barbellata S. K. Chen；髯毛远志(接骨丹)；Barbellate Milkwort，Beard Milkwort ●

307947 Polygala baumii Gürke；鲍姆远志■☆

307948 Polygala bawanglingensis F. W. Xing et Z. X. Li；坝王远志；Bawangling Milkwort ●

307949 Polygala beiliana Eckl. et Zeyh. = Muraltia muraltioides（Eckl. et Zeyh.）Levyns ■☆

307950 Polygala bennae Jacq. -Fél.；本纳远志●☆

307951 Polygala bicornis Burch. ex Chodat = Polygala schinziana Chodat ■☆

307952 Polygala bicornis H. Perrier = Polygala andringitrensis Paiva ■☆

307953 Polygala boissieri Coss.；布瓦西耶远志■☆

307954 Polygala bojeri Chodat = Polygala arvicola Bojer ■☆

307955 Polygala bowkerae Harv.；鲍克远志■☆

307956 Polygala brachyphylla Chodat；短叶远志■☆

307957 Polygala brachystachya Blume = Polygala linarifolia Willd.■

307958 Polygala brachystachya DC. = Polygala arvensis Willd.■

307959 Polygala bracteolata L.；小苞远志■☆

307960 Polygala bracteolata L. var. racemosa Harv. = Polygala bracteolata L.■☆

307961 Polygala bracteolata L. var. umbellata Harv. = Polygala bracteolata L.■☆

307962 Polygala brevifolia Harv. = Polygala brachyphylla Chodat ■☆

307963 Polygala britteniana Chodat；布里滕远志■☆

307964 Polygala buchanani Buch. -Ham. ex D. Don = Polygala persicariifolia DC.■

307965 Polygala buchenavii O. Hoffm. = Polygala schoenlankii O. Hoffm. et Hildebrandt ■☆

307966 Polygala bukobensis Gürke = Polygala capillaris E. Mey. ex Harv.■☆

307967　Polygala burmanii DC. = Polygala scabra L. ■☆

307968　Polygala cabrae Chodat；卡布拉远志●☆

307969　Polygala calcarea J. E. Zetterst. ex Willk. et Lange；灰岩远志；Chalk Milkwort，Milkwort，Shepherd's Blue Thyme，Shepherd's Love，Shepherd's Thyme ■☆

307970　Polygala calcarea J. E. Zetterst. ex Willk. et Lange 'Bulley'；布丽灰岩远志■☆

307971　Polygala calcicola Chodat = Polygala senensis Klotzsch var. calcicola（Chodat）Paiva ■☆

307972　Polygala calycina C. Presl = Polygala bracteolata L. ■☆

307973　Polygala capillaris E. Mey. ex Harv.；发状远志■☆

307974　Polygala capillaris E. Mey. ex Harv. subsp. perrottetiana（Paiva）Paiva；佩罗远志■☆

307975　Polygala capillaris E. Mey. ex Harv. var. angolensis Oliv. = Polygala spicata Chodat ■☆

307976　Polygala capillaris E. Mey. ex Harv. var. bosobolensis E. M. Petit = Polygala capillaris E. Mey. ex Harv. ■☆

307977　Polygala capillaris E. Mey. ex Harv. var. tukpwoensis E. M. Petit = Polygala capillaris E. Mey. ex Harv. ■☆

307978　Polygala cardiocarpa Kurz = Polygala isocarpa Chodat ■

307979　Polygala carnosicaulis W. H. Chen et Y. M. Shui；肉茎远志■

307980　Polygala carrissoana Exell et Mendonça；卡里索远志●☆

307981　Polygala caucasica Rupr.；高加索远志■☆

307982　Polygala caudata Rehder et E. H. Wilson；尾叶远志（毛籽红山桂，毛籽山桂花，木本远志，水黄杨木，乌棒子，野桂花）；Caudate Milkwort，Tailleaf Milkwort ●

307983　Polygala chamaebuxus L.；革叶远志（矮黄杨远志）；Bastard Box，Box-leaved Milkwort，Grondbox Milkwort，Ground Box，Milkwort，Shrubby Milkwort ●☆

307984　Polygala chamaebuxus L. var. grandiflora Gaudich.；紫花革叶远志（大花革叶远志，紫花远志）；Purple Grondbox Milkwort ■☆

307985　Polygala chevalieri Chodat = Polygala lecardii Chodat ●☆

307986　Polygala chinensis L. = Polygala glomerata Lour. ■

307987　Polygala chinensis L. f. arvensis（Willd.）Chodat = Polygala arvensis Willd. ■

307988　Polygala chinensis L. var. brachystachya（Blume）Benn. = Polygala linarifolia Willd. ■

307989　Polygala chinensis L. var. linarifolia（Willd.）Chodat = Polygala linarifolia Willd. ■

307990　Polygala chloroptera Chodat = Polygala serpentaria Eckl. et Zeyh. ■☆

307991　Polygala ciliatifolia Turcz. = Polygala umbellata L. ■☆

307992　Polygala citrina Thulin；柠檬远志■☆

307993　Polygala claessensii Chodat = Polygala paludicola Gürke ■☆

307994　Polygala clarkeana Chodat = Polygala lecardii Chodat ■☆

307995　Polygala colchica Tamamsch.；黑海远志■☆

307996　Polygala comesperma Chodat = Polygala wattersii Hance ●

307997　Polygala comgesta Rehder et E. H. Wilson = Polygala tricornis Gagnep. ●

307998　Polygala comosa Schkuhr；丛毛远志■☆

307999　Polygala comosa Schkuhr var. altaica Chodat = Polygala hybrida DC. ■

308000　Polygala compressa H. Perrier；扁远志●☆

308001　Polygala compressa H. Perrier f. mandrarensis H. Perrier = Polygala compressa H. Perrier ●☆

308002　Polygala comsesperma Chodat = Polygala caudata Rehder et E. H. Wilson ●

308003　Polygala confusa MacOwan = Polygala macowaniana Paiva ■☆

308004　Polygala congoensis Gürke = Polygala nambalensis Gürke ●☆

308005　Polygala conosperma Bojer；束籽远志●☆

308006　Polygala cornuta Kellogg；角状远志；Horned Milkwort，Milkwort ■☆

308007　Polygala costaricensis Chodat ex T. Durand et Pitt.；哥斯达黎加远志■☆

308008　Polygala coursiereana Pomel = Polygala nicaeensis Koch ■☆

308009　Polygala crassiuscula Hayata = Polygala arcuata Hayata ●◇

308010　Polygala cretacea Kotov；白垩远志■☆

308011　Polygala cristata P. Taylor；冠状远志■☆

308012　Polygala crotalarioides Buch. -Ham. ex DC.；西南远志（地花生，翻转红，西藏远志，猪大肠）；SW. China Milkwort，Tibet Milkwort ■

308013　Polygala crotalarioides DC.；猪屎豆远志■☆

308014　Polygala cruciata L.；十字远志；Cross Milkwort，Cross-leaf Milkwort，Cross-leaved Milkwort，Curtiss' Milkwort，Drum-heads，Marsh Milkwort，Polygala ■☆

308015　Polygala cruciata L. var. aquilonia Fernald et B. G. Schub. = Polygala cruciata L. ■☆

308016　Polygala dalmaisiana Bailey；达耳马氏远志（达尔迈远志，玫瑰远志）；Dalmais Milkwort，Sweet Pea Shrub ●☆

308017　Polygala dasyphylla Levyns；毛叶远志■☆

308018　Polygala declinata（Harv.）E. Mey. ex Paiva；外折远志■☆

308019　Polygala decora Sond. = Polygala virgata Thunb. var. decora（Sond.）Harv. ●☆

308020　Polygala densiflora Blume = Polygala glomerata Lour. ■

308021　Polygala dewevrei Exell；德韦远志■☆

308022　Polygala dewevrei Exell var. schmitzii E. M. Petit = Polygala baumii Gürke ■☆

308023　Polygala dictyoptera Boiss.；网翅远志■☆

308024　Polygala didyma C. Y. Wu；肾果远志（肾果山桂花）；Didymous Milkwort，Kidnyfruit Milkwort ●

308025　Polygala discolor Buch. -Ham. ex D. Don = Polygala longifolia Poir. ■

308026　Polygala djalonis A. Chev. = Polygala sparsiflora Oliv. ●☆

308027　Polygala dumosa Poir. = Muraltia dumosa（Poir.）DC. ●☆

308028　Polygala dunniana H. Lév.；贵州远志；Dunn Milkwort，Guizhou Milkwort ●■

308029　Polygala dunniana H. Lév. = Polygala hainanensis Chun et F. C. How ●

308030　Polygala eckloniana C. Presl = Polygala teretifolia L. f. ■☆

308031　Polygala effusa Paiva et Thulin；开展远志■☆

308032　Polygala elegans Wall.；雅致远志；Elegant Milkwort ■

308033　Polygala emirnensis Baker；埃米远志■☆

308034　Polygala emoryi S. F. Blake；喜山远志■☆

308035　Polygala empetrifolia Houtt.；岩高兰叶远志■☆

308036　Polygala engleri Chodat；恩格勒远志●☆

308037　Polygala engleriana Buscal. et Muschl.；恩氏远志■☆

308038　Polygala ephedroides Burch.；麻黄远志■☆

308039　Polygala ericifolia DC.；南非毛叶远志■☆

308040　Polygala ericoides Burm. f. = Muraltia ericoides（Burm. f.）Steud. ●☆

308041　Polygala erioptera DC.；毛翅远志■☆

308042　Polygala erioptera DC. subsp. petraea（Chodat）Paiva；坦桑尼亚毛翅远志■☆

308043　Polygala erioptera var. vahliana（DC.）Chodat = Polygala erioptera DC. ■☆

308044   Polygala erlangeri Gürke ex Chodat;厄兰格远志■☆

308045   Polygala erubescens E. Mey. ex Chodat;变红远志■☆

308046   Polygala esterae Chodat = Polygala gazensis Baker f. ■☆

308047   Polygala exelliana Troupin;埃克塞尔远志■☆

308048   Polygala fallax Hayek = Polygala fallax Hemsl. ex Zahlbr. ●

308049   Polygala fallax Hemsl. = Polygala fallax Hemsl. ex Zahlbr. ●

308050   Polygala fallax Hemsl. ex Zahlbr.;假黄花远志(白马胎,倒吊黄,倒吊黄花,倒吊莲,吊吊黄,吊黄,观音串,黄花参,黄花大远志,黄花倒水莲,黄花吊水莲,黄花远志,黄金印,鸡仔树,木本远志,念健,一身保暖);False-yellowflower Milkwort, False-yellow-flowered Milkwort, Yellowflower Milkwort ●

308051   Polygala fasciculata Poir. = Muraltia ericoides (Burm. f.) Steud. ●☆

308052   Polygala fernandesiana Paiva;费尔南远志●☆

308053   Polygala ficalhoana Exell et Mendonça = Polygala angolensis Chodat ●☆

308054   Polygala filicaulis Baill.;线茎远志■☆

308055   Polygala filifera Chodat = Polygala sphenoptera Fresen. ■☆

308056   Polygala filiformis Thunb. = Muraltia filiformis (Thunb.) DC. ●☆

308057   Polygala fischeri Gürke;菲舍尔远志●☆

308058   Polygala floribunda Dunn = Polygala tricornis Gagnep. ●

308059   Polygala fontqueri Pau = Polygala rupestris Pourr. subsp. fontqueri (Pau) Font Quer ■☆

308060   Polygala forbesii Chodat = Polygala fallax Hemsl. ex Zahlbr. ●

308061   Polygala fragilis Paiva;脆远志■☆

308062   Polygala franciscii Exell;弗朗西斯科远志■☆

308063   Polygala friesii Chodat = Polygala engleriana Buscal. et Muschl. ■☆

308064   Polygala fruticosa P. J. Bergius;灌木状远志■☆

308065   Polygala furcata Royle;叉枝远志(肾果小扁豆,一碗泡);Furcate Milkwort ■

308066   Polygala gagnebiniana Chodat = Polygala sphenoptera Fresen. ■☆

308067   Polygala galpinii Hook. f. = Heterosamara galpinii (Hook. f.) Paiva ●☆

308068   Polygala garcinii DC.;加尔桑远志■☆

308069   Polygala gazensis Baker f.;加兹远志■☆

308070   Polygala genistoides Poir. = Polygala virgata Thunb. ●☆

308071   Polygala genistoides Poir. var. ephedroides (Burch.) DC. = Polygala ephedroides Burch. ■☆

308072   Polygala gerardiana Wall. ex Hassk. = Polygala abyssinica R. Br. ex Fresen. ■☆

308073   Polygala gerrardii Chodat;杰勒德远志■☆

308074   Polygala gilletiana E. M. Petit;吉氏远志■☆

308075   Polygala gilletii Paiva;吉勒特远志■☆

308076   Polygala glaucescens Wall. = Polygala furcata Royle ■

308077   Polygala globulifera Dunn;球冠远志(大金,假指味风,金不换);Ball Denseflower Milkwort, Ball Densi-flowered Milkwort, Miniball Milkwort ●

308078   Polygala globulifera Dunn var. kachinensis (Mukerjee) R. N. Banerjee ex R. N. Banerjee et al. = Polygala globulifera Dunn var. longiracemosa S. K. Chen ●

308079   Polygala globulifera Dunn var. kachinensis (Mukerjee) R. N. Banerjee = Polygala globulifera Dunn var. longiracemosa S. K. Chen ●

308080   Polygala globulifera Dunn var. longiracemosa S. K. Chen;长序球冠远志;Longraceme Milkwort ●

308081   Polygala glomerata Lour.;华南远志(大金不换,大金草,大金牛,大兰青,大蓝青,大叶金不换,肥儿草,甘得,瘠积草,厚皮柑,厚皮桔,金不换,金牛草,金牛远志,坡白草,杀粘,蛇总管,无柄花瓜子金,午时合,银不换,鹧鸪茶,紫背金牛);Chinese Milkwort, S. China Milkwort, South China Milkwort ■

308082   Polygala glomerata Lour. var. pygmaea C. Y. Wu et S. K. Chen;矮华南远志;Dwarf S. China Milkwort, Dwarf South China Milkwort ■

308083   Polygala glomerata Lour. var. villosa C. Y. Wu et S. K. Chen;长毛华南远志;Villose South China Milkwort ■

308084   Polygala goetzei Gürke;格兹远志■☆

308085   Polygala gomesiana Welw. ex Oliv.;戈梅斯远志■☆

308086   Polygala gondarensis Chiov.;贡达尔远志■☆

308087   Polygala gossweileri Exell et Mendonça;戈斯远志■☆

308088   Polygala gracilenta Burtt Davy;细黏远志■☆

308089   Polygala gracilipes Harv.;细梗远志■☆

308090   Polygala grandidieri Baill. = Polygala subglobosa Paiva ■☆

308091   Polygala greveana Baill.;格雷弗远志■☆

308092   Polygala guineensis Willd.;几内亚远志■☆

308093   Polygala gymnoclada MacOwan;裸枝远志■

308094   Polygala gypsophila Thulin;喜钙远志■☆

308095   Polygala hainanensis Chun et F. C. How;海南远志;Hainan Milkwort ●

308096   Polygala hainanensis Chun et F. C. How var. strigosa Chun et F. C. How;粗毛海南远志(毛海南远志);Hispid Hainan Milkwort ●

308097   Polygala hamata Burtt Davy = Polygala uncinata E. Mey. ex Meisn. ■☆

308098   Polygala hasskadii Merr. et Chun = Polygala tricholopha Chodat ●

308099   Polygala heliostigma Chodat = Polygala usafuensis Gürke ■☆

308100   Polygala heterantha H. Perrier;异花毛远志●☆

308101   Polygala hildebrandtii Baill.;希尔德远志●☆

308102   Polygala hispida Burch. ex DC.;硬毛远志■☆

308103   Polygala hispida Busch. ex DC. var. declinata Harv. = Polygala declinata (Harv.) E. Mey. ex Paiva ■☆

308104   Polygala hohenackeriana Fisch. et C. A. Mey.;霍氏远志远志■☆

308105   Polygala hohenackeriana var. stocksiana (Boiss.) Boiss. = Polygala hohenackeriana Fisch. et C. A. Mey. ■☆

308106   Polygala homblei Exell;洪布勒远志■☆

308107   Polygala hondoensis Nakai = Polygala japonica Houtt. ■

308108   Polygala hongkongensis Hemsl. ex Forbes et Hemsl.;香港远志;Hongkong Milkwort ●■

308109   Polygala hongkongensis Hemsl. ex Forbes et Hemsl. var. stenophylla (Hayata) Migo;狭叶香港远志(地丁草,瓜子草,金钥匙,狭叶远志,鸭舌瓜子);Narrowleaf Hongkong Milkwort, Narrowleaf Hongkong Milkwort ●■

308110   Polygala hottentotta C. Presl;豪顿远志●☆

308111   Polygala hottentotta C. Presl var. fleckiana Schinz = Polygala leptophylla Burch. ■☆

308112   Polygala huillensis Welw. ex Oliv.;威拉远志●☆

308113   Polygala humbertii H. Perrier;亨伯特远志●☆

308114   Polygala humbertii H. Perrier var. crinigera H. Perrier = Polygala humbertii H. Perrier ●☆

308115   Polygala humifusa Paiva;平伏远志■☆

308116   Polygala hybrida DC.;新疆远志(远志);Hybrid Milkwort, Xinjiang Milkwort ■

308117   Polygala hyssopifolia Bojer = Polygala schoenlankii O. Hoffm. et Hildebrandt ■☆

308118   Polygala incarnata L.;肉色远志;Pink Milkwort, Procession Flower, Slender Milkwort ■☆

308119   Polygala insularis Chun et F. C. How ex Y. C. Wu et S. K. Chen;海岛远志;Island Milkwort ●■

308120　Polygala irregularis Boiss. ;不对称远志■☆

308121　Polygala isaloensis H. Perrier;伊萨卢远志●☆

308122　Polygala isocarpa Chodat; 心果小扁豆（滑石草）; Heartfruit Milkwort, Isocarp Milkwort ■

308123　Polygala japonica Houtt. ;瓜子金（草鳖黄，产后草，辰砂草，地风消，地莳草，地藤草，二月花，高脚瓜子草，蛤鳖黄，瓜米草，瓜米细辛，瓜子草，瓜子莲，黄瓜仁草，鸡拍翅，歼疟草，接骨红，金牛草，金锁匙，金钥匙，惊风草，苦草，苦远志，拦路枝，蓝花草，柳叶紫花，卵叶远志，女儿红，七寸金，日本远志，散血丹，山黄连，神砂草，铁箭风，铁洗帚，铁线风，通性草，细金不换，小金不换，小金盆，小叶地丁草，小叶瓜子草，小英雄，小远志，银不换，鱼胆草，远志草，竹叶地丁）;Japan Milkwort, Japanese Milkwort ■

308124　Polygala japonica Houtt. f. angustifolia（Koidz.）H. Hara;狭叶瓜子金■☆

308125　Polygala japonica Houtt. f. ciliata Hiyama;缘毛瓜子金■☆

308126　Polygala japonica Houtt. f. virescens Nakai;绿瓜子金■☆

308127　Polygala japonica Houtt. var. angustifolia Koidz. = Polygala japonica Houtt. ■

308128　Polygala juniperifolia Poir. = Muraltia juniperifolia（Poir.）DC. ■☆

308129　Polygala juniperina Cav. = Polygala rupestris Pourr. ■☆

308130　Polygala kagerensis Lebrun et Taton = Polygala transvaalensis Chodat subsp. kagerensis（Lebrun et Taton）Paiva ■☆

308131　Polygala kalaxariensis Schinz;卡拉克萨尔远志■☆

308132　Polygala kasikensis Exell;卡西基远志■☆

308133　Polygala kassasii Chrtek;卡萨斯远志■☆

308134　Polygala katangensis Exell = Polygala myrtillopsis Welw. ex Oliv. ●☆

308135　Polygala kemulariae Tamamsch. ;凯木远志■☆

308136　Polygala khasiana Hassk. ;卡西远志;Khas Milkwort ■

308137　Polygala kinii Courtois = Polygala arvensis Willd. ■

308138　Polygala koi Merr. ;曲江远志（红花倒水莲，一包花）;Ko Milkwort, Qujiang Milkwort ●

308139　Polygala kubangensis Gürke = Polygala kalaxariensis Schinz ■☆

308140　Polygala lacei Craib;思茅远志; Simao Milkwort, Szemao Milkwort ■

308141　Polygala lactiflora Paiva et Brummitt;乳白花远志■☆

308142　Polygala lancilimba Merr. = Polygala tricornis Gagnep. ●

308143　Polygala langebergensis Levyns;朗厄山远志■☆

308144　Polygala lasiosepala Levyns;毛萼远志■☆

308145　Polygala lateriflora Y. K. Yang;侧生花远志（对时接骨草，合叶，合掌草，午时合）■

308146　Polygala laticarpa Doum. = Polygala rupestris Pourr. var. saxatilis（Desf.）Murb. ■☆

308147　Polygala latifolia Ker Gawl. = Polygala fruticosa P. J. Bergius ■☆

308148　Polygala latipetala N. E. Br. ;宽瓣远志■☆

308149　Polygala latouchei Franch. ;大叶金牛（红背兰，天青地紫，岩生远志，一包花）;Largeleaf Milkwort ●

308150　Polygala laxifolia Exell;疏叶远志■☆

308151　Polygala lecardii Chodat;莱卡德远志■☆

308152　Polygala leendertziae Burtt Davy;伦德茨远志■☆

308153　Polygala lehmanniana Eckl. et Zeyh. ;莱曼远志■☆

308154　Polygala lepidota Welw. ex Exell = Polygala welwitschii Chodat subsp. pygmaea（Gürke）Paiva ■☆

308155　Polygala leptalea DC. = Polygala longifolia Poir. ■☆

308156　Polygala leptophylla Burch. ;细叶远志■☆

308157　Polygala leptophylla Burch. var. armata（Chodat）Paiva;具刺细叶远志■☆

308158　Polygala leptorhiza DC. = Polygala erioptera DC. ■☆

308159　Polygala leucocarpa Chodat = Polygala pallida E. Mey. ■☆

308160　Polygala leucothyrsa Woronow;白序远志■☆

308161　Polygala levynsiana Paiva;勒温斯远志■☆

308162　Polygala lhunzeensis C. Y. Wu et S. K. Chen;隆子远志;Longzi Milkwort ■

308163　Polygala lijiangensis C. Y. Wu et S. K. Chen;丽江远志;Lijiang Milkwort ■

308164　Polygala linarifolia Willd. ；金花远志（细叶远志）; Goldenflower Milkwort, Toadflexleaf Milkwort ■

308165　Polygala linearis E. Mey. = Polygala gracilenta Burtt Davy ■☆

308166　Polygala linearis R. Br. = Polygala erioptera DC. ■☆

308167　Polygala lonchophylla Greene = Polygala senega L. ■☆

308168　Polygala longeracemosa H. Perrier;长序远志●☆

308169　Polygala longeracemosa var. retamoides H. Perrier = Polygala longeracemosa H. Perrier ●☆

308170　Polygala longifolia Poir. ；长叶远志（山鼠尾）; Longleaf Milkwort ■

308171　Polygala lourerii Gardner et Champ. = Polygala hongkongensis Hemsl. ex Forbes et Hemsl. ●■

308172　Polygala ludwigiana Eckl. et Zeyh. ;路德维格远志■☆

308173　Polygala lutea L. ;橘黄远志（浅黄远志，糖果草）;Milkwort, Orange Milkwort, Wild Bachelor's Buttons, Yellow Bachelor's Buttons, Yellow Milkwort ■☆

308174　Polygala luteo-viridis Chodat;黄绿远志■☆

308175　Polygala luzoniensis Merr. = Polygala japonica Houtt. ■

308176　Polygala lysimachiifolia Chodat;珍珠菜远志■☆

308177　Polygala macowaniana Paiva;麦克欧文远志■☆

308178　Polygala macradenia A. Gray;大腺远志■☆

308179　Polygala macroptera DC. ;大翅远志●☆

308180　Polygala macrostigma Chodat;大柱头远志■☆

308181　Polygala major Jacq. ;大远志■☆

308182　Polygala mannii Oliv. ;曼氏远志●☆

308183　Polygala mariamae Tamamsch;沼泽远志■☆

308184　Polygala mariesii Hemsl. ex Forbes et Hemsl. = Polygala wattersii Hance ●

308185　Polygala matteiana Pamp. = Polygala senensis Klotzsch ■☆

308186　Polygala melilotoides Chodat;草木犀远志■☆

308187　Polygala mendoncae E. M. Petit;门东萨远志■☆

308188　Polygala meonantha Chodat;细花远志■☆

308189　Polygala meridionalis Levyns;南方远志■☆

308190　Polygala micrantha Perr. et Guillaumin = Polygala capillaris E. Mey. ex Harv. ■☆

308191　Polygala micrantha Thunb. = Muraltia stipulacea Burch. ex DC. ●☆

308192　Polygala microlopha DC. ;小冠远志■☆

308193　Polygala minuta Paiva;微小远志■☆

308194　Polygala mixta L. f. = Muraltia mixta（L. f.）DC. ■☆

308195　Polygala moldavica Kotov;摩尔达瓦远志■☆

308196　Polygala monopetala Cambess. ;单瓣远志;Onepetal Milkwort ■

308197　Polygala monopetala Cambess. = Polygala sibirica L. ■

308198　Polygala mooneyi M. G. Gilbert;穆尼远志■☆

308199　Polygala mossamedensis Paiva;莫萨梅迪远志■☆

308200　Polygala mossii Exell;莫西远志■☆

308201　Polygala mucronata Baker = Polygala ankaratrensis H. Perrier ■☆

308202　Polygala multiflora Mattei = Polygala senensis Klotzsch ■☆

308203　Polygala multiflora Poir. ;多花远志■☆

308204　Polygala multifurcata Mildbr. ;多叉远志■☆

308205　Polygala munbyana Boiss. et Reut. ;芒比远志●☆

308206　Polygala muraltioides Eckl. et Zeyh. = Muraltia muraltioides（Eckl. et Zeyh.）Levyns ■☆

308207　Polygala myriantha Chodat;喀麦隆远志●☆

308208　Polygala myrsinites Royle = Polygala elegans Wall. ■

308209　Polygala myrtifolia L. ;番樱桃叶远志;Myrtle Milkwort,Myrtle-leaf Milkwort ●☆

308210　Polygala myrtifolia L. var. grandiflora Hook. ;美丽番樱桃叶远志;Showy Myrtle Milkwort ■☆

308211　Polygala myrtifolia L. var. pinifolia（Lam. ex Poir.）Paiva;松叶远志☆

308212　Polygala myrtillopsis Welw. ex Oliv. ;黑果越橘远志●☆

308213　Polygala nambalensis Gürke;楠巴莱远志●☆

308214　Polygala natalensis Chodat = Polygala serpentaria Eckl. et Zeyh. ■☆

308215　Polygala neglecta MacOwan = Polygala macowaniana Paiva ■☆

308216　Polygala negrii Chiov. = Polygala steudneri Chodat ■☆

308217　Polygala nematocaulis Levyns;虫茎远志■☆

308218　Polygala nematophylla Exell;蠕虫叶远志■☆

308219　Polygala nemorivaga Pomel = Polygala numidica Pomel ■☆

308220　Polygala nicaeensis Koch;尼赛远志■☆

308221　Polygala nicaeensis Koch subsp. mediterranea Chodat;地中海远志■☆

308222　Polygala nicaeensis Koch subsp. versicolor（Pomel）Batt. = Polygala nicaeensis Koch ■☆

308223　Polygala nicaeensis Koch var. commutata Pomel = Polygala nicaeensis Koch ■☆

308224　Polygala nicaeensis Koch var. courseriana（Pomel）Batt. = Polygala nicaeensis Koch ■☆

308225　Polygala nicaeensis Koch var. nemorivaga（Pomel）Bonnet et Barratte = Polygala numidica Pomel ■☆

308226　Polygala nicaeensis Koch var. obtusata Pomel = Polygala nicaeensis Koch ■☆

308227　Polygala nimborum Dunn = Polygala latouchei Franch. ●

308228　Polygala nodiflora Chodat;节花远志■☆

308229　Polygala nubica Hochst. = Polygala erioptera DC. ■☆

308230　Polygala numidica Pomel;努米底亚远志■☆

308231　Polygala nuttallii Torr. et Gray;纳托尔远志;Nuttall's Milkwort ■☆

308232　Polygala nyikensis Exell;尼卡远志■☆

308233　Polygala obtusata DC. = Polygala erioptera DC. ■☆

308234　Polygala obtusissima Hochst. ex Chodat = Polygala senensis Klotzsch ■☆

308235　Polygala okongavensis Dinter = Polygala pallida E. Mey. ■☆

308236　Polygala oligantha A. Rich. = Polygala erioptera DC. ■☆

308237　Polygala oligophylla DC. = Polygala longifolia Poir. ■

308238　Polygala oligosperma C. Y. Wu;少籽远志;Fewseed Milkwort,Paucity Milkwort,Paucity Seed Milkwort ●

308239　Polygala oliverana Exell et Mendonça;奥里弗远志■☆

308240　Polygala oppositifolia L. = Polygala fruticosa P. J. Bergius ■☆

308241　Polygala oppositifolia L. var. latifolia（Ker Gawl.）Harv. = Polygala fruticosa P. J. Bergius ■☆

308242　Polygala oxycocoides Desf. = Polygala rupestris Pourr. subsp. oxycoccoides（Desf.）Chodat ☆

308243　Polygala pallida E. Mey. ;苍白远志■☆

308244　Polygala paludicola Gürke;沼生远志■☆

308245　Polygala paniculata Forssk. = Polygala erioptera DC. ■☆

308246　Polygala paniculata L. = Polygala paniculata Le Conte ex Torr. et Gray ■

308247　Polygala paniculata Le Conte ex Torr. et Gray;圆锥花远志（圆锥远志）;Panicle Milkwort,Paniculate Milkwort ■

308248　Polygala papilionacea Boiss. ;蝶形远志■☆

308249　Polygala pappeana Eckl. et Zeyh. ;帕珀远志■☆

308250　Polygala parkeri Levyns;帕克远志■☆

308251　Polygala parva Chodat = Polygala seminuda Harv. ■☆

308252　Polygala pauciflora Thunb. = Muraltia pauciflora（Thunb.）DC. ●☆

308253　Polygala paucifolia Willd. ;穗叶远志;Flowering Polygala,Flowering Wintergreen,Flowering-wintergreen,Fringed Milkwort,Fringed Polygala,Gaywings,Gay-wings ■☆

308254　Polygala pearsonii Exell = Polygala pallida E. Mey. ■☆

308255　Polygala peduncularis Burch. ex DC. ;梗花远志■☆

308256　Polygala penaea L. ;佩尼远志■☆

308257　Polygala peplis Baill. ;孛艾远志■☆

308258　Polygala perrottetiana Paiva = Polygala capillaris E. Mey. ex Harv. subsp. perrottetiana（Paiva）Paiva ■☆

308259　Polygala persicariifolia DC. ;蓼叶远志（瓜子金,黄瓜仁草,辣味根,紫饭豆）;Knotweedleaf Milkwort ■

308260　Polygala persicarioides Franch. = Polygala persicariifolia DC. ■

308261　Polygala petitiana A. Rich. ;四萼远志■☆

308262　Polygala petitiana A. Rich. subsp. parviflora（Exell）Paiva;小花四萼远志■☆

308263　Polygala petitiana A. Rich. var. abercornensis Paiva;阿伯康远志■☆

308264　Polygala petitiana A. Rich. var. calceolata Norl. = Polygala petitiana A. Rich. ■☆

308265　Polygala petitiana A. Rich. var. parviflora Exell = Polygala petitiana A. Rich. subsp. parviflora（Exell）Paiva ■☆

308266　Polygala petraea Chodat = Polygala erioptera DC. subsp. petraea（Chodat）Paiva ■☆

308267　Polygala pilosa Baker = Polygala subglobosa Paiva ■☆

308268　Polygala pinifolia L. f. = Polygala teretifolia L. f. ■☆

308269　Polygala pinifolia Lam. ex Poir. = Polygala myrtifolia L. var. pinifolia（Lam. ex Poir.）Paiva ■☆

308270　Polygala pobeguinii A. Chev. et Jacq. -Fél. = Polygala baikiei Chodat subsp. pobeguinii（A. Chev. et Jacq. -Fél.）Paiva ■☆

308271　Polygala podolica DC. ;波多尔远志■☆

308272　Polygala poggei Gürke;波格远志■☆

308273　Polygala polyfolia C. Presl = Polygala arvensis Willd. ■

308274　Polygala polyfolia C. Presl = Polygala glomerata Lour. ■

308275　Polygala polygama Walter; 杂性远志（苦远志）; Bitter Milkwort,Pink Polygala,Purple Milkwort,Racemed Milkwort ■☆

308276　Polygala polygama Walter var. obtusata Chodat;紫苦远志;Bitter Milkwort,Purple Milkwort,Racemed Milkwort ■☆

308277　Polygala polygoniflora Chodat = Polygala sadebeckiana Gürke ■☆

308278　Polygala polyphylla DC. = Muraltia polyphylla（DC.）Levyns ●☆

308279　Polygala pottebergensis Levyns;波太伯格远志■☆

308280　Polygala praetermissa Thulin;疏忽远志■☆

308281　Polygala praticola Chodat;草原远志■☆

308282　Polygala pretzii Pennell = Polygala verticillata L. ■

308283　Polygala producta N. E. Br. ;伸展远志■☆

308284　Polygala pruinosa Boiss. ;粉远志■☆

308285　Polygala pteropoda H. Perrier;翅足远志●☆

308286　Polygala pubiflora Burch. ex DC. ;短毛花远志●☆

308287　Polygala pungens Burch. ;刚毛远志■☆

308288　Polygala pygmaea Gürke = Polygala welwitschii Chodat subsp.

pygmaea（Gürke）Paiva ■☆

308289 Polygala pyramidalis H. Lév. = Polygala longifolia Poir. ■

308290 Polygala quartiniana Quart. -Dill. ex A. Rich. = Polygala sphenoptera Fresen. ■☆

308291 Polygala ramosissima Cav. ;多枝远志■☆

308292 Polygala rarifolia DC. ;稀叶远志■☆

308293 Polygala reflexa Schinz = Polygala kalaxariensis Schinz ■

308294 Polygala refracta DC. ;反折远志■☆

308295 Polygala rehmannii Chodat;拉赫曼远志■☆

308296 Polygala rehmannii Chodat var. gymnoptera ? = Polygala producta N. E. Br. ■☆

308297 Polygala rehmannii Chodat var. latipetala（N. E. Br.）Norl. = Polygala latipetala N. E. Br. ■☆

308298 Polygala rehmannii Chodat var. parviflora ? = Polygala producta N. E. Br. ■☆

308299 Polygala reinii Franch. et Sav. ;黎氏远志(柿叶草)■☆

308300 Polygala reinii Franch. et Sav. f. angustifolia（Makino）Ohwi = Polygala reinii Franch. et Sav. f. stenophylla Yonek. ■☆

308301 Polygala reinii Franch. et Sav. f. angustifolia（T. Ito）K. Iwats. et H. Ohba = Polygala reinii Franch. et Sav. f. stenophylla Yonek. ■☆

308302 Polygala reinii Franch. et Sav. f. stenophylla Yonek. ;狭柿叶草■☆

308303 Polygala resinosa S. K. Chen;斑果远志;Resinose Milkwort ●

308304 Polygala retusa Hochst. = Polygala erioptera DC. ■☆

308305 Polygala rigens Burch. ;硬远志■☆

308306 Polygala riparia Chodat = Polygala usafuensis Gürke ■☆

308307 Polygala riukiuensis Ohwi = Polygala longifolia Poir. ■

308308 Polygala rivularis Gürke;溪边远志■☆

308309 Polygala robsonii Exell;罗布森远志■☆

308310 Polygala robusta Gürke;粗壮远志■☆

308311 Polygala rodrigueana Paiva;纳塔尔远志■☆

308312 Polygala rogersiana Baker f. = Polygala senensis Klotzsch ■☆

308313 Polygala rosea Desf. ;粉红远志■☆

308314 Polygala rosea Desf. subsp. boissieri（Coss.）Maire = Polygala boissieri Coss. ■☆

308315 Polygala rupestris Pourr. ;岩地远志■☆

308316 Polygala rupestris Pourr. subsp. densiflora Braun-Blanq. et Maire = Polygala rupestris Pourr. ■☆

308317 Polygala rupestris Pourr. subsp. fontqueri（Pau）Font Quer;丰特远志■☆

308318 Polygala rupestris Pourr. subsp. oxycoccoides（Desf.）Chodat;红莓苔子远志■☆

308319 Polygala rupestris Pourr. var. canescens（Chodat）Maire = Polygala rupestris Pourr. ■☆

308320 Polygala rupestris Pourr. var. densiflora（Braun-Blanq. et Maire）Maire = Polygala rupestris Pourr. ■☆

308321 Polygala rupestris Pourr. var. desertica（Chodat）Maire = Polygala rupestris Pourr. ■☆

308322 Polygala rupestris Pourr. var. laticarpa（Doum.）Pau et Font Quer = Polygala rupestris Pourr. ■☆

308323 Polygala rupestris Pourr. var. oxycoccoides（Desf.）Chodat = Polygala rupestris Pourr. subsp. oxycoccoides（Desf.）Chodat ■☆

308324 Polygala rupestris Pourr. var. rupicola（Pomel）Batt. = Polygala rupestris Pourr. ■☆

308325 Polygala rupestris Pourr. var. saxatilis（Desf.）Murb. = Polygala rupestris Pourr. ■☆

308326 Polygala rupicola Hochst. et Steud. ex A. Rich. ;岩石远志■☆

308327 Polygala rupicola Pomel = Polygala rupestris Pourr. ■☆

308328 Polygala rutenbergii O. Hoffm. = Polygala schoenlankii O. Hoffm. et Hildebrandt ■☆

308329 Polygala ruwenzoriensis Chodat;鲁文佐里远志■☆

308330 Polygala sadebeckiana Gürke;萨德拜克远志■☆

308331 Polygala sanguinea L. ;血红远志;Blood Milkwort,Blood Polygala, Blood-colored Polygala,Field Milkwort,Purple Milkwort ■☆

308332 Polygala sanguinea L. f. albiflora Millsp. = Polygala sanguinea L. ■☆

308333 Polygala sanguinea L. f. typica Farw. = Polygala sanguinea L. ■☆

308334 Polygala sanguinea L. f. viridescens（L. ）Farw. = Polygala sanguinea L. ■☆

308335 Polygala sansibarensis Gürke;桑给巴尔远志■☆

308336 Polygala saxatilis Desf. = Polygala rupestris Pourr. ■☆

308337 Polygala saxicola Dunn;岩生远志(鸡脚爪,岩生紫金牛); Saxatile Milkwort ●■

308338 Polygala scabra L. ;粗糙远志■☆

308339 Polygala schantziana Dinter = Polygala erioptera DC. ■☆

308340 Polygala schimperi Hassk. = Polygala erioptera DC. ■☆

308341 Polygala schimperi Vatke ex Chodat = Polygala vatkeana Exell ■☆

308342 Polygala schinziana Chodat;欣兹远志■☆

308343 Polygala schlechteri Schinz = Polygala spicata Chodat ■☆

308344 Polygala schoenlankii O. Hoffm. et Hildebrandt;舍恩远志■☆

308345 Polygala schweinfurthii Chodat;施韦远志■☆

308346 Polygala seminuda Harv. ;半裸远志■☆

308347 Polygala senega L. ;美远志;Beautiful Milkwort,Mountain Flax, Rattlesnake-root,Seneca Snakeroot,Senega Root,Senega Snakeroot, Seneka ■☆

308348 Polygala senega L. var. latifolia Torr. et A. Gray;细叶美远志; Seneca Snakeroot ■☆

308349 Polygala senega L. var. latifolia Torr. et A. Gray = Polygala senega L. ■☆

308350 Polygala senega L. var. tenuifolia Pursh;宽叶美远志(日本远志)■☆

308351 Polygala senensis Klotzsch;塞纳远志■☆

308352 Polygala senensis Klotzsch var. calcicola（Chodat）Paiva;钙生远志■☆

308353 Polygala sennenii Pau = Polygala baetica Willk. ■☆

308354 Polygala septemnervia Merr. = Polygala persicariifolia DC. ■

308355 Polygala septentrionalis Troupin;北方远志■☆

308356 Polygala serpentaria Eckl. et Zeyh. ;蛇药远志■☆

308357 Polygala serpyllifolia Wight;百里香叶远志;Common Milkwort, Heath Milkwort ■☆

308358 Polygala shimadai Masam. = Polygala glomerata Lour. ■

308359 Polygala sibirica L. ;西伯利亚远志(辰砂草,大远志,地丁,瓜子草,瓜子金,棘菀,蕀菀,苦远志,宽叶远志,阔叶远志,蓝花地丁,卵叶远志,女儿红,青玉丹草,神砂草,甜远志,万年青,细草,小草,小丁香,小鸡根,小叶远志,蓑,蓑绕,远志,紫花地丁); Siberian Milkwort ■

308360 Polygala sibirica L. var. angustifolia Ledeb. = Polygala tenuifolia Willd. ■

308361 Polygala sibirica L. var. elegans（Wall. ex Royle）Hara = Polygala elegans Wall. ■

308362 Polygala sibirica L. var. japonica（Houtt.）Ito ex Ito et Matsum. = Polygala japonica Houtt. ■

308363 Polygala sibirica L. var. megalopha Franch. ;苦远志(辰砂草,大冠远志,瓜子金,红花地丁,蓝花地丁,蓝花地丁草,四季青,小丁香,小蓝花地丁,小万年青,小细辛,小远志,紫花地丁);Bitter

Milkwort ■

308364　Polygala sibirica L. var. monopetala（Cambess.）Chodat ＝ Polygala monopetala Cambess. ■

308365　Polygala sibirica L. var. monopetala（Cambess.）Chodat ＝ Polygala sibirica L. ■

308366　Polygala sibirica L. var. tenuifolia（Willd.）Backer et Moore ＝ Polygala tenuifolia Willd. ■

308367　Polygala sieboldiana Miq. ＝ Polygala tatarinowii Regel ■

308368　Polygala simadae Masam. ＝ Polygala arvensis Willd. ■

308369　Polygala somaliensis Baker;索马里远志■☆

308370　Polygala sparsiflora Oliv.;稀花远志■☆

308371　Polygala sparsiflora Oliv. f. robustior Chodat ＝ Polygala paludicola Gürke ■☆

308372　Polygala sparsiflora Oliv. var. ukirensis（Gürke）Paiva;乌基尔远志☆

308373　Polygala speciosa Sims ＝ Polygala virgata Thunb. var. speciosa（Sims）Harv. ●☆

308374　Polygala sphenoptera Fresen.;楔翅远志■☆

308375　Polygala sphenoptera Fresen. var. fischeri（Gürke）Petit ＝ Polygala fischeri Gürke ●☆

308376　Polygala spicata Chodat;穗远志■☆

308377　Polygala spinosa L. ＝ Nylandtia spinosa（L.）Dumort. ■☆

308378　Polygala splendens Exell ＝ Polygala macrostigma Chodat ■☆

308379　Polygala squarrosa L. f. ＝ Muraltia squarrosa（L. f.）DC. ●☆

308380　Polygala stanleyana Chodat ＝ Polygala albida Schinz subsp. stanleyana（Chodat）Paiva ■☆

308381　Polygala stanleyana Chodat var. angustifolia ? ＝ Polygala albida Schinz subsp. stanleyana（Chodat）Paiva ■☆

308382　Polygala stanleyana Chodat var. latifolia ? ＝ Polygala albida Schinz subsp. stanleyana（Chodat）Paiva ■☆

308383　Polygala stenopetala Klotzsch;窄瓣远志●☆

308384　Polygala stenophylla Hayata ＝ Polygala hongkongensis Hemsl. ex Forbes et Hemsl. var. stenophylla（Hayata）Migo ●■

308385　Polygala steudneri Chodat;斯托德远志■☆

308386　Polygala stocksiana Boiss.;斯托克斯远志■☆

308387　Polygala stocksiana Boiss. ＝ Polygala hohenackeriana Fisch. et C. A. Mey. ■☆

308388　Polygala suanica Tamamsch.;苏安远志■☆

308389　Polygala subaphylla H. Perrier;亚无叶远志●☆

308390　Polygala subdioica H. Perrier;异株远志●☆

308391　Polygala subglobosa Paiva;亚球形远志■☆

308392　Polygala subopposita S. K. Chen;合叶草（对叶接骨草,合掌草,和合草,排钱金不换,午时合）;Subopposite Milkwort ■

308393　Polygala subspinosa S. Watson;刺远志;Spiny Milkwort ■☆

308394　Polygala supina Schreb.;仰卧远志■☆

308395　Polygala tanganyikensis Troupin ＝ Polygala usafuensis Gürke ■☆

308396　Polygala taquetii H. Lév. ＝ Polygala japonica Houtt. ■

308397　Polygala tatarinowii Regel;小扁豆（苏草,天星吊红,小远志,野豌豆草）;Tatarinow Milkwort ■

308398　Polygala telephioides Willd.;小花远志（金牛草,七寸金,细金草,细金牛草,细叶金不换,小金不换,小金牛草,小兰青,紫背金牛）;Smallflower Milkwort ■

308399　Polygala telephioides Willd. ＝ Polygala arvensis Willd. ■

308400　Polygala tenuicaulis Hook. f.;细茎远志■☆

308401　Polygala tenuicaulis Hook. f. var. tayloriana Paiva;泰勒细茎远志■☆

308402　Polygala tenuifolia Link ＝ Polygala gracilenta Burtt Davy ■☆

308403　Polygala tenuifolia Willd. ;远志（阿只草,草远志,关远志,光棍茶,红籽细辛,棘菀,棘苑,蕀蒬,蕀蒬,苦葽,苦远志,米儿茶,青小草,山茶叶,山胡麻,神砂草,十二月花,细草,细叶远志,线茶,线儿茶,小草,小草根,小鸡根,小鸡棵,小鸡腿,小据,醒心杖,燕子草,葽,葽绕,夷门远志,余粮,蒬葽绕）;Thinleaf Milkwort ■

308404　Polygala tenuis A. Dietr. ＝ Polygala gracilenta Burtt Davy ■☆

308405　Polygala teretifolia Baker f. var. gazensis（Baker f.）Norl. ＝ Polygala gazensis Baker f. ■☆

308406　Polygala teretifolia L. f.;柱叶远志■☆

308407　Polygala tetrasepala Hochst. ex Webb ＝ Polygala petitiana A. Rich. ■☆

308408　Polygala thymifolia Thunb. ＝ Muraltia thymifolia（Thunb.）DC. ■☆

308409　Polygala tinctoria Vahl;染料远志■☆

308410　Polygala tisserantii Jacq. -Fél. ;蒂斯朗特远志■☆

308411　Polygala torrei Exell;托雷远志■☆

308412　Polygala transcaucasica Tamamsch.;外高加索远志■☆

308413　Polygala transvaalensis Chodat;德兰士瓦远志■☆

308414　Polygala transvaalensis Chodat subsp. kagerensis（Lebrun et Taton）Paiva;卡盖拉远志■☆

308415　Polygala tricholopha Chodat;红花远志（藤状远志）;Red Milkwort,Vinelike Milkwort,Vine-like Milkwort ●

308416　Polygala tricornis Gagnep.;密花远志（多花远志,罗氏远志,胖树根,小鸡花）;Denseflower Milkwort,Densiflowered Milkwort ●

308417　Polygala tricornis Gagnep. var. crinita Gagnep. ＝ Polygala tricornis Gagnep. ●

308418　Polygala tricornis Gagnep. var. latifolia Gagnep. ＝ Polygala tricornis Gagnep. ●

308419　Polygala tricornis Gagnep. var. obcordata C. Y. Wu et S. K. Chen;小叶密花远志（少叶远志）;Littleleaf Denseflower Milkwort,Smallleaf Milkwort ●

308420　Polygala trinervia L. f. ＝ Muraltia trinervia（L. f.）DC. ●☆

308421　Polygala triphylla Buch-Ham. ex D. Don ＝ Polygala tatarinowii Regel ■

308422　Polygala triphylla Burm. f. ＝ Polygala furcata Royle ■

308423　Polygala triquetra C. Presl;三棱远志■☆

308424　Polygala tristis Chodat ＝ Polygala sphenoptera Fresen. ■☆

308425　Polygala ukambica Chodat ＝ Polygala sphenoptera Fresen. ■☆

308426　Polygala ukirensis Gürke ＝ Polygala sparsiflora Oliv. var. ukirensis（Gürke）Paiva ■☆

308427　Polygala umbellata L. ;小伞远志■☆

308428　Polygala umbonata Craib;凹籽远志;Umbilicate Milkwort,Umbonate Milkwort ■

308429　Polygala uncinata E. Mey. ex Meisn.;具钩远志■☆

308430　Polygala urarta Tamamsch.;乌拉尔远志■☆

308431　Polygala usafuensis Gürke;乌沙夫远志■☆

308432　Polygala vahliana DC. ＝ Polygala erioptera DC. ■☆

308433　Polygala vatkeana Exell;瓦特凯远志■☆

308434　Polygala vayredae Costa;比利牛斯远志●☆

308435　Polygala verdickii Gürke ＝ Polygala usafuensis Gürke ■☆

308436　Polygala versicolor Pomel ＝ Polygala nicaeensis Koch ■☆

308437　Polygala verticillata L. ;轮生远志;Tall Whorled Milkwort,Whorled Milkwort ■☆

308438　Polygala verticillata L. var. ambigua（Nutt.）A. W. Wood;可疑轮生远志;Whorled Milkwort ■☆

308439　Polygala verticillata L. var. dolichoptera Fernald ＝ Polygala verticillata L. ■☆

308440   Polygala verticillata L. var. isocycla Fernald；相等轮生远志；Short Whorled Milkwort ■☆

308441   Polygala verticillata L. var. isocycla Fernald ＝ Polygala verticillata L. ■☆

308442   Polygala verticillata L. var. sphenostachya Pennell；楔花轮生远志；Whorled Milkwort ■☆

308443   Polygala verticillata L. var. sphenostachya Pennell ＝ Polygala verticillata L. ■☆

308444   Polygala viminalis Gürke ＝ Polygala stenopetala Klotzsch ●☆

308445   Polygala viminalis Gürke var. brachyptera Chodat ＝ Polygala stenopetala Klotzsch ●☆

308446   Polygala virgata Thunb.；好望角远志●☆

308447   Polygala virgata Thunb. var. decora（Sond.）Harv.；装饰远志●☆

308448   Polygala virgata Thunb. var. speciosa（Sims）Harv.；美丽好望角远志●☆

308449   Polygala viridescens L.；绿远志；Field Milkwort, Purple Milkwort ■☆

308450   Polygala viridescens L. ＝ Polygala sanguinea L. ■☆

308451   Polygala vittata Paiva；粗线远志●☆

308452   Polygala volkensii Gürke ＝ Polygala petitiana A. Rich. ■☆

308453   Polygala volubilis Bojer ＝ Polygala macroptera DC. ●☆

308454   Polygala vulgaris L.；普通远志；Christ's Herb, Common Milkwort, Cross Flower, Fairy Soap, Four Sisters, Gang-flower, Hedge Hyssop, Jack-and-the-beanstalk, Milkmaids, Milkwort, Mother Mary's Milk, Procession Flower, Robin's Eye, Robin's Eyes, Rogation Flower, Rogation-flower, Shepherd's Purse, Shepherd's Thyme, Waxworks, Wild Lobelia ■☆

308455   Polygala vulgaris L. ＝ Polygala japonica Houtt. ■

308456   Polygala wallichiana Wight ＝ Polygala persicariifolia DC. ■

308457   Polygala wattersii Hance；长毛远志（长毛籽远志，大毛籽黄山桂，米花子，山桂花，西南远志，细叶远志，油树子）；Longhair Milkwort, Watter Milkwort ●

308458   Polygala wattersii Hance ＝ Polygala caudata Rehder et E. H. Wilson ●

308459   Polygala webbiana Coss.；韦布远志■☆

308460   Polygala welwitschii Chodat；韦氏远志■☆

308461   Polygala welwitschii Chodat subsp. pygmaea（Gürke）Paiva；矮小远志■☆

308462   Polygala wenxianensis Y. S. Zhou et Z. X. Peng；文县远志；Wenxian Milkwort ■

308463   Polygala westii Exell；韦斯特远志■☆

308464   Polygala wilmsii Chodat；维尔姆斯远志■☆

308465   Polygala wistatiifolia Chodat ＝ Polygala arillata Buch. -Ham. ex D. Don ●

308466   Polygala wittebergensis Compton；维特伯格远志■☆

308467   Polygala wittei Exell；维特远志■☆

308468   Polygala wolfgangiana Besser ex Ledeb.；武氏远志；Wolfgang Milkwort ■☆

308469   Polygala woodii Chodat；伍得远志■☆

308470   Polygala xanthina Chodat；黄远志■☆

308471   Polygala youngii Exell；扬氏远志■☆

308472   Polygala yunnanensis Chodat ＝ Polygala tricornis Gagnep. ●

308473   Polygala zambesiaca Paiva；赞比西远志■☆

308474   Polygalaceae Hoffmanns. et Link（1809）（保留科名）；远志科；Milkwort Family ■●

308475   Polygalaceae Juss. ＝ Polygalaceae Hoffmanns. et Link（保留科名）■●

308476   Polygalaceae R. Br. ＝ Polygalaceae Hoffmanns. et Link（保留科名）■●

308477   Polygaloides Haller ＝ Polygala L. ●■

308478   Polygaloides Haller（1768）；拟远志属（假远志属）●☆

308479   Polygaloides chamaebuxus（L.）O. Schwarz；拟远志●☆

308480   Polyglochin Ehrh. ＝ Carex L. ■

308481   Polygonaceae Juss.（1789）（保留科名）；蓼科；Buckwheat Family, Knotweed Family ●■

308482   Polygonanthaceae（Croizat）Croizat ＝ Anisophylleaceae Ridl. ●☆

308483   Polygonanthaceae Croizat ＝ Anisophylleaceae Ridl. ●☆

308484   Polygonanthus Ducke ＝ Polygonanthus Ducke, Baehni et Dans. ●☆

308485   Polygonanthus Ducke, Baehni et Dans.（1932）；蓼花木属●☆

308486   Polygonanthus amazonicus Ducke；蓼花木●☆

308487   Polygonanthus punctulatus Kuhlm.；斑点蓼花木●☆

308488   Polygonastrum Moench（废弃属名）＝ Maianthemum F. H. Wigg.（保留属名）■

308489   Polygonastrum Moench（废弃属名）＝ Smilacina Desf.（保留属名）■

308490   Polygonataceae Salisb.；黄精科■

308491   Polygonataceae Salisb. ＝ Convallariaceae L. ■

308492   Polygonataceae Salisb. ＝ Ruscaceae M. Roem.（保留科名）●

308493   Polygonatum Mill.（1754）；黄精属（玉竹属）；Landpick, Sceau-de-Salomon, Solomon's Seal, Solomon's-Seal, Solomonseal ■

308494   Polygonatum Zinn ＝ Convallaria L. ■

308495   Polygonatum acuminatifolium Kom.；五叶黄精；Acuminateleaf Solomonseal, Fiveleaf Landpick ■

308496   Polygonatum adnatum S. Yun Liang；贴梗黄精；Adnate Solomonseal ■

308497   Polygonatum agglutinatum Hua ＝ Polygonatum kingianum Collett et Hemsl. ■

308498   Polygonatum alte-lobatum Hayata；短筒黄精（阿里山玉竹，台湾黄精）；Shorttube Landpick, Shorttube Solomonseal ■

308499   Polygonatum alternicirrhosum Hand. -Mazz.；互卷黄精；Alternateleaf Solomonseal, Mutualroll Landpick ■

308500   Polygonatum alternicirrhosum Hand. -Mazz. var. piliferum P. Y. Li；粗毛互卷黄精（粗毛黄精）；Pilose Alternateleaf Solomonseal ■

308501   Polygonatum alternicirrhosum Hand. -Mazz. var. piliferum P. Y. Li ＝ Polygonatum hirtellum Hand. -Mazz. ■

308502   Polygonatum amplexicaule DC.；抱茎黄精■☆

308503   Polygonatum anhuiense D. C. Zhang et J. Z. Shao；安徽黄精；Anhui Solomonseal ■

308504   Polygonatum anhuiense D. C. Zhang et J. Z. Shao ＝ Polygonatum zanlanscianense Pamp. ■

308505   Polygonatum anomalum Hua ＝ Polygonatum punctatum Royle ex Kunth ■

308506   Polygonatum arisanense Hayata；阿里黄精（萎蕤）■

308507   Polygonatum arisanense Hayata ＝ Polygonatum odoratum（Mill.）Druce var. pluriflorum（Miq.）Ohwi ■

308508   Polygonatum biflorum（Walter）Elliott；双花黄精；Giant Solomon's Seal, Giant Solomon's-seal, Great Solomon's Seal, Hairy Solomon's Seal, King Solomon's-seal, Small Solomon's Seal, Smooth Solomon's Seal, Smooth Solomon's-seal, Solomon's Seal, Solomon's-seal ☆

308509   Polygonatum biflorum（Walter）Elliott var. commutatum（Schult. f.）Morong ＝ Polygonatum biflorum（Walter）Elliott ■☆

308510   Polygonatum biflorum（Walter）Elliott var. melleum（Farw.）R. P. Ownbey ＝ Polygonatum biflorum（Walter）Elliott ■☆

308511　Polygonatum biflorum Elliott = Polygonatum biflorum（Walter）Elliott ■☆

308512　Polygonatum bodinieri H. Lév. = Disporopsis pernyi（Hua）Diels ■

308513　Polygonatum brachynema Hand. -Mazz. = Polygonatum cyrtonema Hua ■

308514　Polygonatum bulbosum H. Lév. = Polygonatum cirrhifolium（Wall.）Royle ■

308515　Polygonatum canaliculatum（Muhl.）Pursh = Polygonatum biflorum（Walter）Elliott ■☆

308516　Polygonatum canaliculatum Miq. = Polygonatum commutatum A. Dietr. ■☆

308517　Polygonatum cathcarti Baker；棒丝黄精；Cathcart Landpick，Cathcart Solomonseal ■

308518　Polygonatum cavaleriei H. Lév. = Polygonatum kingianum Collett et Hemsl. ■

308519　Polygonatum chinense Kunth = Polygonatum sibiricum Delarb ex Redouté ■

308520　Polygonatum chingshuishanianum S. S. Ying；清水山黄精；Qingshuishan Solomonseal ■

308521　Polygonatum cirrhifoliodes D. M. Liu et W. Z. Zeng = Polygonatum cirrhifolium（Wall.）Royle ■

308522　Polygonatum cirrhifolioides D. M. Liu et W. Z. Zeng；卷叶玉竹；Like Tendrilleaf Solomonseal ■

308523　Polygonatum cirrhifolium（Wall.）Royle；卷叶黄精（白药子，滇钩吻，鄂西黄精，鸡头参，老虎姜，轮叶黄精，裸花黄精，盘龙七，惹涅，山姜，算盘七）；Tendrilleaf Landpick，Tendrilleaf Solomonseal ■

308524　Polygonatum cobrense（Wooton et Standl.）R. R. Gates = Polygonatum biflorum（Walter）Elliott ■☆

308525　Polygonatum commutatum（Schult. f.）A. Dietr. = Polygonatum biflorum（Walter）Elliott ■☆

308526　Polygonatum commutatum（Schult. f.）A. Dietr. f. foliatum H. M. Clarke = Polygonatum biflorum（Walter）Elliott ■☆

308527　Polygonatum commutatum A. Dietr. ；变化黄精（变态广叶黄精，巨黄精）；Giant Solomon's Seal，Great Solomon's Seal，Great Solomon's-seal，Smooth Solomon's Seal，Solomon's Giant Seal，Solomon's Seal ■☆

308528　Polygonatum curvistylum Hua；垂叶黄精（弯花柱黄精）；Curvedstyle Solomonseal，Droopingleaf Landpick ■

308529　Polygonatum cyrtonema Hua；多花黄精（白芨黄精，白及，白及黄精，笔菜，笔管菜，长叶黄精，垂珠，苟格，黄鸡菜，黄精，黄精姜，黄芝，鸡格，鸡头参，姜形黄精，九蒸姜，老虎姜，龙衔，鹿竹，马箭，南黄精，囊丝黄精，山捣臼，山姜，山生姜，生姜，太阳草，甜黄精，土灵芝，兔竹，萎蕤，阳雀蕨，野生姜，野鲜姜，玉竹黄精，重楼，竹姜）；Common Solomon's-seal，Davids Harp，Lady's Seal，Manyflower Landpick，Manyflower Solomonseal ■

308530　Polygonatum darrisii H. Lév. = Polygonatum kingianum Collett et Hemsl. ■

308531　Polygonatum delavayi Hua = Polygonatum prattii Baker ■

308532　Polygonatum desoulavyi Kom. ；长苞黄精；Longbract Landpick，Longbract Solomonseal ■

308533　Polygonatum ensifolium H. Lév. = Disporopsis pernyi（Hua）Diels ■

308534　Polygonatum ensifolium H. Lév. var. didymocarpum H. Lév. = Disporopsis pernyi（Hua）Diels ■

308535　Polygonatum ericoides H. Lév. = Polygonatum kingianum Collett et Hemsl. ■

308536　Polygonatum erythrocarpum Hua = Polygonatum verticillatum（L.）All. ■

308537　Polygonatum esquirolii H. Lév. = Polygonatum kingianum Collett et Hemsl. ■

308538　Polygonatum falcatum A. Gray；镰刀黄精（黄精，老虎姜，马箭，戊己芝）；Falcate Solomonseal ■☆

308539　Polygonatum fargesii Hua = Polygonatum cirrhifolium（Wall.）Royle ■

308540　Polygonatum filipes Merr. ex C. Jeffrey et McEwan；垂丝黄精（长梗黄精）；Longstalk Landpick，Longstalk Solomonseal ■

308541　Polygonatum formosanum（Hayata）Masam. et Shimada = Polygonatum arisanense Hayata ■

308542　Polygonatum formosanum（Hayata）Masam. et Shimada = Polygonatum odoratum（Mill.）Druce var. pluriflorum（Miq.）Ohwi ■

308543　Polygonatum formosanum Masam. et Shimada = Polygonatum odoratum（Mill.）Druce var. pluriflorum（Miq.）Ohwi ■

308544　Polygonatum franchetii Hua；距药黄精（距花黄精）；Franchet Landpick，Franchet Solomonseal ■

308545　Polygonatum fuscum Hua = Polygonatum cirrhifolium（Wall.）Royle ■

308546　Polygonatum gentilianum H. Lév. = Polygonatum prattii Baker ■

308547　Polygonatum giganteum A. Dietr. = Polygonatum biflorum（Walter）Elliott ■☆

308548　Polygonatum giganteum A. Dietr. = Polygonatum commutatum A. Dietr. ■☆

308549　Polygonatum giganteum H. Lév. = Polygonatum cyrtonema Hua ■

308550　Polygonatum ginfushanicum（F. T. Wang et Ts. Tang）F. T. Wang et Ts. Tang = Heteropolygonatum ginfushanicum（F. T. Wang et Ts. Tang）M. N. Tamura，S. Yun Liang et N. J. Turland ■

308551　Polygonatum glaberrimum K. Koch；无毛黄精■☆

308552　Polygonatum gracile P. Y. Li；细根茎黄精；Slenderrhizome Landpick，Slenderrhizome Solomonseal ■

308553　Polygonatum graminifolium Hook. ；禾叶黄精；Grassleaf Landpick，Grassleaf Solomonseal ■

308554　Polygonatum griffithii Baker；三脉黄精■

308555　Polygonatum henryi Diels = Polygonatum cyrtonema Hua ■

308556　Polygonatum hirtellum Hand. -Mazz. ；粗毛黄精；Hirsute Landpick，Hirsute Solomonseal ■

308557　Polygonatum hirtum（Poir.）Pursh = Polygonatum latifolium（Jacq.）Desf. ■☆

308558　Polygonatum hirtum Pursh；毛黄精（宽叶黄精）；Broadleaf Landpick，Broadleaf Solomon's Seal，Broadleaf Solomonseal，Broadleaved Solomon's-seal ■☆

308559　Polygonatum hondoense Nakai ex Koidz. = Polygonatum odoratum（Mill.）Druce ■

308560　Polygonatum hookeri Baker；独花黄精；Hooker Landpick，Hooker Solomonseal ■

308561　Polygonatum huanum H. Lév. = Polygonatum kingianum Collett et Hemsl. ■

308562　Polygonatum humile Fisch. ex Maxim. ；小玉竹；Dwarf Solomon's Seal，Small Landpick，Small Solomonseal ■

308563　Polygonatum humillimum Nakai = Polygonatum humile Fisch. ex Maxim. ■

308564　Polygonatum hybridum Brügger；杂种黄精；David's Harp，Garden Solomon's-seal，Solomon's-seal ■☆

308565　Polygonatum inflatum Kom. ；毛筒玉竹；Inflated Landpick，

Inflated Solomonseal ■

308566　Polygonatum inflatum Kom. var. rotundifolium Hatus. = Polygonatum inflatum Kom. ■

308567　Polygonatum involucratum (Franch. et Sav.) Maxim. ;二苞黄精(小玉竹);Twobract Landpick,Twobract Solomonseal ■

308568　Polygonatum japonicum C. Morren et Decne. = Polygonatum odoratum (Mill.) Druce ■

308569　Polygonatum kalapanum Hand.-Mazz. = Polygonatum stewartianum Diels ■

308570　Polygonatum kansuense Maxim. = Polygonatum verticillatum (L.) All. ■

308571　Polygonatum kansuense Maxim. ex Batalin = Polygonatum verticillatum (L.) All.

308572　Polygonatum kingianum Collett et Hemsl. ;滇黄精(白及,白及黄精,笔菜,笔管菜,垂珠,德保黄精,苟格,黄鸡菜,黄精,黄芝,鸡格,鸡头参,姜形黄精,节节高,金氏黄精,老虎姜,龙衔,鹿竹,马箭,山捣臼,山姜,山生姜,生姜,太阳草,甜黄精,土灵芝,兔竹,萎蕤,西南黄精,仙人饭,阳雀蕨,野生姜,野鲜姜,玉竹黄精,重楼);King Solomonseal,Yunnan Landpick ■

308573　Polygonatum kingianum Collett et Hemsl. var. cavaleriei (H. Lév.) C. Jeffrey et McEwan = Polygonatum kingianum Collett et Hemsl. ■

308574　Polygonatum kingianum Collett et Hemsl. var. ericoideum (H. Lév.) C. Jeffrey et McEwan = Polygonatum kingianum Collett et Hemsl. ■

308575　Polygonatum kingianum Collett et Hemsl. var. grandifolium D. M. Liu et W. Z. Zeng = Polygonatum kingianum Collett et Hemsl. ■

308576　Polygonatum kingianum Collett et Hemsl. var. uncinatum (Diels) C. Jeffrey et McEwan = Polygonatum kingianum Collett et Hemsl. ■

308577　Polygonatum kungii F. T. Wang et Ts. Tang = Polygonatum zanlanscianense Pamp. ■

308578　Polygonatum langyaense D. C. Zhang et J. Z. Shao;琅琊黄精;Langya Solomonseal ■

308579　Polygonatum langyaense D. C. Zhang et J. Z. Shao = Polygonatum odoratum (Mill.) Druce ■

308580　Polygonatum lanuginosum F. T. Wang et Ts. Tang;白芨黄精(绵毛黄精);Woolly Solomonseal ■☆

308581　Polygonatum laoticum Gagnep. = Disporopsis longifolia Craib ■

308582　Polygonatum lasianthum Maxim. ;深山黄精(庐山黄精);Lushan Solomonseal ■

308583　Polygonatum lasianthum Maxim. var. commutatum (A. Dietr.) Baker = Polygonatum commutatum A. Dietr. ■☆

308584　Polygonatum latifolium (Jacq.) Desf. ;宽叶黄精■☆

308585　Polygonatum latifolium Desf. = Polygonatum latifolium (Jacq.) Desf. ■☆

308586　Polygonatum latifolium Desv. = Polygonatum hirtum Pursh ■☆

308587　Polygonatum lebrunii H. Lév. = Polygonatum cirrhifolium (Wall.) Royle ■

308588　Polygonatum leiboense S. C. Chen et D. Q. Liu;雷波黄精;Leibo Landpick,Leibo Solomonseal ■

308589　Polygonatum leveilleanum Fedde = Polygonatum nodosum Hua ■

308590　Polygonatum longipedunculatum S. Yun Liang;长梗黄精(长柄黄精);Longpedicel Landpick,Longpedicel Solomonseal ■

308591　Polygonatum longistylum Y. Wan ex C. Z. Gao;百色黄精;Baise Solomonseal ■

308592　Polygonatum macranthum Koidz. ;大花黄精;Largeflower Solomonseal,Large-leaved Giant Solomon's-seal ■☆

308593　Polygonatum macropodum Turcz. ;热河黄精(大玉竹,多花黄精,小叶珠);Macropodous Landpick,Macropodous Solomonseal ■

308594　Polygonatum mairei H. Lév. = Polygonatum cirrhifolium (Wall.) Royle ■

308595　Polygonatum mairei H. Lév. = Polygonatum nodosum Hua ■

308596　Polygonatum marmoratum H. Lév. = Polygonatum punctatum Royle ex Kunth ■

308597　Polygonatum martinii H. Lév. = Polygonatum cyrtonema Hua ■

308598　Polygonatum maximoviczli F. Schmidt;马氏黄精■

308599　Polygonatum maximowiczii F. Schmidt = Polygonatum odoratum (Mill.) Druce ■

308600　Polygonatum megaphyllum P. Y. Li;大苞黄精;Largeleaf Landpick,Largeleaf Solomonseal ■

308601　Polygonatum melleum Farw. = Polygonatum biflorum (Walter) Elliott ■☆

308602　Polygonatum mengtzense F. T. Wang et Ts. Tang = Polygonatum punctatum Royle ex Kunth ■

308603　Polygonatum minutiflorum H. Lév. = Polygonatum verticillatum (L.) All. ■

308604　Polygonatum multiflorum All. = Polygonatum cyrtonema Hua ■

308605　Polygonatum multiflorum All. var. longifolium Merr. = Polygonatum cyrtonema Hua ■

308606　Polygonatum multiflorum H. Lév. = Polygonatum verticillatum (L.) All. ■

308607　Polygonatum multiflorum Kunth;复花黄精(多花黄精);David's Harp, David's Seal, Eurasian Solomon's-seal, Fraxinell, Jacob's Ladder, Job's Tears, Ladder To Heaven, Lady's Locket, Lady's Seal, Lady's Signet, Lily-of-the-mountain, Many Knees, Sealwort, Solomon's Sale, Solomon's Seal, Solomon's-seal, Sow's Tits, St. Mary's Seal, Vagabond's Friend, Whiteroot, Whitewort ■☆

308608　Polygonatum multiflorum Kunth var. longifolium Merr. = Polygonatum cyrtonema Hua ■

308609　Polygonatum multiflorum sensu All. = Polygonatum cyrtonema Hua ■

308610　Polygonatum nakaianum Tshidoga;小苞黄精■☆

308611　Polygonatum nodosum Hua; 节根黄精; Nodose Landpick, Nodose Solomonseal ■

308612　Polygonatum obtusifolium Miscz. ;钝叶黄精■☆

308613　Polygonatum odoratum (Mill.) Druce;玉竹(白及,百解药,笔管菜,虫蝉,灯笼菜,地管子,地节,黄脚鸡,黄蔓菁,黄芝,节地,句稳草,靠山竹,丽草,连州竹,连竹,铃铛菜,芦莉花,马儿花,马氏黄精,马薰,女草,女萎,青粘,日本黄精,山包米,山姜,山铃子草,山玉竹,十样错,甜草根,娃草,王马,葳参,葳蕤,尾参,委萎,萎,萎蕤,萎香,萎移,乌萎,西竹,香花黄精,香黄精,小笔管菜,药用黄精,萤,玉参,玉术,玉竹参,玉竹面,圆叶玉竹,竹节黄);Angled Solomon's Seal, Angular Solomon's-seal, Drug Solomon's-seal, Fragrant Landpick, Fragrant Solomonseal, Japan Landpick, Japanese Solomonseal, Lesser Solomon's-seal, Medicinal Solomonseal ■

308614　Polygonatum odoratum (Mill.) Druce f. ovalifolium C. Y. Chu = Polygonatum odoratum (Mill.) Druce ■

308615　Polygonatum odoratum (Mill.) Druce f. ovalifolium Y. C. Chu et al. = Polygonatum odoratum (Mill.) Druce ■

308616　Polygonatum odoratum (Mill.) Druce var. pluriflorum (Miq.) Ohwi;多花玉竹(阿里黄精,百解药,长叶玉竹,虫蝉,灯笼菜,地节,豆应仇罗,肥玉竹,黄脚鸡,黄蔓菁,黄芝,节地,靠山竹,丽草,连竹,铃铛菜,铃当菜,芦莉花,马熏,毛筒玉竹,名荧,明玉竹,女草,女萎,青粘,山包米,山姜,山铃子草,山玉竹,十样错,台湾大黄精,甜草根,娃草,王马,威绥,葳参,葳蕤,尾参,委蛇,委萎,萎蕤,萎香,萎薐,乌女,乌萎,西竹,小笔管菜,小玉竹,荧,玉参,玉术,玉竹,蒸玉竹,竹节黄,竹七根);Alishan Landpick,Alishan Solomonseal,Angular Solomon's Seal,Angular Solomon's-seal,Scented Solomon's Seal ■

308617　Polygonatum odoratum (Mill.) Druce var. pluriflorum (Miq.) Ohwi = Polygonatum odoratum (Mill.) Druce ■

308618　Polygonatum officinale All. = Polygonatum odoratum (Mill.) Druce ■

308619　Polygonatum officinale All. var. formosanum Hayata = Polygonatum arisanense Hayata ■

308620　Polygonatum officinale All. var. formosanum Hayata = Polygonatum odoratum (Mill.) Druce var. pluriflorum (Miq.) Ohwi ■

308621　Polygonatum officinale All. var. humile (Fisch. ex Maxim.) Baker = Polygonatum humile Fisch. ex Maxim. ■

308622　Polygonatum officinale All. var. papillosum Franch. = Polygonatum odoratum (Mill.) Druce ■

308623　Polygonatum officinale All. var. pluriflorum Miq. = Polygonatum odoratum (Mill.) Druce var. pluriflorum (Miq.) Ohwi ■

308624　Polygonatum omeiense Z. Y. Zhu;峨眉黄精;Emei Landpick,Emei Solomonseal,Omei Landpick ■

308625　Polygonatum oppositifolium (Wall.) Royle;对叶黄精;Oppositeleaf Landpick,Oppositeleaf Solomonseal ■

308626　Polygonatum ovatum Miscz.;卵叶黄精■☆

308627　Polygonatum parcefolium F. T. Wang et Ts. Tang = Polygonatum punctatum Royle ex Kunth ■

308628　Polygonatum pendulum Z. G. Liu et X. H. Hu = Heteropolygonatum pendulum (Z. G. Liu et X. H. Hu) M. N. Tamura et Ogisu ■

308629　Polygonatum planifilum Kitag. et Hir. Takah. = Polygonatum odoratum (Mill.) Druce ■

308630　Polygonatum platyphyllum Franch. = Polygonatum involucratum (Franch. et Sav.) Maxim. ■

308631　Polygonatum polyanthemum (M. Bieb.) Dietr.;细花黄精■☆

308632　Polygonatum prattii Baker;康定玉竹(大理玉竹,小玉竹);Kangding Landpick,Pratt Solomonseal ■

308633　Polygonatum pubescens (Willd.) Pursh;柔毛黄精;Downy Solomon's-seal,Hairy Solomon's Seal,Hairy Solomon's-seal,Small Solomon's Seal ■☆

308634　Polygonatum pumilum Hua = Polygonatum hookeri Baker ■

308635　Polygonatum punctatum Royle ex Kunth;点花黄精(斑茎黄精,滇钩吻,黄精,蒙自黄精,生扯拢,树刁,树吊,葳参,玉术);Punctated Landpick,Variegated Solomonseal ■

308636　Polygonatum quelpaertense Ohwi = Polygonatum odoratum (Mill.) Druce ■

308637　Polygonatum quinquefolium Kitag. = Polygonatum acuminatifolium Kom. ■

308638　Polygonatum racemosum F. T. Wang et Ts. Tang = Polygonatum alternicirrhosum Hand.-Mazz. ■

308639　Polygonatum roseum (Ledeb.) Kunth;新疆黄精(玫瑰红黄精,玫瑰黄精,玉竹,紫花黄精);Sinkiang Solomonseal,Xinjiang Landpick,Xinjiang Solomonseal ■

308640　Polygonatum sewerzovii Regel;赛氏黄精■☆

308641　Polygonatum sibiricum Delarb ex Redouté;黄精(白及,白及黄精,笔菜,笔管菜,垂珠,高良姜,苟格,黄鸡菜,黄精子,黄芝,鸡格,鸡头参,鸡头黄精,鸡头七,鸡爪参,姜形黄精,老虎姜,良姜,

龙衔,鹿竹,马箭,盘龙七,山捣臼,山姜,山生姜,生姜,太阳草,甜黄精,土灵芝,兔竹,萎蕤,乌鸦七,西伯利亚黄精,阳雀蕻,野生姜,野鲜姜,玉竹黄精,重楼,爪子参);Siberia Landpick,Siberian Solomonseal ■

308642　Polygonatum sibiricum Redouté = Polygonatum sibiricum Delarb ex Redouté ■

308643　Polygonatum simizui Kitag.;春水玉竹■☆

308644　Polygonatum simizui Kitag. = Polygonatum odoratum (Mill.) Druce ■

308645　Polygonatum sinomairei F. T. Wang et Ts. Tang = Polygonatum punctatum Royle ex Kunth ■

308646　Polygonatum souliei Hua = Polygonatum cirrhifolium (Wall.) Royle ■

308647　Polygonatum stenophyllum Maxim.;狭叶黄精;Narrowleaf Landpick,Narrowleaf Solomonseal ■

308648　Polygonatum stewartianum Diels;西南黄精■

308649　Polygonatum strumulosum D. M. Liu et W. Z. Zeng;苦瘤黄精■

308650　Polygonatum strumulosum D. M. Liu et W. Z. Zeng = Polygonatum cirrhifolium (Wall.) Royle ■

308651　Polygonatum tessellatum F. T. Wang et Ts. Tang;格脉黄精(滇竹根七,竹根七);Tessllateleaf Landpick,Tessllateleaf Solomonseal ■

308652　Polygonatum thunbergii C. Morren et Decne. = Polygonatum odoratum (Mill.) Druce ■

308653　Polygonatum tonkinense Gagnep. = Disporopsis longifolia Craib ■

308654　Polygonatum trinerve Hua = Polygonatum cirrhifolium (Wall.) Royle ■

308655　Polygonatum umbellatum Baker = Polygonatum macropodum Turcz. ■

308656　Polygonatum uncinatum Diels;小黄精;Uncinate Solomonseal ■

308657　Polygonatum uncinatum Diels = Polygonatum kingianum Collett et Hemsl. ■

308658　Polygonatum vertichillatum (L.) All. var. stenophyllum (Maxim.) Baker = Polygonatum stenophyllum Maxim. ■

308659　Polygonatum verticillatum (L.) All.;轮叶黄精(臭儿参,地吊,甘肃黄精,红果黄精,拉尼,羊角参,玉竹参);Whorled Solomon's Seal,Whorled Solomon's-seal,Whorledleaf Landpick,Whorledleaf Solomonseal ■

308660　Polygonatum verticillatum (L.) All. var. stenophyllum (Maxim.) Baker = Polygonatum stenophyllum Maxim. ■

308661　Polygonatum virens Nakai = Polygonatum inflatum Kom. ■

308662　Polygonatum vulgare Desf. = Polygonatum odoratum (Mill.) Druce ■

308663　Polygonatum wardii F. T. Wang et Ts. Tang;西藏黄精;Xizang Solomonseal ■

308664　Polygonatum yunnanense H. Lév. = Polygonatum nodosum Hua ■

308665　Polygonatum zanlanscianense Pamp.;湖北黄精(虎其尾,野山姜);Hubei Landpick,Hubei Solomonseal,Hupeh Solomonseal ■

308666　Polygonatum zhejiangensis X. J. Xue et H. Yao;浙江黄精;Zhejiang Landpick,Zhejiang Solomonseal ■

308667　Polygonella Michx. (1803);小蓼属(假蓼属,贴茎蓼属);Jointweed,Wireweed ■☆

308668　Polygonella americana (Fisch. et C. A. Mey.) Small;美洲小蓼;American Jointweed,Jointweed,Southern Jointweed ●☆

308669　Polygonella articulata (L.) Meisn.;节小蓼;Coastal Jointweed,Northern Jointweed,Sand Jointweed ■☆

308670　Polygonella basiramia (Small) G. L. Nesom et V. M. Bates;毛小蓼;Hairy Wireweed ■☆

308671　Polygonella brachystachya Meisn. = Polygonella polygama （Vent.）Engelm. et A. Gray var. brachystachya （Meisn.）Wunderlin ■☆

308672　Polygonella ciliata Meisn. ；缘毛小蓼；Fringed Jointweed ■☆

308673　Polygonella ciliata Meisn. var. basiramia （Small）Horton = Polygonella basiramia （Small）G. L. Nesom et V. M. Bates ■☆

308674　Polygonella croomii Chapm. = Polygonella polygama （Vent.）Engelm. et A. Gray var. croomii （Chapm.）Fernald ■☆

308675　Polygonella cuspidata ？；日本小蓼；Japanese Knotweed ■☆

308676　Polygonella fimbriata （Elliott）Horton；沙丘小蓼；Sandhill Jointweed ■☆

308677　Polygonella fimbriata （Elliott）Horton var. robusta （Small）Horton = Polygonella robusta （Small）G. L. Nesom et V. M. Bates ■☆

308678　Polygonella gracilis Meisn. ；细小蓼；Slender Wireweed ■☆

308679　Polygonella macrophylla Small；大叶小蓼；Large-leaf Wireweed ●☆

308680　Polygonella myriophylla （Small）Horton；木本小蓼；Sandlace, Woody Jointweed ●☆

308681　Polygonella parksii Cory；帕克斯小蓼；Parks' Jointweed ■☆

308682　Polygonella parvifolia Michx. ；小花小蓼■☆

308683　Polygonella polygama （Vent.）Engelm. et A. Gray；十月花小蓼；October-flower ●☆

308684　Polygonella polygama （Vent.）Engelm. et A. Gray var. brachystachya （Meisn.）Wunderlin；短穗小花小蓼■☆

308685　Polygonella polygama （Vent.）Engelm. et A. Gray var. croomii （Chapm.）Fernald；克鲁姆小蓼■☆

308686　Polygonella robusta （Small）G. L. Nesom et V. M. Bates；粗壮小蓼；Stout Jointweed ☆

308687　Polygonifolia Fabr. = Corrigiola L. ■☆

308688　Polygonoidea Ortega = Calligonum L. ●

308689　Polygonon St. -Lag. = Polygonum L. （保留属名）■●

308690　Polygonum L. （1753）（保留属名）；蓼属；Fleece Flower, Jointweed, Knotgrass, Knot-grass, Knotweed, Polygonum, Silver Lace Vine, Smartweed, Tearthumb ■●

308691　Polygonum × bohemicum （Chrtek et Chrtkova）P. F. Zika et Jacobson = Fallopia × bohemica （Chrtek et Chrtkova）J. P. Bailey ●☆

308692　Polygonum abbreviatum Kom. ；缩短蓼■☆

308693　Polygonum acadiense Fernald = Polygonum oxyspermum C. A. Mey. et Bunge ex Ledeb. ■☆

308694　Polygonum acaule Hook. f. = Polygonum hookeri Meisn. ■

308695　Polygonum acerosum Ledeb. ex Meisn. ；松叶萹蓄（松叶蓼）；Pineleaf Knotweed ■

308696　Polygonum acetosellum Klokov；小灰绿蓼■☆

308697　Polygonum acetosum M. Bieb. ；灰绿萹蓄（灰绿蓼，酸蓼）；Greygreen Knotweed ■

308698　Polygonum achoreum S. F. Blake；北美蓼；Beak-seeded Knotweed, Blake's Knotweed, Knotweed, Leathery Knotweed ■☆

308699　Polygonum acidulum Willd. = Polygonum angustifolium Pall. ■

308700　Polygonum acre Kunth；辛辣蓼■☆

308701　Polygonum acre Kunth = Polygonum punctatum Elliott ■☆

308702　Polygonum acre Kunth var. confertiflorum Meisn. = Polygonum punctatum Elliott var. confertiflorum （Meisn.）Fassett ■☆

308703　Polygonum acre Kunth var. leptostachyum Meisn. = Persicaria punctata （Elliott）Small ■☆

308704　Polygonum acre Kunth var. leptostachyum Meisn. = Polygonum punctatum Elliott ■☆

308705　Polygonum acre Kunth var. leptostachyum Meisn. = Polygonum punctatum Elliott var. confertiflorum （Meisn.）Fassett ■☆

308706　Polygonum acuminatum Kunth = Persicaria acuminata （Kunth）M. Gómez ■☆

308707　Polygonum acuminatum Kunth var. microstemon Mart. ex Meisn. = Persicaria acuminata （Kunth）M. Gómez ■☆

308708　Polygonum adenopodum Sam. = Polygonum chinense L. ■

308709　Polygonum aequale Lindm. = Polygonum arenastrum Jord. ex Boreau ■

308710　Polygonum aequale Lindm. = Polygonum aviculare L. subsp. depressum （Meisn.）Arcang. ■☆

308711　Polygonum aequale Lindm. subsp. oedocarpum Lindm. = Polygonum aviculare L. subsp. neglectum （Besser）Arcang. ■☆

308712　Polygonum aequale Lindm. var. platycarpum Koji Ito = Polygonum aviculare L. subsp. neglectum （Besser）Arcang. ■☆

308713　Polygonum affine D. Don；密穗蓼（密穗拳参）；Densespike Knotweed, Dense-spiked Knotweed, Himalayan Fleeceflower ●■

308714　Polygonum affine D. Don = Polygonum macrophyllum D. Don ■

308715　Polygonum afghanicum Meisn. ；阿富汗蓼■☆

308716　Polygonum afromontanum Greenway；非洲山生蓼■☆

308717　Polygonum ajanense （Regel et Tiling）Grig. ；阿扬蓼（阿扬神血宁，高山蓼，兴安蓼）；Ajan Knotweed ■

308718　Polygonum ajanense （Regel et Tiling）Grig. = Aconogonon ajanense （Regel et Tiling）H. Hara ■

308719　Polygonum alatum （Buch. -Ham. ex D. Don）Spreng. = Persicaria nepalensis （Meisn.）H. Gross ■

308720　Polygonum alatum （Buch. -Ham. ex D. Don）Spreng. = Polygonum nepalense Meisn. ■

308721　Polygonum alatum （Buch. -Ham. ex D. Don）Spreng. var. nepalense （Meisn.）Hook. f. = Polygonum nepalense Meisn. ■

308722　Polygonum alatum Buch. -Ham. ex D. Don = Persicaria nepalensis （Meisn.）H. Gross ■

308723　Polygonum alatum Buch. -Ham. ex D. Don = Polygonum nepalense Meisn. ■

308724　Polygonum allocarpum S. F. Blake = Polygonum fowleri B. L. Rob. ■☆

308725　Polygonum allocarpum S. F. Blake = Polygonum ramosissimum Michx. ■☆

308726　Polygonum alopecuroides Turcz. ex Besser；狐尾蓼（狐尾拳参）；Foxtail Knotweed ■

308727　Polygonum alopecuroides Turcz. ex Besser f. pilosum C. F. Fang = Polygonum alopecuroides Turcz. ex Besser ■

308728　Polygonum alopecuroides Turcz. ex Besser f. pilosum W. P. Fang；毛狐尾蓼■

308729　Polygonum alpestre C. A. Mey. ；山地蓼■☆

308730　Polygonum alpinum All. ；高山神血宁■

308731　Polygonum alpinum All. = Persicaria alpina （All.）H. Gross ■

308732　Polygonum alpinum All. var. sinicum Dammer = Polygonum campanulatum Hook. f. var. fulvidum Hook. f. ■

308733　Polygonum alpinum All. var. sinicum Dammer ex Diels = Polygonum campanulatum Hook. f. ■

308734　Polygonum alpinum Willd. var. angustissimum Turcz. = Polygonum angustifolium Pall. ■

308735　Polygonum ambiguum Meisn. = Polygonum amplexicaule D. Don ■☆

308736　Polygonum amblyophyllum （H. Hara）Kitam. = Persicaria amblyophylla H. Hara ■☆

308737　Polygonum ammanioides Jaub. et Spach；水苋菜蓼■☆

308738　Polygonum amoenum Blume = Polygonum orientale L. ■

308739　Polygonum amphibium L. ；两栖蓼（醋柳，木茧，水茧，天蓼，小

黄药）; Amphibia Knotweed, Amphibious Bistort, Amphibious Knotweed, Amphibious Persicaria, Bistort, Dock Flower, Floating Knotweed, Ground Willow, Lakeweed, Redshanks, Shoestring Smartweed, Swamp Smartweed, Water Heart's-ease, Water Lady's-thumb, Water Smartweed, Willow-grass, Willow-weed ■

308740 Polygonum amphibium L. = Persicaria amphibia (L.) Delarbre ■

308741 Polygonum amphibium L. = Persicaria amphibia (L.) Gray ■

308742 Polygonum amphibium L. f. hirtuosum Farw. = Polygonum amphibium L. var. stipulaceum N. Coleman ■

308743 Polygonum amphibium L. f. terrestre (Willd.) S. F. Blake = Polygonum amphibium L. var. emersum Michx. ■☆

308744 Polygonum amphibium L. subsp. laevimarginatum Hultén = Persicaria amphibia (L.) Gray ■

308745 Polygonum amphibium L. subsp. laevimarginatum Hultén = Polygonum amphibium L. var. stipulaceum N. Coleman ■

308746 Polygonum amphibium L. subsp. laevimarginatum Hultén = Polygonum amphibium L. ■

308747 Polygonum amphibium L. var. amurense Korsh. = Polygonum amphibium L. ■

308748 Polygonum amphibium L. var. aquaticum Leyss. = Persicaria amphibia (L.) Gray ■

308749 Polygonum amphibium L. var. coccineum (Muhl. ex Willd.) Farw. = Polygonum amphibium L. var. emersum Michx. ■☆

308750 Polygonum amphibium L. var. emersum Michx.; 长根两栖蓼; Long-root Smartweed, Water Heart's-ease, Water Smartweed ■☆

308751 Polygonum amphibium L. var. emersum Michx. = Persicaria amphibia (L.) Gray ■

308752 Polygonum amphibium L. var. emersum Michx. = Polygonum amphibium L. ■

308753 Polygonum amphibium L. var. hartwrightii (A. Gray) Bissell = Polygonum amphibium L. var. stipulaceum N. Coleman ■

308754 Polygonum amphibium L. var. muhlenbergii Meisn. = Polygonum amphibium L. ■

308755 Polygonum amphibium L. var. natans Michx. = Persicaria amphibia (L.) Gray ■

308756 Polygonum amphibium L. var. natans Michx. = Polygonum amphibium L. ■

308757 Polygonum amphibium L. var. stipulaceum f. fluitans (Eaton) Fernald = Polygonum amphibium L. ■

308758 Polygonum amphibium L. var. stipulaceum N. Coleman; 托叶两栖蓼; Water Heart's-ease, Water Smartweed ■

308759 Polygonum amphibium L. var. stipulaceum N. Coleman = Persicaria amphibia (L.) Gray ■

308760 Polygonum amphibium L. var. stipulaceum N. Coleman = Polygonum amphibium L. ■

308761 Polygonum amphibium L. var. stipulaceum N. Coleman f. fluitans (Eaton) Fernald = Polygonum amphibium L. var. stipulaceum N. Coleman ■

308762 Polygonum amphibium L. var. stipulaceum N. Coleman f. hirtuosum (Farw.) Fernald = Polygonum amphibium L. var. stipulaceum N. Coleman ■

308763 Polygonum amphibium L. var. stipulaceum N. Coleman f. simile Fernald = Polygonum amphibium L. var. stipulaceum N. Coleman ■

308764 Polygonum amphibium L. var. terrestre Leyss.; 干型两栖蓼■

308765 Polygonum amphibium L. var. terrestre Leyss. = Persicaria amphibia (L.) Gray ■

308766 Polygonum amphibium L. var. terrestre Leyss. = Polygonum amphibium L. ■

308767 Polygonum amphibium L. var. terrestre Willd. = Polygonum amphibium L. var. emersum Michx. ■☆

308768 Polygonum amphibium L. var. vestitum Hemsl. = Polygonum amphibium L. ■

308769 Polygonum amplexicaule D. Don; 抱茎蓼（抱茎拳参，墨苹头草）; Amplexicaule Knotweed, Fleece Flower, Mountain Fleece, Red Bistort ■

308770 Polygonum amplexicaule D. Don = Persicaria amplexicaulis (D. Don) Ronse Decr. ■

308771 Polygonum amplexicaule D. Don var. sinense Forbes et Hemsl. = Polygonum amplexicaule D. Don var. sinense Forbes et Hemsl. ex Stewart ■

308772 Polygonum amplexicaule D. Don var. sinense Forbes et Hemsl. ex Stewart; 中华抱茎蓼（倒生莲，红孩儿，红血儿，红血七，鸡心七，鸡血七，荞麦七，蜈蚣七，血三七，中华抱茎拳参）; Chinese Amplexicaule Knotweed ■

308773 Polygonum amplexicaule D. Don var. speciosum (Meisn.) Hook. f.; 华美蓼■☆

308774 Polygonum amplexicaule Meisn. var. speciosum (Meisn.) Hook. f. = Polygonum amplexicaule D. Don ■

308775 Polygonum amurense (Korsh.) Vorosch. = Persicaria amphibia (L.) Delarbre ■

308776 Polygonum anguillanum Koidz. = Persicaria sagittata (L.) H. Gross var. sibirica (Meisn.) Miyabe ■

308777 Polygonum angustifolium Pall.; 狭叶蓼（细叶蓼，狭叶神血宁）; Narrowleaf Knotweed ■

308778 Polygonum angustifolium Pall. var. songaricum (Schrenk) Stewart = Polygonum songaricum Schrenk ■

308779 Polygonum antihaemorrhoidale Mart. f. aquatile Mart. = Polygonum punctatum Elliott var. confertiflorum (Meisn.) Fassett ■☆

308780 Polygonum araraticum Kom.; 亚拉腊蓼■☆

308781 Polygonum arcticum Pall. ex Spreng. = Polygonum sibiricum Laxm. ■

308782 Polygonum arenarium Waldst. et Kit.; 欧洲蓼; European Knotweed, Lesser Red Knot-grass ■☆

308783 Polygonum arenarium Waldst. et Kit. subsp. pulchellum (Loisel.) Thell.; 美丽欧洲蓼■☆

308784 Polygonum arenastrum Boreau = Polygonum arenastrum Jord. ex Boreau ■

308785 Polygonum arenastrum Boreau = Polygonum aviculare L. subsp. depressum (Meisn.) Arcang. ■☆

308786 Polygonum arenastrum Boreau subsp. boreale (Lange) Á. Löve = Polygonum aviculare L. subsp. boreale (Lange) Karlsson ■☆

308787 Polygonum arenastrum Boreau var. platycarpum (Koji Ito) Koji Ito ex Kitag. = Polygonum aviculare L. subsp. neglectum (Besser) Arcang. ■☆

308788 Polygonum arenastrum Jord. ex Boreau; 伏地萹蓄（多脉蓼，伏地蓼）; Adpressed Knotweed, Door-yard Knotweed, Equal-leaved Knotgrass, Knotweed, Oval-leaved Knotweed ■

308789 Polygonum argenteum Skvortsov = Polygonum aviculare L. var. fusco-ochreatum (Kom.) A. J. Li ■

308790 Polygonum argenteum Skvortsov = Polygonum fusco-ochreatum Kom. ■

308791 Polygonum argyrocoleum Steud. ex Kunze; 帚萹蓄（银鞘蓼，帚蓼）; Broom Knotweed, Persian Knotweed, Silversheath Knotweed, Silver-sheathed Knotweed ■

308792　Polygonum arifolium L. = Persicaria arifolia（L.）Haraldson ■

308793　Polygonum arifolium L. = Polygonum thunbergii Siebold et Zucc. ■

308794　Polygonum arifolium L. var. lentiforme Fernald et Griscom = Persicaria arifolia（L.）Haraldson ■☆

308795　Polygonum arifolium L. var. lentiforme Fernald et Griscom = Polygonum arifolium L. ■

308796　Polygonum arifolium L. var. perfoliatum L. = Persicaria perfoliata（L.）H. Gross ■

308797　Polygonum arifolium L. var. perfoliatum L. = Polygonum perfoliatum L. ■

308798　Polygonum arifolium L. var. pubescens（R. Keller）Fernald = Persicaria arifolia（L.）Haraldson ■☆

308799　Polygonum arifolium L. var. pubescens（R. Keller）Fernald = Polygonum arifolium L. ■

308800　Polygonum arifolium Thunb. = Polygonum thunbergii Siebold et Zucc. ■

308801　Polygonum articulatum L. = Polygonella articulata（L.）Meisn. ■☆

308802　Polygonum arussense Chiov. ;阿鲁斯蓼■☆

308803　Polygonum aschersonianum H. Gross,阿塞蓼■☆

308804　Polygonum assamicum Meisn. ;阿萨姆蓼;Assam Knotweed ■

308805　Polygonum atlanticum（B. L. Rob.）E. P. Bicknell = Polygonum ramosissimum Michx. ■☆

308806　Polygonum atraphaxis Thunb. = Polygonum undulatum（L.）P. J. Bergius ■☆

308807　Polygonum atraphaxoides Thunb. = Polygonum undulatum（L.）P. J. Bergius ■☆

308808　Polygonum attenuatum Petr. ex Kom. ;毛叶耳蓼（毛耳叶蓼）; Attenuate Knotweed ■

308809　Polygonum attenuatum Petr. ex Kom. = Polygonum ellipticum Willd. ex Spreng. ■

308810　Polygonum aubertii（L. Henry）Holub = Fallopia aubertii（L. Henry）Holub ●

308811　Polygonum aubertii L. Henry = Fallopia aubertii（L. Henry）Holub ●

308812　Polygonum aubertii L. Henry = Fallopia baldschuanica（Regel）Holub ■☆

308813　Polygonum auriculatum Makino = Polygonum praetermissum Hook. f. ■

308814　Polygonum austiniae Greene;奥氏蓼;Mrs. Austin's knotweed ■☆

308815　Polygonum autumnale Brenckle = Polygonum ramosissimum Michx. ■☆

308816　Polygonum aviculare L. ;萹蓄（百节,百节草,斑鸠台,边血草,编竹,萹蔓,萹蓄草,萹竹,萹竹竹,萹苋,扁瓣,扁蔓,扁畜,扁蓄,扁猪芽,扁猪芽,扁竹,扁竹蓼,扁竹牙,扁苋,残猪草,残竹草,大萹蓄,大铁马鞭,大蓄片,道生草,地萹蓄,地蓼,多茎蓼,藩萹竹,藩水萹,粉节草,俯卧蓼,疳积药,路边草,路柳,蚂蚁草,妹子草,牛鞭草,牛筋草,七星草,姝子草,太阳草,铁锦草,铁片草,王刍,乌蓼,畜瓣,蓄,蓄蔓,野铁扫把,异叶萹蓄,异叶蓼,猪圈草,猪牙草,竹,竹萹,竹萹蓄,竹节草,竹叶草,桌面草）;Allseed, Armstrong, Beggarweed, Bird Knotgrass, Bird's Knotgrass, Bird's Knot-grass, Bird's Tongue, Blackstrap, Bloodwort, Clutch, Common Knot-grass, Common Knotweed, Cow Grass, Crab Grass, Crabweed, Cumberfield, Cumberland, Devil's Lingels, Diverseleaf Knotweed, Doorweed, Dooryard Knotweed, Dooryard Weed, Finzach, Goose Grass, Hogweed, Iron-grass, Knotgrass, Knot-grass, Knotweed, Knotwort, Man-tie, Matgrass, Nine Joints, Ninety Knot, Ninety-knot, Pig Grass, Pig Rush, Pigweed, Pinkweed, Prostrate Knotweed, Red

Legs, Red Robin, Redweed, Snakeweed, Sparrow's Tongue, Surface Twitch, Swine Carse, Swine Grass, Swine's Cress, Swine's Grass, Swine's Grease, Swine's Skir, Swine's Tusker, Tacker-grass, Untrod Den-to-pieces, Untrodden-to-death, Waygrass, Willow-grass, Wiregrass, Wireweed ■

308817　Polygonum aviculare L. = Polygonum arenastrum Jord. ex Boreau ■

308818　Polygonum aviculare L. = Polygonum buxiforme Small ■☆

308819　Polygonum aviculare L. = Polygonum neglectum Besser ■☆

308820　Polygonum aviculare L. f. erectum J. X. Li et F. Q. Zhou;直立萹蓄;Erect Knotgrass ■

308821　Polygonum aviculare L. subsp. aequale（Lindm.）Asch. et Graebn. = Polygonum aviculare L. subsp. depressum（Meisn.）Arcang. ■☆

308822　Polygonum aviculare L. subsp. battandieri Maire et Sennen = Polygonum balansae Boiss. subsp. battandieri（Maire et Sennen）Greuter ■☆

308823　Polygonum aviculare L. subsp. boreale（Lange）Karlsson;北方萹蓄;Northern Knotweed ■☆

308824　Polygonum aviculare L. subsp. buxiforme（Small）Costea et Tardif;美洲萹蓄;American Knotweed, Box Knotweed, Boxwood Knotweed, Knotweed ■☆

308825　Polygonum aviculare L. subsp. buxiforme（Small）Costea et Tardif = Polygonum buxiforme Small ■☆

308826　Polygonum aviculare L. subsp. calcatum（Lindm.）Thell. = Polygonum aviculare L. subsp. depressum（Meisn.）Arcang. ■☆

308827　Polygonum aviculare L. subsp. depressum（Meisn.）Arcang. ;卵叶萹蓄;Common Knotweed, Oval-leaf Knotweed ■☆

308828　Polygonum aviculare L. subsp. heterophyllum Asch. et Graebn. = Polygonum aviculare L. ■

308829　Polygonum aviculare L. subsp. induratum（Asch. et Barbey）Maire = Polygonum balansae Boiss. subsp. induratum（Asch. et Barbey）Greuter ■☆

308830　Polygonum aviculare L. subsp. microspermum（Jord. ex Boreau）Berher = Polygonum aviculare L. subsp. depressum（Meisn.）Arcang. ■☆

308831　Polygonum aviculare L. subsp. monspeliense（Thiéb. -Bern. ex Pers.）Arcang. = Polygonum aviculare L. ■

308832　Polygonum aviculare L. subsp. neglectum（Besser）Arcang. ;窄叶萹蓄;Knotweed, Narrow-leaf Knotweed ■☆

308833　Polygonum aviculare L. subsp. patulum（M. Bieb.）Maire = Polygonum bellardii All. ■☆

308834　Polygonum aviculare L. subsp. rectum Chrtek = Polygonum aviculare L. subsp. neglectum（Besser）Arcang. ■☆

308835　Polygonum aviculare L. subsp. rhizoxylon（Pau et Font Quer）Emb. et Maire = Polygonum balansae Boiss. subsp. rhizoxylon（Pau et Font Quer）Greuter ■☆

308836　Polygonum aviculare L. subsp. rurivagum（Jord. ex Boreau）Berher;玉米地萹蓄;Cornfield Knot-grass, Narrow-leaf Knotweed ■☆

308837　Polygonum aviculare L. var. angustifolium Michx. ;狭叶萹蓄■☆

308838　Polygonum aviculare L. var. angustissimum Meisn. = Polygonum aviculare L. subsp. rurivagum（Jord. ex Boreau）Berher ■☆

308839　Polygonum aviculare L. var. angustissimum Meisn. = Polygonum bellardii All. ■☆

308840　Polygonum aviculare L. var. aphyllum Hayne = Polygonum aviculare L. ■

308841　Polygonum aviculare L. var. arenastrum（Boreau）Rouy = Polygonum arenastrum Jord. ex Boreau ■

308842 Polygonum aviculare L. var. boreale Lange = Polygonum aviculare L. subsp. boreale（Lange）Karlsson ■☆

308843 Polygonum aviculare L. var. buxifolium Ledeb. ；日本萹蓄■☆

308844 Polygonum aviculare L. var. condensatum Becker = Polygonum aviculare L. ■

308845 Polygonum aviculare L. var. crassifolium Lange = Polygonum aviculare L. subsp. buxiforme（Small）Costea et Tardif ■☆

308846 Polygonum aviculare L. var. depressum Meisn. = Polygonum aviculare L. subsp. depressum（Meisn.）Arcang. ■☆

308847 Polygonum aviculare L. var. erectum（L.）Roth ex Meisn. = Polygonum erectum L. ■☆

308848 Polygonum aviculare L. var. erectum（Roth）Hayne = Polygonum aviculare L. ■

308849 Polygonum aviculare L. var. eximium（Lindm.）Asch. et Graebn. = Polygonum aviculare L. ■

308850 Polygonum aviculare L. var. fusco-ochreatum（Kom.）A. J. Li；褐鞘萹蓄（褐鞘蓼）■

308851 Polygonum aviculare L. var. fusco-ochreatum（Kom.）A. J. Li = Polygonum fusco-ochreatum Kom. ■

308852 Polygonum aviculare L. var. heterophyllum Munshi et Javeid = Polygonum aviculare L. ■

308853 Polygonum aviculare L. var. latifolium Coss. et Germ. = Polygonum aviculare L. ■

308854 Polygonum aviculare L. var. littorale（Link）Mert. = Polygonum aviculare L. subsp. buxiforme（Small）Costea et Tardif ■☆

308855 Polygonum aviculare L. var. littorale Koch = Polygonum aviculare L. ■

308856 Polygonum aviculare L. var. minimum Murith = Polygonum aviculare L. ■

308857 Polygonum aviculare L. var. minutiflorum Franch. = Polygonum plebeium R. Br. ■

308858 Polygonum aviculare L. var. neglectum（Besser）Rchb. = Polygonum aviculare L. ■

308859 Polygonum aviculare L. var. parvulum Zapal. = Polygonum aviculare L. ■

308860 Polygonum aviculare L. var. procumbens（Gilib.）Hayne；平铺萹蓄■☆

308861 Polygonum aviculare L. var. rigidum（Skvortsov）H. C. Fu；紧穗萹蓄■

308862 Polygonum aviculare L. var. triviale Rchb. = Polygonum aviculare L. ■

308863 Polygonum aviculare L. var. vegetum Ledeb. ；异叶蓼（萹蓄，多茎萹蓄）■☆

308864 Polygonum aviculare L. var. vegetum Ledeb. = Polygonum aviculare L. ■

308865 Polygonum babingtonii Hance = Polygonum senticosum（Meisn.）Franch. et Sav. ■

308866 Polygonum balansae Boiss. ；巴拉蓼■☆

308867 Polygonum balansae Boiss. subsp. battandieri（Maire et Sennen）Greuter；巴坦巴拉蓼■☆

308868 Polygonum balansae Boiss. subsp. induratum（Asch. et Barbey）Greuter；坚硬巴拉蓼■☆

308869 Polygonum balansae Boiss. subsp. rhizoxylon（Pau et Font Quer）Greuter；木根巴拉蓼■☆

308870 Polygonum balansae Boiss. var. battandieri（Maire et Sennen）Maire = Polygonum balansae Boiss. subsp. battandieri（Maire et Sennen）Greuter ■☆

308871 Polygonum balansae Boiss. var. elongatum Maire et Weiller = Polygonum balansae Boiss. ■☆

308872 Polygonum balansae Boiss. var. induratum（Asch. et Barbey）Maire = Polygonum balansae Boiss. subsp. induratum（Asch. et Barbey）Greuter ■☆

308873 Polygonum balansae Boiss. var. rhizoxylon（Pau et Font Quer）Maire = Polygonum balansae Boiss. subsp. rhizoxylon（Pau et Font Quer）Greuter ■☆

308874 Polygonum balansae Boiss. var. suffruticosum Font Quer = Polygonum balansae Boiss. ■☆

308875 Polygonum balansae Boiss. var. tectifolium Svent. et Kahne = Polygonum balansae Boiss. ■☆

308876 Polygonum baldschuanicum Regel = Fallopia baldschuanica（Regel）Holub ■☆

308877 Polygonum barbatum L. ；毛蓼（红蓼子，蓼子草，拳参，髯毛蓼，水辣蓼，四季青，香草，小蓼子草，小毛蓼）；Hair Knotweed，Hairy Knotweed ■

308878 Polygonum barbatum L. = Persicaria barbata（L.）H. Hara ■

308879 Polygonum barbatum L. var. gracile（Danser）Steward；细刺毛蓼（红蓼子，蓼子草，小蓼子草，小毛蓼）■

308880 Polygonum barbatum L. var. gracile（Danser）Steward = Polygonum longisetum Bruijn var. rotundatum A. J. Li ■

308881 Polygonum bellardii All. ；窄叶蓼；Narrowleaf Knotweed，Narrow-leaved Knotweed ■☆

308882 Polygonum bellardii All. var. gracilius Ledeb. = Polygonum patulum M. Bieb. ■

308883 Polygonum bellardii Blanco = Polygonum bellardii All. ■☆

308884 Polygonum belophyllum Litv. = Polygonum sagittatum L. ■

308885 Polygonum biaristatum Aitch. et Hemsl. ；双芒蓼■☆

308886 Polygonum biconvexum Hayata；双凸戟叶蓼；Biconvex Halbertleaf Knotweed ■

308887 Polygonum biconvexum Hayata = Polygonum thunbergii Siebold et Zucc. ■

308888 Polygonum bicorne Raf. = Persicaria bicornis（Raf.）Nieuwl. ■☆

308889 Polygonum bicorne Raf. = Polygonum pensylvanicum L. ■☆

308890 Polygonum bidwelliae S. Watson；毕氏蓼；Bidwell's Knotweed ■☆

308891 Polygonum biforme Wahlenb. = Persicaria maculata（Raf.）Gray ■

308892 Polygonum birmanicum Gage = Polygonum praetermissum Hook. f. ■

308893 Polygonum bistorta L. ；拳参（草河车，刀剪药，刀箭药，刀枪药，倒根草，地虾，杜蒙，疙瘩参，狗尾巴吊，红蚤休，回头参，马峰七，马蜂七，马行，破伤药，拳参蓼，拳蓼，山柳柳，山虾，山虾子，石蚕，铜罗，虾参，鸢头鸡，重楼，紫参）；Adderwort，Astrologia，Betes，Bistort，Bistort Knotweed，Common Bistort，Dragons，Dragonwort，Easter Giants，Easter Hedges，Easter Ledger，Easter Ledgers，Easter Ledges，Easter Mangiants，Easter Mantgions，Easter Sedge，Eastern Giants，English Serpentary，Esterledges，European Bistort，Gentle Dock，Goose Grass，Great Bistort，Greater Bistort，Meadow Bistort，Meeks，Oisterloit，Ooster-munath-jonnums，Osterick，Passion，Passion Dock，Patience，Patience Dock，Patience-dock，Pencuir Kale，Poor Man's Cabbage，Red Legs，Silver Dock，Smartweed，Snakeroot，Snake's Food，Snakeweed，Sweet Dock，Twice-writhen，Waster Ledges ■

308894 Polygonum bistorta L. = Bistorta officinalis Delarbre ■☆

308895 Polygonum bistorta L. = Persicaria bistorta（L.）Samp. ■☆

308896 Polygonum bistorta L. subsp. ochotense（Petr. ex Kom.）

Vorosch. = Polygonum ochotense Petr. ex Kom. ■

308897　Polygonum bistorta L. subsp. pacificum （Petr. ex Kom.）Vorosch. = Polygonum pacificum Petr. ex Kom. ■

308898　Polygonum bistorta L. subsp. plumosum （Small）Hultén = Bistorta plumosa （Small）Greene ■☆

308899　Polygonum bistorta L. var. ellipticum （Willd. ex Spreng.）Turcz. = Polygonum ellipticum Willd. ex Spreng. ■

308900　Polygonum bistorta L. var. griseum Beck = Persicaria bistorta （L.）Samp. ■☆

308901　Polygonum bistorta L. var. nitens Fisch. et C. A. Mey. = Polygonum ellipticum Willd. ex Spreng. ■

308902　Polygonum bistorta L. var. plumosum （Small）B. Boivin = Bistorta plumosa （Small）Greene ■☆

308903　Polygonum bistortioides Pursh；假拳参；Alpine Smartweed ■☆

308904　Polygonum bistortoides Pursh = Bistorta bistortoides （Pursh）Small ■☆

308905　Polygonum bistortoides Pursh var. linearifolium （S. Watson）Small = Bistorta bistortoides （Pursh）Small ■☆

308906　Polygonum bistortoides Pursh var. oblongifolium （Meisn.）H. St. John = Bistorta bistortoides （Pursh）Small ■☆

308907　Polygonum blumei Meisn.；布氏蓼（马蓼，墨记草）；Blume Knotweed ■

308908　Polygonum blumei Meisn. = Persicaria longiseta （Bruijn）Kitag. ■

308909　Polygonum blumei Meisn. = Polygonum longisetum Bruijn ■

308910　Polygonum blumei Meisn. ex Miq. = Polygonum longisetum Bruijn ■

308911　Polygonum bodinieri H. Lév. et Vaniot = Polygonum strigosum R. Br. ■

308912　Polygonum bohemicum （Chrtek et Chrtková）Zika et Jacobson = Fallopia bohemica （Chrtek et Chrtkova）J. P. Bailey ■☆

308913　Polygonum bolanderi W. H. Brewer；鲍氏蓼；Bolander's Knotweed ■☆

308914　Polygonum bonatii H. Lév. = Fagopyrum gracilipes （Hemsl.）Dammer ex Diels ■

308915　Polygonum boreale （Lange）Small = Polygonum aviculare L. subsp. boreale （Lange）Karlsson ■☆

308916　Polygonum boreale （Lange）Small = Polygonum tatewakianum Koji Ito var. notoroense Koji Ito ■☆

308917　Polygonum brachiatum Poir. = Polygonum chinense L. ■

308918　Polygonum brevifolia Kitag.；短箭叶蓼（短叶箭叶蓼）■

308919　Polygonum buchananii Dammer = Oxygonum buchananii （Dammer）J. B. Gillett ■☆

308920　Polygonum bucharicum Grig. = Polygonum coriarium Grig. ■

308921　Polygonum buisanense Ohki = Polygonum longisetum Bruijn ■

308922　Polygonum bungeanum Turcz.；柳叶刺蓼（本氏蓼）；Bunge's Smartweed，Prickly Smartweed，Willowleaf Knotweed ■

308923　Polygonum bungeanum Turcz. = Persicaria bungeana （Turcz.）Nakai ■☆

308924　Polygonum burnmulleri Litv.；鲍尔蓼 ■☆

308925　Polygonum buxifolium Nutt. ex Bong.；黄杨叶蓼 ■☆

308926　Polygonum buxifolium Nutt. ex Bong. = Polygonum fowleri B. L. Rob. ■☆

308927　Polygonum buxiforme Small = Polygonum aviculare L. subsp. buxiforme （Small）Costea et Tardif ■☆

308928　Polygonum caespitosum Blume = Persicaria posumbu （Buch. -Ham. ex D. Don）H. Gross ■

308929　Polygonum caespitosum Blume = Polygonum posumbu Buch. -Ham. ex D. Don ■

308930　Polygonum caespitosum Blume subsp. yokusaianum （Makino）Danser = Persicaria posumbu （Buch. -Ham. ex D. Don）H. Gross ■

308931　Polygonum caespitosum Blume subsp. yokusaianum （Makino）Danser = Polygonum posumbu Buch. -Ham. ex D. Don ■

308932　Polygonum caespitosum Blume var. laxiflorum Meisn. = Persicaria posumbu （Buch. -Ham. ex D. Don）H. Gross ■

308933　Polygonum caespitosum Blume var. laxiflorum Meisn. = Polygonum posumbu Buch. -Ham. ex D. Don ■

308934　Polygonum caespitosum Blume var. longisetum （Bruijn）Steward = Persicaria longiseta （Bruijn）Kitag. ■

308935　Polygonum caespitosum Blume var. longisetum （Bruijn）Steward = Polygonum longisetum Bruijn ■

308936　Polygonum calcatum Lindm. = Polygonum arenastrum Boreau ■

308937　Polygonum calcatum Lindm. = Polygonum aviculare L. subsp. depressum （Meisn.）Arcang. ■☆

308938　Polygonum californicum Meisn.；加洲蓼 ■☆

308939　Polygonum calostachyum Diels；长梗蓼（美穗拳参）；Clustered Knotweed，Prettyspike Knotweed ■

308940　Polygonum calostachyum Diels = Polygonum griffithii Hook. f. ■

308941　Polygonum campanulatum Hook. f.；钟花蓼（神血宁，钟花神血宁）；Bellflower Knotweed，Bellflower Smartweed，Campanulate Knotweed，Himalayan Knotweed，Lesser Knotweed ■

308942　Polygonum campanulatum Hook. f. = Persicaria campanulata （Hook. f.）Ronse Decr. ■

308943　Polygonum campanulatum Hook. f. var. fulvidum Hook. f.；绒毛钟花蓼（黄花钟花蓼，黄毛神血宁，绒毛钟花神血宁）；Yellow Hairy Knotweed ■

308944　Polygonum campanulatum Hook. f. var. lichiangense （W. W. Sm.）Steward = Polygonum lichiangense W. W. Sm. ■

308945　Polygonum campanulatum Hook. f. var. menmbranifolium Hook. f.；膜叶钟花蓼（腊叶神血草）■☆

308946　Polygonum capitatum Buch. -Ham. ex D. Don；头花蓼（草石椒，红花地丁，红酸杆，回生草，火溜草，惊风草，满地红，石辣蓼，石荞草，石头菜，石头花，水绣球，四季红，太阳草，太阳花，小红草，小红藤，小铜草，绣球草，岩荞麦）；Headflower Knotweed，Knotweed，Pink-head Knotweed，Pinkhead Smartweed ■

308947　Polygonum capitatum Buch. -Ham. ex D. Don = Persicaria capitata （Buch. -Ham. ex D. Don）H. Gross ■

308948　Polygonum careyi Olney；凯里蓼；Carey's Heart's-ease，Carey's Smartweed ■☆

308949　Polygonum careyi Olney = Persicaria careyi （Olney）Greene ■☆

308950　Polygonum carneum K. Koch；肉色蓼 ■☆

308951　Polygonum cascadense W. H. Baker；层叠蓼；Cascade Knotweed ■☆

308952　Polygonum caspicum Kom.；里海蓼 ■☆

308953　Polygonum cathayanum A. J. Li；华蓼（草石椒，华神血宁，酸浆草，太阳花，头花蓼）；Capitate Knotweed，Cathay Knotweed ■

308954　Polygonum caudatum Sam. = Fagopyrum caudatum （Sam.）A. J. Li ■

308955　Polygonum caurianum B. L. Rob. = Polygonum humifusum C. Merck ex K. Koch subsp. caurianum （B. L. Rob.）Costea et Tardif ■☆

308956　Polygonum caurianum B. L. Rob. subsp. hudsonianum S. J. Wolf et McNeill = Polygonum fowleri B. L. Rob. subsp. hudsonianum （S. J. Wolf et McNeill）Costea et Tardif ■☆

308957　Polygonum cavaleriei H. Lév.；卡氏蓼（小蓼花）■☆

308958　Polygonum cavaleriei H. Lév. = Persicaria hastatosagittata

(Makino) Nakai ■

308959 Polygonum cavaleriei H. Lév. = Polygonum hastatosagittatum Makino ■

308960 Polygonum cespitosum Blume;东方蓼;Oriental Lady's-thumb, Smartweed ■☆

308961 Polygonum cespitosum Blume var. longisetum (Bruyn) Stewart;长毛东方蓼;Oriental Lady's-thumb ■☆

308962 Polygonum chanetii H. Lév. = Persicaria bungeana (Turcz.) Nakai ex Mori ■

308963 Polygonum chanetii H. Lév. = Polygonum bungeanum Turcz. ■

308964 Polygonum changii Kitag. = Polygonum plebeium R. Br. ■

308965 Polygonum chinense L.;火炭母(白饭草,白饭藤,赤地利,大叶沙滩子,鸪鹚饭,旱辣蓼,红梅子叶,黄鳝藤,黄泽兰,火炭毛,火炭梅,火炭母草,火炭星,火炭只药,鸡粪蔓,金不换,老鼠蔗,冷饭藤,蓼草,毛甘蔗,清饭藤,鹊糖梅,山荞麦草,水沙柑子,水退痧,乌白饭草,乌饭藤,乌炭子,运药,晕药);China Knotweed, Chinese Knotweed ■

308966 Polygonum chinense L. = Persicaria chinensis (L.) H. Gross ■

308967 Polygonum chinense L. f. hispidum (Hook. f.) Sam. = Polygonum chinense L. var. hispidum Hook. f. ■

308968 Polygonum chinense L. var. brachyatum (Poir.) Meisn.;分枝火炭母■☆

308969 Polygonum chinense L. var. hispidum Hook. f.;硬毛火炭母(粗毛火炭母,大碎草,大碎米草,刚毛火炭母,毛水蓼,土三七,小红人,野辣子草,硬毛垂蓼);Chinese Hispid Knotweed, Hispid Flaccid Knotweed ■

308970 Polygonum chinense L. var. malaicum (Danser) Steven = Polygonum chinense L. var. ovalifolium Meisn. ■

308971 Polygonum chinense L. var. ovalifolium Meisn.;宽叶火炭母(卵叶火炭母);Oval-leaf Knotweed ■

308972 Polygonum chinense L. var. paradoxum (H. Lév.) A. J. Li;窄叶火炭母;Narrowleaf Chinese Knotweed ■

308973 Polygonum chinense L. var. scabrum Meisn.;糙叶火炭母■☆

308974 Polygonum chinense L. var. thunbergianum Meisn.;通贝里火炭母(赤薜荔,赤地利,杠板归,胖根藤,山荞麦,蛇茧,五毒草,五蕺);Thunberg Chinese Knotweed ■

308975 Polygonum chinense L. var. thunbergianum Meisn. = Persicaria chinensis (L.) H. Gross ■

308976 Polygonum chinense L. var. umbellatum Makino;晕药(冷饭藤);Umbellate Chinese Knotweed ■

308977 Polygonum ciliinerve (Nakai) Ohwi = Fallopia ciliinervis (Nakai) K. Hammer ■

308978 Polygonum ciliinerve (Nakai) Ohwi = Fallopia multiflora (Thunb.) Haraldson var. ciliinervis (Nakai) A. J. Li ■

308979 Polygonum cilinode (Michx.) Greene var. laevigatum Fernald = Fallopia cilinodis (Michx.) Holub ■☆

308980 Polygonum cilinode F. Michx. var. erectum (Peck) Fernald = Fallopia cilinodis (Michx.) Holub ■☆

308981 Polygonum cilinode Michx. = Fallopia cilinodis (Michx.) Holub ■☆

308982 Polygonum cilinode Michx. var. erectum Peck = Fallopia cilinodis (Michx.) Holub ■☆

308983 Polygonum cilinode Michx. var. laevigatum Fernald = Fallopia cilinodis (Michx.) Holub ■☆

308984 Polygonum coarctatum Douglas ex Meisn. var. majus Meisn. = Polygonum majus (Meisn.) Piper ■☆

308985 Polygonum coccineum Muhl. = Polygonum coccineum Muhl. ex Willd. ■☆

308986 Polygonum coccineum Muhl. ex Willd.;沼泽蓼;Scarlet Knotweed, Swamp Smartweed, Water Smartweed ■☆

308987 Polygonum coccineum Muhl. ex Willd. = Persicaria amphibia (L.) Gray ■

308988 Polygonum coccineum Muhl. ex Willd. = Polygonum amphibium L. var. emersum Michx. ■☆

308989 Polygonum coccineum Muhl. ex Willd. = Polygonum amphibium L. ■

308990 Polygonum coccineum Muhl. ex Willd. f. natans (Wiegand) Stanford = Polygonum amphibium L. ■

308991 Polygonum coccineum Muhl. ex Willd. f. terrestre (Willd.) Stanford = Polygonum amphibium L. var. emersum Michx. ■☆

308992 Polygonum coccineum Muhl. ex Willd. var. pratincola (Greene) Stanford = Polygonum amphibium L. ■

308993 Polygonum coccineum Muhl. ex Willd. var. pratincola (Greene) Stanford = Persicaria amphibia (L.) Gray ■

308994 Polygonum coccineum Muhl. ex Willd. var. pratincola (Greene) Stanford = Polygonum amphibium L. var. emersum Michx. ■☆

308995 Polygonum coccineum Muhl. ex Willd. var. rigidulum (E. Sheld.) Stanford = Polygonum amphibium L. var. stipulaceum N. Coleman ■

308996 Polygonum coccineum Muhl. ex Willd. var. rigidulum (E. Sheld.) Stanford = Polygonum amphibium L. ■

308997 Polygonum coccineum Muhl. ex Willd. var. rigidulum (E. Sheld.) Stanford = Persicaria amphibia (L.) Gray ■

308998 Polygonum coccineum Muhl. ex Willd. var. terrestre Willd. = Polygonum amphibium L. var. emersum Michx. ■☆

308999 Polygonum cochinchinense (Lour.) Meisn. = Polygonum orientale L. ■

309000 Polygonum cochinchinense Meisn.;印支蓼(红蓼)■☆

309001 Polygonum cognatum Meisn.;岩萹蓄(地皮蓼,近亲蓼,岩蓼,岩生蓼);Indian Knot-grass, Rock Knotweed ■

309002 Polygonum compactum Hook. f. = Fallopia japonica (Houtt.) Ronse Decr. ■

309003 Polygonum compactum Hook. f. = Reynoutria japonica Houtt. ■

309004 Polygonum complexum A. Cunn. = Muehlenbeckia complexa (A. Cunn.) Meisn. ●☆

309005 Polygonum confertiflorum Nutt. ex Piper = Polygonum polygaloides Meisn. subsp. confertiflorum (Nutt. ex Piper) J. C. Hickman ■☆

309006 Polygonum conspicuum (Nakai) Nakai;樱蓼(樱花蓼)■

309007 Polygonum conspicuum (Nakai) Nakai = Persicaria macrantha (Meisn.) Haraldson subsp. conspicua (Nakai) Yonek. ■

309008 Polygonum conspicuum (Nakai) Nakai = Polygonum japonicum Meisn. var. conspicuum Nakai ■

309009 Polygonum constans Cumm. = Polygonum suffultum Maxim. ■

309010 Polygonum convolvulus (L.) H. Lév. = Fallopia convolvulus (L.) Á. Löve ■

309011 Polygonum convolvulus L. = Fallopia convolvulus (L.) Á. Löve ■

309012 Polygonum convolvulus L. var. pauciflorum (Maxim.) Vorosch. = Fallopia dumetora (L.) Holub var. pauciflora (Maxim.) A. J. Li ■

309013 Polygonum convolvulus L. var. subulatum ? = Fallopia convolvulus (L.) Á. Löve ■

309014 Polygonum coriaceum Sam.;革叶蓼(伴蛇莲,革叶参,革叶拳参,鸡爪大王,马蜂七,拳参);Coriaceous Knotweed, Leatherlraf Knotweed ■

309015　Polygonum coriarium Grig. ;白花蓼（白花神血宁，布哈尔蓼，皮蓼）;Whiteflower Knotweed ■

309016　Polygonum corrigioloides Jaub. et Spach;盐生蓼■

309017　Polygonum cretaceum Kom. ;白垩蓼■☆

309018　Polygonum criopolitanum Hance;蓼子草; Herb Knotweed, Japanese Fleeceflower ■

309019　Polygonum cristatum Engelm. et A. Gray = Polygonum scandens L. var. cristatum（Engelm. et A. Gray）Gleason ■☆

309020　Polygonum cristatum Engelm. et A. Gray = Polygonum scandens L. ■☆

309021　Polygonum cuspidatum Siebold et Zucc. = Fallopia japonica（Houtt.）Ronse Decr. ■

309022　Polygonum cuspidatum Siebold et Zucc. = Reynoutria japonica Houtt. ■

309023　Polygonum cuspidatum Siebold et Zucc. f. colorans（Makino）Makino = Fallopia japonica（Houtt.）Ronse Decr. var. compacta（Hook. f.）J. P. Bailey f. colorans（Makino）■☆

309024　Polygonum cuspidatum Siebold et Zucc. f. compactum（Hook. f.）Makino = Fallopia japonica（Houtt.）Ronse Decr. ■

309025　Polygonum cuspidatum Siebold et Zucc. f. compactum（Hook. f.）Makino = Reynoutria japonica Houtt. ■

309026　Polygonum cuspidatum Siebold et Zucc. f. uzenense（Honda）Kitam. = Fallopia japonica（Houtt.）Ronse Decr. var. uzenensis（Honda）Yonek. et H. Ohashi ■☆

309027　Polygonum cuspidatum Siebold et Zucc. var. compactum（Hook. f.）L. H. Bailey = Fallopia japonica（Houtt.）Ronse Decr. ■

309028　Polygonum cuspidatum Siebold et Zucc. var. hachidyoense（Makino）Ohwi = Fallopia japonica（Houtt.）Ronse Decr. var. hachidyoensis（Makino）Yonek. et H. Ohashi ■☆

309029　Polygonum cuspidatum Siebold et Zucc. var. terminale（Honda）Ohwi = Fallopia japonica（Houtt.）Ronse Decr. var. hachidyoensis（Makino）Yonek. et H. Ohashi ■☆

309030　Polygonum cuspidatum Siebold et Zucc. var. uzenense（Honda）Noda f. roseum（Satomi）Satomi = Reynoutria japonica Houtt. var. uzenensis Honda f. rosea Satomi ■☆

309031　Polygonum cuspidatum Siebold et Zucc. var. uzenense（Honda）Noda = Fallopia japonica（Houtt.）Ronse Decr. var. uzenensis（Honda）Yonek. et H. Ohashi ■☆

309032　Polygonum cyanandrum Diels;蓝药蓼（虎杖，黄药子）;Blueanther Knotweed ■

309033　Polygonum cymosum Trevir. = Fagopyrum dibotrys（D. Don）H. Hara ■

309034　Polygonum cynanchoides Hemsl. = Fallopia cynanchoides（Hemsl.）Haraldson ■

309035　Polygonum cynanchoides Hemsl. var. glabriusculum A. J. Li = Fallopia cynanchoides（Hemsl.）Haraldson var. glabriuscula（A. J. Li）A. J. Li ■

309036　Polygonum darrisii H. Lév. ;大箭叶（箭叶蓼）;Big Arrowleaf Knotweed ■

309037　Polygonum debile Meisn. = Persicaria debilis（Meisn.）H. Gross ex W. T. Lee ■☆

309038　Polygonum deciduum Boiss. et Noë;脱落蓼■☆

309039　Polygonum decipiens R. Br. ;迷惑蓼;Swamp Willow-weed ■☆

309040　Polygonum decipiens R. Br. = Persicaria decipiens（R. Br.）K. L. Wilson ■☆

309041　Polygonum deflexipilosum Kitam. = Polygonum molle D. Don var. rude（Meisn.）A. J. Li ■

309042　Polygonum delicatulum Meisn. ;小叶蓼;Smallleaf Knotweed ■

309043　Polygonum densiflorum Blume; 密 花 蓼; Dense-flpwer Knotweed , Kamole , Smartweed ■☆

309044　Polygonum densiflorum Meisn. = Persicaria glabra（Willd.）M. Gómez ■

309045　Polygonum densiflorum Meisn. = Polygonum densiflorum Blume ■☆

309046　Polygonum densiflorum Meisn. = Polygonum glabrum Willd. ■

309047　Polygonum dentatoalatum F. Schmidt = Fallopia dentatoalata（F. Schmidt ex Maxim.）Holub ■

309048　Polygonum dentatoalatum F. Schmidt ex Maxim. = Fallopia dentatoalata（F. Schmidt ex Maxim.）Holub ■

309049　Polygonum denticulatum C. C. Huang = Fallopia denticulata（C. C. Huang）A. J. Li ■

309050　Polygonum dibotrys D. Don = Fagopyrum dibotrys（D. Don）H. Hara ■

309051　Polygonum dichotomum Blume;二歧蓼（箭叶蓼,具梗小蓼花,水红骨蛇）;Dichotomous Knotweed ■

309052　Polygonum dichotomum Blume = Persicaria dichotoma（Blume）Masam. ■

309053　Polygonum dielsii H. Lév. = Polygonum chinense L. var. paradoxum（H. Lév.）A. J. Li ■

309054　Polygonum dissitiflorum Hemsl. ;稀花蓼（红降龙草）;Laxflower Knotweed ■

309055　Polygonum dissitiflorum Hemsl. = Persicaria dissitiflora（Hemsl.）H. Gross ex T. Mori ■

309056　Polygonum divaricatum L. ;叉分蓼（叉分神血宁,叉枝蓼,分叉蓼,分枝蓼,酸不溜,酸姜）;Divaricate Knotweed ■

309057　Polygonum divaricatum L. = Aconogonon divaricatum（L.）Nakai ■

309058　Polygonum divaricatum L. var. angustissimum f. glabrum Meisn. = Polygonum angustifolium Pall. ■

309059　Polygonum divaricatum L. var. limosum Kom. = Polygonum limosum Kom. ■

309060　Polygonum dolichopodum Ohwi = Polygonum persicaria L. ■

309061　Polygonum donianum Spreng. = Polygonum affine D. Don ●■

309062　Polygonum donii Meisn. = Polygonum pubescens Blume ■

309063　Polygonum douglasii Greene;道格拉斯蓼;Douglas' Knotweed ■☆

309064　Polygonum douglasii Greene subsp. austiniae（Greene）E. Murray = Polygonum austiniae Greene ■☆

309065　Polygonum douglasii Greene subsp. engelmannii（Greene）Kartesz et Gandhi = Polygonum engelmannii Greene ■☆

309066　Polygonum douglasii Greene subsp. majus（Meisn.）J. C. Hickman = Polygonum majus（Meisn.）Piper ■☆

309067　Polygonum douglasii Greene subsp. nuttallii（Small）J. C. Hickman = Polygonum nuttallii Small ■☆

309068　Polygonum douglasii Greene subsp. spergulariiforme（Meisn. ex Small）J. C. Hickman = Polygonum spergulariiforme Meisn. ex Small ■☆

309069　Polygonum douglasii Greene var. latifolium（Engelm.）Greene = Polygonum douglasii Greene ■☆

309070　Polygonum douglasii Greene var. utahense（Brenckle et Cottam）S. L. Welsh = Polygonum utahense Brenckle et Cottam ■☆

309071　Polygonum dubium Stein = Polygonum persicaria L. ■

309072　Polygonum duclouxii H. Lév. ex Vaniot = Polygonum campanulatum Hook. f. var. fulvidum Hook. f. ■

309073　Polygonum duclouxii H. Lév. ex Vaniot var. hypoleuca H. Lév. = Polygonum campanulatum Hook. f. var. fulvidum Hook. f. ■

309074　Polygonum dumetorum L. = Fallopia dumetora（L.）Holub ■

309075　Polygonum dumetorum L. = Polygonum convolvulus L. ■

309076　Polygonum dumetorum L. var. scandens（L.）A. Gray = Fallopia scandens（L.）Holub ■☆

309077　Polygonum effusum Meisn. ;开展蓼■☆

309078　Polygonum ellipticum Willd. ex Spreng. ;椭圆叶蓼(亮果蓼,椭圆叶拳参);Ellipticleaf Knotweed ■

309079　Polygonum emaciatum A. Nelson = Polygonum douglasii Greene ■☆

309080　Polygonum emaciatum A. Nelson = Polygonum spergulariiforme Meisn. ex Small ■☆

309081　Polygonum emarginatum Roth = Fagopyrum esculentum Moench ■

309082　Polygonum emersum（Michx.）Britton = Persicaria amphibia（L.）Gray ■

309083　Polygonum emersum（Michx.）Britton = Polygonum amphibium L. ■

309084　Polygonum emodii Meisn. ;匍枝蓼(蔓枝蓼,红藤蓼,水荞,竹叶疏筋,竹叶舒筋);Creeping Knotweed, Himalaya Knotweed ●■

309085　Polygonum emodii Meisn. var. dependens Diels;宽叶匍枝蓼(宽竹叶舒筋,悬垂竹叶舒筋)■

309086　Polygonum engelmannii Greene;恩格尔曼蓼;Engelmann's Knotweed ■☆

309087　Polygonum englerianum H. Gross = Polygonum paronychioides C. A. Mey. ex Hohen. ●

309088　Polygonum equisetiforme Sibth. et Sm. ;木贼蓼■☆

309089　Polygonum equisetiforme Sm. var. graecum Meisn. = Polygonum equisetiforme Sibth. et Sm. ■☆

309090　Polygonum equisetiforme Sm. var. peyerimoffii Batt. et Maire = Polygonum equisetiforme Sibth. et Sm. ■☆

309091　Polygonum equisetiforme Sm. var. spicatum Batt. = Polygonum equisetiforme Sibth. et Sm. ■☆

309092　Polygonum erecto-minus Makino = Persicaria erectominor（Makino）Nakai ■☆

309093　Polygonum erecto-minus Makino var. trigonocarpum（Makino）Kitam. = Persicaria erectominor（Makino）Nakai var. trigonocarpa（Makino）H. Hara ■☆

309094　Polygonum erectum L. ;直立蓼; Erect Knotweed, Russian Knotgrass,Weedy Knotweed ■☆

309095　Polygonum erectum L. subsp. achoreum（S. F. Blake）Á. Löve et D. Löve = Polygonum achoreum S. F. Blake ■☆

309096　Polygonum erectum Roth = Polygonum aviculare L. var. erectum（Roth）Hayne ■

309097　Polygonum esquirolii H. Lév. = Polygonum molle D. Don var. rude（Meisn.）A. J. Li ■

309098　Polygonum excurrens Steward;中轴蓼■

309099　Polygonum excurrens Steward = Polygonum trigonoparpum（Makino）Kudo et Masam. ■

309100　Polygonum excurrens Steward = Polygonum viscoferum Makino ■

309101　Polygonum exsertum Small = Polygonum ramosissimum Michx. ■☆

309102　Polygonum fagopyrum L. = Fagopyrum esculentum Moench ■

309103　Polygonum fastigiatoramosum Makino = Persicaria hydropiper（L.）Spach var. fastigiata（Makino）Araki ■☆

309104　Polygonum fauriei H. Lév. et Vaniot = Polygonum dissitiflorum Hemsl. ■

309105　Polygonum fertile（Maxim.）A. J. Li;青藏蓼;Fertile Knotweed ■

309106　Polygonum filicaule Wall. ex Meisn. ;细茎蓼(高山蓼);Thinstem Knotweed ■

309107　Polygonum filiforme Thunb. = Antenoron filiforme（Thunb.）Rob. et Vautier ■

309108　Polygonum filiforme Thunb. = Persicaria filiformis（Thunb.）Nakai ex W. T. Lee ■☆

309109　Polygonum filiforme Thunb. f. albiflorum Makino = Persicaria filiformis（Thunb.）Nakai ex W. T. Lee f. albiflora（Hiyama）Yonek. ☆

309110　Polygonum filiforme Thunb. subsp. neofiliforme（Nakai）Kitam. = Antenoron filiforme（Thunb.）Rob. et Vautier var. neofiliforme（Nakai）A. J. Li ■

309111　Polygonum filiforme Thunb. subsp. neofiliforme（Nakai）Kitam. = Persicaria neofiliformis（Nakai）Ohki ■

309112　Polygonum filiforme Thunb. var. neofiliforme（Nakai）Ohwi = Antenoron filiforme（Thunb.）Rob. et Vautier var. neofiliforme（Nakai）A. J. Li ■

309113　Polygonum filiforme Thunb. var. neofiliforme（Nakai）Ohwi = Persicaria neofiliformis（Nakai）Ohki ■

309114　Polygonum filiforme Thunb. var. smaragdinum（Nakai ex F. Maek.）Ohwi = Persicaria filiformis（Thunb.）Nakai ex W. T. Lee ■☆

309115　Polygonum fimbriatum Elliott = Polygonella fimbriata（Elliott）Horton ■☆

309116　Polygonum flaccidum Meisn. ;辣蓼(斑蕉草,蝙蝠草,蝙蝠蓼,垂蓼,红辣蓼,辣椒草,辣蓼草,辣柳草,辣马蓼,蓼子草,青蓼);Flaccid Knotweed ■

309117　Polygonum flaccidum Meisn. = Polygonum pubescens Blume ■

309118　Polygonum flaccidum Meisn. var. hispidum（Buch. -Ham. ex D. Don）Hook. f. = Polygonum pubescens Blume ■

309119　Polygonum floribundum Schltdl. ex Meisn. ;繁花蓼■☆

309120　Polygonum fluitans Eaton = Polygonum amphibium L. var. stipulaceum N. Coleman ■

309121　Polygonum foliosum H. Lindb. ;多叶蓼（宜兰蓼）;Leafy Knotweed ■

309122　Polygonum foliosum H. Lindb. = Persicaria foliosa（H. Lindb.）Kitag. ■

309123　Polygonum foliosum H. Lindb. var. nikaii（Makino）Kitam. ;二阶氏多叶蓼■☆

309124　Polygonum foliosum H. Lindb. var. nikaii（Makino）Kitam. = Persicaria foliosa（H. Lindb.）Kitag. var. nikaii（Makino）H. Hara ■☆

309125　Polygonum foliosum H. Lindb. var. paludicola（Makino）Kitam. ;宽基多叶蓼■

309126　Polygonum forbesii Hance = Fallopia forbesii（Hance）Yonek. et H. Ohashi ■

309127　Polygonum forrestii Diels;大铜钱叶蓼(大铜钱叶神血宁,云枝花);Big Copper Knotweed ■

309128　Polygonum forrestii Diels var. pumilio Lingelsh. = Polygonum nummularifolium Meisn. ■

309129　Polygonum fowleri B. L. Rob. ;福勒蓼;Fowler's Knotweed ■☆

309130　Polygonum fowleri B. L. Rob. subsp. hudsonianum（S. J. Wolf et McNeill）Costea et Tardif;赫德森蓼;Hudsonian knotweed ■☆

309131　Polygonum frondosum Meisn. = Polygonum molle D. Don var. frondosum（Meisn.）A. J. Li ●■

309132　Polygonum frutescens L. = Atraphaxis frutescens（L.）Eversm. ●

309133　Polygonum fusco-ochreanum Kom. = Polygonum aviculare L. var. fusco-ochreanum（Kom.）A. J. Li ■

309134　Polygonum fusco-ochreatum（Kom.）A. J. Li f. stans（Kitag.）C. F. Fang = Polygonum aviculare L. var. fusco-ochreanum（Kom.）A. J. Li ■

309135　Polygonum fusco-ochreatum Kom. = Polygonum aviculare L. var. fusco-ochreanum（Kom.）A. J. Li ■

309136　Polygonum fusco-ochreatum Kom. f. starts（Kitag.）C. F. Fang ＝ Polygonum aviculare L. var. fusco-ochreatum（Kom.）A. J. Li ■

309137　Polygonum fusiforme Greene ＝ Persicaria maculosa Gray ■☆

309138　Polygonum fusiforme Greene ＝ Polygonum persicaria L. ■

309139　Polygonum gentilianum（H. Lév.）H. Lév. ＝ Polygonum longisetum Bruijn

309140　Polygonum gilesii Hemsl. ＝ Fagopyrum gilesii（Hemsl.）Hedberg ■

309141　Polygonum giraldii Dammer ＝ Pteroxygonum giraldii Dammer et Diels ●■

309142　Polygonum glabrum Willd. ＝ Persicaria glabra（Willd.）M. Gomez ■

309143　Polygonum glaciale（Meisn.）Hook. f.；冰川蓼（冰雪蓼）；Glacial Knotweed，Glacier Knotweed ■

309144　Polygonum glaciale（Meisn.）Hook. f. var. przewalskii（A. K. Skvortsov et Borodina）A. J. Li；洼点蓼；Przewalsk Glacial Knotweed ■

309145　Polygonum glanduliferum Nakai ＝ Polygonum dissitiflorum Hemsl. ■

309146　Polygonum glandulo-pilosum De Wild. ＝ Persicaria glandulo-pilosa（De Wild.）Soják ■☆

309147　Polygonum glareosum Schischk.；灰蓼●

309148　Polygonum glareosum Schischk. ＝ Polygonum schischkinii N. A. Ivanova ex Borodina ●

309149　Polygonum glaucum Nutt.；海蓼；Sea Knotweed，Seabeach Knotweed ■☆

309150　Polygonum glomeratum Dammer ＝ Persicaria glomerata（Dammer）S. Ortiz et Paiva ■☆

309151　Polygonum gloriosum H. Lév. ＝ Polygonum pinetorum Hemsl. ■

309152　Polygonum glutinosum Meisn. var. capensis ? ＝ Persicaria senegalensis（Meisn.）Soják. f. albotomentosa（R. A. Graham）K. L. Wilson ■☆

309153　Polygonum gracile（Ledeb.）Klokov ＝ Polygonum patulum M. Bieb. ■

309154　Polygonum gracile Nutt. ＝ Polygonella gracilis Meisn. ■☆

309155　Polygonum gracilipes Hemsl. ＝ Fagopyrum gracilipes（Hemsl.）Dammer ex Diels ■

309156　Polygonum gracilipes Hemsl. var. odontopterum（H. Gross）Sam. ＝ Fagopyrum gracilipes（Hemsl.）Dammer ex Diels ■

309157　Polygonum graminifolium Wierzb.；禾叶蓼■☆

309158　Polygonum greenei S. Watson ＝ Polygonum californicum Meisn. ■☆

309159　Polygonum griffithii Hook. f.；长梗拳参■

309160　Polygonum griffithii Hook. f. ＝ Polygonum calostachyum Diels ■

309161　Polygonum grossii H. Lév. ＝ Fagopyrum leptopodum（Diels）Hedberg var. grossii（H. Lév.）Lauener et D. K. Ferguson ■

309162　Polygonum hachidyoense Makino ＝ Fallopia japonica（Houtt.）Ronse Decr. var. hachidyoensis（Makino）Yonek. et H. Ohashi ■☆

309163　Polygonum hangchouense Matsuda ＝ Polygonum jucundum Meisn. ■

309164　Polygonum hartwrightii A. Gray ＝ Persicaria amphibia（L.）Gray ■

309165　Polygonum hartwrightii A. Gray ＝ Polygonum amphibium L. var. stipulaceum N. Coleman ■

309166　Polygonum hartwrightii A. Gray ＝ Polygonum amphibium L. ■

309167　Polygonum hastatoauriculatum Makino ex Nakai ＝ Persicaria praetermissa（Hook. f.）H. Hara ■

309168　Polygonum hastatoauriculatum Makino ex Nakai ＝ Polygonum praetermissum Hook. f. ■

309169　Polygonum hastatosagittatum Makino ＝ Persicaria hastatosagittata（Makino）Nakai ■

309170　Polygonum hastatosagittatum Makino var. latifolium Makino ＝ Polygonum muricatum Meisn. ■

309171　Polygonum hastatotrilobum Makino var. lenticulare Danser ＝ Polygonum biconvexum Hayata ■

309172　Polygonum hastatotrilobum Meisn. ＝ Polygonum thunbergii Siebold et Zucc. ■

309173　Polygonum hastatotrilobum Meisn. var. lenticulare Danser ＝ Polygonum biconvexum Hayata ■

309174　Polygonum hayachinense Makino ＝ Bistorta hayachinensis（Makino）H. Gross ■☆

309175　Polygonum herniarioides Delile ＝ Polygonum plebeium R. Br. ■

309176　Polygonum heterophyllum Lindm. ＝ Polygonum aviculare L. ■

309177　Polygonum heterophyllum Lindm. subsp. boreale（Lange）Á. Löve et D. Löve ＝ Polygonum aviculare L. subsp. boreale（Lange）Karlsson ■☆

309178　Polygonum heterophyllum Lindm. subsp. rurivagum（Jord. ex Boreau）Lindm. ＝ Polygonum aviculare L. subsp. rurivagum（Jord. ex Boreau）Berher ■☆

309179　Polygonum heterophyllum Lindm. var. angustissimum（Meisn.）Lindm. ＝ Polygonum aviculare L. subsp. rurivagum（Jord. ex Boreau）Berher ■☆

309180　Polygonum heterosepalum M. Peck et Ownbey；矮沙蓼；Dwarf Desert Knotweed ■☆

309181　Polygonum hickmanii H. R. Hinds et Rand. Morgan；希克曼蓼；Hickman's Knotweed ■☆

309182　Polygonum himalayense H. Gross ＝ Polygonum paronychioides C. A. Mey. ex Hohen. ●

309183　Polygonum hirsutum Walter ＝ Persicaria hirsuta（Walter）Small ■☆

309184　Polygonum hispidum（Hook. f.）Sam. ＝ Polygonum chinense L. var. hispidum Hook. f. ■

309185　Polygonum hispidum Buch.-Ham. ex D. Don ＝ Polygonum pubescens Blume ■

309186　Polygonum hissaricum Popov；希萨尔蓼■☆

309187　Polygonum honanense H. W. Kung；河南蓼（河南拳参）；Henan Knotweed ■

309188　Polygonum hookeri Meisn.；硬毛蓼（虎克蓼，假大黄，硬毛神血宁）；Hardhair Knotweed，Hooker Knotweed ■

309189　Polygonum huananense A. J. Li；华南蓼；S. China Knotweed ■

309190　Polygonum hubertii Lingelsh. ＝ Polygonum sparsipilosum A. J. Li var. hubertii（Lingelsh.）A. J. Li ■

309191　Polygonum hudsonianum（S. J. Wolf et McNeill）H. R. Hinds ＝ Polygonum fowleri B. L. Rob. subsp. hudsonianum（S. J. Wolf et McNeill）Costea et Tardif ■☆

309192　Polygonum humifusum C. Merck ex K. Koch；普通萹蓄（普通蓼）；Common Knotweed ■

309193　Polygonum humifusum C. Merck ex K. Koch f. yamatutae（Kitag.）C. F. Fang ＝ Polygonum humifusum Merker ex C. Koch ■

309194　Polygonum humifusum C. Merck ex K. Koch subsp. caurianum（B. L. Rob.）Costea et Tardif；阿拉斯加蓼；Alaska Knotweed ■☆

309195　Polygonum humifusum Merker ex C. Koch ＝ Polygonum humifusum C. Merck ex K. Koch ■

309196　Polygonum humifusum Merker ex C. Koch var. mandshurica Skvortsov ＝ Polygonum plebeium R. Br. ■

309197　Polygonum humifusum Pall. ex Ledeb. var. mandshuricum Skvortsov ＝ Polygonum plebeium R. Br. ■

309198　Polygonum humile Meisn. ;矮蓼;Dwarf Knotweed,Low-growing Knotweed Knotweed ■

309199　Polygonum hydropiper L. ;水蓼(白辣蓼,斑蕉草,蔦蓄,蔦竹,川蓼,红辣蓼,红蓼子草,胡辣蓼,假辣蓼,辣蓼,辣蓼草,辣人草,蓼,蓼草,蓼芽菜,柳蓼,蕾,水红花,水辣蓼,疼骨消,痛骨消,小叶辣蓼,药蓼,药蓼子草,虞蓼,泽蓼,竹叶菜);Arsenicke, Arsesmart, Biting Persicaria, Bity Tongue, Bloodwort, Bob Ginger, Ciderage, Common Smartweed, Cow Itch, Culerage, Culrache, Culrage, Curage, Cyderach, Ersmart, Hot Arsesmart, Keliage, Killridge, Lakeweed, Marsh pepper, Marshpepper Knotweed, Marsh-pepper Knotweed, Marshpepper Smartweed, Marsh-pepper Smartweed, Pepper-plant, Red Knees, Red-knees, Redshanks, Redweed, Smartarse, Smartweed, Water Knotweed, Water Pepper, Water Smartweed, Water-pepper,Water-pepper Smartweed,Yes Smart ■

309200　Polygonum hydropiper L. = Persicaria hydropiper (L.) Spach ■

309201　Polygonum hydropiper L. f. purpurascens Makino;紫水蓼■

309202　Polygonum hydropiper L. f. purpurascens Makino = Persicaria hydropiper (L.) Spach f. purpurascens (Makino) Nemoto ■

309203　Polygonum hydropiper L. var. angustifolia (A. Braun) Kitag. ; 狭叶水蓼■

309204　Polygonum hydropiper L. var. diffusa Kitag. ;散枝水蓼■

309205　Polygonum hydropiper L. var. fastigiatum Makino = Persicaria hydropiper (L.) Spach var. fastigiata (Makino) Araki ■☆

309206　Polygonum hydropiper L. var. fastigiatum Makino f. angustissimum Makino = Persicaria hydropiper (L.) Spach var. fastigiata (Makino) Araki f. angustissima (Makino) Araki ■☆

309207　Polygonum hydropiper L. var. fastigiatum Makino f. gramineum (Meisn.) Ohwi = Persicaria hydropiper (L.) Spach var. maximowiczii (Regel) Nemoto ■☆

309208　Polygonum hydropiper L. var. flaccidum (Meisn.) Steward;软水蓼(班蕉草,蝙蝠草,红辣蓼,辣蓼,辣柳草,辣马蓼,青蓼)■

309209　Polygonum hydropiper L. var. flaccidum (Meisn.) Steward = Polygonum pubescens Blume ■

309210　Polygonum hydropiper L. var. gramineum (Meisn.) Ohwi = Persicaria hydropiper (L.) Spach var. maximowiczii (Regel) Nemoto ■☆

309211　Polygonum hydropiper L. var. hispidum (Buch. -Ham. ex D. Don) Steward = Polygonum pubescens Blume ■

309212　Polygonum hydropiper L. var. hispidum (Hook. f.) Steward;刚毛水蓼(水蓼,无辣蓼)■

309213　Polygonum hydropiper L. var. hispidum (Hook. f.) Steward = Polygonum hydropiper L. ■

309214　Polygonum hydropiper L. var. longistachyum Y. L. Chang et S. X. Li;长穗水蓼■

309215　Polygonum hydropiper L. var. longistachyum Y. L. Chang et S. X. Li = Polygonum hydropiper L. ■

309216　Polygonum hydropiper L. var. maximowiczii (Regel) Makino = Persicaria hydropiper (L.) Spach var. maximowiczii (Regel) Nemoto ■☆

309217　Polygonum hydropiper L. var. projectum Stanford = Persicaria hydropiper (L.) Spach ■

309218　Polygonum hydropiper L. var. projectum Stanford = Polygonum hydropiper L. ■

309219　Polygonum hydropiper L. var. vulgare Meisn. = Polygonum hydropiper L. ■

309220　Polygonum hydropiperoides F. Michx. strigosum (Small) Stanford = Persicaria hydropiperoides (Michx.) Small ■☆

309221　Polygonum hydropiperoides Michx. = Persicaria hydropiperoides (Michx.) Small ■☆

309222　Polygonum hydropiperoides Michx. var. adenocalyx (Stanford) Gleason = Polygonum hydropiperoides Michx. ■☆

309223　Polygonum hydropiperoides Michx. var. adenocalyx (Stanford) Gleason = Persicaria hydropiperoides (Michx.) Small ■☆

309224　Polygonum hydropiperoides Michx. var. asperifolium Stanford = Persicaria hydropiperoides (Michx.) Small ■☆

309225　Polygonum hydropiperoides Michx. var. asperifolium Stanford = Polygonum hydropiperoides Michx. ■☆

309226　Polygonum hydropiperoides Michx. var. breviciliatum Fernald = Persicaria hydropiperoides (Michx.) Small ■☆

309227　Polygonum hydropiperoides Michx. var. breviciliatum Fernald = Polygonum hydropiperoides Michx. ■☆

309228　Polygonum hydropiperoides Michx. var. bushianum Stanford = Persicaria hydropiperoides (Michx.) Small ■☆

309229　Polygonum hydropiperoides Michx. var. bushianum Stanford = Polygonum hydropiperoides Michx. ■☆

309230　Polygonum hydropiperoides Michx. var. digitatum Fernald = Persicaria hydropiperoides (Michx.) Small ■☆

309231　Polygonum hydropiperoides Michx. var. digitatum Fernald = Polygonum hydropiperoides Michx. ■☆

309232　Polygonum hydropiperoides Michx. var. euronotorum Fernald = Persicaria hydropiperoides (Michx.) Small ■☆

309233　Polygonum hydropiperoides Michx. var. hydropiperoides f. strigosum (Small) Stanford = Polygonum hydropiperoides Michx. ■☆

309234　Polygonum hydropiperoides Michx. var. opelousanum (Riddell ex Small) W. Stone = Polygonum hydropiperoides Michx. ■☆

309235　Polygonum hydropiperoides Michx. var. opelousanum (Riddell ex Small) W. Stone = Persicaria hydropiperoides (Michx.) Small ■☆

309236　Polygonum hydropiperoides Michx. var. psilostachyum H. St. John = Persicaria hydropiperoides (Michx.) Small ■☆

309237　Polygonum hydropiperoides Michx. var. psilostachyum H. St. John = Polygonum hydropiperoides Michx. ■☆

309238　Polygonum hydropiperoides Michx. var. setaceum (Baldwin ex Elliott) Gleason = Persicaria setacea (Baldwin) Small ■☆

309239　Polygonum hydropiperoides Michx. var. setaceum (Baldwin) Gleason = Persicaria setacea (Baldwin) Small ■☆

309240　Polygonum hydropiperoides Michx. var. strigosum Small = Persicaria hydropiperoides (Michx.) Small ■☆

309241　Polygonum hypoleucum Nakai ex Ohwi = Fallopia multiflora (Thunb.) Haraldson ■

309242　Polygonum hypoleucum Ohwi = Fallopia multiflora (Thunb.) Haraldson var. hypoleuca (Ohwi) Yonek. et H. Ohashi ■

309243　Polygonum hypoleucum Ohwi = Fallopia multiflora (Thunb.) Haraldson ■

309244　Polygonum hypoleucum Ohwi = Polygonum multiflorum Thunb. var. hypoleucum (Ohwi) Tang S. Liu et al. ■

309245　Polygonum hystriculum J. Schust. = Persicaria hystricula (J. Schust.) Soják ■☆

309246　Polygonum ilanense Y. C. Liu et C. H. Ou = Polygonum foliosum H. Lindb. ■

309247　Polygonum imeretinum Kom. ;伊梅里特蓼■☆

309248　Polygonum incanum F. W. Schmidt = Polygonum lapathifolium L. ■

309249　Polygonum incarnatum Elliott = Persicaria lapathifolia (L.) Gray ■

309250　Polygonum incarnatum Elliott ＝ Polygonum lapathifolium L. ■

309251　Polygonum induratum Asch. et Barbey ＝ Polygonum balansae Boiss. subsp. induratum（Asch. et Barbey）Greuter ■☆

309252　Polygonum inflexum Kom. ;内折萹蓄■☆

309253　Polygonum intercedens Petr. ;中间蓼■☆

309254　Polygonum interius Brenckle ＝ Polygonum ramosissimum Michx. ■☆

309255　Polygonum intermedium Nutt. ex S. Watson ＝ Polygonum nuttallii Small ■☆

309256　Polygonum interruptum Bunge ＝ Polygonum longisetum Bruijn ■

309257　Polygonum intramongolicum Borodina;圆叶萹蓄（圆叶蓼）; Roundleaf Knotweed ●

309258　Polygonum intricatum Kom. ;缠结萹蓄■☆

309259　Polygonum inundatum Raf. ＝ Polygonum amphibium L. var. stipulaceum N. Coleman ■

309260　Polygonum islandicum（L.）Hook. f. ＝ Koenigia islandica L. ■

309261　Polygonum janatae Klokov;亚那塔蓼■☆

309262　Polygonum japonicum Meisn. ;蚕茧蓼（蚕茧草,红蓼,红蓼子, 蓼草,蓼子草,水蓼,小蓼子草,樱蓼）;Japan Knotweed, Japanese Knotweed, Low Japanese Fleeceflower ■

309263　Polygonum japonicum Meisn. ＝ Persicaria japonica（Meisn.） Nakai ex Ohki ■

309264　Polygonum japonicum Meisn. var. conspicuum Nakai;显花蓼■

309265　Polygonum japonicum Meisn. var. conspicuum Nakai ＝ Persicaria macrantha（Meisn.）Haraldson subsp. conspicua（Nakai） Yonek. ■

309266　Polygonum japonicum Meisn. var. micranthum Nakai ＝ Polygonum japonicum Meisn. var. conspicuum Nakai ■

309267　Polygonum japonicum Meisn. var. micranthum Nakai f. brevistylum Nakai ＝ Polygonum japonicum Meisn. var. conspicuum Nakai ■

309268　Polygonum jucundum Diels ＝ Polygonum chinense L. var. paradoxum（H. Lév.）A. J. Li ■

309269　Polygonum jucundum Meisn. ;愉悦蓼（紫苞蓼）;Joyful Knotweed, Lovely Knotweed ■

309270　Polygonum jucundum Meisn. ＝ Polygonum chinense L. var. paradoxum（H. Lév.）A. J. Li ■

309271　Polygonum junceum Ledeb. ;灯心草蓼■

309272　Polygonum kawagoeanum Makino ＝ Persicaria tenella（Blume） H. Hara var. kawagoeana（Makino）H. Hara ■

309273　Polygonum kawagoeanum Makino ＝ Polygonum micranthum Boiss. ex Meisn. ■

309274　Polygonum kawagoeanum Makino ＝ Polygonum tenellum Blume var. micranthum（Boiss. ex Meisn.）C. Y. Wu ■

309275　Polygonum kawagoeanum Makino var. densiflorum（H. Hara et I. Ito）Ohwi ＝ Persicaria tenella（Blume）H. Hara ■

309276　Polygonum kelloggii Greene ＝ Polygonum polygaloides Meisn. subsp. kelloggii（Greene）J. C. Hickman ■☆

309277　Polygonum kelloggii Greene var. confertiflorum（Nutt. ex Piper） Dorn ＝ Polygonum polygaloides Meisn. subsp. confertiflorum（Nutt. ex Piper）J. C. Hickman ■☆

309278　Polygonum kemensinum Kingdon-Ward. ＝ Polygonum calostachyum Diels ■

309279　Polygonum kinashii H. Lév. et Vaniot ＝ Polygonum longisetum Bruijn ■

309280　Polygonum kirinense S. X. Li et Y. L. Chang ＝ Polygonum muricatum Meisn. ■

309281　Polygonum kitaibelianum Sadler;基陶蓼■☆

309282　Polygonum komarovii H. Lév. ＝ Polygonum lapathifolium L. ■

309283　Polygonum koreense Nakai ＝ Persicaria erectominor（Makino） Nakai var. koreensis（Nakai）I. Ito ■☆

309284　Polygonum koreense Nakai f. viridiflorum S. X. Li et Y. L. Chang;朝鲜蓼■

309285　Polygonum koreense Nakai f. viridiflorum S. X. Li et Y. L. Chang ＝ Polygonum longisetum Bruijn var. rotundatum A. J. Li ■

309286　Polygonum korshinskianum Makino ＝ Polygonum hastatosagittatum Makino ■

309287　Polygonum korshinskianum Nakai ＝ Polygonum hastatosagittatum Makino ■

309288　Polygonum kotoshoense Ohki ＝ Polygonum barbatum L. ■

309289　Polygonum kuekenthalii H. Lév. ＝ Polygonum viscosum Buch. - Ham. ex D. Don ■

309290　Polygonum labordei H. Lév. et Vaniot ＝ Fagopyrum dibotrys （D. Don）H. Hara ■

309291　Polygonum lacerum Roxb. ;撕裂蓼;Fringed Knotweed ■☆

309292　Polygonum lanatum Roxb. ＝ Persicaria lapathifolia（L.）Gray var. lanata（Roxb.）H. Hara ■

309293　Polygonum lanatum Roxb. ＝ Polygonum lapathifolium L. var. lanatum（Roxb.）Stewart ■

309294　Polygonum lanigerum R. Br. ;绵毛叶蓼（水红花子）■

309295　Polygonum lanigerum R. Br. ＝ Persicaria lanigera（R. Br.） Soják ■☆

309296　Polygonum lanigerum R. Br. var. africanum Meisn. ＝ Persicaria senegalensis（Meisn.）Soják. f. albotomentosa（R. A. Graham）K. L. Wilson ■☆

309297　Polygonum lanigerum R. Br. var. cristatum Hemsl. ＝ Polygonum lapathifolium L. var. lanatum（Roxb.）Stewart ■

309298　Polygonum lapathifolium L. ;酸模叶蓼（白苦柱,白辣蓼,糙叶 蓼,粗糙蓼,大马蓼,旱苗蓼,旱田蓼,红脚马蓼,假辣蓼,蓼吊子, 马蓼,麦蓼,水红花子,小旱苗蓼）;Bale-persicaria Curlytop Knotweed, Curltop Lady's-thumb, Curly-top Knotweed, Dockleaf Knotweed, Dockleaved Knotweed, Dock-leaved Smartweed, Heart's- ease, Pale Persicaria, Pale Smartweed, Pale Willow-weed, Peachwort, Smartweed, Willow Persicaria ■

309299　Polygonum lapathifolium L. ＝ Persicaria lapathifolia（L.）Gray ■

309300　Polygonum lapathifolium L. ＝ Polygonum lapathifolium L. var. lanatum（Roxb.）Steward ■

309301　Polygonum lapathifolium L. subsp. maculatum（Gray）Dyer et Trimen ＝ Persicaria lapathifolia（L.）Gray ■

309302　Polygonum lapathifolium L. subsp. nodosum（Pers.）Fr. ＝ Polygonum lapathifolium L. ■

309303　Polygonum lapathifolium L. subsp. nodosum（Pers.）Weinm. ＝ Polygonum lapathifolium L. ■

309304　Polygonum lapathifolium L. subsp. pallidum（With.）Fr. ＝ Persicaria lapathifolia（L.）Gray var. tomentosa（Schrank）H. Gross ■

309305　Polygonum lapathifolium L. subsp. pallidum（With.）Fr. ＝ Polygonum lapathifolium L. ■

309306　Polygonum lapathifolium L. subsp. verum ? ＝ Persicaria lapathifolia（L.）Gray ■

309307　Polygonum lapathifolium L. var. gibbosum Chabert ＝ Persicaria lapathifolia（L.）Gray ■

309308　Polygonum lapathifolium L. var. glabrum Burtt Davy ＝ Persicaria hystricula（J. Schust.）Soják ■☆

309309　Polygonum lapathifolium L. var. incanum（F. W. Schmidt）W.

D. J. Koch = Polygonum lapathifolium L. ■

309310　Polygonum lapathifolium L. var. incanum Ledeb. = Polygonum lapathifolium L. var. salicifolium Sibth. ■

309311　Polygonum lapathifolium L. var. lanatum (Roxb.) Steward = Persicaria lapathifolia (L.) Gray var. lanata (Roxb.) H. Hara ■

309312　Polygonum lapathifolium L. var. lanatum (Roxb.) Stewart;密毛酸模叶蓼(白苦柱,密毛马蓼);Densehair Dockleaved Knotweed ■

309313　Polygonum lapathifolium L. var. maculatum Dyer et Trimen = Persicaria lapathifolia (L.) Gray ■

309314　Polygonum lapathifolium L. var. nodosum (Pers.) Small = Polygonum lapathifolium L. ■

309315　Polygonum lapathifolium L. var. ovatum A. Braun = Persicaria lapathifolia (L.) Gray ■

309316　Polygonum lapathifolium L. var. ovatum A. Braun = Polygonum lapathifolium L. ■

309317　Polygonum lapathifolium L. var. prostratum Wimm. = Polygonum lapathifolium L. ■

309318　Polygonum lapathifolium L. var. salicifolium Sibth.;绵毛酸模叶蓼(红辣蓼,辣蓼,辣蓼草,柳叶大马蓼,柳叶蓼,绵毛大马蓼,绵毛马蓼);Willowleaf Knotweed ■

309319　Polygonum lapathifolium L. var. salicifolium Sibth. = Persicaria lapathifolia (L.) Gray var. tomentosa (Schrank) H. Gross ■

309320　Polygonum lapathifolium L. var. salicifolium Sibth. = Persicaria lapathifolia (L.) Gray ■

309321　Polygonum lapathifolium L. var. salicifolium Sibth. = Polygonum lapathifolium L. ■

309322　Polygonum lapathifolium L. var. xanthophyllum H. W. Kung;黄斑酸模叶蓼;Yellowspot Dockleaved Knotweed ■

309323　Polygonum lapathifolium L. var. xanthophyllum H. W. Kung = Polygonum lapathifolium L. ■

309324　Polygonum lapidosum (Kitag.) Kitag. = Polygonum bistorta L. ■

309325　Polygonum lapidosum Kitag.;石生蓼■

309326　Polygonum lapidosum Kitag. = Polygonum bistorta L. ■

309327　Polygonum latum Small ex Rydb. = Polygonum ramosissimum Michx. ■☆

309328　Polygonum laxmannii Lepech. = Polygonum ochreatum L. ■

309329　Polygonum lencoranicum Kom.;连科兰蓼■☆

309330　Polygonum leptocarpum B. L. Rob. = Polygonum ramosissimum Michx. ■☆

309331　Polygonum leptopodum Diels = Fagopyrum leptopodum (Diels) Hedberg var. grossii (H. Lév.) Lauener et D. K. Ferguson ■

309332　Polygonum leptopodum Diels = Fagopyrum leptopodum (Diels) Hedberg ■

309333　Polygonum leptopodum Diels var. grossii (H. Lév.) Sam. = Fagopyrum leptopodum (Diels) Hedberg var. grossii (H. Lév.) Lauener et D. K. Ferguson ■

309334　Polygonum liaotungense Kitag.;辽东蓼■☆

309335　Polygonum lichiangense W. W. Sm.;丽江蓼(丽江假虎杖,丽江神血宁);Lijiang Knotweed,Likiang Knotweed ■

309336　Polygonum limbatum Meisn. = Persicaria limbata (Meisn.) H. Hara ■☆

309337　Polygonum limicola Sam.;污泥蓼;Mud Knotweed ■

309338　Polygonum limosum Kom.;谷地蓼(谷地神血宁);Valley Knotweed ■

309339　Polygonum limprichtii Lingelsh. = Polygonum suffultum Maxim. ■

309340　Polygonum lineare Sam. = Fagopyrum lineare (Sam.) Haraldson ■

309341　Polygonum linicola Sutulov = Persicaria lapathifolia (L.) Gray ■

309342　Polygonum linicola Sutulov = Polygonum lapathifolium L. ■

309343　Polygonum litorale Meisn.;滨海蓼■☆

309344　Polygonum littorale Link = Polygonum buxiforme Small ■☆

309345　Polygonum longisetum Bruijn;长鬃蓼(假长尾蓼,睫穗蓼,辣蓼,马蓼,山蓼,水红花);Bristly Lady's-thumb, Longseta Knotweed, Posumbu Knotweed ■

309346　Polygonum longisetum Bruijn = Persicaria longiseta (Bruijn) Kitag. ■

309347　Polygonum longisetum Bruijn f. albiflorum (Honda) Ohwi = Persicaria longiseta (Bruijn) Kitag. f. albiflora (Honda) Masam. ■☆

309348　Polygonum longisetum Bruijn var. rotundatum A. J. Li;圆基长鬃蓼(细刺毛蓼);Roundbase Knotweed ■

309349　Polygonum longisetum Bruyn = Polygonum cespitosum Blume var. longisetum (Bruyn) Stewart ■☆

309350　Polygonum longistylum Small = Persicaria bicornis (Raf.) Nieuwl. ■☆

309351　Polygonum longistylum Small = Polygonum pensylvanicum L. ■☆

309352　Polygonum longistylum Small var. omissum (Greene) Stanford = Polygonum pensylvanicum L. ■☆

309353　Polygonum luxurians Grig.;繁茂蓼■☆

309354　Polygonum luzuloides Jaub. et Spach;地杨梅蓼■☆

309355　Polygonum lyratum Nakai;大头羽裂蓼■☆

309356　Polygonum maackianum Regel;长戟叶蓼(鹿蹄草);Long Harbertleaf Knotweed ■

309357　Polygonum maackianum Regel = Persicaria maackiana (Regel) Nakai ex T. Mori ■

309358　Polygonum macranthum Meisn. = Persicaria macrantha (Meisn.) Haraldson ■☆

309359　Polygonum macranthum Meisn. = Polygonum japonicum Meisn. ■

309360　Polygonum macrophyllum D. Don;圆穗蓼(大叶蓼,蝎子七,圆穗拳参);Macrophyllous Knotweed,Roundspike Knotweed ■

309361　Polygonum macrophyllum D. Don f. tomentosum Kitam. = Polygonum macrophyllum D. Don ■

309362　Polygonum macrophyllum D. Don var. stenophyllum (Meisn.) A. J. Li;狭叶圆穗蓼(狭叶圆穗拳参);Narrow Macrophyllous Knotweed ■

309363　Polygonum madagascariense (Meisn.) Meisn. = Persicaria madagascariensis (Meisn.) S. Ortiz et Paiva ■☆

309364　Polygonum mairei H. Lév. = Fagopyrum urophyllum (Bureau et Franch.) H. Gross ●

309365　Polygonum majanthemifolium (Petr.) Steward = Polygonum suffultum Maxim. ■

309366　Polygonum majus (Meisn.) Piper;细尖蓼;Wiry Knotweed ■☆

309367　Polygonum makinoi Nakai = Persicaria viscofera (Makino) H. Gross var. robusta (Makino) Hiyama ■

309368　Polygonum makinoi Nakai = Polygonum viscoferum Makino ■

309369　Polygonum makinoi Nakai var. laeve Kitag. = Persicaria viscofera (Makino) H. Gross var. robusta (Makino) Hiyama f. laevis (Kitag.) Hiyama ■☆

309370　Polygonum malaicum Danser = Polygonum chinense L. var. ovalifolium Meisn. ■

309371　Polygonum malaicum Denser;傣酸秆■

309372　Polygonum mandshuricum Skvortsov = Polygonum humifusum Merker ex C. Koch ■

309373　Polygonum manshuricola Kitag.;东北蓼■

309374　Polygonum manshuriense Petr. ex Kom.;耳叶蓼(北重楼,耳叶

拳参）；Earleaf Knotweed ■

309375 Polygonum marinense T. R. Mert. et P. H. Raven；马林蓼；Marin Knotweed ■☆

309376 Polygonum maritimum L.；海滨蓼；Sea Knotgrass, Sea Knotgrass ■☆

309377 Polygonum marretii H. Lév. = Polygonum suffultum Maxim. ■

309378 Polygonum martini H. Lév. et Vaniot = Polygonum japonicum Meisn. ■

309379 Polygonum martini H. Lév. et Vaniot = Polygonum longisetum Bruijn ■

309380 Polygonum meeboldii W. W. Sm. = Polygonum palmatum Dunn ■

309381 Polygonum meisnerianum Cham. et Schltdl.；梅氏蓼 ■☆

309382 Polygonum meisnerianum Cham. et Schltdl. = Persicaria meisneriana（Cham. et Schltdl.）M. Gómez ■☆

309383 Polygonum meissnerianum Cham. et Schltdl. var. triangulare Meisn.；三棱蓼 ■☆

309384 Polygonum melaicum Danser = Polygonum chinense L. var. ovalifolium Meisn. ■

309385 Polygonum micranthum Boiss. ex Meisn. = Polygonum tenellum Blume var. micranthum（Boiss. ex Meisn.）C. Y. Wu ■

309386 Polygonum micranthum Meisn. = Persicaria tenella（Blume）H. Hara var. kawagoeana（Makino）H. Hara ■

309387 Polygonum micranthum Meisn. = Polygonum kawagoeanum Makino ■

309388 Polygonum micranthum Meisn. = Polygonum tenellum Blume var. micranthum（Boiss. ex Meisn.）C. Y. Wu ■

309389 Polygonum microcephalum D. Don；小头蓼；Smallhead Knotweed ■

309390 Polygonum microcephalum D. Don var. sphaerocephalum（Wall. ex Meisn.）Murata；腺梗小头蓼 ■

309391 Polygonum microspermum Jord. ex Boreau = Polygonum aviculare L. subsp. depressum（Meisn.）Arcang. ■☆

309392 Polygonum mildbraedii Dammer = Persicaria setosula（A. Rich.）K. L. Wilson ■☆

309393 Polygonum milletii（H. Lév.）H. Lév.；大海蓼（大海拳参,球穗蓼,太白蓼,圆穗蓼）；Sea Knotweed ■

309394 Polygonum minimum S. Watson；阔叶侏儒蓼；Broad-leaf Knotweed, Leafy Dwarf Knotweed ■☆

309395 Polygonum minus Huds.；小蓼；Least Water Pepper, Pygmy Smartweed, Slender Knotgrass, Slender Knot-grass, Small Waterpepper ■

309396 Polygonum minus Huds. = Persicaria minor（Huds.）Opiz ■

309397 Polygonum minus Huds. = Persicaria tenella（Blume）H. Hara var. kawagoeana（Makino）H. Hara ■

309398 Polygonum minus Huds. = Polygonum tenellum Blume var. micranthum（Boiss. ex Meisn.）C. Y. Wu ■

309399 Polygonum minus Huds. f. trigonoparpum Makino = Polygonum trigonoparpum（Makino）Kudo et Masam. ■

309400 Polygonum minus Huds. subsp. decipiens（R. Br.）Danser = Persicaria decipiens（R. Br.）K. L. Wilson ■☆

309401 Polygonum minus Huds. subsp. micranthum（Meisn.）Danser = Persicaria tenella（Blume）H. Hara var. kawagoeana（Makino）H. Hara ■

309402 Polygonum minus Huds. subsp. micranthum（Meisn.）Danser = Polygonum kawagoeanum Makino ■

309403 Polygonum minus Huds. subsp. micranthum（Meisn.）Danser = Polygonum micranthum Boiss. ex Meisn. ■

309404 Polygonum minus Huds. subsp. micranthum（Meisn.）Danser = Polygonum tenellum Blume var. micranthum（Boiss. ex Meisn.）C. Y. Wu ■

309405 Polygonum minus Huds. subsp. procerum Danser = Polygonum kawagoeanum Makino ■

309406 Polygonum minus Huds. var. procerum（Danser）Steward = Polygonum kawagoeanum Makino ■

309407 Polygonum minus Huds. var. subcontinuum（Meisn.）Fernald = Persicaria minor（Huds.）Opiz ■

309408 Polygonum minus Huds. var. subcontinuum（Meisn.）Fernald = Polygonum minus Huds. ■

309409 Polygonum minus Huds. var. subcontinuum（Meisn.）Fernald = Polygonum persicaria L. ■

309410 Polygonum minutissimum L. O. Williams = Polygonum polygaloides Meisn. subsp. kelloggii（Greene）J. C. Hickman ■☆

309411 Polygonum minutissimum Z. Wei et Yun B. Chang；微叶蓼 ■

309412 Polygonum minutulum Makino = Persicaria taquetii（H. Lév.）Koidz. ■

309413 Polygonum minutulum Makino = Polygonum taquetii H. Lév. ■

309414 Polygonum minutum Hayata = Polygonum filicaule Wall. ex Meisn. ■

309415 Polygonum mississippiense Stanford = Polygonum pensylvanicum L. ■☆

309416 Polygonum mississippiense Stanford var. interius Stanford = Polygonum pensylvanicum L. ■☆

309417 Polygonum modosum var. incanum Ledeb. = Polygonum lapathifolium L. var. salicifolium Sibth. ■

309418 Polygonum molle D. Don；绒毛蓼（高山蓼,绒毛神血宁）；Soft Knotweed ●■

309419 Polygonum molle D. Don var. frondosum（Meisn.）A. J. Li；光叶蓼（光叶神血宁,无毛蓼）●■

309420 Polygonum molle D. Don var. rude（Meisn.）A. J. Li；倒毛蓼（倒毛神血宁,地柏,黑酸杆,九牯牛,蓼草,羊耳朵）；Rude Knotweed ■

309421 Polygonum molliiforme Boiss.；丝茎蓼（丝茎萹蓄）；Silkstem Knotweed ■

309422 Polygonum monspeliense Pers. = Polygonum aviculare L. ■

309423 Polygonum monspeliense Thiebaut ex Pers. = Polygonum aviculare L. ■

309424 Polygonum montanum（Small）Greene = Polygonum douglasii Greene ■☆

309425 Polygonum montereyense Brenckle = Polygonum arenastrum Jord. ex Boreau ■

309426 Polygonum montereyense Brenckle = Polygonum aviculare L. subsp. depressum（Meisn.）Arcang. ■☆

309427 Polygonum morrisonense Hayata = Polygonum runcinatum Buch. -Ham. ex D. Don ■

309428 Polygonum muhlenbergii S. Watson；大根蓼；Bigroot Lady's Thumb, Swamp Persicaria, Swamp Smartweed ■☆

309429 Polygonum muhlenbergii S. Watson = Polygonum amphibium L. var. emersum Michx. ■☆

309430 Polygonum muhlenbergii S. Watson f. natans Wiegand = Polygonum amphibium L. var. emersum Michx. ■☆

309431 Polygonum muhlenbergii S. Watson var. terrestre（Willd.）Trel. = Polygonum amphibium L. var. emersum Michx. ■☆

309432 Polygonum multiflorum Thunb. = Fallopia multiflora（Thunb.）Haraldson ■

309433　Polygonum multiflorum Thunb. ex A. Murray = Fallopia multiflora（Thunb.）Haraldson ■

309434　Polygonum multiflorum Thunb. ex A. Murray var. ciliinerve（Nakai）Stewart = Fallopia multiflora（Thunb.）Haraldson var. ciliinervis（Nakai）A. J. Li ■

309435　Polygonum multiflorum Thunb. ex A. Murray var. hypoleucum（Ohwi）Tang S. Liu et al. = Fallopia multiflora（Thunb.）Haraldson ■

309436　Polygonum multiflorum Thunb. var. angulatum S. Y. Liu = Fallopia multiflora（Thunb.）Haraldson ■

309437　Polygonum multiflorum Thunb. var. ciliinerve（Nakai）Steward = Fallopia multiflora（Thunb.）Haraldson var. ciliinervis（Nakai）A. J. Li ■

309438　Polygonum multiflorum Thunb. var. ciliinerve（Nakai）Steward = Fallopia ciliinervis（Nakai）K. Hammer ■

309439　Polygonum multiflorum Thunb. var. hypoleucum（Nakai ex Ohwi）Tang S. Liu et al. = Fallopia multiflora（Thunb.）Haraldson ■

309440　Polygonum multiflorum Thunb. var. hypoleucum（Ohwi）Tang S. Liu et al. = Fallopia multiflora（Thunb.）Haraldson ■

309441　Polygonum multiflorum Thunb. var. hypoleucum（Ohwi）Tang S. Liu et al. = Fallopia multiflora（Thunb.）Haraldson var. hypoleuca（Ohwi）Yonek. et H. Ohashi ■

309442　Polygonum muricatum Meisn.；小花蓼（匍茎蓼，水湿蓼，小蓼花，有刺水湿蓼）；Rough Knotweed ■

309443　Polygonum muricatum Meisn. = Persicaria muricata（Meisn.）Nemoto ■

309444　Polygonum myosurus Franch. = Polygonum japonicum Meisn. ■

309445　Polygonum myriophyllum H. Gross = Polygonum cognatum Meisn. ■

309446　Polygonum myrtillifolium Kom.；黑果越橘蓼■☆

309447　Polygonum nakaii（H. Hara）Ohwi；中井氏蓼■☆

309448　Polygonum nakaii（H. Hara）Ohwi = Aconogonon nakaii（H. Hara）H. Hara ■☆

309449　Polygonum natans（Michx.）Eaton = Persicaria amphibia（L.）Gray ■

309450　Polygonum natans（Michx.）Eaton = Polygonum amphibium L. ■

309451　Polygonum natans Eaton = Polygonum amphibium L. var. stipulaceum N. Coleman ■

309452　Polygonum natans Eaton f. genuinum Stanford = Polygonum amphibium L. var. stipulaceum N. Coleman ■

309453　Polygonum natans Eaton f. hartwrightii（A. Gray）Stanford = Polygonum amphibium L. var. stipulaceum N. Coleman ■

309454　Polygonum neglectum Besser = Polygonum aviculare L. subsp. neglectum（Besser）Arcang. ■☆

309455　Polygonum neglectum Besser = Polygonum aviculare L. ■

309456　Polygonum neglectum Besser = Polygonum bellardii All. ■☆

309457　Polygonum neofiliforme Nakai = Antenoron filiforme（Thunb.）Rob. et Vautier var. neofiliforme（Nakai）A. J. Li ■

309458　Polygonum neofiliforme Nakai = Persicaria neofiliformis（Nakai）Ohki ■

309459　Polygonum nepalense Meisn.；尼泊尔蓼（猫儿眼睛，荞麦苋，水荞麦，头状蓼，小猫眼，野荞麦，野荞麦苗，野荞子）；Nepal Knotweed，Nepal Persicaria，Nepalese Smartweed ■

309460　Polygonum nepalense Meisn. = Persicaria nepalensis（Meisn.）H. Gross ■

309461　Polygonum nepalense Meisn. var. adenothrix Nakai = Polygonum nepalense Meisn. ■

309462　Polygonum ninutum Hayata = Polygonum filicaule Wall. ex Meisn. ■

309463　Polygonum nipponense Makino = Persicaria muricata（Meisn.）Nemoto ■

309464　Polygonum nipponense Makino = Polygonum muricatum Meisn. ■

309465　Polygonum nipponense Makino f. albiflorum Makino；白小花蓼■☆

309466　Polygonum nitens（Fisch. et C. A. Mey.）Petr. ex Kom.；亮果蓼（草河车，短柄蓼，拳参）■

309467　Polygonum nitens（Fisch. et C. A. Mey.）Petr. ex Kom. = Polygonum ellipticum Willd. ex Spreng. ■

309468　Polygonum nodosum Pers.；节蓼（大马蓼，马蓼，曲辣蓼，虾蟆腿，小蓼子草，钟花蓼，猪蓼子草）；Knotted Persicaria, Spotted Persicaria ■

309469　Polygonum nodosum Pers. = Persicaria lapathifolia（L.）Gray ■

309470　Polygonum nodosum Pers. = Polygonum lapathifolium L. ■

309471　Polygonum nodosum Pers. var. amblyophyllum（H. Hara）Ohwi = Persicaria amblyophylla H. Hara ■☆

309472　Polygonum nodosum Pers. var. incanum Ledeb.；绵毛节蓼（白毛垂花蓼，绵毛大马蓼）■☆

309473　Polygonum nodosum Pers. var. incanum Ledeb. = Polygonum lapathifolium L. var. salicifolium Sibth. ■

309474　Polygonum nummularifolium Meisn.；铜钱叶蓼（铜钱叶神血宁）；Copper Knotweed ■

309475　Polygonum nuttallii Small；纳托尔蓼；Nuttall's Knotweed ■☆

309476　Polygonum nyikense Baker = Persicaria setosula（A. Rich.）K. L. Wilson ■☆

309477　Polygonum obtusifolium Täckh. et Boulos = Persicaria obtusifolia（Täckh. et Boulos）Greuter et Burdet ■☆

309478　Polygonum ochotense Petr. ex Kom.；倒根蓼（倒根草，倒根拳参）；Ochotsk Knotweed ■

309479　Polygonum ochreatum L.；白山蓼（白山神血宁）；Ocreate Knotweed ■

309480　Polygonum odontopterum（H. Gross）H. W. Kung = Fagopyrum gracilipes（Hemsl.）Dammer ex Diels ■

309481　Polygonum odontopterum（H. Gross）Kung = Fagopyrum gracilipes（Hemsl.）Dammer ex Diels ■

309482　Polygonum odoratum Lour.；芳香蓼；Asian Mint, Rau Ram, Vietnam Mint ■☆

309483　Polygonum oliganthum Diels = Polygonum muricatum Meisn. ■

309484　Polygonum omerostromum Ohki = Polygonum barbatum L. ■

309485　Polygonum omissum Greene = Persicaria pensylvanica（L.）M. Gómez ■☆

309486　Polygonum omissum Greene = Polygonum pensylvanicum L. ■☆

309487　Polygonum oneillii Brenckle = Polygonum lapathifolium L. ■

309488　Polygonum opacum Sam. = Polygonum persicaria L. var. opacum（Sam.）A. J. Li ■

309489　Polygonum opelousanum Riddell ex Small = Persicaria hydropiperoides（Michx.）Small ■☆

309490　Polygonum opelousanum Riddell ex Small = Polygonum hydropiperoides Michx. ■☆

309491　Polygonum opelousanum Riddell ex Small var. adenocalyx Stanford = Persicaria hydropiperoides（Michx.）Small ■☆

309492　Polygonum opelousanum Riddell ex Small var. adenocalyx Stanford = Polygonum hydropiperoides Michx. ■☆

309493　Polygonum oreophilum（Makino）Ohwi = Persicaria oreophila（Makino）Hiyama ■☆

309494　Polygonum orientale L.；红蓼（八字蓼，冰红花，川蓼，大接骨，大接骨天蓼，大蓼，大毛蓼，丹药头，捣花，东方蓼，狗尾巴花，果麻,何曹花,河蓼,红草,荭,荭草,荭草花,荭蓼,家蓼,九节龙,酒药草,苦苍蓼子曹,辣蓼,辣蓼子,狼尾巴花,蓼花,龙豉,茏古,茏鼓,茏,马蓼,马子银花,山红花,石龙,水红花,水红花秆,水红花子,水红子,水荭,水荭花,水荭子,水辣蓼,水蓬稞,天蓼,游龙,追风草);Gardengate, Kiss Me Over the Garden Gate, Kiss-me-over-the-garden-gate, Lady's Thumb, Prince's Feather, Prince's Feathers, Prince's Plume, Prince's-feather, Prince's-plume' Lady's-thumb, Red Knotweed ■

309495 Polygonum orientale L. = Persicaria orientalis (L.) Spach ■

309496 Polygonum orientale L. var. discolor Benth. = Polygonum orientale L. ■

309497 Polygonum orientale L. var. pilosum (Roxb. ex Meisn.) Meisn. = Polygonum orientale L. ■

309498 Polygonum orientale L. var. pilosum Meisn. = Polygonum orientale L. ■

309499 Polygonum oryzetorum Blume = Polygonum pubescens Blume ■

309500 Polygonum oxianum Kom. ;阿穆达尔蓼■☆

309501 Polygonum oxyphyllum Wall. ex Meisn. = Polygonum amplexicaule D. Don ■

309502 Polygonum oxyspermum C. A. Mey. et Bunge = Polygonum oxyspermum C. A. Mey. et Bunge ex Ledeb. ■☆

309503 Polygonum oxyspermum C. A. Mey. et Bunge ex Ledeb. ;尖籽蓼(雷氏蓼);Ray's Knotgrass, Ray's Knot-grass, Slender Sea Knotgrass ■☆

309504 Polygonum oxyspermum C. A. Mey. et Bunge ex Ledeb. subsp. raii (Bab.) D. A. Webb et Chater;雷氏蓼;Ray's Knotweed ■☆

309505 Polygonum pacificum Petr. ex Kom. ;太平洋蓼(太平洋拳参);Pacific Knotweed ■

309506 Polygonum pacificum Petr. ex Kom. = Bistorta officinalis Delarbre subsp. pacifica (Petr. ex Kom.) Yonek. ■☆

309507 Polygonum paleaceum Wall. ex Hook. f. ;草血竭(草血结,地蜂子,地黑蜂,地马蜂,地蜈蚣,凤凰鸡,弓腰老,拱腰老,回头草,金贵鸡,金黄鸡,老腰弓,蛇疙瘩,土血竭,虾子七,小公公,小么公,血三七,一口血,迂头鸡,紫花根);Herb Dragon's blood, Paleaceous Knotweed ■

309508 Polygonum paleaceum Wall. ex Hook. f. var. pubifolium Sam. ;毛叶草血竭(地七风);Hairy-leaf Knotweed ■

309509 Polygonum pallidum With. = Polygonum lapathifolium L. ■

309510 Polygonum palmatum Dunn;掌叶蓼(猪草);Palmate Leaf Knotweed,Palmleaf Knotweed ■

309511 Polygonum paludicola Makino = Polygonum foliosum H. Lindb. var. paludicola (Makino) Kitam. ■

309512 Polygonum paludosum (Kom.) Kom. = Polygonum sagittatum L. ■

309513 Polygonum paludosum (Kom.) Kom. = Polygonum sieboldii Meisn. ■

309514 Polygonum pamiricum Korsh. = Polygonum sibiricum Laxm. var. thomsonii Meisn. ex Stewart ■

309515 Polygonum pamiroalaicum Kom. ;帕米尔蓼■☆

309516 Polygonum panduriforme H. Lév. et Vaniot = Polygonum runcinatum Buch. -Ham. ex D. Don ■

309517 Polygonum paniculatum Andrz. = Polygonum molle D. Don ●■

309518 Polygonum paniculatum Andrz. var. frondosum (Meisn.) Steward = Polygonum molle D. Don var. frondosum (Meisn.) A. J. Li ●■

309519 Polygonum paniculatum Andrz. var. rude (Meisn.) Steward = Polygonum molle D. Don var. rude (Meisn.) A. J. Li ■

309520 Polygonum paniculatum Blume = Polygonum molle D. Don var. frondosum (Meisn.) A. J. Li ●■

309521 Polygonum paniculatum Blume var. frondosum (Meisn.) Steward = Polygonum molle D. Don var. frondosum (Meisn.) A. J. Li ●■

309522 Polygonum paniculatum Blume var. rude (Meisn.) Steward = Polygonum molle D. Don var. rude (Meisn.) A. J. Li ■

309523 Polygonum paradoxum H. Lév. = Polygonum chinense L. var. paradoxum (H. Lév.) A. J. Li ■

309524 Polygonum paralimicola A. J. Li ;湿地蓼;Marsh Knotweed ◨

309525 Polygonum paronychia Cham. et Schltdl. ; 海滩黑蓼;Beach Knotweed, Black Knotweed, Dune Knotweed ■☆

309526 Polygonum paronychioides C. A. Mey. ex Hohen. ;线叶蓣蓄(线叶蓼);Threadleaf Knotweed, Whilowort-like Knotweed ●

309527 Polygonum parryi Greene;帕里蓼;Parry's Knotweed, Prickly Knotweed ■☆

309528 Polygonum parviflorum Gromov;小果蓼■

309529 Polygonum parviflorum Gromov = Polygonum scabrum Moench ■

309530 Polygonum parviflorum Y. L. Chang et S. H. Li = Polygonum plebeium R. Br. ■

309531 Polygonum patulum M. Bieb. ;展枝蓣蓄(多枝蓼,新疆蓼,展枝蓼);Bellard's Smartweed,Expandtwig Knotweed,Red Knot-grass ■

309532 Polygonum patulum M. Bieb. var. gracilius (Ledeb.) Rouy = Polygonum patulum M. Bieb. ■

309533 Polygonum patulum M. Bieb. var. patulum f. gracilius (Ledeb.) I. Grint = Polygonum patulum M. Bieb. ■

309534 Polygonum patulum M. Bieb. var. virgatum (Loisel.) Rouy = Polygonum bellardii All. ■☆

309535 Polygonum pauciflorum Maxim. = Fallopia dumetora (L.) Holub var. pauciflora (Maxim.) A. J. Li ■

309536 Polygonum pedunculare Wall. = Polygonum dichotomum Blume ■

309537 Polygonum pedunculare Wall. ex Meisn. = Persicaria dichotoma (Blume) Masam. ■

309538 Polygonum pedunculare Wall. ex Meisn. = Polygonum dichotomum Blume ■

309539 Polygonum pedunculare Wall. ex Meisn. var. subsagittatum De Wild. = Polygonum subsagittatum (De Wild.) Park ■☆

309540 Polygonum pedunculare Wall. var. angustissimum Hook. f. = Persicaria strigosa (R. Br.) Nakai ■

309541 Polygonum pensylvanicum Bunge = Polygonum bungeanum Turcz. ■

309542 Polygonum pensylvanicum L. ;宾州蓼;Common Smartweed, Heartseed, Pennsylvania Knotweed, Pennsylvania Persicaria, Pennsylvania Smartweed,Pink Knotweed,Pinkweed ■☆

309543 Polygonum pensylvanicum L. = Persicaria pensylvanica (L.) M. Gómez ■☆

309544 Polygonum pensylvanicum L. subsp. oneillii (Brenckle) Hultén = Polygonum lapathifolium L. ■

309545 Polygonum pensylvanicum L. var. durum Stanford = Persicaria pensylvanica (L.) M. Gómez ■☆

309546 Polygonum pensylvanicum L. var. durum Stanford = Polygonum pensylvanicum L. ■☆

309547 Polygonum pensylvanicum L. var. eglandulosum Myers = Persicaria pensylvanica (L.) M. Gómez ■☆

309548 Polygonum pensylvanicum L. var. eglandulosum Myers = Polygonum pensylvanicum L. ■☆

309549 Polygonum pensylvanicum L. var. genuinum Fernald =

Polygonum pensylvanicum L. ■☆

309550　Polygonum pensylvanicum L. var. laevigatum Fernald = Persicaria pensylvanica (L.) M. Gómez ■☆

309551　Polygonum pensylvanicum L. var. laevigatum Fernald = Polygonum pensylvanicum L. ■☆

309552　Polygonum pensylvanicum L. var. laevigatum Fernald f. albineum Farw. = Polygonum pensylvanicum L. ■☆

309553　Polygonum pensylvanicum L. var. laevigatum Fernald f. pallescens Stanford = Polygonum pensylvanicum L. ■☆

309554　Polygonum pensylvanicum L. var. nesophilum Fernald = Persicaria pensylvanica (L.) M. Gómez ■☆

309555　Polygonum pensylvanicum L. var. nesophilum Fernald = Polygonum pensylvanicum L. ■☆

309556　Polygonum pensylvanicum L. var. oneillii (Brenckle) Hultén = Persicaria lapathifolia (L.) Gray ■

309557　Polygonum pensylvanicum L. var. oneillii (Brenckle) Hultén = Polygonum lapathifolium L. ■

309558　Polygonum pensylvanicum L. var. rosiflorum Norton = Persicaria pensylvanica (L.) M. Gómez ■☆

309559　Polygonum pensylvanicum L. var. rosiflorum Norton = Polygonum pensylvanicum L. ■☆

309560　Polygonum peregrinatoris Paulsen;逆阿落■

309561　Polygonum peregrinatoris Paulsen = Polygonum tortuosum D. Don ■

309562　Polygonum perfoliatum (L.) L. = Persicaria perfoliata (L.) H. Gross ■

309563　Polygonum perfoliatum (L.) L. = Polygonum perfoliatum L. ■

309564　Polygonum perfoliatum L.;杠板归(白大老鸦酸,白笋,穿叶蓼,刺犁头,刺藜头,刺酸浆,刺头草,蒴藜头,大蜣脚,大蜣腿,倒挂紫金钩,倒金钩,地不过,地葡萄,豆干草,方胜板,贯叶蓼,河白草,虎舌草,花头公草,火轮箭,火炭藤,鸡眼睛草,急改索,急解索,扛板归,括耙草,拦路虎,拦蛇风,老虎刺,老虎方,老虎脷,老虎利,烙铁草,雷公藤,犁尖草,犁头草,犁头刺,犁头刺藤,犁头尖,犁头藤,龙仙草,蚂蚱簕,猫公刺,猫牙草,猫爪草,猫爪刺,南蛇风,霹雳木,三角藤,三角藤,三木棉,蛇不过,蛇倒退,蛇牙草,水蓼,水马铃,酸藤,退血草,五毒草,小蓼,有刺粪箕笃,有刺鸹鹕饭,有刺火炭藤,有刺鸠饭草,有刺犁牛草,有刺三角延酸,有笋火炭藤,有笋犁牛草,有簕犁牛草,鱼尾花,鱼牙草,月斑鸠);Asiatic Tearthumb, Devil's-tail Tearthrumb, Giant Climbing Tearthumb, Mile-a-minute Vine, Mile-a-minute Weed, Mile-a-minute-weed, Perfoliate Knotweed, Perforate Fleeceflower ■

309565　Polygonum perforatum L. f. glaciale Meisn. = Polygonum glaciale (Meisn.) Hook. f. ■

309566　Polygonum perforatum Meisn. var. glaciale Meisn. = Polygonum glaciale (Meisn.) Hook. f. ■

309567　Polygonum pergracile Hemsl. = Polygonum suffultum Maxim. var. pergracile (Hemsl.) Sam. ■

309568　Polygonum periginatoris Paulsen = Polygonum tortuosum D. Don ■

309569　Polygonum perpusilum Hook. f.;极小珠芽蓼■

309570　Polygonum persicaria L.;春蓼(蓼,马蓼,牛耳朵菜,山辣蓼,桃叶蓼,野芥菜);Adam's Plaster, Alice, Arsesmart, Blind Withy, Common Persicaria, Crab's Claws, Croneshanks, Devil's Pinch, Fat Hen, Heart's Ease, Heartsease, Heart's-ease, Heartweed, Heartwort, Lady's Thumb, Lakeweed, Lamb's Tongue, Lavender, Lover's Pride, Lovers' Pride, Morub, Peachwort, Persicaria, Persicary, Pig Grass, Pinchweed, Pincushion, Plumbago, Print Pinafore, Red Joints, Red Knees, Red Legs, Redshank, Redshanks, Redweed, Sauchweed, Saucy

Alice, Smartweed, Sourock, Spotted Ladysthumb, Spotted Lady's-thumb, Spotted Notweed, Spotted Persicaria, Spring Knotweed, Useless, Virgin's Pinch, Willow-weed, Yes Smart ■

309571　Polygonum persicaria L. = Persicaria maculata (Raf.) Gray ■

309572　Polygonum persicaria L. = Persicaria maculosa Gray subsp. hirticaulis (Danser) S. Ekman et Knutsson ■

309573　Polygonum persicaria L. = Persicaria maculosa Gray ■☆

309574　Polygonum persicaria L. f. albiflorum Rob. = Persicaria maculata (Raf.) Gray ■

309575　Polygonum persicaria L. f. glabrescens Rob. = Persicaria maculata (Raf.) Gray ■

309576　Polygonum persicaria L. f. humile S. X. Li et Y. L. Chang = Polygonum persicaria L. ■

309577　Polygonum persicaria L. f. jumile S. X. Li et Y. L. Chang;小桃叶蓼;Small Spring Knotweed ■

309578　Polygonum persicaria L. f. latifolium S. X. Li et Y. L. Chang;宽叶桃叶蓼;Broadleaf Spring Knotweed ■

309579　Polygonum persicaria L. f. latifolium S. X. Li et Y. L. Chang = Polygonum persicaria L. ■

309580　Polygonum persicaria L. subsp. hirticaule Danser = Polygonum persicaria L. ■

309581　Polygonum persicaria L. var. amblyophyllum (H. Hara) Ohwi = Persicaria amblyophylla H. Hara ■☆

309582　Polygonum persicaria L. var. angustifolium Beckh. = Persicaria maculosa Gray ■☆

309583　Polygonum persicaria L. var. angustifolium Beckh. = Polygonum persicaria L. ■

309584　Polygonum persicaria L. var. biforme (Wahlenb.) Fr. = Persicaria maculata (Raf.) Gray ■☆

309585　Polygonum persicaria L. var. incanum Roth = Persicaria lapathifolia (L.) Gray ■

309586　Polygonum persicaria L. var. incanum Roth = Polygonum lapathifolium L. var. salicifolium Sibth. ■

309587　Polygonum persicaria L. var. opacum (Sam.) A. J. Li;暗果春蓼(暗果蓼)■

309588　Polygonum persicaria L. var. pubescens Makino = Polygonum persicaria L. ■

309589　Polygonum persicaria L. var. ruderale (Salisb.) Meisn. = Persicaria maculata (Raf.) Gray ■

309590　Polygonum persicaria L. var. ruderale (Salisb.) Meisn. = Polygonum persicaria L. ■

309591　Polygonum persicarioides Kunth = Persicaria hydropiperoides (Michx.) Small ■☆

309592　Polygonum petiolatum D. Don = Polygonum amplexicaule D. Don ■

309593　Polygonum piliferum Tikovsky = Persicaria limbata (Meisn.) H. Hara ■☆

309594　Polygonum pilosum (Maxim.) Forbes et Hemsl. = Polygonum sparsipilosum A. J. Li ■

309595　Polygonum pilosum (Maxim.) Hemsl. = Polygonum sparsipilosum A. J. Li ■

309596　Polygonum pilosum Roxb. = Polygonum orientale L. ■

309597　Polygonum pilosum Roxb. ex Meisn. = Polygonum orientale L. ■

309598　Polygonum pilushanense Y. C. Liu et C. H. Ou;毕禄山蓼;Bilushan Knotweed ■

309599　Polygonum pinetorum Hemsl.;松林蓼(松林神血宁,松荫蓼);Pinewood Knotweed, Pintorum Knotweed ■

309600　Polygonum planum Skvortsov = Polygonum arenastrum Boreau ■

309601　Polygonum platyphyllum S. X. Li et Y. L. Chang;宽叶蓼(宽叶神血宁);Broadleaf Knotweed ■

309602　Polygonum plebeium R. Br.;习见蓼(萹蓄,假萹蓄,节花路蓼,米子蓼,铁马鞭,铁马齿苋,小萹蓄,小叶萹蓄,腋花蓼);Common Knotweed, Joint-flowered Knotweed, Plebian Knotweed, Usual Knotweed ■

309603　Polygonum plebeium R. Br. subsp. changii (Kitag.) Vorosch. = Polygonum plebeium R. Br. ■

309604　Polygonum plumosum Small = Bistorta plumosa (Small) Greene ■☆

309605　Polygonum poiretii Meisn. = Persicaria poiretii (Meisn.) K. L. Wilson ■☆

309606　Polygonum poiretii Meisn. var. madagascariense Meisn. = Persicaria madagascariensis (Meisn.) S. Ortiz et Paiva ■☆

309607　Polygonum polycnemoides Jaub. et Spach;针叶萹蓄(针叶蓼,针枝蓼);Manyleg Knotweed,Needletwig Knotweed ■

309608　Polygonum polygaloides Meisn.;远志蓼;Polygala Knotweed ■☆

309609　Polygonum polygaloides Meisn. subsp. confertiflorum (Nutt. ex Piper) J. C. Hickman;密花远志蓼■☆

309610　Polygonum polygaloides Meisn. subsp. kelloggii (Greene) J. C. Hickman;凯洛格蓼;Kellogg's Knotweed ■☆

309611　Polygonum polygaloides Meisn. var. montanum Brenckle = Polygonum polygaloides Meisn. ■☆

309612　Polygonum polygamum Vent. = Polygonella polygama (Vent.) Engelm. et A. Gray ●☆

309613　Polygonum polymorphum Ledeb. var. ajanense Regel et Tiling = Polygonum ajanense (Regel et Tiling) Grig. ■

309614　Polygonum polymorphum Ledeb. var. angustissimum Korsh. = Polygonum angustifolium Pall. ■

309615　Polygonum polyneuron Franch. et Sav. = Polygonum arenastrum Boreau ■

309616　Polygonum polystachyum Wall. ex Meisn.;多穗蓼(多穗假虎杖,多穗神血宁,辣蓼,菱叶拔毒散,球序蓼,水蓼);Ballspike Knotweed, Cultivated Knotweed, Himalyan Knotweed, Kashmir Plume, Manyspike Knotweed, Polystachous Knotweed, Wallich Knotweed ●■

309617　Polygonum polystachyum Wall. ex Meisn. = Persicaria wallichii Greuter et Burdet ■☆

309618　Polygonum polystachyum Wall. ex Meisn. var. longifolia Hook. f.;长叶多穗蓼(长叶多穗神血宁,长叶假虎杖);Longleaf Manyspike ■

309619　Polygonum polystachyum Wall. ex Meisn. var. pubescens Meisn.;柔毛假虎杖■

309620　Polygonum popovii Borodina;库车萹蓄(库车蓼);Kuche Knotweed ●

309621　Polygonum portoricense Bertero ex Small = Persicaria glabra (Willd.) M. Gómez ■

309622　Polygonum portoricense Bertero ex Small = Polygonum glabrum Willd. ■

309623　Polygonum portoricense Bertol. ex Endl. = Polygonum glabrum Willd. ■

309624　Polygonum posumbu Buch. -Ham. ex D. Don;丛枝蓼(白辣蓼,长尾叶蓼,簇蓼,花蓼,辣蓼,马蓼,水红辣蓼,小辣蓼);Clump Knotweed,Oriental Lady's Thumb, Oriental Ladysthumb, Smartweed, Winkle Knotweed ■

309625　Polygonum posumbu Buch. -Ham. ex D. Don = Persicaria posumbu (Buch. -Ham. ex D. Don) H. Gross ■

309626　Polygonum posumbu Buch. -Ham. ex D. Don var. blumei (Meisn.) Herder = Polygonum longisetum Bruijn ■

309627　Polygonum posumbu Buch. -Ham. ex D. Don var. laxiflorum (Meisn.) Ohwi = Persicaria posumbu (Buch. -Ham. ex D. Don) H. Gross ■

309628　Polygonum posumbu Buch. -Ham. ex D. Don var. stenophyllum (Makino) Murata;窄长尾叶蓼■☆

309629　Polygonum praetermissum Hook. f.;疏蓼(疏忽蓼,细叶犁避,细叶雀翘,遗漏蓼);Omission Knotweed,Omit Knotweed ■

309630　Polygonum praetermissum Hook. f. = Persicaria praetermissa (Hook. f.) H. Hara ■

309631　Polygonum procumbens Gilib. = Polygonum aviculare L. ■

309632　Polygonum procumbens Y. L. Chang et S. X. Li = Polygonum posumbu Buch. -Ham. ex D. Don ■

309633　Polygonum prolificum (Small) B. L. Rob. = Polygonum ramosissimum Michx. subsp. prolificum (Small) Costea et Tardif ■☆

309634　Polygonum prolificum (Small) B. L. Rob. var. autumnale (Brenckle) Brenckle = Polygonum ramosissimum Michx. subsp. prolificum (Small) Costea et Tardif ■☆

309635　Polygonum prolificum (Small) B. L. Rob. var. profusum Brenckle = Polygonum ramosissimum Michx. subsp. prolificum (Small) Costea et Tardif ■☆

309636　Polygonum pronum C. F. Fang = Polygonum posumbu Buch. -Ham. ex D. Don ■

309637　Polygonum propinquum Ledeb. = Polygonum arenastrum Boreau ■

309638　Polygonum prostratum Skvortsov = Polygonum arenastrum Boreau ■

309639　Polygonum provinciale K. Koch = Polygonum bellardii All. ■☆

309640　Polygonum przewalskii A. K. Skvortsov et Borodina = Polygonum glaciale (Meisn.) Hook. f. var. przewalskii (A. K. Skvortsov et Borodina) A. J. Li ■

309641　Polygonum pseudopalmatum G. Hoo;拟掌叶蓼(鸭脚蓼);Palmleaf Knotweed ■

309642　Polygonum pseudopalmatum G. Hoo = Polygonum palmatum Dunn ■

309643　Polygonum pteropus Hance = Polygonum thunbergii Siebold et Zucc. ■

309644　Polygonum pubescens Blume;伏毛蓼(八字蓼,垂蓼,短毛蓼,旱辣蓼,辣蓼,软水蓼,无辣蓼,腺花蓼,腺花毛蓼);Pubescent Knotweed ■

309645　Polygonum pubescens Blume = Persicaria pubescens (Blume) H. Hara ■

309646　Polygonum pulchellum Loisel. = Polygonum arenarium Waldst. et Kit. subsp. pulchellum (Loisel.) Thell. ■☆

309647　Polygonum pulchrum Blume;丽蓼(绒毛蓼);Spiffy Knotweed ■

309648　Polygonum pulchrum Blume = Persicaria attenuata (R. Br.) Soják subsp. pulchra (Blume) K. L. Wilson ■

309649　Polygonum pulchrum Blume = Persicaria lapathifolia (L.) Gray ■

309650　Polygonum pulvinatum Kom.;垫蓼●

309651　Polygonum punctatum Buch. -Ham. ex D. Don = Polygonum nepalense Meisn. ■

309652　Polygonum punctatum Buch. -Ham. ex D. Don = Polygonum punctatum Elliott ■☆

309653　Polygonum punctatum Buch. -Ham. ex D. Don var. alatum Buch. -Ham. ex D. Don = Polygonum nepalense Meisn. ■

309654　Polygonum punctatum Elliott;斑叶蓼;Dotted Smartweed, Smartweed,Water Knotweed,Water Smartweed ■☆

309655　Polygonum punctatum Elliott = Persicaria punctata (Elliott)

Small ■☆

309656 Polygonum punctatum Elliott = Polygonum punctatum Buch. - Ham. ex D. Don ■☆

309657 Polygonum punctatum Elliott var. aquatile（Mart.）Fernald = Polygonum punctatum Elliott ■☆

309658 Polygonum punctatum Elliott var. confertiflorum（Meisn.）Fassett;簇花斑叶蓼;Dotted Smartweed ■☆

309659 Polygonum punctatum Elliott var. confertiflorum（Meisn.）Fassett f. longicollum Fassett = Polygonum punctatum Elliott var. confertiflorum（Meisn.）Fassett ■☆

309660 Polygonum punctatum Elliott var. confertiflorum（Meisn.）Small = Persicaria punctata（Elliott）Small ■☆

309661 Polygonum punctatum Elliott var. ellipticum Fassett = Persicaria punctata（Elliott）Small ■☆

309662 Polygonum punctatum Elliott var. ellipticum Fassett = Polygonum punctatum Elliott ■☆

309663 Polygonum punctatum Elliott var. leptostachyum（Meisn.）Small = Persicaria punctata（Elliott）Small ■☆

309664 Polygonum punctatum Elliott var. leptostachyum（Meisn.）Small = Polygonum punctatum Elliott var. confertiflorum（Meisn.）Fassett ■☆

309665 Polygonum punctatum Elliott var. leptostachyum（Meisn.）Small = Polygonum punctatum Elliott ■☆

309666 Polygonum punctatum Elliott var. littorale Fassett;湿地斑叶蓼;Dotted Smartweed ■☆

309667 Polygonum punctatum Elliott var. majus（Meisn.）Fassett = Persicaria robustior（Small）E. P. Bicknell ■☆

309668 Polygonum punctatum Elliott var. parviflorum Fassett = Persicaria punctata（Elliott）Small ■☆

309669 Polygonum punctatum Elliott var. parviflorum Fassett = Polygonum punctatum Elliott ■☆

309670 Polygonum punctatum Elliott var. parvum Vict. et J. Rousseau = Polygonum punctatum Elliott var. confertiflorum（Meisn.）Fassett ■☆

309671 Polygonum punctatum Elliott var. parvum Vict. et Rousseau = Persicaria punctata（Elliott）Small ■☆

309672 Polygonum punctatum Elliott var. parvum Vict. et Rousseau = Polygonum punctatum Elliott ■☆

309673 Polygonum punctatum Elliott var. robustius Small = Persicaria robustior（Small）E. P. Bicknell ■☆

309674 Polygonum puritanorum Fernald = Persicaria maculosa Gray ■☆

309675 Polygonum puritanorum Fernald = Polygonum persicaria L. ■

309676 Polygonum purpureonervosum A. J. Li;紫脉拳参（紫脉蓼）;Purplevein Knotweed ■

309677 Polygonum pyramidale H. Lév. = Polygonum lapathifolium L. ■

309678 Polygonum quadrifidum Hayata = Polygonum nepalense Meisn. ■

309679 Polygonum quarrei De Wild. = Persicaria setosula（A. Rich.）K. L. Wilson ■☆

309680 Polygonum radicans Hemsl. = Polygonum filicaule Wall. ex Meisn. ■

309681 Polygonum raii Bab. = Polygonum oxyspermum C. A. Mey. et Bunge ex Ledeb. subsp. raii（Bab.）D. A. Webb et Chater ■☆

309682 Polygonum ramosissimum F. Michx. atlanticum B. L. Rob. = Polygonum ramosissimum Michx. ■☆

309683 Polygonum ramosissimum Michx.;密丛蓼;Bushy Knotweed, Long-fruited Knotweed ■☆

309684 Polygonum ramosissimum Michx. = Polygonum prolificum（Small）B. L. Rob. ■☆

309685 Polygonum ramosissimum Michx. subsp. prolificum（Small）Costea et Tardif;多育密丛蓼;Knotweed ■☆

309686 Polygonum ramosissimum Michx. var. prolificum Small = Polygonum ramosissimum Michx. subsp. prolificum（Small）Costea et Tardif ■☆

309687 Polygonum ramosissimum var. prolificum Small = Polygonum ramosissimum Michx. subsp. prolificum（Small）Costea et Tardif ■☆

309688 Polygonum ramuliflorum Kitag. = Polygonum argyrocoleum Steud. ex Kunze ■

309689 Polygonum rayi Bab. = Polygonum oxyspermum C. A. Mey. et Bunge ex Ledeb. ■☆

309690 Polygonum regelianum Kom.;来盖蓼■☆

309691 Polygonum renii L. C. Wang;草原蓼;Ren Knotweed ■

309692 Polygonum renii L. C. Wang = Polygonum viviparum L. ■

309693 Polygonum reynoutria Makino var. ellipticum Koidz. = Fallopia forbesii（Hance）Yonek. et H. Ohashi ■

309694 Polygonum rhizoxylon Pau et Font Quer = Polygonum balansae Boiss. subsp. rhizoxylon（Pau et Font Quer）Greuter ■☆

309695 Polygonum rigidulum E. Sheld. = Polygonum amphibium L. var. emersum Michx. ■☆

309696 Polygonum rigidum Skvortsov;尖果蓼（尖果萹蓄）;Tinefruit Knotweed ■

309697 Polygonum robertii Loisel.;罗伯特蓼■☆

309698 Polygonum robustum（Small）Fernald = Persicaria robustior（Small）E. P. Bicknell ■☆

309699 Polygonum robynsii De Wild.;罗宾斯蓼■☆

309700 Polygonum roseoviride（Kitag.）S. X. Li et Y. L. Chang = Polygonum longisetum Bruijn ■

309701 Polygonum roseoviride（Kitag.）S. X. Li et Y. L. Chang var. manshuricola（Kitag.）C. F. Fang = Polygonum longisetum Bruijn ■

309702 Polygonum roxburghii Meisn.;印度蓼■☆

309703 Polygonum roxburghii Meisn. = Polygonum plebeium R. Br. ■

309704 Polygonum roylei Bab.;罗伊尔蓼■☆

309705 Polygonum rubricaule Cham.;红茎蓼■☆

309706 Polygonum rude Meisn. = Polygonum molle D. Don var. rude（Meisn.）A. J. Li ■

309707 Polygonum rude Meisn. var. sikkimensis Hook. f.;酸藤■

309708 Polygonum ruderale Salisb. = Persicaria maculosa Gray ■☆

309709 Polygonum ruderale Salisb. = Polygonum persicaria L. ■

309710 Polygonum rumicifolium Royle ex Bab. var. oblongum Meisn. = Polygonum campanulatum Hook. f. ■

309711 Polygonum runcinatum Buch. -Ham. ex D. Don;羽叶蓼（草见血,赤胫散,飞蛾七,拐枣七,广川草,红皂药,红泽兰,蝴蝶草,花扁担,花蝴蝶,花脸荞,花脸荞麦,花脸晕药,黄泽兰,鸡脚七,加肿草,九龙盘,苦荞头草,南蛇头,荞黄莲,荞子莲,缺腰叶蓼,散血丹,散血莲,蛇头草,蛇头蓼,田牯七,甜荞莲,土竭力,土三七,小晕药,血当归,亚腰山蓼,玉山蓼,皂药根）;Pinnaleaf Knotweed, Runcinate Knotweed ■

309712 Polygonum runcinatum Buch. -Ham. ex D. Don = Persicaria runcinata（Buch. -Ham. ex D. Don）H. Gross ■

309713 Polygonum runcinatum Buch. -Ham. ex D. Don var. corymbosum C. C. Huang;伞房花赤胫散■

309714 Polygonum runcinatum Buch. -Ham. ex D. Don var. exauriculatum Lingelsh. = Polygonum runcinatum Buch. -Ham. ex D. Don var. sinense Hemsl. ■

309715 Polygonum runcinatum Buch. -Ham. ex D. Don var. exauriculatum Lingelsh.;无耳赤胫散■

309716 Polygonum runcinatum Buch. -Ham. ex D. Don var. sinense Hemsl. ; 赤胫散(花蝴蝶,华赤胫散,华缺腰叶蓼,缺腰叶蓼,蛇头蓼,土血竭,血当归);Chinese Runcinate Knotweed,Redish Powder ■

309717 Polygonum rupestre Kar. et Kir. = Polygonum cognatum Meisn. ■

309718 Polygonum rurivagum Jord. ex Boreau = Polygonum aviculare L. subsp. rurivagum (Jord. ex Boreau) Berher ■☆

309719 Polygonum rurivagum Jord. ex Boreau = Polygonum bellardii All. ■☆

309720 Polygonum ryukyuense Kitag. = Polygonum ramosissimum Michx. ■☆

309721 Polygonum sachalinense F. Schmidt = Fallopia sachalinensis (F. Schmidt) Ronse Decr. ■☆

309722 Polygonum sachalinense F. Schmidt = Reynoutria sachalinensis (F. Schmidt) Nakai ■☆

309723 Polygonum sachalinense F. Schmidt ex Maxim. ;库页蓼;Giant Knotweed,Sakhalian Knotweed ■☆

309724 Polygonum sagittatum L. = Persicaria sagitta (L.) H. Gross ■

309725 Polygonum sagittatum L. = Polygonum sieboldii Meisn. ■

309726 Polygonum sagittatum L. subsp. sieboldii (Meisn.) Vorosch. = Persicaria sagittata (L.) H. Gross var. sibirica (Meisn.) Miyabe ■

309727 Polygonum sagittatum L. subsp. sieboldii (Meisn.) Vorosch. = Polygonum sagittatum L. ■

309728 Polygonum sagittatum L. var. aestivum Makino ex Koidz. = Persicaria sagittata (L.) H. Gross var. sibirica (Meisn.) Miyabe f. aestiva (Ohki) H. Hara ■☆

309729 Polygonum sagittatum L. var. boreale Meisn. = Polygonum sagittatum L. ■

309730 Polygonum sagittatum L. var. gracilentum Fernald = Persicaria sagittata (L.) H. Gross ■

309731 Polygonum sagittatum L. var. gracilentum Fernald = Polygonum sagittatum L. ■

309732 Polygonum sagittatum L. var. paludosum Kom. = Polygonum sieboldii Meisn. ■

309733 Polygonum sagittatum L. var. pubescens R. Keller = Persicaria arifolia (L.) Haraldson ■☆

309734 Polygonum sagittatum L. var. pubescens R. Keller = Polygonum arifolium L. ■

309735 Polygonum sagittatum L. var. sibiricum Meisn. = Persicaria sagittata (L.) H. Gross var. sibirica (Meisn.) Miyabe ■

309736 Polygonum sagittatum L. var. sibiricum Meisn. = Polygonum sagittatum L. ■

309737 Polygonum sagittatum L. var. sibiricum Meisn. f. aestivum (Ohki) Murata = Persicaria sagittata (L.) H. Gross var. sibirica (Meisn.) Miyabe f. aestiva (Ohki) H. Hara ■☆

309738 Polygonum sagittatum L. var. sieboldii (Meisn.) Maxim. ex Kom. = Persicaria sagittata (L.) H. Gross var. sibirica (Meisn.) Miyabe ■

309739 Polygonum sagittatum L. var. sieboldii (Meisn.) Maxim. ex Kom. = Polygonum sieboldii Meisn. ■

309740 Polygonum sagittatum L. var. sieboldii (Meisn.) Maxim. ex Kom. = Polygonum sagittatum L. ■

309741 Polygonum sagittatum L. var. ussurense Regel = Polygonum hastatosagittatum Makino ■

309742 Polygonum sagittifolium H. Lév. et Vaniot ;大箭叶蓼(花蓼子草,戟叶扛板归,箭叶蓼,雀翘,蛇子草);Sagitate Knotweed ■

309743 Polygonum sagittifolium H. Lév. et Vaniot = Polygonum darrisii H. Lév. ■

309744 Polygonum salicifolium Brouss. ex Willd. = Persicaria decipiens (R. Br.) K. L. Wilson ■☆

309745 Polygonum salicifolium Willd. var. serrulatum (Lag.) Maire et Weiller = Persicaria salicifolia (Brouss. ex Willd.) Assenov ■☆

309746 Polygonum salinum A. I. Baranov et Skvortsov = Polygonum patulum M. Bieb. ■

309747 Polygonum salinum A. I. Baranov et Skvortsov ex S. X. Li et Y. L. Chang = Polygonum patulum M. Bieb. ■

309748 Polygonum salinum A. I. Baranov et Skvortsov ex S. X. Li et Y. L. Chang;卤金蓼 ■

309749 Polygonum samarense H. Gross;萨马尔蓼 ■☆

309750 Polygonum sambesiacum J. Schust. = Persicaria senegalensis (Meisn.) Soják ■☆

309751 Polygonum scabrum Moench;糙叶蓼(小早苗蓼) ■

309752 Polygonum scabrum Moench = Persicaria lapathifolia (L.) Gray var. tomentosa (Schrank) H. Gross ■

309753 Polygonum scabrum Moench = Persicaria lapathifolia (L.) Gray ■

309754 Polygonum scabrum Moench = Polygonum lapathifolium L. ■

309755 Polygonum scandens L. = Anredera scandens (L.) Moq. ■☆

309756 Polygonum scandens L. = Fallopia scandens (L.) Holub ■☆

309757 Polygonum scandens L. var. cristatum (Engelm. et A. Gray) Gleason;冠状攀缘首乌;Climbing False Buckwheat ■☆

309758 Polygonum scandens L. var. cristatum (Engelm. et A. Gray) Gleason = Fallopia scandens (L.) Holub ■☆

309759 Polygonum scandens L. var. cristatum (Engelm. et A. Gray) Gleason = Polygonum scandens L. ■☆

309760 Polygonum scandens L. var. dentatoalatum (F. Schmidt) Maxim. ex Franch. et Sav. = Fallopia dentatoalata (F. Schmidt ex Maxim.) Holub ■

309761 Polygonum scandens L. var. dumetorum (L.) Gleason = Fallopia dumetora (L.) Holub ■

309762 Polygonum scandens L. var. dumetorum (L.) Gleason = Polygonum scandens L. ■☆

309763 Polygonum schimperi Vatke ex Engl. ;欣珀蓼 ■☆

309764 Polygonum schinzii C. H. Wright = Persicaria limbata (Meisn.) H. Hara ■☆

309765 Polygonum schinzii J. Schust. = Polygonum hydropiper L. ■

309766 Polygonum schischkinii N. A. Ivanova ex Borodina;新疆萹蓄(新疆蓼);Xinjiang Knotweed ●

309767 Polygonum schugnanicum Kom. ;舒格南蓼 ■☆

309768 Polygonum senegalense Meisn. = Persicaria senegalensis (Meisn.) Soják ■☆

309769 Polygonum senegalense Meisn. f. albotomentosum R. A. Graham = Persicaria senegalensis (Meisn.) Soják. f. albotomentosa (R. A. Graham) K. L. Wilson ■☆

309770 Polygonum senegalense Meisn. subsp. albotomentosum (R. A. Graham) Germish. = Persicaria senegalensis (Meisn.) Soják. f. albotomentosa (R. A. Graham) K. L. Wilson ■☆

309771 Polygonum senegalense Meisn. var. numidicum Maire = Persicaria senegalensis (Meisn.) Soják ■☆

309772 Polygonum senticosum (Meisn.) Franch. et Sav. ;刺蓼(红梗豺狼舌头草,红火老鸦酸草,急解索,廊茵,猫儿草,猫儿刺,猫舌草,南蛇草,蛇不钻);Spine Knotweed ■

309773 Polygonum senticosum (Meisn.) Franch. et Sav. = Persicaria senticosa (Meisn.) H. Gross ■

309774 Polygonum senticosum (Meisn.) Franch. et Sav. var. formosanum Ohwi = Polygonum senticosum (Meisn.) Franch. et

Sav. ■

309775　Polygonum senticosum（Meisn.）Franch. et Sav. var. sagittifolium（H. Lév. et Vaniot）C. W. Park ＝Polygonum darrisii H. Lév. ■

309776　Polygonum senticosum（Meisn.）Franch. et Sav. var. sagittifolium C. W. Park ＝Polygonum darrisii H. Lév. ■

309777　Polygonum sericeum Pall. ;西伯利亚绢毛蓼■☆

309778　Polygonum serratum Poir. ＝ Persicaria poiretii（Meisn.）K. L. Wilson ■☆

309779　Polygonum serrulatoides H. Lindb. ＝Persicaria salicifolia（Brouss. ex Willd.）Assenov ■☆

309780　Polygonum serrulatoides H. Lindb. var. pseudohydropiper（Salzm.）H. Lindb. ＝Persicaria salicifolia（Brouss. ex Willd.）Assenov ■☆

309781　Polygonum serrulatum Lag. ＝Persicaria decipiens（R. Br.）K. L. Wilson ■☆

309782　Polygonum serrulatum Lag. var. salicifolium（Willd.）Ball ＝ Persicaria salicifolia（Brouss. ex Willd.）Assenov ■☆

309783　Polygonum setaceum Baldwin ＝Persicaria setacea（Baldwin）Small ■☆

309784　Polygonum setaceum Baldwin ex Elliott ＝ Persicaria setacea（Baldwin）Small ■☆

309785　Polygonum setaceum Baldwin ex Elliott var. interjectum Fernald ＝ Polygonum setaceum Baldwin ex Elliott ■☆

309786　Polygonum setaceum Baldwin var. interjectum Fernald ＝ Persicaria setacea（Baldwin）Small ■☆

309787　Polygonum setaceum Baldwin var. tonsum Fernald ＝ Persicaria setacea（Baldwin）Small ■☆

309788　Polygonum setosulum A. Rich. ＝ Persicaria setosula（A. Rich.）K. L. Wilson ■☆

309789　Polygonum setosum Jacq. ;刚毛蓼■☆

309790　Polygonum shuchengense Z. Z. Zhou;舒城蓼;Shucheng Knotweed ■

309791　Polygonum shuchengense Z. Z. Zhou ＝Polygonum persicaria L. ■

309792　Polygonum sibiricum Laxm. ;西伯利亚蓼(萹蓄,醋柳,剪刀股,曲玛孜,西伯利亚神血宁);Siberia Knotweed,Siberian Knotweed ■

309793　Polygonum sibiricum Laxm. subsp. thomsonii（Meisn.）Rech. et Schiman-Czeika ＝Polygonum sibiricum Laxm. var. thomsonii Meisn. ex Stewart ■

309794　Polygonum sibiricum Laxm. var. nanum Meisn. ＝Polygonum sibiricum Laxm. var. thomsonii Meisn. ex Stewart ■

309795　Polygonum sibiricum Laxm. var. thomsonii Meisn. ex Stewart;细叶西伯利亚蓼(细叶西伯利亚神血宁);Thomson Siberian Knotweed ■

309796　Polygonum sieboldii Meisn. ;箭叶蓼(长野荞麦草,倒刺林,更生,锯锯草,牛角尖,荞麦刺,去母,雀翘,水红骨蛇,小箭叶蓼,走游草);American Tear-thumb, Arrow Vine, Arrowleaf Knotweed, Arrow-leaved Tear-thumb, Arrow-leaved Tearweed, False Buckwheat, Tear Thumb, Tear-thumb ■

309797　Polygonum sieboldii Meisn. ＝Persicaria sagittata（L.）H. Gross var. sibirica（Meisn.）Miyabe ■

309798　Polygonum sieboldii Meisn. ＝Polygonum brevifolia Kitag. ■

309799　Polygonum sieboldii Meisn. ＝Polygonum sagittatum L. ■

309800　Polygonum sieboldii Meisn. var. aestivum（Ohki）Ohwi ＝Persicaria sagittata（L.）H. Gross var. sibirica（Meisn.）Miyabe f. aestiva（Ohki）H. Hara ■☆

309801　Polygonum sieboldii Meisn. var. pratense Y. L. Chang et S. X. Li;草甸箭叶蓼;Meadow Arrowleaf Knotweed ■

309802　Polygonum sieboldii Meisn. var. pratense Y. L. Chang et S. X. Li ＝ Polygonum sagittatum L. ■

309803　Polygonum sieboldii Meisn. var. pratense Y. L. Chang et S. X. Li ＝

309804　Polygonum sieboldii Meisn.

309804　Polygonum sieboldii Meisn. var. sericeum（Nakai）Nakai ex Ohwi ＝ Persicaria sagittata（L.）H. Gross var. sibirica（Meisn.）Miyabe f. tomentosa（H. Hara）H. Hara ■☆

309805　Polygonum sinense J. F. Gmel. ＝Polygonum chinense L. ■

309806　Polygonum sinicum（Migo）D. Fang et L. Zeng ＝ Polygonum thunbergii Siebold et Zucc. ■

309807　Polygonum sinomontanum Sam. ;翅柄蓼(翅柄拳参,滇拳参,石风丹);Webstalk Knotweed,Yunnan Knotweed ■

309808　Polygonum songaricum Schrenk;准噶尔蓼(准噶尔神血宁);Dzungar Knotweed ■

309809　Polygonum spaethii Dammer ＝Polygonum orientale L. ■

309810　Polygonum sparsipilosum A. J. Li;柔毛蓼;Softhair Knotweed ■

309811　Polygonum sparsipilosum A. J. Li var. hubertii（Lingelsh.）A. J. Li;腺点柔毛蓼■

309812　Polygonum speciosum Meisn. ＝Polygonum amplexicaule D. Don ■

309813　Polygonum spergulariiforme Meisn. ex Small;大竹草蓼;Fall Knotweed,Spurry Knotweed ■☆

309814　Polygonum sphaerocephalum Wall. ex Meisn. ＝ Polygonum microcephalum D. Don var. sphaerocephalum（Wall. ex Meisn.）Murata ■

309815　Polygonum sphaerostachyum Meisn. ＝Polygonum macrophyllum D. Don ■

309816　Polygonum sphaerostachyum Meisn. ＝Polygonum milletii（H. Lév.）H. Lév. ■

309817　Polygonum stans（Kitag.）Kitag. ＝Polygonum aviculare L. var. fusco-ochreatum（Kom.）A. J. Li ■

309818　Polygonum stans Kitag. ＝ Polygonum aviculare L. var. fusco-ochreanum（Kom.）A. J. Li ■

309819　Polygonum statice H. Lév. ＝ Fagopyrum statice（H. Lév.）H. Gross ■

309820　Polygonum staticiflorum Wall. ＝ Persicaria microcephala（D. Don）H. Gross ■

309821　Polygonum stellato-tomentosum W. W. Sm. et Ramas. ? ＝ Polygonum thunbergii Siebold et Zucc. ■

309822　Polygonum stenophyllum Meisn. ＝ Polygonum macrophyllum D. Don var. stenophyllum（Meisn.）A. J. Li ■

309823　Polygonum sterile Nakai ＝ Polygonum japonicum Meisn. var. conspicuum Nakai ■

309824　Polygonum sterile Nakai var. brevistylum（Nakai）Nakai ＝ Polygonum japonicum Meisn. var. conspicuum Nakai ■

309825　Polygonum stevensii Brenckle ＝Polygonum ramosissimum Michx. ■☆

309826　Polygonum stoloniferum F. Schmidt ＝ Polygonum thunbergii Siebold et Zucc. ■

309827　Polygonum striatulum B. L. Rob. ;得州蓼;Texas Knotweed ■☆

309828　Polygonum striatulum B. L. Rob. var. texense（M. C. Johnst.）Costea et Tardif ＝Polygonum striatulum B. L. Rob. ■☆

309829　Polygonum strictum Meisn. var. subcontinuum Meisn. ＝ Polygonum persicaria L. ■

309830　Polygonum strigosum L. var. hastatosagittatum（Makino）Steward ＝Polygonum hastatosagittatum Makino ■

309831　Polygonum strigosum R. Br. ;糙毛蓼(刚毛野蓼花,水湿蓼);Hispid Knotweed,Roughhair Knotweed ■

309832　Polygonum strigosum R. Br. ＝Persicaria strigosa（R. Br.）Nakai ■

309833　Polygonum strigosum R. Br. var. muricatum（Meisn.）Stewart ＝Polygonum muricatum Meisn. ■

309834　Polygonum strigosum R. Br. var. muricatum Meisn. ＝ Polygonum muricatum Meisn. ■

309835　Polygonum strigosum R. Br. var. pedunculare (Wall. ex Meisn.) Stewart = Polygonum dichotomum Blume ■

309836　Polygonum strindbergii J. Schust.；平卧蓼(蔓蓼)；Prostrate Knotweed, Strindberg Knotweed ■

309837　Polygonum subauriculatum Petr.；近耳状蓼■☆

309838　Polygonum subsagittatum (De Wild.) Park；亚箭头蓼■☆

309839　Polygonum subscaposum Diels；大理蓼(抽茎拳参, 大理拳参)；Dali Knotweed, Subscapose Knotweed ■

309840　Polygonum suffultoides A. J. Li；珠芽支柱蓼(珠芽支柱拳参)；Bulbil Knotweed ■

309841　Polygonum suffultum Maxim.；支柱蓼(赶山鞭, 红三七, 鸡血七, 九节雷, 九节犁, 九龙盘, 九牛造, 蓼子七, 螺丝七, 螺丝三七, 扭子七, 伞墩七, 算盘七, 蜈蚣七, 血墩七, 血三七, 支柱拳参)；Strut Knotweed ■

309842　Polygonum suffultum Maxim. = Bistorta suffulta (Maxim.) H. Gross ■

309843　Polygonum suffultum Maxim. var. pergracile (Hemsl.) Sam.；细穗支柱蓼(细穗苷三七, 细穗支柱拳参)■

309844　Polygonum sungareense (Kitag.) Kitag. = Polygonum longisetum Bruijn var. rotundatum A. J. Li ■

309845　Polygonum sungareense (Kitag.) Kitag. f. rubiflorum S. X. Li et Y. L. Chang = Polygonum longisetum Bruijn var. rotundatum A. J. Li ■

309846　Polygonum sungareense Maxim. f. rubriflorum S. X. Li et Y. L. Chang；红花被松江蓼■

309847　Polygonum sungareense Maxim. f. rubriflorum S. X. Li et Y. L. Chang = Polygonum longisetum Bruijn var. rotundatum A. J. Li ■

309848　Polygonum taipaishanense H. W. Kung；太白蓼(大红粉)■

309849　Polygonum taipaishanense H. W. Kung = Polygonum milletii (H. Lév.) H. Lév. ■

309850　Polygonum tairae Ohwi = Persicaria dichotoma (Blume) Masam. ■

309851　Polygonum taliense Lingelsh. = Polygonum subscaposum Diels ■

309852　Polygonum tanganikae J. Schust. = Persicaria senegalensis (Meisn.) Soják ■☆

309853　Polygonum taquetii H. Lév.；细叶蓼；Thinleaf Knotweed ■

309854　Polygonum taquetii H. Lév. = Persicaria taquetii (H. Lév.) Koidz. ■

309855　Polygonum taquetii H. Lév. var. minutulum (Makino) Ohwi = Persicaria taquetii (H. Lév.) Koidz. ■

309856　Polygonum tataricum L. = Fagopyrum tataricum (L.) Gaertn. ■

309857　Polygonum tatewakianum Koji Ito；馆肋蓼■☆

309858　Polygonum tatewakianum Koji Ito var. notoroense Koji Ito；东方馆肋蓼；Northern Knot-grass ■☆

309859　Polygonum tenellum Blume；纤细小蓼；Small Knotweed ■

309860　Polygonum tenellum Blume = Persicaria tenella (Blume) H. Hara ■

309861　Polygonum tenellum Blume var. kawagoeanum (Makino) Murata = Persicaria tenella (Blume) H. Hara var. kawagoeana (Makino) H. Hara ■

309862　Polygonum tenellum Blume var. kawagoeanum (Makino) Murata = Polygonum tenellum Blume var. micranthum (Boiss. ex Meisn.) C. Y. Wu ■

309863　Polygonum tenellum Blume var. micranthum (Boiss. ex Meisn.) C. Y. Wu；柔茎蓼(盘腺蓼)■

309864　Polygonum tenellum Blume var. micranthum (Boiss. ex Meisn.) C. Y. Wu = Persicaria tenella (Blume) H. Hara var. kawagoeana (Makino) H. Hara ■

309865　Polygonum tenellum Blume var. micranthum (Meisn.) C. Y. Wu = Polygonum kawagoeanum Makino ■

309866　Polygonum tenue Michx.；纤细蓼；Pleat-leaf Knotweed, Slender Knotweed ■☆

309867　Polygonum tenue Michx. var. commune Engelm. = Polygonum douglasii Greene ■☆

309868　Polygonum tenue Michx. var. latifolium Engelm. = Polygonum douglasii Greene ■☆

309869　Polygonum tenue Michx. var. protrusum Fernald = Polygonum tenue Michx. ■☆

309870　Polygonum tenuicaule Bisset et S. Moore；日本紫参(细茎蓼, 紫参)■☆

309871　Polygonum tenuicaule Bisset et S. Moore = Bistorta tenuicaulis (Bisset et S. Moore) Nakai ■☆

309872　Polygonum tenuifolium H. W. Kung = Polygonum viviparum L. var. angustum A. J. Li ■

309873　Polygonum tenuifolium H. W. Kung = Polygonum viviparum L. var. tenuifolium (H. W. Kung) Y. L. Liu ■

309874　Polygonum terrestre (Willd.) Britton, Sterns et Poggenb. = Polygonum amphibium L. var. emersum Michx. ■☆

309875　Polygonum tetragonum Blume = Polygonum dichotomum Blume ■

309876　Polygonum texense M. C. Johnst. = Polygonum striatulum B. L. Rob. ■☆

309877　Polygonum thunbergii Siebold et Zucc.；戟叶蓼(藏氏蓼, 鹿蹄草, 水蝴蝶, 水犁避, 水麻, 水麻蓼, 水麻芀, 小青草, 野荞麦)；Halberd-leaved Tear Thumb, Halberd-leaved Tearthumb, Halberd-leaved Tear-thumb, Harbertleaf Knotweed, Hastate Knotgrass, Sickle Grass, Thunberg Knotweed ■

309878　Polygonum thunbergii Siebold et Zucc. = Persicaria thunbergii (Siebold et Zucc.) H. Gross ■

309879　Polygonum thunbergii Siebold et Zucc. f. biconvexum (Hayata) Tang S. Liu, S. S. Ying et M. J. Lai = Polygonum biconvexum Hayata ■

309880　Polygonum thunbergii Siebold et Zucc. f. biconvexum (Hayata) Tang S. Liu, S. S. Ying et M. J. Lai = Polygonum thunbergii Siebold et Zucc. ■

309881　Polygonum thunbergii Siebold et Zucc. var. hastatotrilobum subvar. eciliolatum H. Lév. = Polygonum praetermissum Hook. f. ■

309882　Polygonum thunbergii Siebold et Zucc. var. maackianum (Regel) Maxim. ex Franch. et Sav. = Persicaria maackiana (Regel) Nakai ■

309883　Polygonum thunbergii Siebold et Zucc. var. maackianum (Regel) Maxim. ex Franch. et Sav. = Polygonum maackianum Regel ■

309884　Polygonum thunbergii Siebold et Zucc. var. oreophilum Makino = Persicaria oreophila (Makino) Hiyama ■☆

309885　Polygonum thunbergii Siebold et Zucc. var. spicatum H. Lév. = Polygonum muricatum Meisn. ■

309886　Polygonum thunbergii Siebold et Zucc. var. stoloniferum (F. Schmidt) Makino = Polygonum thunbergii Siebold et Zucc. ■

309887　Polygonum thunbergii Siebold et Zucc. var. stoloniferum (F. Schmidt) Makino = Persicaria thunbergii (Siebold et Zucc.) H. Gross var. stolonifera (F. Schmidt) Nakai ex H. Hara ■

309888　Polygonum thunbergii Siebold et Zucc. var. stoloniferum (F. Schmidt) Makino；沟荞麦■

309889　Polygonum thymifolium Jaub. et Spach；百里香叶蓼(天山蓼)；Tianshan Knotweed ●

309890　Polygonum tianschanicum C. Y. Yang = Polygonum thymifolium Jaub. et Spach ●

309891　Polygonum tibeticum Hemsl. ;西藏蓼（西藏神血宁）；Tibet Knotweed,Xizang Knotweed ●■

309892　Polygonum tiflisiense Kom. ;梯弗里斯蓼■☆

309893　Polygonum tinctorium Aiton;蓼蓝（大青叶,靛,蓝,蓝靛,蓝蓼,蓝实,蓝子,蓼蓝青黛,青板水辣蓼,青黛,小蓝）；Chinese Indigo,Dyer's Knotgrass,Knotweed Blue ■

309894　Polygonum tinctorium Aiton = Persicaria tinctoria（Aiton）Spach ■

309895　Polygonum tinctorium Lour. = Polygonum tinctorium Aiton ■

309896　Polygonum tomentosum Schrank = Persicaria lapathifolia（L.）Gray ■

309897　Polygonum tomentosum Schrank = Polygonum lapathifolium L. ■

309898　Polygonum tomentosum Willd. ;绒毛蓼■

309899　Polygonum tomentosum Willd. = Persicaria attenuata（R. Br.）Soják subsp. pulchra（Blume）K. L. Wilson ■

309900　Polygonum tomentosum Willd. = Persicaria lapathifolia（L.）Gray ■

309901　Polygonum tomentosum Willd. = Polygonum lapathifolium L. ■

309902　Polygonum tomentosum Willd. = Polygonum pulchrum Blume ■

309903　Polygonum torquatum Bruijn = Polygonum orientale L. ■

309904　Polygonum torreyi S. Watson = Polygonum minimum S. Watson ■☆

309905　Polygonum tortuosum（Losinsk.）Lovelius = Polygonum intramongolicum Borodina ●

309906　Polygonum tortuosum D. Don;叉枝蓼（叉枝神血宁,扭转蓼）；Forky Knotweed ■

309907　Polygonum tortuosum D. Don var. tibetanum Meisn. = Polygonum tortuosum D. Don ■

309908　Polygonum triangulum E. P. Bicknell = Polygonum ramosissimum Michx. ■☆

309909　Polygonum trigonocarpum（Makino）Kudo et Masam. ;楔叶蓼（细叶蓼,细叶犬蓼）；Cuneateleaf Knotweed ■

309910　Polygonum trigonocarpum（Makino）Kudo et Masam. = Persicaria erectominor（Makino）Nakai var. trigonocarpa（Makino）H. Hara ■☆

309911　Polygonum tripterocarpum A. Gray;三翅果蓼■☆

309912　Polygonum tristachyum H. Lév. = Fagopyrum dibotrys（D. Don）H. Hara ■

309913　Polygonum tsangschanicum Lingelsh. et Borza = Polygonum molle D. Don var. rude（Meisn.）A. J. Li ■

309914　Polygonum tumidum Delile;肿胀蓼■☆

309915　Polygonum typhoniifolium Hance = Persicaria senticosa（Meisn.）H. Gross ■

309916　Polygonum typhoniifolium Hance = Polygonum senticosum（Meisn.）Franch. et Sav. ■

309917　Polygonum umbellatum（Houtt.）Koidz. = Persicaria chinensis（L.）H. Gross ■

309918　Polygonum umbrosum Sam. ;荫地蓼（林荫蓼）；Shade Knotweed ■

309919　Polygonum undulatum（L.）P. J. Bergius;波蓼■☆

309920　Polygonum undulatum Murray = Polygonum alpinum All. ■

309921　Polygonum uniflorum Y. X. Ma et Y. T. Zhao;丹花蓼■

309922　Polygonum uniflorum Y. X. Ma et Y. T. Zhao = Polygonum muricatum Meisn. ■

309923　Polygonum unifolium Small ex Rydb. = Polygonum polygaloides Meisn. subsp. kelloggii（Greene）J. C. Hickman ■☆

309924　Polygonum urophyllum Bureau et Franch. = Fagopyrum urophyllum（Bureau et Franch.）H. Gross ●

309925　Polygonum ussuriense（Regel）Nakai = Polygonum hastatosagittatum Makino ■

309926　Polygonum ussuriense（Regel）Nakai ex Mori;乌苏里蓼■☆

309927　Polygonum ussuriense Petr. ex Kom. = Persicaria praetermissa（Hook. f.）H. Hara ■

309928　Polygonum ussuriense Petr. ex Kom. = Polygonum hastatoauriculatum Makino ex Nakai ■

309929　Polygonum ussuriense Petr. ex Kom. var. baischanense Y. L. Chang et S. X. Li;细叶乌苏里蓼■

309930　Polygonum utahense Brenckle et Cottam;犹他蓼；Utah Knotweed ■☆

309931　Polygonum utriculosum Tikovsky = Persicaria hystricula（J. Schust.）Soják ■☆

309932　Polygonum uvifera L. = Coccoloba uvifera（L.）L. ●

309933　Polygonum vaccinifolium Wall. ex Meisn. ;乌饭树叶蓼；Blueberryleaf Knotweed ■

309934　Polygonum vaniotianum（H. Lév.）H. Lév. = Polygonum lapathifolium L. ■

309935　Polygonum venosum Steward;多脉蓼■☆

309936　Polygonum virgatum Loisel. = Polygonum bellardii All. ■☆

309937　Polygonum virginatum L. = Antenoron filiforme（Thunb.）Rob. et Vautier ■

309938　Polygonum virginatum L. var. filiforme（Thunb. ex A. Murray）Nakai = Antenoron filiforme（Thunb.）Rob. et Vautier ■

309939　Polygonum virginatum L. var. filiforme（Thunb.）Nakai = Antenoron filiforme（Thunb.）Rob. et Vautier ■

309940　Polygonum virginianum L. ;弗州蓼（人字草）；Bohemian Knotweed, Hybrid Knotweed, Jumpseed, Virginia Knotweed, Woodland Knotweed ■☆

309941　Polygonum virginianum L. = Persicaria virginiana（L.）Gaertn. ■☆

309942　Polygonum virginianum L. f. glabratum Matsuda = Antenoron filiforme（Thunb.）Rob. et Vautier ■

309943　Polygonum virginianum L. var. filiforme（Thunb. ex A. Murray）Nakai = Persicaria filiformis（Thunb.）Nakai ex W. T. Lee ■☆

309944　Polygonum virginianum L. var. filiforme（Thunb.）Nakai = Antenoron filiforme（Thunb.）Rob. et Vautier ■

309945　Polygonum virginianum L. var. glaberrimum（Fernald）Steyerm. = Polygonum virginianum L. ■☆

309946　Polygonum viscoferum Makino;黏蓼（香蓼,黏毛蓼,中轴蓼）；Viscidity Knotweed ■

309947　Polygonum viscoferum Makino = Persicaria viscofera（Makino）H. Gross ■

309948　Polygonum viscoferum Makino subsp. robustum（Makino）Kitam. = Persicaria viscofera（Makino）H. Gross var. robusta（Makino）Hiyama ■

309949　Polygonum viscoferum Makino var. robustum Makino = Persicaria viscofera（Makino）H. Gross var. robusta（Makino）Hiyama ■

309950　Polygonum viscoferum Makino var. robustum Makino = Polygonum viscoferum Makino ■

309951　Polygonum viscosum Buch. -Ham. ex D. Don;香蓼（黏毛蓼）；Aromatic Knotweed ■

309952　Polygonum viscosum Buch. -Ham. ex D. Don = Persicaria viscosa（Buch. -Ham. ex D. Don）H. Gross ex T. Mori ■

309953　Polygonum viscosum Buch. -Ham. ex D. Don var. minus Hook. f. = Polygonum viscosum Buch. -Ham. ex D. Don ■

309954　Polygonum viviparum L. ;珠芽蓼（地黑蜂,核子七,红粉,红三

七,红蝎子七,猴儿七,猴娃七,猴子七,剪刀七,然波,染布子,山高粱,山谷子,蛇疙瘩,蝎子七,野高粱,一口血草,珠芽拳参);Alpine Bistort, Alpine Knotweed, Bulbil Knotweed, Viviparous Bistort, Viviparous Bistort Serpentgrass ■

309955　Polygonum viviparum L. = Bistorta vivipara（L.）Delarbre ■

309956　Polygonum viviparum L. = Persicaria vivipara（L.）Ronse Decr. ■

309957　Polygonum viviparum L. var. angustum A. J. Li = Polygonum viviparum L. var. tenuifolium（H. W. Kung）Y. L. Liu ■

309958　Polygonum viviparum L. var. tenuifolium（H. W. Kung）Y. L. Liu;细叶珠芽蓼(细叶珠芽拳参)■

309959　Polygonum viviparum L. var. tenuifolium（H. W. Kung）Y. L. Liu = Polygonum viviparum L. var. angustum A. J. Li ■

309960　Polygonum viviparum L. var. tenuifolium Y. L. Liu = Polygonum viviparum L. var. tenuifolium（H. W. Kung）Y. L. Liu ■

309961　Polygonum volubile Turcz. = Fagopyrum dibotrys（D. Don）H. Hara ■

309962　Polygonum wallichii Meisn. ;球序蓼■

309963　Polygonum wallichii Meisn. = Polygonum polystachyum Wall. ex Meisn. ●■

309964　Polygonum watsonii Small = Polygonum polygaloides Meisn. subsp. confertiflorum（Nutt. ex Piper）J. C. Hickman ■☆

309965　Polygonum wellensii De Wild. ;韦伦斯蓼■☆

309966　Polygonum weyrichii F. Schmidt;里白蓼;Chinese Knotweed ■☆

309967　Polygonum weyrichii F. Schmidt = Aconogonon weyrichii（F. Schmidt）H. Hara ■☆

309968　Polygonum weyrichii F. Schmidt = Persicaria weyrichii（F. Schmidt）H. Gross ■☆

309969　Polygonum weyrichii F. Schmidt var. alpinum Maxim. ;高山里白蓼■☆

309970　Polygonum weyrichii F. Schmidt var. alpinum Maxim. = Aconogonon weyrichii（F. Schmidt）H. Hara var. alpinum（Maxim.）H. Hara ■☆

309971　Polygonum yamatutae Kitag. = Polygonum humifusum Merker ex C. Koch ■

309972　Polygonum yokusaianum Makino = Persicaria posumbu（Buch.-Ham. ex D. Don）H. Gross ■

309973　Polygonum yokusaianum Makino = Polygonum posumbu Buch.-Ham. ex D. Don ■

309974　Polygonum yokusaianum Makino var. stenophyllum Makino = Persicaria posumbu（Buch.-Ham. ex D. Don）H. Gross var. stenophylla（Makino）Yonek. et H. Ohashi ■☆

309975　Polygonum yunnanense（H. Gross）H. Lév. = Polygonum paleaceum Wall. ex Hook. f. ■

309976　Polygonum yunnanense H. Lév. = Fallopia forbesii（Hance）Yonek. et H. Ohashi ■

309977　Polygonum yunnanense H. Lév. = Reynoutria japonica Houtt. ■

309978　Polygonum zeaense Kom. ;泽蓼■☆

309979　Polygonum zigzag H. Lév. et Vaniot = Polygonum emodii Meisn. var. dependens Diels ■

309980　Polygonum zuccarinii Small = Fallopia japonica（Houtt.）Ronse Decr. ■

309981　Polygyne Phil. = Eclipta L.（保留属名）■

309982　Polylepis Ruiz et Pav.（1794）;多鳞木属(普利勒木属)●☆

309983　Polylepis incana Kunth;灰白多鳞木●☆

309984　Polylepis lanuginosa Kunth;绵毛多鳞木●☆

309985　Polylepis microphylla（Wedd.）Bitter;小叶多鳞木●☆

309986　Polylepis reticulata Kunth;网脉多鳞木●☆

309987　Polylepis sericea Wedd. ;绢毛多鳞木●☆

309988　Polylepis tomentella Wedd. ;毛多鳞木(普利勒木);Quenoa ●☆

309989　Polylepis weberbaueri Pilg. ;韦伯多鳞木●☆

309990　Polylobium Eckl. et Zeyh. = Lotononis（DC.）Eckl. et Zeyh.（保留属名）■

309991　Polylobium angustifolium Eckl. et Zeyh. = Lotononis involucrata（P. J. Bergius）Benth. subsp. peduncularis（E. Mey.）B.-E. van Wyk ■☆

309992　Polylobium brachylobum（E. Mey.）Eckl. et Zeyh. = Lotononis parviflora（P. J. Bergius）D. Dietr. ■☆

309993　Polylobium calycinum（E. Mey.）Benth. = Lotononis calycina（E. Mey.）Benth. ●☆

309994　Polylobium carinatum（E. Mey.）Benth. = Lotononis carinata（E. Mey.）Benth. ■☆

309995　Polylobium corymbosum（E. Mey.）Benth. = Lotononis corymbosa（E. Mey.）Benth. ●☆

309996　Polylobium debile Eckl. et Zeyh. = Lotononis umbellata（L.）Benth. ●☆

309997　Polylobium erubescens（E. Mey.）D. Dietr. = Lotononis pumila Eckl. et Zeyh. ●☆

309998　Polylobium falcatum（E. Mey.）D. Dietr. = Lotononis falcata（E. Mey.）Benth. ●☆

309999　Polylobium fastigiatum（E. Mey.）Eckl. et Zeyh. = Lotononis fastigiata（E. Mey.）B.-E. van Wyk ●☆

310000　Polylobium filiforme Eckl. et Zeyh. = Lotononis umbellata（L.）Benth. ●☆

310001　Polylobium molle（E. Mey.）D. Dietr. = Lotononis mollis（E. Mey.）Benth. ●■☆

310002　Polylobium pallens Eckl. et Zeyh. = Lotononis pallens（Eckl. et Zeyh.）Benth. ●☆

310003　Polylobium sparsiflorum Eckl. et Zeyh. = Lotononis oxyptera（E. Mey.）Benth. ●☆

310004　Polylobium tenellum（E. Mey.）D. Dietr. = Lotononis tenella（E. Mey.）Eckl. et Zeyh. ●☆

310005　Polylobium truncatum（E. Mey.）Eckl. et Zeyh. = Lotononis umbellata（L.）Benth. ●☆

310006　Polylobium umbellatum（L.）Benth. = Lotononis umbellata（L.）Benth. ●☆

310007　Polylophium Boiss.（1844）;多脊草属☆

310008　Polylophium panjutinii Manden et Schischk. ;多脊草☆

310009　Polylychnis Bremek.（1938）;多花爵床属☆

310010　Polylychnis essequibensis Bremek. ;多花爵床☆

310011　Polymeria R. Br.（1810）;澳新旋花属☆

310012　Polymeria angusta F. Muell. ;窄澳新旋花☆

310013　Polymeria longifolia Lindl. ;长叶澳新旋花☆

310014　Polymeria mollis（Benth.）Domin. ;软澳新旋花☆

310015　Polymeria occidentalis F. Muell. ;西方澳新旋花☆

310016　Polymita N. E. Br.（1930）;白玲玉属●☆

310017　Polymita albiflora（L. Bolus）L. Bolus;白花白玲玉■☆

310018　Polymita diutina（L. Bolus）L. Bolus = Polymita albiflora（L. Bolus）L. Bolus ■☆

310019　Polymita pearsonii N. E. Br. = Polymita albiflora（L. Bolus）L. Bolus ■☆

310020　Polymita steenbokensis H. E. K. Hartmann;斯滕白玲玉■☆

310021　Polymnia L.（1753）;杯苞菊属(杯叶菊属);Leaf Cup, Leafcup ■●☆

310022　Polymnia abyssinica L. f. = Guizotia abyssinica（L. f.）Cass. ■☆

310023　Polymnia canadensis L.；加拿大杯苞菊；Pale-flowered Leaf-cup，Small-flowered Leafcup，White-flowered Leaf-cup ■☆

310024　Polymnia canadensis L. f. radiata（A. Gray）Fassett = Polymnia canadensis L. ■☆

310025　Polymnia canadensis L. var. radiata A. Gray = Polymnia canadensis L.；■☆

310026　Polymnia carnosa L. f. = Didelta carnosa（L. f.）W. T. Aiton ■☆

310027　Polymnia caroliniana Poir. = Berlandiera pumila（Michx.）Nutt. ■☆

310028　Polymnia cossatotensis Pittman et V. M. Bates；克撒特杯苞菊；Cossatot Mountain Leafcup ■☆

310029　Polymnia edulis Wedd.；可食杯苞菊；Yacon Strawberry ■☆

310030　Polymnia laevigata Beadle；平滑杯苞菊；Leaf Cup ■☆

310031　Polymnia radiata（A. Gray）Small = Polymnia canadensis L. ■☆

310032　Polymnia sonchifolia Poepp. et Endl.；苦苣菜叶杯苞菊■☆

310033　Polymnia sonchifolia Poepp. et Endl. = Smallanthus sonchifolius（Poeppig et Endl.）H. Rob. ■

310034　Polymnia spinosa L. f. = Didelta spinosa（L. f.）W. T. Aiton ■☆

310035　Polymnia uvedalia（L.）L.；熊脚杯苞菊；Large-flowered Leafcup ■☆

310036　Polymnia uvedalia（L.）L. = Osteospermum uvedalia L. ■☆

310037　Polymnia uvedalia（L.）L. = Smallanthus uvedalius（L.）Mack. ex Small ■

310038　Polymnia uvedalia（L.）L. var. floridana S. F. Blake = Osteospermum uvedalia L. ■☆

310039　Polymniastrum Lam. = Polymnia L. ●● ☆

310040　Polymniastrum Small = Smallanthus Mack. ex Small ■●

310041　Polymorpha Fabr. = Salvia L. ●■

310042　Polyneura Peter = Panicum L. ■

310043　Polyneura squarrosa Peter = Panicum peteri Pilg. ■☆

310044　Polynome Salisb. = Dioscorea L.（保留属名）■

310045　Polyochnella Tiegh. = Ochna L. ■

310046　Polyochnella barteri Tiegh. = Ochna afzelii R. Br. ex Oliv. ●☆

310047　Polyochnella buchneri Tiegh. = Ochna afzelii R. Br. ex Oliv. ●☆

310048　Polyochnella congoensis Tiegh. = Ochna afzelii R. Br. ex Oliv. subsp. congoensis（Tiegh.）N. Robson ●☆

310049　Polyochnella fruticulosa（Gilg）Tiegh. = Ochna leptoclada Oliv. ●☆

310050　Polyochnella gilletiana（Gilg）Tiegh. = Ochna afzelii R. Br. ex Oliv. ●☆

310051　Polyochnella gracilipes（Hiern）Tiegh. = Ochna pygmaea Hiern ●☆

310052　Polyochnella hylophila（Gilg）Tiegh. = Ochna polyneura Gilg ●☆

310053　Polyochnella micrantha（Schweinf. et Gilg）Tiegh. = Ochna micrantha Schweinf. et Gilg ●☆

310054　Polyochnella polyneura（Gilg）Tiegh. = Ochna polyneura Gilg ●☆

310055　Polyochnella welwitschii（Rolfe）Tiegh. = Ochna afzelii R. Br. ex Oliv. subsp. mechowiana（O. Hoffm.）N. Robson ●☆

310056　Polyodon Kunth = Bouteloua Lag.（保留属名）■

310057　Polyodontia Meisn. = Lauro-Cerasus Duhamel ●

310058　Polyodontia Meisn. = Polydontia Blume ●

310059　Polyodontia Meisn. = Pygeum Gaertn. ●

310060　Polyosma Blume（1826）；多香木属；Polyosma ●

310061　Polyosma cambodiana Gagnep.；多香木；Cambodia Polyosma，Cambodian Polyosma ●

310062　Polyosma integrifolia Blume；全缘多香木●☆

310063　Polyosmaceae Blume = Escalloniaceae R. Br. ex Dumort.（保留科名）●

310064　Polyosmaceae Blume = Polyosmataceae Blume ●

310065　Polyosmataceae Blume；多香木科●

310066　Polyosus Lour. = Canthium Lam. + Psychotria L.（保留属名）●

310067　Polyosus Lour. = Polyozus Lour. ●

310068　Polyothyris Koord. = Polyothyrsis Koord. ●★

310069　Polyothyrsis Koord. = Poliothyrsis Oliv. ●★

310070　Polyotidium Garay（1958）；多耳兰属■☆

310071　Polyotidium huebneri（Mansf.）Garay；多耳兰；■☆

310072　Polyotus Nutt. = Asclepias L. ■

310073　Polyouratea Tiegh. = Ouratea Aubl.（保留属名）●

310074　Polyozus Blume = Psychotria L.（保留属名）●

310075　Polyozus Lour. = Canthium Lam. + Psychotria L.（保留属名）●

310076　Polyozus Lour. = Psychotria L.（保留属名）●

310077　Polypappus Less. = Baccharis L.（保留属名）●■☆

310078　Polypappus Nutt. = Tessaria Ruiz et Pav. ●☆

310079　Polypappus sericeus Nutt. = Pluchea sericea（Nutt.）Coville ■☆

310080　Polypara Lour. = Houttuynia Thunb. ■

310081　Polypara cochinchinensis Lour. = Houttuynia cordata Thunb. ■

310082　Polypara cordata Kuntze = Houttuynia cordata Thunb. ■

310083　Polypetalia Hort. = Prunus L. ●

310084　Polyphema Lour. = Artocarpus J. R. Forst. et G. Forst.（保留属名）●

310085　Polyphragmon Desf.（废弃属名）= Timonius DC.（保留属名）●

310086　Polyplethia（Griff.）Tiegh. = Balanophora J. R. Forst. et G. Forst. ■

310087　Polyplethia Tiegh. = Balanophora J. R. Forst. et G. Forst. ■

310088　Polyplethia Tiegh. = Polyplethia（Griff.）Tiegh. ■

310089　Polyplethia kainantensis Yamam. = Balanophora kainantensis Masam. ■

310090　Polyplethia polyaandra（Griff.）Tiegh. = Balanophora polyandra Griff. ■

310091　Polyplethia spicata（Hayata）Nakai = Balanophora laxiflora Hemsl. ■

310092　Polyplethia spicata（Hayata）Nakai = Balanophora spicata Hayata ■

310093　Polypleurella Engl.（1927）；小多脉川苔草属■☆

310094　Polypleurella schmidtiana Engl.；小多脉川苔草■☆

310095　Polypleurum（Taylor ex Tul.）Warm.（1901）；多脉川苔草属■☆

310096　Polypleurum（Tul.）Warm. = Polypleurum（Taylor ex Tul.）Warm. ■☆

310097　Polypleurum acuminatum Warm.；渐尖多脉川苔草■☆

310098　Polypleurum filifolium（Ramam. et J. Joseph）A. S. Rao et Hajra；线形多脉川苔草■☆

310099　Polypleurum longifolium M. Kato；长叶多脉川苔草■☆

310100　Polypleurum orientale Tayl. ex Tul.；东方多脉川苔草■☆

310101　Polypleurum submersum J. B. Hall = Saxicolella submersa（J. B. Hall）C. D. K. Cook et Rutish. ■☆

310102　Polypogon Desf.（1798）；棒头草属；Beard Grass，Beardgrass，Polypogon ■

310103　Polypogon australis Brongn.；澳洲棒头草；Chilean Rabbitsfoot Grass ■☆

310104　Polypogon brachyphyllus E. Fourn. ex Hemsl.；短叶棒头草■☆

310105　Polypogon canadensis E. Fourn.；加拿大棒头草■☆

310106　Polypogon chilensis Pilg.；智利棒头草■☆

310107　Polypogon demissus Steud. = Polypogon fugax Ness ex Steud. ■

310108　Polypogon elongatus Kunth；长棒头草；Elangate Polypogon ■☆

310109　Polypogon fugax Ness ex Steud.；棒头草（沟渠棒头草，麦毛草）；Asia Minor Bluegrass，Ditch Polypogon，Shortlived Beardgrass ■

310110 Polypogon griquensis (Stapf) Gibbs Russ. et L. Fish; 格里夸棒头草■☆

310111 Polypogon higegaweri Steud. = Polypogon fugax Ness ex Steud. ■

310112 Polypogon hissaricus (Roshev.) Bor; 糙毛棒头草■

310113 Polypogon interruptus Kunth; 间断棒头草; Ditch Polypogon ■☆

310114 Polypogon ivanovae Tzvelev; 昆下棒头草(伊凡棒头草)■

310115 Polypogon litoralis (With.) Sm. = Polypogon fugax Ness ex Steud. ■

310116 Polypogon litoralis (With.) Sm. var. hagegaweri (Steud.) Hook. f. = Polypogon fugax Ness ex Steud. ■

310117 Polypogon litoralis Sm. = Polypogon fugax Ness ex Steud. ■

310118 Polypogon littoralis (With.) Sm. var. muticus Hook. f. = Agrostis viridis Gouan ■

310119 Polypogon littoralis (With.) Sm. var. muticus Hook. f. = Polypogon viridis (Gouan) Breistr. ■

310120 Polypogon lutosus (Poir.) Hitchc. = Polypogon fugax Ness ex Steud. ■

310121 Polypogon maritimus Willd.; 裂颖棒头草(地中海棒头草); Mediterranean Polypogon, Mediterranean Rabbitsfoot Grass, Seashore Beardgrass ■

310122 Polypogon maritimus Willd. subsp. subspathaceus (Req.) K. Richt.; 佛焰苞棒头草■☆

310123 Polypogon melillensis Sennen = Polypogon monspeliensis (L.) Desf. ■

310124 Polypogon minutiflorus Pilg.; 微花棒头草■☆

310125 Polypogon monspeliensis (L.) Desf.; 长芒棒头草; Annual Rabbitsfoot Grass, Annual Rabbit's-foot Grass, Beard Grass, Beardgrass, Rabbitfoot Beardgrass, Rabbitfoot Grass, Rabbit-foot Grass, Rabbitfoot Polypogon ■

310126 Polypogon monspeliensis (L.) Desf. var. maritimus (Willd.) Coss. et Durieu = Polypogon maritimus Willd. ■

310127 Polypogon monspeliensis (L.) Desf. var. minor Coss. et Durieu = Polypogon maritimus Willd. ■

310128 Polypogon schimperianus (Hochst. ex Steud.) Cope; 欣珀棒头草■☆

310129 Polypogon semiverticillatus (Forssk.) Hyl.; 半轮棒头草■☆

310130 Polypogon semiverticillatus (Forssk.) Hyl. = Polypogon viridis (Gouan) Breistr. ■

310131 Polypogon strictus Nees; 刚直棒头草■☆

310132 Polypogon subspathaceus Req. = Polypogon maritimus Willd. subsp. subspathaceus (Req.) K. Richt. ■☆

310133 Polypogon tenuis Brongn.; 细棒头草■☆

310134 Polypogon viridis (Gouan) Breistr.; 苔绿棒头草(绿棒头草); Beardless Rabbitsfoot Grass, Beardless Rabbit's-foot Grass, Water Bentgrass ■

310135 Polypogon viridis (Gouan) Breistr. = Agrostis viridis Gouan ■

310136 Polypompholyx Lehm. (1844)(保留属名); 四萼狸藻属■

310137 Polypompholyx Lehm. (保留属名) = Utricularia L. ■

310138 Polypompholyx endlicheri Lehm.; 四萼狸藻■☆

310139 Polypompholyx madecassa H. Perrier = Utricularia scandens Benj. ■

310140 Polyporandra Becc. (1877); 多孔蕊茶萸属●☆

310141 Polyporandra scandens Becc.; 多孔蕊茶萸●☆

310142 Polypremaceae L. Watson ex Doweld et Reveal = Labiatae Juss. (保留科名)●■

310143 Polypremaceae L. Watson ex Doweld et Reveal = Lamiaceae Martinov(保留科名)●■

310144 Polypremaceae L. Watson ex Doweld et Reveal = Tetrachondraceae Skottsb. ex R. W. Sanders et P. D. Cantino ■☆

310145 Polypremum Adans. = Valerianella Mill. ■

310146 Polypremum L. (1763); 美洲四粉草属■☆

310147 Polypremum procumbens L.; 美洲四粉草■☆

310148 Polypsecadium O. E. Schulz(1924); 多碎片芥属■☆

310149 Polypsecadium brasiliense (O. E. Schulz) Al-Shehbaz; 巴西多碎片芥■☆

310150 Polypteris Nutt. = Palafoxia Lag. ■☆

310151 Polypteris integrifolia Nutt. = Palafoxia integrifolia (Nutt.) Torr. et A. Gray ■☆

310152 Polyradicion Garay = Dendrophylax Rchb. f. ■☆

310153 Polyradicion Garay(1969); 多根兰属■☆

310154 Polyradicion lindenii (Lindl.) Garay = Dendrophylax lindenii (Lindl.) Benth. ex Rolfe ■☆

310155 Polyraphis (Trin.) Lindl. = Pappophorum Schreb. ■☆

310156 Polyrhabda C. C. Towns. (1984); 单花苋属●☆

310157 Polyrhabda atriplicifolia C. C. Towns.; 单花苋■☆

310158 Polyrhaphis Lindl. = Pappophorum Schreb. ■☆

310159 Polyrrhiza Pfitzer = Dendrophylax Rchb. f. + Polyradicion Garay ■☆

310160 Polyrrhiza Pfitzer = Dendrophylax Rchb. f. ■☆

310161 Polyrrhiza Pfitzer = Polyradicion Garay ■☆

310162 Polyrrhiza lindenii (Lindl.) Cogn. = Dendrophylax lindenii (Lindl.) Benth. ex Rolfe ■☆

310163 Polyscalia Wall. = Cyathula Blume(保留属名)■

310164 Polyscelis Hook. f. = Polyscalia Wall. ■

310165 Polyschemone Schott, Nyman et Kotschy = Silene L. (保留属名)■

310166 Polyschisma Turcz. = Pelargonium L'Hér. ex Aiton ■■

310167 Polyschistis C. Presl = Pentarrhaphis Kunth ■☆

310168 Polyschistis J. Presl et C. Presl = Pentarrhaphis Kunth ■☆

310169 Polyscias J. R. Forst. et G. Forst. (1775); 南洋参属(福禄桐属, 南洋森属); Polyscias, Umbrella Tree ●

310170 Polyscias aculeata Lowry et G. Plunkett; 皮刺南洋参●☆

310171 Polyscias albersiana Harms; 阿伯斯南洋参●☆

310172 Polyscias amplifolia (Baker) Harms; 大叶南洋参●☆

310173 Polyscias anacardium Bernardi; 腰果南洋参●☆

310174 Polyscias aubrevillei (Bernardi) Bernardi; 奥布南洋参●☆

310175 Polyscias bakeriana (Drake) R. Vig. = Polyscias multibracteata (Baker) Harms ●☆

310176 Polyscias balfouriana (André) L. H. Bailey = Polyscias scutellaria (Burm. f.) Fosberg ●

310177 Polyscias balfouriana L. H. Bailey 'Marginata'; 镶边圆叶南洋参●☆

310178 Polyscias balfouriana L. H. Bailey 'Pennockii'; 白斑南洋参●☆

310179 Polyscias balfouriana L. H. Bailey = Polyscias scutellaria (Burm. f.) Fosberg ●

310180 Polyscias boivinii (Seem.) Bernardi; 博伊文南洋参●☆

310181 Polyscias briquetiana (Bernardi) Lowry et G. Plunkett; 布里凯南洋参●☆

310182 Polyscias chapelieri (Drake) Harms ex R. Vig.; 沙普南洋参●☆

310183 Polyscias compacta Lowry et G. Plunkett; 紧密南洋参●☆

310184 Polyscias confertifolia (Baker) Harms; 密叶南洋参●☆

310185 Polyscias crispata (Bull.) Bull.; 卷叶福禄桐●

310186 Polyscias cumingiana Fern. -Vill.; 线叶南洋参(卡氏南洋参); Malaysian Aralia ●

310187　Polyscias cussonioides (Drake) Bernardi;甘蓝福禄桐●

310188　Polyscias duplicata (Thouars ex Baill.) Lowry et G. Plunkett;成双南洋参●☆

310189　Polyscias elegans (C. Moore et F. Muell.) Harms;美丽南洋参;Celery Wood ●☆

310190　Polyscias elliotii Harms = Polyscias fulva (Hiern) Harms ●☆

310191　Polyscias farinosa (Delile) Harms;被粉南洋参●☆

310192　Polyscias ferruginea (Hiern) Harms = Polyscias fulva (Hiern) Harms ●☆

310193　Polyscias ferruginea Harms ex Hutch. et Dalziel;锈色南洋参●☆

310194　Polyscias filicifolia (Moore ex E. Fourn.) Bailey = Polyscias cumingiana Fern. -Vill. ●

310195　Polyscias filicifolia (Ridl.) Bailey;蕨叶南洋森(线叶南洋参);Fern-leaf Aralia,Fernleaf Polyscias,Fern-leaved Polyscias ●

310196　Polyscias filicifolia (Ridl.) Bailey 'Marginata';银边蕨叶南洋●☆

310197　Polyscias floccosa (Drake) Bernardi;丛卷毛南洋参●☆

310198　Polyscias fraxinifolia Harms;白蜡叶南洋参●☆

310199　Polyscias fruticosa (L.) Harms;南洋参(碎锦福禄桐);Fruticose Polyscias,India Polyscias,Indian Polyscias,Ming Aralia ●

310200　Polyscias fruticosa (L.) Harms var. deleauana N. E. Br.;细叶福禄桐(细裂南洋参)●☆

310201　Polyscias fruticosa (L.) Harms var. plumata Bailey;羽叶南洋参;Pinnate Polyscias,Plumed Fruticose Polyscias,Plumed Indian Polyscias,Plumed Polyscias ●

310202　Polyscias fulva (Hiern) Harms;黄褐南洋参●☆

310203　Polyscias gomphophylla (Baker) Harms = Polyscias amplifolia (Baker) Harms ●☆

310204　Polyscias guilfoylei (Bull. ex Cogn. et J. F. Macbr.) L. H. Bailey;银边南洋参(福禄桐,南洋参,银边南洋参);Black aralia,Coffee Tree,Geranium Aralia,Geranium-leaf Aralia,Guilfoyle Polyscias,Silveredge Polyscias,Wild Coffee,Wild Coffee Tree ●

310205　Polyscias guilfoylei (Bull. ex Cogn. et J. F. Macbr.) L. H. Bailey 'Crispa';皱叶南洋参●

310206　Polyscias guilfoylei (Bull. ex Cogn. et J. F. Macbr.) L. H. Bailey 'Quercifolia';栎叶南洋参●

310207　Polyscias guilfoylei (Bull. ex Cogn. et J. F. Macbr.) L. H. Bailey 'Quinquefolia';五叶南洋参●

310208　Polyscias guilfoylei (Bull. ex Cogn. et J. F. Macbr.) L. H. Bailey 'Victoriae';花边南洋参(碎叶福禄桐)●

310209　Polyscias guilfoylei (Bull. ex Cogn. et J. F. Macbr.) L. H. Bailey var. laciniata L. H. Bailey;狭叶银边南洋参(银边南洋参);Narrow-leaved Guilfoyle Polyscias ●

310210　Polyscias guilfoylei (Bull. ex Cogn. et J. F. Macbr.) L. H. Bailey var. victoriae (Rod. ?) Bailey = Polyscias guilfoylei (Bull. ex Cogn. et J. F. Macbr.) L. H. Bailey 'Victoriae' ●

310211　Polyscias guilfoylei L. H. Bailey = Polyscias guilfoylei (Bull. ex Cogn. et J. F. Macbr.) L. H. Bailey ●

310212　Polyscias hildebrandtii (Drake) Harms = Polyscias nossibensis (Drake) Harms ●☆

310213　Polyscias humbertiana (Bernardi) Lowry et G. Plunkett;亨伯特南洋参●☆

310214　Polyscias kikuyuensis Summerh.;吉库尤南洋参●☆

310215　Polyscias kivuensis Bamps;基伍南洋参●☆

310216　Polyscias lancifolia (Drake) R. Vig.;披针叶南洋参●☆

310217　Polyscias lantzii (Drake) Harms ex R. Vig.;兰兹南洋参●☆

310218　Polyscias leandriana (Bernardi) Lowry et G. Plunkett;利安南洋参●☆

310219　Polyscias lepidota Chiov. = Polyscias farinosa (Delile) Harms ●☆

310220　Polyscias letestui Norman;莱泰斯图南洋参●☆

310221　Polyscias lokobensis (Drake) Harms ex R. Vig. = Polyscias nossibensis (Drake) Harms ●☆

310222　Polyscias madagascariensis (Seem.) Harms;马岛南洋参●☆

310223　Polyscias malosana Harms = Polyscias fulva (Hiern) Harms ●☆

310224　Polyscias maralia (Roem. et Schult.) Bernardi;马拉里南洋参●☆

310225　Polyscias multibracteata (Baker) Harms;多苞南洋参●☆

310226　Polyscias muraltiana Bernardi;厚壁南洋参●☆

310227　Polyscias murrayi (F. Muell.) Harms;狭叶南洋参;Pencil Cedar, Umbrella Tree ●☆

310228　Polyscias myrsine Bernardi;铁仔南洋参●☆

310229　Polyscias nodosa Blanco;结节南洋参●

310230　Polyscias nossibensis (Drake) Harms;诺西波南洋参●☆

310231　Polyscias odorata Blanco = Schefflera odorata (Blanco) Merr. et Rolfe ●

310232　Polyscias pentamera (Baker) Harms;五数南洋参●☆

310233　Polyscias preussii Harms = Polyscias fulva (Hiern) Harms ●☆

310234　Polyscias quintasii Exell;昆塔斯南洋参●☆

310235　Polyscias richardsiae Bamps;理查兹南洋参●☆

310236　Polyscias sambucifolia (DC.) Harms;接骨木叶南洋参;Elederberry Panax ●☆

310237　Polyscias scutellaria (Burm. f.) Fosberg;圆叶南洋参(大叶福禄桐,圆叶福禄桐);Balfour Polyscias, Ming Aralia, Plum Aralia, Roundleaf Polyscias, Shield Aralia ●

310238　Polyscias stuhlmannii Harms;斯图尔曼南洋参●☆

310239　Polyscias stuhlmannii Harms var. inarticulata Tennant = Polyscias stuhlmannii Harms ●☆

310240　Polyscias tennantii Bernardi = Polyscias chapelieri (Drake) Harms ex R. Vig. ●☆

310241　Polyscias terminalia Bernardi;顶生南洋参●☆

310242　Polyscias tripinnata Harms;三羽裂南洋参●☆

310243　Polyscias zanthoxyloides (Baker) Harms;花椒南洋参●☆

310244　Polysolen Rauschert = Indopolysolenia Bennet ■

310245　Polysolen Rauschert = Polysolenia Hook. f. ■

310246　Polysolenia Hook. f. = Indopolysolenia Bennet ■

310247　Polysolenia Hook. f. = Leptomischus Drake ■

310248　Polyspatha Benth. (1849);歧苞草属(多苞鸭跖草属)●☆

310249　Polyspatha hirsuta Mildbr.;毛歧苞草●☆

310250　Polyspatha paniculata Benth.;歧苞草●☆

310251　Polysphaeria Hook. f. (1873);多球茜属■●☆

310252　Polysphaeria acuminata Verdc.;渐尖多球茜■☆

310253　Polysphaeria aethiopica Verdc.;埃塞俄比亚多球茜■☆

310254　Polysphaeria arbuscula K. Schum.;小乔木多球茜■☆

310255　Polysphaeria braunii K. Krause;布劳恩多球茜■☆

310256　Polysphaeria brevifolia K. Krause = Tricalysia coriacea (Benth.) Hiern subsp. nyassae (Hiern) Bridson ●☆

310257　Polysphaeria capuronii Verdc.;凯普伦多球茜■☆

310258　Polysphaeria cleistocalyx Verdc.;闭萼多球茜■☆

310259　Polysphaeria cleistocalyx Verdc. var. pedunculata Verdc.;梗花闭萼多球茜■☆

310260　Polysphaeria congesta (Baill.) Cavaco = Polysphaeria multiflora Hiern ■☆

310261　Polysphaeria grandiflora Cavaco;大花多球茜■☆

310262　Polysphaeria grandis (Baill.) Cavaco;大多球茜■☆

310263　Polysphaeria hirta Verdc.;多毛多球茜■☆

310264　Polysphaeria lagosensis A. Chev. = Polysphaeria arbuscula K. Schum. ■☆

310265　Polysphaeria lanceolata Hiern；披针形多球茜■☆

310266　Polysphaeria lanceolata Hiern subsp. ellipticifolia Verdc.；椭圆叶多球茜■☆

310267　Polysphaeria lanceolata Hiern var. pedata Brenan；鸟足状多球茜■☆

310268　Polysphaeria lepidocarpa Verdc.；鳞果多球茜●☆

310269　Polysphaeria ligustriflora Vatke = Olinia rochetiana Juss.●☆

310270　Polysphaeria macrantha Brenan；小花多球茜●☆

310271　Polysphaeria macrophylla K. Schum.；大叶多球茜■☆

310272　Polysphaeria maxima（Baill.）Cavaco；极大多球茜■☆

310273　Polysphaeria multiflora Hiern；多花多球茜■☆

310274　Polysphaeria multiflora Hiern subsp. pubescens Verdc.；短柔毛多球茜●☆

310275　Polysphaeria neriifolia K. Schum. = Polysphaeria multiflora Hiern ■☆

310276　Polysphaeria ovata Cavaco；卵形多球茜●☆

310277　Polysphaeria parviflora R. D. Good = Chazaliella parviflora（R. D. Good）Verdc.●☆

310278　Polysphaeria parvifolia Hiern；小叶多球茜●☆

310279　Polysphaeria parvifolia Hiern var. glabra ? = Polysphaeria parvifolia Hiern ●☆

310280　Polysphaeria pedunculata K. Schum.；梗花多球茜■☆

310281　Polysphaeria pedunculata K. Schum. var. reducta Verdc.；退缩多球茜■☆

310282　Polysphaeria squarrosa K. Krause = Polysphaeria multiflora Hiern ■☆

310283　Polysphaeria subnudifaux Verdc. subsp. dewevrei Verdc.；德韦多球茜■☆

310284　Polysphaeria tubulosa（Baill.）Cavaco；管状多球茜●☆

310285　Polysphaeria zombensis S. Moore = Polysphaeria lanceolata Hiern var. pedata Brenan ■☆

310286　Polyspora Sweet = Gordonia J. Ellis（保留属名）●

310287　Polyspora acuminata S. X. Yang = Gordonia szechuanensis Hung T. Chang ●

310288　Polyspora acuminata S. X. Yang = Polyspora speciosa（Kochs）B. M. Barthol. et T. L. Ming ●

310289　Polyspora axillaris（Roxb. ex Ker Gawl.）Sweet ex G. Don = Gordonia axillaris（Roxb. ex Ker Gawl.）Endl.●

310290　Polyspora axillaris Sweet = Gordonia axillaris（Roxb. ex Ker Gawl.）Endl.●

310291　Polyspora axillaris var. nantoensis（H. Keng）S. S. Ying = Gordonia axillaris（Roxb. ex Ker Gawl.）Endl.●

310292　Polyspora axillaris var. nantoensis（H. Keng）S. S. Ying = Polyspora axillaris（Roxb. ex Ker Gawl.）Sweet ex G. Don ●

310293　Polyspora balansae（Pit.）Hu = Gordonia hainanensis Hung T. Chang ●

310294　Polyspora chrysandra（Cowan）Hu ex B. M. Barthol. et T. L. Ming = Gordonia chrysandra Cowan ●

310295　Polyspora hainanensis（Hung T. Chang）C. X. Ye ex B. M. Barthol. et T. L. Ming = Gordonia hainanensis Hung T. Chang ●

310296　Polyspora kwangsiensis（Hung T. Chang）C. X. Ye ex S. X. Yang = Gordonia szechuanensis Hung T. Chang ●

310297　Polyspora kwangsiensis（Hung T. Chang）C. X. Ye ex S. X. Yang = Polyspora speciosa（Kochs）B. M. Barthol. et T. L. Ming ●

310298　Polyspora longicarpa（Hung T. Chang）C. X. Ye ex B. M. Barthol. et T. L. Ming = Gordonia longicarpa Hung T. Chang ●

310299　Polyspora shimadae（Ohwi）Ohwi = Gordonia axillaris（Roxb. ex Ker Gawl.）Endl.●

310300　Polyspora shimadae（Ohwi）Ohwi = Polyspora axillaris（Roxb. ex Ker Gawl.）Sweet ex G. Don ●

310301　Polyspora speciosa（Kochs）B. M. Barthol. et T. L. Ming = Gordonia szechuanensis Hung T. Chang ●

310302　Polyspora tagawae（Ohwi）S. S. Ying = Gordonia axillaris（Roxb. ex Ker Gawl.）Endl.●

310303　Polyspora tagawae（Ohwi）S. S. Ying = Polyspora axillaris（Roxb. ex Ker Gawl.）Sweet ex G. Don ●

310304　Polyspora tiantangensis（L. L. Deng et G. S. Fan）S. X. Yang = Gordonia tiantangensis L. L. Deng et G. S. Fan ●

310305　Polyspora tonkinensis（Pit.）B. M. Barthol. et T. L. Ming = Gordonia axillaris（Roxb. ex Ker Gawl.）Endl.●

310306　Polyspora tonkinensis（Pit.）B. M. Barthol. et T. L. Ming = Polyspora axillaris（Roxb. ex Ker Gawl.）Sweet ex G. Don ●

310307　Polyspora yunnanensis Hu = Camellia irrawadiensis P. K. Barua ●

310308　Polyspora yunnanensis Hu = Camellia taliensis（W. W. Sm.）Melch.●

310309　Polystachya Hook.（1824）（保留属名）；多穗兰属；Manyspikeorchis，Polystachya ■

310310　Polystachya acuminata Summerh.；渐尖多穗兰■☆

310311　Polystachya adansoniae Rchb. f.；阿丹松多穗兰；Adanson. Polystachya ■☆

310312　Polystachya adansoniae Rchb. f. var. elongata Summerh.；伸长多穗兰■☆

310313　Polystachya adansoniae Rchb. f. var. stuhlmannii（Kraenzl.）Geerinck；斯图尔曼多穗兰■☆

310314　Polystachya aethiopica P. J. Cribb；埃塞俄比亚多穗兰■☆

310315　Polystachya affinis Lindl.；近缘多穗兰■☆

310316　Polystachya affinis Lindl. var. nana J. B. Hall；矮小近缘多穗兰■☆

310317　Polystachya albescens Ridl.；变白多穗兰■☆

310318　Polystachya albescens Ridl. subsp. angustifolia（Summerh.）Summerh.；窄叶变白多穗兰■☆

310319　Polystachya albescens Ridl. subsp. imbricata（Rolfe）Summerh.；覆瓦变白多穗兰■☆

310320　Polystachya albescens Ridl. subsp. polyphylla（Summerh.）Stévart；多叶变白多穗兰■☆

310321　Polystachya alboviolacea Kraenzl. = Polystachya adansoniae Rchb. f.■☆

310322　Polystachya alpina Lindl.；高山多穗兰■☆

310323　Polystachya anceps Ridl.；马岛多穗兰■☆

310324　Polystachya angularis Rchb. f.；三棱多穗兰■☆

310325　Polystachya angustifolia Summerh.；窄叶多穗兰■☆

310326　Polystachya appendiculata Kraenzl. = Polystachya cultriformis（Thouars）Spreng.■☆

310327　Polystachya aristulifera Rendle = Polystachya simplex Rendle ■☆

310328　Polystachya armeniaca la Croix et P. J. Cribb；亚美尼亚多穗兰■☆

310329　Polystachya ashantensis Kraenzl. = Polystachya tenuissima Kraenzl.■☆

310330　Polystachya asper P. J. Cribb et Podz.；粗糙多穗兰■☆

310331　Polystachya aurantiaca Schltr.；橙色多穗兰■☆

310332　Polystachya babilonii Geerinck；巴比龙多穗兰■☆

310333　Polystachya bancoensis Burg；邦克多穗兰■☆

310334　Polystachya beccarii Martelli = Polystachya steudneri Rchb. f.■☆

310335　Polystachya bella Summerh.；雅致多穗兰■☆

310336　Polystachya bennettiana Rchb. f. ;贝内特多穗兰■☆

310337　Polystachya bequaertii Summerh. ;贝卡尔多穗兰■☆

310338　Polystachya bicalcarata Kraenzl. ;双距多穗兰■☆

310339　Polystachya bicarinata Rendle;双棱多穗兰■☆

310340　Polystachya bicolor Rolfe = Polystachya rosea Ridl. ■☆

310341　Polystachya bifida Lindl. ;双裂多穗兰■☆

310342　Polystachya bracteosa Lindl. = Polystachya affinis Lindl. ■☆

310343　Polystachya brassii Summerh. ;布拉斯多穗兰■☆

310344　Polystachya buchananii Rolfe = Polystachya concreta (Jacq.) Garay et H. R. Sweet ■

310345　Polystachya buchananii Rolfe = Polystachya tessellata Lindl. ■

310346　Polystachya bulbophylloides Rolfe = Bulbophyllum pumilum (Sw.) Lindl. ■☆

310347　Polystachya busseana Kraenzl. = Polystachya dendrobiiflora Rchb. f. ■☆

310348　Polystachya caespitifica Kraenzl. subsp. hollandii (Bolus) P. J. Cribb et Podz. ;霍氏多穗兰■☆

310349　Polystachya caillei Guillaumin = Polystachya adansoniae Rchb. f. ■☆

310350　Polystachya calluniflora Kraenzl. var. hologlossa P. J. Cribb et la Croix = Polystachya hologlossa (P. J. Cribb et la Croix) Szlach. et Olszewski ■☆

310351　Polystachya caloglossa Rchb. f. ;美舌多穗兰■☆

310352　Polystachya calyptrata Kraenzl. ;帽状多穗兰■☆

310353　Polystachya camaridioides Summerh. ;鳃状多穗兰■☆

310354　Polystachya campyloglossa Rolfe;弯舌多穗兰■☆

310355　Polystachya canaliculata Summerh. ;具沟多穗兰■☆

310356　Polystachya candida Kraenzl. ;白多穗兰■☆

310357　Polystachya capensis Sond. ex Harv. = Polystachya ottoniana Rchb. f. ■☆

310358　Polystachya caquetana Schltr. = Polystachya concreta (Jacq.) Garay et H. R. Sweet ■

310359　Polystachya carnea A. Br. = Polystachya rhodoptera Rchb. f. ■☆

310360　Polystachya caudata Summerh. ;尾状多穗兰■☆

310361　Polystachya colombiana Schltr. = Polystachya concreta (Jacq.) Garay et H. R. Sweet ■

310362　Polystachya compereana Geerinck;孔佩尔多穗兰■☆

310363　Polystachya composita Kraenzl. = Polystachya fusiformis (Thouars) Lindl. ■☆

310364　Polystachya composita Kraenzl. = Polystachya superposita Rchb. f. ■☆

310365　Polystachya concreta (Jacq.) Garay et H. R. Sweet;多穗兰;Flavescent Polystachya, Manyspikeorchis, Yellow Spike Orchid ■

310366　Polystachya concreta (Jacq.) Garay et Sweet = Polystachya modesta Rchb. f. ■

310367　Polystachya confusa Rolfe;混乱多穗兰■☆

310368　Polystachya cooperi Summerh. ;库珀多穗兰■☆

310369　Polystachya coriacea Rolfe = Polystachya golungensis Rchb. f. ■☆

310370　Polystachya cornigera Schltr. ;角状多穗兰■☆

310371　Polystachya crassifolia Schltr. = Polystachya mystacioides De Wild. ■☆

310372　Polystachya cucullata (Afzel.) Durieu et Schinz;兜状多穗兰;Hand-shaped Polystachya ■☆

310373　Polystachya cultrata Lindl. = Polystachya cultriformis Lindl. ex Spreng. ■☆

310374　Polystachya cultriformis (Thouars) Lindl. ex Spreng. ;刀片多穗兰■☆

310375　Polystachya cultriformis (Thouars) Spreng. = Polystachya cultriformis (Thouars) Lindl. ex Spreng. ■☆

310376　Polystachya cultriformis (Thouars) Spreng. var. africana Schltr. = Polystachya cultriformis (Thouars) Lindl. ex Spreng. ■☆

310377　Polystachya cultriformis (Thouars) Spreng. var. autogama Schltr. = Polystachya cultriformis (Thouars) Lindl. ex Spreng. ■☆

310378　Polystachya cultriformis (Thouars) Spreng. var. occidentalis Kraenzl. = Polystachya cultriformis (Thouars) Lindl. ex Spreng. ■☆

310379　Polystachya cultriformis Lindl. ex Spreng. = Polystachya cultriformis (Thouars) Lindl. ex Spreng. ■☆

310380　Polystachya cultriformis Lindl. ex Spreng. var. africana Schltr. = Polystachya cultriformis Lindl. ex Spreng. ■☆

310381　Polystachya cultriformis Lindl. ex Spreng. var. autogama Schltr. = Polystachya cultriformis Lindl. ex Spreng. ■☆

310382　Polystachya cultriformis Lindl. ex Spreng. var. humblotii Rchb. f. = Polystachya cultriformis Lindl. ex Spreng. ■☆

310383　Polystachya cultriformis Lindl. ex Spreng. var. occidentalis Kraenzl. = Polystachya cultriformis Lindl. ex Spreng. ■☆

310384　Polystachya cyperacearum A. Chev. = Polystachya microbambusa Kraenzl. ■☆

310385　Polystachya dalzielii Summerh. ;达尔齐尔多穗兰■☆

310386　Polystachya dendrobiiflora Rchb. f. ;枝花多穗兰■☆

310387　Polystachya disticha Rolfe;二列多穗兰■☆

310388　Polystachya dixantha Rchb. f. = Polystachya stricta Rolfe var. laxiflora (Lindl.) Pérez-Vera ■☆

310389　Polystachya dolichophylla Schltr. ;长叶多穗兰■☆

310390　Polystachya dorotheae Rendle = Polystachya concreta (Jacq.) Garay et H. R. Sweet ■

310391　Polystachya dorotheae Rendle = Polystachya mukandaensis De Wild. ■

310392　Polystachya duemmeriana Kraenzl. = Polystachya modesta Rchb. f. ■

310393　Polystachya dusenii Kraenzl. = Polystachya adansoniae Rchb. f. ■☆

310394　Polystachya elastica Lindl. ;弹性多穗兰■☆

310395　Polystachya elegans Rchb. f. ;雅丽多穗兰■☆

310396　Polystachya ellenbeckiana Kraenzl. = Polystachya steudneri Rchb. f. ■☆

310397　Polystachya ensifolia Lindl. = Polystachya rhodoptera Rchb. f. ■☆

310398　Polystachya epidendroides Schltr. ;柱瓣多穗兰■☆

310399　Polystachya erythrocephala Summerh. ;红头多穗兰■☆

310400　Polystachya estrellensis Rchb. f. = Polystachya concreta (Jacq.) Garay et H. R. Sweet ■

310401　Polystachya eurychila Summerh. ;宽多穗兰■☆

310402　Polystachya eusepala Kraenzl. = Polystachya bicarinata Rendle ■☆

310403　Polystachya excelsa Kraenzl. ;高大多穗兰■☆

310404　Polystachya extinctoria Rchb. f. = Polystachya concreta (Jacq.) Garay et H. R. Sweet ■

310405　Polystachya fallax Kraenzl. ;迷惑多穗兰■☆

310406　Polystachya farinosa Kraenzl. = Polystachya bifida Lindl. ■☆

310407　Polystachya fischeri Rchb. f. ex Kraenzl. ;菲舍尔多穗兰■☆

310408　Polystachya flavescens (Blume) J. J. Sm. = Polystachya concreta (Jacq.) Garay et H. R. Sweet ■

310409　Polystachya flexuosa (Rolfe) Schltr. = Polystachya dendrobiiflora Rchb. f. ■☆

310410　Polystachya fusiformis (Thouars) Lindl. ;繁花多穗兰■☆

310411　Polystachya gabonensis Summerh. = Polystachya odorata Lindl. subsp. gabonensis (Summerh.) Stévart ■☆

310412　Polystachya galeata (Sw.) Rchb. f. ;盔形多穗兰■☆

310413　Polystachya galericulata Rchb. f. = Polystachya stricta Rolfe var. laxiflora (Lindl.) Pérez-Vera ■☆

310414　Polystachya geniculata Summerh. ;膝曲多穗兰■☆

310415　Polystachya gerrardii Harv. = Polystachya cultriformis (Thouars) Spreng. ■☆

310416　Polystachya gilletti De Wild. = Polystachya galeata (Sw.) Rchb. f. ■☆

310417　Polystachya glaberrima Schltr. = Polystachya ottoniana Rchb. f. ■☆

310418　Polystachya goetzeana Kraenzl. ;格兹多穗兰■☆

310419　Polystachya golungensis Rchb. f. ;戈龙多穗兰■☆

310420　Polystachya gracilenta Kraenzl. ;细黏多穗兰■☆

310421　Polystachya gracilis De Wild. = Polystachya concreta (Jacq.) Garay et H. R. Sweet ■

310422　Polystachya gracilis De Wild. = Polystachya tessellata Lindl. ■

310423　Polystachya graminoides Kraenzl. = Polystachya stauroglossa Kraenzl. ■☆

310424　Polystachya grandiflora Lindl. = Polystachya galeata (Sw.) Rchb. f. ■☆

310425　Polystachya guerzorum A. Chev. = Polystachya dolichophylla Schltr. ■☆

310426　Polystachya hamiltonii W. W. Sm. = Polystachya dolichophylla Schltr. ■☆

310427　Polystachya hastata Summerh. ;戟形多穗兰■☆

310428　Polystachya heckeliana Schltr. ;赫克尔多穗兰■☆

310429　Polystachya heckmanniana Kraenzl. ;赫克曼多穗兰■☆

310430　Polystachya henrici Schltr. ;昂里克多穗兰■☆

310431　Polystachya hildebrandtii Kraenzl. = Polystachya rosea Ridl. ■☆

310432　Polystachya hislopii Rolfe ;黑氏多穗兰;Hislop Polystachya ■☆

310433　Polystachya hislopii Rolfe = Polystachya zambesiaca Rolfe ■☆

310434　Polystachya hollandii Bolus = Polystachya caespitifica Kraenzl. subsp. hollandii (Bolus) P. J. Cribb et Podz. ■☆

310435　Polystachya holmesiana P. J. Cribb;霍尔梅斯多穗兰■☆

310436　Polystachya holochila Schltr. = Polystachya dendrobiiflora Rchb. f. ■☆

310437　Polystachya hologlossa (P. J. Cribb et la Croix) Szlach. et Olszewski;全舌多穗兰■☆

310438　Polystachya holstii Kraenzl. ;霍尔多穗兰■☆

310439　Polystachya holtzeana Kraenzl. = Neobenthamia gracilis Rolfe ■☆

310440　Polystachya humberti H. Perrier;亨伯特多穗兰■☆

310441　Polystachya huyghei De Wild. = Polystachya concreta (Jacq.) Garay et H. R. Sweet ■

310442　Polystachya huyghei De Wild. = Polystachya mukandaensis De Wild. ■

310443　Polystachya hypocrita Rchb. f. = Polystachya concreta (Jacq.) Garay et H. R. Sweet ■

310444　Polystachya hypocrita Rchb. f. = Polystachya tessellata Lindl. ■

310445　Polystachya imbricata Rolfe = Polystachya albescens Ridl. subsp. imbricata (Rolfe) Summerh. ■☆

310446　Polystachya imbricata Rolfe subsp. angustifolia Summerh. = Polystachya albescens Ridl. subsp. angustifolia (Summerh.) Summerh. ■☆

310447　Polystachya inaperta Guillaumin = Polystachya subulata Finet ■☆

310448　Polystachya inconspicua Rendle = Polystachya tenuissima Kraenzl. ■☆

310449　Polystachya ionocharis Kraenzl. = Polystachya melanantha Schltr. ■☆

310450　Polystachya isochiloides Summerh. ;等舌多穗兰■☆

310451　Polystachya johnsonii Kraenzl. = Polystachya golungensis Rchb. f. ■☆

310452　Polystachya johnstonii Rolfe;约翰斯顿多穗兰■☆

310453　Polystachya johnstonii Rolfe var. roseopurpurea la Croix et P. J. Cribb;紫红约翰斯顿多穗兰■☆

310454　Polystachya jussieuana Rchb. f. = Polystachya concreta (Jacq.) Garay et H. R. Sweet ■

310455　Polystachya kaessneriana Kraenzl. = Polystachya dendrobiiflora Rchb. f. ■☆

310456　Polystachya kaluluensis P. J. Cribb et la Croix;卡卢卢多穗兰■☆

310457　Polystachya kermesina Kraenzl. ;克迈斯多穗兰■☆

310458　Polystachya kerstingii Schltr. ;克斯廷多穗兰■☆

310459　Polystachya kilimanjari Kraenzl. = Polystachya fischeri Rchb. f. ex Kraenzl. ■☆

310460　Polystachya kindtiana De Wild. = Polystachya modesta Rchb. f. ■☆

310461　Polystachya kingii Summerh. ;金多穗兰■☆

310462　Polystachya kirkii Rolfe = Polystachya cultriformis (Thouars) Spreng. ■☆

310463　Polystachya kirkii Rolfe = Polystachya cultriformis Lindl. ex Spreng. ■☆

310464　Polystachya kornasiana Szlach. et Olszewski;科纳斯多穗兰■☆

310465　Polystachya kraenzliana Pabst = Polystachya concreta (Jacq.) Garay et H. R. Sweet ■

310466　Polystachya kupensis P. J. Cribb et B. J. Pollard;库普多穗兰■☆

310467　Polystachya latifolia De Wild. = Polystachya concreta (Jacq.) Garay et H. R. Sweet ■

310468　Polystachya latifolia De Wild. = Polystachya tessellata Lindl. ■

310469　Polystachya laurentii De Wild. ;洛朗多穗兰■☆

310470　Polystachya lawalreana Geerinck;拉瓦尔多穗兰■☆

310471　Polystachya lawrenceana Kuntze;劳伦斯多穗兰; Lawrence Polystachya ■☆

310472　Polystachya laxa R. Schust. ;疏松多穗兰■☆

310473　Polystachya laxiflora Lindl. = Polystachya stricta Rolfe var. laxiflora (Lindl.) Pérez-Vera ■☆

310474　Polystachya lehmbachiana Kraenzl. = Polystachya tessellata Lindl. ■

310475　Polystachya lejolyana Stévart;勒若利多穗兰■☆

310476　Polystachya leonardiana Geerinck;莱奥多穗兰■☆

310477　Polystachya leonensis Rchb. f. ;莱昂多穗兰■☆

310478　Polystachya lepidantha Kraenzl. = Polystachya concreta (Jacq.) Garay et H. R. Sweet ■

310479　Polystachya lepidantha Kraenzl. = Polystachya tessellata Lindl. ■

310480　Polystachya letouzeyana Szlach. et Olszewski;勒图多穗兰■☆

310481　Polystachya lettowiana Kraenzl. = Polystachya concreta (Jacq.) Garay et H. R. Sweet ■

310482　Polystachya lettowiana Kraenzl. = Polystachya tessellata Lindl. ■

310483　Polystachya leucorhoda Kraenzl. = Polystachya poikilantha Kraenzl. var. leucorhoda (Kraenzl.) P. J. Cribb et Podz. ■☆

310484　Polystachya leucosepala P. J. Cribb;白萼多穗兰■☆

310485　Polystachya liberica Rolfe = Polystachya reflexa Lindl. ■☆

310486　Polystachya ligulifolia Summerh. = Polystachya retusiloba Summerh. ■☆

310487　Polystachya lindblomii Schltr. ;林德布卢姆多穗兰■☆

310488　Polystachya lindleyana Harv. = Polystachya pubescens (Lindl.) Rchb. f. ■☆

310489　Polystachya longiscapa Summerh. ;长果多穗兰■☆

310490　Polystachya lujae De Wild. = Polystachya cultriformis Lindl. ex Spreng. ■☆

310491　Polystachya lukwangulensis P. J. Cribb;卢夸古尔多穗兰■☆

310492　Polystachya luteola（Sw.）Hook. ;淡黄多穗兰;Yellowish Polystachya ■☆

310493　Polystachya luteola（Sw.）Hook. = Polystachya concreta（Jacq.）Garay et H. R. Sweet ■

310494　Polystachya macropetala Kraenzl. = Polystachya dendrobiiflora Rchb. f. ■☆

310495　Polystachya macropoda Summerh. ;大足多穗兰■☆

310496　Polystachya maculata P. J. Cribb;斑点多穗兰■☆

310497　Polystachya mafingensis P. J. Cribb;马芬加多穗兰■☆

310498　Polystachya magnibracteata P. J. Cribb;大苞多穗兰■☆

310499　Polystachya mannii Rolfe = Polystachya elegans Rchb. f. ■☆

310500　Polystachya mauritiana Spreng. = Polystachya concreta（Jacq.）Garay et H. R. Sweet ■

310501　Polystachya mayombensis De Wild. = Polystachya golungensis Rchb. f. ■☆

310502　Polystachya megalogenys Summerh. = Polystachya mildbraedii Kraenzl. ■☆

310503　Polystachya melanantha Schltr. ;黑花多穗兰■☆

310504　Polystachya melliodora P. J. Cribb;蜜味多穗兰■☆

310505　Polystachya membranacea A. Rich. = Tropidia polystachya（Sw.）Ames ■☆

310506　Polystachya meyeri P. J. Cribb et Podz. ;迈尔多穗兰■☆

310507　Polystachya microbambusa Kraenzl. ;小多穗兰■☆

310508　Polystachya micropetala（Lindl.）Rolfe = Genyorchis micropetala（Lindl.）Schltr. ■☆

310509　Polystachya mildbraedii Kraenzl. ;米尔德多穗兰■☆

310510　Polystachya mildbraedii Kraenzl. var. angustifolia（Summerh.）Geerinck = Polystachya angustifolia Summerh. ■☆

310511　Polystachya minima Rendle;微小多穗兰■☆

310512　Polystachya minuta（Aubl.）Britton = Polystachya concreta（Jacq.）Garay et H. R. Sweet ■

310513　Polystachya minuta Frapp. et Cordem. = Polystachya concreta（Jacq.）Garay et H. R. Sweet ■

310514　Polystachya minutiflora Ridl. = Polystachya fusiformis（Thouars）Lindl. ■☆

310515　Polystachya miranda Kraenzl. = Polystachya dendrobiiflora Rchb. f. ■☆

310516　Polystachya modesta Rchb. f. = Polystachya concreta（Jacq.）Garay et H. R. Sweet ■

310517　Polystachya monophylla Schltr. ;单叶多穗兰■☆

310518　Polystachya moreauae P. J. Cribb et Podz. ;莫罗多穗兰■☆

310519　Polystachya mukandaensis De Wild. = Polystachya concreta（Jacq.）Garay et H. R. Sweet ■

310520　Polystachya multiflora Ridl. = Polystachya fusiformis（Thouars）Lindl. ■☆

310521　Polystachya mystacioides De Wild. ;触须兰状多穗兰■☆

310522　Polystachya natalensis Rolfe = Polystachya transvaalensis Schltr. ■☆

310523　Polystachya neobenthamia Schltr. = Neobenthamia gracilis Rolfe ■☆

310524　Polystachya nigerica Rendle = Polystachya adansoniae Rchb. f. ■☆

310525　Polystachya nigrescens Rendle = Polystachya transvaalensis Schltr. ■☆

310526　Polystachya nitidula Rchb. f. = Polystachya concreta（Jacq.）Garay et H. R. Sweet ■

310527　Polystachya obanensis Rendle;奥班多穗兰■☆

310528　Polystachya oblanceolata Summerh. ;倒披针形多穗兰■☆

310529　Polystachya odorata Lindl. ;多齿多穗兰■☆

310530　Polystachya odorata Lindl. subsp. gabonensis（Summerh.）Stévart;加蓬多穗兰■☆

310531　Polystachya odorata Lindl. subsp. trilepidis（Summerh.）Stévart;三鳞多齿多穗兰■☆

310532　Polystachya odorata Lindl. var. trilepidis Summerh. = Polystachya odorata Lindl. subsp. trilepidis（Summerh.）Stévart ■☆

310533　Polystachya oligophylla Schltr. = Polystachya albescens Ridl. subsp. imbricata（Rolfe）Summerh. ■☆

310534　Polystachya ottoniana Rchb. f. ;奥托多穗兰;Otton Polystachya ■☆

310535　Polystachya oxychila Schltr. ex Kraenzl. = Polystachya dolichophylla Schltr. ■☆

310536　Polystachya pachyglossa Rchb. f. ;粗舌多穗兰■☆

310537　Polystachya pachyrhiza Kraenzl. = Polystachya simplex Rendle ■☆

310538　Polystachya paniculata（Sw.）Rolfe;圆锥多穗兰■☆

310539　Polystachya parva Summerh. ;瘦小多穗兰■☆

310540　Polystachya parviflora Summerh. ;小花多穗兰■☆

310541　Polystachya perrieri Schltr. ;佩里耶多穗兰■☆

310542　Polystachya pisobulbon Kraenzl. = Polystachya ottoniana Rchb. f. ■☆

310543　Polystachya plehniana Schltr. = Polystachya concreta（Jacq.）Garay et H. R. Sweet ■

310544　Polystachya plehniana Schltr. = Polystachya mukandaensis De Wild. ■

310545　Polystachya pleistantha Kraenzl. = Polystachya concreta（Jacq.）Garay et H. R. Sweet ■

310546　Polystachya pobeguinii（Finet）Rolfe;波别多穗兰■☆

310547　Polystachya pocsii P. J. Cribb;波奇多穗兰■☆

310548　Polystachya poikilantha Kraenzl. var. leucorhoda（Kraenzl.）P. J. Cribb et Podz. ;白红多穗兰■☆

310549　Polystachya polychaete Kraenzl. ;多毛多穗兰■☆

310550　Polystachya polyglossa T. Durand et Schinz = Polystachya pachyglossa Rchb. f. ■☆

310551　Polystachya polyphylla Summerh. = Polystachya albescens Ridl. subsp. polyphylla（Summerh.）Stévart ■☆

310552　Polystachya porphyrochila J. Stewart;紫唇多穗兰■☆

310553　Polystachya praealta Kraenzl. = Polystachya concreta（Jacq.）Garay et H. R. Sweet ■

310554　Polystachya praealta Kraenzl. = Polystachya tessellata Lindl. ■

310555　Polystachya praecipitis Summerh. ;特别多穗兰■☆

310556　Polystachya preussii Kraenzl. = Polystachya alpina Lindl. ■☆

310557　Polystachya principia P. J. Cribb et Stévart;普林西比多穗兰■☆

310558　Polystachya proterantha P. J. Cribb;海神花多穗兰■☆

310559　Polystachya puberula Lindl. ;短柔毛多穗兰;Slightly-hairy Polystachya ■☆

310560　Polystachya pubescens（Lindl.）Rchb. f. ;柔毛多穗兰;Pubescent Polystachya ■☆

310561　Polystachya purpurea Wight = Polystachya concreta（Jacq.）Garay et H. R. Sweet ■

310562　Polystachya purpurea Wight var. lutescens Gagnep. = Polystachya concreta（Jacq.）Garay et H. R. Sweet ■

310563　Polystachya purpureo-alba Kraenzl. = Polystachya gracilenta Kraenzl. ■☆

310564　Polystachya purpureobracteata P. J. Cribb et la Croix;紫苞多穗兰■☆

310565　Polystachya pyramidalis Lindl. ;塔形多穗兰■☆

310566　Polystachya ramulosa Lindl. ;多枝多穗兰■☆

310567　Polystachya reflexa Lindl. ;反折多穗兰■☆

310568　Polystachya reichenbachiana Kraenzl. = Polystachya concreta （Jacq.） Garay et H. R. Sweet ■

310569　Polystachya rendlei Rolfe = Polystachya transvaalensis Schltr. ■☆

310570　Polystachya repens Rolfe = Stolzia repens （Rolfe） Summerh. ■☆

310571　Polystachya reticulata Stévart et Droissart;网状多穗兰■☆

310572　Polystachya retusiloba Summerh. ;微凹多穗兰■☆

310573　Polystachya rhodochila Schltr. ;粉红多穗兰■☆

310574　Polystachya rhodoptera Rchb. f. ;红翅多穗兰■☆

310575　Polystachya ridleyi Rolfe;里德利多穗兰■☆

310576　Polystachya rigidula Rchb. f. = Polystachya concreta （Jacq.） Garay et H. R. Sweet ■

310577　Polystachya rigidula Rchb. f. = Polystachya modesta Rchb. f. ■

310578　Polystachya rivae Schweinf. ;沟多穗兰■☆

310579　Polystachya rolfeana Kraenzl. ;罗尔夫多穗兰■☆

310580　Polystachya rolfeana Kraenzl. = Polystachya excelsa Kraenzl. ■☆

310581　Polystachya rosea Ridl. ;玫瑰多穗兰■☆

310582　Polystachya rosellata Ridl. ;浅红多穗兰■☆

310583　Polystachya rufinula Rchb. f. = Polystachya tessellata Lindl. ■

310584　Polystachya rugosilabia Summerh. ;皱唇多穗兰■☆

310585　Polystachya ruwenzoriensis Rendle;鲁文佐里多穗兰■☆

310586　Polystachya ruwenzoriensis Rendle var. tridentata （Summerh.） Geerinck;三齿多穗兰■☆

310587　Polystachya saccata （Finet） Rolfe;囊状多穗兰■☆

310588　Polystachya sandersonii Harv. ;桑德森多穗兰■☆

310589　Polystachya serpentina P. J. Cribb;蛇形多穗兰■☆

310590　Polystachya seticaulis Rendle;毛茎多穗兰■☆

310591　Polystachya setifera Lindl. ;刚毛多穗兰■☆

310592　Polystachya shirensis Rchb. f. = Polystachya concreta （Jacq.） Garay et H. R. Sweet ■

310593　Polystachya shirensis Rchb. f. = Polystachya modesta Rchb. f. ■

310594　Polystachya siamensis Ridl. = Polystachya concreta （Jacq.） Garay et H. R. Sweet ■

310595　Polystachya similis Rchb. f. = Polystachya concreta （Jacq.） Garay et H. R. Sweet ■

310596　Polystachya similis Rchb. f. = Polystachya modesta Rchb. f. ■

310597　Polystachya simoniana Kraenzl. = Polystachya dolichophylla Schltr. ■☆

310598　Polystachya simplex Rendle;简单多穗兰■☆

310599　Polystachya smytheana Rolfe = Polystachya reflexa Lindl. ■☆

310600　Polystachya spiranthoides Kraenzl. = Polystachya golungensis Rchb. f. ■☆

310601　Polystachya stauroglossa Kraenzl. ;十字舌多穗兰■☆

310602　Polystachya stauroglossa Kraenzl. var. alata Geerinck;具翅十字舌多穗兰■☆

310603　Polystachya steudneri Rchb. f. ;斯托德多穗兰■☆

310604　Polystachya stewartiana Geerinck;斯图尔特多穗兰■☆

310605　Polystachya striata De Wild. = Polystachya odorata Lindl. ■☆

310606　Polystachya stricta Rolfe;刚直多穗兰■☆

310607　Polystachya stricta Rolfe var. laxiflora （Lindl.） Pérez-Vera;疏花刚直多穗兰■☆

310608　Polystachya stuhlmannii Kraenzl. = Polystachya adansoniae Rchb. f. var. stuhlmannii （Kraenzl.） Geerinck ■☆

310609　Polystachya suaveolens P. J. Cribb;芳香多穗兰■☆

310610　Polystachya subcorymbosa Kraenzl. = Polystachya rhodoptera Rchb. f. ■☆

310611　Polystachya subdiphylla Summerh. ;亚二叶多穗兰■☆

310612　Polystachya subulata Finet;钻形多穗兰■☆

310613　Polystachya subumbellata P. J. Cribb et Podz. ;小伞多穗兰■☆

310614　Polystachya sulfurea A. Br. = Polystachya rhodoptera Rchb. f. ■☆

310615　Polystachya superposita Rchb. f. ;后多穗兰■☆

310616　Polystachya supfiana Schltr. ;祖普夫多穗兰■☆

310617　Polystachya talbotii Rolfe = Polystachya alpina Lindl. ■☆

310618　Polystachya tayloriana Rendle = Polystachya dendrobiiflora Rchb. f. ■☆

310619　Polystachya teitensis P. J. Cribb;泰塔多穗兰■☆

310620　Polystachya tenella Summerh. ;细多穗兰■☆

310621　Polystachya tenuissima Kraenzl. ;极细多穗兰■☆

310622　Polystachya tessellata Lindl. = Polystachya concreta （Jacq.） Garay et H. R. Sweet ■

310623　Polystachya tessellata Lindl. var. tricruris （Rchb. f.） Schelpe = Polystachya tessellata Lindl. ■

310624　Polystachya thomensis Summerh. ;爱岛多穗兰■☆

310625　Polystachya transvaalensis Schltr. ;德兰士瓦多穗兰■☆

310626　Polystachya tricruris Rchb. f. = Polystachya concreta （Jacq.） Garay et H. R. Sweet ■

310627　Polystachya tricruris Rchb. f. = Polystachya tessellata Lindl. ■

310628　Polystachya tridentata Summerh. = Polystachya ruwenzoriensis Rendle var. tridentata （Summerh.） Geerinck ■☆

310629　Polystachya troupiniana Geerinck;特鲁皮尼多穗兰■☆

310630　Polystachya tsaratananae H. Perrier;察拉塔纳纳多穗兰■☆

310631　Polystachya tsinjoarivensis H. Perrier;钦祖阿里武多穗兰■☆

310632　Polystachya ugandae Kraenzl. = Polystachya lindblomii Schltr. ■☆

310633　Polystachya ulugurensis P. J. Cribb et Podz. ;乌卢古尔多穗兰■☆

310634　Polystachya undulata P. J. Cribb et Podz. ;波状多穗兰■☆

310635　Polystachya uniflora De Wild. = Polystachya melanantha Schltr. ■☆

310636　Polystachya usambarensis Schltr. = Polystachya odorata Lindl. ■☆

310637　Polystachya vaginata Summerh. ;具鞘多穗兰■☆

310638　Polystachya valentina la Croix et P. J. Cribb;瓦伦特多穗兰（瓦伦蒂多穗兰）■☆

310639　Polystachya victoriae Kraenzl. ;维多利亚多穗兰■☆

310640　Polystachya villosa Rolfe;长柔毛多穗兰■☆

310641　Polystachya virescens Ridl. ;浅绿多穗兰■☆

310642　Polystachya virgata Schltr. ;条纹多穗兰■☆

310643　Polystachya virginea Summerh. ;纯白多穗兰■☆

310644　Polystachya virginea Summerh. var. parva Geerinck;小纯白多穗兰■☆

310645　Polystachya vulcanica Kraenzl. ;火山多穗兰■☆

310646　Polystachya wahisiana De Wild. = Polystachya rhodoptera Rchb. f. ■☆

310647　Polystachya waterlotii Guillaumin;瓦泰洛多穗兰■☆

310648　Polystachya wightii Rchb. f. = Polystachya concreta （Jacq.） Garay et H. R. Sweet ■

310649　Polystachya winkleri Schltr. = Polystachya alpina Lindl. ■☆

310650　Polystachya woosnamii Rendle;沃斯南多穗兰■☆

310651　Polystachya woosnamii Rendle var. nyungweensis Delep. et Lebel;尼永圭多穗兰■☆

310652　Polystachya xerophila Kraenzl. ;旱生多穗兰■☆

310653　Polystachya zambesiaca Rolfe;赞比西多穗兰■☆

310654　Polystachya zanguebarica Rolfe = Polystachya concreta （Jacq.） Garay et H. R. Sweet ■

310655　Polystachya zanguebarica Rolfe = Polystachya tessellata Lindl. ■

310656　Polystachya zeylanica Lindl. = Polystachya concreta（Jacq.）Garay et H. R. Sweet ■

310657　Polystachya zollingeri Rchb. f. = Polystachya concreta（Jacq.）Garay et H. R. Sweet ■

310658　Polystachya zuluensis L. Bolus；祖卢多穗兰■☆

310659　Polystemma Decne.（1844）；多冠萝藦属●☆

310660　Polystemma viridiflora Decne.；多冠萝藦●☆

310661　Polystemon D. Don = Belangera Cambess. ●☆

310662　Polystemon D. Don = Lamanonia Vell. ●☆

310663　Polystemonanthus Harms（1897）；西非多蕊豆属■☆

310664　Polystemonanthus dinklagei Harms；西非多蕊豆■☆

310665　Polystepis Thouars = Epidendrum L.（保留属名）■☆

310666　Polystepis Thouars = Oeoniella Schltr. ☆

310667　Polystigma Meisn. = Byronia Endl. ●

310668　Polystigma Meisn. = Ilex L. ●

310669　Polystorthia Blume = Lauro-Cerasus Duhamel ●

310670　Polystorthia Blume = Polydontia Blume ●

310671　Polystorthia Blume = Pygeum Gaertn. ●

310672　Polystylus Hasselt ex Hassk. = Phalaenopsis Blume ■

310673　Polytaenia DC.（1830）；多带草属■☆

310674　Polytaenia nuttallii DC.；多带草；Nuttall's Prairie-parsley，Prairie Parsley，Prairie-parsley ■☆

310675　Polytaxis Bunge = Jurinea Cass. ●■

310676　Polytaxis Bunge（1843）；肉木香属■

310677　Polytaxis lahmannii Bunge；肉木香■☆

310678　Polytaxis winkleri Iljin；温氏肉木香■☆

310679　Polytemia Raf. = Polytenia Raf. ■☆

310680　Polytenia Raf. = Polytaenia DC. ■☆

310681　Polytepalum Suess. et Beyerle（1938）；多萼木属●☆

310682　Polytepalum angolense Suess. et Beyerle；安哥拉多萼木●☆

310683　Polythecandra Planch. et Triana = Clusia L. ●☆

310684　Polythecanthum Tiegh. = Ochna L. ●

310685　Polythecium Tiegh. = Ochna L. ●

310686　Polythecium carvalhoi（Engl.）Tiegh. = Ochna kirkii Oliv. ●☆

310687　Polythecium citrinum（Gilg）Tiegh. = Ochna citrina Gilg ●☆

310688　Polythecium fischeri（Engl.）Tiegh. = Ochna mossambicensis Klotzsch ●☆

310689　Polythecium hildebrandtii（Engl.）Tiegh. = Ochna thomasiana Engl. et Gilg ex Gilg ●☆

310690　Polythecium kirkii（Oliv.）Tiegh. = Ochna kirkii Oliv. ●☆

310691　Polythecium pulchrum（Hook.）Tiegh. = Ochna pulchra Hook. f. ●☆

310692　Polythecium rehmannii（Szyszyl.）Tiegh. = Ochna pulchra Hook. f. ●☆

310693　Polythecium splendidum（Engl.）Tiegh. = Ochna macrocalyx Oliv. ●☆

310694　Polythecium thomasianum（Engl. et Gilg）Tiegh. = Ochna thomasiana Engl. et Gilg ex Gilg ●☆

310695　Polythrix Nees = Crossandra Salisb. ●

310696　Polythrix stenandrium Nees = Crossandra stenandrium（Nees）Lindau ■☆

310697　Polythysania Hanst. = Alloplectus Mart.（保留属名）●■☆

310698　Polythysania Hanst. = Drymonia Mart. ●☆

310699　Polytoca R. Br.（1838）；多裔草属（多裔黍属）；Polytoca ■

310700　Polytoca barbata（Roxb.）Stapf = Polytoca digitata（L. f.）Druce ■

310701　Polytoca barbata（Roxb.）Stapf ex Hook. f. = Chionachne koenigii（Spreng.）Thwaites ■☆

310702　Polytoca bracteata R. Br. = Polytoca digitata（L. f.）Druce ■

310703　Polytoca digitata（L. f.）Druce；多裔草；Digitate Polytoca ■

310704　Polytoca digitata Druce = Polytoca digitata（L. f.）Druce ■

310705　Polytoca heteroclita（Roxb.）Koord. = Polytoca digitata（L. f.）Druce ■

310706　Polytoca macrophylla Benth.；大叶多裔草■☆

310707　Polytoca massiei（Balansa）Schenck ex Henrard = Chionachne massiei Balansa ■

310708　Polytoca massiei（Balansa）Schenk = Chionachne massiei Balansa ■

310709　Polytoca massii（Balansa）Schenck ex Henrard = Chionachne massiei Balansa ■

310710　Polytoma Lour. ex B. A. Gomes = Bletilla Rchb. f.（保留属名）＋ Aerides Lour. ■

310711　Polytoma Lour. ex B. A. Gomes = Bletilla Rchb. f.（保留属名）●■

310712　Polytrema C. B. Clarke = Ptyssiglottis T. Anderson ■☆

310713　Polytrias Hack.（1889）；单序草属（三穗草属）；Onespikegrass，Three-spikegrass ■☆

310714　Polytrias amaura（Büse）Kuntze = Polytrias indica（Houtt.）Veldkamp ■

310715　Polytrias amaura（Büse）Kuntze var. nana（Keng et S. L. Chen）Keng ex S. L. Chen = Polytrias indica（Houtt.）Veldkamp var. nana（Keng et S. L. Chen）S. M. Phillips et S. L. Chen ■

310716　Polytrias amaura Kuntze = Polytrias indica（Houtt.）Veldkamp ■

310717　Polytrias amaura Kuntze var. nana（Keng et S. L. Chen）S. L. Chen = Polytrias indica（Houtt.）Veldkamp var. nana（Keng et S. L. Chen）S. M. Phillips et S. L. Chen ■

310718　Polytrias diversiflora（Steud.）Nash = Polytrias amaura（Büse）Kuntze ■

310719　Polytrias diversiflora（Steud.）Nash = Polytrias indica（Houtt.）Veldkamp ■

310720　Polytrias indica（Houtt.）Veldkamp；单序草（三穗草）；Onespikegrass ■

310721　Polytrias indica（Houtt.）Veldkamp = Phleum indicum Houtt. ■☆

310722　Polytrias indica（Houtt.）Veldkamp var. nana（Keng et S. L. Chen）S. M. Phillips et S. L. Chen；短毛单序草（矮金茅）；Java Grass，Shorthair Onespikegrass ■

310723　Polytrias praemorsa（Nees ex Steud.）Hack. = Polytrias indica（Houtt.）Veldkamp ■

310724　Polytrias praemorsa（Nees）Hack. = Polytrias amaura（Büse）Kuntze ■

310725　Polytropia C. Presl = Rhynchosia Lour.（保留属名）●■

310726　Polytropis B. D. Jacks. = Polytropia C. Presl ●■

310727　Polytropis C. Presl = Rhynchosia Lour.（保留属名）●■

310728　Polyura Hook. f.（1868）；多尾草属■

310729　Polyura geminata Hook. f.；多尾草■

310730　Polyura geminata Hook. f. = Lerchea micrantha（Drake）H. S. Lo ■

310731　Polyxena Kunth（1843）；外来风信子属■☆

310732　Polyxena angustifolia（L. f.）Baker = Massonia echinata L. f. ■☆

310733　Polyxena bakeri T. Durand et Schinz = Massonia echinata L. f. ■☆

310734　Polyxena calcicola U. Müll. -Doblies et D. Müll. -Doblies；钙生外来风信子■☆

310735　Polyxena comata（Burch. ex Baker）Baker = Daubenya comata（Burch. ex Baker）J. C. Manning et A. M. Van der Merwe ■☆

310736　Polyxena corymbosa（L.）Jessop = Lachenalia corymbosa（L.）

J. C. Manning et Goldblatt ■☆

310737　Polyxena ensifolia（Thunb.）Schönland;剑叶外来风信子■☆

310738　Polyxena haemanthoides Baker = Daubenya marginata（Willd. ex Kunth）J. C. Manning et A. M. Van der Merwe ■☆

310739　Polyxena longituba A. M. Van der Merwe = Lachenalia longituba（Van der Merwe）J. C. Manning et Goldblatt ■☆

310740　Polyxena marginata（Willd. ex Kunth）Baker = Daubenya marginata（Willd. ex Kunth）J. C. Manning et A. M. Van der Merwe ■☆

310741　Polyxena maughanii W. F. Barker = Lachenalia maughanii（W. F. Barker）J. C. Manning et Goldblatt ■☆

310742　Polyxena namaquensis（Schltr.）K. Krause = Daubenya namaquensis（Schltr.）J. C. Manning et Goldblatt ■☆

310743　Polyxena odorata（Hook. f.）Baker = Polyxena ensifolia（Thunb.）Schönland ■☆

310744　Polyxena paucifolia（W. F. Barker）A. M. Van der Merwe et J. C. Manning;寡叶外来风信子■☆

310745　Polyxena pygmaea（Jacq.）Kunth = Polyxena ensifolia（Thunb.）Schönland ■☆

310746　Polyxena rugulosa（Licht. ex Kunth）Baker = Daubenya marginata（Willd. ex Kunth）J. C. Manning et A. M. Van der Merwe ■☆

310747　Polyxena uniflora（Banks ex Baker）Baker;单花外来风信子■☆

310748　Polyzone Endl. = Darwinia Rudge ● ☆

310749　Polyzygus Dalzell（1850）;多轭草属☆

310750　Polyzygus tuberosus Walp.;多轭草☆

310751　Pomaceae Gray = Rosaceae Juss.（保留科名）●■

310752　Pomaderris Labill.（1805）;安匝木属;Anzacwood ● ☆

310753　Pomaderris androsaemifolia A. Cunn. ex Heynh.;显脉安匝木;Veined Anzacwood ● ☆

310754　Pomaderris angustifolia N. A. Wakef.;窄叶安匝木;Narrow-leaved Anzacwood ● ☆

310755　Pomaderris apetala Labill.;无瓣安匝木;Virescent Anzacwood, Virescent Anzac-wood, Virescent Anzae Wood ● ☆

310756　Pomaderris aspera Sieber ex DC.;皱叶安匝木;Wrinkled Anzacwood ● ☆

310757　Pomaderris betulina Hook.;桦叶安匝木;Birchleaf Anzacwood ● ☆

310758　Pomaderris brunnea N. A. Wakef.;褐毛安匝木;Brownish Anzacwood ● ☆

310759　Pomaderris cotoneaster N. A. Wakef.;枸子安匝木;Cotoneaster Anzacwood ● ☆

310760　Pomaderris discolor（Vent.）Poir.;灰毛安匝木;Greyish Anzacwood ● ☆

310761　Pomaderris elliptica Labill.;椭圆叶安匝木;Goldentainui Anzacwood ● ☆

310762　Pomaderris eriocephala N. A. Wakef.;黄毛安匝木;Yellow-hair Anzacwood ● ☆

310763　Pomaderris ferruginea Fenzl;大花安匝木;Bigflower Anzacwood ● ☆

310764　Pomaderris kumeraho A. Cunn. ex Fenzl;丰花安匝木 ● ☆

310765　Pomaderris lanigera（Andréws）Sims;毛序安匝木;Hairy-inflorescence Anzacwood, Woolly Pomaderris ● ☆

310766　Pomaderris ledifolia A. Cunn.;隐脉安匝木;Invisible-veine Anzacwood ● ☆

310767　Pomaderris ligustrina Sieber ex DC.;披针叶安匝木;Lanceolate Anzacwood ● ☆

310768　Pomaderris multiflora Fenzl;多花安匝木;Multiflorous Anzacwood ● ☆

310769　Pomaderris phylicifolia Link;菲利木叶安匝木（灌木状安匝木）;Shrubby Anzacwood ● ☆

310770　Pomaderris prunifolia A. Cunn. ex Fenzl;锈毛安匝木;Rusty-hair Anzacwood ● ☆

310771　Pomaderris rugosa Cheeseman;锈脉安匝木;Rusty-vein Anzacwood ● ☆

310772　Pomaderris sericea N. A. Wakef.;金毛安匝木;Golden-hair Anzacwood ● ☆

310773　Pomaderris sieberiana N. A. Wakef.;单毛安匝木;Simple-hair Anzacwood ● ☆

310774　Pomaderris vellea N. A. Wakef.;密花安匝木;Compact-flower Anzacwood ● ☆

310775　Pomaderris velutina J. H. Willis;绒毛安匝木;Velutinous Anzacwood ● ☆

310776　Pomangium Reinw. = Argostemma Wall. ■

310777　Pomaria Cav. = Caesalpinia L. ●

310778　Pomasterion Miq. = Actinostemma Griff. ■

310779　Pomasterion japonicum Miq. = Actinostemma tenerum Griff. ■

310780　Pomasterium Miq. = Actinostemma Griff. ■

310781　Pomatiderris Roem. et Schult. = Pomaderris Labill. ● ☆

310782　Pomatiderris Roem. et Schult. = Pomatoderris Roem. et Schult. ● ☆

310783　Pomatium C. F. Gaertn. = Bertiera Aubl. ■☆

310784　Pomatium Nees et Mart. ex Lindl. = Ocotea Aubl. ● ☆

310785　Pomatium spicatum C. F. Gaertn. = Bertiera spicata（C. F. Gaertn.）K. Schum. ■☆

310786　Pomatocalpa Breda = Pomatocalpa Breda, Kuhl et Hasselt ■

310787　Pomatocalpa Breda, Kuhl et Hasselt（1829）;鹿角兰属（绣球兰属）;Antlerorchis, Pomatocalpa ■

310788　Pomatocalpa acuminata（Rolfe）Schltr. = Pomatocalpa undulatum（Lindl.）J. J. Sm. subsp. acuminatum（Rolfe）S. Watthana et S. W. Chung ■

310789　Pomatocalpa acuminatum（Rolfe）Schltr. = Pomatocalpa undulatum（Lindl.）J. J. Sm. subsp. acuminatum（Rolfe）S. Watthana et S. W. Chung ■

310790　Pomatocalpa brachybotryum（Hayata）Hayata = Pomatocalpa acuminate（Rolfe）Schltr. ■

310791　Pomatocalpa brachybotryum（Hayata）Schltr. = Pomatocalpa acuminate（Rolfe）Schltr. ■

310792　Pomatocalpa brachybotryum（Hayata）Schltr. = Pomatocalpa undulatum（Lindl.）J. J. Sm. subsp. acuminatum（Rolfe）S. Watthana et S. W. Chung ■

310793　Pomatocalpa breviracemum（Hayata）Hayata = Trichoglottis rosea（Lindl.）Ames ■

310794　Pomatocalpa densiflorum（Lindl.）Ts. Tang et F. T. Wang = Robiquetia spatulata（Blume）J. J. Sm. ■

310795　Pomatocalpa luchuensis（Rolfe）Ts. Tang et F. T. Wang = Staurochilus luchuensis（Rolfe）Fukuy. ■

310796　Pomatocalpa poilanei（Gagnep.）Ts. Tang et F. T. Wang = Smitinandia micrantha（Lindl.）Holttum ■

310797　Pomatocalpa spicatum Breda;鹿角兰（白花鹿角兰）;Antlerorchis, Spikedflower Pomatocalpa, Whiteflower Pomatocalpa ■

310798　Pomatocalpa undulatum（Lindl.）J. J. Sm. subsp. acuminatum（Rolfe）S. Watthana et S. W. Chung;台湾鹿角兰（黄绣球兰）;Acuminate Pomatocalpa, Taiwan Antlerorchis ■

310799　Pomatocalpa virginale（Hance）J. J. Sm. = Robiquetia succisa（Lindl.）Seidenf. et Garay ■

310800　Pomatocalpa vitellinum（Rchb. f.）Ames;蛋黄色鹿角兰;Yolkcolour Pomatocalpa ■☆

310801　Pomatocalpa wendlandorum（Rchb. f.）J. J. Sm.;白花鹿角兰■

310802　Pomatocalpa wendlandorum（Rchb. f.）J. J. Sm. = Pomatocalpa spicatum Breda ■

310803　Pomoderris Roem. et Schult. = Pomaderris Labill. ●☆

310804　Pomoderris Schult. = Pomaderris Labill. ●☆

310805　Pomoealpa brachybotryum（Hayata）Hayata = Pomatocalpa acuminata（Rolfe）Schltr. ■

310806　Pomatosace Maxim.（1881）；羽叶点地梅属；Pomatosace ■★

310807　Pomatosace filicula Maxim.；羽叶点地梅；Common Pomatosace ■

310808　Pomatostoma Stapf = Anerincleistus Korth. ●☆

310809　Pomatotheca F. Muell. = Trianthema L. ■

310810　Pomax Sol. ex DC.（1830）；东亚茜属 ☆

310811　Pomax umbellata（Gaertn.）A. Rich.；东亚茜 ☆

310812　Pomazota Ridl. = Coptophyllum Korth.（保留属名）■☆

310813　Pombalia Vand. = Hybanthus Jacq.（保留属名）■●

310814　Pomelia Durando ex Pomel = Daucus L. ■

310815　Pomelia setifolia（Desf.）Durando = Daucus setifolius Desf. ■☆

310816　Pomelina（Maire）Güemes et Raynaud = Fumana（Dunal）Spach ●☆

310817　Pomereula Dombey ex DC. = Miconia Ruiz et Pav.（保留属名）●☆

310818　Pometia J. R. Forst. et G. Forst.（1775）；番龙眼属；Pometia ●

310819　Pometia Vell. = Neopometia Aubrév. ●☆

310820　Pometia Vell. = Pradosia Liais ●☆

310821　Pometia Willd. = Allophylus L. ●

310822　Pometia acuminata Radlk.；渐尖番龙眼●☆

310823　Pometia pinnata J. R. Forst. et G. Forst.；番龙眼（台东龙眼）；Fiji Longan，Malugay，Pinnate Pometia ●

310824　Pometia pinnata J. R. Forst. et G. Forst. f. tomentosa（Blume）Jacobs = Pometia tomentosa（Blume）Teijsm. et Binn. ●◇

310825　Pometia pinnata J. R. Forst. et G. Forst. f. tomentosa（Blume）Jacobs = Pometia pinnata J. R. Forst. et G. Forst. ●

310826　Pometia tomentosa（Blume）Teijsm. et Binn.；绒毛番龙眼（毛番龙眼，茸毛番龙眼）；Tomentose Pometia ●◇

310827　Pometia tomentosa（Blume）Teijsm. et Binn. = Pometia pinnata J. R. Forst. et G. Forst. ●

310828　Pommereschea Wittm.（1895）；直唇姜属；Pommereschea ■

310829　Pommereschea lackneri Wittm.；直唇姜；Beautiful Pommereschea，Lackner Pommereschea ■

310830　Pommereschea spectabilis（King et Prain）K. Schum.；短柄直唇姜（直唇姜）；Beautiful Pommereschea，Shortstalk Pommereschea ■

310831　Pommereschia T. Durand et Jacks. = Pommereulla L. f. ■☆

310832　Pommereulla L. f.（1779）；单生偏穗草属■☆

310833　Pommereulla cornucopiae L. f.；单生偏穗草■☆

310834　Pommereullia Post et Kuntze = Miconia Ruiz et Pav.（保留属名）●☆

310835　Pommereullia Post et Kuntze = Pomereula Dombey ex DC. ●☆

310836　Pompadoura Buc'hoz ex DC. = Calycanthus L.（保留属名）●

310837　Pomphidea Miers = Ravenia Vell. ●☆

310838　Pompila Noronha = Sterculia L. ●

310839　Ponaea Bubani = Carpesium L. ■

310840　Ponaea Schreb. = Toulicia Aubl. ●☆

310841　Ponapea Becc.（1924）；波纳佩桐属（庞那皮桐属）●☆

310842　Ponapea Becc. = Ptychosperma Labill. ●☆

310843　Ponapea ledermanniana Becc.；波纳佩桐（庞那皮桐）●☆

310844　Ponaria Raf. = Veronica L. ■

310845　Ponceletia R. Br. = Sprengelia Sm. ●☆

310846　Ponceletia Thouars = Psammophila Schult. ■

310847　Ponceletia Thouars = Spartina Schreb. ex J. F. Gmel. ■

310848　Poncirus Raf.（1838）；枳属（枸橘属）；Poncirus，Stockorange，Trifoliate Orange，Trifoliate-orange ●★

310849　Poncirus polyandra S. Q. Ding et al.；富民枳（富民藤）；Fumin Trifoliate-orange，Manystamen Stockorange，Many-stamen Trifoliate-orange，Polyandrous Trifoliate Orange ●◇

310850　Poncirus trifoliata（L.）Raf.；枳（臭橘，臭杞，钢橘子，枸鬃李，枸棘子，枸橘，枸橘李，桔，橘，苦桶子，雀不站，唐橘，铁篱寨，野橙子，野李子，枳壳）；Hardy Orange，Japanese Bitter Orange，Trifoliate Citrus，Trifoliate Orange，Trifoliate Stockorange，Trifoliate-orange ●

310851　Poncirus trifoliata（L.）Raf. = Citrus trifoliata L. ●

310852　Poncirus trifoliata（L.）Raf. f. monstrosa（T. Ito）H. Hara；飞龙枳；Flying Dragon Trifoliate-orange ●☆

310853　Poncirus trifoliata（L.）Raf. f. monstrosa（T. Ito）H. Hara = Poncirus trifoliata（L.）Raf. ●

310854　Poncirus trifoliata（L.）Raf. var. monstrosa（T. Ito）Swingle = Poncirus trifoliata（L.）Raf. f. monstrosa（T. Ito）H. Hara ●☆

310855　Poncirus trifoliata（L.）Raf. var. monstrosa（T. Ito）Swingle = Poncirus trifoliata（L.）Raf. ●

310856　Ponera Lindl.（1831）；波纳兰属（波内兰属）■☆

310857　Ponera affinis（Poepp. et Endl.）Rchb. f.；近缘波纳兰■☆

310858　Ponera australis Cogn.；南方波纳兰●☆

310859　Ponerorchis Rchb. f.（1852）；小红门兰属（少花兰属，小蝶兰属）■

310860　Ponerorchis Rchb. f. = Gymnadenia R. Br. ■

310861　Ponerorchis Rchb. f. = Habenaria Willd. ■

310862　Ponerorchis Rchb. f. = Orchis L. ■

310863　Ponerorchis beesiana（W. W. Sm.）Soó = Ponerorchis chusua（D. Don）Soó ■

310864　Ponerorchis brevicalcarata（Finet）Soó；短距小红门兰（短距红门兰）；Shortspur Orchis ■

310865　Ponerorchis brevicalcarata（Finet）Soó = Orchis brevicalcarata（Finet）Schltr. ■

310866　Ponerorchis chidori（Makino）Ohwi；千岛小红门兰（千岛小蝶兰）■☆

310867　Ponerorchis chidori（Makino）Ohwi f. albiflora（Sugim.）F. Maek.；白花千岛小红门兰（白花千岛小蝶兰）■☆

310868　Ponerorchis chidori（Makino）Ohwi var. curtipes（Ohwi）F. Maek.；短梗千岛小红门兰（短梗千岛小蝶兰）■☆

310869　Ponerorchis chingshuishania S. S. Ying = Orchis takasagomontana Masam. ■

310870　Ponerorchis chingshuishania S. S. Ying = Ponerorchis takasagomontana（Masam.）Ohwi ■

310871　Ponerorchis chrysea（W. W. Sm.）Soó；黄花小红门兰（黄花红门兰）；Yellow Orchis ■

310872　Ponerorchis chrysea（W. W. Sm.）Soó = Orchis chrysea（W. W. Sm.）Schltr. ■

310873　Ponerorchis chusua（D. Don）Soó；广布小红门兰（高山红门兰，广布红门兰，库莎红门兰，千岛兰）；Blazon Orchis，Orchis ■

310874　Ponerorchis chusua（D. Don）Soó = Orchis chusua D. Don ■

310875　Ponerorchis chusua（D. Don）Soó subsp. nana（King et Pantl.）Soó = Orchis chusua D. Don ■

310876　Ponerorchis chusua（D. Don）Soó subsp. nana（King et Pantl.）Soó = Ponerorchis chusua（D. Don）Soó ■

310877　Ponerorchis chusua（D. Don）Soó var. delavayi（Schltr.）Soó = Orchis chusua D. Don ■

310878　Ponerorchis chusua（D. Don）Soó var. delavayi（Schltr.）Soó

= Ponerorchis chusua（D. Don）Soó ■

310879　Ponerorchis chusua（D. Don）Soó var. giraldiana（Kraenzl.）Soó = Orchis chusua D. Don ■

310880　Ponerorchis chusua（D. Don）Soó var. giraldiana（Kraenzl.）Soó = Ponerorchis chusua（D. Don）Soó ■

310881　Ponerorchis chusua（D. Don）Soó var. tenii（Schltr.）Soó = Orchis chusua D. Don ■

310882　Ponerorchis chusua（D. Don）Soó var. tenii（Schltr.）Soó = Ponerorchis chusua（D. Don）Soó ■

310883　Ponerorchis chusua（D. Don）Soó var. unifoliata（Schltr.）Soó = Orchis chusua D. Don ■

310884　Ponerorchis chusua（D. Don）Soó var. unifoliata（Schltr.）Soó = Ponerorchis chusua（D. Don）Soó ■

310885　Ponerorchis crenulata（Schltr.）Soó;齿缘小红门兰（齿缘红门兰）;Crenulate Orchis ■

310886　Ponerorchis crenulata（Schltr.）Soó = Orchis crenulata Schltr. ■

310887　Ponerorchis curtipes（Ohwi）Soó = Ponerorchis chidori（Makino）Ohwi var. curtipes（Ohwi）F. Maek. ■☆

310888　Ponerorchis diantha（Schltr.）Soó = Galearis spathulata（Lindl.）P. F. Hunt ■

310889　Ponerorchis diantha（Schltr.）Soó = Orchis diantha Schltr. ■

310890　Ponerorchis formosensis（S. S. Ying）S. S. Ying = Amitostigma gracile（Blume）Schltr. ■

310891　Ponerorchis graminifolia Rchb. f. ;禾叶小蝶兰■☆

310892　Ponerorchis graminifolia Rchb. f. f. albiflora（Murai）F. Maek. ;白花禾叶小蝶兰■☆

310893　Ponerorchis graminifolia Rchb. f. var. suzukiana（Ohwi）Soó;铃木小蝶兰■☆

310894　Ponerorchis graminifollia Rchb. f. var. kurokamiana（Ohwi et Hatus.）T. Hashim. ;畔上小蝶兰■☆

310895　Ponerorchis hui（Ts. Tang et F. T. Wang）Soó = Orchis limprichtii Schltr. ■

310896　Ponerorchis hui（Ts. Tang et F. T. Wang）Soó = Ponerorchis limprichtii（Schltr.）Soó ■

310897　Ponerorchis joo-iokiana（Makino）Nakai;日本小蝶兰■☆

310898　Ponerorchis kiraishiensis（Hayata）Ohwi;奇莱小红门兰（红小蝶兰）■

310899　Ponerorchis kiraishiensis（Hayata）Ohwi = Orchis kiraishiensis Hayata ■

310900　Ponerorchis kiraishiensis（Hayata）Ohwi var. leucantha（Masam.）A. T. Hsieh = Ponerorchis tominagai（Hayata）H. J. Su et J. J. Chen ■

310901　Ponerorchis kiraishiensis Hayata = Orchis kiraishiensis Hayata ■

310902　Ponerorchis kiraishiensis Hayata = Orchis nanhutashanensis S. S. Ying ■

310903　Ponerorchis kiraishiensis Hayata = Ponerorchis kiraishiensis（Hayata）Ohwi ■

310904　Ponerorchis kiraishiensis Hayata var. leucantha（Masam.）A. T. Hsieh = Orchis tomingai（Hayata）H. J. Su ■

310905　Ponerorchis kiraishiensis Hayata var. leucantha（Masam.）A. T. Hsieh = Ponerorchis tominagai（Hayata）H. J. Su et J. J. Chen ■

310906　Ponerorchis kiraishiensis Hayata var. leucantha（Masam.）A. T. Hsieh = Ponerorchis kuanshanensis（S. S. Ying）S. S. Ying ■

310907　Ponerorchis kuanshanensis（S. S. Ying）S. S. Ying = Orchis tomingai（Hayata）H. J. Su ■

310908　Ponerorchis kuanshanensis（S. S. Ying）S. S. Ying = Ponerorchis tominagai（Hayata）H. J. Su et J. J. Chen ■

310909　Ponerorchis kunihikoana（Masam. et Fukuy.）Soó = Orchis tomingai（Hayata）H. J. Su ■

310910　Ponerorchis kunihikoana（Masam. et Fukuy.）Soó = Ponerorchis tominagai（Hayata）H. J. Su et J. J. Chen ■

310911　Ponerorchis kunihikoana（Masam.）Soó = Orchis tomingai（Hayata）H. J. Su ■

310912　Ponerorchis kunihikoana（Masam.）Soó var. leucantha（Masam.）Soó = Orchis tomingai（Hayata）H. J. Su ■

310913　Ponerorchis kurokamiana（Ohwi et Hatus.）F. Maek. = Ponerorchis graminifollia Rchb. f. var. kurokamiana（Ohwi et Hatus.）T. Hashim. ■☆

310914　Ponerorchis limprichtii（Schltr.）Soó;华西小红门兰（单叶红门兰,华西红门兰,岩一枝箭）;Limpricht Orchis, W. China Orchis ■

310915　Ponerorchis limprichtii（Schltr.）Soó = Orchis limprichtii Schltr. ■

310916　Ponerorchis monophylla（Collett et Hemsl.）Soó;毛轴小红门兰（毛轴红门兰）;Hairaxle Orchis ■

310917　Ponerorchis monophylla（Collett et Hemsl.）Soó = Orchis monophylla（Collett et Hemsl.）Rolfe ■

310918　Ponerorchis nana（King et Pantl.）Soó = Orchis chusua D. Don ■

310919　Ponerorchis nana（King et Pantl.）Soó = Ponerorchis chusua（D. Don）Soó ■

310920　Ponerorchis nanhutashanensis S. S. Ying = Orchis kiraishiensis Hayata ■

310921　Ponerorchis nanhutashanensis S. S. Ying = Orchis nanhutashanensis S. S. Ying ■

310922　Ponerorchis nanhutashanensis S. S. Ying = Ponerorchis kiraishiensis（Hayata）Ohwi ■

310923　Ponerorchis omeishanica（Ts. Tang, F. T. Wang et K. Y. Lang）S. C. Chen, P. J. Cribb et S. W. Gale;峨眉小红门兰（峨眉红门兰）;Emei Orchis, Omei Orchis ■

310924　Ponerorchis pauciflora（Lindl.）Ohwi = Orchis chusua D. Don ■

310925　Ponerorchis pauciflora（Lindl.）Ohwi = Orchis pauciflora（Lindl.）Fisch. ex Schltr. ■☆

310926　Ponerorchis pauciflora（Lindl.）Ohwi = Ponerorchis chusua（D. Don）Soó ■

310927　Ponerorchis pauciflora（Lindl.）Ohwi var. joo-iokiana（Makino）Ohwi = Ponerorchis joo-iokiana（Makino）Nakai ■☆

310928　Ponerorchis pugeensis（K. Y. Lang）S. C. Chen;普格小红门兰（普格红门兰）;Puge Orchis ■

310929　Ponerorchis pulchella（Hand. -Mazz.）Soó = Orchis chusua D. Don ■

310930　Ponerorchis pulchella（Hand. -Mazz.）Soó = Ponerorchis chusua（D. Don）Soó ■

310931　Ponerorchis rotundifolia（Banks ex Pursh）Soó = Amerorchis rotundifolia（Banks ex Pursh）Hultén ■☆

310932　Ponerorchis schlechteri Perner et Y. B. Luo = Orchis crenulata Schltr. ■

310933　Ponerorchis sichuanica（K. Y. Lang）S. C. Chen;四川小红门兰（四川红门兰）;Sichuan Orchis ■

310934　Ponerorchis taitungensis（S. S. Ying）S. S. Ying = Orchis taitungensis S. S. Ying ■

310935　Ponerorchis taitungensis（S. S. Ying）S. S. Ying = Platanthera tipuloides（L. f.）Lindl. ■

310936　Ponerorchis taitungensis（S. S. Ying）S. S. Ying var. alboflorens（S. S. Ying）S. S. Ying = Orchis taitungensis S. S. Ying var.

alboflorens S. S. Ying ■

310937　Ponerorchis taitungensis (S. S. Ying) S. S. Ying var. alboflorens (S. S. Ying) S. S. Ying = Ponerorchis taiwanensis (Fukuy.) Ohwi ■

310938　Ponerorchis taitungensis (S. S. Ying) S. S. Ying var. alboflorens (S. S. Ying) S. S. Ying = Platanthera tipuloides (L. f.) Lindl. ■

310939　Ponerorchis taiwanensis (Fukuy.) Ohwi；台湾小红门兰（台湾红兰，台湾红门兰，台湾兰，台湾小蝶兰）；Taiwan Orchis ■

310940　Ponerorchis taiwanensis (Fukuy.) Ohwi = Orchis taiwanensis Fukuy. ■

310941　Ponerorchis taiwanensis (S. S. Ying) S. S. Ying = Orchis taiwanensis Fukuy. ■

310942　Ponerorchis taiwanensis (S. S. Ying) S. S. Ying var. alboflorens (S. S. Ying) S. S. Ying = Orchis taiwanensis Fukuy. ■

310943　Ponerorchis takasagomontana (Masam.) Ohwi；高山小红门兰（高山红兰，高山红门兰，高山兰，高山小蝶兰）；Alp Orchis ■

310944　Ponerorchis takasagomontana (Masam.) Ohwi = Orchis takasagomontana Masam. ■

310945　Ponerorchis taoloii (S. S. Ying) S. S. Ying = Orchis tomingai (Hayata) H. J. Su ■

310946　Ponerorchis taoloii (S. S. Ying) T. P. Lin = Ponerorchis kunihikoana (Masam. et Fukuy.) Soó ■

310947　Ponerorchis taoloii (S. S. Ying) T. P. Lin = Ponerorchis tominagai (Hayata) H. J. Su et J. J. Chen ■

310948　Ponerorchis tominagai (Hayata) H. J. Su et J. J. Chen；白花小红门兰（白花红门兰，白花兰，大水窟红兰，红斑兰）；White Orchis ■

310949　Pongam Adans. = Pongamia Adans. (保留属名) ●

310950　Pongamia Adans. (1763)(保留属名)；水黄皮属；Pongamia, Waterwampee ●

310951　Pongamia Vent. = Millettia Wight et Arn. (保留属名) ●■

310952　Pongamia Vent. = Pongamia Adans. (保留属名) ●

310953　Pongamia canarensis Dalzell = Paraderris canarensis (Dalzell) Adema ●

310954　Pongamia elliptica Wall. = Derris elliptica (Wall.) Benth. ●

310955　Pongamia elliptica Wall. = Paraderris elliptica (Wall.) Adema ●

310956　Pongamia extensa Wall. = Millettia extensa Benth. ●☆

310957　Pongamia glabra Vent. = Pongamia pinnata (L.) Pierre ex Merr. ●

310958　Pongamia madagascariensis Bojer ex Oliv. = Derris trifoliata Lour. ●

310959　Pongamia mitis (L.) Kurz = Pongamia pinnata (L.) Pierre ●

310960　Pongamia pinnata (L.) Pierre = Millettia pinnata (L.) Panigrahi ●

310961　Pongamia pinnata (L.) Pierre = Pongamia pinnata (L.) Pierre ex Merr. ●

310962　Pongamia pinnata (L.) Pierre ex Merr.；水黄皮（九重吹，水刀豆，水流兵，水流豆，水罗豆，无毛水黄皮，野豆）；Indian Beech, Karanja, Karum Tree, Kemng, Poongaoil Pongamia, Poonga-oil Tree, Thinwin, Waterwampee ●

310963　Pongamia taiwaniana Hayata = Millettia pachycarpa Benth. ●■

310964　Pongamiopsis R. Vig. (1950)；类水黄皮属●☆

310965　Pongamiopsis amygdalina (Baill.) R. Vig.；类水黄皮●☆

310966　Pongamiopsis pervilleana (Baill.) R. Vig.；非洲类水黄皮●☆

310967　Pongamiopsis viguieri Du Puy et Labat；马岛类水黄皮●☆

310968　Pongati Adans. = Sphenoclea Gaertn. (保留属名) ■

310969　Pongatiaceae Endl. = Sphenocleaceae T. Baskerv. (保留科名) ■

310970　Pongatiaceae Endl. ex Meisn. = Sphenocleaceae T. Baskerv. (保留科名) ■

310971　Pongatium Adans. = Pongati Adans. ■

310972　Pongatium Adans. = Sphenoclea Gaertn. (保留属名) ■

310973　Pongatium Juss. = Sphenoclea Gaertn. (保留属名) ■

310974　Pongatium indicum Lam. = Sphenoclea zeylanica Gaertn. ■

310975　Pongatium spongiosum Blanco = Sphenoclea zeylanica Gaertn. ■

310976　Pongelia Raf. (废弃属名) = Dolichandrone (Fenzl) Seem. (保留属名) ●

310977　Pongelion Adans. (废弃属名) = Adenanthera L. ●

310978　Pongelion Adans. (废弃属名) = Ailanthus Desf. (保留属名) ●

310979　Pongelion glandulosum Pierre = Ailanthus altissima (Mill.) Swingle ●

310980　Pongelium Scop. = Pongelion Adans. (废弃属名) ●

310981　Pongonia Grant = Pogonia Juss. ■

310982　Pongonia Griff. ex Grant = Pogonia Juss. ■

310983　Ponista Raf. = Saxifraga L. ●

310984　Ponna Boehm. = Calophyllum L. ●

310985　Pontaletsje Adans. = Hedyotis L. (保留属名) ●■

310986　Pontaletsje Adans. = Poutaletsje Adans. ●■

310987　Pontania Lem. = Brachysema R. Br. ●☆

310988　Pontechium U. -R. Böhle et Hilger(2000)；法国蓝蓟属■☆

310989　Ponteras Hoffmanns. = Pontederia L. ■☆

310990　Pontederia L. (1753)；海寿属（海寿属，美雨久属，梭鱼草属）；Pickerel Weed, Pickerelweed, Pickerel-weed ■☆

310991　Pontederia angustifolia Pursh = Pontederia cordata L. ■☆

310992　Pontederia azurea Sw. = Eichhornia azurea (Sw.) Kunth ■☆

310993　Pontederia cordata L.；海寿花（海寿，梭鱼草）；Pickerel Rush, Pickerel Weed, Pickerelweed, Pickerel-weed, Wampee ■☆

310994　Pontederia cordata L. f. latifolia (Farw.) House = Pontederia cordata L. ■☆

310995　Pontederia cordata L. f. latifolia (Raf.) House = Pontederia cordata L. ■☆

310996　Pontederia cordata L. var. albiflora Short；白花海寿花；Pickerel Weed, White ■☆

310997　Pontederia cordata L. var. angustifolia (Pursh) Torr. et Elliott = Pontederia cordata L. ■☆

310998　Pontederia cordata L. var. lanceolata (Nutt.) Griseb. = Pontederia cordata L. ■☆

310999　Pontederia cordata L. var. lancifolia (Muhl.) Torr. = Pontederia cordata L. ■☆

311000　Pontederia cordata L. var. ovalis (Mart.) Solms；椭圆海寿花■☆

311001　Pontederia crassipes Mart. = Eichhornia crassipes (Mart.) Solms ■

311002　Pontederia dilatata Buch. -Ham. = Monochoria hastata (L.) Solms ■

311003　Pontederia dubia Blume = Hydrocharis dubia (Blume) Backer ■

311004　Pontederia hastata L. = Monochoria hastata (L.) Solms ■

311005　Pontederia lanceolata Nutt. = Pontederia cordata L. ■☆

311006　Pontederia lanceolata Nutt. var. vichadensis Herman = Pontederia cordata L. var. ovalis (Mart.) Solms ■☆

311007　Pontederia lancifolia Muhl. = Pontederia cordata L. ■☆

311008　Pontederia limosa Sw. = Heteranthera limosa (Sw.) Willd. ■☆

311009　Pontederia linearis Hassk. = Monochoria vaginalis (Burm. f.) C. Presl ex Kunth ■

311010　Pontederia natans P. Beauv. = Eichhornia natans (P. Beauv.) Solms ■☆

311011　Pontederia ovalis Mart. = Pontederia cordata L. var. ovalis

（Mart.）Solms ■☆

311012　Pontederia ovata Hook. et Arn. = Monochoria vaginalis（Burm. f.）C. Presl ex Kunth ■

311013　Pontederia ovata L. = Phrynium rheedei Suresh et Nicolson ■

311014　Pontederia pauciflora Blume = Monochoria vaginalis（Burm. f.）C. Presl ex Kunth ■

311015　Pontederia plantaginea Roxb. = Monochoria vaginalis（Burm. f.）C. Presl ex Kunth ■

311016　Pontederia sagittata Roxb. = Monochoria hastata（L.）Solms ■

311017　Pontederia stricta Burm. f. = Onixotis stricta（Burm. f.）Wijnands ■☆

311018　Pontederia vaginalis Burm. f. = Monochoria vaginalis（Burm. f.）C. Presl ex Kunth ■

311019　Pontederiaceae Kunth（1816）（保留科名）；雨久花科；Pickerelweed Family，Pickerel-weed Family ■

311020　Ponthieva R. Br.（1813）；蓬氏兰属■☆

311021　Ponthieva brittoniae Ames；布里顿兰；Mrs. Britton's shadow witch ■☆

311022　Ponthieva glandulosa（Sims）R. Br. = Ponthieva racemosa（Walter）C. Mohr ■☆

311023　Ponthieva racemosa（Walter）C. Mohr；荫地蓬氏兰；Shadow Witch ■☆

311024　Ponthieva racemosa（Walter）C. Mohr var. brittoniae（Ames）Luer = Ponthieva brittoniae Ames ■☆

311025　Pontia Bubani = Chrysanthemum L.（保留属名）■●

311026　Pontia Bubani = Pyrethrum Zinn ■

311027　Pontinia Fries = Eudianthe（Rchb.）Rchb. ■

311028　Pontinia Fries = Silene L.（保留属名）■

311029　Pontopidana Scop. = Couroupita Aubl. ●☆

311030　Pontoppidana Steud. = Pontopidana Scop. ●☆

311031　Pontya A. Chev. = Bosquiea Thouars ex Baill. ●☆

311032　Pontya A. Chev. = Trilepisium Thouars ●☆

311033　Poortmannia Drake = Trianaea Planch. et Linden ●☆

311034　Pootia Dennst. = Canscora Lam. ■

311035　Pootia Miq. = Voacanga Thouars ●

311036　Poponax Raf. = Acacia Mill.（保留属名）●■

311037　Popoviocodonia Fed.（1957）；波氏桔梗属■☆

311038　Popoviocodonia Fed. = Campanula L. ■●

311039　Popoviocodonia stenocarpa（Trautv. et C. A. Mey.）Fed.；狭果波氏桔梗■☆

311040　Popoviocodonia uyemurae（Kudo）Fed.；波氏桔梗■☆

311041　Popoviolimon Lincz.（1971）；简枝补血草属■☆

311042　Popoviolimon turcomanicum（Lincz.）Lincz.；简枝补血草●☆

311043　Popowia Endl.（1839）；嘉陵花属；Popowia ●

311044　Popowia argentea De Wild. = Monanthotaxis poggei Engl. et Diels ●☆

311045　Popowia baillonii（Scott-Elliot）Engl. et Diels = Monanthotaxis mannii（Baill.）Verdc. ●☆

311046　Popowia barteri Baill. = Monanthotaxis barteri（Baill.）Verdc. ●☆

311047　Popowia bequaertii De Wild. = Monanthotaxis littoralis（Bagsh. et Baker f.）Verdc. ●☆

311048　Popowia bicornis Boutique = Monanthotaxis bicornis（Boutique）Verdc. ●☆

311049　Popowia bokoli（De Wild. et T. Durand）Robyns et Ghesq. ex Boutique = Monanthotaxis bokoli（De Wild. et T. Durand）Verdc. ●☆

311050　Popowia buchananii（Engl.）Engl. et Diels = Monanthotaxis buchananii（Engl.）Verdc. ●☆

311051　Popowia buchananii（Engl.）Engl. et Diels var. trichantha Diels = Monanthotaxis stenosepala（Engl. et Diels）Verdc. ●☆

311052　Popowia buchananii（Engl.）Engl. et Diels var. trichantha Diels = Monanthotaxis trichantha（Diels）Verdc. ●☆

311053　Popowia caffra（Sond.）Hook. f. et Thomson ex Benth. = Monanthotaxis caffra（Sond.）Verdc. ●☆

311054　Popowia capea E. G. Camus et A. Camus = Monanthotaxis capea（E. G. Camus et A. Camus）Verdc. ●☆

311055　Popowia caulantha Exell = Monanthotaxis diclina（Sprague）Verdc. ●☆

311056　Popowia cauliflora Chipp = Monanthotaxis cauliflora（Chipp）Verdc. ●☆

311057　Popowia chasei N. Robson = Monanthotaxis chasei（N. Robson）Verdc. ●☆

311058　Popowia congensis（Engl. et Diels）Engl. et Diels = Monanthotaxis laurentii（De Wild.）Verdc. ●☆

311059　Popowia dalzielii Hutch. = Monanthotaxis vogelii（Hook. f.）Verdc. ●☆

311060　Popowia dawei Diels = Monanthotaxis littoralis（Bagsh. et Baker f.）Verdc. ●☆

311061　Popowia diclina Sprague = Monanthotaxis diclina（Sprague）Verdc. ●☆

311062　Popowia dicranantha Diels = Artabotrys brachypetalus Benth. ●☆

311063　Popowia dictyoneura Diels = Monanthotaxis dictyoneura（Diels）Verdc. ●☆

311064　Popowia discolor Diels = Monanthotaxis discolor（Diels）Verdc. ●☆

311065　Popowia djumaensis De Wild. = Monanthotaxis ferruginea（Oliv.）Verdc. ●☆

311066　Popowia djurensis Engl. et Diels = Monanthotaxis buchananii（Engl.）Verdc. ●☆

311067　Popowia elegans Engl. et Diels = Monanthotaxis elegans（Engl. et Diels）Verdc. ●☆

311068　Popowia enghiana Diels = Friesodielsia enghiana（Diels）Verdc. ●☆

311069　Popowia engleriana Exell et Mendonça = Sphaerocoryne gracilis（Engl. et Diels）Verdc. subsp. engleriana（Exell et Mendonça）Verdc. ●☆

311070　Popowia ferruginea（Oliv.）Engl. et Diels = Monanthotaxis ferruginea（Oliv.）Verdc. ●☆

311071　Popowia filamentosa Diels = Monanthotaxis filamentosa（Diels）Verdc. ●☆

311072　Popowia foliosa Engl. et Diels = Monanthotaxis foliosa（Engl. et Diels）Verdc. ●☆

311073　Popowia fornicata Baill. = Monanthotaxis fornicata（Baill.）Verdc. ●☆

311074　Popowia germainii Boutique = Monanthotaxis germainii（Boutique）Verdc. ●☆

311075　Popowia gilletii De Wild. = Monanthotaxis gilletii（De Wild.）Verdc. ●☆

311076　Popowia glomerulata Le Thomas = Monanthotaxis glomerulata（Le Thomas）Verdc. ●☆

311077　Popowia gracilis Engl. et Diels = Sphaerocoryne gracilis（Engl. et Diels）Verdc. ●☆

311078　Popowia gracilis Engl. et Diels subsp. engleriana（Exell et Mendona）N. Robson = Sphaerocoryne gracilis（Engl. et Diels）Verdc. subsp. engleriana（Exell et Mendonça）Verdc. ●☆

311079　Popowia hallei Le Thomas ＝Monanthotaxis letestui Pellegr. var. hallei（Le Thomas）Le Thomas ●☆

311080　Popowia heudelotii Baill. ＝ Monanthotaxis barteri（Baill.）Verdc. ●☆

311081　Popowia iboundjiensis Pellegr. ＝ Monanthotaxis bokoli（De Wild. et T. Durand）Verdc. ●☆

311082　Popowia kirkii Benth. ＝Cleistochlamys kirkii（Benth.）Oliv. ●☆

311083　Popowia klainii Engl. ＝Monanthotaxis klainei（Engl.）Verdc. ●☆

311084　Popowia klainii Engl. var. angustifolia（Boutique）Le Thomas ＝ Monanthotaxis klainei（Engl.）Verdc. var. angustifolia（Boutique）Verdc. ●☆

311085　Popowia klainii Engl. var. lastoursvillensis（Pellegr.）Le Thomas ＝ Monanthotaxis klainei（Engl.）Verdc. var. lastoursvillensis（Pellegr.）Verdc. ●☆

311086　Popowia lastoursvillensis Pellegr. ＝ Monanthotaxis klainei（Engl.）Verdc. var. lastoursvillensis（Pellegr.）Verdc. ●☆

311087　Popowia laurentii De Wild. ＝ Monanthotaxis laurentii（De Wild.）Verdc. ●☆

311088　Popowia letestui Pellegr. ＝ Monanthotaxis pellegrinii Verdc. ●☆

311089　Popowia letouzeyi Le Thomas ＝ Monanthotaxis letouzeyi（Le Thomas）Verdc. ●☆

311090　Popowia littoralis Bagsh. et Baker f. ＝ Monanthotaxis littoralis（Bagsh. et Baker f.）Verdc. ●☆

311091　Popowia louisii Boutique ＝ Monanthotaxis parvifolia（Oliv.）Verdc. ●☆

311092　Popowia lucidula（Oliv.）Engl. et Diels ＝ Monanthotaxis lucidula（Oliv.）Verdc. ●☆

311093　Popowia macrocarpa Engl. et Diels ＝ Sphaerocoryne gracilis（Engl. et Diels）Verdc. subsp. engleriana（Exell et Mendonça）Verdc. ●☆

311094　Popowia malchairi De Wild. ＝ Monanthotaxis filamentosa（Diels）Verdc. ●☆

311095　Popowia mangenotii Sillans ＝ Friesodielsia enghiana（Diels）Verdc. ●☆

311096　Popowia mangenotii Sillans f. concolor ？ ＝ Friesodielsia enghiana（Diels）Verdc. ●☆

311097　Popowia mannii（Oliv.）Engl. et Diels ＝ Monanthotaxis cauliflora（Chipp）Verdc. ●☆

311098　Popowia mannii Baill. ＝Monanthotaxis mannii（Baill.）Verdc. ●☆

311099　Popowia mortehanii De Wild. ＝ Monanthotaxis mortehanii（De Wild.）Verdc. ●☆

311100　Popowia nigritiana Baker f. ＝ Monanthotaxis barteri（Baill.）Verdc. ●☆

311101　Popowia nimbana Schnell ＝ Monanthotaxis nimbana（Schnell）Verdc. ●☆

311102　Popowia obovata（Benth.）Engl. et Diels ＝ Friesodielsia obovata（Benth.）Verdc. ●☆

311103　Popowia ochroleuca Diels ＝ Monanthotaxis schweinfurthii（Engl. et Diels）Verdc. ●☆

311104　Popowia ochroleuca Diels var. keniensis R. E. Fr. ＝ Monanthotaxis schweinfurthii（Engl. et Diels）Verdc. ●☆

311105　Popowia oliverana Exell et Mendonça ＝ Monanthotaxis parvifolia（Oliv.）Verdc. ●☆

311106　Popowia orophila Boutique ＝ Monanthotaxis orophila（Boutique）Verdc. ●☆

311107　Popowia parvifolia（Oliv.）Engl. et Diels ＝ Monanthotaxis parvifolia（Oliv.）Verdc. ●☆

311108　Popowia pisocarpa（Blume）Endl. ；嘉陵花；Peafruit Popowia，Pealikefruit Popowia，Pealike-fruited Popowia ●

311109　Popowia prehensilis A. Chev. ＝ Monanthotaxis whytei（Stapf）Verdc. ●☆

311110　Popowia pynaertii De Wild. ＝ Monanthotaxis diclina（Sprague）Verdc. ●☆

311111　Popowia scamnopetala Exell ＝ Exellia scamnopetala（Exell）Boutique ●☆

311112　Popowia schweinfurthii Engl. et Diels ＝ Monanthotaxis schweinfurthii（Engl. et Diels）Verdc. ●☆

311113　Popowia seretii De Wild. ＝ Monanthotaxis schweinfurthii（Engl. et Diels）Verdc. var. seretii（De Wild.）Verdc. ●☆

311114　Popowia setosa Diels；刚毛嘉陵花☆

311115　Popowia stenosepala Engl. et Diels ＝ Monanthotaxis stenosepala（Engl. et Diels）Verdc. ●☆

311116　Popowia stormsii De Wild. ＝ Friesodielsia obovata（Benth.）Verdc. ●☆

311117　Popowia trichantha（Diels）R. E. Fr. ＝ Monanthotaxis trichantha（Diels）Verdc. ●☆

311118　Popowia trichocarpa Engl. et Diels ＝ Monanthotaxis trichocarpa（Engl. et Diels）Verdc. ●☆

311119　Popowia vogelii（Hook. f.）Baill. ＝ Monanthotaxis vogelii（Hook. f.）Verdc. ●☆

311120　Popowia whytei Stapf ＝Monanthotaxis whytei（Stapf）Verdc. ●☆

311121　Poppia Cam ex Vilm. ＝ Luffa Mill. ■

311122　Poppia Cam ex Vilm. ＝ Poppya Neck. ex M. Roem. ■

311123　Poppigia Hook. et Arn. ＝ Poeppigia Bertero ●☆

311124　Poppigia Hook. et Arn. ＝ Rhaphithamnus Miers ●☆

311125　Poppya Neck. ＝ Luffa Mill. ■

311126　Poppya Neck. ex M. Roem. ＝ Luffa Mill. ■

311127　Populago Mill. ＝ Caltha L. ■

311128　Populina Baill.（1891）；杨爵床属☆

311129　Populina perrieri Benoist；佩里耶杨爵床☆

311130　Populina richardii Baill. ；马岛杨爵床☆

311131　Populus L.（1753）；杨属；Aspen, Cotton Wood, Cottonwood, Poplar ●

311132　Populus 'Eugenei' ＝ Populus × canadensis Moench 'Eugenei' ●

311133　Populus 'Gelrica' ＝ Populus × canadensis Moench 'Gelrica' ●

311134　Populus 'I-214' ＝ Populus × canadensis Moench 'I-214' ●

311135　Populus 'Leipzig' ＝ Populus × canadensis Moench 'Leipzig' ●

311136　Populus 'Marilandica' ＝ Populus × canadensis Moench 'Marilandica' ●

311137　Populus 'Polska A15' ＝ Populus × canadensis Moench 'Polska A15' ●

311138　Populus 'Regenerata' ＝ Populus × canadensis Moench 'Regenerata' ●

311139　Populus 'Sacrau 79' ＝ Populus × canadensis Moench 'Sacrau 79' ●

311140　Populus 'Serotina Aurea'；金杨；Golden Poplar ●☆

311141　Populus 'Serotina' ＝ Populus × canadensis Moench 'Serotina' ●

311142　Populus × acuminata Rydb. ；披针叶杨（尖叶杨）；Lanceleaf Cottonwood, Lanceleaf Poplar, Waxleaf Cottonwood ●

311143　Populus × beijingensis W. Y. Hsu ex Z. Wang et S. L. Tung；北京杨；Beijing Poplar ●

311144　Populus × berolinensis（K. Koch）Dippel；中东杨（柏林杨）；Berlin Poplar, Berolin Poplar, Teretepetiolet Poplar ●

311145　Populus × berolinensis K. Koch ＝ Populus × berolinensis（K.

Koch) Dippel ●

311146　Populus × canadensis Moench；加杨(加拿大白杨,加拿大杨,卡洛林杨,美国大叶白杨,欧美杨)；Canada Poplar, Canadian Poplar, Carolina Poplar, Eugene Poplar, Hybrid Black Poplar, Hybrid Poplar, Railway Poplar, Topola Canadsca ●

311147　Populus × canadensis Moench 'Aurea'；金叶加杨●☆

311148　Populus × canadensis Moench 'Bachelieri'；巴氏加杨●

311149　Populus × canadensis Moench 'Eugenei'；尤金杨(尖叶加杨,欧根杨,柱冠加拿大杨)；Carolina Poplar, Eugene Poplar ●

311150　Populus × canadensis Moench 'Gelrica'；格尔里杨；Gelrica Poplar ●

311151　Populus × canadensis Moench 'Henryana'；毛芽马里兰杨；Henryana Poplar ●

311152　Populus × canadensis Moench 'I-214'；意大利杨(意大利214杨)；Italy I-poplar, Italy Poplar ●

311153　Populus × canadensis Moench 'Leipzig'；莱比锡杨(里普杨)；Leipzig Poplar ●

311154　Populus × canadensis Moench 'Lloydii'；毛柄马里兰杨；Lloydii Poplar ●

311155　Populus × canadensis Moench 'Marilandica'；马里兰杨(马里兰德杨,五月杨)；Marilandica Poplar ●

311156　Populus × canadensis Moench 'Polska A15'；波兰15号杨；Poland15A Poplar ●

311157　Populus × canadensis Moench 'Regenerata'；新生杨；Railway Poplar, Regenerata Poplar ●

311158　Populus × canadensis Moench 'Robusta'；健杨(强壮加拿大杨)；Robust Poplar ●

311159　Populus × canadensis Moench 'Robusta-Naunhof'；隆荷夫健杨●

311160　Populus × canadensis Moench 'Sacrau 79'；沙兰杨(萨克劳)；Sacrau Poplar 79 ●

311161　Populus × canadensis Moench 'Serotina de Selys'；直枝加拿大杨●☆

311162　Populus × canadensis Moench 'Serotina Erecta' = Populus × canadensis Moench 'Serotina de Selys ●☆

311163　Populus × canadensis Moench 'Serotina'；晚花杨(迟叶杨,意大利黑杨)；Black Italian Poplar, Serotina Poplar ●

311164　Populus × canadensis Moench 'Vernirubana'；酱红健杨(施氏杨)●

311165　Populus × canadensis Moench f. eugenei Simon-Louis ex Schelle = Populus × canadensis Moench 'Eugenei' ●

311166　Populus × canadensis Moench var. eugenei (Simon-Louis ex Schelle) Rehder = Populus × canadensis Moench 'Eugenei' ●

311167　Populus × canadensis Moench var. eugenei (Simon-Louis) Schelle = Populus × canadensis Moench 'Eugenei' ●

311168　Populus × canadensis Moench var. gelrica (Houtz.) Geerinck = Populus × canadensis Moench 'Gelrica' ●

311169　Populus × canadensis Moench var. regenerata (C. K. Schneid.) Rehder = Populus × canadensis Moench 'Regenerata' ●

311170　Populus × canadensis Moench var. serotina (Hartig) Rehder = Populus × canadensis Moench 'Serotina' ●

311171　Populus × candicans Aiton；白壳杨(欧洲大叶杨)；Balm of Gilead, Balsam Poplar, Ontario Poplar ●☆

311172　Populus × candicans Aiton 'Aurora'；曙光白壳杨●☆

311173　Populus × eugenei Simon-Louis ex Schelle = Populus × canadensis Moench 'Eugenei' ●

311174　Populus × euramericana (Dode) Guinier 'Grlrica' = Populus

× canadensis Moench 'Gelrica' ●

311175　Populus × euramericana (Dode) Guinier 'I-214' = Populus × canadensis Moench 'I-214' ●

311176　Populus × euramericana (Dode) Guinier 'Lepzig' = Populus × canadensis Moench 'Leipzig' ●

311177　Populus × euramericana (Dode) Guinier 'Polska 15A' = Populus × canadensis Moench 'Polska A15' ●

311178　Populus × euramericana (Dode) Guinier 'Robusta' = Populus × canadensis Moench 'Robusta' ●

311179　Populus × euramericana (Dode) Guinier 'Robusta-Naunhof' = Populus × canadensis Moench 'Robusta-Naunhof' ●

311180　Populus × euramericana (Dode) Guinier 'Sacrau 79' = Populus × canadensis Moench 'Sacrau 79' ●

311181　Populus × euramericana (Dode) Guinier = Populus × canadensis Moench ●

311182　Populus × gansuensis Z. Wang et H. L. Yang；二白杨(二青杨,甘肃杨,青白杨,软白杨)；Gansu Poplar ●

311183　Populus × gelrica Houtz. = Populus × canadensis Moench 'Gelrica' ●

311184　Populus × gileadensis Rouleau = Populus × candicans Aiton ●☆

311185　Populus × hybrida Rchb. = Populus canescens (Aiton) Sm. ●

311186　Populus × jackii Sarg. 杰克杨；Balm-of-gilead, Jack's Poplar ●☆

311187　Populus × jackii Sarg. 'Gileadensis' = Populus × candicans Aiton ●☆

311188　Populus × jrtyschensis Chang Y. Yang；额河杨；Eerjisi Poplar, Ehe Poplar, Ertix River Poplar ●

311189　Populus × marilandica (Poir.) Poir. = Populus × canadensis Moench 'Marilandica' ●

311190　Populus × regenerata Henry ex C. K. Schneid. = Populus × canadensis Moench 'Regenerata' ●

311191　Populus × xiaohei T. S. Huang et Y. Liang；小黑杨；Small Black Poplar, Xiaohei Poplar ●

311192　Populus × xiaozhuanica W. Y. Hsu et C. F. Liang；小钻杨(八里庄杨,白城杨,赤峰杨,大关杨,合作杨,小美杨)；Small Soaring Poplar, Xiaozuan Poplar ●

311193　Populus adenopoda Maxim.；响叶杨(白杨树,风响树,风响杨,山白杨,团叶白杨,团叶杨,腺柄杨,圆叶白杨)；China Aspen, Chinese Aspen ●

311194　Populus adenopoda Maxim. f. cuneata Z. Wang et S. L. Tung；楔叶响叶杨(楔叶杨)；Wedgeleaf Chinese Aspen ●

311195　Populus adenopoda Maxim. f. cuneata Z. Wang et S. L. Tung = Populus adenopoda Maxim. ●

311196　Populus adenopoda Maxim. f. microcarpa Z. Wang et S. L. Tung；小果响叶杨；Small-fruit Chinese Aspen ●

311197　Populus adenopoda Maxim. f. microcarpa Z. Wang et S. L. Tung = Populus adenopoda Maxim. ●

311198　Populus adenopoda Maxim. var. microphylla T. B. Chao；小叶响叶杨；Small-leaf Chinese Aspen ●

311199　Populus adenopoda Maxim. var. nanchaoensis T. B. Chao et C. W. Chiuan；南召响叶杨；Nanzhao Chinese Aspen ●

311200　Populus adenopoda Maxim. var. platyphylla Z. Wang et S. L. Tung；大叶响叶杨；Big-leaf Chinese Aspen ●

311201　Populus adenopoda Maxim. var. platyphylla Z. Wang et S. L. Tung = Populus lasiocarpa Oliv. ●

311202　Populus adenopoda Maxim. var. rotundifolia T. B. Chao；圆叶响叶杨；Round-leaf Chinese Aspen ●

311203　Populus adenopoda Maxim. var. tadenopoda ? = Populus

adenopoda Maxim. ●

311204 Populus afghanica（Aiton et Hemsl.）C. K. Schneid.；阿富汗杨（昆仑杨）；Afghan Poplar ●

311205 Populus afghanica（Aiton et Hemsl.）C. K. Schneid. var. cuneata Z. Wang et Chang Y. Yang；尖叶阿富汗杨●

311206 Populus afghanica（Aiton et Hemsl.）C. K. Schneid. var. tadishikistanica（Kom.）Z. Wang et Chang Y. Yang；喀什阿富汗杨（喀什昆仑杨，毛枝阿富汗杨）●

311207 Populus alaschanica Kom.；阿拉善杨；Alashan Poplar ●

311208 Populus alba L.；银白杨（白背杨，白杨，罗圈杨，银叶杨）；Abbey，Abby，Abeel，Abele，Able-tree，Arbale，Arbeal，Awbel Aubel，Bitterweed，Bolleana Poplar，Downy Poplar，Dutch Arbel，Dutch Beech，Ebble，European White Poplar，Great Aspen，Grey Poplar，Lady Poplar，Pipple，Poplain，Poplar，Poplin，Silberpappel，Silver Leaf Poplar，Silver Poplar，Silverleaf Poplar，Silver-leaf Poplar，Silver-leaved Poplar，Silver-leaved Tree，White Abele，White Asp，White Aspen，White Beech，White Leaf Poplar，White Poplar，Whiteback，Whitewood ●

311209 Populus alba L. 'Nivea'；雪白银白杨●☆

311210 Populus alba L. 'Pendula'；垂枝银白杨●☆

311211 Populus alba L. 'Pyramidalis' = Populus alba L. var. pyramidalis Bunge ●

311212 Populus alba L. 'Raket'；火箭银白杨●☆

311213 Populus alba L. 'Richardii'；理查德银白杨●☆

311214 Populus alba L. 'Rocket' = Populus alba L. 'Raket' ●☆

311215 Populus alba L. f. pyramidalis（Bunge）Dippel = Populus alba L. var. pyramidalis Bunge ●

311216 Populus alba L. subsp. nivea（Willd.）Maire et Weiller = Populus alba L. ●

311217 Populus alba L. subsp. pyramidalis（Bunge）Wettst. = Populus alba L. var. pyramidalis Bunge ●

311218 Populus alba L. var. bachofenii（Wierzb. ex Rchb.）Wesm.；光皮银白杨（巴氏杨）；Bachofen Dutch Beech，Bachofen Poplar ●

311219 Populus alba L. var. blumeana（Lauche）Otto = Populus alba L. var. pyramidalis Bunge ●

311220 Populus alba L. var. bolleana（Lauche）Otto = Populus alba L. var. pyramidalis Bunge ●

311221 Populus alba L. var. bolleana（Lauche）Otto = Populus alba L. ●

311222 Populus alba L. var. canescens Aiton = Populus canescens（Aiton）Sm. ●

311223 Populus alba L. var. hickeliana（Dode）Maire = Populus alba L. ●

311224 Populus alba L. var. integrifolia Ball = Populus alba L. ●

311225 Populus alba L. var. microphylla Maire = Populus alba L. ●

311226 Populus alba L. var. pyramidalis Bunge；新疆杨（圆锥银白杨）；Columnar Poplar，Pyramidal Poplar ●

311227 Populus alba L. var. pyramidalis Bunge = Populus alba L. ●

311228 Populus alba L. var. subintegerrima Lange = Populus alba L. ●

311229 Populus amurensis Kom.；黑龙江杨；Amur Poplar，Heilongjiang Poplar ●

311230 Populus andrewsii Sarg. = Populus jackii Sarg. ●☆

311231 Populus angulata Aiton；棱枝杨；Angle-twig Poplar，Angulate Branchlet Poplar，Angulate Poplar，Carolina Black Poplar ●

311232 Populus angulata Aiton = Populus deltoides Marshall ●

311233 Populus angustifolia James；狭叶杨；Black Cottonwood，Mountain Cottonwood，Narrowleaf Balsam Poplar，Narrowleaf Cottonwood，Narrowleaf Poplar ●

311234 Populus ariana Dode = Populus euphratica Oliv. ●

311235 Populus aurea Tidestr. = Populus tremuloides Michx. ●

311236 Populus bachofenii Wierzb. ex Rchb. = Populus alba L. var. bachofenii（Wierzb. et Rchb.）Wesm. ●

311237 Populus balsamifera L.；脂杨（大叶钻天杨，瑞典香脂杨，树胶杨，香脂白杨，香脂杨）；American Balsam Poplar，Balm，Balm of Gilead，Balsam，Balsam Poplar，Balsam-poplar，Berry-bearing Poplar，Canadian Poplar，Carolina-poplar，Cottonwood，Cotton-wood，Eastern Balsam Poplar，Eastern Cottonwood，Eastern Poplar，Hackmatack，Necklace Poplar，Necklace-bearing Poplar，Northern Cottonwood，Scented Poplar，Southern Poplar，Swedish Balsam Poplar，Tacamahac ●

311238 Populus balsamifera L. = Populus suaveolens Fisch. ●

311239 Populus balsamifera L. var. candicans（Aiton）A. Gray = Populus balsamifera L. ●

311240 Populus balsamifera L. var. candicans（Aiton）A. Gray = Populus candicans Aiton ●

311241 Populus balsamifera L. var. fernaldiana Rouleau = Populus balsamifera L. ●

311242 Populus balsamifera L. var. lanceolata Marshall = Populus balsamifera L. ●

311243 Populus balsamifera L. var. laurifolia Wesm. = Populus laurifolia Ledeb. ●

311244 Populus balsamifera L. var. michauxii（Dode）A. Henry = Populus balsamifera L. ●

311245 Populus balsamifera L. var. simonii（Carrière）Wesm. = Populus simonii Carrière ●

311246 Populus balsamifera L. var. simonii Wesm. = Populus simonii Carrière ●

311247 Populus balsamifera L. var. suaveolens Loudon = Populus suaveolens Fisch. ex Loudon ●

311248 Populus balsamifera L. var. subcordata Hyl. = Populus balsamifera L. ●

311249 Populus balsamifera L. var. subcordata Hyl. = Populus candicans Aiton ●

311250 Populus beijengensis W. Y. Hsu = Populus × beijingensis W. Y. Hsu ex Z. Wang et S. L. Tung ●

311251 Populus bernardii B. Boivin = Populus jackii Sarg. ●☆

311252 Populus besseyana Dode = Populus deltoides Bartram ex Marshall subsp. monilifera（Aiton）Eckenw. ●☆

311253 Populus bolleana Lauche = Populus alba L. var. pyramidalis Bunge ●

311254 Populus bonatii H. Lév. = Populus rotundifolia Griff. var. bonatii（H. Lév.）Z. Wang et S. L. Tung ●

311255 Populus bonnetiana Dode = Populus euphratica Oliv. ●

311256 Populus brabantica（Houtt.）Houtt.；布拉班特杨；Brabantica Poplar ●☆

311257 Populus cana T. Y. Sun；白皮杨（白皮青杨）；White-bark Poplar ●

311258 Populus cana T. Y. Sun = Populus hsinganica Z. Wang et Skvortsov var. trichorachis Z. F. Chen ●

311259 Populus candicans Aiton；欧洲大叶杨（白亮杨）；Balm of Gilead，Balm-of-Gilead，Balm-of-Gilead Poplar，Balsam Poplar，Europe Biglef Poplar，Necklace Poplar，Ontario Poplar ●

311260 Populus candicans Aiton = Populus balsamifera L. ●

311261 Populus canescens（Aiton）Sm.；银灰杨（灰白杨，灰杨）；Curly Poplar，Gray Poplar，Grey Poplar，Silvergray Poplar，Tower Poplar，Vaal Populier ●

311262　Populus canescens（Aiton）Sm.'Macrophylla'；大叶银灰杨；Grey Poplar，Picart's Poplar ●☆

311263　Populus cathayana f. latifolia Z. Wang et C. Y. Yu = Populus cathayana Rehder var. latifolia（Z. Wang et C. Y. Yu）Z. Wang et S. L. Tung ●

311264　Populus cathayana Poljak. = Populus talassica Kom. ●

311265　Populus cathayana Rehder；青杨（大叶白杨，河杨，家白杨，小叶杨）；Cathay Poplar，Green Poplar ●

311266　Populus cathayana Rehder f. latifolia Z. Wang et C. Y. Yu = Populus cathayana Rehder var. latifolia（Z. Wang et C. Y. Yu）Z. Wang et S. L. Tung ●

311267　Populus cathayana Rehder var. latifolia（Z. Wang et C. Y. Yu）Z. Wang et S. L. Tung；宽叶青杨；Broad-leaf Cathay Poplar ●

311268　Populus cathayana Rehder var. pedicellata Z. Wang et S. L. Tung；长果柄青杨（长柄青杨）；Long-pedicel Cathay Poplar ●

311269　Populus cathayana Rehder var. pendula T. B. Chao；垂枝青杨；Dropin Cathay Poplar ●

311270　Populus cathayana Rehder var. schneideri Rehder；云南青杨；Schneider Cathay Poplar ●

311271　Populus cathayana Rehder var. schneideri Rehder = Populus kangdingensis Z. Wang et S. L. Tung var. schneideri（Rehder）N. Chao et J. Liu ●

311272　Populus cathayana Rehder var. schneideri Rehder = Populus schneideri（Rehder）N. Chao ●

311273　Populus cercidiphylla Britton = Populus tremuloides Michx. ●

311274　Populus charbinensis Z. Wang et Skvortsov；哈青杨；Harbin Green Poplar，Harbin Poplar ●

311275　Populus charbinensis Z. Wang et Skvortsov var. pachydermis Z. Wang et S. L. Tung；厚皮哈青杨（隆山杨）；Thick-bark Harbin Poplar ●

311276　Populus ciliata Wall. ex Royle；缘毛杨（睫毛杨）；Bangikat Poplar，Ciliate Poplar ●

311277　Populus ciliata Wall. ex Royle var. aurea C. Marquand et Airy Shaw；金色缘毛杨（金毛杨）；Gold Ciliate Poplar ●

311278　Populus ciliata Wall. ex Royle var. gyirongensis Z. Wang et S. L. Tung；吉隆缘毛杨（吉隆杨）；Jilong Ciliate Poplar ●

311279　Populus ciliata Wall. ex Royle var. weixi Z. Wang et S. L. Tung；维西缘毛杨；Weixi Ciliate Poplar ●

311280　Populus ciupi S. Y. Wang；楸皮杨●

311281　Populus davidiana Dode；山杨（白杨，白杨树，大叶杨，火杨，明杨，山白杨，山小叶杨，响杨）；David Poplar，Wild Poplar ●

311282　Populus davidiana Dode f. foliotardus X. S. Zhang et H. Y. Jiang；晚叶山杨■

311283　Populus davidiana Dode f. laticuneata Nakai；楔叶山杨；Wedgeleaf David Poplar ●

311284　Populus davidiana Dode f. ovata Z. Wang et S. L. Tung；卵叶山杨；Ovateleaf Cottonwood ●

311285　Populus davidiana Dode f. ovata Z. Wang et S. L. Tung = Populus davidiana Dode ●

311286　Populus davidiana Dode f. pendula（Skvortsov）Z. Wang et S. L. Tung；垂枝山杨；Pendentbranch David Poplar ●

311287　Populus davidiana Dode f. pendula（Skvortsov）Z. Wang et S. L. Tung = Populus davidiana Dode ●

311288　Populus davidiana Dode var. longipediolata T. B. Chao；长柄山杨；Long-stalk David Poplar ●

311289　Populus davidiana Dode var. pendula Skvortsov = Populus davidiana Dode f. pendula（Skvortsov）Z. Wang et S. L. Tung ●

311290　Populus davidiana Dode var. pendula Skvortsov = Populus davidiana Dode ●

311291　Populus davidiana Dode var. tomentella（C. K. Schneid.）Nakai；茸毛山杨（毛山杨）；Tomentose David Poplar ●

311292　Populus deltoides Bartram ex Marshall；美洲黑杨（白杨，北美白杨，北美杨，东部杨，东方白杨，棱枝杨，棉白杨，南方棉白杨，三角杨，三角叶杨）；American Black Poplar，Angulata Poplar，Big Cottonwood，Carolina Poplar，Common Cottonwood，Cotton Wood，Cotton-tree，Cottonwood，Eastern Cottonwood，Eastern Poplar，Necklace Poplar，Northern Cottonwood，River Poplar，Southern Cottonwood，Yellow Cottonwood ●

311293　Populus deltoides Bartram ex Marshall 'LUX' I-6955'；鲁克斯杨；LUX Poplar ●

311294　Populus deltoides Bartram ex Marshall subsp. monilifera（Aiton）Eckenw.；得州杨；Plains Cottonwood ●☆

311295　Populus deltoides Bartram ex Marshall var. deltoides f. pilosa（Sarg.）Sudw. = Populus deltoides Bartram ex Marshall ●

311296　Populus deltoides Bartram ex Marshall var. missouriensis（A. Henry）Rehder = Populus deltoides Bartram ex Marshall ●

311297　Populus deltoides Bartram ex Marshall var. occidentalis Rydb. = Populus deltoides Bartram ex Marshall subsp. monilifera（Aiton）Eckenw. ●☆

311298　Populus deltoides Marshall = Populus balsamifera L. ●

311299　Populus deltoides Marshall = Populus deltoides Bartram ex Marshall ●

311300　Populus deltoides Marshall var. missouriensis Henry = Populus balsamifera L. ●

311301　Populus deltoides Marshall var. missouriensis Henry = Populus deltoides Marshall ●

311302　Populus densa Kom. = Populus talassica Kom. ●

311303　Populus dilatata Aiton = Populus nigra L. ●

311304　Populus diversifolia Schrank = Populus euphratica Oliv. ●

311305　Populus duclouxiana Dode = Populus rotundifolia Griff. var. duclouxiana（Dode）Gomb. ●

311306　Populus dutillyi Lepage = Populus jackii Sarg. ●☆

311307　Populus eugenei Simon-Louis = Populus × canadensis Moench 'Eugenei' ●

311308　Populus euphratica Oliv.；胡杨（胡桐，陶来杨，异叶胡杨，异叶发拉底杨，幼发拉底杨）；Diversifolious Poplar，Euphrasia Poplar，Euphrates Aspen，Euphrates Balsam Tree，Euphrates Balsam-tree，Euphrates Poplar，Gharab，Mesopotamian Poplar，Saf-saf ●

311309　Populus euphratica Oliv. var. bonnetiana（Dode）Maire = Populus euphratica Oliv. ●

311310　Populus euphratica Oliv. var. mauritanica（Dode）Maire = Populus euphratica Oliv. ●

311311　Populus euphratica Olivier subsp. denhardtiorum Engl. = Populus ilicifolia（Engl.）Rouleau ●☆

311312　Populus euramericana（Dode）Guinier 'gelrica' = Populus × canadensis Moench 'Gelrica' ●

311313　Populus euramericana（Dode）Guinier 'I-214' = Populus × canadensis Moench 'I-214' ●

311314　Populus euramericana（Dode）Guinier 'leipzig' = Populus × canadensis Moench 'Leipzig' ●

311315　Populus euramericana（Dode）Guinier 'Polska 15A' = Populus × canadensis Moench 'Polska A15' ●

311316　Populus euramericana（Dode）Guinier 'Sacrau 79' = Populus × canadensis Moench 'Sacrau 79' ●

311317　Populus euramericana（Dode）Guinier = Populus × canadensis Moench ●

311318　Populus euramericana Guinier = Populus canadensis Moench ●

311319　Populus eurantiana（Dode）Guinier 'I-214' = Populus × canadensis Moench 'I-214' ●

311320　Populus fangiana N. Chao et J. Liu；方氏杨（方杨）；Fang's Poplar ●

311321　Populus fangiana N. Chao et J. Liu var. microphylla（Z. Wang et S. L. Tung）N. Chao et J. Liu；金沙杨●

311322　Populus fargesii Franch. = Populus lasiocarpa Oliv. ●

311323　Populus fastigiata Poir. = Populus nigra L. var. italica（Moench）Koehne ●

311324　Populus fremontii S. Watson；弗里芒氏杨（弗里蒙特杨，弗氏黑杨，加里福尼亚杨）；Alamillo，Cottonwood，Fremont Cottonwood，Fremont Poplar，Fremont's Cottonwood，Fremont's Poplar，Meseta Cottonwood，Rio Grande Cottonwood，Western Cottonwood ●

311325　Populus gansuensis Z. Wang et H. L. Yang = Populus × gansuensis Z. Wang et H. L. Yang ●

311326　Populus gelrica（Houtz.）Houtz. = Populus × canadensis Moench 'Gelrica' ●

311327　Populus generosa Henry = Populus jackii Sarg. ●☆

311328　Populus gileadensis Rouleau = Populus candicans Aiton ●

311329　Populus gileadensis Rouleau = Populus jackii Sarg. ●☆

311330　Populus girinensis Skvortsov；东北杨（吉林杨）；Jilin Poplar ●

311331　Populus girinensis Skvortsov var. ivaschkevitochii Skvortsov；楔叶东北杨（依氏东北杨）；Wedgeleaf Jilin Poplar ●

311332　Populus glabrata Dode = Populus tomentosa Carrière ●

311333　Populus glauca Haines；灰背杨（灰叶杨）；Glaucous Poplar，Grey-blue Poplar ●

311334　Populus gonggaensis N. Chao et J. R. He；贡嘎杨；Gongga Poplar ●

311335　Populus grandidentata Michx.；大齿杨；Aspen，Bigtooth Aspen，Big-tooth Aspen，Big-toothed Aspen，Canadian Aspen，Largetooth Aspen，Largetoothed Aspen，Large-toothed Aspen ●

311336　Populus grandidentata Michx. var. angustata Vict. = Populus grandidentata Michx. ●

311337　Populus grandidentata Michx. var. meridionalis Tidestr. = Populus grandidentata Michx. ●

311338　Populus grandidentata Michx. var. subcordata Vict. = Populus grandidentata Michx. ●

311339　Populus grandidentata S. Watson = Populus grandidentata Michx. ●

311340　Populus haoana W. C. Cheng et Z. Wang；德钦杨；Deqin Poplar，Hao Poplar ●

311341　Populus haoana W. C. Cheng et Z. Wang = Populus pseudoglauca Z. Wang et P. Y. Fu ●

311342　Populus haoana W. C. Cheng et Z. Wang var. macrocarpa Z. Wang et S. L. Tung = Populus × xiaozhuanica W. Y. Hsu et C. F. Liang ●

311343　Populus haoana W. C. Cheng et Z. Wang var. macrocarpa Z. Wang et S. L. Tung；大果德钦杨；Bigfruit Hao Poplar ●

311344　Populus haoana W. C. Cheng et Z. Wang var. megaphylla Z. Wang et S. L. Tung；大叶德钦杨；Bigleaf Hao Poplar ●

311345　Populus haoana W. C. Cheng et Z. Wang var. microcarpa Z. Wang et S. L. Tung；小果德钦杨；Little Hao Poplar ●

311346　Populus henryana Dode = Populus × canadensis Moench 'Henryana' ●

311347　Populus heterophylla L.；异叶杨（沼生杨）；Black Cottonwood，Downy Poplar，Swamp Cottonwood，Swamp Poplar ●☆

311348　Populus hickeliana Dode = Populus alba L. ●

311349　Populus hopeiensis Hu et H. F. Chow；河北杨（串杨，椴杨）；Hebei Poplar，Hopei Poplar ●

311350　Populus hsinganica Z. Wang et Skvortsov；兴安杨（河杨）；Hsingan Poplar，Xing'an Poplar ●

311351　Populus hsinganica Z. Wang et Skvortsov var. trichorachis Z. F. Chen；毛轴兴安杨；Hairy-rachis Xing'an Poplar ●

311352　Populus hybrida M. Bieb. = Populus canescens（Aiton）Sm. ●

311353　Populus hybrida Rchb. var. berolinensis C. Koch = Populus × berolinensis（C. Koch）Dippel ●

311354　Populus ilicifolia（Engl.）Rouleau；冬青叶杨；Tana River Poplar，Tanariver Poplar ●☆

311355　Populus iliensis Drobow；伊犁杨；Ili Poplar，Yili Poplar ●

311356　Populus intramongolica T. Y. Sun et E. W. Ma；内蒙杨；Inner Mongolian Poplar ●

311357　Populus italica（Du Roi）Moench = Populus nigra L. ●

311358　Populus italica Du Roi = Populus nigra L. ●

311359　Populus italica Moench = Populus nigra L. var. italica（Moench）Koehne ●

311360　Populus jackii Sarg.；雅克杨；Balm-of-gilead，Jack's Poplar ●☆

311361　Populus jesoensis Nakai；虾夷杨●☆

311362　Populus kangdingensis Z. Wang et S. L. Tung；康定杨；Kangding Poplar ●

311363　Populus kangdingensis Z. Wang et S. L. Tung var. lancifolia（N. Chao）N. Chao et J. Liu；瘦叶杨；Lanceolate-leaved Poplar，Lanci-leaf Poplar ●

311364　Populus kangdingensis Z. Wang et S. L. Tung var. schneideri（Rehder）N. Chao et J. Liu = Populus cathayana Rehder var. schneideri Rehder ●

311365　Populus kangdingensis Z. Wang et S. L. Tung var. schneideri（Rehder）N. Chao et J. Liu = Populus schneideri（Rehder）N. Chao ●

311366　Populus kangdingensis Z. Wang et S. L. Tung var. tibetica（C. K. Schneid.）N. Chao et J. Liu = Populus szechuanica C. K. Schneid. var. tibetica C. K. Schneid. ●

311367　Populus keerqinensis T. Y. Sun；科尔沁杨；Kerqin Poplar ●

311368　Populus koreana Rehder；香杨（朝鲜杨，大青杨，黄铁木，皱叶杨）；Korea Poplar，Korean Poplar ●

311369　Populus krauseana Dode；克劳斯杨●☆

311370　Populus lancifolia N. Chao = Populus kangdingensis Z. Wang et S. L. Tung var. lancifolia（N. Chao）N. Chao et J. Liu ●

311371　Populus lasiocarpa Oliv.；大叶杨（大叶泡，大叶响叶杨，水冬瓜）；Bigleaf Poplar，China Poplar，Chinese Necklace Poplar，Chinese Poplar，Western Balsam Poplar ●

311372　Populus lasiocarpa Oliv. var. longiamenta P. Y. Mao et P. X. He；长序大叶杨（镇雄杨）；Long-ament Chinese Poplar ●

311373　Populus lasiocarpa Oliv. var. psilocloda N. Chao et J. Liu；裸枝杨●

311374　Populus lasiocarpa Oliv. var. yiliangensis N. Chao et J. Liu；彝良杨；Yiliang Poplar ●

311375　Populus laurifolia Ledeb.；苦杨；Bitter Poplar，Laurel Poplar，Laurelleaf Poplar，Laurel-leaved Poplar，Russian Poplar ●

311376　Populus laurifolia Ledeb. var. simonii（Carrière）Regel = Populus simonii Carrière ●

311377　Populus laurifolia Ledeb. var. simonii Regel = Populus simonii Carrière ●

311378 Populus liaotungensis Z. Wang et Skvortsov = Populus simonii Carrière var. liaotungensis（Z. Wang et Skvortsov）Z. Wang et S. L. Tung ●

311379 Populus litwinowiana Dode = Populus euphratica Oliv. ●

311380 Populus lloidii Henry = Populus × canadensis Moench 'Lloydii' ●

311381 Populus macranthela H. Lév. et Vaniot = Populus rotundifolia Griff. ●

311382 Populus macranthela H. Lév. et Vaniot = Populus rotundifolia Griff. var. duclouxiana（Dode）Gomb. ●

311383 Populus mainlingensis Z. Wang et S. L. Tung；米林杨；Mainlin Poplar，Milin Poplar ●

311384 Populus mainlingensis Z. Wang et S. L. Tung = Populus pseudoglauca Z. Wang et P. Y. Fu ●

311385 Populus manitobensis Dode = Populus jackii Sarg. ● ☆

311386 Populus manshurica Nakai；热河杨（赤峰杨）；Manshuria Poplar，Manshurian Poplar ●

311387 Populus marilandica Bosc ex Poir. = Populus × canadensis Moench 'Marilandica' ●

311388 Populus mauritanica Dode = Populus euphratica Oliv. ●

311389 Populus maximowiczii A. Henry；辽杨（臭梧桐，马氏杨，日本白杨，杨树）；Doronoki，Japan Poplar，Japanese Poplar，Maximowicz Poplar ●

311390 Populus maximowiczii A. Henry = Populus suaveolens Fisch. ●

311391 Populus maximowiczii A. Henry var. barbinervis Nakai = Populus ussuriensis Kom. ●

311392 Populus michauxii Dode = Populus balsamifera L. ●

311393 Populus microcarpa Hook. f. et Thomson ex Hook. f. = Populus rotundifolia Griff. ●

311394 Populus minhoensis Henry；民和杨 ●

311395 Populus monilifera Aiton = Populus balsamifera L. ●

311396 Populus monilifera Aiton = Populus deltoides Bartram ex Marshall subsp. monilifera（Aiton）Eckenw. ● ☆

311397 Populus nakaii Skvortsov；玉泉杨；Nakai Poplar ●

311398 Populus nigra L.；黑杨（欧亚黑杨，欧洲黑杨，钻天杨）；Aspen，Berry-bearing Poplar，Black Barked，Black Poplar，Catfoot Poplar，Common Black Poplar，Devil's Fingers，Europe Poplar，European Black Poplar，Lady Poplar，Lombardy Poplar，Manchester Poplar，Old English Poplar，Pepilary，Peplar，Peppilary，Popilary，Popillary，Poppilery，Popple，Shiver-tree，Theves Poplar，Water Poplar，Willow Poplar ●

311399 Populus nigra L. 'Gigantea'；巨黑杨；Giant Lombardy Poplar ● ☆

311400 Populus nigra L. 'Italica'；钻天杨（白杨树，笔杨，黑杨，美国白杨，美杨）；Cypress Poplar，Fastigiate Poplar，Italian Poplar，Lady Poplar，Lombardy Poplar，Poplar Pine，Soaring Poplar ●

311401 Populus nigra L. subsp. italica（Du Roi）Asch. et Graebn. = Populus nigra L. ●

311402 Populus nigra L. subsp. thevestina（Dode）Maire = Populus nigra L. ●

311403 Populus nigra L. var. afghanica Aiton et Hemsl. = Populus afghanica（Aiton et Hemsl.）C. K. Schneid. ●

311404 Populus nigra L. var. afghanica Aiton et Hemsl. = Populus nigra L. ●

311405 Populus nigra L. var. betulifolia（Pursh）Torr.；桦叶黑杨；Birch-leaf Black Poplar，Manchester Poplar ● ☆

311406 Populus nigra L. var. italica（Du Roi）Koehne = Populus nigra L. 'Italica' ●

311407 Populus nigra L. var. italica（Moench）Koehne = Populus nigra L. 'Italica' ●

311408 Populus nigra L. var. italica Du Roi = Populus nigra L. 'Italica' ●

311409 Populus nigra L. var. italica Du Roi = Populus nigra L. ●

311410 Populus nigra L. var. italica Moench = Populus nigra L. var. italica（Moench）Koehne ●

311411 Populus nigra L. var. neapolitana Ten.；那不勒斯黑杨；Neapolitan Poplar ● ☆

311412 Populus nigra L. var. pyramidalis（Bork.）Spach = Populus nigra L. 'Italica' ●

311413 Populus nigra L. var. sinensis Carrière = Populus nigra L. var. italica（Moench）Koehne ●

311414 Populus nigra L. var. thevestina（Dode）Bean；箭杆杨（电杆杨）；Arrowshaft Poplar，Greyishbark Poplar ●

311415 Populus ningshanica Z. Wang et S. L. Tung；汉白杨（大白杨，宁陕杨，骚白杨）；Ningshan Poplar，Ningxia Poplar ●

311416 Populus nivea Willd.；雪白杨 ● ☆

311417 Populus nivea Willd. = Populus alba L. ●

311418 Populus occidentalis（Rydb.）Britton ex Rydb. = Populus deltoides Bartram ex Marshall subsp. monilifera（Aiton）Eckenw. ● ☆

311419 Populus palmeri Sarg.；帕氏杨；Palmer Cottonwood ● ☆

311420 Populus pamirica Kom.；帕米尔杨；Pamir Poplar ● ◇

311421 Populus pamirica Kom. var. akqiensis Chang Y. Yang；阿合奇杨 ●

311422 Populus pekinensis L. Henry = Populus tomentosa Carrière ●

311423 Populus pilosa Rehder；柔毛杨；Pilose Poplar ●

311424 Populus pilosa Rehder var. leiocarpa Z. Wang et S. L. Tung；光果柔毛杨；Glabrous-fruit Pilose Poplar ●

311425 Populus pilosa Rehder var. leiocarpa Z. Wang et S. L. Tung = Populus talassica Kom. var. tomortensis Chang Y. Yang ●

311426 Populus platyphylla T. Y. Sun；阔叶青杨（粗枝青杨，二青杨，阔叶杨）；Broadleaf Poplar ●

311427 Populus platyphylla T. Y. Sun var. flaviflora T. Y. Sun；黄花杨；Yellowflower Broadleaf Poplar ●

311428 Populus platyphylla T. Y. Sun var. flaviflora T. Y. Sun = Populus platyphylla T. Y. Sun ●

311429 Populus platyphylla T. Y. Sun var. glauca T. Y. Sun = Populus platyphylla T. Y. Sun ●

311430 Populus platyphylla T. Y. Sun var. glauca T. Y. Sun et Z. F. Chen；青皮杨；Glaucous Broadleaf Poplar ●

311431 Populus platyphylla T. Y. Sun var. glauca T. Y. Sun et Z. F. Chen = Populus platyphylla T. Y. Sun ●

311432 Populus polygonifolia F. G. Bernard = Populus tremuloides Michx. ●

311433 Populus pruinosa Schrenk；灰胡杨（灰杨，灰叶胡杨，灰叶杨）；Bloomy Poplar ● ◇

311434 Populus przewalskii Maxim.；青甘杨（青海杨）；Przewalsk Poplar ●

311435 Populus przewalskii Maxim. f. microphylla Gomb. = Populus przewalskii Maxim. ●

311436 Populus psendo-simonii Kitag.；小青杨；Fake Simon Poplar，False Simon Poplar ●

311437 Populus psendosimonii Kitag. var. patula T. Y. Sun；展枝小青杨；Patulous False Simon Poplar ●

311438 Populus pseudoglauca Z. Wang et P. Y. Fu；长序杨；Fake Glaucous Poplar，False Glaucous Poplar，Longcatkin Poplar ●

311439 Populus pseudoglauca Z. Wang et P. Y. Fu var. weixi（Z. Wang et S. L. Tung）N. Chao et J. Lin = Populus ciliata Wall. ex Royle var. weixi Z. Wang et S. L. Tung ●

311440　Populus pseudoglauca Z. Wang et P. Y. Fu var. yatungensis（Z. Wang et P. Y. Fu）N. Chao ＝ Populus yatungensis（Z. Wang et P. Y. Fu）Z. Wang et S. L. Tung ●

311441　Populus pseudomaximowiczii Z. Wang et S. L. Tung；梧桐杨；False Maximowicz Poplar，Phoenixtree Poplar ●

311442　Populus pseudomaximowiczii Z. Wang et S. L. Tung f. glabrata Z. Wang et S. L. Tung；光果梧桐杨；Glabrous Fake Maximowicz Poplar ●

311443　Populus pseudotomentosa Z. Wang et S. L. Tung；响毛杨；False China Whie Poplar，False Chinese Whie Poplar，False Tomentose Poplar ●

311444　Populus purdomii Rehder；冬瓜杨（太白杨，水冬瓜）；Purdom Poplar，Waxgourd Poplar ●

311445　Populus purdomii Rehder var. rockii（Rehder）C. F. Fang et H. L. Yang；光皮冬瓜杨；Rock Purdom Poplar ●

311446　Populus purdomii Rehder var. rockii（Rehder）C. F. Fang et H. L. Yang ＝ Populus purdomii Rehder ●

311447　Populus pyramidalis Bork. ＝ Populus nigra L. var. italica（Moench）Koehne ●

311448　Populus pyramidalis Moench ＝ Populus nigra L. var. italica（Moench）Koehne ●

311449　Populus pyramidalis Rozier ＝ Populus nigra L. ' Italica ' ●

311450　Populus pyramidalis Rozier ＝ Populus nigra L. var. italica（Moench）Koehne ●

311451　Populus pyramidalis Salisb. ＝ Populus nigra L. var. italica（Moench）Koehne ●

311452　Populus qamdoensis Z. Wang et S. L. Tung；昌都杨；Changdu Poplar ●

311453　Populus qamdoensis Z. Wang et S. L. Tung ＝ Populus kangdingensis Z. Wang et S. L. Tung var. schneideri（Rehder）N. Chao et J. Liu ●

311454　Populus qiongdaoensis T. Hong et P. Luo；琼岛杨；Qiongdao Poplar ●

311455　Populus regenerata C. K. Schneid. ＝ Populus × canadensis Moench ' Regenerata ' ●

311456　Populus regenerata Henry ex Schneid. ＝ Populus × canadensis Moench ' Regenerata ' ●

311457　Populus robusta（Simon-Louis）C. K. Schneid. ＝ Populus × canadensis Moench ' Robusta ' ●

311458　Populus robusta C. K. Schneid. ＝ Populus × canadensis Moench ' Robusta ' ●

311459　Populus rotundifolia Griff. ；圆叶杨（白杨树，团叶杨，响叶杨）；Round-leaf Poplar，Round-leaved Poplar ●

311460　Populus rotundifolia Griff. var. bonatii（H. Lév.）Z. Wang et S. L. Tung ＝ Populus rotundifolia Griff. ●

311461　Populus rotundifolia Griff. var. bonatii（H. Lév.）Z. Wang et S. L. Tung；滇南山杨（白杨树，滇南杨，山白龙，团叶杨，响叶杨，圆叶杨）；South Yunnan Poplar ●

311462　Populus rotundifolia Griff. var. duclouxiana（Dode）Gomb. ；清溪杨；Ducloux Round-leaf Poplar ●

311463　Populus rotundifolia Griff. var. duclouxiana（Dode）Gomb. ＝ Populus rotundifolia Griff. ●

311464　Populus rotundifolia Griff. var. macranthela（H. Lév. et Vaniot）Gomb. ＝ Populus rotundifolia Griff. ●

311465　Populus sargentii Dode；沙氏杨；Great Plains Cottonwood，Plains Poplar，Sargent Cottonwood ●☆

311466　Populus sargentii Dode ＝ Populus deltoides Bartram ex Marshall subsp. monilifera（Aiton）Eckenw. ●☆

311467　Populus sargentii Dode var. texana（Sarg.）Correll ＝ Populus deltoides Bartram ex Marshall subsp. monilifera（Aiton）Eckenw. ●☆

311468　Populus schneideri（Rehder）N. Chao；西南杨；Southwest Poplar ●

311469　Populus schneideri（Rehder）N. Chao ＝ Populus kangdingensis Z. Wang et S. L. Tung var. schneideri（Rehder）N. Chao et J. Liu ●

311470　Populus schneideri（Rehder）N. Chao var. tibetica（C. K. Schneid.）N. Chao ＝ Populus kangdingensis Z. Wang et S. L. Tung var. tibetica（C. K. Schneid.）N. Chao et J. Liu ●

311471　Populus schneideri（Rehder）N. Chao var. tibetica（C. K. Schneid.）N. Chao ＝ Populus szechuanica C. K. Schneid. var. tibetica C. K. Schneid. ●

311472　Populus scytica Dode ＝ Populus nigra L. var. italica（Moench）Koehne ●

311473　Populus serotina Hartig ＝ Populus × canadensis Moench ' Serotina ' ●

311474　Populus serrata T. B. Chao et J. S. Chen；齿叶山杨；Serrate Poplar ●

311475　Populus serrata T. B. Chao et J. S. Chen ＝ Populus ningshanica Z. Wang et S. L. Tung ●

311476　Populus serrata T. B. Chao et J. S. Chen f. acuminati-gemmata T. Hong et J. Zhang ＝ Populus ningshanica Z. Wang et S. L. Tung ●

311477　Populus serrata T. B. Chao et J. S. Chen f. cordata T. Hong et J. Zhang ＝ Populus ningshanica Z. Wang et S. L. Tung ●

311478　Populus serrata T. B. Chao et J. S. Chen f. grosserrata T. Hong et J. Zhang ＝ Populus ningshanica Z. Wang et S. L. Tung ●

311479　Populus shanxiensis Z. Wang et S. L. Tung；青毛杨；Greenhair Poplar，Shaanxi Poplar ●◇

311480　Populus sieboldii Miq. ；日本山杨（日本杨，西氏杨）；Japanese Aspen，Siebeld Aspen ●

311481　Populus silvestrii Pamp. ＝ Populus adenopoda Maxim. ●

311482　Populus simonii Carrière；小叶杨（白杨柳，明杨，南京白杨，青杨，山白杨）；Simon Poplar ●

311483　Populus simonii Carrière f. brachychaeta P. Yu et C. F. Fang；短毛小叶杨；Short-hair Simon Poplar ●

311484　Populus simonii Carrière f. fastigiata C. K. Schneid. ＝ Populus simonii Carrière var. fastigiata C. K. Schneid. ●

311485　Populus simonii Carrière f. fastigiata C. K. Schneid. ＝ Populus simonii Carrière ●

311486　Populus simonii Carrière f. liaotungensis（Z. Wang et Skvortsov）Kitag. ＝ Populus simonii Carrière var. liaotungensis（Z. Wang et Skvortsov）Z. Wang et S. L. Tung ●

311487　Populus simonii Carrière f. pendula C. K. Schneid. ＝ Populus simonii Carrière ●

311488　Populus simonii Carrière f. przewalskii（Maxim.）Rehder ＝ Populus przewalskii Maxim. ●

311489　Populus simonii Carrière f. rhombifolia（Kitag.）Z. Wang et S. L. Tung ＝ Populus simonii Carrière var. rhombifolia Kitag. ●

311490　Populus simonii Carrière f. robusta Z. Wang et S. L. Tung ＝ Populus simonii Carrière ●

311491　Populus simonii Carrière f. robusta Z. Wang et S. L. Tung ＝ Populus simonii Carrière var. robusta Z. Wang et S. L. Tung ●

311492　Populus simonii Carrière var. breviamenta T. Y. Sun；短序小叶杨；Short-ament Simon Poplar ●

311493　Populus simonii Carrière var. breviamenta T. Y. Sun ＝ Populus simonii Carrière var. liaotungensis（Z. Wang et Skvortsov）Z. Wang et S. L. Tung ●

311494　Populus simonii Carrière var. fastigiata C. K. Schneid. ;塔形小叶杨(扫帚小叶杨,塔杨);Broom Simon Poplar ●

311495　Populus simonii Carrière var. griseoalba T. Y. Sun;灰白小叶杨;Alba-grey Simon Poplar ●

311496　Populus simonii Carrière var. griseoalba T. Y. Sun = Populus przewalskii Maxim. ●

311497　Populus simonii Carrière var. latifolia Z. Wang et S. L. Tung;宽叶小叶杨;Brod-leaf Simon Poplar ●

311498　Populus simonii Carrière var. liaotungensis（Z. Wang et Skvortsov）Z. Wang et S. L. Tung;辽东小叶杨(辽东杨);Liaoning Simon Poplar ●

311499　Populus simonii Carrière var. manshurica（Nakai）Kitag. = Populus manshurica Nakai ●

311500　Populus simonii Carrière var. ovata T. Y. Sun;卵叶小叶杨;Ovateleaf Simon Poplar ●

311501　Populus simonii Carrière var. ovata T. Y. Sun = Populus przewalskii Maxim. ●

311502　Populus simonii Carrière var. pendula C. K. Schneid. ;垂枝小叶杨;Pendulous Simon Poplar ●

311503　Populus simonii Carrière var. przewalskii（Maxim.）H. L. Yang = Populus przewalskii Maxim. ●

311504　Populus simonii Carrière var. rhombifolia Kitag. ;菱叶小叶杨(菱形小叶杨);Rhombic-leaf Poplar ●

311505　Populus simonii Carrière var. robusta Z. Wang et S. L. Tung;扎鲁小叶杨;Robuste Simon Poplar ●

311506　Populus simonii Carrière var. rotundifolia S. C. Lu ex Z. Wang et S. L. Tung;圆叶小叶杨;Round-leaf Simon Poplar ●

311507　Populus simonii Carrière var. tsinlingensis Z. Wang et C. Y. Yu;秦岭小叶杨;Qinling Simon Poplar ●

311508　Populus suaveolens Fisch. = Populus suaveolens Fisch. ex Loudon ●

311509　Populus suaveolens Fisch. ex Loudon;甜杨(西伯利亚白杨);Fragrant Poplar,Mongolian Poplar,Sweet Poplar ●

311510　Populus suaveolens Fisch. ex Loudon var. przeuwalskii（Maxim.）C. K. Schneid. = Populus przewalskii Maxim. ●

311511　Populus suaveolens Fisch. subsp. maximowiczii（A. Henry）Tatew. = Populus suaveolens Fisch. ex Loudon ●

311512　Populus suaveolens Fisch. var. przewalskii（Maxim.）C. K. Schneid. = Populus przewalskii Maxim. ●

311513　Populus subintegerrima Lange = Populus alba L. ●

311514　Populus szechuanica C. K. Schneid. ;川杨（四川杨）;Sichuan Poplar,Szechuan Poplar,Szechwan Aspen ●

311515　Populus szechuanica C. K. Schneid. var. rockii Rehder = Populus purdomii Rehder var. rockii（Rehder）C. F. Fang et H. L. Yang ●

311516　Populus szechuanica C. K. Schneid. var. tibetica C. K. Schneid. ;藏川杨（高山杨）;Xizang Sichuan Poplar ●

311517　Populus szechuanica C. K. Schneid. var. tibetica C. K. Schneid. = Populus kangdingensis Z. Wang et S. L. Tung var. tibetica（C. K. Schneid.）N. Chao et J. Liu ●

311518　Populus tacamahacca Mill. = Populus balsamifera L. ●

311519　Populus tacamahacca Mill. var. candicans（Aiton）Stout = Populus balsamifera L. ●

311520　Populus tacamahacca Mill. var. lanceolata（Marshall）Farw. = Populus balsamifera L. ●

311521　Populus tacamahacca Mill. var. michauxii（Dode）Farw. = Populus balsamifera L. ●

311522　Populus tadshikistanica Kom. = Populus afghanica（Aiton et Hemsl.）C. K. Schneid var. tadishikistanica（Kom.）Z. Wang et Chang Y. Yang

311523　Populus tajikistanica Kom. = Populus afghanica（Aiton et Hemsl.）C. K. Schneid var. tadishikistanica（Kom.）Z. Wang et Chang Y. Yang ●

311524　Populus talassica Kom. ;密叶杨;Denseleaf Poplar,Dense-leaved Poplar ●

311525　Populus talassica Kom. var. cordata Chang Y. Yang;心叶密叶杨;Heart-denseleaf Poplar ●

311526　Populus talassica Kom. var. tomortensis Chang Y. Yang;托木尔峰密叶杨;Tuomuer Poplar ●

311527　Populus texana Sarg. = Populus deltoides Bartram ex Marshall subsp. monilifera（Aiton）Eckenw. ● ☆

311528　Populus thevestina Dode = Populus nigra L. var. thevestina（Dode）Bean ●

311529　Populus thevestina Dode = Populus nigra L. ●

311530　Populus tomentosa Carrière;毛白杨(白杨,笨白杨,大叶杨,独摇,响杨,响叶子杨);China White Poplar,Chinese White Poplar,Silver Chinese Poplar ●

311531　Populus tomentosa Carrière f. cordiconeifolia T. B. Chao et J. W. Liu;心楔叶毛白杨 ●

311532　Populus tomentosa Carrière f. deltatifolia T. B. Chao et Z. X. Chen;三角叶毛白杨 ●

311533　Populus tomentosa Carrière f. lerigata T. B. Chao et Z. X. Chen;光皮毛白杨 ●

311534　Populus tomentosa Carrière f. yixianensis H. M. Jiang et J. X. Huang;易县毛白杨 ●

311535　Populus tomentosa Carrière var. fastigiata Z. Wang et S. L. Tung;抱头毛白杨;Broom Chinese White Poplar ●

311536　Populus tomentosa Carrière var. truncata Y. C. Fu et Chung H. Wang;截叶毛白杨;Truncate-leaf Chinese White Poplar ●

311537　Populus tremula L. :欧洲山杨(白杨,火烧杨);Apse,Apsen-tree,Asp,Aspe,Aspen,Aspen Poplar,Berry-bearing Poplar,Ebble,Eps,Esp,Espin,Europe Poplar,European Aspen,Haps Tree,Haspen,Jitterpappel,Mountain Ash,Old Wives' Tongue,Old Wives' Tongues,Owler,Pipple,Quakin Esp,Quaking Ash,Quaking Aspen,Quakiu Asp,Quickbeam,Rattling Asp,Shaking Asp,Shiver-tree,Snapsen,Swedish Aspen,Tremble,Trembling Aspen,Trembling Poplar,Woman's Tongue ●

311538　Populus tremula L. 'Erecta';直枝欧洲山杨 ● ☆

311539　Populus tremula L. 'Pendula';垂枝欧洲山杨;Penddulous European Aspen,Weeping Aspen ●

311540　Populus tremula L. subsp. davidiana（Dode）Hultén = Populus davidiana Dode ●

311541　Populus tremula L. var. adenopoda（Maxim.）Burkill = Populus adenopoda Maxim. ●

311542　Populus tremula L. var. davidiana（Dode）C. K. Schneid. = Populus davidiana Dode ●

311543　Populus tremula L. var. davidiana（Dode）C. K. Schneid. f. tomentella C. K. Schneid. = Populus davidiana Dode var. tomentella（C. K. Schneid.）Nakai ●

311544　Populus tremula L. var. pendula Loudon = Populus tremula L. 'Pendula' ●

311545　Populus tremula L. var. sieboldii（Miq.）H. Ohashi;毛欧洲山

杨;Villose European Aspen ●

311546 Populus tremula Liou et al. = Populus davidiana Dode ●

311547 Populus tremuloides Michx.;美洲山杨(颤杨,美国白杨);American Aspen, Aspen, Golden Aspen, Poplar, Popple, Quaking Aspen, Quaking Asr, Quiver-leaf, Rocky Mountain Aspen, Trembling Aspen, Trembling Poplar, White Poplar ●

311548 Populus tremuloides Michx. var. aurea (Tidestr.) Daniels = Populus tremuloides Michx. ●

311549 Populus tremuloides Michx. var. cercidiphylla (Britton) Sudw. = Populus tremuloides Michx. ●

311550 Populus tremuloides Michx. var. intermedia Vict. = Populus tremuloides Michx. ●

311551 Populus tremuloides Michx. var. magnifica Vict. = Populus tremuloides Michx. ●

311552 Populus tremuloides Michx. var. rhomboidea Vict. = Populus tremuloides Michx. ●

311553 Populus tremuloides Michx. var. vancouveriana (Trel.) Sarg. = Populus tremuloides Michx. ●

311554 Populus trichocarpa Torr. et Gray ex Hook.;毛果杨(美国黑杨);Balm of Gilead, Black Cottonwood, California Poplar, Cotton Wood, Oregon Balsam Poplar, Western Balsam Poplar ●

311555 Populus trinervis Z. Wang et S. L. Tung;三脉青杨(三脉杨);Threevein Poplar, Trinerve Poplar ●

311556 Populus trinervis Z. Wang et S. L. Tung var. shimianica Z. Wang et N. Chao;三脉石棉杨(石棉杨);Shimian Trinerve Poplar ●

311557 Populus tristis Fisch.;褐枝杨;Brown-twig Poplar ●☆

311558 Populus undulata J. Zhang;波叶山杨;Undulate Poplar ●

311559 Populus usbekistanica Kom. 'Afghanica' = Populus afghanica (Aiton et Hemsl.) C. K. Schneid. ●

311560 Populus usbekistanica Kom. = Populus afghanica (Aiton et Hemsl.) C. K. Schneid. ●

311561 Populus usbekistanica Kom. subsp. tadishistanica (Kom.) Bunge = Populus afghanica (Aiton et Hemsl.) C. K. Schneid. var. tadishikistanica (Kom.) Z. Wang et Chang Y. Yang ●

311562 Populus ussuriensis Kom.;大青杨(哈达杨,憨大杨);Ussuri Poplar ●

311563 Populus vancouveriana Trel. = Populus tremuloides Michx. ●

311564 Populus violascens Dode;堇柄杨(紫柄杨);Violet Poplar, Violetish Poplar, Violet-stalked Poplar ●

311565 Populus virginiana Foug.;弗里库尔特杨●☆

311566 Populus wenxianica Z. C. Feng et J. L. Guo ex G. Zhu;文县杨●

311567 Populus wilslizenii Sarg.;得克萨斯杨;Wilslizen Poplar ●☆

311568 Populus wilsonii C. K. Schneid.;椅杨(魏氏杨);Chair Poplar, E. H. Wilson Poplar, Wilson Poplar ●

311569 Populus wilsonii C. K. Schneid. f. brevipetiolata Z. Wang et S. L. Tung;短柄椅杨;Short-stalk E. H. Wilson Poplar ●

311570 Populus wilsonii C. K. Schneid. f. brevipetiolata Z. Wang et S. L. Tung = Populus glauca Haines ●

311571 Populus wilsonii C. K. Schneid. f. brevipetiolata Z. Wang et S. L. Tung = Populus wilsonii C. K. Schneid. ●

311572 Populus wilsonii C. K. Schneid. f. pedicellata Z. Wang et S. L. Tung;长果柄椅杨(长柄椅杨);Longstalk E. H. Wilson Poplar ●

311573 Populus wilsonii C. K. Schneid. f. pedicellata Z. Wang et S. L. Tung = Populus wilsonii C. K. Schneid. ●

311574 Populus wilsonii C. K. Schneid. var. lasioclada N. Chao;毛枝杨●

311575 Populus wuana Z. Wang et S. L. Tung;长叶杨;Longleaf Poplar, Long-leaved Poplar, Wu Poplar ●

311576 Populus wulianensis S. B. Liang et X. W. Li;五莲杨;Wulian Poplar ●

311577 Populus wutanica Mayr = Populus davidiana Dode ●

311578 Populus xiangchengensis Z. Wang et S. L. Tung;乡城杨;Xiangcheng Poplar ●

311579 Populus yatungensis (Z. Wang et P. Y. Fu) Z. Wang et S. L. Tung;亚东杨;Yadong Poplar ●

311580 Populus yatungensis (Z. Wang et P. Y. Fu) Z. Wang et S. L. Tung var. crenata Z. Wang et S. L. Tung;圆齿亚东杨;Crenate Yadong Poplar ●

311581 Populus yatungensis (Z. Wang et P. Y. Fu) Z. Wang et S. L. Tung var. trichorachis Z. Wang et S. L. Tung = Populus schneideri (Rehder) N. Chao ●

311582 Populus yatungensis (Z. Wang et P. Y. Fu) Z. Wang et S. L. Tung var. trichorachis Z. Wang et S. L. Tung;毛轴亚东杨;Hairy-rachis Yadong Poplar ●

311583 Populus yuana Z. Wang et S. L. Tung;五瓣杨;Yu Poplar ●

311584 Populus yunnanensis Dode;滇杨(白泡桐,大叶杨柳,东川杨柳,云南白杨);Yunnan Aspen, Yunnan Poplar ●

311585 Populus yunnanensis Dode var. microphylla Z. Wang et S. L. Tung;小叶滇杨;Littleleaf Yunnan Poplar ●

311586 Populus yunnanensis Dode var. microphylla Z. Wang et S. L. Tung = Populus fangiana N. Chao et J. Liu var. microphylla (Z. Wang et S. L. Tung) N. Chao et J. Liu ●

311587 Populus yunnanensis Dode var. pedicellata Z. Wang et S. L. Tung;长果柄滇杨(长柄滇杨);Long-pedicel Yunnan Poplar ●

311588 Populus yunnanensis Dode var. yatuegensis Z. Wang et P. Y. Fu = Populus yatungensis (Z. Wang et P. Y. Fu) Z. Wang et S. L. Tung ●

311589 Populus yunsiaoshanensis T. B. Chao et C. W. Chiuan;云霄杨;Yunxiaoshan Poplar ●

311590 Poraelia Durando ex Pomel = Daucus L. ■

311591 Porana Burm. f. (1768);翼萼藤属(飞蛾藤属)●■☆

311592 Porana acuminata P. Beauv. = Neuropeltis acuminata (P. Beauv.) Benth. ●☆

311593 Porana brebisepala C. Y. Wu et S. H. Huang = Dinetus dinetoides (C. K. Schneid.) Staples ■●

311594 Porana confertifolia C. Y. Wu = Tridynamia sinensis (Hemsl.) Staples var. delavayi (Gagnep. et Courchet) Staples ●

311595 Porana decora W. W. Sm. = Dinetus decorus (W. W. Sm.) Staples ●

311596 Porana delavayi Gagnep. = Tridynamia sinensis (Hemsl.) Staples var. delavayi (Gagnep. et Courchet) Staples ●

311597 Porana delavayi Gagnep. et Courchet = Tridynamia sinensis (Hemsl.) Staples var. delavayi (Gagnep. et Courchet) Staples ●

311598 Porana densiflora Hallier f. = Metaporana densiflora (Hallier f.) N. E. Br. ●☆

311599 Porana dinetoides C. K. Schneid. = Dinetus dinetoides (C. K. Schneid.) Staples ■

311600 Porana dinetoides C. K. Schneid. var. mienningensis S. H. Huang = Dinetus dinetoides (C. K. Schneid.) Staples ■●

311601 Porana discifera C. K. Schneid. = Poranopsis discifera (C. K. Schneid.) Staples ●

311602 Porana duclouxii Gagnep. et Courchet = Dinetus duclouxii (Gagnep. et Courchet) Staples ■

311603 Porana duclouxii Gagnep. et Courchet var. lasia (C. K. Schneid.) Hand. -Mazz. = Dinetus duclouxii (Gagnep. et Courchet) Staples ■●

311604　Porana esquirolii H. Lév. = Tridynamia sinensis（Hemsl.）Staples ●

311605　Porana gagnepainiana H. Lév. = Dinetus racemosa Ham. ex Sweet ●

311606　Porana grandiflora Wall. = Dinetus grandiflorus（Wall.）Staples ●

311607　Porana henryi Verdc.；白花叶翼萼藤；Chinese Poranopsis，Henry Porana ●

311608　Porana henryi Verdc. = Poranopsis sinensis（Hand.-Mazz.）Staples ●

311609　Porana henryi Verdc. = Poranopsis sinensis（Hemsl.）Staples ●

311610　Porana lobata C. Y. Wu = Dinetus duclouxii（Gagnep. et Courchet）Staples ■●

311611　Porana lobata C. Y. Wu = Porana duclouxii Gagnep. et Courchet var. lasia（C. K. Schneid.）Hand.-Mazz. ■●

311612　Porana lutingensis Lingelsh. = Dinetus duclouxii（Gagnep. et Courchet）Staples ■●

311613　Porana mairei Gagnep. = Dinetus decorus（W. W. Sm.）Staples ●

311614　Porana mairei Gagnep. et Courchet = Dinetus decorus（W. W. Sm.）Staples ●

311615　Porana mairei Gagnep. et Courchet var. holosericea C. Y. Wu = Dinetus decorus（W. W. Sm.）Staples ●

311616　Porana mairei Gagnep. var. holosericea C. Y. Wu = Dinetus decorus（W. W. Sm.）Staples ●

311617　Porana megalantha Merr. = Tridynamia megalantha（Merr.）Staples ●

311618　Porana megathyrsa C. Y. Wu = Dinetus dinetoides（C. K. Schneid.）Staples ■●

311619　Porana megathyrsa C. Y. Wu = Tridynamia megalantha（Merr.）Staples ●

311620　Porana microsepala Hand.-Mazz. = Dinetus decorus（W. W. Sm.）Staples ●

311621　Porana paniculata Roxb.；圆锥翼萼藤；White Corallita ●

311622　Porana paniculata Roxb. = Poranopsis paniculata（Roxb.）Roberty ●

311623　Porana parvifolia K. Afzel. = Metaporana parvifolia（K. Afzel.）Verdc. ●☆

311624　Porana racemosa（Wall.）Sweet var. tomentella C. Y. Wu = Dinetus racemosa Ham. ex Sweet ●

311625　Porana racemosa（Wall.）Sweet var. violacea C. Y. Wu = Dinetus racemosa Ham. ex Sweet ●

311626　Porana racemosa Roxb. = Dinetus racemosa Ham. ex Sweet ●

311627　Porana racemosa Roxb. var. sericocarpa C. Y. Wu = Dinetus racemosa Ham. ex Sweet ●

311628　Porana racemosa Roxb. var. sericocarpa C. Y. Wu = Dinetus truncatus（Kurz）Staples ■

311629　Porana racemosa Roxb. var. sericocarpa C. Y. Wu = Porana truncata Kurz ●

311630　Porana racemosa Roxb. var. tomentella C. Y. Wu = Dinetus racemosa Ham. ex Sweet ●

311631　Porana racemosa Roxb. var. violacea C. Y. Wu = Dinetus racemosa Ham. ex Sweet ●

311632　Porana racemosa Wall. = Dinetus racemosa Ham. ex Sweet ●

311633　Porana sinensis Hemsl. = Poranopsis sinensis（Hemsl.）Staples ●

311634　Porana sinensis Hemsl. = Tridynamia sinensis（Hemsl.）Staples ●

311635　Porana sinensis Hemsl. var. delavayi（Gagnep. et Courchet）Rehder = Tridynamia sinensis（Hemsl.）Staples var. delavayi（Gagnep. et Courchet）Staples ●

311636　Porana speciosa Benth. et Hook. f. = Tridynamia megalantha（Merr.）Staples ●

311637　Porana spectabilis Kurz = Tridynamia megalantha（Merr.）Staples ●

311638　Porana spectabilis Kurz var. megalantha（Merr.）F. C. How = Tridynamia megalantha（Merr.）Staples ●

311639　Porana spectabilis Kurz var. megalantha（Merr.）F. C. How ex H. S. Kiu = Tridynamia megalantha（Merr.）Staples ●

311640　Porana subrotundifolia De Wild. = Paralepistemon shirensis（Oliv.）Lejoly et Lisowski ●☆

311641　Porana triserialis C. K. Schneid. = Dinetus duclouxii（Gagnep. et Courchet）Staples ■●

311642　Porana triserialis C. K. Schneid. var. lasia C. K. Schneid. = Dinetus duclouxii（Gagnep. et Courchet）Staples ■●

311643　Porana triserialis Schneid. = Dinetus duclouxii（Gagnep. et Courchet）Staples ■●

311644　Porana triserialis Schneid. var. lasia Schneid. = Dinetus duclouxii（Gagnep. et Courchet）Staples ■●

311645　Porana truncata Kurz = Dinetus truncatus（Kurz）Staples ●

311646　Poranaceae J. Agardh = Convolvulaceae Juss.（保留科名）●■

311647　Poranaceae J. Agardh；翼萼藤科 ●■

311648　Porandra D. Y. Hong = Amischotolype Hassk. ■

311649　Porandra D. Y. Hong（1974）；孔药花属（孔药藤花属）；Porandra ■

311650　Porandra microphylla Y. Wan；小叶孔药花；Littleleaf Porandra ■

311651　Porandra ramosa D. Y. Hong；孔药花；Brachy Porandra，Porandra ■

311652　Porandra scandens D. Y. Hong；攀缘孔药花；Climbing Porandra ■

311653　Poranopsis Roberty = Porana Burm. f. ●■☆

311654　Poranopsis Roberty（1953）；白花叶属 ●

311655　Poranopsis discifera（C. K. Schneid.）Staples；搭棚藤；Makeawning Porana，Schneider Poranopsis ●

311656　Poranopsis discifera（C. K. Schneid.）Staples = Porana discifera C. K. Schneid. ●

311657　Poranopsis paniculata（Roxb.）Roberty；白花叶（圆锥白花叶，圆锥飞蛾藤，圆锥翼萼藤）；Bridal Bouquet，Christmas-vine，Christmasvine Porana，Christmas-vine Poranopsis，Snow Creeper ●

311658　Poranopsis sinensis（Hand.-Mazz.）Staples = Porana henryi Verdc. ●

311659　Poranopsis sinensis（Hemsl.）Staples；大果飞蛾藤（白花叶，异萼飞蛾藤）；China Porana，Chinese Porana ●

311660　Poranopsis sinensis（Hemsl.）Staples = Porana sinensis Hemsl. ●

311661　Poranthera Raf. = Sorghastrum Nash ■☆

311662　Poranthera Rudge（1811）；孔药大戟属 ■☆

311663　Poranthera alpina Cheeseman ex Hook. f.；孔药大戟 ■☆

311664　Poranthera florosa Halford et R. J. F. Hend.；多花孔药大戟 ■☆

311665　Poranthera glauca Klotzsch；灰绿孔药大戟 ■

311666　Poranthera microphylla Brongn.；小叶孔药大戟 ■☆

311667　Poranthera triandra J. M. Black；三蕊孔药大戟 ■☆

311668　Porantheraceae（Pax）Hurus. = Euphorbiaceae Juss.（保留科名）●■

311669　Porantheraceae（Pax）Hurus. = Phyllanthaceae J. Agardh ●■

311670　Porantheraceae Hurus.；孔药大戟科 ■

311671　Porantheraceae Hurus. = Euphorbiaceae Juss.（保留科名）●■

311672　Porantheraceae Hurus. = Phyllanthaceae J. Agardh ●■

311673　Poraqueiba Aubl.（1775）;巴拿马茱萸属●☆

311674　Poraqueiba guianensis Aubl. ;巴拿马茱萸●☆

311675　Poraresia Gleason ＝ Pogonophora Miers ex Benth. ■☆

311676　Porcelia Pers. ＝ Asimina Adans. ●☆

311677　Porcelia Ruiz et Pav.（1794）;泡泽木属●☆

311678　Porcelia grandiflora（W. Bartram）Pers. ＝ Asimina obovata（Willd.）Nash ●☆

311679　Porcelia grandiflora Pers. ＝ Asimina obovata（Willd.）Nash ●☆

311680　Porcelia macrocarpa R. E. Fr. ;泡泽木●☆

311681　Porcelia microcarpa Donn. Sm. ;小果泡泽木●☆

311682　Porcelia parviflora（Michx.）Pers. ＝ Asimina parviflora（Michx.）Dunal ●☆

311683　Porcelia parviflora Pers. ＝ Asimina parviflora（Michx.）Dunal ●☆

311684　Porcelia pygmaea（W. Bartram）Pers. ＝ Asimina pygmaea（W. Bartram）Dunal ●☆

311685　Porcelia stenopetala Donn. Sm. ;窄瓣泡泽木●☆

311686　Porcelia triloba（L.）Pers. ＝ Asimina triloba（L.）Dunal ●☆

311687　Porcellites Cass. ＝ Hypochaeris L. ■

311688　Porcellites brasiliensis Less. ＝ Hypochaeris brasiliensis Benth. et Hook. f. ex Griseb. ■☆

311689　Porcellites brasiliensis Less. ＝ Hypochaeris chillensis（Kunth）Britton ■☆

311690　Porfiria Boed. ＝ Mammillaria Haw.（保留属名）●

311691　Porfuris Raf. ＝ Nemesia Vent. ■●☆

311692　Porliera Pers. ＝ Porlieria Ruiz et Pav. ●☆

311693　Porlieria Ruiz et Pav.（1794）;长小叶蒺藜属●☆

311694　Porlieria hygrometrs Ruiz et Pav. ;长小叶蒺藜●☆

311695　Porocarpus Gaertn.（废弃属名）＝ Timonius DC.（保留属名）●

311696　Porochna Tiegh. ＝ Ochna L. ●

311697　Porochna antunesii Tiegh. ＝ Ochna pulchra Hook. f. ●☆

311698　Porochna aschersoniana（Schinz）Tiegh. ＝ Ochna pulchra Hook. f. ●☆

311699　Porochna bifolia Tiegh. ＝ Ochna pulchra Hook. f. ●☆

311700　Porochna brunnescens Tiegh. ＝ Ochna pulchra Hook. f. ●☆

311701　Porochna davilliflora Tiegh. ＝ Ochna pulchra Hook. f. ●☆

311702　Porochna hoffmanni-ottonis（Engl.）Tiegh. ＝ Ochna pulchra Hook. f. ●☆

311703　Porochna huillensis Tiegh. ＝ Ochna pulchra Hook. f. ●☆

311704　Porochna membranacea（Oliv.）Tiegh. ＝ Ochna membranacea Oliv. ●☆

311705　Porochna quangensis（Büttner）Tiegh. ＝ Ochna pulchra Hook. f. ●☆

311706　Porochna rubescens（Hiern）Tiegh. ＝ Ochna hiernii（Tiegh.）Exell ●☆

311707　Porocystis Radlk.（1878）;孔囊无患子属●☆

311708　Porocystis toulicioides Radlk. ;孔囊无患子●☆

311709　Porodittia G. Don ＝ Stemotria Wettst. et Harms ex Engl. ●☆

311710　Porodittia G. Don ex Kraenzlin ＝ Stemotria Wettst. et Harms ●☆

311711　Porodittia G. Don ex Kraenzlin ＝ Trianthera Wettst. ●☆

311712　Porolabium Ts. Tang et F. T. Wang（1940）;孔唇兰属;Porolabium ■★

311713　Porolabium biporosum（Maxim.）Ts. Tang et F. T. Wang;孔唇兰;Bipore Porolabium ■

311714　Porophyllum Adans. ＝ Porophyllum Guett. ■●☆

311715　Porophyllum Guett.（1754）;孔叶菊属（点叶菊属）;Poreleaf ■●☆

311716　Porophyllum gracile Benth. ;细孔叶菊;Odora Porophyllum ■☆

311717　Porophyllum greggii A. Gray;格氏孔叶菊■●☆

311718　Porophyllum japonicum（Thunb.）DC. ＝ Gynura japonica（Thunb.）Juel ■

311719　Porophyllum macrocephalum DC. ＝ Porophyllum ruderale（Jacq.）Cass. var. macrocephalum（DC.）Cronquist ■☆

311720　Porophyllum pygmaeum D. J. Keil et Morefield;小孔叶菊■

311721　Porophyllum ruderale（Jacq.）Cass. ;垃圾堆孔叶菊■☆

311722　Porophyllum ruderale（Jacq.）Cass. subsp. macrocephalum（DC.）R. R. Johnson ＝ Porophyllum ruderale（Jacq.）Cass. var. macrocephalum（DC.）Cronquist ■☆

311723　Porophyllum ruderale（Jacq.）Cass. var. macrocephalum（DC.）Cronquist;大头孔叶菊■☆

311724　Porophyllum scoparium A. Gray;帚状孔叶菊;Poreleaf ■☆

311725　Porosectaceae Dulac ＝ Loranthaceae Juss.（保留科名）●

311726　Porospermum F. Muell. ＝ Delarbrea Vieill. ●☆

311727　Porostema Schreb. ＝ Ocotea Aubl. ●☆

311728　Porotheca K. Schum. ＝ Chlaenandra Miq. ●☆

311729　Porothrinax H. Wendl. ex Griseb. ＝ Thrinax L. f. ex Sw. ●☆

311730　Porottddta Walp. ＝ Porodittia G. Don ●☆

311731　Porpa Blume ＝ Triumfetta Plum. ex L. ●■

311732　Porpa repens Blume ＝ Triumfetta repens（Blume）Merr. et Rolfe ■☆

311733　Porpax Lindl.（1845）;盾柄兰属;Porpax ■

311734　Porpax Salisb. ＝ Aspidistra Ker Gawl. ●■

311735　Porpax ustulata（Parl. et Rchb. f.）Rolfe;盾柄兰;Lightblack Porpax ■

311736　Porphyra Lour.（1790）;紫马鞭草属;Laver ●☆

311737　Porphyra Lour. ＝ Callicarpa L. ●

311738　Porphyra dichotoma Lour. ＝ Callicarpa dichotoma（Lour.）K. Koch ●

311739　Porphyranthus Engl. ＝ Panda Pierre ●☆

311740　Porphyrocodon Hook. f. ＝ Cardamine L. ■

311741　Porphyrocoma Scheidw. ＝ Justicia L. ●■

311742　Porphyrocoma Scheidw. ex Hook. ＝ Justicia L. ●■

311743　Porphyrodesme Schltr.（1913）;棕带兰属■☆

311744　Porphyrodesme papuana Schltr. ;棕带兰■☆

311745　Porphyroglottis Ridl.（1896）;紫舌兰属■☆

311746　Porphyroglottis maxwelliae Ridl. ;紫舌兰■☆

311747　Porphyroscias Miq.（1867）;紫花前胡属■

311748　Porphyroscias Miq. ＝ Angelica L. ■

311749　Porphyroscias decursiva Miq. ＝ Angelica decursiva（Miq.）Franch. et Sav. ■

311750　Porphyroscias decursiva Miq. ＝ Peucedanum decursivum（Miq.）Maxim. ■

311751　Porphyroscias decursiva Miq. f. albiflora（Maxim.）Nakai ＝ Angelica decursiva（Miq.）Franch. et Sav. f. albiflora（Maxim.）Nakai ■

311752　Porphyroscias decursiva Miq. var. albiflora（Maxim.）Nakai ＝ Angelica decursiva（Miq.）Franch. et Sav. var. albiflora（Maxim.）Nakai ■

311753　Porphyroscias decursiva Miq. var. albiflora（Maxim.）Nakai ＝ Peucedanum decursivum（Miq.）Maxim. var. albiflora Maxim. ■

311754　Porphyroscias longipedicellata H. Wolff ＝ Angelica longipedicellata（H. Wolff）M. Hiroe ■

311755　Porphyroscias megaphylla H. Boissieu ＝ Angelica megaphylla Diels ■

311756　Porphyrospatha Engl. ＝ Syngonium Schott ■☆

311757　Porphyrostachys Rchb. f.（1854）;紫穗兰属■☆

311758 Porphyrostachys parviflora (C. Schweinf.) Garay；小花紫穗兰■☆

311759 Porphyrostachys pilifera Rchb. f.；紫穗兰■☆

311760 Porphyrostemma Benth. ex Oliv. = Porphyrostemma Grant ex Benth. ex Oliv■☆

311761 Porphyrostemma Grant ex Benth. ex Oliv. (1873)；红脂菊属■☆

311762 Porphyrostemma chevalieri (O. Hoffm.) Hutch. et Dalziel；红脂菊■☆

311763 Porphyrostemma cuanzensis (Welw.) O. Hoffm. = Inula cuanzensis (Welw.) Hiern■☆

311764 Porphyrostemma grantii Benth. ex Oliv.；格朗特红脂菊■☆

311765 Porphyrostemma grantii Benth. ex Oliv. var. chevalieri O. Hoffm. = Porphyrostemma chevalieri (O. Hoffm.) Hutch. et Dalziel■☆

311766 Porphyrostemma grantii Benth. ex Oliv. var. semicalva O. Hoffm. = Porphyrostemma grantii Benth. ex Oliv.■☆

311767 Porphyrostemma monocephala (E. A. Bruce) Leins；单头红脂菊■☆

311768 Porroglossum Schltr. (1920)；葱兰属■☆

311769 Porroglossum echidnum (Rchb. f.) Garay；葱兰■☆

311770 Porrorhachis Garay(1972)；葱刺兰属■☆

311771 Porrorhachis galbina (J. J. Sm.) Garay；葱刺兰■☆

311772 Porrorhachis macrosepala (Schltr.) Garay；大萼葱刺兰■☆

311773 Porroteranthe Steud. = Glyceria R. Br. (保留属名)■

311774 Porrum Mill. = Allium L.■

311775 Porsildia Á. Löve et D. Löve = Minuartia L.■

311776 Porsildia groenlandica (Retz.) Á. Löve et D. Löve = Minuartia groenlandica (Retz.) Ostenf.■☆

311777 Porsildia groenlandica (Retz.) Á. Löve et D. Löve subsp. glabra (Michx.) Á. Löve et D. Löve = Minuartia glabra (Michx.) Mattf.■☆

311778 Portaea Ten. = Juanulloa Ruiz et Pav.●☆

311779 Portalesia Meyen = Nassauvia Comm. ex Juss.●☆

311780 Portea Brongn. ex K. Koch(1856)；星果凤梨属(波提亚属,星果属)■☆

311781 Portea K. Koch = Portea Brongn. ex K. Koch■☆

311782 Portea grandiflora Philcox；大花星果凤梨■☆

311783 Portea kermesina K. Koch；星果凤梨■☆

311784 Portea leptantha Harms；细花星果凤梨■☆

311785 Portea nana Leme et H. Luther；矮星果凤梨■☆

311786 Portenschlagia Tratt. = Elaeodendron J. Jacq.●☆

311787 Portenschlagia Vis. = Portenschlagiella Turin■☆

311788 Portenschlagiella Tutin(1967)；波氏萝卜属■☆

311789 Portenschlagiella ramosissima (Port.) Tutin；波氏萝卜■☆

311790 Porterandia Ridl. (1940)；绢冠茜属；Porterandia●

311791 Porterandia anisophylla (Jack ex Roxb.) Ridl.；异叶绢冠茜●☆

311792 Porterandia annulata (K. Schum.) Keay = Aoranthe annulata (K. Schum.) Somers●☆

311793 Porterandia castaneofulva (S. Moore) Keay = Aoranthe castaneofulva (S. Moore) Somers●☆

311794 Porterandia cladantha (K. Schum.) Keay = Aoranthe cladantha (K. Schum.) Somers●☆

311795 Porterandia nalaensis (De Wild.) Keay = Aoranthe nalaensis (De Wild.) Somers●☆

311796 Porterandia penduliflora (K. Schum.) Keay = Aoranthe penduliflora (K. Schum.) Somers●☆

311797 Porterandia sericantha (W. C. Chen) W. C. Chen；绢冠茜；Sericeousflor Porterandia●

311798 Porteranthus Britton = Gillenia Moench■☆

311799 Porteranthus Small = Gillenia Moench■☆

311800 Porteranthus stipulatus (Muhl. ex Willd.) Britton = Gillenia stipulata (Muhl. ex Willd.) Baill.●☆

311801 Porteranthus trifoliatus (L.) Britton = Gillenia trifoliata (L.) Moench●☆

311802 Porterella Torr. (1872)；波特草属■☆

311803 Porterella Torr. = Laurentia Neck.■☆

311804 Porterella carnosula Hook. et Arn.；波特草■☆

311805 Porteresia Tateoka(1965)；盐稻属■☆

311806 Porteresia coarctata (Roxb.) Tateoka；盐稻■☆

311807 Porteria Hook. = Valeriana L.●■

311808 Portesia Cav. = Trichilia P. Browne(保留属名)●

311809 Portillia Königer = Masdevallia Ruiz et Pav.■☆

311810 Portillia Königer(1996)；厄瓜多尔细瓣兰属■☆

311811 Portlandia Ellis = Gardenia Ellis(保留属名)●

311812 Portlandia L. = Portlandia P. Browne●☆

311813 Portlandia P. Browne(1756)；波特兰木属(波特蓝木属)●☆

311814 Portlandia domingensis Britton；多明戈波特兰木●☆

311815 Portlandia grandiflora L.；大花波特兰木●☆

311816 Portula Hill = Peplis L.●

311817 Portulaca L. (1753)；马齿苋属；Common Purslane, Portulaca, Purslane■

311818 Portulaca amilis Speg.；巴拉圭马齿苋；Paraguayan Purslane■☆

311819 Portulaca anceps A. Rich. = Portulaca quadrifida L.■

311820 Portulaca arachnoides Haw. = Anacampseros arachnoides (Haw.) Sims■☆

311821 Portulaca aurea DC. = Portulaca oleracea L.■

311822 Portulaca australis Endl. = Portulaca pilosa L.■

311823 Portulaca biloba Urb.；双裂马齿苋■☆

311824 Portulaca boninensis Tuyama = Portulaca pilosa L.■

311825 Portulaca bulbifera M. G. Gilbert；球根马齿苋■☆

311826 Portulaca caffra Thunb. = Talinum caffrum (Thunb.) Eckl. et Zeyh.■☆

311827 Portulaca carrissoana (Exell et Mendonça) Nyananyo = Portulaca hereroensis Schinz■☆

311828 Portulaca centrali-africana R. E. Fr.；中非马齿苋■☆

311829 Portulaca collina Dinter；山丘马齿苋■☆

311830 Portulaca commutata M. G. Gilbert；变异马齿苋■☆

311831 Portulaca consanguinea Schltdl. = Portulaca oleracea L.■

311832 Portulaca constricta M. G. Gilbert；缢缩马齿苋■☆

311833 Portulaca coralloides S. M. Phillips；珊瑚状马齿苋■☆

311834 Portulaca coronata Small = Portulaca umbraticola Kunth subsp. coronata (Small) J. F. Matthews et Ketron■☆

311835 Portulaca crocodilorum Poelln. = Portulaca kermesina N. E. Br.■☆

311836 Portulaca cuneifolia Vahl = Talinum portulacifolium (Forssk.) Asch. ex Schweinf.■☆

311837 Portulaca decorticans M. G. Gilbert；脱皮马齿苋■☆

311838 Portulaca decumbens (Forssk.) Vahl = Corbichonia decumbens (Forssk.) Exell■☆

311839 Portulaca diptera Zipp. ex Span. = Portulaca quadrifida L.■

311840 Portulaca eriophora Casar. = Portulaca pilosa L.■

311841 Portulaca erythraeae Schweinf.；浅红马齿苋■☆

311842 Portulaca fascicularis Peter；扁马齿苋■☆

311843 Portulaca filamentosa Haw. = Anacampseros filamentosa (Haw.) Sims■☆

311844 Portulaca fischeri Pax；菲舍尔马齿苋■☆

311845 Portulaca fischeri Pax var. lutea Poelln. = Portulaca kermesina N. E. Br. var. lutea (Poelln.) S. M. Phillips■☆

311846　Portulaca fischeri Pax var. robusta Poelln. = Portulaca foliosa Ker Gawl. ■☆

311847　Portulaca foliosa Ker Gawl. ;多叶马齿苋■☆

311848　Portulaca foliosa Ker Gawl. var. nitidissima Poelln. = Portulaca guineensis Lindl. ex Ker Gawl. ■☆

311849　Portulaca formosana (Hayata) Hayata = Portulaca quadrifida L. ■

311850　Portulaca fruticosa L. = Talinum fruticosum (L.) Juss. ■☆

311851　Portulaca geniculata Royle = Portulaca quadrifida L. ■

311852　Portulaca grandiflora Hook. ;大花马齿苋(半支莲,草杜鹃,打砍不死,佛甲草,金丝杜鹃,龙须牡丹,死不了,松叶牡丹,太阳花,万年草,午时花,洋马齿苋);Bigflower Purslane, Common Portulaca, Eleven o'clock Flower, Largeflower Purslane, Largeflowered Purslane, Moss Rose, Moss-rose, Portulaca, Rose Moss, Rose-moss, Sun Plant ■☆

311853　Portulaca grandiflora Hook. = Portulaca pilosa L. subsp. grandiflora (Hook.) R. Geesink ■

311854　Portulaca grandis Peter;大马齿苋■☆

311855　Portulaca greenwayi M. G. Gilbert;格林韦马齿苋■☆

311856　Portulaca guineensis Lindl. ex Ker Gawl. ;几内亚马齿苋■☆

311857　Portulaca hainanensis Chun et F. C. How;海南马齿苋;Hainan Purslane ■

311858　Portulaca hainanensis Chun et F. C. How = Portulaca pilosa L. ■

311859　Portulaca hainanensis Chun et F. C. How = Portulaca psammotropha Hance ■

311860　Portulaca halimoides L. ;牙买加马齿苋■☆

311861　Portulaca halimoides L. var. brasiliensis Poelln. ;巴西马齿苋■☆

311862　Portulaca halimoides L. var. brevipilosa Poelln. ;短毛牙买加马齿苋■☆

311863　Portulaca hereroensis Schinz;赫雷罗马齿苋■☆

311864　Portulaca heterophylla Peter;互叶马齿苋■☆

311865　Portulaca holosericea Peter = Portulaca kermesina N. E. Br. ☆

311866　Portulaca humilis Peter;低矮马齿苋■☆

311867　Portulaca insularis Hosok. ;小琉球马齿苋■

311868　Portulaca insularis Hosok. = Portulaca oleracea L. ■

311869　Portulaca kermesina N. E. Br. ;克迈斯马齿苋■☆

311870　Portulaca kermesina N. E. Br. var. lutea (Poelln.) S. M. Phillips;黄克迈斯马齿苋■☆

311871　Portulaca laevis Buch-Ham. = Portulaca oleracea L. ■

311872　Portulaca lanata Rich. = Portulaca pilosa L. ■

311873　Portulaca lanceolata Engelm. = Portulaca umbraticola Kunth subsp. lanceolata J. F. Matthews et Ketron ■☆

311874　Portulaca lanceolata Haw. = Anacampseros lanceolata (Haw.) Sweet ■☆

311875　Portulaca lanuginosa Crantz = Portulaca pilosa L. ■

311876　Portulaca linifolia Forssk. = Portulaca quadrifida L. ■

311877　Portulaca marginata Kunth;红边马齿苋■☆

311878　Portulaca massaica S. M. Phillips;马萨马齿苋■☆

311879　Portulaca megalantha Steud. = Portulaca grandiflora Hook. ■

311880　Portulaca mendocinensis Gillies ex Rohrb. = Portulaca grandiflora Hook. ■

311881　Portulaca meridiana L. f. = Portulaca quadrifida L. ■

311882　Portulaca mkatensis Poelln. = Portulaca heterophylla Peter ■☆

311883　Portulaca mundula I. M. Johnst. = Portulaca pilosa L. ■

311884　Portulaca neglecta Mack. et Bush = Portulaca oleracea L. ■

311885　Portulaca neumannii Engl. ex Poelln. ;纽曼马齿苋■☆

311886　Portulaca nogalensis Chiov. ;诺加尔马齿苋■☆

311887　Portulaca oblonga Peter;矩圆马齿苋■☆

311888　Portulaca officinarum Crantz = Portulaca oleracea L. ■

311889　Portulaca okinawensis E. Walker et Tawada;冲绳马齿苋■☆

311890　Portulaca oleracea L. ;马齿苋(安乐菜,长命草,长命苋,长寿菜,地马菜,豆瓣菜,瓜子菜,红胶墙,酱瓣豆草,九头狮子草,麻绳菜,马齿菜,马齿草,马齿龙芽,马齿龙牙,马蜂菜,马凤菜,马马菜,马屈菜,马舌菜,马蛇子菜,马生菜,马食菜,马踏菜,马苋,马苋菜,马子菜,蚂蚁菜,蚂蚱菜,母猪菜,耐旱菜,蛇草,狮岳菜,狮子菜,狮子草,酸菜,酸味菜,酸苋,五方草,五行菜,五行草,猪肥菜,猪母菜,猪母乳,猪母苋);Common Purslane, Green Purslane, Insular Purslane, Little Hogweed, Pigweed, Pot Purslane, Purslane, Pursley, Pusley, Pussley ■

311891　Portulaca oleracea L. subsp. africana Danin et H. G. Baker;非洲马齿苋■☆

311892　Portulaca oleracea L. subsp. grandiflora (Hook.) R. Geesink = Portulaca grandiflora Hook. ■

311893　Portulaca oleracea L. subsp. granulato-stellulata (Poelln.) Danin et H. G. Baker = Portulaca oleracea L. ■

311894　Portulaca oleracea L. subsp. granulato-stellulata (Poelln.) Danin et H. G. Baker;星粒马齿苋■☆

311895　Portulaca oleracea L. subsp. macrantha (Maire) Maire;阿根廷大花马齿苋■☆

311896　Portulaca oleracea L. subsp. nicaraguensis Danin et H. G. Baker = Portulaca oleracea L. ■

311897　Portulaca oleracea L. subsp. nitida Danin et H. G. Baker = Portulaca oleracea L. ■

311898　Portulaca oleracea L. subsp. papillatostellulata Danin et H. G. Baker;乳头星马齿苋■☆

311899　Portulaca oleracea L. subsp. sativa (Haw.) Celak. ;栽培马齿苋; Garden Purslane, Golden Purslane, Pigweed, Porcelane, Puiruellaine, Pursley, Pusley, Pussley ☆

311900　Portulaca oleracea L. subsp. stellata Danin et H. G. Baker;星状马齿苋■☆

311901　Portulaca oleracea L. subsp. sylvestris (DC.) Thell. = Portulaca oleracea L. ■

311902　Portulaca oleracea L. var. granulato-stellulata Poelln. = Portulaca oleracea L. subsp. granulato-stellulata (Poelln.) Danin et H. G. Baker ■☆

311903　Portulaca oleracea L. var. macrantha Maire = Portulaca oleracea L. subsp. macrantha (Maire) Maire ■☆

311904　Portulaca oleracea L. var. opposita Poelln. = Portulaca oleracea L. ■

311905　Portulaca oleracea L. var. sativa DC. = Portulaca oleracea L. subsp. sativa (Haw.) Celak. ■☆

311906　Portulaca oleracea L. var. sylvestris DC. = Portulaca oleracea L. ■

311907　Portulaca olitoria Pall. = Portulaca oleracea L. ■

311908　Portulaca pachyrrhiza Gagnep. = Portulaca psammotropha Hance ■

311909　Portulaca paniculata Jacq. = Talinum paniculatum (Jacq.) Gaertn. ■

311910　Portulaca parviflora Haw. = Portulaca oleracea L. ■

311911　Portulaca parvula A. Gray = Portulaca halimoides L. ■☆

311912　Portulaca patens L. = Talinum paniculatum (Jacq.) Gaertn. ■

311913　Portulaca peteri Poelln. ;彼得马齿苋■☆

311914　Portulaca phaeosperma Urb. = Portulaca rubricaulis Kunth ■☆

311915　Portulaca pilosa L. ;毛马齿苋(半枝莲,多毛马齿苋,禾雀舌,还魂草,日中花);Hair Purslane, Jump-up-and-kiss-me, Purslane, Red Wild Purslane ■

311916　Portulaca pilosa L. subsp. grandiflora (Hook.) R. Geesink =

Portulaca grandiflora Hook. ■

311917　Portulaca pilosa L. subsp. okinawensis（E. Walker et Tawada）R. Geesink ＝Portulaca okinawensis E. Walker et Tawada ■☆

311918　Portulaca pilosa L. var. setacea DC. ＝Portulaca pilosa L. ■

311919　Portulaca portulacastrum L. ＝Sesuvium portulacastrum（L.）L. ■

311920　Portulaca psammotropha Hance；沙生马齿苋；Sandy Purslane ■

311921　Portulaca quadrifida L.；四瓣马齿苋（地锦，四裂马齿苋，小马齿苋）；Chickenweed，Four-cat Purslane，Fourpetal Purslane，Pusley，Ten o'clock Flower ■

311922　Portulaca quadrifida L. var. formosana Hayata ＝Portulaca quadrifida L. ■

311923　Portulaca quadrifida L. var. meridiana DC. ＝Portulaca quadrifida L. ■

311924　Portulaca ramosa Poelln.；分枝马齿苋■☆

311925　Portulaca repens Roxb. ex Wight et Arn. ＝Portulaca quadrifida L. ■

311926　Portulaca retusa Engelm. ＝Portulaca oleracea L. ■

311927　Portulaca rhodesiana R. A. Dyer et E. A. Bruce；罗得西亚马齿苋■☆

311928　Portulaca rubens Haw. ＝Anacampseros arachnoides（Haw.）Sims ■☆

311929　Portulaca rubricaulis Kunth；红茎马齿苋■☆

311930　Portulaca rubriflora Poelln. ＝Portulaca kermesina N. E. Br. ■☆

311931　Portulaca rufescens Haw. ＝Anacampseros rufescens（Haw.）Sweet ■☆

311932　Portulaca sativa Haw. ＝Portulaca oleracea L. var. sativa DC. ■☆

311933　Portulaca saxifragoides Welw. ex Oliv.；虎耳草马齿苋■☆

311934　Portulaca sedoides Spruce ex Rohrb. ＝Portulaca pilosa L. ■

311935　Portulaca sedoides Welw. ex Oliv.；景天马齿苋■☆

311936　Portulaca smallii P. Wilson；斯莫尔马齿苋■☆

311937　Portulaca somalica N. E. Br.；索马里马齿苋■☆

311938　Portulaca stuhlmannii Poelln.；斯图尔曼马齿苋■☆

311939　Portulaca suffrutescens Engelm. ＝Portulaca oleracea L. ■

311940　Portulaca suffruticosa Thwaites ＝Portulaca oleracea L. ■

311941　Portulaca teretifolia Kunth ＝Portulaca pilosa L. ■

311942　Portulaca triangularis Jacq. ＝Talinum fruticosum（L.）Juss. ■☆

311943　Portulaca trianthemoides Bremek.；假海马齿苋■☆

311944　Portulaca tuberosa Roxb.；块根马齿苋■☆

311945　Portulaca umbraticola Kunth；翼梗马齿苋；Wingpod Purslane ■☆

311946　Portulaca umbraticola Kunth subsp. coronata（Small）J. F. Matthews et Ketron；岩地翼梗马齿苋■☆

311947　Portulaca umbraticola Kunth subsp. lanceolata J. F. Matthews et Ketron；剑叶翼梗马齿苋■☆

311948　Portulaca usambarensis Poelln. var. obtusata ？ ＝Portulaca commutata M. G. Gilbert ■☆

311949　Portulaca usambarensis Poelln. var. tuberculata ？ ＝Portulaca commutata M. G. Gilbert ■☆

311950　Portulaca viridis DC. ＝Portulaca oleracea L. ■

311951　Portulaca walteriana Poelln. ＝Portulaca quadrifida L. ■

311952　Portulaca wightiana Wall. ex Wight et Arn.；怀特马齿苋■☆

311953　Portulacaceae Adans. ＝Portulacaceae Juss.（保留科名）■●

311954　Portulacaceae Juss.（1789）（保留科名）；马齿苋科；Blinks Family，Purslane Family ■●

311955　Portulacaria Jacq.（1787）；马齿苋树属（树马齿苋属）；Portulacaria ●☆

311956　Portulacaria afra（L.）Jacq.；马齿苋树（树马齿苋，银公孙树）；Chinese Jade Plant，Elephant Bush，Elephant's Food，Epephant Bush，Jade Plant，Purslane Tree，Spekboom，Tree Purslane ●☆

311957　Portulacaria afra（L.）Jacq. 'Foliisvariegata'；斑叶马齿苋树（斑叶马齿苋，花叶树马齿苋）●☆

311958　Portulacaria afra（L.）Jacq. 'Tricolor'；三色马齿苋树●☆

311959　Portulacaria afra（L.）Jacq. 'Variegata' ＝Portulacaria afra（L.）Jacq. 'Foliisvariegata'●☆

311960　Portulacaria afra Jacq. ＝Portulacaria afra（L.）Jacq. ●☆

311961　Portulacaria namaquensis Sond. ＝Ceraria namaquensis（Sond.）H. Pearson et Stephens ●☆

311962　Portulacaria pygmaea Pillans；矮小马齿苋树●☆

311963　Portulacariaceae Doweld ＝Didiereaceae Radlk.（保留科名）●☆

311964　Portulacariaceae Doweld；马齿苋树科●☆

311965　Portulacastrum Juss. ex Medik. ＝Trianthema L. ■

311966　Portuna Nutt. ＝Pieris D. Don ●

311967　Posadaea Cogn.（1890）；波萨瓜属●☆

311968　Posadaea sphaerocarpa Cogn.；波萨瓜●☆

311969　Posidonia K. D. König（1805）（保留属名）；波喜荡草属（波喜荡属）；Posidonia，Tapeweed ●☆

311970　Posidonia Koniger ＝Posidonia K. D. König（保留属名）■

311971　Posidonia australis Hook. f.；波喜荡（波喜荡草）；Australian Posidonia，Posidonia ■

311972　Posidonia caulinii K. D. König；茎生波喜荡；Mediterranean Tapeweed ■☆

311973　Posidonia oceanica（L.）Delile ＝Posidonia caulinii K. D. König ■☆

311974　Posidonia serrulata（R. Br.）Spreng. ＝Cymodocea serrulata（R. Br.）Asch. et Magnus ■☆

311975　Posidoniaceae Hutch. ＝Posidoniaceae Vines（保留科名）■

311976　Posidoniaceae Lotsy ＝Posidoniaceae Vines（保留科名）■

311977　Posidoniaceae Vines（1895）（保留科名）；波喜荡草科（波喜荡科，海草科，海神草科）■

311978　Posidonion St. -Lag. ＝Posidonia K. D. König（保留属名）■

311979　Poskea Vatke（1882）；密穗球花木属●☆

311980　Poskea africana Vatke；非洲密穗球花木●☆

311981　Poskea newbouldii Braggio ＝Poskea africana Vatke ●☆

311982　Poskea socotrana（Balf. f.）G. Taylor；索科特拉密穗球花木●☆

311983　Posoqueria Aubl.（1775）；波苏茜属（鲍苏栎属）●☆

311984　Posoqueria acuminata Mart.；渐尖特喜无患子●☆

311985　Posoqueria acutifolia Mart.；尖叶特喜无患子●☆

311986　Posoqueria brachyantha Standl.；短花特喜无患子●☆

311987　Posoqueria densiflora Hutch.；密花特喜无患子●☆

311988　Posoqueria fasciculata Roxb.；簇生特喜无患子●☆

311989　Posoqueria floribunda Roxb.；繁花特喜无患子●☆

311990　Posoqueria gracilis（Rudge）Roem. et Schult.；细特喜无患子●☆

311991　Posoqueria grandiflora Standl.；大花特喜无患子●☆

311992　Posoqueria latifolia（Rudge）Roem. et Schult.；宽叶波苏茜（宽叶特喜无患子，阔叶鲍苏栎）；Brazilian Oak ●☆

311993　Posoqueria latifolia Roem. et Schult. ＝Posoqueria latifolia（Rudge）Roem. et Schult. ●☆

311994　Posoqueria laurifolia Mart.；桂叶特喜无患子●☆

311995　Posoqueria longiflora Aubl.；长花波苏茜（长花柏索奎利亚）●☆

311996　Posoqueria longiflora Roxb.；长花特喜无患子●☆

311997　Posoqueria longispina Roxb.；长刺特喜无患子●☆

311998　Posoqueria macrophylla Hemsl.；大叶特喜无患子●☆

311999　Posoqueria multiflora Lemaire；多花特喜无患子●☆

312000　Posoqueria trinitatis DC.；圆裂波苏茜（圆裂鲍苏栎）●☆

312001　Posoria Raf. = Posoqueria Aubl. ●☆

312002　Possira Aubl. (废弃属名) = Swartzia Schreb. (保留属名)●☆

312003　Possiria Steud. = Postia Boiss. et Blanche ●☆

312004　Possura Aubl. ex Steud. = Postia Boiss. et Blanche ●☆

312005　Postia Boiss. et Blanche = Rhanteriopsis Rauschert ●☆

312006　Postiella Kljuykov(1985);波斯特草属■☆

312007　Postiella capillifolia (Post) Kljuykov;波斯特草■☆

312008　Postuera Raf. = Notelaea Vent. ●☆

312009　Potalia Aubl. (1775);龙爪七叶属●☆

312010　Potalia amara Aubl.;龙爪七叶●☆

312011　Potaliaceae Mart.;龙爪七叶科●☆

312012　Potaliaceae Mart. = Gentianaceae Juss. (保留科名)●■

312013　Potameia Thouars(1806);合药樟属(白面柴属,河樟属,马岛樟属,油樟属)●☆

312014　Potameia chinensis (C. K. Allen) Kosterm. = Syndiclis chinensis C. K. Allen ●

312015　Potameia kwangsiensis Kosterm. = Syndiclis kwangsiensis (Kosterm.) H. W. Li ●

312016　Potameia lotungensis (S. K. Lee) Dao = Syndiclis lotungensis S. K. Lee ●

312017　Potameia lucida Kosterm. = Beilschmiedia opposita Kosterm. ●☆

312018　Potameia thouarsiana (Baill.) Capuron;合药樟●☆

312019　Potamica Poiret = Potameia Thouars ●☆

312020　Potamobryon Liebm. = Tristicha Thouars ■☆

312021　Potamobryum Liebm. = Tristicha Thouars ■☆

312022　Potamocharis Rottb. = Mammea L. ●

312023　Potamochloa Griff. = Hygroryza Nees ■

312024　Potamochloa aristata (Retz.) Griff. ex Steud. = Hygroryza aristata (Retz.) Nees ex Wight et Arn. ■

312025　Potamochloa retzii Griff. = Hygroryza aristata (Retz.) Nees ex Wight et Arn. ■

312026　Potamoganos Sandwith(1937);河美紫葳属●☆

312027　Potamoganos microcalyx (G. Mey.) Sandwith;河美紫葳●☆

312028　Potamogeton L. (1753);眼子菜属;Curled Pondweed, Pond Weed, Pondweed, Pondweeds, Water Spike ■

312029　Potamogeton Walter = Myriophyllum L. ■

312030　Potamogeton × apertus Miki;无孔眼子菜■☆

312031　Potamogeton × fauriei (A. Benn.) Miki;福氏眼子菜■☆

312032　Potamogeton × leptocephalus Koidz.;细头眼子菜■☆

312033　Potamogeton × malainoides Miki;拟竹叶眼子菜■☆

312034　Potamogeton × nitens Weber;亮叶眼子菜;Bright-leaved Pondweed ■☆

312035　Potamogeton × orientalis Hagstr.;东方眼子菜■☆

312036　Potamogeton acutifolius Link;单果眼子菜(尖叶眼子菜,柳叶眼子菜);Sharp-leaved Pondweed, Singlefruit Pondweed ■

312037　Potamogeton alpinus Balb.;高山眼子菜;Alpine Pondweed, Red Pondweed, Reddish Pondweed ■☆

312038　Potamogeton alpinus Balb. subsp. tenuifolius (Raf.) Hultén = Potamogeton alpinus Balb. ■☆

312039　Potamogeton alpinus Balb. var. subellipticus (Fernald) Ogden = Potamogeton alpinus Balb. ■☆

312040　Potamogeton alpinus Balb. var. tenuifolius (Raf.) Ogden = Potamogeton alpinus Balb. ■☆

312041　Potamogeton amblyphyllus C. A. Mey. = Stuckenia amblyphylla (C. A. Mey.) Holub ■

312042　Potamogeton americanus Cham. et Schltdl.;美国眼子菜;American Pondweed ■☆

312043　Potamogeton americanus Cham. et Schltdl. = Potamogeton nodosus Poir. ■

312044　Potamogeton amplifolius Tuck.;大叶眼子菜;Big-leaved Pondweed, Broad-leaved Pondweed, Largeleaf Pondweed, Large-leaved Pondweed ■☆

312045　Potamogeton anguillanus Koidz.;细眼子菜■☆

312046　Potamogeton angustifolius Bercht. et C. Presl;狭叶眼子菜;Narrow-leaved Pondweed ■☆

312047　Potamogeton angustifolius Bercht. et C. Presl = Potamogeton illinoensis Morong ■☆

312048　Potamogeton antaicus Hagstr. = Potamogeton pusillus L. ■

312049　Potamogeton applanatus Y. D. Chen = Potamogeton filiformis Pers. var. applanatus (Y. D. Chen) Q. Y. Li ■

312050　Potamogeton applanatus Y. D. Chen = Stuckenia filiformis (Pers.) Börner ■

312051　Potamogeton asiaticus A. Benn. = Potamogeton octandrus Poir. ■

312052　Potamogeton berchtoldii Fieber;纤细眼子菜■

312053　Potamogeton berchtoldii Fieber = Potamogeton pusillus L. subsp. tenuissimus (Mert. et W. D. J. Koch) R. R. Haynes et Hellq. ■☆

312054　Potamogeton berchtoldii Fieber = Potamogeton pusillus L. ■

312055　Potamogeton berchtoldii Fieber var. acuminatus Fieber = Potamogeton pusillus L. subsp. tenuissimus (Mert. et W. D. J. Koch) R. R. Haynes et Hellq. ■☆

312056　Potamogeton berchtoldii Fieber var. colpophilus (Fernald) Fernald = Potamogeton pusillus L. subsp. tenuissimus (Mert. et W. D. J. Koch) R. R. Haynes et Hellq. ■☆

312057　Potamogeton berchtoldii Fieber var. lacunatus (Hagstr.) Fernald = Potamogeton pusillus L. subsp. tenuissimus (Mert. et W. D. J. Koch) R. R. Haynes et Hellq. ■☆

312058　Potamogeton berchtoldii Fieber var. mucronatus Fieber = Potamogeton pusillus L. subsp. tenuissimus (Mert. et W. D. J. Koch) R. R. Haynes et Hellq. ■☆

312059　Potamogeton berchtoldii Fieber var. polyphyllus (Morong) Fernald = Potamogeton pusillus L. subsp. tenuissimus (Mert. et W. D. J. Koch) R. R. Haynes et Hellq. ■☆

312060　Potamogeton berchtoldii Fieber var. tenuissimus (Mert. et W. D. J. Koch) Fernald = Potamogeton pusillus L. subsp. tenuissimus (Mert. et W. D. J. Koch) R. R. Haynes et Hellq. ■☆

312061　Potamogeton berchtoldii Fieber var. tenuissimus (Mert. et W. D. J. Koch) Fernald = Potamogeton berchtoldii Fieber ■

312062　Potamogeton bicupulatus Fernald;螺果眼子菜;Snail-seed Pondweed ■☆

312063　Potamogeton borealis Raf. = Stuckenia filiformis (Pers.) Börner subsp. alpina (Blytt) R. R. Haynes, Les et Král ■☆

312064　Potamogeton bracteatus Y. D. Chen = Potamogeton pectinatus L. ■

312065　Potamogeton bracteatus Y. D. Chen = Stuckenia pectinata (L.) Börner ■

312066　Potamogeton bupleuroides Fernald;美洲眼子菜;Bupleuroid Pondweed ■☆

312067　Potamogeton bupleuroides Fernald = Potamogeton perfoliatus L. ■

312068　Potamogeton capillaceus Poir. = Potamogeton diversifolius Raf. ■☆

312069　Potamogeton capillaceus Poir. var. atripes Fernald = Potamogeton diversifolius Raf. ■☆

312070　Potamogeton chongyangensis W. X. Wang = Potamogeton oxyphyllus Miq. ■

312071　Potamogeton chongyongensis W. X. Wang;崇阳眼子菜;Chongyang Pondweed ■

312072　Potamogeton clystocarpus Fernald；瘤果眼子菜■☆

312073　Potamogeton coloratus Hornem.；车前叶眼子菜；Fen Pondweed，Plantain-leaved Pondweed■☆

312074　Potamogeton coloratus Vahl；着色眼子菜■☆

312075　Potamogeton compressus L.；扁茎眼子菜（大叶藻叶眼子菜，柳叶眼子菜）；Flat-stem Pondweed，Flat-stemmed Pondweed，Grass-wrack Pondweed■

312076　Potamogeton confervoides Rchb.；藻叶眼子菜；Alga Pondweed，Algae-like Pondweed，Algal-leaved Pondweed，Tuckerman's Pondweed■☆

312077　Potamogeton contortus Desf. = Zannichellia contorta（Desf.）Cham. et Schltdl.■☆

312078　Potamogeton cooperi；库珀眼子菜；Cooper's Pondweed■☆

312079　Potamogeton crispus L.；菹草（鹅草，马藻，丝草，虾藻，扎草）；Beginners' Pondweed，Curled Pondweed，Curly Muckweed，Curly Pondweed，Curly-leaf Pondweed，Curly-leaved Pondweed■

312080　Potamogeton crispus L. var. acutifolius Fieber = Potamogeton crispus L.■

312081　Potamogeton crispus L. var. obtusifolius Fieber = Potamogeton crispus L.■

312082　Potamogeton crispus L. var. serrulatus Rchb. = Potamogeton crispus L.■

312083　Potamogeton cristatus Regel et Maack；鸡冠眼子菜（冠果眼子菜，水竹叶，突果眼子菜，小叶眼子菜）；Cockscomb Pondweed，Littleleaf Pondweed■

312084　Potamogeton curtissii Morong = Potamogeton foliosus Raf.■☆

312085　Potamogeton delavayi A. Benn.；牙齿草（滇鸭子草，水案板）■

312086　Potamogeton densus L.；密集眼子菜；Dense Pondweed■☆

312087　Potamogeton densus L. = Groenlandia densa（L.）Fourr.■☆

312088　Potamogeton densus L. var. lancifolius Mert. et Koch = Groenlandia densa（L.）Fourr.■☆

312089　Potamogeton densus L. var. rigidus Opiz = Groenlandia densa（L.）Fourr.■☆

312090　Potamogeton distinctus A. W. Benn.；眼子菜（案板菜，案板芽，滇鸭子草，钉钯七，金梳子草，水案板，鸭吃草，鸭子草，牙齿草，牙拾草，异匙叶藻，扎水板）；Distinct Pondweed，Pond Weed，Pondweed■

312091　Potamogeton diversifolius Raf.；西方异叶眼子菜；Common Snail-seed Pondweed，Waterthread Pondweed，Water-thread Pondweed■☆

312092　Potamogeton diversifolius Raf. = Potamogeton bicupulatus Fernald■☆

312093　Potamogeton diversifolius Raf. var. multidenticulatus（Morong）Asch. et Graebn. = Potamogeton diversifolius Raf.■☆

312094　Potamogeton diversifolius Raf. var. trichophyllus Morong = Potamogeton bicupulatus Fernald■☆

312095　Potamogeton ellipticus Z. S. Diao；圆叶眼子菜；Elliptic Pondweed■

312096　Potamogeton epihydrus Raf.；水生眼子菜；American Pondweed，Emersed Pondweed，Ribbonleaf Pondweed，Ribbon-leaf Pondweed，Ribbon-leaved Pondweed■☆

312097　Potamogeton epihydrus Raf. subsp. nuttallii（Cham. et Schltdl.）Calder et R. L. Taylor = Potamogeton epihydrus Raf.■☆

312098　Potamogeton epihydrus Raf. var. cayugensis（Wiegand）A. Benn. = Potamogeton epihydrus Raf.■☆

312099　Potamogeton epihydrus Raf. var. nuttallii（Cham. et Schltdl.）Fernald = Potamogeton epihydrus Raf.■☆

312100　Potamogeton epihydrus Raf. var. ramosus（Peck）House = Potamogeton epihydrus Raf.■☆

312101　Potamogeton epihydrus Raf. var. typicus Fernald = Potamogeton epihydrus Raf.■☆

312102　Potamogeton erhaiensis Y. D. Chen = Potamogeton pectinatus L.■

312103　Potamogeton erhaiensis Y. D. Chen = Stuckenia pectinata（L.）Börner■

312104　Potamogeton faxonii Morong；法克森眼子菜；Pondweed■☆

312105　Potamogeton fibrillosus Fernald = Potamogeton foliosus Raf. subsp. fibrillosus（Fernald）R. R. Haynes et Hellq.■☆

312106　Potamogeton filiformis Pers. = Stuckenia filiformis（Pers.）Börner■

312107　Potamogeton filiformis Pers. subsp. alpinus（Blytt）Hellq. et R. R. Haynes = Stuckenia filiformis（Pers.）Börner subsp. alpina（Blytt）R. R. Haynes，Les et Král■☆

312108　Potamogeton filiformis Pers. subsp. occidentalis（J. W. Robbins）Hellq. et R. R. Haynes = Stuckenia filiformis（Pers.）Börner subsp. occidentalis（J. W. Robbins）R. R. Haynes，Les et Král■☆

312109　Potamogeton filiformis Pers. var. alpinus（Blytt）Asch. et Graebn. = Stuckenia filiformis（Pers.）Börner subsp. alpina（Blytt）R. R. Haynes，Les et Král■☆

312110　Potamogeton filiformis Pers. var. applanatus（Y. D. Chen）Q. Y. Li；扁茎丝叶眼子菜（扁茎眼子菜）■

312111　Potamogeton filiformis Pers. var. applanatus（Y. D. Chen）Q. Y. Li = Stuckenia filiformis（Pers.）Börner■

312112　Potamogeton filiformis Pers. var. borealis（Raf.）H. St. John = Stuckenia filiformis（Pers.）Börner subsp. alpina（Blytt）R. R. Haynes，Les et Král■☆

312113　Potamogeton filiformis Pers. var. macounii Morong = Stuckenia filiformis（Pers.）Börner subsp. alpina（Blytt）R. R. Haynes，Les et Král■☆

312114　Potamogeton filiformis Pers. var. occidentalis（J. W. Robbins）Morong = Stuckenia filiformis（Pers.）Börner subsp. occidentalis（J. W. Robbins）R. R. Haynes，Les et Král■☆

312115　Potamogeton floridanus Small；佛罗里达眼子菜；Florida Pondweed■☆

312116　Potamogeton fluitans Roth；漂浮眼子菜■☆

312117　Potamogeton foliosus Raf.；多叶眼子菜；Leafy Pondweed■☆

312118　Potamogeton foliosus Raf. subsp. fibrillosus（Fernald）R. R. Haynes et Hellq.；纤维眼子菜■☆

312119　Potamogeton foliosus Raf. var. macellus Fernald = Potamogeton foliosus Raf.■☆

312120　Potamogeton fontigenus Y. H. Guo，X. Z. Sun et H. Q. Wang；泉生眼子菜；Fontinal Pondweed■

312121　Potamogeton fontigenus Y. H. Guo，X. Z. Sun et H. Q. Wang = Potamogeton distinctus A. W. Benn.■

312122　Potamogeton franchetii A. Benn. et Baagoe = Potamogeton distinctus A. W. Benn.■

312123　Potamogeton friesii Rupr.；弗瑞氏眼子菜；Flat-stalked Pondweed，Fries Pondweed，Fries' Pondweed■

312124　Potamogeton gaudichaudii Cham. et Schltdl. = Potamogeton malaianus Miq.■

312125　Potamogeton gemmiparus（J. W. Robbins）Morong = Potamogeton pusillus L. subsp. gemmiparus（J. W. Robbins）R. R. Haynes et Hellq.■☆

312126　Potamogeton gramineus L.；禾叶眼子菜；Gemmiparous Pondweed，Grass-leaved Pondweed，Poaleaf Pondweed，Variable-leaved Pondweed, Various-leaved Pondweed ■

312127　Potamogeton gramineus L. var. graminifolius Fr. = Potamogeton gramineus L. ■

312128　Potamogeton gramineus L. var. heterophyllus（Schreb.）Fr. = Potamogeton heterophyllus Schreb. ■

312129　Potamogeton gramineus L. var. heterophyllus（Schreb.）Fr. = Potamogeton gramineus L. ■

312130　Potamogeton gramineus L. var. maximus Morong = Potamogeton gramineus L. ■

312131　Potamogeton gramineus L. var. maximus Morong ex A. Benn. = Potamogeton gramineus L. ■

312132　Potamogeton gramineus L. var. myriophyllus J. W. Robbins = Potamogeton gramineus L. ■

312133　Potamogeton gramineus L. var. typicus Ogden = Potamogeton gramineus L. ■

312134　Potamogeton griffithii A. Benn.；格氏眼子菜；Griffith's Pondweed ■☆

312135　Potamogeton groenlandicus Hagstr.；格陵兰眼子菜■☆

312136　Potamogeton henningii A. Benn.；亨氏眼子菜；Henning Pondweed ■☆

312137　Potamogeton heterocaulis Dia = Potamogeton gramineus L. ■

312138　Potamogeton heterophyllus Schreb.；异叶眼子菜（牙齿草）；Differleaf Pondweed, Various-leaved Pondweed ■

312139　Potamogeton heterophyllus Schreb. = Potamogeton gramineus L. ■

312140　Potamogeton hillii Morong；希尔眼子菜；Hill's Pondweed ■☆

312141　Potamogeton hubeiensis W. X. Wang, X. Z. Sun et H. Q. Wang；湖北眼子菜；Hubei Pondweed ■

312142　Potamogeton hubeiensis W. X. Wang, X. Z. Sun et H. Q. Wang = Potamogeton octandrus Poir. ■

312143　Potamogeton illinoensis Morong；伊利诺眼子菜；Illinois Pondweed ■☆

312144　Potamogeton indicus Roth ex Roem. et Schult. = Potamogeton nodosus Poir. ■

312145　Potamogeton indicus Roxb. = Potamogeton nodosus Poir. ■

312146　Potamogeton interior Rydb. = Stuckenia filiformis（Pers.）Börner subsp. alpina（Blytt）R. R. Haynes, Les et Král ■☆

312147　Potamogeton interruptus Kit. = Potamogeton intromongolicus Ma ■

312148　Potamogeton interruptus Kit. = Potamogeton pectinatus L. var. interruptus（Kit.）Asch. ■

312149　Potamogeton interruptus Kit. = Stuckenia pectinata（L.）Börner ■

312150　Potamogeton interruptus Kit. = Stuckenia vaginata（Turcz.）Holub ■☆

312151　Potamogeton intortusifolius J. B. He, L. Y. Zhou et H. Q. Wang；扭叶眼子菜；Wringleaf Pondweed ■

312152　Potamogeton intortusifolius J. B. He, L. Y. Zhou et H. Q. Wang = Potamogeton wrightii Morong ■

312153　Potamogeton intramongolicus Ma = Potamogeton pectinatus L. var. interruptus（Kit.）Asch. ■

312154　Potamogeton intramongolicus Ma = Stuckenia pectinata（L.）Börner ■

312155　Potamogeton intromongolicus Ma；内蒙眼子菜（断眼子菜，扎草）；Inner Mongolia Pondweed ■

312156　Potamogeton iriomotensis Masam. = Potamogeton cristatus Regel et Maack ■

312157　Potamogeton japonicus Franch. et Sav. = Potamogeton malaianus Miq. ■

312158　Potamogeton japonicus Franch. et Sav. = Potamogeton wrightii

Morong ■

312159　Potamogeton javanicus Hassk. = Potamogeton octandrus Poir. ■

312160　Potamogeton latifolius（J. W. Robbins）Morong = Stuckenia striata（Ruiz et Pav.）Holub ■☆

312161　Potamogeton leptanthus Y. D. Chen；柔花眼子菜；Softflower Pondweed ■

312162　Potamogeton leptanthus Y. D. Chen = Stuckenia pectinata（L.）Börner ■

312163　Potamogeton limosellifolius Maxim. = Potamogeton octandrus Poir. var. miduhikimo（Makino）H. Hara ■

312164　Potamogeton limosellifolius Maxim. ex Korsh. = Potamogeton octandrus Poir. ■

312165　Potamogeton livingstonei A. Benn. = Potamogeton pectinatus L. ■

312166　Potamogeton longifolius J. Gay ex Poir.；长叶眼子菜；Longleaf Pondweed ■☆

312167　Potamogeton longipetiolatus A. Camus = Potamogeton distinctus A. W. Benn. ■

312168　Potamogeton lucens L.；光叶眼子菜（闪眼子菜）；Brightleaf Pondweed, Glabrousleaf Pondweed, Shining Pondweed ■

312169　Potamogeton lucens L. var. lancifolius Mert. et Koch = Potamogeton lucens L. ■

312170　Potamogeton maackianus A. W. Benn.；微齿眼子菜（黄丝草，马克眼子菜，线叶菹，竹叶草）；Maack Pondweed, Minitooth Pondweed ■

312171　Potamogeton malaianus Miq.；马来眼子菜（匙叶眼子菜，箬叶藻，水龙草，竹草眼子菜，竹叶眼子菜，竹叶藻）；Bambooleaf Pondweed ■

312172　Potamogeton malaianus Miq. = Potamogeton nodosus Poir. ■

312173　Potamogeton manchuriensis（A. Benn.）A. Benn.；东北眼子菜■

312174　Potamogeton marinus L. = Potamogeton pectinatus L. ■

312175　Potamogeton marinus L. f. alpinus Blytt = Stuckenia filiformis（Pers.）Börner subsp. alpina（Blytt）R. R. Haynes, Les et Král ■☆

312176　Potamogeton marinus L. var. alpinus（J. W. Robbins）Morong = Stuckenia filiformis（Pers.）Börner subsp. alpina（Blytt）R. R. Haynes, Les et Král ■☆

312177　Potamogeton marinus L. var. macounii Morong = Stuckenia filiformis（Pers.）Börner subsp. alpina（Blytt）R. R. Haynes, Les et Král ■☆

312178　Potamogeton marinus L. var. occidentalis J. W. Robbins = Stuckenia filiformis（Pers.）Börner subsp. occidentalis（J. W. Robbins）R. R. Haynes, Les et Král ■☆

312179　Potamogeton microstachys Wolfg. ex Schult.；小穗眼子菜；Small-spiked Pondweed ■☆

312180　Potamogeton miduhikimo Makino = Potamogeton octandrus Poir. var. miduhikimo（Makino）H. Hara ■

312181　Potamogeton miduhikimo Makino = Potamogeton octandrus Poir. ■

312182　Potamogeton minatus Y. D. Chen = Potamogeton pectinatus L. ■

312183　Potamogeton miniatus Y. D. Chen = Stuckenia pectinata（L.）Börner ■

312184　Potamogeton morongii A. Benn. = Potamogeton natans L. ■

312185　Potamogeton mucronatus C. Presl = Potamogeton malaianus Miq. ■

312186　Potamogeton mucronatus C. Presl = Potamogeton wrightii Morong ■

312187　Potamogeton mucronatus Schrad. = Potamogeton friesii Rupr. ■

312188　Potamogeton nanus Y. D. Chen；矮眼子菜；Common Pondweed, Dwarf Pondweed, Floating Pondweed, Floating-leaf Pondweed ■

312189　Potamogeton nanus Y. D. Chen = Stuckenia pectinata（L.）

Börner ■

312190 Potamogeton natans L. ;浮叶眼子菜(浮眼子菜,厚叶眼子菜, 活叶眼子菜,水案板,西藏眼子菜); Broad-leaved Pondweed, Common Pondweed, Devil's Spoons, Floating Leaf Pondweed, Floating Pondweed, Floatingleaf Pondweed, Floating-leaf Pondweed, Floatleaf Pondweed, Pickerel Weed, Pond Plantain, Swimming Plantain, Tenchweed, Water Spike ■

312191 Potamogeton natans L. = Potamogeton distinctus A. W. Benn. ■

312192 Potamogeton natans L. = Potamogeton nodosus Poir. ■

312193 Potamogeton nipponicus Makino = Potamogeton × nitens Weber ■☆

312194 Potamogeton nodosus Poir. ;小节眼子菜(节眼子菜); Loddon Pondweed, Long Leaf Pondweed, Longleaf Pondweed, Long-leaf Pondweed, Long-leaved Pondweed, Nodular Pondweed ■

312195 Potamogeton oakesianus J. W. Robbins;奥克斯眼子菜; Oakes Pondweed, Oakes' Pondweed ■☆

312196 Potamogeton oblongus Viv. ;矩圆眼子菜■☆

312197 Potamogeton oblongus Viv. = Potamogeton nodosus Poir. ■

312198 Potamogeton oblongus Viv. = Potamogeton polygonifolius Pourr. ■

312199 Potamogeton obtusifolius Merr. et W. D. J. Koch;钝叶眼子菜; Blunt-leaved Pondweed, Grassy Pondweed, Obtuse Pondweed, Obtuse-leaved Pondweed ■

312200 Potamogeton octandrus Poir. ;八蕊眼子菜(南方眼子菜,眼子菜); Southern Pondweed ■

312201 Potamogeton octandrus Poir. subsp. ethiopicus Lye = Potamogeton octandrus Poir. ■

312202 Potamogeton octandrus Poir. var. limosellifolius (Maxim. ex Korsh.) Tzvelev = Potamogeton octandrus Poir. ■

312203 Potamogeton octandrus Poir. var. miduhikimo (Makino) H. Hara;钝脊眼子菜(南方眼子菜,小浮叶眼子菜); Obtuseridge Pondweed ■

312204 Potamogeton ogdenii Hellq. et R. L. Hilton;奥格登眼子菜; Ogden's Pondweed ■☆

312205 Potamogeton olivaceus Baagoe;优美眼子菜; Graceful Pondweed ■☆

312206 Potamogeton oxyphyllus Miq. ;尖叶眼子菜(线叶藻); Sharpleaf Pondweed ■

312207 Potamogeton pamiricus Baagoe = Stuckenia pamirica (Baagoe) Z. Kaplan ■

312208 Potamogeton pamiricus Baagoe ex Paulsen;帕米尔眼子菜; Pamir Pondweed ■

312209 Potamogeton panormitanus Biv. = Potamogeton pusillus L. ■

312210 Potamogeton panormitanus Biv. var. major G. Fisch. = Potamogeton pusillus L. ■

312211 Potamogeton panormitanus Biv. var. minor Biv. = Potamogeton pusillus L. ■

312212 Potamogeton parvifolius Buchenau = Potamogeton octandrus Poir. ■

312213 Potamogeton pectinatus L. = Stuckenia pectinata (L.) Börner ■

312214 Potamogeton pectinatus L. var. diffusus Hagstr. ;铺散眼子菜; Spread Pondweed ■

312215 Potamogeton pectinatus L. var. interruptus (Kit.) Asch. = Potamogeton intromongolicus Ma ■

312216 Potamogeton pectinatus L. var. interruptus (Kit.) Asch. = Stuckenia pectinata (L.) Börner ■

312217 Potamogeton pectinatus L. var. tenuifolius A. Benn. ;细叶篦齿眼子菜■☆

312218 Potamogeton pectinatus L. var. ungulatus Hagstr. = Potamogeton pectinatus L. ■

312219 Potamogeton pectinatus L. var. vulgaris Cham. = Potamogeton pectinatus L. ■

312220 Potamogeton perfoliatus L. ;穿叶眼子菜(抱茎眼子菜,酸水草,眼子菜); Clasping Leaf Pondweed, Clasping-leaf Pondweed, Crossleaf Pondweed, Perfoliate Pondweed, Red-head-grass, Thorowort Pondweed ■

312221 Potamogeton perfoliatus L. subsp. bupleuroides (Fernald) Hultén = Potamogeton perfoliatus L. ■

312222 Potamogeton perfoliatus L. subsp. richardsonii (A. Benn.) Hultén = Potamogeton richardsonii (A. Benn.) Rydb. ■☆

312223 Potamogeton perfoliatus L. var. bupleuroides (Fernald) Farw. = Potamogeton perfoliatus L. ■

312224 Potamogeton perfoliatus L. var. manchshuriensis A. Benn. = Potamogeton perfoliatus L. ■

312225 Potamogeton perfoliatus L. var. mandchshuriensis A. Benn. = Potamogeton perfoliatus L. ■

312226 Potamogeton perfoliatus L. var. richardsonii A. Benn. = Potamogeton richardsonii (A. Benn.) Rydb. ■☆

312227 Potamogeton perversus A. Benn. = Potamogeton distinctus A. W. Benn. ■

312228 Potamogeton polygonifolius Pourr. ;蓼叶眼子菜(牙齿草,眼子菜); Bog Pondweed, Knotweedleaf Pondweed ■

312229 Potamogeton polygonifolius Pourr. = Potamogeton distinctus A. W. Benn. ■

312230 Potamogeton porsildiorum Fernald = Potamogeton subsibiricus Hagstr. ■☆

312231 Potamogeton porteri Fernald = Potamogeton hillii Morong ■☆

312232 Potamogeton praelongus Wulfen;白茎眼子菜(长眼子菜); Long Pondweed, Long-peduncled Pondweed, Longstalked Pondweed, Whitestem Pondweed, White-stem Pondweed, White-stemmed Pondweed ■

312233 Potamogeton praelongus Wulfen var. angustifolius Graebn. = Potamogeton praelongus Wulfen ■

312234 Potamogeton preussii A. Benn. = Potamogeton octandrus Poir. ■

312235 Potamogeton pulcher Tuck. ;斑点眼子菜; Spotted Pondweed ■☆

312236 Potamogeton pusillus L. ;小眼子菜(柳丝藻,马尾巴草,帕乐幕眼子菜,丝藻,松花草,线叶藻); Baby Pondweed, Broad-leaved Small Pondweed, Lesser Pondweed, Palermo Pondweed, Slender Pondweed, Small Pondweed ■

312237 Potamogeton pusillus L. = Potamogeton oxyphyllus Miq. ■

312238 Potamogeton pusillus L. = Potamogeton panormitanus Biv. ■

312239 Potamogeton pusillus L. subsp. friesii (Rupr.) Hook. f. = Potamogeton friesii Rupr. ■

312240 Potamogeton pusillus L. subsp. gemmiparus (J. W. Robbins) R. R. Haynes et Hellq. ;芽眼子菜■☆

312241 Potamogeton pusillus L. subsp. tenuissimus (Mert. et W. D. J. Koch) R. R. Haynes et Hellq. ;狭叶小眼子菜; Narrow-leaved Small Pondweed, Slender Pondweed, Small Pondweed ■☆

312242 Potamogeton pusillus L. var. africanus A. Benn. = Potamogeton pusillus L. ■

312243 Potamogeton pusillus L. var. berchtoldii (Fieber) Asch. et Graebn. = Potamogeton berchtoldii Fieber ■

312244 Potamogeton pusillus L. var. gemmiparus J. W. Robbins = Potamogeton pusillus L. subsp. gemmiparus (J. W. Robbins) R. R. Haynes et Hellq. ☆

312245 Potamogeton pusillus L. var. minor (Biv.) Fernald et B. G. Schub. = Potamogeton pusillus L. ■

312246　Potamogeton pusillus L. var. mucronatus（Fieber）Graebn. = Potamogeton pusillus L. subsp. tenuissimus（Mert. et W. D. J. Koch）R. R. Haynes et Hellq. ■☆

312247　Potamogeton pusillus L. var. rutiloides（Fernald）B. Boivin = Potamogeton strictifolius A. Benn. ■☆

312248　Potamogeton pusillus L. var. tenuissimus Mert. et W. D. J. Koch = Potamogeton pusillus L. subsp. tenuissimus（Mert. et W. D. J. Koch）R. R. Haynes et Hellq. ■☆

312249　Potamogeton pusillus L. var. tenuissimus Mert. et W. D. J. Koch = Potamogeton berchtoldii Fieber ■

312250　Potamogeton recurvatus Hagstr.；长鞘眼子菜（长鞘菹草）；Longsheath Pondweed ■

312251　Potamogeton recurvatus Hagstr. = Stuckenia pamirica（Baagoe）Z. Kaplan ■

312252　Potamogeton richardii Solms；理查德眼子菜■☆

312253　Potamogeton richardsonii（A. Benn.）Rydb.；理查眼子菜；Richardson's Pondweed ■☆

312254　Potamogeton robbinsii Oakes；罗氏眼子菜；Fern Pondweed, Robbins' Pondweed ■☆

312255　Potamogeton robbinsii Oakes f. cultellatus Fassett = Potamogeton robbinsii Oakes ■☆

312256　Potamogeton rostratus Hagstr. = Stuckenia filiformis（Pers.）Börner ■

312257　Potamogeton rufescens Schrad. = Potamogeton alpinus Balb. ■☆

312258　Potamogeton rufescens Schrad. = Potamogeton heterophyllus Schreb. ■

312259　Potamogeton rutilus Wolfg.；棕眼子菜；Shetland Pondweed ■☆

312260　Potamogeton salicifolius Wolfg.；柳叶眼子菜；Willow-leaved Pondweed ■☆

312261　Potamogeton schweinfurthii A. Benn.；施韦眼子菜■☆

312262　Potamogeton serrulatus Regel et Maack = Potamogeton maackianus A. W. Benn. ■

312263　Potamogeton sibiricus A. Benn. = Potamogeton compressus L. ■

312264　Potamogeton sinicus Migo = Potamogeton lucens L. ■

312265　Potamogeton sparganifolius Laest. ex Fr.；带叶眼子菜；Ribbon-leaved Pondweed ■☆

312266　Potamogeton spathuliformis（J. W. Robbins）Morong；匙状眼子菜■☆

312267　Potamogeton spirillus Tuck.；北方眼子菜；Dimorphous Pondweed, Northern Snailseed Pondweed, Northern Snail-seed Pondweed, Spiral-fruited Pondweed ■☆

312268　Potamogeton stagnorum Hagstr. = Potamogeton nodosus Poir. ■

312269　Potamogeton striatus Ruiz et Pav. = Stuckenia striata（Ruiz et Pav.）Holub ■☆

312270　Potamogeton strictifolius A. Benn.；直叶眼子菜；Narrow-leaved Pondweed, Stiff Pondweed, Straight-leaved Pondweed ■☆

312271　Potamogeton strictifolius A. Benn. var. rutiloides Fernald = Potamogeton strictifolius A. Benn. ■☆

312272　Potamogeton strictifolius A. Benn. var. typicus Fernald = Potamogeton strictifolius A. Benn. ■☆

312273　Potamogeton stylatus Hagstr. = Potamogeton alpinus Balb. ■☆

312274　Potamogeton subretusus Hagstr.；微凹眼子菜■☆

312275　Potamogeton subsibiricus Hagstr.；叶尼塞眼子菜；Yenissei River Pondweed ■☆

312276　Potamogeton suecicus K. Richt.；瑞典眼子菜；Swedish Pondweed ■☆

312277　Potamogeton tennesseensis Fernald；田纳西眼子菜；Tennessee Pondweed ■☆

312278　Potamogeton tenuifolius Raf. = Potamogeton alpinus Balb. ■☆

312279　Potamogeton tenuifolius Raf. var. subellipticus Fernald = Potamogeton alpinus Balb. ■☆

312280　Potamogeton tepperi Benn. = Potamogeton distinctus A. W. Benn. ■

312281　Potamogeton thunbergii Cham. et Schltdl. = Potamogeton nodosus Poir. ■

312282　Potamogeton trichoides Cham. et Schltdl.；毛眼子菜；Hairlike Pondweed, Hair-like Pondweed ■☆

312283　Potamogeton trichoides Cham. et Schltdl. var. tuberculosus Rchb. = Potamogeton trichoides Cham. et Schltdl. ■☆

312284　Potamogeton tuckermanii J. W. Robbins = Potamogeton confervoides Rchb. ■☆

312285　Potamogeton vaginatus Turcz.；大鞘眼子菜；Big-sheath Pondweed, Bigsheath-pondweed, Sheathed Pondweed ■☆

312286　Potamogeton vaginatus Turcz. = Stuckenia vaginata（Turcz.）Holub ■☆

312287　Potamogeton varians Morong = Potamogeton spathuliformis（J. W. Robbins）Morong ■☆

312288　Potamogeton vaseyi J. W. Robbins；瓦齐眼子菜；Vasey's Pondweed ■

312289　Potamogeton vaseyi J. W. Robbins = Potamogeton octandrus Poir. var. miduhikimo（Makino）H. Hara ■

312290　Potamogeton wolfgangii Kihlm.；武氏眼子菜；Wolfgang Pondweed ■☆

312291　Potamogeton wrightii Morong；竹叶眼子菜■

312292　Potamogeton xichangensis Z. S. Diao；西昌眼子菜；Xichang Pondweed ■

312293　Potamogeton xinganensis Ma；兴安眼子菜；Xing'an Pondweed ■

312294　Potamogeton zizii Mert. et W. D. J. Koch；济兹眼子菜；Long-leaved Pondweed ■☆

312295　Potamogeton zosterifolius Schumach. = Potamogeton acutifolius Link ■

312296　Potamogeton zosterifolius Schumach. = Potamogeton compressus L. ■

312297　Potamogeton zosterifolius Schumach. subsp. zosteriformis（Fernald）Hultén = Potamogeton zosteriformis Fernald ■☆

312298　Potamogeton zosterifolius Schumach. var. americanus A. Benn. = Potamogeton zosteriformis Fernald ■☆

312299　Potamogeton zosteriformis Fernald；大叶藻状眼子菜；Flatstem Pondweed, Flat-stem Pondweed, Flat-stemmed Pondweed, Zostera-like Pondweed ■☆

312300　Potamogetonaceae Bercht. et J. Presl（1823）（保留科名）；眼子菜科；Pondweed Family ■

312301　Potamogetonaceae Dumort. = Potamogetonaceae Bercht. et J. Presl（保留科名）■

312302　Potamogetonaceae Rchb. = Potamogetonaceae Bercht. et J. Presl（保留科名）■

312303　Potamogetum Clairv. = Potamogeton L. ■

312304　Potamogiton Raf. = Potamogetum Clairv. ■

312305　Potamophila R. Br.（1810）；小叶河草属■☆

312306　Potamophila Schrank = Microtea Sw. ■☆

312307　Potamophila letestui Koechlin = Maltebrunia letestui（Koechlin）Koechlin ■☆

312308　Potamophila parviflora R. Br.；小叶河草■☆

312309　Potamophila prehensilis（Nees）Benth. = Prosphytochloa

prehensilis (Nees) Schweick. ■☆

312310　Potamophila schliebenii Pilg. = Maltebrunia schliebenii (Pilg.) C. E. Hubb. ■☆

312311　Potamopithys Senb. = Potamopitys Adans. ■

312312　Potamopitys Adans. = Elatine L. ■

312313　Potamopitys Buxb. ex Adans. = Elatine L. ■

312314　Potamotheca Post et Kuntze = Pomatotheca F. Muell. ■

312315　Potamotheca Post et Kuntze = Trianthema L. ■

312316　Potamoxylon Raf. = Tabebuia Gomes ex DC. ●☆

312317　Potaninia Maxim. (1881);绵刺属(蒙古刺属);Cottonspine, Potaninia ●★

312318　Potaninia mongolica Maxim.;绵刺(蒙古包大宁);Mongol Cottonspine,Mongolian Potaninia ●◇

312319　Potarophytum Sandwith(1939);枝序偏穗草属■☆

312320　Potarophytum riparium Sandwith;枝序偏穗草■☆

312321　Potentilla Adans. = Tormentilla L. ■●

312322　Potentilla L. (1753);委陵菜属(翻白草属);Cinquefoil, Five Fingers,Five-finger,Potentil,Potentilla ■●

312323　Potentilla × musashinoana Makino;武藏野委陵菜■☆

312324　Potentilla abyssinica A. Rich. = Potentilla reptans L. ■

312325　Potentilla acaulis L.;星毛委陵菜(无茎委陵菜);Stemless Cinquefoil ■

312326　Potentilla acervata Soják = Potentilla tanacetifolia Willd. ex Schltdl. ■

312327　Potentilla adenophylla Boiss. et Hohen.;腺叶委陵菜■☆

312328　Potentilla adnata Wall. ex Lehm. = Acomastylis elata (Royle) F. Bolle var. humilis (Royle) F. Bolle ■

312329　Potentilla adnata Wall. ex Lehm. = Geum elatum Wall. ex Hook. f. var. humile Franch. ■

312330　Potentilla adpressa (Bunge) Cardot = Sibbaldia adpressa Bunge ■

312331　Potentilla adpressa (Bunge) Cardot var. pumila Hook. f. = Sibbaldia adpressa Bunge ■

312332　Potentilla adpressa (Bunge) Cardot var. sericea Cardot = Sibbaldia sericea (Grubov) Soják ■

312333　Potentilla adscharica Sommier et H. Lév.;阿得扎尔委陵菜■☆

312334　Potentilla aegopodiifolia H. Lév. = Potentilla cryptotaeniae Maxim. ■

312335　Potentilla aemulans Juz. = Potentilla ancistrifolia Bunge ■

312336　Potentilla agrimonioides M. Bieb. = Potentilla strigosa Pall. ex Pursh ■

312337　Potentilla alba L.;白花委陵菜(白委陵菜);White Cinquefoil ■☆

312338　Potentilla albifolia Wall. ex Hook. f. = Sibbaldia micropetala (D. Don) Hand. -Mazz. ■

312339　Potentilla alchemilloides Lapeyr.;羽衣草委陵菜■☆

312340　Potentilla alchemilloides Lapeyr. subsp. atlantica Emb. et Maire;北非委陵菜■☆

312341　Potentilla alchemilloides Lapeyr. var. ghatica Quézel = Potentilla alchemilloides Lapeyr. ■☆

312342　Potentilla alchemilloides Lapeyr. var. iminouakensis Quézel = Potentilla alchemilloides Lapeyr. ■☆

312343　Potentilla alchemilloides Lapeyr. var. maaghalensis Quézel = Potentilla alchemilloides Lapeyr. subsp. atlantica Emb. et Maire ■☆

312344　Potentilla alexeenkoi Lipsky;阿氏委陵菜■☆

312345　Potentilla altaica Bunge = Potentilla virgata Lehm. var. pinnatifida (Lehm.) Te T. Yu et C. L. Li ■

312346　Potentilla ambigua Cambess. = Potentilla cuneata Wall. ex Lehm. ■

312347　Potentilla amurensis Maxim.;小花金梅■

312348　Potentilla amurensis Maxim. = Potentilla heynii Roth ■

312349　Potentilla amurensis Maxim. = Potentilla supina L. var. ternata Peterm. ■

312350　Potentilla anadyrensis Juz.;阿纳代尔委陵菜■☆

312351　Potentilla ancistrifolia Bunge;皱叶委陵菜(钩叶委陵菜,特兰委陵菜);Barbed Leaf Cinquefoil, Wrinkleleaf Cinquefoil ■

312352　Potentilla ancistrifolia Bunge f. lanceolata C. S. Zhu;披针叶皱叶委陵菜;Lanceolate Wrinkleleaf Cinquefoil ■

312353　Potentilla ancistrifolia Bunge var. concolor Liou et C. Y. Li;同色钩叶委陵菜■

312354　Potentilla ancistrifolia Bunge var. dickinsii (Franch. et Sav.) Koidz. = Potentilla dickinsii Franch. et Sav. ■

312355　Potentilla ancistrifolia Bunge var. tomentosa Liou et Y. Y. Li ex Te T. Yu et C. L. Li;白毛皱叶委陵菜;Tomentose Barbed Cinquefoil ■

312356　Potentilla anemonifolia Lehm. = Potentilla kleiniana Wight et Arn. ■

312357　Potentilla anglica Laichard.;英国委陵菜(平铺委陵菜);Creeping Tormentil, English Cinquefoil, Trailing Tormentil ■☆

312358　Potentilla angustifolia DC.;窄叶委陵菜;Narrowleaf Cinquefoil ■☆

312359　Potentilla angustiloba Te T. Yu et C. L. Li;窄裂委陵菜;Narrowlobe Cinquefoil ■

312360　Potentilla anserina L.;鹅绒委陵菜(戳玛,鹅食委陵菜,藩白花,河篦梳,蕨麻,蕨麻委陵菜,莲菜花,莲花菜,莲叶花,曲尖委陵菜,人参果,延寿草,延寿果);Argemone, Argentina, Argentine, Blithran, Bread-and-butter, Bread-and-cheese, Buttercup, Camoroche, Cheese, Cramp-weed, Creeping Tansy, Dog Tansy, Dog's Tansy, Fair Days, Fair Grass, Fern Buttercup, Fernhemp Cinquefoil, Fish Bones, Gander-grass, Golden Sovereigns, Goose Grass, Goose Gray, Goose Tansy, Goosefoot, Gooseweed, Goosewort, Grey Goose, Lamb's Ear, Mash-corns, Midsummer Silver, Moor-grass, Moorhens, Moss-corns, Moss-crop, Prince's Feathers, Scented Buttercup, Silver Feather, Silver Feathers, Silver Fern, Silver Grass, Silver Leaves, Silver Rose, Silver Weed, Silverleaf, Silverweed, Silver-weed, Silverweed Cinquefoil, Silvery Cinquefoil, Spring Carrot, Sweetbread, Swine's Beads, Tansy, Trailing Tansy, Traveller's Ease, Traveller's Leaf, White Tansy, Wild Agrimony, Wild Skirret, Wild Tansy ■

312361　Potentilla anserina L. = Argentina anserina (L.) Rydb. ■

312362　Potentilla anserina L. f. incisa Wolf = Potentilla anserina L. ■

312363　Potentilla anserina L. f. sericea (Hayne) Hayek = Argentina anserina (L.) Rydb. ■

312364　Potentilla anserina L. f. sericea (Hayne) Hayek = Potentilla anserina L. ■

312365　Potentilla anserina L. subsp. pacifica (Howell) Rousi;太平洋委陵菜■☆

312366　Potentilla anserina L. var. concolor Ser. = Argentina anserina (L.) Rydb. ■

312367　Potentilla anserina L. var. concolor Ser. = Potentilla anserina L. ■

312368　Potentilla anserina L. var. grandis Torr. et A. Gray = Potentilla anserina L. subsp. pacifica (Howell) Rousi ■☆

312369　Potentilla anserina L. var. nuda Gaudich.;无毛鹅绒委陵菜(无毛蕨麻);Naked Silverweed Cinquefoil ■

312370　Potentilla anserina L. var. nuda Gaudin = Potentilla anserina L. ■

312371　Potentilla anserina L. var. orientalis Cardot;东方鹅绒委陵菜(东方蕨麻)■

312372　Potentilla anserina L. var. sericea Hayne;灰叶鹅绒委陵菜(灰叶蕨麻);Silky Silverweed Cinquefoil ■

312373　Potentilla anserina L. var. sericea Hayne ＝ Argentina anserina (L.) Rydb. ■

312374　Potentilla anserina L. var. sericea Hayne ＝ Potentilla anserina L. ■

312375　Potentilla anserina L. var. viridis W. D. J. Koch ＝ Potentilla anserina L. ■

312376　Potentilla anserina L. var. yukonensis (Hultén) B. Boivin ＝ Argentina anserina (L.) Rydb. ■

312377　Potentilla anserina L. var. yukonensis (Hultén) B. Boivin ＝ Potentilla anserina L. ■

312378　Potentilla approximata Bunge；近邻委陵菜■

312379　Potentilla approximata Bunge ＝ Potentilla conferta Bunge ■

312380　Potentilla arbuscula D. Don ＝ Potentilla fruticosa L. var. arbuscula (D. Don) Maxim. ●

312381　Potentilla arbuscula D. Don var. albicans (Rehder et E. H. Wilson) Hand. -Mazz. ＝ Potentilla fruticosa L. var. albicans Rehder et E. H. Wilson ●

312382　Potentilla arbuscula D. Don var. albicans Rehder et E. H. Wilson ＝ Pentaphylloides fruticosa (L.) O. Schwarz var. albicans (Rehder et E. H. Wilson) Y. L. Han ●

312383　Potentilla arbuscula D. Don var. albicans Rehder et E. H. Wilson ＝ Potentilla fruticosa L. var. albicans Rehder et E. H. Wilson ●

312384　Potentilla arbuscula D. Don var. bulleyana Balf. f. ex H. R. Fletcher ＝ Potentilla fruticosa L. var. albicans Rehder et E. H. Wilson ●

312385　Potentilla arbuscula D. Don var. pumila (Hook. f.) Hand. -Mazz. ＝ Potentilla fruticosa L. var. pumila Hook. f. ●

312386　Potentilla arbuscula D. Don var. veitchii (E. H. Wilson) Liou ＝ Potentilla glabra Lodd. var. mandshurica (Maxim.) Hand. -Mazz. ●

312387　Potentilla arenaria Borkh. ex Gaertn., Mey. et Scherb.；沙地委陵菜■☆

312388　Potentilla arenosa (Turcz.) Juz.；砂委陵菜■☆

312389　Potentilla argentea L.；银背委陵菜（银白委陵菜，银色委陵菜，银叶委陵菜）；Argent Cinquefoil, Hoary Cinquefoil, Silver Cinquefoil, Silvery Five-fingers, Silvery-leaved Cinquefoil, Siverback Cinquefoil ■

312390　Potentilla argenteaeformis Kauffm. ex Trautv.；拟银色委陵菜■☆

312391　Potentilla arguta Pursh；草地委陵菜；Prairie Cinquefoil, Tall Cinquefoil, Tall Potentilla ■☆

312392　Potentilla argyrophylla Wall.；银光委陵菜；Silverleaf Cinquefoil, Silvery Cinquefoil ■

312393　Potentilla argyrophylla Wall. var. atrosanguinea Hook. f.；紫花银光委陵菜；Dark-bloodcoloured Cinquefoil, Himalaya Cinquefoil, Purpleflower Silvery Cinquefoil ■

312394　Potentilla argyrophylla Wall. var. genuina Hook. f. ＝ Potentilla argyrophylla Wall. ■

312395　Potentilla aristata Soják；多对小叶委陵菜；Manypair Smallleaf Cinquefoil ■

312396　Potentilla articulata Franch.；关节委陵菜；Jointed Cinquefoil ■

312397　Potentilla articulata Franch. var. latipetiolata (C. E. C. Fisch.) Te T. Yu et C. L. Li；宽柄关节委陵菜；Broadpetiole Jointed Cinquefoil ■

312398　Potentilla asiatica (Th. Wolf) Juz. ＝ Potentilla chrysantha Trevir. ■

312399　Potentilla asperrima Turcz.；刚毛委陵菜；Bristle Cinquefoil ■

312400　Potentilla asperrima Turcz. ＝ Potentilla fragarioides L. ■

312401　Potentilla astrachanica Jacq.；阿斯特拉罕委陵菜■☆

312402　Potentilla astragalifolia Bunge；芪叶委陵菜■☆

312403　Potentilla atrosanguinea Lodd. ＝ Potentilla argyrophylla Wall. var. atrosanguinea Hook. f. ■

312404　Potentilla atrosanguinea Lodd. var. argyrophylla (Lehm.) Grierson et D. G. Long ＝ Potentilla argyrophylla Wall. ■

312405　Potentilla atrosanguinea Lodd., G. Lodd. et W. Lodd. ＝ Potentilla argyrophylla Wall. var. atrosanguinea Hook. f. ■

312406　Potentilla aurea L.；金黄委陵菜；Glandular Cinquefoil, Golden Cinquefoil, Tall Cinquefoil ■☆

312407　Potentilla beauvaisii Cardot ＝ Potentilla griffithii Hook. f. f. velutina Cardot ■

312408　Potentilla betonicifolia Poir.；白萼委陵菜（白叶委陵菜，草杜仲，三出委陵菜，三出叶委陵菜）；Betonicleaf Cinquefoil, Whitecalyx Cinquefoil ■

312409　Potentilla biflora Willd. ex Schltdl.；双花委陵菜；Twinflower Cinquefoil, Twoflower Cinquefoil ■

312410　Potentilla biflora Willd. ex Schltdl. var. armerioides (Hook. f.) Hand. -Mazz. ＝ Potentilla articulata Franch. ■

312411　Potentilla biflora Willd. ex Schltdl. var. lahulensis Th. Wolf；五叶双花委陵菜；Fivelaeves Twoflower Cinquefoil ■

312412　Potentilla bifurca L.；二裂委陵菜（叉叶委陵菜，二叉委陵菜，二裂翻白草，二裂叶委陵菜，黄瓜绿草，鸡冠草，鸡冠茶，痔疮草）；Bifurcate Cinquefoil ■

312413　Potentilla bifurca L. ＝ Sibbaldianthe bifurca (L.) Kurtto et T. Erikss. ■

312414　Potentilla bifurca L. subsp. orientalis (Juz.) Soják ＝ Potentilla bifurca L. var. major Ledeb. ■

312415　Potentilla bifurca L. var. canescens Bong. et Mey. ＝ Potentilla imbricata Kar. et Kir. ■

312416　Potentilla bifurca L. var. glabrata Lehm. ＝ Potentilla bifurca L. var. major Ledeb. ■

312417　Potentilla bifurca L. var. humilior Rupr. et Ost. -Sack.；矮生二裂委陵菜；Dwarf Bifurcate Cinquefoil ■

312418　Potentilla bifurca L. var. major Ledeb.；长叶二裂委陵菜（高二裂委陵菜，光叉叶委陵菜，小叉叶委陵菜）；Longleaf Bifurcate Cinquefoil ■

312419　Potentilla bifurca L. var. moocroftii Th. Wolf ＝ Potentilla bifurca L. var. humilior Rupr. et Ost. -Sack. ■

312420　Potentilla bifurca L. var. moorcroftii (Wall. ex Lehm.) Th. Wolf ＝ Potentilla bifurca L. var. humilior Rupr. et Ost. -Sack. ■

312421　Potentilla bifurca L. var. pygmaea Kitag.；微小二裂委陵菜■

312422　Potentilla bifurca L. var. typica Th. Wolf ＝ Potentilla bifurca L. ■

312423　Potentilla bifurca L. var. unijuga Th. Wolf ＝ Sibbaldia adpressa Bunge ■

312424　Potentilla bifurca Willd. ex Schltdl. var. humilior Rupr. et Ost. -Sack. ＝ Potentilla bifurca L. var. humilior Rupr. et Ost. -Sack. ■

312425　Potentilla bifurca Willd. ex Schltdl. var. major Ledeb. ＝ Potentilla bifurca L. var. major Ledeb. ■

312426　Potentilla bodinieri H. Lév. ＝ Potentilla kleiniana Wight et Arn. ■

312427　Potentilla bornmuelleri Borbás；鲍氏委陵菜；Bornmueller Cinquefoil ■☆

312428　Potentilla brachypetala Fisch.；短瓣委陵菜■☆

312429　Potentilla brachystemon Hand. -Mazz. ＝ Sibbaldia perpusilloides (W. W. Sm.) Hand. -Mazz. ■

312430　Potentilla breviscapa Vest；短茎委陵菜■☆

312431　Potentilla bungei Boiss.；邦奇委陵菜；Bunge Cinquefoil ■☆

312432　Potentilla caespitosa Lehm. ＝ Potentilla saundersiana Royle var. caespitosa (Lehm.) Th. Wolf ■

312433   Potentilla callieri（Th. Wolf）Juz.；卡赖氏委陵菜；Callier Cinquefoil ■☆

312434   Potentilla camillae Kolak.；卡米尔委陵菜■☆

312435   Potentilla canadensis L.；加拿大委陵菜（指叶委陵菜）；Canada Cinquefoil, Dwarf Cinquefoil, Five Finger, Running Five-fingers ■☆

312436   Potentilla canescens Besser = Potentilla inclinata Vill. ■

312437   Potentilla cardotiana Hand. -Mazz.；聚伞委陵菜■

312438   Potentilla cardotiana Hand. -Mazz. = Potentilla peduncularis D. Don ■

312439   Potentilla cariandrifolia D. Don = Potentilla coriandrifolia D. Don ■

312440   Potentilla cariandrifolia D. Don var. dumosa Franch. = Potentilla coriandrifolia D. Don var. dumosa Franch. ■

312441   Potentilla caucasica Juz.；高加索委陵菜■☆

312442   Potentilla caulescens L.；灌木状委陵菜；Shrubby White Cinquefoil ■☆

312443   Potentilla caulescens L. subsp. djurdjurae（Chabert）Romo；朱尔朱拉山委陵菜■☆

312444   Potentilla caulescens L. var. djurdjurae Chabert = Potentilla caulescens L. subsp. djurdjurae（Chabert）Romo ■☆

312445   Potentilla caulescens L. var. mesatlantica Maire = Potentilla caulescens L. ■☆

312446   Potentilla centigrana Maxim.；蛇莓委陵菜（日本委陵菜，小翻白草，小蛇莓）；Mockstrawberry Cinquefoil ■☆

312447   Potentilla centigrana Maxim. f. patens Hiyama；匍匐蛇莓委陵菜■☆

312448   Potentilla centigrana Maxim. var. japonica Maxim. = Potentilla centigrana Maxim. ■

312449   Potentilla centigrana Maxim. var. mandshuriva Maxim.；东北蛇莓委陵菜■

312450   Potentilla chinensis Ser.；委陵菜（白头翁，地区草，翻白菜，翻白草，蛤蟆草，根头菜，贵州白头翁，虎爪菜，鸡爪草，老鸦翎，老鸦爪，痢疾草，龙牙草，毛鸡腿子，扑地虎，山萝卜，生血丹，天青地白，五虎嚼血，小毛药，野鸡膀子，野鸠旁花，一白草）；China Cinquefoil, Chinese Cinquefoil ■

312451   Potentilla chinensis Ser. subsp. trigonodonta Hand. -Mazz. = Potentilla chinensis Ser. ■

312452   Potentilla chinensis Ser. var. latifida Koidz. = Potentilla chinensis Ser. ■

312453   Potentilla chinensis Ser. var. lineariloba Franch. et Sav.；细裂委陵菜；Linear-lobe Cinquefoil ■

312454   Potentilla chinensis Ser. var. oligodonta Hand. -Mazz.；疏齿委陵菜■

312455   Potentilla chinensis Ser. var. platyloba Liou et C. Y. Li；薄叶委陵菜；Thinleaf Cinquefoil ■

312456   Potentilla chinensis Ser. var. xerogenes Hand. -Mazz. = Potentilla chinensis Ser. ■

312457   Potentilla chrysantha Trevir.；黄花委陵菜（翻白草，金黄委陵菜，亚洲委陵菜）；Safronyellow Cinquefoil, Yellowflower Cinquefoil ■

312458   Potentilla chrysantha Trevir. var. asiatica Th. Wolf = Potentilla chrysantha Trevir. ■

312459   Potentilla cinerea Chaix ex Vill；灰色委陵菜；Grey Cinquefoil ■☆

312460   Potentilla collina Wibel；山丘委陵菜；Palmleaf Cinquefoil ■☆

312461   Potentilla comarum Nestl. = Comarum palustre L. ●■

312462   Potentilla commutata Lehm. var. polyandra Soják；多蕊委陵菜■

312463   Potentilla compsophylla Hand. -Mazz. = Potentilla potaninii Th. Wolf var. compsophylla（Hand. -Mazz.）Te T. Yu et C. L. Li ■

312464   Potentilla concolor（Franch.）Rolfe = Potentilla macrosepala Cardot ■

312465   Potentilla concolor Rolfe = Potentilla macrosepala Cardot ■

312466   Potentilla concolor Zimm. = Potentilla anserina L. var. sericea Hayne ■

312467   Potentilla conferta Bunge；大萼委陵菜（白毛委陵菜，大头委陵菜，热干巴）；Largecalyx Cinquefoil ■

312468   Potentilla conferta Bunge var. trijuga Te T. Yu et C. L. Li；矮生大萼委陵菜；Dwarf Largecalyx Cinquefoil ■

312469   Potentilla contigua Soják；高山委陵菜■

312470   Potentilla coriandrifolia D. Don；莞叶委陵菜；Coriandrifolious Cinquefoil ■

312471   Potentilla coriandrifolia D. Don var. dumosa Franch.；丛生莞叶委陵菜；Clustered Coriandrifolious Cinquefoil ■

312472   Potentilla crantzii（Crantz）Beck ex Lindm.；克澜氏委陵菜（阿尔卑斯委陵菜）；Alpine Cinquefoil ■☆

312473   Potentilla crassa Tausch ex Opiz；肥厚委陵菜■☆

312474   Potentilla crebridens Juz. = Potentilla nivea L. var. elongata Th. Wolf ■

312475   Potentilla crenulata Te T. Yu et C. L. Li；圆齿委陵菜；Crenulate Cinquefoil, Roundtoothed Cinquefoil ■

312476   Potentilla cristata H. R. Fletcher；鸡冠委陵菜■

312477   Potentilla cristata H. R. Fletcher = Potentilla stenophylla（Franch.）Diels var. cristata（H. R. Fletcher）H. Ikeda et H. Ohba ■

312478   Potentilla cryptotaeniae Maxim.；狼牙委陵菜（播丝草，地蜂子，地蜘蛛，狼牙，三爪金，山蜂子，铁秤砣，铁钮子，铁枕头）；Nippon Cinquefoil ■

312479   Potentilla cryptotaeniae Maxim. var. genuina Kitag. = Potentilla cryptotaeniae Maxim. ■

312480   Potentilla cryptotaeniae Maxim. var. insularis Kitag.；岛生委陵菜■☆

312481   Potentilla cryptotaeniae Maxim. var. obovata Th. Wolf = Potentilla cryptotaeniae Maxim. ■

312482   Potentilla cryptotaeniae Maxim. var. radicans Te T. Yu et C. L. Li；匍行委陵菜（匍行狼牙委陵菜）；Rooting Nippon Cinquefoil ■

312483   Potentilla cuneata Wall. ex Lehm.；楔叶委陵菜；Cuneate Cinquefoil ■

312484   Potentilla curta Soják = Potentilla peduncularis D. Don var. curta（Soják）H. Ikeda et H. Ohba ■

312485   Potentilla dasyphylla Bunge = Potentilla sericea L. ■

312486   Potentilla davidii Franch. = Potentilla eriocarpa Wall. ex Lehm. ●

312487   Potentilla davurica Nestl. = Pentaphylloides fruticosa（L.）O. Schwarz ●

312488   Potentilla davurica Nestl. = Potentilla fruticosa L. ●

312489   Potentilla davurica Nestl. var. mandshurica（Maxim.）Th. Wolf = Potentilla glabra Lodd. var. mandshurica（Maxim.）Hand. -Mazz. ●

312490   Potentilla davurica Nestl. var. mandshurica Th. Wolf = Potentilla glabra Lodd. var. mandshurica（Maxim.）Hand. -Mazz. ●

312491   Potentilla davurica Nestl. var. veitchii Jesson = Potentilla glabra Lodd. var. veitchii（E. H. Wilson）Hand. -Mazz. ●

312492   Potentilla dealbata Bunge = Potentilla virgata Lehm. ■

312493   Potentilla decemjuga Soják = Potentilla commutata Lehm. var. polyandra Soják ■

312494   Potentilla delavayi Franch.；滇西委陵菜；Delavay Cinquefoil ■

312495   Potentilla dentata Forssk.；尖齿委陵菜■☆

312496   Potentilla depressa Willd. ex Schltdl.；压抑委陵菜■☆

312497   Potentilla desertorum Bunge；荒漠委陵菜（草原委陵菜）；

Desert Cinquefoil ■

312498　Potentilla dickinsii Franch. et Sav.；拟薄叶委陵菜；Dickins Barbed Cinquefoil ■

312499　Potentilla dickinsii Franch. et Sav. = Potentilla ancistrifolia Bunge var. dickinsii (Franch. et Sav.) Koidz. ■

312500　Potentilla dickinsii Franch. et Sav. f. simplicifolia (Takeda)；单拟薄叶委陵菜■☆

312501　Potentilla dickinsii Franch. et Sav. var. glabrata Nakai；无毛薄叶委陵菜■☆

312502　Potentilla discolor Bunge；翻白草(白鸡爪草,白漂莲,白头翁,番白草,翻白委陵菜,反白草,茯苓草,湖鸡腿,黄花地丁,鸡脚草,鸡脚爪,鸡距草,鸡腿儿,鸡腿根,鸡腿子,鸡爪参,鸡爪草,鸡爪莲,犄角草,觭角草,金线吊葫芦,兰溪白头翁,千锤打,沙柳草,天藕,天藕儿,天青地白,土菜,土人参,土洋参,委陵菜,乌皮浮儿,细沙扭,鸭脚参,野鸡坝,叶下白,郁苏参)；Discolor Cinquefoil ■

312503　Potentilla discolor Bunge var. formosana (Hance) Franch. = Potentilla discolor Bunge ■

312504　Potentilla discolor Bunge var. formosana Franch. = Potentilla discolor Bunge ■

312505　Potentilla divina Albov；神委陵菜■☆

312506　Potentilla dolichopogon H. Lév. = Potentilla cuneata Wall. ex Lehm. ■

312507　Potentilla dumosa (Franch.) Hand.-Mazz. = Potentilla coriandrifolia D. Don var. dumosa Franch. ■

312508　Potentilla dumosa (Franch.) Hand.-Mazz. subsp. salwinensis Soják = Potentilla coriandrifolia D. Don var. dumosa Franch. ■

312509　Potentilla dumosa (Franch.) Hand.-Mazz. var. stromatodes (Melch.) H. R. Fletcher = Potentilla coriandrifolia D. Don var. dumosa Franch. ■

312510　Potentilla egedii Wormsk.；埃氏委陵菜；Eded Cinquefoil ■☆

312511　Potentilla egedii Wormsk. subsp. grandis (Torr. et A. Gray) Hultén = Potentilla anserina L. subsp. pacifica (Howell) Rousi ■☆

312512　Potentilla egedii Wormsk. subsp. pacifica (Howell) L. A. Sergienko = Potentilla anserina L. subsp. pacifica (Howell) Rousi ■☆

312513　Potentilla egedii Wormsk. subsp. yukonensis (Hultén) Hultén = Argentina anserina (L.) Rydb. ■

312514　Potentilla egedii Wormsk. subsp. yukonensis (Hultén) Hultén = Potentilla anserina L. ■

312515　Potentilla egedii Wormsk. var. grandis (Torr. et A. Gray) J. T. Howell = Potentilla anserina L. subsp. pacifica (Howell) Rousi ■☆

312516　Potentilla elatior Willd. ex Schltdl.；较高委陵菜■☆

312517　Potentilla elegans Cham. et Schltdl.；雅致委陵菜■☆

312518　Potentilla emarginata Pursh；凹沟委陵菜■☆

312519　Potentilla erecta (L.) Hampe = Potentilla erecta (L.) Raeusch. ■☆

312520　Potentilla erecta (L.) Raeusch.；洋委陵菜(洋翻白草,直立委陵菜)；Biscuit, Blood Root, Earth Bark, English Sarsaparilla, Erect Cinquefoil, Ewe Daisy, Ewe-daisy, Five Fingers, Five-finger Grass, Five-fingers, Five-leaved Grass, Flesh-and-blood, Septfoil, Setfoil, Seven-leaf, Sheep's Knapperty, Shepherd's Knot, Snake's Head, Starflower, Thor's Mantle, Tormentil, Tormentilla, Tormentilla Cinquefoil, Tormenting Root, Tormerik, Turmentill, Turmentyne ☆

312521　Potentilla erecta (L.) Raeusch. var. maurorum Maire = Potentilla erecta (L.) Raeusch. ■☆

312522　Potentilla eriocarpa Wall. ex Lehm.；毛果委陵菜(绵毛果委陵菜,小神砂草)；Woollyfruit Cinquefoil ●

312523　Potentilla eriocarpa Wall. ex Lehm. var. cathayana C. K. Schneid. = Potentilla eriocarpa Wall. ex Lehm. ●

312524　Potentilla eriocarpa Wall. ex Lehm. var. dissecta C. Marquand et Airy Shaw = Potentilla eriocarpa Wall. ex Lehm. var. tsarongensis W. E. Evans ●

312525　Potentilla eriocarpa Wall. ex Lehm. var. tsarongensis W. E. Evans；裂叶毛果委陵菜；Schizoleaf Woollyfruit Cinquefoil ●

312526　Potentilla eriocarpoides J. Krause = Potentilla eriocarpa Wall. ex Lehm. var. tsarongensis W. E. Evans ●

312527　Potentilla eriocarpoides J. Krause var. glabrescens J. Krause = Potentilla eriocarpa Wall. ex Lehm. ●

312528　Potentilla etomentosa Rydb.；无毛委陵菜■☆

312529　Potentilla euxantha W. E. Evans = Potentilla hypargyrea Hand.-Mazz. ■

312530　Potentilla eversmanniana Fisch. ex Claus；埃威氏委陵菜；Eversmann Cinquefoil ■☆

312531　Potentilla evestita Th. Wolf；脱绒委陵菜；Hairless Cinquefoil ■

312532　Potentilla exaltata Bunge = Potentilla chinensis Ser. ■

312533　Potentilla fallens Cardot；川滇委陵菜；Sichuan-Yunnan Cinquefoil ■

312534　Potentilla fauriei H. Lév. = Potentilla supina L. ■

312535　Potentilla fedtschenkoana Siegfr. ex Th. Wolf；范氏委陵菜■☆

312536　Potentilla festiva Soják；合耳委陵菜■

312537　Potentilla filipendula Willd. ex Schltdl. = Potentilla tanacetifolia Willd. ex Schltdl. ■

312538　Potentilla flabellata Regel et Schmalh.；扇状委陵菜■☆

312539　Potentilla flabelliformis Lehm. = Potentilla gracilis Douglas ex Hook. var. flabelliformis (Lehm.) Nutt. ex Torr. et A. Gray ■☆

312540　Potentilla flagellaris Willd. ex Schltdl.；匍枝委陵菜(匐枝委陵菜,鸡儿头苗,蔓委陵菜)；Flagellar Cinquefoil, Runnery Cinquefoil ■

312541　Potentilla flagellaris Willd. ex Schltdl. var. oblongifolia Liou et C. Y. Li；宽叶蔓委陵菜；Broadleaf Cinquefoil ■

312542　Potentilla floribunda Pursh；加拿大繁花委陵菜；Golden Hardhack, Shrubby Cinquefoil ●☆

312543　Potentilla floribunda Pursh = Pentaphylloides floribunda (Pursh) Á. Löve ●☆

312544　Potentilla foliosa Sommier et H. Lév.；繁叶委陵菜■☆

312545　Potentilla formosana Hance = Potentilla discolor Bunge ■

312546　Potentilla forrestii W. W. Sm. = Potentilla saundersiana Royle var. jacquemontii Franch. ■

312547　Potentilla forrestii W. W. Sm. var. subpinnata (Hand.-Mazz.) Hand.-Mazz. = Potentilla saundersiana Royle var. subpinnata Hand.-Mazz. ■

312548　Potentilla forrestii W. W. Sm. var. subpinnata Hand.-Mazz. = Potentilla saundersiana Royle var. subpinnata Hand.-Mazz. ■

312549　Potentilla fragariastrum Ehrh. ex Haller f.；脆委陵菜；Barren Strawberry, Strawberry-leaved Potentil ■☆

312550　Potentilla fragarioides L.；莓叶委陵菜(经如草,满山红,毛猴子,瓢о,软梗蛇扭,雉子莚,雉子筵)；Dewberryleaf Cinquefoil ■

312551　Potentilla fragarioides L. var. lancifolia (Honda) H. Hara = Potentilla togasii Ohwi ■☆

312552　Potentilla fragarioides L. var. major Maxim.；大莓叶委陵菜■☆

312553　Potentilla fragarioides L. var. major Maxim. = Potentilla fragarioides L. ■

312554　Potentilla fragarioides L. var. major Maxim. = Potentilla sprengeliana Lehm. ■☆

312555　Potentilla fragarioides L. var. sprengeliana (Lehm.) Maxim. =

Potentilla fragarioides L. ■

312556　Potentilla fragarioides L. var. sprengeliana（Lehm.）Maxim. = Potentilla sprengeliana Lehm. ■☆

312557　Potentilla fragarioides L. var. stononifera f. trifoliola Takeda = Potentilla freyniana Bornm. var. sinica Migo ■

312558　Potentilla fragarioides L. var. ternata Maxim. = Potentilla freyniana Bornm. ■

312559　Potentilla fragarioides L. var. typica Maxim. = Potentilla fragarioides L. ■

312560　Potentilla fragarioides L. var. yamanakae（Naruh.）Naruh. = Potentilla yamanakae（Naruh.）Naruh. ■☆

312561　Potentilla fragiformis Willd.；莓状委陵菜；Strawberry Cinquefoil ■☆

312562　Potentilla fragiformis Willd. ex D. F. K. Schltdl. subsp. megalantha（Takeda）Hultén = Potentilla megalantha Takeda ■☆

312563　Potentilla fragiformis Willd. ex Schltdl. var. gelida（C. A. Mey.）Trautv. = Potentilla gelida C. A. Mey. ■

312564　Potentilla fragiformis Willd. var. gelida Trautv. = Potentilla gelida C. A. Mey. ■

312565　Potentilla freyniana Bornm.；三叶委陵菜（地风子,地蜂子,地蜘蛛,三片风,三叶翻白草,三叶蛇子草,三张叶,三爪金,铁秤砣）；Freyn Cinquefoil,Threeleaf Cinquefoil ■

312566　Potentilla freyniana Bornm. f. miyoshii（Honda et Oishi）Naruh.；三好学三叶委陵菜■☆

312567　Potentilla freyniana Bornm. f. monophylla Kitag.；单叶委陵菜■☆

312568　Potentilla freyniana Bornm. var. grandiflora Th. Wolf = Potentilla rosulifera H. Lév. ■

312569　Potentilla freyniana Bornm. var. grandiflora Th. Wolf = Potentilla yokusaiana Makino ■

312570　Potentilla freyniana Bornm. var. miyoshii Honda et Oishi = Potentilla freyniana Bornm. f. miyoshii（Honda et Oishi）Naruh. ■☆

312571　Potentilla freyniana Bornm. var. sinica Migo；中华三叶委陵菜（白里金梅,地风子,地蜂子,地蜘蛛,独脚金,独脚伞,独脚委陵菜,独立金蛋,蜂子花,蜂子芪,烂苦春,毛猴子,软梗蛇扭,三片风,三叶翻白草,三叶蛇莓,三叶蛇子草,三叶委陵菜,三张叶,三爪金,山蜂子,铁秤砣,铁枕头）；Chinese Freyn Cinquefoil ■

312572　Potentilla fruticosa（L.）Rydb. = Pentaphylloides fruticosa（L.）O. Schwarz ●

312573　Potentilla fruticosa（L.）Rydb. = Potentilla fruticosa L. ●

312574　Potentilla fruticosa L. 'Abbotswood'；阿伯茨伍德金露梅；Bush Cinquefoil,Shrubby Cinquefoil ●☆

312575　Potentilla fruticosa L. 'Beesii'；毕斯金露梅●☆

312576　Potentilla fruticosa L. 'Daydawn'；黎明金露梅（曙光金露梅）●

312577　Potentilla fruticosa L. 'Elizabeth'；伊丽莎白金露梅；Bush Cinquefoil,Shrubby Cinquefoil ●

312578　Potentilla fruticosa L. 'Farrer's White'；法勒白金露梅●☆

312579　Potentilla fruticosa L. 'Friedrichsenii'；弗内德利森金露梅●☆

312580　Potentilla fruticosa L. 'Gold Drop'；金滴金露梅；Gold Drop Shrubby Cinquefoil,Shrubby Cinquefoil ●☆

312581　Potentilla fruticosa L. 'Goldfinger'；金指金露梅；Goldfinger Shrubby Cinquefoil,Shrubby Cinquefoil ●☆

312582　Potentilla fruticosa L. 'Jackman's Variety'；杰克曼金露梅●☆

312583　Potentilla fruticosa L. 'Katherine Dykes'；凯瑟琳堤金露梅●

312584　Potentilla fruticosa L. 'Maaneleys'；马尼斯金露梅●☆

312585　Potentilla fruticosa L. 'Manchu'；满族金露梅●☆

312586　Potentilla fruticosa L. 'Ochroleuca'；淡赭金露梅●

312587　Potentilla fruticosa L. 'Primrose Beauty'；黄美人金露梅●

312588　Potentilla fruticosa L. 'Red Ace'；红王牌金露梅●☆

312589　Potentilla fruticosa L. 'Royal Flush'；品红金露梅●☆

312590　Potentilla fruticosa L. 'Sunset'；日落金露梅●☆

312591　Potentilla fruticosa L. 'Tangerine'；橘红金露梅●

312592　Potentilla fruticosa L. 'Vilmoriniana'；维尔莫林金露梅●☆

312593　Potentilla fruticosa L. = Dasiphora fruticosa（L.）Rydb. ●

312594　Potentilla fruticosa L. = Pentaphylloides fruticosa（L.）O. Schwarz ●

312595　Potentilla fruticosa L. f. mandshurica（Maxim.）Rehder = Dasiphora mandshurica（Maxim.）Juz. ●

312596　Potentilla fruticosa L. f. wardii Rehder = Potentilla parvifolia Fisch. ex Lehm. ●

312597　Potentilla fruticosa L. subsp. floribunda（Pursh）Elkington = Pentaphylloides floribunda（Pursh）Á. Löve ●☆

312598　Potentilla fruticosa L. var. albicans Rehder et E. H. Wilson = Pentaphylloides fruticosa（L.）O. Schwarz var. albicans（Rehder et E. H. Wilson）Y. L. Han ●

312599　Potentilla fruticosa L. var. arbuscula（D. Don）Maxim.；伏毛金露梅；Hairy Bush Cinquefoil ●

312600　Potentilla fruticosa L. var. arbuscula（D. Don）Maxim. = Dasiphora fruticosa（L.）Rydb. ●

312601　Potentilla fruticosa L. var. armerioides Hook. f. = Potentilla articulata Franch. ■

312602　Potentilla fruticosa L. var. dahurica Ser. = Potentilla glabra Lodd. ●

312603　Potentilla fruticosa L. var. dahurica Ser. f. ternata Cardot = Potentilla glabra Lodd. ●

312604　Potentilla fruticosa L. var. davurica（Nestl.）Ser. = Potentilla davurica Nestl. ●

312605　Potentilla fruticosa L. var. grandiflora C. Marquand = Potentilla parvifolia Fisch. ex Lehm. ●

312606　Potentilla fruticosa L. var. leucantha Makino；白花金露梅●

312607　Potentilla fruticosa L. var. leucantha Makino = Dasiphora mandshurica（Maxim.）Juz. ●

312608　Potentilla fruticosa L. var. mandshurica Maxim. = Dasiphora mandshurica（Maxim.）Juz. ●

312609　Potentilla fruticosa L. var. mandshurica Maxim. = Potentilla glabra Lodd. var. mandshurica（Maxim.）Hand.-Mazz. ●

312610　Potentilla fruticosa L. var. mongolica Maxim. = Potentilla glabra Lodd. ●

312611　Potentilla fruticosa L. var. parvifolia（Fisch. ex Lehm.）Th. Wolf = Potentilla parvifolia Fisch. ex Lehm. ●

312612　Potentilla fruticosa L. var. parvifolia Th. Wolf = Potentilla parvifolia Fisch. ex Lehm. ●

312613　Potentilla fruticosa L. var. pumila Hook. f.；垫状金露梅；Cushionshaped Bush Cinquefoil ●

312614　Potentilla fruticosa L. var. purdomii Rehder = Potentilla parvifolia Fisch. ex Lehm. ●

312615　Potentilla fruticosa L. var. rigida（Wall. ex Lehm.）Th. Wolf = Dasiphora fruticosa（L.）Rydb. ●

312616　Potentilla fruticosa L. var. rigida（Wall. ex Lehm.）Th. Wolf = Potentilla fruticosa L. ●

312617　Potentilla fruticosa L. var. subalbicans Hand.-Mazz. = Potentilla glabra Lodd. var. mandshurica（Maxim.）Hand.-Mazz. ●

312618　Potentilla fruticosa L. var. tanguitica Th. Wolf = Potentilla glabra Lodd. ●

312619　Potentilla fruticosa L. var. tenuifolia Lehm. = Pentaphylloides floribunda（Pursh）Á. Löve ●☆

312620 Potentilla fruticosa L. var. veitchii（E. H. Wilson）Bean =
Potentilla glabra Lodd. var. veitchii（E. H. Wilson）Hand. -Mazz. ●

312621 Potentilla fruticosa L. var. veitchii Bean = Potentilla glabra
Lodd. var. veitchii（E. H. Wilson）Hand. -Mazz. ●

312622 Potentilla fruticosa L. var. vilmoriniana Kom. = Potentilla
fruticosa L. var. albicans Rehder et E. H. Wilson ●

312623 Potentilla fulgens Lehm. = Potentilla lineata Trevir. ■

312624 Potentilla fulgens Lehm. var. acutiserrata（Te T. Yu et C. L. Li）
Te T. Yu et C. L. Li = Potentilla lineata Trevir. ■

312625 Potentilla fulgens Lehm. var. intermedia ? = Potentilla fulgens
Wall. ex Hook. ■

312626 Potentilla fulgens Lehm. var. macrophylla Cardot = Potentilla
lineata Trevir. ■

312627 Potentilla fulgens Wall. ex Hook. ;西南委陵菜(白地榆,白头
翁,槟榔仁,地槟榔,地管子,番白草,翻白草,翻白地榆,翻白叶,
翻背白草,管仲,亮叶委陵菜,涩疙瘩,西南香陵菜,银毛委陵菜,
银毛香陵菜);Shiny Cinquefoil ■

312628 Potentilla fulgens Wall. ex Hook. var. acutiserrata（Te T. Yu et
C. L. Li）Te T. Yu et C. L. Li;锐齿西南委陵菜(锐齿川滇委陵
菜);Sharptooth Shiny Cinquefoil,Sharptoothed Cinquefoil ■

312629 Potentilla fulgens Wall. ex Hook. var. macrophylla Cardot =
Potentilla fulgens Wall. ex Hook. ■

312630 Potentilla gelida C. A. Mey. ;耐寒委陵菜(冰委陵菜,绿叶委
陵菜);Frosty Cinquefoil ■

312631 Potentilla gelida C. A. Mey. var. genuina Th. Wolf = Potentilla
gelida C. A. Mey. ■

312632 Potentilla gelida C. A. Mey. var. sericea Te T. Yu et C. L. Li;绢
毛耐寒委陵菜(绢毛绿叶委陵菜);Silky Frosty Cinquefoil ■

312633 Potentilla geoides M. Bieb. ;水杨梅委陵菜■☆

312634 Potentilla glabra Lodd. ;银露梅(白花棍儿茶,银老梅);
Glabrous Cinquefoil ●

312635 Potentilla glabra Lodd. var. longipetala Te T. Yu et C. L. Li;长
瓣银露梅;Longpetal Glabrous Cinquefoil ●

312636 Potentilla glabra Lodd. var. mandshurica（Maxim.）Hand. -
Mazz. ;白毛银露梅(观音茶,华西银蜡梅,华西银露梅);
Manchrian Glabrous Cinquefoil ●

312637 Potentilla glabra Lodd. var. rhodocalyx H. R. Fletcher =
Potentilla glabra Lodd. ●

312638 Potentilla glabra Lodd. var. veitchii（E. H. Wilson）Hand. -
Mazz. ;伏毛银露梅(华西银蜡梅,华西银露梅,银老梅);Veitch
Glabrous Cinquefoil ●

312639 Potentilla glabrata Willd. ex Schltdl. = Potentilla glabra Lodd. ●

312640 Potentilla glabriuscula（Te T. Yu et C. L. Li）Soják;光叶委陵
菜■■

312641 Potentilla glabriuscula（Te T. Yu et C. L. Li）Soják var.
gracilescens Soják = Potentilla turfosa Hand. -Mazz. var. gracilescens
（Soják）H. Ikeda et H. Ohba ■

312642 Potentilla glabriuscula（Te T. Yu et C. L. Li）Soják var.
oligandra（Soják）Soják;多蕊光叶委陵菜■

312643 Potentilla glabriuscula Te T. Yu et C. L. Li var. majuscula Soják
= Potentilla glabriuscula（Te T. Yu et C. L. Li）Soják ■

312644 Potentilla glaucescens Willd. ex Schltdl. ;灰绿委陵菜■☆

312645 Potentilla goldbachii Rupr. ;高氏委陵菜;Goldbach Cinquefoil ■☆

312646 Potentilla gombalana Hand. -Mazz. ;川边委陵菜;Gombalan
Cinquefoil ■

312647 Potentilla gracilis Douglas ex Hook. ;纤弱委陵菜;Northwest
Cinquefoil,Slender Five-finger,Slender Goose Grass ■☆

312648 Potentilla gracilis Douglas ex Hook. var. flabelliformis（Lehm.）
Nutt. ex Torr. et A. Gray;鞭状纤细委陵菜;Comb Five-fingers,
Northwest Cinquefoil ■☆

312649 Potentilla gracillima Te T. Yu et C. L. Li;纤细委陵菜;Slender
Cinquefoil ■

312650 Potentilla grandiflora L. = Potentilla gelida C. A. Mey. ■

312651 Potentilla granulosa Te T. Yu et C. L. Li;腺粒委陵菜;
Gladulose Cinquefoil,Grainy Cinquefoil ■

312652 Potentilla griffithii Hook. f. ;柔毛委陵菜(地蜂,翻白叶,红地
榆,小管仲,小天青);Griffith Cinquefoil ■

312653 Potentilla griffithii Hook. f. f. velutina Cardot;长柔毛委陵菜
(翻白叶,绒毛委陵菜,小管仲,小天青);Velutinous Cinquefoil,
Velvety Cinquefoil ■

312654 Potentilla griffithii Hook. f. var. concolor Franch. = Potentilla
macrosepala Cardot ■

312655 Potentilla griffithii Hook. f. var. pumila（Franch.）Hand. -
Mazz. = Potentilla saundersiana Royle ■

312656 Potentilla heidenreichii Zimm. ;海登氏委陵菜;Heidenreich
Cinquefoil ■☆

312657 Potentilla hemsleyana Th. Wolf = Potentilla reptans L. var.
sericophylla Franch. ■

312658 Potentilla heptaphylla L. ;七叶委陵菜;Seven-leaved
Cinquefoil,Sevenleaves Cinquefoil ■☆

312659 Potentilla heynii Roth;阿穆尔委陵菜(东北委陵菜,小花金
梅);Amur Cinquefoil ■

312660 Potentilla hirsuta Michx. = Potentilla norvegica L. ■☆

312661 Potentilla hirta L. var. afra Pau et Font Quer = Potentilla recta
L. ■

312662 Potentilla hirta L. var. tenuirugis（Pomel）Batt. = Potentilla
recta L. ■

312663 Potentilla hispanica Zimmeter;西班牙委陵菜■☆

312664 Potentilla hispanica Zimmeter var. longepilosa H. Lindb. =
Potentilla hispanica Zimmeter ■☆

312665 Potentilla hispanica Zimmeter var. mesatlantica Dobignard =
Potentilla hispanica Zimmeter ■☆

312666 Potentilla hololeuca Boiss. = Potentilla griffithii Hook. f. f.
velutina Cardot ■

312667 Potentilla hololeuca Boiss. ex Lehm. ;全白委陵菜;Hole-white
Cinquefoil,White Cinquefoil ■

312668 Potentilla holopetala Turcz. ;全瓣委陵菜■☆

312669 Potentilla humifusa Willd. ex Schltdl. ;平伏委陵菜■☆

312670 Potentilla hybrida E. H. L. Krause;杂种委陵菜;Cinquefoil ■☆

312671 Potentilla hypargyrea Hand. -Mazz. ;白背委陵菜;Hypersilvery
Cinquefoil,Whiteback Cinquefoil ■

312672 Potentilla hypargyrea Hand. -Mazz. var. subpinnata Te T. Yu et
C. L. Li;假羽白背委陵菜;Falsepinnate Hypersilvery Cinquefoil ■

312673 Potentilla hypoleuca Turcz. = Potentilla multifida L. ■

312674 Potentilla imbricata Kar. et Kir. ;覆瓦委陵菜(毛二裂委陵
菜);Imbricate Cinquefoil ■

312675 Potentilla impolita Wahlenb. ;连锁委陵菜■☆

312676 Potentilla inclinata Vill. ;薄毛委陵菜(灰白委陵菜,灰毛委陵
菜);Ashy Cinquefoil,Grey Cinquefoil,Inclined Cinquefoil ■

312677 Potentilla indica（Andréws）Th. Wolf = Duchesnea indica
（Andréws）Focke ●■

312678 Potentilla indica（Andréws）Th. Wolf var. wallichii（Franch. et
Sav.）Th. Wolf = Duchesnea chrysantha（Zoll. et Moritzi）Miq. ■

312679 Potentilla indica Andréws = Potentilla reptans L. var.

sericophylla Franch. ■

312680 Potentilla indica Andréws var. wallichii（Franch. et Sav.）Th. Wolf ＝ Duchesnea chrysantha（Zoll. et Moritzi）Miq. ■

312681 Potentilla inglisii Royle ＝ Potentilla biflora Willd. ex Schltdl. ■

312682 Potentilla inquinans Turcz. ＝ Potentilla rupestris L. ■

312683 Potentilla intermedia L.；中型委陵菜；Downy Cinquefoil, Intermediate Cinquefoil, Russian Cinquefoil ■☆

312684 Potentilla intermedia L. var. canescens（Besser）Rupr. ＝ Potentilla inclinata Vill. ■

312685 Potentilla interrupta Te T. Yu et C. L. Li ＝ Potentilla polyphylla Wall. et Lehm. var. interrupta（Te T. Yu et C. L. Li）H. Ikeda et H. Ohba ■

312686 Potentilla italica Lehm. ＝ Potentilla mixta Nolte ex Koch ■☆

312687 Potentilla jacutica Juz.；雅库特委陵菜■☆

312688 Potentilla jailae Juz.；杰里委陵菜■☆

312689 Potentilla japonica Blume ＝ Potentilla sprengeliana Lehm. ■☆

312690 Potentilla kleiniana Wight et Arn.；蛇含委陵菜（地五加,地五甲,地五龙,地五爪,蛇包五披风,蛇含,蛇含草,蛇衔,威蛇,委陵菜,五虎草,五虎下山,五披风,五皮草,五皮风,五皮枫,五匹风,五星草,五叶莓,五叶蛇莓,五爪风,五爪虎,五爪金龙,五爪龙,小龙牙,小龙牙草,银莲花叶委陵菜,紫背草,紫背龙牙）；Klein Cinquefoil ■

312691 Potentilla kleiniana Wight et Arn. ＝ Potentilla griffithii Hook. f. ■

312692 Potentilla kleiniana Wight et Arn. subsp. anemonifolia（Lehm.）Murata ＝ Potentilla anemonifolia Lehm. ■

312693 Potentilla kleiniana Wight et Arn. var. robusta（Franch. et Sav.）Kitag. ＝ Potentilla anemonifolia Lehm. ■

312694 Potentilla kleiniana Wight et Arn. var. robusta（Franch. et Sav.）Kitag.；粗壮蛇含委陵菜■☆

312695 Potentilla komaroviana Th. Wolf；科马罗夫委陵菜■☆

312696 Potentilla kryloviana Th. Wolf；克雷氏委陵菜■☆

312697 Potentilla kulabensis Th. Wolf；库拉波委陵菜■☆

312698 Potentilla kusnetzowii（Gower）Juz.；库氏委陵菜■☆

312699 Potentilla labradorica Lehm. ＝ Potentilla norvegica L. ■☆

312700 Potentilla laeta Rchb.；愉快委陵菜■☆

312701 Potentilla lancinata Cardot；条裂委陵菜；Lance-cleft Cinquefoil, Lanciniate Cinquefoil ■

312702 Potentilla lancinata Cardot var. minor H. R. Fletcher ＝ Potentilla lancinata Cardot ■

312703 Potentilla lapponica（F. Nyl.）Juz.；拉普兰委陵菜■☆

312704 Potentilla latipetiolata C. E. C. Fisch. ＝ Potentilla articulata Franch. var. latipetiolata（C. E. C. Fisch.）Te T. Yu et C. L. Li ■

312705 Potentilla leschenaultiana Ser. ＝ Potentilla griffithii Hook. f. ■

312706 Potentilla leschenaultiana Ser. var. concolor Cardot ＝ Potentilla griffithii Hook. f. f. velutina Cardot ■

312707 Potentilla leschenaultiana Ser. var. concolor Franch. ＝ Potentilla fragarioides L. ■

312708 Potentilla leschenaultiana Ser. var. pumila Franch. ＝ Potentilla griffithii Hook. f. ■

312709 Potentilla leschenaultiana Ser. var. pumila Franch. ＝ Potentilla saundersiana Royle ■

312710 Potentilla leschenaultiana Ser. var. reticulata Franch. ＝ Potentilla griffithii Hook. f. ■

312711 Potentilla leschenaultiana var. concolor Cardot ＝ Potentilla griffithii Hook. f. f. velutina Cardot ■

312712 Potentilla lespedeza H. Lév. ＝ Potentilla fruticosa L. var. arbuscula（D. Don）Maxim. ●

312713 Potentilla leucocarpa Rydb. ＝ Potentilla rivalis Nutt. var. millegrana（Engelm. ex Lehm.）S. Watson ■☆

312714 Potentilla leuconota D. Don；银叶委陵菜（金钱标,锦标草,锦钱镖,进去标,涩草,玉山金梅）；Silverleaf Cinquefoil ■

312715 Potentilla leuconota D. Don var. brachyllaria Cardot；脱毛银叶委陵菜（短叶白斑人参果）；Hairless Silverleaf Cinquefoil ■

312716 Potentilla leuconota D. Don var. corymbosa Cardot ＝ Potentilla cardotiana Hand. -Mazz. ■

312717 Potentilla leuconota D. Don var. corymbosa Cardot ＝ Potentilla peduncularis D. Don ■

312718 Potentilla leuconota D. Don var. morrisonicola Hayata ＝ Potentilla leuconota D. Don ■

312719 Potentilla leuconota D. Don var. omeiensis H. Ikeda et H. Ohba；峨眉银叶委陵菜■

312720 Potentilla leuconota D. Don var. tugitakensis（Masam.）H. L. Li ＝ Potentilla tugitakensis Masam. ■

312721 Potentilla leucophylla Pall.；白叶委陵菜（三出委陵菜）；Whitehair Cinquefoil ■

312722 Potentilla leucophylla Pall. ＝ Potentilla betonicifolia Poir. ■

312723 Potentilla leucophylla Pall. var. morrisonicola Hayata ＝ Potentilla leuconota D. Don ■

312724 Potentilla leucophylla Pall. var. pentaphylla Liou et C. Y. Li；五叶白叶委陵菜；Fiveleaf Whitehair Cinquefoil ■

312725 Potentilla leucophylla Pall. var. tugitakensis（Masam.）C. L. Li ＝ Potentilla tugitakensis Masam. ■

312726 Potentilla leucopolitana P. J. Müll. ex F. Schulz；亮白委陵菜■☆

312727 Potentilla leucotricha Borbás；白毛委陵菜；Whitehair Cinquefoil ■☆

312728 Potentilla limprichtii J. Krause；下江委陵菜（多茎委陵菜,浮尸草）；Limbricht Cinquefoil ■

312729 Potentilla lindenbergii Lehm. ＝ Sibbaldia adpressa Bunge ■

312730 Potentilla lineata Trevir.；西南线委陵菜■

312731 Potentilla lipskyana Th. Wolf；里普委陵菜■☆

312732 Potentilla lomakinii Grossh.；洛马金委陵菜■☆

312733 Potentilla longepetiolata H. Lév. ＝ Potentilla freyniana Bornm. ■

312734 Potentilla longifolia Willd. ex Schltdl.；腺毛委陵菜（委陵菜,粘委陵菜）；Longleaf Cinquefoil, Sticky Cinquefoil ■

312735 Potentilla longifolia Willd. ex Schltdl. var. villosa F. Z. Li；长毛委陵菜■

312736 Potentilla longipes Ledeb.；长花梗委陵菜■☆

312737 Potentilla longipetiolata H. Lév. ＝ Potentilla centigrana Maxim. ■

312738 Potentilla luteopilosa Te T. Yu et C. L. Li；黄毛委陵菜（黄毛小叶委陵菜）；Yellowhair Cinquefoil ■

312739 Potentilla macrantha Ledeb.；大花委陵菜■☆

312740 Potentilla macrosepala Cardot；大瓣委陵菜（大花委陵菜）；Largesepal Cinquefoil ■

312741 Potentilla mairei H. Lév. ＝ Sibbaldia micropetala（D. Don）Hand. -Mazz. ■

312742 Potentilla malacotricha Juz.；软毛委陵菜■☆

312743 Potentilla martinii H. Lév. ＝ Potentilla fulgens Wall. ex Hook. ■

312744 Potentilla martinii H. Lév. ＝ Potentilla lineata Trevir. ■

312745 Potentilla matsumurae Th. Wolf；松村翻白草（翻白草）■

312746 Potentilla matsumurae Th. Wolf f. lasiocarpa（H. Hara）T. Shimizu；毛果翻白草■☆

312747 Potentilla matsumurae Th. Wolf f. paucidentata H. Hara；稀齿翻白草■☆

312748 Potentilla matsumurae Th. Wolf var. apoiensis（Nakai）H. Hara；阿伯伊翻白草■☆

312749　Potentilla matsumurae Th. Wolf var. lasiocarpa H. Hara ＝ Potentilla matsumurae Th. Wolf f. lasiocarpa (H. Hara) T. Shimizu ■☆

312750　Potentilla matsumurae Th. Wolf var. pilosa Koidz. ;高山翻白草■

312751　Potentilla matsumurae Th. Wolf var. yuparensis (Miyabe et Tatew.) Kudo ex H. Hara;北海道翻白草■☆

312752　Potentilla matsuokana Makino ＝ Potentilla nivea L. ■

312753　Potentilla maura Th. Wolf;模糊委陵菜■☆

312754　Potentilla maura Th. Wolf var. brachypetala Emb. ＝ Potentilla maura Th. Wolf ■☆

312755　Potentilla maura Th. Wolf var. glabrescens Emb. et Maire ＝ Potentilla maura Th. Wolf ■☆

312756　Potentilla maura Th. Wolf var. sericea Font Quer ＝ Potentilla maura Th. Wolf ■☆

312757　Potentilla mayeri Boiss. ;迈耶委陵菜;Mayer Cinquefoil ■☆

312758　Potentilla megalantha Takeda;巨花委陵菜(大花莓状委陵菜)■☆

312759　Potentilla megalantha Takeda ＝ Potentilla fragiformis Willd. ex D. F. K. Schltdl. subsp. megalantha (Takeda) Hultén ■☆

312760　Potentilla meifolia Wall. ＝ Potentilla coriandrifolia D. Don ■

312761　Potentilla micrantha DC. ＝ Potentilla micrantha Ramond ex DC. ■

312762　Potentilla micrantha Ramond ex DC. ;小花委陵菜;Smallflower Cinquefoil ■

312763　Potentilla micropetala D. Don ＝ Sibbaldia micropetala (D. Don) Hand. -Mazz. ■

312764　Potentilla microphylla D. Don;小叶委陵菜;Microphyll Cinquefoil,Smallleaf Cinquefoil ■

312765　Potentilla microphylla D. Don var. achilleifolia Hook. f. ;细裂小叶委陵菜;Finelydivided ■

312766　Potentilla microphylla D. Don var. achilleifolia Hook. f. ＝ Potentilla aristata Soják ■

312767　Potentilla microphylla D. Don var. caespitosa Te T. Yu et C. L. Li;丛生小叶委陵菜;Tufted Smallleaf Cinquefoil ■

312768　Potentilla microphylla D. Don var. depressa Wall. ex Lehm. ＝ Potentilla microphylla D. Don ■

312769　Potentilla microphylla D. Don var. glabriuscula Hook. f. ＝ Sibbaldia glabriuscula Te T. Yu et C. L. Li ■

312770　Potentilla microphylla D. Don var. glabriuscula Wall. ;无毛小叶委陵菜;Hairless Smallleaf Cinquefoil ■

312771　Potentilla microphylla D. Don var. glabriuscula Wall. ex Lehm. ＝ Potentilla microphylla D. Don ■

312772　Potentilla microphylla D. Don var. latiloba ? ＝ Potentilla microphylla D. Don ■

312773　Potentilla microphylla D. Don var. luteopilosa (Te T. Yu et C. L. Li) H. Ikeda et H. Ohba ＝ Potentilla luteopilosa Te T. Yu et C. L. Li ■

312774　Potentilla microphylla D. Don var. multijuga Te T. Yu et C. L. Li ＝ Potentilla aristata Soják ■

312775　Potentilla microphylla D. Don var. tapetodes (Soják) H. Ikeda et H. Ohba;地毯小叶委陵菜(丛生小叶委陵菜)■

312776　Potentilla mieheorum Soják ＝ Potentilla commutata Lehm. var. polyandra Soják ■

312777　Potentilla millefolia H. Lév. ＝ Potentilla stenophylla (Franch.) Diels ■

312778　Potentilla millefolium H. Lév. ＝ Potentilla stenophylla (Franch.) Diels ■

312779　Potentilla millegrana Engelm. ex Lehm. ＝ Potentilla rivalis Nutt. ex Torr. et A. Gray ■☆

312780　Potentilla millegrana Engelm. ex Lehm. ＝ Potentilla rivalis Nutt.

var. millegrana (Engelm. ex Lehm.) S. Watson ■☆

312781　Potentilla mixta Nolte ex Koch;意大利委陵菜;Hybrid Cinquefoil ■☆

312782　Potentilla miyabei Makino;日本山地委陵菜(高山委陵菜);Miyabe Cinquefoil ■☆

312783　Potentilla mollissima Lehm. ;柔软委陵菜■☆

312784　Potentilla monspeliensis L. ;多皱委陵菜;Rough Cinquefoil ■☆

312785　Potentilla monspeliensis L. ＝ Potentilla norvegica L. ■☆

312786　Potentilla montana Brot. ;山地委陵菜■☆

312787　Potentilla mooniana Wight ＝ Potentilla polyphylla Wall. et Lehm. ■

312788　Potentilla moorcroftii Wall. ＝ Potentilla bifurca L. var. humilior Rupr. et Ost. -Sack. ■

312789　Potentilla moorcroftii Wall. ex Lehm. ＝ Potentilla bifurca L. var. humilior Rupr. et Ost. -Sack. ■

312790　Potentilla morii Hayata ＝ Potentilla matsumurae Th. Wolf var. pilosa Koidz. ■

312791　Potentilla morrisonicola (Hayata) Hayata ＝ Potentilla leuconota D. Don ■

312792　Potentilla morrisonicola (Hayata) Hayata var. nitida Masam. ＝ Potentilla leuconota D. Don ■

312793　Potentilla morrisonicola (Hayata) Hayata var. sericea Masam. ＝ Potentilla leuconota D. Don ■

312794　Potentilla moupinensis Franch. ＝ Fragaria moupinensis (Franch.) Cardot ■

312795　Potentilla multicaulis Bunge;多茎委陵菜(猫爪子);Manystalk Cinquefoil,Multicaulis Cinquefoil,Prairie Cinquefoil ■

312796　Potentilla multiceps Te T. Yu et C. L. Li;多头委陵菜;Manyhead Cinquefoil,Many-heads Cinquefoil ■

312797　Potentilla multifida L. ;多裂委陵菜(白马肉,翻白草,毛鸡腿,细叶委陵菜);Manycleft Cinquefoil,Staghorn Cinquefoil ■

312798　Potentilla multifida L. f. subpalmata (Krylov) Kitag. ＝ Potentilla multifida L. var. ornithopoda Th. Wolf ■

312799　Potentilla multifida L. var. angustifolia Lehm. ＝ Potentilla multifida L. ■

312800　Potentilla multifida L. var. hypoleuca (Turcz.) Th. Wolf ＝ Potentilla multifida L. ■

312801　Potentilla multifida L. var. minor Ledeb. ＝ Potentilla multifida L. var. nubigena Th. Wolf ■

312802　Potentilla multifida L. var. nubigena Th. Wolf;矮生多裂委陵菜;Dwarf Manycleft Cinquefoil ■

312803　Potentilla multifida L. var. ornithopoda Th. Wolf;掌叶多裂委陵菜(爪细叶委陵菜);Palmate Manycleft Cinquefoil ■

312804　Potentilla multifida L. var. saundersiana (Royle) Hook. f. ＝ Potentilla saundersiana Royle ■

312805　Potentilla multifida L. var. saundersiana Hook. f. ＝ Potentilla saundersiana Royle ■

312806　Potentilla multifida L. var. sericea Bar. et Skvortsov ex Liou ＝ Potentilla multifida L. ■

312807　Potentilla multifida L. var. subpalmata Krylov ＝ Potentilla multifida L. var. ornithopoda Th. Wolf ■

312808　Potentilla multijuga Te T. Yu et C. L. Li;多对委陵菜■

312809　Potentilla nepalensis Hook. 'Miss Willmott';威尔莫特尼泊尔委陵菜■☆

312810　Potentilla nepalensis Hook. f. ;尼泊尔委陵菜■☆

312811　Potentilla nervosa Juz. ;显脉委陵菜(显脉翻白草);Nerved Cinquefoil ■

312812　Potentilla neumanniana Rchb.；春委陵菜（纽曼委陵菜）；Cinquefoil，Spring Cinquefoil，Tabernemontan Cinquefoil ■☆

312813　Potentilla neumanniana Rchb.'Nana'；匍匐春委陵菜（匍匐纽曼委陵菜）；Creeping Potentilla，Spring Cinquefoil ■☆

312814　Potentilla nicolletii（S. Watson）E. Sheld.；尼氏委陵菜；Cinquefoil ■☆

312815　Potentilla niponica Th. Wolf；日本翻白草■

312816　Potentilla nitida L.；光亮委陵菜；Pink Cinquefoil ■☆

312817　Potentilla nivalis Lapeyr. = Potentilla nivea L. ■

312818　Potentilla nivea L.；雪白委陵菜（白里金梅，白委陵菜，假雪委陵菜，雪委陵菜，雪线委陵菜）；Snow Cinquefoil，Snow White Cinquefoil，Snowwhite Cinquefoil，Snowy Cinquefoil ■☆

312819　Potentilla nivea L. var. angustifolia Ledeb. = Potentilla betonicifolia Poir. ■

312820　Potentilla nivea L. var. camtschatica Cham. et Schltdl. = Potentilla nivea L. ■

312821　Potentilla nivea L. var. elongata Th. Wolf；多齿雪白委陵菜；Elongate Snow White Cinquefoil ■

312822　Potentilla nivea L. var. elongata Th. Wolf = Potentilla nervosa Juz. ■

312823　Potentilla nivea L. var. macrantha Ledeb. = Potentilla nivea L. var. elongata Th. Wolf ■

312824　Potentilla nivea L. var. pinnatifida Lehm. = Potentilla virgata Lehm. var. pinnatifida（Lehm.）Te T. Yu et C. L. Li ■

312825　Potentilla nivea L. var. polyphylla Y. Zhang et Z. T. Yin；混叶雪白委陵菜■

312826　Potentilla nivea L. var. yuparensis Miyabe et Tatew. = Potentilla nivea L. ■

312827　Potentilla nordmanniana Ledeb.；诺氏委陵菜■☆

312828　Potentilla norvegica L.；挪威委陵菜；Norwegian Cinquefoil，Rough Cinquefoil，Strawberry-weed，Ternate-leaved Cinquefoil ■☆

312829　Potentilla norvegica L. subsp. hirsuta（Michx.）Hyl. = Potentilla norvegica L. ■☆

312830　Potentilla norvegica L. subsp. monspeliensis（L.）Asch. et Graebn. = Potentilla norvegica L. ■☆

312831　Potentilla norvegica L. var. hirsuta（Michx.）Torr. et A. Gray = Potentilla norvegica L. ■☆

312832　Potentilla norvegica L. var. labradorica（Lehm.）Fernald = Potentilla norvegica L. ■☆

312833　Potentilla nudicaulis Willd. = Potentilla tanacetifolia Willd. ex Schltdl. ■

312834　Potentilla nudicaulis Willd. ex Schltdl.；大委陵菜（红茎委陵菜）■

312835　Potentilla nudicaulis Willd. ex Schltdl. = Potentilla tanacetifolia Willd. ex Schltdl. ■

312836　Potentilla nureusis Boiss. et Hausskn.；努尔委陵菜■☆

312837　Potentilla obscura Willd.；幽委陵菜■☆

312838　Potentilla okuboi Kitag. = Potentilla rupestris L. ■

312839　Potentilla oligandra Soják = Potentilla glabriuscula（Te T. Yu et C. L. Li）Soják var. oligandra（Soják）Soják ■

312840　Potentilla orbiculata Th. Wolf；圆委陵菜■☆

312841　Potentilla orientalis Juz.；东方委陵菜；Oriental Cinquefoil ■

312842　Potentilla orientalis Juz. = Potentilla bifurca Willd. ex Schltdl. var. major Ledeb. ■

312843　Potentilla ornithopoda Tausch = Potentilla multifida L. var. ornithopoda Th. Wolf ■

312844　Potentilla ovalis Lehm. = Fragaria virginiana Duchesne ■☆

312845　Potentilla oxyodonta Soják = Potentilla peduncularis D. Don var. vittata（Soják）H. Ikeda et H. Ohba ■

312846　Potentilla pacifica Howell = Potentilla anserina L. subsp. pacifica（Howell）Rousi ■☆

312847　Potentilla palczowskii Juz. = Potentilla fragarioides L. ■

312848　Potentilla palustris（L.）Scop. = Comarum palustre L. ●■

312849　Potentilla palustris（L.）Scop. var. parvifolia（Raf.）Fernald et B. H. Long = Comarum palustre L. ●■

312850　Potentilla palustris（L.）Scop. var. villosa（Pers.）Lehm. = Comarum palustre L. ●●■

312851　Potentilla pamirica Th. Wolf；帕米尔委陵菜■☆

312852　Potentilla pamiroalaica Juz.；高原委陵菜（帕米尔委陵菜）；Alpine Cinquefoil，Plateau Cinquefoil ■

312853　Potentilla pannifolia Liou et C. Y. Li；假翻白委陵菜■

312854　Potentilla paradoxa Nutt. ex Torr. et A. Gray = Potentilla supina L. ■

312855　Potentilla parviflora Willd. = Sibbaldia cuneata Hornem. ex Kuntze ■

312856　Potentilla parvifolia Fisch. = Pentaphylloides parvifolia（Fisch. ex Lehm.）Soják ●

312857　Potentilla parvifolia Fisch. ex Lehm. = Pentaphylloides parvifolia（Fisch. ex Lehm.）Soják ●

312858　Potentilla parvifolia Fisch. ex Lehm. var. hypoleuca Hand.-Mazz.；白毛小叶金露梅；Whitehair Smallleaf Cinquefoil ●

312859　Potentilla parvipetala B. C. Ding et S. Y. Wang = Potentilla supina L. var. ternata Peterm. ■

312860　Potentilla pectinata Raf.；北美委陵菜；Coast Cinquefoil ■☆

312861　Potentilla pedata Willd.；足裂委陵菜■☆

312862　Potentilla peduncularis D. Don；总梗委陵菜（白地榆，长梗委陵菜，大理人参果，翻白草）；Peduncled Cinquefoil ■

312863　Potentilla peduncularis D. Don var. abbreviata Te T. Yu et C. L. Li；高山总梗委陵菜；Alpine ■

312864　Potentilla peduncularis D. Don var. clarkei Hook. f. = Potentilla contigua Soják ■

312865　Potentilla peduncularis D. Don var. clarkei Hook. f. = Potentilla peduncularis D. Don ■

312866　Potentilla peduncularis D. Don var. curta（Soják）H. Ikeda et H. Ohba；少齿总梗委陵菜■

312867　Potentilla peduncularis D. Don var. elangata Te T. Yu et C. L. Li；疏叶总梗委陵菜（疏叶委陵菜）；Elongate Peduncle Cinquefoil ■

312868　Potentilla peduncularis D. Don var. elongata Te T. Yu et C. L. Li = Potentilla peduncularis D. Don ■

312869　Potentilla peduncularis D. Don var. glabriuscula Te T. Yu et C. L. Li；脱毛总梗委陵菜；Hairless Peduncle Cinquefoil ■

312870　Potentilla peduncularis D. Don var. glabriuscula Te T. Yu et C. L. Li = Potentilla contigua Soják ■

312871　Potentilla peduncularis D. Don var. obscura Hook. f. = Potentilla fallens Cardot ■

312872　Potentilla peduncularis D. Don var. obscura Hook. f. = Potentilla peduncularis D. Don ■

312873　Potentilla peduncularis D. Don var. shweliensis（H. R. Fletcher）H. Ikeda et H. Ohba；多齿总梗委陵菜■

312874　Potentilla peduncularis D. Don var. stenophylla Franch. = Potentilla stenophylla（Franch.）Diels ■

312875　Potentilla peduncularis D. Don var. vittata（Soják）H. Ikeda et H. Ohba；狭叶总梗委陵菜■

312876　Potentilla pendula Te T. Yu et C. L. Li；垂花委陵菜；

Droopingflower Cinquefoil ■

312877　Potentilla pensylvanica L. = Potentilla multicaulis Bunge ■

312878　Potentilla pensylvanica L. subsp. hispanica（Zimmeter）Maire = Potentilla dentata Forssk. ■☆

312879　Potentilla pensylvanica L. var. conferta（Bunge）Ledeb. = Potentilla conferta Bunge ■

312880　Potentilla pensylvanica L. var. conferta Ledeb. = Potentilla conferta Bunge ■

312881　Potentilla pensylvanica L. var. nivea Andr. = Potentilla hispanica Zimmeter ■☆

312882　Potentilla pensylvanica L. var. strigosa（Pall. ex Pursh）Lehm. = Potentilla strigosa Pall. ex Pursh ■

312883　Potentilla pensylvanica L. var. strigosa Lehm. = Potentilla strigosa Pall. ex Pursh ■

312884　Potentilla perpusilloides W. W. Sm. = Sibbaldia perpusilloides（W. W. Sm.）Hand.-Mazz. ■

312885　Potentilla peterae Hand.-Mazz. = Potentilla sischanensis Bunge ex Lehm. var. peterae（Hand.-Mazz.）Te T. Yu et C. L. Li ■

312886　Potentilla pilosa Willd.；疏毛委陵菜；Pilose Cinquefoil ■☆

312887　Potentilla pilosa Willd. = Potentilla recta L. ■

312888　Potentilla pimpinelloides L.；茴芹委陵菜■☆

312889　Potentilla pindicola Hausskn. ex Zimm.；平德斯委陵菜■☆

312890　Potentilla plumosa Te T. Yu et C. L. Li；羽毛委陵菜；Feather Cinquefoil, Feathered Cinquefoil ■

312891　Potentilla plurijuga Hand.-Mazz. = Potentilla multifida L. ■

312892　Potentilla polyphylla Wall. et Lehm.；多叶委陵菜；Manyleaf Cinquefoil, Manyleaves Cinquefoil ■

312893　Potentilla polyphylla Wall. et Lehm. var. interrupta（Te T. Yu et C. L. Li）H. Ikeda et H. Ohba；间断委陵菜；Interrupted Cinquefoil ■

312894　Potentilla polyphylloides H. Ikeda et H. Ohba；似多叶委陵菜■

312895　Potentilla polyschista Boiss. = Potentilla sericea L. var. polyschista Lehm. ■

312896　Potentilla porphyrantha Juz.；紫花委陵菜■☆

312897　Potentilla potaninii Th. Wolf；华西委陵菜；Potanin Cinquefoil ■

312898　Potentilla potaninii Th. Wolf var. compsophylla（Hand.-Mazz.）Te T. Yu et C. L. Li；裂叶华西委陵菜；Clefted-leaf Potanin Cinquefoil ■

312899　Potentilla potaninii Th. Wolf var. subdigitata Th. Wolf = Potentilla saundersiana Royle ■

312900　Potentilla poterioides Willd. ex Schltdl. = Potentilla limprichtii J. Krause ■

312901　Potentilla procumbens（L.）Clairv. = Sibbaldia procumbens L. ■

312902　Potentilla procumbens Sibth. = Potentilla anemonifolia Lehm. ■

312903　Potentilla procumbens Sibth. = Potentilla reptans L. ■

312904　Potentilla pseudosimulatrix W. B. Liao, Si F. Li et Z. Y. Yu；粗齿委陵菜；Broadtooth Cinquefoil ■

312905　Potentilla pulchella R. Br.；美委陵菜■☆

312906　Potentilla pulvinata（Te T. Yu et C. L. Li）Soják = Potentilla coriandrifolia D. Don var. dumosa Franch. ■

312907　Potentilla purdomii N. E. Br. = Coluria longifolia Maxim. ■

312908　Potentilla purpurea（Royle）Hook. f. = Sibbaldia purpurea Royle ■

312909　Potentilla purpurea Royle = Sibbaldia purpurea Royle var. macropetala（Murav.）Te T. Yu et C. L. Li ■

312910　Potentilla querpaertensis Cardot = Potentilla rosulifera H. Lév. ■

312911　Potentilla raddeana Juz.；拉德委陵菜■☆

312912　Potentilla recta L.；直立委陵菜；Erect Cinquefoil, Rough-fruited Cinquefoil, Straight Cinquefoil, Sulphur Cinquefoil, Sulphur Five-fingers ■

312913　Potentilla recta L. ‘Macrantha’ = Potentilla recta L. ‘Warrenii’■☆

312914　Potentilla recta L. ‘Warrenii’；沃伦直立委陵菜■☆

312915　Potentilla recta L. subsp. afra（Pau et Font Quer）Romo；非洲直立委陵菜■☆

312916　Potentilla recta L. var. afra Pau et Font Quer = Potentilla recta L. ■

312917　Potentilla recta L. var. obscura（Nestl.）W. D. J. Koch = Potentilla recta L. ■

312918　Potentilla recta L. var. pilosa（Willd.）Asch. et Graebn. = Potentilla recta L. ■

312919　Potentilla recta L. var. pilosa（Willd.）Ledeb. = Potentilla recta L. ■

312920　Potentilla recta L. var. sulphurea（Lam.）Lam. et DC. = Potentilla recta L. ■

312921　Potentilla recta L. var. sulphurea（Lam.）Peyr. = Potentilla recta L. ■

312922　Potentilla regeliana Th. Wolf；雷格尔委陵菜■☆

312923　Potentilla rehderiana Hand.-Mazz. = Potentilla parvifolia Fisch. ex Lehm. ●

312924　Potentilla remota Soják = Potentilla peduncularis D. Don ■

312925　Potentilla reptans L.；细蔓委陵菜（金棒槌，金金棒，决麻，老瓜咭子花，匍匐委陵菜，匍根委陵菜）；Cinkefield, Cinquefoil, Creeping Cinquefoil, Creeping Jenny, Five Fingers, Five Fingers of Mary, Five-finger Blossom, Five-finger Grass, Five-leaved Grass, Lady's Fingers, Sinkfield, Yellow Strawberry-flower ■

312926　Potentilla reptans L. var. angustiloba Ser. = Potentilla flagellaris Willd. ex Schltdl. ■

312927　Potentilla reptans L. var. argentea Batt. = Potentilla reptans L. ■

312928　Potentilla reptans L. var. incisa Franch. = Potentilla reptans L. var. sericophylla Franch. ■

312929　Potentilla reptans L. var. lanata Lange = Potentilla reptans L. ■

312930　Potentilla reptans L. var. mollis Borbás = Potentilla reptans L. ■

312931　Potentilla reptans L. var. sericophylla Franch.；绢毛细蔓委陵菜（结根草莓，金棒槌，金棒锤，金金棒，绢毛匍匐委陵菜，五爪龙，小五爪龙）；Silky Creeping Cinquefoil ■

312932　Potentilla rhytidocarpa Cardot = Potentilla lancinata Cardot ■

312933　Potentilla rigida Wall. ex Lehm. = Potentilla fruticosa L. ●

312934　Potentilla rigidula Th. Wolf；硬委陵菜■☆

312935　Potentilla riparia Murata；河畔委陵菜■☆

312936　Potentilla rivalis Nutt. ex Torr. et A. Gray；溪畔委陵菜；Brook Cinquefoil, Brook Five-fingers ■☆

312937　Potentilla rivalis Nutt. var. millegrana（Engelm. ex Lehm.）S. Watson；白果溪畔委陵菜；Brook Cinquefoil, Brook Five-fingers ■☆

312938　Potentilla rivalis Nutt. var. millegrana（Engelm. ex Lehm.）S. Watson = Potentilla rivalis Nutt. ex Torr. et A. Gray ■☆

312939　Potentilla robbinsiana Oakes = Potentilla robbinsiana Oakes ex Torr. et Gray ■☆

312940　Potentilla robbinsiana Oakes ex Torr. et Gray；罗宾斯委陵菜；Dwarf Mountain Cinquefoil ■☆

312941　Potentilla rockiana Melch. = Potentilla fallens Cardot ■

312942　Potentilla rosulifera H. Lév.；曲枝委陵菜（匐枝委陵菜，匍枝委陵菜）；Yokusa Cinquefoil ■

312943　Potentilla rosulifera H. Lév. = Potentilla centigrana Maxim. ■

312944　Potentilla rugolosa Kitag.；乌苏里委陵菜；Ussuri Cinquefoil ■

312945　Potentilla rugosa Kitag. = Potentilla ancistrifolia Bunge ■

312946　Potentilla rupestris L. ;石生委陵菜(白花委陵菜,岩生委陵菜); Cliff Cinquefoil, Eliff Cinquefoil, Rocky Cinquefoil, White Cinquefoil ■

312947　Potentilla rupestris L. = Drymocallis rupestris (L.) Soják ■☆

312948　Potentilla ruprechtii Boiss. ;卢普委陵菜■☆

312949　Potentilla sachalinensis Juz. = Potentilla sprengeliana Lehm. ■☆

312950　Potentilla salesoviana Steph. = Comarum salesovianum (Stephan) Asch. et Graebn.●■

312951　Potentilla salesovii Steph. ex Willd. = Comarum salesovianum (Stephan) Asch. et Graebn.●■

312952　Potentilla salwinensis (Soják) Soják = Potentilla coriandrifolia D. Don var. dumosa Franch. ■

312953　Potentilla salwinensis (Soják) Soják var. latiuscula Soják = Potentilla coriandrifolia D. Don var. dumosa Franch. ■

312954　Potentilla salwinensis (Soják) Soják var. parviflora Soják = Potentilla coriandrifolia D. Don var. dumosa Franch. ■

312955　Potentilla sanguisorba Willd. ex Schltdl. ;地榆委陵菜■☆

312956　Potentilla saundersiana Royle;钉柱委陵菜;Saunders Cinquefoil ■

312957　Potentilla saundersiana Royle var. caespitosa (Lehm.) Th. Wolf;丛生钉柱委陵菜;Tufted Saunders Cinquefoil ■

312958　Potentilla saundersiana Royle var. jacquemontii Franch. ;裂萼钉柱委陵菜;Cleftedcalyx Saunders Cinquefoil ■

312959　Potentilla saundersiana Royle var. potaninii (Th. Wolf) Hand. - Mazz. = Potentilla potaninii Th. Wolf ■

312960　Potentilla saundersiana Royle var. subpinnata Hand. -Mazz. ;羽叶钉柱委陵菜;Pinnateleaf Saunders Cinquefoil ■

312961　Potentilla saxosa Lemmon = Potentilla saxosa Lemmon ex Greene ■☆

312962　Potentilla saxosa Lemmon ex Greene;岩生委陵菜;Rock Five-finger ■☆

312963　Potentilla schrenkiana Regel;施伦克委陵菜■☆

312964　Potentilla schurii Fuss ex Zimm. ;苏氏委陵菜;Suchur Cinquefoil ■☆

312965　Potentilla seidlitziana Bien. ;塞德委陵菜■☆

312966　Potentilla semiglabra Juz. = Potentilla bifurca L. var. major Ledeb. ■

312967　Potentilla semiglabrata Juz. ;半光委陵菜■☆

312968　Potentilla semilaciniosa Bornm. ;半裂委陵菜■☆

312969　Potentilla sericea L. ;绢毛委陵菜(白毛小委陵菜,毛叶委陵菜,丝绒委陵菜); Silky Cinquefoil ■

312970　Potentilla sericea L. = Potentilla multicaulis Bunge ■

312971　Potentilla sericea L. var. dasyphylla (Bunge) Ledeb. = Potentilla sericea L. ■

312972　Potentilla sericea L. var. dasyphylla Ledeb. = Potentilla sericea L. ■

312973　Potentilla sericea L. var. multicaulis (Bunge) Lehm. = Potentilla multicaulis Bunge ■

312974　Potentilla sericea L. var. multicaulis Lehm. = Potentilla multicaulis Bunge ■

312975　Potentilla sericea L. var. polyschista Lehm. ;变叶绢毛委陵菜; Varied-leaf Cinquefoil ■

312976　Potentilla shweliensis H. R. Fletcher = Potentilla peduncularis D. Don ■

312977　Potentilla shweliensis H. R. Fletcher = Potentilla peduncularis D. Don var. shweliensis (H. R. Fletcher) H. Ikeda et H. Ohba ■

312978　Potentilla sibbaldai Haller f. = Sibbaldia cuneata Hornem. ex Kuntze ■

312979　Potentilla sibbaldia Griess. = Sibbaldia procumbens L. ■

312980　Potentilla sibbaldia Lehm. = Sibbaldia procumbens L. ■

312981　Potentilla sibbaldii Haller f. = Sibbaldia procumbens L. ■

312982　Potentilla sibirica Th. Wolf = Potentilla strigosa Pall. ex Pursh ■

312983　Potentilla sibirica Th. Wolf var. genuina Th. Wolf = Potentilla strigosa Pall. ex Pursh ■

312984　Potentilla sibirica Th. Wolf. var. longipila Th. Wolf = Potentilla conferta Bunge ■

312985　Potentilla siemersiana Lehm. = Potentilla fulgens Wall. ex Hook. ■

312986　Potentilla siemersiana Lehm. = Potentilla lineata Trevir. ■

312987　Potentilla siemersiana Lehm. var. acutiserrata Te T. Yu et C. L. Li = Potentilla lineata Trevir. ■

312988　Potentilla siemersiana Lehm. var. acutiserrataTe T. Yu et C. L. Li = Potentilla fulgens Wall. ex Hook. var. acutiserrata (Te T. Yu et C. L. Li) Te T. Yu et C. L. Li ■

312989　Potentilla sikkimensis Prain = Sibbaldia sikkimensis (Prain) Chatterjee ■

312990　Potentilla sikkimensis Th. Wolf = Potentilla griffithii Hook. f. ■

312991　Potentilla sikkimensis Th. Wolf = Sibbaldia melinocricha Hand. -Mazz. ■

312992　Potentilla silvestris Neck. = Potentilla erecta (L.) Hampe ■☆

312993　Potentilla simplex Michx. ;简单委陵菜;Common Cinquefoil, Five Finger, Old-field Cinquefoil, Old-field Five-fingers ■☆

312994　Potentilla simplex Michx. var. argyrisma Fernald = Potentilla simplex Michx. ■☆

312995　Potentilla simplex Michx. var. calvescens Fernald = Potentilla simplex Michx. ■☆

312996　Potentilla simplex Michx. var. typica Fernald = Potentilla simplex Michx. ■☆

312997　Potentilla simulatrix Th. Wolf;等齿委陵菜;Equaltoothed Cinquefoil ■

312998　Potentilla simulatrix Th. Wolf var. grossidens Kitag. ;重齿委陵菜■

312999　Potentilla sinonivea Hultén = Potentilla saundersiana Royle var. caespitosa (Lehm.) Th. Wolf ■

313000　Potentilla sischanensis Bunge ex Lehm. ;西山委陵菜;Sishan Cinquefoil, Xishan Cinquefoil ■

313001　Potentilla sischanensis Bunge ex Lehm. var. peterae (Hand. -Mazz.) Te T. Yu et C. L. Li;齿裂西山委陵菜;Toothcleft Sishan Cinquefoil, Toothcleft Xishan Cinquefoil ■

313002　Potentilla smithiana Hand. -Mazz. ;齿萼委陵菜;Smith Cinquefoil ■

313003　Potentilla sommieri Siegfr. et Keller;索姆委陵菜■☆

313004　Potentilla songarica Bunge var. chinensis Bunge = Potentilla sischanensis Bunge ex Lehm. ■

313005　Potentilla soongarica Bunge;准噶尔委陵菜■

313006　Potentilla soongarica Bunge var. chinensis Bunge = Potentilla sischanensis Bunge ex Lehm. ■

313007　Potentilla soraida Zimm. ;污泥委陵菜■☆

313008　Potentilla sordida Klotzsch = Potentilla polyphylla Wall. et Lehm. ■

313009　Potentilla sphenophylla Th. Wolf;欧洲楔叶委陵菜■☆

313010　Potentilla splendens Buch. -Ham. ex Trevir. = Potentilla lineata Trevir. ■

313011　Potentilla splendens DC. = Potentilla montana Brot. ■☆

313012　Potentilla splendens Wall. ex D. Don = Potentilla fulgens Wall. ex Hook. ■

313013　Potentilla sprengeliana Lehm. ;施普伦委陵菜(日本委陵菜)■☆

313014　Potentilla sprengeliana Lehm. = Potentilla fragarioides L. ■

313015　Potentilla sprengeliana Lehm. f. pleniflora (Sugim.) Naruh. ;重瓣日本委陵菜■☆

313016　Potentilla stenophylla (Franch.) Diels；狭叶委陵菜；Narrowleaf Cinquefoil ■

313017　Potentilla stenophylla (Franch.) Diels var. compacta J. Krause = Potentilla stenophylla (Franch.) Diels var. emergens Cardot ■

313018　Potentilla stenophylla (Franch.) Diels var. compacta J. Krause = Potentilla tatsienluensis Th. Wolf ■

313019　Potentilla stenophylla (Franch.) Diels var. cristata (H. R. Fletcher) H. Ikeda et H. Ohba;贡山狭叶委陵菜■

313020　Potentilla stenophylla (Franch.) Diels var. emergens Cardot = Potentilla tatsienluensis Th. Wolf ■

313021　Potentilla stenophylla (Franch.) Diels var. exahata Cardot = Potentilla tatsienluensis Th. Wolf ■

313022　Potentilla stenophylla (Franch.) Diels var. exaltata Cardot = Potentilla stenophylla (Franch.) Diels var. emergens Cardot ■

313023　Potentilla stenophylla (Franch.) Diels var. millefolia Soják = Potentilla stenophylla (Franch.) Diels ■

313024　Potentilla stenophylla (Franch.) Diels var. taliensis (W. W. Sm.) H. Ikeda et H. Ohba;大理委陵菜;Dali Cinquefoil ■

313025　Potentilla sterilis (L.) Garcke;不育委陵菜；Barren Strawberry, Craisey, Lazy-bones, Story-tellers, Strawberry Plant, Strawberry-leaved Potentilla ■☆

313026　Potentilla sterilis Richt. = Potentilla sterilis (L.) Garcke ■☆

313027　Potentilla stipularia L. ;大托叶委陵菜■☆

313028　Potentilla stolonifera Lehm. ex Ledeb. ;匍匐委陵菜■☆

313029　Potentilla strigosa Pall. ex Pursh;茸毛委陵菜(刺毛委陵菜,灰白委陵菜);Hairy Cinquefoil ■

313030　Potentilla strigosa Pall. ex Pursh var. conferta Kitag. = Potentilla tanacetifolia Willd. ex Schltdl. ■

313031　Potentilla strigosa Pall. var. conferta Kitag. = Potentilla tanacetifolia Willd. ex Schltdl. ■

313032　Potentilla stromatodes Melch. = Potentilla coriandrifolia D. Don var. dumosa Franch. ■

313033　Potentilla subacaulis L. = Potentilla acaulis L. ■

313034　Potentilla subdigitata Te T. Yu et C. L. Li;混叶委陵菜；Subdigitate Cinquefoil ■

313035　Potentilla subpalmata Ledeb. ;亚掌委陵菜■☆

313036　Potentilla sulphurea Lam. ;硫黄委陵菜■☆

313037　Potentilla sulphurea Lam. = Potentilla recta L. ■

313038　Potentilla sundaica (Blume) Kuntze var. robusta (Franch. et Sav.) Kitag. = Potentilla anemonifolia Lehm. ■

313039　Potentilla supina L. ;朝天委陵菜(背铺委陵菜,伏委陵菜,鸡毛菜,铺地委陵菜,仰卧委陵菜);Bushy Cinquefoil, Carpet Cinquefoil ■

313040　Potentilla supina L. subsp. paradoxa (Nutt. ex Torr. et A. Gray) Soják = Potentilla supina L. ■

313041　Potentilla supina L. var. campestris Cardot = Potentilla supina L. var. ternata Peterm. ■

313042　Potentilla supina L. var. egibbosa Th. Wolf = Potentilla supina L. ■

313043　Potentilla supina L. var. egibbosa Th. Wolf f. ternata (Peterm.) Wolf = Potentilla supina L. var. ternata Peterm. ■

313044　Potentilla supina L. var. nicollettii (E. Sheld.) Maire = Potentilla supina L. ■

313045　Potentilla supina L. var. paradoxa (Nutt. ex Torr. et A. Gray) Th. Wolf = Potentilla supina L. ■

313046　Potentilla supina L. var. paradoxa (Nutt.) Th. Wolf = Potentilla supina L. ■

313047　Potentilla supina L. var. ternata Peterm. ;三叶朝天委陵菜(三裂朝天委陵菜,三叶委陵菜,小瓣委陵菜);Ternate Carpet Cinquefoil,Threeleaf Cinquefoil ■

313048　Potentilla supina L. var. ternata Peterm. = Potentilla heynii Roth ■

313049　Potentilla sutchuenica Cardot = Potentilla freyniana Bornm. ■

313050　Potentilla svanetica Siegfr. et Keller;斯万涅特委陵菜■☆

313051　Potentilla szovitsii Th. Wolf;绍氏委陵菜■☆

313052　Potentilla tabernemontani Asch. = Potentilla neumanniana Rchb. ■☆

313053　Potentilla taliensis W. W. Sm. = Potentilla stenophylla (Franch.) Diels var. taliensis (W. W. Sm.) H. Ikeda et H. Ohba ■

313054　Potentilla tanacetifolia Willd. ex Schltdl. ;菊叶委陵菜(叉菊委陵菜,蒿叶委陵菜,砂地委陵菜);Taisyleaf Cinquefoil ■

313055　Potentilla tanacetifolia Willd. ex Schltdl. f. decumbens Krylov = Potentilla tanacetifolia Willd. ex Schltdl. ■

313056　Potentilla tanacetifolia Willd. ex Schltdl. f. erecta Krylov = Potentilla tanacetifolia Willd. ex Schltdl. ■

313057　Potentilla tanacetifolia Willd. ex Schltdl. var. decumbens (Krylov) Th. Wolf = Potentilla tanacetifolia Willd. ex Schltdl. ■

313058　Potentilla tanacetifolia Willd. ex Schltdl. var. erecta (Krylov) Th. Wolf = Potentilla tanacetifolia Willd. ex Schltdl. ■

313059　Potentilla tapetodes Soják = Potentilla microphylla D. Don var. tapetodes (Soják) H. Ikeda et H. Ohba ■

313060　Potentilla tapetodes Soják var. decidua Soják = Potentilla microphylla D. Don ■

313061　Potentilla taronensis Y. C. Wu ex Te T. Yu et C. L. Li;大果委陵菜;Bigfruit Cinquefoil ■

313062　Potentilla tatsienluensis Th. Wolf;康定委陵菜; Kangding Cinquefoil ■

313063　Potentilla tatsienluensis Th. Wolf = Potentilla stenophylla (Franch.) Diels var. emergens Cardot ■

313064　Potentilla taurica Willd. ;克里木委陵菜■☆

313065　Potentilla tenuifolia Willd. ex Schltdl. ;细叶金老梅(小叶金露梅)●☆

313066　Potentilla tenuirugis Pomel = Potentilla recta L. ■

313067　Potentilla tephroleuca Th. Wolf;灰白委陵菜■☆

313068　Potentilla ternata K. Koch = Potentilla freyniana Bornm. ■

313069　Potentilla tetrandra (Bunge) Hook. f. = Sibbaldia tetrandra Bunge ■

313070　Potentilla tetrandra Bunge = Dryadanthe tetrandra (Bunge) Juz. ■

313071　Potentilla tetrandra Bunge = Sibbaldia tetrandra Bunge ■

313072　Potentilla thibetica Cardot = Potentilla saundersiana Royle ■

313073　Potentilla thurberi A. Gray ex Lehm. ;瑟伯委陵菜■☆

313074　Potentilla thyrsiflora Kern. ;伞锥花委陵菜■☆

313075　Potentilla tianschanica Th. Wolf;天山委陵菜■☆

313076　Potentilla togasii Ohwi;唐氏委陵菜■☆

313077　Potentilla tollii Trautv. ;托尔委陵菜■☆

313078　Potentilla tonguei W. Baxter;红心委陵菜■☆

313079　Potentilla tormentilla Neck. = Potentilla erecta (L.) Hampe ■☆

313080　Potentilla toyamensis Naruh. et T. Sato;广岛委陵菜■☆

313081　Potentilla transcaspia Th. Wolf;东里海委陵菜■☆

313082　Potentilla tranzschelii Juz. = Potentilla ancistrifolia Bunge ■

313083　Potentilla tridentata Aiton ＝ Sibbaldiopsis tridentata（Aiton） Rydb. ■☆

313084　Potentilla tridentata Aiton f. hirsutifolia Pease ＝ Sibbaldiopsis tridentata（Aiton） Rydb. ■☆

313085　Potentilla tridentata Sol. ;三齿委陵菜;Three-toothed Cinquefoil ■☆

313086　Potentilla tugitakensis Masam. ;台湾委陵菜（雪山翻白草）; Taiwan Cinquefoil ■

313087　Potentilla turfosa Hand. -Mazz. ;簇生委陵菜（白头翁,翻背白草,泥沼人参果,涩疙瘩）;Bog Cinquefoil ■

313088　Potentilla turfosa Hand. -Mazz. var. caudiculata Soják ＝ Potentilla turfosa Hand. -Mazz. var. gracilescens（Soják） H. Ikeda et H. Ohba ■

313089　Potentilla turfosa Hand. -Mazz. var. gracilescens（Soják） H. Ikeda et H. Ohba ;纤细簇生委陵菜 ■

313090　Potentilla umbrosa Steven;喜阴委陵菜■☆

313091　Potentilla uniflora Ledeb. ;单花委陵菜;Uniflower Cinquefoil ■☆

313092　Potentilla ussuriensis Juz. ＝ Potentilla rugolosa Kitag. ■

313093　Potentilla vahliana Lehm. ;瓦氏委陵菜■☆

313094　Potentilla veitchii E. H. Wilson ＝ Potentilla glabra Lodd. var. veitchii（E. H. Wilson） Hand. -Mazz. ●

313095　Potentilla veitchii Hand. -Mazz. ＝ Potentilla glabra Lodd. ●

313096　Potentilla verna L. ＝ Potentilla neumanniana Rchb. ■☆

313097　Potentilla verrucosum G. Don var. teneriffae（Bornm.） Pit. ＝ Sanguisorba verrucosa（Don） Ces. ■☆

313098　Potentilla verticillaris Stephan ex Willd. ;轮叶委陵菜;Whored Cinquefoil,Whorlleaf Cinquefoil ■

313099　Potentilla verticillaris Stephan ex Willd. var. acutipetala Lehm. ＝ Potentilla verticillaris Stephan ex Willd. ■

313100　Potentilla verticillaris Stephan ex Willd. var. condensata Th. Wolf ＝ Potentilla verticillaris Stephan ex Willd. ■

313101　Potentilla verticillaris Stephan ex Willd. var. latisecta Liou et C. Y. Li;宽轮叶委陵菜;Broadtooth Whored Cinquefoil ■

313102　Potentilla verticillaris Stephan ex Willd. var. pedatisecta Liou et C. Y. Li;爪轮叶委陵菜;Petasisecte Whored Cinquefoil ■

313103　Potentilla villosa Pall. ;长绒毛委陵菜■☆

313104　Potentilla virgata Lehm. ;密枝委陵菜;Twiggy Cinquefoil ■

313105　Potentilla virgata Lehm. var. pinnatifida（Lehm.） Te T. Yu et C. L. Li;羽裂密枝委陵菜;Pinnate Twiggy Cinquefoil ■

313106　Potentilla viscosa Donn ex Lehm. ＝ Potentilla longifolia Willd. ex Schltdl. ■

313107　Potentilla viscosa Donn ex Lehm. var. macrophylla Kom. ＝ Potentilla longifolia Willd. ex Schltdl. ■

313108　Potentilla vittata Soják ＝ Potentilla peduncularis D. Don var. vittata（Soják） H. Ikeda et H. Ohba ■

313109　Potentilla vittata Soják var. abbreviata Soják ＝ Potentilla peduncularis D. Don var. vittata（Soják） H. Ikeda et H. Ohba ■

313110　Potentilla vittata Soják var. assidens Soják ＝ Potentilla peduncularis D. Don var. vittata（Soják） H. Ikeda et H. Ohba ■

313111　Potentilla wallichiana Ser. ＝ Duchesnea chrysantha（Zoll. et Moritzi） Miq. ■

313112　Potentilla wenchuensis H. Ikeda et H. Ohba;汶川委陵菜■

313113　Potentilla wibeliana Th. Wolf;韦比氏委陵菜;Wiebel Cinquefoil ■☆

313114　Potentilla xizangensis Te T. Yu et C. L. Li;西藏委陵菜;Xizang Cinquefoil ■

313115　Potentilla yamanakae（Naruh.） Naruh. ;宇和委陵菜■☆

313116　Potentilla yokusaiana Makino ＝ Potentilla rosulifera H. Lév. ■

313117　Potentilla yukonensis Hultén ＝ Argentina anserina（L.） Rydb. ■

313118　Potentilla yukonensis Hultén ＝ Potentilla anserina L. ■

313119　Potentilla zhangbeiensis Yo. Zhang et Z. T. Yin;张北委陵菜; Zhangbei Cinquefoil ■

313120　Potentillaceae Bercht. et J. Presl ＝ Rosaceae Juss. （保留科名）●■

313121　Potentillaceae Bercht. et J. Presl;委陵菜科●■

313122　Potentillaceae Perleb ＝ Rosaceae Juss. （保留科名）●■

313123　Potentillopsis Opiz ＝ Potentilla L. ■●

313124　Poteranthera Bong. （1838）;杯药野牡丹属■☆

313125　Poteranthera pusilla Bong. ;杯药野牡丹■☆

313126　Poteriaceae Raf. ＝ Rosaceae Juss. （保留科名）●■

313127　Poteridium Spach ＝ Sanguisorba L. ■

313128　Poterion St. -Lag. ＝ Poterium L. ■☆

313129　Poterium L. （1753）;肖地榆属■☆

313130　Poterium L. ＝ Sanguisorba L. ■

313131　Poterium alveolosum Spach ＝ Sanguisorba minor Scop. subsp. alveolosa（Spach） Maire ■☆

313132　Poterium anceps Ball ＝ Sanguisorba minor Scop. subsp. maroccana（Coss.） Maire ■☆

313133　Poterium ancistroides Desf. ＝ Sanguisorba ancistroides（Desf.） A. Br. ■☆

313134　Poterium ancistroides Desf. var. parviflorum Pomel ＝ Sanguisorba ancistroides（Desf.） A. Br. ■☆

313135　Poterium crispum Pomel ＝ Sanguisorba minor Scop. subsp. vestita（Pomel） Maire ■☆

313136　Poterium diandrum Hook. f. ＝ Sanguisorba diandra（Hook. f.） Nordborg ■

313137　Poterium duriaei Spach ＝Sanguisorba verrucosa（Don） Ces. ■☆

313138　Poterium filiforme Hook. f. ＝ Sanguisorba filiformis（Hook. f.） Hand. -Mazz. ■

313139　Poterium fontanesii Spach ＝ Sanguisorba mauritanica Desf. ■☆

313140　Poterium lasiocarpum Boiss. et Hausskn. ex Boiss. ;毛果肖地榆■☆

313141　Poterium longifolium（Bertol.） Hook. f. ＝ Sanguisorba officinalis L. var. longifolia（Bertol.） Te T. Yu et C. L. Li ■

313142　Poterium magnolii Spach ＝Sanguisorba verrucosa（Don） Ces. ■☆

313143　Poterium magnolii Spach var. verrucosum（Spach） Batt. ＝ Sanguisorba verrucosa（Don） Ces. ■☆

313144　Poterium maroccanum Batt. ＝ Sanguisorba minor Scop. subsp. maroccana（Coss.） Maire ■☆

313145　Poterium mauritanicum Pomel ＝ Sanguisorba mauritanica Desf. ■☆

313146　Poterium multicaule Boiss. et Reut. ＝ Sanguisorba minor Scop. subsp. balearica（Nyman） Munoz Garm. et C. Navarro ■☆

313147　Poterium muricatum Spach ＝ Sanguisorba minor Scop. subsp. muricata（Spach） Briq. ■☆

313148　Poterium officinale（L.） A. Gray ＝ Sanguisorba officinalis L. ■

313149　Poterium officinale A. Gray ＝ Sanguisorba officinalis L. ■

313150　Poterium polygamum Waldst. et Kit. ＝ Sanguisorba minor Scop. subsp. muricata（Spach） Briq. ■☆

313151　Poterium polyganum Lej. ＝ Poterium sanguisorba L. ■

313152　Poterium polygamum Waldst. et Kit. ＝Sanguisorba officinalis L. ■

313153　Poterium rhodopeum Velen. ＝ Sanguisorba minor Scop. subsp. muricata（Spach） Briq. ■☆

313154　Poterium rupicola Boiss. et Reut. ＝ Sanguisorba rupicola （Boiss. et Reut.） A. Braun et Bouché ■☆

313155　Poterium sanguisorba L. ＝ Sanguisorba minor Scop. ■☆

313156　Poterium sanguisorba L. ＝ Sanguisorba officinalis L. ■

313157　Poterium sanguisorba L. var. verrucosa（G. Don） Ehrenb. ＝ Sanguisorba verrucosa（Don） Ces. ■☆

313158　Poterium sanguisorba L. var. virescens Spach ＝ Sanguisorba minor Scop. ■☆

313159　Poterium sitchense（C. A. Mey.）S. Watson ＝ Sanguisorba stipulata Raf. ■

313160　Poterium spinosum L.；具刺肖地榆；Thorny Burnet ■☆

313161　Poterium spinosum L. ＝ Sarcopoterium spinosum（L.）Spach ●☆

313162　Poterium tenuifolium（Fisch. ex Link）Franch. et Sav. ＝ Sanguisorba tenuifolia Fisch. ex Link ■

313163　Poterium tenuifolium Franch. et Sav. ＝ Sanguisorba tenuifolia Fisch. ex Link ■

313164　Poterium verrucosum G. Don ＝ Sanguisorba verrucosa（Don）Ces. ■☆

313165　Poterium verrucosum G. Don var. magnolii ? ＝ Sanguisorba verrucosa（Don）Ces. ■☆

313166　Poterium vestitum Pomel ＝ Sanguisorba minor Scop. subsp. vestita（Pomel）Maire ■☆

313167　Pothaceae Raf. ＝ Araceae Juss.（保留科名）■●

313168　Pothoaceae Raf. ＝ Araceae Juss.（保留科名）■●

313169　Pothoidium Schott（1856-1857）；假石柑属（假柚叶藤属）；False Pothos，Pothoidium ●■

313170　Pothoidium lobbianum Schott；假石柑（假柚叶藤）；False Pothos，Lobbiana False Pothos，Lobbiana Pothoidium ●■

313171　Pothomorpha Willis ＝ Pothomorphe Miq. ●

313172　Pothomorphe Miq.（1840）；大胡椒属；Pothomorphe，Vinepepper ●

313173　Pothomorphe Miq. ＝ Lepianthes Raf. ●■

313174　Pothomorphe Miq. ＝ Piper L. ●■

313175　Pothomorphe angustifolia Ruiz. et Pav.；狭叶大胡椒●☆

313176　Pothomorphe hirsuta Sw.；硬毛大胡椒●☆

313177　Pothomorphe peltata（L.）Miq.；盾状大胡椒●☆

313178　Pothomorphe subpeltata（Willd.）Miq.；大胡椒（大圆叶胡椒）；Peltatelike Pothomorphe，Peltatelike Vinepepper，Subpeltate-leaf Pepper ●

313179　Pothomorphe subpeltata（Willd.）Miq. ＝ Piper subpeltatum Willd. ●

313180　Pothomorphe subpeltata（Willd.）Miq. ＝ Piper umbellatum L. ●

313181　Pothomorphe tabogana C. DC.；塔博大胡椒●☆

313182　Pothomorphe umbellata（L.）Miq. ＝ Piper umbellatum L. ●

313183　Pothomorphe umbellata（L.）Miq. ＝ Pothomorphe subpeltata（Willd.）Miq. ●

313184　Pothomorphe umbellata L. f. glabra（C. DC.）Steyerm. ＝ Piper umbellatum L. ●

313185　Pothos Adans. ＝ Polianthes L. ■

313186　Pothos L.（1753）；石柑属（石柑子属，柚叶藤属）；Pothos ●■

313187　Pothos L. ＝ Epipremnum Schott ●■

313188　Pothos acaulis Hook. ＝ Anthurium hookeri Kunth ■☆

313189　Pothos angustifolius C. Presl；窄叶石柑（藤橘）；Narrowleaf Pothos ■☆

313190　Pothos aureus Linden et André ＝ Epipremnum pinnatum（L.）Engl. 'Aureum' ●■

313191　Pothos aureus Linden et André ＝ Epipremnum pinnatum（L.）Engl. ●■

313192　Pothos aureus Lindl. et André ＝ Epipremnum aureum（Linden ex André）Bunting ●■

313193　Pothos aureus Lindl. et André ＝ Scindapsus aureus（Lindl. ex André）Engl. et Krause ●■

313194　Pothos balansae Engl.；龙州石柑；Longzhou Pothos ■

313195　Pothos cathcarti Schott；紫苞石柑（石柑子）；Catheart Pothos，Purplebract Pothos ●■

313196　Pothos chapelieri Schott ＝ Pothos scandens L. ●

313197　Pothos chinensis（Raf.）Merr.；石柑子（巴岩姜，巴岩香，百步藤，大疮花，毒蛇上树，风瘫药，柑子菌芋，关刀草，葫芦草，六扑风，落山葫芦，马莲鞍，猛药，爬山虎，千年青，青蒲芦茶，青竹标，上树葫芦，伸筋草，石百足，石柑儿，石葫芦，石蒲藤，石气柑，石上蟾蜍草，石上葫芦茶，石柚，藤橘，铁斑鸠，铁板草，小毛铜钱菜，岩石焦，柚叶藤，竹结草）；China Pothos，Chinese Pothos ■

313198　Pothos chinensis（Raf.）Merr. var. lotinensis C. Y. Wu et H. Li；长柄石柑■

313199　Pothos decursiva Roxb. ＝ Rhaphidophora decursiva（Roxb.）Schott ●■

313200　Pothos kerrii Buchet ex Gagnep.；长梗石柑；Kerr Pothos ■

313201　Pothos loureiri Hook. et Arn. ＝ Pothos repens（Lour.）Druce ■

313202　Pothos malaianus Miq. ＝ Anadendrum montanum（Blume）Schott ●

313203　Pothos peepla Roxb. ＝ Rhaphidophora peepla（Roxb.）Schott ●■

313204　Pothos pilulifer Buchet；地柑（葫芦藤，石蚌寄生）；Globulebearing Pothos ■

313205　Pothos pinnatus L. ＝ Epipremnum pinnatum（L.）Engl. ●■

313206　Pothos repens（Lour.）Druce；百足藤（巴岩姜，百足草，飞天蜈蚣，姜藤，路氏石柑子，落地蜈蚣，神仙对坐草，石上蜈蚣，石蜈蚣，铁斑鸠，蜈蚣草，蜈蚣藤，细蜈蚣草，细叶石柑，细叶藤柑，细叶藤橘，下山蜈蚣，鸭子草）；Creeping Pothos，Loureiro Pothos ■

313207　Pothos scandens L.；螳螂跌打（石柑子，硬骨散）；Climbing Pothos ●

313208　Pothos scandens sensu Lindl. ＝ Pothos chinensis（Raf.）Merr. ■

313209　Pothos seemannii Schott ＝ Pothos chinensis（Raf.）Merr. ■

313210　Pothos warburgii Engl.；台湾石柑；Warburg Pothos ■

313211　Pothos yunnanensis Engl. ＝ Pothos chinensis（Raf.）Merr. ■

313212　Pothuava Gaudich. ＝ Aechmea Ruiz et Pav.（保留属名）■☆

313213　Pothuava Gaudich. ex K. Koch ＝ Hoiriri Adans.（废弃属名）■☆

313214　Potima R. Hedw. ＝ Faramea Aubl. ●☆

313215　Potosia（Schltr.）R. González et Szlach. ex Mytnik ＝ Pelexia Poit. ex Lindl.（保留属名）■

313216　Potosia（Schltr.）R. González et Szlach. ex Mytnik（2003）；美洲肥根兰属■☆

313217　Potoxylon Kosterm.（1978）；婆罗秀樟属（婆罗秀樾属）●☆

313218　Potoxylon melangangai（Symington）Kosterm.；婆罗秀樟●☆

313219　Pottingeria Prain（1898）；托叶假樟属（单室木属）●

313220　Pottingeria acuminata Prain；托叶假樟●

313221　Pottingeria acuminata Prain var. latifolia Airy Shaw；宽叶托叶假樟●☆

313222　Pottingeriaceae Takht.（1987）；托叶假樟科（单室木科）●

313223　Pottingeriaceae Takht. ＝ Celastraceae R. Br.（保留科名）●

313224　Pottsia Hook. et Arn.（1837）；帘子藤属（蒲楲藤属）；Pottsia ●

313225　Pottsia cantonensis Hook. et Arn. ＝ Pottsia laxiflora（Blume）Kuntze ●

313226　Pottsia cantonensis Hook. et Beech ＝ Pottsia laxiflora（Blume）Kuntze ●

313227　Pottsia grandiflora Markgr.；大花帘子藤（乳汁藤）；Bigflowered Pottsia，Largeflower Pottsia ●

313228　Pottsia hookeriana Wight ＝ Pottsia laxiflora（Blume）Kuntze ●

313229　Pottsia laxiflora（Blume）Kuntze；帘子藤（笔须藤，长果胶藤，钩婆藤，红杜仲藤，厚皮藤，花拐藤，黄心泥藤，火烧角，蚂蝗藤，能藤，坭藤母，乳汁藤，山羊角，亚八藤，腰骨藤）；Laxflower

Pottsia，Laxiflowered Pottsia ●

313230　Pottsia laxiflora（Blume）Kuntze var. pubescens（Tsiang）P. T. Li ＝ Pottsia laxiflora（Blume）Kuntze ●

313231　Pottsia laxiflora var. pubescens（Tsiang）P. T. Li ＝ Pottsia laxiflora（Blume）Kuntze ●

313232　Pottsia ovata A. DC. ＝ Pottsia laxiflora（Blume）Kuntze ●

313233　Pottsia ovata DC. ＝ Pottsia laxiflora（Blume）Kuntze ●

313234　Pottsia pubescens Lindl.；毛帘子藤；Hairy Pottsia ●

313235　Pottsia pubescens Lindl. ＝ Pottsia laxiflora（Blume）Kuntze ●

313236　Pottsia pubescens Tsiang ＝ Pottsia laxiflora（Blume）Kuntze ●

313237　Pouchetia A. Rich. ＝ Pouchetia A. Rich. ex DC. ●☆

313238　Pouchetia A. Rich. ex DC.（1830）；普谢茜属●☆

313239　Pouchetia africana A. Rich. ex DC.；非洲普谢茜●☆

313240　Pouchetia africana A. Rich. ex DC. var. aequatorialis N. Hallé；赤道普谢茜●☆

313241　Pouchetia africana A. Rich. ex DC. var. cuneata Hiern ＝ Pouchetia baumanniana Büttner ●☆

313242　Pouchetia baumanniana Büttner；鲍曼普谢茜●☆

313243　Pouchetia confertiflora Mildbr.；密花普谢茜●☆

313244　Pouchetia gilletii De Wild. ＝ Pouchetia baumanniana Büttner ●☆

313245　Pouchetia parviflora Benth.；小花普谢茜●☆

313246　Pouchetia saxifraga Hochst. ＝ Galiniera saxifraga（Hochst.）Bridson ●☆

313247　Poulsenia Eggers（1898）；刺枝桑属●☆

313248　Poulsenia aculeata Eggers；刺枝桑●☆

313249　Pounguia Benoist ＝ Whitfieldia Hook. ■☆

313250　Pounguia purpurata Benoist ＝ Whitfieldia purpurata（Benoist）Heine ■☆

313251　Poupartia Comm. ex Juss.（1789）；波旁漆属（波岛漆属）●☆

313252　Poupartia amazonica Ducke；亚马逊波旁漆●☆

313253　Poupartia axillaris（Roxb.）King et Prain ＝ Choerospondias axillaris（Roxb.）B. L. Burtt et A. W. Hill ●

313254　Poupartia axillaris King et Prain ＝ Choerospondias axillaris（Roxb.）B. L. Burtt et A. W. Hill ●

313255　Poupartia caffra（Sond.）H. Perrier ＝ Sclerocarya birrea（A. Rich.）Hochst. subsp. caffra（Sond.）Kokwaro ●☆

313256　Poupartia chinensis Merr. ＝ Spondias lakonensis Pierre ●

313257　Poupartia fordii Hemsl. ＝ Choerospondias axillaris（Roxb.）B. L. Burtt et A. W. Hill ●

313258　Poupartia fordii Hemsl. var. japonica ？ ＝ Choerospondias axillaris（Roxb.）B. L. Burtt et A. W. Hill ●

313259　Poupartia pinnata（L. f.）Blanco ＝ Spondias pinnata（L. f.）Kurz ●

313260　Poupartiopsis Capuron ex J. D. Mitch. et D. C. Daly（2006）；拟波旁漆属●☆

313261　Poupartiopsis spondiocarpa Capuron ex J. D. Mitch. et D. C. Daly；拟波旁漆●☆

313262　Pourouma Aubl.（1775）；亚马逊葡萄属；Amaxon Grape ●☆

313263　Pourouma cecropiifolia Mart.；亚马逊葡萄；Uvilla ●☆

313264　Pourretia Ruiz et Pav. ＝ Dupuya J. H. Kirkbr. ●☆

313265　Pourretia Willd. ＝ Cavanillesia Ruiz et Pav. ●☆

313266　Pourthiaea Decne.（1874）；鸡丁子属●

313267　Pourthiaea Decne. ＝ Aronia Medik.（保留属名）●☆

313268　Pourthiaea Decne. ＝ Photinia Lindl. ●

313269　Pourthiaea arguta Decne. ＝ Photinia arguta Lindl. ●

313270　Pourthiaea arguta Decne. var. hookeri（Decne.）Hook. f. ＝ Photinia arguta Lindl. var. hookeri（Decne.）Vidal ●

313271　Pourthiaea arguta Decne. var. hookeri Hook. f. ＝ Photinia arguta Lindl. var. hookeri（Decne.）Vidal ●

313272　Pourthiaea arguta Decne. var. wallichii Hook. f. ＝ Photinia arguta Lindl. ●

313273　Pourthiaea beauverdiana（C. K. Schneid.）Hatus. ＝ Photinia beauverdiana C. K. Schneid. ●

313274　Pourthiaea beauverdiana（C. K. Schneid.）Hatus. var. notabilis（C. K. Schneid.）Hatus. ＝ Photinia beauverdiana C. K. Schneid. var. notabilis（C. K. Schneid.）Rehder et E. H. Wilson ●

313275　Pourthiaea beauverdiana（C. K. Schneid.）Hatus. var. notabilis（C. K. Schneid.）Hatus. ＝ Photinia beauverdiana C. K. Schneid. ●

313276　Pourthiaea beauverdiana（C. K. Schneid.）Hatus. var. notabilis（Rehder et E. H. Wilson）Hatus. ＝ Photinia beauverdiana C. K. Schneid. var. notabilis（C. K. Schneid.）Rehder et E. H. Wilson ●

313277　Pourthiaea beauverdiana（C. K. Schneid.）Hatus. var. notabilis（Rehder et E. H. Wilson）Hatus.；台湾老叶儿树●

313278　Pourthiaea beauverdiana（C. K. Schneid.）Migo ＝ Photinia beauverdiana C. K. Schneid. ●

313279　Pourthiaea benthamiana（Hance）Nakai ＝ Photinia benthamiana Hance ●

313280　Pourthiaea benthamina Nakai var. obovata（H. L. Li）Iketani et H. Ohashi ＝ Photinia benthamiana Hance var. obovata H. L. Li ●

313281　Pourthiaea benthamina Nakai var. salicifolia（Cardot）Iketani et H. Ohashi ＝ Photinia benthamiana Hance var. salicifolia Cardot ●

313282　Pourthiaea bergerae（C. K. Schneid.）Iketani et H. Ohashi ＝ Photinia bergerae C. K. Schneid. ●

313283　Pourthiaea blinii（H. Lév.）Iketani et H. Ohashi ＝ Photinia blinii（H. Lév.）Rehder ●

313284　Pourthiaea calleryana Decne. ＝ Photinia benthamiana Hance ●

313285　Pourthiaea calleryana Decne. ＝ Photinia calleryana（Decne.）Cardot ●

313286　Pourthiaea callosa（Chun ex K. C. Kuan）Iketani et H. Ohashi ＝ Photinia callosa Chun ex K. C. Kuan ●

313287　Pourthiaea chingshuiensis T. Shimizu ＝ Photinia chingshuiensis（T. Shimizu）Tang S. Liu et H. J. Su ●

313288　Pourthiaea chingshuiensis T. Shimizu ＝ Photinia parvifolia（Pritz.）C. K. Schneid. var. kankoensis（Hatus.）Te T. Yu et T. C. Kuan ●

313289　Pourthiaea fokienensis（Franch.）Iketani et H. Ohashi ＝ Photinia fokienensis（Franch.）Franch. ex Cardot ●

313290　Pourthiaea formosana（Hance）Koidz. ＝ Photinia lucida（Decne.）C. K. Schneid. ●

313291　Pourthiaea hirsuta（Hand.-Mazz.）Iketani et H. Ohashi ＝ Photinia hirsuta Hand.-Mazz. ●

313292　Pourthiaea hirsuta（Hand.-Mazz.）Iketani et H. Ohashi var. lobulata（Te T. Yu）Iketani et H. Ohashi ＝ Photinia hirsuta Hand.-Mazz. var. lobalata Te T. Yu ●

313293　Pourthiaea hookeri Decne. ＝ Photinia arguta Lindl. var. hookeri（Decne.）Vidal ●

313294　Pourthiaea impressivena（Hayata）Iketani et H. Ohashi ＝ Photinia impressivena Hayata ●

313295　Pourthiaea impressivena（Hayata）Iketani et H. Ohashi var. urceolocarpa（Vidal）Iketani et H. Ohashi ＝ Photinia impressivena Hayata var. urceolocarpa（Vidal）Vidal ●

313296　Pourthiaea kankaoensis Hatus. ＝ Photinia parvifolia（Pritz.）C. K. Schneid. ●

313297　Pourthiaea kankaoensis Hatus. ＝ Photinia parvifolia（Pritz.）C.

K. Schneid. var. kankoensis（Hatus.）Te T. Yu et T. C. Kuan ●

313298 Pourthiaea kankaoensis Hatus. = Photinia villosa（Thunb.）DC. var. sinica Rehder et E. H. Wilson ●

313299 Pourthiaea laevis（Thunb.）Koidz. var. parvifolia（E. Pritz.）Migo = Photinia parvifolia（Pritz.）C. K. Schneid. ●

313300 Pourthiaea lucida Decne. = Photinia lucida（Decne.）C. K. Schneid. ●

313301 Pourthiaea obliqua（Stapf）Iketani et H. Ohashi = Photinia obliqua Stapf ●

313302 Pourthiaea parviflora（Cardot）Iketani et H. Ohashi = Photinia parviflora Cardot ●

313303 Pourthiaea parviflora（Cardot）Iketani et H. Ohashi = Photinia schneideriana Rehder et E. H. Wilson var. parviflora（Cardot）L. T. Lu et C. L. Li ●

313304 Pourthiaea parvifolia E. Pritz. = Photinia parvifolia（Pritz.）C. K. Schneid. ●

313305 Pourthiaea parvifolia E. Pritz. ex Diels = Photinia parvifolia（Pritz.）C. K. Schneid. ●

313306 Pourthiaea pilosicalyx（Te T. Yu）Iketani et H. Ohashi = Photinia pilosicalyx Te T. Yu ●

313307 Pourthiaea podocarpifolia（Te T. Yu）Iketani et H. Ohashi = Photinia podocarpifolia Te T. Yu ●

313308 Pourthiaea salicifolia Decne. = Photinia arguta Lindl. var. salicifolia（Decne.）Vidal ●

313309 Pourthiaea schneideriana（Rehder et E. H. Wilson）Iketani et H. Ohashi = Photinia schneideriana Rehder et E. H. Wilson ●

313310 Pourthiaea tsaii（Rehder）Iketani et H. Ohashi = Photinia tsaii Rehder ●

313311 Pourthiaea variabilis Palib. = Photinia variabilis Hemsl. ex Forbes et Hemsl. ●

313312 Pourthiaea variabilis Palib. = Photinia villosa（Thunb.）DC. ●

313313 Pourthiaea villosa（Thunb. ex A. Murray）Decne. var. chingshuiensis（T. Shimizu）Iketani et H. Ohashi = Photinia chingshuiensis（T. Shimizu）Tang S. Liu et H. J. Su ●

313314 Pourthiaea villosa（Thunb. ex A. Murray）Decne. var. parvifolia（Pritz.）Iketani et H. Ohashi = Photinia parvifolia（Pritz.）C. K. Schneid. ●

313315 Pourthiaea villosa（Thunb.）Decne. = Photinia villosa（Thunb.）DC. ●

313316 Pourthiaea villosa（Thunb.）Decne. f. aurantiaca T. Yamanaka；黄毛叶石楠●☆

313317 Pourthiaea villosa（Thunb.）Decne. var. laevis（Thunb.）Stapf = Pourthiaea villosa（Thunb.）Decne. ●

313318 Pourthiaea villosa（Thunb.）Decne. var. longipes Nakai；长梗毛叶石楠●☆

313319 Pourthiaea villosa（Thunb.）Decne. var. zollingeri（Decne.）Nakai = Pourthiaea villosa（Thunb.）Decne. ●

313320 Pourthiaea villosa Decne. = Photinia villosa（Thunb.）DC. ●

313321 Pourthiaea villosa Decne. var. sinica（Rehder et E. H. Wilson）Migo = Photinia villosa（Thunb.）DC. var. sinica Rehder et E. H. Wilson ●

313322 Pourthiaea villosa Decne. var. tenuipes（P. S. Hsu et L. C. Li）Iketani et H. Ohashi = Photinia villosa（Thunb.）DC. var. tenuipes P. S. Hsu et L. C. Li ●

313323 Pourthiaea zhejiangensis（P. L. Chiu）Iketani et H. Ohashi = Photinia zhejiangensis P. L. Chiu ●

313324 Poutaletsje Adans. = Hedyotis L.（保留属名）●■

313325 Pouteria Aubl.（1775）；桃榄属（山榄属）；Pouteria ●

313326 Pouteria adolfi-friedericii（Engl.）A. Meeuse；阿弗桃榄●☆

313327 Pouteria adolfi-friedericii（Engl.）A. Meeuse subsp. australis（J. H. Hemsl.）L. Gaut.；南方阿弗桃榄●☆

313328 Pouteria adolfi-friedericii（Engl.）A. Meeuse subsp. floccosa（J. H. Hemsl.）L. Gaut.；丛毛阿弗桃榄●☆

313329 Pouteria adolfi-friedericii（Engl.）A. Meeuse subsp. keniensis（R. E. Fr.）L. Gaut.；肯尼亚桃榄●☆

313330 Pouteria adolfi-friedericii（Engl.）A. Meeuse subsp. usambarensis（J. H. Hemsl.）L. Gaut.；乌桑巴拉桃榄●☆

313331 Pouteria afzelii（Engl.）Baehni = Synsepalum afzelii（Engl.）T. D. Penn. ●☆

313332 Pouteria akuedo Baehni = Synsepalum afzelii（Engl.）T. D. Penn. ●☆

313333 Pouteria alnifolia（Baker）Roberty；桤叶桃榄●☆

313334 Pouteria alnifolia（Baker）Roberty var. sacleuxii（Lecomte）L. Gaut.；萨克勒桃榄●☆

313335 Pouteria altissima（A. Chev.）Baehni；高大桃榄●☆

313336 Pouteria androyensis Capuron ex Aubrév. = Capurodendron androyense Aubrév. ●☆

313337 Pouteria ankaranensis Capuron ex Aubrév. = Capurodendron ankaranense Aubrév. ●☆

313338 Pouteria annamensis（Pierre ex Dubard）Baehni；桃榄（大核果树，敏果）；Annam Pouteria ●

313339 Pouteria annamensis（Pierre）Baehni = Pouteria annamensis（Pierre ex Dubard）Baehni ●

313340 Pouteria antongiliensis Capuron ex Aubrév. = Capurodendron antongiliense Aubrév. ●☆

313341 Pouteria antunesii（Engl.）Baehni = Englerophytum magalismontanum（Sond.）T. D. Penn. ●☆

313342 Pouteria apollonioides Capuron ex Aubrév. = Capurodendron apollonioides Aubrév. ●☆

313343 Pouteria aurata（Pierre ex Dubard）Baehni = Eberhardtia aurata（Pierre ex Dubard）Lecomte ●

313344 Pouteria australis（R. Br.）Baehni；黑桃榄；Black Apple，Brush Apple，Buttonwood，Wild Plum ●☆

313345 Pouteria aylmeri（M. B. Scott）Baehni = Neolemonniera clitandrifolia（A. Chev.）Heine ●☆

313346 Pouteria bequaertii（De Wild.）Baehni = Synsepalum msolo（Engl.）T. D. Penn. ●☆

313347 Pouteria boninensis（Nakai）Baehni = Planchonella boninensis（Nakai）Masam. et Yanagih. ●☆

313348 Pouteria brevipes（Baker）Baehni = Synsepalum brevipes（Baker）T. D. Penn. ●☆

313349 Pouteria buluensis（Greves）Baehni = Pachystela buluensis（Greves）Aubrév. et Pellegr. ●☆

313350 Pouteria cainito（Roem. et Schult.）Radlk.；卡米图桃榄；Abiu ●☆

313351 Pouteria cainito Radlk. = Pouteria cainito（Roem. et Schult.）Radlk. ●☆

313352 Pouteria camerounensis（Pierre ex Aubrév. et Pellegr.）Baehni = Synsepalum revolutum（Baker）T. D. Penn. ●☆

313353 Pouteria campechiana（Kunth）Baehni；坎佩切桃榄●☆

313354 Pouteria campechiana（Kunth）Baehni = Lucuma nervosa A. DC. ●

313355 Pouteria cerasifera（Welw.）A. Meeuse = Synsepalum cerasiferum（Welw.）T. D. Penn. ●☆

313356 Pouteria chevalieri（Engl.）Baehni = Synsepalum cerasiferum

（Welw.）T. D. Penn. ●☆

313357　Pouteria clemensii（Lecomte）Baehni = Planchonella clemensii（Lecomte）P. Royen ●

313358　Pouteria congolense（Lecomte）Baehni = Synsepalum congolense Lecomte ●☆

313359　Pouteria costata Capuron ex Aubrév. = Capurodendron costatum Aubrév. ●☆

313360　Pouteria delphinensis Capuron ex Aubrév. = Capurodendron delphinense Aubrév. ●☆

313361　Pouteria densiflora（Baker）Baehni = Synsepalum revolutum（Baker）T. D. Penn. ●☆

313362　Pouteria disaco（Engl.）A. Meeuse = Synsepalum cerasiferum（Welw.）T. D. Penn. ●☆

313363　Pouteria duclitan（Blanco）Baehni = Planchonella duclitan（Blanco）Bakh. f. ●

313364　Pouteria dulcifica（Schumach. et Thonn.）Baehni = Synsepalum dulcificum（Schumach. et Thonn.）Daniell ●☆

313365　Pouteria eluviicola Baehni = Xantolis boniana（Dubard）Royen var. rostrata（Merr.）Royen ●

313366　Pouteria ferruginea Chiov. = Pouteria adolfi-friedericii（Engl.）A. Meeuse ●☆

313367　Pouteria giordanii Chiov. = Pouteria altissima（A. Chev.）Baehni ●☆

313368　Pouteria gracilifolia Capuron ex Aubrév. = Capurodendron gracilifolium Aubrév. ●☆

313369　Pouteria grandifolia（Wall.）Baehni;龙果桃榄（康克桑,龙果,马鸡康）;Big-flowered Pouteria,Largeleaf Pouteria ●

313370　Pouteria guianensis Aubl. ;圭亚那桃榄●☆

313371　Pouteria hainanensis（Merr.）Baehni = Pouteria annamensis（Pierre ex Dubard）Baehni ●

313372　Pouteria hexastemon Baehni;六冠桃榄●☆

313373　Pouteria kaessneri（Engl.）Baehni = Synsepalum kaessneri（Engl.）T. D. Penn. ●☆

313374　Pouteria kemoensis（Dubard）Baehni = Synsepalum kemoense（Dubard）Aubrév. ●☆

313375　Pouteria kerrii（H. R. Fletcher）Baehni = Pouteria grandifolia（Wall.）Baehni ●

313376　Pouteria laurentii（De Wild.）Baehni = Pachystela laurentii（De Wild.）C. M. Evrard ●☆

313377　Pouteria leptosperma Baehni = Breviea sericea Aubrév. et Pellegr. ●☆

313378　Pouteria ligulata Baehni = Synsepalum passargei（Engl.）T. D. Penn. ●☆

313379　Pouteria longecuneata（De Wild.）Baehni = Synsepalum longecuneatum De Wild. ●☆

313380　Pouteria magalismontana（Sond.）A. Meeuse = Englerophytum magalismontanum（Sond.）T. D. Penn. ●☆

313381　Pouteria malchairi（De Wild.）Baehni = Englerophytum oblanceolatum（S. Moore）T. D. Penn. ●☆

313382　Pouteria mandrarensis Capuron ex Aubrév. = Capurodendron mandrarense Aubrév. ●☆

313383　Pouteria msolo（Engl.）A. Meeuse = Synsepalum msolo（Engl.）T. D. Penn. ●☆

313384　Pouteria multiflora Eyma;繁花桃榄;Broad-leaved Lucuma ●☆

313385　Pouteria natalensis（Sond.）A. Meeuse = Englerophytum natalense（Sond.）T. D. Penn. ●☆

313386　Pouteria nodosum Capuron ex Aubrév. = Capurodendron

nodosum Aubrév. ●☆

313387　Pouteria obovata（R. Br.）Baehni = Planchonella obovata（R. Br.）Pierre ●

313388　Pouteria obovata（R. Br.）Baenni var. dubia Koidz. ex H. Hara = Planchonella obovata（R. Br.）Pierre var. dubia（Koidz. ex H. Hara）Hatus. ex T. Yamaz. ●☆

313389　Pouteria ovatostipulata（De Wild.）Baehni = Vincentella ovatostipulata（De Wild.）Aubrév. et Pellegr. ●☆

313390　Pouteria passargei（Engl.）Baehni = Synsepalum passargei（Engl.）T. D. Penn. ●☆

313391　Pouteria pedunculata Baehni = Sinosideroxylon pedunculatum（Hemsl.）H. Chuang ●

313392　Pouteria pierrei（A. Chev.）Baehni;皮埃尔桃榄●☆

313393　Pouteria pseudoracemosa（J. H. Hemsl.）L. Gaut. ;假总花桃榄●☆

313394　Pouteria pseudoterminalia Capuron ex Aubrév. = Capurodendron pseudoterminalia Aubrév. ●☆

313395　Pouteria revoluta（Baker）Baehni = Synsepalum revolutum（Baker）T. D. Penn. ●☆

313396　Pouteria rufescens Capuron ex Aubrév. = Capurodendron rufescens Aubrév. ●☆

313397　Pouteria rufinervis Chiov. = Pouteria adolfi-friedericii（Engl.）A. Meeuse ●☆

313398　Pouteria sakalava Capuron ex Aubrév. = Capurodendron sakalavum Aubrév. ●☆

313399　Pouteria sapota（Jacq.）H. E. Moore et Stearn;山榄桃榄（妈妈果）; Mamey Sapote, Mammee Sapote, Mammee Zapote, Marmalade Plum, Mar-malade Plum, Sapote ●☆

313400　Pouteria seretii（De Wild.）Baehni = Synsepalum seretii（De Wild.）T. D. Penn. ●☆

313401　Pouteria stipulata（Radlk.）Baehni = Synsepalum stipulatum（Radlk.）Engl. ●☆

313402　Pouteria subcordata（De Wild.）Baehni = Synsepalum subcordatum De Wild. ●☆

313403　Pouteria superba（Vermoesen）L. Gaut. ;华丽桃榄（华丽马拉山榄）●☆

313404　Pouteria taiensis（Aubrév. et Pellegr.）Baehni = Aubregrinia taiensis（Aubrév. et Pellegr.）Heine ●☆

313405　Pouteria terminalioides Capuron ex Aubrév. = Capurodendron terminalioides Aubrév. ●☆

313406　Pouteria tridentata Baehni = Synsepalum passargei（Engl.）T. D. Penn. ●☆

313407　Pouteria ulugurensis（Engl.）Baehni = Synsepalum ulugurense（Engl.）Engl. ●☆

313408　Pouteria viridis（Pittier）Cronquist;绿人心果; Green Sapote, Jnjerto, Zapote Jnjerto ●☆

313409　Pouteria zenkeri A. Meeuse = Pachystela robusta Engl. ●☆

313410　Pouzolsia Benth. = Pouzolzia Gaudich. ●■

313411　Pouzolzia Gaudich.（1830）;雾水葛属（水鸡油属）; Pouzolzia, Reekkudzu ●■

313412　Pouzolzia abyssinica（A. Rich.）Blume = Pouzolzia guineensis Benth. ●☆

313413　Pouzolzia andongensis Hiern = Pouzolzia denudata De Wild. et T. Durand ●☆

313414　Pouzolzia angustifolia Wight = Pouzolzia zeylanica（L.）Benn. et R. Br. var. angustifolia（Wight）C. J. Chen ■

313415　Pouzolzia arabica Deflers = Pouzolzia mixta Solms ●☆

313416 Pouzolzia argenteonitida W. T. Wang；银叶雾水葛；Silverleaf Reekkudzu,Silveryleaf Pouzolzia ●

313417 Pouzolzia argenteonitida W. T. Wang = Pouzolzia calophylla W. T. Wang et C. J. Chen ●

313418 Pouzolzia batesii Rendle = Pouzolzia denudata De Wild. et T. Durand ●☆

313419 Pouzolzia bracteosa Friis；多苞片雾水葛●☆

313420 Pouzolzia calophylla W. T. Wang et C. J. Chen；美叶雾水葛；Beautiful-leaf Pouzolzia,Beautyleaf Reekkudzu ●

313421 Pouzolzia conulifera Friis et Jellis = Phenax sonneratii（Poir.）Wedd. ■☆

313422 Pouzolzia cordata Peter = Pouzolzia peteri Friis ●☆

313423 Pouzolzia dewevrei De Wild. et T. Durand = Pouzolzia guineensis Benth. ●☆

313424 Pouzolzia elegans Wedd.；雅致雾水葛（济把燕，水鸡油）；Elegant Pouzolzia,Elegant Reekkudzu,Formosan Elegant Pouzolzia ●

313425 Pouzolzia elegans Wedd. var. delavayi（Gagnep.）W. T. Wang；菱叶雾水葛；Delavay Elegant Pouzolzia,Delavay Elegant Reekkudzu ●

313426 Pouzolzia elegans Wedd. var. delavayi（Gagnep.）W. T. Wang = Pouzolzia elegans Wedd. ●

313427 Pouzolzia elegans Wedd. var. formosana H. L. Li = Pouzolzia elegans Wedd. ●

313428 Pouzolzia elegantula W. W. Sm. = Pouzolzia elegans Wedd. var. delavayi（Gagnep.）W. T. Wang ●

313429 Pouzolzia elegantula W. W. Sm. = Pouzolzia elegans Wedd. ●

313430 Pouzolzia elegantula W. W. Sm. et Jeffrey = Pouzolzia elegans Wedd. ●

313431 Pouzolzia erythraeae Schweinf. = Didymodoxa caffra（Thunb.）Friis et Wilmot-Dear ■☆

313432 Pouzolzia fadenii Friis et Jellis；法登雾水葛●☆

313433 Pouzolzia flaccida A. Rich. = Australina flaccida（A. Rich.）Wedd. ■☆

313434 Pouzolzia fruticosa Engl. = Pouzolzia mixta Solms ●☆

313435 Pouzolzia golungensis Hiern = Pouzolzia guineensis Benth. ●☆

313436 Pouzolzia guineensis Benth.；几内亚雾水葛●☆

313437 Pouzolzia guineensis Benth. var. abyssinica（A. Rich.）Rendle = Pouzolzia guineensis Benth. ●☆

313438 Pouzolzia hirta（Blume）Hassk. = Gonostegia hirta（Blume）Miq. ■

313439 Pouzolzia hirta Blume ex Hassk. = Gonostegia hirta（Blume ex Hassk.）Miq. ■

313440 Pouzolzia huillensis Hiern = Pouzolzia mixta Solms ●☆

313441 Pouzolzia hypericifolia Blume = Gonostegia pentandra（Roxb.）Miq. var. hypericifolia（Blume）Masam. ●

313442 Pouzolzia hypericifolia Blume = Gonostegia pentandra（Roxb.）Miq. ●

313443 Pouzolzia hypoleuca Wedd. = Pouzolzia mixta Solms ●☆

313444 Pouzolzia indica（L.）Gaudich. = Pouzolzia zeylanica（L.）Benn. et R. Br. var. microphylla（Wedd.）W. T. Wang ■

313445 Pouzolzia indica（L.）Gaudich. = Pouzolzia zeylanica（L.）Benn. et R. Br. ●

313446 Pouzolzia indica（L.）Gaudich. var. alienata（L.）Wedd. = Pouzolzia zeylanica（L.）Benn. et R. Br. ■

313447 Pouzolzia indica Gaudich. = Pouzolzia zeylanica（L.）Benn. et R. Br. var. microphylla（Wedd.）W. T. Wang ■

313448 Pouzolzia indica Gaudich. subvar. microphylla Wedd. = Pouzolzia zeylanica（L.）Benn. et R. Br. var. microphylla（Wedd.）W. T. Wang ■

313449 Pouzolzia indica Gaudich. var. alienata（L.）Wedd. subvar. microphylla Wedd. = Pouzolzia zeylanica（L.）Benn. et R. Br. var. microphylla（Wedd.）W. T. Wang ■

313450 Pouzolzia indica Gaudich. var. alienta（L.）Wedd.；全缘叶水鸡油■

313451 Pouzolzia indica Gaudich. var. alienta（L.）Wedd. = Pouzolzia zeylanica（L.）Benn. et R. Br. ■

313452 Pouzolzia indica Gaudich. var. angustifolia（Wight）Wedd. = Pouzolzia angustifolia Wight ■

313453 Pouzolzia matsudae（Yamam.）S. S. Ying = Gonostegia matsudae（Yamam.）Yamam. et Masam. ■☆

313454 Pouzolzia mixta Solms；阿拉伯雾水葛；Tingo Fibre ●☆

313455 Pouzolzia neurocarpa（Yamam.）S. S. Ying = Gonostegia matsudae（Yamam.）Yamam. et Masam. ■☆

313456 Pouzolzia niveotomentosa W. T. Wang；雪毡雾水葛；Snowwhitehair Reekkudzu,Snowy Pouzolzia ●

313457 Pouzolzia ovalis Miq. = Pouzolzia sanguinea（Blume）Merr. ●

313458 Pouzolzia ovalis Miq. var. fulgens Wedd. = Pouzolzia argenteonitida W. T. Wang ●

313459 Pouzolzia ovalis Miq. var. fulgens Wedd. = Pouzolzia calophylla W. T. Wang et C. J. Chen ●

313460 Pouzolzia parasitica（Forssk.）Schweinf.；寄生雾水葛●☆

313461 Pouzolzia parvifolia Wight = Gonostegia parvifolia（Wight）Miq. ●■

313462 Pouzolzia pauciflora（Steud.）A. Rich. = Droguetia iners（Forssk.）Schweinf. ■☆

313463 Pouzolzia pentandra（Roxb.）Benn.；五蕊雾水葛●☆

313464 Pouzolzia pentandra（Roxb.）Benn. = Gonostegia pentandra（Roxb.）Miq. ●

313465 Pouzolzia pentandra（Roxb.）Wedd. var. akoensis（Yamam.）Yamam. et Masam. = Gonostegia pentandra（Roxb.）Miq. ●

313466 Pouzolzia pentandra Benn. = Pouzolzia pentandra（Roxb.）Benn. ●☆

313467 Pouzolzia pentandra Benn. var. hypericifolia（Blume）Masam. = Gonostegia pentandra（Roxb.）Miq. ●

313468 Pouzolzia peteri Friis；彼得雾水葛●☆

313469 Pouzolzia piscicelliana Buscal. et Muschl. = Didymodoxa caffra（Thunb.）Friis et Wilmot-Dear ■☆

313470 Pouzolzia procridioides（Wedd.）Wedd. = Pouzolzia parasitica（Forssk.）Schweinf. ●☆

313471 Pouzolzia sanguinea（Blume）Merr.；红雾水葛（大榄，大黏药，大黏叶，红水麻，接骨灵，青白麻叶，涩叶树，山毛柳，土升麻，小黏榔，嫩麻桐，玄麻，野麻公，黏榔药，黏药根，籽藤）；Red Pouzolzia,Red Reekkudzu ●

313472 Pouzolzia sanguinea（Blume）Merr. var. elegans（Wedd.）Friis,Wilmot-Dear et C. J. Chen = Pouzolzia elegans Wedd. ●

313473 Pouzolzia sanguinea（Blume）Merr. var. fulgens（Wedd.）H. Hara = Pouzolzia argenteonitida W. T. Wang ●

313474 Pouzolzia sanguinea（Blume）Merr. var. fulgens（Wedd.）H. Hara = Pouzolzia calophylla W. T. Wang et C. J. Chen ●

313475 Pouzolzia sanguinea（Blume）Merr. var. nepalensis（Wedd.）H. Hara；尼泊尔雾水葛；Nepal Red Pouzolzia,Nepal Red Reekkudzu ●

313476 Pouzolzia sanguinea（Blume）Merr. var. nepalensis（Wedd.）H. Hara = Pouzolzia sanguinea（Blume）Merr. ●

313477 Pouzolzia shirensis Rendle；希尔雾水葛●☆

313478 Pouzolzia spinosobracteata W. T. Wang;刺苞雾水葛;Spiny-bract Pouzolzia,Thornbract Reekkudzu ●

313479 Pouzolzia spinosobracteata W. T. Wang = Pouzolzia niveotomentosa W. T. Wang ●

313480 Pouzolzia viminea (Wall.) Wedd.;柳枝雾水葛(山麻柳)●

313481 Pouzolzia viminea (Wall.) Wedd. = Pouzolzia sanguinea (Blume) Merr. ●

313482 Pouzolzia zeylanica (L.) Benn. = Pouzolzia zeylanica (L.) Benn. et R. Br. ■

313483 Pouzolzia zeylanica (L.) Benn. et R. Br.;雾水葛(拔脓膏,白石薯,啜脓膏,地消散,杜薯,多枝雾水葛,咄脓膏,脓见消,糯米藤,山参,山三茄,生肉药,石茹,石株,水麻秧,台湾雾水葛,田薯,黏椰根,黏椰果);Ceylan Pouzolzia, Graceful Pouzolzsbush, Reekkudzu ■

313484 Pouzolzia zeylanica (L.) Benn. et R. Br. var. alienata Merr.;台湾雾水葛●

313485 Pouzolzia zeylanica (L.) Benn. et R. Br. var. angustifolia (Wight) C. J. Chen;狭叶雾水葛■

313486 Pouzolzia zeylanica (L.) Benn. et R. Br. var. microphylla (Wedd.) W. T. Wang = Pouzolzia zeylanica (L.) Benn. et R. Br. ■

313487 Pouzolzia zeylanica (L.) Benn. et R. Br. var. microphylla (Wedd.) W. T. Wang;多枝雾水葛(强盗草,石珠,石珠子)■

313488 Povedadaphne W. C. Burger = Ocotea Aubl. ●☆

313489 Povedadaphne W. C. Burger(1988);肖绿心樟属●☆

313490 Povedadaphne quadriporata W. C. Burger;肖绿心樟●☆

313491 Pozoa Hook. f. = Schizeilema (Hook. f.) Domin ●☆

313492 Pozoa Lag. (1816);波索草属■☆

313493 Pozoa coriacea Lag.;波索草■☆

313494 Pozoopsis Benth. = Pozopsis Hook. ■☆

313495 Pozopsis Hook. = Huanaca Cav. ■☆

313496 Pradosia Liais(1872);普拉榄属●☆

313497 Pradosia spinosa Ewango et Breteler;普拉榄●☆

313498 Praealstonia Miers = Symplocos Jacq. ●

313499 Praecereus Buxb. = Cereus Mill. ●

313500 Praecitrullus Pangalo(1944);印度瓜属■☆

313501 Praecitrullus fistulosus (Stocks) Pangalo;印度瓜■☆

313502 Praecoxanthus Hopper et A. P. Br. (2000);澳洲无叶兰属■☆

313503 Praecoxanthus Hopper et A. P. Br. = Caladenia R. Br. ■☆

313504 Praenanthes Hook. = Prenanthes L. ■

313505 Praetoria Baill. = Pipturus Wedd. ●

313506 Prageluria N. E. Br. = Telosma Coville ●

313507 Pragmatropa Pierre = Euonymus L. (保留属名)●

313508 Pragmatropa Pierre = Vyenomus C. Presl ●

313509 Pragmotessara Pierre = Euonymus L. (保留属名)●

313510 Pragmotessera ilicifolia Pierre = Glyptopetalum ilicifolium (Franch.) C. Y. Cheng et Q. S. Ma ●◇

313511 Pragmotropa Pierre = Euonymus L. (保留属名)●

313512 Prainea King ex Hook. f. (1888);陷毛桑属●☆

313513 Prainea scandens King;陷毛桑●☆

313514 Pranceacanthus Wassh. (1984);巴西刺爵床属 ☆

313515 Pranceacanthus coccineus Wassh.;巴西刺爵床 ☆

313516 Prangos Lindl. (1825);栓翅芹属■☆

313517 Prangos acaulis (DC.) Bornm.;无茎栓翅芹■☆

313518 Prangos bucharica B. Fedtsch.;布赫栓翅芹■☆

313519 Prangos cachroides (Schrenk) Pimenov et V. N. Tikhom.;毛栓翅芹■

313520 Prangos cylindrocarpa Korovin;柱果栓翅芹■☆

313521 Prangos didyma (Regel) Pimenov et V. N. Tikhom.;双生栓翅芹■

313522 Prangos fedtschenkoi (Regel et Schmalh.) Korovin;范氏栓翅芹■☆

313523 Prangos herderi (Regel) Herrnst. et Heyn subsp. xinjiangensis X. Y. Chen et Q. X. Liu;新疆栓翅芹■

313524 Prangos latiloba Korovin;栓翅芹(阔裂栓翅芹,牧草栓翅芹)■☆

313525 Prangos ledebourii Herrnst. et Heyn;大果栓翅芹■

313526 Prangos lipskyi Korovin;利普斯基栓翅芹■☆

313527 Prangos lophotera Boiss.;冠饰栓翅芹■☆

313528 Prangos pabularia Lindl.;裂栓翅芹■;Hay Plant ■☆

313529 Prangos tschimganica B. Fedtsch.;契穆干栓翅芹■☆

313530 Prangos uloptera DC.;卷栓翅芹■☆

313531 Prantleia Mez = Orthophytum Beer ■☆

313532 Praravinia Korth. (1842);菲岛茜属●☆

313533 Praravinia affinis (Merr.) Bremek.;近缘菲岛茜●☆

313534 Praravinia densiflora Korth.;密花菲岛茜●☆

313535 Praravinia microphylla (Merr.) Bremek.;小叶菲岛茜●☆

313536 Praravinia parviflora Bremek.;小花菲岛茜●☆

313537 Prasanthea (DC.) Decne. = Codonophora Lindl. ●☆

313538 Prasanthea (DC.) Decne. = Paliavana Vell. ex Vand. ●☆

313539 Prasanthea Decne. = Codonophora Lindl. ●☆

313540 Prasanthea Decne. = Paliavana Vell. ex Vand. ●☆

313541 Prascoenum Post et Kuntze = Allium L. ●

313542 Prascoenum Post et Kuntze = Praskoinon Raf. ●

313543 Prasiteles Salisb. = Narcissus L. ●

313544 Prasium L. (1753);葱草属●☆

313545 Prasium majus L.;葱草●☆

313546 Prasium majus L. subsp. neglectum Bég. et Vacc. = Prasium majus L. ●☆

313547 Prasium majus L. var. liparitanum (Tod.) = Prasium majus L. ●☆

313548 Praskoinon Raf. = Allium L. ●

313549 Prasopepon Naudin = Cucurbitella Walp. ■☆

313550 Prasophyllum R. Br. (1810);韭兰属;Leek Orchid ■☆

313551 Prasoxylon M. Roem. = Dysoxylum Blume ●

313552 Pratia Gaudich. (1825);铜锤玉带草属;Pratia,Purplehammer ■

313553 Pratia Gaudich. = Lobelia L. ●■

313554 Pratia angulata (G. Forst.) Hook. f. = Lobelia angulata G. Forst. ■☆

313555 Pratia angulata Hook. f.;角棱铜锤玉带草;Blue Star Creeper, Creeping Pratia,Fragrant Carpet ■☆

313556 Pratia begonifolia (Wall.) Lindl. = Lobelia nummularia Lam. ■

313557 Pratia begonifolia (Wall.) Lindl. = Pratia nummularia (Lam.) A. Br. et Asch. ■

313558 Pratia begoniifolia (Wall.) Lindl. = Lobelia angulata G. Forst. ■☆

313559 Pratia brevisepala Y. S. Lian = Lobelia brevisepala (Y. S. Lian) Lammers ■

313560 Pratia fangiana E. Wimm. = Lobelia fangiana (E. Wimm.) S. Y. Hu ■

313561 Pratia gilletii (De Wild.) E. Wimm. = Lobelia gilletii De Wild. ■☆

313562 Pratia montana (Reinw. ex Blume) Hassk.;山紫锤草(地钮子,小铜锤);Montane Pratia, Wild Purplehammer ■

313563 Pratia montana (Reinw. ex Blume) Hassk. = Lobelia montana Reinw. ex Blume ■

313564 Pratia nummularia (Lam.) A. Br. et Asch.;铜锤玉带草(地浮萍,地扣子,地纽子,地茄子,地茄子草,地石榴,红头带,金钱草,扣子草,马金钟,马莲草,骂补神,米汤果,普刺特草,铜锤草,乌

金钟，小铜锤，鳖子草）；Common Pratia, Copperhammer and Jadebelt ■

313565 Pratia nummularia（Lam.）A. Braun et Asch. = Lobelia angulata G. Forst. ■☆

313566 Pratia nummularia（Lam.）A. Braun et Asch. = Lobelia nummularia Lam. ■

313567 Pratia ovata Elmer = Lobelia zeylanica L. ■

313568 Pratia pedunculata（R. Br.）Benth.；具梗铜锤玉带草；Blue Star Creeper ■☆

313569 Pratia pedunculata（R. Br.）Benth. 'County Park'；野趣园具梗铜锤玉带草■☆

313570 Pratia purpurascens（R. Br.）E. Wimm.；淡紫铜锤玉带草（白根紫锤草）；White Root ■☆

313571 Pratia radicans G. Don = Lobelia chinensis Lour. ■

313572 Pratia reflexa Y. S. Lian = Lobelia reflexisepala Lammers ■

313573 Pratia thunbergii G. Don = Lobelia chinensis Lour. ■

313574 Pratia torricellensis Lauterb. = Lobelia zeylanica L. ■

313575 Pratia wollastonii S. Moore；广西铜锤草（土半边莲）；Guangxi Purplehammer, Wallaston Pratia ■

313576 Pratia zeylanica Hassk. = Pratia nummularia（Lam.）A. Br. et Asch. ■

313577 Praticola Ehrh. = Thalictrum L. ■

313578 Pravinaria Bremek.（1940）普拉茜属☆

313579 Pravinaria endertii Bremek.；普拉茜☆

313580 Pravinaria leucocarpa Bremek.；白果普拉茜☆

313581 Praxeliopsis G. M. Barroso(1949)；寡毛假臭草属■☆

313582 Praxeliopsis mattogrossensis G. M. Barroso；寡毛假臭草■☆

313583 Praxelis Cass.（1826）；假臭草属（南美蓟属）■●

313584 Praxelis clematidea R. M. King et H. Rob.；假臭草●■

313585 Preauxia Sch. Bip. = Argyranthemum Webb ex Sch. Bip. ●

313586 Preauxia Sch. Bip. = Chrysanthemum L.（保留属名）■●

313587 Preauxia canariensis Sch. Bip. = Argyranthemum adauctum（Link）Humphries ●☆

313588 Preauxia dugourii Bolle = Argyranthemum adauctum（Link）Humphries subsp. dugourii（Bolle）Humphries ●☆

313589 Preauxia jacobifolia Sch. Bip. = Argyranthemum adauctum（Link）Humphries subsp. jacobiifolium（Sch. Bip.）Humphries ●☆

313590 Preauxia perraudieri Sch. Bip. = Argyranthemum adauctum（Link）Humphries ●☆

313591 Precopiania Gusul. = Symphytum L. ■

313592 Preissia Opiz = Avena L. ■

313593 Premna L.（1771）（保留属名）；豆腐柴属（臭黄荆属，臭娘子属，臭鱼木属，腐婢属）；Musk 'Maple', Premna ●■

313594 Premna acuminatissima Merr. = Premna chevalieri Dop ●

313595 Premna acuminatissima Merr. = Premna octonervia Merr. et F. P. Metcalf ●

313596 Premna acutata W. W. Sm.；尖齿豆腐柴（尖齿大青，尖叶臭黄荆，尖叶臭牡丹）；Sharpleaf Premna, Sharp-leaved Premna, Sharptooth Premna ●

313597 Premna ambongensis Moldenke；安邦豆腐柴●☆

313598 Premna angolensis Gürke；安哥拉豆腐柴●☆

313599 Premna angolensis Gürke var. cuneata De Wild. = Premna angolensis Gürke ●☆

313600 Premna angustifolia Hung T. Chang = Callicarpa hungtaii C. P' ei et S. L. Chen ●

313601 Premna arborea（Roxb.）Roth = Gmelina arborea Roxb. ●◇

313602 Premna barbata Wall. ex Schauer；髯毛豆腐柴●☆

313603 Premna bequaertii Moldenke；贝卡尔豆腐柴●☆

313604 Premna bodinieri H. Lév. = Premna puberula Pamp. var. bodinieri（H. Lév.）C. Y. Wu et S. Y. Pao ●

313605 Premna bracteata Wall.；苞序豆腐柴（苞序大青）；Bracteate Premna ●

313606 Premna cavaleriei H. Lév.；黄药豆腐柴；Cavaierie Premna ●

313607 Premna chevalieri Dop；尖叶豆腐柴（尖叶大青）；Chevalier. Premna ●

313608 Premna chrysoclada（Bojer）Gürke；黄枝豆腐柴●☆

313609 Premna claessensii De Wild. = Premna angolensis Gürke ●☆

313610 Premna colorata Hiern = Vitex sulphurea Baker ●☆

313611 Premna confinis C. P' ei et S. L. Chen ex C. Y. Wu；滇桂豆腐柴（滇桂大青，滇南豆腐柴，太阳木）；Adjoining Premna ●

313612 Premna congolensis Moldenke；刚果豆腐柴●☆

313613 Premna congolensis Moldenke var. integrifolia ? = Premna congolensis Moldenke ●☆

313614 Premna cordifolia Wight = Premna corymbosa（Burm. f.）Rottl. et Willd. ●

313615 Premna corymbosa（Burm. f.）Rottl. et Willd. = Premna serratifolia L. ●

313616 Premna corymbosa（Burm. f.）Rottl. et Willd. var. obtusifolia（R. Br.）Flecher = Premna serratifolia L. ●

313617 Premna corymbosa A. Meeuse = Premna obtusifolia R. Br. ●

313618 Premna corymbosa A. Meeuse = Premna serratifolia L. ●

313619 Premna corymbosa Rottl. et Willd. = Premna serratifolia L. ●

313620 Premna corymbosa Rottl. et Willd. var. obtusifolia（R. Br.）H. R. Fletcher = Premna obtusifolia R. Br. ●

313621 Premna crassa Hand. -Mazz.；石山豆腐柴（黄皮树，石山大青）；Thick Premna ●

313622 Premna crassa Hand. -Mazz. var. yui Moldenke；风庆豆腐柴（风庆大青）；Fengqing Premna ●

313623 Premna decaryi Moldenke；德卡里豆腐柴●☆

313624 Premna discolor Verdc.；异色豆腐柴●☆

313625 Premna dopi C. P' ei = Premna tapintzeana Dop ●

313626 Premna elskensii De Wild. = Premna angolensis Gürke ●☆

313627 Premna esquirolii H. Lév. = Viburnum congestum Rehder ●

313628 Premna ferruginea A. Rich.；锈色豆腐柴●☆

313629 Premna flavescens Buch. -Ham. ex Wall.；淡黄豆腐柴（淡黄大青）；Yellowish Premna ●

313630 Premna fohaiensis C. P' ei et S. L. Chen ex C. Y. Wu；勐海豆腐柴（勐海大青）；Menghai Premna ●

313631 Premna fordii Dunn et Tutcher；长序臭黄荆；Ford Premna ●

313632 Premna fordii Dunn et Tutcher var. glabra S. L. Chen；无毛臭黄荆；Glabrous Ford Premna ●

313633 Premna formosana Maxim. = Premna microphylla Turcz. ●

313634 Premna fortunati Dop = Premna fulva Craib ●

313635 Premna fulva Craib；黄毛豆腐柴（斑鸠占，黄毛大青，神仙豆腐柴，战骨）；Yellow-haired Premna, Yellowhairy Premna ●

313636 Premna fulva Merr. = Premna crassa Hand. -Mazz. ●

313637 Premna glandulosa C. P' ei = Premna henryana（Hand. -Mazz.）C. Y. Wu ●

313638 Premna glandulosa Hand. -Mazz.；腺叶豆腐柴（腺叶大青）；Glandleaf Premna, Glandular Premna ●

313639 Premna gracilis A. Chev. = Premna lucens A. Chev. ●☆

313640 Premna gracillima Verdc.；纤细豆腐柴●☆

313641 Premna grandifolia A. Meeuse；大叶豆腐柴●☆

313642 Premna hainanensis Chun et F. C. How；海南臭黄荆（海南腐

313715   Premna polita Hiern;亮豆腐柴●☆

313716   Premna puberula Hand.-Mazz. = Premna puberula Pamp.●

313717   Premna puberula Pamp.;狐臭柴(斑鸠叶豆腐柴,斑鸠占,长柄臭黄荆,臭黄荆,臭树,跌打王,毛豆婢,木臭牡丹,神仙豆腐柴,水白蜡,土常山,小青树);Puberulent Premna●

313718   Premna puberula Pamp. var. bodinieri (H. Lév.) C. Y. Wu et S. Y. Pao;毛狐臭柴(臭叶草);Bodinier Puberulent Premna●

313719   Premna puerensis Y. Y. Qian;普洱豆腐柴;Puer Premna●

313720   Premna punicea C. Y. Wu;玫花豆腐柴(玫花大青);Darkred Premna,Redpurple Premna,Scarlet Premna●

313721   Premna pygmaea Wall. = Pygmaeopremna herbacea (Roxb.) Moldenke●

313722   Premna pyramidata Wall. ex Schauer;塔序豆腐柴(塔序大青);Pyramidal Premna●

313723   Premna quadrifolia Schumach. et Thonn.;四叶豆腐柴●☆

313724   Premna quadrifolia Schumach. et Thonn. var. subglabra Moldenke = Premna angolensis Gürke●☆

313725   Premna quadrifolia Schumach. et Thonn. var. warneckeana Moldenke = Premna quadrifolia Schumach. et Thonn.●☆

313726   Premna racemosa Wall.;总序豆腐柴(总序大青);Racemose Premna●

313727   Premna remnapygmaes Wall. = Pygmaeopremna herbacea (Roxb.) Moldenke●

313728   Premna resinosa (Hochst.) Schauer;胶豆腐柴●☆

313729   Premna resinosa (Hochst.) Schauer subsp. holstii (Gürke) Verdc.;霍尔豆腐柴●☆

313730   Premna resinosa (Hochst.) Schauer. f. grossedentata Moldenke = Premna resinosa (Hochst.) Schauer●☆

313731   Premna richardsiae Moldenke;理查兹豆腐柴●☆

313732   Premna rotundifolia C. P'ei = Premna tenii C. P'ei●

313733   Premna rubroglandulosa C. Y. Wu;红腺豆腐柴(红腺大青);Redgland Premna,Red-glanded Premna●

313734   Premna scandens Bojer = Premna corymbosa (Burm. f.) Rottl. et Willd.●

313735   Premna scandens Roxb.;藤豆腐柴(藤大青);Scandent Premna●

313736   Premna schimperi Engl.;欣氏豆腐柴●☆

313737   Premna schliebenii Werderm.;施利本豆腐柴●☆

313738   Premna scoriarum W. W. Sm.;腾冲豆腐柴(腾冲大青);Tengchong Premna●

313739   Premna senensis Klotzsch;塞纳豆腐柴●☆

313740   Premna serratifolia L.;伞序豆腐柴(臭娘子,伞序臭黄荆);Corymbose Premna,Headache Tree,Serratifoliate Premna●

313741   Premna serratifolia L. = Premna corymbosa (Burm. f.) Rottl. et Willd.●

313742   Premna somaliensis Baker = Premna oligotricha Baker●☆

313743   Premna spinosa Roxb. = Premna corymbosa (Burm. f.) Rottl. et Willd.●

313744   Premna stenantha Merr. = Premna fordii Dunn et Tutcher●

313745   Premna steppicola Hand.-Mazz.;草坡豆腐柴(草坡大青);Steppe Premna,Steppeliving Premna●

313746   Premna steppicola Hand.-Mazz. var. henryana Hand.-Mazz. = Premna henryana (Hand.-Mazz.) C. Y. Wu●

313747   Premna straminicaulis C. Y. Wu;草黄枝豆腐柴(草黄枝大青);Straw-coloured Premna,Strawyellow Premna,Yellowbranch Premna●

313748   Premna suaveolens Chiov. = Clerodendrum glabrum E. Mey.●☆

313749   Premna subcapitata Rehder;近头状豆腐柴(近头状大青,头序臭黄荆);Subcapitate Premna●

313750   Premna subcordata Nakai = Premna puberula Pamp.●

313751   Premna subscandens Merr.;攀缘豆腐柴(攀缘臭黄荆,攀缘大青);Climbing Premna,Subclimbing Premna●

313752   Premna sulphurea (Baker) Gürke = Vitex sulphurea Baker●☆

313753   Premna sunyiensis C. P'ei;塘虱角(大蛇药,牛尾鸟);Catfishhorn,Sunyi Premna●

313754   Premna szemaoensis C. P'ei;思茅豆腐柴(戳皮树,接骨树,类梧桐,绿泽兰,蚂蚁鼓堆树,思茅大青,思茅腐柴);Simao Premna●◇

313755   Premna taitensis Schauer;塔岛豆腐柴●☆

313756   Premna tanganyikensis Moldenke;坦噶尼喀豆腐柴●☆

313757   Premna tapintzeana Dop;大坪子豆腐柴(大坪子大青);Dapingzi Premna●

313758   Premna tenii C. P'ei;圆叶豆腐柴;Roundleaf Premna,Round-leaved Premna●

313759   Premna urticifolia Rehder;麻叶豆腐柴(筋骨散,麻叶大青,麻叶腐婢,荨麻叶臭黄荆,荨麻叶豆腐柴);Nettle-leaf Premna,Nettle-leaved Premna●

313760   Premna valbruyi H. Lév. = Viburnum foetidum Wall. var. ceanothoides (C. H. Wright) Hand.-Mazz.●

313761   Premna velutina C. Y. Wu;黄绒豆腐柴(黄绒大青);Velvet-like Premna,Velvety Premna,Yellownappy Premna●

313762   Premna velutina Gürke;短绒毛豆腐柴●☆

313763   Premna venulosa Moldenke;细脉豆腐柴●☆

313764   Premna vestita Schauer = Premna odorata Blanco●

313765   Premna viburnoides A. Rich. = Premna schimperi Engl.●☆

313766   Premna viburnoides A. Rich. var. genuina Pic. Serm. = Premna schimperi Engl.●☆

313767   Premna viburnoides A. Rich. var. schimperi (Engl.) Pic. Serm. = Premna schimperi Engl.●☆

313768   Premna viburnoides Kurz = Premna latifolia Roxb. var. cuneata C. B. Clarke●

313769   Premna viburnoides Kurz = Premna latifolia Roxb.●

313770   Premna viburnoides Wall. ex Schauer = Premna latifolia Roxb. var. cuneata C. B. Clarke●

313771   Premna yunnanensis Dop = Premna tapintzeana Dop●

313772   Premna yunnanensis W. W. Sm.;云南豆腐柴(虎珀,云南大青);Yunnan Premna●

313773   Premna zanzibarensis Vatke = Premna chrysoclada (Bojer) Gürke●☆

313774   Premna zenkeri Gürke = Premna angolensis Gürke●☆

313775   Prenanthella Rydb. (1906);小福王草属■☆

313776   Prenanthella exigua (A. Gray) Rydb.;小福王草;Brightwhite, Desert Prenanthella■☆

313777   Prenanthella exigua Rydb. = Prenanthella exigua (A. Gray) Rydb.■☆

313778   Prenanthes L. (1753);福王草属(盘果菊属);Cankerweed, Gall-of-the-earth,Rattlesnake Root,Rattlesnakeroot■

313779   Prenanthes abietina (Boiss. et Balansa) Kirp.;冷杉福王草■☆

313780   Prenanthes acaulis Roxb. = Launaea acaulis (Roxb.) Babc. ex Kerr■

313781   Prenanthes acerifolia (Maxim.) Matsum.;日本福王草■☆

313782   Prenanthes acerifolia (Maxim.) Matsum. f. heterophylla Matsum. et Koidz.;异叶日本福王草■☆

313783   Prenanthes acerifolia (Maxim.) Matsum. f. nipponica (Franch. et Sav.) Matsum. et Koidz.;本州福王草■☆

313784   Prenanthes acerifolia (Maxim.) Matsum. var. nipponica

（Franch. et Sav.）Makino ＝ Prenanthes acerifolia（Maxim.）Matsum. f. nipponica（Franch. et Sav.）Matsum. et Koidz. ■☆

313785 Prenanthes alata（Hook.）D. Dietr.；翅福王草；Western Rattlesnakeroot，Western White Lettuce，White Lettuce，Wing-leaved Rattlesnakeroot ■☆

313786 Prenanthes alata（Hook.）D. Dietr. var. sagittata A. Gray ＝ Prenanthes sagittata（A. Gray）A. Nelson ■☆

313787 Prenanthes alba L.；白福王草（白盘果菊）；Lion's Foot，Rattlesnake Root，Rattlesnake-root，White Canker Weed，White Lettuce，White Rattlesnakeroot，White Snakeroot，White-lettuce ■☆

313788 Prenanthes altissima L.；极高盘果菊；Rattlesnake Root，Tall Rattlesnakeroot，Tall White Lettuce，Tall White-lettuce ■☆

313789 Prenanthes altissima L. var. cinnamomea Fernald ＝ Prenanthes altissima L. ■☆

313790 Prenanthes altissima L. var. hispidula Fernald ＝ Prenanthes altissima L. ■☆

313791 Prenanthes angustifolia Boulos；狭叶福王草■☆

313792 Prenanthes angustiloba C. Shih；细裂福王草；Narrowlobed Rattlesnakeroot ■

313793 Prenanthes aphylla Nutt. ＝ Lygodesmia aphylla（Nutt.）DC. ■☆

313794 Prenanthes aspera Michx. ＝ Chondrilla aspera（Schrad. ex Willd.）Poir. ■

313795 Prenanthes aspera Michx. ＝ Prenanthes aspera Schrad. ex Willd. ■

313796 Prenanthes aspera Schrad. ex Willd.；粗糙福王草；Rattlesnake Root，Rough Rattlesnakeroot，Rough Rattlesnake-root，Rough White Lettuce，Rough White-lettuce ■

313797 Prenanthes aspera Schrad. ex Willd. ＝ Chondrilla aspera（Schrad. ex Willd.）Poir. ■

313798 Prenanthes autumnalis Walter；纤细福王草；Slender Rattlesnakeroot ■☆

313799 Prenanthes barbata（Torr. et A. Gray）Milstead ex Cronquist；刺福王草；Barbed Rattlesnakeroot ■☆

313800 Prenanthes blinii（H. Lév.）Kitag. ＝ Nabalus ochroleucus Maxim. ■

313801 Prenanthes bootii（DC.）D. Dietr.；布特福王草；Boot's Rattlesnakeroot ■☆

313802 Prenanthes cacaliifolia P. Beauv.；蟹甲草叶福王草■☆

313803 Prenanthes carrii Singhurst，O'Kennon et W. C. Holmes；卡尔福王草；Carr's Rattlesnakeroot ■☆

313804 Prenanthes cavaleriei（H. Lév.）Stebbins ＝ Faberia cavaleriei H. Lév. ■

313805 Prenanthes cavaleriei（H. Lév.）Stebbins ex Lauener ＝ Faberia cavaleriei H. Lév. ■

313806 Prenanthes chaffanjoni H. Lév. ＝ Crepis napifera（Franch.）Babc. ■

313807 Prenanthes chinensis Thunb. ＝ Ixeridium chinense（Thunb.）Tzvelev ■

313808 Prenanthes chinensis Thunb. ＝ Ixeris chinensis（Thunb.）Nakai ■

313809 Prenanthes crepidinea Michx.；俯垂福王草；Great White Lettuce，Midwestern White-lettuce，Nodding Rattlesnakeroot，Nodding Rattlesnake-root，Rattlesnake Root ■☆

313810 Prenanthes cylindrica（Small）E. L. Braun ＝ Prenanthes roanensis（Chick.）Chick. ■☆

313811 Prenanthes debilis Thunb. ＝ Ixeris debilis（Thunb.）A. Gray ■

313812 Prenanthes debilis Thunb. ＝ Ixeris japonica（Burm. f.）Nakai ■

313813 Prenanthes debilis Thunb. ex Murray ＝ Ixeris debilis（Thunb.）A. Gray ■

313814 Prenanthes dentata Thunb. ＝ Ixeridium dentatum（Thunb.）Tzvelev ■

313815 Prenanthes dentata Thunb. ＝ Ixeris dentata（Thunb.）Nakai ■

313816 Prenanthes denticulata Houtt. ＝ Paraixeris denticulata（Houtt.）Nakai ■

313817 Prenanthes diversifolia（Vaniot）C. C. Chang ＝ Paraprenanthes sororia（Miq.）C. Shih ■

313818 Prenanthes diversifolia Ledeb. ex Spreng. ＝ Youngia diversifolia（Ledeb. ex Spreng.）Ledeb. ■

313819 Prenanthes exigua A. Gray ＝ Prenanthella exigua Rydb. ■☆

313820 Prenanthes faberi Hemsl. ex Forbes et Hemsl.；狭锥福王草（薯蓣叶福王草）；Faber Rattlesnakeroot ■

313821 Prenanthes fastigiata Blume ＝ Youngia japonica（L.）DC. ■

313822 Prenanthes formosana Kitam.；台湾福王草■

313823 Prenanthes formosana Kitam. ＝ Notoseris formosana（Kitam.）C. Shih ■

313824 Prenanthes glandulosa Dunn；腺毛福王草■

313825 Prenanthes glandulosa Dunn ＝ Notoseris glandulosa（Dunn）C. Shih ■

313826 Prenanthes glomerata Decne. ＝ Soroseris glomerata（Decne.）Stebbins ■

313827 Prenanthes glomerata Decne. ex Jacq. ＝ Soroseris glomerata（Decne.）Stebbins ■

313828 Prenanthes graciliflora Wall. ＝ Stenoseris graciliflora（Wall. ex DC.）C. Shih ■

313829 Prenanthes graminea Fisch. ＝ Ixeridium gramineum（Fisch.）Tzvelev ■

313830 Prenanthes graminifolia Vaniot et H. Lév. ＝ Senecio wightii（DC. ex Wight）Benth. ex C. B. Clarke ■

313831 Prenanthes hastata Thunb. ＝ Paraixeris denticulata（Houtt.）Nakai ■

313832 Prenanthes henryi Dunn ＝ Notoseris gracilipes C. Shih ■

313833 Prenanthes henryi Dunn ＝ Notoseris henryi（Dunn）C. Shih ■

313834 Prenanthes hieracifolia H. Lév. ＝ Pterocypsela elata（Hemsl.）C. Shih ■

313835 Prenanthes humilis Thunb. ＝ Lapsana humilis（Thunb.）Makino ■

313836 Prenanthes integra Thunb. ＝ Crepidiastrum lanceolatum（Houtt.）Nakai ■

313837 Prenanthes integrifolia（Cass.）Small ＝ Prenanthes serpentaria Pursh ■☆

313838 Prenanthes japonica L. ＝ Youngia japonica（L.）DC. ■

313839 Prenanthes juncea Pursh ＝ Lygodesmia juncea（Pursh）D. Don ex Hook. ■☆

313840 Prenanthes laciniata Houtt. ＝ Pterocypsela laciniata（Houtt.）C. Shih ■

313841 Prenanthes laevigata Blume ＝ Ixeridium laevigatum（Blume）C. Shih ■

313842 Prenanthes laevigata Blume ＝ Ixeris laevigata（Blume）Sch. Bip. ex Maxim. ■

313843 Prenanthes lanceolata Houtt. ＝ Crepidiastrum lanceolatum（Houtt.）Nakai ■

313844 Prenanthes leptantha C. Shih；细花福王草；Finehead Rattlesnakeroot，Smallflower Rattlesnakeroot ■

313845 Prenanthes lessingii Hultén ＝ Prenanthes alata（Hook.）D.

Dietr. ■☆

313846　Prenanthes lyrata Thunb. = Youngia pseudosenecio (Vaniot) C. Shih ■

313847　Prenanthes lyrata Thunb. ex Houtt. = Youngia pseudosenecio (Vaniot) C. Shih ■

313848　Prenanthes macilentas Vaniot et H. Lév.；钝叶福王草(长柄福王草,穗花福王草)■

313849　Prenanthes macrophylla Franch.；多裂福王草(大叶福王草,大叶盘果菊)；Largeleaf Rattlesnakeroot ■

313850　Prenanthes maximowiczii Kirp.；马氏福王草■

313851　Prenanthes maximowiczii Kirp. = Nabalus ochroleucus Maxim. ■

313852　Prenanthes multiflora Thunb. = Youngia japonica (L.) DC. ■

313853　Prenanthes muralis L. = Mycelis muralis Dumort. ■☆

313854　Prenanthes ochroleuca (Maxim.) Hemsl. = Nabalus ochroleucus Maxim. ■

313855　Prenanthes ochroleuca (Maxim.) Hemsl. = Prenanthes blinii (H. Lév.) Kitag. ■

313856　Prenanthes ochroleuca (Maxim.) Hemsl. var. tanakae (Franch. et Sav. ex Y. Tanaka et Ono) Koidz. = Prenanthes tanakae (Franch. et Sav. ex Y. Tanaka et Ono) Koidz. ■☆

313857　Prenanthes pauciflora Torr. = Stephanomeria pauciflora (Torr.) A. Nelson ■☆

313858　Prenanthes pendula Sch. Bip. = Sonchus pendulus (Sch. Bip.) Sennikov ■☆

313859　Prenanthes pendula Sch. Bip. subsp. flaccida Svent. = Sonchus pendulus (Sch. Bip.) Sennikov subsp. flaccidus (Svent.) N. Kilian et Greuter ■☆

313860　Prenanthes pinnata L. f. = Sonchus leptocephalus Cass. ■☆

313861　Prenanthes polymorpha Ledeb. var. flaccida Ledeb. = Crepis nana Richardson ■

313862　Prenanthes polymorpha Ledeb. var. flexuosa Ledeb. = Crepis flexuosa (Ledeb.) C. B. Clarke ■

313863　Prenanthes polymorpha Ledeb. var. pygmaea (Ledeb.) Ledeb. subvar. integrifolia Ledeb. = Crepis nana Richardson ■

313864　Prenanthes pontica (Boiss.) Leskov；蓬特福王草■☆

313865　Prenanthes procumbens Roxb. = Paramicrorhynchus procumbens (Roxb.) Kirp. ■

313866　Prenanthes purpurea L.；紫福王草(紫花盘果菊)；Purple Lettuce, Purple Rattlesnake Root, Purple Rattlesnake-root, Purplehead Rattlesnakeroot ■☆

313867　Prenanthes pygmaea Ledeb. = Crepis nana Richardson ■

313868　Prenanthes pyramidalis C. Shih = Prenanthes tatarinowii Maxim. ■

313869　Prenanthes quinqueloba Wall. ex DC. = Parasenecio quinquelobus (Wall. ex DC.) Y. L. Chen ■

313870　Prenanthes racemiformis C. Shih = Prenanthes tatarinowii Maxim. ■

313871　Prenanthes racemosa Michx.；蓝绿福王草；Glaucous Rattlesnakeroot, Glaucous White Lettuce, Glaucous White-lettuce, Purple Rattlesnakeroot, Purple Rattlesnake-root, Racemose White-lettuce, Rattlesnake Root ■☆

313872　Prenanthes racemosa Michx. subsp. multiflora Cronquist = Prenanthes racemosa Michx. ■☆

313873　Prenanthes racemosa Michx. var. multiflora (Cronquist) Dorn = Prenanthes racemosa Michx. ■☆

313874　Prenanthes racemosa Michx. var. pinnatifida A. Gray = Prenanthes racemosa Michx. ■☆

313875　Prenanthes repens L. = Chorisis repens (L.) DC. ■

313876　Prenanthes repens L. = Ixeris repens (L.) A. Gray ■

313877　Prenanthes roanensis (Chick.) Chick.；杂色山地福王草；Roan Mountain Rattlesnakeroot ■☆

313878　Prenanthes sagittata (A. Gray) A. Nelson；窄叶福王草；Arrowleaf Snakeroot ■☆

313879　Prenanthes sarmentosa Willd. = Launaea sarmentosa (Willd.) Kuntze ■

313880　Prenanthes scandens Hook. f. ex Benth. et Hook. f.；藤本福王草■

313881　Prenanthes serpentaria Pursh；蛇足盘果菊；Butterweed, Cankerweed, Gall-of-the-earth, Lion's-foot ■☆

313882　Prenanthes serpentaria Pursh var. barbata (Torr. et A. Gray) A. Gray = Prenanthes barbata (Torr. et A. Gray) Milstead ex Cronquist ■☆

313883　Prenanthes sikkimensis Hook. f. = Cicerbita sikkimensis (Hook. f.) C. Shih ■

313884　Prenanthes sinensis (Hemsl.) Stebbins ex Babc. = Faberia sinensis Hemsl. ■

313885　Prenanthes somaliensis C. Jeffrey；索马里福王草■☆

313886　Prenanthes sonchifolia Bunge = Ixeridium sonchifolium (Maxim.) C. Shih ■

313887　Prenanthes sonchifolia Willd. = Ixeridium sonchifolium (Maxim.) C. Shih ■

313888　Prenanthes sonchifolia Willd. = Launaea intybacea (Jacq.) Beauverd ■☆

313889　Prenanthes spathulata Turcz. ex Herder = Youngia stenoma (Turcz.) Ledeb. ■

313890　Prenanthes squarosa Thunb. = Pterocypsela indica (L.) C. Shih ■

313891　Prenanthes stricta Blume = Youngia japonica (L.) DC. ■

313892　Prenanthes stricta Greene = Rainiera stricta (Greene) Greene ■☆

313893　Prenanthes subpeltata Stebbins；亚盾状福王草■☆

313894　Prenanthes tanakae (Franch. et Sav. ex Y. Tanaka et Ono) Koidz.；田中氏福王草；Tanaka's Rattlesnakeroot ■☆

313895　Prenanthes tatarinowii Maxim.；福王草(卵叶福王草,盘果菊,锥序福王草,总序福王草)；Panicle Rattlesnakeroot, Pyramid Rattlesnakeroot, Raceme Rattlesnakeroot, Racemose Rattlesnakeroot, Tatarinow Rattlesnakeroot ■

313896　Prenanthes tatarinowii Maxim. subsp. macrantha Stebbins ex Walker = Prenanthes macrophylla Franch. ■

313897　Prenanthes tatarinowii Maxim. var. divisa (Nakai et Kitag.) Kitag. = Prenanthes macrophylla Franch. ■

313898　Prenanthes tenuifolia Torr. = Stephanomeria tenuifolia (Raf.) H. M. Hall ■☆

313899　Prenanthes thirionni H. Lév. = Paraprenanthes sororia (Miq.) C. Shih ■

313900　Prenanthes triflora (Hemsl.) C. C. Chang = Notoseris triflora (Hemsl.) C. Shih ■

313901　Prenanthes trifoliolata (Cass.) Fernald；三小叶福王草；Gall-of-the-earth, Ratdesnake-root, Tall Rattlesnake-root, Threeleaved Rattlesnakeroot, Three-leaved White-lettuce ■☆

313902　Prenanthes trifoliolata (Cass.) Fernald var. nana (Bigelow) Fernald = Prenanthes trifoliolata (Cass.) Fernald ■☆

313903　Prenanthes trifoliata Fernald = Prenanthes trifoliolata (Cass.) Fernald ■☆

313904　Prenanthes violifolia Decne. = Paraprenanthes sororia (Miq.) C. Shih ■

313905　Prenanthes virgata Michx. = Prenanthes autumnalis Walter ■☆

313906　Prenanthes vitifolia Diels；葡萄叶福王草(三叶福王草,圆齿福

王草)■

313907　Prenanthes wilsonii C. C. Chang ＝ Notoseris wilsonii (C. C. Chang) C. Shih ■

313908　Prenanthes yakoensis Jeffrey ex Diels;云南福王草(垭口盘果菊);Yunnan Rattlesnakeroot ■

313909　Prenia N. E. Br. (1925);花姬属●☆

313910　Prenia N. E. Br. ＝ Phyllobolus N. E. Br. ●☆

313911　Prenia englishiae (L. Bolus) Gerbaulet;英格利希花姬■☆

313912　Prenia olivacea (Schltr.) H. Jacobsen ＝ Phyllobolus tenuiflorus (Jacq.) Gerbaulet ●☆

313913　Prenia pallens (Aiton) N. E. Br. ;苍白花姬■☆

313914　Prenia pallens (Aiton) N. E. Br. subsp. lancea (Thunb.) Gerbaulet;披针状苍白花姬●☆

313915　Prenia pallens (Aiton) N. E. Br. subsp. lutea L. Bolus;黄苍白花姬■☆

313916　Prenia pallens (Aiton) N. E. Br. subsp. namaquensis Gerbaulet;纳马夸花姬■☆

313917　Prenia radicans (L. Bolus) Gerbaulet;辐射花姬●☆

313918　Prenia relaxata (Willd.) N. E. Br. ＝ Prenia pallens (Aiton) N. E. Br. ■☆

313919　Prenia sladeniana (L. Bolus) L. Bolus;斯莱登花姬■☆

313920　Prenia tetragona (Thunb.) Gerbaulet;四角花姬●☆

313921　Prenia vanrensburgii L. Bolus;范伦花姬■☆

313922　Preonanthus (DC.) Schur ＝ Anemone L. (保留属名)■

313923　Preonanthus Ehrh. ＝ Anemone L. (保留属名)■

313924　Prepodesma N. E. Br. ＝ Aloinopsis Schwantes ■☆

313925　Prepodesma orpenii (N. E. Br.) N. E. Br. ＝ Mesembryanthemum orpenii N. E. Br. ■☆

313926　Prepodesma uncipetala N. E. Br. ＝ Hereroa wilmaniae L. Bolus ■☆

313927　Preptanthe Rchb. f. ＝ Calanthe R. Br. (保留属名)■

313928　Prepusa Mart. (1827);显龙胆属■☆

313929　Prepusa alata Porto et Brade;翅显龙胆■☆

313930　Prepusa montana Mart. ;山地显龙胆■☆

313931　Prepusa viridiflora Brade;绿花显龙胆■☆

313932　Prescotia Lindl. ＝ Prescottia Lindl. ■☆

313933　Prescottia Lindl. (1824)('Prescotia');普雷兰属■☆

313934　Prescottia gracilis Schltr. ＝ Prescottia oligantha (Sw.) Lindl. ■☆

313935　Prescottia myosurus Rchb. ex Griseb. ＝ Prescottia oligantha (Sw.) Lindl. ■☆

313936　Prescottia oligantha (Sw.) Lindl. ;普雷兰■☆

313937　Prescottia panamensis Schltr. ＝ Prescottia oligantha (Sw.) Lindl. ■☆

313938　Preslaea Mart. ＝ Heliotropium L. ●■

313939　Preslea G. Don ＝ Preslia Opiz ■

313940　Preslea Spreng. ＝ Heliotropium L. ●■

313941　Preslea Spreng. ＝ Preslaea Mart. ●■

313942　Preslia Opiz ＝ Mentha L. ●●

313943　Preslia cervina (L.) Fresen. ＝ Mentha cervina L. ■☆

313944　Presliophytum (Urb. et Gilg) Weigend(1997);爪瓣刺莲花属●☆

313945　Presliophytum arequipensis Weigend;爪瓣刺莲花●☆

313946　Prestelia Sch. Bip. ＝ Prestelia Sch. Bip. ex Benth. et Hook. f. ●☆

313947　Prestelia Sch. Bip. ex Benth. et Hook. f. (1865);无茎灯头菊属●☆

313948　Prestelia Sch. Bip. ex Benth. et Hook. f. ＝ Eremanthus Less. ●☆

313949　Prestelia eriopus Sch. Bip. ;无茎灯头菊■☆

313950　Prestinaria Sch. Bip. ex Hochst. ＝ Coreopsis L. ●■

313951　Prestinaria bidentoides Sch. Bip. ＝ Bidens prestinaria (Sch. Bip. ex Walp.) Cufod. ■☆

313952　Prestoea Hook. f. (1883)(保留属名);山甘蓝椰属(不列思多棕属,粉轴椰属,派斯斯托桐属);Prestoea ●☆

313953　Prestoea acuminata (Willd.) H. E. Moore;渐尖山甘蓝椰●☆

313954　Prestoea brachyclada (Burret) R. Bernal, Galeano et A. J. Hend. ;短枝山甘蓝椰●☆

313955　Prestoea integrifolia de Nevers et A.J.Hend. ;全缘山甘蓝椰●☆

313956　Prestonia R. Br. (1810)(保留属名);五角木属;Prestonia ●☆

313957　Prestonia Scop. (废弃属名)＝ Abutilon Mill. ●■

313958　Prestonia Scop. (废弃属名)＝ Lass Adans. (废弃属名)●■☆

313959　Prestonia Scop. (废弃属名)＝ Prestonia R. Br. (保留属名)●☆

313960　Prestonia acutifolia (Benth. ex Müll. Arg.) K. Schum. ;尖叶五角木●☆

313961　Prestonia amazonica J. F. Macbr. ;亚马逊五角木●☆

313962　Prestonia boliviana J. F. Morales et A. Fuentes;玻利维亚五角木●☆

313963　Prestonia cordifolia Woodson;心叶五角木●☆

313964　Prestonia discolor Woodson;异色五角木●☆

313965　Prestonia gracilis Rusby;细五角木●☆

313966　Prestonia grandiflora L. O. Williams;大花五角木●☆

313967　Prestonia laxa Rusby ex Woodson;松散五角木●☆

313968　Prestonia longifolia (Sessé et Moç.) J. F. Morales;长叶五角木●☆

313969　Prestonia lutescens Müll. Arg. ;浅黄五角木●☆

313970　Prestonia macrocarpa Hemsl. ;大果五角木●☆

313971　Prestonia macroneura (Müll. Arg.) Woodson;粗脉五角木●☆

313972　Prestonia macrophylla Woodson;大叶五角木●☆

313973　Prestonia mexicana A. DC. ;墨西哥五角木●☆

313974　Prestonia mucronata Rusby;钝尖五角木●☆

313975　Prestonia parviflora Baill. ;小花五角木●☆

313976　Prestonia quinquangularis Spreng. ;五角木;Fiveangular Prestonia ●☆

313977　Prestonia rotundifolia K. Schum. ex Woodson;圆叶五角木●☆

313978　Prestonia tomentosa Seem. ;毛五角木●☆

313979　Prestonia trifida (Poepp.) Woodson;三裂五角木●☆

313980　Prestoniopsis Müll. Arg. (1860);类五角木属●☆

313981　Prestoniopsis Müll. Arg. ＝ Dipladenia A. DC. ●

313982　Prestoniopsis pubescens Müll. Arg. ;类五角木●☆

313983　Pretrea J. Gay ＝ Dicerocaryum Bojer ■☆

313984　Pretrea J. Gay ex Meisn. ＝ Dicerocaryum Bojer ■☆

313985　Pretrea artemisiifolia Klotzsch ＝ Dicerocaryum senecioides (Klotzsch) Abels ■☆

313986　Pretrea bojeriana Decne. ＝ Dicerocaryum zanguebarium (Lour.) Merr. ■☆

313987　Pretrea eriocarpa Decne. ＝ Dicerocaryum eriocarpum (Decne.) Abels ■☆

313988　Pretrea forbesii Decne. ＝ Dicerocaryum forbesii (Decne.) A. E. van Wyk ■☆

313989　Pretrea loasifolia Klotzsch ＝ Dicerocaryum senecioides (Klotzsch) Abels ■☆

313990　Pretrea senecioides Klotzsch ＝ Dicerocaryum senecioides (Klotzsch) Abels ■☆

313991　Pretrea zanguebaria (Lour.) J. Gay ex DC. ＝ Dicerocaryum zanguebarium (Lour.) Merr. ■☆

313992　Pretreothamnus Engl. ＝ Josephinia Vent. ●■☆

313993　Pretreothamnus africanus (Vatke) B. Fedtsch. ＝ Josephinia africana Vatke ●☆

313994　Pretreothamnus rosaceus Engl. ＝ Josephinia africana Vatke ●☆

313995　Preussiella Gilg(1897);普罗野牡丹属☆

313996 Preussiella chevalieri Jacq. -Fél. = Preussiella kamerunensis Gilg ☆

313997 Preussiella gabonensis Jacq. -Fél. ;非洲普罗野牡丹 ☆

313998 Preussiella kamerunensis Gilg;普罗野牡丹 ☆

313999 Preussiodora Keay(1958);普罗茜属 ● ☆

314000 Preussiodora sulphurea (K. Schum.) Keay;普罗茜 ● ☆

314001 Prevoita Steud. = Cerastium L. ■

314002 Prevoita Steud. = Prevotia Adans. ■

314003 Prevostea Choisy = Calycobolus Willd. ex Schult. ● ☆

314004 Prevostea acuminata Pilg. = Calycobolus acuminatus (Pilg.) Heine ● ☆

314005 Prevostea acuta Pilg. = Calycobolus acutus (Pilg.) Heine ● ☆

314006 Prevostea africana (G. Don) Benth. = Calycobolus africanus (G. Don) Heine ● ☆

314007 Prevostea alternifolia (Planch.) Hallier f. = Calycobolus africanus (G. Don) Heine ● ☆

314008 Prevostea cabrae De Wild. et T. Durand = Calycobolus heudelotii (Baker ex Oliv.) Heine ● ☆

314009 Prevostea campanulata K. Schum. ex Hallier f. = Calycobolus campanulatus (K. Schum. ex Hallier f.) Heine ● ☆

314010 Prevostea claessensii De Wild. = Calycobolus claessensii (De Wild.) Heine ● ☆

314011 Prevostea cordata Hallier f. = Bonamia semidigyna (Roxb.) Hallier f. ● ☆

314012 Prevostea cordata Hallier f. = Calycobolus cordatus (Hallier f.) Heine ● ☆

314013 Prevostea gilgiana Pilg. =Calycobolus gilgianus (Pilg.) Heine ● ☆

314014 Prevostea heudelotii (Baker ex Oliv.) Hallier f. = Calycobolus heudelotii (Baker ex Oliv.) Heine ● ☆

314015 Prevostea heudelotii (Baker ex Oliv.) Hallier f. var. minor Rendle = Calycobolus heudelotii (Baker ex Oliv.) Heine ● ☆

314016 Prevostea insignis Rendle = Calycobolus insignis (Rendle) Heine ● ☆

314017 Prevostea klaineana Pierre ex Pellegr. = Calycobolus klaineanus (Pierre ex Pellegr.) Heine ● ☆

314018 Prevostea lucida R. D. Good = Calycobolus heudelotii (Baker ex Oliv.) Heine ● ☆

314019 Prevostea mayombensis Pellegr. = Calycobolus mayombensis (Pellegr.) Heine ● ☆

314020 Prevostea mayumbensis R. D. Good = Calycobolus goodii Heine ● ☆

314021 Prevostea morthehanii De Wild. = Calycobolus mortehanii (De Wild.) Heine ● ☆

314022 Prevostea mossambicensis Klotzsch = Bonamia mossambicensis (Klotzsch) Hallier f. ● ☆

314023 Prevostea nigerica Rendle = Calycobolus africanus (G. Don) Heine ● ☆

314024 Prevostea oddonii De Wild. = Calycobolus campanulatus (K. Schum. ex Hallier f.) Heine subsp. oddonii (De Wild.) Lejoly et Lisowski ● ☆

314025 Prevostea parviflora Mangenot = Calycobolus parviflorus (Mangenot) Heine ● ☆

314026 Prevostea racemosa R. D. Good = Calycobolus racemosus (R. D. Good) Heine ● ☆

314027 Prevotia Adans. = Cerastium L. ■

314028 Priamosia Urb. (1919);海地木属 ● ☆

314029 Priamosia domingensis Urb. ;海地木 ● ☆

314030 Prianthes Pritz. = Prionanthes Schrank ■● ☆

314031 Prianthes Pritz. = Trixis P. Browne ■● ☆

314032 Prickothamnus Nutt. ex Baill. = Pickeringia Nutt. (保留属名) ● ☆

314033 Pridania Gagnep. = Pycnarrhena Miers ex Hook. f. et Thomson ●

314034 Pridania petelotii Gagnep. = Pycnarrhena poilanei (Gagnep.) Forman ●

314035 Priestleya DC. (1825);普里豆属 ■ ☆

314036 Priestleya DC. = Liparia L. ■● ☆

314037 Priestleya Moc. et Sessé ex DC. = Montanoa Cerv. ■● ☆

314038 Priestleya angustifolia Eckl. et Zeyh. = Liparia angustifolia (Eckl. et Zeyh.) A. L. Schutte ■ ☆

314039 Priestleya boucheri E. G. H. Oliv. et Fellingham = Liparia boucheri (E. G. H. Oliv. et Fellingham) A. L. Schutte ■ ☆

314040 Priestleya calycina L. Bolus = Liparia calycina (L. Bolus) A. L. Schutte ■ ☆

314041 Priestleya capitata (Thunb.) DC. = Liparia capitata Thunb. ■ ☆

314042 Priestleya capitata (Thunb.) DC. var. pilosa (E. Mey.) Harv. = Liparia capitata Thunb. ■ ☆

314043 Priestleya cephalotes E. Mey. = Liparia umbellifera Thunb. ■ ☆

314044 Priestleya cephalotes E. Mey. var. angustifolia ? = Liparia umbellifera Thunb. ■ ☆

314045 Priestleya elliptica DC. = Xiphotheca elliptica (DC.) A. L. Schutte et B. -E. van Wyk ■ ☆

314046 Priestleya ericifolia (L.) DC. = Amphithalea ericifolia (L.) Eckl. et Zeyh. ■ ☆

314047 Priestleya glauca T. M. Salter = Xiphotheca lanceolata (E. Mey.) Eckl. et Zeyh. ■ ☆

314048 Priestleya graminifolia (L.) DC. = Liparia graminifolia L. ■ ☆

314049 Priestleya guthriei L. Bolus = Xiphotheca guthriei (L. Bolus) A. L. Schutte et B. -E. van Wyk ■ ☆

314050 Priestleya hirsuta (Thunb.) DC. = Liparia hirsuta Thunb. ■ ☆

314051 Priestleya hirsuta (Thunb.) DC. var. subenervia Meisn. = Liparia hirsuta Thunb. ■ ☆

314052 Priestleya hirsuta (Thunb.) DC. var. trinervia Meisn. = Liparia hirsuta Thunb. ■ ☆

314053 Priestleya laevigata (L.) DC. = Liparia laevigata (L.) Thunb. ■ ☆

314054 Priestleya laevigata (L.) DC. var. glabra E. Mey. = Liparia capitata Thunb. ■ ☆

314055 Priestleya laevigata (L.) DC. var. pilosa E. Mey. = Liparia capitata Thunb. ■ ☆

314056 Priestleya laevigata (L.) Druce = Liparia laevigata (L.) Thunb. ■ ☆

314057 Priestleya lanceolata E. Mey. = Xiphotheca lanceolata (E. Mey.) Eckl. et Zeyh. ■ ☆

314058 Priestleya latifolia Benth. = Liparia latifolia (Benth.) A. L. Schutte ■ ☆

314059 Priestleya leiocarpa Eckl. et Zeyh. = Liparia myrtifolia Thunb. ■ ☆

314060 Priestleya meyeri Meisn. = Amphithalea imbricata (L.) Druce ■ ☆

314061 Priestleya myrtifolia (Thunb.) DC. = Liparia myrtifolia Thunb. ■ ☆

314062 Priestleya rotundifolia (Eckl. et Zeyh.) Walp. = Xiphotheca tecta (Thunb.) A. L. Schutte et B. -E. van Wyk ■ ☆

314063 Priestleya schlechteri L. Bolus = Xiphotheca canescens (Thunb.) A. L. Schutte et B. -E. van Wyk ■ ☆

314064 Priestleya stokoei L. Bolus = Xiphotheca tecta (Thunb.) A. L. Schutte et B. -E. van Wyk ■ ☆

314065 Priestleya tecta (Thunb.) DC. = Xiphotheca tecta (Thunb.) A. L. Schutte et B. -E. van Wyk ■ ☆

314066 Priestleya teres (Thunb.) DC. = Liparia genistoides (Lam.)

A. L. Schutte ■☆

314067　Priestleya thunbergii Benth. = Liparia laevigata（L.）Thunb. ■☆

314068　Priestleya thunbergii Benth. var. villosa Harv. = Liparia laevigata（L.）Thunb. ■☆

314069　Priestleya tomentosa（L.）Druce = Liparia vestita Thunb. ■☆

314070　Priestleya umbellifera（Thunb.）DC. = Liparia umbellifera Thunb. ■☆

314071　Priestleya umbellifera Eckl. et Zeyh. = Liparia laevigata（L.）Thunb. ■☆

314072　Priestleya vestita（Thunb.）DC. = Liparia vestita Thunb. ■☆

314073　Priestleya villosa（Thunb.）Druce = Liparia angustifolia（Eckl. et Zeyh.）A. L. Schutte ■☆

314074　Priestleya villosa DC. = Xiphotheca fruticosa（L.）A. L. Schutte et B. -E. van Wyk ■☆

314075　Prieurea DC. = Ludwigia L. ●■

314076　Prieurea senegalensis DC. = Ludwigia senegalensis（DC.）Troch. ☆

314077　Prieurella Pierre = Bumelia Sw.（保留属名）●☆

314078　Prieurella Pierre = Chrysophyllum L. ●

314079　Prieuria Benth. et Hook. f. = Ludwigia L. ●■

314080　Prieuria Benth. et Hook. f. = Prieurea DC. ●■

314081　Prieuria DC. = Ludwigia L. ●■

314082　Primula Kuntze = Androsace L. ■

314083　Primula L.（1753）；报春花属（报春属，樱草属）；Cowslip, Freckled Face，Primrose ■

314084　Primula × kewensis W. Watson；邱园报春；Kew Primrose ■☆

314085　Primula × polyantha L. H. Bailey；多花报春；Elatior Hybrid Primroses，Primrose ■☆

314086　Primula abchasia Sosn. ；阿伯哈斯报春■☆

314087　Primula acaulis（L.）L. ；无茎报春■☆

314088　Primula acaulis（L.）L. subsp. atlantica（Maire et Wilczek）Greuter et Burdet；北非无茎报春■☆

314089　Primula acaulis（L.）L. var. atlantica Maire et Wilczek = Primula acaulis（L.）L. subsp. atlantica（Maire et Wilczek）Greuter et Burdet ■☆

314090　Primula acaulis Hill = Primula vulgaris（L.）Huds. ■☆

314091　Primula adenantha Balf. f. et Cooper = Primula bellidifolia King ex Hook. f. ■

314092　Primula adenophora Blatt. = Primula denticulata Sm. ■

314093　Primula advena W. W. Sm. ；折瓣雪山报春；Flexpetal Primrose ■

314094　Primula advena W. W. Sm. var. argentata W. W. Sm. = Primula advena W. W. Sm. ■

314095　Primula advena W. W. Sm. var. concolor W. W. Sm. = Primula advena W. W. Sm. ■

314096　Primula advena W. W. Sm. var. euprepes（W. W. Sm.）F. H. Chen et C. M. Hu；紫折瓣报春■

314097　Primula aemula Balf. f. ；粗葶报春■

314098　Primula aequalis Craib = Primula denticulata Sm. ■

314099　Primula aequipila Craib = Primula ovalifolia Franch. ■

314100　Primula aerinantha Balf. f. et Purdom；裂瓣穗花报春（裂瓣穗状报春）■

314101　Primula agleniana Balf. f. et Forrest；乳黄雪山报春；Milky-yellow Primrose ■

314102　Primula agleniana Balf. f. et Forrest var. alba Forrest = Primula agleniana Balf. f. et Forrest ■

314103　Primula agleniana Balf. f. et Forrest var. atrocrocea Kingdon-Ward = Primula agleniana Balf. f. et Forrest ■

314104　Primula aitchisonii Pax = Primula macrophylla D. Don ■☆

314105　Primula alchemilloides（Franch.）Derganc = Androsace alchemilloides Franch. ■

314106　Primula algida Adam ex Weber et Mohr；寒地报春（冷报春，冷地报春）；Algid Primrose ■

314107　Primula aliciae G. Taylor ex W. W. Sm. ；西藏缺裂报春；Incised Primrose ■

314108　Primula allionii Loisel. ；单花报春■☆

314109　Primula alpicola（W. W. Sm.）Stapf；杂色钟报春（顶花报春，高山报春）；Variegate Primrose ■

314110　Primula alpicola（W. W. Sm.）Stapf subsp. luna Stapf = Primula alpicola（W. W. Sm.）Stapf ■

314111　Primula alpicola（W. W. Sm.）Stapf subsp. violacea Stapf = Primula alpicola（W. W. Sm.）Stapf ■

314112　Primula alpicola（W. W. Sm.）Stapf var. alba W. W. Sm. = Primula alpicola（W. W. Sm.）Stapf ■

314113　Primula alsophila Balf. f. et Farrer；蔓茎报春；Creeping Primrose ■

314114　Primula alta Balf. f. et Forrest = Primula denticulata Sm. subsp. sinodenticulata（Balf. f. et Forrest）W. W. Sm. et Forrest ■

314115　Primula amabilis Balf. f. et Forrest = Primula diantha Bureau et Franch. ■

314116　Primula ambita Balf. f. ；圆迴报春（黄花鄂报春，圆回报春，圆迦报春）；Cycle Primrose ■

314117　Primula amethystina Franch. ；紫晶报春（紫花报春）；Amethyst Primrose，Purpleflower Primrose ■

314118　Primula amethystina Franch. subsp. argutidens（Franch.）W. W. Sm. et H. R. Fletcher；尖齿紫晶报春；Sharptooth ■

314119　Primula amethystina Franch. subsp. brevifolia（Forrest）W. W. Sm. et Forrest；短叶紫晶报春■

314120　Primula amoena M. Bieb. ；秀丽报春■☆

314121　Primula androsacea Pax = Primula forbesii Franch. ■

314122　Primula anisodora Balf. f. et Forrest；茴香灯台报春（茴香报春，异味报春）；Anise Primrose，Fennel Primrose ■

314123　Primula annulata Balf. f. et Kingdon-Ward；单花小报春；Singleflower Primrose ■

314124　Primula apoclita Balf. f. et Forrest = Primula pinnatifida Franch. ■

314125　Primula arctica Koidz. ；北极报春■

314126　Primula argutidens Franch. = Primula amethystina Franch. subsp. argutidens（Franch.）W. W. Sm. et H. R. Fletcher ■

314127　Primula arizonica（A. Gray）Derganc = Androsace occidentalis Pursh ■☆

314128　Primula aromatica W. W. Sm. et Forrest；香花报春（香报春）；Fragrantflower Primrose ■

314129　Primula articulata W. W. Sm. = Primula bracteata Franch. ■

314130　Primula articulata W. W. Sm. var. sublinearis W. W. Sm. = Primula bracteata Franch. ■

314131　Primula asarifolia H. R. Fletcher；细辛叶报春（心叶鄂报春）；Wild-gingerleaf Primrose ■

314132　Primula asperulata N. P. Balakr. = Primula blinii H. Lév. ■

314133　Primula atricapilla Balf. f. et Cooper = Primula bellidifolia King ex Hook. f. ■

314134　Primula atrodentata W. W. Sm. ；白心球花报春（黑齿报春）；Blacktoothed Primrose ■

314135　Primula atrodentata W. W. Sm. subsp. orestora（Craib et Cooper）W. W. Sm. et Forrest = Primula atrodentata W. W. Sm. ■

314136　Primula atrotubata W. W. Sm. et Forrest = Primula malvacea Franch. ■

314137　Primula atroviolacea Jacq. ex Duby = Primula macrophylla D. Don ■

314138　Primula atuntzuensis Balf. f. et Forrest = Primula minor Balf. f. et Kingdon-Ward ■

314139　Primula aurantiaca W. W. Sm. et Forrest;橙红灯台报春(橙红报春);Orange Primrose ■

314140　Primula aurantiaca W. W. Sm. et Forrest = Primula prenantha Balf. f. et W. W. Sm. ■

314141　Primula aurantiaca W. W. Sm. et Forrest subsp. morsheandiana F. H. Chen et C. M. Hu = Primula prenantha Balf. f. et W. W. Sm. subsp. morsheadiana (Kingdon-Ward) F. H. Chen et C. M. Hu ■

314142　Primula aureata H. R. Fletcher;尼泊尔黄报春■☆

314143　Primula auricula L. ;熊耳报春;Auricula,Bazier Basier,Bear's Ear, Bear's Ears,Bearsear,Bezors,Boar's Ear,Boar's Ears,Bore's Ear,Bore's Ears,Cat's Lugs,Cowslip,Dusty Miller. , French Cowslip, Mountain Cowslip, Primmily, Racalus, Rackeler Rackliss, Reckless, Rigglers, Tanner's Apron ■☆

314144　Primula auriculata Lam. ;耳叶报春(耳状报春,耳状报春花)■☆

314145　Primula auriculata Lam. var. polyphylla Franch. = Primula pseudodenticulata Pax ■

314146　Primula baileyana Kingdon-Ward;圆叶报春;Bailey Primrose ■

314147　Primula baldshuanica B. Fedtsch. ;巴尔德报春■☆

314148　Primula balfourii H. Lév. = Primula spicata Franch. ■

314149　Primula baokongensis F. H. Chen et C. M. Hu = Primula neurocalyx Franch. ■

314150　Primula barbatula W. W. Sm. ;紫球毛小报春;Bearded Primrose ■

314151　Primula barbeyana Petitm. = Primula forbesii Franch. ■

314152　Primula barbicalyx C. H. Wright;毛萼鄂报春;Haircalyx Primrose ■

314153　Primula barnardoana W. W. Sm. et Kingdon-Ward = Primula elongata Watt var. barnardoana (W. W. Sm. et Kingdon-Ward) C. M. Hu ■

314154　Primula barybotrys Hand. -Mazz. = Primula malvacea Franch. ■

314155　Primula bathangensis Petitm. et Hand. -Mazz. ;巴塘报春;Batang Primrose,Patang Primrose ■

314156　Primula bayerni Rupr. ;巴氏报春■

314157　Primula beesiana Forrest;霞红灯台报春(霞红报春,鱼肠草);Bees Primrose ☆

314158　Primula begoniiformis Petitm. = Primula obconica Hance subsp. begoniiformis (Petitm.) W. W. Sm. et Forrest ■

314159　Primula bella Franch. ;山丽报春;Beautiful Primrose ■

314160　Primula bella Franch. subsp. bonattana (Petitm.) W. W. Sm. et Forrest = Primula bella Franch. ■

314161　Primula bella Franch. subsp. cyclostegia (Hand. -Mazz.) W. W. Sm. = Primula bella Franch. ■

314162　Primula bella Franch. subsp. cyclostegia (Hand. -Mazz.) W. W. Sm. et Forrest = Primula bella Franch. ■

314163　Primula bella Franch. subsp. moschophora (Balf. f. et Forrest) W. W. Sm. et Forrest = Primula moschophora Balf. f. et Forrest ■

314164　Primula bella Franch. subsp. nanobella (Balf. f. et Forrest) W. W. Sm. et Forrest = Primula bella Franch. ■

314165　Primula bellidifolia King ex Hook. f. ;菊叶穗花报春(雏菊叶报春);Daisyleaf Primrose ■

314166　Primula bhutanica H. R. Fletcher = Primula whitei W. W. Sm. ■

314167　Primula biondiana Petitm. = Primula stenocalyx Maxim. ■

314168　Primula biserrata Forrest = Primula serratifolia Franch. ■

314169　Primula blattariformis Franch. ;地黄叶报春(毛蕊草报春);Hairystamen Primrose ■

314170　Primula blattariformis Franch. subsp. tenana (Bonati ex Balf. f.) W. W. Sm. et Forrest = Primula blattariformis Franch. ■

314171　Primula blattariformis Franch. var. duclouxii Bonati = Primula blattariformis Franch. ■

314172　Primula blinii H. Lév. ;糙毛报春(羽叶报春);Scabroushair Primrose ■

314173　Primula bomiensis F. H. Chen et C. M. Hu;波密脆蒴报春;Bomi Primrose ■

314174　Primula bonatiana Petitm. = Primula bella Franch. ■

314175　Primula bonatii R. Knuth = Primula obconica Hance ■

314176　Primula boothii Craib = Primula bracteosa Craib ■

314177　Primula borealis Duby;北方报春■☆

314178　Primula boreio-calliantha Balf. f. et Cooper;木里报春;Muli Primrose ■

314179　Primula brachystoma W. W. Sm. = Primula prenantha Balf. f. et W. W. Sm. ■

314180　Primula bracteata Franch. ;小苞报春(鳞茎报春);Bracteate Primrose ■

314181　Primula bracteosa Craib;叶苞脆蒴报春;Manybract Primrose ■

314182　Primula brevicula Balf. f. et Forrest = Primula diantha Bureau et Franch. ■

314183　Primula brevifolia Forrest = Primula amethystina Franch. subsp. brevifolia (Forrest) W. W. Sm. et Forrest ■

314184　Primula breviscapa Franch. ;短葶报春(葶花卵叶报春);Shortstalk Primrose ■

314185　Primula bryophila Balf. f. et Farrer = Primula calliantha Franch. subsp. bryophila (Balf. f. et Farrer) W. W. Sm. et Forrest ■

314186　Primula bullata Franch. ;皱叶报春(黄葵报春);Bullate Primrose ■

314187　Primula bullata Franch. = Primula forrestii Balf. f. ■

314188　Primula bullata Franch. var. rufa (Balf. f.) W. W. Sm. et H. R. Fletcher = Primula forrestii Balf. f. ■

314189　Primula bulleyana Forrest;橘红灯台报春(橘红报春);Bulley Primrose,Primrose ■

314190　Primula burmanica Balf. f. et Kingdon-Ward = Primula beesiana Forrest ■

314191　Primula buryana Balf. f. ;珠峰垂花报春;Jolmolungma Primrose ■

314192　Primula cachemeriana Munro = Primula denticulata Sm. ■

314193　Primula calcicola Balf. f. et Forrest = Primula yunnanensis Franch. ■

314194　Primula calciphila Hutch. = Primula rupestris Balf. f. et Forrest ■

314195　Primula caldaria W. W. Sm. et Forrest;匍枝粉报春;Creepingbract Primrose ■

314196　Primula caldaria W. W. Sm. et Forrest var. nana W. W. Sm. et Forrest = Primula caldaria W. W. Sm. et Forrest ■

314197　Primula calderiana Balf. f. et Cooper;暗紫脆蒴报春(卡德报春);Calder Primrose ■

314198　Primula calderiana Balf. f. et Cooper f. alba (W. W. Sm.) H. Hara = Primula calderiana Balf. f. et Cooper ■

314199　Primula calderiana Balf. f. et Cooper subsp. strumosa (Balf. f. et Cooper) A. J. Richards = Primula strumosa Balf. f. et Cooper ■

314200　Primula calderiana Balf. f. et Cooper var. alba (W. W. Sm.) W. W. Sm. et H. R. Fletcher = Primula calderiana Balf. f. et Cooper ■

314201　Primula calliantha Franch. ;美花报春(雪山厚叶报春,紫鹃报春);Beautifulflower Primrose,Thickleaf Primrose ■

314202　Primula calliantha Franch. subsp. bryophila (Balf. f. et Farrer) W. W. Sm. et Forrest;黛粉美花报春(黛粉雪山报春)■

314203　Primula calliantha Franch. subsp. kiuchiangensis (Balf. f. et Forrest) W. W. Sm. et Forrest = Primula diantha Bureau et Franch. ■

314204　Primula calliantha Franch. subsp. mishmiensis (Kingdon-Ward) C. M. Hu;黄美花报春(亮叶雪山报春)■

314205　Primula calliantha Franch. var. albiflos W. W. Sm. et Forrest = Primula calliantha Franch. subsp. bryophila (Balf. f. et Farrer) W. W. Sm. et Forrest ■

314206　Primula calliantha Franch. var. nuda Farrer ex W. W. Sm. = Primula calliantha Franch. subsp. bryophila (Balf. f. et Farrer) W. W. Sm. et Forrest ■

314207　Primula calthifolia W. W. Sm.;驴蹄草叶报春;Marshmarigoldleaf Primrose ■

314208　Primula calyptrata X. Gong et R. C. Fang;帽盖报春■☆

314209　Primula cana Balf. f. et Cave = Primula caveana W. W. Sm. ■

314210　Primula candicans W. W. Sm.;亮白小报春;Bright Primrose ■

314211　Primula candidissima W. W. Sm. et Forrest = Primula sinuata Franch. ■

314212　Primula capitata Hook.;头序报春(头状樱草);Capitate Primrose,Round-headed Mealy Primrose ■

314213　Primula capitata Hook. subsp. craibeana (Balf. f. et W. W. Sm.) W. W. Sm. et Forrest = Primula capitata Hook. subsp. lacteocapitata (Balf. f. et W. W. Sm.) W. W. Sm. et Forrest ■

314214　Primula capitata Hook. subsp. crispata (Balf. f. et W. W. Sm.) W. W. Sm. et Forrest = Primula capitata Hook. ■

314215　Primula capitata Hook. subsp. lacteocapitata (Balf. f. et W. W. Sm.) W. W. Sm. et Forrest;黄粉头序报春■

314216　Primula capitata Hook. subsp. mooreana (Balf. f. et W. W. Sm.) W. W. Sm. et Forrest = Primula capitata Hook. ■

314217　Primula capitata Hook. subsp. sphaerocephala (Balf. f. et W. W. Sm.) W. W. Sm. et Forrest;无粉头序报春(大回心草,俯垂球花报春,水白菜)■

314218　Primula capitellata Boiss.;小头报春■☆

314219　Primula cardiophylla Balf. f. et W. W. Sm. = Primula rotundifolia Wall. ex Roxb. ■

314220　Primula carnosula Balf. f. et Forrest = Primula gemmifera Batalin var. amoena F. H. Chen ■

314221　Primula carpathica Fuss;卡尔帕索斯报春■☆

314222　Primula cavaleriei Petitm.;黔西报春(贵州报春);Cavalearie Primrose ■

314223　Primula caveana W. W. Sm.;短葶圆叶报春;Cavea Primrose ■

314224　Primula cawdoriana Kingdon-Ward;条裂垂花报春(芳香垂花报春);Cawdor Primrose ■

314225　Primula celsiaeformis Franch.;显脉报春(青葙报春);Mulleinlike Primrose ■

314226　Primula cephalantha Balf. f. = Primula pinnatifida Franch. ■

314227　Primula cerina H. R. Fletcher;蜡黄报春;Waxyellow Primrose ■

314228　Primula cernua Franch.;垂花穗状报春(垂花穗花报春,米伞花,斜倾报春,斜倾报春花,野洋参);Nodding Primrose,Nutantspike Primrose ■

314229　Primula chamaedoron W. W. Sm.;单花脆蒴报春(矮茎报春);Singleflower Primrose ■

314230　Primula chamaethauma W. W. Sm.;异葶脆蒴报春(高原报春,缅藏报春);Variegatestalk Primrose ■

314231　Primula chamaethauma W. W. Sm. var. chiukiangensis F. H. Chen = Primula chamaethauma W. W. Sm. ■

314232　Primula chapaensis Gagnep. ;马关报春;Maguan Primrose ■

314233　Primula chartacea Franch. ;草叶报春(纸叶报春);Leathryleaf Primrose ■

314234　Primula cheniana W. P. Fang;时雅报春;Chen Primrose ■

314235　Primula cheniana W. P. Fang = Primula epilosa Craib ■

314236　Primula chienii W. P. Fang;青城报春(雨农报春);Chien Primrose,Qingcheng Primrose ■

314237　Primula chionantha Balf. f. et Forrest;玉葶报春(紫花雪山报春);Snowblossom Primrose ■

314238　Primula chionata W. W. Sm.;裂叶脆蒴报春;Splitleaf Primrose ■

314239　Primula chionata W. W. Sm. var. violacea W. W. Sm. ;蓝花裂叶脆蒴报春(蓝花裂叶报春)■

314240　Primula chionogenes H. R. Fletcher;粗齿脆蒴报春;Bigtoothed Primrose ■

314241　Primula chlorodryas W. W. Sm. = Primula dryadifolia Franch. subsp. chlorodryas (W. W. Sm.) F. H. Chen et C. M. Hu ■

314242　Primula chrysochlora Balf. f. et Kingdon-Ward;腾冲灯台报春;Yellow Primrose ■

314243　Primula chrysopa Balf. f. et Forrest = Primula gemmifera Batalin var. amoena F. H. Chen ■

314244　Primula chrysophylla Balf. f. et Forrest = Primula dryadifolia Franch. ■

314245　Primula chumbiensis W. W. Sm. ;厚叶钟报春;Thickleaf Primrose ■

314246　Primula chungensis Balf. f. et Kingdon-Ward;中甸灯台报春(中甸报春);Chungien Primrose,Zhongdian Primrose ■

314247　Primula cicutariifolia Pax;毛茛叶报春;Crowflower-leaf Primrose,Waterhemlockleaf Primrose ■

314248　Primula cinerascens Franch. ;灰绿报春(灰白报春);Greyish Primrose ■

314249　Primula cinerascens Franch. subsp. sinomollis (Balf. f. et Forrest) W. W. Sm. et Forrest = Primula sinomollis Balf. f. et Forrest ■

314250　Primula cinerascens Franch. subsp. sylvicola (Hutch.) W. W. Sm. et Forrest = Primula sinomollis Balf. f. et Forrest ■

314251　Primula cinerascens Franch. subsp. violodora (Dunn) W. W. Sm. et Forrest = Primula cinerascens Franch. ■

314252　Primula citrina Balf. f. et Purdom = Primula flava Maxim. ■

314253　Primula clarkei Watt et Stokes;克拉克报春(克什米尔报春);C. B. Clarke Primrose ■☆

314254　Primula clusiana Tausch;白斑红报春■☆

314255　Primula clutterbuckii Kingdon-Ward;短茎粉报春(克鲁特报春);Shortstem Primrose ■

314256　Primula cockburniana Hemsl.;鹅黄灯台报春(川西报春,鹅黄报春);Cockbum Primrose ■

314257　Primula coerulea Forrest;蓝花大叶报春(天蓝报春);Lightblue Primrose ■

314258　Primula cognata Duthie = Primula stenocalyx Maxim. ■

314259　Primula comata H. R. Fletcher;镇康报春(团丛报春);Tasselled Primrose ■

314260　Primula compsantha Balf. f. et Forrest = Primula pulchella Franch. ■

314261　Primula concholoba Stapf et Sealy;短筒穗花报春;Shorttube Primrose ■

314262　Primula concinna Watt;雅洁粉报春(美观樱草);Graceful Primrose,Pretty Primrose ■

314263　Primula congestifolia Forrest = Primula dryadifolia Franch. ■

314264　Primula conica Balf. f. et Forrest = Primula deflexa Duthie ■

314265　Primula consocia W. W. Sm. ＝ Primula littledalei Balf. f. et Watt ■

314266　Primula conspersa Balf. f. et Purdom;散布报春(灯台报春,灯台大苞报春,密布);Scattered Primrose ■

314267　Primula cordata Balf. f. ex W. W. Sm. et Forrest ＝ Primula rotundifolia Wall. ex Roxb. ■

314268　Primula cordifolia Pax ＝ Primula rotundifolia Wall. ex Roxb. ■

314269　Primula cordifolia Rupr. ;喜马拉雅心叶报春■☆

314270　Primula cordifolia Schur ＝ Primula officinalis Jacq. ■☆

314271　Primula cortusoides L. ;晚报春(翠兰花,拟报春)■☆

314272　Primula cortusoides L. var. lichiangensis Forrest ＝ Primula polyneura Franch. ■

314273　Primula coryana Balf. f. et Forrest ex W. W. Sm. ＝ Primula boreio-calliantha Balf. f. et Cooper ■

314274　Primula craibeana Balf. f. et W. W. Sm. ＝ Primula capitata Hook. subsp. lacteocapitata (Balf. f. et W. W. Sm.) W. W. Sm. et Forrest ■

314275　Primula crassa Hand. -Mazz. ;毛卵叶报春■

314276　Primula crassa Hand. -Mazz. ＝ Primula ovalifolia Franch. ■

314277　Primula crispa Balf. f. et W. W. Sm. ＝ Primula glomerata Pax ■

314278　Primula crispata Balf. f. et W. W. Sm. ＝ Primula capitata Hook. ■

314279　Primula crocifolis Pax et K. Hoffm. ;番红报春(红花报春);Crocusleaf Primrose ■

314280　Primula cuneifolia Ledeb. ;北海道小报春;Yezo Primrose ■☆

314281　Primula cuneifolia Ledeb. f. leucantha H. Hara;白花北海道小报春■☆

314282　Primula cuneifolia Ledeb. subsp. hakusanensis (Franch.) W. W. Sm. et Forrest ＝ Primula cuneifolia Ledeb. var. hakusanensis (Franch.) Makino ■☆

314283　Primula cuneifolia Ledeb. subsp. heterodonta (Makino) W. W. Sm. et Forrest ＝ Primula heterodonta Franch. ■☆

314284　Primula cuneifolia Ledeb. var. hakusanensis (Franch.) Makino;白山小报春■☆

314285　Primula cuneifolia Ledeb. var. hakusanensis (Franch.) Makino f. alba H. Hara;白花白山小报春■☆

314286　Primula cuneifolia Ledeb. var. heterodona Makino f. nivea H. Hara;雪白异齿北海道小报春■☆

314287　Primula cuneifolia Ledeb. var. heterodonta Makino ＝ Primula heterodonta Franch. ■☆

314288　Primula cunninghamii King et Craib;小脆蒴报春(康氏报春);Cunningham Primrose ■

314289　Primula cyanantha Balf. f. et Forrest ＝ Primula watsonii Dunn ■

314290　Primula cyancephala Balf. f. ＝ Primula denticulata Sm. subsp. sinodenticulata (Balf. f. et Forrest) W. W. Sm. et Forrest ■

314291　Primula cyclaminifolia Franch. ex Petitm. ＝ Primula partschiana Pax ■

314292　Primula cycliophylla Balf. f. et Farrer ＝ Primula dryadifolia Franch. ■

314293　Primula cyclostegia Hand. -Mazz. ＝ Primula bella Franch. ■

314294　Primula cylindriflora Hand. -Mazz. ＝ Primula faberi Oliv. et Hand. -Mazz. ■

314295　Primula davidii Franch. ;大叶宝兴报春(大卫报春);David Primrose ■

314296　Primula debilis Bonati ＝ Primula pellucida Franch. ■

314297　Primula decurva Balf. f. et Forrest ＝ Primula szechuanica Pax ■

314298　Primula deflexa Duthie;穗花报春(俯垂报春);Deflexed Primrose ■

314299　Primula delavayi Franch. ＝ Omphalogramma delavayi (Franch.) Franch. ■

314300　Primula deleiensis Kingdon-Ward ＝ Primula firmipes Balf. f. et Forrest ■

314301　Primula delicata Forrest ＝ Primula spicata Franch. ■

314302　Primula delicata Petitm. ＝ Primula malacoides Franch. ■

314303　Primula densa Balf. f. ;小叶鄂报春;Dense Primrose ■

314304　Primula denticulata Sm. ＝ Primula elliptica Royle ■

314305　Primula denticulata Sm. subsp. alta (Balf. f. et Forrest) W. W. Sm. et H. R. Fletcher ＝ Primula denticulata Sm. subsp. sinodenticulata (Balf. f. et Forrest) W. W. Sm. et Forrest ■

314306　Primula denticulata Sm. subsp. cyanocephala (Balf. f.) W. W. Sm. et Forrest ＝ Primula denticulata Sm. subsp. sinodenticulata (Balf. f. et Forrest) W. W. Sm. et Forrest ■

314307　Primula denticulata Sm. subsp. erythrocarpa (Craib) W. W. Sm. et Forrest ＝ Primula erythrocarpa Craib ■

314308　Primula denticulata Sm. subsp. sinodenticulata (Balf. f. et Forrest) W. W. Sm. et Forrest;滇北球花报春(米伞花,球花报春,野洋参);Sessile Primrose ■

314309　Primula denticulata Sm. subsp. stolanifera (Balf. f.) W. W. Sm. et Forrest ＝ Primula pseudodenticulata Pax ■

314310　Primula denticulata Wight;球花报春(球序报春,喜马报春);Denticulate Primrose, Drumstick Primrose, Drumstick Primula, Himalayan Primrose, Indian Primrose, Smalltoothed Primrose ■

314311　Primula diantha Bureau et Franch. ; 双花报春;Dualflower Primrose ■

314312　Primula dickieana Watt;展瓣紫晶报春(第克报春,漏斗报春,绿心报春);Dicky Primrose, Funnel Primrose ■

314313　Primula dickieana Watt var. chlorops W. W. Sm. et Forrest ＝ Primula dickieana Watt ■

314314　Primula dickieana Watt var. gouldii H. R. Fletcher ＝ Primula kingii Watt ■

314315　Primula dickieana Watt var. pantlingii (King) W. W. Sm. et Forrest ＝ Primula dickieana Watt ■

314316　Primula dielsii Petitm. ＝ Primula tongolensis Franch. ■

314317　Primula dissecta (Franch.) Derganc ＝ Androsace dissecta (Franch.) Franch. ■

314318　Primula divaricata F. H. Chen et C. M. Hu;叉梗报春;Manyforked Primrose ■

314319　Primula doshongensis W. W. Sm. ＝ Primula glabra Klatt subsp. genestieriana (Hand. -Mazz.) C. M. Hu ■

314320　Primula dryadifolia Franch. ;石岩报春(岩报春);Dryadleaf Primrose ■

314321　Primula dryadifolia Franch. subsp. chlorodryas (W. W. Sm.) F. H. Chen et C. M. Hu;黄花岩报春■

314322　Primula dryadifolia Franch. subsp. chrysophylla (Balf. f. et Forrest) W. W. Sm. et Forrest ＝ Primula dryadifolia Franch. ■

314323　Primula dryadifolia Franch. subsp. congestifolia (Forrest) W. W. Sm. et Forrest ＝ Primula dryadifolia Franch. ■

314324　Primula dryadifolia Franch. subsp. jonardunii (W. W. Sm.) F. H. Chen et C. M. Hu;翅柄岩报春■

314325　Primula drymophila Craib ＝ Primula sonchifolia Franch. ■

314326　Primula dubernardiana Forrest ＝ Primula bracteata Franch. ■

314327　Primula duclouxii Petitm. ;曲柄报春(杜氏报春);Curvedstipe Primrose ■

314328　Primula dumicola W. W. Sm. et Forrest;灌丛报春(矮林报春);Shrub Primrose ■

314329　Primula eburnea Balf. f. et Cooper;乳白垂花报春;Ivary

Primrose ■

314330　Primula edgeworthii Pax = Primula nana Wall. ■☆

314331　Primula efarinosa Pax;无粉报春(偷偷还阳,云苔草);Epruinose Primrose ■

314332　Primula effusa W. W. Sm. et Forrest;散花报春(散生报春); Scatterflower Primrose ■

314333　Primula elatior (L.) Hill;高报春;Bardfield Oxlip, Oxlip, Oxlip Primrose,Paigle ■☆

314334　Primula elegans Duby = Primula rosea Royle ■☆

314335　Primula elizabethae Ludlow ex W. W. Sm.;卵叶雪山报春;Elizabeth Primrose ■

314336　Primula elliptica Royle;椭圆形报春;Elliptic Primrose ■☆

314337　Primula elongata Watt;黄齿雪山报春(长樱草);Elongate Primrose ■

314338　Primula elongata Watt var. barnardoana (W. W. Sm. et Kingdon-Ward) C. M. Hu;黄花圆叶报春 ■

314339　Primula elwesiana King ex Watt = Omphalogramma elwesianum (King ex Watt) Franch. ■☆

314340　Primula engleri R. Knuth = Omphalogramma vincaeflorum (Franch.) Franch. ■

314341　Primula epilithica F. H. Chen et C. M. Hu;石面报春;Stoneface Primrose ■

314342　Primula epilosa Craib;二郎山报春(川南报春,豆叶参,鄂西粗叶报春,西南报春);Hairless Primrose,Primrose ■

314343　Primula epilosa Craib = Primula tardiflora (C. M. Hu) C. M. Hu ■

314344　Primula erodioides Schltr.;安徽报春;Anhui Primrose, Anhwei Primrose ■

314345　Primula erodioides Schltr. = Primula cicutariifolia Pax ■

314346　Primula erosa (Wall. ex Duby) Regel = Primula glomerata Pax ■

314347　Primula erosa Wall.;啮齿状报春;Erose Primrose ■☆

314348　Primula erratica W. W. Sm.;甘南报春(奇特报春,野报春); Gannan Primrose,Wondeful Primrose ■

314349　Primula erratica W. W. Sm. = Primula gemmifera Batalin ■

314350　Primula erythrocarpa Craib;黄心球花报春;Redfruit Primrose ■

314351　Primula esquirolii Petitm.;贵州卵叶报春(艾氏报春); Esquirol Primrose ■

314352　Primula eucyclia W. W. Sm. et Forrest = Primula vaginata Watt subsp. eucyclia (W. W. Sm. et Forrest) F. H. Chen et C. M. Hu ■

314353　Primula eugeniae Fed.;欧氏报春 ■☆

314354　Primula euosma Craib;绿眼报春(紫心报春);Greeneye Primrose ■

314355　Primula euosma Craib var. puralba W. W. Sm. = Primula taliensis Forrest ■

314356　Primula eximia Greene;优异报春 ■☆

314357　Primula exscapa F. H. Chen et C. M. Hu;无葶脆蒴报春; Stalkleaf Primrose ■

314358　Primula faberi Oliv. et Hand.-Mazz.;峨眉报春(峨山雪莲花);Faber Primrose ■

314359　Primula fagosa Balf. f. et Craib;城口报春(川东粗叶报春,蜀报春);Chengkou Primrose ■

314360　Primula falcifolia Kingdon-Ward;镰叶雪山报春;Sickleleaf Primrose ■

314361　Primula falcifolia Kingdon-Ward var. farinifera C. M. Hu;波密镰叶雪山报春(波密镰叶报春);Bomi Sickleleaf Primrose ■

314362　Primula fangii F. H. Chen et C. M. Hu;金川粉报春;Fang Primrose ■

314363　Primula fangingensis F. H. Chen et C. M. Hu;梵净报春;

Fanjing Primrose ■

314364　Primula fargesii Franch. = Primula nutantiflora Hemsl. ■

314365　Primula farinifolia Rupr.;粉叶报春 ■☆

314366　Primula farinosa L.;粉报春(长白山报春,红花粉叶报春,黄报春,适度报春,优雅报春,有粉报春);Biddy's Eyes, Bird's Eye Primrose, Bird's Eye, Bird's-eye Primrose, Bird's-eye Primula, Bogbean, Bonny Bird Een, Bug Bean, Farinose Primrose, Mealy Primrose,Powdered Bean ■

314367　Primula farinosa L. = Primula algida Adam ex Weber et Mohr ■

314368　Primula farinosa L. subsp. fauriei (Franch.) Murata = Primula fauriai Franch. ■☆

314369　Primula farinosa L. subsp. fistulosa (Turkev.) W. W. Sm. et Forrest = Primula fistulosa Turkev. ■

314370　Primula farinosa L. subsp. modesta (Bisset et S. Moore) Pax;长白山报春 ■☆

314371　Primula farinosa L. subsp. modesta (Bisset et S. Moore) Pax var. fauriei (Franch.) Miyabe = Primula fauriai Franch. ■☆

314372　Primula farinosa L. subsp. modesta (Bisset et S. Moore) Pax var. fauriei (Franch.) Miyabe f. leucantha (H. Hara) T. Yamaz.;白花法氏报春 ■☆

314373　Primula farinosa L. subsp. modesta (Bisset et S. Moore) Pax var. macrantha Murata;大花长白山报春 ■☆

314374　Primula farinosa L. subsp. modesta (Bisset et S. Moore) Pax var. matsumurae (Petitm.) T. Yamaz. = Primula matsumurae Petitm. ■☆

314375　Primula farinosa L. subsp. modesta (Bisset et S. Moore) Pax var. matsumurae (Petitm.) T. Yamaz. f. alba (H. Hara) T. Yamaz.;白花松村报春 ■☆

314376　Primula farinosa L. subsp. modesta (Bisset et S. Moore) Pax var. matsumurae (Petitm.) T. Yamaz.;松村报春 ■☆

314377　Primula farinosa L. subsp. modesta (Bisset et S. Moore) Pax var. modesta (Bisset et S. Moore) Makino;适度粉报春 ■☆

314378　Primula farinosa L. subsp. xanthophylla (Trautv. et C. A. Mey.) Kitag.;黄叶粉报春 ■☆

314379　Primula farinosa L. subsp. xanthophylla (Trautv. et C. A. Mey.) Kitag. = Primula farinosa L. ■

314380　Primula farinosa L. subsp. yuparensis (Takeda) Kitam. = Primula yuparensis Takeda ■☆

314381　Primula farinosa L. var. concinna (Watt) Pax = Primula concinna Watt ■

314382　Primula farinosa L. var. denudata Koch;裸报春;Naked Farinose Primrose ■

314383　Primula farinosa L. var. xanthophylla (Trautv. et C. A. Mey.) Trautv. et C. A. Mey. = Primula farinosa L. subsp. xanthophylla (Trautv. et C. A. Mey.) Kitag. ■☆

314384　Primula farinosa L. var. xanthophylla Trautv. et C. A. Mey. = Primula farinosa L. ■

314385　Primula farrerisna Balf. f.;大通报春(法瑞报春,甘肃厚叶报春);Farrer Primrose ■

314386　Primula fasciculata Balf. f. et Kingdon-Ward;束花粉报春(束花报春);Fascicleflower Primrose ■

314387　Primula fauriai Franch.;法氏报春 ■☆

314388　Primula fedschenkoi Regel;范氏报春 ■☆

314389　Primula fernaldiana W. W. Sm.;雅东粉报春;Fernald Primrose ■

314390　Primula filchnerae Knuth;陕西羽叶报春;Pinnateleaf Primrose ■

314391　Primula filipes Watt;线柄报春;Thread-stalked Primrose ■☆

314392　Primula finmarchica Jacq. = Primula sibirica Jacq. ■

314393　Primula finno-marchica Georgi ＝Primula finmarchica Jacq. ■

314394　Primula firmipes Balf. f. et Forrest；葶立钟报春（柔枝报春，葶立报春）；Firmstip Primrose,Flexiblebranch Primrose ■

314395　Primula firmipes Balf. f. et Forrest subsp. flexilipes（Balf. f. et Forrest）W. W. Sm. et Forrest ＝Primula firmipes Balf. f. et Forrest ■

314396　Primula fistulosa Turkev.；箭报春；Arrow Primrose ■

314397　Primula fistulosa Turkev. var. breviscapa P. H. Huang et L. H. Zhuo；短葶箭报春（短葶报春）■

314398　Primula flabellifera W. W. Sm.；扇叶垂花报春；Fanlike Primrose ■

314399　Primula flaccida N. P. Balakr.；垂花报春；Weak Primrose ■

314400　Primula flava Maxim.；黄花粉叶报春（黄花报春）；Yellowflower Primrose ■

314401　Primula flavescens Derganc ＝Androsace flavescens Maxim. ■

314402　Primula flavicans Hand. -Mazz. ＝Primula ambita Balf. f. ■

314403　Primula flexilipes Balf. f. et Forrest ＝Primula firmipes Balf. f. et Forrest ■

314404　Primula flexuosa Turkev.；之字报春■☆

314405　Primula floribunda Wall.；繁花报春；Buttercup Primrose, Profusely Flowering Primrose ■☆

314406　Primula florida Balf. f. et Forrest ＝Primula blinii H. Lév. ■

314407　Primula florindae Kingdon-Ward；巨伞钟报春（藏报春）；Florind Primrose, Giant Cowslip, Himalayan Cowslip, Primrose, Tibetan Cowslip ■

314408　Primula forbesii Franch.；小报春（佛贝氏樱草，癫痫头花，痫头花）；Baby Primrose,Forbes Primrose ■

314409　Primula forbesii Franch. subsp. androsacea（Pax）W. W. Sm. et Forrest ＝Primula forbesii Franch. ■

314410　Primula forbesii Franch. subsp. delicata（Petitm.）W. W. Sm. et Forrest ＝Primula malacoides Franch. ■

314411　Primula forbesii Franch. subsp. duclouxii（Petitm.）W. W. Sm. et Forrest ＝Primula duclouxii Petitm. ■

314412　Primula forbesii Franch. subsp. hypoleuca（Hand. -Mazz.）W. W. Sm. ＝Primula hypoleuca Hand. -Mazz. ■

314413　Primula forbesii Franch. subsp. hypoleuca（Hand. -Mazz.）W. W. Sm. et Forrest ＝Primula hypoleuca Hand. -Mazz. ■

314414　Primula forbesii Franch. var. brevipes Bonati ＝Primula duclouxii Petitm. ■

314415　Primula forrestii Balf. f.；灰岩皱叶报春（黄葵报春,鹿角七,石瘘参,松打七,岩笃米花,猪尾七）；Forrest Primrose ■

314416　Primula frae Pax et K. Hoffm. ＝Primula amethystina Franch. subsp. argutidens（Franch.）W. W. Sm. et H. R. Fletcher ■

314417　Primula fragilis Balf. f. et Kingdon-Ward ＝Primula yunnanensis Franch. ■

314418　Primula franchetii Pax ＝Omphalogramma souliei Franch. ■

314419　Primula frondosa Janka；巴尔干报春；Leafy Primrose ■☆

314420　Primula gagnepainiana Hand. -Mazz. ＝Primula szechuanica Pax ■

314421　Primula gagnepainii Petitm. ＝Primula heucherifolia Franch. ■

314422　Primula gambeliana Watt；长蒴圆叶报春（冈伯尔报春）；Gambel Primrose,Longcapsule Primrose ■

314423　Primula gemmifera Batalin；苞芽报春（苞芽粉报春,多芽报春）；Budding Primrose ■

314424　Primula gemmifera Batalin var. amoena F. H. Chen；厚叶苞芽报春■

314425　Primula gemmifera Batalin var. licentii（W. W. Sm. et Forrest）W. W. Sm. et H. R. Fletcher ＝Primula conspersa Balf. f. et Purdom ■

314426　Primula gemmifera Batalin var. monantha（W. W. Sm. et Forrest）W. W. Sm. et H. R. Fletcher ＝Primula gemmifera Batalin var. amoena F. H. Chen ■

314427　Primula gemmifera Batalin var. rupestris（Pax et K. Hoffm.）W. W. Sm. et H. R. Fletcher ＝Primula gemmifera Batalin var. amoena F. H. Chen ■

314428　Primula gemmifera Batalin var. zambalensis（Petitm.）W. W. Sm. et H. R. Fletcher ＝Primula gemmifera Batalin var. amoena F. H. Chen ■

314429　Primula genestieriana Hand. -Mazz. ＝Primula glabra Klatt subsp. genestieriana（Hand. -Mazz.）C. M. Hu ■

314430　Primula gentianoides W. W. Sm. et Kingdon-Ward ＝Primula tongolensis Franch. ■

314431　Primula geraldinae W. W. Sm. ＝Primula rhodochroa W. W. Sm. var. geraldinae（W. W. Sm.）F. H. Chen et C. M. Hu ■

314432　Primula geraniifolia Hook. f.；滇藏掌叶报春（老鹳草叶报春）；Xizang Geraniumleaf Primrose ■

314433　Primula gigantea Jacq. ＝Primula farinosa L. var. denudata Koch ■

314434　Primula giraldiana Pax；太白山紫穗报春（季氏报春）；Girald Primrose ■

314435　Primula glabra Klatt；光叶粉报春（无毛樱草）；Glabrous Primrose,Psilate Primrose ■

314436　Primula glabra Klatt subsp. genestieriana（Hand. -Mazz.）C. M. Hu；纤葶粉报春■

314437　Primula glabra Klatt subsp. kongboensis（Kingdon-Ward）Halda ＝Primula kongboensis Kingdon-Ward ■

314438　Primula glacialis Franch. ＝Primula diantha Bureau et Franch. ■

314439　Primula glomerata Pax；立花头序报春；Congregate Primrose ■

314440　Primula glycyosma Petitm. ＝Primula wilsonii Dunn ■

314441　Primula gracilenta Dunn；长瓣穗花报春（细柳报春）；Slender Primrose ■

314442　Primula gracilipes Craib；纤柄脆蒴报春；Slenderlongstipe Primrose ■

314443　Primula graminifolia Pax et K. Hoffm.；禾叶报春；Grassleaf Primrose,Poaleaf Primrose ■

314444　Primula grandis Trautv ＝Sredinskya grandis（Trautv.）Fed. ■☆

314445　Primula gratissima Forrest ＝Primula sonchifolia Franch. ■

314446　Primula griffithii（Watt）Pax；高葶脆蒴报春；Griffith Primrose ■

314447　Primula hakonensis Nakai ＝Primula reinii Franch. et Sav. ■☆

314448　Primula hakusanensis Franch. var. heterodonta ？ ＝Primula heterodonta Franch. ■☆

314449　Primula halleri J. F. Gmel.；哈氏报春■☆

314450　Primula handeliana W. W. Sm. et Forrest；陕西报春（山西报春,西北厚叶报春）；Shaanxi Primrose ■

314451　Primula harroviana Balf. f. et Cooper ＝Primula eburnea Balf. f. et Cooper ■

314452　Primula harsukhii Craib ＝Primula denticulata Sm. ■

314453　Primula helodoxa Balf. f.；泽地灯台报春（晋台报春,篩沼报春,指状报春）；Mashy Primrose ■

314454　Primula helodoxa Balf. f. ＝Primula prolifera Wall. ■☆

314455　Primula helodoxa Balf. f. subsp. chrysochlora（Balf. f. et Kingdon-Ward）W. W. Sm. et Forrest ＝Primula chrysochlora Balf. f. et Kingdon-Ward ■

314456　Primula helvenacea Balf. f. et Kingdon-Ward ＝Primula minor Balf. f. et Kingdon-Ward ■

314457　Primula henrici Bureau et Franch.；鳞茎报春；Bulbous Primrose ■☆

314458　Primula henrici Bureau et Franch. ＝Primula bracteata Franch. ■

314459　Primula henryi（Hemsl.）Pax；滇南报春（亨利报春）；S.

Yunnan Primrose ■

314460　Primula heterochroma Stapf;异色报春■☆

314461　Primula heterodonta Franch.;异齿报春■☆

314462　Primula heucherifolia Franch.;宝兴掌叶报春;Baoxing Primrose,Lobedleaf Primrose ■

314463　Primula heucherifolia Franch. subsp. humicola（Balf. f. et Forrest）W. W. Sm. et Forrest = Primula geraniifolia Hook. f. ■

314464　Primula heydei Watt = Primula minutissima Jacquem. ex Duby ■

314465　Primula hidakana Miyabe et Kudo ex Nakai;日高报春■☆

314466　Primula hidakana Miyabe et Kudo ex Nakai f. kamuiana（Miyabe et Tatew.）T. Yamaz.;久武报春■☆

314467　Primula hidakana Miyabe et Kudo ex Nakai var. kamuiana（Miyabe et Tatew.）H. Hara = Primula hidakana Miyabe et Kudo ex Nakai f. kamuiana（Miyabe et Tatew.）T. Yamaz. ■☆

314468　Primula hilaris W. W. Sm.;大花脆蒴报春（脐点报春）;Bigflower Primrose ■

314469　Primula hirsuta Rchb. ex Nyman;红花报春■☆

314470　Primula hoffmanniana W. W. Sm.;川北脆蒴报春（川北苣叶报春）;N. Sichuan Primrose ■

314471　Primula hoi W. P. Fang;单伞长柄报春（何氏报春花）;Ho Primrose ■

314472　Primula homogama F. H. Chen et C. M. Hu;峨眉缺裂报春;Equallyflower Primrose ■

314473　Primula hondoensis Nakai et Kitag. = Primula jesoana Miq. ■☆

314474　Primula hookeri Watt;春花脆蒴报春（胡克报春）;Hooker. Primrose ■

314475　Primula hookeri Watt var. violacea（W. W. Sm.）C. M. Hu;蓝春花报春■

314476　Primula hopeana Balf. f. et Cooper = Primula ioessa W. W. Sm. ■

314477　Primula hopeana Balf. f. et Cooper = Primula sikkimensis Hook. ■

314478　Primula hsiungiana W. P. Fang;济华报春;Jihua Primrose ■

314479　Primula hsiungiana W. P. Fang = Primula walshii Craib ■

314480　Primula huana W. W. Sm. = Primula chapaensis Gagnep. ■

314481　Primula huashanensis F. H. Chen et C. M. Hu;华山报春;Huashan Primrose ■

314482　Primula humicola Balf. f. et Forrest = Primula geraniifolia Hook. f. ■

314483　Primula humilis Pax et K. Hoffm.;矮葶缺裂报春;Short Primrose ■

314484　Primula hupehensis Craib = Primula odontocalyx（Franch.）Pax ■

314485　Primula hyacinthina W. W. Sm. = Primula bellidifolia King ex Hook. f. ■

314486　Primula hylobia W. W. Sm.;亮叶报春（海螺报春）;Brightleaf Primrose ■

314487　Primula hylophilla Balf. f. et Farrer = Primula odontocalyx（Franch.）Pax ■

314488　Primula hymenophylla Balf. f. et Forrest = Primula polyneura Franch. ■

314489　Primula hypoleuca Hand. -Mazz.;白背小报春■

314490　Primula iljinskii Fed.;伊尔报春■☆

314491　Primula incisa Franch.;羽叶报春（窄裂叶报春）;Pinnatipartite Primrose ■

314492　Primula incisa Franch. = Primula blinii H. Lév. ■

314493　Primula incisa Franch. subsp. pectinata（Balf. f. et Forrest）W. W. Sm. et Forrest = Primula blinii H. Lév. ■

314494　Primula indobella Balf. f. et W. W. Sm. = Primula tenuiloba

（Watt）Pax ■

314495　Primula inflata Lehm. subsp. macrocalyx（Bunge）O. Schwarz = Primula veris L. subsp. macrocalyx（Bunge）Ludi ■

314496　Primula ingens W. W. Sm. et Forrest = Primula chionantha Balf. f. et Forrest ■

314497　Primula inopinata H. R. Fletcher;迷离报春;Puzzle Primrose ■

314498　Primula intercedens Fernald = Primula mistassinica Michx. ■☆

314499　Primula interjacens F. H. Chen;景东报春;Chingtung Primrose,Jingdong Primrose ■

314500　Primula interjacens F. H. Chen var. epilosa C. M. Hu;光叶景东报春;Epilose Jingdong Primrose ■

314501　Primula intermedia Sims;中型报春;Intermediate Primrose ■☆

314502　Primula intermedia Sims. = Primula longiscapa Ledeb. ■

314503　Primula involucrata Wall. ex Duby;花苞报春（大总苞报春）;Involucral Primrose,Pinnacle Primrose ■

314504　Primula involucrata Wall. ex Duby subsp. yargongensis（Petitm.）W. W. Sm. et Forrest;雅江报春;Yajiang Primrose ■

314505　Primula ioessa W. W. Sm.;缺叶钟报春（缺叶报春）;Incise Primrose ■

314506　Primula ionantha Pax et K. Hoffm. = Primula russeola Balf. f. et Forrest ■

314507　Primula jaffreyana King;藏南粉报春（喜岭报春）;Jaffrey Primrose ■

314508　Primula japonica A. Gray;日本报春（九轮草,七重草,日本樱草）;Japan Primrose,Japanese Cowslip,Japanese Primrose ■

314509　Primula japonica A. Gray 'Miller's Crimson';米勒绯红日本报春■☆

314510　Primula japonica A. Gray 'Postford White';波斯特福特白日本报春■☆

314511　Primula japonica A. Gray f. albiflora Hort.;白花日本报春■☆

314512　Primula japonica A. Gray f. robusta Hemsl. = Primula pulverulenta Duthie ■

314513　Primula japonica A. Gray var. angustidens Franch. = Primula stenodonta Balf. f. ex W. W. Sm. et H. R. Fletcher ■

314514　Primula jesoana Miq.;鸭绿报春;Yalu River Primrose ■☆

314515　Primula jesoana Miq. = Primula loeseneri Kitag. ■

314516　Primula jesoana Miq. f. leucantha H. Hara;白花鸭绿报春■☆

314517　Primula jesoana Miq. f. pubescens Takeda = Primula jesoana Miq. var. pubescens（Takeda）Takeda et H. Hara ■

314518　Primula jesoana Miq. subsp. pubescens（Takeda）Kitam. = Primula jesoana Miq. var. pubescens（Takeda）Takeda et H. Hara ■

314519　Primula jesoana Miq. var. glabra Takeda et H. Hara;无毛鸭绿报春■☆

314520　Primula jesoana Miq. var. glabra Takeda et H. Hara = Primula jesoana Miq. ■☆

314521　Primula jesoana Miq. var. pubescens（Takeda）Takeda et H. Hara;毛鸭绿报春（肾叶报春,心叶报春）;Cordatelaef Primrose,Kidneyleaf Primrose ■

314522　Primula jesoana Miq. var. pubescens（Takeda）Takeda et H. Hara f. albiflora Tatew. ex H. Hara;白花毛鸭绿报春■☆

314523　Primula jesoana Miq. var. pubescens（Takeda）Takeda et H. Hara f. nudiuscula（Nakai et Kitag.）H. Hara;稍裸毛鸭绿报春■☆

314524　Primula jonardunii W. W. Sm. = Primula dryadifolia Franch. subsp. jonardunii（W. W. Sm.）F. H. Chen et C. M. Hu ■

314525　Primula jucunda W. W. Sm.;山南脆蒴报春（欢喜报春）;Joyful Primrose ■

314526　Primula jucunda W. W. Sm. var. ponticula W. W. Sm. = Primula jucunda W. W. Sm. ■

314527 Primula juliae Kusn. ;尤里报春■☆

314528 Primula junior Balf. f. et Forrest = Primula calliantha Franch. subsp. bryophila (Balf. f. et Farrer) W. W. Sm. et Forrest ■

314529 Primula kanseana Pax et K. Hoffm. = Primula stenocalyx Maxim. ■

314530 Primula kaufmanniana Regel;考夫报春■☆

314531 Primula kawasimae H. Hara;川岛报春■☆

314532 Primula kewensis W. Watson = Primula × kewensis W. Watson ■☆

314533 Primula kialensis Franch. ;等梗报春（康边报春）; Equallypedicel Primrose,Kang Primrose ■

314534 Primula kialensis Franch. subsp. brevituba C. M. Hu;短筒等梗报春■

314535 Primula kichanensis Franch. ex Petitm. = Primula yunnanensis Franch. ■

314536 Primula kingii Watt;高葶紫晶报春;King Primrose ■

314537 Primula kisoana Miq. ;木曾报春■☆

314538 Primula kisoana Miq. var. shikokiana Makino = Primula kisoana Miq. ■☆

314539 Primula kiuchiangensis Balf. f. et Forrest = Primula diantha Bureau et Franch. ■

314540 Primula klattii N. P. Balakr. ;单朵垂花报春;Klatt Primrose ■

314541 Primula klaveriana Forrest;云南卵叶报春（克氏报春）;Klaver Primrose ■

314542 Primula knorringiana Fed. ;克诺林报春■

314543 Primula knuthiana Pax;阔萼粉报春（克努报）■■

314544 Primula knuthiana Pax var. brevipes Pax = Primula knuthiana Pax ■

314545 Primula knuthiana Pax var. major Pax =Primula knuthiana Pax ■

314546 Primula komarovii Losinsk. ;科马罗夫报春■☆

314547 Primula kongboensis Kingdon-Ward;工布报春（康波报春）; Gongbu Primrose ■

314548 Primula kusnetzovii Fed. ;库氏报春■☆

314549 Primula kwangtungensis W. W. Sm. ;广东报春; Guangdong Primrose,Kwangtung Primrose ■

314550 Primula kweichouensis W. W. Sm. ;贵州报春; Guizhou Primrose,Kweichou Primrose ■

314551 Primula kweichouensis W. W. Sm. var. venulosa F. H. Chen et C. M. Hu;多脉贵州报春;Multivein Guizhou Primrose ■

314552 Primula laberi Oliv. ;峨山雪莲花（峨眉报春）;Emei Primrose, Omei Primrose ■

314553 Primula lacei Hemsl. et Watt = Dionysia lacei (Hemsl. et Watt) Clay ■☆

314554 Primula lacerata W. W. Sm. ;瓣脆蒴报春;Tesselpetal Primrose ■

314555 Primula laciniata Pax et K. Hoffm. ;条裂叶报春（大苞裂叶报春）;Laciniateleaf Primrose,Lacinose Primrose ■

314556 Primula lacteocapitata Balf. f. et W. W. Sm. = Primula capitata Hook. subsp. lacteocapitata (Balf. f. et W. W. Sm.) W. W. Sm. et Forrest ■

314557 Primula lactucoides F. H. Chen et C. M. Hu;襄谦报春; Lettucelike Primrose ■

314558 Primula laeta W. W. Sm. =Primula calderiana Balf. f. et Cooper ■

314559 Primula lanata Pax et K. Hoffm. = Primula heucherifolia Franch. ■

314560 Primula lancifolia Pax et K. Hoffm. =Primula russeola Balf. f. et Forrest ■

314561 Primula langkongensis Forrest = Primula malvacea Franch. ■

314562 Primula latisecta W. W. Sm. ;宽裂掌叶报春; Broadsplit Primrose ■

314563 Primula laurentiana Fernald;粉状报春; Bird's-eye Primrose, Mealy Primrose ■☆

314564 Primula laxiuscula W. W. Sm. ;疏序球花报春; Spareflower Primrose ■

314565 Primula lecomtei Petitm. = Primula faberi Oliv. et Hand. -Mazz. ■

314566 Primula legendrei Bonati = Primula souliei Franch. ■

314567 Primula leimonophila Balf. f. = Primula virginis H. Lév. ■

314568 Primula lepta Balf. f. et Forrest = Primula pinnatifida Franch. ■

314569 Primula leptophylla Craib;薄叶长柄报春;Thinleaf Primrose ■

314570 Primula leptopoda Bureau et Franch. = Primula stenocalyx Maxim. ■

314571 Primula leucantha Balf. f. et Forrest = Primula beesiana Forrest ■

314572 Primula leucochnoa Hand. -Mazz. = Primula melanops W. W. Sm. et Kingdon-Ward ■

314573 Primula leucophylla Pax;白叶报春■☆

314574 Primula leucops W. W. Sm. et Kingdon-Ward = Primula diantha Bureau et Franch. ■

314575 Primula leucops W. W. Sm. et Kingdon-Ward var. anopa Hand. -Mazz. = Primula diantha Bureau et Franch. ■

314576 Primula levicalyx C. M. Hu et Z. R. Xu;光萼报春;Smoothcalyx Primrose ■

314577 Primula lhasaensis Balf. f. et W. W. Sm. = Primula jaffreyana King ■

314578 Primula licentii W. W. Sm. et Forrest = Primula conspersa Balf. f. et Purdom ■

314579 Primula lichiangensis (Forrest) Forrest = Primula polyneura Franch. ■

314580 Primula lichiangensis (Forrest) Forrest var. halapa Balf. f. et Forrest = Primula polyneura Franch. ■

314581 Primula limbata Balf. f. et Forrest;匙叶雪山报春（镶边报春）; Spoonleaf Primrose ■

314582 Primula limnoica Craib = Primula denticulata Sm. subsp. sinodenticulata (Balf. f. et Forrest) W. W. Sm. et Forrest ■

314583 Primula limprichtii Pax = Primula ovalifolia Franch. ■

314584 Primula limprichtii Pax et K. Hoffm. = Primula ovalifolia Franch. ■

314585 Primula listeri Forbes et Hemsl. var. glabrescens Franch. = Primula obconica Hance subsp. begoniiformis (Petitm.) W. W. Sm. et Forrest ■

314586 Primula listeri King = Primula sinolisteri Balf. f. ■

314587 Primula listeri King ex Hook. f. ;里斯特报春;Lister Primrose ■☆

314588 Primula listeri King ex Hook. f. = Primula obconica Hance subsp. begoniiformis (Petitm.) W. W. Sm. et Forrest ■

314589 Primula listeri King ex Hook. f. var. rotundifolia Franch. = Primula obconica Hance subsp. begoniiformis (Petitm.) W. W. Sm. et Forrest ■

314590 Primula lithophila F. H. Chen et C. M. Hu;习水报春;Xishui Primrose ■

314591 Primula littledalei Balf. f. et Watt;白粉圆叶报春（白粉叶报春）;Pale Primrose ■

314592 Primula littoniana Forrest = Primula vialii Delavay ex Franch. ■

314593 Primula littoniana Forrest var. robusta Forrest = Primula vialii Delavay ex Franch. ■

314594 Primula loeseneri Kitag. = Primula jesoana Miq. var. pubescens

（Takeda）Takeda et H. Hara ■

314595　Primula longipes Freyn et Sint.；长梗报春■☆

314596　Primula longipetiolata Pax et K. Hoffm.；长柄雪山报春；
Longstipe Primrose ■

314597　Primula longipetiolata Pax et K. Hoffm. = Primula optata Farrer
ex Balf. f. ■

314598　Primula longipinnatifida F. H. Chen = Primula blinii H. Lév. ■

314599　Primula longiscapa Ledeb.；长葶报春；Longstalk Primrose ■

314600　Primula longituba Forrest = Primula membranifolia Franch. ■

314601　Primula lungchiensis W. P. Fang；龙池报春（龙溪报春）；
Longchi Primrose，Longxi Primrose，Lungchi Primrose ■

314602　Primula luteola Rupr.；浅黄报春■☆

314603　Primula macrocalyx Bunge = Primula veris L. subsp. macrocalyx
（Bunge）Ludi ■

314604　Primula macrocarpa Maxim.；姬报春（姬小樱）■

314605　Primula macrophylla D. Don；大叶报春；Largeleaf Primrose ■

314606　Primula macrophylla D. Don var. atra W. W. Sm. et H. R.
Fletcher；黄粉大叶报春；Yellow Largeleaf Primrose ■

314607　Primula macrophylla D. Don var. macrocarpa（Wall.）W. W.
Sm. et H. R. Fletcher = Primula megalocarpa H. Hara ■

314608　Primula macrophylla D. Don var. moorcroftiana（Wall. ex Klatt）
W. W. Sm. et H. R. Fletcher；长苞大叶报春；Longbract Largeleaf
Primrose ■

314609　Primula macrophylla D. Don var. ninguida（W. W. Sm.）W. W.
Sm. et H. R. Fletcher = Primula ninguida W. W. Sm. ■

314610　Primula macropoda Craib = Primula ovalifolia Franch. ■

314611　Primula maikhaensis Balf. f. et Forrest；怒江报春；Nujiang
Primrose ■

314612　Primula mairei H. Lév. = Primula pinnatifida Franch. ■

314613　Primula malacoides Franch.；报春花；Baby Primrose，Fairy
Primrose ■

314614　Primula malacoides Franch. subsp. pseudomalacoides（Stewart）
W. W. Sm. et Forrest = Primula malacoides Franch. ■

314615　Primula mallophylla Balf. f.；川东灯台报春（近东报春，毛叶
报春）；Softhairleaf Primrose ■

314616　Primula malvacea Franch.；葵叶报春；Mallowleaf Primrose ■

314617　Primula malvacea Franch. suhsp. rosthornii（Diels）W. W. Sm.
et Forrest = Primula neurocalyx Franch. ■

314618　Primula malvacea Franch. var. alba Forrest = Primula malvacea
Franch. ■

314619　Primula malvacea Franch. var. intermedia W. W. Sm. et Forrest
= Primula malvacea Franch. ■

314620　Primula mandarina Hoffm. = Primula sinensis Sabine ex Lindl. ■

314621　Primula marginata Curtis；齿缘报春■☆

314622　Primula marginata Curtis 'Prichard's Variety'；普氏齿缘报春■☆

314623　Primula matsumurae Petitm. = Primula farinosa L. subsp.
modesta（Bisset et S. Moore）Pax var. matsumurae（Petitm.）T.
Yamaz. ☆

314624　Primula maximowiczii Regel；胭脂花（段报春，假报春，陕甘报
春花，胭脂报春）；Maximowicz Primrose，Rougeflower Primrose ■

314625　Primula maximowiczii Regel var. brevifolia Pax = Primula
maximowiczii Regel ■

314626　Primula maximowiczii Regel var. dielsiana Pax = Primula
maximowiczii Regel ■

314627　Primula maximowiczii Regel var. euprepes W. W. Sm. = Primula
advena W. W. Sm. var. euprepes（W. W. Sm.）F. H. Chen et C. M.
Hu ■

314628　Primula maximowiczii Regel var. flaviflorida D. Z. Lu；黄花胭脂
花（黄胭脂花）；Yellowflower Maximowicz Primrose ■

314629　Primula meeboldii Pax = Primula macrophylla D. Don var.
moorcroftiana（Wall. ex Klatt）W. W. Sm. et H. R. Fletcher ■

314630　Primula megalocarpa H. Hara；大果报春（凹瓣大叶报春）；
Bigfruit Primrose ■

314631　Primula meiotera（W. W. Sm. et H. R. Fletcher）C. M. Hu；深
齿小报春；Deeptooth Primrose ■

314632　Primula melanantha（Franch.）C. M. Hu；深紫报春■

314633　Primula melanodonta W. W. Sm.；芒齿灯台报春（藏黄报春）；
Awntooth Primrose ■

314634　Primula melanops W. W. Sm. et Kingdon-Ward；粉葶报春（粉
茎报春）；Farinosestalk Primrose ■

314635　Primula membranifolia Franch.；薄叶粉报春（膜叶报春）；
Membraanousleaf Primrose ■

314636　Primula menziesiana Balf. f. et W. W. Sm. = Primula bellidifolia
King ex Hook. f. ■

314637　Primula merrilliana Schltr.；安徽羽叶报春；Merril Primrose ■

314638　Primula meyeri Rupr.；迈氏报春■☆

314639　Primula microdonta Franch. ex Petitm. = Primula sikkimensis
Hook. ■

314640　Primula microdonta Franch. ex Petitm. var. alpicola W. W. Sm.
= Primula alpicola（W. W. Sm.）Stapf ■

314641　Primula microdonta Franch. ex Petitm. var. alpicola W. W. Sm.
f. micromeres W. W. Sm. = Primula sikkimensis Hook. ■

314642　Primula microdonta Franch. ex Petitm. var. alpicola W. W. Sm.
f. micromeres W. W. Sm. et Ward = Primula sikkimensis Hook. ■

314643　Primula microloma Hand. -Mazz. = Primula prenantha Balf. f. et
W. W. Sm. ■

314644　Primula micropetala Balf. f. et Cooper = Primula bellidifolia
King ex Hook. f. ■

314645　Primula microstachys Balf. f. et Forrest = Primula blattariformis
Franch. ■

314646　Primula minima L.；微小报春；Least Primrose ■☆

314647　Primula minkwitziae W. W. Sm.；明氏报春■☆

314648　Primula minor Balf. f. et Kingdon-Ward；雪山小报春（小报
春）；Morelittle Primrose ■

314649　Primula minutiflora Forrest = Androsace umbellata（Lour.）
Merr. ■

314650　Primula minutissima Jacquem. ex Duby；高峰小报春（海地报
春，微小报春）；Heyde Primrose，Least Primrose，Minute Primrose ■

314651　Primula mishmiemsis Kingdon-Ward = Primula calliantha
Franch. subsp. mishmiemsis（Kingdon-Ward）C. M. Hu ■

314652　Primula mistassinica Michx.；奎伯克报春；Bird's-eye Primrose，
Dwarf Canadian Primrose，Lake Mistassini Primrose，Mistassini
Primrose ■☆

314653　Primula mistassinica Michx. var. intercedens（Fernald）B.
Boivin = Primula mistassinica Michx. ■☆

314654　Primula mistassinica Michx. var. noveboracensis Fernald =
Primula mistassinica Michx. ■☆

314655　Primula miyabeana Ito et Kawak.；玉山灯台报春（玉山樱草）；
Yushan Primrose ■

314656　Primula modesta Bisset et S. Moore = Primula farinosa L. subsp.
modesta（Bisset et S. Moore）Pax ■☆

314657　Primula modesta Bisset et S. Moore = Primula farinosa L. ■

314658　Primula modesta Bisset et S. Moore subsp. fauriei（Franch.）
W. W. Sm. et Forrest = Primula farinosa L. subsp. modesta（Bisset et

S. Moore) Pax var. fauriei (Franch.) Miyabe ■☆

314659 Primula modesta Bisset et S. Moore var. fauriei (Franch.) Takeda f. leucantha H. Hara = Primula farinosa L. subsp. modesta (Bisset et S. Moore) Pax var. fauriei (Franch.) Miyabe f. leucantha (H. Hara) T. Yamaz. ■☆

314660 Primula modesta Bisset et S. Moore var. fauriei (Franch.) Takeda = Primula farinosa L. subsp. modesta (Bisset et S. Moore) Pax var. fauriei (Franch.) Miyabe ■☆

314661 Primula modesta Bisset et S. Moore var. matsumurae (Petitm.) Takeda f. alba H. Hara = Primula farinosa L. subsp. modesta (Bisset et S. Moore) Pax var. matsumurae (Petitm.) T. Yamaz. f. alba (H. Hara) T. Yamaz. ■☆

314662 Primula modesta Bisset et S. Moore var. matsumurae (Petitm.) Takeda = Primula farinosa L. subsp. modesta (Bisset et S. Moore) Pax var. matsumurae (Petitm.) T. Yamaz. ■☆

314663 Primula modesta Bisset et S. Moore var. shikoku-montana ? = Primula modesta Bisset et S. Moore ■☆

314664 Primula mollis Nutt. ex Hook. ;灰毛报春(绒毛报春,柔毛报春);Flamingo Primrose,Softhair Primrose,Soft-hairy Primrose ■

314665 Primula mollis Nutt. ex Hook. subsp. seclusa (Balf. f. et Forrest) W. W. Sm. et Forrest = Primula mollis Nutt. ex Hook. ■

314666 Primula monantha W. W. Sm. et Forrest = Primula gemmifera Batalin var. amoena F. H. Chen ■

314667 Primula monbeigii Balf. f. = Primula bracteata Franch. ■

314668 Primula monticola (Hand.-Mazz.) F. H. Chen et C. M. Hu;中甸海水仙;Mountain Primrose ■

314669 Primula moorcroftiana Wall. = Primula macrophylla D. Don var. moorcroftiana (Wall. ex Klatt) W. W. Sm. et H. R. Fletcher ■

314670 Primula moorcroftiana Wall. ex Klatt ;穆氏报春■

314671 Primula moorcroftiana Wall. ex Klatt = Primula macrophylla D. Don var. moorcroftiana (Wall. ex Klatt) W. W. Sm. et H. R. Fletcher ■

314672 Primula mooreana Balf. f. et W. W. Sm. = Primula capitata Hook. ■

314673 Primula morsheadiana Kingdon-Ward = Primula aurantiaca W. W. Sm. et Forrest subsp. morsheandiana F. H. Chen et C. M. Hu ■

314674 Primula morsheadiana Kingdon-Ward = Primula prenantha Balf. f. et W. W. Sm. subsp. morsheandiana (Kingdon-Ward) F. H. Chen et C. M. Hu ■

314675 Primula moschophora Balf. f. et Forrest;麝香美报春;Musk Primrose ■

314676 Primula moupinensis Franch. ;宝兴报春;Baoxing Primrose, Paohsing Primrose ■

314677 Primula moupioensis Franch. subsp. barkamensis C. M. Hu;马尔康报春■

314678 Primula moystrophylla Balf. f. et Forrest = Primula dryadifolia Franch. ■

314679 Primula mulans Delavay ex Franch. = Primula flaccida N. P. Balakr. ■

314680 Primula muliensis Hand.-Mazz. = Primula boreio-calliantha Balf. f. et Cooper ■

314681 Primula multicaulis Petitm. = Primula forbesii Franch. ■

314682 Primula munroi Lindl. = Primula involucrata Wall. ex Duby ■

314683 Primula mupinensis Pax = Primula moupinensis Franch. ■

314684 Primula muscarioides Hemsl. ;麝草报春(麝香穗花报春);Flylike Primrose,Primrose ■

314685 Primula muscarioides Hemsl. subsp. conica (Balf. f. et Forrest) W. W. Sm. et Forrest = Primula deflexa Duthie ■

314686 Primula muscoides Hook. f. ex Watt;苔状小报春;Flylike Primrose,Mosslike Primrose ■

314687 Primula muscoides Hook. f. ex Watt var. tenuiloba Watt = Primula tenuiloba (Watt) Pax ■

314688 Primula mystrophylla Balf. f. et Forrest = Primula dryadifolia Franch. ■

314689 Primula nana Wall. ;矮报春■☆

314690 Primula nanobella Balf. f. et Forrest = Primula bella Franch. ■

314691 Primula nasturtiifolia F. H. Chen et C. M. Hu = Primula runcinata W. W. Sm. et H. R. Fletcher ex C. M. Hu ■

314692 Primula nemoralis Balf. f. = Primula sinuata Franch. ■

314693 Primula neurocalyx Franch. ;保康报春;Baokang Primrose, Veincalyx Primrose ■

314694 Primula neurocalyx Franch. subsp. riparia (Balf. f. et Farrer) W. W. Sm. et Forrest = Primula cinerascens Franch. ■

314695 Primula ninguida W. W. Sm. ;林芝报春(尖萼大叶报春);Linzhi Primrose ■

314696 Primula nipponica Yatabe;本州报春■☆

314697 Primula nivalis Pall. ;雪山报春(雪报春);Snowmountain Primrose ■

314698 Primula nivalis Pall. = Primula sinopurpurea Balf. f. ex Hutch. ■

314699 Primula nivalis Pall. var. colorata Regel = Primula nivalis Pall. var. farinosa Schrenk ■

314700 Primula nivalis Pall. var. farinosa Schrenk;准噶尔报春(新疆报春);Dzungar Primrose ■

314701 Primula nivalis Pall. var. longifolia Regel = Primula nivalis Pall. var. farinosa Schrenk ■

314702 Primula nivalis Pall. var. macrocarpa (Watt) Pax = Primula megalocarpa H. Hara ■

314703 Primula nivalis Pall. var. macrophylla (D. Don) Pax = Primula macrophylla D. Don ■

314704 Primula nivalis Pall. var. melanantha Franch. = Primula melanantha (Franch.) C. M. Hu ■

314705 Primula nivalis Pall. var. moorcroftiana (Wall. ex Klatt) Pax = Primula macrophylla D. Don var. moorcroftiana (Wall. ex Klatt) W. Sm. et H. R. Fletcher ■

314706 Primula nivalis Pall. var pumila Ledeb. ;偃伏雪山报春■☆

314707 Primula nivalis Pall. var. purpurea Franch. = Primula sinopurpurea Balf. f. ex Hutch. ■

314708 Primula nivalis Pall. var. sinensis Pax = Primula sinoplantaginea Balf. f. ■

314709 Primula nivalis Pall. var. turkestanica Regel = Primula nivalis Pall. var. farinosa Schrenk ■

314710 Primula nomaniana Kingdon-Ward = Primula vaginata Watt subsp. normaniana (Kingdon-Ward) F. H. Chen et C. M. Hu ■

314711 Primula norwegica Retz. ;挪威报春■☆

314712 Primula nutans Delavay ex Franch. = Primula flaccida N. P. Balakr. ■

314713 Primula nutans Georgi;天山报春(垂花报春,伞报春,西伯利亚报春);Drooping Primrose,Tianshan Primrose ■

314714 Primula nutans Georgi = Primula sibirica Jacq. ■

314715 Primula nutantiflora Hemsl. ;俯垂粉报春(垂花报春,管状报春花);Noddingflower,Nutantflower Primrose ■

314716 Primula nutantiflora Hemsl. = Primula homogama F. H. Chen et C. M. Hu ■

314717 Primula obconica Hance;鄂报春(四季报春,四季樱草,仙鹤莲,鲜荷莲报春,鲜鹤莲报春,岩丸子);Hubei Primrose,Poison

Primrose, Top Primrose ■

314718 Primula obconica Hance subsp. barbicalyx（C. H. Wright）W. W. Sm. = Primula barbicalyx C. H. Wright ■

314719 Primula obconica Hance subsp. begoniiformis（Balf. f.）W. W. Sm. et Forrest = Primula obconica Hance subsp. begoniiformis（Petitm.）W. W. Sm. et Forrest ■

314720 Primula obconica Hance subsp. begoniiformis（Petitm.）W. W. Sm. et Forrest；海棠叶报春 ■

314721 Primula obconica Hance subsp. densa（Balf. f.）W. W. Sm. et Forrest = Primula densa Balf. f. ■

314722 Primula obconica Hance subsp. fujianensis C. M. Hu et G. S. He；福建报春；Fujian Primrose ■

314723 Primula obconica Hance subsp. nigraglandulosa（W. W. Sm. et H. R. Fletcher）C. M. Hu；黑腺鄂报春 ■

314724 Primula obconica Hance subsp. parva（Balf. f.）W. W. Sm. et Forrest；小型报春 ■

314725 Primula obconica Hance subsp. parva W. W. Sm. et Forrest = Primula obconica Hance subsp. begoniiformis（Petitm.）W. W. Sm. et Forrest ■

314726 Primula obconica Hance subsp. petitmengini（Bonati）W. W. Sm. et Forrest = Primula obconica Hance ■

314727 Primula obconica Hance subsp. sinolisteri（Balf. f.）W. W. Sm. et Forrest = Primula sinolisteri Balf. f. ■

314728 Primula obconica Hance subsp. vilmoriniana（Petitm.）W. W. Sm. et Forrest = Primula vilmoriniana Petitm. ■

314729 Primula obconica Hance subsp. werringtonensis（Forrest）W. W. Sm. et Forrest；薄叶鄂报春 ■

314730 Primula obconica Hance var. hispida Franch. = Primula obconica Hance ■

314731 Primula obconica Hance var. nigroglandulosa W. W. Sm. et H. R. Fletcher = Primula obconica Hance subsp. nigraglandulosa（W. W. Sm. et H. R. Fletcher）C. M. Hu ■

314732 Primula obconica Hance var. parva Balf. f. = Primula obconica Hance subsp. nigraglandulosa（W. W. Sm. et H. R. Fletcher）C. M. Hu ■

314733 Primula obconica Hance var. rotundifolia Franch. = Primula obconica Hance subsp. begoniiformis（Petitm.）W. W. Sm. et Forrest ■

314734 Primula obconica Hance var. werrigtonensis（Forrest）W. W. Sm. et H. R. Fletcher = Primula obconica Hance ■

314735 Primula obconica Hance var. werringtonensis（Forrest）W. W. Sm. et H. R. Forrest = Primula obconica Hance subsp. werringtonensis（Forrest）W. W. Sm. et Forrest ■

314736 Primula oblanceolata Balf. f. = Primula wilsonii Dunn ■

314737 Primula obliqua W. W. Sm.；斜花雪山报春；Oblique Primrose ■

314738 Primula obovata（Hemsl.）Pax；蒙自报春；Obovateleaf Primrose ■

314739 Primula obovata（Hemsl.）Pax = Primula rugosa N. P. Balakr. ■

314740 Primula obsessa W. W. Sm.；肥满报春；Fat Primrose ■

314741 Primula obtusifolia Royle；钝叶报春（钝叶缨子草）；Obtuse-leaved Primrose ■☆

314742 Primula obtusifolia Royle = Primula calderiana Balf. f. et Cooper ■

314743 Primula obtusifolia Royle var. griffithii Watt = Primula griffithii（Watt）Pax ■

314744 Primula occlusa W. W. Sm.；扇叶小报春；Fanleaf Primrose ■

314745 Primula ochracea Pax et K. Hoffm. = Primula orbicularis Hemsl. ■

314746 Primula oculata Duthie = Primula heucherifolia Franch. ■

314747 Primula odontica W. W. Sm.；粗齿紫晶报春；Tooth Primrose ■

314748 Primula odontocalyx（Franch.）Pax；齿萼报春（湖北报春）；Hubei Primrose, Hupeh Primrose, Toothedcalyx Primrose ■

314749 Primula odontophylla Wall. = Primula rotundifolia Wall. ex Roxb. ■

314750 Primula officinalis（L.）Hill var. macrocalyx（Bunge）C. Koch = Primula veris L. subsp. macrocalyx（Bunge）Ludi ■

314751 Primula officinalis Hill = Primula veris L. ■

314752 Primula officinalis Jacq.；药樱草（樱草花）■☆

314753 Primula officinalis Jacq. = Primula veris L. ■

314754 Primula officinalis Jacq. var. macrocalyx（Bunge）C. Koch = Primula veris L. subsp. macrocalyx（Bunge）Ludi ■

314755 Primula okamotoi Koidz. = Primula reinii Franch. et Sav. ■☆

314756 Primula olgae Regel；奥氏报春 ■☆

314757 Primula operculata R. Knuth = Primula cockburniana Hemsl. ■

314758 Primula optata Farrer ex Balf. f.；心愿报春（报春花，甘肃高葶雪山报春）；Dream Primrose ■

314759 Primula orbicularis Hemsl.；圆瓣黄花报春（圆叶报春）；Orbicular Primrose, Orbiculate Primrose ■

314760 Primula oreina Balf. f. et Cooper = Primula dryadifolia Franch. subsp. jonardunii（W. W. Sm.）F. H. Chen et C. M. Hu ■

314761 Primula oreocharis Hance = Primula maximowiczii Regel ■

314762 Primula oreodoxa Franch.；迎阳报春；Mountainglary Primrose, Mountainglory Primrose ■

314763 Primula oresbia Balf. f. = Primula blinii H. Lév. ■

314764 Primula orestora Craib et Cooper = Primula atrodentata W. W. Sm. ■

314765 Primula ossetica Kusn.；骨质报春 ■☆

314766 Primula ovalifolia Franch.；卵叶报春（粗莛报春）；Ovalleaf Primrose, Thick Primrose ■

314767 Primula ovalifolia Franch. subsp. tardiflora C. M. Hu = Primula tardiflora（C. M. Hu）C. M. Hu ■

314768 Primula oxygraphdifolia W. W. Sm. et Kingdon-Ward；鸭跖花叶报春（四川粉报春）；Oxygraphisleaf Primrose ■

314769 Primula palinuri Petagna；意大利报春 ■☆

314770 Primula pallasii Lehm.；帕拉氏报春（阿尔泰报春）■☆

314771 Primula palmata Hand.-Mazz.；掌叶报春；Palmateleaf Primrose ■

314772 Primula pamirica Fed.；帕米尔报春 ■☆

314773 Primula pantlingii King = Primula dickieana Watt ■

314774 Primula partschiana Pax；心叶报春；Heartleaf Primrose ■

314775 Primula parva Balf. f. = Primula obconica Hance subsp. parva（Balf. f.）W. W. Sm. et Forrest ■

314776 Primula parva Balf. f. = Primula obconica Hance ■

314777 Primula parvula Pax et K. Hoffm. = Primula souliei Franch. ■

314778 Primula patens（Turcz.）E. Busch = Primula sieboldii E. Morren ■

314779 Primula patens Turcz. = Primula sieboldii E. Morren ■

314780 Primula patens Turcz. var. genuina Skvortzov = Primula sieboldii E. Morren ■

314781 Primula patens Turcz. var. manshurica Skvortzov = Primula sieboldii E. Morren ■

314782 Primula paucifolia（Hook. f.）Watt ex Craib = Primula denticulata Sm. ■

314783 Primula paucifolia Watt ex Craib = Primula denticulata Sm. ■

314784 Primula pauliana W. W. Sm. et Forrest；总序报春；Paul Primrose ■

314785 Primula pauliana W. W. Sm. et Forrest var. huiliensis F. H. Chen et C. M. Hu；会理总序报春（会理报春）；Huili Primrose ■

314786　Primula paxiana Kuntze ＝ Primula loeseneri Kitag. ■

314787　Primula pectinata Balf. f. et Forrest；篦齿报春；Pectinate Primrose ■

314788　Primula pectinata Balf. f. et Forrest ＝ Primula blinii H. Lév. ■

314789　Primula pedemontana E. Thomas；皮埃蒙特报春；Piedmont Primula ■☆

314790　Primula pellucida Franch.；钻齿报春（光叶报春）；Glabrousleaf Primrose ■

314791　Primula penduliflora Franch. ex Petitm. ＝ Primula flaccida N. P. Balakr. ■

314792　Primula petiolaris Wall.；有柄报春（齿瓣报春）；Petiolate Primrose ■☆

314793　Primula petiolaris Wall. var. odontocalyx Franch. ＝ Primula odontocalyx（Franch.）Pax ■

314794　Primula petiolaris Wall. var. scapigera Hook. f. ＝ Primula scapigera（Hook. f.）Craib ■

314795　Primula petiolaris Wall. var. setschwanica Pax et K. Hoffm. ＝ Primula hoffmanniana W. W. Sm. ■

314796　Primula petitmenginii Bonati ＝ Primula obconica Hance ■

314797　Primula petraea Balf. f. et Forrest ＝ Primula minor Balf. f. et Kingdon-Ward ■

314798　Primula petrocallis F. H. Chen et C. M. Hu；饰岩报春；Rockface Primrose ■

314799　Primula petrocallis F. H. Chen et C. M. Hu var. glabrata F. H. Chen et C. M. Hu；无毛饰岩报春（光叶饰岩报春）■

314800　Primula petrocharis Pax et K. Hoffm. ＝ Primula walshii Craib ■

314801　Primula petrophyes Balf. f. ＝ Primula virginis H. Lév. ■

314802　Primula philoresia Balf. f. ＝ Primula dryadifolia Franch. ■

314803　Primula philoresia Balf. f. et Kingdon-Ward ＝ Primula dryadifolia Franch. ■

314804　Primula pinnata Popov et Fed.；羽状报春 ■☆

314805　Primula pinnatifida Franch.；羽叶穗花报春（裂叶报春，羽裂报春，羽裂叶报春）；Pinnatifid Primrose ■

314806　Primula pinnatifida Franch. ＝ Primula blinii H. Lév. ■

314807　Primula pinnatifida Franch. subsp. apoclita（Balf. f. et Forrest）W. W. Sm. et Forrest ＝ Primula pinnatifida Franch. ■

314808　Primula pinnatifida Franch. suhsp. cephalantha（Balf.）W. W. Sm. et Forrest ＝ Primula pinnatifida Franch. ■

314809　Primula pintchouanensis Petitm. ＝ Primula bathangensis Petitm. et Hand. -Mazz. ■

314810　Primula pirolifolia H. Lév. ＝ Primula veitchiana Petitm. ■

314811　Primula planiflora Hand. -Mazz. ＝ Primula poissonii Franch. ■

314812　Primula platycrana Craib ＝ Primula denticulata Sm. ■

314813　Primula plebeia Balf. f. ＝ Primula sinuata Franch. ■

314814　Primula poculiformis Hook. f. ＝ Primula obconica Hance ■

314815　Primula poissonii Franch.；海仙报春（海仙花，平瓣报春）；Poisson Primrose ■

314816　Primula poissonii Franch. ＝ Primula wilsonii Dunn ■

314817　Primula poissonii Franch. subsp. anguistidens（Franch.）Pax ex W. W. Sm. et Forrest ＝ Primula stenodonta Balf. f. ex W. W. Sm. et H. R. Fletcher ■

314818　Primula poissonii Franch. subsp. wilsonii（Dunn）W. W. Sm. et Forrest ＝ Primula wilsonii Dunn ■

314819　Primula polia Craib ＝ Primula ovalifolia Franch. ■

314820　Primula polyantha Mill.；丛花报春；English Primrose, Polyanthus, Polyanthus Primrose ■☆

314821　Primula polyneura Franch.；多脉报春；Manynerve Primrose,

314822　Primula polyneura Franch. subsp. hymenophylla（Balf. f. et Forrest）W. W. Sm. et Forrest ＝ Primula polyneura Franch. ■

314823　Primula polyneura Franch. subsp. lichiangensis（Forrest）W. W. Sm. et Forrest ＝ Primula polyneura Franch. ■

314824　Primula polyneura Franch. subsp. sataniensis（Balf. f. et Farrer）W. W. Sm. et Forrest ＝ Primula polyneura Franch. ■

314825　Primula polyneura Franch. subsp. sikuensis（Balf. f. et Farrer）W. W. Sm. et Forrest ＝ Primula polyneura Franch. ■

314826　Primula polyneura Franch. subsp. veitchii（Duthie）W. W. Sm. et Forrest ＝ Primula polyneura Franch. ■

314827　Primula polyphylla（Franch.）Petitm. ＝ Primula pseudodenticulata Pax ■

314828　Primula polyphylla（Franch.）Petitm. var. monticola Hand. -Mazz. ＝ Primula monticola（Hand. -Mazz.）F. H. Chen et C. M. Hu ■

314829　Primula polyphylla Franch.；无粉海仙花；Manyleaf Primrose ■☆

314830　Primula praeflorens F. H. Chen et C. M. Hu；早花脆蒴报春；Earlybloom Primrose ■

314831　Primula praenitens Ker Gawl. ＝ Primula sinensis Sabine ex Lindl. ■

314832　Primula praetermissa W. W. Sm.；匙叶小报春；Spoonleaf Primrose ■

314833　Primula praticola Craib ＝ Primula taliensis Forrest ■

314834　Primula prattii Hemsl.；雅砻黄报春；Pratt Primrose ■

314835　Primula prenantha Balf. f. et W. W. Smith.；小花灯台报春；Droopflower Primrose ■

314836　Primula prenantha Balf. f. et W. W. Sm. subsp. morsheadiana（Kingdon-Ward）F. H. Chen et C. M. Hu；朗贡灯台报春（朗贡报春）■

314837　Primula prevernalis F. H. Chen et C. M. Hu；云龙报春；Earlyspring Primrose ■

314838　Primula primulina（Spreng.）Hara；球毛小报春（细羽樱草）；Insignificant Primrose, Mini Primrose ■

314839　Primula primulina Spreng. ＝ Primula primulina（Spreng.）Hara ■

314840　Primula primuloides D. Don ＝ Primula primulina（Spreng.）Hara ■

314841　Primula prionotes Balf. f. et Watt ＝ Primula waltonii Watt ex Balf. f. ■

314842　Primula proba Balf. f. et Forrest ＝ Primula calliantha Franch. subsp. bryophila（Balf. f. et Farrer）W. W. Sm. et Forrest ■

314843　Primula prolifera Wall.；多育报春（灯台报春）；Proliferous Primrose ■☆

314844　Primula propinqua Balf. f. et Forrest ＝ Primula boreio-calliantha Balf. f. et Cooper ■

314845　Primula pseudobracteata Petitm.；假苞报春（点地梅，假具色报春）■☆

314846　Primula pseudobracteata Petitm. ＝ Primula bracteata Franch. ■

314847　Primula pseudobracteata Petitm. var. polyphylla（Franch.）W. W. Sm.；多叶假苞报春 ■☆

314848　Primula pseudocapitata Kingdon-Ward ＝ Primula capitata Hook. subsp. sphaerocephala（Balf. f. et W. W. Sm.）W. W. Sm. et Forrest ■

314849　Primula pseudodenticulata Pax；滇海水仙花（海水仙）；Falsedenticulate Primrose ■

314850　Primula pseudodenticulata Pax subsp. polyphylla（Franch.）W. W. Sm. et Forrest ＝ Primula pseudodenticulata Pax ■

314851　Primula pseudoglabra Hand. -Mazz.；松潘报春；False Psilate

Primrose ■

314852　Primula pseudomalacoides L. B. Stewart　= Primula malacoides Franch. ■

314853　Primula pseudomalacoides L. B. Stewart ex Balf. f.　= Primula malacoides Franch. ■

314854　Primula pseudomalacoides Smart　=Primula malacoides Franch. ■

314855　Primula pseudosikkimensis Forrest　=Primula sikkimensis Hook. ■

314856　Primula pubescens Jacq. ;细毛报春;Garden Primrose ■☆

314857　Primula pudibunda W. W. Sm.　= Primula sikkimensis Hook. ■

314858　Primula pulchella Franch. ;丽花报春;Beautifulflower Primrose ■

314859　Primula pulchelloides Kingdon-Ward　= Primula pulchella Franch. ■

314860　Primula pulchra Watt;美丽缨草■

314861　Primula pulinata Balf. f. et Forrest;垫状报春■

314862　Primula pulverea Fed. ;多粉报春■☆

314863　Primula pulverulenta Duthie;粉被灯台报春(川东报春,粉被报春,粉莛报春);Powder Primrose, Pulverulent Primrose, Silverdust Primrose ■

314864　Primula pulverulenta Duthie 'Bartley';巴特尼尔粉被灯台报春■☆

314865　Primula pulvinata Balf. f. et Kingdon-Ward　= Primula bracteata Franch. ■

314866　Primula pumilio Maxim. ;柔小粉报春(侏儒报春)■

314867　Primula purdomii Craib;紫罗兰报春(朴氏报春);Purdom Primrose ■

314868　Primula purdomii Craib　= Primula woodwardii Balf. f. ■

314869　Primula purpurea Royle　= Primula macrophylla D. Don ■

314870　Primula pusilla Wall.　= Primula primulina (Spreng.) Hara ■

314871　Primula pusilla Wall. var. flabellata W. W. Sm.　= Primula primulina (Spreng.) Hara ■

314872　Primula pycnoloba Bureau et Franch. ;密裂报春(长萼报春);Denselobate Primrose ■

314873　Primula pygmaeorum Balf. f. et W. W. Sm.　= Primula pumilio Maxim. ■

314874　Primula qinghaiensis F. H. Chen et C. M. Hu;青海报春;Qinghai Primrose ■

314875　Primula racemosa Bonati　= Primula bathangensis Petitm. et Hand. -Mazz. ■

314876　Primula racemosa H. Lév.　= Primula celsiaeformis Franch. ■

314877　Primula ragotiana H. Lév.　= Primula sinuata Franch. ■

314878　Primula ranunculoides F. H. Chen　= Primula cicutariifolia Pax ■

314879　Primula ranunculoides F. H. Chen var. minor F. H. Chen　= Primula cicutariifolia Pax ■

314880　Primula redolens Balf. f. et Kingdon-Ward　= Primula forrestii Balf. f. ■

314881　Primula reflexa Petitm. ;嫩黄报春■

314882　Primula refracta Hand. -Mazz.　= Primula duclouxii Petitm. ■

314883　Primula reginella Balf. f.　= Primula fasciculata Balf. f. et Kingdon-Ward ■

314884　Primula reidii Duthie;雷氏报春■☆

314885　Primula reinii Franch. et Sav. ;赖因报春■☆

314886　Primula reinii Franch. et Sav. f. albiflora Makino;白花赖因报春■☆

314887　Primula reinii Franch. et Sav. var. brachycarpa (H. Hara) Ohwi　= Primula tosaensis Yatabe var. brachycarpa (H. Hara) Ohwi ■☆

314888　Primula reinii Franch. et Sav. var. kitadakensis (H. Hara) Ohwi;信州北岳报春■☆

314889　Primula reinii Franch. et Sav. var. okamotoi (Koidz.) Murata =

Primula reinii Franch. et Sav. ■☆

314890　Primula reinii Franch. et Sav. var. ovatifolia Ohwi;卵叶赖因报春■☆

314891　Primula reinii Franch. et Sav. var. rhodotricha (Nakai et F. Maek.) T. Yamaz. ;粉毛赖因报春■☆

314892　Primula renifolia Volgunov;肾叶报春■☆

314893　Primula reptans Hook. f. ex Watt;匍匐报春;Reptant Primrose ■☆

314894　Primula reticulata Wall. ; 网叶钟报春; Netleaf Primrose, Reticulate Primrose ■

314895　Primula rhodochroa W. W. Sm. ;密丛小报春;Netleaf Primrose ■

314896　Primula rhodochroa W. W. Sm. var. geraldinae (W. W. Sm.) F. H. Chen et C. M. Hu;洛拉小报春■

314897　Primula rhodochroa W. W. Sm. var. meiotera W. W. Sm. et H. R. Fletcher　= Primula meiotera (W. W. Sm. et H. R. Fletcher) C. M. Hu ■

314898　Primula rhodotricha Nakai et F. Maek.　= Primula reinii Franch. et Sav. var. rhodotricha (Nakai et F. Maek.) T. Yamaz. ■☆

314899　Primula riae Pax et K. Hoffm.　= Primula amethystina Franch. subsp. argutidens (Franch.) W. W. Sm. et H. R. Fletcher ■

314900　Primula rigida Balf. f. et Forrest　= Primula diantha Bureau et Franch. ■

314901　Primula rimicola W. W. Sm. ;岩生小报春;Rimose Primrose ■

314902　Primula riparia Balf. f. et Farrer　= Primula cinerascens Franch. ■

314903　Primula rockii W. W. Sm. ;纤柄皱叶报春(川藏皱叶报春);Rock Primrose ■

314904　Primula rosea Royle;玫瑰色报春(玫红报春);Rose Primrose ■☆

314905　Primula rosea Royle var. elegans ?　= Primula rosea Royle ■☆

314906　Primula rosthornii Diels;短葶叶报春;Rosthorn Primrose ■☆

314907　Primula rosthornii Diels　= Primula neurocalyx Franch. ■

314908　Primula rotundifolia Wall.　= Primula rotundifolia Wall. ex Roxb. ■

314909　Primula rotundifolia Wall. ex Roxb. ;大圆叶报春(圆叶报春);Bigroundleaf Primrose, Round-leaved Primrose ■

314910　Primula roxburghii N. P. Balakr.　= Primula rotundifolia Wall. ex Roxb. ■

314911　Primula roylei Balf. f. et W. W. Sm.　= Primula calderiana Balf. f. et Cooper ■

314912　Primula roylei Balf. f. et W. W. Sm. subsp. calderiana (Balf. f. et Cooper) W. W. Sm. et Forrest　= Primula calderiana Balf. f. et Cooper ■

314913　Primula roylei Balf. f. et W. W. Sm. var. alba W. W. Sm.　= Primula calderiana Balf. f. et Cooper ■

314914　Primula rubicunda H. R. Fletcher;深红小报春;Deepred Primrose ■

314915　Primula rubifolia C. M. Hu;莓叶报春;Blackberryleaf Primrose ■

314916　Primula rufa Balf. f.　= Primula forrestii Balf. f. ■

314917　Primula rugosa N. P. Balakr. ;倒卵叶报春(蒙自报春);Wrinkleleaf Primrose ■

314918　Primula runcinata W. W. Sm. et H. R. Fletcher ex C. M. Hu;芥叶报春;Runcinate Primrose ■

314919　Primula rupestris Balf. f. et Forrest;巴蜀报春■

314920　Primula rupestris Pax et K. Hoffm.　= Primula gemmifera Batalin var. amoena F. H. Chen ■

314921　Primula rupicola Balf. f. et Forrest;黄粉缺裂报春(石报春);Yellowpowder ■

314922　Primula rupicola Balf. f. et Forrest var. albicolor W. W. Sm. et H. R. Fletcher　= Primula rupicola Balf. f. et Forrest ■

314923　Primula ruprechtii Kusn. ;鲁普雷希特报春■☆

314924　Primula russeola Balf. f. et Forrest;黑萼报春(黑花报春,红花

雪山报春）;Blackcalyx Primrose ■

314925 Primula sachalinensis Nakai;库页报春■

314926 Primula sandemaniana W. W. Sm. ;粉萼垂花报春;Sandeman Primrose ■

314927 Primula sapphirina Hook. f. et Thomson ex Watt;小垂花报春; Verdant Primrose ■

314928 Primula sataniensis Balf. f. et Forrest = Primula polyneura Franch.

314929 Primula saturata W. W. Sm. et H. R. Fletcher;黄葵叶报春; Muskmallowleaf Primrose ■

314930 Primula saxatilis Kom. ;岩生报春;Saxicolous Primrose ■

314931 Primula saxatilis Kom. var. pubescens Pax et K. Hoffm. = Primula polyneura Franch. ■

314932 Primula scapigera（Hook. f.）Craib;葶花脆蒴报春;Scapose Primrose ■

314933 Primula scopulorum Balf. f. et Farrer;米仓山报春;Micangshan Primrose ■

314934 Primula scullyi Craib = Primula gracilipes Craib ■

314935 Primula seclusa Balf. f. et Forrest = Primula mollis Nutt. ex Hook. ■

314936 Primula secundiflora Franch.;偏花报春（报春,带叶报春,偏花叶报春,偏花钟报春,条纹报春）;Excentricflower Primrose ■

314937 Primula semperflorens Loisel. ex Steud. = Primula sinensis Sabine ex Lindl.

314938 Primula septemloba Franch.;七指报春（七裂报春）; Sevenlobate Primrose ■

314939 Primula septemloba Franch. = Primula heucherifolia Franch. ■

314940 Primula septemloba Franch. var. minor Kingdon-Ward;小七指报春■

314941 Primula serratifolia Franch.;齿叶灯台报春（齿叶报春,多齿叶报春,烂泥蒿）;Toothedleaf Primrose ■

314942 Primula sertulosa Kickx f. = Primula sinensis Sabine ex Lindl. ■

314943 Primula sertulum Franch.;小伞报春（岷山苞花报春）; Smallumbrella Primrose ■

314944 Primula sherriffae W. W. Sm.;长管垂花报春;Sherriff Primrose ■

314945 Primula shihmienensis W. P. Fang;石棉报春; C. Shihmien Primrose,Shimian Primrose ■

314946 Primula shihmienensis W. P. Fang = Primula pulverulenta Duthie ■

314947 Primula shwelicalliantha Balf. f et Forrest = Primula calliantha Franch. subsp. bryophila（Balf. f. et Farrer）W. W. Sm. et Forrest ■

314948 Primula sibirica Jacq.;西伯利亚报春（天山报春）;Siberian Primrose ■

314949 Primula sibirica Jacq. = Primula nutans Georgi ■

314950 Primula sibirica Jacq. var. kashmiriana Hook. f. = Primula nutans Georgi ■

314951 Primula sibthorpi Hoffm. ;西氏报春■☆

314952 Primula sieboldii E. Morren;樱草（翠兰花,翠蓝报春,翠蓝草,翠南报春,野白菜）;Siebold Primrose,Siebold's Primrose ■

314953 Primula sieboldii E. Morren 'Sumina';大花樱草■☆

314954 Primula sieboldii E. Morren 'Wine Lady';醉女樱草■☆

314955 Primula sieboldii E. Morren f. albiflora H. Hara = Primula sieboldii E. Morren f. lactiflora（Nakai）H. Hara ■☆

314956 Primula sieboldii E. Morren f. incisa（Miq.）Makino;锐裂樱草■☆

314957 Primula sieboldii E. Morren f. lactiflora（Nakai）H. Hara;乳花樱草■☆

314958 Primula sieboldii E. Morren f. patens（Turcz.）Kitag. = Primula sieboldii E. Morren ■

314959 Primula sikangensis F. H. Chen = Primula amethystina Franch. subsp. brevifolia（Forrest）W. W. Sm. et Forrest ■

314960 Primula sikkimensis Hook. = Primula sikkimensis Hook. f. ■

314961 Primula sikkimensis Hook. f. ;钟花报春（黄花报春,锡金报春）;Himalayan Cowslip,Sikkim Cowslip,Sikkim Primrose ■

314962 Primula sikkimensis Hook. f. subsp. pseudosikkimensis（Forrest）W. W. Sm. et Forrest = Primula sikkimensis Hook. ■

314963 Primula sikkimensis Hook. f. subsp. pudibunda（W. W. Sm.）W. W. Sm. et Forrest = Primula sikkimensis Hook. f. ■

314964 Primula sikkimensis Hook. f. subsp. subpinnatifida W. W. Sm. = Primula ioessa W. W. Sm. ■

314965 Primula sikkimensis Hook. f. var. hookeri Stapf = Primula sikkimensis Hook. f. ■

314966 Primula sikkimensis Hook. f. var. lorifolia W. W. Sm. = Primula sikkimensis Hook. f. ■

314967 Primula sikkimensis Hook. f. var. microdonta Stapf = Primula waltonii Watt ex Balf. f. ■

314968 Primula sikkimensis Hook. f. var. pudibunda（W. W. Sm.）W. W. Sm. et Forrest = Primula sikkimensis Hook. f. ■

314969 Primula sikkimensis Hook. f. var. pudibunda（W. W. Sm.）W. W. Sm. et H. R. Fletcher = Primula sikkimensis Hook. f. ■

314970 Primula sikkimensis Hook. f. var. subpinnatifida W. W. Sm. = Primula ioessa W. W. Sm. ■

314971 Primula sikuensis Balf. f. et Farrer = Primula polyneura Franch. ■

314972 Primula silaensis Petitm.;贡山紫晶报春（贡山报春）; Gongshan Primrose ■

314973 Primula silenantha Pax et K. Hoffm. = Primula tangutica Pax ■

314974 Primula simensis Hochst. = Primula verticillata Forssk. subsp. simensis（Hochst.）W. W. Sm. et Forrest ■☆

314975 Primula sinensis Sabine ex Lindl.;藏报春（报春花,藏报春花,华报春,年景花,四季报春）;China Primrose, Chinese Primrose, Chinese Primula ■

314976 Primula sinodenticulata Balf. f. et Forrest = Primula denticulata Sm. subsp. sinodenticulata（Balf. f. et Forrest）W. W. Sm. et Forrest ■

314977 Primula sinolisteri Balf. f.;铁梗报春（铁丝报春）;Sinolister Primrose ■

314978 Primula sinolisteri Balf. f. var. aspera W. W. Sm. et H. R. Fletcher;糙叶铁梗报春■

314979 Primula sinolisteri Balf. f. var. longicalyx D. W. Xue et C. Q. Zhang;长萼铁梗报春■

314980 Primula sinomollis Balf. f. et Forrest;华柔毛报春;Velvety Primrose ■

314981 Primula sinomollis Balf. f. et Forrest var. alba Balf. f. et Forrest = Primula sinomollis Balf. f. et Forrest ■

314982 Primula sinonivalis Balf. f. et Forrest = Primula limbata Balf. f. et Forrest ■

314983 Primula sinoplantaginea Balf. f.;车前叶报春（窄叶报春花,中华车前叶报春）;Narrowleaf Primrose,Plantainleaf Primrose ■

314984 Primula sinoplantaginea Balf. f. subsp. graminifolia（Pax et K. Hoffm.）W. W. Sm. et Forrest = Primula graminifolia Pax et K. Hoffm. ■

314985 Primula sinoplantaginea Balf. f. var. fengxiangiana W. L. Zheng; 凤翔报春;Fengxiang Primrose ■

314986 Primula sinoplantaginea Balf. f. var. graminifolia（Pax et K. Hoffm.）W. W. Sm. et H. R. Fletcher = Primula graminifolia Pax et K. Hoffm. ■

314987 Primula sinopurpurea Balf. f. = Primula longipetiolata Pax et K. Hoffm. ■

314988 Primula sinopurpurea Balf. f. ex Hutch. ;紫花雪山报春（车前草叶报春，华蓝报春花，华紫报春，三月花，中华紫报春）; Sinopurple Primrose，Violetflower Primrose ■

314989 Primula sinopurpurea Balf. f. ex Hutch. = Primula chionantha Balf. f. et Forrest ■

314990 Primula sinuata Franch. ;波缘报春（齿裂苣叶报春）; Sinuate Primrose ■

314991 Primula smithiana Craib;亚东灯台报春; Smith Primrose ■

314992 Primula socialis F. H. Chen et C. M. Hu;群居粉报春; Gather Primrose ■

314993 Primula soldanelloides Watt;低温花状报春; Soldanellalike Primrose ■☆

314994 Primula sonchifolia Franch. ;苣叶报春（峨山雪莲花，苣叶脆蒴报春，苦苣叶报春花）;Sowthistleleaf Primrose ■

314995 Primula sonchifolia Franch. subsp. emeiensis C. M. Hu;峨眉苣叶报春;Emei Sowthistleleaf Primrose ■

314996 Primula sonchifolia Franch. var. atrocoerulea Forrest = Primula sonchifolia Franch. ■

314997 Primula soongii F. H. Chen et C. M. Ha;滋圃报春; Soong Primrose ■

314998 Primula souliei Franch. ;缺裂报春（苏氏报春）; Soulie Primrose ■

314999 Primula souliei Franch. subsp. florida（Balf. f. et Forrest）W. W. Sm. et Forrest = Primula blinii H. Lév. ■

315000 Primula souliei Franch. subsp. humilis（Pax et K. Hoffm. ）W. W. Sm. et Forrest = Primula humilis Pax et K. Hoffm. ■

315001 Primula souliei Franch. subsp. legendrei（Bonati）W. W. Sm. et Forrest = Primula souliei Franch. ■

315002 Primula souliei Franch. subsp. oresbia（Balf. f. ）W. W. Sm. et Forrest = Primula blinii H. Lév. ■

315003 Primula spathulacea Jacquem. ex Duby = Primula elliptica Royle ■☆

315004 Primula speluncicola Petitm. = Primula pellucida Franch. ■

315005 Primula sphaerocephala Balf. f. et Forrest = Primula capitata Hook. subsp. sphaerocephala（Balf. f. et W. W. Sm. ）W. W. Sm. et Forrest ■

315006 Primula spicata Franch. ;穗状垂花报春（穗花报春）; Spike Primrose ■

315007 Primula stenocalyx Maxim. ;狭萼报春;Narrowcalyx Primrose ■

315008 Primula stenocalyx Maxim. var. luteofarinosa W. W. Sm. = Primula stenocalyx Maxim. ■

315009 Primula stenodonta Balf. f. ex W. W. Sm. et H. R. Fletcher;凉山灯台报春;Liangshan Primrose ■

315010 Primula stephanocalyx Hand. -Mazz. = Primula bathangensis Petitm. et Hand. -Mazz. ■

315011 Primula stirtoniana Watt;斯提氏报春;Stirton Primrose ■☆

315012 Primula stolonifera Balf. f. = Primula pseudodenticulata Pax ■

315013 Primula stracheyi Hook. f. et Thomson ex Watt = Primula reptans Hook. f. ex Watt ■☆

315014 Primula stragulata Balf. f. et Forrest = Primula bella Franch. ■

315015 Primula stricta Hornem. ;直报春■☆

315016 Primula strumosa Balf. f. et Cooper;金黄脆蒴报春（瘤报春）; Tumorous Primrose ■

315017 Primula strumosa Balf. f. et Cooper subsp. emeiensis C. M. Hu;峨眉金黄报春;Emei Tumorous Primrose ■

315018 Primula strumosa Balf. f. et Cooper subsp. tenuipes C. M. Hu;矩

圆金黄报春■

315019 Primula stuartii Wall. ;斯氏报春;Stuart Primrose ■☆

315020 Primula stuartii Wall. var. macrocarpa Watt = Primula megalocarpa H. Hara ■

315021 Primula stuartii Wall. var. moorcroftiana（Wall. ex Klatt）Watt = Primula macrophylla D. Don var. moorcroftiana（Wall. ex Klatt）W. W. Sm. et H. R. Fletcher ■

315022 Primula stuartii Wall. var. purpurea（Royle）Watt = Primula macrophylla D. Don ■

315023 Primula subtropica Hand. -Mazz. = Primula vilmoriniana Petitm. ■

315024 Primula subularia W. W. Sm. ;线叶小报春;Linear Primrose ■

315025 Primula sulphurea Pax et K. Hoffm. = Primula prattii Hemsl. ■

315026 Primula sulphurea Pax et K. Hoffm. var. rosea Pax et K. Hoffm. = Primula pulchella Franch. ■

315027 Primula sylvicola Hutch. = Primula sinomollis Balf. f. et Forrest ■

315028 Primula szechuanica Pax;四川报春（偷筋草）; Sichuan Primrose，Szechuan Primrose ■

315029 Primula takedana Tatew. ;武田氏报春■☆

315030 Primula taliensis Forrest;大理报春;Dali Primrose ■

315031 Primula taliensis Forrest subsp. procera C. M. Hu;金粉大理报春■

315032 Primula tangutica Duthie = Primula tangutica Pax ■

315033 Primula tangutica Pax;甘青报春（唐古特报春）; Tangut Primrose ■

315034 Primula tangutica Pax var. flavescens F. H. Chen et C. M. Hu;黄甘青报春■

315035 Primula tangutica Pax var. serrata W. W. Sm. et H. R. Fletcher = Primula tangutica Pax ■

315036 Primula tanneri King;心叶脆蒴报春（谭氏报春）; Cordate Primrose ■

315037 Primula tanneri King subsp. tsariensis（W. W. Sm. ）A. J. Richards = Primula tsariensis W. W. Sm. ■

315038 Primula tanupoda Balf. f. et W. W. Sm. = Primula tibetica Watt ■

315039 Primula tapeina Balf. f. et Forrest = Primula bracteata Franch. ■

315040 Primula tapetodes（Bunge）Kuntze = Dionysia tapetodes Bunge ■☆

315041 Primula taraxacoides Balf. f. = Primula sonchifolia Franch. ■

315042 Primula tardiflora（C. M. Hu）C. M. Hu;晚花报春（晚花卵叶报春）;Late Ovalleaf Primrose ■

315043 Primula tayloriana H. R. Fletcher;淡粉报春;Taylor Primrose ■

315044 Primula tenana Bonati ex Balf. f. = Primula blattariformis Franch. ■

315045 Primula tenella King ex Hook. f. ;匍茎小报春（细柔樱草）; Lithe Primrose，Slender Primrose ■

315046 Primula tenuiloba（Watt）Pax;细裂小报春; Narrowlobate Primrose ■

315047 Primula tenuipes F. H. Chen et C. M. Hu;纤柄报春; Narrowstipe Primrose ■

315048 Primula tenuissima Pax = Primula odontocalyx（Franch. ）Pax ■

315049 Primula tibetica Watt;西藏报春（藏东报春）; Tibet Primrose，Xizang Primrose ■

315050 Primula tongolensis Franch. ;东俄洛报春;Tongol Primrose ◪

315051 Primula tosaensis Yatabe;土佐报春■☆

315052 Primula tosaensis Yatabe f. albiflora Mizuno;白花土佐报春（白花东俄洛报春）■☆

315053 Primula tosaensis Yatabe f. brachycarpa H. Hara = Primula tosaensis Yatabe var. brachycarpa（H. Hara）Ohwi ■☆

315054 Primula tosaensis Yatabe var. brachycarpa（H. Hara）Ohwi;短

果土佐报春■☆

315055 Primula tosaensis Yatabe var. rhodotricha（Nakai et F. Maek.）Ohwi = Primula reinii Franch. et Sav. var. rhodotricha（Nakai et F. Maek.）T. Yamaz. ■☆

315056 Primula tournefortii Rupr.；图尔报春■☆

315057 Primula tribola Balf. f. et Forrest = Primula calliantha Franch. subsp. bryophila（Balf. f. et Farrer）W. W. Sm. et Forrest ■

315058 Primula tridentife F. H. Chen et C. M. Hu；三齿卵叶报春（三齿脆蒴报春）；Threetooth Primrose ■

315059 Primula triloba Balf. f. et Forrest；三裂叶报春；Threelobe Primrose ■

315060 Primula tsariensis W. W. Sm.；察日脆蒴报春；Chari Primrose ■

315061 Primula tsariensis W. W. Sm. var. porrecta W. W. Sm.；大察日报春；Big Chari Primrose ■

315062 Primula tsarongensis Balf. f. et Forrest = Primula muscarioides Hemsl. ■

315063 Primula tsiangiae W. P. Fang = Primula longipetiolata Pax et K. Hoffm. ■

315064 Primula tsiangii W. W. Sm.；绒毛报春（蒋氏报春）；Tsiang Primrose ■

315065 Primula tsongpenii H. R. Fletcher；丛毛岩报春；Thickhair Primrose ■

315066 Primula turkestanica（Regel）E. White = Primula nivalis Pall. var. farinosa Schrenk ■

315067 Primula tyoseniana Nakai ex Kitag. = Primula loeseneri Kitag. ■

315068 Primula tzetsouensis Petitm.；心叶黄花报春；Tzetsou Primrose ■

315069 Primula ulophylla Hand. -Mazz. = Primula bracteata Franch. ■

315070 Primula umbellata（Lour.）Bentv. = Androsace umbellata（Lour.）Merr. ■

315071 Primula umbrella Forrest = Primula yunnanensis Franch. ■

315072 Primula uniflora Klatt；单花樱草；Uniflorous Primrose ■

315073 Primula uniflora Klatt = Primula klattii N. P. Balakr. ■

315074 Primula urticifolia Maxim.；荨麻叶报春；Nettleleaf Primrose ■

315075 Primula vaginata Watt；鞘柄掌叶报春（叶鞘报春）；Sheathed Primrose，Sheathpalm Primrose ■

315076 Primula vaginata Watt subsp. eucyclia（W. W. Sm. et Forrest）F. H. Chen et C. M. Hu；圆叶鞘柄报春（圆叶报春，正轮掌叶报春）■

315077 Primula vaginata Watt subsp. normaniana（Kingdon-Ward）F. H. Chen et C. M. Hu；短梗鞘柄报春■

315078 Primula valentiniana Hand. -Mazz.；暗红紫晶报春（紫红报春）；Velentin Primrose ■

315079 Primula variabilis Bastard；多变报春；Polyanthus Primrose ■☆

315080 Primula veitchiana Petitm.；川西 瓣报春（维奇报春花，维氏报春）；Veitch Primrose ■

315081 Primula veitchii Duthie = Primula polyneura Franch. ■

315082 Primula veris L.；黄花九轮草（标准报春，标准报春花，黄花九轮樱，药用报春）；Beagle，Boys-and-girls，Bunch of Keys，Bunch-of-keys，Butter Rose，Carslope，Cooslip，Cove Keys，Cove-keys，Cow Paigle，Cow Peggle，Cow Slip，Cow Stripling，Cow Stropple，Cow Strupple，Cow's Mouth，Cower-slop，Cowflop，Cowslip，Cowslip Primrose，Cowslop，Cowslop Cowslap，Creivel，Crewel，Crows，Cruel，Cuckoo，Culverkeys，Fairy Basin，Fairy Bells，Fairy Cups，Fairy's Basin，Fairy's Cup，Fastenblume，Freckled Face，Galligaskins，Gaskins，Gelbe Zeitlose，Gichtblume，Golden Bells，Golden Drops，Herb Paralysy，Herb Peter，Hodrod，Holrod，Horse Buckle，Key-flower，Keys of Heaven，Keys-of-heaven，Keywort，Lady Cues，Lady's Bunch of Keys，Lady's Bunch of-keys，Lady's Fingers，Lady's Keys，Long Legs，Long-legs，Mayflower，May-flower，Milk Maidens，Nine-wheel Primrose，Oddrod，Paggle，Pagle，Paigle，Palsywort，Palsy-wort，Pea Gull，Peagle，Peeps，Peggle，Peterkin，Petty Mullein，Piggle，Pips，Plaggis，Primrose，Racconals，Slap，Slop，St. Peter's Herb，St. Peter's Keys，Symyl，Tisty-tosty，Tosty ■

315083 Primula veris L. subsp. macrocalyx（Bunge）Ludi；硕萼报春（大萼报春）■

315084 Primula vernicosa Kingdon-Ward = Primula hookeri Watt ■

315085 Primula vernicosa Kingdon-Ward ex Balf. f. = Primula hookeri Watt ■

315086 Primula vernicosa Kingdon-Ward var. violacea W. W. Sm. = Primula hookeri Watt var. violacea（W. W. Sm.）C. M. Hu ■

315087 Primula verticillata Forssk.；阿拉伯报春；Arabian Primrose ■☆

315088 Primula verticillata Forssk. subsp. simensis（Hochst.）W. W. Sm. et Forrest；锡米报春■☆

315089 Primula vialii Delavay ex Franch.；高穗报春花（高穗花报春）；Vial Primrose ■

315090 Primula vilmoriniana Petitm.；毛叶鄂报春■

315091 Primula vinciflora Franch. = Omphalogramma vinciflora（Franch.）Franch. ■

315092 Primula vinosa Stapf = Primula waltonii Watt ex Balf. f. ■

315093 Primula violacea W. W. Sm. et Kingdon-Ward；紫穗报春（堇叶报春）；Violetspike Primrose ■

315094 Primula viola-grandis Farrer et Purdom = Omphalogramma vinciflorum（Franch.）Franch. ■

315095 Primula viola-grandis Farrer et Purdom ex Balf. f. = Omphalogramma vinciflorum（Franch.）Franch. ■

315096 Primula violaris W. W. Sm. et H. R. Fletcher；堇菜报春；Violetlike Primrose ■

315097 Primula violodora Dunn = Primula cinerascens Franch. ■

315098 Primula virginis H. Lév.；乌蒙紫晶报春（处女报春，纯白报春）；Wumeng Primrose ■

315099 Primula vittata Bureau et Franch.；带叶报春（报春花，条纹报春）；Beltleaf Primrose ■

315100 Primula vittata Bureau et Franch. = Primula secundiflora Franch. ■

315101 Primula vulgaris（L.）Huds.；欧洲樱草（单花樱草，欧樱草，普通报春，无茎报春，无茎樱草）；Auntie Polly，Boys-and-girls，Buckie-faalie，Butter Rose，Common Primrose，Darling April，Darling Ofapril，Early Rose，Easter Rose，English Primrose，First Rose，Golden Rose，Golden Star，Lent Rose，Mary Spink，May-flower，May-spink，Paigle，Palsy Plant，Peagle，Pegyll，Pimmerose，Pimrose，Pink Primrose，Plumrocks，Primarose，Primmy Rose，Primrose，Simmeren，Simmerin，Sumach，Sumark ■☆

315102 Primula vulgaris（L.）Huds.‘Alba Plena’；白重瓣欧洲樱草 ■☆

315103 Primula vulgaris（L.）Huds.‘Gigha White’；吉阿白欧洲樱草 ■☆

315104 Primula vulgaris（L.）Huds. = Primula acaulis（L.）L. ■☆

315105 Primula vulgaris（L.）Huds. subsp. atlantica（Maire et Wilczek）Greuter et Burdet = Primula acaulis（L.）L. subsp. atlantica（Maire et Wilczek）Greuter et Burdet ■☆

315106 Primula vulgaris（L.）Huds. var. elatior ?；高欧洲樱草；Auntie Polly，Bullslop，Cow Sinkin，Five Fingers，Five-fingers，Great Cowslip，Jack-in-the-green，Lady's Candlesticks，None-so-pretty，

Polant, Polly Andréws, Polly Ann, Pollyandice, Polyanthus, Pug-in-a-pinner, Spring-flower ■☆

315107 Primula waddellii Balf. f. et W. W. Sm. ; 窄筒小报春; Waddell Primrose ■

315108 Primula walshii Craib; 腺毛小报春; Walsh Primrose ■

315109 Primula waltonii Watt ex Balf. f. ; 紫钟报春（瓦顿报春）; Walton Primrose ■

315110 Primula waltonii Watt ex Balf. f. subsp. prionotes (Balf. f. et Watt) W. W. Sm. et Forrest = Primula waltonii Watt ex Balf. f. ■

315111 Primula wangii F. H. Chen et C. M. Hu; 广南报春; Wang Primrose ■

315112 Primula wardii Balf. f. = Primula involucrata Wall. ex Duby subsp. yargongensis (Petitm.) W. W. Sm. et Forrest ■

315113 Primula warshenewskiana B. Fedtsch.; 瓦尔报春■☆

315114 Primula warshenewskiana B. Fedtsch. subsp. rhodantha ?; 粉红花瓦尔报春■☆

315115 Primula watsonii Dunn; 靛蓝穗花报春（短柄紫花报春, 瓦震报春）; Watson Primrose ■

315116 Primula watsonii Dunn subsp. cyanantha (Balf. f. et Forrest) W. W. Sm. et Forrest = Primula watsonii Dunn ■

315117 Primula wenshanensis F. H. Chen et G. M. Hu; 滇南脆蒴报春（白环报春）; Wenshan Primrose ■

315118 Primula werringtonensis Forrest = Primula obconica Hance subsp. werringtonensis (Forrest) W. W. Sm. et Forrest ■

315119 Primula werringtonensis Forrest = Primula obconica Hance var. werrigtonensis (Forrest) W. W. Sm. et H. R. Fletcher ■

315120 Primula werringtonensis Forrest = Primula obconica Hance ■

315121 Primula whitei W. W. Sm.; 鹃林脆蒴报春（不丹报春, 怀特报春）; White Primrose ■

315122 Primula willmottiae Petitm. = Primula forbesii Franch. ■

315123 Primula wilsonii Dunn; 香海仙报春（威氏报春花）; E. H. Wilson Primrose ■

315124 Primula wollastonii Balf. f. ; 钟状垂花报春; Wollaston Primrose ■

315125 Primula woodwardii Balf. f. ; 岷山报春（西藏紫花报春）; Woodward Primrose ■

315126 Primula woonyoungiana W. P. Fang; 焕镛报春; Woonyoung Primrose ■

315127 Primula woronovii Losinsk.; 沃氏报春■☆

315128 Primula xanthobasis Fed.; 黄基报春■☆

315129 Primula yargongensis Petitm. = Primula involucrata Wall. ex Duby subsp. yargongensis (Petitm.) W. W. Sm. et Forrest ■

315130 Primula yargongensis Petitm. var. liensis W. P. Fang = Primula involucrata Wall. ex Duby subsp. yargongensis (Petitm.) W. W. Sm. et Forrest ■

315131 Primula youngeriana W. W. Sm.; 展萼雪山报春; Younger Primrose ■

315132 Primula younghusbandiana Balf. f. = Primula caveana W. W. Sm. ■

315133 Primula yuana F. H. Chen = Primula tzetsouensis Petitm. ■

315134 Primula yunnanensis Franch.; 云南报春; Yunnan Primrose ■

315135 Primula yunnanensis Franch. subsp. fragilis (Balf. f. et Kingdon-Ward) W. W. Sm. et Forrest = Primula yunnanensis Franch. ■

315136 Primula yuparensis Takeda; 北海道报春■☆

315137 Primula zambalensis Petitm. = Primula gemmifera Batalin var. amoena F. H. Chen ■

315138 Primulaceae Batsch ex Borkh. (1797)（保留科名）; 报春花科; Primrose Family, Primula Family ●■

315139 Primulaceae Vent. = Primulaceae Batsch ex Borkh. (保留科

名)●■

315140 Primularia Brenan = Cincinnobotrys Gilg ■☆

315141 Primularia pulchella Brenan = Cincinnobotrys pulchella (Brenan) Jacq. -Fél. ■☆

315142 Primulidium Spach = Auganthus Link ■

315143 Primulidium Spach = Primula L. ■

315144 Primulidium sinense (Sabine ex Lindl.) Spach = Primula sinense Sabine ex Lindl. ■

315145 Primulidium sinense Spach = Primula sinensis Sabine ex Lindl. ■

315146 Primulina Hance(1883); 报春苣苔属; Primulina ★

315147 Primulina sinensis Hook. f. = Primulina tabacum Hance ■

315148 Primulina tabacum Hance; 报春苣苔（石烟）; Primulina, Tabaccoodored Primulina ■

315149 Princea Dubard et Dop = Triainolepis Hook. f. ■☆

315150 Principina Uittien = Hypolytrum Rich. ex Pers. ■

315151 Principina Uittien(1935); 松莎属■☆

315152 Principina grandis Uittien = Hypolytrum grande (Uittien) Koyama ■

315153 Pringlea T. Anderson ex Hook. f. (1845); 普林芥属■☆

315154 Pringlea antiscorbutica R. Br. ex Hook. f. ; 普林芥; Kerguelen Cabbage ■☆

315155 Pringleochloa Scribn. (1896); 喜钙匍茎草属■☆

315156 Pringleochloa stolonifera Scribn. ; 喜钙匍茎草■☆

315157 Pringleophytum A. Gray = Berginia Harv. ■☆

315158 Pringleophytum A. Gray = Holographis Nees ■☆

315159 Prinodia Griseb. = Prinos Gronov. ex L. ●

315160 Prinoides (DC.) Willis = Prinos Gronov. ex L. ●

315161 Prinos Gronov. ex L. = Ilex L. ●

315162 Prinos L. = Ilex L. ●

315163 Prinos asprellus Hook. et Arn. = Ilex asprella (Hook. et Arn.) Champ. ex Benth. ●

315164 Prinos godajam Colebr. = Ilex godajam Colebr. ex Wall. ●

315165 Prinos godajam Colebr. ex Wall. = Ilex godajam Colebr. ex Wall. ●

315166 Prinos integer Hook. et Arn. = Ilex integra Thunb. ●◇

315167 Prinos laurinus Thury = Ilex mitis (L.) Radlk. ●☆

315168 Prinos padifolius Willd. = Ilex verticillata (L.) A. Gray ●☆

315169 Prinos verticillatus L. = Ilex verticillata (L.) A. Gray ●☆

315170 Prinos verticillatus L. var. tenuifolius Torr. = Ilex verticillata (L.) A. Gray ●☆

315171 Prinsepia Royle (1835); 扁核木属（假皂荚属, 蕤核属）; Cherry Prinsepia, Prinsepia ●

315172 Prinsepia chinensis Oliv. ex Kom. = Prinsepia sinensis (Oliv.) Kom. ●

315173 Prinsepia nanhutashanense S. S. Ying; 南湖大山扁核木●

315174 Prinsepia scandens Hayata; 台湾扁核木（假皂荚）; Climbing Prinsepia ●

315175 Prinsepia sinensis (Oliv.) Kom. = Prinsepia sinensis (Oliv.) Oliv. ex Bean ●

315176 Prinsepia sinensis (Oliv.) Oliv. ex Bean; 东北扁核木(扁担胡子, 东北蕤核, 华北扁核木, 辽东扁核木, 辽宁扁核木, 中华扁核木); Cherry Prinsepia ●

315177 Prinsepia uniflora Batalin; 扁核木(白桜, 打枪果, 打油果, 单花扁核木, 鸡蛋糕, 李子蕤, 马茹, 梅花刺, 蒙自扁核木, 牛奶锤, 炮筒果, 枪子果, 青刺尖, 茹茹, 蕤核, 蕤李子, 蕤子, 山桃, 椹, 孙奶子, 小马茹, 械); Hedge Prinsepia, Prinsepia ●

315178 Prinsepia uniflora Batalin var. serrata Rehder; 齿叶扁核木;

Serrate Hedge Prinsepia ●

315179 Prinsepia utilis Royle;总花扁核木(扁核木,打枪果,打油果,鸡蛋果,梅花刺果,枪刺果,青刺尖);Himalayan Prinsepia,Himalayas Prinsepia ●

315180 Prinsepia utilis Royle = Prinsepia scandens Hayata ●

315181 Printzia Cass. (1825)(保留属名);尾药菀属■●☆

315182 Printzia aromatica (L.) Less. ;芳香尾药菀■☆

315183 Printzia asteroides Schltr. ex Bews = Lepidostephium asteroides (Bolus et Schltr.) Kroner ■

315184 Printzia auriculata Harv. ;耳形尾药菀■☆

315185 Printzia bergii Cass. = Printzia polifolia (L.) Hutch. ■☆

315186 Printzia cernua (P. J. Bergius) Druce = Printzia polifolia (L.) Hutch. ■☆

315187 Printzia densifolia J. M. Wood et M. S. Evans = Printzia auriculata Harv. ■☆

315188 Printzia huttoni Harv. ;赫顿尾药菀■☆

315189 Printzia laxa N. E. Br. = Printzia auriculata Harv. ■☆

315190 Printzia nutans (Bolus) Leins;俯垂尾药菀■☆

315191 Printzia polifolia (L.) Hutch. ;灰叶尾药菀■☆

315192 Printzia pyrifolia Less. ;梨叶尾药菀■☆

315193 Prinus Post et Kuntze = Ilex L. ●

315194 Prinus Post et Kuntze = Prinos L. ●

315195 Priogymnanthus P. S. Green(1994);裸锯花属●☆

315196 Priogymnanthus apertus (B. Stahl) P. S. Green;无翅裸锯花●☆

315197 Priogymnanthus hasslerianus (Chodat) P. S. Green;裸锯花●☆

315198 Prionachne Nees = Prionanthium Desv. ■☆

315199 Prionachne ecklonii Nees = Prionanthium ecklonii (Nees) Stapf ■☆

315200 Prionanthes Schrank = Trixis P. Browne ■●☆

315201 Prionanthium Desv. (1831);锯花禾属■☆

315202 Prionanthium dentatum (L. f.) Henrard;尖齿锯花禾■☆

315203 Prionanthium ecklonii (Nees) Stapf;埃氏锯花禾■☆

315204 Prionanthium pholiuroides Stapf;鳞尾草锯花禾■☆

315205 Prionanthium rigidum Desv. = Prionanthium dentatum (L. f.) Henrard ☆

315206 Prionantium Desv. = Prionanthium Desv. ■☆

315207 Prioniaceae S. L. Munro et H. P. Linder = Thurniaceae Engl. ■☆

315208 Prioniaceae S. L. Munro et H. P. Linder;南非灯心草科■

315209 Prionites Pritz. = Prionotes R. Br. ●☆

315210 Prionitis Adans. = Falcaria Fabr. (保留属名)■

315211 Prionitis Oerst. = Barleria L. ●■

315212 Prionium E. Mey. (1832);南非灯心草属■☆

315213 Prionium palmita E. Mey. = Prionium serratum (L. f.) Drège ex E. Mey. ■☆

315214 Prionium serratum (L. f.) Drège ex E. Mey. ;南非灯心草;Palmier ■☆

315215 Prionolepis Poepp. et Endl. = Liabum Adans. ■●☆

315216 Prionophyllum C. Koch = Dyckia Schult. et Schult. f. ■☆

315217 Prionophyllum K. Koch = Dyckia Schult. et Schult. f. ■☆

315218 Prionoplectus Oerst. = Alloplectus Mart. (保留属名)●■☆

315219 Prionopsis Nutt. (1840);锯菊属■☆

315220 Prionopsis Nutt. = Haplopappus Cass. (保留属名)■●☆

315221 Prionopsis chapmanii Torr. et A. Gray = Eurybia eryngiifolia (Torr. et A. Gray) G. L. Nesom ☆

315222 Prionopsis ciliata (Nutt.) Nutt. = Grindelia ciliata (Nutt.) Spreng. ☆

315223 Prionopsis ciliata (Nutt.) Nutt. = Grindelia papposa G. L. Nesom et Y. B. Suh ■☆

315224 Prionoschoenus (Rchb.) Kuntze = Prionium E. Mey. ■☆

315225 Prionosciadium S. Watson(1888);锯伞芹属■☆

315226 Prionosciadium acuminatum Robinson ex J. M. Coult. et Rose;渐尖锯伞芹■☆

315227 Prionosciadium cuneatum J. M. Coult. et Rose;楔形锯伞芹■☆

315228 Prionosciadium filifolium J. M. Coult. et Rose;线叶锯伞芹■☆

315229 Prionosciadium megacarpum J. M. Coult. et Rose;大果锯伞芹■☆

315230 Prionosciadium mexicanum S. Watson;墨西哥锯伞芹■☆

315231 Prionosepalum Steud. = Chaetanthus R. Br. ■☆

315232 Prionosepalum Steud. = Chrysanthemum L. (保留属名)■●

315233 Prionostachys Hassk. = Aneilema R. Br. ■☆

315234 Prionostachys Hassk. = Murdannia Royle(保留属名)■

315235 Prionostemma Miers(1872);普瑞木属(齿梗木属)●☆

315236 Prionostemma aspera Miers;普瑞木●☆

315237 Prionostemma delagoensis (Loes.) N. Hallé;迪拉果普瑞木●☆

315238 Prionostemma delagoensis (Loes.) N. Hallé var. ritschardii (R. Wilczek) N. Hallé;里恰德普瑞木●☆

315239 Prionostemma fimbriata (Exell) N. Hallé;流苏普瑞木●☆

315240 Prionostemma unguiculata (Loes.) N. Hallé;爪状普瑞木●☆

315241 Prionotaceae Hutch. = Epacridaceae R. Br. (保留科名)●☆

315242 Prionotaceae Hutch. = Ericaceae Juss. (保留科名)●

315243 Prionotes R. Br. (1810);电珠石南属●☆

315244 Prionotes cerinthoides (Labill.) R. Br. ;电珠石南●☆

315245 Prionotis Benth. et Hook. f. = Prionotes R. Br. ●☆

315246 Prionotrichon Botsch. et Vved. (1948);齿毛芥属(锯毛芥属)■☆

315247 Prionotrichon Botsch. et Vved. = Rhammatophyllum O. E. Schulz ■☆

315248 Prionotrichon afghanicum (Rech. f.) Botsch. ;阿富汗齿毛芥■☆

315249 Prionotrichon erysimoides (Kar. et Kir.) Botsch. et Vved. ;齿毛芥■☆

315250 Prionotrichon erysimoides (Kar. et Kir.) Botsch. et Vved. = Arabis erysimoides Kar. et Kir. ■☆

315251 Priopetalon Raf. = Alstroemeria L. ■☆

315252 Prioria Griseb. (1860);脂苏木属(普疗木属)●☆

315253 Prioria balsamifera (Vermoesen) Breteler = Gossweilerodendron balsamiferum (Vermoesen) Harms ●☆

315254 Prioria buchholzii (Harms) Breteler;布赫脂苏木●☆

315255 Prioria copaifera Griseb. ;脂苏木(普疗木);Cativo ●☆

315256 Prioria gilbertii (J. Léonard) Breteler;吉尔伯特脂苏木●☆

315257 Prioria joveri (Normand ex Aubrév.) Breteler;约瓦里脂苏木●☆

315258 Prioria mannii (Baill.) Breteler;曼氏脂苏木(门哈威豆,门氏印度苏木)●☆

315259 Prioria msoo (Harms) Breteler;热非脂苏木●☆

315260 Prioria oxyphylla (Harms) Breteler;尖叶脂苏木●☆

315261 Priotropis Wight et Arn. (1834); 黄雀儿属; Priotropis,Siskinling ●

315262 Priotropis Wight. et Arn. = Crotalaria L. ●■

315263 Priotropis cydisoides (Roxb. ex DC.) Wight et Arn. ;黄雀儿(小扁豆);Broom-like Priotropis,Siskinling ●

315264 Priotropis cytisoides (Roxb. ex DC.) Wight et Arn. = Crotalaria cytisoides Roxb. ex DC. ■

315265 Priotropis cytisoides (Roxb. ex DC.) Wight et Arn. = Crotalaria psoralioides D. Don ■

315266 Priotropis inopinata Harms = Crotalaria inopinata (Harms) Polhill ■☆

315267 Prisciana Raf. = Carpopodium (DC.) Eckl. et Zeyh. ●■☆

315268 Prisciana Raf. = Heliophila Burm. f. ex L. ●■☆

315269 Prismanthus Hook. et Arn. = Siphonostegia Benth. ■

315270　Prismatanthus Hook. et Arn. = Siphonostegia Benth. ■

315271　Prismatocarpus L'Hér. (1789)(保留属名);棱果桔梗属●■■☆

315272　Prismatocarpus acerosus Schinz = Prismatocarpus sessilis Eckl. ex A. DC. ●☆

315273　Prismatocarpus alpinus（Bond）Adamson;高山棱果桔梗●☆

315274　Prismatocarpus brevilobus A. DC.;短裂棱果桔梗●☆

315275　Prismatocarpus burchellii Vatke = Prismatocarpus tenerrimus H. Buek ●☆

315276　Prismatocarpus campanuloides（L. f.）Sond.;钟棱果桔梗●☆

315277　Prismatocarpus campanuloides（L. f.）Sond. var. dentatus Adamson;尖齿棱果桔梗●☆

315278　Prismatocarpus candolleanus Cham.;康多勒棱果桔梗●☆

315279　Prismatocarpus cliffortioides Adamson;可利果棱果桔梗●☆

315280　Prismatocarpus crispus L'Hér.;皱波棱果桔梗●☆

315281　Prismatocarpus debilis Adamson;弱小棱果桔梗●☆

315282　Prismatocarpus debilis Adamson var. elongatus ?;伸长弱小棱果桔梗■☆

315283　Prismatocarpus decurrens Adamson;下延棱果桔梗●☆

315284　Prismatocarpus diffusus（L. f.）A. DC.;松散棱果桔梗●☆

315285　Prismatocarpus ecklonii A. DC.;埃氏棱果桔梗●☆

315286　Prismatocarpus falcatus Ten. = Legousia falcata（Ten.）Janch. ●☆

315287　Prismatocarpus fastigiatus C. Presl ex A. DC.;帚状棱果桔梗●☆

315288　Prismatocarpus fruticosus L'Hér.;灌丛棱果桔梗●☆

315289　Prismatocarpus hildebrandtii Vatke;希尔德棱果桔梗●☆

315290　Prismatocarpus hookeri Sweet = Prismatocarpus nitidus L'Hér. ●☆

315291　Prismatocarpus implicatus Adamson;纠缠棱果桔梗●☆

315292　Prismatocarpus interruptus L'Hér. = Prismatocarpus pedunculatus（P. J. Bergius）A. DC. ●☆

315293　Prismatocarpus junceus H. Buek = Wahlenbergia juncea（H. Buek）Lammers ■☆

315294　Prismatocarpus lasiophyllus Adamson;毛叶棱果桔梗●☆

315295　Prismatocarpus lycioides Adamson;枸杞状棱果桔梗●☆

315296　Prismatocarpus lycopodioides A. DC. = Prismatocarpus subulatus（Thunb.）A. DC. var. pauciflorus Sond. ●☆

315297　Prismatocarpus lycopodioides A. DC. var. hispidus Adamson;硬毛棱果桔梗●☆

315298　Prismatocarpus nitidus L'Hér.;光亮棱果桔梗●☆

315299　Prismatocarpus nitidus L'Hér. var. ovatus Adamson = Prismatocarpus debilis Adamson ●☆

315300　Prismatocarpus paniculatus L'Hér. = Prismatocarpus pedunculatus（P. J. Bergius）A. DC. ●☆

315301　Prismatocarpus pauciflorus Adamson;少花棱果桔梗●☆

315302　Prismatocarpus pedunculatus（P. J. Bergius）A. DC.;梗花棱果桔梗●☆

315303　Prismatocarpus pilosus Adamson;疏毛棱果桔梗●☆

315304　Prismatocarpus rhodesicus Adamson = Gunillaea rhodesica（Adamson）Thulin ●☆

315305　Prismatocarpus roelloides（L. f.）Sond. = Prismatocarpus pedunculatus（P. J. Bergius）A. DC. ●☆

315306　Prismatocarpus rogersii Fourc.;罗杰斯棱果桔梗●☆

315307　Prismatocarpus schinzianus Markgr. = Namacodon schinzianum（Markgr.）Thulin ●☆

315308　Prismatocarpus schlechteri Adamson;施莱棱果桔梗●☆

315309　Prismatocarpus sessilis Eckl. ex A. DC.;无柄棱果桔梗●☆

315310　Prismatocarpus sessilis Eckl. ex A. DC. var. macrocarpus Adamson;大果无柄棱果桔梗●☆

315311　Prismatocarpus speculum L. = Legousia speculum-veneris（L.）Chaix ●☆

315312　Prismatocarpus spinosus Adamson;具刺棱果桔梗●☆

315313　Prismatocarpus strictus A. DC. = Prismatocarpus campanuloides（L. f.）Sond. ●☆

315314　Prismatocarpus subulatus（Thunb.）A. DC. var. pauciflorus Sond. = Prismatocarpus lycopodioides A. DC. ●☆

315315　Prismatocarpus tenellus Oliv.;柔弱棱果桔梗●☆

315316　Prismatocarpus tenerrimus H. Buek;极细棱果桔梗●☆

315317　Prismatocarpus virgatus Fourc.;条纹棱果桔梗●☆

315318　Prismatomeris Thwaites（1856）;南山花属（三角瓣花属）; Prismatomeris ●

315319　Prismatomeris albiflora Thwaites = Prismatomeris tetrandra（Roxb.）K. Schum. ●

315320　Prismatomeris bravipes Hutch. = Damnacanthus henryi（H. Lév.）H. S. Lo ●

315321　Prismatomeris connata Y. Z. Ruan;南山花（黄根,三角瓣花）; Joined Prismatomeris ●

315322　Prismatomeris connata Y. Z. Ruan subsp. hainanensis Y. Z. Ruan;海南三角瓣花;Hainan Joined Prismatomeris ●

315323　Prismatomeris henryi（H. Lév.）Rehder = Damnacanthus henryi（H. Lév.）H. S. Lo ●

315324　Prismatomeris labordei（H. Lév.）Merr.;柳叶山花●

315325　Prismatomeris labordei（H. Lév.）Merr. = Damnacanthus labordei（H. Lév.）H. S. Lo ●

315326　Prismatomeris labordei（H. Lév.）Merr. ex Rehder = Damnacanthus labordei（H. Lév.）H. S. Lo ●

315327　Prismatomeris linearis Hutch. = Damnacanthus henryi（H. Lév.）H. S. Lo ●

315328　Prismatomeris malayana Ridl.;马来西亚黄根;Malaya Prismatomeris ●

315329　Prismatomeris multiflora Ridl. = Prismatomeris tetrandra（Roxb.）K. Schum. subsp. multiflora（Ridl.）Y. Z. Ruan ●

315330　Prismatomeris multiflora Ridl. = Prismatomeris tetrantra（Roxb.）K. Schum. ●

315331　Prismatomeris subsessilis King et Gamble = Gentingia subsessilis（King et Gamble）J. T. Johansson et K. M. Wong ●☆

315332　Prismatomeris tetrandra（Roxb.）K. Schum.;四蕊三角瓣花（狗骨木,黄根,南山花,三角瓣花）;Fourstamen Prismatomeris, Four-stamened Prismatomeris,Prismatomeris ●

315333　Prismatomeris tetrandra（Roxb.）K. Schum. = Prismatomeris connata Y. Z. Ruan ●

315334　Prismatomeris tetrandra（Roxb.）K. Schum. subsp. multiflora（Ridl.）Y. Z. Ruan;多花三角瓣花（黄根,南山花,三角瓣花,四蕊三角瓣花）;Manyflower Fourstamen Prismatomeris ●

315335　Prismatomeris tetrantra（Roxb.）K. Schum. subsp. multiflora（Ridl.）Y. Z. Ruan = Prismatomeris tetrantra（Roxb.）K. Schum. ●

315336　Prismocarpa Raf. = Prismatocarpus L'Hér.（保留属名）●■☆

315337　Prismophylis Thouars = Bulbophyllum Thouars(保留属名）■

315338　Pristidia Thwaites = Gaertnera Lam. ●

315339　Pristiglottis Cretz. et J. J. Sm.（1934）;双丸兰属■

315340　Pristiglottis Cretz. et J. J. Sm. = Odontochilus Blume ■

315341　Pristiglottis bisaccata（Hayata）K. Nakaj. = Odontochilus lanceolatus（Lindl.）Blume ■

315342　Pristiglottis bisaccata（Hayata）K. Nakej. = Anoectochilus lanceolatus Lindl. ■

315343　Pristiglottis humilis（Fukuy.）Fukuy. = Kuhlhasseltia yakushimensis（Yamam.）Ormerod ■

315344 Pristiglottis humilis (Fukuy.) Fukuy. = Vexillabium yakushimense (Yamam.) F. Maek. ■

315345 Pristiglottis integra Fukuy. = Kuhlhasseltia yakushimensis (Yamam.) Ormerod ■

315346 Pristiglottis integra Fukuy. = Vexillabium yakushimense (Yamam.) F. Maek. ■

315347 Pristiglottis saprophytica Aver. = Odontochilus saprophyticus (Aver.) Ormerod ■

315348 Pristiglottis tashiroi (Maxim.) Cretz. et J. J. Sm. = Odontochilus tashiroi (Maxim.) Makino ex Kuroiwa ■☆

315349 Pristiglottis torta (King et Pantl.) Aver. = Anoectochilus tortus (King et Pantl.) King et Pantl. ■

315350 Pristiglottis yakushimense (Yamam.) Masam. = Vexillabium yakushimense (Yamam.) F. Maek. ■

315351 Pristiglottis yakushimensis (Yamam.) Masam. = Kuhlhasseltia yakushimensis (Yamam.) Ormerod ■

315352 Pristimera Miers(1872);扁蒴藤属(扁蒴果属);Pristimera ●

315353 Pristimera andongensis (Welw. ex Oliv.) N. Hallé;安东扁蒴藤●☆

315354 Pristimera andongensis (Welw. ex Oliv.) N. Hallé var. cinerascens N. Hallé;灰色扁蒴藤●☆

315355 Pristimera andongensis (Welw. ex Oliv.) N. Hallé var. volkensii (Loes.) N. Hallé;福尔扁蒴藤●☆

315356 Pristimera arborea (Roxb.) A. C. Sm.;二籽扁蒴藤;Biseeded Pristimera, Twoseed Pristimera ●

315357 Pristimera bojeri (Tul.) N. Hallé;博耶尔扁蒴藤●☆

315358 Pristimera breteleri N. Hallé;布勒泰尔扁蒴藤●☆

315359 Pristimera cambodiana (Pierre) A. C. Sm.;风车扁蒴藤(风车果);Cambodia Pristimera ●

315360 Pristimera graciliflora (Welw. ex Oliv.) N. Hallé;细花扁蒴藤●☆

315361 Pristimera indica (Willd.) A. C. Sm.;扁蒴藤(印度扁蒴藤);India Pristimera, Indian Pristimera ●

315362 Pristimera indica (Willd.) A. C. Sm. = Reissantia indica (Willd.) N. Hallé ●☆

315363 Pristimera longipetiolata (Oliv.) N. Hallé;长梗扁蒴藤●☆

315364 Pristimera luteoviridis (Exell) N. Hallé;黄绿扁蒴藤●☆

315365 Pristimera malifolia (Baker) N. Hallé;苹果叶扁蒴藤●☆

315366 Pristimera mouilensis (N. Hallé) N. Hallé;莫伊拉扁蒴藤●☆

315367 Pristimera paniculata (Vahl) N. Hallé;圆锥扁蒴藤●☆

315368 Pristimera plumbea (Blakelock et R. Wilczek) N. Hallé;普拉姆扁蒴藤●☆

315369 Pristimera polyantha (Loes.) N. Hallé;多花扁蒴藤●☆

315370 Pristimera preussii (Loes.) N. Hallé;普罗伊斯扁蒴藤●☆

315371 Pristimera setulosa A. C. Sm.;毛扁蒴藤;Hairy Pristimera ●

315372 Pristimera tetramera (H. Perrier) N. Hallé;四数扁蒴藤●☆

315373 Pristocarpha E. Mey. ex DC. = Athanasia L. ●☆

315374 Pristocarpha capitata E. Mey. ex DC. = Athanasia capitata (L.) L. ●☆

315375 Pritaelago Kuntze = Hornungia Rchb. ■

315376 Pritchardia Seem. et H. Wendl. (1862)(保留属名);太平洋棕属(卜力查得棕属,布氏榈属,金棕属,普权榈属,普利加德属,夏威夷葵属,夏威夷棕属);Fan Palm, Pritchardia, Pritchardia Palm ●☆

315377 Pritchardia Seem. et H. Wendl. ex H. Wendl. = Pritchardia Seem. et H. Wendl. (保留属名)●☆

315378 Pritchardia Unger ex Endl. (废弃属名) = Pritchardia Seem. et H. Wendl. (保留属名)●☆

315379 Pritchardia beccarianaRock;贝氏太平洋棕●☆

315380 Pritchardia filamentosa Franceschi = Washingtonia filifera (Linden ex André) H. Wendl. ex de Bary ●

315381 Pritchardia filamentosa H. Wendl. = Washingtonia filifera (Linden ex André) H. Wendl. ex de Bary ●

315382 Pritchardia filamentosa H. Wendl. ex Franceschi = Washingtonia filifera (Linden ex André) H. Wendl. ex de Bary ●

315383 Pritchardia filifera Linden = Washingtonia filifera (Linden ex André) H. Wendl. ex de Bary ●

315384 Pritchardia filifera Linden ex André = Washingtonia filifera (Linden ex André) H. Wendl. ex de Bary ●

315385 Pritchardia maideniana Becc.;少女金棕●☆

315386 Pritchardia pacifica Seem. et H. Wendl.;太平洋棕(斐济榈);Fiji Fan-palm, Sakiki Palm ●☆

315387 Pritchardia thurstonii F. Muell. et Drude;红果太平洋棕●☆

315388 Pritchardia wrightii Becc.;赖特太平洋棕●☆

315389 Pritchardiopsis Becc. (1910);脊果棕属(大果棕属)●☆

315390 Pritchardiopsis jennencyi Becc.;脊果棕(大果棕)●☆

315391 Pritzelago Kuntze = Hornungia Rchb. ■

315392 Pritzelago Kuntze(1891);岩羚羊芥属■☆

315393 Pritzelago alpina (L.) Kuntze;岩羚羊芥■☆

315394 Pritzelago alpina (L.) Kuntze = Hornungia alpina (L.) Appel ■☆

315395 Pritzelago alpina (L.) Kuntze subsp. fontqueri (Sauvage) Greuter et Burdet = Hornungia alpina (L.) Appel subsp. fontqueri (Sauvage) Appel ■☆

315396 Pritzelia F. Muell. = Hetaeria Endl. ■☆

315397 Pritzelia F. Muell. = Philydrella Caruel ■☆

315398 Pritzelia Klotzsch = Begonia L. ●■

315399 Pritzelia Schauer = Scholtzia Schauer ●☆

315400 Pritzelia Walp. = Trachymene Rudge ■☆

315401 Priva Adans. (1763);异柱马鞭草属■☆

315402 Priva abyssinica Jaub. et Spach = Priva adhaerens (Forssk.) Chiov. ■☆

315403 Priva abyssinica Jaub. et Spach = Priva cordifolia (L. f.) Druce ■☆

315404 Priva adhaerens (Forssk.) Chiov.;附着异柱马鞭草■☆

315405 Priva adhaerens (Forssk.) Chiov. var. forskaolaei (Vahl) Chiov. = Priva adhaerens (Forssk.) Chiov. ■☆

315406 Priva africana Moldenke;非洲异柱马鞭草■☆

315407 Priva angolensis Moldenke = Priva auricoccea A. Meeuse ■☆

315408 Priva auricoccea A. Meeuse;安哥拉异柱马鞭草■☆

315409 Priva cordifolia (L. f.) Druce;心叶异柱马鞭草■☆

315410 Priva cordifolia (L. f.) Druce var. abyssinica (Jaub. et Spach) Moldenke = Priva adhaerens (Forssk.) Chiov. ■☆

315411 Priva cordifolia (L. f.) Druce var. flabelliformis Moldenke = Priva flabelliformis (Moldenke) R. Fern. ■☆

315412 Priva dentata Juss. = Priva adhaerens (Forssk.) Chiov. ■☆

315413 Priva flabelliformis (Moldenke) R. Fern.;扇状异柱马鞭草■☆

315414 Priva forskaolaei (Vahl) Jaub. et Spach = Priva adhaerens (Forssk.) Chiov. ■☆

315415 Priva lappulacea (L.) Pers.;鹤虱异柱马鞭草■☆

315416 Priva leptostachya Juss. = Priva cordifolia (L. f.) Druce ■☆

315417 Priva meyeri Jaub. et Spach;迈尔异柱马鞭草■☆

315418 Priva tenax Verdc.;黏异柱马鞭草■☆

315419 Proatriplex (W. A. Weber) Stutz et G. L. Chu = Atriplex L. ■●

315420 Proatriplex Stutz et G. L. Chu = Atriplex L. ■●

315421 Proatriplex pleiantha (W. A. Weber) Stutz et G. L. Chu = Atriplex pleiantha W. A. Weber ■☆

315422 Probatea Raf. = Asarina Mill. ■☆

315423 Problastes Reinw. = Lumnitzera Willd. ●

315424 Problastes cuneifolia Reinw. = Lumnitzera racemosa Willd. ●

315425 Probletostemon K. Schum. = Tricalysia A. Rich. ex DC. ●

315426 Probletostemon elliotii K. Schum. = Tricalysia elliotii（K. Schum.）Hutch. et Dalziel ●☆

315427 Proboscella Tiegh. = Ochna L. ●

315428 Proboscella emarginata Tiegh. = Ochna pygmaea Hiern ●☆

315429 Proboscella hoepfneri Tiegh. = Ochna pygmaea Hiern ●☆

315430 Proboscidea Schmidel（1763）；长角胡麻属（单角胡麻属，角胡麻属）；Devil's Claws，Devil's-claws，Proboscidea，Unicorn Plant ■

315431 Proboscidea althaeifolia Decne. ；蜀葵叶角胡麻；Devil's Horn，Devil's-claw，Unicorn-plant ■☆

315432 Proboscidea fragrans（Lindl.）Decne. = Proboscidea louisianica（Mill.）Thell. subsp. fragrans（Lindl.）Bretting ■☆

315433 Proboscidea jussieui A. Keller = Proboscidea louisiana（Mill.）Wooton et Standl. ■

315434 Proboscidea louisiana（Mill.）Wooton et Standl. ；长角胡麻（单角胡麻，角胡麻）；Common Devil's Claws，Common Unicorn Flower，Devil's Claw，Proboscidea，Proboscis Flower，Ram's Horn，Unicorn Plant ■

315435 Proboscidea louisianica（Mill.）Thell. = Proboscidea louisiana（Mill.）Wooton et Standl. ■

315436 Proboscidea louisianica（Mill.）Thell. subsp. fragrans（Lindl.）Bretting；香长角胡麻（香单角胡麻）；Sweet Devil's-claws ■☆

315437 Proboscidea lutea（Lindl.）Stapf = Ibicella lutea（Lindl.）Van Eselt. ■☆

315438 Proboscidea parviflora（Wooton）Wooton et Standl. ；小花长角胡麻；Devil's Claw ■☆

315439 Proboscidea parviflora Wooton et Standl. = Proboscidea parviflora（Wooton）Wooton et Standl. ■☆

315440 Proboscidea triloba（Schltdl. et Cham.）Decne. ；墨西哥长角胡麻■☆

315441 Proboscidia Rich. ex DC. = Rhynchanthera DC.（保留属名）●☆

315442 Prosciphora Neck. = Rhynchocorys Griseb.（保留属名）■☆

315443 Procephalium Post et Kuntze = Proscephaleium Korth. ☆

315444 Prochnyanthes S. Watson（1887）；跪花龙舌兰属■☆

315445 Prochnyanthes mexicans（Zucc.）Rose；跪花龙舌兰■☆

315446 Prochynanthes Baker = Prochnyanthes S. Watson ■☆

315447 Prockia P. Browne ex L.（1759）；普罗椴属●☆

315448 Prockia flava H. Karst. ；黄普罗椴●☆

315449 Prockia glabra Briq. ；光普罗椴●☆

315450 Prockia grandiflora Herzog；大花普罗椴●☆

315451 Prockia theiformis（Vahl）Willd. = Aphloia theiformis（Vahl）Benn. ●☆

315452 Prockiaceae Bertuch = Flacourtiaceae Rich. ex DC.（保留科名）●

315453 Prockiaceae D. Don = Flacourtiaceae Rich. ex DC.（保留科名）●

315454 Prockiopsis Baill.（1886）；普罗木属●☆

315455 Prockiopsis calcicola G. E. Schatz et Lowry；岩地普罗木●☆

315456 Prockiopsis hildebrandtii Baill. ；普罗木●☆

315457 Prockiopsis orientalis Capuron ex G. E. Schatz et Lowry；东方普罗木●☆

315458 Proclesia Klotzsch = Cavendishia Lindl.（保留属名）●☆

315459 Procopiana Gusul. = Symphytum L. ■

315460 Procrassula Griseb. = Sedum L. ●■

315461 Procris Comm. ex Juss.（1789）；藤麻属（望北京属，乌来麻属）；Procris，Vinenettle ●■

315462 Procris Juss. = Procris Comm. ex Juss. ●

315463 Procris acuminata Poir. = Elatostema acuminatum（Poir.）Brongn. ●

315464 Procris boninensis Tuyama；小笠原藤麻■☆

315465 Procris crenata C. B. Rob. ；藤麻（石羊草，乌来麻，虾公菜，下山连，眼睛草，一支林，一支麻，一枝林，圆齿藤麻）；Crenate Procris，Wight Vinenettle ●■

315466 Procris cyrtandrifolia Zoll. = Elatostema cyrtandrifolium（Zoll. et Moritzi）Miq. ■

315467 Procris cyrtandrifolia Zoll. et Moritzi = Elatostema cyrtandrifolium（Zoll. et Moritzi）Miq. ■

315468 Procris diversifolia Wall. = Elatostema monandrum（D. Don）Hara ■

315469 Procris elegans Wall. = Elatostema monandrum（D. Don）Hara ■

315470 Procris ficoidea Wall. = Elatostema ficoides Wedd. ■

315471 Procris gibbosa Wall. = Pellionia repens（Lour.）Merr. ■

315472 Procris heyneana Wall. = Pellionia heyneana Wedd. ●■

315473 Procris humblotii H. Schroet. = Procris pedunculata（J. R. Forst. et G. Forst.）Wedd. ■

315474 Procris hypoleuca Steud. = Debregeasia saeneb（Forssk.）Hepper et J. R. I. Wood ●

315475 Procris integrifolia D. Don = Elatostema integrifolium（D. Don）Wedd. ●■

315476 Procris laevigata Blume；台湾藤麻（平滑楼梯草，藤麻，乌来草，乌来麻）；Procris，Vinenettle ●

315477 Procris laevigata Blume = Procris wightiana Wall. ex Wedd. ■

315478 Procris laevigata Miq. = Procris crenata C. B. Rob. ●■

315479 Procris latifolia Blume = Pellionia latifolia（Blume）Boerl. ■

315480 Procris monandra D. Don = Elatostema monandrum（D. Don）Hara ■

315481 Procris obtusa Royle = Lecanthus peduncularis（Wall. ex Royle）Wedd. ■

315482 Procris parva Blume = Elatostema parvum（Blume）Miq. ■

315483 Procris peduncularis Royle = Lecanthus peduncularis（Wall. ex Royle）Wedd. ■

315484 Procris peduncularis Wall. = Lecanthus peduncularis（Wall. ex Royle）Wedd. ■

315485 Procris peduncularis Wall. ex Royle = Lecanthus peduncularis（Wall. ex Royle）Wedd. ■

315486 Procris pedunculata（J. R. Forst. et G. Forst.）Wedd. ；短花梗藤麻●

315487 Procris pedunculata（J. R. Forst. et G. Forst.）Wedd. var. humblotii（H. Schroet.）Léandri = Procris pedunculata（J. R. Forst. et G. Forst.）Wedd. ●

315488 Procris racemosa Royle = Pilea racemosa（Royle）Tuyama ■

315489 Procris radicans Siebold et Zucc. = Pellionia radicans（Siebold et Zucc.）Wedd. ■

315490 Procris rupestris Buch. -Ham. = Elatostema macintyrei Dunn ●■

315491 Procris rupestris Buch. -Ham. = Elatostema rupestre（Buch. -Ham.）Wedd. ●■

315492 Procris sesquifolia Reinw. ex Blume = Elatostema integrifolium（D. Don）Wedd. ●■

315493 Procris wightiana Wall. ex Wedd. = Procris crenata C. B. Rob. ■

315494 Proctoria Luer = Pleurothallis R. Br. ■☆

315495 Proctoria Luer（2004）；开曼兰属■☆

315496 Proineta Ehrh. = Aira L.（保留属名）■☆

315497 Proiphys Herb.（1821）；初泡石蒜属■☆

315498 Proiphys Herb. = Eurycles Salisb. ■☆

315499 Proiphys amboinensis（L.）Herb.；假玉簪；Brisbane Lily，Cardwell Lily ■☆

315500 Proiphys amboinensis（L.）Herb. = Eurycles amboinensis（L.）Lindl. ■☆

315501 Proiphys cunninghamii（Lindl.）Mabb.；初泡石蒜；Brisbane Lily ■☆

315502 Prolobus R. M. King et H. Rob.（1982）；黏叶柄泽兰属●☆

315503 Prolobus nitidulus（Baker）R. M. King et H. Rob.；黏叶柄泽兰 ☆

315504 Prolongoa Boiss.（1840）（保留属名）；长莛菊属☆

315505 Prolongoa Boiss. = Chrysanthemum L.（保留属名）■●

315506 Prolongoa hispanica G. López et C. E. Jarvis；长莛菊■☆

315507 Prolongoa macrocarpa（Sch. Bip.）Alavi = Endopappus macrocarpus Sch. Bip. ■☆

315508 Prolongoa pseudanthemis Kuntze = Hymenostemma pseudanthemis（Kuntze）Willk. ■☆

315509 Promenaea Lindl.（1843）；普罗兰属☆

315510 Promenaea acuminata Schltr.；渐尖普罗兰■☆

315511 Promenaea florida Rchb. f.；佛罗里达普罗兰■☆

315512 Promenaea graminea Lindl.；禾状普罗兰■☆

315513 Promenaea microptera Rchb. f.；小翅普罗兰■☆

315514 Prometheum（A. Berger）H. Ohba（1978）；普罗景天属■☆

315515 Pronacron Cass. = Melampodium L. ■●

315516 Pronaya Hügel = Pronaya Hügel ex Endl. ●☆

315517 Pronaya Hügel ex Endl.（1837）；普罗桐属；Pronaya ●☆

315518 Pronaya fraseri（Hook.）E. M. Benn.；普罗桐●☆

315519 Prosanerpis S. F. Blake = Clidemia D. Don ●☆

315520 Prosartema Gagnep. = Trigonostemon Blume（保留属名）●

315521 Prosartema stellaris Gagnep. = Trigonostemon thyrsoideus Stapf ●

315522 Prosartes D. Don = Disporum Salisb. ex D. Don ■

315523 Prosartes D. Don（1839）；附加百合属■☆

315524 Prosartes hookeri Torr.；胡克附加百合；Fairy-bells ■☆

315525 Prosartes lanuginosa（Michx.）D. Don；黄附加百合；Hairy Mandarin，Yellow Mandarin ■☆

315526 Prosartes maculata（Buckley）A. Gray；斑点附加百合；Nodding Mandarin ■☆

315527 Prosartes oregana S. Watson = Prosartes hookeri Torr. ■☆

315528 Prosartes ovalis（Ohwi）M. N. Tamura = Streptopus ovalis（Ohwi）F. T. Wang et Y. C. Tang ■

315529 Prosartes parvifolia S. Watson = Prosartes hookeri Torr. ■☆

315530 Prosartes smithii（Hook.）Utech；史密斯附加百合；Fairy-lantern，Large-flowered Fairy-bells ■☆

315531 Prosartes trachycarpa S. Watson；糙果附加百合；Rough-fruited Fairy-bells，Rough-fruited Mandarin ■☆

315532 Prosartes viridescens（Maxim.）Regel = Disporum viridescens（Maxim.）Nakai ■

315533 Proscephaleium Korth.（1851）；顶头茜属☆

315534 Proscephaleium javanicum Korth.；顶头茜☆

315535 Proscephalium Benth. et Hook. f. = Proscephaleium Korth. ☆

315536 Proselia D. Don = Chaetanthera Ruiz et Pav. ●☆

315537 Proselias Steven = Astragalus L. ●■

315538 Proserpinaca L.（1753）；葡萄仙草属■☆

315539 Proserpinaca palustris L.；葡萄仙草；Marsh Mermaid-weed，Mermaid Weed，Mermaidweed ●☆

315540 Proserpinaca palustris L. var. crebra Fernald et Griscom；普通葡萄仙草；Common Mermaid-weed，Marsh Mermaid-weed ●☆

315541 Prosopanche de Bary（1868）；牧豆寄生属■☆

315542 Prosopanche clavata Chodat；棍棒牧豆寄生■☆

315543 Prosopanche minor Chodat；小牧豆寄生■☆

315544 Prosopia Rchb. = Pedicularis L. ■

315545 Prosopidastrum Burkart（1964）；小牧豆树属（球形牧豆树属）●☆

315546 Prosopidastrum mexicanum（Dressler）Burkart；小牧豆树 ●☆

315547 Prosopis L.（1767）；牧豆树属；Algaroba，Mesquita，Mesquite ●

315548 Prosopis africana（Guillaumin et Perr.）Taub.；非洲牧豆树；Ironwood ●☆

315549 Prosopis africana Taub. = Prosopis africana（Guillaumin et Perr.）Taub. ●☆

315550 Prosopis alba Griseb.；阿根廷牧豆树；Argentine Mesquite ●

315551 Prosopis chilensis（Molina）Stuntz；智利牧豆树；Algaroba，Chilean Mesquite ●☆

315552 Prosopis chilensis Stuntz = Prosopis chilensis（Molina）Stuntz ●☆

315553 Prosopis cineraria（L.）Druce；牧豆树（瓜叶菊牧豆树，瓜叶牧豆树）；Common Mesquite，Melonleaf Mesquite，Mesquite ●☆

315554 Prosopis farcta（Banks et Sol.）J. F. Macbr.；叙利亚牧豆树（实心牧豆树）；Syrian Mesquite ●☆

315555 Prosopis farcta J. Macbr. = Prosopis farcta（Banks et Sol.）J. F. Macbr. ●☆

315556 Prosopis fischeri Taub. = Pseudoprosopis fischeri（Taub.）Harms ●☆

315557 Prosopis glandulosa Torr.；腺牧豆树（蜜豆灌木，普通豆灌木，天鹅绒豆灌木，甜豆，腺质牧豆树）；Algaroba，Common Mesquite，Honey Mesquite，Mesquite，Texas Mesquite，Velvet Mesquite ●☆

315558 Prosopis glandulosa Torr. subsp. torreyana（L. D. Benson）A. E. Murray；托里腺牧豆树（托雷亚腺牧豆树）●☆

315559 Prosopis glandulosa Torr. var. torreyana（L. D. Benson）M. C. Johnst. = Prosopis glandulosa Torr. subsp. torreyana（L. D. Benson）A. E. Murray ●☆

315560 Prosopis juliflora（Sw.）DC.；柔荑花牧豆树（结亚木，墨西哥合欢，牧豆树）；Algarreba，Algarrobo，Common Mesquite，Honey Mesquite，Kiawe Bean，Mesquite ●

315561 Prosopis juliflora（Sw.）DC. var. torreyana L. D. Benson = Prosopis glandulosa Torr. var. torreyana（L. D. Benson）M. C. Johnst. ●☆

315562 Prosopis juliflora（Sw.）DC. var. torreyana L. D. Benson = Prosopis glandulosa Torr. ●☆

315563 Prosopis juliflora DC. = Prosopis juliflora（Sw.）DC. ●

315564 Prosopis kuntzei Harms et Hassl.；康氏牧豆树●☆

315565 Prosopis laevigata（Humb. et Bonpl. ex Willd.）M. C. Johnst.；平滑牧豆树（无毛牧豆树）；Mesquite，Smooth Mesquite ●☆

315566 Prosopis nigra（Griseb.）Hieron.；黑牧豆树；Algarrobo Negro，Black Mesquite ●☆

315567 Prosopis oblonga Benth. = Prosopis africana（Guillaumin et Perr.）Taub. ●☆

315568 Prosopis pallida（Humb. et Bonpl. ex Willd.）Kunth；灰叶牧豆（绿花牧豆树）；Algaroba，Kiawe，Mesquite ●☆

315569 Prosopis pubescens Benth.；柔毛牧豆树；Screw Bean Mesquite，Screw Mesquite，Screwbean，Screwbean Mesquite ●☆

315570 Prosopis spicigera L. = Prosopis cineraria（L.）Druce ●☆

315571 Prosopis stephaniana（M. Bieb.）Kunth ex Spreng. = Prosopis farcta（Banks et Sol.）J. F. Macbr. ●☆

315572 Prosopis stephaniana（M. Bieb.）Spreng. = Prosopis farcta（Banks et Sol.）J. F. Macbr. ●☆

315573 Prosopis strombulifera Benth.；螺旋牧豆树；Argentine Screwbean ●☆

315574 Prosopis tamarugo F. Phil. ;北方牧豆树●☆

315575 Prosopis velutina Wooton;亚利桑那牧豆树;Arizona Mesquite, Velvet Mesquite ●☆

315576 Prosopostelma Baill. (1890);假冠萝藦属●☆

315577 Prosopostelma aculeatum Desc. = Cynanchum aculeatum (Desc.) Liede et Meve ●☆

315578 Prosopostelma grandiflorum Choux = Cynanchum floriferum Liede et Meve ■☆

315579 Prosopostelma madagascariense Jum. et H. Perrier;假冠萝藦●☆

315580 Prosopostelma madagascariense Jum. et H. Perrier = Cynanchum toliari Liede et Meve ●☆

315581 Prosorus Dalzell = Margaritaria L. f. ●

315582 Prosorus indicus Dalzell = Margaritaria indica (Dalzell) Airy Shaw ●

315583 Prospero Salisb. (1866);无苞风信子属■☆

315584 Prospero Salisb. = Scilla L. ■

315585 Prospero autumnale (L.) Speta;秋无苞风信子■☆

315586 Prospero fallax (Steinh.) Speta;迷惑无苞风信子■☆

315587 Prospero obtusifolium (Poir.) Speta;钝叶无苞风信子■☆

315588 Prosphyais Dulac = Festuca L. ■

315589 Prosphysis Dulac = Nardurus (Bluff, Nees et Schauer) Rchb. ■

315590 Prosphysis Dulac = Vulpia C. C. Gmel. ■

315591 Prosphytochloa Schweick. (1961);南非攀高草属■☆

315592 Prosphytochloa prehensilis (Nees) Schweick. ;南非攀高草■☆

315593 Prosporus Thwaites = Margaritaria L. f. ●

315594 Prosporus Thwaites = Prosorus Dalzell ●

315595 Prostanthera Labill. (1806);薄荷木属(木薄荷属);Australian Mint, Mint Bush, Mintbush, Mint-bush ●☆

315596 Prostanthera aspalathoides A. Cunn. ex Benth. ;猩红薄荷木;Scarlet Mint Bush ●☆

315597 Prostanthera cuneata Benth. ;高山薄荷木;Alpine Mint Bush ●☆

315598 Prostanthera denticulata R. Br. ;细齿薄荷木;Mint Bush ●☆

315599 Prostanthera incana A. Cunn. ex Benth. ;毛薄荷木●☆

315600 Prostanthera incisa R. Br. ;深裂叶薄荷木;Cut-leafed Mint Bush ●☆

315601 Prostanthera lasianthos Labill. ;艳丽薄荷木;Christmas Bush, Victorian Christmas-bush, Victorian Mint Bush ●☆

315602 Prostanthera linearis R. Br. ;薄荷木;Mint Bush, Narrow-leaf Mint Bush ●☆

315603 Prostanthera magnifica C. A. Gardner;壮丽薄荷木;Magnificent Mint Bush, Splendid Mint Bush ●☆

315604 Prostanthera nivea A. Cunn. ;雪白薄荷木;Snowy Mint Bush ●☆

315605 Prostanthera ovalifolia R. Br. ;紫花薄荷木(卵叶木薄荷);Mint Bush, Oval-leaf Mint-bush, Purple Mint Bush ●☆

315606 Prostanthera rotundifolia R. Br. ;圆叶薄荷木(圆叶薄荷);Round-leaved Mint Bush, Round-leaved Mint-bush ●☆

315607 Prostanthera saxicola R. Br. ;细枝薄荷木●☆

315608 Prostanthera striatiflora F. Muell. ;大花薄荷木;Mint Bush, Streaked Mint Bush ●☆

315609 Prostanthera walteri F. Muell. ;紫斑薄荷木;Blotchy Mint Bush ●☆

315610 Prosteo Cambess. = Pometia J. R. Forst. et G. Forst. ●

315611 Prosthechea Knowles et Wastc. = Epidendrum L. (保留属名)■☆

315612 Prosthechea Knowles et Westc. (1838);附属物兰属■☆

315613 Prosthechea boothiana (Lindl.) W. E. Higgins;布思附属物兰■☆

315614 Prosthechea boothiana (Lindl.) W. E. Higgins var. erythronioides (Small) W. E. Higgins;佛罗里达附属物兰;Dollar Orchid, Florida Dollar-orchid ■☆

315615 Prosthechea cochleata (L.) W. E. Higgins;美洲附属物兰(螺壳附属物兰)■☆

315616 Prosthechea cochleata (L.) W. E. Higgins var. triandra (Ames) W. E. Higgins;三蕊附属物兰;Clam-shell Orchid, Florida Clamshell Orchid ■☆

315617 Prosthechea glauca Knowles et Westc. ;灰附属物兰■☆

315618 Prosthechea pygmaea (Hook.) W. E. Higgins;微小附属物兰;Dwarf Butterfly Orchid, Dwarf Encyclia, Dwarf Epidendrum ■☆

315619 Prosthecidiscus Dorm. Sm. (1898);附盘萝摩属■☆

315620 Prosthecidiscus guatemalensis Dorm. Sm. ;附盘萝摩■☆

315621 Prosthesia Blume = Rinorea Aubl. (保留属名)●

315622 Protamomum Ridl. = Orchidantha N. E. Br. ■

315623 Protarum Engl. (1901);原始南星属■☆

315624 Protarum sechellarum Engl. ;原始南星■☆

315625 Protasparagus Oberm. = Asparagus L. ■

315626 Protasparagus acocksii (Jessop) Oberm. = Asparagus acocksii Jessop ■☆

315627 Protasparagus aethiopicus (L.) Oberm. = Asparagus aethiopicus L. ■

315628 Protasparagus africanus (Lam.) Oberm. = Asparagus africanus Lam. ■☆

315629 Protasparagus aggregatus Oberm. = Asparagus aggregatus (Oberm.) Fellingham et N. L. Mey. ■☆

315630 Protasparagus angusticladus (Jessop) Oberm. = Asparagus angusticladus (Jessop) J. -P. Lebrun et Stork ■☆

315631 Protasparagus aspergillus (Jessop) Oberm. = Asparagus aspergillus Jessop ■☆

315632 Protasparagus bayeri Oberm. = Asparagus bayeri (Oberm.) Fellingham et N. L. Mey. ■☆

315633 Protasparagus bechuanicus (Baker) Oberm. = Asparagus bechuanicus Baker ■☆

315634 Protasparagus biflorus Oberm. = Asparagus biflorus (Oberm.) Fellingham et N. L. Mey. ■☆

315635 Protasparagus buchananii (Baker) Oberm. = Asparagus buchananii Baker ■☆

315636 Protasparagus burchellii (Baker) Oberm. = Asparagus burchellii Baker ■☆

315637 Protasparagus capensis (L.) Oberm. ;好望角原始南星■☆

315638 Protasparagus capensis (L.) Oberm. var. litoralis (Suess. et Karl) Oberm. = Asparagus capensis L. var. litoralis Suess. et Karl ■☆

315639 Protasparagus clareae Oberm. = Asparagus clareae (Oberm.) Fellingham et N. L. Mey. ■☆

315640 Protasparagus coddii Oberm. = Asparagus coddii (Oberm.) Fellingham et N. L. Mey. ■☆

315641 Protasparagus concinnus (Baker) Oberm. et Immelman = Asparagus concinnus (Baker) Kies ■☆

315642 Protasparagus confertus (K. Krause) Oberm. = Asparagus confertus K. Krause ■☆

315643 Protasparagus cooperi (Baker) Oberm. = Asparagus cooperi Baker ■☆

315644 Protasparagus crassicladus (Jessop) Oberm. = Asparagus crassicladus Jessop ■☆

315645 Protasparagus densiflorus (Kunth) Oberm. 'Myers';迈尔原始南星;Foxtail Fern ■☆

315646 Protasparagus densiflorus (Kunth) Oberm. = Asparagus densiflorus (Kunth) Jessop ■

315647 Protasparagus denudatus (Kunth) Oberm. = Asparagus denudatus

（Kunth）Baker ■☆

315648　Protasparagus divaricatus Oberm. = Asparagus divaricatus（Oberm.）Fellingham et N. L. Mey. ■

315649　Protasparagus edulis Oberm. = Asparagus edulis（Oberm.）J. -P. Lebrun et Stork ■☆

315650　Protasparagus exsertus Oberm. = Asparagus exsertus（Oberm.）Fellingham et N. L. Mey. ■☆

315651　Protasparagus exuvialis（Burch.）Oberm. = Asparagus exuvialis Burch. ■☆

315652　Protasparagus exuvialis（Burch.）Oberm. f. ecklonii（Baker）Oberm. = Asparagus exuvialis Burch. f. ecklonii（Baker）Fellingham et N. L. Mey. ■☆

315653　Protasparagus falcatus（L.）Oberm. = Asparagus falcatus L. ■☆

315654　Protasparagus filicladus Oberm. = Asparagus filicladus（Oberm.）Fellingham et N. L. Mey. ■☆

315655　Protasparagus flavicaulis Oberm. ；黄茎原始南星■☆

315656　Protasparagus flavicaulis Oberm. subsp. flavicaulis = Asparagus flavicaulis（Oberm.）Fellingham et N. L. Mey. ■☆

315657　Protasparagus fouriei Oberm. = Asparagus fourei（Oberm.）Fellingham et N. L. Mey. ■☆

315658　Protasparagus fractiflexus Oberm. = Asparagus fractiflexus（Oberm.）Fellingham et N. L. Mey. ■☆

315659　Protasparagus glaucus（Kies）Oberm. = Asparagus glaucus Kies ■☆

315660　Protasparagus graniticus Oberm. = Asparagus graniticus（Oberm.）Fellingham et N. L. Mey. ■☆

315661　Protasparagus humilis（Engl.）B. Mathew = Asparagus humilis Engl. ■☆

315662　Protasparagus intricatus Oberm. = Asparagus intricatus（Oberm.）Fellingham et N. L. Mey. ■☆

315663　Protasparagus krebsianus（Kunth）Oberm. = Asparagus krebsianus（Kunth）Jessop ■☆

315664　Protasparagus laricinus（Burch.）Oberm. = Asparagus laricinus Burch. ■☆

315665　Protasparagus lignosus（Burm. f.）Oberm. = Asparagus lignosus Burm. f. ■☆

315666　Protasparagus longicladus（N. E. Br.）B. Mathew = Asparagus longicladus N. E. Br. ■☆

315667　Protasparagus macowanii（Baker）Oberm. = Asparagus macowanii Baker ■☆

315668　Protasparagus mariae Oberm. = Asparagus mariae（Oberm.）Fellingham et N. L. Mey. ■☆

315669　Protasparagus meyeri ? = Protasparagus densiflorus（Kunth）Oberm. ‘Myers’■☆

315670　Protasparagus meyersii ? = Protasparagus densiflorus（Kunth）Oberm. ‘Myers’■☆

315671　Protasparagus microraphis（Kunth）Oberm. = Asparagus microraphis（Kunth）Baker ■☆

315672　Protasparagus minutiflorus（Kunth）Oberm. = Asparagus minutiflorus（Kunth）Baker ■☆

315673　Protasparagus mollis Oberm. = Asparagus mollis（Oberm.）Fellingham et N. L. Mey. ■☆

315674　Protasparagus mucronatus（Jessop）Oberm. = Asparagus mucronatus Jessop ■☆

315675　Protasparagus multiflorus（Baker）Oberm. = Asparagus multiflorus Baker ■☆

315676　Protasparagus natalensis（Baker）Oberm. = Asparagus natalensis（Baker）J. -P. Lebrun et Stork ■☆

315677　Protasparagus nelsii（Schinz）Oberm. = Asparagus nelsii Schinz ■☆

315678　Protasparagus nodulosus Oberm. = Asparagus nodulosus（Oberm.）J. -P. Lebrun et Stork ■☆

315679　Protasparagus oliveri Oberm. = Asparagus oliveri（Oberm.）Fellingham et N. L. Mey. ■☆

315680　Protasparagus oxyacanthus（Baker）Oberm. = Asparagus oxyacanthus Baker ■☆

315681　Protasparagus pearsonii（Kies）Oberm. = Asparagus pearsonii Kies ■☆

315682　Protasparagus pendulus Oberm. = Asparagus pendulus（Oberm.）J. -P. Lebrun et Stork ☆

315683　Protasparagus plumosus（Baker）Oberm. = Asparagus setaceus（Kunth）Jessop ■

315684　Protasparagus racemosus（Willd.）Oberm. = Asparagus racemosus Willd. ■

315685　Protasparagus recurvispinus Oberm. = Asparagus recurvispinus（Oberm.）Fellingham et N. L. Mey. ■☆

315686　Protasparagus retrofractus（L.）Oberm. = Asparagus retrofractus L. ■☆

315687　Protasparagus rigidus（Jessop）Oberm. = Asparagus rigidus Jessop ■☆

315688　Protasparagus rubicundus（P. J. Bergius）Oberm. = Asparagus rubicundus P. J. Bergius ■☆

315689　Protasparagus schroederi（Engl.）Oberm. = Asparagus schroederi Engl. ■☆

315690　Protasparagus sekukuniensis Oberm. = Asparagus sekukuniensis（Oberm.）Fellingham et N. L. Mey. ■☆

315691　Protasparagus setaceus（Kunth）Oberm. = Asparagus setaceus（Kunth）Jessop ■

315692　Protasparagus spinescens（Steud. ex Roem. et Schult.）Oberm. = Asparagus spinescens Steud. ex Roem. et Schult. ■☆

315693　Protasparagus stellatus（Baker）Oberm. = Asparagus stellatus Baker ■☆

315694　Protasparagus stipulaceus（Lam.）Oberm. = Asparagus stipulaceus Lam. ■☆

315695　Protasparagus striatus（L. f.）Oberm. = Asparagus striatus（L. f.）Thunb. ■☆

315696　Protasparagus suaveolens（Burch.）Oberm. = Asparagus suaveolens Burch. ■☆

315697　Protasparagus subulatus（Thunb.）Oberm. = Asparagus subulatus Thunb. ■☆

315698　Protasparagus transvaalensis Oberm. = Asparagus transvaalensis（Oberm.）Fellingham et N. L. Mey. ■☆

315699　Protasparagus virgatus（Baker）Oberm. = Asparagus virgatus Baker ■☆

315700　Protea L.（1753）（废弃属名）= Leucadendron R. Br.（保留属名）●

315701　Protea L.（1753）（废弃属名）= Protea L.（1771）（保留属名）●☆

315702　Protea L.（1771）（保留属名）；海神木属（布罗特属，帝王花属，多变花属，鳞果山龙眼属，帕洛梯属，普罗梯亚木属，山龙眼属）；Protea, Sugar Bush ●☆

315703　Protea R. Br. = Protea L.（保留属名）●☆

315704　Protea abyssinica Willd. = Protea gaguedii J. F. Gmel. ●☆

315705　Protea abyssinica Willd. var. adolphi-friderici Engl. = Protea welwitschii Engl. ●☆

315706　Protea abyssinica Willd. var. brevifolia Engl. = Protea welwitschii Engl. ●☆

315707　Protea acaulis（L.）Reichard var. cockscombensis Archibald =

Protea tenax（Salisb.）R. Br. ●☆

315708　Protea acaulis Thunb. =Protea acaulos（L.）Reichard ●☆

315709　Protea acaulos（L.）Reichard;无茎海神木●☆

315710　Protea acerosa R. Br. = Protea subulifolia（Salisb. ex Knight）Rourke ●☆

315711　Protea acuminata Sims;渐尖海神木●☆

315712　Protea adscendens Lam. =Serruria adscendens（Lam.）R. Br. ●☆

315713　Protea alba Thunb. =Leucadendron album（Thunb.）Fourc. ●☆

315714　Protea albida De Wild. = Protea micans Welw. ●☆

315715　Protea alpina Salisb. ex Knight = Vexatorella alpina（Salisb. ex Knight）Rourke ●☆

315716　Protea amplexicaulis（Salisb.）R. Br. ;抱茎海神木●☆

315717　Protea angolensis Burtt Davy et Hoyle = Protea madiensis Oliv. subsp. occidentalis（Beard）Chisumpa et Brummitt ●☆

315718　Protea angolensis Welw. ;安哥拉海神木（安哥拉帕洛梯）;Angola Protea,Sugar Bush ●☆

315719　Protea angolensis Welw. f. trichanthera（Baker）Beard = Protea angolensis Welw. var. trichanthera（Baker）Brummitt ●☆

315720　Protea angolensis Welw. var. albiflora Engl. = Protea angolensis Welw. ●☆

315721　Protea angolensis Welw. var. divaricata（Engl. et Gilg）Beard;叉开海神木●☆

315722　Protea angolensis Welw. var. glabribracteata Hauman = Protea angolensis Welw. ●☆

315723　Protea angolensis Welw. var. roseola Chisumpa et Brummitt;粉红安哥拉海神木●☆

315724　Protea angolensis Welw. var. trichanthera（Baker）Brummitt;毛花安哥拉海神木●☆

315725　Protea angustata R. Br. ;狭海神木●☆

315726　Protea arborea Houtt. = Protea nitida Mill. ●☆

315727　Protea arcuata Lam. = Leucadendron arcuatum（Lam.）I. Williams ●☆

315728　Protea argentea L. = Leucadendron argenteum（L.）R. Br. ●☆

315729　Protea argyrea Hauman;银色海神木●☆

315730　Protea argyrea Hauman subsp. zambiana Chisumpa et Brummitt;赞比亚海神木●☆

315731　Protea argyrophaea Hutch. = Protea madiensis Oliv. ●☆

315732　Protea aristata E. Phillips;松叶海神木（松叶帕洛梯）;Christmas Protea,Pine Sugar Protea ●☆

315733　Protea aspera E. Phillips;粗糙海神木●☆

315734　Protea aurea（Burm. f.）Rourke;多型海神木（多型帕洛梯）●☆

315735　Protea auriculata Tausch = Protea eximia（Salisb. ex Knight）Fourc. ●☆

315736　Protea barbigera Meisn. = Protea magnifica Andréws ●☆

315737　Protea baumii Engl. et Gilg;鲍姆海神木●☆

315738　Protea baumii Engl. et Gilg subsp. robusta Chisumpa et Brummitt;粗壮鲍姆海神木●☆

315739　Protea bequaertii De Wild. = Protea madiensis Oliv. ●☆

315740　Protea bequaertii De Wild. var. pilosa Hauman = Protea madiensis Oliv. ●☆

315741　Protea bolusii E. Phillips = Protea caffra Meisn. ●☆

315742　Protea burchellii Stapf;狭叶海神木（美丽海神木,美丽帕洛梯,狭叶帕洛梯）;Pretty Protea ●☆

315743　Protea caespitosa Andréws;丛生海神木●☆

315744　Protea caffra Meisn. ;球花海神木（球花帕洛梯）●☆

315745　Protea caffra Meisn. subsp. gazensis（Beard）Chisumpa et Brummitt;加兹海神木●☆

315746　Protea caffra Meisn. subsp. kilimandscharica（Engl.）Chisumpa et Brummitt;基利海神木●☆

315747　Protea caffra Meisn. subsp. mafingensis Chisumpa et Brummitt;马芬加海神木●☆

315748　Protea caffra Meisn. subsp. nyasae（Rendle）Chisumpa et Brummitt;尼亚萨海神木●☆

315749　Protea caffra Sim subsp. falcata（Beard）Lötter;镰形海神木●☆

315750　Protea calocephala Meisn. = Protea obtusifolia H. Buek ex Meisn. ●☆

315751　Protea canaliculata Andréws;具沟海神木●☆

315752　Protea candicans Thunb. = Paranomus candicans（Thunb.）Kuntze ●☆

315753　Protea caudata Thunb. =Spatalla caudata（Thunb.）R. Br. ●☆

315754　Protea cedromontana Schltr. = Protea acuminata Sims ●☆

315755　Protea chionantha Engl. et Gilg = Protea angolensis Welw. ●☆

315756　Protea chionantha Engl. et Gilg var. divaricata ? = Protea angolensis Welw. var. divaricata（Engl. et Gilg）Beard ●☆

315757　Protea chrysolepis Engl. et Gilg = Protea gaguedii J. F. Gmel. ●☆

315758　Protea cinerea Sol. ex Aiton = Leucadendron cinereum（Sol. ex Aiton）R. Br. ●☆

315759　Protea comigera Stapf = Protea laurifolia Thunb. ●☆

315760　Protea comosa Thunb. = Leucadendron comosum（Thunb.）R. Br. ●☆

315761　Protea compacta R. Br. ;红宝石海神木（红宝石帕洛梯）;Prince Protea,River Protea,Ruby Protea ●☆

315762　Protea comptonii Beard;康普顿海神木●☆

315763　Protea concava Lam. = Diastella thymelaeoides（P. J. Bergius）Rourke ●☆

315764　Protea congensis Engl. = Protea welwitschii Engl. ●☆

315765　Protea conica Lam. = Leucadendron conicum（Lam.）I. Williams ●☆

315766　Protea conifera L. = Leucadendron coniferum（L.）Meisn. ●☆

315767　Protea conocarpa Thunb. = Leucospermum conocarpodendron（L.）H. Buek ●☆

315768　Protea convexa E. Phillips;弯曲海神木●☆

315769　Protea cordata Thunb. ;心形海神木●☆

315770　Protea coronata Lam. ;长梗海神木（长梗帕洛梯）●☆

315771　Protea crassulifolia Salisb. ex Knight = Leucadendron arcuatum（Lam.）I. Williams ●☆

315772　Protea crinita Beard = Protea wentzeliana Engl. ●☆

315773　Protea crinita Thunb. = Leucospermum oleifolium（P. J. Bergius）R. Br. ●☆

315774　Protea curvata N. E. Br. ;内折海神木●☆

315775　Protea cuspidata Beard = Protea kibarensis Hauman subsp. cuspidata（Beard）Chisumpa et Brummitt ●☆

315776　Protea cynaroides（L.）L. ;巨大海神木（蓟状多变花,巨大帕洛梯,帕洛梯王）;Giant Protea,Honeypot Protea,King Protea ●☆

315777　Protea cynaroides L. = Protea cynaroides（L.）L. ●☆

315778　Protea daphnoides Thunb. = Leucadendron daphnoides（Thunb.）Meisn. ●☆

315779　Protea decumbens Thunb. = Serruria decumbens（Thunb.）R. Br. ●☆

315780　Protea decurrens E. Phillips;下延海神木●☆

315781　Protea dekindtiana Engl. ;德金海神木●☆

315782　Protea denticulata Rourke;细齿海神木●☆

315783　Protea doddii E. Phillips = Protea simplex E. Phillips ●☆

315784　Protea dracomontana Beard;德拉科海神木●☆

315785　Protea dykei E. Phillips = Protea rupicola Mund ex Meisn. ●☆

315786　Protea echinulata Meisn. = Protea restionifolia (Salisb. ex Knight) Rycroft ●☆

315787　Protea echinulata Meisn. var. minor E. Phillips = Protea restionifolia (Salisb. ex Knight) Rycroft ●☆

315788　Protea effusa E. Mey. ex Meisn.；开展海神木●☆

315789　Protea eickii Engl. = Protea welwitschii Engl. ●☆

315790　Protea elliottii C. H. Wright = Protea madiensis Oliv. ●☆

315791　Protea elliottii C. H. Wright var. angustifolia Keay = Protea madiensis Oliv. ☆

315792　Protea elliptica Thunb. = Leucospermum cuneiforme (Burm. f.) Rourke ●☆

315793　Protea enervis Wild；无脉海神木●☆

315794　Protea eximia (Knight) Fourc. = Protea eximia (Salisb. ex Knight) Fourc. ●☆

315795　Protea eximia (Salisb. ex Knight) Fourc.；阔叶海神木(阔叶帕洛梯)；Duchess Protea, Ray-flowered Protea ●☆

315796　Protea flanaganii E. Phillips = Protea simplex E. Phillips ●☆

315797　Protea flavopilosa Beard；褐毛海神木●☆

315798　Protea florida Thunb. = Serruria florida (Thunb.) Salisb. ex Knight ●☆

315799　Protea foliosa Rourke；多叶海神木●☆

315800　Protea formosa Andréws = Leucospermum formosum (Andréws) Sweet ●☆

315801　Protea fulva Tausch = Protea lepidocarpodendron (L.) L. ●☆

315802　Protea fusciflora Jacq. = Leucadendron linifolium (Jacq.) R. Br. ●☆

315803　Protea gaguedii J. F. Gmel.；格氏海神木(格哥特帕洛梯)；Gagued Protea ●☆

315804　Protea gaguedii J. F. Gmel. subsp. laetans (L. E. Davidson) Beard = Protea laetans L. E. Davidson ●☆

315805　Protea gazensis Beard = Protea caffra Meisn. subsp. gazensis (Beard) Chisumpa et Brummitt ●☆

315806　Protea glabra Thunb.；光滑海神木●☆

315807　Protea glaucophylla Salisb. = Protea acaulos (L.) Reichard ●☆

315808　Protea globosa Kenn. ex Andréws = Leucadendron globosum (Kenn. ex Andréws) I. Williams ●☆

315809　Protea goetzeana Engl. = Protea welwitschii Engl. ●☆

315810　Protea grandiceps Tratt.；桃叶海神木(桃叶帕洛梯)；Peach Protea, Princes Protea, Red Sugarbush ●☆

315811　Protea grandiflora Thunb. = Protea nitida Mill. ●☆

315812　Protea haemantha Engl. et Gilg = Protea poggei Engl. subsp. haemantha (Engl. et Gilg) Chisumpa et Brummitt ●☆

315813　Protea haemantha Engl. et Gilg subsp. vernicosa (Hauman) Beard = Protea poggei Engl. ●☆

315814　Protea harmeri E. Phillips = Protea canaliculata Andréws ●☆

315815　Protea heckmanniana Engl.；赫克曼海神木●☆

315816　Protea heckmanniana Engl. subsp. angustifolia (Hauman) Chisumpa et Brummitt；窄叶海神木●☆

315817　Protea heckmanniana Engl. var. angustifolia Hauman = Protea heckmanniana Engl. subsp. angustifolia (Hauman) Chisumpa et Brummitt ●☆

315818　Protea heterophylla Thunb. = Leucospermum heterophyllum (Thunb.) Rourke ●☆

315819　Protea hirta Klotzsch = Protea welwitschii Engl. ●☆

315820　Protea hirta Klotzsch subsp. glabrescens Beard = Protea welwitschii Engl. ●☆

315821　Protea holosericea (Salisb. ex Knight) Rourke；全毛海神木●☆

315822　Protea homblei De Wild. = Protea angolensis Welw. var. trichanthera (Baker) Brummitt ●☆

315823　Protea humiflora Andréws；平花海神木●☆

315824　Protea humifusa Beard；平伏海神木●☆

315825　Protea hypophylla Thunb. = Leucospermum hypophyllocarpodendron (L.) Druce ●☆

315826　Protea ignota E. Phillips = Protea longifolia Andréws ●☆

315827　Protea imbricata Thunb. = Sorocephalus imbricatus (Thunb.) R. Br. ●☆

315828　Protea incana Meisn. = Protea roupelliae Meisn. ●☆

315829　Protea incompta R. Br. = Protea coronata Lam. ●☆

315830　Protea incompta R. Br. var. susannae E. Phillips = Protea coronata Lam. ●☆

315831　Protea incurva Thunb. = Spatalla incurva (Thunb.) R. Br. ●☆

315832　Protea intonsa Rourke；须毛海神木●☆

315833　Protea inyanganiensis Beard；伊尼海神木●☆

315834　Protea kibarensis Hauman；基巴拉海神木●☆

315835　Protea kibarensis Hauman subsp. cuspidata (Beard) Chisumpa et Brummitt；骤尖海神木●☆

315836　Protea kilimandscharica Engl. = Protea caffra Meisn. subsp. kilimandscharica (Engl.) Chisumpa et Brummitt ●☆

315837　Protea kingaensis Engl. = Protea rubrobracteata Engl. ●☆

315838　Protea kirkii C. H. Wright = Protea welwitschii Engl. ●☆

315839　Protea lacticolor Salisb.；厚叶海神木(厚叶帕洛梯)●☆

315840　Protea lacticolor Salisb. var. angustata E. Phillips = Protea subvestita N. E. Br. ●☆

315841　Protea lacticolor Salisb. var. orientalis (Sim) E. Phillips = Protea subvestita N. E. Br. ●☆

315842　Protea laetans L. E. Davidson；愉悦海神木●☆

315843　Protea laevis R. Br.；平滑海神木●☆

315844　Protea lagopus Thunb. = Paranomus lagopus (Thunb.) Salisb. ●☆

315845　Protea lanata Thunb. = Sorocephalus lanatus (Thunb.) R. Br. ●☆

315846　Protea lanceolata E. Mey. ex Meisn.；披针形海神木●☆

315847　Protea latifolia R. Br. = Protea eximia (Salisb. ex Knight) Fourc. ●☆

315848　Protea latifolia R. Br. var. auriculata (Tausch) Kuntze = Protea eximia (Salisb. ex Knight) Fourc. ●☆

315849　Protea laureola Lam. = Leucadendron laureolum (Lam.) Fourc. ●☆

315850　Protea laurifolia Thunb.；月桂叶海神木●☆

315851　Protea lemairei De Wild. = Protea micans Welw. subsp. lemairei (De Wild.) Chisumpa et Brummitt ●☆

315852　Protea lepidocarpodendron (L.) L.；黑海神木(黑帕洛梯)；Black Protea ●☆

315853　Protea lepidocarpodendron (L.) L. var. villosa E. Phillips = Protea lepidocarpodendron (L.) L. ●☆

315854　Protea lepidocarpodendron L. = Protea lepidocarpodendron (L.) L. ●☆

315855　Protea leucoblepharis (Hiern) Baker = Protea welwitschii Engl. ●☆

315856　Protea ligulifolia (Salisb. ex Knight) Sweet = Protea longifolia Andréws ●☆

315857　Protea linearifolia Engl.；线叶海神木●☆

315858　Protea linifolia Jacq. = Leucadendron linifolium (Jacq.) R. Br. ●☆

315859　Protea longicaulis Salisb. ex Knight = Leucadendron linifolium (Jacq.) R. Br. ●☆

315860　Protea longiflora Lam.；长花海神木(长花帕洛梯)；Longflower

Protea, Sir Lowry's Pass ●☆

315861   Protea longiflora Lam. = Protea aurea (Burm. f.) Rourke ●☆

315862   Protea longiflora Lam. var. ovalis E. Phillips = Protea aurea (Burm. f.) Rourke ●☆

315863   Protea longifolia Andr. var. minor E. Phillips = Protea longifolia Andréws ●☆

315864   Protea longifolia Andréws;长叶海神木●☆

315865   Protea loranthifolia Salisb. ex Knight = Leucadendron loranthifolium (Salisb. ex Knight) I. Williams ●☆

315866   Protea lorifolia (Salisb. ex Knight) Fourc. ;纽叶海神木●☆

315867   Protea macrocephala Thunb. = Protea coronata Lam. ●☆

315868   Protea macrophylla R. Br. = Protea lorifolia (Salisb. ex Knight) Fourc. ●☆

315869   Protea madiensis Oliv. ;马地海神木●☆

315870   Protea madiensis Oliv. f. pilosa Engl. = Protea madiensis Oliv. ●☆

315871   Protea madiensis Oliv. subsp. occidentalis (Beard) Chisumpa et Brummitt;西方海神木●☆

315872   Protea madiensis Oliv. var. angustifolia (Keay) Beard = Protea madiensis Oliv. ●☆

315873   Protea madiensis Oliv. var. claessensii De Wild. = Protea madiensis Oliv. ●☆

315874   Protea madiensis Oliv. var. elliottii (C. H. Wright) Beard = Protea madiensis Oliv. ●☆

315875   Protea madiensis Oliv. var. pilosa (Engl.) Hauman = Protea madiensis Oliv. ●☆

315876   Protea magnifica Andréws;壮丽海神木(芒须多变花,壮丽帕洛梯);Bearded Protea, Mearded Protea, Queen Protea ●☆

315877   Protea manikensis De Wild. = Protea gaguedii J. F. Gmel. ●☆

315878   Protea marginata Thunb. = Protea laurifolia Thunb. ●☆

315879   Protea marginata Willd. = Leucadendron spissifolium (Salisb. ex Knight) I. Williams ●☆

315880   Protea marlothii E. Phillips = Protea effusa E. Mey. ex Meisn. ●☆

315881   Protea mellifera Thunb. = Protea repens (L.) L. ●☆

315882   Protea mellifera Thunb. var. albiflora Andréws = Protea repens (L.) L. ●☆

315883   Protea melliodora Engl. et Gilg = Protea welwitschii Engl. ●☆

315884   Protea micans Welw. ;弱光泽海神木●☆

315885   Protea micans Welw. subsp. lemairei (De Wild.) Chisumpa et Brummitt;勒迈尔海神木●☆

315886   Protea micans Welw. subsp. suffruticosa (Beard) Chisumpa et Brummitt;亚灌木海神木●☆

315887   Protea micans Welw. subsp. trichophylla (Engl. et Gilg) Chisumpa et Brummitt;毛叶海神木●☆

315888   Protea minima Hauman;微小海神木●☆

315889   Protea minor (E. Phillips) Compton = Protea longifolia Andréws ●☆

315890   Protea montana E. Mey. ex Meisn. ;山地海神木●☆

315891   Protea mucronifolia Salisb. ;微凸海神木●☆

315892   Protea multibracteata E. Phillips = Protea caffra Meisn. ●☆

315893   Protea mundii Klotzsch;美叶海神木(美叶帕洛梯)●☆

315894   Protea myrsinifolia Engl. et Gilg = Protea welwitschii Engl. ●☆

315895   Protea myrtifolia Thunb. = Diastella myrtifolia (Thunb.) Salisb. ex Knight ●☆

315896   Protea namaquana Rourke;纳马夸海神木●☆

315897   Protea nana (P. J. Bergius) Thunb. ;小海神木(小帕洛梯);Montain Rose ●☆

315898   Protea nana Lam. = Protea nana (P. J. Bergius) Thunb. ●☆

315899   Protea neocrinita Beard = Protea wentzeliana Engl. ●☆

315900   Protea neriifolia R. Br. ;夹竹桃叶海神木(夹竹桃叶多变花,夹竹桃叶帕洛梯);Blue Sugarbush, Oleander-leafed Protea, Oleander-leaved Protea, Pink Mink ●☆

315901   Protea nitens Thunb. = Mimetes argenteus Salisb. ex Knight ●☆

315902   Protea nitida Mill. ;光亮海神木●☆

315903   Protea nubigena Rourke;云雾海神木●☆

315904   Protea nyasae Rendle = Protea caffra Meisn. subsp. nyasae (Rendle) Chisumpa et Brummitt ●☆

315905   Protea obtusata Thunb. = Vexatorella obtusata (Thunb.) Rourke ●☆

315906   Protea obtusifolia De Wild. = Protea welwitschii Engl. ●☆

315907   Protea obtusifolia H. Buek ex Meisn. ;钝叶海神木(钝叶帕洛梯);Bluntleaf Protea, Bush Protea ●☆

315908   Protea occidentalis Beard = Protea madiensis Oliv. subsp. occidentalis (Beard) Chisumpa et Brummitt ●☆

315909   Protea odorata Thunb. ;芳香海神木●☆

315910   Protea oleracea Guthrie = Protea caespitosa Andréws ●☆

315911   Protea orientalis Sim = Protea subvestita N. E. Br. ●☆

315912   Protea paludosa (Hiern) Engl. ;沼泽海神木●☆

315913   Protea paludosa (Hiern) Engl. subsp. secundifolia (Hauman) Chisumpa et Brummitt;单侧叶沼泽海神木●☆

315914   Protea parvula Beard;较小海神木●☆

315915   Protea patens R. Br. = Protea holosericea (Salisb. ex Knight) Rourke ●☆

315916   Protea pedunculata Lam. = Serruria pedunculata (Lam.) R. Br. ●☆

315917   Protea pendula R. Br. ;下垂海神木●☆

315918   Protea petiolaris (Hiern) Baker et C. H. Wright;柄生海神木(柄生帕洛梯);Petiolebore Protea ●☆

315919   Protea petiolaris (Hiern) Baker et C. H. Wright subsp. elegans Chisumpa et Brummitt;雅致柄生海神木●☆

315920   Protea petiolaris Welw. = Protea petiolaris (Hiern) Baker et C. H. Wright ●☆

315921   Protea plumosa Aiton = Leucadendron rubrum Burm. f. ●☆

315922   Protea poggei Engl. ;波格海神木●☆

315923   Protea poggei Engl. subsp. haemantha (Engl. et Gilg) Chisumpa et Brummitt;血红花海神木●☆

315924   Protea poggei Engl. subsp. mwinilungensis Chisumpa et Brummitt;穆维尼海神木●☆

315925   Protea praticola Engl. ;草原海神木●☆

315926   Protea prolifera Thunb. = Spatalla prolifera (Thunb.) Salisb. ex Knight ●☆

315927   Protea prostrata Thunb. = Leucospermum prostratum (Thunb.) Stapf ●☆

315928   Protea pruinosa Rourke;白粉海神木●☆

315929   Protea pubera L. = Leucospermum calligerum (Salisb. ex Knight) Rourke ●☆

315930   Protea pudens Rourke;地被海神木(地被帕洛梯);Ground Protea ●☆

315931   Protea pulchella Andréws = Protea burchellii Stapf ●☆

315932   Protea pulchella Andréws var. undulata E. Phillips = Protea burchellii Stapf ●☆

315933   Protea pulchra Rycroft = Protea burchellii Stapf ●☆

315934   Protea punctata Meisn. ;斑点海神木●☆

315935   Protea purpurea L. = Diastella proteoides (L.) Druce ●☆

315936   Protea recondita H. Buek ex Meisn. ;隐蔽海神木●☆

315937   Protea repens (L.) L. ;匍匐海神木(布罗特,多变花,蜜花帕

洛梯,蜜帕洛梯,蜜味帕洛梯);Honey Flower,Honey Protea, Honeyflower,Honey-flower,Sugar Bush,Sugar Protea,Sugarbush ●☆

315938　Protea repens L. =Protea repens (L.) L. ●☆

315939　Protea restionifolia (Salisb. ex Knight) Rycroft;帚灯草叶海神木●☆

315940　Protea revoluta R. Br. ;外卷海神木●☆

315941　Protea rhodantha Hook. f. =Protea caffra Meisn. ●☆

315942　Protea rhodantha Hook. f. var. falcata Beard =Protea caffra Sim subsp. falcata (Beard) Lötter ●☆

315943　Protea rosacea L. =Protea nana (P. J. Bergius) Thunb. ●☆

315944　Protea roupelliae Meisn. ;细毛海神木(细毛帕洛梯)●☆

315945　Protea roupelliae Meisn. subsp. hamiltonii Beard ex Rourke;汉密尔顿海神木●☆

315946　Protea rubrobracteata Engl. ;红苞海神木●☆

315947　Protea rubropilosa Beard;红毛海神木●☆

315948　Protea rupestris R. E. Fr. ;岩地海神木●☆

315949　Protea rupicola Mund ex Meisn. ;岩生海神木(岩生帕洛梯); Protea ●☆

315950　Protea salicifolia Mildbr. =Protea heckmanniana Engl. ●☆

315951　Protea scabra R. Br. ;糙海神木●☆

315952　Protea scabriuscula E. Phillips;略粗糙海神木●☆

315953　Protea sceptrum-gustavianus Sparrm. =Paranomus sceptrum-gustavianus (Sparrm.) Hyl. ●☆

315954　Protea scolymocephala (L.) Reichard;绿苞海神木(绿苞帕洛梯,朱蓟头状帕洛梯); Green Button Protea, Green Protea, Greenbract Protea,Mini Protea ●☆

315955　Protea scolymocephala Reichard =Protea scolymocephala (L.) Reichard ●☆

315956　Protea secundiflora Hauman =Protea paludosa (Hiern) Engl. subsp. secundifolia (Hauman) Chisumpa et Brummitt ●☆

315957　Protea sericea Thunb. =Leucadendron sericeum (Thunb.) R. Br. ●☆

315958　Protea simplex E. Phillips;简单海神木●☆

315959　Protea spathulata Thunb. =Paranomus spathulatus (Thunb.) Kuntze ●☆

315960　Protea speciosa (L.) L. ;美丽海神木(褐毛海神木,美丽帕洛梯); Brown-bearded Sugarbush ●☆

315961　Protea speciosa (L.) L. var. angustata Meisn. =Protea speciosa (L.) L. ●☆

315962　Protea speciosa L. =Protea speciosa (L.) L. ●☆

315963　Protea stellaris Sims =Leucadendron stellare (Sims) Sweet ●☆

315964　Protea stokoei E. Phillips;斯托克海神木●☆

315965　Protea strobilina L. =Leucadendron strobilinum (L.) Druce ●☆

315966　Protea subpulchella Stapf =Protea burchellii Stapf ●☆

315967　Protea subulifolia (Salisb. ex Knight) Rourke;球叶海神木●☆

315968　Protea subvestita N. E. Br. ;包被海神木●☆

315969　Protea suffruticosa Beard =Protea micans Welw. subsp. suffruticosa (Beard) Chisumpa et Brummitt ●☆

315970　Protea susannae E. Phillips;萨氏海神木(萨珊娜帕洛梯); Pink Protea ●☆

315971　Protea swynnertonii S. Moore =Protea welwitschii Engl. ●☆

315972　Protea tenax (Salisb.) R. Br. ;黏海神木●☆

315973　Protea tenax (Salisb.) R. Br. var. latifolia Meisn. =Protea foliosa Rourke ●☆

315974　Protea tenuifolia R. Br. =Protea scabra R. Br. ●☆

315975　Protea teretifolia Andréws =Leucadendron teretifolium (Andréws) I. Williams ●☆

315976　Protea thymifolia Salisb. ex Knight =Leucadendron thymifolium (Salisb. ex Knight) I. Williams ●☆

315977　Protea tomentosa Thunb. =Leucospermum tomentosum (Thunb.) R. Br. ●☆

315978　Protea transvaalensis E. Phillips =Protea simplex E. Phillips ●☆

315979　Protea triandra Schltr. =Protea compacta R. Br. ●☆

315980　Protea trichanthera Baker =Protea angolensis Welw. var. trichanthera (Baker) Brummitt ●☆

315981　Protea trichophylla Engl. et Gilg =Protea micans Welw. subsp. trichophylla (Engl. et Gilg) Chisumpa et Brummitt ●☆

315982　Protea trigona E. Phillips =Protea gaguedii J. F. Gmel. ●☆

315983　Protea triternata Thunb. =Serruria triternata (Thunb.) R. Br. ●☆

315984　Protea truncata Thunb. =Leucadendron cinereum (Sol. ex Aiton) R. Br. ●☆

315985　Protea turbiniflora (Salisb.) R. Br. =Protea caespitosa Andréws ●☆

315986　Protea uhehensis Engl. =Protea welwitschii Engl. ●☆

315987　Protea umbellata Thunb. =Aulax umbellata (Thunb.) R. Br. ●☆

315988　Protea umbonalis (Salisb. ex Knight) Sweet =Protea longifolia Andréws ●☆

315989　Protea urundinensis Hauman =Protea angolensis Welw. ●☆

315990　Protea venusta Compton;雅致海神木●☆

315991　Protea vernicosa Hauman =Protea poggei Engl. ●☆

315992　Protea verticillata Thunb. =Leucadendron verticillatum (Thunb.) Meisn. ●☆

315993　Protea vestita Lam. =Leucospermum vestitum (Lam.) Rourke ●☆

315994　Protea villosa Lam. =Serruria villosa (Lam.) R. Br. ●☆

315995　Protea welwitschii Engl. ;韦尔海神木●☆

315996　Protea welwitschii Engl. subsp. adolphi-friderici (Engl.) Beard =Protea welwitschii Engl. ●☆

315997　Protea welwitschii Engl. subsp. glabrescens (Beard) Beard =Protea welwitschii Engl. ●☆

315998　Protea welwitschii Engl. subsp. goetzeana (Engl.) Beard =Protea welwitschii Engl. ●☆

315999　Protea welwitschii Engl. subsp. hirta Beard =Protea welwitschii Engl. ●☆

316000　Protea welwitschii Engl. subsp. melliodora (Engl. et Gilg) Beard =Protea welwitschii Engl. ●☆

316001　Protea welwitschii Engl. subsp. mocoensis Beard =Protea welwitschii Engl. ●☆

316002　Protea welwitschii Engl. var. goetzeana (Engl.) Beard =Protea welwitschii Engl. ●☆

316003　Protea welwitschii Engl. var. melliodora (Engl. et Gilg) Beard =Protea welwitschii Engl. ●☆

316004　Protea wentzeliana Engl. ;文策尔海神木●☆

316005　Protea wentzeliana Engl. var. prostrata Beard =Protea humifusa Beard ●☆

316006　Protea xanthoconus Kuntze =Leucadendron xanthoconus (Kuntze) K. Schum. ●☆

316007　Proteaceae Juss. (1789)(保留科名);山龙眼科;Protea Family ●■

316008　Proteinia (Ser.) Rchb. =Saponaria L. ■

316009　Proteinia Rchb. =Saponaria L. ■

316010　Proteinophallus Hook. f. =Amorphophallus Blume ex Decne. (保留属名)■●

316011　Proteinophallus vivieri Hook. f. =Amorphophallus rivieri Durieu ex Carrière ■

316012　Proteocarpus Börner =Carex L. ■

316013　Proteopsis Mart. et Zucc. ex DC. (1863);尖苞灯头菊属■☆

316014　Proteopsis Mart. et Zucc. ex Sch. Bip. = Proteopsis Mart. et Zucc. ex DC. ■☆

316015　Proteopsis argentea Mart. et Zucc. ex Sch. Bip. ;尖苞灯头菊■☆

316016　Proterpia Raf. = Tabebuia Gomes ex DC. ●☆

316017　Protionopsis Blume = Commiphora Jacq. (保留属名)●

316018　Protium Burm. f. (1768)(保留属名);马蹄果属(白蹄果属);Hooffruit,Protium,Resin Tree,Resintree ●

316019　Protium Wight et Arn. = Commiphora Jacq. (保留属名)●

316020　Protium africanum Harv. = Commiphora harveyi (Engl.) Engl. ●☆

316021　Protium copal (Schltdl. et Cham.) Engl. ;科帕马蹄果●☆

316022　Protium guianense Marchand;圭亚那马蹄果;Cayenne Incense ●☆

316023　Protium heptaphyllum (Aubl.) Marchand;七叶马蹄果;Brazilian Elemi,Incense Tree ●☆

316024　Protium insigne Engl. ;显著马蹄果●☆

316025　Protium javanicum Burm. f. ;爪哇马蹄果●☆

316026　Protium madagascariense Engl. ;马岛马蹄果●☆

316027　Protium mossambicense Oliv. = Commiphora mossambicensis (Oliv.) Engl. ●☆

316028　Protium panamense (Rose) I. M. Johnst. ;巴拿马马蹄果●☆

316029　Protium robustum (Swart) D. M. Porter;粗壮马蹄果●☆

316030　Protium rubrum Cuatrec. ;红马蹄果●☆

316031　Protium serratum (Wall. ex Colebr.) Engl. ;马蹄果(白蹄果);Hooffruit,Serrate Protium ●

316032　Protium serratum (Wall. ex Colebr.) Engl. = Bursera serrata Wall. ex Colebr. ●

316033　Protium tenuifolium Engl. ;细叶马蹄果●☆

316034　Protium unifoliolatum Engl. ;单叶马蹄果●☆

316035　Protium yunnanense (Hu) Kalkman;滇马蹄果;Yunnan Hooffruit,Yunnan Protium ●◇

316036　Protocamusia Gand. = Buphthalmum L. ■

316037　Protoceras Joseph et Vajr. = Pteroceras Hasselt ex Hassk. ■

316038　Protocyrtandra Hosok. = Cyrtandra J. R. Forst. et G. Forst. ●■

316039　Protogabunia Boiteau = Tabernaemontana L. ●

316040　Protogabunia latifolia Boiteau = Tabernaemontana letestui (Pellegr.) Pichon ●☆

316041　Protogabunia letestui (Pellegr.) Boiteau = Tabernaemontana letestui (Pellegr.) Pichon ●☆

316042　Protohopea Miers = Hopea Roxb. (保留属名)●

316043　Protohopea Miers = Symplocos Jacq. ●

316044　Protolepis Steud. = Proteopsis Mart. et Zucc. ex DC. ■☆

316045　Protoliriaceae Makino = Melanthiaceae Batsch ex Borkh. (保留科名)■

316046　Protoliriaceae Makino = Petrosaviaceae Hutch. (保留科名)■

316047　Protolirion Ridl. = Petrosavia Becc. ■

316048　Protolirion miyoshia-sakuraii (Makino) Makino = Petrosavia sakurai (Makino) J. J. Sm. ex Steenis ■

316049　Protolirion miyoshia-sakuraii Makino = Petrosavia sakurai (Makino) J. J. Sm. ex Steenis ■

316050　Protolirion sakurai (Makino) Dandy = Petrosavia sakurai (Makino) J. J. Sm. ex Steenis ■

316051　Protolirion sinii K. Krause = Petrosavia sakurai (Makino) J. J. Sm. ex Steenis ■

316052　Protolirion sinii K. Krause = Petrosavia sinii (K. Krause) Gagnep. ■

316053　Protomegabaria Hutch. (1911);平舟大戟属●☆

316054　Protomegabaria macrophylla (Pax) Hutch. ;大叶平舟大戟●☆

316055　Protomegabaria meiocarpa J. Léonard;平舟大戟●☆

316056　Protomegabaria stapfiana (Beille) Hutch. ;斯氏平舟大戟●☆

316057　Protonopsis Pfeiff. = Commiphora Jacq. (保留属名)●

316058　Protonopsis Pfeiff. = Protionopsis Blume ●

316059　Protorhus Engl. (1881);原始漆木属●☆

316060　Protorhus longifolia (Bernh.) Engl. ;原始漆木●☆

316061　Protorhus namaquensis Sprague = Ozoroa namaquensis (Sprague) Von Teichman et A. E. vanWyk ●☆

316062　Protosavia miyoshia-sakuraii (Makino) Makino = Petrosavia sakurai (Makino) J. J. Sm. ex Steenis ■

316063　Protoschwenckia Soler. = Protoschwenkia Soler. ■☆

316064　Protoschwenkia Soler. (1898);原始施文克茄属■☆

316065　Protoschwenkia mandonii Soler. ;原始施文克茄■☆

316066　Prototulbaghia Vosa(2007);南非紫瓣花属■☆

316067　Proustia Lag. (1807) = Actinotus Labill. ●■☆

316068　Proustia Lag. (1811);刺枝钝柱菊属●☆

316069　Proustia Lag. ex DC. (1830) = Actinotus Labill. ●■☆

316070　Proustia angustifolia Wedd. ;窄叶刺枝钝柱菊●☆

316071　Proustia cinerea Phil. ;灰刺枝钝柱菊●☆

316072　Proustia cuneifolia D. Don;楔叶刺枝钝柱菊●☆

316073　Provancheria B. Boivin = Cerastium L. ■

316074　Provencheria cerastoides (L.) B. Boivin = Cerastium cerastoides (L.) Britton ■

316075　Provenzalia Adans. = Calla L. ■

316076　Prozetia Neck. = Pouteria Aubl. ●

316077　Prozopsis C. Muell. = Huanaca Cav. ■☆

316078　Prozopsis C. Muell. = Pozopsis Hook. ■☆

316079　Prumnopityaceae A. V. Bobrov et Melikyan = Podocarpaceae Endl. (保留科名)●

316080　Prumnopityaceae Melikian et A. V. Bobrov = Podocarpaceae Endl. (保留科名)●

316081　Prumnopitys Phil. (1861);异罗汉松属(鳞梗杉属);Matai,Plum-fir ●☆

316082　Prumnopitys Phil. = Podocarpus Pers. (保留属名)●

316083　Prumnopitys amara (Blume) de Laub. ;阿玛拉异罗汉松●☆

316084　Prumnopitys andina (Poepp. ex Endl.) de Laub. ;安第斯异罗汉松(安第斯鳞梗杉,智利罗汉松);Andes Podocarpus,Andes Yaccatree,Chile Yaccatree,Plum Yew,Plum-fruited Yew,Plum-yew ●☆

316085　Prumnopitys ferruginea (D. Don) de Laub. = Prumnopitys ferruginea (G. Benn. ex D. Don) de Laub. ●☆

316086　Prumnopitys ferruginea (G. Benn. ex D. Don) de Laub. ;亮红果异罗汉松(新西兰罗汉松,锈色异罗汉松);Miro,Rusty Podocarp ●☆

316087　Prumnopitys ladei (Bailey) de Laub. ;蕨叶异罗汉松;Black Pine ●☆

316088　Prumnopitys taxifolia (D. Don) de Laub. = Prumnopitys taxifolia (Sol. ex D. Don) de Laub. ●☆

316089　Prumnopitys taxifolia (Sol. ex D. Don) de Laub. ;紫杉叶异罗汉松;Matai,New Zealand Black Pine ●☆

316090　Prunaceae Bercht. et J. Presl = Rosaceae Juss. (保留科名)●■

316091　Prunaceae Burnett = Amygdalaceae Bartl. ●

316092　Prunaceae Burnett = Rosaceae Juss. (保留科名)●■

316093　Prunaceae Burnett;李科(樱科)●

316094　Prunaceae Martinov = Rosaceae Juss. (保留科名)●■

316095　Prunella L. (1753);夏枯草属;Selfheal,Self-heal ■☆

316096　Prunella afriquena Pau et Font Quer = Prunella laciniata (L.) L. ■☆

316097　Prunella asiatica Nakai;山菠菜(长冠夏枯草,大花夏枯草,灯

笼草,灯笼头,东北夏枯草,麦穗夏枯草,山苏子,夏枯草);Asia Selfheal,Asian Selfheal,Wild Selfheal ■

316098　Prunella asiatica Nakai = Prunella vulgaris L. subsp. asiatica (Nakai) H. Hara ■

316099　Prunella asiatica Nakai = Prunella vulgaris L. ■

316100　Prunella asiatica Nakai var. albiflora（Koidz.）Nakai;白花山菠菜（白花东北夏枯草）■

316101　Prunella asiatica Nakai var. albiflora（Koidz.）Nakai = Prunella asiatica Nakai ■

316102　Prunella asiatica Nakai var. nanhutashanensis S. S. Ying;高山夏枯草■

316103　Prunella asiatica Nakai var. taiwaniana S. S. Ying = Prunella asiatica Nakai ■

316104　Prunella grandiflora（L.）Jacq.;大花夏枯草;Bigflower Selfheal,Big-flowered Selfheal,Large Selfheal,Large Self-heal ■

316105　Prunella grandiflora（L.）Jacq. ' Loveliness';丽人大花夏枯草■☆

316106　Prunella grandiflora（L.）Jacq. ' Pink Loveliness';粉丽人大花夏枯草■☆

316107　Prunella grandiflora（L.）Jacq. ' White Loveliness';白丽人大花夏枯草■☆

316108　Prunella grandiflora（L.）Moench = Prunella grandiflora（L.）Jacq. ■

316109　Prunella hispida Benth.;硬毛夏枯草（刚毛夏枯草）;Hispid Selfheal ■

316110　Prunella hyssopifolia L.;神香草叶夏枯草■☆

316111　Prunella indica Burm. f. = Acrocephalus indicus（Burm. f.）Kuntze ■

316112　Prunella intermedia Link;中夏枯草;Intermediate Selfheal ■☆

316113　Prunella japonica Makino = Prunella vulgaris L. subsp. asiatica （Nakai）H. Hara ■

316114　Prunella japonica Makino = Prunella vulgaris L. subsp. asiatica （Nakai）H. Hara var. aleutica Fernald ■☆

316115　Prunella japonica Makino = Prunella vulgaris L. ■

316116　Prunella laciniata（L.）L.;条裂夏枯草（白花夏枯草）;Cutleaf Self Heal,Cutleaf Selfheal,Cut-leaved Selfheal,Cut-leaved Self-heal,Heal All,Laciniate Selfheal ■☆

316117　Prunella laciniata（L.）L. var. macrostachya Pau et Font Quer = Prunella laciniata（L.）L. ■☆

316118　Prunella laciniata（L.）L. var. subintegra Ham. = Prunella laciniata（L.）L. ■☆

316119　Prunella laciniata L. = Prunella laciniata（L.）L. ■☆

316120　Prunella lanceolata W. P. C. Barton = Prunella vulgaris L. subsp. lanceolata（W. P. C. Barton）Hultén ■

316121　Prunella parviflora Gilib. = Prunella vulgaris L. ■

316122　Prunella pennsylvanica Willd. var. lanceolata W. P. C. Barton = Prunella vulgaris L. var. lanceolata（W. P. C. Barton）Fernald ■

316123　Prunella prunelliformis（Maxim.）Makino;山生夏枯草（高山夏枯草）■☆

316124　Prunella prunelliformis（Maxim.）Makino f. albiflora M. Mizush.;白花高山夏枯草■☆

316125　Prunella prunelliformis（Maxim.）Makino f. lilacina（Nakai）Honda;淡紫高山夏枯草■☆

316126　Prunella stolonifera H. Lév. et Giraudias = Prunella hispida Benth. ■

316127　Prunella vulgaris L.;夏枯草（棒槌草,棒头花,滁州夏枯草,大头花,灯笼草,古牛草,牯牛岭,牯牛头,金疮小草,槊头草,锣

锤草,麦穗夏枯草,麦夏枯,毛虫药,乃东,牛低头,欧夏枯草,丝线吊铜钟,铁色草,铁线夏枯,铁线夏枯草,土枇杷,夕句,夏枯菜,夏枯头,小本蛇药草,燕面,羊蹄尖）;Black Man's Flower,Blaw-weary,Blue Curls,Brownwort,Brunel,Brunell,Bumble-bee,Bummeltykite,Carpenter Grass,Carpenter's Herb,Common Selfheal,Common Self-heal,Fly Flower,Framlington Clover,Heal-all,Heart-o'-the-earth,Heartsease,Hedley's Jamie Clover,Hercules'Woundwort,Hook-heal,Hookweed,Jamie Hedley's Clover,Lawn Prunella,London Bottle,London Bottles,Oak of Mamre,Pansy,Pickpocket,Poverty Pink,Poverty-pink,Prince's Feathers,Proud Carpenter,Prunell,Scotch Granfer-Griggles,Self Heal,Selfheal,Self-heal,Slough-heal,Sloughwort,Snake's Meat,Soldier Buttons,Soldier's Buttons,Touch-and-heal,Unsavoury Marjoram,Wood Sage ■

316128　Prunella vulgaris L. = Prunella asiatica Nakai ■

316129　Prunella vulgaris L. subsp. asiatica（Nakai）H. Hara = Prunella asiatica Nakai ■

316130　Prunella vulgaris L. subsp. asiatica（Nakai）H. Hara f. albiflora Nakai = Prunella asiatica Nakai var. albiflora（Koidz.）Nakai ■

316131　Prunella vulgaris L. subsp. asiatica（Nakai）H. Hara var. aleutica Fernald = Prunella vulgaris L. var. aleutica Fernald ■☆

316132　Prunella vulgaris L. subsp. asiatica（Nakai）H. Hara var. lilacina Nakai = Prunella vulgaris L. var. lilacina Nakai ■☆

316133　Prunella vulgaris L. subsp. asiatica（Nakai）H. Hara var. taiwaniana T. Yamaz.;台湾夏枯草■

316134　Prunella vulgaris L. subsp. lanceolata（W. P. C. Barton）Hultén = Prunella vulgaris L. var. lanceolata（W. P. C. Barton）Fernald ■

316135　Prunella vulgaris L. var. albiflora Koidz. = Prunella asiatica Nakai ■

316136　Prunella vulgaris L. var. aleutica Fernald;阿留申夏枯草（阿留夏枯草）;Aleutian Selfheal ■☆

316137　Prunella vulgaris L. var. atropurpurea Fernald = Prunella vulgaris L. ■

316138　Prunella vulgaris L. var. calvescens Fernald = Prunella vulgaris L. ■

316139　Prunella vulgaris L. var. elongata Benth. = Prunella vulgaris L. var. lanceolata（W. P. C. Barton）Fernald ■

316140　Prunella vulgaris L. var. elongata Benth. = Prunella vulgaris L. ■

316141　Prunella vulgaris L. var. elongata Makino = Prunella vulgaris L. ■

316142　Prunella vulgaris L. var. grandiflora L. = Prunella grandiflora（L.）Jacq. ■

316143　Prunella vulgaris L. var. hispida（Benth.）Benth. = Prunella hispida Benth. ■

316144　Prunella vulgaris L. var. hispida Benth. = Prunella vulgaris L. ■

316145　Prunella vulgaris L. var. japonica Kudo = Prunella vulgaris L. ■

316146　Prunella vulgaris L. var. lanceolata（W. P. C. Barton）Fernald;狭叶夏枯草（剑叶夏枯草,狭夏枯草）;Lance Self-heal,Narrowleaf Selfheal ■

316147　Prunella vulgaris L. var. lanceolata（W. P. C. Barton）Fernald f. iodocalyx Fernald = Prunella vulgaris L. var. lanceolata（W. P. C. Barton）Fernald ■

316148　Prunella vulgaris L. var. leucantha Schrad.;白花夏枯草;Whiteflower Selfheal ■

316149　Prunella vulgaris L. var. leucantha Schrad. = Prunella vulgaris L. ■

316150　Prunella vulgaris L. var. lilacina Nakai;日本夏枯草■☆

316151　Prunella vulgaris L. var. lilacina Nakai = Prunella vulgaris L. subsp. asiatica（Nakai）H. Hara ■

316152 Prunella vulgaris L. var. major Batt. = Prunella vulgaris L. ■

316153 Prunella vulgaris L. var. minor Sm. = Prunella vulgaris L. ■

316154 Prunella vulgaris L. var. nahutashanense S. S. Ying;南湖大山夏枯草(高山夏枯草)■

316155 Prunella vulgaris L. var. nana Clute = Prunella vulgaris L. ■

316156 Prunella vulgaris L. var. parviflora (Gilib.) J. W. Moore = Prunella vulgaris L. ■

316157 Prunella vulgaris L. var. rouleauiana Vict. = Prunella vulgaris L. ■

316158 Prunella vulgaris L. var. vulgaris Benth. = Prunella vulgaris L. ■

316159 Prunella webbiana N. Taylor;韦布夏枯草;Self-heal ■☆

316160 Prunellopsis Kudo = Prunella L. ■

316161 Prunellopsis Kudo(1920);拟夏枯草属■☆

316162 Prunellopsis prunelliformis Kudo;拟夏枯草■☆

316163 Prunellopsis prunelliformis Kudo = Prunella prunelliformis (Maxim.) Makino ■☆

316164 Prunophora Neck. = Prunus L. ●

316165 Prunopsis André = Louiseania Carrière ●

316166 Prunus L. (1753);李属(梅属,樱桃属,樱属);Apricot, Bird Cherry, Cherry, Cherry Plum, Laurel Cherry, Peach, Plum, Prune ●

316167 Prunus Mill. = Prunus L. ●

316168 Prunus Ser. = Prunus L. ●

316169 Prunus 'Snow fountain';雪泉李;Weeping Cherry ●☆

316170 Prunus × affinis Makino = Prunus furuseana Ohwi ●

316171 Prunus × bukosanensis Moriya = Cerasus × yuyamae (Sugim.) H. Ohba nothovar. bukosanensis (Moriya) H. Ohba ●☆

316172 Prunus × chichibuensis Kubota et Moriya = Cerasus × chichibuensis (Kubota et Moriya) H. Ohba ●☆

316173 Prunus × chichibuensis Kubota et Moriya nothovar. aizuensis Kawas. = Cerasus × chichibuensis (Kubota et Moriya) H. Ohba nothovar. aizuensis (Kawas.) Yonek. ●☆

316174 Prunus × chichibuensis Kubota et Moriya var. uyekii Kubota = Cerasus × chichibuensis (Kubota et Moriya) H. Ohba var. uyekii (Kubota) H. Ohba ●☆

316175 Prunus × cistena ?;紫叶沙梨;Purple Leaf Sand Cherry, Purple-leaf Sand Cherry, Purple-leaved Sand Cherry ●☆

316176 Prunus × gondouini (Poit. et Turpin) Rehder = Cerasus × gondouini Poit. et Turpin ●☆

316177 Prunus × juddii E. S. Anderson = Cerasus × juddii (E. S. Anderson) H. Ohba ●☆

316178 Prunus × kubotana Kawas. = Cerasus × kubotana (Kawas.) H. Ohba ●☆

316179 Prunus × mitsuminensis Moriya = Cerasus × mitsuminensis (Moriya) H. Ohba ●☆

316180 Prunus × miyasakana H. Kubota = Cerasus × miyasakana (H. Kubota) H. Ohba ●☆

316181 Prunus × miyoshii Ohwi = Cerasus × miyoshii (Ohwi) H. Ohba ●☆

316182 Prunus × mochizukiana Nakai = Cerasus × mochizukiana (Nakai) H. Ohba ●☆

316183 Prunus × moniwana Kawas. = Cerasus × moniwana (Kawas.) Yonek. ●☆

316184 Prunus × oneyamensis Hayashi = Cerasus × oneyamensis (Hayashi) H. Ohba ●☆

316185 Prunus × pseudaffinis Kawas. = Cerasus × furuseana (Ohwi) H. Ohba nothovar. pseudaffinis (Kawas.) H. Ohba ●☆

316186 Prunus × pseudoverecunda H. Kubota et Funatsu = Cerasus × tschoniskii (Maxim.) H. Ohba nothovar. pseudoverecunda (H. Kubota et Moriya) H. Ohba ●☆

316187 Prunus × sacra Miyoshi = Cerasus × sacra (Miyoshi) H. Ohba ●☆

316188 Prunus × sakabae Makino;坂场李●☆

316189 Prunus × schmittii Rehder;亮皮樱桃●☆

316190 Prunus × syodoi Nakai = Cerasus × syodoi (Nakai) H. Ohba ●☆

316191 Prunus × takasawana Kubota et Funatsu = Cerasus × oneyamensis (Hayashi) H. Ohba nothovar. takasawana (Kubota et Funatsu) H. Ohba ●☆

316192 Prunus × takinoensis Kawas. = Cerasus × takinoensis (Kawas.) Yonek. ●☆

316193 Prunus × tschonoskii Koehne = Cerasus × tschonoskii (Koehne) H. Ohba ●☆

316194 Prunus × yanashimana H. Kubota et Moriya = Cerasus × yanashimana (H. Kubota et Moriya) H. Ohba ●☆

316195 Prunus × yuyamae Sugim. = Cerasus × yuyamae (Sugim.) H. Ohba ●☆

316196 Prunus acuminata (Wall.) D. Dietr. = Laurocerasus undulata (Buch. -Ham. ex D. Don) M. Roem. ●

316197 Prunus acuminata (Wall.) D. Dietr. f. elongata Koehne = Laurocerasus undulata (Buch. -Ham. ex D. Don) M. Roem. ●

316198 Prunus acuminata (Wall.) D. Dietr. f. microbotrys (Koehne) Koehne = Laurocerasus undulata (Buch. -Ham. ex D. Don) M. Roem. ●

316199 Prunus acuminata (Wall.) Dietr. = Laurocerasus undulata (D. Don) M. Roem. ●

316200 Prunus acuminata Hook. f. f. elongata Koehne = Laurocerasus undulata (D. Don) M. Roem. f. elongata (Koehne) Te T. Yu et L. T. Lu ●

316201 Prunus acuminata Hook. f. f. microbotrys Koehne = Laurocerasus undulata (D. Don) M. Roem. f. microbotrys (Koehne) Te T. Yu et L. T. Lu ●

316202 Prunus acuminata Hook. f. f. microbotrys Koehne = Prunus microbotrys Koehne ●

316203 Prunus africana (Hook. f.) Kalkman;非洲李;African Cherry, Red Stinkwood ●☆

316204 Prunus alleghaniensis Porter;阿根廷李(艾利盖尼李);Alleghany Plum, Alleghany Sloe, Sloe Plum ●☆

316205 Prunus alleghaniensis Porter var. davisii W. Wight;达维氏阿根廷李(达维艾利盖尼李);Davis Alleghany Plum ●☆

316206 Prunus americana Marshall;美国李(美国刺李,美洲李);American Plum, American Red Plum, American Wild Plum, August Plum, Canada Plum, Goose Plum, Hog Plum, Horse Plum, Prairie Plum, Red Plum, River Plum, Sloe, Thorn Plum, Wild Plum, Yellow Plum ●

316207 Prunus americana Marshall var. lanata Sudw. = Prunus americana Marshall ●

316208 Prunus americana Marshall var. lanata Sudw. = Prunus mexicana S. Watson ●☆

316209 Prunus americana Marshall var. nigra (Aiton) Waugh = Armeniaca dasycarpa (Ehrh.) Borkh. ●

316210 Prunus amygdalo-persica Rehder;扁桃叶李(杏桃);Flowering Almond ●☆

316211 Prunus amygdalo-persica Rehder 'Pollardii';波拉里杏桃●☆

316212 Prunus amygdalus Batsch = Amygdalus communis L. ●

316213 Prunus amygdalus Batsch = Prunus dulcis (Mill.) D. A. Webb ●

316214 Prunus amygdalus Batsch var. amara (DC.) Focke =

Amygdalus communis L. var. amara Ludw. ex DC. ●

316215　Prunus amygdalus Batsch var. fragilis（Borkh.）Focke ＝ Amygdalus communis L. var. fragilis（Borkh.）Ser. ●

316216　Prunus amygdalus Batsch var. satina Focke ＝ Amygdalus communis L. var. dulcis Borkh. ex DC. ●

316217　Prunus amygdalus Stokes ＝ Amygdalus communis L. ●

316218　Prunus andersonii Hook. f. ＝ Laurocerasus andersonii（Hook. f.）Te T. Yu et L. T. Lu ●

316219　Prunus angustifolia Marshall；狭叶李（奇克索李）；Chickasaw Plum，Mountain Cherry，Sand Plum ●☆

316220　Prunus angustifolia Marshall var. watsonii Waugh；沙李；Chickasaw Plum，Sand Chickasaw Plum ●☆

316221　Prunus ansu（Maxim.）Kom. ＝ Amygdalus communis L. var. ansu Ludw. ex DC. ●

316222　Prunus ansu（Maxim.）Kom. ＝ Armeniaca vulgaris Lam. var. ansu（Maxim.）Te T. Yu et L. T. Lu ●

316223　Prunus ansu（Maxim.）Kom. ＝ Prunus armeniaca L. var. ansu Maxim. ●

316224　Prunus ansu Kom. ＝ Armeniaca vulgaris Lam. var. ansu（Maxim.）Te T. Yu et L. T. Lu ●

316225　Prunus anzygdalus ? ＝ Prunus dulcis（Mill.）D. A. Webb ●

316226　Prunus apetala（Siebold et Zucc.）Franch. et Sav. ＝ Cerasus apetala（Siebold et Zucc.）Ohle ex H. Ohba ●☆

316227　Prunus apetala（Siebold et Zucc.）Franch. et Sav. subsp. pilosa（Koidz.）H. Ohba ＝ Cerasus apetala（Siebold et Zucc.）Ohle ex H. Ohba var. pilosa（Koidz.）H. Ohba ●☆

316228　Prunus apetala（Siebold et Zucc.）Franch. et Sav. var. pilosa（Koidz.）E. H. Wilson ＝ Cerasus apetala（Siebold et Zucc.）Ohle ex H. Ohba var. pilosa（Koidz.）H. Ohba ●☆

316229　Prunus apetala（Siebold et Zucc.）Franch. et Sav. var. pilosa（Koidz.）E. H. Wilson f. multipetala Kawas. ＝ Cerasus apetala（Siebold et Zucc.）Ohle ex H. Ohba var. pilosa（Koidz.）H. Ohba f. multipetala（Kawas.）H. Ohba ●☆

316230　Prunus armeniaca L. ；亚美尼亚杏；Apricot ●☆

316231　Prunus armeniaca L. ＝ Armeniaca vulgaris Lam. var. ansu（Maxim.）Te T. Yu et L. T. Lu ●

316232　Prunus armeniaca L. ＝ Armeniaca vulgaris Lam. ●

316233　Prunus armeniaca L. var. ansu Maxim. ＝ Armeniaca vulgaris Lam. var. ansu（Maxim.）Te T. Yu et L. T. Lu ●

316234　Prunus armeniaca L. var. dasycarpa（Ehrh.）K. Koch ＝ Armeniaca dasycarpa（Ehrh.）Borkh. ●

316235　Prunus armeniaca L. var. dasycarpa K. Koch ＝ Armeniaca dasycarpa（Ehrh.）Borkh. ●

316236　Prunus armeniaca L. var. holosericea Batalin ＝ Armeniaca holosericea（Batalin）Kostina ●◇

316237　Prunus armeniaca L. var. mandshurica Maxim. ＝ Armeniaca mandshurica（Maxim.）Skvortsov ●

316238　Prunus armeniaca L. var. mandshurica Maxim. ＝ Prunus mandshurica（Maxim.）Koehne ●

316239　Prunus armeniaca L. var. sibirica（L.）K. Koch ＝ Armeniaca sibirica（L.）Lam. ●

316240　Prunus armeniaca L. var. toypica Maxim. ＝ Armeniaca vulgaris Lam. ●

316241　Prunus aspera Thunb. ＝ Aphananthe aspera（Thunb. ex A. Murray）Planch. ●

316242　Prunus aspera Thunb. ex A. Murray ＝ Aphananthe aspera（Thunb. ex A. Murray）Planch. ●

316243　Prunus avium（L.）L. ‘Asplenifolia’；铁角蕨叶尖尾樱桃●☆

316244　Prunus avium（L.）L. ‘Cavalier’；骑士尖尾樱桃●☆

316245　Prunus avium（L.）L. ‘Multiplex’ ＝ Prunus avium L. ‘Plena’●☆

316246　Prunus avium（L.）L. ‘Pendula’；垂枝尖尾樱桃●☆

316247　Prunus avium（L.）L. ‘Plena’；重瓣尖尾樱桃（重瓣欧洲甜樱桃）；Double Gean，Double-fowered Cherry ●☆

316248　Prunus avium（L.）L. ‘Rubrifolia’；红叶尖尾樱桃●☆

316249　Prunus avium（L.）L. ＝ Cerasus avium（L.）Moench ●

316250　Prunus avium（L.）L. var. duracina L. ＝ Prunus avium（L.）L. ●

316251　Prunus avium（L.）L. var. juliana L. ＝ Prunus avium（L.）L. ●

316252　Prunus avium（L.）L. var. sylvestris（Kirschl.）Mart. et ＝ Prunus avium（L.）L. ●

316253　Prunus avium（L.）L. var. tazekkensis Sauvage ＝ Prunus avium（L.）L. ●

316254　Prunus avium L. ＝ Cerasus avium（L.）Moench ●

316255　Prunus balansae Koehne ＝ Laurocerasus fordiana（Dunn）Browicz ●

316256　Prunus balansae Koehne ＝ Prunus fordiana Dunn ●

316257　Prunus balfourii Cardot ＝ Laurocerasus spinulosa（Siebold et Zucc.）C. K. Schneid. ●

316258　Prunus balfourii Cardot ＝ Prunus spinulosa Siebold et Zucc. ●

316259　Prunus batalinii（C. K. Schneid.）Koehne ＝ Cerasus tomentosa（Thunb.）Wall. ●

316260　Prunus besseyi L. H. Bailey ‘Black Beaty’；黑美人沙樱桃●☆

316261　Prunus besseyi L. H. Bailey ‘Hansen’；汉森沙樱桃●☆

316262　Prunus besseyi L. H. Bailey ＝ Cerasus besseyi（L. H. Bailey）Smyth ●☆

316263　Prunus bicolor Koehne；双色稠李●

316264　Prunus botan André ＝ Prunus salicina Lindl. ●

316265　Prunus brachypoda Batalin ＝ Padus brachypoda（Batalin）C. K. Schneid. ●

316266　Prunus brachypoda Batalin var. eglandulosa W. C. Cheng；无腺橉木（山莓梨）●

316267　Prunus brachypoda Batalin var. eglandulosa W. C. Cheng ＝ Padus brachypoda（Batalin）C. K. Schneid. ●

316268　Prunus brachypoda Batalin var. hwasiensis Te T. Yu et C. L. Li；褐毛短柄稠李；Brownhair Bird Cherry ●

316269　Prunus brachypoda Batalin var. microdonta Koehne ＝ Padus brachypoda（Batalin）C. K. Schneid. var. microdonta（Koehne）Te T. Yu et T. C. Ku ●

316270　Prunus brachypoda Batalin var. pseudossiori Koehne ＝ Padus brachypoda（Batalin）C. K. Schneid. ●

316271　Prunus brigantina Nyman ＝ Prunus brigantiaca Vill. ●☆

316272　Prunus brigantina Vill. ＝ Armeniaca brigantiaca Pers. ●☆

316273　Prunus buergeriana Miq. ＝ Padus buergeriana（Miq.）Te T. Yu et T. C. Ku ●

316274　Prunus buergeriana Miq. var. nudiuscula Koehne ＝ Padus buergeriana（Miq.）Te T. Yu et T. C. Ku ●

316275　Prunus buergeriana Miq. var. nudiuscula Koehne ＝ Prunus buergeriana Miq. ●

316276　Prunus buergeriana Miq. var. stellipila（Koehne）Te T. Yu et C. L. Li ＝ Padus stellipila（Koehne）Te T. Yu et T. C. Ku ●

316277　Prunus campanulata Maxim. ＝ Cerasus campanulata（Maxim.）A. V. Vassil. ●

316278　Prunus canescens Vilm. et Bois ＝ Cerasus canescens（Bois）S. Y. Sokolov ●

316279 Prunus cantabrigiensis Stapf;中国酸樱桃;Chinese Sour Cherry ●☆

316280 Prunus capollin Zucc. ;墨西哥卡李;Capulin ●☆

316281 Prunus capuli Cav. ex Spreng. ;卡帕李●☆

316282 Prunus carcharis Koehne;南川樱桃●

316283 Prunus carmesina H. Hara ＝ Cerasus cerasoides （D. Don）Sokoloff var. rubea （Ingram）Te T. Yu et C. L. Li ●

316284 Prunus carmesina H. Hara ＝ Cerasus cerasoides （D. Don）Sokoloff ●

316285 Prunus caroliniana （Mill.）Aiton;卡罗里纳李（卡罗来纳樱桃,野橙）;American Cherry Laurel,Carolina Cherry,Carolina Cherry Laurel, Carolina Cherrylaurel, Carolina Laurel Cherry, Carolina Laurelcherry, Carolina Laurel-cherry, Cherry Laurel, Laurel Cherry, Wild Orande,Wild Peach ●☆

316286 Prunus caudata （Hance）Koidz. ＝ Cerasus pogonostyla （Maxim.）Te T. Yu et C. L. Li ●

316287 Prunus caudata （Hance）Koidz. var. globosa （Koehne）F. P. Metcalf ＝ Cerasus pogonostyla （Maxim.）Te T. Yu et C. L. Li ●

316288 Prunus caudata （Hance）Koidz. var. obovata （Koehne）F. P. Metcalf ＝ Cerasus pogonostyla （Maxim.）Te T. Yu et C. L. Li var. obovata （Koehne）Te T. Yu et C. L. Li ●

316289 Prunus caudata Franch. ＝ Cerasus caudata （Franch.）Te T. Yu et C. L. Li ●

316290 Prunus caudata Franch. ＝ Cerasus pogonostyla （Maxim.）Te T. Yu et C. L. Li ●

316291 Prunus caudata Franch. var. globosa （Koehne）Metcalf ＝ Cerasus pogonostyla （Maxim.）Te T. Yu et C. L. Li ●

316292 Prunus caudata Franch. var. obovata （Koehne）Metcalf ＝ Cerasus pogonostyla （Maxim.）Te T. Yu et C. L. Li var. obovata （Koehne）Te T. Yu et C. L. Li ●

316293 Prunus caudata Franch. var. obovata （Koehne）Metcalf ＝ Prunus pogonostyla Maxim. var. obovata Koehne ●

316294 Prunus cerasifera Ehrh. ;樱桃李（樱李）;Balkan Plum,Cherry Plum,Cherry-plum,Flowering Plum,Mirabelle,Myrobalan,Myrobalan Plum,Myrobalans,Purple Leaf Plum,Roblum,White Sprite Cherry-plum,Wild Myrobalan Plum ●◇

316295 Prunus cerasifera Ehrh. 'Hessei';哈塞樱桃李（哈塞樱李）●☆

316296 Prunus cerasifera Ehrh. 'Lindsayae';林德萨亚樱桃李（林德萨亚樱李）●☆

316297 Prunus cerasifera Ehrh. 'Newport';纽波特樱桃李（纽波特樱李）●☆

316298 Prunus cerasifera Ehrh. 'Nigra';黑叶樱桃李（黑叶樱李）●☆

316299 Prunus cerasifera Ehrh. 'Pendula';垂枝樱桃李（垂枝樱李）●☆

316300 Prunus cerasifera Ehrh. 'Pissardii';皮萨尔迪樱桃李（紫叶樱桃李）●☆

316301 Prunus cerasifera Ehrh. 'Thundercloud';雷云樱桃李（雷云樱李）●☆

316302 Prunus cerasifera Ehrh. f. atropurpurea （Jacq.）Rehder;紫叶樱桃李（紫叶李）;Flowering Plum, Purple Cherry Plum, Purpleleaf Cherry Plum,Purple-leaved Plum ●

316303 Prunus cerasifera Ehrh. subsp. divaricata （Ledeb.）C. K. Schneid. ＝ Prunus cerasifera Ehrh. ●◇

316304 Prunus cerasifera Ehrh. subsp. myrobalana （L.）C. K. Schneid. ＝ Prunus cerasifera Ehrh. ●◇

316305 Prunus cerasifera Ehrh. subsp. myrobalana C. K. Schneid. ＝ Prunus cerasifera Ehrh. ●◇

316306 Prunus cerasifera Ehrh. var. campanulata （Maxim.）Koidz. ＝ Cerasus campanulata （Maxim.）Te T. Yu et C. L. Li ●

316307 Prunus cerasifera Ehrh. var. campanulata （Maxim.）Koidz. ＝ Prunus campanulata Maxim. ●

316308 Prunus cerasifera Ehrh. var. stropurpurea Jacq. ;红叶樱桃李（红叶李）;Redleaf Cherry Plum ●

316309 Prunus cerasoides Buch. -Ham. ex D. Don ＝ Cerasus cerasoides （D. Don）Sokoloff ●

316310 Prunus cerasoides Buch. -Ham. ex D. Don var. campanulata （Maxim.）Koidz. ＝ Cerasus campanulata （Maxim.）A. V. Vassil. ●

316311 Prunus cerasoides Buch. -Ham. ex D. Don var. majestica （Koehne）Ingram ＝ Cerasus cerasoides （D. Don）Sokoloff ●

316312 Prunus cerasoides Buch. -Ham. ex D. Don var. rubea Ingram ＝ Cerasus cerasoides （D. Don）Sokoloff ●

316313 Prunus cerasoides Buch. -Ham. ex D. Don var. tibetica （Batalin）C. K. Schneid. ＝ Cerasus serrula （Franch.）Te T. Yu et C. L. Li ●

316314 Prunus cerasoides D. Don ＝ Cerasus cerasoides （D. Don）Sokoloff ●

316315 Prunus cerasoides D. Don ＝ Prunus cerasoides Buch. -Ham. ex D. Don ●

316316 Prunus cerasoides D. Don var. campanulata （Maxim.）Koidz. ＝ Cerasus campanulata （Maxim.）A. V. Vassil. ●

316317 Prunus cerasoides Ehrh. var. rubea Ingram ＝ Prunus cerasoides Buch. -Ham. ex D. Don var. rubea Ingram ●

316318 Prunus cerasus L. 'Marasca';马拉斯基樱桃;Marasca Sour Cherry,Maraschino Cherry,Maraschino Marasco,Morello ●☆

316319 Prunus cerasus L. ＝ Cerasus vulgaris Mill. ●

316320 Prunus cerasus L. var. austera L. ;涩欧樱;Morello ●☆

316321 Prunus cerasus L. var. avium L. ＝ Cerasus avium （L.）Moench ●

316322 Prunus cerasus L. var. marasca ? ＝ Prunus cerasus L. 'Marasca' ●☆

316323 Prunus cerasus L. var. subhirtella ? ＝ Cerasus subhirtella （Miq.）Sokoloff ●

316324 Prunus cerasus L. var. subhirtella ? ＝ Prunus subhirtella Miq. ●

316325 Prunus cerasus L. var. typica C. K. Schneid. ＝ Cerasus vulgaris Mill. ●

316326 Prunus ceylanica Miq. ＝ Pygeum zeylanicum Gaertn. ●

316327 Prunus chamaecerasus Jacq. ＝ Cerasus fruticosa （Pall.）Woronow ●

316328 Prunus chamaecerasus Jacq. ＝ Prunus fruticosa Pall. ●

316329 Prunus chicasa Michx. ＝ Prunus angustifolia Marshall ●☆

316330 Prunus chikusiensis Koidz. ＝ Cerasus jamasakura （Siebold ex Koidz.）H. Ohba var. chikusiensis （Koidz.）H. Ohba ●☆

316331 Prunus cinerascens Franch. ＝ Cerasus tomentosa （Thunb.）Wall. ●

316332 Prunus clarofolia C. K. Schneid. ＝ Cerasus clarofolia （C. K. Schneid.）Te T. Yu et C. L. Li ●

316333 Prunus claudiana Poir. ;克劳迪樱桃;Greengage,Reine Claude ●☆

316334 Prunus claviculata Te T. Yu et C. L. Li;长腺樱;Claviculate Cherry ●

316335 Prunus cocomilia Ten. ;椰粟李●☆

316336 Prunus communis （L.）Arcang. ＝ Prunus dulcis （Mill.）D. A. Webb ●

316337 Prunus communis （L.）Fritsch ＝ Amygdalus communis L. ●

316338 Prunus communis （L.）Fritsch f. amara Schneid. ＝ Amygdalus communis L. var. amara Ludw. ex DC. ●

316339 Prunus communis （L.）Fritsch f. dulcis Schneid. ＝ Amygdalus communis L. var. dulcis Borkh. ex DC. ●

316340 Prunus communis （L.）Fritsch f. fragilis Arcang. ＝ Amygdalus communis L. var. fragilis (Borkh.) Ser. ●

316341　Prunus communis (L.) Fritsch f. sativa Asch. et Graebn. = Amygdalus communis L. var. dulcis Borkh. ex DC. ●

316342　Prunus communis Huds. = Prunus domestica L. ●

316343　Prunus communis Huds. = Prunus salicina Lindl. ●

316344　Prunus compressa P. Beauv. = Amygdalus vulgaris Mill. var. compressa Loudon ●

316345　Prunus compta (Koidz.) Tatew. = Cerasus × compta (Koidz.) H. Ohba ●☆

316346　Prunus conadenia Koehne;锥腺樱(锥腺樱桃);Subulate Cherry, Subulate Gland Cherry ●

316347　Prunus conadenia Koehne = Cerasus conadenia (Koehne) Te T. Yu et C. L. Li ●

316348　Prunus conradinae Koehne = Cerasus conradinae (Koehne) Te T. Yu et C. L. Li ●

316349　Prunus cornuta (Wall. ex Royle) Steud. = Padus cornuta (Wall. ex Royle) Carrière ●

316350　Prunus cornuta (Wall. ex Royle) Steud. f. villosa (Hara) Hara;毛华中樱桃●☆

316351　Prunus cornuta (Wall. ex Royle) Steud. var. integrifolia Te T. Yu = Padus integrifolia Te T. Yu et T. C. Ku ●

316352　Prunus cornuta (Wall. ex Royle) Steud. var. villosa Hara = Prunus cornuta (Wall. ex Royle) Steud. f. villosa (Hara) Hara ●☆

316353　Prunus crassifolia (Hauman) Kalkman;厚叶樱桃●☆

316354　Prunus crataegifolia Hand. -Mazz. = Cerasus crataegifolia (Hand. -Mazz.) Te T. Yu et C. L. Li ●

316355　Prunus cuneata Raf. = Prunus pumila L. var. susquehanae (Willd.) H. Jaeger ●☆

316356　Prunus curdica ?;亚美尼亚樱桃;Armenian Plum ●☆

316357　Prunus cyclamina Koehne = Cerasus cyclamina (Koehne) Te T. Yu et C. L. Li ●

316358　Prunus cyclamina Koehne var. biflora Koehne = Cerasus cyclamina (Koehne) Te T. Yu et C. L. Li var. biflora (Koehne) Te T. Yu et C. L. Li ●

316359　Prunus damascena Ehrh. = Prunus domestica L. ●

316360　Prunus dasycarpa Ehrh. = Armeniaca dasycarpa (Ehrh.) Borkh. ●

316361　Prunus davidiana (Carrière) Franch. = Amygdalus davidiana (Carrière) de Voss ex Henry ●

316362　Prunus davidiana (Carrière) Franch. var. potaninii Rehder = Amygdalus davidiana (Carrière) de Voss ex Henry var. potaninii (Batalin) Te T. Yu et A. M. Lu ●

316363　Prunus dehiscens Koehne = Amygdalus tangutica (Batalin) Korsh. ●

316364　Prunus dehiscens Koehne = Prunus tangutica (Batalin) Koehne ●

316365　Prunus demissa (Nutt.) D. Dietr. = Cerasus demissa Nutt. ●☆

316366　Prunus demissa D. Dietr. = Cerasus demissa Nutt. ●☆

316367　Prunus depressa Lieg. = Prunus domestica L. ●

316368　Prunus depressa Pursh = Prunus pumila L. var. depressa (Pursh) Bean ●☆

316369　Prunus depressa Pursh = Prunus pumila L. ●☆

316370　Prunus dictyoneura Diels = Cerasus dictyoneura (Diels) Te T. Yu ●

316371　Prunus dielsiana C. K. Schneid. = Cerasus dielsiana (C. K. Schneid.) Te T. Yu et C. L. Li ●

316372　Prunus dielsiana C. K. Schneid. var. abbreviata Cardot = Cerasus dielsiana (C. K. Schneid.) Te T. Yu et C. L. Li var. abbreviana (Cardot) Te T. Yu et C. L. Li ●

316373　Prunus dielsiana C. K. Schneid. var. conferta Koehne = Cerasus

dielsiana (C. K. Schneid.) Te T. Yu et C. L. Li ●

316374　Prunus dielsiana C. K. Schneid. var. conferta Koehne = Prunus dielsiana C. K. Schneid. ●

316375　Prunus dielsiana C. K. Schneid. var. laxa Koehne = Cerasus dielsiana (C. K. Schneid.) Te T. Yu et C. L. Li ●

316376　Prunus dielsiana C. K. Schneid. var. laxa Koehne = Prunus dielsiana C. K. Schneid. ●

316377　Prunus discadenia Koehne = Cerasus discadenia (Koehne) S. Y. Jiang et C. L. Li ●

316378　Prunus discadenia Koehne = Cerasus szechuanica (Batalin) Te T. Yu et C. L. Li ●

316379　Prunus discadenia Koehne = Prunus szechuanica Batalin ●

316380　Prunus divaricata Ledeb. = Prunus cerasifera Ehrh. ●◇

316381　Prunus dolichadenia Cardot = Cerasus dolichadenia (Cardot) S. Y. Jiang et C. L. Li ●

316382　Prunus domestica L. ;欧洲李(西洋李,杏梅,洋李);Ballam, Bellum, Bullace Plum, Common Plum, Cultivated Plum, Damson Plum, Europe Plum, European Plum, Garden Plum, Plum, Prune Tree, Quetschen Plum, Quetscheplum, Wild Plum ●

316383　Prunus domestica L. ' Angelina Burdett ';安吉利那·布尔德特欧洲李●☆

316384　Prunus domestica L. ' Buhlerfruhwetsch ';布勒尔福鲁万奇欧洲李●☆

316385　Prunus domestica L. ' Coe's Golden Drop ';落金欧洲李●☆

316386　Prunus domestica L. ' Greengage ';绿盖奇欧洲李●☆

316387　Prunus domestica L. ' Mount Royal ';蒙罗亚尔欧洲李●☆

316388　Prunus domestica L. ' President ';总统欧洲李●☆

316389　Prunus domestica L. = Prunus salicina Lindl. ●

316390　Prunus domestica L. subsp. insititia (L.) C. K. Schneid. = Prunus insititia L. ●

316391　Prunus domestica L. subsp. italica (Borkh.) Gams ex Hegi;意大利李;Greengage, Reine Claude ●☆

316392　Prunus domestica L. subsp. oeconomica (Borkh.) C. K. Schneid. = Prunus domestica L. ●

316393　Prunus domestica L. subsp. oeconomica Schneid. = Prunus domestica L. ●

316394　Prunus domestica L. subsp. syriaca (Borkh.) Janch. ex Mansf. ;叙利亚李;Mirabelle, Mirabelle Plum ●☆

316395　Prunus domestica L. var. damascena L. = Prunus domestica L. ●

316396　Prunus domestica L. var. insititia (L.) B. Boivin = Prunus insititia L. ●

316397　Prunus domestica L. var. insititia (L.) Fiori et Paol. = Prunus insititia L. ●

316398　Prunus domestica L. var. myrobalana L. = Prunus cerasifera Ehrh. ●◇

316399　Prunus domestica L. var. spinosa (L.) Kuntze = Prunus spinosa L. ●

316400　Prunus duclouxii Koehne;西南樱(西南樱桃);Ducloux Cherry ●

316401　Prunus duclouxii Koehne = Cerasus duclouxii (Koehne) Te T. Yu et C. L. Li ●

316402　Prunus duclouxii Koehne = Cerasus yunnanensis (Franch.) Te T. Yu et C. L. Li ●

316403　Prunus dulcis (Mill.) D. A. Webb ' Alba Plena ';重瓣白扁桃(重瓣白甜杏)●☆

316404　Prunus dulcis (Mill.) D. A. Webb ' Macrocarpa ';大果扁桃(大果甜杏)●☆

316405　Prunus dulcis (Mill.) D. A. Webb ' Roseoplena ';重瓣玫瑰扁

桃（粉重瓣扁桃，重瓣玫瑰甜杏）；Double Almond ●☆

316406 Prunus dulcis（Mill.）D. A. Webb = Amygdalus communis L. ●

316407 Prunus dulcis（Mill.）D. A. Webb = Prunus amygdalus Batsch ●

316408 Prunus dulcis（Mill.）D. A. Webb var. amara（DC.）Buchheim = Amygdalus communis L. var. amara Ludw. ex DC. ●

316409 Prunus dunniana H. Lév. = Padus wilsonii C. K. Schneid. ●

316410 Prunus elliptica Thunb. = Elaeocarpus sylvestris（Lour.）Poir. ●

316411 Prunus emarginata（Douglas ex Hook.）Walp.；苦樱；Bitter Cherry，Quinine Cherry，Wild Cherry ●☆

316412 Prunus eriogyna Mason；毛蕊李；Desert Apricot ●☆

316413 Prunus fasciculata A. Gray；簇生李；Desert Range Almond，Wild Almond ●☆

316414 Prunus fasciculata A. Gray subsp. punctata（Jeps.）E. Murray；斑点簇生李●☆

316415 Prunus ferganensis（Kostina et Rjabov）Y. Y. Yao = Amygdalus ferganensis（Kostina et Rjabov）Te T. Yu et L. T. Lu ●

316416 Prunus ferganica Lincz.；费尔干李●☆

316417 Prunus fordiana Dunn = Laurocerasus fordiana（Dunn）Browicz ●

316418 Prunus fordiana Dunn var. balansae（Koehne）J. E. Vidal = Laurocerasus fordiana（Dunn）Browicz ●

316419 Prunus formosana Matsum. = Cerasus pogonostyla（Maxim.）Te T. Yu et C. L. Li ●

316420 Prunus formosana Matsum. = Prunus pogonostyla Maxim. ●

316421 Prunus fremontii S. Watson = Amygdalus fremontii（S. Watson）Abrams ●☆

316422 Prunus fruticosa Pall. = Cerasus fruticosa（Pall.）Woronow ●

316423 Prunus glabra（Pamp.）Koehne = Cerasus conradinae（Koehne）Te T. Yu et C. L. Li ●

316424 Prunus glabra（Pamp.）Koehne = Cerasus glabra（Pamp.）Te T. Yu et C. L. Li ●

316425 Prunus glandulosa Thunb.'Alba Plena'；重瓣白麦李（白重瓣麦李）●☆

316426 Prunus glandulosa Thunb.'Rosea Plena' = Prunus glandulosa Thunb.'Sinensis'●☆

316427 Prunus glandulosa Thunb.'Sinensis'；中国麦李（粉重瓣麦李）●☆

316428 Prunus glandulosa Thunb. = Cerasus glandulosa（Thunb.）Loisel. ●

316429 Prunus glandulosa Thunb. var. albiplena（Koehne）Nakai = Cerasus glandulosa（Thunb.）Loisel. ●

316430 Prunus glandulosa Thunb. var. salicifolia（Kom.）Koehne = Cerasus humilis（Bunge）Sokoloff ●

316431 Prunus glandulosa Thunb. var. trichostyla Koehne = Cerasus glandulosa（Thunb.）Loisel. var. trichostyla（Koehne）J. X. Yang ●☆

316432 Prunus glyptocarya Koehne = Prunus trichostoma Koehne ●

316433 Prunus gracilifolia Koehne = Cerasus setulosa（Batalin）Te T. Yu et C. L. Li ●

316434 Prunus gracilis Engelm. et A. Gray；俄克拉何马李；Oklahoma Plum，Prairie Cherry ●☆

316435 Prunus grayana Maxim. = Padus grayana（Maxim.）C. K. Schneid. ●

316436 Prunus grisea（Blume ex C. H. Müll.）Kalkman；兰屿野樱花（布列氏野樱桃，柿叶野樱，台湾臀果木）●

316437 Prunus henryi（C. K. Schneid.）Koehne = Cerasus henryi（C. K. Schneid.）Te T. Yu et C. L. Li ●

316438 Prunus herincquina Koehne = Cerasus subhirtella（Miq.）Sokoloff ●

316439 Prunus hiberniflora Koidz.；霜月樱●☆

316440 Prunus hirtifolia Koehne = Cerasus yunnanensis（Franch.）Te T. Yu et C. L. Li ●

316441 Prunus hirtipes Hemsl. var. glabra Pamp. = Cerasus conradinae（Koehne）Te T. Yu et C. L. Li ●

316442 Prunus hirtipes Hemsl. var. glabra Pamp. = Cerasus glabra（Pamp.）Te T. Yu et C. L. Li ●

316443 Prunus hisauchiana Koidz. ex Hisauti = Cerasus × hisauchiana（Koidz. ex Hisauti）H. Ohba ●☆

316444 Prunus hortulana L. H. Bailey；好图兰李；Hortulan Plum，Hortulana Plum，Wild Goose Plum，Wild-goose Plum ●☆

316445 Prunus humilis Bunge = Cerasus humilis（Bunge）Sokoloff ●

316446 Prunus humilis Bunge var. villosula Bunge = Cerasus dictyoneura（Diels）Te T. Yu ●

316447 Prunus hypotricha Rehder = Laurocerasus hypotricha（Rehder）Te T. Yu et L. T. Lu ●

316448 Prunus hypotricha Rehder = Prunus zippeliana Miq. var. puberifolia（Koehne）Te T. Yu et C. L. Li ●

316449 Prunus hypotrichodes Cardot = Armeniaca hypotrichodes（Cardot）L. C. Li et S. Y. Jiang ●

316450 Prunus hypotrichodes Cardot = Laurocerasus hypotricha（Rehder）Te T. Yu et L. T. Lu ●

316451 Prunus ichangana C. K. Schneid. = Prunus salicina Lindl. ●

316452 Prunus ilicifolia（Hook. et Arn.）D. Dietr.；冬青叶李；Evergreen Cherry，Hollyleaf Cherry，Holly-leaf Cherry，Holly-leafed Cherry，Islay ●

316453 Prunus imanishii Kitam. = Cerasus trichantha（Koehne）S. Y. Jiang et C. L. Li ●

316454 Prunus immanishi Kitam. = Cerasus rufa Wall. var. trichantha（Koehne）Te T. Yu et C. L. Li ●

316455 Prunus incisa Thunb.'February Pink'；二月粉豆樱●☆

316456 Prunus incisa Thunb. = Cerasus incisa（Thunb.）Loisel. ●

316457 Prunus incisa Thunb. f. plenissima S. Watan. = Cerasus incisa（Thunb.）Loisel. f. chrysantha H. Ohba ●☆

316458 Prunus incisa Thunb. f. yamadae（Makino）Ohwi = Cerasus incisa（Thunb.）Loisel. f. yamadae（Makino）H. Ohba ●☆

316459 Prunus incisa Thunb. subsp. kinkiensis（Koidz.）Kitam. = Cerasus incisa（Thunb.）Loisel. var. kinkiensis（Koidz.）H. Ohba ●☆

316460 Prunus incisa Thunb. subsp. kinkiensis（Koidz.）Kitam. f. plena（Satomi）Sugim. = Cerasus incisa（Thunb.）Loisel. var. kinkiensis（Koidz.）H. Ohba f. plena（Satomi）H. Ohba ●☆

316461 Prunus incisa Thunb. var. alpina（Koidz.）Kitam. = Cerasus nipponica（Matsum.）Ohle ex H. Ohba var. alpina（Koidz.）H. Ohba ●☆

316462 Prunus incisa Thunb. var. bukosanensis（Honda）H. Hara = Cerasus incisa（Thunb.）Loisel. var. bukosanensis（Honda）H. Ohba ●☆

316463 Prunus incisa Thunb. var. globosa Kawas. = Cerasus incisa（Thunb.）Loisel. var. incisa f. globosa（Kawas.）H. Ohba ●☆

316464 Prunus incisa Thunb. var. kinkiensis（Koidz.）Ohwi = Cerasus incisa（Thunb.）Loisel. var. kinkiensis（Koidz.）H. Ohba ●☆

316465 Prunus incisa Thunb. var. shikokuensis Moriya = Cerasus shikokuensis（Moriya）H. Ohba ●☆

316466 Prunus incisa Thunb. var. tomentosa Koidz.；毛豆樱●☆

316467 Prunus incisa Thunb. var. tomentosa Koidz. = Cerasus × hisauchiana（Koidz. ex Hisauti）H. Ohba ●☆

316468 Prunus incisa Thunb. var. urceolata Koidz. = Cerasus incisa

（Thunb.）Loisel. f. urceolata（Koidz.）H. Ohba ●☆

316469　Prunus incisa Thunb. var. yamadae Makino = Cerasus incisa（Thunb.）Loisel. f. yamadae（Makino）H. Ohba ●☆

316470　Prunus insititia L. ;西洋李（欧亚野李,乌荆子李）;Bolas, Bollas, Bullace, Bullace Plum, Bullers, Bullesse, Bullies, Bullins, Bullions, Bullison, Bulloe, Bullums, Bullum-tree, Bully-blooms, Christian, Christlings, Crex, Cricks, Cricksey, Crislings, Cristens, Crystal, Custin, Damsel, Damsil, Damson, Damson Plum, Eurasian Wild Plum, European Plum Gristlings, Kerslins, Keslings, Kix, Krislings, Mirabelle, Scad, Scad-tree, Skeg, Slath, Wild Damson, Wild Plum ●

316471　Prunus insititia L. = Prunus domestica L. ●

316472　Prunus insititia L. var. italica（Borkh.）Asch. et Graebn. = Prunus insititia L. ●

316473　Prunus insititia L. var. nigra（Rchb.）Asch. et Graebn. = Prunus insititia L. ●

316474　Prunus involucrata Koehne = Cerasus pseudocerasus（Lindl.）G. Don ●

316475　Prunus italica Borkh. = Prunus domestica L. subsp. italica（Borkh.）Gams ex Hegi ●☆

316476　Prunus italica Borkh. = Prunus insititia L. ●

316477　Prunus itosakura Siebold var. taiwaniana（Hayata）Kudo et Masam. = Prunus taiwaniana Hayata ●

316478　Prunus jacquemontii（Edgew.）Hook. f. ;西藏樱桃●

316479　Prunus jacquemontii Hook. f. ;印度矮扁桃●☆

316480　Prunus jamasakura Siebold ex Koidz. = Cerasus jamasakura（Siebold ex Koidz.）H. Ohba ●☆

316481　Prunus jamasakura Siebold ex Koidz. f. pubescens（Makino）Ohwi = Cerasus jamasakura（Siebold ex Koidz.）H. Ohba f. pubescens（Makino）H. Ohba ●☆

316482　Prunus jamasakura Siebold ex Koidz. var. chikusiensis（Koidz.）Ohwi = Cerasus jamasakura（Siebold ex Koidz.）H. Ohba var. chikusiensis（Koidz.）H. Ohba ●☆

316483　Prunus japonica Thunb. = Cerasus japonica（Thunb.）Loisel. ●

316484　Prunus japonica Thunb. var. glandulosa（Thunb.）Maxim. = Cerasus glandulosa（Thunb.）Loisel. ●

316485　Prunus japonica Thunb. var. multiplex Makino;西洋李花（垂瓣郁李,多叶郁李）●

316486　Prunus japonica Thunb. var. nakaii（H. Lév.）Rehder = Cerasus japonica（Thunb.）Loisel. var. nakaii（H. Lév.）Te T. Yu et C. L. Li ●

316487　Prunus japonica Thunb. var. salicifolia Kom. = Cerasus humilis（Bunge）Sokoloff ●

316488　Prunus japonica Thunb. var. zhejiangensis Yun B. Chang;浙江郁李;Zhejiang Bushcherry ●

316489　Prunus jenkinsii（Hook. f.）Te T. Yu et A. M. Lu;坚核桂樱（阿萨姆稠李）;Assam Cherry-laurel ●

316490　Prunus jenkinsii（Hook. f.）Te T. Yu et A. M. Lu = Laurocerasus jenkinsii（Hook. f.）Te T. Yu et L. T. Lu ●

316491　Prunus jenkinsii Hook. f. et Thomson ex Hook. f. = Laurocerasus jenkinsii（Hook. f.）Te T. Yu et L. T. Lu ●

316492　Prunus kanehirai Hayata ex Hisauti = Laurocerasus zippeliana（Miq.）Te T. Yu et L. T. Lu ●

316493　Prunus kanehirai Hayata ex Hisauti = Prunus zippeliana Miq. ●

316494　Prunus kansuensis Rehder = Amygdalus kansuensis（Rehder）Skeels ●

316495　Prunus kinkiensis Koidz. = Cerasus incisa（Thunb.）Loisel.

var. kinkiensis（Koidz.）H. Ohba ●☆

316496　Prunus koshiensis Koidz. = Cerasus spachiana Lavalée ex H. Otto var. koshiensis（Koidz.）H. Ohba ●☆

316497　Prunus kurilensis（Miyabe）Miyabe ex Takeda = Cerasus nipponica（Matsum.）Ohle ex H. Ohba var. kurilensis（Miyabe）H. Ohba ●☆

316498　Prunus lannesiana（Carrière）E. H. Wilson = Cerasus lannesiana Carrière ●

316499　Prunus lannesiana（Carrière）E. H. Wilson = Cerasus serrulata（Lindl.）G. Don ex Loudon var. lannesiana（Carrière）Makino ●

316500　Prunus lannesiana（Carrière）E. H. Wilson var. speciosa（Koidz.）Makino = Cerasus speciosa（Koidz.）H. Ohba ●☆

316501　Prunus lannesiana E. H. Wilson = Cerasus serrulata（Lindl.）G. Don ex Loudon var. lannesiana（Carrière）Makino ●

316502　Prunus latidentata Koehne = Cerasus trichostoma（Koehne）Te T. Yu et C. L. Li ●

316503　Prunus latidentata Koehne = Prunus trichostoma Koehne ●

316504　Prunus latidentata Koehne var. trichostoma（Koehne）C. K. Schneid. = Cerasus trichostoma（Koehne）Te T. Yu et C. L. Li ●

316505　Prunus laurocerasus L. 'Etna';埃特纳桂樱●☆

316506　Prunus laurocerasus L. 'Otto Luyken';密枝桂樱;Otto Luken Cherry Laurel ●☆

316507　Prunus laurocerasus L. 'Schipkaensis';展枝桂樱●☆

316508　Prunus laurocerasus L. 'Zabeliana';狭叶桂樱（扎贝利安桂樱）;Zabel's Cherry Laurel ●☆

316509　Prunus laurocerasus L. = Laurocerasus officinalis M. Roem. ●

316510　Prunus laxiflora Koehne = Padus laxiflora（Koehne）T. C. Ku ●

316511　Prunus leveilleana Koehne = Cerasus serrulata（Lindl.）G. Don ex Loudon var. pubescens（Makino）Te T. Yu et C. L. Li ●

316512　Prunus limbata Cardot = Laurocerasus spinulosa（Siebold et Zucc.）C. K. Schneid. ●

316513　Prunus limbata Cardot = Prunus spinulosa Siebold et Zucc. ●

316514　Prunus litiginosa C. K. Schneid. ;湖北樱桃●

316515　Prunus litiginosa C. K. Schneid. var. abbreviata Koehne = Prunus litiginosa C. K. Schneid. ●

316516　Prunus lobulata Koehne = Cerasus trichostoma（Koehne）Te T. Yu et C. L. Li ●

316517　Prunus lobulata Koehne = Prunus trichostoma Koehne ●

316518　Prunus lusitanica L. ;葡萄牙稠李（葡萄牙桂樱）;Cherry Bay, Cherry Laurel, Cherry-laurel, Laurel, Portugal Laurel, Portuguese Cherrylaurel, Portuguese Laurel, Portuguese Laurel Cherry, Portuguese Laurel-cherry ●☆

316519　Prunus lusitanica L. 'Variegata';银边葡萄牙桂樱●☆

316520　Prunus maackii Rupr. = Padus maackii（Rupr.）Kom. ●

316521　Prunus macrodenia Koehne = Cerasus conadenia（Koehne）Te T. Yu et C. L. Li ●

316522　Prunus macrodenia Koehne = Prunus conadenia Koehne ●

316523　Prunus macrophylla Siebold et Zucc = Laurocerasus zippeliana（Miq.）Browicz ●

316524　Prunus macrophylla Siebold et Zucc. = Laurocerasus zippeliana（Miq.）Te T. Yu et L. T. Lu ●

316525　Prunus macrophylla Siebold et Zucc. = Prunus zippeliana Miq. ●

316526　Prunus macrophylla Siebold et Zucc. var. crassistyla Cardot = Laurocerasus zippeliana（Miq.）Browicz ●

316527　Prunus macrophylla Siebold et Zucc. var. crassistyla Cardot = Laurocerasus zippeliana（Miq.）Te T. Yu et L. T. Lu var. crassistyla（Cardot）Te T. Yu et L. T. Lu ●

316528　Prunus macrophylla Siebold et Zucc. var. puberifolia Koehne = Laurocerasus hypotricha (Rehder) Te T. Yu et L. T. Lu ●

316529　Prunus macrophylla Siebold et Zucc. var. sphaerocarpa Nakai;台湾黄土树●

316530　Prunus mahaleb L. 'Aurea';金叶圆叶樱桃●☆

316531　Prunus mahaleb L. 'Bommii';波米圆叶樱桃●☆

316532　Prunus mahaleb L. 'Xanthocarpa';黄果圆叶樱桃●☆

316533　Prunus mahaleb L. = Cerasus mahaleb (L.) Mill. ●

316534　Prunus mairei H. Lév. = Symplocos paniculata (Thunb.) Miq. ●

316535　Prunus majestica (Koehne) Ingram = Cerasus cerasoides (D. Don) Sokoloff ●

316536　Prunus majestica Koehne;冬花樱(大樱,冬樱花,樱桃)●

316537　Prunus majestica Koehne = Cerasus cerasoides (D. Don) Sokoloff ●

316538　Prunus mandshurica (Maxim.) Koehne = Armeniaca mandshurica (Maxim.) Skvortsov ●

316539　Prunus mandshurica (Maxim.) Koehne var. glabra Nakai = Armeniaca mandshurica (Maxim.) Skvortsov var. glabra (Nakai) Te T. Yu et L. T. Lu ●

316540　Prunus marasca Rchb. = Prunus cerasus L. 'Marasca' ●☆

316541　Prunus marginata Dunn = Laurocerasus marginata (Dunn) Te T. Yu et L. T. Lu ●

316542　Prunus marginata Dunn = Prunus spinulosa Siebold et Zucc. ●

316543　Prunus maritima Marshall;海滨李;Beach Plum,Sand Plum ●☆

316544　Prunus maritima Wangenh. 'Eastham';伊珊海滨李●☆

316545　Prunus maritima Wangenh. 'Hancock';汉考克海滨李●☆

316546　Prunus maritima Wangenh. = Prunus maritima Marshall ●☆

316547　Prunus marrupialis Kalkman = Cerasus glandulosa (Thunb.)Loisel. ●

316548　Prunus masu Koehne = Prunus salicina Lindl. ●

316549　Prunus matuurai Sasaki;太平山樱(太平山樱花);Taipingshan Plum ●

316550　Prunus maximowiczii Rupr. = Cerasus maximowiczii (Rupr.) Kom. ●

316551　Prunus media Miyoshi;蒲樱●☆

316552　Prunus melanocarpa (A. Nelson) Rydb. ;黑果李●☆

316553　Prunus mexicana S. Watson;墨西哥李;Big Tree Plum, Bigtree Plum,Big-tree Plum Inch Plum,Mexican Plum,Wild Plum ●☆

316554　Prunus mexicana S. Watson sensu Gleason et Cronquist = Prunus americana Marshall ●

316555　Prunus microbotrys Koehne = Laurocerasus undulata (Buch. -Ham.) M. Roem. ●

316556　Prunus microbotrys Koehne = Laurocerasus undulata (Buch. -Ham.) M. Roem. f. microbotrys (Koehne) Te T. Yu et L. T. Lu ●

316557　Prunus microbotrys Koehne = Prunus wallichii Steud. ●

316558　Prunus microlepis Koehne = Cerasus subhirtella (Miq.) Sokoloff ●

316559　Prunus mira Koehne = Amygdalus mira (Koehne) Te T. Yu et A. M. Lu ●

316560　Prunus mongolica Maxim. = Amygdalus mongolica (Maxim.) Ricker ●◇

316561　Prunus mugus Hand. -Mazz. = Cerasus mugus (Hand. -Mazz.) Te T. Yu et C. L. Li ●

316562　Prunus multiglandulosa Cav. = Prunus lusitanica L. ●☆

316563　Prunus multipunctata Cardot = Laurocerasus fordiana (Dunn) Browicz ●

316564　Prunus multipunctata Cardot = Prunus fordiana Dunn ●

316565　Prunus mume (Siebold) Siebold et Zucc. 'Alboplena';重瓣白梅●☆

316566　Prunus mume (Siebold) Siebold et Zucc. 'Benichidori';班尼斯道利梅(红千鸟梅)●☆

316567　Prunus mume (Siebold) Siebold et Zucc. 'Beni-shidori' = Prunus mume (Siebold) Siebold et Zucc. 'Benichido'●☆

316568　Prunus mume (Siebold) Siebold et Zucc. 'Dawn';黎明梅●☆

316569　Prunus mume (Siebold) Siebold et Zucc. 'Geisha';艺妓梅●☆

316570　Prunus mume (Siebold) Siebold et Zucc. 'Microcarpa';小果梅●

316571　Prunus mume (Siebold) Siebold et Zucc. 'Omoi-no-mama';顺心梅●☆

316572　Prunus mume (Siebold) Siebold et Zucc. 'Omoi-no-wac' = Prunus mume (Siebold) Siebold et Zucc. 'Omoi-no-mama'●☆

316573　Prunus mume (Siebold) Siebold et Zucc. 'Pendula';垂枝梅●☆

316574　Prunus mume (Siebold) Siebold et Zucc. = Armeniaca mume (Siebold et Zucc.) de Vriese ●

316575　Prunus mume (Siebold) Siebold et Zucc. f. viridicalyx (Makino) T. Y. Chen = Armeniaca mume (Siebold et Zucc.) de Vriese f. viridicalyx (Makino) T. Y. Chen ●

316576　Prunus mume (Siebold) Siebold et Zucc. var. cernua Franch. = Armeniaca mume (Siebold et Zucc.) de Vriese var. cernua (Franch.) Te T. Yu et L. T. Lu ●

316577　Prunus mume (Siebold) Siebold et Zucc. var. flvescens Makino;黄梅●☆

316578　Prunus mume (Siebold) Siebold et Zucc. var. formosana Masam. ex Kudo et Masam. = Prunus mume (Siebold) Siebold et Zucc. ●

316579　Prunus mume (Siebold) Siebold et Zucc. var. microcarpa Makino = Prunus mume (Siebold) Siebold et Zucc. 'Microcarpa' ●

316580　Prunus mume (Siebold) Siebold et Zucc. var. pallescens Franch. = Armeniaca mume (Siebold et Zucc.) de Vriese var. pallescens (Franch.) Te T. Yu et L. T. Lu ●

316581　Prunus mume (Siebold) Siebold et Zucc. var. pleiocarpa Maxim. ;品字梅●☆

316582　Prunus mume Siebold et Zucc. = Armeniaca mume (Siebold et Zucc.) de Vriese ●

316583　Prunus munsoniana W. Wight et Hedrick;芒森李(野天鹅李);Munson Plum,Wild Goose Plum,Wildgoose Plum ●☆

316584　Prunus mutabilis Miyoshi = Prunus serrulata Lindl. var. spontanea (Maxim.) E. H. Wilson ●

316585　Prunus myrobalana Loisel. = Prunus cerasifera Ehrh. ●◇

316586　Prunus nakaii H. Lév. = Cerasus japonica (Thunb.) Loisel. var. nakaii (H. Lév.) Te T. Yu et C. L. Li ●

316587　Prunus nana (L.) Stokes = Amygdalus nana L. ●

316588　Prunus napaulensis (Ser.) Steud. = Padus napaulensis (Ser.) C. K. Schneid. ●

316589　Prunus napaulensis (Ser.) Steud. var. sericea Batalin = Padus wilsonii C. K. Schneid. ●

316590　Prunus neglecta Koehne. = Cerasus henryi (C. K. Schneid.) Te T. Yu et C. L. Li ●

316591　Prunus nigra Aiton;加拿大黑李;Canada Plum ●☆

316592　Prunus nigra Aiton = Armeniaca dasycarpa (Ehrh.) Borkh. ●

316593　Prunus nikaii (Honda) Koidz. = Cerasus nikaii (Honda) H. Ohba ●☆

316594　Prunus nipponica Matsum. 'Kursar';库萨尔日本樱(库萨尔本州樱桃)●☆

316595　Prunus nipponica Matsum. 'Kursar' = Prunus nipponica Matsum. f. kurilensis (Miyabe) Hiroe ●☆

316596   Prunus nipponica Matsum. = Cerasus nipponica （Matsum.） Ohle ex H. Ohba ●☆

316597   Prunus nipponica Matsum. f. kurilensis （Miyabe） Hiroe；千岛樱桃●☆

316598   Prunus nipponica Matsum. var. alpina （Koidz.） Sugim. = Cerasus nipponica （Matsum.） Ohle ex H. Ohba var. alpina （Koidz.） H. Ohba ●☆

316599   Prunus nipponica Matsum. var. kurilensis （Miyabe） E. H. Wilson = Cerasus nipponica （Matsum.） Ohle ex H. Ohba var. kurilensis （Miyabe） H. Ohba ●☆

316600   Prunus obtusata Koehne = Padus obtusata （Koehne） Te T. Yu et T. C. Ku ●

316601   Prunus odontocalyx H. Lév. = Cerasus serrula （Franch.） Te T. Yu et C. L. Li ●

316602   Prunus odontocalyx H. Lév. = Cerasus serrulata （Lindl.） G. Don ex Loudon ●

316603   Prunus odontocalyx H. Lév. = Prunus serrula Franch. ●

316604   Prunus ohwii Kaneh. et Hatus. = Padus obtusata （Koehne） Te T. Yu et T. C. Ku ●

316605   Prunus ohwii Kaneh. et Hatus. ex Kaneh. = Padus obtusata （Koehne） Te T. Yu et T. C. Ku ●

316606   Prunus ohwii Kaneh. et Hatus. ex Kaneh. = Prunus obtusata Koehne ●

316607   Prunus oxycarpa （Hance） Maxim. = Laurocerasus zippeliana （Miq.） Browicz ●

316608   Prunus oxycarpa （Hance） Maxim. = Laurocerasus zippeliana （Miq.） Te T. Yu et L. T. Lu ●

316609   Prunus oxycarpa （Hance） Maxim. = Prunus zippeliana Miq. ●

316610   Prunus padus L. ‘Aucubifolia’；桃叶珊瑚叶稠李●☆

316611   Prunus padus L. ‘Colorata’；多彩稠李（染色稠李）●☆

316612   Prunus padus L. ‘Grandiflora’ = Prunus padus L. ‘Watereri’●☆

316613   Prunus padus L. ‘Pendula’；垂枝稠李●☆

316614   Prunus padus L. ‘Plena’；重瓣稠李●☆

316615   Prunus padus L. ‘Stricta’；直立稠李●☆

316616   Prunus padus L. ‘Watereri’；沃特尔稠李●☆

316617   Prunus padus L. = Padus avium Mill. ●

316618   Prunus padus L. = Padus cornuta （Wall. ex Royle） Carrière ●

316619   Prunus padus L. = Padus racemosa （Lam.） Gilib. ●

316620   Prunus padus L. f. pubescens （Regel） Kitag. = Padus avium Mill. ●

316621   Prunus padus L. var. japonica Miq. = Padus grayana （Maxim.） C. K. Schneid. ●

316622   Prunus padus L. var. japonica Miq. = Prunus grayana Maxim. ●

316623   Prunus padus L. var. pubescens Regel = Padus avium Mill. ●

316624   Prunus padus L. var. pubescens Regel et Tiling = Padus avium Mill. var. pubescens （Regel et Tiling） T. C. Ku et B. M. Barthol. ●

316625   Prunus padus L. var. pubescens Regel et Tiling = Padus racemosa （Lam.） Gilib. var. pubescens （Regel et Tiling） C. K. Schneid. ●

316626   Prunus padus L. var. pubescens Regel et Tiling f. purdomii Koehne = Padus racemosa （Lam.） Gilib. var. pubescens （Regel et Tiling） C. K. Schneid. ●

316627   Prunus paniculata Thunb. = Cerasus pseudocerasus （Lindl.） G. Don ●

316628   Prunus paniculata Thunb. = Symplocos paniculata （Thunb.） Miq. ●

316629   Prunus paracerasus Koehne = Cerasus yedoensis （Matsum.） A. V. Vassil. ●

316630   Prunus paracerasus Koehne = Prunus yedoensis Matsum. ●

316631   Prunus parvifolia （Matsum.） Koehne = Cerasus parvifolia （Matsum.） H. Ohba ●☆

316632   Prunus patentipila Hand. -Mazz. = Cerasus patentipila （Hand. - Mazz.） Te T. Yu et C. L. Li ●

316633   Prunus pauciflora Bunge = Cerasus pseudocerasus （Lindl.） G. Don ●

316634   Prunus pauciflora Bunge = Prunus pseudocerasus Lindl. ●

316635   Prunus pedunculata （Pall.） Maxim. = Amygdalus pedunculata Pall. ●◇

316636   Prunus pendula Maxim.；垂枝李（软条海棠）；Double Weeping Cherry，Pendulous Peach ●☆

316637   Prunus pendula Maxim. f. ascendens （Makino） Ohwi = Cerasus spachiana Lavalée ex H. Otto f. ascendens （Makino） H. Ohba ●

316638   Prunus pendula Maxim. var. koshiensis （Koidz.） Ohwi = Cerasus spachiana Lavalée ex H. Otto var. koshiensis （Koidz.） H. Ohba ●☆

316639   Prunus pensylvanica L. f. = Cerasus pensylvanica （L. f.） Loisel. ●

316640   Prunus perrulata Koehne = Prunus buergeriana Miq. ●

316641   Prunus persica （L.） Batsch ‘Alboplena’；重瓣白桃●☆

316642   Prunus persica （L.） Batsch ‘Cresthaven’；科雷塞文桃●☆

316643   Prunus persica （L.） Batsch ‘Jerseyglo’；杰斯格罗桃●☆

316644   Prunus persica （L.） Batsch ‘Klara Mey.’；克拉拉·麦伊桃 （克拉拉·迈尔桃）●☆

316645   Prunus persica （L.） Batsch ‘Nana’；矮小桃●☆

316646   Prunus persica （L.） Batsch ‘Prince Charming’；魅力王子桃●☆

316647   Prunus persica （L.） Batsch ‘Russel’s Red’；鲁塞尔红桃●☆

316648   Prunus persica （L.） Batsch ‘Texstar’；泰斯塔桃●☆

316649   Prunus persica （L.） Batsch ‘Versicolor’；杂色桃●☆

316650   Prunus persica （L.） Batsch = Amygdalus persica L. ●

316651   Prunus persica （L.） Batsch f. aganopersica （Reichard） Voss = Amygdalus persica L. var. aganopersica Reichard ●

316652   Prunus persica （L.） Batsch f. compressa （Loudon） Rehder = Amygdalus persica L. var. compressa （Loudon） Te T. Yu et A. M. Lu ●

316653   Prunus persica （L.） Batsch f. scleropersica （Reichard） Voss = Amygdalus persica L. var. scleropersica （Rchb.） Te T. Yu et L. T. Lu ●

316654   Prunus persica （L.） Batsch subsp. davidiana （Carrière） D. Rivera，Obón，S. Ríos，Selma，F. Méndez，Verde et F. Cano = Persica davidiana Carrière ●

316655   Prunus persica （L.） Batsch subsp. davidiana （Carrière） D. Rivera，Obón，S. Ríos，Selma，F. Méndez，Verde et F. Cano = Amygdalus davidiana （Carrière） de Vos ex Henry ●

316656   Prunus persica （L.） Batsch subsp. domestica （Risso） D. Rivera，Obón，S. Ríos，Selma，F. Méndez，Verde et F. Cano = Persica domestica Risso ●☆

316657   Prunus persica （L.） Batsch subsp. ferganensis Kostina et Rjabov = Amygdalus ferganensis （Kostina et Rjabov） Te T. Yu et L. T. Lu ●

316658   Prunus persica （L.） Batsch subsp. nucipersica （Suckow） Dippel = Amygdalus persica L. var. nucipersica Suckow ●☆

316659   Prunus persica （L.） Batsch subsp. platycarpa （Decne.） D. Rivera，Obón，S. Ríos，Selma，F. Méndez，Verde et F. Cano = Persica platycarpa Decne. ●

316660   Prunus persica （L.） Batsch subsp. platycarpa （Decne.） D.

Rivera, Obón, S. Ríos, Selma, F. Méndez, Verde et F. Cano ＝ Prunus persica Siebold et Zucc. var. compressa (Loudon) Bean ●

316661　Prunus persica (L.) Batsch var. aganopersica (Reichard) Voss ＝ Amygdalus persica L. var. aganopersica Reichard ●

316662　Prunus persica (L.) Batsch var. compressa Bean ＝ Amygdalus persica L. var. compressa (Loudon) Te T. Yu et A. M. Lu ●

316663　Prunus persica (L.) Batsch var. davidiana Maxim. ＝ Amygdalus davidiana (Carrière) de Vos ex Henry ●

316664　Prunus persica (L.) Batsch var. nectarina (W. T. Aiton) Maxim. ＝ Amygdalus persica L. var. nectarina W. T. Aiton ●

316665　Prunus persica (L.) Batsch var. nectarina Maxim. ＝ Prunus persica (L.) Batsch var. nucipersica C. K. Schneid. ●☆

316666　Prunus persica (L.) Batsch var. nectarina Maxim. ＝ Prunus simonii Carrière ●

316667　Prunus persica (L.) Batsch var. nucifera ?;坚果桃;Nectarine ●☆

316668　Prunus persica (L.) Batsch var. nucipersica C. K. Schneid. ＝ Amygdalus persica L. var. nucipersica Suckow ●☆

316669　Prunus persica (L.) Batsch var. nucipersica C. K. Schneid. f. aganonucipersica (Schübl. et Martel) Rehder ＝ Amygdalus persica L. var. aganopersica Reichard ●

316670　Prunus persica (L.) Batsch var. nucipersica C. K. Schneid. f. scleronucipersica (Schübl. et Martel) Rehder ＝ Amygdalus persica L. var. scleropersica (Rchb.) Te T. Yu et L. T. Lu ●

316671　Prunus persica (L.) Batsch var. platycarpa (Decne.) Bailey ＝ Amygdalus persica L. var. compressa (Loudon) Te T. Yu et A. M. Lu ●

316672　Prunus persica (L.) Batsch var. potaninii Batalin ＝ Amygdalus davidiana (Carrière) de Voss ex Henry var. potaninii (Batalin) Te T. Yu et A. M. Lu ●

316673　Prunus persica (L.) Stokes ＝ Amygdalus persica L. ●

316674　Prunus persica Siebold et Zucc. ＝ Prunus persica (L.) Batsch ●

316675　Prunus persica Siebold et Zucc. subsp. ferganensis Kostina et Rjabov ＝ Amygdalus ferganensis (Kostina et Rjabov) Te T. Yu et A. M. Lu ●

316676　Prunus persica Siebold et Zucc. var. compressa (Loudon) Bean ＝ Amygdalus persica L. var. compressa (Loudon) Te T. Yu et A. M. Lu ●

316677　Prunus persica Siebold et Zucc. var. duplex Rehder ＝ Amygdalus persica L. f. duplex Rehder ●

316678　Prunus persica Siebold et Zucc. var. nectarina (W. T. Aiton) Maxim. ＝ Amygdalus persica L. var. nectarina W. T. Aiton ●

316679　Prunus persica Siebold et Zucc. var. nucipersica (Borkh.) C. K. Schneid. f. aganonucipersica (Schübeler et M. Martens) Rehder ＝ Amygdalus persica L. var. aganonucipersica (Schübl. et M. Martens) Te T. Yu et L. T. Lu ●

316680　Prunus persica Siebold et Zucc. var. nucipersica (Borkh.) C. K. Schneid. f. scleronucipersica (Schübeler et M. Martens) Rehder ＝ Prunus persica Siebold et Zucc. var. nucipersica (Borkh.) C. K. Schneid. f. scleronucipersica (Schübeler et M. Martens) Te T. Yu et A. M. Lu ●

316681　Prunus persica Siebold et Zucc. var. nucipersica (Borkh.) C. K. Schneid. f. scleronucipersica (Schübeler et M. Martens) Te T. Yu et A. M. Lu ＝ Amygdalus persica L. var. scleronucipersica (Schübl. et M. Martens) Te T. Yu et L. T. Lu ●

316682　Prunus persica Stokes ＝ Prunus persica (L.) Batsch ●

316683　Prunus perulata Koehne ＝ Padus perulata (Koehne) Te T. Yu et T. C. Ku ●

316684　Prunus phaeosticta (Hance) Maxim. ＝ Laurocerasus phaeosticta (Hance) C. K. Schneid. ●

316685　Prunus phaeosticta (Hance) Maxim. f. ciliospinosa Chun ＝ Laurocerasus phaeosticta (Hance) C. K. Schneid. f. ciliospinosa Chun ex Te T. Yu et L. T. Lu ●

316686　Prunus phaeosticta (Hance) Maxim. f. dentigera Rehder;粗齿桂樱;Roughtooth Cherry-laurel, Rough-tooth Cherry-laurel ●

316687　Prunus phaeosticta (Hance) Maxim. f. dentigera Rehder ＝ Laurocerasus phaeosticta (Hance) C. K. Schneid. ●

316688　Prunus phaeosticta (Hance) Maxim. f. dentigera Rehder ＝ Laurocerasus phaeosticta (Hance) C. K. Schneid. f. dentigera (Rehder) Te T. Yu et L. T. Lu ●

316689　Prunus phaeosticta (Hance) Maxim. f. lasioclada Rehder ＝ Laurocerasus phaeosticta (Hance) C. K. Schneid. f. lasioclada (Rehder) Te T. Yu et L. T. Lu ●

316690　Prunus phaeosticta (Hance) Maxim. var. ancylocarpa J. E. Vidal ＝ Laurocerasus fordiana (Dunn) Browicz ●

316691　Prunus phaeosticta (Hance) Maxim. var. dimorphophylla J. E. Vidal ＝ Laurocerasus fordiana (Dunn) Browicz ●

316692　Prunus phaeosticta (Hance) Maxim. var. ilicifolia Yamam.;冬青叶桃仁●

316693　Prunus phaeosticta (Hance) Maxim. var. promeccocarpa Cardot ＝ Laurocerasus fordiana (Dunn) Browicz ●

316694　Prunus phaeosticta f. lasioclada Rehder ＝ Laurocerasus phaeosticta (Hance) C. K. Schneid. ●

316695　Prunus pilosa (Turcz.) Maxim.;戈壁桃;Gobi Peach ●

316696　Prunus pilosa (Turcz.) Maxim. ＝ Amygdalus pedunculata Pall. ●◇

316697　Prunus pilosiuscula (C. K. Schneid.) Koehne ＝ Cerasus clarofolia (C. K. Schneid.) Te T. Yu et C. L. Li ●

316698　Prunus pilosiuscula (C. K. Schneid.) Koehne ＝ Prunus clarofolia C. K. Schneid. ●

316699　Prunus pilosiuscula (C. K. Schneid.) Koehne var. barbata Koehne ＝ Prunus litiginosa C. K. Schneid. ●

316700　Prunus pilosiuscula (C. K. Schneid.) Koehne var. media Koehne ＝ Cerasus clarofolia (C. K. Schneid.) Te T. Yu et C. L. Li ●

316701　Prunus pilosiuscula (C. K. Schneid.) Koehne var. subvestita Koehne ＝ Cerasus clarofolia (C. K. Schneid.) Te T. Yu et C. L. Li ●

316702　Prunus pleiocerasus Koehne ＝ Cerasus pleiocerasus (Koehne) Te T. Yu et C. L. Li ●

316703　Prunus pleuroptera Koehne ＝ Cerasus trichostoma (Koehne) Te T. Yu et C. L. Li ●

316704　Prunus pleuroptera Koehne ＝ Prunus trichostoma Koehne ●

316705　Prunus pogonostyla Maxim.;毛柱樱(高岭梅花,毛柱郁李,庭梅);Hairystyle Cherry, Hairy-styled Cherry ●

316706　Prunus pogonostyla Maxim. ＝ Cerasus pogonostyla (Maxim.) Te T. Yu et C. L. Li ●

316707　Prunus pogonostyla Maxim. var. globosa Koehne ＝ Cerasus pogonostyla (Maxim.) Te T. Yu et C. L. Li ●

316708　Prunus pogonostyla Maxim. var. obovata Koehne ＝ Cerasus pogonostyla (Maxim.) Te T. Yu et C. L. Li var. obovata (Koehne) Te T. Yu et C. L. Li ●

316709　Prunus polytricha Koehne ＝ Cerasus polytricha (Koehne) Te T. Yu et C. L. Li ●

316710　Prunus prostrata Labill.;平卧樱;Rock Cherry ●☆

316711　Prunus prostrata Labill. var. concolor (Boiss.) Lipsky ＝ Cerasus tianschanica Pojark. ●

316712　Prunus prostrata Labill. var. discolor Raulin ＝ Prunus prostrata

Labill. ●☆

316713   Prunus prostrata Labill. var. glabrifolia Moris = Prunus prostrata Labill. ●☆

316714   Prunus prostrata Labill. var. incana Litard. et Maire = Prunus prostrata Labill. ●☆

316715   Prunus pseudocerasus Lindl. = Cerasus pseudocerasus (Lindl.) G. Don ●

316716   Prunus pseudocerasus Lindl. var. borealis Makino = Cerasus sargentii (Rehder) H. Ohba ●☆

316717   Prunus pseudocerasus Lindl. var. jamasakura (Siebold et Zucc.) Makino subvar. pubescens Makino = Cerasus serrulata (Lindl.) G. Don ex Loudon var. pubescens (Makino) Te T. Yu et C. L. Li ●

316718   Prunus pseudocerasus Lindl. var. jamasakura Makino = Cerasus jamasakura (Siebold ex Koidz.) H. Ohba ●☆

316719   Prunus pseudocerasus Lindl. var. jamasakura Makino subvar. pubescens Makino = Cerasus serrulata (Lindl.) G. Don ex Loudon var. pubescens (Makino) Te T. Yu et C. L. Li ●

316720   Prunus pubigera (C. K. Schneid.) Koehne = Padus obtusata (Koehne) Te T. Yu et T. C. Ku ●

316721   Prunus pubigera (C. K. Schneid.) Koehne var. longifolia Cardot = Padus obtusata (Koehne) Te T. Yu et T. C. Ku ●

316722   Prunus pubigera (C. K. Schneid.) Koehne var. obovata Koehne = Padus obtusata (Koehne) Te T. Yu et T. C. Ku ●

316723   Prunus pubigera (C. K. Schneid.) Koehne var. ohwii (Kaneh. et Hatus. ex Kaneh.) S. S. Ying = Padus obtusata (Koehne) Te T. Yu et T. C. Ku ●

316724   Prunus pubigera (C. K. Schneid.) Koehne var. potaninii Koehne = Padus obtusata (Koehne) Te T. Yu et T. C. Ku ●

316725   Prunus pubigera (C. K. Schneid.) Koehne var. prattii Koehne = Padus obtusata (Koehne) Te T. Yu et T. C. Ku ●

316726   Prunus pubigera Koehne = Prunus obtusata Koehne ●

316727   Prunus pubigera Koehne var. longifolia Cardot = Padus obtusata (Koehne) Te T. Yu et T. C. Ku ●

316728   Prunus pubigera Koehne var. obovata Koehne = Padus obtusata (Koehne) Te T. Yu et T. C. Ku ●

316729   Prunus pubigera Koehne var. prattii Koehne = Padus obtusata (Koehne) Te T. Yu et T. C. Ku ●

316730   Prunus puddum (Roxb. ex Ser.) Brandis = Cerasus cerasoides (D. Don) Sokoloff ●

316731   Prunus puddum Miq. = Cerasus campanulata (Maxim.) Te T. Yu et C. L. Li ●

316732   Prunus puddum Roxb. ex Brandis = Cerasus cerasoides (D. Don) Sokoloff ●

316733   Prunus puddum Roxb. ex Wall. var. tibetica Batalin = Cerasus serrula (Franch.) Te T. Yu et C. L. Li ●

316734   Prunus puddum Roxb. ex Wall. var. tibetica Batalin = Cerasus serrulata (Lindl.) G. Don ex Loudon ●

316735   Prunus pumila L. ;沙地矮樱桃(矮樱桃);Dwarf Cherry, Great Lakes Sand Cherry, Sand Cherry ●☆

316736   Prunus pumila L. subsp. susquehanae (Willd.) R. T. Clausen = Prunus pumila L. var. susquehanae (Willd.) H. Jaeger ●☆

316737   Prunus pumila L. var. besseyi (L. H. Bailey) Gleason = Prunus besseyi L. H. Bailey ●☆

316738   Prunus pumila L. var. cuneata (Raf.) L. H. Bailey = Prunus pumila L. var. susquehanae (Willd.) H. Jaeger ●☆

316739   Prunus pumila L. var. depressa (Pursh) Bean;匍匐矮樱桃(法

国李);Dwarf Cherry, Great Lakes Sand Cherry, Sand Cherry ●☆

316740   Prunus pumila L. var. susquehanae (Willd.) H. Jaeger;苏斯科维汉矮樱桃;Sand Cherry, Susquehana Sand Cherry ●☆

316741   Prunus pumila L. var. typica Groh et H. Senn = Prunus pumila L. ●☆

316742   Prunus punctata Hook. f. = Laurocerasus phaeosticta (Hance) C. K. Schneid. ●

316743   Prunus punctata Hook. f. et Thomson = Laurocerasus phaeosticta (Hance) C. K. Schneid. ●

316744   Prunus pusilliflora Cardot = Cerasus pusilliflora (Cardot) Te T. Yu et C. L. Li ●

316745   Prunus pygeoides Koehne = Laurocerasus andersonii (Hook. f.) Te T. Yu et L. T. Lu ●

316746   Prunus quelpaertensis Nakai = Cerasus jamasakura (Siebold ex Koidz.) H. Ohba f. pubescens (Makino) H. Ohba ●☆

316747   Prunus racemosa Lam. = Padus avium Mill. ●

316748   Prunus racemosa Lam. = Padus racemosa (Lam.) Gilib. ●

316749   Prunus racemosa Lam. = Prunus padus L. ●

316750   Prunus rehderiana Koehne = Prunus litiginosa C. K. Schneid. ●

316751   Prunus reverchonii Sarg. ;美国矮李;Hog Plum ●☆

316752   Prunus rufa (Wall.) Hook. f. = Cerasus rufa Wall. ●

316753   Prunus rufa (Wall.) Hook. f. var. trichantha (Koehne) Hara = Cerasus rufa Wall. var. trichantha (Koehne) Te T. Yu et C. L. Li ●

316754   Prunus rufa Hook. f. var. trichantha (Koehne) H. Hara = Cerasus trichantha (Koehne) S. Y. Jiang et C. L. Li ●

316755   Prunus rufa Steud. = Cerasus rufa Wall. ●

316756   Prunus rufoides C. K. Schneid. = Cerasus dielsiana (C. K. Schneid.) Te T. Yu et C. L. Li ●

316757   Prunus rufomicans Koehne = Padus wilsonii C. K. Schneid. ●

316758   Prunus rufomicans Koehne = Prunus napaulensis (Ser.) Steud. var. sericea Batalin ●

316759   Prunus sachalinensis (F. Schmidt) Koidz. = Prunus sargentii Rehder ●☆

316760   Prunus salicifolia Kunth;柳叶野黑樱;Capulin, Capulin Black Cherry, Capulin Cherry, Mexican Bird Cherry ●☆

316761   Prunus salicina Lindl. ;李(嘉庆子,嘉应子,李仔,李子,山李子,玉皇李,中国李);Californian Plum, China Plum, Chinese Plum, Japan Plum, Japanese Plum, Plum, Santa Rosa Plum ●

316762   Prunus salicina Lindl. 'Methley';麦斯李 ●☆

316763   Prunus salicina Lindl. 'Red Heart';红心李 ●☆

316764   Prunus salicina Lindl. var. cordata Y. He et J. Y. Zhang;柰李 ●

316765   Prunus salicina Lindl. var. mandshurica (Skvortsov) Skvortsov et A. I. Baranov = Prunus ussuriensis Kov. et Kostina ●

316766   Prunus salicina Lindl. var. mandshurica (Skvortsov) Skvortsov et A. I. Baranov = Prunus salicina Lindl. ●

316767   Prunus salicina Lindl. var. pubipes (Koehne) L. H. Bailey;毛梗李 ●

316768   Prunus sargentii Rehder 'Accolade';荣誉萨金特樱桃 ●☆

316769   Prunus sargentii Rehder = Cerasus sargentii (Rehder) H. Ohba ●☆

316770   Prunus sargentii Rehder f. albida (Miyoshi) H. Hara = Cerasus sargentii (Rehder) H. Ohba f. albida (Miyoshi) H. Ohba ●☆

316771   Prunus sargentii Rehder f. compta (Koidz.) Ohwi = Cerasus × compta (Koidz.) H. Ohba ●☆

316772   Prunus sargentii Rehder f. nagaokae Tatew. ex Koji Ito = Cerasus sargentii (Rehder) H. Ohba f. pendula (Honda) Yonek. ●☆

316773   Prunus sargentii Rehder f. pendula (Honda) Okuyama = Cerasus sargentii (Rehder) H. Ohba f. pendula (Honda) Yonek. ●☆

316774 Prunus sargentii Rehder f. pubescens（Tatew.）Ohwi＝Cerasus sargentii（Rehder）H. Ohba f. pubescens（Tatew.）H. Ohba ●☆

316775 Prunus sargentii Rehder f. pubescens（Tatew.）Ohwi＝Prunus sargentii Rehder var. pubescens（Tatew.）Ohwi ●☆

316776 Prunus sargentii Rehder var. compta（Koidz.）H. Hara＝Cerasus × compta（Koidz.）H. Ohba ●☆

316777 Prunus sargentii Rehder var. pubescens（Tatew.）Ohwi；萨金特氏毛樱 ●☆

316778 Prunus sativa Rouy et Camus subsp. domestica（L.）Rouy et E. G. Camus＝Prunus domestica L. ●

316779 Prunus schneideriana Koehne＝Cerasus schneideriana（Koehne）Te T. Yu et C. L. Li ●

316780 Prunus scopulorum Koehne＝Cerasus pseudocerasus（Lindl.）G. Don ●

316781 Prunus scopulorum Koehne＝Cerasus scopulorum（Koehne）Te T. Yu et C. L. Li ●

316782 Prunus semiarmillata Koehne＝Laurocerasus andersonii（Hook. f.）Te T. Yu et L. T. Lu ●

316783 Prunus sericea（Batalin）Koehne＝Padus wilsonii C. K. Schneid. ●

316784 Prunus sericea（Batalin）Koehne＝Prunus napaulensis（Ser.）Steud. var. sericea Batalin ●

316785 Prunus sericea（Batalin）Koehne var. batalinii（Batalin）Koehne＝Padus wilsonii C. K. Schneid. ●

316786 Prunus sericea（Batalin）Koehne var. brevifolia Koehne＝Padus wilsonii C. K. Schneid. ●

316787 Prunus serotina Ehrh.；晚熟樱桃；American Cherry，Black Cherry，Cabinet Cherry，Capulin，Rum Cherry ●☆

316788 Prunus serotina Ehrh. 'Cartilaginia'；软骨黑果樱桃 ●☆

316789 Prunus serotina Ehrh. 'Pendula'；垂枝黑果樱桃 ●☆

316790 Prunus serotina Ehrh.＝Cerasus serotina（Ehrh.）Loisel. ●

316791 Prunus serotina Ehrh. subsp. virens（Wooton et Standl.）McVaugh；西南野黑樱；Southwestern Chokecherry ●☆

316792 Prunus serotina Ehrh. var. salicifolia Koehne；柳叶晚熟樱桃（柳叶野黑樱）；Capulin Black Cherry ●☆

316793 Prunus serrula Franch.＝Cerasus serrula（Franch.）Te T. Yu et C. L. Li ●

316794 Prunus serrula Franch. var. tibetica（Batalin）Koehne＝Cerasus serrula（Franch.）Te T. Yu et C. L. Li ●

316795 Prunus serrula Franch. var. tibetica Koehne＝Cerasus serrula（Franch.）Te T. Yu et C. L. Li ●

316796 Prunus serrulata Lindl. 'Sachalinensis'＝Prunus sargentii Rehder ●☆

316797 Prunus serrulata Lindl.＝Cerasus serrulata（Lindl.）G. Don ex Loudon ●

316798 Prunus serrulata Lindl. f. purpurascens Miyoshi；寒山樱 ●

316799 Prunus serrulata Lindl. var. lannesiana（Carrière）Makino＝Cerasus serrulata（Lindl.）G. Don ex Loudon var. lannesiana（Carrière）Makino ●

316800 Prunus serrulata Lindl. var. pubescens（Makino）E. H. Wilson＝Cerasus serrulata（Lindl.）G. Don ex Loudon var. pubescens（Makino）Te T. Yu et C. L. Li ●

316801 Prunus serrulata Lindl. var. pubescens E. H. Wilson＝Prunus serrulata Lindl. var. pubescens（Makino）E. H. Wilson ●

316802 Prunus serrulata Lindl. var. spontanea（Maxim.）E. H. Wilson＝Cerasus serrulata（Lindl.）G. Don ex Loudon ●

316803 Prunus serrulata Lindl. var. spontanea（Maxim.）E. H. Wilson＝

Prunus jamasakura Siebold ex Koidz. ●☆

316804 Prunus serrulata Lindl. var. spontanea E. H. Wilson＝Cerasus serrulata（Lindl.）G. Don ex Loudon ●

316805 Prunus setulosa Batalin＝Cerasus setulosa（Batalin）Te T. Yu et C. L. Li ●

316806 Prunus shikokuensis（Moriya）Kubota＝Cerasus shikokuensis（Moriya）H. Ohba ●☆

316807 Prunus sibirica L.＝Armeniaca sibirica（L.）Lam. ●

316808 Prunus sibirica L. var. pubescens（Kostina）Nakai＝Armeniaca sibirica（L.）Lam. var. pubescens Kostina ●

316809 Prunus sieboldii（Carrière）Wittm.；茶碗樱（亮枝樱桃）●☆

316810 Prunus sieboldii（Carrière）Wittm. 'Caespitosa'；簇生茶碗樱（亮枝樱桃簇生）●☆

316811 Prunus sieboldii（Carrière）Wittm.＝Cerasus sieboldii Carrière ●☆

316812 Prunus sieboldii（Carrière）Wittm. var. parvifolia（Matsum.）E. H. Wilson＝Cerasus parvifolia（Matsum.）H. Ohba ●☆

316813 Prunus simonii Carrière；杏李（红李，鸡血李，苦李，秋根李，秋根子）；Apricot Plum，Simon Plum ●

316814 Prunus sogdiana Vassilcz.＝Prunus cerasifera Ehrh. ●◇

316815 Prunus spachiana（Laval ex H. Otto）Kitam.＝Cerasus spachiana Lavalée ex H. Otto ●☆

316816 Prunus spachiana（Laval ex H. Otto）Kitam. f. ascendens（Makino）Kitam.＝Cerasus spachiana Lavalée ex H. Otto f. ascendens（Makino）H. Ohba ●

316817 Prunus spachiana（Laval ex H. Otto）Kitam. var. koshiensis（Koidz.）Kitam.＝Cerasus spachiana Lavalée ex H. Otto var. koshiensis（Koidz.）H. Ohba ●☆

316818 Prunus speciosa（Koidz.）Nakai＝Cerasus speciosa（Koidz.）H. Ohba ●☆

316819 Prunus spinosa L.；黑刺李（刺李，欧洲黑刺李，乌荆子）；Black Thorn，Blackthorn，Blackthorn May，Blackthorn Plum，Bolloms，Buckthorn，Bullem，Bullies，Bullins，Bullison，Bullister，Bullums，Castings，Damsil，Egg-peg Bush，Hedge Picks，Hedge Pigs，Hedge Speak，Hedgepegs，Hedgepicks，Hedgespeaks，Hedgespecks，Hegpeg，Heg-peg Bush，Horse Gog，Horse-gog，Irish Tea，Kex，Picks，Pig-in-the-hedge，Scrog，Siam，Skeg，Slaa Thorn，Slaa-thorn，Slaa-tree，Slacen-bush，Slae，Slag，Slaigh，Slaun-bush，Slaun-tree，Slaw，Slawnes，Slea，Sloanes，Sloe，Sloe-bush，Sloethorn，Slon，Slon-bush，Slone，Slon-tree，Sloon，Slow，Slue，Snag-bush，Snags，Snegs，Spiny Plum，Wayside Beauty，Winter Kecksies，Winterpicks ●

316820 Prunus spinosa L. 'Plena'；重瓣黑刺李 ●☆

316821 Prunus spinosa L. 'Purpurea'；紫黑刺李 ●☆

316822 Prunus spinosa L. subsp. fruticans（Weihe）Nyman；灌木黑刺李 ●☆

316823 Prunus spinosa L. var. typica C. K. Schneid.＝Prunus spinosa L. ●

316824 Prunus spinulosa Siebold et Zucc.；刺叶樱（刺叶桂樱，刺叶野樱，櫏木）；Spinulose Leaf Cherry，Spinuloseleaf Cherry，Spinyleaf Cherry，Spinyleaf Cherrylaurel ●

316825 Prunus spinulosa Siebold et Zucc.＝Laurocerasus spinulosa（Siebold et Zucc.）C. K. Schneid. ●

316826 Prunus spinulosa Siebold et Zucc. var. pubiflora Koehne＝Laurocerasus spinulosa（Siebold et Zucc.）C. K. Schneid. ●

316827 Prunus spinulosa Siebold et Zucc. var. pubiflora Koehne＝Prunus spinulosa Siebold et Zucc. ●

316828 Prunus ssiori F. Schmidt＝Padus brachypoda（Batalin）C. K. Schneid. ●

316829 Prunus ssiori F. Schmidt = Padus ssiori (F. Schmidt) C. K. Schneid. ●

316830 Prunus stellipila Koehne = Padus stellipila (Koehne) Te T. Yu et T. C. Ku ●

316831 Prunus stellipila Koehne = Prunus buergeriana Miq. var. stellipila (Koehne) Te T. Yu et C. L. Li ●

316832 Prunus stipulacea Maxim. = Cerasus stipulacea (Maxim.) Te T. Yu et C. L. Li ●

316833 Prunus subcordata Benth.；太平洋李；Klamath Plum, Pacific Plum, Sierra Plum ●

316834 Prunus subhirtella Miq. 'Autumnalis' = Cerasus 'Autumnalis' ●☆

316835 Prunus subhirtella Miq. 'Hally Jolivette'；哈利·菊利维特大叶早樱 ●☆

316836 Prunus subhirtella Miq. 'Pendula Rosea'；垂枝玫瑰红大叶早樱；Weeping Spring Cherry ●☆

316837 Prunus subhirtella Miq. 'Pendula Rubra'；红垂枝彼岸樱 ●☆

316838 Prunus subhirtella Miq. 'Pendula'；垂枝彼岸樱（垂枝大叶早樱）●

316839 Prunus subhirtella Miq. 'Stellata'；星形大叶早樱（星花彼岸樱）●☆

316840 Prunus subhirtella Miq. = Cerasus subhirtella (Miq.) Sokoloff ●

316841 Prunus subhirtella Miq. f. autumnalis (Makino) Koehne = Cerasus 'Autumnalis' ●☆

316842 Prunus subhirtella Miq. var. ascendens (Makino) E. H. Wilson = Cerasus spachiana Lavalée ex H. Otto f. ascendens (Makino) H. Ohba ●

316843 Prunus subhirtella Miq. var. ascendens E. H. Wilson = Cerasus subhirtella (Miq.) Sokoloff ●

316844 Prunus subhirtella Miq. var. autumnalis Makino = Cerasus 'Autumnalis' ●☆

316845 Prunus subhirtella Miq. var. koshiensis (Koidz.) Ohwi = Cerasus spachiana Lavalée ex H. Otto var. koshiensis (Koidz.) H. Ohba ●☆

316846 Prunus subhirtella Miq. var. pendula (Maxim.) Tanaka f. ascendens (Makino) Ohwi = Cerasus spachiana Lavalée ex H. Otto f. ascendens (Makino) H. Ohba ●

316847 Prunus subhirtella Miq. var. pendula Tanaka = Cerasus subhirtella (Miq.) Sokoloff var. pendula (Tanaka) Te T. Yu et C. L. Li ●

316848 Prunus sundaica Miq. = Laurocerasus spinulosa (Siebold et Zucc.) C. K. Schneid. ●

316849 Prunus sundaica Miq. = Prunus spinulosa Siebold et Zucc. ●

316850 Prunus susquehanae Willd. = Prunus pumila L. var. susquehanae (Willd.) H. Jaeger ●☆

316851 Prunus szechuanica Batalin = Cerasus szechuanica (Batalin) Te T. Yu et C. L. Li ●

316852 Prunus taiwaniana Hayata；雾社山樱花（雾社樱花）；Wusheh Cherry ●

316853 Prunus taiwaniana Hayata = Cerasus subhirtella (Miq.) Sokoloff var. pendula (Tanaka) Te T. Yu et C. L. Li ●

316854 Prunus takasago-montana Sasaki；山白樱（白花山樱）●

316855 Prunus tangutica (Batalin) Koehne = Amygdalus tangutica (Batalin) Korsh. ●

316856 Prunus tatsienensis Batalin = Cerasus tatsieenensis (Batalin) Te T. Yu et C. L. Li ●

316857 Prunus tatsienensis Batalin var. pilosiuscula C. K. Schneid. = Cerasus clarofolia (C. K. Schneid.) Te T. Yu et C. L. Li ●

316858 Prunus tatsienensis Batalin var. stenadenia Koehne = Cerasus pleiocerasus (Koehne) Te T. Yu et C. L. Li ●

316859 Prunus tenella Batsch；俄罗斯柔弱扁桃；Dwarf Russian Almond, Russian Dwarf Almond ●☆

316860 Prunus tenella Rehder 'Fire Hill'；火丘矮扁桃 ●☆

316861 Prunus tenella Rehder = Amygdalus nana L. ●

316862 Prunus tenella Rehder = Prunus nana (L.) Stokes ●

316863 Prunus tianschanica (Pojark.) Te T. Yu et C. L. Li = Cerasus tianschanica Pojark. ●

316864 Prunus tiliifolia Salisb. = Armeniaca vulgaris Lam. ●

316865 Prunus tomentosa Thunb. = Cerasus tomentosa (Thunb.) Wall. ●

316866 Prunus tomentosa Thunb. var. batalinii C. K. Schneid. = Cerasus tomentosa (Thunb.) Wall. ●

316867 Prunus tomentosa Thunb. var. breviflora Koehne = Cerasus tomentosa (Thunb.) Wall. ●

316868 Prunus tomentosa Thunb. var. endotricha Koehne = Cerasus tomentosa (Thunb.) Wall. ●

316869 Prunus tomentosa Thunb. var. heteromera Koehne = Cerasus tomentosa (Thunb.) Wall. ●

316870 Prunus tomentosa Thunb. var. kashkarovii Koehne = Cerasus tomentosa (Thunb.) Wall. ●

316871 Prunus tomentosa Thunb. var. souliei Koehne = Cerasus tomentosa (Thunb.) Wall. ●

316872 Prunus tomentosa Thunb. var. trichocarpa (Bunge) Koehne = Cerasus tomentosa (Thunb.) Wall. ●

316873 Prunus tomentosa Thunb. var. tsuluensis Koehne = Cerasus tomentosa (Thunb.) Wall. ●

316874 Prunus transarisanensis Hayata；阿里山樱花（阿里山山樱）；Arisan Cherry ●

316875 Prunus transarisanensis Hayata f. matuurai Suzuki = Prunus matuurai Sasaki ●

316876 Prunus trichantha Koehne = Cerasus rufa Wall. var. trichantha (Koehne) Te T. Yu et C. L. Li ●

316877 Prunus trichantha Koehne = Cerasus trichantha (Koehne) S. Y. Jiang et C. L. Li ●

316878 Prunus trichantha Koehne = Prunus rufa (Wall.) Hook. f. var. trichantha (Koehne) Hara ●

316879 Prunus trichocarpa Bunge = Cerasus tomentosa (Thunb.) Wall. ●

316880 Prunus trichostoma Koehne = Cerasus trichostoma (Koehne) Te T. Yu et C. L. Li ●

316881 Prunus triflora Roxb. = Prunus salicina Lindl. ●

316882 Prunus triflora Roxb. var. mandshurica Skvortsov = Prunus ussuriensis Kov. et Kostina ●

316883 Prunus triflora Roxb. var. pubipes Koehne = Prunus salicina Lindl. var. pubipes (Koehne) L. H. Bailey ●

316884 Prunus triloba Lindl. 'Multiplex'；重瓣榆叶梅；Doubleflower Flowering Plum ●

316885 Prunus triloba Lindl. = Amygdalus triloba (Lindl.) Ricker ●

316886 Prunus triloba Lindl. f. plena Dippel = Prunus triloba Lindl. 'Multiplex' ●

316887 Prunus triloba Lindl. var. plena Dippel = Prunus triloba Lindl. f. plena Dippel ●

316888 Prunus triloba Lindl. var. truncata Kom. = Amygdalus triloba (Lindl.) Ricker var. truncata (Kom.) S. Q. Nie ●

316889 Prunus ulmifolia Franch. = Amygdalus triloba (Lindl.) Ricker ●☆

316890 Prunus umbellata Elliott；小伞李；Black Sloe, Hog Plum, Sloe ●☆

316891 Prunus undulata Buch.-Ham. = Laurocerasus undulata (Buch.-

Ham. ex D. Don）M. Roem. ●

316892　Prunus undulata Buch. -Ham. = Prunus undulata Buch. -Ham. ex D. Don ●

316893　Prunus undulata Buch. -Ham. = Prunus wallichii Steud. ●

316894　Prunus undulata Buch. -Ham. ex D. Don = Laurocerasus undulata（Buch. -Ham. ex D. Don）M. Roem. ●

316895　Prunus undulata Buch. -Ham. ex D. Don = Prunus wallichii Steud. ●

316896　Prunus undulata Buch. -Ham. ex D. Don f. venosa Koehne = Padus buergeriana（Miq.）Te T. Yu et T. C. Ku ●

316897　Prunus undulata Buch. -Ham. ex D. Don var. stellipila（Koehne）Te T. Yu et C. L. Li = Prunus buergeriana Miq. var. stellipila（Koehne）Te T. Yu et C. L. Li ●

316898　Prunus undulata Buch. -Ham. f. venosa Koehne = Padus buergeriana（Miq.）Te T. Yu et T. C. Ku ●

316899　Prunus undulata Buch. -Ham. var. stellipila（Koehne）Te T. Yu et C. L. Li = Prunus buergeriana Miq. var. stellipila（Koehne）Te T. Yu et C. L. Li ●

316900　Prunus ussuriensis Kov. et Kostina；东北李（乌苏里李）；Ussuri Plum，Ussurian Plum ●

316901　Prunus ussuriensis Kov. et Kostina = Prunus salicina Lindl. ●

316902　Prunus vaniotii H. Lév. = Padus obtusata（Koehne）Te T. Yu et T. C. Ku ●

316903　Prunus vaniotii H. Lév. = Prunus obtusata Koehne ●

316904　Prunus vaniotii H. Lév. var. obovata（Koehne）Rehder = Padus obtusata（Koehne）Te T. Yu et T. C. Ku ●

316905　Prunus vaniotii H. Lév. var. potanini（Koehne）Rehder = Padus obtusata（Koehne）Te T. Yu et T. C. Ku ●

316906　Prunus variabilis Koehne = Prunus litiginosa C. K. Schneid. ●

316907　Prunus veitchii Koehne = Cerasus serrulata（Lindl.）G. Don ex Loudon var. pubescens（Makino）Te T. Yu et C. L. Li ●

316908　Prunus velutina Batalin = Padus velutina（Batalin）C. K. Schneid. ●

316909　Prunus venosa Koehne = Padus buergeriana（Miq.）Te T. Yu et T. C. Ku ●

316910　Prunus venosa Koehne = Prunus buergeriana Miq. ●

316911　Prunus venusta Koehne = Cerasus clarofolia（C. K. Schneid.）Te T. Yu et C. L. Li ●

316912　Prunus venusta Koehne = Prunus clarofolia C. K. Schneid. ●

316913　Prunus verecunda（Koidz.）Koehne；羞怯樱（华东山樱）；Keyamasakura ●

316914　Prunus verecunda（Koidz.）Koehne = Cerasus serrulata（Lindl.）G. Don ex Loudon var. pubescens（Makino）Te T. Yu et C. L. Li ●

316915　Prunus verecunda（Koidz.）Koehne f. tomentella（Nakai）Kubota et Moriya = Cerasus serrulata（Lindl.）G. Don ex Loudon var. pubescens（Makino）Te T. Yu et C. L. Li ●

316916　Prunus verecunda（Koidz.）Koehne var. pendula H. Hara = Cerasus leveilleana（Koehne）H. Ohba f. pendula（H. Hara）H. Ohba ●☆

316917　Prunus verecunda（Koidz.）Koehne var. pubipes H. Hara = Cerasus serrulata（Lindl.）G. Don ex Loudon var. pubescens（Makino）Te T. Yu et C. L. Li ●

316918　Prunus virginiana L. ；北美稠李（弗吉尼亚樱桃，维州稠李）；American Bird Cherry，Capuli，Choke Cherry，Chokecherry，Common Choke Cherry，Eastern Chokecherry，Red Chokecherry，Rum Cherry，Virginia Bird-cherry，Virginian Bird Cherry，Western Chokecherry，

Wild Black Cherry ●☆

316919　Prunus virginiana L. ‘Shubert’；紫叶稠李●☆

316920　Prunus virginiana L. f. deamii G. N. Jones = Prunus virginiana L. ●☆

316921　Prunus vulgaris（Mill.）Schur. = Cerasus vulgaris Mill. ●

316922　Prunus waleri（Wight）Kalkman；瓦尔克李●☆

316923　Prunus wallichii Steud. = Laurocerasus undulata（Buch. -Ham. ex D. Don）M. Roem. ●

316924　Prunus wallichii Steud. var. crenulata F. P. Metcalf = Laurocerasus undulata（Buch. -Ham. ex D. Don）M. Roem. ●

316925　Prunus watsoni Sarg. = Prunus angustifolia Marshall var. watsonii Waugh ●☆

316926　Prunus wilsonii（C. K. Schneid.）Koehne = Padus wilsonii C. K. Schneid. ●

316927　Prunus wilsonii（C. K. Schneid.）Koehne var. leiobotrys Koehne = Padus wilsonii C. K. Schneid. ●

316928　Prunus wilsonii（Diels）Koehne = Padus wilsonii C. K. Schneid. ●

316929　Prunus wilsonii Diels ex Koehne = Padus wilsonii C. K. Schneid. ●

316930　Prunus wilsonii Diels ex Koehne var. leiobotrys Koehne = Padus wilsonii C. K. Schneid. ●

316931　Prunus xerocarpa Hemsl. = Laurocerasus phaeosticta（Hance）C. K. Schneid. ●

316932　Prunus yedoensis Matsum. ‘Akebono’；阿科布诺日本樱花（阿科布诺东京樱花）●☆

316933　Prunus yedoensis Matsum. ‘Ivensii’；艾文汜日本樱花（东京樱花）；Akebono Flowered Cherry ●☆

316934　Prunus yedoensis Matsum. ‘Shidare-yoshino’；西达雷姚西诺日本樱花（东京樱花）●☆

316935　Prunus yedoensis Matsum. = Cerasus yedoensis（Matsum.）A. V. Vassil. ●

316936　Prunus yedoensis Matsum. var. candida Kawas. ；纯白花樱桃●☆

316937　Prunus yedoensis Matsum. var. nikaii Honda = Cerasus nikaii（Honda）H. Ohba ●☆

316938　Prunus yedoensis Matsum. var. nudiflora Koehne = Cerasus yedoensis（Matsum.）A. V. Vassil. ●

316939　Prunus yedoensis Matsum. var. rubriflora Kawas. ；红花樱桃●☆

316940　Prunus yunnanensis Franch. = Cerasus yunnanensis（Franch.）Te T. Yu et C. L. Li ●

316941　Prunus yunnanensis Franch. var. henryi C. K. Schneid. = Cerasus henryi（C. K. Schneid.）Te T. Yu et C. L. Li ●

316942　Prunus yunnanensis Franch. var. henryi C. K. Schneid. = Prunus henryi（C. K. Schneid.）Koehne ●

316943　Prunus yunnanensis Franch. var. polybotrys Koehne；多花樱桃（多花云南樱桃）；Manyflower Yunnan Cherry ●

316944　Prunus yunnanensis Franch. var. polybotrys Koehne = Cerasus yunnanensis（Franch.）Te T. Yu et C. L. Li var. polybotrys（Koehne）Te T. Yu et C. L. Li ●

316945　Prunus zippeliana Miq. = Laurocerasus zippeliana（Miq.）Browicz ●

316946　Prunus zippeliana Miq. f. angustifolia Te T. Yu et L. T. Lu = Laurocerasus zippeliana（Miq.）Browicz ●

316947　Prunus zippeliana Miq. f. infravelutina（Makino）Sugim. = Laurocerasus zippeliana（Miq.）Browicz ●

316948　Prunus zippeliana Miq. var. crassistyla（Cardot）J. E. Vidal = Laurocerasus zippeliana（Miq.）Te T. Yu et L. T. Lu var. crassistyla

（Cardot）Te T. Yu et L. T. Lu ●

316949　Prunus zippeliana Miq. var. crassistyla （Cardot）J. E. Vidal ＝ Laurocerasus zippeliana （Miq.）Browicz ●

316950　Prunus zippeliana Miq. var. puberifolia （Koehne）Te T. Yu et L. T. Lu；柔毛大叶桂樱；Pubescent Bigleaf Laurel Cherry ●

316951　Prunus-Cerasus Weston ＝ Cerasus Mill. ●

316952　Pryona Miq. ＝ Crudia Schreb.（保留属名）●☆

316953　Przewalskia Maxim.（1882）；马尿泡属；Horsebladder, Przewalskia ■☆

316954　Przewalskia roborowskii Batalin ＝ Przewalskia tangutica Maxim. ■

316955　Przewalskia shebbearei （C. E. C. Fisch.）Grubov ＝ Przewalskia tangutica Maxim. ■

316956　Przewalskia tangutica Maxim.；马尿泡（矮莨菪,青海矮莨菪,唐古特马尿泡,羊尿泡）；Tangut Horsebladder,Tangut Przewalskia ■

316957　Psacadocalymma Bremek. ＝ Justicia L. ●■

316958　Psacadocalymma Bremek. ＝ Stethoma Raf. ●■

316959　Psacadocalymma pectorale （Jacq.）Bremek. ＝ Justicia pectoralis Jacq. ■☆

316960　Psacadopaepale Bremek. ＝ Strobilanthes Blume ●■

316961　Psacaliopsis H. Rob. et Brettell(1974)；类粒菊属（类印第安菊属）■☆

316962　Psacaliopsis pudica （Standl. et Steyerm.）H. Rob. et Brettell；类粒菊■☆

316963　Psacalium Cass.（1826）；粒菊属（印第安菊属）■☆

316964　Psacalium decompositum （A. Gray）H. Rob. et Brettell；多裂粒菊（印第安菊）；Matarique ■☆

316965　Psacalium peltatum Cass.；盾状粒菊■☆

316966　Psacalium strictum Greene ＝ Rainiera stricta （Greene）Greene ■☆

316967　Psalina Raf. ＝ Gentiana L. ■

316968　Psamma P. Beauv. ＝ Ammophila Host ■☆

316969　Psammagrostis C. A. Gardner et C. E. Hubb.（1938）；沙剪股颖属■☆

316970　Psammagrostis wiseana C. A. Gardner et C. E. Hubb.；沙剪股颖■☆

316971　Psammanthe Hance ＝ Sesuvium L. ■

316972　Psammanthe Rchb. ＝ Minuartia L. ■

316973　Psammanthe Rchb. ＝ Rhodalsine J. Gay ■

316974　Psammetes Hepper(1962)；马岛沙玄参属■☆

316975　Psammetes madagascariensis （Bonati）Eb. Fisch. et Hepper；马岛沙玄参■☆

316976　Psammetes nigerica Hepper ＝ Psammetes madagascariensis （Bonati）Eb. Fisch. et Hepper ■☆

316977　Psammisia Klotzsch(1851)；杞莓属●☆

316978　Psammisia alpicola Klotzsch；高山杞莓●☆

316979　Psammisia amazonica Luteyn；亚马孙杞莓●☆

316980　Psammisia bicolor Klotzsch；二色杞莓●☆

316981　Psammisia breviflora Klotzsch；短花杞莓●☆

316982　Psammisia elegans Rusby；雅致杞莓●☆

316983　Psammisia elliptica （Rusby）A. C. Sm.；椭圆杞莓●☆

316984　Psammisia falcata Klotzsch；镰叶杞莓●☆

316985　Psammisia ferruginea A. C. Sm.；锈色杞莓●☆

316986　Psammisia glabra Klotzsch；光杞莓●☆

316987　Psammisia grandiflora Hoerold；大花杞莓●☆

316988　Psammisia leucostoma Benth.；白口杞莓●☆

316989　Psammisia longicaulis A. C. Sm.；长茎杞莓●☆

316990　Psammisia longifolia Klotzsch；长叶杞莓●☆

316991　Psammisia macrocalyx A. C. Sm.；大萼杞莓●☆

316992　Psammisia macrophylla Klotzsch；小叶杞莓●☆

316993　Psammisia micrantha （A. C. Sm.）Luteyn；小花杞莓●☆

316994　Psammisia montana Luteyn；山地杞莓●☆

316995　Psammisia nitida Klotzsch；亮杞莓●☆

316996　Psammisia occidentalis A. C. Sm.；西方杞莓●☆

316997　Psammisia orientalis Luteyn；东方杞莓●☆

316998　Psammisia pauciflora Griseb.；少花杞莓●☆

316999　Psammisia pterocalyx Luteyn；翅萼杞莓●☆

317000　Psammochloa Hitchc.（1927）；沙鞭属（沙茅属）；Psammochloa,Sandwhip ■

317001　Psammochloa Hitchc. et Bor ＝ Psammochloa Hitchc. ■

317002　Psammochloa mongolica （Hitchc.）Roshev. ＝ Psammochloa villosa （Trin.）Bor ■

317003　Psammochloa mongolica Hitchc. ＝ Psammochloa villosa （Trin.）Bor ■

317004　Psammochloa villosa （Trin.）Bor；沙鞭（沙竹）；Common Sandwhip,Mongolian Psammochloa ■

317005　Psammocorchorus Rchb. ＝ Corchorus L. ■●

317006　Psammogeton Edgew.(1845)；沙地芹属■☆

317007　Psammogeton biternatum Edgew.；三出沙地芹■☆

317008　Psammogeton biternatum Edgew. var. villosa C. B. Clarke ＝ Psammogeton canescens （DC.）Vatke ■☆

317009　Psammogeton canescens （DC.）Vatke；灰白沙地芹■☆

317010　Psammogeton canescens （DC.）Vatke subsp. biternatus （Edgew.）Waga ＝ Psammogeton biternatum Edgew. ■☆

317011　Psammogeton crinata Boiss. ＝ Psammogeton canescens （DC.）Vatke ■☆

317012　Psammogeton setifolium Boiss.；毛叶沙地芹■☆

317013　Psammogeton setifolium Boiss. ＝ Cuminum setifolium （Boiss.）Koso-Pol. ■☆

317014　Psammogeton stocksii （Boiss.）Nasir；斯托克斯沙地芹■☆

317015　Psammogonum Nieuwi. ＝ Polygonella Michx. ■☆

317016　Psammomoya Diels et Loes.（1904）；沙莫亚卫矛属●☆

317017　Psammomoya choretroides （F. Muell.）Diels et Loes.；沙莫亚卫矛●☆

317018　Psammophila Fourr. ＝ Gypsophila L. ■●

317019　Psammophila Fourr. ex Ikonn. ＝ Psammophiliella Ikonn. ■●

317020　Psammophila Ikonn. ＝ Gypsophila L. ■●

317021　Psammophila Schult. ＝ Ponceletia Thouars ■

317022　Psammophila Schult. ＝ Spartina Schreb. ex J. F. Gmel. ■

317023　Psammophila muralis （L.）Fourr. ＝ Gypsophila muralis L. ■

317024　Psammophiliella Ikonn. ＝ Gypsophila L. ■●

317025　Psammophiliella muralis （L.）Ikonn. ＝ Gypsophila muralis L. ■

317026　Psammophora Dinter et Schwantes(1926)；藏沙玉属●☆

317027　Psammophora herrei L. Bolus ＝ Psammophora longifolia L. Bolus ●☆

317028　Psammophora longifolia L. Bolus；长叶藏沙玉●☆

317029　Psammophora modesta （Dinter et A. Berger）Dinter et Schwantes；莫德斯托藏沙玉●☆

317030　Psammophora nissenii （Dinter）Dinter et Schwantes；尼森藏沙玉●☆

317031　Psammophora pillansii L. Bolus ＝ Arenifera pillansii （L. Bolus）Herre ■☆

317032　Psammophora saxicola H. E. K. Hartmann；岩生藏沙玉●☆

317033　PsammopyrumÁ. Löve ＝ Elymus L. ■☆

317034　Psammoseris Boiss. et Reut. ＝ Crepis L. ■

317035　Psammosilene W. C. Wu et C. Y. Wu（1945）；金铁锁属；Golden Ironlock,Psammosilene ■☆

317036 Psammosilene tunicoides W. C. Wu et C. Y. Wu;金铁锁(白马分鬃,百步杨,穿石甲,独丁子,独钉子,独定子,独鹿角姜,对叶七,金丝矮陀陀,昆明沙参,麻参,麦方草,土人参,蜈蚣七,夷方草);Tunica-like Psammosilene,Tuniclike Golden Ironlock ■

317037 Psammostachys C. Presl = Striga Lour. ■

317038 Psammotropha Eckl. et Zeyh. (1836);沙粟草属●☆

317039 Psammotropha alternifolia Killick;互叶沙粟草●☆

317040 Psammotropha androsacea Fenzl = Psammotropha mucronata (Thunb.) Fenzl ●☆

317041 Psammotropha androsacea Fenzl var. enervis ? = Psammotropha mucronata (Thunb.) Fenzl ●☆

317042 Psammotropha breviscapa Burtt Davy = Psammotropha myriantha Sond. ●☆

317043 Psammotropha diffusa Adamson;松散沙粟草●☆

317044 Psammotropha frigida Schltr.;硬沙粟草●☆

317045 Psammotropha marginata (Thunb.) Druce;具边沙粟草●☆

317046 Psammotropha mucronata (Thunb.) Fenzl;短尖沙粟草●☆

317047 Psammotropha mucronata (Thunb.) Fenzl var. foliosa Adamson;多叶沙粟草●☆

317048 Psammotropha mucronata (Thunb.) Fenzl var. marginata Adamson;具边短尖沙粟草●☆

317049 Psammotropha myriantha Sond.;多花沙粟草●☆

317050 Psammotropha obovata Adamson;倒卵沙粟草●☆

317051 Psammotropha obtusa Adamson;钝沙粟草●☆

317052 Psammotropha quadrangularis (L. f.) Fenzl;棱角沙粟草●☆

317053 Psammotropha rigida (Bartl.) Fenzl = Adenogramma rigida (Bartl.) Sond. ■☆

317054 Psammotropha spicata Adamson;长穗沙粟草●☆

317055 Psammotropha stipulacea F. M. Leight. = Polpoda stipulacea (Leight.) Adamson ●☆

317056 Psammotrophe Benth. et Hook. f. = Psammotropha Eckl. et Zeyh. ●☆

317057 Psanacetum (Neck. ex Less.) Spach = Tanacetum L. ■●

317058 Psanacetum Neck. = Tanacetum L. ■●

317059 Psanchum Neck. = Cynanchum L. ●■

317060 Psatherips Raf. = Salix L. (保留属名)●

317061 Psathura Comm. ex Juss. (1789);脆茜属■☆

317062 Psathura fryeri Hemsl. = Triainolepis africana Hook. f. subsp. hildebrandtii (Vatke) Verdc. ■☆

317063 Psathura lancifolia Bremek.;披针叶脆茜■☆

317064 Psathura lutescens Bremek.;淡黄脆茜■☆

317065 Psathura myriantha Bremek.;多花脆茜■☆

317066 Psathurochaeta DC. = Melanthera Rohr ■●☆

317067 Psathurochaeta dregei DC. = Melanthera scandens (Schumach. et Thonn.) Roberty subsp. dregei (DC.) Wild ■☆

317068 Psathyra Spreng. = Psathura Comm. ex Juss. ■☆

317069 Psathyranthus Ule = Psittacanthus Mart. ●☆

317070 Psathyrochaeta Post et Kuntze = Melanthera Rohr ■●☆

317071 Psathyrochaeta Post et Kuntze = Psathurochaeta DC. ■●☆

317072 Psathyrodes Willis = Psathyrotes (Nutt.) A. Gray ■☆

317073 Psathyrostachys (Boiss.) Nevski(1934);新麦草属(华新麦草属);Newstraw,Psathyrostachys ■

317074 Psathyrostachys Hevski = Hordeum L. ■

317075 Psathyrostachys Nevski = Psathyrostachys (Boiss.) Nevski ■

317076 Psathyrostachys Nevski ex Roshev. = Psathyrostachys (Boiss.) Nevski ■

317077 Psathyrostachys daghestanica (Alex.) Nevski;达赫斯坦新麦草■☆

317078 Psathyrostachys fragilis (Boiss.) Nevski;纤细新麦草■☆

317079 Psathyrostachys huashanica Keng ex P. C. Kuo;华山新麦草;Huashan Newstraw, Huashan Psathyrostachys ■

317080 Psathyrostachys hyalantha (Rupr.) Tzvelev = Psathyrostachys juncea (Fisch.) Nevski subsp. hyalantha (Rupr.) Tzvelev ■

317081 Psathyrostachys juncea (Fisch.) Nevski;新麦草(灯心草野麦,苏联宾麦);Newstraw, Rush-like Psathyrostachys, Russian Wild Rye, Russian Wildrye ■

317082 Psathyrostachys juncea (Fisch.) Nevski subsp. hyalantha (Rupr.) Tzvelev;紫药新麦草■■

317083 Psathyrostachys kronenburgii (Hack.) Nevski;单花新麦草(大麦新麦草,单穗新麦草,克罗氏新麦草,新疆新麦草);Kronenburg Newstraw, Kronenburg Psathyrostachys ■

317084 Psathyrostachys lanuginosa (Trin.) Nevski;毛穗新麦草;Hair Spike Newstraw, Woolly Psathyrostachys ■

317085 Psathyrostachys perennis Keng = Psathyrostachys juncea (Fisch.) Nevski ■

317086 Psathyrostachys rupestris (Alex.) Nevski;岩地新麦草■☆

317087 Psathyrostachys stoloniformis Baden;匍茎新麦草■

317088 Psathyrotes (Nutt.) A. Gray(1853);龟背菊属■☆

317089 Psathyrotes A. Gray = Psathyrotes (Nutt.) A. Gray ■☆

317090 Psathyrotes annua (Nutt.) A. Gray;一年龟背菊■☆

317091 Psathyrotes incisa A. Gray = Trichoptilium incisum (A. Gray) A. Gray ■☆

317092 Psathyrotes pilifera A. Gray;纤毛龟背菊■☆

317093 Psathyrotes ramosissima (Torr.) A. Gray;龟背菊;Desert Velvet, Turtleback ■☆

317094 Psathyrotes scaposa A. Gray = Psathyrotopsis scaposa (A. Gray) H. Rob. ■☆

317095 Psathyrotopsis Rydb. (1927);类龟背菊属■☆

317096 Psathyrotopsis Rydb. = Psathyrotes (Nutt.) A. Gray ■☆

317097 Psathyrotopsis scaposa (A. Gray) H. Rob.;类龟背菊■☆

317098 Psatura Poir. = Psathura Comm. ex Juss. ■☆

317099 Psectra (Endl.) P. Tomšovic = Echinops L. ■☆

317100 Psectra (Endl.) P. Tomšovic(1997);削刀蓝刺头属■☆

317101 Psedera Neck. = Parthenocissus Planch. (保留属名)●

317102 Psedera Neck. ex Greene = Ampelopsis Michx. ●

317103 Psedera Neck. ex Greene = Parthenocissus Planch. (保留属名)●

317104 Psedera henryana (Hemsl.) C. K. Schneid. = Parthenocissus henryana (Hemsl.) Diels et Gilg ●

317105 Psedera himalayana (Royle) C. K. Schneid. = Parthenocissus semicordata (Wall.) Planch. ●

317106 Psedera quinquefolia (L.) Greene = Parthenocissus quinquefolius (L.) Planch. ●

317107 Psedera thomsonii (M. A. Lawson) Stuntz = Yua thomsonii (M. A. Lawson) C. L. Li ●

317108 Psedera thunbergii (Siebold et Zucc.) Nakai = Parthenocissus tricuspidata (Siebold et Zucc.) Planch. ●

317109 Psedera thunbergii Nakai = Parthenocissus tricuspidatus (Siebold et Zucc.) Planch. ●

317110 Psedera tricuspidata (Siebold et Zucc.) Rehder = Parthenocissus tricuspidatus (Siebold et Zucc.) Planch. ●

317111 Psedera tricuspidata Rehder = Parthenocissus tricuspidatus (Siebold et Zucc.) Planch. ●

317112 Psedera vitacea (Knerr) Greene = Parthenocissus vitacea (Knerr) Hitchc. ●☆

317113　Psednotrichia Hiern(1898);丝莲菊属■☆

317114　Psednotrichia australis Alston = Felicia australis（Alston）E. Phillips ■☆

317115　Psednotrichia newtonii（O. Hoffm.）Anderb. et P. O. Karis;纽敦丝莲菊■☆

317116　Psednotrichia tenella Hiern = Psednotrichia xyridopsis（O. Hoffm.）Anderb. et P. O. Karis ■☆

317117　Psednotrichia xyridopsis（O. Hoffm.）Anderb. et P. O. Karis;丝莲菊■☆

317118　Psedomelia Neck. = Bromelia L. ■☆

317119　Pseliaceae Raf. = Menispermaceae Juss.（保留科名）●■

317120　Pselium Lour.（废弃属名）= Pericampylus Miers + Stephania Lour. ●

317121　Pselium Lour.（废弃属名）= Pericampylus Miers（保留属名）●

317122　Psephellus Cass.（1826）;膜片菊属●■☆

317123　Psephellus Cass. = Centaurea L.（保留属名）●■

317124　Psephellus calocephalus Cass.;膜片菊●■☆

317125　Pseuclomorus Bureau = Streblus Lour. ●

317126　Pseudabutilon R. E. Fr.（1908）;假苘麻属●☆

317127　Pseudabutilon scabrum（C. Presl）R. E. Fr.;假苘麻■☆

317128　Pseudabutilon stuckertii R. E. Fr.;斯图假苘麻;Velvetleaf Indian Mallow ■☆

317129　Pseudacacia Moench = Pseudo-Acacia Duhamel ●

317130　Pseudacacia Moench = Robinia L. ●

317131　Pseudacanthopale Benoist = Strobilanthopsis S. Moore ●☆

317132　Pseudacoridium Ames(1922);假足柱兰属■☆

317133　Pseudacoridium woodianum Ames;假足柱兰■☆

317134　Pseudactis S. Moore = Emilia（Cass.）Cass. ■

317135　Pseudaechmanthera Bremek.（1944）;假尖药草属（假尖蕊属,假尖药花属,毛叶草属）;Pseudaechmanthera ■

317136　Pseudaechmanthera Bremek. = Strobilanthes Blume ●■

317137　Pseudaechmanthera glutinosa（Nees）Bremek.;假尖药草（假尖蕊,毛叶草,黏毛假尖蕊）;Pseudaechmanthera ■

317138　Pseudaechmanthera glutinosa（Nees）Bremek. = Strobilanthes glutinosa Nees ■☆

317139　Pseudaechmea L. B. Sm. et Read(1982);假光萼荷属■☆

317140　Pseudaechmea ambigua L. B. Sm. et Read;假光萼荷■☆

317141　Pseudaegiphila Rusby = Aegiphila Jacq. ●■☆

317142　Pseudaegle Miq. = Poncirus Raf. ●☆

317143　Pseudaegle sepiaria Miq. = Poncirus trifoliata（L.）Raf. ●

317144　Pseudaegle trifoliata Makino = Poncirus trifoliata（L.）Raf. ●

317145　Pseudagrostistachys Pax et K. Hoffm.（1912）;假田穗戟属■☆

317146　Pseudagrostistachys africana（Müll. Arg.）Pax et K. Hoffm.;非洲假田穗戟■☆

317147　Pseudagrostistachys africana（Müll. Arg.）Pax et K. Hoffm. subsp. humbertii（Lebrun）J. Léonard;亨氏假田穗戟■☆

317148　Pseudagrostistachys humbertii Lebrun = Pseudagrostistachys africana（Müll. Arg.）Pax et K. Hoffm. subsp. humbertii（Lebrun）J. Léonard ■☆

317149　Pseudagrostistachys ugandensis（Hutch.）Pax et K. Hoffm.;假田穗戟■☆

317150　Pseudaidia Tirveng.（1987）;假茜树属●☆

317151　Pseudaidia speciosa（Bedd.）Tirveng.;假茜树●☆

317152　Pseudais Decne. = Phaleria Jack ●☆

317153　Pseudalangium F. Muell. = Alangium Lam.（保留属名）●

317154　Pseudalbizzia Britton et Rose = Albizia Durazz. ●

317155　Pseudaleia Thouars = Olax L. ●

317156　Pseudaleioides Thouars = Olax L. ●

317157　Pseudaleioides Thouars = Pseudaleia Thouars ●☆

317158　Pseudaleioides thouarsii DC. = Olax thouarsii（DC.）Valeton ●☆

317159　Pseudaleiopsis Rchb. = Olax L. ●

317160　Pseudaleiopsis Rchb. = Pseudaleia Thouars ●☆

317161　Pseudalepyrum Dandy = Centrolepis Labill. ■

317162　Pseudalomia Zoll. et Moritzi = Ethulia L. f. ■

317163　Pseudalthenia（Graebn.）Nakai(1943);拟加利亚草属■☆

317164　Pseudalthenia（Graetn.）Nakai = Zannichellia L. ■

317165　Pseudalthenia Nakai = Pseudalthenia（Graebn.）Nakai ■☆

317166　Pseudalthenia aschersoniana（Graebn.）Hartog;拟加利亚草■☆

317167　Pseudammi H. Wolff = Seseli L. ■

317168　Pseudammi ehrenbergii H. Wolff = Seseli strictum Ledeb. ■

317169　Pseudanamomis Kausel = Myrtus L. ●

317170　Pseudanamomis Kausel(1956);假繁花桃金娘属●☆

317171　Pseudanamomis umbellulifera（Kunth）Kausel;假繁花桃金娘;Ciruelas ●☆

317172　Pseudananas（Hassl.）Harms = Pseudananas Hassl. ex Harms ■☆

317173　Pseudananas Hassl. = Pseudananas Hassl. ex Harms ■☆

317174　Pseudananas Hassl. ex Harms(1930);假凤梨属■☆

317175　Pseudananas sagenarius（Arruda）Camargo;假凤梨■☆

317176　Pseudanastatica（Boiss.）Lemee = Clypeola L. ■☆

317177　Pseudanastatica（Boiss.）Lemee = Pseudoanastatica（Boiss.）Grossh. ☆

317178　Pseudanchusa（A. DC.）Kuntze = Lindelofia Lehm. ■

317179　Pseudannona（Baill.）Saff. = Annona L. ●

317180　Pseudannona（Baill.）Saff. = Xylopia L.（保留属名）●

317181　Pseudannona Saff. = Annona L. ●

317182　Pseudannona Saff. = Xylopia L.（保留属名）●

317183　Pseudanthaceae Endl.;假花大戟科■

317184　Pseudanthaceae Endl. = Arecaceae Bercht. et J. Presl（保留科名）●

317185　Pseudanthaceae Endl. = Euphorbiaceae Juss.（保留科名）●■

317186　Pseudanthaceae Endl. = Micrantheaceae J. Agardh ●

317187　Pseudanthaceae Endl. = Palmae Juss.（保留科名）●

317188　Pseudanthaceae Endl. = Picrodendraceae Small（保留科名）●☆

317189　Pseudanthistiria Hook. f. = Pseudanthistria（Hack.）Hook. f. ■

317190　Pseudanthistiria emeinica S. L. Chen et T. D. Zhuang = Themeda villosa（Poir.）A. Camus ■

317191　Pseudanthistiria hispida Hook. f. = Pseudanthistria heteroclita（Roxb.）Hook. f. ■

317192　Pseudanthistria（Hack.）Hook. f.（1896）;假铁秆草属（假铁秆蒿属,伪铁秆草属）;False Ironculm,Pseudanthistria ■

317193　Pseudanthistria emeiica S. L. Chen et T. D. Zhuang;峨眉假铁秆草;Emei False Ironculm,Emei Pseudanthistria ■

317194　Pseudanthistria heteroclita（Roxb.）Hook. f.;假铁秆草;Curious Pseudanthistria,Ironculm ■

317195　Pseudanthus Sieber ex A. Spreng.（1827）;假花大戟属■☆

317196　Pseudanthus Sieber ex A. Spreng. = Nothosaerva Wight ■☆

317197　Pseudanthus Wight = Nothosaerva Wight ■☆

317198　Pseudanthus axillaris（A. S. George）Radcl. -Sm.;腋生假花大戟■☆

317199　Pseudanthus brachiatus（L.）Wight = Nothosaerva brachiata（L.）Wight ■☆

317200　Pseudanthus brachyphyllus F. Muell.;短叶假花大戟■☆

317201　Pseudanthus nitidus Müll. Arg.;亮假花大戟■☆

317202　Pseudanthus orientalis F. Muell. ;东方假花大戟■☆

317203　Pseudanthus ovalifolius F. Muell. ;卵叶假花大戟■☆

317204　Pseudanthus pauciflorus Halford et R. J. F. Hend. ;少花假花大戟■☆

317205　Pseudarabidella O. E. Schulz ＝ Arabidella（F. Muell.）O. E. Schulz ■☆

317206　Pseudarrhenatherum Rouy ＝ Arrhenatherum P. Beauv. ■

317207　Pseudarrhenatherum Rouy（1921）;肖燕麦草属■☆

317208　Pseudarrhenatherum longifolium（Thore）Rouy;长叶肖燕麦草■☆

317209　Pseudartabotrys Pellegr.（1920）;非洲鹰爪属（拟鹰爪属）●☆

317210　Pseudartabotrys letestui Pellegr. ;非洲鹰爪花（拟鹰爪花）●☆

317211　Pseudarthria Wight et Arn.（1834）;假节豆属■☆

317212　Pseudarthria alba A. Chev. ＝ Pseudarthria hookeri Wight et Arn. ●☆

317213　Pseudarthria capitata Hassk. ＝ Desmodium styracifolium （Osbeck）Merr. ●■

317214　Pseudarthria confertiflora（A. Rich.）Baker;密花假节豆●☆

317215　Pseudarthria cordata Klotzsch ＝ Desmodium velutinum （Willd.）DC. ●

317216　Pseudarthria crenata Welw. ex Hiern;圆齿假节豆●☆

317217　Pseudarthria hookeri Wight et Arn. ;胡克假节豆●☆

317218　Pseudarthria hookeri Wight et Arn. var. argyrophylla Verdc. ;银叶胡克假节豆●☆

317219　Pseudarthria macrophylla Welw. ex Baker;大叶假节豆●☆

317220　Pseudarthria robusta（E. Mey.）Schltr. ＝ Pseudarthria hookeri Wight et Arn. ●☆

317221　Pseudatalaya Baill. ＝ Atalaya Blume ●☆

317222　Pseudechinolaena Stapf（1919）; 钩毛黍属; Barkgrass, Pseudochinolaena ■

317223　Pseudechinolaena madagascariensis（A. Camus）Bosser;马岛钩毛黍（马岛钩毛草）■☆

317224　Pseudechinolaena moratii Bosser;莫拉特钩毛黍（莫拉特钩毛草）■☆

317225　Pseudechinolaena perrieri A. Camus;佩里耶钩毛黍（佩里耶钩毛草）■☆

317226　Pseudechinolaena polystachya（Kunth）Stapf;钩毛草（马耳朵草）; Manyspliked Barkgrass, Manyspliked Pseudochinolaena ■

317227　Pseudechinolaena tenuis Bosser;细钩毛黍（细钩毛草）■☆

317228　Pseudechinopepon（Cogn.）Kuntze ＝ Vaseyanthus Cogn. ■☆

317229　Pseudehretia Turcz. ＝ Ilex L. ●

317230　Pseudehretia umbellulata（Wall.）Turcz. ＝ Ilex umbellulata （Wall.）Loes. ●

317231　Pseudelephantopus Rohr（1792）（'Pseudo-Elephantopus'）（保留属名）;假地胆草属; Fake Earthgall, False Elephant's Foot, Pseudelephantopus, Pseudolephantopus ■

317232　Pseudelephantopus Rohr（保留属名）＝ Elephantopus L. ■

317233　Pseudelephantopus spicatus（Juss. ex Aubl.）Gleason;假地胆草（穗花地胆草）; Dog's-tongue, Fake Earthgall, Spicate Pseudelephantopus, Spicate Pseudolephantopus ■

317234　Pseudelleanthus Brieger ＝ Elleanthus C. Presl ■☆

317235　Pseudellipanthus G. Schellenb. ＝ Ellipanthus Hook. f. ●

317236　Pseudeminia Verdc.（1970）;热非草属■☆

317237　Pseudeminia benguellensis（Torre）Verdc. ;本格拉热非草豆■☆

317238　Pseudeminia comosa（Baker）Verdc. ;簇毛热非草豆■☆

317239　Pseudeminia mendoncae（Torre）Verdc. ;门东萨热非草豆■☆

317240　Pseudencyclia Chiron et V. P. Castro ＝ Epidendrum L.（保留属名）■☆

317241　Pseudencyclia Chiron et V. P. Castro（2003）;假围柱兰属■☆

317242　Pseudephedranthus Aristeg.（1969）;假麻黄花属●☆

317243　Pseudephedranthus fragrans（R. E. Fr.）Aristeg. ;假麻黄花●☆

317244　Pseudepidendrura Rchb. f. ＝ Epidendrum L.（保留属名）■☆

317245　Pseuderanthemum Radlk.（1895）;钩粉草属（拟美花属,山壳骨属）; Pseuderanthemum ●■

317246　Pseuderanthemum Radlk. ex Lindau ＝ Pseuderanthemum Radlk. ●■

317247　Pseuderanthemum alatum Nees;翼柄钩粉草;Chocolate Plant ■☆

317248　Pseuderanthemum albocoeruleum Champl. ;白蓝钩粉草■☆

317249　Pseuderanthemum albocoeruleum Champl. subsp. robustum Champl. ;粗壮白蓝钩粉草■☆

317250　Pseuderanthemum atropurpureum L. H. Bailey;紫叶钩粉草（喜花草,紫叶拟美花）●☆

317251　Pseuderanthemum bicolor（Schrank）Radlk. ;双色钩粉草■

317252　Pseuderanthemum campylosiphon Mildbr. ;弯管钩粉草■☆

317253　Pseuderanthemum carruthersii（Seem.）Guillaumin;拟美花; Carruthers' Falseface ■☆

317254　Pseuderanthemum carruthersii（Seem.）Guillaumin var. atropurpureum（W. Bull）Fosberg;紫叶拟美花■☆

317255　Pseuderanthemum connatum Lindau ＝ Rhinacanthus nasutus （L.）Kuntze ●■

317256　Pseuderanthemum couderci Benoist;狭叶钩粉草（钩毛草）; Narrowleaf Pseuderanthemum ●■

317257　Pseuderanthemum decurrens（Hochst. ex Nees）Radlk. ＝ Ruspolia decurrens（Hochst. ex Nees）Milne-Redh. ●☆

317258　Pseuderanthemum dichotomum Lindau ＝ Rhinacanthus dichotomus（Lindau）I. Darbysh. ●☆

317259　Pseuderanthemum graciliflorum（Nees）Ridl. ;云南山壳骨; Malacca Pseuderanthemum ●

317260　Pseuderanthemum haikangense C. Y. Wu et H. S. Lo;海康钩粉草（兰心草）;Haikang Pseuderanthemum ●

317261　Pseuderanthemum hildebrandtii Lindau;希尔德钩粉草■☆

317262　Pseuderanthemum hypocrateriforme（Vahl）Radlk. ＝ Ruspolia hypocrateriformis（Vahl）Milne-Redh. ●☆

317263　Pseuderanthemum hypocrateriforme（Vahl）Radlk. var. breviflorum S. Moore ＝ Pseuderanthemum kewense L. H. Bailey et E. Z. Bailey ■☆

317264　Pseuderanthemum katangense Champl. ;加丹加拟美花■☆

317265　Pseuderanthemum kewense L. H. Bailey ＝ Pseuderanthemum kewense L. H. Bailey et E. Z. Bailey ■☆

317266　Pseuderanthemum kewense L. H. Bailey et E. Z. Bailey;黑叶拟美花■☆

317267　Pseuderanthemum latifolium（Vahl）B. Hansen;山壳骨（钩粉草,小驳骨）;Palatebearing Pseuderanthemum ■

317268　Pseuderanthemum ludovicianum（Büttner）Lindau;吕多维克拟美花■☆

317269　Pseuderanthemum malaccense（C. B. Clarke）Lindau ＝ Pseuderanthemum graciliflorum（Nees）Ridl. ●

317270　Pseuderanthemum metallicum Hallier;亮泽拟美花■☆

317271　Pseuderanthemum nigritianum（T. Anderson）Radlk. ＝ Pseuderanthemum tunicatum（Afzel.）Milne-Redh. ■☆

317272　Pseuderanthemum palatiferum（Nees）Radlk. ＝ Pseuderanthemum latifolium（Vahl）B. Hansen ■

317273　Pseuderanthemum palatiferum（Wall.）Radlk. ex Lindau ＝ Pseuderanthemum latifolium（Vahl）B. Hansen ■

317274　Pseuderanthemum polyanthum（C. B. Clarke ex Oliv.）Merr. ＝

Pseuderanthemum polyanthum（C. B. Clarke）Merr. ■

317275　Pseuderanthemum polyanthum（C. B. Clarke）Merr.；多花山壳骨■

317276　Pseuderanthemum reticulatum（Bull.）Radlk.；金叶拟美花■

317277　Pseuderanthemum senense（Klotzsch）Radlk. = Ruspolia decurrens（Hochst. ex Nees）Milne-Redh. ●☆

317278　Pseuderanthemum seticalyx（C. B. Clarke）Stapf = Ruspolia seticalyx（C. B. Clarke）Milne-Redh. ●☆

317279　Pseuderanthemum shweliense（W. W. Sm.）C. Y. Wu et C. C. Hu；瑞丽山壳骨；Ruili Pseuderanthemum ●■

317280　Pseuderanthemum sinuatum（Vahl）Radlk.；深波状钩粉草；Sinuate Pseuderanthemum ■

317281　Pseuderanthemum subviscosum（C. B. Clarke）Stapf；黏拟美花■☆

317282　Pseuderanthemum tapingense（W. W. Sm.）C. Y. Wu et C. C. Hu；太平山壳骨；Taipingshan Pseuderanthemum ●■

317283　Pseuderanthemum tapingense（W. W. Sm.）C. Y. Wu et H. S. Lo ex C. C. Hu = Pseuderanthemum tapingense（W. W. Sm.）C. Y. Wu et C. C. Hu ●■

317284　Pseuderanthemum teysmannii（C. B. Clarke）Ridl.；红河山壳骨●

317285　Pseuderanthemum teysmannii Ridl. = Pseuderanthemum teysmannii（C. B. Clarke）Ridl. ●

317286　Pseuderanthemum teysmnnioides（C. B. Clarke）Merr. = Pseuderanthemum teysmannii Ridl. ●

317287　Pseuderanthemum tunicatum（Afzel.）Milne-Redh.；着衣拟美花■☆

317288　Pseuderanthemum tunicatum（Afzel.）Milne-Redh. var. infundibuliformis Champl.；漏斗状钩粉草■☆

317289　Pseuderanthemum variabile（R. Br.）Radlk.；多变钩粉草；Night and Afternoon ■☆

317290　Pseuderemostachys Popov（1941）；假沙穗属■☆

317291　Pseuderemostachys sewerzovii（Herd.）Popov；假沙穗■☆

317292　Pseuderia Schltr.（1912）；假毛兰属■☆

317293　Pseuderia brevifolia J. J. Sm.；短叶假毛兰■☆

317294　Pseuderia floribunda Schltr.；多花假毛兰■☆

317295　Pseuderiopsis Rchb. f. = Eriopsis Lindl. ■☆

317296　Pseudernestia Post et Kuntze = Pseudoernestia（Cogn.）Krasser ☆

317297　Pseuderucaria（Boiss.）O. E. Schulz（1916）；假芝麻芥属■☆

317298　Pseuderucaria O. E. Schulz = Pseuderucaria（Boiss.）O. E. Schulz ■☆

317299　Pseuderucaria clavata（Boiss. et Reut.）O. E. Schulz；珊瑚假芝麻芥■☆

317300　Pseuderucaria clavata（Boiss. et Reut.）O. E. Schulz subsp. tourneuxii（Coss.）Maire；图尔假芝麻芥■☆

317301　Pseuderucaria clavata（Boiss. et Reut.）O. E. Schulz var. ozendae Leredde = Pseuderucaria clavata（Boiss. et Reut.）O. E. Schulz ■☆

317302　Pseuderucaria teretifolia（Desf.）O. E. Schulz；假芝麻芥■☆

317303　Pseuderucaria teretifolia（Desf.）O. E. Schulz var. grandiflora O. E. Schulz = Pseuderucaria teretifolia（Desf.）O. E. Schulz ■☆

317304　Pseuderucaria teretifolia（Desf.）O. E. Schulz var. parviflora Batt. = Pseuderucaria teretifolia（Desf.）O. E. Schulz ■☆

317305　Pseudetalon Raf. = Pseudopetalon Raf. ●

317306　Pseudetalon Raf. = Zanthoxylum L. ●

317307　Pseudeugenia D. Legrand et Mattos（1966）；巴西番樱桃属●☆

317308　Pseudeugenia Post et Kuntze = Pseudoeugenia Scort. ●

317309　Pseudeugenia Post et Kuntze = Syzygium R. Br. ex Gaertn.（保留属名）●

317310　Pseudevax DC. ex Pomel = Evax Gaertn. ■☆

317311　Pseudevax DC. ex Steud. = Evax Gaertn. ■☆

317312　Pseudevax Pomel = Evax Gaertn. ■☆

317313　Pseudevax mauritanica Pomel = Filago mauritanica（Pomel）Dobignard ■☆

317314　Pseudibatia Malme（1900）；假伊巴特萝藦属●☆

317315　Pseudibatia australis Malme；澳洲假伊巴特萝藦●☆

317316　Pseudibatia boliviensis Schltr.；玻利维亚假伊巴特萝藦●☆

317317　Pseudibatia ciliata（E. Fourn.）Malme；睫毛假伊巴特萝藦●☆

317318　Pseudima Radlk.（1878）；中南美无患子属●☆

317319　Pseudima pallidum Radlk.；中南美无患子●☆

317320　Pseudiosma A. Juss. = Zanthoxylum L. ●

317321　Pseudiosma DC.（1824）；假逸香木属●☆

317322　Pseudiosma asiatica G. Don；假逸香木●☆

317323　Pseudipomoea Roberty = Ipomoea L.（保留属名）●■

317324　Pseudiris Chukr et A. Gil（2008）；假鸢尾属■☆

317325　Pseudiris Post et Kuntze = Iris L. ■

317326　Pseudiris Post et Kuntze = Pseudo-Iris Medik. ■

317327　Pseuditea Hassk. = Pittosporum Banks ex Gaertn.（保留属名）●☆

317328　Pseudixora Miq. = Anomanthodia Hook. f. ●☆

317329　Pseudixora Miq. = Randia L. ●☆

317330　Pseudixus Hayata = Korthalsella Tiegh. ●

317331　Pseudixus japonicus（Thunb.）Hayata = Korthalsella japonica（Thunb.）Engl. ●

317332　Pseudoacacia Duhamel = Robinia L. ●

317333　Pseudo-Acacia Duhamel = Robinia L. ●

317334　Pseudoacanthocereus F. Ritter = Acanthocereus（Engelm. ex A. Berger）Britton et Rose ●☆

317335　Pseudoacanthocereus F. Ritter（1979）；假刺萼柱属■☆

317336　Pseudoacanthocereus brasiliensis（Britton et Rose）F. Ritter；假刺萼柱■☆

317337　Pseudoampelopsis Planch. = Ampelopsis Michx. ●

317338　Pseudoampelopsis Planch. = Ampelopsis Planch. ●

317339　Pseudoanastatica（Boiss.）Grossh.（1930）；假复活草属■☆

317340　Pseudoanastatica（Boiss.）Grossh. = Clypeola L. ■☆

317341　Pseudoanastatica Grossh. = Clypeola L. ■☆

317342　Pseudoanastatica dichotoma（Boiss.）Grossh. = Clypeola dichotoma Boiss ■☆

317343　Pseudoarabidopsis Al-Shehbaz, O'Kane et R. A. Price（1999）；假鼠耳芥属■

317344　Pseudoarabidopsis toxophylla（M. Bieb.）Al-Shehbaz, O'Kane et R. A. Price；假鼠耳芥（毒大蒜芥, 弓叶鼠耳芥, 箭叶大蒜芥）；Arched Leaf Mouseear Cress, Archedleaf Mouseear Cress ■

317345　Pseudoarrenatherum Holub = Pseudarrhenatherum Rouy ■☆

317346　Pseudobaccharis Cabrera = Baccharis L.（保留属名）●■☆

317347　Pseudobaccharis Cabrera = Psila Phil. ●■☆

317348　Pseudobaccharis Cabrera（1944）；假种棉木属●☆

317349　Pseudobaccharis spartioides（Hook. et Arn. ex DC.）Cabrera；假种棉木●☆

317350　Pseudobaccharis spartioides（Hook. et Arn. ex DC.）Cabrera = Heterothalamus spartioides Hook. et Arn. ex DC. ●☆

317351　Pseudobaccharis spartioides（Hook. et Arn.）Cabrera = Pseudobaccharis spartioides（Hook. et Arn. ex DC.）Cabrera ●☆

317352　Pseudobaeckea Nied.（1891）；假岗松属●☆

317353　Pseudobaeckea africana（Burm. f.）Pillans；非洲假岗松●☆

317354　Pseudobaeckea cordata（Burm. f.）Nied.；心形假岗松●☆

317355 Pseudobaeckea gracilis Dümmer = Pseudobaeckea cordata（Burm. f.）Nied. ●☆

317356 Pseudobaeckea pinifolia（L. f.）Nied. = Pseudobaeckea africana（Burm. f.）Pillans ●☆

317357 Pseudobaeckea stokoei Pillans;斯托克假岗松●☆

317358 Pseudobaeckea teres（Oliv.）Dümmer;圆柱假岗松●☆

317359 Pseudobaeckea thymeleoides Schltr. = Pseudobaeckea cordata（Burm. f.）Nied. ●☆

317360 Pseudobaeckea virgata Nied. = Raspalia virgata（Brongn.）Pillans ●☆

317361 Pseudobahia（A. Gray）Rydb.（1915）;旭日菊属（假黄羽菊属）;Sunburst ■☆

317362 Pseudobahia Rydb. = Pseudobahia（A. Gray）Rydb. ■☆

317363 Pseudobahia bahiifolia（Benth.）Rydb.;哈氏旭日菊;Hartweg's Golden Sunburst ■☆

317364 Pseudobahia heermannii（Durand）Rydb.;希氏旭日菊;Brittlestem,Foothill Sunburst ■☆

317365 Pseudobahia peirsonii Munz;皮尔逊旭日菊;San Joaquin Adobe Sunburst ■☆

317366 Pseudobambusa T. Q. Nguyen（1991）;假竹属●☆

317367 Pseudobambusa schizostachyoides（Kurz）T. Q. Nguyen;假竹●☆

317368 Pseudobarleria Oerst. = Barleria L. ●■

317369 Pseudo-Barleria Oerst. = Barleria L. ●■

317370 Pseudobarleria T. Anderson = Petalidium Nees ■☆

317371 Pseudobarleria T. Anderson（1863）;肖扁爵床属■☆

317372 Pseudobarleria canescens Engl. = Petalidium canescens（Engl.）C. B. Clarke ■☆

317373 Pseudobarleria coccinea（S. Moore）Lindau = Petalidium coccineum S. Moore ■☆

317374 Pseudobarleria currorii Lindau = Petalidium currorii（Lindau）Benth. ex S. Moore ■☆

317375 Pseudobarleria engleriana Schinz = Petalidium englerianum（Schinz）C. B. Clarke ■☆

317376 Pseudobarleria glandulifera Lindau = Petalidium rautanenii Schinz ■☆

317377 Pseudobarleria glandulosa（S. Moore）Lindau = Petalidium glandulosum S. Moore ■☆

317378 Pseudobarleria glutinosa Engl. = Petalidium variabile（Engl.）C. B. Clarke ■☆

317379 Pseudobarleria halimoides（Nees）Lindau = Petalidium halimoides（Nees）S. Moore ■☆

317380 Pseudobarleria hirsuta T. Anderson;肖扁爵床■☆

317381 Pseudobarleria hirsuta T. Anderson = Petalidium currorii（Lindau）Benth. ex S. Moore ■☆

317382 Pseudobarleria lanata Engl. = Petalidium lanatum（Engl.）C. B. Clarke ■☆

317383 Pseudobarleria latifolia Schinz = Petalidium englerianum（Schinz）C. B. Clarke ■☆

317384 Pseudobarleria lindaui Dewèvre = Onus submuticus（C. B. Clarke）Gilli ■☆

317385 Pseudobarleria linifolia（T. Anderson）Lindau = Petalidium linifolium T. Anderson ■☆

317386 Pseudobarleria loranthifolia（S. Moore）Lindau = Petalidium halimoides（Nees）S. Moore ■☆

317387 Pseudobarleria ovata Schinz = Petalidium englerianum（Schinz）C. B. Clarke ■☆

317388 Pseudobarleria physaloides（S. Moore）Lindau = Petalidium physaloides S. Moore ■☆

317389 Pseudobarleria rupestre（S. Moore）Lindau = Petalidium rupestre S. Moore ■☆

317390 Pseudobarleria variabilis Engl. = Petalidium variabile（Engl.）C. B. Clarke ■☆

317391 Pseudobarleria variabilis Engl. var. incana ? = Petalidium variabile（Engl.）C. B. Clarke ■☆

317392 Pseudobarleria welwitschii（S. Moore）Lindau = Petalidium welwitschii S. Moore ■☆

317393 Pseudobartlettia Rydb.（1927）;假巴氏菊属■☆

317394 Pseudobartlettia Rydb. = Psathyrotes（Nutt.）A. Gray ■☆

317395 Pseudobartlettia Rydb. = Psathyrotopsis Rydb. ■☆

317396 Pseudobartlettia scaposa（A. Gray）Rydb.;假巴氏菊■☆

317397 Pseudobartlettia scaposa（A. Gray）Rydb. = Psathyrotopsis scaposa（A. Gray）H. Rob. ■☆

317398 Pseudobartsia D. Y. Hong = Parentucellia Viv. ■☆

317399 Pseudobartsia D. Y. Hong（1979）;五齿萼属（五齿草属）;Palmcalyx,Pseudobartsia ■☆

317400 Pseudobartsia yunnanensis D. Y. Hong;五齿萼（具腺松蒿,五齿草）;Glandular Phtheirospermum, Yunnan Palmcalyx, Yunnan Pseudobartsia ■

317401 Pseudobartsia yunnanensis D. Y. Hong = Phtheirospermum glandulosum Benth. et Hook. f. ■

317402 Pseudobasilicum Steud. = Picramnia Sw.（保留属名）●☆

317403 Pseudobastardia Hassl. = Herissantia Medik. ●■

317404 Pseudobastardia crispa（L.）Hassl. = Abutilon crispum（L.）Medik. ■

317405 Pseudobastardia crispa（L.）Hassl. = Herissantia crispa（L.）Brizicky ■

317406 Pseudoberlinia P. A. Duvign.（1950）;假鞋木豆属●☆

317407 Pseudoberlinia P. A. Duvign. = Julbernardia Pellegr. ●☆

317408 Pseudoberlinia baumii（Harms）P. A. Duvign. = Julbernardia paniculata（Benth.）Troupin ●☆

317409 Pseudoberlinia globiflora（Benth.）P. A. Duvign. = Julbernardia globiflora（Benth.）Troupin ●☆

317410 Pseudoberlinia paniculata（Benth.）P. A. Duvign. = Julbernardia paniculata（Benth.）Troupin ●☆

317411 Pseudobersama Verdc.（1956）;莫桑比克棟属●☆

317412 Pseudobersama mossambicensis（Sim）Verdc.;莫桑比克棟●☆

317413 Pseudobesleria Oerst. = Besleria L. ●■☆

317414 Pseudobetckea（Höck）Lincz.（1958）;贝才草属■☆

317415 Pseudobetckea caucasica（Boiss.）Lincz.;贝才草■☆

317416 Pseudoblepharis Baill. = Sclerochiton Harv. ●☆

317417 Pseudoblepharis boivinii Baill. = Sclerochiton boivinii（Baill.）C. B. Clarke ●☆

317418 Pseudoblepharis coerulea Lindau = Sclerochiton coeruleus（Lindau）S. Moore ●☆

317419 Pseudoblepharis dusenii Lindau = Crossandrella dusenii（Lindau）S. Moore ●☆

317420 Pseudoblepharis grandidieri（Baill.）Lindau = Crossandra grandidieri（Baill.）Benoist ●☆

317421 Pseudoblepharis heinsenii Lindau = Sclerochiton boivinii（Baill.）C. B. Clarke ●☆

317422 Pseudoblepharis holstii Lindau = Sclerochiton vogelii（Nees）T. Anderson subsp. holstii（Lindau）Napper ●☆

317423 Pseudoblepharis insignis Mildbr. = Sclerochiton insignis（Mildbr.）Vollesen ●☆

317424　Pseudoblepharis nitida（S. Moore）Lindau = Sclerochiton nitidus（S. Moore）C. B. Clarke ●☆

317425　Pseudoblepharis obtusisepala（C. B. Clarke）Lindau = Sclerochiton obtusisepalus C. B. Clarke ●☆

317426　Pseudoblepharis preussii Lindau = Sclerochiton preussii（Lindau）C. B. Clarke ●☆

317427　Pseudoblepharispermum J. -P. Lebrun et Stork（1982）；假睑子菊属●☆

317428　Pseudoblepharispermum bremeri J. -P. Lebrun et Stork；假睑子菊■☆

317429　Pseudoblepharispermum mudugense Beentje et D. J. N. Hind；穆杜杜格假睑子菊■☆

317430　Pseudoboivinella Aubrév. et Pellegr. = Englerophytum K. Krause ●☆

317431　Pseudoboivinella laurentii（De Wild.）Aubrév. et Pellegr. = Pachystela laurentii（De Wild.）C. M. Evrard ●☆

317432　Pseudoboivinella oblanceolata（S. Moore）Aubrév. et Pellegr. = Englerophytum oblanceolatum（S. Moore）T. D. Penn. ●☆

317433　Pseudoboivinella subverticillata（E. A. Bruce）Aubrév. et Pellegr. = Synsepalum subverticillatum（E. A. Bruce）T. D. Penn. ●☆

317434　Pseudobombax Dugand（1943）；假木棉属●☆

317435　Pseudobombax ellipticum（Kunth）Dugand；假木棉；Shaving Brush Tree ●☆

317436　Pseudobotrys Moeser（1912）；总状荼黄属●☆

317437　Pseudobotrys cauliflora（Pulle）Sleumer；茎花总状荼黄●☆

317438　Pseudobotrys dorae Moeser；总状荼黄●☆

317439　Pseudobrachiaria Launert = Brachiaria（Trin.）Griseb. ■

317440　Pseudobrachiaria Launert = Urochloa P. Beauv. ■

317441　Pseudobrachiaria deflexa（Schumach.）Launert = Brachiaria deflexa（Schumach.）C. E. Hubb. ex Robyns ■☆

317442　Pseudo-brasilium Adans.（废弃属名）= Picramnia Sw.（保留属名）●☆

317443　Pseudobrasilium Adans. = Picramnia Sw.（保留属名）●☆

317444　Pseudobrasilium Plum. ex Adans. = Picramnia Sw.（保留属名）●☆

317445　Pseudobrassaiopsis R. N. Banerjee = Brassaiopsis Decne. et Planch. ●

317446　Pseudobrassaiopsis alpina（C. B. Clarke）R. N. Banerjee = Merrilliopanax alpinus（C. B. Clarke）C. B. Shang ●

317447　Pseudobrassaiopsis hainla（Buch. -Ham.）R. N. Banerjee = Brassaiopsis hainla（Buch. -Ham.）Seem. ●

317448　Pseudobrassaiopsis hispida（Seem.）R. N. Banerjee = Brassaiopsis hispida Seem. ●

317449　Pseudobrassaiopsis polyacantha（Wall.）R. N. Banerjee = Brassaiopsis polyacantha（Wall.）R. N. Banerjee ●

317450　Pseudobravoa Rose = Polianthes L. ■

317451　Pseudobravoa Rose（1899）；假布拉沃兰属■☆

317452　Pseudobravoa densiflora（B. L. Rob. et Fernald）Rose；假布拉沃兰■☆

317453　Pseudobraya Korsh. = Draba L. ■

317454　Pseudobraya kizylarti Korsh. = Draba oreades Schrenk ■

317455　Pseudobrazzeia Engl. = Brazzeia Baill. ●☆

317456　Pseudobrickellia R. M. King et H. Rob.（1972）；线叶肋泽兰属●☆

317457　Pseudobrickellia angustissima（Spreng. ex Baker）R. M. King et H. Rob.；窄线叶肋泽兰●☆

317458　Pseudobrickellia brasiliensis（Spreng.）R. M. King et H. Rob.；巴西线叶肋泽兰●☆

317459　Pseudobromus K. Schum.（1895）；肖羊茅属■☆

317460　Pseudobromus K. Schum. = Festuca L. ■

317461　Pseudobromus africanus（Hack.）Stapf = Festuca africana（Hack.）Clayton ■☆

317462　Pseudobromus brassii C. E. Hubb. = Pseudobromus engleri（Pilg.）Clayton ■☆

317463　Pseudobromus engleri（Pilg.）Clayton；肖羊茅■☆

317464　Pseudobromus silvaticus K. Schum. = Festuca africana（Hack.）Clayton ■☆

317465　Pseudobrownanthus Ihlenf. et Bittrich（1985）；假褐花属（坚果露花树属）●☆

317466　Pseudobrownanthus nucifer Ihlenf. et Bittrich；假褐花●☆

317467　Pseudobrownanthus nucifer Ihlenf. et Bittrich = Brownanthus nucifer（Ihlenf. et Bittrich）S. M. Pierce et Gerbaulet ●☆

317468　Pseudobulbostylis Nutt. = Brickellia Elliott（保留属名）■●

317469　Pseudocadia Harms = Xanthocercis Baill. ●■☆

317470　Pseudocadia anomala（Vatke）Harms = Xanthocercis madagascariensis Baill. ●☆

317471　Pseudocadia zambesiaca（Baker）Harms = Xanthocercis zambesiaca（Baker）Dumaz-le-Grand ■☆

317472　Pseudocadiscus Lisowski = Stenops B. Nord. ■☆

317473　Pseudocadiscus Lisowski（1987）；假水漂菊属■☆

317474　Pseudocadiscus zairensis Lisowski = Stenops zairensis（Lisowski）B. Nord. ■☆

317475　Pseudocalymma A. Samp. et Kuhlm.（1933）；假绿苞草属●☆

317476　Pseudocalymma A. Samp. et Kuhlm. = Mansoa DC. ●☆

317477　Pseudocalymma alliaceum（Lam.）Sandwith；假绿苞草（张氏紫葳，紫铃藤）；Garlic Vine，Garlic-scented Vine，Trumpet Flower ●☆

317478　Pseudocalyx Radlk.（1883）；假萼爵床属■☆

317479　Pseudocalyx africanus S. Moore = Pseudocalyx saccatus Radlk. ■☆

317480　Pseudocalyx aurantiacus Benoist；橙色假萼爵床■☆

317481　Pseudocalyx libericus Breteler；利比里亚假萼爵床■☆

317482　Pseudocalyx macrophyllus McPherson et Louis；大叶假萼爵床■☆

317483　Pseudocalyx ochraceus Champl.；淡黄褐假萼爵床■☆

317484　Pseudocalyx saccatus Radlk.；假萼爵床■☆

317485　Pseudocamelina（Boiss.）N. Busch（1928）；假亚麻荠属■☆

317486　Pseudocamelina N. Busch = Pseudocamelina（Boiss.）N. Busch ■☆

317487　Pseudocamelina szovitsii（Boiss.）N. Busch；假亚麻荠（假山茶）●☆

317488　Pseudocampanula Kolak. = Campanula L. ■●

317489　Pseudocannaboides B. -E. van Wyk（1999）；假大麻属■☆

317490　Pseudocannaboides andringitrensis（Humbert）B. -E. van Wyk；假大麻■☆

317491　Pseudocapsicum Medik. = Solanum L. ●■

317492　Pseudocarapa Hemsl.（1884）；假酸渣树属●☆

317493　Pseudocarapa Hemsl. = Dysoxylum Blume ●

317494　Pseudocarapa championii Hemsl.；假酸渣树●☆

317495　Pseudocarapa nitidula（Benth.）Merr. et L. M. Perry；光亮假酸渣树●☆

317496　Pseudocarex Miq. = Carex L. ■

317497　Pseudocarpidium Millsp.（1906）；假果马鞭草属●☆

317498　Pseudocarpidium avicennioides（A. Rich.）Millsp.；假果马鞭草●☆

317499　Pseudocarpidium rigens Britton；硬假果马鞭草●☆

317500　Pseudocarum C. Norman（1924）；假葛缕子属●☆

317501　Pseudocarum clematidifolium C. Norman = Pseudocarum eminii（Engl.）H. Wolff ■☆

317502　Pseudocarum eminii（Engl.）H. Wolff；假葛缕子■☆

317503　Pseudocarum laxiflorum（Baker）B. -E. van Wyk；疏花假葛缕

子■☆

317504 Pseudocaryophyllus O. Berg = Pimenta Lindl. ●☆

317505 Pseudocaryophyllus O. Berg(1856);假蒲桃属(假石竹属)●

317506 Pseudocaryophyllus jaccondii Mattos;假蒲桃●☆

317507 Pseudocaryopteris(Briq.)P. D. Cantino(1999);假莸属●

317508 Pseudocaryopteris bicolor(Hardw.)P. D. Cantino;二色假莸●☆

317509 Pseudocaryopteris foetida(D. Don)P. D. Cantino;臭假莸●☆

317510 Pseudocaryopteris paniculata(C. B. Clarke)P. D. Cantino;圆锥假莸●☆

317511 Pseudocassia Britton et Rose = Senna Mill. ●■

317512 Pseudocassine Bredell = Crocoxylon Eckl. et Zeyh. ●☆

317513 Pseudocassine transvaalensis(Burtt Davy)Bredell = Elaeodendron transvaalense(Burtt Davy)R. H. Archer ●☆

317514 Pseudocatalpa A. H. Gentry(1973);假梓属●☆

317515 Pseudocatalpa caudiculata(Standl.)A. H. Gentry;假梓●☆

317516 Pseudocedrela Harms(1895);假洋椿属●☆

317517 Pseudocedrela caudata Sprague = Entandrophragma caudatum(Sprague)Sprague ●☆

317518 Pseudocedrela chevalieri C. DC. = Pseudocedrela kotschyi(Schweinf.)Harms ●☆

317519 Pseudocedrela cylindrica Sprague = Entandrophragma cylindricum(Sprague)Sprague ●☆

317520 Pseudocedrela excelsa Dawe et Sprague = Entandrophragma excelsum(Dawe et Sprague)Sprague ●☆

317521 Pseudocedrela kotschyi(Schweinf.)Harms;假洋椿●☆

317522 Pseudocedrela utilis Dawe et Sprague = Entandrophragma utile(Dawe et Sprague)Sprague ●☆

317523 Pseudocentema Chiov. = Centema Hook. f. ■●☆

317524 Pseudocentema angolensis(Hook. f.)Chiov. = Centema angolensis Hook. f. ■☆

317525 Pseudocentrum Lindl.(1858);假心兰属■☆

317526 Pseudocentrum macrostachyum Lindl. ;假心兰■☆

317527 Pseudocentrum minus Benth. ;小假心兰■☆

317528 Pseudocerastium C. Y. Wu,X. H. Guo et X. P. Zhang(1998);假卷耳属;Pseudocerastium ■☆

317529 Pseudocerastium stellarioides C. Y. Wu, X. H. Guo et X. P. Zhang;假卷耳;Pseudocerastium ■

317530 Pseudochaenomeles Carrière = Chaenomeles Lindl.(保留属名)●

317531 Pseudo-chaenomeles Carrière = Chaenomeles Lindl.(保留属名)●

317532 Pseudochaetochloa Hitchc.(1924);澳洲假刚毛草属■☆

317533 Pseudochaetochloa australiensis Hitchc. ;澳洲假刚毛草■☆

317534 Pseudochamaesphacos Parsa(1946);假矮刺苏属■☆

317535 Pseudochamaesphacos spinosa Parsa;假矮刺苏■☆

317536 Pseudochimarrhis Ducke = Chimarrhis Jacq. ■☆

317537 Pseudochirita W. T. Wang(1983);异裂苣苔属;Falsechirita ■☆

317538 Pseudochirita guangxiensis(S. Z. Huang)W. T. Wang;异裂苣苔(广西唇柱苣苔, 两面稠);Guangxi Chirita, Guangxi Falsechirita,Kwangsi Falsechirita ■

317539 Pseudochirita guangxiensis(S. Z. Huang)W. T. Wang var. glauca Y. G. Wei et Yan Liu;粉绿异裂苣苔;Glaucous Guangxi Falsechirita ■

317540 Pseudochrosia Blume = Ochrosia Juss. ●

317541 Pseudocimum Bremek. = Endostemon N. E. Br. ●■☆

317542 Pseudocimum tenuiflorus Bremek. = Endostemon tenuiflorus(Benth.)M. Ashby ■☆

317543 Pseudocinchona A. Chev. = Corynanthe Welw. ●☆

317544 Pseudocinchona A. Chev. ex Perrot = Pausinystalia Pierre ex Beille ●☆

317545 Pseudocinchona africana A. Chev. ex Perrot = Corynanthe pachyceras K. Schum. ●☆

317546 Pseudocinchona johimbe(K. Schum.)A. Chev. = Pausinystalia johimbe(K. Schum.)Pierre ex Beille ●☆

317547 Pseudocinchona mayumbensis(R. D. Good)Raym. -Hamet = Corynanthe mayumbensis(R. D. Good)Raym. -Hamet ex N. Hallé ●☆

317548 Pseudocinchona moebiusii(W. Brandt)A. Chev. = Pausinystalia talbotii Wernham ●☆

317549 Pseudocinchona pachyceras(K. Schum.)A. Chev. = Corynanthe pachyceras K. Schum. ●☆

317550 Pseudocione Mart. ex Engl. = Thyrsodium Salzm. ex Benth. ●☆

317551 Pseudocladia Pierre = Lucuma Molina ●

317552 Pseudocladia Pierre = Pouteria Aubl. ●

317553 Pseudoclappia Rydb.(1923);假盐菊属●☆

317554 Pseudoclappia arenaria Rydb.;假盐菊●☆

317555 Pseudoclappia watsonii A. M. Powell et B. L. Turner;沃森假盐菊●☆

317556 Pseudoclausena T. P. Clark = Clausena Burm. f. ●

317557 Pseudoclausena T. P. Clark(1994);假黄皮属●

317558 Pseudoclausia Popov(1955);假香芥属■

317559 Pseudoclausia gracillima(Popov)A. Vassil. ;细长假香芥■☆

317560 Pseudoclausia hispida(Regel)Popov;柔毛假香芥■☆

317561 Pseudoclausia mollissima(Lipsky)A. Vassil. ;柔软假香芥■☆

317562 Pseudoclausia olgae(Regel et Schmalh.)Botsch. ;奥氏假香芥■☆

317563 Pseudoclausia papillosa(Vassilcz.)A. Vassil. ;乳头假香芥■☆

317564 Pseudoclausia tschimganica(Popov)A. Vassil. ;契穆干假香芥■☆

317565 Pseudoclausia turkestanica(Lipsky)A. Vassilicz. ;突厥假香芥(假香芥, 土耳其斯坦香花芥, 腺果香芥);Glandularfruit Aromcress,Glandularfruit Clausia ■

317566 Pseudoclausia vvedenskyi Pachom. ;韦氏假香芥■☆

317567 Pseudoclinium Kuntze = Leptoclinium(Nutt.)Benth. et Hook. f. ●☆

317568 Pseudocoeloglossum(Szlach. et Olszewski)Szlach.(2003);拟凹舌兰属■☆

317569 Pseudocoix A. Camus(1924);假薏苡竹属●☆

317570 Pseudocoix perrieri A. Camus;假薏苡竹●☆

317571 Pseudocoix perrieri A. Camus = Hickelia africana S. Dransf. ●☆

317572 Pseudoconnarus Radlk.(1887);假牛栓藤属●☆

317573 Pseudoconnarus macrophyllus Radlk. ;大叶假牛栓藤●☆

317574 Pseudoconnarus reticulatus G. Schellenb. ;网脉假牛栓藤●☆

317575 Pseudoconyza Cuatrec.(1961);尾酒草属●☆

317576 Pseudoconyza Cuatrec. = Laggera Sch. Bip. ex Benth. ■

317577 Pseudoconyza lyrata(Kunth)Cuatrec. = Pseudoconyza viscosa(Mill.)D' Arcy ■☆

317578 Pseudoconyza viscosa(Mill.)D' Arcy;尾酒草;Clammy False Oxtongue ■☆

317579 Pseudoconyza viscosa(Mill.)D' Arcy var. lyrata(Kunth)D' Arcy = Pseudoconyza viscosa(Mill.)D' Arcy ■☆

317580 Pseudocopaiva Britton = Pseudocopaiva Britton et P. Wilson ●☆

317581 Pseudocopaiva Britton et P. Wilson = Copaifera L.(保留属名)●☆

317582 Pseudocopaiva Britton et P. Wilson = Guibourtia Benn. ●☆

317583 Pseudocorchorus Capuron(1963);假黄麻属■☆

317584 Pseudocorchorus rostratus(Danguy)Mabb. ;喙假黄麻■☆

317585 Pseudocranichis Garay(1982);假宝石兰属■☆

317586 Pseudocranichis thysanochila（Robins. et Greenm.）Garay；假宝石兰■☆

317587 Pseudocroton Müll. Arg.（1872）；假巴豆属●☆

317588 Pseudocroton tinctorius Müll. Arg.；假巴豆■☆

317589 Pseudocrupina Velen.（1923）；假半毛菊属●☆

317590 Pseudocrupina Velen. = Leysera L. ■●☆

317591 Pseudocrupina arabica Velen.；假半毛菊●☆

317592 Pseudocrupina arabica Velen. = Leysera leyseroides（Desf.）Maire ●☆

317593 Pseudocryptocarya Teschner = Cryptocarya R. Br.（保留属名）●

317594 Pseudoctomeria Kraenzl. = Pleurothallis R. Br. ■

317595 Pseudocunila Brade = Hedeoma Pers. ■●☆

317596 Pseudocyclanthera Mart. Crov.（1954）；假小雀瓜属■☆

317597 Pseudocyclanthera australis（Cogn.）Mart. Crov.；假小雀瓜■☆

317598 Pseudocydonia（C. K. Schneid.）C. K. Schneid.（1906）；木瓜属（假榅桲属，木李属）●☆

317599 Pseudocydonia（C. K. Schneid.）C. K. Schneid. = Cydonia Mill. ●

317600 Pseudocydonia C. K. Schneid. = Chaenomeles Lindl.（保留属名）●

317601 Pseudocydonia C. K. Schneid. = Pseudocydonia（C. K. Schneid.）C. K. Schneid. ●☆

317602 Pseudocydonia sinensis（Thouin）C. K. Schneid.；木瓜（光木瓜，光皮木瓜，海棠，假榅桲，蛮楂，榠楂，榠樝，木梨，木李，瘙楂，铁脚梨，土木瓜，香瓜）；Chinese False-quince，Chinese Flowering Quince，Chinese Floweringquince，Chinese Flowering-quince，Chinese Quince，Chinese-quince，Quince ●

317603 Pseudocydonia sinensis（Thouin）C. K. Schneid. = Chaenomeles sinensis（Thouin）Koehne ●

317604 Pseudocydonia sinensis C. K. Schneid. = Chaenomeles sinensis（Thouin）Koehne ●

317605 Pseudocymopterus J. M. Coult. et Rose（1888）；假聚散翼属■☆

317606 Pseudocymopterus montanus J. M. Coult. et Rose；山地假聚散翼■☆

317607 Pseudocymopterus purpureus（J. M. Coult. et Rose）Rydb.；紫假聚散翼■☆

317608 Pseudocymopterus purpureus Rydb. = Pseudocymopterus montanus J. M. Coult. et Rose ■☆

317609 Pseudocymopterus tidestromii J. M. Coult. et Rose；假聚散翼■☆

317610 Pseudocynometra（Wight et Arn.）Kuntze = Maniltoa Scheff. ■☆

317611 Pseudocynometra Kuntze = Maniltoa Scheff. ■☆

317612 Pseudocyperus Steud. = Fimbristylis Vahl（保留属名）■

317613 Pseudocytisus Kuntze = Vella L. ●☆

317614 Pseudocytisus integrifolius（Salisb.）Rehder = Vella pseudocytisus L. ●☆

317615 Pseudocytisus integrifolius（Salisb.）Rehder subsp. glabrescens（Coss.）Litard. et Maire = Vella pseudocytisus L. subsp. glabrata Greuter ●☆

317616 Pseudocytisus integrifolius（Salisb.）Rehder subsp. iberica Litard. et Maire = Vella pseudocytisus L. ●☆

317617 Pseudocytisus mairei（Humbert）Maire = Vella mairei Humbert ●☆

317618 Pseudodacryodes R. Pierlot（1997）；假蜡烛木属●☆

317619 Pseudodanthonia Bor et C. E. Hubb.（1958）；假扁芒草属（假三蕊草属）■☆

317620 Pseudodanthonia himalaica（Hook. f.）Bor et C. E. Hubb.；假扁芒草■☆

317621 Pseudodanthonia trigyna（Keng）Clayton；三蕊假扁芒草■☆

317622 Pseudodatura Zijp = Brugmansia Pers. ●

317623 Pseudodesmos Spruce ex Engl. = Moronobea Aubl. ●☆

317624 Pseudodichanthium Bor（1940）；假双花草属■☆

317625 Pseudodichanthium cookei（Stapf ex Cooke）M. R. Almeida；假双花草■☆

317626 Pseudodicliptera Benoist（1939）；假狗肝菜属●■☆

317627 Pseudodicliptera coursii Benoist；库尔斯假狗肝菜■☆

317628 Pseudodicliptera humilis Benoist；矮小假狗肝菜■☆

317629 Pseudodicliptera longifolia（Benoist）Benoist；长叶假狗肝菜■☆

317630 Pseudodicliptera sulfureolilacina Benoist；假狗肝菜■☆

317631 Pseudodictamnus Fabr. = Ballota L. ●■☆

317632 Pseudodigera Chiov. = Digera Forssk. ■☆

317633 Pseudodigera pollaccii Chiov. = Digera muricata（L.）Mart. var. trinervis C. C. Towns. ■☆

317634 Pseudodiphryllum Nevski = Platanthera Rich.（保留属名）■

317635 Pseudodiphryllum Nevski（1935）；假对叶兰属■☆

317636 Pseudodiphryllum chorisianum（Cham.）Nevski；假对叶兰■☆

317637 Pseudodiphryllum chorisianum（Cham.）Nevski = Platanthera chorisiana（Cham.）Rchb. f. ■☆

317638 Pseudodiplospora Deb = Diplospora DC. ●

317639 Pseudodiplospora Deb（2001）；假狗骨柴属●☆

317640 Pseudodissochaeta M. P. Nayar = Medinilla Gaudich. ex DC. ●

317641 Pseudodissochaeta M. P. Nayar（1969）；假双毛藤属●☆

317642 Pseudodissochaeta assamica（C. B. Clarke）M. P. Nayar = Medinilla assamica（C. B. Clarke）C. Chen ●

317643 Pseudodissochaeta assamica（C. B. Clarke）Nayar；假双毛藤●☆

317644 Pseudodissochaeta lanceata M. P. Nayar = Medinilla lanceata（M. P. Nayar）C. Chen ●

317645 Pseudodissochaeta roseus（Guillaumin）J. F. Maxwell；粉红假双毛藤●☆

317646 Pseudodissochaeta septentrionalis（W. W. Sm.）M. P. Nayar = Medinilla septentrionalis（W. W. Sm.）H. L. Li ●

317647 Pseudodissochaeta spirei（Guillaumin）Veldkamp et J. F. Maxwell = Medinilla assamica（C. B. Clarke）C. Chen ●

317648 Pseudodissochaeta subsessilis（Craib）M. P. Nayar = Medinilla assamica（C. B. Clarke）C. Chen ●

317649 Pseudodracontium N. E. Br.（1882）；假小龙南星属■☆

317650 Pseudodracontium lacourii N. E. Br.；假小龙南星■☆

317651 Pseudodracontium lanceolatum Serebryanyi；披针叶假小龙南星■☆

317652 Pseudodracontium latifolium Serebryanyi；宽叶假小龙南星■☆

317653 Pseudodracontium macrophyllum Gagnep. ex Serebryanyi；大叶假小龙南星■☆

317654 Pseudodrimys Doweld（2000）；假辛酸木属●☆

317655 Pseudoechinocereus Buining = Borzicactus Riccob. ■☆

317656 Pseudoechinocereus Buining = Cleistocactus Lem. ●☆

317657 Pseudoechinopepon（Cogn.）Cockerell = Vaseyanthus Cogn. ■☆

317658 Pseudoehretia umbellulata（Wall.）Turcz. = Ilex umbellulata（Wall.）Loes. ●

317659 Pseudoelephantopus Rohr = Pseudelephantopus Rohr（保留属名）■

317660 Pseudo-Elephantopus Rohr = Pseudelephantopus Rohr（保留属名）■

317661 Pseudo-elephantopus Steud. = Pseudelephantopus Rohr（保留属名）■

317662 Pseudoelephantopus spicatus（Juss. ex Aubl.）C. F. Baker = Elephantopus spicatus Juss. ex Aubl. ■

317663 Pseudoelephantopus spicatus（Juss. ex Aubl.）Gleason =

Pseudelephantopus spicatus （Juss. ex Aubl.）Gleason ■

317664 Pseudoelephantopus spicatus （Juss.）Rohr ＝ Pseudelephantopus spicatus （Juss. ex Aubl.）Gleason ■

317665 Pseudoentada Britton et Rose ＝ Adenopodia C. Presl ■☆

317666 Pseudoentada Britton et Rose ＝ Entada Adans.（保留属名）●

317667 Pseudoentada rotundifolia （Harms）Guinet ＝ Adenopodia rotundifolia （Harms）Brenan ■☆

317668 Pseudoentada scelerata （A. Chev.）Guinet ＝ Adenopodia scelerata （A. Chev.）Brenan ■☆

317669 Pseudo-eranthemum Radlk. ＝ Pseuderanthemum Radlk. ●■

317670 Pseudoeremostachys Popov ＝ Pseuderemostachys Popov ●☆

317671 Pseudoeremostachys sewertzowii Popov ＝ Pseuderemostachys sewerzovii （Herd.）Popov ■☆

317672 Pseudoeria （Schltr.）Schltr. ＝ Pseuderia Schltr. ■☆

317673 Pseudoeriosema Hauman（1955）；假鸡头薯属■●☆

317674 Pseudoeriosema andongense （Baker）Hauman；安东假鸡头薯（安东鸡头薯）■☆

317675 Pseudoeriosema andongense （Baker）Hauman subsp. bequaertii （De Wild.）Verdc. ＝ Pseudoeriosema andongense （Baker）Hauman ■☆

317676 Pseudoeriosema bequaertii （De Wild.）Hauman ＝ Pseudoeriosema andongense （Baker）Hauman ■☆

317677 Pseudoeriosema borianii （Schweinf.）Hauman；博里假鸡头薯（博里鸡头薯）■☆

317678 Pseudoeriosema borianii （Schweinf.）Hauman subsp. longipedunculatum Verdc. ；长柄假鸡头薯■☆

317679 Pseudoeriosema homblei （De Wild.）Hauman；洪布勒假鸡头薯（洪布勒鸡头薯）■☆

317680 Pseudoeriosema homblei （De Wild.）Hauman var. latistipulatum Hauman ＝ Pseudoeriosema longipes （Harms）Hauman ●☆

317681 Pseudoeriosema longipes （Harms）Hauman；长梗假鸡头薯●☆

317682 Pseudoernestia （Cogn.）Krasser（1893）；拟欧内野牡丹属☆

317683 Pseudoernestia Krasser ＝ Pseudoernestia （Cogn.）Krasser ☆

317684 Pseudoernestia cordifolia （O. Berg ex Triana）Krasser；拟欧内野牡丹☆

317685 Pseudoespostoa Backeb.（1933）；肖老乐柱属；Catton-ball，Catton Ball ■

317686 Pseudoespostoa Backeb. ＝ Espostoa Britton et Rose ●

317687 Pseudoespostoa melanostele （Vaupel）Backeb. ；幻乐（假长毛柱）●☆

317688 Pseudoespostoa melanostele （Vaupel）Backeb. var. rubrispina F. Ritter；梦幻乐（赤刺幻乐，红寿乐）●☆

317689 Pseudoespostoa nana （F. Ritter）Backeb. ；白宫殿●☆

317690 Pseudoeugenia Scort. ＝ Syzygium R. Br. ex Gaertn.（保留属名）●

317691 Pseudoeugonia Scort. ＝ Aphanomyrtus Miq. ●

317692 Pseudoeugonia Scort. ＝ Syzygium R. Br. ex Gaertn.（保留属名）●

317693 Pseudoeurya Yamam. ＝ Eurya Thunb. ●

317694 Pseudoeurya crenatifolia Yamam. ＝ Eurya crenatifolia （Yamam.）Kobuski ●

317695 Pseudoeurystyles Hoehne ＝ Eurystyles Wawra ■☆

317696 Pseudoeverardia Gilly ＝ Everardia Ridl. ■☆

317697 Pseudofortuynia Hedge（1968）；假曲序芥属■☆

317698 Pseudofortuynia esfandiarii Hedge；假曲序芥■☆

317699 Pseudofumaria Medik.（废弃属名）＝ Corydalis DC.（保留属名）■

317700 Pseudo-fumaria Medik. ＝ Corydalis DC.（保留属名）■

317701 Pseudofumaria alba （Mill.）Lidén ＝ Corydalis ochroleuca Koch ■☆

317702 Pseudofumaria lutea Medik. ＝ Corydalis lutea （L.）DC. ■☆

317703 Pseudofumaria ochroleuca （W. D. J. Koch）Holub ＝ Pseudofumaria alba （Mill.）Lidén ■☆

317704 Pseudogaillonia Lincz.（1973）；假加永茜属■☆

317705 Pseudogaillonia Lincz. ＝ Gaillonia A. Rich. ex DC. ■☆

317706 Pseudogaillonia hymenostephana （Jaub. et Spach）Lincz. ；假加永茜■☆

317707 Pseudogaltonia （Kuntze）Engl.（1888）；假夏风信子属■☆

317708 Pseudogaltonia （Kuntze）Engl. ＝ Neogaillonia Lincz. ■☆

317709 Pseudogaltonia Kuntze ＝ Pseudogaltonia （Kuntze）Engl. ■☆

317710 Pseudogaltonia clavata （Mast.）E. Phillips；假夏风信子■☆

317711 Pseudogaltonia pechuelii （Kuntze）Engl. ＝ Pseudogaltonia clavata （Mast.）E. Phillips ■☆

317712 Pseudogaltonia subspicata Baker ＝ Pseudogaltonia clavata （Mast.）E. Phillips ■☆

317713 Pseudogardenia Keay（1958）；假栀子属●☆

317714 Pseudogardenia kalbreyeri （Hiern）Keay ＝ Adenorandia kalbreyeri （Hiern）Robbr. et Bridson ●☆

317715 Pseudogardneria Racib. ＝ Gardneria Wall. ex Roxb. ●

317716 Pseudogardneria angustifolia （Wall.）Racib. ＝ Gardneria angustifolia Wall. ●

317717 Pseudogardneria multiflora （Makino）Pamp. ＝ Gardneria multiflora Makino ●

317718 Pseudogardneria nutans （Siebold et Zucc.）Racib. ＝ Gardneria angustifolia Wall. ●

317719 Pseudogardneria nutans （Siebold et Zucc.）Racib. ＝ Gardneria nutans Siebold et Zucc. ●

317720 Pseudoglochidion Gamble ＝ Glochidion J. R. Forst. et G. Forst.（保留属名）●

317721 Pseudoglossanthis P. P. Poljakov ＝ Glossanthis P. P. Poljakov ■☆

317722 Pseudoglossanthis P. P. Poljakov ＝ Trichanthemis Regel et Schmalh. ■●☆

317723 Pseudoglossanthis P. P. Poljakov（1967）；假毛春黄菊属■●☆

317724 Pseudoglycine F. J. Herm. ＝ Ophrestia H. M. L. Forbes ●■

317725 Pseudoglycine lyallii （Benth.）F. J. Herm. ＝ Ophrestia lyallii （Benth.）Verdc. ■☆

317726 Pseudognaphalium Kirp.（1950）；假鼠麴草属■☆

317727 Pseudognaphalium Kirp. ＝ Gnaphalium L. ■

317728 Pseudognaphalium affine （D. Don）Anderb. ＝ Gnaphalium affine D. Don ■

317729 Pseudognaphalium arizonicum （A. Gray）Anderb. ；亚利桑那假鼠麴草；Arizona Rabbit-tobacco ■☆

317730 Pseudognaphalium attenuatum （DC.）Anderb. ；锥形假鼠麴草；Tapered Cudweed ■☆

317731 Pseudognaphalium austrotexanum G. L. Nesom；南得州假鼠麴草；South Texas Rabbit-tobacco ■☆

317732 Pseudognaphalium beneolens （Davidson）Anderb. ；香假鼠麴草；Fragrant Rabbit-tobacco ■☆

317733 Pseudognaphalium biolettii Anderb. ；比奥莱特假鼠麴草；Bioletti's Rabbit-tobacco ■☆

317734 Pseudognaphalium californicum （DC.）Anderb. ；加州假鼠麴草；California Rabbit-tobacco ■☆

317735 Pseudognaphalium canescens （DC.）Anderb. ；赖特假鼠麴草；Wright's Rabbit-tobacco ■☆

317736 Pseudognaphalium canescens （DC.）Anderb. subsp. beneolens （Davidson）Kartesz ＝ Pseudognaphalium beneolens （Davidson）Anderb. ■☆

317737 Pseudognaphalium canescens （DC.）Anderb. subsp.

microcephalum （Nutt.） Kartesz = Pseudognaphalium microcephalum （Nutt.） Anderb. ■☆

317738 Pseudognaphalium canescens （DC.） Anderb. subsp. thermale （E. E. Nelson） Kartesz = Pseudognaphalium thermale （E. E. Nelson） G. L. Nesom ■☆

317739 Pseudognaphalium chrysocephalum （Franch.） Hilliard et B. L. Burtt = Gnaphalium chrysanthum Y. S. Chen ■

317740 Pseudognaphalium elegans （Kunth） Kartesz；雅致假鼠麴草；Royal Cudweed ■☆

317741 Pseudognaphalium flavescens （Kitam.） Anderb. = Gnaphalium flavescens Kitam. ■

317742 Pseudognaphalium helleri （Britton） Anderb.；黑勒假鼠麴草；Cat's-foot，Cudweed，Heller's Rabbit-tobacco，Heller's Cudweed ■☆

317743 Pseudognaphalium helleri （Britton） Anderb. = Gnaphalium helleri Britton ■☆

317744 Pseudognaphalium helleri （Britton） Anderb. subsp. micradenium （Weath.） Kartesz = Pseudognaphalium micradenium （Weath.） G. L. Nesom ■☆

317745 Pseudognaphalium hypoleucum （DC.） Hilliard et B. L. Burtt = Gnaphalium hypoleucum DC. ■

317746 Pseudognaphalium jaliscense （Greenm.） Anderb.；哈利斯科假鼠麴草；Jalisco Rabbit-tobacco ■☆

317747 Pseudognaphalium leucocephalum （A. Gray） Anderb.；白头假鼠麴草；White Rabbit-tobacco ■☆

317748 Pseudognaphalium luteoalbum （L.） Hilliard et B. L. Burtt；泽西假鼠麴草；Jersey Cudweed，Jersey Rabbit-tobacco，Red-tip Rabbit-tobacco ■☆

317749 Pseudognaphalium luteoalbum （L.） Hilliard et B. L. Burtt = Gnaphalium luteoalbum L. ■

317750 Pseudognaphalium luteoalbum （L.） Hilliard et B. L. Burtt subsp. affine （D. Don） Hilliard et B. L. Burtt = Gnaphalium affine D. Don ■

317751 Pseudognaphalium luteoalbum （L.） Hilliard et B. L. Burtt subsp. affine （D. Don） Hilliard et B. L. Burtt = Pseudognaphalium affine （D. Don） Anderb. ■

317752 Pseudognaphalium luteoalbum （L.） Hilliard et B. L. Burtt subsp. affine （D. Don） Hilliard et B. L. Burtt；近缘黄白假鼠麴草 ■☆

317753 Pseudognaphalium macounii （Greene） Kartesz；马昆假鼠麴草；Clammy Cudweed，Clammy Everlasting，Macoun's Rabbit-tobacco，Western Cudweed ■☆

317754 Pseudognaphalium macounii （Greene） Kartesz = Gnaphalium macounii Greene ■☆

317755 Pseudognaphalium melanosphaerum （A. Rich.） Hilliard；黑球假鼠麴草 ■☆

317756 Pseudognaphalium micradenium （Weath.） G. L. Nesom；小巧假鼠麴草；Cat's-foot，Cudweed，Delicate Rabbit-tobacco，Everlasting，Heller's Cudweed ■☆

317757 Pseudognaphalium micradenium （Weath.） G. L. Nesom = Gnaphalium helleri Britton var. micradenium （Weath.） Mahler ■☆

317758 Pseudognaphalium micradenium （Weath.） Nesom；小腺假鼠麴草 ■☆

317759 Pseudognaphalium microcephalum （Nutt.） Anderb.；小头假鼠麴草；San Diego Rabbit-tobacco ■☆

317760 Pseudognaphalium microcephalum （Nutt.） Anderb. var. thermale （E. E. Nelson） Dorn = Pseudognaphalium thermale （E. E. Nelson） G. L. Nesom ■☆

317761 Pseudognaphalium obtusifolium （L.） Hilliard et B. L. Burtt；钝

叶假鼠麴草；Catfoot，Eastern Rabbit-tobacco，Fragrant Cudweed，Old-field Balsam，Sweet Everlasting ■☆

317762 Pseudognaphalium obtusifolium （L.） Hilliard et B. L. Burtt = Gnaphalium obtusifolium L. ■☆

317763 Pseudognaphalium obtusifolium （L.） Hilliard et B. L. Burtt var. saxicola （Fassett） Kartesz = Pseudognaphalium saxicola （Fassett） H. E. Ballard et Feller ■☆

317764 Pseudognaphalium oligandrum （DC.） Hilliard et B. L. Burtt；寡蕊假鼠麴草 ■☆

317765 Pseudognaphalium petitianum （A. Rich.） Mesfin；佩蒂蒂假鼠麴草 ■☆

317766 Pseudognaphalium pringlei （A. Gray） Anderb.；普林格尔假鼠麴草；Pringle's Rabbit-tobacco ■☆

317767 Pseudognaphalium ramosissimum （Nutt.） Anderb.；粉红假鼠麴草；Pink Rabbit-tobacco ■☆

317768 Pseudognaphalium richardianum （Cufod.） Hilliard et B. L. Burtt；理氏假鼠麴草 ■☆

317769 Pseudognaphalium roseum （Kunth） Anderb.；玫瑰假鼠麴草；Rosy Cudweed，Rosy Rabbit-tobacco ■☆

317770 Pseudognaphalium saxicola （Fassett） H. E. Ballard et Feller；峭壁假鼠麴草；Cliff Cudweed，Rabbit-tobacco ■☆

317771 Pseudognaphalium saxicola （Fassett） H. E. Ballard et Feller = Gnaphalium saxicola Fassett ■☆

317772 Pseudognaphalium stramineum （Kunth） Anderb.；禾秆色假鼠麴草；Cotton-batting-plant ■☆

317773 Pseudognaphalium thermale （E. E. Nelson） G. L. Nesom；西北假鼠麴草；Northwestern Rabbit-tobacco ■☆

317774 Pseudognaphalium undulatum （L.） Hilliard et B. L. Burtt；波状假鼠麴草 ■☆

317775 Pseudognaphalium viscosum （Kunth） Anderb.；黏假鼠麴草；Sticky Rabbit-tobacco ■☆

317776 Pseudognidia E. Phillips = Gnidia L. ●☆

317777 Pseudognidia anomala （Meisn.） E. Phillips = Gnidia anomala Meisn. ●☆

317778 Pseudogomphrena R. E. Fr. （1920）；假千日红属 ■☆

317779 Pseudogomphrena scandens R. E. Fr.；假千日红 ■☆

317780 Pseudogonocalyx Bisse et Berazain = Schoepfia Schreb. ●

317781 Pseudogonocalyx Bisse et Berazain（1984）；假棱萼杜鹃属 ■☆

317782 Pseudogonocalyx paradoxa Bisse et R. Berazaín；假棱萼杜鹃 ●☆

317783 Pseudogoodyera Schltr. （1920）；假斑叶兰属 ■☆

317784 Pseudogoodyera wrightii （Rchb. f.） Schltr.；假斑叶兰 ■☆

317785 Pseudo-gunnera Oerst. = Gunnera L. ■☆

317786 Pseudogynoxys （Greenm.） Cabrera（1950）；蔓黄金菊属 ■●☆

317787 Pseudogynoxys chenopodioides （Kunth） Cabrera；蔓黄金菊；Mexican Flamevine ■☆

317788 Pseudohamelia Wernham（1912）；假长隔木属 ●☆

317789 Pseudohamelia hirsuta Wernham；假长隔木 ●☆

317790 Pseudohandelia Tzvelev（1961）；假天山蓍属（蓍菊属，腺果菊属）■

317791 Pseudohandelia umbellifera （Boiss.） Tzvelev；假天山蓍（腺果菊）■

317792 Pseudohemipilia Szlach. （2003）；莫桑比克兰属 ■☆

317793 Pseudohexadesmia Brieger = Hexadesmia Brongn. ■☆

317794 Pseudohexadesmia Brieger = Scaphyglottis Poepp. et Endl. （保留属名）☆

317795 Pseudohomalomena A. D. Hawkes = Zantedeschia Spreng. （保留属名）■

317796　Pseudohydrosme Engl.（1892）;假魔芋属■☆

317797　Pseudohydrosme buettneri Engl.;假魔芋■☆

317798　Pseudohydrosme gabunensis Engl.;加蓬假魔芋■☆

317799　Pseudo-Iris Medik. = Iris L. ■

317800　Pseudojacobaea（Hook. f.）R. Mathur = Senecio L. ■●

317801　Pseudojacobaea lavandulifolius（DC.）R. Mathur = Senecio lavandulifolius Wall. ■☆

317802　Pseudokyrsteniopsis R. M. King et H. Rob.（1973）;假展毛修泽兰属■☆

317803　Pseudokyrsteniopsis perpetiolata R. M. King et H. Rob.;假展毛修泽兰■☆

317804　Pseudolabatia Aubrév. et Pellegr. = Pouteria Aubl. ●

317805　Pseudolachnostylis Pax（1899）;假毛柱大戟属●☆

317806　Pseudolachnostylis bussei Pax ex Hutch. = Pseudolachnostylis maprouneifolia Pax var. glabra（Pax）Brenan ●☆

317807　Pseudolachnostylis dekindtii Pax = Pseudolachnostylis maprouneifolia Pax var. dekindtii（Pax）Radcl. -Sm.●☆

317808　Pseudolachnostylis dekindtii Pax var. glabra Pax = Pseudolachnostylis maprouneifolia Pax var. glabra（Pax）Brenan ●☆

317809　Pseudolachnostylis glauca（Hiern）Hutch. = Pseudolachnostylis maprouneifolia Pax var. glabra（Pax）Brenan ●☆

317810　Pseudolachnostylis maprouneifolia Pax;马龙戟;Duiker Tree ●☆

317811　Pseudolachnostylis maprouneifolia Pax var. dekindtii（Pax）Radcl. -Sm.;德金假毛柱大戟●☆

317812　Pseudolachnostylis maprouneifolia Pax var. glabra（Pax）Brenan;光滑假毛柱大戟●☆

317813　Pseudolachnostylis maprouneifolia Pax var. polygyna（Pax et K. Hoffm.）Radcl. -Sm.;多雌蕊假毛柱大戟●☆

317814　Pseudolachnostylis polygyna Pax et K. Hoffm. = Pseudolachnostylis maprouneifolia Pax var. polygyna（Pax et K. Hoffm.）Radcl. -Sm.●☆

317815　Pseudolachnostylis verdickii De Wild. = Pseudolachnostylis maprouneifolia Pax var. glabra（Pax）Brenan ●☆

317816　Pseudolaelia Porto = Pseudolaelia Porto et Brade ■☆

317817　Pseudolaelia Porto et Brade = Schomburgkia Lindl. ■☆

317818　Pseudolaelia Porto et Brade（1935）;假蕾丽兰属■☆

317819　Pseudolaelia corcovadensis Porto et Brade;假蕾丽兰■☆

317820　Pseudolariacis（A. Camus）A. Camus = Lasiacis（Griseb.）Hitchc. ■☆

317821　Pseudolarix Gordon（1858）;金钱松属;False-larch, Golden Larch, Goldenlarch, Golden-larch, Goldlarch ●★

317822　Pseudolarix amabilis（J. Nelson）Rehder;金钱松（金松,金叶松,荆皮树,荆树,山松,水树,水松,天枞,土槿,土荆）;China Golden Larch, Chinese Golden Larch, Chinese Larch, Coin Larch, Golden Larch, Golden Pine, Golden-larch, Lovely Golden Larch, Lovely Golden-larch ●◇

317823　Pseudolarix amabilis（J. Nelson）Rehder = Pseudolarix kaempferi（Lindl.）Gordon ●◇

317824　Pseudolarix fortunei Mayr = Pseudolarix amabilis（J. Nelson）Rehder ●◇

317825　Pseudolarix fortunei Mayr = Pseudolarix kaempferi（Lindl.）Gordon ●◇

317826　Pseudolarix kaempferi（Lindl.）Gordon = Pseudolarix amabilis（J. Nelson）Rehder ●◇

317827　Pseudolarix kaempferi Gordon = Pseudolarix amabilis（J. Nelson）Rehder ●◇

317828　Pseudolarix pourteti Ferre = Pseudolarix amabilis（J. Nelson）Rehder ●◇

317829　Pseudolarix pourteti Ferre = Pseudolarix kaempferi（Lindl.）Gordon ●◇

317830　Pseudolasiacis（A. Camus）A. Camus = Lasiacis（Griseb.）Hitchc. ■☆

317831　Pseudolasiacis（A. Camus）A. Camus（1945）;假毛尖草属■☆

317832　Pseudolasiacis bathiei A. Camus;假毛尖草■☆

317833　Pseudolepanthes（Luer）Archila = Trichosalpinx Luer ■☆

317834　Pseudolepanthes（Luer）Archila（2000）;假鳞花兰属■☆

317835　Pseudolephantopus Rohr = Pseudelephantopus Rohr（保留属名）■

317836　Pseudolephantopus spicatus（Juss. ex Aubl.）Gleason = Pseudelephantopus spicatus（Juss. ex Aubl.）Gleason ■

317837　Pseudolgsimachion Opiz = Veronica L. ■

317838　Pseudoligandra Dillon et Sagast. = Chionolaena DC. ■☆

317839　Pseudolinosyris Novopokr.（1918）;假麻菀属■☆

317840　Pseudolinosyris Novopokr. = Linosyris Cass. ■

317841　Pseudolinosyris capusi Novopokr.;假麻菀■☆

317842　Pseudolinosyris grimmii Novopokr.;格氏假麻菀（格氏类麻菀）■☆

317843　Pseudolinosyris microcephala（Novopokr.）Tamamsch.;小头假麻菀（小头类麻菀）■☆

317844　Pseudolinosyris sintenisii（Bornm.）Tamamsch.;西氏假麻菀（西氏类麻菀）■☆

317845　Pseudoliparis Finet = Crepidium Blume ■

317846　Pseudoliparis Finet = Malaxis Sol. ex Sw. ■

317847　Pseudoliparis ramosii（Ames）Marg. et Szlach. = Crepidium ramosii（Ames）Szlach. ■

317848　Pseudolitchi Danguy et Choux = Stadmannia Lam. ●☆

317849　Pseudolithos P. R. O. Bally（1965）;假石萝藦属■☆

317850　Pseudolithos cubiformis（P. R. O. Bally）P. R. O. Bally;管状假石萝藦■☆

317851　Pseudolithos cubiformis（P. R. O. Bally）P. R. O. Bally var. viridiflorus Horw. = Pseudolithos cubiformis（P. R. O. Bally）P. R. O. Bally ■☆

317852　Pseudolithos dodsonianus（Lavranos）Bruyns et Meve;多德森假石萝藦■☆

317853　Pseudolithos gigas Dioli;巨大假石萝藦■☆

317854　Pseudolithos horwoodii P. R. O. Bally et Lavranos;霍伍得假石萝藦■☆

317855　Pseudolithos migiurtinus（Chiov.）P. R. O. Bally;米朱蒂假石萝藦■☆

317856　Pseudolithos sphaericus（P. R. O. Bally）P. R. O. Bally = Pseudolithos migiurtinus（Chiov.）P. R. O. Bally ■☆

317857　Pseudolitsea Yen C. Yang = Litsea Lam.（保留属名）●

317858　Pseudolitsea tsaii Yen C. Yang = Litsea pedunculata（Diels）Yen C. Yang et P. H. Huang ●

317859　Pseudolmedia H. Karat. = Olmediophaena H. Karst. ●☆

317860　Pseudolmedia Trécul（1847）;假牛筋属●☆

317861　Pseudolmedia alnifolia Rusby;桤叶假牛筋树●☆

317862　Pseudolmedia boliviana C. C. Berg et Villav.;玻利维亚假牛筋树●☆

317863　Pseudolmedia ferruginea Trécul;锈色假牛筋树●☆

317864　Pseudolmedia laevis J. F. Macbr.;平滑假牛筋树●☆

317865　Pseudolmedia macrophylla Trécul;大叶假牛筋树●☆

317866　Pseudolmedia multinervis Mildbr.;多脉假牛筋树●☆

317867　Pseudolmedia rigida（Klotzsch et H. karst.）Cuatrec.;硬假牛

筋树●☆

317868　Pseudolobelia A. Chev. = Torenia L. ■

317869　Pseudolobivia（Backeb.）Backeb.（1942）；肖丽花属■☆

317870　Pseudolobivia（Backeb.）Backeb. = Echinopsis Zucc. ●

317871　Pseudolobivia Backeb. = Echinopsis Zucc. ●

317872　Pseudolobivia Backeb. = Pseudolobivia（Backeb.）Backeb. ■☆

317873　Pseudolobivia ancistrophora（Speg.）Backeb. ex Krainz = Echinopsis ancistrophora Speg. ■☆

317874　Pseudolobivia aurea（Britton et Rose）Backeb. = Echinopsis aurea Britton et Rose ■☆

317875　Pseudolobivia aurea（Britton et Rose）Backeb. var. elegans Backeb. = Lobivia aurea Backeb. var. elegans Backeb. ■☆

317876　Pseudolobivia aurea（Britton et Rose）Backeb. var. fallax（Oehme）Backeb. = Echinopsis aurea Britton et Rose var. fallax（Oehme）J. Ullmann ■☆

317877　Pseudolobivia aurea（Britton et Rose）Backeb. var. fallax（Oehme）Backeb. = Lobivia fallax Oehme ■☆

317878　Pseudolobivia ferox（Britton et Rose）Backeb. = Echinopsis ferox（Britton et Rose）Backeb. ■☆

317879　Pseudolobivia hamatacantha（Backeb.）Backeb. = Echinopsis haematantha（Speg.）D. R. Hunt ■☆

317880　Pseudolobivia kermesina Krainz = Echinopsis kermesina（Krainz）Krainz ■☆

317881　Pseudolobivia kratochviliana（Backeb.）Backeb. = Echinopsis kratochviliana Backeb. ■☆

317882　Pseudolobivia leucorhodantha（Backeb.）Backeb. ex Krainz = Echinopsis leucorhodantha Backeb. ■☆

317883　Pseudolobivia longispina（Britton et Rose）Backeb. ex Krainz = Echinopsis longispina（Britton et Rose）Werderm. ■☆

317884　Pseudolobivia obrepanda（Salm-Dyck）Backeb. ex Krainz = Echinopsis obrepanda（Salm-Dyck）K. Schum. ■☆

317885　Pseudolobivia pelecyrhachis（Backeb.）Backeb. ex Krainz = Echinopsis pelecyrhachis Backeb. ■☆

317886　Pseudolobivia polyancistra（Backeb.）Backeb. = Echinopsis polyancistra Backeb. ■☆

317887　Pseudolobivia potosina（Werderm.）Backeb. ex Krainz = Echinopsis potosina Werderm. ■☆

317888　Pseudolobivia rojasii（Cárdenas）Backeb. = Echinopsis rojasii Cárdenas ■☆

317889　Pseudolopezia Rose = Lopezia Cav. ■☆

317890　Pseudolophanthus Kuprian. = Pseudolophanthus Levin ■

317891　Pseudolophanthus Levin = Marmorites Benth. ■

317892　Pseudolophanthus Levin = Phyllophyton Kudo ■★

317893　Pseudolophanthus complanatus（Dunn）Levin = Marmoritis complanata（Dunn）A. L. Budantzev ■

317894　Pseudolophanthus decolorans（Hemsl.）Levin = Marmoritis decolorans（Hemsl.）H. W. Li ■

317895　Pseudolophanthus nivalis（Jacq. ex Benth.）Levin = Marmoritis nivalis（Jacq. ex Benth.）Hedge ■

317896　Pseudolophanthus nivalis（Jacq.）Levin = Marmoritis nivalis（Jacq. ex Benth.）Hedge ■

317897　Pseudolophanthus nivalis（Jacq.）Levin = Phyllophyton nivale（Jacq. ex Benth.）C. Y. Wu ■

317898　Pseudolophanthus pharicus（Prain）Kupr. = Marmoritis pharica（Prain）A. L. Budantzev ■

317899　Pseudolophanthus tibeticus（Jacq. ex Benth.）Kupr. = Marmoritis rotundifolia Benth. ■

317900　Pseudolophanthus tibeticus（Jacq.）Kupr. = Marmoritis rotundifolia Benth. ■

317901　Pseudolotus Rech. f. = Lotus L. ■

317902　Pseudolotus makranicus（Rech. f. et Esfand.）Rech. f. = Lotus makranicus Rech. f. et Esfand. ■☆

317903　Pseudoludovia Harling = Sphaeradenia Harling ■☆

317904　Pseudolysimachion（W. D. J. Koch）Opiz（1852）；穗花属（水萝卜属,水蔓菁属,兔尾草属）■

317905　Pseudolysimachion Opiz = Pseudolysimachion（W. D. J. Koch）Opiz ■

317906　Pseudolysimachion Opiz = Veronica L. ■

317907　Pseudolysimachion alatavicum（Popov）Holub；阿拉套穗花（阿拉套婆婆纳）；Alatao Speedwell, Alataw Speedwell ■

317908　Pseudolysimachion alatavicum（Popov）Holub = Veronica alatavica Popov ■

317909　Pseudolysimachion dahuricum（Steven）Holub；大穗花（大婆婆纳,心叶婆婆纳）；Dahur Speedwell, Dahurian Speedwell, Heartleaf Speedwell ■

317910　Pseudolysimachion dahuricum（Steven）Holub = Veronica dahurica Steven ■

317911　Pseudolysimachion galactites（Hance）Holub = Pseudolysimachion linariifolium（Pall. ex Link）Holub subsp. dilatatum（Nakai et Kitag.）D. Y. Hong ■

317912　Pseudolysimachion incanum（L.）Holub；白兔儿尾苗（白婆婆纳,毛叶水苦荬）；Silver Speedwell, White Speedwell, Woolly Speedwell ■

317913　Pseudolysimachion incanum（L.）Holub = Veronica incana L. ■

317914　Pseudolysimachion kiusianum（Furumi）Holub = Pseudolysimachion ovatum（Nakai）T. Yamaz. subsp. kiusianum（Furumi）T. Yamaz. ■☆

317915　Pseudolysimachion kiusianum（Furumi）Holub subsp. maritimum（Nakai）T. Yamaz. = Pseudolysimachion ovatum（Nakai）T. Yamaz. subsp. maritimum（Nakai）T. Yamaz. ■☆

317916　Pseudolysimachion kiusianum（Furumi）Holub subsp. maritimum（Nakai）T. Yamaz. var. canescens（Satake）T. Yamaz. = Pseudolysimachion ovatum（Nakai）T. Yamaz. subsp. maritimum（Nakai）T. Yamaz. f. canescens（Satake）T. Yamaz. ■☆

317917　Pseudolysimachion kiusianum（Furumi）Holub subsp. miyabei（Nakai et Honda）T. Yamaz. = Pseudolysimachion ovatum（Nakai）T. Yamaz. subsp. miyabei（Nakai et Honda）T. Yamaz. ■☆

317918　Pseudolysimachion kiusianum（Furumi）Holub subsp. miyabei（Nakai et Honda）T. Yamaz. var. villosum（Furumi）T. Yamaz. = Pseudolysimachion ovatum（Nakai）T. Yamaz. subsp. miyabei（Nakai et Honda）T. Yamaz. var. villosum（Furumi）T. Yamaz. ■☆

317919　Pseudolysimachion kiusianum（Furumi）Holub subsp. miyabei（Nakai et Honda）T. Yamaz. var. japonicum（Miq.）T. Yamaz. = Pseudolysimachion ovatum（Nakai）T. Yamaz. subsp. miyabei（Nakai et Honda）T. Yamaz. var. japonicum（Miq.）T. Yamaz. ■☆

317920　Pseudolysimachion kiusianum（Furumi）Holub var. kitadakemontanum（T. Yamaz.）T. Yamaz. = Pseudolysimachion ovatum（Nakai）T. Yamaz. subsp. kiusianum（Furumi）T. Yamaz. var. kitadakemontanum（T. Yamaz.）T. Yamaz. ■☆

317921　Pseudolysimachion kiusianum（Furumi）T. Yamaz.；长毛穗花（长毛婆婆纳）；Longhairy Speedwell ■

317922　Pseudolysimachion kiusianum（Furumi）T. Yamaz. = Veronica kiusiana Furumi ■

317923　Pseudolysimachion laetum（Kar. et Kir.）Holub =

Pseudolysimachion pinnatum (L.) Holub ■

317924　Pseudolysimachion lineariifolium (Pall. ex Link) Holub;细叶穗花(高山婆婆纳,勒马回,蜈蚣草,细叶婆婆纳,一支香,斩龙剑,追风草);Linearleaf Speedwell ■

317925　Pseudolysimachion lineariifolium (Pall. ex Link) Holub subsp. dilatatum (Nakai et Kitag.) D. Y. Hong;水蔓菁(宽叶婆婆纳,气管炎草,蜈蚣草,细叶婆婆纳,细叶穗花,一枝香,斩龙剑,追风草);Dilatate Linearleaf Speedwell ■

317926　Pseudolysimachion longifolium (L.) Opiz;长叶穗花■☆

317927　Pseudolysimachion longifolium (L.) Opiz;兔儿尾苗(长尾婆婆纳,长叶婆婆纳,四方麻,兔耳尾苗);Beach Speedwell, Clump Speedwell, Garden Speedwell, Garden Veronica, Longleaf Speedwell, Long-leaved Speedwell, Rabbitleaf Speedwell, Seaside Speedwell, Speedwell ■

317928　Pseudolysimachion longifolium (L.) Opiz = Veronica longifolia L. ■

317929　Pseudolysimachion miyabei (Nakai et Honda) Holub = Pseudolysimachion ovatum (Nakai) T. Yamaz. subsp. miyabei (Nakai et Honda) T. Yamaz. ■☆

317930　Pseudolysimachion ogurae T. Yamaz.;御座穗花■☆

317931　Pseudolysimachion ornatum (Monjuschko) Holub;装饰穗花■☆

317932　Pseudolysimachion ovatum (Nakai) T. Yamaz.;卵叶穗花■☆

317933　Pseudolysimachion ovatum (Nakai) T. Yamaz. subsp. kiusianum (Furumi) T. Yamaz.;九州穗花■☆

317934　Pseudolysimachion ovatum (Nakai) T. Yamaz. subsp. kiusianum (Furumi) T. Yamaz. var. kitadakemontanum (T. Yamaz.) T. Yamaz.;信州北岳穗花■☆

317935　Pseudolysimachion ovatum (Nakai) T. Yamaz. subsp. maritimum (Nakai) T. Yamaz.;滨海穗花■☆

317936　Pseudolysimachion ovatum (Nakai) T. Yamaz. subsp. maritimum (Nakai) T. Yamaz. f. canescens (Satake) T. Yamaz.;灰滨海穗花■☆

317937　Pseudolysimachion ovatum (Nakai) T. Yamaz. subsp. miyabei (Nakai et Honda) T. Yamaz.;宫部氏穗花■☆

317938　Pseudolysimachion ovatum (Nakai) T. Yamaz. subsp. miyabei (Nakai et Honda) T. Yamaz. var. villosum (Furumi) T. Yamaz.;毛宫部氏穗花■☆

317939　Pseudolysimachion ovatum (Nakai) T. Yamaz. subsp. miyabei (Nakai et Honda) T. Yamaz. var. japonicum (Miq.) T. Yamaz.;日本毛宫部氏穗花■☆

317940　Pseudolysimachion pinnatum (L.) Holub;羽叶穗花(羽裂婆婆纳,羽叶婆婆纳);Pinnate Speedwell ■

317941　Pseudolysimachion pinnatum (L.) Holub = Veronica pinnata L. ■

317942　Pseudolysimachion rotundum (Nakai) Holub var. petiolatum (Nakai) T. Yamaz. f. albiflorum (H. Hara) T. Yamaz.;白花具柄羽叶穗花■☆

317943　Pseudolysimachion rotundum (Nakai) Holub var. petiolatum (Nakai) T. Yamaz.;具柄羽叶穗花■☆

317944　Pseudolysimachion rotundum (Nakai) Holub var. subintegrum (Nakai) T. Yamaz. f. petiolatum (Nakai) T. Yamaz. = Pseudolysimachion rotundum (Nakai) Holub var. petiolatum (Nakai) T. Yamaz. ■☆

317945　Pseudolysimachion rotundum (Nakai) T. Yamaz.;无柄穗花(无柄婆婆纳);Sessile Speedwell ■

317946　Pseudolysimachion rotundum (Nakai) T. Yamaz. = Veronica rotunda Nakai ■

317947　Pseudolysimachion rotundum (Nakai) T. Yamaz. var. coreanum (Nakai) D. Y. Hong;朝鲜穗花(朝鲜婆婆纳);Korea Speedwell, Korean Speedwell ■

317948　Pseudolysimachion rotundum (Nakai) T. Yamaz. var. coreanum (Nakai) D. Y. Hong = Veronica rotunda Nakai var. coreana (Nakai) T. Yamaz. ■

317949　Pseudolysimachion rotundum (Nakai) T. Yamaz. var. subintegrum (Nakai) D. Y. Hong;东北穗花(东北婆婆纳);NE. China Speedwell ■

317950　Pseudolysimachion rotundum (Nakai) T. Yamaz. var. subintegrum (Nakai) D. Y. Hong = Veronica rotunda Nakai var. subintegra (Nakai) T. Yamaz. ■

317951　Pseudolysimachion schmidtianum (Regel) T. Yamaz.;斯氏穗花■☆

317952　Pseudolysimachion schmidtianum (Regel) T. Yamaz. = Veronica schmidtiana Regel ■☆

317953　Pseudolysimachion schmidtianum (Regel) T. Yamaz. f. albiflorum (Sugaw.) T. Yamaz.;白花斯氏穗花■☆

317954　Pseudolysimachion schmidtianum (Regel) T. Yamaz. f. candidum (T. Yamaz.) T. Yamaz.;纯白斯氏穗花■☆

317955　Pseudolysimachion schmidtianum (Regel) T. Yamaz. subsp. senanense (Maxim.) T. Yamaz. f. bandaianum (Makino) T. Yamaz. = Astilbe odontophylla Miq. var. bandaica (Honda) H. Hara ■☆

317956　Pseudolysimachion schmidtianum (Regel) T. Yamaz. subsp. senanense (Maxim.) T. Yamaz. f. daisenense (Makino) T. Yamaz.;大山斯氏穗花■☆

317957　Pseudolysimachion schmidtianum (Regel) T. Yamaz. subsp. senanense (Maxim.) T. Yamaz. f. tomentosum (T. Yamaz.) T. Yamaz.;绒毛斯氏穗花■☆

317958　Pseudolysimachion schmidtianum (Regel) T. Yamaz. subsp. senanense (Maxim.) T. Yamaz.;信浓斯氏穗花■☆

317959　Pseudolysimachion schmidtianum (Regel) T. Yamaz. subsp. senanense (Maxim.) T. Yamaz. var. bandaianum (Makino) T. Yamaz.;班代穗花■☆

317960　Pseudolysimachion schmidtianum (Regel) T. Yamaz. var. yezoalpinum (Koidz. ex H. Hara) T. Yamaz. f. exiguum (Takeda) T. Yamaz.;弱小斯氏穗花■☆

317961　Pseudolysimachion schmidtianum (Regel) T. Yamaz. var. yezoalpinum (Koidz. ex H. Hara) T. Yamaz.;北海道山地穗花■☆

317962　Pseudolysimachion schmidtianum (Regel) T. Yamaz. var. yezoalpinum (Koidz. ex H. Hara) T. Yamaz. f. pubescens (H. Hara) T. Yamaz.;短柔毛斯氏穗花■☆

317963　Pseudolysimachion sieboldianum (Miq.) Holub;西氏穗花■☆

317964　Pseudolysimachion spicatum (L.) Opiz;穗花(密穗水苦荬,穗花婆婆纳,穗花水苦荬,穗婆婆纳);Bastard Speedwell, Speedwell, Spike Speedwell, Spiked Speedwell, Upright Speedwell ■

317965　Pseudolysimachion spicatum (L.) Opiz = Veronica spicata L. ■

317966　Pseudolysimachion spurium (L.) Rauschert;轮叶穗花(狗日巴花,狼尾拉花,轮叶婆婆纳,气管炎草,水蔓菁,一枝香);Bastard Speedwell ■

317967　Pseudolysimachion spurium (L.) Rauschert = Veronica spuria L. ■

317968　Pseudolysimachion subsessile (Miq.) Holub;近无柄穗花(无柄婆婆纳)■☆

317969　Pseudolysimachion subsessile (Miq.) Holub f. albiflorum T. Yamaz.;白花近无柄穗花■☆

317970　Pseudolysimachion subsessile (Miq.) Holub var. ibukiense T. Yamaz.;伊吹穗花■☆

317971　Pseudo-Lysimachium（W. D. J. Koch）Opiz ＝ Pseudolysimachion（W. D. J. Koch）Opiz ■

317972　Pseudo-Lysimachium（W. D. J. Koch）Opiz ＝ Veronica L. ■

317973　Pseudomachaerium Hassl.（1906）;拟军刀豆属☆

317974　Pseudomachaerium Hassl. ＝ Nissolia Jacq.（保留属名）■☆

317975　Pseudomachaerium rojasianum Hassl.;拟军刀豆☆

317976　Pseudomacodes Rolfe ＝ Macodes（Blume）Lindl. ■☆

317977　Pseudomacrolobium Hauman(1952);刚果异萼豆属●☆

317978　Pseudomacrolobium mengei（De Wild.）Hauman;刚果异萼豆☆

317979　Pseudomalachra H. Monteiro ＝ Sida L. ●■

317980　Pseudomalmea Chatrou ＝ Malmea R. E. Fr. ●☆

317981　Pseudomalmea Chatrou（1998）;假马尔木属●☆

317982　Pseudomammillaria Buxb. ＝ Mammillaria Haw.（保留属名）●

317983　Pseudomantalania J. -F. Leroy（1973）;拟曼塔茜属☆

317984　Pseudomantalania macrophylla J. -F. Leroy;大叶拟曼塔茜☆

317985　Pseudomariscus Rauschert ＝ Courtoisina Soják ■

317986　Pseudomariscus cyperoides（Roxb.）Rauschert ＝ Courtoisina cyperoides（Roxb.）Soják ■

317987　Pseudomarrubium Popov(1940);类欧夏至草属■☆

317988　Pseudomarrubium eremostachydioides Popov;类欧夏至草■☆

317989　Pseudomarsdenia Baill.;类牛奶菜属●☆

317990　Pseudomarsdenia Baill. ＝ Marsdenia R. Br.（保留属名）●

317991　Pseudomarsdenia bourgeana Baill.;类牛奶菜●☆

317992　Pseudomaxillaria Hoehne ＝ Maxillaria Ruiz et Pav. ■☆

317993　Pseudomaxillaria Hoehne ＝ Ornithidium R. Br. ■☆

317994　Pseudomelasma Eb. Fisch.（1996）;假黑蒴属■☆

317995　Pseudomelasma pedicularioides（Baker）Eb. Fisch.;假黑蒴☆

317996　Pseudomelissitus Ovcz., Rassulova et Kinzik. ＝ Medicago L.（保留属名）●■

317997　Pseudomelissitus Ovcz.,Rassulova et Kinzik. ＝ Radiata Medik. ●■

317998　Pseudomertensia Riedl ＝ Oreocharis（Decne.）Lindl.（废弃属名）■

317999　Pseudomertensia Riedl(1967);假滨紫草属■

318000　Pseudomertensia chitralensis（Riedl）Riedl;吉德拉尔假滨紫草■☆

318001　Pseudomertensia drummondii Kazmi;德拉蒙德假滨紫草■☆

318002　Pseudomertensia echioides（Benth.）Riedl;刺假滨紫草■☆

318003　Pseudomertensia efornicata（Rech. f. et Riedl）Riedl;无拱长假滨紫草■☆

318004　Pseudomertensia elongata（Decne.）Riedl;伸长假滨紫草■☆

318005　Pseudomertensia moltkioides（Royle ex Benth.）Kazmi;弯果假滨紫草■☆

318006　Pseudomertensia moltkioides（Royle ex Benth.）Kazmi var. primuloides ?;报春假滨紫草■☆

318007　Pseudomertensia moltkioides（Royle ex Benth.）Kazmi var. tanneri ?;坦纳假滨紫草■☆

318008　Pseudomertensia nemorosa（DC.）R. R. Stewart et Kazmi;森林假滨紫草■☆

318009　Pseudomertensia parvifolia（Decne.）Riedl;小叶假滨紫草■☆

318010　Pseudomertensia primuloides（Decne.）Riedl ＝ Pseudomertensia moltkioides（Royle ex Benth.）Kazmi var. primuloides ? ■☆

318011　Pseudomertensia racemosa（Royle）Kazmi;总花假滨紫草■☆

318012　Pseudomertensia rosulata（Ovcz. et Czukav.）Ovcz. et Czukav. ＝ Scapicephalus rosulatus Ovcz. et Czukav. ■☆

318013　Pseudomertensia sericophylla（Riedl）Y. J. Nasir;绢毛叶假滨紫草■☆

318014　Pseudomertensia trollii（Melch.）R. R. Stewart et Kazmi;特洛尔假滨紫草■☆

318015　Pseudomertensia trollii（Melch.）R. R. Stewart et Kazmi var. harrissii ?;哈里斯假滨紫草■☆

318016　Pseudomiltemia Borhidi ＝ Kohleria Regel ●■☆

318017　Pseudomiltemia Borhidi ＝ Omiltemia Standl. ●☆

318018　Pseudomiltemia Borhidi(2004);假奥米茜属●☆

318019　Pseudomisopates Güemes ＝ Misopates Raf. ■☆

318020　Pseudomisopates Güemes(1997);假劣属■☆

318021　Pseudomitrocereus Bravo et Buxb. ＝ Neobuxbaumia Backeb. ●☆

318022　Pseudomitrocereus Bravo et F. Buxb. ＝ Pachycereus（A. Berger）Britton et Rose ●

318023　Pseudomonotes A. C. Londoño, E. Alvarez et Forero（1995）;假单列木属●☆

318024　Pseudomonotes tropenbosii A. C. Londono, E. Alvarez et Forero;假单列木●☆

318025　Pseudomorus Bureau ＝ Streblus Lour. ●

318026　Pseudomuscari Garbari et Greuter ＝ Muscari Mill. ■☆

318027　Pseudomuscari Garbari et Greuter(1970);假葡萄风信子属■☆

318028　Pseudomuscari acutifolium（Boiss.）Garbari;假葡萄风信子■☆

318029　Pseudomussaenda Wernham(1916);假玉叶金花属■☆

318030　Pseudomussaenda angustifolia Troupin et E. M. Petit;窄叶假玉叶金花■☆

318031　Pseudomussaenda flava Verdc.;黄色假玉叶金花■☆

318032　Pseudomussaenda gossweileri Wernham ＝ Pseudomussaenda monteroi（Wernham）Wernham ■☆

318033　Pseudomussaenda lanceolata（Forssk.）Wernham ＝ Pentas lanceolata（Forssk.）K. Schum. ●■

318034　Pseudomussaenda lanceolata Wernham ＝ Pseudomussaenda flava Verdc. ■☆

318035　Pseudomussaenda monteroi（Wernham）Wernham;假玉叶金花■☆

318036　Pseudomussaenda mozambicensis Verdc.;莫桑比克假玉叶金花■☆

318037　Pseudomussaenda stenocarpa（Hiern）E. M. Petit;窄果假玉叶金花■☆

318038　Pseudomyrcianthes Kausel ＝ Myrcianthes O. Berg ●☆

318039　Pseudonemacladus McVaugh(1943);假丝枝参属■☆

318040　Pseudonemacladus oppositifolius（B. L. Rob.）McVaugh;假丝枝参■☆

318041　Pseudonephelium Radlk. ＝ Dimocarpus Lour. ●

318042　Pseudonephelium confine F. C. How et G. Hoo ＝ Dimocarpus confinis（F. C. How et G. Hoo）H. S. Lo ●

318043　Pseudonesohedyotis Tennant（1965）;假美耳茜属☆

318044　Pseudonesohedyotis bremekampii Tennant;假美耳茜☆

318045　Pseudonopalxochia Backeb.（1958）;假令箭荷花属●☆

318046　Pseudonopalxochia Backeb. ＝ Disocactus Lindl. ●☆

318047　Pseudonopalxochia Backeb. ＝ Nopalxochia Britton et Rose ■☆

318048　Pseudonopalxochia conzattiana（T. M. MacDoug.）Backeb.;假令箭荷花●☆

318049　Pseudonoseris H. Rob. et Brettell(1974);红安菊属■☆

318050　Pseudonoseris discolor（Muschl.）H. Rob. et Brettell;异色红安菊■☆

318051　Pseudonoseris striatum（Cuatrec.）H. Rob. et Brettell;直红安菊■☆

318052　Pseudonoseris szyszylowiczii（Hieron.）H. Rob. et Brettell;红安菊■☆

318053　Pseudoorleanesia Rauschert ＝ Orleanesia Barb. Rodr. ■☆

318054 Pseudopachystela Aubrév. et Pellegr. (1961);假粗柱山榄属●☆

318055 Pseudopachystela Aubrév. et Pellegr. = Synsepalum（A. DC.）Daniell ●☆

318056 Pseudopachystela lastoursvillensis Aubrév. et Pellegr.;拉斯图维尔假粗柱山榄●☆

318057 Pseudopachystela oyemensis Aubrév. et Pellegr.;假粗柱山榄●☆

318058 Pseudopaegma Urb. = Anemopaegma Mart. ex Meisn.（保留属名）●☆

318059 Pseudopanax C. Koch. = Pseudopanax K. Koch ●■

318060 Pseudopanax K. Koch = Nothopanax Miq. ●

318061 Pseudopanax K. Koch(1859);假人参属（新树参属）●■

318062 Pseudopanax arboreus（L. f. et Murray）Philipson;乔木假人参（乔木新树参,五指树）;Five Fingers,Five-fingers ●☆

318063 Pseudopanax crassifolius（Cunn.）K. Koch;厚叶假人参（矛木）;Horoeka,Lancewood ●☆

318064 Pseudopanax crassifolius K. Koch = Pseudopanax crassifolius（Cunn.）K. Koch ●☆

318065 Pseudopanax davidii（Franch.）Philipson = Metapanax davidii（Franch.）Frodin ex J. Wen et Frodin ●

318066 Pseudopanax delavayi（Franch.）Philipson = Metapanax delavayi（Franch.）Frodin ex J. Wen et Frodin ●

318067 Pseudopanax ferox Kirk;粗齿假人参（齿叶矛木）;Toothed Lancewood ●☆

318068 Pseudopanax laetus（Kirk）Philipson;美丽假人参（亮叶新树参）●☆

318069 Pseudopanax lessonii K. Koch;雷苏假人参;Houpara ●☆

318070 Pseudopanax simplex（G. Forst.）K. Koch;新西兰假人参●☆

318071 Pseudopanax simplex K. Koch = Pseudopanax simplex（G. Forst.）K. Koch ●☆

318072 Pseudopancovia Pellegr.（1955）;假潘考夫无患子属●☆

318073 Pseudopancovia heteropetala Pellegr.;假潘考夫无患子●☆

318074 Pseudoparis H. Perrier(1936);假重楼属■☆

318075 Pseudoparis cauliflora H. Perrier;茎花假重楼■☆

318076 Pseudoparis monandra H. Perrier;假重楼■☆

318077 Pseudopavonia Hassl. = Pavonia Cav.（保留属名）●■☆

318078 Pseudopectinaria Lavranos = Echidnopsis Hook. f. ■☆

318079 Pseudopectinaria Lavranos(1971);假苦瓜掌属■☆

318080 Pseudopectinaria malum Lavranos;假苦瓜掌■☆

318081 Pseudopentameris Conert(1971);假五部芒属（假五数草属）●■☆

318082 Pseudopentameris brachyphylla（Stapf）Conert;短叶假五数草■☆

318083 Pseudopentameris caespitosa N. P. Barker;假五数草■☆

318084 Pseudopentameris macrantha（Schrad.）Conert;大花假五数草■☆

318085 Pseudopentameris obtusifolia（Hochst.）N. P. Barker;钝叶假五数草■☆

318086 Pseudopentatropis Costantin(1912);假朱砂莲属■☆

318087 Pseudopentatropis oblongifolia Costantin;假朱砂莲■☆

318088 Pseudopeponidium Homolle ex Arènes = Pyrostria Comm. ex Juss. ●☆

318089 Pseudoperistylus（P. F. Hunt）Szlach. et Olszewski = Habenaria Willd. ■

318090 Pseudoperistylus（P. F. Hunt）Szlach. et Olszewski(1998);假阔蕊兰属■☆

318091 Pseudoperistylus aethiopicus Szlach. et Olszewski = Montolivaea aethiopica（Szlach. et Olszewski）Szlach. ■☆

318092 Pseudoperistylus attenuatus（Hook. f.）Szlach. et Olszewski = Habenaria attenuata Hook. f. ■☆

318093 Pseudoperistylus bequaertii（Summerh.）Szlach. et Olszewski = Habenaria bequaertii Summerh. ■☆

318094 Pseudoperistylus bracteosus（Hochst. ex A. Rich.）Szlach. et Olszewski = Habenaria bracteosa Hochst. ex A. Rich. ■☆

318095 Pseudoperistylus ituriensis Szlach. et Olszewski;假阔蕊兰■☆

318096 Pseudoperistylus lefebureanus（A. Rich.）Szlach. et Olszewski = Habenaria lefebureana（A. Rich.）T. Durand et Schinz ■☆

318097 Pseudoperistylus microceras（Hook. f.）Szlach. et Olszewski = Habenaria microceras Hook. f. ■☆

318098 Pseudoperistylus montolivaea（Kraenzl. ex Engl.）Szlach. et Olszewski = Habenaria montolivaea Kraenzl. ex Engl. ■☆

318099 Pseudoperistylus petitianus（A. Rich.）Szlach. et Olszewski = Habenaria petitiana（A. Rich.）T. Durand et Schinz ■☆

318100 Pseudopetalon Raf. = Zanthoxylum L. ●

318101 Pseudophacelurus A. Camus = Phacelurus Griseb. ■

318102 Pseudophacelurus A. Camus = Rottboellia L. f.（保留属名）■

318103 Pseudophacelurus A. Camus(1921);假束尾草属■

318104 Pseudophacelurus latifolius A. Camus;宽叶假束尾草■☆

318105 Pseudophacelurus speciosus（Steud.）A. Camus = Phacelurus speciosus（Steud.）C. E. Hubb. ■☆

318106 Pseudophacelurus speciosus（Steud.）C. E. Hubb.;假束尾草■☆

318107 Pseudophacelurus speciosus A. Camus = Phacelurus speciosus（Steud.）C. E. Hubb. ■☆

318108 Pseudophleum Dogan = Phleum L. ■

318109 Pseudophleum Dogan(1982);类梯牧草属■☆

318110 Pseudophleum gibbum（Boiss.）Dogan;类梯牧草■☆

318111 Pseudophoeniaceae O. F. Cook = Arecaceae Bercht. et J. Presl（保留科名）●

318112 Pseudophoeniaceae O. F. Cook = Palmae Juss.（保留科名）●

318113 Pseudophoenicaceae O. F. Cook = Arecaceae Bercht. et J. Presl（保留科名）●

318114 Pseudophoenicaceae O. F. Cook = Palmae Juss.（保留科名）●

318115 Pseudophoenix H. Wendl. = Pseudophoenix H. Wendl. et Drude ex Drude ●☆

318116 Pseudophoenix H. Wendl. et Drude = Pseudophoenix H. Wendl. et Drude ex Drude ●☆

318117 Pseudophoenix H. Wendl. et Drude ex Drude = Pseudophoenix H. Wendl. ex Sarg.（废弃属名）●☆

318118 Pseudophoenix H. Wendl. et Drude ex Drude(1888);假刺葵属（葫芦椰子属,假海枣属,樱桃椰属）●☆

318119 Pseudophoenix H. Wendl. ex Sarg.（废弃属名）= Sargentia S. Watson（保留属名）●☆

318120 Pseudophoenix ekmanii Burret;埃氏假刺葵●☆

318121 Pseudophoenix lediniana Read;莱氏假刺葵●☆

318122 Pseudophoenix sargenti H. Wendl. = Sargentia aricocca H. Wendl. et Drude ex Salomon ●☆

318123 Pseudophoenix vinifera（Mart.）Becc. = Euterpe vinifera Mart. ●☆

318124 Pseudopholidia A. DC. = Pholidia R. Br. ●☆

318125 Pseudophyllanthus（Müll. Arg.）Voronts. et Petra Hoffm.(2008);肖叶下珠属●☆

318126 Pseudophyllanthus（Müll. Arg.）Voronts. et Petra Hoffm. = Phyllanthus L. ●■

318127 Pseudopilocereus Buxb. = Pilosocereus Byles et G. D. Rowley ●☆

318128 Pseudopimpinella F. Ghahrem.,Khajepiri et Mozaff.(2010);假茴芹属●☆

318129 Pseudopimpinella F. Ghahrem., Khajepiri et Mozaff. = Pimpinella L. ■

318130 Pseudopinanga Burret = Pinanga Blume ●

318131 Pseudopinanga Burret(1936);假山槟榔属●☆

318132 Pseudopinanga albescens (Becc.) Burret;灰白假山槟榔●☆

318133 Pseudopinanga maculata (Porte ex Lem.) Burret;斑点假山槟榔●☆

318134 Pseudopinanga multisecta Burret;多刚毛假山槟榔●☆

318135 Pseudopinanga pilosa Burret;毛假山槟榔●☆

318136 Pseudopinanga tashiroi (Hayata) Burret = Pinanga tashiroi Hayata ●

318137 Pseudopinanga urosperma (Becc.) Burret;尾籽假山槟榔●☆

318138 Pseudopiptadenia Rauschert(1982);假落腺豆属■☆

318139 Pseudopiptadenia leptostachya (Benth.) Rauschert;假落腺豆■☆

318140 Pseudopiptocarpha H. Rob. (1994);假落苞菊属●☆

318141 Pseudopiptocarpha elaeagnoides (Kunth) H. Rob.;假落苞菊●☆

318142 Pseudopipturus Skottsb. = Nothocnide Blume ex Chew ●☆

318143 Pseudoplantago Suess. (1934);车前苋属■☆

318144 Pseudoplantago friesii Suess.;车前苋■☆

318145 Pseudopodospermum (Lipsch. et Krasch.) A. I. Kuth. = Scorzonera L. ■

318146 Pseudopogonatherum A. Camus = Eulalia Kunth ■

318147 Pseudopogonatherum A. Camus (1921);假金发草属(笔草属);Pengrass, Pseudopogonatherum, Shamgoldquitch ■

318148 Pseudopogonatherum capilliphyllum S. L. Chen = Pseudopogonatherum filifolium (S. L. Chen) H. Yu, Y. F. Deng et N. X. Zhao ■

318149 Pseudopogonatherum collinum (Balansa) A. Camus = Pseudopogonatherum contortum (Brongn.) A. Camus ■

318150 Pseudopogonatherum contortum (Brongn.) A. Camus;笔草(刺叶笔草,刺叶假金发草,刺叶金茅,刺叶荸草,荸草);Bent Eulalia, Pen Shamgoldquitch, Pengrass, Spineleaf Shamgoldquitch, Spiny Pseudopogonatherum ■

318151 Pseudopogonatherum contortum (Brongn.) A. Camus var. linearifolia S. L. Chen;线叶笔草;Linearleaf Pseudopogonatherum, Threadleaf Eulalia ■

318152 Pseudopogonatherum contortum (Brongn.) A. Camus var. linearifolium (Keng) P. C. Keng = Pseudopogonatherum contortum (Brongn.) A. Camus var. linearifolia S. L. Chen ■

318153 Pseudopogonatherum contortum (Brongn.) A. Camus var. sinense (Keng ex S. L. Chen) Keng ex S. L. Chen;中华笔草;China Pseudopogonatherum ■

318154 Pseudopogonatherum filifolium (S. L. Chen) H. Yu, Y. F. Deng et N. X. Zhao;假金发草(线叶金茅);Linearifolious Shamgoldquitch ■

318155 Pseudopogonatherum filifolium S. L. Chen = Pseudopogonatherum capilliphyllum S. L. Chen ■

318156 Pseudopogonatherum koretrostachys (Trin.) Henrard;刺叶假金发草■

318157 Pseudopogonatherum koretrostachys (Trin.) Henrard = Pseudopogonatherum contortum (Brongn.) A. Camus ■

318158 Pseudopogonatherum quadrinerve (Hack.) Ohwi = Eulalia quadrinervis (Hack.) Kuntze ■

318159 Pseudopogonatherum setifolium (Nees) A. Camus = Pseudopogonatherum contortum (Brongn.) A. Camus ■

318160 Pseudopogonatherum setifolium (Nees) A. Camus = Pseudopogonatherum koretrostachys (Trin.) Henrard ■

318161 Pseudopogonatherum speciosum (Debeaux) Ohwi = Eulalia speciosa (Debeaux) Kuntze ■

318162 Pseudopogonatherum trispicatum (Schult.) Ohwi = Eulalia trispicata (Schult.) Henrard ■

318163 Pseudoponera Brieger = Ponera Lindl. ■☆

318164 Pseudoprimula (Pax) O. Schwarz;假报春花属■☆

318165 Pseudoprimula (Pax) O. Schwarz = Primula L. ■

318166 Pseudoprosopis Harms(1902);假牧豆树属●☆

318167 Pseudoprosopis bampsiana Lisowski;邦氏假牧豆树●☆

318168 Pseudoprosopis claessensii (De Wild.) G. C. C. Gilbert et Boutique;克莱森假牧豆树●☆

318169 Pseudoprosopis euryphylla Harms;宽叶假牧豆树●☆

318170 Pseudoprosopis fischeri (Taub.) Harms;菲舍尔假牧豆树●☆

318171 Pseudoprosopis gilletii (De Wild.) Villiers;吉莱假牧豆树●☆

318172 Pseudoprosopis sericea (Hutch. et Dalziel) Brenan;绢毛假牧豆树●☆

318173 Pseudoprosopis uncinata Evrard;具钩假牧豆树●☆

318174 Pseudoprospero Speta(1998);双珠风信子属■☆

318175 Pseudoprospero firmifolium (Baker) Speta;双珠风信子■☆

318176 Pseudoprotorhus H. Perrier = Filicium Thwaites ex Benth. ●☆

318177 Pseudoprotorhus longifolia H. Perrier = Filicium longifolium (H. Perrier) Capuron ●☆

318178 Pseudopteris Baill. (1874);假翼无患子属●☆

318179 Pseudopteris ankaranensis Capuron;假翼无患子●☆

318180 Pseudopteris arborea Capuron;树状假翼无患子●☆

318181 Pseudopteris decipiens Baill.;马岛假翼无患子●☆

318182 Pseudopteryxia Rydb. = Pseudocymopterus J. M. Coult. et Rose ■☆

318183 Pseudopyxis Miq. (1867);假盖果草属;Pseudopyxis ■

318184 Pseudopyxis depressa Miq.;假盖果草■☆

318185 Pseudopyxis depressa Miq. f. angustiloba (Makino) H. Hara;狭裂假盖果草■☆

318186 Pseudopyxis depressa Miq. f. variegata (Makino) H. Hara;多变假盖果草■☆

318187 Pseudopyxis heterophylla (Miq.) Maxim.;异叶假盖果草;Diverseleaf Pseudopyxis ■

318188 Pseudoraachaerium Hassl. = Nissolia Jacq. (保留属名)■☆

318189 Pseudorachicallis Post et Kuntze = Arcytophyllum Willd. ex Schult. et Schult. f. ●☆

318190 Pseudorachicallis Post et Kuntze = Mallostoma H. Karst. ●☆

318191 Pseudorachicallis Post et Kuntze = Pseudrachicallis H. Karst. ●☆

318192 Pseudoraphis Griff. = Pseudoraphis Griff. ex R. Pilger ■

318193 Pseudoraphis Griff. ex R. Pilger(1928);伪针茅属(大伪针茅属,蚰蜒草属);Fake Needlegrass, Pseudoraphis ■

318194 Pseudoraphis balansae Henrard;长稃伪针茅(长稃伪针草);Longlemma Fake Needlegrass, Longlemma Pseudoraphis ■

318195 Pseudoraphis brunoniana (Wall. et Griff.) Griff. = Pseudoraphis spinescens (R. Br.) Vickery ■

318196 Pseudoraphis brunoniana (Wall. et Griff.) Pilg.;伪针茅■

318197 Pseudoraphis depauperata (Nees ex Hook. f.) Keng = Pseudoraphis sordida (Thwaites) S. M. Phillips et S. L. Chen ■

318198 Pseudoraphis depauperata (Nees) Keng = Pseudoraphis spinescens (R. Br.) Vickery var. depauperata (Nees) Bor ■

318199 Pseudoraphis longipaleacea L. C. Chia = Pseudoraphis balansae Henrard ■

318200 Pseudoraphis simaoensis Y. Y. Qian;思茅伪针茅;Simao Pseudoraphis ■

318201 Pseudoraphis sordida (Thwaites) S. M. Phillips et S. L. Chen;瘦瘠伪针茅;Depauperate Fake Needlegrass, Depauperate Pseudoraphis ■

318202 Pseudoraphis spinescens (R. Br.) Vickery;大伪针茅(糙伪针茅,伪针茅);Spiny Fake Needlegrass, Spiny Pseudoraphis, Squarrose Pseudoraphis ■

318203　Pseudoraphis spinescens（R. Br.）Vickery var. depauperata（Nees ex Hook. f.）Bor = Pseudoraphis sordida（Thwaites）S. M. Phillips et S. L. Chen ■

318204　Pseudoraphis spinescens（R. Br.）Vickery var. depauperata（Nees）Bor = Pseudoraphis sordida（Thwaites）S. M. Phillips et S. L. Chen ■

318205　Pseudoraphis squarrosa（L. f.）Chase = Pseudoraphis spinescens（R. Br.）Vickery ■

318206　Pseudoraphis squarrosa（L. f.）Chase var. depauperata（Nees）Senaratna = Pseudoraphis spinescens（R. Br.）Vickery var. depauperata（Nees）Bor ■

318207　Pseudoraphis ukishiba Ohwi = Pseudoraphis sordida（Thwaites）S. M. Phillips et S. L. Chen ■

318208　Pseudorchis Gray = Liparis Rich.（保留属名）■

318209　Pseudorchis Ség.（1754）;拟红门兰属（白兰属）;Leucorchis, Small White Orchid ■☆

318210　Pseudorchis albida（L.）Á. Löve et D. Löve;拟红门兰（白兰）;Small White Orchid, White Fragrant Orchid, White Frog Orchid, White Mountain Orchid ■☆

318211　Pseudorchis albida（L.）Á. Löve et D. Löve subsp. straminea（Fernald）Á. Löve et D. Löve;禾秆色白兰■☆

318212　Pseudorchis frivaldii（Hampe ex Griseb.）P. F. Hunt;弗氏拟红门兰;Frivald's Frog Orchid ■☆

318213　Pseudoreoxis Rydb. = Pseudocymopterus J. M. Coult. et Rose ■☆

318214　Pseudorhachicallis Benth. et Hook. f. = Mallostoma H. Karst. ●☆

318215　Pseudorhachicallis Benth. et Hook. f. = Pseudorachicallis H. Karst. ●☆

318216　Pseudorhachicallis Hook. f. = Arcytophyllum Willd. ex Schult. et Schult. f. ●☆

318217　Pseudorhipsalis Britton et Rose = Disocactus Lindl. ●☆

318218　Pseudorhipsalis Britton et Rose（1923）;假丝苇属（假苇属,假仙人棒属）●☆

318219　Pseudorhipsalis acuminata Cufod.;尖假丝苇（尖假仙人棒）●☆

318220　Pseudorhipsalis alata（Sw.）Britton et Rose;翅假丝苇（翅假仙人棒）●☆

318221　Pseudorhipsalis amazonica（K. Schum.）Ralf Bauer;亚马孙假丝苇（亚马孙假仙人棒）●☆

318222　Pseudorhipsalis macrantha Alexander;大花假丝苇（大花假仙人棒）●☆

318223　Pseudorhipsalis ramulosa（Salm-Dyck）Barthlott;多枝假丝苇●☆

318224　Pseudoridolfia Reduron, Mathez et S. R. Downie（2009）;假里多尔菲草属■☆

318225　Pseudorlaya（Murb.）Murb.（1897）;假奥尔雷草属●☆

318226　Pseudorlaya Murb. = Pseudorlaya（Murb.）Murb. ●☆

318227　Pseudorlaya maritima（L.）Murb. = Pseudorlaya pumila（L.）Parl. ●☆

318228　Pseudorlaya minuscula（Pau）M. Lainz;小假奥尔雷草■☆

318229　Pseudorlaya pumila（L.）Grande var. breviaculeata（Boiss.）Zohary = Pseudorlaya pumila（L.）Parl. ●☆

318230　Pseudorlaya pumila（L.）Parl.;偃俯假奥尔雷草●☆

318231　Pseudorlaya pycnacantha H. Lindb. = Pseudorlaya pumila（L.）Parl. ●☆

318232　Pseudorleanesia Rauschert = Orleanesia Barb. Rodr. ■☆

318233　Pseudornelissitus Ovcz., Rassulova et Kinzik. = Medicago L.（保留属名）●■

318234　Pseudorobanche Rouy = Alectra Thunb. ■

318235　Pseudoroegneria（Nevski）Á. Löve = Elymus L. ■

318236　Pseudoroegneria（Nevski）Á. Löve = Elytrigia Desv. ■

318237　Pseudoroegneria（Nevski）Á. Löve（1980）;假鹅观草属■

318238　Pseudoroegneria cognata（Hack.）Á. Löve;假鹅观草■

318239　Pseudoroegneria elytrigioides（C. Yen et J. L. Yang）B. Rong Lu = Elymus elytrigioides（C. Yen et J. L. Yang）S. L. Chen ■

318240　Pseudoroegneria strigosa（M. Bieb.）Á. Löve subsp. aegilopoides（Drobow）Á. Löve = Elytrigia gmelinii（Trin.）Nevski ■

318241　Pseudorontium（A. Gray）Rothm.（1943）;假金棒芋属■☆

318242　Pseudorontium cyathiferum（Benth.）Rothm.;假金棒芋■☆

318243　Pseudosularia Gurgen. = Prometheum（A. Berger）H. Ohba ■☆

318244　Pseudoruellia Benoist（1962）;拟芦莉草属■☆

318245　Pseudoruellia perrieri（Benoist）Benoist;拟芦莉草■☆

318246　Pseudoryza Griff. = Leersia Sw.（保留属名）■

318247　Pseudosabicea N. Hallé（1963）;假萨比斯茜属●☆

318248　Pseudosabicea arborea（K. Schum.）N. Hallé;树状假萨比斯茜●☆

318249　Pseudosabicea arborea（K. Schum.）N. Hallé subsp. bequaertii（De Wild.）Verdc.;贝卡尔假萨比斯茜●☆

318250　Pseudosabicea arborea（K. Schum.）N. Hallé var. bequaertii（De Wild.）Verdc. = Pseudosabicea arborea（K. Schum.）N. Hallé subsp. bequaertii（De Wild.）Verdc. ●☆

318251　Pseudosabicea aurifodinae N. Hallé var. crystallina N. Hallé;水晶假萨比斯茜●☆

318252　Pseudosabicea batesii（Wernham）N. Hallé;贝茨假萨比斯茜●☆

318253　Pseudosabicea becquetii N. Hallé;贝凯假萨比斯茜●☆

318254　Pseudosabicea floribunda（K. Schum.）N. Hallé;繁花假萨比斯茜●☆

318255　Pseudosabicea mildbraedii（Wernham）N. Hallé;米尔德假萨比斯茜●☆

318256　Pseudosabicea mildbraedii（Wernham）N. Hallé var. letestui N. Hallé;莱泰斯图假萨比斯茜●☆

318257　Pseudosabicea mitisphaera N. Hallé;软球假萨比斯茜●☆

318258　Pseudosabicea nobilis（R. D. Good）N. Hallé;名贵假萨比斯茜●☆

318259　Pseudosabicea pedicellata（Wernham）N. Hallé;梗花假萨比斯茜●☆

318260　Pseudosabicea sanguinosa N. Hallé;血红假萨比斯茜●☆

318261　Pseudosabicea segregata（Hiern）N. Hallé;隔离假萨比斯茜●☆

318262　Pseudosagotia Secco（1985）;假萨戈大戟属☆

318263　Pseudosagotia brevipetiolata Secco;假萨戈大戟☆

318264　Pseudosalacia Codd（1972）;假五层龙属●☆

318265　Pseudosalacia streyi Codd;假五层龙●☆

318266　Pseudosamanea Harms = Albizia Durazz. ●

318267　Pseudosamanea Harms（1930）;拟雨树属●☆

318268　Pseudosamanea guachapele（Kunth）Harms;拟雨树●☆

318269　Pseudosantalum Kuntze = Osmoxylon Miq. ●

318270　Pseudosantalum Mill. = Caesalpinia L. ●

318271　Pseudosantalum Rumph. = Caesalpinia L. ●

318272　Pseudosantalum Rumph. ex Kuntze = Caesalpinia L. ●

318273　Pseudosaponaria（F. N. Williams）Ikonn. = Gypsophila L. ●●

318274　Pseudosarcolobus Costantin = Gymnema R. Br. ●

318275　Pseudosarcopera Gir. -Cañas（2007）;假穗状附生藤属●☆

318276　Pseudosasa Makino = Sasa Makino et Shibata ●

318277　Pseudosasa Makino ex Nakai（1925）;茶秆竹属（假箬竹属,箭竹属,青篱竹属,矢竹属）;Arrow Bamboo, Bamboo, Pseudosasa ●

318278　Pseudosasa Nakai = Sasa Makino et Shibata ●

318279　Pseudosasa acutivagina T. H. Wen et S. C. Chen;尖箨茶秆竹（尖箨茶竿竹）;Sharp-sheath Pseudosasa ●

318280　Pseudosasa acutivagina T. H. Wen et S. C. Chen ＝ Acidosasa nanunica (McClure) C. S. Chao et G. Y. Yang ●

318281　Pseudosasa aeria T. H. Wen;空心竹(空心苦);Aerial Pseudosasa,Empty Pseudosasa ●

318282　Pseudosasa altiligulata T. H. Wen ＝ Acidosasa nanunica (McClure) C. S. Chao et G. Y. Yang ●

318283　Pseudosasa amabilis (McClure) P. C. Keng;茶秆竹(茶竿竹,苦竹,青篱竹,沙白竹);Lavable Pseudosasa,Teastick Bamboo,Tonkin Bamboo,Tonkin Cane,Tonkin Canebrake,Tonkin Pseudosasa ●

318284　Pseudosasa amabilis (McClure) P. C. Keng var. convexa Z. P. Wang et G. H. Ye;福建茶秆竹(福建茶竿竹);Fujian Tonkin Pseudosasa ●

318285　Pseudosasa amabilis (McClure) P. C. Keng var. farinosa C. S. Chao ex S. L. Chen et G. Y. Sheng;厚粉茶秆竹(厚粉茶竿竹); Mealy Tonkin Pseudosasa ●

318286　Pseudosasa amabilis (McClure) P. C. Keng var. tenuis S. L. Chen et G. Y. Sheng ＝ Pseudosasa amabilis (McClure) P. C. Keng var. convexa Z. P. Wang et G. H. Ye ●

318287　Pseudosasa amabilis (McClure) P. C. Keng var. tenuis S. L. Chen et G. Y. Sheng;薄箨茶秆竹(薄箨茶竿竹);Thin Tonkin Pseudosasa ●

318288　Pseudosasa aureovagina W. T. Lin;金箨茶秆竹(金箨茶竿竹)●

318289　Pseudosasa aureovagina W. T. Lin ＝ Pseudosasa hindsii (Munro) S. L. Chen et G. Y. Sheng ex T. G. Liang ●

318290　Pseudosasa baiyunensis W. T. Lin;白云矢竹●

318291　Pseudosasa baiyunensis W. T. Lin ＝ Pseudosasa hindsii (Munro) S. L. Chen et G. Y. Sheng ex T. G. Liang ●

318292　Pseudosasa brevivaginata G. H. Lai;短鞘茶秆竹●

318293　Pseudosasa cantorii (Munro) P. C. Keng ＝ Pseudosasa cantorii (Munro) P. C. Keng ex S. L. Chen et al. ●

318294　Pseudosasa cantorii (Munro) P. C. Keng ex S. L. Chen et al.; 托竹(篙竹,拖竹);Cantor Pseudosasa,Hinds Pseudosasa ●

318295　Pseudosasa flexuosa T. P. Yi et X. M. Zhou ＝ Oligostachyum scabriflorum (McClure) Z. P. Wang et G. H. Ye ●

318296　Pseudosasa gracilis S. L. Chen et G. Y. Sheng;纤细茶秆竹(纤细茶竿竹);Slender Pseudosasa ●

318297　Pseudosasa guanxianensis T. P. Yi;笔竿竹;Guangxi Pseudosasa,Kwangsi Pseudosasa ●

318298　Pseudosasa guanxianensis T. P. Yi ＝ Indocalamus longiauritus Hand. -Mazz. ●

318299　Pseudosasa hainanensis G. A. Fu;海南茶秆竹(海南茶竿竹)●

318300　Pseudosasa hainanensis G. A. Fu ＝ Pseudosasa cantorii (Munro) P. C. Keng ex S. L. Chen et al. ●

318301　Pseudosasa hamadae Hatus. ＝ Indocalamus tessellatus (Munro) P. C. Keng ●

318302　Pseudosasa hindsii (Munro) C. D. Chu et C. S. Chao ＝ Pleioblastus hindsii (Munro) Nakai ●

318303　Pseudosasa hindsii (Munro) S. L. Chen et G. Y. Sheng ex T. G. Liang;箬竹(寒山竹,篙竹,四时竹,邢氏苦竹,箬竹);Hinds Cane,Hinds' Pseudosasa,Ramrod Bamboo ●

318304　Pseudosasa hindsii (Munro) S. L. Chen et G. Y. Sheng ex T. G. Liang ＝ Pleioblastus hindsii (Munro) Nakai ●

318305　Pseudosasa hirta S. L. Chen et G. Y. Sheng;庐山茶秆竹(庐山茶竿竹);Hairy Pseudosasa,Lushan Pseudosasa ●

318306　Pseudosasa hirta S. L. Chen et G. Y. Sheng ＝ Indocalamus latifolius (Keng) McClure ●

318307　Pseudosasa japonica (Siebold et Zucc. ex Steud.) Makino ex Nakai;矢竹(箭竹,日本赤竹,日本矢竹);Arrow Bamboo,Bamboo, Japan Pseudosasa,Japanese Arrow Bamboo,Japanese Sasa,Metake ●

318308　Pseudosasa japonica (Siebold et Zucc. ex Steud.) Makino ex Nakai 'Flavo-variegata';黄斑矢竹●☆

318309　Pseudosasa japonica (Siebold et Zucc. ex Steud.) Makino ex Nakai 'Tsutsumiana';绿矢竹;Green Onion Bamboo ●☆

318310　Pseudosasa japonica (Siebold et Zucc. ex Steud.) Makino ex Nakai var. tsutsumiana Yanagita ＝ Pseudosasa japonica (Siebold et Zucc. ex Steud.) Makino ex Nakai 'Tsutsumiana' ●☆

318311　Pseudosasa japonica (Siebold et Zucc. ex Steud.) Nakai ＝ Pseudosasa japonica (Siebold et Zucc. ex Steud.) Makino ex Nakai ●

318312　Pseudosasa japonica (Siebold et Zucc.) Makino f. flavovariegata Makino ＝ Pseudosasa japonica (Siebold et Zucc. ex Steud.) Makino ex Nakai 'Flavo-variegata' ●☆

318313　Pseudosasa japonica (Siebold et Zucc.) Makino f. variegata Muroi;白斑矢竹●☆

318314　Pseudosasa japonica (Siebold et Zucc.) Makino var. usawai (Hayata) Muroi ＝ Pseudosasa usawai (Hayata) Makino et Nemoto ●

318315　Pseudosasa jiangleensis N. X. Zhao et N. H. Xia;将乐茶秆竹(将乐茶竿竹);Jiangle Pseudosasa ●

318316　Pseudosasa kunishii (Hayata) Makino et Nemoto ＝ Gelidocalamus kunishii (Hayata) P. C. Keng et T. H. Wen ●

318317　Pseudosasa longiligula T. H. Wen;广竹;Guang Pseudosasa, Long-ligule Pseudosasa ●

318318　Pseudosasa longiligulata (McClure) Koidz. ＝ Sasa longiligulata McClure ●

318319　Pseudosasa longivaginata H. R. Zhao et Y. L. Yang;长鞘茶秆竹(长鞘茶竿竹);Long-sheath Pseudosasa ●

318320　Pseudosasa longivaginata H. R. Zhao et Y. L. Yang ＝ Indocalamus tessellatus (Munro) P. C. Keng ●

318321　Pseudosasa maculifera J. L. Lu;鸡公山茶秆竹(鸡公山茶竿竹);Jigongshan Pseudosasa ●

318322　Pseudosasa maculifera J. L. Lu var. hirsuta S. L. Chen et G. Y. Sheng;毛箨茶秆竹(毛箨茶竿竹);Hispid Jigongshan Pseudosasa ●

318323　Pseudosasa magilaminaria B. M. Yang;江永茶秆竹(江永茶竿竹);Jiangyong Pseudosasa ●

318324　Pseudosasa multifloscula (W. T. Lin) W. T. Lin ＝ Pseudosasa hindsii (Munro) S. L. Chen et G. Y. Sheng ex T. G. Liang ●

318325　Pseudosasa nabeshimana (Koidz.) Koidz.;九州矢竹●☆

318326　Pseudosasa naibunensis (Hayata) Makino et Nemoto ＝ Ampelocalamus naibunensis (Hayata) T. H. Wen ●

318327　Pseudosasa naibunensis (Hayata) Makino et Nemoto ＝ Drepanostachyum naibunense (Hayata) P. C. Keng ●

318328　Pseudosasa nanunica (McClure) Z. P. Wang et G. H. Ye;长舌茶秆竹(长舌茶竿竹);Longtogue Pseudosasa ●

318329　Pseudosasa nanunica (McClure) Z. P. Wang et G. H. Ye ＝ Acidosasa nanunica (McClure) C. S. Chao et G. Y. Yang ●

318330　Pseudosasa nanunica (McClure) Z. P. Wang et G. H. Ye var. angustifolia S. L. Chen et G. Y. Sheng;狭叶长舌茶秆竹(狭叶茶秆竹,狭叶长舌茶竿竹);Narrowleaf Pseudosasa ●

318331　Pseudosasa nigrinodis G. A. Fu ＝ Pseudosasa hindsii (Munro) S. L. Chen et G. Y. Sheng ex T. G. Liang ●

318332　Pseudosasa notata Z. P. Wang et G. H. Ye;斑箨茶秆竹(斑箨茶竿竹);Marked Pseudosasa,Spotsheath Pseudosasa ●

318333　Pseudosasa oiwakensis (Hayata) Makino ＝ Yushania niitakayamensis (Hayata) P. C. Keng ●

318334　Pseudosasa oiwakensis (Hayata) Makino et Nemoto ＝ Yushania niitakayamensis（Hayata）P. C. Keng ●

318335　Pseudosasa orthotropa S. L. Chen et T. H. Wen；面秆竹（面竿竹）；Erect Pseudosasa

318336　Pseudosasa owatarii（Makino）Makino ex Nakai；大渡茶秆竹●☆

318337　Pseudosasa pallidiflora（McClure）S. L. Chen et G. Y. Sheng；少花茶秆竹（少花茶竿竹）；Fewflower Pseudosasa, Poorflower Pseudosasa ●

318338　Pseudosasa pallidi-flora（McClure）S. L. Chen et G. Y. Sheng = Pseudosasa pubiflora（Keng）P. C. Keng ex D. Z. Li et L. M. Gao ●

318339　Pseudosasa parilis T. P. Yi et D. H. Hu；抽展茶秆竹（抽展茶竿竹）●

318340　Pseudosasa parilis T. P. Yi et D. H. Hu = Pseudosasa pubiflora（Keng）P. C. Keng ex D. Z. Li et L. M. Gao ●

318341　Pseudosasa pubiflora（Keng）P. C. Keng = Pseudosasa pubiflora（Keng）P. C. Keng ex D. Z. Li et L. M. Gao ●

318342　Pseudosasa pubiflora（Keng）P. C. Keng ex D. Z. Li et L. M. Gao；毛花茶秆竹（毛花茶竿竹，毛花青篱竹）；Pubescent Cane ●

318343　Pseudosasa pubioicatrix W. T. Lin；毛痕矢竹●

318344　Pseudosasa subsolida S. L. Chen et G. Y. Sheng；近实心茶秆竹（近实心茶竿竹）；Bearsolid Pseudosasa, Subsolid Pseudosasa ●

318345　Pseudosasa taiwanensis Masam. et Mori = Gelidocalamus kunishii（Hayata）P. C. Keng et T. H. Wen ●

318346　Pseudosasa tessellata（Munro）Hatus. = Indocalamus tessellatus（Munro）P. C. Keng ●

318347　Pseudosasa truncatula S. L. Chen et G. Y. Sheng；平截茶秆竹（平截茶竿竹）；Truncate Pseudosasa ●

318348　Pseudosasa truncatula S. L. Chen et G. Y. Sheng = Indocalamus latifolius（Keng）McClure ●

318349　Pseudosasa usawae（Hayata）Makino et Nemoto = Pseudosasa japonica（Siebold et Zucc. ex Steud.）Makino ex Nakai ●

318350　Pseudosasa usawai（Hayata）Makino et Nemoto；矢竹仔（包箨箭竹，包箨矢竹）；Usawa Pseudosasa ●

318351　Pseudosasa vicina（Keng）T. Q. Nguyen = Fargesia vicina（Keng）T. P. Yi ●

318352　Pseudosasa victorialis（P. C. Keng）T. P. Yi = Indocalamus victorialis P. C. Keng ●

318353　Pseudosasa viridula S. L. Chen et G. Y. Sheng；笔竹；Greenish Pseudosasa, Virescent Pseudosasa ●

318354　Pseudosasa vittata B. M. Yang = Indocalamus longiauritus Hand. -Mazz. ●

318355　Pseudosasa vulgata（W. T. Lin et X. B. Ye）W. T. Lin = Indocalamus longiauritus Hand. -Mazz. ●

318356　Pseudosasa wuyiensis S. L. Chen et G. Y. Sheng；武夷山茶秆竹（武夷山茶竿竹）；Wuyisha Pseudosasa, Wuyishan Pseudosasa ●

318357　Pseudosasa yangshanensis（W. T. Lin）T. P. Yi；阳山茶秆竹●

318358　Pseudosasa yuelushanensis B. M. Yang；岳麓山茶秆竹（岳麓山茶竿竹）；Yuelushan Pseudosasa ●

318359　Pseudosassafras Lecomte = Sassafras J. Presl ●

318360　Pseudosassafras Lecomte（1912）；肖檫木属（假檫木属）●

318361　Pseudosassafras laxiflora（Hemsl.）Nakai = Sassafras tsumu（Hemsl.）Hemsl. ●

318362　Pseudosassafras laxiflora（Hemsl.）Nakai var. randaiensis（Hayata）Nakai = Sassafras randaiense（Hayata）Rehder ●

318363　Pseudosassafras laxiflorum（Hemsl.）Nakai = Sassafras tsumu（Hemsl.）Hemsl. ●

318364　Pseudosassafras laxiflorum（Hemsl.）Nakai var. randaiensis（Hayata）Nakai = Sassafras randaiense（Hayata）Rehder ●

318365　Pseudosassafras tsumu（Hemsl.）Lecomte = Sassafras tsumu（Hemsl.）Hemsl. ●

318366　Pseudosbeckia A. Fern. et R. Fern.（1956）；假金锦香属●☆

318367　Pseudosbeckia swynnertonii（Baker f.）A. Fern. et R. Fern.；假金锦香●☆

318368　Pseudosbeckia swynnertonii（Baker f.）A. Fern. et R. Fern. = Dissotis swynnertonii（Baker f.）A. Fern. et R. Fern.●☆

318369　Pseudoscabiosa Devesa = Scabiosa L. ●■

318370　Pseudoscabiosa Devesa（1984）；假蓝盆草属●■☆

318371　Pseudoscabiosa africana（Font Quer）Romo et al.；非洲假蓝盆花■☆

318372　Pseudoscabiosa grosii（Font Quer）Devesa var. africana（Font Quer）Romo = Pseudoscabiosa africana（Font Quer）Romo et al.■☆

318373　Pseudoschoenus（C. B. Clarke）Oteng-Yeb.（1974）；假赤箭莎属■☆

318374　Pseudoschoenus（C. B. Clarke）Oteng-Yeb. = Scirpus L.（保留属名）■

318375　Pseudoschoenus inanis（Thunb.）Oteng-Yeb.；假赤箭莎■☆

318376　Pseudosciadium Baill.（1878）；假伞五加属●☆

318377　Pseudosciadium balansae Baill.；假伞五加●☆

318378　Pseudosclerochloa Tzvelev（2004）；假硬草属（耿氏假硬草属）■

318379　Pseudosclerochloa kengiana（Ohwi）Tzvelev；耿氏假硬草■

318380　Pseudoscolopia E. Phillips = Pseudoscolopia Gilg ●☆

318381　Pseudoscolopia Gilg（1917）；假箣柊属●☆

318382　Pseudoscolopia Phillips = Pseudoscolopia Gilg ●☆

318383　Pseudoscolopia fraseri E. Phillips = Pseudoscolopia polyantha Gilg ●☆

318384　Pseudoscolopia polyantha Gilg；假箣柊●☆

318385　Pseudoscordum Herb. = Nothoscordum Kunth（保留属名）■☆

318386　Pseudosecale（Godr.）Degen = Dasypyrum（Coss. et Durieu）T. Durand ■☆

318387　Pseudosecale（Godr.）Degen = Haynaldia Schur ■☆

318388　Pseudosedum（Boiss.）A. Berger（1930）；合景天属（假景天属，六瓣景天属）；Falsestonecrop ●■

318389　Pseudosedum A. Berger = Pseudosedum（Boiss.）A. Berger ●■

318390　Pseudosedum affine（Schrenk）A. Berger；白花合景天（白花假景天，白花景天）；Whiteflower Falsestonecrop ●■

318391　Pseudosedum bucharicum Boriss.；布哈尔合景天●☆

318392　Pseudosedum campanuliforum Boriss.；风铃草景天■☆

318393　Pseudosedum condensatum Boriss.；密集合景天■☆

318394　Pseudosedum fedtschenkoanum Boriss.；费氏合景天■☆

318395　Pseudosedum ferganense Boriss.；费尔干合景天■☆

318396　Pseudosedum karatavicum Boriss.；卡拉塔夫合景天■☆

318397　Pseudosedum lievenii（Ledeb.）A. Berger；合景天（假景天，六瓣景天）；Lieven Falsestonecrop ●■

318398　Pseudosedum longidentatum Boriss.；长齿合景天■☆

318399　Pseudosedum multicaule（Boiss. et Buhse）Boriss.；多茎合景天■☆

318400　Pseudoselago Hilliard = Selago L. ●☆

318401　Pseudoselago Hilliard（1995）；假塞拉玄参属●☆

318402　Pseudoselago arguta（E. Mey.）Hilliard；亮假塞拉玄参●☆

318403　Pseudoselago ascendens（E. Mey.）Hilliard；上升假塞拉玄参●☆

318404　Pseudoselago bella Hilliard；雅致假塞拉玄参●☆

318405　Pseudoselago burmannii（Choisy）Hilliard；布尔曼假塞拉玄参●☆

318406　Pseudoselago caerulescens Hilliard；浅蓝假塞拉玄参●☆

318407　Pseudoselago candida Hilliard；纯白假塞拉玄参●☆

318408　Pseudoselago densifolia（Hochst.）Hilliard；密叶假塞拉玄参●☆

318409　Pseudoselago diplotricha Hilliard；双毛假塞拉玄参●☆

318410　Pseudoselago gracilis Hilliard；纤细假塞拉玄参●☆

318411　Pseudoselago guttata（E. Mey.）Hilliard；油点假塞拉玄参●☆

318412　Pseudoselago humilis（Rolfe）Hilliard；低矮假塞拉玄参●☆

318413　Pseudoselago langebergensis Hilliard；朗厄山塞拉玄参●☆

318414　Pseudoselago outeniquensis Hilliard；南非假塞拉玄参●☆

318415　Pseudoselago parvifolia Hilliard；小叶假塞拉玄参●☆

318416　Pseudoselago prolixa Hilliard；伸展塞拉玄参●☆

318417　Pseudoselago prostrata Hilliard；平卧假塞拉玄参●☆

318418　Pseudoselago pulchra Hilliard；美丽假塞拉玄参●☆

318419　Pseudoselago quadrangularis（Choisy）Hilliard；棱角假塞拉玄参●☆

318420　Pseudoselago rapunculoides（L.）Hilliard；小钟假塞拉玄参●☆

318421　Pseudoselago recurvifolia Hilliard；反曲叶假塞拉玄参●☆

318422　Pseudoselago serrata（P. J. Bergius）Hilliard；具齿假塞拉玄参●☆

318423　Pseudoselago similis Hilliard；相似假塞拉玄参●☆

318424　Pseudoselago spuria（L.）Hilliard；可疑假塞拉玄参；Blue Haze ●☆

318425　Pseudoselago subglabra Hilliard；近光假塞拉玄参●☆

318426　Pseudoselago verbenacea（L. f.）Hilliard；马鞭草假塞拉玄参●☆

318427　Pseudoselago violacea Hilliard；堇色塞拉玄参☆

318428　Pseudoselinum C. Norman（1929）；假亮蛇床属■☆

318429　Pseudoselinum angolense（C. Norman）C. Norman；假亮蛇床■☆

318430　Pseudosempervivum（Boiss.）Grossh.（1930）；假长生草属■☆

318431　Pseudosempervivum（Boiss.）Grossh. = Cochlearia L. ■

318432　Pseudosempervivum karsianum（N. Busch）Grossh.；假长生草■☆

318433　Pseudosenefeldera Esser（2001）；假塞内大戟属■☆

318434　Pseudosericocoma Cavaco（1962）；假绢毛苋属●☆

318435　Pseudosericocoma pungens（Fenzl）Cavaco；假绢毛苋●☆

318436　Pseudoseris Baill. = Gerbera L.（保留属名）■

318437　Pseudoseris Baill. = Gerbera L. ex Cass. ■

318438　Pseudoseris grandidieri Baill. = Gerbera bojeri（DC.）Sch. Bip. ■☆

318439　Pseudoseris rutenbergii Baill. = Gerbera piloselloides（L.）Cass. ■

318440　Pseudosicydium Harms（1927）；肖野胡瓜属■☆

318441　Pseudosicydium acariaeanthum Harms；肖野胡瓜■☆

318442　Pseudosindora Symington = Copaifera L.（保留属名）●☆

318443　Pseudosindora Symington（1944）；假油楠属●☆

318444　Pseudosindora palustris Symington；假油楠●☆

318445　Pseudosmelia Sleumer（1954）；假香木属●☆

318446　Pseudosmelia moluccana Sleumer；假香木●☆

318447　Pseudosmilax Hayata = Heterosmilax Kunth ●■

318448　Pseudosmilax Hayata（1920）；假菝葜属（假土茯苓属）；Pseudosmilax ●■★

318449　Pseudosmilax hogoensis Hayata = Heterosmilax seisuiensis（Hayata）F. T. Wang et Ts. Tang ●■

318450　Pseudosmilax seisuiensis Hayata；假菝葜（假土茯苓，台湾肖菝葜，台中假土茯苓）；Pseudosmilax，Taiwan Heterosmilax ●■

318451　Pseudosmilax seisuiensis Hayata = Heterosmilax seisuiensis（Hayata）F. T. Wang et Ts. Tang ●■

318452　Pseudosmodingium Engl.（1881）；假肿漆属●☆

318453　Pseudosmodingium multifolium Rose；假肿漆●☆

318454　Pseudosolisia Y. Ito = Neolloydia Britton et Rose ●☆

318455　Pseudosophora（DC.）Sweet = Radiusia Rchb. ●■

318456　Pseudosophora（DC.）Sweet = Sophora L. ●■

318457　Pseudosophora Sweet = Radiusia Rchb. ●■

318458　Pseudosophora Sweet = Sophora L. ●■

318459　Pseudosopubia Engl.（1897）；假短冠草属■●☆

318460　Pseudosopubia delamerei S. Moore；德拉米尔假短冠草■☆

318461　Pseudosopubia elata Hemsl. = Pseudosopubia hildebrandtii（Vatke）Engl. ■☆

318462　Pseudosopubia hildebrandtii（Vatke）Engl.；希氏假短冠草■☆

318463　Pseudosopubia kituiensis（Vatke）Engl.；假短冠草■☆

318464　Pseudosopubia obtusifolia Engl. = Pseudosopubia hildebrandtii（Vatke）Engl. ■☆

318465　Pseudosopubia polemonioides Chiov. = Pseudosopubia hildebrandtii（Vatke）Engl. ■☆

318466　Pseudosopubia procumbens Hemsl.；平铺假短冠草■☆

318467　Pseudosopubia verruculosa Chiov. = Pseudosopubia procumbens Hemsl. ■☆

318468　Pseudosorghum A. Camus（1921）；肖高粱属（假蜀黍属，拟高粱属）■

318469　Pseudosorghum fasciculare（Roxb.）A. Camus；肖高粱●

318470　Pseudosorghum zollingeri（Steud.）A. Camus = Pseudosorghum fasciculare（Roxb.）A. Camus ■

318471　Pseudosorocea Baill. = Sorocea A. St. -Hil. + Acanthinophyllum M. Allemão ●☆

318472　Pseudosorocea Baill. = Sorocea A. St. -Hil. ●☆

318473　Pseudospermum Gray = Physospermum Cusson ex Juss. ■☆

318474　Pseudospermum Spreng. ex Gray = Physospermum Cusson ex Juss. ■☆

318475　Pseudospigelia W. Klett = Spigelia L. ■☆

318476　Pseudospigelia W. Klett（1923）；假驱虫草属■☆

318477　Pseudospigelia polystachya W. Klett；假驱虫草■☆

318478　Pseudospondias Engl.（1883）；假槟榔青属●☆

318479　Pseudospondias gigantea A. Chev. = Ganophyllum giganteum（A. Chev.）Hauman ●☆

318480　Pseudospondias longifolia Engl.；长叶假槟榔青●☆

318481　Pseudospondias luxurians A. Chev. = Trichoscypha lucens Oliv. ●☆

318482　Pseudospondias microcarpa（A. Rich.）Engl.；小果假槟榔青●☆

318483　Pseudosrnilax Hayata = Heterosmilax Kunth ●■

318484　Pseudostachyum Munro = Schizostachyum Nees ●

318485　Pseudostachyum Munro（1868）；泡竹属（假穗竹属，小薄竹属）；Bubblebamboo，Thinestwalled Bamboon，Thinest-walled Bamboon ●

318486　Pseudostachyum compactiflorum Kurz = Melocalamus compactiflorus（Kurz）Benth. et Hook. f. ●

318487　Pseudostachyum polymorphum Munro；泡竹（假穗竹）；Bubblebamboo，Thinestwalled Bamboo，Thinest-walled Bamboon，Thin-walled Bamboo ●

318488　Pseudostelis Schltr. = Pleurothallis R. Br. ■☆

318489　Pseudostellaria Pax（1934）；孩儿参属（假繁缕属，太子参属）；Childseng，False Chickweed，Falsestarwort，Sticky Starwort ■

318490　Pseudostellaria cashmiriana Schaeftl. = Pseudostellaria heterantha（Maxim.）Pax var. himalaica（Franch.）Ohwi ■

318491　Pseudostellaria cashmiriana Schaeftl. = Pseudostellaria himalaica（Franch.）Pax ■

318492　Pseudostellaria dalaolingensis Z. E. Zhou et J. Q. Wu = Pseudostellaria himalaica（Franch.）Pax ■

318493　Pseudostellaria davidii（Franch.）Pax；蔓孩儿参（蔓假繁缕）；David Pseudostellaria，Tendril Childseng ■

318494　Pseudostellaria eritrichoides（Diels）Ohwi = Pseudostellaria heterantha（Maxim.）Pax ■

318495　Pseudostellaria helanshanensis W. Z. Di et Y. Ren;贺兰山孩儿参;Helanshan Childseng,Helanshan Pseudostellaria ■

318496　Pseudostellaria heterantha（Maxim.）Pax;异花孩儿参（假繁缕,异花假繁缕）;Heteroflower Childseng,Heteroflower Pseudostellaria,Maximowicz Pseudostellaria ■

318497　Pseudostellaria heterantha（Maxim.）Pax var. himalaica（Franch.）Ohwi = Pseudostellaria himalaica（Franch.）Pax ■

318498　Pseudostellaria heterantha（Maxim.）Pax var. himalaica Ohwi = Pseudostellaria himalaica（Franch.）Pax ■

318499　Pseudostellaria heterantha（Maxim.）Pax var. linearifolia（Takeda）Nemoto;线叶异花孩儿参■☆

318500　Pseudostellaria heterantha（Maxim.）Pax var. tibetica（Ohwi）Kozhevn. = Pseudostellaria tibetica Ohwi ■

318501　Pseudostellaria heterophylla（Miq.）Pax;孩儿参（假繁缕,双批七,四叶菜,太子参,童参,异叶假繁缕）;Childseng,Different Leaves Pseudostellaria ■

318502　Pseudostellaria heterophylla（Miq.）Pax var. stenopetala Kitag.;辽宁孩儿参■

318503　Pseudostellaria himalaica（Franch.）Pax;须弥孩儿参;Himalayas Childseng,Himalayas Pseudostellaria ■

318504　Pseudostellaria himalaica（Franch.）Pax = Pseudostellaria heterantha（Maxim.）Pax ■

318505　Pseudostellaria jamesiana（Torr.）W. A. Weber et R. L. Hartm.;美洲孩儿参（詹姆斯孩儿参）■☆

318506　Pseudostellaria japonica（Korsh.）Pax;毛脉孩儿参（毛孩儿参,毛假繁缕）;Hairyvein Childseng,Hairyvein Pseudostellaria ■

318507　Pseudostellaria maximowicziana（Franch. et Sav.）Pax;矮小孩儿参（棒棒草,孩儿参,假繁缕,米太子参,双批七,太子参,童参,异叶假繁缕）;Dwarf Childseng,Dwarf Pseudostellaria ■

318508　Pseudostellaria maximowicziana（Franch. et Sav.）Pax = Pseudostellaria heterantha（Maxim.）Pax ■

318509　Pseudostellaria musashiensis Hiyama = Pseudostellaria heterantha（Maxim.）Pax var. linearifolia（Takeda）Nemoto ■☆

318510　Pseudostellaria oxyphylla（B. L. Rob.）R. L. Hartm. et Rabeler;尖叶孩儿参;Robinson's Starwort ■☆

318511　Pseudostellaria palibiniana（Takeda）Ohwi;帕利宾孩儿参■☆

318512　Pseudostellaria rhaphanorrhiza（Hemsl.）Pax = Pseudostellaria heterophylla（Miq.）Pax ■

318513　Pseudostellaria rupestris（Turcz.）Pax;石生孩儿参（石假繁缕）;Pseudostellaria,Saxicolous Childseng ■

318514　Pseudostellaria sierrae Rabeler et R. L. Hartm.;齿叶孩儿参;Sierra Starwort ■☆

318515　Pseudostellaria sylvatica（Maxim.）Pax;细叶孩儿参（疙瘩七,林生孩儿参,森林假繁缕,细叶假繁缕,狭叶孩儿参,狭叶假繁缕）;Narrowleaf seudostellaria,Thinleaf Childseng,Thinleaf seudostellaria ■

318516　Pseudostellaria terminalis W. Z. Di et Y. Ren = Pseudostellaria rupestris（Turcz.）Pax ■

318517　Pseudostellaria tibetica Ohwi;西藏孩儿参;Xizang Childseng,Xizang Pseudostellaria ■

318518　Pseudostellaria zhejiangensis X. F. Jin et B. Y. Ding;浙江孩儿参■

318519　Pseudostenomesson Velarde（1949）;假狭管石蒜属■☆

318520　Pseudostenomesson morrisonii（Vargas）Velarde;假狭管石蒜■☆

318521　Pseudostenosiphonium Lindau = Gutzlaffia Hance ●■

318522　Pseudostenosiphonium Lindau = Strobilanthes Blume ●■

318523　Pseudostifftia H. Rob.（1979）;假亮毛菊属●☆

318524　Pseudostifftia kingii H. Rob. ;假亮毛菊■☆

318525　Pseudostonium Kuntze = Pseudostenosiphonium Lindau ●■

318526　Pseudostreblus Bureau = Streblus Lour. ●

318527　Pseudostreblus Bureau（1873）;类鹊肾树属●☆

318528　Pseudostreblus candatus Ridl. ;类鹊肾树●☆

318529　Pseudostreblus indicus Bureau = Streblus indicus（Bureau）Corner ●

318530　Pseudostreptogyne A. Camus = Streblochaete Hochst. ex Pilg. ■☆

318531　Pseudostriga Bonati（1911）;假独脚金属■☆

318532　Pseudostriga cambodiana Bonati;假独脚金■☆

318533　Pseudostrophis T. Durand et B. D. Jacks. = Pseudotrophis Warb. ●

318534　Pseudostrophis T. Durand et B. D. Jacks. = Streblus Lour. ●

318535　Pseudotaenidia Mack.（1903）;假太尼草属☆

318536　Pseudotaenidia Mack. = Taenidia（Torr. et A. Gray）Drude ☆

318537　Pseudotaenidia montana Mack. ;假太尼草☆

318538　Pseudotaxus W. C. Cheng（1947）;白豆杉属;White Aril Yew,Whitearil Yew,White-aril Yew ●★

318539　Pseudotaxus chienii（W. C. Cheng）W. C. Cheng;白豆杉（短水松）;False Yew,White Aril Yew,Whitearil Yew,White-aril Yew ●

318540　Pseudotaxus liana Silba = Pseudotaxus chienii（W. C. Cheng）W. C. Cheng ●

318541　Pseudotenanthera R. B. Majumdar = Pseudoxytenanthera Soderstr. et R. P. Ellis ●☆

318542　Pseudotenanthera albociliata（Munro）R. B. Majumdar = Gigantochloa albociliata（Munro）Kurz ●

318543　Pseudotephrocactus Frič = Opuntia Mill. ●

318544　Pseudotephrocactus Frič et Schelle = Opuntia Mill. ●

318545　Pseudotigandra Dillon et Sagast. = Chionolaena DC. ●☆

318546　Pseudotrachydium（Kljuykov, Pimenov et V. N. Tikhom.）Pimenov et Kljuykov（2000）;假瘤果芹属■☆

318547　Pseudotragia Pax = Pterococcus Hassk.（保留属名）●☆

318548　Pseudotragia scandens Pax = Plukenetia africana Sond. ●☆

318549　Pseudotragia schinzii Pax = Plukenetia africana Sond. ●☆

318550　Pseudotreculia（Baill.）B. D. Jacks. = Treculia Decne. ex Trécul ●☆

318551　Pseudotreculia Baill. = Treculia Decne. ex Trécul ●☆

318552　Pseudotrewia Miq. = Wetria Baill. ●☆

318553　Pseudotrillium S. B. Farmer = Trillium L. ■

318554　Pseudotrillium S. B. Farmer（2002）;假延龄草属■☆

318555　Pseudotrimezia R. C. Foster（1945）;假枝端花属■☆

318556　Pseudotrimezia barretoi R. C. Foster;假枝端花■☆

318557　Pseudotrimezia fulva Ravenna;黄假枝端花■☆

318558　Pseudotrimezia gracilis Chukr;细假枝端花■☆

318559　Pseudotrimezia monticola（Klatt）Ravenna;山地假枝端花■☆

318560　Pseudotrophis Warb. = Streblus Lour. ●

318561　Pseudotrophis laxiflora Warb. = Streblus ilicifolius（S. Vidal）Corner ●

318562　Pseudotsuga Carrière（1867）;黄杉属;Douglas Fir,Douglas-fir,Hongcone-fir ●

318563　Pseudotsuga argyrophylla（Chun et Kuang）Greguss = Cathaya argyrophylla Chun et Kuang ●◇

318564　Pseudotsuga brevifolia W. C. Cheng et L. K. Fu;短叶黄杉;Shortleaf Douglas Fir,Short-leaved Douglas Fir ●◇

318565　Pseudotsuga californica Flous = Pseudotsuga macrocarpa（Vasey）Mayr ●◇

318566　Pseudotsuga davidiana Bertrand = Keteleeria davidiana（Bertrand）Beissn. ●

318567 Pseudotsuga douglasii （Lindl.） Carrière = Pseudotsuga menziesii （Mirb.） Franco ●

318568 Pseudotsuga douglasii （Lindl.） Carrière var. glauca Mayr = Pseudotsuga menziesii （Mirb.） Franco var. glauca （Mayr） Franco ●

318569 Pseudotsuga douglasii （Lindl.） Carrière var. macrocarpa （Vasey） Engelm. = Pseudotsuga macrocarpa （Vasey） Mayr ●◇

318570 Pseudotsuga douglasii （Sabine ex D. Don） Carrière = Pseudotsuga menziesii （Mirb.） Franco ●

318571 Pseudotsuga forrestii Craib；澜沧黄杉（湄公黄杉）；Forrest Douglas Fir，Yunnan Douglas Fir ●

318572 Pseudotsuga gaussenii Flous；华东黄杉（浙皖黄杉）；Gaussen Douglas Fir ●◇

318573 Pseudotsuga gaussenii Flous = Pseudotsuga sinensis Dode ●◇

318574 Pseudotsuga glauca Mayr；灰色黄杉；Blue Douglas Fir，Rocky Mountain Douglas Fir ●☆

318575 Pseudotsuga japonica （Shiras.） Beissn.；日本黄杉；Japan Douglas Fir，Japanese Douglas Fir ●☆

318576 Pseudotsuga japonica Beissn. = Pseudotsuga sinensis Dode var. wilsoniana （Hayata） L. K. Fu et Nan Li ●◇

318577 Pseudotsuga jezoensis （Carri） Bertrand = Keteleeria fortunei （A. Murray） Carrière ●

318578 Pseudotsuga macrocarpa （Torr.） Mayr = Pseudotsuga macrocarpa （Vasey） Mayr ●◇

318579 Pseudotsuga macrocarpa （Vasey） Mayr；大果黄杉（北美大果黄杉，大果花旗松）；Bigcone Douglas Fir，Big-cone Douglas Fir，Bigcone Douglas-fir，Bigcone Spruce，Bigcone-spruce，Large-cone Douglas Fir，Large-coned Douglas Fir ●◇

318580 Pseudotsuga menziesii （Mirb.） Franco；花旗松（北美黄杉，落基山黄杉，美国黄杉）；Blue Douglas Fir，British Columbian Pine，Coast Douglas Fir，Coast Douglas-fir，Columbian British Pine，Columbian Pine，Common Douglas Fir，Douglas Fir，Douglas Spruce，Douglas Tree，Douglas-fir，Douglas-spruce，False Spruce，Green Douglas Fir，Norway Fir，Oregon Fir，Oregon Pine，Oregon-pine，Red Fir，Red Pine，Yellow Fir ●

318581 Pseudotsuga menziesii （Mirb.） Franco 'Densa'；密集花旗松 ●☆

318582 Pseudotsuga menziesii （Mirb.） Franco 'Fletcheri'；富莱切利花旗松（矮生花旗松）●☆

318583 Pseudotsuga menziesii （Mirb.） Franco 'Fretsii'；短叶花旗松 ●☆

318584 Pseudotsuga menziesii （Mirb.） Franco var. glauca （Boiss.） Franco = Pseudotsuga menziesii （Mirb.） Franco var. glauca （Mayr） Franco ●

318585 Pseudotsuga menziesii （Mirb.） Franco var. glauca （Mayr） Franco；蓝色花旗松（粉绿花旗松，蓝花旗松）；Blue Douglas Fir，Colorado Douglas Fir，Rocky Mountain Douglas-fir ●

318586 Pseudotsuga mucronata （Raf.） Sudw. = Pseudotsuga menziesii （Mirb.） Franco ●

318587 Pseudotsuga salvadori Flous = Pseudotsuga sinensis Dode var. wilsoniana （Hayata） L. K. Fu et Nan Li ●◇

318588 Pseudotsuga salvadori Flous = Pseudotsuga wilsoniana Hayata ●◇

318589 Pseudotsuga shaanxiensis S. Z. Qu et K. Y. Wang；陕西黄杉；Shaanxi Douglas Fir ●

318590 Pseudotsuga shaanxiensis S. Z. Qu et K. Y. Wang = Pseudotsuga sinensis Dode ●◇

318591 Pseudotsuga sinensis Dode；黄杉；China Douglas Fir，Chinese Douglas Fir ●◇

318592 Pseudotsuga sinensis Dode var. brevifolia （W. C. Cheng et L. K. Fu） Farjon et Silba = Pseudotsuga brevifolia W. C. Cheng et L. K. Fu ●◇

318593 Pseudotsuga sinensis Dode var. forrestii （Craib） Silba = Pseudotsuga forrestii Craib ●◇

318594 Pseudotsuga sinensis Dode var. gaussenii （Flous） Silba = Pseudotsuga sinensis Dode ●◇

318595 Pseudotsuga sinensis Dode var. wilsoniana （Hayata） L. K. Fu et Nan Li = Pseudotsuga wilsoniana Hayata ●◇

318596 Pseudotsuga taxifolia （Lamb.） Britton = Pseudotsuga menziesii （Mirb.） Franco ●

318597 Pseudotsuga taxifolia （Poir.） Britton ex Sudw. = Pseudotsuga menziesii （Mirb.） Franco ●

318598 Pseudotsuga trifoliata ？；三小叶黄杉 ●☆

318599 Pseudotsuga wilsoniana Hayata；台湾黄杉（萨尔瓦多黄杉，威氏帝杉）；Taiwan Douglas Fir，E. H. Wilson Douglas Fir，Formosan Douglas Fir，Salvador Douglas Fir ●◇

318600 Pseudotsuga wilsoniana Hayata = Pseudotsuga forrestii Craib ●◇

318601 Pseudotsuga wilsoniana Hayata = Pseudotsuga sinensis Dode var. wilsoniana （Hayata） L. K. Fu et Nan Li ●◇

318602 Pseudotsuga xichangensis C. T. Kuan et L. J. Zhou；西昌黄杉；Xichang Douglas Fir ●

318603 Pseudotsuga xichangensis C. T. Kuan et L. J. Zhou = Pseudotsuga sinensis Dode ●◇

318604 Pseudoturritis Al-Shehbaz = Arabis L. ●■

318605 Pseudoturritis Al-Shehbaz（2005）；意大利南芥属 ■☆

318606 Pseudoturritis turrita （L.） Al-Shehbaz；意大利芥 ■☆

318607 Pseudourceolina Vargas = Urceolina Rchb.（保留属名）■☆

318608 Pseudovanilla Garay（1986）；假香荚兰属 ■☆

318609 Pseudovanilla foliata （F. Muell.） Garay；假香荚兰 ■☆

318610 Pseudovesicaria （Boiss.） Rupr.（1869）；膀胱菜属 ■☆

318611 Pseudovesicaria Boiss. = Pseudovesicaria （Boiss.） Rupr. ■☆

318612 Pseudovesicaria digitata （C. A. Mey.） Rupr.；膀胱菜 ■☆

318613 Pseudovigna （Harms） Verdc.（1970）；假豇豆属 ■☆

318614 Pseudovigna argentea （Willd.） Verdc.；假豇豆 ■☆

318615 Pseudovigna puerarioides Ern；非洲假豇豆 ■☆

318616 Pseudovossia A. Camus = Phacelurus Griseb. ■

318617 Pseudovouapa Britton et Killip = Macrolobium Schreb.（保留属名）●☆

318618 Pseudovouapa Britton et Rose = Macrolobium Schreb.（保留属名）●☆

318619 Pseudoweinmannia Engl.（1930）；假万灵木属 ●☆

318620 Pseudoweinmannia lachnocarpa （F. Muell.） Engl.；假万灵木；Marara ●☆

318621 Pseudo-willughbeia Markgr. = Melodinus J. R. Forst. et G. Forst. ●

318622 Pseudowillughbeia Markgr. = Melodinus J. R. Forst. et G. Forst. ●

318623 Pseudowintera Dandy（1933）；假林仙属（哈罗皮图木属）●☆

318624 Pseudowintera axillaris （J. R. Forst. et G. Forst.） Dandy；假林仙（哈罗皮图木）；Haropito，Pepper-tree，Pepper Tree ●☆

318625 Pseudowintera colorata （Raoul） Dandy；彩叶假林仙（彩色哈罗皮图木）；Haropito，Pepper Tree ●☆

318626 Pseudowolffia Hartog et Plas = Wolffiella （Hegelm.） Hegelm. ■☆

318627 Pseudowolffia Hartog et Plas（1970）；假芜萍属 ■☆

318628 Pseudowolffia hyalina （Delile） Hartog et Plas = Wolffiella hyalina （Delile） Monod ■☆

318629 Pseudowolffia monodii （Ast） Hartog et Plas = Wolffiella hyalina （Delile） Monod ■☆

318630 Pseudowolffia repanda （Hegelm.） Hartog et Plas = Wolffiella repanda （Hegelm.） Monod ■☆

318631 Pseudoxalis Rose = Oxalis L. ■●

318632　Pseudoxandra R. E. Fr.（1937）;假剑木属（假酸蕊花属）●☆

318633　Pseudoxandra angustifolia Maas;窄叶假剑木●☆

318634　Pseudoxandra cauliflora Maas;茎花假剑木●☆

318635　Pseudoxandra guianensis（R. E. Fr.）R. E. Fr.;圭亚那假剑木●☆

318636　Pseudoxandra leiophylla（Diels）R. E. Fr.;光叶假剑木●☆

318637　Pseudoxandra longipes Maas;长梗假剑木●☆

318638　Pseudoxandra parvifolia Maas;小叶假剑木●☆

318639　Pseudoxandra pilosa Maas;毛假剑木●☆

318640　Pseudoxytenanthera Soderstr. et R. P. Ellis ＝ Schizostachyum Nees ●

318641　Pseudoxytenanthera Soderstr. et R. P. Ellis(1988);假锐药竹属●☆

318642　Pseudoxytenanthera albociliata（Munro）T. Q. Nguyen ＝ Gigantochloa albociliata（Munro）Kurz ●

318643　Pseudoxytenanthera monadelpha（Thwaites）Soderstr. et R. P. Ellis;锡兰假锐药竹;Ceylon Bamboo ●☆

318644　Pseudoxytenanthera nigrociliata（Büse）T. Q. Nguyen ＝ Gigantochloa nigrociliata（Büse）Kurz ●

318645　Pseudoxythece Aubrév. ＝ Pouteria Aubl. ●

318646　Pseudoyoungia D. Maity et Maiti(2010);假黄鹌菜属■

318647　Pseudoyoungia angustifolia（Tzvelev）D. Maity et Maiti;窄叶假黄鹌菜■

318648　Pseudoyoungia angustifolia（Tzvelev）D. Maity et Maiti ＝ Tibetoseris angustifolia Tzvelev ■

318649　Pseudoyoungia conjuctiva（Babc. et Stebbins）D. Maity et Maiti ＝ Tibetoseris conjunctiva（Babc. et Stebbins）Sennikov ■

318650　Pseudoyoungia cristata（C. Shih et C. Q. Cai）D. Maity et Maiti ＝ Tibetoseris cristata（C. Shih et C. Q. Cai）Sennikov ■

318651　Pseudoyoungia gracilipes（Hook. f.）D. Maity et Maiti ＝ Tibetoseris gracilipes（Hook. f.）Sennikov ■

318652　Pseudoyoungia ladyginii（Tzvelev）D. Maity et Maiti ＝ Tibetoseris ladyginii Tzvelev ■

318653　Pseudoyoungia parva（Babc. et Stebbins）D. Maity et Maiti ＝ Tibetoseris parva（Babc. et Stebbins）Sennikov ■

318654　Pseudoyoungia sericea（C. Shih）D. Maity et Maiti ＝ Tibetoseris sericeus（C. Shih）Sennikov ■

318655　Pseudoyoungia simulatrix（Babc.）D. Maity et Maiti ＝ Tibetoseris simulatrix（Babc.）Sennikov ■

318656　Pseudoyoungia tianshanica（C. Shih）D. Maity et Maiti ＝ Tibetoseris tianshanica（C. Shih）Tzvelev ■

318657　Pseudozoysia Chiov.（1928）;假结缕属（卷曲刺毛叶草属）■☆

318658　Pseudozoysia sessilis Chiov.;假结缕（卷曲刺毛叶草）■☆

318659　Pseudozygocactus Backeb. ＝ Hatiora Britton et Rose ●

318660　Pseudrachicallis H. Karst. ＝ Mallostoma H. Karst. ●☆

318661　Pseuduvaria Miq.（1858）;金钩花属（假紫玉盘属）;Goldenhook,Pseuduvaria ●

318662　Pseuduvaria indochinensis Merr.;金钩花;Indochina Goldenhook,Indochina Pseuduvaria ●

318663　Pseusmagenetus Ruschenb. ＝ Marsdenia R. Br.（保留属名）●

318664　Pseva Raf. ＝ Chimaphila Pursh ●■

318665　Psiadia Jacq.（1803）;黄胶菊属●☆

318666　Psiadia agathaeoides（Cass.）Drake;费利菊状黄胶菊■☆

318667　Psiadia alticola Humbert;高原黄胶菊●☆

318668　Psiadia altissima（DC.）Drake;高大黄胶菊●☆

318669　Psiadia altissima（DC.）Drake subsp. angustifolia Humbert ＝ Psiadia angustifolia（Humbert）Humbert ●☆

318670　Psiadia altissima（DC.）Drake subsp. coarctata Humbert ＝ Psiadia coarctata（Humbert）Humbert ●☆

318671　Psiadia altissima（DC.）Drake subsp. serrata Humbert ＝ Psiadia serrata（Humbert）Humbert ●☆

318672　Psiadia altissima（DC.）Drake var. boinensis Humbert ＝ Psiadia altissima（DC.）Drake ●☆

318673　Psiadia altissima（DC.）Drake var. cloiselii Humbert ＝ Psiadia angustifolia（Humbert）Humbert ●☆

318674　Psiadia altissima（DC.）Drake var. latifolia Humbert ＝ Psiadia glutinosa Jacq. ●☆

318675　Psiadia angustifolia（Humbert）Humbert;窄叶黄胶菊●☆

318676　Psiadia aparine Muschl. ＝ Psiadia punctulata（DC.）Vatke ●☆

318677　Psiadia arabica Jaub. et Spach ＝ Psiadia punctulata（DC.）Vatke ●☆

318678　Psiadia auriculata Baker ＝ Psiadia hispida（DC.）Benth. et Hook. f. ●☆

318679　Psiadia cacuminum（Humbert）Humbert;渐尖黄胶菊●☆

318680　Psiadia catatii Drake ＝ Psiadia salviifolia Baker ●☆

318681　Psiadia coarctata（Humbert）Humbert;密集黄胶菊●☆

318682　Psiadia coursii Humbert;库尔斯黄胶菊●☆

318683　Psiadia cuspidifera Baker ＝ Conyza ageratoides DC. ●☆

318684　Psiadia decaryi Humbert;德卡里黄胶菊●☆

318685　Psiadia decurrens Klatt ＝ Psiadia altissima（DC.）Drake ●☆

318686　Psiadia depauperata Humbert;萎缩黄胶菊●☆

318687　Psiadia dimorpha Humbert;二型黄胶菊●☆

318688　Psiadia flavocinerea Humbert;黄灰黄胶菊●☆

318689　Psiadia glutinosa Jacq.;黏性黄胶菊●☆

318690　Psiadia gnaphaliopsis Schweinf. et Volkens ＝ Conyza boranensis（S. Moore）Cufod. ■☆

318691　Psiadia godotiana Humbert;戈多黄胶菊●☆

318692　Psiadia grandidentata Steetz;大齿黄胶菊●☆

318693　Psiadia grevei Baill. ＝ Pluchea grevei（Baill.）Humbert ●☆

318694　Psiadia hendersoniae S. Moore;亨德森黄胶菊●☆

318695　Psiadia hispida（DC.）Benth. et Hook. f.;硬毛黄胶菊●☆

318696　Psiadia inaequidentata Humbert;不等齿黄胶菊●☆

318697　Psiadia incana Oliv. et Hiern;灰毛黄胶菊●☆

318698　Psiadia inuloides O. Hoffm. ＝ Conyza vernonioides（Sch. Bip. ex A. Rich.）Wild ■☆

318699　Psiadia leucophylla（Baker）Humbert;白叶黄胶菊●☆

318700　Psiadia leucophylla（Baker）Humbert subsp. cacuminum Humbert ＝ Psiadia cacuminum（Humbert）Humbert ●☆

318701　Psiadia leucophylla（Baker）Humbert subsp. marojejyensis Humbert ＝ Psiadia marojejyensis（Humbert）Humbert ●☆

318702　Psiadia lucida（Cass.）Drake;光亮黄胶菊●☆

318703　Psiadia lycioides Hiern ＝ Pluchea lycioides（Hiern）Merxm. ●☆

318704　Psiadia madagascariensis（Lam.）DC. ＝ Psiadia lucida（Cass.）Drake ●☆

318705　Psiadia marojejyensis（Humbert）Humbert;马鲁杰黄胶菊●☆

318706　Psiadia minor Steetz;较小黄胶菊●☆

318707　Psiadia modesta Baker ＝ Pluchea bojeri（DC.）Humbert ●☆

318708　Psiadia mollissima O. Hoffm.;柔软黄胶菊●☆

318709　Psiadia nigrescens Humbert;变黑黄胶菊●☆

318710　Psiadia pseudonigrescens Buscal. et Muschl.;假变黑黄胶菊●☆

318711　Psiadia punctulata（DC.）Vatke;小斑黄胶菊●☆

318712　Psiadia quartziticola Humbert;阔茨黄胶菊●☆

318713　Psiadia resiniflua（Hochst. et Steud. ex DC.）Sch. Bip. ＝ Psiadia punctulata（DC.）Vatke ●☆

318714　Psiadia salviifolia Baker;鼠尾草叶黄胶菊●☆

318715　Psiadia serrata（Humbert）Humbert;具齿黄胶菊●☆

318716　Psiadia tanala Humbert;塔纳尔黄胶菊●☆

318717　Psiadia tortuosa Klatt = Psiadia lucida (Cass.) Drake ●☆

318718　Psiadia trinervia Willd.;三脉黄胶菊●☆

318719　Psiadia tsaratananensis Humbert;察拉塔纳纳黄胶菊●☆

318720　Psiadia urticifolia Baker = Conyza urticifolia (Baker) Humbert ■☆

318721　Psiadia vernicosa Schinz = Psiadia punctulata (DC.) Vatke ●☆

318722　Psiadia vernicosa Steetz;光泽黄胶菊●☆

318723　Psiadia vestita Humbert;包被黄胶菊●☆

318724　Psiadia volubilis (DC.) Baill. ex Humbert;缠绕黄胶菊●☆

318725　Psiadiella Humbert(1923);单脉黄胶菊属●☆

318726　Psiadiella humilis Humbert;单脉黄胶菊●☆

318727　Psiadiella humilis Humbert subvar. sciaphila Humbert = Psiadiella humilis Humbert ●☆

318728　Psidiastrum Bello = Eugenia L. ●

318729　Psidiomyrtus Guillaumin = Rhodomyrtus (DC.) Rchb. ●

318730　Psidiopsis O. Berg = Psidium L. ●

318731　Psidiopsis O. Berg(1856);类番石榴属●☆

318732　Psidiopsis moritziana O. Berg;类番石榴●☆

318733　Psidium L. (1753);番石榴属(兽石榴属);Guava ●

318734　Psidium araca Raddi;巴西番石榴;Brazilian Guava ●☆

318735　Psidium araca Raddi = Psidium guineense Sw. ●☆

318736　Psidium cattleianum Sabine = Psidium cattleyanum Sabine ●

318737　Psidium cattleyanum Sabine;草莓番石榴(长梗草莓番石榴,海滨番石榴);Cattley Guava, Cherry Guava, Pineapple Guava, Purple Guava,Strawberry Guava,Yellow Guava ●

318738　Psidium cattleyanum Sabine var. coriaceum ?;革质番石榴●☆

318739　Psidium cattleyanum Sabine var. lucidum Hort.;黄果草莓番石榴(黄果番石榴);Yellow Strawberry Guava ●☆

318740　Psidium friedrichsthalianum (O. Berg) Nied.;弗氏番石榴;Costa Rican Guava ●☆

318741　Psidium friedrichsthalianum Nied. = Psidium friedrichsthalianum (O. Berg) Nied. ●☆

318742　Psidium guajava L.;番石榴(拔仔,拔子,百子树,番鬼子,番稔,番桃,番桃树,蕃石榴,广东石榴,花稔,鸡矢茶,鸡失果,鸡屎果,鸡头果,交趾果,郊树,胶桃,胶子果,篮拔,梨仔菝,莉仔芰,缅桃,木八仔,木八子,那拔,秋果);Common Guava, Common Guava Goyave, Guava, Guava Tree, Jambu Batu, Jambu Biji, Red Malaysian Guava,Tropical Guava ●

318743　Psidium guianense Pers.;圭亚那番石榴●☆

318744　Psidium guineense Sw.;几内亚番石榴(巴西番石榴);Brazilian Guava,Guisaro,Wild Guava ●☆

318745　Psidium humile Vell.;矮番石榴(倭番石榴,小叶番石榴)●☆

318746　Psidium inermis L.;菲岛番石榴●☆

318747　Psidium laurifolium Berg.;樟叶番石榴;Costa Rican Guava ●☆

318748　Psidium laurifolium Berg. = Psidium friedrichsthalianum Nied. ●☆

318749　Psidium littorale Raddi 'Lucidum';柠檬番石榴;Lemon Guava ●☆

318750　Psidium littorale Raddi = Psidium cattleyanum Sabine ●

318751　Psidium littorale Raddi var. longipes (O. Berg ex Mart.) Fosberg = Psidium littorale Raddi ●

318752　Psidium longifolium Schumach.;长叶番石榴●☆

318753　Psidium pomiferum L. = Psidium guajava L. ●

318754　Psidium pyriferum L. = Psidium guajava L. ●

318755　Psidium sartorianum (O. Berg) Nied.;萨托番石榴●☆

318756　Psidium variabile O. Berg = Psidium cattleyanum Sabine ●

318757　Psiguria Arn. = Psiguria Neck. ex Arn. ■☆

318758　Psiguria Neck. = Psiguria Neck. ex Arn. ■☆

318759　Psiguria Neck. ex Arn. (1841);热美葫芦属■☆

318760　Psiguria triphylla (Miq.) C. Jeffrey;三叶热美葫芦■☆

318761　Psila Phil. = Baccharis L.(保留属名)●■☆

318762　Psilachaenia Post et Kuntze = Psilachenia Benth. ■

318763　Psilachenia Benth. = Crepis L. ■

318764　Psilachenia Benth. = Psilochenia Nutt. ■

318765　Psilactis A. Gray(1849);裸冠菀属■☆

318766　Psilactis asteroides A. Gray;裸冠菀■☆

318767　Psilactis brevilingulata Sch. Bip. ex Hemsl.;短舌裸冠菀■☆

318768　Psilactis gentryi (Standl.) D. R. Morgan;金特里裸冠菀■☆

318769　Psilactis heterocarpa (R. L. Hartm. et M. A. Lane) D. R. Morgan;异果裸冠菀■☆

318770　Psilactis leptos Shinners = Psilactis asteroides A. Gray ■☆

318771　Psilactis tenuis S. Watson;小裸冠菀■☆

318772　Psilaea Miq. = Linostoma Wall. ex Endl. ●☆

318773　Psilantha (C. Koch) Tzvelev = Eragrostis Wolf ■

318774　Psilantha (K. Koch) Tzvelev = Boriskellera Terechov ■

318775　Psilantha (K. Koch) Tzvelev = Eragrostis Wolf ■

318776　Psilanthele Lindau(1897);裸花爵床属☆

318777　Psilanthele eggersii Lindau;裸花爵床☆

318778　Psilanthopsis A. Chev. = Coffea L. ●

318779　Psilanthopsis kapakata A. Chev. = Coffea kapakata (A. Chev.) Bridson ●☆

318780　Psilanthus (DC.) Juss ex M. Roem.(废弃属名) = Psilanthus Hook. f.(保留属名)●☆

318781　Psilanthus (DC.) M. Roem. = Passiflora L. ●■

318782　Psilanthus (DC.) M. Roem. = Psilanthus Hook. f.(保留属名)●☆

318783　Psilanthus Hook. f. (1873)(保留属名);光花咖啡属●☆

318784　Psilanthus Juss. = Psilanthus Hook. f. (保留属名)●☆

318785　Psilanthus Juss. ex M. Roem. = Psilanthus Hook. f.(保留属名)●☆

318786　Psilanthus comoensis Pierre ex De Wild. = Psilanthus mannii Hook. f. ●☆

318787　Psilanthus ebracteolatus Hiern = Coffea ebracteolata (Hiern) Brenan ●☆

318788　Psilanthus epiphyticus Mildbr. ex A. Chev. = Chassalia petitiana Piesschaert ●☆

318789　Psilanthus jasminoides Hutch. et Dalziel = Argocoffeopsis rupestris (Hiern) Robbr. ●☆

318790　Psilanthus lebrunianus (R. Germ. et Kesler) Leroy ex Bridson;勒布伦光花咖啡●☆

318791　Psilanthus ledermannii A. Chev.;莱德光花咖啡●☆

318792　Psilanthus leroyi Bridson;勒罗伊光花咖啡●☆

318793　Psilanthus mannii Hook. f.;曼氏光花咖啡●☆

318794　Psilanthus melanocarpus (Welw. ex Hiern) J. -F. Leroy;黑果光花咖啡●☆

318795　Psilanthus minor A. Chev.;小光花咖啡●☆

318796　Psilanthus sapini De Wild.;萨潘光花咖啡●☆

318797　Psilanthus semsei Bridson;塞姆光花咖啡●☆

318798　Psilarabis Fourr. = Arabis L. ●■

318799　Psilathera Link = Sesleria Scop. ■☆

318800　Psilobium Jack(废弃属名) = Acranthera Arn. ex Meisn.(保留属名)●

318801　Psilocarphus Nutt. (1840);绵石菊属;Woolly Marbles, Woollyheads ■☆

318802　Psilocarphus brevissimus Nutt.;小绵石菊;Dwarf Woollyheads, Short Woollyheads ■☆

318803　Psilocarphus brevissimus Nutt. var. multiflorus Cronquist;多花小绵石菊;Delta Woolly Marbles ■☆

318804　Psilocarphus caulescens Benth. = Hesperevax caulescens（Benth.）A. Gray ■☆

318805　Psilocarphus chilensis A. Gray；智利绵石菊；Round Woolly Marbles ■☆

318806　Psilocarphus elatior（A. Gray）A. Gray；高绵石菊；Meadow Woollyheads, Tall Woollyheads ■☆

318807　Psilocarphus globiferus Nutt. = Psilocarphus brevissimus Nutt. ■☆

318808　Psilocarphus oregonus Nutt.；俄勒冈绵石菊；Oregon Woolly Marbles, Oregon Woollyheads ■☆

318809　Psilocarphus oregonus Nutt. var. elatior A. Gray = Psilocarphus elatior（A. Gray）A. Gray ■☆

318810　Psilocarphus tenellus Nutt.；纤细绵石菊；Slender Woolly Marbles ■☆

318811　Psilocarphus tenellus Nutt. var. globiferus（Bertero ex DC.）Morefield = Psilocarphus chilensis A. Gray ■☆

318812　Psilocarphus tenellus Nutt. var. tenuis（Eastw.）Cronquist = Psilocarphus chilensis A. Gray ■☆

318813　Psilocarpus Pritz. = Psophocarpus Neck. ex DC.（保留属名）■

318814　Psilocarya Torr. = Rhynchospora Vahl（保留属名）■

318815　Psilocarya candida Nees = Rhynchospora candida（Nees）Boeck. ■☆

318816　Psilocarya nitens（Vahl）A. W. Wood = Rhynchospora nitens（Vahl）A. Gray ■☆

318817　Psilocarya rhynchosporoides Torr. = Rhynchospora nitens（Vahl）A. Gray ■☆

318818　Psilocarya schiedeana（Nees）Liebm. = Rhynchospora eximia（Nees）Boeck. ■☆

318819　Psilocarya scirpoides Torr. = Rhynchospora scirpoides（Torr.）Griseb. ■☆

318820　Psilocaulon N. E. Br.（1925）；裸茎日中花属 ■☆

318821　Psilocaulon absimile N. E. Br. = Psilocaulon coriarium（Burch. ex N. E. Br.）N. E. Br. ■☆

318822　Psilocaulon acutisepalum（A. Berger）N. E. Br. = Psilocaulon junceum（Haw.）Schwantes ■☆

318823　Psilocaulon annuum L. Bolus = Psilocaulon articulatum（Thunb.）N. E. Br. ■☆

318824　Psilocaulon arenosum（Schinz）L. Bolus = Brownanthus arenosus（Schinz）Ihlenf. et Bittrich ●☆

318825　Psilocaulon articulatum（Thunb.）N. E. Br.；关节裸茎日中花 ■☆

318826　Psilocaulon asperulum N. E. Br. = Psilocaulon articulatum（Thunb.）N. E. Br. ■☆

318827　Psilocaulon baylissii L. Bolus = Psilocaulon dinteri（Engl.）Schwantes ■☆

318828　Psilocaulon bicorne（Sond.）Schwantes；双角裸茎日中花 ■☆

318829　Psilocaulon bijliae N. E. Br. = Psilocaulon junceum（Haw.）Schwantes ■☆

318830　Psilocaulon bryantii L. Bolus = Psilocaulon articulatum（Thunb.）N. E. Br. ■☆

318831　Psilocaulon caducum（Aiton）N. E. Br. = Mesembryanthemum nodiflorum L. ■☆

318832　Psilocaulon calvinianum L. Bolus = Psilocaulon junceum（Haw.）Schwantes ■☆

318833　Psilocaulon candidum L. Bolus = Psilocaulon junceum（Haw.）Schwantes ■☆

318834　Psilocaulon ciliatum（Aiton）Friedrich = Brownanthus vaginatus（Lam.）Chess. et M. Pignal ●☆

318835　Psilocaulon clavulatum（A. Berger）N. E. Br. = Psilocaulon subnodosum（A. Berger）N. E. Br. ■☆

3!8836　Psilocaulon corallinum（Thunb.）Schwantes = Brownanthus corallinus（Thunb.）Ihlenf. et Bittrich ●☆

318837　Psilocaulon coriarium（Burch. ex N. E. Br.）N. E. Br.；革质裸茎日中花 ■☆

318838　Psilocaulon dejagerae L. Bolus = Psilocaulon articulatum（Thunb.）N. E. Br. ■☆

318839　Psilocaulon delosepalum L. Bolus = Psilocaulon junceum（Haw.）Schwantes ■☆

318840　Psilocaulon densum N. E. Br.；密集裸茎日中花 ■☆

318841　Psilocaulon dimorphum（Welw. ex Oliv.）N. E. Br.；二型裸茎日中花 ■☆

318842　Psilocaulon dinteri（Engl.）Schwantes；丁特裸茎日中花 ■☆

318843　Psilocaulon distinctum N. E. Br. = Aptenia geniculiflora（L.）Bittrich ex Gerbaulet ■☆

318844　Psilocaulon diversipapillosum（A. Berger）N. E. Br. = Brownanthus arenosus（Schinz）Ihlenf. et Bittrich ●☆

318845　Psilocaulon duthiae L. Bolus = Psilocaulon articulatum（Thunb.）N. E. Br. ■☆

318846　Psilocaulon fasciculatum N. E. Br. = Psilocaulon dinteri（Engl.）Schwantes ■☆

318847　Psilocaulon filipetalum L. Bolus = Psilocaulon subnodosum（A. Berger）N. E. Br. ■☆

318848　Psilocaulon fimbriatum L. Bolus = Psilocaulon salicornioides（Pax）Schwantes ■☆

318849　Psilocaulon foliolosum L. Bolus；多小叶裸茎日中花 ■☆

318850　Psilocaulon framesii L. Bolus = Psilocaulon junceum（Haw.）Schwantes ■☆

318851　Psilocaulon gessertianum（Dinter et A. Berger）Dinter et Schwantes；格氏裸茎日中花 ■☆

318852　Psilocaulon glareosum（A. Berger）Dinter et Schwantes = Psilocaulon salicornioides（Pax）Schwantes ■☆

318853　Psilocaulon godmaniae L. Bolus；戈德曼裸茎日中花 ■☆

318854　Psilocaulon godmaniae L. Bolus var. godmaniae ？ = Psilocaulon dinteri（Engl.）Schwantes ■☆

318855　Psilocaulon godmaniae L. Bolus var. gracile ？ = Psilocaulon dinteri（Engl.）Schwantes ■☆

318856　Psilocaulon granulicaule（Haw.）Schwantes；颗粒裸茎日中花 ■☆

318857　Psilocaulon gymnocladum（Schltr. et Diels）Dinter et Schwantes = Brownanthus arenosus（Schinz）Ihlenf. et Bittrich ●☆

318858　Psilocaulon herrei L. Bolus = Psilocaulon dinteri（Engl.）Schwantes ■☆

318859　Psilocaulon hirtellum L. Bolus = Psilocaulon articulatum（Thunb.）N. E. Br. ■☆

318860　Psilocaulon imitans L. Bolus = Psilocaulon junceum（Haw.）Schwantes ■☆

318861　Psilocaulon implexum N. E. Br. = Psilocaulon parviflorum（Jacq.）Schwantes ■☆

318862　Psilocaulon inachabense L. Bolus = Psilocaulon salicornioides（Pax）Schwantes ■☆

318863　Psilocaulon inconspicuum L. Bolus = Psilocaulon salicornioides（Pax）Schwantes ■☆

318864　Psilocaulon inconstrictum L. Bolus = Psilocaulon subnodosum（A. Berger）N. E. Br. ■☆

318865　Psilocaulon junceum（Haw.）Schwantes；灯心草裸茎日中花 ■☆

318866　Psilocaulon kuntzei（Schinz）Dinter et Schwantes = Brownanthus kuntzei（Schinz）Ihlenf. et Bittrich ●☆

318867　Psilocaulon laxiflorum L. Bolus ＝ Psilocaulon junceum（Haw.）Schwantes ■☆

318868　Psilocaulon leightoniae L. Bolus ＝ Psilocaulon junceum（Haw.）Schwantes ■☆

318869　Psilocaulon levynsiae N. E. Br. ＝ Psilocaulon junceum（Haw.）Schwantes ■☆

318870　Psilocaulon lewisiae L. Bolus ＝ Psilocaulon junceum（Haw.）Schwantes ■☆

318871　Psilocaulon liebenbergii L. Bolus ＝ Psilocaulon articulatum（Thunb.）N. E. Br. ■☆

318872　Psilocaulon lindequistii（Engl.）Schwantes ＝ Aridaria noctiflora（L.）Schwantes ●☆

318873　Psilocaulon littlewoodii L. Bolus ＝ Psilocaulon dinteri（Engl.）Schwantes ■☆

318874　Psilocaulon littlewoodii L. Bolus. f. laxum ? ＝ Psilocaulon dinteri（Engl.）Schwantes ■☆

318875　Psilocaulon luteum L. Bolus ＝ Psilocaulon gessertianum（Dinter et A. Berger）Dinter et Schwantes ■☆

318876　Psilocaulon marlothii（Pax）Friedrich ＝ Brownanthus marlothii（Pax）Schwantes ●☆

318877　Psilocaulon melanospermum N. E. Br. ＝ Aptenia geniculiflora（L.）Bittrich ex Gerbaulet ■☆

318878　Psilocaulon mentiens（A. Berger）N. E. Br. ＝ Psilocaulon coriarium（Burch. ex N. E. Br.）N. E. Br. ■☆

318879　Psilocaulon micranthon L. Bolus ＝ Psilocaulon parviflorum（Jacq.）Schwantes ■☆

318880　Psilocaulon mucronulatum（Dinter）N. E. Br. ＝ Psilocaulon articulatum（Thunb.）N. E. Br. ■☆

318881　Psilocaulon namaquense（Sond.）Schwantes ＝ Psilocaulon junceum（Haw.）Schwantes ■☆

318882　Psilocaulon namibense（Marloth）Friedrich ＝ Brownanthus namibensis（Marloth）Bullock ●☆

318883　Psilocaulon oculatum L. Bolus ＝ Psilocaulon junceum（Haw.）Schwantes ■☆

318884　Psilocaulon otzenianum（Dinter）L. Bolus ＝ Lampranthus otzenianus（Dinter）Friedrich ■☆

318885　Psilocaulon pageae L. Bolus ＝ Psilocaulon dinteri（Engl.）Schwantes ■☆

318886　Psilocaulon pageae L. Bolus var. grandiflorum ? ＝ Psilocaulon dinteri（Engl.）Schwantes ■☆

318887　Psilocaulon parviflorum（Jacq.）Schwantes；小花裸茎日中花■☆

318888　Psilocaulon pauciflorum（Sond.）Schwantes ＝ Psilocaulon junceum（Haw.）Schwantes ■☆

318889　Psilocaulon pauper L. Bolus ＝ Psilocaulon granulicaule（Haw.）Schwantes ■☆

318890　Psilocaulon peersii L. Bolus ＝ Brownanthus corallinus（Thunb.）Ihlenf. et Bittrich ●☆

318891　Psilocaulon pfeilii（Engl.）Schwantes ＝ Prenia tetragona（Thunb.）Gerbaulet ■☆

318892　Psilocaulon pillansii（L. Bolus）Friedrich ＝ Brownanthus pubescens（N. E. Br. ex C. A. Maass）Bullock ●☆

318893　Psilocaulon planisepalum L. Bolus ＝ Psilocaulon junceum（Haw.）Schwantes ■☆

318894　Psilocaulon planum L. Bolus ＝ Psilocaulon salicornioides（Pax）Schwantes ■☆

318895　Psilocaulon pomeridianum L. Bolus ＝ Mesembryanthemum stenandrum（L. Bolus）L. Bolus ■☆

318896　Psilocaulon puberulum Dinter ＝ Psilocaulon articulatum（Thunb.）N. E. Br. ■☆

318897　Psilocaulon pubescens N. E. Br. ＝ Psilocaulon articulatum（Thunb.）N. E. Br. ■☆

318898　Psilocaulon rogersiae L. Bolus ＝ Psilocaulon junceum（Haw.）Schwantes ■☆

318899　Psilocaulon roseoalbum L. Bolus ＝ Psilocaulon articulatum（Thunb.）N. E. Br. ■☆

318900　Psilocaulon salicornioides（Pax）Schwantes；盐角草裸茎日中花■☆

318901　Psilocaulon schlichtianum（Sond.）Schwantes ＝ Brownanthus arenosus（Schinz）Ihlenf. et Bittrich ●☆

318902　Psilocaulon semilunatum L. Bolus ＝ Psilocaulon junceum（Haw.）Schwantes ■☆

318903　Psilocaulon simile（Sond.）Schwantes ＝ Psilocaulon junceum（Haw.）Schwantes ■☆

318904　Psilocaulon simulans L. Bolus ＝ Psilocaulon junceum（Haw.）Schwantes ■☆

318905　Psilocaulon sinus-redfordiani Dinter ex Range ＝ Psilocaulon salicornioides（Pax）Schwantes ■☆

318906　Psilocaulon squamifolium N. E. Br. ＝ Psilocaulon junceum（Haw.）Schwantes ■☆

318907　Psilocaulon stayneri L. Bolus ＝ Psilocaulon junceum（Haw.）Schwantes ■☆

318908　Psilocaulon stenopetalum L. Bolus ＝ Psilocaulon coriarium（Burch. ex N. E. Br.）N. E. Br. ■☆

318909　Psilocaulon subintegrum L. Bolus ＝ Psilocaulon junceum（Haw.）Schwantes ■☆

318910　Psilocaulon subnodosum（A. Berger）N. E. Br.；多节日中花■☆

318911　Psilocaulon tenue（Haw.）Schwantes ＝ Psilocaulon parviflorum（Jacq.）Schwantes ■☆

318912　Psilocaulon trothai（Engl.）Schwantes ＝ Psilocaulon salicornioides（Pax）Schwantes ■☆

318913　Psilocaulon uncinatum L. Bolus ＝ Psilocaulon coriarium（Burch. ex N. E. Br.）N. E. Br. ■☆

318914　Psilocaulon utile L. Bolus ＝ Psilocaulon junceum（Haw.）Schwantes ■☆

318915　Psilocaulon variabile L. Bolus ＝ Psilocaulon dinteri（Engl.）Schwantes ■☆

318916　Psilocaulon woodii L. Bolus ＝ Psilocaulon salicornioides（Pax）Schwantes ■☆

318917　Psilochenia Nutt. ＝ Crepis L. ■

318918　Psilochenia occidentalis（Nutt.）Nutt. ＝ Crepis occidentalis Nutt. ■☆

318919　Psilochilus Barb. Rodr.（1882）；弱唇兰属■☆

318920　Psilochilus Barb. Rodr. ＝ Pogonia Juss. ■

318921　Psilochilus macrophyllus Ames；大叶弱唇兰■☆

318922　Psilochilus mollis Garay；柔软弱唇兰■☆

318923　Psilochlaena Walp. ＝ Crepis L. ■

318924　Psilochlaena Walp. ＝ Psilochenia Nutt. ■

318925　Psilochloa Launert ＝ Panicum L. ■

318926　Psilochloa pilgeriana（Schweick.）Launert ＝ Panicum pilgerianum（Schweick.）Clayton ■☆

318927　Psilodigera Suess. ＝ Saltia R. Br. ex Moq. ●☆

318928　Psilodigera spicata Suess. ＝ Psilotrichum spicatum（Suess.）Cavaco ■☆

318929　Psiloesthes Benoist ＝ Peristrophe Nees ■

318930　Psilogyne DC. = Vitex L. ●

318931　Psilolaemus I. M. Johnst. (1954) ; 裸喉紫草属■☆

318932　Psilolaemus revolutus (B. L. Rob.) I. M. Johnst. ; 裸喉紫草■☆

318933　Psilolemma S. M. Phillips(1974) ; 光颖草属■☆

318934　Psilolemma jaegeri (Pilg.) S. M. Phillips ; 光颖草■☆

318935　Psilolepis C. Presl = Aspalathus L. ●☆

318936　Psilolepus bracteatus (Thunb.) C. Presl = Aspalathus bracteata Thunb. ●☆

318937　Psilolepus lanatus (E. Mey.) C. Presl = Aspalathus lanata E. Mey. ●☆

318938　Psilolepus pedunculatus (L'Hér.) C. Presl = Aspalathus bracteata Thunb. ●☆

318939　Psilonema C. A. Mey. = Alyssum L. ■●

318940　Psilonema alyssoides (L.) Heideman = Alyssum alyssoides (L.) L. ■

318941　Psilonema calycinum (L.) C. A. Mey. = Alyssum alyssoides (L.) L. ■

318942　Psilonema dasycarpum (Steph. ex Willd.) C. A. Mey. = Alyssum dasycarpum Stephan ex Willd. ■

318943　Psilonema homalocarpum Fisch. et C. A. Mey. = Alyssum homalocarpum (Fisch. et C. A. Mey.) Boiss. ■☆

318944　Psilonema minimum Schur. = Alyssum desertorum Stapf ■

318945　Psilopeganum Hemsl. (1886) ; 裸芸香属(臭草属, 臭节草属, 拟芸香属, 山麻黄属) ; Nakerue, Psilopeganum ■★

318946　Psilopeganum Hemsl. ex Forb. et Hemsl. = Psilopeganum Hemsl. ■★

318947　Psilopeganum sinense Hemsl. ; 裸芸香(臭草, 臭节草, 大蓼, 拟芸香, 千垂乌, 千垂乌, 山麻黄, 蛇皮草, 蛇咬药, 虱子草) ; Chinese Psilopeganum, Nakerue ■

318948　Psilopogon Hochst. = Arthraxon P. Beauv. ■

318949　Psilopogon Hochst. = Microstegium Nees ■

318950　Psilopogon Phil. = Picrosia D. Don ■☆

318951　Psilopogon capensis Hochst. = Microstegium nudum (Trin.) A. Camus ■

318952　Psilo-pogon schimperi Hochst. ex A. Rich. = Arthraxon lanceolatus (Roxb.) Hochst. ■

318953　Psilopsis Neck. = Lamium L. ■

318954　Psilorhegma (Vogel) Britton et Rose = Cassia L. (保留属名)●■

318955　Psilorhegma Britton et Rose = Cassia L. (保留属名)●■

318956　Psilosanthus Neck. = Liatris Gaertn. ex Schreb. (保留属名)■☆

318957　Psilosiphon Welw. ex Baker = Lapeirousia Pourr. ■☆

318958　Psilosolena C. Presl = Peddiea Harv. ex Hook. ●☆

318959　Psilostachys Hochst. = Poechia Endl. ●■

318960　Psilostachys Hochst. = Psilotrichum Blume ●■

318961　Psilostachys Steud. = Dimeria R. Br. ■

318962　Psilostachys Turcz. = Cleidion Blume ●

318963　Psilostachys boiviniana Baill. = Psilotrichum sericeum (Jos. König ex Roxb.) Dalzell ●☆

318964　Psilostachys filiformis Dalzell et A. Gibson = Dimeria ornithopoda Trin. ■

318965　Psilostachys filipes Baill. = Psilotrichum sericeum (Jos. König ex Roxb.) Dalzell ●☆

318966　Psilostachys gnaphalobrya Hochst. = Psilotrichum gnaphalobryum (Hochst.) Schinz ■☆

318967　Psilostachys kirkii Baker = Psilotrichum sericeum (Jos. König ex Roxb.) Dalzell ●☆

318968　Psilostachys nervulosa Baill. = Psilotrichum sericeum (Jos. König ex Roxb.) Dalzell ●☆

318969　Psilostachys sericea (J. König ex Roxb.) Hook. f. = Psilotrichum sericeum (Jos. König ex Roxb.) Dalzell ●☆

318970　Psilostemon DC. = Trachystemon D. Don ●☆

318971　Psilostoma Klotzsch = Canthium Lam. ●

318972　Psilostoma Klotzsch = Plectronia L. ●☆

318973　Psilostoma Klotzsch ex Eckl. et Zeyh. = Canthium Lam. ●

318974　Psilostoma Klotzsch ex Eckl. et Zeyh. = Plectronia L. ●☆

318975　Psilostoma ciliata Klotzsch ex Eckl. et Zeyh. = Canthium ciliatum (Klotzsch ex Eckl. et Zeyh.) Kuntze ●☆

318976　Psilostrophe DC. (1838) ; 纸花菊属 ; Paperflower ■☆

318977　Psilostrophe bakeri Greene ; 贝克纸花菊■☆

318978　Psilostrophe cooperi (A. Gray) Greene ; 库珀纸花菊 ; Cooper's Paperflower, Paper Flower, Paperflower ■☆

318979　Psilostrophe gnaphalodes DC. ; 鼠麹草状纸花菊■☆

318980　Psilostrophe sparsiflora (A. Gray) A. Nelson ; 散花纸花菊■☆

318981　Psilostrophe tagetina (Nutt.) Greene ; 毛纸花菊 ; Woolly Paperflower ■☆

318982　Psilostrophe tagetina (Nutt.) Greene var. grandiflora (Rydb.) Heiser = Psilostrophe tagetina (Nutt.) Greene ■☆

318983　Psilostrophe tagetina (Nutt.) Greene var. lanata A. Nelson = Psilostrophe tagetina (Nutt.) Greene ■☆

318984　Psilostrophe villosa Rydb. ex Britton ; 长柔毛纸花菊■☆

318985　Psilothamnus DC. = Gamolepis Less. ■☆

318986　Psilothamnus DC. = Steirodiscus Less. ■☆

318987　Psilothamnus adpressifolius DC. = Euryops ericifolius (Bél.) B. Nord. ■☆

318988　Psilothamnus ericifolius (Bél.) DC. = Euryops ericifolius (Bél.) B. Nord. ■☆

318989　Psilothonna (E. Mey. ex DC.) E. Phillips = Gamolepis Less. ■☆

318990　Psilothonna E. Mey. ex DC. = Steirodiscus Less. ■☆

318991　Psilothonna capillacea (L. f.) E. Phillips = Steirodiscus capillaceus (L. f.) Less. ■☆

318992　Psilothonna schlechteri (Bolus ex Schltr.) E. Phillips = Steirodiscus schlechteri Bolus ex Schltr. ■☆

318993　Psilothonna speciosa (Pillans) E. Phillips = Steirodiscus speciosus (Pillans) B. Nord. ■☆

318994　Psilothonna tagetes (L.) E. Mey. ex DC. = Steirodiscus tagetes (L.) Schltr. ■☆

318995　Psilotrichium Hassk. = Psilotrichum Blume ●■

318996　Psilotrichopsis C. C. Towns. (1974) ; 青花苋属(假林地苋属, 类林地苋属)■

318997　Psilotrichopsis curtisii (Oliv.) C. C. Towns. ; 青花苋■☆

318998　Psilotrichopsis curtisii (Oliv.) C. C. Towns. var. hainanensis (F. C. How) H. S. Kiu ; 海南青花苋(青花苋)■

318999　Psilotrichopsis hainanensis (F. C. How) C. C. Towns. = Psilotrichopsis curtisii (Oliv.) C. C. Towns. ■☆

319000　Psilotrichum Blume (1826) ; 林地苋属(裸被苋属) ; Psilotrichum ●■

319001　Psilotrichum africanum Oliv. = Psilotrichum scleranthum Thwaites ●☆

319002　Psilotrichum africanum Oliv. f. intermedium Suess. = Psilotrichum majus Peter ■☆

319003　Psilotrichum africanum Oliv. var. pilosum Suess. = Psilotrichum elliotii Baker et C. B. Clarke ■☆

319004　Psilotrichum amplum Suess. ; 膨大林地苋■☆

319005　Psilotrichum angustifolium Gilg = Psilotrichum schimperi Engl. ■☆

319006　Psilotrichum aphyllum C. C. Towns. ;无叶林地苋■☆

319007　Psilotrichum axillare C. B. Clarke = Psilotrichum sericeum（Jos. König ex Roxb.）Dalzell ●☆

319008　Psilotrichum axilliflorum Suess. ;腋花林地苋■☆

319009　Psilotrichum boivinianum（Baill.）Cavaco = Psilotrichum sericeum（Jos. König ex Roxb.）Dalzell ●☆

319010　Psilotrichum calceolatum Hook. = Psilotrichum elliotii Baker et C. B. Clarke ■☆

319011　Psilotrichum camporum Lebrun et L. Touss. ex Hauman = Psilotrichum schimperi Engl. ■☆

319012　Psilotrichum concinnum Baker = Psilotrichum scleranthum Thwaites ●☆

319013　Psilotrichum confertum（Schinz）C. B. Clarke = Centemopsis conferta（Schinz）Suess. ■☆

319014　Psilotrichum cordatum Moq. = Psilotrichum gnaphalobryum（Hochst.）Schinz ■☆

319015　Psilotrichum cyathuloides Suess. et Launert;杯苋状林地苋■☆

319016　Psilotrichum debile Baker = Pandiaka welwitschii（Schinz）Hiern ■☆

319017　Psilotrichum densiflorum Lopr. = Achyropsis leptostachya（E. Mey. ex Meisn.）Baker et C. B. Clarke ■☆

319018　Psilotrichum edule C. B. Clarke = Psilotrichum sericeum（Jos. König ex Roxb.）Dalzell ●☆

319019　Psilotrichum elliotii Baker et C. B. Clarke;爱利林地苋■☆

319020　Psilotrichum erythrostachyum Gagnep. ;茑叶林地苋■

319021　Psilotrichum fallax C. C. Towns. ;迷惑林地苋■☆

319022　Psilotrichum ferrugineum（Roxb.）Moq. ;林地苋;Rusty Psilotrichum ■

319023　Psilotrichum ferrugineum（Roxb.）Moq. var. ximengense Y. Y. Qian;西盟林地苋;Ximeng Psilotrichum ■

319024　Psilotrichum filiforme E. A. Bruce = Centemopsis filiformis（E. A. Bruce）C. C. Towns. ■☆

319025　Psilotrichum gloveri Suess. ;格洛韦尔林地苋■☆

319026　Psilotrichum gnaphalobryum（Hochst.）Schinz;鼠麹林地苋■☆

319027　Psilotrichum gracilentum（Hiern）C. B. Clarke = Centemopsis gracilenta（Hiern）Schinz ■☆

319028　Psilotrichum gracilipes Hutch. et E. A. Bruce;细梗林地苋■☆

319029　Psilotrichum gramineum Suess. = Psilotrichum schimperi Engl. ■☆

319030　Psilotrichum kirkii（Baker）C. B. Clarke = Psilotrichum sericeum（Jos. König ex Roxb.）Dalzell ●☆

319031　Psilotrichum lanatum C. C. Towns. ;绵毛林地苋■☆

319032　Psilotrichum leptostachys C. C. Towns. ;细穗林地苋■☆

319033　Psilotrichum majus Peter;大林地苋■☆

319034　Psilotrichum mildbraedii Schinz = Psilotrichum elliotii Baker et C. B. Clarke ■☆

319035　Psilotrichum moquinianum Abeyw. = Psilotrichum elliotii Baker et C. B. Clarke ■☆

319036　Psilotrichum ochradenoides Chiov. = Pleuropterantha revoilii Franch. ●☆

319037　Psilotrichum ovatum（Moq.）Hauman = Psilotrichum elliotii Baker et C. B. Clarke ■☆

319038　Psilotrichum ovatum Peter = Psilotrichum elliotii Baker et C. B. Clarke ■☆

319039　Psilotrichum peterianum Suess. = Psilotrichum elliotii Baker et C. B. Clarke ■☆

319040　Psilotrichum robecchii Lopr. = Neocentema robecchii（Lopr.）Schinz ■☆

319041　Psilotrichum rubellum Baker = Centemopsis biflora（Schinz）Schinz ■☆

319042　Psilotrichum ruspolii Lopr. = Lopriorea ruspolii（Lopr.）Schinz ■☆

319043　Psilotrichum schimperi Engl. ;欣珀林地苋■☆

319044　Psilotrichum schimperi Engl. var. gramineum（Suess.）Suess. = Psilotrichum schimperi Engl. ■☆

319045　Psilotrichum scleranthum Thwaites;硬花林地苋●☆

319046　Psilotrichum sericeovillosum Chiov. = Psilotrichum sericeum（Jos. König ex Roxb.）Dalzell ●☆

319047　Psilotrichum sericeum（Jos. König ex Roxb.）Dalzell;绢毛林地苋●☆

319048　Psilotrichum spicatum（Suess.）Cavaco;穗状林地苋■☆

319049　Psilotrichum stenanthum C. C. Towns. ;狭花林地苋■☆

319050　Psilotrichum suffruticosum C. C. Towns. ;亚灌木林地苋●☆

319051　Psilotrichum tomentosum Chiov. ;绒毛林地苋■☆

319052　Psilotrichum trichophyllum Baker = Psilotrichum scleranthum Thwaites ●☆

319053　Psilotrichum trichotomum Blume = Psilotrichum ferrugineum（Roxb.）Moq. ■

319054　Psilotrichum villosiflorum Lopr. = Psilotrichum gnaphalobryum（Hochst.）Schinz ■☆

319055　Psilotrichum virgatum C. C. Towns. ;条纹林地苋■☆

319056　Psilotrichum vollesenii C. C. Towns. ;福勒森林地苋■☆

319057　Psilotrichum yunnanense D. D. Tao;云南林地苋■

319058　Psiloxylaceae Croizat = Myrtaceae Juss.（保留科名）●

319059　Psiloxylaceae Croizat（1960）;亮皮树科（裸木科）●☆

319060　Psiloxylon Thouars ex Tul.（1856）;亮皮树属（裸木属）●☆

319061　Psiloxylon mauritianum Thouars ex Benth. ;亮皮树●☆

319062　Psilurus Trin.（1820）;内曲草属;Psilurus ●☆

319063　Psilurus aristatus（L.）Duval-Jouve = Psilurus incurvus（Gouan）Schinz et Thell. ■☆

319064　Psilurus incurvus（Gouan）Schinz et Thell. ;内曲草;Incurvate Psilurus ■☆

319065　Psilurus incurvus（Gouan）Schinz et Thell. var. hirtellus（Simonk.）Asch. et Graebn. = Psilurus incurvus（Gouan）Schinz et Thell. ■☆

319066　Psilurus nardoides Trin. = Psilurus incurvus（Gouan）Schinz et Thell. ■☆

319067　Psilurus rottboelloides Griff. = Psilurus incurvus（Gouan）Schinz et Thell. ■☆

319068　Psistina Raf. = Helianthemum Mill. ●■

319069　Psistus Neck. = Psistina Raf. ●■

319070　Psithyrisma Herb. = Phaiophleps Raf. ■☆

319071　Psithyrisma Herb. = Symphyostemon Miers ex Klatt ■☆

319072　Psittacanthaceae Nakai = Loranthaceae Juss.（保留科名）●

319073　Psittacanthus Mart.（1830）;鹦鹉刺属●☆

319074　Psittacanthus Mart. = Loranthus L.（废弃属名）●

319075　Psittacanthus acuminatus Kuijt;渐尖鹦鹉刺●☆

319076　Psittacanthus allenii Woodson et Schery;阿伦鹦鹉刺●☆

319077　Psittacanthus amazonicus（Ule）Kuijt;亚马孙鹦鹉刺●☆

319078　Psittacanthus angustifolius Kuijt;窄叶鹦鹉刺●☆

319079　Psittacanthus brachypodus Kuijt;短梗鹦鹉刺●☆

319080　Psittacanthus brasiliensis Blume;巴西鹦鹉刺●☆

319081　Psittacanthus crassipes Kuijt;成梗鹦鹉刺●☆

319082　Psittacanthus gigas Kuijt;巨大鹦鹉刺●☆

319083　Psittacanthus grandiflorus G. Don;大花鹦鹉刺●☆

319084　Psittacanthus grandifolius Mart. ;大叶鹦鹉刺●☆

319085　Psittacanthus lasianthus Sandwith；毛花鹦鹉刺●☆

319086　Psittacanthus macrantherus Eichler；大药鹦鹉刺●☆

319087　Psittacanthus mexicanus（Presl ex Schult. f.）G. Don；墨西哥鹦鹉刺●☆

319088　Psittacanthus microphyllus Kuijt；小叶鹦鹉刺●☆

319089　Psittacanthus obovatus Eichler；倒卵鹦鹉刺●☆

319090　Psittacanthus ovatus Kuijt；卵形鹦鹉刺●☆

319091　Psittacanthus salicifolius Kuijt；柳叶鹦鹉刺●☆

319092　Psittacanthus stenanthus Rizzini；窄花鹦鹉刺●☆

319093　Psittacaria Fabr. = Amaranthus L. ■

319094　Psittacoglossum La Llave et Lex. = Maxillaria Ruiz et Pav. ■☆

319095　Psittacoschoenus Nees = Gahnia J. R. Forst. et G. Forst. ■

319096　Psittaglossum Post et Kuntze = Maxillaria Ruiz et Pav. ■☆

319097　Psittaglossum Post et Kuntze = Psittacoglossum La Llave et Lex. ■☆

319098　Psolanum Neck. = Solanum L. ●■

319099　Psophiza Raf. = Aristolochia L. ■●

319100　Psophocarpus Neck. = Psophocarpus Neck. ex DC.（保留属名）■

319101　Psophocarpus Neck. ex DC.（1825）（保留属名）；四棱豆属；Goabean ■

319102　Psophocarpus golungensis Welw. ex Romariz = Psophocarpus scandens（Endl.）Verdc. ■☆

319103　Psophocarpus grandiflorus R. Wilczek；大花四棱豆■☆

319104　Psophocarpus lancifolius Harms；剑叶四棱豆■☆

319105　Psophocarpus lecomtei Tisser.；勒孔特四棱豆■☆

319106　Psophocarpus longipedunculatus Hassk. = Psophocarpus scandens（Endl.）Verdc. ■☆

319107　Psophocarpus lukafuensis（De Wild.）R. Wilczek；卢卡夫四棱豆■☆

319108　Psophocarpus mabala Welw. = Psophocarpus scandens（Endl.）Verdc. ■☆

319109　Psophocarpus monophyllus Harms；单叶四棱豆■☆

319110　Psophocarpus obovalis Tisser.；倒卵四棱豆■☆

319111　Psophocarpus palustris Desv.'Wondo Surprise'Westphal = Psophocarpus grandiflorus R. Wilczek ■☆

319112　Psophocarpus paustris Desv.；豆菜■☆

319113　Psophocarpus scandens（Endl.）Verdc.；攀缘四棱豆■☆

319114　Psophocarpus tetragonolobus（L.）DC.；四棱豆（翅豆，四翅豆，四角豆，杨桃豆，翼豆）；Asparagus Bean，Asparagus Pea，Dambala，Goa Bean，Goabean，Indies Goa Bean，Indies Goabean，Indies Goa-bean，Manila Bean，Prince's Pea，Princess Bean，Winged Bean，Winged Pea ■

319115　Psora Hill = Centaurea L.（保留属名）●■

319116　Psoralea L.（1753）；补骨脂属（南非补骨脂属）；Scurf Pea，Scurfpea，Scurfy Pea，Scurfy-pea ●■

319117　Psoralea abbottii C. H. Stirt.；阿巴特补骨脂■☆

319118　Psoralea aculeata L.；皮刺补骨脂■☆

319119　Psoralea aculeata Thunb. = Otholobium fruticans（L.）C. H. Stirt. ●☆

319120　Psoralea acuminata Lam. = Otholobium acuminatum（Lam.）C. H. Stirt. ●☆

319121　Psoralea affinis Eckl. et Zeyh.；近缘补骨脂■☆

319122　Psoralea alata（Thunb.）T. M. Salter；具翅补骨脂■☆

319123　Psoralea albicans Eckl. et Zeyh. = Otholobium argenteum（Thunb.）C. H. Stirt. ●☆

319124　Psoralea algoensis Eckl. et Zeyh. = Otholobium bracteolatum（Eckl. et Zeyh.）C. H. Stirt. ●☆

319125　Psoralea americana L.；美洲补骨脂；Scurfy-pea ■☆

319126　Psoralea americana L. = Cullen americanum（L.）Rydb. ●☆

319127　Psoralea americana L. var. integrifolia Batt. = Cullen americanum（L.）Rydb. ●☆

319128　Psoralea americana L. var. polystachya（Poir.）Cout. = Cullen americanum（L.）Rydb. ●☆

319129　Psoralea andongensis Baker = Pseudoeriosema andongense（Baker）Hauman ■☆

319130　Psoralea angustifolia Jacq.；窄叶补骨脂■☆

319131　Psoralea aphylla L.；无叶补骨脂■☆

319132　Psoralea arborea Sims；树状补骨脂■☆

319133　Psoralea argentea Thunb. = Otholobium argenteum（Thunb.）C. H. Stirt. ●☆

319134　Psoralea argophylla Pursh；银叶补骨脂；Scurf Pea，Scurfy Pea，Silvedeaf Scurfpea ■☆

319135　Psoralea argophylla Pursh = Pediomelum argophyllum（Pursh）J. W. Grimes ■☆

319136　Psoralea axillaris Eckl. et Zeyh. = Psoralea glaucina Harv. ■☆

319137　Psoralea axillaris L.；腋生补骨脂■☆

319138　Psoralea biflora Harv. = Cullen biflorum（Harv.）C. H. Stirt. ■☆

319139　Psoralea bituminosa L.；树脂补骨脂（阿拉伯补骨脂）；Pitch Trefofi ■☆

319140　Psoralea bituminosa L. = Bituminaria bituminosa（L.）C. H. Stirt. ■☆

319141　Psoralea bituminosa L. var. angustifolia（Guss.）Strobl = Bituminaria bituminosa（L.）C. H. Stirt. ■☆

319142　Psoralea bituminosa L. var. atropurpurea Maire = Bituminaria bituminosa（L.）C. H. Stirt. ■☆

319143　Psoralea bituminosa L. var. communis Webb et Berthel. = Bituminaria bituminosa（L.）C. H. Stirt. ■☆

319144　Psoralea bituminosa L. var. decumbens Sennen = Bituminaria bituminosa（L.）C. H. Stirt. ■☆

319145　Psoralea bituminosa L. var. humilis Bég. et Vacc. = Bituminaria bituminosa（L.）C. H. Stirt. ■☆

319146　Psoralea bituminosa L. var. latifolia Moris = Bituminaria bituminosa（L.）C. H. Stirt. ■☆

319147　Psoralea bituminosa L. var. laxa Maire et Weiller = Bituminaria bituminosa（L.）C. H. Stirt. ■☆

319148　Psoralea bituminosa L. var. palestina Webb et Berthel. = Bituminaria bituminosa（L.）C. H. Stirt. ■☆

319149　Psoralea bituminosa L. var. plumosa（Rchb.）Rchb. = Bituminaria bituminosa（L.）C. H. Stirt. ■☆

319150　Psoralea bituminosa L. var. rotundata Maire = Bituminaria bituminosa（L.）C. H. Stirt. ■☆

319151　Psoralea bolusii H. M. L. Forbes = Otholobium bolusii（H. M. L. Forbes）C. H. Stirt. ●☆

319152　Psoralea bowieana Harv. = Otholobium bowieanum（Harv.）C. H. Stirt. ●☆

319153　Psoralea bracteata L. = Otholobium fruticans（L.）C. H. Stirt. ●☆

319154　Psoralea bracteolata Eckl. et Zeyh. = Otholobium bracteolatum（Eckl. et Zeyh.）C. H. Stirt. ●☆

319155　Psoralea caffra Eckl. et Zeyh. = Otholobium caffrum（Eckl. et Zeyh.）C. H. Stirt. ●☆

319156　Psoralea candicans Eckl. et Zeyh. = Otholobium candicans（Eckl. et Zeyh.）C. H. Stirt. ●☆

319157　Psoralea canescens Eckl. et Zeyh. = Otholobium canescens（Eckl. et Zeyh.）C. H. Stirt. ●☆

319158　Psoralea capitata L. f. = Psoralea ensifolia（Houtt.）Merr. ■☆

319159　Psoralea carnea E. Mey. = Otholobium carneum（E. Mey.）C. H. Stirt. ●☆

319160　Psoralea cephalotes E. Mey. = Otholobium uncinatum（Eckl. et Zeyh.）C. H. Stirt. ●☆

319161　Psoralea cinerea Lindl. ;灰补骨脂■☆

319162　Psoralea collina Rydb. = Pediomelum argophyllum（Pursh）J. W. Grimes ■☆

319163　Psoralea cordata（L.）T. M. Salter = Psoralea monophylla（L.）C. H. Stirt. ■☆

319164　Psoralea corylifolia L. = Cullen corylifolium（L.）Medik. ■

319165　Psoralea cuneifolia Dum. Cours. = Otholobium fruticans（L.）C. H. Stirt. ●☆

319166　Psoralea cytisoides L. = Indigofera cytisoides（L.）L. ●☆

319167　Psoralea decidua P. J. Bergius = Psoralea aphylla L. ■☆

319168　Psoralea decumbens Aiton = Otholobium virgatum（Burm. f.）C. H. Stirt. ■☆

319169　Psoralea decumbens Willd. = Otholobium mundianum（Eckl. et Zeyh.）C. H. Stirt. ●☆

319170　Psoralea densa E. Mey. = Otholobium acuminatum（Lam.）C. H. Stirt. ●☆

319171　Psoralea dentata DC. ;齿叶补骨脂■☆

319172　Psoralea dentata DC. = Cullen americanum（L.）Rydb. ●☆

319173　Psoralea diffusa Eckl. et Zeyh. = Psoralea repens L. ■☆

319174　Psoralea drupacea Bunge;核果补骨脂（核状补骨脂）■☆

319175　Psoralea eckloniana Otto = Otholobium striatum（Thunb.）C. H. Stirt. ■☆

319176　Psoralea eglandulosa Elliott;无腺补骨脂■☆

319177　Psoralea ensifolia（Houtt.）Merr. ;剑叶补骨脂■☆

319178　Psoralea esculenta Pursh;食用补骨脂; Breadroot, Indian Breadroot, Pomme Blanche, Prairie Apple, Prairie Potato, Prairie Turnip ■☆

319179　Psoralea esculenta Pursh = Pediomelum esculentum（Pursh）Rydb. ■☆

319180　Psoralea fascicularis DC. ;扁补骨脂■☆

319181　Psoralea filifolia Thunb. ;丝叶补骨脂■☆

319182　Psoralea foliosa Oliv. = Otholobium foliosum（Oliv.）C. H. Stirt. ●☆

319183　Psoralea foliosa Oliv. var. gazensis Baker f. = Otholobium foliosum（Oliv.）C. H. Stirt. subsp. gazense（Baker f.）Verdc. ●☆

319184　Psoralea fruticans（L.）Druce = Otholobium fruticans（L.）C. H. Stirt. ●☆

319185　Psoralea glabra E. Mey. ;光滑补骨脂■☆

319186　Psoralea glandulosa L. ;库蓝补骨脂（具腺补骨脂）; Culen, Jesuit's Tea ●☆

319187　Psoralea glaucescens Eckl. et Zeyh. ;浅绿补骨脂■☆

319188　Psoralea glaucina Harv. ;灰绿补骨脂■☆

319189　Psoralea gueinzii Harv. ;吉内斯补骨脂■☆

319190　Psoralea hamata Harv. = Otholobium hamatum（Harv.）C. H. Stirt. ●☆

319191　Psoralea heterosepala Fourc. = Otholobium heterosepalum（Fourc.）C. H. Stirt. ■☆

319192　Psoralea hilaris Eckl. et Zeyh. = Otholobium racemosum（Thunb.）C. H. Stirt. ●☆

319193　Psoralea hirta L. = Otholobium hirtum（L.）C. H. Stirt. ■☆

319194　Psoralea holubii Burtt Davy = Cullen holubii（Burtt Davy）C. H. Stirt. ■☆

319195　Psoralea imbricata（L. f.）T. M. Salter;覆瓦补骨脂■☆

319196　Psoralea implexa C. H. Stirt. ;错乱补骨脂■☆

319197　Psoralea involucrata Thunb. = Polhillia involucrata（Thunb.）B. -E. van Wyk et A. L. Schutte ●☆

319198　Psoralea keetii Schönland;克特补骨脂■☆

319199　Psoralea lanceolata Pursh;黄补骨脂;Yellow Scurf Pea ■☆

319200　Psoralea latifolia（Harv.）C. H. Stirt. = Psoralea arborea Sims ■☆

319201　Psoralea laxa T. M. Salter;疏松补骨脂■☆

319202　Psoralea linearis Burm. f. = Aspalathus linearis（Burm. f.）R. Dahlgren ●☆

319203　Psoralea macradenia Harv. = Otholobium macradenium（Harv.）C. H. Stirt. ■☆

319204　Psoralea macrostachya DC. ;革根补骨脂■☆

319205　Psoralea mephitica S. Watson = Psoralea mephitica S. Watson ex Palmer ■☆

319206　Psoralea mephitica S. Watson ex Palmer;犹他补骨脂;Utah Breadroot ■☆

319207　Psoralea mexicana（L. f.）Vail;墨西哥补骨脂■☆

319208　Psoralea monophylla（L.）C. H. Stirt. ;单叶补骨脂■☆

319209　Psoralea multicaulis Jacq. = Psoralea ensifolia（Houtt.）Merr. ■☆

319210　Psoralea mundiana Eckl. et Zeyh. = Otholobium mundianum（Eckl. et Zeyh.）C. H. Stirt. ■☆

319211　Psoralea obliqua E. Mey. ;偏斜补骨脂■☆

319212　Psoralea obliqua E. Mey. = Otholobium obliquum（E. Mey.）C. H. Stirt. ■☆

319213　Psoralea obtusifolia DC. = Cullen tomentosum（Thunb.）J. W. Grimes ■☆

319214　Psoralea odorata Blatt. et Hallb. = Psoralea plicata Delile ■☆

319215　Psoralea odoratissima Jacq. ;极香补骨脂■☆

319216　Psoralea oligophylla Eckl. et Zeyh. ;寡叶补骨脂■☆

319217　Psoralea onobrychis Nutt. = Orbexilum onobrychis（Nutt.）Rydb. ■☆

319218　Psoralea ononoides Lam. = Otholobium virgatum（Burm. f.）C. H. Stirt. ■☆

319219　Psoralea oreophila Schltr. ;喜山补骨脂■☆

319220　Psoralea parviflora E. Mey. = Otholobium parviflorum（E. Mey.）C. H. Stirt. ■☆

319221　Psoralea pattersoniae Schönland = Cullen corylifolium（L.）Medik. ■

319222　Psoralea pedunculata Ker Gawl. = Otholobium sericeum（Poir.）C. H. Stirt. ■☆

319223　Psoralea pedunculata Vail;萨氏补骨脂;Sampson's Snakeroot ■☆

319224　Psoralea pinnata L. ;羽叶补骨脂; African Scurf Pea, African Scurf-pea, Blue Pea, Blue Pea Bush ●☆

319225　Psoralea pinnata L. var. glabra（E. Mey.）Harv. = Psoralea glabra E. Mey. ■☆

319226　Psoralea pinnata L. var. latifolia Harv. = Psoralea arborea Sims ■☆

319227　Psoralea plauta C. H. Stirt. ;澳非补骨脂■☆

319228　Psoralea plicata Delile;折补骨脂■☆

319229　Psoralea plicata Delile = Cullen plicatum（Delile）C. H. Stirt. ■☆

319230　Psoralea plumosa Rchb. ;羽毛补骨脂■☆

319231　Psoralea plumosa Rchb. = Bituminaria bituminosa（L.）C. H. Stirt. ■☆

319232　Psoralea polyphylla Eckl. et Zeyh. = Otholobium polyphyllum（Eckl. et Zeyh.）C. H. Stirt. ●☆

319233　Psoralea polystachya Poir. = Cullen americanum（L.）Rydb. ●☆

319234　Psoralea polysticta Benth. ex Harv. = Otholobium polystictum（Benth. ex Harv.）C. H. Stirt. ■☆

319235 Psoralea psoralioides（Walter）Cory ＝ Orbexilum pedunculatum（Mill.）Rydb. ■☆

319236 Psoralea psoralioides（Walter）Cory var. eglandulosa（Elliott）F. L. Freeman ＝ Orbexilum pedunculatum（Mill.）Rydb. ■☆

319237 Psoralea racemosa Thunb. ＝ Otholobium racemosum（Thunb.）C. H. Stirt. ●☆

319238 Psoralea repens L. ；匍匐补骨脂■☆

319239 Psoralea restioides Eckl. et Zeyh. ；绳补骨脂■☆

319240 Psoralea rotundifolia L. f. ＝ Otholobium rotundifolium（L. f.）C. H. Stirt. ●☆

319241 Psoralea royffei H. M. L. Forbes ＝ Otholobium caffrum（Eckl. et Zeyh.）C. H. Stirt. ●☆

319242 Psoralea rupicola Eckl. et Zeyh. ＝ Otholobium striatum（Thunb.）C. H. Stirt. ■☆

319243 Psoralea sericea Poir. ＝ Otholobium sericeum（Poir.）C. H. Stirt. ■☆

319244 Psoralea spathulata E. Mey. ＝ Otholobium mundianum（Eckl. et Zeyh.）C. H. Stirt. ■☆

319245 Psoralea speciosa Eckl. et Zeyh. ；美丽补骨脂■☆

319246 Psoralea spicata L. ＝ Otholobium spicatum（L.）C. H. Stirt. ■☆

319247 Psoralea stachydis L. f. ＝ Otholobium hirtum（L.）C. H. Stirt. ■☆

319248 Psoralea stachyera Eckl. et Zeyh. ＝ Otholobium stachyerum（Eckl. et Zeyh.）C. H. Stirt. ■☆

319249 Psoralea stachyos Thunb. ＝ Otholobium hirtum（L.）C. H. Stirt. ■☆

319250 Psoralea stricta Thunb. ＝ Otholobium striatum（Thunb.）C. H. Stirt. ■☆

319251 Psoralea tenuifolia L. ；细花补骨脂；Indian Turnip ■☆

319252 Psoralea tenuifolia Thunb. ＝ Psoralea fascicularis DC. ■☆

319253 Psoralea tenuissima E. Mey. ；极细补骨脂■☆

319254 Psoralea tetragonoloba L. ＝ Cyamopsis tetragonoloba（L.）Taub. ■

319255 Psoralea thomii Harv. ＝ Otholobium thomii（Harv.）C. H. Stirt. ■☆

319256 Psoralea thunbergiana Eckl. et Zeyh. ＝ Psoralea fascicularis DC. ■☆

319257 Psoralea tomentosa Thunb. ＝ Otholobium sericeum（Poir.）C. H. Stirt. ■☆

319258 Psoralea triantha E. Mey. ＝ Otholobium trianthum（E. Mey.）C. H. Stirt. ●☆

319259 Psoralea triflora Poir. ＝ Otholobium trianthum（E. Mey.）C. H. Stirt. ●☆

319260 Psoralea triflora Thunb. ；三花补骨脂■☆

319261 Psoralea trullata C. H. Stirt. ；杓补骨脂■☆

319262 Psoralea uncinata Eckl. et Zeyh. ＝ Otholobium uncinatum（Eckl. et Zeyh.）C. H. Stirt. ●☆

319263 Psoralea venusta Eckl. et Zeyh. ＝ Otholobium venustum（Eckl. et Zeyh.）C. H. Stirt. ●☆

319264 Psoralea verrucosa Willd. ；多疣补骨脂■☆

319265 Psoralea wilmsii Harms ＝ Otholobium wilmsii（Harms）C. H. Stirt. ■☆

319266 Psoralea zeyheri Harv. ＝ Otholobium zeyheri（Harv.）C. H. Stirt. ■☆

319267 Psoralidium Rydb. （1919）；小痂豆属■☆

319268 Psoralidium Rydb. ＝ Orbexilum Raf. ■☆

319269 Psoralidium argophyllum（Pursh）Rydb. ＝ Pediomelum argophyllum（Pursh）J. W. Grimes ■☆

319270 Psoralidium tenuiflorum（Pursh）Rydb. ；细叶小痂豆；Gray Scurf-pea，Slimflower Scurfpea ■☆

319271 Psorobates Willis ＝ Psorobatus Rydb. ●☆

319272 Psorobatus Rydb. ＝ Dalea L. （保留属名）●■☆

319273 Psorobatus Rydb. ＝ Errazurizia Phil. ●☆

319274 Psorodendron Rydb. ＝ Dalea L. （保留属名）●■☆

319275 Psorodendron Rydb. ＝ Psorothamnus Rydb. ●☆

319276 Psorophytum Spach ＝ Hypericum L. ■●

319277 Psorospermum Spach ＝ Harungana Lam. ■☆

319278 Psorospermum Spach（1836）；普梭木属（普梭草属）●■☆

319279 Psorospermum albidum（Oliv.）Engl. ＝ Psorospermum febrifugum Spach ■☆

319280 Psorospermum alternifolium Hook. f. ；互叶普梭木●☆

319281 Psorospermum angolense Exell ＝ Psorospermum mechowii Engl. ●☆

319282 Psorospermum angustifolium Spirlet ＝ Psorospermum febrifugum Spach ■☆

319283 Psorospermum aurantiacum Engl. ；橙色普梭木●☆

319284 Psorospermum baumannii Engl. ＝ Psorospermum febrifugum Spach ■☆

319285 Psorospermum baumii Engl. ；鲍姆普梭木●☆

319286 Psorospermum bracteolatum Spirlet ＝ Vismia affinis Oliv. ●☆

319287 Psorospermum campestre Engl. ＝ Psorospermum febrifugum Spach ■☆

319288 Psorospermum chariense A. Chev. ＝ Psorospermum febrifugum Spach ■☆

319289 Psorospermum chevalieri Hochr. ；舍瓦利耶普梭木●☆

319290 Psorospermum corymbiferum Hochr. ；伞序普梭木●☆

319291 Psorospermum corymbiferum Hochr. var. doeringii（Engl.）Keay et Milne-Redh. ；多林普梭木●☆

319292 Psorospermum corymbiferum Hochr. var. kerstingii（Engl.）Keay et Milne-Redh. ＝ Psorospermum corymbiferum Hochr. var. doeringii（Engl.）Keay et Milne-Redh. ●☆

319293 Psorospermum corymbosellum Spirlet ＝ Psorospermum febrifugum Spach ■☆

319294 Psorospermum corymbosum Spirlet ＝ Psorospermum febrifugum Spach ■☆

319295 Psorospermum cuneifolium Hochr. ＝ Psorospermum tenuifolium Hook. f. ●☆

319296 Psorospermum densipunctatum Engl. ；密斑普梭木●☆

319297 Psorospermum discolor Spirlet ＝ Psorospermum febrifugum Spach ■☆

319298 Psorospermum ellipticum Spirlet ＝ Psorospermum febrifugum Spach ■☆

319299 Psorospermum febrifugum Spach；普梭木（普梭草）■☆

319300 Psorospermum febrifugum Spach var. albida Oliv. ＝ Psorospermum febrifugum Spach ■☆

319301 Psorospermum febrifugum Spach var. ferrugineum ？；锈色普梭木（血红普梭草，血红普梭木）；Rhodesian Holly ■☆

319302 Psorospermum ferrugineum Hook. f. ＝ Psorospermum febrifugum Spach ■☆

319303 Psorospermum floribundum Hutch. et Dalziel ＝ Psorospermum febrifugum Spach ■☆

319304 Psorospermum gillardinii Spirlet ＝ Psorospermum febrifugum Spach ■☆

319305 Psorospermum glaberrimum Hochr. ；无毛普梭木●☆

319306 Psorospermum glaucum Engl. ；灰绿普梭木●☆

319307 Psorospermum gracile Spirlet ＝ Psorospermum tenuifolium Hook. f. ●☆

319308　Psorospermum hundtii Exell ex Mendonca ＝ Psorospermum mechowii Engl. ●☆

319309　Psorospermum kaniamae Spirlet ＝ Psorospermum febrifugum Spach ■☆

319310　Psorospermum kerstingii Engl. ＝ Psorospermum corymbiferum Hochr. var. doeringii（Engl.）Keay et Milne-Redh. ●☆

319311　Psorospermum kisantuense Spirlet ＝ Psorospermum febrifugum Spach ■☆

319312　Psorospermum kivuense Spirlet ＝ Psorospermum tenuifolium Hook. f. ●☆

319313　Psorospermum lanatum Hochr.；绵毛普梭木●☆

319314　Psorospermum lanceolatum Spirlet ＝ Psorospermum febrifugum Spach ■☆

319315　Psorospermum laurifolium（Pellegr.）Bamps；月桂叶普梭木●☆

319316　Psorospermum laxiflorum（Engl.）Engl. ＝ Psorospermum glaberrimum Hochr. ●☆

319317　Psorospermum ledermannii Engl. var. doeringii ？ ＝ Psorospermum corymbiferum Hochr. var. doeringii（Engl.）Keay et Milne-Redh. ●☆

319318　Psorospermum leopoldvilleanum Spirlet ＝ Psorospermum febrifugum Spach ■☆

319319　Psorospermum macrophyllum Spirlet ＝ Psorospermum febrifugum Spach ■☆

319320　Psorospermum magniflorum Spirlet ＝ Psorospermum febrifugum Spach ■☆

319321　Psorospermum mahagiense Spirlet ＝ Psorospermum febrifugum Spach ■☆

319322　Psorospermum mechowii Engl.；安哥拉普梭木●☆

319323　Psorospermum membranaceum C. H. Wright；膜质普梭木●☆

319324　Psorospermum microphyllum A. Chev. ＝ Psorospermum febrifugum Spach ■☆

319325　Psorospermum mossoense Spirlet ＝ Psorospermum febrifugum Spach ■☆

319326　Psorospermum nigrum Spirlet ＝Psorospermum febrifugum Spach ■☆

319327　Psorospermum niloticum Kotschy ex Schweinf. et Asch. ＝ Psorospermum febrifugum Spach ■☆

319328　Psorospermum orbiculare Spirlet ＝ Psorospermum febrifugum Spach ■☆

319329　Psorospermum ovatum Spirlet ＝Psorospermum febrifugum Spach ■☆

319330　Psorospermum parviflorum Engl.；小花普梭木●☆

319331　Psorospermum pauciflorum Spirlet ＝ Psorospermum febrifugum Spach ■☆

319332　Psorospermum pectinatum Spirlet ＝ Psorospermum febrifugum Spach ■☆

319333　Psorospermum pubescens Spirlet ＝ Psorospermum febrifugum Spach ■☆

319334　Psorospermum rigidum Spirlet ＝ Psorospermum tenuifolium Hook. f. ●☆

319335　Psorospermum robynsii Spirlet ＝ Psorospermum tenuifolium Hook. f. ●☆

319336　Psorospermum rotundatifolium Spirlet ＝ Psorospermum febrifugum Spach ■☆

319337　Psorospermum rubescens Spirlet ＝ Psorospermum tenuifolium Hook. f. ●☆

319338　Psorospermum salicifolium Engl. ＝ Psorospermum febrifugum Spach ■☆

319339　Psorospermum senegalense Spach；塞内加尔普梭木●☆

319340　Psorospermum staneranum Spirlet ＝ Psorospermum febrifugum Spach ■☆

319341　Psorospermum staudtii Engl.；施陶普梭木●☆

319342　Psorospermum stuhlmannii Engl. ＝ Psorospermum febrifugum Spach ■☆

319343　Psorospermum suffruticosum Engl.；亚灌木普梭木●☆

319344　Psorospermum tenuifolium Hook. f.；细叶普梭木●☆

319345　Psorospermum tenuifolium Hook. f. var. laxiflorum Engl. ＝ Psorospermum glaberrimum Hochr. ●☆

319346　Psorospermum thompsonii Hutch. et Dalziel ＝ Psorospermum corymbiferum Hochr. var. doeringii（Engl.）Keay et Milne-Redh. ●☆

319347　Psorospermum troupinii Spirlet ＝ Psorospermum tenuifolium Hook. f. ●☆

319348　Psorospermum uelense Spirlet ＝ Psorospermum febrifugum Spach ■☆

319349　Psorospermum umbellatum Hutch. et Dalziel ＝ Psorospermum glaberrimum Hochr. ●☆

319350　Psorospermum victoranum Spirlet ＝ Psorospermum febrifugum Spach ■☆

319351　Psorothamnus Rydb.（1919）；癣豆属●☆

319352　Psorothamnus Rydb. ＝ Dalea L.（保留属名）●■☆

319353　Psorothamnus arborescens（A. Gray）Barneby；乔木癣豆●☆

319354　Psorothamnus arborescens（A. Gray）Barneby var. minutifolia（Parish）Barneby；小叶乔木癣豆●☆

319355　Psorothamnus dentatus Rydb.；齿叶癣豆●☆

319356　Psorothamnus fremontii（Torr.）Barneby；弗氏癣豆●☆

319357　Psorothamnus kingii（S. Watson）Barneby；金氏癣豆●☆

319358　Psorothamnus scoparius Rydb.；帚状癣豆；Broom Dalea ●☆

319359　Psorothamnus spinosus（A. Gray）Barneby；多刺癣豆；Smoke Tree ●☆

319360　Psorothamnus subnudus Rydb.；近裸癣豆●☆

319361　Psorothamnus tinctorius Rydb.；着色癣豆●☆

319362　Pstathura Raf. ＝ Psathura Comm. ex Juss. ■☆

319363　Psychanthus（K. Schum.）Ridl. ＝ Pleuranthodium（K. Schum.）R. M. Sm. ■☆

319364　Psychanthus Raf. ＝ Polygala L. ●■

319365　Psychanthus Ridl. ＝ Alpinia Roxb.（保留属名）■

319366　Psychechilos Breda ＝Zeuxine Lindl.（保留属名）■

319367　Psychilis Raf.（1838）；蝶唇兰属■☆

319368　Psychilis Raf. ＝ Epidendrum L.（保留属名）■☆

319369　Psychilis amena Raf.；蝶唇兰■☆

319370　Psychilis atropurpurea（Willd.）Sauleda；暗紫蝶唇兰■☆

319371　Psychilis bifida（Aubl.）Sauleda；二裂蝶唇兰■☆

319372　Psychilis truncata（Cogniaux）Sauleda；平截蝶唇兰■☆

319373　Psychilus Raf. ＝ Epidendrum L.（保留属名）■☆

319374　Psychine Desf.（1798）；蝶荠属■☆

319375　Psychine stylosa Desf.；多柱蝶荠（蝶荠）■☆

319376　Psychine stylosa Desf. var. auriculata Font Quer ＝ Psychine stylosa Desf. ☆

319377　Psychine stylosa Desf. var. maroccana Murb. ＝ Psychine stylosa Desf. ☆

319378　Psychochilus Post et Kuntze ＝ Psychechilos Breda ■

319379　Psychochilus Post et Kuntze ＝Zeuxine Lindl.（保留属名）■

319380　Psychodendron Walp. ex Voigt ＝ Bischofia Blume ●

319381　Psycholobium Blume ex Burck ＝Mucuna Adans.（保留属名）●■

319382　Psychopsiella Lückel et Braem（1982）；赛蝶唇兰属■☆

319383　Psychopsiella limminghei（E. Morren ex Lindl.）Lückel et

Braem;赛蝶唇兰■☆

319384　Psychopsiella limminghei（Lindl.）Lückel et Braem ＝ Psychopsiella limminghei（E. Morren ex Lindl.）Lückel et Braem ■☆

319385　Psychopsis Nutt. ex Greene ＝ Hosackia Douglas ex Benth. ■☆

319386　Psychopsis Raf.（1838）;拟蝶唇兰属;Butterfly Orchid ●☆

319387　Psychopsis Raf. ＝ Oncidium Sw.（保留属名）■☆

319388　Psychopsis papilio（Lindl.）H. G. Jones;拟蝶唇兰●☆

319389　Psychopterys W. R. Anderson et S. Corso（2007）;蝶翅藤属●☆

319390　Psychosperma Dumort. ＝ Ptychosperma Labill. ●☆

319391　Psychothria L. ＝ Psychotria L.（保留属名）●

319392　Psychotria L.（1759）（保留属名）;九节属;Ninenode, Psychotria, Wild Coffee, Wild-coffee ●

319393　Psychotria abouabouensis（Schnell）Verdc. ;阿布九节■☆

319394　Psychotria abrupta Hepper ＝ Chazaliella cupulicalyx Verdc. ●☆

319395　Psychotria abrupta Hiern ＝ Chazaliella abrupta（Hiern）E. M. Petit et Verdc. ●☆

319396　Psychotria aemulans K. Schum. ;匹敌九节●☆

319397　Psychotria afzelii Hiern ＝Chassalia afzelii（Hiern）K. Schum. ■☆

319398　Psychotria alatipes Wernham;翅梗九节●☆

319399　Psychotria albicaulis Scott-Elliot;白茎九节●☆

319400　Psychotria albidocalyx K. Schum. ＝ Psychotria amboniana K. Schum. ●☆

319401　Psychotria albidocalyx K. Schum. var. angustifolia S. Moore ＝ Psychotria amboniana K. Schum. ●☆

319402　Psychotria albidocalyx K. Schum. var. mosambicensis E. M. Petit ＝ Psychotria amboniana K. Schum. subsp. mosambicensis（E. M. Petit）Verdc. ●☆

319403　Psychotria albidocalyx K. Schum. var. subumbellata E. M. Petit ＝ Psychotria pumila Hiern var. subumbellata（E. M. Petit）Verdc. ●☆

319404　Psychotria albidocalyx K. Schum. var. velutina E. M. Petit ＝ Psychotria amboniana K. Schum. var. velutina（E. M. Petit）Verdc. ●☆

319405　Psychotria albifaux K. Schum. ex Hutch. et Dalziel ＝ Chazaliella sciadephora（Hiern）E. M. Petit et Verdc. ●☆

319406　Psychotria albiflora（K. Krause）De Wild. ＝ Chassalia albiflora K. Krause ■☆

319407　Psychotria aledjoensis De Wild. ＝ Psychotria mannii Hiern ●☆

319408　Psychotria aledjoensis De Wild. var. glabra ? ＝ Psychotria brachyantha Hiern ●☆

319409　Psychotria amboniana K. Schum. ;安博九节●☆

319410　Psychotria amboniana K. Schum. subsp. mosambicensis（E. M. Petit）Verdc. ;莫桑比克九节●☆

319411　Psychotria amboniana K. Schum. var. velutina（E. M. Petit）Verdc. ;短绒毛九节●☆

319412　Psychotria anetoclada Hiern ＝ Psychotria subobliqua Hiern ●☆

319413　Psychotria anetoclada Hiern var. angustifolia A. Chev. ＝ Psychotria subobliqua Hiern ●☆

319414　Psychotria anomovenosa R. D. Good ＝ Psychotria dermatophylla（K. Schum.）E. M. Petit ●☆

319415　Psychotria ansellii Hiern ＝ Chassalia laxiflora Benth. ■☆

319416　Psychotria arabica Klotzsch ＝ Pentas lanceolata（Forssk.）K. Schum. ●■

319417　Psychotria arborea Hiern;树状九节●☆

319418　Psychotria arnoldiana De Wild. ;阿诺德九节●☆

319419　Psychotria articulata（Hiern）E. M. Petit;关节九节●☆

319420　Psychotria asiatica L. ;亚洲九节●

319421　Psychotria asiatica Wall. ＝ Psychotria calocarpa Kurz ●

319422　Psychotria assimilis Bremek. ;相似肖九节●☆

319423　Psychotria bacteriophila Valeton ＝Psychotria punctata Vatke ●☆

319424　Psychotria bagshawei E. M. Petit;巴格肖九节●☆

319425　Psychotria bangweana K. Schum. ;邦韦肖九节●☆

319426　Psychotria batangana K. Schum. ;巴坦加肖九节●☆

319427　Psychotria beniensis De Wild. ＝ Psychotria kirkii Hiern var. mucronata（Hiern）Verdc. ●☆

319428　Psychotria benthamiana Hiern ＝ Chassalia kolly（Schumach.）Hepper ■☆

319429　Psychotria bequaertii De Wild. ＝ Psychotria lauracea（K. Schum.）E. M. Petit ●☆

319430　Psychotria biaurita（Hutch. et Dalziel）Verdc. ;双耳九节●☆

319431　Psychotria bicarinata Mildbr. ＝ Psychotria recurva Hiern ●☆

319432　Psychotria bidentata（Thunb. ex Roem. et Schult.）Hiern;双齿九节●☆

319433　Psychotria bieleri De Wild. ＝ Gaertnera bieleri（De Wild.）E. M. Petit ●☆

319434　Psychotria bifaria Hiern;二列九节●☆

319435　Psychotria bifaria Hiern var. pauridiantha（Hiern）E. M. Petit;寡花二列九节●☆

319436　Psychotria boivinii Bremek. ;博伊文九节●☆

319437　Psychotria boninensis Nakai;小笠原九节●☆

319438　Psychotria brachyantha Hiern;短花九节●☆

319439　Psychotria brachyanthoides De Wild. ;假短花九节●☆

319440　Psychotria brachythamnus K. Schum. et K. Krause ＝ Psychotria pumila Hiern ●☆

319441　Psychotria bracteosa Hiern ＝ Peripeplus bracteosus（Hiern）E. M. Petit ●☆

319442　Psychotria brassii Hiern;布拉斯九节●☆

319443　Psychotria brenanii Hepper ＝ Psychotria minuta E. M. Petit ●☆

319444　Psychotria brevicaulis K. Schum. ;短茎九节●☆

319445　Psychotria brevipaniculata De Wild. ;短圆锥九节●☆

319446　Psychotria brevipuberula E. M. Petit;短毛九节●☆

319447　Psychotria brevistipulata De Wild. ＝ Chazaliella sciadephora（Hiern）E. M. Petit et Verdc. ●☆

319448　Psychotria brieyi De Wild. ;布里九节●☆

319449　Psychotria brucei Verdc. ;布鲁斯肖九节●☆

319450　Psychotria brunnea Schweinf. ex Hiern;褐色九节●☆

319451　Psychotria buchwaldii（K. Schum.）De Wild. ＝ Chassalia buchwaldii K. Schum. ■☆

319452　Psychotria bukobensis K. Schum. ;布科巴九节●☆

319453　Psychotria bullulata Bremek. ;小泡九节●☆

319454　Psychotria butayei De Wild. ;布塔耶九节●☆

319455　Psychotria butayei De Wild. var. glabra（R. D. Good）E. M. Petit;光滑九节●☆

319456　Psychotria butayei De Wild. var. simplex E. M. Petit;简单九节●☆

319457　Psychotria buzica S. Moore ＝ Psychotria pumila Hiern var. buzica（S. Moore）E. M. Petit ●☆

319458　Psychotria cabrae De Wild. ＝ Psychotria dermatophylla（K. Schum.）E. M. Petit ●☆

319459　Psychotria caduciflora De Wild. ＝ Psychotria subobliqua Hiern ●☆

319460　Psychotria callensii E. M. Petit;卡伦斯九节●☆

319461　Psychotria calocarpa Kurz;美果九节（花叶九节,花叶九节木,美果九节木,小功劳,牙齿硬）;Prettyfruit Ninenode, Prettyfruit Psychotria, Pretty-fruited Psychotria ●

319462　Psychotria calocarpa Kurz ＝ Psychotria errafica Hook. f. ●

319463　Psychotria calva Hiern;光秃九节●☆

319464　Psychotria calva Hiern ＝ Uragoga thonningii Kuntze ●☆

319465　Psychotria camerunensis E. M. Petit;喀麦隆九节●☆

319466　Psychotria camptopus Verdc.;曲梗九节●☆

319467　Psychotria capensis（Eckl.）Vatke;好望角九节;Wild Coffee●☆

319468　Psychotria capensis（Eckl.）Vatke subsp. riparia（K. Schum. et K. Krause）Verdc.;河岸九节●☆

319469　Psychotria capensis（Eckl.）Vatke var. puberula（E. M. Petit）Verdc.;微毛好望角九节●☆

319470　Psychotria capensis（Eckl.）Vatke var. pubescens（Sond.）E. M. Petit;柔毛好望角九节九节●☆

319471　Psychotria capensis Schönland = Psychotria capensis（Eckl.）Vatke ●☆

319472　Psychotria capitellata A. Chev. ex De Wild. = Psychotria longituba A. Chev. ex De Wild. ●☆

319473　Psychotria castaneifolia E. M. Petit;栗叶九节●☆

319474　Psychotria cataractorum K. Schum. = Psychotria ebensis K. Schum. ●☆

319475　Psychotria catetensis（Hiern）E. M. Petit;卡特蒂九节●☆

319476　Psychotria cephalidantha K. Schum.;头花九节●☆

319477　Psychotria cephalophora Merr.;兰屿九节（兰屿九节木）;Lanyu Ninenode, Lanyu Psychotria, Philippine Psychotria, Philippine Wild Coffee ●

319478　Psychotria chalconeura（K. Schum.）E. M. Petit var. montana E. M. Petit = Psychotria lebrunii Cheek ●☆

319479　Psychotria chrysoclada K. Schum.;金枝九节●☆

319480　Psychotria ciliata De Wild. = Psychotria ebensis K. Schum. ●☆

319481　Psychotria ciliatocostata Cufod. = Psychotria kirkii Hiern var. nairobiensis（Bremek.）Verdc.●☆

319482　Psychotria cinerea De Wild.;灰色九节●☆

319483　Psychotria coaetanea K. Schum. = Chazaliella abrupta（Hiern）E. M. Petit et Verdc.●☆

319484　Psychotria coeruleo-violacea K. Schum. = Psychotria ebensis K. Schum. ●☆

319485　Psychotria coffeosperma K. Schum. = Chazaliella coffeosperma（K. Schum.）Verdc.●☆

319486　Psychotria comperei E. M. Petit;孔佩尔九节●☆

319487　Psychotria copeensis De Wild. =Psychotria subobliqua Hiern ●☆

319488　Psychotria coralloides A. Chev. ex De Wild. = Chassalia subherbacea（Hiern）Hepper ■☆

319489　Psychotria cornuta Hiern = Psychotria humilis Hiern var. cornuta（Hiern）E. M. Petit ●☆

319490　Psychotria coursii Bremek.;库尔斯九节●☆

319491　Psychotria crassicalyx K. Krause;厚萼九节●☆

319492　Psychotria crassipetala E. M. Petit;厚瓣九节●☆

319493　Psychotria crispa Hiern =Psychotria latistipula Benth.●☆

319494　Psychotria cristata Hiern =Chassalia cristata（Hiern）Bremek.■☆

319495　Psychotria cryptogrammata E. M. Petit;隐九节●☆

319496　Psychotria cyanopharynx K. Schum.;蓝色九节●☆

319497　Psychotria dalzielii Hutch. = Psychotria eminiana（Kuntze）E. M. Petit ●☆

319498　Psychotria decaryi Bremek.;德卡里九节●☆

319499　Psychotria decolor Drake ex Bremek.;褪色九节●☆

319500　Psychotria densa W. C. Chen;密集脉九节;Densevein Ninenode, Densevein Psychotria ●

319501　Psychotria densinervia（K. Krause）Verdc.;密脉九节●☆

319502　Psychotria denticulata Wall.;齿九节●☆

319503　Psychotria dermatophylla（K. Schum.）E. M. Petit;革叶九节●☆

319504　Psychotria dewevrei De Wild. = Psychotria eminiana（Kuntze）E. M. Petit ●☆

319505　Psychotria dimorphophylla K. Schum. = Psychotria psychotrioides（DC.）Roberty ●☆

319506　Psychotria diploneura（K. Schum.）Bridson et Verdc.;二脉九节●☆

319507　Psychotria distegia K. Schum. = Psychotria brevicaulis K. Schum. ●☆

319508　Psychotria diversinodula（Verdc.）Verdc.;异节九节●☆

319509　Psychotria djumaensis De Wild.;朱马九节●☆

319510　Psychotria djumaensis De Wild. var. zambesiaca E. M. Petit;赞比西九节●☆

319511　Psychotria doniana Benth. = Chassalia doniana（Benth.）G. Taylor ■☆

319512　Psychotria dorotheae Wernham;多罗特娅九节●☆

319513　Psychotria dusenii K. Schum.;杜森九节●☆

319514　Psychotria dusenii Standl. = Psychotria vogeliana Benth. ●☆

319515　Psychotria ealaensis De Wild.;埃阿拉九节●☆

319516　Psychotria ebensis K. Schum.;埃贝九节●☆

319517　Psychotria eckloniana F. Muell. = Psychotria capensis（Eckl.）Vatke ●☆

319518　Psychotria elachistacantha K. Schum. = Psychotria elachistantha（K. Schum.）E. M. Petit ●☆

319519　Psychotria elachistantha（K. Schum.）E. M. Petit;微花九节●☆

319520　Psychotria elliotii Bremek.;埃利九节●☆

319521　Psychotria elliptica Ker Gawl. = Psychotria asiatica L. ●

319522　Psychotria elliptica Ker Gawl. =Psychotria rubra（Lour.）Poir.●

319523　Psychotria elongatosepala（De Wild.）E. M. Petit;长萼九节●☆

319524　Psychotria emetica L. f.;大沟九节（大沟吐根）●☆

319525　Psychotria emetica L. f. = Psychotria yapoensis（Schnell）Verdc.●☆

319526　Psychotria eminiana（Kuntze）E. M. Petit;埃明九节●☆

319527　Psychotria eminiana（Kuntze）E. M. Petit var. heteroclada E. M. Petit;互枝埃明九节●☆

319528　Psychotria eminiana（Kuntze）E. M. Petit var. stolzii（K. Krause）E. M. Petit = Psychotria eminiana（Kuntze）E. M. Petit ●☆

319529　Psychotria eminiana（Kuntze）E. M. Petit var. tenuifolia Verdc.;细叶埃明九节●☆

319530　Psychotria engleri K. Krause =Chassalia parvifolia K. Schum. ■☆

319531　Psychotria epiphytica Mildbr. =Chassalia petitiana Piesschaert ●☆

319532　Psychotria erratica Hook. f.;西藏九节;Xizang Ninenode, Xizang Psychotria ●

319533　Psychotria erythrocarpa K. Krause = Psychotria lauracea（K. Schum.）E. M. Petit ●☆

319534　Psychotria erythropus K. Schum.;红足九节●☆

319535　Psychotria esquirolii H. Lév. = Psychotria asiatica L. ●

319536　Psychotria esquirolii H. Lév. = Psychotria rubra（Lour.）Poir.●

319537　Psychotria evrardiana E. M. Petit;埃夫拉尔九节●☆

319538　Psychotria exellii R. Alves, Figueiredo et A. P. Davis;埃克塞尔九节●☆

319539　Psychotria expansissima K. Schum.;扩展九节●☆

319540　Psychotria farmari Hutch. et Dalziel = Psychotria schweinfurthii Hiern ●☆

319541　Psychotria fernandopoensis E. M. Petit;费尔南九节●☆

319542　Psychotria ficoidea K. Krause = Psychotria mahonii C. H. Wright var. puberula（E. M. Petit）Verdc.●☆

319543　Psychotria fimbriata A. Chev. ex Hutch. et Dalziel = Bertiera fimbriata（A. Chev. ex Hutch. et Dalziel）Hepper ■☆

319544　Psychotria fimbriatifolia R. D. Good;流苏叶九节●☆

319545　Psychotria fleuryana E. M. Petit;弗勒里九节●☆

319546　Psychotria fleuryi De Wild. = Psychotria fleuryana E. M. Petit ●☆

319547　Psychotria floribunda De Wild. = Psychotria calva Hiern ●☆

319548　Psychotria fluviatilis Chun ex W. C. Chen;溪边九节;Fluvial Ninenode,Fluvial Psychotria,Riverside Psychotria ●

319549　Psychotria foliosa Hiern;密叶九节●☆

319550　Psychotria fractinervata E. M. Petit;碎脉九节●☆

319551　Psychotria furcellata Baill. ex Vatke = Gaertnera furcellata (Baill. ex Vatke) Malcomber et A. P. Davis ●☆

319552　Psychotria gabonica Hiern;加蓬九节●☆

319553　Psychotria garrettii K. Schum. = Psychotria bidentata (Thunb. ex Roem. et Schult.) Hiern ●☆

319554　Psychotria gilletii De Wild.;吉勒特九节●☆

319555　Psychotria giorgii De Wild. = Psychotria succulenta (Schweinf. ex Hiern) E. M. Petit ●☆

319556　Psychotria globiceps K. Schum.;球头九节●☆

319557　Psychotria globosa Hiern;球形九节●☆

319558　Psychotria globosa Hiern var. ciliata (Hiern) E. M. Petit;睫毛球形九节●☆

319559　Psychotria globuloso-baccata (De Wild.) E. M. Petit;小球九节●☆

319560　Psychotria goetzei (K. Schum.) E. M. Petit;格兹九节●☆

319561　Psychotria goetzei (K. Schum.) E. M. Petit var. meridiana E. M. Petit = Psychotria zombamontana (Kuntze) E. M. Petit ●☆

319562　Psychotria goetzei (K. Schum.) E. M. Petit var. platyphylla ?;宽叶九节●☆

319563　Psychotria gossweileri E. M. Petit;戈斯九节●☆

319564　Psychotria gracilescens De Wild. = Psychotria cinerea De Wild. ●☆

319565　Psychotria graciliflora Benth. ex Oerst.;细花九节●☆

319566　Psychotria griseola K. Schum.;灰九节●☆

319567　Psychotria guerkeana K. Schum.;盖尔克九节●☆

319568　Psychotria guillotii Hochr. = Gaertnera inflexa Boivin ex Baill. ●☆

319569　Psychotria hainanensis H. L. Li;海南九节;Hainan Ninenode,Hainan Psychotria ●

319570　Psychotria hallei Aké Assi et Bouton = Psychotria ivorensis De Wild. ●☆

319571　Psychotria hamata De Wild.;顶钩九节●☆

319572　Psychotria haplantha Bremek.;单花九节●☆

319573　Psychotria hawaiiensis (A. Gray) Fosberg;夏威夷九节●☆

319574　Psychotria hemsleyi Verdc.;昂斯莱九节●☆

319575　Psychotria henriquesiana K. Schum. = Psychotria lucens Hiern ●☆

319576　Psychotria henryi H. Lév.;滇南九节;Henry Ninenode,Henry Psychotria ●

319577　Psychotria henryi H. Lév. = Psychotria hainanensis H. L. Li ●

319578　Psychotria herbacea Jacq. = Geophila herbacea (Jacq.) K. Schum. ■

319579　Psychotria herbacea Jacq. = Geophila repens (L.) I. M. Johnst. ■☆

319580　Psychotria herbacea L. = Geophila repens (L.) I. M. Johnst. ■☆

319581　Psychotria heterosticta E. M. Petit;异点九节●☆

319582　Psychotria heterosticta E. M. Petit var. acuminata ?;渐尖九节●☆

319583　Psychotria heterosticta E. M. Petit var. plurinervata ?;寡脉异点九节●☆

319584　Psychotria heterosticta E. M. Petit var. pubescens Verdc.;短柔毛异点九节●☆

319585　Psychotria himanthophylla Bremek.;带叶九节●☆

319586　Psychotria hirtella Oliv. = Psychotria kirkii Hiern var. hirtella (Oliv.) Verdc.■☆

319587　Psychotria holtzii (K. Schum.) E. M. Petit;霍尔茨九节●☆

319588　Psychotria holtzii (K. Schum.) E. M. Petit var. pubescens Verdc.;短柔毛九节●☆

319589　Psychotria homalosperma A. Gray = Psychotria manillensis Bartl. ex DC. ●

319590　Psychotria homolleae Bremek.;奥莫勒九节●☆

319591　Psychotria huae De Wild. = Psychotria calva Hiern ●☆

319592　Psychotria humbertii Bremek.;亨伯特九节●☆

319593　Psychotria humilis Hiern;低矮九节●☆

319594　Psychotria humilis Hiern var. cornuta (Hiern) E. M. Petit;角状九节●☆

319595　Psychotria humilis Hiern var. maior E. M. Petit;大株低矮九节●☆

319596　Psychotria humilis Hutch. = Psychotria eminiana (Kuntze) E. M. Petit ●☆

319597　Psychotria hypoleuca K. Schum.;白背九节●☆

319598　Psychotria ilendensis K. Krause;伊伦德九节●☆

319599　Psychotria infundibularis Hiern;漏斗状九节●☆

319600　Psychotria ionantha K. Schum. = Psychotria globosa Hiern ●☆

319601　Psychotria ipecacuanha Stokes = Cephaelis ipecacuatha (Brot.) A. Rich. ●

319602　Psychotria iringensis Verdc.;伊林加九节●☆

319603　Psychotria ischnophylla K. Schum. = Chassalia ischnophylla (K. Schum.) Hepper ■☆

319604　Psychotria ituriensis De Wild. ex E. M. Petit;伊图里九节●☆

319605　Psychotria ivorensis De Wild.;伊沃里九节●☆

319606　Psychotria ixoprides Bartl.;大叶九节;Big-leaved Psychotria ●

319607　Psychotria ixoroides Bartl. ex DC. = Psychotria serpens L. ●

319608　Psychotria jasminiflora (Linden et André) Mast.;素馨花九节●☆

319609　Psychotria kassneri Bremek. = Psychotria kirkii Hiern var. volkensii (K. Schum.) Verdc. ●☆

319610　Psychotria kilimandscharica Engl.;基利九节●☆

319611　Psychotria kimuenzae De Wild.;基姆扎九节●☆

319612　Psychotria kirkii Hiern;柯克九节●☆

319613　Psychotria kirkii Hiern var. diversinodula Verdc. = Psychotria diversinodula (Verdc.) Verdc. ●☆

319614　Psychotria kirkii Hiern var. hirtella (Oliv.) Verdc.;多毛九节■☆

319615　Psychotria kirkii Hiern var. mucronata (Hiern) Verdc.;钝尖柯克九节●☆

319616　Psychotria kirkii Hiern var. nairobiensis (Bremek.) Verdc.;内罗比九节●☆

319617　Psychotria kirkii Hiern var. swynnertonii (Bremek.) Verdc.;斯温纳顿九节●☆

319618　Psychotria kirkii Hiern var. volkensii (K. Schum.) Verdc.;福尔九节●☆

319619　Psychotria kisantuensis De Wild. = Psychotria calva Hiern ●☆

319620　Psychotria klainei Schnell;克莱恩九节●☆

319621　Psychotria kolly Schumach. = Chassalia kolly (Schumach.) Hepper ■☆

319622　Psychotria konguensis Hiern;孔古九节●☆

319623　Psychotria konkourensis Schnell = Psychotria rufipilis De Wild. ●☆

319624　Psychotria kotoensis Hayata = Psychotria cephalophora Merr. ●

319625　Psychotria kwangsiensis H. L. Li = Psychotria yunnanensis Hutch. ●

319626　Psychotria laevis (Benth.) K. Schum. = Psychotria calva Hiern ●☆

319627　Psychotria lagenocarpa K. Schum. = Psychotria subobliqua Hiern ●☆

319628　Psychotria lamprophylla K. Schum.;亮叶九节●☆

319629　Psychotria lanceifolia K. Schum.;披针九节●☆

319630 Psychotria latistipula Benth. ;宽托叶九节●☆

319631 Psychotria laui Merr. et F. P. Metcalf = Cephaelis laui（Merr. et F. P. Metcalf）F. C. How et W. C. Ko ●

319632 Psychotria lauracea（K. Schum.）E. M. Petit;劳拉九节;Laura Ninenode ●☆

319633 Psychotria laurentii De Wild. ;洛朗九节●☆

319634 Psychotria lebrunii Cheek;勒布伦九节●☆

319635 Psychotria leonardiana E. M. Petit;莱奥九节●☆

319636 Psychotria leptophylla Hiern;非洲细叶九节●☆

319637 Psychotria letouzeyi E. M. Petit;勒图九节●☆

319638 Psychotria leucocentron K. Schum. ;白距九节●☆

319639 Psychotria leuconeura K. Schum. et K. Krause = Psychotria pumila Hiern var. leuconeura（K. Schum. et K. Krause）E. M. Petit ●☆

319640 Psychotria leucopoda E. M. Petit;白梗九节●☆

319641 Psychotria leucothyrsa K. Krause = Gaertnera leucothyrsa（K. Krause）E. M. Petit ●☆

319642 Psychotria liberica Hepper;利比里亚九节●☆

319643 Psychotria limba Scott-Elliot;具边九节●☆

319644 Psychotria linderi Hepper;林德九节●☆

319645 Psychotria linearifolia Bremek. ;线叶九节●☆

319646 Psychotria linearisepala E. M. Petit;线萼九节●☆

319647 Psychotria linearisepala E. M. Petit var. subobtusa Verdc. ;钝线萼九节●☆

319648 Psychotria liukiuensis Hatus. = Psychotria manillensis Bartl. ex DC. ●

319649 Psychotria lomiensis K. Krause = Psychotria cyanopharynx K. Schum. ●☆

319650 Psychotria longevaginalis Schweinf. ex Hiern = Gaertnera longevaginalis（Schweinf. ex Hiern）E. M. Petit ●☆

319651 Psychotria longistylis Hiern = Chazaliella longistylis（Hiern）E. M. Petit et Verdc. ●☆

319652 Psychotria longituba A. Chev. ex De Wild. ;长管九节●☆

319653 Psychotria lophoclada Hiern = Chazaliella lophoclada（Hiern）E. M. Petit et Verdc. ●☆

319654 Psychotria louisii E. M. Petit;路易斯九节●☆

319655 Psychotria lovettii Borhidi et Verdc. ;洛维特九节●☆

319656 Psychotria lubutuensis De Wild. ;卢布图九节●☆

319657 Psychotria lucens Hiern;光亮九节●☆

319658 Psychotria lucens Hiern var. minor E. M. Petit;大株光亮九节●☆

319659 Psychotria lucidula Baker;明亮九节●☆

319660 Psychotria macrodiscus（K. Schum.）De Wild. = Chassalia macrodiscus K. Schum. ■☆

319661 Psychotria macrophylla Ruiz et Pav. = Psychotria boninensis Nakai ●☆

319662 Psychotria maculata S. Moore = Psychotria kirkii Hiern var. mucronata（Hiern）Verdc. ●☆

319663 Psychotria madandensis S. Moore = Chazaliella abrupta（Hiern）E. M. Petit et Verdc. ●☆

319664 Psychotria magnisepala Bremek. ;大萼九节●☆

319665 Psychotria mahonii C. H. Wright;马洪九节●☆

319666 Psychotria mahonii C. H. Wright var. puberula（E. M. Petit）Verdc. = Psychotria mahonii C. H. Wright ●☆

319667 Psychotria mahonii C. H. Wright var. pubescens（Robyns）Verdc. = Psychotria mahonii C. H. Wright ●☆

319668 Psychotria maliensis Schnell = Psychotria multinervis De Wild. ●☆

319669 Psychotria mangenotii（Aké Assi）Verdc. ;芒热诺九节●☆

319670 Psychotria manillensis Bartl. = Psychotria manillensis Bartl. ex DC. ●

319671 Psychotria manillensis Bartl. ex DC. ;琉球九节;Liuqiu Ninenode ●

319672 Psychotria mannii Hiern;曼氏九节●☆

319673 Psychotria mannii Hiern var. nigrescens E. M. Petit;黑曼氏九节●☆

319674 Psychotria marcgravii Spreng. ;马氏九节●☆

319675 Psychotria marginata Bremek. = Psychotria kirkii Hiern var. nairobiensis（Bremek.）Verdc. ●☆

319676 Psychotria marojejensis Bremek. ;马鲁杰九节●☆

319677 Psychotria megalopus Verdc. ;粗梗九节●☆

319678 Psychotria megistosticta（S. Moore）E. M. Petit = Psychotria mahonii C. H. Wright var. puberula（E. M. Petit）Verdc. ●☆

319679 Psychotria megistosticta（S. Moore）E. M. Petit var. puberula E. M. Petit = Psychotria mahonii C. H. Wright var. puberula（E. M. Petit）Verdc. ●☆

319680 Psychotria megistosticta（S. Moore）E. M. Petit var. punicea ? = Psychotria mahonii C. H. Wright var. puberula（E. M. Petit）Verdc. ●☆

319681 Psychotria melanosticta K. Schum. = Psychotria punctata Vatke ●☆

319682 Psychotria membranifolia Bartl. ex DC. ;膜叶九节●

319683 Psychotria meridiano-montana E. M. Petit = Psychotria zombamontana（Kuntze）E. M. Petit ●☆

319684 Psychotria meridiano-montana E. M. Petit var. angustifolia ? = Psychotria zombamontana（Kuntze）E. M. Petit ●☆

319685 Psychotria meridiano-montana E. M. Petit var. glabra ? = Psychotria zombamontana（Kuntze）E. M. Petit ●☆

319686 Psychotria meridiano-montana E. M. Petit var. meridiana ? = Psychotria zombamontana（Kuntze）E. M. Petit ●☆

319687 Psychotria micheliana J. -G. Adam;米歇尔九节●☆

319688 Psychotria microdon Urb. ;尖齿九节●☆

319689 Psychotria microgrammata Bremek. ;小九节●☆

319690 Psychotria microthyrsa E. M. Petit;小序九节●☆

319691 Psychotria minima R. D. Good;微小九节●☆

319692 Psychotria minimicalyx K. Schum. ;小萼九节●☆

319693 Psychotria minuta E. M. Petit;弱小九节●☆

319694 Psychotria molleri K. Schum. ;默勒九节●☆

319695 Psychotria monticola Hiern = Psychotria nubicola G. Taylor ●☆

319696 Psychotria morindoides Hutch. ;聚果九节（假巴戟）;Indian-mulberryoid Psychotria ●

319697 Psychotria mortehanii De Wild. ;莫特汉九节●☆

319698 Psychotria moseskemei Cheek;莫塞九节●☆

319699 Psychotria mucronata Hiern = Psychotria kirkii Hiern var. mucronata（Hiern）Verdc. ●☆

319700 Psychotria multiflora Schumach. et Thonn. = Keetia multiflora（Schumach. et Thonn.）Bridson ●☆

319701 Psychotria multinervis De Wild. ;多脉九节●☆

319702 Psychotria mushiticola E. M. Petit = Psychotria mahonii C. H. Wright ●☆

319703 Psychotria mwinilungae Verdc. ;穆维尼九节●☆

319704 Psychotria nairobiensis Bremek. = Psychotria kirkii Hiern var. nairobiensis（Bremek.）Verdc. ●☆

319705 Psychotria nebulosa（Dwyer）C. M. Taylor = Psychotria nebulosa K. Krause ●☆

319706 Psychotria nebulosa K. Krause;星云九节●☆

319707 Psychotria nervosa Benth. ;约翰九节;St. John's Bush ●☆

319708 Psychotria nervosa Benth. = Psychotria denticulata Wall. ●☆

319709 Psychotria neurodictyon K. Schum. = Hymenocoleus neurodictyon

（K. Schum.）Robbr.■☆

319710　Psychotria nigerica Hepper = Psychotria globosa Hiern var. ciliata（Hiern）E. M. Petit ●☆

319711　Psychotria nigrescens De Wild. = Psychotria albicaulis Scott-Elliot ●☆

319712　Psychotria nigrifolia Gilli = Rutidea fuscescens Hiern ●☆

319713　Psychotria nigropunctata Hiern;黑斑九节☆

319714　Psychotria nimbana Schnell = Psychotria rufipilis De Wild. ●☆

319715　Psychotria nimbana Schnell var. djalonensis ? = Psychotria rufipilis De Wild. ●☆

319716　Psychotria nimbana Schnell var. gaidensis ? = Psychotria rufipilis De Wild. ●☆

319717　Psychotria nimbana Schnell. f. vallicola ? = Psychotria rufipilis De Wild. ●☆

319718　Psychotria nubica Delile;云雾九节●☆

319719　Psychotria nubicola G. Taylor;云生九节●☆

319720　Psychotria obanensis Wernham = Chazaliella obanensis（Wernham）E. M. Petit et Verdc. ●☆

319721　Psychotria oblanceolata（R. D. Good）Ruhsam;倒披针形九节●☆

319722　Psychotria obovatifolia De Wild. ;倒卵叶九节●☆

319723　Psychotria obscura Benth. = Psychotria schweinfurthii Hiern ●☆

319724　Psychotria obtusifolia Poir. ;钝叶九节●☆

319725　Psychotria obvallata Schumach. = Geophila obvallata（Schumach.）Didr. ■☆

319726　Psychotria oddonii De Wild. = Chazaliella oddonii（De Wild.）E. M. Petit et Verdc. ●☆

319727　Psychotria ogowensis De Wild. = Psychotria brieyi De Wild. ●☆

319728　Psychotria oligocarpa K. Schum. ;寡果九节●☆

319729　Psychotria ombrophila（Schnell）Verdc. ;喜雨九节●☆

319730　Psychotria orophila E. M. Petit;喜山九节●☆

319731　Psychotria owariensis（P. Beauv.）Hiern;尾张九节●☆

319732　Psychotria pachyclada K. Schum. et K. Krause = Psychotria punctata Vatke ●☆

319733　Psychotria pachygrammata Bremek. ;粗纹九节●☆

319734　Psychotria palustris E. M. Petit;沼泽九节●☆

319735　Psychotria pandurata Verdc. ;琴形九节●☆

319736　Psychotria parkeri Baker;帕克九节●☆

319737　Psychotria parvifolia（K. Schum.）De Wild. = Chassalia parvifolia K. Schum. ■☆

319738　Psychotria parvistipulata E. M. Petit;小托叶九节●☆

319739　Psychotria pauciflora De Wild. = Psychotria subobliqua Hiern ●☆

319740　Psychotria pauridiantha Hiern = Psychotria bifaria Hiern var. pauridiantha（Hiern）E. M. Petit ●☆

319741　Psychotria peduncularis（Salisb.）Steyerm. ;梗花九节●☆

319742　Psychotria peduncularis（Salisb.）Steyerm. var. angustibracteata Verdc. ;窄苞九节●☆

319743　Psychotria peduncularis（Salisb.）Steyerm. var. ciliato-stipulata Verdc. ;睫毛托叶九节●☆

319744　Psychotria peduncularis（Salisb.）Steyerm. var. guineensis（Schnell）Verdc. ;几内亚九节●☆

319745　Psychotria peduncularis（Salisb.）Steyerm. var. hypsophila（K. Schum. et K. Krause）Verdc. ;喜高九节●☆

319746　Psychotria peduncularis（Salisb.）Steyerm. var. ivorensis（Schnell）Verdc. ;伊沃里梗花九节●☆

319747　Psychotria peduncularis（Salisb.）Steyerm. var. nyassana（K. Krause）Verdc. ;尼亚萨九节●☆

319748　Psychotria peduncularis（Salisb.）Steyerm. var. palmetorum

（DC.）Verdc. ;棕榈九节■☆

319749　Psychotria peduncularis（Salisb.）Steyerm. var. semlikiensis Verdc. ;塞姆利基九节■☆

319750　Psychotria peduncularis（Salisb.）Steyerm. var. suaveolens（Hiern）Verdc. ;香梗花九节●☆

319751　Psychotria peduncularis（Salisb.）Steyerm. var. tabouensis（Schnell）Verdc. ;塔布九节●☆

319752　Psychotria perrieri Bremek. ;佩里耶九节●☆

319753　Psychotria peteri E. M. Petit;彼得九节●☆

319754　Psychotria petitii Verdc. ;佩蒂蒂九节●☆

319755　Psychotria petroxenos K. Schum. = Psychotria kirkii Hiern ●☆

319756　Psychotria pilifera Hutch. = Psychotria vogeliana Benth. ●☆

319757　Psychotria pilifera Hutch. et Dalziel = Psychotria vogeliana Benth. ●☆

319758　Psychotria pilosula De Wild. = Chazaliella oddonii（De Wild.）E. M. Petit et Verdc. ●☆

319759　Psychotria plantaginoidea E. M. Petit;车前状九节●☆

319760　Psychotria pleuroneura K. Schum. ;侧脉九节●☆

319761　Psychotria pocsii Borhidi et Verdc. ;波克斯九节●☆

319762　Psychotria pocsii Borhidi et Verdc. subsp. ferruginea Borhidi et Verdc. ;锈色波克斯九节●☆

319763　Psychotria podocarpa E. M. Petit;柄果九节●☆

319764　Psychotria poggei K. Schum. = Chazaliella poggei（K. Schum.）E. M. Petit et Verdc. ●☆

319765　Psychotria polygrammata Bremek. ;多纹九节●☆

319766　Psychotria polyphylla Bremek. ;多叶九节●☆

319767　Psychotria porphyroclada K. Schum. ;紫枝九节●☆

319768　Psychotria portonoversis De Wild. = Psychotria calva Hiern ●☆

319769　Psychotria potamogetonoides Wernham = Psychotria recurva Hiern ●☆

319770　Psychotria potamophila K. Schum. ;河生九节●☆

319771　Psychotria prainii H. Lév. ;驳骨九节(百样花,茶山虫,花叶九节,毛九节,小功劳);Prain's Ninenode, Prain's Psychotria ●

319772　Psychotria principensis G. Taylor;普林西比九节●☆

319773　Psychotria pseudoplatyphylla E. M. Petit;假宽叶九节●☆

319774　Psychotria psychotrioides（DC.）Roberty;药用九节●☆

319775　Psychotria psychotrioides（Schnell）Schnell = Psychotria rufipilis De Wild. ●☆

319776　Psychotria pteropetala K. Schum. = Chassalia pteropetala（K. Schum.）Cheek ■☆

319777　Psychotria pubescens Bartl. ex DC. ;柔毛九节●☆

319778　Psychotria pubifolia De Wild. = Psychotria kirkii Hiern var. mucronata（Hiern）Verdc. ●☆

319779　Psychotria pumila Hiern;偃伏九节●☆

319780　Psychotria pumila Hiern var. buzica（S. Moore）E. M. Petit;布兹九节●☆

319781　Psychotria pumila Hiern var. leuconeura（K. Schum. et K. Krause）E. M. Petit;白脉九节●☆

319782　Psychotria pumila Hiern var. puberula E. M. Petit;微毛偃伏九节●☆

319783　Psychotria pumila Hiern var. subumbellata（E. M. Petit）Verdc. ;小伞九节●☆

319784　Psychotria punctata Vatke;斑点九节;Dotted Wild Coffee ●☆

319785　Psychotria punctata Vatke var. hirtella Chiov. = Psychotria kirkii Hiern var. nairobiensis（Bremek.）Verdc. ●☆

319786　Psychotria punctata Vatke var. minor E. M. Petit;小偃伏九节●☆

319787　Psychotria punctata Vatke var. tenuis E. M. Petit;细九节●☆

319788　Psychotria pygmaeodendron K. Schum. ;矮小九节●☆

319789　Psychotria recurva Hiern;反曲九节●☆

319790　Psychotria reducta De Wild. = Psychotria verschuerenii De Wild. var. reducta E. M. Petit ☆

319791　Psychotria reevesii Wall. = Psychotria asiatica L. ●

319792　Psychotria reevesii Wall. = Psychotria rubra (Lour.) Poir. ●

319793　Psychotria reevesii Wall. var. pilosa Pit. = Psychotria asiatica L. ●

319794　Psychotria reevesii Wall. var. pilosa Pit. = Psychotria rubra (Lour.) Poir. var. pilosa (Pit.) W. C. Chen ●

319795　Psychotria refractiflora K. Schum. 曲花九节●☆

319796　Psychotria refractiloba K. Schum. = Psychotria subobliqua Hiern ●☆

319797　Psychotria refractistipula De Wild. = Psychotria cyanopharynx K. Schum. ●☆

319798　Psychotria repens (L.) L. ;匍匐九节●☆

319799　Psychotria reptans Benth. ;俯卧九节●☆

319800　Psychotria retiphlebia Baker;网脉九节●☆

319801　Psychotria rhizomatosa De Wild. ;根茎九节●☆

319802　Psychotria rhizomatosa De Wild. var. minor E. M. Petit;较小根茎九节●☆

319803　Psychotria riparia (K. Schum. et K. Krause) E. M. Petit = Psychotria capensis (Eckl.) Vatke subsp. riparia (K. Schum. et K. Krause) Verdc. ●☆

319804　Psychotria riparia (K. Schum. et K. Krause) E. M. Petit var. puberula E. M. Petit = Psychotria capensis (Eckl.) Vatke var. puberula (E. M. Petit) Verdc. ●☆

319805　Psychotria robynsiana E. M. Petit = Psychotria mahonii C. H. Wright var. pubescens (Robyns) Verdc. ●☆

319806　Psychotria robynsiana E. M. Petit var. glabra ? = Psychotria mahonii C. H. Wright var. pubescens (Robyns) Verdc. ●☆

319807　Psychotria robynsiana E. M. Petit var. pauciorinervata ? = Psychotria mahonii C. H. Wright var. pubescens (Robyns) Verdc. ●☆

319808　Psychotria rotundifolia R. D. Good = Chazaliella rotundifolia (R. D. Good) E. M. Petit et Verdc. ●☆

319809　Psychotria rowlandii Hutch. et Dalziel = Psychotria gabonica Hiern ●☆

319810　Psychotria rubra (Lour.) Poir. ;九节(暗山公,暗山谷,暗山香,吹筒管,吹筒树,大丹叶,大罗伞,大退七,刀斧伤,刀枪木,刀伤木,火筒树,假木竹,金鸡爪,九节木,牛屎乌,喷筒,青龙吐雾,散血丹,山打大刀,山大刀,山大颜,血丝罗伞);Red Ninenode, Red Psychotria, Wild Coffee ●

319811　Psychotria rubra (Lour.) Poir. var. pilosa (Pit.) W. C. Chen;毛叶九节;Pilose Red Ninenode,Pilose Red Psychotria ●

319812　Psychotria rubra (Lour.) Poir. var. pilosa (Pit.) W. C. Chen = Psychotria asiatica L. ●

319813　Psychotria rubristipulata R. D. Good;红托叶九节●☆

319814　Psychotria rubropilosa De Wild. ;红柔毛九节●☆

319815　Psychotria rufipila A. Chev. = Psychotria rufipilis De Wild. ●☆

319816　Psychotria rufipilis De Wild. ;红毛九节●☆

319817　Psychotria rufipilis De Wild. var. konkourensis (Schnell) Hepper = Psychotria rufipilis De Wild. ●☆

319818　Psychotria rutshuruensis De Wild. = Psychotria kirkii Hiern var. mucronata (Hiern) Verdc. ●☆

319819　Psychotria sabukaensis De Wild. = Psychotria mannii Hiern ●☆

319820　Psychotria sadebeckiana K. Schum. ;萨德拜克九节●☆

319821　Psychotria sadebeckiana K. Schum. var. elongata E. M. Petit;伸长九节●☆

319822　Psychotria sakaleonensis Bremek. ;萨卡莱乌纳九节●☆

319823　Psychotria sambiranensis Bremek. ;桑比朗九节●☆

319824　Psychotria sangalkamensis (Schnell) Schnell = Psychotria bidentata (Thunb. ex Roem. et Schult.) Hiern ●☆

319825　Psychotria scabrida Bremek. ;微糙九节●☆

319826　Psychotria scandens Hook. et Arn. = Psychotria serpens L. ●☆

319827　Psychotria scheffleri K. Schum. et K. Krause;谢夫勒九节●☆

319828　Psychotria schliebenii E. M. Petit;施利本九节●☆

319829　Psychotria schliebenii E. M. Petit var. parvipaniculata ?;小圆锥九节●☆

319830　Psychotria schnellii (Aké Assi) Verdc. ;施内尔九节●☆

319831　Psychotria schweinfurthii Hiern;施韦九节●☆

319832　Psychotria sciadephora Hiern = Chazaliella sciadephora (Hiern) E. M. Petit et Verdc. ●☆

319833　Psychotria serpens L. ;蔓九节(白花风不动,白珠藤,穿根藤,春根藤,风不动,风不动藤,广东络石藤,络石藤,木头痧,匍匐九节,拎壁龙,拎树龙,上木蛇,伸筋藤,石邦子,松筋藤,崧根藤,崧筋藤,蜈蚣藤,银珠果);Creeping Ninenode, Creeping Psychotria ●

319834　Psychotria serpens L. var. macrophylla Koidz. = Psychotria boninensis Nakai ●☆

319835　Psychotria setacea Hiern = Psychotria leptophylla Hiern ●☆

319836　Psychotria setistipulata (R. D. Good) E. M. Petit;毛托叶九节●☆

319837　Psychotria siamica (Craib) Hutch. ;毛九节(驳骨草,花叶九节,小功劳);Siam Ninenode ●

319838　Psychotria siamica (Craib) Hutch. = Psychotria prainii H. Lév. ●

319839　Psychotria sidamensis Cufod. = Psychotria orophila E. M. Petit ●☆

319840　Psychotria simplex (K. Krause) De Wild. = Chassalia simplex K. Krause ■☆

319841　Psychotria sodifera De Wild. = Psychotria schweinfurthii Hiern ●☆

319842　Psychotria solfiana K. Krause = Psychotria bifaria Hiern ●☆

319843　Psychotria soyauxii Hiern = Psychotria schweinfurthii Hiern ●☆

319844　Psychotria sparsipila Bremek. ;疏毛九节●☆

319845　Psychotria spathacea (Hiern) Verdc. ;佛焰苞九节●☆

319846　Psychotria sphaerocarpa (Hiern) Hutch. et Dalziel = Psychotria fernandopoensis E. M. Petit ●☆

319847　Psychotria spithamea S. Moore;距九节●☆

319848　Psychotria stictophylla Hiern = Gaertnera stictophylla (Hiern) E. M. Petit ●☆

319849　Psychotria stigmatophylla K. Schum. ;点叶九节●☆

319850　Psychotria straminea Hutch. ;黄脉九节(草绿九节);Yellownerve Ninenode, Yellownerve Psychotria, Yellow-nerved Psychotria ●

319851　Psychotria subcapitata Bremek. ;亚头状九节●☆

319852　Psychotria subcordatifolia De Wild. = Chassalia subcordatifolia (De Wild.) Piessch. ■☆

319853　Psychotria subglabra De Wild. ;近光九节●☆

319854　Psychotria subglabroides Schnell = Psychotria subglabra De Wild. ●☆

319855　Psychotria subherbacea Hiern = Chassalia subherbacea (Hiern) Hepper ■☆

319856　Psychotria subhirtella K. Schum. = Psychotria kirkii Hiern var. volkensii (K. Schum.) Verdc. ●☆

319857　Psychotria subnuda Hiern = Chassalia subnuda (Hiern) Hepper ■☆

319858　Psychotria subobliqua Hiern;偏斜九节●☆

319859　Psychotria subochreata De Wild. = Chassalia subochreata (De Wild.) Robyns ■☆

319860　Psychotria subpunctata Hiern;小斑九节●☆

319861　Psychotria succulenta (Schweinf. ex Hiern) E. M. Petit;多汁九

节●☆

319862　Psychotria swynnertonii Bremek. = Psychotria kirkii Hiern var. swynnertonii（Bremek.）Verdc.●☆

319863　Psychotria sycophylla（K. Schum.）E. M. Petit；无花果叶九节●☆

319864　Psychotria symplocifolia Kurz；山矾叶九节；Symplocosleaf Ninenode，Symplocosleaf Psychotria，Symplocos-leaved Psychotria ●

319865　Psychotria taitensis Verdc.；泰塔九节●☆

319866　Psychotria talbotii Wernham；塔尔博特九节●☆

319867　Psychotria tanganyicensis Verdc.；坦噶尼喀九节●☆

319868　Psychotria tanganyicensis Verdc. subsp. longipes Verdc.；长梗坦噶尼喀九节●☆

319869　Psychotria tanganyicensis Verdc. var. ferruginea Verdc.；锈色坦噶尼喀九节●☆

319870　Psychotria tenuifolia Sw.；多明细叶九节●☆

319871　Psychotria tenuipetiolata Verdc.；细柄九节●☆

319872　Psychotria tenuissima E. M. Petit；极细九节●☆

319873　Psychotria ternata Bremek.；三出九节●☆

319874　Psychotria thomensis G. Taylor；爱岛九节●☆

319875　Psychotria trachystyla Hiern = Gaertnera trachystyla（Hiern）E. Petit ●☆

319876　Psychotria trichanthera K. Schum.；毛花九节●☆

319877　Psychotria trichopleura Mildbr. = Psychotria globosa Hiern var. ciliata（Hiern）E. M. Petit ●☆

319878　Psychotria triclada E. M. Petit；三枝九节●☆

319879　Psychotria triflora Thonn. = Cremaspora triflora（Thonn.）K. Schum.●☆

319880　Psychotria tutcheri Dunn；假九节（小叶九节）；Small-leaved Psychotria，Tutcher Ninenode，Tutcher Psychotria ●

319881　Psychotria umbellata Thonn. = Psychotria calva Hiern ●☆

319882　Psychotria umbellifera E. M. Petit；伞花九节●☆

319883　Psychotria umbraticola（Vatke）Hiern = Chassalia umbraticola Vatke ■☆

319884　Psychotria umbraticola Williams = Chassalia umbraticola Vatke ■☆

319885　Psychotria usambarensis Verdc.；乌桑巴拉九节●☆

319886　Psychotria vaginalis G. Don = Gaertnera vaginans（DC.）Merr.●☆

319887　Psychotria vaginans DC. = Gaertnera vaginans（DC.）Merr.●☆

319888　Psychotria vanderystii De Wild. = Chassalia vanderystii（De Wild.）Verdc. ■☆

319889　Psychotria variopunctulata De Wild. = Psychotria mannii Hiern ●☆

319890　Psychotria velutipes K. Schum. = Psychotria thomensis G. Taylor ●☆

319891　Psychotria venosa（Hiern）E. M. Petit；毛脉九节●☆

319892　Psychotria verdcourtii Borhidi；韦尔德九节●☆

319893　Psychotria verschuerenii De Wild.；费许伦九节●☆

319894　Psychotria verschuerenii De Wild. var. reducta E. M. Petit；退缩九节●☆

319895　Psychotria virens Hiern = Chassalia hiernii（Kuntze）G. Taylor ■☆

319896　Psychotria viridicalyx R. D. Good = Chazaliella viridicalyx（R. D. Good）Verdc.●☆

319897　Psychotria viticoides Wernham；葡萄九节●☆

319898　Psychotria vogeliana Benth.；沃格尔九节（百样花，驳骨草，小功劳）；Piliferous Psychotria，Siam Ninenode，Siam Psychotria ●☆

319899　Psychotria vogeliana Benth. var. bipindensis Schnell = Psychotria vogeliana Benth. ●☆

319900　Psychotria vogeliana Benth. var. chariensis Schnell = Psychotria vogeliana Benth. ●☆

319901　Psychotria vogeliana Benth. var. korhogoensis Schnell = Psychotria vogeliana Benth. ●☆

319902　Psychotria vogeliana Benth. var. letestui Schnell = Psychotria vogeliana Benth. ●☆

319903　Psychotria volkensii K. Schum. = Psychotria kirkii Hiern var. volkensii（K. Schum.）Verdc.●☆

319904　Psychotria walikalensis E. M. Petit；瓦利卡莱九节●☆

319905　Psychotria wallichiana Spreng. = Psychotria denticulata Wall. ●☆

319906　Psychotria warneckei K. Schum. et K. Krause = Chassalia kolly（Schumach.）Hepper ■☆

319907　Psychotria wauensis K. Krause = Psychotria lauracea（K. Schum.）E. M. Petit ●☆

319908　Psychotria welwitschii（Hiern）Bremek.；韦尔九节●☆

319909　Psychotria wildemaniana T. Durand ex De Wild. = Chazaliella wildemaniana（T. Durand ex De Wild.）E. M. Petit et Verdc.●☆

319910　Psychotria williamsii Hutch. et Dalziel = Psychotria calva Hiern ●☆

319911　Psychotria yabaensis De Wild. = Psychotria subobliqua Hiern ●☆

319912　Psychotria yapoensis（Schnell）Verdc.；亚波九节●☆

319913　Psychotria yorubensis（K. Schum.）De Wild. = Chassalia kolly（Schumach.）Hepper ■☆

319914　Psychotria yunnanensis Hutch.；云南九节（滇九节）；Yunnan Ninenode，Yunnan Psychotria ●

319915　Psychotria zambesiana Hiern = Psychotria capensis（Eckl.）Vatke ●☆

319916　Psychotria zanguebarica Hiern = Chassalia umbraticola Vatke ■☆

319917　Psychotria zenkeri K. Schum. = Psychotria subobliqua Hiern ●☆

319918　Psychotria zombamontana（Kuntze）E. M. Petit；宗巴九节●☆

319919　Psychotriaceae F. Rudolphi = Rubiaceae Juss.（保留科名）●■

319920　Psychotriaceae F. Rudolphi；九节科●

319921　Psychotrophum P. Browne（废弃属名）= Psychotria L.（保留属名）●

319922　Psychridium Steven = Astragalus L.●■

319923　Psychrobatia Greene = Ametron Raf.●■

319924　Psychrobatia Greene = Rubus L.●■

319925　Psychrogeton Boiss.（1875）；寒蓬属（寒菊属）；Psychrogeton ■

319926　Psychrogeton andryaloides Novopokr. ex Krasch.；印度寒蓬■☆

319927　Psychrogeton andryaloides Novopokr. ex Krasch. var. poncinsii（Franch.）Grierson = Psychrogeton poncinsii（Franch.）Y. Ling et Y. L. Chen ■

319928　Psychrogeton cabulicus Boiss.；喀布尔寒蓬；Kabul Psychrogeton ■☆

319929　Psychrogeton nigromontanus（Boiss. et Buhse）Grierson；黑山寒蓬；Black Mountain Psychrogeton ■

319930　Psychrogeton poncinsii（Franch.）Y. Ling et Y. L. Chen；藏寒蓬；Xizang Psychrogeton ■

319931　Psychrophila（DC.）Bercht. et J. Presl = Caltha L.■

319932　Psychrophila（DC.）Bercht. et J. Presl（1823）；寒金盏花属■☆

319933　Psychrophila Bercht. et J. Presl = Caltha L.■

319934　Psychrophila Bercht. et J. Presl = Psychrophila（DC.）Bercht. et J. Presl ■☆

319935　Psychrophila leptosepala（DC.）W. Weber = Caltha leptosepala DC.■☆

319936　Psychrophila sagittata Bercht. et J. Presl；寒金盏花●☆

319937　Psychrophyton Beauverd = Raoulia Hook. f. ex Raoul ■☆

319938　Psychrophyton Beauverd（1910）；喜寒菊属；Vegetable Sheep ■☆

319939　Psychrophyton eximium Beauverd；优异喜寒菊■☆

319940　Psychrophyton grandiflorum Beauverd；大花喜寒菊■☆

319941　Psychrophyton mammillare Beauverd；喜寒菊；Vegetable Sheep ■☆

319942　Psychrophyton rubrum Beauverd；红喜寒菊■☆

319943　Psycothria L. = Psychotria L.（保留属名）●

319944　Psycrophila Raf. = Caltha L. ■

319945　Psycrophila Raf. = Psychrophila（DC.）Bercht. et J. Presl ■☆

319946　Psydaranta Neck. = Calathea G. Mey. ■

319947　Psydaranta Neck. ex Raf. = Calathea G. Mey. ■

319948　Psydarantha Steud. = Psydaranta Neck. ex Raf. ■

319949　Psydax Steud. = Psydrax Gaertn. ●☆

319950　Psydrax Gaertn.（1788）；疱茜属●☆

319951　Psydrax Gaertn. = Canthium Lam. ●

319952　Psydrax acutiflora（Hiern）Bridson；尖花疱茜●☆

319953　Psydrax arnoldiana（De Wild. et T. Durand）Bridson；阿诺德疱茜●☆

319954　Psydrax austro-orientalis（Cavaco）A. P. Davis et Bridson；东南疱茜●☆

319955　Psydrax bathieana（Cavaco）A. P. Davis et Bridson；巴西疱茜●☆

319956　Psydrax dicoccos Gaertn. = Canthium dicoccum（Gaertn.）Teijsm. et Binn. ●

319957　Psydrax faulknerae Bridson；福克纳疱茜●☆

319958　Psydrax fragrantissima（K. Schum.）Bridson；脆疱茜●☆

319959　Psydrax gilletii（De Wild.）Bridson；吉勒特疱茜●☆

319960　Psydrax graniticola（Chiov.）Bridson；格兰特疱茜●☆

319961　Psydrax horizontalis（Schumach.）Bridson；平展疱茜●☆

319962　Psydrax kaessneri（S. Moore）Bridson；卡斯纳疱茜●☆

319963　Psydrax kibuwae Bridson；基布瓦疱茜●☆

319964　Psydrax kraussioides（Hiern）Bridson；拟克劳斯疱茜●☆

319965　Psydrax livida（Hiern）Bridson；铅色疱茜●☆

319966　Psydrax locuples（K. Schum.）Bridson；繁茂疱茜●☆

319967　Psydrax lynesii Bullock et Bridson；莱恩斯疱茜●☆

319968　Psydrax manensis（Aubrév. et Pellegr.）Bridson；热非疱茜●☆

319969　Psydrax martinii（Dunkley）Bridson；马丁疱茜●☆

319970　Psydrax micans（Bullock）Bridson；弱光泽疱茜●☆

319971　Psydrax moandensis Bridson；莫安达疱茜●☆

319972　Psydrax moggii Bridson；莫格疱茜●☆

319973　Psydrax obovata（Klotzsch ex Eckl. et Zeyh.）Bridson；倒卵疱茜●☆

319974　Psydrax obovata（Klotzsch ex Eckl. et Zeyh.）Bridson subsp. elliptica Bridson；椭圆疱茜●☆

319975　Psydrax occidentalis（Cavaco）A. P. Davis et Bridson；西方疱茜●☆

319976　Psydrax palma（K. Schum.）Bridson；掌疱茜●☆

319977　Psydrax parviflora（Afzel.）Bridson；小花疱茜●☆

319978　Psydrax parviflora（Afzel.）Bridson subsp. chapmanii Bridson；查普曼疱茜●☆

319979　Psydrax parviflora（Afzel.）Bridson subsp. melanophengos（Bullock）Bridson；黑光小花疱茜●☆

319980　Psydrax parviflora（Afzel.）Bridson subsp. rubrocostata（Robyns）Bridson；红脉小花疱茜●☆

319981　Psydrax polhillii Bridson；普尔疱茜●☆

319982　Psydrax recurvifolia（Bullock）Bridson；反曲叶疱茜●☆

319983　Psydrax richardsiae Bridson；理查兹疱茜●☆

319984　Psydrax robertsoniae Bridson；罗伯逊疱茜●☆

319985　Psydrax sambiranensis（Cavaco）A. P. Davis et Bridson；桑比朗疱茜●☆

319986　Psydrax schimperiana（A. Rich.）Bridson；欣珀疱茜●☆

319987　Psydrax schimperiana（A. Rich.）Bridson subsp. occidentalis Bridson；西方欣珀疱茜●☆

319988　Psydrax splendens（K. Schum.）Bridson；光亮疱茜●☆

319989　Psydrax subcordata（DC.）Bridson；亚心形疱茜●☆

319990　Psydrax subcordata（DC.）Bridson var. connata（De Wild. et T. Durand）Bridson；合生疱茜●☆

319991　Psydrax virgata（Hiern）Bridson；条纹疱茜●☆

319992　Psydrax whitei Bridson；瓦特疱茜●☆

319993　Psygmorchis Dodson et Dressler（1972）；扇兰属■☆

319994　Psygmorchis gnomus（Kraenzl.）Dodson et Dressler；扇兰■☆

319995　Psylliaceae Horan. = Plantaginaceae Juss.（保留科名）■

319996　Psylliostachys（Jaub. et Spach）Nevski（1937）；长筒补血草属；Statice ■☆

319997　Psylliostachys anceps（Regel）Roshkova；二棱长筒补血草■☆

319998　Psylliostachys beludshistanica Roshkova；巴基斯坦长筒补血草■☆

319999　Psylliostachys hymenostegia Rech. f. et Koeie = Psylliostachys beludshistanica Roshkova ■☆

320000　Psylliostachys leptostachya（Boiss.）Roshkova；细穗长筒补血草■☆

320001　Psylliostachys spicata（Willd.）Nevski；长筒补血草■☆

320002　Psylliostachys suvorovii（Regel）Roshkova；土耳其长筒补血草（俄国矶松，苏沃补血草，索罗补血草）；Candlewick Statice, Statice, Suworow Statice ■☆

320003　Psylliostachys suworowii（Regel）Roshkova = Psylliostachys suvorovii（Regel）Roshkova ■☆

320004　Psyllium Juss. = Plantago L. ■●

320005　Psyllium Mill. = Plantago L. ■●

320006　Psyllium Tourn. ex Juss. = Plantago L. ■●

320007　Psyllium arenarium（Waldst. et Kit.）Mirb. = Plantago arenaria Waldst. et Kit. ■

320008　Psyllium indicum（L.）Dumont = Plantago arenaria Waldst. et Kit. ■

320009　Psyllium mauritanicum（Boiss. et Reut.）Soják = Plantago mauritanica Boiss. et Reut. ■☆

320010　Psyllium squalidum（Salisb.）Soják = Plantago afra L. ■☆

320011　Psyllocarpus Mart. = Psyllocarpus Mart. et Zucc. ■☆

320012　Psyllocarpus Mart. et Zucc.（1824）；蚤茜属■☆

320013　Psyllocarpus Mart. et Zucc. = Psyllocarpus Mart. ■☆

320014　Psyllocarpus Pohl ex DC. = Declieuxia Kunth ■☆

320015　Psyllocarpus foliosus Pohl；多叶蚤茜■☆

320016　Psyllocarpus glaber Pohl；光蚤茜■☆

320017　Psyllophora Ehrh. = Carex L. ■

320018　Psyllophora Heuffel = Carex L. ■

320019　Psyllothamnus Oliv. = Sphaerocoma T. Anderson ●☆

320020　Psyllothamnus beevori Oliv. = Sphaerocoma hookeri T. Anderson ●☆

320021　Psylostachys Oerst. = Chamaedorea Willd.（保留属名）●☆

320022　Psyloxylon Thouars ex Gaudich. = Psiloxylon Thouars ex Tul. ●☆

320023　Psythirhisma Herb. ex Lindl. = Psithyrisma Herb. ■☆

320024　Psythirhisma Herb. ex Lindl. = Symphyostemon Miers ex Klatt ■☆

320025　Psythirhisma Lindl. = Psithyrisma Herb. ■☆

320026　Psythirhisma Lindl. = Symphyostemon Miers ex Klatt ■☆

320027　Ptacoseia Ehrh. = Carex L. ■

320028　Ptaeroxylaceae J.-F. Leroy = Rutaceae Juss.（保留科名）●■

320029　Ptaeroxylaceae J.-F. Leroy（1960）；喷嚏木科（嚏树科）●☆

320030　Ptaeroxylaceae Sander = Ptaeroxylaceae J.-F. Leroy ●☆

320031　Ptaeroxylaceae Sander = Rutaceae Juss.（保留科名）●■

320032　Ptaeroxylon Eckl. et Zeyh.（1835）；喷嚏木属（喷嚏树属，嚏树属）；Sneezewood ●☆

320033　Ptaeroxylon obliquum（Thunb.）Radlk.；喷嚏木；Sneezewood ●☆

320034　Ptaeroxylon utile Eckl. et Zeyh. = Ptaeroxylon obliquum

（Thunb.）Radlk. ●☆

320035　Ptarmica Mill.（1754）；长舌蓍属（假蓍属）■●☆

320036　Ptarmica Mill. = Achillea L. ■

320037　Ptarmica Neck. = Achillea L. ■

320038　Ptarmica acuminata Ledeb. = Achillea acuminata（Ledeb.）Sch. Bip. ■

320039　Ptarmica acuminata Ledeb. = Achillea ptarmica L. var. acuminata（Ledeb.）Heimerl ■

320040　Ptarmica alpina（L.）DC. = Achillea alpina L. ■

320041　Ptarmica alpina DC. = Achillea alpina L. ■

320042　Ptarmica alpina DC. = Achillea ledebouri Heimerl ■

320043　Ptarmica camtschatica（Heimerl）Kom. = Achillea alpina L. subsp. camtschatica（Heimerl）Kitam. ■☆

320044　Ptarmica clavennae DC. = Achillea clavennae L. ■☆

320045　Ptarmica grandiflora DC. ；大花长舌蓍■☆

320046　Ptarmica impatiens（L.）DC. = Achillea impatiens L. ■

320047　Ptarmica impatiens DC. = Achillea impatiens L. ■

320048　Ptarmica japonica（Heimerl）Vorosch. = Achillea alpina L. subsp. japonica（Heimerl）Kitam. ■☆

320049　Ptarmica macrophylla DC. ；大叶长舌蓍■☆

320050　Ptarmica mongolica（Fisch. ex Spreng.）DC. ；蒙古长舌蓍■☆

320051　Ptarmica mongolica（Fisch. ex Spreng.）DC. = Achillea alpina L. ■

320052　Ptarmica mongolica DC. = Ptarmica mongolica（Fisch. ex Spreng.）DC. ■☆

320053　Ptarmica moschata DC. = Achillea moschata Jacq. ■☆

320054　Ptarmica multiflora（Hook.）Tzvelev；多花长舌蓍■☆

320055　Ptarmica nana DC. = Achillea nana L. ■☆

320056　Ptarmica ptarmicoides（Maxim.）Vorosch. = Achillea alpina L. var. discoidea（Regel）Kitam. ■☆

320057　Ptarmica ptarmicoides（Maxim.）Vorosch. = Achillea ptarmicoides Maxim. ■

320058　Ptarmica salicifolia（Besser）Myrzakulov；柳叶长舌菊■☆

320059　Ptarmica sibirica（Ledeb.）Ledeb. = Achillea alpina L. var. longiligulata H. Hara ■

320060　Ptarmica sibirica（Ledeb.）Ledeb. = Achillea alpina L. ■

320061　Ptarmica sibirica Ledeb. = Achillea alpina L. ■

320062　Ptarmica speciosa DC. ；美丽长舌蓍■☆

320063　Ptarmica tenuifolia Schur；细叶长舌蓍■☆

320064　Ptelandra Triana = Macrolenes Naudin ex Miq. ●☆

320065　Ptelea L.（1753）；榆橘属（翅果椒属）；Hop Tree, Hoptree, Hop-tree, Wafer Ash ●

320066　Ptelea baldwinii Torr. et A. Gray；细毛榆橘●☆

320067　Ptelea baldwinii Torr. et A. Gray = Ptelea trifoliata L. ●

320068　Ptelea lutescens Greene；黄榆橘；Yellow Hoptree ●☆

320069　Ptelea microcarpa Small = Ptelea trifoliata L. ●

320070　Ptelea serrata Small = Ptelea trifoliata L. ●

320071　Ptelea trifoliata L. ；榆橘（三叶椒）；Common Hop Tree, Common Hoptree, Common Hop-tree, Hop Tree, Hoptree, Shrubby Trefoil, Stinking Ash, Stinkingash, Stinking-ash, Swamp Dogwood, Three-leaved Hop Tree, Water Ash, Water-ash, Wingseed ●

320072　Ptelea trifoliata L. 'Aurea'；金叶榆橘●☆

320073　Ptelea trifoliata L. var. deamiana Nieuwl. = Ptelea trifoliata L. ●

320074　Ptelea viscosa L. = Dodonaea angustifolia L. f. ●

320075　Ptelea viscosa L. = Dodonaea viscosa（L.）Jacq. ●

320076　Pteleaceae Kunth = Rutaceae Juss.（保留科名）●■

320077　Pteleaceae Kunth；榆橘科●

320078　Pteleocarpa Oilv.（1873）；翅果紫草属●☆

320079　Pteleocarpa lamponga（Miq.）Heyne；翅果紫草●☆

320080　Pteleodendron K. Schum. = Pleodendron Tiegh. ●☆

320081　Pteleopsis Engl.（1895）；假榆橘属●☆

320082　Pteleopsis albidiflora De Wild. = Pteleopsis hylodendron Mildbr. ●☆

320083　Pteleopsis anisoptera（Welw. ex M. A. Lawson）Engl. et Diels；不等翅假榆橘●☆

320084　Pteleopsis apetala Vollesen；无瓣假榆橘●☆

320085　Pteleopsis barbosae Exell；巴尔博萨假榆橘●☆

320086　Pteleopsis bequaertii De Wild. = Pteleopsis hylodendron Mildbr. ●☆

320087　Pteleopsis diptera（Welw.）Engl. et Diels；双翅假榆橘●☆

320088　Pteleopsis hylodendron Mildbr. ；林生假榆橘●☆

320089　Pteleopsis kerstingii Gilg ex Engl. ；克斯廷假榆橘●☆

320090　Pteleopsis ledermannii Engl. et Gilg；莱德曼假榆橘●☆

320091　Pteleopsis myrtifolia（M. A. Lawson）Engl. et Diels；番樱桃叶假榆橘●☆

320092　Pteleopsis myrtifolia Engl. et Diels = Pteleopsis myrtifolia（M. A. Lawson）Engl. et Diels ●☆

320093　Pteleopsis obovata Hutch. = Pteleopsis myrtifolia（M. A. Lawson）Engl. et Diels ●☆

320094　Pteleopsis pteleopsoides（Exell）Vollesen；普通假榆橘●☆

320095　Pteleopsis ritschardii De Wild. = Pteleopsis anisoptera（Welw. ex M. A. Lawson）Engl. et Diels ●☆

320096　Pteleopsis stenocarpa Engl. et Diels = Pteleopsis myrtifolia（M. A. Lawson）Engl. et Diels ●☆

320097　Pteleopsis suberosa Engl. et Diels；木栓假榆橘●☆

320098　Pteleopsis tetraptera Wickens；四翅假榆橘●☆

320099　Pteleopsis variifolia Engl. = Pteleopsis myrtifolia（M. A. Lawson）Engl. et Diels ●☆

320100　Ptelidium Thouars(1804)；榆橘卫矛属●☆

320101　Ptelidium integrifolium J. St. -Hil. ；全缘榆橘卫矛●☆

320102　Ptelidium ovatum Poir. ；倒卵形榆橘卫矛●☆

320103　Pteracanthus（Nees）Bremek.（1944）；翅柄马蓝属（对节叶属，马蓝属）●■

320104　Pteracanthus（Nees）Bremek. = Strobilanthes Blume ●■

320105　Pteracanthus aenobarbus（W. W. Sm.）C. Y. Wu et C. C. Hu；铜毛马蓝（刚毛紫云菜，十三年花，铜毛紫云菜，铜色紫云菜）■

320106　Pteracanthus alatiramosus（H. S. Lo et D. Fang）C. Y. Wu et C. C. Hu；翅枝马蓝■

320107　Pteracanthus alatus（Nees）Bremek. = Pteracanthus alatus（Wall. ex Nees）Bremek. ●■

320108　Pteracanthus alatus（Wall. ex Nees）Bremek. ；翅柄马蓝（翅柄马兰，对节叶，三花马蓝）；Triflorous Conehead ●■

320109　Pteracanthus alatus（Wall. ex Nees）Bremek. = Strobilanthes wallichii Nees ●■

320110　Pteracanthus alatus（Wall.）Bremek. = Pteracanthus alatus（Wall. ex Nees）Bremek. ●■

320111　Pteracanthus botryanthus（D. Fang et H. S. Lo）C. Y. Hu；串花马蓝（多穗马蓝）●

320112　Pteracanthus calycinus（Nees）Bremek. ；曲序马蓝；Bowedinflorescense Conehead ●

320113　Pteracanthus clavicalatus（C. B. Clarke ex W. W. Sm.）C. Y. Wu；棒果马蓝；Styckyfruit Conehead ■

320114　Pteracanthus claviculatus（C. B. Clarke ex W. W. Sm.）H. P. Tsui = Pteracanthus clavicalatus（C. B. Clarke ex W. W. Sm.）C. Y. Wu ■

320115　Pteracanthus cognatus（Benoist）C. Y. Wu et C. C. Hu；奇瓣马

蓝■

320116 Pteracanthus congestus (Terao) C. Y. Wu et C. C. Hu;密序马
蓝●

320117 Pteracanthus cyphanthus (Diels) C. Y. Wu et C. C. Hu;弯花马
蓝(弯花紫云菜);Bentflower Conehead ●

320118 Pteracanthus dryadum (C. B. Clarke ex Benoist) C. Y. Wu et
C. C. Hu;林马蓝(林紫云菜);Forest Conehead ■

320119 Pteracanthus duclouxii (C. B. Clarke ex Benoist) C. Y. Wu et
C. C. Hu;高原马蓝(高原紫云菜);Highland Conehead ■

320120 Pteracanthus extensus (Nees) Bremek.;展翅马蓝●■

320121 Pteracanthus flexus (Benoist) C. Y. Wu et C. C. Hu;城口马蓝■

320122 Pteracanthus forrestii (Diels) C. Y. Wu;腺毛马蓝(味牛膝);
Forrest Conehead ●■

320123 Pteracanthus forrestii (Diels) C. Y. Wu = Strobilanthes forrestii
Diels ●■

320124 Pteracanthus gongshanensis H. P. Tsui;贡山马蓝■

320125 Pteracanthus grandissimus (H. P. Tsui) C. Y. Wu et C. C. Hu;
大叶马蓝(大叶金足草)■

320126 Pteracanthus guangxiensis (S. Z. Huang) C. Y. Wu et C. C. Hu;
广西马蓝;Guangxi Conehead,Kwangsi Conehead ■

320127 Pteracanthus hygrophiloides (C. B. Clarke ex W. W. Sm.) H.
W. Li;假水蓑衣●

320128 Pteracanthus inflatus (T. Anderson) Bremek.;锡金马蓝■

320129 Pteracanthus lamius (C. B. Clarke ex W. W. Sm.) C. Y. Wu et
C. C. Hu;野芝麻马蓝(灯笼草,灯笼花)■

320130 Pteracanthus leucotrichus (Benoist) C. Y. Wu et C. C. Hu;白毛
马蓝(白毛紫云菜);Whitehair Conehead ■

320131 Pteracanthus mekongensis (W. W. Sm.) C. Y. Wu et C. C. Hu;
澜沧马蓝(澜沧紫云菜);Lancang Conehead ●

320132 Pteracanthus nemorosus (Benoist) C. Y. Wu et C. C. Hu;森林
马蓝(牛膝马蓝);Nemoros Conehead ■

320133 Pteracanthus oresbius (W. W. Sm.) C. Y. Wu et C. C. Hu;山
马蓝(山紫云菜);Montane Conehead ■

320134 Pteracanthus panduratus (Hand.-Mazz.) C. Y. Wu et C. C.
Hu;琴叶马蓝(木里叉花草,琴叶紫云菜);Pandurate Conehead ■

320135 Pteracanthus pinnatifidus (C. Z. Zheng) C. Y. Wu et C. C. Hu;
羽裂马蓝;Pinnatifid Conehead ■

320136 Pteracanthus rotundifolius (D. Don) Bremek.;圆叶马蓝(圆叶
紫云菜);Roundleaf Conehead ■

320137 Pteracanthus tibeticus (J. R. I. Wood) C. Y. Wu et C. C. Hu;西
藏马蓝●

320138 Pteracanthus urophyllus (Nees) Bremek.;尾叶马蓝●

320139 Pteracanthus urticifolius (Kuntze) Bremek.;荨麻叶马蓝●

320140 Pteracanthus urticifolius (Kuntze) Bremek. = Strobilanthes
urticifolia Wall. ex Kuntze ●

320141 Pteracanthus versicolor (Diels) H. W. Li;变色马蓝●■

320142 Pteracanthus versicolor (Diels) Hand.-Mazz. ex C. Y. Wu =
Pteracanthus versicolor (Diels) H. W. Li ●■

320143 Pteracanthus yunnanensis (Diels) C. Y. Wu et C. C. Hu;云南
马蓝(滇紫云英,云南马兰);Yunnan Conehead ●

320144 Pterachaenia (Benth. et Hook. f.) Lipsch. = Pterachenia
(Benth.) Lipsch. ■☆

320145 Pterachaenia (Benth.) Lipsch. = Pterachenia (Benth.)
Lipsch. ■☆

320146 Pterachenia (Benth.) Lipsch. (1939);三翅苣属■☆

320147 Pterachenia stewartii (Hook. f.) R. R. Stewart;三翅苣■☆

320148 Pterachne Schrad. ex Nees = Ascolepis Nees ex Steud. (保留属

名)■☆

320149 Pteralyxia K. Schum. (1895);大翅夹竹桃属●☆

320150 Pteralyxia macrocarpa K. Schum.;大翅夹竹桃●☆

320151 Pterandra A. Juss. (1833);翼雄花属●☆

320152 Pterandra coerulescens Jack;天蓝翼雄花●☆

320153 Pteranthera Blume = Vatica L. ●

320154 Pteranthera sinensis Blume = Vatica mangachapoi Blanco ●◇

320155 Pteranthus Forssk. (1775);翼萼裸果草属(翅萼指甲草属,翼
萼裸果木属,翼花裸果木属)■☆

320156 Pteranthus dichotomus Forssk.;翼萼裸果木(翼花裸果木)●☆

320157 Pteranthus dichotomus Forssk. var. trigynus (Caball.) Maire =
Pteranthus dichotomus Forssk. ●☆

320158 Pteranthus echinatus Desf. = Pteranthus dichotomus Forssk. ●☆

320159 Pteranthus trigynus Caball. = Pteranthus dichotomus Forssk. ●☆

320160 Pteraton Raf. = Bupleurum L. ●■

320161 Pterichis Lindl. (1840);翼兰属■☆

320162 Pterichis Lindl. = Acraea Lindl. ■☆

320163 Pterichis acuminata Schltr.;尖翼兰■☆

320164 Pterichis boliviana Schltr.;玻利维亚翼兰■☆

320165 Pterichis macroptera Schltr.;大翅翼兰■☆

320166 Pterichis multiflora (Lindl.) Schltr.;多花翼兰■☆

320167 Pterichis parvifolia (Lindl.) Schltr.;小叶翼兰■☆

320168 Pterichis pauciflora Schltr.;少花翼兰■☆

320169 Pteridocalyx Wernham(1911);翼萼茜属☆

320170 Pteridocalyx appunii Wernham;翼萼茜☆

320171 Pteridocalyx minor Wernham;小翼萼茜☆

320172 Pteridophyilaceae Reveal et Hoogland = Pteridophyllaceae
(Murbeck) Sugiura ex Nakai ■☆

320173 Pteridophyllaceae (Murbeck) Sugiura ex Nakai(1943);蕨叶草
科(蕨罂粟科)■☆

320174 Pteridophyllaceae Nakai ex Reveal et Hoogland =
Pteridophyllaceae (Murbeck) Sugiura ex Nakai ■☆

320175 Pteridophyllaceae Sugiura ex Nakai = Papaveraceae Juss. (保留
科名)●■

320176 Pteridophyllaceae Sugiura ex Nakai = Pteridophyllaceae
(Murbeck) Sugiura ex Nakai ■☆

320177 Pteridophyllaceae Sugiura ex Nakai = Pterostemonaceae Small
(保留科名)●☆

320178 Pteridophyllum Siebold et Zucc. (1843);蕨叶草属(蕨罂粟
属)■☆

320179 Pteridophyllum Thwaites = Filicium Thwaites ex Benth. ●☆

320180 Pteridophyllum decipiens Thwaites = Pteridophyllum racemosum
Siebold et Zucc. ■☆

320181 Pteridophyllum racemosum Siebold et Zucc.;蕨叶草■☆

320182 Pterigeron (DC.) Benth. = Allopterigeron Dunlop ■☆

320183 Pterigeron (DC.) Benth. = Oliganthemum F. Muell. ■☆

320184 Pterigeron (DC.) Benth. = Streptoglossa Steetz ex F. Muell. ■●☆

320185 Pterigeron A. Gray = Oliganthemum F. Muell. ■☆

320186 Pterigium Corrêa = Dipterocarpus C. F. Gaertn. ●

320187 Pterigostachyum Nees ex Steud. = Dimeria R. Br. ■

320188 Pterigostachyum Nees ex Steud. = Pterygostachyum Nees ex
Steud. ■

320189 Pterilema Reinw. = Engelhardia Lesch. ex Blume ●

320190 Pterilema aceriflorum Reinw. = Engelhardtia aceriflora
(Reinw.) Blume ●

320191 Pteriphis Raf. = Aristolochia L. ■●

320192 Pterisanthaceae J. Agardh = Vitaceae Juss. (保留科名)●■

320193 Pterisanthes Blume(1825);翼花藤属●☆

320194 Pterisanthes cissioides Blume;翼花藤●☆

320195 Pterisanthes glabra Ridl.;光翼花藤●☆

320196 Pterisanthes heterantha M. A. Lawson;异花翼花藤●☆

320197 Pterium Desv. = Lamarckia Moench(保留属名)■☆

320198 Pternandra Jack(1822);翼药花属;Pternandra ●

320199 Pternandra caerulescens Jack;翼药花;Coerulescent Pternandra ●

320200 Pternandra cordata Baill.;心形翼药花●☆

320201 Pternix Hill = Carduus L. ■

320202 Pternix Raf. = Silybum Vaill.(保留属名)■

320203 Pternopetalum Franch.(1885);囊瓣芹属(肿瓣芹属);Cystopetal ■

320204 Pternopetalum affine(H. Wolff)C. Y. Wu = Pternopetalum delicatulum(H. Wolff)Hand. -Mazz. ■

320205 Pternopetalum affine(H. Wolff)M. Hiroe = Pternopetalum delicatulum(H. Wolff)Hand. -Mazz. ■

320206 Pternopetalum botrychioides(Dunn)Hand. -Mazz.;散血芹(散血草,水芹花);Graperfern-like Cystopetal ■

320207 Pternopetalum botrychioides(Dunn)Hand. -Mazz. var. latipinnulatum R. H. Shan;宽叶散血芹;Broadleaf Graperfern-like Cystopetal ■

320208 Pternopetalum brevium(R. H. Shan et F. T. Pu)K. T. Fu = Pternopetalum longicaule R. H. Shan var. humile R. H. Shan et F. T. Pu ■

320209 Pternopetalum caespitosum R. H. Shan;丛枝囊瓣芹;Caespitose Cystopetal ■

320210 Pternopetalum cardiocarpum(Franch.)Hand. -Mazz.;心果囊瓣芹;Hearfruit Cystopetal ■

320211 Pternopetalum cartilagineum C. Y. Wu;骨缘囊瓣芹;Boneedge Cystopetal ■

320212 Pternopetalum confusum C. Norman = Pternopetalum leptophyllum(Dunn)Hand. -Mazz. ■

320213 Pternopetalum cuneifolium(H. Wolff)Hand. -Mazz.;楔叶囊瓣芹;Wedgeleaf Cystopetal ■

320214 Pternopetalum cuneifolium(H. Wolff)Hand. -Mazz. = Pternopetalum molle(Franch.)Hand. -Mazz. ■

320215 Pternopetalum davidii Franch.;囊瓣芹(水芹菜);David Cystopetal ■

320216 Pternopetalum decipiens(C. Norman)M. Hiroe = Pternopetalum trichomanifolium(Franch.)Hand. -Mazz. ■

320217 Pternopetalum delavayi(Franch.)Hand. -Mazz.;澜沧囊瓣芹(洱源囊瓣芹);Delavay Cystopetal ■

320218 Pternopetalum delicatulum(H. Wolff)Hand. -Mazz.;嫩弱襄瓣芹;Tender Cystopetal ■

320219 Pternopetalum filicinum(Franch.)Hand. -Mazz.;羊齿囊瓣芹;Fernlike Cystopetal ■

320220 Pternopetalum gracillimum(H. Wolff)Hand. -Mazz.;纤细襄瓣芹■

320221 Pternopetalum heterophyllum Hand. -Mazz.;异叶囊瓣芹;Diversefolius Cystopetal ■

320222 Pternopetalum kiangsiense(H. Wolff)Hand. -Mazz.;江西囊瓣芹;Jiangxi Cystopetal ■

320223 Pternopetalum kiangsiense(H. Wolff)Hand. -Mazz. = Pternopetalum trichomanifolium(Franch.)Hand. -Mazz. ■

320224 Pternopetalum lamellosociliare K. T. Fu;片毛囊瓣芹;Lamellosehair Cystopetal ■

320225 Pternopetalum lamellosociliare K. T. Fu = Pternopetalum gracillimum(H. Wolff)Hand. -Mazz. ■

320226 Pternopetalum leptophyllum(Dunn)Hand. -Mazz.;薄叶囊瓣芹(水中芹);Thinleaf Cystopetal ■

320227 Pternopetalum longicaule R. H. Shan;长茎囊瓣芹;Longstem Cystopetal ■

320228 Pternopetalum longicaule R. H. Shan var. brevium R. H. Shan et F. T. Pu = Pternopetalum longicaule R. H. Shan var. humile R. H. Shan et F. T. Pu ■

320229 Pternopetalum longicaule R. H. Shan var. humile R. H. Shan et F. T. Pu;短茎囊瓣芹(矮茎囊瓣芹,矮型长茎囊瓣芹);Dwarf Longstem Cystopetal ■

320230 Pternopetalum mairei(Diels)Hand. -Mazz.;东川囊瓣芹;Maire Cystopetal ■

320231 Pternopetalum molle(Franch.)Hand. -Mazz.;洱源囊瓣芹(柔软囊瓣芹);Eryuan Cystopetal ■

320232 Pternopetalum molle(Franch.)Hand. -Mazz. var. crenulatum R. H. Shan et F. T. Pu;圆齿囊瓣芹(圆齿柔软囊瓣芹);Roundtooth Eryuan Cystopetal ■

320233 Pternopetalum molle(Franch.)Hand. -Mazz. var. crenulatum R. H. Shan et F. T. Pu = Pternopetalum molle(Franch.)Hand. -Mazz. ■

320234 Pternopetalum molle(Franch.)Hand. -Mazz. var. dissectum R. H. Shan et F. T. Pu;裂叶囊瓣芹(裂叶柔软囊瓣芹);Dissecte Eryuan Cystopetal ■

320235 Pternopetalum nudicaule(H. Boissieu)Hand. -Mazz.;裸茎囊瓣芹(药芹菜);Nakestem Cystopetal ■

320236 Pternopetalum nudicaule(H. Boissieu)Hand. -Mazz. var. esetosum Hand. -Mazz. = Pternopetalum nudicaule(H. Boissieu)Hand. -Mazz. ■

320237 Pternopetalum nudicaule(H. Boissieu)Hand. -Mazz. var. esetosum Hand. -Mazz.;光滑囊瓣芹;Smooth Nakestem Cystopetal ■

320238 Pternopetalum rosthornii(Diels)Hand. -Mazz.;川鄂囊瓣芹;Rosthorn Cystopetal ■

320239 Pternopetalum sinense(Franch.)Hand. -Mazz.;华囊瓣芹;Chinese Cystopetal ■

320240 Pternopetalum subalpinum Hand. -Mazz.;高山囊瓣芹;Alpine Cystopetal ■

320241 Pternopetalum tanakae(Franch. et Sav.)Hand. -Mazz.;东亚囊瓣芹;Tanaka Cystopetal ■

320242 Pternopetalum tanakae(Franch. et Sav.)Hand. -Mazz. f. conforme ? = Pternopetalum tanakae(Franch. et Sav.)Hand. -Mazz. ■

320243 Pternopetalum tanakae(Franch. et Sav.)Hand. -Mazz. f. lineare ? = Pternopetalum tanakae(Franch. et Sav.)Hand. -Mazz. ■

320244 Pternopetalum tanakae(Franch. et Sav.)Hand. -Mazz. var. fulcrantum Y. H. Zhang;假苞囊瓣芹■

320245 Pternopetalum trichomanifolium(Franch.)Hand. -Mazz.;膜蕨囊瓣芹(细沙毛,细叶囊瓣芹);Trichomanefolious Cystopetal ■

320246 Pternopetalum trifoliatum R. H. Shan et F. T. Pu;鹧鸪山囊瓣芹;Zhegushan Cystopetal ■

320247 Pternopetalum viride(C. Norman)Hand. -Mazz. = Pternopetalum leptophyllum(Dunn)Hand. -Mazz. ■

320248 Pternopetalum vulgare(Dunn)Hand. -Mazz.;五匹青(囊瓣芹,五匹青囊瓣芹,肿瓣芹,踵瓣芹,紫金沙,紫金砂);Common Cystopetal ■

320249 Pternopetalum vulgare(Dunn)Hand. -Mazz. var. acuminatum C. Y. Wu;尖叶五匹青(尖叶囊瓣芹,刷把草);Sharpleaf Cystopetal ■

320250　Pternopetalum vulgare（Dunn）Hand.-Mazz. var. foliosum R. H. Shan et F. T. Pu；钝叶五匹青(多叶五匹青)；Obtuseleaf Cystopetal ■

320251　Pternopetalum vulgare（Dunn）Hand.-Mazz. var. strigosum R. H. Shan et F. T. Pu；毛叶五匹青；Hairleaf Cystopetal ■

320252　Pternopetalum vulgare var. foliosum R. H. Shan et F. T. Pu = Pternopetalum vulgare（Dunn）Hand.-Mazz. ■

320253　Pternopetalum wangianum Hand.-Mazz.；天全囊瓣芹；Wang Cystopetal ■

320254　Pternopetalum wangianum Hand.-Mazz. = Pternopetalum gracillimum（H. Wolff）Hand.-Mazz. ■

320255　Pternopetalum wolffianum（Fedde）Hand.-Mazz.；滇西囊瓣芹(腾冲囊瓣芹)；Wolff Cystopetal ■

320256　Pternopetalum yiliangense R. H. Shan et F. T. Pu；宜良囊瓣芹(彝良囊瓣芹)；Yiliang Cystopetal ■

320257　Pterobesleria C. V. Morton = Besleria L. ●■☆

320258　Pterocactus K. Schum.（1897）；翅子掌属(真翅仙人掌属)；Wing Cactus ●☆

320259　Pterocactus hickenii Britton et Rose；怒黄龙■☆

320260　Pterocactus tuberosus（Pfeiff.）Britton et Rose；黑龙■☆

320261　Pterocalymma Bernh. et Hook. f. = Pterocalymma Turcz. ●

320262　Pterocalymma Turcz. = Lagerstroemia L. ●

320263　Pterocalyx Schrenk = Alexandra Bunge ■☆

320264　Pterocariaceae Nakai = Juglandaceae DC. ex Perleb(保留科名)●

320265　Pterocarpos St.-Lag. = Pterocarpus Jacq.（保留属名）●

320266　Pterocarpus Bergius = Dalbergia L. f.（保留属名）●

320267　Pterocarpus Bergius = Ecastaphyllum P. Browne(废弃属名)●

320268　Pterocarpus Burm. = Brya P. Browne ●☆

320269　Pterocarpus Jacq.（1763）（保留属名）；紫檀属；Amboyna Wood，Blood Wood，Padauk，Padouk，Sandalwood，Vermilion Wood ●

320270　Pterocarpus Kuntze = Derris Lour.（保留属名）●

320271　Pterocarpus L.（废弃属名）= Derris Lour.（保留属名）●

320272　Pterocarpus L.（废弃属名）= Pterocarpus Jacq.（保留属名）●

320273　Pterocarpus P. J. Bergius = Dalbergia L. f.（保留属名）●

320274　Pterocarpus P. J. Bergius = Ecastaphyllum P. Browne(废弃属名)●

320275　Pterocarpus abyssinicus Hochst. ex A. Rich. = Pterocarpus lucens Lepr. ex Guillaumin et Perr. ●☆

320276　Pterocarpus adansonii DC. = Pterocarpus erinaceus Poir. ●☆

320277　Pterocarpus amazonicus Huber = Pterocarpus santalinoides L'Hér. ex DC. ●☆

320278　Pterocarpus angolensis DC.；安哥拉紫檀(北方红柝,得兰士瓦紫檀,东部红柝,非洲紫檀,红柝,灰橡木,加拿大红柝,罗得西亚紫檀,美国红柝,南非吉纳木,山地红柝,血檀)；American Red Oak，Barwood，Bloodwood，Bloodwood Tree，Brown African Padauk，Camwood，Canadian Red Oak，Eastern Red Oak，Gray Oak，Kiaat，Mountain Red Oak，Muninga，Northern Red Oak，Red Oak，Rhodesian Bloodwood，Transvaal Teak ●☆

320279　Pterocarpus antunesii（Taub.）Harms = Pterocarpus lucens Lepr. ex Guillaumin et Perr. subsp. antunesii（Taub.）Rojo ●☆

320280　Pterocarpus brenanii Barbosa et Torre；布雷南紫檀●☆

320281　Pterocarpus buchananii Schinz = Pterocarpus rotundifolius（Sond.）Druce ●☆

320282　Pterocarpus bussei Harms = Pterocarpus angolensis DC. ●☆

320283　Pterocarpus cabrae De Wild. = Pterocarpus tinctorius Welw. ●☆

320284　Pterocarpus cambodianus Pierre；越柬紫檀●☆

320285　Pterocarpus casteelsii De Wild. = Pterocarpus soyauxii Taub. ●☆

320286　Pterocarpus casteelsii De Wild. var. ealaensis Hauman =

Pterocarpus soyauxii Taub. ●☆

320287　Pterocarpus chrysothrix Taub. = Pterocarpus tinctorius Welw. ●☆

320288　Pterocarpus dalbergioides Roxb.；安达曼紫檀(黄檀状紫檀,印度紫檀)；Andaman Padauk，East Indian Mahogany，Padauk，Padouk ●☆

320289　Pterocarpus dekindtianus Harms = Pterocarpus angolensis DC. ●☆

320290　Pterocarpus delevoyi De Wild. = Pterocarpus tinctorius Welw. ●☆

320291　Pterocarpus draco L.；龙血紫檀；Dragonblood，Dragonblood Padauk，Dragon's-blood，Dragon's-blood Padauk，Grenadillo ●

320292　Pterocarpus elisabethvillensis De Wild. = Dalbergia nitidula Baker ●☆

320293　Pterocarpus erinaceus Poir.；西非紫檀(刺猬紫檀,刺紫檀,美果紫檀)；African Kino，African Rosewood，African Teak，Barwood，Gambian Kino，Senegal Rosewood，West African Kino，West African Rosewood ●☆

320294　Pterocarpus esculentus Schumach. et Thonn. = Pterocarpus santalinoides L'Hér. ex DC. ●☆

320295　Pterocarpus gilletii De Wild. = Pterocarpus officinalis Jacq. subsp. gilletii（De Wild.）Rojo ●☆

320296　Pterocarpus gilletii De Wild. var. angustifolius Hauman = Pterocarpus officinalis Jacq. subsp. gilletii（De Wild.）Rojo ●☆

320297　Pterocarpus grandiflorus Micheli = Craibia grandiflora（Micheli）Baker f. ●☆

320298　Pterocarpus grandis Cowan = Pterocarpus santalinoides L'Hér. ex DC. ●☆

320299　Pterocarpus hockii De Wild. = Pterocarpus tinctorius Welw. ●☆

320300　Pterocarpus holtzii Harms = Pterocarpus tinctorius Welw. ●☆

320301　Pterocarpus indicus Willd.；紫檀(檗木,赤檀,赤血树,红木,花榈,花榈木,黄柏木,榈木,蘗木,蔷薇木,青龙木,胜沉香,印度紫檀,羽叶檀,紫檀香,紫栴木,紫真檀)；Amboyna Wood，Andaman Redwood，Burma Coast Padauk，Burma Coast Padouk，Burmacoast Padauk，Burma-coast Padauk，Burmese Rosewood，Burmese Rose-wood，Manila Padauk，New Guinea Rosewood，Padauk，Papua New Guinea Rosewood，Rose Wood ●

320302　Pterocarpus kaessneri Harms = Pterocarpus tinctorius Welw. ●☆

320303　Pterocarpus lucens Lepr. ex Guillaumin et Perr.；光亮紫檀●☆

320304　Pterocarpus lucens Lepr. ex Guillaumin et Perr. subsp. antunesii（Taub.）Rojo；安图光亮紫檀●☆

320305　Pterocarpus lunatus L. f. = Machaerium lunatum（L. f.）Ducke ●☆

320306　Pterocarpus macrocarpus Kurz；缅甸大果紫檀(大果紫檀,花梨,花榈木,缅甸紫檀)；Burma Padauk，Maidu，Padauk，Padouk ●☆

320307　Pterocarpus marsupium Roxb.；吉纳紫檀(花榈木,吉纳檀,马拉巴紫檀,囊状紫檀)；Bastard Teak，Bustard Teak，Eucalyptus Indian Kino，Gamalu，Kino，Maidu，Malabar Kino，Narra Padauk，Vengai Padauk ●

320308　Pterocarpus martinii Dunkley = Pterocarpus rotundifolius（Sond.）Druce subsp. martinii（Dunkley）Lock ●☆

320309　Pterocarpus megalocarpus Harms = Pterocarpus tinctorius Welw. ●☆

320310　Pterocarpus melliferus Welw. ex Baker = Pterocarpus rotundifolius（Sond.）Druce ●☆

320311　Pterocarpus michelii Britton = Pterocarpus santalinoides L'Hér. ex DC. ●☆

320312　Pterocarpus mildbraedii Harms；米尔德紫檀●☆

320313　Pterocarpus mildbraedii Harms subsp. usambarensis（Verdc.）Polhill = Pterocarpus mildbraedii Harms ●☆

320314　Pterocarpus odoratus De Wild. = Pterocarpus tinctorius Welw. ●☆

320315　Pterocarpus officinalis Jacq.；药用紫檀●☆

320316　Pterocarpus officinalis Jacq. subsp. gilletii（De Wild.）Rojo；吉勒特紫檀●☆

320317　Pterocarpus osun Craib；娥孙紫檀●☆

320318　Pterocarpus pedatus Pierre；鸟足紫檀●☆

320319　Pterocarpus peltaria DC. = Wiborgia fusca Thunb. ■☆

320320　Pterocarpus podocarpus S. F. Blake；柄果紫檀●☆

320321　Pterocarpus polyanthus Harms = Pterocarpus rotundifolius（Sond.）Druce subsp. polyanthus（Harms）Mendonça et E. C. Sousa ●☆

320322　Pterocarpus rotundifolius（Sond.）Druce；圆叶紫檀●☆

320323　Pterocarpus rotundifolius（Sond.）Druce subsp. martinii（Dunkley）Lock；马丁紫檀●☆

320324　Pterocarpus rotundifolius（Sond.）Druce subsp. polyanthus（Harms）Mendonça et E. C. Sousa；多花圆叶紫檀●☆

320325　Pterocarpus rotundifolius（Sond.）Druce var. martinii（Dunkley）Mendonça et E. C. Sousa = Pterocarpus rotundifolius（Sond.）Druce subsp. martinii（Dunkley）Lock ●☆

320326　Pterocarpus santalinoides L'Hér. ex DC.；非洲紫檀●☆

320327　Pterocarpus santalinus L. f.；檀香紫檀（赤檀，红木，红檀，酸枝树，檀，香紫檀，正紫檀，紫檀，紫檀香，紫榆）；Red Sandalwood, Red Sandal-wood, Red Sanders, Red Saunders, Rubywood, Sandalwood Padauk, Sandal-wood Padauk, Sanders, Sanderswood, Saunders, Saunderswood ●☆

320328　Pterocarpus sericeus Benth. = Pterocarpus rotundifolius（Sond.）Druce ●☆

320329　Pterocarpus simplicifolius Baker = Pterocarpus lucens Lepr. ex Guillaumin et Perr. ●☆

320330　Pterocarpus soyauxii Taub.；苏姚紫檀（非洲珊瑚木，非洲紫檀，索约紫檀，紫檀）；African Coralwood, African Coral-wood, African Padauk, African Padouk, Barwood, Camwood, Comwood, Padauk, Padouk, Redwood, W. African Padauk ●☆

320331　Pterocarpus stevensonii Burtt Davy = Pterocarpus lucens Lepr. ex Guillaumin et Perr. subsp. antunesii（Taub.）Rojo ●☆

320332　Pterocarpus stolzii Harms = Pterocarpus tinctorius Welw. ●☆

320333　Pterocarpus tessmannii Harms；泰斯曼紫檀●☆

320334　Pterocarpus tinctorius Welw.；染色紫檀●☆

320335　Pterocarpus tinctorius Welw. var. chrysothrix（Taub.）Hauman = Pterocarpus tinctorius Welw. ●☆

320336　Pterocarpus tinctorius Welw. var. odoratus（De Wild.）Hauman = Pterocarpus tinctorius Welw. ●☆

320337　Pterocarpus ulei Harms；乌里紫檀●☆

320338　Pterocarpus usambarensis Verdc. = Pterocarpus mildbraedii Harms ●☆

320339　Pterocarpus velutinus De Wild. = Pterocarpus tinctorius Welw. ●☆

320340　Pterocarpus vidalianus Rolfe；菲律宾紫檀（八重山紫檀）；Philippine Padauk ●☆

320341　Pterocarpus violaceus Vogel；卵叶紫檀●☆

320342　Pterocarpus wallichii Wight et Arn. = Pterocarpus indicus Willd. ●

320343　Pterocarpus zenkeri Harms；岑克尔紫檀●☆

320344　Pterocarpus zimmermannii Harms = Pterocarpus tinctorius Welw. ●☆

320345　Pterocarpus zollingeri Miq. = Pterocarpus indicus Willd. ●

320346　Pterocarya Kunth（1824）；枫杨属；Chinese Wing-nut, Wing Nut, Wingnut ●

320347　Pterocarya Nutt. ex Moq. = Atriplex L. ■●

320348　Pterocarya caucasica C. A. Mey. = Pterocarya fraxinifolia（Lam.）Spach ●☆

320349　Pterocarya chinensis Lavaleé = Pterocarya stenoptera C. DC. ●

320350　Pterocarya delavayi Franch.；云南枫杨；Delavay Wingnut ●

320351　Pterocarya delavayi Franch. = Pterocarya macroptera Batalin var. delavayi（Franch.）W. E. Manning ●

320352　Pterocarya esquirolii H. Lév. = Pterocarya stenoptera C. DC. ●

320353　Pterocarya forrestii W. W. Sm. = Pterocarya delavayi Franch. ●

320354　Pterocarya forrestii W. W. Sm. = Pterocarya macroptera Batalin var. delavayi（Franch.）W. E. Manning ●

320355　Pterocarya forrestii W. W. Sm. ex Hand. -Mazz. = Pterocarya delavayi Franch. ●

320356　Pterocarya fraxinifolia（Lam.）Spach；白蜡叶枫杨（梣叶枫杨，高加索枫杨）；Caucasian Wing Nut, Caucasian Wingnut, Caucasus Wingnut, Wingnut ●☆

320357　Pterocarya fraxinifolia（Poir.）Spach = Pterocarya fraxinifolia（Lam.）Spach ●☆

320358　Pterocarya hupehensis V. Naray.；湖北枫杨（枫杨，枫杨柳，麻柳，麻柳树，山柳树，山麻柳）；Hubei Wingnut, Hupeh Wingnut ●

320359　Pterocarya insignis Rehder et E. H. Wilson；华西枫杨（棘枫杨，山麻柳，瓦山水胡桃，野椿）；Insignis Wingnut ●

320360　Pterocarya insignis Rehder et E. H. Wilson = Pterocarya macroptera Batalin var. insignis（Rehder et E. H. Wilson）W. E. Manning ●

320361　Pterocarya japonica Dippel = Pterocarya stenoptera C. DC. ●

320362　Pterocarya japonica Lavaleé = Pterocarya stenoptera C. DC. ●

320363　Pterocarya laevigata Lavaleé = Pterocarya stenoptera C. DC. ●

320364　Pterocarya macroptera Batalin；甘肃枫杨；Largewing Wingnut, Large-wing Wingnut ●

320365　Pterocarya macroptera Batalin var. delavayi（Franch.）W. E. Manning = Pterocarya delavayi Franch. ●

320366　Pterocarya macroptera Batalin var. insignis（Rehder et E. H. Wilson）W. E. Manning = Pterocarya insignis Rehder et E. H. Wilson ●

320367　Pterocarya micropaliurus P. C. Tsoong = Cyclocarya paliurus（Batalin）Iljinsk. ●

320368　Pterocarya nanjiangensis T. P. Yi；南江枫杨；Nanjiang Wingnut ●

320369　Pterocarya paliurus Batalin = Cyclocarya paliurus（Batalin）Iljinsk. ●

320370　Pterocarya pterocarpa（Michx.）Kunth = Pterocarya fraxinifolia（Lam.）Spach ●☆

320371　Pterocarya rehderiana C. K. Schneid.；雷氏枫杨（阿诺德枫杨）；Rehder Wing Nut, Rehder's Wingnut ●☆

320372　Pterocarya rhoifolia Siebold et Zucc.；水胡桃（华西枫杨，日本枫杨）；Japanese Wing Nut, Japanese Wingnut, Water Wingnut ●

320373　Pterocarya rhoifolia Siebold et Zucc. = Pterocarya macroptera Batalin var. insignis（Rehder et E. H. Wilson）W. E. Manning ●

320374　Pterocarya sorbifolia Siebold et Zucc. = Pterocarya fraxinifolia（Lam.）Spach ●☆

320375　Pterocarya sorbifolia Siebold et Zucc. = Pterocarya rhoifolia Siebold et Zucc. ●

320376　Pterocarya sprengeri Pamp. = Pterocarya hupehensis V. Naray. ●

320377　Pterocarya stenoptera C. DC.；枫杨（臭树柳，大叶柳，枫柳，枸树，鬼柳，柜柳，榉，榉柳，魁柳，柳丝子，麻柳，麻柳树，胖柳，平柳，平阳柳，杞柳，嵌宝枫，嵌实枫，嵌实树，水槐树，水柳树，水麻柳，蜈蚣柳，溪钩树，溪榉，溪口树，溪柳，溪麻柳，溪杨，燕子柳，燕子树，元宝枫，元宝树）；China Wingnut, Chinese Ash, Chinese Wing Nut, Chinese Wingnut ●

320378 Pterocarya stenoptera C. DC. var. brevialata Pamp. = Pterocarya stenoptera C. DC. ●

320379 Pterocarya stenoptera C. DC. var. kouitchensis Franch. = Pterocarya stenoptera C. DC. ●

320380 Pterocarya stenoptera C. DC. var. sinensis Graebn. = Pterocarya stenoptera C. DC. ●

320381 Pterocarya stenoptera C. DC. var. tonkinensis Franch. = Pterocarya tonkinensis (Franch.) Dode ●

320382 Pterocarya stenoptera C. DC. var. typica Franch. = Pterocarya stenoptera C. DC. ●

320383 Pterocarya stenoptera C. DC. var. zhijiangensis Z. E. Chao et C. J. Zheng;枝江枫杨;Zhijiang Wingnut ●

320384 Pterocarya tonkinensis (Franch.) Dode;越南枫杨(滇桂枫杨, 东京枫杨,麻柳,妹宗);Tonkin Wingnut,Tonkin Wing-nut,Vietnam Wingnut ●

320385 Pterocaryaceae Nakai = Juglandaceae DC. ex Perleb(保留科名)●

320386 Pterocassia Britton et Rose = Cassia L.(保留属名)●■

320387 Pterocassia Britton et Rose = Senna Mill. ●

320388 Pterocaulon Elliott = Neojeffreya Cabrera ■☆

320389 Pterocaulon Elliott(1823);翼茎草属(翅茎菊属);Pterocaulon ■

320390 Pterocaulon bojeri Baker = Neojeffreya decurrens (L.) Cabrera ■☆

320391 Pterocaulon cylindrostachyum C. B. Clarke = Pterocaulon redolens (G. Forst.) Fern.-Vill. ■

320392 Pterocaulon decurrens (L.) S. Moore = Neojeffreya decurrens (L.) Cabrera ■☆

320393 Pterocaulon pycnostachyum (Michx.) Elliott;虎尾翼茎草;Coastal Blackroot,Fox-tail Blackroot ■☆

320394 Pterocaulon redolens (G. Forst.) Fern.-Vill. = Pterocaulon redolens (Willd.) Fern.-Vill. ■

320395 Pterocaulon redolens (Willd.) Fern.-Vill.;翼茎草(黏叶子草);Fragrant Pterocaulon ■

320396 Pterocaulon serrulatum Guillaumin;锯齿翅茎菊■☆

320397 Pterocaulon sphacelatum (Labill.) F. Muell.;枯翅茎菊■☆

320398 Pterocaulon undulatum C. Mohr;波状翼茎草■

320399 Pterocaulon undulatum C. Mohr = Pterocaulon pycnostachyum (Michx.) Elliott ■☆

320400 Pterocaulon virgatum (L.) DC.;棒状翼茎草;Wand Blackroot ■☆

320401 Pterocelastrus Meisn.(1837);翅蛇藤属●☆

320402 Pterocelastrus dregeanus (C. Presl) Sond.;德雷翅蛇藤●☆

320403 Pterocelastrus echinatus N. E. Br.;具刺翅蛇藤●☆

320404 Pterocelastrus galpinii Loes.;盖尔翅蛇藤●☆

320405 Pterocelastrus litoralis Walp. = Pterocelastrus tricuspidatus (Lam.) Walp. ●☆

320406 Pterocelastrus rehmannii Davison = Pterocelastrus echinatus N. E. Br. ●☆

320407 Pterocelastrus rostratus (Thunb.) Walp.;喙状翅蛇藤●☆

320408 Pterocelastrus stenopterus Walp. = Pterocelastrus tricuspidatus (Lam.) Walp. ●☆

320409 Pterocelastrus tetrapterus Walp. = Pterocelastrus tricuspidatus (Lam.) Walp. ●☆

320410 Pterocelastrus tricuspidatus (Lam.) Walp.;骤尖翅蛇藤●☆

320411 Pteroceltis Maxim.(1873);青檀属(青朴属,翼朴属);Whinghack Berry,Whinghackberry,Whing-hackberry,Wing Celtis,Wing-cactus,Wingceltis,Winged Hackberry ●★

320412 Pteroceltis tartarinowii Maxim.;青檀(毛青檀,毛翼朴,青壳椰树,檀,檀树,摇钱树,翼朴);Pubescent Whing-hackberry,Tara

Wingceltis,Whinghack Berry,Whinghackberry,Whing-hackberry,Wingceltis ●

320413 Pteroceltis tatarinowii Maxim. var. pubescens Hand.-Mazz. = Pteroceltis tartarinowii Maxim. ●

320414 Pterocephalidium G. López = Pterocephalus Vaill. ex Adans. ●■

320415 Pterocephalidium G. López(1987);小头翅续断属■☆

320416 Pterocephalidium diandrum (Lag.) G. López;小头翅续断■☆

320417 Pterocephalodes V. Mayer et Ehrend.(2000);拟翼首花属●■

320418 Pterocephalodes bretschneideri (Batalin) V. Mayer et Ehrend. = Pterocephalus bretschneideri (Batalin) Pritz. ex Diels ●■

320419 Pterocephalodes hookeri (C. B. Clarke) V. Mayer et Ehrend. = Pterocephalus hookeri (C. B. Clarke) Höck ●■

320420 Pterocephalodes siamensis (Craib) V. Mayer et Ehrend.;缅甸拟翼首花●■☆

320421 Pterocephalus Adans.(1763);翼首花属;Whinghead,Whinghead Flower ●■

320422 Pterocephalus Vaill. ex Adans. = Pterocephalus Adans. ●■

320423 Pterocephalus afghanica (Aitch. et Hemsl.) Boiss.;阿富汗翼首花■☆

320424 Pterocephalus batangensis Pax ex K. Hoffm. = Pterocephalus hookeri (C. B. Clarke) Höck ●■

320425 Pterocephalus bretschneideri (Batalin) Pritz. = Pterocephalus bretschneideri (Batalin) Pritz. ex Diels ●■

320426 Pterocephalus bretschneideri (Batalin) Pritz. ex Diels;裂叶翼首花(棒子头,大树小黑牛,爬岩夕,狮子草,小铜锤,岩七);Lobedleaf Whinghead,Lobedleaf Whinghead flower ●■

320427 Pterocephalus bretschneideri (Batalin) Pritz. ex Diels = Pterocephalodes bretschneideri (Batalin) V. Mayer et Ehrend. ●■

320428 Pterocephalus brevis Coult.;短翼首花●☆

320429 Pterocephalus depressus Coss. et Balansa;凹陷翼首花●☆

320430 Pterocephalus depressus Coss. et Balansa subsp. rifanus (Emb. et Maire) Dobignard;里夫翼首花■☆

320431 Pterocephalus dumetorus (Willd.) Coult.;灌丛翼首花■☆

320432 Pterocephalus frutescens Hochst. ex A. Rich.;灌木翼首花●☆

320433 Pterocephalus frutescens Hochst. ex A. Rich. var. dentatus Chiov. = Pterocephalus frutescens Hochst. ex A. Rich. ●☆

320434 Pterocephalus frutescens Hochst. ex A. Rich. var. tomentellus Beck = Pterocephalus frutescens Hochst. ex A. Rich. ●☆

320435 Pterocephalus hookeri (C. B. Clarke) Höck;匙叶翼首花(帮子毒鸟,棒子头,狮子草,土苦参,翼首草,翼首花);Hooker Whinghead,Hooker Whinghead Flower,Hooker's Whing Head ●■

320436 Pterocephalus hookeri (C. B. Clarke) Höck = Pterocephalodes hookeri (C. B. Clarke) V. Mayer et Ehrend. ●■

320437 Pterocephalus kunkelianus Peris et Romo et Stübing = Pterocephalus depressus Coss. et Balansa subsp. rifanus (Emb. et Maire) Dobignard ■☆

320438 Pterocephalus lasiospermus Buch;毛籽翼首花■☆

320439 Pterocephalus papposus (L.) Coult. = Pterocephalus brevis Coult. ●☆

320440 Pterocephalus parnassi Spreng. = Pterocephalus perennis DC. ■☆

320441 Pterocephalus perennis DC.;翼首花(希腊翼首花)■☆

320442 Pterocephalus plumosus (L.) Coult.;羽状翼首花■☆

320443 Pterocephalus plumosus Coult. = Pterocephalus plumosus (L.) Coult. ■☆

320444 Pterocephalus porphyranthus Svent.;紫花翼首花■☆

320445 Pterocephalus virens Berthel.;绿翼首花■☆

320446 Pteroceras Hasselt ex Hassk.(1842);翅足兰属(长脚兰属);

Longlegorchis，Pteroceras ■

320447　Pteroceras Hassk. = Pteroceras Hasselt ex Hassk. ■

320448　Pteroceras appendiculatum（Blume）Holttum；翅足兰（长脚兰，长足兰）■☆

320449　Pteroceras asperatum（Schltr.）P. F. Hunt；毛莛翅足兰（毛莛长足兰）；Hairscape Longlegorchis，Hairscape Pteroceras ■

320450　Pteroceras elobe Seidenf. = Parapteroceras elobe（Seidenf.）Aver. ■

320451　Pteroceras leopardinum（Parl. et Rchb. f.）Seidenf. et Sm.；双臂翅足兰（长足兰，双臂长足兰）；Leopard Longlegorchis，Leopard Pteroceras ■

320452　Pteroceras loratum（Rolfe ex Downie）Seidenf. = Staurochilus loratus（Rolfe ex Downie）Seidenf. ■

320453　Pteroceras loratum（Rolfe ex Downie）Seidenf. et Smitinand = Staurochilus loratus（Rolfe ex Downie）Seidenf. ■

320454　Pteroceras pricei（Rolfe）Aver. = Thrixspermum formosanum（Hayata）Schltr. ■

320455　Pterocereus T. MacDoug. et Miranda = Pachycereus（A. Berger）Britton et Rose ●

320456　Pterocereus T. MacDoug. et Miranda（1954）；翼柱属（有翼柱属）●☆

320457　Pterocereus foetidus T. MacDoug. et Miranda；翼柱●☆

320458　Pterochaeta Boiss. = Pulicaria Gaertn. ■●

320459　Pterochaeta Steetz = Waitzia J. C. Wendl. ■☆

320460　Pterochaeta Steetz（1845）；彩苞金绒草属■☆

320461　Pterochaeta paniculata Steetz；彩苞金绒草■☆

320462　Pterochaete Arn. ex Boeck. = Rhynchospora Vahl（保留属名）■

320463　Pterochaete Boiss. = Platychaeta Boiss. ■●

320464　Pterochaete Boiss. = Pulicaria Gaertn. ■●

320465　Pterochaete glutinosa Boiss. = Pulicaria glutinosa（Boiss.）Jaub. et Spach ■☆

320466　Pterochilus Hook. et Arn. = Crepidium Blume ■

320467　Pterochilus Hook. et Arn. = Maxillaria Ruiz et Pav. ■☆

320468　Pterochiton Tore. = Atriplex L. ■●

320469　Pterochiton Torr. et Frém. = Atriplex L. ■●

320470　Pterochlaena Chiov. = Alloteropsis J. Presl ex C. Presl ■

320471　Pterochlaena catangensis Chiov. = Alloteropsis semialata（R. Br.）Hitchc. ■

320472　Pterochlamys Fisch. ex Endl. = Panderia Fisch. et C. A. Mey. ■

320473　Pterochlamys Roberty = Hildebrandtia Vatke ex A. Braun. ●☆

320474　Pterochlamys somalensis（Engl.）Roberty = Hildebrandtia somalensis Engl. ●☆

320475　Pterochloris（A. Camus）A. Camus = Chloris Sw. ●■

320476　Pterochloris A. Camus = Chloris Sw. ●■

320477　Pterochrosia Baill. = Cerberiopsis Vieill. ex Pancher et Sebert ●☆

320478　Pterocissus Urb. et Ekman = Cissus L. ●

320479　Pterocissus Urb. et Ekman（1926）；翅春藤属●☆

320480　Pterocissus mirabilis Urb. et Ekman；翅春藤●☆

320481　Pterocladis Lamb. ex G. Don = Baccharis L.（保留属名）●■☆

320482　Pterocladon Hook. f. = Miconia Ruiz et Pav.（保留属名）●☆

320483　Pterocladum Triana = Pterocladon Hook. f. ●☆

320484　Pterococcus Hassk.（1842）（保留属名）；翼果大戟属●☆

320485　Pterococcus Pall.（废弃属名）= Calligonum L. ●

320486　Pterococcus Pall.（废弃属名）= Pterococcus Hassk.（保留属名）●☆

320487　Pterococcus africanus（Sond.）Pax et K. Hoffm. = Plukenetia africana Sond. ●☆

320488　Pterococcus aphyllus Pall. = Calligonum aphyllum（Pall.）Gürke ●

320489　Pterococcus glaberrimus Hassk.；翼果大戟●☆

320490　Pterococcus leucocladus Schrenk = Calligonum leucocladum（Schrenk）Bunge ●

320491　Pterococcus procumbens（Prain）Pax et K. Hoffm.；平铺翼果大戟●☆

320492　Pterococcus songaricus C. A. Mey. var. rubicundus C. A. Mey. = Calligonum rubicundum Bunge ●

320493　Pterocoelion Turcz. = Berrya Roxb.（保留属名）●

320494　Pterocoellion Turcz. = Berrya Roxb.（保留属名）●

320495　Pterocyclus Klotzsch = Pleurospermum Hoffm. ■

320496　Pterocyclus Klotzsch（1862）；滇羌活属（翼轮芹属）■

320497　Pterocyclus angelicoides（Wall. ex DC.）Klotzsch = Pleurospermum angelicoides（Wall. ex DC.）Benth. ex C. B. Clarke ■

320498　Pterocyclus angelicoides Klotzsch = Pleurospermum angelicoides（Wall. ex DC.）Benth. ex C. B. Clarke ■

320499　Pterocyclus forrestii（Diels）Pimenov et Kljuykov. = Pleurospermum angelicoides（Wall. ex DC.）Benth. ex C. B. Clarke ■

320500　Pterocyclus rivulorum（Diels）H. Wolff = Pleurospermum rivulorum（Diels）K. T. Fu et Y. C. Ho ■

320501　Pterocyclus rotundatus（DC.）Pimenov et Kljuykov = Pleurospermum rotundatum（DC.）C. B. Clarke ■

320502　Pterocyclus wolffianus Fedde ex H. Wolff = Pleurospermum longicarpum R. H. Shan et Z. H. Pan ■

320503　Pterocymbium R. Br.（1844）；翅梧桐属●☆

320504　Pterocymbium javanicum R. Br.；翅梧桐●☆

320505　Pterocymbium parviflorum Merr.；小花翅梧桐●☆

320506　Pterocyperus（Peterm.）Opiz = Cyperus L. ■

320507　Pterocyperus Opiz = Cyperus L. ■

320508　Pterocypsela C. Shih（1988）；翅果菊属（刺果菊属）；Pterocypsela，Samaradaisy ■

320509　Pterocypsela × mansuensis（Hayata）C. I. Peng；恒春翅果菊（恒春山苦荬）■

320510　Pterocypsela auriculiformis C. Shih = Paraprenanthes sororia（Miq.）C. Shih ■

320511　Pterocypsela elata（Hemsl.）C. Shih；高大翅果菊（高莴苣，高株山莴苣，剪刀草，老蛇药，山苦菜，水紫菀，野苦麻，野洋烟）；Tall Pterocypsela，Tall Samaradaisy ■

320512　Pterocypsela elata（Hemsl.）C. Shih = Lactuca raddiana Maxim. var. elata（Hemsl.）Kitam. ■

320513　Pterocypsela formosana（Maxim.）C. Shih；台湾翅果菊（八楞麻，八楞木，叉头草，丁萝卜，蛾子草，高脚蒲公英，花苦菜，灰地菜，九刀参，苦丁，龙喳口，乳浆草，台湾山苦荬，台湾山莴苣，台湾莴苣，小山萝卜，野苦菜，野苦麻，野莴苣）；Formosan Lettuce，Taiwan Pterocypsela，Taiwan Samaradaisy ■

320514　Pterocypsela formosana（Maxim.）C. Shih = Lactuca formosana Maxim. ■

320515　Pterocypsela indica（L.）C. Shih；翅果菊（白龙头，长椭圆叶鹅仔草，鹅仔草，蝴蝶菜，苦菜，苦芥菜，苦马菜，苦马地丁，苦莴苣，驴干粮，山马草，山莴苣，土莴苣，鸭子食，野大烟，野生菜，野莴苣，印度鹅仔草，猪人参）；Common Pterocypsela，India Lettuce，Samaradaisy，Wild Lettuce ■

320516　Pterocypsela indica（L.）C. Shih = Lactuca indica L. ■

320517　Pterocypsela indica（L.）C. Shih var. laciniata（Houtt.）H. C. Fu = Pterocypsela laciniata（Houtt.）C. Shih ■

320518　Pterocypsela laciniata（Houtt.）C. Shih；多裂翅果菊；

Manylobed Samaradaisy, Pinnate Pterocypsela ■

320519　Pterocypsela laciniata (Houtt.) C. Shih = Lactuca indica L. ■

320520　Pterocypsela raddeana (Maxim.) C. Shih;毛脉翅果菊(高株山莴苣,老蛇药,毛脉山莴苣,山苦菜,水紫菀,野洋烟); Hairyvein Lettuce, Hairyvein Pterocypsela, Hairyvein Samaradaisy ■

320521　Pterocypsela raddeana (Maxim.) C. Shih = Lactuca raddiana Maxim. ■

320522　Pterocypsela sonchus (H. Lév. et Vaniot) C. Shih;细喙翅果菊;Finerostrate Pterocypsela, Sowthinstlelike Samaradaisy ■

320523　Pterocypsela triangulata (Maxim.) C. Shih;翼柄翅果菊(翼柄山莴苣); Triangul Lettuce, Triangular Pterocypsela, Wingstipe Samaradaisy ■

320524　Pterocypsela triangulata (Maxim.) C. Shih = Lactuca triangulata Maxim. ■

320525　Pterodiscus Hook. (1844);翅盘麻属■☆

320526　Pterodiscus angustifolius Engl. ;窄叶翅盘麻■☆

320527　Pterodiscus aurantiacus Welw. ;橙色翅盘麻;Pterodiscus ■☆

320528　Pterodiscus coeruleus Chiov. ;青蓝翅盘麻■☆

320529　Pterodiscus elliottii Baker ex Stapf;埃利翅盘麻■☆

320530　Pterodiscus gayi Decne. ;盖伊翅盘麻■☆

320531　Pterodiscus heterophyllus Stapf = Pterodiscus saccatus S. Moore ■☆

320532　Pterodiscus intermedius Engl. ;间型翅盘麻■☆

320533　Pterodiscus kellerianus Schinz;凯勒翅盘麻■☆

320534　Pterodiscus luridus Hook. f. ;灰黄翅盘麻■☆

320535　Pterodiscus ngamicus N. E. Br. ex Stapf;恩加姆翅盘麻■☆

320536　Pterodiscus purpureus Chiov. ;紫翅盘麻■☆

320537　Pterodiscus ruspolii Engl. ;索马里翅盘麻■☆

320538　Pterodiscus ruspolii Engl. var. roseus Chiov. = Pterodiscus undulatus Baker f. ■☆

320539　Pterodiscus saccatus S. Moore;囊状翅盘麻■☆

320540　Pterodiscus somaliensis Baker ex Stapf = Pterodiscus ruspolii Engl. ■☆

320541　Pterodiscus speciosus Hook. ;美丽翅盘麻;Pterodiscus ■☆

320542　Pterodiscus undulatus Baker f. ;波状翅盘麻■☆

320543　Pterodiscus wellbyi Stapf = Pterodiscus ruspolii Engl. ■☆

320544　Pterodon Vogel(1837);翼齿豆属(翅齿豆属)■☆

320545　Pterodon apparicioi Pedersoli;翼齿豆■☆

320546　Pterodon pubescens Benth. ;短毛翼齿豆■☆

320547　Pterogaillonia Lincz. (1973);翅加永茜属■☆

320548　Pterogaillonia calycoptera (Decne.) Lincz. ;翅加永茜■☆

320549　Pterogaillonia stscherbinovskii Lincz. = Pterogaillonia calycoptera (Decne.) Lincz. ■☆

320550　Pterogastra Naudin(1850);翅果野牡丹属●■☆

320551　Pterogastra divaricata Naudin;翅果野牡丹●■☆

320552　Pteroglossa Schltr. (1920);南美翼舌兰属■☆

320553　Pteroglossa Schltr. = Stenorrhynchos Rich. ex Spreng. ■☆

320554　Pteroglossa macrantha Schltr. ;大花南美翼舌兰■☆

320555　Pteroglossaspis Rchb. f. (1878);翼舌兰属■☆

320556　Pteroglossaspis carsonii Rolfe = Eulophia ruwenzoriensis Rendle ■☆

320557　Pteroglossaspis clandestina Börge Pett. ;隐匿翼舌兰■☆

320558　Pteroglossaspis corymbosa G. Will. ;伞序翼舌兰■☆

320559　Pteroglossaspis distans Summerh. ;远离翼舌兰■☆

320560　Pteroglossaspis ecristata (Fernald) Rolfe;翼舌兰■☆

320561　Pteroglossaspis engleriana Kraenzl. = Eulophia eustachya (Rchb. f.) Geerinck ■☆

320562　Pteroglossaspis eustachya Rchb. f. = Eulophia eustachya (Rchb. f.) Geerinck ■☆

320563　Pteroglossaspis ruwenzoriensis (Rendle) Rolfe = Eulophia ruwenzoriensis Rendle ■☆

320564　Pteroglossaspis stricta Schltr. = Eulophia ruwenzoriensis Rendle ■☆

320565　Pteroglossis Miers = Reyesia Clos ■☆

320566　Pteroglossis Miers = Salpiglossis Ruiz et Pav. ■☆

320567　Pterogonum H. Gross = Eriogonum Michx. ●■☆

320568　Pterogonum H. Gross(1913);翼胚木属●☆

320569　Pterogonum alatum (Torr.) H. Gross = Eriogonum alatum Torr. ■☆

320570　Pterogonum atrorubens (Engelm.) H. Gross;翼胚木●☆

320571　Pterogonum atrorubens (Engelm.) H. Gross = Eriogonum atrorubens Engelm. ■☆

320572　Pterogonum hieracifolium (Benth.) H. Gross = Eriogonum heracleoides Nutt. ■☆

320573　Pterogyne Schrad. ex Nees = Ascolepis Nees ex Steud. (保留属名)■☆

320574　Pterogyne Tul. (1843);翅雌豆属(翼豆属)■☆

320575　Pterogyne nitens Tul. ;翅雌豆(南美荚豆,翼豆)■☆

320576　Pterolepis (DC.) Miq. (1840)(保留属名);翼鳞野牡丹属●■☆

320577　Pterolepis Endl. = Osbeckia L. ●■

320578　Pterolepis Miq. = Pterolepis (DC.) Miq. (保留属名)●■☆

320579　Pterolepis Schrad. (废弃属名) = Pterolepis (DC.) Miq. (保留属名)●■☆

320580　Pterolepis Schrad. (废弃属名) = Scirpus L. (保留属名)■

320581　Pterolepis alpestris Triana;高山翼鳞野牡丹●☆

320582　Pterolepis boliviensis Cogn. ;玻利维亚翼鳞野牡丹●☆

320583　Pterolepis filiformis Triana;线形翼鳞野牡丹●☆

320584　Pterolepis fragilis L. O. Williams;脆翼鳞野牡丹●☆

320585　Pterolepis macranthera Triana;大花翼鳞野牡丹●☆

320586　Pterolepis nitida (Graham) Triana;光亮翼鳞野牡丹●☆

320587　Pterolepis pauciflora Triana;少花翼鳞野牡丹●☆

320588　Pterolepis scirpoides Schrad. = Schoenoplectus scirpoideus (Schrad.) Browning ■☆

320589　Pterolepis stenophylla Gleason;窄叶翼鳞野牡丹●☆

320590　Pterolobium Andrz. = Pterolobium R. Br. ex Wight et Arn. (保留属名)●

320591　Pterolobium Andrz. ex C. A. Mey. (废弃属名) = Pterolobium R. Br. ex Wight et Arn. (保留属名)●

320592　Pterolobium Andrz. ex DC. = Pachyphragma (DC.) Rchb. ■☆

320593　Pterolobium R. Br. = Pterolobium R. Br. ex Wight et Arn. (保留属名)●

320594　Pterolobium R. Br. ex Wight et Arn. (1834)(保留属名);老虎刺属(雀不踏属,崖颊簕属);Pterolobium, Tigerthorn ●

320595　Pterolobium exosum (J. F. Gmel.) Baker f. = Pterolobium stellatum (Forssk.) Brenan ●☆

320596　Pterolobium indicum A. Rich. ;印度老虎刺●☆

320597　Pterolobium indicum A. Rich. = Pterolobium punctatum Hemsl. ex Forbes et Hemsl. ●

320598　Pterolobium indicum A. Rich. var. macropterum (Kurz) Baker = Pterolobium macropterum Kurz ●

320599　Pterolobium indicum A. Rich. var. macropterum Baker = Pterolobium macropterum Kurz ●

320600　Pterolobium indicum Hance = Pterolobium punctatum Hemsl. ●

320601　Pterolobium kantuffa Wight et Arn. ex Steud. = Pterolobium stellatum (Forssk.) Brenan ●☆

320602　Pterolobium lacerans R. Br. = Pterolobium stellatum (Forssk.) Brenan ●☆

320603　Pterolobium macropterum Kurz;大翅老虎刺(大翅果老虎刺,

老母狗果,小花老虎刺);Bigwing Pterolobium, Bigwing Tigerthorn, Big-winged Pterolobium ●

320604　Pterolobium micranthum Gagnep.；小花老虎刺；Little-flower Pterolobium, Little-flower Tigerthorn ●☆

320605　Pterolobium micranthum Gagnep. = Pterolobium macropterum Kurz ●

320606　Pterolobium punctatum Hemsl. = Pterolobium punctatum Hemsl. ex Forbes et Hemsl. ●

320607　Pterolobium punctatum Hemsl. ex Forbes et Hemsl.；老虎刺(蝉翼豆,倒钩藤,倒爪刺,老虎树,老鹰刺,牛王刺藤,雀不踏,石龙花,崖婆箭,蚰蛇利);Variegated Pterolobium, Variegated Tigerthorn ●

320608　Pterolobium rosthornii Harms = Pterolobium punctatum Hemsl. ex Forbes et Hemsl. ●

320609　Pterolobium sinense J. E. Vidal = Pterolobium macropterum Kurz ●

320610　Pterolobium stellatum (Forssk.) Brenan；小星老虎刺●☆

320611　Pterolobium subvestitum Hance = Caesalpinia millettii Hook. et Arn. ●

320612　Pteroloma Desv. ex Benth. (1852);蔓茎葫芦茶属(葫芦茶属)●☆

320613　Pteroloma Desv. ex Benth. = Desmodium Desv. (保留属名)●■

320614　Pteroloma Desv. ex Benth. = Tadehagi H. Ohashi ■

320615　Pteroloma Hochst. et Steud. = Dipterygium Decne. ●☆

320616　Pteroloma pseudotriquetrum (DC.) Schindl. = Tadehagi pseudotriquetra (A. DC.) H. Ohashi ●

320617　Pteroloma triquetrum (L.) Desv. = Desmodium triquetrum (L.) DC. ●

320618　Pteroloma triquetrum (L.) Desv. ex Benth. = Tadehagi triquetra (L.) H. Ohashi ●

320619　Pteroloma triquetrum (L.) Desv. ex Benth. subsp. pseudotriquetrum (DC.) H. Ohashi = Tadehagi pseudotriquetra (DC.) H. Ohashi ●

320620　Pterolophus Cass. = Centaurea L. (保留属名)●■

320621　Pteromarathrum W. D. J. Koch ex DC. = Prangos Lindl. ■☆

320622　Pteromimosa Britton = Mimosa L. ●■

320623　Pteromischus Pichon = Crescentia L. ●

320624　Pteromonnina B. Eriksen(1993);裂柱远志属■☆

320625　Pteromonnina boliviana (Chodat) B. Eriksen;玻利维亚裂柱远志■☆

320626　Pteromonnina cardiocarpa (A. St. -Hil.) B. Eriksen;心果裂柱远志■☆

320627　Pteromonnina dictyocarpa (Griseb.) B. Eriksen;指果裂柱远志■☆

320628　Pteromonnina filifolia (Chodat) B. Eriksen;线叶裂柱远志■☆

320629　Pteromonnina leptostachya (Benth.) B. Eriksen;细穗裂柱远志■☆

320630　Pteromonnina macrocarpa (Chodat) B. Eriksen;大果裂柱远志■☆

320631　Pteromonnina pterocarpa Ruiz et Pav.;翅果裂柱远志■☆

320632　Pteromonnina stenophylla (A. St. -Hil.) B. Eriksen;窄叶裂柱远志■☆

320633　Pteronema Pierre = Spondias L. ●

320634　Pteroneuron Meisn. = Pteroneurum DC. ■

320635　Pteroneurum DC. = Cardamine L. ■

320636　Pteronia L. (1763)(保留属名);橙菀属●☆

320637　Pteronia acerosa DC. = Pteronia teretifolia (Thunb.) Fourc. ●☆

320638　Pteronia acuminata DC.;渐尖橙菀●☆

320639　Pteronia acuta Muschl.;锐利橙菀●☆

320640　Pteronia adenocarpa Harv.;腺果橙菀●☆

320641　Pteronia aizoides Muschl. = Eremothamnus marlothianus O.

Hoffm. ●☆

320642　Pteronia anisata B. Nord.;异型橙菀●☆

320643　Pteronia anisata Dinter ex Merxm. = Pteronia glabrata L. f. ●☆

320644　Pteronia aspalatha DC.;芳香木橙菀●☆

320645　Pteronia aspera Thunb. = Pteronia camphorata (L.) L. ●☆

320646　Pteronia baccharoides Less. = Pteronia teretifolia (Thunb.) Fourc. ●☆

320647　Pteronia beckeoides DC.;斯诺登橙菀●☆

320648　Pteronia bolusii E. Phillips;博卢斯橙菀●☆

320649　Pteronia bromoides S. Moore = Pteronia lucilioides DC. ●☆

320650　Pteronia callosa DC.;硬皮橙菀●☆

320651　Pteronia camphorata (L.) L.;樟脑橙菀●☆

320652　Pteronia camphorata (L.) L. var. armata Harv.;具刺橙菀●☆

320653　Pteronia camphorata (L.) L. var. aspera (Thunb.) Harv. = Pteronia camphorata (L.) L. ●☆

320654　Pteronia camphorata (L.) L. var. laevigata Harv.;光滑樟脑橙菀●☆

320655　Pteronia camphorata (L.) L. var. longifolia Harv.;长叶樟脑橙菀●☆

320656　Pteronia camphorata (L.) L. var. stricta (Aiton) Harv. = Pteronia stricta Aiton ●☆

320657　Pteronia candollei Harv. = Pteronia glauca Thunb. ●☆

320658　Pteronia canescens DC. = Pteronia cinerea L. f. ●☆

320659　Pteronia carnosa Muschl. = Pteronia acuminata DC. ●☆

320660　Pteronia centauroides DC.;矢车菊橙菀●☆

320661　Pteronia chlorolepis Dinter = Pteronia sordida N. E. Br. ●☆

320662　Pteronia chrysocomifolia Poir. = Pteronia stricta Aiton ●☆

320663　Pteronia ciliata Thunb.;缘毛橙菀●☆

320664　Pteronia cinerea L. f.;灰色橙菀●☆

320665　Pteronia connata DC. = Pteronia flexicaulis L. f. ●☆

320666　Pteronia cylindracea DC.;柱形橙菀●☆

320667　Pteronia decumbens Banks ex S. Moore = Pteronia fastigiata Thunb. ●☆

320668　Pteronia dinteri S. Moore = Pteronia mucronata DC. ●☆

320669　Pteronia diosmifolia Brusse;逸香木橙菀●☆

320670　Pteronia divaricata (P. J. Bergius) Less.;叉开橙菀●☆

320671　Pteronia echinata Thunb. = Felicia echinata (Thunb.) Nees ■☆

320672　Pteronia eenii S. Moore;埃恩橙菀●☆

320673　Pteronia elata B. Nord.;高橙菀●☆

320674　Pteronia elegans Sch. Bip. ex Walp. = Pteronia tenuifolia DC. ●☆

320675　Pteronia elongata Thunb.;伸长橙菀●☆

320676　Pteronia empetrifolia DC.;岩高兰叶橙菀●☆

320677　Pteronia engleriana Muschl. = Amphiglossa tomentosa (Thunb.) Harv. ■☆

320678　Pteronia erythrochaeta DC.;红毛橙菀●☆

320679　Pteronia fasciculata L. f.;簇生橙菀●☆

320680　Pteronia fastigiata Thunb.;帚状橙菀●☆

320681　Pteronia feddeana Muschl. = Pteronia acuminata DC. ●☆

320682　Pteronia feldtmanniana Dinter ex Merxm. = Pteronia eenii S. Moore ●☆

320683　Pteronia flexicaulis DC. = Pteronia paniculata Thunb. ●☆

320684　Pteronia flexicaulis L. f.;曲茎橙菀●☆

320685　Pteronia foleyi Hutch. et E. Phillips;福莱橙菀●☆

320686　Pteronia geigerioides Muschl. ex Dinter = Athanasia minuta (L. f.) Källersjö ●☆

320687　Pteronia glabrata DC. = Pteronia glauca Thunb. ●☆

320688　Pteronia glabrata L. f.;光滑橙菀●☆

320689 Pteronia glauca Thunb. ;灰绿橙菀●☆

320690 Pteronia glaucescens DC. ;粉绿橙菀●☆

320691 Pteronia glomerata L. f. ;小叶橙菀●☆

320692 Pteronia gymnocline E. Mey. = Pteronia lucilioides DC. ●☆

320693 Pteronia heterocarpa DC. ;异果橙菀●☆

320694 Pteronia hirsuta L. f. ;粗毛橙菀●☆

320695 Pteronia hutchinsoniana Compton;哈钦森橙菀●☆

320696 Pteronia incana（Burm.）DC. ;灰毛橙菀●☆

320697 Pteronia inflexa Thunb. ex L. f. ;内折橙菀●☆

320698 Pteronia intermedia Hutch. et E. Phillips;间型橙菀●☆

320699 Pteronia kingesii Merxm. = Pteronia polygalifolia O. Hoffm. ●☆

320700 Pteronia laricina Houtt. ex DC. = Pteronia camphorata（L.）L. ●☆

320701 Pteronia latisquama DC. = Pteronia glauca Thunb. ●☆

320702 Pteronia leptospermoides DC. ;薄子木橙菀●☆

320703 Pteronia leucoclada Turcz. ;白枝橙菀●☆

320704 Pteronia leucoloma DC. ;白边橙菀●☆

320705 Pteronia lucilioides DC. ;长毛紫绒草橙菀●☆

320706 Pteronia lucilioides DC. var. sparsifolia Harv. = Pteronia lucilioides DC. ●☆

320707 Pteronia lupulina DC. = Pteronia inflexa Thunb. ex L. f. ●☆

320708 Pteronia lycioides Muschl. ex Dinter = Pteronia scariosa L. f. ●☆

320709 Pteronia marlothiana（O. Hoffm.）Dinter = Eremothamnus marlothianus O. Hoffm. ●☆

320710 Pteronia membranacea L. f. ;膜质橙菀●☆

320711 Pteronia microphylla DC. = Pteronia glomerata L. f. ●☆

320712 Pteronia minuta L. f. = Athanasia minuta（L. f.）Källersjö ●☆

320713 Pteronia mooreiana Hutch. ;穆尔橙菀●☆

320714 Pteronia mucronata DC. ;短尖橙菀●☆

320715 Pteronia mucronata DC. subsp. dinteri（S. Moore）Merxm. = Pteronia mucronata DC. ●☆

320716 Pteronia oblanceolata E. Phillips;倒披针形橙菀●☆

320717 Pteronia onobromoides DC. ;红花橙菀●☆

320718 Pteronia oppositifolia L. ;对叶橙菀●☆

320719 Pteronia ovalifolia DC. ;卵叶橙菀●☆

320720 Pteronia pallens L. f. ;变苍白橙菀●☆

320721 Pteronia paniculata Thunb. ;圆锥橙菀●☆

320722 Pteronia paniculata Thunb. var. fastigiata（Thunb.）Harv. = Pteronia fastigiata Thunb. ●☆

320723 Pteronia pauciflora Sims = Syncarpha staehelina（L.）B. Nord. ■☆

320724 Pteronia pillansii Hutch. ;皮朗斯橙菀●☆

320725 Pteronia polygalifolia O. Hoffm. ;远志叶橙菀●☆

320726 Pteronia pomonae Merxm. ;波莫纳橙菀●☆

320727 Pteronia punctata E. Phillips;斑点橙菀●☆

320728 Pteronia quadrifaria Dinter = Pteronia lucilioides DC. ●☆

320729 Pteronia quinquecostata Dinter = Pteronia polygalifolia O. Hoffm. ●☆

320730 Pteronia quinqueflora DC. ;五花橙菀●☆

320731 Pteronia rangei Muschl. ;朗格橙菀●☆

320732 Pteronia retorta L. f. = Pteronia hirsuta L. f. ●☆

320733 Pteronia roesemaniana Dinter ex Merxm. = Pteronia lucilioides DC. ●☆

320734 Pteronia scabra Harv. ;粗糙橙菀●☆

320735 Pteronia scariosa L. f. ;干膜质橙菀●☆

320736 Pteronia sesuviifolia DC. = Pteronia glabrata L. f. ●☆

320737 Pteronia sordida N. E. Br. ;污浊橙菀●☆

320738 Pteronia spinulosa E. Phillips;细刺橙菀●☆

320739 Pteronia stricta Aiton;刚直橙菀●☆

320740 Pteronia stricta Aiton var. longifolia E. Phillips;长叶橙菀●☆

320741 Pteronia succulenta Thunb. ;多汁橙菀●☆

320742 Pteronia tenuifolia DC. ;细叶橙菀●☆

320743 Pteronia teretifolia（Thunb.）Fourc. ;四叶橙菀●☆

320744 Pteronia thymifolia Muschl. et Dinter = Pteronia glauca Thunb. ●☆

320745 Pteronia tricephala DC. ;三头橙菀●☆

320746 Pteronia trigona E. Phillips = Pteronia teretifolia（Thunb.）Fourc. ●☆

320747 Pteronia turbinata DC. = Pteronia ciliata Thunb. ●☆

320748 Pteronia uncinata DC. ;具钩橙菀●☆

320749 Pteronia undulata DC. ;波状橙菀●☆

320750 Pteronia unguiculata S. Moore;爪状橙菀●☆

320751 Pteronia uniflora Poir. = Pteronia hirsuta L. f. ●☆

320752 Pteronia utilis Hutch. ;有用橙菀●☆

320753 Pteronia verticillata DC. = Pteronia uncinata DC. ●☆

320754 Pteronia villosa L. f. ;长柔毛橙菀●☆

320755 Pteronia viscosa DC. = Pteronia adenocarpa Harv. ●☆

320756 Pteronia viscosa Thunb. ;黏橙菀●☆

320757 Pteronia xantholepis DC. = Pteronia incana（Burm.）DC. ●☆

320758 Pteroon Luer = Masdevallia Ruiz et Pav. ■☆

320759 Pteropappus Pritz. = Pterygopappus Hook. f. ■☆

320760 Pteropavonia Mattei = Pavonia Cav.（保留属名）●■☆

320761 Pteropentacoilanthus F. Rappa et Camarrone = Mesembryanthemum L.（保留属名）■●

320762 Pteropentacoilanthus fastigiatus（Dinter）Rappa et Camorrone = Mesembryanthemum fastigiatum Thunb. ■☆

320763 Pteropentacoilanthus hypertrophicus（Dinter）Rappa et Camarrone = Mesembryanthemum hypertrophicum Dinter ■☆

320764 Pteropepon（Cogn.）Cogn.（1916）;翼瓠果属■☆

320765 Pteropepon Cogn. = Pteropepon（Cogn.）Cogn. ■☆

320766 Pteropepon deltoideus Cogn. ;翼瓠果■☆

320767 Pteropetalum Pax = Euadenia Oliv. ●☆

320768 Pteropetalum klingii Pax = Euadenia trifoliata（Schumach. et Thonn.）Oliv. ●☆

320769 Pterophacos Rydb. = Astragalus L. ●■

320770 Pterophora Harv. = Dregea E. Mey.（保留属名）●

320771 Pterophora L.（废弃属名）= Bigelowia DC.（保留属名）●☆

320772 Pterophora L.（废弃属名）= Pteronia L.（保留属名）●☆

320773 Pterophora L.（废弃属名）= Spermacoce L. ●■

320774 Pterophora dregea Harv. = Dregea floribunda E. Mey. ●☆

320775 Pterophorus Boehm = Pteronia L.（保留属名）●☆

320776 Pterophorus Vaill.（废弃属名）= Pteronia L.（保留属名）●☆

320777 Pterophorus Vaill. ex Adans. = Pteronia L.（保留属名）●☆

320778 Pterophylla D. Don = Weinmannia L.（保留属名）●☆

320779 Pterophyllus J. Nelson = Ginkgo L. ●★

320780 Pterophyton Cass. = Actinomeris Nutt.（保留属名）●■☆

320781 Pteropodium DC. = Jacaranda Juss. ●

320782 Pteropodium DC. ex Meisn. = Jacaranda Juss. ●

320783 Pteropodium Steud. = Calamagrostis Adans. ■

320784 Pteropodium Willd. ex Steud. = Deyeuxia Clarion ■

320785 Pteropogon A. Cunn. ex DC. = Helipterum DC. ex Lindl. ■☆

320786 Pteropogon A. Cunn. ex DC. = Rhodanthe Lindl. ●■☆

320787 Pteropogon DC. = Helipterum DC. ex Lindl. ■☆

320788 Pteropogon DC. = Rhodanthe Lindl. ●■☆

320789 Pteropogon Fenzl = Facelis Cass. ■☆

320790 Pteropogon Neck. = Scabiosa L. ●■

320791 Pteroptychia Bremek.（1944）;假蓝属;Pteroptychia ●■

320792　Pteroptychia Bremek. = Strobilanthes Blume ●■

320793　Pteroptychia dalziellii（Sm.）H. S. Lo；曲枝假蓝（蓝靛，曲枝黄猄草）；Dalziel Pteroptychia ●■

320794　Pteropyrum Jaub. et Spach（1844）；中亚翅果蓼属（翅蓼属）●☆

320795　Pteropyrum aucheri Jaub. et Spach；奥切翅蓼■☆

320796　Pteropyrum olivieri Jaub. et Spach；中亚翅果蓼■☆

320797　Pterorhachis Harms（1895）；刺翅楝属●☆

320798　Pterorhachis le-testui Pellegr.；刺翅楝●☆

320799　Pterorhachis zenkeri Harms；非洲刺翅楝●☆

320800　Pteroscleria Nees = Diplacrum R. Br. ■

320801　Pteroscleria longifolia Griseb. = Diplacrum longifolium（Griseb.）C. B. Clarke ■☆

320802　Pteroselinum（Rchb.）Rchb.（1832）；翅蛇床属■☆

320803　Pteroselinum（Rchb.）Rchb. = Peucedanum L. ■

320804　Pteroselinum Rchb. = Peucedanum L. ■

320805　Pteroselinum Rchb. = Pteroselinum（Rchb.）Rchb. ■☆

320806　Pteroselinum alsaticum Rchb.；翅蛇床■☆

320807　Pterosenecio Sch. Bip. ex Baker = Senecio L. ■●

320808　Pterosicyos Brandegee = Sechiopsis Naudin ■

320809　Pterosiphon Turcz. = Cedrela P. Browne ●

320810　Pterospartum（Spach.）K. Koch = Chamaespartium Adans. ●

320811　Pterospartum（Spach.）K. Koch = Genista L. ●

320812　Pterospartum（Spach.）K. Koch（1853）；肖染料木属●

320813　Pterospartum K. Koch = Chamaespartium Adans. ●

320814　Pterospartum K. Koch = Genista L. ●

320815　Pterospartum K. Koch = Pterospartum（Spach.）K. Koch ●

320816　Pterospartum Willk. = Genista L. ●

320817　Pterospartum sagittale Willk. = Genista sagittalis L. ●☆

320818　Pterospartum tridentatum（L.）Willk.；三齿肖染料木●☆

320819　Pterospartum tridentatum（L.）Willk. subsp. lasianthum（Spach）Talavera et Gibbs；毛花金雀花●☆

320820　Pterospartum tridentatum（L.）Willk. subsp. rhiphaeum（Pau et Font Quer）Talavera et P. E. Gibbs；山金雀花●☆

320821　Pterospartum tridentatum（L.）Willk. var. gomaricum（Emb. et Maire）Sauvage = Pterospartum tridentatum（L.）Willk. ●☆

320822　Pterospartum tridentatum（L.）Willk. var. lasianthum（Spach）Sauvage = Pterospartum tridentatum（L.）Willk. ●☆

320823　Pterospartum tridentatum（L.）Willk. var. rhiphaeum（Pau et Font Quer）Sauvage = Pterospartum tridentatum（L.）Willk. ●☆

320824　Pterospermadendron Kuntze = Pterospermum Schreb.（保留属名）●

320825　Pterospermopsis Arch. = Macarisia Thouars ●☆

320826　Pterospermopsis Arènes（1949）；拟翅子树属●☆

320827　Pterospermum Schreb.（1791）（保留属名）；翅子树属；Pterospermum，Wingseed Tree，Wingseedtree，Wing-seed-tree ●

320828　Pterospermum acerifolium（L.）Willd.；翅子树（白桐，翅子木，大巴巴叶，大钩藤叶，大毛红花，大毛红叶，槭叶翅子树）；Baynt Tree，Maple-leaf Bayur，Mapleleaf Wingseedtree，Maple-leaved Pterospermum，Maple-leaved Wing-seed-tree ●

320829　Pterospermum acerifolium Benth. = Pterospermum heterophyllum Hance ●

320830　Pterospermum acerifolium Willd. = Pterospermum acerifolium（L.）Willd. ●

320831　Pterospermum diversifolium Blume；异叶翅子树；Diverseleaf Wingseedtree ●☆

320832　Pterospermum diversifolium Blume = Pterospermum acerifolium Willd. ●

320833　Pterospermum formosanum Matsum. = Pterospermum niveum S. Vidal ●

320834　Pterospermum glabrescens Wight et Arn.；光翅子树；Glabrous Wingseedtree ●☆

320835　Pterospermum grande Craib；大翅子树（大巴巴叶，大钩藤叶，大红毛叶，大毛红花）；Big Wingseedtree ●

320836　Pterospermum heterophyllum Hance；翻白叶树（白背枫，半边枫荷，半梧桐，大叶半枫荷，番张麻，红半枫荷，米纸，异叶半枫荷，异叶翅子树，阴阳叶）；Diversifolious Pterospermum，Heterophyllous Wingseedtree，Heterophyllous Wing-seed-tree ●

320837　Pterospermum heyneanum Wall.；海纳翅子树；Heyne Wingseedtree ●☆

320838　Pterospermum jackianum Wall.；札克翅子树；Jack Wingseedtree ●☆

320839　Pterospermum kingtungense C. Y. Wu ex H. H. Hsue；景东翅子树（大巴巴叶，大钩藤叶，大红毛叶，大毛红花）；Jingdong Wingseedtree，Jingdong Wing-seed-tree ●◇

320840　Pterospermum lanceifolium Roxb. et DC.；窄叶半枫荷（翅子树，火草树，假棉木，剑叶翅子树，窄叶翅子树）；Lanceleaf Wingseedtree，Lance-leaved Wing-seed-tree ●

320841　Pterospermum levinei Merr. = Pterospermum heterophyllum Hance ●

320842　Pterospermum menglunense H. H. Hsue；勐仑翅子树；Menglun Wingseedtree，Menglun Wing-seed-tree ●◇

320843　Pterospermum niveum S. Vidal；台湾翅子树（翅子树，里白翅子木，里白翅子树）；Lanyu Pterospermum，Taiwan Wingseedtree，Taiwan Wing-seed-tree ●

320844　Pterospermum obtusifolium Wight ex Mast.；钝叶翅子树；Obtuseleaf Wingseedtree ●☆

320845　Pterospermum proteus Burkill；变叶翅子树；Variable-leaved Wing-seed-tree，Variantleaf Wingseedtree ●

320846　Pterospermum reticulatum Wight et Arn.；网脉翅子树；Reticulate Wingseedtree ●☆

320847　Pterospermum rubiginosum K. Heyne et G. Don；褐赤翅子树；Red Wingseedtree ●☆

320848　Pterospermum semisagittatum Buch. -Ham. ex Roxb.；半箭形翅子树；Halfsagittate Wingseedtree ●☆

320849　Pterospermum suberifolium Lam.；栓叶翅子树；Corkleaf Wingseedtree ●☆

320850　Pterospermum truncatolobatum Gagnep.；截裂翅子树（缺叶翻白树）；Truncatelobed Wingseedtree，Truncatelobed Wing-seed-tree ●

320851　Pterospermum yunnanense H. H. Hsue；云南翅子树；Yunnan Wingseedtree，Yunnan Wing-seed-tree ●◇

320852　Pterospora Nutt.（1818）；松滴兰属（翅孢属，松球属）■☆

320853　Pterospora andromedea Nutt.；松滴兰（大鸟巢翅孢）；Giant Bird's-nest，Giant Pinedrops，Pine-drops，Woodland Pinedrops ■☆

320854　Pterosporopsis Kellogg = Sarcodes Torr. ●☆

320855　Pterostegia Fisch. et C. A. Mey.（1836）；翅苞蓼属；Woodland Threadstem ■☆

320856　Pterostegia drymarioides Fisch. et C. A. Mey.；翅苞蓼■☆

320857　Pterostelma Wight = Hoya R. Br. ●

320858　Pterostemma Kraenzl.（1899）；翅冠兰属■☆

320859　Pterostemma Lehm. et Kraenzl. = Pterostemma Kraenzl. ■☆

320860　Pterostemma antioquiense F. Lehm. et Kraenzl. ex Kraenzl.；翅冠兰■☆

320861　Pterostemma calceolaris Garay = Sarmenticola calceolaris（Garay）Senghas et Garay ■☆

320862　Pterostemon Schauer（1847）；翼蕊木属（齿蕊属）；Green Hoods，Hood Orchids ■☆

320863　Pterostemon mexicanus Schauer；翼蕊木（齿蕊）■☆

320864　Pterostemonaceae Small（1905）（保留科名）；翼蕊木科（齿蕊科）●☆

320865　Pterostemonaceae Small（保留科名）= Grossulariaceae DC.（保留科名）●

320866　Pterostephanus Kellogg = Anisocoma Torr. et A. Gray ☆

320867　Pterostephus C. Presl = Spermacoce L. ●■

320868　Pterostigma Benth. = Adenosma R. Br. ■

320869　Pterostigma capitatum Benth. = Adenosma indiana（Lour.）Merr. ■

320870　Pterostigma grandiflorum Benth. = Adenosma glutinosa（L.）Druce ■

320871　Pterostylis R. Br.（1810）（保留属名）；翅柱兰属；Green Hoods，Greenhood，Greenhoods，Hood Orchids ■☆

320872　Pterostylis bansii R. Br.；斑克翅柱兰；Banks Greenhoods ■☆

320873　Pterostylis baptisii Fitzg.；巴波翅柱兰；Baptist Greenhoods ■☆

320874　Pterostylis curta R. Br.；短翅柱兰；Short Greenhoods ■☆

320875　Pterostylis longifolia R. Br.；长叶翅柱兰；Longleaf Greenhoods ■☆

320876　Pterostylis nutans R. Br.；垂花翅柱兰；Droopingflower Greenhoods，Nodding Greenhood ■☆

320877　Pterostylis trullifolia Hook. f.；翅柱兰；Common Greenhoods ■☆

320878　Pterostyrax Siebold et Zucc.（1839）；白辛树属；Epaulette Tree，Epaulettetree，Epaulette-tree ●

320879　Pterostyrax cavaleriei Guillaumin = Pterostyrax psilophyllus Diels ex Perkins ●

320880　Pterostyrax corymbosus Siebold et Zucc.；小叶白辛树；Corymbose Epaulettetree，Corymbose Epaulette-tree，Epaulette Tree，Little Epaulettetree ●

320881　Pterostyrax henryi Dümmer = Sinojackia henryi（Dümmer）Merr. ●

320882　Pterostyrax hispidus Rehder et E. H. Wilson = Pterostyrax psilophyllus Diels ex Perkins ●

320883　Pterostyrax hispidus Siebold et Zucc. = Pterostyrax psilophyllus Diels ex Perkins ●

320884　Pterostyrax leveillei（Fedde ex H. Lév.）Chun = Pterostyrax psilophyllus Diels ex Perkins ●

320885　Pterostyrax micranthus Siebold et Zucc.；小花白辛树；Littleflower Epaulettetree，Smallflower Epaulettetree ●☆

320886　Pterostyrax psilophyllus Diels ex Perkins；白辛树（大叶白辛树，鄂西野茉莉，刚毛白辛树，肩饰白辛树，裂叶白辛树）；Epaulette Tree，Epaulettetree，Fragrant Epaulette Tree，Glabrousleaf Epaulettetree，Glabrous-leaved Epaulette-tree ●

320887　Pterostyrax psilophyllus Diels ex Perkins var. leveillei（Fedde）Hara = Pterostyrax psilophyllus Diels ex Perkins ●

320888　Pterota P. Browne（废弃属名）= Fagara L.（保留属名）●

320889　Pterotaberna Stapf = Tabernaemontana L. ●

320890　Pterotaberna inconspicua（Stapf）Stapf = Tabernaemontana inconspicua Stapf ●☆

320891　Pterotheca C. Presl = Rhynchospora Vahl（保留属名）■

320892　Pterotheca Cass. = Crepis L. ■

320893　Pterothrix DC.（1838）；羽冠帚鼠麹属●☆

320894　Pterothrix engleriana（Muschl.）Hutch. et E. Phillips = Amphiglossa tomentosa（Thunb.）Harv. ■☆

320895　Pterothrix flaccida Schltr. ex Hutch.；软羽冠帚鼠麹●☆

320896　Pterothrix flaccida Schltr. ex Hutch. et E. Phillips = Amphiglossa tomentosa（Thunb.）Harv. ■☆

320897　Pterothrix perotrichoides（DC.）Harv. = Amphiglossa perotrichoides DC. ■☆

320898　Pterothrix spinescens DC.；羽冠帚鼠麹●☆

320899　Pterothrix spinescens DC. = Amphiglossa triflora DC. ■☆

320900　Pterothrix tecta Brusse = Amphiglossa tecta（Brusse）Koekemoer ■☆

320901　Pterothrix tomentosa（Thunb.）DC.；毛羽冠帚鼠麹●☆

320902　Pterothrix tomentosa（Thunb.）DC. = Amphiglossa tomentosa（Thunb.）Harv. ■☆

320903　Pterothrix tomentosa DC. = Pterothrix tomentosa（Thunb.）DC. ●☆

320904　Pterotrichis Rchb. = Lachnostoma Kunth ●☆

320905　Pterotrichis Rchb. = Pherotrichis Decne. ●☆

320906　Pterotropia Hillebr. = Dipanax Seem. ●☆

320907　Pterotropia W. F. Hillebr. = Tetraplasandra A. Gray ●☆

320908　Pterotropis（DC.）Fourr. = Thlaspi L. ■

320909　Pterotropis Fourr. = Thlaspi L. ■

320910　Pteroxygonum Dammer et Diels = Fagopyrum Mill.（保留属名）●■

320911　Pteroxygonum Dammer et Diels（1905）；翼蓼属（红药子属）；Pteroxygonum，Wingknotweed ●■★

320912　Pteroxygonum giraldii Dammer et Diels；翼蓼（白药子，红药子，红要子，金荞仁，荞麦七，石天荞）；Girald Knotweed，Girald Pteroxygonum，Wingknotweed ●■

320913　Pteroxylon Hook. f. = Ptaeroxylon Eckl. et Zeyh. ●☆

320914　Pterygiella Oliv.（1896）；翅茎草属（马松蒿属）；Pterygiella ●■★

320915　Pterygiella bartschioides Hand. -Mazz. = Xizangia bartschioides（Hand. -Mazz.）C. Y. Wu et D. D. Tao ■

320916　Pterygiella cylindrica P. C. Tsoong；圆茎翅茎草（利胆草）；Cylindrical Pterygiella ■

320917　Pterygiella duclouxii Franch.；疏毛翅茎草（草连翘，翅茎草，杜氏翅茎草，疳积药，鬼见羽，松毛杆，牙痛草，翼茎草，鱼邦草，紫茎牙痛草）；Ducloux Pterygiella ■

320918　Pterygiella nigrescens Oliv.；黑翅茎草（翅茎草）；Pterygiella ■

320919　Pterygiella suffruticosa D. Y. Hong；川滇翅茎草 ■

320920　Pterygiosperma O. E. Schulz（1924）；翅籽芥属■☆

320921　Pterygiosperma tehuelches O. E. Schulz；翅籽芥■☆

320922　Pterygium Endl. = Dipterocarpus C. F. Gaertn. ●

320923　Pterygium Endl. = Pterigium Corrêa ●

320924　Pterygocalyx Maxim.（1859）；翼萼蔓属；Pterygocalyx ■

320925　Pterygocalyx Maxim. = Crawfurdia Wall. ■

320926　Pterygocalyx Maxim. = Gentiana L. ■

320927　Pterygocalyx volubilis Maxim.；翼萼蔓（翼萼蔓龙胆）；Twining Pterygocalyx ■

320928　Pterygocarpus Hochst. = Dregea E. Mey.（保留属名）●

320929　Pterygocarpus abyssinicus Hochst. = Dregea abyssinica（Hochst.）K. Schum. ●☆

320930　Pterygodium Sw.（1800）；非洲兰属（非兰属）■☆

320931　Pterygodium acutifolium Lindl.；尖叶非洲兰■☆

320932　Pterygodium alare（L. f.）Druce = Pterygodium catholicum（L.）Sw. ■☆

320933　Pterygodium alatum（Thunb.）Sw.；具翅非洲兰■☆

320934　Pterygodium atratum（L.）Sw. = Ceratandra atrata（L.）T. Durand et Schinz ■☆

320935　Pterygodium bicolorum（Thunb.）Schltr. = Corycium bicolorum（Thunb.）Sw. ■☆

320936　Pterygodium bifidum（Sond.）Schltr. = Corycium bifidum Sond. ■☆

320937　Pterygodium caffrum (L.) Sw.;开菲尔非洲兰■☆

320938　Pterygodium carnosum Lindl. = Corycium carnosum (Lindl.) Rolfe ■☆

320939　Pterygodium catholicum (L.) Sw.;普通非洲兰■☆

320940　Pterygodium connivens Schelpe;靠合非洲兰■☆

320941　Pterygodium cooperi Rolfe;库珀非洲兰■☆

320942　Pterygodium crispum (Thunb.) Schltr. = Corycium crispum (Thunb.) Sw.■☆

320943　Pterygodium deflexum Bolus = Corycium deflexum (Bolus) Rolfe ■☆

320944　Pterygodium excisum (Lindl.) Schltr. = Corycium excisum Lindl.■☆

320945　Pterygodium flanaganii Bolus = Corycium flanaganii (Bolus) Kurzweil et H. P. Linder ■☆

320946　Pterygodium hallii (Schelpe) Kurzweil et H. P. Linder;霍尔非洲兰■☆

320947　Pterygodium hastatum Bolus;戟形非洲兰■☆

320948　Pterygodium inversum (Thunb.) Sw.;倒垂非洲兰■☆

320949　Pterygodium leucanthum Bolus;白花非洲兰■☆

320950　Pterygodium macloughlinii L. Bolus = Pterygodium cooperi Rolfe ■☆

320951　Pterygodium magnum Rchb. f.;大非洲兰■☆

320952　Pterygodium microglossum (Lindl.) Schltr. = Corycium microglossum Lindl.■☆

320953　Pterygodium mundii Schltr. = Corycium bicolorum (Thunb.) Sw.■☆

320954　Pterygodium newdigatae Bolus;纽迪盖特非洲兰■☆

320955　Pterygodium newdigatae Bolus var. cleistogamum ?;封闭非洲兰■☆

320956　Pterygodium nigrescens (Sond.) Schltr. = Corycium nigrescens Sond. ■☆

320957　Pterygodium orobanchoides (L. f.) Schltr. = Corycium orobanchoides (L. f.) Sw. ■☆

320958　Pterygodium patersoniae Schltr. = Corycium carnosum (Lindl.) Rolfe ■☆

320959　Pterygodium pentherianum Schltr.;彭泰尔非洲兰■☆

320960　Pterygodium platypetalum Lindl.;阔瓣非洲兰■☆

320961　Pterygodium rubiginosum Sond. ex Bolus = Evotella rubiginosa (Sond. ex Bolus) Kurzweil et H. P. Linder ■☆

320962　Pterygodium sulcatum Roxb. = Pecteilis susannae (L.) Raf. ■

320963　Pterygodium sulcatum Roxb. = Zeuxine strateumatica (L.) Schltr. ■

320964　Pterygodium tricuspidatum (Bolus) Schltr. = Corycium tricuspidatum Bolus ■☆

320965　Pterygodium ukingense Schltr.;尤金非洲兰■☆

320966　Pterygodium venosum Lindl. = Ceratandra venosa (Lindl.) Schltr. ■☆

320967　Pterygodium vestitum (Sw.) Schltr. = Corycium orobanchoides (L. f.) Sw. ■☆

320968　Pterygodium volucris (L. f.) Sw.;翼非洲兰■☆

320969　Pterygolepis Rchb. = Pterolepis Schrad.(废弃属名)●■☆

320970　Pterygolepis Rchb. = Scirpus L.(保留属名)■

320971　Pterygoloma Hanst. = Alloplectus Mart. (保留属名)●■☆

320972　Pterygoloma Hanst. = Columnea L.●■☆

320973　Pterygopappus Hook. f. (1874);尖叶藓菊属■☆

320974　Pterygopappus lawrencii Hook. f.;尖叶藓菊■☆

320975　Pterygopleurum Kitag. (1937);翅棱芹属(翅肋芹属,凤尾参属);Pterygopleurum ■

320976　Pterygopleurum neurophyllum (Maxim.) Kitag.;翅棱芹(凤尾参,脉叶翅棱芹);Nervedleaf Pterygopleurum ■

320977　Pterygopodium Harms = Oxystigma Harms ●☆

320978　Pterygopodium balsamiferum Vermoesen = Prioria balsamifera (Vermoesen) Breteler ●☆

320979　Pterygopodium oxyphyllum Harms = Prioria oxyphylla (Harms) Breteler ●☆

320980　Pterygostachyum Nees ex Steud. = Dimeria R. Br. ■

320981　Pterygostachyum Steud. = Dimeria R. Br. ■

320982　Pterygostemon V. V. Botsch. (1977);翅蕊芥属■☆

320983　Pterygostemon V. V. Botsch. = Fibigia Medik. ■☆

320984　Pterygostemon spathulatus (Kar. et Kir.) V. V. Boch.;翅蕊芥■☆

320985　Pterygota Schott et Endl. (1832);翅苹婆属(翅子桐属);Pterygota ●

320986　Pterygota adolfi-friederici Engl. et K. Krause;弗里德里西翅苹婆●☆

320987　Pterygota alata (Roxb.) R. Br.;翅苹婆(翅果苹婆,海南苹婆,绿花翅苹婆);Buddha's Coconut, Pterygota, Winged Pterygota ●

320988　Pterygota alata K. Schum. = Pterygota schumanniana Engl. ●☆

320989　Pterygota bequaertii De Wild.;贝卡尔翅苹婆●☆

320990　Pterygota kamerunensis K. Schum. et Engl.;喀麦隆翅苹婆●☆

320991　Pterygota macrocarpa K. Schum.;大果翅苹婆●☆

320992　Pterygota mildbraedii Engl.;米尔德翅苹婆●☆

320993　Pterygota roxburghii Schott et Endl. = Pterygota alata (Roxb.) R. Br. ●

320994　Pterygota schumanniana Engl.;舒曼翅苹婆●☆

320995　Pterygota schweinfurthii Engl.;施韦翅苹婆●☆

320996　Pterypodium Rchb. f. = Pterygodium Sw. ●☆

320997　Pteryxia (Nutt.) J. M. Coult. et Rose = Pteryxia (Torr. et A. Gray) J. M. Coult. et Rose ■☆

320998　Pteryxia (Torr. et A. Gray) J. M. Coult. et Rose(1900);北美芹属■☆

320999　Pteryxia Nutt. = Pteryxia (Torr. et A. Gray) J. M. Coult. et Rose ■☆

321000　Pteryxia Nutt. ex Torr. et A. Gray = Cymopterus Raf. ■☆

321001　Pteryxia terebinthina (Hook.) J. M. Coult. et Rose;北美芹(乳香北美芹)■☆

321002　Ptichochilus Benth. = Ptychochilus Schauer ■

321003　Ptichochilus Benth. = Tropidia Lindl. ■

321004　Ptilagrostis Griseb. (1852);细柄茅属(剪股颖属,细柄草属);Ptilagrostis ■

321005　Ptilagrostis Griseb. = Stipa L. ■

321006　Ptilagrostis alpina (F. Schmidt) Sipliv. = Eragrostis virescens J. Presl ■☆

321007　Ptilagrostis alpina (F. Schmidt) Sipliv. = Stipa alpina (F. Schmidt) Petr. ■☆

321008　Ptilagrostis concinna (Hook. f.) Roshev.;太白细柄茅(优雅细柄茅);Elegant Ptilagrostis ■

321009　Ptilagrostis concinna (Hook. f.) Roshev. subsp. schischkiana Tzvelev = Ptilagrostis concinna (Hook. f.) Roshev. ■

321010　Ptilagrostis dichotoma Keng = Ptilagrostis dichotoma Keng ex Tzvelev ■

321011　Ptilagrostis dichotoma Keng ex Tzvelev;双叉细柄茅;Bifork Ptilagrostis, Dichotomous Ptilagrostis ■

321012　Ptilagrostis dichotoma Keng ex Tzvelev var. roshevitsiana Tzvelev;小花细柄茅;Roshvits Ptilagrostis ■

321013　Ptilagrostis junatovii Grubov;窄穗细柄茅;Junatov Ptilagrostis ■

321014　Ptilagrostis luquensis P. M. Peterson et al.;短花细柄茅■

321015　Ptilagrostis macrospicula L. B. Cai = Ptilagrostis yadongensis P.

C. Keng et J. S. Tang ■

321016 Ptilagrostis mongholica (Turcz. ex Trin.) Griseb.;细柄茅(蒙古细柄茅);Mongolian Ptilagrostis ■

321017 Ptilagrostis mongholica (Turcz. ex Trin.) Griseb. = Stipa mongholica Turcz. ex Trin. ■

321018 Ptilagrostis pelliotii (Danguy) Grubov;中亚细柄茅;Pelliot Ptilagrostis ■

321019 Ptilagrostis purpurea (Griseb.) Roshev. = Stipa purpurea Griseb. ■

321020 Ptilagrostis subsessiliflora (Rupr.) Roshev. = Stipa subsessiliflora (Rupr.) Roshev. ■

321021 Ptilagrostis subsessilifolia (Rupr.) Grubov = Stipa subsessiliflora (Rupr.) Roshev. ■

321022 Ptilagrostis tibetica (Mez) Tzvelev = Ptilagrostis mongholica (Turcz. ex Trin.) Griseb. ■

321023 Ptilagrostis yadongensis P. C. Keng et J. S. Tang;大穗细柄茅■

321024 Ptilanthelium Steud. = Schoenus L. ■

321025 Ptilanthus Gleason = Graffenrieda DC. ●☆

321026 Ptilepida Raf. = Actinea Juss. ■

321027 Ptilepida Raf. = Actinella Pers. ■

321028 Ptilepida Raf. = Helenium L. ■

321029 Ptileris Raf. = Erechtites Raf. ■

321030 Ptilimnium Raf. (1825);沼毛草属■☆

321031 Ptilimnium capillaceum (Michx.) Raf.;乳突沼毛草;Mock Bishop's Weed,Mock Bishop's-weed ■☆

321032 Ptilimnium costatum (Elliott) Raf.;沼毛草;Mock Bishop's Weed ■☆

321033 Ptilimnium nuttallii (DC.) Britton;纳他尔沼毛草;Mock Bishop's Weed ■☆

321034 Ptilina Nutt. ex Torr. et A. Gray = Didiplis Raf. ●■

321035 Ptilina Nutt. ex Torr. et A. Gray = Lythrum L. ●■

321036 Ptilium Pers. = Fritillaria L. ■

321037 Ptilium Pers. = Petilium Ludw. ■

321038 Ptilocalais A. Gray ex Greene = Microseris D. Don ■☆

321039 Ptilocalais Torrey ex Greene = Microseris D. Don ■☆

321040 Ptilocalyx Torr. et A. Gray = Coldenia L. ■

321041 Ptilocalyx Torr. et A. Gray(1857);羽萼紫草属■☆

321042 Ptilocalyx greggii Torr. et A. Gray;羽萼紫草■☆

321043 Ptilochaeta Nees(废弃属名) = Ptilochaeta Turcz.(保留属名)●☆

321044 Ptilochaeta Nees(废弃属名) = Rhynchospora Vahl(保留属名)●☆

321045 Ptilochaeta Turcz.(1843)(保留属名);翼木属●☆

321046 Ptilochaeta densiflora Nied.;密花翼毛木●☆

321047 Ptilochaeta elegans Nied.;雅致翼毛木●☆

321048 Ptilochaeta glabra Nied.;光翼毛木●☆

321049 Ptilochaeta nudipes Griseb.;裸梗翼毛木●☆

321050 Ptilocnema D. Don = Pholidota Lindl. ex Hook. ■

321051 Ptilocnema bracteata D. Don = Pholidota bracteata (D. Don) Seidenf. ■

321052 Ptilocnema bracteata D. Don = Pholidota imbricata Hook. ■

321053 Ptilomeria Nutt. = Baeria Fisch. et C. A. Mey. ■☆

321054 Ptilomeris coronaria Nutt. = Lasthenia coronaria (Nutt.) Ornduff ■☆

321055 Ptiloneilema Steud. = Melanocenchris Nees ■☆

321056 Ptiloneilema plumosum Steud. = Melanocenchris abyssinica (R. Br. ex Fresen.) Hochst. ■☆

321057 Ptilonella Nutt. = Blepharipappus Hook.(废弃属名)●☆

321058 Ptilonella Nutt. = Lebetanthus Endl.(保留属名)●☆

321059 Ptilonema Hook. f. = Melanocenchris Nees ■☆

321060 Ptilonema Hook. f. = Ptiloneilema Steud. ■☆

321061 Ptilonilema Post et Kuntze = Melanocenchris Nees ■☆

321062 Ptilonilema Post et Kuntze = Ptilonema Hook. f. ■☆

321063 Ptilophora (Torr. et A. Gray ex Hook. f.) A. Gray = Microseris D. Don ■☆

321064 Ptilophora (Torr. et A. Gray ex Hook. f.) A. Gray = Ptilocalais A. Gray ex Greene ■☆

321065 Ptilophora A. Gray = Microseris D. Don ■☆

321066 Ptilophora A. Gray = Ptilocalais A. Gray ex Greene ■☆

321067 Ptilophyllum (Nutt.) Rchb. = Burshia Raf. ■

321068 Ptilophyllum Raf. = Myriophyllum L. ■

321069 Ptiloria Raf.(废弃属名) = Stephanomeria Nutt.(保留属名)●■☆

321070 Ptiloria pleurocarpa Greene = Stephanomeria virgata Benth. subsp. pleurocarpa (Greene) Gottlieb ■☆

321071 Ptiloria tenuifolia Raf. = Stephanomeria tenuifolia (Raf.) H. M. Hall ■☆

321072 Ptilosciadlum Steud. = Rhynchospora Vahl(保留属名)■

321073 Ptilosla Tausch = Picris L. ■

321074 Ptilostemon Cass. (1816);卵果蓟属(羽蕊菊属)■☆

321075 Ptilostemon abylensis (Pau et Font Quer) Greuter;阿比尔卵果蓟■☆

321076 Ptilostemon casabonae (L.) Greuter;卡萨卵果蓟■☆

321077 Ptilostemon dyricola (Maire) Greuter;荒地卵果蓟■☆

321078 Ptilostemon gnaphaloides (Cirillo) Soják;鼠麴卵果蓟■☆

321079 Ptilostemon leptophyllus (Pau et Font Quer) Greuter;互叶卵果蓟■☆

321080 Ptilostemon muticus Cass.;卵果蓟(羽蕊菊)■☆

321081 Ptilostemon rhiphaeus (Pau et Font Quer) Greuter;山地卵果蓟■☆

321082 Ptilostemon rhiphaeus (Pau et Font Quer) Greuter var. tetouanensis (Font Quer) Greuter = Ptilostemon rhiphaeus (Pau et Font Quer) Greuter ■☆

321083 Ptilostemum Steud. = Ptilostemon Cass. ■☆

321084 Ptilostephium Kunth = Tridax L. ■●

321085 Ptilothrix K. L. Wilson(1994);羽毛莎属■●☆

321086 Ptilothrix deusta (R. Br.) K. L. Wilson;羽毛莎■●☆

321087 Ptilotrichum C. A. Mey. (1831);燥原荠属(节毛芥属,节毛荠属);Ptilotrichum ●■

321088 Ptilotrichum C. A. Mey. = Alyssum L. ■●

321089 Ptilotrichum canescens (DC.) C. A. Mey.;燥原荠(灰毛庭荠);Canescent Ptilotrichum,Ptilotrichum ●■

321090 Ptilotrichum canescens (DC.) C. A. Mey. = Alyssum canescens DC. ■

321091 Ptilotrichum canescens (DC.) C. A. Mey. subsp. tenuifolium (Steph. ex Willd.) Hanelt et Davazamc = Alyssum tenuifolium Steph. ex Willd. ■

321092 Ptilotrichum cretaceum (Adams) Ledeb. = Alyssum canescens DC. ■

321093 Ptilotrichum elongamm (DC.) C. A. Mey. = Alyssum canescens DC. var. elongatum DC. ■

321094 Ptilotrichum elongamm (DC.) C. A. Mey. = Alyssum tenuifolium Steph. ex Willd. ■

321095 Ptilotrichum elongatum C. A. Mey. = Alyssum canescens DC. ■

321096 Ptilotrichum spinosum (L.) Boiss. = Hormatophylla spinosa (L.) P. Küpfer ■☆

321097 Ptilotrichum tenuifolium (Steph. ex Willd.) C. A. Mey. = Alyssum tenuifolium Steph. ex Willd. ■

321098　Ptilotrichum tenuifolium C. A. Mey.；薄叶燥原荠；Thinleaf Ptilotrichum ●

321099　Ptilotrichum wageri Jafri；西藏燥原荠；Tibet Ptilotrichum，Xizang Ptilotrichum ■

321100　Ptilotrichum wageri Jafri ＝ Draba winterbottomii（Hook. f. et Thomson）Pohle ■

321101　Ptilotum Dulac ＝ Dryas L. ●■

321102　Ptilotus R. Br.（1810）；澳洲苋属；Multa-mulla，Pussy Tail ■●☆

321103　Ptilotus manglesii（Lindl.）F. Muell.；曼格澳洲苋■☆

321104　Ptilotus ovatus Moq. ＝ Psilotrichum elliotii Baker et C. B. Clarke ■☆

321105　Ptilurus D. Don ＝ Leuceria Lag. ■☆

321106　Ptosimopappus Boiss. ＝ Centaurea L.（保留属名）●■

321107　Ptyas Salisb. ＝ Aloe L. ●■

321108　Ptyas Salisb. ＝ Kumara Medik. ●■

321109　Ptyanthera Decne.（1844）；褶药萝藦属☆

321110　Ptyanthera berteroi Decne.；褶药萝藦☆

321111　Ptychandra Scheff.（1876）；襞蕊桐属●☆

321112　Ptychandra Scheff. ＝ Heterospathe Scheff. ●☆

321113　Ptychandra glabra Burret；光襞蕊桐●☆

321114　Ptychandra glauca Scheff.；襞蕊桐●☆

321115　Ptychandra montana Burret；山地襞蕊桐●☆

321116　Ptychanthera Post et Kuntze ＝ Ptycanthera Decne. ☆

321117　Ptychocarpa（R. Br.）Spach ＝ Grevillea R. Br. ex Knight（保留属名）●

321118　Ptychocarpa Spach ＝ Grevillea R. Br. ex Knight（保留属名）●

321119　Ptychocarpus Hils. ex Sieber ＝ Melochia L.（保留属名）●■

321120　Ptychocarpus Kuhlm. ＝ Neoptychocarpus Buchheim ●☆

321121　Ptychocarya R. Br. ＝ Ptychocarya R. Br. ex Wall. ■☆

321122　Ptychocarya R. Br. ex Wall. ＝ Scirpodendron Zipp. ex Kurz ■☆

321123　Ptychocaryum Kuntze ex H. Pfeiff. ＝ Ptychocarya R. Br. ex Wall. ■☆

321124　Ptychocaryum Kuntze ex H. Pfeiff. ＝ Scirpodendron Zipp. ex Kurz ■☆

321125　Ptychocentrum（Wight et Arn.）Benth. ＝ Rhynchosia Lour.（保留属名）●■

321126　Ptychocentrum Benth. ＝ Rhynchosia Lour.（保留属名）●■

321127　Ptychochilus Schauer ＝ Tropidia Lindl. ■

321128　Ptychococcus Becc.（1885）；皱果片棕属（襞果桐属，襞果椰属，襞实桐属，折果椰子属，皱果椰属）；Ptychococcus ●☆

321129　Ptychococcus arecinus Becc.；皱果片棕●☆

321130　Ptychococcus elatus Becc.；大皱果片棕●☆

321131　Ptychodea Willd. ex Cham. ＝ Sipanea Aubl. ●■☆

321132　Ptychodea Willd. ex Cham. et Schltdl. ＝ Sipanea Aubl. ●■☆

321133　Ptychodon（Klotzsch ex Endl.）Rchb. ＝ Lafoensia Vand. ●

321134　Ptychodon Klotzsch ex Rchb. ＝ Lafoensia Vand. ●

321135　Ptychogyne Pfitzer ＝ Coelogyne Lindl. ■

321136　Ptycholepis Griseb. ＝ Blepharodon Decne. ■☆

321137　Ptycholepis Griseb. ex Lechler ＝ Blepharodon Decne. ■☆

321138　Ptycholobium Harms（1915）；异灰毛豆属■☆

321139　Ptycholobium biflorum（E. Mey.）Brummitt；双花异灰毛豆■☆

321140　Ptycholobium biflorum（E. Mey.）Brummitt subsp. angolensis（Baker）Brummitt；安哥拉异灰毛豆■☆

321141　Ptycholobium contortum（N. E. Br.）Brummitt；旋异灰毛豆■☆

321142　Ptycholobium plicatum（Oliv.）Harms；折扇异灰毛豆■☆

321143　Ptychomeria Benth. ＝ Gymnosiphon Blume ■

321144　Ptychopetalum Benth.（1843）；褶瓣树属（巴西椇椇木属）●☆

321145　Ptychopetalum acuminatissimum Engl. ＝ Ptychopetalum

petiolatum Oliv. ●☆

321146　Ptychopetalum alliaceum De Wild. ＝ Olax gambecola Baill. ●☆

321147　Ptychopetalum anceps Oliv.；二棱褶瓣树●☆

321148　Ptychopetalum cuspidatum R. E. Fr. ＝ Rhaphiostylis beninensis（Hook. f. ex Planch.）Planch. ex Benth. ●☆

321149　Ptychopetalum laurentii De Wild. ＝ Olax subscorpioidea Oliv. ●☆

321150　Ptychopetalum nigricans De Wild.；黑褶瓣树●☆

321151　Ptychopetalum olacoides Benth.；褶瓣树（巴西椇椇木）●☆

321152　Ptychopetalum petiolatum Oliv.；柄叶褶瓣树●☆

321153　Ptychopetalum petiolatum Oliv. var. paniculatum Engl.；圆锥褶瓣树●☆

321154　Ptychopetalum uncinatum Anselmino；沟褶瓣树●☆

321155　Ptychopyxis Miq.（1861）；皱果大戟属●☆

321156　Ptychopyxis angustifolia Gage；窄叶皱果大戟●☆

321157　Ptychopyxis grandis Airy Shaw；大皱果大戟●☆

321158　Ptychoraphis Becc. ＝ Rhopaloblaste Scheff. ●☆

321159　Ptychosema Benth.（1839）；异荚属■☆

321160　Ptychosema pusillum Benth.；异荚豆■☆

321161　Ptychosperma Labill.（1809）；绉子棕属（襞籽椰属，海桃椰子属，麦加绉子棕属，射叶椰子属，射叶椰子属，绉籽椰属，绉子棕属，皱子棕属）；Ptychosperma，Solitaire Palm ●☆

321162　Ptychosperma angustifolium Blume；细叶绉子棕（射叶椰子）●☆

321163　Ptychosperma burretianum Essig；布氏绉子棕（巴提青棕）●☆

321164　Ptychosperma elegans（R. Br.）Blume；优雅皱子棕（海桃椰子）；Alexander Palm，Alexander's Palm，Australian Feather Palm，Princess Palm，Solitaire Palm，Solitary Palm ●☆

321165　Ptychosperma hosinoi（Kaneh.）H. E. Moore et Fosberg；星夜氏皱子棕●☆

321166　Ptychosperma hospitum（Burret）Burret；奇异皱子棕●☆

321167　Ptychosperma laccospadix Benth. ＝ Laccospadix australasica H. Wendl. et Drude ●☆

321168　Ptychosperma macarthuri（H. Wendl.）G. Nicholson ＝ Ptychosperma macarthurii（H. Wendl. ex Veitch）G. Nicholson ●☆

321169　Ptychosperma macarthuri（H. Wendl.）H. Wendl. ex Hook. f. ＝ Ptychosperma macarthurii（H. Wendl. ex Veitch）G. Nicholson ●☆

321170　Ptychosperma macarthurii（H. Wendl. ex Veitch）G. Nicholson；马卡氏皱子棕（马氏射叶椰子，青棕）；Cluster Palm，Hurricane Palm，MacArthur Feather Palm，Macarthur Palm ●☆

321171　Ptychosperma microcarpum（Burret）Burret；小果皱子棕●☆

321172　Ptychosperma nicolai Burret；尼古拉绉子棕●☆

321173　Ptychosperma propinquum Becc. ex Martelli；洋皱子棕●☆

321174　Ptychosperma sanderianum Ridl.；桑德皱子棕（所罗门绉子棕，新几内亚绉子棕）；Sander Ptychosperma ●☆

321175　Ptychosperma scheffierii Becc. ex Martelli；穴穗棕●☆

321176　Ptychosperma waitianum Essig；威提亚皱子棕●☆

321177　Ptychostigma Hochst. ＝ Galiniera Delile ●☆

321178　Ptychostigma saxifraga（Hochst.）Hochst. ＝ Galiniera saxifraga（Hochst.）Bridson ●☆

321179　Ptychostoma Post et Kuntze ＝ Lonchostoma Wikstr.（保留属名）●☆

321180　Ptychostoma Post et Kuntze ＝ Ptyxostoma Vahl（废弃属名）■☆

321181　Ptychostylus Tiegh. ＝ Loranthus Jacq.（保留属名）●

321182　Ptychostylus Tiegh. ＝ Struthanthus Mart.（保留属名）●☆

321183　Ptychotis W. D. J. Koch ＝ Carum L. ■

321184　Ptychotis W. D. J. Koch（1824）；褶耳草属■☆

321185　Ptychotis achilleifolia DC. ＝ Meeboldia achilleifolia（DC.）P. K. Mukh. et Constance ■

321186　Ptychotis ajowan DC. = Trachyspermum ammi（L.）Sprague ■

321187　Ptychotis ammoides W. D. J. Koch；褶耳草■☆

321188　Ptychotis ammoides W. D. J. Koch　= Ammoides pusilla（Brot.）Breistr. ■☆

321189　Ptychotis aspera Pomel = Ammoides pusilla（Brot.）Breistr. ■☆

321190　Ptychotis atlantica Coss. et Durieu = Ammoides atlantica（Coss. et Durieu）H. Wolff ■☆

321191　Ptychotis coptica（L.）DC. = Trachyspermum ammi（L.）Sprague ■

321192　Ptychotis didyma Sond. = Stoibrax capense（Lam.）B. L. Burtt ■☆

321193　Ptychotis hispida（Thunb.）Sond. = Sonderina hispida（Thunb.）H. Wolff ■☆

321194　Ptychotis involucrata（Roxb.）Lindl. = Trachyspermum roxburghianum（DC.）H. Wolff ■

321195　Ptychotis meisneri Sond. = Sonderina humilis（Meisn.）H. Wolff ■☆

321196　Ptychotis puberula DC. = Pimpinella puberula（DC.）H. Boissieu ■

321197　Ptychotis roxburghiana DC. = Trachyspermum roxburghianum（DC.）H. Wolff ■

321198　Ptychotis tenuis Sond. = Sonderina tenuis（Sond.）H. Wolff ■☆

321199　Ptyssiglottis T. Anderson（1860）；折舌爵床属■☆

321200　Ptyssiglottis flava Ridl.；黄折舌爵床■☆

321201　Ptyssiglottis lanceolata Hallier f. ；披针叶折舌爵床■☆

321202　Ptyssiglottis laxa（Lindau）Benoist；松散折舌爵床■☆

321203　Ptyssiglottis leptoneura Hallier f. ；细脉折舌爵床■☆

321204　Ptyssiglottis leptostachya S. Moore；细穗折舌爵床■☆

321205　Ptyssiglottis obovata S. Moore；倒卵形折舌爵床■☆

321206　Ptyssiglottis parviflora Ridl. ；小花折舌爵床■☆

321207　Ptyxostoma Vahl（废弃属名）= Lonchostoma Wikstr.（保留属名）●☆

321208　Ptyxostoma monogyna Vahl = Lonchostoma monogynum（Vahl）Pillans ●☆

321209　Ptyxostoma myrtoides Vahl = Lonchostoma myrtoides（Vahl）Pillans ●☆

321210　Ptyxostoma quadrifidum Kuntze = Campylostachys cernua（L. f.）Kunth ●☆

321211　Puberula Rydb. = Johanneshowellia Reveal ■☆

321212　Pubeta L.（废弃属名）= Duroia L. f.（保留属名）●☆

321213　Pubilaria Raf. = Simethis Kunth（保留属名）●☆

321214　Pubistylus Thoth.（1966）；柔毛茜属●☆

321215　Pubistylus andamanensis Thoth. ；柔毛茜●☆

321216　Pubistylus andamanensis Thoth. = Diplospora andamanensis（Thoth.）M. Gangop. et Chakrab. ●☆

321217　Publicaria Deflers = Pulicaria Gaertn. ■●

321218　Pucara Ravenna（1972）；北秘鲁石蒜属■☆

321219　Pucara leucantha Ravenna；北秘鲁石蒜■☆

321220　Puccinellia Parl.（1848）（保留属名）；碱茅属（卜氏草属,铺茅属,盐茅属）；Alkali Grass, Alkaligrass, Alkali-grass ■

321221　Puccinellia adpressa Ohwi；匍匐碱茅■

321222　Puccinellia adpressa Ohwi = Puccinellia kurilensis（Takeda）Honda ■

321223　Puccinellia adpressa Ohwi = Puccinellia nipponica Ohwi ■

321224　Puccinellia airoides（Nutt.）S. Watson et J. M. Coult. = Puccinellia nuttalliana（Schult.）Hitchc. ■☆

321225　Puccinellia alascana Scribn. et Merr. = Puccinellia tenella（Lange）Holmb. ex A. E. Porsild ■

321226　Puccinellia altaica Tzvelev；阿尔泰碱茅■

321227　Puccinellia angusta（Nees）C. A. Sm. et C. E. Hubb.；狭碱茅■☆

321228　Puccinellia angustata（R. Br.）E. L. Rand et Redfield；侧序碱茅■

321229　Puccinellia anisoclada（V. I. Krecz.）Parsa = Puccinellia gigantea（Grossh.）Grossh. ■

321230　Puccinellia anisoclada V. I. Krecz. ；异枝碱茅■

321231　Puccinellia arjinshanensis D. F. Cui；阿尔金山碱茅；Arjinshan Alkaligrass ■

321232　Puccinellia borealis Swallen = Puccinellia distans（Jacq.）Parl. subsp. borealis（Holmb.）W. E. Hughes

321233　Puccinellia bulbosa（Grossh.）Grossh. ；鳞茎碱茅■

321234　Puccinellia capillaris（Lilj.）R. K. Jansen；细穗碱茅■

321235　Puccinellia capillaris（Lilj.）R. K. Jansen subsp. pulvinata（Fr.）Tzvelev = Puccinellia pulvinata（Fr.）V. I. Krecz.

321236　Puccinellia chinampoensis Ohwi；朝鲜碱茅（铺茅）；Korea Alkaligrass ■

321237　Puccinellia choresmica V. I. Krecz. ；短生碱茅■

321238　Puccinellia coarctata Fernald et Weath. = Puccinellia distans（Jacq.）Parl. subsp. borealis（Holmb.）W. E. Hughes ■

321239　Puccinellia coarctata Fernald et Weath. var. pseudofasciculata T. J. Sorensen = Puccinellia distans（Jacq.）Parl. subsp. borealis（Holmb.）W. E. Hughes

321240　Puccinellia convoluta（Hornem.）Fourr. ；卷叶碱茅■

321241　Puccinellia coreensis（Hack.）Honda；高丽碱茅■

321242　Puccinellia coreensis（Hack.）Honda var. asperifolia Kitag. = Puccinellia coreensis（Hack.）Honda ■

321243　Puccinellia coreensis Hack. ex Honda var. asperifolia Kitag. = Puccinellia coreensis（Hack.）Honda ■

321244　Puccinellia cusickii Weath. = Puccinellia nuttalliana（Schult.）Hitchc. ■☆

321245　Puccinellia degeensis L. Liou；德格碱茅■

321246　Puccinellia diffusa（V. I. Krecz.）V. I. Krecz. ex Drobow；展穗碱茅■

321247　Puccinellia diffusa V. I. Krecz. = Puccinellia diffusa（V. I. Krecz.）V. I. Krecz. ex Drobow ■

321248　Puccinellia distans（Jacq.）Parl. ；碱茅（铺茅）；Alkali Grass, Alkaligrass, European Alkali Grass, Lax Puccinellia, Weeping Alkali Grass, Weeping Alkaligrass ■

321249　Puccinellia distans（Jacq.）Parl. subsp. borealis（Holmb.）W. E. Hughes；北方碱茅；European Alkali Grass, Reflexed Meadow-grass, Weeping Alkali Grass ■

321250　Puccinellia distans（Jacq.）Parl. var. angustifolia（Blytt）Holmb. = Puccinellia distans（Jacq.）Parl. subsp. borealis（Holmb.）W. E. Hughes ■

321251　Puccinellia distans（Jacq.）Parl. var. tenuis（Uechtr.）Fernald et Weath. = Puccinellia distans（Jacq.）Parl. ■

321252　Puccinellia distans（L.）Parl. f. robusta Roshev. = Puccinellia anisoclada V. I. Krecz. ■

321253　Puccinellia distans（L.）Parl. subsp. borealis（Holmb.）W. E. Hughes = Puccinellia distans（Jacq.）Parl. subsp. borealis（Holmb.）W. E. Hughes ■

321254　Puccinellia distans（L.）Parl. subsp. convoluta（Hornem.）Maire et Weiller = Puccinellia convoluta（Hornem.）Fourr. ■

321255　Puccinellia distans（L.）Parl. subsp. embergeri（H. Lindb.）Maire et Weiller = Puccinellia festuciformis（Host）Parl. ■

321256　Puccinellia distans（L.）Parl. subsp. festuciformis（Host）

Maire et Weiller ＝Puccinellia festuciformis（Host）Parl. ■

321257　Puccinellia distans（L.）Parl. subsp. fontqueri Maire ＝ Puccinellia festuciformis（Host）Parl. ■

321258　Puccinellia distans（L.）Parl. subsp. glauca（Regel）V. I. Krecz. ＝Puccinellia distans（Jacq.）Parl. ■

321259　Puccinellia distans（L.）Parl. subsp. sevangensis（Grossh.）Tzvelev ＝Puccinellia sevangensis Grossh. ■

321260　Puccinellia distans（L.）Parl. subsp. tenuifolia（Boiss. et Reut.）Maire et ＝Puccinellia tenuifolia（Boiss. et Reut.）H. Lindb. ■☆

321261　Puccinellia distans（L.）Parl. var. fallax Maire ＝Puccinellia festuciformis（Host）Parl. ■

321262　Puccinellia distans（L.）Parl. var. halophila（Trab.）Emb. et Maire ＝Puccinellia fasciculata（Torr.）E. P. Bicknell ■☆

321263　Puccinellia distans（L.）Parl. var. micrandra Keng ＝ Puccinellia micrandra（Keng）Keng et S. L. Chen ■

321264　Puccinellia distans（L.）Parl. var. permixta（Guss.）Trab. ＝ Puccinellia fasciculata（Torr.）E. P. Bicknell ■☆

321265　Puccinellia distans（L.）Parl. var. poiformis Emb. et Maire ＝ Puccinellia fasciculata（Torr.）E. P. Bicknell ■☆

321266　Puccinellia distans（L.）Parl. var. salina Fuss ＝Puccinellia festuciformis（Host）Parl. ■

321267　Puccinellia distans（L.）Parl. var. vulgaris Coss. et Durieu ＝ Puccinellia fasciculata（Torr.）E. P. Bicknell subsp. pseudodistans（Crép.）Kerguélen ■☆

321268　Puccinellia distans（Wahlb.）Parl. ＝Puccinellia distans（L.）Parl. ■

321269　Puccinellia distans（Wahlb.）Parl. subsp. glauca（Regel）Tzvelev ＝Puccinellia distans（Jacq.）Parl. ■

321270　Puccinellia dolicholepis（V. I. Krecz.）Pavlov；毛稃碱茅（灰绿碱茅）■

321271　Puccinellia dolicholepis V. I. Krecz. ＝Puccinellia dolicholepis（V. I. Krecz.）Pavlov ■

321272　Puccinellia dolicholepis V. I. Krecz. var. paradosa Serg. ＝ Puccinellia altaica Tzvelev ■

321273　Puccinellia embergeri（H. Lindb.）H. Lindb. ＝ Puccinellia festuciformis（Host）Parl. ■

321274　Puccinellia expansa（Crép.）Julià et J. M. Monts.；扩展碱茅■☆

321275　Puccinellia fasciculata（Torr.）E. P. Bicknell；簇生碱茅；Borrer's Saltmarsh Grass ■☆

321276　Puccinellia fasciculata（Torr.）E. P. Bicknell subsp. pseudodistans（Crép.）Kerguélen；拟碱茅■☆

321277　Puccinellia fasciculata（Torr.）E. P. Bicknell var. caespitosa Allan et Jansen ＝Puccinellia fasciculata（Torr.）E. P. Bicknell ■☆

321278　Puccinellia fernaldii（Hitchc.）E. G. Voss；弗氏碱茅；Fernald's False Manna Grass ■☆

321279　Puccinellia festuciformis（Host）Parl.；羊茅状碱茅■

321280　Puccinellia festuciformis（Host）Parl. subsp. convoluta（Hornem.）W. E. Hughes ＝Puccinellia convoluta（Hornem.）Fourr. ■

321281　Puccinellia festuciformis（Host）Parl. subsp. intermedia（Schrad.）Hughes ＝Puccinellia intermedia（Schrad.）Janch. ■

321282　Puccinellia festuciformis（Host）Parl. subsp. tenuifolia（Boiss. et Reut.）W. E. Hughes ＝Puccinellia tenuifolia（Boiss. et Reut.）H. Lindb. ■☆

321283　Puccinellia filifolia（Trin.）Tzvelev；线叶碱茅■

321284　Puccinellia filiformis Keng ＝Puccinellia distans（Jacq.）Parl. ■

321285　Puccinellia florida D. F. Cui；玫花碱茅（多花碱茅）；Manyflower Alkaligrass ■

321286　Puccinellia geniculata V. I. Krecz.；膝曲碱茅■

321287　Puccinellia gigantea（Grossh.）Grossh.；大碱茅■

321288　Puccinellia gigantea（Grossh.）Grossh. subsp. bulbosa（Grossh.）Tzvelev ＝Puccinellia bulbosa（Grossh.）Grossh. ■

321289　Puccinellia glauca（Regel）V. I. Krecz. ex Drobow；灰绿碱茅■

321290　Puccinellia grossheimiana（V. I. Krecz.）V. I. Krecz.；格海碱茅■

321291　Puccinellia gyirongensis L. Liou；吉隆碱茅■

321292　Puccinellia gyirongensis L. Liou ＝Puccinellia tianschanica（Tzvelev）Ikonn. ■

321293　Puccinellia hackeliana（Roshev.）Pers. ＝Puccinellia hackeliana（V. I. Krecz.）V. I. Krecz. ex Drobow ■

321294　Puccinellia hackeliana（Roshev.）Pers. subsp. humilis（Litv. ex V. I. Krecz.）Tzvelev ＝Puccinellia humilis Litv. ex V. I. Krecz. ■

321295　Puccinellia hackeliana（V. I. Krecz.）V. I. Krecz. ex Drobow；高山碱茅■

321296　Puccinellia hauptiana（Trin.）V. I. Krecz.；鹤甫碱茅（短药碱茅,小林碱茅）；Kobayashi Alkaligrass ■

321297　Puccinellia himalaica Tzvelev；喜马拉雅碱茅；Ximalaya Alkaligrass ■

321298　Puccinellia humilis（Litv. ex V. I. Krecz.）Bor；矮碱茅（喀什碱茅）■

321299　Puccinellia humilis Litv. ex V. I. Krecz. ＝Puccinellia hackeliana（Roshev.）Pers. subsp. humilis（Litv. ex V. I. Krecz.）Tzvelev ■

321300　Puccinellia humilis Litv. ex V. I. Krecz. ＝Puccinellia humilis（Litv. ex V. I. Krecz.）Bor ■

321301　Puccinellia iberica（Wolley-Dod）Tzvelev ＝Puccinellia festuciformis（Host）Parl. ■

321302　Puccinellia iliensis（V. I. Krecz.）Serg.；伊犁碱茅■

321303　Puccinellia iliensis Tzvelev ＝Puccinellia iliensis（V. I. Krecz.）Serg. ■

321304　Puccinellia intermedia（Schrad.）Janch.；中间碱茅■

321305　Puccinellia jeholensis Kitag.；热河碱茅■

321306　Puccinellia jeholensis Kitag. ＝Puccinellia macranthera V. I. Krecz. ■

321307　Puccinellia jenisseiensis（Roshev.）Tzvelev；热尼斯碱茅■☆

321308　Puccinellia kackeliana subsp. humilis（Litv. ex V. I. Krecz.）Tzvelev ＝Puccinellia humilis Litv. ex V. I. Krecz. ■

321309　Puccinellia kamtschatica（Holmb.）V. I. Krecz.；堪察加碱茅；Kamtschatka Alkaligrass ■

321310　Puccinellia kamtschatica（Holmb.）V. I. Krecz. var. asperula Holmb. ＝Puccinellia kamtschatica（Holmb.）V. I. Krecz. ■

321311　Puccinellia kamtschatica（Holmb.）V. I. Krecz. var. sublaevis Holmb. ＝Puccinellia kurilensis（Takeda）Honda ■

321312　Puccinellia kamtschatica Holmb. var. sublaevis Holmb. ＝ Puccinellia kurilensis（Takeda）Honda ■

321313　Puccinellia kanashiroi Ohwi；金城碱茅；Kanashira Alkaligrass ■

321314　Puccinellia kashmiriana Bor；克什米尔碱茅■

321315　Puccinellia kengiana Ohwi ＝Pseudosclerochloa kengiana（Ohwi）Tzvelev ■

321316　Puccinellia kengiana Ohwi ＝Sclerochloa kengiana（Ohwi）Tzvelev ■

321317　Puccinellia kobayashii Ohwi；小林碱茅■

321318　Puccinellia kobayashii Ohwi ＝Puccinellia hauptiana（Trin.）V. I. Krecz. ■

321319 Puccinellia kobayashii Ohwi = Puccinellia macranthera V. I. Krecz. ■

321320 Puccinellia koeieana（Grossh.）Grossh. = Puccinellia koeieana Melderis ■

321321 Puccinellia koeieana Melderis；科氏碱茅■

321322 Puccinellia kulundensis Serg. = Puccinellia manshuriensis Ohwi ■

321323 Puccinellia kunlunica Tzvelev；昆仑碱茅■

321324 Puccinellia kurilensis（Takeda）Honda；千岛碱茅；Dwarf Puccinellia ■

321325 Puccinellia ladakhensis（H. Hartm.）Dickoré；拉达克碱茅■

321326 Puccinellia ladyginii K. V. Ivanova ex Tzvelev；布达尔碱茅■

321327 Puccinellia laeviuscula V. I. Krecz. = Puccinellia tenella（Lange）Holmb. ex A. E. Porsild ■

321328 Puccinellia leiolepis L. Liou；光稃碱茅■

321329 Puccinellia letroflexa var. pulvinata Holmb. ex Linden. = Puccinellia pulvinata（Fr.）V. I. Krecz. ■

321330 Puccinellia limosa（Schur）Holmb.；沼泞碱茅■

321331 Puccinellia lucida Fernald et Weath.；光亮碱茅；Bright Puccinellia ■☆

321332 Puccinellia macranthera（V. I. Krecz.）Norl.；大药碱茅（热河碱茅）；Rehe Alkaligrass，Reho Alkaligrass ■

321333 Puccinellia macranthera V. I. Krecz. = Puccinellia macranthera（V. I. Krecz.）Norl. ■

321334 Puccinellia macropus V. I. Krecz.；大足碱茅■☆

321335 Puccinellia manchuriensis Ohwi；东北碱茅（柔枝碱茅）；Northeast Alkaligrass ■

321336 Puccinellia maritima（Huds.）Parl.；海滨碱茅；Sea Mannagrass，Sea Meadow-grass，Seashore Alkali Grass，Seashore Alkaligrass，Seashore Alkali-grass，Seaside Alkaligrass ■

321337 Puccinellia micrandra（Keng）Keng et S. L. Chen；微药碱茅；Little-anther Alkaligrass ■

321338 Puccinellia micranthera D. F. Cui；小药碱茅；Small-anther Alkaligrass ■

321339 Puccinellia minuta Bor；侏碱茅■

321340 Puccinellia mongolica（Norl.）Bubnova = Puccinellia tenuiflora（Griseb.）Scribn. et Merr. ■

321341 Puccinellia multiflora L. Liou；多花碱茅■

321342 Puccinellia nipponica Ohwi；日本碱茅■

321343 Puccinellia nudiflora（Hack.）Tzvelev；裸花碱茅■

321344 Puccinellia nuttalliana（Schult.）Hitchc.；纳托尔碱茅；Nuttall's Alkali Grass ■☆

321345 Puccinellia pallida（Torr.）R. T. Clausen；苍白碱茅；Pale False Manna Grass ■☆

321346 Puccinellia pallida（Torr.）R. T. Clausen = Torreyochloa pallida（Torr.）G. L. Church ■☆

321347 Puccinellia pallida（Torr.）R. T. Clausen = Windsoria pallida Torr. ■☆

321348 Puccinellia pallida（Torr.）T. Koyama = Windsoria pallida Torr. ■☆

321349 Puccinellia pallida（Torr.）T. Koyama subsp. natans（Kom.）T. Koyama = Torreyochloa natans（Kom.）Church ■☆

321350 Puccinellia pallida（Torr.）T. Koyama subsp. viridis（Honda）T. Koyama = Torreyochloa viridis（Honda）Church ■☆

321351 Puccinellia palustris（Seenus）Grossh. subsp. jeholensis（Kitag.）Norl. = Puccinellia jeholensis Kitag. ■

321352 Puccinellia palustris（Seenus）Hayek = Puccinellia festuciformis（Host）Parl. ■

321353 Puccinellia palustris（Seenus）Hayek subsp. festuciformis（Host）Briq. = Puccinellia festuciformis（Host）Parl. ■

321354 Puccinellia palustris（Seenus）Hayek subsp. tenuifolia（Boiss. et Reut.）Emb. et Maire = Puccinellia tenuifolia（Boiss. et Reut.）H. Lindb. ■☆

321355 Puccinellia palustris Grossh. subsp. jeholensis（Kitag.）Norl. = Puccinellia jeholensis Kitag. ■

321356 Puccinellia pamirica（Roshev.）V. I. Krecz. ex Ovcz. et Czukav. subsp. vachanica（Ovcz. et Czukav.）Tzvelev = Puccinellia vachanica Ovcz. et Czukav. ■

321357 Puccinellia pamirica（Roshev.）V. I. Krecz. ex Roshev. et Czukav.；帕米尔碱茅■

321358 Puccinellia pauciramea（Hack.）V. I. Krecz. ex Ovcz. et Czukav.；少枝碱茅■

321359 Puccinellia pauciramea（Hack.）V. I. Krecz. ex Ovcz. et Czukav. = Puccinellia nudiflora（Hack.）Tzvelev ■

321360 Puccinellia paupercula Fernald et Weath. = Puccinellia pumila（Vassilcz.）Hitchc. ■

321361 Puccinellia phryganodes（Trin.）Scribn. et Merr.；佛利碱茅■☆

321362 Puccinellia phryganodes（Trin.）Scribn. et Merr. subsp. geniculata（V. I. Krecz.）Tzvelev = Puccinellia geniculata V. I. Krecz. ■

321363 Puccinellia platyglumis L. Liou = Poa platyglumis（L. Liou）L. Liou ■

321364 Puccinellia platyglumis L. Liou = Poa pseudamoena Bor ■

321365 Puccinellia platyglumis L. Liou = Puccinellia minuta Bor ■

321366 Puccinellia poaeoides Keng；莓系碱茅；Bluegrass-shaped Alkaligrass ■

321367 Puccinellia poaeoides Keng = Puccinellia macranthera V. I. Krecz. ■

321368 Puccinellia poecilantha（C. Koch）V. I. Krecz. = Puccinellia poecilantha（K. Koch）V. I. Krecz. ■

321369 Puccinellia poecilantha（K. Koch）V. I. Krecz.；斑秤碱茅■

321370 Puccinellia przewalskii Tzvelev；勃氏碱茅■

321371 Puccinellia pseudodistans（Crép.）Jansen et Wacht. = Puccinellia fasciculata（Torr.）E. P. Bicknell subsp. pseudodistans（Crép.）Kerguélen ■☆

321372 Puccinellia pulvinata（Fr.）V. I. Krecz.；腋枕碱茅■

321373 Puccinellia pumila（Vassilcz.）Hitchc. = Puccinellia kurilensis（Takeda）Honda ■

321374 Puccinellia qinghaica Tzvelev；青海碱茅■

321375 Puccinellia retroflexa（Curtis）Holmb. = Puccinellia distans（Jacq.）Parl. ■

321376 Puccinellia roborovskyi Tzvelev；疏穗碱茅■

321377 Puccinellia roshevitsiana（Schischk.）V. I. Krecz. ex Tzvelev；西域碱茅■

321378 Puccinellia rupestris（With.）Fernald et Weath.；英国碱茅；British Alkaligrass，Proeumbent Meadow-grass ■

321379 Puccinellia saclinaria（Sim.）Holmb. = Puccinellia intermedia（Schrad.）Janch. ■

321380 Puccinellia schischkinii Tzvelev；希施碱茅（斯碱茅）■

321381 Puccinellia sclerodes（V. I. Krecz.）V. I. Krecz. ex Drobow = Puccinellia gigantea（Grossh.）Grossh. ■

321382 Puccinellia sclerodes V. I. Krecz.；硬碱茅■

321383 Puccinellia sevangensis Grossh.；塞文碱茅■

321384 Puccinellia shuanghuensis L. Liou；双湖碱茅；Shuanghu Alkaligrass ■

321385 Puccinellia sibirica Holmb. ;西伯利亚碱茅■

321386 Puccinellia sibirica Holmb. = Puccinellia distans（Jacq.）Parl. subsp. borealis（Holmb.）W. E. Hughes ■

321387 Puccinellia stapfiana R. R. Stewart;藏北碱茅■

321388 Puccinellia stenophylla Kerguélen = Puccinellia tenuifolia （Boiss. et Reut.）H. Lindb. ■☆

321389 Puccinellia stricta Keng = Pseudosclerochloa kengiana（Ohwi）Tzvelev ■

321390 Puccinellia stricta Keng =Sclerochloa kengiana（Ohwi）Tzvelev ■

321391 Puccinellia strictura L. Liou;坚碱茅■

321392 Puccinellia subspicata V. I. Krecz. = Puccinellia subspicata V. I. Krecz. ex Ovcz. et Czukav. ■

321393 Puccinellia subspicata V. I. Krecz. ex Ovcz. et Czukav. ;穗序碱茅■

321394 Puccinellia suecica Holmb. ;瑞典碱茅■☆

321395 Puccinellia suksdorfii H. St. John = Puccinellia distans（Jacq.）Parl. ■

321396 Puccinellia tenella（Lange）Holmb. ex A. E. Porsild;细雅碱茅■

321397 Puccinellia tenuiflora（Griseb.）Scribn. et Merr. ; 星星草; Fineflowered Alkaligrass ■

321398 Puccinellia tenuiflora（Griseb.）Scribn. et Merr. subsp. tianschanica Tzvelev = Puccinellia tianschanica（Tzvelev）Ikonn. ■

321399 Puccinellia tenuiflora（Griseb.）Scribn. et Merr. var. mongolica Norl. = Puccinellia tenuiflora（Griseb.）Scribn. et Merr. ■

321400 Puccinellia tenuifolia（Boiss. et Reut.）H. Lindb. ;细叶碱茅■☆

321401 Puccinellia tenuissima（Litv. ex V. I. Krecz.）Litv. ex Pavlov; 纤细碱茅■

321402 Puccinellia tenuissima Litv. ex V. I. Krecz. = Puccinellia tenuissima（Litv. ex V. I. Krecz.）Litv. ex Pavlov ■

321403 Puccinellia thomsonii（Stapf）R. R. Stewart;长穗碱茅■

321404 Puccinellia tianschanica（Tzvelev）Ikonn. ;天山碱茅;Tianshan Alkaligrass ■

321405 Puccinellia vachanica Ovcz. et Czukav. ;文昌碱茅■

321406 Puccionia Chiov.（1929）;普奇尼南星属■☆

321407 Puccionia macradenia Chiov. ;普奇尼南星■☆

321408 Pucedanum Hill = Peucedanum L. ■

321409 Puebloa Doweld = Pediocactus Britton et Rose ●☆

321410 Puelia Franch.（1887）;皮埃尔禾属（珀尔筷属）■☆

321411 Puelia acuminata Pilg. = Puelia ciliata Franch. ■☆

321412 Puelia ciliata Franch. ;缘毛皮埃尔禾■☆

321413 Puelia coriacea Clayton;革质皮埃尔禾■☆

321414 Puelia dewevrei De Wild. et T. Durand;德韦皮埃尔禾■☆

321415 Puelia olyriformis（Franch.）Clayton;刚果皮埃尔禾■☆

321416 Puelia schumanniana Clayton = Puelia schumanniana Pilg. ■☆

321417 Puelia schumanniana Pilg. ;舒曼皮埃尔禾■☆

321418 Puelia subsessilis Pilg. = Puelia ciliata Franch. ■☆

321419 Pueraria DC.（1825）;葛属（葛藤属）;Kudzu Bean, Kudzu Vine,Kudzubean,Kudzuvine,Pueraria ●■

321420 Pueraria alopecuroides Craib;密花葛（狐尾葛）;Alopeculus-like Kudzuvine,Denseflower Kudzuvine,Foxtaail-like Kudzuvine ●

321421 Pueraria anabaptis Kurz = Shuteria hirsuta Baker ■

321422 Pueraria argyi H. Lév. et Vaniot = Pueraria lobata（Willd.）Ohwi ●■

321423 Pueraria bicalcarata Gagnep. = Pueraria edulis Pamp. ■

321424 Pueraria bodinieri H. Lév. et Vaniot = Pueraria lobata（Willd.）Ohwi ●■

321425 Pueraria brachycarpa Kurz = Pueraria stricta Kurz ●■

321426 Pueraria calycina Franch. ;黄毛尊葛（黄毛尊葛藤,黄毛葛）; Bigcalyx Kudzuvine,Yellow-hairy Calyx Kudzuvine ●

321427 Pueraria chinensis（Benth.）Ohwi;中国葛（华葛）●☆

321428 Pueraria chinensis Benth. = Pueraria lobata（Willd.）Ohwi subsp. thomsonii（Benth.）H. Ohashi et Tateishi ●

321429 Pueraria coerulea H. Lév. et Vaniot = Pueraria lobata（Willd.）Ohwi ●■

321430 Pueraria collettii Prain = Pueraria stricta Kurz ●■

321431 Pueraria edulis Pamp. ;食用葛（粉葛,甘葛,葛根,葛藤,食用葛藤）;Edible Kudzuvine,Edible Sandcress ■

321432 Pueraria elegans F. T. Wang and Ts. Tang;丽花葛●☆

321433 Pueraria ficifolia（Benth.）L. Bolus = Neorautanenia ficifolia （Benth. ex Harv.）C. A. Sm. ■☆

321434 Pueraria forrestii Evans = Pueraria calycina Franch. ●

321435 Pueraria hirsuta（Thunb.）C. K. Schneid. = Pueraria montana （Lour.）Merr. var. lobata（Willd.）Maesen et S. M. Almeida ex Sanjappa et Predeep ●■

321436 Pueraria hirsuta C. K. Schneid. = Pueraria lobata（Willd.）Ohwi ●■

321437 Pueraria hirsuta Kurz = Pueraria stricta Kurz ●■

321438 Pueraria hochstetteri Chiov. = Neorautanenia mitis（A. Rich.）Verdc. ■☆

321439 Pueraria javanica（Benth.）Benth. = Pueraria phaseoloides （Roxb.）Benth. var. javanica（Benth.）Baker ■☆

321440 Pueraria koten H. Lév. et Vaniot = Pueraria lobata（Willd.）Ohwi ●■

321441 Pueraria lobata（Willd.）Ohwi;葛藤（大葛藤,粉葛,粉葛藤,甘葛,甘葛藤,葛,葛胆,葛根,葛根条,葛花,葛麻茹,葛麻藤,葛条,葛子,黄葛根,黄葛藤,黄斤,黄芹,鸡齐根,鸡脐根,苦葛,鹿豆,鹿豆忠,鹿藿,浅裂葛藤,田葛藤,希绤草,野扁根,野葛,刘头茹）;Japan Arrowroot,Japanese Arrowroot,Kudzu,Kudzu Vine,Kudzuvine,Kudzu-vine,Lobed Kudzuvine,Thunberg Kudzu Bean,Thunberg Kudzu Vine,Thunberg Kudzu-bean,Thunberg Kudzu-vine ●■

321442 Pueraria lobata（Willd.）Ohwi = Pueraria montana（Lour.）Merr. var. lobata（Willd.）Maesen et S. M. Almeida ex Sanjappa et Predeep ●■

321443 Pueraria lobata（Willd.）Ohwi f. alborosea（Makino）Okuyama;粉白葛藤●☆

321444 Pueraria lobata（Willd.）Ohwi f. leucostachya（Honda）Okuyama;白花葛藤（大葛藤）●☆

321445 Pueraria lobata（Willd.）Ohwi subsp. thomsonii（Benth.）H. Ohashi et Tateishi;汤氏葛藤（粉葛）●

321446 Pueraria lobata（Willd.）Ohwi var. chinensis Benth. = Pueraria lobata（Willd.）Ohwi subsp. thomsonii（Benth.）H. Ohashi et Tateishi ●

321447 Pueraria lobata（Willd.）Ohwi var. insularis M. Mizush. ;海岛葛藤●☆

321448 Pueraria lobata（Willd.）Ohwi var. montana（Lour.）Maesen = Pueraria lobata（Willd.）Ohwi ●■

321449 Pueraria lobata（Willd.）Ohwi var. montana（Lour.）Maesen = Pueraria montana（Lour.）Merr. ●■

321450 Pueraria lobata（Willd.）Ohwi var. thomsonii（Benth.）Maesen = Pueraria lobata（Willd.）Ohwi subsp. thomsonii（Benth.）H. Ohashi et Tateishi ●

321451 Pueraria longicarpa Thuan = Pueraria stricta Kurz ●■

321452 Pueraria mirifica Airy Shaw et Suvat. ;泰国野葛（奇葛）●☆

321453 Pueraria montana（Lour.）Merr. ;山葛（葛麻姆,台湾葛,越南葛

藤,越南野葛);Kudzu,Montane Kudzuvine,Vietnam Kudzuvine ●■

321454　Pueraria montana (Lour.) Merr. = Pueraria lobata (Willd.) Ohwi var. montana (Lour.) Maesen ●■

321455　Pueraria montana (Lour.) Merr. = Pueraria lobata (Willd.) Ohwi ●■

321456　Pueraria montana (Lour.) Merr. var. lobata (Willd.) Maesen et S. M. Almeida ex Sanjappa et Predeep = Pueraria lobata (Willd.) Ohwi ■

321457　Pueraria omeiensis F. T. Wang et Ts. Tang;峨眉葛藤(苦葛,苦葛花);Emei Kudzuvine,Omei Kudzuvine ■

321458　Pueraria omeiensis F. T. Wang et Ts. Tang = Pueraria montana (Lour.) Merr. ●■

321459　Pueraria peduncularis (Graham ex Benth.) Benth.;苦葛(白苦葛,红苦葛,苦葛藤,云南葛,云南葛藤);Bitter Kudzuvine, Yunnan Kudzuvine ■

321460　Pueraria phaseoloides (Roxb.) Benth.;三裂叶野葛(假菜豆,热带葛藤,三裂叶葛藤);Trilobedleaf Kudzuvine, Trilobeleaf Kudzuvine, Tropical Kudzu,Tropical Kudzubean ■

321461　Pueraria phaseoloides (Roxb.) Benth. var. javanica (Benth.) Baker;爪哇三裂叶野葛■☆

321462　Pueraria pseudohirsuta Ts. Tang et F. T. Wang = Pueraria lobata (Willd.) Ohwi ●■

321463　Pueraria rogersii L. Bolus = Neorautanenia brachypus (Harms) C. A. Sm. ■☆

321464　Pueraria stricta Kurz;小花野葛;Small Flowered Kudzuvine, Strict Kudzuvine ●■

321465　Pueraria subspicata Benth. = Pueraria phaseoloides (Roxb.) Benth. ■

321466　Pueraria thomsonii Benth.;甘葛藤(粉葛,甘葛);Thomson Kudzuvine ■

321467　Pueraria thunbergiana (Siebold et Zucc.) Benth. = Pueraria lobata (Willd.) Ohwi ●■

321468　Pueraria thunbergiana (Siebold et Zucc.) Benth. = Pueraria montana (Lour.) Merr. var. lobata (Willd.) Maesen et S. M. Almeida ex Sanjappa et Predeep ●■

321469　Pueraria tonkinensis Gagnep.;台湾葛藤(乾葛);Taiwan Kudzubean,Tonkin Kudzuvine ■

321470　Pueraria tonkinensis Gagnep. = Pueraria montana (Lour.) Merr. ●■

321471　Pueraria tuberosa (Roxb. ex Willd.) DC.;块茎葛(块根状葛根,块茎葛藤)■☆

321472　Pueraria tuberosa DC. = Pueraria tuberosa (Roxb. ex Willd.) DC. ■☆

321473　Pueraria wallichii DC.;须弥葛(思茅葛,瓦氏葛藤,喜马拉雅葛藤,须弥菜,紫梗藤,紫铆树小豆花);Wallich Kudzuvine ●■

321474　Pueraria yunnanensis Franch. = Pueraria peduncularis (Graham ex Benth.) Benth. ■

321475　Pugetia (Gand.) Gand. = Rosa L. ●

321476　Pugetia Gand. = Rosa L. ●

321477　Pugionella Salisb. = Strumaria Jacq. ■☆

321478　Pugionium Gaertn. (1791);沙芥属(漠芥属);Pugionium, Sandcress ■

321479　Pugionium calcaratum Kom.;距果沙芥(距花沙芥,距沙芥);Spur Pugionium,Spur Sandcress ■

321480　Pugionium calcaratum Kom. = Pugionium dolabratum Maxim. ■

321481　Pugionium cornutum (L.) Gaertn.;沙芥(沙白菜,沙盖,沙芥菜,沙萝卜,山盖,山萝卜,山羊沙芥);Cornuted Pugionium, Cornuted Sandcress ■

321482　Pugionium cristatum Kom.;鸡冠沙芥;Crest Sandcress, Crested

Pugionium ■

321483　Pugionium cristatum Kom. = Pugionium dolabratum Maxim. ■

321484　Pugionium dolabratum Maxim.;斧翅沙芥(斧形沙芥,宽翅沙芥,绵羊沙芥);Axe Sandcress, Axe-shaped ugionium ■

321485　Pugionium dolabratum Maxim. var. latipterum H. L. Yang = Pugionium pterocarpum Kom. ■

321486　Pugionium dolabratum Maxim. var. platypterum H. L. Yang = Pugionium dolabratum Maxim. ■

321487　Pugionium dolabratum Maxim. var. platypterum H. L. Yang = Pugionium pterocarpum Kom. ■

321488　Pugionium pterocarpum Kom.;宽翅沙芥(翅果沙芥);Broadwing Axe-shaped Pugionium ■

321489　Pugiopappus A. Gray = Coreopsis L. ●■

321490　Pugiopappus A. Gray ex Torr. = Coreopsis L. ●■

321491　Pugiopappus bigelovii A. Gray = Coreopsis bigelovii (A. Gray) Voss ■☆

321492　Puja Molina = Puya Molina ■☆

321493　Pukanthus Raf. = Grabowskia Schltdl. ●☆

321494　Pukateria Raoul = Griselinia J. R. Forst. et G. Forst. ●☆

321495　Pulassarium Kuutae = Alyxia Banks ex R. Br. (保留属名)●

321496　Pulassarium Rumph. = Alyxia Banks ex R. Br. (保留属名)●

321497　Pulassarium Rumph. ex Kuntze = Alyxia Banks ex R. Br. (保留属名)●

321498　Pulassarium madagascariense (A. DC.) Kuntze = Petchia erythrocarpa (Vatke) Leeuwenb. ●☆

321499　Pulchea hirsuta Less. = Blumea clarkei Hook. f. ■

321500　Pulcheria Comm. ex Moewes = Polycardia Juss. ●☆

321501　Pulcherta Noronha = Kadsura Kaempf. ex Juss. ●

321502　Pulchia Steud. = Diclidanthera Mart. ●■☆

321503　Pulchia Steud. = Pluchia Vell. ●■☆

321504　Pulchranthus V. M. Baum, Reveal et Nowicke(1983);美花爵床属●■☆

321505　Pulchranthus surinamensis (Bremek.) V. M. Baum, Reveal et Nowicke;美花爵床●☆

321506　Pulegium Mill. = Mentha L. ■●

321507　Pulegium Ray ex Mill. = Mentha L. ■●

321508　Pulegium vulgare Mill. = Mentha pulegium L. ■

321509　Pulicaria Gaertn. (1791);蚤草属(臭蚤草属);False Fleabane,Fleabane,Fleaweed,Pulicaria ■●

321510　Pulicaria adenophora Franch. = Pulicaria hildebrandtii Vatke ■☆

321511　Pulicaria alata E. Phillips = Pentatrichia alata S. Moore ■☆

321512　Pulicaria albida E. Gamal-Eldin;微白蚤草■☆

321513　Pulicaria alveolosa Batt. et Trab.;热非蚤草■☆

321514　Pulicaria antiatlantica Förther et Podlech = Pulicaria glandulosa Caball. ■☆

321515　Pulicaria arabica (L.) Cass.;阿拉伯蚤草;Ladies' False Fleabane ■☆

321516　Pulicaria arabica (L.) Cass. subsp. hispanica (Boiss.) Murb. = Pulicaria paludosa Link ■☆

321517　Pulicaria arabica (L.) Cass. subsp. inuloides (Poir.) Maire = Pulicaria inuloides (Poir.) DC. ■☆

321518　Pulicaria arabica (L.) Cass. subsp. longifolia Ball = Pulicaria arabica (L.) Cass. subsp. inuloides (Poir.) Maire ■☆

321519　Pulicaria arabica (L.) Cass. var. herteri Sennen = Pulicaria arabica (L.) Cass. ■☆

321520　Pulicaria arabica (L.) Cass. var. paludosa (Link) Pau et Font Quer = Pulicaria paludosa Link ■☆

321521　Pulicaria arabica Bourg. ex Nyman ＝Pulicaria arabica（L.） Cass. ■☆

321522　Pulicaria areysiana Deflers ＝Pulicaria argyrophylla Franch. ■☆

321523　Pulicaria argyrophylla Franch.；银叶蚤草■☆

321524　Pulicaria aromatica R. Br. ＝Pulicaria incisa（Lam.） DC. ■☆

321525　Pulicaria aspera Pomel ＝Pulicaria arabica（L.） Cass. subsp. inuloides（Poir.） Maire ■☆

321526　Pulicaria attenuata Hutch. et B. L. Burtt；渐狭蚤草■☆

321527　Pulicaria aylmeri Baker；艾梅蚤草■☆

321528　Pulicaria burchardii Hutch.；伯查德蚤草■☆

321529　Pulicaria burchardii Hutch. subsp. longifolia E. Gamal-Eldin；长叶蚤草■☆

321530　Pulicaria canariensis Bolle；加那利蚤草■☆

321531　Pulicaria canariensis Bolle subsp. lanata（Font Quer et Svent.） Bramwell et G. Kunkel；绵毛加那利蚤草■☆

321532　Pulicaria canariensis Bolle var. lanata Font Quer et Svent. ＝Pulicaria canariensis Bolle subsp. lanata（Font Quer et Svent.） Bramwell et G. Kunkel ■☆

321533　Pulicaria capensis DC. ＝Pulicaria scabra（Thunb.） Druce ■☆

321534　Pulicaria capensis DC. var. erigeroides（DC.） Harv. ＝Pulicaria scabra（Thunb.） Druce ■☆

321535　Pulicaria chrysantha（Diels） Y. Ling；金仙草（齿叶旋覆花，金花蚤草，金仙花）；Yellowflower Fleawee，Yellowflower Pulicaria ■

321536　Pulicaria chrysantha（Diels） Y. Ling var. oligochaeta Y. Ling；少毛金仙草■

321537　Pulicaria chrysopsidoides Schweinf.；金菊蚤草■☆

321538　Pulicaria chudaei Batt. et Trab.；朱丹蚤草■☆

321539　Pulicaria clausonis Pomel；克劳森蚤草■☆

321540　Pulicaria confertifolia Klatt ex Merxm. ＝Pentatrichia avasmontana Merxm. ■☆

321541　Pulicaria confusa E. Gamal-Eldin；混乱蚤草■☆

321542　Pulicaria crispa（Forssk.） Benth. ex Oliv. ＝Pulicaria undulata（L.） C. A. Mey. ■☆

321543　Pulicaria crispa（Forssk.） Benth. ex Oliv. subsp. argyrophylla E. Gamal-Eldin ＝Pulicaria undulata（L.） C. A. Mey. subsp. argyrophylla（E. Gamal-Eldin） D. J. N. Hind et Boulos ■☆

321544　Pulicaria crispa（Forssk.） Benth. ex Oliv. subsp. candidissima（Maire） E. Gamal-Eldin ＝Pulicaria undulata（L.） C. A. Mey. subsp. candidissima（Maire） D. J. N. Hind et Boulos ■☆

321545　Pulicaria crispa（Forssk.） Benth. ex Oliv. subsp. fogensis E. Gamal-Eldin ＝Pulicaria diffusa（Shuttlew. ex S. Brunner） B. Peterson ■☆

321546　Pulicaria crispa（Forssk.） Benth. ex Oliv. subsp. tomentosa E. Gamal-Eldin ＝Pulicaria undulata（L.） C. A. Mey. subsp. tomentosa（E. Gamal-Eldin） D. J. N. Hind et Boulos ■☆

321547　Pulicaria crispa（Forssk.） Benth. ex Oliv. var. candidissima Maire ＝Pulicaria undulata（L.） C. A. Mey. subsp. candidissima（Maire） D. J. N. Hind et Boulos ■☆

321548　Pulicaria crispa（Forssk.） Benth. ex Oliv. var. gracillima Maire ＝Pulicaria undulata（L.） C. A. Mey. ■☆

321549　Pulicaria crispa（Forssk.） Benth. ex Oliv. var. virescens Maire ＝Pulicaria undulata（L.） C. A. Mey. subsp. candidissima（Maire） D. J. N. Hind et Boulos ■☆

321550　Pulicaria decumbens（Litard. et Maire） Greuter ＝Pulicaria vulgaris Gaertn. ■

321551　Pulicaria desertorum DC. ＝Pulicaria incisa（Lam.） DC. subsp. candolleana E. Gamal-Eldin ■☆

321552　Pulicaria diffusa（Shuttlew. ex S. Brunner） B. Peterson；铺散蚤草■☆

321553　Pulicaria discoidea（Chiov.） N. Kilian；盘状蚤草■☆

321554　Pulicaria dysenterica（L.） Gaertn.；止痢蚤草；Antidy Senteric Pulicaria, Antidysenteric Fleaweed, Cammock, Common Fleabane, Fleabane, Fleabane-mullet, Harvest Flower, Herb Christopher, Job's Tears, Mare's Fat, Meadow False Fleabane, Middle Fleabane, Pig Daisy, Pig's Daisy, Wild Marigold, Yellow Fleabane ■

321555　Pulicaria ehrenbergiana Sch. Bip. ex Schweinf. ＝Pulicaria schimperi DC. ■☆

321556　Pulicaria erigeroides DC. ＝Pulicaria scabra（Thunb.） Druce ■☆

321557　Pulicaria filaginoides Pomel；絮菊蚤草■☆

321558　Pulicaria glandulosa Caball.；具腺蚤草■☆

321559　Pulicaria glutinosa（Boiss.） Jaub. et Spach；黏蚤草■☆

321560　Pulicaria glutinosa（Boiss.） Jaub. et Spach subsp. somalensis E. Gamal-Eldin；索马里黏蚤草■☆

321561　Pulicaria gnaphaloides（Vent.） Boiss.；鼠麹蚤草；Cudweedlike Fleaweed, Cudweedlike Pulicaria ■

321562　Pulicaria grantii Oliv. et Hiern；格兰特蚤草■☆

321563　Pulicaria hesperia Maire et al. ＝Pulicaria glandulosa Caball. ■☆

321564　Pulicaria hildebrandtii Vatke；希尔德蚤草■☆

321565　Pulicaria hispanica（Boiss.） Boiss. ＝Pulicaria paludosa Link ■☆

321566　Pulicaria incisa（Lam.） DC.；锐裂蚤草■☆

321567　Pulicaria incisa（Lam.） DC. subsp. candolleana E. Gamal-Eldin；康氏蚤草■☆

321568　Pulicaria incisa（Lam.） DC. subsp. denticulata E. Gamal-Eldin；细齿锐裂蚤草■☆

321569　Pulicaria incisa（Lam.） DC. subsp. suffrutescens E. Gamal-Eldin；亚灌木蚤草●☆

321570　Pulicaria insignis J. R. Drumm. ex Dunn；臭蚤草（金花旋覆，山葵花，虱草花）；Insignis Pulicaria, Stink Fleaweed ■

321571　Pulicaria inuloides（Poir.） DC.；旋覆花蚤草■☆

321572　Pulicaria involucrata R. Br. ＝Pulicaria schimperi DC. ■☆

321573　Pulicaria jaubertii E. Gamal-Eldin；若贝尔蚤草■☆

321574　Pulicaria kouyangensis Vaniot ＝Synotis nagensis（C. B. Clarke） C. Jeffrey et Y. L. Chen ■

321575　Pulicaria laciniata（Coss. et Durieu） Thell.；撕裂蚤草■☆

321576　Pulicaria leucophylla Baker ＝Pulicaria argyrophylla Franch. ■☆

321577　Pulicaria longifolia（Wagenitz et E. Gamal-Eldin） N. Kilian ＝Pulicaria uniseriata N. Kilian ■☆

321578　Pulicaria longifolia Boiss. ＝Pulicaria inuloides（Poir.） DC. ■☆

321579　Pulicaria longifolia Boiss. var. herteri Sennen ＝Pulicaria arabica（L.） Cass. subsp. inuloides（Poir.） Maire ■☆

321580　Pulicaria lozanoi Caball. ＝Pulicaria burchardii Hutch. ■☆

321581　Pulicaria marsahitensis Buscal. et Muschl. ＝Pulicaria somalensis O. Hoffm. ■☆

321582　Pulicaria mauritanica Batt.；毛里塔尼亚蚤草■☆

321583　Pulicaria monocephala Franch.；单头蚤草■☆

321584　Pulicaria odora（L.） Rchb.；芬芳蚤草■☆

321585　Pulicaria odora（L.） Rchb. var. atlantica（Pau） Maire ＝Pulicaria odora（L.） Rchb. ■☆

321586　Pulicaria odora（L.） Rchb. var. lanata Maire et Sennen ＝Pulicaria odora（L.） Rchb. ■☆

321587　Pulicaria odora（L.） Rchb. var. macrocephala Ball ＝Pulicaria odora（L.） Rchb. ■☆

321588　Pulicaria orientalis Jaub. et Spach ＝Pulicaria jaubertii E. Gamal-Eldin ■☆

321589 Pulicaria paludosa Link;西班牙蚤草;Spanish False Fleabane ■☆

321590 Pulicaria paludosa Link subsp. inuloides（Poir.）Valdés = Pulicaria arabica（L.）Cass. subsp. inuloides（Poir.）Maire ■☆

321591 Pulicaria paludosa Steud. = Pulicaria paludosa Link ■☆

321592 Pulicaria petiolaris Jaub. et Spach;柄叶蚤草■☆

321593 Pulicaria phillipsiae S. Moore = Iphiona phillipsiae（S. Moore）Anderb. ■☆

321594 Pulicaria pomeliana Faure et Maire = Pulicaria vulgaris Gaertn. subsp. pomeliana Faure et Maire ■☆

321595 Pulicaria prostrata（Gilib.）Asch. ;蚤草;Fleaweed, Prostrate Pulicaria ■

321596 Pulicaria renschiana Vatke;伦施蚤草■☆

321597 Pulicaria rueppellii Sch. Bip. ex A. Rich. = Pulicaria schimperi DC. ■☆

321598 Pulicaria salviifolia Bunge;鼠尾蚤草; Sageleaf Fleaweed, Sageleaf Pulicaria ■

321599 Pulicaria scabra（Thunb.）Druce;粗糙蚤草■☆

321600 Pulicaria schimperi DC. ;欣珀蚤草■☆

321601 Pulicaria sericea E. Gamal-Eldin;绢毛蚤草■☆

321602 Pulicaria sicula（L.）Moris;西西里蚤草■☆

321603 Pulicaria sicula（L.）Moris var. radiata（DC.）Batt. = Pulicaria sicula（L.）Moris ■☆

321604 Pulicaria sicula（L.）Moris var. virescens Batt. = Pulicaria arabica（L.）Cass. ■☆

321605 Pulicaria somalensis O. Hoffm. ;索马里蚤草■☆

321606 Pulicaria uliginosa Hoffmanns. et Link = Pulicaria paludosa Link ■☆

321607 Pulicaria uliginosa Stev. ;沼泽蚤草■☆

321608 Pulicaria undulata（L.）C. A. Mey. ;波状蚤草■☆

321609 Pulicaria undulata（L.）C. A. Mey. subsp. argyrophylla（E. Gamal-Eldin）D. J. N. Hind et Boulos;银色波状蚤草■☆

321610 Pulicaria undulata（L.）C. A. Mey. subsp. candidissima（Maire）D. J. N. Hind et Boulos;白波状蚤草■☆

321611 Pulicaria undulata（L.）C. A. Mey. subsp. fogensis（E. Gamal-Eldin）A. Hansen et Sunding = Pulicaria diffusa（Shuttlew. ex S. Brunner）B. Peterson ■☆

321612 Pulicaria undulata（L.）C. A. Mey. subsp. tomentosa（E. Gamal-Eldin）D. J. N. Hind et Boulos;绒毛波状蚤草■☆

321613 Pulicaria undulata（L.）C. A. Mey. var. alveolosa（Batt. et Trab.）Maire = Pulicaria alveolosa Batt. et Trab. ■☆

321614 Pulicaria undulata（L.）C. A. Mey. var. candidissima Maire = Pulicaria undulata（L.）C. A. Mey. subsp. candidissima（Maire）D. J. N. Hind et Boulos ■☆

321615 Pulicaria undulata Mey. = Pulicaria prostrata（Gilib.）Asch. ■

321616 Pulicaria uniseriata N. Kilian;单丝蚤草■☆

321617 Pulicaria villosa（Vahl ex Hornem.）Link = Pulicaria arabica（L.）Cass. ■☆

321618 Pulicaria volkonskyana Maire;沃尔孔斯基蚤草■☆

321619 Pulicaria vulgaris Gaertn. ;欧洲蚤草; Small Fleabane, Small Pulicaria ■

321620 Pulicaria vulgaris Gaertn. = Pulicaria prostrata（Gilib.）Asch. ■

321621 Pulicaria vulgaris Gaertn. subsp. dentata Batt. = Pulicaria vulgaris Gaertn. ■

321622 Pulicaria vulgaris Gaertn. subsp. graeca（Sch. Bip.）Quézel et Santa = Pulicaria clausonis Pomel ■☆

321623 Pulicaria vulgaris Gaertn. subsp. pomeliana Faure et Maire;波梅尔蚤草■☆

321624 Pulicaria vulgaris Gaertn. var. decumbens Litard. et Maire =

Pulicaria vulgaris Gaertn. ■

321625 Pulicaria zimbabwensis S. Moore = Philyrophyllum schinzii O. Hoffm. ■☆

321626 Puliculum Haines = Eulalia Kunth ■

321627 Puliculum Stapf ex Haines = Pseudopogonatherum A. Camus ■

321628 Pullea Schltr.（1914）;普莱木属●☆

321629 Pullea Schltr. = Codia J. R. Forst. et G. Forst. ●☆

321630 Pullea mollis Schltr. ;普莱木●☆

321631 Pullipes Raf. = Caucalis L. ■☆

321632 Pullipuntu Ruiz = Phytelephas Ruiz et Pav. + Yarina O. F. Cook ●☆

321633 Pulmonaria L.（1753）;肺草属;Lungwort ■

321634 Pulmonaria angustifolia L. = Pulmonaria longifolia Bastard ex Bor ■☆

321635 Pulmonaria davurica Sims = Mertensia davurica（Sims）G. Don ■

321636 Pulmonaria longifolia Bastard ex Bor;长叶肺草（狭叶肺草,窄叶肺草）;Adam-and-eve, Blue Cowslip, Cowslip Lungwort, Joseph-and-Mary, Lungwort, Narrowleaf Lungwort, Narrow-leaved Jerusalem Cowslip, Narrow-leaved Lungwort, Snake's Flower ■☆

321637 Pulmonaria maritima L. ;海滨肺草; Officinal Bugloss, Sea Lungwort ■☆

321638 Pulmonaria mollis H. Wolff ex Heller;软肺草（软紫草）■☆

321639 Pulmonaria mollissima J. Kern. ;腺毛肺草（软肺草,肺草）; Glandularhair Lungwort ■

321640 Pulmonaria montana Lej. ;山地肺草;Mountain Lungwort ■☆

321641 Pulmonaria obscura Dumort. ;暗色肺草（暗色紫草,黑肺草）; Unspotted Lungwort ■☆

321642 Pulmonaria obscura Dumort. = Pulmonaria officinalis L. ■☆

321643 Pulmonaria officinalis L. ;药用肺草（肺草,蓝花肺草,疗肺草）;Abraham, Adam-and-eve, Bedlam Cowslip, Beelzebub, Beggar's Basket, Bethlehem Cowslip, Bethlehem Sage, Blue Lungwort, Bottle-of-all-sorts, Bugloss Cowslip, Children of Israel, Children-of-israel, Common Lungwort, Cow's Parsley, Crayfery, Fool Gooseberry, French Paigle, Good Friday Plant, Herb of Mary, Hope And Charity Faith, Hundreds-and-thousands, Jersalem Cowslip, Jerusalem Cowslip, Jerusalem Primrose, Jerusalem Sage, Jerusalem Seeds, Joseph-and-Mary, Joseph's Coat, Kwort, Lady Mary's Tears, Lady's Mii, Lady's Milksile, Lady's Milkwort, Lady's Pincushion, Liverwort, Lungwort, Mary's Tears, Mountain Sage, Soldier-and-his-wife, Soldiers-and-sailors, Spared Bugloss, Spotted Comfrey, Spotted Dog, Spotted Mary, Spotted Virgin, Thunder-and-lightning, Today-and-tomorrow, Twelve Apostles, Virgin Mary, Virgin Mary's Cowslip, Virgin Mary's Honeysuckle, Virgin Mary's Milkdrops, Virgin Mary's Tears, Virginian Cowslip, Wild Comfrey, William-and-Mary ■☆

321644 Pulmonaria officinalis L. 'Sissiinghurst White';白花药用肺草■☆

321645 Pulmonaria officinalis L. obscura ? = Pulmonaria obscura Dumort. ■☆

321646 Pulmonaria rubra Schott;红花肺草（红肺草）;Red Lungwort ■☆

321647 Pulmonaria saccharata Mill. ;白斑叶肺草（甘肺草）;Bethlehem Lungwort, Bethlehem Sage, Lungwort ■☆

321648 Pulmonaria saccharata Mill. 'Sissiinghurst White' = Pulmonaria officinalis L. 'Sissiinghurst White' ■☆

321649 Pulmonaria sibirica L. = Mertensia sibirica（L.）G. Don ■

321650 Pulmonaria vallarsae A. Kern. ;肺草■☆

321651 Pulmouaria filarszkyana Jav. ;菲拉尔肺草■☆

321652 Pulpaceae Dulac = Grossulariaceae DC.（保留科名）●

321653 Pulsatilla Mill.（1754）;白头翁属;Anemone, European Pasqueflower,Pasque Flower,Pasqueflower,Pulsatilla,Windflower ■

321654   Pulsatilla Mill. = Anemone L. (保留属名)■

321655   Pulsatilla ajanensis Regel;阿赞白头翁■☆

321656   Pulsatilla albana (Stev.) Bercht. et J. Presl;阿尔班白头翁■☆

321657   Pulsatilla albana (Stev.) Bercht. et J. Presl var. campanella Fisch. ex Regel et Tiling = Pulsatilla campanella Fisch. ex Regel et Tiling ■

321658   Pulsatilla alpina (L.) Delarbre;高山白头翁;Alpine Anemone, Devil's Bane,Devil's Beard ■☆

321659   Pulsatilla ambigua (Turcz. ex Pritz.) Juz.;蒙古白头翁(白头翁,北白头翁,高山白头翁);Mongol Pulsatilla,Mongolian Pulsatilla ■

321660   Pulsatilla ambigua (Turcz. ex Pritz.) Juz. var. barbata J. G. Liou;髯毛蒙古白头翁(拟蒙古白头翁);Barbate Mongolian Pulsatilla ■

321661   Pulsatilla armena (Boiss.) Rupr.;亚美尼亚白头翁■☆

321662   Pulsatilla aurea (N. Busch) Juz.;金黄色白头翁;Golden Pulsatilla ■

321663   Pulsatilla bungeana C. A. Mey.;布恩白头翁■☆

321664   Pulsatilla caffra Eckl. et Zeyh. = Anemone caffra (Eckl. et Zeyh.) Harv. ■☆

321665   Pulsatilla campanella Fisch. ex Regel et Tiling;钟萼白头翁(阿尔泰白头翁,白头翁,小花白头翁);Bellcalyx Pulsatilla ■

321666   Pulsatilla cernua (Thunb. ex Murray) Bercht. et C. Presl = Pulsatilla cernua (Thunb.) Bercht. et C. Presl ■

321667   Pulsatilla cernua (Thunb.) Bercht. et C. Presl;朝鲜白头翁(白头翁,伏垂银莲花,姑朵花,毛骨朵花);Korea Pulsatilla, Korean Pulsatilla ■

321668   Pulsatilla cernua (Thunb.) Bercht. et C. Presl f. flava Y. N. Lee;黄花朝鲜白头翁;Yellowflower Korean Pulsatilla ■

321669   Pulsatilla cernua (Thunb.) Bercht. et C. Presl f. plumbea J. X. Ji et Y. T. Zhao;灰花白头翁;Grayflower Korean Pulsatilla ■

321670   Pulsatilla cernua (Thunb.) Bercht. et C. Presl var. koreana (Y. Yabe ex Nakai) Y. N. Lee = Pulsatilla cernua (Thunb.) Bercht. et C. Presl ■

321671   Pulsatilla cernua Thunb. var. koreana (Y. Yabe ex Nakai) Y. N. Lee = Pulsatilla cernua (Thunb.) Bercht. et C. Presl ■

321672   Pulsatilla chinensis (Bunge) Regel;白头翁(白头翁草,白头公,大将军草,大碗花,粉草,粉乳草,毫笔花,耗子花,耗子尾巴花,胡王使者,犄角花,将军草,菊菊苗,老白发,老白毛,老公花,老姑草,老姑子花,老观花,老冠花,老和尚头,老婆子花,老人发,老翁发,老翁花,老翁须,猫古都,猫头花,猫爪子花,毛姑朵花,奈何草,山棉花,头痛棵,耋草,细叶白头翁,羊胡子花,野丈人,注之花);China Pulsatilla,Chinese Pulsatilla ■

321673   Pulsatilla chinensis (Bunge) Regel f. alba D. K. Zang;白花白头翁;Whiteflower Chinese Pulsatilla ■

321674   Pulsatilla chinensis (Bunge) Regel f. plurisepala D. K. Zang;多萼白头翁;Manysepal Chinese Pulsatilla ■

321675   Pulsatilla chinensis (Bunge) Regel var. kissii (Mandl) S. H. Li et Y. Huei Huang;金县白头翁;Jinxian Pulsatilla ■

321676   Pulsatilla dahurica (Fisch.) Spreng.;兴安白头翁(白头翁,达呼尔白头翁,姑朵花,毛骨朵花);Dahur Pulsatilla, Dahurian Pulsatilla ■

321677   Pulsatilla dahurica (Fisch.) Spreng. f. alba H. W. Jen ex D. Z. Lu;白花兴安白头翁;Whiteflower Dahurian Pulsatilla ■

321678   Pulsatilla flavescens (Zucc.) Juz. = Pulsatilla patens (L.) Mill. subsp. flavescens (Zucc.) Zämelis ■

321679   Pulsatilla georgica Rupr.;乔治白头翁■☆

321680   Pulsatilla glaucifolia (Franch.) Huth = Anemoclema glaucifolium (Franch.) W. T. Wang ■

321681   Pulsatilla grandis Wender.;大白头翁;Big Pulsatilla ■☆

321682   Pulsatilla halleri (All.) Willd.;哈勒氏白头翁(哈理氏白头翁,堇花白头翁);Haller Pulsatilla ■☆

321683   Pulsatilla hirsutissima (Pursh) Britton = Anemone patens L. var. multifida Pritz. ■

321684   Pulsatilla hirsutissima (Pursh) Britton = Pulsatilla patens (L.) Mill. subsp. multifida (Pritz.) Zämelis ■

321685   Pulsatilla kissi Mandl = Pulsatilla chinensis (Bunge) Regel var. kissi (Mandl) S. H. Li et Y. Huei Huang ■

321686   Pulsatilla koreana (Y. Yabe ex Nakai) Nakai ex Mori = Pulsatilla cernua (Thunb.) Bercht. et C. Presl ■

321687   Pulsatilla koreana Nakai ex Mori = Pulsatilla cernua (Thunb.) Bercht. et C. Presl ■

321688   Pulsatilla kostyczewii (Korsh.) Juz.;紫蕊白头翁;Purplestamen Pulsatilla ■

321689   Pulsatilla ludoviciana (Nutt.) A. Heller = Anemone patens L. var. multifida Pritz. ■

321690   Pulsatilla ludoviciana (Nutt.) A. Heller = Pulsatilla patens (L.) Mill. subsp. multifida (Pritz.) Zämelis ■

321691   Pulsatilla millefolium (Hemsl. et E. H. Wilson) Ulbr.;西南白头翁(川滇白头翁,千叶白头翁);Manyleaf Pulsatilla ■

321692   Pulsatilla montana (Hooper) Rchb.;山白头翁■☆

321693   Pulsatilla morii (Yamam.) Masam. = Ranunculus morii (Yamam.) Ohwi ■

321694   Pulsatilla multiceps Greene = Anemone multiceps (Greene) Standl. ■☆

321695   Pulsatilla multifida (Pritz.) Juz. = Pulsatilla patens (L.) Mill. subsp. multifida (Pritz.) Zämelis ■

321696   Pulsatilla nigricans Stoerck ex DC.;黑色白头翁(黑白头翁)■☆

321697   Pulsatilla nipponica (Takeda) Ohwi;日本白头翁■☆

321698   Pulsatilla nuttalliana (DC.) Spreng. = Anemone patens L. var. multifida Pritz. ■

321699   Pulsatilla nuttalliana (DC.) Spreng. = Pulsatilla patens (L.) Mill. subsp. multifida (Pritz.) Zämelis ■

321700   Pulsatilla occidentalis (S. Watson) Freyn;西洋白头翁■☆

321701   Pulsatilla occidentalis (S. Watson) Freyn = Anemone occidentalis S. Watson ■☆

321702   Pulsatilla occidentalis Freyn = Pulsatilla occidentalis (S. Watson) Freyn ■☆

321703   Pulsatilla patens (L.) Mill.;肾叶白头翁(白头翁,伸展白头翁,展形白头翁);American Pasqueflower, Eastern Pasque Flower, Eastern Pasque-flower, Kidneyleaf Pulsatilla, Prairie-smoke, Spreading Pulsatilla ■

321704   Pulsatilla patens (L.) Mill. = Anemone patens L. ■

321705   Pulsatilla patens (L.) Mill. subsp. asiatica Krylov et Serg. = Anemone patens L. var. multifida Pritz. ■

321706   Pulsatilla patens (L.) Mill. subsp. asiatica Krylov et Serg. = Pulsatilla patens (L.) Mill. subsp. multifida (Pritz.) Zämelis ■

321707   Pulsatilla patens (L.) Mill. subsp. flavescens (Zucc.) Zämelis;黄色白头翁(发黄白头翁);Yellow Pulsatilla ■

321708   Pulsatilla patens (L.) Mill. subsp. multifida (Pritz.) Zämelis;多裂白头翁(多裂草原银莲花,掌叶白头翁);American Pasqueflower, Multifid Pulsatilla, Pasqueflower, Prairie-crocus, Prairie-smoke,Pulsatille ■

321709   Pulsatilla patens (L.) Mill. subsp. multifida (Pritz.) Zämelis = Anemone patens L. var. multifida Pritz. ■

321710　Pulsatilla patens（L.）Mill. var. multifida（Pritz.）S. H. Li et Y. Huei Huang ＝Anemone patens L. var. multifida Pritz. ■

321711　Pulsatilla patens（L.）Mill. var. multifida（Pritz.）S. H. Li et Y. Huei Huang ＝Pulsatilla patens（L.）Mill. subsp. multifida（Pritz.）Zämelis ■

321712　Pulsatilla pratensis（L.）Mill. ＝Anemone pratensis L. ■☆

321713　Pulsatilla sachalinensis Hara；库页白头翁■☆

321714　Pulsatilla sukaczewii Juz.；黄花白头翁（白头翁）；Yellow Pulsatilla，Yellowflower Pulsatilla ■

321715　Pulsatilla tenuiloba（Hayek）Juz.；细裂白头翁■

321716　Pulsatilla turczaninovii Krylov et Serg.；细叶白头翁（白头翁）；Slenderleaf Pulsatilla ■

321717　Pulsatilla turczaninovii Krylov et Serg. f. albiflora Y. Z. Zhao；白花细叶白头翁；Whiteflower Slenderleaf Pulsatilla ■

321718　Pulsatilla turczaninovii Krylov et Serg. f. albiflora Y. Z. Zhao ＝Pulsatilla turczaninovii Krylov et Serg. ■

321719　Pulsatilla vernalis（L.）Mill. ＝Anemone vernalis L. ■☆

321720　Pulsatilla violacea Rupr.；蓝紫色白头翁；Violet Pulsatilla ■☆

321721　Pulsatilla vulgaris Mill.；欧洲白头翁（白头翁状银莲花，欧白头翁,铺散草原银莲花,普通白头翁）；Bastard Anemone，Blue Money，Bluemony，Coral Bells，Coventry Bells，Dane's Blood，Dane's Flower，Dane's Weed，Devil's Bane，Devil's Beard，Easter Flower，European Pasque Flower，European Pasqueflower，Flaw-flower，Jupiter's Beard，Laughing Parsley，Long-sheathed Anemone，Meadow Anemone，Meadow Anemony，Paschal Flower，Pasque Flower，Pasqueflower，Pasque-flower，Passe-flower，Prairie Crocus，Spreading Anemone，Spreading Pasqueflower，Wind Flower，Windflower ■☆

321722　Pulsatilla vulgaris Mill. ＝Anemone pulsatilla L. ■☆

321723　Pulsatilloides（DC.）Starod.（1991）；拟白头翁属■

321724　Pulsatilloides（DC.）Starod. ＝Anemone L.（保留属名)■

321725　Pulsatilloides begoniifolia（H. Lév. et Vaniot）Starod.；拟白头翁■☆

321726　Pultenaea Sm.（1794）；灌木豆属（普尔特木属）；Bush-pea ●☆

321727　Pultenaea altissima F. Muell. ex Benth.；高大灌木豆（高普尔特木）；Tall Bush-pea ●☆

321728　Pultenaea cunninghamii（Benth.）H. B. Will.；杉灌木豆（杉普尔特木）●☆

321729　Pultenaea flexilis Sm.；优美灌木豆（优美普尔特木）；Graceful Bush-pea ●☆

321730　Pultenaea pedunculata Hook.；地被灌木豆（地被普尔特木）；Matted Pea Bush ●☆

321731　Pultenaea rosmarinifolia Sieber ex DC.；黄灌木豆（黄普尔特木）；Yellow Pea ●☆

321732　Pultenaea scabra R. Br.；密毛灌木豆（密毛普尔特木）●☆

321733　Pultenaea stipularis Sm.；松叶灌木豆（松叶普尔特木）●☆

321734　Pultenaea villosa Willd.；毛灌木豆（毛普尔特木）●☆

321735　Pultenea A. St. -Hil. ＝Pultenaea Sm. ●☆

321736　Pulteneja Hoffmanns. ＝Pultenaea Sm. ●☆

321737　Pulteneya Hoffmanns. ＝Pultenaea Sm. ●☆

321738　Pulteneya Post et Kuntze ＝Pultenaea Sm. ●☆

321739　Pultnaea Graham ＝Pultenaea Sm. ●☆

321740　Pultoria Raf. ＝Ilex L. ●

321741　Pultoria Raf. ＝Paltoria Ruiz et Pav. ●

321742　Pulvinaria E. Fourn. ＝Lhotzkyella Rauschert ●☆

321743　Pumilea P. Browne ＝Turnera L. ●■☆

321744　Pumilo Schltdl. ＝Rutidosis DC. ■☆

321745　Puna R. Kiesling ＝Opuntia Mill. ●

321746　Puncticularia N. E. Br. ex Lemee ＝Pleiospilos N. E. Br. ■☆

321747　Puncticularia N. E. Br. ex Lemee ＝Punctillaria N. E. Br. ■☆

321748　Punctilaria Lemee ＝Pleiospilos N. E. Br. ■☆

321749　Punctilaria Lemee ＝Punctillaria N. E. Br. ■☆

321750　Punctillaria N. E. Br. ＝Pleiospilos N. E. Br. ■☆

321751　Punctillaria cana L. Bolus ＝Pleiospilos compactus（Aiton）Schwantes subsp. canus（Haw.）H. E. K. Hartmann et Liede ■☆

321752　Punctillaria compacta（Aiton）N. E. Br. ＝Pleiospilos compactus（Aiton）Schwantes ■☆

321753　Punctillaria dekenahi N. E. Br. ＝Pleiospilos compactus（Aiton）Schwantes subsp. canus（Haw.）H. E. K. Hartmann et Liede ■☆

321754　Punctillaria magnipunctata（Haw.）N. E. Br. ＝Pleiospilos compactus（Aiton）Schwantes subsp. canus（Haw.）H. E. K. Hartmann et Liede ■☆

321755　Punctillaria optata（N. E. Br.）N. E. Br. ＝Pleiospilos compactus（Aiton）Schwantes ■☆

321756　Punctillaria roodiae N. E. Br. ＝Tanquana prismatica（Schwantes）H. E. K. Hartmann et Liede ■☆

321757　Punctillaria sesquiuncialis N. E. Br. ＝Pleiospilos compactus（Aiton）Schwantes subsp. canus（Haw.）H. E. K. Hartmann et Liede ■☆

321758　Punctillaria sororia（N. E. Br.）N. E. Br. ＝Pleiospilos compactus（Aiton）Schwantes subsp. sororius（N. E. Br.）H. E. K. Hartmann et Liede ■☆

321759　Punduana Steetz ＝Vernonia Schreb.（保留属名)●■

321760　Puneeria Stocks ＝Withania Pauquy（保留属名)●■

321761　Puneeria coagulans Stocks ＝Withania coagulans（Stocks）Dunal ●■☆

321762　Pungamia Lam. ＝Pongamia Adans.（保留属名)●

321763　Punica L.（1753）；石榴属（安石榴属）；Pomegranate ●

321764　Punica granatum L.；石榴（安榴,安石榴,丹若,海石榴,花石榴,金罂,若榴,若榴木,山力叶,天浆,榭榴）；Balustine Flowers，Carthaginian Apple，Common Pomegranate，Delima，Gerneter，Granada，Grenadine，Pomegarnet，Pomegranate，Pound-garnet，Wild Pomegranate ●

321765　Punica granatum L. 'Albescens'；白石榴；White Pomegranate ●

321766　Punica granatum L. 'Chico'；重瓣橙红石榴●

321767　Punica granatum L. 'Flavescens'；黄石榴；Yellow Pomegranate ●

321768　Punica granatum L. 'Legrellei'；玛瑙石榴；Agate Pomegranate ●

321769　Punica granatum L. 'Multiplex'；重瓣白石榴（复瓣白石榴,千瓣白石榴,重瓣白花石榴）；Doublepetalous White Pomegranate ●

321770　Punica granatum L. 'Nana Plena'；重瓣矮石榴●☆

321771　Punica granatum L. 'Nana'；月季石榴（矮生石榴,矮石榴）；Dwarf Pomegranate ●

321772　Punica granatum L. 'Nochi Shibari'；诺奇·西巴利石榴●☆

321773　Punica granatum L. 'Wonderful'；奇妙石榴●☆

321774　Punica granatum L. var. nana（L.）Pers. ＝Punica granatum L. 'Nana' ●

321775　Punica granatum L. var. plena Voss；重瓣月季石榴●

321776　Punica granatum L. var. pleniflora Hayata；重瓣石榴（千瓣大红,千瓣红,千瓣红石榴,重瓣红石榴）；Double-flower Pomegranate，Granada ●

321777　Punicaceae Bercht. et J. Presl（1825）（保留科名）；石榴科（安石榴科）；Pomegranate Family ●

321778　Punicaceae Bercht. et J. Presl（保留科名）＝Lythraceae J. St. -Hil.（保留科名)■●

321779　Punicaceae Horan. ＝Punicaceae Bercht. et J. Presl（保留科名)●

321780　Punicaceae Horan. ＝Putranjivaceae Endl. ●

321781　Punicella Turcz. ＝Balaustion Hook. ●☆

321782　Punjuba Britton et Rose ＝Pithecellobium Mart.（保留属名）●

321783　Punjuba Britton et Rose（1928）;热美围涎树属●☆

321784　Punjuba racemiflora（Donn. Sm.）Britton et Rose;热美围涎树●☆

321785　Puntia Hedge ＝Endostemon N. E. Br. ●■☆

321786　Puntia Hedge（1983）;蓬特草属●☆

321787　Puntia stenocaulis Hedge ＝Endostemon stenocaulis（Hedge）Ryding et A. J. Paton et Thulin ■☆

321788　Pupal Adans.（废弃属名）＝Pupalia Juss.（保留属名）■☆

321789　Pupalia Adans. ＝Pupalia Juss.（保留属名）■☆

321790　Pupalia Juss.（1803）（保留属名）;钩刺苋属（非洲苋属,钩牛膝属）■☆

321791　Pupalia affinis Engl. ＝Pupalia lappacea（L.）A. Juss. var. velutina（Moq.）Hook. f. ■☆

321792　Pupalia alopecurus Fenzl ＝Cyathula cylindrica Moq. ■☆

321793　Pupalia atropurpurea（Lam.）Moq. ＝Pupalia lappacea（L.）A. Juss. ■☆

321794　Pupalia brachystachys Peter ＝Pupalia lappacea（L.）A. Juss. var. velutina（Moq.）Hook. f. ■☆

321795　Pupalia distantiflora A. Rich. ＝Pupalia lappacea（L.）A. Juss. ■☆

321796　Pupalia erecta Suess. ＝Cyathula orthacantha（Hochst. ex Asch.）Schinz ■☆

321797　Pupalia grandiflora Peter;大花钩刺苋■☆

321798　Pupalia huillensis Hiern ＝Cyathula cylindrica Moq. ■☆

321799　Pupalia lappacea（L.）A. Juss. ;钩刺苋■☆

321800　Pupalia lappacea（L.）A. Juss. var. argyrophylla C. C. Towns. ;银叶钩刺苋■☆

321801　Pupalia lappacea（L.）A. Juss. var. glabrescens C. C. Towns. ;渐光钩刺苋■☆

321802　Pupalia lappacea（L.）A. Juss. var. grandiflora（Peter）Suess. ＝Pupalia grandiflora Peter ■☆

321803　Pupalia lappacea（L.）A. Juss. var. tomentosa（Peter）Suess. ＝Pupalia lappacea（L.）A. Juss. var. velutina（Moq.）Hook. f. ■☆

321804　Pupalia lappacea（L.）A. Juss. var. velutina（Moq.）Hook. f. ;短绒毛钩刺苋■☆

321805　Pupalia lappacea（L.）A. Juss. var. velutina Hook. f. ＝Pupalia lappacea（L.）A. Juss. ■☆

321806　Pupalia micrantha Hauman;小花钩刺苋■☆

321807　Pupalia mollis（Thonn.）Moq. ＝Pupalia lappacea（L.）A. Juss. var. velutina（Moq.）Hook. f. ■☆

321808　Pupalia natalensis Sond. ＝Cyathula natalensis Sond. ■☆

321809　Pupalia orthacantha Hochst. ex Asch. ＝Cyathula orthacantha（Hochst. ex Asch.）Schinz ■☆

321810　Pupalia prostrata（L.）C. Mart. ＝Cyathula prostrata（L.）Blume ■

321811　Pupalia prostrata（L.）Mart. ＝Cyathula prostrata（L.）Blume ■

321812　Pupalia psilotrichoides Suess. ＝Pupalia micrantha Hauman ■☆

321813　Pupalia remotiflora（Hook.）Lopr. ＝Sericorema remotiflora（Hook.）Lopr. ●☆

321814　Pupalia remotiflora Moq. ＝Sericorema remotiflora（Hook.）Lopr. ●☆

321815　Pupalia robecchii Lopr. ;罗贝克钩刺苋■☆

321816　Pupalia sericea Fiori ＝Pupalia lappacea（L.）A. Juss. var. velutina（Moq.）Hook. f. ■☆

321817　Pupalia subfusca Moq. ＝Centema subfusca（Moq.）T. Cooke ■☆

321818　Pupalia thonningii（Schumach.）Moq. ＝Pupalia lappacea

321819　Pupalia tomentosa Peter ＝Pupalia lappacea（L.）A. Juss. var. velutina（Moq.）Hook. f. ■☆

321820　Pupalia velutina Moq. ＝Pupalia lappacea（L.）A. Juss. var. velutina（Moq.）Hook. f. ■☆

321821　Pupartia Post et Kuntze ＝Poupartia Comm. ex Juss. ●☆

321822　Pupilla Rizzini ＝Justicia L. ●■

321823　Puraria Wall. ＝Pueraria DC. ●■

321824　Purchia Dumort. ＝Onosmodium Michx. ■☆

321825　Purchia Dumort. ＝Purshia Spreng. ■☆

321826　Purdiaea Planch.（1846）;宽萼桤叶树属●☆

321827　Purdiaea angustifolia C. Wright;窄叶宽萼桤叶树●☆

321828　Purdiaea cubensis Urb. ;古巴宽萼桤叶树●☆

321829　Purdiaea cubensis Urb. var. albosepala Vict. ;古巴白宽萼桤叶树●☆

321830　Purdiaea parvifolia（Vict.）J. L. Thomas;小叶宽萼桤叶树●☆

321831　Purdieanthus Gilg ＝Lehmanniella Gilg ■☆

321832　Purga Schiede ex Zucc. ＝Exogonium Choisy ■☆

321833　Purgosea Haw. ＝Crassula L. ●■☆

321834　Purgosea alooides（Dryand.）Haw. ＝Crassula hemisphaerica Thunb. ■☆

321835　Purgosea alpestris（Thunb.）G. Don ＝Crassula alpestris Thunb. ■☆

321836　Purgosea barbata（Thunb.）G. Don ＝Crassula barbata Thunb. ■☆

321837　Purgosea capitella（Thunb.）Sweet ＝Crassula capitella Thunb. ■☆

321838　Purgosea cephalophora（Thunb.）G. Don ＝Crassula nudicaulis L. ■☆

321839　Purgosea ciliata（L.）Sweet ＝Crassula ciliata L. ■☆

321840　Purgosea conspicua（Haw.）Sweet ＝Crassula tomentosa Thunb. ■☆

321841　Purgosea corymbulosa（Link et Otto）Sweet ＝Crassula capitella Thunb. subsp. thyrsiflora（Thunb.）Toelken ■☆

321842　Purgosea cotyledonis（Thunb.）Sweet ＝Crassula cotyledonis Thunb. ■☆

321843　Purgosea crenulata（Thunb.）G. Don ＝Crassula crenulata Thunb. ■☆

321844　Purgosea debilis（Thunb.）G. Don ＝Crassula thunbergiana Schult. ■☆

321845　Purgosea dentata（Thunb.）G. Don ＝Crassula dentata Thunb. ■☆

321846　Purgosea hemisphaerica（Thunb.）G. Don ＝Crassula hemisphaerica Thunb. ■☆

321847　Purgosea hirta（Thunb.）G. Don ＝Crassula nudicaulis L. ■☆

321848　Purgosea lingulifolia（Haw.）Haw. ＝Crassula tomentosa Thunb. ■☆

321849　Purgosea lingulifolia（Haw.）Sweet ＝Crassula tomentosa Thunb. ■☆

321850　Purgosea minima（Thunb.）G. Don ＝Crassula dentata Thunb. ■☆

321851　Purgosea montana（Thunb.）G. Don ＝Crassula montana Thunb. ●☆

321852　Purgosea obovata（Haw.）Haw. ＝Crassula obovata Haw. ■☆

321853　Purgosea pertusa（Haw.）Haw. ＝Crassula capitella Thunb. subsp. thyrsiflora（Thunb.）Toelken ■☆

321854　Purgosea pertusula Haw. ＝Crassula capitella Thunb. subsp. thyrsiflora（Thunb.）Toelken ■☆

321855　Purgosea pyramidalis（Thunb.）G. Don ＝Crassula pyramidalis Thunb. ●☆

321856　Purgosea sediflora Eckl. et Zeyh. ＝Crassula sediflora（Eckl. et

Zeyh. ） Endl. et Walp. ■☆

321857　Purgosea spicata （Thunb.） G. Don ＝ Crassula capitella Thunb. ■☆

321858　Purgosea tecta （Thunb.） G. Don ＝ Crassula tecta Thunb. ■☆

321859　Purgosea thyrsiflora （Thunb.） Sweet ＝ Crassula capitella Thunb. subsp. thyrsiflora （Thunb.） Toelken ■☆

321860　Purgosea tomentosa （Thunb.） Haw. ＝ Crassula tomentosa Thunb. ■☆

321861　Purgosea turrita （Thunb.） Sweet ＝ Crassula capitella Thunb. subsp. thyrsiflora （Thunb.） Toelken ■☆

321862　Purgosea turrita （Thunb.） Sweet var. alba Sweet ＝ Crassula capitella Thunb. ■☆

321863　Purgosta G. Don ＝ Purgosea Haw. ●■☆

321864　Puria N. C. Nair ＝ Cissus L. ●

321865　Purkayasthaea Purkayastha ＝ Beilschmiedia Nees ●

321866　Purkinjia C. Presl ＝ Ardisia Sw. （保留属名）●■

321867　Purpurabenis Thouars ＝ Cynorkis Thouars ■☆

321868　Purpurabenis Thouars ＝ Habenaria Willd. ■

321869　Purpurella Naudin ＝ Tibouchina Aubl. ●■☆

321870　Purpureostemon Gugerli（1939）；紫蕊桃金娘属●☆

321871　Purpureostemon ciliatus （J. R. Forst. et G. Forst.） Gugerli；紫蕊桃金娘●☆

321872　Purpurocynis Thouars ＝ Cynorkis Thouars ■☆

321873　Purpurocynis Thouars ＝ Cynosorchis Thouars ■☆

321874　Purpusia Brandegee ＝ Potentilla L. ■●

321875　Purpusia Brandegee（1899）；普尔蔷薇属●☆

321876　Purpusia saxosa Brandegee；普尔蔷薇●☆

321877　Purshia DC. ＝ Purshia DC. ex Poir. ●☆

321878　Purshia DC. ex Poir. （1816）；珀什蔷薇属；Antelope Bush ●☆

321879　Purshia Dennst. ＝ Centranthera R. Br. ■

321880　Purshia Poir. ＝ Purshia DC. ex Poir. ●☆

321881　Purshia Raf. ＝ Burshia Raf. ■

321882　Purshia Raf. ＝ Myriophyllum L. ■

321883　Purshia Spreng. ＝ Onosmodium Michx. ■☆

321884　Purshia humilis Raf. ＝ Myriophyllum humile Morong ■

321885　Purshia mexicana （D. Don） S. L. Welsh；墨西哥珀什蔷薇；Mexican Cliffrose ●☆

321886　Purshia tridentata （Pursh） DC. ；三齿珀什蔷薇；Antelope Brush ●☆

321887　Puruma J. St. -Hil. ＝ Pourouma Aubl. ●☆

321888　Pusaetha Kuntze ＝ Entada Adans. （保留属名）●

321889　Pusaetha africana （Guillaumin et Perr.） Kuntze ＝ Entada africana Guillaumin et Perr. ●☆

321890　Pusaetha stuhlmannii Taub. ＝ Entada stuhlmannii （Taub.） Harms ●☆

321891　Pusaetha wahlbergii （Harv.） Kuntze ＝ Entada wahlbergii Harv. ●☆

321892　Puschkinia Adams （1805）；蚁播花属；Lebanon Squill, Puschkinia ■☆

321893　Puschkinia hyacinthoides Baker；风信子状蚁播花■☆

321894　Puschkinia libanotica Zucc. ；黎巴嫩蚁播花；Lebanon Squill, Striped Squill ■☆

321895　Puschkinia libanotica Zucc. ＝ Puschkinia scilloides Adams ■☆

321896　Puschkinia scilloides Adams；蚁播花；Lebanon Squill, Striped Squill ■☆

321897　Puschkinia sicula Van Houtte ＝ Puschkinia scilloides Adams ■☆

321898　Pusillanthus Kuijt ＝ Phthirusa Mart. ●☆

321899　Pusillanthus Kuijt（2008）；弱花桑寄生属●☆

321900　Pusiphylis Thouars ＝ Bulbophyllum Thouars（保留属名）■

321901　Putoria Pers. （1805）；臭茜属；Bitterbrush ●☆

321902　Putoria brevifolia Coss. et Durieu ex Batt. ；短叶臭茜●☆

321903　Putoria brevifolia Coss. et Durieu ex Batt. var. typica Maire ＝ Putoria brevifolia Coss. et Durieu ex Batt. ●☆

321904　Putoria brevifolia Coss. et Durieu var. demnatensis Litard. et Maire ＝ Putoria brevifolia Coss. et Durieu ex Batt. ●☆

321905　Putoria brevifolia Coss. et Durieu var. dyris Jahand. et Maire ＝ Putoria brevifolia Coss. et Durieu ex Batt. ●☆

321906　Putoria brevifolia Coss. et Durieu var. melillensis Maire et Sennen ＝ Putoria brevifolia Coss. et Durieu ex Batt. ●☆

321907　Putoria brevifolia Coss. et Durieu var. microphylla （Pomel） Batt. ＝ Putoria brevifolia Coss. et Durieu ex Batt. ●☆

321908　Putoria brevifolia Coss. et Durieu var. tenella （Pomel） Batt. ＝ Putoria brevifolia Coss. et Durieu ex Batt. ●☆

321909　Putoria calabrica （L. f.） Pers. ；臭茜；Stinking Madder ●☆

321910　Putoria calabrica （L. f.） Pers. subsp. hispanica （Boiss. et Reut.） Nyman ＝ Putoria calabrica （L. f.） Pers. ●☆

321911　Putoria calabrica （L. f.） Pers. var. atlantis Litard. et Maire ＝ Putoria calabrica （L. f.） Pers. ●☆

321912　Putoria calabrica （L. f.） Pers. var. hispanica （Boiss. et Reut.） Jahand. et Maire ＝ Putoria calabrica （L. f.） Pers. ●☆

321913　Putoria calabrica Pers. ＝ Putoria calabrica （L. f.） Pers. ●☆

321914　Putoria cymosa Pomel ＝ Putoria calabrica （L. f.） Pers. ●☆

321915　Putoria hispanica Boiss. et Reut. ＝ Putoria calabrica （L. f.） Pers. ●☆

321916　Putoria microphylla Pomel ＝ Putoria brevifolia Coss. et Durieu ex Batt. ●☆

321917　Putoria tenella Pomel ＝ Putoria brevifolia Coss. et Durieu ex Batt. ●☆

321918　Putranjiva Wall. （1826）；羽柱果属●

321919　Putranjiva Wall. ＝ Drypetes Vahl ●

321920　Putranjiva formosana Kaneh. et Sasaki ex Shimada ＝ Drypetes formosana （Kaneh. et Sasaki ex Shimada） Kaneh. ●

321921　Putranjiva integerrima Koidz. ＝ Drypetes integerrima （Koidz.） Hosok. ●☆

321922　Putranjiva matsumurae Koidz. ＝ Drypetes matsumurae （Koidz.） Kaneh. ●

321923　Putranjiva roxburghii Wall. ＝ Drypetes roxburghii （Wall.） Hurus. ●

321924　Putranjivaceae Endl. ＝ Euphorbiaceae Juss. （保留科名）●■

321925　Putranjivaceae Endl. ＝ Putranjivaceae Meisn. ●

321926　Putranjivaceae Meisn. （1842）；羽柱果科●

321927　Putterlickia Endl. （1840）；普氏卫矛属（普特里开亚属,普特木属）●☆

321928　Putterlickia pyracantha （L.） Szyszyl. ；火棘普氏卫矛（火刺普特里开亚,火刺普特木）●☆

321929　Putterlickia pyracantha Endl. ＝ Putterlickia pyracantha （L.） Szyszyl. ●☆

321930　Putterlickia retrospinosa A. E. van Wyk et Mostert；弯刺普氏卫矛●☆

321931　Putterlickia saxatilis （Burch.） Jordaan；岩生普氏卫矛●☆

321932　Putterlickia verrucosa （E. Mey. ex Sond.） Szyszyl. ；多痣普氏卫矛（多痣普特里开亚,多痣普特木）●☆

321933　Putterlickia verrucosa Sim ＝ Putterlickia verrucosa （E. Mey. ex Sond.） Szyszyl. ●☆

321934　Putzeysia Klotzsch ＝ Begonia L. ●■

321935　Putzeysia Planch. et Linden　＝Aesculus L. ●

321936　Puya Molina(1782)；普亚凤梨属（安第斯凤梨属，粗茎凤梨属，火星草属，蒲雅凤梨属，蒲亚属，普雅属，普亚属，普椰属）；Puya ■☆

321937　Puya alpestris（Poepp.）Gay；高山普亚凤梨（亚高山火星草，亚高山普亚）；Alpine Puya ■☆

321938　Puya berteroniana Mez；贝氏普亚凤梨；Puya ■☆

321939　Puya chilensis Molina；火星草（普亚）■☆

321940　Puya coerulea Miers；灰普亚■☆

321941　Puya gigas André；大普亚■☆

321942　Puya raimondii Harms；芮氏普亚凤梨；Raimond Puya ■☆

321943　Pyankovia Akhani et Roalson　＝Salsola L. ●■

321944　Pyankovia Akhani et Roalson(2007)；皮氏猪毛菜属●☆

321945　Pycanthus Post et Kuntze　＝Grabowskia Schltdl. ●☆

321946　Pycanthus Post et Kuntze　＝Pukanthus Raf. ●☆

321947　Pychnanthemum G. Don　＝Pycnanthemum Michx.（保留属名）■☆

321948　Pychnostachys G. Don　＝Pycnostachys Hook. ●■☆

321949　Pycnandra Benth.（1876）；密蕊榄属●☆

321950　Pycnandra benthamii Baill.；本氏密蕊榄●☆

321951　Pycnandra elegans Vink；雅致密蕊榄●☆

321952　Pycnanthemum Michx.（1803）（保留属名）；山薄荷属；Mountain Mint，Mountain-mint ■☆

321953　Pycnanthemum albescens Torr. et A. Gray；白山薄荷；White Mountain Mint ■☆

321954　Pycnanthemum decurrens Blanco　＝Hyptis rhomboides M. Martens et Galeotti ■

321955　Pycnanthemum elongatum Blanco　＝Hyptis spicigera Lam. ■

321956　Pycnanthemum flexuosum（Walter）Britton, Sterns et Poggenb.；反卷山薄荷■☆

321957　Pycnanthemum incanum（L.）Michx.；灰色山薄荷；Hoary Mountain-mint，Mountain-mint ■☆

321958　Pycnanthemum muticum（Michx.）Pers.；钝山薄荷；Mountain-mint，Short-toothed Mountain Mint ■☆

321959　Pycnanthemum muticum Pers.　＝Pycnanthemum muticum（Michx.）Pers. ■☆

321960　Pycnanthemum pilosum Nutt.；毛山薄荷；Hairy Leaf Mountain Mint，Hairy Mountain Mint ■☆

321961　Pycnanthemum tenuifolium Schrad.；细山薄荷；Narrow Leaf Mountain Mint，Narrow-leafed Mountain-mint，Narrow-leaved Mountain Mint，Slender Mountain Mint ■☆

321962　Pycnanthemum torrei Benth.；托雷山薄荷；Mountain Mint ■☆

321963　Pycnanthemum verticillatum Pers.；轮状山薄荷；Torrey's Mountain-mint ■☆

321964　Pycnanthemum virginianum（L.）T. Durand et B. D. Jacks. ex B. L. Rob. et Fernald；普通山薄荷（弗州山薄荷）；Common Mountain Mint，Mountain Mint，Virgdnia Mountain-mint，Virginia Mountain Mint ■☆

321965　Pycnanthes Raf.　＝Pycnanthemum Michx.（保留属名）■☆

321966　Pycnanthus Warb.（1895）；密花属（密花楠属）●☆

321967　Pycnanthus angolensis（Welw.）Warb.；安哥拉密花；Akomu，Eteng，Ilomba，Kombo False Nutmeg ●☆

321968　Pycnanthus angolensis（Welw.）Warb. subsp. schweinfurthii（Warb.）Verdc.；施韦密花●☆

321969　Pycnanthus angolensis（Welw.）Warb. var. amarantifolius Compère　＝Pycnanthus angolensis（Welw.）Warb. ●☆

321970　Pycnanthus dinklagei Warb.；丁克密花●☆

321971　Pycnanthus kombo（Baill.）Warb.　＝Pycnanthus angolensis

（Welw.）Warb. ●☆

321972　Pycnanthus kombo（Baill.）Warb. var. angolensis（Welw.）Warb.　＝Pycnanthus angolensis（Welw.）Warb. ●☆

321973　Pycnanthus kombo Warb.；野密花；False Nutmeg，White Cedar，Wild Nutmeg ●☆

321974　Pycnanthus mechowii Warb.　＝Pycnanthus angolensis（Welw.）Warb. ●☆

321975　Pycnanthus schweinfurthii Warb.　＝Pycnanthus angolensis（Welw.）Warb. subsp. schweinfurthii（Warb.）Verdc. ●☆

321976　Pycnarrhena Miers　＝Pycnarrhena Miers ex Hook. f. et Thomson ●

321977　Pycnarrhena Miers ex Hook. f. et Thomson(1855)；密花藤属；Pycnarrhena ●

321978　Pycnarrhena calocarpa（Kurz）Diels　＝Pycnarrhena lucida（Teijsm. et Binn.）Miq. ●

321979　Pycnarrhena fasciculata（Miers）Diels　＝Pycnarrhena lucida（Teijsm. et Binn.）Miq. ●

321980　Pycnarrhena longifolia（Decne. ex Miq.）Becc.；长叶密花藤；Longleaf Pycnarrhena ●☆

321981　Pycnarrhena lucida（Teijsm. et Binn.）Miq.；密花藤；Fasciculate Pycnarrhena，Pycnarrhena ●

321982　Pycnarrhena macrocarpa Diels　＝Eleutharrhena macrocarpa（Diels）Forman ●◇

321983　Pycnarrhena poilanei（Gagnep.）Forman；硬骨藤；Poilane Pycnarrhena ●

321984　Pycnobolus Willd. ex O. E. Schulz　＝Eudema Humb. et Bonpl. ■☆

321985　Pycnobotrya Benth.（1876）；密穗夹竹桃属●☆

321986　Pycnobotrya multiflora K. Schum. ex Stapf　＝Pycnobotrya nitida Benth. ●☆

321987　Pycnobotrya nitida Benth.；光亮密穗夹竹桃●☆

321988　Pycnobregma Baill.（1890）；密额萝藦属☆

321989　Pycnobregma funckii Baill.；密额萝藦☆

321990　Pycnocephalum（Less.）DC.（1836）；密头菊属●☆

321991　Pycnocephalum DC.　＝Eremanthus Less. ●☆

321992　Pycnocephalum angustifolium（Gardner）MacLeish；窄叶密头菊●☆

321993　Pycnocephalum pinnatifidum（Philipson）MacLeish；羽裂密头菊●☆

321994　Pycnocephalum spathulifolium DC.；匙叶密头菊●☆

321995　Pycnocoma Benth.（1849）；密毛大戟属●☆

321996　Pycnocoma angustifolia Prain；窄叶密毛大戟●☆

321997　Pycnocoma bampsiana J. Léonard；邦氏密毛大戟●☆

321998　Pycnocoma beillei A. Chev. ex Hutch. et Dalziel　＝Pycnocoma angustifolia Prain ●☆

321999　Pycnocoma brachystachya Pax；短穗密毛大戟●☆

322000　Pycnocoma chevalieri Beille；舍瓦利耶密毛大戟●☆

322001　Pycnocoma cornuta Müll. Arg.；角状密毛大戟●☆

322002　Pycnocoma dentata Hiern；尖齿密毛大戟●☆

322003　Pycnocoma devredii J. Léonard；德夫雷大戟☆

322004　Pycnocoma hirsuta Prain　＝Argomuellera macrophylla Pax ●☆

322005　Pycnocoma hutchinsonii Beille　＝Argomuellera macrophylla Pax ●☆

322006　Pycnocoma insularum J. Léonard；海岛密毛大戟●☆

322007　Pycnocoma laurentii De Wild.　＝Argomuellera macrophylla Pax ●☆

322008　Pycnocoma littoralis Pax；滨海密毛大戟●☆

322009　Pycnocoma longipes Pax　＝Pycnocoma thonneri Pax ex De Wild. et T. Durand ●☆

322010　Pycnocoma louisii J. Léonard；路易斯密毛大戟●☆

322011　Pycnocoma lucida Pax et K. Hoffm.　＝Pycnocoma chevalieri

Beille ●☆

322012 Pycnocoma macrantha Pax;大花密毛大戟●☆

322013 Pycnocoma macrophylla Benth.;大叶密毛大戟●☆

322014 Pycnocoma macrophylla Benth. var. genuina Pax et K. Hoffm. = Pycnocoma macrophylla Benth. ●☆

322015 Pycnocoma macrophylla Benth. var. longicornuta J. Léonard = Pycnocoma macrophylla Benth. ●☆

322016 Pycnocoma macrophylla Benth. var. microsperma Pax et K. Hoffm.;小籽密毛大戟●☆

322017 Pycnocoma minor Müll. Arg.;小密毛大戟●☆

322018 Pycnocoma mortehanii De Wild. = Pycnocoma reygaertii De Wild. ●☆

322019 Pycnocoma parviflora Pax = Argomuellera macrophylla Pax ●☆

322020 Pycnocoma reygaertii De Wild.;赖氏密毛大戟●☆

322021 Pycnocoma sapinii De Wild. = Argomuellera macrophylla Pax ●☆

322022 Pycnocoma sassandrae Beille = Argomuellera macrophylla Pax ●☆

322023 Pycnocoma subflava J. Léonard;浅黄密毛大戟●☆

322024 Pycnocoma thollonii Prain;托伦密毛大戟●☆

322025 Pycnocoma thonneri Pax ex De Wild. et T. Durand;托内密毛大戟●☆

322026 Pycnocoma trilobata De Wild. = Pycnocoma thonneri Pax ex De Wild. et T. Durand ●☆

322027 Pycnocoma zenkeri Pax;岑克尔密毛大戟●☆

322028 Pycnocomon Hoffmanns. et Link = Scabiosa L. ●■

322029 Pycnocomon Hoffmanns. et Link(1820);密毛续断属■☆

322030 Pycnocomon St. -Lag. = Cirsium Mill. ■

322031 Pycnocomon St. -Lag. = Picnomon Adans. ■☆

322032 Pycnocomon Wallr. = Cephalaria Schrad. (保留属名)■

322033 Pycnocomon montanum Pomel = Pycnocomon rutifolium (Vahl) Hoffmanns. et Link ■☆

322034 Pycnocomon rutifolium (Vahl) Hoffmanns. et Link;芸香叶密毛续断■☆

322035 Pycnocomon rutifolium (Vahl) Hoffmanns. et Link var. montanum (Pomel) Maire = Pycnocomon rutifolium (Vahl) Hoffmanns. et Link ■☆

322036 Pycnocomum Link = Pycnocomon Hoffmanns. et Link ■☆

322037 Pycnocomus Hill = Calcitrapoides Fabr. ●■

322038 Pycnocomus Hill = Centaurea L. (保留属名)●■

322039 Pycnocycla Lindl. (1835);密环草属☆

322040 Pycnocycla aucheriana Decne. ex Boiss.;奥切尔密环草☆

322041 Pycnocycla caespitosa Boiss. et Hausskn. ex Boiss.;丛生密环草☆

322042 Pycnocycla glauca Lindl.;灰绿密环草☆

322043 Pycnocycla ledermannii H. Wolff;莱德密环草●☆

322044 Pycnolachne Turcz. = Lachnostachys Hook. ●☆

322045 Pycnoneurum Decne. (1838);密脉萝藦属■☆

322046 Pycnoneurum Decne. = Cynanchum L. ●■

322047 Pycnoneurum junciforme Decne. = Cynanchum junciforme (Decne.) Liede ■☆

322048 Pycnoneurum sessiliflorum Decne. = Cynanchum sessiliflorum (Decne.) Liede ■☆

322049 Pycnonia L. A. S. Johnson et B. G. Briggs(1975);澳洲山龙眼属●☆

322050 Pycnonia teretifolia (R. Br.) L. A. S. Johnson et B. G. Briggs;澳洲山龙眼●☆

322051 Pycnophyllopsis Skottsb. (1916);类密叶花属■☆

322052 Pycnophyllopsis muscosa Skottsb.;类密叶花■☆

322053 Pycnophyllum J. Rémy(1846);密叶花属■☆

322054 Pycnophyllum filiforme Mattf.;线形密叶花■☆

322055 Pycnophyllum macropetalum Mattf.;大瓣密叶花■☆

322056 Pycnophyllum macrophyllum Muschl.;大叶密叶花■☆

322057 Pycnoplinthopsis Jafri(1972);假簇芥属■

322058 Pycnoplinthopsis bhuatanica Jafri;假簇芥■

322059 Pycnoplinthopsis minor Jafri = Pycnoplinthopsis bhuatanica Jafri ■

322060 Pycnoplinthus O. E. Schulz(1924);簇芥属;Pycnoplinthus ■

322061 Pycnoplinthus uniflorus (Hook. f. et Thomson) O. E. Schulz;簇芥;Oneflower Pycnoplinthus ■

322062 Pycnorhachis Benth. (1876);密刺萝藦属☆

322063 Pycnorhachis benthamiana Baill.;密刺萝藦☆

322064 Pycnosandra Blume = Drypetes Vahl ●

322065 Pycnosorus Benth. (1837);密头彩鼠麹属■☆

322066 Pycnosorus Benth. = Craspedia G. Forst. ■☆

322067 Pycnosorus chrysanthus (Schldl.) Sond.;金花密头彩鼠麹■☆

322068 Pycnospatha Thorel ex Gagnep. (1941);密苞南星属■☆

322069 Pycnospatha palmata Gagnep. 密苞南星■☆

322070 Pycnosphace (Benth.) Rydb. = Salvia L. ●■

322071 Pycnosphace Rydb. = Salvia L. ●■

322072 Pycnosphaera Gilg(1903);密球龙胆属■☆

322073 Pycnosphaera buchananii (Baker) N. E. Br.;密球龙胆■☆

322074 Pycnosphaera quarrei De Wild. = Pycnosphaera buchananii (Baker) N. E. Br. ■☆

322075 Pycnosphaera vanderysti De Wild. = Pycnosphaera buchananii (Baker) N. E. Br. ■☆

322076 Pycnospora R. Br. ex Wight et Arn. (1834);密子豆属;Seedybean, Pycnospora ●■

322077 Pycnospora hedysaroides R. Br. ex Wight et Arn. = Pycnospora lutescens (Poir.) Schindl. ●■

322078 Pycnospora lutescens (Poir.) Schindl.;密子豆(假地豆,假番豆);Lutescent Pycnospora, Lutescent Seedybean, Three Goldenpoint ●■

322079 Pycnospora nervosa Wight et Arn. = Pycnospora lutescens (Poir.) Schindl. ●■

322080 Pycnostachys Hook. (1826);密穗花属(密穗木属)●■☆

322081 Pycnostachys abyssinica Fresen.;阿比西尼亚密穗花(阿比西尼亚密穗木)●☆

322082 Pycnostachys affinis Gürke = Pycnostachys speciosa Gürke ■☆

322083 Pycnostachys angolensis G. Taylor;安哥拉密穗花(安哥拉密穗木)●☆

322084 Pycnostachys ballotoides Perkins = Pycnostachys pseudospeciosa Buscal. et Muschl. ■☆

322085 Pycnostachys batesii Baker;贝茨密穗花■☆

322086 Pycnostachys bequaertii De Wild. = Pycnostachys stuhlmannii Gürke ■☆

322087 Pycnostachys bowalensis A. Chev. = Pycnostachys meyeri Gürke ■☆

322088 Pycnostachys brevipetiolata De Wild. = Pycnostachys coerulea Hook. ■☆

322089 Pycnostachys bussei Gürke = Pycnostachys orthodonta Gürke ■☆

322090 Pycnostachys butaguensis De Wild. = Pycnostachys elliotii S. Moore ■☆

322091 Pycnostachys carigensis Gürke ex De Wild. = Pycnostachys kassneri De Wild. ■☆

322092 Pycnostachys chevalieri Briq.;舍瓦利耶密穗花■☆

322093 Pycnostachys ciliata Bramley;缘毛密穗花■☆

322094 Pycnostachys cinerascens Robyns et Lebrun = Pycnostachys elliotii S. Moore ■☆

322095 Pycnostachys clinodon Mildbr. = Pycnostachys ruandensis De

Wild. ■☆

322096 Pycnostachys coerulea Hook. ;天蓝密穗花■☆

322097 Pycnostachys congensis Gürke;刚果密穗花■☆

322098 Pycnostachys cyanea Gürke ＝Pycnostachys orthodonta Gürke ■☆

322099 Pycnostachys dawei N. E. Br. ;头状密穗花■☆

322100 Pycnostachys dawei N. E. Br. ＝Pycnostachys speciosa Gürke ■☆

322101 Pycnostachys decussata Baker ＝ Pycnostachys niamniamensis Gürke ■☆

322102 Pycnostachys deflexifolia Baker;外折叶密穗花■☆

322103 Pycnostachys deflexifolia Baker var. gattii Fiori ＝Pycnostachys deflexifolia Baker ■☆

322104 Pycnostachys descampsii Briq. ;德康密穗花■☆

322105 Pycnostachys de-wildemaniana Robyns et Lebrun;德怀尔德曼密穗花■☆

322106 Pycnostachys elliotii S. Moore;埃里密穗花■☆

322107 Pycnostachys eminii Gürke;埃明密穗花■☆

322108 Pycnostachys erici-rosenii R. E. Fr. ;欧石南密穗花■☆

322109 Pycnostachys goetzenii Gürke;格氏密穗花■☆

322110 Pycnostachys gracilis R. D. Good;纤细密穗花■☆

322111 Pycnostachys graminifolia Perkins;禾叶密穗花■☆

322112 Pycnostachys hanningtonii Baker ＝ Pycnostachys orthodonta Gürke ■☆

322113 Pycnostachys holophylla Briq. ＝ Pycnostachys reticulata (E. Mey.) Benth. ■☆

322114 Pycnostachys kaessneri Perkins ＝Pycnostachys perkinsii E. A. Bruce ☆

322115 Pycnostachys kassneri De Wild. ;卡斯纳密穗花■☆

322116 Pycnostachys kirkii Baker ＝Pycnostachys reticulata (E. Mey.) Benth. ■☆

322117 Pycnostachys lancifolia Bramley;披针叶密穗花■☆

322118 Pycnostachys lavanduloides Perkins ＝Pycnostachys linifolia Gürke ■☆

322119 Pycnostachys leptophylla Baker ＝ Pycnostachys orthodonta Gürke ■☆

322120 Pycnostachys lindblomii T. C. E. Fr. ＝ Pycnostachys niamniamensis Gürke ■☆

322121 Pycnostachys linifolia Gürke;亚麻叶密穗花■☆

322122 Pycnostachys longebracteata De Wild. ＝ Pycnostachys meyeri Gürke ■☆

322123 Pycnostachys longiacuminata Perkins ＝Pycnostachys abyssinica Fresen. ●☆

322124 Pycnostachys longifolia De Wild. ＝ Pycnostachys stuhlmannii Gürke ■☆

322125 Pycnostachys mausaensis Gürke ex De Wild. ＝ Pycnostachys pseudospeciosa Buscal. et Muschl. ■☆

322126 Pycnostachys meyeri Gürke;迈尔密穗花■☆

322127 Pycnostachys micrantha Gürke ＝Pycnostachys coerulea Hook. ■☆

322128 Pycnostachys mildbraedii Perkins ＝Pycnostachys elliotii S. Moore ■☆

322129 Pycnostachys nepetifolia Baker;荆芥叶密穗花■☆

322130 Pycnostachys niamniamensis Gürke;尼亚密穗花■☆

322131 Pycnostachys oblongifolia Baker ＝Pycnostachys meyeri Gürke ■☆

322132 Pycnostachys orthodonta Gürke;直齿密穗花■☆

322133 Pycnostachys ovoideo-conica De Wild. ＝ Pycnostachys meyeri Gürke ■☆

322134 Pycnostachys pallide-caerulea Perkins;白蓝密穗花■☆

322135 Pycnostachys parvifolia Baker;小叶密穗花■☆

322136 Pycnostachys perkinsii E. A. Bruce;珀金斯密穗花■☆

322137 Pycnostachys petherickii Baker ＝ Pycnostachys niamniamensis Gürke ■☆

322138 Pycnostachys prittwitzii Perkins;普里特密穗花■☆

322139 Pycnostachys pseudospeciosa Buscal. et Muschl. ;假美丽密穗花■☆

322140 Pycnostachys pubescens Gürke ＝Pycnostachys urticifolia Hook. ●☆

322141 Pycnostachys purpurascens Briq. ＝Pycnostachys reticulata (E. Mey.) Benth. ■☆

322142 Pycnostachys recurvata Ryding;反曲密穗花■☆

322143 Pycnostachys remotifolia Baker ＝ Pycnostachys stuhlmannii Gürke ■☆

322144 Pycnostachys reticulata (E. Mey.) Benth. ;网状密穗花■☆

322145 Pycnostachys reticulata (E. Mey.) Benth. var. angustifolia Benth. ＝ Pycnostachys reticulata (E. Mey.) Benth. ■☆

322146 Pycnostachys rotundato-dentata De Wild. ＝ Pycnostachys eminii Gürke ■☆

322147 Pycnostachys ruandensis De Wild. ;卢旺达密穗花■☆

322148 Pycnostachys ruwenzoriensis Baker ＝Pycnostachys eminii Gürke ■☆

322149 Pycnostachys schlechteri Briq. ＝ Pycnostachys reticulata (E. Mey.) Benth. ■☆

322150 Pycnostachys schliebenii Mildbr. ;施利本密穗花■☆

322151 Pycnostachys schweinfurthii Briq. ;施韦密穗花■☆

322152 Pycnostachys speciosa Gürke;美丽密穗花■☆

322153 Pycnostachys sphaerocephala Baker;球头密穗花■☆

322154 Pycnostachys stenostachys Baker ＝ Pycnostachys coerulea Hook. ■☆

322155 Pycnostachys stuhlmannii Gürke;斯图尔曼密穗花■☆

322156 Pycnostachys togoensis Perkins ＝ Pycnostachys schweinfurthii Briq. ■☆

322157 Pycnostachys uliginosa Gürke ＝ Pycnostachys reticulata (E. Mey.) Benth. ■☆

322158 Pycnostachys umbrosa (Vatke) Perkins;耐荫密穗花■☆

322159 Pycnostachys urticifolia Hook. ;荨麻叶密穗木(荨麻叶密穗花);Blue Boys ●☆

322160 Pycnostachys urticifolia Hook. var. pubescens Gürke ＝ Pycnostachys urticifolia Hook. ●☆

322161 Pycnostachys verticillata Baker;轮生密穗花■☆

322162 Pycnostachys volkensii Gürke ＝Pycnostachys meyeri Gürke ■☆

322163 Pycnostachys vulcanicola Lebrun et L. Touss. ;火山密穗花■☆

322164 Pycnostachys whytei Baker;怀特密穗花■☆

322165 Pycnostelma Bunge ex Decne. (1844);徐长卿属■

322166 Pycnostelma Bunge ex Decne. ＝Cynanchum L. ●■

322167 Pycnostelma Decne. ＝Pycnostelma Bunge ex Decne. ■

322168 Pycnostelma chinense Bunge ex Decne. ＝ Cynanchum paniculatum (Bunge) Kitag. ex H. Hara ■

322169 Pycnostelma lateriflorum Hemsl. ＝ Cynanchum mongolicum (Maxim.) Hemsl. ■

322170 Pycnostelma leucanthum Kitag. ＝ Cynanchum paniculatum (Bunge) Kitag. ex H. Hara ■

322171 Pycnostelma paniculatum (Bunge) K. Schum. ＝ Cynanchum paniculatum (Bunge) Kitag. ex H. Hara ■

322172 Pycnostelma paniculatum K. Schum. ＝ Cynanchum paniculatum (Bunge) Kitag. ex H. Hara ■

322173 Pycnostylis Pierre ＝Triclisia Benth. ●☆

322174 Pycnostylis sacleuxii Pierre ＝Triclisia sacleuxii (Pierre) Diels ●☆

322175 Pycnothryx M. E. Jones ＝Drudeophytum J. M. Coult. et Rose ■☆

322176 Pycnothryx M. E. Jones = Tauschia Schltdl. (保留属名)■☆

322177 Pycnothymus (Benth.) Small = Piloblephis Raf. ●☆

322178 Pycnothymus Small = Satureja L. ●■

322179 Pycreus P. Beauv. (1816);扁莎属(侧扁莎属);Flatsedge,Pycreus ■

322180 Pycreus P. Beauv. = Cyperus L. ■

322181 Pycreus acaulis Nelmes;无茎扁莎■☆

322182 Pycreus acuticarinatus (Kük.) Cherm.;尖棱扁莎■☆

322183 Pycreus afrozonatus Lye = Cyperus zonatus Kük. ■☆

322184 Pycreus albomarginatus Mart. et Schrad. ex Nees = Cyperus flavicomus Michx. ■☆

322185 Pycreus albomarginatus Nees = Pycreus macrostachyos (Lam.) J. Raynal ■☆

322186 Pycreus angulatus (Nees) Nees = Cyperus unioloides R. Br. ■

322187 Pycreus angulatus (Nees) Nees = Pycreus unioloides (R. Br.) Urb. ■

322188 Pycreus ater (C. B. Clarke) Cherm.;黑色扁莎■☆

322189 Pycreus atribulbus (Kük.) Napper;黑球根扁莎■☆

322190 Pycreus atronervatus (Boeck.) C. B. Clarke;暗脉扁莎■☆

322191 Pycreus atrorubidus Nelmes;暗红扁莎■☆

322192 Pycreus baoulensis A. Chev. = Cyperus baoulensis Kük. ■☆

322193 Pycreus bipartitus (Torr.) C. B. Clarke = Cyperus bipartitus Torr. ■☆

322194 Pycreus capillifolius (A. Rich.) C. B. Clarke;毛叶扁莎■☆

322195 Pycreus cataractarum C. B. Clarke;瀑布群高原扁莎■☆

322196 Pycreus chekiangensis Ts. Tang et F. T. Wang;浙江扁莎;Chekiang Pycreus,Zhejiang Flatsedge,Zhejiang Pycreus ■

322197 Pycreus chrysanthus (Boeck.) C. B. Clarke;金花扁莎■☆

322198 Pycreus colchicus C. Koch;黑海扁莎■☆

322199 Pycreus cooperi C. B. Clarke;库珀扁莎■☆

322200 Pycreus cuanzensis (Ridl.) C. B. Clarke = Pycreus smithianus (Ridl.) C. B. Clarke ■☆

322201 Pycreus debilissimus C. B. Clarke = Pycreus flavescens (L.) P. Beauv. ex Rchb. subsp. tanaensis (Kük.) Lye ■☆

322202 Pycreus delavayi C. B. Clarke;黑鳞扁莎;Delavay Flatsedge,Delavay Pycreus ■

322203 Pycreus densespicatus Hayata = Cyperus imbricatus Retz. var. elongatus (Boeck.) L. K. Dai ■

322204 Pycreus densus (Link) C. B. Clarke = Pycreus lanceolatus (Poir.) C. B. Clarke ■☆

322205 Pycreus dewildeorum J. Raynal;德维尔德扁莎■☆

322206 Pycreus diander (Torr.) C. B. Clarke = Cyperus diandrus Torr. ■☆

322207 Pycreus diaphanus (Schrad. ex Roem. et Schult.) S. S. Hooper et T. Koyama ex S. S. Hooper = Cyperus diaphanus Schrad. ex Roem. et Schult. ■

322208 Pycreus diaphanus (Schrad. ex Roem. et Schult.) S. S. Hooper et T. Koyama = Cyperus diaphanus Schrad. ex Roem. et Schult. ■

322209 Pycreus diaphanus (Schrad. ex Roem. et Schult.) S. S. Hooper et T. Koyama subsp. setiformis (Korsh.) T. Koyama = Cyperus diaphanus Schrad. ex Roem. et Schult. ■

322210 Pycreus diaphanus (Schrad. ex Roem. et Schult.) S. S. Hooper et T. Koyama ex S. S. Hooper;宽穗扁莎;Broad Spike Pycreus,Broadspike Flatsedge ■

322211 Pycreus diaphanus (Schrad. ex Schult.) S. S. Hooper et T. Koyama = Pycreus diaphanus (Schrad. ex Roem. et Schult.) S. S. Hooper et T. Koyama ex S. S. Hooper ■

322212 Pycreus divulsus Ridl. subsp. africanus S.S. Hooper;非洲扁莎■☆

322213 Pycreus djalonis A. Chev. = Cyperus pustulatus Vahl ■☆

322214 Pycreus dwarkensis (Sahni et H. B. Naithani) S. S. Hooper;杜瓦尔卡扁莎■☆

322215 Pycreus elegantulus (Steud.) C. B. Clarke = Pycreus niger (Ruiz et Pav.) Cufod. subsp. elegantulus (Steud.) Lye ■☆

322216 Pycreus eragrostis Lam. f. melanocephalus (Miq.) Suringar = Pycreus sanguinolentus (Vahl) Nees ex C. B. Clarke ■

322217 Pycreus eragrostis Lam. var. humilis Miq. = Pycreus sanguinolentus (Vahl) Nees ex C. B. Clarke ■

322218 Pycreus esculentus (L.) Hayek = Cyperus esculentus L. ■☆

322219 Pycreus felicis J. Raynal;多育扁莎■☆

322220 Pycreus ferrugineus (Poir.) C. B. Clarke = Pycreus intactus (Vahl) J. Raynal ■☆

322221 Pycreus ferrugineus (Poir.) C. B. Clarke var. baroni (C. B. Clarke) Cherm. = Pycreus intactus (Vahl) J. Raynal ■☆

322222 Pycreus fibrillosus (Kük.) Cherm.;须毛扁莎■☆

322223 Pycreus flavescens (L.) P. Beauv. ex Rchb.;浅黄扁莎■☆

322224 Pycreus flavescens (L.) P. Beauv. ex Rchb. subsp. intermedius (Steud.) Lye = Pycreus intermedius (Steud.) C. B. Clarke ■☆

322225 Pycreus flavescens (L.) P. Beauv. ex Rchb. subsp. laevinux Lye = Pycreus overlaetii Cherm. ex S. S. Hooper et J. Raynal ■☆

322226 Pycreus flavescens (L.) P. Beauv. ex Rchb. subsp. microglumis Lye;小颖浅黄扁莎■☆

322227 Pycreus flavescens (L.) P. Beauv. ex Rchb. subsp. tanaensis (Kük.) Lye;塔纳扁莎■☆

322228 Pycreus flavescens (L.) P. Beauv. ex Rchb. var. castaneus Lye = Pycreus flavescens (L.) P. Beauv. ex Rchb. ■☆

322229 Pycreus flavescens (L.) P. Beauv. ex Rchb. var. rehmannianus (C. B. Clarke) Kük. = Pycreus flavescens (L.) P. Beauv. ex Rchb. ■☆

322230 Pycreus flavescens (L.) Rchb. = Cyperus flavescens L. ■☆

322231 Pycreus flavescens (L.) Rchb. = Pycreus flavescens (L.) P. Beauv. ex Rchb. ■☆

322232 Pycreus flavicomus (Michx.) C. D. Adams = Cyperus flavicomus Michx. ■☆

322233 Pycreus flavidus (Retz.) T. Koyama;球穗扁莎(带黄扁莎,飞天蜈蚣,梳子草);Ballspike Flatsedge,Globular Spike Pycreus ■

322234 Pycreus flavidus (Retz.) T. Koyama = Cyperus flavidus Retz. ■

322235 Pycreus flavidus (Retz.) T. Koyama f. atroferrugineus (Steud.) C. Y. Wu;深锈扁莎;Deeptust Ballspike Flatsedge ■

322236 Pycreus flavidus (Retz.) T. Koyama f. fuscoater (Meinsh.) T. Koyama = Cyperus flavidus Retz. ■

322237 Pycreus flavidus (Retz.) T. Koyama f. pauperior (Boeck.) C. Y. Wu;瘦弱扁莎■

322238 Pycreus flavidus (Retz.) T. Koyama var. minimus (Kük.) L. K. Dai = Pycreus globosus (All.) Rchb. var. minimvs (Kük.) Ts. Tang et F. T. Wang ■

322239 Pycreus flavidus (Retz.) T. Koyama var. nilagiricus (Hochst. ex Steud.) C. Y. Wu ex Karthik. = Cyperus flavidus Retz. ■

322240 Pycreus flavidus (Retz.) T. Koyama var. nilagiricus (Hochst. ex Steud.) Karthik. = Cyperus flavidus Retz. ■

322241 Pycreus flavidus (Retz.) T. Koyama var. nilagiricus (Hochst. ex Steud.) Karthik. = Pycreus flavidus (Retz.) T. Koyama var. nilagiricus (Hochst. ex Steud.) C. Y. Wu ex Karthik. ■

322242 Pycreus flavidus (Retz.) T. Koyama var. nilagiricus (Hochst. ex Steud.) C. Y. Wu ex Karthik.;小球穗扁莎;Small Globular Spike Pycreus ■

322243 Pycreus flavidus (Retz.) T. Koyama var. nilaguricus (Hochst. ex Steud.) C. Y. Wu = Pycreus flavidus (Retz.) T. Koyama var. nilagiricus (Hochst. ex Steud.) C. Y. Wu ex Karthik. ■

322244 Pycreus flavidus（Retz.）T. Koyama var. strictus（Roxb.）C. Y. Wu ex Karthik.；少花扁莎（直球穗扁莎）；Strict Globular Spike Pycreus ■

322245 Pycreus flavidus（Retz.）T. Koyama var. strictus（Roxb.）C. Y. Wu = Pycreus flavidus（Retz.）T. Koyama var. strictus（Roxb.）C. Y. Wu ex Karthik. ■

322246 Pycreus fluminalis（Ridl.）Troupin；河流扁莎■☆

322247 Pycreus fuscus f. virescens（Hoffm.）Vahl = Cyperus fuscus L. ■

322248 Pycreus globosus（All.）Rchb. = Cyperus flavidus Retz. ■

322249 Pycreus globosus（All.）Rchb. = Pycreus flavidus（Retz.）T. Koyama ■

322250 Pycreus globosus（All.）Rchb. f. atroferrugineus（Steud.）Kük. = Pycreus flavidus（Retz.）T. Koyama f. atroferrugineus（Steud.）C. Y. Wu ■

322251 Pycreus globosus（All.）Rchb. f. fuscoater（Meinsh.）T. Koyama = Cyperus flavidus Retz. ■

322252 Pycreus globosus（All.）Rchb. f. pauperior（Boeck.）Kük. = Pycreus flavidus（Retz.）T. Koyama f. pauperior（Boeck.）C. Y. Wu ■

322253 Pycreus globosus（All.）Rchb. var. minimvs（Kük.）Ts. Tang et F. T. Wang；矮球穗扁莎；Dwarf Globular Spike Pycreus ■

322254 Pycreus globosus（All.）Rchb. var. nilagiricus（Hochst.）C. B. Clarke = Pycreus flavidus（Retz.）T. Koyama var. nilaguricus（Hochst. ex Steud.）C. Y. Wu ■

322255 Pycreus globosus（All.）Rchb. var. strictus（Roxb.）C. B. Clarke = Pycreus flavidus（Retz.）T. Koyama var. strictus（Roxb.）C. Y. Wu ■

322256 Pycreus globosus All. var. strictus（Roxb.）C. B. Clarke = Pycreus flavidus（Retz.）T. Koyama var. strictus（Roxb.）C. Y. Wu ex Karthik. ■

322257 Pycreus globosus Rchb. = Pycreus flavidus（Retz.）T. Koyama ■

322258 Pycreus globosus Rchb. var. minimus（Kük.）Ts. Tang et F. T. Wang = Pycreus flavidus（Retz.）T. Koyama var. minimus（Kük.）L. K. Dai ■

322259 Pycreus globosus Rchb. var. nilagiricus（Hochst. ex Steud.）C. B. Clarke = Pycreus flavidus（Retz.）T. Koyama var. nilagiricus（Hochst. ex Steud.）C. Y. Wu ex Karthik. ■

322260 Pycreus globosus Rchb. var. strictus（Roxb.）C. B. Clarke = Pycreus flavidus（Retz.）T. Koyama var. strictus（Roxb.）C. Y. Wu ex Karthik. ■

322261 Pycreus gracillimus Chiov.；细长扁莎■☆

322262 Pycreus hildebrandtii（K. Schum.）C. B. Clarke；希尔德德扁莎■☆

322263 Pycreus humboldtianus（Schult.）Cufod.；洪堡扁莎■☆

322264 Pycreus hyalinus（Vahl）Domin = Queenslandiella hyalina（Vahl）F. Ballard ■☆

322265 Pycreus imbricatus subsp. elongatus（Boeck.）T. Koyama = Cyperus imbricatus Retz. var. elongatus（Boeck.）L. K. Dai ■

322266 Pycreus imbricatus var. densespicatus（Hayata）Ohwi = Cyperus imbricatus Retz. var. elongatus（Boeck.）L. K. Dai ■

322267 Pycreus intactus（Vahl）J. Raynal；洁净扁莎■☆

322268 Pycreus intermedius（Steud.）C. B. Clarke；间型扁莎■☆

322269 Pycreus intermedius（Steud.）C. B. Clarke f. tenuis（Boeck.）Cufod.；细间型扁莎■☆

322270 Pycreus katangensis Cherm. = Pycreus fibrillosus（Kük.）Cherm. ■☆

322271 Pycreus korshinskyi（Meinsh.）V. I. Krecz.；槽鳞扁莎（红鳞扁莎）■

322272 Pycreus korshinskyi（Meinsh.）V. I. Krecz. = Pycreus sangninolentus（Vahl）Nees ex C. B. Clarke ■

322273 Pycreus korshinskyi Meinsh. = Pycreus sanguinolentus（Vahl）Nees ex C. B. Clarke ■

322274 Pycreus laevigatus（L.）Nees = Cyperus laevigatus L. ■☆

322275 Pycreus lanceolatus（Poir.）C. B. Clarke；披针形扁莎■☆

322276 Pycreus lanceolatus（Poir.）C. B. Clarke subsp. ugandensis Lye = Pycreus mortonii S. S. Hooper ■☆

322277 Pycreus lanceus（Thunb.）Turrill = Pycreus nitidus（Lam.）J. Raynal ■☆

322278 Pycreus lanceus（Thunb.）Turrill var. melanopus（Boeck.）Troupin = Pycreus nitidus（Lam.）J. Raynal ■☆

322279 Pycreus latespicatus（Boeck.）C. B. Clarke = Pycreus diaphanus（Schrad. ex Roem. et Schult.）S. S. Hooper et T. Koyama ex S. S. Hooper ■

322280 Pycreus latespicatus（Boeck.）C. B. Clarke = Pycreus diaphanus（Schrad. ex Schult.）S. S. Hooper et T. Koyama ■

322281 Pycreus latespicatus（Boeck.）C. B. Clarke subsp. setiformis（Korsh.）T. Koyama = Cyperus diaphanus Schrad. ex Roem. et Schult. ■

322282 Pycreus lijiangensis L. K. Dai；丽江扁莎；Lijiang Flatsedge, Lijiang Pycreus ■

322283 Pycreus limosus（Maxim.）Schischk. = Juncellus limosus（Maxim.）C. B. Clarke ■

322284 Pycreus longistolon（Peter et Kük.）Napper；长葡匐枝扁莎■☆

322285 Pycreus longistolon（Peter et Kük.）Napper subsp. atrofuscus Lye；暗褐扁莎■☆

322286 Pycreus longus（L.）Hayek = Cyperus longus L. ■☆

322287 Pycreus macranthus（Boeck.）C. B. Clarke；大花扁莎■☆

322288 Pycreus macranthus（Boeck.）C. B. Clarke var. angustifolius（Ridl.）C. B. Clarke ex Rendle；窄叶扁莎■☆

322289 Pycreus macrostachyos（Lam.）J. Raynal；大穗扁莎■☆

322290 Pycreus macrostachyos（Lam.）J. Raynal subsp. tremulus（Poir.）Lye = Pycreus macrostachyos（Lam.）J. Raynal ■☆

322291 Pycreus macrostachyos（Lam.）J. Raynal var. tenuis（Boeck.）Wickens；细大穗扁莎■☆

322292 Pycreus malangensis Meneses；马兰加扁莎■☆

322293 Pycreus melanacme Nelmes；黑边扁莎■☆

322294 Pycreus melanocephalus Miq. = Pycreus sanguinolentus（Vahl）Nees ex C. B. Clarke ■

322295 Pycreus minimus（K. Schum.）C. B. Clarke；小扁莎■☆

322296 Pycreus mortonii S. S. Hooper；莫顿扁莎■☆

322297 Pycreus mundtii Nees；蒙特扁莎■☆

322298 Pycreus mundtii Nees var. uniceps（C. B. Clarke）Napper = Pycreus mundtii Nees ■☆

322299 Pycreus muricatus（Kük.）Napper；粗糙扁莎■☆

322300 Pycreus niger（Ruiz et Pav.）Cufod.；黑扁莎■☆

322301 Pycreus niger（Ruiz et Pav.）Cufod. subsp. elegantulus（Steud.）Lye；雅致黑扁莎■☆

322302 Pycreus nigricans（Steud.）C. B. Clarke；变黑扁莎■☆

322303 Pycreus nigricans（Steud.）C. B. Clarke var. firmior（Kük.）Cherm.；密变黑扁莎■☆

322304 Pycreus nilagiricus（Hochst. ex Steud.）E. G. Camus；黑紫鳞扁莎■

322305 Pycreus nilagiricus（Hochst. ex Steud.）E. G. Camus = Cyperus flavidus Retz. ■

322306 Pycreus nitens Nees = Pycreus pumilus（L.）Domin ■

322307　Pycreus nitidus（Lam.）J. Raynal；光亮扁莎■☆

322308　Pycreus nyasensis C. B. Clarke；尼亚斯扁莎■☆

322309　Pycreus oakfortensis C. B. Clarke = Pycreus permutatus（Boeck.）Napper■☆

322310　Pycreus okavangensis Podlech；奥卡万戈扁莎■☆

322311　Pycreus overlaetii Cherm. ex S. S. Hooper et J. Raynal；奥弗莱特扁莎■☆

322312　Pycreus pagotii J. Raynal；帕戈扁莎■☆

322313　Pycreus patens（Vahl）Cherm. = Pycreus pumilus（L.）Domin■

322314　Pycreus pauper（A. Rich.）C. B. Clarke；贫乏扁莎■☆

322315　Pycreus pennatus Lam. = Mariscus javanicus（Houtt.）Merr. et F. P. Metcalf■

322316　Pycreus permutatus（Boeck.）Napper；全变扁莎■☆

322317　Pycreus polystachyos（Rottb.）P. Beauv.；多枝扁莎■

322318　Pycreus polystachyos（Rottb.）P. Beauv. = Cyperus polystachyus（Rottb.）P. Beauv.■

322319　Pycreus polystachyos（Rottb.）P. Beauv. var. brevispiculatus（F. C. How）L. K. Dai；短穗多枝扁莎；Short-spike Branchy Pycreus■

322320　Pycreus polystachyos（Rottb.）P. Beauv. var. laxiflorus（Benth.）C. B. Clarke；疏花扁莎■☆

322321　Pycreus polystachyus（Rottb.）P. Beauv. = Cyperus polystachyos Rottb.■

322322　Pycreus polystachyus（Rottb.）P. Beauv. var. brevispiculatus F. C. How = Pycreus polystachyos（Rottb.）P. Beauv. var. brevispiculatus（F. C. How）L. K. Dai■

322323　Pycreus pratorum（Korotky）Schischk.；普拉塔扁莎■☆

322324　Pycreus propinquus Nees = Pycreus lanceolatus（Poir.）C. B. Clarke■☆

322325　Pycreus pseudodiaphanus S. S. Hooper；假透明扁莎■☆

322326　Pycreus pseudodiaphanus S. S. Hooper var. occidentalis ?；西方扁莎■☆

322327　Pycreus pseudolatespicatus L. K. Dai；拟宽穗扁莎（似宽穗扁莎）；False Broadspike Flatsedge，False Broadspike Pycreus■

322328　Pycreus pubescens Turrill；短柔毛扁莎■☆

322329　Pycreus pumilus（L.）Domin；矮扁莎；Dwarf Flatsedge，Dwarf Pycreus，Low Flatsedge■

322330　Pycreus pumilus（L.）Domin = Cyperus pumilus L.■

322331　Pycreus pumilus（L.）Domin subsp. patens（Vahl）Podlech = Pycreus pumilus（L.）Domin■

322332　Pycreus pumilus（L.）Nees = Pycreus pumilus（L.）Domin■

322333　Pycreus radians Nees et Meyen var. floribundus（E. G. Camus）Kük. = Mariscus radians（Nees et Meyen）Ts. Tang et F. T. Wang var. floribundus（E. G. Camus）S. M. Huang■

322334　Pycreus radiatus Vahl = Cyperus imbricatus Retz.■

322335　Pycreus rehmannianus C. B. Clarke = Pycreus flavescens（L.）P. Beauv. ex Rchb.■☆

322336　Pycreus rehmannii（Boiss.）Palla ex Grossh.；拉赫曼扁莎■☆

322337　Pycreus rivularis（Kunth）Palla = Cyperus bipartitus Torr.■☆

322338　Pycreus sabulosus Mart. et Schrad. ex Nees = Cyperus flavicomus Michx.■☆

322339　Pycreus sanguineo-squamatus Van der Veken；血鳞扁莎■☆

322340　Pycreus sanguinolentus（Vahl）Nees = Cyperus sanguinolentus Vahl■

322341　Pycreus sanguinolentus（Vahl）Nees = Pycreus sanguinolentus（Vahl）Nees ex C. B. Clarke■

322342　Pycreus sanguinolentus（Vahl）Nees ex C. B. Clarke；红鳞扁莎（荸荠草，红扁鳞莎草，红鳞扁莎草，路州莎草，水花毛）；

Louisiana Flatsedge，Pale Galingale，Redscale Flatsedge，Red-scaled Flat-sedge，Sanguineous Scale Pycreus，Sanguineous-scale Galingale，Sanguineous-scale Pycreus■

322343　Pycreus sanguinolentus（Vahl）Nees ex C. B. Clarke = Cyperus sanguinolentus Vahl■

322344　Pycreus sanguinolentus（Vahl）Nees ex C. B. Clarke f. flaccidus（Boeck.）Cufod；柔软红鳞扁莎■☆

322345　Pycreus sanguinolentus（Vahl）Nees ex C. B. Clarke f. humilis（Miq.）L. K. Dai = Pycreus sanguinolentus（Vahl）Nees■

322346　Pycreus sanguinolentus（Vahl）Nees ex C. B. Clarke f. humilis（Miq.）L. K. Dai；矮红鳞扁莎；Dwarf Sanguineous-scale Pycreus■

322347　Pycreus sanguinolentus（Vahl）Nees ex C. B. Clarke f. melanocephalus（Miq.）Kük. = Pycreus sanguinolentus（Vahl）Nees ex C. B. Clarke■

322348　Pycreus sanguinolentus（Vahl）Nees ex C. B. Clarke f. melanocephalus（Miq.）L. K. Dai = Pycreus sanguinolentus（Vahl）Nees ex C. B. Clarke■

322349　Pycreus sanguinolentus（Vahl）Nees ex C. B. Clarke f. melanocephalus（Miq.）L. K. Dai；黑头扁莎（黑扁莎）；Black Sanguineous-scale Pycreus■

322350　Pycreus sanguinolentus（Vahl）Nees ex C. B. Clarke f. neurotropis（Steud.）Cufod.；凸脉扁莎■☆

322351　Pycreus sanguinolentus（Vahl）Nees ex C. B. Clarke f. nipponicus（Ohwi）T. Koyama = Cyperus sanguinolentus Vahl■

322352　Pycreus sanguinolentus（Vahl）Nees ex C. B. Clarke f. rubromarginatus（Schrenk）Kük. = Pycreus sanguinolentus（Vahl）Nees ex C. B. Clarke■

322353　Pycreus sanguinolentus（Vahl）Nees ex C. B. Clarke f. rubromarginatus（Schrenk）L. K. Dai = Pycreus sanguinolentus（Vahl）Nees ex C. B. Clarke■

322354　Pycreus sanguinolentus（Vahl）Nees ex C. B. Clarke f. rubromarginatus（Schrenk）L. K. Dai；红边扁莎；Red-margin Sanguineous-scale Pycreus■

322355　Pycreus sanguinolentus（Vahl）Nees ex C. B. Clarke var. spectabilis（Makino）T. Koyama = Cyperus sanguinolentus Vahl f. spectabilis（Makino）Ohwi■☆

322356　Pycreus sanguinolentus（Vahl）Nees f. flaccidus（Boeck.）Cufod. = Pycreus sanguinolentus（Vahl）Nees ex C. B. Clarke f. flaccidus（Boeck.）Cufod■☆

322357　Pycreus sanguinolentus（Vahl）Nees f. neurotropis（Steud.）Cufod. = Pycreus sanguinolentus（Vahl）Nees ex C. B. Clarke f. neurotropis（Steud.）Cufod.■☆

322358　Pycreus sanguinolentus（Vahl）Nees subsp. nairobiensis Lye = Pycreus sanguinolentus（Vahl）Nees subsp. nairobiensis Lye ex C. B. Clarke■☆

322359　Pycreus sanguinolentus（Vahl）Nees subsp. nairobiensis Lye ex C. B. Clarke；内罗比扁莎■☆

322360　Pycreus sanguinolentus Vahl f. humilis（Miq.）Kük. = Pycreus sanguinolentus（Vahl）Nees ex C. B. Clarke■

322361　Pycreus scaettae Cherm. = Pycreus fibrillosus（Kük.）Cherm.■☆

322362　Pycreus scaettae Cherm. var. katangensis ? = Pycreus fibrillosus（Kük.）Cherm.■☆

322363　Pycreus segmentatus C. B. Clarke；短裂扁莎■☆

322364　Pycreus serotinus Rottb. f. depauperatus Kük. = Juncellus serotinus（Rottb.）C. B. Clarke■

322365　Pycreus setiformis（Korsh.）Nakai = Cyperus diaphanus Schrad. ex Roem. et Schult.■

322366 Pycreus sinensis Debeaux = Mariscus radians (Nees et Meyen) Ts. Tang et F. T. Wang ■

322367 Pycreus smithianus (Ridl.) C. B. Clarke;史密斯扁莎■☆

322368 Pycreus spissiflorus C. B. Clarke;密花扁莎■☆

322369 Pycreus substellatus E. G. Camus = Pycreus sulcinux (C. B. Clarke) C. B. Clarke ■

322370 Pycreus subtrigonus C. B. Clarke;亚三角扁莎■☆

322371 Pycreus sulcinux (C. B. Clarke) C. B. Clarke;槽果扁莎(槽果莎草,垦丁扁莎);Sulcatefruit Flatsedge, Sulcatefruit Galingale, Sulcate-fruit Pycreus ■

322372 Pycreus testui Cherm.;泰斯蒂扁莎■☆

322373 Pycreus tremulus (Poir.) C. B. Clarke = Pycreus macrostachyos (Lam.) J. Raynal ■☆

322374 Pycreus tremulus (Poir.) Clarke;颤扁莎■☆

322375 Pycreus umbrosus Nees = Pycreus nitidus (Lam.) J. Raynal ■☆

322376 Pycreus unioloides (R. Br.) Urb.;禾状扁莎(牛露草,水社扁莎);Grasslike Flatsedge, Grass-like Flat-sedge, Unida-like Pycreus, Uniola-like Galingale ■

322377 Pycreus unioloides (R. Br.) Urb. = Cyperus unioloides R. Br. ■

322378 Pycreus unioloides (R. Br.) Urb. var. wightii (C. B. Clarke) C. Y. Wu;扁莎;Wright Pycreus ■

322379 Pycreus vanderystii Cherm.;范德扁莎■☆

322380 Pycreus vavavatensis Cherm. = Pycreus nigricans (Steud.) C. B. Clarke ■☆

322381 Pycreus vicinus Cherm. = Pycreus intermedius (Steud.) C. B. Clarke ■☆

322382 Pycreus virescens Hoffm. = Cyperus fuscus L. ■

322383 Pycreus waillyi Cherm.;瓦伊扁莎■☆

322384 Pycreus xantholepis Nelmes;黄鳞扁莎■☆

322385 Pycreus zonatus Cherm.;环带扁莎■☆

322386 Pygeum Gaertn. (1788);臀果木属(肾果木属,臀形果属,臀形木属,野樱属);Pygeum ●

322387 Pygeum Gaertn. = Lauro-Cerasus Duhamel ●

322388 Pygeum Gaertn. = Prunus L. ●

322389 Pygeum africanum Hook. f.;非洲臀果木(红臭木);Red Stinkwood ●☆

322390 Pygeum africanum Hook. f. = Prunus africana (Hook. f.) Kalkman ●☆

322391 Pygeum andersonii Hook. f. = Laurocerasus andersonii (Hook. f.) Te T. Yu et L. T. Lu ●

322392 Pygeum arboreum Endl. ex Kurz;楝桃●☆

322393 Pygeum caudatum Merr. = Pygeum lancilimbum Merr. ●

322394 Pygeum crassifolium Hauman = Prunus africana (Hook. f.) Kalkman ●☆

322395 Pygeum crassifolium Hauman = Prunus crassifolia (Hauman) Kalkman ●☆

322396 Pygeum griseum Blume ex C. H. Müll. = Prunus grisea (Blume ex C. H. Müll.) Kalkman ●

322397 Pygeum henryi Dunn;云南臀果木;Henry Pygeum, Yunnan Pygeum, Yunnan Stinkwood ●☆

322398 Pygeum lancilimbum Merr.;滇臀果木●

322399 Pygeum latifolium Miq. var. macrocarpum (Te T. Yu et A. M. Lu) C. Y. Wu et H. Chu = Pygeum macrocarpum Te T. Yu et A. M. Lu ●

322400 Pygeum laxiflorum Merr. ex H. L. Li;疏花臀果木;Laxiflowered Pygeum, Loose-flower Pygeum ●

322401 Pygeum macrocarpum Te T. Yu et A. M. Lu;大果臀果木; Bigfruit Pygeum, Big-fruited Pygeum, Largefruit Pygeum ●

322402 Pygeum megaphyllum Merr. ex Elmer;大叶臀果木(大叶野樱桃);Bigleaf Pygeum ●

322403 Pygeum montanum Hook. f.;喜马拉雅臀果木;Ximalaya Pygeum ●

322404 Pygeum oblongum Te T. Yu et A. M. Lu;长圆果臀果木(长圆臀果木);Oblong Pygeum, Oblongifruited Pygeum ●

322405 Pygeum oxycarpum Hance = Laurocerasus zippeliana (Miq.) Browicz ●

322406 Pygeum oxycarpum Hance = Pygeum zeylanicum Gaertn. ●

322407 Pygeum phaeosticta Hance = Laurocerasus phaeosticta (Hance) C. K. Schneid. ●

322408 Pygeum phaeostictum Hance = Laurocerasus phaeosticta (Hance) C. K. Schneid. ●

322409 Pygeum phaeostictum Hance = Prunus phaeosticta (Hance) Maxim. ●

322410 Pygeum preslii Merr. = Prunus grisea (Blume ex C. H. Müll.) Kalkman ●

322411 Pygeum tokangpengii Merr. = Pygeum topengii Merr. ●

322412 Pygeum topengii Merr.;臀果木(荷包李,鹿角,木虱檎,木虱罗,臀形果);Pygeum, Topeng Pygeum ●

322413 Pygeum wilsonii Koehne;西南臀果木;E. H. Wilson Pygeum, Wilson Pygeum ●

322414 Pygeum wilsonii Koehne var. macrophyllum L. D. Lu;大叶西南臀果木;Bigleaf E. H. Wilson Pygeum ●

322415 Pygeum zeylanicum Gaertn.;锡兰臀果木;Ceylon Pygeum ●

322416 Pygmaea Hook. f. (1895);侏儒婆婆纳属■☆

322417 Pygmaea Jacks. = Chionohebe B. G. Briggs et Ehrend. ●☆

322418 Pygmaea Jacks. = Pygmaea Hook. f. ■☆

322419 Pygmaeocereus J. H. Johnson et Backeb. (1957);矮小天轮柱属●☆

322420 Pygmaeocereus J. H. Johnson et Backeb. = Echinopsis Zucc. ●

322421 Pygmaeopremna Merr. (1910);千解草属;Pygmaeopremna ●

322422 Pygmaeopremna Merr. = Premna L. (保留属名)●■

322423 Pygmaeopremna herbacea (Roxb.) Moldenke;千解草(草臭黄荆,灰叶树,土巴戟,细八棱麻,细三对节,小八棱麻,小常山);Herbaceous Pygmaeopremna ●

322424 Pygmaeopremna herbacea (Roxb.) Moldenke = Premna herbacea Roxb. ●

322425 Pygmaeopremna humilis Merr. = Premna herbacea Roxb. ●

322426 Pygmaeopremna nana (Collett et Hemsl.) Moldenke = Premna herbacea Roxb. ●

322427 Pygmaeorchis Brade (1939);侏儒兰属■☆

322428 Pygmaeorchis brasiliensis Brade;侏儒兰■☆

322429 Pygmaeothamnus Robyns (1928);矮灌茜属●☆

322430 Pygmaeothamnus chamaedendrum (Kuntze) Robyns;矮灌茜●☆

322431 Pygmaeothamnus chamaedendrum (Kuntze) Robyns var. setulosus Robyns;小刚毛矮灌茜●☆

322432 Pygmaeothamnus longipes Robyns = Pygmaeothamnus chamaedendrum (Kuntze) Robyns var. setulosus Robyns ●☆

322433 Pygmaeothamnus pilosus Robyns = Pygmaeothamnus chamaedendrum (Kuntze) Robyns var. setulosus Robyns ●☆

322434 Pygmaeothamnus zeyheri (Sond.) Robyns;泽赫矮灌茜●☆

322435 Pygmaeothamnus zeyheri (Sond.) Robyns var. livingstonianus Robyns = Pygmaeothamnus zeyheri (Sond.) Robyns var. rogersii Robyns ●☆

322436 Pygmaeothamnus zeyheri (Sond.) Robyns var. oatesii (Rolfe)

Robyns = Pygmaeothamnus zeyheri (Sond.) Robyns ●☆

322437　Pygmaeothamnus zeyheri (Sond.) Robyns var. rogersii Robyns;罗杰斯矮灌茜●☆

322438　Pygmea Hook. f. = Chionohebe B. G. Briggs et Ehrend. ●☆

322439　Pygmea J. Buchanan = Pygmaea Hook. f. ■☆

322440　Pylostachya Raf. = Polygala L. ●■

322441　Pynaertia De Willd. = Anopyxis (Pierre) Engl. ●☆

322442　Pynaertia ealaensis De Wild. = Anopyxis klaineana (Pierre) Engl. ●☆

322443　Pynaertia occidentalis A. Chev. = Anopyxis klaineana (Pierre) Engl. ●☆

322444　Pynaertiodendron De Wild. = Cryptosepalum Benth. ●☆

322445　Pynaertiodendron congolanum De Wild. = Cryptosepalum congolanum (De Wild.) J. Léonard ●☆

322446　Pynaertiodendron pellegrinianum J. Léonard = Cryptosepalum pellegrinianum (J. Léonard) J. Léonard ●☆

322447　Pynanthemum Raf. = Pycnanthemum Michx. (保留属名)■☆

322448　Pyracantha M. Roem. (1847);火棘属(蔷若属,火把果属,火刺木属);Fire Thorn, Firethorn ●

322449　Pyracantha 'Mohave';莫哈维火棘;Mohave Firethorn ●☆

322450　Pyracantha angustifolia (Franch.) C. K. Schneid.;窄叶火棘;Egyptian Thorn, Firethorn, Narrowleaf Fire Thorn, Narrowleaf Firethorn, Narrowleafed Firethorn, Narrow-leaved Fire Thorn, Narrow-leaved Firethorn, Orange Fire Thorn, Orange Firethorn, Orange Fire-thorn ●

322451　Pyracantha atalantioides (Hance) Stapf;全缘火棘(救军粮,木瓜刺);Chinese Firethorn, Entire Firethorn, Gibbs Firethorn ●

322452　Pyracantha atalantioides (Hance) Stapf 'Aurea';金果全缘火棘●☆

322453　Pyracantha chinensis M. Roem. = Pyracantha crenulata (D. Don) M. Roem. ●

322454　Pyracantha coccinea M. Roem.;欧洲火棘(暗红火棘,火棘木,猩红火棘);European Firethorn, Everlasting Fire Thorn, Everlasting Thorn, Fiery Thorn, Fire Bush, Fire Thorn, Firethorn, Pyracantha, Red Fire Thorn, Red Fire-thorn, Scarlet Fire Thorn, Scarlet Firethorn ●☆

322455　Pyracantha coccinea M. Roem. 'Lalandei';莱兰欧洲火棘(拉兰德暗红火棘)●☆

322456　Pyracantha crenatoserrata (Hance) Rehder;圆锯齿火棘●

322457　Pyracantha crenatoserrata (Hance) Rehder = Pyracantha fortuneana (Maxim.) H. L. Li ●

322458　Pyracantha crenulata (D. Don) M. Roem.;细圆齿火棘(蔷若,红子,火把果,火棘,圆锯齿火棘);Crenate Firethorn, Himalayan Firethorn, Nepal Firethorn, Nepalese Firethorn, Nepalese White-thorn ●

322459　Pyracantha crenulata (D. Don) M. Roem. var. kansuensis Rehder;甘肃细圆齿火棘(甘肃火棘,细叶细圆齿火棘);Gansu Crenate Firethorn, Kansu Crenate Firethorn ●

322460　Pyracantha crenulata (D. Don) M. Roem. var. rogersiana A. B. Jacks. = Pyracantha fortuneana (Maxim.) H. L. Li ●

322461　Pyracantha crenulata (D. Don) M. Roem. var. yunnanensis M. Vilm. ex Mottet = Pyracantha fortuneana (Maxim.) H. L. Li ●

322462　Pyracantha densiflora Te T. Yu;密花火棘;Denseflower Firethorn, Densiflowered Firethorn ●

322463　Pyracantha discolor Rehder = Pyracantha atalantioides (Hance) Stapf ●

322464　Pyracantha formosana Kaneh. = Pyracantha koidzumii (Hayata) Rehder ●

322465　Pyracantha fortuneana (Maxim.) H. L. Li;火棘(赤果,赤阳

子,纯阳子,豆金娘,钝阳子,红子,火把果,救兵粮,救军粮,救命粮,水搓子,水沙子);Chinese Firethorn, Fire Thorn, Firethorn, Fortune Firethorn, Pygeum, Yunnan Firethorn ●

322466　Pyracantha gibbsii A. B. Jacks. = Pyracantha atalantioides (Hance) Stapf ●

322467　Pyracantha gibbsii A. B. Jacks. = Pyracantha fortuneana (Maxim.) H. L. Li ●

322468　Pyracantha gibbsii A. B. Jacks. var. yunnanensis (M. Vilm. ex Mottet) Osborn = Pyracantha fortuneana (Maxim.) H. L. Li ●

322469　Pyracantha heterophylla T. B. Chao et Z. X. Chen;异型叶火棘;Diversileaf Firethorn ●

322470　Pyracantha inermis Vidal;澜沧火棘;Lancang Firethorn, Prickless Firethorn ●

322471　Pyracantha koidzumii (Hayata) Rehder;台湾火棘(台东火刺木,台湾火刺木);Formosa Firethorn, Formosa Pyracantha, Koidzum Firethorn, Taitung Firethorn, Taiwan Firethorn ●

322472　Pyracantha koidzumii (Hayata) Rehder var. taitoensis (Hayata) Masam. = Pyracantha koidzumii (Hayata) Rehder ●

322473　Pyracantha koidzumii (Hayata) Rehder var. taitoensis Masam. = Pyracantha koidzumii (Hayata) Rehder ●

322474　Pyracantha koidzumii Hayata 'Victory';胜利台湾火棘;Victory Pyracantha ●

322475　Pyracantha koidzumii Hayata var. taitoensis (Hayata) Masam. = Pyracantha koidzumii (Hayata) Rehder ●

322476　Pyracantha loureiroi (Kostel.) Merr. = Pyracantha atalantioides (Hance) Stapf ●

322477　Pyracantha mekongensis Te T. Yu = Pyracantha inermis Vidal ●

322478　Pyracantha pauciflora André = Pyracantha coccinea M. Roem. ●☆

322479　Pyracantha rogersiana (A. B. Jacks.) L. H. Bailey = Pyracantha crenulata (D. Don) M. Roem. var. rogersiana A. B. Jacks. ●

322480　Pyracantha rogersiana Coltm. -Rog.;罗杰斯火棘(罗氏火棘);Asian Firethorn, Rogers Firethorn ●☆

322481　Pyracantha rogersiana Coltm. -Rog. 'Flava';黄果罗杰斯火棘●☆

322482　Pyracantha stoloniformis T. B. Chao et Z. X. Chen;匍匐火棘;Stoloniform Firethorn ●

322483　Pyracantha yunnanensis (M. Vilm. ex Mottet) Chitt. = Pyracantha fortuneana (Maxim.) H. L. Li ●

322484　Pyracantha yunnanensis Chitt. = Pyracantha fortuneana (Maxim.) H. L. Li ●

322485　Pyraceae Burnett = Pomaceae Gray ●■

322486　Pyraceae Burnett = Rosaceae Juss. (保留科名)●■

322487　Pyraceae Vest = Rosaceae Juss. (保留科名)●■

322488　Pyragma Noronha = Stelechocarpus Hook. f. et Thomson ●☆

322489　Pyragma Noronha = Uvaria L. ●

322490　Pyragra Bremek. (1958);皮拉茜属■☆

322491　Pyragra ankarensis Bremek.;皮拉茜■☆

322492　Pyragra obtusifolia Bremek.;钝叶皮拉茜■☆

322493　Pyramia Cham. = Cambessedesia DC. (保留属名)●■☆

322494　Pyramidanthe Miq. (1865);长瓣银帽花属●☆

322495　Pyramidanthe prismatica (Hook. f. et Thomson) Sinclair;长瓣银帽花●☆

322496　Pyramidium Boiss. = Veselskya Opiz ■☆

322497　Pyramidocarpus Oliv. = Dasylepis Oliv. ●☆

322498　Pyramidocarpus blackii Oliv. = Dasylepis blackii (Oliv.) Chipp ●☆

322499　Pyramidocarpus petiolaris Pierre ex A. Chev. = Oncoba mannii Oliv. ●☆

322500　Pyramidoptera Boiss. (1856);锥翅草属☆

322501 Pyramidoptera cabulica Boiss. ;锥翅草☆

322502 Pyramidostylium Mart. ex Peyr. = Salacia L. (保留属名)●

322503 Pyramidostylum Mart. = Salacia L. (保留属名)●

322504 Pyranthus Du Puy et Labat(1995);脓花豆属●☆

322505 Pyranthus ambatoana (Baill.) Du Puy et Labat;安巴托脓花豆●☆

322506 Pyranthus lucens (R. Vig.) Du Puy et Labat;光亮脓花豆●☆

322507 Pyranthus monantha (Baker) Du Puy et Labat;单花脓花豆●☆

322508 Pyranthus pauciflora (Baker) Du Puy et Labat;少花脓花豆●☆

322509 Pyrarda Cass. = Grangea Adans. ■

322510 Pyrarda ceruanoides Cass. = Grangea ceruanoides Cass. ■

322511 Pyrecnia Noronha = Pyreenia Noronha ●■

322512 Pyrecnia Noronha ex Hassk. = Pyreenia Noronha ●■

322513 Pyreenia Noronha = Laportea Gaudich. (保留属名)●■

322514 Pyrenacantha Hook. = Pyrenacantha Wight(保留属名)●

322515 Pyrenacantha Hook. ex Wight = Pyrenacantha Wight(保留属名)●

322516 Pyrenacantha Hook. f. = Pyrenacantha Wight(保留属名)●

322517 Pyrenacantha Wight (1830) (保留属名);刺核藤属;Pyrenacantha ●

322518 Pyrenacantha acuminata Engl. ;渐尖刺核藤●☆

322519 Pyrenacantha ambrensis Labat,El-Achkar et R. Rabev. ;昂布尔刺核藤●☆

322520 Pyrenacantha andapensis Labat,El-Achkar et R. Rabev. ;安达帕刺核藤●☆

322521 Pyrenacantha brevipes Engl. = Desmostachys brevipes (Engl.) Sleumer ■☆

322522 Pyrenacantha canaliculata Pierre = Pyrenacantha vogeliana Baill. ●☆

322523 Pyrenacantha capitata H. Perrier;头状刺核藤●☆

322524 Pyrenacantha chlorantha Baker;绿花刺核藤●☆

322525 Pyrenacantha chlorantha Baker f. opulenta H. Perrier = Pyrenacantha perrieri Labat,El-Achkar et R. Rabev. ●☆

322526 Pyrenacantha chlorantha Baker var. sambiranensis H. Perrier = Pyrenacantha perrieri Labat,El-Achkar et R. Rabev. ●☆

322527 Pyrenacantha cordata Villiers = Pyrenacantha cordicula Villiers ●☆

322528 Pyrenacantha cordicula Villiers;心形刺核藤●☆

322529 Pyrenacantha dinklagei Engl. = Pyrenacantha vogeliana Baill. ●☆

322530 Pyrenacantha fissistigma Sleumer = Pyrenacantha laetevirens Sleumer ●☆

322531 Pyrenacantha gabonica Breteler et Villiers;加蓬刺核藤●☆

322532 Pyrenacantha glabrescens (Engl.) Engl. ;渐光刺核藤●☆

322533 Pyrenacantha globosa Engl. = Pyrenacantha malvifolia Engl. ●☆

322534 Pyrenacantha grandiflora Baill. ;大花刺核藤●☆

322535 Pyrenacantha grandifolia Engl. ;大叶刺核藤●☆

322536 Pyrenacantha humblotii (Baill. ex Grandid.) Sleumer;洪布刺核藤●☆

322537 Pyrenacantha kamassana Baill. = Pyrenacantha kaurabassana Baill. ●☆

322538 Pyrenacantha kaurabassana Baill. ;考拉刺核藤●☆

322539 Pyrenacantha kirkii Baill. ;柯克刺核藤●☆

322540 Pyrenacantha klaineana Pierre ex Exell et Mendonça;克莱恩刺核藤●☆

322541 Pyrenacantha klaineana Pierre ex Exell et Mendonça var. congolana Boutique;刚果刺核藤●☆

322542 Pyrenacantha laetevirens Sleumer;鲜绿刺核藤●☆

322543 Pyrenacantha lebrunii Boutique;勒布伦刺核藤●☆

322544 Pyrenacantha longirostrata Villiers;长喙刺核藤●☆

322545 Pyrenacantha malvifolia Engl. ;锦葵叶刺核藤●☆

322546 Pyrenacantha mangenotiana J. Miège = Pyrenacantha glabrescens (Engl.) Engl. ●☆

322547 Pyrenacantha menyharthii Schinz = Pyrenacantha kaurabassana Baill. ●☆

322548 Pyrenacantha perrieri Labat,El-Achkar et R. Rabev. ;佩里耶刺核藤●☆

322549 Pyrenacantha puberula Boutique;微毛刺核藤●☆

322550 Pyrenacantha rakotozafyi Labat,El-Achkar et R. Rabev. ;拉库图扎菲刺核藤●☆

322551 Pyrenacantha ruspolii Engl. = Pyrenacantha malvifolia Engl. ●☆

322552 Pyrenacantha scandens Planch. ex Harv. ;攀缘刺核藤●☆

322553 Pyrenacantha staudtii (Engl.) Engl. ;斯氏刺核藤●☆

322554 Pyrenacantha staudtii Engl. = Pyrenacantha staudtii (Engl.) Engl. ●☆

322555 Pyrenacantha sylvestris S. Moore;林地刺核藤●☆

322556 Pyrenacantha taylori Engl. ;泰勒刺核藤●☆

322557 Pyrenacantha tessmanii Engl. = Pyrenacantha cordicula Villiers ●☆

322558 Pyrenacantha ugandensis Hutch. et Robyns = Pyrenacantha staudtii (Engl.) Engl. ●☆

322559 Pyrenacantha undulata Engl. ;波状刺核藤●☆

322560 Pyrenacantha vitifolia Engl. = Pyrenacantha kaurabassana Baill. ●☆

322561 Pyrenacantha vogeliana Baill. ;沃格尔刺核藤●☆

322562 Pyrenacantha volubilis Wight;刺核藤;Pyrenacantha, Twining Pyrenacantha ●

322563 Pyrenaceae Vent. = Verbenaceae J. St. -Hil. (保留科名)●■

322564 Pyrenaria Blume(1827);核果茶属(雕果茶属,乌皮茶属);Drupetea, Pyrenaria ●

322565 Pyrenaria albiflora Hung T. Chang;白色核果茶;White-flowered Pyrenaria ●

322566 Pyrenaria brevisepala Hung T. Chang;短萼核果茶;Shortcalyx Pyrenaria,Shortsepal Drupetea, Short-sepaled Pyrenaria ●

322567 Pyrenaria brevisepala Hung T. Chang = Pyrenaria diospyricarpa Kurz ●

322568 Pyrenaria buisanensis (Sasaki) C. F. Hsieh et al. = Pyrenaria microcarpa (Dunn) H. Keng var. ovalifolia (H. L. Li) T. L. Ming et S. X. Yang ●

322569 Pyrenaria burmanica T. K. Paul et Nayar = Pyrenaria diospyricarpa Kurz ●

322570 Pyrenaria camellioides Hu = Camellia yunnanensis (Pit. ex Diels) Cohen-Stuart var. camellioides (Hu) T. L. Ming ●

322571 Pyrenaria camellioides Hu = Camellia yunnanensis (Pit. ex Diels) Cohen-Stuart ●

322572 Pyrenaria championii (Nakai) H. Keng = Tutcheria championi Nakai ●

322573 Pyrenaria championii (Nakai) H. Keng = Tutcheria spectabilis Dunn ●

322574 Pyrenaria championii H. Keng = Pyrenaria spectabilis (Champ.) C. Y. Wu et S. X. Yang ●

322575 Pyrenaria cheliensis Hu;景洪核果茶(景宏核果茶);Jinghong Drupetea, Jinghong Pyrenaria ●

322576 Pyrenaria cheliensis Hu = Pyrenaria diospyricarpa Kurz ●

322577 Pyrenaria chmponi (Nakai) H. Keng = Tutcheria championi Nakai ●

322578 Pyrenaria diospyricarpa Kurz;叶萼核果茶●

322579 Pyrenaria garrettiana Craib;短叶核果茶;Shortleaf Drupetea, Shortleaf Pyrenaria,Short-leaved Pyrenaria ●

322580 Pyrenaria garrettiana Craib = Pyrenaria diospyricarpa Kurz ●

322581 Pyrenaria grandiflora (Y. C. Wu) T. L. Ming et S. X. Yang = Pyrenaria hirta (Hand. -Mazz.) H. Keng ●

322582 Pyrenaria greeniae (Chun) H. Keng = Pyrenaria spectabilis (Champ.) C. Y. Wu et S. X. Yang var. greeniae (Chun) S. X. Yang ●

322583 Pyrenaria greeniae (Chun) H. Keng = Tutcheria greeniae Chun ●

322584 Pyrenaria greeniae H. Keng = Tutcheria greeniae Chun ●

322585 Pyrenaria hirta (Hand. -Mazz.) H. Keng;粗毛核果茶(粗毛石笔木,毛石笔木);Hairy-leaf Slatepentree, Hairy-leaf Tutcheria, Hisute Tutcheria, Shag Slatepentree ●

322586 Pyrenaria hirta (Hand. -Mazz.) H. Keng = Tutcheria hirta (Hand. -Mazz.) H. L. Li ●

322587 Pyrenaria hirta (Hand. -Mazz.) H. Keng var. cordatula (H. L. Li) S. X. Yang et T. L. Ming = Tutcheria hirta (Hand. -Mazz.) H. L. Li var. cordatula H. L. Li ●

322588 Pyrenaria hirta (Hand. -Mazz.) H. Keng var. cordatula (H. L. Li) S. X. Yang et T. L. Ming;心叶核果茶(心叶石笔木); Cordateleaf Slatepentree, Cordateleaf Tutcheria ●

322589 Pyrenaria hirta H. Keng = Tutcheria hirta (Hand. -Mazz.) H. L. Li ●

322590 Pyrenaria jonquieriana Pierre ex Lanessan subsp. multisepala (Merr. et Chun) S. X. Yang;多萼核果茶●

322591 Pyrenaria khasiana R. N. Paul;印藏核果茶●

322592 Pyrenaria kwangsiensis Hung T. Chang;广西核果茶(广西石笔木);Guangxi Slatepentree, Kwangsi Slatepentree ●

322593 Pyrenaria kwangsiensis Hung T. Chang = Tutcheria kwangsiensis (Hung T. Chang) Hung T. Chang et C. X. Ye ●

322594 Pyrenaria maculatoclada (Y. K. Li) S. X. Yang;斑支核果茶●

322595 Pyrenaria menglaensis G. D. Tao;勐腊核果茶;Mengla Drupetea, Mengla Pyrenaria ●◇

322596 Pyrenaria microcarpa (Dunn) H. Keng;小果核果茶(狭叶石笔木,小果石笔木);Littlefruit Tutcheria, Little-fruited Tutcheria, Smallfruit Slatepentree ●

322597 Pyrenaria microcarpa (Dunn) H. Keng = Tutcheria microcarpa Dunn ●

322598 Pyrenaria microcarpa (Dunn) H. Keng var. ovalifolia (H. L. Li) T. L. Ming et S. X. Yang = Tutcheria ovalifolia H. L. Li ●

322599 Pyrenaria microcarpa (Dunn) H. Keng var. ovalifolia (H. L. Li) T. L. Ming et S. X. Yang;卵叶核果茶(卵叶石笔木);Eggleaf Slatepentree, Ovateleaf Tutcheria, Ovate-leaved Tutcheria ●

322600 Pyrenaria microcarpa (Dunn) H. Keng var. shinkoensis (Hayata) T. L. Ming et S. X. Yang = Pyrenaria microcarpa (Dunn) H. Keng ●

322601 Pyrenaria microcarpa (Dunn) H. Keng var. tenuifolia (Hung T. Chang) T. L. Ming et S. X. Yang = Pyrenaria microcarpa (Dunn) H. Keng ●

322602 Pyrenaria microcarpa H. Keng = Tutcheria microcarpa Dunn ●

322603 Pyrenaria multisepala (Merr. et Chun) H. Keng = Parapyrenaria multisepala (Merr. et Chun) Hung T. Chang ●◇

322604 Pyrenaria multisepala (Merr. et Chun) H. Keng = Pyrenaria jonquieriana Pierre ex Lanessan subsp. multisepala (Merr. et Chun) S. X. Yang ●

322605 Pyrenaria oblongicarpa Hung T. Chang;长核果茶;Long Drupetea, Longfruit Pyrenaria, Oblongifruited Pyrenaria ●

322606 Pyrenaria ovalifolia (H. L. Li) H. Keng = Pyrenaria microcarpa (Dunn) H. Keng var. ovalifolia (H. L. Li) T. L. Ming et S. X. Yang ●

322607 Pyrenaria ovalifolia (H. L. Li) H. Keng = Tutcheria ovalifolia H. L. Li ●

322608 Pyrenaria ovalifolia H. Keng = Tutcheria ovalifolia H. L. Li ●

322609 Pyrenaria pingpienensis (Hung T. Chang) S. X. Yang et T. L. Ming;屏边核果茶(屏边石笔木);Hairy-ribbed Tutcheria, Pingbian Slatepentree, Pingbian Tutcheria ●

322610 Pyrenaria pingpienensis (Hung T. Chang) S. X. Yang et T. L. Ming = Tutcheria pingpienensis Hung T. Chang ●

322611 Pyrenaria rostrata S. X. Yang et T. L. Ming = Pyrenaria spectabilis (Champ.) C. Y. Wu et S. X. Yang var. greeniae (Chun) S. X. Yang ●

322612 Pyrenaria shinkoensis (Hayata) H. Keng;乌皮茶(水冬瓜,台湾石笔木,圆果石笔木);Ballfruit Slatepentree, Round-fruit Tutcheria, Taiwan Tutcheria ●

322613 Pyrenaria shinkoensis (Hayata) H. Keng = Pyrenaria microcarpa (Dunn) H. Keng ●

322614 Pyrenaria shinkoensis (Hayata) H. Keng = Tutcheria shinkoensis (Hayata) Nakai ●

322615 Pyrenaria shinkoensis H. Keng = Tutcheria shinkoensis (Hayata) Nakai ●

322616 Pyrenaria sophiae (Hu) S. X. Yang et T. L. Ming;云南核果茶(云南石笔木);Yunnan Drupetea, Yunnan Pyrenaria, Yunnan Slatepentree, Yunnan Tutcheria ●

322617 Pyrenaria spectabilis (Champ.) C. Y. Wu et S. X. Yang;大果核果茶●

322618 Pyrenaria spectabilis (Champ.) C. Y. Wu et S. X. Yang var. greeniae (Chun) S. X. Yang = Pyrenaria greeniae (Chun) H. Keng ●

322619 Pyrenaria spectabilis (Champ.) C. Y. Wu et S. X. Yang var. greeniae (Chun) S. X. Yang = Tutcheria greeniae Chun ●

322620 Pyrenaria spectabilis (Champ.) C. Y. Wu et S. X. Yang var. greeniae (Chun) S. X. Yang;长柱核果茶(薄瓣石笔木,长柄石笔木);Longstalk Slatepentree, Longstalk Tutcheria, Long-stalked Tutcheria ●

322621 Pyrenaria symplocifolia (Merr. et F. P. Metcalf) H. Keng = Pyrenaria microcarpa (Dunn) H. Keng var. ovalifolia (H. L. Li) T. L. Ming et S. X. Yang ●

322622 Pyrenaria symplocifolia (Merr. et F. P. Metcalf) H. Keng = Tutcheria symplocifolia Merr. et Chun ●

322623 Pyrenaria symplocifolia H. Keng = Tutcheria symplocifolia Merr. et Chun ●

322624 Pyrenaria tibetana Hung T. Chang;西藏核果茶;Tibet Pyrenaria, Xizang Drupetea, Xizang Pyrenaria ●

322625 Pyrenaria tibetana Hung T. Chang = Pyrenaria khasiana R. N. Paul ●

322626 Pyrenaria turbinata S. X. Yang = Pyrenaria spectabilis (Champ.) C. Y. Wu et S. X. Yang var. greeniae (Chun) S. X. Yang ●

322627 Pyrenaria virgata H. Keng = Pyrenaria microcarpa (Dunn) H. Keng ●

322628 Pyrenaria wuana (Hung T. Chang) S. X. Yang;长萼核果茶(长萼石笔木,长毛石笔木,吴氏石笔木);Longcalyx Slatepentree, Wu Tutcheria ●

322629 Pyrenaria yunnanensis Hu = Pyrenaria diospyricarpa Kurz ●

322630 Pyrenia Clairv. = Pyrus L. ●

322631 Pyrenocarpa Hung T. Chang et R. H. Miao = Decaspermum J. R. Forst. et G. Forst. ●

322632 Pyrenocarpa Hung T. Chang et R. H. Miao(1975);多核果属;Pyrenocarpa ●★

322633 Pyrenocarpa hainanensis (Merr.) Hung T. Chang et R. H.

Miao;多核果;Hainan Pyrenocarpa ●◇

322634　Pyrenocarpa teretis Hung T. Chang et R. H. Miao;圆枝多核果;
Cylindric-branch Pyrenocarpa,Cylindric-branched Pyrenocarpa ●◇

322635　Pyrenoglyphis H. Karst. = Bactris Jacq. ex Scop. ●

322636　Pyrethraria Pers. ex Steud. = Cotula L. ■

322637　Pyrethrum Medik. = Spilanthes Jacq. ■

322638　Pyrethrum Zinn = Chrysanthemum L.（保留属名）■●

322639　Pyrethrum Zinn = Tanacetum L. ■●

322640　Pyrethrum Zinn(1757)（'Pyrethum'）;匹菊属(除虫菊属,菊
属,小黄菊属）;Painted Daisy,Pyrethrum ■

322641　Pyrethrum abrotanifolium Bunge ex Ledeb.;丝叶匹菊;Filarious
Pyrethrum,Silkleaf Pyrethrum ■

322642　Pyrethrum acheilifolium M. Bieb. var. discoideum Kar. et Kir. =
Tanacetum barclayanum DC. ■

322643　Pyrethrum achilleifolium Bunge ex Ledeb. var. discoideum Kar.
et Kir. = Tanacetum barclayanum DC. ■

322644　Pyrethrum alatavicum (Herder) O. Fedtsch. et B. Fedtsch.;新
疆匹菊;Sinkiang Pyrethrum,Xinjiang Pyrethrum ■

322645　Pyrethrum alatavicum (Herder) O. Fedtsch. et B. Fedtsch.
subsp. krylovianum (Krasch.) N. M. Boldyreva = Pyrethrum
krylovianum Krasch. ■

322646　Pyrethrum alpinum (L.) Schrank;山地匹菊;Alpine
Chrysanthemum ■☆

322647　Pyrethrum ambiguum Ledeb. = Tripleurospermum ambiguum
(Ledeb.) Franch. et Sav. ■

322648　Pyrethrum arassanicum (C. Winkl.) O. Fedtsch. et B. Fedtsch.
= Pyrethrum pyrethroides (Kar. et Kir.) B. Fedtsch. ex Krasch. ■

322649　Pyrethrum arcticum (L.) Stankov;北除虫菊(北极除虫菊,除
虫菊,新疆除虫菊)■☆

322650　Pyrethrum arrasanicum (C. Winkl.) O. Fedtsch. et B.
Fedtsch.;光滑匹菊(光叶匹菊);Smootjlraf Pyrethrum ■

322651　Pyrethrum arundanum Boiss. = Rhodanthemum arundanum
(Boiss.) B. H. Wilcox, K. Bremer et Humphries ●☆

322652　Pyrethrum arundanum Boiss. var. depressum (Ball) Pau =
Rhodanthemum arundanum (Boiss.) B. H. Wilcox, K. Bremer et
Humphries ●☆

322653　Pyrethrum atkinsonii (C. B. Clarke) Y. Ling et C. Shih;藏匹
菊;Atkinson Pyrethrum,Xizang Pyrethrum ■

322654　Pyrethrum aucherianum DC.;奥切尔匹菊■☆

322655　Pyrethrum balsamitoides (Nábelek) Tzvelev;拟菊蒿匹菊■☆

322656　Pyrethrum balsamitum (L.) Willd.;菊蒿匹菊■☆

322657　Pyrethrum balsamitum (L.) Willd. var. tanacetoides Boiss. =
Balsamita major Desf. ■☆

322658　Pyrethrum carneum M. Bieb.;肉色除虫菊(除虫菊)■☆

322659　Pyrethrum caucasicum M. Bieb. = Chrysanthemum caucasicum
Pers. ■☆

322660　Pyrethrum cinerariifolium Trevir.;除虫菊（白花除虫菊);
Dalmatian Insect Flower,Dalmatian Pyrethrum,Insect Powder Plant,
Pyrethrum Flower ■

322661　Pyrethrum cinerariifolium Trevir. = Chrysanthemum cinerariifolium
(Trevir.) Vis. ■

322662　Pyrethrum cinerariifolium Trevir. = Tanacetum cinerariifolium
(Trevir.) Sch. Bip. ■

322663　Pyrethrum clausonis Pomel = Mauranthemum paludosum (Poir.)
Vogt et Oberpr. ■☆

322664　Pyrethrum clusii Fisch. ex Rchb.;克氏匹菊■☆

322665　Pyrethrum coccineum (Willd.) Vorosch.;红花除虫菊(除虫

菊,粉蒿蒿）; Caucasian Pyrethrum, Common Pyrethrum, Florists
Pyrethrum,Florist's Pyrethrum, Florists' Pyrethrum, Painted Daisy,
Painted Lady, Persian Pellitory, Pyrethrum, Pyrethum Daisy,
Scarletflower Pyrethrum ■

322666　Pyrethrum coccineum (Willd.) Vorosch. = Tanacetum coccineum
(Willd.) Grierson ■

322667　Pyrethrum coccineum (Willd.) Vorosch. = Pyrethrum coccineum
(Willd.) Vorosch. ■

322668　Pyrethrum corymbiferum (L.) Schrank = Tanacetum corymbosum
(L.) Sch. Bip. ■☆

322669　Pyrethrum corymbiforme Tzvelev;匹菊;Common Pyrethrum ■

322670　Pyrethrum corymbosum (L.) Scop. = Tanacetum corymbosum
(L.) Sch. Bip. ■☆

322671　Pyrethrum corymbosum (L.) Scop. subsp. achilleae (L.)
Murb. = Tanacetum corymbosum (L.) Sch. Bip. ■☆

322672　Pyrethrum corymbosum (L.) Scop. var. achilleae (L.) Chabert
= Tanacetum corymbosum (L.) Sch. Bip. ■☆

322673　Pyrethrum corymbosum Link = Chrysanthemum anethifolium
Brouss. ex Willd. ■☆

322674　Pyrethrum crassipes Stschegl. = Tanacetum crassipes
(Stschegl.) Tzvelev ■

322675　Pyrethrum daghestanicum (Rupr. ex Boiss.) Sosn. et Manden.;
达赫斯坦匹菊■☆

322676　Pyrethrum demnatense (Murb.) Pau = Rhodanthemum
gayanum (Coss. et Durieu) B. H. Wilcox, K. Bremer et Humphries
subsp. demnatense (Murb.) Vogt ■☆

322677　Pyrethrum deserticola Murb. = Chrysanthoglossum deserticola
(Murb.) B. H. Wilcox, K. Bremer et Humphries ■☆

322678　Pyrethrum discoideum Ledeb. = Cancrinia discoides (Ledeb.)
Poljakov ex Tzvelev ■

322679　Pyrethrum djilgense (Franch.) Tzvelev;裂齿匹菊■

322680　Pyrethrum ferulaceum Webb et Berthel. = Tanacetum ferulaceum
(Webb) Sch. Bip. ■☆

322681　Pyrethrum foeniculaceum Willd. = Argyranthemum
foeniculaceum (Willd.) Webb ex Sch. Bip. ●☆

322682　Pyrethrum frutescens (L.) Willd. = Argyranthemum frutescens
(L.) Sch. Bip. ●

322683　Pyrethrum fruticulosum Fenzl ex Boiss.;小灌木匹菊●☆

322684　Pyrethrum fuscatum (Desf.) Bonnet = Heteromera fuscata
(Desf.) Pomel ■☆

322685　Pyrethrum galae Popov;嘎氏匹菊■☆

322686　Pyrethrum gayanum Coss. = Rhodanthemum gayanum (Coss. et
Durieu) B. H. Wilcox, K. Bremer et Humphries ■☆

322687　Pyrethrum gayanum Coss. var. nanum Batt. = Rhodanthemum
gayanum (Coss. et Durieu) B. H. Wilcox, K. Bremer et Humphries ■☆

322688　Pyrethrum glanduliferum Sommier et H. Lév.;腺体匹菊■☆

322689　Pyrethrum grossheimii Sosn.;格罗氏匹菊■☆

322690　Pyrethrum heldreichianum Fenzl;海氏匹菊■☆

322691　Pyrethrum hissaricum Krasch.;希萨尔匹菊■☆

322692　Pyrethrum indicum (L.) Cass. = Dendranthema indicum (L.)
Des Moul. ■

322693　Pyrethrum karelinii Krasch.;卡氏匹菊■☆

322694　Pyrethrum karelinii Krasch. = Pyrethrum richterioides (C.
Winkl.) Krasn. ■

322695　Pyrethrum kasakhstanicum Krasch. = Tanacetum santolina C.
Winkl. ■

322696　Pyrethrum kaschgharicum Krasch.;托毛匹菊;Kaschghar

Pyrethrum ■

322697 Pyrethrum kelleri (Krylov et Plotn.) Krasch. ;凯勒匹菊■☆

322698 Pyrethrum kotschyi Boiss. ;考氏匹菊■☆

322699 Pyrethrum krylovianum Krasch. ; 黑苞匹菊; Blackbract Pyrethrum ■

322700 Pyrethrum kubense Grossh. ;库班匹菊■☆

322701 Pyrethrum lanuginosum (Sch. Bip. et Herder) Tzvelev;多毛匹菊■☆

322702 Pyrethrum lavandulifolium Fisch. ex Trautv. = Dendranthema lavandulifolium (Fisch. ex Trautv.) Kitam. ■

322703 Pyrethrum lavandulifolium Fisch. ex Trautv. = Dendranthema lavandulifolium (Fisch. ex Trautv.) Y. Ling et C. Shih ■

322704 Pyrethrum leptopodium (Winkl.) Tzvelev;火绒匹菊■

322705 Pyrethrum leucanthemum (L.) Franch. = Leucanthemum vulgare Lam. ■

322706 Pyrethrum leucanthemum (L.) Sch. Bip. = Leucanthemum vulgare Lam. ■

322707 Pyrethrum macrocephalum (Viv.) Coss. et Durieu = Chrysanthoglossum trifurcatum (Desf.) B. H. Wilcox, K. Bremer et Humphries ■☆

322708 Pyrethrum macrophyllum Willd. = Chrysanthemum macrophyllum Waldst. et Kit. ■☆

322709 Pyrethrum majus (Desf.) Tzvelev;大匹菊■☆

322710 Pyrethrum majus (Desf.) Tzvelev = Balsamita major Desf. ■☆

322711 Pyrethrum majus (Desf.) Tzvelev = Tanacetum balsamita L. ■☆

322712 Pyrethrum maresii Coss. = Rhodanthemum maresii (Coss.) B. H. Wilcox, K. Bremer et Humphries ■☆

322713 Pyrethrum marionii Albov;马氏匹菊■☆

322714 Pyrethrum mikeschinii Tzvelev;米氏匹菊■☆

322715 Pyrethrum millefoliatum Willd. = Tanacetum tanacetoides (DC.) Tzvelev ■

322716 Pyrethrum myconnis (L.) Moench. = Coleostephus myconis (L.) Cass. ■

322717 Pyrethrum neglectum Tzvelev;疏忽匹菊■☆

322718 Pyrethrum olivieri Chabert;奥氏匹菊■☆

322719 Pyrethrum pallasianum (Fisch. ex Besser) Maxim. = Ajania pallasiana (Fisch. ex Besser) Poljakov ■

322720 Pyrethrum pallidum (Mill.) Pau = Leucanthemopsis pallida (Mill.) Heywood ■☆

322721 Pyrethrum pallidum (Mill.) Pau var. longipectinatum Pau = Leucanthemopsis longipectinata (Pau) Heywood ■☆

322722 Pyrethrum pallidum (Mill.) Pau var. radicans (Cav.) Pau = Leucanthemopsis pectinata (L.) G. López et Jarvis ■☆

322723 Pyrethrum parthenifolium Willd. ; 伞房匹菊; Corymbose Pyrethrum ■

322724 Pyrethrum parthenium (L.) Sm.;短舌匹菊(艾菊,短舌菊蒿, 菊蒿,玲珑菊,小白菊);Arsesmart, Bachelor's Buttons, Bertram, Bothen, Burtons, Devil Daisy, Devil's Daisy, Featherbow, Featherfall, Featherfew, Featherfoe, Featherfoil, Featherfold, Featherfowl, Featherfoy, Featherfull, Featherwheelie, Feathyfew, Fedderfew, Fetferfoe, Feverfew, Feverfew Chrysanthemum, Feverfoullie, Field Daisy, Flirtwort, Hen-and-chickens, Madron, Maghet, Maids, Maithes, Mayweed, Michaelmas Daisy, Midsummer Daisy, Nosebleed, Old Maid's Scent, Pellitory, Shortligule Pyrethrum, St. Peter's Wort, Stink Daisy, Vether Vaw, Vether-vo, Vethervow, Vivvervaw, Vivvyvew, Whitewort, Yard Daisy ■

322725 Pyrethrum parthenium (L.) Sm. =Chrysanthemum parthenium

322726 Pyrethrum parthenium (L.) Sm. = Tanacetum karelinii Tzvelev ■

322727 Pyrethrum parthenium (L.) Sm. = Tanacetum parthenium (L.) Sch. Bip. ■

322728 Pyrethrum petrareum C. Shih;岩匹菊;Rocky Pyrethrum ■

322729 Pyrethrum peucedanifolium (Sosn.) Manden. ;前胡叶匹菊■☆

322730 Pyrethrum pulchrum Ledeb. ; 美丽匹菊(小黄菊,小匹菊); Little Pyrethrum ■

322731 Pyrethrum punctatum Bordz. ;斑点匹菊■☆

322732 Pyrethrum pyrethroides (Kar. et Kir.) B. Fedtsch. ex Krasch. ; 灰叶匹菊;Greyleaf Pyrethrum ■

322733 Pyrethrum radicans Cav. = Leucanthemopsis pectinata (L.) G. López et Jarvis ■☆

322734 Pyrethrum richterioides (C. Winkl.) Krasn. ;单头匹菊; Singlehead Pyrethrum ■

322735 Pyrethrum roseum Adam. = Pyrethrum coccineum (Willd.) Vorosch. ■

322736 Pyrethrum roseum Lindl. = Pyrethrum coccineum (Willd.) Vorosch. ■

322737 Pyrethrum roseum Lindl. var. adami Trautv. = Pyrethrum coccineum (Willd.) Vorosch. ■

322738 Pyrethrum saxatile Kar. et Kir. = Tanacetum karelinii Tzvelev ■

322739 Pyrethrum scopulorum Krasch. = Tanacetum scopulorum (Krasch.) Tzvelev ■

322740 Pyrethrum segetum (L.) Moech. =Chrysanthemum segetum L. ■

322741 Pyrethrum semenowii C. Winkl. ex O. Fedtsch. et B. Fedtsch. ; 谢氏匹菊■☆

322742 Pyrethrum sericeum (Adam) M. Bieb. ;绢毛匹菊■☆

322743 Pyrethrum seticuspe Maxim. = Dendranthema lavandulifolium (Fisch. ex Trautv.) Kitam. ■

322744 Pyrethrum sinense (Sabine) DC. = Dendranthema morifolium (Ramat.) Tzvelev ■

322745 Pyrethrum songoricum Tzvelev;准噶尔匹菊■

322746 Pyrethrum sorbifolium Boiss. ex Boiss. ;花楸匹菊■☆

322747 Pyrethrum tanacetoides DC. = Tanacetum tanacetoides (DC.) Tzvelev ■

322748 Pyrethrum tatsienense (Bureau et Franch.) Y. Ling et C. Shih; 川西小黄菊(鞑靼菊,鞑新菊,打箭菊);W. Sichuan Pyrethrum, West Szechuan Pyrethrum ■

322749 Pyrethrum tatsienense (Bureau et Franch.) Y. Ling et C. Shih var. tanacetopsis (W. W. Sm.) Y. Ling et C. Shih;无舌川西小黄 菊;Tongueless Pyrethrum ■

322750 Pyrethrum tianschanicum Krasn. ; 天山匹菊; Tianshan Pyrethrum ■

322751 Pyrethrum transiliense (Herder) Regel et Schmalh. ;白花匹菊; White Pyrethrum, Whiteflower Pyrethrum ■

322752 Pyrethrum transiliense (Herder) Regel et Schmalh. var. subvillosum Regel et Schmalh. = Pyrethrum pyrethroides (Kar. et Kir.) B. Fedtsch. ex Krasch. ■

322753 Pyrethrum transiliense (Herder) Regel et Schmalh. var. subvilosum Regel et Schmalh. = Pyrethrum transiliense (Herder) Regel et Schmalh. ■

322754 Pyrethrum trasiliense (Herder) Regel et Schmalh. = Pyrethrum pyrethroides (Kar. et Kir.) B. Fedtsch. ex Krasch. ■

322755 Pyrethrum tricholobum Sosn. ex Manden. ;毛片匹菊■☆

322756 Pyrethrum trifurcatum (Desf.) Willd. = Chrysanthoglossum trifurcatum (Desf.) B. H. Wilcox, K. Bremer et Humphries ■☆

322757 Pyrethrum turlanicum Pavlov = Tanacetum barclayanum DC. ■

322758 Pyrethrum uvlgare (L.) Boiss. = Tanacetum vulgare L. ■

322759 Pyrethrum webbianum Coss. = Tanacetum corymbosum (L.) Sch. Bip. ■☆

322760 Pyrethrum zawadskii (Herb.) Nyman = Dendranthema zawadskii (Herb.) Tzvelev ■

322761 Pyrgophyllum (Gagnep.) T. L. Wu et Z. Y. Chen(1989);苞叶姜属;Pyrgophyllum ■★

322762 Pyrgophyllum yunnanense (Gagnep.) T. L. Wu et Z. Y. Chen;苞叶姜(大苞姜,滇姜三七,姜三七,曲蕊姜,云南曲蕊姜);Pyrgophyllum, Yunnan Bigbractginger ■

322763 Pyrgosea Eckl. et Zeyh. = Crassula L. ●■☆

322764 Pyrgosea Eckl. et Zeyh. = Purgosea Haw. ●■☆

322765 Pyrgosea Eckl. et Zeyh. = Turgosea Haw. ●■☆

322766 Pyrgus Lour. = Ardisia Sw. (保留属名)●■

322767 Pyriluma (Baill.) Aubrév. = Pouteria Aubl. ●

322768 Pyriluma Baill. = Pouteria Aubl. ●

322769 Pyriluma Baill. ex Aubrév. = Pouteria Aubl. ●

322770 Pyro-cydonia Guillaumin = Pirocydonia H. K. A. Winkl. ex L. L. Daniel ●☆

322771 Pyrocydonia Rehder = Pirocydonia H. K. A. Winkl. ex L. L. Daniel ●☆

322772 Pyrogennema J. Lunell = Chamaenerion Adans. ■

322773 Pyrogennema J. Lunell = Epilobium L. ■

322774 Pyrola Alef. = Orthilia Raf. ■

322775 Pyrola L. (1753);鹿蹄草属;Pyrola, Shinleaf, Winter green, Wintergreen, Winter-green ●■

322776 Pyrola alba Andres = Pyrola decorata Andres var. alba (Andres) T. L. Chou et R. C. Zho ●

322777 Pyrola alba Andres = Pyrola decorata Andres ●

322778 Pyrola alba Andres var. viridiflora Andres = Pyrola decorata Andres ●

322779 Pyrola alboreticulata Hayata;花叶鹿蹄草(阿里山鹿蹄草,白脉鹿蹄草,斑纹鹿蹄草);Garishleaf Pyrola, Whitenetvein Pyrola ●

322780 Pyrola alpina Andres;高山鹿蹄草●☆

322781 Pyrola alpina Andres f. rosea Sugim. ;粉色高山鹿蹄草●☆

322782 Pyrola americana Sweet;美洲鹿蹄草; American Round-leaved Wintergreen, American Wintergreen ●☆

322783 Pyrola americana Sweet = Pyrola rotundifolia L. subsp. americana (Sweet) R. T. Clausen ●☆

322784 Pyrola americana Sweet var. dahurica Andres = Pyrola dahurica (Andres) Kom. ●

322785 Pyrola andresii Krisa = Pyrola calliantha Andréws var. tibetana (Andres) Y. L. Chou ●

322786 Pyrola andresii Krisa = Pyrola calliantha Andréws ●

322787 Pyrola asarifolia Michx. ;细辛叶鹿蹄草(粉鹿蹄草,细叶鹿蹄草); Asarum-leaved Wintergreen, Bog Wintergreen, Liver-leaf Wintergreen, Pink Pyrola, Pink Shin-leaf ●☆

322788 Pyrola asarifolia Michx. subsp. americana (Sweet) Krisa = Pyrola rotundifolia L. subsp. americana (Sweet) R. T. Clausen ●☆

322789 Pyrola asarifolia Michx. subsp. incarnata (DC.) E. Murray = Pyrola incarnata (DC.) Fisch. ex Kom. ●

322790 Pyrola asarifolia Michx. subsp. incarnata (DC.) Haber et Hir. Takah. ;红花鹿蹄草(红鹿衔草);Redflower Pyrola ●

322791 Pyrola asarifolia Michx. var. incarnata (DC.) Fernald = Pyrola asarifolia Michx. ●☆

322792 Pyrola asarifolia Michx. var. incarnata (DC.) Fernald = Pyrola

322792 asarifolia Michx. subsp. incarnata (DC.) Haber et Hir. Takah. ●

322793 Pyrola asarifolia Michx. var. japonica (Klenze ex Alef.) Miq. = Pyrola japonica Klenze ex Alef. ●

322794 Pyrola asarifolia Michx. var. japonica (Klenze) Miq. = Pyrola japonica Klenze ex Alef. ●

322795 Pyrola asarifolia Michx. var. ovata Farw. = Pyrola asarifolia Michx. ●☆

322796 Pyrola asarifolia Michx. var. purpurea (Bunge) Fernald = Pyrola asarifolia Michx. subsp. incarnata (DC.) Haber et Hir. Takah. ●

322797 Pyrola asarifolia Michx. var. purpurea (Bunge) Fernald = Pyrola asarifolia Michx. ●☆

322798 Pyrola atropurpurea Franch. ;紫背鹿蹄草(鹿蹄草,鹿衔草,深紫鹿蹄草);Purpleback Pyrola ●

322799 Pyrola atropurpurea Franch. var. gracilis Andres = Pyrola atropurpurea Franch. ●

322800 Pyrola californica Krisa = Pyrola asarifolia Michx. ●☆

322801 Pyrola calliantha Andréws;鹿蹄草(白鹿寿草,常绿茶,川北鹿蹄草,大肺金草,大肺筋草,河北鹿蹄草,红肺筋草,鹿安茶,鹿含草,鹿寿草,鹿寿茶,鹿衔草,罗汉茶,美花鹿蹄草,破血丹,紫背金牛草);Pyrola, Chinese Pyrola ●

322802 Pyrola calliantha Andréws var. tibetana (Andres) Y. L. Chou;西藏鹿蹄草;Tibet Pyrola, Xizang Pyrola ●

322803 Pyrola calliantha Andréws var. tibetana (Andres) Y. L. Chou = Pyrola calliantha Andréws ●

322804 Pyrola chlorantha Sw. ;绿花鹿蹄草(绿色鹿蹄草,索伦鹿蹄草);Green Pyrola, Greenflower Pyrola, Green-flowered Wintergreen, Greenish Wintergreen, Greenish-flowered Pyrola, Shin-leaf, Yellow Wintergreen ●

322805 Pyrola chlorantha Sw. subsp. fallax Krisa = Pyrola dahurica (Andres) Kom. ●

322806 Pyrola chlorantha Sw. var. convoluta (W. P. C. Barton) Fernald = Pyrola chlorantha Sw. ●

322807 Pyrola chlorantha Sw. var. paucifolia Fernald = Pyrola chlorantha Sw. ●

322808 Pyrola chlorantha Sw. var. revoluta Jenn. = Pyrola chlorantha Sw. ●

322809 Pyrola chlorantha Sw. var. saximontana Fernald = Pyrola chlorantha Sw. ●

322810 Pyrola chouana Chang Y. Yang;阿尔泰鹿蹄草;Altai Pyrola ●

322811 Pyrola compacta Jenn. = Pyrola elliptica Nutt. ●☆

322812 Pyrola convoluta W. P. C. Barton = Pyrola chlorantha Sw. ●

322813 Pyrola corbieri H. Lév. ;贵阳鹿蹄草;Corbier Pyrola, Guiyang Pyrola ●

322814 Pyrola dahurica (Andres) Kom. ;兴安鹿蹄草;Dahur Pyrola, Dahurian Pyrola ●

322815 Pyrola decorata Andres;普通鹿蹄草(斑纹鹿蹄草,大肺金草,大肺筋草,红肺筋草,鹿安茶,鹿含草,鹿寿草,鹿寿茶,鹿蹄草,鹿衔草,卵叶鹿蹄草,破血丹,山美人鹿蹄草,雅美鹿蹄草,紫背金牛草);Common Pyrola ●

322816 Pyrola decorata Andres = Pyrola alboreticulata Hayata ●

322817 Pyrola decorata Andres var. alba (Andres) T. L. Chou et R. C. Zho;白花鹿蹄草(白鹿蹄草);Whiteflower Common Pyrola ●

322818 Pyrola decorata Andres var. alba (Andres) Y. L. Chou et R. C. Zhou = Pyrola decorata Andres ●☆

322819 Pyrola elata Nutt. = Pyrola asarifolia Michx. ●☆

322820 Pyrola elegantula Andres;长叶鹿蹄草;Longleaf Pyrola ●

322821 Pyrola elegantula Andres var. jiangxiensis Y. L. Chou et R. C.

Zhou;江西长叶鹿蹄草;Jiangxi Longleaf Pyrola ●

322822 Pyrola elegantula Andres var. jiangxiensis Y. L. Chou et R. C. Zhou = Pyrola elegantula Andres ●

322823 Pyrola elliptica Nutt. ;亮叶鹿蹄草(鹿蹄草,鹿衔草,破血丹,椭圆叶鹿蹄草);Elliptic Shin-leaf,Large-leaved Shin-leaf,Shinleaf,Wax-flower Shin-leaf,Wild Lily of the Velley ●☆

322824 Pyrola elliptica Nutt. var. morrisonensis Hayata = Pyrola morrisonensis (Hayata) Hayata ●

322825 Pyrola faurieana Andres;法氏鹿蹄草●☆

322826 Pyrola forrestiana Andres;大理鹿蹄草(西南鹿蹄草);Forrest Pyrola ●

322827 Pyrola forrestiana Andres subsp. dahurica (Andres) Krisa = Pyrola dahurica (Andres) Kom. ●

322828 Pyrola gracilis (Andres) Andres = Pyrola atropurpurea Franch. ●

322829 Pyrola gracilis Andres = Pyrola atropurpurea Franch. ●

322830 Pyrola grandiflora Radius;大花鹿蹄草(冬绿鹿蹄草,高山鹿蹄草);Arctic Pyrola,Arctic Wintergreen ●☆

322831 Pyrola handeliana Andres = Pyrola decorata Andres ●

322832 Pyrola hopeiensis Nakai = Pyrola calliantha Andréws ●

322833 Pyrola incarnata (DC.) Fisch. ex Kom. = Pyrola asarifolia Michx. subsp. incarnata (DC.) Haber et Hir. Takah. ●

322834 Pyrola incarnata (DC.) Fisch. ex Kom. subsp. dahurica (Andres) Krisa = Pyrola dahurica (Andres) Kom. ●

322835 Pyrola incarnata (DC.) Fisch. ex Kom. var. japonica (Klenze ex Alef.) Koidz. = Pyrola japonica Klenze ex Alef. ●

322836 Pyrola incarnata (DC.) Fisch. ex Kom. var. japonica (Klenze) Koidz. = Pyrola japonica Klenze ex Alef. ●

322837 Pyrola incarnata (DC.) Fisch. ex Kom. var. ovatifolia Y. Z. Zhao;卵叶红花鹿蹄草■

322838 Pyrola incarnata (DC.) Fisch. ex Kom. var. ovatifolia Y. Z. Zhao = Pyrola asarifolia Michx. subsp. incarnata (DC.) Haber et Hir. Takah. ●

322839 Pyrola incarnata (DC.) Fisch. ex Kom. var. subaphylla Satomi;亚无叶红花鹿蹄草●☆

322840 Pyrola incarnata (DC.) Freyn = Pyrola asarifolia Michx. subsp. incarnata (DC.) Haber et Hir. Takah. ●

322841 Pyrola japonica Klenze = Pyrola dahurica (Andres) Kom. ●

322842 Pyrola japonica Klenze ex Alef. ;日本鹿蹄草(鹿寿草,鹿寿茶,鹿衔草);Japan Pyrola,Japanese Pyrola ●

322843 Pyrola japonica Klenze ex Alef. f. rosiflora H. Hara;粉红花鹿蹄草☆

322844 Pyrola japonica Klenze ex Alef. f. subaphylla (Maxim.) Ohwi = Pyrola japonica Klenze ex Alef. ●

322845 Pyrola japonica Klenze ex Alef. var. subaphylla (Maxim.) Andres = Pyrola japonica Klenze ex Alef. ●

322846 Pyrola japonica Klenze ex Alef. var. subaphylla (Maxim.) Hara = Pyrola subaphylla (Maxim.) Ohwi ●

322847 Pyrola japonica Klenze ex Alef. var. subaphylla (Maxim.) Hara = Pyrola subaphylla Maxim. ●

322848 Pyrola japonica Siebold = Pyrola japonica Klenze ex Alef. ●

322849 Pyrola macrocalyx Ohwi;长萼鹿蹄草;Longcalyx Pyrola ●

322850 Pyrola markonica Y. L. Chou et R. C. Zhou;马尔康鹿蹄草;Markang Pyrola ●

322851 Pyrola mattfeldiana Andres;贵州鹿蹄草;Guizhou Pyrola ●

322852 Pyrola media Sw. ;小叶鹿蹄草(天葵,紫背天葵);Greater Wintergreen,Intermediate Wintergreen,Littleleaf Pyrola ●

322853 Pyrola minor L. ;短柱鹿蹄草;Common Wintergreen, Lesser Wintergreen, Little Shin-leaf, Small Shin-leaf, Snowline Pyrola, Wintergreen ●

322854 Pyrola minor L. = Pyrola sororia Andres ●

322855 Pyrola minor L. subsp. faurieana (Andres) Vorosch. ex Khokhr. = Pyrola faurieana Andres ●☆

322856 Pyrola minor L. var. parviflora B. Boivin = Pyrola minor L. ●

322857 Pyrola monophylla Y. L. Chou et R. C. Zhou;单叶鹿蹄草;Monoleaf Pyrola,Singleleaf Pyrola ●

322858 Pyrola morrisonensis (Hayata) Hayata;台湾鹿蹄草(新高山鹿蹄草,玉山鹿蹄草);Taiwan Pyrola ●

322859 Pyrola nephrophylla (Andres) Andres;圆肾叶鹿蹄草(肾叶鹿蹄草)●☆

322860 Pyrola nephrophylla (Andres) Andres f. rosea Sugim. ;粉花肾叶鹿蹄草●☆

322861 Pyrola norvegica Knaben;挪威鹿蹄草;Norwegian Wintergreen ☆

322862 Pyrola nummularia (Rupr.) Rupr. ex Kom. = Orthilia obtusata (Turcz.) H. Hara ■

322863 Pyrola nummularia Rupr. ex Kom. = Orthilia obtusata (Turcz.) H. Hara ■

322864 Pyrola obovata Bertol. = Pyrola rotundifolia L. subsp. americana (Sweet) R. T. Clausen ●☆

322865 Pyrola obtusata (Turcz.) Pavlov = Orthilia obtusata (Turcz.) H. Hara ■

322866 Pyrola obtusata (Turcz.) Turcz. ex Kom. = Orthilia obtusata (Turcz.) H. Hara ■

322867 Pyrola obtusata Turcz. ex Kom. = Orthilia obtusata (Turcz.) H. Hara ■

322868 Pyrola oreodoxa Andres = Pyrola decorata Andres ●

322869 Pyrola picta Andres;彩色鹿蹄草;White-veined Shinleaf ●☆

322870 Pyrola renifolia Maxim. ;肾叶鹿蹄草;Kidneyleaf Pyrola ●

322871 Pyrola rockii Krisa = Pyrola calliantha Andréws ●

322872 Pyrola rotundifolia L. ;圆叶鹿蹄草(白花鹿蹄草,鹿含草,鹿蹄草,鹿衔草);Canker Lettuce, Dollar-leaf, European Pyrola, False Wintergreen, Larger Wintergreen, Pear-leaf Wintergreen, Round-leafed Pyrola, Round-leaved Shin-leaf, Round-leaved Wintergreen, Round-leaved Winter-green, Shin-leaf, Wild Lily of the Valley, Wild Lily-of-the-valley ●

322873 Pyrola rotundifolia L. = Pyrola morrisonensis (Hayata) Hayata ●

322874 Pyrola rotundifolia L. f. subaphylla (Maxim.) Makino = Pyrola japonica Klenze ex Alef. ●

322875 Pyrola rotundifolia L. subsp. americana (Sweet) R. T. Clausen;美国圆叶鹿蹄草;Round-leaved Shin-leaf ●☆

322876 Pyrola rotundifolia L. subsp. chinensis (Andres) Andres = Pyrola calliantha Andréws ●

322877 Pyrola rotundifolia L. subsp. chinensis Andres = Pyrola calliantha Andréws ●

322878 Pyrola rotundifolia L. subsp. chinensis Andres var. communre Andres = Pyrola calliantha Andréws ●

322879 Pyrola rotundifolia L. subsp. chinensis Andres var. laurifolia Andres = Pyrola calliantha Andréws ●

322880 Pyrola rotundifolia L. subsp. chinensis Andres var. sphaeroides Andres = Pyrola calliantha Andréws ●

322881 Pyrola rotundifolia L. subsp. dahurica (Andres) Andres = Pyrola dahurica (Andres) Kom. ●

322882 Pyrola rotundifolia L. subsp. incarnata (DC.) Krylov = Pyrola asarifolia Michx. subsp. incarnata (DC.) Haber et Hir. Takah. ●

322883 Pyrola rotundifolia L. subsp. incarnata (Fisch.) Krylov = Pyrola incarnata（DC.）Fisch. ex Kom. ●

322884　Pyrola rotundifolia L. subsp. tibetana（Andres）Andres ＝ Pyrola calliantha Andréws ●

322885　Pyrola rotundifolia L. subsp. tibetana Andres ＝ Pyrola calliantha Andréws var. tibetana（Andres）Y. L. Chou ●

322886　Pyrola rotundifolia L. var. albiflora Maxim. ＝ Pyrola japonica Klenze ex Alef. ●

322887　Pyrola rotundifolia L. var. americana（Sweet）Fernald ＝ Pyrola rotundifolia L. subsp. americana（Sweet）R. T. Clausen ●☆

322888　Pyrola rotundifolia L. var. chinensis Andres ＝ Pyrola calliantha Andréws ●

322889　Pyrola rotundifolia L. var. communis Andres ＝ Pyrola calliantha Andréws ●

322890　Pyrola rotundifolia L. var. grandiflora DC. ＝ Pyrola rotundifolia L. ●

322891　Pyrola rotundifolia L. var. incarnata（Fisch.）DC. ＝ Pyrola incarnata（DC.）Fisch. ex Kom. ●

322892　Pyrola rotundifolia L. var. incarnata（Fisch.）DC. f. subaphylla（Maxim.）Makino ＝ Pyrola subaphylla（Maxim.）Ohwi ●

322893　Pyrola rotundifolia L. var. incarnata DC. ＝ Pyrola asarifolia Michx. ●☆

322894　Pyrola rotundifolia L. var. incarnata DC. ＝ Pyrola asarifolia Michx. subsp. incarnata（DC.）Haber et Hir. Takah. ●

322895　Pyrola rotundifolia L. var. incarnata DC. ＝ Pyrola incarnata（DC.）Fisch. ex Kom. ●

322896　Pyrola rotundifolia L. var. laurifolia Andres ＝ Pyrola calliantha Andréws ●

322897　Pyrola rotundifolia L. var. purpurea Bunge ＝ Pyrola asarifolia Michx. subsp. incarnata（DC.）Haber et Hir. Takah. ●

322898　Pyrola rotundifolia L. var. purpurea Bunge ＝ Pyrola asarifolia Michx. ●☆

322899　Pyrola rotundifolia L. var. purpurea Bunge ＝ Pyrola incarnata（DC.）Fisch. ex Kom. ●

322900　Pyrola rotundifolia L. var. sphaeroides Andres ＝ Pyrola calliantha Andréws ●

322901　Pyrola rotundifolia L. var. tibetana Andres ＝ Pyrola calliantha Andréws ●

322902　Pyrola rugosa Andres;皱叶鹿蹄草;Wrinkledleaf Pyrola ●

322903　Pyrola secunda L. ＝ Orthilia secunda（L.）House ■

322904　Pyrola secunda L. subsp. obtusata（Turcz.）Hultén ＝ Orthilia obtusata（Turcz.）H. Hara ■

322905　Pyrola secunda L. subsp. obtusata（Turcz.）Hultén ＝ Orthilia secunda（L.）House ■

322906　Pyrola secunda L. var. nummularia Rupr. ＝ Orthilia obtusata（Turcz.）H. Hara ■

322907　Pyrola secunda L. var. obtusata Turcz. ＝ Orthilia secunda（L.）House ■

322908　Pyrola secunda L. var. vulgaris Turcz. ＝ Orthilia secunda（L.）House ■

322909　Pyrola shanxiensis Y. L. Chou et R. C. Zhou;山西鹿蹄草;Shanxi Pyrola ●

322910　Pyrola sikkimensis Krisa;锡金鹿蹄草;Sikkim Pyrola ●

322911　Pyrola sikkimensis Krisa ＝ Pyrola sororia Andres ●

322912　Pyrola soldanellifolia Andres ＝ Pyrola renifolia Maxim. ●

322913　Pyrola solunica S. D. Zhao ＝ Pyrola chlorantha Sw. ●

322914　Pyrola sororia Andres;珍珠鹿蹄草（群生鹿蹄草）;Pearl Pyrola ●

322915　Pyrola subaphylla（Maxim.）Ohwi;鳞叶鹿蹄草;Scaleleaf Pyrola ●

322916　Pyrola subaphylla Maxim. ＝ Pyrola japonica Klenze ex Alef. f. subaphylla（Maxim.）Ohwi ●

322917　Pyrola subaphylla Maxim. ＝ Pyrola japonica Klenze ex Alef. ●

322918　Pyrola subaphylla Maxim. ＝ Pyrola subaphylla（Maxim.）Ohwi ●

322919　Pyrola szechuanica Andres;四川鹿蹄草;Sichuan Pyrola ●

322920　Pyrola tibetana Andres ＝ Pyrola calliantha Andréws var. tibetana（Andres）Y. L. Chou ●

322921　Pyrola tschanbaischanica Y. L. Chou et Y. L. Chang;长白鹿蹄草（长白山鹿蹄草）;Changbai Pyrola, Changbaishan Pyrola ●

322922　Pyrola uliginosa Torr. et A. Gray;北美鹿蹄草;Bog Wintergreen ●☆

322923　Pyrola uliginosa Torr. et A. Gray ex Torr. ＝ Pyrola asarifolia Michx. ●☆

322924　Pyrola uliginosa Torr. et A. Gray ex Torr. var. gracilis Jenn. ＝ Pyrola asarifolia Michx. ●☆

322925　Pyrola umbellata L. ＝ Chimaphila umbellata（L.）W. P. C. Barton ■

322926　Pyrola uniflora L. ＝ Moneses uniflora（L.）A. Gray ■

322927　Pyrola virens Schweigg. ＝ Pyrola chlorantha Sw. ●

322928　Pyrola virens Schweigg. f. paucifolia（Fernald）Fernald ＝ Pyrola chlorantha Sw. ●

322929　Pyrola virens Schweigg. var. convoluta（W. P. C. Barton）Fernald ＝ Pyrola chlorantha Sw. ●

322930　Pyrola virens Schweigg. var. saximontana（Fernald）Fernald ＝ Pyrola chlorantha Sw. ●

322931　Pyrola virescens N. Busch ＝ Pyrola chlorantha Sw. ●

322932　Pyrola xinjiangensis Y. L. Chou et R. C. Zhou;新疆鹿蹄草;Xinjiang Pyrola ●

322933　Pyrolaceae Dumort. ＝ Ericaceae Juss.（保留科名）●

322934　Pyrolaceae Dumort. ＝ Pyrolaceae Lindl.（保留科名）●■

322935　Pyrolaceae Lindl.（1829）（保留科名）;鹿蹄草科;Pyrola Family, Wintergreen Family ●■

322936　Pyrolirion Herb.（1821）;火石蒜属■☆

322937　Pyrolirion albicans Herb.;白火石蒜■☆

322938　Pyrolirion aureum Herb.;黄火石蒜■☆

322939　Pyrolirion boliviense（Baker）Sealy;玻利维亚火石蒜■☆

322940　Pyrolirion tubiflorum M. Roem.;管花火石蒜■☆

322941　Pyrophorum DC. ＝ Pirophorum Neck. ●

322942　Pyrophorum DC. ＝ Pyrus L. ●

322943　Pyropsis Hort. ex Fisch. Mey. et Avé-Lall. ＝ Madia Molina ●☆

322944　Pyrorchis D. L. Jones et M. A. Clem.（1995）;澳火兰属■☆

322945　Pyrorchis D. L. Jones et M. A. Clem. ＝ Lyperanthus R. Br. ●☆

322946　Pyrospermum Miq. ＝ Bhesa Buch. -Ham. ex Arn. ●

322947　Pyrospermum Miq. ＝ Kurrimia Wall. ex Thwaites ●

322948　Pyrostegia C. Presl(1845);炮仗藤属(火把果属,炮仗花属);Crackflower, Pyrostegia ●

322949　Pyrostegia ignea（Vell.）C. Presl ＝ Pyrostegia venusta（Ker Gawl.）Miers ●

322950　Pyrostegia venusta（Ker Gawl.）Miers;炮仗藤(黄花炮仗藤,黄鳝藤,炮仗花);Beautiful Pyrostegia, Chinese Cracker Flower, Crackflower, Flame Flower, Flame Vine, Flamecoloured Pyrostegia, Flamevine, Golden Shower, Hame Flower, Hame Vine, Orange Trumpet Vine ●

322951　Pyrostoma G. Mey. ＝ Vitex L. ●

322952　Pyrostria Comm. ex Juss.（1789）;火畦茜属●☆

322953　Pyrostria Roxb. ＝ Timonius DC.（保留属名）●

322954　Pyrostria affinis（Robyns）Bridson;近缘火畦茜●☆

322955 Pyrostria analamazaotrensis Arènes ex Cavaco;阿纳拉马火畔茜●☆

322956 Pyrostria angustifolia (A. Rich. ex DC.) Cavaco;窄叶火畔茜●☆

322957 Pyrostria ankazobeensis Arènes ex Cavaco;阿卡祖贝火畔茜●☆

322958 Pyrostria bibracteata (Baker) Cavaco;双小苞火畔茜●☆

322959 Pyrostria bispathacea (Mildbr.) Bridson;双苞火畔茜●☆

322960 Pyrostria buxifolia Hochr.;黄杨叶火畔茜●☆

322961 Pyrostria chapmanii Bridson;查普曼火畔茜●☆

322962 Pyrostria hystrix (Bremek.) Bridson;豪猪火畔茜●☆

322963 Pyrostria lobulata Bridson;小裂片火畔茜●☆

322964 Pyrostria madagascariensis Lecomte;马岛火畔茜●☆

322965 Pyrostria major (A. Rich. ex DC.) Cavaco;大火畔茜●☆

322966 Pyrostria media (A. Rich. ex DC.) Cavaco;中间火畔茜●☆

322967 Pyrostria pendula Lantz,Klack. et Razafim.;下垂火畔茜●☆

322968 Pyrostria phyllanthoidea (Baill.) Bridson;叶花火畔茜●☆

322969 Pyrostria pseudocommersonii Cavaco;假科梅逊火畔茜●☆

322970 Pyrostria serpentina Lantz,Klack. et Razafim.;蛇形火畔茜●☆

322971 Pyrostria syringifolia Hochr.;丁香叶火畔茜●☆

322972 Pyrostria urschii Arènes ex Cavaco;乌尔施火畔茜●☆

322973 Pyrostria uzungwaensis Bridson;乌尊季沃火畔茜●☆

322974 Pyrostria variistipula Arènes ex Cavaco;杂托叶火畔茜●☆

322975 Pyrostria verdcourtii (Cavaco) Razafim,Lantz et B. Bremer = Neoleroya verdcourtii Cavaco ●☆

322976 Pyrotheca Steud. = Gyrotheca Salisb. ■☆

322977 Pyrotheca Steud. = Lachnanthes Elliott(保留属名)■☆

322978 Pyrrhanthera Zotov(1963);侏儒草属(小侏儒草属)■☆

322979 Pyrrhanthera exigua (Kirk) Zotov;侏儒草■☆

322980 Pyrrhanthus Jack = Lumnitzera Willd. ●

322981 Pyrrhanthus littoreus Jack = Lumnitzera littorea (Jacq.) Voigt ●◇

322982 Pyrrheima Hassk. = Siderasis Raf. ■☆

322983 Pyrrheima loddigesii Hassk. var. reginae (L. Linden et Rodigas) Bonstedt = Dichorisandra reginae (L. Linden et Rodigas) W. Mill. ■☆

322984 Pyrrhocactus (A. Berger) Backeb. = Neoporteria Britton et Rose ●■

322985 Pyrrhocactus (A. Berger) Backeb. et F. M. Knuth = Neoporteria Britton et Rose ●■

322986 Pyrrhocactus A. Berger(1929);焰刺球属■☆

322987 Pyrrhocactus Backeb. = Pyrrhocactus A. Berger ■☆

322988 Pyrrhocactus Backeb. et F. M. Knuth = Neoporteria Britton et Rose ●■

322989 Pyrrhocactus catamarcensis (F. A. C. Weber) Backeb.;铁心球(铁心丸)■☆

322990 Pyrrhocactus dubius Backeb.;怪魔玉■☆

322991 Pyrrhocactus strausianus (K. Schum.) A. Berger;吼熊球(吼熊丸)■☆

322992 Pyrrhocactus umadeave (Werderm.) Backeb.;寒鬼玉■☆

322993 Pyrrhocoma Walp. = Haplopappus Cass. (保留属名)●☆

322994 Pyrrhocoma Walp. = Pyrrocoma Hook. ■☆

322995 Pyrrhopappus A. Rich. = Lactuca L. ■

322996 Pyrrhopappus DC. (1838)(保留属名);火红苣属■☆

322997 Pyrrhopappus carolinianus (Walter) DC.;加州火红苣;False Dandelion ■☆

322998 Pyrrhopappus carolinianus (Walter) DC. var. georgianus (Shinners) H. E. Ahles = Pyrrhopappus carolinianus (Walter) DC. ■☆

322999 Pyrrhopappus geiseri Shinners = Pyrrhopappus pauciflorus (D. Don) DC. ■☆

323000 Pyrrhopappus georgianus Shinners = Pyrrhopappus carolinianus (Walter) DC. ■☆

323001 Pyrrhopappus grandiflorus (Nutt.) Nutt.;大花火红苣■☆

323002 Pyrrhopappus hochstetteri A. Rich. = Lactuca inermis Forssk. ■☆

323003 Pyrrhopappus humilis A. Rich. = Lactuca inermis Forssk. ■☆

323004 Pyrrhopappus multicaulis DC. = Pyrrhopappus pauciflorus (D. Don) DC. ■☆

323005 Pyrrhopappus multicaulis DC. var. geiseri (Shinners) North. = Pyrrhopappus pauciflorus (D. Don) DC. ■☆

323006 Pyrrhopappus pauciflorus (D. Don) DC.;疏花火红苣■☆

323007 Pyrrhopappus rothrockii A. Gray;罗思罗克火红苣■☆

323008 Pyrrhosa (Blume) Eudl. = Horsfieldia Willd. ●

323009 Pyrrhosa Eudl. = Horsfieldia Willd. ●

323010 Pyrrhotrichia Wight et Arn. = Eriosema (DC.) Desv. (保留属名)●■

323011 Pyrrocoma Hook. (1833);红毛菀属;Goldenweed ■☆

323012 Pyrrocoma Hook. = Haplopappus Cass. (保留属名)●●☆

323013 Pyrrocoma apargioides (A. Gray) Greene;高山红毛菀;Alpine-flames ■☆

323014 Pyrrocoma carthamoides Hook.;大花红毛菀;Large-flower Goldenweed ■☆

323015 Pyrrocoma carthamoides Hook. var. cusickii (A. Gray) Kartesz et Gandhi;库西克红毛菀■☆

323016 Pyrrocoma carthamoides Hook. var. subsquarrosa (Greene) G. K. Br. et D. J. Keil;糠秕红毛菀■☆

323017 Pyrrocoma clementis Rydb.;卷须红毛菀;Tranquil Goldenweed ■☆

323018 Pyrrocoma clementis Rydb. var. villosa (Rydb.) Mayes ex G. K. Br. et D. J. Keil;长柔毛红毛菀■☆

323019 Pyrrocoma crocea (A. Gray) Greene;弯头红毛菀;Curly-head Goldenweed ■☆

323020 Pyrrocoma crocea (A. Gray) Greene var. genuflexa (Greene) Mayes ex G. K. Br. et D. J. Keil;膝曲红毛菀■☆

323021 Pyrrocoma demissa Greene = Pyrrocoma apargioides (A. Gray) Greene ■☆

323022 Pyrrocoma foliosa A. Gray = Oonopsis foliosa (A. Gray) Greene ■☆

323023 Pyrrocoma genuflexa Greene = Pyrrocoma crocea (A. Gray) Greene var. genuflexa (Greene) Mayes ex G. K. Br. et D. J. Keil ■☆

323024 Pyrrocoma gossypina Greene = Pyrrocoma uniflora (Hook.) Greene var. gossypina (Greene) Kartesz et Gandhi ■☆

323025 Pyrrocoma grindelioides DC. = Hazardia squarrosa (Hook. et Arn.) Greene var. grindelioides (DC.) W. D. Clark ●☆

323026 Pyrrocoma hirta (A. Gray) Greene;粗毛红毛菀;Tacky Goldenweed ■☆

323027 Pyrrocoma hirta (A. Gray) Greene var. lanulosa (Greene) Mayes ex G. K. Br. et D. J. Keil;绵毛红毛菀■☆

323028 Pyrrocoma hirta (A. Gray) Greene var. sonchifolia (Greene) Kartesz et Gandhi;苦苣菜叶红毛菀■☆

323029 Pyrrocoma howellii (A. Gray) Greene = Pyrrocoma uniflora (Hook.) Greene ■☆

323030 Pyrrocoma insecticruris (L. F. Hend.) A. Heller;蚂蚱腿红毛菀;Bug-leg Goldenweed ■☆

323031 Pyrrocoma integrifolia (Porter ex A. Gray) Greene;全缘叶红毛菀;Smooth Goldenweed ■☆

323032 Pyrrocoma lanceolata (Hook.) Greene;剑叶红毛菀;Lance-leaf Goldenweed ■☆

323033 Pyrrocoma lanceolata (Hook.) Greene var. subviscosa (Greene) Mayes ex G. K. Br. et D. J. Keil;亚黏红毛菀■☆

323034 Pyrrocoma lanulosa Greene = Pyrrocoma hirta (A. Gray) Greene var. lanulosa (Greene) Mayes ex G. K. Br. et D. J. Keil ■☆

323035 Pyrrocoma liatriformis Greene;蛇鞭菊红毛菀;Palouse

Goldenweed ■☆

323036　Pyrrocoma linearis（D. D. Keck）Kartesz et Gandhi；湿地红毛菀；Marsh Goldenweed ■☆

323037　Pyrrocoma lucida（D. D. Keck）Kartesz et Gandhi；亮红毛菀；Sticky Goldenweed ■☆

323038　Pyrrocoma menziesii Hook. et Arn. = Isocoma menziesii（Hook. et Arn.）G. L. Nesom ■☆

323039　Pyrrocoma racemosa（Nutt.）Torr. et A. Gray；丛生红毛菀；Clustered Goldenweed ■☆

323040　Pyrrocoma racemosa（Nutt.）Torr. et A. Gray subsp. duriuscula（Greene）H. M. Hall = Pyrrocoma racemosa（Nutt.）Torr. et A. Gray var. paniculata（Nutt.）Kartesz et Gandhi ■☆

323041　Pyrrocoma racemosa（Nutt.）Torr. et A. Gray subsp. glomeratus（Nutt.）H. M. Hall = Pyrrocoma racemosa（Nutt.）Torr. et A. Gray var. paniculata（Nutt.）Kartesz et Gandhi ■☆

323042　Pyrrocoma racemosa（Nutt.）Torr. et A. Gray subsp. halophilus（Greene）H. M. Hall = Pyrrocoma racemosa（Nutt.）Torr. et A. Gray var. paniculata（Nutt.）Kartesz et Gandhi ■☆

323043　Pyrrocoma racemosa（Nutt.）Torr. et A. Gray subsp. prionophyllus（Greene）H. M. Hall = Pyrrocoma racemosa（Nutt.）Torr. et A. Gray var. paniculata（Nutt.）Kartesz et Gandhi ■☆

323044　Pyrrocoma racemosa（Nutt.）Torr. et A. Gray var. congesta（Greene）Mayes ex G. K. Br. et D. J. Keil；红毛菀■☆

323045　Pyrrocoma racemosa（Nutt.）Torr. et A. Gray var. glomerellus A. Gray = Pyrrocoma racemosa（Nutt.）Torr. et A. Gray var. paniculata（Nutt.）Kartesz et Gandhi ■☆

323046　Pyrrocoma racemosa（Nutt.）Torr. et A. Gray var. paniculata（Nutt.）Kartesz et Gandhi；圆锥红毛菀■☆

323047　Pyrrocoma racemosa（Nutt.）Torr. et A. Gray var. pinetorum（D. D. Keck）Kartesz et Gandhi；松林红毛菀■☆

323048　Pyrrocoma racemosa（Nutt.）Torr. et A. Gray var. prionophyllus（Greene）S. L. Welsh = Pyrrocoma racemosa（Nutt.）Torr. et A. Gray var. paniculata（Nutt.）Kartesz et Gandhi ■☆

323049　Pyrrocoma racemosa（Nutt.）Torr. et A. Gray var. sessiliflora（Greene）Mayes ex G. K. Br. et D. J. Keil；无梗丛生红毛菀■☆

323050　Pyrrocoma radiata Nutt.；斯内克红毛菀；Snake River Goldenweed ■☆

323051　Pyrrocoma sessiliflora Greene = Pyrrocoma racemosa（Nutt.）Torr. et A. Gray var. sessiliflora（Greene）Mayes ex G. K. Br. et D. J. Keil ■☆

323052　Pyrrocoma sonchifolia Greene = Pyrrocoma hirta（A. Gray）Greene var. sonchifolia（Greene）Kartesz et Gandhi ■☆

323053　Pyrrocoma subsquarrosa Greene = Pyrrocoma carthamoides Hook. var. subsquarrosa（Greene）G. K. Br. et D. J. Keil ■☆

323054　Pyrrocoma subviscosa Greene = Pyrrocoma lanceolata（Hook.）Greene var. subviscosa（Greene）Mayes ex G. K. Br. et D. J. Keil ■☆

323055　Pyrrocoma uniflora（Hook.）Greene；单花红毛菀；Plantain Goldenweed ■☆

323056　Pyrrocoma uniflora（Hook.）Greene var. gossypina（Greene）Kartesz et Gandhi；棉花红毛菀■☆

323057　Pyrrocoma villosa Rydb. = Pyrrocoma clementis Rydb. var. villosa（Rydb.）Mayes ex G. K. Br. et D. J. Keil ■☆

323058　Pyrrorhiza Maguire et Wurdack（1957）；红根血草属■☆

323059　Pyrrorhiza neblinae Maguire et Wurdack；红根血草■☆

323060　Pyrrothrix Bremek.（1944）；红毛蓝属■

323061　Pyrrothrix Bremek. = Strobilanthes Blume ●■

323062　Pyrrothrix heterochrous（Hand.-Mazz.）C. Y. Wu et C. C. Hu；异色红毛蓝（异色紫云菜）；Discolor Conehead ■

323063　Pyrrothrix hossei（C. B. Clarke）C. Y. Wu et C. C. Hu；泰北红毛蓝■

323064　Pyrrothrix rufohirtus（C. B. Clarke）C. Y. Wu et C. C. Hu；红毛蓝（红毛紫云菜）；Conehead ■

323065　Pyrsonota Ridl. = Sericolea Schltr. ●☆

323066　Pyrularia Michx.（1803）；檀梨属（冠梨属，油葫芦属）；Oilnut，Oil-nut ●

323067　Pyrularia bullata P. C. Tam；泡叶檀梨；Bladdery Oil-nut，Bullate Oilnut ●

323068　Pyrularia bullata P. C. Tam = Pyrularia edulis（Wall.）A. DC. ●

323069　Pyrularia edulis（Wall.）A. DC.；檀梨（冠梨，鹿子果，油葫芦）；Edible Oilnut，Edible Oil-nut ●

323070　Pyrularia inermis S. S. Chien；无刺檀梨（四川檀梨）；Spineless Oilnut，Spineless Oil-nut ●

323071　Pyrularia inermis S. S. Chien = Pyrularia edulis（Wall.）A. DC. ●

323072　Pyrularia pubera Michx.；美洲檀梨（北美檀梨）；American Oil-nut，Buffalo Nut，Elk Nut ●

323073　Pyrularia sinensis Y. C. Wu；华檀梨（中华檀梨）；China Oilnut，Chinese Oilnut，Chinese Oil-nut ●

323074　Pyrularia sinensis Y. C. Wu = Pyrularia edulis（Wall.）A. DC. ●

323075　Pyrularia zeylanica A. DC. = Scleropyrum wallichianum（Wight et Arn.）Arn. ●

323076　Pyrus L.（1753）；梨属（棠梨属）；Pear ●

323077　Pyrus alnifolia（Siebold et Zucc.）Franch. et Sav. = Sorbus alnifolia（Siebold et Zucc.）K. Koch ●

323078　Pyrus alnifolia Lindl. = Aria alnifolia（Siebold et Zucc.）Decne. ●

323079　Pyrus alnifolia Lindl. = Sorbus alnifolia（Siebold et Zucc.）K. Koch ●

323080　Pyrus americana（Marshall）DC. = Sorbus americana Marshall ●☆

323081　Pyrus americana（Marshall）DC. var. decora Sarg. = Sorbus decora（Sarg.）C. K. Schneid. ●

323082　Pyrus americana（Marshall）DC. var. microcarpa（Pursh）Torr. et A. Gray = Sorbus americana Marshall ●☆

323083　Pyrus americana（Marshall）Spreng.；美洲梨；American Mountain-ash，Dogwood，Missy-moosey，Mountain Ash，Roundwood，Witchwood ●☆

323084　Pyrus americana（Marshall）Spreng. var. microcarpa（Pursh）Torr. et A. Gray；小果美洲梨●☆

323085　Pyrus amoena Koidz.；涩梨●☆

323086　Pyrus amygdaliformis Vill.；扁桃梨（杏梨）；Almond Pear，Almond-leafed Pear，Almond-shaped Pear Free，Almond-shaped Pear-tree ●☆

323087　Pyrus angustifolia Aiton = Malus angustifolia（Aiton）Michx. ●☆

323088　Pyrus angustifolia Aiton var. spinosa（Rehder）L. H. Bailey = Malus angustifolia（Aiton）Michx. ●☆

323089　Pyrus anthyllidifolia Sm. = Osteomeles anthyllidifolia（Sm.）Lindl. ●

323090　Pyrus arbutifolia（L. f.）Pers.；乔鹃梨；Red Chokeberry ●☆

323091　Pyrus arbutifolia（L. f.）Pers. var. atropurpurea（Britton）B. L. Rob. = Aronia prunifolia（Marshall）Rehder ●☆

323092　Pyrus arbutifolia（L. f.）Pers. var. nigra Willd. = Aronia melanocarpa（Michx.）Elliott ●☆

323093　Pyrus armeniacifolia Te T. Yu；杏叶梨（野梨）；Apricotleaf Pear，Apricot-leaved Pear ●

323094　Pyrus aromatica Kikuchi;岩手山梨●☆

323095　Pyrus asiae-mediae Popov;中亚梨●☆

323096　Pyrus astateria Cardot ＝ Sorbus astateria（Cardot）Hand.-Mazz. ●

323097　Pyrus atropurpurea（Britton）L. H. Bailey;黑紫梨;Purple Crab ●☆

323098　Pyrus atrovirens Hort. ex K. Koch;墨绿紫梨●☆

323099　Pyrus aucuparia（L.）Gaertn. ＝ Sorbus aucuparia L. ●

323100　Pyrus aucuparia（L.）Gaertn. var. randaiensis Hayata ＝ Sorbus randaiensis（Hayata）Koidz. ●

323101　Pyrus aucuparia（L.）Gaertn. var. trilocularis Hayata ＝ Sorbus randaiensis（Hayata）Koidz. ●

323102　Pyrus aucuparia Ehrh. ＝ Sorbus aucuparia L. ●

323103　Pyrus aucuparia L. ＝ Sorbus pohuashanensis（Hance）Hedl. ●

323104　Pyrus baccata L. ＝ Malus baccata（L.）Borkh. ●

323105　Pyrus baccata L. var. himalaica Maxim. ＝ Malus rockii Rehder ●

323106　Pyrus baccata L. var. mandshurica Maxim. ＝ Malus mandshurica（Maxim.）Kom. ex Juz. ●

323107　Pyrus balansae Decne. ;巴兰萨梨●☆

323108　Pyrus barbulata Michx. ;沙盖花;Flowering Moss ●

323109　Pyrus bartramiana Tausch ＝ Amelanchier bartramiana（Tausch）M. Roem. ●☆

323110　Pyrus betulifolia Bunge;杜梨(白棠,北支豆梨,赤棠,杜,杜棠,甘棠,海棠梨,灰梨,满洲豆梨,唐豆梨,棠,棠梨,土梨,野梨,野梨子);Birch-leaf,Birchleaf Pear,Birch-leaf Pear,Birch-leaved Pear ●

323111　Pyrus bodinieri H. Lév. ＝ Decaspermum parviflorum（Lam.）A. J. Scott ●

323112　Pyrus boissieriana Buhse;布瓦西耶梨●☆

323113　Pyrus bourgaeana Decne. ;布尔加梨●☆

323114　Pyrus bretschneideri Rehder;白梨(白挂梨,白罐梨,白桂梨,罐梨,果宗,快果,蜜父,玉乳);Bretschneider Pear,White Pear ●

323115　Pyrus bucharica Litv. ;布哈尔梨●☆

323116　Pyrus calleryana Decne. ;豆梨(车头梨,赤罗,杜梨,梨丁子,梨宁子,鹿梨,罗,鸟梨,山梨,鼠梨,树梨,酸梨,檖,糖梨,阳檖,杨檖,野梨);Bean Pear,Bradford Pear,Callery Pear,Flowering Pear ●

323117　Pyrus calleryana Decne. 'Bradford';布拉德福豆梨●☆

323118　Pyrus calleryana Decne. 'Chanticleer';雄鸡豆梨;Columnar Ornament Pear ●☆

323119　Pyrus calleryana Decne. f. tomentolla Rehder;绒毛豆梨(毛豆梨,绒毛柳叶豆梨);Tomentose Callery Pear ●

323120　Pyrus calleryana Decne. var. calleryana f. tomentella Rehder ＝ Pyrus calleryana Decne. ●

323121　Pyrus calleryana Decne. var. dimorphophylla（Makino）Koidz. ＝ Pyrus calleryana Decne. ●

323122　Pyrus calleryana Decne. var. integrifolia Te T. Yu;全缘叶豆梨;Entire Callery Pear ●

323123　Pyrus calleryana Decne. var. koehnei（C. K. Schneid.）Te T. Yu;楔叶豆梨;Koehne Callery Pear ●

323124　Pyrus calleryana Decne. var. lanceolata Rehder;柳叶豆梨(披针叶豆梨);Lanceolate Callery Pear ●

323125　Pyrus caloneura（Stapf）Bean ＝ Sorbus caloneura（Stapf）Rehder ●

323126　Pyrus caloneura Bean ＝ Sorbus caloneura（Stapf）Rehder ●

323127　Pyrus candidissima Chev. ＝ Sorbus granulosa（Bertol.）Rehder ●

323128　Pyrus cathayensis Hemsl. ＝ Chaenomeles sinensis（Thouin）Koehne ●

323129　Pyrus caucasica Fed. ;高加索梨●☆

323130　Pyrus cavaleriei H. Lév. ＝ Stranvaesia davidiana Decne. ●

323131　Pyrus chinensis Spreng. ＝ Chaenomeles sinensis（Thouin）Koehne ●

323132　Pyrus chinensis Spreng. ＝ Pseudocydonia sinensis C. K. Schneid. ●

323133　Pyrus communis L. ;西洋梨(梨,沙梨,洋梨);Callery Pear,Choke Pear,Common Pear,European Pear,Garden Pear,Occidental Pear,Pear,Pyrrie,West Pear,Wild Pear ●

323134　Pyrus communis L. 'Bartlett' ＝ Pyrus communis L. 'Williams' Bon Chretien' ●☆

323135　Pyrus communis L. 'Beech Hill';红叶西洋梨●☆

323136　Pyrus communis L. 'Beurre d'Anjou';布勒德阿娇西洋梨●☆

323137　Pyrus communis L. 'Cascade';瀑布西洋梨●☆

323138　Pyrus communis L. 'Clapp's Favorite' ＝ Pyrus communis L. 'Clapp's Liebling' ●☆

323139　Pyrus communis L. 'Clapp's Liebling';科莱普列波灵西洋梨(科莱普的最爱西洋梨)●☆

323140　Pyrus communis L. 'Comice' ＝ Pyrus communis L. 'Doyenne du Comice' ●☆

323141　Pyrus communis L. 'Conference';集会西洋梨●☆

323142　Pyrus communis L. 'Doyenne du Comice';多尼世纪西洋梨●☆

323143　Pyrus communis L. 'Gellerts Butterbine';布特比伦及勒兹西洋梨●☆

323144　Pyrus communis L. 'Red Bartlett';红巴特利特西洋梨●☆

323145　Pyrus communis L. 'Williams' Bon Chretien';威廉斯伯克雷廷西洋梨●☆

323146　Pyrus communis L. subsp. gharbiana（Trab.）Maire ＝ Pyrus cordata Desv. ●☆

323147　Pyrus communis L. subsp. longipes（Coss. et Durieu）Maire ＝ Pyrus cordata Desv. ●☆

323148　Pyrus communis L. subsp. mamorensis（Trab.）Maire ＝ Pyrus bourgaeana Decne. ●☆

323149　Pyrus communis L. subsp. sativa（Lam. et DC.）Asch. et Graebn. ＝ Pyrus communis L. ●

323150　Pyrus communis L. var. brevipes Emb. et Maire ＝ Pyrus communis L. ●

323151　Pyrus communis L. var. sativa（DC.）DC. ;洋梨(西洋梨,栽培西洋梨)●

323152　Pyrus cordata Desv. ;心叶梨;Plymouth Pear ●☆

323153　Pyrus coronaria L. ＝ Malus coronaria（L.）Mill. ●☆

323154　Pyrus coronaria L. var. lancifolia（Rehder）Fernald ＝ Malus coronaria（L.）Mill. ●☆

323155　Pyrus coronata Cardot ＝ Sorbus coronata（Cardot）Te T. Yu et H. T. Tsai ●

323156　Pyrus corymbifera Nakai;和尚梨●

323157　Pyrus cossonii Rehder;长柄梨●☆

323158　Pyrus crenata Lindl. ＝ Sorbus cuspidata（Spach）Hedl. ●

323159　Pyrus cydonia L. ＝ Cydonia oblonga Mill. ●

323160　Pyrus davidiana Decne. var. formosana（Cardot）H. Ohashi et H. Iketami ＝ Stranvaesia davidiana Decne. ●

323161　Pyrus decora（Sarg.）Hyl. ＝ Sorbus decora（Sarg.）C. K. Schneid. ●

323162　Pyrus delavayi Franch. ＝ Docynia delavayi（Franch.）C. K. Schneid. ●

323163　Pyrus discolor Maxim. ＝ Sorbus discolor（Maxim.）Maxim. ●

323164　Pyrus doumeri Bois ＝ Malus doumeri（Bois）A. Chev. ●

323165　Pyrus elaeagnifolia Pall. ;胡颓子叶梨(野橄榄梨);Elaeagnus

Pear, Oleaster Pear, Pleaster Pear ●☆

323166　Pyrus esquirolii H. Lév. = Malus sieboldii（Regel）Rehder ●

323167　Pyrus feddei H. Lév. = Stranvaesia amphidoxa C. K. Schneid. ●

323168　Pyrus ferruginea（Wenz.）Hook. f. = Sorbus ferruginea（Wenz.）Rehder ●

323169　Pyrus ferruginea Hook. f. = Sorbus ferruginea（Wenz.）Rehder ●

323170　Pyrus floribunda（Siebold）Kirchn.；繁花梨；Purple Chokeberry ●☆

323171　Pyrus floribunda（Siebold）Kirchn. = Malus floribunda Siebold ex Van Houtte ●☆

323172　Pyrus floribunda Lindl. = Aronia prunifolia（Marshall）Rehder ●☆

323173　Pyrus folgneri（C. K. Schneid.）C. K. Schneid. ex Bean = Sorbus folgneri（C. K. Schneid.）Rehder ●

323174　Pyrus folgneri Bean = Sorbus folgneri（C. K. Schneid.）Rehder ●

323175　Pyrus foliolosa Wall. = Sorbus foliolosa（Wall.）Spach ●

323176　Pyrus foliolosa Wall. var. ambigua Cardot = Sorbus foliolosa（Wall.）Spach ●

323177　Pyrus foliolosa Wall. var. subglabra Cardot = Sorbus poteriifolia Hand. -Mazz. ●

323178　Pyrus formosana Kawak. et Koidz. ex Hayata = Malus doumeri（Bois）A. Chev. ●

323179　Pyrus gharbiana Trab. = Pyrus cordata Desv. ●☆

323180　Pyrus glabrescens Cardot = Sorbus oligodonta（Cardot）Hand. -Mazz. ●

323181　Pyrus glomerulata（Koehne）Bean = Sorbus glomerulata Koehne ●

323182　Pyrus glomerulata Bean = Sorbus glomerulata Koehne ●

323183　Pyrus gramulosa Bertol. = Sorbus granulosa（Bertol.）Rehder ●

323184　Pyrus granulosa Bertol. = Sorbus corymbifera（Miq.）T. H. Nguyên et Yakovlev ●

323185　Pyrus grossheimii Fed. ；格罗梨 ●☆

323186　Pyrus halliana（Koehne）Voss = Malus halliana Koehne ●

323187　Pyrus halliana Voss = Malus halliana Koehne ●

323188　Pyrus harrowiana Balf. f. et W. W. Sm. = Sorbus harrowiana（Balf. f. et W. W. Sm.）Rehder ●

323189　Pyrus harrowiana Balf. f. et W. W. Sm. = Sorbus insignis（Hook. f.）Hedl. ●

323190　Pyrus hedlundii Lacaita = Sorbus hedlundii C. K. Schneid. ●☆

323191　Pyrus hepehensis Bean = Sorbus hupehensis C. K. Schneid. ●

323192　Pyrus heterophylla Regel et Schmalh. = Pyrus regelii Rehder ●☆

323193　Pyrus hondoensis Nakai et Kikuchi = Pyrus ussuriensis Maxim. var. hondoensis（Nakai et Kikuchi）Rehder ●☆

323194　Pyrus hopeiensis Te T. Yu；河北梨（黑丁子，红丁子，红杜梨）；Hebei Pear, Hopei Pear ●

323195　Pyrus hopeiensis Te T. Yu var. peninsula D. K. Zang et W. D. Peng；半岛梨；Peninsula Hebei Pear ●

323196　Pyrus hopeiensis Te T. Yu var. peninsula D. K. Zang et W. D. Peng = Pyrus hopeiensis Te T. Yu ●

323197　Pyrus hupehensis（C. K. Schneid.）Bean = Sorbus hupehensis C. K. Schneid. ●

323198　Pyrus hupehensis Pamp. = Malus hupehensis（Pamp.）Rehder ●

323199　Pyrus hypoglauca Cardot = Sorbus rehderiana Koehne ●

323200　Pyrus indica Wall. = Docynia indica（Wall.）Decne. ●

323201　Pyrus insignis Hook. f. = Sorbus insignis（Hook. f.）Hedl. ●

323202　Pyrus insueta Koidz. ；异常梨 ●☆

323203　Pyrus ioensis（A. W. Wood）L. H. Bailey = Malus ioensis（A. W. Wood）Britton ●☆

323204　Pyrus japonica Thunb. = Chaenomeles japonica（Thunb.）

Lindl. ex Spach ●

323205　Pyrus kansuensis Batalin = Malus kansuensis（Batalin）C. K. Schneid. ●

323206　Pyrus karensium Kurz = Sorbus granulosa（Bertol.）Rehder ●

323207　Pyrus kawakamii Hayata；台湾梨（台湾野梨）；Evergreen Pear, Taiwan Pear ●

323208　Pyrus kawakamii Hayata = Pyrus calleryana Decne. ●

323209　Pyrus keissleri H. Lév. = Sorbus keissleri（C. K. Schneid.）Rehder ●

323210　Pyrus kikuchii Nakai；菊池梨 ●☆

323211　Pyrus koehneana（C. K. Schneid.）Cardot = Sorbus koehneana C. K. Schneid. ●

323212　Pyrus koehnei C. K. Schneid. = Pyrus calleryana Decne. var. koehnei（C. K. Schneid.）Te T. Yu ●

323213　Pyrus kolupana C. K. Schneid. ；陕西木梨 ●

323214　Pyrus korshinskyi Litv. ；考尔梨 ●☆

323215　Pyrus laosensis Cardot = Malus doumeri（Bois）A. Chev. ●

323216　Pyrus lasyogyna Koidz. ；三郎梨 ●☆

323217　Pyrus lindleyi Rehder；岭南梨（鸟梨）；Lindley. Pear ●

323218　Pyrus longipes Coss. et Durieu = Pyrus cordata Desv. ●☆

323219　Pyrus malus Aiton = Malus prunifolia（Willd.）Borkh. ●

323220　Pyrus malus L. = Malus pumila Mill. ●

323221　Pyrus malus L. = Malus sylvestris Mill. ●☆

323222　Pyrus malus L. var. pumila Henry = Malus pumila Mill. ●

323223　Pyrus mamorensis Trab. ；摩洛哥梨；Mamora Pear ●☆

323224　Pyrus mamorensis Trab. = Pyrus bourgaeana Decne. ●☆

323225　Pyrus mamorensis Trab. var. brevipes Emb. et Maire = Pyrus bourgaeana Decne. ●☆

323226　Pyrus matsumurae（Koidz.）Cardot = Malus asiatica Nakai ●

323227　Pyrus matsumurae Cardot = Malus asiatica Nakai ●

323228　Pyrus maulei Mast. = Chaenomeles japonica（Thunb.）Lindl. ex Spach ●

323229　Pyrus megalocarpa（Rehder）Bean = Sorbus megalocarpa Rehder ●

323230　Pyrus megalocarpa Bean = Sorbus megalocarpa Rehder ●

323231　Pyrus melanocarpa（Michx.）Willd. = Aronia melanocarpa（Michx.）Elliott ●☆

323232　Pyrus meliosmifolia（Rehder）Bean = Sorbus meliosmifolia Rehder ●

323233　Pyrus meliosmifolia Bean = Sorbus meliosmifolia Rehder ●

323234　Pyrus melliana Hand. -Mazz. = Malus doumeri（Bois）A. Chev. ●

323235　Pyrus mesogea Cardot = Sorbus hupehensis C. K. Schneid. ●

323236　Pyrus microcarpa（Pursh）DC. = Sorbus americana Marshall ●☆

323237　Pyrus micromalus（Makino）Makino = Malus micromalus Makino ●

323238　Pyrus micromalus Bailey = Malus micromalus Makino ●

323239　Pyrus microphylla Wall. ex Hook. f. = Sorbus microphylla Wenz. ●

323240　Pyrus mikado Koidz. ；黄涩梨 ●☆

323241　Pyrus miyabei Sarg. = Sorbus alnifolia（Siebold et Zucc.）K. Koch ●

323242　Pyrus monbeigii Cardot = Sorbus monbeigii（Cardot）Te T. Yu ●

323243　Pyrus montana Poit. et Turpin = Pyrus pyrifolia（Burm. f.）Nakai ●

323244　Pyrus nepalensis Decne. = Pyrus pashia Buch. -Ham. ex D. Don ●

323245　Pyrus nepalensis Hortorum ex Decne. = Pyrus pashia Buch. -Ham. ex D. Don ●

323246   Pyrus nivalis Jacq. ；南欧雪梨(雪梨)；Snow Pear ●☆

323247   Pyrus nussia Buch. -Ham. ex D. Don = Stranvaesia nussia (Buch. -Ham. ex D. Don) Decne. ●

323248   Pyrus obsoletidentata Cardot = Sorbus obsoletidentata (Cardot) Te T. Yu ●

323249   Pyrus oligodonta Cardot = Sorbus oligodonta (Cardot) Hand. -Mazz. ●

323250   Pyrus ovoidea Rehder；香水梨(卵果梨) ●

323251   Pyrus pashia Buch. -Ham. ex D. Don；川梨(山里红,棠梨,棠梨刺,野山查)；Himalayan Pear, Indian Pear, Indian Wild Pear, Pashi Pear, Pashia Pear, Sichuan Pear ●

323252   Pyrus pashia Buch. -Ham. ex D. Don var. grandiflora Cardot；大花川梨；Largeflower Pashi Pear ●

323253   Pyrus pashia Buch. -Ham. ex D. Don var. kumaoni Stapf；无毛川梨(光梨)；Hairless Pashi Pear ●

323254   Pyrus pashia Buch. -Ham. ex D. Don var. kumaoni Stapf = Pyrus pashia Buch. -Ham. ex D. Don ●

323255   Pyrus pashia Buch. -Ham. ex D. Don var. obtusata Cardot；钝叶川梨；Obtuse Pashi Pear ●

323256   Pyrus pashia Buch. -Ham. ex D. Don var. sikkimensis Wenz. = Malus sikkimensis (Wenz.) Koehne ●

323257   Pyrus pekiennsis Cardot = Sorbus discolor (Maxim.) Maxim. ●

323258   Pyrus pekinensis (Koehne) Cardot = Sorbus discolor (Maxim.) Maxim. ●

323259   Pyrus phaeocarpa Rehder；褐梨(杜梨,棠杜梨)；Dusky Pear ●

323260   Pyrus pobuashanensis Hance = Sorbus pohuashanensis (Hance) Hedl. ●

323261   Pyrus prattii Hemsl. = Malus ombrophila Hand. -Mazz. ●

323262   Pyrus prattii Hemsl. = Malus prattii (Hemsl.) C. K. Schneid. ●

323263   Pyrus prunifolia Steud. = Malus prunifolia (Willd.) Borkh. ●

323264   Pyrus prunifolia Willd. = Malus prunifolia (Willd.) Borkh. ●

323265   Pyrus pseudopashia Te T. Yu；滇梨；Yunnan Pear ●

323266   Pyrus pumila (Mill.) K. Koch = Malus pumila Mill. ●

323267   Pyrus pyraster Medik. ；野梨；Wild Pear ●☆

323268   Pyrus pyrifolia (Burm. f.) Nakai；沙梨(白梨,淡水梨,果宗,快果,梨,梨树,梨仔,橘,麻安梨,蜜父,山梨,水梨,酸梨子树,棠杜梨,糖罐梨,糖梨,玉乳,日本山梨)；Asian Pear, China Pear, Chinese Apple Pear, Chinese Pear, Chinese Pear Apple, Chinese Pear-apple, Chinese Sand Pear, Japan Pear, Japanese Pear, Nashi Pear, Oriental Pear, Sand Pear, Sand Pear Tree, Sandy Pear ●

323269   Pyrus pyrifolia (Burm. f.) Nakai 'Chojuro'；乔举拉沙梨 ●☆

323270   Pyrus pyrifolia (Burm. f.) Nakai 'Shinko'；辛考沙梨 ●☆

323271   Pyrus pyrifolia (Burm. f.) Nakai var. culta (Makino) Nakai = Pyrus pyrifolia (Burm. f.) Nakai var. culta Rehder ●

323272   Pyrus pyrifolia (Burm. f.) Nakai var. culta Rehder；酥梨(库尔塔中国梨,梨,乳梨,栽培梨) ●

323273   Pyrus pyrifolia (Burm. f.) Nakai var. montana Nakai = Pyrus pyrifolia (Burm. f.) Nakai ●

323274   Pyrus raddeaua Woronow；拉德梨 ●☆

323275   Pyrus reducta W. W. Sm. = Sorbus poteriifolia Hand. -Mazz. ●

323276   Pyrus regelii Rehder；雷氏梨 ●☆

323277   Pyrus rehderiana (Koehne) Cardot = Sorbus rehderiana Koehne ●

323278   Pyrus rhamnoides (Decne.) Hook. f. = Sorbus rhamnoides (Decne.) Rehder ●

323279   Pyrus ringo Wenz. = Malus asiatica Nakai ●

323280   Pyrus rufifolia H. Lév. = Docynia indica (Wall.) Decne. ●

323281   Pyrus salicifolia Pall. ；柳叶梨；Silver Pear, Willowleaf Pear, Willow-leafed Pear ●☆

323282   Pyrus salicifolia Pall. 'Pendula'；垂枝柳叶梨；Silver Weeping Pear, Weeping Silver Pear ●☆

323283   Pyrus salvifolia DC. ；鼠尾草叶梨 ●☆

323284   Pyrus sanguinea Pursh = Amelanchier sanguinea (Pursh) DC. ●

323285   Pyrus sargentiana (Koehne) Bean = Sorbus sargentiana Koehne ●

323286   Pyrus sargentiana Bean = Sorbus sargentiana Koehne ●

323287   Pyrus sativa DC. = Pyrus communis L. var. sativa (DC.) DC. ●

323288   Pyrus sativa DC. = Pyrus communis L. ●

323289   Pyrus scabrifolia Franch. = Crataegus scabrifolia (Franch.) Rehder ●

323290   Pyrus scalaris (Koehne) Bean = Sorbus scalaris Koehne ●

323291   Pyrus scalaris Bean = Sorbus scalaris Koehne ●

323292   Pyrus scandica Asch. ；法国梨；French Hales ●☆

323293   Pyrus serotina Rehder = Pyrus pyrifolia (Burm. f.) Nakai ●

323294   Pyrus serrulata Rehder；麻梨(黄皮梨,麻梨子)；Serrulate Pear ●

323295   Pyrus sieboldii Regel = Malus sieboldii (Regel) Rehder ●

323296   Pyrus sieversii Ledeb. = Malus sieversii (Ledeb.) Roem. ●◇

323297   Pyrus sikkimensis (Wenz.) Hook. f. = Malus sikkimensis (Wenz.) Koehne ●

323298   Pyrus simonii Carrière = Pyrus ussuriensis Maxim. ●

323299   Pyrus sinensis (Thouin) Poir. = Chaenomeles sinensis (Thouin) Koehne ●

323300   Pyrus sinensis Lindl. = Pyrus pyrifolia (Burm. f.) Nakai ●

323301   Pyrus sinensis Lindl. var. ussuriensis Makino = Pyrus ussuriensis Maxim. ●

323302   Pyrus sinensis Poir. = Chaenomeles sinensis (Thouin) Koehne ●

323303   Pyrus sinensis Poir. = Pyrus lindleyi Rehder ●

323304   Pyrus sinensis Poir. = Pyrus pyrifolia (Burm. f.) Nakai ●

323305   Pyrus sinensis Poir. = Pyrus ussuriensis Maxim. ●

323306   Pyrus sinkiangensis Te T. Yu；新疆梨；Sinkiang Pear, Xinjiang Pear ●

323307   Pyrus sosnovskii Fed. ；锁斯诺夫斯基梨 ●☆

323308   Pyrus spectabilis Aiton = Malus halliana Koehne ●

323309   Pyrus spectabilis Aiton = Malus spectabilis (Aiton) Borkh. ●

323310   Pyrus subcrataegifolia H. Lév. = Malus sieboldii (Regel) Rehder ●

323311   Pyrus syriaca Boiss. ；雪梨 ●☆

323312   Pyrus taihangshanensis S. Y. Wang et C. L. Chang；太行梨；Taihangshan Pear ●

323313   Pyrus taihangshanensis S. Y. Wang et C. L. Chang = Pyrus xerophila Te T. Yu ●

323314   Pyrus taiwanensis Iketani et H. Ohashi；台湾野梨 ●

323315   Pyrus taquetii H. Lév. = Amelanchier asiatica (Siebold et Zucc.) Endl. ex Walp. ●

323316   Pyrus theifera L. H. Bailey = Malus hupehensis (Pamp.) Rehder ●

323317   Pyrus thibetica Cardot = Sorbus thibetica (Cardot) Hand. -Mazz. ●

323318   Pyrus thomsonii King ex Hook. f. = Sorbus thomsonii (King) Rehder ●

323319   Pyrus tianschanica (Rupr.) Franch. = Sorbus tianschanica Rupr. ●

323320   Pyrus tianschanica Franch. = Sorbus tianschanica Rupr. ●

323321   Pyrus toringo (K. Koch) Miq. = Malus sieboldii (Regel) Rehder ●

323322   Pyrus toringo Siebold ex Miq. ；日本棠梨 ●☆

323323　Pyrus toringoides（Rehder）Osborn ＝ Malus toringoides（Rehder）Hughes ●

323324　Pyrus toringoides Osborn ＝ Malus toringoides（Rehder）Hughes ●

323325　Pyrus transitoria Batalin ＝ Malus transitoria（Batalin）C. K. Schneid. ●

323326　Pyrus transitoria Batalin var. toringoides（Rehder）Bailey ＝ Malus toringoides（Rehder）Hughes ●

323327　Pyrus transitoria Batalin var. toringoides Bailey ＝ Malus toringoides（Rehder）Hughes ●

323328　Pyrus trilocularis D. K. Zang et P. C. Huang；崂山梨；Laoshan Pear ●

323329　Pyrus turcomanica Maleev；土库曼梨●☆

323330　Pyrus ussuriensis Maxim. ；秋子梨（果宗，花盖梨，快果，蜜父，青梨，沙果梨，山梨，酸梨，野梨，玉乳）；Chinese Pear，Mongolian Pear，Sand Pear，Ussuri Pear，Ussurian Pear ●

323331　Pyrus ussuriensis Maxim. var. aromatica（Kikuchi et Nakai）Rehder ＝ Pyrus ussuriensis Maxim. ●

323332　Pyrus ussuriensis Maxim. var. hondoensis（Nakai et Kikuchi）Rehder；绿果秋子梨（苦梨，青梨，仙顶梨）●☆

323333　Pyrus ussuriensis Maxim. var. ovoidea（Rehder）Rehder ＝ Pyrus ovoidea Rehder ●

323334　Pyrus vaniotii H. Lév. ＝ Amelanchier asiatica（Siebold et Zucc.）Endl. ex Walp. ●

323335　Pyrus variolosa Wall. ＝ Pyrus pashia Buch. -Ham. ex D. Don ●

323336　Pyrus variolosa Wall. ex G. Don ＝ Pyrus pashia Buch. -Ham. ex D. Don ●

323337　Pyrus veitchii Hort. ＝ Malus yunnanensis（Franch.）C. K. Schneid. var. veitchii（Veitch）Rehder ●

323338　Pyrus vestita Wall. ＝ Sorbus cuspidata（Spach）Hedl. ●

323339　Pyrus vestita Wall. ex Hook. f. ＝ Sorbus cuspidata（Spach）Hedl. ●

323340　Pyrus vestita Wall. var. khasiana Hook. ＝ Sorbus coronata（Cardot）Te T. Yu et H. T. Tsai ●

323341　Pyrus wallichii Hook. f. ＝ Sorbus foliolosa（Wall.）Spach ●

323342　Pyrus wallichii Hook. f. ＝ Sorbus wallichii（Hook. f.）Te T. Yu ●

323343　Pyrus wilsoniana（C. K. Schneid.）Cardot ＝ Sorbus wilsoniana C. K. Schneid. ●

323344　Pyrus wilsoniana C. K. Schneid. ＝ Sorbus oligodonta（Cardot）Hand. -Mazz. ●

323345　Pyrus xanthoneura（Rehder）Cardot ＝ Sorbus hemsleyi（C. K. Schneid.）Rehder ●

323346　Pyrus xerophila Te T. Yu；木梨（大梨，酸梨，棠梨，野梨）；Arid Pear，Woody Pear ●

323347　Pyrus yamatensis Koidz. ；奈良短柄梨●☆

323348　Pyrus yoshinoi Koidz. ；吉野梨●☆

323349　Pyrus yunnanensis Franch. ＝ Malus yunnanensis（Franch.）C. K. Schneid. ●

323350　Pyrus yunnanensis Franch. var. veitchii Osbom ＝ Malus yunnanensis（Franch.）C. K. Schneid. var. veitchii（Veitch）Rehder ●

323351　Pyrus yunnanensis Franch. var. veitchii Osborn ＝ Malus yunnanensis（Franch.）C. K. Schneid. var. veitchii（Osborn）Rehder ●

323352　Pyrus zangezura Maleev；赞格祖尔梨●☆

323353　Pyrus-cydonia Weston ＝ Cydonia Mill. ●

323354　Pythagorea Lour. ＝ Homalium Jacq. ●

323355　Pythagorea Raf. ＝ Lythrum L. ●■

323356　Pythion Mart. （废弃属名）＝ Amorphophallus Blume ex Decne. （保留属名）■●

323357　Pythius B. D. Jacks. ＝ Euphorbia L. ●■

323358　Pythonium Schott ＝ Amorphophallus Blume ex Decne. （保留属名）■●

323359　Pythonium Schott ＝ Thomsonia Wall. （废弃属名）■●

323360　Pythonium hookeri Kunth ＝ Anchomanes difformis（Blume）Engl. ■☆

323361　Pytinicarpa G. L. Nesom（1994）；锐托菀属■☆

323362　Pytinicarpa neocaledonica（Guillaumin）G. L. Nesom；锐托菀■☆

323363　Pyxa Noronha ＝ Costus L. ■

323364　Pyxidanthera Michx. （1803）；岩樱属●☆

323365　Pyxidanthera Muehlenbeck ＝ Lepuropetalon Elliott ■☆

323366　Pyxidanthera barbulata Michx. ；岩樱；Flowering Moss，Pyxie ●☆

323367　Pyxidanthus Naudin ＝ Blakea P. Browne ■☆

323368　Pyxidaria Kuntze ＝ Lindernia All. ●

323369　Pyxidaria Schott ＝ ? Lecythis Loefl. ●☆

323370　Pyxidaria crustacea（L.）Kuntze ＝ Lindernia crustacea（L.）F. Muell. ■

323371　Pyxidaria diffusa（L.）Kuntze ＝ Lindernia diffusa（L.）Wettst. ■☆

323372　Pyxidaria nummulariifolia（D. Don）Kuntze ＝ Lindernia nummulariifolia（D. Don）Wettst. ■

323373　Pyxidaria senegalensis（Benth.）Kuntze ＝ Lindernia senegalensis（Benth.）V. Naray. ■☆

323374　Pyxidiaceae Dulac ＝ Plantaginaceae Juss. （保留科名）■

323375　Pyxidium Moench ex Moq. ＝ Amaranthus L. ■

323376　Pyxidium Moq. ＝ Amaranthus L. ■

323377　Pyxipoma Fenzl ＝ Sesuvium L. ■

323378　Qaeria Raf. ＝ Queria L. ■

323379　Qaisera Omer（1989）；夸伊泽龙胆属■

323380　Qiongzhuea（T. H. Wen et Ohrnb. ）J. R. Xue et T. P. Yi ＝ Chimonobambusa Makino ●

323381　Qiongzhuea（T. H. Wen et Ohrnb. ）J. R. Xue et T. P. Yi（1980）；筇竹属；Qiongzhu，Swollennoded Cane ●★

323382　Qiongzhuea J. R. Xue et T. P. Yi ＝ Chimonobambusa Makino ●

323383　Qiongzhuea J. R. Xue et T. P. Yi ＝ Qiongzhuea（T. H. Wen et Ohrnb. ）J. R. Xue et T. P. Yi ●★

323384　Qiongzhuea communis J. R. Xue et T. P. Yi ＝ Chimonobambusa communis（J. R. Xue et T. P. Yi）T. H. Wen et Ohrnb. ex Ohrnb. ●

323385　Qiongzhuea intermedia J. R. Xue et D. Z. Li ＝ Chimonobambusa hsuehiana D. Z. Li et H. Q. Yang ●

323386　Qiongzhuea luzhiensis J. R. Xue et T. P. Yi ＝ Chimonobambusa luzhiensis（J. R. Xue et T. P. Yi）T. H. Wen et Ohrnb. ex Ohrnb. ●

323387　Qiongzhuea macrophylla J. R. Xue et T. P. Yi ＝ Chimonobambusa macrophylla（J. R. Xue et T. P. Yi）T. H. Wen et Ohrnb. ex Ohrnb. ●

323388　Qiongzhuea macrophylla J. R. Xue et T. P. Yi f. leiboensis J. R. Xue et D. Z. Li ＝ Chimonobambusa macrophylla（J. R. Xue et T. P. Yi）T. H. Wen et Ohrnb. ex Ohrnb. f. leiboensis（J. R. Xue et T. P. Yi）T. H. Wen et Ohrnb. ex Ohrnb. ●

323389　Qiongzhuea macrophylla J. R. Xue et T. P. Yi var. leiboensis（J. R. Xue et D. Z. Li）J. R. Xue et D. Z. Li ＝ Chimonobambusa macrophylla（J. R. Xue et T. P. Yi）T. H. Wen et Ohrnb. ex Ohrnb. f. leiboensis（J. R. Xue et T. P. Yi）T. H. Wen et Ohrnb. ex Ohrnb. ●

323390　Qiongzhuea maculata T. H. Wen ＝ Chimonobambusa opienensis（J. R. Xue et T. P. Yi）T. H. Wen et Ohrnb. ex Ohrnb. ●

323391　Qiongzhuea montigena T. P. Yi ＝ Chimonobambusa montigena

（T. P. Yi）Ohrnb. ●

323392　Qiongzhuea opienensis（J. R. Xue et T. P. Yi）D. Z. Li et J. R. Xue = Chimonobambusa opienensis（J. R. Xue et T. P. Yi）T. H. Wen et Ohrnb. ex Ohrnb. ●

323393　Qiongzhuea opienensis J. R. Xue et T. P. Yi = Chimonobambusa opienensis（J. R. Xue et T. P. Yi）T. H. Wen et Ohrnb. ex Ohrnb. ●

323394　Qiongzhuea puberula J. R. Xue et T. P. Yi = Chimonobambusa puberula（J. R. Xue et T. P. Yi）T. H. Wen et Ohrnb. ex Ohrnb. ●

323395　Qiongzhuea rigidula（J. R. Xue et T. P. Yi）T. H. Wen et Ohrnb. = Qiongzhuea rigidula J. R. Xue et T. P. Yi ●

323396　Qiongzhuea rigidula J. R. Xue et T. P. Yi = Chimonobambusa rigidula（J. R. Xue et T. P. Yi）T. H. Wen et Ohrnb. ex Ohrnb. ●

323397　Qiongzhuea tumidinoda（Ohrnb.）J. R. Xue et T. P. Yi = Chimonobambusa tumidissinoda J. R. Xue et T. P. Yi ex Ohrnb. ●◇

323398　Qiongzhuea tumidissinoda（J. R. Xue et T. P. Yi ex Ohrnb.）J. R. Xue et T. P. Yi = Chimonobambusa tumidissinoda J. R. Xue et T. P. Yi ex Ohrnb. ●◇

323399　Qiongzhuea unifolia T. P. Yi = Chimonobambusa unifolia（T. P. Yi）T. H. Wen et Ohrnb. ●

323400　Qiongzhuea verruculosa T. P. Yi = Chimonobambusa verruculosa（T. P. Yi）T. H. Wen et Ohrnb. ●

323401　Quadrangula Baum. -Bod. = Gymnostoma L. A. S. Johnson ●☆

323402　Quadrania Noronha = ? Kopsia Blume（保留属名）●

323403　Quadrasia Elmer = Claoxylon A. Juss. ●

323404　Quadrella（DC.）J. Presl = Capparis L. ●

323405　Quadrella J. Presl = Capparis L. ●

323406　Quadria Mutis = Vismia Vand.（保留属名）●☆

323407　Quadria Rniz et Pav. = Gevuina Molina ●☆

323408　Quadriala Siebold et Zucc. = Buckleya Torr.（保留属名）●

323409　Quadriala lanceolata Siebold et Zucc. = Buckleya lanceolata（Siebold et Zucc.）Miq. ●

323410　Quadricasaea Woodson = Tabernaemontana L. ●

323411　Quadricosta Dulac = Isnardia L. ●■

323412　Quadricosta Dulac = Ludwigia L. ●■

323413　Quadrifaria Manetti ex Gordon = Araucaria Juss. ●

323414　Quadripterygium Tardieu = Euonymus L.（保留属名）●

323415　Quaiacum Scop. = Guaiacum L.（保留属名）●

323416　Qualea Aubl.（1775）；腺托囊萼花属●☆

323417　Qualea acuminata Spruce ex Warm. ；尖腺托囊萼花●☆

323418　Qualea albiflora Warm. ；白花腺托囊萼花●☆

323419　Qualea brasiliana Stafleu et Marc. -Berti；巴西腺托囊萼花●☆

323420　Qualea fasciculata Spreng. ；簇生腺托囊萼花●☆

323421　Qualea ferruginea Steyerm. ；锈色腺托囊萼花●☆

323422　Qualea glauca Warm. ；灰绿腺托囊萼花●☆

323423　Qualea grandiflora Mart. ；大花腺托囊萼花●☆

323424　Qualea intermedia Warm. ；间型腺托囊萼花●☆

323425　Qualea maliformis Link ex Warm. ；苹果腺托囊萼花●☆

323426　Qualea megalocarpa Stafleu；大果腺托囊萼花●☆

323427　Qualea microphylla Warm. ；小叶腺托囊萼花●☆

323428　Qualea minor Spreng. ；小腺托囊萼花●☆

323429　Qualea multiflora Mart. ；多花腺托囊萼花●☆

323430　Qualea nitida Stafleu；光亮腺托囊萼花●☆

323431　Qualea parviflora Mart. ；小花腺托囊萼花●☆

323432　Qualea pilosa Warm. ；毛腺托囊萼花●☆

323433　Qualea rigida Stafleu；硬腺托囊萼花●☆

323434　Quamasia Raf. = Camassia Lindl.（保留属名）■☆

323435　Quamasia angusta（Engelm. et A. Gray）Piper = Camassia

angusta（Engelm. et A. Gray）Blank. ■☆

323436　Quamasia azurea A. Heller = Camassia quamash（Pursh）Greene subsp. azurea（A. Heller）Gould ■☆

323437　Quamasia cusickii（S. Watson）Coville = Camassia cusickii S. Watson ■☆

323438　Quamasia howellii（S. Watson）Coville = Camassia howellii S. Watson ■☆

323439　Quamasia hyacinthina（Raf.）Britton = Camassia scilloides（Raf.）Cory ■☆

323440　Quamasia leichtlinii（Baker）Coville = Camassia leichtlinii（Baker）S. Watson ■☆

323441　Quamasia suksdorfii（Greenm.）Piper = Camassia leichtlinii（Baker）S. Watson subsp. suksdorfii（Greenm.）Gould ■☆

323442　Quamasia walpolei Piper = Camassia quamash（Pursh）Greene subsp. walpolei（Piper）Gould ■☆

323443　Quamassia B. D. Jacks. = Quamasia Raf. ■☆

323444　Quamoclidion Choisy（1849）；山豆茉莉属■☆

323445　Quamoclidion angulatum Choisy；山豆茉莉■☆

323446　Quamoclidion froebelii（Behr）Standl. = Mirabilis multiflora（Torr.）A. Gray var. pubescens S. Watson ■☆

323447　Quamoclidion multiflorum（Torr.）Torr. ex A. Gray = Mirabilis multiflora（Torr.）A. Gray ■☆

323448　Quamoclidion multiflorum（Torr.）Torr. ex A. Gray subsp. glandulosum Standl. = Mirabilis multiflora（Torr.）A. Gray var. glandulosa（Standl.）J. F. Macbr. ■☆

323449　Quamoclidion oxybaphoides A. Gray = Mirabilis oxybaphoides（A. Gray）A. Gray ■☆

323450　Quamoclit Mill.（1754）；茑萝属；Cypress Vine，Star Glory，Star-glory ■

323451　Quamoclit Mill. = Ipomoea L.（保留属名）●■

323452　Quamoclit Moench = Ipomoea L.（保留属名）●■

323453　Quamoclit Tourn. ex Moench = Ipomoea L.（保留属名）●■

323454　Quamoclit × sloteri House = Ipomoea × sloteri（House）Ooststr. ■☆

323455　Quamoclit angulata（Lam.）Bojer = Ipomoea hederifolia L. ■

323456　Quamoclit angulata Bojer = Quamoclit coccinea（L.）Moench ■

323457　Quamoclit cardinalis Hort. ；心叶茑萝（葵叶茑萝，槭叶茑萝，心脏叶茑萝，羽衣留红草）；Cardinal Climber，Cardinal Creeper，Hearts And Honey Vine，Hearts-and-honey，Hearts-and-honey Vine，Sweet Potato Vine ■☆

323458　Quamoclit cardinalis Hort. = Ipomoea × multifida（Raf.）Shinners ■☆

323459　Quamoclit coccinea（L.）Moench；橙红茑萝（红茑萝，圆叶茑萝）；Orange Cypress Vine，Red Morning Glory，Redstar，Scarlet Ipomoea，Scarlet Morning-glory，Star Ipomoea，Star-glory ■

323460　Quamoclit coccinea（L.）Moench = Ipomoea coccinea L. ■☆

323461　Quamoclit coccinea（L.）Moench = Ipomoea quamoclit L. ■☆

323462　Quamoclit grandiflora G. Don；大花茑萝；Large-flower Star Glory ■☆

323463　Quamoclit lobata（Cerv.）House = Mina lobata Cerv. ■

323464　Quamoclit mina G. Don = Mina lobata Cerv. ■

323465　Quamoclit pennata（Desr.）Bojer；茑萝（翠翎草，金凤毛，金丝线，锦屏封，鸟萝，茑萝番薯，茑萝松，女萝，五角星花）；Cardinal Climber，Cypress Vine，Cypressvine，Ipomea Quamoclit，Star Glory ■

323466　Quamoclit pennata（Desr.）Bojer = Ipomoea quamoclit L. ■

323467　Quamoclit phoenicea（Roxb.）Choisy = Ipomoea hederifolia L. ■

323468　Quamoclit pinnata（Desr.）Bojer = Ipomoea quamoclit L. ■

323469　Quamoclit pinnata Bojer ＝Ipomoea quamoclit L. ■

323470　Quamoclit sloteri House；葵叶茑萝（槭叶茑萝，杂种茑萝，掌叶茑萝）；Cardinal Climber, Cardinal Star Glory, Cardinal Star Ipomoea , Palmlaef Cypress Vine ■

323471　Quamoclit sloteri House ＝Ipomoea × sloteri（House）Ooststr. ■☆

323472　Quamoclit vulgaris Choisy ＝Ipomoea quamoclit L. ■

323473　Quamoclit vulgaris Choisy ＝Quamoclit pennata（Desr.）Bojer ■

323474　Quamoclita Raf. ＝Quamoclit Mill. ■

323475　Quamoclitia Lowe ＝Quamoclit Mill. ■

323476　Quamoclitium Post et Kuntze ＝Quamoclidion Choisy ■☆

323477　Quamoctita Raf. ＝Ipomoea L.（保留属名）●■

323478　Quamoctita Raf. ＝Quamoclit Mill. ■

323479　Quapoja Batsch ＝Quapoya Aubl. ●☆

323480　Quapoya Aubl.（1775）；秘鲁藤黄属●☆

323481　Quapoya acuminata Kuntze；渐尖秘鲁藤黄●☆

323482　Quapoya longipes（Ducke）Maguire；长梗秘鲁藤黄●☆

323483　Quapoya microphylla Klotzsch；小叶秘鲁藤黄●☆

323484　Quapoya robusta Klotzsch ex Engl.；粗壮秘鲁藤黄●☆

323485　Quapoya sulphurea Poepp. et Endl.；硫色秘鲁藤黄●☆

323486　Quaqua N. E. Br.（1879）；南非萝藦属■☆

323487　Quaqua N. E. Br. ＝Caralluma R. Br. ■

323488　Quaqua acutiloba（N. E. Br.）Bruyns；尖浅裂南非萝藦■☆

323489　Quaqua arenicola（N. E. Br.）Plowes；沙地南非萝藦■☆

323490　Quaqua arenicola（N. E. Br.）Plowes subsp. pilifera（Bruyns）Bruyns；纤毛沙地南非萝藦■☆

323491　Quaqua arida（Masson）Bruyns；旱生南非萝藦■☆

323492　Quaqua armata（N. E. Br.）Bruyns；具刺南非萝藦■☆

323493　Quaqua armata（N. E. Br.）Bruyns subsp. arenicola（N. E. Br.）Bruyns ＝Quaqua arenicola（N. E. Br.）Plowes ■☆

323494　Quaqua armata（N. E. Br.）Bruyns subsp. maritima Bruyns；滨海南非萝藦■☆

323495　Quaqua armata（N. E. Br.）Bruyns subsp. pilifera Bruyns ＝Quaqua arenicola（N. E. Br.）Plowes subsp. pilifera（Bruyns）Bruyns ■☆

323496　Quaqua aurea（C. A. Lückh.）Plowes；黄南非萝藦■☆

323497　Quaqua bayeriana（Bruyns）Plowes；巴耶尔南非萝藦■☆

323498　Quaqua cincta（C. A. Lückh.）Bruyns；围绕南非萝藦■☆

323499　Quaqua confusa Plowes ＝Quaqua parviflora（Masson）Bruyns subsp. confusa（Plowes）Bruyns ■☆

323500　Quaqua dependens（N. E. Br.）Plowes ＝Quaqua parviflora（Masson）Bruyns subsp. dependens（N. E. Br.）Bruyns ■☆

323501　Quaqua framesii（Pillans）Bruyns；弗雷斯南非萝藦■☆

323502　Quaqua gracilis（C. A. Lückh.）Plowes ＝Quaqua parviflora（Masson）Bruyns subsp. gracilis（C. A. Lückh.）Bruyns ■☆

323503　Quaqua hottentotorum N. E. Br. ＝Quaqua incarnata（L. f.）Bruyns subsp. hottentotorum（N. E. Br.）Bruyns ■☆

323504　Quaqua incarnata（L. f.）Bruyns；肉色南非萝藦■☆

323505　Quaqua incarnata（L. f.）Bruyns subsp. aurea（C. A. Lückh.）Bruyns ＝Quaqua aurea（C. A. Lückh.）Plowes ■☆

323506　Quaqua incarnata（L. f.）Bruyns subsp. hottentotorum（N. E. Br.）Bruyns；霍屯督南非萝藦■☆

323507　Quaqua incarnata（L. f.）Bruyns subsp. tentaculata（Bruyns）Bruyns；触角南非萝藦■☆

323508　Quaqua incarnata（L. f.）Bruyns var. tentaculata Bruyns ＝Quaqua incarnata（L. f.）Bruyns subsp. tentaculata（Bruyns）Bruyns ■☆

323509　Quaqua inversa（N. E. Br.）Bruyns；倒垂南非萝藦■☆

323510　Quaqua inversa（N. E. Br.）Bruyns var. cincta（C. A. Lückh.）Bruyns ＝Quaqua cincta（C. A. Lückh.）Bruyns ■☆

323511　Quaqua linearis（N. E. Br.）Bruyns；线状南非萝藦■☆

323512　Quaqua mammillaris（L.）Bruyns；乳突南非萝藦■☆

323513　Quaqua maritima（Bruyns）Plowes ＝Quaqua armata（N. E. Br.）Bruyns subsp. maritima Bruyns ■☆

323514　Quaqua marlothii（N. E. Br.）Bruyns；马洛斯南非萝藦■☆

323515　Quaqua multiflora（R. A. Dyer）Bruyns；繁花南非萝藦■☆

323516　Quaqua pallens Bruyns；变苍白南非萝藦■☆

323517　Quaqua parviflora（Masson）Bruyns；小花南非萝藦■☆

323518　Quaqua parviflora（Masson）Bruyns subsp. bayeriana Bruyns ＝Quaqua bayeriana（Bruyns）Plowes ■☆

323519　Quaqua parviflora（Masson）Bruyns subsp. confusa（Plowes）Bruyns；混乱南非萝藦■☆

323520　Quaqua parviflora（Masson）Bruyns subsp. dependens（N. E. Br.）Bruyns；悬垂南非萝藦■☆

323521　Quaqua parviflora（Masson）Bruyns subsp. gracilis（C. A. Lückh.）Bruyns；纤细小花南非萝藦■☆

323522　Quaqua parviflora（Masson）Bruyns subsp. pulchra Bruyns ＝Quaqua pulchra（Bruyns）Plowes ■☆

323523　Quaqua parviflora（Masson）Bruyns subsp. swanepoelii（Lavranos）Bruyns；斯旺南非萝藦■☆

323524　Quaqua pilifera（Bruyns）Plowes ＝Quaqua arenicola（N. E. Br.）Plowes subsp. pilifera（Bruyns）Bruyns ■☆

323525　Quaqua pillansii（N. E. Br.）Bruyns；皮氏南非萝藦■☆

323526　Quaqua pruinosa（Masson）Bruyns；白粉南非萝藦■☆

323527　Quaqua pulchra（Bruyns）Plowes；美丽南非萝藦■☆

323528　Quaqua radiata Plowes ＝Quaqua incarnata（L. f.）Bruyns ■☆

323529　Quaqua ramosa（Masson）Bruyns；分枝南非萝藦■☆

323530　Quaqua swanepoelii（Lavranos）Plowes ＝Quaqua parviflora（Masson）Bruyns subsp. swanepoelii（Lavranos）Bruyns ■☆

323531　Quaqua tentaculata（Bruyns）Plowes ＝Quaqua incarnata（L. f.）Bruyns subsp. tentaculata（Bruyns）Bruyns ■☆

323532　Quararibea Aubl.（1775）；夸拉木属●☆

323533　Quararibea cordata（Bonpl.）Vischer；夸拉木●☆

323534　Quararibea fieldii Millsp.；搅棒夸拉木（搅棒树）●☆

323535　Quararibea funebris（La Llave）Vischer；墓地夸拉木●☆

323536　Quararibea pumila Alverson；矮夸拉木●☆

323537　Quarena Raf. ＝Cordia L.（保留属名）●

323538　Quartinia A. Rich. ＝Pterolobium R. Br. ex Wight et Arn.（保留属名）●

323539　Quartinia Endl. ＝Rotala L. ■

323540　Quartinia abyssinica A. Rich. ＝Pterolobium stellatum（Forssk.）Brenan ●☆

323541　Quartinia turfosa A. Rich. ＝Rotala repens（Hochst.）Koehne ■☆

323542　Quassia L.（1762）；类苦木属（苦木属）；Quassia ●☆

323543　Quassia africana（Baill.）Baill.；非洲苦木（非洲苦香木，苦木）●☆

323544　Quassia africana Baill. ＝Quassia africana（Baill.）Baill. ●☆

323545　Quassia amara L. ＝Simarouba amara Aubl. ●☆

323546　Quassia cedron Baill.；塞庄苦木；Cedron ●☆

323547　Quassia gabonensis Pierre ＝Odyendyea gabonensis（Pierre）Engl. ●☆

323548　Quassia glauca Spreng.；灰绿类苦木；Paradise Tree ●☆

323549　Quassia grandifolia（Engl.）Noot. ＝Pierreodendron africanum（Hook. f.）Little ●☆

323550　Quassia indica（Gaertn.）Noot. ＝Samadera indica Gaertn. ●☆

323551　Quassia klaineana Pierre ＝ Quassia undulata （Guillaumin et Perr.） F. Dietr. ●☆

323552　Quassia schweinfurthii （Oliv.） Noot. ;施韦类苦木●☆

323553　Quassia simarouba L. f. ;希马类苦木;Aceituna ●☆

323554　Quassia undulata （Guillaumin et Perr.） F. Dietr. ;波纹类苦木（波叶哈诺苦木,波叶苦木）●☆

323555　Quassiaceae Bertol. ＝ Simaroubaceae DC.（保留科名）●

323556　Quaternella Ehrh. ＝ Moenchia Ehrh.（保留属名）■☆

323557　Quaternella Pedersen(1990);四数苋属■☆

323558　Quaternella confusa Pedersen;四数苋■☆

323559　Quebitea Aubl. ＝ Piper L. ●■

323560　Quebrachia Griseb.（1874）＝ Loxopterygium Hook. f. ●☆

323561　Quebrachia Griseb.（1879）;破斧木属;Red Quebracho ●☆

323562　Quebrachia Griseb.（1879）＝ Schinopsis Engl. ●☆

323563　Quebrachia lorentzii Griseb. ;破斧木（红坚木,洛伦破斧木）;Lorentz Red Quebracho ,Quebracho ●☆

323564　Quebrachia lorentzii Griseb. ＝ Schinopsis lorentzii Engl. ●☆

323565　Quechualia H. Rob.（1993）;毛喉斑鸠菊属●☆

323566　Quechualia fulta （Griseb.） H. Rob. ;毛喉斑鸠菊●☆

323567　Quechualia smithii H. Rob. ;史密斯毛喉斑鸠菊●☆

323568　Queenslandiella Domin(1915);昆士兰莎草属■☆

323569　Queenslandiella hyalina （Vahl） F. Ballard;昆士兰莎草;Queensland Sedge ■☆

323570　Queenslandiella hyalina （Vahl） F. Ballard ＝ Cyperus hyalinus Vahl ■☆

323571　Queenslandiella mira Domin ＝ Queenslandiella hyalina （Vahl） F. Ballard ■☆

323572　Quekettia Lindl.（1839）;快特兰属■☆

323573　Quekettia microscopica Lindl. ;快特兰■☆

323574　Quekettia vermeuleniana Determann ＝ Scolopendrogyne vermeuleniana （Determann） Szlach. et Mytnik ■☆

323575　Quelchia N. E. Br.（1901）;寡枝菊属●☆

323576　Quelchia conferta N. E. Br. ;寡枝菊●☆

323577　Queltia Salisb. ＝ Narcissus L. ■

323578　Quelusia Vand. ＝ Fuchsia L. ●■

323579　Quercaceae Martinov ＝ Fagaceae Dumort.（保留科名）●

323580　Quercus L.（1753）;栎属（麻栎属,橡属）;Oak ●

323581　Quercus × angustilepidota Nakai;狭鳞栎●☆

323582　Quercus × bebbiana C. K. Schneid. ;拜比栎;Bebb's Oak ,Oak ●☆

323583　Quercus × bushii Sarg. ;布西栎;Bush's Oak ,Oak ●☆

323584　Quercus × crispuloserrata （Sugim.） M. Kikuchi;皱齿栎●☆

323585　Quercus × deamii Trel. ;迪米栎;Deam's Oak ●☆

323586　Quercus × hawkinsii Sudw. ;霍氏栎;Hawkins' Oak ●☆

323587　Quercus × jackiana C. K. Schneid. ;杰克栎●☆

323588　Quercus × kiusiana Nakai;九州栎●☆

323589　Quercus × leana Nutt. ;利纳栎●☆

323590　Quercus × major Nakai;大栎●☆

323591　Quercus × nipponica Koidz. ;本州栎●☆

323592　Quercus × serratoides Uyeki;齿栎●☆

323593　Quercus × serratoides Uyeki nothovar. crispuloserrata Sugim. ＝ Quercus × crispuloserrata （Sugim.） M. Kikuchi ●☆

323594　Quercus × takaoyamensis Makino;高尾山栎●☆

323595　Quercus acerifolia （E. J. Palmer） Stoynoff et W. J. Hess;枫叶栎;Maple-leaf Oak ●☆

323596　Quercus acrodonta Seemen;岩栎;Cliff Oak ●

323597　Quercus acroglandis Kellogg ＝ Quercus agrifolia Née ●☆

323598　Quercus acuminata （Michx.） Sarg. ＝ Quercus muchlenbergii

Engelm. ●☆

323599　Quercus acuta Thunb. ;日本常绿栎(红材栎,尖叶栎,血槠）;Japanese Evergreen Oak ,Japanese Red Oak ●☆

323600　Quercus acuta Thunb. ex A. Murray ＝ Quercus acuta Thunb. ●☆

323601　Quercus acuta Thunb. f. lanceolata Hatus. ;剑叶日本常绿栎●☆

323602　Quercus acuta Thunb. var. acutiformis Nakai;锐栎●☆

323603　Quercus acuta Thunb. var. megaphylla （Hayashi） ＝ Quercus acuta Thunb. ●☆

323604　Quercus acuta Thunb. var. yanagitae Makino ＝ Quercus acuta Thunb. ●☆

323605　Quercus acuta Thunb. var. yokohamensis （Makino） Nakai ＝ Quercus acuta Thunb. var. acutiformis Nakai ●☆

323606　Quercus acuta Thunb. var. yokohamensis （Makino） Nakai ＝ Quercus acuta Thunb. ●☆

323607　Quercus acutidentata （Maxim. ex Shirai） Koidz. ＝ Quercus aliena Blume var. acuteserrata Maxim. ex Wenz. ●

323608　Quercus acutidentata （Maxim. ex Shirai） Koidz. var. latifolia Liou ＝ Quercus aliena Blume var. acuteserrata Maxim. ex Wenz. ●

323609　Quercus acutidentata Koidz. ＝ Quercus aliena Blume var. acuteserrata Maxim. ex Wenz. ●

323610　Quercus acutidentata Koidz. var. latifolia Liou ＝ Quercus aliena Blume var. acuteserrata Maxim. ●

323611　Quercus acutissima Carruth. ;麻栎（北方麻栎,黄麻栎,栃子,栎,栎木,栎子栎,茅栗,青刚,青杠转,柞,橡栎,橡栗,橡树,橡椀树,橡碗树,橡子树,栩,皂斗,杼斗,紫栎,柞栎,柞树）;Japanese Chestnut Oak ,Japanese Oak ,Low-level Oak ,Low-leveled Oak ,North Sawtooth Oak ,Sawtooth Oak ●

323612　Quercus acutissima Carruth. subsp. chenii （Nakai） A. Camus ＝ Quercus chenii Nakai ●

323613　Quercus acutissima Carruth. subsp. euacutissima A. Camus ＝ Quercus acutissima Carruth. ●

323614　Quercus acutissima Carruth. var. brevipetiolata G. Hoo ＝ Quercus chenii Nakai ●

323615　Quercus acutissima Carruth. var. chenii （Nakai） Menitsky ＝ Quercus chenii Nakai ●

323616　Quercus acutissima Carruth. var. depressinucata H. Wei Jen et R. Q. Gao ＝ Quercus acutissima Carruth. ●

323617　Quercus acutissima Carruth. var. deptessinucata H. Wei Jen et R. Q. Gao ;扁果麻栎;Flatfruit Sawtooth Oak ●

323618　Quercus acutissima Carruth. var. macrocarpa X. W. Li et Y. Q. Zhu;大果麻栎;Big-fruit Sawtooth Oak ●

323619　Quercus acutissima Carruth. var. septentrionalis Liou ＝ Quercus acutissima Carruth. ●

323620　Quercus aegilops L. ＝ Quercus macrolepis L. ●☆

323621　Quercus agrifolia Née;加州栎（糙叶栎,加州野栎）;California Live Oak ,Coast Live Oak ,Encina ●☆

323622　Quercus agrifolia Née var. berberidifolia （Liebm.） Wenz. ＝ Quercus berberidifolia Liebm. ●☆

323623　Quercus agrifolia Née var. oxyadenia （Torr.） J. T. Howell ＝ Quercus agrifolia Née ●☆

323624　Quercus ajoensis C. H. Mull. ;阿霍栎;Ajo Mountain Scrub Oak ●☆

323625　Quercus alba L. ;美国白栎（白栎,白色栎,东部白栎,欧洲橡树,山脊白栎,桶栎,西部白橡树）;American Oak ,American White Oak ,Eastern White Oak ,Quebec Oak ,Ridge White Oak ,Stave Oak ,Stone Oak ,Tanner's Oak ,White Oak ●☆

323626　Quercus alba L. var. subcaerulea Pickens ＝ Quercus alba L. ●☆

323627　Quercus alba L. var. subflavea Pickens ＝ Quercus alba L. ●☆

323628 Quercus albicaulis Chun et W. C. Ko = Cyclobalanopsis albicaulis (Chun et W. C. Ko) Y. C. Hsu et H. Wei Jen ●◇

323629 Quercus alexanderi Britton = Quercus muchlenbergii Engelm. ●☆

323630 Quercus aliena Blume;槲栎(白皮栎,大槲树,大叶栎柴,青冈,青刚树,青岗,青心子,细皮青冈,橡碗树,小槲栎树);Oriental Oak,Oriental White Oak ●

323631 Quercus aliena Blume f. alticupuliformis (H. Wei Jen et L. M. Wang) H. Wei Jen et L. M. Wang = Quercus aliena Blume var. alticupuliformis (H. Wei Jen et L. M. Wang) H. Wei Jen et L. M. Wang ●

323632 Quercus aliena Blume f. pellucida (Blume) Kitam. = Quercus aliena Blume var. pellucida Blume ●

323633 Quercus aliena Blume var. acuteserrata Maxim. = Quercus aliena Blume var. acuteserrata Maxim. ex Wenz. ●

323634 Quercus aliena Blume var. acutidentata Maxim. ex Shirai = Quercus aliena Blume var. acutiserrata Maxim. ex Wenz. ●

323635 Quercus aliena Blume var. acutiserrata Maxim. ex Wenz.;锐齿槲栎(宇字栎,尖齿槲栎,锐齿宇字栎,锐齿栎);Borbor Oak, Sharptooth Oak ●

323636 Quercus aliena Blume var. alticupuliformis (H. Wei Jen et L. M. Wang) H. Wei Jen et L. M. Wang = Quercus aliena Blume var. pekingensis Schottky ●

323637 Quercus aliena Blume var. alticupuliformis H. Wei Jen et L. M. Wang = Quercus aliena Blume var. pellucida Blume ●

323638 Quercus aliena Blume var. alticupuliformis H. Wei Jen et L. M. Wang = Quercus aliena Blume var. pekingensis Schottky ●

323639 Quercus aliena Blume var. griffithii (Hook. f. et Thomson ex Miq.) Schottky = Quercus griffithii Hook. f. et Thomson ●

323640 Quercus aliena Blume var. griffithii Schottky = Quercus griffithii Hook. f. et Thomson ●

323641 Quercus aliena Blume var. jeholensis (Liou et S. X. Li) H. Wei Jen et L. M. Wang = Quercus aliena Blume var. pekingensis Schottky ●

323642 Quercus aliena Blume var. jeholensis Liou et S. X. Li = Quercus aliena Blume var. pellucida Blume ●

323643 Quercus aliena Blume var. jeholensis Liouet S. X. Li = Quercus aliena Blume var. pekingensis Schottky ●

323644 Quercus aliena Blume var. pekingensis Schottky = Quercus aliena Blume var. pellucida Blume ●

323645 Quercus aliena Blume var. pekingensis Schottky f. jeholensis (Liou et S. X. Li) H. Wei Jen et L. M. Wang = Quercus aliena Blume var. pellucida Blume ●

323646 Quercus aliena Blume var. pekingensis Schottky f. jeholensis (Liou et S. X. Li) H. Wei Jen et L. M. Wang = Quercus aliena Blume var. pekingensis Schottky ●

323647 Quercus aliena Blume var. pellucida Blume;北京槲栎(高壳槲栎);Alticupula Oriental Oak,Beijing Oak,Peking Oak ●

323648 Quercus aliena Blume var. urticifolia V. Naray. = Quercus yunnanensis Franch. ●

323649 Quercus alnifolia Poech;赤杨叶栎(桤叶栎,塞浦路斯黄栎);Cyprian Golden Oak,Golden Oak of Cyprus,Golden Oak-of-cyprus ●☆

323650 Quercus ambigua F. Michx. = Quercus rubra L. ●☆

323651 Quercus amygdalifolia V. Naray. = Lithocarpus amygdalifolius (V. Naray. ex Forbes et Hemsl.) Hayata ●

323652 Quercus amygdalifolia V. Naray. ex Forbes et Hemsl. = Lithocarpus amygdalifolius (V. Naray. ex Forbes et Hemsl.) Hayata ●

323653 Quercus annulata Buckley = Quercus sinuata Walter var. breviloba (Torr.) C. H. Mull. ●☆

323654 Quercus annulata Sm. = Cyclobalanopsis annulata (Sm.) Oerst. ●

323655 Quercus annulata Sm. = Quercus glauca Thunb. ●

323656 Quercus apucidentata Franch. ex Nakai = Cyclobalanopsis sessilifolia (Blume) Schottky ●

323657 Quercus aquifolioides Rehder et E. H. Wilson;巴郎栎(巴郎山栎,川滇高山栎,高山栎);Hollyleaf Oak,Hollyleaf-like Oak ●

323658 Quercus aquifolioides Rehder et E. H. Wilson = Quercus semecarpifolia Sm. ●

323659 Quercus aquifolioides Rehder et E. H. Wilson var. rufescens (Franch.) Rehder et E. H. Wilson = Quercus guyavifolia H. Lév. ●

323660 Quercus arcaulis Buch.-Ham. ex Spreng. = Lithocarpus arcuala (Spreng.) C. C. Huang et Y. T. Chang ●

323661 Quercus arcuala D. Don = Lithocarpus arcuala (Spreng.) C. C. Huang et Y. T. Chang ●

323662 Quercus arcuala Spreng. = Lithocarpus arcuala (Spreng.) C. C. Huang et Y. T. Chang ●

323663 Quercus argyi H. Lév. = Castanopsis chinensis Hance ●

323664 Quercus argyrotricha A. Camus = Cyclobalanopsis argyrotricha (A. Camus) Chun et Y. T. Chang ex Y. C. Hsu et H. Wei Jen ●◇

323665 Quercus ariifolia Trel. = Quercus rugosa Née ●☆

323666 Quercus arisanensis Hayata = Lithocarpus hancei (Benth.) Rehder ●

323667 Quercus arisanensis Hayata = Pasania hancei (Benth.) Schottky var. arisanensis (Hayata) J. C. Liao ●

323668 Quercus arizonica Sarg.;亚利桑那栎;Arizona Oak, Arizona White Oak ●☆

323669 Quercus arkansana Sarg.;阿肯色栎;Arkansas Oak ●☆

323670 Quercus asakii Kaneh. = Cyclobalanopsis glauca (Thunb. ex A. Murray) Oerst. ●

323671 Quercus atropurpurea Hort. ex K. Koch;紫栎●☆

323672 Quercus attenuata V. Naray. = Lithocarpus attenuatus (V. Naray.) Rehder ●

323673 Quercus augustinii (V. Naray.) Schottky = Cyclobalanopsis angustinii (V. Naray.) Schottky ●

323674 Quercus augustinii (V. Naray.) Schottky var. angustifolia A. Camus = Cyclobalanopsis angustinii (V. Naray.) Schottky ●

323675 Quercus augustinii (V. Naray.) Schottky var. rockiana A. Camus = Cyclobalanopsis angustinii (V. Naray.) Schottky ●

323676 Quercus augustinii Skan = Cyclobalanopsis augustinii (Skan) Schottky ●

323677 Quercus augustinii Skan var. angustifolia A. Camus = Cyclobalanopsis augustinii (Skan) Schottky ●

323678 Quercus augustinii Skan var. rockiana A. Camus. = Cyclobalanopsis augustinii (Skan) Schottky ●

323679 Quercus augustinii V. Naray. = Cyclobalanopsis angustinii (V. Naray.) Schottky ●

323680 Quercus augustinii V. Naray. var. angustifolia A. Camus = Cyclobalanopsis augustinii Schottky ●

323681 Quercus augustinii V. Naray. var. rockiana A. Camus = Cyclobalanopsis augustinii Schottky ●

323682 Quercus aurea Hort. ex K. Koch;金黄栎●☆

323683 Quercus aurtro-glauca Y. T. Chang = Cyclobalanopsis austro-glauca Y. T. Chang ex Y. C. Hsu et H. Wei Jen ●

323684 Quercus austrina Small;北美栎●☆

323685 Quercus austrocochinchinensis Hickel = Cyclobalanopsis austrocochinchinensis (Hickel et A. Camus) Hjelmq. ●

323686 Quercus austrocochinchinensis Hickel et A. Camus =

Cyclobalanopsis austrocochinchinensis （Hickel et A. Camus） Hjelmq. ●

323687　Quercus austroglauca （Y. T. Chang ex Y. C. Hsu et H. Wei Jen） Y. T. Chang ＝Cyclobalanopsis austroglauca Y. T. Chang ex Y. C. Hsu et H. Wei Jen ●

323688　Quercus austro-glauca （Y. T. Chang） Y. T. Chang ＝ Cyclobalanopsis austroglauca Y. T. Chang ex Y. C. Hsu et H. Wei Jen ●

323689　Quercus balansae Drake ＝Lithocarpus balansae （Drake） A. Camus ●

323690　Quercus balla Chun et Tsiang ＝Cyclobalanopsis bella （Chun et Tsiang） Chun ex Y. C. Hsu et H. Wei Jen ●

323691　Quercus baloot Griff. ；阿富汗栎●☆

323692　Quercus bambusifolia Fortune ＝Cyclobalanopsis myrsinifolia （Blume） Oerst. ●

323693　Quercus bambusifolia Hance ＝Cyclobalanopsis bambusifolia （Hance） Chun ex Y. C. Hsu et H. Wei Jen ●

323694　Quercus bambusifolia Hance ＝Cyclobalanopsis neglecta Schottky ●

323695　Quercus baronii V. Naray. ；橿子栎（巴氏铁橿，垂枝黄橿，多毛橿子栎，多毛橿子树，黄橿子，橿子树，老黄橿，青冈子，栀子树）；Baron Oak，Capilatus Baron Oak，Pendulate Baron Oak ●

323696　Quercus baronii V. Naray. f. capillata Kosl. ＝Quercus baronii V. Naray. var. capillata （Kosl.） Liou ●

323697　Quercus baronii V. Naray. f. capillata Kosl. ＝Quercus baronii V. Naray. ●

323698　Quercus baronii V. Naray. var. capillata （Kosl.） Liou ＝ Quercus baronii V. Naray. ●

323699　Quercus baronii V. Naray. var. pendula S. Y. Wang et C. L. Chang ＝Quercus baronii V. Naray. ●

323700　Quercus basellata Chun et W. C. Ko；基座栎；Basellate Oak ●

323701　Quercus basellata Chun et W. C. Ko ＝Cyclobalanopsis phanera （Chun） Y. C. Hsu et H. Wei Jen ●

323702　Quercus baviensis Drake ＝Lithocarpus truncatus （King） Rehder var. baviensis （Drake） A. Camus ●

323703　Quercus bawanglingensis C. C. Huang, Z. X. Li et F. W. Xing；坝王栎（霸王栎）；Bawang Oak，Bawangling Oak ●

323704　Quercus bebbiana C. K. Schneid. ；贝波栎；Bebb's Oak ●☆

323705　Quercus bella Chun et Tsiang ＝Cyclobalanopsis bella （Chun et Tsiang） Chun ex Y. C. Hsu et H. Wei Jen ●

323706　Quercus berberidifolia Liebm. ；加州胭脂栎；California Scrub Oak ●☆

323707　Quercus bicolor Willd. ；二色栎；Swamp Oak，Swamp White Oak ●☆

323708　Quercus bicolor Willd. var. angustifolia Dippel ＝Quercus bicolor Willd. ●☆

323709　Quercus bicolor Willd. var. lyrata （Walter） Dippel ＝Quercus lyrata Watt ●☆

323710　Quercus bicolor Willd. var. platanoides A. DC. ＝Quercus bicolor Willd. ●☆

323711　Quercus blakei V. Naray. ＝Cyclobalanopsis blakei （R. H. Shan） Schottky ●

323712　Quercus blakei V. Naray. var. parvifolia Merr. ＝Cyclobalanopsis blakei （R. H. Shan） Schottky ●

323713　Quercus bocoynensis C. H. Mull. ＝Quercus depressipes Trel. ●☆

323714　Quercus borealis Michx. ＝Quercus rubra L. ●☆

323715　Quercus borealis Michx. var. maxima （Marshall） Ashe ＝ Quercus rubra L. ●☆

323716　Quercus boyntonii Beadle；鲍伊栎；Boynton Oak ●☆

323717　Quercus brantii Lindl. ；布兰迪栎●☆

323718　Quercus brayi Small ＝Quercus muchlenbergii Engelm. ●☆

323719　Quercus brevicaudata V. Naray. ＝Lithocarpus brevicaudatus （V. Naray.） Hayata ●

323720　Quercus breviloba （Torr.） Sarg. ＝Quercus sinuata Walter var. breviloba （Torr.） C. H. Mull. ●☆

323721　Quercus breviradiata （W. C. Cheng） C. C. Huang ＝ Cyclobalanopsis breviradiata W. C. Cheng ●

323722　Quercus broteroi Cout. ；布罗特罗栎●☆

323723　Quercus brunnea H. Lév. ＝Castanopsis hystrix Hook. f. et Thomson ex A. DC. ●

323724　Quercus brunnea H. Lév. ＝Castanopsis hystrix Miq. ●

323725　Quercus buckleyi Nixon et Dorr；得州红栎；Buckley Oak，Buckley's Oak，Texas Red Oak ●☆

323726　Quercus bullata Seemen ＝Quercus spinosa A. David ex Franch. ●

323727　Quercus bungeana F. B. Forbes ＝Quercus variabilis Blume ●

323728　Quercus calathiformis V. Naray. ＝Castanopsis calathiformis （V. Naray.） Rehder et P. Wilson ●

323729　Quercus californica （Torr.） Cooper ＝Quercus kelloggii Newb. ●☆

323730　Quercus camusae Trel. ex Hickel et A. Camus ＝Cyclobalanopsis camusae （Trel. ex Hickel et A. Camus） Y. C. Hsu et H. Wei Jen ●

323731　Quercus camusiae Trel. ex Hickel et A. Camus ＝ Cyclobalanopsis camusae （Trel. ex Hickel et A. Camus） Y. C. Hsu et H. Wei Jen ●

323732　Quercus canariensis Willd. ；加那利栎（加纳利栎）；Algerian Fir Oak，Algerian Oak，Canary Oak，Mirbeck Oak，Mirbeck's Oak ●☆

323733　Quercus canbyi Cory et Parks ＝Quercus graciliformis C. H. Mull. ●☆

323734　Quercus caput-rivuli Ashe ＝Quercus arkansana Sarg. ●☆

323735　Quercus carlesii Hemsl. ＝Castanopsis carlesii （Hemsl.） Hayata ●

323736　Quercus carolinae V. Naray. ＝Lithocarpus carolinae （V. Naray.） Rehder ●

323737　Quercus castaneifolia C. A. Mey. ；栗叶栎；Chestnut-leafed Oak Chestnut-leaf Oak ●☆

323738　Quercus castanopsifolia Hayata ＝Lithocarpus lepidocarpus （Hayata） Hayata ●

323739　Quercus castanopsis H. Lév. ＝Castanopsis eyrei （Champ. ex Benth.） Tutcher ●

323740　Quercus catesbaei Michx. ＝Quercus laevis Walter ●☆

323741　Quercus cathayana Seemen ＝Lithocarpus truncatus （King） Rehder ●

323742　Quercus caudatilimba Merr. ＝Lithocarpus caudatilimbus （Merr.） A. Camus ●

323743　Quercus cavaleriei H. Lév. et Vaniot ＝Castanopsis eyrei （Champ. ex Benth.） Tutcher ●

323744　Quercus cepifera H. Lév. ＝Castanopsis eyrei （Champ. ex Benth.） Tutcher ●

323745　Quercus cerris L. ；土耳其栎；Adriatic Oak，European Turkey Oak，Manna Oak，Massy-cup Oak，Turkey Oak，Wainscot Oak ●☆

323746　Quercus cerris L. 'Argenteovariegata'；银边土耳其栎（镶边土耳其栎）●☆

323747　Quercus cerris L. 'Laciniata'；条裂土耳其栎●☆

323748　Quercus championii Benth. ＝Cyclobalanopsis championii （Benth.） Oerst. ex Schottky ●

323749　Quercus chapensis Hickel et A. Camus ＝Cyclobalanopsis chapensis （Hickel et A. Camus） Y. C. Hsu et H. Wei Jen ●

323750　Quercus chapmanii Sarg. ；查普曼栎；Chapman Oak，Chapman

White Oak ●☆

323751　Quercus chenii Nakai;小叶栎(苍落,杜木,黄栎树,临安栎,铁栎柴);Chen's Oak,Linan Chen's Oak ●

323752　Quercus chenii Nakai var. linanensis M. C. Liu et X. L. Shen = Quercus chenii Nakai ●

323753　Quercus chevalieri Hickel et A. Camus = Cyclobalanopsis chevalieri (Hickel et A. Camus) Y. C. Hsu et H. Wei Jen ●

323754　Quercus chihuahuensis Trel.;奇瓦瓦栎;Chihuahua Oak,Felt Oak ●☆

323755　Quercus chinensis Abel = Castanopsis sclerophylla (Lindl.) Schottky ●

323756　Quercus chinensis Bunge = Quercus variabilis Blume ●

323757　Quercus chingii F. P. Metcalf = Cyclobalanopsis sessilifolia (Blume) Schottky ●

323758　Quercus chingsiensis Y. T. Chang = Cyclobalanopsis chingsiensis (Y. T. Chang) Y. T. Chang ●

323759　Quercus chingsiensis Y. T. Chang = Cyclobalanopsis thorelii (Hickel et A. Camus) Hu ●

323760　Quercus chisosensis (Sarg.) C. H. Mull. = Quercus gravesii Sudw. ●☆

323761　Quercus chrysocalyx Hickel et A. Camus = Cyclobalanopsis chrysocalyx (Hickel et A. Camus) Hjelmq. ●

323762　Quercus chrysolepis Liebm.;黄鳞栎;Californian Line Oak,Canyon Live Oak,Canyon Oak,Goldcup Oak,Goldencup Oak,Live Oak,Maul Oak ●☆

323763　Quercus chrysolepis Liebm. var. nana (Jeps.) Jeps. = Quercus chrysolepis Liebm. ●☆

323764　Quercus chungii F. P. Metcalf = Cyclobalanopsis chungii (F. P. Metcalf) Y. C. Hsu et H. Wei Jen ex Q. F. Zheng ●

323765　Quercus ciliaris C. C. Huang et Y. T. Chang = Cyclobalanopsis gracilis (Rehder et E. H. Wilson) W. C. Cheng et T. Hong ●

323766　Quercus cinerea Michx. = Quercus incana Roxb. ●☆

323767　Quercus cleistocarpa Seemen = Lithocarpus cleistocarpus (Seemen) Rehder et E. H. Wilson ●

323768　Quercus clentatoides Liou = Quercus yunnanensis Franch. ●

323769　Quercus coccifera L.;胭脂红栎(虫瘿橡,大红栎,胭脂虫栎);Berry Bearing Oak,Grain Tree,Kermes Oak ●☆

323770　Quercus coccifera L. subsp. calliprinos (Webb) Holmboe = Quercus coccifera L. ●☆

323771　Quercus coccifera L. var. angustifolia E. Laguna = Quercus coccifera L. ●☆

323772　Quercus coccifera L. var. auzendei (Gren. et Godr.) A. DC. = Quercus coccifera L. ●☆

323773　Quercus coccifera L. var. imbricata A. DC. = Quercus coccifera L. ●☆

323774　Quercus coccifera L. var. integrifolia Boiss. = Quercus coccifera L. ●☆

323775　Quercus coccifera L. var. pseudococcifera (Desf.) A. DC. = Quercus coccifera L. ●☆

323776　Quercus coccifera L. var. stenocarpa Pamp. = Quercus coccifera L. ●☆

323777　Quercus cocciferoides Hand.-Mazz.;铁橡栎;Berrylike Oak,Ciccifer-like Oak ●

323778　Quercus cocciferoides Hand.-Mazz. var. taliensis (A. Camus) Y. C. Hsu et H. Wei Jen;大理栎;Dali Oak ●

323779　Quercus cocciferoides Hand.-Mazz. var. taliensis (A. Camus) Y. C. Hsu et H. Wei Jen = Quercus cocciferoides Hand.-Mazz. ●

323780　Quercus coccinea Münchh.;大红栎(绯红栎,绯栎);Scarlet Oak ●☆

323781　Quercus coccinea Münchh. 'Splendens';华美绯栎●☆

323782　Quercus coccinea Münchh. var. tuberculata Sarg. = Quercus coccinea Münchh. ●☆

323783　Quercus conduplicans Chun = Cyclobalanopsis pachyloma (Seemen) Schottky ●

323784　Quercus conferta Kit. = Quercus frainetto Ten. ●☆

323785　Quercus conocarpa Oudem.;新加坡栎;Singapore Oak ●☆

323786　Quercus conspersa Benth.;散布栎●☆

323787　Quercus cornea Lour. = Lithocarpus corneus (Lour.) Rehder ●

323788　Quercus cornea Lour. = Pasania cornea (Lour.) J. C. Liao ●

323789　Quercus cornea Lour. var. konishii (Hayata) Hayata = Lithocarpus konishii (Hayata) Hayata ●

323790　Quercus cornea Lour. var. konishii Hayata = Lithocarpus konishii (Hayata) Hayata ●

323791　Quercus cornelius-mulleri Nixon et K. P. Steele;马勒栎;Muller Oak ●☆

323792　Quercus crassifolia Humb. et Bonpl.;厚叶栎●☆

323793　Quercus crassipes Humb. et Bonpl.;粗柄栎●☆

323794　Quercus crispula Blume = Quercus mongolica Fisch. ex Ledeb. ●

323795　Quercus crispula Blume = Quercus mongolica Fisch. ex Turcz. var. grosserrata (Blume) Rehder et E. H. Wilson ●

323796　Quercus crispula Blume f. lomgifolia (Nakai) M. Kikuchi = Quercus crispula Blume ●

323797　Quercus crispula Blume var. horikawae H. Ohba;堀川栎●☆

323798　Quercus crispula Blume var. manshurica Koidz. = Quercus mongolica Fisch. ex Turcz. ●

323799　Quercus cryptoneuron H. Lév. = Castanopsis fargesii Franch. ●

323800　Quercus cuspidata (Thunb.) Schottky = Castanopsis carlesii (Hemsl.) Hayata ●

323801　Quercus cuspidata Thunb. = Castanopsis sieboldii (Makino) Hatus. ex T. Yamaz. et Mashiba ●☆

323802　Quercus cuspidata Thunb. var. sinensis A. DC. = Castanopsis sclerophylla (Lindl.) Schottky ●

323803　Quercus cyrtocarpa Drake = Lithocarpus cyrtocarpus (Drake) A. Camus ●

323804　Quercus daimingshanensis (S. K. Lee) C. C. Huang = Cyclobalanopsis damingshanensis S. K. Lee ●

323805　Quercus dalechampii Ten.;戴尔查栎;Durmast Oak ●☆

323806　Quercus dalicatula Chun et Tsiang = Cyclobalanopsis delicatula (Chun et Tsiang) Y. C. Hsu et H. Wei Jen ●

323807　Quercus damingsbanensis (S. K. Lee) C. C. Huang = Cyclobalanopsis damingshanensis S. K. Lee ●

323808　Quercus dealbata Hook. f. et Thomson ex DC. = Lithocarpus dealbatus (Hook. f. et Thomson ex DC.) Rehder ●

323809　Quercus delavayi Franch. = Cyclobalanopsis delavayi (Franch.) Schottky ●

323810　Quercus delicatula Chun et Tsiang = Cyclobalanopsis delicatula (Chun et Tsiang) Y. C. Hsu et H. Wei Jen ●

323811　Quercus densiflora Hook. et Arn.;密花栎;Tan Oak ●☆

323812　Quercus densiflora Hook. et Arn. = Lithocarpus densiflorus (Hook. et Arn.) Rehder ●☆

323813　Quercus dentaoides Liou = Quercus yunnanensis Franch. ●

323814　Quercus dentata Thunb.;波罗栎(波罗树,波罗叶,薄罗,槲罗树,大叶栎,大叶柞,槲,槲栎,槲若,槲树,金鸡菊,朴檄,青冈,青刚树,橡树,小波拉叶,柞,柞栎);Daimio Oak,Daimyo Oak ●

323815　Quercus dentata Thunb. 'Aurea'；黄波罗栎；Golden Daimyo Oak ●☆

323816　Quercus dentata Thunb. f. angustifolia (Ito) Hayashi = Quercus dentata Thunb. ●

323817　Quercus dentata Thunb. f. erectisquamosa (Nakai) Hayashi = Quercus dentata Thunb. ●

323818　Quercus dentata Thunb. f. grandifolia (Koidz.) Hayashi = Quercus dentata Thunb. ●

323819　Quercus dentata Thunb. f. laciniata (Makino) Kitam. et T. Horik.；撕裂波罗栎●☆

323820　Quercus dentata Thunb. f. pinnatifida (Franch. et Sav.) Kitam. et T. Horik.；羽裂波罗栎●☆

323821　Quercus dentata Thunb. f. pinnatiloba (Makino) Kitam. et T. Horik.；羽裂栎●☆

323822　Quercus dentata Thunb. fvar. grandifolia Koidz. = Quercus dentata Thunb. ●

323823　Quercus dentata Thunb. subsp. eudentata A. Camus = Quercus dentata Thunb. ●

323824　Quercus dentata Thunb. subsp. stewardii (Rehder) A. Camus = Quercus stewardii Rehder ●

323825　Quercus dentata Thunb. subsp. yunnanensis (Franch.) Menitsky = Quercus yunnanensis Franch. ●

323826　Quercus dentata Thunb. var. oxyloba Franch. = Quercus yunnanensis Franch. ●

323827　Quercus dentata Thunb. var. stewardii (Rehder)? = Quercus stewardii Rehder ●

323828　Quercus dentatoides Liou = Quercus dentata Thunb. var. oxyloba Franch. ●

323829　Quercus dentatoides Liou = Quercus yunnanensis Franch. ●

323830　Quercus depressipes Trel. ；匍匐栎；Depressed Oak ●☆

323831　Quercus digitata Sudw. = Quercus falcata Michx. ●☆

323832　Quercus dinghuensis Chun C. Huang = Cyclobalanopsis dinghuensis (C. C. Huang) Y. C. Hsu et H. Wei Jen ●

323833　Quercus disciformis Chun et Tsiang = Cyclobalanopsis disciformis (Chun et Tsiang) Y. C. Hsu et H. Wei Jen ●

323834　Quercus dispar Chun et Tsiang = Cyclobalanopsis kerrii (Craib) Hu ●

323835　Quercus diversicolor Trel. = Quercus rugosa Née ●☆

323836　Quercus djiringensis A. Camus = Cyclobalanopsis xanthotricha (A. Camus) Y. C. Hsu et H. Wei Jen ●

323837　Quercus dodonaeifolia Hayata = Lithocarpus dodonaeifolius (Hayata) Hayata ●

323838　Quercus doichangensis A. Camus = Formanodendron doichangensis (A. Camus) Nixon et Crepet ●◇

323839　Quercus doichangensis A. Camus = Trigonobalanus dolichangensis (A. Camus) Forman ●

323840　Quercus dolicholepis A. Camus var. elliptica (Y. C. Hsu et H. Wei Jen) Y. C. Hsu et H. Wei Jen；丽江栎；Lijiang Oak ●

323841　Quercus dolicholepis A. Camus var. elliptica (Y. C. Hsu et H. Wei Jen) Y. C. Hsu et H. Wei Jen = Quercus dolicholepis A. Camus ●

323842　Quercus dolichostyla A. Camus；匙叶栎(青橿)；Spoonleaf Oak，Spoon-leaved Oak，Spoonshapeleaf Oak ●

323843　Quercus dolichostyla A. Camus = Quercus engleriana Seemen ●

323844　Quercus dongfangensis C. C. Huang, F. W. Xing et Z. X. Li = Cyclobalanopsis dongfangensis (C. C. Huang, F. W. Xing et Z. X. Li) Y. T. Chang ●

323845　Quercus douglasii Hook. et Arn. ；道格拉斯栎；Blue Oak ●☆

323846　Quercus douglasii Hook. et Arn. var. gambelii (Nutt.) A. DC. = Quercus gambelii Nutt. ●☆

323847　Quercus douglasii Hook. et Arn. var. neaei (Liebm.) A. DC. = Quercus garryana Douglas ex Hook. ●☆

323848　Quercus douglasii Hook. et Arn. var. ransomii (Kellogg) Beissn. = Quercus douglasii Hook. et Arn. ●☆

323849　Quercus douglassi Hook. et Arn. ；美国铁橡树；Blue Oak，California Blue Oak，California Scrub Oak，Iron Oak，Scrub Oak ●☆

323850　Quercus dumosa Nutt. ；胭脂栎；Coastal Sage Scrub Oak，Scrub Oak ●☆

323851　Quercus dumosa Nutt. var. bullata Engelm. = Quercus durata Jeps. ●☆

323852　Quercus dumosa Nutt. var. munita Greene = Quercus berberidifolia Liebm. ●☆

323853　Quercus dumosa Nutt. var. polycarpa Greene = Quercus pacifica Nixon et C. H. Mull. ●☆

323854　Quercus dumosa Nutt. var. revoluta Sarg. = Quercus durata Jeps. ●☆

323855　Quercus dumosa Nutt. var. turbinella (Greene) Jeps. = Quercus turbinella Greene ●☆

323856　Quercus dunniana H. Lév. = Sageretia rugosa Hance ●

323857　Quercus dunnii Kellogg；邓恩栎；Dunn Oak，Palmer Oak ●☆

323858　Quercus dunnii Kellogg ex Curran = Quercus palmeri Engelm. ●☆

323859　Quercus durandii Buckley；杜氏栎；Bluff Oak，Durand Oak，Pin Oak，White Oak ●☆

323860　Quercus durandii Buckley = Quercus sinuata Walter ●☆

323861　Quercus durandii Buckley var. austrina (Small) E. J. Palmer = Quercus austrina Small ●☆

323862　Quercus durandii Buckley var. breviloba (Torr.) E. J. Palmer = Quercus sinuata Walter var. breviloba (Torr.) C. H. Mull. ●☆

323863　Quercus durandii Buckley var. san-sabeana (Buckley ex M. J. Young) Buckley = Quercus sinuata Walter var. breviloba (Torr.) C. H. Mull. ●☆

323864　Quercus durangensis Trel. = Quercus rugosa Née ●☆

323865　Quercus durata Jeps. ；革栎；Leather Oak ●☆

323866　Quercus echinoides R. Br. = Lithocarpus densiflorus (Hook. et Arn.) Rehder var. echinoides (R. Br.) Abrams ●☆

323867　Quercus edithae V. Naray. = Cyclobalanopsis edithae (V. Naray.) Schottky ●

323868　Quercus elaeagnifolia Seemen = Lithocarpus elaeagnifolius (Seemen) Chun ●

323869　Quercus elevaticostata (Q. F. Zheng) C. C. Huang = Cyclobalanopsis elevaticostata Q. F. Zheng ●

323870　Quercus elizabethiae Tutcher = Lithocarpus elizabethae (Tutcher) Rehder ●

323871　Quercus ellipsoidalis E. J. Hill；美国北方栎(椭圆果栎，椭圆栎)；Black Oak，Hills Oak，Hill's Oak，Jack Oak，Northern Pin Oak ●☆

323872　Quercus ellipsoidalis E. J. Hill var. kaposianensis J. W. Moore = Quercus ellipsoidalis E. J. Hill ●☆

323873　Quercus emoryi Torr. ；矮栎；Black Oak，Blackjack Oak，Emory Oak，Live Oak ●☆

323874　Quercus engelmannii Greene；恩氏栎(平顶栎)；Engelmann Oak，Evergreen White Oak，Mesa Oak ●☆

323875　Quercus engleriana Seemen；巴东栎(橡实，小青冈，小叶青冈)；Engler Oak ●

323876　Quercus erucifolia Stev. ；芥叶栎●☆

323877　Quercus essilifoliaBlume = Cyclobalanopsis sessilifolia (Blume)

Schottky ●

323878　Quercus eyrei Champ. = Castanopsis eyrei（Champ. ex Benth.） Tutcher ●

323879　Quercus eyrei Champ. ex Benth. = Castanopsis eyrei（Champ. ex Benth.）Tutcher ●

323880　Quercus eyrei Champ. ex Benth. = Lithocarpus attenuatus（V. Naray.）Rehder ●

323881　Quercus fabri Hance；白栎（白柴蒲树，白栎蔀，白栗，白皮栎，金刚栎，枥柴，栎，青冈栎，青冈树，青岗，青杠，小白栎，小白栎青冈树，小白皮栎，皂斗，泽栗，柞子柴）；Faber Oak，White Oak ●

323882　Quercus faginea Lam. = Quercus lusitanica Lam. ●☆

323883　Quercus faginea Lam. subsp. alpestris（Boiss.）Maire = Quercus faginea Lam. ●☆

323884　Quercus faginea Lam. subsp. baetica（Webb）Maire = Quercus canariensis Willd. ●☆

323885　Quercus faginea Lam. subsp. broteroi（Cout.）A. Camus = Quercus broteroi Cout. ●☆

323886　Quercus faginea Lam. subsp. lusitanica（Lam.）Maire = Quercus lusitanica Lam. ●☆

323887　Quercus faginea Lam. subsp. tlemcenensis（A. DC.）Greuter et Burdet = Quercus faginea Lam. subsp. broteroi（Cout.）A. Camus ●☆

323888　Quercus faginea Lam. var. fagifolia Trab. = Quercus faginea Lam. ●☆

323889　Quercus faginea Lam. var. maroccana（Braun-Blanq. et Maire）Maire = Quercus faginea Lam. ●☆

323890　Quercus faginea Lam. var. microphylla（Trab.）Maire = Quercus faginea Lam. ●☆

323891　Quercus faginea Lam. var. mirbeckii（Durieu）Emb. = Quercus canariensis Willd. ●☆

323892　Quercus faginea Lam. var. spinosa Maire et Trab. = Quercus faginea Lam. ●☆

323893　Quercus faginea Ten. = Quercus robur L. ●☆

323894　Quercus falcata Michx.；南部红栎（西班牙红栎，西班牙栎）；American Red Oak，Red Oak，Southern Red Oak，Spanish Oak，Swamp Red Oak ●☆

323895　Quercus falcata Michx. var. leucophylla（Ashe）E. J. Palmer et Steyerm. = Quercus pagoda Raf. ●☆

323896　Quercus falcata Michx. var. pagodifolia（Elliott）E. Murray；樱皮栎；Cherrybark Oak ●☆

323897　Quercus falcata Michx. var. pagodifolia Elliott = Quercus pagoda Raf. ●☆

323898　Quercus falcata Michx. var. triloba（Michx.）Nutt. = Quercus falcata Michx. ●☆

323899　Quercus fangshanensis Liou；房山栎；Fangshan Oak ●

323900　Quercus fargesii Franch. = Cyclobalanopsis oxyodon（Miq.）Oerst. ●

323901　Quercus farinulenta Hance = Lithocarpus farinulentus（Hance）A. Camus ●

323902　Quercus farnetto Ten. = Quercus frainetto Ten. ●☆

323903　Quercus fenestrata Roxb. = Lithocarpus fenestratus（Roxb.）Rehder ●

323904　Quercus fengchengensis H. Wei Jen；凤城栎；Fengcheng Oak ●

323905　Quercus fenzeliana（A. Camus）Merr. = Lithocarpus fenzelianus A. Camus ●

323906　Quercus ferox Roxb. = Castanopsis ferox Spach ●

323907　Quercus fimbriata Y. C. Hsu et H. Wei Jen；长苞高山栎；Longbract Alpine Oak，Longbract Oak，Long-bracted Oak ●

323908　Quercus fissa Champ. = Castanopsis fissa（Champ. ex Benth.）Rehder et E. H. Wilson ●

323909　Quercus fissa Champ. ex Benth. = Castanopsis fissa（Champ. ex Benth.）Rehder et E. H. Wilson ●

323910　Quercus fleuryi Hickel et A. Camus = Cyclobalanopsis fleuryi（Hickel et A. Camus）Chun ●

323911　Quercus fokienensis Nakai = Quercus phillyraeoides A. Gray ●

323912　Quercus fooningensis Hu et W. C. Cheng = Quercus phillyraeoides A. Gray ●

323913　Quercus fordiana Hemsl. = Lithocarpus fordianus（Hemsl.）Chun ●

323914　Quercus formosana V. Naray. = Lithocarpus formosanus（V. Naray. ex Forbes et Hemsl.）Hayata ●

323915　Quercus formosanus V. Naray. ex Forbes et Hemsl. = Pasania formosana（V. Naray. ex Forbes et Hemsl.）Schottky ●

323916　Quercus fragifera Franch. = Lithocarpus cleistocarpus（Seemen）Rehder et E. H. Wilson ●

323917　Quercus frainetto Ten.；匈牙利栎；Farnetto，Hungarian Oak，Italian Oak ●☆

323918　Quercus franchetiana H. Lév. = Castanopsis tibetana Hance ●

323919　Quercus franchetiana H. Lév. ex A. Camus = Castanopsis tibetana Hance ●

323920　Quercus franchetii V. Naray.；锥连栎（川滇山栎，黄栗）；Franchet Oak ●

323921　Quercus fruticosa Brot. = Quercus lusitanica Lam. ●☆

323922　Quercus fuhsingensis Y. T. Chang = Cyclobalanopsis xanthotricha（A. Camus）Y. C. Hsu et H. Wei Jen ●

323923　Quercus fuliginosa Chun et W. C. Ko = Cyclobalanopsis fuliginosa（Chun et W. C. Ko）Y. Y. Luo et R. J. Wang ●

323924　Quercus fulviseriaca（Y. C. Hsu et D. M. Wang）Z. K. Zhou；黄枝青冈；Yellow-branch Oak，Yellow-branched Cyclobalanopsis，Yellowtwig Qinggang ●

323925　Quercus fulviseriaca（Y. C. Hsu et D. M. Wang）Z. K. Zhou = Cyclobalanopsis lungmaiensis Hu ●

323926　Quercus fusiformis Small；高原栎；Plateau Oak，Texas Live Oak ●☆

323927　Quercus gambelii Nutt.；落基山白栎（加姆贝尔栎）；Blue Oak，Gambel Oak，Gambel's Oak，Oregon White Oak，Rocky Mountain White Oak，Shin Oak，Utah White Oak ●☆

323928　Quercus gambelii Nutt. var. gunnisonii Wenz. = Quercus fusiformis Small ●☆

323929　Quercus gambleana A. Camus = Cyclobalanopsis gambleana（A. Camus）Y. C. Hsu et H. Wei Jen ●

323930　Quercus garrettiana Carib = Lithocarpus garrettianus（Craib）A. Camus ●

323931　Quercus garryana Douglas ex Hook.；俄勒冈白栎（俄勒冈栎）；Brewer Oak，Garry Oak，Oregon Oak，Oregon White Oak，Post Oak，White Oak ●☆

323932　Quercus garryana Douglas ex Hook. var. jacobi（R. Br.）Zabel = Quercus garryana Douglas ex Hook. ●☆

323933　Quercus geminata Hickel et A. Camus = Cyclobalanopsis camusae（Trel. ex Hickel et A. Camus）Y. C. Hsu et H. Wei Jen ●

323934　Quercus geminata Small；沙栎；Sand Live Oak ●☆

323935　Quercus georgiana M. A. Curtis；乔治亚栎；Georgia Oak ●☆

323936　Quercus gilliana Rehder et E. H. Wilson；川西栎（野青冈）；Gill Oak，W. Sichuan Oak，West Sichuan Oak，West Szechwan Oak ●

323937　Quercus gilliana Rehder et E. H. Wilson = Quercus spinosa A. David ex Franch. ●

323938 Quercus gilva Blume = Cyclobalanopsis gilva (Blume) Oerst. ●

323939 Quercus glabra Thunb. = Lithocarpus glaber (Thunb.) Nakai ●

323940 Quercus glabra Thunb. ex A. Murray = Lithocarpus glaber (Thunb.) Nakai ●

323941 Quercus glabra Thunb. ex A. Murray = Pasania glabra (Thunb.) Oerst. ●

323942 Quercus glandulifera Blume;枹栎(白皮栎,字落树,勃落树,大叶青冈,枹,枹树,桴栎,青冈树,青刚树,橡子树,小橡树,柞木);Glandular Oak, Shortpetiole Oak, Shortpetiole Serrate Oak ●

323943 Quercus glandulifera Blume = Quercus serrata Murray ●

323944 Quercus glandulifera Blume var. brevipetiolata (A. DC.) Nakai;短柄枹栎(菠萝,柴树,青冈柳,青栲栎,思茅楮栎,柞树);Shortstalk Oak, Short-stypes Oak ●

323945 Quercus glandulifera Blume var. brevipetiolata (A. DC.) Nakai = Quercus serrata Thunb. ●

323946 Quercus glandulifera Blume var. stellatopilosa W. H. Zhang = Quercus serrata Murray ●

323947 Quercus glandulifera Blume var. stellatopilosa W. H. Zhang = Quercus serrata Thunb. var. tomentosa (B. C. Ding et T. B. Chao) Y. C. Hsu et H. Wei Jen ●

323948 Quercus glandulifera Blume var. tomentosa B. C. Ding et T. B. Chao = Quercus serrata Murray ●

323949 Quercus glandulifera Blume var. tomentosa W. H. Zhang = Quercus serrata Thunb. var. tomentosa (B. C. Ding et T. B. Chao) Y. C. Hsu et H. Wei Jen ●

323950 Quercus glandulifera Blume var. tomentosa W. H. Zhang = Quercus serrata Thunb. ●

323951 Quercus glauca (Thunb. ex A. Murray) Oerst. var. kuyuensis J. C. Liao = Cyclobalanopsis glauca (Thunb. ex A. Murray) Oerst. ●

323952 Quercus glauca Thunb. ' Fastigiata';帚状栎●☆

323953 Quercus glauca Thunb. = Cyclobalanopsis glauca (Thunb. ex A. Murray) Oerst. ●

323954 Quercus glauca Thunb. ex A. Murray = Cyclobalanopsis glauca (Thunb. ex A. Murray) Oerst. ●

323955 Quercus glauca Thunb. ex A. Murray var. annulata (Sm.) A. Camus = Cyclobalanopsis annulata (Sm.) Oerst. ●

323956 Quercus glauca Thunb. f. gracilis Rehder et E. H. Wilson = Cyclobalanopsis gracilis (Rehder et E. H. Wilson) W. C. Cheng et T. Hong ●

323957 Quercus glauca Thunb. f. lacera (Blume) Kitam.;撕裂青冈(撕裂栎)●☆

323958 Quercus glauca Thunb. f. latifolia (Nakai) Hiyama = Quercus glauca Thunb. ●

323959 Quercus glauca Thunb. f. stenocarpa Honda = Quercus glauca Thunb. ●

323960 Quercus glauca Thunb. subsp. annulata (Sm.) A. Camus = Cyclobalanopsis annulata (Sm.) Oerst. ●

323961 Quercus glauca Thunb. subsp. euglauca A. Camus = Quercus glauca Thunb. ●

323962 Quercus glauca Thunb. var. annulata (Sm.) A. Camus = Cyclobalanopsis annulata (Sm.) Oerst. ●

323963 Quercus glauca Thunb. var. gracilis f. subintegrifolia Ling = Cyclobalanopsis myrsinifolia (Blume) Oerst. ●

323964 Quercus glauca Thunb. var. hyparyrea Seemen = Cyclobalanopsis multinervis W. C. Cheng et T. Hong ●

323965 Quercus glauca Thunb. var. kuyuensis J. C. Liao = Cyclobalanopsis glauca (Thunb. ex A. Murray) Oerst. var. kuyuensis (J. C. Liao) J. C. Liao ●

323966 Quercus glauca Thunb. var. kuyuensis J. C. Liao = Cyclobalanopsis glauca (Thunb. ex A. Murray) Oerst. ●

323967 Quercus glauca Thunb. var. salicina (Blume) Menitsky = Quercus salicina Blume ●

323968 Quercus glaucoides (Schottky) Koidz. = Cyclobalanopsis glaucoides Schottky ●

323969 Quercus globosa (W. F. Lin et T. Liu) J. C. Liao = Cyclobalanopsis globosa W. F. Lin et T. Liu ●

323970 Quercus globosa (W. F. Lin et T. Liu) J. C. Liao f. chiapautaiensis J. C. Liao = Cyclobalanopsis globosa W. F. Lin et T. Liu f. chiapautaiensis (J. C. Liao) J. C. Liao ●

323971 Quercus gracilenta Chun = Cyclobalanopsis pachyloma (Seemen) Schottky ●

323972 Quercus graciliformis C. H. Mull.;纤细栎;Chisos Oak ●☆

323973 Quercus graciliformis C. H. Mull. var. parvilobata C. H. Mull. = Quercus graciliformis C. H. Mull. ●☆

323974 Quercus gracilis (Rehder et E. H. Wilson) Wuzhi = Cyclobalanopsis gracilis (Rehder et E. H. Wilson) W. C. Cheng et T. Hong ●

323975 Quercus graeca Kotschy = Quercus macrolepis L. ●☆

323976 Quercus grandifolia D. Don = Lithocarpus elegans (Blume) Hatus. ex Soepadmo ●☆

323977 Quercus grandifolia D. Don = Lithocarpus grandifolius (D. Don) S. N. Biswas ●

323978 Quercus gravesii Sudw.;格雷夫斯栎;Chisos Red Oak, Graves Oak ●☆

323979 Quercus griffithii Hook. f. et Thomson;大叶栎;Bigleaf Oak, Griffith Oak ●

323980 Quercus griffithii Hook. f. et Thomson ex Miq. var. urticifolia Franch. = Quercus yunnanensis Franch. ●

323981 Quercus griffithii Hook. f. et Thomson var. urticifolia Franch. = Quercus malacotricha A. Camus ●

323982 Quercus grisea Liebm.;北美灰栎;Gray Oak, Scrub Oak, Shun Oak ●☆

323983 Quercus grosserrata Blume = Quercus mongolica Fisch. ex Turcz. var. grosserrata (Blume) Rehder et E. H. Wilson ●

323984 Quercus grosseserrata Blume = Quercus crispula Blume ●

323985 Quercus grosseserrata Blume = Quercus mongolica Fisch. ex Ledeb. ●

323986 Quercus guyavifolia H. Lév.;帽斗栎;Capcupule Oak, Cupula Oak ●

323987 Quercus hainanensis C. C. Huang et Y. T. Zhang = Cyclobalanopsis litoralis Chun et P. C. Tam ●

323988 Quercus hainanensis Merr. = Lithocarpus corneus (Lour.) Rehder var. hainanensis (Merr.) C. C. Huang et Y. T. Chang ●

323989 Quercus hainanica C. C. Huang et Y. T. Chang = Cyclobalanopsis litoralis Chun et P. C. Tam ex Y. C. Hsu et H. Wei Jen ●

323990 Quercus hancei Benth. = Lithocarpus hancei (Benth.) Rehder ●

323991 Quercus handeliana A. Camus = Quercus acrodonta Seemen ●

323992 Quercus harlandii Hance = Pasania harlandii (Hance) Oerst. ●

323993 Quercus harlandii Hance ex Walp. = Lithocarpus harlandii (Hance) Rehder ●

323994 Quercus harlandii Hance var. integrifolia Dunn = Lithocarpus harlandii (Hance) Rehder ●

323995 Quercus hartmanii Trel. = Quercus toumeyi Sarg. ●☆

323996　Quercus hartwissiana Stev. ;哈特维斯栎●☆

323997　Quercus hastata Liebm. = Quercus emoryi Torr. ●☆

323998　Quercus havardii Rydb. ;阿瓦尔栎;Havard Oak,Shinnery Shin ●☆

323999　Quercus helferiana A. DC. = Cyclobalanopsis helferiana (A. DC.) Oerst. ●

324000　Quercus hemisphaerica Drake = Lithocarpus corneus (Lour.) Rehder var. zonatus C. C. Huang et Y. T. Chang ●

324001　Quercus hemisphaerica Endl. = Quercus hemisphaerica W. Bartram ex Willd. ●☆

324002　Quercus hemisphaerica W. Bartram ex Willd. ;月桂栎; Darlington Oak,Laurel Oak ●☆

324003　Quercus hennongii C. C. Huang et S. H. Fu = Cyclobalanopsis shennongii (C. C. Huang et S. H. Fu) Y. C. Hsu et H. Wei Jen ●

324004　Quercus henryi Seemen = Lithocarpus henryi (Seemen) Rehder et E. H. Wilson ●

324005　Quercus heterophylla Michx. ;异叶栎;Bartram's Oak ●☆

324006　Quercus hinckleyi C. H. Mull. ;欣克利栎;Hinckley Oak ●☆

324007　Quercus hindsii Benth. = Quercus lobata Née ●☆

324008　Quercus hingjenensis Y. T. Chang = Cyclobalanopsis disciformis (Chun et Tsiang) Y. C. Hsu et H. Wei Jen ●

324009　Quercus hirsulata Blume = Quercus aliena Blume ●

324010　Quercus hispanica Lam. ;西班牙栎;Hispania Oak,Spanish Oak ●☆

324011　Quercus hispanica Lam. ' Lucombeana ';大叶西班牙栎; Spanish Oak ●☆

324012　Quercus hopeiensis Liou;河北栎;Hebei Oak,Hopei Oak ●

324013　Quercus houstoniana C. H. Mull. = Quercus michauxii Nutt. ●☆

324014　Quercus hsiensiui Chun et W. C. Ko;先骕栎;Hu's Oak ●

324015　Quercus hsiensiui Chun et W. C. Ko = Cyclobalanopsis thorelii (Hickel et A. Camus) Hu ●

324016　Quercus hui Chun = Cyclobalanopsis hui (Chun) Chun ex Y. C. Hsu et H. Wei Jen ●

324017　Quercus hunanensis Hand. -Mazz. = Cyclobalanopsis gilva (Blume) Oerst. ●

324018　Quercus hypargyrea (Seemen ex Diels) C. C. Huang et Y. T. Chang = Cyclobalanopsis multinervis W. C. Cheng et T. Hong ●

324019　Quercus hypargyrea (Seemen) C. C. Huang et Y. T. Chang = Cyclobalanopsis multinervis W. C. Cheng et T. Hong ●

324020　Quercus hypochrysa Stev. ;金背栎●☆

324021　Quercus hypoleuca Engelm. ;白背叶栎（银叶栎）;Silverleaf Oak,Whiteleaf Oak ●☆

324022　Quercus hypoleuca Engelm. = Quercus hypoleucoides A. Camus ●☆

324023　Quercus hypoleucoides A. Camus;假白背叶栎;Silverleaf Oak ●☆

324024　Quercus hypophaea Hayata = Cyclobalanopsis hypophaea (Hayata) Kudo ●

324025　Quercus iberica Stev. ex M. Bieb. ;伊比利亚栎●☆

324026　Quercus ieboldiana Blume = Lithocarpus glaber (Thunb.) Nakai ●

324027　Quercus ilex L. ;圣栎（冬青栎）;Evergreen Oak,Grain Tree, Holly Oak,Holm Oak,Hulver Oak,Ilex Oak,Kerm Oak,Kermes Oak,Scarlet Holm Oak,Scarlet Oak,Spanish Oak,Winter Oak ●

324028　Quercus ilex L. subsp. ballota (Desf.) Samp. = Quercus ilex L. ●

324029　Quercus ilex L. subsp. rotundifolia (Lam.) Tab. Morais = Quercus ilex L. subsp. ballota (Desf.) Samp. ●

324030　Quercus ilex L. var. acrodonta (Seemen) V. Naray. = Quercus acrodonta Seemen ●

324031　Quercus ilex L. var. ballota (Desf.) A. DC. = Quercus ilex L. subsp. ballota (Desf.) Samp. ●

324032　Quercus ilex L. var. phillyraeoides Franch. = Quercus phillyraeoides A. Gray ●

324033　Quercus ilex L. var. phillyreoides (A. Gray) Franch. = Quercus phillyraeoides A. Gray ●

324034　Quercus ilex L. var. rufescens Franch. = Quercus guyavifolia H. Lév. ●

324035　Quercus ilex L. var. rutescens Franch. = Quercus pannosa Hand. -Mazz. ●

324036　Quercus ilex L. var. spinosa (A. David ex Franch.) Franch. = Quercus spinosa A. David ex Franch. ●

324037　Quercus ilex L. var. spinosa Franch. = Quercus spinosa A. David ex Franch. ●

324038　Quercus ilicifolia Griff. = Quercus baloot Griff. ●☆

324039　Quercus ilicifolia Wangenh. ;矮橡树（冬青叶栎）;Bear Oak, Scrub Oak ●☆

324040　Quercus ilvicolarum Hance = Lithocarpus silvicolarum (Hance) Chun ●

324041　Quercus imbricaria Michx. ;覆瓦栎（单果栎，短柄栎，沙栎）; Laurel Oak,Shingle Oak ●☆

324042　Quercus imbricata Buch. -Ham. ex D. Don = Quercus lamellosa Sm. ●

324043　Quercus imeretina Stev. ex Woronow;伊梅里特栎●☆

324044　Quercus impressivena Hayata = Lithocarpus brevicaudatus (V. Naray.) Hayata ●

324045　Quercus incana Roxb. ;灰栎;Blue Jack Oak,Bluejack Oak ●☆

324046　Quercus incana W. Bartram = Quercus incana Roxb. ●☆

324047　Quercus incana W. Bartram = Quercus leucotrichophora A. Camus ●☆

324048　Quercus indica (Roxb. ex Lindl.) Drake = Castanopsis indica (Roxb. ex Lindl.) A. DC. ●

324049　Quercus indica (Roxb.) Drake = Castanopsis indica (Roxb.) A. DC. ●

324050　Quercus infectoria Oliv. ;塞浦路斯栎（刺齿栎，没石子，没食子槲，没食子栎，没食子树，墨石子，染色栎，无石子，无食子）; Aleppo Oak,Dyer's Oak,Gall Oak,Nutgall Oak ●☆

324051　Quercus infectoria Oliv. = Quercus pubescens Willd. ●☆

324052　Quercus infralutea Trel. = Quercus chihuahuensis Trel. ●☆

324053　Quercus inpressivena Hayata = Lithocarpus brevicaudatus (V. Naray.) Hayata ●

324054　Quercus insularis Chun et P. C. Tam;南岛椆●

324055　Quercus insularis Chun et P. C. Tam = Cyclobalanopsis phanera (Chun) Y. C. Hsu et H. Wei Jen ●

324056　Quercus intricata Trel. ;缠结栎;Intricate Oak ●☆

324057　Quercus irwinii Hance = Lithocarpus irwinii (Hance) Rehder ●◇

324058　Quercus iteaphylla Hance = Lithocarpus iteaphyllus (Hance) Rehder ●

324059　Quercus ithaburensis Decne. ;塔布尔山栎;Vallonea Oak ●☆

324060　Quercus ithaburensis Decne. = Quercus macrolepis L. ●☆

324061　Quercus ithaburensis Decne. subsp. macrolepis (Kotschy) Hedge et Yalt. = Quercus macrolepis L. ●☆

324062　Quercus jacobi R. Br. = Quercus garryana Douglas ex Hook. ●☆

324063　Quercus jalicensis Trel. = Quercus chihuahuensis Trel. ●☆

324064　Quercus jenkinsii Benth. = Lithocarpus jenkinsii (Benth.) C. C. Huang et Y. T. Chang ●

324065　Quercus jenseniana Hand. -Mazz. = Cyclobalanopsis jenseniana (Hand. -Mazz.) W. C. Cheng et T. Hong ex Q. F. Zheng ●

324066　Quercus jinpinensis (Y. C. Hsu et H. Wei Jen) C. C. Huang =

Cyclobalanopsis jinpinensis Y. C. Hsu et H. Wei Jen ●

324067　Quercus john-tuckeri Nixon et C. H. Mull. ;褶栎; Desert Scrub Oak , Tucker Oak ●☆

324068　Quercus junghuhnii Miq. = Castanopsis carlesii（Hemsl.）Hayata ●

324069　Quercus kawakamii Hayata = Lithocarpus kawakamii（Hayata）Hayata ●

324070　Quercus kawakamii Hayata = Pasania kawakamii（Hayata）Schottky ●

324071　Quercus kelloggii Newb. ;加州黑栎; Balck Oak , California Black Oak , Californian Black Oak , Kellogg Oak ●☆

324072　Quercus kerrii Craib = Cyclobalanopsis kerrii（Craib）Hu ●

324073　Quercus kingiana Craib;澜沧栎(薄叶高山栎); Lancang Oak , Thinleaf Oak , Thin-leaved Oak ●◇

324074　Quercus kirinensis Nakai = Quercus mongolica Fisch. ex Ledeb. ●

324075　Quercus kiukiangensis（Y. T. Chang ex Y. C. Hsu et H. Wei Jen）Y. T. Chang = Cyclobalanopsis kiukiangensis Y. T. Chang ex Y. C. Hsu et H. Wei Jen ●

324076　Quercus kiukiangensis（Y. T. Chang）Y. T. Chang = Cyclobalanopsis kiukiangensis Y. T. Chang ex Y. C. Hsu et H. Wei Jen ●

324077　Quercus kodaihoensis Hayata = Lithocarpus corneus（Lour.）Rehder ●

324078　Quercus kodaihoensis Hayata = Pasania cornea（Lour.）J. C. Liao ●

324079　Quercus kongshanensis Y. C. Hsu et H. Wei Jen;贡山栎; Gongshan Oak , Kungshan Oak ●

324080　Quercus kongshanensis Y. C. Hsu et H. Wei Jen = Quercus engleriana Seemen ●

324081　Quercus konishii Hayata = Lithocarpus konishii（Hayata）Hayata ●

324082　Quercus konishii Hayata = Pasania konishii（Hayata）Schottky ●

324083　Quercus kontumensis A. Camus = Cyclobalanopsis saravanensis（A. Camus）Hjelmq. ●

324084　Quercus kouangsiensis A. Camus = Cyclobalanopsis kouangsiensis（A. Camus）Y. C. Hsu et H. Wei Jen ●◇

324085　Quercus kozloviana Liou = Quercus baronii V. Naray. ●

324086　Quercus kozlowskyi Woronow ex Grossh. ;科兹罗夫斯基栎●☆

324087　Quercus laevis Walter;土耳其光栎（光滑栎）; American Turkey Oak , Catesby Oak , Scrub Oak , Turkey Oak ●☆

324088　Quercus lamelloides C. C. Huang = Cyclobalanopsis lamelloides（C. C. Huang）Y. T. Chang ●

324089　Quercus lamelloides C. C. Huang = Cyclobalanopsis lamellosa（Sm.）Oerst. ●

324090　Quercus lamellosa Sm. = Cyclobalanopsis lamellosa（Sm.）Oerst. ●

324091　Quercus lanata Sm. ;通麦栎●

324092　Quercus lancefolia Roxb. ;披针叶栎●☆

324093　Quercus lanceolata S. Z. Qu et W. H. Zhang;青树栎; Lanceolate Oak ●

324094　Quercus lanceolata S. Z. Qu et W. H. Zhang = Quercus engleriana Seemen ●

324095　Quercus laurifolia Michx. ;月桂叶栎; Darlington Oak , Diamond Leaf Oak , Diamondleaf Oak , Laurel Oak , Swamp Laurel Oak ●☆

324096　Quercus lepidocarpa Hayata = Lithocarpus lepidocarpus（Hayata）Hayata ●

324097　Quercus lesueuri C. H. Mull. = Quercus fusiformis Small ●☆

324098　Quercus leucophylla Ashe = Quercus pagoda Raf. ●☆

324099　Quercus leucotrichophora A. Camus = Quercus lanata Sm. ●

324100　Quercus liaotungensis（Koidz.）Nakai;辽东栎(柴忽拉,柴树,杠木,辽东柞,青冈柳,青杠,小叶青冈); East-Liaoning Oak ●

324101　Quercus liaotungensis（Koidz.）Nakai = Quercus mongolica Fisch. ex Turcz. ●

324102　Quercus liaotungensis（Koidz.）Nakai = Quercus wutaishanica Mayr ●

324103　Quercus liaotungensis Koidz. = Quercus mongolica Fisch. ex Ledeb. ●

324104　Quercus libani Oliv. ;黎巴嫩栎; Lebanon Oak ●☆

324105　Quercus liboensis Z. K. Zhou = Cyclobalanopsis gracilis（Rehder et E. H. Wilson）W. C. Cheng et T. Hong ●

324106　Quercus lichuanensis W. C. Cheng = Quercus phillyraeoides A. Gray ●

324107　Quercus lineata Blume var. grandifolia V. Naray. = Cyclobalanopsis oxyodon（Miq.）Oerst. ●

324108　Quercus lineata Blume var. lobbii Hook. f. et Thomson ex Wenz. = Cyclobalanopsis lobbii（Hook. f. et Thomson ex Wenz.）Y. C. Hsu et H. Wei Jen ●

324109　Quercus lineata Blume var. oxyodon（Miq.）Wenz. = Cyclobalanopsis oxyodon（Miq.）Oerst. ●

324110　Quercus liouana W. C. Cheng et T. Hong = Quercus aliena Blume var. acuteserrata Maxim. ex Wenz. ●

324111　Quercus listeri King = Lithocarpus listeri（King）Grierson et D. G. Long ●

324112　Quercus litseifolia Hance = Lithocarpus litseifolius（Hance）Chun ●

324113　Quercus litseoides Dunn = Cyclobalanopsis litseoides（Dunn）Y. C. Hsu et H. Wei Jen ●

324114　Quercus lobata Née;谷地栎(澳洲栎,谷栎,加州白栎); Bur Oak , California Oak , California Valley Oak , California White Oak , Valley Oak , Valley White Oak , White Oak ●☆

324115　Quercus lobata Née var. breweri（Engelm.）Wenz. = Quercus garryana Douglas ex Hook. ●☆

324116　Quercus lobata Née var. hindsii（Benth.）Wenz. = Quercus lobata Née ●☆

324117　Quercus lobbii（Hook. f. et Thomson ex Wenz.）A. Camus = Cyclobalanopsis lobbii（Hook. f. et Thomson ex Wenz.）Y. C. Hsu et H. Wei Jen ●

324118　Quercus lobbii Etting = Cyclobalanopsis lobbii（Etting）Y. C. Hsu et H. Wei Jen ●

324119　Quercus lodicosa E. F. Warb. ;西藏栎; Tibet Oak , Xizang Oak ●

324120　Quercus longicaudata Hayata = Castanopsis carlesii（Hemsl.）Hayata ●

324121　Quercus longifolia K. Koch;长叶栎●☆

324122　Quercus longiglanda Frém. = Quercus lobata Née ●☆

324123　Quercus longinux Hayata = Cyclobalanopsis longinux（Hayata）Schottky ●

324124　Quercus longinux Hayata var. kanehirai（Nakai）J. C. Liao = Cyclobalanopsis longinux（Hayata）Schottky var. kuoi J. C. Liao ●

324125　Quercus longinux Hayata var. kuoi（J. C. Liao）J. C. Liao = Cyclobalanopsis longinux（Hayata）Schottky var. kuoi J. C. Liao ●

324126　Quercus longinux Hayata var. lativiolaciifolia（J. C. Liao）J. C. Liao = Cyclobalanopsis longinux（Hayata）Schottky var. lativiolaciifolia J. C. Liao ●

324127　Quercus longipes Hu;长柄栎; Long-stalk Oak ●

324128 Quercus longipes Hu = Cyclobalanopsis glauca（Thunb. ex A. Murray）Oerst. ●

324129 Quercus longispica（Hand.-Mazz.）A. Camus；长穗高山栎；Longspike Oak，Long-spiked Oak ●☆

324130 Quercus longispica（Hand.-Mazz.）A. Camus = Quercus rehderiana Hand.-Mazz. ●

324131 Quercus lotungensis Chun et W. C. Ko；乐东栎；Ledong Oak ●

324132 Quercus lunglingensis Hu = Quercus acutissima Carruth. ●

324133 Quercus lungmaiensis（Hu）C. C. Huang et Y. T. Chang = Cyclobalanopsis lungmaiensis Hu ●

324134 Quercus lusitanica Lam.；葡萄牙栎（山毛榉栎）；Cyprus Oak，Dead Sea Apple，Lusitanian Oak，Portuguese Oak，Swamp Laurel Oak ●☆

324135 Quercus lycoperdon V. Naray. = Lithocarpus lycoperdon（V. Naray.）A. Camus ●

324136 Quercus lyoniifolia W. C. Cheng = Quercus engleriana Seemen ●

324137 Quercus lyrata Watt；琴叶栎；Overcup Oak，Swamp Oak，Swamp Post Oak ●☆

324138 Quercus macedonaldii Greene；麦克唐纳栎；Island Scrub Oak，McDonald Oak ●☆

324139 Quercus macedonica A. DC. = Quercus trojana Webb ●☆

324140 Quercus macranthera Fisch. et C. A. Mey. ex Hohen.；高加索栎（大花栎）；Caucasian Oak，Persian Oak ●☆

324141 Quercus macrocarpa F. Michx.；大果栎；Blue Oak，Bur Oak，Burr Oak，Mossy Cup Oak，Mossy-cup Oak ●☆

324142 Quercus macrocarpa F. Michx. subsp. olivaeformis（F. Michx.）A. Camus = Quercus macrocarpa F. Michx. ●☆

324143 Quercus macrocarpa F. Michx. var. depressa（Nutt.）Engelm. = Quercus macrocarpa F. Michx. ●☆

324144 Quercus macrocarpa F. Michx. var. olivaeformis（F. Michx.）A. Gray = Quercus macrocarpa F. Michx. ●☆

324145 Quercus macrocarpa F. Michx. var. olivaeformis（F. Michx.）Trel. = Quercus macrocarpa F. Michx. ●☆

324146 Quercus macrolepis L.；大鳞栎（大鳞片栎，大鳞塔布尔山栎）；Palamut，Turkish Oak，Valonea，Valonia Oak，Velanidi Oak ●☆

324147 Quercus maingayi（Schottky）Burkill；迈氏栎；Maingay's Oak ●☆

324148 Quercus mairei（Schottky）H. Lév. = Lithocarpus mairei（Schottky）Rehder ●

324149 Quercus mairei H. Lév. = Lithocarpus megalophyllus Rehder et E. H. Wilson ●

324150 Quercus malacotricha A. Camus；毛叶槲栎（软毛槲栎，污毛栎）；Hairleaf Oak，Softhaired Oak，Soft-haired Oak ●

324151 Quercus malacotricha A. Camus = Quercus yunnanensis Franch. ●

324152 Quercus mandanensis Rydb. = Quercus macrocarpa F. Michx. ●☆

324153 Quercus margarettae（Ashe）Small；矮沙栎；Dwarf Post Oak，Sand Post Oak ●☆

324154 Quercus mariakii Hayata = Lithocarpus silvicolarum（Hance）Chun ●

324155 Quercus marilandica Münchh.；马里兰德栎（马里兰栎，马利兰栎）；Black Jack Oak，Blackjack，Blackjack Oak，Jack Oak ●☆

324156 Quercus marlipoensis Hu et W. C. Cheng；麻栗坡栎（大叶高山栎）；Malipo Oak ●

324157 Quercus marshii C. H. Mull. = Quercus fusiformis Small ●☆

324158 Quercus maxima（Marshall）Ashe；北部红栎；Champion Oak，Northern Red Oak，Red Oak ●☆

324159 Quercus maxima（Marshall）Ashe = Quercus rubra L. ●☆

324160 Quercus maxima Ashe = Quercus rubra L. ●☆

324161 Quercus meihuashanensis（Q. F. Zheng）C. C. Huang = Cyclobalanopsis meihuashanensis Q. F. Zheng ●

324162 Quercus meihuashanensis（Q. F. Zheng）C. C. Huang = Cyclobalanopsis obovatifolia（C. C. Huang）Y. C. Hsu et H. Wei Jen ●

324163 Quercus meridionalis Liou = Quercus aliena Blume var. acuteserrata Maxim. ex Wenz. ●

324164 Quercus meridionalis Liou var. chungnanensis Liou = Quercus aliena Blume var. acuteserrata Maxim. ex Wenz. ●

324165 Quercus meridionalis Liou var. gnanensis Liou = Quercus aliena Blume var. acuteserrata Maxim. ex Wenz. ●

324166 Quercus michauxii Nutt.；湿地栎（米楚栎，湿生栎）；Basket Oak，Chestnut Oak，Cow Oak，Dwarf-oak，Michaux. Oak，Rock Chestnut Oak，Rock Oak，Swamp Chestnut Oak，Swamp Chestnut-oak ●☆

324167 Quercus minima（Sarg.）Small；极小栎；Minimal Oak ●☆

324168 Quercus minor（Marshall）Sarg. = Quercus stellata Wangenh. ●☆

324169 Quercus minor（Marshall）Sarg. var. margarettae Ashe = Quercus margarettae（Ashe）Small ●☆

324170 Quercus mirbeckii Durieu = Quercus canariensis Willd. ●☆

324171 Quercus miyagii Koidz.；宫木栎●☆

324172 Quercus mohriana Buckley ex Rydb.；莫尔栎；Mohr Oak，Scrub Oak，Shin Oak ●☆

324173 Quercus mongolica Fisch. ex Ledeb.；蒙古栎（粗齿蒙古栎，蒙古柞，蒙栎，青风栎，青冈栎，青刚木，青岗，青栎，橡子树，小叶槲树，小叶柞，小叶柞树，柞，柞栎，柞木，柞树）；Japanese Oak，Mongol Oak，Mongolian Oak ●

324174 Quercus mongolica Fisch. ex Ledeb. f. lomgifolia（Nakai）Kitam. et T. Horik. = Quercus crispula Blume ●

324175 Quercus mongolica Fisch. ex Ledeb. f. macrocarpa（Nakai）Kitam. et T. Horik. = Quercus crispula Blume ●

324176 Quercus mongolica Fisch. ex Ledeb. subsp. crispula（Blume）Menitsky = Quercus mongolica Fisch. ex Ledeb. ●

324177 Quercus mongolica Fisch. ex Ledeb. subsp. crispula（Blume）Menitsky = Quercus crispula Blume ●

324178 Quercus mongolica Fisch. ex Ledeb. var. grosseserrata（Blume）Rehder et E. H. Wilson f. laciniata Hayashi；撕裂栎●☆

324179 Quercus mongolica Fisch. ex Ledeb. var. grosseserrata（Blume）Rehder et E. H. Wilson = Quercus mongolica Fisch. ex Ledeb. ●

324180 Quercus mongolica Fisch. ex Ledeb. var. grosseserrata（Blume）Rehder et E. H. Wilson = Quercus crispula Blume ●

324181 Quercus mongolica Fisch. ex Ledeb. var. kirinensis（Nakai）Kitag. = Quercus mongolica Fisch. ex Ledeb. ●

324182 Quercus mongolica Fisch. ex Ledeb. var. liaotungensis（Koidz.）Nakai = Quercus mongolica Fisch. ex Ledeb. ●

324183 Quercus mongolica Fisch. ex Ledeb. var. liaotungensis（Koidz.）Nakai = Quercus liaotungensis（Koidz.）Nakai ●

324184 Quercus mongolica Fisch. ex Ledeb. var. macrocarpa H. Wei Jen et L. M. Wang = Quercus mongolica Fisch. ex Ledeb. ●

324185 Quercus mongolica Fisch. ex Ledeb. var. manschurica（Koidz.）Nakai = Quercus mongolica Fisch. ex Ledeb. ●

324186 Quercus mongolica Fisch. ex Ledeb. var. undulatifolia Kitam. et T. Horik. = Quercus crispula Blume var. horikawae H. Ohba ●☆

324187 Quercus mongolica Fisch. ex Turcz. = Quercus mongolica Fisch. ex Ledeb. ●

324188 Quercus mongolica Fisch. ex Turcz. subsp. crispula（Blume）Menitsky = Quercus mongolica Fisch. ex Turcz. ●

324189 Quercus mongolica Fisch. ex Turcz. subsp. crispula（Blume）Menitsky = Quercus mongolica Fisch. ex Turcz. var. grosserrata（Blume）Rehder et E. H. Wilson ●

324190 Quercus mongolica Fisch. ex Turcz. var. grosserrata（Blume）Rehder et E. H. Wilson；粗齿蒙古栎；Coarse-serrate Mongolian Oak，Coarseserrate Oak，Shallow-cup Mongolian Oak ●

324191 Quercus mongolica Fisch. ex Turcz. var. grosserrata（Blume）Rehder et E. H. Wilson = Quercus mongolica Fisch. ex Turcz. ●

324192 Quercus mongolica Fisch. ex Turcz. var. grosserrata Rehder et E. H. Wilson = Quercus mongolica Fisch. ex Turcz. var. grosserrata（Blume）Rehder et E. H. Wilson ●

324193 Quercus mongolica Fisch. ex Turcz. var. kirinensis（Nakai）Kitag. = Quercus mongolica Fisch. ex Turcz. ●

324194 Quercus mongolica Fisch. ex Turcz. var. liaotungensis（Koidz.）Nakai = Quercus liaotungensis（Koidz.）Nakai ●

324195 Quercus mongolica Fisch. ex Turcz. var. macrocarpa H. Wei Jen et L. M. Wang；大果蒙古栎；Bigfruit Mongolian Oak ●

324196 Quercus mongolica Fisch. ex Turcz. var. macrocarpa H. Wei Jen et L. M. Wang = Quercus mongolica Fisch. ex Turcz. ●

324197 Quercus mongolica Fisch. ex Turcz. var. manshurica（Koidz.）Nakai = Quercus mongolica Fisch. ex Turcz. ●

324198 Quercus mongolico-dentata Nakai；柞槲栎；Mongol-Daimyo Oak，Mongolian-Daimyo Oak ●

324199 Quercus monimotricha Hand. -Mazz.；矮山栎（矮高山栎）；Dwarf Alpine Oak，Dwarf Oak ●

324200 Quercus monnula Y. C. Hsu et H. Wei Jen；长叶枹栎；Agreeable Oak，Longleaf Oak，Long-leaved Oak ●

324201 Quercus montana Willd.；山地栎（栗栎）；Chestnut Oak，Mountain Chestnut Oak，Rock Chestnut Oak，Rock Oak ●☆

324202 Quercus morii Hayata = Cyclobalanopsis morii（Hayata）Schottky ●

324203 Quercus motuoensis C. C. Huang = Cyclobalanopsis motuoensis（C. C. Huang）Y. C. Hsu et H. Wei Jen ●

324204 Quercus muchlenbergii Engelm.；黄栎（黄栗栎，米伦贝格栎）；Chestnut Oak，Chinkapin，Chinkapin Oak，Chinquapin，Chinquapin Oak，Rock Oak，Yellow Chestnut Oak，Yellow Oak ●☆

324205 Quercus muchlenbergii Engelm. f. alexanderi（Britton）Trel. = Quercus muchlenbergii Engelm. ●☆

324206 Quercus muliensis Hu = Quercus seneacens Hand. -Mazz. var. muliensis（Hu）Y. C. Hsu et H. Wei Jen ●

324207 Quercus myricifolia Hu = Quercus phillyraeoides A. Gray ●

324208 Quercus myricifolia Hu et W. C. Cheng = Quercus phillyraeoides A. Gray ●

324209 Quercus myrsinifolia Blume = Cyclobalanopsis myrsinifolia（Blume）Oerst. ●

324210 Quercus myrtifolia Willd.；香桃叶栎；Myrtle Oak，Scrub Oak ●☆

324211 Quercus naiadarum Hance = Lithocarpus naiadarum（Hance）Chun ●

324212 Quercus nana（Marshall）Sarg. = Quercus ilicifolia Wangenh. ●☆

324213 Quercus nana Willd. = Quercus nigra L. ●☆

324214 Quercus nanchuanica C. C. Huang；南川青冈；Nanchuan Oak ●

324215 Quercus nanchuanica C. C. Huang = Cyclobalanopsis gambleana（A. Camus）Y. C. Hsu et H. Wei Jen ●

324216 Quercus nantoensis Hayata = Lithocarpus nantoensis（Hayata）Hayata ●

324217 Quercus nantoensis Hayata = Pasania nantoensis（Hayata）Schottky ●

324218 Quercus nariakii Hayata = Lithocarpus silvicolarum（Hance）Chun ●

324219 Quercus neaei Liebm. = Quercus garryana Douglas ex Hook. ●☆

324220 Quercus neglecta（Schottky）Koidz. = Cyclobalanopsis bambusifolia（Hance）Chun ex Y. C. Hsu et H. Wei Jen ●

324221 Quercus neglecta（Schottky）Koidz. = Cyclobalanopsis neglecta Schottky ●

324222 Quercus nemoralis Chun = Cyclobalanopsis kouangsiensis（A. Camus）Y. C. Hsu et H. Wei Jen ●◇

324223 Quercus neoglandulifera Nakai；拟枹栎 ●☆

324224 Quercus neriifolia Seemen = Lithocarpus naiadarum（Hance）Chun ●

324225 Quercus nigra L.；美国黑栎（黑栎，水栎）；Blackjack Oak，Duck Oak，Possum Oak，Spotted Oak，Water Oak ●☆

324226 Quercus nigra L. var. tridentifera Sarg. = Quercus nigra L. ●☆

324227 Quercus ningangensis（W. C. Cheng et Y. C. Hsu）C. C. Huang = Cyclobalanopsis ningangensis W. C. Cheng et Y. C. Hsu ●

324228 Quercus ningqiangensis S. Z. Qu et W. H. Zhang；宁强栎；Ningqiang Oak ●

324229 Quercus ningqiangensis S. Z. Qu et W. H. Zhang = Quercus serrata Thunb. ●

324230 Quercus novomexicana Rydb. = Quercus fusiformis Small ●☆

324231 Quercus nubium Hand. -Mazz. = Cyclobalanopsis sessilifolia（Blume）Schottky ●

324232 Quercus nuttallii E. J. Palmer；美国南方栎；Nuttal Oak，Pin Oak，Red Oak ●☆

324233 Quercus nuttallii E. J. Palmer = Quercus texana Buckley ●☆

324234 Quercus obconicus Y. C. Hsu ex Z. K. Zhou = Cyclobalanopsis litoralis Chun et P. C. Tam ex Y. C. Hsu et H. Wei Jen ●

324235 Quercus oblongifolia Torr.；墨西哥蓝栎；Mexican Blue Oak，Sonoran Blue Oak ●☆

324236 Quercus oboconicus Y. C. Hsu ex Z. K. Zhou = Cyclobalanopsis litoralis Chun et P. C. Tam ●

324237 Quercus obovata Bunge = Quercus dentata Thunb. ●

324238 Quercus obovatifolia C. C. Huang = Cyclobalanopsis obovatifolia（C. C. Huang）Y. C. Hsu et H. Wei Jen ●

324239 Quercus obscura Seemen = Quercus engleriana Seemen ●

324240 Quercus obtusa（Willd.）Ashe；钝栎；Diamondleaf Oak，Diamond-leaf Oak，Laurel Oak ●☆

324241 Quercus obtusa（Willd.）Ashe = Quercus laurifolia Michx. ●☆

324242 Quercus obtusifolia D. Don = Quercus semicarpifolia Sm. ●

324243 Quercus obtusifolia D. Don var. breviloba Torr. = Quercus sinuata Walter var. breviloba（Torr.）C. H. Mull. ●☆

324244 Quercus obtusifolia Michx.；钝叶栎；Iron Oak，Post Oak ●☆

324245 Quercus obtusiloba Michx. = Quercus stellata Wangenh. ●☆

324246 Quercus obtusiloba Michx. var. depressa Nutt. = Quercus macrocarpa F. Michx. ●☆

324247 Quercus occidentalis J. Gay；西方栎；Biennial Cork Oak ●☆

324248 Quercus oglethorpensis Duncan；奥格栎；Oglethorpe Oak ●☆

324249 Quercus olivaeformis F. Michx. = Quercus macrocarpa F. Michx. ●☆

324250 Quercus oxyodon Miq. = Cyclobalanopsis oxyodon（Miq.）Oerst. ●

324251 Quercus oxyphylla（E. H. Wilson）Hand. -Mazz.；尖叶栎（铁槲树）；Sharpleaf Oak，Sharp-leaved Oak ●

324252 Quercus pachyloma Seemen = Cyclobalanopsis pachyloma（Seemen）Schottky ●

324253 Quercus pachyloma Seemen var. mubianensis（Y. C. Hsu et H. Wei Jen）C. C. Huang = Cyclobalanopsis pachyloma（Seemen）Schottky ●

324254 Quercus pachyloma Seemenvar. mubianensis（Y. C. Hsu et H.

Wei Jen) C. C. Huang ＝ Cyclobalanopsis pachyloma (Seemen) Schottky var. mubianensis Y. C. Hsu et H. Wei Jen ●

324255　Quercus pachyphylla Kurz ＝ Lithocarpus pachyphyllus (Kurz) Rehder ●

324256　Quercus pachyphylla Kurz var. fruticosa G. Watt ex King ＝ Lithocarpus pachyphyllus (Kurz) Rehder var. fruticosus (G. Watt ex King) A. Camus ●

324257　Quercus pachyphylla Kurz var. fruticosus Watt ex King ＝ Lithocarpus pachyphyllus (Kurz) Rehder var. fruticosus (G. Watt ex King) A. Camus ●

324258　Quercus pacifica Nixon et C. H. Mull. ;多果栎●☆

324259　Quercus pagoda Raf. ＝ Quercus falcata Michx. var. pagodifolia (Elliott) E. Murray ●☆

324260　Quercus pagodifolia (Elliott) Ashe ＝ Quercus pagoda Raf. ●☆

324261　Quercus palmeri Engelm. ;帕默栎;Palmer Oak ●☆

324262　Quercus palustris Münchh. ;沼生栎(美国沼地栎);Marsh Oak,Paludal Oak,Pin Oak,Spanish Oak,Swamp Oak ●☆

324263　Quercus pannonica Endl. ＝ Quercus frainetto Ten. ●☆

324264　Quercus pannosa Hand. -Mazz. ;黄背栎(椶白槲);Yellowback Oak,Yellow-backed Oak,Yellowleafback Oak ●

324265　Quercus pannosa Hand. -Mazz. ＝ Quercus guyavifolia H. Lév. ●

324266　Quercus paohangii Chun et Tsiang ＝ Castanopsis uraiana (Hayata) Kaneh. et Hatus. ●

324267　Quercus parvifolia Hand. -Mazz. ＝ Quercus acrodonta Seemen ●

324268　Quercus parvula Greene var. shrevei (C. H. Mull.) Nixon ＝ Quercus wislizenii A. DC. ●☆

324269　Quercus parvula Greene var. tamalpaisensis S. Langer ＝ Quercus wislizenii A. DC. ●☆

324270　Quercus patelliformis Chun ＝ Cyclobalanopsis patelliformis (Chun) Y. C. Hsu et H. Wei Jen ●

324271　Quercus paucidentata Franch. ＝ Cyclobalanopsis sessilifolia (Blume) Schottky ●

324272　Quercus paucidentata Franch. ex Nakai ＝ Cyclobalanopsis sessilifolia (Blume) Schottky ●

324273　Quercus peduncularis Née ＝ Quercus robur L. ●

324274　Quercus pedunculata Ehrh. ＝ Quercus robur L. ●

324275　Quercus pedunculiflora K. Koch;垂花栎●☆

324276　Quercus penduculata Ehrh. ＝ Quercus robur L. ●

324277　Quercus pentacycla Y. T. Chang ＝ Cyclobalanopsis pentacycla (Y. T. Chang) Y. T. Chang ex Y. C. Hsu et H. Wei Jen ●

324278　Quercus petraea (Matt.) Liebl. ;无梗栎(欧岩栎,无柄花栎,无柄叶栎,无梗花栎,岩生栎);Bay Oak,Durmast Oak,English Oak,French Oak,Malden Oak,Oak,Polish Oak,Sessile Oak,Short-stalked Oak,Slavonian Oak,White Oak ●☆

324279　Quercus petraea Liebl. 'Columna';柱冠欧岩栎●☆

324280　Quercus petraea Liebl. ＝ Quercus petraea (Matt.) Liebl. ●☆

324281　Quercus phanera Chun ＝ Cyclobalanopsis phanera (Chun) Y. C. Hsu et H. Wei Jen ●

324282　Quercus phellos L. ;柳叶栎(柳栎);Peach Oak,Peachleaf Oak,Pin Oak,Willow Oak,Willow-leafed Oak ●☆

324283　Quercus phillyraeoides A. Gray;乌冈栎(九冈树,九刚树,姥芽栎,楞芽栎,青冈树,石楠柴,乌岗山栎);Mocketprivet-like Oak,Ubame Oak,Wugang Oak ●

324284　Quercus phillyraeoides A. Gray f. crispa (Matsum.) Kitam. et T. Horik. ;皱波乌冈栎●☆

324285　Quercus phillyraeoides A. Gray f. subcrispa (Matsum.) Kitam. et T. Horik. ;微皱乌冈栎●☆

324286　Quercus phillyraeoides A. Gray f. wrightii (Nakai) Makino;赖氏乌冈栎●☆

324287　Quercus phillyraeoides A. Gray subsp. fokienensis (Nakai) Menitsky ＝ Quercus phillyraeoides A. Gray ●

324288　Quercus phullata Buch. -Ham. ex D. Don ＝ Quercus glauca Thunb. ●

324289　Quercus picata Sm. ＝ Lithocarpus henryi (Seemen) Rehder et E. H. Wilson ●

324290　Quercus picata Sm. var. collettii Hook. f. ＝ Lithocarpus collettii (King ex Hook. f.) A. Camus ●

324291　Quercus picata Sm. var. collettii Kiger ＝ Lithocarpus fohaiensis (Hu) A. Camus ●

324292　Quercus picata Sm. var. gracilipes Hook. f. ＝ Lithocarpus himalaicus C. C. Huang et Y. T. Chang ●

324293　Quercus pileata Hu et W. C. Cheng ＝ Quercus guyavifolia H. Lév. ●

324294　Quercus pinbianensis (Y. C. Hsu et H. Wei Jen) C. C. Huang et Y. T. Chang ＝ Cyclobalanopsis jenseniana (Hand. -Mazz.) W. C. Cheng et T. Hong ex Q. F. Zheng ●

324295　Quercus pinbianensis (Y. C. Hsu et H. Wei Jen) C. C. Huang et Y. T. Chang ＝ Cyclobalanopsis pinbianensis Y. C. Hsu et H. Wei Jen ●

324296　Quercus pinfaensis H. Lév. ＝ Castanopsis fargesii Franch. ●

324297　Quercus platanoides (Lam.) Sudw. ＝ Quercus bicolor Willd. ●☆

324298　Quercus poilanei Hickel et A. Camus ＝ Cyclobalanopsis poilanei (Hickel et A. Camus) Hjelmq. ●

324299　Quercus polymorpha Schltdl. et Cham. ;多形栎;Coahuila Oak,Net-leaf White Oak ●☆

324300　Quercus polystachya V. Naray. ＝ Lithocarpus litseifolius (Hance) Chun ●

324301　Quercus pontica K. Koch;亚美尼亚栎;Armenian Oak,Pontic Oak,Pontine Oak ●☆

324302　Quercus prainiana H. Lév. ＝ Cyclobalanopsis helferiana (A. DC.) Oerst. ●

324303　Quercus pricei Sudw. ＝ Quercus agrifolia Née ●☆

324304　Quercus prinoides Willd. ;粉栎;Dwarf Chestnut Oak,Dwarf Chinkapin,Dwarf Chinquapin,Scrub Chestnut Oak,Scrub Oak ●☆

324305　Quercus prinoides Willd. var. acuminata (Michx.) Gleason ＝ Quercus muchlenbergii Engelm. ●☆

324306　Quercus prinoides Willd. var. rufescens Rehder ＝ Quercus prinoides Willd. ●☆

324307　Quercus prinus L. ＝ Quercus michauxii Nutt. ●☆

324308　Quercus prinus L. var. acuminata Michx. ＝ Quercus muchlenbergii Engelm. ●☆

324309　Quercus pseudococcifera Labill. ;亚伯拉罕栎;Abraham's Oak,Oak of Mamre ●☆

324310　Quercus pseudoglauca Z. K. Zhou ＝ Cyclobalanopsis pseudoglauca Y. K. Li et X. M. Wang ●

324311　Quercus pseudomyrsinifolia Hayata ＝ Cyclobalanopsis longinux (Hayata) Schottky ●

324312　Quercus pseudosemicarpifolia A. Camus ＝ Quercus rehderiana Hand. -Mazz. ●

324313　Quercus pseudoserrata Liou ＝ Quercus baronii V. Naray. ●

324314　Quercus pubescens Willd. ;柔毛栎(密毛栎,棉栎);Aleppo Galls,Downy Oak,Levant Galls,Mecca Galls,Pubescent Oak,Turkish Galls,White Oak ●☆

324315　Quercus pumila Walter;铺散栎;Runner Oak ●☆

324316　Quercus pungens Liebm. ;锐尖栎;Holly Oak,Pungent Oak,

Scrub Oak ●☆

324317　Quercus pungens Liebm. var. vaseyana（Buckley）C. H. Mull. = Quercus vaseyana Buckley ●☆

324318　Quercus pypargyrea（Seemen）C. C. Huang et Y. T. Chang = Cyclobalanopsis multinervis W. C. Cheng et T. Hong ●

324319　Quercus pyrenaica Willd.；比利牛斯栎；Pyrenean Oak, Pyrenees Oak ●☆

324320　Quercus randaiensis Hayata = Castanopsis uraiana（Hayata）Kaneh. et Hatus. ●

324321　Quercus ransomii Kellogg = Quercus douglasii Hook. et Arn. ●☆

324322　Quercus rehderiana Hand. -Mazz.；毛脉栎（高山栎, 光叶高山栎, 光叶山栎, 毛脉高山栎）；False Alpine Oak, Hairvein Alpine Oak, High-moutain Oak, Rehder Oak, Smoothleaf Alpine Oak ●

324323　Quercus reinwardtii Drake = Lithocarpus pseudoreinwardtii（Drake）A. Camus ●

324324　Quercus repandifolia J. C. Liao = Cyclobalanopsis glauca（Thunb. ex A. Murray）Oerst. ●

324325　Quercus repandifolia J. C. Liao = Cyclobalanopsis repandifolia（J. C. Liao）J. C. Liao ●

324326　Quercus reticulata Engelm.；网叶栎；Net-leaf Oak ●☆

324327　Quercus reticulata Humb. et Bonpl. = Quercus rugosa Née ●☆

324328　Quercus rex Hemsl. = Cyclobalanopsis rex（Hemsl.）Schottky ● ◇

324329　Quercus rhodophlebia Trel. = Quercus rugosa Née ●☆

324330　Quercus rhombica Sarg. = Quercus laurifolia Michx. ●☆

324331　Quercus rhombocarpa Hayata = Lithocarpus taitoensis（Hayata）Hayata ●

324332　Quercus rhombocarpa Hayata = Lithocarpus taitoensis C. C. Huang et Y. T. Chang ●

324333　Quercus rhombocarpa Hayata = Pasania synbalanos（Hance）Schottky ●

324334　Quercus rigida K. Koch ex A. DC.；硬栎●☆

324335　Quercus robur L.；英国栎（欧洲白栎, 欧洲栎, 夏栎, 夏橡, 夏橡树, 橡树, 柞栎, 柞树）；Achorn, Acorn, Acorn Tree, Aik, Aitchorn, Akyr, Archarde, Atchern, Ayk, Black Oak, Common Oak, Cup-and-ladle, Eacor, Eak, Eike-tree, Eke-tree, Ekkern, English Oak, European Oak, French Oak, Geanucanach's Pipes, Golden Oak, Hakerne, Hatch-horn, Jove's Nut, Knappers, Mace, Macey, Macey-tree, Mask, Mass, Mess, Oak, Oak Atchern, Oak Nut, Oakberry, Oak-macey, Ovest, Pedunculate Oak, Polish Oak, Robur Oak, Shick-shack Tree, Slavonian Oak, Sussex Weed, Tom Paine, Truffle Oak, Wuk, Yak, Yakkron, Yeaker, Yek, Yik ●

324336　Quercus robur L. 'Argenteovariegata'；银边欧洲栎●☆

324337　Quercus robur L. 'Concordia'；肯考迪亚欧洲栎（黄叶夏栎）；Golden Oak ●☆

324338　Quercus robur L. 'Fastigiata'；帚状欧洲栎（密枝夏栎）；Columnar English Oak, Cypress Oak ●☆

324339　Quercus robur L. 'Pendula'；垂枝欧洲栎；Weeping Oak ●☆

324340　Quercus robur L. subsp. eu-robur A. Camus = Quercus robur L. ●

324341　Quercus robur L. var. fastigiata ? = Quercus robur L. 'Fastigiata' ●☆

324342　Quercus robusta C. H. Mull.；粗壮栎●☆

324343　Quercus rotundifolia Lam.；圆叶栎；Round-leaved Oak ●☆

324344　Quercus rubra L.；美国红栎（赤栎, 红槲栎, 红栎）；American Red Oak, Champion Oak, Gray Oak, Northern Red Oak, Red Oak, Southern Red Oak, Spanish Oak ●☆

324345　Quercus rubra L. 'Aurea'；金叶红栎●☆

324346　Quercus rubra L. var. ambigua（F. Michx.）Fernald = Quercus rubra L. ●☆

324347　Quercus rubra L. var. ambigua（F. Michx.）Houba = Quercus rubra L. ●☆

324348　Quercus rubra L. var. borealis（F. Michx.）Farw. = Quercus rubra L. ●☆

324349　Quercus rubra L. var. maxima Marshall = Quercus rubra L. ●☆

324350　Quercus rubra L. var. pagodifolia（Elliott）Ashe；沼地红栎；Swamp Red Oak ●☆

324351　Quercus rubra L. var. texana（Buckley）Buckley = Quercus texana Buckley ●☆

324352　Quercus rufescens Hook. f. et Thunb. = Castanopsis wattii（King）A. Camus ●

324353　Quercus rugosa Née；皱皮栎（网叶栎）；Née Tree, Netleaf Oak ●☆

324354　Quercus runcinata Engelm.；倒齿栎；Botton Oak ●☆

324355　Quercus sacame Trel. = Quercus arizonica Sarg. ●☆

324356　Quercus sadleriana R. Br. ter.；鹿栎；Deer Oak ●☆

324357　Quercus salicina Blume = Cyclobalanopsis salicina（Blume）Oerst. ●

324358　Quercus salicina Blume f. angustata（Nakai）H. Ohba；窄柳叶青冈●☆

324359　Quercus salicina Blume f. latifolia（Nakai）H. Ohba；阔柳叶青冈●☆

324360　Quercus salicina Blume var. stenophylla（Blume）Hatus. = Quercus salicina Blume ●

324361　Quercus salicina Blume var. stenophylloides（Hayata）S. S. Ying = Cyclobalanopsis stenophylloides（Hayata）Kudo et Masam. ex Kudo ●

324362　Quercus sansabeana Buckley ex M. J. Young = Quercus sinuata Walter var. breviloba（Torr.）C. H. Mull. ●☆

324363　Quercus santaclarensis C. H. Mull. = Quercus chihuahuensis Trel. ●☆

324364　Quercus saravanensis A. Camus = Cyclobalanopsis saravanensis（A. Camus）Hjelmq. ●

324365　Quercus sasakii Kaneh. = Cyclobalanopsis glauca（Thunb. ex A. Murray）Oerst. ●

324366　Quercus schneckii Britton = Quercus shumardii Buckley ●☆

324367　Quercus schottkyana Rehder et E. H. Wilson = Cyclobalanopsis glaucoides Schottky ●

324368　Quercus sclerophylla Lindl. = Castanopsis sclerophylla（Lindl.）Schottky ●

324369　Quercus semecarpifolia Sm. var. glabra Franch. = Quercus rehderiana Hand. -Mazz. ●

324370　Quercus semecarpifolia Sm. var. longispica Hand. -Mazz. = Quercus rehderiana Hand. -Mazz. ●

324371　Quercus semecarpifolia Sm. var. rufescens（Franch.）Schottky = Quercus guyavifolia H. Lév. ●

324372　Quercus semecarpifolia Sm. var. spinosa（A. David ex Franch.）Schottky = Quercus spinosa A. David ex Franch. ●

324373　Quercus semicarpifolia Sm.；高山栎；Alpine Oak ●

324374　Quercus semicarpifolia Sm. var. glabra Franch. = Quercus rehderiana Hand. -Mazz. ●

324375　Quercus semicarpifolia Sm. var. longispica Hand. -Mazz. = Quercus longispica（Hand. -Mazz.）A. Camus ●☆

324376　Quercus semicarpifolia Sm. var. longispica Hand. -Mazz. = Quercus rehderiana Hand. -Mazz. ●

324377　Quercus semicarpifolia Sm. var. rufescens Schottky = Quercus monimotricha Hand. -Mazz. ●

324378　Quercus semicarpifolia Sm. var. spinosa Schottky ＝ Quercus spinosa A. David ex Franch. ●

324379　Quercus semiserrata Roxb. ＝ Cyclobalanopsis semiserrata (Roxb.) Oerst. ●

324380　Quercus semiserratoides (Y. C. Hsu et H. Wei Jen) C. C. Huang et Y. T. Chang ＝ Cyclobalanopsis semiserrata (Roxb.) Oerst. ●

324381　Quercus semiserratoides (Y. C. Hsu et H. Wei Jen) C. C. Huang et Y. T. Zhang ＝ Cyclobalanopsis semiserratoides Y. C. Hsu et H. Wei Jen ●

324382　Quercus senescens Hand. -Mazz.；灰背栎（灰背高山栎）；Greyback Oak，Grey-dorsal Oak，Greyleafback Oak ●

324383　Quercus senescens Hand. -Mazz. var. muliensis (Hu) Y. C. Hsu et H. Wei Jen；木里栎；Muli Oak ●

324384　Quercus serrata Murray；枹栎（枹栎，槲橡，栎，青刚树，土骨皮，橡椀树）；Oak，Serrate Oak ●

324385　Quercus serrata Murray f. concolor (Sugim.) H. Ohba；同色枹栎●☆

324386　Quercus serrata Murray f. dependens Nakai；悬垂枹栎●☆

324387　Quercus serrata Murray f. donarium (Nakai) Kitam.；神地枹栎●☆

324388　Quercus serrata Murray f. longicarpa (Uyeki) Kitam. et T. Horik.；长果枹栎●☆

324389　Quercus serrata Murray f. pinnata (Koidz.) Kitam. et T. Horik.；羽状枹栎●☆

324390　Quercus serrata Murray f. pinnatifida (Koidz.) Kitam. et T. Horik.；羽裂枹栎●☆

324391　Quercus serrata Murray f. suberosa Hayashi ＝ Quercus serrata Murray ●

324392　Quercus serrata Murray var. brevipetiolata (A. DC.) Nakai ＝ Quercus glandulifera Blume var. brevipetiolata (A. DC.) Nakai ●

324393　Quercus serrata Murray var. brevipetiolata (A. DC.) Nakai ＝ Quercus serrata Murray ●

324394　Quercus serrata Murray var. concolor Sugim. ＝ Quercus serrata Murray f. concolor (Sugim.) H. Ohba ●☆

324395　Quercus serrata Murray var. pseudovariabilis Nakai；易变枹栎●☆

324396　Quercus serrata Murray var. tomentosa (B. C. Ding et T. B. Chao) Y. C. Hsu et W. Jen ＝ Quercus serrata Murray ●

324397　Quercus serrata Siebold et Zucc. ＝ Quercus acutissima Carruth. ●

324398　Quercus serrata Thunb. ＝ Quercus serrata Murray ●

324399　Quercus serrata Thunb. var. brevipetiolata (A. DC.) Nakai ＝ Quercus glandulifera Blume var. brevipetiolata (A. DC.) Nakai ●

324400　Quercus serrata Thunb. var. brevipetiolata (A. DC.) Nakai ＝ Quercus serrata Murray ●

324401　Quercus serrata Thunb. var. tomentosa (B. C. Ding et T. B. Chao) Y. C. Hsu et H. Wei Jen；绒毛枹栎；Tomentose Serrate Oak ●

324402　Quercus serrata Thunb. var. tomentosa (B. C. Ding et T. B. Chao) Y. C. Hsu et H. Wei Jen ＝ Quercus serrata Thunb. ●

324403　Quercus serrata Thunb. var. variabilis (Blume) Matsum. ＝ Quercus variabilis Blume ●

324404　Quercus serratoides Uyeki；齿状栎●☆

324405　Quercus sessiliflora Salisb. ＝ Quercus petraea (Matt.) Liebl. ●☆

324406　Quercus sessiliflora Salisb. var. mongolica (Fisch. ex Ledeb.) Franch. ＝ Quercus mongolica Fisch. ex Ledeb. ●

324407　Quercus sessiliflora Salisb. var. mongolica Franch. ＝ Quercus mongolica Fisch. ex Turcz. ●

324408　Quercus sessilifolia Blume ＝ Cyclobalanopsis sessilifolia (Blume) Schottky ●

324409　Quercus setulosa Hickel et A. Camus；富宁栎；Funing Oak，Setose Oak，Ssetulose Oak，Thinbristle Oak ●

324410　Quercus shangxiensis Z. K. Zhou ＝ Quercus engleriana Seemen ●

324411　Quercus shangxiensis Z. K. Zhou ＝ Quercus lanceolata S. Z. Qu et W. H. Zhang ●

324412　Quercus shennongii C. C. Huang et S. H. Fu ＝ Cyclobalanopsis gracilis (Rehder et E. H. Wilson) W. C. Cheng et T. Hong ●

324413　Quercus shennongii C. C. Huang et S. H. Fu ＝ Cyclobalanopsis shennongii (C. C. Huang et S. H. Fu) Y. C. Hsu et H. Wei Jen ●

324414　Quercus shingjenensis Y. T. Chang ＝ Cyclobalanopsis disciformis (Chun et Tsiang) Y. C. Hsu et H. Wei Jen ●

324415　Quercus shumardii Buckley；舒马栎（苏麻德栎）；Shumard Oak，Shumard Red Oak，Shumard's Oak，Shumard's Red Oak，Spotted Oak，Swamp Red Oak，Texas Red Oak，Water Oak ●☆

324416　Quercus shumardii Buckley var. acerifolia E. J. Palmer ＝ Quercus acerifolia (E. J. Palmer) Stoynoff et W. J. Hess ●☆

324417　Quercus shumardii Buckley var. microcarpa (Torr.) Shinners ＝ Quercus gravesii Sudw. ●☆

324418　Quercus shumardii Buckley var. texana (Buckley) Ashe ＝ Quercus texana Buckley ●☆

324419　Quercus sichourensis (Hu) C. C. Huang et Y. T. Chang ＝ Cyclobalanopsis sichourensis Hu ●◇

324420　Quercus sieboldiana Blume ＝ Lithocarpus glaber (Thunb.) Nakai ●

324421　Quercus silvicolarum Hance ＝ Lithocarpus silvicolarum (Hance) Chun ●

324422　Quercus similis Ashe；相似栎；Delta Post Oak，Swamp Post Oak ●☆

324423　Quercus singuliflora (H. Lév.) A. Camus ＝ Quercus phillyraeoides A. Gray ●

324424　Quercus singuliflora A. Camus ＝ Quercus phillyraeoides A. Gray ●

324425　Quercus sinii Chun ＝ Quercus setulosa Hickel et A. Camus ●

324426　Quercus sinuata Walter；深波栎；Bastard Oak，Bastard White Oak，Durand Oak ●☆

324427　Quercus sinuata Walter var. breviloba (Torr.) C. H. Mull.；浅波栎●☆

324428　Quercus skaniana Dunn ＝ Lithocarpus skanianus (Dunn) Rehder ●

324429　Quercus spathulata Seemen ＝ Quercus dolicholepis A. Camus ●

324430　Quercus spathulata Seemen var. elliptica Y. C. Hsu et H. Wei Jen ＝ Quercus dolicholepis A. Camus var. elliptica (Y. C. Hsu et H. Wei Jen) Y. C. Hsu et H. Wei Jen ●

324431　Quercus spathulata Seemen var. elliptica Y. C. Hsu et H. Wei Jen ＝ Quercus dolichostyla A. Camus ●

324432　Quercus spathulata Seemen var. oxyphylla E. H. Wilson ＝ Quercus oxyphylla (E. H. Wilson) Hand. -Mazz. ●

324433　Quercus spicata sensu Sm. ＝ Lithocarpus grandifolius (D. Don) S. N. Biswas ●

324434　Quercus spicata Sm. ＝ Lithocarpus grandifolius (D. Don) S. N. Biswas ●

324435　Quercus spicata Sm. var. collettii King ex Hook. f. ＝ Lithocarpus collettii (King ex Hook. f.) A. Camus ●

324436　Quercus spinosa A. David ex Franch.；刺栎（刺青冈，刺叶高山栎，刺叶山栎，高山栎，九刚斧，铁将，铁匠树，铁青冈，铁橡，铁橡树，铁橡子树）；Spineleaf Alpine Oak，Spineleaf Oak，Spineleaved Oak ●

324437　Quercus spinosa A. David ex Franch. var. miyabei Hayata ＝ Quercus spinosa A. David ex Franch. ●

324438　Quercus spinosa A. David ex Franch. var. miyabei Hayata f.

rugosa（Masam.）J. C. Liao ＝ Quercus tatakaensis Tomiya ●

324439 Quercus spinosa A. David ex Franch. var. miyabei Hayata f. tatakaensis（Tomiya）J. C. Liao ＝ Quercus tatakaensis Tomiya ●

324440 Quercus spinosa A. David ex Franch. var. monimotricha Hand. - Mazz. ＝ Quercus monimotricha Hand. -Mazz. ●

324441 Quercus squamata Roxb. ＝ Lithocarpus elegans（Blume）Hatus. ex Soepadmo ●☆

324442 Quercus squamata Roxb. ＝ Lithocarpus grandifolius（D. Don）S. N. Biswas ●

324443 Quercus stellata Wangenh.；星毛栎；Delta Post Oak，Iron Oak，Post Oak，Sand Post Oak ●☆

324444 Quercus stellata Wangenh. var. boyntonii（Beadle）Sarg. ＝ Quercus boyntonii Beadle ●☆

324445 Quercus stellata Wangenh. var. margarettae（Ashe）Sarg. ＝ Quercus margarettae（Ashe）Small ●☆

324446 Quercus stellata Wangenh. var. paludosa Sarg. ＝ Quercus similis Ashe ●☆

324447 Quercus stellipila（Sarg.）Parks ex Cory ＝ Quercus gravesii Sudw. ●☆

324448 Quercus stenophylla（Blume）Makino var. stenophylloides（Hayata）A. Camus ＝ Cyclobalanopsis stenophylloides（Hayata）Kudo et Masam. ex Kudo ●

324449 Quercus stenophylla Makino；狭叶栎；Narrowleaf Oak ●☆

324450 Quercus stenophylloides Hayata ＝ Cyclobalanopsis stenophylloides（Hayata）Kudo et Masam. ex Kudo ●

324451 Quercus stewardiana A. Camus ＝ Cyclobalanopsis stewardiana（A. Camus）Y. C. Hsu et H. Wei Jen ●

324452 Quercus stewardii Rehder；黄山栎（黄山槲栎）；Huangshan Oak，Steward Oak ●

324453 Quercus suber L.；欧洲栓皮栎（栓皮槠，西班牙栓皮栎）；Cork Oak，Cork Tree，Cork-tree ●

324454 Quercus subhinoidea Chun et W. C. Ko ＝ Cyclobalanopsis subhinoidea（Chun et W. C. Ko）Y. C. Hsu et H. Wei Jen ex Y. T. Chang ●

324455 Quercus subreticulata Hayata ＝ Lithocarpus hancei（Benth.）Rehder ●

324456 Quercus subturbinella Trel. ＝ Quercus turbinella Greene ●☆

324457 Quercus sutchuenensis Franch. ＝ Quercus engleriana Seemen ●

324458 Quercus synbalanos Hance ＝ Lithocarpus litseifolius（Hance）Chun ●

324459 Quercus synbalanos Hance ＝ Pasania synbalanos（Hance）Schottky ●

324460 Quercus taichuensis Hayata ＝ Cyclobalanopsis longinux（Hayata）Schottky ●

324461 Quercus taitoensis Hayata ＝ Lithocarpus taitoensis（Hayata）Hayata ●

324462 Quercus taitoensis Hayata ＝ Pasania taitoensis（Hayata）J. C. Liao ●

324463 Quercus taiyuensis Y. Ling ＝ Quercus spinosa A. David ex Franch. ●

324464 Quercus takatorensis Makino；小槲树●☆

324465 Quercus taliensis A. Camus ＝ Quercus cocciferoides Hand. -Mazz. var. taliensis（A. Camus）Y. C. Hsu et H. Wei Jen ●

324466 Quercus taliensis A. Camus ＝ Quercus cocciferoides Hand. -Mazz. ●

324467 Quercus tarokoensis Hayata；台湾栎（太鲁阁栎）；Tailuger Oak，Taiwan Oak，Taroko Oak ●

324468 Quercus tatakaensis Tomiya；锐叶高山栎●

324469 Quercus tatakaensis Tomiya ＝ Quercus spinosa A. David ex Franch. ●

324470 Quercus taxana Buckley var. chisosensis Sarg. ＝ Quercus gravesii Sudw. ●☆

324471 Quercus taxana Buckley var. stellapila Sarg. ＝ Quercus gravesii Sudw. ●☆

324472 Quercus tenuicupula（Y. C. Hsu et H. Wei Jen）C. C. Huang ＝ Cyclobalanopsis tenuicupula Y. C. Hsu et H. Wei Jen ●

324473 Quercus tenuicupula（Y. C. Hsu et H. Wei Jen）C. C. Huang et Y. T. Chang ＝ Cyclobalanopsis tenuicupula Y. C. Hsu et H. Wei Jen ●

324474 Quercus tephrocarpa Drake ＝ Lithocarpus tephrocarpus（Drake）A. Camus ●

324475 Quercus tephrosis Chun et W. C. Ko ＝ Cyclobalanopsis edithae（V. Naray.）Schottky ●

324476 Quercus ternaticupula Hayata ＝ Lithocarpus hancei（Benth.）Rehder ●

324477 Quercus ternaticupula Hayata ＝ Pasania hancei（Benth.）Schottky var. ternaticupula（Hayata）J. C. Liao ●

324478 Quercus tetris ?；暗淡栎；Manna Oak ●☆

324479 Quercus texana Buckley；得克萨斯栎；Nuttall Oak，Nuttall's Oak，Taxan Oak，Taxas Oak ●☆

324480 Quercus thalassica Hance ＝ Lithocarpus glaber（Thunb.）Nakai ●

324481 Quercus thalassica Hance var. obtusiglans Dunn ＝ Lithocarpus glaber（Thunb.）Nakai ●

324482 Quercus thalassica Hance var. vestita Franch. ＝ Lithocarpus dealbatus（Hook. f. et Thomson ex DC.）Rehder ●

324483 Quercus thomsonii Miq. ＝ Lithocarpus thomsonii（Miq.）Rehder ●

324484 Quercus thorelii Hickel et A. Camus ＝ Cyclobalanopsis thorelii（Hickel et A. Camus）Hu ●

324485 Quercus tiaoloshanica Chun et W. C. Ko ＝ Cyclobalanopsis tiaoloshanica（Chun et W. C. Ko）Y. C. Hsu et H. Wei Jen ●

324486 Quercus tinctoria W. Bartram ＝ Quercus velutina Lam. ●☆

324487 Quercus tinctoria W. Bartram var. californica Torr. ＝ Quercus kelloggii Newb. ●☆

324488 Quercus tipitata Hayata ex Koidz. ＝ Castanopsis carlesii（Hemsl.）Hayata ●

324489 Quercus tomentella Engelm.；岛栎（细毛栎）；Channel Island Oak，Island Live Oak，Island Oak ●☆

324490 Quercus tomentosicupula Hayata；毛壳栎●

324491 Quercus tomentosicupula Hayata ＝ Cyclobalanopsis pachyloma（Seemen）Schottky ●

324492 Quercus tomentosinervis（Y. C. Hsu et H. Wei Jen）C. C. Huang ＝ Cyclobalanopsis tomentosinervis Y. C. Hsu et H. Wei Jen ●

324493 Quercus toumeyi Sarg.；图米栎；Toumey Oak ●☆

324494 Quercus toza Griseb. ＝ Quercus pyrenaica Willd. ●☆

324495 Quercus tribuloides Lindl. ＝ Castanopsis tribuloides（Lindl.）A. DC. ●

324496 Quercus tribuloides Sm. ＝ Castanopsis tribuloides（Lindl.）A. DC. ●

324497 Quercus trinervis H. Lév. ＝ Castanopsis eyrei（Champ.）Tutcher ●

324498 Quercus trojana Webb；马其顿栎；Macedonian Oak ●☆

324499 Quercus truncata King ＝ Lithocarpus truncatus（King）Rehder ●

324500 Quercus truncata King ex Hook. f. ＝ Lithocarpus truncatus（King）Rehder ●

324501　Quercus tsinglingensis Liou ex S. Z. Qu et W. H. Zhang;秦岭栎;Qinling Oak ●

324502　Quercus tsinglingensis Liou ex S. Z. Qu et W. H. Zhang = Quercus aliena Blume var. acuteserrata Maxim. ex Wenz. ●

324503　Quercus tsoi Chun = Cyclobalanopsis fleuryi（Hickel et A. Camus）Chun ●

324504　Quercus tsoi Chun ex Menitsky = Cyclobalanopsis fleuryi（Hickel et A. Camus）Chun ●

324505　Quercus tungmaiensis Y. T. Chang;通卖栎（通麦栎）;Tongmai Oak ●

324506　Quercus tungmaiensis Y. T. Chang = Quercus lanata Sm. ●

324507　Quercus tunkinensis Drake = Castanopsis fissa（Champ. ex Benth.）Rehder et E. H. Wilson ●

324508　Quercus turbinata Roxb. = Lithocarpus thomsonii（Miq.）Rehder ●

324509　Quercus turbinella Greene;灌木栎;Scrub Oak, Shrub Live Oak, Sonoran Scrub Oak, Turbinella Oak ●☆

324510　Quercus turbinella Greene subsp. ajoensis（C. H. Mull.）Felger et C. H. Lowe = Quercus ajoensis C. H. Mull. ●☆

324511　Quercus turbinella Greene subsp. californica J. M. Tucker = Quercus john-tuckeri Nixon et C. H. Mull. ●☆

324512　Quercus turneri A. DC. ;图纳栎（聚果栎）●☆

324513　Quercus turneri A. DC. 'Paeudoturneri';假图纳栎●☆

324514　Quercus uliginosa Wangenh. = Quercus nigra L. ●☆

324515　Quercus undulata Torr. ;常绿栎;Evergreen Oak, Mountain Rocky Oak, Rocky Mountain Oak, Scrub Oak, Wavyleaf Oak ●☆

324516　Quercus undulata Torr. var. gambelii（Nutt.）Engelm. = Quercus fusiformis Small ●☆

324517　Quercus undulata Torr. var. grisea（Liebm.）Engelm. = Quercus grisea Liebm. ●☆

324518　Quercus undulata Torr. var. pungens（Liebm.）Engelm. = Quercus pungens Liebm. ●☆

324519　Quercus undulata Torr. var. vaseyana（Buckley）Rydb. = Quercus vaseyana Buckley ●☆

324520　Quercus uraiana Hayata = Castanopsis uraiana（Hayata）Kaneh. et Hatus. ●

324521　Quercus urbanii Trel. ;尤本栎●☆

324522　Quercus urticifolia Blume var. brevipetiolata A. DC. = Quercus glandulifera Blume var. brevipetiolata（A. DC.）Nakai ●

324523　Quercus urticifolia Blume var. brevipetiolata A. DC. = Quercus serrata Thunb. var. brevipetiolata（A. DC.）Nakai ●

324524　Quercus urticifolia Blume var. brevipetiolata A. DC. = Quercus serrata Murray ●

324525　Quercus utanensis Rydb. ;犹他州栎;Utah Oak ●☆

324526　Quercus utahensis Rydb. = Quercus fusiformis Small ●☆

324527　Quercus utchuanensis Franch. = Quercus engleriana Seemen ●

324528　Quercus utilis Hu et W. C. Cheng;炭栎;Charcoal Oak, Utility Oak ●

324529　Quercus uvariifolia Hance = Lithocarpus uvariifolius（Hance）Rehder ●

324530　Quercus vaccinifolia Kellogg ex Curran;越橘栎;Huckleberry Oak ●☆

324531　Quercus vaniotii H. Lév. = Cyclobalanopsis glauca（Thunb. ex A. Murray）Oerst. ●

324532　Quercus variabilis Blume;栓皮栎（白麻栎,粗皮栎,粗皮青冈,大叶橡,花栎,花栎木,黄划栎,老栎,栗壳,青杠碗,软木栎,校力,柞）;Chinese Cork Oak, Cork Oak, Oriental Oak ●

324533　Quercus variabilis Blume var. megaphylla T. B. Chao = Quercus variabilis Blume ●

324534　Quercus variabilis Blume var. pyramidalis T. B. Chao, Z. I. Chang et W. C. Li;塔形栓皮栎●

324535　Quercus variabilis Blume var. pyramidalis T. B. Chao, Z. I. Chang et W. C. Li = Quercus variabilis Blume ●

324536　Quercus variolosa Franch. = Lithocarpus variolosus（Franch.）Chun ●

324537　Quercus vaseyana Buckley;瓦齐栎;Vasey Oak ●☆

324538　Quercus vellifera Trel. = Quercus rugosa Née ●☆

324539　Quercus velutina Lam. ;美洲黑栎（短绒栎,黑栎,美国黑栎）;Black Oak, Dyer's Oak, Dyer's-oak, Quercitron Oak, Smoothbark Oak, Yellow Bark Oak, Yellow Oak, Yellowbark Oak, Yellow-bark Oak ●☆

324540　Quercus velutina Lam. f. missouriensis（Sarg.）Trel. = Quercus velutina Lam. ●☆

324541　Quercus velutina Lam. var. missouriensis Sarg. = Quercus velutina Lam. ●☆

324542　Quercus virens Aiton = Quercus virginiana Mill. ●☆

324543　Quercus virens Aiton var. dentata Chapm. = Quercus minima（Sarg.）Small ●☆

324544　Quercus virginiana Mill. ;弗吉尼亚栎（强生栎）;Live Oak, Southern Live Oak, Virginia Oak ●☆

324545　Quercus virginiana Mill. var. dentata（Chapm.）Sarg. = Quercus minima（Sarg.）Small ●☆

324546　Quercus virginiana Mill. var. eximea Sarg. = Quercus virginiana Mill. ●☆

324547　Quercus virginiana Mill. var. fusiformis（Small）Sarg. = Quercus fusiformis Small ●☆

324548　Quercus virginiana Mill. var. geminata（Small）Sarg. = Quercus geminata Small ●☆

324549　Quercus virginiana Mill. var. minima Sarg. = Quercus minima（Sarg.）Small ●☆

324550　Quercus wallichiana Lindl. ;沃利克栎;Wallich's Oak ●☆

324551　Quercus wangii Hu et W. C. Cheng;王氏栎;Wang's Oak ●

324552　Quercus wangii Hu et W. C. Cheng = Lithocarpus pachylepis A. Camus ●

324553　Quercus warburgii A. Camus;剑桥栎;Cambridge Oak ●☆

324554　Quercus wilcoxii Rydb. = Quercus chrysolepis Liebm. ●☆

324555　Quercus wilsonii Seemen = Lithocarpus cleistocarpus（Seemen）Rehder et E. H. Wilson ●

324556　Quercus wislizenii A. DC. ;高地栎（内陆野栎）;Desert Oak, Highland Live Oak, Interior Live Oak, Large Scrub Oak, Sierra Live Oak ●☆

324557　Quercus wislizenii A. DC. var. frutescens Engelm. = Quercus wislizenii A. DC. ●☆

324558　Quercus woronowii Maleev;沃氏栎●☆

324559　Quercus wutaishanica Mayr;五台栎（辽东栎）;East-Liaoning Oak, Wutaishan Oak ●

324560　Quercus wutaishanica Mayr = Quercus mongolica Fisch. ex Turcz. ●

324561　Quercus xanthotricha A. Camus = Cyclobalanopsis xanthotricha（A. Camus）Y. C. Hsu et H. Wei Jen ●

324562　Quercus xizangensis（Y. C. Hsu et H. Wei Jen）C. C. Huang et Y. T. Chang = Cyclobalanopsis kiukiangensis Y. T. Chang ex Y. C. Hsu et H. Wei Jen ●

324563　Quercus xizangensis（Y. C. Hsu et H. Wei Jen）C. C. Huang et

Y. T. Zhang = Cyclobalanopsis xizangensis Y. C. Hsu et H. Wei Jen ●

324564 Quercus xylocarpa Kurz = Lithocarpus xylocarpus (Kurz) Markgr. ●

324565 Quercus yaeyamensis Koidz. = Quercus miyagii Koidz. ●☆

324566 Quercus yiwuensis C. C. Huang ex Y. C. Hsu et H. Wei Jen;易武栎;Yiwu Oak ●

324567 Quercus yonganensis L. G. Lin et C. C. Huang = Cyclobalanopsis yonganensis (L. Lin et C. C. Huang) Y. C. Hsu et H. Wei Jen ●

324568 Quercus yongchuanana Z. K. Zhou = Cyclobalanopsis longifolia Y. C. Hsu et Q. Z. Dong ●

324569 Quercus yongchuanana Z. K. Zhou = Cyclobalanopsis lungmaiensis Hu ●

324570 Quercus yui Liou = Quercus dentata Thunb. var. oxyloba Franch. ●

324571 Quercus yui Liou = Quercus yunnanensis Franch. ●

324572 Quercus yunnanensis Franch.;云南波罗栎(锐齿波罗栎,云南柞栎);Yunnan Daimyo Oak,Yunnan Oak,Yunnan Pineapple Oak ●

324573 Querezia L. = Queria L. ■

324574 Queria L. = Minuartia L. ■

324575 Queria Loefl. = Queria L. ■

324576 Queria trichotoma Thunb. = Diplomorpha trichotoma (Thunb.) Nakai ●

324577 Quesnelia Gaudich. (1842);豪华菠萝属(龟甲凤梨属,魁氏凤梨属)■☆

324578 Quesnelia arvensis Mez;豪华菠萝(豪华凤梨)■☆

324579 Quesnelia humilis Mez;小豪华菠萝■☆

324580 Quesnelia lateralis Wawra;侧生豪华菠萝■☆

324581 Quesnelia liboniana (De Jonghe) Mez;里氏豪华菠萝■☆

324582 Quesnelia marmorata (Lem.) Read;大理石状豪华菠萝■☆

324583 Quesnelia morreniana (Baker) Mez;默氏豪华菠萝■☆

324584 Quesnelia quesneliana (Brongn.) L. B. Sm.;凯氏豪华菠萝■☆

324585 Quesnelia testudo Lindm.;龟甲状豪华菠萝■☆

324586 Queteletia Blume = Orchipedum Breda ■☆

324587 Quetia Gand. = Chaerophyllum L. ■

324588 Quetzalia Lundell(1970);奎茨卫矛属●☆

324589 Quetzalia areolata (Lundell) Lundell;奎茨卫矛●☆

324590 Quetzalia occidentalis (Loes.) Lundell;西方奎茨卫矛●☆

324591 Quetzalia pauciflora Lundell;小花奎茨卫矛●☆

324592 Quezelia H. Scholz = Quezeliantha H. Scholz ex Rauschert ■☆

324593 Quezeliantha H. Scholz = Quezeliantha H. Scholz ex Rauschert ■☆

324594 Quezeliantha H. Scholz ex Rauschert(1982);撒哈拉芥属■☆

324595 Quezeliantha tibestica (Scholz) Rauschert;撒哈拉芥■☆

324596 Quiabentia Britton et Rose(1923);顶花麒麟掌属(奎阿本特属)●☆

324597 Quiabentia chacoensis Backeb.;叶仙人掌●☆

324598 Quidproquo Greuter et Burdet = Raphanus L. ■

324599 Quiducia Gagnep. = Silvianthus Hook. f. ●

324600 Quiducia tonkinensis Gagnep. = Silvianthus tonkinensis (Gagnep.) Ridsdale ●■

324601 Quiina Aubl. (1775);绒子树属●☆

324602 Quiina albiflora A. C. Sm.;白花绒子树●☆

324603 Quiina gracilis A. C. Sm.;细绒子树●☆

324604 Quiina lanceolata Dusén ex Ducke;披针叶绒子树●☆

324605 Quiina leptoclada Tul.;细枝绒子树●☆

324606 Quiina longifolia Spruce ex Planch. et Triana;长叶绒子树●☆

324607 Quiina lucida Glaz.;亮叶绒子树●☆

324608 Quiina macrophylla Tul.;大叶绒子树●☆

324609 Quiina macrostachya Tul.;大穗绒子树●☆

324610 Quiina multiflora Cuatrec.;多花绒子树●☆

324611 Quiina parvifolia Lanj. et Heerdt;小叶绒子树●☆

324612 Quiina rigidifolia Pires;硬叶绒子树●☆

324613 Quiina silvatica Pulle;林地绒子树●☆

324614 Quiinaceae Choisy = Quiinaceae Choisy ex Engl. (保留科名)●☆

324615 Quiinaceae Choisy ex Engl. (1888)(保留科名);绒子树科(羽叶树科)●☆

324616 Quiinaceae Engl. = Quiinaceae Choisy ex Engl. (保留科名)●☆

324617 Quilamum Blanco = Crypteronia Blume ●

324618 Quilesia Blanco = Dichapetalum Thouars ●

324619 Quiliusa Hook. f. = Fuchsia L. ●■

324620 Quiliusa Hook. f. = Quelusia Vand. ●■

324621 Quillaia Molina = Quillaja Molina ●☆

324622 Quillaiaceae D. Don = Rosaceae Juss. (保留科名)●■

324623 Quillaja Molina (1782);皂树属(肥皂树属,奎拉雅属);Soapbark Tree ●☆

324624 Quillaja saponaria Molina;皂树(肥皂树,石碱木);Quillai,Soapbark Tree,Soap-bark Tree ●☆

324625 Quillaja smegmadermis St. -Lag.;黏膜皂树●☆

324626 Quillajaceae D. Don = Rosaceae Juss. (保留科名)●■

324627 Quillajaceae D. Don(1831);皂树科(皂皮树科)●☆

324628 Quinaria Lour. = Clausena Burm. f. ●

324629 Quinaria Raf. = Parthenocissus Planch. (保留属名)●

324630 Quinaria hederacea Raf. = Parthenocissus quinquefolius (L.) Planch. ●

324631 Quinaria lansium Lour. = Clausena lansium (Lour.) Skeels ●

324632 Quinaria tricuspidata (Siebold et Zucc.) Koehne = Parthenocissus tricuspidatus (Siebold et Zucc.) Planch. ●

324633 Quinaria tricuspidata Koehne = Parthenocissus tricuspidatus (Siebold et Zucc.) Planch. ●

324634 Quinaria veitchii Koehne = Parthenocissus tricuspidatus (Siebold et Zucc.) Planch. ●

324635 Quinasis Raf. = Polylepis Ruiz et Pav. ●☆

324636 Quinata Medik. (废弃属名) = Machaerium Pers. (保留属名)●☆

324637 Quinchamala Willd. = Quinchamalium Molina(保留属名)●☆

324638 Quinchamalium Juss. = Quinchamalium Molina(保留属名)●☆

324639 Quinchamalium Molina(1782)(保留属名);智利檀●☆

324640 Quinchamalium majus Brongn.;五月智利檀●☆

324641 Quincula Raf. (1832);北美茄属■☆

324642 Quincula Raf. = Physalis L. ■

324643 Quincula lobata Raf.;北美茄;Purple Groundcherry ☆

324644 Quinetia Cass. (1830);紫鼠麴属■☆

324645 Quinetia urvillei Cass.;紫鼠麴■☆

324646 Quinio Schltdl. = Cocculus DC. (保留属名)●

324647 Quinquedula Noronha = Litsea Lam. (保留属名)●

324648 Quinquefolium Ség. = Potentilla L. ■●

324649 Quinquelobus Benj. = Bacopa Aubl. (保留属名)■

324650 Quinquelobus Benj. = Benjaminia Mart. ex Benj. ■☆

324651 Quinquelocularia C. Koch = Campanula L. ■●

324652 Quinquelocularia K. Koch = Campanula L. ■●

324653 Quinqueremulus Paul G. Wilson(1987);线鼠麴属■☆

324654 Quinqueremulus linearis Paul G. Wilson;线鼠麴●☆

324655 Quinquina Boehm. = Cinchona L. ■●

324656 Quinquina Condam. = Quinquina Boehm. ■●

324657 Quinsonia Montrouz. = Pittosporum Banks ex Gaertn. (保留属

名)●

324658　Quintilia Endl. = Anomorhegmia Meisn. ■

324659　Quintilia Endl. = Stauranthera Benth. ■

324660　Quintinia A. DC. (1830);昆廷树属(昆亭尼亚属,昆亭树属);Quintinia ●☆

324661　Quintinia serrata A. Cunn.;新西兰昆廷树(昆亭尼亚);New Zeyland Quintinia ●☆

324662　Quintinia sieberi A. DC.;昆廷树;Popossum Wood,Possumwood ●☆

324663　Quintiniaceae Doweld;昆廷树科●☆

324664　Quiotania Zarucchi(1991);哥伦比亚夹竹桃属●☆

324665　Quiotania colombiana Zarucchi;哥伦比亚夹竹桃●☆

324666　Quirina Raf. = Cuphea Adans. ex P. Browne ●■

324667　Quirivelia Poir. = Ichnocarpus R. Br. (保留属名)●■

324668　Quirosia Blanco = Crotalaria L. ●■

324669　Quisqualis L. (1762);使君子属;Quisqualis ●

324670　Quisqualis L. = Combretum Loefl. (保留属名)●

324671　Quisqualis caudata Craib;小花使君子;Caudate Quisqualis, Small-flower Quisqualis ●

324672　Quisqualis conferta (Jack) Exell;密花使君子(小花使君子);Denseflower Quisqualis ●

324673　Quisqualis densiflora Wall. ex Miq. = Quisqualis conferta (Jack) Exell ●

324674　Quisqualis fructa Welw. ex Hiern;多果使君子●☆

324675　Quisqualis glabra Burm. f. = Quisqualis indica L. ●

324676　Quisqualis grandiflora Miq. = Quisqualis indica L. ●

324677　Quisqualis indica L.;使君子(病柑子,冬均子,留求子,山羊屎,史君子,水君子,四君子,索子果,五棱子,削痄子);Akar Dani,Drunken Sailor,Rangoon Creeper,Rangooncreeper ●

324678　Quisqualis indica L. var. oxypetala Kurz = Quisqualis indica L. ●

324679　Quisqualis indica L. var. villosa (Roxb.) C. B. Clarke;毛叶使君子(毛使君子,手君子,西蜀使君子);Villose Rangoon Creepet ●

324680　Quisqualis indica L. var. villosa (Roxb.) C. B. Clarke = Quisqualis indica L. ●

324681　Quisqualis indica L. var. villosa C. B. Clarke = Quisqualis indica L. ●

324682　Quisqualis longiflora C. Presl = Quisqualis indica L. ●

324683　Quisqualis loureiroi G. Don = Quisqualis indica L. ●

324684　Quisqualis malabarica Bedd. ;马拉巴使君子●☆

324685　Quisqualis obovata Schumach. et Thon ning = Quisqualis indica L. ●

324686　Quisqualis pubescens Burm. f. = Quisqualis indica L. ●

324687　Quisqualis sinensis Lindl. = Quisqualis indica L. ●

324688　Quisqualis spinosa Blanco = Quisqualis indica L. ●

324689　Quisqualis villosa Roxb. = Quisqualis indica L. ●

324690　Quisqueya Dod(1979);海地兰属■☆

324691　Quisqueya ekmanii Dod;海地兰■☆

324692　Quisqueya rosea (Schltr.) Dod;粉红海地兰■☆

324693　Quisumbingia Merr. (1936);奎苏萝藦属●☆

324694　Quisumbingia merrillii (Schltr.) Merr.;奎苏萝藦●☆

324695　Quivisia Cav. = Turraea L. ●

324696　Quivisia Comm. ex Juss. = Turraea L. ●

324697　Quivisiantha Willis = Quivisianthe Baill. ●☆

324698　Quivisianthe Baill. (1893);基维棣属●☆

324699　Quivisianthe papinae Baill. ;基维棣●☆

324700　Quoya Gaudich. = Pityrodia R. Br. ●☆

324701　Qveria L. = Minuartia L. ■

324702　Qveria L. = Queria L. ■

324703　Raaltema Mus. Lugd. ? ex C. B. Clarke = Boea Comm. ex Lam. ■

324704　Rabarbarum Post et Kuntze = Rhabarbarum Adans. ■

324705　Rabarbarum Post et Kuntze = Rheum L. ■

324706　Rabdadenia Post et Kuntze = Rhabdadenia Müll. Arg. ●☆

324707　Rabdia Post et Kuntze = Rhabdia Mart. ●

324708　Rabdia Post et Kuntze = Rotula Lour. ●

324709　Rabdochloa P. Beauv. = Leptochloa P. Beauv. ■

324710　Rabdochloa mucronata (Michx.) P. Beauv. = Dactyloctenium aegyptium (L.) Willd. ■

324711　Rabdochloa vulpiastrum De Not. = Leptocaryon vulpiastrum (De Not.) Stapf ■☆

324712　Rabdosia (Blume) Hassk.(1842);小香茶菜属(回菜花属);Rabdosia ●■

324713　Rabdosia (Blume) Hassk. = Isodon (Schrad. ex Benth.) Spach ●■

324714　Rabdosia Hassk. = Isodon (Schrad. ex Benth.) Spach ●■

324715　Rabdosia Hassk. = Plectranthus L' Hér. (保留属名)●■

324716　Rabdosia × inamii (Murata) Murata = Isodon × inamii Murata ■☆

324717　Rabdosia × ohwii (Okuyama) H. Hara = Isodon × ohwii Okuyama ■☆

324718　Rabdosia × togashii (Okuyama) H. Hara = Isodon × togashii Okuyama ■☆

324719　Rabdosia adenantha (Diels) H. Hara = Isodon adenanthus (Diels) Kudo ■

324720　Rabdosia adenoloma (Hand. -Mazz.) H. Hara = Isodon adenolomus (Hand. -Mazz.) H. Hara ●

324721　Rabdosia albopilosa C. Y. Wu et H. W. Li = Isodon albopilosus (C. Y. Wu et H. W. Li) H. Hara ■

324722　Rabdosia alborubra C. Y. Wu;粉红香茶菜(粉白香茶菜);Pink Rabdosia ■

324723　Rabdosia alborubra C. Y. Wu = Isodon alborubrus (C. Y. Wu) H. Hara ■

324724　Rabdosia alborubra C. Y. Wu = Isodon sculponeatus (Vaniot) Kudo ■

324725　Rabdosia amethystoides (Benth.) H. Hara = Isodon amethystoides (Benth.) H. Hara ■

324726　Rabdosia angustifolia (Dunn) H. Hara = Isodon angustifolius (Dunn) Kudo ■

324727　Rabdosia angustifolia (Dunn) H. Hara var. glabrescens C. Y. Wu et H. W. Li = Isodon angustifolius (Dunn) H. Hara var. glabrescens (C. Y. Wu et H. W. Li) H. W. Li ■

324728　Rabdosia anisochila C. Y. Wu;异唇香茶菜;Anisolip Rabdosia, Differentlip Rabdosia, Unequallip Rabdosia ●■

324729　Rabdosia anisochila C. Y. Wu = Isodon coetsus (Buch. -Ham. ex D. Don) Kudo ●■

324730　Rabdosia arakii (Murata) H. Hara = Isodon arakii Murata ■☆

324731　Rabdosia bifidocalyx (Dunn) H. Hara = Isodon macrocalyx (Dunn) Kudo ■

324732　Rabdosia brachythyrsa C. Y. Wu et H. W. Li;短锥香茶菜;Shortpanicle Rabdosia,Short-thyrse Rabdosia ■

324733　Rabdosia brachythyrsa C. Y. Wu et H. W. Li = Isodon muliensis (W. W. Sm.) Kudo ●

324734　Rabdosia brevicalcarata C. Y. Wu et H. W. Li = Isodon brevicalcaratus (C. Y. Wu et H. W. Li) H. Hara ■

324735　Rabdosia brevifolia (Hand. -Mazz.) H. Hara = Isodon brevifolius (Hand. -Mazz.) H. W. Li ●

324736　Rabdosia bulleyana (Diels) H. Hara = Isodon bulleyanus (Diels) Kudo ●

324737　Rabdosia bulleyana（Diels）H. Hara var. foliosa C. Y. Wu ＝ Isodon bulleyanus（Diels）Kudo ●

324738　Rabdosia calcicola（Hand.-Mazz.）H. Hara ＝ Isodon calcicola（Hand.-Mazz.）H. Hara ■

324739　Rabdosia calcicola（Hand.-Mazz.）H. Hara var. subcalva（Hand.-Mazz.）C. Y. Wu et H. W. Li ＝ Isodon calcicola（Hand.-Mazz.）H. Hara var. subcalva（Hand.-Mazz.）H. W. Li ■

324740　Rabdosia calycina（Benth.）Codd ＝ Rabdosiella calycina（Benth.）Codd ●■

324741　Rabdosia chionantha C. Y. Wu；雪花香茶菜；Snowflake Rabdosia,Snowflower Rabdosia,Snow-flowered Rabdosia ●

324742　Rabdosia chionantha C. Y. Wu ＝ Isodon muliensis（W. W. Sm.）Kudo ●

324743　Rabdosia coetsa（Buch.-Ham. ex D. Don）H. Hara ＝ Isodon coetsus（Buch.-Ham. ex D. Don）Kudo ●■

324744　Rabdosia coetsa（Buch.-Ham. ex D. Don）H. Hara var. cavaleriei（H. Lév.）C. Y. Wu et H. W. Li ＝ Isodon coetsus（Buch.-Ham. ex D. Don）H. Hara var. cavaleriei（H. Lév.）H. W. Li ●■

324745　Rabdosia coetsoides C. Y. Wu；假细锥香茶菜（野坝子）；False Smallpanicl Rabdosia ■

324746　Rabdosia coetsoides C. Y. Wu ＝ Isodon coetsus（Buch.-Ham. ex D. Don）Kudo ●■

324747　Rabdosia daitonensis（Hayata）H. Hara；大屯延命草（台湾香茶菜）■

324748　Rabdosia daitonensis（Hayata）H. Hara ＝ Isodon amethystoides（Benth.）H. Hara ■

324749　Rabdosia daitonensis（Hayata）H. Hara ＝ Isodon daitonensis（Hayata）Kudo ■

324750　Rabdosia daitonensis（Hayata）H. Hara ＝ Rabdosia amethystoides（Benth.）H. Hara ■

324751　Rabdosia dawoensis（Hand.-Mazz.）H. Hara ＝ Isodon dawoensis（Hand.-Mazz.）H. Hara ●

324752　Rabdosia dhankutana（Murata）Hara；尼泊尔香茶菜■☆

324753　Rabdosia dichromophylla（Diels）H. Hara ＝ Isodon rubescens（Hemsl.）H. Hara ●

324754　Rabdosia dichromophylla（Diels）H. Hara ＝ Rabdosia rubescens（Hemsl.）H. Hara ●

324755　Rabdosia drogotschiensis（Hand.-Mazz.）H. Hara ＝ Isodon barbeyanus（H. Lév.）H. W. Li ■

324756　Rabdosia drogotschiensis（Hand.-Mazz.）H. Hara ＝ Isodon drogotschiensis（Hand.-Mazz.）H. Hara ●

324757　Rabdosia effusa（Maxim.）H. Hara；疏展香茶菜（疏展毛莨）；Effuse Rabdosia ■☆

324758　Rabdosia effusa（Maxim.）H. Hara ＝ Isodon effusus（Maxim.）H. Hara ■☆

324759　Rabdosia enanderiana（Hand.-Mazz.）H. Hara ＝ Isodon enanderianus（Hand.-Mazz.）H. W. Li ●

324760　Rabdosia eriocalyx（Dunn）H. Hara ＝ Isodon eriocalyx（Dunn）Kudo ●■

324761　Rabdosia eriocalyx（Dunn）H. Hara var. laxiflora C. Y. Wu et H. W. Li ＝ Isodon eriocalyx（Dunn）Kudo ●■

324762　Rabdosia excisa（Maxim.）H. Hara ＝ Isodon excisus（Maxim.）Kudo ●

324763　Rabdosia excisoides（Y. Z. Sun ex C. H. Hu）C. Y. Wu et H. W. Li ＝ Isodon excisoides（Y. Z. Sun ex C. H. Hu）H. Hara ■

324764　Rabdosia fangii（Y. Z. Sun）H. Hara ＝ Isodon lophanthoides（Buch.-Ham. ex D. Don）H. Hara ■

324765　Rabdosia flabelliformis C. Y. Wu ＝ Isodon flabelliformis（C. Y. Wu）H. Hara ■

324766　Rabdosia flavida（Hand.-Mazz.）H. Hara ＝ Isodon flavidus（Hand.-Mazz.）H. Hara ■

324767　Rabdosia flexicaulis C. Y. Wu et H. W. Li ＝ Isodon flexicaulis（C. Y. Wu et H. W. Li）H. Hara ●

324768　Rabdosia forrestii（Diels）H. Hara ＝ Isodon forrestii（Diels）Kudo ■

324769　Rabdosia forrestii（Diels）H. Hara var. intermedia C. Y. Wu et H. W. Li；居间紫萼香茶菜；Intermediate Purplesepal Rabdosia ■

324770　Rabdosia forrestii（Diels）H. Hara var. intermedia C. Y. Wu et H. W. Li ＝ Isodon forrestii（Diels）Kudo ■

324771　Rabdosia gesneroides（J. Sinclair）H. Hara ＝ Isodon gesneroides（J. Sinclair）H. Hara ●

324772　Rabdosia gibbosa C. Y. Wu et H. W. Li ＝ Isodon gibbosus（C. Y. Wu et H. W. Li）H. Hara ■

324773　Rabdosia glutinosa C. Y. Wu et H. W. Li ＝ Isodon glutinosus（C. Y. Wu et H. W. Li）H. Hara ●

324774　Rabdosia grandifolia（Hand.-Mazz.）H. Hara ＝ Isodon grandifolius（Hand.-Mazz.）H. Hara ●

324775　Rabdosia grandifolia（Hand.-Mazz.）H. Hara var. atuntzeensis C. Y. Wu ＝ Isodon grandifolius（Hand.-Mazz.）H. Hara var. atunzensis（C. Y. Wu）H. W. Li ●

324776　Rabdosia grandifolia（Hand.-Mazz.）H. Hara var. atunzensis C. Y. Wu ＝ Isodon grandifolius（Hand.-Mazz.）H. Hara var. atunzensis（C. Y. Wu）H. W. Li ●

324777　Rabdosia grosseserrata（Dunn）H. Hara ＝ Isodon grosseserratus（Dunn）Kudo ■

324778　Rabdosia henryi（Hemsl.）H. Hara ＝ Isodon henryi（Hemsl.）Kudo ■

324779　Rabdosia hirtella（Hand.-Mazz.）H. Hara ＝ Isodon hirtellus（Hand.-Mazz.）H. Hara ●

324780　Rabdosia hispida（Benth.）H. Hara ＝ Isodon hispidus（Benth.）Murata ■

324781　Rabdosia incana（Link）H. Hara；灰毛香茶菜（灰毛毛莨）；Greywhitehair Plectranthus ■☆

324782　Rabdosia inflexa（Thunb.）H. Hara ＝ Isodon inflexus（Thunb.）Kudo ■

324783　Rabdosia inflexa（Thunb.）H. Hara var. macrophylla（Maxim.）H. Hara ＝ Isodon inflexus（Thunb.）Kudo ■

324784　Rabdosia interrupta C. Y. Wu et H. W. Li ＝ Isodon interruptus（C. Y. Wu et H. W. Li）H. Hara ●

324785　Rabdosia irrorata（Forrest ex Diels）H. Hara ＝ Isodon irroratus（Forrest ex Diels）Kudo ●

324786　Rabdosia irrorata（Forrest ex Diels）H. Hara var. crenata C. Y. Wu et H. W. Li；圆齿露珠香茶菜（圆齿香茶菜）；Crenate Dew Rabdosia ●

324787　Rabdosia irrorata（Forrest ex Diels）H. Hara var. crenata C. Y. Wu et H. W. Li ＝ Isodon irroratus（Forrest ex Diels）Kudo ●

324788　Rabdosia irrorata（Forrest ex Diels）H. Hara var. longipes C. Y. Wu et H. W. Li ＝ Isodon irroratus（Forrest ex Diels）Kudo ●

324789　Rabdosia irrorata（Forrest ex Diels）H. Hara var. rungshiaensis C. Y. Wu et H. W. Li；绒辖香茶菜（绒辖露珠香茶菜）；Rongxia Rabdosia ●

324790　Rabdosia irrorata（Forrest ex Diels）H. Hara var. rungshiaensis

C. Y. Wu et H. W. Li = Isodon irroratus（Forrest ex Diels）Kudo ●

324791　Rabdosia japonica（Burm. f.）H. Hara = Isodon japonicus（Burm. f.）H. Hara ■

324792　Rabdosia japonica（Burm. f.）H. Hara var. glaucocalyx（Maxim.）H. Hara = Isodon japonicus（Burm. f.）H. Hara var. glaucocalyx（Maxim.）H. W. Li ■

324793　Rabdosia kangtingensis C. Y. Wu et H. W. Li = Isodon flabelliformis（C. Y. Wu）H. Hara ■

324794　Rabdosia kangtingensis C. Y. Wu et H. W. Li = Isodon kangtingensis（C. Y. Wu et H. W. Li）H. Hara ■

324795　Rabdosia koroensis（Kudo）H. Hara = Isodon amethystoides（Benth.）H. Hara ■

324796　Rabdosia koroensis（Kudo）H. Hara = Isodon koroensis Kudo ■

324797　Rabdosia koroensis（Kudo）H. Hara = Rabdosia amethystoides（Benth.）H. Hara ■

324798　Rabdosia kunmingensis C. Y. Wu et H. W. Li = Isodon interruptus（C. Y. Wu et H. W. Li）H. Hara ●

324799　Rabdosia kunmingensis C. Y. Wu et H. W. Li = Isodon kunmingensis（C. Y. Wu et H. W. Li）H. Hara ●

324800　Rabdosia lasiocarpa（Hayata）H. Hara = Isodon lasiocarpus（Hayata）Kudo ■

324801　Rabdosia lasiocarpa（Hayata）H. Hara = Isodon serrus（Maxim.）Kudo ■

324802　Rabdosia latiflora C. Y. Wu et H. W. Li；宽花香茶菜；Broadflower Rabdosia ■

324803　Rabdosia latiflora C. Y. Wu et H. W. Li = Isodon scoparius（C. Y. Wu et H. W. Li）H. Hara ●

324804　Rabdosia latiflora C. Y. Wu et H. W. Li = Isodon scrophularioides（Wall. ex Benth.）Murata ■

324805　Rabdosia latifolia C. Y. Wu et H. W. Li = Isodon latifolius（C. Y. Wu et H. W. Li）H. Hara ■

324806　Rabdosia leucophylla（Dunn）H. Hara = Isodon leucophyllus（Dunn）Kudo ●

324807　Rabdosia liangshanica C. Y. Wu et H. W. Li = Isodon liangshanicus（C. Y. Wu et H. W. Li）H. Hara ■

324808　Rabdosia lihsienensis C. Y. Wu et H. W. Li = Isodon lihsienensis（C. Y. Wu et H. W. Li）H. ■

324809　Rabdosia longituba（Miq.）H. Hara = Isodon longitubus（Miq.）Kudo ■

324810　Rabdosia lophanthoides（Buch. -Ham. ex D. Don）H. Hara = Isodon lophanthoides（Buch. -Ham. ex D. Don）H. Hara ■

324811　Rabdosia lophanthoides（Buch. -Ham. ex D. Don）H. Hara var. gerardiana（Benth.）H. Hara = Isodon lophanthoides（Buch. -Ham. ex D. Don）H. Hara var. gerardiana（Benth.）H. Hara ■

324812　Rabdosia lophanthoides（Buch. -Ham. ex D. Don）H. Hara var. graciliflora（Benth.）H. Hara = Isodon lophanthoides（Buch. -Ham. ex D. Don）H. Hara var. graciliflora（Benth.）H. Hara ■

324813　Rabdosia lophanthoides（Buch. -Ham. ex D. Don）H. Hara var. micrantha C. Y. Wu = Isodon lophanthoides（Buch. -Ham. ex D. Don）H. Hara var. micrantha（C. Y. Wu）H. W. Li ■

324814　Rabdosia loxothyrsa（Hand. -Mazz.）H. Hara = Isodon loxothyrsus（Hand. -Mazz.）H. Hara ●

324815　Rabdosia lungshengensis C. Y. Wu et H. W. Li = Isodon lungshengensis（C. Y. Wu et H. W. Li）H. Hara ■

324816　Rabdosia macrantha（Hook. f.）H. Hara = Siphocranion macranthum（Hook. f.）C. Y. Wu ■

324817　Rabdosia macrocalyx（Dunn）H. Hara = Isodon macrocalyx（Dunn）Kudo ■

324818　Rabdosia macrophylla（Migo）C. Y. Wu et H. W. Li = Isodon macrophyllus（Migo）H. Hara ●■

324819　Rabdosia medilungensis C. Y. Wu et H. W. Li = Isodon medilungensis（C. Y. Wu et H. W. Li）H. Hara ●

324820　Rabdosia megathyrsa（Diels）H. Hara = Isodon megathyrsus（Diels）H. W. Li ■

324821　Rabdosia megathyrsa（Diels）H. Hara var. strigosissima C. Y. Wu et H. W. Li = Isodon megathyrsus（Diels）H. W. Li var. strigosissimus（C. Y. Wu et H. W. Li）H. W. Li ■

324822　Rabdosia megathyrsoides H. W. Li；拟锥香茶菜；False Bigthyrse Rabdosia ■

324823　Rabdosia megathyrsoides H. W. Li = Isodon coetsus（Buch. -Ham. ex D. Don）Kudo ●■

324824　Rabdosia megathyrsoides H. W. Li = Isodon melissiformis（C. Y. Wu）H. Hara ■

324825　Rabdosia megathyrsoides H. W. Li = Isodon melissoides（Benth.）H. Hara ■

324826　Rabdosia melissiformis C. Y. Wu = Isodon melissiformis（C. Y. Wu）H. Hara ■

324827　Rabdosia melissiformis C. Y. Wu = Isodon melissoides（Benth.）H. Hara ■

324828　Rabdosia melissoides（Benth.）H. Hara = Isodon melissoides（Benth.）H. Hara ■

324829　Rabdosia mucronata C. Y. Wu et H. W. Li = Isodon mucronatus（C. Y. Wu et H. W. Li）H. Hara ●■

324830　Rabdosia muliensis（W. W. Sm.）H. Hara = Isodon muliensis（W. W. Sm.）Kudo ●

324831　Rabdosia nervosa（Hemsl.）C. Y. Wu et H. W. Li = Isodon nervosus（Hemsl.）Kudo ■

324832　Rabdosia oresbia（W. W. Sm.）H. Hara = Isodon oresbius（W. W. Sm.）Kudo ●

324833　Rabdosia pachythyrsa（Hand. -Mazz.）H. Hara = Isodon leucophyllus（Dunn）Kudo ●

324834　Rabdosia pantadenia（Hand. -Mazz.）H. Hara = Isodon pantadenius（Hand. -Mazz.）H. W. Li ■

324835　Rabdosia parvifolia（Batalin）H. Hara = Isodon parvifolius（Batalin）H. Hara ●

324836　Rabdosia pharica（Prain）H. Hara = Isodon pharicus（Prain）Murata ●

324837　Rabdosia phyllopoda（Diels）H. Hara = Isodon phyllopodus（Diels）Kudo ●■

324838　Rabdosia phyllostachys（Diels）H. Hara = Isodon phyllostachys（Diels）Kudo ●

324839　Rabdosia phyllostachys（Diels）H. Hara var. leptophylla C. Y. Wu et H. W. Li = Isodon phyllostachys（Diels）Kudo ●

324840　Rabdosia phyllostachys（Diels）Kudo = Isodon phyllostachys（Diels）Kudo ●

324841　Rabdosia phyllostachys（Diels）Kudo var. leptophylla C. Y. Wu et H. W. Li = Isodon phyllostachys（Diels）Kudo ●

324842　Rabdosia pleiophylla（Diels）C. Y. Wu et H. W. Li = Isodon pleiophyllus（Diels）Kudo ●

324843　Rabdosia pleiophylla（Diels）C. Y. Wu et H. W. Li var. dolichodens C. Y. Wu et H. W. Li = Isodon pleiophyllus（Diels）Kudo var. dolichodens（C. Y. Wu et H. W. Li）H. W. Li ●

324844　Rabdosia pleiophylla（Diels）H. Hara var. dolichodens C. Y. Wu et H. W. Li = Isodon pleiophyllus（Diels）Kudo var. dolichodens

（C. Y. Wu et H. W. Li）H. W. Li ●

324845　Rabdosia pluriflora C. Y. Wu et H. W. Li；多花香茶菜；Flowery Rabdosia，Manyflower Rabdosia ■

324846　Rabdosia pluriflora C. Y. Wu et H. W. Li = Isodon coetsus（Buch. -Ham. ex D. Don）Kudo ●■

324847　Rabdosia polystachys（Y. Z. Sun ex C. H. Hu）C. Y. Wu et H. W. Li；多穗香茶菜；Manyspike Rabdosia，Polystachous Rabdosia ●■

324848　Rabdosia polystachys（Y. Z. Sun ex C. H. Hu）C. Y. Wu et H. W. Li var. phyllodioides C. Y. Wu；排钱多穗香茶菜 ●■

324849　Rabdosia polystachys（Y. Z. Sun ex C. H. Hu）C. Y. Wu et H. W. Li var. phyllodioides C. Y. Wu = Isodon coetsus（Buch. -Ham. ex D. Don）Kudo ●■

324850　Rabdosia polystachys（Y. Z. Sun ex C. H. Hu）C. Y. Wu et H. W. Li = Isodon coetsus（Buch. -Ham. ex D. Don）Kudo ●■

324851　Rabdosia provicarii（H. Lév.）H. Hara；白龙香茶菜；Provicar Rabdosia，Whitedragon Rabdosia ●■

324852　Rabdosia provicarii（H. Lév.）H. Hara = Isodon bulleyanus（Diels）Kudo ●

324853　Rabdosia pseudoirrorata C. Y. Wu = Isodon pharicus（Prain）Murata ●

324854　Rabdosia pseudoirrorata C. Y. Wu var. centellifolia C. Y. Wu；马蹄叶香茶菜（阔叶川藏香茶菜）●

324855　Rabdosia pseudoirrorata C. Y. Wu var. centellifolia C. Y. Wu = Isodon pharicus（Prain）Murata ●

324856　Rabdosia pseudoirrorata C. Y. Wu var. pleiophylla（Diels）C. Y. Wu et H. W. Li；多叶香茶菜；Leafy Rabdosia，Manyleaf Rabdosia，Manyleaf Sichuan-Xizang Rabdosia，Pleiophyllous Rabdosia ●

324857　Rabdosia racemosa（Hemsl.）H. Hara = Isodon racemosus（Hemsl.）H. W. Li ■

324858　Rabdosia ricinisperma（Pamp.）H. Hara = Isodon rubescens（Hemsl.）H. Hara ●

324859　Rabdosia ricinisperma（Pamp.）H. Hara = Rabdosia rubescens（Hemsl.）H. Hara ●

324860　Rabdosia rosthornii（Diels）H. Hara = Isodon rosthornii（Diels）Kudo ■

324861　Rabdosia rubescens（Hemsl.）H. Hara = Isodon rubescens（Hemsl.）H. Hara ●

324862　Rabdosia rubescens（Hemsl.）H. Hara f. lushanensis Z. Y. Gao et Y. R. Li；鲁山香茶菜（鲁山冬凌草）；Lushan Blushred Rabdosia ■

324863　Rabdosia rubescens（Hemsl.）H. Hara var. lushiensis Z. Y. Gao et Y. R. Li；卢氏香茶菜；Lushi Blushred Rabdosia ■

324864　Rabdosia rubescens（Hemsl.）H. Hara var. taihangensis Z. Y. Gao et Y. R. Li；冬凌草；Taihang Blushred Rabdosia ■

324865　Rabdosia rugosa（Wall. ex Benth.）H. Hara = Isodon rugosa（Wall. ex Benth.）Codd ●☆

324866　Rabdosia rugosiformis（Hand. -Mazz.）H. Hara = Isodon rugosiformis（Hand. -Mazz.）H. Hara ●

324867　Rabdosia scoparia C. Y. Wu et H. W. Li = Isodon scoparius（C. Y. Wu et H. W. Li）H. Hara ●

324868　Rabdosia scrophularioides（Wall. ex Benth.）H. Hara = Isodon scrophularioides（Wall. ex Benth.）Murata ■

324869　Rabdosia sculponeata（Vaniot）H. Hara = Isodon sculponeatus（Vaniot）Kudo ■

324870　Rabdosia secundiflora C. Y. Wu = Isodon secundiflorus（C. Y. Wu）H. Hara ●

324871　Rabdosia serra（Maxim.）H. Hara = Isodon serrus（Maxim.）Kudo ■

324872　Rabdosia setschwanensis（Hand. -Mazz.）H. Hara = Isodon setschwanensis（Hand. -Mazz.）H. Hara ●

324873　Rabdosia setschwanensis（Hand. -Mazz.）H. Hara var. yungshengensis C. Y. Wu et H. W. Li = Isodon setschwanensis（Hand. -Mazz.）H. Hara ●

324874　Rabdosia shikokiana（Makino）H. Hara = Isodon shikokianus（Makino）H. Hara ■☆

324875　Rabdosia shikokiana（Makino）H. Hara var. intermedia（Kudo）H. Hara = Isodon shikokianus（Makino）H. Hara var. intermedius（Kudo）Murata ■☆

324876　Rabdosia shikokiana（Makino）H. Hara var. occidentalis（Murata）H. Hara = Isodon shikokianus（Makino）H. Hara var. occidentalis Murata ■☆

324877　Rabdosia shimizuana Murata = Isodon hispidus（Benth.）Murata ■

324878　Rabdosia silvatica C. Y. Wu et H. W. Li = Isodon silvaticus（C. Y. Wu et H. W. Li）H. W. Li ●

324879　Rabdosia sinuolata C. Y. Wu et H. W. Li；波齿香茶菜；Sinuate Rabdosia，Sinuolate Rabdosia，Sinuoustooth Rabdosia ●

324880　Rabdosia sinuolata C. Y. Wu et H. W. Li = Isodon pharicus（Prain）Murata ●

324881　Rabdosia smithiana（Hand. -Mazz.）H. Hara = Isodon smithianus（Hand. -Mazz.）H. Hara ●

324882　Rabdosia stenodonta C. Y. Wu et H. W. Li；狭齿香茶菜；Narrowtooth Rabdosia ■

324883　Rabdosia stenodonta C. Y. Wu et H. W. Li = Isodon angustifolius（Dunn）Kudo ■

324884　Rabdosia stenophylla（Migo）Hara = Isodon nervosus（Hemsl.）Kudo ■

324885　Rabdosia stracheyi（Benth. ex Hook. f.）H. Hara = Isodon stracheyi（Benth. ex Hook. f.）Kudo ■

324886　Rabdosia stracheyi（Benth. ex Hook. f.）H. Hara = Isodon walkeri（Arn.）H. Hara ■

324887　Rabdosia taiwanensis（Masam.）H. Hara；台湾延命草 ■

324888　Rabdosia taiwanensis（Masam.）H. Hara = Isodon macrocalyx（Dunn）Kudo ■

324889　Rabdosia taiwanensis（Masam.）H. Hara = Rabdosia macrocalyx（Dunn）H. Hara ■

324890　Rabdosia taliensis C. Y. Wu；大理香茶菜；Dali Rabdosia，Tali Rabdosia ●

324891　Rabdosia taliensis C. Y. Wu = Isodon setschwanensis（Hand. -Mazz.）H. Hara ●

324892　Rabdosia tenuifolia（W. W. Sm.）H. Hara = Isodon tenuifolius（W. W. Sm.）Kudo ●

324893　Rabdosia ternifolia（D. Don）H. Hara = Isodon ternifolius（D. Don）Kudo ●■

324894　Rabdosia thiothyrsa（Hand. -Mazz.）H. Hara = Isodon leucophyllus（Dunn）Kudo ●

324895　Rabdosia trichocarpa（Maxim.）H. Hara = Isodon trichocarpus（Maxim.）Kudo ■☆

324896　Rabdosia trichocarpa（Maxim.）H. Hara f. crythrantha（Ikegami）H. Hara = Isodon trichocarpus（Maxim.）Kudo f. crythranthus Ikegami ■☆

324897　Rabdosia umbrosa（Maxim.）H. Hara = Isodon umbrosus（Maxim.）H. Hara ■☆

324898　Rabdosia umbrosa（Maxim.）H. Hara var. hakusanensis（Kudo）H. Hara = Isodon umbrosus（Maxim.）H. Hara var.

hakusanensis（Kudo）K. Asano ■☆

324899 Rabdosia umbrosa（Maxim.）H. Hara var. komaensis（Okuyama）H. Hara ＝ Isodon umbrosus（Maxim.）H. Hara var. latifolius Okuyama ■☆

324900 Rabdosia umbrosa（Maxim.）H. Hara var. latifolia（Okuyama）H. Hara ＝ Isodon umbrosus（Maxim.）H. Hara var. latifolius Okuyama ■☆

324901 Rabdosia umbrosa（Maxim.）H. Hara var. leucantha（Murai）H. Hara f. kameba（Okuyama ex Ohwi）H. Hara ＝ Isodon umbrosus（Maxim.）H. Hara var. leucanthus（Murai）K. Asano f. kameba（Okuyama ex Ohwi）K. Asano ■☆

324902 Rabdosia umbrosa（Maxim.）H. Hara var. leucantha（Murai）H. Hara ＝ Isodon umbrosus（Maxim.）H. Hara f. leucanthus（Murai）K. Asano ■☆

324903 Rabdosia wardii（C. Marquand et Airy Shaw）H. Hara ＝ Isodon wardii（C. Marquand et Airy Shaw）H. Hara ●

324904 Rabdosia websteri（Hemsl.）H. Hara ＝ Isodon websteri（Hemsl.）Kudo ■

324905 Rabdosia weisiensis C. Y. Wu ＝ Isodon weisiensis（C. Y. Wu）H. Hara ■

324906 Rabdosia wikstroemioides（Hand.-Mazz.）H. Hara ＝ Isodon wikstroemioides（Hand.-Mazz.）H. Hara ●

324907 Rabdosia xerophila C. Y. Wu et H. W. Li ＝ Isodon xerophilus（C. Y. Wu et H. W. Li）H. Hara ●

324908 Rabdosia yunnanensis（Hand.-Mazz.）H. Hara ＝ Isodon yunnanensis（Hand.-Mazz.）H. Hara ●

324909 Rabdosiella Codd ＝ Plectranthus L' Hér.（保留属名）●■

324910 Rabdosiella Codd（1984）；肖香茶菜属 ●■

324911 Rabdosiella calycina（Benth.）Codd ＝ Plectranthus calycinus Benth. ●■

324912 Rabdosiella leemanni N. Hahn；肖香茶菜 ●☆

324913 Rabdosiella ternifolia（D. Don）Codd ＝ Plectranthus ternifolius D. Don ●■

324914 Rabdosiella ternifolia（D. Don）Codd ＝ Rabdosia ternifolia（D. Don）H. Hara ●■

324915 Rabelaisia Planch. ＝ Lunasia Blanco ●☆

324916 Rabenhorstia Rchb. ＝ Berzelia Brongn. ●☆

324917 Rabenhorstia Rchb. ＝ Heterodon Meisn. ●☆

324918 Rabiea N. E. Br.（1930）；旭波属 ■☆

324919 Rabiea albinota（Haw.）N. E. Br.；旭波 ■☆

324920 Rabiea albinota（Haw.）N. E. Br. var. longipetala L. Bolus ＝ Rabiea albinota（Haw.）N. E. Br. ■☆

324921 Rabiea albinota（Haw.）N. E. Br. var. microstigma L. Bolus ＝ Rabiea albinota（Haw.）N. E. Br. ■☆

324922 Rabiea albipuncta（Haw.）N. E. Br.；静波 ■☆

324923 Rabiea albipuncta（Haw.）N. E. Br. var. major L. Bolus ＝ Rabiea albipuncta（Haw.）N. E. Br. ■☆

324924 Rabiea comptonii（L. Bolus）L. Bolus；康普顿旭波 ■☆

324925 Rabiea difformis（L. Bolus）L. Bolus；参差旭波 ■☆

324926 Rabiea jamesii（L. Bolus）L. Bolus；詹姆斯旭波 ■☆

324927 Rabiea lesliei N. E. Br.；莱斯利旭波 ■☆

324928 Rabiea tersa N. E. Br. ＝ Prepodesma orpenii（N. E. Br.）N. E. Br. ■☆

324929 Racapa M. Roem. ＝ Carapa Aubl. ●☆

324930 Racaria Aubl. ＝ Talisia Aubl. ●☆

324931 Racemaria Raf. ＝ Smilacina Desf.（保留属名）■

324932 Racemobambos Holttum（1956）；总序竹属；Racemobambos ●

324933 Racemobambos prainii（Gamble）P. C. Keng et T. H. Wen；总序竹（西藏新小竹，新小竹）；Common Neomicrocalamus，Racemobambos，Small-leaf Neomicrocalamus ●

324934 Racemobambos prainii（Gamble）P. C. Keng et T. H. Wen ＝ Neomicrocalamus prainii（Gamble）P. C. Keng ●

324935 Racemobambos yunnanensis T. H. Wen；云南总序竹（云南新小竹）；Yunnan Racemobambos ●

324936 Racemobambos yunnanensis T. H. Wen ＝ Melocalamus yunnanensis（T. H. Wen）T. P. Yi ●

324937 Racemobambos yunnanensis T. H. Wen ＝ Neomicrocalamus yunnanensis（T. H. Wen）Ohrnb. ●

324938 Rachea DC. ＝ Crassula L. ●■☆

324939 Racheella Pax ＝ Lyallia Hook. f. ■☆

324940 Rachelia J. M. Ward et Breitw.（1997）；腋头紫绒草属 ■☆

324941 Rachelia glaria J. M. Ward et Breitw.；腋头紫绒草 ■☆

324942 Rachia Klotzsch ＝ Begonia L. ●■

324943 Rachicallis DC. ＝ Arcytophyllum Willd. ex Schult. et Schult. f. ●☆

324944 Rachicallis DC. ＝ Rhachicallis DC. ●☆

324945 Raciborskanthos Szlach.（1995）；拉氏兰属 ■☆

324946 Raciborskanthos Szlach. ＝ Ascochilus Ridl. ■☆

324947 Racinaea M. A. Spencer et L. B. Sm.（1993）；拉西纳凤梨属（拉辛铁兰属）■☆

324948 Racinaea blassii（L. B. Sm.）M. A. Spencer et L. B. Sm.；拉西纳凤梨 ■☆

324949 Racka J. F. Gmel. ＝ Avicennia L. ●

324950 Raclathris Raf. ＝ Rochelia Rchb.（保留属名）■

324951 Racletia Adans. ＝ ? Reaumuria L. ●

324952 Racoma Willd. ex Steud. ＝ Rocama Forssk. ■

324953 Racoma Willd. ex Steud. ＝ Trianthema L. ■

324954 Racosperma（DC.）Mart. ＝ Acacia Mill.（保留属名）●■

324955 Racosperma Mart. ＝ Acacia Mill.（保留属名）●■

324956 Racosperma aneurum（F. Muell.）Pedley ＝ Acacia aneura F. Muell. ●☆

324957 Racosperma auriculiforme（A. Cunn. ex Benth.）Pedley ＝ Acacia auriculiformis A. Cunn. ex Benth. ●

324958 Racosperma baileyanum（F. Muell.）Pedley ＝ Acacia baileyana F. Muell. ●☆

324959 Racosperma confusum（Merr.）Pedley ＝ Acacia confusa Merr. ●

324960 Racosperma cultriforme（A. Cunn. ex G. Don）Pedley ＝ Acacia cultriformis A. Cunn. ex G. Don ●☆

324961 Racosperma dealbatum（Link）Pedley ＝ Acacia dealbata Link ●

324962 Racosperma decurrens（Willd.）Pedley ＝ Acacia decurrens（J. C. Wendl.）Willd. ●

324963 Racosperma elata（A. Cunn. ex Benth.）Pedley ＝ Acacia elata A. Cunn. ex Benth. ●☆

324964 Racosperma kempeanum（F. Muell.）Pedley ＝ Acacia kempeana F. Muell. ●☆

324965 Racosperma mearnsii（De Wild.）Pedley ＝ Acacia mearnsii De Wild. ●

324966 Racosperma melanoxylon（R. Br.）Maitland ＝ Acacia melanoxylon R. Br. ●

324967 Racosperma peuce（F. Muell.）Pedley ＝ Acacia peuce F. Muell. ●☆

324968 Racosperma podalyriifolium（A. Cunn. ex G. Don）Pedley ＝ Acacia podalyriifolia A. Cunn. ex G. Don ●☆

324969 Racosperma saligna（Labill.）Pedley ＝ Acacia saligna（Labill.）H. L. Wendl. ●☆

324970 Racoubea Aubl. ＝ Homalium Jacq. ●

324971　Racua J. F. Gruel　= Avicennia L. ●

324972　Racua J. F. Gruel　= Racka J. F. Gmel. ●

324973　Radamaea Benth. (1846);马岛林列当属●☆

324974　Radamaea montana Benth. ;马岛林列当●☆

324975　Radcliffea Petra Hoffm. et K. Wurdack(2006);拉德大戟属●☆

324976　Radcliffea smithii Petra Hoffm. et K. Wurdack;拉德大戟●☆

324977　Radcliffea smithii Petra Hoffm. et K. Wurdack = Rapanea erythroxyloides (Thouars ex Roem. et Schult.) Mez ●☆

324978　Raddia Bertol. (1819);雷迪禾属■☆

324979　Raddia DC. ex Miers = Raddisia Leandro ●

324980　Raddia DC. ex Miers = Salacia L. (保留属名)●

324981　Raddia Mazziari　= Crypsis Aiton(保留属名)■

324982　Raddia Miers　= Raddisia Leandro ●

324983　Raddia Miers　= Salacia L. (保留属名)●

324984　Raddia Pieri　= Crypsis Aiton(保留属名)■

324985　Raddia Post et Kuntze = Barbacenia Vand. ■☆

324986　Raddia Post et Kuntze = Radia A. Rich. ex Kunth ■☆

324987　Raddia brasiliensis Bertol. ;雷迪禾■☆

324988　Raddiella Swallen(1948);小雷迪禾属■☆

324989　Raddiella nana (Döll) Swallen;小雷迪禾■☆

324990　Raddisia Leandro = Salacia L. (保留属名)●

324991　Rademachia Steud. = Artocarpus J. R. Forst. et G. Forst. (保留属名)●

324992　Rademachia Steud. = Radermachia Thunb. ●

324993　Radermachera Zoll. et Moritzi(1855);菜豆树属(山菜豆属); Belltree,Bell Tree ●

324994　Radermachera alata Dop　= Pauldopia ghorta (Buch. -Ham. ex G. Don) Steenis ●

324995　Radermachera bipinnata (Collett et Hemsl.) Steenis ex Chatterjee = Pauldopia ghorta (Buch. -Ham. ex G. Don) Steenis ●

324996　Radermachera frondosa Chun et F. C. How;美叶菜豆树(红花树,美丽菜豆树,牛尾连);Beautifulleaf Belltree, Leafy Belltree, Leafy Bell-tree ●

324997　Radermachera glandulosa (Blume) Miq. ;广西菜豆树; Glandulose Belltree, Guangxi Bell Tree, Guangxi Belltree, Kwangsi Belltree ●

324998　Radermachera hainanensis Merr. ;海南菜豆树(大叶牛尾连, 大叶牛尾林,牛尾林);Hainan Bell Tree, Hainan Belltree ●

324999　Radermachera ignea (Kurz) Steenis = Mayodendron igneum (Kurz) Kurz ●

325000　Radermachera microcalyx C. Y. Wu et W. C. Yin;小萼菜豆树; Littlecalyx Belltree, Smallcalyx Belltree, Small-calyxed Bell Tree ●

325001　Radermachera pentandra Hemsl. ;豇豆树;Cowpeatree, Five-stamened Belltree, Pentandrous Bell Tree ●

325002　Radermachera sinica (Hance) Hemsl. ;菜豆树(白鹤参,朝阳花,大朝阳,大朗伞,跌死猫树,豆角木,豆角树,鸡豆木,豇豆树,接骨凉伞,苦苓舅,辣椒树,牛尾豆,牛尾木,牛尾树,森木郎伞,森木凉伞,山菜豆,山苦楝,山苦苓,蛇树,蛇子豆);Asia Bell Tree, Asia Belltree, Asian Bell, Asian Bell Tree ●

325003　Radermachera sinica (Hance) Hemsl. = Radermachera frondosa Chun et F. C. How ●

325004　Radermachera sinica (Hance) Hemsl. = Radermachera yunnanensis C. Y. Wu et W. C. Yin ●

325005　Radermachera tonkinensis Dop = Radermachera sinica (Hance) Hemsl. ●

325006　Radermachera xylocarpa (Roxb.) K. Schum. ;木果菜豆树●☆

325007　Radermachera yunnanensis C. Y. Wu et W. C. Yin;滇菜豆树 (豇豆树,蛇尾树,土厚朴);Yunnan Bell Tree, Yunnan Belltree ●

325008　Radermachia B. D. Jacks. = Radermachera Zoll. et Moritzi ●

325009　Radermachia Thunb. = Artocarpus J. R. Forst. et G. Forst. (保留属名)●

325010　Radermachia incisa Thunb. = Artocarpus altilis (Parkinson) Fosberg ●

325011　Radermachia incisa Thunb. = Artocarpus communis J. R. Forst. et G. Forst. ●

325012　Radermachia incisa Thunb. = Artocarpus incisus (Thunb.) L. f. ●

325013　Radia A. Rich. = Barbacenia Vand. ■☆

325014　Radia A. Rich. ex Kunth = Barbacenia Vand. ■☆

325015　Radia Noronha = Mimusops L. ●☆

325016　Radiana Raf. = Cypselea Turpin ■☆

325017　Radiana Raf. ex DC. = Cypselea Turpin ■☆

325018　Radiata Medik. = Medicago L. (保留属名)●■

325019　Radiaxaceae Dulac = Cornaceae Bercht. et J. Presl(保留科名)●■

325020　Radicula Dill. ex Moench = Rorippa Scop. ■

325021　Radicula Hill = Nasturtium W. T. Aiton(保留属名)■

325022　Radicula Hill = Rorippa Scop. ■

325023　Radicula Moench = Rorippa Scop. ■

325024　Radicula aquatica (Eaton) B. L. Rob. = Armoracia lacustris (A. Gray) Al-Shehbaz et V. M. Bates ■☆

325025　Radicula armoracia (L.) B. L. Rob. = Armoracia rusticana (Lam.) Gaertn. ,B. Mey. et Scherb. ■

325026　Radicula hispida (Desv.) Britton = Rorippa palustris (L.) Besser subsp. hispida (Desv.) Jonsell ■☆

325027　Radicula montana (Wall. ex Hook. f. et Thomson) Hu ex C. P'ei = Rorippa indica (L.) Hiern ■

325028　Radicula nasturtium-aquaticum (L.) Britten et Rendle = Nasturtium officinale R. Br. ■

325029　Radicula sessiliflora (Nutt.) Greene = Rorippa sessiliflora (Nutt. ex Torr. et A. Gray) Hitchc. ■☆

325030　Radicula sinuata (Nutt.) Greene = Rorippa sinuata (Nutt. ex Torr. et A. Gray) Hitchc. ■☆

325031　Radicula sylvestris (L.) Druce = Rorippa sylvestris (L.) Besser ■

325032　Radinocion Ridl. = Aerangis Rchb. f. ■☆

325033　Radinocion flexuosa Ridl. = Aerangis flexuosa (Ridl.) Schltr. ■☆

325034　Radinosiphon N. E. Br. (1932);细管鸢尾属■☆

325035　Radinosiphon cameronii N. E. Br. = Radinosiphon leptostachyis (Baker) N. E. Br. ■☆

325036　Radinosiphon leptosiphon (F. Bolus) N. E. Br. = Gladiolus leptosiphon F. Bolus ■☆

325037　Radinosiphon leptostachyis (Baker) N. E. Br. ;细穗细管鸢尾■☆

325038　Radinosiphon lomatensis (N. E. Br.) N. E. Br. ;细管鸢尾■☆

325039　Radiola Hill(1756);射线亚麻属;Allseed, Flaxseed ■☆

325040　Radiola Roth = Radiola Hill ■☆

325041　Radiola linoides Roth;射线亚麻;Allseed, Flaxseed, Thyme-leaved Flaxseed ■☆

325042　Radiola millegrana Sm. = Radiola linoides Roth ■☆

325043　Radiola multiflora (Lam.) Asch. = Radiola linoides Roth ■☆

325044　Radiusia Rchb. = Sophora L. ●■

325045　Radlkofera Gilg(1897);拉氏无患子属●☆

325046　Radlkofera calodendron Gilg;拉氏无患子●☆

325047　Radlkoferella Pierre = Lucuma Molina ●

325048　Radlkoferella Pierre = Pouteria Aubl. ●

325049 Radlkoferotoma Kuntze(1891);玫菊木属●☆

325050 Radlkoferotoma cistifolium Kuntze;玫菊木●☆

325051 Radojitskya Turcz. = Lachnaea L. ●☆

325052 Radyera Bullock(1957);拉迪锦葵属●☆

325053 Radyera urens (L. f.) Bullock;拉迪锦葵●☆

325054 Raffenaldia Godr. (1853);北非芥属■☆

325055 Raffenaldia platycarpa (Coss.) Stapf;宽果北非芥■☆

325056 Raffenaldia primuloides Godr.;北非芥■☆

325057 Raffenaldia primuloides Godr. subsp. riphaensis J. M. Monts.;山地北非芥■☆

325058 Raffenaldia primuloides Godr. var. lutea Maire = Raffenaldia primuloides Godr. ■☆

325059 Raffenaldia primuloides Godr. var. violacea Maire = Raffenaldia primuloides Godr. ■☆

325060 Rafflesia R. Br. = Rafflesia R. Br. ex Gray ■☆

325061 Rafflesia R. Br. ex Gray(1821);大花草属;Monster Flower, Monsterflower, Rafflesia ■☆

325062 Rafflesia arnoldii R. Br.;大花草■☆

325063 Rafflesia kerrii Meijer;泰国大花草■☆

325064 Rafflesia pricei Meijer;婆罗洲大花草■☆

325065 Rafflesiaceae Dumort. (1829)(保留科名);大花草科;Monsterflower Family, Rafflesia Family ■

325066 Rafflesiaceae Dumort. (保留科名) = Cytinaceae Brongn. ☆

325067 Rafia Bory = Raphia P. Beauv. ●

325068 Rafinesquia Nutt. (1841)(保留属名);雪苣属;Rafinesqui's Chicory ■☆

325069 Rafinesquia Raf. (废弃属名) = Clinopodium L. ■●

325070 Rafinesquia Raf. (废弃属名) = Diodeilis Raf. ■●

325071 Rafinesquia Raf. (废弃属名) = Hosackia Douglas ex Benth. ●☆

325072 Rafinesquia Raf. (废弃属名) = Jacaranda Juss. ●

325073 Rafinesquia Raf. (废弃属名) = Rafinesquia Nutt. (保留属名)■☆

325074 Rafinesquia californica Nutt.;加州雪苣;California Chicory, California Plumeseed ■☆

325075 Rafinesquia neomexicana A. Gray;雪苣;Desert Chicory, New Mexico Plumseed ■☆

325076 Rafnia Thunb. (1800);拉菲豆属(雷夫豆属)■☆

325077 Rafnia acuminata (E. Mey.) G. J. Campb. et B. -E. van Wyk;渐尖拉菲豆■☆

325078 Rafnia affinis Harv. = Rafnia elliptica Thunb. ■☆

325079 Rafnia alata G. J. Campb. et B. -E. van Wyk;具翅拉菲豆■☆

325080 Rafnia alpina Eckl. et Zeyh. = Rafnia triflora (L.) Thunb. ■☆

325081 Rafnia amplexicaulis (L.) Thunb.;抱茎拉菲豆■☆

325082 Rafnia angulata Thunb.;棱角拉菲豆■☆

325083 Rafnia angulata Thunb. subsp. ericifolia (T. M. Salter) G. J. Campb. et B. -E. van Wyk;毛叶棱角拉菲豆■☆

325084 Rafnia angulata Thunb. subsp. humilis (Eckl. et Zeyh.) G. J. Campb. et B. -E. van Wyk;低矮棱角拉菲豆■☆

325085 Rafnia angulata Thunb. subsp. montana G. J. Campb. et B. -E. van Wyk;山地棱角拉菲豆■☆

325086 Rafnia angulata Thunb. subsp. thunbergii (Harv.) G. J. Campb. et B. -E. van Wyk;通贝里拉菲豆■☆

325087 Rafnia angulata Thunb. var. angustifolia (Thunb.) E. Mey. = Rafnia angulata Thunb. ■☆

325088 Rafnia angulata Thunb. var. filifolia (Thunb.) E. Mey. = Rafnia angulata Thunb. ■☆

325089 Rafnia angulata Thunb. var. latifolia Harv. = Rafnia angulata Thunb. ■☆

325090 Rafnia angustifolia Thunb. = Rafnia angulata Thunb. ■☆

325091 Rafnia axillaris Thunb. = Rafnia elliptica Thunb. ■☆

325092 Rafnia capensis (L.) Schinz;好望角拉菲豆■☆

325093 Rafnia capensis (L.) Schinz subsp. calycina G. J. Campb. et B. -E. van Wyk;萼状拉菲豆■☆

325094 Rafnia capensis (L.) Schinz subsp. carinata G. J. Campb. et B. -E. van Wyk;龙骨状拉菲豆■☆

325095 Rafnia capensis (L.) Schinz subsp. dichotoma (Eckl. et Zeyh.) G. J. Campb. et B. -E. van Wyk;二歧好望角拉菲豆■☆

325096 Rafnia capensis (L.) Schinz subsp. ovata (P. J. Bergius) G. J. Campb. et B. -E. van Wyk;卵形好望角拉菲豆■☆

325097 Rafnia capensis (L.) Schinz subsp. pedicellata G. J. Campb. et B. -E. van Wyk;梗花拉菲豆■☆

325098 Rafnia cordata (L.) Mart. = Rafnia triflora (L.) Thunb. ■☆

325099 Rafnia cordata Eckl. et Zeyh. = Rafnia capensis (L.) Schinz subsp. ovata (P. J. Bergius) G. J. Campb. et B. -E. van Wyk ■☆

325100 Rafnia corymbosa (E. Mey.) Walp. = Rafnia capensis (L.) Schinz ■☆

325101 Rafnia crassifolia Harv.;厚叶拉菲豆■☆

325102 Rafnia crispa C. H. Stirt.;皱波拉菲豆■☆

325103 Rafnia cuneifolia Thunb. = Rafnia capensis (L.) Schinz subsp. ovata (P. J. Bergius) G. J. Campb. et B. -E. van Wyk ■☆

325104 Rafnia cuneifolia Thunb. var. lanceolata Harv. = Rafnia capensis (L.) Schinz subsp. ovata (P. J. Bergius) G. J. Campb. et B. -E. van Wyk ■☆

325105 Rafnia cuneifolia Thunb. var. obovata Harv. = Rafnia capensis (L.) Schinz subsp. ovata (P. J. Bergius) G. J. Campb. et B. -E. van Wyk ■☆

325106 Rafnia cuneifolia Thunb. var. rhomboidea (E. Mey.) Harv. = Rafnia capensis (L.) Schinz subsp. ovata (P. J. Bergius) G. J. Campb. et B. -E. van Wyk ■☆

325107 Rafnia dichotoma Eckl. et Zeyh. = Rafnia capensis (L.) Schinz subsp. dichotoma (Eckl. et Zeyh.) G. J. Campb. et B. -E. van Wyk ■☆

325108 Rafnia diffusa Eckl. et Zeyh. = Rafnia triflora (L.) Thunb. ■☆

325109 Rafnia diffusa Thunb.;松散拉菲豆■☆

325110 Rafnia elliptica Thunb.;椭圆拉菲豆■☆

325111 Rafnia elliptica Thunb. var. acuminata Harv. = Rafnia elliptica Thunb. ■☆

325112 Rafnia elliptica Thunb. var. erecta Harv. = Rafnia elliptica Thunb. ■☆

325113 Rafnia elliptica Thunb. var. intermedia (Vogel ex Walp.) Harv. = Rafnia elliptica Thunb. ■☆

325114 Rafnia ericifolia T. M. Salter = Rafnia angulata Thunb. subsp. ericifolia (T. M. Salter) G. J. Campb. et B. -E. van Wyk ■☆

325115 Rafnia fastigiata Eckl. et Zeyh. = Rafnia triflora (L.) Thunb. ■☆

325116 Rafnia filifolia Thunb. = Rafnia angulata Thunb. ■☆

325117 Rafnia gibba (E. Mey.) Druce = Rafnia capensis (L.) Schinz subsp. dichotoma (Eckl. et Zeyh.) G. J. Campb. et B. -E. van Wyk ■☆

325118 Rafnia globosa G. J. Campb. et B. -E. van Wyk;球形拉菲豆■☆

325119 Rafnia humilis Eckl. et Zeyh. = Rafnia angulata Thunb. subsp. humilis (Eckl. et Zeyh.) G. J. Campb. et B. -E. van Wyk ■☆

325120 Rafnia inaequalis G. J. Campb. et B. -E. van Wyk;不等拉菲豆■☆

325121 Rafnia intermedia Benth. = Rafnia elliptica Thunb. ■☆

325122 Rafnia intermedia Vogel ex Walp. = Rafnia elliptica Thunb. ■☆

325123 Rafnia lancea (Thunb.) DC.;披针状拉菲豆■☆

325124 Rafnia lancifolia C. Presl;剑叶拉菲豆■☆

325125 Rafnia meyeri Schinz = Rafnia ovata E. Mey. ■☆

325126　Rafnia myrtifolia C. Presl;香桃木叶拉菲豆■☆

325127　Rafnia opposita (L.) Thunb. = Rafnia capensis (L.) Schinz ■☆

325128　Rafnia ovata (P. J. Bergius) Schinz = Rafnia capensis (L.) Schinz subsp. ovata (P. J. Bergius) G. J. Campb. et B. -E. van Wyk ☆

325129　Rafnia ovata E. Mey.;卵形拉菲豆■☆

325130　Rafnia pauciflora Eckl. et Zeyh. = Rafnia capensis (L.) Schinz ■☆

325131　Rafnia perfoliata (Thunb.) E. Mey. = Rafnia acuminata (E. Mey.) G. J. Campb. et B. -E. van Wyk ■☆

325132　Rafnia racemosa Eckl. et Zeyh.;总拉菲豆■☆

325133　Rafnia racemosa Eckl. et Zeyh. subsp. pumila G. J. Campb. et B. -E. van Wyk;矮拉菲豆■☆

325134　Rafnia retroflexa Thunb. = Rafnia capensis (L.) Schinz ■☆

325135　Rafnia rhomboidea (E. Mey.) Walp. = Rafnia capensis (L.) Schinz subsp. ovata (P. J. Bergius) G. J. Campb. et B. -E. van Wyk ■☆

325136　Rafnia rostrata G. J. Campb. et B. -E. van Wyk;喙状拉菲豆■☆

325137　Rafnia rostrata G. J. Campb. et B. -E. van Wyk subsp. pluriflora G. J. Campb. et B. -E. van Wyk;多花拉菲豆■☆

325138　Rafnia schlechteriana Schinz;施莱拉菲豆■☆

325139　Rafnia spicata Eckl. et Zeyh. = Rafnia capensis (L.) Schinz ■☆

325140　Rafnia spicata Thunb.;穗花拉菲豆■☆

325141　Rafnia thunbergii Harv. = Rafnia angulata Thunb. subsp. thunbergii (Harv.) G. J. Campb. et B. -E. van Wyk ■☆

325142　Rafnia triflora (L.) Thunb.;三花拉菲豆■☆

325143　Rafnia virens E. Mey. = Rafnia amplexicaulis (L.) Thunb. ■☆

325144　Rafnia vlokii G. J. Campb. et B. -E. van Wyk;弗劳克拉菲豆■☆

325145　Ragadiolus Post et Kuntze = Rhagadiolus Vaill.(保留属名)■☆

325146　Ragala Pierre = Chrysophyllum L. ●

325147　Ragenium Gand. = Geranium L. ■●

325148　Rahowardiana D'Arcy = Markea Rich. ●☆

325149　Rahowardiana D'Arcy(1974);拉氏茄属●☆

325150　Rahowardiana wardiana D'Arcy;拉氏茄☆

325151　Raiania Scop. = Rajania L. ■☆

325152　Raillarda Endl. = Raillardia Spreng. ■☆

325153　Raillarda Endl. = Railliardia Gaudich. ●■☆

325154　Raillardella (A. Gray) Benth. (1873);小轮菊属■☆

325155　Raillardella Benth. = Raillardella (A. Gray) Benth. ■☆

325156　Raillardella argentea (A. Gray) A. Gray;银色小轮菊■☆

325157　Raillardella muirii A. Gray = Carlquistia muirii (A. Gray) B. G. Baldwin ■☆

325158　Raillardella paniculata Greene = Arnica viscosa A. Gray ■☆

325159　Raillardella scaposa (A. Gray) A. Gray;小轮菊■☆

325160　Raillardia Gaudich. = Dubautia Gaudich. ●■☆

325161　Raillardia Spreng. = Railliardia Gaudich. ●■☆

325162　Raillardiopsis Rydb. (1927);拟轮菊属■☆

325163　Raillardiopsis Rydb. = Raillardella (A. Gray) Benth. ■☆

325164　Raillardiopsis muirii (A. Gray) Rydb. = Carlquistia muirii (A. Gray) B. G. Baldwin ■☆

325165　Raillardiopsis scabrida (Eastw.) Rydb. = Anisocarpus scabridus (Eastw.) B. G. Baldwin ■☆

325166　Railliarda DC. = Railliardia Gaudich. ●■☆

325167　Railliardia Gaudich. = Dubautia Gaudich. ●■☆

325168　Railliardia argentea A. Gray = Raillardella argentea (A. Gray) A. Gray ■☆

325169　Railliardia scaposa A. Gray = Raillardella scaposa (A. Gray) A. Gray ■☆

325170　Raimannia Rose = Oenothera L. ●■

325171　Raimannia Rose ex Britton et A. Br. = Oenothera L. ●■

325172　Raimannia grandis (Britton) Rose = Oenothera grandis (Britton) Smyth ■☆

325173　Raimannia laciniata (Hill) Rose = Oenothera laciniata Hill ■

325174　Raimannia laciniata (Hill) Rose ex Britton et A. Br. = Oenothera laciniata Hill ■

325175　Raimannia rhombipetala (Nutt. ex Torr. et A. Gray) Rose = Oenothera rhombipetala Nutt. ex Torr. et A. Gray ■☆

325176　Raimondia Saff. (1913);拉伊木属(雷蒙木属)●☆

325177　Raimondia monoica Saff.;山地拉伊木●☆

325178　Raimondia tenuiflora (Mart.) R. E. Fr.;细花拉伊木●☆

325179　Raimondianthus Harms = Chaetocalyx DC. ■☆

325180　Raimundochloa A. M. Molina = Koeleria Pers. ■

325181　Rainiera Greene = Luina Benth. ■●☆

325182　Rainiera Greene(1898);长序蟹甲草属■☆

325183　Rainiera stricta (Greene) Greene;长序蟹甲草■☆

325184　Raja Burm. = Rajania L. ■☆

325185　Rajania L. (1753);闭果薯蓣属;Tuber Vine ■☆

325186　Rajania Walter = Brunnichia Banks ex Gaertn. ●☆

325187　Rajania cordata L.;闭果薯蓣;Bihi, Carib Yam ■☆

325188　Rajania hastata L.;戟叶闭果薯蓣■☆

325189　Rajania ovata Walter = Brunnichia ovata (Walter) Shinners ■☆

325190　Rajania quinata Houtt. = Akebia quinata (Thunb.) Decne. ●

325191　Rajania quinquefolia L.;五叶闭果薯蓣■☆

325192　Raleighia Gardner = Abatia Ruiz et Pav. ●☆

325193　Ramangis Thouars = Angraecum Bory ■

325194　Ramatuela Kunth = Terminalia L.(保留属名)●

325195　Ramatuella Poir. = Terminalia L.(保留属名)●

325196　Ramelia Baill. = Bocquillonia Baill. ●☆

325197　Rameya Baill. = Triclisia Benth. ●☆

325198　Rameya loucoubensis Baill. = Triclisia loucoubensis Baill. ●☆

325199　Rameya macrocarpa Baill. = Triclisia macrocarpa (Baill.) Diels ●☆

325200　Ramirezella Rose = Vigna Savi(保留属名)■

325201　Ramirezia A. Rich. = Poeppigia C. Presl ■☆

325202　Ramischia Opiz = Orthilia Raf. ■

325203　Ramischia Opiz ex Garcke = Orthilia Raf. ■

325204　Ramischia elatior Rydb. = Orthilia secunda (L.) House ■

325205　Ramischia obtusata (Turcz.) Freyn = Orthilia obtusata (Turcz.) H. Hara ■

325206　Ramischia secunda (L.) Garcke = Orthilia secunda (L.) House ■

325207　Ramischia secunda (L.) Garcke subsp. obtusata (Turcz.) Andres = Orthilia obtusata (Turcz.) H. Hara ■

325208　Ramischia secunda (L.) Garcke var. nummularia Rupr. = Orthilia obtusata (Turcz.) H. Hara ■

325209　Ramischia secunda (L.) Garcke var. obtusata (Turcz.) House = Orthilia obtusata (Turcz.) H. Hara ■

325210　Ramischia secunda (L.) Garcke var. pumila Cham. = Orthilia obtusata (Turcz.) H. Hara ■

325211　Ramischia secundiflora Opiz = Orthilia secunda (L.) House ■

325212　Ramisia Glaz. = Ramisia Glaz. ex Baill. ●☆

325213　Ramisia Glaz. ex Baill. (1887);巴西茉莉属●☆

325214　Ramisia brasiliensis Glaz.;巴西茉莉●☆

325215　Ramium Kuntze = Boehmeria Jacq. ●

325216　Ramium Rumph. = Boehmeria Jacq. ●

325217　Ramium Rumph. ex Kuntze = Boehmeria Jacq. ●

325218　Ramium niveum (L.) Small = Boehmeria nivea (L.) Gaudich. ●

325219　Ramona Greene ＝ Audibertia Benth. ●■

325220　Ramona Greene ＝ Salvia L. ●■

325221　Ramonda Caruel ＝ Ramonda Rich. （保留属名）■☆

325222　Ramonda Pers. ＝ Ramonda Rich. （保留属名）■☆

325223　Ramonda Rich. （1805）（保留属名）；欧洲苣苔属（拉蒙达花属，拉蒙苣苔属）；Pyrenean-violet, Ramonda, Ramondia ■☆

325224　Ramonda Rich. ex Pers. ＝ Ramonda Rich. （保留属名）■☆

325225　Ramonda myconi （L.） Rchb. ；欧洲苣苔, Pyrenean-violet, Rosette Mullein ■☆

325226　Ramonda nathaliae Pancic et Petrovic；那塔利欧洲苣苔（巴尔干苣苔，那塔利拉蒙苣苔）；Nathalia Ramonda ■☆

325227　Ramonda pyrenaica ? ＝ Ramonda myconi （L.） Rchb. ■☆

325228　Ramonda serbica Pancic；塞尔维亚欧洲苣苔（塞尔维亚拉蒙苣苔，紫药欧苣苔）；Servian Ramonda ■☆

325229　Ramondaceae Godr. ＝ Ramondaceae Godr. et Gren. ex Godr. ■

325230　Ramondaceae Godr. et Gren. ＝ Gesneriaceae Rich. et Juss. （保留科名）■●

325231　Ramondaceae Godr. et Gren. ＝ Ramondaceae Godr. et Gren. ex Godr. ■

325232　Ramondaceae Godr. et Gren. ex Godr. ；欧洲苣苔科■

325233　Ramondaceae Godr. et Gren. ex Godr. ＝ Gesneriaceae Rich. et Juss. （保留科名）■●

325234　Ramondia J. St. -Hil. ＝ Ramonda Rich. （保留属名）■☆

325235　Ramondia Mirb. （废弃属名）＝ Ramonda Rich. （保留属名）■☆

325236　Ramondia Rich. ＝ Ramonda Rich. （保留属名）■☆

325237　Ramonia Post et Kuntze ＝ Ramona Greene ●■

325238　Ramonia Post et Kuntze ＝ Salvia L. ●■

325239　Ramonia Schltr. ＝ Hexadesmia Brongn. ■☆

325240　Ramonia Schltr. ＝ Scaphyglottis Poepp. et Endl. （保留属名）■☆

325241　Ramorinoa Speg. （1924）；拉莫豆属■☆

325242　Ramorinoa girolae Speg. ；拉莫豆■☆

325243　Ramosia Merr. ＝ Centotheca Desv. （保留属名）■

325244　Ramosmania Tirveng. et Verdc. （1982）；拉氏茜属●☆

325245　Ramosmania heterophylla （Balf. f.） Tirveng. et Verdc. ；互叶拉氏茜●☆

325246　Ramosmania rodriguesii Tirveng. ；拉氏茜●☆

325247　Ramostigmaceae Dulac ＝ Empetraceae Hook. et Lindl. （保留科名）●

325248　Ramotha Raf. ＝ Xyris L. ■

325249　Ramphicarpa Rchb. ＝ Rhamphicarpa Benth. ■☆

325250　Ramphidia Miq. ＝ Hetaeria Blume（保留属名）■

325251　Ramphidia Miq. ＝ Myrmechis （Lindl.） Blume ■

325252　Ramphidia Miq. ＝ Rhamphidia Lindl. ■

325253　Ramphidia mannii Rchb. f. ＝ Zeuxine mannii （Rchb. f.） Geerinck ■☆

325254　Ramphocarpus Neck. ＝ Geranium L. ■●

325255　Rampholepis Stapf ＝ Rhampholepis Stapf ■

325256　Rampholepis Stapf ＝ Sacciolepis Nash ■

325257　Ramphospermum Andrz. ex Rchb. ＝ Rhamphospermum Rchb. ■

325258　Ramphospermum Andrz. ex Rchb. ＝ Sinapis L. ■

325259　Rampinia C. B. Clarke ＝ Herpetospermum Wall. ex Hook. f. ■

325260　Ramsaia W. Anderson ex R. Br. ＝ Bauera Banks ex Andréws ●☆

325261　Ramsdenia Britton ＝ Phyllanthus L. ●■

325262　Ramspekia Scop. ＝ Posoqueria Aubl. ●☆

325263　Ramtilla DC. ＝ Guizotia Cass. （保留属名）■●

325264　Ramusia E. Mey. ＝ Asystasia Blume ●■

325265　Ramusia Nees ＝ Peristrophe Nees ■

325266　Ramusia tridentata （E. Mey.） Nees ＝ Peristrophe tridentata （E. Mey.） Baill. ■☆

325267　Ramusia tridentata Nees ＝ Isoglossa hypoestiflora Lindau ■☆

325268　Ranalisma Stapf（1900）；毛茛泽泻属；Ranalisma ■

325269　Ranalisma humile （Rich.） Hutch. ；小毛茛泽泻■☆

325270　Ranalisma rostratum Stapf；长喙毛茛泽泻；Longbeak Ranalisma ■

325271　Ranapalus Kellogg ＝ Bacopa Aubl. （保留属名）■

325272　Ranapalus Kellogg ＝ Herpestis C. F. Gaertn. ■

325273　Ranaria Cham. ＝ Bacopa Aubl. （保留属名）■

325274　Rancagua Poepp. et Endl. ＝ Lasthenia Cass. ■☆

325275　Randalia Desv. ＝ Eriocaulon L. ■

325276　Randalia P. Beauv. ex Desv. ＝ Eriocaulon L. ■

325277　Randalia decangularis P. Beauv. ＝ Eriocaulon decangulare L. ■☆

325278　Randia L. （1753）；山黄皮属（鸡爪簕属，茜草树属）；Randia ●

325279　Randia L. ＝ Hyperacanthus E. Mey. ex Bridson ●☆

325280　Randia accedens Hance ＝ Fagerlindia scandens （Thunb.） Tirveng. ●

325281　Randia aculeata L. ；尖山黄皮（刺茜树）●☆

325282　Randia acuminata （G. Don） Benth. ＝ Massularia acuminata （G. Don） Bullock ex Hoyle ●☆

325283　Randia acuminatissima Merr. ＝ Aidia pycnantha （Drake） Tirveng. ●

325284　Randia acutidens Hemsl. et E. H. Wilson ＝ Aidia cochinchinensis Lour. ●

325285　Randia acutidens Hemsl. et E. H. Wilson ＝ Randia henryi E. Pritz. ●

325286　Randia adolfi-friederici K. Krause ＝ Aoranthe nalaensis （De Wild.） Somers ●☆

325287　Randia africana G. Don；非洲山黄皮●☆

325288　Randia amaralioides K. Schum. ＝ Sherbournia streptocaulon （K. Schum.） Hepper ●☆

325289　Randia amaralioides K. Schum. ex Hutch. et Dalziel ＝ Sherbournia calycina （G. Don） Hua ●☆

325290　Randia andongensis Hiern ＝ Aulacocalyx jasminiflora Hook. f. ●☆

325291　Randia angolensis Hutch. ＝ Leptactina angolensis （Hutch.） Bullock ex I. Nogueira ●☆

325292　Randia annulata K. Schum. ＝ Aoranthe annulata （K. Schum.） Somers ●☆

325293　Randia barteri （Hook. f. ex Hiern） K. Schum. ＝ Mitriostigma barteri Hook. f. ex Hiern ●☆

325294　Randia bellatula K. Schum. ＝ Rothmannia capensis Thunb. ●☆

325295　Randia bispinosa （Griff.） Craib；弯刺山黄皮；Bispine Randia ●

325296　Randia bowieana A. Cunn. ex Hook. ＝ Euclinia longiflora Salisb. ●☆

325297　Randia brachythamnus K. Schum. ＝ Gardenia brachythamnus （K. Schum.） Launert ●☆

325298　Randia bruneelii De Wild. ＝ Rothmannia lateriflora （K. Schum.） Keay ●☆

325299　Randia buchananii Oliv. ＝ Rothmannia manganjae （Hiern） Keay ●☆

325300　Randia cacaocarpa Wernham ＝ Massularia acuminata （G. Don） Bullock ex Hoyle ●☆

325301　Randia canthioides Champ. ex Benth. ；台北茜草树（香楠）●

325302　Randia canthioides Champ. ex Benth. ＝ Aidia canthioides （Champ. ex Benth.） Masam. ●

325303　Randia castaneofulva S. Moore ＝ Aoranthe castaneofulva （S. Moore） Somers ●☆

325304　Randia caudata Hiern ＝Aulacocalyx caudata（Hiern）Keay ●☆

325305　Randia caudatifolia Merr. ＝Aidia cochinchinensis Lour. ●

325306　Randia chloroleuca K. Schum. ＝ Pleiocoryne fernandense（Hiern）Rauschert ●☆

325307　Randia chromocarpa K. Krause;色果山黄皮●☆

325308　Randia cladantha K. Schum. ＝ Aoranthe cladantha（K. Schum.）Somers ●☆

325309　Randia cochinchinensis（Lour.）Merr.;茜草树（龙虾,山桂花,山黄皮）●

325310　Randia cochinchinensis（Lour.）Merr. ＝Aidia cochinchinensis Lour. ●

325311　Randia congestiflora K. Krause ＝Aidia micrantha（K. Schum.）F. White var. zenkeri（S. Moore）E. M. Petit ●☆

325312　Randia congolana De Wild. et T. Durand ＝Aidia micrantha（K. Schum.）F. White var. congolana（De Wild.）E. M. Petit ●☆

325313　Randia coriacea Benth. ＝Tricalysia coriacea（Benth.）Hiern ●☆

325314　Randia coriacea K. Schum. ex Hutch. et Dalziel ＝Rothmannia lujae（De Wild.）Keay ●☆

325315　Randia cunliffeae Wernham ＝Rothmannia octomera（Hook.）Fagerl. ●☆

325316　Randia curvipes Wernham ＝Sherbournia curvipes（Wernham）N. Hallé ●☆

325317　Randia cuvelieriana De Wild. ＝Rothmannia whitfieldii（Lindl.）Dandy ●☆

325318　Randia densiflora（Wall.）Benth. ＝Aidia cochinchinensis Lour. ●

325319　Randia depauperata Drake ＝Fagerlindia depauperata（Drake）Tirveng. ●

325320　Randia devoniana（Lindl.）Benth. et Hook. f. ex B. D. Jacks. ＝Euclinia longiflora Salisb. ●☆

325321　Randia discolor K. Krause;异色山黄皮●☆

325322　Randia doniana Benth. ＝Sherbournia calycina（G. Don）Hua ●☆

325323　Randia dorothea Wernham ＝Aulacocalyx jasminiflora Hook. f. ●☆

325324　Randia dumetorum（Retz.）Lam. ＝Catunaregam spinosa（Thunb.）Tirveng. ●

325325　Randia echinocarpa Moc. et Sessé;刺果山黄皮●☆

325326　Randia eetveldiana De Wild. et T. Durand ＝Rothmannia whitfieldii（Lindl.）Dandy ●☆

325327　Randia eetveldiana De Wild. et T. Durand var. elongata De Wild. ＝Rothmannia whitfieldii（Lindl.）Dandy ●☆

325328　Randia engleriana K. Schum. ＝Rothmannia engleriana（K. Schum.）Keay ●☆

325329　Randia evenosa Hutch. ＝Oxyceros evenosa（Hutch.）T. Yamaz. ●

325330　Randia exserta K. Schum. ＝Preussiodora sulphurea（K. Schum.）Keay ●☆

325331　Randia fischeri K. Schum. ＝Rothmannia fischeri（K. Schum.）Bullock ●☆

325332　Randia fischeri K. Schum. var. major ? ＝Rothmannia fischeri（K. Schum.）Bullock subsp. verdcourtii Bridson ●☆

325333　Randia formosa（Jacq.）Schum. ;美丽茜树;Jasmin De Rosa ●☆

325334　Randia forrestii J. Anthony ＝Oxyceros griffithii（Hook. f.）W. C. Chen ●

325335　Randia fratrum K. Krause ＝Rothmannia manganjae（Hiern）Keay ●☆

325336　Randia galtonii Wernham ＝Rothmannia octomera（Hook.）Fagerl. ●☆

325337　Randia griffithii Hook. f. ＝Oxyceros griffithii（Hook. f.）W. C. Chen ●

325338　Randia hainanensis Merr. ＝Oxyceros griffithii（Hook. f.）W. C. Chen ●

325339　Randia hapalophylla Wernham ＝Sherbournia hapalophylla（Wernham）Hepper ●☆

325340　Randia heinsioides Schweinf. ex Hua ＝Sherbournia bignoniiflora（Welw.）Hua ●☆

325341　Randia henryi E. Pritz. ;鄂西茜树（西南茜树,西南香楠）;Henry Randia ●

325342　Randia hispida K. Schum. ＝Rothmannia hispida（K. Schum.）Fagerl. ●☆

325343　Randia hockii De Wild. ＝Aulacocalyx laxiflora E. M. Petit ●☆

325344　Randia homblei De Wild. ＝Rothmannia whitfieldii（Lindl.）Dandy ●☆

325345　Randia immanifolia Wernham ＝Schumanniophyton magnificum（K. Schum.）Harms ●☆

325346　Randia jasminodora K. Krause;茉莉茜树●☆

325347　Randia katentaniae De Wild. ＝Rothmannia engleriana（K. Schum.）Keay ●☆

325348　Randia kerstingii K. Krause;克斯廷茜树●☆

325349　Randia kraussii Harv. ＝Catunaregam obovata（Hochst.）A. E. Gonc. ●☆

325350　Randia kuhniana F. Hoffm. et K. Schum. ＝Rothmannia engleriana（K. Schum.）Keay ●☆

325351　Randia lacourtiana De Wild. ＝Rothmannia engleriana（K. Schum.）Keay ●☆

325352　Randia lane-poolei Hutch. et Dalziel ＝Rothmannia munsae（Schweinf. ex Hiern）E. M. Petit subsp. megalostigma（Wernham）Somers ●☆

325353　Randia lasiophylla K. Krause;毛叶茜树●☆

325354　Randia lemairei De Wild. ＝Rothmannia engleriana（K. Schum.）Keay ●☆

325355　Randia lemblinii A. Chev. ＝Argocoffeopsis lemblinii（A. Chev.）Robbr. ●☆

325356　Randia leptactinoides（K. Schum.）Hutch. et Dalziel ＝Aulacocalyx caudata（Hiern）Keay ●☆

325357　Randia letestui Pellegr. ＝Aoranthe annulata（K. Schum.）Somers ●☆

325358　Randia leucocarpa Champ. ex Benth. ＝Alleizettella leucocarpa（Champ. ex Benth.）Tirveng. ●

325359　Randia lichiangensis W. W. Sm. ＝Himalrandia lichiangensis（W. W. Sm.）Tirveng. ●

325360　Randia liebrechtsiana De Wild. et T. Durand ＝Rothmannia liebrechtsiana（De Wild. et T. Durand）Keay ●☆

325361　Randia longiflora（Salisb.）T. Durand et Schinz ＝Rothmannia longiflora Salisb. ●☆

325362　Randia longiflora Salisb. ;长花茜树●☆

325363　Randia longiflora Salisb. ＝Euclinia longiflora Salisb. ●☆

325364　Randia longipedicellata K. Schum. ＝Atractogyne bracteata（Wernham）Hutch. et Dalziel ●☆

325365　Randia longistyla DC. ＝Macrosphyra longistyla（DC.）Hiern ●☆

325366　Randia lucida A. Chev. ＝Gardenia sokotensis Hutch. ●☆

325367　Randia lucidula Hiern ＝Aidia micrantha（K. Schum.）F. White ●☆

325368　Randia lujae De Wild. ＝Rothmannia lujae（De Wild.）Keay ●☆

325369　Randia macrantha（Schult.）DC. ＝Euclinia longiflora Salisb. ●☆

325370 Randia macrocarpa Hiern = Rothmannia macrocarpa （Hiern） Keay ●☆

325371 Randia macrosiphon K. Schum. ex Engl. = Rothmannia macrosiphon （K. Schum. ex Engl.） Bridson ●☆

325372 Randia madagascariensis （Lam.） DC. = Hyperacanthus madagascariensis （Lam.） Rakotonas. et A. P. Davis ●☆

325373 Randia malleifera （Hook.） Hook. f. = Rothmannia whitfieldii （Lindl.） Dandy ●☆

325374 Randia malleiflora Walp. = Rothmannia whitfieldii （Lindl.） Dandy ●☆

325375 Randia mayumbensis R. D. Good = Rothmannia mayumbensis （R. D. Good） Keay ●☆

325376 Randia megalostigma Wernham = Rothmannia munsae （Schweinf. ex Hiern） E. M. Petit subsp. megalostigma （Wernham） Somers ●☆

325377 Randia merrillii Chun;柳叶山黄皮;Merrill Randia, Willow-leaved Randia ●☆

325378 Randia micrantha K. Schum. = Aidia micrantha （K. Schum.） F. White ●☆

325379 Randia micrantha K. Schum. var. poggeana ? = Aidia micrantha （K. Schum.） F. White ●☆

325380 Randia micrantha K. Schum. var. zenkeri S. Moore = Aidia micrantha （K. Schum.） F. White var. zenkeri （S. Moore） E. M. Petit ●☆

325381 Randia microphylla K. Schum. = Hyperacanthus microphyllus （K. Schum.） Bridson ●☆

325382 Randia monteiroae K. Schum. = Catunaregam obovata （Hochst.） A. E. Gonc. ●☆

325383 Randia mossica A. Chev. = Gardenia sokotensis Hutch. ●☆

325384 Randia myrmecophylla De Wild. = Rothmannia macrocarpa （Hiern） Keay ●☆

325385 Randia myrmecophylla De Wild. var. glabra ? = Rothmannia macrocarpa （Hiern） Keay ●☆

325386 Randia myrmecophylla De Wild. var. subglabra ? = Rothmannia macrocarpa （Hiern） Keay ●☆

325387 Randia myrmecophylla De Wild. var. typica ? = Rothmannia macrocarpa （Hiern） Keay ●☆

325388 Randia nalaensis De Wild. = Aoranthe nalaensis （De Wild.） Somers ●☆

325389 Randia nalaensis De Wild. var. rotundata Vermoesen ex De Wild. = Aoranthe castaneofulva （S. Moore） Somers ●☆

325390 Randia naucleoides S. Moore = Bertiera naucleoides （S. Moore） Bridson ■☆

325391 Randia nilotica Stapf;尼罗河山黄皮●☆

325392 Randia nilotica Stapf = Catunaregam nilotica （Stapf） Tirveng. ●☆

325393 Randia nipponensis Makino;日本山黄皮;Japanese Randia ●☆

325394 Randia ochroleuca K. Schum. = Aidia ochroleuca （K. Schum.） E. M. Petit ●☆

325395 Randia octomera （Hook.） Hook. f. = Rothmannia octomera （Hook.） Fagerl. ●☆

325396 Randia oligoneura K. Schum. = Aulacocalyx talbotii （Wernham） Keay ●☆

325397 Randia oppositifolia （Roxb.） Koord. = Aidia cochinchinensis Lour. ●

325398 Randia oxydonta Drake = Aidia oxydonta （Drake） T. Yamaz. ●

325399 Randia pallens Hiern = Aulacocalyx pallens （Hiern） Bridson et Figueiredo ●☆

325400 Randia parvifolia Harv. = Coddia rudis （E. Mey. ex Harv.） Verdc. ●☆

325401 Randia penduliflora K. Schum. = Aoranthe penduliflora （K. Schum.） Somers ●☆

325402 Randia physophylla K. Schum. = Gardenia imperialis K. Schum. subsp. physophylla （K. Schum.） L. Pauwels ●☆

325403 Randia pierrei A. Chev. = Aoranthe cladantha （K. Schum.） Somers ●☆

325404 Randia psychotrioides K. Schum. = Pleiocoryne fernandense （Hiern） Rauschert ●☆

325405 Randia purpureo-maculata C. H. Wright = Adenorandia kalbreyeri （Hiern） Robbr. et Bridson ●☆

325406 Randia pycnantha Drake = Aidia pycnantha （Drake） Tirveng. ●

325407 Randia pynaertii De Wild. = Rothmannia hispida （K. Schum.） Fagerl. ●☆

325408 Randia quintasii K. Schum. = Aidia quintasii （K. Schum.） G. Taylor ●☆

325409 Randia ravae Chiov. = Rothmannia ravae （Chiov.） Bridson ●☆

325410 Randia rectispina Merr. = Oxyceros rectispina （Merr.） T. Yamaz. ●

325411 Randia refractiloba K. Krause = Aidia rubens （Hiern） G. Taylor ●☆

325412 Randia reticulata Benth. = Tricalysia reticulata （Benth.） Hiern ●☆

325413 Randia rubens Hiern = Aidia rubens （Hiern） G. Taylor ●☆

325414 Randia rudis E. Mey. ex Harv. = Coddia rudis （E. Mey. ex Harv.） Verdc. ●☆

325415 Randia salicifolia H. L. Li = Aidia salicifolia （H. L. Li） T. Yamaz. ●

325416 Randia sapinii De Wild. = Rothmannia longiflora Salisb. ●☆

325417 Randia scabra Chiov. = Oxyanthus zanguebaricus （Hiern） Bridson ●☆

325418 Randia seretii De Wild. = Gardenia vogelii Hook. f. ex Planch. var. seretii （De Wild.） L. Pauwels ●☆

325419 Randia sericantha K. Schum. = Aoranthe penduliflora （K. Schum.） Somers ●☆

325420 Randia sericantha W. C. Chen = Porterandia sericantha （W. C. Chen） W. C. Chen ●

325421 Randia sherbourniae （Hook.） Hook. = Sherbournia calycina （G. Don） Hua ●☆

325422 Randia shweliensis J. Anthony = Aidia shweliensis （J. Anthony） W. C. Chen ●

325423 Randia shweliensis J. Anthony = Fosbergia shweliensis （J. Anthony） Tirveng. et Sastre ●

325424 Randia siamensis Craib;泰国山黄皮●☆

325425 Randia sinensis （Lour.） Roem. et Schult. = Oxyceros sinensis Lour. ●

325426 Randia sinensis （Lour.） Schult. = Oxyceros sinensis Lour. ●

325427 Randia sootepensis Craib;泰北山黄皮●☆

325428 Randia spathacea De Wild. = Rothmannia longiflora Salisb. ●☆

325429 Randia spathicalyx De Wild. = Rothmannia urcelliformis （Hiern） Robyns ●☆

325430 Randia spathulifolia R. D. Good = Brenania brieyi （De Wild.） E. M. Petit ●☆

325431 Randia sphaerocoryne K. Schum. = Aidia micrantha （K. Schum.） F. White var. zenkeri （S. Moore） E. M. Petit ●☆

325432 Randia spinosa （Thunb.） Blume = Catunaregam spinosa （Thunb.） Tirveng. ●

325433 Randia spinosa （Thunb.） Poir. = Catunaregam spinosa （Thunb.） Tirveng. ●

325434 Randia stanleyana （Hook. ex Lindl.） Walp. = Rothmannia longiflora Salisb. ●☆

325435　Randia stenophylla K. Krause ＝ Rothmannia urcelliformis（Hiern）Robyns ●☆

325436　Randia stolzii K. Schum. et K. Krause ＝ Rothmannia whitfieldii（Lindl.）Dandy ●☆

325437　Randia streptocaulon K. Schum. ＝ Sherbournia streptocaulon（K. Schum.）Hepper ●☆

325438　Randia stricta（Roxb. ex Roem. et Schult.）Roxb. ＝ Hyptianthera stricta（Roxb.）Wight et Arn. ●

325439　Randia stricta Roxb. ＝ Hyptianthera stricta（Roxb.）Wight et Arn. ●

325440　Randia submontana K. Krause ＝ Rutidea smithii Hiern subsp. submontana（K. Krause）Bridson ●☆

325441　Randia suishaensis Hayata ＝ Aidia cochinchinensis Lour. ●

325442　Randia sulphurea K. Schum. ＝ Preussiodora sulphurea（K. Schum.）Keay ●☆

325443　Randia swynnertonii S. Moore ＝ Catunaregam swynnertonii（S. Moore）Bridson ●☆

325444　Randia talbotii Wernham ＝ Rothmannia talbotii（Wernham）Keay ●☆

325445　Randia taylorii S. Moore ＝ Catunaregam taylorii（S. Moore）Bridson ●☆

325446　Randia ternifolia Ficalho et Hiern ＝ Rothmannia engleriana（K. Schum.）Keay ●☆

325447　Randia tetrasperma（Roxb.）Benth. et Hook. f. ex Brandis；四子山黄皮●☆

325448　Randia thomasii Hutch. et Dalziel ＝ Rothmannia longiflora Salisb. ●☆

325449　Randia torulosa K. Krause ＝ Gardenia ternifolia Schumach. et Thonn. var. goetzei（Stapf et Hutch.）Verdc. ●☆

325450　Randia troposepala K. Schum. ；棱萼山黄皮●☆

325451　Randia tubaeformis Pellegr. ＝ Rothmannia talbotii（Wernham）Keay ●☆

325452　Randia urcelliformis（Hiern）Eggeling ＝ Rothmannia urcelliformis（Hiern）Robyns ●☆

325453　Randia urcelliformis（Hiern）Schweinf. ex Hiern ＝ Rothmannia urcelliformis（Hiern）Robyns ●☆

325454　Randia vestita S. Moore；南非山黄皮●☆

325455　Randia vestita S. Moore ＝ Catunaregam taylorii（S. Moore）Bridson ●☆

325456　Randia violascens Hiern ＝ Rothmannia urcelliformis（Hiern）Robyns ●☆

325457　Randia walkeri Pellegr. ＝ Brenania brieyi（De Wild.）E. M. Petit ●☆

325458　Randia wallichii Hook. f. ＝ Tarennoidea wallichii（Hook. f.）Tirveng. ●

325459　Randia yunnanensis Hutch. ＝ Aidia yunnanensis（Hutch.）T. Yamaz. ●

325460　Randiaceae Martinov ＝ Rubiaceae Juss.（保留科名）●■

325461　Randiaceae Martinov；山黄皮科●

325462　Randonia Coss.（1859）；朗东木犀草属●☆

325463　Randonia africana Coss.；非洲朗东木犀草●☆

325464　Randonia somalensis Schinz ＝ Ochradenus somalensis Baker f. ●☆

325465　Ranevea L. H. Bailey ＝ Ravenea H. Wendl. ex C. D. Bouché ●☆

325466　Ranevea L. H. Bailey（1902）；拉内棕属（拉昵棕属）；Ranevea ●☆

325467　Ranevea hildebrandtii L. H. Bailey；拉内棕（拉昵棕）●☆

325468　Rangaeris（Schltr.）Summerh.（1936）；朗加兰属■☆

325469　Rangaeris amaniensis（Kraenzl.）Summerh. ；阿马尼朗加兰■☆

325470　Rangaeris biglandulosa Summerh. ＝ Cribbia brachyceras（Summerh.）Senghas ■☆

325471　Rangaeris brachyceras（Summerh.）Summerh. ＝ Cribbia brachyceras（Summerh.）Senghas ■☆

325472　Rangaeris longicaudata（Rolfe）Summerh. ；长尾朗加兰■☆

325473　Rangaeris muscicola（Rchb. f.）Summerh. ；苔藓朗加兰■☆

325474　Rangaeris rhipsalisocia（Rchb. f.）Summerh. ；莱普朗加兰■☆

325475　Rangaeris schliebenii（Mansf.）P. J. Cribb；施利本朗加兰■☆

325476　Rangaeris trilobata Summerh. ；三裂朗加兰■☆

325477　Rangia Griseb. ＝ Randia L. ●

325478　Rangium Juss. ＝ Forsythia Vahl（保留属名）●

325479　Rangium mandshuricum（Uyeki）Uyeki et Kitag. ＝ Forsythia mandschurica Uyeki ●

325480　Rangium ovatum（Nakai）Ohwi ＝ Forsythia ovata Nakai ●

325481　Rangium suspensum（Thunb.）Ohwi ＝ Forsythia ovata Nakai ●

325482　Rangium suspensum（Thunb.）Ohwi ＝ Forsythia suspensa（Thunb.）Vahl ●

325483　Rangium viridissimum（Lindl.）Ohwi ＝ Forsythia viridissima Lindl. ●

325484　Ranisia Salisb. ＝ Gladiolus L. ■

325485　Ranisia Salisb. ＝ Lilio-gladiolus Trew ■

325486　Ranopisoa J. -F. Leroy（1977）；拉诺玄参属●☆

325487　Ranopisoa rakotosonii（Capuron）J. -F. Leroy；拉诺玄参●☆

325488　Ranorchis D. L. Jones et M. A. Clem.（2002）；拉诺兰属■☆

325489　Ranorchis D. L. Jones et M. A. Clem. ＝ Pterostylis R. Br.（保留属名）■☆

325490　Ranugia（Schltdl.）Post et Kuntze ＝ Dieudonnaea Cogn. ■☆

325491　Ranugia（Schltdl.）Post et Kuntze ＝ Gurania（Schltdl.）Cogn. ■☆

325492　Ranula Fourr. ＝ Ranunculus L. ■

325493　Ranunculaceae Adans. ＝ Ranunculaceae Juss.（保留科名）●■

325494　Ranunculaceae Juss.（1789）（保留科名）；毛茛科；Buttercup Family，Crowfoot Family ■●

325495　Ranunculastrum Fabr. ＝ Trollius L. ■

325496　Ranunculastrum Fourr. ＝ Ranunculus L. ■

325497　Ranunculastrum Heist. ex Fabr. ＝ Trollius L. ■

325498　Ranunculus L.（1753）；毛茛属；Butter Cup，Buttercup，Craw Foot，Crowfoot，Frogflower，Paigle，Ranunculus ■

325499　Ranunculus × bachii Wirtg. ；贝奇毛茛；Wirtgen's Water-crowfoot ■☆

325500　Ranunculus × kelchoensis S. D. Webster；凯尔索毛茛；Kelso Water-crowfoot ■☆

325501　Ranunculus × levenensis Druce ex Gornall；利文毛茛；Loch Leven Spearwort ■☆

325502　Ranunculus × novae-forestae S. D. Webster；新林毛茛；New Forest Crowfoot ■☆

325503　Ranunculus abaensis W. T. Wang ＝ Ranunculus indivisus（Maxim.）Hand. -Mazz. var. abaensis（W. T. Wang）W. T. Wang ■

325504　Ranunculus abchasicus Freyn；阿伯哈斯毛茛■☆

325505　Ranunculus aberdaricus Ulbr. ；肯尼亚毛茛■☆

325506　Ranunculus abortivus L. ；北美小花毛茛；Early Wood Buttercup，Little-leaf Buttercup，Small-flowered Buttercup，Small-flowered Crowfoot ■☆

325507　Ranunculus abortivus L. subsp. acrolasius（Fernald）B. M. Kapoor et Á. Löve ＝ Ranunculus abortivus L. ■☆

325508　Ranunculus abortivus L. var. acrolasius Fernald ＝ Ranunculus abortivus L. ■☆

325509　Ranunculus abortivus L. var. eucyclus Fernald ＝ Ranunculus abortivus L. ■☆

325510　Ranunculus abortivus L. var. harveyi A. Gray ＝ Ranunculus harveyi (A. Gray) Britton ■☆

325511　Ranunculus abortivus L. var. indivisus Fernald ＝ Ranunculus abortivus L. ■☆

325512　Ranunculus abortivus L. var. typicus Fernald ＝ Ranunculus abortivus L. ■☆

325513　Ranunculus abyssinicus Schube ex Engl. ;阿比西尼亚毛茛 ■☆

325514　Ranunculus acer B. Fedtsch. ＝ Ranunculus laetus Wall. ■

325515　Ranunculus acer L. ;辛辣毛茛 (槭叶毛茛,五裂毛茛); Mapleleaf Buttercup ■

325516　Ranunculus aconitifolius L. ; 欧洲乌头叶毛茛; Aconite Buttercup, Aconite-leaved Buttercup, Bachelor's Buttons, Double White Crowfoot, Fair Maid of France, Fair Maid of Kent, Fair Maids of France, Fair Maids of Kent, Fair Maids-of-france, White Bachelor's Buttons, White Buttercup ■☆

325517　Ranunculus aconitifolius L. ' Plore Pleno' ;重瓣乌头叶毛茛■☆

325518　Ranunculus acris L. ;高毛茛 (毛茛,欧毛茛); Bachelor's Buttons, Bassinet, Billy Buttons, Blister Cup, Blister Plant, Buster Cup, Butter Churn, Butter Cress, Butter Daisy, Butter Flower, Butter Rose, Butter-and-cheese, Butterbumps, Buttercup, Buttercup Crowfoot, Caltrop, Clovewort, Common Buttercup, Common Meadow Buttercup, Cowslip, Craw-taes, Crazy, Crazy Bet, Crazy Bets, Crazy Weed, Crow Flower, Crowfoot, Crowtoes, Cuckoo-buds, Dale-cup, Dellcup, Dew Cup, Dewcup, Dill-cup, Double Buttercup, Eggs-and-butter, Fairy Basin, Fairy's Basin, Giant Buttercup, Gil Cup, Gilcup, Gildcup, Gilted Cup, Gilty Cup, Gilty-cup, Glennies, Gllted Cup, Go-cup, Gold Cup, Gold Knob, Gold Knop, Gold Knots, Gold-crap, Goldcrop, Golden Cup, Goldweed, Goldy, Goldy Knob, Golland, Gowan, Guilty Cup, Guilty-cup, Gulty Cup, Gulty-cup, Hunger-weed, King Cob, Kingcup, King's Clover, King's Cob, King's Knob, King's Knobs, Lady's Slipper, Lawyer-weed, Marybud, Maybuds, Meadow Buttercup, Meadow Crowfoot, Old Man's Buttons, Paigle, Queen's Button, Ram's Claws, Ram's Glass, Showy Buttercup, Sit-siccar, Sit-sicker, Soldier Buttons, Soldier's Buttons, Tall Buttercup, Tall Crowfoot, Tall Field Buttercup, Teacups, Upright Meadow, Upright Meadow Crowfoot, Yellow Bachelor's Buttons, Yellow Caul, Yellow Creams, Yellow Cup, Yellow Gollan, Yellow Gowan ■☆

325519　Ranunculus acris L. ' Fore Pleno' ;重瓣高毛茛; Double Meadow Buttercup, Yellow Bachelor's-buttons ■☆

325520　Ranunculus acris L. subsp. atlanticus (Ball) Ball ＝ Ranunculus granatensis Boiss. ■☆

325521　Ranunculus acris L. subsp. granatensis (Boiss.) Nyman ＝ Ranunculus granatensis Boiss. ■☆

325522　Ranunculus acris L. subsp. japonicus (Thunb.) Hultén ＝ Ranunculus japonicus Thunb. ■

325523　Ranunculus acris L. subsp. strigulosus (Schur) Hyl. ＝ Ranunculus acris L. ■☆

325524　Ranunculus acris L. var. atlanticus (Ball) Maire ＝ Ranunculus granatensis Boiss. ■☆

325525　Ranunculus acris L. var. japonicus (Thunb.) Maxim. ＝ Ranunculus japonicus Thunb. ■

325526　Ranunculus acris L. var. japonicus (Thunb.) Maxim. ex Makino et al. ＝ Ranunculus japonicus Thunb. ■

325527　Ranunculus acris L. var. latisectus Beck ＝ Ranunculus acris L. ■☆

325528　Ranunculus acris L. var. monticola (Kitag.) Tajmura ＝ Ranunculus paishanensis Kitag. ■

325529　Ranunculus acris L. var. propinquus (C. A. Mey.) Maxim. ＝ Ranunculus japonicus Thunb. var. propinquus (C. A. Mey.) W. T. Wang ■

325530　Ranunculus acris L. var. schizophyllus H. Lév. ＝ Ranunculus japonicus Thunb. ■

325531　Ranunculus acris L. var. stevenii (Andr.) Regel ＝ Ranunculus japonicus Thunb. var. propinquus (C. A. Mey.) W. T. Wang ■

325532　Ranunculus acris L. var. stevenii (Andrz. ex Besser) Lange ＝ Ranunculus acris L. ■☆

325533　Ranunculus acris L. var. typicus Beck ＝ Ranunculus acris L. ■☆

325534　Ranunculus acris L. var. villosus (Drabble) S. M. Coles ＝ Ranunculus acris L. ■☆

325535　Ranunculus acutidentatus Rupr. ;尖齿毛茛■☆

325536　Ranunculus acutilobus Ledeb. ;尖裂毛茛■☆

325537　Ranunculus adoneus A. Gray;阿东尼斯毛茛■☆

325538　Ranunculus adoneus A. Gray var. alpinus (S. Watson) L. D. Benson ＝ Ranunculus adoneus A. Gray ■☆

325539　Ranunculus adoxifolius Hand. -Mazz. ;五福花叶毛茛■

325540　Ranunculus affinis R. Br. ＝ Ranunculus pedatifidus Sm. var. affinis (R. Br.) L. D. Benson ■☆

325541　Ranunculus affinis R. Br. ＝ Ranunculus pedatifidus Sm. ■

325542　Ranunculus affinis R. Br. var. capillaceus Franch. ＝ Ranunculus nematolobus Hand. -Mazz. ■

325543　Ranunculus affinis R. Br. var. filiformis Finet et Gagnep. ＝ Ranunculus nematolobus Hand. -Mazz. ■

325544　Ranunculus affinis R. Br. var. flabellatus Franch. ＝ Ranunculus felixii H. Lév. ■

325545　Ranunculus affinis R. Br. var. indivisus Maxim. ＝ Ranunculus indivisus (Maxim.) Hand. -Mazz. ■

325546　Ranunculus affinis R. Br. var. stracheyanus Maxim. ＝ Ranunculus popovii Ovcz. var. stracheyanum (Maxim.) W. T. Wang ■

325547　Ranunculus affinis R. Br. var. tanguticus Maxim. ＝ Ranunculus tanguticus (Maxim.) Ovcz. ■

325548　Ranunculus affinis R. Br. var. ternatus Diels ＝ Ranunculus hirtellus Royle ■

325549　Ranunculus affinis R. Br. var. ternatus Franch. ＝ Ranunculus tanguticus (Maxim.) Ovcz. ■

325550　Ranunculus affinis R. Br. var. tibeticus Maxim. ＝ Ranunculus nephelogenes Edgew. ■

325551　Ranunculus afghanicus Aitch. et Hemsl. ;阿富汗毛茛■☆

325552　Ranunculus africanus Hort. ;非洲毛茛■☆

325553　Ranunculus ageri Bertol. ;阿氏毛茛■☆

325554　Ranunculus ailaoshanicus W. T. Wang;哀劳山毛茛■

325555　Ranunculus alaiensis Ostenf. ;阿赖毛茛■

325556　Ranunculus alaschanicus Y. Z. Zhao;贺兰山毛茛 (长茎毛茛); Helanshan Buttercup ■

325557　Ranunculus alaschanicus Y. Z. Zhao ＝ Ranunculus membranaceus Royle var. pubescens (W. T. Wang) W. T. Wang ■

325558　Ranunculus albertii Regel et Schmalh. ;宽瓣毛茛; Albert Buttercup ■

325559　Ranunculus alleghenensis Britton. ;北美山地毛茛; Mountain Crowfoot ■☆

325560　Ranunculus allemannii Braun-Blanq. ;阿鲁氏毛茛 (阿勒曼毛茛); Alleman Buttercup ■☆

325561 Ranunculus allenii B. L. Rob. ;阿伦毛茛■☆

325562 Ranunculus alpestris L. ; 亚高山毛茛; Alpine Buttercup, Moraine Buttercup ■☆

325563 Ranunculus altaicus Laxm. ;阿尔泰毛茛;Altai Buttercup ■

325564 Ranunculus altaicus Laxm. var. fraternus (Schrenk) Trautv. = Ranunculus fraternus Schrenk ■

325565 Ranunculus altaicus Laxm. var. sulphureus Finet et Gagnep. = Ranunculus nematolobus Hand. -Mazz. ■

325566 Ranunculus alveolatus A. M. Carter = Ranunculus bonariensis Poir. var. trisepalus (Gillies ex Hook. et Arn.) Lourteig ■☆

325567 Ranunculus ambigens S. Watson;漫生毛茛; Water-plantain Spearwort ■☆

325568 Ranunculus amieri H. Lév. = Ranunculus yunnanensis Franch. ■

325569 Ranunculus amphibius James = Ranunculus aquatilis L. var. diffusus With. ■☆

325570 Ranunculus amplexicaulis L. ;抱茎毛茛;Pyrenean Buttercup ■☆

325571 Ranunculus amurensis Kom. ;长叶毛茛(披针毛茛,披针叶毛茛);Amur Buttercup ■

325572 Ranunculus andersonii (A. Gray) Jeps. var. juniperinus (M. E. Jones) S. L. Welsh = Ranunculus andersonii A. Gray ■☆

325573 Ranunculus andersonii (A. Gray) Jeps. var. tenellus S. Watson = Ranunculus andersonii A. Gray ■☆

325574 Ranunculus andersonii A. Gray;安德森毛茛■☆

325575 Ranunculus andersonii A. Gray var. juniperinus (M. E. Jones) S. L. Welsh = Ranunculus andersonii A. Gray ■☆

325576 Ranunculus andersonii A. Gray var. tenellus S. Watson = Ranunculus andersonii A. Gray ■☆

325577 Ranunculus anemoneus F. Muell. ; 澳洲毛茛; Anemone Buttercup ■☆

325578 Ranunculus anemonifolius DC. ;银莲花叶毛茛■

325579 Ranunculus angustisepalus W. T. Wang;狭萼毛茛■

325580 Ranunculus aquatilis L. ;欧洲水毛茛(梅花藻);Bacon-and-eggs, Common Water-crowfoot, Cow-weed, Eelweed Eelware, Lodewort, Pickerel-weed, Rait, Ram's Foot, Rawheads, Reate, Water Anemone, Water Crowfoot, Water Lily, Water Liverwort, Water Nemony, Water Snow-cup, Water Snow-cups, White Crowfoot, White Water Crowfoot ■☆

325581 Ranunculus aquatilis L. = Batrachium bungei (Steud.) L. Liou ■

325582 Ranunculus aquatilis L. subsp. baudotii (Godr.) Ball = Ranunculus peltatus Schrank subsp. baudotii (Godr.) C. D. K. Cook ■☆

325583 Ranunculus aquatilis L. subsp. minoriflorus (Pau) Maire;小花欧洲水毛茛■☆

325584 Ranunculus aquatilis L. subsp. triphyllus (Wallr.) Maire = Ranunculus peltatus Schrank ■☆

325585 Ranunculus aquatilis L. var. baudotii (Godr.) Batt. et Trab. = Ranunculus peltatus Schrank subsp. baudotii (Godr.) C. D. K. Cook ■☆

325586 Ranunculus aquatilis L. var. calvescens (W. B. Drew) L. D. Benson = Ranunculus aquatilis L. var. diffusus With. ■☆

325587 Ranunculus aquatilis L. var. capillaceus (Thuill.) DC. = Ranunculus aquatilis L. var. diffusus With. ■☆

325588 Ranunculus aquatilis L. var. confusus (Gren. et Godr.) Batt. et Trab. = Ranunculus peltatus Schrank subsp. baudotii (Godr.) C. D. K. Cook ■☆

325589 Ranunculus aquatilis L. var. diffusus With. ;白欧洲水毛茛; White Water Crowfoot ■☆

325590 Ranunculus aquatilis L. var. elegans Chabert = Ranunculus aquatilis L. ■☆

325591 Ranunculus aquatilis L. var. eradicatus Laest. = Batrachium eradicatum (Laest.) Fr. ■

325592 Ranunculus aquatilis L. var. eradicatus Laest. = Ranunculus aquatilis L. var. diffusus With. ■☆

325593 Ranunculus aquatilis L. var. harrisii L. D. Benson = Ranunculus aquatilis L. var. diffusus With. ■☆

325594 Ranunculus aquatilis L. var. heleophyllus (Arv. -Touv.) Beck = Ranunculus peltatus Schrank ■☆

325595 Ranunculus aquatilis L. var. heterophyllus (Weber) DC. = Ranunculus aquatilis L. ■☆

325596 Ranunculus aquatilis L. var. hispidulus Drew = Ranunculus aquatilis L. ■☆

325597 Ranunculus aquatilis L. var. lalondei L. D. Benson = Ranunculus aquatilis L. var. diffusus With. ■☆

325598 Ranunculus aquatilis L. var. leontinensis (Freyn) Maire = Ranunculus peltatus Schrank ■☆

325599 Ranunculus aquatilis L. var. longirostris (Godr.) Lawson = Ranunculus aquatilis L. var. diffusus With. ■☆

325600 Ranunculus aquatilis L. var. porteri (Britton) L. D. Benson = Ranunculus aquatilis L. var. diffusus With. ■☆

325601 Ranunculus aquatilis L. var. pseudofluitans (Syme) Freyn = Ranunculus penicillatus (Dumort.) Bab. ■☆

325602 Ranunculus aquatilis L. var. pseudoleontinensis Maire = Ranunculus peltatus Schrank ■☆

325603 Ranunculus aquatilis L. var. renifolius Pau et Font Quer = Ranunculus aquatilis L. ■☆

325604 Ranunculus aquatilis L. var. rhiphaeus Pau et Font Quer = Ranunculus aquatilis L. ■☆

325605 Ranunculus aquatilis L. var. saniculifolius (Viv.) Batt. = Ranunculus peltatus Schrank subsp. saniculifolius (Viv.) C. D. K. Cook ■☆

325606 Ranunculus aquatilis L. var. subrigidus (W. B. Drew) Breitung = Ranunculus aquatilis L. var. diffusus With. ■☆

325607 Ranunculus aquatilis L. var. trichophyllus (Chaix) A. Gray = Ranunculus aquatilis L. var. diffusus With. ■☆

325608 Ranunculus arachnoideus C. A. Mey. ;蛛网毛茛■☆

325609 Ranunculus arcuans S. S. Chien = Ranunculus sieboldii Miq. ■

325610 Ranunculus arizonicus Lemmon ex A. Gray;亚利桑那毛茛■☆

325611 Ranunculus arizonicus Lemmon ex A. Gray var. subaffinis A. Gray = Ranunculus inamoenus Greene var. subaffinis (A. Gray) L. D. Benson ■☆

325612 Ranunculus arvensis L. ; 田野毛茛(野毛茛);Buttercup, Clench, Cogweed, Corn Buttercup, Corn Crowfoot, Cornfield Crowfoot, Crow Claws, Crow Peck, Devil-on-all-sides, Devil-on-both-sides, Devil's Claws, Devils Coachwheel, Devil's Coachwheel, Devil's Coach-wheel, Devils Currycomb, Devil's Curry-comb, Devil's Fingers, Dill-cup, Eggs-and-bacon, Goldweed, Gye, Hardine, Hedehog, Hedgehogs, Hellweed, Horse Gold, Hungerweed, Hunger-weed, Jack-o '-both-sides, Jack-o '-twosides, Jackweed, Joy, Kennin Herb, Kenning Heal, King Cob, King-cob, King's Cob, Lady's Fingers, Peagle, Prickleback, Scratch Bur, Scratch-bur, Starve-acre, Urchin Crowfoot, Watch Wheels, Yellow Cup ■

325613 Ranunculus arvensis L. var. tuberculatus DC. = Ranunculus arvensis L. ■

325614 Ranunculus asiaticus L. ;花毛茛(波斯毛茛,陆莲花);Garden Ranunculus, Garden-crowfoot, Persian Buttercup, Persian Kingcup, Persian King-cup, Persian Ranunculus, Scarlet Crowfoot, Turban

Buttercup ■

325615　Ranunculus asiaticus L. var. bereniceus Pamp. = Ranunculus asiaticus L. ■

325616　Ranunculus asiaticus L. var. bicolor Pamp. = Ranunculus asiaticus L. ■

325617　Ranunculus asiaticus L. var. flavus Sickenb. = Ranunculus asiaticus L. ■

325618　Ranunculus asiaticus L. var. grandiflorus Bég. et Vacc. = Ranunculus asiaticus L. ■

325619　Ranunculus asiaticus L. var. sanguineus DC. = Ranunculus asiaticus L. ■

325620　Ranunculus asiaticus L. var. variegatus Sickenb. = Ranunculus asiaticus L. ■

325621　Ranunculus atlanticus Ball = Ranunculus granatensis Boiss. ■☆

325622　Ranunculus atlanticus Pomel = Ranunculus peltatus Schrank subsp. baudotii（Godr.）C. D. K. Cook ■☆

325623　Ranunculus aurasiacus Pomel；奥拉斯毛茛■☆

325624　Ranunculus aurasiacus Pomel var. ayachicus（Emb.）Emb. = Ranunculus aurasiacus Pomel ■☆

325625　Ranunculus aurasiacus Pomel var. djurdjurae Chabert = Ranunculus aurasiacus Pomel ■☆

325626　Ranunculus aurasiacus Pomel var. pseudodemissus Chabert = Ranunculus aurasiacus Pomel ■☆

325627　Ranunculus aureopetalus Kom.；黄瓣毛茛■☆

325628　Ranunculus auricomus L.；金发状毛茛（毛茛，茅根）；Golden-haired Crowfoot, Goldilocks, Goldilocks Buttercup, Sweet Crowfoot, Wood Crowfoot, Wood Goldilocks ■☆

325629　Ranunculus auricomus L. subsp. boecheri Fagerstr. et G. Kvist = Ranunculus auricomus L. ■☆

325630　Ranunculus auricomus L. subsp. glabratus（Lynge）Fagerstr. et G. Kvist = Ranunculus auricomus L. ■☆

325631　Ranunculus auricomus L. subsp. hartzii Fagerstr. et G. Kvist = Ranunculus auricomus L. ■☆

325632　Ranunculus auricomus L. subsp. sibiricus Korsh. = Ranunculus monophyllus Ovcz. ■

325633　Ranunculus auricomus L. var. glabratus Lynge = Ranunculus auricomus L. ■☆

325634　Ranunculus auricomus L. var. sibiricus Glehn = Ranunculus monophyllus Ovcz. ■

325635　Ranunculus balangshanicus W. T. Wang；巴郎山毛茛；Balangshan Buttercup ■

325636　Ranunculus baldschuanicus Regel ex Kom.；巴尔德毛茛■☆

325637　Ranunculus balikunensis J. G. Liou；巴里坤毛茛；Balikun Buttercup ■

325638　Ranunculus banguoensis L. Liou；班戈毛茛；Bange Buttercup, Banguo Buttercup ■

325639　Ranunculus banguoensis L. Liou var. grandiflorus W. T. Wang；普兰毛茛；Pulan Buttercup ■

325640　Ranunculus batrachioides Pomel；水毛茛状毛茛■☆

325641　Ranunculus batrachioides Pomel subsp. brachypodus G. López = Ranunculus batrachioides Pomel ■☆

325642　Ranunculus batrachioides Pomel var. mesatlanticus Maire = Ranunculus batrachioides Pomel ■☆

325643　Ranunculus batrachioides Pomel var. xantholeucos（Coss. et Durieu）Maire = Ranunculus batrachioides Pomel ■☆

325644　Ranunculus baudotii Godr.；海滨毛茛；Brackish Water Crowfoot, Brackish-water Crowfoot, Seaside Water Crowfoot ■☆

325645　Ranunculus baudotii Godr. = Ranunculus peltatus Schrank subsp. baudotii（Godr.）C. D. K. Cook ■☆

325646　Ranunculus baudotii Godr. var. submersum Gren. et Godr. = Ranunculus peltatus Schrank subsp. baudotii（Godr.）C. D. K. Cook ■☆

325647　Ranunculus baurii MacOwan；巴利毛茛■☆

325648　Ranunculus bequaertii De Wild. ；贝卡尔北美毛茛■☆

325649　Ranunculus blepharicarpos Boiss. = Ranunculus spicatus Desf. subsp. blepharicarpos（Boiss.）Grau ■☆

325650　Ranunculus bloomeri S. Watson = Ranunculus orthorhynchus Hook. var. bloomeri（S. Watson）L. D. Benson ■☆

325651　Ranunculus bonariensis Poir.；北美毛茛■☆

325652　Ranunculus bonariensis Poir. var. trisepalus（Gillies ex Hook. et Arn.）Lourteig；三萼北美毛茛■☆

325653　Ranunculus bonatianus Ulbr. = Ranunculus ficariifolius H. Lév. et Vaniot ■

325654　Ranunculus bongardii Greene = Ranunculus uncinatus D. Don ■☆

325655　Ranunculus bongardii Greene var. tenellus（A. Gray）Greene = Ranunculus uncinatus D. Don ■☆

325656　Ranunculus borealis Trautv.；北毛茛■

325657　Ranunculus boreanus Jord. = Ranunculus acris L. ■☆

325658　Ranunculus brachylobus Boiss. et Hohen.；短裂毛茛■☆

325659　Ranunculus brachyrhynchus S. S. Chien = Ranunculus cantoniensis DC. ■

325660　Ranunculus brotherusii Freyn = Ranunculus tanguticus（Maxim.）Ovcz. ■

325661　Ranunculus brotherusii Freyn var. dasycarpus（Maxim.）Hand. -Mazz. = Ranunculus tanguticus（Maxim.）Ovcz. var. dasycarpus（Maxim.）L. Liou ■

325662　Ranunculus brotherusii Freyn var. tanguticus Tamura = Ranunculus brotherusii Freyn ■

325663　Ranunculus brotherusii Freyn var. tanguticus Tamura = Ranunculus tanguticus（Maxim.）Ovcz. ■

325664　Ranunculus brutius Ten.；布吕特毛茛■☆

325665　Ranunculus buhsei Boiss.；布塞毛茛■☆

325666　Ranunculus bulbosus L.；鳞茎毛茛（鳞茎状毛茛）；Arnicks, Bachelor's Buttons, Bulb Buttercup, Bulbous Buttercup, Bulbous Crowfoot, Butter Churn, Butter Daisy, Butter Flower, Butter Rose, Butter-and-cheese, Butterbumps, Buttercreese, Caltrop, Churn, Cowslip, Cranops, Craw, Craw Crowfoot, Crazy, Crazy Bet, Crazy Bets, Crazy Weed, Crow Bells, Crow Packle, Crow Pightle, Crowfoot, Crow's Foot, Crowtoes, Cuckoo-buds, Dale-cup, Dew Cup, Dewcup, Dill-cup, Eggs-and-bacon, Eggs-and-butter, Fair Grass, Fairy Basin, Fairy's Basin, Frog's Foot, Gilcup, Gildcup, Gilted Cup, Gilty Cup, Gilty-cup, Glennies, Gllted Cup, Gold Cup, Gold Knop, Gold-crap, Goldcup, Goldweed, Goldy, Golland, Horse Gold, King Cob, Kingcup, King's Clover, King's Cob, King's Knob, King's Knobs, King's Nobs, Knobbed Crowfoot, Lawyer-weed, Maidentile-meadow, Maid-in-the-meadow, Marybud, Maybuds, Old Man's Buttons, Onion-rooted Crowfoot, Paigle, Rape Crowfoot, Round-rooted Crowfoot, Sit-siccar, Sit-sicker, Soldier Buttons, Soldier's Buttons, St. Anthony's Rape, St. Anthony's Rope, St. Anthony's Turnip, Teacups, Yellow Caul, Yellow Craw, Yellow Creams, Yellow Crees, Yellow Cup, Yellow Gollan ■☆

325667　Ranunculus bulbosus L. var. apiifolius Pau et Font Quer = Ranunculus bulbosus L. ■☆

325668　Ranunculus bulbosus L. var. giganteus Ball = Ranunculus bulbosus L. ■☆

325669 Ranunculus bulbosus L. var. glabrescens Maire = Ranunculus bulbosus L. ■☆

325670 Ranunculus bulbosus L. var. hirtus Briq. = Ranunculus bulbosus L. ■☆

325671 Ranunculus bulbosus L. var. hispanicus Freyn = Ranunculus bulbosus L. ■☆

325672 Ranunculus bulbosus L. var. leiopodus Briq. = Ranunculus bulbosus L. ■☆

325673 Ranunculus bulbosus L. var. neapolitanus (Ten.) Coss. = Ranunculus bulbosus L. ■☆

325674 Ranunculus bulbosus L. var. petiolulatus E. H. L. Krause = Ranunculus bulbosus L. ■☆

325675 Ranunculus bulbosus L. var. pomelianus Maire = Ranunculus bulbosus L. ■☆

325676 Ranunculus bulbosus L. var. radicosus Maire = Ranunculus bulbosus L. ■☆

325677 Ranunculus bulbosus L. var. valdepubens (Jord.) Briq. = Ranunculus bulbosus L. ■☆

325678 Ranunculus bullatus L. ;泡叶毛茛■☆

325679 Ranunculus bullatus L. subsp. cyrenaicus (Pamp.) Maire = Ranunculus cytheraceus Halácsy ■☆

325680 Ranunculus bullatus L. subsp. cytheraceus (Halácsy) Vierh. = Ranunculus cytheraceus Halácsy ■☆

325681 Ranunculus bullatus L. var. cyrenaicus Pamp. = Ranunculus cytheraceus Halácsy ■☆

325682 Ranunculus bullatus L. var. plantagineus (Jord. et Fourr.) Maire = Ranunculus bullatus L. ■☆

325683 Ranunculus bullatus L. var. prolifer Gennari = Ranunculus cytheraceus Halácsy ■☆

325684 Ranunculus bullatus L. var. supranudus (Jord. et Fourr.) Maire = Ranunculus bullatus L. ■☆

325685 Ranunculus bungei Steud. = Batrachium bungei (Steud.) L. Liou ■

325686 Ranunculus buschii Ovcz. ;布什毛茛■☆

325687 Ranunculus calandrinioides (L.) Garcke ;岩马齿苋毛茛■☆

325688 Ranunculus calandrinioides (L.) Garcke var. glaberrimus Maire = Ranunculus calandrinioides (L.) Garcke ■☆

325689 Ranunculus calandrinioides Oliv. ;车前叶毛茛;Golden Knob, Pissabed ■☆

325690 Ranunculus californicus Benth. ; 加 州 毛 茛; Californian Buttercup, Yellow-blossom ■☆

325691 Ranunculus californicus Benth. var. austromontanus L. D. Benson = Ranunculus californicus Benth. ■☆

325692 Ranunculus californicus Benth. var. canus (Benth.) W. H. Brewer et S. Watson = Ranunculus canus Benth. ■☆

325693 Ranunculus californicus Benth. var. gratus Jeps. = Ranunculus californicus Benth. ■☆

325694 Ranunculus californicus Benth. var. ludovicianus (Greene) K. C. Davis = Ranunculus canus Benth. var. ludovicianus (Greene) L. D. Benson ■☆

325695 Ranunculus californicus Benth. var. rugulosus (Greene) L. D. Benson = Ranunculus californicus Benth. ■☆

325696 Ranunculus cangshanicus W. T. Wang;苍山毛茛;Cangshan Buttercup ■

325697 Ranunculus cantoniensis DC. ;禹毛茛(点草,猴蒜,黄花虎掌草,回回蒜,假芹菜,鹿蹄草,毛建草,毛茛,毛芹菜,毛水虎掌草,千里光,水辣菜,水芹菜,天灸,田芹菜,小虎掌草,小回回蒜,鸭足板,野芹菜,禹毛茛,自扣草,自炙,自炙草);Canton Buttercup

325698 Ranunculus cantoniensis DC. subsp. tachiroei (Franch. et Sav.) Kitam. = Ranunculus tachiroei Franch. et Sav. ■

325699 Ranunculus cantoniensis DC. var. sieboldii (Miq.) Kitam. ex Hatus. = Ranunculus sieboldii Miq. ■

325700 Ranunculus canus Benth. ;苍白毛茛■☆

325701 Ranunculus canus Benth. var. laetus (Greene) L. D. Benson = Ranunculus canus Benth. ■☆

325702 Ranunculus canus Benth. var. ludovicianus (Greene) L. D. Benson;卢氏毛茛■☆

325703 Ranunculus capensis Thunb. ;好望角毛茛■☆

325704 Ranunculus capillaceus Thuill. = Ranunculus trichophyllus Chaix ■☆

325705 Ranunculus cardiophyllus Hook. ;心叶毛茛■☆

325706 Ranunculus cardiophyllus Hook. var. coloradensis L. D. Benson = Ranunculus cardiophyllus Hook. ■☆

325707 Ranunculus cardiophyllus Hook. var. subsagittatus (A. Gray) L. D. Benson = Ranunculus cardiophyllus Hook. ■☆

325708 Ranunculus caricetorum Greene = Ranunculus hispidus Michx. var. caricetorum (Greene) T. Duncan ■☆

325709 Ranunculus carolinianus DC. ;卡罗林毛茛;Carolina Buttercup ■☆

325710 Ranunculus carolinianus DC. = Ranunculus hispidus Michx. var. nitidus (Chapm.) T. Duncan ■☆

325711 Ranunculus carolinianus DC. var. villicaulis Shinners = Ranunculus hispidus Michx. var. nitidus (Chapm.) T. Duncan ■☆

325712 Ranunculus cassius Finet et Gagnep. = Ranunculus distans Wall. et Royle ■

325713 Ranunculus cassius Finet et Gagnep. = Ranunculus laetus Wall. ■

325714 Ranunculus cassubicus L. ;扭曲毛茛■☆

325715 Ranunculus caucasicus M. Bieb. ;高加索毛茛;Caucasia ■☆

325716 Ranunculus changpingensis W. T. Wang;昌平毛茛■

325717 Ranunculus cheirophyllus Hayata;掌叶毛茛;Palmleaf Buttercup ■

325718 Ranunculus chinensis Bunge;茴茴蒜(鹅巴掌,黄花草,黄花虎掌草,回回蒜,回回蒜毛茛,辣辣草,糯虎掌,青果草,绒毛犬脚迹,山辣椒,石龙芮,水胡椒,水虎掌草,水杨梅,土细辛,小虎掌草,小回回蒜,小桑子,小桑子,蝎虎草,鸭脚板,野大蒜,野桑椹);China Buttercup, Chinese Buttercup ■

325719 Ranunculus chinghoensis L. Liou;青 河 毛 茛; Chingho Buttercup, Qinghe Buttercup ■

325720 Ranunculus chius DC. ;岛生毛茛■☆

325721 Ranunculus chondrodes Pomel = Ranunculus paludosus Poir. ■☆

325722 Ranunculus chuanchingensis L. Liou;川青毛茛; Chuanching Buttercup, Chuanqing Buttercup ■

325723 Ranunculus circinatus Sibth. ;卷须毛茛■☆

325724 Ranunculus circinatus Sibth. = Batrachium foeniculaceum (Gilib.) Krecz. ■

325725 Ranunculus circinatus Sibth. var. rodiei Maire = Ranunculus peltatus Schrank subsp. saniculifolius (Viv.) C. D. K. Cook ■☆

325726 Ranunculus circinatus Sibth. var. subrigidus (W. B. Drew) L. D. Benson = Ranunculus aquatilis L. var. diffusus With. ■☆

325727 Ranunculus coenosus Guss. = Ranunculus hederaceus L. ■☆

325728 Ranunculus confervoides (Fr.) Fr. = Ranunculus aquatilis L. var. diffusus With. ■☆

325729 Ranunculus confervoides Franch. = Batrachium eradicatum (Laest.) Fr. ■

325730 Ranunculus constantinopolitanus d'Urv. ;康士坦丁堡毛茛(丽毛茛)■☆

325731　Ranunculus constantinopolitanus d'Urv. 'Plenus'; 重瓣丽毛茛■☆

325732　Ranunculus cooleyae Vasey et Rose ex Rose; 库利毛茛■☆

325733　Ranunculus cooperi Oliv. = Ranunculus baurii MacOwan ■☆

325734　Ranunculus cornutus DC. var. trachycarpus Coss. = Ranunculus sardous Crantz ■

325735　Ranunculus cortusifolius Willd.; 假报春叶毛茛■☆

325736　Ranunculus cortusifolius Willd. var. maroccanus Coss. = Ranunculus spicatus Desf. subsp. maroccanus (Coss.) Greuter et Burdet ■☆

325737　Ranunculus cortusifolius Willd. var. rupestris Webb et Berthel. = Ranunculus cortusifolius Willd. ■☆

325738　Ranunculus cortusifolius Willd. var. silvaticus Webb et Berthel. = Ranunculus cortusifolius Willd. ■☆

325739　Ranunculus cortusifolius Willd. var. villosus Pit. = Ranunculus cortusifolius Willd. ■☆

325740　Ranunculus crassifolius (Rupr.) Grossh.; 厚叶毛茛■☆

325741　Ranunculus crenatus Waldst. et Kit.; 莲座毛茛(圆叶毛茛)■☆

325742　Ranunculus creticus L.; 克里特毛茛■☆

325743　Ranunculus cuneifolius Maxim.; 楔叶毛茛; Cuneateleaf Buttercup ■

325744　Ranunculus cuneifolius Maxim. var. latisectus S. H. Li et Y. Huei Huang; 宽楔叶毛茛; Broad Cuneateleaf Buttercup ■

325745　Ranunculus cuneifolius Maxim. var. latisectus S. H. Li et Y. Huei Huang = Ranunculus cuneifolius Maxim. ■

325746　Ranunculus cuneilobus A. Rich.; 楔裂毛茛■☆

325747　Ranunculus cyclocarpus Pamp.; 环果毛茛■☆

325748　Ranunculus cymbalaria (Pursh) Green = Halerpestes sarmentosa (Adans.) Kom. et Aliss. ■

325749　Ranunculus cymbalaria Pursh; 海岸毛茛; Alkali Buttercup, Seaside Crowfoot, Shore Buttercup ■☆

325750　Ranunculus cymbalaria Pursh = Halerpestes cymbalaria (Pursh) Green ■

325751　Ranunculus cymbalaria Pursh = Halerpestes sarmentosa (Adans.) Kom. et Aliss. ■

325752　Ranunculus cymbalaria Pursh f. hebecaulis Fernald = Ranunculus cymbalaria Pursh ■☆

325753　Ranunculus cymbalaria Pursh f. multisectus S. H. Li et Y. H. Huang = Halerpestes sarmentosa (Adans.) Kom. et Aliss. var. multisecta (S. H. Li et Y. Huei Huang) W. T. Wang ■

325754　Ranunculus cymbalaria Pursh subsp. sarmentosus (Adams) Kitag. = Halerpestes sarmentosa (Adans.) Kom. et Aliss. ■

325755　Ranunculus cymbalaria Pursh subsp. sarmentosus Kitag. = Halerpestes sarmentosa (Adans.) Kom. et Aliss. ■

325756　Ranunculus cymbalaria Pursh subsp. saximontanus (Fernald) Thorne = Ranunculus cymbalaria Pursh ■☆

325757　Ranunculus cymbalaria Pursh var. alpinus Hook. = Halerpestes cymbalaria (Pursh) Green ■

325758　Ranunculus cymbalaria Pursh var. alpinus Hook. = Ranunculus cymbalaria Pursh ■☆

325759　Ranunculus cymbalaria Pursh var. saximontanus Fernald = Halerpestes cymbalaria (Pursh) Green ■

325760　Ranunculus cymbalaria Pursh var. saximontanus Fernald = Ranunculus cymbalaria Pursh ■☆

325761　Ranunculus cymbalaria Pursh var. typicus L. D. Benson = Ranunculus cymbalaria Pursh ■☆

325762　Ranunculus cytheraceus Halácsy; 布袋兰毛茛■☆

325763　Ranunculus decipiens Pomel = Ranunculus macrophyllus Desf. ■☆

325764　Ranunculus delphinifolius Fries ex Steud.; 翠雀花叶毛茛; Yellow Water-crowfoot ■☆

325765　Ranunculus delphinifolius Torr. = Ranunculus flabellaris Raf. ■☆

325766　Ranunculus delphiniifolius Torr. ex Eaton = Ranunculus flabellaris Raf. ■☆

325767　Ranunculus delphiniifolius Torr. ex Eaton var. terrestris Farw. = Ranunculus flabellaris Raf. ■☆

325768　Ranunculus densiciliatus W. T. Wang; 睫毛毛茛; Denseciliate Buttercup ■

325769　Ranunculus densiciliatus W. T. Wang var. glabrescens W. L. Cheng; 变裸毛茛; Glabrous Denseciliate Buttercup ■

325770　Ranunculus densiciliatus W. T. Wang var. glabrescens W. L. Zheng = Ranunculus densiciliatus W. T. Wang ■

325771　Ranunculus densiciliatus W. T. Wang var. nyingchiensis W. L. Cheng; 林芝毛茛; Linzhi Denseciliate Buttercup ■

325772　Ranunculus densiciliatus W. T. Wang var. nyingchiensis W. L. Cheng = Ranunculus densiciliatus W. T. Wang ■

325773　Ranunculus dichotomus (Schmalh.) Orlova; 二歧毛茛■☆

325774　Ranunculus dielsianus Ulbr.; 康定毛茛; Kangding Buttercup ■

325775　Ranunculus dielsianus Ulbr. var. leiogynus W. T. Wang; 大通毛茛; Smooth Kangding Buttercup ■

325776　Ranunculus dielsianus Ulbr. var. longipilosus W. T. Wang; 长毛康定毛茛; Longhair Kangding Buttercup ■

325777　Ranunculus dielsianus Ulbr. var. suprasericeus Hand.-Mazz.; 丽江毛茛(毛叶毛茛); Woollyleaf Buttercup ■

325778　Ranunculus dielsianus Ulbr. var. suprasericeus Hand.-Mazz. = Ranunculus suprasericeus (Hand.-Mazz.) L. Liou ■

325779　Ranunculus diffusus DC.; 铺散毛茛; Spreading Buttercup ■

325780　Ranunculus diffusus DC. f. mollis (Wall.) Diels = Ranunculus diffusus DC. ■

325781　Ranunculus diffusus DC. f. subpinnatus Forbes et Hemsl. = Ranunculus cantoniensis DC. ■

325782　Ranunculus dilatatus Ovcz.; 膨大毛茛■☆

325783　Ranunculus dingjieensis L. Liou; 定结毛茛; Dingjie Buttercup ■

325784　Ranunculus dissectus M. Bieb.; 分节毛茛■☆

325785　Ranunculus distans Wall. et Royle; 黄毛茛; Yellow Buttercup ■

325786　Ranunculus distrias Steud. ex A. Rich. = Ranunculus cuneilobus A. Rich. ■☆

325787　Ranunculus divaricatus Schrank = Batrachium divaricatum (Schrank) Schur ■

325788　Ranunculus dolosus Fisch. et C. A. Mey.; 假毛茛■☆

325789　Ranunculus dongrergensis Hand.-Mazz.; 圆裂毛茛; Roundlobe Buttercup ■

325790　Ranunculus dongrergensis Hand.-Mazz. var. altifidus W. T. Wang; 深圆裂毛茛; Deepfid Roundlobe Buttercup ■

325791　Ranunculus ducloxii Finet et Gagnep. = Ranunculus ficariifolius H. Lév. et Vaniot ■

325792　Ranunculus dyris (Maire) H. Lindb.; 荒地毛茛■☆

325793　Ranunculus dyris (Maire) H. Lindb. var. ayachicus Emb. = Ranunculus aurasiacus Pomel ■☆

325794　Ranunculus dyris (Maire) H. Lindb. var. pseudobulbosus Maire = Ranunculus dyris (Maire) H. Lindb. ■☆

325795　Ranunculus eastwoodianus L. D. Benson = Ranunculus pedatifidus Sm. var. affinis (R. Br.) L. D. Benson ■☆

325796　Ranunculus elegans K. Koch; 雅致毛茛■☆

325797　Ranunculus elgonensis Ulbr. = Ranunculus volkensii Engl. ■

325798　Ranunculus ellipticus Greene = Ranunculus glaberrimus Hook.

var. ellipticus（Greene）Greene ■☆

325799 Ranunculus eradicatus （ Laest. ） Johans. = Batrachium eradicatum（Laest.）Fr. ■

325800 Ranunculus eschscholtzii Schltdl.；埃绍毛茛；Subalpine Buttercup ■☆

325801 Ranunculus eschscholtzii Schltdl. var. adoneus（A. Gray）C. L. Hitchc. = Ranunculus adoneus A. Gray ■☆

325802 Ranunculus eschscholtzii Schltdl. var. alpinus（S. Watson）C. L. Hitchc. = Ranunculus adoneus A. Gray ■☆

325803 Ranunculus eschscholtzii Schltdl. var. eximius（Greene）L. D. Benson；高山草甸毛茛 ■☆

325804 Ranunculus eschscholtzii Schltdl. var. suksdorfii （A. Gray）L. D. Benson；苏克毛茛 ■☆

325805 Ranunculus eschscholtzii Schltdl. var. trisectus（B. L. Rob.）L. D. Benson；三刚毛埃绍毛茛 ■☆

325806 Ranunculus eximius Greene = Ranunculus eschscholtzii Schltdl. var. eximius（Greene）L. D. Benson ■☆

325807 Ranunculus extensus （ Hook. f. ） Schube ex Engl. = Ranunculus multifidus Forssk. ■☆

325808 Ranunculus extensus Hook. f. = Ranunculus multifidus Forssk. ■☆

325809 Ranunculus extorris Hance = Ranunculus ternatus Thunb. ■

325810 Ranunculus falcatus L. = Ceratocephala falcata （L.）Pers. ■

325811 Ranunculus falcatus L. subsp. incurvus （Steven）Maire et Weiller = Ceratocephalus falcatus （ L. ） Pers. subsp. incurvus （Steven）Chrtek et Chrtková ■☆

325812 Ranunculus fascicularis Muhl. ex J. M. Bigelow；早毛茛；Early Buttercup，Prairie Buttercup，Thick-root Buttercup ■☆

325813 Ranunculus fascicularis Muhl. ex J. M. Bigelow var. apricus （Greene）Fernald = Ranunculus fascicularis Muhl. ex J. M. Bigelow ■☆

325814 Ranunculus fascicularis Muhl. ex J. M. Bigelow var. cuneiformis （Small）L. D. Benson = Ranunculus macranthus Scheele ■☆

325815 Ranunculus fascicularis Muhl. ex J. M. Bigelow var. typicus L. D. Benson = Ranunculus fascicularis Muhl. ex J. M. Bigelow ■☆

325816 Ranunculus felixii H. Lév.；扇叶毛茛；Fanleaf Buttercup ■

325817 Ranunculus felixii H. Lév. var. forrestii Hand. -Mazz.；心基扇叶毛茛；Forrest Buttercup ■

325818 Ranunculus felixii H. Lév. var. forrestii Hand. -Mazz. = Ranunculus felixii H. Lév. ■

325819 Ranunculus fibrosus Pomel = Ranunculus paludosus Poir. ■☆

325820 Ranunculus ficaria L.；无花果毛茛（倭毛茛）；Bright，Bright Eye，Brighteye，Butter，Butter-and-cheese，Butterchops，Buttercup Ficaria，Celandine，Cheese-and-butter，Cheesecups，Crain，Crazy，Crazy Bet，Crazy Bets，Crazy Cup，Cream-and-butter，Crow Pightle，Dill-cup，Fig Buttercup，Figroot Buttercup，Figwort，Foalfoot，Fogwort，Foxwort，Frog's Foot，Gentlemen's Cap-and-frills，Gilcup，Gilt Cup，Gilty Cup，Gilty-cup，Go-cup，Golden Cap-and-frills，Golden Cup，Golden Daisy，Golden Drinking Cup，Golden Guinea，Golden Star，Golding Cup，Goldy Knob，Kingcup，King's Evil，Legwort，Lesser Celandine，Marsh Pilewort，Pilewort，Power-wort，Purple Leaved Lesser Celandine，Smaillwort，Spring Messenger，Star-flower，Swallow-wort，Tetter，Yellow Crain，Yellow Crane ■☆

325821 Ranunculus ficaria L. 'Albus'；白花倭毛茛 ■☆

325822 Ranunculus ficaria L. 'Aurantiacus'；橙花倭毛茛 ■☆

325823 Ranunculus ficaria L. 'Brazen Hussy'；铜色倭毛茛 ■☆

325824 Ranunculus ficaria L. 'Flore Pleno'；重瓣倭毛茛 ■☆

325825 Ranunculus ficaria L. subsp. bulbifer Lawalrée = Ranunculus ficaria L. ■☆

325826 Ranunculus ficaria L. subsp. bulbilifer Lambinon = Ranunculus ficaria L. ■☆

325827 Ranunculus ficaria L. subsp. calthifolius （Rchb.）Arcang. = Ranunculus ficaria L. ■☆

325828 Ranunculus ficaria L. subsp. ficariiformis Rouy et Foucaud；大无花果毛茛 ■☆

325829 Ranunculus ficaria L. subsp. nudicaulis （ A. Kern. ） Rouy et Foucaud = Ranunculus ficaria L. subsp. calthifolius （ Rchb. ） Arcang. ☆

325830 Ranunculus ficaria L. var. bulbifera Albert = Ranunculus ficaria L. ■☆

325831 Ranunculus ficaria L. var. intermedius Ball = Ranunculus ficaria L. ■☆

325832 Ranunculus ficariifolius H. Lév. et Vaniot；西南毛茛（卵叶毛茛）；Ficarialeaf Buttercup，SW. China Buttercup ■

325833 Ranunculus ficariifolius H. Lév. et Vaniot var. crenatus H. Lév. = Ranunculus ficariifolius H. Lév. et Vaniot ■

325834 Ranunculus ficariifolius H. Lév. et Vaniot var. erythrosepalus H. Lév. = Ranunculus ficariifolius H. Lév. et Vaniot ■

325835 Ranunculus ficariifolius H. Lév. et Vaniot var. ovalifolius H. Lév. = Ranunculus ficariifolius H. Lév. et Vaniot ■

325836 Ranunculus filiformis Michx. = Ranunculus flammula L. var. reptans （L.）Schltdl. ■☆

325837 Ranunculus filiformis Michx. var. ovalis J. M. Bigelow = Ranunculus flammula L. var. ovalis （J. M. Bigelow）L. D. Benson ■☆

325838 Ranunculus flabellaris Raf.；黄水毛茛；Fan-leaved Buttercup，Jersey Butrercup，Yellow Water Buttercup，Yellow Water Crowfoot，Yellow Water-buttercup ■☆

325839 Ranunculus flabellaris Raf. f. riparius Fernald = Ranunculus flabellaris Raf. ■☆

325840 Ranunculus flabellatus Desf. = Ranunculus paludosus Poir. ■☆

325841 Ranunculus flabellatus Desf. subsp. fibrosus Pomel = Ranunculus paludosus Poir. ■☆

325842 Ranunculus flabellatus Desf. subsp. granulatus （Pomel）Sennen = Ranunculus paludosus Poir. var. granulatus （Pomel）Emb. et Maire ■☆

325843 Ranunculus flabellatus Desf. subsp. robustus Pomel = Ranunculus paludosus Poir. var. robustus （Pomel）Emb. et Maire ■☆

325844 Ranunculus flabellatus Desf. var. chondrodes （Pomel）Batt. et Trab. = Ranunculus ficaria L. ■☆

325845 Ranunculus flabellatus Desf. var. faurei Maire = Ranunculus paludosus Poir. ■☆

325846 Ranunculus flabellatus Desf. var. fibrosus （Pomel）Batt. et Trab. = Ranunculus ficaria L. ■☆

325847 Ranunculus flabellatus Desf. var. flabellatus （Desf.）Batt. et Trab. = Ranunculus ficaria L. ■☆

325848 Ranunculus flabellatus Desf. var. granulatus （Pomel）Batt. et Trab. = Ranunculus ficaria L. ■☆

325849 Ranunculus flabellatus Desf. var. leucothrix （Ball）Maire = Ranunculus paludosus Poir. ■☆

325850 Ranunculus flabellatus Desf. var. manzanoanus Sennen et Mauricio = Ranunculus paludosus Poir. ■☆

325851 Ranunculus flabellatus Desf. var. nervosus Ducell. et Maire = Ranunculus paludosus Poir. ■☆

325852 Ranunculus flabellatus Desf. var. nidulans （Pomel）Batt. et Trab. = Ranunculus ficaria L. ■☆

325853 Ranunculus flabellatus Desf. var. paludosus （Poir.）Batt. et

Trab. = Ranunculus ficaria L. ■☆

325854　Ranunculus flabellatus Desf. var. pseudomonspeliacus Maire = Ranunculus paludosus Poir. ■☆

325855　Ranunculus flabellatus Desf. var. reesei Sennen = Ranunculus paludosus Poir. ■☆

325856　Ranunculus flabellatus Desf. var. robustus（Pomel）Batt. et Trab. = Ranunculus ficaria L. ■☆

325857　Ranunculus flabellatus Desf. var. rufulus Brot. = Ranunculus ficaria L. ■☆

325858　Ranunculus flabellatus Desf. var. termieri Sennen et Mauricio = Ranunculus paludosus Poir. ■☆

325859　Ranunculus flaccidus Hook. f. et Thomson = Ranunculus ficariifolius H. Lév. et Vaniot ■

325860　Ranunculus flaccidus Pers. = Batrachium trichophyllum（Chaix）Bosch ■

325861　Ranunculus flaccidus Pers. = Ranunculus aquatilis L. var. diffusus With. ■☆

325862　Ranunculus flaccidus Pers. var. rionii（Lagger）Hegi = Batrachium rionii（Lagger）Nyman ■

325863　Ranunculus flagellifolius Nakai = Ranunculus reptans L. ■

325864　Ranunculus flammula L.；焰毛茛；Banebind，Cow Grass，Creeping Spearwort，Crowfoot，Flame-leaved Crowfoot，Golden Butter，Goose Tongue，Goose-tongue，Lesser Spearwort，Miniature Spearwort，Small Spearwort，Snake's Tongue，Spear Crowfoot，Speargrass，Spearwort，Spearwort Buttercup，Spirewort，Spurwood，Water Buttercup，Wilfire，Yellow Crain，Yellow Crane ■☆

325865　Ranunculus flammula L. subsp. reptans Turcz. = Ranunculus reptans L. ■

325866　Ranunculus flammula L. var. alismifolius Vill. = Ranunculus flammula L. ■☆

325867　Ranunculus flammula L. var. angustifolius Wallr. = Ranunculus flammula L. ■☆

325868　Ranunculus flammula L. var. filiformis（Michx.）Hook. = Ranunculus flammula L. var. reptans（L.）Schltdl. ■☆

325869　Ranunculus flammula L. var. numidicus Maire = Ranunculus flammula L. ■☆

325870　Ranunculus flammula L. var. ovalis（J. M. Bigelow）L. D. Benson；卵形焰毛茛；Creeping Spearwort，Spearwort Buttercup ■☆

325871　Ranunculus flammula L. var. reptans（L.）E. Mey. = Ranunculus flammula L. var. reptans（L.）Schltdl. ■☆

325872　Ranunculus flammula L. var. reptans（L.）Piper et Beattie = Ranunculus flammula L. var. reptans（L.）Schltdl. ■☆

325873　Ranunculus flammula L. var. reptans（L.）Schltdl. = Ranunculus reptans L. ■☆

325874　Ranunculus flammula L. var. reptans Fleisch. = Ranunculus reptans L. ■

325875　Ranunculus flammula L. var. samolifolius（Greene）L. D. Benson = Ranunculus flammula L. var. ovalis（J. M. Bigelow）L. D. Benson ■☆

325876　Ranunculus flavidus（Hand. -Mazz.）C. D. K. Cook. = Batrachium bungei（Steud.）L. Liou var. flavidum（Hand. -Mazz.）L. Liou ■

325877　Ranunculus fluitans Lam. = Batrachium fluitans Wimm. ■☆

325878　Ranunculus foeniculaceus Gilib. = Batrachium foeniculaceum（Gilib.）Krecz. ■

325879　Ranunculus formosa-montanus Ohwi；蓬莱毛茛（南湖毛茛，疏花毛茛）；Nanhu Buttercup，Taiwan Montane Buttercup ■

325880　Ranunculus formosanus Masam. = Ranunculus ternatus Thunb. ■

325881　Ranunculus forskoehlii DC. = Ranunculus multifidus Forssk. ■☆

325882　Ranunculus franchetii H. Boissieu；深山毛茛；Franchet. Buttercup ■

325883　Ranunculus fraternus Schrenk；团叶毛茛 ■

325884　Ranunculus furcatifidus W. T. Wang；叉裂毛茛；Furcatefid Buttercup ■

325885　Ranunculus garianicus Borzí et Mattei = Ranunculus asiaticus L. ■

325886　Ranunculus gelidus Kar. et Kir. ；冷地毛茛；Cold Buttercup ■

325887　Ranunculus gelidus Kar. et Kir. subsp. grayi（Britton）Hultén = Ranunculus gelidus Kar. et Kir. ■

325888　Ranunculus geranifolium Hayata = Ranunculus taisanensis Hayata ■

325889　Ranunculus geraniifolius Pourr. subsp. aurasiacus（Pomel）Maire = Ranunculus aurasiacus Pomel ■☆

325890　Ranunculus geraniifolius Pourr. subsp. dyris Maire = Ranunculus dyris（Maire）H. Lindb. ■☆

325891　Ranunculus geraniifolius Pourr. var. ayachicus（Emb.）Maire = Ranunculus aurasiacus Pomel ■☆

325892　Ranunculus geraniifolius Pourr. var. mesatlanticus Maire = Ranunculus aurasiacus Pomel ■☆

325893　Ranunculus glaberrimus Hook. var. ellipticus（Greene）Greene；椭圆毛茛 ■☆

325894　Ranunculus glaberrimus Hook. var. reconditus L. D. Benson = Ranunculus triternatus A. Gray ■☆

325895　Ranunculus glabricaulis（Hand. -Mazz.）L. Liou；甘藏毛茛；Gan-Zang Buttercup，Glabrousstemmed Buttercup ■

325896　Ranunculus glabricaulis（Hand. -Mazz.）L. Liou var. viridisepalus W. T. Wang；绿萼甘藏毛茛；Greenaepal Glabrousstemmed Buttercup ■

325897　Ranunculus glacialiformis Hand. -Mazz. ；宿萼毛茛；Glacialform Buttercup，Persistent Buttercup ■

325898　Ranunculus glacialis L. ；冰河毛茛；Glacier Buttercup，Glacier Crowfoot，Snow Buttercup ■☆

325899　Ranunculus glacialis L. var. gelidus（Kar. et Kir.）Finet et Gagnep. = Ranunculus gelidus Kar. et Kir. ■

325900　Ranunculus glareosus Hand. -Mazz. ；砾地毛茛；Glareosous Buttercup ■

325901　Ranunculus gmelinii DC. ；小掌叶毛茛（格麦氏毛茛，小叶毛茛）；Gmelin Buttercup，Gmelin's Buttercup，Small Yellow Water-crowfoot ■

325902　Ranunculus gmelinii DC. subsp. purshii（Richardson）Hultén = Ranunculus gmelinii DC. ■

325903　Ranunculus gmelinii DC. var. hookeri（D. Don）L. D. Benson = Ranunculus gmelinii DC. ■

325904　Ranunculus gmelinii DC. var. limosus（Nutt.）H. Hara = Ranunculus gmelinii DC. ■

325905　Ranunculus gmelinii DC. var. prolificus（Fernald）H. Hara = Ranunculus gmelinii DC. ■

325906　Ranunculus gmelinii DC. var. purshii（Richardson）H. Hara = Ranunculus gmelinii DC. ■

325907　Ranunculus gmelinii DC. var. radicans（C. A. Mey.）Krylov = Ranunculus radicans C. A. Mey. ■

325908　Ranunculus gmelinii DC. var. radicans Krylov = Ranunculus radicans C. A. Mey. ■

325909　Ranunculus gmelinii DC. var. terrestris（Ledeb.）L. D. Benson = Ranunculus gmelinii DC. ■

325910　Ranunculus gmelinii DC. var. terrestris（Ledeb.）L. D. Benson f. purshii（Richardson）Fassett ＝ Ranunculus gmelinii DC. ■

325911　Ranunculus gmelinii DC. var. typicus L. D. Benson ＝ Ranunculus gmelinii DC. ■

325912　Ranunculus gramineus L. ;禾叶毛茛;Grass-leaved Buttercup, Ranunkel ■☆

325913　Ranunculus gramineus L. var. luzulifolius Boiss. ＝ Ranunculus gramineus L. ■☆

325914　Ranunculus granatensis Boiss. ;格拉毛茛■☆

325915　Ranunculus grandifolius C. A. Mey. ;大叶毛茛;Largeleaf Buttercup ■

325916　Ranunculus grandis Honda;大毛茛;Great Buttercup ■

325917　Ranunculus grandis Honda var. manshuricus Hara;帽儿山毛茛;Maorshan Great Buttercup ■

325918　Ranunculus granulatus Pomel ＝ Ranunculus paludosus Poir. ■☆

325919　Ranunculus grayi Britton ＝ Ranunculus gelidus Kar. et Kir. ■☆

325920　Ranunculus gymnadenus Sommier et H. Lév. ;裸腺毛茛■☆

325921　Ranunculus hamiensis J. G. Liu;哈密毛茛（毛瓣毛茛）;Hami Buttercup ■

325922　Ranunculus harveyi（A. Gray）Britton;哈尔维毛茛;Harvey's Buttercup ■☆

325923　Ranunculus hederaceus L. ;常春藤叶毛茛;Ivy Crowfoot, Ivyleaved Crowfoot,Ivy-leaved Crowfoot,Ivy-leaved Water Buttercup, Water Anemone ■☆

325924　Ranunculus hederaceus L. var. coenosus（Guss.）Coss. ＝ Ranunculus hederaceus L. ■☆

325925　Ranunculus hejingensis W. T. Wang;和静毛茛;Hejing Buttercup ■

325926　Ranunculus helenae Albov;海伦娜毛茛■☆

325927　Ranunculus helogeton Ulbr. ＝ Ranunculus volkensii Engl. ■

325928　Ranunculus heterophyllus F. H. Wigg. ＝ Ranunculus aquatilis L. ■☆

325929　Ranunculus heterophyllus Sm. ;异叶毛茛;Eggs-and-bacon, Various-leaved Crowfoot ■☆

325930　Ranunculus hetianensis L. Liou;和田毛茛;Hetian Buttercup ■

325931　Ranunculus hexasepalus（L. D. Benson）L. D. Benson ＝ Ranunculus occidentalis Nutt. var. hexasepalus L. D. Benson ■☆

325932　Ranunculus hirsutus Curtis;毛茎毛茛■☆

325933　Ranunculus hirtellus Royle;吉隆三裂毛茛（基隆毛茛,三裂毛茛）;Trilobe Buttercup ■

325934　Ranunculus hirtellus Royle ＝ Ranunculus dielsianus Ulbr. var. suprasericeus Hand. -Mazz. ■

325935　Ranunculus hirtellus Royle var. glabrescens W. L. Cheng;光柄毛茛;Glabrousstalk Trilobe Buttercup ■

325936　Ranunculus hirtellus Royle var. glabrescens W. L. Cheng ＝ Ranunculus hirtellus Royle var. orientalis W. T. Wang ■

325937　Ranunculus hirtellus Royle var. glabricaulis Hand. -Mazz. ＝ Ranunculus glabricaulis（Hand. -Mazz.）L. Liou ■

325938　Ranunculus hirtellus Royle var. humilis W. T. Wang;小基隆毛茛;Dwarf Trilobe Buttercup ■

325939　Ranunculus hirtellus Royle var. orientalis W. T. Wang;东方三裂毛茛（三裂毛茛）;Oriental Trilobe Buttercup ■

325940　Ranunculus hirtellus Royle var. sigyiaicus W. L. Cheng;色季拉毛茛;Sejila Trilobe Buttercup ■

325941　Ranunculus hirtellus Royle var. sigyiaicus W. L. Cheng ＝ Ranunculus hirtellus Royle var. orientalis W. T. Wang ■

325942　Ranunculus hispidus Michx. ;刚毛毛茛;Bristly Buttercup, Hispid Buttercup,Rough Buttercup,Swamp Buttercup ■☆

325943　Ranunculus hispidus Michx. var. caricetorum（Greene）T. Duncan;苔地毛茛;Bristly Buttercup, Hispid Buttercup, Rough Buttercup ■☆

325944　Ranunculus hispidus Michx. var. eurylobus L. D. Benson ＝ Ranunculus hispidus Michx. ■☆

325945　Ranunculus hispidus Michx. var. falsus Fernald ＝ Ranunculus hispidus Michx. ■☆

325946　Ranunculus hispidus Michx. var. greenmanii L. D. Benson ＝ Ranunculus hispidus Michx. ■☆

325947　Ranunculus hispidus Michx. var. marilandicus（Poir.）L. D. Benson ＝ Ranunculus hispidus Michx. ■☆

325948　Ranunculus hispidus Michx. var. nitidus（Chapm.）T. Duncan;光亮刚毛毛茛;Bristly Buttercup, Hispid Buttercup, Rough Buttercup ■☆

325949　Ranunculus hispidus Michx. var. typicus L. D. Benson ＝ Ranunculus hispidus Michx. ■☆

325950　Ranunculus holophyllus Hance ＝ Ranunculus sceleratus L. ■

325951　Ranunculus homoeophyllus Ten. ＝ Ranunculus hederaceus L. ■☆

325952　Ranunculus hsinganensis Kitag. ＝ Ranunculus japonicus Thunb. var. hsinganensis（Kitag.）W. T. Wang ■

325953　Ranunculus humillimus W. T. Wang;低毛茛■

325954　Ranunculus hybridus Biria;杂种毛茛;Persian Buttercup ■☆

325955　Ranunculus hydrocharis Spenn. f. bungei（Steud.）Hiern ＝ Batrachium bungei（Steud.）L. Liou ■

325956　Ranunculus hydrocharis Spenn. f. lobbii Hiern ＝ Ranunculus lobbii（Hiern）A. Gray ■☆

325957　Ranunculus hydrocharoides A. Gray;水鳖毛茛■☆

325958　Ranunculus hydrocharoides A. Gray var. stolonifer（Hemsl.）L. D. Benson ＝ Ranunculus hydrocharoides A. Gray ■☆

325959　Ranunculus hydrophilus Bunge ＝ Batrachium bungei（Steud.）L. Liou ■

325960　Ranunculus hyperboreus Rottb. ;北方毛茛■☆

325961　Ranunculus hyperboreus Rottb. ＝ Ranunculus pseudopygmaeus Hand. -Mazz. ■

325962　Ranunculus hyperboreus Rottb. subsp. arnellii Scheutz ＝ Ranunculus hyperboreus Rottb. ■☆

325963　Ranunculus hyperboreus Rottb. subsp. intertextus（Greene）B. M. Kapoor et Á. Löve ＝ Ranunculus hyperboreus Rottb. ■☆

325964　Ranunculus hyperboreus Rottb. var. natans Regel;北方矮毛茛■☆

325965　Ranunculus hyperboreus Rottb. var. natans Regel ＝ Ranunculus natans C. A. Mey. ■

325966　Ranunculus hyperboreus Rottb. var. radicans Hook. f. ＝ Ranunculus radicans C. A. Mey. ■

325967　Ranunculus hyperboreus Rottb. var. samojedorum（Rupr.）Perfil. ＝ Ranunculus hyperboreus Rottb. ■☆

325968　Ranunculus hyperboreus Rottb. var. tricrenatus Rupr. ＝ Ranunculus hyperboreus Rottb. ■☆

325969　Ranunculus hyperboreus Rottb. var. turquetilianus Polunin ＝ Ranunculus hyperboreus Rottb. ■☆

325970　Ranunculus hystriculus A. Gray;豪猪毛茛■☆

325971　Ranunculus illyricus L. ;依赖利毛茛;Illyrian Buttercup ■☆

325972　Ranunculus inamoenus Greene;不悦毛茛■☆

325973　Ranunculus inamoenus Greene var. alpeophilus（A. Nelson）L. D. Benson ＝ Ranunculus inamoenus Greene ■☆

325974　Ranunculus inamoenus Greene var. subaffinis（A. Gray）L. D. Benson;近缘不悦毛茛■☆

325975 Ranunculus indivisus（Maxim.）Hand.-Mazz.；圆叶毛茛；Roundleaf Buttercup ■

325976 Ranunculus indivisus（Maxim.）Hand.-Mazz. var. abaensis（W. T. Wang）W. T. Wang；阿坝毛茛；Aba Buttercup ■

325977 Ranunculus intromongolicus Y. Z. Zhao；内蒙古毛茛■

325978 Ranunculus involucratus Maxim. = Ranunculus similis Hemsl. ■

325979 Ranunculus involucratus Maxim. var. minor L. Liou = Ranunculus minor（L. Liou）W. T. Wang ■

325980 Ranunculus jacuticus Ovcz.；雅库特毛茛■☆

325981 Ranunculus japonicus Thunb.；毛茛（大脚迹,鹤膝草,猴蒜,回回蒜,火筒青,瞌睡草,辣辣草,辣子草,烂肺草,老虎草,老虎脚底板,老虎脚迹,老虎脚迹草,老虎脚爪草,老虎须,老鼠脚底板,毛建,毛建草,毛堇,毛芹菜,毛田菜,起泡草,千里光,日本毛茛,三脚虎,山辣椒,水茛,水芹菜,天灸,五虎草,鸭脚板,野芹菜,鱼疗草,自灸）；Japan Buttercup,Japanese Buttercup ■

325982 Ranunculus japonicus Thunb. f. latissimus（Kitag.）Kitag. = Ranunculus japonicus Thunb. ■

325983 Ranunculus japonicus Thunb. f. pleniflorus（Makino）Honda；重瓣毛茛；Doubleflower Japan Buttercup ■

325984 Ranunculus japonicus Thunb. var. hsinganensis（Kitag.）W. T. Wang；银叶毛茛；Silverleaf Japanese Buttercup ■

325985 Ranunculus japonicus Thunb. var. latissimus Kitag. = Ranunculus japonicus Thunb. ■

325986 Ranunculus japonicus Thunb. var. monticola Kitag. = Ranunculus paishanensis Kitag. ■

325987 Ranunculus japonicus Thunb. var. pratensis Kitag.；草地毛茛；Grassland Japanese Buttercup ■

325988 Ranunculus japonicus Thunb. var. pratensis Kitag. = Ranunculus japonicus Thunb. var. propinquus（C. A. Mey.）W. T. Wang ■

325989 Ranunculus japonicus Thunb. var. propinquus（C. A. Mey.）W. T. Wang；伏毛毛茛；Prostratehair Japanese Buttercup ■

325990 Ranunculus japonicus Thunb. var. smirnovii（Ovcz.）L. Liou = Ranunculus smirnovii Ovcz. ■

325991 Ranunculus japonicus Thunb. var. ternatifolius L. Liao；三小叶毛茛（三叶毛茛）；Threeleaf Japanese Buttercup ■

325992 Ranunculus jilongensis L. Liou = Ranunculus hirtellus Royle ■

325993 Ranunculus jingyuanensis W. T. Wang；靖远毛茛■

325994 Ranunculus jovis A. Nelson；朱庇特毛茛■☆

325995 Ranunculus junipericola Ohwi；桧林毛茛（高山毛茛）；Juniperforest Buttercup,Juniperwoods Buttercup ■

325996 Ranunculus juniperinus M. E. Jones = Ranunculus andersonii A. Gray ■☆

325997 Ranunculus kamtschaticus DC.；勘察加毛茛■☆

325998 Ranunculus kamtschaticus DC. = Oxygraphis glacialis（Fisch. ex DC.）Bunge ■

325999 Ranunculus kaufmannii Clerc ex Trautv. = Batrachium kauffmannii（Clerc）V. I. Krecz. ■

326000 Ranunculus kawakamii Hayata = Ranunculus cheirophyllus Hayata ■

326001 Ranunculus keniensis Milne-Redh. et Turrill = Ranunculus aberdaricus Ulbr. ■☆

326002 Ranunculus komarovii Freyn；科马罗夫毛茛■

326003 Ranunculus kopetdaghensis Litv.；科佩特毛茛■☆

326004 Ranunculus kotschyi Boiss.；考奇毛茛■☆

326005 Ranunculus krasnovii Ovcz.；克拉毛茛■☆

326006 Ranunculus krylovii Ovcz. = Ranunculus monophyllus Ovcz. ■

326007 Ranunculus kunlunensis J. G. Liu；昆仑毛茛；Kunlun Buttercup ■

326008 Ranunculus kunmingensis W. T. Wang；昆明毛茛；Kunming Buttercup ■

326009 Ranunculus kunmingensis W. T. Wang f. leipoensis（L. Liou）W. T. Wang；雷波毛茛；Leibo Kunming Buttercup ■

326010 Ranunculus kunmingensis W. T. Wang f. leipoensis（L. Liou）W. T. Wang = Ranunculus kunmingensis W. T. Wang ■

326011 Ranunculus kunmingensis W. T. Wang var. hispidus W. T. Wang；展毛昆明毛茛（展尾昆明毛茛）；Hispid Kunming Buttercup ■

326012 Ranunculus labordei H. Lév. et Vaniot = Ranunculus japonicus Thunb. ■

326013 Ranunculus laetus Royle = Ranunculus distans Wall. et Royle ■

326014 Ranunculus laetus Wall. = Ranunculus distans Wall. et Royle ■

326015 Ranunculus laetus Wall. var. leipoensis L. Liou = Ranunculus kunmingensis W. T. Wang f. leipoensis（L. Liou）W. T. Wang ■

326016 Ranunculus lancifolius Bertero = Halerpestes lancifolia（Bertol.）Hand.-Mazz. ■

326017 Ranunculus lanuginosiformis Selin ex J. Fellm.；毛毛茛■☆

326018 Ranunculus lanuginosus L.；绒毛茛（棉毛茛）；Lanuginous Buttercup,Woolly Buttercup ■☆

326019 Ranunculus lappaceus Sm.；钩毛茛；Grassland Buttercup ■☆

326020 Ranunculus lapponicus L.；拉普兰毛茛；Lapland Buttercup ■☆

326021 Ranunculus lasiocarpus C. A. Mey.；光果毛茛■☆

326022 Ranunculus lateriflorus DC. = Batrachium lateriflorus（DC.）Ovcz. ■☆

326023 Ranunculus laxicaulis Darby；疏茎毛茛；Spearwort,Water Plantain,Water Spearwort ■☆

326024 Ranunculus leiocladus Hayata = Ranunculus ternatus Thunb. ■

326025 Ranunculus lenormandii F. W. Schultz = Ranunculus omiophyllus Ten. ■☆

326026 Ranunculus leontinensis Freyn = Ranunculus peltatus Schrank ■☆

326027 Ranunculus leptorrhynchus Aitch. et Hemsl.；细嘴毛茛■☆

326028 Ranunculus leucothrix（Ball）Ball = Ranunculus paludosus Poir. ■☆

326029 Ranunculus limosus Nutt. = Ranunculus gmelinii DC. ■

326030 Ranunculus limprichtii Ulbr.；纺锤毛茛；Limpricht Buttercup ■

326031 Ranunculus limprichtii Ulbr. var. flavus Hand.-Mazz.；狭瓣纺锤毛茛■

326032 Ranunculus limprichtii Ulbr. var. flavus Hand.-Mazz. = Ranunculus limprichtii Ulbr. ■

326033 Ranunculus linearilobus Bunge；线裂毛茛■☆

326034 Ranunculus lingua L.；条叶毛茛（长叶毛茛,舌状毛茛）；Banebind,Greater Spearwort,Longleaf Buttercup,Sparrow-weed,Spear Crowfoot,Spearwort,Tongue Buttercup,Tongue-leaved Crowfoot ■

326035 Ranunculus lingua L. 'Grandiflorus'；大花长叶毛茛■☆

326036 Ranunculus lobatus Jacq.；浅裂毛茛；Lobate Buttercup ■

326037 Ranunculus lobbii（Hiern）A. Gray；洛布毛茛■☆

326038 Ranunculus lojkae Sommier et H. Lév.；洛伊毛茛■☆

326039 Ranunculus lomatocarpus Fisch. et C. A. Mey.；缘果毛茛■☆

326040 Ranunculus longicaulis C. A. Mey. = Ranunculus pulchellus C. A. Mey. var. longicaulis Trautv. ■

326041 Ranunculus longicaulis C. A. Mey. var. geniculatus（Hand.-Mazz.）L. Liou = Ranunculus nephelogenes Edgew. var. geniculatus（Hand.-Mazz.）W. T. Wang ■

326042 Ranunculus longicaulis C. A. Mey. var. nephelogenes（Edgew.）L. Liou = Ranunculus nephelogenes Edgew. ■

326043 Ranunculus longilobus Ovcz.；长裂毛茛■☆

326044 Ranunculus longipetalus Hand. -Mazz. ＝ Ranunculus micronivalis Hand. -Mazz. ■

326045 Ranunculus longirostris Godr. ；长喙毛茛；Long-beaked Water-crowfoot，White Water Crowfoot，White Water-buttercup ■☆

326046 Ranunculus longirostris Godr. ＝ Ranunculus aquatilis L. var. diffusus With. ■☆

326047 Ranunculus ludovicianus Greene ＝ Ranunculus canus Benth. var. ludovicianus （Greene） L. D. Benson ■☆

326048 Ranunculus luoergaiensis L. Liou；若尔盖毛茛；Luoergai Buttercup，Rorgai Buttercup ■

326049 Ranunculus lyalli Rydb. ；大圆叶毛茛；Giant Buttercup，Mount Cook Lily ■☆

326050 Ranunculus macauleyi A. Gray；麦考毛茛■☆

326051 Ranunculus macounii Britton；马昆毛茛■☆

326052 Ranunculus macounii Britton var. oreganus （A. Gray） K. C. Davis ＝ Ranunculus macounii Britton ■☆

326053 Ranunculus macranthus Scheele；大花毛茛■☆

326054 Ranunculus macrophyllus Desf. ；欧美大叶毛茛■☆

326055 Ranunculus macrophyllus Desf. var. corsicus （DC.） Briq. ＝ Ranunculus macrophyllus Desf. ■☆

326056 Ranunculus macrophyllus Desf. var. procerus （Moris） Freyn ＝ Ranunculus macrophyllus Desf. ■☆

326057 Ranunculus mainlingensis W. T. Wang；米林毛茛；Milin Buttercup ■

326058 Ranunculus mairei H. Lév. ＝ Ranunculus ficariifolius H. Lév. et Vaniot ■

326059 Ranunculus mairei H. Lév. ＝ Ranunculus yunnanensis Franch. ■

326060 Ranunculus manshuricus S. H. Li ＝ Ranunculus rigescens Turcz. ex Ost. -Sack. et Rupr. ■

326061 Ranunculus marginatus d'Urv. ；花边毛茛；Margined Buttercup，St. Martin's Buttercup ■☆

326062 Ranunculus marginatus d'Urv. subsp. trachycarpus （Fisch. et C. A. Mey.） Elenevsky et T. G. Derviz-Sokolova；粗果花边毛茛；Margined Buttercup ■☆

326063 Ranunculus matsudai Hayata ex Masam. ；疏花毛茛；Laxflower Buttercup ■

326064 Ranunculus matsudai Hayata ex Masam. ＝ Ranunculus formosa-montanus Ohwi ■

326065 Ranunculus mauritanicus Pomel ＝ Ranunculus hederaceus L. ■☆

326066 Ranunculus maximowiczii Pamp. ＝ Ranunculus similis Hemsl. ■

326067 Ranunculus megacarpus W. Koch；大果毛茛；Bigfruit Buttercup，Largefruit Buttercup ■☆

326068 Ranunculus meifolius Pomel ＝ Ranunculus millefoliatus Vahl ■☆

326069 Ranunculus meinshausenii Schrenk；梅氏毛茛■☆

326070 Ranunculus melanogynus W. T. Wang；黑果毛茛；Blackfruit Buttercup ■

326071 Ranunculus membranaceus Fresen. ＝ Ranunculus multifidus Forssk. ■☆

326072 Ranunculus membranaceus Royle；绵毛茛（绢毛毛茛，棉毛茛）；Woolly Buttercup ■

326073 Ranunculus membranaceus Royle var. floribundus W. T. Wang；多花绵毛茛（多花毛茛，多花柔毛茛）；Flowery Woolly Buttercup ■

326074 Ranunculus membranaceus Royle var. pubescens （W. T. Wang） W. T. Wang；柔毛茛（柔毛云生毛茛）；Pubescent Cloudy Buttercup，Pubescent Woolly Buttercup ■

326075 Ranunculus mengyuanensis W. T. Wang；门源毛茛；Menyuan Buttercup ■

326076 Ranunculus meyeri Harv. ；迈尔毛茛■☆

326077 Ranunculus meyeri Harv. var. transvaalensis Szyszyl. ＝ Ranunculus meyeri Harv. ■☆

326078 Ranunculus meyerianus Rupr. ；短喙毛茛；Shortbeak Buttercup ■

326079 Ranunculus micranthus Nutt. ＝ Ranunculus abortivus L. ■☆

326080 Ranunculus micranthus Nutt. var. cymbalistes （Greene） Fernald ＝ Ranunculus micranthus Nutt. ■☆

326081 Ranunculus micranthus Nutt. var. delitescens （Greene） Fernald ＝ Ranunculus micranthus Nutt. ■☆

326082 Ranunculus micronivalis Hand. -Mazz. ；窄瓣毛茛；Narrowpetal Buttercup ■

326083 Ranunculus micronivalis Hand. -Mazz. var. platypetalus Hand. -Mazz. ＝ Ranunculus platypetalus （Hand. -Mazz.） Hand. -Mazz. ■

326084 Ranunculus microphyllus Hand. -Mazz. ；小叶毛茛；Littleleaf Buttercup ■

326085 Ranunculus microphyllus Hand. -Mazz. ＝ Ranunculus ficariifolius H. Lév. et Vaniot ■

326086 Ranunculus millefoliatus Vahl；蜜小叶毛茛■☆

326087 Ranunculus millefoliatus Vahl var. inor Pamp. ＝ Ranunculus millefoliatus Vahl ■☆

326088 Ranunculus millefoliatus Vahl var. meifolius （Pomel） Batt. et Trab. ＝ Ranunculus millefoliatus Vahl ■☆

326089 Ranunculus millefolius Banks et Sol. ；粟草叶毛茛■☆

326090 Ranunculus minor （L. Liou） W. T. Wang；小苞毛茛；Smallbract Buttercup ■

326091 Ranunculus mississippiensis Small ＝ Ranunculus laxicaulis Darby ■☆

326092 Ranunculus moellendorffii Hance ＝ Anemone rivularis Buch. -Ham. var. flore-minore Maxim. ■

326093 Ranunculus monophyllus Ovcz. ；单叶毛茛（齿裂毛茛）；Krylov Buttercup，Unifolious Buttercup ■

326094 Ranunculus monophyllus Ovcz. f. latisectus Ovcz. ；齿裂毛茛；Broadtooth Unifolious Buttercup ■

326095 Ranunculus monophyllus Ovcz. f. latisectus Ovcz. ＝ Ranunculus monophyllus Ovcz. ■

326096 Ranunculus montanus Willd. ；山毛茛；Mountain Buttercup ■☆

326097 Ranunculus montanus Willd. 'Molten Gold'；金花山毛茛■☆

326098 Ranunculus montanus Willd. subsp. aurasiacus （Pomel） Maire ＝ Ranunculus aurasiacus Pomel ■☆

326099 Ranunculus montanus Willd. var. aurasiacus （Pomel） Maire；奥拉斯山毛茛■☆

326100 Ranunculus montanus Willd. var. ayachicus Emb. ＝ Ranunculus dyris （Maire） H. Lindb. ■☆

326101 Ranunculus montanus Willd. var. djurdjurae Chabert ＝ Ranunculus aurasiacus Pomel ■☆

326102 Ranunculus montanus Willd. var. mesatlanticus Maire ＝ Ranunculus aurasiacus Pomel ■☆

326103 Ranunculus morii （Yamam.） Ohwi；森氏毛茛（长柄毛茛）■

326104 Ranunculus multifidus Forssk. ；多裂毛茛；Rhenoster，Wild Buttercup ■☆

326105 Ranunculus multifidus Robyns ＝ Ranunculus volkensii Engl. ■

326106 Ranunculus munroanus J. R. Drumm. ex Dunn；荏弱毛茛（藏西毛茛）；Weak Buttercup ■

326107 Ranunculus munroanus J. R. Drumm. ex Dunn var. minor Tamura ＝ Ranunculus munroanus J. R. Drumm. ex Dunn ■

326108 Ranunculus muricatus L. ；刺果毛茛（粗糙毛茛，具刺毛茛，野元宵）；Rough-fruited Buttercup，Scilly Buttercup，Spinefruit

Buttercup ■

326109 Ranunculus muricatus L. var. graecus Heldr. et Sart. = Ranunculus muricatus L. ■

326110 Ranunculus muricatus L. var. pygmaea Pit. = Ranunculus muricatus L. ■

326111 Ranunculus muscigenus W. T. Wang;薛丛毛茛■

326112 Ranunculus nankotaizanus Ohwi;南湖毛茛■

326113 Ranunculus napellifolius DC. ;乌头叶毛茛■

326114 Ranunculus natans C. A. Mey. ;浮毛茛;Floating Buttercup ■

326115 Ranunculus natans C. A. Mey. var. intertextus (Greene) L. D. Benson = Ranunculus hyperboreus Rottb. ■☆

326116 Ranunculus neapolitanus Ten. ;那不勒斯毛茛■☆

326117 Ranunculus nematolobus Hand. -Mazz. ;丝叶毛茛(线裂毛茛);Filiformleaf Buttercup ■

326118 Ranunculus nemorosus DC. ;栎林毛茛(荫蔽毛茛);Wildwood Buttercup,Woods Buttercup ■☆

326119 Ranunculus neopolitanus Ten. ;新光毛茛■☆

326120 Ranunculus nephelogenes Edgew. ;云生毛茛;Cloudy Buttercup ■

326121 Ranunculus nephelogenes Edgew. var. geniculatus (Hand. -Mazz.) W. T. Wang;曲升毛茛;Geniculate Buttercup ■

326122 Ranunculus nephelogenes Edgew. var. longicaulis (Trautv.) W. T. Wang = Ranunculus alaschanicus Y. Z. Zhao ■

326123 Ranunculus nephelogenes Edgew. var. longicaulis (Trautv.) W. T. Wang = Ranunculus pulchellus C. A. Mey. var. longicaulis Trautv. ■

326124 Ranunculus nephelogenes Edgew. var. pseudohirculus (Trautv.) J. G. Liu = Ranunculus pseudohirculus (Trautv.) J. G. Liu ■☆

326125 Ranunculus nephelogenes Edgew. var. pubescens W. T. Wang = Ranunculus membranaceus Royle var. pubescens (W. T. Wang) W. T. Wang ■

326126 Ranunculus nidulans Pomel = Ranunculus paludosus Poir. ■☆

326127 Ranunculus nitidus Muhl. ex Elliott = Ranunculus hispidus Michx. var. nitidus (Chapm.) T. Duncan ■☆

326128 Ranunculus nivalis L. ;雪线毛茛;Snow Buttercup ■☆

326129 Ranunculus nivalis L. var. tianschanicus Rupr. = Ranunculus transiliensis Popov ■

326130 Ranunculus novae-forestae S. D. Webster;新林地毛茛;New Forest Crowfoot ■☆

326131 Ranunculus nyalamensis W. T. Wang;聂拉木毛茛;Nielamu Buttercup ■

326132 Ranunculus nyalamensis W. T. Wang var. angustipetalus W. T. Wang;浪卡子毛茛;Langkazi Buttercup ■

326133 Ranunculus obesus Trautv. ;肥胖毛茛■☆

326134 Ranunculus oblongifolius Elliot = Ranunculus pusillus Poir. ■☆

326135 Ranunculus occidentalis Nutt. ;西方毛茛■☆

326136 Ranunculus occidentalis Nutt. subsp. insularis Hultén = Ranunculus occidentalis Nutt. var. brevistylis Greene ■☆

326137 Ranunculus occidentalis Nutt. subsp. nelsonii (DC.) Hultén = Ranunculus occidentalis Nutt. var. nelsonii (DC.) L. D. Benson ■☆

326138 Ranunculus occidentalis Nutt. var. brevistylis Greene;短柱西方毛茛■☆

326139 Ranunculus occidentalis Nutt. var. dissectus L. F. Hend. ;深裂西方毛茛■☆

326140 Ranunculus occidentalis Nutt. var. eisenii (Kellogg) A. Gray = Ranunculus occidentalis Nutt. ■☆

326141 Ranunculus occidentalis Nutt. var. hexasepalus L. D. Benson;六瓣毛茛■☆

326142 Ranunculus occidentalis Nutt. var. howellii Greene;豪厄尔毛茛■☆

326143 Ranunculus occidentalis Nutt. var. nelsonii (DC.) L. D. Benson;纳尔逊毛茛■☆

326144 Ranunculus occidentalis Nutt. var. rattanii A. Gray = Ranunculus occidentalis Nutt. ■☆

326145 Ranunculus occidentalis Nutt. var. turneri (Greene) L. D. Benson = Ranunculus turneri Greene ■☆

326146 Ranunculus oligocarpos Hochst. ex A. Rich. ;寡果毛茛■☆

326147 Ranunculus omiophyllus Ten. ; 勒氏毛茛;Lenormand's Water Crowfoot,Round-leaved Crowfoot ■☆

326148 Ranunculus ophioglossifolius Vill. ; 舌花毛茛;Adder's Tongue Spearwort, Adder's-tongue Spearwort, Badgeworth Buttercup, Serpent's Tongue Spearwort, Snaketongue Crowfoot, Snake-tongue Crowfoot ■☆

326149 Ranunculus ophioglossifolius Vill. var. dentatus Rouy et Foucaud = Ranunculus ophioglossifolius Vill. ■☆

326150 Ranunculus ophioglossifolius Vill. var. laevis Chabert = Ranunculus ophioglossifolius Vill. ■☆

326151 Ranunculus ophioglossifolius Vill. var. rhoedifolius (DC.) Webb = Ranunculus ophioglossifolius Vill. ■☆

326152 Ranunculus oreionannos Marquand et Airy Shaw;花葶毛茛■

326153 Ranunculus oreogenes Greene = Ranunculus glaberrimus Hook. var. ellipticus (Greene) Greene ■☆

326154 Ranunculus oreophilus M. Bieb. ;山地毛茛■☆

326155 Ranunculus oreophytus Delile var. stolonifer Ulbr. ;匍匐毛茛■☆

326156 Ranunculus oresterus L. D. Benson;东方毛茛■☆

326157 Ranunculus orthorhynchus Hook. ;直喙毛茛■☆

326158 Ranunculus orthorhynchus Hook. subsp. alaschensis (L. D. Benson) Hultén = Ranunculus orthorhynchus Hook. ■☆

326159 Ranunculus orthorhynchus Hook. var. alaschensis L. D. Benson = Ranunculus orthorhynchus Hook. ■☆

326160 Ranunculus orthorhynchus Hook. var. bloomeri (S. Watson) L. D. Benson;布卢默毛茛■☆

326161 Ranunculus orthorhynchus Hook. var. hallii Jeps. = Ranunculus orthorhynchus Hook. ■☆

326162 Ranunculus orthorhynchus Hook. var. platyphyllus A. Gray;宽叶直喙毛茛■☆

326163 Ranunculus oryzetorum Bunge = Ranunculus sceleratus L. ■

326164 Ranunculus osseticus Ovcz. ;骨质毛茛■☆

326165 Ranunculus ovalis Raf. = Ranunculus rhomboideus Goldie ■☆

326166 Ranunculus oxyspermus M. Bieb. ;尖果毛茛■☆

326167 Ranunculus pacificus (Hultén) L. D. Benson;太平洋毛茛■☆

326168 Ranunculus paishanensis Kitag. ; 白山毛茛;Baishan, Paishan Japanese Buttercup ■

326169 Ranunculus paishanensis Kitag. f. oreodoxa (Kitag.) Kitag. = Ranunculus paishanensis Kitag. ■

326170 Ranunculus paishanensis Kitag. var. oreodoxa Kitag. = Ranunculus paishanensis Kitag. ■

326171 Ranunculus palifolius Dunn = Halerpestes lancifolia (Bertol.) Hand. -Mazz. ■

326172 Ranunculus pallasii Schltdl. ;帕拉氏毛茛;Pallas Buttercup ■☆

326173 Ranunculus palmatus Elliott = Ranunculus hispidus Michx. var. nitidus (Chapm.) T. Duncan ■☆

326174 Ranunculus paludosus Poir. ;沼生毛茛;Jersey Buttercup ■☆

326175 Ranunculus paludosus Poir. var. acinacilobus Freyn = Ranunculus paludosus Poir. ■☆

326176 Ranunculus paludosus Poir. var. acutifolius (Freyn) Emb. et Maire = Ranunculus paludosus Poir. ■☆

326177 Ranunculus paludosus Poir. var. amphicarpus Pamp. =

Ranunculus paludosus Poir. ■☆

326178　Ranunculus paludosus Poir. var. chondrodes（Pomel）Batt. = Ranunculus paludosus Poir. ■☆

326179　Ranunculus paludosus Poir. var. cinerascens（Freyn）Emb. et Maire ＝ Ranunculus paludosus Poir. ■☆

326180　Ranunculus paludosus Poir. var. confertus Freyn ＝ Ranunculus paludosus Poir. ■☆

326181　Ranunculus paludosus Poir. var. faurei Maire ＝ Ranunculus paludosus Poir. ■☆

326182　Ranunculus paludosus Poir. var. fibrosus（Pomel）Emb. et Maire ＝ Ranunculus paludosus Poir. ■☆

326183　Ranunculus paludosus Poir. var. flabellatus（Desf.）DC. ＝ Ranunculus paludosus Poir. ■☆

326184　Ranunculus paludosus Poir. var. flavescens（Freyn）Emb. et Maire ＝ Ranunculus paludosus Poir. ■☆

326185　Ranunculus paludosus Poir. var. granulatus（Pomel）Emb. et Maire ＝ Ranunculus paludosus Poir. ■☆

326186　Ranunculus paludosus Poir. var. leucothrix（Ball）Maire ＝ Ranunculus paludosus Poir. ■☆

326187　Ranunculus paludosus Poir. var. nervosus（Ducell. et Maire）Emb. et Maire ＝ Ranunculus paludosus Poir. ■☆

326188　Ranunculus paludosus Poir. var. nidulans（Pomel）Batt. ＝ Ranunculus paludosus Poir. ■☆

326189　Ranunculus paludosus Poir. var. ovatus（Freyn）Emb. et Maire ＝ Ranunculus paludosus Poir. ■☆

326190　Ranunculus paludosus Poir. var. pseudomonspeliacus（Maire）Emb. et Maire ＝ Ranunculus paludosus Poir. ■☆

326191　Ranunculus paludosus Poir. var. robustus（Pomel）Emb. et Maire ＝ Ranunculus paludosus Poir. ■☆

326192　Ranunculus paludosus Poir. var. subcinerascens Maire et Weiller ＝ Ranunculus paludosus Poir. ■☆

326193　Ranunculus pamiri Korsh. ；帕米尔毛茛■☆

326194　Ranunculus parviflorus L. ；小花毛茛；Smallflower Buttercup，Small-flowered Buttercup，Small-flowered Crowfoot ■☆

326195　Ranunculus parviflorus L. var. acutilobus DC. ＝ Ranunculus parviflorus L. ■☆

326196　Ranunculus parvulus L. ＝ Ranunculus sardous Crantz ■

326197　Ranunculus paucidentatus Schrenk；少齿毛茛■☆

326198　Ranunculus paucistamineus Tausch ＝ Batrachium eradicatum（Laest.）Fr. ■

326199　Ranunculus pauperculus Ovcz. ；贫乏毛茛■☆

326200　Ranunculus pectinatilobus W. T. Wang；栉节毛茛（栉裂毛茛）；Pectinatilobe Buttercup ■

326201　Ranunculus pedatifidus Sm. ；裂叶毛茛；Pedatifid Buttercup ■

326202　Ranunculus pedatifidus Sm. ＝ Ranunculus rigescens Turcz. ex Ost. -Sack. et Rupr. ■

326203　Ranunculus pedatifidus Sm. subsp. affinis（R. Br.）Hultén ＝ Ranunculus pedatifidus Sm. var. affinis（R. Br.）L. D. Benson ■☆

326204　Ranunculus pedatifidus Sm. var. affinis（R. Br.）L. D. Benson；近缘裂叶毛茛■☆

326205　Ranunculus pedatifidus Sm. var. cardiophyllus（Hook.）Britton ＝ Ranunculus cardiophyllus Hook. ■☆

326206　Ranunculus pedatus Waldst. et Kit. ；鸟足毛茛■☆

326207　Ranunculus pedicellatus Hand. -Mazz. ；长梗毛茛（具梗毛茛）；Longpedicel Buttercup ■

326208　Ranunculus pegaeus Hand. -Mazz. ；爬地毛茛（泉毛茛）；Prostrate Buttercup ■

326209　Ranunculus pekinensis（L. Liou）Luferov ＝ Batrachium pekinense L. Liou ■

326210　Ranunculus peltatus Schrank；盾状毛茛；Pond Water-crowfoot ■☆

326211　Ranunculus peltatus Schrank subsp. baudotii（Godr.）C. D. K. Cook；博多盾状毛茛■☆

326212　Ranunculus peltatus Schrank subsp. fucoides（Freyn）Munoz Garm. ＝ Ranunculus peltatus Schrank subsp. saniculifolius（Viv.）C. D. K. Cook ■☆

326213　Ranunculus peltatus Schrank subsp. saniculifolius（Viv.）C. D. K. Cook；变豆菜毛茛■☆

326214　Ranunculus peltatus Schrank subsp. sphaerospermus（Boiss. et Blanche）Meikle；球籽毛茛■☆

326215　Ranunculus peltatus Schrank var. rhiphaeus Pau et Font Quer ＝ Ranunculus peltatus Schrank ■☆

326216　Ranunculus penicillatus（Dumort.）Bab. ；帚状毛茛；Stream Water-crowfoot ■☆

326217　Ranunculus pensylvanicus L. f. ；宾州毛茛；Bristly Buttercup，Bristly Crowfoot，Pennsylvania Buttercup ■☆

326218　Ranunculus pensylvanicus L. f. ＝ Ranunculus chinensis Bunge ■

326219　Ranunculus pensylvanicus L. f. ＝ Ranunculus trigonus Hand. -Mazz. ■

326220　Ranunculus pensylvanicus L. f. var. chinensis Maxim. ＝ Ranunculus chinensis Bunge ■

326221　Ranunculus pensylvanicus L. f. var. sieboldii Ito ＝ Ranunculus sieboldii Miq. ■

326222　Ranunculus petiolaris Kunth ex DC. var. arsenei（L. D. Benson）T. Duncan ＝ Ranunculus fascicularis Muhl. ex J. M. Bigelow ■☆

326223　Ranunculus petrogeiton Ulbr. et Hand. -Mazz. ；太白山毛茛；Taibai Buttercup ■

326224　Ranunculus peucedanoides Desf. ＝ Ranunculus trichophyllus Chaix ■☆

326225　Ranunculus philonotis Ehrh. ＝ Ranunculus sardous Crantz ■

326226　Ranunculus philonotis Ehrh. var. intermedius（Poir.）Coss. ＝ Ranunculus sardous Crantz ■

326227　Ranunculus philonotis Ehrh. var. trilobus（Desf.）Loisel. ＝ Ranunculus sardous Crantz ■

326228　Ranunculus pimpinelloides D. Don ＝ Callianthemum pimpinlloides（D. Don）Hook. f. et Thomson ■

326229　Ranunculus pinnatisectus Popov；重齿毛茛■☆

326230　Ranunculus pinnatus Poir. ＝ Ranunculus multifidus Forssk. ■☆

326231　Ranunculus pinnatus Poir. var. extensus Hook. f. ＝ Ranunculus multifidus Forssk. ■☆

326232　Ranunculus pinnatus Poir. var. hermannii DC. ＝ Ranunculus multifidus Forssk. ■☆

326233　Ranunculus plantaginifolius Murray ＝ Halerpestes ruthenica（Jacq.）Ovcz. ■

326234　Ranunculus platensis Spreng. ；草原毛茛；Prairie Buttercup ■☆

326235　Ranunculus platypetalus（Hand. -Mazz.）Hand. -Mazz. ；大瓣毛茛（宽瓣毛茛）；Broadpetal Buttercup ■

326236　Ranunculus platypetalus（Hand. -Mazz.）Hand. -Mazz. var. macranthus W. T. Wang；硕花大瓣毛茛；Bigflower Broadpetal Buttercup ■

326237　Ranunculus platyspermus Fisch. ；宽翅毛茛（宽果毛茛）；Broadseed Buttercup ■

326238　Ranunculus plebeius DC. ＝ Ranunculus plebeius R. Br. ex DC. ■☆

326239　Ranunculus plebeius R. Br. ex DC. ；普通澳洲毛茛；Common Australian Buttercup ■☆

326240　Ranunculus podocarpus W. T. Wang；柄果毛茛；Stalkfruit Buttercup ■

326241　Ranunculus polii Franch. ex Forbes et Hemsl. ；肉根毛茛（上海毛茛）；Fleshyroot Buttercup ■

326242　Ranunculus polyanthemus L. ；多花毛茛；Manyflower Buttercup ■

326243　Ranunculus polypetalus Royle ＝ Oxygraphis endicheri（Walp. ）Bennet et S. Chandra ■

326244　Ranunculus polyphyllus Waldst. et Kit. ex Willd. ；多叶毛茛■☆

326245　Ranunculus polyrhizus Stephan ex Willd. ；多根毛茛（密根毛茛）；Manyroot Buttercup ■

326246　Ranunculus polyrhizus Stephan ex Willd. var. major Maxim. ＝ Ranunculus franchetii H. Boissieu ■

326247　Ranunculus polyrhzus Steph ＝ Ranunculus franchetii H. Boissieu ■

326248　Ranunculus popovii Ovcz. ；天山毛茛；Tianshan Buttercup ■

326249　Ranunculus popovii Ovcz. var. stracheyanum（Maxim. ）W. T. Wang；深齿毛茛■

326250　Ranunculus potaninii Kom. ；川滇毛茛（洱源毛茛）；Potanin Buttercup ■

326251　Ranunculus propinquus C. A. Mey. ＝ Ranunculus japonicus Thunb. var. propinquus（C. A. Mey. ）W. T. Wang ■

326252　Ranunculus propinquus C. A. Mey. ＝ Ranunculus japonicus Thunb. ■

326253　Ranunculus pseudobulbosus Schur；假鳞茎毛茛■☆

326254　Ranunculus pseudoflaccidus Petunn. ；柔叶毛茛■☆

326255　Ranunculus pseudohirculus（Trautv. ）J. G. Liu；沼泽毛茛■

326256　Ranunculus pseudolaetus Tamura ＝ Ranunculus distans Wall. et Royle ■

326257　Ranunculus pseudolobatus L. Liou；大金毛茛；Dajin Buttercup，False Lobate Buttercup ■

326258　Ranunculus pseudoparviflorus H. Lév. ＝ Ranunculus meyerianus Rupr. ■

326259　Ranunculus pseudopygmaeus Hand. -Mazz. ；拟矮毛茛（矮毛茛）；Pygmy Buttercup，Pygmy-like Buttercup ■

326260　Ranunculus pubescens Thunb. ＝ Ranunculus multifidus Forssk. ■☆

326261　Ranunculus pulchellus C. A. Mey. ；美丽毛茛；Beautiful Buttercup ■

326262　Ranunculus pulchellus C. A. Mey. var. geniculatus Hand. -Mazz. ＝ Ranunculus nephelogenes Edgew. var. geniculatus（Hand. -Mazz. ）W. T. Wang ■

326263　Ranunculus pulchellus C. A. Mey. var. longicaulis Trautv. ；长茎毛茛；Longstem Buttercup ■

326264　Ranunculus pulchellus C. A. Mey. var. membranaceus（Royle）Mukerjee ＝ Ranunculus membranaceus Royle ■

326265　Ranunculus pulchellus C. A. Mey. var. potaninii（Kom. ）Hand. -Mazz. ＝ Ranunculus potaninii Kom. ■

326266　Ranunculus pulchellus C. A. Mey. var. sericeus Hook. f. et Thomson ＝ Ranunculus membranaceus Royle ■

326267　Ranunculus pulchellus C. A. Mey. var. stracheyanus（Maxim. ）Hand. -Mazz. ＝ Ranunculus popovii Ovcz. var. stracheyanum（Maxim. ）W. T. Wang ■

326268　Ranunculus pulchellus C. A. Mey. var. yinshanicus Y. Z. Zhao ＝ Ranunculus yinshanicus（Y. Z. Zhao）Y. Z. Zhao ■

326269　Ranunculus purshii Richardson；珀什毛茛；Pursh's Crowfoot ■☆

326270　Ranunculus purshii Richardson ＝ Ranunculus gmelinii DC. ■

326271　Ranunculus pusillus Poir. ；侏儒毛茛；Low Spearwort ■☆

326272　Ranunculus pusillus Poir. var. angustifolius（Engelm. ex

Engelm. et A. Gray）L. D. Benson ＝ Ranunculus pusillus Poir. ■☆

326273　Ranunculus pygmaeus Wahlenb. ；矮毛茛（小毛茛）；Dwarf Buttercup，Pygmy Buttercup，Pyrenean Buttercup ■☆

326274　Ranunculus pygmaeus Wahlenb. subsp. sabinei（R. Br. ）Hultén ＝ Ranunculus sabinei R. Br. ■

326275　Ranunculus pygmaeus Wahlenb. var. langeana Nathorst ＝ Ranunculus pygmaeus Wahlenb. ■☆

326276　Ranunculus quelpaertensis Nakai；日本回回蒜（回回蒜）■

326277　Ranunculus raddeanus Regel；拉德毛茛■☆

326278　Ranunculus radicans C. A. Mey. ；沼地毛茛；Marshy Buttercup ■

326279　Ranunculus ranunculinus（Nutt. ）Rydb. ＝ Cyrtorrhyncha ranunculina Nutt. ■☆

326280　Ranunculus rectirostris Coss. et Durieu；阿尔及利亚毛茛■☆

326281　Ranunculus recurvatus Poir. ；钩毛茛；Blisterwort，Hooked Buttercup，Hooked Crowfoot ■☆

326282　Ranunculus recurvatus Poir. f. laevicaulis Weath. ＝ Ranunculus recurvatus Poir. ■☆

326283　Ranunculus recurvatus Poir. var. adpressipilis Weath. ＝ Ranunculus recurvatus Poir. ■☆

326284　Ranunculus recurvatus Poir. var. nelsonii DC. ＝ Ranunculus occidentalis Nutt. var. nelsonii（DC. ）L. D. Benson ■☆

326285　Ranunculus recurvatus Poir. var. typicus L. D. Benson ＝ Ranunculus recurvatus Poir. ■☆

326286　Ranunculus regelianus Ovcz. ；扁果毛茛；Flatfruit Buttercup ■

326287　Ranunculus repens L. ；匍枝毛茛（伏生毛茛，蔓枝毛茛）；Bachelor's Buttons，Bar Crowfoot，Bur Crowfoot，Butter Churn，Butter Daisy，Butter Flower，Butter Rose，Butter-and-cheese，Butterbumps，Buttercreese，Caltrop，Cat's Claws，Churn，Cowslip，Cranops，Crawfoot，Craw-taes，Crazy，Crazy Bet，Crazy Bets，Crazy Weed，Crazymar，Crazy-moir，Crazy-more，Creeping Buds，Creeping Buttercup，Creeping Crazy，Creeping Crowfoot，Crow Claws，Crow Toe，Crow Toes，Crowtoes，Cuckoo-buds，Dale-cup，Deity Cup，Delty-cup，Devil's Guts，Dew Cup，Dewcup，Dill-cup，Fairy Basin，Fairy's Basin，Gilcup，Gildcup，Gilty-cup，Glennies，Gold Cup，Gold Knob，Gold Knop，Goldweed，Goldy，Golland，Granny Threads，Granny-threads，Gye，Hod-the-rake，King Cob，Kingcup，King's Clover，King's Cob，Lantern-leaves，Lawyer-weed，Many-feet，Marybud，Maybuds，Meg Many-feet，Meg-many-feet，Old Man's Buttons，Old Wife's Threads，Paigle，Pickpocket，Ram's Claws，Raven's Claws，Sitfast，Sit-siccar，Sit-sicker，Soldier Buttons，Soldier's Buttons，Tangle-grass，Teacups，Tether Toad，Tether-toad，Toad Tether，Yellow Caul，Yellow Crees，Yellow Cup，Yellow Gollan ■

326288　Ranunculus repens L. 'Buttered Popcorn'；奶油色匍枝毛茛；Creeping Buttercup ■☆

326289　Ranunculus repens L. 'Flore-Pleno' ＝ Ranunculus repens L. 'Pleniflorus' ■

326290　Ranunculus repens L. 'Pleniflorus'；重瓣匍枝毛茛；Creeping Buttercup，Double Buttercup Daisy ■

326291　Ranunculus repens L. f. polypetalus S. H. Li ＝ Ranunculus repens L. ■

326292　Ranunculus repens L. var. brevistylus Maxim. ＝ Ranunculus repens L. ■

326293　Ranunculus repens L. var. erectus DC. ＝ Ranunculus repens L. ■

326294　Ranunculus repens L. var. glabratus DC. ＝ Ranunculus repens L. ■

326295　Ranunculus repens L. var. linearilobus DC. ＝ Ranunculus repens L. ■

326296　Ranunculus repens L. var. loponensis H. Lév. = Ranunculus ficariifolius H. Lév. et Vaniot ■

326297　Ranunculus repens L. var. major Nakai = Ranunculus repens L. ■

326298　Ranunculus repens L. var. nitidus Chapm. = Ranunculus hispidus Michx. var. nitidus (Chapm.) T. Duncan ■☆

326299　Ranunculus repens L. var. pleniflorus Fernald = Ranunculus repens L. ■

326300　Ranunculus repens L. var. typicus Beck = Ranunculus repens L. ■

326301　Ranunculus repens L. var. villosus Lamotte = Ranunculus repens L. ■

326302　Ranunculus reptabundus Rupr. ;爬行毛茛■☆

326303　Ranunculus reptans L. ;松叶毛茛（匍匐焰毛茛，匍枝毛茛，掌裂毛茛）; Creeping Crowfoot, Creeping Spearwort, Flam Buttercup, Palmatelobe Buttercup, Pineleaf Buttercup, Spearwort Buttercup ■

326304　Ranunculus reptans L. = Ranunculus flammula L. var. reptans (L.) Schltdl. ■☆

326305　Ranunculus reptans L. var. filiformis (Michx.) DC. = Ranunculus flammula L. var. reptans (L.) Schltdl. ■☆

326306　Ranunculus reptans L. var. flagellifolius (Nakai) Ohwi = Ranunculus reptans L. ■

326307　Ranunculus reptans L. var. intermedius (Hook.) Torr. et A. Gray = Ranunculus flammula L. var. reptans (L.) Schltdl. ■☆

326308　Ranunculus reptans L. var. ovalis (J. M. Bigelow) Torr. et A. Gray = Ranunculus flammula L. var. ovalis (J. M. Bigelow) L. D. Benson ■☆

326309　Ranunculus rhomboideus Goldie;菱形毛茛; Prairie Buttercup, Prairie Crowfoot ■☆

326310　Ranunculus rigescens Turcz. ex Ost. -Sack. et Rupr. ;掌裂毛茛;Palmtelobe Buttercup ■

326311　Ranunculus rigescens Turcz. ex Ost. -Sack. et Rupr. var. leiocarpus Kitag. = Ranunculus rigescens Turcz. ex Ost. -Sack. et Rupr. ■

326312　Ranunculus rionii Lagger = Batrachium rionii (Lagger) Nyman ■

326313　Ranunculus rivularis Banks et Sol. ex DC. ;溪畔毛茛;Waoriki ■☆

326314　Ranunculus robustus Pomel = Ranunculus paludosus Poir. ■☆

326315　Ranunculus rodiei Maire = Ranunculus peltatus Schrank subsp. saniculifolius (Viv.) C. D. K. Cook ■☆

326316　Ranunculus rodiei Maire var. illudens ? = Ranunculus peltatus Schrank subsp. saniculifolius (Viv.) C. D. K. Cook ■☆

326317　Ranunculus rubrocalyx Regel ex Kom. ;红萼毛茛; Redsepal Buttercup ■

326318　Ranunculus rufosepalus Franch. ;棕萼毛茛■

326319　Ranunculus rufosepalus Franch. var. parviflorus Kom. = Ranunculus rubrocalyx Regel ex Kom. ■

326320　Ranunculus ruthenicus Jacq. = Halerpestes ruthenica (Jacq.) Ovcz. ■

326321　Ranunculus ruthenicus Jacq. f. multidentatus S. H. Li et Y. H. Huang = Halerpestes ruthenica (Jacq.) Ovcz. ■

326322　Ranunculus sabinei R. Br. ;萨比娜毛茛■☆

326323　Ranunculus salsuginosus Pall. = Halerpestes sarmentosa (Adans.) Kom. et Aliss. ■

326324　Ranunculus samojedorum Rupr. ;涅涅茨基毛茛■☆

326325　Ranunculus saniculifolius Viv. = Ranunculus peltatus Schrank subsp. saniculifolius (Viv.) C. D. K. Cook ■☆

326326　Ranunculus sardous Crantz;欧毛茛（沙丁毛茛）;Buttercup, Hairy Buttercup, Pale-leaved Crowfoot ■

326327　Ranunculus sardous Crantz subsp. intermedius (Poir.) Jahand. et Maire = Ranunculus sardous Crantz ■

326328　Ranunculus sardous Crantz subsp. philonotis (Ehrh.) Briq. = Ranunculus sardous Crantz ■

326329　Ranunculus sardous Crantz subsp. trilobus (Desf.) Rouy et Foucaud = Ranunculus trilobus Desf. ■☆

326330　Ranunculus sardous Crantz subsp. xatartii (Lapeyr.) Rouy et Foucaud = Ranunculus sardous Crantz ■

326331　Ranunculus sardous Crantz var. cossoniana Maire = Ranunculus sardous Crantz ■

326332　Ranunculus sardous Crantz var. fontanesii Maire = Ranunculus sardous Crantz ■

326333　Ranunculus sardous Crantz var. macrocarpus Freyn = Ranunculus sardous Crantz ■

326334　Ranunculus sardous Crantz var. monanthos Finet et Gagnep. = Ranunculus sieboldii Miq. ■

326335　Ranunculus sardous Crantz var. rhoeadifolius (DC.) Webb = Ranunculus sardous Crantz ■

326336　Ranunculus sardous Crantz var. trilobus (Desf.) Burnat = Ranunculus sardous Crantz ■

326337　Ranunculus sardous Crantz var. tuberculatus Celak. = Ranunculus sardous Crantz ■

326338　Ranunculus sarmentosus Adams = Halerpestes sarmentosa (Adans.) Kom. et Aliss. ■

326339　Ranunculus sartorianus Boiss. et Heldr. ;萨尔毛茛■☆

326340　Ranunculus sceleratus L. ;石龙芮(打锣锤，地椹，鹊孙头草，鬼见愁，和尚菜，胡椒菜，胡椒草，黄花菜，黄爪草，回回蒜，鸡脚爬草，姜苔，堇菜，堇葵，苦堇，鲁果能，彭根，清香草，生堇，石龙芮毛茛，石能，水胡椒，水虎掌草，水黄瓜香，水姜苔，水堇，水毛茛，水芹菜，天豆，铜锤，无毛野芹菜，小木杨梅，小水杨梅，鸭巴掌，野堇菜，野芹菜，油灼灼，治寇草）; Banebind, Blister Buttercup, Blisterwort, Celery-leaf Buttercup, Celery-leaved Buttercup, Celery-leaved Crowfoot, Cloffing, Cursed Crowfoot, Devil's Buttercup, German-and-English, Hurtful Crowfoot, Loveache, Marsh Crowfoot, May Blob, May-blob, Poison Cup, Poisonous Buttercup, Water Blob, Water Celery ■

326341　Ranunculus sceleratus L. f. natans Glück = Ranunculus sceleratus L. ■

326342　Ranunculus sceleratus L. subsp. multifidus (Nutt.) Hultén = Ranunculus sceleratus L. var. multifidus Nutt. ■☆

326343　Ranunculus sceleratus L. var. globosus (Freyn) Batt. = Ranunculus sceleratus L. ■

326344　Ranunculus sceleratus L. var. multifidus Nutt. ;多裂石龙芮; Blister Buttercup, Cursed Crowfoot ■☆

326345　Ranunculus sceleratus L. var. sinensis H. Lév. = Ranunculus sceleratus L. ■

326346　Ranunculus sceleratus L. var. typicus L. D. Benson = Ranunculus sceleratus L. ■

326347　Ranunculus sceleratus L. var. umbellatus ? = Ranunculus sceleratus L. ■

326348　Ranunculus schaftoanus (Aitch. et Hemsl.) Boiss. ;沙夫毛茛■☆

326349　Ranunculus schimperianus Hochst. ex A. Rich. = Ranunculus simensis Fresen. ■☆

326350　Ranunculus seguieri Vill. ;塞氏毛茛;Sequier's Buttercup ■☆

326351　Ranunculus septentrionalis De Bray ex Fleisch. et Lindem. ;北方沼毛茛;Northern Crowfoot, Swamp Buttercup ■☆

326352　Ranunculus septentrionalis Poir. = Ranunculus hispidus Michx.

var. nitidus（Chapm.）T. Duncan ■☆

326353 Ranunculus septentrionalis Poir. = Ranunculus hispidus Michx. ■☆

326354 Ranunculus septentrionalis Poir. subsp. pacificus Hultén = Ranunculus pacificus（Hultén）L. D. Benson ■☆

326355 Ranunculus septentrionalis Poir. var. caricetorum（Greene）Fernald = Ranunculus hispidus Michx. var. caricetorum（Greene）T. Duncan ■☆

326356 Ranunculus septentrionalis Poir. var. caricetorum（Greene）Fernald = Ranunculus hispidus Michx. ■☆

326357 Ranunculus septentrionalis Poir. var. nitidus Chapm. = Ranunculus hispidus Michx. var. nitidus（Chapm.）T. Duncan ■☆

326358 Ranunculus septentrionalis Poir. var. pterocarpus L. D. Benson = Ranunculus hispidus Michx. var. nitidus（Chapm.）T. Duncan ■☆

326359 Ranunculus septentrionalis Poir. var. pterocarpus L. D. Benson = Ranunculus hispidus Michx. ■☆

326360 Ranunculus sericeus Banks et Sol. ;绢毛毛茛■☆

326361 Ranunculus sewerzovii Regel;塞沃毛茛■☆

326362 Ranunculus shuichengensis L. Liao;水城毛茛;Shuicheng Buttercup ■

326363 Ranunculus sibiricus Adams = Ranunculus monophyllus Ovcz. ■

326364 Ranunculus siciformis Mack. et Bush ex Rydb. = Ranunculus hispidus Michx. var. caricetorum（Greene）T. Duncan ■☆

326365 Ranunculus sieboldi Miq. ;扬子毛茛（大本水芹菜,地胡椒,鹅脚板,瞎睡果子,辣子草,毛茛,起泡草,水辣菜,西氏毛茛,新疆毛茛,鸭脚板草,野芹菜）;Cantonese Buttercup,Siebold Buttercup,Songaria Buttercup ■

326366 Ranunculus sieboldii Miq. var. arcuans（S. S. Chien）H. Hara = Ranunculus sieboldii Miq. ■

326367 Ranunculus silerifolius H. Lév. ;钩柱毛茛■

326368 Ranunculus silerifolius H. Lév. = Ranunculus cantoniensis DC. ■

326369 Ranunculus silerifolius H. Lév. var. dolicanthus L. Liao;长花毛茛■

326370 Ranunculus simensis Fresen. ;锡米毛茛■☆

326371 Ranunculus simensis Fresen. var. stagnalis（Hochst. ex A. Rich.）Oliv. = Ranunculus stagnalis Hochst. ex A. Rich. ■☆

326372 Ranunculus similis Hemsl. ;苞毛茛;Involucrate Buttercup ■

326373 Ranunculus similis Hemsl. = Ranunculus involucratus Maxim. ■

326374 Ranunculus sinovaginatus W. T. Wang;褐鞘毛茛;Brownsheath Buttercup ■

326375 Ranunculus smirnovii Ovcz. ;兴安毛茛（大叶毛茛）;Smirnov Japanese Buttercup ■

326376 Ranunculus sommieri Albov;索米毛茛■☆

326377 Ranunculus songaricus Schrenk;新疆毛茛（准噶尔毛茛）;Dzungar Buttercup ■

326378 Ranunculus songaricus Schrenk var. lasiopetalus Maxim. = Ranunculus songaricus Schrenk ■

326379 Ranunculus songaricus Schrenk var. partitus Rupr. = Ranunculus trautvetterianus Regel ex Ovcz. ■

326380 Ranunculus sphaerospermus Boiss. et Blanche = Ranunculus peltatus Schrank subsp. sphaerospermus（Boiss. et Blanche）Meikle ■☆

326381 Ranunculus sphaerospermus Boiss. et Blanche var. illudens Maire = Ranunculus peltatus Schrank subsp. sphaerospermus（Boiss. et Blanche）Meikle ■☆

326382 Ranunculus sphaerospermus Boiss. et Blanche var. rodieri（Maire）Maire = Ranunculus peltatus Schrank subsp. sphaerospermus（Boiss. et Blanche）Meikle ■☆

326383 Ranunculus spicatus Desf. ;长穗毛茛■☆

326384 Ranunculus spicatus Desf. subsp. blepharicarpos（Boiss.）Grau;脉果长穗毛茛■☆

326385 Ranunculus spicatus Desf. subsp. fontqueri Romo;丰特长穗毛茛■☆

326386 Ranunculus spicatus Desf. subsp. maroccanus（Coss.）Greuter et Burdet;摩洛哥长穗毛茛■☆

326387 Ranunculus spicatus Desf. subsp. rupestris（Guss.）Maire;岩生长穗毛茛■☆

326388 Ranunculus spicatus Desf. var. blepharicarpos（Boiss.）Ball = Ranunculus spicatus Desf. subsp. blepharicarpos（Boiss.）Grau ■☆

326389 Ranunculus spicatus Desf. var. cortusoides Maire = Ranunculus spicatus Desf. subsp. maroccanus（Coss.）Greuter et Burdet ■☆

326390 Ranunculus spicatus Desf. var. maroccanus（Coss.）Maire = Ranunculus spicatus Desf. subsp. maroccanus（Coss.）Greuter et Burdet ■☆

326391 Ranunculus spicatus Desf. var. transiens Emb. et Maire = Ranunculus spicatus Desf. ■☆

326392 Ranunculus stagnalis Hochst. ex A. Rich. ;喜沼毛茛■☆

326393 Ranunculus stenopetalus Ovcz. ;狭瓣毛茛■☆

326394 Ranunculus stenorhynchus Franch. ;宝兴毛茛■

326395 Ranunculus stenorhynchus Franch. = Ranunculus hirtellus Royle ■

326396 Ranunculus stevenii Andrz. = Ranunculus japonicus Thunb. var. propinquus（C. A. Mey.）W. T. Wang ■

326397 Ranunculus stevenii Andrz. ex Besser;司梯文氏毛茛;Steven's Buttercup ■☆

326398 Ranunculus striatus Hochst. ex A. Rich. = Ranunculus multifidus Forssk. ■☆

326399 Ranunculus strigillosus Boiss. et Huet;硬毛毛茛■☆

326400 Ranunculus subcordatus E. O. Beal = Ranunculus laxicaulis Darby ■☆

326401 Ranunculus subcorymbosus Kom. subsp. grandis（Honda）Tamura = Ranunculus grandis Honda ■

326402 Ranunculus subcorymbosus Kom. subsp. grandis（Honda）Tamura var. ovczimikovii Tamura = Ranunculus grandis Honda ■

326403 Ranunculus subcorymbosus Kom. var. grandis（Honda）Kitag. = Ranunculus grandis Honda ■

326404 Ranunculus subcorymbosus Kom. var. manshuricus（H. Hara）Kitag. = Ranunculus grandis Honda var. manshuricus Hara ■

326405 Ranunculus submarginatus Ovcz. ；棱边毛茛; Ribedge Buttercup,Submarginate Buttercup ■

326406 Ranunculus subrigidus W. B. Drew = Ranunculus aquatilis L. var. diffusus With. ■☆

326407 Ranunculus subsimilis Printz. = Halerpestes cymbalaria（Pursh）Green ■

326408 Ranunculus subtilis Trautv. ;纤细毛茛■☆

326409 Ranunculus suksdorfii A. Gray = Ranunculus eschscholtzii Schltdl. var. suksdorfii（A. Gray）L. D. Benson ■☆

326410 Ranunculus sulphureus Sol. = Ranunculus sulphureus Sol. ex Phipps ■☆

326411 Ranunculus sulphureus Sol. ex Phipps;硫黄毛茛■☆

326412 Ranunculus sulphureus Sol. ex Phipps var. albertii Maxim. = Ranunculus albertii Regel et Schmalh. ■

326413 Ranunculus sulphureus Sol. ex Phipps var. altaica Trautv. = Ranunculus altaicus Laxm. ■

326414 Ranunculus sulphureus Sol. ex Phipps var. intercedens Hultén = Ranunculus sulphureus Sol. ex Phipps ■☆

326415 Ranunculus sulphureus Sol. var. intercedens Hultén =

Ranunculus sulphureus Sol. ex Phipps ■☆

326416 Ranunculus suprasericeus（Hand.-Mazz.）L. Liou ＝Ranunculus dielsianus Ulbr. var. suprasericeus Hand.-Mazz. ■

326417 Ranunculus tachiroei Franch. et Sav. ;长嘴毛茛(长喙毛茛,长咀毛茛);Longbeak Buttercup ■

326418 Ranunculus tachiroei Franch. et Sav. var. tripartitus Ohwi ＝Ranunculus tachiroei Franch. et Sav. ■

326419 Ranunculus taisanensis Hayata;鹿场毛茛(台湾毛茛)■

326420 Ranunculus taiwanensis Hayata;台湾毛茛;Taiwan Buttercup ■

326421 Ranunculus taizanensis Yamam. ＝Ranunculus morii（Yamam.）Ohwi ■

326422 Ranunculus tanguticus（Maxim.）Ovcz. ;高原毛茛(结察,辣子草,唐古特毛茛);Birdfoot Buttercup,Pleteau Buttercup ■

326423 Ranunculus tanguticus（Maxim.）Ovcz. var. capillaceus（Franch.）L. Liou;丝叶高原毛茛(丝叶毛茛)■

326424 Ranunculus tanguticus（Maxim.）Ovcz. var. capillaceus（Franch.）L. Liou ＝Ranunculus nematolobus Hand.-Mazz. ■

326425 Ranunculus tanguticus（Maxim.）Ovcz. var. dasycarpus（Maxim.）L. Liou;毛果高原毛茛(毛果毛茛);Hairyfruit Buttercup ■

326426 Ranunculus tanguticus（Maxim.）Ovcz. var. xinglongshanicus Z. X. Peng et Y. J. Zhang;兴隆山毛茛;Xinglongshan Buttercup ■

326427 Ranunculus tembensis Fresen. ;滕博毛茛■☆

326428 Ranunculus tener C. Mohr ＝Ranunculus pusillus Poir. ■☆

326429 Ranunculus tenuilobus Regel ex Kom. ;细裂毛茛■☆

326430 Ranunculus tenuirostris Pomel ＝Ranunculus rectirostris Coss. et Durieu ■☆

326431 Ranunculus ternatus Thunb. ;猫爪草(茨栢,回回蒜,金花草,猫爪,三散草,小毛茛);Catclaw Buttercup ■

326432 Ranunculus ternatus Thunb. var. dissectissimus（Migo）Hand.-Mazz. ;细裂猫爪草;Thinlobed Catclaw Buttercup ■

326433 Ranunculus ternatus Thunb. var. hirsutus H. Boissieu ＝Ranunculus silerifolius H. Lév. ■

326434 Ranunculus testiculatus Crantz ＝Ceratocephala testiculata（Crantz）Roth ■

326435 Ranunculus testiculatus Crantz ＝Consolida pubescens（DC.）Soó ■☆

326436 Ranunculus testiculatus M. Bieb. ＝Ceratocephala testiculata（Crantz）Roth ■

326437 Ranunculus tetrandrus W. T. Wang;四蕊毛茛;Bur Buttercup,Curve-seed Butterwort,Fourstamen Buttercup ■

326438 Ranunculus texensis Engelm. ex Engelm. et A. Gray ＝Ranunculus laxicaulis Darby ■☆

326439 Ranunculus thora L. ;陶氏毛茛;Thora Buttercup,Thore's But Tercup,Thore's Buttercup ■☆

326440 Ranunculus trachycarpus Fisch. et C. A. Mey. ;疣果毛茛(糙果毛茛);Roughfruit Buttercup ■

326441 Ranunculus transiliensis Popov;截叶毛茛;Truncateleaf Buttercup ■

326442 Ranunculus trautvetterianus Regel ex Ovcz. ;毛托毛茛(毛茛);Hairreceptacle Buttercup,Trautvetter Buttercup ■

326443 Ranunculus triangularis W. T. Wang; 三角叶毛茛; Threeleves Buttercup ■

326444 Ranunculus trichocarpus Boiss . et Kotschy ex Boiss. ;毛果毛茛■☆

326445 Ranunculus trichophyllus（Chaix）Bosch ＝Ranunculus aquatilis L. var. diffusus With. ■☆

326446 Ranunculus trichophyllus（Chaix）Bosch subsp. lutulentus（Perrier et Songeon）Vierh. ＝Ranunculus aquatilis L. var. diffusus With. ■☆

326447 Ranunculus trichophyllus（Chaix）Bosch var. calvescens W. B.

Drew ＝Ranunculus aquatilis L. var. diffusus With. ■☆

326448 Ranunculus trichophyllus（Chaix）Bosch var. eradicatus（Laest.）W. B. Drew ＝Ranunculus aquatilis L. var. diffusus With. ■☆

326449 Ranunculus trichophyllus（Chaix）Bosch var. hispidulus（Drew）W. B. Drew ＝Ranunculus aquatilis L. ■☆

326450 Ranunculus trichophyllus Chaix;毛叶毛茛■☆

326451 Ranunculus trichophyllus Chaix ＝Ranunculus aquatilis L. var. diffusus With. ■☆

326452 Ranunculus trichophyllus Chaix ex Vill. ＝Batrachium trichophyllum（Chaix）Bosch ■

326453 Ranunculus trichophyllus Chaix ex Vill. subsp. rionii（Lagger）Soó ＝Batrachium rionii（Lagger）Nyman ■

326454 Ranunculus trichophyllus Chaix ex Vill. var. chanetii H. Lév. ＝Batrachium bungei（Steud.）L. Liou ■

326455 Ranunculus trichophyllus Chaix ex Vill. var. terrestris Gren. et Godr. ＝Batrachium eradicatum（Laest.）Fr. ■

326456 Ranunculus trichophyllus Chaix subsp. capillaceus（Thuill.）Maire ＝Ranunculus trichophyllus Chaix ■☆

326457 Ranunculus trichophyllus Chaix subsp. drouetii（Gren. et Godr.）Maire ＝Ranunculus trichophyllus Chaix var. drouetii（Gren. et Godr.）Batt. et Trab. ■☆

326458 Ranunculus trichophyllus Chaix subsp. lutulentus（Perrier et Songeon）Vierh. ＝Ranunculus aquatilis L. var. diffusus With. ■☆

326459 Ranunculus trichophyllus Chaix var. calvescens W. B. Drew ＝Ranunculus aquatilis L. var. diffusus With. ■☆

326460 Ranunculus trichophyllus Chaix var. chanetii H. Lév. ＝Batrachium bungei（Steud.）L. Liou ■

326461 Ranunculus trichophyllus Chaix var. drouetii（Gren. et Godr.）Batt. et Trab. ＝Ranunculus trichophyllus Chaix ■☆

326462 Ranunculus trichophyllus Chaix var. eradicatus（Laest.）W. B. Drew ＝Ranunculus aquatilis L. var. diffusus With. ■☆

326463 Ranunculus trichophyllus Chaix var. heterophyllus Freyn ＝Ranunculus trichophyllus Chaix ■☆

326464 Ranunculus trichophyllus Chaix var. peucedanoides（Desf.）Batt. et Trab. ＝Ranunculus trichophyllus Chaix ■☆

326465 Ranunculus trichophyllus Chaix var. trichophylloides（Hunm.）Hegi ＝Ranunculus trichophyllus Chaix ■☆

326466 Ranunculus trichophyllus Chaix var. tripartitus（DC.）Batt. et Trab. ＝Ranunculus trichophyllus Chaix ■☆

326467 Ranunculus trichophyllus Chaix var. typicus W. B. Drew ＝Ranunculus aquatilis L. var. diffusus With. ■☆

326468 Ranunculus tricuspis Maxim. ＝Halerpestes tricuspis（Maxim.）Hand.-Mazz. ■

326469 Ranunculus tricuspis Maxim. var. lancifolius（Bertol.）H. Hara ＝Halerpestes lancifolia（Bertol.）Hand.-Mazz. ■

326470 Ranunculus trigonus Hand.-Mazz. ;棱喙毛茛(三角毛茛,黄花虎掌草);Ribbeak Buttercup,Triangular Buttercup ■

326471 Ranunculus trigonus Hand.-Mazz. var. strigosus W. T. Wang;伏毛棱喙毛茛;Strigose Triangular Buttercup ■

326472 Ranunculus trilobus Desf. ;三裂毛茛;Threelobe Buttercup ■☆

326473 Ranunculus trilobus Desf. var. fontanesii Webb et Berthel. ＝Ranunculus trilobus Desf. ■☆

326474 Ranunculus trilobus Desf. var. rhoeadifolius（DC.）Webb et Berthel. ＝Ranunculus trilobus Desf. ■☆

326475 Ranunculus tripartitus DC. ;三裂水毛茛;Mud Crowfoot,Three-lobed Crowfoot,Three-lobed Water Crowfoot ■☆

326476 Ranunculus tripartitus DC. var. fluitans Gren. et Godr. ＝

Ranunculus hederaceus L. ■☆

326477　Ranunculus trisectilis Ovcz. ;三齿毛茛■☆

326478　Ranunculus trisectus Eastw. ex B. L. Rob. = Ranunculus eschscholtzii Schltdl. var. trisectus (B. L. Rob.) L. D. Benson ■☆

326479　Ranunculus trisepalus Gillies ex Hook. et Arn. = Ranunculus bonariensis Poir. var. trisepalus (Gillies ex Hook. et Arn.) Lourteig ■☆

326480　Ranunculus triternatus A. Gray;重三出毛茛■☆

326481　Ranunculus turkestanicus Franch. ;土耳其斯坦毛茛■☆

326482　Ranunculus turneri Greene;特纳毛茛■☆

326483　Ranunculus uncinatus D. Don;钩刺毛茛■☆

326484　Ranunculus uncinatus D. Don var. earlei (Greene) L. D. Benson = Ranunculus uncinatus D. Don ■☆

326485　Ranunculus uncinatus D. Don var. parviflorus (Torr.) L. D. Benson = Ranunculus uncinatus D. Don ■☆

326486　Ranunculus usneoides Greene = Ranunculus aquatilis L. var. diffusus With. ■☆

326487　Ranunculus ussuriensis Kom. ;乌苏里毛茛;Ussuri Buttercup ■☆

326488　Ranunculus ussuriensis Kom. = Ranunculus franchetii H. Boissieu ■

326489　Ranunculus vaginatus Hand. -Mazz. = Ranunculus sinovaginatus W. T. Wang ■

326490　Ranunculus vaniotii H. Lév. = Ranunculus ficariifolius H. Lév. et Vaniot ■

326491　Ranunculus verecundus B. L. Rob. ex Piper = Ranunculus gelidus Kar. et Kir. ■

326492　Ranunculus vernyi Franch. et Sav. ;光毛茛(毛茛)■

326493　Ranunculus vernyi Franch. et Sav. var. japonicus Nakai;日本光毛茛■

326494　Ranunculus villosus DC. ;长柔毛毛茛;Villose Buttercup ■

326495　Ranunculus volkensii Engl. ;沃尔毛茛■☆

326496　Ranunculus walteri Regel ex Freyn;瓦尔毛茛■☆

326497　Ranunculus wangianus Q. E. Yang;文采毛茛;Wang Buttercup ■

326498　Ranunculus xinningensis W. T. Wang;新宁毛茛;Xinning Buttercup ■

326499　Ranunculus yanshanensis W. T. Wang;砚山毛茛;Yanshan Buttercup ■

326500　Ranunculus yaoanus W. T. Wang;姚氏毛茛;Yao Buttercup ■

326501　Ranunculus yechengensis W. T. Wang;叶城毛茛;Yecheng Buttercup ■

326502　Ranunculus yinshanicus (Y. Z. Zhao) Y. Z. Zhao;阴山毛茛(阴山美丽毛茛);Yinshan Buttercup ■

326503　Ranunculus yunnanensis Franch. ;云南毛茛;Yunnan Buttercup ■

326504　Ranunculus zhungdianensis W. T. Wang;中甸毛茛;Zhongdian Buttercup ■

326505　Ranunculus zuccarinii Miq. = Ranunculus ternatus Thunb. ■

326506　Ranunculus zuccarinii Miq. var. dissectissimus Migo = Ranunculus ternatus Thunb. var. dissectissimus (Migo) Hand. -Mazz. ■

326507　Ranzania T. Ito(1888);草檗属(兰山草属)■☆

326508　Ranzania japonica (T. Ito ex Maxim.) T. Ito;草檗(兰山草,日本鬼臼)■☆

326509　Ranzaniaceae Takht. (1994);草檗科■☆

326510　Ranzaniaceae Takht. = Berberidaceae Juss. (保留科名)●■

326511　Ranzaniaceae Takht. = Rapateaceae Dumort. (保留科名)●☆

326512　Raoulia Hook. f. = Raoulia Hook. f. ex Raoul ■☆

326513　Raoulia Hook. f. ex Raoul (1846);鲜菊属(薜菊属);Mat Daisy , New Zealand Pincushion , Raoulia , Vegetable Sheep ■☆

326514　Raoulia australis Hook. f. ex Raoul;南方鲜菊;Raoulia ■☆

326515　Raoulia eximia Hook. f. ;超凡鲜菊;Vegetable Sheep ■☆

326516　Raoulia glabra Hook. f. ;光鲜菊■☆

326517　Raoulia haastii Hook. f. ;三脉鲜菊■☆

326518　Raoulia hookeri Allan;胡克鲜菊■☆

326519　Raoulia rubra Buchanan;红鲜菊■☆

326520　Raouliopsis S. F. Blake(1938);类薜菊属■☆

326521　Raouliopsis seifrizii S. F. Blake;类薜菊■☆

326522　Rapa Mill. = Brassica L. ■●

326523　Rapanea Aubl. (1775);密花树属(酸金牛属);Rapanea ●

326524　Rapanea Aubl. = Myrsine L. ●

326525　Rapanea affinis (A. DC.) Mez = Myrsine affinis A. DC. ●

326526　Rapanea aurea H. Lév. = Eurya aurea (H. Lév.) Hu et L. K. Ling ●

326527　Rapanea cicatricosa C. Y. Wu et C. Chen = Myrsine cicatricosa (C. Y. Wu et C. Chen) Pipoly et C. Chen ●

326528　Rapanea erythroxyloides (Thouars ex Roem. et Schult.) Mez;马岛密花树●☆

326529　Rapanea faberi Mez = Myrsine faberi (Mez) Pipoly ●

326530　Rapanea ferruginea (Ruiz et Pav.) Mez;锈色密花树●☆

326531　Rapanea gilliana (Sond.) Mez;吉尔密花树●☆

326532　Rapanea kwangsiensis E. Walker = Myrsine kwanfsiensis (E. Walker) Pipoly et C. Chen ●

326533　Rapanea kwangsiensis E. Walker var. lanceolata C. Y. Wu et C. Chen = Myrsine kwanfsiensis (E. Walker) Pipoly et C. Chen ●

326534　Rapanea linearis (Lour.) S. Moore = Myrsine linearis (Lour.) Poir. ●

326535　Rapanea maximowiczii Koidz. = Myrsine maximowiczii (Koidz.) E. Walker ●☆

326536　Rapanea maximowiczii Koidz. var. okabeana (Tuyama) T. Yamaz. = Myrsine maximowiczii (Koidz.) E. Walker ●☆

326537　Rapanea melanophleos (L.) Mez;海角密花树(密花树);Cape Beech ●☆

326538　Rapanea neriifolia (Siebold et Zucc.) Mez = Myrsine seguinii H. Lév. ●

326539　Rapanea neriifolia (Siebold et Zucc.) Mez var. yunnanensis (Mez) E. Walker = Myrsine seguinii H. Lév. ●

326540　Rapanea neriifolia (Siebold et Zucc.) Mez var. yunnanensis (Mez) E. Walker = Rapanea neriifolia (Siebold et Zucc.) Mez ●

326541　Rapanea neriifolia (Siebold et Zucc.) Mez var. yunnanensis Walker = Myrsine sequinii H. Lév. ●

326542　Rapanea neriifolia Mez = Myrsine seguinii H. Lév. ●

326543　Rapanea neriifolia Mez var. yunnanensis (Mez) E. Walker = Myrsine seguinii H. Lév. ●

326544　Rapanea okabeana Tuyama = Myrsine maximowiczii (Koidz.) E. Walker ●☆

326545　Rapanea playfairii (Hemsl.) Mez = Myrsine linearis (Lour.) Poir. ●

326546　Rapanea playfairii (Hemsl.) Mez = Myrsine seguinii H. Lév. ●

326547　Rapanea playfairii Mez = Myrsine linearis (Lour.) Poir. ●

326548　Rapanea stolonifera (Koidz.) Nakai = Myrsine stolonifera (Koidz.) E. Walker ●

326549　Rapanea stolonifera Nakai = Myrsine stolonifera (Koidz.) E. Walker ●

326550　Rapanea usambarensis Gilg et G. Schellenb. ;东非密花树●☆

326551　Rapanea verruculosa C. Y. Wu = Myrsine verruculosa (C. Y. Wu et C. Chen) Pipoly et C. Chen ●

326552　Rapanea verruculosa C. Y. Wu et C. Chen = Myrsine verruculosa (C. Y. Wu et C. Chen) Pipoly et C. Chen ●

326553　Rapanea walkeriana Hand. -Mazz. = Myrsine seguinii H. Lév. ●

326554　Rapanea yunnanensis Mez = Myrsine seguinii H. Lév. ●

326555　Rapanea yunnanensis Mez = Rapanea melanophleos (L.) Mez ●☆

326556　Raparia F. K. Mey. = Thlaspi L. ■

326557　Rapatea Aubl. (1775);偏穗草属(雷巴第属,瑞碑题雅属)■☆

326558　Rapatea angustifolia Spruce ex Körn.;窄叶偏穗草■☆

326559　Rapatea flava (Link) Kunth;黄偏穗草■☆

326560　Rapatea gracilis Seub. ;细偏穗草■☆

326561　Rapatea linearis Gleason;线形偏穗草■☆

326562　Rapatea longipes Spruce ex Körn.;长梗偏穗草■☆

326563　Rapatea membranacea Maguire;膜质偏穗草■☆

326564　Rapatea parviflora Nees ex Seub. ;小花偏穗草■☆

326565　Rapateaceae Dumort. (1829) (保留科名);偏穗草科(雷巴第科,瑞碑题雅科)■☆

326566　Raphanaceae Horan. = Brassicaceae Burnett(保留科名)■●

326567　Raphanaceae Horan. = Cruciferae Juss. (保留科名)■●

326568　Raphanaceae Horan. = Rapateaceae Dumort. (保留科名)■☆

326569　Raphanis Dod. ex Moench = Armoracia P. Gaertn. , B. Mey. et Scherb. (保留属名)■

326570　Raphanis Moench(废弃属名) = Armoracia P. Gaertn. , B. Mey. et Scherb. (保留属名)■

326571　Raphanistrocarpus (Baill.) Pax = Momordica L. ■

326572　Raphanistrocarpus Baill. = Momordica L. ■

326573　Raphanistrum Mill. = Raphanus L. ■

326574　Raphanocarpus Hook. f. = Momordica L. ■

326575　Raphanopsis Welw. = Oxygonum Burch. ex Campd. ●■☆

326576　Raphanorhyncha Rollins(1976);萝卜秧属■☆

326577　Raphanorhyncha crassa Rollins;萝卜秧■☆

326578　Raphanus L. (1753);萝卜属(莱菔属);Radish ■

326579　Raphanus acanthiformis J. M. Morel ex L. Sisley;滨莱菔■

326580　Raphanus acanthiformis J. M. Morel ex L. Sisley = Raphanus sativus L. ■

326581　Raphanus acanthiformis J. M. Morel ex L. Sisley = Raphanus sativus L. var. hortensis Backer ■

326582　Raphanus acanthiformis J. M. Morel ex L. Sisley f. raphanistroides (Makino) H. Hara = Raphanus sativus L. var. hortensis Backer f. raphanistroides Makino ■

326583　Raphanus acanthiformis J. M. Morel ex L. Sisley f. raphanistroides (Makino) H. Hara = Raphanus sativus L. ■

326584　Raphanus acanthiformis J. M. Morel ex L. Sisley var. giganthissimus Nakai;樱岛萝卜■

326585　Raphanus acanthiformis J. M. Morel ex L. Sisley var. horyo Nakai;和寮萝卜■

326586　Raphanus acanthiformis J. M. Morel ex L. Sisley var. miyashige Nakai;宫重萝卜■

326587　Raphanus acanthiformis J. M. Morel ex L. Sisley var. nerima Nakai;练马萝卜■

326588　Raphanus acanthiformis J. M. Morel ex L. Sisley var. sempervinus Nakai;四季萝卜■

326589　Raphanus acanthiformis J. M. Morel ex L. Sisley var. shogoin Nakai;圣护院萝卜■

326590　Raphanus caudatus L. = Raphanus sativus L. var. caudatus (L.) Hook. f. et T. Anderson ■☆

326591　Raphanus chanetii H. Lév. = Orychophragmus violaceus (L.) O. E. Schulz ■

326592　Raphanus chinensis (L.) Crantz = Brassica rapa L. var. chinensis (L.) Kitam. ■

326593　Raphanus chinensis Mill. = Raphanus sativus L. ■

326594　Raphanus courtoisii H. Lév. = Orychophragmus violaceus (L.) O. E. Schulz ■

326595　Raphanus junceus (L.) Crantz = Brassica juncea (L.) Czern. ■

326596　Raphanus laevigatus M. Bieb. = Goldbachia laevigata (M. Bieb.) DC. ■

326597　Raphanus landra Moretti ex DC. = Raphanus maritimus Sm. subsp. landra (Moretti ex DC.) Rivas Mart. ■☆

326598　Raphanus lyratus Forssk. = Enarthrocarpus lyratus (Forssk.) DC. ■☆

326599　Raphanus macropodus H. Lév. = Raphanus sativus L. ■

326600　Raphanus macropodus H. Lév. var. spontaneus Nakai = Raphanus sativus L. ■

326601　Raphanus maritimus Sm. ;海滨萝卜;Sea Radish ■☆

326602　Raphanus maritimus Sm. subsp. landra (Moretti ex DC.) Rivas Mart. ;巨根萝卜;Mediterranean Radish ■☆

326603　Raphanus monnetii H. Lév. = Chorispora tenella (Pall.) DC. ■

326604　Raphanus niger Mill. = Raphanus sativus L. ■

326605　Raphanus rapa (L.) Crantz. = Brassica rapa L. ■

326606　Raphanus raphanistroides (Makino) Nakai = Raphanus sativus L. var. raphanistroides (Makino) Makino ■

326607　Raphanus raphanistroides (Makino) Nakai = Raphanus sativus L. ■

326608　Raphanus raphanistroides Nakai = Raphanus sativus L. ■

326609　Raphanus raphanistrum L. ;野萝卜;Cadlock, Charlock, Cranops, Curlock, Field Wallflower, Jointed Charlock, Kedlock, Kellock, Ketlock, Radishweed, Runch, Skeldick, Skeldock, Skellock, Skillock, White Charlock, White Runch, Wild Mustard, Wild Radish ■

326610　Raphanus raphanistrum L. maritimus ? = Raphanus maritimus Sm. ■☆

326611　Raphanus raphanistrum L. subsp. landra (DC.) Bonnier et Layens = Raphanus maritimus Sm. subsp. landra (Moretti ex DC.) Rivas Mart. ■☆

326612　Raphanus raphanistrum L. var. sativus (L.) Beck = Raphanus sativus L. ■

326613　Raphanus raphanistrum L. var. sativus (L.) Domin = Raphanus sativus L. ■

326614　Raphanus rostratus DC. ;喙萝卜■☆

326615　Raphanus satiuus L. ' Longipinnatus ' = Raphanus sativus L. var. hortensis Backer ■

326616　Raphanus sativus L. ;萝卜(白萝卜,雹荽,菜头,楚菘,春莲花,大根,地灯笼,地枯萝,地骷髅,菲,红萝卜,红色大萝卜,胡萝卜,黄萝卜,枯萝卜,莱菔,老萝卜头,老人头,芦菔,芦葍,罗服,萝白,萝购,萝菖,满阳花,荠,荠根,秦菘,寿星头,葵,葵子,土酥,温菘,薏菜,仙人骨,仙人头,紫花菘,紫菘);Alman Radice, Cultivated Radish, Daikon, Garden Radish, Rabone, Radish, Rafort, Rape Radice, Rawbone, Reefort, Ryfart, Small Radish, Summer Radish, Wild Radish ■

326617　Raphanus sativus L. f. raphanistroides Makino = Raphanus sativus L. ■

326618　Raphanus sativus L. f. raphanistroides Makino = Raphanus sativus L. var. raphanistroides (Makino) Makino ■

326619　Raphanus sativus L. var. acanthiformis Nakai;菜头(萝卜)■

326620　Raphanus sativus L. var. alba DC. ;白萝卜;White Radish ■

326621　Raphanus sativus L. var. caudatus (L.) Hook. f. et T.

Anderson;尾萝卜;Rat-tailed Radish ■☆

326622　Raphanus sativus L. var. hortensis Backer;长羽裂萝卜(长羽叶萝卜);Chinese Radish, Daikon, Japanese Radish, Longpinnate Radish, Mooli, Mula, Mull ■

326623　Raphanus sativus L. var. hortensis Backer f. raphanistroides Makino = Raphanus sativus L. ■

326624　Raphanus sativus L. var. longipinnatus L. H. Bailey = Raphanus sativus L. var. hortensis Backer ■

326625　Raphanus sativus L. var. macropodus (H. Lév.) Makino = Raphanus sativus L. ■

326626　Raphanus sativus L. var. nigra Pers. ;黑萝卜■☆

326627　Raphanus sativus L. var. oleiferus Makino;茹菜■

326628　Raphanus sativus L. var. radicula DC. ;大菜■

326629　Raphanus sativus L. var. raphanistroides (Makino) Makino;蓝花子(滨莱菔,冬子菜,茹菜);Raphanistrum-like Radish ■

326630　Raphanus sativus L. var. raphanistroides (Makino) Makino = Raphanus sativus L. ■

326631　Raphanus sibiricus L. = Chorispora sibirica (L.) DC. ■

326632　Raphanus strictus Fisch. ex M. Bieb. = Diptychocarpus strictus (Fisch. ex M. Bieb.) Trautv. ■

326633　Raphanus taquetii H. Lév. = Raphanus sativus L. ■

326634　Raphanus tenellus Pall. = Chorispora tenella (Pall.) DC. ■

326635　Raphanus violaceus (L.) Crantz = Orychophragmus violaceus (L.) O. E. Schulz ■

326636　Raphanus violaceus L. = Orychophragmus violaceus (L.) O. E. Schulz ■

326637　Raphelingia Dumort. = Ornithogalum L. ■

326638　Raphia P. Beauv. (1806);酒椰属(酒椰子属,拉非椰子属,拉菲亚椰子属,拉菲棕属,罗非亚椰子属,棕竹属);Raffia, Raffia Palm, Raphia, Raphia Palm ●

326639　Raphia farinifera (Gaertn.) Hyl. ;罗非亚椰子(酒椰);Madagascar Raffia Palm, Madagascar Raffia-palm, Raffia, Raffia Palm, Raphia, Roffia Palm ●

326640　Raphia hookeri G. Mann et H. Wendl. ;胡克酒椰(胡克罗非亚椰子);African Piassava, Hooker Raffia Palm ●

326641　Raphia pedunculata P. Beauv. = Raphia ruffia Mart. ●

326642　Raphia ruffia (Jacq.) Mart. = Raphia farinifera (Gaertn.) Hyl. ●

326643　Raphia ruffia Mart. = Raphia farinifera (Gaertn.) Hyl. ●

326644　Raphia ruwenzorica Otedoh;鲁文佐里酒椰●☆

326645　Raphia sassandrensis A. Chev. = Raphia hookeri G. Mann et H. Wendl. ●

326646　Raphia sudanica A. Chev. ;苏丹酒椰●☆

326647　Raphia taedigera (Mart.) Mart. ;胶酒椰●☆

326648　Raphia textilis Welw. ;编织酒椰●☆

326649　Raphia vinifera P. Beauv. ;酒椰(酒罗非亚椰子,酒椰子,西非酒棕,竹棕);Bamboo Palm, Jupati Palm, Pharaoh's Date Palm, Raffia, Raffia Palm, Roffia Palm, Wine Palm ●

326650　Raphia vinifera P. Beauv. var. nigerica Otedoh;尼日利亚酒椰●☆

326651　Raphia welwitschii H. Wendl. = Raphia textilis Welw. ●☆

326652　Raphia wendlandi Becc. = Raphia vinifera P. Beauv. ●

326653　Raphiacme K. Schum. = Raphionacme Harv. ■☆

326654　Raphiacme angolensis K. Schum. = Raphionacme angolensis (K. Schum.) N. E. Br. ■☆

326655　Raphiacme globosa K. Schum. = Raphionacme globosa (K. Schum.) K. Schum. ■☆

326656　Raphiacme linearis K. Schum. = Raphionacme linearis (K.

Schum.) K. Schum. ■☆

326657　Raphiacme longifolia K. Schum. = Raphionacme longifolia (K. Schum.) N. E. Br. ■☆

326658　Raphidiocystis Hook. f. (1867);针囊葫芦属■☆

326659　Raphidiocystis brachypoda Baker;短足针囊葫芦■☆

326660　Raphidiocystis caillei Hutch. et Dalziel = Raphidiocystis chrysocoma (Schumach.) C. Jeffrey ■☆

326661　Raphidiocystis chrysocoma (Schumach.) C. Jeffrey;金针囊葫芦■☆

326662　Raphidiocystis jeffreyana R. Fern. et A. Fern. ;杰弗里针囊葫芦■☆

326663　Raphidiocystis mannii Hook. f. ;曼氏针囊葫芦■☆

326664　Raphidiocystis phyllocalyx C. Jeffrey et Rabenant. ;叶萼针囊葫芦■☆

326665　Raphidiocystis sakalavensis Baker = Raphidiocystis brachypoda Baker ■☆

326666　Raphidiocystis ugandensis Rolfe;乌干达针囊葫芦■☆

326667　Raphidiocystis welwitschii Hook. f. = Raphidiocystis chrysocoma (Schumach.) C. Jeffrey ■☆

326668　Raphidophora Hassk. = Rhaphidophora Hassk. ●■

326669　Raphidophyllum Hochst. (1841);针叶玄参属;Needle Palm ■☆

326670　Raphidophyllum Hochst. = Sopubia Buch. -Ham. ex D. Don ■

326671　Raphidophyllum ramosum Hochst. = Sopubia ramosa (Hochst.) Hochst. ■☆

326672　Raphidophyllum simplex Hochst. = Sopubia simplex (Hochst.) Hochst. ■☆

326673　Raphidospora Rchb. = Justicia L. ●■

326674　Raphidospora Rchb. = Rhaphidospora Nees ●■

326675　Raphinastrum Mill. = Raphanus L. ●■

326676　Raphiocarpus Chun = Didissandra C. B. Clarke(保留属名)●■

326677　Raphiocarpus Chun(1946);针果苣苔属(细蒴苣苔属)■☆

326678　Raphiocarpus sinicus Chun = Didissandra sinica (Chun) W. T. Wang ●■

326679　Raphiodon Benth. = Hyptis Jacq. (保留属名)●■

326680　Raphiodon Benth. = Rhaphiodon Schauer ■☆

326681　Raphiolepis Lindl. = Rhaphiolepis Lindl. (保留属名)●

326682　Raphiolepis cheniana F. P. Metcalf = Rhaphiolepis salicifolia Lindl. ●

326683　Raphiolepis delacouri André = Rhaphiolepis delacouri André ●☆

326684　Raphiolepis ferruginea F. P. Metcalf = Rhaphiolepis ferruginea F. P. Metcalf ●

326685　Raphiolepis ferruginea F. P. Metcalf var. serrata F. P. Metcalf = Rhaphiolepis ferruginea F. P. Metcalf var. serrata F. P. Metcalf ●

326686　Raphiolepis gracilis Nakai = Rhaphiolepis indica (L.) Lindl. ex Ker ●

326687　Raphiolepis hainanensis Metcalf = Rhaphiolepis lanceolata Hu ●

326688　Raphiolepis hiiranensis Kaneh. = Rhaphiolepis indica (L.) Lindl. ex Ker var. shilanensis Y. P. Yang et H. Y. Liu ●

326689　Raphiolepis impressivena Masam. = Rhaphiolepis impressivena Masam. ●

326690　Raphiolepis indica (L.) Lindl. = Rhaphiolepis indica (L.) Lindl. ex Ker ●

326691　Raphiolepis indica (L.) Lindl. ex Ker = Rhaphiolepis indica (L.) Lindl. ex Ker ●

326692　Raphiolepis indica (L.) Lindl. var. angustifolia Cardot = Rhaphiolepis lanceolata Hu ●

326693　Raphiolepis indica (L.) Lindl. var. grandifolia Franch. = Rhaphiolepis major Cardot ●

326694 Raphiolepis indica (L.) Lindl. var. hiiranensis (Kaneh.) H. L. Li = Rhaphiolepis indica (L.) Lindl. ex Ker ●

326695 Raphiolepis indica (L.) Lindl. var. tashiroi Hayata ex Matsum. et Hayata = Rhaphiolepis indica (L.) Lindl. ex Ker var. tashiroi Hayata ex Matsum. et Hayata ●

326696 Raphiolepis indica (L.) Lindl. var. umbellata (Thunb.) Ohashi = Rhaphiolepis umbellata (Thunb.) Makino ●

326697 Raphiolepis integerrima Hook. et Arn. = Rhaphiolepis indica (L.) Lindl. ex Ker var. umbellata (Thunb.) H. Ohashi ●

326698 Raphiolepis integerrima Hook. et Arn. = Rhaphiolepis integerrima Hook. et Arn. ●

326699 Raphiolepis integerrima Hook. et Arn. var. mertensii (Siebold et Zucc.) Makino ex Koidz. = Rhaphiolepis integerrima Hook. et Arn. ●

326700 Raphiolepis integerrima Hook. et Arn. var. mertensii (Siebold et Zucc.) Makino ex Koidz. = Rhaphiolepis indica (L.) Lindl. ex Ker var. umbellata (Thunb. ex Murray) H. Ohashi ●

326701 Raphiolepis japonica Siebold et Zucc. = Rhaphiolepis umbellata (Thunb.) Makino ●

326702 Raphiolepis japonica Siebold et Zucc. var. integerrima Hook. f. = Rhaphiolepis indica (L.) Lindl. ex Ker var. umbellata (Thunb.) H. Ohashi ●

326703 Raphiolepis kwangsiensis Hu = Rhaphiolepis salicifolia Lindl. ●

326704 Raphiolepis lanceolata Hu = Rhaphiolepis lanceolata Hu ●

326705 Raphiolepis liukiuensis Nakai = Rhaphiolepis liukiuensis Nakai ●

326706 Raphiolepis major Cardot = Rhaphiolepis indica (L.) Lindl. ex Ker ●

326707 Raphiolepis major Cardot = Rhaphiolepis major Cardot ●

326708 Raphiolepis mertensii Siebold et Zucc. = Rhaphiolepis integerrima Hook. et Arn. ●

326709 Raphiolepis parvibracteolata Mem = Rhaphiolepis indica (L.) Lindl. ex Ker ●

326710 Raphiolepis rubra (Lour.) Lindl. = Rhaphiolepis indica (L.) Lindl. ex Ker ●

326711 Raphiolepis rubra Lindl. = Rhaphiolepis indica (L.) Lindl. ex Ker ●

326712 Raphiolepis rugosa Nakai = Rhaphiolepis indica (L.) Lindl. ex Ker ●

326713 Raphiolepis salicifolia Lindl. = Rhaphiolepis salicifolia Lindl. ●

326714 Raphiolepis sinensis Roem. = Rhaphiolepis indica (L.) Lindl. ex Ker ●

326715 Raphiolepis umbellata (Thunb. ex Murray) Makino = Rhaphiolepis indica (L.) Lindl. ex Ker var. umbellata (Thunb.) H. Ohashi ●

326716 Raphiolepis umbellata (Thunb. ex Murray) Makino = Rhaphiolepis umbellata (Thunb.) Makino ●

326717 Raphiolepis umbellata (Thunb. ex Murray) Ohashi f. integerrima Rehder = Rhaphiolepis umbellata (Thunb.) Makino ●

326718 Raphiolepis umbellata (Thunb. ex Murray) Ohashi f. ovata (Briot) Schneid. = Rhaphiolepis umbellata (Thunb.) Makino ●

326719 Raphiolepis umbellata (Thunb. ex Murray) Ohashi var. integerrima (Hook. et Arn.) Masam. = Rhaphiolepis indica (L.) Lindl. ex Ker var. umbellata (Thunb.) H. Ohashi ●

326720 Raphiolepis umbellata (Thunb. ex Murray) Ohashi var. mertensii (Siebold et Zucc.) Makino = Rhaphiolepis integerrima Hook. et Arn. ●

326721 Raphiolepis umbellata (Thunb. ex Murray) Ohashi var. mertensii Makino = Rhaphiolepis integerrima Hook. et Arn. ●

326722 Raphionacme Harv. (1842) ;澳非萝藦属■☆

326723 Raphionacme abyssinica Chiov. = Schlechterella abyssinica (Chiov.) Venter et R. L. Verh. ■☆

326724 Raphionacme angolensis (K. Schum.) N. E. Br. ;安哥拉澳非萝藦■☆

326725 Raphionacme arabica A. G. Mill. et Biagi;阿拉伯澳非萝藦■☆

326726 Raphionacme bagshawei S. Moore;巴格肖澳非萝藦■☆

326727 Raphionacme baguirmiensis A. Chev. = Brachystelma mortonii Walker ■☆

326728 Raphionacme bingeri (A. Chev.) Lebrun et Stork;宾格澳非萝藦■☆

326729 Raphionacme borenensis Venter et M. G. Gilbert;北方澳非萝藦■☆

326730 Raphionacme brownii Scott-Elliot;布朗澳非萝藦■☆

326731 Raphionacme brownii Scott-Elliot var. longifolia A. Chev. = Raphionacme brownii Scott-Elliot ■☆

326732 Raphionacme burkei N. E. Br. = Raphionacme velutina Schltr. ■☆

326733 Raphionacme caerulea E. A. Bruce = Pentagonanthus caeruleus (E. A. Bruce) Bullock ■☆

326734 Raphionacme chimanimaniana Venter et R. L. Verh. ;奇马尼曼澳非萝藦■☆

326735 Raphionacme daronii Berhaut = Raphionacme bingeri (A. Chev.) Lebrun et Stork ■☆

326736 Raphionacme decolor Schltr. ;褪色澳非萝藦■☆

326737 Raphionacme denticulata N. E. Br. ;细齿澳非萝藦■☆

326738 Raphionacme dinteri Schltr. ex Schinz = Raphionacme velutina Schltr. ■☆

326739 Raphionacme divaricata Harv. = Raphionacme hirsuta (E. Mey.) R. A. Dyer ■☆

326740 Raphionacme divaricata Harv. var. glabra N. E. Br. = Raphionacme hirsuta (E. Mey.) R. A. Dyer ■☆

326741 Raphionacme dyeri Retief et Venter;戴尔澳非萝藦■☆

326742 Raphionacme elata N. E. Br. = Raphionacme galpinii Schltr. ■☆

326743 Raphionacme engleriana Schltr. ex Dinter;恩格尔澳非萝藦■☆

326744 Raphionacme excisa Schltr. = Raphionacme brownii Scott-Elliot ■☆

326745 Raphionacme flanaganii Schltr. ;弗拉纳根澳非萝藦■☆

326746 Raphionacme galpinii Schltr. ;盖尔澳非萝藦■☆

326747 Raphionacme globosa (K. Schum.) K. Schum. ;球形澳非萝藦■☆

326748 Raphionacme gossweileri S. Moore;戈斯澳非萝藦■☆

326749 Raphionacme grandiflora N. E. Br. = Pentagonanthus grandiflorus (N. E. Br.) Bullock ■☆

326750 Raphionacme grandiflora N. E. Br. subsp. glabrescens Bullock = Pentagonanthus grandiflorus (N. E. Br.) Bullock subsp. glabrescens (Bullock) Bullock ■☆

326751 Raphionacme hirsuta (E. Mey.) R. A. Dyer;粗毛澳非萝藦■☆

326752 Raphionacme hirsuta (E. Mey.) R. A. Dyer var. glabra (N. E. Br.) R. A. Dyer = Raphionacme hirsuta (E. Mey.) R. A. Dyer ■☆

326753 Raphionacme inconspicua H. Huber;显著澳非萝藦■☆

326754 Raphionacme jurensis N. E. Br. = Raphionacme brownii Scott-Elliot ■☆

326755 Raphionacme keayii Bullock;凯伊澳非萝藦■☆

326756 Raphionacme kubangensis S. Moore;古邦澳非萝藦■☆

326757 Raphionacme lanceolata Schinz;披针形澳非萝藦■☆

326758 Raphionacme lanceolata Schinz var. latifolia N. E. Br. = Raphionacme lanceolata Schinz ■☆

326759 Raphionacme linearis (K. Schum.) K. Schum. ;线状澳非萝藦■☆

326760 Raphionacme linearis (K. Schum.) K. Schum. var. glabra K. Schum. = Raphionacme linearis (K. Schum.) K. Schum. ■☆

326761 Raphionacme loandae Schltr. et Rendle;罗安达澳非萝藦■☆

326762 Raphionacme lobulata Venter et R. L. Verh.;小裂片澳非萝藦■☆

326763 Raphionacme longifolia (K. Schum.) N. E. Br.;长叶澳非萝藦■☆

326764 Raphionacme longituba E. A. Bruce;长管澳非萝藦■☆

326765 Raphionacme lucens Venter et R. L. Verh.;光亮澳非萝藦■☆

326766 Raphionacme macrorrhiza Schltr. = Raphionacme galpinii Schltr. ■☆

326767 Raphionacme madiensis S. Moore;马迪澳非萝藦■☆

326768 Raphionacme michelii De Wild.;米歇尔澳非萝藦■☆

326769 Raphionacme monteiroae (Oliv.) N. E. Br.;蒙泰鲁澳非萝藦■☆

326770 Raphionacme moyalicus Venter et R. L. Verh.;莫亚莱澳非萝藦■☆

326771 Raphionacme namibiana Venter et R. L. Verh.;纳米比亚澳非萝藦■☆

326772 Raphionacme obovata Turcz. = Raphionacme hirsuta (E. Mey.) R. A. Dyer ■☆

326773 Raphionacme pachyodon K. Schum. ex Schinz;粗齿澳非萝藦■☆

326774 Raphionacme palustris Venter et R. L. Verh.;沼泽澳非萝藦■☆

326775 Raphionacme procumbens Schltr.;平铺澳非萝藦■☆

326776 Raphionacme pubescens (Hochst.) Hochst. = Raphionacme hirsuta (E. Mey.) R. A. Dyer ■☆

326777 Raphionacme pulchella Venter et R. L. Verh.;美丽澳非萝藦■☆

326778 Raphionacme purpurea Harv. = Raphionacme hirsuta (E. Mey.) R. A. Dyer ■☆

326779 Raphionacme scandens N. E. Br. = Raphionacme flanaganii Schltr. ■☆

326780 Raphionacme splendens Schltr.;亮澳非萝藦■☆

326781 Raphionacme sudanica A. Chev. ex Hutch. et Dalziel = Pentagonanthus caeruleus (E. A. Bruce) Bullock ■☆

326782 Raphionacme sylvicola Venter et R. L. Verh.;森林澳非萝藦■☆

326783 Raphionacme utilis N. E. Br. et Stapf;有用澳非萝藦;Bitinga Rubber ■☆

326784 Raphionacme velutina Schltr.;短绒毛澳非萝藦■☆

326785 Raphionacme verdickii De Wild.;韦尔迪克澳非萝藦■☆

326786 Raphionacme vignei Bruce;维涅澳非萝藦■☆

326787 Raphionacme virgultorum S. Moore;条纹澳非萝藦■☆

326788 Raphionacme volubilis Schltr. = Buckollia volubilis (Schltr.) Venter et R. L. Verh. ●☆

326789 Raphionacme welwitschii Schltr. et Rendle;韦尔澳非萝藦■☆

326790 Raphionacme wilczekiana R. Germ.;维尔切克澳非萝藦■☆

326791 Raphionacme zeyheri Harv.;泽赫澳非萝藦■☆

326792 Raphione Salisb. = Allium L. ■

326793 Raphiostyles Benth. et Hook. f. = Rhaphiostylis Planch. ex Benth. ●☆

326794 Raphis P. Beauv. = Chrysopogon Trin.(保留属名)■

326795 Raphis P. Beauv. = Rhaphis Lour.(废弃属名)■

326796 Raphisanthe Lilja = Blumenbachia Schrad.(保留属名)■☆

326797 Raphisanthe Lilja = Caiophora C. Presl ■☆

326798 Raphistemma Wall.(1831);大花藤属;Raphistemma ●

326799 Raphistemma brevipedunculatum Y. Wan = Raphistemma hooperianum (Blume) Decne. ●

326800 Raphistemma hooperianum (Blume) Decne.;广西大花藤;Guangxi Raphistemma ●

326801 Raphistemma pulchellum (Roxb.) Wall.;大花藤;Common Raphistemma ●

326802 Raphithamnus Dalli Torre et Harms = Rhaphithamnus Miers ●☆

326803 Rapicactus Buxb. et Oehme = Turbinicarpus (Backeb.) Buxb.

326804 et Backeb. ■☆

326804 Rapinia Lour. = Sphenoclea Gaertn.(保留属名)■

326805 Rapinia Montrouz. = Neorapinia Moldenke ●☆

326806 Rapinia herbacea Lour. = Sphenoclea zeylanica Gaertn. ■

326807 Rapis L. f. ex Aiton = Rhapis L. f. ex Aiton ●

326808 Rapistrella Pomel = Rapistrum Crantz(保留属名)■☆

326809 Rapistrella Pomel(1860);小匕果芥属■☆

326810 Rapistrella ramosissima Pomel;小匕果芥■☆

326811 Rapistrum Bergeret = Calepina Adans. ■☆

326812 Rapistrum Crantz(1769)(保留属名);匕果芥属(小萝卜属);Cabbage,Rapistrum ■☆

326813 Rapistrum Fabr. = Neslia Desv.(保留属名)■

326814 Rapistrum Haller f. = Neslia Desv.(保留属名)■

326815 Rapistrum Medik. = Crambe L. ■

326816 Rapistrum Mill. = Sinapis L. ■

326817 Rapistrum R. Br. = Ochthodium DC. ■☆

326818 Rapistrum Scop.(废弃属名) = Neslia Desv.(保留属名)■

326819 Rapistrum Scop.(废弃属名) = Rapistrum Crantz(保留属名)■☆

326820 Rapistrum Tourn. ex Medik. = Crambe L. ■

326821 Rapistrum aegyptium (L.) Coss. = Didesmus aegyptius (L.) Desv. ■☆

326822 Rapistrum bipinnatum (Desf.) Coss. et Kralik = Didesmus bipinnatus (Desf.) DC. ■☆

326823 Rapistrum confusum Pomel = Rapistrum rugosum (L.) All. subsp. orientale (L.) Arcang. ■☆

326824 Rapistrum conoideum Pomel = Rapistrum rugosum (L.) All. ■☆

326825 Rapistrum cordylocarpum (Coss. et Durieu) Pomel = Kremeriella cordylocarpa (Coss. et Durieu) Maire ■☆

326826 Rapistrum hispanicum (L.) Crantz = Rapistrum rugosum (L.) All. subsp. linnaeanum (Boiss. et Reut.) Rouy et Foucaud ■☆

326827 Rapistrum hispanicum (L.) Crantz = Rapistrum rugosum (L.) All. ■☆

326828 Rapistrum linnaeanum Boiss. et Reut. = Rapistrum rugosum (L.) All. subsp. linnaeanum (Boiss. et Reut.) Rouy et Foucaud ■☆

326829 Rapistrum microcarpum Jord. ex Loret = Rapistrum rugosum (L.) All. ■☆

326830 Rapistrum orientale (L.) DC. = Rapistrum rugosum (L.) All. subsp. orientale (L.) Arcang. ■☆

326831 Rapistrum orientale (L.) DC. = Rapistrum rugosum (L.) All. ■☆

326832 Rapistrum pereene L.;西伯利亚匕果芥(西伯利亚小萝卜);Steppe Cabbage ■☆

326833 Rapistrum rugosum (L.) All.;匕果芥(小萝卜);Annual Bastard Cabbage,Annual Bastard-cabbage,Bastard Cabbage,Oriental Bastard-cabbage, Rugose Rapistrum, Turnip Weed, Turnip-weed, Wild Rape,Wild Turnip ■☆

326834 Rapistrum rugosum (L.) All. subsp. linnaeanum (Boiss. et Reut.) Rouy et Foucaud;林奈匕果芥(林奈小萝卜)■☆

326835 Rapistrum rugosum (L.) All. subsp. orientale (L.) Arcang.; 东方匕果芥(东方小萝卜);Annual Bastard-cabbage, Oriental Bastard-cabbage,Wild Rape ■☆

326836 Rapistrum rugosum (L.) All. subsp. orientale (L.) Arcang. = Rapistrum rugosum (L.) All. ■☆

326837 Rapistrum rugosum (L.) All. var. conoideum (Pomel) Maire et Weiller = Rapistrum rugosum (L.) All. ■☆

326838 Rapistrum rugosum (L.) All. var. glabrum Cariot = Rapistrum rugosum (L.) All. ■☆

326839 Rapistrum rugosum (L.) All. var. linnaeanum (Boiss. et

Reut.）Coss. = Rapistrum rugosum（L.）All. subsp. linnaeanum（Boiss. et Reut.）Rouy et Foucaud ■☆

326840 Rapistrum rugosum（L.）All. var. linnaeanum（Boiss. et Reut.）Coss. = Rapistrum rugosum（L.）All. ■☆

326841 Rapistrum rugosum（L.）All. var. microcarpum（Loret）Rouy et Foucaud = Rapistrum rugosum（L.）All. ■☆

326842 Rapistrum rugosum（L.）All. var. microcarpum Thell.；小果匕果芥(小果小萝卜)■☆

326843 Rapistrum rugosum（L.）All. var. orientale（L.）Coss. = Rapistrum rugosum（L.）All. ■☆

326844 Rapistrum rugosum（L.）All. var. scabrum（Rouy et Foucaud）Pit. et Proust = Rapistrum rugosum（L.）All. ■☆

326845 Rapistrum rugosum（L.）All. var. strictissimum（Pomel）Batt. = Rapistrum rugosum（L.）All. ■☆

326846 Rapistrum rugosum（L.）All. var. venosum（Pers.）DC. = Rapistrum rugosum（L.）All. ■☆

326847 Rapistrum rugosum Berger var. scabrum Rouy et Foucaud = Rapistrum rugosum（L.）All. ■☆

326848 Rapistrum strictissimum Pomel = Rapistrum rugosum（L.）All. ■☆

326849 Rapolocarpus Bojer = Ropalocarpus Bojer ●☆

326850 Rapona Baill.（1890）；椴叶旋花属●☆

326851 Rapona madagascariensis Baill. = Rapona tiliifolia（Baker）Verdc. ●☆

326852 Rapona tiliifolia（Baker）Verdc.；椴叶旋花●☆

326853 Rapourea Rchb. = Diospyros L. ●

326854 Rapourea Rchb. = Ropourea Aubl. ●

326855 Raptostylus T. Post et Kuntze = Heisteria Jacq.（保留属名）●☆

326856 Raptostylus T. Post et Kuntze = Rhaptostylum Humb. et Bonpl. ●☆

326857 Rapum Hill = Brassica L. ■●

326858 Rapum Hill = Rapa Mill. ■●

326859 Rapunculus Fourr. = Campanula L. ■●

326860 Rapunculus Mill. = Phyteuma L. ■☆

326861 Rapunculus Tourn. ex Mill. = Phyteuma L. ■☆

326862 Rapuntia Chevall. = Campanula L. ■●

326863 Rapuntium Mill. = Lobelia L. ●■

326864 Rapuntium Post et Ktmtze = Campanula L. ■●

326865 Rapuntium Post et Kuntze = Rapuntia Chevall. ■

326866 Rapuntium Tourn. ex Mill. = Lobelia L. ●■

326867 Rapuntium acutangulum C. Presl = Lobelia flaccida（C. Presl）A. DC. ■☆

326868 Rapuntium affine C. Presl = Lobelia zeylanica L. ■

326869 Rapuntium angulatum（G. Forst.）C. Presl = Lobelia angulata G. Forst. ■☆

326870 Rapuntium arabideum C. Presl = Wimmerella arabidea（C. Presl）Serra, M. B. Crespo et Lammers ■☆

326871 Rapuntium caespitosum C. Presl = Lobelia chinensis Lour. ■

326872 Rapuntium campanuloides C. Presl = Lobelia chinensis Lour. ■

326873 Rapuntium capillifolium C. Presl = Lobelia capillifolia（C. Presl）A. DC. ■☆

326874 Rapuntium ceratophyllum C. Presl = Lobelia chamaepitys Lam. var. ceratophylla（C. Presl）E. Wimm ■☆

326875 Rapuntium chamaedryfolium C. Presl = Lobelia chamaedryfolia（C. Presl）A. DC. ■☆

326876 Rapuntium chinense C. Presl = Lobelia chinensis Lour. ■

326877 Rapuntium coloratura C. Presl = Lobelia colorata Wall. ■

326878 Rapuntium dregeanum C. Presl = Lobelia dregeana（C. Presl）A. DC. ■☆

326879 Rapuntium ecklonianum C. Presl = Lobelia eckloniana（C. Presl）A. DC. ■☆

326880 Rapuntium ericoides C. Presl = Monopsis lutea（L.）Urb. ■☆

326881 Rapuntium euphrasioides C. Presl = Monopsis lutea（L.）Urb. ■☆

326882 Rapuntium flaccidum C. Presl = Lobelia flaccida（C. Presl）A. DC. ■☆

326883 Rapuntium flavum C. Presl = Monopsis flava（C. Presl）E. Wimm. ■☆

326884 Rapuntium genistoides C. Presl = Lobelia patula L. f. ■☆

326885 Rapuntium incisum C. Presl = Lobelia pubescens Aiton var. incisa（C. Presl）Sond. ■☆

326886 Rapuntium kamtschaticum C. Presl = Lobelia sessilifolia Lamb. ■

326887 Rapuntium lasianthum C. Presl = Lobelia linearis Thunb. ■☆

326888 Rapuntium linarioides C. Presl = Lobelia flaccida（C. Presl）A. DC. ■☆

326889 Rapuntium maculare C. Presl = Lobelia tomentosa L. f. ■☆

326890 Rapuntium maculare C. Presl var. procerum？ = Lobelia tomentosa L. f. ■☆

326891 Rapuntium microdon C. Presl = Lobelia comosa L. var. microdon（C. Presl）E. Wimm. ■☆

326892 Rapuntium nummularium（Lam.）C. Presl = Lobelia angulata G. Forst. ■☆

326893 Rapuntium nummularium C. Presl = Pratia nummularia（Lam.）A. Br. et Asch. ■

326894 Rapuntium ottonianum C. Presl = Lobelia anceps L. f. ■☆

326895 Rapuntium ovatum C. Presl var. hirsutum？ = Lobelia cuneifolia Link et Otto var. hirsuta（C. Presl）E. Wimm. ■☆

326896 Rapuntium procumbens C. Presl = Lobelia erinus L. ■

326897 Rapuntium pteropodum C. Presl = Lobelia pteropoda（C. Presl）A. DC. ■☆

326898 Rapuntium pyramidale C. Presl = Lobelia pyramidalis Wall. ■

326899 Rapuntium radicans C. Presl = Lobelia chinensis Lour. ■

326900 Rapuntium reinwardtiamum C. Presl = Lobelia zeylanica L. ■

326901 Rapuntium scabripes C. Presl = Lobelia flaccida（C. Presl）A. DC. ■☆

326902 Rapuntium spartioides C. Presl = Lobelia linearis Thunb. ■☆

326903 Rapuntium succulentum C. Presl = Lobelia zeylanica L. ■

326904 Rapuntium wallichianum C. Presl = Lobelia pyramidalis Wall. ■

326905 Rapuntium zeylanicum C. Presl = Lobelia zeylanica L. ■

326906 Raputia Aubl.（1775）；拉普芸香属●☆

326907 Raputia alba（Nees et Mart.）Engl.；拉普芸香●☆

326908 Raputia alba Engl. = Raputia alba（Nees et Mart.）Engl. ●☆

326909 Raputia amazonica（Huber）Kallunki；亚马孙拉普芸香●☆

326910 Raputia aromatica Aubl. = Raputia amazonica（Huber）Kallunki ●☆

326911 Raputia heptaphylla Pittier；异叶拉普芸香●☆

326912 Raputia heterophylla DC. = Tabebuia heterophylla（DC.）Britton ●☆

326913 Raputiarana Emmerich = Raputia Aubl. ●☆

326914 Raram Adans. = Cenchrus L. ■

326915 Raritebe Wernham（1917）；哥伦比亚茜属●☆

326916 Raritebe axillare C. M. Taylor；腋生哥伦比亚茜●☆

326917 Raritebe blumii（Dwyer）Dwyer；布氏哥伦比亚茜●☆

326918 Raritebe palicoureoides Wernham；哥伦比亚茜●☆

326919 Raritebe trifoliatum（Dwyer et M. V. Hayden）Dwyer；三小叶哥伦比亚茜●☆

326920 Raspailia C. Presl = Polypogon Desf. ■

326921　Raspailia Endl. = Raspalia Brongn. ●☆

326922　Raspailia J. Presl et C. Presl = Polypogon Desf. ■

326923　Raspalia Brongn.（1826）;南非鳞叶树属●☆

326924　Raspalia affinis Nied.;近缘南非鳞叶树●☆

326925　Raspalia angulata（Sond.）Nied.;棱角南非鳞叶树●☆

326926　Raspalia aspera E. Mey. = Nebelia aspera（Sond.）Kuntze●☆

326927　Raspalia barnardii Pillans;巴纳德南非鳞叶树●☆

326928　Raspalia dregeana（Sond.）Nied.;德雷南非鳞叶树●☆

326929　Raspalia globosa（Lam.）Pillans;球形南非鳞叶树●☆

326930　Raspalia microphylla（Thunb.）Brongn.;小叶南非鳞叶树●☆

326931　Raspalia oblongifolia Pillans;矩圆叶南非鳞叶树●☆

326932　Raspalia palustris（Schltr. ex Kirchn.）Pillans;沼泽鳞叶树●☆

326933　Raspalia passerinoides（Schltdl.）C. Presl = Raspalia phylicoides（Thunb.）Arn. ●☆

326934　Raspalia phylicoides（Thunb.）Arn.;菲利木鳞叶树●☆

326935　Raspalia sacculata（Bolus ex Kirchn.）Pillans;小囊南非鳞叶树●☆

326936　Raspalia schlechteri Dümmer;施莱南非鳞叶树●☆

326937　Raspalia squalida Dümmer = Raspalia globosa（Lam.）Pillans●☆

326938　Raspalia staavioides（Sond.）Pillans;斯塔南非鳞叶树●☆

326939　Raspalia stokoei Pillans;斯托克南非鳞叶树●☆

326940　Raspalia struthioloides C. Presl;花束南非鳞叶树●☆

326941　Raspalia trigyna（Schltr.）Dümmer;三蕊南非鳞叶树●☆

326942　Raspalia variabilis Pillans;易变南非鳞叶树●☆

326943　Raspalia villosa C. Presl;长柔毛南非鳞叶树●☆

326944　Raspalia virgata（Brongn.）Pillans;条纹南非鳞叶树●☆

326945　Rassia Neck. = Gentiana L. ■

326946　Rastrophyllum Wild et G. V. Pope（1977）;耙叶菊属（锄叶菊属）■☆

326947　Rastrophyllum apiifolium M. G. Gilbert;耙叶菊■☆

326948　Rastrophyllum pinnatipartitum Wild et G. V. Pope;羽裂耙叶菊■☆

326949　Ratabida Loudon = Ratibida Raf. ■☆

326950　Ratanhia Raf. = Krameria L. ex Loefl. ●■☆

326951　Rathbunia Britton et Rose（废弃属名）= Stenocereus（A. Berger）Riccob.（保留属名）●☆

326952　Rathbunia thurberi（Engelm.）P. V. Heath = Stenocereus thurberi（Engelm.）Buxb. ●☆

326953　Rathea H. Karst.（1858）= Synechanthus H. Wendl. ●☆

326954　Rathea H. Karst.（1860）= Passiflora L. ●■

326955　Rathkea Schumch. = Ormocarpum P. Beauv.（保留属名）●

326956　Rathkea Schumch. et Thonn. = Ormocarpum P. Beauv.（保留属名）●

326957　Ratibida Raf.（1817）;草原松果菊属（草光菊属,拉提比达菊属,松果菊属）;Mexican Hat, Mexican-hat, Prairie Coneflower, Prairie Voneflower, Ratibida ■☆

326958　Ratibida columnaris（Sims）D. Don = Ratibida columnifera（Nutt.）Wooton et Standl. ■☆

326959　Ratibida columnaris（Sims）D. Don var. pulcherrima（DC.）D. Don = Ratibida columnifera（Nutt.）Wooton et Standl. ■☆

326960　Ratibida columnifera（Nutt.）Wooton et Standl.;草原松果菊;Columnar Coneflower, Coneflower, Longhead Prairie Coneflower, Long-headed Coneflower, Mexican Hat, Prairie Coneflower, Redspike Mexican-hat, Upright Prairie Coneflower ■☆

326961　Ratibida mexicana（Watson）Sharp;墨西哥草原松果菊（拉提比达菊）■☆

326962　Ratibida peduncularis（Torr. et A. Gray）Barnhart;裸草原松果菊;Naked Mexican-hat, Naked Prairie Coneflower ■☆

326963　Ratibida peduncularis（Torr. et A. Gray）Barnhart var. picta（A. Gray）W. M. Sharp;着色草原松果菊■☆

326964　Ratibida pinnata（Vent.）Barnhart;羽状草原松果菊;Drooping Coneflower, Globular Coneflower, Grayhead Prairie Coneflower, Gray-headed Coneflower, Pinnate Prairie Coneflower, Prairie Coneflower, Yellow Coneflower ■☆

326965　Ratibida pinnata Barnhart = Ratibida pinnata（Vent.）Barnhart ■☆

326966　Ratibida tagetes（E. James）Barnhart;绿草原松果菊;Green Mexican-hat, Prairie Coneflower, Shortray Prairie Coneflower ■☆

326967　Ratonia DC. = Matayba Aubl. ●☆

326968　Ratopitys Carrière = Cunninghamia R. Br.（保留属名）●★

326969　Rattraya J. B. Phipps = Danthoniopsis Stapf ■☆

326970　Rattraya anomala（C. E. Hubb. et Schweick.）Butzin = Danthoniopsis ramosa（Stapf）Clayton ■☆

326971　Rattraya petiolata J. B. Phipps = Danthoniopsis petiolata（J. B. Phipps）Clayton ■☆

326972　Rattraya ramosa（Stapf）Butzin = Danthoniopsis ramosa（Stapf）Clayton ■☆

326973　Rattraya simulans（C. E. Hubb.）Butzin = Danthoniopsis simulans（C. E. Hubb.）Clayton ■☆

326974　Ratzeburgia Kunth（1831）;极美草属■☆

326975　Ratzeburgia pulcherrima Kunth;极美草■☆

326976　Rauhia Traub（1957）;劳氏石蒜属■☆

326977　Rauhia megistophylla（Kraenzl.）Traub;劳氏石蒜■☆

326978　Rauhia multiflora（Kunth）Ravenna;多花劳氏石蒜■☆

326979　Rauhia occidentalis Ravenna;西方劳氏石蒜■☆

326980　Rauhiella Pabst et Braga（1978）;劳兰属■☆

326981　Rauhiella brasiliensis Pabst et Braga;劳兰■☆

326982　Rauhocereus Backeb. = Weberbauerocereus Backeb. ●☆

326983　Rauia Nees et Mart. = Angostura Roem. et Schult. ●☆

326984　Raukana Seem. = Pseudopanax K. Koch ●■

326985　Raukaua Seem. = Pseudopanax K. Koch ●■

326986　Raulinoa R. S. Cowan（1960）;多刺芸香属●☆

326987　Raulinoa echinata R. S. Cowan;多刺芸香●☆

326988　Raulinoreitzia R. M. King et H. Rob.（1971）;簇泽兰属●☆

326989　Raulinoreitzia crenulata（Spreng.）R. M. King et H. Rob.;簇泽兰●☆

326990　Rausslnia Neck. = Pachira Aubl. ●

326991　Rautanenia Buchenau = Burnatia Micheli ■☆

326992　Rautanenia Buchenau（1897）;劳氏泽泻属（劳坦宁泻属）■☆

326993　Rautanenia schinzii（Buchenau）Buchenau = Burnatia enneandra Micheli ■☆

326994　Rautanenia schinzii Buchenau;劳氏泽泻■☆

326995　Rauvolfia L.（1753）;萝芙木属（矮青木属,萝芙藤属）;Devil Pepper, Devilpepper, Devil-pepper, Rauvolfia, Rauwolfia ●

326996　Rauvolfia Plum. ex L. = Rauvolfia L. ●

326997　Rauvolfia altodiscifera R. H. Miao;高盘萝芙木●

326998　Rauvolfia altodiscifera R. H. Miao = Rauvolfia verticillata（Lour.）Baill. ●

326999　Rauvolfia brevistyla Tsiang;矮青木;Dwarf Devilpepper, Dwarf Greenwood, Shortstyle Devilpepper ●

327000　Rauvolfia brevistyla Tsiang = Rauvolfia verticillata（Lour.）Baill. ●

327001　Rauvolfia caffra Sond.;卡夫拉萝芙木（卡弗萝芙木）;Caffra Devilpepper ●☆

327002　Rauvolfia caffra Sond. var. natalensis Stapf ex Hiern = Rauvolfia caffra Sond. ●☆

327003　Rauvolfia cambodiana Pierre ex Pit. = Rauvolfia verticillata (Lour.) Baill. ●

327004　Rauvolfia canescens L. = Rauvolfia tetraphylla L. ●

327005　Rauvolfia capuronii Markgr. ;凯普伦萝芙木●☆

327006　Rauvolfia cardiocarpa K. Schum. = Rauvolfia mannii Stapf ●☆

327007　Rauvolfia celastrifolia Baker = Stephanostegia hildebrandtii Baill. ●☆

327008　Rauvolfia chinensis (Spreng.) Hemsl. = Rauvolfia verticillata (Lour.) Baill. ●

327009　Rauvolfia chinensis Hemsl. = Rauvolfia verticillata (Lour.) Baill. ●

327010　Rauvolfia congolana De Wild. et T. Durand = Rauvolfia vomitoria Afzel. ●

327011　Rauvolfia cubana A. DC. ;古巴萝芙木;Cuba Devilpepper, Cuba Devil-pepper ●

327012　Rauvolfia cumminsii Stapf ;卡明萝芙木;Cummins Devilpepper ●☆

327013　Rauvolfia cumminsii Stapf = Rauvolfia mannii Stapf ●☆

327014　Rauvolfia densiflora Benth. ex Hook. f. ;密花萝芙木●☆

327015　Rauvolfia dichotoma K. Schum. = Rauvolfia vomitoria Afzel. ●

327016　Rauvolfia faucium Engl. = Rauvolfia volkensii (K. Schum.) Stapf ●☆

327017　Rauvolfia goetzei Stapf = Rauvolfia caffra Sond. ●☆

327018　Rauvolfia heterophylla Willd. ex Roem. et Schult. ;异叶萝芙木●☆

327019　Rauvolfia inebrians K. Schum. = Rauvolfia caffra Sond. ●☆

327020　Rauvolfia ivorensis A. Chev. = Rauvolfia mannii Stapf ●☆

327021　Rauvolfia lamarkii A. DC. ;拉马萝芙木●☆

327022　Rauvolfia latifrons Tsiang ;阔叶萝芙木(大叶萝芙木,风湿木);Broadleaf Devilpepper, Rheumatism Wood ●

327023　Rauvolfia latifrons Tsiang = Rauvolfia verticillata (Lour.) Baill. ●

327024　Rauvolfia letouzeyi Leeuwenb. ;勒图萝芙木●☆

327025　Rauvolfia leucopoda K. Schum. ex De Wild. et T. Durand = Rauvolfia caffra Sond. ●☆

327026　Rauvolfia liberiensis Stapf = Rauvolfia mannii Stapf ●☆

327027　Rauvolfia ligustrina Willd. ex Schult. ;女贞萝芙木●☆

327028　Rauvolfia littoralis Rusby ;海滨萝芙木●☆

327029　Rauvolfia longiacuminata De Wild. et T. Durand = Rauvolfia mannii Stapf ●☆

327030　Rauvolfia macrocarpa Standl. ;大果萝芙木●☆

327031　Rauvolfia macrophylla Ruiz et Pav. ;大叶萝芙木●☆

327032　Rauvolfia macrophylla Stapf = Rauvolfia caffra Sond. ●☆

327033　Rauvolfia mannii Stapf;曼氏萝芙木●☆

327034　Rauvolfia mayombensis Pellegr. = Rauvolfia caffra Sond. ●☆

327035　Rauvolfia media Pichon;中间萝芙木●☆

327036　Rauvolfia mombasiana Stapf;蒙巴萨萝芙木●☆

327037　Rauvolfia monopyrena K. Schum. = Rauvolfia mombasiana Stapf ●☆

327038　Rauvolfia nana E. A. Bruce;矮小萝芙木●☆

327039　Rauvolfia natalensis Sond. ;奎宁树●☆

327040　Rauvolfia natalensis Sond. = Rauvolfia caffra Sond. ●☆

327041　Rauvolfia nitida Jacq. ;光亮萝芙木;Shining Devilpepper ●☆

327042　Rauvolfia obliquinervis Stapf = Rauvolfia caffra Sond. ●☆

327043　Rauvolfia obscura K. Schum. ;刚果萝芙木●☆

327044　Rauvolfia obscura K. Schum. = Rauvolfia mannii Stapf ●☆

327045　Rauvolfia obtusiflora A. DC. ;钝花萝芙木●☆

327046　Rauvolfia ochrosioides K. Schum. = Rauvolfia caffra Sond. ●☆

327047　Rauvolfia oreogiton Markgr. = Rauvolfia volkensii (K. Schum.) Stapf ●☆

327048　Rauvolfia oxyphylla Stapf = Rauvolfia caffra Sond. ●☆

327049　Rauvolfia paraensis Ducke;巴拉萝芙木;Para Devilpepper ●☆

327050　Rauvolfia perakensis King et Gamble;霹雳萝芙木; Perak Devilpepper ●

327051　Rauvolfia perakensis King et Gamble = Rauvolfia verticillata (Lour.) Baill. ●

327052　Rauvolfia pleiosiadica K. Schum. = Rauvolfia vomitoria Afzel. ●

327053　Rauvolfia preussii K. Schum. = Rauvolfia mannii Stapf ●☆

327054　Rauvolfia rosea K. Schum. = Rauvolfia mannii Stapf ●☆

327055　Rauvolfia sambesiaca Schinz = Rauvolfia mombasiana Stapf ●☆

327056　Rauvolfia senegambiae A. DC. = Rauvolfia vomitoria Afzel. ●

327057　Rauvolfia serpentina (L.) Benth. = Rauvolfia serpentina (L.) Benth. ex Kurz ●

327058　Rauvolfia serpentina (L.) Benth. ex Kurz;印度萝芙木(蛇根木,印度蛇根草,印度蛇根木,印度蛇木);Devil Pepper, Indian Snakeroot, Java Devilpepper, Java Devil-pepper, Snake Wood ●

327059　Rauvolfia serpentina (L.) Kurz = Rauvolfia serpentina (L.) Benth. ex Kurz ●

327060　Rauvolfia stuhlmannii K. Schum. = Rauvolfia vomitoria Afzel. ●

327061　Rauvolfia sumatrana Jack; 苏门答腊萝芙木; Sumatra Devilpepper, Sumatra Devil-pepper ●

327062　Rauvolfia superaxillaris P. T. Li et S. Z. Huang;腋生花萝芙木; Axillaryflower Devilpepper ●

327063　Rauvolfia superaxillaris P. T. Li et S. Z. Huang = Rauvolfia verticillata (Lour.) Baill. ●

327064　Rauvolfia taiwanensis Tsiang;台湾萝芙木;Taiwan Devilpepper ●

327065　Rauvolfia taiwanensis Tsiang = Rauvolfia verticillata (Lour.) Baill. ●

327066　Rauvolfia tchibangensis Pellegr. = Rauvolfia caffra Sond. ●☆

327067　Rauvolfia tetraphylla L. ;四叶萝芙木(灰白萝芙木,灰萝芙木,灰毛萝芙木,卡涅斯萝芙木,异叶萝芙木);Fourleaf Devilpepper, Four-leaved Devil-pepper ●

327068　Rauvolfia tiaolushanensis Tsiang;吊罗山萝芙木;Diaoluoshan Devilpepper ●

327069　Rauvolfia verticillata (Lour.) Baill. ;萝芙木(矮青木,白花丹,白花莲,刀伤药,地郎伞,毒狗药,甘榕木,红果萝芙木,红果木,火烙木,鸡眼子,假辣椒,假鱼胆,阔叶萝芙木,辣椒树,萝芙藤,麻木端,麻三端,麻桑瑞,青辣椒,染布子,三叉虎,三叉叶,山胡椒,山辣椒,山马蹄,蛇根木,十八爪,塘婢粘,通骨消,万药归宗,羊姆奶,羊屎子,药用萝芙木,野辣椒,鱼胆木);Common Devilpepper, Common Devil-pepper, Devilpepper, Taiwan Devilpepper ●

327070　Rauvolfia verticillata (Lour.) Baill. var. hainanensis Tsiang;海南萝芙木(海南红果萝芙木,红果萝芙木,山番椒,山马蹄根);Hainan Devilpepper ●

327071　Rauvolfia verticillata (Lour.) Baill. var. hainanensis Tsiang = Rauvolfia verticillata (Lour.) Baill. ●

327072　Rauvolfia verticillata (Lour.) Baill. var. oblanceolata Tsiang;倒披针叶萝芙木;Oblanceleaf Devilpepper, Oblanceolate-leaf Devilpepper ●

327073　Rauvolfia verticillata (Lour.) Baill. var. oblanceolata Tsiang = Rauvolfia verticillata (Lour.) Baill. ●

327074　Rauvolfia verticillata (Lour.) Baill. var. officinalis Tsiang;药用萝芙木(大叶萝芙木,奎宁树);Medicinal Devilpepper ●

327075　Rauvolfia verticillata (Lour.) Baill. var. officinalis Tsiang = Rauvolfia verticillata (Lour.) Baill. ●

327076　Rauvolfia verticillata (Lour.) Baill. var. rubrocarpa Tsiang;红

果萝芙木;Redfruit Devilpepper ●

327077 Rauvolfia verticillata A. Chev. = Rauvolfia caffra Sond. ●☆

327078 Rauvolfia volkensii (K. Schum.) Stapf;福尔萝芙木●☆

327079 Rauvolfia vomitoria Afzel. = Rauvolfia vomitoria Afzel. ex Spreng. ●

327080 Rauvolfia vomitoria Afzel. ex Spreng.;催吐萝芙木;Emetic Devilpepper,Emetic Devil-pepper,Poison Devil's-pepper,Swizzlestick Tree ●

327081 Rauvolfia welwitschii Stapf = Rauvolfia caffra Sond. ●☆

327082 Rauvolfia yunnanensis Tsiang;云南萝芙木(辣多,篊毒); Yunnan Devilpepper ●

327083 Rauvolfia yunnanensis Tsiang = Rauvolfia verticillata (Lour.) Baill. ●

327084 Rauvolfia yunnanensis Tsiang var. angustifolia C. Y. Wu;狭叶萝芙木;Narrowleaf Yunnan Devilpepper ●

327085 Rauwenhoffia Scheff. (1885);爪瓣玉盘属●☆

327086 Rauwenhoffia Scheff. = Melodorum Lour. ●☆

327087 Rauwenhoffia oligocarpa Diels;寡果爪瓣玉盘●☆

327088 Rauwenhoffia uvarioides Scheff.;爪瓣玉盘●☆

327089 Rauwolfia L. = Rauvolfia L. ●

327090 Rauwolfia Ruiz et Pav. = Citharexylum Mill. ●☆

327091 Rauwolfia serpentina Benth. = Rauvolfia serpentina (L.) Benth. ex Kurz ●

327092 Rauwolfia verticillata (Lour.) Baill. = Rauvolfia verticillata (Lour.) Baill. ●

327093 Ravenala Adans. (1763);旅人蕉属;Traveler's-tree, Travelerstree,Yunnan Devilpepper ●■

327094 Ravenala madagascariensis Adans. = Ravenala madagascariensis J. F. Gmel. ●■

327095 Ravenala madagascariensis J. F. Gmel.;旅人蕉(旅人木,扇芭蕉);Madagascar Travelerstree, Madagascar Traveler's-tree, Pilgrim's Tree, Ravenala, Traveler's Fan, Traveler's Palm, Traveler's Tree, Traveler's-tree, Travelerstree, Traveller Palm, Traveller Tree, Traveller's Palm ●■

327096 Ravenala madagascariensis Sonn. = Ravenala madagascariensis J. F. Gmel. ●■

327097 Ravenea C. D. Bouché = Ravenea H. Wendl. ex C. D. Bouché ●☆

327098 Ravenea C. D. Bouché ex H. Wendl. = Ravenea H. Wendl. ex C. D. Bouché ●☆

327099 Ravenea H. Wendl. = Ravenea H. Wendl. ex C. D. Bouché ●☆

327100 Ravenea H. Wendl. ex C. D. Bouché(1878);国王椰子属(国王椰属,溪棕属)●☆

327101 Ravenea L. H. Bailey = Ravenea H. Wendl. ex C. D. Bouché ●☆

327102 Ravenea albicans (Jum.) Beentje;白国王椰子●☆

327103 Ravenea amara Jum. = Ravenea sambiranensis Jum. et H. Perrier ●☆

327104 Ravenea dransfieldii Beentje;德兰斯菲尔德国王椰子●☆

327105 Ravenea glauca Jum. et H. Perrier;银色国王椰子●☆

327106 Ravenea latisecta Jum.;宽裂国王椰子●☆

327107 Ravenea louvelii Beentje;卢韦尔国王椰子●☆

327108 Ravenea madagascariensis Becc.;马岛国王椰子●☆

327109 Ravenea madagascariensis Becc. var. monticola Jum. = Ravenea madagascariensis Becc. ●☆

327110 Ravenea nana Beentje;矮国王椰子●☆

327111 Ravenea rivularis Jum. et H. Perrier;国王椰子(溪棕);Majesty Palm ●☆

327112 Ravenea robustior Jum. et H. Perrier;粗壮国王椰子●☆

327113 Ravenea sambiranensis Jum. et H. Perrier;桑比朗国王椰子●☆

327114 Ravenea xerophila Jum.;旱生国王椰子●

327115 Ravenia Vell. (1829);拉氏芸香属●☆

327116 Ravenia spectabilis Engl.;拉氏芸香●☆

327117 Raveniopsis Gleason(1939);拟拉氏芸香属●☆

327118 Raveniopsis abyssicola R. S. Cowan;阿比西尼亚拟拉氏芸香●☆

327119 Raveniopsis capitata R. S. Cowan;头状拟拉氏芸香●☆

327120 Raveniopsis linearis (Gleason) R. S. Cowan;线形拟拉氏芸香●☆

327121 Raveniopsis tomentosa Gleason;毛拟拉氏芸香●☆

327122 Raveniopsis trifoliolata R. S. Cowan;三小叶拟拉氏芸香●☆

327123 Ravensara Sonn. (废弃属名) = Agathophyllum Juss. ●☆

327124 Ravensara Sonn. (废弃属名) = Cryptocarya R. Br. (保留属名)●

327125 Ravensara aromatica Sonn. = Cryptocarya agathophylla van der Werff ●☆

327126 Ravensara parvifolia Scott-Elliot = Aspidostemon parvifolium (Scott-Elliot) van der Werff ●☆

327127 Ravia Schult. = Rauia Nees et Mart. ●☆

327128 Ravinia Post et Kuntze = Ravenia Vell. ●☆

327129 Ravnia Oerst. (1852);拉夫恩茜属●☆

327130 Ravnia triflora Oerst.;拉夫恩茜●☆

327131 Rawsonia Harv. et Sond. (1860);罗森木属●☆

327132 Rawsonia burtt-davyi (Edlin) F. White;伯特·戴维罗森木●☆

327133 Rawsonia lucida Harv. et Sond.;光亮罗森木●☆

327134 Rawsonia reticulata Gilg = Rawsonia lucida Harv. et Sond. ●☆

327135 Rawsonia schlechteri Gilg = Rawsonia lucida Harv. et Sond. ●☆

327136 Rawsonia spinidens (Hiern) Mendonça et Sleumer = Rawsonia lucida Harv. et Sond. ●☆

327137 Rawsonia transjubensis Chiov. = Rawsonia lucida Harv. et Sond. ●☆

327138 Rawsonia ugandensis Dawe et Sprague = Rawsonia lucida Harv. et Sond. ●☆

327139 Rawsonia uluguruensis Sleumer = Rawsonia lucida Harv. et Sond. ●☆

327140 Rawsonia usambarensis Engl. et Gilg = Rawsonia lucida Harv. et Sond. ●☆

327141 Raxamaris Raf. = Soulamea Lam. ●☆

327142 Raxopitys J. Nelson = Belis Salisb. (废弃属名)●★

327143 Raxopitys J. Nelson = Cunninghamia R. Br. (保留属名)●★

327144 Raxopitys cunninghamii J. Nelson = Cunninghamia lanceolata (Lamb.) Hook. ●

327145 Rayania Meisn. = Brunnichia Banks ex Gaertn. ●☆

327146 Rayania Meisn. = Rajania L. ■☆

327147 Rayania Raf. = Rajania L. ■☆

327148 Raycadenco Dodson(1851);厄瓜多尔兰属■☆

327149 Raycadenco ecuadorensis Dodson;厄瓜多尔兰■☆

327150 Rayera Gaudich. = Nolana L. ex L. f. ■☆

327151 Rayeria Gaudich. = Alona Lindl. ■☆

327152 Rayjacksonia R. L. Hartm. et M. A. Lane(1996);樟雏菊属■☆

327153 Rayjacksonia annua (Rydb.) R. L. Hartm. et M. A. Lane;黏樟雏菊;Viscid Camphor-daisy ■☆

327154 Rayjacksonia aurea (A. Gray) R. L. Hartm. et M. A. Lane;黄樟雏菊;Houston camphor-daisy ■☆

327155 Rayjacksonia phyllocephala (DC.) R. L. Hartm. et M. A. Lane;海湾樟雏菊;Gulf Coast camphor-daisy ■☆

327156 Rayleya Cristóbal(1981);雷梧桐属●☆

327157 Rayleya bahiensis Cristóbal;雷梧桐●☆

327158　Raynalia Soják = Alinula J. Raynal ■☆

327159　Raynalia lipocarphioides （Kük.） Soják = Alinula lipocarphioides （Kük.） J. Raynal ■☆

327160　Raynaudetia Bubani = Telephium L. ■☆

327161　Raynia Raf. = Rajania L. ■☆

327162　Razafimandimbisonia Kainul. et B. Bremer = Alberta E. Mey. ●☆

327163　Razafimandimbisonia Kainul. et B. Bremer(2009);拉扎木属●☆

327164　Razisea Oerst. (1854);拉齐爵床属☆

327165　Razisea breviflora D. N. Gibson;短花拉齐爵床☆

327166　Razisea spicata Oerst.;拉齐爵床☆

327167　Razisea villosa Gómez-Laur. et Hammel;毛拉齐爵床☆

327168　Razoumofskia Hoffm. = Arceuthobium M. Bieb.（保留属名）●

327169　Razoumofskya Hoffm.（废弃属名）= Arceuthobium M. Bieb.（保留属名）●

327170　Razoumofskya caucasica Hoffm. = Arceuthobium oxycedri （DC.） M. Bieb. ●

327171　Razoumofskya oxycedri （DC.） F. W. Schultz ex Nyman = Arceuthobium oxycedri （DC.） M. Bieb. ●

327172　Razoumofskya oxycedri Schultz = Arceuthobium oxycedri （DC.） M. Bieb. ●

327173　Razoumofskya pusilla （Peck） Kuntze = Arceuthobium pusillum Peck ●☆

327174　Razoumowskia Hoffm. ex M. Bieb. = Arceuthobium M. Bieb.（保留属名）●

327175　Razoumowskya Hoffm.（废弃属名）= Arceuthobium M. Bieb.（保留属名）●

327176　Razulia Raf. = Angelica L. ■

327177　Razumovia Spreng. (1805) = Humea Sm. ●☆

327178　Razumovia Spreng. (1807) = Centranthera R. Br. ■

327179　Razumovia Spreng. ex Juss. = Calomeria Vent. ■●☆

327180　Razumovia cochinchinensis （Lour.） Merr. = Centranthera cochinchinensis （Lour.） Merr. ■

327181　Razumovia cochinchinensis （Lour.） Merr. var. lutea H. Hara = Centranthera cochinchinensis （Lour.） Merr. subsp. lutea （H. Hara） T. Yamaz. ■

327182　Razumovia cochinchinensis （Lour.） Merr. var. nepalensis （D. Don） Merr. = Centranthera cochinchinensis （Lour.） Merr. var. nepalensis （D. Don） Merr. ■

327183　Razumovia grandiflora （Benth.） Merr. = Centranthera grandiflora Benth. ■

327184　Razumovia longiflora Merr. = Centranthera cochinchinensis （Lour.） Merr. ■

327185　Razumovia longiflora Merr. = Centranthera cochinchinensis （Lour.） Merr. subsp. lutea （H. Hara） T. Yamaz. ■

327186　Razumovia tonkinensis （Bonati） Merr. = Centranthera tranquebarica （Spreng.） Merr. ■

327187　Razumovia tranquebarica Spreng. = Centranthera tranquebarica （Spreng.） Merr. ■

327188　Rea Bertero ex Decne. = Dendroseris D. Don ●☆

327189　Readea Gillespie(1930);里德茜属●☆

327190　Readea membranacea Gillespie;里德茜●☆

327191　Reana Brign. = Euchlaena Schrad. ■

327192　Reana Brign. = Zea L. ■

327193　Reaumurea Steud. = Reaumuria L. ●

327194　Reaumuria L. (1759);红砂柳属（红沙属,红砂属,枇杷柴属,琵琶柴属,瑞茉花属）;Reaumuria,Redsandplant ●

327195　Reaumuria alternifolia （Labill.） Britton;互叶红砂柳（互叶红砂）;Alternateleaf Redsandplant,Alternate-leaved Redsandplant ●

327196　Reaumuria alternifolia （Labill.） Britton subsp. panjgurica （Blatt. et Hallb.） Qaiser;伊朗红砂柳●☆

327197　Reaumuria alternifolia （Labill.） Grande = Reaumuria alternifolia （Labill.） Britton ●

327198　Reaumuria alternifolia （Labill.） Grande subsp. panjgurica （Blatt. et Hallb.） Qaiser = Reaumuria alternifolia （Labill.） Britton subsp. panjgurica （Blatt. et Hallb.） Qaiser ●☆

327199　Reaumuria badghysi Korovin;巴德红砂柳●☆

327200　Reaumuria billarderi Jaub. et Spach = Reaumuria alternifolia （Labill.） Britton ●

327201　Reaumuria cistoides Adam;岩蔷薇红砂柳●☆

327202　Reaumuria cystoides Adam = Reaumuria alternifolia （Labill.） Britton ●

327203　Reaumuria desertorum Hausskn. ex Bornm. = Reaumuria alternifolia （Labill.） Britton ●

327204　Reaumuria fruticosa Boiss.;灌丛红砂柳●☆

327205　Reaumuria gabrielae Bornm. = Reaumuria stocksii Boiss. ●☆

327206　Reaumuria hirtella Jaub. et Spach;多毛红砂柳●☆

327207　Reaumuria hypericoides Willd. = Reaumuria alternifolia （Labill.） Britton ●

327208　Reaumuria hypericoides Willd. var. angustifolia （M. Bieb.） Trautv. = Reaumuria alternifolia （Labill.） Britton ●

327209　Reaumuria hypericoides Willd. var. cystoides （Adam） Regel et Mlok. = Reaumuria alternifolia （Labill.） Britton ●

327210　Reaumuria hypericoides Willd. var. cystoides （Adam） Regel et Mlok. f. hyrcanica （Jaub. et Spach） Regel et Mlok. = Reaumuria alternifolia （Labill.） Britton ●

327211　Reaumuria hyrcanica Jaub. et Spach = Reaumuria alternifolia （Labill.） Britton ●

327212　Reaumuria kaschgarica Rupr.;五柱红砂柳（长柱红砂,五柱红砂,五柱枇杷柴）;Kaschgar Reaumuria,Kaschgar Redsandplant ●

327213　Reaumuria kaschgarica Rupr. var. nanschanica Maxim. = Reaumuria kaschgarica Rupr. ●

327214　Reaumuria kaschgarica Rupr. var. przewalskii Maxim. = Reaumuria kaschgarica Rupr. ●

327215　Reaumuria korovinii Lincz. = Reaumuria alternifolia （Labill.） Britton ●

327216　Reaumuria kuznetzovii Sosn. et Manden.;库氏红砂柳●☆

327217　Reaumuria kuznetzovii Sosn. et Manden. = Reaumuria alternifolia （Labill.） Britton ●

327218　Reaumuria linifolia Salisb. = Reaumuria alternifolia （Labill.） Britton ●

327219　Reaumuria minfengensis D. F. Cui et M. J. Zhong;民丰枇杷柴;Minfeng Reaumuria,Minfeng Redsandplant ●

327220　Reaumuria mucronata Jaub. et Spach = Reaumuria vermiculata L. ●☆

327221　Reaumuria oxiana Boiss.;阿穆达尔红砂柳●☆

327222　Reaumuria palaestina Boiss. var. acuminate Blatt. et Hallb.;渐尖红砂柳●☆

327223　Reaumuria panjgurica Blatt. et Hallb. = Reaumuria alternifolia （Labill.） Grande subsp. panjgurica （Blatt. et Hallb.） Qaiser ●☆

327224　Reaumuria persica Boiss.;波斯红砂柳●☆

327225　Reaumuria reflexa Lipsky;反折红砂柳●☆

327226　Reaumuria soongarica （Pall.） Maxim.;红砂柳（海葫芦根,红沙,红砂,红虮,杷杷柴,枇杷柴,琵琶柴,杉柳）;Dzungar Redsandplant,Songory Reaumuria ●

327227　Reaumuria squarrosa Jaub. et Spach = Reaumuria alternifolia（Labill.）Britton ●

327228　Reaumuria stocksii Boiss.；斯托克斯红砂柳●☆

327229　Reaumuria tatarica Jaub. et Spach = Reaumuria alternifolia（Labill.）Britton ●

327230　Reaumuria trigyna Maxim.；长叶红砂柳（长叶红沙，长叶红砂，黄花红砂，黄花枇杷柴，牛板筋，三柱枇杷柴）；Yellowflower Reaumuria, Yellowflower Redsandplant, Yellow-flowered Reaumuria ●

327231　Reaumuria turkestanica Gorschk.；突厥红砂柳（突厥红砂，土耳其斯坦枇杷柴）●

327232　Reaumuria vermiculata L.；虫状红砂柳●☆

327233　Reaumuria vermuculata-angustifolia M. Bieb. = Reaumuria alternifolia（Labill.）Britton ●

327234　Reaumuria zakirovii Gorschk.；扎氏红砂柳●☆

327235　Reaumuriaceae Ehrenb. = Reaumuriaceae Ehrenb. ex Lindl. ●

327236　Reaumuriaceae Ehrenb. = Tamaricaceae Link（保留科名）●■

327237　Reaumuriaceae Ehrenb. ex Lindl.；红砂柳科

327238　Reaumuriaceae Ehrenb. ex Lindl. = Tamaricaceae Link（保留科名）●■

327239　Rebentischia Opiz = Trisetum Pers. ■

327240　Rebis Spach = Ribes L. ●

327241　Reboudia Coss. et Durieu = Erucaria Gaertn. ■☆

327242　Reboudia Coss. et Durieu（1856）；粉花芥属■☆

327243　Reboudia erucarioides Coss. et Durieu = Erucaria erucarioides（Coss. et Durieu）C. Müll. ■☆

327244　Reboudia pinnata（Viv.）O. E. Schulz = Erucaria microcarpa Boiss. ■☆

327245　Reboulea Kunth = Sphenopholis Scribn. ■☆

327246　Rebsamenia Conz. = Robinsonella Rose et Baker f. ●☆

327247　Rebulobivia Frič = Echinopsis Zucc. ●

327248　Rebulobivia Frič = Rebutia K. Schum. ●

327249　Rebulobivia Fri č. et Schelle ex Backeb. et F. M. Knuth = Lobivia Britton et Rose ■

327250　Rebutia K. Schum.（1895）；子孙球属（宝山属，翁宝属）；Crown Cactus, Crowncactus ●

327251　Rebutia albiflora F. Ritter et Buining；白花子孙球●☆

327252　Rebutia albipilosa F. Ritter；白毛子孙球●☆

327253　Rebutia albopectinata Rausch；白篦子孙球●☆

327254　Rebutia almeyeri W. Heinr.；桃红花子孙球●☆

327255　Rebutia arenacea Cardenas；砂地球（黄花沟宝山）●☆

327256　Rebutia aureiflora Backeb.；黄丽球（黄丽丸，金黄花子孙球）●☆

327257　Rebutia aureiflora Backeb. subsp. elegans（Backeb.）Donald；优雅球●☆

327258　Rebutia binnewaldiana W. Heinr.；猩红花子孙球●☆

327259　Rebutia calliantha Bewer.；白象球（白象丸）●☆

327260　Rebutia candiae Cárdenas；扁球沟宝山●☆

327261　Rebutia canigueralii Cárdenas；小球沟宝山●☆

327262　Rebutia chrysacantha Backeb.；锦宝球（锦宝丸）；Yellowspine Crowncactus ●

327263　Rebutia chrysacantha Backeb. var. elegans（Backeb.）Backeb.；美宝球（美宝丸）●

327264　Rebutia deminuta Britton et Rose；丽盛球（丽盛丸）●☆

327265　Rebutia diersiana Rausch；黛西子孙球●☆

327266　Rebutia einsteinii Frič；华宝子孙球●☆

327267　Rebutia eos Rausch；艾欧斯子孙球●☆

327268　Rebutia euanthema（Backeb.）Buining et Donald；春黄子孙球（真春黄菊子孙球）●☆

327269　Rebutia fiebrigii（Gurke）Britton et Rose；新玉●☆

327270　Rebutia fulviseta Rausch；褐刚毛子孙球●☆

327271　Rebutia glomeriseta Cardenas；金黄花子孙球（团集刚毛沟宝山）●☆

327272　Rebutia glomerispina Cárdenas；聚刺沟宝山●☆

327273　Rebutia graciliflora Backeb.；细长花子孙球●☆

327274　Rebutia grandiflora Backeb.；伟宝球（大花细仙人球，伟宝丸）；Scarlet Crowncactus ●

327275　Rebutia heliosa Rausch；红宝山●☆

327276　Rebutia hoffmannii Diers et Rausch；如月球（赫富曼子孙球）●☆

327277　Rebutia hyalacantha（Backeb.）Backeb.；透明刺子孙球●☆

327278　Rebutia krainziana Kesselr.；绯宝球（绯宝丸，绯宝玉，克氏沟宝山，艳红细仙人球）●☆

327279　Rebutia krugerae（Cárdenas）Backeb.；圆筒球沟宝山●☆

327280　Rebutia kupperiana Boed = Aylostera kupperiana（Boed.）Backeb. ●☆

327281　Rebutia marsoneri Werderm.；金簪球（金簪丸）●☆

327282　Rebutia mentosa（F. Ritter）Donald；短筒沟宝山●☆

327283　Rebutia minuscula K. Schum.；子孙球（宝山，细仙人球）；Red Crown Cactus ●

327284　Rebutia minuscula K. Schum. f. kariusiana（Wess.）Donald；红花宝山●☆

327285　Rebutia minuscula K. Schum. subsp. grandiflora（Backeb.）Donald = Rebutia grandiflora Backeb. ●☆

327286　Rebutia minuscula K. Schum. subsp. violaciflora（Backeb.）Donald；紫宝球●☆

327287　Rebutia muscula F. Ritter et Thiele；橙花子孙球●☆

327288　Rebutia neocumingii（Backeb.）D. R. Hunt = Weingartia neocumingii Backeb. ■☆

327289　Rebutia nigricans（Wessner）Pilbeam = Mediolobivia nigricans（Wessner）Krainz ●☆

327290　Rebutia perplexa Donald；缠结子孙球●☆

327291　Rebutia pseudodeminuta Backeb = Aylostera pseudodeminuta（Backeb.）Backeb. ●☆

327292　Rebutia pulchra Cárdenas；黑丽球●☆

327293　Rebutia pulvinosa F. Ritter et Buining；枕突子孙球●☆

327294　Rebutia pygmaea（R. E. Fr.）Britton et Rose；白宫球（白宫丸）●☆

327295　Rebutia rauschii Zecher；劳氏球●☆

327296　Rebutia ritteri（Wessner）Buining et Donald；里特子孙球（里特里子孙球）●☆

327297　Rebutia senilis Backeb.；翁宝球（翁宝丸）；Fire Crowncactus ●

327298　Rebutia senilis Backeb. var. kesselringiana Bewer.；金蝶球（金蝶丸）●☆

327299　Rebutia senilis Backeb. var. sieperdaiana（Buining）Backeb.；橙蝶球（橙蝶丸）●☆

327300　Rebutia senilis Backeb. var. stuemeri Backeb.；鹤宝球（鹤宝丸）●☆

327301　Rebutia spegazziniana Backeb.；壮丽球（壮丽丸）●☆

327302　Rebutia spinosissima Backeb.；多刺子孙球●☆

327303　Rebutia spiralisepala（Schütz）Šída；旋萼子孙球●☆

327304　Rebutia steinmannii（Solms）Britton et Rose；施氏子孙球●☆

327305　Rebutia tiraquensis Cárdenas；大扁平球沟宝山●☆

327306　Rebutia tiraquensis Cardenas；紫环球●☆

327307　Rebutia torquata F. Ritter et Buining；珠节子孙球●☆

327308　Rebutia violaciflora Backeb. = Rebutia minuscula K. Schum. sutsp. violaciflora（Backeb.）Donald ●☆

327309　Rebutia violaciflora Backeb. = Rebutia minuscula K. Schum. ●☆

327310　Rebutia violaciflora Backeb. var. knuthiana（Backeb.）Donald；妍宝球（妍宝丸）●☆

327311　Rebutia wessneriana Bewer.；银宝球（银宝丸）●☆

327312　Rebutia xanthocarpa Backeb. ;熏宝球（熏宝丸）; Yellowfruit Crowncactus ●

327313　Rebutia xanthocarpa Backeb. var. citricarpa Frič ex Backeb. ;橙宝球（橙宝丸）●☆

327314　Rebutia xanthocarpa Backeb. var. coerulescens Backeb. ;丽宝球（丽宝丸）●☆

327315　Rebutia xanthocarpa Backeb. var. dasyphrissa（Werderm.）Backeb. ;鲜宝球（鲜宝丸）●☆

327316　Rebutia xanthocarpa Backeb. var. luteiorosea Backeb. ;晚宝球●☆

327317　Rebutia xanthocarpa Backeb. var. salmonea Backeb. ;桃宝球（桃宝丸）●☆

327318　Recchia Moc. et Sessé ex DC.（1817）;短梗苦木属●☆

327319　Recchia mexicana Moc. et Sessé;短梗苦木●☆

327320　Receveura Vell. = Hypericum L. ■●

327321　Rechsteineria Kuntze = Rechsteineria Regel（保留属名）■☆

327322　Rechsteineria Regel（1848）（保留属名）;月宴属（月之宴属）; Gesnera, Gesneria ■☆

327323　Rechsteineria cardinalis Kuntze;绯红月之宴■☆

327324　Rechsteineria leucotricha Hoehne ex Morton;月之宴; Brazilian Edelweiss ●☆

327325　Rechsteineria lineata Hjelmq. ;线纹月之宴■☆

327326　Rechsteineria macropoda（Sprague）C. H. Curtis;长柄月之宴■☆

327327　Rechsteineria warszewiczii（C. D. Bouché et Hanst.）Kuntze;瓦氏月之宴■☆

327328　Recordia Moldenke（1934）;雷科德草属●☆

327329　Recordia boliviana Moldenke;雷科德草●☆

327330　Recordoxylon Ducke（1934）;雷豆木属（记木豆属）●☆

327331　Recordoxylon amazonicum（Ducke）Ducke;雷豆木●☆

327332　Rectangis Thouars = Angraecum Bory ■

327333　Rectangis Thouars = Jumellea Schltr. ■☆

327334　Rectanthera O. Deg. = Callisia Loefl. ■☆

327335　Rectomitra Blume = Kibessia DC. ●

327336　Rectomitra Blume = Pternandra Jack ●

327337　Rectophylis Thouars = Bulbophyllum Thouars（保留属名）■

327338　Rectophyllum Post et Kuntze = Rhektophyllum N. E. Br. ■☆

327339　Redfieldia Vasey（1887）;雷德禾属（毛之枝草属）■☆

327340　Redfieldia flexuosa Vasey;雷德禾■☆

327341　Redia Casar. = Cleidion Blume ●

327342　Redoutea Vent. = Cienfuegosia Cav. ■●☆

327343　Redovskia Cham. et Schltdl. = Redowskia Cham. et Schltdl. ■☆

327344　Redowskia Cham. et Schltdl.（1826）;播娘蒿叶芥属●☆

327345　Redowskia sophiifolia Cham. et Schltdl. ;播娘蒿叶芥■☆

327346　Redutea Pers. = Cienfuegosia Cav. ■●☆

327347　Redutea Pers. = Redoutea Vent. ■●☆

327348　Redutea Vent. = Redoutea Vent. ■●☆

327349　Reederochloa Soderstr. et H. F. Decker（1964）;啮蚀草属■☆

327350　Reederochloa eludens Soderstr. et H. F. Decker;啮蚀草■☆

327351　Reedia F. Muell.（1859）;里德莎草属●☆

327352　Reedia spathacea F. Muell. ;里德莎草■☆

327353　Reedrollinsia J. W. Walker = Stenanona Standl. ●☆

327354　Reedrollinsia J. W. Walker（1971）;雷德木属●☆

327355　Reedrollinsia cauliflora J. W. Walker;雷德木●☆

327356　Reesia Ewart = Polycarpaea Lam.（保留属名）■●

327357　Reevesia Lindl.（1827）;梭罗树属（梭罗属）; Reevesia, Suoluo ●

327358　Reevesia Walp. = Reevesia Lindl. ●

327359　Reevesia botingensis H. H. Hsue;保亭梭罗; Baoting Reevesia, Baoting Suoluo ●

327360　Reevesia cavaleriei H. Lév. et Vaniot = Reevesia pubescens Mast. ●

327361　Reevesia formosana Hayata = Reevesia formosana Sprague ●

327362　Reevesia formosana Sprague;台湾梭罗（台湾梭罗树）; Formosan Reevesia, Taiwan Reevesia, Taiwan Suoluo ●

327363　Reevesia glaucophylla H. H. Hsue;瑶山梭罗（九层皮）; Paleleaf Reevesia, Paleleaf Suoluo, Pale-leaved Reevesia ●

327364　Reevesia lancilolia H. L. Li;剑叶梭罗（剑叶梭罗木）; Lance-leaved Reevesia, Lanceoleaf Reevesia, Swordleaf Suoluo ●

327365　Reevesia lofouensis Chun et H. H. Hsue;罗浮梭罗; Lofu Reevesia, Lofu Suoluo, Luofu Reevesia ●

327366　Reevesia longipetiolata Merr. et Chun;长柄梭罗（长柄梭罗木，海南梭罗树，细棉木，硬壳果树）; Longpetiole Reevesia, Long-petioled Reevesia, Longstalk Suoluo ●

327367　Reevesia lunglingensis H. H. Hsue ex S. J. Xu;隆林梭罗; Longlin Reevesia, Longlin Suoluo ●

327368　Reevesia megaphylla Hu = Reevesia pubescens Mast. ●

327369　Reevesia membranacea H. H. Hsue = Reevesia pubescens Mast. ●

327370　Reevesia orbicularifolia H. H. Hsue;圆叶梭罗; Roundleaf Reevesia, Roundleaf Suoluo, Round-leaved Reevesia ●◇

327371　Reevesia pubescens Mast. ;梭罗树（毛叶梭罗树）; Common Reevesia, Common Suoluo ●

327372　Reevesia pubescens Mast. var. kwangsiensis H. H. Hsue;广西梭罗; Guangxi Common Reevesia, Guangxi Suoluo, Kwangsi Suoluo ●

327373　Reevesia pubescens Mast. var. siamensis（Craib）J. Anthony;泰梭罗; Siam Reevesia, Siam Suoluo ●

327374　Reevesia pubescens Mast. var. xuefengensis C. J. Qi = Reevesia xuefengensis（C. J. Qi）C. J. Qi ●

327375　Reevesia pycnantha W. T. Ling;密花梭罗; Denseflower Reevesia, Denseflower Suoluo, Densiflowered Reevesia ●

327376　Reevesia rotundifolia Chun;粗齿梭罗（岭南梭罗树，圆叶梭罗）; Bigtooth Suoluo, Roundleaf Reevesia, Round-leaved Reevesia ●◇

327377　Reevesia rubronervia H. H. Hsue;红脉梭罗; Red-nerved Reevesia, Redvein Reevesia, Redvein Suoluo ●

327378　Reevesia shangszeensis H. H. Hsue;上思梭罗; Shangsi Reevesia, Shangsi Suoluo ●

327379　Reevesia siamensis Craib = Reevesia pubescens Mast. var. siamensis（Craib）J. Anthony ●

327380　Reevesia sinica E. H. Wilson = Reevesia pubescens Mast. ●

327381　Reevesia taiwanensis Chun et H. H. Hsue = Reevesia formosana Sprague ●

327382　Reevesia thyrsoidea Lindl. ;两广梭罗（脆皮树，两广梭罗树，密花梭罗树，牛关麻，细叶马甲，油在麻，油在树）; Bunchlike Suoluo, Bunch-like Reevesia ●

327383　Reevesia thyrsoidea Lindl. = Reevesia pubescens Mast. ●

327384　Reevesia tomentosa H. L. Li;绒果梭罗; Tomentose Reevesia, Tomentose Suoluo ●

327385　Reevesia wallichii R. Br. ;瓦立克梭罗; Wallich Tomentose Reevesia ●☆

327386　Reevesia wallichii R. Br. f. pubescens（Mast.）Malick = Reevesia pubescens Mast. ●

327387　Reevesia xuefengensis（C. J. Qi）C. J. Qi;雪峰山梭罗; Xuefengshan Reevesia, Xuefengshan Suoluo ●

327388　Reevesia xuefengensis（C. J. Qi）C. J. Qi = Reevesia pubescens Mast. var. xuefengensis C. J. Qi ●

327389　Regalia Luer = Masdevallia Ruiz et Pav. ■☆

327390　Regalia Luer（2006）;雷加尔兰属■☆

327391 Regelia (Lem.) Lindm. = Neoregelia L. B. Sm. ■☆

327392 Regelia H. Wendl. = Verschaffeltia H. Wendl. ●☆

327393 Regelia Hort. ex H. Wendl. = Verschaffeltia H. Wendl. ●☆

327394 Regelia Lem. = Aregelia Kuntze ■☆

327395 Regelia Lindm. = Neoregelia L. B. Sm. ■☆

327396 Regelia Schauer(1843);雷格尔木属●☆

327397 Regelia ciliata Schauer;雷格尔木●☆

327398 Reggeria Raf. = Gagea Salisb. ■

327399 Regia Loudon ex DC. = Juglans L. ●

327400 Regina Buc'hoz = Bontia L. ●☆

327401 Registaniella Rech. f. (1987);勒吉斯坦草属☆

327402 Registaniella hapaxlegomena Rech. f.;勒吉斯坦草☆

327403 Regmus Dulac = Circaea L. ■

327404 Regnaldia Baill. = Chaetocarpus Thwaites(保留属名)●

327405 Regnellia Barb. Rodr. (1877);伦内尔兰属■☆

327406 Regnellia Barb. Rodr. = Bletia Ruiz et Pav. ■☆

327407 Regnellia purpurea Barb. Rodr. ;伦内尔兰■☆

327408 Rehdera Moldenke(1935);雷德尔草属●☆

327409 Rehdera trinervis (Blake) Moldenke;雷德尔草●☆

327410 Rehderodendron Hu (1932);木瓜红属;Rehder Tree, Rehdertree,Rehder-tree ●

327411 Rehderodendron conostyle C. Y. Wu;屏边木瓜红;Pingbian Rehdertree ●

327412 Rehderodendron fengii Hu = Rehderodendron indochinense H. L. Li ●

327413 Rehderodendron gongshanense Y. C. Tang;贡山木瓜红; Gongshan Rehdertree,Gongshan Rehder-tree ●

327414 Rehderodendron hui Chun = Rehderodendron kwangtungense Chun ●

327415 Rehderodendron indochinense H. L. Li;越南木瓜红(滇南木瓜红); Indo-China Rehder-tree, Vietnam Rehdertree, Yunnan Rehdertree ●

327416 Rehderodendron kwangtungense Chun;广东木瓜红(奥芮德木,红木冬瓜木,红木木瓜红,揭阳木瓜红,岭南木瓜红); Guangdong Rehdertree, Guangdong Rehder-tree, Kwangtung Rehdertree ●

327417 Rehderodendron kweichowense Hu;贵州木瓜红(滇芮德木,蒋氏芮德木,矩圆果芮德木,毛果木瓜红,云南木瓜红);Guizhou Rehdertree,Guizhou Rehder-tree,Hairyfruit Rehdertree ●

327418 Rehderodendron macrocarpum Hu;木瓜红(大果芮德木,马边木瓜红,硕果芮德木,野草果);Big-fruited Rehder-tree,Largefruit Rehder Tree,Largefruit Rehdertree ●◇

327419 Rehderodendron mapienense Hu = Rehderodendron macrocarpum Hu ●◇

327420 Rehderodendron membranifolium C. Y. Wu;膜叶木瓜红; Membrana-leaf Rehdertree ●

327421 Rehderodendron microcarpum K. M. Feng;小果木瓜红;Small-fruit Rehdertree ●

327422 Rehderodendron praeteritum Sleumer = Rehderodendron kweichowense Hu ●

327423 Rehderodendron tsiangii Hu et W. C. Cheng = Rehderodendron kweichowense Hu ●

327424 Rehderodendron yunnanensis Hu = Rehderodendron kweichowense Hu ●

327425 Rehderophoenix Burret = Drymophloeus Zipp. ●☆

327426 Rehia Fijten(1975);珠芽禾属●☆

327427 Rehia nervata (Swallen) Fijten;珠芽禾■☆

327428 Rehmannia Libosch. = Rehmannia Libosch. ex Fisch. et C. A. Mey.(保留属名)■★

327429 Rehmannia Libosch. ex Fisch. et C. A. Mey. (1835)(保留属名);地黄属;Chinese Foxglove,Rehmannia ■★

327430 Rehmannia angulata (Oliv.) Hemsl. = Rehmannia piasezkii Maxim. ■

327431 Rehmannia angulata Hemsl. = Rehmannia piasezkii Maxim. ■

327432 Rehmannia chinensis Libosch. ex Fisch. et C. A. Mey. = Rehmannia glutinosa (Gaertn.) Libosch. ex Fisch. et C. A. Mey. ■

327433 Rehmannia chingii H. L. Li;天目地黄(浙地黄,紫花地黄); Ching Rehmannia,Tianmu Rehmannia ■

327434 Rehmannia elata N. E. Br. ex Prain;高地黄(棱地黄);China Foxglove,Chinese Fox Glove,Chinese Foxglove,Tall Rehmannia ■

327435 Rehmannia glutinosa (Gaertn.) Libosch. ex Fisch. et C. A. Mey.;地黄(次生地,大生地,地髓,蜂糖罐,干生地,蛤蟆草,根生地,狗奶子,胡面莽,苄,怀庆地黄,怀庆地黄,怀生地,酒壶花,酒盅盅花,蜜罐棵,蜜蜜罐,蜜蜜罐棵,牛奶子,炮掌,婆婆奶,婆婆妮,苣,山白菜,山旱烟,山烟根,山烟棵,生地,生地黄,生地炭,甜酒棵,细生地,小生地,阳精,淮生地,原生地);Adhesive Rehmannia,Sticky Rehmannia ■

327436 Rehmannia glutinosa (Gaertn.) Libosch. ex Fisch. et C. A. Mey. f. huechingensis (Chao et Shih) P. G. Hsiao = Rehmannia glutinosa (Gaertn.) Libosch. ex Fisch. et C. A. Mey. ■

327437 Rehmannia glutinosa (Gaertn.) Libosch. ex Fisch. et C. A. Mey. f. lutea (Maxim.) Matsuda;筑桥地黄■☆

327438 Rehmannia glutinosa (Gaertn.) Libosch. ex Fisch. et C. A. Mey. f. purpurea Matsuda = Rehmannia glutinosa (Gaertn.) Libosch. ex Fisch. et C. A. Mey. ■

327439 Rehmannia glutinosa (Gaertn.) Libosch. ex Fisch. et C. A. Mey. var. angulata Oliv. = Rehmannia piasezkii Maxim. ■

327440 Rehmannia glutinosa (Gaertn.) Libosch. ex Fisch. et C. A. Mey. var. hemsleyana Diels = Rehmannia glutinosa (Gaertn.) Libosch. ex Fisch. et C. A. Mey. ■

327441 Rehmannia glutinosa (Gaertn.) Libosch. ex Fisch. et C. A. Mey. var. huechingensis Chao et Shih = Rehmannia glutinosa (Gaertn.) Libosch. ex Fisch. et C. A. Mey. ■

327442 Rehmannia glutinosa (Gaertn.) Libosch. ex Fisch. et C. A. Mey. var. japonica Thunb. = Rehmannia japonica (Thunb.) Makino ex T. Yamaz. ☆

327443 Rehmannia glutinosa (Gaertn.) Libosch. ex Fisch. et C. A. Mey. var. lutea Makino = Rehmannia glutinosa (Gaertn.) Libosch. ex Fisch. et C. A. Mey. f. lutea (Maxim.) Matsuda ■☆

327444 Rehmannia glutinosa (Gaertn.) Libosch. ex Fisch. et C. A. Mey. var. makinoi Matsuda = Rehmannia japonica (Thunb.) Makino ex T. Yamaz. ■☆

327445 Rehmannia glutinosa (Gaertn.) Libosch. ex Fisch. et C. A. Mey. var. piasezkii Diels = Rehmannia piasezkii Maxim. ■

327446 Rehmannia glutinosa (Gaertn.) Libosch. ex Fisch. et C. A. Mey. var. purpurea Makino;紫地黄■☆

327447 Rehmannia glutinosa (Gaertn.) Libosch. ex Fisch. et C. A. Mey. var. typica f. purpurea Matsuda = Rehmannia glutinosa (Gaertn.) Libosch. ex Fisch. et C. A. Mey. ■

327448 Rehmannia henryi N. E. Br.;湖北地黄(鄂地黄,岩白菜); Henry Rehmannia,Hubei Rehmannia ■

327449 Rehmannia integra H. L. Li = Triaenophora integra (H. L. Li) Ivanina ■

327450 Rehmannia japonica (Thunb.) Makino ex T. Yamaz. ;日本地

黄■☆

327451　Rehmannia lutea Maxim. ;深黄地黄■☆

327452　Rehmannia lutea Maxim. = Rehmannia glutinosa（Gaertn.）
Libosch. ex Fisch. et C. A. Mey. f. lutea（Maxim.）Matsuda ■☆

327453　Rehmannia oldhamii Hemsl. = Titanotrichum oldhamii
（Hemsl.）Soler. ■

327454　Rehmannia piasezkii Maxim. ;裂叶地黄;Piasezk Rehmannia ■☆

327455　Rehmannia rupestris Hemsl. = Triaenophora rupestris（Hemsl.）
Soler. ■

327456　Rehmannia solanifolia P. C. Tsoong et T. L. Chin;茄叶地黄;
Eggplantleaf Rehmannia,Nightshadeleaf Rehmannia ■

327457　Rehmanniaceae G. Kunkel;地黄科■

327458　Rehsonia Stritch = Wisteria Nutt.（保留属名）●

327459　Reichantha Luer = Masdevallia Ruiz et Pav. ☆

327460　Reichantha Luer(2006);赖克兰属■☆

327461　Reichardia Dennst. = Tabernaemontana L. ●

327462　Reichardia Roth(1787);直梗栓果菊（赖卡菊属）■☆

327463　Reichardia Roth(1800) = Maurandya Ortega ■☆

327464　Reichardia Roth(1821) = Pterolobium R. Br. ex Wight et Arn.
（保留属名）●☆

327465　Reichardia crystallina（Sch. Bip.）Bramwell;水晶直梗栓果菊■☆

327466　Reichardia decapetala Roth = Caesalpinia decapetala（Roth）
Alston ●

327467　Reichardia dichotoma（Vahl）Freyn;二歧直梗栓果菊■☆

327468　Reichardia gaditana（Willk.）Cout. ;加迪特直梗栓果菊■☆

327469　Reichardia intermedia（Sch. Bip.）Cout. ;间型直梗栓果菊■☆

327470　Reichardia laciniata Klatt = Lactuca petrensis Hiern ☆

327471　Reichardia ligulata（Vent.）G. Kunkel et Sunding;舌状直梗栓
果菊■☆

327472　Reichardia orientalis（L.）Hochr. = Reichardia tingitana（L.）
Roth ■☆

327473　Reichardia picroides（L.）Roth;直梗栓果菊（赖卡菊）;
Common Brighteyes ■☆

327474　Reichardia picroides（L.）Roth subsp. intermedia（Sch. Bip.）
Batt. = Reichardia intermedia（Sch. Bip.）Cout. ■☆

327475　Reichardia picroides（L.）Roth var. intermedium Hochr. =
Reichardia picroides（L.）Roth ■☆

327476　Reichardia picroides（L.）Roth var. maritima（Batt.）Maire =
Reichardia picroides（L.）Roth ■☆

327477　Reichardia picroides Roth = Reichardia picroides（L.）Roth ■☆

327478　Reichardia tingitana（L.）Roth;假苦苣菜;False Sowthistle ■☆

327479　Reichardia tingitana（L.）Roth subsp. discolor（Pomel）Batt. ;
异色直梗栓果菊■☆

327480　Reichardia tingitana（L.）Roth var. arabica（Hochst. et）Asch.
et Schweinf. = Reichardia tingitana（L.）Roth ■☆

327481　Reichardia tingitana（L.）Roth var. mauritii Sennen =
Reichardia tingitana（L.）Roth ☆

327482　Reichardia tingitana（L.）Roth var. orientalis（L.）Pamp. =
Reichardia tingitana（L.）Roth subsp. discolor（Pomel）Batt. ■☆

327483　Reichardia tingitana（L.）Roth var. pinnatifida（Lag.）Pau et
Font Quer = Reichardia tingitana（L.）Roth ■☆

327484　Reichardia tingitana（L.）Roth var. subintegra（Boiss.）
Pamp. = Reichardia tingitana（L.）Roth ■☆

327485　Reichea Kausel = Myrcianthes O. Berg ●☆

327486　Reicheella Pax(1900);藓缀属■☆

327487　Reicheella andicola（Phil.）Pax;藓缀■☆

327488　Reicheia Kausel = Myrcianthes O. Berg ●☆

327489　Reichelea A. W. Benn. = Reichelia Schreb. ■

327490　Reichelia Schreb. = Hydrolea L.（保留属名）■

327491　Reichembachanthus B. D. Jacks. = Reichenbachanthus Barb.
Rodr. ■☆

327492　Reichenbachanthus Barb. Rodr.（1882）;赖兴巴赫兰属■☆

327493　Reichenbachanthus Barb. Rodr. = Fractiunguis Schltr. ■☆

327494　Reichenbachanthus modestus Barb. Rodr. ;赖兴巴赫兰■☆

327495　Reichenbachia Spreng.（1823）;管花茉莉属●☆

327496　Reichenbachia hirsuta Spreng. ;管花茉莉●☆

327497　Reichenheimia Klotzsch = Begonia L. ●■

327498　Reicheocactus Backeb.（1942）;螺棱球属■☆

327499　Reicheocactus Backeb. = Pyrrhocactus（A. Berger）Backeb. et
F. M. Knuth ●■

327500　Reicheocactus Backeb. = Rebutia K. Schum. ●

327501　Reicheocactus floribundus Backeb. ;多花螺棱球■☆

327502　Reicheocactus neoreichei（Backeb.）Backeb. ;螺棱球■☆

327503　Reicheocactus pseudoreicheanus Backeb. ;黑斜子■☆

327504　Reichertia H. Karst. = Schultesia Mart.（保留属名）■☆

327505　Reidia Wight = Eriococcus Hassk. ●■

327506　Reidia Wight = Phyllanthus L. ●■

327507　Reidia gracilipes Miq. = Phyllanthus gracilipes（Miq.）Müll.
Arg. ●

327508　Reidia gracilis（Hassk.）Miq. = Phyllanthus gracilipes（Miq.）
Müll. Arg. ●

327509　Reifferscheidia C. Presl = Dillenia L. ●

327510　Reigera Opiz = Bolboschoenus（Asch.）Palla ■

327511　Reilia Steud.（1855）;阿根廷草属☆327512　Reimaria Flüggé
= Paspalum L. ■

327513　Reimaria Humb. et Bonpl. ex Flüggé = Paspalum L. ■

327514　Reimarochloa Hitchc.（1909）;沼生雀稗属■☆

327515　Reimarochloa Hitchc. = Reimaria Flüggé ■

327516　Reimarochloa acuta Hitchc. ;尖沼生雀稗■☆

327517　Reimarochloa brasiliensis Hitchc. ;巴西沼生雀稗■☆

327518　Reimarochloa oligostachya Hitchc. ;沼生雀稗■☆

327519　Reimbolea Debeaux = Echinaria Desf.（保留属名）■☆

327520　Reimbolea Debeaux(1890);雷穗草属■☆

327521　Reimbolea spicata Debeaux;雷穗草■☆

327522　Reimbolea spicata Debeaux = Echinaria spicata Debeaux ■☆

327523　Reineckea H. Karst. = Synechanthus H. Wendl. ●☆

327524　Reineckea Kunth(1844)（保留属名）;吉祥草属;Reineckia ■

327525　Reineckea carnea（Andréws）Kunth;吉祥草（长春草,滇吉祥
草,佛顶珠,观音草,广东万年青,结实兰,解晕草,九节连,九节
莲,米腊参,七厘麻,软筋藤,松寿兰,西边兰,细叶万年青,小九
龙盘,小青胆,小叶万年青,洋吉祥草,玉带草,真武草,中叶麦
冬,竹根七,竹节伤,竹叶草,竹叶青,紫袍玉带草）;Pink
Reineckea,Pink Reineckia,Reineckea ■

327526　Reineckea carnea（Andréws）Kunth var. rubra H. Lév. =
Reineckea carnea（Andréws）Kunth ■

327527　Reineckea ovata Z. Y. Zhu = Reineckea carnea（Andréws）
Kunth ■

327528　Reineckea ovata Z. Y. Zhu et Z. R. Chen;卵果吉祥草;
Ovatefruit Reineckea ■

327529　Reineckea ovata Z. Y. Zhu et Z. R. Chen = Reineckea carnea
（Andréws）Kunth ■

327530　Reineckea yunnanensis W. W. Sm. = Reineckea carnea
（Andréws）Kunth ■

327531　Reineckia H. Karst. = Synechanthus H. Wendl. ●☆

327532 Reineckia Kunth = Reineckea Kunth(保留属名)■

327533 Reineckia carnea（Andréws）Kunth = Reineckea carnea（Andréws）Kunth ■

327534 Reineckia ovata Z. Y. Zhu et Z. R. Chen = Reineckea carnea（Andréws）Kunth ■

327535 Reineckia ovata Z. Y. Zhu et Z. R. Chen = Reineckea ovata Z. Y. Zhu et Z. R. Chen ■

327536 Reinera Dennst. = Leptadenia R. Br. ●☆

327537 Reineria Moench（废弃属名）= Tephrosia Pers.（保留属名）●■

327538 Reinhardtia Liebm.（1849）;窗孔椰属（来哈特椰属,鸢氏椰子属,美兰葵属,芮哈德桐属）;Reinhardt Palm ●☆

327539 Reinhardtia gracilis（H. Wendl.）Dammer = Reinhardtia gracilis Burret ●☆

327540 Reinhardtia gracilis Burret;窗孔椰（窗孔椰子,来哈特棕,美兰葵）;Little Window-palm ●☆

327541 Reinhardtia gracilis Burret var. gracilior（Burret）H. E. Moore;小窗孔椰（小来哈特棕）●☆

327542 Reinhardtia gracilis Burret var. simplex Burret = Reinhardtia simplex（H. Wendl.）Burret ●☆

327543 Reinhardtia simplex（H. Wendl.）Burret;单干窗孔椰（单干来哈特棕）●☆

327544 Reinhardtia simplex Burret = Reinhardtia simplex（H. Wendl.）Burret ●☆

327545 Reinia Franch. = Itea L. ●

327546 Reinia Franch. et Sav. = Itea L. ●

327547 Reinwardtia Blume ex Nees = Saurauia Willd.（保留属名）●

327548 Reinwardtia Dumort.（1822）;石海椒属（过山青属）;Stonecayenne,Yellow Flax,Yellowflax,Yellow-flax ●

327549 Reinwardtia Korth. = Ternstroemia Mutis ex L. f.（保留属名）●

327550 Reinwardtia Spreng. = Breweria R. Br. ●☆

327551 Reinwardtia Spreng. = Dufourea Gren. ■

327552 Reinwardtia Spreng. = Prevostea Choisy ●☆

327553 Reinwardtia indica Dumort.;石海椒（过山青,黄花香草,黄亚麻,迎春柳）;Common Yellowflax, Common Yellow-flax, India Stonecayenne, Indian Yellowflax, Stonecayenne, Yellow Flax, Yellow Flax Bush ●

327554 Reinwardtia sinensis Hemsl. = Tirpitzia sinensis（Hemsl.）Hallier f. ●

327555 Reinwardtia trigyna（Roxb.）Planch. = Reinwardtia indica Dumort. ●

327556 Reinwardtiodendron Koord.（1898）;雷棟属;Thundermelia ●

327557 Reinwardtiodendron celebicum Koord.;西里伯斯雷棟●☆

327558 Reinwardtiodendron dubium（Merr.）X. M. Chen;雷棟;Dubious Thundermelia ●

327559 Reissantia N. Hallé(1958);星刺卫矛属（星刺属）●

327560 Reissantia angustipetala（H. Perrier）N. Hallé;窄瓣星刺●☆

327561 Reissantia astericantha N. Hallé;星刺●

327562 Reissantia astericantha N. Hallé = Reissantia indica（Willd.）N. Hallé var. astericantha（N. Hallé）N. Hallé ●☆

327563 Reissantia buchananii（Loes.）N. Hallé;布氏星刺●☆

327564 Reissantia indica（Willd.）N. Hallé;印度星刺●☆

327565 Reissantia indica（Willd.）N. Hallé var. astericantha（N. Hallé）N. Hallé;菊花星刺●☆

327566 Reissantia indica（Willd.）N. Hallé var. loeseneriana（Hutch. et M. B. Moss）N. Hallé;勒泽纳星刺●☆

327567 Reissantia indica（Willd.）N. Hallé var. orientalis N. Hallé et B. Mathew;东方星刺●☆

327568 Reissantia parviflora（N. E. Br.）N. Hallé;小花星刺●☆

327569 Reissantia parvifolia（Oliv.）N. Hallé = Elachyptera parvifolia（Oliv.）N. Hallé ●☆

327570 Reissekia Endl.（1840）;赖斯鼠李属●☆

327571 Reissekia cordifolia Endl.;赖斯鼠李●☆

327572 Reissipa Steud. ex Klotzsch = Monotaxis Brongn. ■☆

327573 Reitzia Swallen(1956);赖茨禾属■☆

327574 Reitzia smithii Swallen;赖茨禾■☆

327575 Rejoua Gaudich.（1828）;假金橘属;Rejoua ●

327576 Rejoua Gaudich. = Tabernaemontana L. ●

327577 Rejoua dichotoma（Roxb.）Gamble;假金橘（革叶山辣椒,革叶山马茶,红头三友花,兰屿马蹄花,兰屿山马茶）;Dichotomous Tabernaemontana, Eve's Apple, Forbidden Fruit, Lanyu Tabernaemontana,Rejoua,Twofork Rejoua ●

327578 Rejoua dichotoma（Roxb.）Gamble = Tabernaemontana dichotoma Roxb. ex Wall. ●

327579 Relbunium（Endl.）Benth. et Hook. f.（1873）;肖拉拉藤属■☆

327580 Relbunium（Endl.）Benth. et Hook. f. = Galium L. ■●

327581 Relbunium（Endl.）Hook. f. = Relbunium（Endl.）Benth. et Hook. f. ■☆

327582 Relbunium Benth. et Hook. f. = Relbunium（Endl.）Benth. et Hook. f. ■☆

327583 Relbunium nitidum（Kunth）K. Schum.;肖拉拉藤■☆

327584 Relbunium nitidum K. Schum. = Relbunium nitidum（Kunth）K. Schum. ■☆

327585 Relchardia Roth = Maurandya Ortega ■☆

327586 Relchela Steud.（1854）;黍状凌风草属■☆

327587 Relchela Steud. = Briza L. ■

327588 Relchela panicoides Steud.;黍状凌风草■☆

327589 Reldia Wiehler(1977);雷尔苣苔属■☆

327590 Reldia alternifolia Wiehler;异叶雷尔苣苔■☆

327591 Reldia grandiflora L. P. Kvist et L. E. Skog;大花雷尔苣苔■☆

327592 Reldia minutiflora（L. Skog）L. P. Kvist et L. E. Skog;小花雷尔苣苔■☆

327593 Reldia multiflora L. P. Kvist et L. E. Skog;多花雷尔苣苔■☆

327594 Relhamia J. F. Gmel. = Curtisia Aiton（保留属名）●☆

327595 Relhania L' Hér.（1789）（保留属名）;寡头鼠麴木属●☆

327596 Relhania acerosa（DC.）K. Bremer;针状寡头鼠麴木●☆

327597 Relhania acicularis Desf. ex Cass. = Relhania pungens L'Hér. ●☆

327598 Relhania affinis Sond. ex Harv. = Oedera viscosa（L'Hér.）Anderb. et K. Bremer ●☆

327599 Relhania apiculata（DC.）Harv. = Relhania calycina（L. f.）L'Hér. subsp. apiculata（DC.）K. Bremer ●☆

327600 Relhania biennis（Jacq.）K. Bremer = Nestlera biennis（Jacq.）Spreng. ■☆

327601 Relhania calycina（L. f.）L'Hér.;萼状寡头鼠麴木●☆

327602 Relhania calycina（L. f.）L'Hér. subsp. apiculata（DC.）K. Bremer;细尖寡头鼠麴木●☆

327603 Relhania calycina（L. f.）L'Hér. subsp. lanceolata K. Bremer;披针形寡头鼠麴木●☆

327604 Relhania centauroides（L.）Harv. = Athanasia crenata（L.）L. ●☆

327605 Relhania conferta Hutch. = Oedera conferta（Hutch.）Anderb. et K. Bremer ●☆

327606 Relhania corymbosa（Bolus）K. Bremer;伞序寡头鼠麴木●☆

327607 Relhania cuneata L'Hér. = Oedera uniflora（L. f.）Anderb. et K. Bremer ●☆

327608　Relhania cuneata L'Hér. var. virgata（L'Hér.）S. Moore = Oedera uniflora（L. f.）Anderb. et K. Bremer ●☆

327609　Relhania decussata L'Hér. ;对生寡头鼠麴木●☆

327610　Relhania dieterlenii（E. Phillips）K. Bremer;迪特尔寡头鼠麴木●☆

327611　Relhania ericoides（P. J. Bergius）Cass. = Relhania fruticosa（L.）K. Bremer ●☆

327612　Relhania foveolata K. Bremer = Oedera foveolata（K. Bremer）Anderb. et K. Bremer ●☆

327613　Relhania fruticosa（L.）K. Bremer;灌丛寡头鼠麴木●☆

327614　Relhania garnotii（Less.）K. Bremer;加尔诺寡头鼠麴木●☆

327615　Relhania genistifolia（L.）L'Hér. = Oedera genistifolia（L.）Anderb. et K. Bremer ●☆

327616　Relhania genistifolia（L.）L'Hér. var. angustifolia Harv. = Oedera genistifolia（L.）Anderb. et K. Bremer ●☆

327617　Relhania genistifolia（L.）L'Hér. var. discoidea Harv. = Oedera genistifolia（L.）Anderb. et K. Bremer ●☆

327618　Relhania genistifolia（L.）L'Hér. var. glutinosa（DC.）Harv. = Oedera viscosa（L'Hér.）Anderb. et K. Bremer ●☆

327619　Relhania lanata Compton = Rosenia spinescens DC. ●☆

327620　Relhania lateriflora L'Hér. = Rhynchopsidium sessiliflorum（L. f.）DC. ■☆

327621　Relhania latifolia Compton = Oedera squarrosa（L.）Anderb. et K. Bremer ●☆

327622　Relhania laxa L'Hér. = Rhynchopsidium pumilum（L. f.）DC. ■☆

327623　Relhania microphylla L'Hér. = Oedera genistifolia（L.）Anderb. et K. Bremer ●☆

327624　Relhania multipunctata DC. = Oedera multipunctata（DC.）Anderb. et K. Bremer ●☆

327625　Relhania nordenstamii K. Bremer = Oedera nordenstamii（K. Bremer）Anderb. et K. Bremer ●☆

327626　Relhania passerinoides L'Hér. = Oedera genistifolia（L.）Anderb. et K. Bremer ●☆

327627　Relhania patersoniae L. Bolus = Relhania decussata L'Hér. ●☆

327628　Relhania pedunculata（DC.）Harv. = Rhynchopsidium pumilum（L. f.）DC. ■☆

327629　Relhania pumila（L. f.）Thunb. = Rhynchopsidium pumilum（L. f.）DC. ■☆

327630　Relhania pungens L'Hér. ;锐尖寡头鼠麴木●☆

327631　Relhania pungens L'Hér. subsp. angustifolia（DC.）K. Bremer; 窄叶寡头鼠麴木●☆

327632　Relhania pungens L'Hér. subsp. trinervis（Thunb.）K. Bremer; 三脉寡头鼠麴木●☆

327633　Relhania quinquenervis Thunb. = Relhania calycina（L. f.）L'Hér. ●☆

327634　Relhania reflexa Thunb. = Nestlera biennis（Jacq.）Spreng. ■☆

327635　Relhania relhanioides（Schltr.）K. Bremer;普通寡头鼠麴木●☆

327636　Relhania resinifera K. Bremer = Oedera resinifera（K. Bremer）Anderb. et K. Bremer ●☆

327637　Relhania rigida Hoffm. et Muschl. = Relhania calycina（L. f.）L'Hér. ●☆

327638　Relhania rotundifolia Less. ;圆叶寡头鼠麴木●☆

327639　Relhania sedifolia（DC.）Harv. = Oedera sedifolia（DC.）Anderb. et K. Bremer ●☆

327640　Relhania sessiliflora（L. f.）Thunb. = Rhynchopsidium sessiliflorum（L. f.）DC. ■☆

327641　Relhania silicicola K. Bremer = Oedera silicicola（K. Bremer）Anderb. et K. Bremer ●☆

327642　Relhania spathulifolia K. Bremer;苞叶寡头鼠麴木●☆

327643　Relhania speciosa（DC.）Harv. ;美丽寡头鼠麴木●☆

327644　Relhania speciosa（DC.）Harv. var. schizolepis？= Relhania speciosa（DC.）Harv. ●☆

327645　Relhania squarrosa（L.）L'Hér. = Oedera squarrosa（L.）Anderb. et K. Bremer ●☆

327646　Relhania squarrosa（L.）L'Hér. var. brevifolia Harv. = Oedera squarrosa（L.）Anderb. et K. Bremer ●☆

327647　Relhania styphelioides（DC.）Harv. = Relhania speciosa（DC.）Harv. ●☆

327648　Relhania tomentosa Thunb. = Amphiglossa tomentosa（Thunb.）Harv. ■☆

327649　Relhania tricephala（DC.）K. Bremer;三头寡头鼠麴木●☆

327650　Relhania trinervis Thunb. = Relhania pungens L'Hér. subsp. trinervis（Thunb.）K. Bremer ●☆

327651　Relhania tuberosa K. Bremer = Comborhiza virgata（N. E. Br.）Anderb. et K. Bremer ●☆

327652　Relhania uniflora（L. f.）Druce = Oedera uniflora（L. f.）Anderb. et K. Bremer ●☆

327653　Relhania virgata L'Hér. = Oedera uniflora（L. f.）Anderb. et K. Bremer ●☆

327654　Relhania viscosa L'Hér. = Oedera viscosa（L'Hér.）Anderb. et K. Bremer ●☆

327655　Rellesta Turcz. = Swertia L. ■

327656　Remaclea C. Morren = Trimeza Salisb. ■☆

327657　Rembertia Adans. = Diapensia L. ●

327658　Reme Adans. = Trianthema L. ■

327659　Remijia DC. (1829);铜色树属●☆

327660　Remijia macrophylla Benth. et Hook. f. ex Flueck. ;大叶铜色树●☆

327661　Remijia pedunculata André;铜色树;Cuprea Bark ●☆

327662　Remijia purdieana Wedd. ;普通铜色树;Cuprea Bark ●☆

327663　Remirea Aubl. (1775);海滨莎属（海莎草属）;Remirea ■

327664　Remirea maritima Aubl. ; 海滨莎（海滩莎草）;Beach Star, Remirea, Seashore Remirea ■

327665　Remirea maritima Aubl. = Cyperus pedunculatus（R. Br.）J. Kern ■

327666　Remirea maritima Aubl. = Mariscus pedunculatus（R. Br.）T. Koyama ■

327667　Remirea pedunculata R. Br. = Remirea maritima Aubl. ■

327668　Remirema Kerr(1943);雷米花属☆

327669　Remirema bracteata Kerr;雷米花☆

327670　Remusatia Schott(1832);岩芋属（零余芋属,目贼芋属,曲苞芋属）;Rocktaro ■

327671　Remusatia bulbifera Vilm. = Remusatia vivipara（Lodd.）Schott ■

327672　Remusatia formosana Hayata;台湾岩芋(台湾目贼芋);Taiwan Rocktaro ■

327673　Remusatia garrettii Gagnep. = Gonatanthus pumilus（D. Don）Engl. et Krause ■

327674　Remusatia hookeriana Schott;早花岩芋;Hooker. Rocktaro ■

327675　Remusatia ornata（Schott）H. Li et Q. F. Guo;秀丽岩芋■

327676　Remusatia vivipara（Lodd.）Schott;岩芋（红半夏,红天椒,红岩芋,红芋,红芋头,零余芋,台湾目贼芋,野木芋）;Viviparous Rocktaro ■

327677　Remusatia vivipara（Roxb.）Schott = Remusatia vivipara（Lodd.）Schott ■

327678 Remya Hillebr. ex Benth. = Remya Hillebr. ex Benth. et Hook. f. ●☆

327679 Remya Hillebr. ex Benth. et Hook. f. (1873);黄绒菀属●☆

327680 Remya kauiensis Hillebr.;黄绒菀■☆

327681 Renanthera Lour. (1790);火焰兰属(龙爪兰属,肾药兰属); Flameorchis,Renanthera ■

327682 Renanthera bilinguis Rchb. f. = Arachnis labrosa (Lindl. et Paxton) Rchb. f. ■

327683 Renanthera citrina Aver.;中华火焰兰;Chinese Renanthera ■

327684 Renanthera citrina Aver. var. sinica (Z. J. Liu et S. C. Chen) R. Rice = Renanthera citrina Aver. ■

327685 Renanthera coccinea Lour.;火焰兰(绯红火焰兰,红珊瑚,山 观带,山裙带);Flameorchis,Scarlet Renanthera ■

327686 Renanthera elongata Lindl.;长茎火焰兰;Longstem Renanthera ■☆

327687 Renanthera hybrida Hort.;杂种火焰兰(杂交肾药兰)■☆

327688 Renanthera imschootiana Rolfe;云南火焰兰(矮兰,火焰兰); Common Renanthera,Yunnan Flameorchis,Yunnan Renanthera ■☆

327689 Renanthera labrosa (Lindl. et Paxton) Rchb. f. = Arachnis labrosa (Lindl. et Paxton) Rchb. f. ■

327690 Renanthera labrosa (Lindl.) Rchb. f. = Renanthera labrosa (Lindl. et Paxton) Rchb. f. ■

327691 Renanthera leptantha Fukuy. = Renanthera labrosa (Lindl.) Rchb. f. ■

327692 Renanthera matutina (Blume) Lindl.;晨花火焰兰(豹斑红珊 瑚);Morningflower Renanthera ■☆

327693 Renanthera monachica Ames;冬花火焰兰(豹纹肾药兰); Winterflowering Renanthera ■☆

327694 Renanthera pulchella Rolfe;美丽火焰兰;Beautiful Renanthera ■☆

327695 Renanthera rohaniana Rchb. f.;罗汉氏火焰兰■☆

327696 Renanthera sinica Z. J. Liu et S. C. Chen = Renanthera citrina Aver. ■

327697 Renanthera storiei Rchb. f.;斯托里火焰兰;Storie Renanthera ■☆

327698 Renantherella Ridl. (1896);小火焰兰属■☆

327699 Renantherella histrionica Ridl.;小火焰兰■☆

327700 Renarda Regel = Hymenolaena DC. ■

327701 Renardia (Regel et Schmalh.) Kuntze = Schmalhausenia C. Winkl. ■★

327702 Renardia Moc. et Sessé ex DC. = Rhynchanthera DC. (保留属 名)●☆

327703 Renardia Turcz. = Trimeria Harv. ●☆

327704 Renardia lejocarpa Turcz. = Trimeria grandifolia (Hochst.) Warb. ●☆

327705 Renata Ruschi = Pseudolaelia Porto et Brade ■☆

327706 Rendlia Chiov. = Microchloa R. Br. ■

327707 Rendlia altera (Rendle) Chiov. = Microchloa altera (Rendle) Stapf ■☆

327708 Rendlia annua Kupicha et Cope = Microchloa annua (Kupicha et Cope) Cope ■☆

327709 Rendlia cupricola P. A. Duvign. = Microchloa altera (Rendle) Stapf ■☆

327710 Rendlia nelsonii (Stapf) Chiov. = Microchloa altera (Rendle) Stapf ■☆

327711 Rendlia obtusifolia Chiov. = Microchloa altera (Rendle) Stapf ■☆

327712 Rendlia pseudoharpechloa Chiov. = Harpochloa pseudoharpechloa (Chiov.) Clayton ■☆

327713 Renealmia Houtt. = Villarsia Vent. (保留属名)●☆

327714 Renealmia L. (废弃属名) = Renealmia L. f. (保留属名)■☆

327715 Renealmia L. (废弃属名) = Tillandsia L. ■☆

327716 Renealmia L. f. (1782)(保留属名);雷内姜属(润尼花属)■☆

327717 Renealmia R. Br. = Libertia Spreng. (保留属名)■☆

327718 Renealmia africana (K. Schum.) Benth.;非洲雷内姜(非洲润 尼花)■☆

327719 Renealmia albo-rosea K. Schum.;白红雷内姜■☆

327720 Renealmia alpinia (Rottb.) Maas;高山雷内姜■☆

327721 Renealmia batangana K. Schum.;巴坦加雷内姜■☆

327722 Renealmia battenbergiana Cummins ex Baker;巴氏雷内姜■☆

327723 Renealmia brachythyrsa Loes.;短序雷内姜■☆

327724 Renealmia bracteata De Wild. et T. Durand;具苞雷内姜■☆

327725 Renealmia cabrae De Wild. et T. Durand;卡布拉雷内姜■☆

327726 Renealmia capensis Houtt. = Villarsia capensis (Houtt.) Merr. ■☆

327727 Renealmia cincinnata (K. Schum.) Baker;卷毛雷内姜■☆

327728 Renealmia congoensis Gagnep.;刚果雷内姜■☆

327729 Renealmia congolana De Wild. et T. Durand;康戈尔雷内姜■☆

327730 Renealmia dewevrei De Wild. et T. Durand ex Gagnep.;德韦雷 内姜■☆

327731 Renealmia engleri K. Schum.;恩格勒雷内姜■☆

327732 Renealmia erythroneura Gagnep. = Renealmia cabrae De Wild. et T. Durand ■☆

327733 Renealmia exaltata L. f.;极高雷内姜■☆

327734 Renealmia fischeri K. Schum.;菲舍尔雷内姜■☆

327735 Renealmia guianensis Mass;圭亚那雷内姜■☆

327736 Renealmia ivorensis A. Chev. = Renealmia longifolia K. Schum. ■☆

327737 Renealmia laxa K. Schum.;疏松雷内姜■☆

327738 Renealmia longifolia K. Schum.;长叶雷内姜■☆

327739 Renealmia macrocolea K. Schum.;大雷内姜■☆

327740 Renealmia maculata Stapf;斑点雷内姜■☆

327741 Renealmia mannii Hook. f.;曼氏雷内姜■☆

327742 Renealmia monostachia L. = Guzmania monostachia (L.) Rusby ex Mez ■☆

327743 Renealmia polyantha K. Schum.;多花雷内姜■☆

327744 Renealmia polypus Gagnep.;多足雷内姜■☆

327745 Renealmia recurvata L. = Tillandsia recurvata (L.) L. ■☆

327746 Renealmia sancti-thomae I. M. Turner;非洲山姜■☆

327747 Renealmia sessilifolia Gagnep.;无柄叶雷内姜■☆

327748 Renealmia stenostachya K. Schum.;狭穗雷内姜■☆

327749 Renealmia talbotii Hutch. = Renealmia africana (K. Schum.) Benth. ■☆

327750 Renealmia thyesoidea (Ruiz et Pav.) Poepp. et Endl.;锥花雷 内姜(锥花润尼花)■☆

327751 Renealmia usneoides L.;西班牙雷内姜(西班牙润尼花); Spanish Moss ■☆

327752 Renealmia usneoides L. = Tillandsia usuneoides (L.) L. ■☆

327753 Renggeria Meisn. (1837);伦格藤属●☆

327754 Renggeria comans (Mart.) Meisn. ex Engl.;伦格藤●☆

327755 Rengifa Poepp. et Endl. = Quapoya Aubl. ●☆

327756 Renia Noronha = Inocarpus J. Forst. et G. Forst. (保留属名)●☆

327757 Renistipula Borhidi = Rondeletia L. ●

327758 Renistipula Borhidi(2004);肾叶木属●

327759 Rennellia Korth. (1851);伦内尔茜属●☆

327760 Rennellia elliptica Korth.;椭圆伦内尔茜●☆

327761 Rennellia microcephala (Ridl.) K. M. Wong;小头伦内尔茜●☆

327762 Rennellia ovalis Korth.;卵形伦内尔茜●☆

327763 Rennera Merxm. (1957);皱果菊属■☆

327764 Rennera eenii (S. Moore) Källersjö;埃恩皱果菊■☆

327765　Rennera laxa (Bremek. et Oberm.) Källersjö;舒松皱果菊■☆

327766　Rennera limnophila Merxm. ;皱果菊■☆

327767　Rennera stellata Herman;星皱果菊■☆

327768　Renschia Vatke(1881);伦施木属●☆

327769　Renschia heteroptypica (S. Moore) Vatke;伦施木●☆

327770　Renschia mirabilis Bullock = Tinnea mirabilis (Bullock) Vollesen ●☆

327771　Rensonia S. F. Blake(1923);稻翅菊属●☆

327772　Rensonia salvadorica S. F. Blake;稻翅菊☆

327773　Rensselaeria Beck = Peltandra Raf. (保留属名)■☆

327774　Renvoizea Zuloaga et Morrone = Panicum L. ■

327775　Renvoizea Zuloaga et Morrone(2008);赖恩草属■☆

327776　Renzorchis Szlach. et Olszewski(1998);伦兰属■☆

327777　Renzorchis pseudoplatycoryne Szlach. et Olszewski;伦兰■☆

327778　Repandra Lindl. = Disa P. J. Bergius ■☆

327779　Rephesis Raf. = Ficus L. ●

327780　Rephesis Raf. = Urostigma Gasp. ●

327781　Reptonia A. DC. = Monotheca A. DC. ●☆

327782　Reptonia A. DC. = Sideroxylon L. ●☆

327783　Reptonia buxifolia (Falc.) A. DC. = Monotheca buxifolia (Falc.) A. DC. ●☆

327784　Reptonia buxifolia (Falc.) A. DC. = Sideroxylon mascatense (A. DC.) T. D. Penn. ●☆

327785　Reptonia laurina Benth. = Sarcosperma laurinum (Benth.) Hook. f. ●

327786　Reptonia mascatensis (A. DC.) Radlk. ex O. Schwartz = Monotheca buxifolia (Falc.) A. DC. ●☆

327787　Reptonia mascatensis (A. DC.) Radlk. ex O. Schwartz = Sideroxylon mascatense (A. DC.) T. D. Penn. ●☆

327788　Requienia DC. (1825);勒基灰毛豆属■☆

327789　Requienia DC. = Tephrosia Pers. (保留属名)●■

327790　Requienia obcordata (Lam. ex Poir.) DC. ;勒基灰毛豆■☆

327791　Requienia pseudosphaerosperma (Schinz) Brummitt;假球籽勒基灰毛豆■☆

327792　Requienia sphaerosperma DC. ;球籽勒基灰毛豆■☆

327793　Reseda L. (1753);木犀草属;Mignonette, Reseda ■

327794　Reseda Tourn. ex L. = Reseda L. ■

327795　Reseda alba L. ;白木犀草(白色木犀草);Upright Mignonette, White Mignonette,White Upright Mignonette ■

327796　Reseda alba L. subsp. decursiva (Forssk.) Maire = Reseda decursiva Forssk. ■☆

327797　Reseda alba L. subsp. gayana (Boiss.) Maire = Reseda undata L. ■☆

327798　Reseda alba L. subsp. maritima Maire = Reseda alba L. ■☆

327799　Reseda alba L. subsp. myriosperma (Murb.) Maire;多籽白木犀草■☆

327800　Reseda alba L. subsp. paui Sennen;波氏木犀草■☆

327801　Reseda alba L. subsp. tricuspis (Coss.) Maire = Reseda alba L. subsp. trigyna (Batt.) Greuter et Burdet ■☆

327802　Reseda alba L. subsp. trigyna (Batt.) Greuter et Burdet;三蕊白木犀草■☆

327803　Reseda alba L. var. eremophila (Boiss.) Maire = Reseda decursiva Forssk. ■☆

327804　Reseda alba L. var. firma Müll. Arg. = Reseda alba L. ■

327805　Reseda alba L. var. laetevirens Müll. Arg. = Reseda alba L. ■

327806　Reseda alba L. var. myriosperma (Murb.) Abdallah et de Wit = Reseda alba L. ■

327807　Reseda alba L. var. propinqua (R. Br.) Maire = Reseda decursiva Forssk. ■☆

327808　Reseda alba L. var. subtrimera Maire et Sam. = Reseda alba L. ■

327809　Reseda alba L. var. trigyna (Batt.) Maire = Reseda alba L. subsp. trigyna (Batt.) Greuter et Burdet ■☆

327810　Reseda amblycarpa Fresen. ;钝果木犀草■☆

327811　Reseda amblycarpa Fresen. var. somala Chiov. = Reseda amblycarpa Fresen. ■☆

327812　Reseda arabica Boiss. ;阿拉伯木犀草■☆

327813　Reseda arabica Boiss. subsp. moroccana Abdallah et de Wit;摩洛哥木犀草■☆

327814　Reseda arabica Boiss. var. integrifolia Abdallah et de Wit = Reseda arabica Boiss. ■☆

327815　Reseda attenuata (Ball) Ball = Reseda fruticulosa L. subsp. attenuata (Ball) Maire ■☆

327816　Reseda aucheri Boiss. ;奥切木犀草■☆

327817　Reseda balansae Müll. Arg. ;巴氏木犀草(巴兰氏木犀草)■☆

327818　Reseda battandieri Pit. ;巴坦木犀草■☆

327819　Reseda battandieri Pit. var. limicola (Maire et Sam.) Abdallah et de Wit = Reseda battandieri Pit. ■☆

327820　Reseda battandieri Pit. var. tuberculata (Batt. et Jahand.) Maire = Reseda battandieri Pit. ■☆

327821　Reseda bracteata Boiss. = Reseda pruinosa Delile ■☆

327822　Reseda brevipedunculata N. Busch;短梗木犀草■☆

327823　Reseda bucharica Litv. ;布哈尔木犀草■☆

327824　Reseda capensis Thunb. = Oligomeris dipetala (Aiton) Turcz. ■☆

327825　Reseda carmensylvae Volkens et Schweinf. = Reseda amblycarpa Fresen. ■☆

327826　Reseda collina Müll. Arg. = Reseda phyteuma L. subsp. collina (Müll. Arg.) Batt. ■☆

327827　Reseda confusa Pomel = Reseda phyteuma L. ■☆

327828　Reseda crystallina Webb et Berthel. = Reseda lancerotae Delile ■☆

327829　Reseda crystallina Webb et Berthel. var. graciosae Pit. = Reseda lancerotae Delile ■☆

327830　Reseda decursiva Forssk. ;下延木犀草■☆

327831　Reseda decursiva Forssk. var. eremophila (Boiss.) Maire = Reseda decursiva Forssk. ■☆

327832　Reseda decursiva Forssk. var. propinqua (R. Br.) Maire = Reseda decursiva Forssk. ■☆

327833　Reseda decursiva Forssk. var. trigyna (Batt.) Maire = Reseda decursiva Forssk. ■☆

327834　Reseda diffusa (Ball) Ball;松散木犀草■☆

327835　Reseda dimerocarpa Rouy et Foucaud = Reseda luteola L. ■

327836　Reseda dipetala Aiton = Oligomeris dipetala (Aiton) Turcz. ■☆

327837　Reseda dregeana C. Presl = Oligomeris dregeana (Müll. Arg.) Müll. Arg. ■☆

327838　Reseda duriaeana Müll. Arg. ;乳突木犀草■☆

327839　Reseda duriaeana Müll. Arg. var. papillosa (Müll. Arg.) Abdallah et de Wit = Reseda duriaeana Müll. Arg. ■☆

327840　Reseda dutoitii Sennen et Mauricio = Reseda lutea L. ■

327841　Reseda elata Müll. Arg. ;高木犀草■☆

327842　Reseda elata Müll. Arg. var. malvalii (Maire) Maire = Reseda elata Müll. Arg. ■☆

327843　Reseda elata Müll. Arg. var. villosa Maire = Reseda elata Müll. Arg. ■☆

327844　Reseda ellenbeckii Perkins;埃伦木犀草■☆

327845　Reseda eremophylla Boiss. = Reseda alba L. ■

327846　Reseda fruticulosa L. ;小灌木木犀草●☆

327847　Reseda fruticulosa L. subsp. attenuata（Ball）Maire；渐狭木犀草■☆

327848　Reseda fruticulosa L. subsp. gayana（Boiss.）Maire ＝ Reseda undata L. ■☆

327849　Reseda fruticulosa L. var. attenuata Ball ＝ Reseda fruticulosa L. subsp. attenuata（Ball）Maire ■☆

327850　Reseda gayana Boiss. ＝ Reseda undata L. ■☆

327851　Reseda gayana Boiss. subsp. attenuata Ball ＝ Reseda undata L. ■☆

327852　Reseda gayana Boiss. var. trigyna Batt. ＝ Reseda undata L. ■☆

327853　Reseda gilgiana Perkins ＝ Reseda amblycarpa Fresen. ■☆

327854　Reseda gilgiana Perkins var. nogalensis（Chiov.）Abdallah et de Wit ＝ Reseda amblycarpa Fresen. ■☆

327855　Reseda globulosa Fisch. et C. A. Mey. ；小球木犀草■☆

327856　Reseda gussonei Boiss. ＝ Reseda luteola L. ■

327857　Reseda inodora Rchb. ；无味木犀草■☆

327858　Reseda lanceolata Lag. ；披针形木犀草■☆

327859　Reseda lanceolata Lag. subsp. constricta（Lange）Valdés Berm. ；缢缩木犀草■☆

327860　Reseda lanceolata Lag. var. constricta（Lange）Ball ＝ Reseda lanceolata Lag. subsp. constricta（Lange）Valdés Berm. ■☆

327861　Reseda lanceolata Lag. var. maura Maire ＝ Reseda lanceolata Lag. ■☆

327862　Reseda lanceolata Lag. var. trifida Pau et Font Quer ＝ Reseda lanceolata Lag. ■☆

327863　Reseda lancerotae Delile；兰斯罗特木犀草■☆

327864　Reseda linifolia Vahl ＝ Oligomeris linifolia（Vahl）J. F. Macbr. ■

327865　Reseda linifolia Vahl ex Hornem. ＝ Oligomeris linifolia（Vahl ex Hornem.）J. F. Macbr. ■

327866　Reseda lutea L. ；黄木犀草（黄色木犀草，细叶木犀草）；Crambling Rocket, Cut-leaf Mignonette, Dyer's-rocket, Italian Rocket, Lest-we-forget, Little Darling, Mignonette, Wild Mignonette, Yellow Mignonette ■

327867　Reseda lutea L. subsp. neglecta（Müll. Arg.）Abdallah et de Wit；忽视黄木犀草■☆

327868　Reseda lutea L. subsp. neglecta（Müll. Arg.）Ball ＝ Reseda lutea L. subsp. neglecta（Müll. Arg.）Abdallah et de Wit ■☆

327869　Reseda lutea L. var. australis Müll. Arg. ＝ Reseda luteola L. ■

327870　Reseda lutea L. var. crispata（Link）Müll. Arg. ＝ Reseda luteola L. ■

327871　Reseda lutea L. var. gussonei（Boiss. et Reut.）Müll. Arg. ＝ Reseda luteola L. ■

327872　Reseda lutea L. var. maritima Müll. Arg. ＝ Reseda lutea L. ■

327873　Reseda lutea L. var. mucronulata（Tineo）Müll. Arg. ＝ Reseda lutea L. ■

327874　Reseda lutea L. var. muelleri（Boiss.）Abdallah et de Wit；米勒木犀草■☆

327875　Reseda lutea L. var. nutans Boiss. ；俯垂木犀草■☆

327876　Reseda lutea L. var. pulchella Müll. Arg. ＝ Reseda lutea L. ■

327877　Reseda lutea L. var. stricta Müll. Arg. ＝ Reseda lutea L. ■

327878　Reseda lutea L. var. subreflexa Maire ＝ Reseda lutea L. ■

327879　Reseda luteola L. ；淡黄木犀草（彩色木犀草，木犀草）；Base Rocket, Dutch Pink Plant, Dyer's Greenweed, Dyer's Rocket, Dyer's Weed, Dyer's Weld, Dyer's Yellow, Greenweed, Jerusalem Ash, Weld, Weld Mignonette, Wild Rocket, Wild Woad, Woad, Wold, Yellow Weed ■

327880　Reseda luteola L. var. australis Müll. Arg. ＝ Reseda luteola L. ■

327881　Reseda luteola L. var. crispata（Link）Ball ＝ Reseda luteola L. ■

327882　Reseda luteola L. var. gussonei（Boiss.）Müll. Arg. ＝ Reseda luteola L. ■

327883　Reseda luteola L. var. hispanica Kuntze ＝ Reseda luteola L. ■

327884　Reseda malvalii Maire ＝ Reseda elata Müll. Arg. ■☆

327885　Reseda massae Chiov. ；马萨木犀草■☆

327886　Reseda media Lag. ；中间木犀草■☆

327887　Reseda microcarpa Müll. Arg. ；小果木犀草■☆

327888　Reseda microphylla C. Presl；小叶木犀草■☆

327889　Reseda microphylla C. Presl ＝ Oligomeris dipetala（Aiton）Turcz. ■☆

327890　Reseda muricata C. Presl；粗糙木犀草■☆

327891　Reseda myriosperma Murb. ＝ Reseda alba L. subsp. myriosperma（Murb.）Maire ■☆

327892　Reseda nainii Maire；奈恩木犀草■☆

327893　Reseda nainii Maire subsp. integrifolia Maire et Weiller ＝ Reseda nainii Maire ■☆

327894　Reseda nilgherrensis Müll. Arg. ＝ Reseda odorata L. ■

327895　Reseda nogalensis Chiov. ＝ Reseda amblycarpa Fresen. ■☆

327896　Reseda odorata L. ；木犀草（草木犀，香草）；Common Mignonette, Garden Mignonette, Mignonette, Scented Mignonette, Sweet Mignonette, Sweet-scented Mignonette ■

327897　Reseda oligomeroides Schinz ＝ Reseda amblycarpa Fresen. ■☆

327898　Reseda orientalis（Müll. Arg.）Boiss. ；东方木犀草■☆

327899　Reseda orientalis Boiss. ＝ Reseda orientalis（Müll. Arg.）Boiss. ■☆

327900　Reseda pampaniniana Maire et Weiller ＝ Reseda urnigera Webb ■☆

327901　Reseda papillosa Müll. Arg. ＝ Reseda duriaeana Müll. Arg. ■☆

327902　Reseda papillosa Müll. Arg. var. duriaeana（Müll. Arg.）Batt. ＝ Reseda duriaeana Müll. Arg. ■☆

327903　Reseda phyteuma Kralik ex Müll. Arg. ＝ Reseda phyteuma L. ■☆

327904　Reseda phyteuma L. ；牧根木犀草；Common Mignonette, Corn Mignonette, Rampion Mignonette ■☆

327905　Reseda phyteuma L. subsp. collina（Müll. Arg.）Batt. ；山丘木犀草■☆

327906　Reseda phyteuma L. subsp. diffusa Ball ＝ Reseda diffusa（Ball）Ball ■☆

327907　Reseda phyteuma L. subsp. media（Lag.）Ball ＝ Reseda media Lag. ■☆

327908　Reseda phyteuma L. var. fragrans Texidor ＝ Reseda phyteuma L. ■☆

327909　Reseda phyteuma L. var. integrifolia Texidor ＝ Reseda phyteuma L. ■☆

327910　Reseda propinqua R. Br. ＝ Reseda decursiva Forssk. ■☆

327911　Reseda propinqua R. Br. var. decursiva（Forssk.）Batt. ＝ Reseda decursiva Forssk. ■☆

327912　Reseda pruinosa Delile；白粉木犀草■☆

327913　Reseda pseudovirens Friv. ＝ Reseda luteola L. ■

327914　Reseda scoparia Willd. ；帚状木犀草■☆

327915　Reseda sessilifolia Thulin；无柄叶木犀草■☆

327916　Reseda somalensis Baker f. ；索马里木犀草■☆

327917　Reseda stenobotrys Maire et Sam. ；窄穗木犀草■☆

327918　Reseda stricta Pers. ＝ Reseda lutea L. ■

327919　Reseda stricta Pers. var. reuteriana Müll. Arg. ＝ Reseda stricta Pers. ■

327920　Reseda subulata Delile ＝ Oligomeris linifolia（Vahl ex Hornem.）J. F. Macbr. ■

327921　Reseda telephiifolia（Chiov.）Abdallah et de Wit；耳托指甲草

状木犀草■☆

327922 Reseda tomentosa Boiss. ;绒毛木犀草■☆

327923 Reseda tricuspis Coss. = Reseda alba L. subsp. trigyna（Batt.）Greuter et Burdet ■☆

327924 Reseda undata L. ;钝波木犀草■☆

327925 Reseda urnigera Webb;北非木犀草■☆

327926 Reseda villosa Coss. ;长柔毛木犀草■☆

327927 Reseda villosa Coss. var. garamantum Maire = Reseda villosa Coss. ■☆

327928 Reseda villosa Coss. var. glabrescens Maire;渐光木犀草■☆

327929 Reseda villosa Coss. var. typica Maire = Reseda villosa Coss. ■☆

327930 Resedaceae Bercht. et J. Presl = Resedaceae Martinov（保留科名）■●

327931 Resedaceae DC. ex Gray = Astrocarpaceae A. Kern. ■●

327932 Resedaceae DC. ex Gray = Resedaceae Martinov（保留科名）■●

327933 Resedaceae Gray = Resedaceae Martinov（保留科名）■●

327934 Resedaceae Martinov（1820）（保留科名）;木犀草科;Mignonette Family ■●

327935 Resedella Webb et Berthel. = Oligomeris Cambess.（保留属名）■●

327936 Resedella dipetala（Aiton）Webb et Berthel. ;小木犀草■☆

327937 Resedella dipetala（Aiton）Webb et Berthel. = Oligomeris dipetala（Aiton）Turcz. ■☆

327938 Resedella dipetala Webb et Berthel. = Resedella dipetala（Aiton）Webb et Berthel. ■☆

327939 Resedella dregeana Müll. Arg. = Oligomeris dregeana（Müll. Arg.）Müll. Arg. ■☆

327940 Resia H. E. Moore(1962);雷斯芭苔属■●☆

327941 Resia nimbicola H. E. Moore;雷斯芭苔■●☆

327942 Resinanthus（Borhidi）Borhidi = Antirhea Comm. ex Juss. ●

327943 Resinanthus（Borhidi）Borhidi = Stenostomum C. F. Gaertn. ●

327944 Resinanthus（Borhidi）Borhidi(2007);脂花木属●☆

327945 Resinaria Comm. ex Lam. = Terminalia L.（保留属名）●

327946 Resinocaulon Lunell = Silphium L. ■

327947 Resnova Van der Merwe = Drimiopsis Lindl. et Paxton ■☆

327948 Resnova Van der Merwe(1946);肖辛酸木属■☆

327949 Resnova humifusa（Baker）U. Müll. -Doblies et D. Müll. -Doblies;肖辛酸木■☆

327950 Resnova lachenalioides（Baker）Van der Merwe;立金花肖辛酸木■☆

327951 Resnova maxima Van der Merwe;大肖辛酸木■☆

327952 Resnova minor Van der Merwe;小肖辛酸木■☆

327953 Resnova pilosa Van der Merwe;疏毛肖辛酸木■☆

327954 Resnova schlechteri（Baker）Van der Merwe = Resnova humifusa（Baker）U. Müll. -Doblies et D. Müll. -Doblies ■☆

327955 Resnova transvaalensis Van der Merwe = Resnova humifusa（Baker）U. Müll. -Doblies et D. Müll. -Doblies ■☆

327956 Restella Pobed.（1941）;中美瑞香属●☆

327957 Restella Pobed. = Wikstroemia Endl.（保留属名）●

327958 Restella alberti（Regel）Pobed. ;中美瑞香●☆

327959 Restiaria Kuntze = Commersonia J. R. Forst. et G. Forst. + Rulingia R. Br.（保留属名）●☆

327960 Restiaria Lour. = Uncaria Schreb.（保留属名）●

327961 Restio L.（废弃属名）= Restio Rottb.（保留属名）■☆

327962 Restio L.（废弃属名）= Thamnochortus P. J. Bergius ■☆

327963 Restio Rottb.（1772）（保留属名）;绳草属;Rope Grass, Rope-grass ■☆

327964 Restio acockii Pillans;阿科科绳草■☆

327965 Restio acuminatus Kunth = Cannomois parviflora（Thunb.）Pillans ■☆

327966 Restio acuminatus Thunb. = Chondropetalum nudum Rottb. ■☆

327967 Restio alboaristatus Nees = Hypodiscus alboaristatus（Nees）Mast. ■☆

327968 Restio alticola Pillans;高原绳草■☆

327969 Restio ambiguus Mast. ;可疑绳草■☆

327970 Restio ameles Steud. = Ischyrolepis subverticillata Steud. ■☆

327971 Restio anceps（Mast.）Pillans = Platycaulos anceps（Mast.）H. P. Linder ■☆

327972 Restio araneosus Mast. = Ischyrolepis triflora（Rottb.）H. P. Linder ■☆

327973 Restio arcuatus Mast. ;弓绳草■☆

327974 Restio argenteus Thunb. = Hypodiscus argenteus（Thunb.）Mast. ■☆

327975 Restio aridus Pillans = Ischyrolepis arida（Pillans）H. P. Linder ■☆

327976 Restio aristatus Thunb. = Hypodiscus aristatus（Thunb.）C. Krauss ■☆

327977 Restio articulatus Retz. = Lepironia articulata（Retz.）Domin ■

327978 Restio aspericaulis Pillans = Platycaulos anceps（Mast.）H. P. Linder ■☆

327979 Restio asperiflorus Nees = Elegia asperiflora（Nees）Kunth ■☆

327980 Restio bifarius Mast. ;二行绳草■☆

327981 Restio bifidus Thunb. ;二裂绳草■☆

327982 Restio bigeminus Mast. = Restio micans Nees ■☆

327983 Restio bolusii Pillans;博卢斯绳草■☆

327984 Restio brachiatus（Mast.）Pillans;短绳草■☆

327985 Restio brownianus（Mast.）Pillans = Restio debilis Nees ■☆

327986 Restio brunneus Pillans;褐色绳草■☆

327987 Restio burchellii Pillans;伯切尔绳草■☆

327988 Restio callistachyus Kunth = Platycaulos callistachyus（Kunth）H. P. Linder ■☆

327989 Restio capillaris Kunth;发状绳草■☆

327990 Restio cascadensis Pillans = Platycaulos cascadensis（Pillans）H. P. Linder ■☆

327991 Restio cernuus L. f. = Staberoha cernua（L. f.）T. Durand et Schinz ■☆

327992 Restio chondropetalum Nees = Chondropetalum deustum Rottb. ■☆

327993 Restio cincinnatus Mast. = Ischyrolepis cincinnata（Mast.）H. P. Linder ■☆

327994 Restio cirratus Mast. = Ischyrolepis gaudichaudiana（Kunth）H. P. Linder ■☆

327995 Restio communis Pillans;普通绳草■☆

327996 Restio comosus N. E. Br. = Rhodocoma gigantea（Kunth）H. P. Linder ■☆

327997 Restio compressus Rottb. = Platycaulos compressus（Rottb.）H. P. Linder ■☆

327998 Restio compressus Rottb. var. major Mast. = Platycaulos major（Mast.）H. P. Linder ■☆

327999 Restio concolor Steud. = Platycaulos callistachyus（Kunth）H. P. Linder ■☆

328000 Restio confusus Pillans;混乱绳草■☆

328001 Restio consimilis Mast. = Ischyrolepis sieberi（Kunth）H. P. Linder ■☆

328002 Restio conspicuus（Mast.）Pillans = Restio dispar Mast. ■☆

328003 Restio corneolus Esterh. ;小角绳草■☆

328004　Restio crinalis Mast. = Anthochortus crinalis (Mast.) H. P. Linder ■☆

328005　Restio cuspidatus Thunb. = Ischyrolepis capensis (L.) H. P. Linder ■☆

328006　Restio cymosus (Mast.) Pillans；聚伞绳草 ■☆

328007　Restio debilis Nees；弱小绳草 ■☆

328008　Restio debilis Nees var. subulatus (Mast.) Pillans = Restio debilis Nees ■☆

328009　Restio decipiens (N. E. Br.) H. P. Linder；迷惑绳草 ■☆

328010　Restio degenerans Pillans；不纯绳草 ■☆

328011　Restio depauperatus Kunth = Platycaulos depauperatus (Kunth) H. P. Linder ■☆

328012　Restio dichotomus L. = Ischyrolepis capensis (L.) H. P. Linder ■☆

328013　Restio digitatus Thunb. = Mastersiella digitata (Thunb.) Gilg-Ben. ●☆

328014　Restio dimorphostachyus Mast. = Restio pachystachyus Kunth ■☆

328015　Restio dispar Mast. ；异型绳草 ■☆

328016　Restio distachyos Rottb. = Staberoha distachyos (Rottb.) Kunth ■☆

328017　Restio distans Pillans；远离绳草 ■☆

328018　Restio distichus Rottb. ；二列绳草 ■☆

328019　Restio distractus Mast. = Ischyrolepis distracta (Mast.) H. P. Linder ■☆

328020　Restio divaricatus Mast. = Ischyrolepis sieberi (Kunth) H. P. Linder ■☆

328021　Restio dodii Pillans；多德绳草 ■☆

328022　Restio dodii Pillans var. purpureus？ = Restio dodii Pillans ■☆

328023　Restio echinatus Kunth；具刺绳草 ■☆

328024　Restio ecklonii Mast. = Calopsis viminea (Rottb.) H. P. Linder ■☆

328025　Restio egregius Hochst. ；优秀绳草 ■☆

328026　Restio egregius Hochst. var. nutans Mast. = Restio egregius Hochst. ■☆

328027　Restio elatus Mast. = Ischyrolepis gaudichaudiana (Kunth) H. P. Linder ■☆

328028　Restio elegans Poir. = Cannomois virgata (Rottb.) Steud. ■☆

328029　Restio elegia Murray = Elegia juncea L. ■☆

328030　Restio eleocharis Mast. = Ischyrolepis eleocharis (Mast.) H. P. Linder ■☆

328031　Restio elongatus Thunb. = Thamnochortus erectus (Thunb.) Mast. ●☆

328032　Restio erectus Thunb. = Thamnochortus erectus (Thunb.) Mast. ●☆

328033　Restio esterhuyseniae Pillans = Ischyrolepis esterhuyseniae (Pillans) H. P. Linder ■☆

328034　Restio exilis Mast. ；瘦小绳草 ■☆

328035　Restio fastigiatus Mast. = Platycaulos callistachyus (Kunth) H. P. Linder ■☆

328036　Restio ferruginosus Kunth = Ischyrolepis gaudichaudiana (Kunth) H. P. Linder ■☆

328037　Restio festuciformis Mast. ；羊茅状绳草 ■☆

328038　Restio filicaulis Pillans；线茎绳草 ■☆

328039　Restio filiformis Poir. ；线形绳草 ■☆

328040　Restio filiformis Poir. var. monostachyus (Mast.) Mast. = Restio filiformis Poir. ■☆

328041　Restio filiformis Poir. var. oligostachyus (Mast.) Mast. = Restio filiformis Poir. ■☆

328042　Restio foliosus N. E. Br. = Rhodocoma gigantea (Kunth) H. P. Linder ■☆

328043　Restio fourcadei Pillans；富尔卡德绳草 ■☆

328044　Restio fragilis Esterh. ；脆绳草 ■☆

328045　Restio fraternus Kunth = Ischyrolepis fraterna (Kunth) H. P. Linder ■☆

328046　Restio fruticosus Thunb. = Rhodocoma fruticosa (Thunb.) H. P. Linder ■☆

328047　Restio fuirenoides Kunth = Ischyrolepis setiger (Kunth) H. P. Linder ■☆

328048　Restio furcatus Mast. = Restio pachystachyus Kunth ■☆

328049　Restio fuscidulus Pillans = Ischyrolepis fuscidula (Pillans) H. P. Linder ■☆

328050　Restio fusiformis Pillans；纺锤形绳草 ■☆

328051　Restio galpinii Pillans；盖尔绳草 ■☆

328052　Restio garnotianus Kunth = Restio filiformis Poir. ■☆

328053　Restio garnotianus Kunth var. monostachyus Mast. = Restio filiformis Poir. ■☆

328054　Restio garnotianus Kunth var. oligostachyus Mast. = Restio filiformis Poir. ■☆

328055　Restio gaudichaudianus Kunth；戈迪绍绳草 ■☆

328056　Restio gaudichaudianus Kunth = Ischyrolepis gaudichaudiana (Kunth) H. P. Linder ■☆

328057　Restio gaudichaudianus Kunth var. luxurians Pillans = Ischyrolepis gaudichaudiana (Kunth) H. P. Linder ■☆

328058　Restio gaudichaudianus Kunth var. microstachyus Mast. = Ischyrolepis gaudichaudiana (Kunth) H. P. Linder ■☆

328059　Restio giganteus (Kunth) N. E. Br. = Rhodocoma gigantea (Kunth) H. P. Linder ■☆

328060　Restio glomeratus Thunb. = Willdenowia glomerata (Thunb.) H. P. Linder ■☆

328061　Restio graminifolius Kunth = Anthochortus graminifolius (Kunth) H. P. Linder ■☆

328062　Restio grandis Nees = Elegia grandis (Nees) Kunth ■☆

328063　Restio harveyi Mast. ；哈维绳草 ■☆

328064　Restio helenae Mast. = Ischyrolepis helenae (Mast.) H. P. Linder ■☆

328065　Restio humilis Pillans = Ischyrolepis wallichii (Mast.) H. P. Linder ■☆

328066　Restio hystrix Mast. = Ischyrolepis hystrix (Mast.) H. P. Linder ■☆

328067　Restio imbricatus Thunb. = Staberoha distachyos (Rottb.) Kunth ■☆

328068　Restio implexus Mast. = Restio capillaris Kunth ■☆

328069　Restio implicatus Esterh. ；纠缠绳草 ■☆

328070　Restio impolitus Kunth = Calopsis impolita (Kunth) H. P. Linder ■☆

328071　Restio inconspicuus Esterh. ；显著绳草 ■☆

328072　Restio incurvatus Pillans = Calopsis viminea (Rottb.) H. P. Linder ■☆

328073　Restio incurvatus Thunb. = Willdenowia incurvata (Thunb.) H. P. Linder ■☆

328074　Restio insignis Pillans；标志绳草 ■☆

328075　Restio intermedius Kunth = Ischyrolepis sieberi (Kunth) H. P. Linder ■☆

328076　Restio intermedius Steud. = Elegia intermedia (Steud.) Pillans ☆

328077　Restio intricatus Mast. = Restio patens Mast. ■☆

328078　Restio involutus Pillans；内卷绳草 ■☆

328079　Restio junceus (L.) Nees = Elegia juncea L. ■☆

328080　Restio kunthii Steud. = Ischyrolepis triflora (Rottb.) H. P.

Linder ■☆

328081　Restio laniger Kunth = Ischyrolepis laniger（Kunth）H. P. Linder ■☆

328082　Restio laniger Kunth var. distractus（Mast.）Pillans = Ischyrolepis distracta（Mast.）H. P. Linder ■☆

328083　Restio leptoclados Mast. = Ischyrolepis leptoclados（Mast.）H. P. Linder ■☆

328084　Restio leptostachyus Kunth；细穗绳草■☆

328085　Restio luceanus Kunth = Ischyrolepis gaudichaudiana（Kunth）H. P. Linder ■☆

328086　Restio lucens Poir. = Thamnochortus lucens（Poir.）H. P. Linder ●■☆

328087　Restio lucens Poir. var. minor Mast. = Thamnochortus lucens（Poir.）H. P. Linder ●☆

328088　Restio ludwigii Steud. = Ischyrolepis tenuissima（Kunth）H. P. Linder ■☆

328089　Restio macer Kunth = Ischyrolepis macer（Kunth）H. P. Linder ■☆

328090　Restio macowanii Pillans = Rhodocoma fruticosa（Thunb.）H. P. Linder ■☆

328091　Restio madagascariensis Cherm. = Restio mahonii（N. E. Br.）Pillans ■☆

328092　Restio mahonii（N. E. Br.）Pillans；马洪绳草■☆

328093　Restio major（Mast.）Pillans = Platycaulos major（Mast.）H. P. Linder ■☆

328094　Restio marlothii Pillans = Ischyrolepis marlothii（Pillans）H. P. Linder ■☆

328095　Restio marlothii Pillans var. parviflorus ? = Ischyrolepis marlothii（Pillans）H. P. Linder ■☆

328096　Restio mastersii F. Muell. = Platycaulos callistachyus（Kunth）H. P. Linder ■☆

328097　Restio membranaceus Nees = Elegia intermedia（Steud.）Pillans ■☆

328098　Restio micans Nees；弱光泽绳草■☆

328099　Restio miser Kunth；贫弱绳草■☆

328100　Restio mlanjiensis H. P. Linder；梅兰杰绳草■☆

328101　Restio monanthos Mast. = Ischyrolepis monanthos（Mast.）H. P. Linder ■☆

328102　Restio monostachyus Steud. = Restio filiformis Poir. ■☆

328103　Restio montanus Esterh. ；山地绳草■☆

328104　Restio mucronatus Nees = Chondropetalum mucronatum（Nees）Pillans ■☆

328105　Restio multicurvus N. E. Br. = Restio filiformis Poir. ■☆

328106　Restio multiflorus Spreng. ；多花绳草■☆

328107　Restio multiflorus Spreng. var. tuberculatus Pillans = Restio multiflorus Spreng. ☆

328108　Restio neesii Mast. = Ischyrolepis sieberi（Kunth）H. P. Linder ■☆

328109　Restio nodosus Pillans；多节绳草■☆

328110　Restio nudus（Rottb.）Nees = Chondropetalum nudum Rottb. ■☆

328111　Restio nutans Steud. = Ischyrolepis tenuissima（Kunth）H. P. Linder ■☆

328112　Restio nutans Thunb. = Thamnochortus nutans（Thunb.）Pillans ●☆

328113　Restio nuwebergensis Esterh. ；纽沃绳草■☆

328114　Restio oblongus Mast. = Anthochortus crinalis（Mast.）H. P. Linder ■☆

328115　Restio obscurus Pillans；隐匿绳草■☆

328116　Restio obtusissimus Steud. = Nevillea obtusissima（Steud.）H.

328117　Restio occultus（Mast.）Pillans；隐蔽绳草■☆

328118　Restio oligostachyus Kunth = Restio filiformis Poir. ■☆

328119　Restio pachystachyus Kunth；粗穗绳草■☆

328120　Restio paludosus Pillans = Ischyrolepis paludosa（Pillans）H. P. Linder ■☆

328121　Restio paniculatus Rottb. = Calopsis paniculata（Rottb.）Desv. ■☆

328122　Restio pannosus Mast. = Restio triticeus Rottb. ■☆

328123　Restio papyraceus Pillans；纸质绳草■☆

328124　Restio parviflorus Thunb. = Cannomois parviflora（Thunb.）Pillans ■☆

328125　Restio patens Mast.；铺展绳草■☆

328126　Restio pauciflorus Poir. = Calopsis viminea（Rottb.）H. P. Linder ■☆

328127　Restio peculiaris Esterh.；特殊绳草■☆

328128　Restio pedicellatus Mast.；梗花绳草■☆

328129　Restio penicillatus Mast. = Ischyrolepis leptoclados（Mast.）H. P. Linder ■☆

328130　Restio perplexus Kunth；缠结绳草■☆

328131　Restio perplexus Kunth var. gracilis Mast. = Restio capillaris Kunth ■☆

328132　Restio pillansii H. P. Linder；皮朗斯绳草■☆

328133　Restio polystachyus Kunth = Platycaulos callistachyus（Kunth）H. P. Linder ■☆

328134　Restio pondoensis Mast. = Calopsis paniculata（Rottb.）Desv. ■☆

328135　Restio praefixus Mast. = Platycaulos compressus（Rottb.）H. P. Linder ■☆

328136　Restio procurrens Mast. = Ischyrolepis helenae（Mast.）H. P. Linder ■☆

328137　Restio productus Mast. = Ischyrolepis helenae（Mast.）H. P. Linder ■☆

328138　Restio propinquus Nees = Elegia juncea L. ■☆

328139　Restio protractus Mast. = Restio multiflorus Spreng. ■☆

328140　Restio pseudoleptocarpus Kunth = Restio bifidus Thunb. ■☆

328141　Restio pulvinatus Esterh. ；叶枕绳草■☆

328142　Restio pumilus Esterh. ；矮绳草■☆

328143　Restio punctulatus Mast. = Restio distichus Rottb. ■☆

328144　Restio purpurascens Mast. ；紫绳草■☆

328145　Restio pusillus Pillans = Restio leptostachyus Kunth ■☆

328146　Restio pycnostachyus Mast. = Ischyrolepis sieberi（Kunth）H. P. Linder ■☆

328147　Restio pygmaeus Pillans = Ischyrolepis pygmaea（Pillans）H. P. Linder ■☆

328148　Restio quadratus Mast. ；四方形绳草■☆

328149　Restio quartziticola H. P. Linder；阔茨绳草■☆

328150　Restio quinquefarius Nees；五列绳草■☆

328151　Restio racemosus Poir. = Elegia racemosa（Poir.）Pers. ■☆

328152　Restio ramiflorus Nees = Calopsis paniculata（Rottb.）Desv. ■☆

328153　Restio rarus Esterh. ；稀少绳草■☆

328154　Restio rhodocoma Mast. = Rhodocoma capensis Steud. ■☆

328155　Restio rottboellioides Kunth = Ischyrolepis rottboellioides（Kunth）H. P. Linder ■☆

328156　Restio rupicola Esterh. ；岩生绳草■☆

328157　Restio sabulosus Pillans = Ischyrolepis sabulosa（Pillans）H. P. Linder ■☆

328158　Restio sarocladus Mast. ；柔枝绳草■☆

328159　Restio scaber Mast. ；粗糙绳草■☆

328160　Restio scaberulus N. E. Br. ;略粗糙绳草■☆

328161　Restio scariosus Thunb. = Thamnochortus fruticosus P. J. Bergius ●☆

328162　Restio schlechteri（Mast.）Pillans = Restio occultus（Mast.）Pillans ■☆

328163　Restio schoenoides Kunth = Ischyrolepis schoenoides（Kunth）H. P. Linder ■☆

328164　Restio scopa Thunb. = Cannomois virgata（Rottb.）Steud. ■☆

328165　Restio scoparius Kunth = Ischyrolepis sieberi（Kunth）H. P. Linder ■☆

328166　Restio scopula Mast. = Restio perplexus Kunth ■☆

328167　Restio secundus（Pillans）H. P. Linder;单侧绳草■☆

328168　Restio setiger Kunth = Ischyrolepis setiger（Kunth）H. P. Linder ■☆

328169　Restio sieberi Kunth = Ischyrolepis sieberi（Kunth）H. P. Linder ■☆

328170　Restio sieberi Kunth var. schoenoides（Kunth）Pillans = Ischyrolepis schoenoides（Kunth）H. P. Linder ■☆

328171　Restio sieberi Kunth var. venustulus（Kunth）Pillans = Ischyrolepis sieberi（Kunth）H. P. Linder ■☆

328172　Restio similis Pillans;相似绳草■☆

328173　Restio singularis Esterh. ;单一绳草■☆

328174　Restio sonderianus Mast. = Restio pedicellatus Mast. ■☆

328175　Restio sparsus Mast. = Restio strobilifer Kunth ■☆

328176　Restio spicifer Poir. = Staberoha vaginata（Thunb.）Pillans ■☆

328177　Restio spicigerus Thunb. = Thamnochortus spicigerus（Thunb.）Spreng. ●☆

328178　Restio spiculatus Mast. = Ischyrolepis sieberi（Kunth）H. P. Linder ■☆

328179　Restio spinulosus Kunth = Platycaulos callistachyus（Kunth）H. P. Linder ■☆

328180　Restio stereocaulis Mast. ;扭茎绳草■☆

328181　Restio stokoei Pillans;斯托克绳草■☆

328182　Restio strictus N. E. Br. ;刚直绳草■☆

328183　Restio strobilifer Kunth;球果绳草■☆

328184　Restio subcompressus Pillans = Platycaulos subcompressus（Pillans）H. P. Linder ■☆

328185　Restio subfalcatus Mast. = Ischyrolepis sieberi（Kunth）H. P. Linder ■☆

328186　Restio subtilis Mast. ;纤细绳草■☆

328187　Restio subulatus Mast. = Restio debilis Nees ■☆

328188　Restio subverticillatus（Steud.）Mast. = Ischyrolepis subverticillata Steud. ■☆

328189　Restio sulcatus Kunth = Hypodiscus willdenowia（Nees）Mast. ■☆

328190　Restio synchroolepis Steud. = Hypodiscus synchroolepis（Steud.）Mast. ■☆

328191　Restio tabularis Pillans = Restio sarocladus Mast. ■☆

328192　Restio tectorum L. f. = Chondropetalum tectorum（L. f.）Raf. ■☆

328193　Restio tenuissimus Kunth = Ischyrolepis tenuissima（Kunth）H. P. Linder ■☆

328194　Restio tetragonus Thunb. ;四角绳草■☆

328195　Restio tetraphyllus Labill. ;四叶绳草■☆

328196　Restio tetraphyllus Labill. subsp. meiostachyus ?;小穗四叶绳草;Tassel Cord Rush ■☆

328197　Restio tetrasepalus Steud. = Staberoha aemula（Kunth）Pillans ■☆

328198　Restio thamnochortus Thunb. = Thamnochortus lucens（Poir.）H. P. Linder ●☆

328199　Restio thyrsifer Rottb. = Elegia thyrsifera（Rottb.）Pers. ■☆

328200　Restio trichocaulis Mast. = Restio perplexus Kunth ■☆

328201　Restio triflorus Rottb. = Ischyrolepis triflora（Rottb.）H. P. Linder ■☆

328202　Restio triticeus Rottb. ;澳非绳草■☆

328203　Restio tuberculatus Pillans;多疣绳草■☆

328204　Restio umbellatus Thunb. = Staberoha distachyos（Rottb.）Kunth ■☆

328205　Restio vaginatus Thunb. = Staberoha vaginata（Thunb.）Pillans ■☆

328206　Restio venustulus Kunth = Ischyrolepis sieberi（Kunth）H. P. Linder ■☆

328207　Restio verrucosus Esterh. ;密疣绳草■☆

328208　Restio versatilis H. P. Linder;丁字绳草■☆

328209　Restio verticillaris L. f. = Elegia capensis（Burm. f.）Schelpe ■☆

328210　Restio vilis Kunth = Ischyrolepis vilis（Kunth）H. P. Linder ■☆

328211　Restio vimineus Rottb. = Calopsis viminea（Rottb.）H. P. Linder ■☆

328212　Restio virgatus Rottb. = Cannomois virgata（Rottb.）Steud. ■☆

328213　Restio virgeus Mast. = Ischyrolepis virgea（Mast.）H. P. Linder ■☆

328214　Restio wallichii Mast. = Ischyrolepis wallichii（Mast.）H. P. Linder ■☆

328215　Restio xyridioides Kunth = Restio quinquefarius Nees ■☆

328216　Restio zuluensis H. P. Linder;祖卢绳草■☆

328217　Restio zwartbergensis Pillans;茨瓦特伯格绳草■☆

328218　Restionaceae R. Br.（1810）(保留科名);帚灯草科;Ropegrass Family ■

328219　Restrepia Kunth(1816);雷氏兰属■☆

328220　Restrepia biflora Regel;双花雷氏兰■☆

328221　Restrepia maculata Lindl. ;斑点雷氏兰■☆

328222　Restrepia parvifolia Lindl. ;小叶雷氏兰■☆

328223　Restrepiella Garay et Dunst.（1966）;小雷氏兰属■☆

328224　Restrepiella ophiocephala（Lindl.）Garay et Dunst. ;小雷氏兰■☆

328225　Restrepiopsis Luer(1978);拟雷氏兰属■☆

328226　Restrepiopsis microptera（Schltr.）Luer;小翅拟雷氏兰■☆

328227　Resupinaria Raf. = Agati Adans.（废弃属名)●■

328228　Resuptnaria Raf. = Sesbania Scop.（保留属名)●■

328229　Retalaceae Dulac = Pyrolaceae Lindl.（保留科名)●■

328230　Retama Boiss. = Genista L. ●

328231　Retama Boiss. = Lygos Adans.（废弃属名)●☆

328232　Retama Raf.（1838)（保留属名）;勒塔木属（杜松豆属）;Retam ●☆

328233　Retama atlantica Pomel = Retama sphaerocarpa（L.）Boiss. ●☆

328234　Retama bovei Spach = Retama raetam（Forssk.）Webb subsp. bovei（Spach）Talavera et Gibbs ●☆

328235　Retama dasycarpa Coss. ;毛果勒塔木●☆

328236　Retama duriaei Spach = Retama raetam（Forssk.）Webb ●☆

328237　Retama monosperma（L.）Boiss. ;紫萼勒塔木;Bridal Broom ●☆

328238　Retama monosperma（L.）Boiss. subsp. bovei（Spach）Maire = Retama raetam（Forssk.）Webb subsp. bovei（Spach）Talavera et Gibbs ●☆

328239　Retama monosperma（L.）Boiss. var. bovei（Spach）Pau = Retama raetam（Forssk.）Webb subsp. bovei（Spach）Talavera et Gibbs ●☆

328240　Retama monosperma（L.）Boiss. var. hipponensis（Webb）Maire = Retama raetam（Forssk.）Webb subsp. bovei（Spach）Talavera et Gibbs ●☆

328241　Retama monosperma（L.）Boiss. var. webbii（Spach）Maire =

Retama monosperma （L.） Boiss. ●☆

328242   Retama monosperma Boiss. = Retama monosperma （L.） Boiss. ●☆

328243   Retama raetam （Forssk.） Webb；毛萼勒塔木（杜松豆）；Retam, White Broom ●☆

328244   Retama raetam （Forssk.） Webb subsp. bovei （Spach） Talavera et Gibbs；包氏毛萼勒塔木●☆

328245   Retama raetam （Forssk.） Webb var. duriaei （Spach） Letourn. = Retama raetam （Forssk.） Webb ●☆

328246   Retama raetam （Forssk.） Webb var. pallens L. Chevall. = Retama raetam （Forssk.） Webb ●☆

328247   Retama raetam （Forssk.） Webb var. rigidula DC. = Retama raetam （Forssk.） Webb ●☆

328248   Retama sphaerocarpa （L.） Boiss.；球果勒塔木（球果杜松豆）●☆

328249   Retama sphaerocarpa （L.） Boiss. var. atlantica （Pomel） Batt. = Retama sphaerocarpa （L.） Boiss. ●☆

328250   Retamilia Miers = Retanilla （DC.） Brongn. ●☆

328251   Retanilla （DC.） Brongn. （1826）；南美鼠李属（瑞大尼拉木属）●☆

328252   Retanilla Brongn. = Retanilla （DC.） Brongn. ●☆

328253   Retanilla ephedra Brongn.；南美鼠李（瑞大尼拉木）●☆

328254   Retinaria Gaertn. = Gouania Jacq. ●

328255   Retinaria scandens Gaertn. = Gouania scandens （Gaertn.） R. B. Drumm. ●☆

328256   Retiniphyllum Bonpl. （1806）；脂叶茜属●☆

328257   Retiniphyllum Humb. et Bonpl. = Retiniphyllum Bonpl. ●☆

328258   Retiniphyllum secundiflorum Humb. et Bonpl.；脂叶茜☆

328259   Retinispora Siebold et Zucc. = Chamaecyparis Spach ●

328260   Retinispora filicoides Syme = Chamaecyparis obtusa （Siebold et Zucc.） Siebold et Zucc. ex Endl. 'Filicoides' ●

328261   Retinispora obtusa Siebold et Zucc. = Chamaecyparis obtusa （Siebold et Zucc.） Siebold et Zucc. ex Endl. ●

328262   Retinispora pisifera Siebold et Zucc. = Chamaecyparis pisifera （Siebold et Zucc.） Siebold et Zucc. ex Endl. ●

328263   Retinispora squarrosa Zucc. = Chamaecyparis pisifera （Siebold et Zucc.） Siebold et Zucc. ex Endl. 'Squarrosa' ●

328264   Retinispora tetragona Syme = Chamaecyparis obtusa （Siebold et Zucc.） Siebold et Zucc. ex Endl. 'Tetragona' ●

328265   Retinodendron Korth. = Vatica L. ●

328266   Retinodendropsis Heim = Retinodendron Korth. ●

328267   Retinodendropsis Heim = Vatica L. ●

328268   Retinophleum Benth. et Hook. f. = Cercidium Tul. ●☆

328269   Retinophleum Benth. et Hook. f. = Rhetinophloeum H. Karst. ●☆

328270   Retinospora Carrière = Retinispora Siebold et Zucc. ●

328271   Retispatha J. Dransf. （1980）；网苞藤属●☆

328272   Retispatha dumefosa J. Dransf.；网苞藤●☆

328273   Retrophyllum C. N. Page = Podocarpus Pers. （保留属名）●

328274   Retrophyllum C. N. Page（1989）；扭叶罗汉松属●☆

328275   Retrophyllum comptonii （J. Buchholz） C. N. Page；扭叶罗汉松●☆

328276   Retrophyllum minor （Carrière） C. N. Page；小扭叶罗汉松●☆

328277   Retrosepalaceae Dulac = Violaceae Batsch（保留科名）●■

328278   Retrulobivia Frič = Echinopsis Zucc. ●

328279   Rettbergia Raddi = Chusquea Kunth ●☆

328280   Retzia Thunb. （1776）；异轮叶属（轮叶木属,轮叶属）●☆

328281   Retzia capensis Thunb.；异轮叶●☆

328282   Retziaceae Bartl. （1830）；异轮叶科（轮叶科,轮叶木科）●☆

328283   Retziaceae Bartl. = Stilbaceae Kunth（保留科名）●☆

328284   Retziaceae Choisy = Retziaceae Bartl. ●☆

328285   Reussia Dennst. = Paederia L. （保留属名）●■

328286   Reussia Endl. （1836）（保留属名）；罗伊斯花属■☆

328287   Reussia Endl. = Pontederia L. ■☆

328288   Reussia rotundifolia （L. f.） Castell.；罗伊斯花■☆

328289   Reutealis Airy Shaw（1967）；菲律宾大戟属●☆

328290   Reutealis trisperma （Blanco） Airy Shaw；菲律宾大戟●☆

328291   Reutera Boiss. = Pimpinella L. ■

328292   Reutera acuminate Edgew. = Pimpinella acuminata （Edgew.） C. B. Clarke ■

328293   Reutera aurea （DC.） Boiss. = Pimpinella aurea DC. ■☆

328294   Reutera fontanesii Boiss. = Pimpinella lutea Desf. ■☆

328295   Reutera lutea （Desf.） Maire = Pimpinella lutea Desf. ■☆

328296   Revatophyllum Roehl. = Ceratophyllum L. ■

328297   Revealia R. M. King et H. Rob. （1976）；短芒菊属●☆

328298   Revealia R. M. King et H. Rob. = Carphochaete A. Gray ■☆

328299   Revealia macrocephala （Paray） R. M. King etH. Rob.；短芒菊■☆

328300   Reveesia Walp. = Reevesia Lindl. ●

328301   Reverchonia A. Gray（1880）；勒韦雄大戟属☆

328302   Reverchonia Gand. = Armeria Willd. （保留属名）■☆

328303   Reverchonia arenaria A. Gray；勒韦雄大戟■☆

328304   Reya Kuntze = Burchardia R. Br. （保留属名）●☆

328305   Reyemia Hilliard（1993）；雷耶玄参属■☆

328306   Reyemia chasmanthiflora Hilliard；非洲雷耶玄参■☆

328307   Reyemia nemesioides （Diels） Hilliard；雷耶玄参■☆

328308   Reyesia Clos = Salpiglossis Ruiz et Pav. ■☆

328309   Reyesia Gay（1849）；雷耶斯茄属☆

328310   Reyesia chilensis Gay；智利雷耶斯茄■☆

328311   Reyesia laxa （Miers） D' Arcy；松散雷耶斯茄■☆

328312   Reyesia parviflora （Phil.） Hunz.；小花雷耶斯茄■☆

328313   Reymondia H. Karst. = Pleurothallis R. Br. ■☆

328314   Reymondia H. Karst. et Kuntze = Myoxanthus Poepp. et Endl. ■☆

328315   Reynandia B. D. Jacks. = Reynaudia Kunth ■☆

328316   Reynaudia Kunth（1829）；丝形草属■☆

328317   Reynaudia filiformis （Spreng.） Kunth；丝形草■☆

328318   Reynoldsia A. Gray（1854）；雷诺五加属●☆

328319   Reynoldsia sandwicensis A. Gray；雷诺五加●☆

328320   Reynosia Griseb. （1866）；雷诺木属（瑞诺木属）●☆

328321   Reynosia affinis Urb. et Ekman；近缘雷诺木☆

328322   Reynosia barbatula M. C. Johnst. et Lundell；髯毛雷诺木●☆

328323   Reynosia cuneifolia Urb. et Ekman；楔叶雷诺木●☆

328324   Reynosia latifolia Griseb.；宽叶雷诺木●☆

328325   Reynosia orbiculata Urb.；圆雷诺木●☆

328326   Reynosia reticulata Urb.；网脉雷诺木●☆

328327   Reynosia truncata Urb.；平截雷诺木●☆

328328   Reynoutria Houtt. （1777）；虎杖属；Japanes Knotweed, Reynoutria, Tigerstick ■

328329   Reynoutria Houtt. = Fallopia Adans. ●■

328330   Reynoutria aubertii （L. Henry） Moldenke = Fallopia aubertii （L. Henry） Holub ●

328331   Reynoutria baldschuanica （Regel） Moldenke = Fallopia baldschuanica （Regel） Holub ■☆

328332   Reynoutria baldschuanica （Regel） Shinners = Polygonum baldschuanicum Regel ■☆

328333   Reynoutria bohemica Chrtek et Chrtková = Fallopia bohemica （Chrtek et Chrtková） J. P. Bailey ■☆

328334   Reynoutria campanulata （Hook. f.） Moldenke = Polygonum campanulatum Hook. f. ■

328335　Reynoutria campanulata（Hook. f.）Moldenke ＝ Polygonum campanulatum Hook. f. var. fulvidum Hook. f. ■

328336　Reynoutria ciliinervis（Nakai）Moldenke ＝ Fallopia ciliinervis（Nakai）K. Hammer ■

328337　Reynoutria cilinodis（Michx.）Shinners ＝ Fallopia cilinodis（Michx.）Holub ■☆

328338　Reynoutria convolvulus（L.）Shinners ＝Fallopia convolvulus（L.）Á. Löve ■

328339　Reynoutria convolvulus（L.）Shinners ＝ Polygonum convolvulus L. ■

328340　Reynoutria elliptica（Koidz.）Migo ex Nakai ＝ Fallopia forbesii（Hance）Yonek. et H. Ohashi ■

328341　Reynoutria forbesii（Hance）T. Yamaz. ＝ Fallopia forbesii（Hance）Yonek. et H. Ohashi ■

328342　Reynoutria hachidyoensis（Makino）Honda ＝ Fallopia japonica（Houtt.）Ronse Decr. var. hachidyoensis（Makino）Yonek. et H. Ohashi ■☆

328343　Reynoutria hachidyoensis（Makino）Honda var. terminalis Honda ＝ Fallopia japonica（Houtt.）Ronse Decr. var. hachidyoensis（Makino）Yonek. et H. Ohashi ■☆

328344　Reynoutria henryi Nakai ＝ Reynoutria japonica Houtt. ■

328345　Reynoutria japonica Houtt.；虎杖（班草，班根，班杖根，般倒甑，斑草，斑根，斑龙紫，斑杖，斑杖根，斑庄，斑庄根，川筋龙，大虫杖，大接骨，大叶蛇总管，端阳，干烟，刚药台，高粱笋子，号筒草，红贯脚，红三七，猴竹根，花斑竹，黄地榆，黄药子，活血丹，金锁王，金杨草，九龙根，苦杖，马龙鱼，鸟不踏，胖官头，日本何首乌，散血草，山大黄，山茄子，蛇总管，酸杆，酸榴根，酸汤杆，酸汤梗，酸通，酸桶芦，酸桶笋，酸筒杆，酸筒根，蒤，土地榆，雄黄连，血藤，野黄连，阴阳莲，竹节参，紫金龙）；Donkey's Rhubarb, Giant Knotweed, Hancock's Curse, Japan Fleeceflower, Japanese Bamboo, Japanese Fleece Flower, Japanese Fleeceflower, Japanese Fleece-flower, Japanese Knotweed, Mexican Bamboo, Tigen Stick, Tigerstick ■

328346　Reynoutria japonica Houtt. ＝ Fallopia japonica（Houtt.）Ronse Decr. ■

328347　Reynoutria japonica Houtt. ＝ Polygonum cuspidatum Siebold et Zucc. ■

328348　Reynoutria japonica Houtt. f. colorans（Makino）Nemoto ＝ Fallopia japonica（Houtt.）Ronse Decr. var. compacta（Hook. f.）J. P. Bailey f. colorans（Makino）■☆

328349　Reynoutria japonica Houtt. f. compacta（Hook. f.）Nemoto ＝ Fallopia japonica（Houtt.）Ronse Decr. ■

328350　Reynoutria japonica Houtt. f. compacta（Hook. f.）Nemoto ＝ Reynoutria japonica Houtt. ■

328351　Reynoutria japonica Houtt. f. elata（Nakai）Hiyama ＝ Fallopia japonica（Houtt.）Ronse Decr. f. elata（Nakai）■☆

328352　Reynoutria japonica Houtt. var. compacta（Hook. f.）Hiyama ＝ Fallopia japonica（Houtt.）Ronse Decr. ■

328353　Reynoutria japonica Houtt. var. compacta（Hook. f.）Hiyama ＝ Reynoutria japonica Houtt. ■

328354　Reynoutria japonica Houtt. var. hachidyoensis（Makino）M. Mizush. ＝ Fallopia japonica（Houtt.）Ronse Decr. var. hachidyoensis（Makino）Yonek. et H. Ohashi ■☆

328355　Reynoutria japonica Houtt. var. terminalis Honda ＝ Fallopia japonica（Houtt.）Ronse Decr. var. hachidyoensis（Makino）Yonek. et H. Ohashi ☆

328356　Reynoutria japonica Houtt. var. uzenensis Honda ＝ Fallopia japonica（Houtt.）Ronse Decr. var. uzenensis（Honda）Yonek. et H. Ohashi ■☆

328357　Reynoutria japonica Houtt. var. uzenensis Honda f. rosea Satomi；粉红羽前虎杖■☆

328358　Reynoutria lichiangensis（W. W. Sm.）Moldenke ＝ Polygonum lichiangense W. W. Sm. ■

328359　Reynoutria multiflora（Thunb.）Moldenke ＝ Fallopia multiflora（Thunb.）Haraldson ■

328360　Reynoutria polystachya（Wall. ex Meisn.）Moldenke ＝ Persicaria wallichii Greuter et Burdet ●☆

328361　Reynoutria polystachya（Wall. ex Meisn.）Moldenke ＝ Polygonum polystachyum Wall. ex Meisn. ●■

328362　Reynoutria sachalinensis（F. Schmidt）Nakai ＝ Fallopia sachalinensis（F. Schmidt）Ronse Decr. ■☆

328363　Reynoutria sachalinensis（F. Schmidt）Nakai var. intermedia（Tatew.）Tatew. ex Miyabe et Kudo ＝ Fallopia sachalinensis（F. Schmidt）Ronse Decr. var. intermedia（Tatew.）Yonek. et H. Ohashi ■☆

328364　Reynoutria sachalinensis（F. W. Schmidt ex Maxim.）Nakai ＝ Polygonum sachalinense F. Schmidt ex Maxim. ■☆

328365　Reynoutria scandens（L.）Shinners ＝ Fallopia scandens（L.）Holub ■☆

328366　Reynoutria scandens（L.）Shinners var. cristata（Engelm. et A. Gray）Shinners ＝ Polygonum scandens L. var. cristatum（Engelm. et A. Gray）Gleason ■☆

328367　Reynoutria scandens（L.）Shinners var. cristata（Engelm. et A. Gray）Shinners ＝ Fallopia scandens（L.）Holub ■☆

328368　Reynoutria scandens（L.）Shinners var. dumetorum（L.）Shinners ＝ Fallopia dumetora（L.）Holub ■

328369　Reynoutria uzenensis（Honda）Honda ＝ Fallopia japonica（Houtt.）Ronse Decr. var. uzenensis（Honda）Yonek. et H. Ohashi ■☆

328370　Reynoutria yunnanensis（H. Lév.）Nakai ex Migo ＝ Fallopia forbesii（Hance）Yonek. et H. Ohashi ■

328371　Rhabarbarum Adans. ＝ Rheum L. ■

328372　Rhabarbarum Fabr. ＝ Rheum L. ■

328373　Rhabarbarum Tourn. ex Adans. ＝ Rheum L. ■

328374　Rhabdadenia Müll. Arg.（1860）；杆腺木属●☆

328375　Rhabdadenia biflora（Jacq.）Müll. Arg.；二花杆腺木（双花杆腺木）；Mangrove Rubber Vine ●☆

328376　Rhabdadenia biflora Müll. Arg. ＝ Rhabdadenia biflora（Jacq.）Müll. Arg. ●☆

328377　Rhabdadenia latifolia Malme；宽叶杆腺木●☆

328378　Rhabdadenia macrantha Donn. Sm.；大花杆腺木●☆

328379　Rhabdadenia polyneura Urb.；密脉杆腺木●☆

328380　Rhabdia Mart.（1827）；杆紫草属●

328381　Rhabdia Mart. ＝ Rotula Lour. ●

328382　Rhabdia lycioides Mart. ＝ Rotula aquatica Lour. ●

328383　Rhabdia viminea（Wall.）Dalziel et Gibson ＝ Rotula aquatica Lour. ●

328384　Rhabdia viminea Wall. ex Dalziel ＝ Rotula aquatica Lour. ●

328385　Rhabdocalyx（DC.）Lindl. ＝ Cordia L.（保留属名）●

328386　Rhabdocalyx Lindl. ＝ Cordia L.（保留属名）●

328387　Rhabdocaulon（Benth.）Epling（1936）；棒茎草属●■☆

328388　Rhabdocaulon coccineum（Benth.）Epling；灰棒茎草■☆

328389　Rhabdocaulon erythrostachys Epling；红穗棒茎草■☆

328390　Rhabdocaulon gracile（Benth.）Epling；细棒茎草■☆

328391　Rhabdochloa Kunth ＝ Leptochloa P. Beauv. ■

328392　Rhabdochloa Kunth ＝ Rabdochloa P. Beauv. ■

328393　Rhabdochloa virgata（Sw.）P. Beauv. ＝ Chloris virgata Sw. ■

328394　Rhabdocrinum Rchb. ＝ Lloydia Salisb. ex Rchb.（保留属名）■

328395 Rhabdodendraceae（Huber）Prance（1968）;棒状木科（棒木科）●☆

328396 Rhabdodendraceae Prance ＝ Rhabdodendraceae （ Huber ） Prance ●☆

328397 Rhabdodendron Gilg et Pilg.（1905）;棒状木属●☆

328398 Rhabdodendron columnare Gilg et Pilg.;棒状木●☆

328399 Rhabdolosperma Hartl ＝ Verbascum L. ■●

328400 Rhabdophyllum Tiegh.（1902）;棒叶金莲木属●☆

328401 Rhabdophyllum Tiegh. ＝ Gomphia Schreb. ●

328402 Rhabdophyllum acutissimum（Gilg）Tiegh. ＝ Rhabdophyllum affine（Hook. f.）Tiegh. subsp. acutissimum（Gilg）Farron ●☆

328403 Rhabdophyllum affine（Hook. f.）Tiegh.;近缘棒叶金莲木●☆

328404 Rhabdophyllum affine（Hook. f.）Tiegh. subsp. acutissimum（Gilg）Farron;尖近缘棒叶金莲木●☆

328405 Rhabdophyllum affine（Hook. f.）Tiegh. subsp. monanthum（Gilg ex Engl.）Farron;单花近缘棒叶金莲木●☆

328406 Rhabdophyllum affine（Hook. f.）Tiegh. subsp. myrioneurum（Gilg）Farron;多脉近缘棒叶金莲木●☆

328407 Rhabdophyllum affine（Hook. f.）Tiegh. subsp. pauciflorum（Tiegh.）Farron;少花近缘棒叶金莲木●☆

328408 Rhabdophyllum arnoldianum（De Wild. et T. Durand）Tiegh.;阿尔棒叶金莲木●☆

328409 Rhabdophyllum arnoldianum（De Wild. et T. Durand）Tiegh. var. quintasii（Tiegh.）Farron;昆塔斯棒叶金莲木●☆

328410 Rhabdophyllum arnoldianum（De Wild. et T. Durand）Tiegh. var. staudtii（Tiegh.）Farron;施陶棒叶金莲木●☆

328411 Rhabdophyllum barteri Tiegh. ＝ Rhabdophyllum arnoldianum（De Wild. et T. Durand）Tiegh. ●☆

328412 Rhabdophyllum bracteolatum（Gilg）Farron;多苞片棒叶金莲木●☆

328413 Rhabdophyllum calophyllum（Hook. f.）Tiegh.;美叶棒叶金莲木●☆

328414 Rhabdophyllum leptoneurum（Gilg）Tiegh. ＝ Rhabdophyllum welwitschii Tiegh. ●☆

328415 Rhabdophyllum letestui Farron;莱泰斯图棒叶金莲木●☆

328416 Rhabdophyllum longipes Tiegh. ＝ Rhabdophyllum arnoldianum（De Wild. et T. Durand）Tiegh. ●☆

328417 Rhabdophyllum myrioneurum（Gilg）Tiegh. ＝ Rhabdophyllum affine（Hook. f.）Tiegh. subsp. myrioneurum（Gilg）Farron ●☆

328418 Rhabdophyllum nutans Tiegh.;俯垂棒叶金莲木●☆

328419 Rhabdophyllum pauciflorum Tiegh. ＝ Rhabdophyllum affine（Hook. f.）Tiegh. subsp. pauciflorum（Tiegh.）Farron ●☆

328420 Rhabdophyllum pedicellatum Tiegh. ＝ Rhabdophyllum welwitschii Tiegh. ●☆

328421 Rhabdophyllum preussii Tiegh. ＝ Rhabdophyllum calophyllum（Hook. f.）Tiegh. ●☆

328422 Rhabdophyllum quintasii Tiegh. ＝ Rhabdophyllum arnoldianum（De Wild. et T. Durand）Tiegh. var. quintasii（Tiegh.）Farron ●☆

328423 Rhabdophyllum refractum（De Wild. et T. Durand）Tiegh.;反折棒叶金莲木●☆

328424 Rhabdophyllum rigidum（De Wild.）Farron;硬棒叶金莲木●☆

328425 Rhabdophyllum staudtii Tiegh. ＝ Rhabdophyllum arnoldianum（De Wild. et T. Durand）Tiegh. var. staudtii（Tiegh.）Farron ●☆

328426 Rhabdophyllum stenorachis（Gilg）Tiegh. ＝ Rhabdophyllum affine（Hook. f.）Tiegh. subsp. myrioneurum（Gilg）Farron ●☆

328427 Rhabdophyllum stenorhachis（Gilg）Tiegh. ＝ Rhabdophyllum affine（Hook. f.）Tiegh. subsp. myrioneurum（Gilg）Farron ●☆

328428 Rhabdophyllum thollonii Tiegh. ＝ Rhabdophyllum arnoldianum（De Wild. et T. Durand）Tiegh. ●☆

328429 Rhabdophyllum thonneri（De Wild.）Farron;托内棒叶金莲木●☆

328430 Rhabdophyllum viancinii Tiegh. ＝ Rhabdophyllum arnoldianum（De Wild. et T. Durand）Tiegh. ●☆

328431 Rhabdophyllum welwitschii Tiegh.;韦尔棒叶金莲木●☆

328432 Rhabdosciadium Boiss.（1844）;棒伞芹属■☆

328433 Rhabdosciadium microcalycinum Hand. -Mazz.;小萼棒伞芹■☆

328434 Rhabdosciadium oligocarpum（Post ex Boiss.）Hedge et Lamond;寡果棒伞芹■☆

328435 Rhabdosciadium stenophyllum Boiss. et Hausskn.;窄叶棒伞芹■☆

328436 Rhabdostigma Hook. f. ＝ Kraussia Harv. ●☆

328437 Rhabdostigma kirkii Hook. f. ＝ Kraussia kirkii（Hook. f.）Bullock ●☆

328438 Rhabdostigma schlechteri K. Schum. ＝ Kraussia floribunda Harv. ●☆

328439 Rhabdothamnopsis Hemsl.（1903）; 长 冠 苣 苔 属; Rhabdothamnopsis ●★

328440 Rhabdothamnopsis chinensis（Franch.）Hand. -Mazz. ＝ Rhabdothamnopsis sinensis Hemsl. ■

328441 Rhabdothamnopsis chinensis（Franch.）Hand. -Mazz. var. ochroleuca（W. W. Sm.）Hand. -Mazz. ＝ Rhabdothamnopsis sinensis Hemsl. ■

328442 Rhabdothamnopsis limprichtiana Lingelsh. et Borza ＝ Rhabdothamnopsis sinensis Hemsl. ■

328443 Rhabdothamnopsis sinensis Hemsl.; 长 冠 苣 苔; China Rhabdothamnopsis,Chinese Rhabdothamnopsis ■

328444 Rhabdothamnopsis sinensis Hemsl. var. ochroleuca W. W. Sm. ＝ Rhabdothamnopsis sinensis Hemsl. ■

328445 Rhabdothamnus A. Cunn.（1838）;杆丛苣苔属;Rhabdothamus ●☆

328446 Rhabdothamnus solandri A. Cunn.;杆丛苣苔;New Zealand Gloxinia,Rhabdothamus,Waiuatua ●☆

328447 Rhabdotheca Cass. ＝ Launaea Cass. ■

328448 Rhabdotheca angustifolia（Desf.）Pomel ＝ Launaea angustifolia（Desf.）Kuntze ■☆

328449 Rhabdotheca brunneri Webb ＝ Launaea brunneri（Webb）Amin ex Boulos ■☆

328450 Rhabdotheca chondrilloides Webb ＝ Launaea nudicaulis（L.）Hook. f. ■☆

328451 Rhabdotheca korovinii（Popov）Kirp. ＝ Launaea korovinii Popov ■☆

328452 Rhabdotheca longiloba（Boiss. et Reut.）Pomel ＝ Launaea fragilis（Asso）Pau ■☆

328453 Rhabdotheca nudicaulis（L.）Pomel ＝ Launaea nudicaulis（L.）Hook. f. ■☆

328454 Rhabdotheca pumila（Cav.）Pomel ＝ Launaea pumila（Cav.）Kuntze ■☆

328455 Rhabdotheca quercifolia（Desf.）Pomel ＝ Launaea quercifolia（Desf.）Pamp. ■☆

328456 Rhabdotheca resedifolia Pomel ＝ Launaea fragilis（Asso）Pau ■☆

328457 Rhabdotheca squarrosa Pomel ＝ Launaea angustifolia（Desf.）Kuntze ■☆

328458 Rhabdotheca tenuiloba（Boiss.）Pomel ＝ Launaea fragilis（Asso）Pau ■☆

328459 Rhabdotosperma Hartl ＝ Verbascum L. ■●

328460 Rhabdotosperma Hartl（1977）;棒籽花属●☆

328461 Rhabdotosperma brevipedicellata（Engl.）Hartl;短梗棒籽花●☆

328462　Rhabdotosperma densifolia（Hook. f.）Hartl;密花棒籽花●☆

328463　Rhabdotosperma keniensis（Murb.）Hartl;肯尼亚棒籽花●☆

328464　Rhabdotosperma ledermannii（Murb.）Hartl;莱德棒籽花●☆

328465　Rhabdotosperma schimperi（V. Naray.）Hartl;欣珀棒籽花●☆

328466　Rhabdotosperma scrophulariifolia（Hochst. ex A. Rich.）Hartl;玄参叶棒籽花●☆

328467　Rhabdotosperma scrophulariifolia（Hochst. ex A. Rich.）Hartl subsp. foliosa（Chiov.）Hartl;密玄参叶棒籽花●☆

328468　Rhachicallis DC.;美刺茜属●☆

328469　Rhachicallis DC. = Arcythophyllum Willd. ex Schltdl. ●☆

328470　Rhachicallis Spach = Arcythophyllum Willd. ex Schltdl. ●☆

328471　Rhachicallis americana Kuntze;美洲美刺茜●☆

328472　Rhachicallis maritima Schum.;美刺茜●☆

328473　Rhachidospermum Vasey = Jouvea E. Fourn. ■☆

328474　Rhacodiscus Lindau = Justicia L. ●■

328475　Rhacoma Adans. = Leuzea DC. ●☆

328476　Rhacoma L. = Crossopetalum P. Browne ●☆

328477　Rhacoma P. Browne ex L. = Crossopetalum P. Browne ●☆

328478　Rhadamanthopsis（Oberm.）Speta = Rhadamanthus Salisb. ■☆

328479　Rhadamanthopsis（Oberm.）Speta（1998）;拟细花风信子属■☆

328480　Rhadamanthopsis karooicus（Oberm.）Speta;拟细花风信子■☆

328481　Rhadamanthus Salisb.（1866）;细花风信子属■☆

328482　Rhadamanthus albiflorus B. Nord. = Drimia albiflora（B. Nord.）J. C. Manning et Goldblatt ■☆

328483　Rhadamanthus arenicola B. Nord. = Drimia arenicola（B. Nord.）J. C. Manning et Goldblatt ■☆

328484　Rhadamanthus convallarioides（L. f.）Baker = Drimia convallarioides（L. f.）J. C. Manning et Goldblatt ■☆

328485　Rhadamanthus cyanelloides Baker = Drimia cyanelloides（Baker）J. C. Manning et Goldblatt ■☆

328486　Rhadamanthus fasciatus B. Nord. = Drimia fasciata（B. Nord.）J. C. Manning et Goldblatt ■☆

328487　Rhadamanthus involutus J. C. Manning et Snijman = Drimia involuta（J. C. Manning et Snijman）J. C. Manning et Goldblatt ■☆

328488　Rhadamanthus karrooicus Oberm. = Drimia karooica（Oberm.）J. C. Manning et Goldblatt ■☆

328489　Rhadamanthus montanus B. Nord. = Drimia convallarioides（L. f.）J. C. Manning et Goldblatt ■☆

328490　Rhadamanthus namibensis Oberm. = Drimia namibensis（Oberm.）J. C. Manning et Goldblatt ■☆

328491　Rhadamanthus platyphyllus B. Nord. = Drimia platyphylla（B. Nord.）J. C. Manning et Goldblatt ■☆

328492　Rhadamanthus secundus B. Nord. = Drimia secunda（B. Nord.）J. C. Manning et Goldblatt ■☆

328493　Rhadamanthus urantherus R. A. Dyer = Drimia uranthera（R. A. Dyer）J. C. Manning et Goldblatt ■☆

328494　Rhadinocarpus Vogel = Chaetocalyx DC. ■☆

328495　Rhadinopus S. Moore（1930）;细足茜属☆

328496　Rhadinopus papuanus S. Moore;细足茜☆

328497　Rhadinothamnus Paul G. Wilson（1971）;柔灌芸香属●☆

328498　Rhadinothamnus euphemiae（F. Muell.）Paul G. Wilson;柔灌芸香●☆

328499　Rhadiola Savi = Radiola Hill ■☆

328500　Rhaeo C. B. Clarke = Rhoeo Hance ■

328501　Rhaeo C. B. Clarke = Tradescantia L. ■

328502　Rhaesteria Summerh.（1966）;东非兰属■☆

328503　Rhaesteria eggelingii Summerh.;东非兰■☆

328504　Rhagadiolus Juss. = Rhagadiolus Vaill.（保留属名）■☆

328505　Rhagadiolus Scop. = Rhagadiolus Vaill.（保留属名）■☆

328506　Rhagadiolus Tourn. ex Scop. = Rhagadiolus Vaill.（保留属名）■☆

328507　Rhagadiolus Vaill.（1789）（保留属名）;双苞苣属（线苞果属）■☆

328508　Rhagadiolus Zinn（废弃属名）= Hedypnois Mill. ■☆

328509　Rhagadiolus Zinn（废弃属名）= Rhagadiolus Vaill.（保留属名）■☆

328510　Rhagadiolus angulosus（Jaub. et Spach）Kupicha = Garhadiolus angulosus Jaub. et Spach ■☆

328511　Rhagadiolus arenarius（Schousb.）Ball = Hedypnois arenaria（Schousb.）DC. ■☆

328512　Rhagadiolus edulis Gaertn.;可食双苞苣（可食拟小疮菊）■☆

328513　Rhagadiolus hebelaenus（DC.）Vassilcz.;柔毛双苞苣■☆

328514　Rhagadiolus hedypnois All. = Hedypnois rhagadioloides（L.）F. W. Schmidt ■☆

328515　Rhagadiolus koelpinia Willd. = Koelpinia linearis Pall. ■

328516　Rhagadiolus lampsanoides Desf. = Rhagadiolus edulis Gaertn. ■☆

328517　Rhagadiolus papposus Kuntze = Garhadiolus papposus Boiss. et Buhse ■

328518　Rhagadiolus pendulus Benth. et Hook. f. = Hedypnois rhagadioloides（L.）F. W. Schmidt ■☆

328519　Rhagadiolus rigidus Pomel = Rhagadiolus edulis Gaertn. ■☆

328520　Rhagadiolus stellatus（L.）Gaertn.;星状双苞苣（拟小疮菊）;Endive Daisy ■☆

328521　Rhagadiolus stellatus（L.）Gaertn. var. eriocarpus Faure et Maire = Rhagadiolus stellatus（L.）Gaertn. ■☆

328522　Rhagadiolus stellatus（L.）Gaertn. var. hebelaenus DC. = Rhagadiolus stellatus（L.）Gaertn. ■☆

328523　Rhagadiolus stellatus（L.）Gaertn. var. intermedius（Ten.）DC. = Rhagadiolus stellatus（L.）Gaertn. ■☆

328524　Rhagadiolus stellatus（L.）Gaertn. var. leiocarpus DC. = Rhagadiolus stellatus（L.）Gaertn. ■☆

328525　Rhagadiolus stellatus Gaertn. = Rhagadiolus stellatus（L.）Gaertn. ■☆

328526　Rhagadiolus zacintha（L.）All. = Crepis zacintha（L.）Babc. ■☆

328527　Rhaganus E. Mey. = Bersama Fresen. ●☆

328528　Rhagodia R. Br.（1810）;肉被藜属（假葡萄属）●☆

328529　Rhagodia hastata R. Br.;戟形肉被藜●

328530　Rhagodia latifolia（Benth.）Paul G. Wilson;宽叶肉被藜●☆

328531　Rhagodia linifolia R. Br.;线叶肉被藜●☆

328532　Rhagodia nutans R. Br. = Einadia nutans（R. Br.）A. J. Scott ●☆

328533　Rhagodia obovata Moq.;倒卵肉被藜●☆

328534　Rhagodia parvifolia Moq.;小叶肉被藜●☆

328535　Rhagodia spinescens R. Br.;刺肉被藜;Spiny Saltbush ●☆

328536　Rhamindium Sarg. = Rhamnidium Reissek ●☆

328537　Rhammatophyllum O. E. Schulz = Erysimum L. ■●

328538　Rhammatophyllum O. E. Schulz（1933）;假糖芥属■☆

328539　Rhammatophyllum afghanicum（Rech. f.）Kamelin;阿富汗假糖芥■☆

328540　Rhammatophyllum pachyrrhizum（Kar. et Kir.）O. E. Schulz;假糖芥■☆

328541　Rham-Mloluma Baill. = Lucuma Molina ●

328542　Rham-Mloluma Baill. = Rhamnoluma Baill. ●

328543　Rhamnaceae Juss.（1789）（保留科名）;鼠李科;Buckthorn Family ●■

328544　Rhamnella Miq.（1867）;猫乳属（长叶绿柴属,假鼠李属）;

Rhamnella ●

328545　Rhamnella berchemiifolia Makino ＝ Berchemiella berchemiifolia（Makino）Nakai ●

328546　Rhamnella caudata Merr. et Chun；尾叶猫乳；Caudate Rhamnella，Tail Rhamnell，Tailleaf Rhamnella ●

328547　Rhamnella crenulata（Hand.-Mazz.）T. Yamaz. ＝ Chaydaia rubrinervis（H. Lév.）C. Y. Wu ●

328548　Rhamnella crenulata（Hand.-Mazz.）T. Yamaz. ＝ Rhamnella rubrinervis（H. Lév.）Rehder ●

328549　Rhamnella forrestii W. W. Sm.；川滇猫乳(清西绿柴)；Forrest Rhamnella ●

328550　Rhamnella franguloides（Maxim.）Weberb.；猫乳（长叶绿柴，糯米牙，山黄，鼠矢枣，鼠屎枣）；Frangula-like Rhamnella，Rhamnella ●

328551　Rhamnella franguloides（Maxim.）Weberb. var. inaequilatera（Ohwi）Hatus.；异边猫乳●☆

328552　Rhamnella gilgitica Mansf. et Melch.；西藏猫乳（森等，升登，生等）；Gilgit Rhamnella，Xizang Rhamnella ●

328553　Rhamnella hainanensis Merr. ＝ Chaydaia rubrinervis（H. Lév.）C. Y. Wu ●

328554　Rhamnella hainanensis Merr. ＝ Rhamnella rubrinervis（H. Lév.）Rehder ●

328555　Rhamnella inaequilatera Ohwi ＝ Rhamnella franguloides（Maxim.）Weberb. var. inaequilatera（Ohwi）Hatus. ●☆

328556　Rhamnella japonica Miq. ＝ Rhamnella franguloides（Maxim.）Weberb. ●

328557　Rhamnella julianae C. K. Schneid.；毛背猫乳；Hairyback Rhamnella，Hairy-back Rhamnella ●

328558　Rhamnella laui Chun ＝ Rhamnus henryi C. K. Schneid. ●

328559　Rhamnella longifolia H. T. Tsai et K. M. Feng ＝ Chaydaia rubrinervis（H. Lév.）C. Y. Wu ●

328560　Rhamnella longifolia H. T. Tsai et K. M. Feng ＝ Rhamnella rubrinervis（H. Lév.）Rehder ●

328561　Rhamnella mairei C. K. Schneid. ＝ Rhamnella martinii（H. Lév.）C. K. Schneid. ●

328562　Rhamnella martinii（H. Lév.）C. K. Schneid.；多脉猫乳（秤杆木，香叶树）；Martin Rhamnella ●

328563　Rhamnella obovalis C. K. Schneid. ＝ Rhamnella franguloides（Maxim.）Weberb. ●

328564　Rhamnella rubrinervis（H. Lév.）Rehder ＝ Chaydaia rubrinervis（H. Lév.）C. Y. Wu ●

328565　Rhamnella wilsonii C. K. Schneid.；卵叶猫乳（小叶猫乳）；Ovateleaf Rhamnella，Ovate-leaved Rhamnella ●

328566　Rhamnicastrum Kuntze ＝ Scolopia Schreb.（保留属名）●

328567　Rhamnicastrum ecklonii（Nees）Kuntze ＝ Scolopia zeyheri（Nees）Harv. ●☆

328568　Rhamnicastrum mundtii（Eckl. et Zeyh.）Kuntze ＝ Scolopia mundtii（Eckl. et Zeyh.）Warb. ●☆

328569　Rhamnicastrum zeyheri（Nees）Kuntze ＝ Scolopia zeyheri（Nees）Harv. ●☆

328570　Rhamnidium Reissek（1861）；肖鼠李属●☆

328571　Rhamnidium acuminatum Urb.；渐尖肖鼠李●☆

328572　Rhamnidium bicolor Britton et P. Wilson；二色肖鼠李●☆

328573　Rhamnidium brevifolium Borhidi；短叶肖鼠李●☆

328574　Rhamnidium cubense Britton et P. Wilson；古巴肖鼠李●☆

328575　Rhamnidium dictyophyllum Urb.；指叶肖鼠李●☆

328576　Rhamnidium ellipticum Britton et P. Wilson；椭圆肖鼠李●☆

328577　Rhamnidium glabrum Reissek；光肖鼠李●☆

328578　Rhamnidium molle Reissek；柔软肖鼠李●☆

328579　Rhamnidium oblongifolium Britton et P. Wilson；矩圆叶肖鼠李●☆

328580　Rhamnobrina H. Perrier ＝ Colubrina Rich. ex Brongn.（保留属名）●

328581　Rhamnoides Mill. ＝ Hippophae L. ●

328582　Rhamnoides Tourn. ex Moench ＝ Hippophae L. ●

328583　Rhamnoides hippophae Moench ＝ Hippophae rhamnoides L. ●

328584　Rhamnoluma Baill. ＝ Lucuma Molina ●

328585　Rhamnoluma Baill. ＝ Pichonia Pierre ●☆

328586　Rhamnoneuron Gilg（1894）；鼠皮树属；Rhamnoneuron ●

328587　Rhamnoneuron balansae（Drake）Gilg；鼠皮树（剥皮树，粗皮树，红花鼠皮树）；Red-flower Rhamnoneuron，Vietnam Rhamnoneuron ●

328588　Rhamnoneuron balansae Drake ＝ Rhamnoneuron balansae（Drake）Gilg ●

328589　Rhamnoneuron rubriflorum C. Y. Wu ex S. C. Huang ＝ Rhamnoneuron balansae（Drake）Gilg ●

328590　Rhamnopsis Rchb. ＝ Flacourtia Comm. ex L'Hér. ●

328591　Rhamnos St.-Lag. ＝ Rhamnus L. ●

328592　Rhamnus L.（1753）；鼠李属；Alder Buckthorn，Buckthorn，Chinese Green Indigo ●

328593　Rhamnus acuminatifolia Hayata ＝ Rhamnus crenata Siebold et Zucc. ●

328594　Rhamnus affinis Blume ＝ Sageretia hamosa（Wall. ex Roxb.）Brongn. ●

328595　Rhamnus alaternus L.；意大利鼠李；Buckthorn，Evergreen Buckthorn，Italian Buckthorn，Italian Buck-thorn，Mediterranean Buckthorn ●☆

328596　Rhamnus alaternus L. 'Argenteovariegata'；银边意大利鼠李；Variegata Italian Buckth ●☆

328597　Rhamnus alaternus L. 'Variegata' ＝ Rhamnus alaternus L. 'Argenteovariegata' ●☆

328598　Rhamnus alaternus L. subsp. myrtifolia（Willk.）Maire ＝ Rhamnus myrtifolia Willk. ●☆

328599　Rhamnus alaternus L. subsp. pendula（Pamp.）Jafri；下垂意大利鼠李●☆

328600　Rhamnus alaternus L. var. angustifolia（Mill.）Aiton；窄叶意大利鼠李；Narrow-leaf Italian Buckthorn ●☆

328601　Rhamnus alaternus L. var. gebelensis Andr. ＝ Rhamnus alaternus L. ●☆

328602　Rhamnus alaternus L. var. prostrata Boiss. ＝ Rhamnus myrtifolia Willk. ●☆

328603　Rhamnus alnifolia L'Hér.；桤叶鼠李（矮鼠李）；Alder Buckthorn，Alder-leaf Buckthorn，Alder-leaved Buckthorn，Arrow-wood，Black Alder，Black Aller，Black Dogwood，Butcher's Prick-tree，Butcher's Prickwood，Dogwood，European Alder Buckthorn，Glossy Buckthorn，Stinking Roger ●☆

328604　Rhamnus alpina L.；高山鼠李(无刺鼠李)；Alpine Buckthorn ●☆

328605　Rhamnus alpina L. var. kabylica Maire ＝ Rhamnus alpina L. ●☆

328606　Rhamnus alpina L. var. libanotica（Boiss.）Batt. ＝ Rhamnus alpina L. ●☆

328607　Rhamnus amygdalina Desf. ＝ Rhamnus lycioides L. subsp. oleoides（L.）Jahand. et Maire ●☆

328608　Rhamnus amygdalina Desf. var. hirsuta Debeaux ＝ Rhamnus lycioides L. subsp. oleoides（L.）Jahand. et Maire ●☆

328609　Rhamnus arguta Maxim.；锐齿鼠李（火李，尖齿鼠李，老乌眼，

牛李子,照家茶);Sharptooth Buckthorn, Sharp-tooth Buckthorn, Sharp-toothed Buckthorn ●

328610　Rhamnus arguta Maxim. var. betulifolia Liou et C. Y. Li = Rhamnus arguta Maxim. ●

328611　Rhamnus arguta Maxim. var. cuneafolia F. T. Wang et H. L. Li = Rhamnus arguta Maxim. ●

328612　Rhamnus arguta Maxim. var. nakaharae Hayata = Rhamnus nakaharai (Hayata) Hayata ●

328613　Rhamnus arguta Maxim. var. nakaharai Hayata = Rhamnus nakaharai (Hayata) Hayata ●

328614　Rhamnus arguta Maxim. var. rotundifolia F. T. Wang et H. L. Li = Rhamnus arguta Maxim. ●

328615　Rhamnus arguta Maxim. var. velutina Hand. -Mazz.;毛背锐齿鼠李;Hairyback Sharptooth Buckthorn ●

328616　Rhamnus atlanticus (Murb.) Stübing et Peris et Figuerola = Rhamnus lycioides L. subsp. atlantica (Murb.) Jahand. et Maire ●☆

328617　Rhamnus aurea Heppeler;云南鼠李(铁马鞭);Horsewhip Buckthorn, Horse-whip Buckthorn, Ironwhip Buckthorn ●

328618　Rhamnus baldschuanica Grubov;巴尔德鼠李●☆

328619　Rhamnus betulifolia Greene;桦叶鼠李;Birchleaf Buckthorn, Birch-leaf Buckthorn ●☆

328620　Rhamnus betulifolia Greene var. obovata Kearney et Peebles;倒卵叶鼠李;Obovate-leaf Buckthorn ●☆

328621　Rhamnus blinii (H. Lév.) Rehder = Rhamnus hemsleyana C. K. Schneid. ●

328622　Rhamnus blinii (H. Lév.) Rehder var. sargentiana (C. K. Schneid.) Rehder = Rhamnus sargentiana C. K. Schneid. ●

328623　Rhamnus bodinieri H. Lév.;陷脉鼠李(地马桑,鼠李);Bodinier Buckthorn ●

328624　Rhamnus bodinieri H. Lév. f. silvicola C. K. Schneid. = Rhamnus bodinieri H. Lév. ●

328625　Rhamnus brachypoda C. Y. Wu ex Y. L. Chen;山绿柴;Buckthorn, Short-stalked Buckthorn ●

328626　Rhamnus bungeana J. J. Vassil.;卵叶鼠李(麻李,小叶鼠李);Bunge Buckthorn ●

328627　Rhamnus buxifolia Poir. = Rhamnus lycioides L. ●☆

328628　Rhamnus calcicola Hatus.;石生鼠李;Calicicolous Buckthorn, Rock Buckthorn ●

328629　Rhamnus calicicola Q. H. Chen = Rhamnus calcicola Hatus. ●

328630　Rhamnus californica Eschsch.;加州鼠李;California Buckthorn, Coffee Berry, Coffeebarry, Pigeon-berry, Redberry ●☆

328631　Rhamnus californica Eschsch. 'Ed Holm';艾德·胡尔姆加州鼠李●☆

328632　Rhamnus californica Eschsch. 'Eve Case';伊夫·凯斯加州鼠李●☆

328633　Rhamnus californica Eschsch. 'Mound San Bruno';圣布鲁诺堤加州鼠李●☆

328634　Rhamnus californica Eschsch. 'Sea View';海景加州鼠李●☆

328635　Rhamnus californica Eschsch. var. crassifolia Jeps.;多毛加州鼠李;California Buckthorn ●☆

328636　Rhamnus californica Eschsch. var. occidentalis (Howell) Jeps.;黄叶加州鼠李;Yellowish-green California Buckthorn ●☆

328637　Rhamnus californica Eschsch. var. tomentella (Benth.) W. H. Brewer et S. Watson;短毛加州鼠李;Shorthair California Buckthorn ●☆

328638　Rhamnus californica Eschsch. var. ursina (Greene) McMinn;厄辛山加州鼠李;California Buckthorn ●☆

328639　Rhamnus californica Eschsch. var. viridula Jeps.;齿叶加州鼠李;California Buckthorn ●☆

328640　Rhamnus cambodiana Pierre ex Pit. = Rhamnus crenata Siebold et Zucc. ●

328641　Rhamnus caroliniana Walter;美国鼠李(卡罗林纳鼠李,卡罗琳鼠李);Carolina Buckthorn, Indian Cherry, Indian-cherry, Yellow Buckthorn ●☆

328642　Rhamnus cathartica L.;药鼠李(泻鼠李);Buckler Thorn, Buckthorn, Carolina Buckthorn, Common Buckthom, European Buckthorn, European Waythorn, French-berries, Hart's-thorn, Hart's Thorn, Hartshorn, Hartsthorn, Laxative Ram, Medicinal Buckthorn, Purging Buckthorn, Purging Thorn, Rainberry Thorn, Rainberry-thorn, Ramsthorn, Rhine Berry, Rhineberry, Sapgreen, Waythorn ●

328643　Rhamnus cathartica L. var. dahurica Maxim. = Rhamnus ussuriensis J. J. Vassil. ●

328644　Rhamnus cathartica L. var. davurica Pall. = Rhamnus davurica Pall. ●

328645　Rhamnus cathartica L. var. intermedia Maxim. = Rhamnus ussuriensis J. J. Vassil. ●

328646　Rhamnus cathartica L. var. nipponica ? = Rhamnus davurica Pall. ●

328647　Rhamnus cathartica L. var. pubescens Rehder = Rhamnus cathartica L. ●

328648　Rhamnus cavaleriei H. Lév. = Rhamnus heterophylla Oliv. ●

328649　Rhamnus cavaleriei H. Lév. = Rhamnus rosthornii E. Pritz. ex Diels ●

328650　Rhamnus chekiangensis W. C. Cheng = Rhamnus rugulosa Hemsl. ex Forbes et Hemsl. var. chekiangensis (W. C. Cheng) Y. L. Chen et P. K. Chou ●

328651　Rhamnus chiennanensis Z. R. Xu;黔南鼠李;South Guizhou Buckthorn ●

328652　Rhamnus chingshuiensis Shimizu;清水鼠李;Qingshui Buckthorn ●

328653　Rhamnus chingshuiensis Shimizu var. tashanensis Y. C. Liu et C. M. Wang;塔山鼠李●

328654　Rhamnus chlorophora Decne. = Rhamnus globosa Bunge ●

328655　Rhamnus chugokuensis Hatus. = Rhamnus yoshinoi Makino ●

328656　Rhamnus coriacea (Regel) Kom.;革质鼠李●☆

328657　Rhamnus coriaceifolia H. Lév. = Sinosideroxylon wightianum (Hook. et Arn.) Aubrév. ●

328658　Rhamnus coriophylla Hand. -Mazz.;革叶鼠李(硬叶鼠李);Leatherleaf Buckthorn, Leather-leaved Buckthorn ●

328659　Rhamnus coriophylla Hand. -Mazz. var. acutidens Y. L. Chen et P. K. Chou;锐齿革叶鼠李(疏齿革叶鼠李);Acuteserrate Leatherleaf Buckthorn ●

328660　Rhamnus costata Maxim.;黑桦叶鼠李●☆

328661　Rhamnus costata Maxim. f. nambuana (Honda) H. Hara = Rhamnus costata Maxim. ●☆

328662　Rhamnus costata Maxim. f. pubescens Hiyama = Rhamnus costata Maxim. ●☆

328663　Rhamnus costata Maxim. var. nambuana ? = Rhamnus costata Maxim. ●☆

328664　Rhamnus costata Miq. = Sageretia hamosa (Wall. ex Roxb.) Brongn. ●

328665　Rhamnus crenata Siebold et Zucc.;长叶冻绿(长叶绿柴,长叶鼠李,冻绿,冻绿树,钝齿鼠李,过路黄,红点秤,黄药,尖尾叶鼠李,苦李根,癞痢柴,黎辣根,六厘柴,绿篱,绿篱柴,马灵仙,拿蒴,拿柳,三黄,山黑子,山黄,山六厘,山绿,山绿篱,水冻绿,铁

包金，土黄柏，亡药，一扫光，圆齿鼠李，掌牛仔）；Oriental Buckthorn ●

328666　Rhamnus crenata Siebold et Zucc. var. cambodiana（Pierre ex Pit.）Tardieu ＝ Rhamnus crenata Siebold et Zucc. ●

328667　Rhamnus crenata Siebold et Zucc. var. discolor Rehder；两色冻绿；Bicolor Oriental Buckthorn ●

328668　Rhamnus crenata Siebold et Zucc. var. ilicifolia（Keller）Greene；冬青叶冻绿（冬青叶鼠李）；Hollyleaf Buckthorn ●☆

328669　Rhamnus crenata Siebold et Zucc. var. oreigenes（Hance）Tardieu ＝ Rhamnus crenata Siebold et Zucc. ●

328670　Rhamnus crenata Siebold et Zucc. var. stenophylla ? ＝ Rhamnus davurica Pall. ●

328671　Rhamnus crenata Siebold et Zucc. var. yakushimensis Makino ex Masam.；屋久岛鼠李●☆

328672　Rhamnus crenata Siebold et Zucc. var. yakushimensis Makino ex Masam. ＝ Rhamnus crenata Siebold et Zucc. ●

328673　Rhamnus crenulata Aiton；细圆齿鼠李●☆

328674　Rhamnus crocea Nutt.；冬青叶红鼠李（红鼠李）；Indian Cherry，Redberry，Redberry Buckthorn ●☆

328675　Rhamnus dahurica Pall. ＝ Rhamnus davurica Pall. ●

328676　Rhamnus dahurica Pall. var. liukiuensis E. H. Wilson ＝ Rhamnus liukiuensis（E. H. Wilson）Koidz. ●

328677　Rhamnus dahurica Pall. var. nipponica Makino ＝ Rhamnus ussuriensis J. J. Vassil. ●

328678　Rhamnus dahuricus Lawson ＝ Rhamnus virgata Roxb. ●

328679　Rhamnus dalianensis S. Y. Li et Z. H. Ning；大连鼠李；Dali Buckthorn ●

328680　Rhamnus davurica Pall.；鼠李（臭李子，楮李，大绿子，冻绿，冻绿柴，冻绿皮，禾镰子，红皮绿树，虎梓，苦楸，老鸹眼，老鹳眼，老乌眼，老攸言，録子，绿子，牛筋子，牛李，牛李子，牛消子，牛皂子，女儿茶，椑，山李子，鼠梓，乌槎，乌槎树，乌槎子，乌巢子，乌冈子，乌罡子，无实子，羊史子，梗，皂子，蘸子，赵子）；Dahurian Buckthorn，Dahursk Buckthorn，Davur Buckthorn，Davurian Buckthorn ●

328681　Rhamnus davurica Pall. var. hirsuta（Wight et Arn.）M. A. Lawson ＝ Rhamnus virgata Roxb. var. hirsuta（Wight et Arn.）Y. L. Chen et P. K. Chou ●

328682　Rhamnus davurica Pall. var. hirsuta M. A. Lawson ＝ Rhamnus virgata Roxb. var. hirsuta（Wight et Arn.）Y. L. Chen et P. K. Chou ●

328683　Rhamnus davurica Pall. var. liukiuensis E. H. Wilson ＝ Rhamnus liukiuensis（E. H. Wilson）Koidz. ●

328684　Rhamnus davurica Pall. var. nipponica Makino；东鼠李（本州鼠李，乌苏里鼠李）；Dahurian Buckthorn，Japanese Dahurian Buckthorn ●

328685　Rhamnus davurica Pall. var. nipponica Makino f. pubescens H. Hara；毛东鼠李●☆

328686　Rhamnus davurica Pall. var. nipponica Makino f. pubescens H. Hara ＝ Rhamnus davurica Pall. ●

328687　Rhamnus deflersii Fiori ＝ Rhamnus staddo A. Rich. ●☆

328688　Rhamnus depressa Grubov；凹陷鼠李●☆

328689　Rhamnus diamantiaca Nakai；金钢鼠李（金刚鼠李，老鸹眼）；Diamond Mountain Buckthorn ●

328690　Rhamnus dolichophylla Gontsch.；长叶鼠李●☆

328691　Rhamnus dumetorum C. K. Schneid.；叫李子（刺鼠李）；Hedge Buckthorn ●

328692　Rhamnus dumetorum C. K. Schneid. var. crenoserrata Rehder et E. H. Wilson；圆齿刺鼠李（圆叶刺鼠李）；Crenate Hedge Buckthorn ●

328693　Rhamnus erythroxyloides Hoffmanns.；假柳叶鼠李；Pallas' Buckthorn ●☆

328694　Rhamnus erythroxylon Pall.；柳叶鼠李（黑疙瘩，黑格兰，黑格铃，红木鼠李，窄叶鼠李）；Willowleaf Buckthorn，Willow-leaved Buckthorn ●

328695　Rhamnus esquirolii H. Lév.；贵州鼠李（铁滚子，无刺鼠李，紫棍柴）；Esquirol Buckthorn，Guizhou Buckthorn ●

328696　Rhamnus esquirolii H. Lév. var. glabrata Y. L. Chen et P. K. Chou；木子花；Glabrousleaf Esquirol Buckthorn ●

328697　Rhamnus fallax Boiss.；假鼠李（肉质鼠李）；Carniolan Buckthorn ●☆

328698　Rhamnus filiformis Roth ＝ Sageretia filiformis（Roth）G. Don ●☆

328699　Rhamnus flavescens Y. L. Chen et P. K. Chou；淡黄鼠李；Paleyellow Buckthorn，Pale-yellow Buckthorn ●

328700　Rhamnus formosana Matsum.；台湾鼠李（桶钩藤）；Formosan Buckthorn，Taiwan Buckthorn ●

328701　Rhamnus frangula L.；欧鼠李（弗郎鼠李，泻鼠李，药绿柴，药炭鼠李）；Alder Buckthorn，Berry-bearing Glossy Buckthorn，European Alder Buckthorn，Glossy Buckthorn ●

328702　Rhamnus frangula L. 'Fine Line'；蕨叶欧鼠李；Fern-leaf Buckthorn ●

328703　Rhamnus frangula L. ＝ Frangula alnus Mill. ●

328704　Rhamnus frangula L. f. angustifolia（Loudon）Schelle ＝ Rhamnus frangula L. ●

328705　Rhamnus frangula L. var. angustifolia Loudon ＝ Rhamnus frangula L. ●

328706　Rhamnus fulvo-tincta F. P. Metcalf；黄鼠李（紫背药）；Yellow Buckthorn ●

328707　Rhamnus gilgiana Heppeler；川滇鼠李（刺绿皮，金沙鼠李）；Chuan-Dian Buckthorn，Gilg Buckthorn ●

328708　Rhamnus glabra（Nakai）Nakai ＝ Rhamnus schneideri H. Lév. et Vaniot ●

328709　Rhamnus glabra Nakai ＝ Rhamnus schneideri H. Lév. et Vaniot ●

328710　Rhamnus glabra Nakai ＝ Rhamnus yoshinoi Makino ●

328711　Rhamnus glabra Nakai var. manshurica Nakai ＝ Rhamnus schneideri H. Lév. et Vaniot var. manshurica Nakai ●

328712　Rhamnus glandulosa Aiton；具腺鼠李●☆

328713　Rhamnus globosa Bunge；圆叶鼠李（臭李子，冻绿，冻绿刺，洞皮树，黑旦子，老鹳眼，偶栗子，山绿柴，鸭屎树，野苦楝子）；Lokao Buckthorn，Roundleaf Buckthorn ●

328714　Rhamnus globosa Bunge var. glabra Nakai ＝ Rhamnus schneideri H. Lév. et Vaniot ●

328715　Rhamnus globosa Bunge var. glabra Nakai ＝ Rhamnus yoshinoi Makino ●

328716　Rhamnus globosa Bunge var. meyeri（C. K. Schneid.）S. Y. Li et Z. H. Ning ＝ Rhamnus globosa Bunge ●

328717　Rhamnus globosa Bunge var. ziziphifolia Ts. Tang ＝ Rhamnus parvifolia Bunge ●

328718　Rhamnus grandiflora C. Y. Wu ex Y. L. Chen；大花鼠李；Big-flowered Buckthorn，Largeflower Buckthorn ●

328719　Rhamnus grandifolia（Fisch. et C. A. Mey.）Ledeb.；大叶鼠李●☆

328720　Rhamnus hainanensis Merr. et Chun；海南鼠李；Hainan Buckthorn ●

328721　Rhamnus hamatidens H. Lév. ＝ Rhamnus lamprophylla C. K. Schneid. ●

328722　Rhamnus hemsleyana C. K. Schneid.；亮叶鼠李；Hemsley Buckthorn，Shinyleaf Buckthorn ●

328723　Rhamnus hemsleyana C. K. Schneid. var. paucinervata G. S. Fan et L. L. Deng = Rhamnus hemsleyana C. K. Schneid. var. yunnanensis C. Y. Wu ex Y. L. Chen ●

328724　Rhamnus hemsleyana C. K. Schneid. var. yunnanensis C. Y. Wu ex Y. L. Chen;高山亮叶鼠李;Yunnan Hemslet Buckthorn, Yunnan Shinyleaf Buckthorn ●

328725　Rhamnus henryi C. K. Schneid. ; 毛叶鼠李(黄柴, 绉棉藤); Henry Buckthorn ●

328726　Rhamnus heterophylla Oliv. ;异叶鼠李(黄茶根, 女儿茶, 女儿红, 崖枣树, 岩果紫, 岩枣树); Diversifolious Buckthorn, Variable-leaved Buckthorn ●

328727　Rhamnus heterophylla Oliv. var. oblongifolia E. Pritz. = Rhamnus heterophylla Oliv. ●

328728　Rhamnus heterophylla Oliv. var. oblongifolia Pfitzer = Rhamnus heterophylla Oliv. ●

328729　Rhamnus hirsuta Wight et Arn. = Rhamnus virgata Roxb. var. hirsuta (Wight et Arn.) Y. L. Chen et P. K. Chou ●

328730　Rhamnus holstii Engl. = Rhamnus staddo A. Rich. ●☆

328731　Rhamnus hupehensis C. K. Schneid. ; 湖北鼠李; Hubei Buckthorn ●

328732　Rhamnus hypochrysa C. K. Schneid. = Rhamnus utilis Decne. var. hypochrysa (C. K. Schneid.) Rehder ●

328733　Rhamnus iguanaea Jacq. = Celtis iguanaea (Jacq.) Sarg. ●☆

328734　Rhamnus ilicifolia Kellogg; 冬青叶鼠李; Hollyleaf Red Berry, Hollyleaf Redberry Buckthorn ●☆

328735　Rhamnus imeretina Dippel;利比里亚鼠李(高加索鼠李)●☆

328736　Rhamnus inconspicua Grubov = Rhamnus leptophylla C. K. Schneid. ●

328737　Rhamnus infectoria L. ; 南欧鼠李; Avignon Berry, French Berries, Persian Berries, Persian Berry Buckthorn, Yellow Berries, Yellow Berry ●☆

328738　Rhamnus integrifolia DC. ;全缘叶鼠李●☆

328739　Rhamnus iteinophylla C. K. Schneid. ;桃叶鼠李(冻绿树); Peachleaf Buckthorn, Peach-leaved Buckthorn ●

328740　Rhamnus japonica Maxim. ; 日本鼠李(鼠李); Japanese Buckthorn ●☆

328741　Rhamnus japonica Maxim. var. angustifolia Nakai; 狭叶日本鼠李;Narrowleaf Japanese Buckthorn ●☆

328742　Rhamnus japonica Maxim. var. decipiens Maxim. ;迷惑鼠李●☆

328743　Rhamnus japonica Maxim. var. decipiens Maxim. f. chlorocarpa (Honda) H. Hara;绿果日本鼠李●☆

328744　Rhamnus japonica Maxim. var. decipiens Maxim. f. senanensis (Koidz.) H. Hara;信浓鼠李●☆

328745　Rhamnus japonica Maxim. var. microphylla H. Hara;小叶日本鼠李●☆

328746　Rhamnus japonica Maxim. var. parvifolia (Honda) Sugim. = Rhamnus japonica Maxim. var. decipiens Maxim. f. senanensis (Koidz.) H. Hara ●☆

328747　Rhamnus japonica Maxim. var. parvifolia Sugim. ;小花日本鼠李;Smallflower Japanese Buckthorn ●☆

328748　Rhamnus japonica Maxim. var. senanensis (Koidz.) H. Ohashi = Rhamnus japonica Maxim. var. decipiens Maxim. f. senanensis (Koidz.) H. Hara ●☆

328749　Rhamnus jujuba L. = Ziziphus mauritiana Lam. ●

328750　Rhamnus kanagusukii Makino;变叶鼠李●

328751　Rhamnus kiusiana Hatus. = Rhamnus yoshinoi Makino ●

328752　Rhamnus koraiensis C. K. Schneid. ;朝鲜鼠李(老乌眼籽); Korea Buckthorn, Korean Buckthorn ●

328753　Rhamnus kwangsiensis Y. L. Chen et P. K. Chou;广西鼠李; Guangxi Buckthorn ●

328754　Rhamnus lamprophylla C. K. Schneid. ;钩齿鼠李(鹿角刺); Hooktooth Buckthorn, Shiningleaf Buckthorn, Shiny-leaved Buckthorn ●

328755　Rhamnus lanceolata Pursh;披针叶鼠李; Buckthorn, Lance-leaf Buckthorn, Lance-leaved Buckthorn ●☆

328756　Rhamnus lanceolata Pursh var. glabrata Gleason;光披针叶鼠李(无毛披针叶鼠李); Lance-leaved Buckthorn, Smooth Lance-leaf Buckthorn ●☆

328757　Rhamnus laoshanensis D. K. Zang; 崂山鼠李; Laoshan Buckthorn ●

328758　Rhamnus leptacantha C. K. Schneid. ;纤花鼠李; Fineflower Buckthorn, Fine-flowered Buckthorn ●

328759　Rhamnus leptacantha Schneid. = Rhamnus gilgiana Heppeler ●

328760　Rhamnus leptophylla C. K. Schneid. ;薄叶鼠李(白赤木, 白色木, 打枪子, 黑龙须, 黑枣子, 绛耳木, 绛梨木, 郊李子, 叫耳母子, 叫梨子, 叫铃子, 金钱子, 嚼连木, 蜡子树, 雷震子, 鹿角刺, 铁包金, 乌槎子, 乌苕子刺, 细叶鼠李, 叶铃子, 震天雷); Thinleaf Buckthorn, Thin-leaved Buckthorn ●

328761　Rhamnus leptophylla C. K. Schneid. var. milensis C. K. Schneid. = Rhamnus virgata Roxb. ●

328762　Rhamnus leptophylla C. K. Schneid. var. scabrella Rehder = Rhamnus tangutica J. J. Vassil. ●

328763　Rhamnus leptophylla C. K. Schneid. var. villosissima C. Y. Wu; 毛薄叶鼠李●

328764　Rhamnus leveilleana Fedde = Rhamnus rosthornii E. Pritz. ex Diels ●

328765　Rhamnus libanotica Boiss. ;黎巴嫩鼠李;Lebanon Buckthorn ●☆

328766　Rhamnus lineata L. = Berchemia lineata (L.) DC. ●

328767　Rhamnus liukiuensis (E. H. Wilson) Koidz. ;琉球鼠李;Liukiu Buckthorn, Liuqiu Buckthorn ●

328768　Rhamnus longipes Merr. et Chun; 长柄鼠李; Longstalk Buckthorn, Long-stalked Buckthorn ●

328769　Rhamnus lotus L. = Ziziphus lotus (L.) Lam. ●☆

328770　Rhamnus lycioides L. ;枸杞鼠李●☆

328771　Rhamnus lycioides L. subsp. atlantica (Murb.) Jahand. et Maire;北非鼠李●☆

328772　Rhamnus lycioides L. subsp. oleoides (L.) Jahand. et Maire;齐墩果鼠李●☆

328773　Rhamnus lycioides L. subsp. velutina (Boiss.) Nyman; 短绒毛鼠李●☆

328774　Rhamnus lycioides L. var. amygdalina (Desf.) Jahand. et Maire = Rhamnus lycioides L. ●☆

328775　Rhamnus lycioides L. var. angustifolia (Lange) Maire = Rhamnus lycioides L. ●☆

328776　Rhamnus lycioides L. var. angustissima Maire = Rhamnus lycioides L. ●☆

328777　Rhamnus lycioides L. var. faureliana Maire = Rhamnus lycioides L. ●☆

328778　Rhamnus lycioides L. var. latifolia (Lange) Jahand. et Maire = Rhamnus lycioides L. ●☆

328779　Rhamnus mairei Schneid. = Rhamnella martinii (H. Lév.) C. K. Schneid. ●

328780　Rhamnus martinii H. Lév. = Rhamnella martinii (H. Lév.) C. K. Schneid. ●

328781　Rhamnus maximovicz, Rhamnus maximovicziana J. J. Vassil. ;黑桦鼠李(黑桦树);

Maximovicz Buckthorn ●

328782　Rhamnus maximovicziana J. J. Vassil. var. oblongifolia Y. L. Cheng et P. K. Chou；矩叶黑桦鼠李（矩叶黑桦树）；Oblongleaf Maximovicz Buckthorn ●

328783　Rhamnus meyeri C. K. Schneid. ；山东鼠李●

328784　Rhamnus meyeri C. K. Schneid. = Rhamnus globosa Bunge ●

328785　Rhamnus micranthus L. = Trema micrantha（L.）Blume ●☆

328786　Rhamnus microcarpa Boiss. ；小果鼠李●

328787　Rhamnus mildbraedii Engl. ；米尔德鼠李●☆

328788　Rhamnus minuta Grubov；矮小鼠李；Dwarf Buckthorn ●

328789　Rhamnus myrtifolia Willk. ；香桃木叶鼠李●☆

328790　Rhamnus myrtillus H. Lév. = Myrsine africana L. ●

328791　Rhamnus myrtina Burm. f. = Scutia myrtina（Burm. f.）Kurz ●

328792　Rhamnus mystacinus Aiton = Helinus mystacinus（Aiton）E. Mey. ex Steud. ●☆

328793　Rhamnus nakaharai（Hayata）Hayata；台中鼠李（中原氏鼠李）；Nakahara Buckthorn，Taizhong Buckthorn ●

328794　Rhamnus nepalensis（Wall.）M. A. Lawson；尼泊尔鼠李（尼布鲁鼠李，染布叶，纤序鼠李）；Nepal Buckthorn ●

328795　Rhamnus nigricans Hand. -Mazz. ；黑背鼠李；Blackish Buckthorn，Nigrescent Buckthorn ●

328796　Rhamnus ninglangensis Y. L. Chen；宁蒗鼠李●

328797　Rhamnus nipponica（Makino）Grubov = Rhamnus davurica Pall. var. nipponica Makino ●

328798　Rhamnus nipponica（Makino）Grubov = Rhamnus davurica Pall. ●

328799　Rhamnus obovatilimba Merr. et F. P. Metcalf = Rhamnus rugulosa Hemsl. ex Forbes et Hemsl. ●

328800　Rhamnus oenopolia L. = Ziziphus oenopolia（L.）Mill. ●

328801　Rhamnus oiwakensis Hayata = Rhamnus kanagusukii Makino ●

328802　Rhamnus oiwakensis Hayata = Rhamnus parvifolia Bunge ●

328803　Rhamnus oleoides L. = Rhamnus lycioides L. subsp. oleoides（L.）Jahand. et Maire ●☆

328804　Rhamnus oleoides L. subsp. atlantica Murb. = Rhamnus lycioides L. subsp. atlantica（Murb.）Jahand. et Maire ●☆

328805　Rhamnus oleoides L. subsp. libyca（Asch. et Schweinf.）Brullo et Furnari = Rhamnus lycioides L. subsp. oleoides（L.）Jahand. et Maire ●☆

328806　Rhamnus oleoides L. var. amygdalina（Desf.）Ball = Rhamnus lycioides L. subsp. oleoides（L.）Jahand. et Maire ●☆

328807　Rhamnus oleoides L. var. libyca Asch. et Schweinf. = Rhamnus lycioides L. subsp. oleoides（L.）Jahand. et Maire ●☆

328808　Rhamnus oreigenes Hance = Rhamnus crenata Siebold et Zucc. ●

328809　Rhamnus owiakensis Hayata = Rhamnus parvifolia Bunge ●

328810　Rhamnus palaestina Boiss. ；燥地鼠李；Palestine Buckthorn ●☆

328811　Rhamnus paliurus L. = Paliurus spina-christi Mill. ●

328812　Rhamnus pallasii Fisch. et C. A. Mey. ；帕拉斯鼠李；Pallas' Buckthorn ●☆

328813　Rhamnus pallasii Fisch. et C. A. Mey. = Rhamnus erythroxyloides Hoffmanns. ●☆

328814　Rhamnus paniculiflora C. K. Schneid. = Rhamnus nepalensis（Wall.）M. A. Lawson ●

328815　Rhamnus paniculiflorus Schneid. = Rhamnus nepalensis（Wall.）M. A. Lawson ●

328816　Rhamnus parvifolia Bunge；小叶鼠李（臭李子，大绿，黑格铃，黑格令，叫驴子，叫驴子刺，琉璃枝，驴子刺，麻绿，挠胡子，雅西勒）；Littleleaf Buckthorn，Small-leaved Buckthorn ●

328817　Rhamnus parvifolia Bunge var. tumetica（Grubov）E. W. Ma；土默特鼠李；Tumote Buckthorn ●

328818　Rhamnus parvifolia Bunge var. tumetica（Grubov）N. W. Ma = Rhamnus parvifolia Bunge ●

328819　Rhamnus pasteurii H. Lév. = Gardneria multiflora Makino ●

328820　Rhamnus pauciflora A. Rich. = Rhamnus prinoides L'Hér. ●☆

328821　Rhamnus persica Boiss. ；波斯鼠李●☆

328822　Rhamnus persica Lawson = Rhamnus prostrata Jacq. ex Parker ●

328823　Rhamnus petiolaris Boiss. et Balansa；叶柄鼠李●☆

328824　Rhamnus pianensis Kaneh. = Rhamnus kanagusukii Makino ●

328825　Rhamnus pianensis Kaneh. = Rhamnus parvifolia Bunge ●

328826　Rhamnus pilushanensis Y. C. Liu et C. M. Wang；毕禄山鼠李；Bilushan Buckthorn ●

328827　Rhamnus polymorpha Turcz. = Rhamnus parvifolia Bunge ●

328828　Rhamnus polymorphus Turcz. = Rhamnus parvifolia Bunge ●

328829　Rhamnus potaninii J. J. Vassil. = Rhamnus tangutica J. J. Vassil. ●

328830　Rhamnus prinoides L'Hér. ；南非鼠李；Southern Africcan Dogwood ●☆

328831　Rhamnus procumbens Edgew. ；蔓生鼠李；Procumbent Buckthorn ●

328832　Rhamnus prostrata Jacq. = Rhamnus prostrata Jacq. ex Parker ●

328833　Rhamnus prostrata Jacq. ex Parker；平卧鼠李（旱鼠李）；Prostrate Buckthorn ●

328834　Rhamnus pseudofrangula H. Lév. = Rhamnus crenata Siebold et Zucc. ●

328835　Rhamnus pumila L. ；矮鼠李；Dwarf Buckthorn ●☆

328836　Rhamnus pumila Turra = Rhamnus pumila L. ●☆

328837　Rhamnus purpurea Edgew. ；紫鼠李●☆

328838　Rhamnus purshiana DC. ；北美鼠李（波斯鼠李，美鼠李，珀希鼠李，药鼠李）；Bearberry，Californian Buckthorn，Cascara，Cascara Buckthorn，Cascara Sagrada，Cascara Tree，Shittim Bark ●☆

328839　Rhamnus qianweiensis Z. Y. Zhu；犍为鼠李；Jianwei Buckthorn ●

328840　Rhamnus rhodesicus Suess. = Rhamnus staddo A. Rich. ●☆

328841　Rhamnus rhododendriphylla Y. L. Chen et P. K. Chou；杜鹃叶鼠李；Azalealeaf Buckthorn，Rhododendronleaf Buckthorn，Rhododendron-leaved Buckthorn ●

328842　Rhamnus rosthornii E. Pritz. ex Diels；小冻绿树；Rosthorns Buckthorn ●

328843　Rhamnus rugulosa Hemsl. ex Forbes et Hemsl. ；皱叶鼠李；Wrinkledleaf Buckthorn，Wrinkle-leaved Buckthorn ●

328844　Rhamnus rugulosa Hemsl. ex Forbes et Hemsl. var. chekiangensis（W. C. Cheng）Y. L. Chen et P. K. Chou；浙江鼠李；Zhejiang Buckthorn ●

328845　Rhamnus rugulosa Hemsl. ex Forbes et Hemsl. var. glabrata Y. L. Chen et P. K. Chou；脱毛皱叶鼠李；Glabrate Wrinkledleaf Buckthorn ●

328846　Rhamnus rupestris Scop. ；岩地鼠李●☆

328847　Rhamnus sanguinea Pers. = Rhamnus frangula L. ●

328848　Rhamnus sargentiana C. K. Schneid. ；多脉鼠李（西南鼠李）；Sargent Buckthorn ●

328849　Rhamnus saxatilis Jacq. ；岩生鼠李；Avignon Berry，Persian Berry，Rock Buckthorn，Saxatile Buckthorn，Yellow Berry ●☆

328850　Rhamnus saxatilis X. H. Song = Rhamnus saxatilis Jacq. ●☆

328851　Rhamnus schneideri H. Lév. et Vaniot；长梗鼠李；Schneider Buckthorn ●

328852　Rhamnus schneideri H. Lév. et Vaniot = Rhamnus yoshinoi

Makino ●

328853 Rhamnus schneideri H. Lév. et Vaniot var. manshurica Nakai；东北长梗鼠李（东北鼠李）；Manchurian Schneider Buckthorn，NE. China Schneider Buckthorn ●

328854 Rhamnus serphillifolia H. Lév.；滇东鼠李●

328855 Rhamnus serpyllacea Greuter et Burdet；百里香鼠李●☆

328856 Rhamnus serpyllifolia Emb. et Maire ＝ Rhamnus serpyllacea Greuter et Burdet ●☆

328857 Rhamnus shozyoensis Nakai var. glabrata ？ ＝ Rhamnus yoshinoi Makino ●

328858 Rhamnus sieboldiana Makino ＝ Rhamnus utilis Decne. ●

328859 Rhamnus sintenisii Rech. f.；西恩鼠李●☆

328860 Rhamnus smithii Greene；史密斯鼠李；Smith Buckthorn ●☆

328861 Rhamnus songorica Gontsch.；新疆鼠李（土茶叶）；Dzungar Buckthorn，Songory Buckthorn ●

328862 Rhamnus spathulifolia Fisch. et C. A. Mey.；匙叶鼠李●☆

328863 Rhamnus spiciflora A. Rich. ＝ Sageretia thea（Osbeck）M. C. Johnst. ●

328864 Rhamnus spina-christi L. ＝Ziziphus spina-christi（L.）Desf. ●☆

328865 Rhamnus staddo A. Rich.；施塔多鼠李●☆

328866 Rhamnus staddo A. Rich. var. deflersii（Fiori）Engl. ＝ Rhamnus staddo A. Rich. ●☆

328867 Rhamnus subapetala Merr.；紫背鼠李；Purpleback Buckthorn，Purple-backed Buckthorn ●

328868 Rhamnus tangutica J. J. Vassil.；甘青鼠李（粗叶鼠李，冻绿）；Tangut Buckthorn ●

328869 Rhamnus tetragona L. f. ＝ Lauridia tetragona（L. f.）R. H. Archer ●☆

328870 Rhamnus thea Osbeck ＝Sageretia thea（Osbeck）M. C. Johnst. ●

328871 Rhamnus theezans L. ＝ Sageretia thea（Osbeck）M. C. Johnst. ●

328872 Rhamnus tibetica Y. L. Chen et P. K. Chou；藏鼠李；Tibet Buckthorn，Xizang Buckthorn ●

328873 Rhamnus tinctoria Hemsl. ＝ Rhamnus globosa Bunge ●

328874 Rhamnus tinctoria Waldst. et Kit.；染色鼠李●☆

328875 Rhamnus tonkinensis Pit.；越南鼠李；Tonkin Buckthorn，Vietnam Buckthorn ●☆

328876 Rhamnus tonkinensis Pit. ＝ Rhamnus nepalensis（Wall.）M. A. Lawson ●

328877 Rhamnus tripartita Ucria ＝ Rhus tripartita（Ucria）Grande ●☆

328878 Rhamnus triquetra（Wall.）Brandis；三棱鼠李●☆

328879 Rhamnus tumetica Grubov ＝ Rhamnus parvifolia Bunge ●

328880 Rhamnus tzekweiensis Y. L. Chen et P. K. Chou；鄂西鼠李；Tzekwei Buckthorn，W. Hubei Buckthorn，West Buckthorn ●

328881 Rhamnus uhligii Engl. ＝ Rhamnus staddo A. Rich. ●☆

328882 Rhamnus ussuriensis J. J. Vassil.；乌苏里鼠李（臭李子，老鸹眼）；Ussuri Buckthorn ●

328883 Rhamnus utilis Decne.；冻绿（剥皮刺，搭绿皮，大脑头，冻绿柴，冻绿鼠李，冻绿树，冻木刺，冻木树，狗李，过路黄，黑狗丹，红冻，鹿蹄根，绿皮刺，鼠李，小黄，油葫芦子）；China Buckthorn，Chinese Buckthorn ●

328884 Rhamnus utilis Decne. f. glaber Rehder ＝ Rhamnus utilis Decne. ●

328885 Rhamnus utilis Decne. var. hypochrysa（C. K. Schneid.）Rehder；毛冻绿（黑刺，毛叶冻绿）；Hairy China Buckthorn，Hairy Chinese Buckthorn ●

328886 Rhamnus utilis Decne. var. multinervis Y. Q. Zhu et D. K. Zang；多脉毛冻绿（多脉鼠李）；Many-veins Chinese Buckthorn ●

328887 Rhamnus utilis Decne. var. multinervis Y. Q. Zhu et D. K. Zang ＝ Rhamnus utilis Decne. ●

328888 Rhamnus utilis Decne. var. szechuanensis Y. L. Chen et P. K. Chou；高山冻绿；Alpine Chinese Buckthorn ●

328889 Rhamnus velutina J. Anthony；毡毛鼠李；Velvet-like Buckthorn ●

328890 Rhamnus velutina J. Anthony ＝ Rhamnus ninglangensis Y. L. Chen ●

328891 Rhamnus virgata Maxim. ＝ Rhamnus davurica Pall. ●

328892 Rhamnus virgata Roxb.；帚枝鼠李（老攸言，牛李子，细叶鼠李，小叶冻绿）；Twiggy Buckthorn ●

328893 Rhamnus virgata Roxb. var. aprica Maxim. ＝ Rhamnus maximovicziana J. J. Vassil. ●

328894 Rhamnus virgata Roxb. var. hirsuta（Wight et Arn.）Y. L. Chen et P. K. Chou；糙毛帚枝鼠李；Hairy Twiggy Buckthorn ●

328895 Rhamnus virgata Roxb. var. mongolica Maxim. ＝ Rhamnus maximovicziana J. J. Vassil. ●

328896 Rhamnus virgata Roxb. var. parvifolia Maxim. ＝ Rhamnus tangutica J. J. Vassil. ●

328897 Rhamnus virgata Roxb. var. sylvestris Maxim. ＝ Rhamnus diamantiaca Nakai ●

328898 Rhamnus viridifolia Liou；金县鼠李；Jinxian Buckthorn ●

328899 Rhamnus vitisidaea Burm. f. ＝ Breynia vitis-idaea（Burm. f.）C. E. C. Fisch. ●

328900 Rhamnus wilsonii C. K. Schneid.；山鼠李（冻绿，郊李子，庐山鼠李）；E. H. Wilson Buckthorn，Wild Buckthorn，Wilson Buckthorn ●

328901 Rhamnus wilsonii Roxb. var. pilosa Rehder；毛山鼠李；Hairy E. H. Wilson Buckthorn，Hairy Wild Buckthorn ●

328902 Rhamnus wumingensis Y. L. Chen et P. K. Chou；武鸣鼠李；Wuming Buckthorn ●

328903 Rhamnus xizangensis Y. L. Chen et P. K. Chou；西藏鼠李；Xizang Buckthorn ●

328904 Rhamnus yoshinoi Makino；吉野氏鼠李（东北鼠李）；Yoshino Buckthorn ●

328905 Rhamnus yoshinoi Makino var. velvetina T. Shimizu ＝ Rhamnus chugokuensis Hatus. ●

328906 Rhamnus yoshinoi Makino var. velvetina T. Shimizu ＝ Rhamnus yoshinoi Makino ●

328907 Rhamnus yunnanensis Heppeler ＝ Rhamnella martinii（H. Lév.）C. K. Schneid. ●

328908 Rhamnus zeyberi Sond.；红象牙木；Red Ebony，Red Ivory，Red Ivory Wood ●☆

328909 Rhamnus zeyheri Sond. ＝ Berchemia zeyheri（Sond.）Grubov ●☆

328910 Rhamnus zizyphus L. ＝ Ziziphus jujuba Mill. ●

328911 Rhamnus zizyphus L. ＝ Ziziphus zizyphus（L.）H. Karst. ●

328912 Rhamophidia Lindl. ＝ Hetaeria Blume（保留属名）■

328913 Rhamophidia elongata（Lindl.）Lindl. ＝ Hetaeria elongata（Lindl.）Hook. f. ■

328914 Rhamophidia elongata Lindl. ＝ Hetaeria elongata（Lindl.）Hook. f. ■

328915 Rhamophidia japonica Rchb. f. ＝ Myrmechis japonica（Rchb. f.）Rolfe ■

328916 Rhamophidia rubens（Lindl.）Lindl. ＝ Hetaeria rubens（Lindl.）Benth. et Hook. f. ■

328917 Rhamphicarpa Benth.（1836）；钩果列当属（钩果玄参属）■☆

328918 Rhamphicarpa albersii（Engl.）V. Naray. ＝ Cycnium veronicifolium（Vatke）Engl. subsp. suffruticosum（Engl.）O. J. Hansen ●☆

328919　Rhamphicarpa angolense Engl. = Cycnium angolense (Engl.) O. J. Hansen ■☆

328920　Rhamphicarpa aquatica (Engl.) V. Naray. = Cycnium tubulosum (L. f.) Engl. ■☆

328921　Rhamphicarpa asperrima (Engl.) V. Naray. = Cycnium volkensii Engl. ■☆

328922　Rhamphicarpa brevifolia (De Wild.) Staner = Cycnium tubulosum (L. f.) Engl. subsp. montanum (N. E. Br.) O. J. Hansen ■☆

328923　Rhamphicarpa brevipedicellata O. J. Hansen；短梗钩果列当■☆

328924　Rhamphicarpa cameroniana Oliv. = Cycnium cameronianum (Oliv.) Engl. ■☆

328925　Rhamphicarpa capillacea A. Raynal；细毛钩果列当■☆

328926　Rhamphicarpa claessensii (De Wild.) Staner = Cycnium tubulosum (L. f.) Engl. subsp. montanum (N. E. Br.) O. J. Hansen ☆

328927　Rhamphicarpa curviflora Benth. = Cycnium tubulosum (L. f.) Engl. ■☆

328928　Rhamphicarpa ellenbeckii (Engl.) V. Naray. = Cycnium volkensii Engl. ■☆

328929　Rhamphicarpa elliotii S. Moore = Cycnium veronicifolium (Vatke) Engl. ■☆

328930　Rhamphicarpa elongata (Hochst.) O. J. Hansen；伸长钩果列当■☆

328931　Rhamphicarpa elskensii (De Wild.) Staner = Cycnium tubulosum (L. f.) Engl. subsp. montanum (N. E. Br.) O. J. Hansen ■☆

328932　Rhamphicarpa filicalyx E. A. Bruce = Cycnium filicalyx (E. A. Bruce) O. J. Hansen ■☆

328933　Rhamphicarpa fistulosa (Hochst.) Benth. ；管钩果列当■☆

328934　Rhamphicarpa gallaensis (Engl.) Cufod. = Cycnium volkensii Engl. ■☆

328935　Rhamphicarpa hamata (Engl. et Gilg) V. Naray. = Cycnium tubulosum (L. f.) Engl. ■☆

328936　Rhamphicarpa herzfeldiana Vatke = Cycnium herzfeldianum (Vatke) Engl. ■☆

328937　Rhamphicarpa herzfeldiana Vatke var. subauriculata ? = Cycnium herzfeldianum (Vatke) Engl. ■☆

328938　Rhamphicarpa heuglinii Hochst. ex Schweinf. = Cycnium tubulosum (L. f.) Engl. ■☆

328939　Rhamphicarpa humilis Hochst. = Cycniopsis humifusa (Forssk.) Engl. ■☆

328940　Rhamphicarpa humilis Hochst. ex Benth. = Cycniopsis humilis (Hochst. ex Benth.) Backlund, Asfaw Hunde et E. Langström ■☆

328941　Rhamphicarpa jamesii V. Naray. = Cycnium jamesii (V. Naray.) O. J. Hansen ■☆

328942　Rhamphicarpa montana N. E. Br. = Cycnium tubulosum (L. f.) Engl. subsp. montanum (N. E. Br.) O. J. Hansen ■☆

328943　Rhamphicarpa multicaulis V. Naray. = Cycnium tubulosum (L. f.) Engl. subsp. montanum (N. E. Br.) O. J. Hansen ■☆

328944　Rhamphicarpa neghellensis Fiori = Cycnium tubulosum (L. f.) Engl. subsp. montanum (N. E. Br.) O. J. Hansen ■☆

328945　Rhamphicarpa paucidentata (Engl.) Fiori = Cycnium tubulosum (L. f.) Engl. subsp. montanum (N. E. Br.) O. J. Hansen ■☆

328946　Rhamphicarpa paucidentata (Engl.) Fiori. f. subintegra Engl. = Cycnium tubulosum (L. f.) Engl. subsp. montanum (N. E. Br.) O. J. Hansen ■☆

328947　Rhamphicarpa recurva Oliv. = Cycnium recurvum (Oliv.) Engl. ■☆

328948　Rhamphicarpa spicata (Engl.) V. Naray. = Cycnium volkensii Engl. ■☆

328949　Rhamphicarpa suffruticosa (Engl.) V. Naray. = Cycnium veronicifolium (Vatke) Engl. subsp. suffruticosum (Engl.) O. J. Hansen ●☆

328950　Rhamphicarpa tenuisecta Standl. = Cycnium recurvum (Oliv.) Engl. ■☆

328951　Rhamphicarpa tubulosa (L. f.) Benth. = Cycnium tubulosum (L. f.) Engl. ■☆

328952　Rhamphicarpa tubulosa (L. f.) Benth. var. curviflora (Benth.) Chiov. = Cycnium tubulosum (L. f.) Engl. ■☆

328953　Rhamphicarpa veronicifolium Vatke = Cycnium veronicifolium (Vatke) Engl. ■☆

328954　Rhamphicarpa volkensii (Engl.) V. Naray. = Cycnium volkensii Engl. ■☆

328955　Rhamphicarpa volkensii (Engl.) V. Naray. var. keniensis R. E. Fr. = Cycnium volkensii Engl. ■☆

328956　Rhamphidia (Lindl.) Lindl. = Hetaeria Blume(保留属名)■

328957　Rhamphidia Lindl. = Hetaeria Blume(保留属名)■

328958　Rhamphidia discoidea Rchb. f. = Hetaeria oblongifolia Blume ■

328959　Rhamphidia elongata (Lindl.) Lindl. = Hetaeria finlaysoniana Seidenf. ■

328960　Rhamphidia japonica Rchb. f. = Myrmechis japonica (Rchb. f.) Rolfe ■

328961　Rhamphidia rubens (Lindl.) Lindl. = Hetaeria affinis (Griff.) Seidenf. et Ormerod ■

328962　Rhamphidia rubicunda (Blume) F. Muell. = Goodyera rubicunda (Rchb. f.) J. J. Sm. ■

328963　Rhamphidia rubicunda (Rchb. f.) Rchb. f. = Hetaeria oblongifolia Blume ■

328964　Rhamphidia tenuis Lindl. = Hetaeria oblongifolia Blume ■

328965　Rhamphocarya Kuang = Annamocarya A. Chev. ●

328966　Rhamphocarya Kuang = Carya Nutt. (保留属名)●

328967　Rhamphocarya integrifolialata Kuang = Annamocarya sinensis (Dode) Leroy ●◇

328968　Rhamphogyne S. Moore(1914)；喙果菀属■☆

328969　Rhamphogyne rhynchocarpa S. Moore；喙果菀■☆

328970　Rhampholepis Stapf = Sacciolepis Nash ■

328971　Rhamphorhynchus Garay(1977)；新钩喙兰属■☆

328972　Rhamphorhynchus mendoncae (Brade et Pabst) Garay；新钩喙兰■☆

328973　Rhamphospermum Andrz. ex Besser = Sinapis L. ■

328974　Rhamphospermum Rchb. = Sinapis L. ■

328975　Rhanteriopsis Rauschert(1982)；隆脉菊属●☆

328976　Rhanteriopsis microcephala (Boiss.) Rauschert；小头隆脉菊●☆

328977　Rhanteriopsis puberula (Boiss. et Hausskn.) Rauschert；毛隆脉菊●☆

328978　Rhanterium Desf. (1799)；外包菊属●☆

328979　Rhanterium adpressum Coss. et Durieu；匍匐外包菊●☆

328980　Rhanterium adpressum Coss. et Durieu subsp. intermedium Pomel = Rhanterium adpressum Coss. et Durieu ●☆

328981　Rhanterium suaveolens Desf. ；芳香外包菊●☆

328982　Rhanterium suaveolens Desf. subsp. adpressum (Coss. et Durieu) Quézel et Santa = Rhanterium adpressum Coss. et Durieu ●☆

328983　Rhaphedosera Wight = Justicia L. ●■

328984　Rhaphedosera Wight = Rhaphidospora Nees ●■

328985　Rhaphiacme K. Schum. = Raphionacme Harv. ■☆

328986　Rhaphidanthe Hiern ex Gürke = Diospyros L. ●

328987 Rhaphidanthe obliquifolia Hiern ex Gürke = Diospyros obliquifolia（Hiern ex Gürke）F. White ●☆

328988 Rhaphidanthe soyauxii Stapf = Diospyros obliquifolia（Hiern ex Gürke）F. White ●☆

328989 Rhaphidiocystis Hook. f. = Raphidiocystis Hook. f. ■☆

328990 Rhaphidophora Hassk.（1842）;崖角藤属（莉牟芋属,针房藤属）;Rhaphidophora ●■

328991 Rhaphidophora africana N. E. Br.;非洲崖角藤●☆

328992 Rhaphidophora aurea（Lindau ex André）Birdsey;黄金藤●■

328993 Rhaphidophora aurea（Lindau ex André）Birdsey = Epipremnum aureum（Linden ex André）Bunting ●■

328994 Rhaphidophora aurea（Linden et André）Birdsey = Epipremnum pinnatum（L.）Engl. ●■

328995 Rhaphidophora australasica F. M. Bailey;澳洲崖角藤●☆

328996 Rhaphidophora crassicaulis Engl. et Krause;粗茎崖角藤;Thickstem Rhaphidophora ●■

328997 Rhaphidophora decursiva（Roxb.）Schott;爬树龙（大过山龙,大憨毒,大青龙,大青蛇,大青竹标,当年见,过岗龙,过江龙,过山标,过山龙,金草箍,老蛇藤,裂叶崖角藤,爬山虎,麒麟叶,青竹标,山包谷,上木蜈蚣,石莲藕,石蛇,万丈洁,下延崖角藤,鸭绿江,羽叶崖角滕）;Climbing Rhaphidophora, Decurrent Rhaphidophora,Treeclimbing Rhaphidophora ●■

328998 Rhaphidophora dulongensis H. Li;独龙崖角藤;Dulong Rhaphidophora ●

328999 Rhaphidophora dunniana H. Lév. = Amydrium sinense（Engl.）H. Li ●

329000 Rhaphidophora hongkongensis Schott;狮子尾（百足草,大南苏,大青龙,大青蛇,大软筋藤,大蛇翁,过山龙,厚叶藤,金竹标,九龙上调,密脉崖角藤,蜜腺崖角藤,爬树龙,爬树蜈蚣,爬岩龙,青竹标,青竹丝,上木蜈蚣,石壁枫,石风,水底龙,水底蜈蚣,水蜈蚣,顽纠占,香港崖角藤,香港针房藤,小过山龙,小南苏,小上石百足,崖角藤,岩角藤）;Hongkong Rhaphidophora, Liontail Rhaphidophora ●■

329001 Rhaphidophora hookeri Schott;毛过山龙（大百步还阳,大岩藤,大叶崖角藤,过山龙,龙嘴草,爬树龙）;Hooker Rhaphidophora ●■

329002 Rhaphidophora laichouensis Gagnep.;莱州崖角藤（粗茎崖角藤）;Laizhou Rhaphidophora ●■

329003 Rhaphidophora lancifolia Schott;上树蜈蚣（过山龙,小青龙）;Lanceolate Rhaphidophora,Treeclimbing Centipede ●■

329004 Rhaphidophora liukiuensis Hatus.;琉球针房藤（爬树龙,针房藤）●

329005 Rhaphidophora luchunensis H. Li;禄春崖角藤（绿春崖角藤）;Luchun Rhaphidophora ●■

329006 Rhaphidophora maclurei Merr. = Scindapsus maclurei（Merr.）Merr. et F. P. Metcalf ■

329007 Rhaphidophora megaphylla H. Li;大叶崖角藤（软筋藤）;Largeleaf Rhaphidophora,Longleaf Rhaphidophora ●■

329008 Rhaphidophora ovoidea A. Chev. = Rhaphidophora africana N. E. Br. ●☆

329009 Rhaphidophora peepla（Roxb.）Schott;大叶南苏（过江龙,金竹标,爬山虎,爬树龙,青竹标,软筋藤,万年青,小过山龙,小南苏）;Largeleaf Rhaphidophora ●■

329010 Rhaphidophora peepla Schott. = Rhaphidophora hongkongensis Schott ●■

329011 Rhaphidophora perkinsiae Engl.;针房藤（爬树龙）●■

329012 Rhaphidophora pinnata（L.）Schott = Epipremnum pinnatum（L.）Engl. ●■

329013 Rhaphidophora pinnata Schott = Epipremnum pinnatum（L.）Engl. ●■

329014 Rhaphidophora pusilla N. E. Br.;微小崖角藤●☆

329015 Rhaphidophora tonkinensis Engl. et K. Krause = Rhaphidophora hongkongensis Schott ●■

329016 Rhaphidophyllum Benth. = Raphidophyllum Hochst. ■☆

329017 Rhaphidophyllum Benth. = Sopubia Buch. -Ham. ex D. Don ■

329018 Rhaphidophyton Iljin = Noaea Moq. ■●☆

329019 Rhaphidophyton Iljin（1936）;硬叶蓬属●☆

329020 Rhaphidophyton regelii（Bunge）Iljin;雷格尔硬叶蓬■☆

329021 Rhaphidorhynchus Finet = Aerangis Rchb. f. ■☆

329022 Rhaphidorhynchus Finet = Beclardia A. Rich. ■☆

329023 Rhaphidorhynchus Finet = Calyptrochilum Kraenzl. ■☆

329024 Rhaphidorhynchus Finet = Microcoelia Lindl. ■☆

329025 Rhaphidorhynchus Finet = Tridactyle Schltr. ■☆

329026 Rhaphidorhynchus articulatus（Rchb. f.）Poiss. = Aerangis articulata（Rchb. f.）Schltr. ☆

329027 Rhaphidorhynchus batesii（Rolfe）Finet = Aerangis arachnopus（Rchb. f.）Schltr. ☆

329028 Rhaphidorhynchus citratus（Thouars）Finet = Aerangis citrata（Thouars）Schltr. ■☆

329029 Rhaphidorhynchus curnowianus Finet = Aerangis monantha Schltr. ■☆

329030 Rhaphidorhynchus fastuosus（Rchb. f.）Finet = Aerangis fastuosa（Rchb. f.）Schltr. ■☆

329031 Rhaphidorhynchus kotschyi（Rchb. f.）Finet = Aerangis kotschyana（Rchb. f.）Schltr. ■☆

329032 Rhaphidorhynchus macrostachys（Thouars）Finet = Beclardia macrostachya（Thouars）A. Rich. ■☆

329033 Rhaphidorhynchus modestus（Hook. f.）Finet = Aerangis modesta（Hook. f.）Schltr. ■☆

329034 Rhaphidorhynchus modestus（Hook. f.）Finet var. sanderianus（Rchb. f.）Poiss. = Aerangis modesta（Hook. f.）Schltr. ■☆

329035 Rhaphidorhynchus moloneyi（Rolfe）Finet = Calyptrochilum christyanum（Rchb. f.）Summerh. ■☆

329036 Rhaphidorhynchus rohlfsianus（Kraenzl.）Finet = Aerangis brachycarpa（A. Rich.）T. Durand et Schinz ■☆

329037 Rhaphidorhynchus spiculatus Finet = Aerangis spiculata（Finet）Senghas ■☆

329038 Rhaphidorhynchus stylosus Finet = Aerangis cryptodon（Rchb. f.）Schltr. ■☆

329039 Rhaphidorhynchus umbonatus Finet = Aerangis fuscata（Rchb. f.）Schltr. ■☆

329040 Rhaphidosperma G. Don = Rhaphidospora Nees ●■

329041 Rhaphidospora Nees = Justicia L. ●■

329042 Rhaphidospora Nees（1832）;针子草属;Needlegrass, Needlegrass ●■

329043 Rhaphidospora abyssinica Nees = Justicia glabra J. König ex Roxb. ●☆

329044 Rhaphidospora anisophylla Mildbr. = Justicia anisophylla（Mildbr.）Brummitt ■☆

329045 Rhaphidospora cordata（Nees）Nees = Justicia cordata（Nees）T. Anderson ■☆

329046 Rhaphidospora glabra（J. König ex Roxb.）Nees = Justicia glabra J. König ex Roxb. ●☆

329047 Rhaphidospora leptantha Nees = Justicia leptantha（Nees）T. Anderson ■☆

329048　Rhaphidospora oblongifolia Lindau ＝ Justicia oblongifolia（Lindau）M. E. Steiner ●☆

329049　Rhaphidospora vagabunda（Benoist）C. Y. Wu ex C. C. Hu；针子草；Common Needlegrass, Common Needle-grass ●■

329050　Rhaphidura Bremek.（1940）；针尾茜属 ●☆

329051　Rhaphidura lowii（Ridl.）Bremek.；针尾茜 ●☆

329052　Rhaphiodon Schauer ＝ Hyptis Jacq.（保留属名）●■

329053　Rhaphiodon Schauer（1844）；针齿草属 ■☆

329054　Rhaphiodon echinus Schauer；针齿草 ■☆

329055　Rhaphiolepis Ker Gawl. ＝ Rhaphiolepis Lindl.（保留属名）●

329056　Rhaphiolepis Lindl.（1820）（'Raphiolepis'）（保留属名）；石斑木属（车轮梅属,春花属）；Hawthorn, Indian Hawthorn, Raphiolepis ●

329057　Rhaphiolepis Lindl. ex Ker Gawl. ＝ Rhaphiolepis Lindl.（保留属名）●

329058　Rhaphiolepis Poir. ＝ Rhaphiolepis Lindl.（保留属名）●

329059　Rhaphiolepis cheniana F. P. Metcalf ＝ Rhaphiolepis salicifolia Lindl. ●

329060　Rhaphiolepis delacouri André；戴氏石斑木；Indian Hawthorn ●☆

329061　Rhaphiolepis ferruginea F. P. Metcalf；锈毛石斑木；Rustyhair Raphiolepis, Rusty-haired Raphiolepis ●

329062　Rhaphiolepis ferruginea F. P. Metcalf var. serrata F. P. Metcalf；齿叶锈毛石斑木（齿叶石斑木）；Serrate Rustyhair Raphiolepis ●

329063　Rhaphiolepis gracilis Nakai ＝ Rhaphiolepis indica（L.）Lindl. ex Ker ●

329064　Rhaphiolepis hainanensis F. P. Metcalf ＝ Rhaphiolepis lanceolata Hu ●

329065　Rhaphiolepis hiiranensis Kaneh. ＝ Rhaphiolepis indica（L.）Lindl. ex Ker var. shilanensis Y. P. Yang et H. Y. Liu ●

329066　Rhaphiolepis impressivena Masam.；刻脉石斑木 ●

329067　Rhaphiolepis indica（L.）Lindl ＝ Rhaphiolepis indica（L.）Lindl. ex Ker ●

329068　Rhaphiolepis indica（L.）Lindl. ex Ker；石斑木（白杏花,车轮梅,春花,春花木,雷公树,伞状春花,山花木,石桂,石棠木,铁里木,印度石斑木,凿角）；Hongkong Hawthorn, Hongkong Raphiolepis, Indian Hawthorn, Yeddo Hawthorn ●

329069　Rhaphiolepis indica（L.）Lindl. ex Ker subsp. umbellata（Thunb.）Hatus. ＝ Rhaphiolepis indica（L.）Lindl. ex Ker var. umbellata（Thunb.）H. Ohashi ●

329070　Rhaphiolepis indica（L.）Lindl. ex Ker var. integerrima（Hook. et Arn.）Kitam. ＝ Rhaphiolepis indica（L.）Lindl. ex Ker var. umbellata（Thunb.）H. Ohashi ●

329071　Rhaphiolepis indica（L.）Lindl. ex Ker var. liukiuensis（Koidz.）Kitam. ＝ Rhaphiolepis indica（L.）Lindl. ex Ker var. shilanensis Y. P. Yang et H. Y. Liu ●

329072　Rhaphiolepis indica（L.）Lindl. ex Ker var. minor（Makino）Kitam. ＝ Rhaphiolepis indica（L.）Lindl. ex Ker var. umbellata（Thunb.）H. Ohashi f. minor（Makino）H. Ohashi ●☆

329073　Rhaphiolepis indica（L.）Lindl. ex Ker var. shilanensis Y. P. Yang et H. Y. Liu；恒春石斑木（琉球春花）；Hiiranshan Hawthorn, Hirane Indian Hawthorn ●

329074　Rhaphiolepis indica（L.）Lindl. ex Ker var. shilanensis Y. P. Yang et H. Y. Liu ＝ Rhaphiolepis indica（L.）Lindl. ex Ker ●

329075　Rhaphiolepis indica（L.）Lindl. ex Ker var. tashiroi Hayata ex Matsum. et Hayata；毛序石斑木（石斑木,田代氏石斑木）；Tashiro Indian Hawthorn ●

329076　Rhaphiolepis indica（L.）Lindl. ex Ker var. umbellata（Thunb. ex Murray）H. Ohashi ＝ Rhaphiolepis indica（L.）Lindl. ex Ker

329077　Rhaphiolepis indica（L.）Lindl. ex Ker var. umbellata（Thunb.）H. Ohashi f. minor（Makino）H. Ohashi；小石斑木（小伞状春花）●☆

329078　Rhaphiolepis indica（L.）Lindl. ex Ker var. umbellata（Thunb.）H. Ohashi ＝ Rhaphiolepis umbellata（Thunb.）Makino ●

329079　Rhaphiolepis indica（L.）Lindl. f. umbellata（Thunb.）Hatus. ＝ Rhaphiolepis umbellata（Thunb.）Makino ●

329080　Rhaphiolepis indica（L.）Lindl. var. angustifolia Cardot ＝ Rhaphiolepis lanceolata Hu ●

329081　Rhaphiolepis indica（L.）Lindl. var. grandifolia Franch. ＝ Rhaphiolepis major Cardot ●

329082　Rhaphiolepis indica（L.）Lindl. var. hiiranensis（Kaneh.）H. L. Li ＝ Rhaphiolepis indica（L.）Lindl. ex Ker var. shilanensis Y. P. Yang et H. Y. Liu ●

329083　Rhaphiolepis indica（L.）Lindl. var. shilanensis Y. P. Yang et H. Y. Liu ＝ Rhaphiolepis indica（L.）Lindl. ex Ker ●

329084　Rhaphiolepis indica（L.）Lindl. var. tashiroi Hayata ex Matsum. et Hayata ＝ Rhaphiolepis indica（L.）Lindl. ex Ker var. tashiroi Hayata ex Matsum. et Hayata ●

329085　Rhaphiolepis indica（L.）Lindl. var. umbellata（Thunb.）Ohashi ＝ Rhaphiolepis umbellata（Thunb.）Makino ●

329086　Rhaphiolepis integerrima Hook. et Arn.；全缘石斑木（革叶石斑木,厚叶石斑木）；Entireleaf Raphiolepis, Entire-leaved Raphiolepis, Wholen-leaf Hawthorn ●

329087　Rhaphiolepis integerrima Hook. et Arn. ＝ Rhaphiolepis indica（L.）Lindl. ex Ker var. umbellata（Thunb.）H. Ohashi ●

329088　Rhaphiolepis integerrima Hook. et Arn. var. mertensii（Siebold et Zucc.）Makino ex Koidz. ＝ Rhaphiolepis integerrima Hook. et Arn. ●

329089　Rhaphiolepis integerrima Hook. et Arn. var. mertensis（Siebold et Zucc.）Makino ex Koidz. ＝ Rhaphiolepis indica（L.）Lindl. ex Ker var. umbellata（Thunb.）H. Ohashi ●

329090　Rhaphiolepis integerrima Hook. et Arn. var. mertensis（Siebold et Zucc.）Makino ex Koidz. ＝ Rhaphiolepis integerrima Hook. et Arn. ●

329091　Rhaphiolepis japonica Siebold et Zucc. ＝ Rhaphiolepis umbellata（Thunb.）Makino ●

329092　Rhaphiolepis japonica Siebold et Zucc. var. integerrima Hook. f. ＝ Rhaphiolepis umbellata（Thunb.）Makino ●

329093　Rhaphiolepis japonica Siebold et Zucc. var. integerrima Hook. f. ＝ Rhaphiolepis indica（L.）Lindl. ex Ker var. umbellata（Thunb.）H. Ohashi ●

329094　Rhaphiolepis kwangsiensis Hu ＝ Rhaphiolepis salicifolia Lindl. ●

329095　Rhaphiolepis lanceolata Hu；细叶石斑木；Lanceolate Raphiolepis ●

329096　Rhaphiolepis liukiuensis Nakai；琉球石斑木；Liuqiu Raphiolepis ●

329097　Rhaphiolepis major Cardot；大叶石斑木；Largeleaf Raphiolepis, Large-leaved Raphiolepis ●

329098　Rhaphiolepis major Cardot ＝ Rhaphiolepis indica（L.）Lindl. ex Ker ●

329099　Rhaphiolepis mertensii Siebold et Zucc. ＝ Rhaphiolepis integerrima Hook. et Arn. ●

329100　Rhaphiolepis ovata Briot ＝ Rhaphiolepis umbellata（Thunb.）Makino ●

329101　Rhaphiolepis parvibracteolata Merr. ＝ Rhaphiolepis indica（L.）Lindl. ex Ker ●

329102　Rhaphiolepis rubra（Lour.）Lindl. = Rhaphiolepis indica（L.）Lindl. ex Ker ●

329103　Rhaphiolepis rubra Lindl. = Rhaphiolepis indica（L.）Lindl. ex Ker ●

329104　Rhaphiolepis rugosa Nakai = Rhaphiolepis indica（L.）Lindl. ex Ker ●

329105　Rhaphiolepis salicifolia Lindl.；柳叶石斑木；Willowleaf Raphiolepis，Willow-leaved Raphiolepis ●

329106　Rhaphiolepis sinensis M. Roem. = Rhaphiolepis indica（L.）Lindl. ex Ker ●

329107　Rhaphiolepis umbellata（Thunb. ex Murray）Makino = Rhaphiolepis indica（L.）Lindl. ex Ker var. umbellata（Thunb. ex Murray）H. Ohashi ●

329108　Rhaphiolepis umbellata（Thunb. ex Murray）Makino = Rhaphiolepis umbellata（Thunb.）Makino ●

329109　Rhaphiolepis umbellata（Thunb. ex Murray）Ohashi f. integerrima Rehder = Rhaphiolepis umbellata（Thunb. ex Murray）Makino ●

329110　Rhaphiolepis umbellata（Thunb. ex Murray）Ohashi f. ovata（Briot）Schneid. = Rhaphiolepis umbellata（Thunb. ex Murray）Makino ●

329111　Rhaphiolepis umbellata（Thunb. ex Murray）Ohashi var. integerrima（Hook. et Arn.）Masam. = Rhaphiolepis indica（L.）Lindl. ex Ker var. umbellata（Thunb.）H. Ohashi ●

329112　Rhaphiolepis umbellata（Thunb. ex Murray）Ohashi var. mertensii（Siebold et Zucc.）Makino = Rhaphiolepis integerrima Hook. et Arn. ●

329113　Rhaphiolepis umbellata（Thunb. ex Murray）Ohashi var. mertensii Makino = Rhaphiolepis integerrima Hook. et Arn. ●

329114　Rhaphiolepis umbellata（Thunb.）Makino；厚叶石斑木（伞花石斑木，水木犀，指甲花）；Yeddo Raphiolepis，Yeddo-hawthorn ●

329115　Rhaphiolepis umbellata（Thunb.）Makino = Rhaphiolepis indica（L.）Lindl. ex Ker var. umbellata（Thunb.）H. Ohashi ●

329116　Rhaphiolepis umbellata（Thunb.）Makino f. integerrima（Hook. f.）Rehder = Rhaphiolepis umbellata（Thunb.）Makino ●

329117　Rhaphiolepis umbellata（Thunb.）Makino f. ovata（Briot）C. K. Schneid. = Rhaphiolepis umbellata（Thunb.）Makino ●

329118　Rhaphiolepis umbellata（Thunb.）Makino var. integerrima（Hook. et Arn.）Masam. = Rhaphiolepis indica（L.）Lindl. ex Ker var. umbellata（Thunb.）H. Ohashi ●

329119　Rhaphiolepis umbellata（Thunb.）Makino var. integerrima（Hook. et Arn.）Masam. = Rhaphiolepis integerrima Hook. et Arn. ●

329120　Rhaphiolepis umbellata（Thunb.）Makino var. liukiuensis Koidz. = Rhaphiolepis indica（L.）Lindl. ex Ker var. liukiuensis（Koidz.）Kitam. ●

329121　Rhaphiolepis umbellata（Thunb.）Makino var. mertensii（Siebold et Zucc.）Makino = Rhaphiolepis integerrima Hook. et Arn. ●

329122　Rhaphiolepis umbellata（Thunb.）Makino var. minor Makino = Rhaphiolepis indica（L.）Lindl. ex Ker var. umbellata（Thunb.）H. Ohashi f. minor（Makino）H. Ohashi ●☆

329123　Rhaphionacme C. Muell. = Raphionacme Harv. ■☆

329124　Rhaphionacme decolor Schltr. = Raphionacme decolor Schltr. ■☆

329125　Rhaphiophallus Schott = Amorphophallus Blume ex Decne.（保留属名）■●

329126　Rhaphiostylis Planch. ex Benth.（1849）；针柱茱萸属●☆

329127　Rhaphiostylis beninensis（Hook. f. ex Planch.）Planch. ex Benth.；贝宁针柱茱萸●☆

329128　Rhaphiostylis cordifolia Hutch. et Dalziel；心叶针柱茱萸●☆

329129　Rhaphiostylis elegans Engl.；雅致针柱茱萸●☆

329130　Rhaphiostylis ferruginea Engl.；锈色针柱茱萸●☆

329131　Rhaphiostylis ferruginea Engl. var. parvifolia S. Moore = Rhaphiostylis parvifolia（S. Moore）Exell ●☆

329132　Rhaphiostylis ferruginea Engl. var. villosa（Pellegr.）Villiers；长柔毛针柱茱萸●☆

329133　Rhaphiostylis fusca（Pierre）Pierre；棕色针柱茱萸●☆

329134　Rhaphiostylis fusca（Pierre）Pierre var. villosa Pellegr. = Rhaphiostylis ferruginea Engl. var. villosa（Pellegr.）Villiers ●☆

329135　Rhaphiostylis jollyana Pierre = Rhaphiostylis beninensis（Hook. f. ex Planch.）Planch. ex Benth. ●☆

329136　Rhaphiostylis latifolia Pierre = Rhaphiostylis beninensis（Hook. f. ex Planch.）Planch. ex Benth. ●☆

329137　Rhaphiostylis ovatifolia Engl. ex Sleumer；卵叶针柱茱萸●☆

329138　Rhaphiostylis parvifolia（S. Moore）Exell；小叶针柱茱萸●☆

329139　Rhaphiostylis poggei Engl.；波格针柱茱萸●☆

329140　Rhaphiostylis preussi Engl.；普罗伊斯针柱茱萸●☆

329141　Rhaphiostylis scandens Engl. = Rhaphiostylis beninensis（Hook. f. ex Planch.）Planch. ex Benth. ●☆

329142　Rhaphiostylis stuhlmannii Engl. = Rhaphiostylis beninensis（Hook. f. ex Planch.）Planch. ex Benth. ●☆

329143　Rhaphiostylis subsessilifolia Engl.；近无柄针柱茱萸●☆

329144　Rhaphiostylis zenkeri Engl. = Rhaphiostylis beninensis（Hook. f. ex Planch.）Planch. ex Benth. ●☆

329145　Rhaphis Lour.（废弃属名）= Chrysopogon Trin.（保留属名）■

329146　Rhaphis Walp. = Rhapis L. f. ex Aiton ●

329147　Rhaphis acicularis（Retz.）Desv. = Chrysopogon aciculatus（Retz. ex Roem. et Schult.）Trin. ■

329148　Rhaphis aciculata（Retz. ex Roem. et Schult.））Honda = Chrysopogon aciculatus（Retz. ex Roem. et Schult.）Trin. ■

329149　Rhaphis aciculata（Retz.）Honda = Lecanorchis triloba J. J. Sm. ■☆

329150　Rhaphis arundinacea Desv. = Sorghum arundinaceum（Desv.）Stapf ■☆

329151　Rhaphis echinulata Nees = Chrysopogon echinulatus（Steud.）W. Watson ■

329152　Rhaphis echinulata Nees = Chrysopogon gryllus（L.）Trin. subsp. echinulatus（Nees）Cope ■☆

329153　Rhaphis echinulata Nees ex Royle = Chrysopogon echinulatus（Steud.）W. Watson ■

329154　Rhaphis gryllus（L.）Trin. = Chrysopogon echinulatus（Steud.）W. Watson ■

329155　Rhaphis gryllus（L.）Trin. = Chrysopogon gryllus Trin. ■☆

329156　Rhaphis microstachya Nees = Capillipedium parviflorum（R. Br.）Stapf ■

329157　Rhaphis orientalis Desv. = Chrysopogon orientalis（Desv.）A. Camus ■

329158　Rhaphis trivalvis Lour. = Chrysopogon aciculatus（Retz. ex Roem. et Schult.）Trin. ■

329159　Rhaphis villosula（Nees ex Steud.）Jacks. = Capillipedium parviflorum（R. Br.）Stapf ■

329160　Rhaphispermum Benth.（1846）；针子参属●☆

329161　Rhaphistemma Meisn. = Raphistemma Wall. ●

329162　Rhaphistemum Walp. = Raphistemma Wall. ●

329163　Rhaphitamnus B. D. Jacks. = Rhaphithamnus Miers ●☆

329164　Rhaphithamnus Miers(1870);刺番樱桃属●☆

329165　Rhaphithamnus cyanocarpus Miers;蓝果刺番樱桃●☆

329166　Rhaphithamnus spinosus（Juss.）Moldenke;刺番樱桃;Prickly Myrtle ●☆

329167　Rhapidaceae Bercht. et J. Presl ＝ Arecaceae Bercht. et J. Presl（保留科名）●

329168　Rhapidaceae Bercht. et J. Presl ＝ Palmae Juss.（保留科名）●

329169　Rhapidophyllum H. Wendl. et Drude(1876);棕叶椰属（发棕榈属,针棕属）;Needle Palm ●☆

329170　Rhapidophyllum hystrix（Pursh）H. Wendl. et Drude;棕叶椰;Needle Palm ●☆

329171　Rhapidospora Rchb. ＝ Rhaphidospora Nees ●■

329172　Rhapis L. f. ＝ Rhapis L. f. ex Aiton ●

329173　Rhapis L. f. ex Aiton(1789);棕竹属（观音棕竹属,竹棕属）;China Cane,Lady Palm,Ladypalm,Ladypalms,Patridge Cane,Rhapis ●

329174　Rhapis acaulis Willd. ＝ Sabal minor（Jacq.）Pers. ●

329175　Rhapis cochinchinensis（Lour.）Mart.;老挝棕竹●☆

329176　Rhapis exselsa（Thunb.）Henry ex Rehder;棕竹（观音竹,观音棕竹,虎散竹,筋头竹,美三,树棕）;Bamboo Palm, Dwarf Ground Rattan, Dwarf Ground-rattan, Fan Palm, Ground Rattan-cane, Japanese Peace Palm, Kan-non-chiku, Lady Palm, Ladypalms, Large Lady Palm, Slender Lady Palm ●

329177　Rhapis exselsa（Thunb.）Rehder ＝ Rhapis exselsa（Thunb.）Henry ex Rehder ●

329178　Rhapis filiformis Burret;丝状棕竹;Filiform Ladypalm, Filiform Ladypalms, Silky Ladypalms ●

329179　Rhapis filiformis Burret ex F. N. Wei ＝ Guihaia grossefibrosa（Gagnep.）J. Dransf.,S. K. Lee et F. N. Wei ●

329180　Rhapis flabellateis L'Her. ex Aiton ＝ Rhapis excelsa（Thunb.）Henry ex Rehder ●

329181　Rhapis flabelliformis L'Hér. ex Aiton ＝ Rhapis exselsa（Thunb.）Henry ex Rehder ●

329182　Rhapis gracilis Burret;细棕竹（细叶棕竹）;Slender China Cane, Thin Ladypalm, Thin Ladypalms ●

329183　Rhapis grossefibrosa Gagnep. ＝ Guihaia grossefibrosa（Gagnep.）J. Dransf.,S. K. Lee et F. N. Wei ●

329184　Rhapis humilis（Thunb.）Blume;矮棕竹（观音棕竹,筋头竹,朴竹,竹棕,棕榈竹,棕竹,樱桐竹,樱竹）;Dwarf Ladypalm, Dwarf Ladypalms, Lady Palm, Low Ground Rattan, Low Ground-rattan, Low Ground-rattan Palm, Rhapis Palm, Slender Lady Palm, Slender Ladypalm, Slender Lady-palm ●

329185　Rhapis humilis Blume ＝ Rhapis humilis（Thunb.）Blume ●

329186　Rhapis multifida Burret;多裂棕竹;Disected Ladypalms, Many-lobed Ladypalms, Multifid Ladypalm ●

329187　Rhapis orientalis Desv. ＝ Chrysopogon orientalis（Desv.）A. Camus ■

329188　Rhapis robusta Burret;粗棕竹（龙州棕竹）;Robust China Cane, Robust Ladypalm, Robust Ladypalms, Thick Ladypalms ●

329189　Rhapis subtilis Becc.;薄叶棕竹（姬达摩）●

329190　Rhapontica Hill ＝ Rhaponticum Ludw. ■

329191　Rhaponticoides Vaill.(1754);缘膜菊属■☆

329192　Rhaponticoides africana（Lam.）M. V. Agab. et Greuter;非洲缘膜菊■☆

329193　Rhaponticoides eriosiphon（Emb. et Maire）M. V. Agab. et Greuter;缘膜菊■☆

329194　Rhaponticum Adans. ＝ Rhaponticum Ludw. ■

329195　Rhaponticum Haller ＝ Centaurea L.（保留属名）●■

329196　Rhaponticum Hill ＝ Stemmacantha Cass. ■

329197　Rhaponticum Ludw.（1757）;漏芦属（祁州漏芦属,洋漏芦属）;Rhapontic, Swiss Centaury ■

329198　Rhaponticum Ludw. ＝ Centaurea L.（保留属名）●■

329199　Rhaponticum Ludw. ＝ Jacea Mill. ●■

329200　Rhaponticum Ludw. ＝ Leuzea DC. ■☆

329201　Rhaponticum acaule（L.）DC.;无茎漏芦■☆

329202　Rhaponticum acaule（L.）DC. var. ochroleucum Maire ＝ Rhaponticum acaule（L.）DC. ■☆

329203　Rhaponticum acaule（L.）DC. var. purpureum Emb. et Maire ＝ Rhaponticum acaule（L.）DC. ■☆

329204　Rhaponticum atriplicifolium（Trevis.）DC. ＝ Synurus deltoides（Aiton）Nakai ■

329205　Rhaponticum aulieatense Iljin;奥利漏芦■☆

329206　Rhaponticum berardioides（Batt.）Dobignard;双绵菊漏芦■☆

329207　Rhaponticum carthamoides（Willd.）Iljin ＝ Stemmacantha carthamoides（Willd.）Dittrich ■

329208　Rhaponticum caulescens Coss. et Balansa ＝ Rhaponticum cossonianum（Ball）Greuter ■☆

329209　Rhaponticum coniferum（L.）Greuter;球果漏芦■☆

329210　Rhaponticum coniferum（L.）Greuter subsp. berardioides（Batt.）Greuter ＝ Rhaponticum berardioides（Batt.）Dobignard ■☆

329211　Rhaponticum cossonianum（Ball）Greuter;科森漏芦■☆

329212　Rhaponticum cynaroides Less.;菜蓟漏芦■☆

329213　Rhaponticum dahuricum（Bunge）Turcz. ＝ Rhaponticum uniflorum（L.）DC. ■

329214　Rhaponticum dahuricum（Bunge）Turcz. ＝ Stemmacantha uniflora（L.）Dittrich ■

329215　Rhaponticum exaltatum（Willk.）Greuter;极高漏芦■☆

329216　Rhaponticum imatongensis（Philipson）Soják ＝ Ochrocephala imatongensis（Philipson）Dittrich ■☆

329217　Rhaponticum integrifolium C. Winkl.;全缘叶漏芦■☆

329218　Rhaponticum karatavicum Regel et Schmalh.;卡拉塔夫漏芦■☆

329219　Rhaponticum longifolium（Hoffmanns. et Link）Dittrich;长叶漏芦■☆

329220　Rhaponticum longifolium（Hoffmanns. et Link）Dittrich subsp. ericeticola（Font Quer）Greuter;石南漏芦■☆

329221　Rhaponticum longifolium（Hoffmanns. et Link）Dittrich var. ericetifolia（Font Quer）Dittrich ＝ Rhaponticum longifolium（Hoffmanns. et Link）Dittrich subsp. ericeticola（Font Quer）Greuter ■☆

329222　Rhaponticum lyratum C. Winkl. ex Iljin;大头羽裂漏芦■☆

329223　Rhaponticum monanthum（Georgi）Vorosch. ＝ Stemmacantha uniflora（L.）Dittrich ■

329224　Rhaponticum namansanicum Iljin;纳曼干漏芦■☆

329225　Rhaponticum nanum Lipsky;矮小漏芦■☆

329226　Rhaponticum nitidum Fisch.;光亮漏芦■☆

329227　Rhaponticum pulchrum Fisch. et C. A. Mey.;美丽漏芦■☆

329228　Rhaponticum pungens Franch. et Sav. ＝ Synurus pungens（Franch. et Sav.）Kitam. ☆

329229　Rhaponticum satzyperovii Soskov ＝ Rhaponticum uniflorum（L.）DC. ■

329230　Rhaponticum satzyperovii Soskov ＝ Stemmacantha uniflora（L.）Dittrich ■

329231　Rhaponticum serratuloides（Georgi）Bobrov;齿状漏芦■☆

329232　Rhaponticum uniflorum（L.）DC. ＝ Stemmacantha uniflora（L.）Dittrich ■

329233　Rhaponticum uniflorum DC. = Centaurea monantha Georgi ■☆

329234　Rhaponticum uniflorum DC. = Stemmacantha uniflora（L.）Dittrich ■

329235　Rhaptocalymma Börner = Carex L. ■

329236　Rhaptocarpus Miers = Echites P. Browne ●☆

329237　Rhaptomeris Miers = Cyclea Arn. ex Wight ●■

329238　Rhaptonema Miers（1867）；枭丝藤属●☆

329239　Rhaptonema bakeriana Diels；贝克枭丝藤●☆

329240　Rhaptonema cancellata Miers；枭丝藤●☆

329241　Rhaptonema densiflora（Baker）Diels；密花枭丝藤●☆

329242　Rhaptonema glabrifolia Diels；光叶枭丝藤●☆

329243　Rhaptonema latifolia Diels；宽叶枭丝藤●☆

329244　Rhaptonema swinglei Kundu et S. Guha，斯温格尔枭丝藤●☆

329245　Rhaptonema thouarsiana（Baill.）Diels；图氏枭丝藤●☆

329246　Rhaptopetalaceae Pierre ex Tiegh. = Lecythidaceae A. Rich.（保留科名）●

329247　Rhaptopetalaceae Pierre ex Tiegh. = Scytopetalaceae Engl.（保留科名）●☆

329248　Rhaptopetalaceae Tiegh. = Lecythidaceae A. Rich.（保留科名）●

329249　Rhaptopetalaceae Tiegh. = Scytopetalaceae Engl.（保留科名）●☆

329250　Rhaptopetalum Oliv.（1864）；带梗革瓣花属●☆

329251　Rhaptopetalum beguei Mangenot；贝格带梗革瓣花（贝格枭丝藤）●☆

329252　Rhaptopetalum breteleri Letouzey；布勒泰尔带梗革瓣花●☆

329253　Rhaptopetalum coriaceum Oliv.；革质带梗革瓣花●☆

329254　Rhaptopetalum depressum Letouzey；凹陷带梗革瓣花●☆

329255　Rhaptopetalum evrardii R. Germ.；埃夫拉尔带梗革瓣花●☆

329256　Rhaptopetalum pachyphyllum（Gürke）Engl. = Diospyros pachyphylla Gürke ●☆

329257　Rhaptopetalum sessilifolium Engl.；无柄叶带梗革瓣花●☆

329258　Rhaptopetalum soyauxii Oliv. = Brazzeia soyauxii（Oliv.）Tiegh. ●☆

329259　Rhaptostylum Bonpl. = Heisteria Jacq.（保留属名）●☆

329260　Rhaptostylum Humb. et Bonpl. = Heisteria Jacq.（保留属名）●☆

329261　Rhazya Decne.（1835）；拉兹草属（瑞兹草属）；Rhazya ■☆

329262　Rhazya greissii Täckh. et Boulos = Rhazya stricta Decne. ■☆

329263　Rhazya orientalis（Decne.）A. DC.；东方拉兹草（东方瑞兹亚，东方水甘草）；Oriental Rhazya ■☆

329264　Rhazya orientalis A. DC. = Amsonia orientalis Decne. ■☆

329265　Rhazya orientalis A. DC. = Rhazya orientalis（Decne.）A. DC. ■☆

329266　Rhazya stricta Decne.；拉兹草■☆

329267　Rhea Endl. = Rea Bertero ex Decne. ●☆

329268　Rheedia L.（1753）；瑞氏木属（瑞地木亚属，瑞地木属）；Rheedia ●☆

329269　Rheedia L. = Garcinia L. ●

329270　Rheedia americana Christm.；美洲瑞氏木●☆

329271　Rheedia brasiliensis Planch. et Triana；巴西瑞氏木；Bakupari，Bakupari Rheedia ●☆

329272　Rheedia brevipes Britton；短梗瑞氏木●☆

329273　Rheedia floribunda Planch. et Triana；多花瑞氏木●☆

329274　Rheedia lateriflora L.；普通瑞氏木；Halstand Rheedia ●☆

329275　Rheedia macrophylla Planch. et Triana；大叶瑞氏木●☆

329276　Rheedia madruno（Kunth）Planch. et Triana；马都瑞氏木（马都瑞地木）●☆

329277　Rheedia madruno（Kunth）Planch. et Triana var. bituberculata Pittier；双瘤马都瑞氏木（马都瑞地木）●☆

329278　Rheedia parviflora Planch. et Triana；小花瑞氏木●☆

329279　Rheedia smeathmannii Planch. et Triana = Garcinia smeathmannii（Planch. et Triana）Oliv. ●☆

329280　Rheedia tenuifolia Engl.；细叶瑞氏木●☆

329281　Rheithrophyllum Hassk. = Aeschynanthus Jack（保留属名）●■

329282　Rhektophyllum N. E. Br.（1882）；肖网纹芋属；Rhektophyllum ■☆

329283　Rhektophyllum N. E. Br. = Cercestis Schott ■☆

329284　Rhektophyllum camerunense Ntepe-Nyame = Cercestis camerunensis（Ntepe-Nyame）Bogner ■☆

329285　Rhektophyllum congense De Wild. et T. Durand = Cercestis mirabilis（N. E. Br.）Bogner ■☆

329286　Rhektophyllum mirabile N. E. Br.；肖网纹芋；Wonderful Rhektophyllum ■☆

329287　Rhektophyllum mirabile N. E. Br. = Cercestis mirabilis（N. E. Br.）Bogner ■☆

329288　Rhenactina Less. = Krylovia Schischk. ■

329289　Rheo Hance = Tradescantia L. ■

329290　Rheochloa Filg., P. M. Peterson et Y. Herrera（1999）；新巴西禾属■☆

329291　Rheome Goldblatt = Moraea Mill.（保留属名）■

329292　Rheome Goldblatt（1980）；南非鸢尾属■☆

329293　Rheome maximiliani（Schltr.）Goldblatt = Moraea maximiliani（Schltr.）Goldblatt et J. C. Manning ■☆

329294　Rheome umbellata（Thunb.）Goldblatt = Moraea umbellata Thunb. ■☆

329295　Rhesa Walp. = Bhesa Buch.-Ham. ex Arn. ●

329296　Rhetinantha M. A. Blanco = Maxillaria Ruiz et Pav. ■☆

329297　Rhetinantha M. A. Blanco（2007）；脂花兰属■☆

329298　Rhetinocarpha Paul G. Wilson et M. A. Wilson = Myriocephalus Benth. ■☆

329299　Rhetinocarpha Paul G. Wilson et M. A. Wilson（2006）；澳洲万头菊属■☆

329300　Rhetinodendron Meisn. = Robinsonia DC.（保留属名）●☆

329301　Rhetinolepis Coss.（1856）；微囊菊属■☆

329302　Rhetinolepis Coss. = Anthemis L. ■

329303　Rhetinolepis lonadioides Coss.；微囊菊■☆

329304　Rhetinophloeum H. Karst. = Cercidium Tul. ●☆

329305　Rhetinosperma Radlk. = Chisocheton Blume ●

329306　Rheum L.（1753）；大黄属；Rhubarb ●☆

329307　Rheum × cultorum Thorsrud et Reisaeter = Rheum hybridum Murray ■☆

329308　Rheum × hybridum Murray；杂种大黄；Garden Rhubarb，Rhubarb ■☆

329309　Rheum acuminatum Hook. f. et Thomson；心叶大黄（渐尖大黄，曲礼）；Acuminate Rhubarb，Heartleaf Rhubarb，Rhubarb ■

329310　Rheum acuminatum Hook. f. et Thomson ex Hook. = Rheum acuminatum Hook. f. et Thomson ■

329311　Rheum alexandrae Batalin；苞叶大黄（苞叶，葛叶大黄，水大黄，水黄）；Alexander Rhubarb，Bractleaf Rhubarb ■

329312　Rheum altaicum Losinsk.；阿尔泰大黄；Altai Rhubarb ■

329313　Rheum aplostachyum Kar. et Kir. = Rheum rhizostachyum Schrenk ■

329314　Rheum australe D. Don；南方大黄（藏边大黄）；Austral Rhubarb，Himalayan Rhubarb，Indian Rhubarb ■

329315　Rheum cacaliformis H. Lév. = Rheum kialense Franch. ■

329316　Rheum caspicum Pall. = Rheum tataricum L. f. ■

329317　Rheum collinianum Baill.；山地大黄■

329318　Rheum compactum L.；密序大黄（密穗大黄，新疆大黄）；

Densespike Rhubarb ■

329319　Rheum cordatum Losinsk.；心形大黄■☆

329320　Rheum coreanum Nakai；朝鲜大黄；Korea Rhubarb ■☆

329321　Rheum crispum G. Don；牛舌大黄(火风棠,牛耳大黄,四季菜根,土大黄,羊蹄,羊蹄草,羊蹄根,皱叶酸模,皱叶羊蹄)；Crispate Rhubarb ■

329322　Rheum cruentum Siev. ex Pall. = Rheum nanum Siev. ex Pall. ■

329323　Rheum darvasicum Titov；达尔瓦斯大黄■☆

329324　Rheum delavayi Franch.；滇边大黄(白小黄,牛尾七,沙七,小大黄)；Delavay Rhubarb ■

329325　Rheum dentatum L.；齿果大黄(齿果酸模,牛舌草)；Dentate Rhubarb ■

329326　Rheum emodi Wall. ex Meisn. = Rheum australe D. Don ■

329327　Rheum emodii Wall.；藏边大黄(白牛尾七,大岩七,牛尾七,双曲,土大黄,印边大黄)；Himalayan Rhubarb ■

329328　Rheum emodii Wall. = Rheum likiangense Sam. ■

329329　Rheum fedtschenkoi Maxim. ex Regel；范氏大黄■☆

329330　Rheum forrestii Diels；牛尾七(红马蹄鸟,红牛尾七,小大黄,小黄,雪三七)；Forrest Rhubarb,Oxtail Rhubarb ■

329331　Rheum franzenbachii Münter；华北大黄(波叶大黄,河北大黄,山大黄,台黄,唐大黄,土大黄,峪黄,籽黄,子黄)；Franzenbach Rhubarb,N. China Rhubarb ■

329332　Rheum franzenbachii Münter = Rheum rhabarbarum L. ■

329333　Rheum franzenbachii Münter var. mongolicum Münter = Rheum franzenbachii Münter ■

329334　Rheum franzenbachii Münter var. mongolium Münter = Rheum rhabarbarum L. ■

329335　Rheum glabricaule Sam.；光茎大黄；Nakedstem Rhubarb ■

329336　Rheum glabricaule Sam. f. brevilobatum Sam.；短叶光茎大黄■

329337　Rheum globulosum Gagnep.；头序大黄(头花大黄)；Head Rhubarb ■

329338　Rheum hastatum D. Don；戟叶大黄(戟叶酸模,酸浆草,太阳草,土大黄)；Hastate Rhubarb ■

329339　Rheum hirsutifolium Losinsk. = Polygonum hookeri Meisn. ■

329340　Rheum hirsutum Maxim. ex Franch. = Polygonum hookeri Meisn. ■

329341　Rheum hissaricum Losinsk.；希萨尔大黄■☆

329342　Rheum hotaoense C. Y. Cheng et T. C. Kao；河套大黄(波叶大黄,土大黄)；Hetao Rhubarb ■

329343　Rheum hybridum Murray = Rheum × hybridum Murray ■☆

329344　Rheum inopinatum Prain；红脉大黄；Redvein Rhubarb ■

329345　Rheum kialense Franch.；疏枝大黄(康定大黄,蟹甲状酸模)；Cacaliform Rhubarb,Kangding Rhubarb,Loosetwig Rhubarb ■

329346　Rheum korshinskyi Titov；考尔大黄■☆

329347　Rheum laciniatum Prain；条裂大黄；Laciniate Rhubarb ■

329348　Rheum leucorrhizum Pall. = Rheum nanum Siev. ex Pall. ■

329349　Rheum lhasaense A. J. Li et P. K. Hsiao；拉萨大黄(伊波佐大黄)；Lasa Rhubarb ■

329350　Rheum likiangense Sam.；丽江大黄(黑七,雪三七)；Lijiang Rhubarb,Likiang Rhubarb ■

329351　Rheum lobatum Litv. ex Losinsk.；浅裂大黄■☆

329352　Rheum lucidum Losinsk.；光亮大黄■☆

329353　Rheum macrocarpum Losinsk.；大果大黄■☆

329354　Rheum maculatum C. Y. Cheng et T. C. Kao；斑茎大黄；Spotstem Rhubarb ■

329355　Rheum maritimum L.；海滨羊蹄(假菠菜,假大黄,酸模)；Maritime Rhubarb ■☆

329356　Rheum maximowiczii Losinsk.；马氏大黄；Maximowicz Rhubarb ■☆

329357　Rheum micranthum Sam. = Rheum kialense Franch. ■

329358　Rheum moorcroftianum Royle；卵果大黄(长穗大黄)；Eggfruit Rhubarb ■

329359　Rheum nanum Lingelsh. = Polygonum hookeri Meisn. ■

329360　Rheum nanum Lingelsh. ex Limpr. = Polygonum hookeri Meisn. ■

329361　Rheum nanum Siev. ex Pall.；矮大黄；Short Rhubarb ■

329362　Rheum nepalensis Spreng.；土大黄(金不换,牛耳大黄,羊蹄根)；Nepal Rhubarb ■

329363　Rheum nobile Hook. f. et Thomson；高山大黄(曲玛孜,塔黄)；Noble Rhubarb ■

329364　Rheum nutans Pall. = Rheum compactum L. ■

329365　Rheum obtusifolium L.；钝叶大黄(钝叶酸模,金不换,救命王,土大黄,吐血草,癣药,血当)；Blunt-leaf Rhubarb ■

329366　Rheum officinale Baill.；药用大黄(川军,大大黄,大黄,肤如,黄良,火参,将军,锦纹大黄,峻,马蹄大黄,马蹄黄,南大黄,生军,西大黄,香大黄)；Chinese Rhubarb,Drug Rhubarb,Medicinal Rhubarb,Russian Rhubarb,Shensi Rhubarb,Turkey Rhubarb ■

329367　Rheum orieni-xizangense Y. K. Yang,J. K. Wu et Gasangs.；藏东大黄；E. Xizang Rhubarb ■

329368　Rheum orientale Losinsk.；东方大黄■☆

329369　Rheum orientali-xizangense Y. K. Yang,J. K. Wu et Gasangs. = Rheum acuminatum Hook. f. et Thomson ■

329370　Rheum ovatum C. Y. Cheng et T. C. Kao；卵叶大黄；Ovateleaf Rhubarb ■

329371　Rheum ovatum C. Y. Cheng et T. C. Kao = Rheum likiangense Sam. ■

329372　Rheum palmatum L.；掌叶大黄(北大黄,波叶大黄,川锦纹,川军,川纹,大黄,大黄炭,大王,蛋吉,肤如,狗头大黄,河州大黄,黑大黄,黄良,火参,鸡爪大黄,将军,金大黄,锦文,锦纹,锦纹大黄,锦庄生军,锦庄黄,酒大黄,酒制军,峻,苦大黄,块黄,葵叶大黄,凉黄,凉周大黄,马蹄大黄,马蹄黄,马蹄金,岷县大黄,南大黄,牛舌大黄,片吉,破门,清水大黄,铨水大黄,山大黄,上广军,上湘黄,上湘军,生大黄,生锦纹,生军,熟大黄,熟军,四川大黄,苏吉,唐古特大黄,天水大黄,土蕃大黄,文县大黄,无声虎,西大黄,西吉,西开片,西宁大黄,香大黄,香结,雅黄,药用大黄,制锦纹,制军,庄黄,庄浪大黄)；Chinese Rhubarb,Chinghai Rhubarb,East Indian Rhubarb,Ornamental Rhubarb,Palmleaf Rhubarb,Palmleaved Rhubarb,Rhubarb,Sorrel Rhubarb,Turkish Rhubarb ■

329373　Rheum palmatum L. 'Atrosanguineum'；深紫掌叶大黄■☆

329374　Rheum palmatum L. subsp. dissect Stapf = Rheum tanguticum Maxim. ex Balf. ■

329375　Rheum palmatum L. subsp. dissectum Stapf f. rubiflora Stapf = Rheum tanguticum Maxim. ex Balf. ■

329376　Rheum palmatum L. var. tanguticum Maxim. ex Regel = Rheum tanguticum Maxim. ex Balf. ■

329377　Rheum plicatum Losiask.；折叠大黄■☆

329378　Rheum potaninii Losinsk. = Rheum palmatum L. ■

329379　Rheum przewalskyi Losinsk.；歧穗大黄(歧序大黄)；Forkspike Rhubarb ■

329380　Rheum pumilum Maxim.；小大黄(矮大黄,曲马孜)；Dwarf Rhubarb ■

329381　Rheum qinlingense Y. K. Yang,D. K. Zhang et J. K. Wu；秦岭大黄；Qinling Rhubarb ■

329382　Rheum qinlingense Y. K. Yang,D. K. Zhang et J. K. Wu = Rheum palmatum L. ■

329383　Rheum racemiferum Maxim.；总序大黄；Dwarf Rhubarb ■

329384　Rheum remotiflorum Sam.；稀花酸模；Remote Rhubarb ■☆

329385　Rheum reticulatum Losinsk.；网脉大黄；Reticulate Rhubarb ■

329386　Rheum rhabarbarum L.；波叶大黄（大黄，山大黄，台黄，唐大黄，田园大黄，土大黄，峪黄，籽黄）；Garden Rhubarb, Pie Plant, Pie-plant, Rhubarb, Waveleaf Rhubarb, Wavy Rhubarb, Wine-plant ■

329387　Rheum rhaponticum Herder ＝ Rheum altaicum Losinsk. ■

329388　Rheum rhaponticum L.；食用大黄（土大黄，圆叶大黄）；English Rhubarb, Garden Rhubarb, Pie Plant, Pie-plant, Rhubarb, Tart Rhubarb, Wine Plant ■

329389　Rheum rhizostachyum Schrenk；直穗大黄；Shootspike Rhubarb ■

329390　Rheum rhomboideum Losinsk.；菱叶大黄 ■

329391　Rheum ribes L.；茶藨大黄■☆

329392　Rheum rupestre Litv. ex Losinsk.；岩生大黄■☆

329393　Rheum scaberrimum Lingelsh. ＝ Rheum przewalskyi Losinsk. ■

329394　Rheum scaberrimum Lingelsh. ＝ Rheum spiciforme Royle ■

329395　Rheum scaberrimum Lingelsh. ex Limpr. ＝ Rheum spiciforme Royle ■

329396　Rheum songaricum Schrenk ＝ Rheum tataricum L. f. ■

329397　Rheum spiciforme Royle；穗序大黄（歧穗大黄，双曲，穗花大黄，亚大黄）；Spicate Rhubarb ■

329398　Rheum strictum Franch. ＝ Rheum delavayi Franch. ■

329399　Rheum subacaule Sam.；垂枝大黄；Droopy Rhubarb ■

329400　Rheum sublanbeolatum C. Y. Cheng et T. C. Kao；窄叶大黄；Narrowleaf Rhubarb ■

329401　Rheum tanguticum Maxim. ex Balf.；鸡爪大黄（川军，肤如，黄良，火参，将军，锦纹大黄，峻，马蹄黄，生军，唐古特大黄，香大黄）；Claw Rhubarb ■

329402　Rheum tanguticum Maxim. ex Balf. var. liupanshanense C. Y. Cheng et T. C. Kao；六盘山大黄（六盘山鸡爪大黄）；Liupanshan Claw Rhubarb ■

329403　Rheum tanguticum Maxim. ex Balf. var. viridiflorum Y. K. Yang et D. K. Zhang；绿花唐古特大黄；Greenflower Claw Rhubarb ■

329404　Rheum tanguticum Maxim. ex Regel ＝ Rheum palmatum L. ■

329405　Rheum tanguticum Maxim. ex Regel var. viridiflorum Y. K. Yang et D. K. Zhang ＝ Rheum tanguticum Maxim. ex Balf. ■

329406　Rheum tataricum L. f.；圆叶大黄（鞑靼大黄）；Roundleaf Rhubarb ■

329407　Rheum tibeticum Maxim. ex Hook. f.；西藏大黄；Xizang Rhubarb ■

329408　Rheum turkestauicum Janisch.；土耳其斯坦大黄■☆

329409　Rheum undalatum L. ＝ Rheum franzenbachii Münter ■

329410　Rheum undulatum L. ＝ Rheum rhabarbatum L. ■

329411　Rheum undulatum L. var. longifolium C. Y. Cheng et T. C. Kao；长波叶大黄；Longleaf Waveleaf Rhubarb ■

329412　Rheum undulatum L. var. longifolium C. Y. Cheng et T. C. Kao ＝ Rheum rhabarbarum L. ■

329413　Rheum uninerve Maxim.；单脉大黄；Monovein Rhubarb ■

329414　Rheum webbianum Royle；喜马拉雅大黄（须弥大黄）；Himalayas Rhubarb ■

329415　Rheum wittrockii Lundstr.；天山大黄（新疆大黄）；Tianshan Rhubarb ■

329416　Rheum yunginigensis Sam. ＝ Rumex yungningensis Sam. ■

329417　Rheum yungningense Sam.；永宁大黄（永宁酸模）；Yongning Rhubarb ■

329418　Rheum yunnanense Sam.；云南大黄（滇大黄）；Yunnan Rhubarb ■

329419　Rhexia L. (1753)；鹿草属（瑞克希阿木属，瑞克希阿属）；Meadow Beauty ●■☆

329420　Rhexia alifanus Walter；萨瓦纳鹿草；Savannah Meadow-beauty ■

329421　Rhexia interior Pennell ＝ Rhexia mariana L. ■☆

329422　Rhexia mariana L.；苍白鹿草；Maryland Meadow Beauty, Meadow Beauty, Pale Meadow-beauty ■☆

329423　Rhexia princeps Kunth ＝ Dissotis princeps (Kunth) Triana ●☆

329424　Rhexia purshii Spreng. ＝ Rhexia virginica L. ■☆

329425　Rhexia septemnervia Walter ＝ Rhexia virginica L. ■☆

329426　Rhexia stricta Pursh ＝ Rhexia virginica L. ■☆

329427　Rhexia virginica L.；弗州鹿草；Deer-grass, Handsome-harry, Meadow Beauty, Virginia Meadow Beauty, Virginia Meadow-beauty, Wing-stem Meadow-pitchers ■☆

329428　Rhexia virginica L. var. purshii (Spreng.) C. W. James ＝ Rhexia virginica L. ■☆

329429　Rhexia virginica L. var. septemnervia (Walter) Pursh ＝ Rhexia virginica L. ■☆

329430　Rhexiaceae Dumort. ＝ Melastomataceae Juss. (保留科名)●■

329431　Rhigicarya Miers ＝ Rhigiocarya Miers ●☆

329432　Rhigiocarya Miers (1864)；硬果藤属●☆

329433　Rhigiocarya chevalieri Hutch. et Dalziel ＝ Kolobopetalum chevalieri (Hutch. et Dalziel) Troupin ●☆

329434　Rhigiocarya nervosa (Miers) A. Chev. ＝ Rhigiocarya racemifera Miers ●☆

329435　Rhigiocarya peltata J. Miège；盾状硬果藤●☆

329436　Rhigiocarya racemifera Miers；总序硬果藤●☆

329437　Rhigiophyllum Hochst. (1842)；霜叶桔梗属●☆

329438　Rhigiophyllum squarrosum Hochst.；霜叶桔梗●☆

329439　Rhigiothamnus Spach ＝ Dicoma Cass. ●☆

329440　Rhigospira Miers ＝ Tabernaemontana L. ●

329441　Rhigospira Miers (1878)；南美夹竹桃属●☆

329442　Rhigospira paucifolia Miers；南美夹竹桃●☆

329443　Rhigozum Burch. (1822)；五蕊簇叶木属●☆

329444　Rhigozum angolense Bamps ＝ Lycium tetrandrum Thunb. ●☆

329445　Rhigozum brevispinosum Kuntze；短刺五蕊簇叶木●☆

329446　Rhigozum linifolium S. Moore ＝ Rhigozum brevispinosum Kuntze ●☆

329447　Rhigozum madagascariense Drake；马岛刺五蕊簇叶木●☆

329448　Rhigozum obovatum Burch.；倒卵刺五蕊簇叶木●☆

329449　Rhigozum somalense Hallier f.；索马里五蕊簇叶木●☆

329450　Rhigozum spinosum Burch. ex Sprague ＝ Rhigozum brevispinosum Kuntze ●☆

329451　Rhigozum trichotomum Burch.；毛片五蕊簇叶木●☆

329452　Rhigozum virgatum Merxm. et A. Schreib.；条纹五蕊簇叶木●☆

329453　Rhigozum zambesiacum Baker；赞比西五蕊簇叶木●☆

329454　Rhinacanthus Nees (1832)；白鹤灵芝属（老鼠簕属，灵芝草属，灵枝草属）；Rhinacanthus ●■

329455　Rhinacanthus angulicaulis I. Darbysh.；棱茎白鹤灵芝●☆

329456　Rhinacanthus beesianus Diels；滇灵枝草●■

329457　Rhinacanthus breviflorus Benoist；短花白鹤灵芝●☆

329458　Rhinacanthus carcaratus (Wall.) Nees；距白鹤灵芝（滑液灵枝草，滑液树）●

329459　Rhinacanthus chiovendae Fiori；基奥文达白鹤灵芝●☆

329460　Rhinacanthus chiovendae Rendle ＝ Ecbolium gymnostachyum (Nees) Milne-Redh. ●☆

329461　Rhinacanthus communis C. B. Clarke ＝ Rhinacanthus gracilis Klotzsch ●☆

329462　Rhinacanthus communis Nees ＝ Rhinacanthus nasutus (L.)

Kurz ●■

329463　Rhinacanthus dewevrei De Wild. et T. Durand ＝ Rhinacanthus virens（Nees）Milne-Redh. ●☆

329464　Rhinacanthus dichotomus（Lindau）I. Darbysh.；二歧白鹤灵芝●☆

329465　Rhinacanthus dichotomus（Lindau）I. Darbysh. var. emaculatus I. Darbysh.；无斑二歧白鹤灵芝●☆

329466　Rhinacanthus gracilis Klotzsch；纤细白鹤灵芝●☆

329467　Rhinacanthus gracilis Klotzsch var. latilabiatus K. Balkwill ＝ Rhinacanthus latilabiatus（K. Balkwill）I. Darbysh. ●☆

329468　Rhinacanthus kaokoensis K. Balkwill et S. D. Will.；卡奥科白鹤灵芝●☆

329469　Rhinacanthus latilabiatus（K. Balkwill）I. Darbysh.；宽唇白鹤灵芝●☆

329470　Rhinacanthus minimus S. Moore；微小白鹤灵芝●☆

329471　Rhinacanthus nasutus（L.）Kuntze；白鹤灵芝（白鹤灵芝草，假红蓝，灵芝草，灵枝草，仙鹤草，仙鹤灵芝草，癣草）；Bignose Rhinacanthus, Big-nosed Rhinacanthus, Ringworm Root ●■

329472　Rhinacanthus nasutus（L.）Kurz ＝ Rhinacanthus nasutus（L.）Kuntze ●■

329473　Rhinacanthus nasutus（L.）Lindau ＝ Rhinacanthus nasutus（L.）Kuntze ●■

329474　Rhinacanthus ndorensis Schweinf. ex Engl.；恩多罗白鹤灵芝●☆

329475　Rhinacanthus oblongus Nees；矩圆白鹤灵芝●☆

329476　Rhinacanthus obtusifolius（Heine）I. Darbysh.；钝叶白鹤灵芝●☆

329477　Rhinacanthus osmospermus Bojer ex Nees ＝ Rhinacanthus nasutus（L.）Kuntze ●■

329478　Rhinacanthus parviflorus T. Anderson ex De Wild. et T. Durand ＝ Rhinacanthus virens（Nees）Milne-Redh. ●☆

329479　Rhinacanthus perrieri Benoist；佩里耶白鹤灵芝●☆

329480　Rhinacanthus pulcher Milne-Redh.；美丽白鹤灵芝●☆

329481　Rhinacanthus rotundifolius C. B. Clarke；圆叶白鹤灵芝●☆

329482　Rhinacanthus scoparius Balf. f.；帚状白鹤灵芝●☆

329483　Rhinacanthus selousensis I. Darbysh.；塞卢斯白鹤灵芝●☆

329484　Rhinacanthus subcaudatus C. B. Clarke ＝ Rhinacanthus virens（Nees）Milne-Redh. ●☆

329485　Rhinacanthus submontanus T. M. Harris et I. Darbysh.；亚山地白鹤灵芝●☆

329486　Rhinacanthus tenuipes（S. Moore）Aké Assi ＝ Justicia tenuipes S. Moore ■☆

329487　Rhinacanthus virens（Nees）Milne-Redh.；绿叶白鹤灵芝●☆

329488　Rhinacanthus virens（Nees）Milne-Redh. var. obtusifolius Heine ＝ Rhinacanthus obtusifolius（Heine）I. Darbysh. ●☆

329489　Rhinacanthus xerophilus A. Meeuse；旱生白鹤灵芝●☆

329490　Rhinactina Less. ＝ Krylovia Schischk. ■

329491　Rhinactina Less. ＝ Rhinactinidia Novopokr. ■

329492　Rhinactina Willd. ＝ Jungia L. f.（保留属名）■●☆

329493　Rhinactina limoniifolia Less. ＝ Krylovia limoniifolia（Less.）Schischk. ■

329494　Rhinactina uniflora Bunge ＝ Krylovia eremophila（Bunge）Schischk. ■

329495　Rhinactina uniflora Bunge ex DC. ＝ Krylovia eremophila（Bunge）Schischk. ■

329496　Rhinactinidia Novopokr. ＝ Aster L. ●■

329497　Rhinactinidia Novopokr. ＝ Krylovia Schischk. ■

329498　Rhinanthaceae Vent.；鼻花科●■

329499　Rhinanthaceae Vent. ＝ Orobanchaceae Vent.（保留科名）●■

329500　Rhinanthaceae Vent. ＝ Scrophulariaceae Juss.（保留科名）●■

329501　Rhinanthera Blume ＝ Scolopia Schreb.（保留属名）●

329502　Rhinanthera Blume（1827）；鼻药红木属●☆

329503　Rhinanthera odoratissima Blume；鼻药红木●☆

329504　Rhinanthus L.（1753）；鼻花属；Rattlebox, Rattleweed, Yellow Rattle ■

329505　Rhinanthus aestivalis（N. W. Zinger）Schischk. et Serg.；夏鼻花■☆

329506　Rhinanthus alata Gilib.；翅鼻花■☆

329507　Rhinanthus alpinus Baumg.；高山鼻花■☆

329508　Rhinanthus altaicus Nutt.；阿尔泰鼻花■☆

329509　Rhinanthus amplexicaulis Benth.；单茎鼻花■☆

329510　Rhinanthus angustifolius J. C. Gmel.；狭叶鼻花（迟鼻花）；Greater Yellow Rattle, Greater Yellow-rattle, Late-flowering Yellowrattle ■☆

329511　Rhinanthus armeniacus Bordz.；亚美尼亚鼻花■☆

329512　Rhinanthus borealis（Stern.）Druce；北方鼻花；Northern Rattle ■☆

329513　Rhinanthus borealis Druce subsp. kyrolliae（Chabert）Pennell ＝ Rhinanthus minor L. ■☆

329514　Rhinanthus canescens Bong.；灰鼻花■☆

329515　Rhinanthus capensis L. ＝ Bartsia trixago L. ■☆

329516　Rhinanthus charadzeae Kern.；哈氏鼻花■☆

329517　Rhinanthus chloranthus Kotschy et Boiss.；绿花鼻花■☆

329518　Rhinanthus chrysanthus Jaub. et Spach；黄花鼻花■☆

329519　Rhinanthus colchicus Vassilcz.；黑海鼻花■☆

329520　Rhinanthus cretaceus Vassilcz.；白垩鼻花■☆

329521　Rhinanthus crista-galli L.；小鼻花；Baby's Rattle, Bull's Pease, Clock, Cock-grass, Cock's Comb, Cockscomb Rattleweed, Common Rattle, Cow-wheat, Dog's Pennies, Dog's Siller, Fiddle-cases, Gowk's Shillings, Gowk's Siller, Gowk's Sixpences, Hay Rattle, Hemper, Hen's Comb, Honeysuckle, Horse Pen, Horse Pennies, Lamb's Tongue, Locusts, Meadow Rattle, Money, Money Grass, New Mown Hay, Pence, Pennies-and-happenies, Penny Box, Penny Girse, Penny Grass, Penny Ratile, Penny Rattle, Pennyweed, Pepper-box, Peter's Grass, Poverty-weed, Purse, Rattle Jack, Rattle Penny, Rattle-bags, Rattle-basket, Rattle-box, Rattle-caps, Rattle-grass, Rattle-Jack, Rattle-penny, Rattle-traps, Rattlewort, Rochlis, Rottle-penny, Shackle-bag, Shackle-basket, Shackle-cap, Shacklers, Shackles, Shekel Basket, Shekel Box, Shepherd's Purse, Snaffles, St. Peter's Flower, White Rattle, Yellow Ratdebox, Yellow Rattle ■☆

329522　Rhinanthus crista-galli L. var. fallax（Wimm. et Grab.）Druce ＝ Rhinanthus minor L. ■☆

329523　Rhinanthus czernjakowskiana B. Fedtsch.；克氏鼻花■☆

329524　Rhinanthus fedtschenkoi Gorschk.；费氏鼻花■☆

329525　Rhinanthus ferganensis Vassilcz.；费尔干鼻花■☆

329526　Rhinanthus frigidus Boiss.；冷鼻花☆

329527　Rhinanthus glaber Lam.；鼻花（春鼻花，大鼻花）；Glabrous Rattlebox, Rattlebox ■

329528　Rhinanthus goldeana Juz.；高氏鼻花■☆

329529　Rhinanthus grayana Maxim. ex Kom.；格拉鼻花■☆

329530　Rhinanthus greenlandicus（Ostenf.）Chabert；绿地鼻花■☆

329531　Rhinanthus haematanthus Boiss. et Heldr.；血花鼻花■☆

329532　Rhinanthus integrlfolius Pavlov；全叶鼻花■☆

329533　Rhinanthus kotschyanus Benth.；考氏鼻花■☆

329534　Rhinanthus kyrolliae Chabert ＝ Rhinanthus minor L. ■☆

329535　Rhinanthus lateriflorus Trautv.；阔叶鼻花■☆

329536　Rhinanthus litwinowii B. Fedtsch.；利特氏鼻花■☆

329537 Rhinanthus major（Sterneck）Fedtsch. = Rhinanthus glaber Lam. ■

329538 Rhinanthus major Ehrh. = Rhinanthus glaber Lam. ■

329539 Rhinanthus mandshuricus Maxim. ;东北鼻花■☆

329540 Rhinanthus maximowiczii Gorschk. ;马氏鼻花■☆

329541 Rhinanthus maximus Lam. = Parentucellia viscosa（L.）Caruel ■☆

329542 Rhinanthus mediterraneus（Stern）Adamov;地中海鼻花■☆

329543 Rhinanthus minor L. ; 微 小 鼻 花; Cock's-comb Rhinanthus, European Yellowrattle, Hayrattle, Little Yellow-rattle, Yellow Ratde, Yellow Rattle ■☆

329544 Rhinanthus mollis Sommier et H. Lév. ;柔软鼻花■☆

329545 Rhinanthus montanus Saut. ;山地鼻花■☆

329546 Rhinanthus multicaulis Turcz. ;多茎鼻花■☆

329547 Rhinanthus nigricans Meinsh. ;黑鼻花■☆

329548 Rhinanthus nikitinii Gorschk. ;尼氏鼻花■☆

329549 Rhinanthus olgae Grossh. ;奥氏鼻花■☆

329550 Rhinanthus orientalis（L.）Benth. ;东方鼻花■☆

329551 Rhinanthus patulus（Stern.）Thell. et Schinz;张开鼻花■☆

329552 Rhinanthus pectinatus（Behrend）Vassilcz. ;篦状鼻花■☆

329553 Rhinanthus ponticus（Sterneck）Vassilcz. ;蓬特鼻花■☆

329554 Rhinanthus pruinosus Boiss. ;粉鼻花■☆

329555 Rhinanthus rigidus Chabert = Rhinanthus minor L. ■☆

329556 Rhinanthus rumelicus Velen. ;鲁默尔鼻花■☆

329557 Rhinanthus rupestris M. Bieb. ex Willd. ;岩地鼻花■☆

329558 Rhinanthus ruprechtii Boiss. ;鲁氏鼻花■☆

329559 Rhinanthus rusticulus（Chabert）Druce;乡村鼻花■☆

329560 Rhinanthus sachalineasis Vassilcz. ;库页鼻花■☆

329561 Rhinanthus scaber Thunb. = Alectra sessiliflora（Vahl）Kuntze ■☆

329562 Rhinanthus schischkinii Vassilcz. ;希施鼻花■☆

329563 Rhinanthus scopolinus Schult. ;斯氏鼻花■☆

329564 Rhinanthus serotinus（Schönh.）Oborny = Rhinanthus angustifolius J. G. Gmel. ■☆

329565 Rhinanthus songaricus（Sterneck）B. Fedtsch. ;准噶尔鼻花■☆

329566 Rhinanthus songaricus（Sterneck）B. Fedtsch. = Rhinanthus glaber Lam. ■

329567 Rhinanthus stenophyllus（Schur）Schinz et Thell. = Rhinanthus minor L. ■☆

329568 Rhinanthus strictus K. Koch;直鼻花■☆

329569 Rhinanthus subulatus（Sterneck）Soó;钻形鼻花■☆

329570 Rhinanthus vernalis（N. W. Zinger）Schischk. = Rhinanthus glaber Lam. ■

329571 Rhinanthus vernalis（N. W. Zinger）Schischk. et Serg. = Rhinanthus glaber Lam. ■

329572 Rhinanthus versicolor Lam. = Bartsia trixago L. ■☆

329573 Rhinanthus verticillatus Gontsch. et Grig. ;轮叶鼻花■☆

329574 Rhinathus songaricus（Stern.）B. Fedtsch. = Rhinanthus glaber Lam. ■

329575 Rhinchoglossum Blume = Rhynchoglossum Blume（保留属名）■

329576 Rhinchosia Zoll. et Moritzi = Rhynchosia Lour.（保留属名）●■

329577 Rhincospora Gaudich. = Rhynchospora Vahl（保留属名）■

329578 Rhinephyllum N. E. Br.（1927）;鼻叶草属（鼻叶花属）●☆

329579 Rhinephyllum broomii L. Bolus;鼻叶草（稚儿祭）●☆

329580 Rhinephyllum comptonii L. Bolus;飞行玉●☆

329581 Rhinephyllum frithii（L. Bolus）L. Bolus = Peersia frithii（L. Bolus）L. Bolus ●☆

329582 Rhinephyllum inaequale L. Bolus;不等鼻叶草●☆

329583 Rhinephyllum inaequale L. Bolus var. latipetalum？= Rhinephyllum inaequale L. Bolus ●☆

329584 Rhinephyllum luteum（L. Bolus）L. Bolus;黄鼻叶草●☆

329585 Rhinephyllum macradenium（L. Bolus）L. Bolus;大齿玉●☆

329586 Rhinephyllum macradenium（L. Bolus）L. Bolus = Peersia macradenia（L. Bolus）L. Bolus ●☆

329587 Rhinephyllum muiri N. E. Br. ;缪尔玉●☆

329588 Rhinephyllum obliquum L. Bolus;偏斜鼻叶草●☆

329589 Rhinephyllum parvifolium L. Bolus;小叶鼻叶草●☆

329590 Rhinephyllum pillansii N. E. Br. ;皮朗斯鼻叶草●☆

329591 Rhinephyllum rouxii（L. Bolus）L. Bolus = Chasmatophyllum rouxii L. Bolus ●☆

329592 Rhinephyllum vanheerdei L. Bolus = Peersia vanheerdei（L. Bolus）H. E. K. Hartmann ●☆

329593 Rhinephyllum willowmorense L. Bolus = Chasmatophyllum willowmorense（L. Bolus）L. Bolus ●☆

329594 Rhinerrhiza Rupp（1951）;锉根兰属■☆

329595 Rhinerrhiza divitiflora（F. Muell. ex Benth.）Rupp;锉根兰■☆

329596 Rhiniachne Hochst. ex Steud. = Thelepogon Roth ex Roem. et Schult. ■☆

329597 Rhiniachne Steud. = Thelepogon Roth ex Roem. et Schult. ■☆

329598 Rhinium Schreb. = Tetracera L. ●

329599 Rhinium Schreb. = Tigarea Aubl. ●

329600 Rhinocarpus Bertero ex Kunth = Anacardium L. ●

329601 Rhinocarpus Bertero ex Kunth（1824）;鼻果漆属●☆

329602 Rhinocarpus Bertero et Balbis ex Kunth = Rhinocarpus Bertero ex Kunth ●☆

329603 Rhinocarpus excelsa Bert. ex Kunth;鼻果漆●☆

329604 Rhinocerotidium Szlach. = Oncidium Sw.（保留属名）■☆

329605 Rhinocidium Baptista = Oncidium Sw.（保留属名）■☆

329606 Rhinoglossum Pritz. = Rhynchoglossum Blume（保留属名）■

329607 Rhinolobium Arn. = Lagarinthus E. Mey. ■☆

329608 Rhinolobium Arn. = Schizoglossum E. Mey. ■☆

329609 Rhinolobium lineare Decne. = Aspidoglossum gracile（E. Mey.）Kupicha ■☆

329610 Rhinolobium tenue Arn. = Aspidoglossum gracile（E. Mey.）Kupicha ■☆

329611 Rhinopetalum Fisch. ex Alex. = Fritillaria L. ■☆

329612 Rhinopetalum arianum Loz.-Losinsk. et Vved. = Fritillaria ariana（Loz.-Lozinsk. et Vved.）Rix ■☆

329613 Rhinopetalum bucharicum（Regel）Losinsk. = Fritillaria bucharica Regel. ■☆

329614 Rhinopetalum gibbosum（Boiss.）Losinsk. et Vved. = Fritillaria gibbosa Boiss. ■☆

329615 Rhinopetalum karelinii Fisch. = Fritillaria karelinii（Fisch. ex D. Don）Baker ■

329616 Rhinopetalum karelinii Fisch. ex D. Don = Fritillaria karelinii（Fisch. ex D. Don）Baker ■

329617 Rhinopetalum stenantherum Regel = Fritillaria stenanthera（Regel）Regel ■☆

329618 Rhinopterys Nied. = Acridocarpus Guill. et Perr.（保留属名）●☆

329619 Rhinopterys angustifolia Sprague = Acridocarpus spectabilis（Nied.）Doorn-Hoekm. ●☆

329620 Rhinopterys kerstingii（Engl.）Nied. = Acridocarpus spectabilis（Nied.）Doorn-Hoekm. ●☆

329621 Rhinopterys spectabilis Nied. = Acridocarpus spectabilis（Nied.）Doorn-Hoekm. ●☆

329622 Rhinopterys spectabilis Nied. subsp. angustifolia（Sprague）

Nied. = Acridocarpus spectabilis (Nied.) Doorn-Hoekm. ●☆

329623　Rhinopteryx Nied. = Rhinopterys Nied. ●☆

329624　Rhinostegia Turcz. = Thesium L. ■

329625　Rhinostigma Miq. = Garcinia L. ●

329626　Rhiphidosperma G. Don = Justicia L. ●■

329627　Rhipidantha Bremek. (1940);扇花茜属●☆

329628　Rhipidantha chlorantha (K. Schum.) Bremek. ;扇花茜●☆

329629　Rhipidia Markgr. = Condylocarpon Desf. ●☆

329630　Rhipidocladum McClure(1973);扇枝竹属●☆

329631　Rhipidocladum geminatum (McClure) McClure;扇枝竹●☆

329632　Rhipidodendron Spreng. = Aloe L. ●■

329633　Rhipidodendron dichotomum (Masson) Willd. = Aloe dichotoma Masson ●☆

329634　Rhipidodendron distichum (Medik.) Willd. = Aloe plicatilis (L.) Mill. ●☆

329635　Rhipidodendron plicatile (L.) Haw. = Aloe plicatilis (L.) Mill. ●☆

329636　Rhipidodendrum Willd. = Aloe L. ●■

329637　Rhipidoglossum Schltr. (1918);扇舌兰属■☆

329638　Rhipidoglossum Schltr. = Diaphananthe Schltr. ■☆

329639　Rhipidoglossum bilobatum (Summerh.) Szlach. et Olszewski;二裂片扇舌兰■☆

329640　Rhipidoglossum brevifolium Summerh. = Diaphananthe brevifolia (Summerh.) Summerh. ■☆

329641　Rhipidoglossum cuneatum (Summerh.) Garay = Diaphananthe cuneata Summerh. ■☆

329642　Rhipidoglossum curvatum (Rolfe) Garay;内折扇舌兰■☆

329643　Rhipidoglossum densiflorum Summerh. ;密花扇舌兰■☆

329644　Rhipidoglossum gerrardii (Rchb. f.) Schltr. = Diaphananthe xanthopollinia (Rchb. f.) Summerh. ■☆

329645　Rhipidoglossum globulosocalcaratum (De Wild.) Summerh. ;小球扇舌兰■☆

329646　Rhipidoglossum kamerunense (Schltr.) Garay;喀麦隆扇舌兰■☆

329647　Rhipidoglossum laxiflorum Summerh. = Diaphananthe laxiflora (Summerh.) Summerh. ■☆

329648　Rhipidoglossum longicalcar Summerh. = Diaphananthe longicalcar (Summerh.) Summerh. ■☆

329649　Rhipidoglossum magnicalcar Szlach. et Olszewski;大扇舌兰■☆

329650　Rhipidoglossum microphyllum Summerh. = Diaphananthe microphylla (Summerh.) Summerh. ■☆

329651　Rhipidoglossum mildbraedii (Kraenzl.) Garay;米尔德扇舌兰■☆

329652　Rhipidoglossum montanum (Piers) Senghas = Diaphananthe montana (Piers) P. J. Cribb et J. Stewart ■☆

329653　Rhipidoglossum obanense (Rendle) Summerh. ;奥班扇舌兰■☆

329654　Rhipidoglossum ochyrae Szlach. et Olszewski;奥吉拉扇舌兰■☆

329655　Rhipidoglossum orientalis (Mansf.) Szlach. et Olszewski;东方扇舌兰■☆

329656　Rhipidoglossum ovale (Summerh.) Garay;椭圆扇舌兰■☆

329657　Rhipidoglossum paucifolium D. Johanss. ;少花扇舌兰■☆

329658　Rhipidoglossum peglerae (Bolus) Schltr. = Diaphananthe xanthopollinia (Rchb. f.) Summerh. ■☆

329659　Rhipidoglossum polyanthum (Kraenzl.) Szlach. et Olszewski;多花扇舌兰■☆

329660　Rhipidoglossum polydactylum (Kraenzl.) Garay;多指扇舌兰■☆

329661　Rhipidoglossum pulchellum (Summerh.) Garay;美丽扇舌兰■☆

329662　Rhipidoglossum pulchellum (Summerh.) Garay var. geniculatum (Summerh.) Szlach. et Olszewski;膝曲美丽扇舌兰■☆

329663　Rhipidoglossum rutilum (Rchb. f.) Schltr. ;橙红扇舌兰■☆

329664　Rhipidoglossum stellatum (P. J. Cribb) Szlach. et Olszewski;星状扇舌兰■☆

329665　Rhipidoglossum tanneri (P. J. Cribb) Senghas = Diaphananthe tanneri P. J. Cribb ■☆

329666　Rhipidoglossum xanthopollinium (Rchb. f.) Schltr. = Diaphananthe xanthopollinia (Rchb. f.) Summerh. ■☆

329667　Rhipidorchis D. L. Jones et M. A. Clem. (2004);小扇兰属■☆

329668　Rhipidorchis D. L. Jones et M. A. Clem. = Oberonia Lindl. (保留属名)■

329669　Rhipidostigma Hassk. = Diospyros L. ●

329670　Rhipogonaceae Conran. et Clifford = Ripogonaceae Conran. et Clifford ●☆

329671　Rhipogonum J. R. Forst. et G. Forst. (1776);无须菝葜属(菝葜藤属,无须藤属)●☆

329672　Rhipogonum J. R. Forst. et G. Forst. = Ripogonum J. R. Forst. et G. Forst. ●☆

329673　Rhipogonum Spreng. = Ripogonum J. R. Forst. et G. Forst. ●☆

329674　Rhipogonum album R. Br. ;白无须菝葜●☆

329675　Rhipogonum brevifolium Conran et Clifford;短叶无须菝葜●☆

329676　Rhipogonum discolor F. Muell. ;异色无须菝葜●☆

329677　Rhipogonum parviflorum R. Br. ;小花无须菝葜藤;Supple Jack,Supple Jack Vine ●☆

329678　Rhipsalidopsis Britton et Rose = Hatiora Britton et Rose ●

329679　Rhipsalidopsis Britton et Rose(1923);假仙人棒属(假昙花属);Rhipsalidopsis ●

329680　Rhipsalidopsis gaertneri (Regel) Lindig = Schlumbergera gaertneri (Regel) Britton et Rose ■☆

329681　Rhipsalidopsis rosea (Lagerh.) Britton et Rose;假仙人棒(假昙花,落花之舞,肖蟹爪);Easter Cactus, Pink Rhipsalidopsis, Rose Pink ●

329682　Rhipsalidopsis rosea (Lagerh.) Britton et Rose = Hatiora rosea (Lagerh.) Barthlott ●

329683　Rhipsalis Gaertn. (1788)(保留属名);仙人棒属(丝苇属);Mistletoe Cactus, Rhipsalis, Wickerware Cactus ●

329684　Rhipsalis baccifera (J. S. Muell.) Stearn;浆果丝苇;Mistletoe Cactus ●☆

329685　Rhipsalis baccifera (J. S. Muell.) Stearn subsp. erythrocarpa (K. Schum.) Barthlott;红果仙人棒●☆

329686　Rhipsalis baccifera (J. S. Muell.) Stearn subsp. mauritiana (DC.) Barthlott;毛里求斯仙人棒●☆

329687　Rhipsalis baccifera (Sol. ex Mill.) Stearn = Rhipsalis baccifera (J. S. Muell.) Stearn ●☆

329688　Rhipsalis capilliformis F. A. C. Weber;松风■☆

329689　Rhipsalis cassutha Gaertn. ;丝苇;Mistletoe Cactus ■☆

329690　Rhipsalis cassytha Gaertn. = Rhipsalis baccifera (J. S. Muell.) Stearn ●☆

329691　Rhipsalis cassytha Gaertn. var. mauritiana DC. = Rhipsalis baccifera (J. S. Muell.) Stearn subsp. mauritiana (DC.) Barthlott ●☆

329692　Rhipsalis cereuscula Haw. ex Phil. ;青柳(龙爪);Coral Cactus ■☆

329693　Rhipsalis clavata F. A. C. Weber = Hatiora clavata (A. A. Weber) Moran ●☆

329694　Rhipsalis cribrata Sweet ex N. E. Br. ;苇仙人棒(浪花苇)■

329695　Rhipsalis crispata (Haw.) Pfeiff. ;窗梅(窗之梅);Curled Wickerware Cactus ■

329696　Rhipsalis erythrocarpa K. Schum. = Rhipsalis baccifera (J. S. Muell.) Stearn subsp. erythrocarpa (K. Schum.) Barthlott ●☆

329697　Rhipsalis floccosa Salm-Dyck ex Pfeiff. ;绵苇(筑羽苇)■☆

329698　Rhipsalis guineensis A. Chev. = Rhipsalis baccifera (J. S. Muell.) Stearn ●☆

329699　Rhipsalis houlletiana Lem. ;花柳■☆

329700　Rhipsalis leucorhaphis K. Schum. ;清姬■☆

329701　Rhipsalis lumbricoides Lem. ;蚯蚓苇■☆

329702　Rhipsalis macrocarpa Miq. = Epiphyllum phyllanthus (L.) Haw. ☆

329703　Rhipsalis mesembryanthemoides Haw. ;拟真昼丝苇■☆

329704　Rhipsalis pachyptera Pfeiff. ;厚翼丝苇■☆

329705　Rhipsalis paradoxa Salm-Dyck;玉 柳; Chain Cactus, Chain Rhipsalis, Link Plant ■

329706　Rhipsalis pentaptera A. Dietr. ;五翼丝苇■☆

329707　Rhipsalis phyllanthus (L.) K. Schum. = Epiphyllum phyllanthus (L.) Haw. ■☆

329708　Rhipsalis regnellii G. Lindb. ;丽人柳; Regnell Wickerware Cactus ■

329709　Rhipsalis rhombea Pfeiff. ;黄仙人棒(黄梅)■☆

329710　Rhipsalis russellii Britton et Rose;红珍珠■☆

329711　Rhipsalis shaferi Britton et Rose;横笛■☆

329712　Rhipsalis tonduzii F. A. C. Weber;东天红■☆

329713　Rhipsalis virgata F. A. C. Weber;未摘花☆

329714　Rhipsalis zanzibarica A. Weber = Rhipsalis baccifera (J. S. Muell.) Stearn ●☆

329715　Rhizaeris Raf. = Laguncularia C. F. Gaertn. ●☆

329716　Rhizakenia Raf. = Limnobium Rich. ■☆

329717　Rhizanota Lour. ex Gomes = Corchorus L. ■●

329718　Rhizanthella R. S. Rogers(1928) ;小根花兰属■☆

329719　Rhizanthella gardneri R. S. Rogers;小根花兰■☆

329720　Rhizanthemum Tiegh. = Amyema Tiegh. ■☆

329721　Rhizanthes Dumort. (1829) ;根生花属■☆

329722　Rhizanthes zippelii (Blume) Spach;根生花■☆

329723　Rhizemys Raf. = Dioscorea L. (保留属名)■

329724　Rhizemys Raf. = Testudinaria Salisb. ■☆

329725　Rhizirideum (G. Don) Fourr. = Allium L. ■

329726　Rhizirideum Fourr. = Allium L. ■

329727　Rhizium Dulac = Elatine L. ■

329728　Rhizobolaceae DC. = Caryocaraceae Voigt(保留科名)●☆

329729　Rhizobolus Gaertn. ex Schreb. = Caryocar F. Allam. ex L. ●☆

329730　Rhizobolus Gaertn. ex Schreb. = Pekea Aubl. ●☆

329731　Rhizobotrya Tausch(1836) ;奥地利山芥属■☆

329732　Rhizobotrya alpina Tausch;奥地利山芥■☆

329733　Rhizocephalum Wedd. (1858) ;安第斯桔梗属■☆

329734　Rhizocephalum brachysiphonium Zahlbr. ;短管安第斯桔梗■☆

329735　Rhizocephalum candollei Wedd. ;安第斯桔梗■☆

329736　Rhizocephalum gracile E. Wimm. ;细安第斯桔梗■☆

329737　Rhizocephalum pumilum Wedd. ;小安第斯桔梗■☆

329738　Rhizocephalus Boiss. (1844) ;微秆草属■☆

329739　Rhizocephalus turkestanicus (Litv.) Roshev. ;微秆草■☆

329740　Rhizocorallon Gagnebin = Corallorhiza Gagnebin(保留属名)■

329741　Rhizogum Rchb. = Rhigozum Burch. ●☆

329742　Rhizomonanthes Danser(1933) ;根单花寄生属●☆

329743　Rhizomonanthes curvifolia (K. Krause) Danser;根单花寄生●☆

329744　Rhizophora L. (1753) ;红树属(红茄苳属) ;Mangrove ●

329745　Rhizophora apiculata Blume;红树(鸡笼笞,尖叶红树,五足驴) ;Kongkang, Sharpleaf Mangrove, Sharp-leaved Mangrove ●

329746　Rhizophora candel L. = Kandelia candel (L.) Druce ●

329747　Rhizophora candelaria DC. = Rhizophora apiculata Blume ●

329748　Rhizophora caryophylloides Burm. f. = Bruguiera cylindrica (L.) Blume ●

329749　Rhizophora caseolaris L. = Sonneratia caseolaris (L.) Engl. ●

329750　Rhizophora conjugata L. = Bruguiera gymnorhiza (L.) Savigny ●

329751　Rhizophora conjugata L. = Rhizophora apiculata Blume ●

329752　Rhizophora conjugata sensu Arn. = Rhizophora apiculata Blume ●

329753　Rhizophora corniculata L. = Aegiceras corniculatum (L.) Blanco ●

329754　Rhizophora cylindrica L. = Bruguiera cylindrica (L.) Blume ●

329755　Rhizophora gymnorhiza L. = Bruguiera gymnorhiza (L.) Savigny ●

329756　Rhizophora gymnorrhiza L. = Bruguiera gymnorhiza (L.) Lam. ●

329757　Rhizophora harrisonii Leechm. ;哈里斯红树●☆

329758　Rhizophora latifolia Miq. ;宽叶红树●☆

329759　Rhizophora longissima Blanco = Rhizophora mucronata Poir. ●

329760　Rhizophora macrorrhiza Griff. = Rhizophora mucronata Poir. ●

329761　Rhizophora mangle L. ;美洲红树(美国红树) ;American Mangrove, Mangrove, Red Mangrove ●☆

329762　Rhizophora mangle Roxb. = Rhizophora mucronata Poir. ●

329763　Rhizophora mucronata Lam. = Rhizophora mucronata Poir. ●

329764　Rhizophora mucronata Lam. var. stylosa (Griff.) Schimp. = Rhizophora stylosa Griff. ●

329765　Rhizophora mucronata Poir. ;红茄苳(红茄,红茄冬,红树,厚皮,栲皮,木考皮,茄藤,五跤梨,五梨跤) ;Boriti Poles, Fourpetale Mangrove, Fourpetaled Mangrove, Mangrove ●

329766　Rhizophora mucronata Poir. var. stylosa Schimp. = Rhizophora stylosa Griff. ●

329767　Rhizophora parviflora Roxb. ;小花红树●☆

329768　Rhizophora pauciflora Griff. ;少花红树●☆

329769　Rhizophora racemosa G. Mey. ;总花红树●☆

329770　Rhizophora sexangula Lour. = Bruguiera sexangula (Lour.) Poir. ●

329771　Rhizophora stylosa Griff. ;红海兰(红海榄,厚皮,鸡爪榄) ;Stylose Mangrove ●

329772　Rhizophora stylosa Griff. = Rhizophora mucronata Lam. ●

329773　Rhizophora tagal Perr. = Ceriops tagal (Perr.) C. B. Rob. ●

329774　Rhizophora timoriensis DC. = Ceriops tagal (Perr.) C. B. Rob. ●

329775　Rhizophora tinoriensis DC. = Ceriops tagal (Perr.) C. B. Rob. ●

329776　Rhizophoraceae Pers. (1806) (保留科名) ;红树科; Mangrove Family ●

329777　Rhizophoraceae R. Br. = Rhizophoraceae Pers. (保留科名)●

329778　Rhoaceae Spreng. ex J. Sadler = Anacardiaceae R. Br. (保留科名)●

329779　Rhoanthus Raf. = Mentzelia L. ●■☆

329780　Rhodactinea Gardner = Barnadesia Mutis ex L. f. ●☆

329781　Rhodactinia Benth. et Hook. f. = Rhodactinea Gardner ●☆

329782　Rhodaetinia Hook. f. = Rhodactinea Gardner ●☆

329783　Rhodalix Raf. = Spiraea L. ●

329784　Rhodalsine J. Gay = Arenaria L. ■

329785　Rhodalsine J. Gay = Minuartia L. ■

329786　Rhodalsine gayana (Christ) Holub;盖伊肖米努草■☆

329787　Rhodalsine geniculata (Poir.) F. N. Williams;膝曲肖米努草■☆

329788　Rhodalsine geniculata (Poir.) F. N. Williams var. fontqueri (Maire) Dobignard = Rhodalsine geniculata (Poir.) F. N. Williams ■☆

329789　Rhodalsine geniculata (Poir.) F. N. Williams var. maroccana (Batt.) Dobignard = Rhodalsine geniculata (Poir.) F. N. Williams ■☆

329790　Rhodalsine geniculata (Poir.) F. N. Williams var. procumbens

（Vahl）Dubuis = Rhodalsine geniculata（Poir.）F. N. Williams ■☆

329791 Rhodalsine procumbens（Vahl）J. Gay = Rhodalsine geniculata（Poir.）F. N. Williams ■☆

329792 Rhodalsine senneniana（Maire et Mauricio）Greuter et Burdet；塞奈尼肖米努草■☆

329793 Rhodamnia Jack（1822）；玫瑰木属；Rhodamnia ●

329794 Rhodamnia dumetorum（Poir.）Merr. et L. M. Perry；玫瑰木；Bushy Rhodamnia，Rhodamnia，Shrubby Rhodamnia ●

329795 Rhodamnia dumetorum（Poir.）Merr. et L. M. Perry var. hainanensis Merr. et L. M. Perry；海南玫瑰木（河南三脉木，山大尼，小叶山齐，银稔）；Hainan Rhodamnia ●

329796 Rhodamnia siamensis Craib = Rhodamnia dumetorum（Poir.）Merr. et L. M. Perry ●

329797 Rhodanthe Lindl.（1834）；鳞托菊属；Australian Everlasting，Strawflower ●■☆

329798 Rhodanthe Lindl. = Helipterum DC. ex Lindl. ■☆

329799 Rhodanthe chlorocephala（Turcz.）Paul G. Wilson；绿头鳞托菊；Rosy Sunray ■☆

329800 Rhodanthe chlorocephala（Turcz.）Paul G. Wilson subsp. rosea（Hook.）Paul G. Wilson；粉舌鳞托菊■☆

329801 Rhodanthe manglesii Lindl.；鳞托菊（鳞托花）；Mangles Everlasting，Paper Daisy，Swan River Everlasting ■☆

329802 Rhodanthe manglesii Lindl. = Helipterum manglesii（Lindl.）F. Muell. ex Benth. ■☆

329803 Rhodanthemum（Vogt）K. Bremer et Humphries = Rhodanthemum B. H. Wilcox，K. Bremer et Humphries ●■☆

329804 Rhodanthemum B. H. Wilcox，K. Bremer et Humphries（1993）；假匹菊属●■☆

329805 Rhodanthemum arundanum（Boiss.）B. H. Wilcox，K. Bremer et Humphries；苇状假匹菊●☆

329806 Rhodanthemum arundanum（Boiss.）B. H. Wilcox，K. Bremer et Humphries var. minutum（Emb. et Maire）Dobignard；小苇状假匹菊■☆

329807 Rhodanthemum atlanticum（Ball）B. H. Wilcox，K. Bremer et Humphries；大西洋假匹菊■☆

329808 Rhodanthemum atlanticum（Ball）B. H. Wilcox，K. Bremer et Humphries subsp. gelidum（Maire）Dobignard；寒地假匹菊■☆

329809 Rhodanthemum briquetii（Maire）B. H. Wilcox，K. Bremer et Humphries；布里凯假匹菊■☆

329810 Rhodanthemum depressum（Ball）B. H. Wilcox，K. Bremer et Humphries；凹陷假匹菊■☆

329811 Rhodanthemum floccosa Pomel = Andryala laxiflora DC. ■☆

329812 Rhodanthemum gayanum（Coss. et Durieu）B. H. Wilcox，K. Bremer et Humphries；盖伊假匹菊■☆

329813 Rhodanthemum gayanum（Coss. et Durieu）B. H. Wilcox，K. Bremer et Humphries subsp. antiatlanticum（Emb. et Maire）Vogt et Greuter；安蒂假匹菊■☆

329814 Rhodanthemum gayanum（Coss. et Durieu）B. H. Wilcox，K. Bremer et Humphries subsp. demnatense（Murb.）Vogt；代姆纳特假匹菊■☆

329815 Rhodanthemum gayanum（Coss. et Durieu）B. H. Wilcox，K. Bremer et Humphries subsp. fallax（Maire et Weiller）Vogt；迷惑盖伊假匹菊■☆

329816 Rhodanthemum gayanum（Coss. et Durieu）B. H. Wilcox，K. Bremer et Humphries var. maroccanum（Batt.）Dobignard = Rhodanthemum gayanum（Coss. et Durieu）B. H. Wilcox，K. Bremer et Humphries ■☆

329817 Rhodanthemum hosmariense（Ball）B. H. Wilcox，K. Bremer et Humphries；假匹菊■☆

329818 Rhodanthemum ifniense（Font Quer）Ibn Tattou；伊夫尼假匹菊■☆

329819 Rhodanthemum maresii（Coss.）B. H. Wilcox，K. Bremer et Humphries；马雷斯假匹菊■☆

329820 Rhodanthemum maroccanum（Batt.）B. H. Wilcox，K. Bremer et Humphries = Rhodanthemum gayanum（Coss. et Durieu）B. H. Wilcox，K. Bremer et Humphries ■☆

329821 Rhodanthemum redieri（Maire）B. H. Wilcox，K. Bremer et Humphries；雷迪尔假匹菊■☆

329822 Rhodanthemum redieri（Maire）B. H. Wilcox，K. Bremer et Humphries subsp. humberti Gómiz = Rhodanthemum redieri（Maire）B. H. Wilcox，K. Bremer et Humphries ■☆

329823 Rhodax Spach = Helianthemum Mill. ●■

329824 Rhodea Endl. = Rohdea Roth ■

329825 Rhodia Adans. = Rhodiola L. ■

329826 Rhodiola L.（1753）；红景天属；Rhodiola ■

329827 Rhodiola L. = Sedum L. ●■

329828 Rhodiola Lour. = Cardiospermum L. ■

329829 Rhodiola algida（Ledeb.）Fisch. et C. A. Mey.；喜冷红景天■☆

329830 Rhodiola algida（Ledeb.）Fisch. et C. A. Mey. var. jenisense（Maxim.）S. H. Fu；叶尼塞喜冷红景天（喜冷红景天）■

329831 Rhodiola algida（Ledeb.）Fisch. et C. A. Mey. var. tangutica（Maxim.）S. H. Fu = Rhodiola tangutica（Maxim.）S. H. Fu ■

329832 Rhodiola algida Ledeb. var. tangutica（Maxim.）S. H. Fu = Rhodiola tangutica（Maxim.）S. H. Fu ■

329833 Rhodiola alsia（Fröd.）S. H. Fu；西川红景天；W. Sichuan Rhodiola，West Szechuan Rhodiola ■

329834 Rhodiola alsia（Fröd.）S. H. Fu subsp. kawaguchii H. Ohba；河口红景天■

329835 Rhodiola alterna S. H. Fu；互生红景天；Alternate Rhodiola ■

329836 Rhodiola angusta Nakai；长白红景天（长白红景天，乌苏里景天）；Narrow Rhodiola ■

329837 Rhodiola aporontica（Fröd.）S. H. Fu；大苞红景天；Largebract Rhodiola ■

329838 Rhodiola aporontica（Fröd.）S. H. Fu = Rhodiola atuntsuensis（Praeger）S. H. Fu ■

329839 Rhodiola arctica Boriss.；北极红景天■☆

329840 Rhodiola atropurpurea（Turcz.）Trautv. et C. A. Mey. = Rhodiola rosea L. ■

329841 Rhodiola atsaensis（Fröd.）H. Ohba；亚查红景天（柴胡红景天）；Atsa Rhodiola ■

329842 Rhodiola atuntsuensis（Praeger）S. H. Fu；德钦红景天（优美红景天）；Atuntsuen Rhodiola，Deqin Rhodiola，Elegant Rhodiola ■

329843 Rhodiola balfouri（Raym. -Hamet）S. H. Fu = Ohbaea balfourii（Raym. -Hamet）Byalt et I. V. Sokolova ■

329844 Rhodiola bhutanica（Praeger）S. H. Fu = Rhodiola bupleuroides（Wall. ex Hook. f. et Thomson）S. H. Fu ■

329845 Rhodiola borealis Boriss.；北方红景天■☆

329846 Rhodiola brevipetiolata（Fröd.）S. H. Fu；短柄红景天；Short Petiole Rhodiola，Shortpetiole Rhodiola ■

329847 Rhodiola brevipetiolata（Fröd.）S. H. Fu = Rhodiola atuntsuensis（Praeger）S. H. Fu ■

329848 Rhodiola bupleuroides（Wall. ex Hook. f. et Thomson）S. H. Fu；柴胡红景天（不丹红景天，柴胡景天，伸长红景天）；Thorowaxlike Rhodiola ■

329849 Rhodiola calliantha（H. Ohba）H. Ohba；美花红景天；Beautifulflower Rhodiola ■

329850 Rhodiola chrysanthemifolia（H. Lév.）S. H. Fu；菊叶红景天（菊叶景天，苘菜景天）；Chrysanthemumleaf Rhodiola ■

329851 Rhodiola chrysanthemifolia（H. Lév.）S. H. Fu subsp. sacra（Prain ex Raym. -Hamet）H. Ohba = Rhodiola sacra（Prain ex Raym. -Hamet）S. H. Fu ■

329852 Rhodiola coccinea（Royle）Boriss. = Rhodiola quadrifida（Pall.）Fisch. et C. A. Mey. ■

329853 Rhodiola coccinea（Royle）Boriss. subsp. scabrida（Franch.）H. Ohba；粗糙红景天（糙红景天，糙景天，雪松）；Scabrous Rhodiola ■

329854 Rhodiola concinna（Praeger）S. H. Fu；优美红景天 ■

329855 Rhodiola concinna（Praeger）S. H. Fu = Rhodiola atuntsuensis（Praeger）S. H. Fu ■

329856 Rhodiola crassipes（Hook. f. et Thomson）Boriss. = Rhodiola wallichiana（Hook.）S. H. Fu ■

329857 Rhodiola crassipes（Wall. ex Hook. f. et Thomson）A. Berger = Rhodiola wallichiana（Hook.）S. H. Fu ■

329858 Rhodiola crassipes（Wall. ex Hook. f. et Thomson）A. Berger var. cretinii（Raym. -Hamet）H. Jacobsen = Rhodiola cretinii（Raym. -Hamet）H. Ohba ■

329859 Rhodiola crassipes var. cretinii（Raym. -Hamet）H. Jacobsen = Rhodiola cretinii（Raym. -Hamet）H. Ohba ■

329860 Rhodiola crenulata（Hook. f. et Thomson）H. Ohba；大花红景天（大和七，大红七，宽瓣红景天，宽叶景天，圆景天）；Bigflower Rhodiola ■

329861 Rhodiola cretinii（Raym. -Hamet）H. Ohba；根出红景天；Cretin Rhodiola ■

329862 Rhodiola cretinii（Raym. -Hamet）H. Ohba subsp. sino-alpina（Fröd.）H. Ohba；高山红景天（高山蔷薇景天）；Alpine Cretin Rhodiola ■

329863 Rhodiola dielsiana（H. Limpr.）S. H. Fu；川西红景天；Diels Rhodiola ■

329864 Rhodiola dielsiana（H. Limpr.）S. H. Fu = Rhodiola chrysanthemifolia（H. Lév.）S. H. Fu ■

329865 Rhodiola discolor（Franch.）S. H. Fu；异色红景天（异色柴胡景天）；Discolor Rhodiola ■

329866 Rhodiola dumulosa（Franch.）S. H. Fu；小丛红景天（凤凰草，凤尾草，凤尾七，雾灵景天，香景天）；Shrubbery Rhodiola ■

329867 Rhodiola durisii（Raym. -Hamet）S. H. Fu = Rosularia alpestris（Kar. et Kir.）Boriss. ■

329868 Rhodiola elongata（Ledeb.）Fisch. et C. A. Mey. = Rhodiola rosea L. ■

329869 Rhodiola elongata（Ledeb.）Fisch. et C. A. Mey. = Rhodiola sachalinensis Boriss. ■

329870 Rhodiola eurycarpa（Fröd.）S. H. Fu；宽果红景天（前进景天）；Broadfruit Rhodiola ■

329871 Rhodiola eurycarpa（Fröd.）S. H. Fu = Rhodiola macrocarpa（Praeger）S. H. Fu ■

329872 Rhodiola euryphylla（Fröd.）S. H. Fu = Rhodiola crenulata（Hook. f. et Thomson）H. Ohba ■

329873 Rhodiola fastigiata（Hook. f. et Thomson）S. H. Fu；长鞭红景天（大理景天，宽叶红景天，竖枝景天）；Longwhip Rhodiola ■

329874 Rhodiola fastigiata（Hook. f. et Thomson）S. H. Fu var. gelida（Schrenk）H. Jacobsen = Rhodiola gelida Schrenk ■

329875 Rhodiola forrestii（Raym. -Hamet）S. H. Fu；长圆红景天（川滇景天，少花云南景天）；Forrest Rhodiola ■

329876 Rhodiola gannanica K. T. Fu；甘南红景天（甘肃红景天）；Gannan Rhodiola ■

329877 Rhodiola gelida Schrenk；长鳞红景天（冻结红景天，红景天鬼灯檠，小叶红景天）；Cold Rhodiola ■

329878 Rhodiola handelii H. Ohba；小株红景天；Handel Rhodiola ■

329879 Rhodiola henryi（Diels）S. H. Fu；菱叶红景天（白三七，打不死，豆瓣七，还阳参，还阳参景天，姜皮矮陀陀，接骨丹，接骨七，三步接骨丹，三面七，三匹七，水三七，豌豆七，血三七，岩还阳，岩活阳，岩见血参，岩老鼠，岩田三七，岩豌豆，一代宗，玉蝴蝶）；Henry Rhodiola，Pea Seven ■

329880 Rhodiola henryi（Diels）S. H. Fu = Rhodiola yunnanensis（Franch.）S. H. Pu ■

329881 Rhodiola heterodonta（Hook. f. et Thomson）Boriss.；异齿红景天；Differenttooth Rhodiola ■

329882 Rhodiola himalensis（D. Don）S. H. Fu；喜马拉雅红景天（喜马红景天）；Himalayas Rhodiola ■

329883 Rhodiola himalensis（D. Don）S. H. Fu var. taohoensis（S. H. Fu）H. Ohba；洮河红景天；Taohe Rhodiola，Taoho Rhodiola ■

329884 Rhodiola hobsonii（Prain ex Raym. -Hamet）S. H. Fu；背药红景天（高峰景天，贺氏红景天，西藏景天）；Hobson Rhodiola ■

329885 Rhodiola hookeri S. H. Fu = Rhodiola bupleuroides（Wall. ex Hook. f. et Thomson）S. H. Fu ■

329886 Rhodiola humilis（Hook. f. et Thomson）S. H. Fu；矮生红景天（矮景天，二型叶景天，卡倍红景天）；Dwarf Rhodiola ■

329887 Rhodiola imbricata Edgew.；覆瓦红景天■☆

329888 Rhodiola integrifolia Raf.；全缘叶红景天■☆

329889 Rhodiola ishidae（Miyabe et Kudo）H. Hara；石田氏红景天■☆

329890 Rhodiola junggarica Chang Y. Yang et N. R. Cui；准噶尔红景天；Dzungar Rhodiola ■

329891 Rhodiola junggarica Chang Y. Yang et N. R. Cui = Rhodiola rosea L. ■

329892 Rhodiola juparensis（Fröd.）S. H. Fu；圆丛红景天；Jupar Rhodiola ■

329893 Rhodiola juparensis（Fröd.）S. H. Fu = Rhodiola coccinea（Royle）Boriss. ■

329894 Rhodiola kansuensis（Pröd.）S. H. Fu；甘肃红景天；Gansu Rhodiola，Kansu Rhodiola ■

329895 Rhodiola karpelesae（Raym. -Hamet）S. H. Fu = Rhodiola humilis（Hook. f. et Thomson）S. H. Fu ■

329896 Rhodiola kaschgarica Boriss.；喀什红景天；Kaschgar Rhodiola ■

329897 Rhodiola kirilowii Regel ex Maxim.；狭叶红景天（长茎红景天，大株鬼灯檠，大株红景天，高壮景天，景天三七，九莲花，九头狮子七，涩疙瘩，涩疙疸，狮子草，狮子七，狮子头，石菜兰，条叶红景天，土三七，线叶红景天）；Kirilow Rhodiola，Linearleaf Rhodiola，Narrowleaf Rhodiola ■

329898 Rhodiola kirilowii Regel ex Maxim. var. latifolia S. H. Fu；宽狭叶红景天；Broad Kirilow Rhodiola ■

329899 Rhodiola kirilowii Regel ex Maxim. var. latifolia S. H. Fu = Rhodiola kirilowii Regel ex Maxim. ■

329900 Rhodiola komarovii Boriss. = Rhodiola angusta Nakai ■

329901 Rhodiola liciae（Raym. -Hamet）S. H. Fu；昆明红景天（丽西红景天，扇叶景天）；Kunming Rhodiola ■

329902 Rhodiola likiangensie（Fröd.）S. H. Fu；丽江红景天（丽江景天）；Lijiang Rhodiola，Likiang Rhodiola ■

329903 Rhodiola likiangensie（Fröd.）S. H. Fu = Rhodiola coccinea（Royle）Boriss. subsp. scabrida（Franch.）H. Ohba ■

329904　Rhodiola linearifolia Boriss. = Rhodiola kirilowii Regel ex Maxim. ■

329905　Rhodiola litvinowii Boriss.；黄萼红景天（多花红景天）；Litwinow Rhodiola, Yellowcalyx Rhodiola ■

329906　Rhodiola longicaulis（Praeger）S. H. Fu = Rhodiola kirilowii Regel ex Maxim. ■

329907　Rhodiola macrocarpa（Praeger）S. H. Fu；大果红景天；Bigfruit Rhodiola ■

329908　Rhodiola macrolepis（Franch.）S. H. Fu；大鳞红景天；Bigscale Rhodiola ■

329909　Rhodiola macrolepis（Franch.）S. H. Fu = Rhodiola kirilowii Regel ex Maxim. ■

329910　Rhodiola megalophylla（Fröd.）S. H. Fu；大叶红景天（大叶景天）；Bigleaf Rhodiola ■

329911　Rhodiola megalophylla（Fröd.）S. H. Fu = Rhodiola crenulata（Hook. f. et Thomson）H. Ohba ■

329912　Rhodiola nobilis（Franch.）S. H. Fu；优秀红景天（贵景天）；Noticeable Rhodiola ■

329913　Rhodiola nobilis（Franch.）S. H. Fu subsp. atuntsuensis（Praeger）H. Ohba = Rhodiola atuntsuensis（Praeger）S. H. Fu ■

329914　Rhodiola ovatisepela（Raym.-Hamet）S. H. Fu；卵萼红景天；Ovatesepal Rhodiola ■

329915　Rhodiola ovatisepela（Raym.-Hamet）S. H. Fu var. chingii S. H. Fu；线萼红景天；Linearsepal Rhodiola ■

329916　Rhodiola pachyclados（Aitch. et Hemsl.）H. Ohba；粗枝红景天 ■☆

329917　Rhodiola pamiro-alaica Boriss.；帕米红景天；Pamir Rhodiola ■

329918　Rhodiola papillocarpa（Fröd.）S. H. Fu；肿果红景天；Teatfruit Rhodiola ■

329919　Rhodiola papillocarpa（Fröd.）S. H. Fu = Rhodiola yunnanensis（Franch.）S. H. Pu ■

329920　Rhodiola petiolata（Fröd.）S. H. Fu；有柄红景天；Stalked Rhodiola ■

329921　Rhodiola petiolata（Fröd.）S. H. Fu = Rhodiola prainii（Raym.-Hamet）H. Ohba ■

329922　Rhodiola phariensis（H. Ohba）S. H. Fu = Rhodiola purpureoviridis（Praeger）S. H. Fu subsp. phariensis（H. Ohba）H. Ohba ■

329923　Rhodiola pinnatifida Boriss.；羽裂红景天（羽齿红景天）；Pinnatifid Rhodiola ■

329924　Rhodiola pleurogynantha（Hand.-Mazz.）S. H. Fu = Rhodiola primuloides（Franch.）S. H. Fu ■

329925　Rhodiola prainii（Raym.-Hamet）H. Ohba；四轮红景天；Prain Rhodiola ■

329926　Rhodiola primuloides（Franch.）S. H. Fu；报春红景天（报春景天,侧花红景天,侧花景天）；Primrose Rhodiola, Primrose-like Rhodiola ■

329927　Rhodiola primuloides（Franch.）S. H. Fu subsp. kongboensis H. Ohba；工布红景天；Gongbu Rhodiola ■

329928　Rhodiola purpureoviridis（Praeger）S. H. Fu；紫绿红景天（紫绿景天）；Purplegreen Rhodiola ■

329929　Rhodiola purpureoviridis（Praeger）S. H. Fu subsp. phariensis（H. Ohba）H. Ohba；帕里红景天；Phari Rhodiola ■

329930　Rhodiola quadrifida（Pall.）Fisch. et C. A. Mey.；四裂红景天（大红红景天,高山红景天,罗迪拉草,四裂景天,圆丛红景天）；Foursplit Rhodiola ■

329931　Rhodiola ramosa Nakai = Rhodiola angusta Nakai ■

329932　Rhodiola recticaulis Boriss.；直茎红景天；Erectstem Rhodiola, Straightstem Rhodiola ■

329933　Rhodiola robusta（Praeger）S. H. Fu；壮健红景天；Robust Rhodiola ■

329934　Rhodiola robusta（Praeger）S. H. Fu = Rhodiola kirilowii Regel ex Maxim. ■

329935　Rhodiola rosea L.；红景天（大紫红景天,蔷薇红,蔷薇景天,深紫蔷薇景天）；Blackpurple Rhodiola, Jealousy, Lus Na Laoch, Midsummer Men, Priest's Pintel, Priest's Pintle, Rose Root, Roseroot, Rose-root, Roseroot Stonecrop, Rose-root Stonecrop, Snowdon Rose ■

329936　Rhodiola rosea L. = Sedum rosea（L.）Scop. ■☆

329937　Rhodiola rosea L. f. purpurascens Y. Meng et J. J. Tian；塞北红景天；Purple Rose-root ■

329938　Rhodiola rosea L. var. elongata（Ledeb.）H. Jacobsen = Rhodiola rosea L. ■

329939　Rhodiola rosea L. var. microphylla（Fröd.）S. H. Fu；小叶红景天；Smallleaf Rose-root ■

329940　Rhodiola rosea var. elongata（Ledeb.）H. Jacobsen = Rhodiola rosea L. ■

329941　Rhodiola rotundata（Hemsl.）S. H. Fu = Rhodiola crenulata（Hook. f. et Thomson）H. Ohba ■

329942　Rhodiola rotundifolia（Fröd.）S. H. Fu；圆叶红景天（山蚕豆）；Roundleaf Rhodiola ■

329943　Rhodiola rotundifolia（Fröd.）S. H. Fu = Rhodiola yunnanensis（Franch.）S. H. Pu ■

329944　Rhodiola sachalinensis Boriss.；库页红景天（高山鬼灯檠,高山红景天,鬼灯檠,红景天）；Kuye Rhodiola, Sachalin Rhodiola ■

329945　Rhodiola sacra（Prain ex Raym.-Hamet）S. H. Fu；圣地红景天（红景天,全瓣红景天,扫罗玛尔布,圣景天）；Integripetal Rhodiola ■

329946　Rhodiola sacra（Prain ex Raym.-Hamet）S. H. Fu var. tsuiana（S. H. Fu）S. H. Fu；长毛圣地红景天（友文红景天）■

329947　Rhodiola sangpotibetana（Fröd.）S. H. Fu = Rhodiola smithii（Raym.-Hamet）S. H. Fu ■

329948　Rhodiola saxifragoides（Fröd.）H. Ohba；岩生红景天 ■☆

329949　Rhodiola scabrida（Franch.）S. H. Fu = Rhodiola coccinea（Royle）Boriss. subsp. scabrida（Franch.）H. Ohba ■

329950　Rhodiola semenovii（Regel et Herder）Boriss.；柱花红景天；Semenov Rhodiola ■

329951　Rhodiola serrata H. Ohba；齿叶红景天；Serrate Rhodiola ■

329952　Rhodiola sexifolia S. H. Fu；六叶红景天；Sixleaf Rhodiola, Sixleaves Rhodiola ■

329953　Rhodiola sherriffii H. Ohba；小杯红景天；Smallcup Rhodiola ■

329954　Rhodiola sinica（Diels）Jacobsen = Rhodiola yunnanensis（Franch.）S. H. Pu ■

329955　Rhodiola sinoalpina（Fröd.）S. H. Fu = Rhodiola cretinii（Raym.-Hamet）H. Ohba subsp. sino-alpina（Fröd.）H. Ohba ■

329956　Rhodiola sinuata（Royle ex Edgew.）S. H. Fu；裂叶红景天（深波线叶景天）；Sinuate Rhodiola ■

329957　Rhodiola smithii（Raym.-Hamet）S. H. Fu；异鳞红景天（藏布红景天,史密红景天）；Smith Rhodiola ■

329958　Rhodiola staminea（Paulsen）S. H. Fu；长蕊红景天（长蕊景天）；Longstamen Rhodiola ■

329959　Rhodiola staminea（Paulsen）S. H. Fu = Rhodiola alsia（Fröd.）S. H. Fu ■

329960　Rhodiola stapfii（Raym.-Hamet）S. H. Fu；托花红景天（史塔红景天,印边景天）；Stapf Rhodiola ■

329961　Rhodiola stephanii（Cham.）Trautv. et C. A. Mey. ;兴安红景天;Stephan Rhodiola ■

329962　Rhodiola supposita（Maxim.）Jacobsen;对叶红景天;Oppositeleaf Rhodiola ■

329963　Rhodiola tangutica（Maxim.）S. H. Fu;唐古特红景天;Tangut Rhodiola ■

329964　Rhodiola taohoensis S. H. Fu ＝ Rhodiola himalensis（D. Don）S. H. Fu var. taohoensis（S. H. Fu）H. Ohba ■

329965　Rhodiola telephioides（Maxim.）S. H. Fu;东疆红景天;E. Xinjiang Rhodiola ■

329966　Rhodiola telephioides（Maxim.）S. H. Fu ＝ Rhodiola rosea L. ■

329967　Rhodiola tibetica（Hook. f. et Thomson）S. H. Fu;西藏红景天;Tibet Rhodiola,Xizang Rhodiola ■

329968　Rhodiola tieghemii（Raym. -Hamet）S. H. Fu;巴塘红景天（藏南红景天）;Batang Rhodiola,Tieghem Rhodiola ■

329969　Rhodiola tsuiana S. H. Fu ＝ Rhodiola sacra（Prain ex Raym. -Hamet）S. H. Fu var. tsuiana（S. H. Fu）S. H. Fu ■

329970　Rhodiola venusta（Praeger）S. H. Fu ＝ Rhodiola atuntsuensis（Praeger）S. H. Fu ■

329971　Rhodiola venusta（Praeger）S. H. Fu ＝ Rhodiola fastigiata（Hook. f. et Thomson）S. H. Fu ■

329972　Rhodiola viridum Boriss. ;绿叶红景天■

329973　Rhodiola wallichiana（Hook.）S. H. Fu;粗茎红景天（豺帚,粗干景天,大株鬼灯檠,大株红景天）;Wallich Rhodiola ■

329974　Rhodiola wallichiana（Hook.）S. H. Fu var. cholaensis（Praeger）S. H. Fu;大株粗茎红景天（大株鬼灯檠,大株红景天）;Big Wallich Rhodiola ■

329975　Rhodiola yunnanensis（Franch.）S. H. Pu;云南红景天（白三七,蚕豆七,豆叶狼毒,豆叶七,胡豆莲,胡豆七,还阳草,黄花参,姜皮矮陀陀,金剪刀,绿豆莲,三匹七,三台观音,铁脚莲,豌豆七,玉蝴蝶,云南景天）;Yunnan Rhodiola ■

329976　Rhodiola yunnanensis（Franch.）S. H. Pu var. oblanceolata（Fröd.）S. H. Fu;倒披针叶红景天;Oblongate Yunnan Rhodiola ■

329977　Rhodiolaceae Martinov ＝ Crassulaceae J. St. -Hil.（保留科名）●■

329978　Rhodiolaceae Martinov;红景天科■

329979　Rhodocactus（A. Berger）F. M. Knuth ＝ Pereskia Mill. ●

329980　Rhodocactus F. M. Knuth ＝ Pereskia Mill. ●

329981　Rhodocactus bleo（Kunth）F. M. Knuth ＝ Pereskia bleo（Kunth）DC. ●☆

329982　Rhodocactus grandifolius（Haw.）F. M. Knuth ＝ Pereskia grandifolia Haw. ■

329983　Rhodocalyx Müll. Arg.（1860）;红萼夹竹桃属●☆

329984　Rhodocalyx rotundifolius Müll. Arg. ;红萼夹竹桃●☆

329985　Rhodochiton Zucc. ＝ Rhodochiton Zucc. ex Otto et A. Dietr. ■☆

329986　Rhodochiton Zucc. ex Otto et A. Dietr.（1834）;缠柄花属■☆

329987　Rhodochiton atrosanguineum（Zucc.）Rothm. ;缠柄花;Purple Bell Vine ■☆

329988　Rhodochlaena Spreng. ＝ Rhodolaena Thouars ●☆

329989　Rhodochlamys S. Schauer ＝ Salvia L. ●■

329990　Rhodocistus Spach ＝ Cistus L. ●

329991　Rhodoclada Baker ＝ Asteropeia Thouars ●☆

329992　Rhodoclada rhopaloides Baker ＝ Asteropeia rhopaloides（Baker）Baill. ●☆

329993　Rhodococcum（Rupr.）Avrorin ＝ Vaccinium L. ●

329994　Rhodococcum minus（Lodd.）Avrorin ＝ Vaccinium vitis-idaea L. ●

329995　Rhodococcum vitis-idaea（L.）Avrorin ＝ Vaccinium vitis-idaea L. ●

329996　Rhodocodon Baker ＝ Rhadamanthus Salisb. ■☆

329997　Rhodocodon Baker(1881);红钟风信子属■☆

329998　Rhodocodon urgineoides Baker;红钟风信子■☆

329999　Rhodocolea Baill.（1887）;红鞘紫葳属●☆

330000　Rhodocolea boivinii（Baill.）H. Perrier;博伊文红鞘紫葳●☆

330001　Rhodocolea involucrata（Bojer ex DC.）H. Perrier;总苞红鞘紫葳●☆

330002　Rhodocolea linearis H. Perrier;线状红鞘紫葳●☆

330003　Rhodocolea nobilis Baill. ;名贵红鞘紫葳●☆

330004　Rhodocolea perrieri Capuron;佩里耶红鞘紫葳●☆

330005　Rhodocolea racemosa（Lam.）H. Perrier;总花红鞘紫葳●☆

330006　Rhodocolea schatzii A. H. Gentry;沙茨红鞘紫葳●☆

330007　Rhodocolea telfairiae（Bojer ex Hook.）H. Perrier;马达加斯加红鞘紫葳●☆

330008　Rhodocoma Nees ＝ Restio Rottb.（保留属名）■☆

330009　Rhodocoma Nees(1836);红毛帚灯草属■☆

330010　Rhodocoma alpina H. P. Linder et Vlok;高山红毛帚灯草■☆

330011　Rhodocoma arida H. P. Linder et Vlok;旱生红毛帚灯草■☆

330012　Rhodocoma capensis Steud. ;好望角红毛帚灯草■☆

330013　Rhodocoma fruticosa（Thunb.）H. P. Linder;灌丛红毛帚灯草■☆

330014　Rhodocoma gigantea（Kunth）H. P. Linder;大红毛帚灯草;Rhodocoma ■☆

330015　Rhodocoma gracilis H. P. Linder et Vlok;纤细红毛帚灯草■☆

330016　Rhododendraceae Juss. ＝ Ericaceae Juss.（保留科名）●

330017　Rhododendron L.（1753）;杜鹃花属（杜鹃属）;Azalea, Rhododendron, Rose Bay, Rosebay ●

330018　Rhododendron aberconwayi Cowan;蝶花杜鹃（滇东杜鹃,碟花杜鹃）; Aberconway Rhododendron, Butterflyflower Azalea, Saucershaped Flowers Rhododendron ●

330019　Rhododendron aberrans Tagg et Forrest ＝ Rhododendron traillanum Forrest et W. W. Sm. ●

330020　Rhododendron achroanthum Balf. f. et W. W. Sm. ;多色杜鹃 ●

330021　Rhododendron achroanthum Balf. f. et W. W. Sm. ＝ Rhododendron rupicola W. W. Sm. ●

330022　Rhododendron acraium Balf. f. et W. W. Sm. ＝ Rhododendron primuliflorum Bureau et Franch. ●

330023　Rhododendron adamsii Rehder;阿氏杜鹃●☆

330024　Rhododendron adenanthum M. Y. He;腺花杜鹃;Gland-flower Azalea, Gland-flower Rhododendron, Glandular-flowered Rhododendron ●

330025　Rhododendron adenobracteum X. F. Gao et Y. L. Peng;腺苞杜鹃●

330026　Rhododendron adenogynum Diels;腺房杜鹃（被腺杜鹃）; Gland-bearing Azalea, Gland-bearing Rhododendron, Glandpistil Azalea, Glandular-ovared Rhododendron, Glandular-ovary Rhododendron ●

330027　Rhododendron adenophorum Balf. f. et W. W. Sm. ＝ Rhododendron adenogynum Diels ●

330028　Rhododendron adenopodum Franch. ;腺枝杜鹃（弯尖杜鹃,弯尖叶杜鹃）;Curve Leaf-tip Rhododendron,Flextine Azalea,Glandular Pedicel Rhododendron,Glandular-pediceled Rhododendron ●

330029　Rhododendron adenostemonum Balf. f. et W. W. Sm. ＝ Rhododendron irroratum Franch. subsp. pogonostylum（Balf. f. et W. W. Sm.）D. F. Chamb. ex Cullen et D. F. Chamb. ●

330030　Rhododendron adenostylum W. P. Fang et M. Y. He;腺柱杜鹃;Glandular Atyle Azalea,Glandular Atyle Rhododendron ●☆

330031 Rhododendron adenpsum Davidian；木里杜鹃（枯鲁杜鹃）；Kulu Azalea，Kulu Rhododendron，Muli Azalea，Muli Rhododendron ●

330032 Rhododendron admirabile Balf. f. et Forrest = Rhododendron lukiangense Franch. ●

330033 Rhododendron adoxum Balf. f. et Forrest = Rhododendron vernicosum Franch. ●

330034 Rhododendron adroserum Balf. f. et Forrest = Rhododendron lukiangense Franch. ●

330035 Rhododendron aechmophyllum Balf. f. et Forrest；矛头杜鹃●

330036 Rhododendron aechmophyllum Balf. f. et Forrest = Rhododendron yunnanense Franch. ●

330037 Rhododendron aemulorum Balf. f. = Rhododendron mallotum Balf. f. et Kingdon-Ward ●

330038 Rhododendron aeruginosum Hook. f. = Rhododendron campanulatum D. Don subsp. aeruginosum（Hook. f.）D. F. Chamb. ●

330039 Rhododendron afghanicum Aitch. et Hemsl.；阿富汗杜鹃；Afghan Azalea，Afghan Rhododendron ●

330040 Rhododendron aganniphum Balf. f. et Kingdon-Ward；雪山杜鹃（白雪杜鹃，海绵杜鹃，灰背杜鹃，软雪杜鹃，闪亮杜鹃）；Jokul Azalea，Shining Covering Rhododendron，Snowy Rhododendron，Whitish Indumentum Rhododendron，Whitish Rhododendron ●

330041 Rhododendron aganniphum Balf. f. et Kingdon-Ward var. flavorufum（Balf. f. et Forrest）D. F. Chamb.；黄毛雪山杜鹃（淡黄毛杜鹃，黄红杜鹃，黄红毛杜鹃）；Red-yellow Hairs Rhododendron，Yellowhair Jokul Azalea ●

330042 Rhododendron aganniphum Balf. f. et Kingdon-Ward var. glaucopeplum（Balf. f. et Forrest）T. L. Ming = Rhododendron aganniphum Balf. f. et Kingdon-Ward ●

330043 Rhododendron aganniphum Balf. f. et Kingdon-Ward var. schizopeplum（Balf. f. et Forrest）T. L. Ming；裂毛雪山杜鹃●

330044 Rhododendron agapetum Balf. f. et Kingdon-Ward；腺梗杜鹃；Glandular Branches Rhododendron ●☆

330045 Rhododendron agapetum Balf. f. et Kingdon-Ward = Rhododendron kyawi Lace et W. W. Sm. ●

330046 Rhododendron agastum Balf. f. et W. W. Sm.；迷人杜鹃（垂俯杜鹃，羽脉杜鹃）；Bewitch Azalea，Charming Rhododendron，Pinnately Azalea，Pinnately Veined Rhododendron ●

330047 Rhododendron agastum Balf. f. et W. W. Sm. var. pennivenium（Balf. f. et W. W. Sm.）T. L. Ming；光柱迷人杜鹃；Smoothstyle Bewitch Azalea，Smoothstyle Charming Rhododendron ●

330048 Rhododendron agetum Balf. f. et Forrest = Rhododendron neriiflorum Franch. var. agetum（Balf. f. et Forrest）T. L. Ming ●

330049 Rhododendron agglutinatum Balf. f. et Forrest = Rhododendron phaeochrysum Balf. f. et W. W. Sm. var. agglutinatum（Balf. f. et Forrest）D. F. Chamb. ●

330050 Rhododendron aiolosalpinx Balf. f. et Forrest = Rhododendron stewartianum Diels ●

330051 Rhododendron aiolpeplum Balf. f. et Forrest = Rhododendron phaeochrysum Balf. f. et W. W. Sm. var. levistratum（Balf. f. et Forrest）D. F. Chamb. ●

330052 Rhododendron aischropeplum Balf. f. et Forrest = Rhododendron roxieanum Forrest ex W. W. Sm. ●

330053 Rhododendron alabamense Rehder；阿拉巴马杜鹃；Alabama Azalea，Alabama Rhododendron ●☆

330054 Rhododendron albertsenianum Forrest ex Balf. f.；亮红杜鹃（怒江杜鹃）；Albertson's Rhododendron，Brightred Azalea，M. O. Albertson's Rhododendron ●

330055 Rhododendron albicaule H. Lév. = Rhododendron decorum Franch. ●

330056 Rhododendron albicaule H. Lév. = Rhododendron fortunei Lindl. ●

330057 Rhododendron albiflorum Hook.；白花杜鹃；Cascade Azalea，White Flowers Azalea，White Flowers Rhododendron，White Rose-bay ●

330058 Rhododendron albrechtii Maxim.；极美杜鹃（加拿大映山红，樱色杜鹃）；Dr. M. Albrecht's Azalea ●☆

330059 Rhododendron albrechtii Maxim. f. albiflorum T. Yamaz.；白花极美杜鹃●☆

330060 Rhododendron albrechtii Maxim. f. canescens Sugim.；灰色极美杜鹃●☆

330061 Rhododendron albrechtii Maxim. f. hypoleucum（Honda）H. Hara；里白极美杜鹃●☆

330062 Rhododendron album Blume；白花树状杜鹃；White-flower Tree Azalea，White-flower Tree Rhododendron ●

330063 Rhododendron album Blume = Rhododendron arboreum Sm. subsp. roseum Lindl. ●

330064 Rhododendron album Buch.-Ham. ex D. Don = Rhododendron arboreum Sm. f. roseum Sweet ●

330065 Rhododendron alpicola Rehder et E. H. Wilson；深山杜鹃；Mountain Dwarf Rhododendron ●

330066 Rhododendron alpicola Rehder et E. H. Wilson = Rhododendron nivale Hook. f. subsp. boreale M. N. Philipson et Philipson ●

330067 Rhododendron alpicola Rehder et E. H. Wilson var. strictum Rehder et E. H. Wilson = Rhododendron nivale Hook. f. subsp. boreale M. N. Philipson et Philipson ●

330068 Rhododendron alutaceum Balf. f. et W. W. Sm.；棕背杜鹃（革质杜鹃，瘤枝杜鹃，软革叶杜鹃）；Bearing Gnaled Branches Rhododendron，Brownback Azalea，Soft Leather Leaves Rhododendron，Soft-leather-leaved Rhododendron ●

330069 Rhododendron alutaceum Balf. f. et W. W. Sm. var. iodes（Balf. f. et Forrest）D. F. Chamb.；毛枝棕背杜鹃（蓝紫杜鹃，锈色杜鹃）；Rust-coloured Rhododendron ●

330070 Rhododendron alutaceum Balf. f. et W. W. Sm. var. russotinctum（Balf. f. et Forrest）D. F. Chamb.；腺房棕背杜鹃（亮叶杜鹃，染褐杜鹃，染黑杜鹃，三带杜鹃）；Polished Leaves Rhododendron，Red-tinged Rhododendron，Triple Moles Rhododendron ●

330071 Rhododendron amagianum（Makino）Makino ex Nemoto；阿马基山杜鹃（伊豆杜鹃）；Mt. Amagi Azalea ●☆

330072 Rhododendron amakusaense（K. Takada ex T. Yamaz.）T. Yamaz.；天草杜鹃●☆

330073 Rhododendron amamiense Ohwi = Rhododendron latoucheae Franch. var. amamiense（Ohwi）T. Yamaz. ●☆

330074 Rhododendron amamiense Ohwi = Rhododendron latoucheae Franch. ●

330075 Rhododendron amandum Cowan；细枝杜鹃；Slender Twigs Rhododendron，Slender-twigged Rhododendron，Thintwig Azalea ●

330076 Rhododendron amanoi Ohwi；天农杜鹃●☆

330077 Rhododendron amanoi Ohwi var. glandulistylum Hatus. = Rhododendron amanoi Ohwi ●☆

330078 Rhododendron amaurophyllum Balf. f. et Forrest = Rhododendron saluenense Franch. ●

330079 Rhododendron ambiguum Hemsl.；问客杜鹃；Doubtful Azalea，Doubtful Rhododendron ●

330080 Rhododendron ambiguum Hemsl. = Rhododendron wongii Hemsl. et E. H. Wilson ●

330081　Rhododendron amesiae Rehder et E. H. Wilson；紫花杜鹃；Ames' Rhododendron, M. S. Ames' Rhododendron, Purpleflower Azalea ●

330082　Rhododendron amoenum（Lindl.）Planch.；可爱杜鹃（钝叶杜鹃，石岩）；Beloved Rhododendron, Blunt Leaves Azalea, Hiryu Azalea ●

330083　Rhododendron amundsonianum Hand. -Mazz.；暗叶杜鹃（西昌杜鹃）；Amundson's Rhododendron, Darkleaf Azalea ●

330084　Rhododendron annae Franch.；桃叶杜鹃（哈定杜鹃，红毛杜鹃）；H. I. Harding's Rhododendron, Peachleaf Azalea, Peachleaf Rhododendron, Peach-leaved Rhododendron ●

330085　Rhododendron annae Franch. subsp. laxiflorum（Balf. f. et Forrest）T. L. Ming；滇西桃叶杜鹃（疏花杜鹃）；Loose-flower Peachleaf Rhododendron, Loose-flowered Rhododendron, W. Yunnan Azalea ●

330086　Rhododendron annamense Rehder；越南杜鹃；Viet Nam Rhododendron, Vietnam Azalea ●

330087　Rhododendron anthopogon D. Don；髯毛杜鹃（髯花杜鹃）；Bearded Flowers Rhododendron, Beardflower Azalea, Beard-flowered Rhododendron, Hairyflower Rhododendron ●

330088　Rhododendron anthopogon D. Don = Rhododendron hypenanthum Balf. f. ●

330089　Rhododendron anthopogon D. Don subsp. hypenanthum（Balf. f.）Cullen = Rhododendron hypenanthum Balf. f. ●

330090　Rhododendron anthopogon D. Don var. album Davidian = Rhododendron anthopogon D. Don ●

330091　Rhododendron anthopogon D. Don var. haemonium（Balf. f. et R. E. Cooper）Cowan et Davidian = Rhododendron anthopogon D. Don ●

330092　Rhododendron anthopogonoides Maxim.；烈香杜鹃（白香柴，黄花杜鹃，似髯花杜鹃，小叶枇杷）；Savoury Rhododendron, Strong Aromatic Azalea, Strong Fragrance Rhododendron, Strong Fragrant Rhododendron ●

330093　Rhododendron anthopogonoides Maxim. subsp. hoi（W. P. Fang）W. P. Fang et Z. X. Xiong = Rhododendron hoi W. P. Fang ●

330094　Rhododendron anthosphaerum Diels；团花杜鹃（桃叶杜鹃，重奖杜鹃）；Groupflower Azalea, Highly Prized Rhododendron, Round-flowered Rhododendron ●

330095　Rhododendron anthosphaerum Diels subsp. hylothreptum（Balf. f. et W. W. Sm.）Tagg = Rhododendron anthosphaerum Diels ●

330096　Rhododendron anthosphaerum Diels var. eritimum（Balf. f. et W. W. Sm.）Davidian = Rhododendron anthosphaerum Diels ●

330097　Rhododendron anwheiense E. H. Wilson；黄山杜鹃（安徽杜鹃）；Huangshan Azalea, Huangshan Rhododendron ●

330098　Rhododendron anwheiense E. H. Wilson = Rhododendron maculiferum Franch. subsp. anwheiense（E. H. Wilson）D. F. Chamb. ex Cullen et D. F. Chamb. ●

330099　Rhododendron aperantum Balf. f. et Kingdon-Ward；宿鳞杜鹃（无限杜鹃）；Limitless Rhododendron, Persistentscale Azalea ●

330100　Rhododendron apiculatum Rehder et E. H. Wilson = Rhododendron concinnum Hemsl. ●

330101　Rhododendron apodectum Balf. f. et W. W. Sm. = Rhododendron dichroanthum Diels et Cowan var. apodectum（Balf. f. et W. W. Sm.）Cowan ●

330102　Rhododendron apodectum Balf. f. et W. W. Sm. = Rhododendron dichroanthum Diels et Cowan subsp. apodectum（Balf. f. et W. W. Sm.）Cowan ●

330103　Rhododendron apricum P. C. Tam = Rhododendron rufulum P. C. Tam ●

330104　Rhododendron apricum P. C. Tam var. falcinellum P. C. Tam = Rhododendron apricum P. C. Tam ●

330105　Rhododendron apricum P. C. Tam var. falcinellum P. C. Tam = Rhododendron falcinellum P. C. Tam ●

330106　Rhododendron apricum P. C. Tam var. falcinellum P. C. Tam = Rhododendron rufulum P. C. Tam ●

330107　Rhododendron araiophyllum Balf. f. et W. W. Sm.；窄叶杜鹃（秀雅杜鹃）；Narrowleaf Azalea, Narrow-leaved Rhododendron ●

330108　Rhododendron araiophyllum Balf. f. et W. W. Sm. subsp. lapidosum（T. L. Ming）M. Y. Fang；石生杜鹃；Lapidicolous Narrow-leaved Rhododendron ●

330109　Rhododendron araliiforme Balf. f. et Forrest = Rhododendron vernicosum Franch. ●

330110　Rhododendron arborescens（Pursh）Torr.；乔木状杜鹃（乔状杜鹃，甜香杜鹃）；Smooth Azalea, Sweet Azalea, Tree Rhododendron ●

330111　Rhododendron arboreum Sm.；树形杜鹃（树状杜鹃，树状杜鹃花）；Flame Tree, Large Rhododendron, Tree Azalea, Tree Rhododendron, Treelike Rhododendron, Tree-like Rhododendron ●

330112　Rhododendron arboreum Sm. f. album Sm. = Rhododendron album Blume ●

330113　Rhododendron arboreum Sm. f. roseum Sweet；粉红树形杜鹃（红花杜鹃，树形蔷薇色杜鹃）；Rosy Treelike Rhododendron ●

330114　Rhododendron arboreum Sm. subsp. campbelliae（Hook. f.）Tagg = Rhododendron arboreum Sm. subsp. cinnammomeum Wall. ex Lindl. ●

330115　Rhododendron arboreum Sm. subsp. campbelliae（Hook. f.）Tagg = Rhododendron arboreum Sm. var. cinnamomeum（Wall. ex G. Don）Lindl. ●

330116　Rhododendron arboreum Sm. subsp. cinnammomeum（Lindl.）Tagg = Rhododendron arboreum Sm. subsp. cinnammomeum Wall. ex Lindl. ●

330117　Rhododendron arboreum Sm. subsp. cinnammomeum Wall. ex Lindl. = Rhododendron arboreum Sm. var. cinnamomeum（Wall. ex G. Don）Lindl. ●

330118　Rhododendron arboreum Sm. subsp. cinnamomeum（Wall. ex G. Don）Tagg = Rhododendron arboreum Sm. var. cinnamomeum（Wall. ex G. Don）Lindl. ●

330119　Rhododendron arboreum Sm. subsp. delavayi（Franch.）Chamb. ex Cullen et Chamb. = Rhododendron delavayi Franch. ●

330120　Rhododendron arboreum Sm. subsp. delavayi（Franch.）Chamb. ex Cullen et Chamb. var. peramoenum（Balf. f. et Forrest）Chamb. ex Cullen et Chamb. = Rhododendron delavayi Franch. var. peramoenum（Balf. f. et Forrest）D. F. Chamb. ●

330121　Rhododendron arboreum Sm. subsp. delavayi（Franch.）D. F. Chamb. = Rhododendron delavayi Franch. ●

330122　Rhododendron arboreum Sm. subsp. peramoenum（Balf. f. et Forrest）D. F. Chamb. = Rhododendron delavayi Franch. var. peramoenum（Balf. f. et Forrest）D. F. Chamb. ●

330123　Rhododendron arboreum Sm. subsp. roseum Lindl. = Rhododendron arboreum Sm. f. roseum Sweet ●

330124　Rhododendron arboreum Sm. subsp. zeylanicum（Booth）Tagg；斯里兰卡树形杜鹃 ●☆

330125　Rhododendron arboreum Sm. var. album Wall. = Rhododendron arboreum Sm. subsp. roseum Lindl. ●

330126　Rhododendron arboreum Sm. var. album Wall. = Rhododendron arboreum Sm. f. roseum Sweet ●

330127　Rhododendron arboreum Sm. var. cinnamomeum（Wall. ex G. Don）Lindl.；棕色树形杜鹃（樟叶杜鹃，紫钟杜鹃）；Cinnamomum Leaves Rhododendron，Marschall. Campbell's Rhododendron ●

330128　Rhododendron arboreum Sm. var. cinnamrnomeum（Wall. ex Lindl.）W. K. Hu＝Rhododendron arboreum Sm. subsp. cinnammomeum Wall. ex Lindl. ●

330129　Rhododendron arboreum Sm. var. peramoenum（Balf. f. et Forrest）D. F. Chamb.＝Rhododendron delavayi Franch. var. peramoenum（Balf. f. et Forrest）D. F. Chamb. ●

330130　Rhododendron arboreum Sm. var. roseum（Lindl.）W. K. Hu＝Rhododendron arboreum Sm. subsp. roseum Lindl. ●

330131　Rhododendron argenteum Hook. f.＝Rhododendron grande Wight ●

330132　Rhododendron argipeplum Balf. f. et R. E. Cooper＝Rhododendron smithii Nutt. ●

330133　Rhododendron argyi H. Lév.＝Rhododendron mucronatum（Blume）G. Don ●

330134　Rhododendron argyrophyllum Franch.；银叶杜鹃（羊角花）；Silverleaf Azalea，Silverleaf Rhododendron，Silver-leaved Rhododendron ●

330135　Rhododendron argyrophyllum Franch. subsp. hejiangense W. P. Fang＝Rhododendron hejiangense M. Y. He ●

330136　Rhododendron argyrophyllum Franch. subsp. hejiangense W. P. Fang＝Rhododendron insigne Hemsl. et E. H. Wilson var. hejiangense（W. P. Fang）M. Y. Fang ●

330137　Rhododendron argyrophyllum Franch. subsp. hypoglaucum（Hemsl.）D. F. Chamb.；粉白杜鹃（灰蓝叶背杜鹃）；Blue-grey Underneath Rhododendron，Limy Rhododendron，Whiteback Azalea ●

330138　Rhododendron argyrophyllum Franch. subsp. hypoglaucum（Hemsl.）D. F. Chamb.＝Rhododendron hypoglaucum Hemsl. ●

330139　Rhododendron argyrophyllum Franch. subsp. hypogluacum（Hemsl.）D. F. Chamb. ex Cullen et Chamb.＝Rhododendron hypoglaucum Hemsl. ●

330140　Rhododendron argyrophyllum Franch. subsp. nankingense（Cowan）D. F. Chamb.；黔东银叶杜鹃；Nanjing Azalea，Nanjing Rhododendron ●

330141　Rhododendron argyrophyllum Franch. subsp. omeiense（Rehder et E. H. Wilson）D. F. Chamb.；峨眉银叶杜鹃；Emei Azalea，Emei Rhododendron ●

330142　Rhododendron argyrophyllum Franch. var. cupulare Rehder et E. H. Wilson＝Rhododendron argyrophyllum Franch. ●

330143　Rhododendron argyrophyllum Franch. var. glabriovarium M. Y. He；光房银叶杜鹃；Glabrous Silverleaf Azalea，Glabrous Silverleaf Rhododendron ●

330144　Rhododendron argyrophyllum Franch. var. hejiangense W. P. Fang＝Rhododendron insigne Hemsl. et E. H. Wilson var. hejiangense（W. P. Fang）M. Y. Fang ●

330145　Rhododendron argyrophyllum Franch. var. leiadrum Hutch.＝Rhododendron argyrophyllum Franch. subsp. nankingense（Cowan）D. F. Chamb. ●

330146　Rhododendron argyrophyllum Franch. var. nankingense Cowan＝Rhododendron argyrophyllum Franch. subsp. nankingense（Cowan）D. F. Chamb. ●

330147　Rhododendron argyrophyllum Franch. var. omeiense Rehder et E. H. Wilson＝Rhododendron argyrophyllum Franch. subsp. omeiense（Rehder et E. H. Wilson）D. F. Chamb. ●

330148　Rhododendron arizelum Balf. f. et Forrest；夺目杜鹃（斜钟杜鹃）●

330149　Rhododendron arizelum Balf. f. et Forrest＝Rhododendron rex H. Lév. subsp. arizelum（Balf. f. et Forrest）D. F. Chamb. ●

330150　Rhododendron artosquameum Balf. f. et Forrest；微心杜鹃（女山神杜鹃）；Compressed Scales Rhododendron，Cordate Rhododendron，Mountain Bred Rhododendron ●

330151　Rhododendron artosquameum Balf. f. et Forrest＝Rhododendron oreotrephes W. W. Sm. ●

330152　Rhododendron asmenistum Balf. f. et Forrest＝Rhododendron sanguineum Franch. var. cloiophorum（Balf. f. et Forrest）D. F. Chamb. ●

330153　Rhododendron asperulum Hutch. et Kingdon-Ward；瘤枝杜鹃；Nubbletwig Azalea，Rough Branches Rhododendron，Rough-branched Rhododendron ●

330154　Rhododendron asteium Balf. f. et Forrest＝Rhododendron eudoxum Balf. f. et Forrest var. mesopolium（Balf. f. et Forrest）D. F. Chamb. ●

330155　Rhododendron asterochnoum Diels；汶川星毛杜鹃（川西杜鹃，星毛杜鹃）；Stellate Hairs Rhododendron，Stellate-haired Rhododendron，Wenchuan Starhair Azalea ●

330156　Rhododendron asterochnoum Diels var. brevipediallatum W. K. Hu；短梗星毛杜鹃；Short-pedicel Stellate Hairs Rhododendron ●

330157　Rhododendron astrocalyx Balf. f. et Forrest＝Rhododendron wardii W. W. Sm. ●

330158　Rhododendron atentsiense Hand. -Mazz.＝Rhododendron dendricola Hutch. ●

330159　Rhododendron atjehense Sleumer；苏门答腊杜鹃；Sumatra Azalea，Sumatra Rhododendron ●

330160　Rhododendron atlanticum（Ashe）Rehder；海岸杜鹃（矮杜鹃，亚特兰杜鹃）；Coast Azalea，Coastal Azalea，Dwarf Azalea，Dwarf Rhododendron ●

330161　Rhododendron atropunicum H. P. Yang；暗紫杜鹃；Darkpurple Azalea，Dark-purple Rhododendron ●

330162　Rhododendron atrovirens Franch.；大关杜鹃（暗绿杜鹃）；Dark Green Rhododendron，Darkgreen Azalea，Dark-green Rhododendron ●

330163　Rhododendron aucklandii Hook. f.＝Rhododendron griffithianum Wight ●

330164　Rhododendron aucubifolium Hemsl.＝Rhododendron stamineum Franch. ●

330165　Rhododendron augustinii Hemsl.；毛肋杜鹃（奥氏杜鹃，褐斑杜鹃，张口杜鹃）；Augustine Henry's Rhododendron，Augustine Rhododendron，Hairrib Azalea，Vilmorin Rhododendron ●

330166　Rhododendron augustinii Hemsl. f. grandifolium Franch.＝Rhododendron augustinii Hemsl. subsp. chasmanthum（Diels）Cullen ●

330167　Rhododendron augustinii Hemsl. f. hardyi（Davidian）R. C. Fang＝Rhododendron augustinii Hemsl. subsp. chasmanthum（Diels）Cullen ●

330168　Rhododendron augustinii Hemsl. f. subglabra Franch.＝Rhododendron augustinii Hemsl. subsp. chasmanthum（Diels）Cullen ●

330169　Rhododendron augustinii Hemsl. subsp. chasmanthum（Diels）Cullen；张口杜鹃（绿斑杜鹃）；Caping Flowers Rhododendron，Olivespot Rhododendron，Open Azalea ●

330170　Rhododendron augustinii Hemsl. subsp. chasmanthum（Diels）Cullen f. hardyi（Davidian）R. C. Fang；白花张口杜鹃●

330171　Rhododendron augustinii Hemsl. subsp. chasmanthum（Diels）Cullen f. rubrum（Davidian）R. C. Fang；红花张口杜鹃●

330172　Rhododendron augustinii Hemsl. subsp. hardyi（Davidian）Cullen＝Rhododendron augustinii Hemsl. subsp. chasmanthum（Diels）Cullen f. hardyi（Davidian）R. C. Fang ●

330173 Rhododendron augustinii Hemsl. subsp. hardyi (Davidian) Cullen = Rhododendron augustinii Hemsl. subsp. chasmanthum (Diels) Cullen ●

330174 Rhododendron augustinii Hemsl. subsp. rubrum (Davidian) Cullen = Rhododendron augustinii Hemsl. subsp. chasmanthum (Diels) Cullen f. rubrum (Davidian) R. C. Fang ●

330175 Rhododendron augustinii Hemsl. subsp. rubrum (Davidian) Cullen = Rhododendron augustinii Hemsl. subsp. chasmanthum (Diels) Cullen ●

330176 Rhododendron augustinii Hemsl. var. chasmanthum (Diels) R. C. Fang = Rhododendron augustinii Hemsl. subsp. chasmanthum (Diels) Cullen ●

330177 Rhododendron augustinii Hemsl. var. rubrum Davidian = Rhododendron augustinii Hemsl. subsp. chasmanthum (Diels) Cullen f. rubrum (Davidian) R. C. Fang ●

330178 Rhododendron augustinii Hemsl. var. rubrum Davidian = Rhododendron augustinii Hemsl. subsp. chasmanthum (Diels) Cullen ●

330179 Rhododendron augustinii Hemsl. var. yui W. P. Fang = Rhododendron augustinii Hemsl. ●

330180 Rhododendron augustinii var. yui W. P. Fang = Rhododendron augustinii Hemsl. ●

330181 Rhododendron aureum Franch. = Rhododendron xanthostephanum Merr. ●

330182 Rhododendron aureum Georgi;牛皮杜鹃(冬桃,牛皮茶);Cowskin Azalea, Golden Rhododendron, Golden-flower Rhododendron, Goldmat Azalea ●◇

330183 Rhododendron aureum Georgi = Rhododendron xanthostephanum Merr. ●

330184 Rhododendron aureum Georgi f. albiflorum S. Watan.;白花牛皮杜鹃●☆

330185 Rhododendron aureum Georgi f. senanense Hara;重瓣牛皮杜鹃;Doubleflower Golden Rhododendron ●☆

330186 Rhododendron auriculatum Hemsl.;耳叶杜鹃;Auriculate Azalea, Auriculate Rhododendron, Auriculate-leaved Rhododendron, Earleaf Azalea, Ear-shape-leaved Rhododendron ●

330187 Rhododendron aurigeranum Sleumer;二色杜鹃●☆

330188 Rhododendron auritum Tagg;折萼杜鹃(长耳杜鹃);Foldcalyx Azalea, Long-ear Rhododendron, Long-eared Rhododendron ●

330189 Rhododendron australe Balf. f. et Forrest = Rhododendron leptothrium Balf. f. et W. W. Sm. ●

330190 Rhododendron austrinum (Small) Rehder;南方杜鹃(弗州之火杜鹃);Florida Azalea, Florida Flame Azalea, Florida Rhododendron, Southern Flame Azalea ●☆

330191 Rhododendron austrokiusianum Hatus. = Rhododendron obtusum (Lindl.) Planch. ●

330192 Rhododendron axium Balf. f. et Forrest = Rhododendron selense Franch. ●

330193 Rhododendron bachii H. Lév.;腺萼马银花(石壁杜鹃);Bach Rhododendron, Glandcalyx Azalea, Glandular Calyx Rhododendron ●

330194 Rhododendron baileyi Balf. f.;辐花杜鹃(贝蕾杜鹃);Bailey Rhododendron, Spokeflower Azalea ●

330195 Rhododendron bainbridgeanum Tagg et Forrest;毛萼杜鹃(白黄杜鹃);Haircalyx Azalea, Mainbridge's Rhododendron, Mr. Mainbridge's Rhododendron ●

330196 Rhododendron bakeri (W. P. Lemmon et McKay) H. H. Hume;坎伯兰杜鹃;Cumberland Azalea ●☆

330197 Rhododendron balangense W. P. Fang;巴朗杜鹃;Balang Azalea, Balang Rhododendron ●◇

330198 Rhododendron balfourianum Diels;粉钟杜鹃(巴氏杜鹃,大理腺萼杜鹃,大理腺蕊杜鹃,腺萼杜鹃);Balfour Azalea, Balfour's Rhododendron ●

330199 Rhododendron balfourianum Diels var. aganniphoides Tagg et Forrest;白毛粉钟杜鹃;White-haired Balfour's Rhododendron ●

330200 Rhododendron balsaminiflorum (Carrière) Nicholson;香膏杜鹃;Balsam Rhododendron ●☆

330201 Rhododendron bamaense Z. J. Zhao;班玛杜鹃(斑玛杜鹃);Banma Azalea, Banma Rhododendron ●

330202 Rhododendron barbatum Wall. = Rhododendron barbatum Wall. ex G. Don ●

330203 Rhododendron barbatum Wall. ex G. Don;硬刺杜鹃(髯毛杜鹃);Bristle Rhododendron, Hardspine Azalea, Hard-spinous Rhododendron ●

330204 Rhododendron barkamense D. F. Chamb.;马尔康杜鹃(巴尔康杜鹃);Barkam's Rhododendron, Maerkang Azalea ●

330205 Rhododendron basilicum Balf. f. et W. W. Sm.;粗枝杜鹃(大叶杜鹃,王杜鹃);Royal Rhododendron, Thicktwig Azalea ●

330206 Rhododendron batangense Balf. f. = Rhododendron nivale Hook. f. subsp. boreale M. N. Philipson et Philipson ●

330207 Rhododendron batangense Balf. f. et Forrest = Rhododendron nivale Hook. f. subsp. boreale M. N. Philipson et Philipson ●

330208 Rhododendron bathyphyllum Balf. f. et Forrest;多叶杜鹃;Leafy Azalea, Thickly Leaved Rhododendron, Thickly-leaved Rhododendron ●

330209 Rhododendron bauhiniiflorum Watt ex Hutch.;羊蹄甲杜鹃(羊蹄甲花杜鹃);Bauhinia-like Rhododendron ●☆

330210 Rhododendron beanianum Cowan;刺枝杜鹃(边氏杜鹃,短果杜鹃);Bean's Rhododendron, Spinebranchlet Rhododendron, Spinetwig Azalea, W. J. Bean's Rhododendron ●

330211 Rhododendron beesianum Diels;宽钟杜鹃(毕氏杜鹃);Bees Rhododendron, Broadbell Azalea ●

330212 Rhododendron beesianum Diels var. compactum Cowan = Rhododendron piercei Davidian ●

330213 Rhododendron beimaense Balf. f. et Forrest = Rhododendron erythrocalyx Balf. f. et Forrest ●

330214 Rhododendron bellissimum D. F. Chamb.;美鳞杜鹃(美丽杜鹃);Belle Rhododendron, Finescale Azalea ●

330215 Rhododendron bellum H. P. Yang = Rhododendron bellissimum D. F. Chamb. ●

330216 Rhododendron bellum W. P. Fang = Rhododendron simsii Planch. ●

330217 Rhododendron bellum W. P. Fang et G. Z. Li = Rhododendron simsii Planch. ●

330218 Rhododendron benthamianum Hemsl. = Rhododendron concinnum Hemsl. ●

330219 Rhododendron bergii Davidian = Rhododendron augustinii Hemsl. subsp. chasmanthum (Diels) Cullen f. rubrum (Davidian) R. C. Fang ●

330220 Rhododendron bergii Davidian = Rhododendron augustinii Hemsl. subsp. chasmanthum (Diels) Cullen ●

330221 Rhododendron beyerinckianum Koord.;拜氏杜鹃;Prof. Beijerinck's Rhododendron ●☆

330222 Rhododendron bhotanicum C. B. Clarke = Rhododendron lindleyi T. Moore ●

330223 Rhododendron bicolor P. C. Tam;双色杜鹃;Twocolour Azalea, Two-colour Rhododendron, Two-coloured Rhododendron ●

330224 Rhododendron bicolor P. C. Tam = Rhododendron simsii Planch. ●

330225　Rhododendron bicorniculatum P. C. Tam；双角杜鹃；Two-horn Rhododendron，Two-horns Azalea，Two-horns Rhododendron ●

330226　Rhododendron bicorniculatum P. C. Tam ＝Rhododendron mariae Hance ●

330227　Rhododendron bijiangense T. L. Ming；碧江杜鹃；Bijiang Azalea，Bijiang Rhododendron ●

330228　Rhododendron bivelatum Balf. f.；双被杜鹃(二包被杜鹃)；Double-perianth Azalea，Two-velate Rhododendron，Two-velates Rhododendron ●

330229　Rhododendron blandulum Balf. f. et W. W. Sm. ＝Rhododendron selense Franch. subsp. jucundum （Balf. f. et W. W. Sm.） D. F. Chamb. ●◇

330230　Rhododendron blepharocalyx Franch. ＝Rhododendron intricatum Franch. ●

330231　Rhododendron blinii H. Lév. ＝Rhododendron lutescens Franch. ●

330232　Rhododendron bodinieri Franch. ＝Rhododendron yunnanense Franch. ●

330233　Rhododendron boninense Nakai；小笠原杜鹃（博宁杜鹃）；Bonin Azalea，Bonin Rhododendron ●☆

330234　Rhododendron bonvalotii Bureau et Franch.；折多杜鹃（短柄杜鹃）；Bonvalot Azalea，Bonvalot Rhododendron ●

330235　Rhododendron boothii Nutt.；黄花花杜鹃（北印杜鹃，布氏杜鹃，柠檬杜鹃）；Assam Rhododendron，Booth's Rhododendron，Yellow Spotty Azalea ●

330236　Rhododendron brachyandrum Balf. f. et Forrest ＝Rhododendron eclecteum Balf. f. et Forrest ●

330237　Rhododendron brachyanthum Franch.；短花杜鹃（短柱杜鹃）；Sapgreen Rhododendron，Shortflor Azalea，Shortflower Rhododendron，Short-flowered Rhododendron ●

330238　Rhododendron brachyanthum Franch. subsp. hypolepidotum （Franch.） Cullen；绿柱短花杜鹃（绿柱杜鹃）；Green-styled Shortflower Rhododendron ●

330239　Rhododendron brachyanthum Franch. var. hypolepidotum Franch. ＝Rhododendron brachyanthum Franch. subsp. hypolepidotum （Franch.） Cullen ●

330240　Rhododendron brachycarpum D. Don ex G. Don；短果杜鹃（短花杜鹃，福氏杜鹃，富士山杜鹃）；Fujiyama Rhododendron，Pere L. F. Faurie's Rhododendron，Shortfruited Rhododendron ●☆

330241　Rhododendron brachycarpum D. Don ex G. Don f. fauriei （Franch.） Murata；法氏短果杜鹃●☆

330242　Rhododendron brachycarpum D. Don ex G. Don f. nematoanum Makino；重瓣短果杜鹃；Doubleflower Shortfruited Rhododendron ●☆

330243　Rhododendron brachycarpum D. Don ex G. Don subsp. fauriei （Franch.） D. F. Chamb. ＝Rhododendron brachycarpum D. Don ex G. Don f. fauriei （Franch.） Murata ●☆

330244　Rhododendron brachycarpum D. Don ex G. Don var. nemotoanum Makino f. normale （H. Hara） Kitam. ＝Rhododendron brachycarpum D. Don ex G. Don f. fauriei （Franch.） Murata ●☆

330245　Rhododendron brachycarpum D. Don ex G. Don var. roseum Koidz.；红花短果杜鹃●☆

330246　Rhododendron brachycarpum D. Don ex G. Don var. roseum Koidz. ＝Rhododendron japonoheptamerum Kitam. var. hondoense （Nakai） Kitam. f. leucanthum （Nakai） T. Yamaz. ●☆

330247　Rhododendron brachypodum W. P. Fang et P. S. Liu；短梗杜鹃；Short-podium Rhododendron，Shortstalk Azalea ●

330248　Rhododendron brachysiphon Balf. f. ex Hutch.；短管杜鹃；Short-tubed Rhododendron ●☆

330249　Rhododendron brachysiphon Balf. f. ex Hutch. ＝Rhododendron maddenii Hook. f. ●

330250　Rhododendron brachystylum Balf. f. et Kingdon-Ward ＝Rhododendron trichocladum Franch. ●

330251　Rhododendron bracteatum Rehder et E. H. Wilson；苞叶杜鹃（红点杜鹃）；Bract Furnished Rhododendron，Bractleaf Azalea，Red-spotted Rhododendron ●

330252　Rhododendron bretii Hemsl. et E. H. Wilson ＝Rhododendron longesquamatum C. K. Schneid. ●

330253　Rhododendron brevicaudatum R. C. Fang et S. S. Chang；短尾杜鹃；Short-caudate Rhododendron，Shorttail Azalea ●

330254　Rhododendron brevinerve Chun et W. P. Fang；短脉杜鹃；Shortvein Azalea，Shortvein Rhododendron，Shortveined Rhododendron ●

330255　Rhododendron breviperulatum Hayata；短鳞芽杜鹃（短鳞杜鹃，短芽鳞杜鹃，南澳杜鹃，埔里杜鹃）；Nan-au Azalea，Short Bud Scales Rhododendron，Short-perula Rhododendron ●

330256　Rhododendron brevipetiolatum M. Y. Fang；短柄杜鹃；Shortpetiole Azalea，Shortstalk Rhododendron，Short-stalked Rhododendron ●

330257　Rhododendron brevistylum Franch.；短柱杜鹃●

330258　Rhododendron brevistylum Franch. ＝Rhododendron heliolepis Franch. ●

330259　Rhododendron brevitubum Balf. f. et R. E. Cooper ＝Rhododendron maddenii Hook. f. ●

330260　Rhododendron breynii Planch. ＝Rhododendron indicum （L.） Sweet ●

330261　Rhododendron brookeanum H. Low ex Lindl.；布鲁克杜鹃；Sir. J. Brooke's Rhododendron ●☆

330262　Rhododendron brunneifolium Balf. f. et Forrest ＝Rhododendron eudoxum Balf. f. et Forrest var. bruneifolium （Balf. f. et Forrest） D. F. Chamb. ●

330263　Rhododendron bullatum Franch. ＝Rhododendron edgeworthii Hook. f. ●

330264　Rhododendron bulu Hutch.；蜿蜒杜鹃（散鳞杜鹃）；Loose Scaled Rhododendron，Loose-scaled Rhododendron，Wriggle Azalea ●

330265　Rhododendron bureavii Franch.；锈红毛杜鹃（布鲁杜鹃，厚毛杜鹃，似锈红杜鹃，锈红杜鹃，血红大理杜鹃）；Bureav's Rhododendron，E. Bureav's Rhododendron，Like E. Bureau's Rhododendron，Red Dali Rhododendron，Rust Azalea ●

330266　Rhododendron bureavioides Balf. f. ＝Rhododendron bureavii Franch. ●

330267　Rhododendron burmanicum Hutch.；缅甸杜鹃；Burma Rhododendron ●☆

330268　Rhododendron burrifolium Balf. f. et Forrest ＝Rhododendron diphrocalyx Balf. f. ●

330269　Rhododendron butyricum Kingdon-Ward ＝Rhododendron chrysodoron Tagg ex Hutch. ●

330270　Rhododendron caendeoglaucum Balf. f. et Forrest ＝Rhododendron campylogynum Franch. ●

330271　Rhododendron caeruleum H. Lév. ＝Rhododendron rigidum Franch. ●

330272　Rhododendron caesium Hutch.；蓝灰糙毛杜鹃（灰蓝叶杜鹃，蓝灰杜鹃）；Bluegrey Azalea，Dullish Blue Leaves Rhododendron，Dullish Blue-leaved Rhododendron ●

330273　Rhododendron caespitulum P. C. Tam；丛枝杜鹃；Fascicled Branches Rhododendron，Fascicled-branched Rhododendron ●

330274　Rhododendron caespitulum P. C. Tam ＝Rhododendron myrsinifolium Ching ex W. P. Fang et M. Y. He ●

330275　Rhododendron calciphilum Hutch. et Kingdon-Ward ＝

Rhododendron calostrotum Balf. f. et Kingdon-Ward var. caliphilum (Hutch. et Kingdon-Ward) Davidian ●

330276　Rhododendron calendulaceum（Michx.）Torr.；火焰杜鹃（黄焰杜鹃）；Flame Azalea，Yellow Azalea，Yellow Honeysuckle ●☆

330277　Rhododendron californicum Hook.；西岸杜鹃（加州杜鹃）；California，California Rhododendron，California Rose-bay，Californian Rose Bay，West Coast Rhododendron ●☆

330278　Rhododendron calleryi Planch. = Rhododendron simsii Planch. ●

330279　Rhododendron callimorphum Balf. f. et W. W. Sm.；卵叶杜鹃；Lovely-shaped Rhododendron，Ovateleaf Azalea，Ovateleaf Rhododendron ●

330280　Rhododendron callimorphum Balf. f. et W. W. Sm. var. myiagrum（Balf. f. et Forrest）D. F. Chamb. ex Cullen et D. F. Chamb.；白花卵叶杜鹃（捕蝇杜鹃，黏梗杜鹃）；Fly-catcher Rhododendron，Whiteflower Lovely-shaped Rhododendron，Whiteflower Ovateleaf Azalea ●

330281　Rhododendron calophyllum Nutt.；美叶杜鹃；Beautiful-leaves Rhododendron ●☆

330282　Rhododendron calophyllum Nutt. = Rhododendron maddenii Hook. f. ●

330283　Rhododendron calophytum Franch.；美容杜鹃（大叶杜鹃）；Beautiful Rhododendron，Bigleaf Rhododendron，Spiffy Azalea ●

330284　Rhododendron calophytum Franch. subsp. jingfuense W. P. Fang = Rhododendron calophytum Franch. var. jinfuense W. P. Fang et W. K. Hu ●

330285　Rhododendron calophytum Franch. var. jinfuense W. P. Fang et W. K. Hu；金佛山美容杜鹃；Jinfoshan Azalea，Jinfoshan Beautiful Rhododendron ●

330286　Rhododendron calophytum Franch. var. openshawianum（Rehder et E. H. Wilson）D. F. Chamb.；尖叶美容杜鹃（川尖叶杜鹃，川西尖叶杜鹃）；Sharp-leaf Beautiful Azalea，Sharp-leaf Beautiful Rhododendron，West Sichuan Point Leaves Rhododendron ●

330287　Rhododendron calophytum Franch. var. pauciflorum W. K. Hu；疏花美容杜鹃；Loose-flower Beautiful Azalea，Loose-flower Beautiful Rhododendron ●

330288　Rhododendron calophytum Balf. f. et Forrest subsp. jinfuense W. P. Fang ex M. Y. Fang = Rhododendron calophytum Franch. var. jinfuense W. P. Fang et W. K. Hu ●

330289　Rhododendron calostrotum Balf. f. et Forrest subsp. keleticum（Balf. f. et Forrest）Cullen = Rhododendron keleticum Nutt. ●

330290　Rhododendron calostrotum Balf. f. et Forrest subsp. riparioides Cullen = Rhododendron riparioides（Cullen）Cubey ●

330291　Rhododendron calostrotum Balf. f. et Forrest var. riparioides（Cullen）R. C. Fang = Rhododendron riparioides（Cullen）Cubey ●

330292　Rhododendron calostrotum Balf. f. et Kingdon-Ward；美被杜鹃；Beautiful-covering Rhododendron，Beautiful-perianth Azalea，Beautiful-perianth Rhododendron，Beautiful-perianthed Rhododendron ●

330293　Rhododendron calostrotum Balf. f. et Kingdon-Ward 'Gigha'；杰格哈美被杜鹃●☆

330294　Rhododendron calostrotum Balf. f. et Kingdon-Ward subsp. keleticum Cullen = Rhododendron keleticum Nutt. ●

330295　Rhododendron calostrotum Balf. f. et Kingdon-Ward subsp. riparioides Cullen = Rhododendron calostrotum Balf. f. et Kingdon-Ward var. riparioides（Cullen）R. C. Fang ●

330296　Rhododendron calostrotum Balf. f. et Kingdon-Ward var. caliphilum（Hutch. et Kingdon-Ward）Davidian；小叶美被杜鹃●

330297　Rhododendron calostrotum Balf. f. et Kingdon-Ward var. riparioides（Cullen）R. C. Fang；雪龙美被杜鹃●

330298　Rhododendron calostrotum Franch. subsp. riparium（Kingdon-Ward）Cullen = Rhododendron calostrotum Balf. f. et Kingdon-Ward ●

330299　Rhododendron caloxanthum Balf. f. et Forrest = Rhododendron campylocarpum Hook. f. subsp. caloxanthum（Balf. f. et Forrest）D. F. Chamb. ●

330300　Rhododendron calvescens Balf. f. et Forrest；变光杜鹃（光秃杜鹃）；Calvous Rhododendron，Shedhair Azalea ●

330301　Rhododendron calvescens Balf. f. et Forrest var. duseimatum（Balf. f. et Forrest）D. F. Chamb.；长梗变光杜鹃；Longstalk Calvous Rhododendron，Longstalk Shedhair Azalea ●

330302　Rhododendron camelliiflorum Hook. f.；茶花杜鹃；Camellia-like Rhododendron，Teaflower Azalea ●

330303　Rhododendron campanulatum D. Don；钟花杜鹃；Bellflower Azalea，Bellflower Rhododendron，Bell-flowered Rhododendron，Bell-shaped Rhododendron ●

330304　Rhododendron campanulatum D. Don subsp. aeruginosum（Hook. f.）D. F. Chamb.；铜叶钟花杜鹃；Cupi-coloured Rhododendron ●

330305　Rhododendron campanulatum D. Don var. aeruginosum（Hook. f.）Cowan et Davidian = Rhododendron campanulatum D. Don subsp. aeruginosum（Hook. f.）D. F. Chamb. ●

330306　Rhododendron campanulatum D. Don var. wallichii（Hook. f.）Hook. = Rhododendron wallichii Hook. f. ●

330307　Rhododendron campanulatum D. Don var. wallichii Hook. f. = Rhododendron wallichii Hook. f. ●

330308　Rhododendron campbelliae Hook. f. = Rhododendron arboreum Sm. subsp. cinnammomeum Wall. ex Lindl. ●

330309　Rhododendron campbelliae Hook. f. = Rhododendron arboreum Sm. var. cinnamomeum（Wall. ex G. Don）Lindl. ●

330310　Rhododendron campylocarpum Franch. var. orbiculare Demoly = Rhododendron campylocarpum Hook. f. subsp. caloxanthum（Balf. f. et Forrest）D. F. Chamb. ●

330311　Rhododendron campylocarpum Hook. f.；弯果杜鹃（黄花杜鹃，蜜钟杜鹃）；Bent Fruits Rhododendron，Bowfruit Rhododendron，Bow-fruited Rhododendron，Curvefruit Azalea，Honey-bell Rhododendron ●

330312　Rhododendron campylocarpum Hook. f. subsp. caloxanthum（Balf. f. et Forrest）D. F. Chamb.；美丽弯果杜鹃（美黄杜鹃，诱人杜鹃）；Beautiful Yellow Rhododendron，Beutiful Curvefruit Azalea，Conspicuous Rhododendron ●

330313　Rhododendron campylocarpum Hook. f. subsp. telopeum（Balf. f. et Forrest）D. F. Chamb. = Rhododendron campylocarpum Hook. f. subsp. caloxanthum（Balf. f. et Forrest）D. F. Chamb. ●

330314　Rhododendron campylogynum Franch.；弯柱杜鹃（弯雄蕊杜鹃）；Bent Ovary Rhododendron，Curvestyle Azalea，Duskybloom Rhododendron，Dusky-bloomed Rhododendron ●

330315　Rhododendron campylogynum Franch. var. celsum Davidian = Rhododendron campylogynum Franch. ●

330316　Rhododendron campylogynum Franch. var. cremastum（Balf. f. et Forrest）Davidian = Rhododendron campylogynum Franch. ●

330317　Rhododendron campylogynum Franch. var. myrtilloides（Balf. f. et Kingdon-Ward）Davidian = Rhododendron campylogynum Franch. ●

330318　Rhododendron camtschaticum Pall. = Therorhodion camtschaticum（Pall.）Small ●☆

330319　Rhododendron camtschaticum Pall. subsp. glandulosum（Standl.）Hultén = Therorhodion camtschaticum（Pall.）Small var. pumilum（E. A. Busch）T. Yamaz. ●☆

330320　Rhododendron camtschaticum Pall. var. barbatum（Nakai）Tatew.

= Therorhodion camtschaticum (Pall.) Small var. pumilum (E. A. Busch) T. Yamaz. ● ☆

330321　Rhododendron canadense (L.) Torr.；加拿大迎红杜鹃；Canada Azalea, Rhodora ● ☆

330322　Rhododendron canescens (Michx.) Sweet；西方灰杜鹃；Florida Pinxter, Hoary Azalea, Mountain Azalea, Piedmont Azalea, Pinxter Flower, Wild Azalea ● ☆

330323　Rhododendron cantabile Balf. f. ex Hutch. = Rhododendron russatum Balf. f. et Forrest ●

330324　Rhododendron capitatum Maxim.；头花杜鹃（黑香柴，头状杜鹃，小头花杜鹃，小叶杜鹃）；Capitate Azalea, Capitate Rhododendron, Head Flowers Rhododendron ●

330325　Rhododendron cardiobasis Sleumer = Rhododendron orbiculare Decne. subsp. cardiobasis (Sleumer) D. F. Chamb. ●

330326　Rhododendron carneum Hutch.；肉色杜鹃；Fleshcolor Azalea, Fleshcolored Rhododendron, Flesh-colored Rhododendron ●

330327　Rhododendron carolinianum Rehder；卡罗来纳杜鹃（卡罗莱纳杜鹃）；Carolina Rhododendron ● ☆

330328　Rhododendron caryophyllum Hayata = Rhododendron rubropilosum Hayata ●

330329　Rhododendron catacosmum Balf. f. ex Tagg；瓣萼杜鹃（装饰杜鹃，尊敬杜鹃）；Adorned Rhododendron, Petalcalyx Azalea ●

330330　Rhododendron catapastum Balf. f. et Forrest = Rhododendron rubiginosum Franch. ●

330331　Rhododendron catawbiense Michx.；酒红杜鹃（北美杜鹃，山夹竹桃）；Catawba Rhododendron, Catawba River Rhododendron, Mountain Rose Bary, Mountain Rosebay, Mountain Rosebay Rhododendron, Purple Rhododendron ● ☆

330332　Rhododendron catawbiense Michx. ‘Album’；白花酒红杜鹃 ● ☆

330333　Rhododendron catawbiense Michx. ‘English Roseum’；英国玫瑰酒红杜鹃 ● ☆

330334　Rhododendron caucasicum Pall.；高加索杜鹃；Caucasian Rhododendron ● ☆

330335　Rhododendron cavaleriei H. Lév.；多花杜鹃（羊角杜鹃）；Cavalerie Rhododendron, Flowery Azalea ●

330336　Rhododendron cavaleriei H. Lév. var. chaffanjonii H. Lév. = Rhododendron stamineum Franch. ●

330337　Rhododendron cephalanthoides Balf. f. et W. W. Sm. = Rhododendron primuliflorum Bureau et Franch. var. cephalanthoides (Balf. f. et W. W. Sm.) Cowan et Davidian ●

330338　Rhododendron cephalanthum Franch.；毛喉杜鹃（小叶杜鹃，小叶枇杷）；Hairthroat Azalea, Headed Flowers Rhododendron, Whitebutton Azalea, Whitebutton Rhododendron, White-button Rhododendron ●

330339　Rhododendron cephalanthum Franch. subsp. platyphyllum (Franch. ex Balf. f. et W. W. Sm.) Cullen = Rhododendron platyphyllum Franch. ex Balf. f. et W. W. Sm. ●

330340　Rhododendron cephalanthum Franch. var. platyphyllum Franch. ex Diels = Rhododendron platyphyllum Franch. ex Balf. f. et W. W. Sm. ●

330341　Rhododendron ceraceum Balf. f. et W. W. Sm. = Rhododendron lukiangense Franch. ●

330342　Rhododendron cerasiflorum Kingdon-Ward = Rhododendron campylogynum Franch. ●

330343　Rhododendron cerasinum Tagg；樱花杜鹃；Cherry-coloured Rhododendron, Cherryflower Azalea ●

330344　Rhododendron cerinum Balf. f. et Forrest = Rhododendron sulphureum Franch. ● ◇

330345　Rhododendron cerochitum Balf. f. et Forrest = Rhododendron tanastylum Balf. f. et Kingdon-Ward ●

330346　Rhododendron cerochitum Tagg = Rhododendron tanastylum Balf. f. et Kingdon-Ward ●

330347　Rhododendron chaetomallum Balf. f. et Forrest = Rhododendron haematodes Franch. subsp. chaetomallum (Balf. f. et Forrest) Cowan ●

330348　Rhododendron chaetomallum Balf. f. et Forrest var. glaucescens Tagg et Forrest = Rhododendron haematodes Franch. subsp. chaetomallum (Balf. f. et Forrest) Cowan ●

330349　Rhododendron chaffanjonii H. Lév. = Rhododendron stamineum Franch. ●

330350　Rhododendron chalarocladum Balf. f. et Forrest = Rhododendron selense Franch. ●

330351　Rhododendron chamaethomsonii (Tagg et Forrest) Cowan et Davidian；云雾杜鹃（矮汤姆逊杜鹃，矮小杜鹃）；Dwarf Thomson Rhododendron, Nabilous Azalea ●

330352　Rhododendron chamaethomsonii (Tagg et Forrest) Cowan et Davidian var. chamaedoron (Tagg et Forrest) D. F. Chamb. ex Cullen et D. F. Chamb.；毛背云雾杜鹃；Hairback Nabilous Azalea ●

330353　Rhododendron chamaethomsonii (Tagg et Forrest) Cowan et Davidian var. chamaethauma (Tagg et Forrest) Cowan et Davidian；短萼云雾杜鹃；Shortcalyx Nabilous Azalea ●

330354　Rhododendron chamaethomsonii (Tagg et Forrest) Cowan et Davidian var. chamaedoron (Tagg et Forrest) D. F. Chamb. ex Cullen et D. F. Chamb. = Rhododendron chamaethomsonii (Tagg et Forrest) Cowan et Davidian var. chamaethauma (Tagg et Forrest) Cowan et Davidian ●

330355　Rhododendron chamaetortum Balf. f. et Kingdon-Ward = Rhododendron cephalanthum Franch. ●

330356　Rhododendron chamaezelum Balf. f. et Forrest；向地杜鹃；Groud Seeking Rhododendron ● ☆

330357　Rhododendron chameunum Balf. f. et Forrest = Rhododendron saluenense Franch. var. prostratum (W. W. Sm.) R. C. Fang ●

330358　Rhododendron championae Hook.；刺毛杜鹃（短柄杜鹃，短柄马银花，狗脚骨，牛舌柴，瘦石榴，太平杜鹃）；Champion Azalea, Champion's Rhododendron, Short Petioles Azalea, Short Petioles Rhododendron ●

330359　Rhododendron championae Hook. var. ovatifolium P. C. Tam；山荷桃；Ovate-leaf Rhododendron ●

330360　Rhododendron championiae Hook. var. ovatifolium P. C. Tam = Rhododendron championae Hook. ●

330361　Rhododendron champmanii A. Gray；康氏杜鹃（堪普曼杜鹃）；Champman Azalea, Champman's Rhododendron ● ☆

330362　Rhododendron changii (W. P. Fang) W. P. Fang；树枫杜鹃；Chang Azalea, Chang Rhododendron ●

330363　Rhododendron chaoanense P. C. Tam et T. C. Wu；潮安杜鹃；Chao' an Azalea, Chao' an Rhododendron ●

330364　Rhododendron chapaense Dop = Rhododendron maddenii Hook. f. subsp. crassum (Franch.) Cullen ●

330365　Rhododendron chapmanii A. Gray；查普曼杜鹃；Chapman's Rhododendron ● ☆

330366　Rhododendron charianthum Hutch.；淑花杜鹃（凹叶杜鹃，美花杜鹃）；Elegentflower Azalea, Elegentflower Rhododendron, Elegent-flowered Rhododendron, Graceful Flowers Rhododendron ●

330367　Rhododendron charianthum Hutch. = Rhododendron davidsonianum Rehder et E. H. Wilson ●

330368　Rhododendron charidotes Balf. f. et Forrest = Rhododendron saluenense Franch. var. prostratum (W. W. Sm.) R. C. Fang ●

330369　Rhododendron charitopes Balf. f. et Forrest;雅容杜鹃;Beautiful Lady Rhododendron,Elegent Azalea,Elegent Rhododendron ●

330370　Rhododendron charitopes Balf. f. et Forrest subsp. tsangpoense (Kingdon-Ward) Cullen;藏布雅容杜鹃(沧波河杜鹃,藏布杜鹃);Tsangpo River Rhododendron,Zangbu Elegent Azalea,Zangbu Elegent Rhododendron ●

330371　Rhododendron charitostreptum Balf. f. et Kingdon-Ward = Rhododendron brachyanthum Franch. subsp. hypolepidotum (Franch.) Cullen ●

330372　Rhododendron charopoeum Balf. f. et Forrest = Rhododendron campylogynum Franch. ●

330373　Rhododendron chartophyllum Franch. ;纸叶杜鹃●

330374　Rhododendron chartophyllum Franch. = Rhododendron yunnanense Franch. ●

330375　Rhododendron chartophyllum Franch. f. praecox Diels = Rhododendron yunnanense Franch. ●

330376　Rhododendron chasmanthoides Balf. f. et Forrest = Rhododendron augustinii Hemsl. subsp. chasmanthum (Diels) Cullen ●

330377　Rhododendron chasmanthum Diels = Rhododendron augustinii Hemsl. subsp. chasmanthum (Diels) Cullen ●

330378　Rhododendron chawchiense Balf. f. et Forrest = Rhododendron anthosphaerum Diels ●

330379　Rhododendron cheilanthum Balf. f. et Forrest = Rhododendron cuneatum W. W. Sm. ●

330380　Rhododendron chengianum W. P. Fang = Rhododendron hemsleyanum E. H. Wilson var. chengianum W. P. Fang ex Ching ●

330381　Rhododendron chengshienianum W. P. Fang;拟黄花杜鹃;Cengshien Rhododendron ●

330382　Rhododendron chengshienianum W. P. Fang = Rhododendron ambiguum Hemsl. ●

330383　Rhododendron chienianum W. P. Fang = Rhododendron longipes Rehder et E. H. Wilson var. chienianum (W. P. Fang) D. F. Chamb. ex Cullen et D. F. Chamb. ●

330384　Rhododendron chihsinianum Chun et W. P. Fang;红滩杜鹃(济新杜鹃,龙胜杜鹃);Chihsin Azalea,Chihsin Rhododendron,Jixin Rhododendron ●◇

330385　Rhododendron chionanthum Tagg et Forrest;高山白花杜鹃(雪花杜鹃);Alp White Azalea,Snow Flower Rhododendron,Snow-flowered Rhododendron ●

330386　Rhododendron chionophyllum Diels = Rhododendron argyrophyllum Franch. ●

330387　Rhododendron chlanidotum Balf. f. et Forrest = Rhododendron citriniflorum Balf. f. et Forrest ●

330388　Rhododendron chloranthum Balf. f. et Forrest;黄绿杜鹃●

330389　Rhododendron chloranthum Balf. f. et Forrest = Rhododendron mekongense Franch. var. melinanthum (Balf. f. et Kingdon-Ward) Cullen ●

330390　Rhododendron chlorops Cowan;绿心杜鹃;Green Eye Rhododendron ●☆

330391　Rhododendron chrysanthum Matsum. et Hayata = Rhododendron pseudochrysanthum Hayata ●

330392　Rhododendron chrysanthum Pall. = Rhododendron aureum Georgi ●◇

330393　Rhododendron chryseum Balf. f. et Kingdon-Ward = Rhododendron rupicola W. W. Sm. var. chryseum (Balf. f. et Kingdon-Ward) M. N. Philipson et Philipson ●

330394　Rhododendron chrysocalyx H. Lév. et Vaniot;金萼杜鹃(羊老毡);Goldencalyx Azalea, Goldencalyx Rhododendron, Golden-calyxed Rhododendron ●

330395　Rhododendron chrysocalyx H. Lév. et Vaniot var. xiushanense (W. P. Fang) M. Y. He;秀山金萼杜鹃;Xiushan Goldencalyx Rhododendron,Xiushan Rhododendron ●

330396　Rhododendron chrysodoron Tagg ex Hutch. ;金黄杜鹃(纯黄杜鹃,金礼杜鹃);Pureyellow Azalea, Pureyellow Rhododendron, Pure-yellow Rhododendron ●

330397　Rhododendron chrysolepis Hutch. et Kingdon-Ward;金鳞杜鹃(椿年杜鹃);Golden Scales Rhododendron ●

330398　Rhododendron chunii W. P. Fang;宿柱杜鹃(龙山杜鹃);Chun Azalea,Chun Rhododendron,Persistent Styles Rhododendron ●

330399　Rhododendron chunnienii Chun et W. P. Fang;椿年杜鹃(花坪杜鹃,金鳞杜鹃);Chunnian Azalea,Chunnien Rhododendron ●

330400　Rhododendron chunnienii Chun et W. P. Fang = Rhododendron liliiflorum H. Lév. ●

330401　Rhododendron ciliatopedicellatum Hayata = Rhododendron henryi Hance ●

330402　Rhododendron ciliatum Hook. f. ;睫毛杜鹃(纤毛杜鹃);Ciliate Azalea,Ciliate Rhododendron,Fringed Rhododendron ●

330403　Rhododendron ciliicalyx Franch. ;睫毛萼杜鹃;Ciliatecalyx Azalea, Ciliatecalyx Rhododendron, Ciliate-calyxed Rhododendron, Fringed Calyx Rhododendron ●

330404　Rhododendron ciliicalyx Franch. subsp. lyi (H. Lév.) R. C. Fang;长柱睫萼杜鹃(长柱杜鹃,李氏杜鹃);Longstyle Ciliatecalyx Azalea,Ly's Rhododendron ●

330405　Rhododendron ciliipes Hutch. ;香花白杜鹃(香花杜鹃);Fragrant Azalea, Fragrant Flowers Rhododendron, Fragrant-flowered Rhododendron ●

330406　Rhododendron cinereoserratum P. C. Tam;灰齿杜鹃;Pale Serrate Azalea,Pale Serrate Leaves Rhododendron ●

330407　Rhododendron cinereoserratum P. C. Tam = Rhododendron mariesii Hemsl. et E. H. Wilson ●

330408　Rhododendron cinereum Balf. f. = Rhododendron cuneatum W. W. Sm. ●

330409　Rhododendron cinnabarinum Hook. f. ;朱红杜鹃(朱砂杜鹃);Cinnabar-red Rhododendron,Vermilion Azalea,Vermilion Rhododendron ●

330410　Rhododendron cinnabarinum Hook. f. 'Mount Everest';珠穆朗玛峰朱砂杜鹃●☆

330411　Rhododendron cinnabarinum Hook. f. subsp. tamaense (Davidian) Cullen;龙江朱砂杜鹃●

330412　Rhododendron cinnabarinum Hook. f. subsp. xanthocodon (Hutch.) Cullen = Rhododendron xanthocodon Hutch. ●

330413　Rhododendron cinnabarinum Hook. f. var. pallidum Hook. = Rhododendron xanthocodon Hutch. ●

330414　Rhododendron cinnabarinum Hook. f. var. purpurellum Cowan = Rhododendron xanthocodon Hutch. ●

330415　Rhododendron cinnabarinum Wall. var. purpureum Cowan;紫色朱砂杜鹃;Purple Cinnabar-red Rhododendron, Purple Vermilion Azalea ●

330416　Rhododendron cinnabarinum Wall. var. roylei (Hook. f.) Hutch. ;深红朱砂杜鹃;Deepred Vermilion Azalea ●

330417　Rhododendron cinnammomeum Wall. ex G. Don = Rhododendron arboreum Sm. subsp. cinnammomeum Wall. ex Lindl. ●

330418　Rhododendron cinnamomeum Wall. = Rhododendron arboreum Sm. subsp. cinnammomeum Wall. ex Lindl. ●

330419   Rhododendron cinnamomeum Wall. ex G. Don = Rhododendron arboreum Sm. var. cinnamomeum (Wall. ex G. Don) Lindl. ●

330420   Rhododendron circinnatum Cowan et Kingdon-Ward；卷毛杜鹃；Coilhair Azalea，Curvihaired Rhododendron，Kurly Hairs Rhododendron ●

330421   Rhododendron circinnatum Cowan et Kingdon-Ward subsp. aureolum Cowan = Rhododendron citriniflorum Balf. f. et Forrest var. horaeum (Balf. f. et Forrest) D. F. Chamb. ●

330422   Rhododendron citriniflorum Balf. f. et Forrest；橙黄杜鹃（香橼花杜鹃）；Lemonyellow Flowered Rhododendron，Orange Azalea，Orange Rhododendron ●

330423   Rhododendron citriniflorum Balf. f. et Forrest subsp. aureolum Cowan = Rhododendron citriniflorum Balf. f. et Forrest var. horaeum (Balf. f. et Forrest) D. F. Chamb. ●

330424   Rhododendron citriniflorum Balf. f. et Forrest subsp. horaeum (Balf. f. et Forrest) Cowan = Rhododendron citriniflorum Balf. f. et Forrest var. horaeum (Balf. f. et Forrest) D. F. Chamb. ●

330425   Rhododendron citriniflorum Balf. f. et Forrest subsp. horaeum Cowan = Rhododendron citriniflorum Balf. f. et Forrest var. horaeum (Balf. f. et Forrest) D. F. Chamb. ●

330426   Rhododendron citriniflorum Balf. f. et Forrest var. horaeum (Balf. f. et Forrest) D. F. Chamb.；美艳橙黄杜鹃；Beautiful Orange Rhododendron ●

330427   Rhododendron clementinae Forrest ex W. W. Sm.；麻点杜鹃；Clementine's Rhododendron，Dotty Azalea ●

330428   Rhododendron clementinae Forrest ex W. W. Sm. subsp. aureodorsale W. P. Fang；金背麻点杜鹃（金背杜鹃，金背枇杷）；Golden-bark Clementine's Rhododendron，Golden-bark Dotty Azalea ●

330429   Rhododendron clivicola Balf. f. et W. W. Sm. = Rhododendron primuliflorum Bureau et Franch. ●

330430   Rhododendron cloiophorum Balf. f. et Forrest = Rhododendron sanguineum Franch. var. cloiophorum (Balf. f. et Forrest) D. F. Chamb. ●

330431   Rhododendron cloiophorum Balf. f. et Forrest subsp. asmenistum (Balf. f. et Forrest) Tagg = Rhododendron sanguineum Franch. var. cloiophorum (Balf. f. et Forrest) D. F. Chamb. ●

330432   Rhododendron cloiophorum Balf. f. et Forrest subsp. leucopetalum (Balf. f. et Forrest) Tagg = Rhododendron sanguineum Franch. var. cloiophorum (Balf. f. et Forrest) D. F. Chamb. ●

330433   Rhododendron cloiophorum Balf. f. et Forrest subsp. mannophorum (Balf. f. et Forrest) Tagg = Rhododendron sanguineum Franch. var. didymoides Tagg et Forrest ●

330434   Rhododendron cloiophorum Balf. f. et Forrest subsp. roseotinctum (Balf. f. et Forrest) Tagg = Rhododendron sanguineum Franch. var. didymoides Tagg et Forrest ●

330435   Rhododendron cloiophorum Balf. f. et Forrest var. leucopetalum (Balf. f. et Forrest) Davidian = Rhododendron sanguineum Franch. var. cloiophorum (Balf. f. et Forrest) D. F. Chamb. ●

330436   Rhododendron cloiophorum Balf. f. et Forrest var. mannophorum (Balf. f. et Forrest) Davidian = Rhododendron sanguineum Franch. var. didymoides Tagg et Forrest ●

330437   Rhododendron cloiophorum Balf. f. et Forrest var. roseotinctum (Balf. f. et Forrest) Davidian = Rhododendron sanguineum Franch. var. didymoides Tagg et Forrest ●

330438   Rhododendron coccinopeplum Balf. f. et Forrest = Rhododendron roxieanum Forrest ex W. W. Sm. var. cucullatum (Hand. -Mazz.) D. F. Chamb. ex Cullen et D. F. Chamb. ●

330439   Rhododendron codonanthum Balf. f. et Forrest；腺蕊杜鹃（黄花脉杜鹃，黄花脉柱杜鹃，黄花腺蕊杜鹃）；Gland Styles Yellow Rhododendron，Glandanther Azalea，Glandular-styled Yellow Rhododendron ●

330440   Rhododendron coelicum Balf. f. et Forrest；滇缅杜鹃（亮红杜鹃，天堂杜鹃）；Heavenly Rhododendron，Yunnan-Burma Azalea ●

330441   Rhododendron coeloneurum Diels；麻叶杜鹃（粗脉杜鹃，空脉杜鹃）；Coelomate Veins Rhododendron，Thickvein Azalea，Thick-veined Rhododendron ●

330442   Rhododendron coeruleum H. Lév.；堇蓝杜鹃（红基杜鹃，基毛杜鹃）；Blue Rhododendron，Stiff Rhododendron ●☆

330443   Rhododendron collettianum Aitch. et Hemsl.；柯来特杜鹃（阿富汗杜鹃）；H. Collett's Rhododendron ●☆

330444   Rhododendron colletum Balf. f. et Forrest = Rhododendron beesianum Diels ●

330445   Rhododendron comisteum Balf. f. et Forrest；砾石杜鹃（锈毛杜鹃）；Gravel Azalea，Rusty-haired Rhododendron，Rusty-hairs Rhododendron ●

330446   Rhododendron commodum Balf. f. et Forrest = Rhododendron sulphureum Franch. ●◇

330447   Rhododendron compactum Hutch.；紧密杜鹃；Compact Rhododendron ●☆

330448   Rhododendron compactum Hutch. = Rhododendron polycladum Franch. ●

330449   Rhododendron complexum Balf. f. et W. W. Sm.；锈红杜鹃（环绕杜鹃，交绕杜鹃）；Complex Rhododendron，Interwoven Rhododendron，Rusty Azalea ●

330450   Rhododendron concatenans Hutch. = Rhododendron xanthocodon Hutch. ●

330451   Rhododendron concinnoides Hutch. et Kingdon-Ward；似秀雅杜鹃；Like R. concinnum Rhododendron ●☆

330452   Rhododendron concinnum Hemsl.；秀雅杜鹃（臭枇杷，尖细杜鹃，枇杷杜鹃，优雅杜鹃，紫蕊杜鹃）；Excellent Azalea，Neat Rhododendron，Purplenymph Rhododendron，Purple-stamened Rhododendron，Sharp Apices Rhododendron ●

330453   Rhododendron concinnum Hemsl. f. laetevirens Cowan = Rhododendron concinnum Hemsl. ●

330454   Rhododendron concinnum Hemsl. var. benthamianum (Hemsl.) Davidian = Rhododendron concinnum Hemsl. ●

330455   Rhododendron concinnum Hemsl. var. lepidanthum (Rehder et E. H. Wilson) Rehder = Rhododendron concinnum Hemsl. ●

330456   Rhododendron concinnum Hemsl. var. pseudoyanthinum (Balf. f. ex Hutch.) Davidian = Rhododendron concinnum Hemsl. ●

330457   Rhododendron confertissimum Nakai；毛毡杜鹃 ●

330458   Rhododendron confertissimum Nakai = Rhododendron lapponicum (L.) Wahlenb. ●

330459   Rhododendron cookeanum Davidian = Rhododendron sikangense W. P. Fang ●◇

330460   Rhododendron coombense Hemsl. = Rhododendron concinnum Hemsl. ●

330461   Rhododendron cordatum H. Lév. = Rhododendron souliei Franch. ●

330462   Rhododendron coriaceum Franch.；革叶杜鹃；Leatherleaf Azalea，Leatherleaf Rhododendron，Leather-leaved Rhododendron ●

330463   Rhododendron coryanum Tagg et Forrest；光蕊杜鹃；Bared Stamens Rhododendron，Bare-stamened Rhododendron，Nakedstamen Azalea ●

330464   Rhododendron coryphaeum Balf. f. et Forrest = Rhododendron praestans Balf. f. et W. W. Sm. ●

330465　Rhododendron cosmetum Balf. f. et Forrest ＝ Rhododendron saluenense Franch. var. prostratum (W. W. Sm.) R. C. Fang ●

330466　Rhododendron costulatum Franch. ＝ Rhododendron lutescens Franch. ●

330467　Rhododendron cowanianum Davidian;尼泊尔杜鹃(郭万里杜鹃,毛缘叶杜鹃);Cowan's Rhododendron,Dr. Cowan's Rhododendron ●☆

330468　Rhododendron crassimedium P. C. Tam;棒柱杜鹃;Claviform-styled Rhododendron,Stickstyle Azalea,Thick Styles Rhododendron ●

330469　Rhododendron crassistylum M. Y. He;粗柱杜鹃;Thickstyle Azalea,Thick-styled Rhododendron ●

330470　Rhododendron crassum Franch. ＝ Rhododendron maddenii Hook. f. subsp. crassum (Franch.) Cullen ●

330471　Rhododendron cremastum Balf. f. et Forrest ＝ Rhododendron campylogynum Franch. ●

330472　Rhododendron cremnastes Balf. f. et Forrest ＝ Rhododendron lepidotum Wall. et G. Don ●

330473　Rhododendron cremnophilum Balf. f. et W. W. Sm. ＝ Rhododendron primuliflorum Bureau et Franch. ●

330474　Rhododendron crenatum H. Lév. ＝ Rhododendron racemosum Franch. ●

330475　Rhododendron crenulatum Hutch. ex Sleumer;细圆齿杜鹃;Crenulate Leaves Rhododendron ●☆

330476　Rhododendron cretaceum P. C. Tam;白枝杜鹃;WhitebranchAzalea, White-branched Rhododendron ●

330477　Rhododendron crinigerum Franch. ;长粗毛杜鹃;Bearing Hairs Rhododendron,Early-flushed Rhododendron,Longshag Azalea ●

330478　Rhododendron crinigerum Franch. var. euadnium Tagg et Forrest;腺背长粗毛杜鹃●

330479　Rhododendron croceum Balf. f. et W. W. Sm. ＝ Rhododendron wardii W. W. Sm. ●

330480　Rhododendron cruentum H. Lév. ＝ Rhododendron bureavii Franch. ●

330481　Rhododendron cubittii Hutch. ; 邱比特杜鹃; Cubitt Rhododendron ●☆

330482　Rhododendron cucullatum Hand. -Mazz. ＝ Rhododendron roxieanum Forrest ex W. W. Sm. var. cucullatum (Hand. -Mazz.) D. F. Chamb. ex Cullen et D. F. Chamb. ●

330483　Rhododendron cuffeanum Craib ex Hutch. ;库菲杜鹃;W. Cuffe's Rhododendron ●☆

330484　Rhododendron cumberlandense E. L. Braun;昆伯兰杜鹃;Cumberland Azalea ●☆

330485　Rhododendron cuneatum W. W. Sm. ;楔叶杜鹃(灰色杜鹃,岩生蔷薇杜鹃);Cliffrose Azalea, Cliffrose Rhododendron, Cuneateleaf Azalea, Grey Rhododendron, Wedge-leaved Rhododendron, Wedge-shaped Leaves Rhododendron ●

330486　Rhododendron cupressens Nitz. ＝ Rhododendron phaeochrysum Balf. f. et W. W. Sm. ●

330487　Rhododendron curvistylum Kingdon-Ward ＝ Rhododendron charitopes Balf. f. et Forrest subsp. tsangpoense (Kingdon-Ward) Cullen ●

330488　Rhododendron cyanocarpum (Franch.) Franch. et W. W. Sm. ; 蓝果杜鹃;Blue Fruits Rhododendron, Bluefruit Azalea, Bluepodded Rhododendron ●◇

330489　Rhododendron cyanocarpum (Franch.) Franch. et W. W. Sm. var. eriphyllum (Balf. f. et W. W. Sm.) Tagg ＝ Rhododendron cyanocarpum (Franch.) Franch. et W. W. Sm. ●◇

330490　Rhododendron cyclium Balf. f. et Forrest ＝ Rhododendron callimorphum Balf. f. et W. W. Sm. ●

330491　Rhododendron cymbomorphum Balf. f. et Forrest ＝ Rhododendron erythrocalyx Balf. f. et Forrest ●

330492　Rhododendron dabanshanense W. P. Fang et S. X. Wang ＝ Rhododendron przewalskii Maxim. ●

330493　Rhododendron dachengense G. Z. Li;大橙杜鹃;Dacheng Azalea,Dacheng Rhododendron ●

330494　Rhododendron dachengense G. Z. Li var. scopulum G. Z. Li;圣堂杜鹃;Shengtangshan Rhododendron ●

330495　Rhododendron dahuricum L. ;兴安杜鹃(达乌里杜鹃,达子香,鞑子香,东北满山红,红杜鹃,靠山红,满山红,山崩子,兴安杜鹃花,野杜鹃花,迎山红,映山红);Dahurian Rhododendron, Pink Azalea,Xing'an Azalea ●

330496　Rhododendron dahuricum L. f. albiflorum S. Watan. ＝ Rhododendron dauricum L. var. albiflorum Turcz. ●

330497　Rhododendron dahuricum L. var. albiflorum Turcz. ;白花兴安杜鹃●

330498　Rhododendron daiyuenshanicum P. C. Tam ＝ Rhododendron mariesii Hemsl. et E. H. Wilson ●

330499　Rhododendron daiyunicum P. C. Tam ＝ Rhododendron mariesii Hemsl. et E. H. Wilson ●

330500　Rhododendron daiyunshanicum P. C. Tam ＝ Rhododendron mariesii Hemsl. et E. H. Wilson ●

330501　Rhododendron dalhousiei Hook. f. ;长药杜鹃(棒杜鹃,红条杜鹃);Dalhousiae's Rhododendron, Longanther Azalea, Red Striped Rhododendron ●

330502　Rhododendron dalhousiae Hook. f. var. rhabdotum (Balf. f. et R. E. Cooper) Cullen;红线长药杜鹃●

330503　Rhododendron dalhousiae Hook. f. var. rhabdotum (Balf. f. et R. E. Cooper) Cullen ＝ Rhododendron dalhousiae Hook. f. ●

330504　Rhododendron damascenum Balf. f. et Forrest ＝ Rhododendron campylogynum Franch. ●

330505　Rhododendron danbaense L. C. Hu;丹巴杜鹃;Danba Azalea, Danba Rhododendron ●

330506　Rhododendron danielsianum Planch. ＝ Rhododendron indicum (L.) Sweet ●

330507　Rhododendron daphniflorum Diels ＝ Rhododendron rufescens Franch. ●

330508　Rhododendron dasycladoides Hand. -Mazz. ;漏斗杜鹃(似毛枝杜鹃);Filler Azalea,Hairy-twigged Rhododendron ●

330509　Rhododendron dasycladum Balf. f. et W. W. Sm. ＝ Rhododendron selense Franch. subsp. dasycladum (Balf. f. et W. W. Sm.) D. F. Chamb. ●

330510　Rhododendron dasypetalum Balf. f. et Forrest;毛瓣杜鹃(厚瓣杜鹃);Hairpetal Azalea, Hairy Petals Rhododendron, Hairy-petaled Rhododendron ●

330511　Rhododendron dauricum L. ;西伯利亚杜鹃(鞑子香)●☆

330512　Rhododendron dauricum L. subsp. mucronulatum (Turcz.) Vorosch. ＝ Rhododendron mucronulatum Turcz. ●

330513　Rhododendron dauricum L. var. albiflorum Turcz. ;白花西伯利亚杜鹃●☆

330514　Rhododendron dauricum L. var. mucronulatum (Turcz.) Maxim. ＝ Rhododendron mucronulatum Turcz. ●

330515　Rhododendron davidii Franch. ;腺果杜鹃;David Azalea, David's Rhododendron ●

330516　Rhododendron davidsonianum Rehder et E. H. Wilson;凹叶杜鹃(淑花杜鹃);Davidson Azalea, Davidson Rhododendron, Retuse

Rhododendron ●

330517　Rhododendron dawoense H. P. Yang；达沃杜鹃（道孚杜鹃）；Daofu Azalea，Dawo Rhododendron ●

330518　Rhododendron dayiense M. Y. He；大邑杜鹃；Dayi Azalea，Dayi Rhododendron ●

330519　Rhododendron decandrum（Makino）Makino = Rhododendron dilatatum Miq. var. decandrum Makino ●☆

330520　Rhododendron decandrum（Makino）Makino f. lasiocarpum H. Hara = Rhododendron dilatatum Miq. var. lasiocarpum Koidz. ex H. Hara ●☆

330521　Rhododendron decandrum（Makino）Makino var. albiflorum Honda = Rhododendron dilatatum Miq. var. decandrum Makino f. albiflorum（Honda）Hatus. ●☆

330522　Rhododendron decipiens Lacaita；飘渺杜鹃；Deceptive Rhododendron ●☆

330523　Rhododendron declivatum Ching et H. P. Yang；陡生杜鹃；SlopeAzalea，Steepy Rhododendron ●

330524　Rhododendron decorum Franch.；大白花杜鹃（白映山红，大白杜鹃，大白花，美丽杜鹃）；Ornamental Rhododendron，Sweetshell Azalea，Sweetshell Rhododendron，Sweet-shelled Rhododendron ●

330525　Rhododendron decorum Franch. subsp. cordatum W. K. Hu；心基大白杜鹃（心叶大白杜鹃）；Cordate Ornamental Rhododendron，Cordate Sweetshell Azalea ●

330526　Rhododendron decorum Franch. subsp. diaprepes（Balf. f. et W. W. Sm.）T. L. Ming；高尚大白杜鹃●

330527　Rhododendron decorum Franch. subsp. parvistigmaticum W. K. Hu；小头大白杜鹃（小柱大白杜鹃）；Small-headed Sweetshell Rhododendron ●

330528　Rhododendron decorum Franch. var. cordatum W. K. Hu = Rhododendron decorum Franch. ●

330529　Rhododendron decorum Franch. var. diaprepes（Balf. f. et W. W. Sm.）T. L. Ming = Rhododendron diaprepes Balf. f. et W. W. Sm. ●

330530　Rhododendron decumbens D. Don = Rhododendron indicum（L.）Sweet ●

330531　Rhododendron decumbens D. Don ex G. Don = Rhododendron indicum（L.）Sweet ●

330532　Rhododendron degronianum Carrière；五裂杜鹃；Degron's Rhododendron，Five-partite Rhododendron，M. Degron's Rhododendron ●☆

330533　Rhododendron degronianum Carrière = Rhododendron metternichii Siebold et Zucc. var. pentamerum T. Yamaz. ●☆

330534　Rhododendron degronianum Carrière f. album Sugim.；白花五裂杜鹃；Whitflower M. Degron's Rhododendron ●☆

330535　Rhododendron degronianum Carrière f. leucanthum（Makino）H. Hara；日本白花五裂杜鹃●☆

330536　Rhododendron degronianum Carrière f. nakai Hara；深红五裂杜鹃；Nakai M. Degron's Rhododendron ●☆

330537　Rhododendron degronianum Carrière f. variegatum Nakai；黄斑五裂杜鹃；Variegate M. Degron's Rhododendron ●☆

330538　Rhododendron degronianum Carrière subsp. heptamerum（Maxim.）H. Hara；七出杜鹃●☆

330539　Rhododendron degronianum Carrière subsp. heptamerum（Maxim.）H. Hara var. kyomaruense（T. Yamaz.）H. Hara f. amagianum（T. Yamaz.）H. Hara = Rhododendron degronianum Carrière var. amagianum（T. Yamaz.）T. Yamaz. ●☆

330540　Rhododendron degronianum Carrière subsp. heptamerum（Maxim.）H. Hara var. hondoense（Nakai）H. Hara = Rhododendron japonoheptamerum Kitam. var. hondoense（Nakai）Kitam. ●☆

330541　Rhododendron degronianum Carrière subsp. yakushimanum（Nakai）H. Hara var. intermedium（Sugim.）H. Hara = Rhododendron yakushimanum Nakai var. intermedium（Sugim.）T. Yamaz. ●☆

330542　Rhododendron degronianum Carrière subsp. yakushimanum（Nakai）H. Hara = Rhododendron yakushimanum Nakai ●☆

330543　Rhododendron degronianum Carrière var. amagianum（T. Yamaz.）T. Yamaz.；天城杜鹃●☆

330544　Rhododendron degronianum Carrière var. yakushimanum（Nakai）Kitam. = Rhododendron yakushimanum Nakai ●☆

330545　Rhododendron dekatanum Cowan；隆子杜鹃；Dekata Rhododendron，Longzi Azalea ●

330546　Rhododendron delavayi Franch.；马缨杜鹃（苍山杜鹃，杜鹃，狗血花，红杜鹃，红山茶，麻力光，马银花，马缨花，密筒花，蜜筒花，映山红）；Delavay Azalea，Delavay Rhododendron ●

330547　Rhododendron delavayi Franch. var. peramoenum（Balf. f. et Forrest）D. F. Chamb.；狭叶马缨杜鹃（狭叶马缨花，悦人杜鹃，窄叶马缨花）；Very Pleasing Rhododendron ●

330548　Rhododendron delavayi Franch. var. pilostylum K. M. Feng；毛柱马缨花；Hairy-style Delavay Rhododendron ●

330549　Rhododendron deleiense Hutch. et Kingdon-Ward = Rhododendron tephropeplum Balf. f. et Forrest ●

330550　Rhododendron dendricola Hutch.；附生杜鹃；Dweller on Trees Rhododendron，Epiphytic Azalea，Tree-attached Rhododendron ●

330551　Rhododendron dendritrichum Balf. f. et Forrest = Rhododendron uvariifolium Diels ●

330552　Rhododendron dendrocharis Franch.；树生杜鹃；Growing on Tree Rhododendron，Treelived Rhododendron，Tree-lived Rhododendron，Treephilous Azalea ●◇

330553　Rhododendron densifolium K. M. Feng；密叶杜鹃；Denseleaf Azalea，Denseleaf Rhododendron，Dense-leaved Rhododendron ●

330554　Rhododendron dentampullum Chun et P. C. Tam；齿萼杜鹃；Dentate Calyx Rhododendron，Dentate-calyxed Rhododendron，Toothcalyx Azalea ●

330555　Rhododendron dentampullum Chun et P. C. Tam = Rhododendron cavaleriei H. Lév. ●

330556　Rhododendron denudatum H. Lév.；皱叶杜鹃（裸露杜鹃）；Babbleleaf Azalea，Denudate Rhododendron，Wrinkledleaf Rhododendron ●

330557　Rhododendron depile Balf. f. et Forrest = Rhododendron oreotrephes W. W. Sm. ●

330558　Rhododendron desquamatum Balf. f. et Forrest = Rhododendron rubiginosum Franch. ●

330559　Rhododendron detersile Franch.；干净杜鹃（洁净杜鹃）；Clean Azalea，Clean Rhododendron ●

330560　Rhododendron detonsum Balf. f. et Forrest；落毛杜鹃（云南红杜鹃）；Downhair Azalea，Shorn Hairs Rhododendron，Yunnan Rose Rhododendron ●

330561　Rhododendron diacritum Balf. f. et W. W. Sm. = Rhododendron telmateium Balf. f. et W. W. Sm. ●

330562　Rhododendron diaprepes Balf. f. et W. W. Sm.；高尚杜鹃（高尚大白杜鹃）；Dietinguished Rhododendron ●

330563　Rhododendron diaprepes Balf. f. et W. W. Sm. = Rhododendron decorum Franch. var. diaprepes（Balf. f. et W. W. Sm.）T. L. Ming ●

330564 Rhododendron diaprepes Balf. f. et W. W. Sm. = Rhododendron decorum Franch. subsp. diaprepes (Balf. f. et W. W. Sm.) T. L. Ming ●

330565 Rhododendron dichroanthum Diels et Cowan;两色杜鹃;Bicolor Rhododendron,Twocolor Azalea,Two-colored Rhododendron ●

330566 Rhododendron dichroanthum Diels et Cowan subsp. apodectum (Balf. f. et W. W. Sm.) Cowan;可喜杜鹃;Acceptable Rhododendron ●

330567 Rhododendron dichroanthum Diels et Cowan subsp. herpesticum (Balf. f. et Kingdon-Ward) Cowan = Rhododendron dichroanthum Diels et Cowan subsp. scyphocalyx (Balf. f. et Forrest) Cowan ●

330568 Rhododendron dichroanthum Diels et Cowan subsp. scyphocalyx (Balf. f. et Forrest) Cowan;杯萼两色杜鹃(杯萼杜鹃);Cap-shaped Rhododendron,Cup-calyx Two-colored Rhododendron ●

330569 Rhododendron dichroanthum Diels et Cowan subsp. septentrionale Cowan;腺梗两色杜鹃●

330570 Rhododendron dichroanthum Diels et Cowan var. apodectum (Balf. f. et W. W. Sm.) T. L. Ming = Rhododendron dichroanthum Diels et Cowan subsp. apodectum (Balf. f. et W. W. Sm.) Cowan ●

330571 Rhododendron dichroanthum Diels et Cowan var. scyphocalyx (Balf. f. et Forrest) T. L. Ming = Rhododendron dichroanthum Diels et Cowan subsp. scyphocalyx (Balf. f. et Forrest) Cowan ●

330572 Rhododendron dichroanthum Diels et Cowan var. septentHonale (Cowan) T. L. Ming = Rhododendron dichroanthum Diels et Cowan subsp. septentrionale Cowan ●

330573 Rhododendron dichropeplum Balf. f. et Forrest = Rhododendron phaeochrysum Balf. f. et W. W. Sm. var. levistratum (Balf. f. et Forrest) D. F. Chamb. ●

330574 Rhododendron dictyotum Balf. f. ex Tagg = Rhododendron traillanum Forrest et W. W. Sm. var. dictyotum (Balf. f. ex Tagg) D. F. Chamb. ex Cullen et D. F. Chamb. ●

330575 Rhododendron didymum Balf. f. et Forrest = Rhododendron sanguineum Franch. var. didymum (Balf. f. et Forrest) T. L. Ming ●

330576 Rhododendron dignabile Cowan;疏毛杜鹃(脱毛杜鹃);Nonscales Rhododendron,Scaleless Rhododendron,Scatterhair Azalea ●

330577 Rhododendron dilatatum Miq.;扩展杜鹃●☆

330578 Rhododendron dilatatum Miq. f. hypopilosum Sa. Kurata;背毛扩展杜鹃●☆

330579 Rhododendron dilatatum Miq. f. leucanthum Sugim.;白花扩展杜鹃●☆

330580 Rhododendron dilatatum Miq. var. boreale Sugim.;北方扩展杜鹃●☆

330581 Rhododendron dilatatum Miq. var. decandrum Makino;十雄扩展杜鹃●☆

330582 Rhododendron dilatatum Miq. var. decandrum Makino f. albiflorum (Honda) Hatus.;白花十雄扩展杜鹃●☆

330583 Rhododendron dilatatum Miq. var. decandrum Makino f. lasiocarpum (H. Hara) Hatus. = Rhododendron dilatatum Miq. var. lasiocarpum Koidz. ex H. Hara ●☆

330584 Rhododendron dilatatum Miq. var. glaucum Hatus. = Rhododendron osuzuyamense T. Yamaz. ●☆

330585 Rhododendron dilatatum Miq. var. lasiocarpum (H. Hara) T. Yamaz. = Rhododendron dilatatum Miq. var. lasiocarpum Koidz. ex H. Hara ●☆

330586 Rhododendron dilatatum Miq. var. lasiocarpum Koidz. ex H. Hara;毛果扩展杜鹃●☆

330587 Rhododendron dilatatum Miq. var. pilosum H. Hara =

330588 Rhododendron dilatatum Miq. var. decandrum Makino ●☆

330588 Rhododendron dilatatum Miq. var. satsumense T. Yamaz.;萨摩扩展杜鹃●☆

330589 Rhododendron dilatatum Miq. var. viscistylum (Nakai) Hatus. = Rhododendron viscistylum Nakai ●☆

330590 Rhododendron dimitrium Balf. f. et Forrest;重套杜鹃(苍山杜鹃,大萼杜鹃);Cangshan Azalea,Cangshan Rhododendron,Large Calyx Rhododendron,Large-calyxed Rhododendron ●

330591 Rhododendron diphrocalyx Balf. f.;腾冲杜鹃(长萼杜鹃,宽萼杜鹃);Tengchong Azalea,Tengchong Rhododendron,Wide Calyx Rhododendron,Wide-calyxed Rhododendron ●

330592 Rhododendron discolor Franch.;喇叭杜鹃(广福杜鹃,豪氏杜鹃,马缨花,小叶野枇杷,异色杜鹃);Bugle Azalea,G. Houlston's Rhododendron,G. Houlston's Rhododendron Azalea,Guangfu Azalea,Guangfu Rhododendron,Kwangfu Rhododendron,Mandarin Rhododendron,Various Colours Rhododendron ●

330593 Rhododendron discolor Franch. = Rhododendron decorum Franch. var. diaprepes (Balf. f. et W. W. Sm.) T. L. Ming ●

330594 Rhododendron dityotum Balf. f. et Tagg;网脉杜鹃;Ned-veins Rhododendron ●☆

330595 Rhododendron diversipilosum (Nakai) Harmaja = Ledum palustre L. subsp. diversipilosum (Nakai) H. Hara ●

330596 Rhododendron dolerum Balf. f. et Forrest = Rhododendron selense Franch. subsp. dasycladum (Balf. f. et W. W. Sm.) D. F. Chamb. ●

330597 Rhododendron doshongense Tagg;得胜杜鹃(墨脱杜鹃);Dosheng La Rhododendron ●

330598 Rhododendron doshongense Tagg = Rhododendron aganniphum Balf. f. et Kingdon-Ward var. schizopeplum (Balf. f. et Forrest) T. L. Ming ●

330599 Rhododendron drumonium Balf. f. et Kingdon-Ward = Rhododendron telmateium Balf. f. et W. W. Sm. ●

330600 Rhododendron dryophyllum Balf. f. et Forrest = Rhododendron phaeochrysum Balf. f. et W. W. Sm. ●

330601 Rhododendron dryophyllum Balf. f. et Forrest = Rhododendron sigillatum Balf. f. et Forrest ●

330602 Rhododendron duclouxii H. Lév.;粉红爆杖花(昆明杜鹃);Kunming Rhododendron,Pink Firecracker-flower Azalea ●

330603 Rhododendron dumicola Tagg et Forrest;灌丛杜鹃;Bushy Rhododendron,Dweller in Thicketts Rhododendron,Shrub Azalea ●

330604 Rhododendron dumosulum Balf. f. et Forrest = Rhododendron phaeochrysum Balf. f. et W. W. Sm. ●

330605 Rhododendron dumulosum Balf. f. et Forrest = Rhododendron phaeochrysum Balf. f. et W. W. Sm. var. agglutinatum (Balf. f. et Forrest) D. F. Chamb. ●

330606 Rhododendron dunnii E. H. Wilson = Rhododendron henryi Hance var. dunnii (E. H. Wilson) M. Y. He ●

330607 Rhododendron duseimatum Balf. f. et Forrest = Rhododendron calvescens Balf. f. et Forrest var. duseimatum (Balf. f. et Forrest) D. F. Chamb. ●

330608 Rhododendron ebianense M. Y. Fang;峨边杜鹃;Ebian Azalea,Ebian Rhododendron ●

330609 Rhododendron eclecteum Balf. f. et Forrest;杂色杜鹃;Diversicolored Rhododendron,Muchcolor Rhododendron,Shosen Out Rhododendron,Variegate Azalea ●

330610 Rhododendron eclecteum Balf. f. et Forrest var. bellatulum Balf. f. ex Tagg;长柄杂色杜鹃;Longstalk Variegate Azalea,Long-stalked

Diversicolored Rhododendron ●

330611　Rhododendron eclecteum Balf. f. et Forrest var. brachyandrum（Balf. f. et Forrest）Coean = Rhododendron eclecteum Balf. f. et Forrest ●

330612　Rhododendron edgarianum Rehder et E. H. Wilson;埃氏杜鹃(埃德加杜鹃);J. H. Edgar Azalea,J. H. Edgar's Rhododendron ●☆

330613　Rhododendron edgeworthii Hook. f.;泡泡叶杜鹃(密花香杜鹃,皱叶杜鹃);Bubbleleaf Azalea, Edgeworth's Rhododendron, M. P. Edgeworth's Rhododendron,Ruckerleaf Rhododendron ●

330614　Rhododendron elaeagnoides Hook. f. = Rhododendron lepidotum Wall. et G. Don ●

330615　Rhododendron elegantulum Tagg et Forrest;金江杜鹃(雅洁杜鹃);Elegant Rhododendron,Jinjiang Azalea ●

330616　Rhododendron elliottii Watt ex Brandis;印度红杜鹃;Mr. Elliott's Rhododendron ●☆

330617　Rhododendron ellipticum Maxim.;西施花(大屯杜鹃,光柄杜鹃,光脚杜鹃,青紫花,青紫木,台长叶杜鹃,台湾长叶杜鹃,台湾西施花,椭圆叶杜鹃);Bared Foot Rhododendron, Elliptic Leaves Rhododendron,Taiwan Rhododendron,Xishi Azalea ●

330618　Rhododendron ellipticum Maxim. = Rhododendron latoucheae Franch. ●

330619　Rhododendron ellipticum Maxim. var. leptosanthum（Hayata）S. S. Ying = Rhododendron ellipticum Maxim. ●

330620　Rhododendron ellipticum Maxim. var. leptosanthum（Hayata）S. S. Ying = Rhododendron latoucheae Franch. ●

330621　Rhododendron emaculatum Balf. f. et Forrest = Rhododendron beesianum Diels ●

330622　Rhododendron emarginatum Hemsl. et E. H. Wilson;缺顶杜鹃（匍枝杜鹃,卫矛叶杜鹃）;Emarginate Azalea, Emarginate Rhododendron, Euonymusleaf Azalea, Euonymusleaf Rhododendron, Euonymus-leaved Rhododendron, Like Euonymus Leaves Rhododendron ●

330623　Rhododendron emarginatum Hemsl. et E. H. Wilson var. erioacarpum K. M. Feng;毛果缺顶杜鹃;Hairy-fruit Emarginate Azalea,Hairy-fruit Emarginate Rhododendron ●

330624　Rhododendron epapillatum Balf. f. et R. E. Cooper = Rhododendron papillatum Balf. f. et R. E. Cooper ●

330625　Rhododendron epipastum Balf. f. et Forrest = Rhododendron eudoxum Balf. f. et Forrest var. mesopolium（Balf. f. et Forrest）D. F. Chamb. ●

330626　Rhododendron erastum Balf. f. et Forrest;匍匐杜鹃(心爱杜鹃);Beloved Rhododendron,Creeping Azalea,Lovely Rhododendron ●

330627　Rhododendron eriandrum H. Lév. ex Hutch. = Rhododendron rigidum Franch. ●

330628　Rhododendron erileucum Balf. f. et Forrest = Rhododendron zaleucum Balf. f. et W. W. Sm. ●

330629　Rhododendron eriocarpum（Hayata）Nakai;毛果皋月杜鹃 ●☆

330630　Rhododendron eriocarpum（Hayata）Nakai var. tawadae Ohwi;田和代氏杜鹃 ●☆

330631　Rhododendron eriogynum Balf. f. et W. W. Sm. = Rhododendron facetum Balf. f. et Kingdon-Ward ●

330632　Rhododendron eriphydum Balf. f. et Forrest = Rhododendron cyanocarpum（Franch.）Franch. et W. W. Sm. ●◇

330633　Rhododendron eritimum Balf. f. et W. W. Sm. = Rhododendron anthosphaerum Diels ●

330634　Rhododendron eritimum Balf. f. et W. W. Sm. subsp. chawchiense（Balf. f. et Forrest）Tagg = Rhododendron anthosphaerum Diels ●

330635　Rhododendron eritimum Balf. f. et W. W. Sm. subsp. gymnogynum（Balf. f. et Forrest）Tagg = Rhododendron anthosphaerum Diels ●

330636　Rhododendron eritimum Balf. f. et W. W. Sm. subsp. heptamerum（Balf. f.）Tagg = Rhododendron anthosphaerum Diels ●

330637　Rhododendron eritimum Balf. f. et W. W. Sm. subsp. persicinum（Hand. -Mazz.）Tagg = Rhododendron anthosphaerum Diels ●

330638　Rhododendron erosum Cowan;啮蚀杜鹃(霸地杜鹃,侵蚀杜鹃);Eaten Away Rhododendron,Erose Azalea,Erose Rhododendron ●

330639　Rhododendron erubescens Hutch. = Rhododendron oreodoxa Franch. var. fargesii（Franch.）D. F. Chamb. ●

330640　Rhododendron erythrocalyx Balf. f. et Forrest;显萼杜鹃(赤萼杜鹃,红萼杜鹃);Red Calyx Rhododendron,Redcalyx Azalea,Red-calyxed Rhododendron ●

330641　Rhododendron erythrocalyx Balf. f. et Forrest subsp. beimaense（Balf. f. et Forrest）Tagg = Rhododendron erythrocalyx Balf. f. et Forrest ●

330642　Rhododendron erythrocalyx Balf. f. et Forrest subsp. docimum Balf. f. ex Tagg = Rhododendron erythrocalyx Balf. f. et Forrest ●

330643　Rhododendron erythrocalyx Balf. f. et Forrest subsp. eucallum（Balf. f. et Forrest）Tagg = Rhododendron erythrocalyx Balf. f. et Forrest ●

330644　Rhododendron esetulosum Balf. f. et Forrest;喙尖杜鹃(薄被杜鹃,少毛杜鹃,无毛杜鹃);Bill Azalea, Hairless Rhododendron, Loose Covering Rhododendron ●

330645　Rhododendron esquirolii H. Lév.;滇黔杜鹃(滇北长果杜鹃);North Yunnan Long Fruit Rhododendron ●

330646　Rhododendron esquirolii H. Lév. = Rhododendron stamineum Franch. ●

330647　Rhododendron euanthum Balf. f. et W. W. Sm. = Rhododendron vernicosum Franch. ●

330648　Rhododendron eucallum Balf. f. et Forrest = Rhododendron erythrocalyx Balf. f. et Forrest ●

330649　Rhododendron euchaites Balf. f. et Forrest = Rhododendron neriiflorum Franch. ●

330650　Rhododendron euchroum Balf. f. et Kingdon-Ward;滇西杜鹃(滇缅杜鹃);Beautiful Colored Rhododendron, Beautiful-colored Rhododendron,W. Yunnan Azalea ●

330651　Rhododendron eudoxum Balf. f. et Forrest;华丽杜鹃(喜报杜鹃,真红杜鹃);Braveness Azalea, Good Report Rhododendron, Gorgeous Rhododendron ●

330652　Rhododendron eudoxum Balf. f. et Forrest subsp. asteium（Balf. f. et Forrest）Tagg = Rhododendron eudoxum Balf. f. et Forrest var. mesopolium（Balf. f. et Forrest）D. F. Chamb. ●

330653　Rhododendron eudoxum Balf. f. et Forrest subsp. bruneifolium（Balf. f. et Forrest）Tagg = Rhododendron eudoxum Balf. f. et Forrest var. bruneifolium（Balf. f. et Forrest）D. F. Chamb. ●

330654　Rhododendron eudoxum Balf. f. et Forrest subsp. epipastum（Balf. f. et Forrest）Tagg = Rhododendron eudoxum Balf. f. et Forrest var. mesopolium（Balf. f. et Forrest）D. F. Chamb. ●

330655　Rhododendron eudoxum Balf. f. et Forrest subsp. glaphyrum（Balf. f. et Forrest）Tagg = Rhododendron temenium Balf. f. et Forrest var. dealbatum（Cowan）D. F. Chamb. ●

330656　Rhododendron eudoxum Balf. f. et Forrest subsp. mesopolium（Balf. f. et Forrest）Tagg = Rhododendron eudoxum Balf. f. et Forrest var. mesopolium（Balf. f. et Forrest）D. F. Chamb. ●

330657　Rhododendron eudoxum Balf. f. et Forrest subsp. pothinum（Balf. f. et Forrest）Tagg = Rhododendron temenium Balf. f. et Forrest ●

330658　Rhododendron eudoxum Balf. f. et Forrest subsp. temenium（Balf. f. et Forrest）Tagg = Rhododendron temenium Balf. f. et Forrest ●

330659　Rhododendron eudoxum Balf. f. et Forrest subsp. trichomiscum（Balf. f. et Forrest）Tagg = Rhododendron eudoxum Balf. f. et Forrest var. bruneifolium（Balf. f. et Forrest）D. F. Chamb. ●

330660　Rhododendron eudoxum Balf. f. et Forrest subsp. trichomiscum（Balf. f. et Forrest）Tagg = Rhododendron eudoxum Balf. f. et Forrest ●

330661　Rhododendron eudoxum Balf. f. et Forrest var. bruneifolium（Balf. f. et Forrest）D. F. Chamb.；褐叶华丽杜鹃（褐叶杜鹃）；Brown-leaf Good Report Rhododendron，Brown-leaves Rhododendron ●

330662　Rhododendron eudoxum Balf. f. et Forrest var. mesopolium（Balf. f. et Forrest）D. F. Chamb.；白毛华丽杜鹃；White Hair Good Report Rhododendron ●

330663　Rhododendron euonymifolium H. Lév. = Rhododendron emarginatum Hemsl. et E. H. Wilson ●

330664　Rhododendron eurysiphon Tagg et Forrest；宽筒杜鹃（宽管杜鹃）；Broad Tube Rhododendron，Broadtube Azalea，Broad-tubed Rhododendron ●

330665　Rhododendron exasperatum Tagg；粗糙叶杜鹃（粗糙杜鹃，美叶杜鹃）；Rough Azalea，Rough Rhododendron ●

330666　Rhododendron excellens Hemsl. et E. H. Wilson；大喇叭杜鹃；Big Trumpet Rhododendron，Excellent Rhododendron，Largebugel Azalea ●

330667　Rhododendron excelsum A. Chev.；高大杜鹃（越南杜鹃）；Vietnam，Vietnam Taller Rhododendron ●☆

330668　Rhododendron eximium Nutt.；最优杜鹃（超级大叶杜鹃，超级杜鹃）；Exvellent Rhododendron ●☆

330669　Rhododendron exquisitum Hutch. = Rhododendron oreotrephes W. W. Sm. ●

330670　Rhododendron exquisitum Hutch. = Rhododendron sikangense W. P. Fang var. exquistum（T. L. Ming）T. L. Ming ●

330671　Rhododendron exquisitum T. L. Ming = Rhododendron sikangense W. P. Fang var. exquistum（T. L. Ming）T. L. Ming ●

330672　Rhododendron faberi Hemsl.；金顶杜鹃（费氏杜鹃，拟金顶杜鹃）；Faber Rhododendron，Jinding Azalea，Like R. Faber's Rhododendron ●◇

330673　Rhododendron faberi Hemsl. subsp. prattii（Franch.）D. F. Chamb. ex Cullen et D. F. Chamb.；大叶金顶杜鹃（康定杜鹃，李民杜鹃，普拉杜鹃，紫腺杜鹃）；Pratt's Rhododendron，Purple Glands Azalea，Purple Glands Ovary Rhododendron ●

330674　Rhododendron faberioides Balf. f. = Rhododendron faberi Hemsl. ●◇

330675　Rhododendron facetum Balf. f. et Kingdon-Ward；绵毛房杜鹃（长毛子房杜鹃，文雅杜鹃）；Dwarf Indica Azalea，Elegant Rhododendron，Lanoseovary Azalea，Woolly Ovary Rhododendron ●

330676　Rhododendron faithae Chun；大云锦杜鹃（信宜杜鹃）；Big Clouds Azalea，Bigger R. Fourtunei Rhododendron，Faith Rhododendron ●

330677　Rhododendron falcinellum（P. C. Tam）P. C. Tam = Rhododendron apricum P. C. Tam ●

330678　Rhododendron falcinellum P. C. Tam；镰叶杜鹃；Falcate-leaved Rhododendron，Sickle-shap Azalea，Sickle-shap Leaves Rhododendron ●

330679　Rhododendron falcinellum P. C. Tam = Rhododendron rufulum P. C. Tam ●

330680　Rhododendron falconeri Hook. f.；大叶杜鹃（杯毛杜鹃，福坎杜鹃）；Falconer's Rhododendron，H. Falconer's Rhododendron ●

330681　Rhododendron falconeri Hook. f. subsp. eximium（Nutt.）D. F. Chamb. = Rhododendron eximium Nutt. ●☆

330682　Rhododendron fangchengense P. C. Tam；防城杜鹃；Fangcheng Azalea，Fangcheng Rhododendron ●

330683　Rhododendron fangchengense P. C. Tam var. minor P. C. Tam；小南边杜鹃；Small Fangcheng Rhododendron ●

330684　Rhododendron fangchengense P. C. Tam var. setistylum P. C. Tam；糙叶防城杜鹃；Rough Fangcheng Rhododendron ●

330685　Rhododendron fargesii Franch. = Rhododendron oreodoxa Franch. var. fargesii（Franch.）D. F. Chamb. ●

330686　Rhododendron farinosum H. Lév.；钝头杜鹃（苍白杜鹃）；Farinose Azalea，Mealy Rhododendron ●

330687　Rhododendron farrerae Tate ex Sweet；丁香杜鹃（法耳杜鹃，华丽杜鹃）；Clove Azalea，Farrer Rhododendron，Mrs. Farrer's Rhododendron ●

330688　Rhododendron farrerae Tate ex Sweet var. mediocre Diels = Rhododendron mariesii Hemsl. et E. H. Wilson ●

330689　Rhododendron farrerae Tate ex Sweet var. typicum Diels = Rhododendron farrerae Tate ex Sweet ●

330690　Rhododendron farrerae Tate ex Sweet var. weyrichii Diels = Rhododendron mariesii Hemsl. et E. H. Wilson ●

330691　Rhododendron fastigiatum Franch.；密枝杜鹃（小枇杷）；Autumnpurple Rhododendron，Autumn-purpled Rhododendron，Densebranch Azalea，Erect Rhododendron ●

330692　Rhododendron faucium Cowan；喉斑杜鹃（猴斑杜鹃）；Monkeyblotch Azalea，Throat Spot Rhododendron，Throat-spoted Rhododendron ●

330693　Rhododendron fauriei Franch. = Rhododendron brachycarpum D. Don ex G. Don f. fauriei（Franch.）Murata ●☆

330694　Rhododendron fauriei Franch. = Rhododendron brachycarpum D. Don ex G. Don ●☆

330695　Rhododendron ferrugineum L.；高山玫瑰杜鹃（矮杜鹃，锈色杜鹃）；Alpenrose，Alpine Rose，Rock Rhododendron，Rusty-colored Rhododendron ●☆

330696　Rhododendron ferrugineum L. 'Album Coccineum'；深红高山玫瑰杜鹃●☆

330697　Rhododendron ferrugineum L. 'Glenarn'；格勒纳高山玫瑰杜鹃●☆

330698　Rhododendron fictolacteum Balf. f. = Rhododendron rex H. Lév. subsp. fictolacteum（Balf. f.）D. F. Chamb. ●◇

330699　Rhododendron fictolacteum Balf. f. var. miniforme Davidian = Rhododendron rex H. Lév. subsp. fictolacteum（Balf. f.）D. F. Chamb. ●◇

330700　Rhododendron fimbriatum Hutch. = Rhododendron hippophaeroides Balf. f. et W. W. Sm. ●

330701　Rhododendron fissotectum Balf. f. et Forrest = Rhododendron aganniphum Balf. f. et Kingdon-Ward var. schizopeplum（Balf. f. et Forrest）T. L. Ming ●

330702　Rhododendron flammeum（Michx.）Sarg.；火焰黄杜鹃；Oconee Azalea ●☆

330703　Rhododendron flavantherum Hutch. ex Kingdon-Ward；黄药杜鹃；Yellow Anthers Rhododendron，Yellowanther Azalea，Yellowanthered Rhododendron ●

330704　Rhododendron flavidum Franch. ;川西淡黄杜鹃(淡黄杜鹃,琥珀色杜鹃); Amberbloom Rhododendron, Lightyellow Azalea, Somewhat Yellow Rhododendron,Yellow-flowered Rhododendron ●

330705　Rhododendron flavidum Franch. var. psilostylum Rehder et E. H. Wilson;光柱淡黄杜鹃●

330706　Rhododendron flavoflorum T. L. Ming;淡黄杜鹃(淡黄花杜鹃,黄杜鹃,黄花杜鹃,泸水杜鹃); Yellow Azalea, Yellow Flowered Rhododendron ●

330707　Rhododendron flavorufum Balf. f. et Forrest = Rhododendron aganniphum Balf. f. et Kingdon-Ward var. flavorufum (Balf. f. et Forrest) D. F. Chamb. ●

330708　Rhododendron fletcherianum Davidian;翅柄杜鹃;Fletcher's Rhododendron,H. R. Fletcher's Rhododendron,Wingstalk Azalea ●

330709　Rhododendron flinckii Davidian = Rhododendron lanatum Hook. f. ●

330710　Rhododendron floccigerum Franch. ;绵毛杜鹃(深红杜鹃); Cottony Azalea,Woolly Rhododendron ●

330711　Rhododendron floccigerum Franch. subsp. appropinquans (Tagg et Forrest) Chamb. ex Cullen et Chamb. = Rhododendron neriiflorum Franch. var. appropinquans (Tagg et Forrest) W. K. Hu ●

330712　Rhododendron floccigerum Franch. subsp. appropinquans (Tagg et Forrest) D. F. Chamb. = Rhododendron neriiflorum Franch. var. appropinquans (Tagg et Forrest) W. K. Hu ●

330713　Rhododendron floccigerum Franch. var. appropinquans Tagg et Forrest = Rhododendron neriiflorum Franch. var. appropinquans (Tagg et Forrest) W. K. Hu ●

330714　Rhododendron floribundum Franch. ;繁花杜鹃(多花杜鹃); Free Flowering Rhododendron, Purplequeen Azalea, Purplequeen Rhododendron ●

330715　Rhododendron floribundum Franch. ' Swinhoe';斯维豪繁花杜鹃●☆

330716　Rhododendron florulentum P. C. Tam;龙岩杜鹃(小花杜鹃); Small-flowered Rhododendron ●

330717　Rhododendron flosculum W. P. Fang et G. Z. Li;子花杜鹃(小花杜鹃); Floweret Azalea, Floweret Rhododendron, Small Flowers Rhododendron ●

330718　Rhododendron flumineum W. P. Fang et M. Y. He;河边杜鹃; River Side Azalea,Riverside Rhododendron ●

330719　Rhododendron fokienense Franch. = Rhododendron simiarum Hance ●

330720　Rhododendron fongkaiense C. N. Wu et P. C. Tam = Rhododendron kwangtungense Merr. et Chun ●

330721　Rhododendron fordii Hemsl. = Rhododendron simiarum Hance ●

330722　Rhododendron formosanum Hemsl. ;台湾杜鹃(红条杜鹃,天目杜鹃,云锦杜鹃); Red-stripe Rhododendron, Red-striped Rhododendron,Taiwan Azalea,Taiwan Rhododendron ●

330723　Rhododendron formosum Wall. ;美丽杜鹃(红条杜鹃); Beautiful Rhododendron ●

330724　Rhododendron forrestii Balf. f. ex Diels;紫背杜鹃(匍匐杜鹃); Creeping Azalea,Creeping Rhododendron,Forrest Rhododendron,G. Forrest's Rhododendron,Purpleback Azalea ●

330725　Rhododendron forrestii Balf. f. ex Diels subsp. papillatum Cowan;乳突紫背杜鹃;Papillate Forrest Rhododendron ●

330726　Rhododendron forrestii Balf. f. ex Diels var. repens (Balf. f. et Forrest) Cowan et Davidian = Rhododendron forrestii Balf. f. ex Diels ●

330727　Rhododendron fortunei Lindl. ;云锦杜鹃(山枇杷,天目杜鹃,

野枇杷,云南杜鹃);Chinese Rhododendron,Clouds Azalea,Fortune Rhododendron ●

330728　Rhododendron fortunei Lindl. subsp. discolor (Franch.) D. F. Chamb. = Rhododendron discolor Franch. ●

330729　Rhododendron fortunei Lindl. subsp. discolor (Franch.) D. F. Chamb. ex Cullen et D. F. Chamb. = Rhododendron discolor Franch. ●

330730　Rhododendron fortunei Lindl. var. houlstonii (Hemsl. et E. H. Wilson) Rehder et E. H. Wilson = Rhododendron discolor Franch. ●

330731　Rhododendron fortunei Lindl. var. houlstonii Rehder et E. H. Wilson = Rhododendron discolor Franch. ●

330732　Rhododendron fortunei Lindl. var. kwangfuense (Chun et W. P. Fang) G. Z. Li = Rhododendron kwangfuense Chun et W. P. Fang ●

330733　Rhododendron fortunei Lindl. var. kwangfuense (Chun et W. P. Fang) G. Z. Li = Rhododendron discolor Franch. ●

330734　Rhododendron foveolatum Rehder et E. H. Wilson = Rhododendron coriaceum Franch. ●

330735　Rhododendron frachetianum H. Lév. = Rhododendron decorum Franch. ●

330736　Rhododendron fragariflorum Kingdon-Ward;草莓花杜鹃; Strawberryflower Azalea,Strawberry-flowered Rhododendron ●

330737　Rhododendron fragrans sensu Franch. = Rhododendron trichostomum Franch. ●

330738　Rhododendron franchetianum H. Lév. = Rhododendron decorum Franch. ●

330739　Rhododendron fuchsiiflorum H. Lév. = Rhododendron spinuliferum Franch. ●

330740　Rhododendron fuchsiifolium H. Lév. ;贵定杜鹃(倒挂金钟花杜鹃); Fuchsia Flowered Rhododendron, Fuchsia-flowered Rhododendron,Guiding Azalea ●

330741　Rhododendron fulgens Hook. f. ;猩红杜鹃(光亮杜鹃,亮叶杜鹃); Pompom Rhododendron, Scarlet Azalea, Shining-leaved Rhododendron ●

330742　Rhododendron fulvastrum Balf. f. et Forrest;茶色杜鹃(黄褐杜鹃);Somewhat Tawny Rhododendron ●☆

330743　Rhododendron fulvastrum Balf. f. et Forrest subsp. epipastum (Balf. f. et Forrest) Cowan = Rhododendron eudoxum Balf. f. et Forrest var. mesopolium (Balf. f. et Forrest) D. F. Chamb. ●

330744　Rhododendron fulvastrum Balf. f. et Forrest subsp. mesopolium (Balf. f. et Forrest) Cowan = Rhododendron eudoxum Balf. f. et Forrest var. mesopolium (Balf. f. et Forrest) D. F. Chamb. ●

330745　Rhododendron fulvastrum Balf. f. et Forrest subsp. trichomiscum (Balf. f. et Forrest) Cowan = Rhododendron eudoxum Balf. f. et Forrest ●

330746　Rhododendron fulvastrum Balf. f. et Forrest var. gilvum (Cowan) Davidian = Rhododendron temenium Balf. f. et Forrest var. gilvutum (Cowan) D. F. Chamb. ●

330747　Rhododendron fulvoides Balf. f. et Forrest = Rhododendron fulvum Balf. f. et W. W. Sm. ●

330748　Rhododendron fulvoides Balf. f. et Forrest = Rhododendron fulvum Balf. f. et W. W. Sm. subsp. fulvoides (Balf. f. et Forrest) D. F. Chamb. ●

330749　Rhododendron fulvum Balf. f. et W. W. Sm. ;镰果杜鹃(肉桂色杜鹃);Cinnamon Rhododendron, Sicklefruit Azalea, Twony-flowered Rhododendron ●

330750　Rhododendron fulvum Balf. f. et W. W. Sm. subsp. fulvoides (Balf. f. et Forrest) D. F. Chamb. ;棕叶镰果杜鹃●

330751　Rhododendron fumidum Balf. f. et W. W. Sm. = Rhododendron heliolepis Franch. var. fumidum（Balf. f. et W. W. Sm.）R. C. Fang ●

330752　Rhododendron fuscipilum M. Y. He；棕毛杜鹃；Brown-haired Rhododendron，Darkbrown-hairy Rhododendron，Palmhair Azalea ●

330753　Rhododendron fuyuanense Z. H. Yang；富源杜鹃；Fuyuan Azalea，Fuyuan Rhododendron ●

330754　Rhododendron galactinum Balf. f. ex Tagg；乳黄叶杜鹃（乳白杜鹃，乳黄叶背杜鹃）；Milk Rhododendron，Milkleaf Azalea，Milky Rhododendron ●

330755　Rhododendron gannanense Z. C. Feng et X. G. Sun；甘南杜鹃；Gannan Azalea，Gannan Rhododendron ●

330756　Rhododendron gemmiferum M. N. Philipson et Philipson；大芽杜鹃；Big-budded Rhododendron，Large Buds Rhododendron，Largebud Azalea ●

330757　Rhododendron genestierianum Forrest；灰白杜鹃；Greywhite Rhododendron，Grey-white Rhododendron，Pale Azalea，Pere A. Gregestier's Rhododendron ●

330758　Rhododendron giganteum Forrest ＝Rhododendron protistum Balf. f. et Forrest var. giganteum（Forrest ex Tagg）D. F. Chamb. ●◇

330759　Rhododendron giganteum Forrest ex Tagg ＝Rhododendron protistum Balf. f. et Forrest var. giganteum（Forrest ex Tagg）D. F. Chamb. ●◇

330760　Rhododendron giganteum Forrest ex Tagg var. seminudum Tagg et Forrest ＝Rhododendron protistum Balf. f. et Forrest ●

330761　Rhododendron giganteum Forrest var. seminudum Tagg et Forrest ＝Rhododendron protistum Balf. f. et Forrest ●

330762　Rhododendron giraudiasii H. Lév. ＝Rhododendron decorum Franch. ●

330763　Rhododendron glanduliferum Franch.；大果杜鹃（腺体杜鹃）；Big-fruited Rhododendron，Glandular Rhododendron，Largefruit Azalea ●

330764　Rhododendron glandulostylum W. P. Fang et M. Y. He ＝Rhododendron guizhongense G. Z. Li ●

330765　Rhododendron glandulosum Standl. ex Small；多腺杜鹃；Many Glands Rhododendron ●☆

330766　Rhododendron glandulosum Standl. ex Small ＝Rhododendron kamtschaticum Pall. ●☆

330767　Rhododendron glaphyrum Balf. f. et Forrest ＝Rhododendron temenium Balf. f. et Forrest var. dealbatum（Cowan）D. F. Chamb. ●

330768　Rhododendron glaphyrum Balf. f. et Forrest var. dealbatum（Cowan）Davidian ＝Rhododendron temenium Balf. f. et Forrest var. dealbatum（Cowan）D. F. Chamb. ●

330769　Rhododendron glaucoaureum Balf. f. et Forrest ＝Rhododendron campylogynum Franch. ●

330770　Rhododendron glaucopeplum Balf. f. et Forrest ＝Rhododendron aganniphum Balf. f. et Kingdon-Ward ●

330771　Rhododendron glaucophyllum Rehder ＝Rhododendron tubiforme（Cowan et Davidian）Davidian ●

330772　Rhododendron glaucophyllum Rehder subsp. tubiforme（Cowan et Davidian）D. G. Long ＝Rhododendron tubiforme（Cowan et Davidian）Davidian ●

330773　Rhododendron glaucophyllum Rehder var. tubiforme Cowan et Davidian ＝Rhododendron tubiforme（Cowan et Davidian）Davidian ●

330774　Rhododendron glischroides Tagg et Forrest；似黏毛杜鹃；Resembling R. Glischrum Rhododendron ●☆

330775　Rhododendron glischrum Balf. f. et W. W. Sm.；黏毛杜鹃（黏叶杜鹃）；Stickinghair Azalea，Stickyleaf Rhododendron，Sticky-leaved Rhododendron ●

330776　Rhododendron glischrum Balf. f. et W. W. Sm. subsp. rude（Tagg et Forrest）Cowan；红黏毛杜鹃（粗毛杜鹃，红粗毛杜鹃，硬毛杜鹃）；Red Stickyleaf Rhododendron，Rough Azalea，Rough Rhododendron ●

330777　Rhododendron glischrum Balf. f. et W. W. Sm. var. adenosum Cowan et Davidian ＝Rhododendron adenosum Davidian ●

330778　Rhododendron globigerum Balf. f. et Forrest ＝Rhododendron alutaceum Balf. f. et W. W. Sm. ●

330779　Rhododendron gloeoblastum Balf. f. et Forrest ＝Rhododendron wardii W. W. Sm. ●

330780　Rhododendron glomerulatum Hutch. ＝Rhododendron yungningense Balf. f. ex Hutch. ●

330781　Rhododendron gnaphalocarpum Hayata ＝Rhododendron mariesii Hemsl. et E. H. Wilson ●

330782　Rhododendron gologense C. J. Xu et Z. J. Zhao；果洛杜鹃；Guoluo Azalea，Guoluo Rhododendron ●

330783　Rhododendron gonggashanense W. K. Hu；贡嘎山杜鹃；Gonggashan Azalea，Gonggashan Rhododendron ●

330784　Rhododendron gonggashanense W. K. Hu ＝Rhododendron vernicosum Franch. ●

330785　Rhododendron gongshanense T. L. Ming；贡山杜鹃；Gongshan Azalea，Gongshan Rhododendron ●

330786　Rhododendron gracilescens（Nakai）F. Maek. ＝Rhododendron nudipes Nakai var. nagasakianum（Nakai）T. Yamaz. ●☆

330787　Rhododendron gracilipes Franch. ＝Rhododendron argyrophyllum Franch. subsp. hypoglaucum（Hemsl.）D. F. Chamb. ●

330788　Rhododendron gracilipes Franch. ＝Rhododendron hypoglaucum Hemsl. ●

330789　Rhododendron grande Wight；巨魁杜鹃（阿根廷杜鹃，大叶杜鹃）；Argentina Rhododendron，Giant Azalea，Giant Rhododendron，Large Rhododendron ●

330790　Rhododendron gratiosum P. C. Tam；多姿杜鹃；Like Lovely Rhododendron ●

330791　Rhododendron gratiosum P. C. Tam ＝Rhododendron mariae Hance ●

330792　Rhododendron gratum T. L. Ming ＝Rhododendron rex H. Lév. subsp. gratum（T. L. Ming）M. Y. Fang ●

330793　Rhododendron griersonianum Balf. f. et Forrest；朱红大杜鹃；Grierson's Rhododendron，R. C. Grierson's Rhododendron，Vermeil Azalea ●

330794　Rhododendron griffithianum Wight；不丹杜鹃（奥克兰杜鹃，格瑞杜鹃，腺柱杜鹃）；Auckland Azalea，Auckland Rhododendron，Bhutan Azalea，Griffith Rhododendron ●

330795　Rhododendron griffithianum Wight var. aucklandii（Hook. f.）Hook. ＝Rhododendron griffithianum Wight ●

330796　Rhododendron groenlandicum（Oeder）Kron et Judd ＝Ledum groenlandicum Oeder ■☆

330797　Rhododendron guangnanense R. C. Fang；广南杜鹃；Guangnan Azalea，Guangnan Rhododendron ●

330798　Rhododendron guihainianum G. Z. Li；桂海杜鹃；Guihai Azalea ●

330799　Rhododendron guizhongense G. Z. Li；桂中杜鹃（腺柱杜鹃）；Glandstyle Azalea，Glandular Styles Rhododendron，Glandular-styled Rhododendron ●

330800　Rhododendron guizhouense M. Y. Fang；贵州杜鹃；Guizhou Azalea，Guizhou Rhododendron ●

330801　Rhododendron gymnanthum Diels ＝Rhododendron lukiangense Franch. ●

330802　Rhododendron gymnocarpum Balf. f. ex Tagg ＝Rhododendron microgynum Balf. f. et Forrest ●

330803　Rhododendron gymnogynum Balf. f. et Forrest = Rhododendron anthosphaerum Diels ●

330804　Rhododendron gymnomiscum Balf. f. et Kingdon-Ward = Rhododendron primuliflorum Bureau et Franch. ●

330805　Rhododendron habrotrichum Balf. f. et W. W. Sm.；粗毛杜鹃（柔毛杜鹃）；Rough-haired Rhododendron, Shag Azalea, Soft Hairs Rhododendron ●

330806　Rhododendron haemaleum Balf. f. et Forrest = Rhododendron sanguineum Franch. var. haemaleum（Balf. f. et Forrest）D. F. Chamb. ●

330807　Rhododendron haematocheilum Craib = Rhododendron oreodoxa Franch. ●

330808　Rhododendron haematodes Franch.；似血杜鹃；Bloodlike Azalea, Blood-like Rhododendron ●◇

330809　Rhododendron haematodes Franch. subsp. chaetomallum（Balf. f. et Forrest）Cowan；绢毛杜鹃（厚毛杜鹃，羊毛叶杜鹃）；Fleeceleaf Azalea, Fleeceleaf Rhododendron, Foeecy Hair Rhododendron ●

330810　Rhododendron haematodes Franch. subsp. chaetomallum（Balf. f. et Forrest）Cowan var. calycinum Franch. = Rhododendron haematodes Franch. subsp. chaetomallum（Balf. f. et Forrest）Cowan ●

330811　Rhododendron haematodes Franch. subsp. chaetomallum（Balf. f. et Forrest）Cowan var. hypoteucum Franch. = Rhododendron haematodes Franch. subsp. chaetomallum（Balf. f. et Forrest）Cowan ●

330812　Rhododendron haematodes Franch. var. calycinum Franch. = Rhododendron haematodes Franch. ●◇

330813　Rhododendron haematodes Franch. var. hypoleucum Franch. = Rhododendron haematodes Franch. ●◇

330814　Rhododendron haemonium Balf. f. et R. E. Cooper = Rhododendron anthopogon D. Don ●

330815　Rhododendron hainanense Merr.；海南杜鹃；Hainan Azalea, Hainan Rhododendron ●

330816　Rhododendron hanceanum Hemsl.；疏叶杜鹃（汉氏杜鹃）；H. F. Hance's Rhododendron, Hance's Rhododendron, Poorleaf Azalea ●

330817　Rhododendron hancockii Hemsl.；滇南杜鹃（汉克杜鹃）；Hancock Azalea, Hancock Rhododendron ●

330818　Rhododendron hancockii Hemsl. var. longisepalum R. C. Fang et C. H. Yang；长萼滇南杜鹃；Longsepal Hancock Azalea, Long-sepal Hancock Rhododendron ●

330819　Rhododendron hangnoense Nakai = Rhododendron indicum（L.）Sweet ●

330820　Rhododendron hangzhouense W. P. Fang et M. Y. He；杭州杜鹃（杭州马银花）；Hangzhou Azalea, Hangzhou Rhododendron ●

330821　Rhododendron hangzhouense W. P. Fang et M. Y. He = Rhododendron bachii H. Lév. ●

330822　Rhododendron hangzhouense W. P. Fang et M. Y. He var. pubescens K. M. Feng；柔毛弯蒴杜鹃；Pubescent Hangzhou Rhododendron ●

330823　Rhododendron hannoense Nakai = Rhododendron indicum（L.）Sweet ●

330824　Rhododendron haofui Chun et W. P. Fang；光枝杜鹃（灏富杜鹃，红岩杜鹃）；Bared Branches Rhododendron, Nakedtwig Azalea ●

330825　Rhododendron hardingii Forrest ex Tagg = Rhododendron annae Franch. ●

330826　Rhododendron hardingii Forrest ex Tagg = Rhododendron annae Franch. subsp. laxiflorum（Balf. f. et Forrest）T. L. Ming ●

330827　Rhododendron hardyi Davidian = Rhododendron augustinii Hemsl. subsp. chasmanthum（Diels）Cullen f. hardyi（Davidian）R. C. Fang ●

330828　Rhododendron hardyi Davidian = Rhododendron augustinii Hemsl. subsp. chasmanthum（Diels）Cullen ●

330829　Rhododendron harrovianum Hemsl. = Rhododendron polylepis Franch. ●

330830　Rhododendron headfortianum Hutch.；佛得角杜鹃（独花杜鹃）；Headfort Rhododendron ●☆

330831　Rhododendron headfortianum Hutch. = Rhododendron taggianum Hutch. ●

330832　Rhododendron hedythammim Hutch. var. eglandulosum Hand.-Mazz. = Rhododendron callimorphum Balf. f. et W. W. Sm. ●

330833　Rhododendron hedythamnum Balf. f. et Forrest = Rhododendron callimorphum Balf. f. et W. W. Sm. ●

330834　Rhododendron hedythamnum Balf. f. et Forrest = Rhododendron selense Franch. subsp. jucundum（Balf. f. et W. W. Sm.）D. F. Chamb. ●◇

330835　Rhododendron hedythamnum Balf. f. et Forrest var. eglandulosum Hand.-Mazz. = Rhododendron cyanocarpum（Franch.）Franch. et W. W. Sm. ●◇

330836　Rhododendron heftii Davidian = Rhododendron wallichii Hook. f. ●

330837　Rhododendron heishuiense W. P. Fang；黑水杜鹃；Heishui Azalea, Heishui Rhododendron ●

330838　Rhododendron heishuiense W. P. Fang = Rhododendron tatsienense Franch. ●

330839　Rhododendron heizhugouense M. Y. He et L. C. Hu；黑竹沟杜鹃；Heizhugou Azalea, Heizhugou Rhododendron ●

330840　Rhododendron hejiangense M. Y. He；合江杜鹃；Hejiang Azalea, Hejiang Rhododendron ●

330841　Rhododendron heliolepis Franch.；亮鳞杜鹃（短柱杜鹃）；Glittering Scales Rhododendron, Glitter-scaled Rhododendron, Shining-lepidote Rhododendron, Shinscale Azalea, Shortstyle Azalea, Short-style Rhododendron, Short-styled Rhododendron ●

330842　Rhododendron heliolepis Franch. var. brevistylum（Franch.）Cullen = Rhododendron heliolepis Franch. ●

330843　Rhododendron heliolepis Franch. var. fumidum（Balf. f. et W. W. Sm.）R. C. Fang；灰褐亮鳞杜鹃（灰杜鹃，灰褐杜鹃）；Smoke-coloured Rhododendron, Smoky Shining-lepidote Rhododendron ●

330844　Rhododendron heliolepis Franch. var. fumidum（Balf. f. et W. W. Sm.）R. C. Fang = Rhododendron fumidum Balf. f. et W. W. Sm. ●

330845　Rhododendron heliolepis Franch. var. oporinum（Balf. f. et Kingdon-Ward）A. L. Chang ex R. C. Fang；毛冠亮鳞杜鹃（毛蕊壳鳞杜鹃，秋花杜鹃）；Autumn Flowering Rhododendron ●

330846　Rhododendron helvolum Balf. f. et Forrest = Rhododendron phaeochrysum Balf. f. et W. W. Sm. var. levistratum（Balf. f. et Forrest）D. F. Chamb. ●

330847　Rhododendron hemidartum Balf. f. ex Tagg = Rhododendron pocophorum Balf. f. ex Tagg var. hemidartum（Tagg）D. F. Chamb. ●

330848　Rhododendron hemitrichotum Balf. f. et Forrest；粉背碎米花；Brick-budded Rhododendron, Half-hairy Leaved Rhododendron, Paleback Azalea ●

330849　Rhododendron hemsleyanum E. H. Wilson；波叶杜鹃（亨氏杜鹃）；Hemsley Azalea, Hemsley's Rhododendron, W. B. Hemsley's Rhododendron ●

330850　Rhododendron hemsleyanum E. H. Wilson var. chengianum W.

P. Fang ex Ching；无腺杜鹃；Cheng Azalea, Cheng Hemsley's Rhododendron ●

330851 Rhododendron henanense W. P. Fang；河南杜鹃；Henan Azalea, Henan Rhododendron ●◇

330852 Rhododendron henanense W. P. Fang subsp. lingbaoense W. P. Fang；灵宝杜鹃；Lingbao Azalea, Lingbao Henan Rhododendron, Lingbao Rhododendron ●

330853 Rhododendron henryi Hance；弯蒴杜鹃；Henry Azalea, Henry Rhododendron ●

330854 Rhododendron henryi Hance var. dunnii（E. H. Wilson）M. Y. He；秃房弯蒴杜鹃（东南杜鹃，杜恩杜鹃，秃房杜鹃）；Dunn Azalea, Dunn's Henry Azalea, Dunn's Henry Rhododendron, Dunn's Rhododendron ●

330855 Rhododendron henryi Hance var. pubescens K. M. Feng et A. L. Chang = Rhododendron cavaleriei H. Lév. ●

330856 Rhododendron hepaticum P. C. Tam = Rhododendron florulentum P. C. Tam ●

330857 Rhododendron hepaticum P. C. Tam = Rhododendron rufulum P. C. Tam ●

330858 Rhododendron heptamerum Balf. f. = Rhododendron anthosphaerum Diels ●

330859 Rhododendron herpesticum Balf. f. et Kingdon-Ward = Rhododendron dichroanthum Diels et Cowan subsp. scyphocalyx（Balf. f. et Forrest）Cowan ●

330860 Rhododendron hesperium Balf. f. et Forrest = Rhododendron rigidum Franch. ●

330861 Rhododendron heteroclitum H. P. Yang；异常杜鹃；Heteroclite Rhododendron ●

330862 Rhododendron hexamerum Hand.-Mazz. = Rhododendron vernicosum Franch. ●

330863 Rhododendron hidaense Makino ex H. Hara；斐太杜鹃●☆

330864 Rhododendron hidakanum H. Hara = Rhododendron dilatatum Miq. var. boreale Sugim. ●☆

330865 Rhododendron hillieri Davidian；察瓦龙杜鹃（海氏杜鹃，山地杜鹃）；Chawalong Azalea, Chawalong Rhododendron, Hillier Rhododendron ●

330866 Rhododendron himertum Balf. f. et Forrest = Rhododendron sanguineum Franch. var. himertum（Balf. f. et Forrest）T. L. Ming ●

330867 Rhododendron himertum Balf. f. et Forrest subsp. nebrites（Balf. f. et Forrest）Tagg = Rhododendron sanguineum Franch. var. himertum（Balf. f. et Forrest）T. L. Ming ●

330868 Rhododendron himertum Balf. f. et Forrest subsp. nebrites（Balf. f. et Forrest）Tagg = Rhododendron sanguineum Franch. var. haemaleum（Balf. f. et Forrest）D. F. Chamb. ●

330869 Rhododendron himertum Balf. f. et Forrest subsp. poliopelum（Balf. f. et Forrest）Tagg = Rhododendron sanguineum Franch. var. haemaleum（Balf. f. et Forrest）D. F. Chamb. ●

330870 Rhododendron himertum Balf. f. et Forrest subsp. poliopeplum（Balf. f. et Forrest）Tagg = Rhododendron sanguineum Franch. var. himertum（Balf. f. et Forrest）T. L. Ming ●

330871 Rhododendron hippophaeroides Balf. f. et W. W. Sm.；灰背杜鹃（沙棘状杜鹃，细穗杜鹃）；Greyback Azalea, Minutely Fringed Rhododendron, Resembling Hippophae Rhododendron, Sea-buckthorn Rhododendron, Seabuckthorn-like Rhododendron ●

330872 Rhododendron hippophaeroides Balf. f. et W. W. Sm. var. occidentale M. N. Philipson et Philipson；长柱灰被杜鹃●

330873 Rhododendron hirsuticostatum Hand.-Mazz. = Rhododendron

augustinii Hemsl. subsp. chasmanthum（Diels）Cullen ●

330874 Rhododendron hirsutipetiolatum A. L. Chang et R. C. Fang；凸脉杜鹃（粗毛叶柄杜鹃）；Convexed Veins Rhododendron, Convex-veined Rhododendron, Hirsute Petiole Rhododendron, Hirsutepetiole Azalea ●

330875 Rhododendron hirsutum L.；中欧毛杜鹃（欧洲高山杜鹃）；Garland Rhododendron, Hairy Alpen Rose, Hairy Rhododendron ●☆

330876 Rhododendron hirtipes Tagg；硬毛杜鹃（西藏杜鹃，硬毛柄杜鹃）；Hispid Azalea, Shaggy-footed Rhododendron, Xizang Azalea, Xizang Rhododendron ●

330877 Rhododendron hodgsonii Hook. f.；多裂杜鹃；B. H. Hedgson's Rhododendron, Hodgson Azalea, Hodgson Rhododendron ●

330878 Rhododendron hoi W. P. Fang；川北杜鹃（何氏杜鹃）；Ho's Rhododendron ●

330879 Rhododendron hongkongense Hutch.；香港杜鹃（白马杜鹃，白马银花）；Hongkong Azaleastrum, White Azalea ●

330880 Rhododendron hookeri Nutt.；串珠杜鹃（虎克杜鹃）；Hooker Azalea, Hooker Rhododendron, Hooker's Rhododendron ●

330881 Rhododendron horaeum Balf. f. et Forrest = Rhododendron citriniflorum Balf. f. et Forrest var. horaeum（Balf. f. et Forrest）D. F. Chamb. ●

330882 Rhododendron hormophorum Balf. f. et Forrest = Rhododendron yunnanense Franch. ●

330883 Rhododendron hortense Nakai；紫琉球杜鹃●☆

330884 Rhododendron houlstonii Hemsl. et E. H. Wilson = Rhododendron discolor Franch. ●

330885 Rhododendron huadingense B. Y. Ding et Y. Y. Fang；华顶杜鹃；Huading Rhododendron ●

330886 Rhododendron huanum W. P. Fang；凉山杜鹃；Dr. Hu's Rhododendron, Hu's Rhododendron, Liangshan Azalea ●

330887 Rhododendron huguangense P. C. Tam；大鳞杜鹃（湖广杜鹃）；Bigscale Azalea, Hunan-Guangdong Rhododendron, Hunan-Guangxi Rhododendron ●

330888 Rhododendron huidongense T. L. Ming；会东杜鹃；Huidong Azalea, Huidong Rhododendron ●

330889 Rhododendron huiyangense W. P. Fang et M. Y. He = Rhododendron tingwuense P. C. Tam ●

330890 Rhododendron hukwangense P. C. Tam = Rhododendron huguangense P. C. Tam ●

330891 Rhododendron hunanense Chun ex P. C. Tam；湖南杜鹃；Hunan Azalea, Hunan Rhododendron ●

330892 Rhododendron hunanense Chun ex P. C. Tam var. mangshanicum P. C. Tam = Rhododendron hunanense Chun ex P. C. Tam ●

330893 Rhododendron hunnewellianum Rehder et E. H. Wilson；岷江杜鹃（粉点杜鹃，汶川杜鹃，羊闹花）；Minjiang Azalea, Pink Spotted Rhododendron, Pink-spotted Rhododendron ●

330894 Rhododendron hunnewellianum Rehder et E. H. Wilson subsp. rockii（E. H. Wilson）D. F. Chamb.；黄毛岷江杜鹃；Rock Pink Spotted Rhododendron, Yellowhair Minjiang Azalea, Yellow-haired Pink Spotted Rhododendron ●

330895 Rhododendron hutchinsonianum W. P. Fang；二郎山杜鹃●

330896 Rhododendron hutchinsonianum W. P. Fang = Rhododendron concinnum Hemsl. ●

330897 Rhododendron hylaeum Balf. f. et Forrest；粉果杜鹃（林间杜鹃）；Farinosefruit Azalea, Forest Belonging Rhododendron ●

330898 Rhododendron hylothreptum Balf. f. et W. W. Sm. = Rhododendron anthosphaerum Diels ●

330899　Rhododendron hymenanthes Makino ＝ Rhododendron degronianum Carrière ●☆

330900　Rhododendron hypenanthum Balf. f.；毛花杜鹃（髯毛黄杜鹃）；Bearded Flowers Rhododendron, Hairflower Azalea, Hairyflower Rhododendron, Hairy-flowered Rhododendron ●

330901　Rhododendron hyperythrum Hayata；微笑杜鹃（红点背面杜鹃，红点叶背杜鹃，红星杜鹃，南湖杜鹃，台湾山地杜鹃，台中杜鹃，小西氏杜鹃）；Reddish Punctulation Rhododendron, Reddish-punctulate Rhododendron, Red-spotted Rhododendron, Smile Azalea, Taiwan Montane Azalea, Thick-flowered Rhododendron, Thick-flowers Rhododendron ●

330902　Rhododendron hypoblematosum P. C. Tam；背绒杜鹃（黄绒杜鹃）；Flossback Azalea, Hair Beneath Rhododendron, Hypotomentose Rhododendron ●

330903　Rhododendron hypoglaucum Hemsl. ＝ Rhododendron argyrophyllum Franch. subsp. hypoglaucum（Hemsl.）D. F. Chamb. ●

330904　Rhododendron hypolepidotum（Franch.）Balf. f. et Forrest；绿柱杜鹃●

330905　Rhododendron hypolepidotum（Franch.）Balf. f. et Forrest ＝ Rhododendron brachyanthum Franch. subsp. hypolepidotum（Franch.）Cullen ●

330906　Rhododendron hypophaeum Balf. f. et Forrest ＝ Rhododendron tatsiense Franch. ●

330907　Rhododendron hypopitys Pojark.；松下杜鹃●☆

330908　Rhododendron hypotrichotum Balf. f. et Forrest ＝ Rhododendron oreotrephes W. W. Sm. ●

330909　Rhododendron hyugaense（T. Yamaz.）T. Yamaz.；日向杜鹃●☆

330910　Rhododendron idoneum Balf. f. et W. W. Sm. ＝ Rhododendron telmateium Balf. f. et W. W. Sm. ●

330911　Rhododendron igneum Cowan；肉红杜鹃；Fleshcolor Azalea, Igneous Coloured Rhododendron, Igneous-coloured Rhododendron ●

330912　Rhododendron imberbe Hutch.；无须杜鹃；Not Bearded Rhododendron ●☆

330913　Rhododendron impeditum Balf. f. et W. W. Sm.；粉紫杜鹃（粉紫矮杜鹃，易混杜鹃，云界杜鹃）；Cloudland Rhododendron, Dwarf Purple Rhododendron, Tangled Rhododendron, Whitepurple Azalea ●

330914　Rhododendron imperator Hutch. et Kingdon-Ward；帝王杜鹃；Emperator Rhododendron ●☆

330915　Rhododendron imperator Hutch. et Kingdon-Ward ＝ Rhododendron uniflorum P. C. Tam var. imperator（Kingdon-Ward）Cullen ●

330916　Rhododendron inaequale Hutch.；极香杜鹃；Unequal Size Rhododendron, Very Fragrant Rhododendron ●☆

330917　Rhododendron indicum（L.）Sweet；皋月杜鹃（大花杂种杜鹃，杜鹃花，皋月，夏鹃，印度杜鹃，映山红，踯躅）；India Azalea, Indian Rhododendron, Indica Azalea, Macranthum Azalea, Southern Indica Hybrid Azalea ●

330918　Rhododendron indicum（L.）Sweet 'Balsaminiflorum'；凤仙花皋月杜鹃●☆

330919　Rhododendron indicum（L.）Sweet 'Macranthum'；大花皋月杜鹃●☆

330920　Rhododendron indicum（L.）Sweet ＝ Rhododendron simsii Planch. ●

330921　Rhododendron indicum（L.）Sweet var. formosanum Hayata ＝ Rhododendron simsii Planch. ●

330922　Rhododendron indicum（L.）Sweet var. ignescens Sweet ＝ Rhododendron simsii Planch. ●

330923　Rhododendron indicum（L.）Sweet var. pulchrum（Sweet）G. Don ＝ Rhododendron pulchrum Sweet ●

330924　Rhododendron indicum（L.）Sweet var. pulchrum G. Don ＝ Rhododendron pulchrum Sweet ●

330925　Rhododendron indicum（L.）Sweet var. puniceum Sweet ＝ Rhododendron simsii Planch. ●

330926　Rhododendron indicum（L.）Sweet var. simsii（Planch.）Maxim. ＝ Rhododendron simsii Planch. ●

330927　Rhododendron indicum（L.）Sweet var. simsii Maxim. ＝ Rhododendron simsii Planch. ●

330928　Rhododendron indicum（L.）Sweet var. smithii Sweet ＝ Rhododendron pulchrum Sweet ●

330929　Rhododendron inobeanum Honda ＝ Rhododendron dilatatum Miq. var. lasiocarpum Koidz. ex H. Hara ●☆

330930　Rhododendron inopinum Balf. f. ex Tagg；短尖杜鹃（意外杜鹃）；Shorttine Azalea, Unexpected Rhododendron ●

330931　Rhododendron insculptum Hutch. et Kingdon-Ward；雕纹杜鹃；Sculptured Rhododendron ●

330932　Rhododendron insigne Hemsl. et E. H. Wilson；不凡杜鹃（显丽杜鹃）；Notable Azalea, Remarkable Rhododendron ●◇

330933　Rhododendron insigne Hemsl. et E. H. Wilson var. hejiangense（W. P. Fang）M. Y. Fang；合江银叶杜鹃；Hejiang Rhododendron ●

330934　Rhododendron intortum Balf. f. et Forrest ＝ Rhododendron phaeochrysum Balf. f. et W. W. Sm. var. levistratum（Balf. f. et Forrest）D. F. Chamb. ●

330935　Rhododendron intricatum Franch.；隐蕊杜鹃（错综杜鹃，睫萼杜鹃）；Bluet Rhododendron, Blurred Rhododendron, Fringed-calyx Azalea, Fringed-calyx Rhododendron, Latentpistil Azalea ●

330936　Rhododendron invictum Balf. f. et Forrest；绝伦杜鹃；Extremely Azalea, Extremely Rhododendron, Invincible Rhododendron ●

330937　Rhododendron ioanthum Balf. f. ＝ Rhododendron siderophyllum Franch. ●

330938　Rhododendron iodes Balf. f. et Forrest ＝ Rhododendron alutaceum Balf. f. et W. W. Sm. var. iodes（Balf. f. et Forrest）D. F. Chamb. ●

330939　Rhododendron irroratum Franch.；露珠杜鹃；Dew Azalea, Dewcovered Rhododendron, Dew-covered Rhododendron ●

330940　Rhododendron irroratum Franch. 'Polka Dot'；普尔卡点露珠杜鹃●☆

330941　Rhododendron irroratum Franch. subsp. pogonostylum（Balf. f. et W. W. Sm.）D. F. Chamb. ex Cullen et D. F. Chamb.；红花露珠杜鹃（须柱杜鹃）；Bearded Style Rhododendron, Redflower Dew Azalea ●

330942　Rhododendron irroratum Franch. subsp. pogonostylum（Balf. f. et W. W. Sm.）D. F. Chamb. ex Cullen ＝ Rhododendron agastum Balf. f. et W. W. Sm. var. pennivenium（Balf. f. et W. W. Sm.）T. L. Ming ●

330943　Rhododendron iteophyllum Hutch.；柳叶杜鹃；Willow-leaved Rhododendron ●☆

330944　Rhododendron ixeuticum Balf. f. et W. W. Sm. ＝ Rhododendron crinigerum Franch. ●

330945　Rhododendron jahandiezii H. Lév. ＝ Rhododendron siderophyllum Franch. ●

330946　Rhododendron jangtzowense Balf. f. et Forrest ＝ Rhododendron dichroanthum Diels et Cowan subsp. apodectum（Balf. f. et W. W. Sm.）Cowan ●

330947　Rhododendron japonicum（A. Gray）Suringar ＝ Rhododendron molle（Blume）G. Don subsp. japonicum（A. Gray）Kron ●☆

330948　Rhododendron japonicum（A. Gray）Suringar f. canescens

Sugim. = Rhododendron molle G. Don subsp. japonicum（A. Gray）Kron f. canescens（Sugim.）Yonek. ●☆

330949　Rhododendron japonicum（A. Gray）Suringar f. flavum（Miyoshi）Nakai = Rhododendron molle G. Don subsp. japonicum（A. Gray）Kron f. flavum（Miyoshi）Yonek. ●☆

330950　Rhododendron japonicum（A. Gray）Suringar f. glaucophyllum（Nakai）H. Hara = Rhododendron molle G. Don subsp. japonicum（A. Gray）Kron f. glaucophyllum（Nakai）Yonek. ●☆

330951　Rhododendron japonicum（A. Gray）Suringar f. multifidum Nakai = Rhododendron molle G. Don subsp. japonicum（A. Gray）Kron f. multifidum（Nakai）Yonek. ●☆

330952　Rhododendron japonicum（Blume）C. K. Schneid. var. pentamerum（Maxim.）Hutch.；五数日本杜鹃●☆

330953　Rhododendron japonoheptamerum Kitam. var. hondoense（Nakai）Kitam.；本岛杜鹃（本岛石南）●☆

330954　Rhododendron japonoheptamerum Kitam. var. hondoense（Nakai）Kitam. f. leucanthum（Nakai）T. Yamaz.；白花本岛杜鹃●☆

330955　Rhododendron jasminiflorum Hook.；茉莉杜鹃；Iasmine-flowered Rhododendron ●☆

330956　Rhododendron jasminoides M. Y. He；素馨杜鹃；Jasninelike Azalea, Jasnine-like Rhododendron ●

330957　Rhododendron javanicum（Blume）Benn.；爪哇杜鹃；Javanese Azalea ●☆

330958　Rhododendron jenkinsii Nutt. = Rhododendron maddenii Hook. f. ●

330959　Rhododendron jinchangense Z. H. Yang；金厂杜鹃；Jinchang Azalea, Jinchang Rhododendron ●

330960　Rhododendron jingangshanicum P. C. Tam；井冈山杜鹃；Jinggangshan Azalea, Jinggangshan Rhododendron ●

330961　Rhododendron jinpingense W. P. Fang et M. Y. He；金平杜鹃；Jinping Azalea, Jinping Rhododendron ●

330962　Rhododendron jinxiuense W. P. Fang et M. Y. He；金秀杜鹃；Jinxiu Azalea, Jinxiu Rhododendron ●

330963　Rhododendron johnstoneanum Watt ex Hutch.；鳞瓣杜鹃（约翰斯顿杜鹃）；Mrs. Johnston's Rhododendron ●☆

330964　Rhododendron joniense Ching et H. P. Yang；卓尼杜鹃；Joni Azalea, Joni Rhododendron, Zhuoni Rhododendron ●

330965　Rhododendron jucundum Balf. f. et W. W. Sm. = Rhododendron selense Franch. subsp. jucundum（Balf. f. et W. W. Sm.）D. F. Chamb. ●◇

330966　Rhododendron kaempferi Planch.；山踯躅（红踯躅，火把杜鹃，火炬杜鹃，堪氏杜鹃）；Kaempfer's Rhododendron, Kaempferi Azalea, Torch Azalea ●☆

330967　Rhododendron kaempferi Planch. f. album Nakai；白花山踯躅●☆

330968　Rhododendron kaempferi Planch. f. angustifolium Nakai；狭叶山踯躅●☆

330969　Rhododendron kaempferi Planch. f. cryptopetalum Makino et Nakai；无瓣山踯躅●☆

330970　Rhododendron kaempferi Planch. f. cryptopetalum Makino et Nemoto；蕊瓣让步山杜鹃●☆

330971　Rhododendron kaempferi Planch. f. komatsui（Nakai）H. Hara；重瓣山踯躅●☆

330972　Rhododendron kaempferi Planch. f. komatsui Hara = Rhododendron kaempferi Planch. f. komatsui（Nakai）H. Hara ●☆

330973　Rhododendron kaempferi Planch. f. latifolium H. Hara；阔叶山踯躅●☆

330974　Rhododendron kaempferi Planch. f. macrogemmum Nakai =

Rhododendron macrogemma Nakai ●☆

330975　Rhododendron kaempferi Planch. f. mikawanum（Makino）H. Hara；紫花山踯躅●☆

330976　Rhododendron kaempferi Planch. f. mikawanum（Makino）H. Hara = Rhododendron kaempferi Planch. var. mikawanum（Makino）Makino ●☆

330977　Rhododendron kaempferi Planch. f. mikawanum Makino = Rhododendron kaempferi Planch. f. mikawanum（Makino）H. Hara ●☆

330978　Rhododendron kaempferi Planch. f. multicolor Makino et Nakai；杂色山踯躅●☆

330979　Rhododendron kaempferi Planch. f. semperflorens H. Hara；常花山踯躅●☆

330980　Rhododendron kaempferi Planch. f. tubiflorum（Komatsu）H. Hara；筒花山踯躅●☆

330981　Rhododendron kaempferi Planch. f. tubiflorum（Komatsu）H. Hara = Rhododendron kaempferi Planch. var. tubiflorum Komatsu ●☆

330982　Rhododendron kaempferi Planch. var. lucidusculum（Nakai）Sugim. = Rhododendron kaempferi Planch. ●☆

330983　Rhododendron kaempferi Planch. var. macrogemma Nakai = Rhododendron macrogemma Nakai ●☆

330984　Rhododendron kaempferi Planch. var. macrostemon Makino；厚冠山踯躅●☆

330985　Rhododendron kaempferi Planch. var. mikawanum（Makino）Makino = Rhododendron kaempferi Planch. f. mikawanum（Makino）H. Hara ●☆

330986　Rhododendron kaempferi Planch. var. saikaiense（T. Yamaz.）T. Yamaz.；西海杜鹃●☆

330987　Rhododendron kaempferi Planch. var. tubiflorum Komatsu = Rhododendron kaempferi Planch. f. tubiflorum（Komatsu）H. Hara ●☆

330988　Rhododendron kaliense W. P. Fang et M. Y. He = Rhododendron westlandii Hemsl. ●

330989　Rhododendron kamtschaticum Pall.；勘察加杜鹃●☆

330990　Rhododendron kanehirae E. H. Wilson；台北杜鹃（金平杜鹃，乌来杜鹃）；Kanehira Azalea, Taibei Azalea, Taibei Rhododendron ●

330991　Rhododendron kangdingense Z. J. Zhao = Rhododendron tatsienense Franch. ●

330992　Rhododendron kasoense Hutch. et Kingdon-Ward；黄管杜鹃；Kaso Rhododendron, Yellowtube Azalea, Yellow-tube Rhododendron ●

330993　Rhododendron kawakamii Hayata；着生杜鹃（川上氏杜鹃，附生杜鹃，著生杜鹃）；Kawakami Azalea, Kawakami Rhododendron ●

330994　Rhododendron kawakamii Hayata var. flaviflorum Tang S. Liu et T. I. Chuang；黄色着生杜鹃（黄花着生杜鹃）；Yellow Kawakami Azalea, Yellow Kawakami Rhododendron ●

330995　Rhododendron keiskei Miguel = Rhododendron keiskei Miq. ●☆

330996　Rhododendron keiskei Miguel var. cordifolium Masam. = Rhododendron keiskei Miguel ●☆

330997　Rhododendron keiskei Miguel var. hypoglaucum Suto et Suzuki；里白伊东杜鹃●☆

330998　Rhododendron keiskei Miq.；伊藤杜鹃（心叶伊东杜鹃）；Ito Keisuke's Rhododendron ●☆

330999　Rhododendron keiskei Miq. var. hypoglaucum Suto et T. Suzuki；里白伊藤杜鹃●☆

331000　Rhododendron keiskei Miq. var. ozawae T. Yamaz.；小泽杜鹃●☆

331001　Rhododendron keleticum Nutt.；独龙杜鹃（娇美杜鹃，茎根杜鹃，凯勒迪库美被杜鹃，拉萨杜鹃，闪光杜鹃）；Charming Rhododendron, Dulong Azalea, Lahsa Rhododendron, Rooting Branches Rhododendron, Shining Rhododendron ●

331002　Rhododendron kendrickii Nutt.；多斑杜鹃(不丹杜鹃,潘吉拉杜鹃)；Dr. Kendrick's Rhododendron, H. Shepherd's Rhododendron, Kendrick Azalea,Kendrick's Rhododendron,Pakim La Rhododendron ●

331003　Rhododendron keysii Nutt.；管花杜鹃（凯氏杜鹃）；Keys' Rhododendron, Mr. Keys' Rhododendron ●

331004　Rhododendron keysii Nutt. var. unicolor Hutch. ex Steam = Rhododendron keysii Nutt. ●

331005　Rhododendron kialense Franch. = Rhododendron przewalskii Maxim. ●

331006　Rhododendron kiangsiense W. P. Fang；江西杜鹃；Jiangxi Azalea,Jiangxi Rhododendron ●

331007　Rhododendron kingdonii Merr. = Rhododendron calostrotum Balf. f. et Kingdon-Ward ●

331008　Rhododendron kirkii Millais = Rhododendron discolor Franch. ●

331009　Rhododendron kisoanum Okuhara；木曽杜鹃●☆

331010　Rhododendron kiusianum Makino；九州山杜鹃（九州岛杜鹃,九州杜鹃, 久留米杜鹃）；Gilbralter Azalea, Kyushu Azalea, Mt. Kyushu Rhododendron ●☆

331011　Rhododendron kiusianum Makino f. albiflorum Honda et Maeda；白花九州山杜鹃●☆

331012　Rhododendron kiusianum Makino var. sataense（Nakai）D. F. Chamb. et Rae = Rhododendron obtusum（Lindl.）Planch. ●

331013　Rhododendron klossii Ridl. = Rhododendron moulmainense Hook. ●

331014　Rhododendron komatsui T. Yamaz.；小松氏杜鹃■☆

331015　Rhododendron komiyamae Makino；笼山杜鹃●☆

331016　Rhododendron kongboense Hutch.；工布杜鹃；Gongbu Azalea, Gongbu Rhododendron,Kongbo Rhododendron ●

331017　Rhododendron konori Becc.；灰叶杜鹃●☆

331018　Rhododendron kotschyi Simonk.；纤枝矮杜鹃；Kotschy's Azalea,Kotschy's Rhododendron ●☆

331019　Rhododendron kouytchense H. Lév. = Rhododendron chrysocalyx H. Lév. et Vaniot ●

331020　Rhododendron kuluense Chamb. ex Cullen et Chamb. = Rhododendron adenosum Davidian ●

331021　Rhododendron kuluense D. F. Chamb. = Rhododendron adenosum Davidian ●

331022　Rhododendron kuratanum S. Watan.；仓田杜鹃●☆

331023　Rhododendron kurohimense Arakawa；黑姬山杜鹃●☆

331024　Rhododendron kwangfuense Chun et W. P. Fang = Rhododendron discolor Franch. ●

331025　Rhododendron kwangsiense Hu = Rhododendron kwangsiense Hu ex P. C. Tam ●

331026　Rhododendron kwangsiense Hu ex P. C. Tam；广西杜鹃；Guangxi Azalea,Guangxi Rhododendron,Kwangsi Azalea ●

331027　Rhododendron kwangsiense Hu ex P. C. Tam var. obovatifolium P. C. Tam；钝圆杜鹃；Ovate-leaf Guangxi Azalea,Ovate-leaf Guangxi Rhododendron ●

331028　Rhododendron kwangsiense Hu ex P. C. Tam var. salicinum P. C. Tam = Rhododendron kwangsiense Hu ex P. C. Tam ●

331029　Rhododendron kwangsiense Hu ex P. C. Tam var. subfalcatum P. C. Tam = Rhododendron kwangsiense Hu ex P. C. Tam ●

331030　Rhododendron kwangsiense Hu ex W. P. Fang = Rhododendron kwangsiense Hu ex P. C. Tam ●

331031　Rhododendron kwangtungense Merr. et Chun；广东杜鹃(封开杜鹃)；Fengkai Azalea, Fengkai Rhododendron, Guangdong Azalea, Guangdong Rhododendron,Kwangtung Rhododendron ●

331032　Rhododendron kyawi Lace et W. W. Sm.；星毛杜鹃(考氏杜鹃,腺梗杜鹃)；Kyaw's Rhododendron,Starhair Azalea ●

331033　Rhododendron labolengense Ching et H. P. Yang；拉卜楞杜鹃；Labuleng Azalea, Labuleng Rhododendron ●

331034　Rhododendron lacteum Franch.；乳黄杜鹃(乳白杜鹃)；Gream Rhododendron, Milky Rhododendron, Milkyellow Azalea ●

331035　Rhododendron lacteum Franch. = Rhododendron rex H. Lév. subsp. fictolacteum（Balf. f.）D. F. Chamb. ●◇

331036　Rhododendron lacteum Franch. var. macrophyllum Franch. = Rhododendron rex H. Lév. subsp. fictolacteum（Balf. f.）D. F. Chamb. ●◇

331037　Rhododendron laetevirens Rehder；鲜绿杜鹃(威氏杜鹃)；Bright-green Rhododendron ●☆

331038　Rhododendron laetum J. J. Sm.；肉花杜鹃●☆

331039　Rhododendron lagopus Nakai；兔足杜鹃●☆

331040　Rhododendron lagopus Nakai var. niphophilum（T. Yamaz.）T. Yamaz. f. albiflorum（Satomi et Fukushima）T. Yamaz.；白花兔足杜鹃●☆

331041　Rhododendron lagopus Nakai var. tokushimense（T. Yamaz.）T. Yamaz. = Rhododendron tsurugisanense（T. Yamaz.）T. Yamaz. ●☆

331042　Rhododendron lampropeplum Balf. f. et Forrest = Rhododendron proteoides Balf. f. et W. W. Sm. ●

331043　Rhododendron lamprophyllum Hayata = Rhododendron ovatum Planch. var. lamprophyllum（Hayata）Y. C. Liu,F. Y. Lu et C. H. Ou ●

331044　Rhododendron lanatoides Cowan；淡钟杜鹃(似黄钟杜鹃,碗状杜鹃)；False Woolly-leaved Rhododendron, Lightcolorbell Azalea, Resembling R. Lanatum Rhododendron ●

331045　Rhododendron lanatum Hook. f.；黄钟杜鹃(林生杜鹃,绵毛杜鹃)；Woolly-leaved Rhododendron, Yellowbell Azalea ●

331046　Rhododendron lanatum Hook. f. var. luciferum Cowan = Rhododendron lanatum Hook. f. ●

331047　Rhododendron lancifolium Hook. f. = Rhododendron barbatum Wall. ex G. Don ●

331048　Rhododendron lanigerum Tagg；绵绒杜鹃(褐芽杜鹃,林生杜鹃)；Cottony Azalea, Russet Bud Rhododendron, Russet-budded Rhododendron ●

331049　Rhododendron lanigerum Tagg 'Chpal Wood'；教条木绵绒杜鹃●☆

331050　Rhododendron lanigerum Tagg 'Round Wood'；圆木绵绒杜鹃●☆

331051　Rhododendron lanigerum Tagg 'Silvia'；塞尔维亚绵绒杜鹃●☆

331052　Rhododendron lanigerum Tagg 'Stonehurst'；斯通胡斯特绵绒杜鹃●☆

331053　Rhododendron lanigerum Tagg var. silvaticum（Cowan）Davidian = Rhododendron lanigerum Tagg ●

331054　Rhododendron laojunense T. L. Ming = Rhododendron laojunshanense M. Y. Fang ●

331055　Rhododendron laojunshanense M. Y. Fang；老君山杜鹃；Laojunshan Azalea,Laojunshan Rhododendron ●

331056　Rhododendron lapidosum T. L. Ming = Rhododendron araiophyllum Balf. f. et W. W. Sm. subsp. lapidosum（T. L. Ming）M. Y. Fang ●

331057　Rhododendron lapponicum（L.）Wahlenb.；高山杜鹃(千岛小叶杜鹃,小杜鹃花,小叶杜鹃)；Arctic Rhododendron, Lapland Azalea, Lapland Rhododendron, Lapland Rose Bay, Lapland Rosebay,Lapland Rose-bay ●

331058　Rhododendron lapponicum（L.）Wahlenb. subsp. parvifolium（Adams）T. Yamaz.；小叶高山杜鹃●

331059　Rhododendron lapponicum（L.）Wahlenb. subsp. parvifolium（Adams）T. Yamaz. = Rhododendron lapponicum（L.）Wahlenb. ●

331060　Rhododendron lapponicum（L.）Wahlenb. var. parvifolium（Adams）Herder = Rhododendron lapponicum（L.）Wahlenb. subsp. parvifolium（Adams）T. Yamaz. ●☆

331061　Rhododendron lasiopodum Hutch. = Rhododendron roseatum Hutch. ●

331062　Rhododendron lasiostylum Hayata；毛花柱杜鹃（毛柱杜鹃，埔里杜鹃）；Hairstyle Azalea, Woollystyle Rhododendron, Woolly-styled Rhododendron ●

331063　Rhododendron lateriflorum R. C. Fang et A. L. Zhang；侧花杜鹃；Lateral Flowered Rhododendron, Lateral-flowered Rhododendron, Laterflower Azalea ●

331064　Rhododendron lateritium Planch. = Rhododendron indicum（L.）Sweet ●

331065　Rhododendron latoucheae Franch.；鹿角杜鹃（湖北单花杜鹃，鹿角杜鹃，老虎花，西施花，岩杜鹃，紫蓝花杜鹃）；Deeppurple Mrs. E. H. Wilson's Azalea, Deerhorn Azalea, Deerhorn Rhododendron, Deer-horn Rhododendron, Mrs. E. H. Wilson's Azalea, Mrs. E. H. Wilson's Rhododendron ●

331066　Rhododendron latoucheae Franch. var. amamiense（Ohwi）T. Yamaz.；天见杜鹃●☆

331067　Rhododendron latoucheae Franch. var. amamiense（Ohwi）T. Yamaz. = Rhododendron latoucheae Franch. ●

331068　Rhododendron latoucheae Franch. var. ionanthum（W. P. Fang）G. Z. Li = Rhododendron latoucheae Franch. ●

331069　Rhododendron laudandum Cowan；毛冠杜鹃（赞美杜鹃）；Haircorol Azalea, Praiseworthy Rhododendron ●

331070　Rhododendron laudandum Cowan var. temoense Kingdon-Ward ex Cowan et Davidian；疏毛冠杜鹃；Praiseworthy Rhododendron, Tomentose Rhododendron ●

331071　Rhododendron laxiflorum Balf. f. et Forrest = Rhododendron annae Franch. subsp. laxiflorum（Balf. f. et Forrest）T. L. Ming ●

331072　Rhododendron leclerei H. Lév. = Rhododendron rubiginosum Franch. var. leclerei（H. Lév.）R. C. Fang ●

331073　Rhododendron ledebouri Pojark.；赖氏杜鹃●☆

331074　Rhododendron ledifolium（Hook.）G. Don = Rhododendron mucronatum（Blume）G. Don ●

331075　Rhododendron ledifolium G. Don = Rhododendron mucronatum（Blume）G. Don ●

331076　Rhododendron ledoides Balf. f. et W. W. Sm. = Rhododendron trichostomum Franch. var. ledoides（Balf. f. et W. W. Sm.）Cowan et Davidian ●

331077　Rhododendron leei W. P. Fang；紫腺杜鹃●

331078　Rhododendron leei W. P. Fang = Rhododendron faberi Hemsl. subsp. prattii（Franch.）D. F. Chamb. ex Cullen et D. F. Chamb. ●

331079　Rhododendron leiboense Z. J. Zhao；雷波杜鹃；Leibo Azalea, Leibo Rhododendron ●

331080　Rhododendron leilungense Balf. f. et Forrest = Rhododendron tatsienense Franch. ●

331081　Rhododendron leiopodum Hayata；光脚杜鹃●

331082　Rhododendron leiopodum Hayata = Rhododendron ellipticum Maxim. ●

331083　Rhododendron leiopodum Hayata = Rhododendron latoucheae Franch. ●

331084　Rhododendron leiopodum Hayata var. amamiense（Ohwi）Ohwi = Rhododendron latoucheae Franch. var. amamiense（Ohwi）T. Yamaz. ●☆

331085　Rhododendron leishanicum W. P. Fang et X. S. Zhang ex D. F. Chamb.；雷山杜鹃；Leishan Azalea, Leishan Rhododendron ●

331086　Rhododendron lemeei H. Lév. = Rhododendron lutescens Franch. ●

331087　Rhododendron lepidanthum Balf. f. et W. W. Sm. = Rhododendron primuliflorum Bureau et Franch. var. lepidanthum（Balf. f. et W. W. Sm.）Cowan et Davidian ●

331088　Rhododendron lepidostylum Balf. f. et W. W. Sm.；常绿糙毛杜鹃（常绿杜鹃，鳞柱杜鹃）；Evergreen Azalea, Scaly Styles Rhododendron, Scaly-styled Rhododendron ●

331089　Rhododendron lepidotum Wall. ex G. Don；鳞腺杜鹃（柳叶杜鹃，小鳞杜鹃）；Fleshy Scales Rhododendron, Scaly Azalea, Willowleaf Rhododendron, Willow-leaved Rhododendron ●

331090　Rhododendron leptanthum Hayata = Rhododendron ellipticum Maxim. ●

331091　Rhododendron leptanthum Hayata = Rhododendron latoucheae Franch. ●

331092　Rhododendron leptocarpum Nutt.；异鳞杜鹃●

331093　Rhododendron leptocladon Dop；细枝林生杜鹃（金平林生杜鹃）●

331094　Rhododendron leptopeplum Balf. f. et Forrest；腺绒杜鹃（薄腺杜鹃）；Glandfloss Azalea, Thin Glands Rhododendron, Thin-glanded Rhododendron ●

331095　Rhododendron leptosanthum Hayata = Rhododendron ellipticum Maxim. ●

331096　Rhododendron leptosanthum Hayata = Rhododendron latoucheae Franch. ●

331097　Rhododendron leptothrium Balf. f. et W. W. Sm.；薄叶马银花（薄叶杜鹃）；Thinleaf Azalea, Thinleaf Rhododendron, Thin-leaved Rhododendron ●

331098　Rhododendron leucandrum H. Lév. = Rhododendron siderophyllum Franch. ●

331099　Rhododendron leucaspis Tagg；白背杜鹃（白盾杜鹃）；White Shield Rhododendron, Whiteback Azalea, White-backed Rhododendron ●

331100　Rhododendron leucobotrys Ridl. = Rhododendron moulmainense Hook. ●

331101　Rhododendron leucolasium Diels = Rhododendron hunnewellianum Rehder et E. H. Wilson ●

331102　Rhododendron leucopetalum Balf. f. et Forrest = Rhododendron sanguineum Franch. var. cloiophorum（Balf. f. et Forrest）D. F. Chamb. ●

331103　Rhododendron levinei Merr.；南岭杜鹃（北江杜鹃）；Levine Rhododendron, Nanling Azalea ●

331104　Rhododendron levistratum Balf. f. et Forrest = Rhododendron phaeochrysum Balf. f. et W. W. Sm. var. levistratum（Balf. f. et Forrest）D. F. Chamb. ●

331105　Rhododendron liaoxiense S. L. Tung et Z. Lu；辽西杜鹃；Liaoxi Azalea, Liaoxi Rhododendron ●

331106　Rhododendron liboense Zheng R. Chen et K. M. Lan；荔波杜鹃●

331107　Rhododendron liliiflorum H. Lév.；百合花杜鹃；Lily Azalea, Lily-like Rhododendron ●

331108　Rhododendron limprichtii Diels = Rhododendron oreodoxa Franch. ●

331109　Rhododendron lindleyi T. Moore；大花杜鹃（林氏杜鹃）；J. Lindley's Rhododendron, Lindley Azalea, Lindley's Rhododendron ●

331110　Rhododendron linearicalyx T. L. Ming；线裂杜鹃；Linear Calyx

Azalea, Linear Calyx Rhododendron ●

331111   Rhododendron linearicupulare P. C. Tam;横县杜鹃(线杯杜鹃); Hengxian Azalea, Hengxian Rhododendron, Linear Cupulate Rhododendron ●

331112   Rhododendron linearicupulare P. C. Tam = Rhododendron westlandii Hemsl. ●

331113   Rhododendron linearifolium Siebold et Zucc.;线叶杜鹃;Linear Leaves Rhododendron ●

331114   Rhododendron linearifolium Siebold et Zucc. = Rhododendron macrosepalum Maxim. ●☆

331115   Rhododendron linearilobum R. C. Fang et A. L. Chang;线萼杜鹃;Linear Lobed Calyx Azalea, Linear Lobed Calyx Rhododendron, Linear-sepaled Rhododendron ●

331116   Rhododendron lingii Chun ex Ching = Rhododendron rhuyuenense Chun ex P. C. Tam ●

331117   Rhododendron linguiense G. Z. Li;临桂杜鹃;Lingui Rhododendron ●

331118   Rhododendron liratum Balf. f. et Forrest = Rhododendron dichroanthum Diels et Cowan subsp. apodectum (Balf. f. et W. W. Sm.) Cowan ●

331119   Rhododendron litangense Balf. f. et Hutch. = Rhododendron impeditum Balf. f. et W. W. Sm. ●

331120   Rhododendron litchiifolium T. C. Wu et P. C. Tam;荔枝叶杜鹃(荔叶杜鹃);Litceeleaf Azalea, Litchi Leaves Rhododendron, Litchi-leaved Rhododendron ●

331121   Rhododendron lithophilum Balf. f. et Kingdon-Ward = Rhododendron trichocladum Franch. ●

331122   Rhododendron litiense Balf. f. et Forrest = Rhododendron wardii W. W. Sm. ●

331123   Rhododendron lochae F. Muell.;澳洲杜鹃(罗氏杜鹃); Australian Rhododendron ●☆

331124   Rhododendron lochmium Balf. f.;矮丛杜鹃;Coppice Rhododendron ●☆

331125   Rhododendron longesquamatum C. K. Schneid.;长鳞杜鹃;Long Scales Rhododendron, Longscale Azalea, Long-scaled Rhododendron ●

331126   Rhododendron longesquamatum C. K. Schneid. var. glabristylum Y. Y. Geng et Z. L. Zhao = Rhododendron longesquamatum C. K. Schneid. ●

331127   Rhododendron longesquamatum C. K. Schneid. var. glabristylum Y. Y. Geng et Z. L. Zhao;光柱长鳞杜鹃;Smoothstyle Longscale Azalea ●

331128   Rhododendron longicalyx M. Y. Fang;长萼杜鹃;Long Calyx Rhododendron, Longcalyx Azalea, Long-calyxed Rhododendron ●

331129   Rhododendron longifalcatum P. C. Tam;长尖杜鹃(长镰杜鹃); Long Falcate-fruit Rhododendron, Longifalcate-fruited Rhododendron, Longsickle Azalea ●

331130   Rhododendron longiflorum Nutt. =Rhododendron grande Wight ●

331131   Rhododendron longilobum L. M. Gao et D. Z. Li, Novon;凸纹杜鹃●

331132   Rhododendron longiperulatum Hayata;长鳞芽杜鹃(大屯杜鹃,大屯满山红);Long-perula Rhododendron ●

331133   Rhododendron longiperulatum Hayata = Rhododendron simsii Planch. ●

331134   Rhododendron longipes Rehder et E. H. Wilson;长柄杜鹃(长梗杜鹃);Longstalk Azalea, Long-stalk Rhododendron, Long-stalked Rhododendron ●◇

331135   Rhododendron longipes Rehder et E. H. Wilson var. chienianum

(W. P. Fang) D. F. Chamb. ex Cullen et D. F. Chamb.;金山杜鹃; Jinshan Azalea, Jinshan Long-stalked Rhododendron, Jinshan Rhododendron ●

331136   Rhododendron longistylum Rehder et E. H. Wilson;长柱杜鹃(长花柱杜鹃,长轴杜鹃);Longstyle Azalea, Longstyle Rhododendron, Long-styled Rhododendron ●

331137   Rhododendron longistylum Rehder et E. H. Wilson subsp. decumbense R. C. Fang;平卧长轴杜鹃;Decumbent Longstyle Azalea, Decumbent Longstyle Rhododendron ●

331138   Rhododendron loniceriflorum P. C. Tam;忍冬杜鹃;Honeysuckle Azalea, Lonicera Flowers Rhododendron, Lonicera-flowered Rhododendron ●

331139   Rhododendron lophogynum Balf. f. et Forrest ex Hutch.;鸡冠子房杜鹃;Crested Ovary Rhododendron ●☆

331140   Rhododendron lophogynum Balf. f. et Forrest ex Hutch. = Rhododendron trichocladum Franch. ●

331141   Rhododendron lophophorum Balf. f. et Forrest = Rhododendron phaeochrysum Balf. f. et W. W. Sm. var. agglutinatum (Balf. f. et Forrest) D. F. Chamb. ●

331142   Rhododendron lopsangianum Cowan;达赖杜鹃;Lopsang Dalai Rhododendron ●☆

331143   Rhododendron lopsangianum Cowan = Rhododendron thomsonii Hook. f. subsp. lopsangianum (Cowan) D. F. Chamb. ●

331144   Rhododendron loranthiflorum Sleumer;桑寄生叶杜鹃●☆

331145   Rhododendron lowndesii Davidian;罗恩杜鹃;Lowndes Rhododendron ●☆

331146   Rhododendron lucidum Franch. = Rhododendron vernicosum Franch. ●

331147   Rhododendron lucidum Nutt. = Rhododendron camelliiflorum Hook. f. ●

331148   Rhododendron lucidum Nutt. = Rhododendron vernicosum Franch. ●

331149   Rhododendron luciferum (Cowan) Cowan = Rhododendron lanatum Hook. f. ●

331150   Rhododendron ludlowii Cowan;广口杜鹃(路氏杜鹃);Ludlow Azalea, Ludlow's Rhododendron ●

331151   Rhododendron luhuoense H. P. Yang;炉霍杜鹃;Luhuo Azalea, Luhuo Rhododendron ●

331152   Rhododendron lukiangense Franch.;蜡叶杜鹃;Lukiang Rhododendron, Waxleaf Azalea ●

331153   Rhododendron lukiangense Franch. subsp. admirabile (Balf. f. et Forrest) Tagg = Rhododendron lukiangense Franch. ●

331154   Rhododendron lukiangense Franch. subsp. adroserum (Balf. f. et Forrest) Tagg = Rhododendron lukiangense Franch. ●

331155   Rhododendron lukiangense Franch. subsp. ceraceum (Balf. f. et W. W. Sm.) Tagg = Rhododendron lukiangense Franch. ●

331156   Rhododendron lukiangense Franch. subsp. gymnanthum (Diels) Tagg = Rhododendron lukiangense Franch. ●

331157   Rhododendron lulangense L. C. Hu et Tateishi;鲁浪杜鹃; Lulang Azalea, Lulang Rhododendron ●◇

331158   Rhododendron lungchiense W. P. Fang;龙溪杜鹃;Longxi Azalea, Longxi Rhododendron ●

331159   Rhododendron lutescens Franch.;黄花杜鹃(变黄杜鹃); Becoming Yellow Rhododendron, Canary Rhododendron, Yellow Azalea ●

331160   Rhododendron lutescens Franch. f. shimianense (W. P. Fang et P. S. Liu) Y. Y. Geng = Rhododendron shimianense W. P. Fang et

P. S. Liu ●

331161　Rhododendron luteum Sweet;纯黄杜鹃(黄杜鹃,黄花杜鹃,黄香杜鹃,黄香杜鹃花,欧洲黄杜鹃);Amber-bloom Rhododendron, Common Yellow Azalea, Flame Azalea, Golden Rhododendron, Pontic Azalea, Yellow Azalea, Yellow Rhododendron ●

331162　Rhododendron lyi H. Lév. = Rhododendron ciliicalyx Franch. subsp. lyi (H. Lév.) R. C. Fang ●

331163　Rhododendron lysolepis Hutch.;疏鳞杜鹃;Loose Scales Rhododendron ●☆

331164　Rhododendron macabeanum Watt ex Balf. f.;麦卡杜鹃(印度黄花杜鹃);McCabe's Rhododendron ●☆

331165　Rhododendron macgregoria F. Muell.;窄冠杜鹃●☆

331166　Rhododendron mackenzianum Forrest;长蒴杜鹃●

331167　Rhododendron mackenzianum Forrest = Rhododendron stenaulum Balf. f. et W. W. Sm. ●

331168　Rhododendron macranthum (Bunge) G. Don = Rhododendron indicum (L.) Sweet ●

331169　Rhododendron macranthum Griff. = Rhododendron maddenii Hook. f. ●

331170　Rhododendron macrogemma Nakai;大鳞山踯躅●☆

331171　Rhododendron macrophyllum D. Don ex G. Don;太平洋杜鹃; Pacific Rhododendron ●☆

331172　Rhododendron macrophyllum D. Don ex G. Don = Rhododendron californicum Hook. ●☆

331173　Rhododendron macrosepalum Maxim.;日本大萼杜鹃●☆

331174　Rhododendron macrosepalum Maxim. ' Linearifolium ';线叶日本大萼杜鹃●☆

331175　Rhododendron macrosepalum Maxim. f. albiflorum (Honda) H. Hara;白花日本大萼杜鹃●☆

331176　Rhododendron macuiferum Franch.;麻花杜鹃(斑点杜鹃,黑斑杜鹃);Blotchflor Azalea, Duskbloch Rhododendron, Maculate Rhododendron, Spotted Rhododendron ●

331177　Rhododendron maculiferum Franch. subsp. anwheiense (E. H. Wilson) D. F. Chamb. ex Cullen et D. F. Chamb.;安徽杜鹃(黄山杜鹃);Anhui Rhododendron ●

331178　Rhododendron maculiferum Franch. subsp. anwheiense (E. H. Wilson) D. F. Chamb. = Rhododendron maculiferum Franch. subsp. anwheiense (E. H. Wilson) D. F. Chamb. ex Cullen et D. F. Chamb. ●

331179　Rhododendron maddenii Hook. f.;隐脉杜鹃(马登杜鹃,麦登杜鹃);Madden Rhododendron, Maior Madden's Rhododendron ●

331180　Rhododendron maddenii Hook. f. subsp. crassum (Franch.) Cullen;滇藏隐脉杜鹃(滇隐脉杜鹃,肥厚杜鹃,厚叶杜鹃);Fleshy Rhododendron ●

331181　Rhododendron maddenii Hook. f. var. longiflorum W. Watson = Rhododendron maddenii Hook. f. ●

331182　Rhododendron magnificum Kingdon-Ward;强壮杜鹃(矮壮杜鹃,大花杜鹃,雄壮杜鹃);Magnific Rhododendron, Strong Azalea ●

331183　Rhododendron magniflorum W. K. Hu;贵州大花杜鹃;Big-flowered Rhododendron, Guizhou Bigflower Azalea, Guizhou Rhododendron ●

331184　Rhododendron maguanense K. M. Feng;马关杜鹃;Maguan Azalea, Maguan Rhododendron ●

331185　Rhododendron mainlingense S. H. Huang et R. C. Fang;米林杜鹃;Milin Azalea, Milin Rhododendron ●

331186　Rhododendron mairei H. Lév. = Rhododendron lacteum Franch. ●

331187　Rhododendron makinoi Tagg ex Nakai;牧野杜鹃;Makino's Rhododendron ●☆

331188　Rhododendron makinoi Tagg ex Nakai f. leucanthum (Makino) Sugim.;白花牧野杜鹃;Whiteflower Makino's Rhododendron ●☆

331189　Rhododendron makinoi Tagg ex Nakai f. muranoanum (Makino) H. Hara;五裂牧野杜鹃;Fivelobed Makino's Rhododendron ●☆

331190　Rhododendron makinoi Tagg ex Nakai f. plenum Sugim.;重瓣牧野杜鹃●☆

331191　Rhododendron makinoi Tagg ex Nakai var. basirosaceum Makino;基红牧野杜鹃●☆

331192　Rhododendron makinoi Tagg ex Nakai var. roseum Makino;红花牧野杜鹃;Rose Makino's Rhododendron ●☆

331193　Rhododendron malayanum Jack;马来杜鹃;Malaya Rhododendron ●☆

331194　Rhododendron malipoense M. Y. He;麻栗坡杜鹃;Malipo Azalea, Malipo Rhododendron ●

331195　Rhododendron mallotum Balf. f. et Kingdon-Ward;羊毛杜鹃(软毛杜鹃);Fleecy Rhododendron, Wool Azalea, Woolly Rhododendron ●

331196　Rhododendron mandarinorum Diels = Rhododendron discolor Franch. ●

331197　Rhododendron manipurense Balf. f. et Watt;曼尼坡杜鹃;Manipur Azalea, Manipur Rhododendron ●☆

331198　Rhododendron manopeplum Balf. f. et Forrest = Rhododendron esetulosum Balf. f. et Forrest ●

331199　Rhododendron manophorum Balf. f. et Forrest = Rhododendron sanguineum Franch. var. didymoides Tagg et Forrest ●

331200　Rhododendron maoerense W. P. Fang et G. Z. Li;猫儿山杜鹃;Maoershan Azalea, Maoershan Rhododendron ●

331201　Rhododendron maowenense Ching et H. P. Yang;茂汶杜鹃;Maowen Azalea, Maowen Rhododendron ●

331202　Rhododendron mariae Hance;岭南杜鹃(广东紫花杜鹃,土牡丹花,紫杜鹃,紫花杜鹃);Lingnan Azalea, Southeast China Rhododendron ●

331203　Rhododendron mariesii Hemsl. et E. H. Wilson;满山红(戴云山杜鹃,马礼士杜鹃,玛丽杜鹃,守城满山红);Daiyunshan Rhododendron, Maries' Azalea, Maries' Rhododendron, Redhillall Azalea ●

331204　Rhododendron mariesii Hemsl. et E. H. Wilson var. albescens B. Y. Ding et G. R. Chen;白花满山红;Whiteflower Maries' Azalea ●

331205　Rhododendron martinianum Balf. f. et Forrest;少花杜鹃(马丁杜鹃);Martin's Rhododendron, Poorflower Azalea ●☆

331206　Rhododendron maximowiczianum H. Lév. = Rhododendron irroratum Franch. ●

331207　Rhododendron maximum L.;极大杜鹃(最大杜鹃);American Great Laurel, American Great Rhododendron, Great American Laurel, Great Laurel, Great Laurel Rhododendron, Rose Bay, Rosebaby Rhododendron, Rosebay, Very Large Rhododendron, White Laurel ●☆

331208　Rhododendron maximum L. ' Summertime ';夏日时光极大杜鹃●☆

331209　Rhododendron meddianum Forrest;深红杜鹃(红萼杜鹃,米德杜鹃);G. Medd's Rhododendron, Redcalyx Azalea, Red-calyxed Rhododendron ●

331210　Rhododendron meddianum Forrest var. atrokermesinum Tagg;腺房红萼杜鹃●

331211　Rhododendron medoense W. P. Fang et M. Y. He;墨脱马银花(墨脱杜鹃);Medog Rhododendron, Motuo Azalea, Motuo Rhododendron ●

331212　Rhododendron megacalyx Balf. f. et Kingdon-Ward;大萼杜鹃;

Bigcalyx Rhododendron, Large Calyx Rhododendron, Large Rhododendron, Largecalyx Azalea ●

331213 Rhododendron megalanthum M. Y. Fang;墨脱大花杜鹃(大花杜鹃,墨脱杜鹃,西藏杜鹃);Motuo Rhododendron ●

331214 Rhododendron megaphyllum Balf. f. et Forrest = Rhododendron basilicum Balf. f. et W. W. Sm. ●

331215 Rhododendron megeratum Balf. f. et Forrest;招展杜鹃;Flaunt Azalea, Outspread Rhododendron, Passing Lovely Rhododendron ●

331216 Rhododendron mekongense Franch. ;弯月杜鹃(湄公河杜鹃);Mekong River Rhododendron, Newmoon Azalea ●

331217 Rhododendron mekongense Franch. var. longipilosum (Cowan) Cullen;长毛弯月杜鹃(蜜花弯月杜鹃);Long-haired Mekong River Rhododendron ●

331218 Rhododendron mekongense Franch. var. melinanthum (Balf. f. et Kingdon-Ward) Cullen;密花弯月杜鹃(黄绿杜鹃,密花杜鹃,弯月杜鹃); Greenflower Azalea, Half-crescent-shaped Rhododendron, Honeyflower Azalea, Honey-flower Rhododendron, Newmoon Rhododendron ●

331219 Rhododendron mekongense Franch. var. rubrolineatum (Balf. f. et Forrest) Cullen;红线弯月杜鹃(红线杜鹃);Red-linear Mekong River Rhododendron, Red-lined Rhododendron ●

331220 Rhododendron melinanthum Balf. f. et Kingdon-Ward = Rhododendron mekongense Franch. var. melinanthum (Balf. f. et Kingdon-Ward) Cullen ●

331221 Rhododendron mengtszense Balf. f. et W. W. Sm. ;蒙自杜鹃;Mengzi Azalea, Mengzi Rhododendron ●

331222 Rhododendron meridionale P. C. Tam;南边杜鹃(南方杜鹃);Nanbian Azalea, Southern Rhododendron ●

331223 Rhododendron meridionale P. C. Tam var. minor P. C. Tam;狭叶南边杜鹃(狭叶南方杜鹃);Narrow-leaved Southern Rhododendron ●

331224 Rhododendron meridionale P. C. Tam var. setistylum P. C. Tam;糙柱南边杜鹃(糙柱杜鹃);Setistyled Southern Rhododendron ●

331225 Rhododendron mesopolium Balf. f. et Forrest = Rhododendron eudoxum Balf. f. et Forrest var. mesopolium (Balf. f. et Forrest) D. F. Chamb. ●

331226 Rhododendron metrium Balf. f. et Forrest = Rhododendron selense Franch. ●

331227 Rhododendron metternichii Siebold et Zucc. ;麦特杜鹃(梅特尼氏杜鹃,七数杜鹃,石南);Metterich Rhododendron ●☆

331228 Rhododendron metternichii Siebold et Zucc. = Rhododendron degronianum Carrière ●☆

331229 Rhododendron metternichii Siebold et Zucc. = Rhododendron degronianum Carrière subsp. heptamerum (Maxim.) H. Hara ●☆

331230 Rhododendron metternichii Siebold et Zucc. f. leucanthum Nakai;白花麦特杜鹃(白花石南)●☆

331231 Rhododendron metternichii Siebold et Zucc. subsp. pentamerum (Maxim.) Sugim. = Rhododendron japonicum (Blume) C. K. Schneid. var. pentamerum (Maxim.) Hutch. ●☆

331232 Rhododendron metternichii Siebold et Zucc. subsp. yakushimanum (Nakai) Sugim. = Rhododendron yakushimanum Nakai ●☆

331233 Rhododendron metternichii Siebold et Zucc. var. hondoense Nakai = Rhododendron japonoheptamerum Kitam. var. hondoense (Nakai) Kitam. ●☆

331234 Rhododendron metternichii Siebold et Zucc. var. pentamerum Maxim. = Rhododendron japonicum (Blume) C. K. Schneid. var. pentamerum (Maxim.) Hutch. ●☆

331235 Rhododendron metternichii Siebold et Zucc. var. pentamerum T. Yamaz. = Rhododendron japonicum (Blume) C. K. Schneid. var. pentamerum (Maxim.) Hutch. ●☆

331236 Rhododendron metternichii Siebold et Zucc. var. yakushimanum (Nakai) Ohwi = Rhododendron yakushimanum Nakai ●☆

331237 Rhododendron metternichii Siebold et Zucc. var. yakusimanum (Nakai) Ohwi;屋久岛杜鹃;Yakusima Rhododendron ●☆

331238 Rhododendron mianningense Z. J. Zhao;冕宁杜鹃;Mianning Azalea, Mianning Rhododendron ●

331239 Rhododendron micranthum Turcz. ;照山白(白花杜鹃,白镜子,达里,铁石茶,万斤,万经棵,小白花杜鹃,小杜鹃,照白杜鹃); Manchurian Rhododendron, Small Flowered Rhododendron, White-hill Azalea ●

331240 Rhododendron microgynum Balf. f. et Forrest;短蕊杜鹃(光果杜鹃,具鳞杜鹃,鳞芽杜鹃,小花杜鹃,小子房杜鹃);Naked Fruits Rhododendron, Scaly Buds Rhododendron, Shortpistil Azalea, Small Ovary Rhododendron, Small-ovared Rhododendron ●

331241 Rhododendron micromeres Tagg;匍匐小杜鹃(异鳞杜鹃); Differscale Azalea, Micromered Rhododendron, Small Straggly Rhododendron ●

331242 Rhododendron micromeres Tagg = Rhododendron leptocarpum Nutt. ●

331243 Rhododendron microphyton Franch. ;亮毛杜鹃(金瓶花,酒瓶花,小杜鹃); Brighthair Azalea, Pinkflush Rhododendron, Pink-flushed Rhododendron, Small Azalea, Small Rhododendron ●

331244 Rhododendron microphyton Franch. var. trichanthum A. L. Chang ex R. C. Fang;碧江亮毛杜鹃;Bijiang Azalea, Bijiang Pinkflush Rhododendron ●

331245 Rhododendron mimetes Tagg et Forrest;小萼杜鹃(模拟杜鹃,优异杜鹃);Excellent Azalea, Imitative Rhododendron ●

331246 Rhododendron mimetes Tagg et Forrest var. simulans Tagg et Forrest = Rhododendron simulans (Tagg et Forrest) D. F. Chamb. ●

331247 Rhododendron miniatum Cowan;焰红杜鹃;Fire-red Azalea, Fleme Red Rhododendron, Fleme-red Rhododendron ●

331248 Rhododendron minus Michx. ;较小杜鹃;Dwarf Azalea, Dwarf Rhododendron, Piedmont Azalea, Piedmont Rhododendron, Smaller Rhododendron ●☆

331249 Rhododendron minutiflorum Hu;小花杜鹃(细花杜鹃); Miniflower Azalea, Minute-flowered Rhododendron, Minuteflower Rhododendron, Smallest Flowers Rhododendron ●

331250 Rhododendron minyaense M. N. Philipson et Philipson;黄褐杜鹃(斑鳞杜鹃);Spoted Scales Rhododendron, Yellow-brown Azalea ●

331251 Rhododendron mirabile Kingdon-Ward = Rhododendron genestierianum Forrest ●

331252 Rhododendron mishmiense Hutch. et Kingdon-Ward = Rhododendron boothii Nutt. ●

331253 Rhododendron missionarium H. Lév. = Rhododendron ciliicalyx Franch. ●

331254 Rhododendron mitriforme P. C. Tam;头巾马银花(冠形杜鹃,头巾杜鹃,兴安马银花);Coif Azalea, Mitriform Rhododendron ●

331255 Rhododendron mitriforme P. C. Tam var. setaceum P. C. Tam;腺刺马银花(腺刺杜鹃);Setaceous Mitriform Rhododendron ●

331256 Rhododendron miyiense W. K. Hu;米易杜鹃;Miyi Azalea, Miyi Rhododendron ●

331257 Rhododendron molle (Blume) G. Don;羊踯躅(八厘麻,八里麻,巴山虎,豹狗花,出山彪,大叶株标,黄稻节柴,黄杜鹃,黄杜鹃花,黄牯牛花,黄花花,黄喇叭花,黄色映山红,黄蛇豹花,黄踯躅,黄株标,惊羊花,老虎花,老鸦花,六轴子,毛老虎,闷头花,南天竺草,闹牛花,闹羊花,三钱三,山枇杷,山芝麻,石菊花,石六轴,石棠花,水兰花,搜山虎,天芝麻,土连翘,羊不吃草,羊不食草,羊踯躅花,一杯倒,一杯醉,影山黄,映山黄,玉支,踯躅,踯躅花,中国杜鹃,坐山虎);Chinese Azalea, Chinese Rhododendron, Deciduous Azalea, Sheeploitered Azalea ●

331258 Rhododendron molle (Blume) G. Don subsp. japonicum (A. Gray) Kron;日本杜鹃(闹羊花,日本杜鹃花,日本羊踯躅);Japanese Azalea ●☆

331259 Rhododendron molle (Blume) G. Don var. glabrius Miq. = Rhododendron molle (Blume) G. Don subsp. japonicum (A. Gray) Kron ●☆

331260 Rhododendron molle G. Don subsp. japonicum (A. Gray) Kron f. canescens (Sugim.) Yonek.;灰色日本杜鹃(灰色日本羊踯躅)●☆

331261 Rhododendron molle G. Don subsp. japonicum (A. Gray) Kron f. flavum (Miyoshi) Yonek.;黄花日本杜鹃(黄杜鹃,黄色日本羊踯躅);Yellow Japanese Azalea ●☆

331262 Rhododendron molle G. Don subsp. japonicum (A. Gray) Kron f. glaucophyllum (Nakai) Yonek.;里白日本杜鹃(蓝绿叶日本羊踯躅)●☆

331263 Rhododendron molle G. Don subsp. japonicum (A. Gray) Kron f. multifidum (Nakai) Yonek.;多裂日本杜鹃(多裂日本羊踯躅)●☆

331264 Rhododendron mollianum Cowan et Davidian = Rhododendron montroseanum Davidian ●

331265 Rhododendron mollianum Koord. = Rhododendron montroseanum Davidian ●

331266 Rhododendron mollicomum Balf. f. et W. W. Sm.;柔毛碎米花(毛叶杜鹃);Pubescent-leaved Rhododendron, Softhair Azalea, Soft-haired Leaves Rhododendron ●

331267 Rhododendron mollicomum Balf. f. et W. W. Sm. var. rockii Tagg = Rhododendron mollicomum Balf. f. et W. W. Sm. ●

331268 Rhododendron mollyanum Cowan et Davidian = Rhododendron montroseanum Davidian ●

331269 Rhododendron mombeigii Rehder et E. H. Wilson = Rhododendron uvarifolium Diels ●

331270 Rhododendron monanthum Balf. f. et W. W. Sm.;一朵花杜鹃(独花杜鹃);Oneflower Azalea, Oneflower Rhododendron, Uniflorous Rhododendron ●

331271 Rhododendron monbeigii Rehder et E. H. Wilson = Rhododendron uvarifolium Diels ●

331272 Rhododendron monosematum Hutch. = Rhododendron strigillosum Franch. var. monosematum (Hutch.) T. L. Ming ●

331273 Rhododendron montigenum T. L. Ming;山地杜鹃;Montane Azalea, Monticule Rhododendron ●

331274 Rhododendron montroseanum Davidian;慕氏杜鹃(墨脱杜鹃);Molly's Rhododendron, Montro Rhododendron, Motuo Azalea, Motuo Rhododendron ●

331275 Rhododendron moriakianum T. Suzuki;铃木氏杜鹃;Noriake Azalea, Noriake Rhododendron ●☆

331276 Rhododendron morii Hayata;玉山杜鹃(森氏杜鹃)●

331277 Rhododendron morii Hayata = Rhododendron pseudochrysanthum Hayata subsp. morii (Hayata) T. Yamaz. ●

331278 Rhododendron morii Hayata var. taitunense (T. Yamaz.) D. F. Chamb. = Rhododendron morii Hayata ●

331279 Rhododendron motsouense H. Lév. = Rhododendron racemosum Franch. ●

331280 Rhododendron moulmainense Hook.;毛棉杜鹃(白花木,黄白管花杜鹃,尖叶杜鹃,毛棉杜鹃花,南海杜鹃,丝线吊芙蓉,韦氏杜鹃);Maomian Azalea, Moulmain's Rhododendron, Pointed Azalea, Pointed Leaves Rhododendron, Short Stamen Rhododendron, Westland Azalea, Westland Rhododendron ●

331281 Rhododendron moulmainense Hook. var. calcaratum G. Z. Li;距药毛棉杜鹃;Calcarate Moulmain's Rhododendron ●

331282 Rhododendron moulmainense Hook. var. calcaratum G. Z. Li = Rhododendron westlandii Hemsl. ●

331283 Rhododendron moupinense Franch.;宝兴杜鹃(穆坪杜鹃,穆坪杜鹃花);Baoxing Azalea, Mouping Rhododendron, Paohsing Rhododendron ●

331284 Rhododendron mucronatum (Blume) G. Don;钝白花杜鹃(白杜鹃,白杜鹃花,白花杜鹃花,白花迎山红,白花映山红,白艳山红,白映山红,尖叶杜鹃,满山红,毛白杜鹃,迎红杜鹃,映山红,照山白);Snow Azalea, Snow Rhododendron, White Azalea ●

331285 Rhododendron mucronatum (Blume) G. Don 'Narcissiflorum';水仙花杜鹃●☆

331286 Rhododendron mucronatum (Blume) G. Don 'Plenum';重瓣白花杜鹃●☆

331287 Rhododendron mucronatum (Blume) G. Don var. ripense (Makino) E. H. Wilson = Rhododendron ripense Makino ●☆

331288 Rhododendron mucronatum G. Don = Rhododendron mucronatum (Blume) G. Don ●

331289 Rhododendron mucronulatum Turcz. 迎红杜鹃(尖叶杜鹃,蓝荆子,满山红,迎山红,映山红);Korean Azalea, Korean Rhododendron, Point Leaves Rhododendron ●

331290 Rhododendron mucronulatum Turcz. 'Alba';白花迎红杜鹃●☆

331291 Rhododendron mucronulatum Turcz. 'Cornell Pink';康奈尔粉迎红杜鹃●☆

331292 Rhododendron mucronulatum Turcz. 'Crater's Edge';碗边迎红杜鹃●☆

331293 Rhododendron mucronulatum Turcz. 'Mahogany Red';红褐迎红杜鹃●☆

331294 Rhododendron mucronulatum Turcz. f. albiflorum Okuyama = Rhododendron mucronulatum Turcz. 'Alba' ●☆

331295 Rhododendron mucronulatum Turcz. f. ciliatum (Nakai) Kitag. = Rhododendron mucronulatum Turcz. var. ciliatum Nakai ●☆

331296 Rhododendron mucronulatum Turcz. var. ciliatum Nakai;粗毛迎红杜鹃(毛叶迎红杜鹃)●☆

331297 Rhododendron mucronulatum Turcz. var. ciliatum Nakai f. leucanthum T. Yamaz.;白花粗毛迎红杜鹃●☆

331298 Rhododendron muliense Balf. f. et Forrest = Rhododendron rupicola W. W. Sm. var. muliense (Balf. f. et Forrest) M. N. Philipson et Philipson ●

331299 Rhododendron mussoti Franch. = Rhododendron wardii W. W. Sm. ●

331300 Rhododendron myiagrum Balf. f. et Forrest = Rhododendron callimorphum Balf. f. et W. W. Sm. var. myiagrum (Balf. f. et Forrest) D. F. Chamb. ex Cullen et D. F. Chamb. ●

331301 Rhododendron myrsinifolium Ching ex W. P. Fang et M. Y. He;铁仔叶杜鹃(丛枝杜鹃,铁仔杜鹃);Myrsine-like Azalea, Myrsine-like Rhododendron ●

331302 Rhododendron myrtilloides Balf. f. et Kingdon-Ward = Rhododendron campylogynum Franch. ●

331303 Rhododendron naamkwanense Merr.;南昆杜鹃(溪岩杜鹃);Nankun Azalea, Nankun Rhododendron, Stream Stone Side Rhododendron, Stream-side Rhododendron ●

331304 Rhododendron naamkwanense Merr. var. cryptonerve P. C. Tam；紫薇春；Cryptonerved Nankun Rhododendron ●

331305 Rhododendron nakaharai Hayata；大河口杜鹃（百里香叶杜鹃，大屯山杜鹃，那克哈杜鹃，中村杜鹃，中原氏杜鹃）；Datunshan Rhododendron, G. Nakahara's Rhododendron, Nakahara Azalea, Tatunsan Rhododendron, Thyme Leaves Azalea, Wildthyme Azalea ●

331306 Rhododendron nakaii Komatsu = Rhododendron degronianum Carrière ●☆

331307 Rhododendron nakotiltum Balf. f. et Forrest；德钦杜鹃（间断杜鹃）；Deqin Azalea, Interrupted Rhododendron ●

331308 Rhododendron nanfumontanum Hayata；南湖杜鹃；Nanhu Azalea, Nanhu Rhododendron ●

331309 Rhododendron nanjianense K. M. Feng et Z. H. Yang；南涧杜鹃；Nanjian Azalea, Nanjian Rhododendron ●

331310 Rhododendron nankingense (Cowan) D. F. Chamb.；梵净山杜鹃；Fanjingshan Azalea, Fanjingshan Rhododendron ●

331311 Rhododendron nankotaisanense Hayata；南口大山杜鹃（玉山杜鹃）；Mt. Nankotaisan Rhododendron ●

331312 Rhododendron nankotaisanense Hayata = Rhododendron pseudochrysanthum Hayata ●

331313 Rhododendron nanothamnum Balf. f. et Forrest = Rhododendron selense Franch. ●

331314 Rhododendron nanpingense P. C. Tam；南平杜鹃；Nanping Azalea, Nanping Rhododendron ●

331315 Rhododendron nanum H. Lév. = Rhododendron farrerae Tate ex Sweet ●

331316 Rhododendron nanum H. Lév. = Rhododendron fastigiatum Franch. ●

331317 Rhododendron nebrites Balf. f. et Forrest = Rhododendron sanguineum Franch. var. himertum (Balf. f. et Forrest) T. L. Ming ●

331318 Rhododendron nebrities Balf. f. et Forrest = Rhododendron sanguineum Franch. var. haemaleum (Balf. f. et Forrest) D. F. Chamb. ●

331319 Rhododendron nematocalyx Balf. f. et W. W. Sm. = Rhododendron moulmainense Hook. ●

331320 Rhododendron nemorosum R. Z. Fang；金平林生杜鹃（林生杜鹃）；Forest Rhododendron, Jinping Forest Azalea ●

331321 Rhododendron neriiflorum Franch.；火红杜鹃（夹竹桃杜鹃，美毛杜鹃）；Beautiful Hairs Rhododendron, Oleander Flowered Rhododendron, Oleander Rhododendron ●

331322 Rhododendron neriiflorum Franch. subsp. agetum (Balf. f. et Forrest) Tagg = Rhododendron neriiflorum Franch. var. agetum (Balf. f. et Forrest) T. L. Ming ●

331323 Rhododendron neriiflorum Franch. subsp. euchaites (Balf. f. et Forrest) Tagg = Rhododendron neriiflorum Franch. ●

331324 Rhododendron neriiflorum Franch. subsp. phaedropum (Balf. f. et Forrest) Tagg = Rhododendron neriiflorum Franch. var. appropinquans (Tagg et Forrest) W. K. Hu ●

331325 Rhododendron neriiflorum Franch. subsp. phoenicodum (Balf. f. et Forrest) Tagg = Rhododendron neriiflorum Franch. ●

331326 Rhododendron neriiflorum Franch. var. agetum (Balf. f. et Forrest) T. L. Ming；网眼火红杜鹃●

331327 Rhododendron neriiflorum Franch. var. appropinquans (Tagg et Forrest) W. K. Hu；腺房火红杜鹃●

331328 Rhododendron neriiflorum Franch. var. euchaites (Balf. f. et Forrest) Davidian = Rhododendron neriiflorum Franch. ●

331329 Rhododendron neriiflorum Franch. var. phaedropum (Balf. f. et Forrest) T. L. Ming = Rhododendron neriiflorum Franch. var. appropinquans (Tagg et Forrest) W. K. Hu ●

331330 Rhododendron neriiflorum Franch. var. phaedropum (Balf. f. et Forrest) T. L. Ming = Rhododendron neriiflorum Franch. var. appropinquans (Tagg et Forrest) W. K. Hu ●

331331 Rhododendron nigroglandulosum Nitz.；大炮山杜鹃（黑腺杜鹃）；Black Glands Rhododendron, Blackgland Azalea ●

331332 Rhododendron nigropunctatum Bureau et Franch.；黑斑杜鹃（黑鳞杜鹃）；Black Spots Rhododendron, Blackscale Azalea ●

331333 Rhododendron nigropunctatum Bureau et Franch. = Rhododendron nivale Hook. f. subsp. boreale M. N. Philipson et Philipson ●

331334 Rhododendron nikoense (Komatsu) Nakai = Rhododendron pentaphyllum Maxim. var. nikoense Komatsu ●☆

331335 Rhododendron nikomontanum Nakai；日光山地杜鹃●☆

331336 Rhododendron nilagiricum Zenker；尼山杜鹃●☆

331337 Rhododendron ningyuenense Hand.-Mazz.；宁云杜鹃；Ningyun Azalea, Ningyun Rhododendron ●

331338 Rhododendron ningyuenense Hand.-Mazz. = Rhododendron irroratum Franch. ●

331339 Rhododendron niphargum Balf. f. et Kingdon-Ward = Rhododendron uvarifolium Diels ●

331340 Rhododendron nipholobum Balf. f. et Forrest = Rhododendron stewartianum Diels ●

331341 Rhododendron nipponicum Matsum.；东瀛杜鹃（大叶杜鹃）；Japanese Rhododendron ●☆

331342 Rhododendron nitens Hutch. = Rhododendron keleticum Nutt. ●

331343 Rhododendron nitidulum Rehder et E. H. Wilson；光亮杜鹃（堇花杜鹃）；Nitid Azalea, Violetbloom Rhododendron, Violet-bloom Rhododendron, Violet-purple Rhododendron ●

331344 Rhododendron nitidulum Rehder et E. H. Wilson var. nubigenum Rehder et E. H. Wilson = Rhododendron nitidulum Rehder et E. H. Wilson ●

331345 Rhododendron nitidulum Rehder et E. H. Wilson var. omeiense M. N. Philipson et Philipson；峨眉光亮杜鹃（光亮峨眉杜鹃）；Azalea, Emei Violetbloom Rhododendron, Omei Violetbloom Rhododendron ●

331346 Rhododendron nivale Hook. f.；雪层杜鹃（雪白杜鹃，沼泽杜鹃）；Glacier Rhododendron, Marshy Rhododendron, Snowlayer Azalea, Snowy Rhododendron ●

331347 Rhododendron nivale Hook. f. subsp. australe M. N. Philipson et Philipson；南方雪层杜鹃；South Snowy Rhododendron ●

331348 Rhododendron nivale Hook. f. subsp. boreale M. N. Philipson et Philipson；北方雪层杜鹃（点叶杜鹃，多枝杜鹃，紫丁杜鹃）；North Snowy Rhododendron, Spotted-leaves Rhododendron, Very Branched Rhododendron, Violet-colored Rhododendron ●

331349 Rhododendron niveum Hook. f.；西藏毛脉杜鹃（毛脉杜鹃，如雪杜鹃）；Snowlike Rhododendron, Tibet Rhododendron, Xizang Hairvein Azalea ●

331350 Rhododendron noriakianum T. Suzuki；南湖大山杜鹃（北部高山红花杜鹃，细叶杜鹃，志佳阳杜鹃）●☆

331351 Rhododendron notatum Hutch. = Rhododendron dendricola Hutch. ●

331352 Rhododendron nudiflorum (L.) Torr.；裸花杜鹃；Honeysuckle, Naked Flowers Rhododendron, Pink Azalea, Pinksterflower, Pinxter Flower, Pinxterbloom, Pinxterbloom Azalea, Pinxterflower, Purple Azalea ●☆

331353 Rhododendron nudiflorum Torr. = Rhododendron nudiflorum (L.) Torr. ●☆

331354 Rhododendron nudipes Nakai；斋国三叶杜鹃●☆

331355 Rhododendron nudipes Nakai subsp. niphophilum T. Yamaz. var. lagopus (Nakai) T. Yamaz. = Rhododendron lagopus Nakai ●☆

331356 Rhododendron nudipes Nakai subsp. niphophilum T. Yamaz. var. tsurugisanense T. Yamaz. = Rhododendron tsurugisanense (T. Yamaz.) T. Yamaz. ●☆

331357 Rhododendron nudipes Nakai var. gracilescens (Nakai) H. Hara = Rhododendron nudipes Nakai var. nagasakianum (Nakai) T. Yamaz. ●☆

331358 Rhododendron nudipes Nakai var. nagasakianum (Nakai) T. Yamaz.；长木杜鹃●☆

331359 Rhododendron nudipes Nakai var. tokushimense T. Yamaz. = Rhododendron tsurugisanense (T. Yamaz.) T. Yamaz. ●☆

331360 Rhododendron numorosum R. C. Fang；林生杜鹃（裸生杜鹃花）；Forest Rhododendron ●☆

331361 Rhododendron nuttallii Booth；木兰杜鹃（大果杜鹃）；Magnolia Azalea，Nuttall Rhododendron，T. Nuttall's Rhododendron ●

331362 Rhododendron nuttallii Booth var. stellatum Hutch.；小花木兰杜鹃●

331363 Rhododendron nyingchiense R. C. Fang et S. H. Huang；林芝杜鹃；Linzhi Azalea，Linzhi Rhododendron ●

331364 Rhododendron nymphaeoides W. K. Hu；睡莲叶杜鹃；Nymphae-leaf Rhododendron，Sleepinglily Azalea，Waterlily-leaved Rhododendron ●

331365 Rhododendron oblancifolium M. Y. Fang；倒矛杜鹃；Oblanceolate-leaved Rhododendron，Oblanceoleaf Azalea，Oblanceoleaf Rhododendron ●

331366 Rhododendron oblongifolium (Small) Millais；长圆叶杜鹃；Oblong Leaves Azalea ●☆

331367 Rhododendron oblongum Griff. = Rhododendron griffithianum Wight ●

331368 Rhododendron obovatum Hook. f. = Rhododendron lepidotum Wall. et G. Don ●

331369 Rhododendron obscurum Franch. ex Balf. f. = Rhododendron siderophyllum Franch. ●

331370 Rhododendron obtusum (Lindl.) Planch.；钝叶杜鹃（本雾岛，飞龙，密毛杜鹃，麒麟杜鹃，雾岛杜鹃，雾岛踯躅）；Blunt Rhododendron，Hiryu Azalea，Japanese Azalea，Kirishima Azalea，Kurume Azalea，Obtuseleaf Azalea，Obtuse-leaf Rhododendron ●

331371 Rhododendron obtusum (Lindl.) Planch. f. album (Rehder) C. K. Schneid.；白花钝叶杜鹃●☆

331372 Rhododendron obtusum (Lindl.) Planch. f. honkirishima Komatsu = Rhododendron obtusum (Lindl.) Planch. ●

331373 Rhododendron obtusum (Lindl.) Planch. var. japonicum (Maxim.) Kitam. = Rhododendron kiusianum Makino ●☆

331374 Rhododendron obtusum (Lindl.) Planch. var. kaempferi (Planch.) E. H. Wilson f. albiflorum E. H. Wilson = Rhododendron kaempferi Planch. f. album Nakai ●☆

331375 Rhododendron obtusum (Lindl.) Planch. var. kaempferi (Planch.) E. H. Wilson f. komatsui (Nakai) E. H. Wilson = Rhododendron kaempferi Planch. f. komatsui (Nakai) H. Hara ●☆

331376 Rhododendron obtusum (Lindl.) Planch. var. kaempferi (Planch.) E. H. Wilson = Rhododendron kaempferi Planch. ●☆

331377 Rhododendron obtusum (Lindl.) Planch. var. macrogemma (Nakai) Kitam. = Rhododendron macrogemma Nakai ●☆

331378 Rhododendron obtusum (Lindl.) Planch. var. mikawanum (Makino) T. Yamaz. = Rhododendron kaempferi Planch. var. mikawanum (Makino) Makino ●☆

331379 Rhododendron obtusum (Lindl.) Planch. var. saikaiense T. Yamaz. = Rhododendron kaempferi Planch. var. saikaiense (T. Yamaz.) T. Yamaz. ●☆

331380 Rhododendron obtusum (Lindl.) Planch. var. sakamotoi Komatsu；久留米杜鹃●☆

331381 Rhododendron obtusum (Lindl.) Planch. var. tosaense (Makino) Kitam. = Rhododendron tosaense Makino ●☆

331382 Rhododendron obtusum (Lindl.) Planch. var. tubiflorum (Komatsu) T. Yamaz. = Rhododendron kaempferi Planch. var. tubiflorum Komatsu ●☆

331383 Rhododendron occidentale A. Gray；西方杜鹃（西洋杜鹃）；Western Azalea，Western Rhododendron ●☆

331384 Rhododendron ochraceum Rehder et E. H. Wilson；峨马杜鹃；Ema Azalea，Ochreyellow Rhododendron，Ochre-yellow Rhododendron，Oma Rhododendron ●

331385 Rhododendron ochraceum Rehder et E. H. Wilson var. brevicarpum W. K. Hu；短果峨马杜鹃（短果杜鹃）；Shortfruit Ema Azalea，Short-fruit Oma Rhododendron ●

331386 Rhododendron octandrum M. Y. He；八蕊杜鹃；Eight-males Rhododendron，Eightstamen Azalea，Eight-stamened Rhododendron ●

331387 Rhododendron odoriferum Hutch.；香甜杜鹃（香花杜鹃）；Fragrant Rhododendron ●

331388 Rhododendron odoriferum Hutch. = Rhododendron maddenii Hook. f. subsp. crassum (Franch.) Cullen ●

331389 Rhododendron odoriferum Hutch. = Rhododendron maddenii Hook. f. ●

331390 Rhododendron officinale Salisb. = Rhododendron aureum Georgi ●◇

331391 Rhododendron oldhamii Maxim.；砖红杜鹃（奥氏杜鹃，金毛杜鹃）；Brick-red Azalea，Oldham Rhododendron，Oldham's Rhododendron ●

331392 Rhododendron oldhamii Maxim. var. glandulosum Hayata = Rhododendron oldhamii Maxim. ●

331393 Rhododendron oleifolium Franch.；油叶杜鹃●

331394 Rhododendron oleifolium Franch. = Rhododendron virgatum Hook. f. ●

331395 Rhododendron oligocarpum W. P. Fang et X. S. Zhang；稀果杜鹃；Few-fruit Rhododendron，Oligrant Rhododendron，Poorfruit Azalea ●

331396 Rhododendron ombrachares Balf. f. et Kingdon-Ward = Rhododendron tanastylum Balf. f. et Kingdon-Ward ●

331397 Rhododendron oomurasakii Makino；大紫杜鹃●☆

331398 Rhododendron openshawianum Rehder et E. H. Wilson = Rhododendron calophytum Franch. var. openshawianum (Rehder et E. H. Wilson) D. F. Chamb. ●

331399 Rhododendron oporinum Balf. f. et Kingdon-Ward = Rhododendron heliolepis Franch. var. oporinum (Balf. f. et Kingdon-Ward) A. L. Chang ex R. C. Fang ●

331400 Rhododendron orbiculare Decne.；团叶杜鹃（圆叶杜鹃）；Globe Rhododendron，Orbicular-leaved Rhododendron，Roundleaf Azalea ●

331401 Rhododendron orbiculare Decne. subsp. cardiobasis (Sleumer) D. F. Chamb.；心基杜鹃（心形杜鹃）；Cordiform-based Rhododendron，Heart-based Globe Rhododendron ●

331402 Rhododendron orbiculare Decne. subsp. maolingense G. Z. Li；猫岭杜鹃；Maoling Roundleaf Azalea ●

331403　Rhododendron orbiculare Decne. subsp. oblongum W. K. Hu；长圆团叶杜鹃（长圆叶杜鹃）；Oblong Orbicular-leaved Rhododendron

331404　Rhododendron orbiculatum Ridl. ；圆厚叶杜鹃●☆

331405　Rhododendron oreinum Balf. f. = Rhododendron nivale Hook. f. subsp. boreale M. N. Philipson et Philipson ●

331406　Rhododendron oreodoxa Franch.；山光杜鹃（光背杜鹃，山景杜鹃）；Glory of the Mountains Rhododendron, Montaneleader Azalea, Mountainglory Rhododendron, Mountain-glory Rhododendron ●

331407　Rhododendron oreodoxa Franch. var. adenostylosum W. P. Fang et W. K. Hu；腺柱山光杜鹃；Gland-style Rhododendron ●

331408　Rhododendron oreodoxa Franch. var. fargesii（Franch.）D. F. Chamb.；粉红杜鹃（法氏杜鹃，粉红山景杜鹃，红润杜鹃，红晕杜鹃）；Blushing Azalea, Blushing Rhododendron, Farges Rhododendron, Pere Farges Rhododendron, Pink Azalea ●

331409　Rhododendron oreodoxa Franch. var. shensiense D. F. Chamb.；陕西山光杜鹃（陕西杜鹃）；Shaanxi Rhododendron ●

331410　Rhododendron oreogenum L. C. Hu；藏东杜鹃；E. Xizang Azalea, East Xizang Rhododendron ●

331411　Rhododendron oreotrephes W. W. Sm.；山生杜鹃（精美杜鹃，可敬杜鹃，山育杜鹃，优美杜鹃）；Exquisite Rhododendron, Graceful Azalea, Montane Azalea, Oread Rhododendron, Threeflower Azalea, Threeflower Rhododendron ●

331412　Rhododendron oresbium Balf. f. et Kingdon-Ward = Rhododendron nivale Hook. f. subsp. boreale M. N. Philipson et Philipson ●

331413　Rhododendron oresterum Balf. f. et Forrest = Rhododendron wardii W. W. Sm. ●

331414　Rhododendron orthocladum Balf. f. et Forrest；直枝杜鹃（直茎杜鹃）；Straight Twigs Rhododendron, Straighttwig Azalea, Straight-twigged Rhododendron ●

331415　Rhododendron orthocladum Balf. f. et Forrest var. longistylum M. N. Philipson et Philipson；长柱直枝杜鹃；Long-style Straight Twigs Rhododendron, Longstyle Straighttwig Azalea ●

331416　Rhododendron osmerum Balf. f. et Forrest = Rhododendron russatum Balf. f. et Forrest ●

331417　Rhododendron osuzuyamense T. Yamaz. ；尾铃山杜鹃●☆

331418　Rhododendron oulotrichum Balf. f. et Forrest = Rhododendron trichocladum Franch. ●

331419　Rhododendron ovatosepalum Yamam. = Rhododendron oldhamii Maxim. ●

331420　Rhododendron ovatum Planch. ；马银花（卵叶杜鹃）；Azalea, Eggleaf Rhododendron, Egg-shaped Leaves Rhododendron, Ovate-leaved Rhododendron ●

331421　Rhododendron ovatum Planch. var. lamprophyllum (Hayata) Y. C. Liu, F. Y. Lu et C. H. Ou；长卵叶马银花（长卵叶杜鹃）●

331422　Rhododendron ovatum Planch. var. prismatum P. C. Tam；尖萼马银花（尖萼杜鹃）；Sharpcalyx Azalea ●

331423　Rhododendron ovatum Planch. var. prismatum P. C. Tam = Rhododendron ovatum Planch. ●

331424　Rhododendron ovatum Planch. var. setuliferum M. Y. He；刚毛马银花；Setose Azalea, Setose Eggleaf Rhododendron ●

331425　Rhododendron oxyphyllum Franch. = Rhododendron moulmainense Hook. ●

331426　Rhododendron pachyphyllum W. P. Fang；厚叶杜鹃；Thick-leaved Rhododendron, Thick-leaves Rhododendron ●

331427　Rhododendron pachypodum Balf. f. et W. W. Sm.；粗柄杜鹃（白豆花，波瓣杜鹃，红鳞杜鹃，云上杜鹃）；Above the Clouds Rhododendron, B. Scott's Rhododendron, Mach Red Scales Rhododendron, Scott Azalea, Scott Rhododendron, Thick-footed Rhododendron, Thickstalk Azalea, West Yunnan Azalea, West Yunnan Rhododendron ●

331428　Rhododendron pachysanthum Hayata；台湾山地杜鹃（台中杜鹃）●

331429　Rhododendron pachysanthum Hayata = Rhododendron hyperythrum Hayata ●

331430　Rhododendron pachytrichum Franch.；绒毛杜鹃（毛梗杜鹃）；Floss Azalea, Hairystalk Rhododendron, Thick Hairs Rhododendron, Thick-haired Rhododendron ●

331431　Rhododendron pachytrichum Franch. var. monosematum (Hutch.) D. F. Chamb. = Rhododendron strigillosum Franch. var. monosematum (Hutch.) T. L. Ming ●

331432　Rhododendron pachytrichum Franch. var. tenuistylum W. K. Hu；瘦柱绒毛杜鹃；Tenuistyle Thick Hairs Rhododendron ●

331433　Rhododendron pagophilum Balf. f. et Kingdon-Ward = Rhododendron selense Franch. ●

331434　Rhododendron pallescens Hutch. ；苍白杜鹃；Becoming Paler Rhododendron ●☆

331435　Rhododendron paludosum Hutch. et Kingdon-Ward = Rhododendron nivale Hook. f. ●

331436　Rhododendron palustre (L.) Kron et Judd = Ledum palustre L. ●

331437　Rhododendron palustre Turcz. = Rhododendron lapponicum (L.) Wahlenb. ●

331438　Rhododendron pankimense Cowan et Kingdon-Ward = Rhododendron kendrickii Nutt. ●

331439　Rhododendron panteumorphum Balf. f. et W. W. Sm. ；全真杜鹃；Pantomorphic Rhododendron ●

331440　Rhododendron panteumorphum Balf. f. et W. W. Sm. = Rhododendron erythrocalyx Balf. f. et Forrest ●

331441　Rhododendron papillatum Balf. f. et R. E. Cooper；乳突杜鹃（无乳突杜鹃）；Nipple Azalea, No Papillary Twigs Rhododendron, Papillary Rhododendron ●

331442　Rhododendron papyrociliare P. C. Tam；青留杜鹃（纸毛杜鹃）；Qingliu Azalea, Qingliu Rhododendron ●

331443　Rhododendron papyrociliare P. C. Tam = Rhododendron mariae Hance ●

331444　Rhododendron paradoxum Balf. f. ex Tagg；奇异杜鹃；Oddity Azalea, Unexpected Rhododendron ●

331445　Rhododendron parishii C. B. Clarke, Lace et W. W. Sm. ；毛淡绵杜鹃；Moulmein Rhododendron, Parish Rhododendron ●☆

331446　Rhododendron parmulatum Cowan；盘萼杜鹃（小盾杜鹃）；Dishcalyx Azalea, Small Shield Rhododendron, Small-shielded Rhododendron ●

331447　Rhododendron parryae Hutch. ；帕瑞杜鹃；Mrs. Parry's Rhododendron ●☆

331448　Rhododendron parviflorum F. Schmidt = Rhododendron lapponicum (L.) Wahlenb. ●

331449　Rhododendron parvifolium Adams；小叶杜鹃；Tundra Azalea, Tundra Rhododendron ●

331450　Rhododendron parvifolium Adams = Rhododendron lapponicum (L.) Wahlenb. subsp. parvifolium (Adams) T. Yamaz. ●

331451　Rhododendron parvifolium Adams = Rhododendron lapponicum (L.) Wahlenb. ●

331452　Rhododendron parvifolium Adams subsp. confertissimum (Nakai) A. P. Khokhr. = Rhododendron lapponicum (L.)

Wahlenb. ●

331453 Rhododendron parvifolium Adams var. albiflorum Herder;白花小叶杜鹃●☆

331454 Rhododendron patulum Kingdon-Ward = Rhododendron pemakoense Kingdon-Ward ●

331455 Rhododendron pectinatum Hutch.;篦齿杜鹃;Comb Toothed Rhododendron ●☆

331456 Rhododendron pectinatum Hutch. = Rhododendron moulmainense Hook. ●

331457 Rhododendron pemakoense Kingdon-Ward;假单花杜鹃(东藏杜鹃,展枝杜鹃);Falseoneflower Azalea, Falseoneflower Rhododendron, False-uniflorous Rhododendron, Pemako Rhododendron, Spreading Rhododendron ●

331458 Rhododendron pendulum Hook. f.;凸叶杜鹃(垂俯杜鹃,悬垂杜鹃);Hanging Rhododendron,Pendent Azalea ●

331459 Rhododendron pennivenium Balf. f. et Forrest = Rhododendron agastum Balf. f. et W. W. Sm. var. pennivenium (Balf. f. et W. W. Sm.) T. L. Ming ●

331460 Rhododendron pennivenium Balf. f. et Forrest = Rhododendron agastum Balf. f. et W. W. Sm. ●

331461 Rhododendron pentamerum (Maxim.) Matsum. et Nakai = Rhododendron degronianum Carrière ●☆

331462 Rhododendron pentaphyllum Maxim.;日本五叶杜鹃(五叶杜鹃,崖生五叶杜鹃);Fiveleaf Azalea,Five-leaved Rhododendron ●☆

331463 Rhododendron pentaphyllum Maxim. var. nikoense Komatsu;日光杜鹃●☆

331464 Rhododendron pentaphyllum Maxim. var. nikoense Komatsu f. albiflorum Asai et E.Torii;白花日光杜鹃●☆

331465 Rhododendron pentaphyllum Maxim. var. shikokianum T. Yamaz.;四国杜鹃●☆

331466 Rhododendron pentaphyllum Maxim. var. villosum Koidz. = Rhododendron pentaphyllum Maxim. ●☆

331467 Rhododendron peramabile Hutch. = Rhododendron intricatum Franch. ●

331468 Rhododendron peramoenum Balf. f. et Forrest = Rhododendron delavayi Franch. var. peramoenum (Balf. f. et Forrest) D. F. Chamb. ●

331469 Rhododendron peregrinum Tagg;外来杜鹃;Foreign Rhododendron ●☆

331470 Rhododendron periclymenoides (Michx.) Shinners = Rhododendron nudiflorum (L.) Torr. ●☆

331471 Rhododendron persicinum Hand. -Mazz. = Rhododendron anthosphaerum Diels ●

331472 Rhododendron perulatum Balf. f. et Forrest = Rhododendron microgynum Balf. f. et Forrest ●

331473 Rhododendron petium P. C. Tam = Rhododendron simsii Planch. ●

331474 Rhododendron petrocharis Diels;饰石杜鹃(密鳞杜鹃);Dense Scales Rhododendron,Stone-ornaments Azalea ●

331475 Rhododendron phaedropum Balf. f. et Forrest = Rhododendron neriiflorum Franch. var. appropinquans (Tagg et Forrest) W. K. Hu ●

331476 Rhododendron phaeochlorum Balf. f. et Forrest = Rhododendron oreotrephes W. W. Sm. ●

331477 Rhododendron phaeochrysum Balf. f. et W. W. Sm.;褐黄杜鹃(矮丛杜鹃,栎叶杜鹃);Dark Golden Rhododendron, Dark-golden Rhododendron,Oakleaf Azalea,Small Bushy Rhododendron ●

331478 Rhododendron phaeochrysum Balf. f. et W. W. Sm. f. smithii E. H. Wilson = Rhododendron pulchrum Sweet ●

331479 Rhododendron phaeochrysum Balf. f. et W. W. Sm. var. agglutinatum (Balf. f. et Forrest) D. F. Chamb.;凝毛杜鹃(凝花杜鹃);Agglutinated Azalea, Stuck Hairs Rhododendron ●

331480 Rhododendron phaeochrysum Balf. f. et W. W. Sm. var. levistratum (Balf. f. et Forrest) D. F. Chamb.;毡毛栎叶杜鹃(斑点杜鹃花);Dapplebloom Rhododendron, Dotty Rhododendron ●

331481 Rhododendron phoeniceum G. Donf. smithii (Sweet) E. H. Wilson = Rhododendron pulchrum Sweet ●

331482 Rhododendron phoenicodum Balf. f. et Forrest = Rhododendron neriiflorum Franch. ●

331483 Rhododendron pholidotum Balf. f. et W. W. Sm.;苍山杜鹃;Scaly Azalea,Scaly Rhododendron ●

331484 Rhododendron pholidotum Balf. f. et W. W. Sm. = Rhododendron heliolepis Franch. ●

331485 Rhododendron piceum P. C. Tam = Rhododendron florulentum P. C. Tam ●

331486 Rhododendron piceum P. C. Tam = Rhododendron rufulum P. C. Tam ●

331487 Rhododendron piercei Davidian;察隅杜鹃;Chayu Azalea, Chayu Rhododendron,Pierce Rhododendron ●

331488 Rhododendron pilicalyx Hutch. = Rhododendron pachypodum Balf. f. et W. W. Sm. ●

331489 Rhododendron pilostylum W. K. Hu;金平毛柱杜鹃(毛柱杜鹃);Jinping Hairstyle Azalea,Jinping Rhododendron ●

331490 Rhododendron pilovittatum Balf. f. et W. W. Sm. = Rhododendron delavayi Franch. ●

331491 Rhododendron pinetorum P. C. Tam;松树杜鹃;Pineforest Azalea, Pinewoods Rhododendron,Under Pine Tree Rhododendron ●

331492 Rhododendron pingbianense M. Y. Fang;屏边杜鹃;Pingbian Azalea, Pingbian Rhododendron ●

331493 Rhododendron pingianum W. P. Fang;海绵杜鹃(粉背杜鹃);Sponge Azalea,Sponge Rhododendron ●◇

331494 Rhododendron pittosporifolium Hemsl. = Rhododendron stamineum Franch. ●

331495 Rhododendron planetum Balf. f.;阔口杜鹃(漂泊杜鹃);Walf Azalea, Walf Rhododendron,Wandering Rhododendron ●

331496 Rhododendron platyphyllum Franch. ex Balf. f. et W. W. Sm.;阔叶杜鹃(宽叶杜鹃);Broadleaf Azalea,Broad-leaved Rhododendron, Broad-leaves Rhododendron ●

331497 Rhododendron platypodum Diels;阔柄杜鹃;Broad Petioles Rhododendron,Broadstalk Azalea,Broad-stalked Rhododendron ●◇

331498 Rhododendron plebeium Balf. f. et W. W. Sm. = Rhododendron fulvum Balf. f. et W. W. Sm. ●

331499 Rhododendron plebeium Balf. f. et W. W. Sm. = Rhododendron heliolepis Franch. ●

331500 Rhododendron pleianthum Sleumer;新几内亚多花杜鹃(多花杜鹃)●☆

331501 Rhododendron pleistanthum Balf. f. ex Hutch.;极多花杜鹃●

331502 Rhododendron pleistanthum Balf. f. ex Hutch. = Rhododendron yunnanense Franch. ●

331503 Rhododendron pocophorum Balf. f. ex Tagg;杯萼杜鹃(雪毛杜鹃,羊毛杜鹃);Cupcalyx Azalea, Cupular-calyxed Rhododendron, Fleece-bearing Rhododendron ●☆

331504 Rhododendron pocophorum Balf. f. ex Tagg var. hemidartum (Tagg) D. F. Chamb.;腺柄杯萼杜鹃(碎毛被叶杜鹃,腺柄杜鹃);Patchy Indumentum Rhododendron ●

331505 Rhododendron poecilodermum Balf. f. et Forrest =

Rhododendron roxieanum Forrest ex W. W. Sm. ●

331506　Rhododendron pogonostylum Balf. f. et W. W. Sm. = Rhododendron irroratum Franch. subsp. pogonostylum (Balf. f. et W. W. Sm.) D. F. Chamb. ex Cullen et D. F. Chamb. ●

331507　Rhododendron polifolium Franch. = Rhododendron thymifolium Maxim. ●

331508　Rhododendron poliopeplum Balf. f. et Forrest = Rhododendron sanguineum Franch. var. himertum (Balf. f. et Forrest) T. L. Ming ●

331509　Rhododendron poliopeplum Balf. f. et Forrest = Rhododendron sanguineum Franch. var. haemaleum (Balf. f. et Forrest) D. F. Chamb. ●

331510　Rhododendron poluandrum Hutch. ;多蕊杜鹃;Many-stamens Rhododendron ●☆

331511　Rhododendron polyandrum Hutch. = Rhododendron maddenii Hook. f. ●

331512　Rhododendron polycladum Franch. ;多枝杜鹃(石堆杜鹃);Branchy Azalea, Many-branched Rhododendron, Multi-branched Rhododendron, Polyclade Rhododendron, Sparkling Rhododendron ●

331513　Rhododendron polyfolium Franch. ;西方多叶杜鹃;Many-leaved Rhododendron ●☆

331514　Rhododendron polylepis Franch. ;多鳞杜鹃(巴塘杜鹃);Batang Rhododendron, ManyscaleAzalea, Many-scales Rhododendron ●

331515　Rhododendron polyraphidoideum P. C. Tam;千针叶杜鹃(千针叶);Kiloneedleleaf Azalea, More Needle Like Leaves Rhododendron, Polyraphioid Rhododendron ●

331516　Rhododendron polyraphidoideum P. C. Tam = Rhododendron chunii W. P. Fang ●

331517　Rhododendron polyraphidoideum P. C. Tam var. montanum P. C. Tam;岭上杜鹃;Mountain Polyraphioid Rhododendron ●

331518　Rhododendron polytrichum W. P. Fang;多毛杜鹃;Hairy Azalea, Many-hairs Rhododendron, Polytrichous Rhododendron ●

331519　Rhododendron pomense Cowan et Davidian;波密杜鹃;Bomi Azalea, Bomi Rhododendron, Pome Rhododendron ●

331520　Rhododendron ponticum L. ;长序杜鹃(本都山杜鹃,黑海杜鹃,秋花杜鹃);Common Rhododendron, Mt. Pontus Rhododendron, Pontic Rhododendron, Rhododendron, Rosydandrum ●

331521　Rhododendron ponticum L. 'Silver Edge';银边本都山杜鹃●☆

331522　Rhododendron ponticum L. 'Variegatum';斑叶本都山杜鹃●☆

331523　Rhododendron populare Cowan;蜜腺杜鹃;Nectary Azalea, Popular Rhododendron ●

331524　Rhododendron porphyroblastum Balf. f. et Forrest = Rhododendron roxieanum Forrest ex W. W. Sm. var. cucullatum (Hand. -Mazz.) D. F. Chamb. ex Cullen et D. F. Chamb. ●

331525　Rhododendron porphyrophyllum Balf. f. et Forrest = Rhododendron erastum Balf. f. et Forrest ●

331526　Rhododendron porrosquameum Balf. f. et Forrest = Rhododendron heliolepis Franch. ●

331527　Rhododendron porrosquameum Balf. f. et Forrest = Rhododendron polylepis Franch. ●

331528　Rhododendron potaninii Batalin;甘肃杜鹃;Gansu Azalea, Gansu Rhododendron, Potanin Rhododendron ●

331529　Rhododendron pothinum Balf. f. et Forrest = Rhododendron temenium Balf. f. et Forrest ●

331530　Rhododendron poukanense Maxim. ;布干山杜鹃;Mt. Poukhan Azalea, Yodogawa Azalea ●☆

331531　Rhododendron poxophorum Balf. f. et Tagg;雪毛杜鹃●☆

331532　Rhododendron praestans Balf. f. et W. W. Sm. ;优秀杜鹃(魁斗杜鹃); Excellence Azalea, Excellent Rhododendron, Great Bear Azalea, Leading Size Rhododendron ●

331533　Rhododendron praeteritum Hutch. ;鄂西杜鹃(五裂杜鹃);Passed Over Rhododendron, W. Hubei Azalea, Western Hubei Rhododendron ●

331534　Rhododendron praeteritum Hutch. var. hirsutum W. K. Hu;毛房杜鹃;Hirsute Passed Over Rhododendron ●

331535　Rhododendron praevernum Hutch. ;早春杜鹃(二月杜鹃);Early Spring Rhododendron, Earlyspring Azalea, February Rhododendron ●

331536　Rhododendron prasinocalyx Balf. f. et Forrest = Rhododendron wardii W. W. Sm. ●

331537　Rhododendron prattii Franch. = Rhododendron faberi Hemsl. subsp. prattii (Franch.) D. F. Chamb. ex Cullen et D. F. Chamb. ●

331538　Rhododendron preptum Balf. f. et Forrest;复毛杜鹃(显异杜鹃);Ditinguished Rhododendron, Duplicatehair Azalea ●

331539　Rhododendron primuliflorum Bureau et Franch. ;樱草杜鹃(报春花杜鹃);Primroseflower Azalea, Primrose-flower Rhododendron, Primrose-flowered Rhododendron ●

331540　Rhododendron primuliflorum Bureau et Franch. var. cephalanthoides (Balf. f. et W. W. Sm.) Cowan et Davidian;微毛樱草杜鹃(微毛杜鹃);Puberulent Azalea, Puberulent Rhododendron ●

331541　Rhododendron primuliflorum Bureau et Franch. var. lepidanthum (Balf. f. et W. W. Sm.) Cowan et Davidian;鳞花樱草杜鹃(鳞花杜鹃); Scale-flower Puberulent Azalea, Scale-flower Puberulent Rhododendron ●

331542　Rhododendron primulinum Hemsl. = Rhododendron flavidum Franch. ●

331543　Rhododendron principis Bureau et Franch. ;藏南杜鹃(紫斑杜鹃); Purple Spots Rhododendron, Purple-spoted Rhododendron, S. Xizang Azalea ●

331544　Rhododendron principis Bureau et Franch. = Rhododendron vellereum Hutch. ex Tagg ●

331545　Rhododendron principis Bureau et Franch. var. vellereum (Hutch. ex Tagg) T. L. Ming = Rhododendron vellereum Hutch. ex Tagg ●

331546　Rhododendron prinophyllum (Small) Millais;冬青叶杜鹃;Azalea, Early Azalea, Election Pink, Election Pink Azalea, Mountain Azalea, Mountain Pink, Piedmont Azalea, Rosehell Azalea, Wild Honeysuckle ●☆

331547　Rhododendron prinophyllum (Small) Millais = Rhododendron roseum (Loisel.) Rehder ●☆

331548　Rhododendron pritzelianum Diels = Rhododendron micranthum Turcz. ●

331549　Rhododendron probum Balf. f. et Forrest = Rhododendron selense Franch. ●

331550　Rhododendron procumbens (L.) A. W. Wood = Loiseleuria procumbens (L.) Desv. ex Loisel. ●☆

331551　Rhododendron pronum W. W. Sm. ;平卧杜鹃;Prostrate Azalea, Prostrate Rhododendron ●

331552　Rhododendron prophantum Balf. f. et Forrest = Rhododendron kyawi Lace et W. W. Sm. ●

331553　Rhododendron propinquum Tagg = Rhododendron rupicola W. W. Sm. ●

331554　Rhododendron prostratum W. W. Sm. = Rhododendron pronum W. W. Sm. ●

331555　Rhododendron prostratum W. W. Sm. = Rhododendron saluenense Franch. var. prostratum (W. W. Sm.) R. C. Fang ●

331556　Rhododendron proteoides Balf. f. et W. W. Sm.；矮生杜鹃（似山龙眼杜鹃，异花杜鹃）；Dwarf Azalea，Dwarf Rhododendron，Proteoid Rhododendron，Resembling Protea Rhododendron ●

331557　Rhododendron protistum Balf. f. et Forrest；翘首杜鹃（魁首杜鹃）；First of the First Rhododendron，Raisehead Azalea ●

331558　Rhododendron protistum Balf. f. et Forrest var. giganteum（Forrest ex Tagg）D. F. Chamb.；大树杜鹃；Gigantic Rhododendron，Great First of the First Rhododendron，Large-tree Rhododendron ●◇

331559　Rhododendron pruniflorum Hutch. et Kingdon-Ward；桃花杜鹃；Peach-flowered Rhododendron ●

331560　Rhododendron prunifolium（Small）Millais；李叶杜鹃；Plum Leaf Azalea，Plumleaf Azalea，Plum-like Leaves Rhododendron ●☆

331561　Rhododendron przewalskii Maxim.；陇蜀杜鹃（大坂山杜鹃，光背枇杷，金背杜鹃，金背枇杷，青海杜鹃，野枇杷）；Dabanshan Azalea，Dabanshan Rhododendron，Przewalsk Azalea，Przewalsk Rhododendron ●

331562　Rhododendron przewalskii Maxim. subsp. chrysophyllum W. P. Fang et S. X. Wang；金背陇蜀杜鹃；Golden-back Przewalsk Azalea，Golden-back Przewalsk Rhododendron ●

331563　Rhododendron przewalskii Maxim. subsp. huzhuense W. P. Fang et S. X. Wang；互助陇蜀杜鹃（互助杜鹃）；Huzhu Azalea，Huzhu Rhododendron ●

331564　Rhododendron przewalskii Maxim. subsp. yushuense W. P. Fang et S. X. Wang；玉树陇蜀杜鹃（玉树杜鹃花）；Yushu Azalea，Yushu Rhododendron ●

331565　Rhododendron pseudochrysanthum Hayata；阿里山杜鹃（红斑杜鹃，红点杜鹃，假牛皮杜鹃，满山红，莫尔杜鹃，森氏杜鹃，森氏满山红，台高山杜鹃，台湾红斑杜鹃，玉山杜鹃）；Alishan Azalea，Like R. Chrysanthum Rhododendron，Mori Rhododendron，Redspot Azalea，Red-spots Rhododendron，Taiwan Alpine Rhododendron，Yushan Azalea，Yushan Rhododendron ●

331566　Rhododendron pseudochrysanthum Hayata f. rufovelutinum T. Yamaz. = Rhododendron hyperythrum Hayata ●

331567　Rhododendron pseudochrysanthum Hayata subsp. morii（Hayata）T. Yamaz. = Rhododendron morii Hayata ●

331568　Rhododendron pseudochrysanthum Hayata var. nankotaisanense（Hayata）T. Yamaz. = Rhododendron pseudochrysanthum Hayata ●

331569　Rhododendron pseudochrysanthum Hayata var. taitunense T. Yamaz. = Rhododendron morii Hayata ●

331570　Rhododendron pseudociliicalyx Hutch. = Rhododendron ciliicalyx Franch. ●

331571　Rhododendron pseudociliipes Cullen；褐叶杜鹃（福川杜鹃）；Brownleaf Azalea，Fuchuan Rhododendron ●

331572　Rhododendron pseudoyanthinum Balf. f. ex Hutch.；紫蕊杜鹃●

331573　Rhododendron pseudoyanthinum Balf. f. ex Hutch. = Rhododendron concinnum Hemsl. ●

331574　Rhododendron psilostylum（Rehder et E. H. Wilson）Balf. f. = Rhododendron flavidum Franch. var. psilostylum Rehder et E. H. Wilson ●

331575　Rhododendron pubescens Balf. f. et Forrest；柔毛杜鹃（狭叶米碎花）；Downy Rhododendron，Pubescent Azalea，Pubescent Rhododendron ●

331576　Rhododendron pubicostatum T. L. Ming；毛脉杜鹃（滇毛脉杜鹃）；Hairvein Azalea，Yunnan Hairy-veined Rhododendron，Yunnan Hairy-veins Rhododendron ●

331577　Rhododendron pubigerum Balf. f. et Forrest = Rhododendron oreotrephes W. W. Sm. ●

331578　Rhododendron pudorosum Cowan；羞怯杜鹃；Bushful Rhododendron，Shyness Azalea ●

331579　Rhododendron pugeense L. C. Hu；普格杜鹃；Puge Azalea，Puge Rhododendron ●

331580　Rhododendron pulchroides Chun et W. P. Fang；美艳杜鹃；Colorful Azalea，False-beautiful Rhododendron，Resembling R. Pulcherum Rhododendron ●◇

331581　Rhododendron pulchrum Sweet；锦绣杜鹃（艳紫杜鹃）；Beautiful Rhododendron，Lovely Azalea，Lovely Rhododendron，Samite Azalea ●

331582　Rhododendron pulchrum Sweet 'Speciosum'；美丽锦绣杜鹃●☆

331583　Rhododendron pulchrum Sweet var. phoeniceum（G. Don）Rehder；凤凰杜鹃（紫花杜鹃）●

331584　Rhododendron pumilum Hook. f.；矮小杜鹃；Dwarf Rhododendron，Dwarfish Rhododendron，Low Rhododendron，Mini Azalea ●

331585　Rhododendron punctifolium L. C. Hu；斑叶杜鹃；Blotchleaf Azalea，Spot-leaved Rhododendron，Spot-leaves Rhododendron ●

331586　Rhododendron puniceum Roxb. = Rhododendron arboreum Sm. ●

331587　Rhododendron puralbum Balf. f. et W. W. Sm. = Rhododendron wardii W. W. Sm. var. puralbum（Balf. f. et W. W. Sm.）D. F. Chamb. ●

331588　Rhododendron purdomii Rehder et E. H. Wilson；太白杜鹃（金背枇杷，陕西杜鹃，药枇杷）；Chinling Rhododendron，Purdom Rhododendron，Shaanxi Azalea，Taibai Rhododendron ●

331589　Rhododendron purdomii Rehder et E. H. Wilson var. villosum L. H. Wu；毛叶太白杜鹃；Villose Taibai Rhododendron ●

331590　Rhododendron pycnocladum Balf. f. et W. W. Sm. = Rhododendron telmateium Balf. f. et W. W. Sm. ●

331591　Rhododendron qianyangense M. Y. He；黔阳杜鹃；Qianyang Azalea，Qianyang Rhododendron ●

331592　Rhododendron qinghaiense Ching ex W. Y. Wang；青海杜鹃；Qinghai Azalea，Qinghai Rhododendron ●

331593　Rhododendron quinquefolium Bisset et S. Moore；五叶杜鹃（五叶杜鹃花）；Five-leaves Azalea，Five-leaves Rhododendron ●☆

331594　Rhododendron quinquefolium Bisset et S. Moore 'Five Arrows'；五箭杜鹃●☆

331595　Rhododendron quinquefolium Bisset et S. Moore f. speciosum N. Yonez.；美丽五叶杜鹃●☆

331596　Rhododendron racemosum Franch.；腋花杜鹃（总状杜鹃）；Manyflower Azalea，Manyflower Rhododendron，Raceme Flowers Rhododendron，Racemose Rhododendron ●

331597　Rhododendron racemosum Franch. 'Forest'；福雷斯特腋花杜鹃●☆

331598　Rhododendron racemosum Franch. 'Glendoick'；格兰道伊科腋花杜鹃●☆

331599　Rhododendron racemosum Franch. 'Rock Rose'；沙漠座莲腋花杜鹃●☆

331600　Rhododendron racemosum Franch. var. rigidum（Franch.）Rehnelt = Rhododendron rigidum Franch. ●

331601　Rhododendron radendum W. P. Fang；毛叶杜鹃（康定刚毛杜鹃）；Hairleaf Azalea，Kangding Bristly Rhododendron ●

331602　Rhododendron radicans Balf. f. et Forrest = Rhododendron keleticum Nutt. ●

331603　Rhododendron radinum Balf. f. et W. W. Sm. = Rhododendron trichostomum Franch. var. radinum（Balf. f. et W. W. Sm.）Cowan et

Davidian ●

331604　Rhododendron ramipilosum T. L. Ming;腺裂杜鹃(线裂杜鹃);
Branchhair Azalea, Linear-lobed Rhododendron, Ramipilose
Rhododendron ●

331605　Rhododendron ramosissimum Franch. = Rhododendron nivale
Hook. f. subsp. boreale M. N. Philipson et Philipson ●

331606　Rhododendron ramsdenianum Cowan;长枝杜鹃(长轴杜鹃,阮
氏杜鹃);L. Ramsden's Rhododendron, Longaxis Azalea, Ramsden's
Rhododendron ●

331607　Rhododendron randaiense Hayata = Rhododendron rubropilosum
Hayata ●

331608　Rhododendron rarile Balf. f. et W. W. Sm. = Rhododendron
decorum Franch. var. diaprepes (Balf. f. et W. W. Sm.) T. L. Ming ●

331609　Rhododendron rarosquameum Balf. f. = Rhododendron rigidum
Franch. ●

331610　Rhododendron rasile Balf. f. et W. W. Sm. = Rhododendron
decorum Franch. subsp. diaprepes (Balf. f. et W. W. Sm.) T. L.
Ming ●

331611　Rhododendron ravum Balf. f. et W. W. Sm. = Rhododendron
cuneatum W. W. Sm. ●

331612　Rhododendron recurvoides Tagg et Kingdon-Ward;下弯杜鹃;
Resembling R. Recurvum Rhododendron ●☆

331613　Rhododendron recurvum Balf. f. et Forrest = Rhododendron
roxieanum Forrest ex W. W. Sm. ●

331614　Rhododendron recurvum Balf. f. et Forrest var. oreonastes Balf.
f. et Forrest = Rhododendron roxieanum Forrest ex W. W. Sm. var.
oreonastes (Balf. f. et Forrest) T. L. Ming ●

331615　Rhododendron redowskianum Maxim.;叶状苞杜鹃●◇

331616　Rhododendron redowskianum Maxim. = Therorhodion
redowskianum (Maxim.) Hutch. ●◇

331617　Rhododendron regale Balf. f. et Kingdon-Ward = Rhododendron
basilicum Balf. f. et W. W. Sm. ●

331618　Rhododendron reginaldii Balf. f. = Rhododendron oreodoxa
Franch. ●

331619　Rhododendron repens Balf. f. et Forrest = Rhododendron forrestii
Balf. f. ex Diels ●

331620　Rhododendron repens Balf. f. et Forrest var. chamaethauma Tagg
= Rhododendron chamaethomsonii (Tagg et Forrest) Cowan et
Davidian var. chamaethauma (Tagg et Forrest) Cowan et Davidian ●

331621　Rhododendron repens Balf. f. et Forrest var. chamaethomsonii
Tagg et Forrest = Rhododendron chamaethomsonii (Tagg et Forrest)
Cowan et Davidian ●

331622　Rhododendron repens Balf. f. et Forrest var. chamaethomsonii
Tagg = Rhododendron chamaethomsonii (Tagg et Forrest) Cowan et
Davidian ●

331623　Rhododendron reticulatum D. Don ex G. Don;网纹杜鹃(网纹
杜鹃花);Net-like Rhododendron ●☆

331624　Rhododendron reticulatum D. Don ex G. Don f. albiflorum
(Makino) Makino ex H. Hara;白花网纹杜鹃(白花三叶杜鹃)●☆

331625　Rhododendron reticulatum D. Don ex G. Don f. bifolium (T.
Yamaz.) T. Yamaz. ;双叶网纹杜鹃●☆

331626　Rhododendron reticulatum D. Don ex G. Don f. ciliatum
(Nakai) Sugim. ;睫毛网纹杜鹃(粗毛网纹杜鹃)●☆

331627　Rhododendron reticulatum D. Don ex G. Don f. glabrescens
(Nakai et H. Hara) T. Yamaz. ;渐光网纹杜鹃●☆

331628　Rhododendron reticulatum D. Don ex G. Don f. parvifolium (T.
Yamaz.) T. Yamaz. ;小叶网纹杜鹃●☆

331629　Rhododendron reticulatum D. Don ex G. Don f. versicolor
(Nakai) H. Hara;斑叶网纹杜鹃●☆

331630　Rhododendron reticulatum D. Don ex G. Don var. bifolium T.
Yamaz. = Rhododendron reticulatum D. Don ex G. Don f. bifolium
(T. Yamaz.) T. Yamaz. ●☆

331631　Rhododendron reticulatum D. Don ex G. Don var. ciliatum Nakai
= Rhododendron reticulatum D. Don ex G. Don f. ciliatum (Nakai)
Sugim. ●☆

331632　Rhododendron reticulatum D. Don ex G. Don var. nudipes
(Nakai) Hatus. = Rhododendron nudipes Nakai ●☆

331633　Rhododendron reticulatum D. Don ex G. Don var. parvifolium T.
Yamaz. = Rhododendron reticulatum D. Don ex G. Don f. parvifolium
(T. Yamaz.) T. Yamaz. ●☆

331634　Rhododendron retusum A. W. Benn. ;爪哇钝叶杜鹃;Blunt-
leaves Java Rhododendron ●☆

331635　Rhododendron rex H. Lév. ;大王杜鹃;King Azalea, King
Rhododendron ●◇

331636　Rhododendron rex H. Lév. subsp. arizelum (Balf. f. et Forrest)
D. F. Chamb. ex Cullen et Chamb. ;斜钟杜鹃(夺目杜鹃);
Apparent Rhododendron, Brilliant Azalea, Notable Rhododendron ●

331637　Rhododendron rex H. Lév. subsp. arizelum (Balf. f. et Forrest)
D. F. Chamb. = Rhododendron rex H. Lév. subsp. arizelum (Balf. f.
et Forrest) D. F. Chamb. ex Cullen et Chamb. ●

331638　Rhododendron rex H. Lév. subsp. arizelum (Balf. f. et Forrest)
D. F. Chamb. ex Cullen et Chamb. = Rhododendron arizelum Balf. f.
et Forrest ●

331639　Rhododendron rex H. Lév. subsp. arizelum (Balf. f. et Forrest)
D. F. Chamb. = Rhododendron arizelum Balf. f. et Forrest ●

331640　Rhododendron rex H. Lév. subsp. fictolacteum (Balf. f.) D. F.
Chamb. ;假乳黄杜鹃(假乳黄叶杜鹃,科里杜鹃,棕背杜鹃);
Corrietree Azalea, Corrietree Rhododendron, Corrie-tree
Rhododendron, False Lacteum Rhododendron ●◇

331641　Rhododendron rex H. Lév. subsp. gratum (T. L. Ming) M. Y.
Fang;可爱大王杜鹃(可爱杜鹃);Lovely Rhododendron, Pleasing
King Azalea, Pleasing King Rhododendron ●

331642　Rhododendron rex subsp. arizelum (Balf. f. et Forrest) D. F.
Chamb. = Rhododendron arizelum Balf. f. et Forrest ●

331643　Rhododendron rhabdotum Balf. f. et R. E. Cooper =
Rhododendron dalhousiae Hook. f. ●

331644　Rhododendron rhabdotum Balf. f. et R. E. Cooper =
Rhododendron dalhousiae Hook. f. var. rhabdotum (Balf. f. et R. E.
Cooper) Cullen ●

331645　Rhododendron rhaibocarpum Balf. f. et W. W. Sm. =
Rhododendron selense Franch. subsp. dasycladum (Balf. f. et W. W.
Sm.) D. F. Chamb. ●

331646　Rhododendron rhantum Balf. f. et W. W. Sm. = Rhododendron
vernicosum Franch. ●

331647　Rhododendron rhodanthum M. Y. He;淡红杜鹃;Lightred
Azalea, Pink-flower Rhododendron, Reddish-flower Rhododendron ●

331648　Rhododendron rhombicum Diels = Rhododendron reticulatum
D. Don ex G. Don ●☆

331649　Rhododendron rhombicum Miq. = Rhododendron mariesii
Hemsl. et E. H. Wilson ●

331650　Rhododendron rhombifolium R. C. Fang;菱形叶杜鹃(菱形杜
鹃,菱叶杜鹃);Rhombic Leaves Rhododendron, Rhombic-leaved
Rhododendron, Rhombleaf Azalea ●

331651　Rhododendron rhuyuenense Chun ex P. C. Tam;乳源杜鹃(林

氏杜鹃）；Ling's Azalea, Ling's Rhododendron, Ruyuan Azalea, Ruyuan Rhododendron ●

331652　Rhododendron rigidum Franch.；基毛杜鹃；Rigid Azalea, Rigidde Rhododendron ●

331653　Rhododendron ripaecola P. C. Tam = Rhododendron naamkwanense Merr. ●

331654　Rhododendron riparioides（Cullen）Cubey = Rhododendron calostrotum Balf. f. et Kingdon-Ward var. riparioides（Cullen）R. C. Fang ●

331655　Rhododendron riparium A. Wang et P. C. Tam = Rhododendron naamkwanense Merr. ●

331656　Rhododendron riparium Kingdon-Ward = Rhododendron calostrotum Balf. f. et Kingdon-Ward ●

331657　Rhododendron ripense Makino；岸生杜鹃●☆

331658　Rhododendron ripense Makino f. leucanthum Sugim.；白花岸生杜鹃●☆

331659　Rhododendron ripicola P. C. Tam = Rhododendron naamkwanense Merr. ●

331660　Rhododendron ririei Hemsl. et E. H. Wilson；大钟杜鹃（里氏杜鹃）；B. Ririe's Rhododendron, Bigbell Azalea, Ririe's Rhododendron ●◇

331661　Rhododendron rivulare Hand. -Mazz.；溪畔杜鹃（贵州杜鹃）；Brooklet Azalea, Brooklet Rhododendron, River Side Rhododendron, River-side Rhododendron ●

331662　Rhododendron rivulare Hand.-Mazz. = Rhododendron calostrotum Balf. f. et Kingdon-Ward ●

331663　Rhododendron rivulare Hand. -Mazz. = Rhododendron kwangtungense Merr. et Chun ●

331664　Rhododendron rockii E. H. Wilson = Rhododendron hunnewellianum Rehder et E. H. Wilson subsp. rockii（E. H. Wilson）D. F. Chamb. ●

331665　Rhododendron roseatum Hutch.；红晕杜鹃（毛足杜鹃,玫瑰杜鹃）；Rose-like Rhododendron, Woolly-footed Rhododendron ●

331666　Rhododendron roseotinctum Balf. f. et Forrest = Rhododendron sanguineum Franch. var. didymoides Tagg et Forrest ●

331667　Rhododendron roseum（Loisel.）Rehder；蔷薇杜鹃（蔷薇杜鹃花）；Hoary Azalea, Mountain Azalea, Roselike Rhododendron, Rosy Rhododendron ●☆

331668　Rhododendron roseum（Loisel.）Rehder = Rhododendron prinophyllum（Small）Millais ●☆

331669　Rhododendron roseum（Loisel.）Rehder f. albidum Steyerm. = Rhododendron prinophyllum（Small）Millais ●☆

331670　Rhododendron rosmarinifolium（Burm. f.）Dippel = Rhododendron mucronatum（Blume）G. Don ●

331671　Rhododendron rosthornii Diels = Rhododendron micranthum Turcz. ●

331672　Rhododendron rothschildii Davidian；宽柄杜鹃；Broadstalk Azalea, Rothschild Rhododendron ●

331673　Rhododendron rotundifolium David = Rhododendron orbiculare Decne. ●

331674　Rhododendron rotundifolium David ex Franch. = Rhododendron orbiculare Decne. ●

331675　Rhododendron roxieanum Forrest ex W. W. Sm.；卷叶杜鹃（兜叶杜鹃）；Coilleaf Azalea, Convolute-leaved Rhododendron, Mrs. Roxie's Rhododendron, Rolledleaf Rhododendron, Roxie's Rhododendron ●

331676　Rhododendron roxieanum Forrest ex W. W. Sm. var. cucullatum（Hand. -Mazz.）D. F. Chamb. ex Cullen et D. F. Chamb.；兜尖卷叶杜鹃；Cucullate Roxie's Rhododendron ●

331677　Rhododendron roxieanum Forrest ex W. W. Sm. var. globigerum（Balf. f. et Forrest）Chamb. ex Cullen et D. F. Chamb. = Rhododendron roxieanum Forrest ex W. W. Sm. ●

331678　Rhododendron roxieanum Forrest ex W. W. Sm. var. globigerum（Balf. f. et Forrest）D. F. Chamb. = Rhododendron alutaceum Balf. f. et W. W. Sm. ●

331679　Rhododendron roxieanum Forrest ex W. W. Sm. var. oreonastes（Balf. f. et Forrest）T. L. Ming；线形卷叶杜鹃；Linear Roxie's Rhododendron ●

331680　Rhododendron roxieanum Forrest ex W. W. Sm. var. recurvum（Balf. f. et Forrest）Davidian = Rhododendron roxieanum Forrest ex W. W. Sm. ●

331681　Rhododendron roxieoides D. F. Chamb.；巫山杜鹃（似卷叶杜鹃）；False Roxie's Rhododendron, Like R. Roxieanum Rhododendron, Wushan Azalea, Wushan Rhododendron ●

331682　Rhododendron roylei Hook. f. = Rhododendron cinnabarinum Wall. var. roylei（Hook. f.）Hutch. ●

331683　Rhododendron rubiginosum Franch.；红棕杜鹃（茶花杜鹃,茶花叶杜鹃,无鳞杜鹃,锈色杜鹃）；Bereft Scales Rhododendron, Mauvequeen Rhododendron, Redbrown Azalea, Reddish-brown Rhododendron, Rusty Rhododendron ●

331684　Rhododendron rubiginosum Franch. var. leclerei（H. Lév.）R. C. Fang；洁净红棕杜鹃●

331685　Rhododendron rubiginosum Franch. var. ptilostylum R. C. Fang；毛柱红棕杜鹃；Hairy-style Rusty Rhododendron ●

331686　Rhododendron rubriflorum Kingdon-Ward = Rhododendron campylogynum Franch. ●

331687　Rhododendron rubrolineatum Balf. f. et Forrest = Rhododendron mekongense Franch. var. rubrolineatum（Balf. f. et Forrest）Cullen ●

331688　Rhododendron rubroluteum Davidian = Rhododendron mekongense Franch. ●

331689　Rhododendron rubropilosum Hayata；赤毛杜鹃（红毛杜鹃,台红毛杜鹃）；Red-haired Rhododendron, Red-hairy Rhododendron, Red-pilose Rhododendron, Taiwan Redhair Azalea ●

331690　Rhododendron rubropilosum Hayata var. breviperulatum（Hayata）T. Yamaz. = Rhododendron breviperulatum Hayata ●

331691　Rhododendron rubropilosum Hayata var. taiwanalpinum（Ohwi）S. Y. Lu, Yuen P. Yang et Y. H. Tseng；台湾高山杜鹃●

331692　Rhododendron rubropilosum Hayata var. taiwanalpinum（Ohwi）S. Y. Lu, Yuen P. Yang et Y. H. Tseng = Rhododendron taiwanalpinum Ohwi ●

331693　Rhododendron rubropuctatum Hayata = Rhododendron pseudochrysanthum Hayata ●

331694　Rhododendron rubropunctatum H. Lév. et Vaniot = Rhododendron siderophyllum Franch. ●

331695　Rhododendron rubropunctatum Hayata = Rhododendron hyperythrum Hayata ●

331696　Rhododendron rubropunctatum T. L. Ming；红点杜鹃；Red-spotted Rhododendron ●

331697　Rhododendron rubropunctatum T. L. Ming = Rhododendron tanastylum Balf. f. et Kingdon-Ward var. lingzhiense M. Y. Fang ●

331698　Rhododendron rubropunctatumm Hayata = Rhododendron hyperythrum Hayata ●

331699　Rhododendron rude Tagg et Forrest = Rhododendron glischrum Balf. f. et W. W. Sm. subsp. rude（Tagg et Forrest）Cowan ●

331700 Rhododendron rufescens Franch. ;红背杜鹃(茶绒杜鹃,瑞香杜鹃花); Becaming Red Rhododendron, Daphne Flowered Rhododendron, Daphne Rhododendron, Redback Azalea, Red-back Rhododendron ●

331701 Rhododendron rufescens P. C. Tam = Rhododendron apricum P. C. Tam ●

331702 Rhododendron rufescens P. C. Tam = Rhododendron rufulum P. C. Tam ●

331703 Rhododendron rufohirtum Hand. -Mazz. ;滇红毛杜鹃(滇红杜鹃); Red-haired Rhododendron, Red-haired Yunnan Azalea, Yunnan Red Azalea ●

331704 Rhododendron rufosquamosum Hutch. = Rhododendron pachypodum Balf. f. et W. W. Sm. ●

331705 Rhododendron rufulum P. C. Tam;茶绒杜鹃(褐色杜鹃,黑叶杜鹃,上杭杜鹃); Brow-coloured Rhododendron, Dark-brown Leaves Azalea, Dark-brown Leaves Rhododendron, Dark-leaved Rhododendron, Light-red Rhododendron, Redish Azalea, Shanghang Azalea, Shanghang Rhododendron, Shang-hang Rhododendron ●

331706 Rhododendron rufulum P. C. Tam = Rhododendron apricum P. C. Tam ●

331707 Rhododendron rufum Batalin;黄毛杜鹃(深红斑杜鹃,魏氏杜鹃); Crimson Spotted Rhododendron, Crimson-spotted Rhododendron, S. M. Weld Rhododendron, Talispot Rhododendron, Yellowhair Azalea, Yellow-hair Rhododendron, Yellow-haired Rhododendron ●

331708 Rhododendron rufum Batalin var. glandulosum G. H. Wang;腺房黄毛杜鹃;Gland Yellowhair Azalea ●

331709 Rhododendron rufum Batalin var. pachysanthum (Hayata) S. S. Ying = Rhododendron hyperythrum Hayata ●

331710 Rhododendron rupicola W. W. Sm. ;岩生杜鹃(多色杜鹃); Cliff-plum Rhododendron, Ill-coloured Azalea, Ill-coloured Rhododendron, Multicolor Azalea, Stony Place Dwellers Rhododendron ●

331711 Rhododendron rupicola W. W. Sm. var. chryseum (Balf. f. et Kingdon-Ward) M. N. Philipson et Philipson;金黄多色杜鹃(金黄杜鹃);Golden Azalea, Golden Rhododendron ●

331712 Rhododendron rupicola W. W. Sm. var. muliense (Balf. f. et Forrest) M. N. Philipson et Philipson;木里多色杜鹃(金黄杜鹃,木里杜鹃,木里金黄杜鹃);Muli Azalea, Muli Rhododendron ●

331713 Rhododendron russatum Balf. f. et Forrest;紫蓝杜鹃;Bluish-purple Rhododendron, Purpleblue Azalea, Purpleblue Rhododendron, Purple-blue Rhododendron ●

331714 Rhododendron russatum Balf. f. et Forrest 'Collingwood Ingram';科灵伍德·英格拉姆紫蓝杜鹃●☆

331715 Rhododendron russatum Balf. f. et Forrest 'Keillour';凯罗尔紫蓝杜鹃●☆

331716 Rhododendron russatum Balf. f. et Forrest 'Maryborough';玛栗鲍鲁紫蓝杜鹃●☆

331717 Rhododendron russatum Balf. f. et Forrest 'Night Editor';夜编辑紫蓝杜鹃●☆

331718 Rhododendron russotinctum Balf. f. et Forrest = Rhododendron alutaceum Balf. f. et W. W. Sm. var. russotinctum (Balf. f. et Forrest) D. F. Chamb. ●

331719 Rhododendron salignum Hook. f. = Rhododendron lepidotum Wall. et G. Don ●

331720 Rhododendron saluenense Franch. ;怒江杜鹃;Nujiang Azalea, Salwen Rhododendron ●

331721 Rhododendron saluenense Franch. subsp. chameunum (Balf. f. et Forrest) Cullen = Rhododendron saluenense Franch. var. prostratum (W. W. Sm.) R. C. Fang ●

331722 Rhododendron saluenense Franch. var. prostratum (W. W. Sm.) R. C. Fang;平卧怒江杜鹃(伏地杜鹃);Lying Dround Rhododendron, Prostrate Nujiang Azalea, Prostrate Salwen Rhododendron ●

331723 Rhododendron sanctum Nakai;神宫杜鹃●☆

331724 Rhododendron sanctum Nakai var. lasiogynum Nakai ex H. Hara;毛蕊神宫杜鹃●☆

331725 Rhododendron sanctum Nakai var. lasiogynum Nakai ex H. Hara f. albiflorum Sugim. ;白花毛蕊神宫杜鹃●☆

331726 Rhododendron sanguineum Franch. ;血红杜鹃;Bloodred Azalea, Blood-red Rhododendron, Tibetblood Rhododendron ●

331727 Rhododendron sanguineum Franch. subsp. aizoides Cowan = Rhododendron sanguineum Franch. var. himertum (Balf. f. et Forrest) T. L. Ming ●

331728 Rhododendron sanguineum Franch. subsp. cloiophorum (Balf. f. et Forrest) Cowan = Rhododendron sanguineum Franch. var. cloiophorum (Balf. f. et Forrest) D. F. Chamb. ●

331729 Rhododendron sanguineum Franch. subsp. consanguineum Cowan = Rhododendron sanguineum Franch. var. didymoides Tagg et Forrest ●

331730 Rhododendron sanguineum Franch. subsp. didymoides (Tagg et Forrest) Cowan = Rhododendron sanguineum Franch. var. didymoides Tagg et Forrest ●

331731 Rhododendron sanguineum Franch. subsp. didymum (Balf. f. et Forrest) Cowan = Rhododendron sanguineum Franch. var. didymum (Balf. f. et Forrest) T. L. Ming ●

331732 Rhododendron sanguineum Franch. subsp. haemaleum (Balf. f. et Forrest) Cowan = Rhododendron sanguineum Franch. var. haemaleum (Balf. f. et Forrest) D. F. Chamb. ●

331733 Rhododendron sanguineum Franch. subsp. himertum (Balf. f. et Forrest) Cowan = Rhododendron sanguineum Franch. var. himertum (Balf. f. et Forrest) T. L. Ming ●

331734 Rhododendron sanguineum Franch. subsp. leucopetalum (Balf. f. et Forrest) Cowan = Rhododendron sanguineum Franch. var. cloiophorum (Balf. f. et Forrest) D. F. Chamb. ●

331735 Rhododendron sanguineum Franch. subsp. melleum Cowan = Rhododendron sanguineum Franch. var. himertum (Balf. f. et Forrest) T. L. Ming ●

331736 Rhododendron sanguineum Franch. subsp. meozaeum Balf. f. ex Cowan = Rhododendron sanguineum Franch. var. haemaleum (Balf. f. et Forrest) D. F. Chamb. ●

331737 Rhododendron sanguineum Franch. subsp. roseotinctum (Balf. f. et Forrest) Cowan = Rhododendron sanguineum Franch. var. didymoides Tagg et Forrest ●

331738 Rhododendron sanguineum Franch. subsp. sanguineoides Cowan = Rhododendron sanguineum Franch. ●

331739 Rhododendron sanguineum Franch. var. cloiophorum (Balf. f. et Forrest) D. F. Chamb. ;褪色血红杜鹃●

331740 Rhododendron sanguineum Franch. var. consanguineum (Cowan) Davidian = Rhododendron sanguineum Franch. var. didymoides Tagg et Forrest ●

331741 Rhododendron sanguineum Franch. var. didymoides Tagg et Forrest;变色血红杜鹃●

331742 Rhododendron sanguineum Franch. var. didymum (Balf. f. et Forrest) T. L. Ming;黑红杜鹃(二褶杜鹃,黑红血红杜鹃,黑色血

红杜鹃）；Mandarin Azalea, Mandarin Rhododendron, Two-fold Rhododendron ●

331743　Rhododendron sanguineum Franch. var. haemaleum（Balf. f. et Forrest）D. F. Chamb.；紫血杜鹃（血红杜鹃，紫血血红杜鹃）；Blood-red Rhododendron ●

331744　Rhododendron sanguineum Franch. var. himertum（Balf. f. et Forrest）T. L. Ming；蜜黄血红杜鹃●

331745　Rhododendron sanguineum Franch. var. mesaeum（Balf. f. ex Cowan）Davidian ＝ Rhododendron sanguineum Franch. var. haemaleum（Balf. f. et Forrest）D. F. Chamb. ●

331746　Rhododendron sanguineum Franch. var. sanguineoides（Cowan）Davidian ＝ Rhododendron sanguineum Franch. ●

331747　Rhododendron sanidodeum P. C. Tam；桂东杜鹃；East Guangxi Rhododendron, Guidong Rhododendron ●

331748　Rhododendron sanidodeum P. C. Tam ＝ Rhododendron bachii H. Lév. ●

331749　Rhododendron sargentianum Rehder et E. H. Wilson；水仙杜鹃（峨眉杜鹃）；Daffodil Azalea, Sargent's Rhododendron ●

331750　Rhododendron sasakii E. H. Wilson ＝ Rhododendron breviperulatum Hayata ●

331751　Rhododendron sasakii E. H. Wilson ＝ Rhododendron lasiostylum Hayata ●

331752　Rhododendron sataense Nakai ＝ Rhododendron obtusum（Lindl.）Planch. ●

331753　Rhododendron satsumense Hatus. ;萨摩杜鹃●☆

331754　Rhododendron sawuense H. P. Yang；道孚杜鹃；Daofu Azalea, Daofu Rhododendron, Dawu Rhododendron ●

331755　Rhododendron saxatile B. Y. Ding et Y. Y. Fang；崖壁杜鹃；Cliff Azalea, Saxatile Rhododendron, Saxicolous Rhododendron ●

331756　Rhododendron scabrifolium Franch.；糙叶杜鹃；Coalseleaf Azalea, Rough-leaves Rhododendron, Scabrousleaf Rhododendron, Scabrous-leaved Rhododendron ●

331757　Rhododendron scabrifolium Franch. var. pauciflorum Franch.；疏花糙叶杜鹃（疏毛杜鹃花）；Loose-flowers Coalseleaf Azalea, Loose-flowers Rough-leaves Rhododendron ●

331758　Rhododendron scabrifolium Franch. var. spiciferum（Franch.）Cullen ＝ Rhododendron spiciferum Franch. ●

331759　Rhododendron scabrum G. Don；琉球杜鹃（火红杜鹃，计良间杜鹃，琉球杜鹃花）；Liuqiu Azalea, Luchu Azalea, Rough Azalea ●☆

331760　Rhododendron scabrum G. Don f. albiflorum Hatus.；白花琉球杜鹃●☆

331761　Rhododendron scabrum G. Don f. coccineum E. H. Wilson ＝ Rhododendron scabrum G. Don ●☆

331762　Rhododendron scabrum G. Don f. roseum Hatus. et Kudaka；粉花琉球杜鹃●☆

331763　Rhododendron scabrum G. Don subsp. amanoi（Ohwi）D. F. Chamb. et Rae ＝ Rhododendron amanoi Ohwi ●☆

331764　Rhododendron scabrum G. Don var. yakuinsulare（Masam.）T. Yamaz. ＝ Rhododendron yakuinsulare Masam. ●☆

331765　Rhododendron schistocalyx Balf. f. et Forrest；裂萼杜鹃；Schizo-calyxed Rhododendron, Split Calyx Rhododendron, Splitcalyx Azalea ●

331766　Rhododendron schizopeplum Balf. f. et Forrest ＝ Rhododendron aganniphum Balf. f. et Kingdon-Ward var. schizopeplum（Balf. f. et Forrest）T. L. Ming ●

331767　Rhododendron schizopeplum Balf. f. et Forrest ＝ Rhododendron schistocalyx Balf. f. et Forrest ●

331768　Rhododendron schlippenbachii Maxim.；大字杜鹃（大字香，新

杜鹃）；Royal Azalea, Royal Rhododendron ●

331769　Rhododendron sciaphilum Balf. f. et Kingdon-Ward ＝ Rhododendron edgeworthii Hook. f. ●

331770　Rhododendron scintillans Balf. f. et W. W. Sm. ＝ Rhododendron polycladum Franch. ●

331771　Rhododendron sclerocladum Balf. f. et Forrest ＝ Rhododendron cuneatum W. W. Sm. ●

331772　Rhododendron scopulorum Hutch.；石峰杜鹃（峭壁杜鹃）；Crags Rhododendron, Stonepeak Azalea, Stonepeak Rhododendron, Stone-peak Rhododendron ●

331773　Rhododendron scottianum Hutch. ;波瓣杜鹃●

331774　Rhododendron scottianum Hutch. ＝ Rhododendron pachypodum Balf. f. et W. W. Sm. ●

331775　Rhododendron scyphocalyx Balf. f. et Forrest ＝ Rhododendron dichroanthum Diels et Cowan subsp. scyphocalyx（Balf. f. et Forrest）Cowan ●

331776　Rhododendron searsiae Rehder et E. H. Wilson；绿点杜鹃（红点杜鹃，席氏杜鹃）；G. Sears' Rhododendron, Greendot Azalea, Sears Rhododendron ●

331777　Rhododendron seguini H. Lév. ＝ Rhododendron yunnanense Franch. ●

331778　Rhododendron seinghkuense Kingdon-Ward；辛库杜鹃（黄花泡泡叶杜鹃，黄花泡叶杜鹃）；Seinghku Rhododendron, Yellow Bubbleleaf Azalea ●

331779　Rhododendron selense Franch.；多变杜鹃（滇西杜鹃）；Sie-la Rhododendron, Variable Rhododendron, Varied Azalea, Varied Rhododendron ●

331780　Rhododendron selense Franch. subsp. dasycladum（Balf. f. et W. W. Sm.）D. F. Chamb.；毛枝多变杜鹃●

331781　Rhododendron selense Franch. subsp. duseimatum（Balf. f. et Forrest）Tagg ＝ Rhododendron calvescens Balf. f. et Forrest var. duseimatum（Balf. f. et Forrest）D. F. Chamb. ●

331782　Rhododendron selense Franch. subsp. jucundum（Balf. f. et W. W. Sm.）D. F. Chamb.；粉背多变杜鹃（和蔼杜鹃，愉快杜鹃）；Pleasant Azalea, Pleasant Rhododendron ●◇

331783　Rhododendron selense Franch. subsp. setiferum（Balf. f. et Forrest）Chamb. ex Cullen et Chamb. ＝ Rhododendron setiferum Balf. f. et Forrest ●

331784　Rhododendron selense Franch. subsp. setiferum（Balf. f. et Forrest）D. F. Chamb. ＝ Rhododendron setiferum Balf. f. et Forrest ●

331785　Rhododendron selense Franch. var. dasycladum（Balf. f. et W. W. Sm.）T. L. Ming ＝ Rhododendron selense Franch. subsp. dasycladum（Balf. f. et W. W. Sm.）D. F. Chamb. ●

331786　Rhododendron selense Franch. var. duseimatum（Balf. f. et Forrest）Cowan et Davidian ＝ Rhododendron calvescens Balf. f. et Forrest var. duseimatum（Balf. f. et Forrest）D. F. Chamb. ●

331787　Rhododendron selense Franch. var. jucundum（Balf. f. et W. W. Sm.）T. L. Ming ＝ Rhododendron selense Franch. subsp. jucundum（Balf. f. et W. W. Sm.）D. F. Chamb. ●◇

331788　Rhododendron selense Franch. var. pagophilum（Balf. f. et Kingdon-Ward）Cowan et Davidian ＝ Rhododendron selense Franch. ●

331789　Rhododendron selense Franch. var. probum（Balf. f. et Forrest）Cowan et Davidian ＝ Rhododendron selense Franch. ●

331790　Rhododendron semanteum Balf. f. ＝ Rhododendron impeditum Balf. f. et W. W. Sm. ●

331791　Rhododendron semibarbatum Maxim.；半硬剌杜鹃；Partially Bearded Rhododendron ●☆

331792 Rhododendron semilunatum Balf. f. et Forrest = Rhododendron mekongense Franch. var. melinanthum（Balf. f. et Kingdon-Ward）Cullen ●

331793 Rhododendron semnoides Tagg et Forrest;圆头杜鹃（圆头叶杜鹃）;Resembling R. Semnum Rhododendron, Roundhead Azalea, Roundtop Rhododendron,Round-topped Rhododendron ●

331794 Rhododendron semnum Balf. f. et Forrest = Rhododendron praestans Balf. f. et W. W. Sm. ●

331795 Rhododendron seniavinii Maxim.;毛果杜鹃（福建杜鹃,满山白,小花满山白）;Hairfruit Azalea, Seniavin Rhododendron ●

331796 Rhododendron seniavinii Maxim. var. crassifolium P. C. Tam;厚叶毛果杜鹃（厚叶照山白）;Thickleaf Hairfruit Azalea ●

331797 Rhododendron seniavinii Maxim. var. crassifolium P. C. Tam = Rhododendron seniavinii Maxim. ●

331798 Rhododendron sermum Balf. f. et Forrest = Rhododendron praestans Balf. f. et W. W. Sm. ●

331799 Rhododendron serotinum Hutch.;晚花杜鹃●☆

331800 Rhododendron serpens Balf. f. et Forrest = Rhododendron erastum Balf. f. et Forrest ●

331801 Rhododendron serpyllifolium（A. Gray）Miq.;百里香叶杜鹃●☆

331802 Rhododendron serpyllifolium（A. Gray）Miq. = Rhododendron nakaharai Hayata ●

331803 Rhododendron serpyllifolium（A. Gray）Miq. f. albiflorum（Makino）H. Hara = Rhododendron serpyllifolium（A. Gray）Miq. var. albiflorum Makino ●☆

331804 Rhododendron serpyllifolium（A. Gray）Miq. f. album T. Yamaz.;白百里香叶杜鹃●☆

331805 Rhododendron serpyllifolium（A. Gray）Miq. var. albiflorum Makino;白花百里香叶杜鹃●☆

331806 Rhododendron serpyllifolium Hayata = Rhododendron nakaharai Hayata ●

331807 Rhododendron serrulatum（Small）Millais;细齿杜鹃（细齿杜鹃花）;Small Tooth Leaves Rhododendron ●☆

331808 Rhododendron setiferum Balf. f. et Forrest;刚刺杜鹃（被背杜鹃, 具髯杜鹃）; Bristle-bearing Rhododendron, Bristly Rhododendron,Clothed Rhododendron,Setiferous Azalea ●

331809 Rhododendron setosum D. Don;刚毛杜鹃（枝杜鹃）;Bristly Branchlets Rhododendron,Setose Azalea,Setose Rhododendron ●

331810 Rhododendron shaanxiense W. P. Fang et Z. J. Zhao = Rhododendron yunnanense Franch. ●

331811 Rhododendron shanii W. P. Fang;都支杜鹃; Shan Azalea, Shan's Rhododendron ●◇

331812 Rhododendron sheltoniae Hemsl. et E. H. Wilson = Rhododendron vernicosum Franch. ●

331813 Rhododendron sheltonii Hemsl. et E. H. Wilson =Rhododendron vernicosum Franch. ●

331814 Rhododendron shensiense R. C. Ching = Rhododendron oreodoxa Franch. var. shensiense D. F. Chamb. ●

331815 Rhododendron shepherdii Nutt. = Rhododendron kendrickii Nutt. ●

331816 Rhododendron sherriffii Cowan;红钟杜鹃; Redbell Azalea, Sherriff's Rhododendron ●

331817 Rhododendron shimenense Q. X. Liu et C. M. Zhang;石门杜鹃; Shimen Azalea,Shimen Rhododendron ●

331818 Rhododendron shimianense W. P. Fang et P. S. Liu;石棉杜鹃; Shimian Azalea,Shimian Rhododendron ●

331819 Rhododendron shiwandashanense P. C. Tam;十万大山杜鹃; Shiwandashan Azalea,Shiwandashan Rhododendron ●

331820 Rhododendron shiwandashanense P. C. Tam = Rhododendron henryi Hance ●

331821 Rhododendron shweliense Balf. f. et Forrest;瑞丽杜鹃;Ruili Azalea,Ruili Rhododendron,Shweli-river Rhododendron ●

331822 Rhododendron siamense Diels = Rhododendron moulmainense Hook. ●

331823 Rhododendron sichotense Pojark.;锡浩特杜鹃●☆

331824 Rhododendron sidereum Balf. f.;银灰杜鹃（特异杜鹃）; Argenteous Azalea, Excellent Rhododendron, Silver-grey Rhododendron ●

331825 Rhododendron siderophylloides Huwh.？ = Rhododendron oreotrephes W. W. Sm. ●

331826 Rhododendron siderophyllum Franch.;锈叶杜鹃（小白花）; Rustleaf Azalea, Rusty Coated Leaves Rhododendron, Rusty-leaved Rhododendron ●

331827 Rhododendron sigillatum Balf. f. et Forrest = Rhododendron phaeochrysum Balf. f. et W. W. Sm. var. levistratum（Balf. f. et Forrest）D. F. Chamb. ●

331828 Rhododendron sikangense W. P. Fang;川西杜鹃（库克杜鹃,西康杜鹃）; R. B. Cooke's Rhododendron, Xikang Azalea, Xikang Rhododendron ●◇

331829 Rhododendron sikangense W. P. Fang var. exquistum（T. L. Ming）T. L. Ming;优美杜鹃;Beautiful Xikang Azalea, Beautiful Xikang Rhododendron ●

331830 Rhododendron sikayotaizanense Masam.;细叶杜鹃（细叶杜鹃花, 志佳阳杜鹃）; Fine Leaves Rhododendron, Sikayotaizan Rhododendron ●

331831 Rhododendron silvaticum Cowan = Rhododendron lanigerum Tagg ●

331832 Rhododendron simiarum Hance;猴头杜鹃（福建杜鹃,南华杜鹃, 西米杜鹃）; Fujian Azalea, Fujian Rhododendron, Fukien Rhododendron, Monkeyhead Azalea, Monkeyhead Rhododendron, Monkey-head Rhododendron,Monkeys Rhododendron ●

331833 Rhododendron simiarum Hance subsp. youngae（W. P. Fang）Chamb. ex Cullen et Chamb. = Rhododendron adenopodum Franch. ●

331834 Rhododendron simiarum Hance subsp. youngiae（W. P. Fang）D. F. Chamb. = Rhododendron adenopodum Franch. ●

331835 Rhododendron simiarum Hance var. deltoideum P. C. Tam;折角杜鹃●

331836 Rhododendron simiarum Hance var. deltoideum P. C. Tam = Rhododendron simiarum Hance ●

331837 Rhododendron simiarum Hance var. grandifolium G. Z. Li;大叶南华杜鹃●

331838 Rhododendron simiarum Hance var. versicolor（Chun et W. P. Fang）M. Y. Fang;变色杜鹃;Changing-coloured Rhododendron, Variable-coloured Rhododendron,Versicoloured Rhododendron ●

331839 Rhododendron simsii Planch.;杜鹃（报春花,长鳞芽杜鹃,长条杜鹃,虫鸟花,大屯杜鹃,大屯满山红,灯盏红花,杜鹃花,翻山虎,红柴片花,红杜鹃,红花杜鹃,红踯躅,怀春花,荚蒾叶杜鹃,满山红,清明花,山茶花,山归来,山石榴,山踯躅,搜山虎,唐杜鹃,艳红山,艳山花,应春花,迎山红,映山红,照山红）; Azalea, Chinese Evergreen Azalea, Indian Azalea,Indian Rhododendron,Long Marked Azalea, Long Marked Rhododendron, Long-marked Rhododendron,Longperula Azalea, Long-perula Rhododendron, Sims Azalea,Sims' Azalea, Sims' Rhododendron, Sims' Indian Azalea, Viburnum-leaved Rhododendron ●

331840　Rhododendron simsii Planch. = Rhododendron hainanense Merr. ●

331841　Rhododendron simsii Planch. var. albiflorum R. L. Liu;白花映山红;White-flower Indian Rhododendron ●

331842　Rhododendron simsii Planch. var. strigoso-stylum G. Z. Li;糙柱映山红;Rough-style Rhododendron ●

331843　Rhododendron simsii Planch. var. tamurae（Makino）Kaneh. et Hatus.;大花映山红●

331844　Rhododendron simsii Planch. var. tamurae（Makino）Kaneh. et Hatus. = Rhododendron eriocarpum（Hayata）Nakai ●

331845　Rhododendron simsii Planch. var. tawadae（Ohwi）Kaneh. et Hatus. = Rhododendron eriocarpum（Hayata）Nakai var. tawadae Ohwi ●☆

331846　Rhododendron simsii Planch. var. yakuinsulare（Masam.）T. Yamaz. = Rhododendron yakuinsulare Masam. ●☆

331847　Rhododendron simulans（Tagg et Forrest）D. F. Chamb. = Rhododendron sinosimulans D. F. Chamb. ●

331848　Rhododendron sinense Sweet = Rhododendron molle（Blume）G. Don ●

331849　Rhododendron sinensis（Lodd.）Sweet = Rhododendron molle（Blume）G. Don ●

331850　Rhododendron sino-falconeri Balf. f.;宽杯杜鹃（华杯毛杜鹃,圆叶杜鹃,中国大叶杜鹃）;Broadcup Azalea, Chinese Falconer Rhododendron, Chinese R. Falconer Rhododendron ●

331851　Rhododendron sinogrande Balf. f. et W. W. Sm.;凸尖杜鹃（大叶杜鹃,凸头杜鹃,凸叶杜鹃,中国大叶杜鹃）;Chinese R. Grande Rhododendron, Convextine Azalea, Greatleaf Rhododendron, Great-leaved Rhododendron ●

331852　Rhododendron sinogrande Balf. f. et W. W. Sm. var. boreale Tagg et Forrest = Rhododendron sinogrande Balf. f. et W. W. Sm. ●

331853　Rhododendron sinolepidotum Balf. f. = Rhododendron lepidotum Wall. ex G. Don ●

331854　Rhododendron sinomuttallii Balf. f. et Forrest = Rhododendron excellens Hemsl. et E. H. Wilson ●

331855　Rhododendron sinonuttallii Balf. f. et Forrest;华木兰杜鹃（大果杜鹃,中国木兰杜鹃,中国木兰杜鹃花）;Bigfruit Azalea, Chinese Nuttall Rhododendron, Chinese R. Nuttall Rhododendron, Chinese Rhododendron ●

331856　Rhododendron sinosimulans D. F. Chamb.;裂毛杜鹃;Imitative Rhododendron, Splithair Azalea ●

331857　Rhododendron sinovaccinioides Balf. f. et Forrest = Rhododendron vaccinioides Hook. f. ●

331858　Rhododendron smirnowii Trautv. ex Regel;斯密尔杜鹃（斯密尔杜鹃花）;M. Smirnow's Rhododendron, Turkish Rhododendron ●☆

331859　Rhododendron smithii Nutt.;毛枝杜鹃（不丹杜鹃,斯密斯杜鹃）;Bhudanese Rhododendron, Hairtwig Azalea, J. E. Smith's Rhododendron, Smith Rhododendron ●

331860　Rhododendron sordidum Hutch. = Rhododendron pruniflorum Hutch. et Kingdon-Ward ●

331861　Rhododendron souliei Franch.;白碗杜鹃（索氏杜鹃）;Soulie Rhododendron, Whitebowl Azalea ●

331862　Rhododendron spadiceum P. C. Tam;蔗黄杜鹃;Cany Yellow Rhododendron, Cany-yellow Rhododendron, Sugarcane-yellow Azalea ●

331863　Rhododendron spanotrichum Balf. f. et W. W. Sm.;红花杜鹃（光柱杜鹃,光柱杜鹃花）;Bare Styles Rhododendron, Bare-styled Rhododendron, Redflower Azalea ●

331864　Rhododendron sparsifolium W. P. Fang;川南杜鹃（疏叶杜鹃）; Chuannan Rhododendron, S. Sichuan Azalea, South Sichuan Rhododendron ●

331865　Rhododendron speciosum（Willd.）Sweet;火黄杜鹃;Golden-looking Azalea, Oconee Azalea ●☆

331866　Rhododendron sperabile Balf. f. et Forrest;纯红杜鹃（缅甸红杜鹃）;Burmaflame Rhododendron, Purered Azalea, Upper Burma Red Rhododendron ●

331867　Rhododendron sperabile Balf. f. et Forrest var. weihsiense Tagg et Forrest;维西纯红杜鹃;Weisi Burmaflame Rhododendron ●

331868　Rhododendron sperabiloides Tagg et Forrest;糠秕杜鹃;Chaff Azalea, False Burmaflame Rhododendron, Like R. Sperabile Rhododendron ●

331869　Rhododendron sphaeranthum Balf. f. et W. W. Sm. = Rhododendron trichostomum Franch. ●

331870　Rhododendron sphaeroblastum Balf. f. et Forrest;宽叶杜鹃（粉红芽杜鹃,圆芽杜鹃）;Broadleaf Azalea, Pink-bud Rhododendron, Rounded-buds Rhododendron ●

331871　Rhododendron sphaeroblastum Balf. f. et Forrest var. wumengense K. M. Feng;乌蒙宽叶杜鹃（乌蒙杜鹃）;Wumeng Rounded-buds Rhododendron ●

331872　Rhododendron spiciferum Franch.;碎米花（毛叶杜鹃,穗花杜鹃）;Breaking Spikes Rhododendron, Brokenriceflower Azalea, Spiked Rhododendron ●

331873　Rhododendron spiciferum Franch. var. album K. M. Feng;白碎米花;White Spiked Rhododendron ●

331874　Rhododendron spilanthum Hutch. = Rhododendron thymifolium Maxim. ●

331875　Rhododendron spilotum Balf. f. et Forrest;染红杜鹃;Stained Crimson Rhododendron ●☆

331876　Rhododendron spinigerum H. Lév. = Rhododendron chrysocalyx H. Lév. et Vaniot ●

331877　Rhododendron spinuliferum Franch.;爆仗花杜鹃（爆仗花,爆杖花,炮仗花,细刺杜鹃）;Bearing Spines Rhododendron, Firecracker-flower Azalea, Torch Rhododendron ●

331878　Rhododendron spinuliferum Franch. var. duclouxii H. Lév.;粉红爆仗花;Pink Torch Rhododendron ●

331879　Rhododendron spinuliferum Franch. var. glabrescens K. M. Feng ex R. C. Fang;少毛爆仗花杜鹃（少花爆杖花,少毛爆仗花）;Glabrescent Firecracker-flower Azalea ●

331880　Rhododendron spodopeplum Balf. f. et Farter = Rhododendron tephropeplum Balf. f. et Forrest ●

331881　Rhododendron spooneri Hemsl. et E. H. Wilson = Rhododendron decorum Franch. ●

331882　Rhododendron stamineum Franch.;长蕊杜鹃（多蕊杜鹃,六骨筋）;Longstamen Azalea, Longstamen Rhododendron, Long-stamened Rhododendron, Many-stamens Rhododendron ●

331883　Rhododendron stamineum Franch. var. lasiocarpum R. C. Fang et C. H. Yang;毛果长蕊杜鹃;Hairyfruit Longstamen Azalea, Hairyfruit Longstamen Rhododendron ●

331884　Rhododendron stenaulum Balf. f. et W. W. Sm.;细管杜鹃（长蒴杜鹃）;Longcapsule Azalea, Narrow-grooved Rhododendron ●

331885　Rhododendron stenaulum Balf. f. et W. W. Sm. = Rhododendron moulmainense Hook. ●

331886　Rhododendron stenopetalum（Hogg）Mabb.;蜘蛛杜鹃;Spider Azalea ●☆

331887　Rhododendron stenophyllum Balf. f. et W. W. Sm. = Rhododendron makinoi Tagg ex Nakai ●☆

331888   Rhododendron stenoplastum Balf. f. et Forrest = Rhododendron rubiginosum Franch. ●

331889   Rhododendron stereophyllum Balf. f. et W. W. Sm. = Rhododendron tatsiense Franch. ●

331890   Rhododendron stewartianum Diels；多趣杜鹃；L. B. Stewart's Rhododendron，Mach-interest Azalea，Stewart Rhododendron ●

331891   Rhododendron stewartianum Diels var. aiolosalpinx（Balf. f. et Forrest）Cowan et Davidian = Rhododendron stewartianum Diels ●

331892   Rhododendron stewartianum Diels var. tantulum Cowan et Davidian = Rhododendron stewartianum Diels ●

331893   Rhododendron stictophyllum Balf. f. = Rhododendron nivale Hook. f. subsp. boreale M. N. Philipson et Philipson ●

331894   Rhododendron strigillosum Franch.；芒刺杜鹃（刺芒杜鹃，大羊角树）；Awnspine Azalea，Scarletball Rhododendron，Scarlet-balled Rhododendron，Stiff Bristless Rhododendron ●

331895   Rhododendron strigillosum Franch. var. monosematum（Hutch.）T. L. Ming；紫斑杜鹃；One Blotch Rhododendron，Purplespot Azalea，Purple-spotted Scarletball Rhododendron ●

331896   Rhododendron strigosum R. L. Liu；伏毛杜鹃；Strigose Rhododendron ●

331897   Rhododendron subarcticum Harmaja = Ledum palustre L. var. decumbens Aiton ●

331898   Rhododendron subcerinum P. C. Tam；蜡黄杜鹃；Waxy Yellow Rhododendron，Waxy-yellow Rhododendron ●

331899   Rhododendron subenerve P. C. Tam = Rhododendron tsoi Merr. ●

331900   Rhododendron subenerve P. C. Tam var. nudistylum P. C. Tam；细石榴花 ●

331901   Rhododendron subenerve P. C. Tam var. nudistylum P. C. Tam = Rhododendron tsoi Merr. ●

331902   Rhododendron suberinum P. C. Tam；秆黄杜鹃 ●

331903   Rhododendron suberosum Balf. f. et Forrest = Rhododendron yunnanense Franch. ●

331904   Rhododendron subespitatum Chun ex P. C. Tam = Rhododendron championae Hook. ●

331905   Rhododendron subestipetatum P. C. Tam = Rhododendron championae Hook. ●

331906   Rhododendron subflumineum P. C. Tam；涧上杜鹃；Gully Azalea，Ravine Rhododendron ●

331907   Rhododendron sulfureum Franch. = Rhododendron monanthum Balf. f. et W. W. Sm. ●

331908   Rhododendron sulphureum Franch.；硫黄杜鹃；Sulphur Azalea，Sulphur Rhododendron，Sulphur-coloured Rhododendron ●◇

331909   Rhododendron supranubium Hutch. = Rhododendron pachypodum Balf. f. et W. W. Sm. ●

331910   Rhododendron surasianum Balf. f. et Craib；泰国杜鹃（泰国杜鹃花）；Saim Rhododendron ●☆

331911   Rhododendron sutchuenense Franch.；四川杜鹃（大叶羊角）；Sichuan Azalea，Sichuan Rhododendron，Szechwan Rhododendron ●

331912   Rhododendron sycnanthum Balf. f. et W. W. Sm. = Rhododendron rigidum Franch. ●

331913   Rhododendron syncollum Balf. f. et Forrest = Rhododendron phaeochrysum Balf. f. et W. W. Sm. var. agglutinatum（Balf. f. et Forrest）D. F. Chamb. ●

331914   Rhododendron taggianum Hutch.；白喇叭杜鹃；Tagg's Rhododendron，Whitebugle Azalea ●

331915   Rhododendron taibaiense Ching et H. P. Yang；陕西杜鹃（平卧太白杜鹃，太白杜鹃）；Taibai Rhododendron ●

331916   Rhododendron taipaoense T. C. Wu et P. C. Tam；大埔杜鹃；Dapu Azalea，Dapu Rhododendron ●

331917   Rhododendron taishunense B. Y. Ding et Y. Y. Fang；泰顺杜鹃；Taishun Azalea ●

331918   Rhododendron taiwanalpinum Ohwi = Rhododendron rubropilosum Hayata var. taiwanalpinum（Ohwi）S. Y. Lu，Yuen P. Yang et Y. H. Tseng ●

331919   Rhododendron taiwanensis S. S. Ying = Rhododendron kawakamii Hayata ●

331920   Rhododendron taiwanianum S. S. Ying = Rhododendron kawakamii Hayata var. flaviflorum Tang S. Liu et T. I. Chuang ●

331921   Rhododendron taliense Franch.；大理杜鹃；Dali Azalea，Dali Rhododendron，Tali-range Rhododendron ●

331922   Rhododendron tamaense Davidian = Rhododendron cinnabarinum Hook. f. subsp. tamaense（Davidian）Cullen ●

331923   Rhododendron tamurae（Makino）Masam. = Rhododendron eriocarpum（Hayata）Nakai ●

331924   Rhododendron tamurae（Makino）Masam. = Rhododendron simsii Planch. var. tamurae（Makino）Kaneh. et Hatus. ●

331925   Rhododendron tanakae Hayata = Rhododendron latoucheae Franch. ●

331926   Rhododendron tanakai Hayata；阿里山石楠（阿里山杜鹃）；Alishan Azalea ●

331927   Rhododendron tanakai Hayata = Rhododendron ellipticum Maxim. ●

331928   Rhododendron tanastylum Balf. f. et Kingdon-Ward；光柱杜鹃（长柱杜鹃，粉蜡杜鹃）；Long-styled Rhododendron，Naked Styles Azalea，Naked Styles Rhododendron，Nakedstyle Azalea，Waxy Pink Rhododendron ●

331929   Rhododendron tanastylum Balf. f. et Kingdon-Ward var. lingzhiense M. Y. Fang；林芝光柱杜鹃（红点杜鹃）；Linzhi Azalea，Linzhi Rhododendron ●

331930   Rhododendron tanastylum Balf. f. et Kingdon-Ward var. pennivenium（Balf. f. et Forrest）Cahb. ex Cullen et Chamb. = Rhododendron agastum Balf. f. et W. W. Sm. var. pennivenium（Balf. f. et W. W. Sm.）T. L. Ming ●

331931   Rhododendron tapeinum Balf. f. et Forrest = Rhododendron megeratum Balf. f. et Forrest ●

331932   Rhododendron tapelouense H. Lév. = Rhododendron tatsiense Franch. ●

331933   Rhododendron tapetiforme Balf. f. et Kingdon-Ward；狭萼杜鹃（单色杜鹃，地毯杜鹃）；Carpet-like Rhododendron，Purecolor Azalea ●

331934   Rhododendron taronense Hutch.；薄皮杜鹃（独龙杜鹃）；Taron Azalea，Taron Rhododendron ●

331935   Rhododendron taronense Hutch. = Rhododendron dendricola Hutch. ●

331936   Rhododendron tashiroi Maxim.；大武杜鹃；Tashiro Azalea，Tashiro Rhododendron ●

331937   Rhododendron tashiroi Maxim. f. leucanthum Masam.；白花大武杜鹃 ●

331938   Rhododendron tashiroi Maxim. var. lasiophyllum Hatus. ex T. Yamaz.；光叶大武杜鹃 ●☆

331939   Rhododendron tatsiense Franch.；硬叶杜鹃（打箭炉杜鹃，灰背杜鹃）；Grey Beneath Rhododendron，Hardleaf Azalea，Stereo-leaves Rhododendron，Tatsienlu Rhododendron ●

331940   Rhododendron tatsiense Franch. var. nudatum R. C. Fang；丽江硬

叶杜鹃;Naked Hardleaf Azalea,Naked Tatsienlu Rhododendron ●

331941　Rhododendron tawangense K. C. Sahni et H. B. Naithani = Rhododendron neriiflorum Franch. var. appropinquans（Tagg et Forrest）W. K. Hu ●

331942　Rhododendron tawangense Sahni et Naithani = Rhododendron neriiflorum Franch. var. appropinquans（Tagg et Forrest）W. K. Hu ●

331943　Rhododendron tectum Koidz. ;屋顶杜鹃●☆

331944　Rhododendron telmateium Balf. f. et W. W. Sm. ;草原杜鹃(白喉杜鹃,豆叶杜鹃,蓝紫杜鹃,林下杜鹃,适宜杜鹃,沼泽杜鹃);Grassland Azalea, Marsh Rhododendron, Marshes Rhododendron, Suitable Rhododendron,White-throat Rhododendron,Woods Rhododendron ●

331945　Rhododendron telmateium Balf. f. et W. W. Sm. subsp. albipetalum Cowan = Rhododendron eudoxum Balf. f. et Forrest ●

331946　Rhododendron telmateium Balf. f. et W. W. Sm. subsp. chrysanthemum Cowan = Rhododendron temenium Balf. f. et Forrest var. gilvutum（Cowan）D. F. Chamb. ●

331947　Rhododendron telmateium Balf. f. et W. W. Sm. subsp. dealbatum Cowan = Rhododendron temenium Balf. f. et Forrest var. dealbatum（Cowan）D. F. Chamb. ●

331948　Rhododendron telmateium Balf. f. et W. W. Sm. subsp. gilvum Cowan = Rhododendron temenium Balf. f. et Forrest var. gilvutum（Cowan）D. F. Chamb. ●

331949　Rhododendron telmateium Balf. f. et W. W. Sm. subsp. glaphyrum（Balf. f. et Forrest）Cowan = Rhododendron temenium Balf. f. et Forrest var. dealbatum（Cowan）D. F. Chamb. ●

331950　Rhododendron telmateium Balf. f. et W. W. Sm. subsp. pothinum（Balf. f. et Forrest）Cowan = Rhododendron temenium Balf. f. et Forrest ●

331951　Rhododendron telmateium Balf. f. et W. W. Sm. subsp. rhodanthum Cowan = Rhododendron eudoxum Balf. f. et Forrest ●

331952　Rhododendron telopeum Balf. f. et Forrest = Rhododendron campylocarpum Hook. f. subsp. caloxanthum（Balf. f. et Forrest）D. F. Chamb. ●

331953　Rhododendron temenium Balf. f. et Forrest;滇藏杜鹃(矮红杜鹃,圣地杜鹃);Sacred Place Rhododendron,Yun-Zang Azalea ●

331954　Rhododendron temenium Balf. f. et Forrest subsp. albipetalum Cowan = Rhododendron eudoxum Balf. f. et Forrest ●

331955　Rhododendron temenium Balf. f. et Forrest subsp. chrysanthemum Cowan = Rhododendron temenium Balf. f. et Forrest var. gilvutum（Cowan）D. F. Chamb. ●

331956　Rhododendron temenium Balf. f. et Forrest subsp. dealbatum Cowan = Rhododendron temenium Balf. f. et Forrest var. dealbatum（Cowan）D. F. Chamb. ●

331957　Rhododendron temenium Balf. f. et Forrest subsp. gilvum Cowan = Rhododendron temenium Balf. f. et Forrest var. gilvutum（Cowan）D. F. Chamb. ●

331958　Rhododendron temenium Balf. f. et Forrest subsp. glaphyrum（Balf. f. et Forrest）Cowan = Rhododendron temenium Balf. f. et Forrest var. dealbatum（Cowan）D. F. Chamb. ●

331959　Rhododendron temenium Balf. f. et Forrest subsp. pothinum（Balf. f. et Forrest）Cowan = Rhododendron temenium Balf. f. et Forrest ●

331960　Rhododendron temenium Balf. f. et Forrest subsp. rhodanthum Cowan = Rhododendron eudoxum Balf. f. et Forrest ●

331961　Rhododendron temenium Balf. f. et Forrest var. dealbatum（Cowan）D. F. Chamb. ;粉红滇藏杜鹃;Pink Yun-Zang Azalea ●

331962　Rhododendron temenium Balf. f. et Forrest var. gilvutum（Cowan）D. F. Chamb. ; 黄花滇藏杜鹃; Yellow-flower Sacred Place

Rhododendron,Yellow-flower Yun-Zang Azalea ●

331963　Rhododendron tenue Ching ex W. P. Fang et M. Y. He;细瘦杜鹃;Thin Rhododendron,Weak Azalea ●

331964　Rhododendron tenuifolium R. C. Fang et S. H. Huang;薄叶朱砂杜鹃(薄叶杜鹃,薄叶管花杜鹃);Thinleaf Rhododendron, Thin-leaved Rhododendron ●

331965　Rhododendron tenuilaminare P. C. Tam;薄片杜鹃;Film Azalea,Thin Laminae Rhododendron,Thin-laminated Rhododendron ●

331966　Rhododendron tephropeplum Balf. f. et Forrest;灰鳞杜鹃(灰背杜鹃, 灰被杜鹃); Ash-grey Covered Rhododendron, Ashrobe Azalea,Ashrobe Rhododendron ●

331967　Rhododendron tetramerum（Komatsu）Nakai = Rhododendron tschonoskii Maxim. var. tetramerum Komatsu ●☆

331968　Rhododendron thayerianum Rehder et E. H. Wilson;反边杜鹃;Thayer Azalea,Thayer's Rhododendron ●

331969　Rhododendron theiochroum Balf. f. et W. W. Sm. = Rhododendron sulphureum Franch. ●◇

331970　Rhododendron theiophyllum Balf. f. et Forrest = Rhododendron phaeochrysum Balf. f. et W. W. Sm. var. levistratum（Balf. f. et Forrest）D. F. Chamb. ●

331971　Rhododendron thomsonii Hook. f. ;半圆叶杜鹃(汤氏杜鹃);T. Thomson's Rhododendron, Thomson Azalea, Thomson's Rhododendron ●

331972　Rhododendron thomsonii Hook. f. subsp. lopsangianum（Cowan）D. F. Chamb. ;小半圆叶杜鹃;Small Thomson Azalea ●

331973　Rhododendron thomsonii Hook. f. var. cyanocarpum Franch. = Rhododendron cyanocarpum（Franch. ）Franch. et W. W. Sm. ●◇

331974　Rhododendron thomsonii Hook. f. var. lopsangianum（Cowan）T. L. Ming = Rhododendron thomsonii Hook. f. subsp. lopsangianum（Cowan）D. F. Chamb. ●

331975　Rhododendron thymifolium Maxim. ;千里香叶杜鹃(百里香杜鹃,百里香叶杜鹃,斑花杜鹃,黑香柴,千里香杜鹃,麝香草叶杜鹃,小叶杜鹃);Spotted Flowers Rhododendron, Thymeleaf Azalea, Thyme-leaf Rhododendron, Thyme-leaved Rhododendron, Thyme-leaves Rhododendron ●

331976　Rhododendron thyodocum Balf. f. et R. E. Cooper = Rhododendron baileyi Balf. f. ●

331977　Rhododendron tianlinense P. C. Tam;田林马银花(田林杜鹃);Tianlin Azalea,Tianlin Rhododendron ●

331978　Rhododendron tiantangense G. Z. Li;天堂杜鹃;Tiantangshan Azalea ●

331979　Rhododendron timeteum Balf. f. et Forrest = Rhododendron oreotrephes W. W. Sm. ●

331980　Rhododendron tingwuense P. C. Tam;鼎湖杜鹃(惠阳杜鹃);Dinghu Azalea, Dinghu Rhododendron, Dinghushan Rhododendron, Huiyang Azalea,Huiyang Rhododendron,Labrador Tea ●

331981　Rhododendron tomentosum Harmaja = Ledum palustre L. ●

331982　Rhododendron tomentosum Harmaja subsp. subarcticum（Harmaja）G. D. Wallace. = Ledum palustre L. var. decumbens Aiton ●

331983　Rhododendron torquatum Balf. f. et Forrest = Rhododendron dichroanthum Diels et Cowan subsp. scyphocalyx（Balf. f. et Forrest）Cowan ●

331984　Rhododendron torquatum L. C. Hu;曲枝杜鹃; Coiled Rhododendron,Twisted Rhododendron,Wringchain Azalea ●

331985　Rhododendron tosaense Makino;土佐杜鹃;Tosa Azalea ●☆

331986　Rhododendron tosaense Makino f. albiflorum T. Yamaz. ;白花土

佐杜鹃●☆

331987 Rhododendron traillanum Forrest et W. W. Sm.；川滇杜鹃（特雷氏杜鹃,异常杜鹃）；G. W. Traill's Rhododendron, Traill Azalea, Traill Rhododendron, Unusual Azalea, Unusual Rhododendron ●

331988 Rhododendron traillanum Forrest et W. W. Sm. var. dictyotum（Balf. f. ex Tagg）D. F. Chamb. ex Cullen et D. F. Chamb.；棕背川滇杜鹃（长叶川滇杜鹃）；Long-leaved Traill Rhododendron ●

331989 Rhododendron transiens Nakai；间型杜鹃●☆

331990 Rhododendron transtylum Balf. f. et Kingdon-Ward；腊质杜鹃（长柱杜鹃,腊质杜鹃花）；Long Style Rhododendron ●☆

331991 Rhododendron trichanthum Rehder；长毛杜鹃（红毛杜鹃,毛花杜鹃）；Hairy Flower Rhododendron, Longhair Azalea, Villous Rhododendron ●

331992 Rhododendron trichocladum Franch.；糙毛杜鹃（卷发杜鹃,卷毛杜鹃）；Curly Hairs Rhododendron, Hairy-twig Rhododendron, Hairy-twigged Rhododendron, Roughhair Azalea ●

331993 Rhododendron trichocladum Franch. var. longipilosum Cowan = Rhododendron mekongense Franch. var. longipilosum（Cowan）Cullen ●

331994 Rhododendron trichogynum L. C. Hu；理县杜鹃；Lixian Azalea, Lixian Rhododendron ●

331995 Rhododendron trichomiscum Balf. f. et Forrest = Rhododendron eudoxum Balf. f. et Forrest ●

331996 Rhododendron trichophlebium Balf. f. et Forrest = Rhododendron eudoxum Balf. f. et Forrest ●

331997 Rhododendron trichophorum E. H. Wilson；毛梗杜鹃（粉紫杜鹃）；Bearing Hairs Rhododendron ●☆

331998 Rhododendron trichopodum Balf. f. et Forrest = Rhododendron oreotrephes W. W. Sm. ●

331999 Rhododendron trichostomum Franch.；毛嘴杜鹃；Hairrostra Azalea, Hairy Mouthed Rhododendron, Hairy-stomatic Rhododendron ●

332000 Rhododendron trichostomum Franch. var. ledoides（Balf. f. et W. W. Sm.）Cowan et Davidian；筒花毛嘴杜鹃（喇叭茶状杜鹃,筒花杜鹃）；Resembling Ledum Rhododendron, Tube-flowered Hairy-stomatic Rhododendron ●

332001 Rhododendron trichostomum Franch. var. radinum（Balf. f. et W. W. Sm.）Cowan et Davidian；鳞斑毛嘴杜鹃（高山蔷薇杜鹃,厚鳞毛嘴杜鹃）；Sensly Hairy Mouthed Rhododendron ●

332002 Rhododendron triflorum Hook. f.；三花杜鹃；Three Flowered Rhododendron, Triflor Azalea, Triflorous Rhododendron ●

332003 Rhododendron triflorum Hook. f. subsp. multiflorum R. C. Fang；云南三花杜鹃；Yunnan Three Flowered Rhododendron, Yunnan Triflor Azalea ●

332004 Rhododendron trilectorum Cowan；朗贡杜鹃；Langgong Azalea, Langgong Rhododendron ●

332005 Rhododendron trinerve Franch. ex H. Boissieu = Rhododendron tschonoskii Maxim. subsp. trinerve（Franch. ex H. Boissieu）Kitam. ●☆

332006 Rhododendron triplonaevium Balf. f. et Forrest = Rhododendron alutaceum Balf. f. et W. W. Sm. var. russotinctum（Balf. f. et Forrest）D. F. Chamb. ●

332007 Rhododendron tritifolium Balf. f. et Forrest = Rhododendron alutaceum Balf. f. et W. W. Sm. var. russotinctum（Balf. f. et Forrest）D. F. Chamb. ●

332008 Rhododendron truncatulum Balf. f. et Forrest = Rhododendron erythrocalyx Balf. f. et Forrest ●

332009 Rhododendron tsaii W. P. Fang；昭通杜鹃（蔡氏杜鹃,大海杜鹃）；H. T. Tsai's Rhododendron, Tsai's Rhododendron, Zhaotong Azalea ●

332010 Rhododendron tsangpoense Hutch. et Kingdon-Ward = Rhododendron charitopes Balf. f. et Forrest subsp. tsangpoense（Kingdon-Ward）Cullen ●

332011 Rhododendron tsangpoense Hutch. et Kingdon-Ward var. curvistylum（Kingdon-Ward）Cowan et Davidian = Rhododendron charitopes Balf. f. et Forrest subsp. tsangpoense（Kingdon-Ward）Cullen ●

332012 Rhododendron tsangpoense Hutch. et Kingdon-Ward var. pruniflorum（Hutch.）Cowan et Davidian = Rhododendron pruniflorum Hutch. et Kingdon-Ward ●

332013 Rhododendron tsangpoense Kingdon-Ward = Rhododendron charitopes Balf. f. et Forrest subsp. tsangpoense（Kingdon-Ward）Cullen ●

332014 Rhododendron tsangpoense Kingdon-Ward var. curvistylum Kingdon-Ward ex Cowan et Davidian = Rhododendron charitopes Balf. f. et Forrest subsp. tsangpoense（Kingdon-Ward）Cullen ●

332015 Rhododendron tsangpoense Kingdon-Ward var. pruniflorum（Hutch. et Kingdon-Ward）Cowan et Davidian = Rhododendron pruniflorum Hutch. et Kingdon-Ward ●

332016 Rhododendron tsariense Cowan；察里杜鹃（白钟杜鹃）；Tsari Rhododendron, Whitebell Azalea ●

332017 Rhododendron tsarongense Balf. f. et Forrest = Rhododendron primuliflorum Bureau et Franch. ●

332018 Rhododendron tschonoskii Maxim.；须川氏杜鹃（冲诺杜鹃）；Tschonoski Azalea ●☆

332019 Rhododendron tschonoskii Maxim. f. roseum Sugim.；粉花须川氏杜鹃●☆

332020 Rhododendron tschonoskii Maxim. subsp. trinerve（Franch. ex H. Boissieu）Kitam.；三脉须川氏杜鹃（三脉冲诺杜鹃）●☆

332021 Rhododendron tschonoskii Maxim. var. tetramerum Komatsu；四数须川氏杜鹃●☆

332022 Rhododendron tschonoskii Maxim. var. trinerve（Franch. ex H. Boissieu）Makino = Rhododendron tschonoskii Maxim. subsp. trinerve（Franch. ex H. Boissieu）Kitam. ●☆

332023 Rhododendron tschonoskii Maxim. var. trinerve（Franch.）Makino = Rhododendron trinerve Franch. ex H. Boissieu ●☆

332024 Rhododendron tsinlingense W. P. Fang；秦岭杜鹃；Qinling Azalea, Qinling Rhododendron ●

332025 Rhododendron tsoi Merr.；两广杜鹃（灌阳杜鹃,细石榴树,亚隐脉杜鹃,隐脉杜鹃,增城杜鹃）；Invisible-veined Rhododendron, Latentvein Azalea, Tso Azalea, Tso's Rhododendron ●

332026 Rhododendron tsurugisanense（T. Yamaz.）T. Yamaz.；剑山杜鹃●☆

332027 Rhododendron tsurugisanense（T. Yamaz.）T. Yamaz. var. nudipetiolatum T. Yamaz.；裸柄杜鹃●☆

332028 Rhododendron tsusiophyllum Sugim. = Tsusiophyllum tanakae Maxim. ●☆

332029 Rhododendron tuba Sleumer；管杜鹃（喇叭杜鹃）●☆

332030 Rhododendron tubiforme（Cowan et Davidian）Davidian；灰蓝叶杜鹃（苍白杜鹃）；Blush-grey Leaves Rhododendron, Blush-grey-leaved Rhododendron, Pale Azalea, Tube-form Rhododendron ●

332031 Rhododendron tubulosum Ching ex W. Y. Wang；长管杜鹃；Longtube Azalea, Long-tube Rhododendron ●

332032 Rhododendron tutcherae Hemsl. et E. H. Wilson；香缅树杜鹃（蒙自杜鹃,香缅杜鹃）；Tucher Azalea, Tucher Rhododendron ●

332033 Rhododendron tutcherae Hemsl. et E. H. Wilson var. glabrifolium

L. M. Gao et D. Z. Li;光叶香缅树杜鹃;Glabrous-leaf Tucher Azalea,Glabrous-leaf Tucher Rhododendron ●

332034　Rhododendron tutcherae Hemsl. et E. H. Wilson var. gymnocatpum A. L. Chang ex R. C. Fang;光果香缅树杜鹃●

332035　Rhododendron tutcherae Hemsl. et E. H. Wilson var. gymnocatpum A. L. Chang ex R. C. Fang = Rhododendron tutcherae Hemsl. et E. H. Wilson ●

332036　Rhododendron umbelliferum H. Lév. = Rhododendron mariesii Hemsl. et E. H. Wilson ●

332037　Rhododendron unciferum P. C. Tam;垂钩杜鹃;Hook-shap Flowers Rhododendron,Hookship-flowered Rhododendron,Nutanthook Azalea ●

332038　Rhododendron ungernii Trautv.;耐寒杜鹃;Ungern Rhododendron ●☆

332039　Rhododendron uniflorum Hutch. et Kingdon-Ward;单花杜鹃;Solitary-flowered Rhododendron, Uniflor Azalea, Uniflorous Rhododendron ●

332040　Rhododendron uniflorum P. C. Tam var. imperator（Kingdon-Ward）Cullen;尖叶单花杜鹃●

332041　Rhododendron urophyllum W. P. Fang;尾叶杜鹃;Caudate-leaf Rhododendron,Caudate-leaved Rhododendron,Tailleaf Azalea ●

332042　Rhododendron uvarifolium Diels;紫玉盘杜鹃(紫玉盘叶杜鹃）; Uvarialeaf Azalea, Uvarialeaf Rhododendron, Uvaria-leaved Rhododendron, Uvaria-like Rhododendron ●

332043　Rhododendron uvarifolium Diels var. griseum Cowan = Rhododendron uvarifolium Diels ●

332044　Rhododendron uvariifolium var. griseum Cowan = Rhododendron uvarifolium Diels ●

332045　Rhododendron uwaense H. Hara et T. Yamanaka;宇和杜鹃●☆

332046　Rhododendron vaccinioides Hook. f.;越橘杜鹃(似越橘杜鹃,越桔杜鹃);Blueberry Azalea, Blueberry-like Rhododendron, Like Vaccinium Rhododendron ●

332047　Rhododendron valentinianum Forrest ex Hutch.;毛柄杜鹃(情念杜鹃, 瓦氏杜鹃）; Hairstalk Azalea, Pere Valentin's Rhododendron,Valentin's Rhododendron ●

332048　Rhododendron valentinianum Forrest ex Hutch. var. changii W. P. Fang = Rhododendron changii（W. P. Fang）W. P. Fang ●

332049　Rhododendron valentinianum Forrest ex Hutch. var. oblongilobatum R. C. Fang;滇南毛柄杜鹃;S. Yunnan Pere Valentin's Rhododendron, South Yunnan Pere Valentin's Rhododendron ●

332050　Rhododendron vaseyi A. Gray;粉壳杜鹃;Pinkshell Azalea,Pink-shell Azalea ●☆

332051　Rhododendron vaseyi A. Gray f. album Rehder;白粉壳杜鹃;White Pinkshell Azalea ●☆

332052　Rhododendron veitchianum Hook.;维奇杜鹃;Veitch's Rhododendron ●☆

332053　Rhododendron vellereum Hutch. ex Tagg;白毛杜鹃(柔白杜鹃);Fleecy Rhododendron,Whitehair Azalea ●

332054　Rhododendron venator Tagg;毛柱杜鹃(猎手杜鹃,猩红杜鹃); Haircolumn Azalea, Scarlet-flowered Rhododendron, Scarlet-flowers Rhododendron ●

332055　Rhododendron vernicosum Franch.;亮叶杜鹃;Brightleaf Azalea,Shiny Rhododendron,Vernicose Rhododendron ●

332056　Rhododendron verruciferum W. K. Hu;疣梗杜鹃;Warts Rhododendron,Wartstalk Azalea,Warty Rhododendron ●

332057　Rhododendron verruculosum Rehder et E. H. Wilson;小疣杜鹃;Small Warts Rhododendron ●☆

332058　Rhododendron versicolor Chun et W. P. Fang = Rhododendron simiarum Hance var. versicolor（Chun et W. P. Fang）M. Y. Fang ●

332059　Rhododendron vesiculiferum Franch.;小囊杜鹃(泡毛杜鹃); Bearing-vesicle Rhododendron,Vesicular-haired Rhododendron ●

332060　Rhododendron vestitum Tagg et Forrest = Rhododendron setiferum Balf. f. et Forrest ●

332061　Rhododendron vetchianum Hook.;维琪杜鹃;Veitchia Family's Rhododendron ●☆

332062　Rhododendron vialii Delavay et Franch.;红马银花;Red Axillary Flowered Rhododendron,Red Azalea,Vial Rhododendron ●◇

332063　Rhododendron viburnifolium W. P. Fang = Rhododendron simsii Planch. ●

332064　Rhododendron vicarium Balf. f. = Rhododendron nivale Hook. f. subsp. boreale M. N. Philipson et Philipson ●

332065　Rhododendron vicinum Balf. f. et Forrest = Rhododendron phaeochrysum Balf. f. et W. W. Sm. var. levistratum（Balf. f. et Forrest）D. F. Chamb. ●

332066　Rhododendron villosum Hemsl. et E. H. Wilson = Rhododendron trichanthum Rehder ●

332067　Rhododendron villosum Roth = Rhododendron trichanthum Rehder ●

332068　Rhododendron vilmorinianum Balf. f. = Rhododendron augustinii Hemsl. ●

332069　Rhododendron violaceum Rehder et E. H. Wilson = Rhododendron nivale Hook. f. subsp. boreale M. N. Philipson et Philipson ●

332070　Rhododendron virescens Hutch.;变绿杜鹃;Becoming Green Rhododendron ●☆

332071　Rhododendron virgatum Hook. f.;柳条杜鹃(橄榄叶杜鹃,油叶杜鹃); Olive-like Foliage Rhododendron, Willowshoot Azalea, Willowshoot Rhododendron, Willow-shooted Rhododendron, Willowy Twigs Rhododendron ●

332072　Rhododendron virgatum Hook. f. subsp. oleifolium（Franch.）Cullen = Rhododendron virgatum Hook. f. ●

332073　Rhododendron virgatum Hook. f. var. glabriflorum K. M. Feng;少毛柳条杜鹃;Glabriflorous Willowshoot Rhododendron ●

332074　Rhododendron virgatum Hook. f. var. glabriflorum K. M. Feng = Rhododendron virgatum Hook. f. ●

332075　Rhododendron virgatum Hook. f. var. glabriflorum K. M. Feng ex R. C. Fang = Rhododendron virgatum Hook. f. ●

332076　Rhododendron virgatum Hook. f. var. oleifolium（Franch.）Demoly = Rhododendron virgatum Hook. f. ●

332077　Rhododendron viridescens Hutch.;显绿杜鹃;Green Azalea,Virescent Rhododendron ●

332078　Rhododendron viscidifolium Davidian;铜色杜鹃;Coppercolor Azalea,Viscoid Leaves Rhododendron,Viscoid-leaved Rhododendron ●

332079　Rhododendron viscidum C. Z. Guo et Z. H. Liu;黏质杜鹃(黏灰杜鹃); Viscid Rhododendron, Viscidity Azalea, Viscosus Rhododendron ●

332080　Rhododendron viscigemmatum P. C. Tam;黏芽杜鹃;Adhibitgemma Azalea, Viscid-buds Rhododendron, Viscose-budded Rhododendron ●

332081　Rhododendron viscigemmatum P. C. Tam = Rhododendron chunii W. P. Fang ●

332082　Rhododendron viscistylum Nakai;黏柱杜鹃●☆

332083　Rhododendron viscistylum Nakai var. amakusaense K. Takada ex T. Yamaz.;天草黏柱杜鹃●☆

332084　Rhododendron viscistylum Nakai var. glaucum（Hatus.）Sugim. = Rhododendron osuzuyamense T. Yamaz. ●☆

332085　Rhododendron viscistylum Nakai var. hyugaense T. Yamaz. = Rhododendron hyugaense（T. Yamaz.）T. Yamaz. ●☆

332086　Rhododendron viscosum（L.）Torr.；沼生杜鹃（黏杜鹃）；Beggar's Buttons, Burdock, Clammy Azalea, Great Burdock, Pig's Rhubarb, Snake's Rhubarb, Swamp Azalea, Swamp Honey Suckle, Swamp Honeysuckle, White Azalea, White Swamp Azalea ●☆

332087　Rhododendron viscosum（L.）Torr. var. serrulatum（Small）H. E. Ahles；小齿沼生杜鹃；Hammock Sweet Azalea ●☆

332088　Rhododendron vupivalleculatum P. C. Tam；岩谷杜鹃；Rock Vallecula Rhododendron, Rocky-valley Rhododendron, Saxicolous Azalea ●

332089　Rhododendron wadanum Makino；东国三叶踯躅●☆

332090　Rhododendron wadanum Makino f. leucanthum（Makino）H. Hara；白花东国三叶踯躅●☆

332091　Rhododendron wallichii Hook. f.；簇毛杜鹃；N. Wallich's Rhododendron, Wallich Rhododendron ●

332092　Rhododendron walongense Kingdon-Ward；瓦弄杜鹃；Wanong Azalea, Wanong Rhododendron ●

332093　Rhododendron wardii W. W. Sm.；黄杯杜鹃（立地坪杜鹃, 沃氏杜鹃, 星萼杜鹃）；Kingdon-ward Rhododendron, Li-ti-ping Rhododendron, Stary Calyx Rhododendron, Ward Azalea, Ward's Rhododendron ●

332094　Rhododendron wardii W. W. Sm. var. puralbum（Balf. f. et W. W. Sm.）D. F. Chamb.；纯白杜鹃（紫白杜鹃）；Pure White Azalea, Pure White Rhododendron, White Ward's Rhododendron ●

332095　Rhododendron wasonii Hemsl. et E. H. Wilson；褐毛杜鹃（异色杜鹃）；Brownhair Azalea, Variable Coloured Rhododendron, Wason Rhododendron ●

332096　Rhododendron wasonii Hemsl. et E. H. Wilson var. wenchuanense L. C. Hu；汶川褐毛杜鹃；Wenchuan Azalea, Wenchuan Variable Coloured Rhododendron ●

332097　Rhododendron watsonii Hemsl. et E. H. Wilson；无柄杜鹃；Sessile Azalea, Watson Rhododendron ●

332098　Rhododendron websterianum Rehder et E. H. Wilson；毛蕊杜鹃；F. G. Webster Rhododendron, Webster Azalea, Webster Rhododendron ●

332099　Rhododendron websterianum Rehder et E. H. Wilson var. yulongense Philipson et M. N. Philipson；黄花毛蕊杜鹃；Yulong Webster Rhododendron ●

332100　Rhododendron weldianum Rehder et E. H. Wilson = Rhododendron rufum Batalin ●

332101　Rhododendron wenshanense K. M. Feng；文山杜鹃；Wenshan Rhododendron ●

332102　Rhododendron westlandii Hemsl.；凯里杜鹃（丝线吊芙蓉）；Kaili Azalea, Kaili Rhododendron ●

332103　Rhododendron westlandii Hemsl. = Rhododendron moulmainense Hook. ●

332104　Rhododendron weyrichi Maxim.；朝鲜杜鹃（朝鲜杜鹃花, 威氏杜鹃）；Dr. Weych Azalea ●☆

332105　Rhododendron weyrichii Maxim. f. albiflorum T. Yamaz.；白花朝鲜杜鹃●☆

332106　Rhododendron weyrichii Maxim. f. purpuriflorum T. Yamaz.；紫花朝鲜杜鹃●☆

332107　Rhododendron weyrichii Maxim. var. psilostylum Nakai；光柱朝鲜杜鹃●☆

332108　Rhododendron weyrichii Maxim. var. sanctum（Nakai）Hatus. = Rhododendron sanctum Nakai ●☆

332109　Rhododendron wightii Hook. f.；宏钟杜鹃；Wight Azalea, Wight Rhododendron ●

332110　Rhododendron williamsianum Rehder et E. H. Wilson；圆叶杜鹃（威廉氏杜鹃, 惟丽杜鹃）；J. C. Williams's Rhododendron, Williams Azalea, Williams Rhododendron ●

332111　Rhododendron wilsonae Hemsl. et E. H. Wilson = Rhododendron latoucheae Franch. ●

332112　Rhododendron wilsonae Hemsl. et E. H. Wilson var. ionanthum W. P. Fang = Rhododendron latoucheae Franch. ●

332113　Rhododendron wiltonii Hemsl. et E. H. Wilson；皱皮杜鹃（毡皮杜鹃, 皱叶杜鹃）；Green-felt Twigs Rhododendron, Wilton Azalea, Wilton Rhododendron ●

332114　Rhododendron windsorii Nutt. = Rhododendron arboreum Sm. ●

332115　Rhododendron wolongense W. K. Hu；卧龙杜鹃；Wolong Azalea, Wolong Rhododendron ●

332116　Rhododendron wongii Hemsl. et E. H. Wilson；康南杜鹃；Kangnan Azalea, Wong Rhododendron ●

332117　Rhododendron wuense Balf. f. = Rhododendron faberi Hemsl. ●◇

332118　Rhododendron wumengshanense T. L. Ming；乌蒙杜鹃；Wumengshan Rhododendron ●

332119　Rhododendron wumingense W. P. Fang；武鸣杜鹃；Wuming Azalea, Wuming Rhododendron ●◇

332120　Rhododendron wuyishanicum L. K. Ling；武夷山杜鹃；Wuyishan Azalea ●

332121　Rhododendron xantheneuron H. Lév. = Rhododendron denudatum H. Lév. ●

332122　Rhododendron xanthinum Balf. f. et W. W. Sm. = Rhododendron trichocladum Franch. ●

332123　Rhododendron xanthocodon Hutch.；黄铃杜鹃（黄钟杜鹃, 蓝幼叶杜鹃）；Linking Together Rhododendron, Yellow Bell Rhododendron, Yellowbell Azalea, Yellowcodded Rhododendron ●

332124　Rhododendron xanthoneuron H. Lév. = Rhododendron denudatum H. Lév. ●

332125　Rhododendron xanthostephanum Merr.；鲜黄杜鹃；Freshyellow Azalea, Yellowcoralla Rhododendron, Yellow-corallate Rhododendron ●

332126　Rhododendron xiaoxidongense W. K. Hu；小溪洞杜鹃；Xiaoxidong Azalea, Xiaoxidong Rhododendron ●

332127　Rhododendron xichangense Z. J. Zhao；西昌杜鹃；Xichang Azalea, Xichang Rhododendron ●

332128　Rhododendron xichangense Z. J. Zhao = Rhododendron davidsonianum Rehder et E. H. Wilson ●

332129　Rhododendron xiguense Ching et H. P. Yang；西固杜鹃；Xigu Azalea, Xigu Rhododendron ●

332130　Rhododendron xinganense G. Z. Li；兴安马银花；Xing'an Azalea ●

332131　Rhododendron xiushanense W. P. Fang = Rhododendron chrysocalyx H. Lév. et Vaniot var. xiushanense（W. P. Fang）M. Y. He ●

332132　Rhododendron xizangense W. P. Fang et W. K. Hu ex Q. Z. Yu = Rhododendron hirtipes Tagg ●

332133　Rhododendron yakuinsulare Masam.；屋久岛生杜鹃●☆

332134　Rhododendron yakumontanum（T. Yamaz.）T. Yamaz.；屋久岛山地杜鹃●☆

332135　Rhododendron yakushimanum Nakai = Rhododendron metternichii Siebold et Zucc. var. yakusimanum（Nakai）Ohwi ●☆

332136　Rhododendron yakushimanum Nakai subsp. makinoi（Tagg ex

Nakai）D. F. Chamb. = Rhododendron makinoi Tagg ex Nakai ●☆

332137　Rhododendron yakushimanum Nakai var. intermedium（Sugim.）T. Yamaz. ;间型屋久岛山地杜鹃●☆

332138　Rhododendron yangmingshanense P. C. Tam;阳明山杜鹃;Yangmingshan Azalea, Yangmingshan Rhododendron ●

332139　Rhododendron yanthinum Bureau et Franch. = Rhododendron concinnum Hemsl. ●

332140　Rhododendron yanthinum Bureau et Franch. var. lepidanthum Rehder et E. H. Wilson = Rhododendron concinnum Hemsl. ●

332141　Rhododendron yaogangxianense Q. X. Liu;瑶岗仙杜鹃;Yaogangxiang Rhododendron ●

332142　Rhododendron yaoshanicum W. P. Fang et M. Y. He;瑶山杜鹃;Yaoshan Azalea, Yaoshan Rhododendron ●

332143　Rhododendron yaragongense Balf. f. = Rhododendron nivale Hook. f. subsp. boreale M. N. Philipson et Philipson ●

332144　Rhododendron yedoense Maxim. ex Regel;东京杜鹃（江户杜鹃）;Tokyo Azalea ●☆

332145　Rhododendron yedoense Maxim. ex Regel f. poukhanense（H. Lév.）Sugim. ex T. Yamaz. ;北汉山杜鹃●☆

332146　Rhododendron yedoense Maxim. ex Regel var. poukhanense（H. Lév.）Nakai = Rhododendron yedoense Maxim. ex Regel f. poukhanense（H. Lév.）Sugim. ex T. Yamaz. ●☆

332147　Rhododendron yizhangense Q. X. Liu;宜章杜鹃;Yizhang Azalea ●

332148　Rhododendron youngae W. P. Fang;弯尖杜鹃●

332149　Rhododendron youngae W. P. Fang = Rhododendron adenopodum Franch. ●

332150　Rhododendron yuefengense G. Z. Li;越峰杜鹃;Yuefeng Rhododendron ●

332151　Rhododendron yulingense W. P. Fang;榆林杜鹃;Yulin Rhododendron ●

332152　Rhododendron yulingense W. P. Fang = Rhododendron nitidulum Rehder et E. H. Wilson ●

332153　Rhododendron yungchangense Cullen;少鳞杜鹃;Poorscale Azalea, Yunchang Rhododendron ●

332154　Rhododendron yungningense Balf. f. ex Hutch. ;永宁杜鹃（小花序杜鹃）;Small Clusters Rhododendron, Yongning Azalea, Yongning Rhododendron, Yungning Rhododendron ●

332155　Rhododendron yunnanense Franch. ;云南杜鹃（多蕊杜鹃,丰花杜鹃,矛头杜鹃,矛叶杜鹃,木栓杜鹃,陕西杜鹃,项链杜鹃,云贵杜鹃,纸叶杜鹃）;Apearleaf Rhododendron, Bearing Necklace Rhododendron, Cork Bark Rhododendron, More Flowers Rhododendron, Paperleaf Azalea, Paper-leaved Rhododendron, Shaanxi Azalea, Shaanxi Rhododendron, Spearleaf Azalea, Yunnan Azalea, Yunnan Rhododendron ●

332156　Rhododendron yushuense Z. J. Zhao;玉树杜鹃;Yushu Azalea, Yushu Rhododendron ●

332157　Rhododendron zaleucum Balf. f. et W. W. Sm. ;白面杜鹃（白背杜鹃, 丝白背杜鹃）;Greywhite Rhododendron, Grey-white Rhododendron, White Lower Surface Rhododendron, Whiteface Azalea ●

332158　Rhododendron zaleucum Balf. f. et W. W. Sm. var. pubilifolium R. C. Fang;毛叶白面杜鹃;Hairy-leaf Greywhite Rhododendron, Hairy-leaf Whiteface Azalea ●

332159　Rhododendron zeylanicum Balf. f. et W. W. Sm. ;锡兰杜鹃（白背杜鹃,白面杜鹃）;Ceylon Azalea, Ceylon Rhododendron ●☆

332160　Rhododendron zheguense Ching et H. P. Yang;鹧鸪杜鹃;Zhegu Azalea, Zhegu Rhododendron ●

332161　Rhododendron zhekoense Y. D. Sun et Z. J. Zhao;泽库杜鹃;Zeku Azalea, Zeku Rhododendron ●

332162　Rhododendron zhongdianense L. C. Hu;中甸杜鹃;Zhongdian Azalea, Zhongdian Rhododendron ●

332163　Rhododendron ziyuanense P. C. Tam var. pachyphyllum（W. P. Fang）G. Z. Li = Rhododendron pachyphyllum W. P. Fang ●

332164　Rhododendron ziyuenense P. C. Tam;资源杜鹃;Ziyuan Azalea, Ziyuan Rhododendron ●

332165　Rhododendros zoelleri Warb. ;佐乐杜鹃●☆

332166　Rhododendros Adans. = Andromeda + Kalmia + Chamaedaphne + Rhododendron 等●

332167　Rhododendrum L. = Rhododendron L. ●

332168　Rhododon Epling（1939）;红齿草属■☆

332169　Rhododon angulatus（Tharp）B. L. Turner;窄红齿草■☆

332170　Rhododon ciliatus（Benth.）Epling;红齿草■☆

332171　Rhodogeron Griseb.（1866）;红蓬属■☆

332172　Rhodogeron Griseb. = Sachsia Griseb. ■☆

332173　Rhodogeron coronopifolius Griseb. ;红蓬■☆

332174　Rhodognaphalon（Ulbr.）Roberty = Bombax L.（保留属名）●

332175　Rhodognaphalon（Ulbr.）Roberty = Pachira Aubl. ●

332176　Rhodognaphalon brevicuspe（Sprague）Roberty = Bombax brevicuspe Sprague ●☆

332177　Rhodognaphalon lukayense（De Wild. et T. Durand）A. Robyns = Bombax lukayense De Wild. et T. Durand ●☆

332178　Rhodognaphalon mossambicense（A. Robyns）A. Robyns = Bombax rhodognaphalon K. Schum. var. tomentosum A. Robyns ●☆

332179　Rhodognaphalon schumannianum A. Robyns = Bombax rhodognaphalon K. Schum. ●☆

332180　Rhodognaphalon stolzii（Ulbr.）A. Robyns = Bombax rhodognaphalon K. Schum. var. tomentosum A. Robyns ●☆

332181　Rhodognaphalon tanganyikense A. Robyns = Bombax rhodognaphalon K. Schum. ●☆

332182　Rhodognaphalopsis A. Robyns = Bombax L.（保留属名）●

332183　Rhodognaphalopsis A. Robyns = Pachira Aubl. ●

332184　Rhodohypoxis Nel（1914）;红金梅草属（红星花属,樱茅属）;Oxblood Lilies, Rhodohypoxis ■☆

332185　Rhodohypoxis baurii（Baker）Nel;红金梅草（红星花,樱茅）;Red Star Grass ■☆

332186　Rhodohypoxis baurii（Baker）Nel var. platypetala ?;窄瓣红金梅草■☆

332187　Rhodohypoxis baurii Nel = Rhodohypoxis baurii（Baker）Nel ■☆

332188　Rhodohypoxis deflexa Hilliard et B. L. Burtt;外折红金梅草■☆

332189　Rhodohypoxis incompta Hilliard et B. L. Burtt;装饰红金梅草■☆

332190　Rhodohypoxis milloides（Baker）Hilliard et B. L. Burtt;澳非红金梅草■☆

332191　Rhodohypoxis rubella（Baker）Nel;微红红金梅草■☆

332192　Rhodolaena Thouars（1805）;红被花属●☆

332193　Rhodolaena acutifolia Baker;尖叶红被花●☆

332194　Rhodolaena bakeriana Baill. ;贝克红被花●☆

332195　Rhodolaena coriacea G. E. Schatz, Lowry et A. -E. Wolf;革质红被花●☆

332196　Rhodolaena echinata H. Perrier = Schizolaena exinvolucrata Baker ●☆

332197　Rhodolaena humblotii Baill. ;洪布红被花●☆

332198　Rhodolaena leroyana G. E. Schatz, Lowry et A. -E. Wolf;勒罗伊红被花●☆

332199　Rhodolaena macrocarpa G. E. Schatz, Lowry et A. -E. Wolf;大果红被花●☆

332200　Rhodolaena parviflora F. Gérard ＝ Schizolaena parviflora (F. Gérard) H. Perrier ●☆

332201　Rhodolaena rotundifolia F. Gérard ＝ Eremolaena rotundifolia (F. Gérard) Danguy ●☆

332202　Rhodolaenaceae Bullock ＝ Ericaceae Juss. (保留科名)●

332203　Rhodolaenaceae Bullock ＝ Sarcolaenaceae Caruel(保留科名)●☆

332204　Rhodoleia Champ. ex Hook. (1850);红花荷属(红苞荷属,红苞木属,红花木荷属);Rhodoleia ●

332205　Rhodoleia championii Hook. f. ;红花荷(吊钟王,广卵叶红花荷,红苞荷,红苞木,红花木荷,萝多木);Champion Rhodoleia, Hong Kong Rose,Rhodoleia ●

332206　Rhodoleia forrestii Chun ex Exell;绒毛红花荷(滇西红花荷);Floss Rhodoleia, Forrest Rhodoleia ●

332207　Rhodoleia henryi S. Q. Tong;小脉红花荷(滇南红花荷,红花树,山茶花,显脉红花荷);Henry Rhodoleia, Showvein Rhodoleia ●

332208　Rhodoleia latiovatifolia G. A. Fu ＝ Rhodoleia championii Hook. f. ●

332209　Rhodoleia macrocarpa Hung T. Chang;大果红花荷(大果红苞木);Bigfruit Rhodoleia, Big-fruited Rhodoleia ●

332210　Rhodoleia parvipetala S. Q. Tong;小花红花荷(红苞木,红花,红花荷,红花树);Little Flower Rhodoleia, Littleflower Rhodoleia ●

332211　Rhodoleia stenopetala Hung T. Chang;海南红花荷(海南红苞木,窄瓣红花荷);Hainan Rhodoleia, Narrowpetal Rhodoleia, Stenopetalous Rhodoleia ●

332212　Rhodoleia teysmannii Miq. ;特氏红花荷●☆

332213　Rhodoleiaceae Nakai ＝ Hamamelidaceae R. Br. (保留科名)●

332214　Rhodoleiaceae Nakai ＝ Rhoipteleaceae Hand. -Mazz. (保留科名)●

332215　Rhodoleiaceae Nakai(1943);红花荷科(红花木荷科)●

332216　Rhodolirion Phil. ＝ Hippeastrum Herb. (保留属名)■

332217　Rhodolirium Phil. ＝ Rhodophiala C. Presl ■☆

332218　Rhodomyrtus (DC.) Rchb. (1841);桃金娘属;Rose Myrtle, Rosemyrtle, Rose-myrtle ●

332219　Rhodomyrtus Rchb. ＝ Rhodomyrtus (DC.) Rchb. ●

332220　Rhodomyrtus macrocarpa Benth. ;大果桃金娘;Finger Cherry ●☆

332221　Rhodomyrtus tomentosa (Aiton) Hassk. ;桃金娘(当梨,倒黏子,豆稔,豆稔干,多莲,岗枔花,岗拈,岗稔,江稔,金丝桃,麦,木刀莲,哖仔,稔子,山旦仔,山东稔,山多奶,山稔,山稔子,水刀莲,苏园子,唐莲,桃娘,乌肚子);Downy Rose Myrtle, Downy Rosemyrtle, Downy Rose-myrtle, Hill Gosseberry, Rose Myrtle, Rosemyrtle ●

332222　Rhodopentas Kårehed et B. Bremer(2007);红星花属☆

332223　Rhodophiala C. Presl ＝ Hippeastrum Herb. (保留属名)■

332224　Rhodophiala C. Presl(1845);红瓶兰属■☆

332225　Rhodophiala advena (Ker Gawl.) Traub;小红瓶兰■☆

332226　Rhodophora Neck. ＝ Rosa L. ●

332227　Rhodopis Urb. (1900)(保留属名);玫瑰豆属■☆

332228　Rhodopsis (Endl.) Rchb. ＝ Rosa L. ●

332229　Rhodopsis (Ledeb.) Dippel ＝ Hulthemia Dumort. ●☆

332230　Rhodopsis Lilja(废弃属名) ＝ Calandrinia Kunth(保留属名)■☆

332231　Rhodopsis Lilja(废弃属名) ＝ Rhodopis Urb. (保留属名)■☆

332232　Rhodopsis Lilja(废弃属名) ＝ Tegneria Lilja ■☆

332233　Rhodopsis Rchb. ＝ Rosa L. ●

332234　Rhodoptera Raf. ＝ Rumex L. ■●

332235　Rhodora L. ＝ Rhododendron L. ●

332236　Rhodora deflexa Griff. ＝ Enkianthus deflexus (Griff.) C. K. Schneid. ●

332237　Rhodoraceae Vent. ＝ Ericaceae Juss. (保留科名)●

332238　Rhodorhiza Webb ＝ Convolvulus L. ■●

332239　Rhodormis Raf. ＝ Salvia L. ●■

332240　Rhodorrhiza Webb et Berthel. ＝ Convolvulus L. ■●

332241　Rhodorrhiza Webb et Berthel. ＝ Rhodorhiza Webb ■●

332242　Rhodosciadium S. Watson(1889);红伞芹属■☆

332243　Rhodosciadium glaucum J. M. Coult. et Rose;灰绿红伞芹■☆

332244　Rhodosciadium macrophyllum Mathias et Constance;大叶红伞芹■☆

332245　Rhodosciadium montanum (J. M. Coult. et Rose) Mathias et Constance;山地红伞芹■☆

332246　Rhodosciadium nudicaule Drude;裸茎红伞芹■☆

332247　Rhodosciadium purpureum (Rose) Mathias et Constance;紫红伞芹■☆

332248　Rhodosepala Baker ＝ Dissotis Benth. (保留属名)●☆

332249　Rhodosepala erecta Cogn. ＝ Dissotis senegambiensis Triana ●☆

332250　Rhodosepala pauciflora Baker ＝ Dissotis senegambiensis Triana ●☆

332251　Rhodosepala procumbens Cogn. ＝ Dissotis senegambiensis Triana ●☆

332252　Rhodoseris Turcz. ＝ Onoseris Willd. ●■☆

332253　Rhodospatha Poepp. (1845);红匙南星属■☆

332254　Rhodospatha latifolia Poepp. et Endl. ;阔叶红匙南星●☆

332255　Rhodosphaera Engl. (1881);红球漆属●☆

332256　Rhodosphaera rhodanthema (F. Muell.) Engl. ;红球漆;Deep Yellow-wood ●☆

332257　Rhodostachys Phil. ＝ Ochagavia Phil. ■☆

332258　Rhodostegiella (Pobed.) C. Y. Wu et D. Z. Li ＝ Cynanchum L. ●■

332259　Rhodostegiella (Pobed.) C. Y. Wu et D. Z. Li(1990);地梢瓜属(雀瓢属)●■

332260　Rhodostegiella C. Y. Wu et D. Z. Li ＝ Cynanchum L. ●■

332261　Rhodostegiella D. Z. Li ＝ Cynanchum L. ●■

332262　Rhodostegiella sibirica (L.) C. Y. Wu et D. Z. Li ＝ Cynanchum thesioides (Freyn) K. Schum. ■

332263　Rhodostegiella sibirica (L.) C. Y. Wu et D. Z. Li ＝ Vincetoxicum sibiricum (L.) Decne. ■

332264　Rhodostegiella sibirica (L.) C. Y. Wu et D. Z. Li var. australis (Maxim.) C. Y. Wu et D. Z. Li ＝ Cynanchum thesioides (Freyn) K. Schum. ■

332265　Rhodostegiella sibirica (L.) C. Y. Wu et D. Z. Li var. australis (Maxim.) C. Y. Wu et D. Z. Li ＝ Vincetoxicum sibiricum Decne. var. australe Maxim. ■

332266　Rhodostegiella sibirica (L.) C. Y. Wu et D. Z. Li var. australis (Maxim.) C. Y. Wu et D. Z. Li ＝ Cynanchum thesioides (Freyn) K. Schum. var. australe (Maxim.) Tsiang et P. T. Li ●■

332267　Rhodostemonodaphne Rohwer et Kubitzki(1985);红蕊樟属●☆

332268　Rhodostemonodaphne anomala (Mez) Rohwer;红蕊樟●☆

332269　Rhodostemonodaphne grandis (Mez) Rohwer;大红蕊樟●☆

332270　Rhodostemonodaphne laxa (Meisn.) Rohwer;松散红蕊樟●☆

332271　Rhodostemonodaphne longiflora Madrinan;长花红蕊樟●☆

332272　Rhodostemonodaphne macrocalyx (Meisn.) Rohwer ex Madrinan;大萼红蕊樟●☆

332273　Rhodostemonodaphne ovatifolia Madrinan;卵叶红蕊樟●☆

332274　Rhodostoma Scheidw. ＝ Palicourea Aubl. ●☆

332275　Rhodothamnus Lindl. et Paxton ＝ Rhododendron L. ●

332276　Rhodothamnus Lindl. et Paxton ＝ Therorhodion （ Maxim. ）
Small ●

332277　Rhodothamnus Rchb. （1827）（保留属名）；伏石花属●☆

332278　Rhodothamnus chamaecistus （L.） Rchb.；伏石花；Miniature
Rhododendron ●☆

332279　Rhodothamnus chamaecistus Rchb. ＝ Rhodothamnus
chamaecistus （L.） Rchb. ●☆

332280　Rhodotypaceae J. Agardh ＝ Rosaceae Juss.（保留科名）●■

332281　Rhodotypaceae J. Agardh；鸡麻科●

332282　Rhodotypos Siebold et Zucc.（1841）；鸡麻属；Black Jetbead,
Jetbead ●

332283　Rhodotypos kerrioides Siebold et Zucc. ＝ Rhodotypos scandens
（Thunb.） Makino ●

332284　Rhodotypos scandens （Thunb.） Makino；鸡麻（白棣棠,山葫
芦子,双珠母,水葫芦杆）；Black Jetbead, Black Jet-bead, Jetbead,
Jetberry Bush, White Jew's-mallow, White Kerria ●

332285　Rhodotypos tetrapetala （ Siebold ） Makino ＝ Rhodotypos
scandens （Thunb.） Makino ●

332286　Rhodotypus Endl. ＝ Rhodotypos Siebold et Zucc. ●

332287　Rhodoxylon Raf. ＝ Convolvulus L. ●■

332288　Rhodusia Vasilch. ＝ Medicago L.（保留属名）●■

332289　Rhodusia arborea （L.） Vassilcz. ＝ Medicago arborea L. ●

332290　Rhoea St. -Lag. ＝ Punica L. ●

332291　Rhoeidlum Greeue ＝ Rhus L. ●

332292　Rhoeo Hance ＝ Tradescantia L. ■

332293　Rhoeo Hance（1852）；紫万年青属（紫背万年青属）；Oyster
Plant, Rhoeo ■

332294　Rhoeo discolor （L' Hér.） Hance ＝ Rhoeo spathacea （Sw.）
Stearn ■

332295　Rhoeo discolor （L' Hér.） Hance ＝Tradescantia spathacea Sw. ■

332296　Rhoeo discolor （L' Hér.） Hance ex Walp. ＝ Tradescantia
spathacea Sw. ■

332297　Rhoeo spathacea （Sw.） Stearn ' Compacta '；小蚌花■☆

332298　Rhoeo spathacea （Sw.） Stearn ' Dwarf Variegata '；条纹小蚌
花■☆

332299　Rhoeo spathacea （Sw.） Stearn ' Vittata '；金线紫万年青（花
叶紫万年青,金线蚌花）■☆

332300　Rhoeo spathacea （Sw.） Stearn ＝ Tradescantia spathacea Sw. ■

332301　Rhoiacarpos A. DC.（1857）；榴果檀香属●☆

332302　Rhoiacarpos capensis （Harv.） A. DC.；榴果檀香●☆

332303　Rhoicarpos B. D. Jacks. ＝ Rhoiacarpos A. DC. ●☆

332304　Rhoicissus Planch.（1887）；菱叶藤属●☆

332305　Rhoicissus capensis （Willd.） Planch. ＝ Rhoicissus tomentosa
（Lam.） Wild et R. B. Drumm. ●☆

332306　Rhoicissus capensis Planch. ＝ Rhoicissus tomentosa （Lam.）
Wild et R. B. Drumm. ●☆

332307　Rhoicissus cirrhiflora （L. f.） Gilg et M. Brandt ＝ Rhoicissus
digitata （L. f.） Gilg et M. Brandt ●☆

332308　Rhoicissus cuneifolia （Eckl. et Zeyh.） Planch. ＝ Rhoicissus
tridentata （L. f.） Wild et R. B. Drumm. subsp. cuneifolia （Eckl. et
Zeyh.） Urton ●☆

332309　Rhoicissus cymbifoliua C. A. Sm. ＝Rhoicissus revoilii Planch. ●☆

332310　Rhoicissus digitata （L. f.） Gilg et M. Brandt；指裂菱叶藤●☆

332311　Rhoicissus dimidiata （Thunb.） Gilg et Brandt；半片菱叶藤●☆

332312　Rhoicissus edulis De Wild. ＝ Ampelocissus edulis （De Wild.）
Gilg et M. Brandt ●☆

332313　Rhoicissus erythrodes （ Fresen. ） Planch. ＝ Rhoicissus

332314　Rhoicissus holstii Engl.；霍尔菱叶藤●☆

332315　Rhoicissus kougabergensis Retief et Van Jaarsv.；科加伯格菱叶
藤●☆

332316　Rhoicissus laetans Retief；愉悦菱叶藤●☆

332317　Rhoicissus microphylla （Turcz.） Gilg et M. Brandt；小叶菱叶
藤●☆

332318　Rhoicissus revoilii Planch.；雷瓦尔菱叶藤●☆

332319　Rhoicissus rhomboidea （E. Mey. ex Harv.） Planch. ＝ Vitis
rhombifolia Khakhlov ●☆

332320　Rhoicissus sapinii De Wild. ＝ Ampelocissus sapinii （De
Wild.） Gilg et M. Brandt ●☆

332321　Rhoicissus schlechteri Gilg et M. Brandt ＝ Rhoicissus revoilii
Planch. ●☆

332322　Rhoicissus sekhukhuniensis Retief, Siebert et A. E. van Wyk；塞
库菱叶藤●☆

332323　Rhoicissus sessilifolia Retief；无柄叶菱叶藤●☆

332324　Rhoicissus tomentosa （Lam.） Wild et R. B. Drumm.；好望角
菱叶藤；Cape Grape, Evergreen Grape ●☆

332325　Rhoicissus tridentata （L. f.） Wild et R. B. Drumm.；三齿菱叶
藤●☆

332326　Rhoicissus tridentata （L. f.） Wild et R. B. Drumm. subsp.
cuneifolia （Eckl. et Zeyh.） Urton；楔叶菱叶藤●☆

332327　Rhoicissus usambarensis Gilg ＝ Rhoicissus tridentata （L. f.）
Wild et R. B. Drumm. ●☆

332328　Rhoiptelea Diels et Hand. -Mazz.（1932）；马尾树属（漆榕
属）；Horsetailtree, Rhoiptelea ●

332329　Rhoiptelea chiliantha Diels et Hand. -Mazz.；马尾树（马尾花,马尾
丝,漆榆,穗果木）；China Horsetailtree, Chinese Rhoiptelea ●◇

332330　Rhoipteleaceae Hand. -Mazz.（1932）（保留科名）；马尾树科；
Horsetailtree Family, Rhoiptelea Family ●

332331　Rhombifolium Rich. ex DC. ＝ Clitoria L. ●

332332　Rhombochlamys Lindau（1897）；菱被爵床属☆

332333　Rhombochlamys rosulata Lindau；菱被爵床☆

332334　Rhomboda Lindl.（1857）；菱兰属■☆

332335　Rhomboda Lindl. ＝ Hetaeria Blume（保留属名）■

332336　Rhomboda Lindl. ＝ Zeuxine Lindl.（保留属名）■

332337　Rhomboda abbreviata （Lindl.） Ormerod；小片菱兰（翻唇兰,
小片齿唇兰）；Common Hetaeria, Shrink Forkliporchis ■

332338　Rhomboda abbreviata （ Lindl. ） Ormerod ＝ Anoectochilus
abbreviatus （Lindl.） Seidenf. ■

332339　Rhomboda fanjingensis Ormerod；贵州菱兰■

332340　Rhomboda moulmeinensis （E. C. Parish et Rchb. f.） Ormerod；
艳丽菱兰（艳丽齿唇兰,艳丽开唇兰）；Colorful Forkliporchis ■

332341　Rhomboda moulmeinensis （Parish et Rchb. f.） Ormerod ＝
Anoectochilus moulmeinensis （Parish, Rchb. f. et Sineref.） Seidenf.
et Smitinand ■

332342　Rhomboda pogonorrhyncha （ Hand. -Mazz. ） Ormerod ＝
Rhomboda tokioi （Fukuy.） Ormerod ■

332343　Rhomboda taiwaniana （S. S. Ying） Ormerod ＝ Habenaria lucida
Wall. ex Lindl. ■

332344　Rhomboda tokioi （Fukuy.） Ormerod；白肋菱兰（白点伴兰）■☆

332345　Rhombolobium Rich. ex Kunth ＝ Clitoria L. ●

332346　Rhombolythrum Airy Shaw ＝ Rhombolytrum Link ■☆

332347　Rhombolythrum Link ＝ Rhombolytrum Link ■☆

332348　Rhombolytrum Link（1833）；石坡草属■☆

332349　Rhombolytrum albescens Nash；白石坡草■☆

332350　Rhombolytrum rhomboideum Link；石坡草■☆

332351　Rhombonema Schltr. = Parapodium E. Mey.●☆

332352　Rhombonema luridum Schltr. = Parapodium costatum E. Mey.●☆

332353　Rhombophyllum（Schwantes）Schwantes（1927）；棱叶属（快刀乱麻属，菱叶草属）●☆

332354　Rhombophyllum Schwantes = Rhombophyllum（Schwantes）Schwantes ●☆

332355　Rhombophyllum albanense（L. Bolus）H. E. K. Hartmann；阿尔邦棱叶●☆

332356　Rhombophyllum dolabriforme（L.）Schwantes；银鲜●☆

332357　Rhombophyllum dolabriforme Schwantes = Rhombophyllum dolabriforme（L.）Schwantes ●☆

332358　Rhombophyllum dyeri（L. Bolus）H. E. K. Hartmann；戴尔棱叶●☆

332359　Rhombophyllum nelii Schwantes；快刀乱麻；Elkhorns ●☆

332360　Rhombophyllum rhomboideum（Salm-Dyck）Schwantes；棱叶（青崖）●☆

332361　Rhombophyllum rhomboideum（Salm-Dyck）Schwantes var. groppiorum Heinrich = Rhombophyllum rhomboideum（Salm-Dyck）Schwantes ●☆

332362　Rhombophyllum rhomboideum Schwantes = Rhombophyllum rhomboideum（Salm-Dyck）Schwantes ●☆

332363　Rhombospora Korth. = Greenea Wight et Arn. ●☆

332364　Rhoogeton Leeuwenb.（1958）；莱文苣苔属■☆

332365　Rhoogeton cyclophyllus Leeuwenb.；莱文苣苔■☆

332366　Rhopala Schreb. = Roupala Aubl. ●☆

332367　Rhopalandria Stapf = Dioscoreophyllum Engl. ●☆

332368　Rhopalandria cumminsii Stapf = Dioscoreophyllum volkensii Engl. ●☆

332369　Rhopalandria lobatum C. H. Wright = Dioscoreophyllum volkensii Engl. ●☆

332370　Rhopalephora Hassk.（1864）；毛果竹叶菜属（钩毛子草属）■

332371　Rhopalephora Hassk. = Aneilema R. Br. ■☆

332372　Rhopalephora Hassk. = Dictyospermum Wight ■

332373　Rhopalephora rugosa（H. Perrier）Faden；马岛毛果竹叶菜☆

332374　Rhopalephora scaberrima（Blume）Faden；毛果竹叶菜（糙叶水竹叶，大水竹叶，钩毛子草，毛果网籽草）；Hairfruit Netseed，Hairyfruit Netseedgrass ■

332375　Rhopaloblaste Scheff.（1876）；杖花棕属（棒果芽椰属，棒椰属，垂叶椰属，垂羽椰属，拟槟榔椰属，手杖棕属）；Rhopaloblaste ●☆

332376　Rhopaloblaste ceramica Burret；印尼杖花棕（印尼垂叶椰）●☆

332377　Rhopaloblaste hexandra Scheff.；杖花棕；Molucca Palm ●☆

332378　Rhopalobrachium Schltr. et K. Krause（1908）；短棒茜属 ☆

332379　Rhopalobrachium fragrans Schltr. et K. Krause；短棒茜 ☆

332380　Rhopalocarpaceae Hemsl. = Rhopalocarpaceae Hemsl. ex Takht. ●☆

332381　Rhopalocarpaceae Hemsl. = Sphaerosepalaceae Tiegh. ex Bullock. ●☆

332382　Rhopalocarpaceae Hemsl. ex Takht.；棒果树科（刺果树科）●☆

332383　Rhopalocarpaceae Hemsl. ex Takht. = Sphaerosepalaceae Tiegh. ex Bullock. ●☆

332384　Rhopalocarpus Bojer（1846）；棒果树属（刺果树属）●☆

332385　Rhopalocarpus Teijsm. et Binn. ex Miq. = Anaxagorea A. St. -Hil. ●

332386　Rhopalocarpus alternifolius（Baker）Capuron；互叶棒果树（球萼树）●☆

332387　Rhopalocarpus alternifolius（Baker）Capuron var. sambiranensis Capuron = Rhopalocarpus alternifolius（Baker）Capuron ●☆

332388　Rhopalocarpus binervius Capuron；双脉棒果树●☆

332389　Rhopalocarpus coriaceus（Scott-Elliot）Capuron；革质棒果树●☆

332390　Rhopalocarpus coriaceus（Scott-Elliot）Capuron var. crassinervius Capuron = Rhopalocarpus crassinervius（Capuron）G. E. Schatz, Lowry et A. -E. Wolf ●☆

332391　Rhopalocarpus coriaceus（Scott-Elliot）Capuron var. trichopetalus Capuron = Rhopalocarpus alternifolius（Baker）Capuron ●☆

332392　Rhopalocarpus crassinervius（Capuron）G. E. Schatz, Lowry et A. -E. Wolf；粗脉棒果树●☆

332393　Rhopalocarpus excelsus Capuron；高大棒果树●☆

332394　Rhopalocarpus longipetiolatus Hemsl.；长柄棒果树●☆

332395　Rhopalocarpus louvelii（Danguy）Capuron；卢韦尔棒果树●☆

332396　Rhopalocarpus louvelii（Danguy）Capuron var. parvifolius Capuron = Rhopalocarpus parvifolius（Capuron）G. E. Schatz, Lowry et A. -E. Wolf ●☆

332397　Rhopalocarpus lucidus Bojer；光亮棒果树●☆

332398　Rhopalocarpus macrorhamnifolius Capuron；大叶棒果树●☆

332399　Rhopalocarpus macrorhamnifolius Capuron f. occidentalis Capuron = Rhopalocarpus louvelii（Danguy）Capuron ●☆

332400　Rhopalocarpus mollis G. E. Schatz et Lowry；柔软棒果树●☆

332401　Rhopalocarpus parvifolius（Capuron）G. E. Schatz, Lowry et A. -E. Wolf；小叶棒果树●☆

332402　Rhopalocarpus pseudothouarsianus Capuron = Rhopalocarpus thouarsianus Baill. ●☆

332403　Rhopalocarpus similis Hemsl.；相似棒果树●☆

332404　Rhopalocarpus similis Hemsl. subsp. velutinus Capuron = Rhopalocarpus similis Hemsl. ●☆

332405　Rhopalocarpus suarezensis Capuron ex Bosser；苏亚雷斯棒果树●☆

332406　Rhopalocarpus thouarsianus Baill.；图氏棒果树●☆

332407　Rhopalocarpus triplinervius Baill.；三脉棒果树●☆

332408　Rhopalocarpus undulatus Capuron；波状棒果树●☆

332409　Rhopalocnemis Jungh.（1841）；盾片蛇菰属（大蛇菰属，鬼笔蛇菰属，双柱蛇菰属）；Rhopalocnemis ■

332410　Rhopalocnemis malagasica Jum. et H. Perrier = Ditepalanthus malagasicus（Jum. et H. Perrier）Fagerl. ■☆

332411　Rhopalocnemis phalloides Jungh.；盾片蛇菰（鬼笔蛇菰，双柱蛇菰）；Rhopalocnemis ■

332412　Rhopalocyclus Schwantes = Leipoldtia L. Bolus ●☆

332413　Rhopalocyclus herrei Schwantes = Leipoldtia schultzei（Schltr. et Diels）Friedrich ●☆

332414　Rhopalocyclus nelii Schwantes = Leipoldtia schultzei（Schltr. et Diels）Friedrich ●☆

332415　Rhopalocyclus weigangianus（Dinter）Dinter et Schwantes = Leipoldtia weigangiana（Dinter）Dinter et Schwantes ●☆

332416　Rhopalopilia Pierre（1896）；毛杖木属●☆

332417　Rhopalopilia altescandens Engl.；攀缘毛杖木●☆

332418　Rhopalopilia bequaertii（De Wild.）J. Léonard = Rhopalopilia altescandens Engl. ●☆

332419　Rhopalopilia hallei Villiers；哈勒毛杖木●☆

332420　Rhopalopilia marquesii Engl. = Pentarhopalopilia marquesii（Engl.）Hiepko ●☆

332421　Rhopalopilia pallens Pierre；变苍白毛杖木●☆

332422　Rhopalopilia pallens Pierre var. glabriflora J. Léonard；光花毛杖木●☆

332423　Rhopalopilia poggei Engl. = Rhopalopilia pallens Pierre var.

glabriflora J. Léonard ●☆

332424 Rhopalopilia poggei Engl. var. bequaertii De Wild. = Rhopalopilia altescandens Engl. ●☆

332425 Rhopalopilia soyauxii Engl. = Pentarhopalopilia marquesii（Engl.）Hiepko ●☆

332426 Rhopalopilia ubanghensis A. Chev. = Urobotrya sparsiflora（Engl.）Hiepko ●☆

332427 Rhopalopilia umbellulata（Baill.）Engl. = Pentarhopalopilia umbellulata（Baill.）Hiepko ●☆

332428 Rhopalopilia verdickii De Wild. = Pentarhopalopilia marquesii（Engl.）Hiepko ●☆

332429 Rhopalopodium Ulbr. = Ranunculus L. ■

332430 Rhopalosciadium Rech. f.（1952）;波斯伞芹属■☆

332431 Rhopalosciadium stereocalyx Rech. f.；波斯伞芹■☆

332432 Rhopalostigma Phil. = Phrodus Miers ■☆

332433 Rhopalostigma Schott = Asterostigma Fisch. et C. A. Mey. ■☆

332434 Rhopalostylis H. Wendl. et Drude(1875);棒柱桐（棒花棕属,胡刷椰子属,细叶椰子属,香棕属,新西兰椰子属）;Rhopalostylis ●☆

332435 Rhopalostylis Klotzsch ex Baill. = Dalechampia L. ●

332436 Rhopalostylis baueri H. Wendl. et Druce;鲍氏棒柱桐（鲍氏棒花棕,诺福克香棕）; Bauer Rhopalostylis, Norfolk Betel Palm, Norfolk Island Palm, Norfolk Palm ●☆

332437 Rhopalostylis cheesemanii Becc.;齐思曼棒柱桐（齐思曼棒花棕）;Cheeseman Rhopalostylis ●☆

332438 Rhopalostylis sapida H. Wendl. et Druce;美味曼棒柱桐棕（美味曼棒花棕,尼卡椰子,刷子椰子,香棕）; Nikau Palm, Nikau Rhopalostylis, Shaving Brush Palm ●☆

332439 Rhopalota N. E. Br. = Crassula L. ●■☆

332440 Rhopalota aphylla（Schönland et Baker f.）N. E. Br. = Crassula aphylla Schönland et Baker f. ■☆

332441 Rhophostemon Wittst. = Nervilia Comm. ex Gaudich.（保留属名）■

332442 Rhophostemon Wittst. = Roptrostemon Blume ■

332443 Rhopium Schreb. = Meborea Aubl. ●■

332444 Rhopium Schreb. = Phyllanthus L. ●■

332445 Rhoradendron Griseb. = Phoradendron Nutt. ●☆

332446 Rhouancou Augier = Rouhamon Aubl. ●

332447 Rhouancou Augier = Strychnos L. ●

332448 Rhuacophila Blume = Dianella Lam. ex Juss. ●■

332449 Rhuacophila Blume(1827);线脐籽属■☆

332450 Rhuacophila javanica Blume;线脐籽■☆

332451 Rhus（Tourn.）L. = Rotala L. ■

332452 Rhus L.（1753）;盐肤木属（漆树属）;Stag's-horn Sumach, Sumac, Sumch ●

332453 Rhus abyssinica Hochst. ex Oliv. = Rhus glutinosa Hochst. ex A. Rich. subsp. abyssinica（Hochst. ex Oliv.）M. G. Gilbert ●☆

332454 Rhus acocksii Moffett;阿氏盐肤木●☆

332455 Rhus acuminata DC. = Toxicodendron acuminatum（DC.）C. Y. Wu et T. L. Ming ●

332456 Rhus acuminata E. Mey. = Rhus chirindensis Baker f. ●☆

332457 Rhus acuminatissima R. Fern. et A. Fern.;渐尖盐肤木●☆

332458 Rhus africana Mill. = Rhus lucida L. ●☆

332459 Rhus africana Mill. var. macrophylla Sond. = Rhus scytophylla Eckl. et Zeyh. ●☆

332460 Rhus ailanthoides Bunge = Picrasma quassioides（D. Don）A. W. Benn. ●

332461 Rhus alatum Thunb. = Hippobromus pauciflorus（L. f.）Radlk. ■☆

332462 Rhus albida Schousb.;淡白盐肤木●☆

332463 Rhus albomarginata Sond.;白边盐肤木●☆

332464 Rhus ambigua H. Lév. ex Dippel = Toxicodendron orientale Greene ●

332465 Rhus amboensis Schinz = Rhus tenuinervis Engl. ●☆

332466 Rhus amerina Meikle = Rhus leptodictya Diels ●☆

332467 Rhus amharica Pic. Serm. = Rhus glutinosa Hochst. ex A. Rich. ●☆

332468 Rhus ampla Engl.;膨大盐肤木●☆

332469 Rhus anchietae Ficalho ex Hiern. f. suffruticosa（Meikle）R. Fern. et A. Fern.;亚灌木盐肤木●☆

332470 Rhus angolensis Engl.;安哥拉盐肤木●☆

332471 Rhus angolensis Engl. f. glabrescens R. Fern.;渐光盐肤木●☆

332472 Rhus angustifolia L.;窄叶盐肤木●☆

332473 Rhus apiculata Engl.;细尖盐肤木●☆

332474 Rhus arenaria Engl.;沙地盐肤木●☆

332475 Rhus argentea A. Chev.;银色盐肤木●

332476 Rhus argyi H. Lév. = Pistacia chinensis Bunge ●

332477 Rhus aromatica Aiton;香漆（香葛藤,香漆树,香枝漆）; Aromatic Sumac, Aromatic Sumach, Fragrant Sumac, Fragrant Sumach, Lemon Sumac, Lemon Sumach, Polecat Bush, Squaw-berry, Squaw-bush, Sumac, Sweet Sumach, White Sumach ●☆

332478 Rhus aromatica Aiton 'Grow-Low';格鲁矮香漆; Dwarf Fragrant Sumac, Fragrant Sumac ●☆

332479 Rhus aromatica Aiton 'Laciniata';条裂香漆●☆

332480 Rhus aromatica Aiton subsp. serotina（Greene）R. E. Brooks = Rhus aromatica Aiton var. serotina（Greene）Rehder ●☆

332481 Rhus aromatica Aiton var. flabelliformis Shinners;臭漆;Skunk-bumash Sumac ●☆

332482 Rhus aromatica Aiton var. illinoensis（Greene）Rehder = Rhus aromatica Aiton ●☆

332483 Rhus aromatica Aiton var. serotina（Greene）Rehder;迟香漆; Fragrant Sumac, Squaw-bush ●☆

332484 Rhus atomica Jacq. = Rhus laevigata L. ●☆

332485 Rhus baurii Schönland = Rhus pyroides Burch. ●☆

332486 Rhus bequaertii Robyns et Lawalrée = Rhus ruspolii Engl. ●☆

332487 Rhus blanda Meikle;光滑盐肤木●☆

332488 Rhus blanda Meikle. f. exelliana（Meikle）R. Fern. et A. Fern.;埃克塞尔盐肤木●☆

332489 Rhus blinii H. Lév. = Cipadessa cinerascens（Pell.）Hand.-Mazz. ●

332490 Rhus bodinieri H. Lév. = Choerospondias axillaris（Roxb.）B. L. Burtt et A. W. Hill ●

332491 Rhus bofillii H. Lév. = Meliosma oldhamii Siebold et Zucc. ●

332492 Rhus bolusii Sond. ex Engl.;博卢斯盐肤木●☆

332493 Rhus borealis（Britton）Gleason = Rhus pulvinata Greene ●☆

332494 Rhus borealis Greene = Rhus glabra L. ●☆

332495 Rhus brenanii Kokwaro;布雷南盐肤木●☆

332496 Rhus buettneri Engl.;比特纳盐肤木●☆

332497 Rhus burchellii Sond. ex Engl.;伯切尔盐肤木●☆

332498 Rhus burchellii Sond. ex Engl. var. tricrenata Engl. = Rhus burchellii Sond. ex Engl. ●☆

332499 Rhus burkeana Sond. = Rhus magalismontana Sond. ●☆

332500 Rhus cacodendron Ehrh. = Ailanthus altissima（Mill.）Swingle ●

332501 Rhus calophylla Greene = Rhus glabra L. ●☆

332502 Rhus canadensis Marshall;香枝漆;Fragrant Sumac ●☆

332503 Rhus canadensis Marshall var. illinoensis（Greene）Fernald = Rhus aromatica Aiton ●☆

332504　Rhus carnosula Schönland；肉质盐肤木●☆

332505　Rhus carnosula Schönland var. parvifolia ? = Rhus carnosula Schönland ●☆

332506　Rhus caustica Hook. et Arn. ；苛性盐肤木（苛性葛）●☆

332507　Rhus cavaleriei H. Lév. = Eurycorymbus cavaleriei （H. Lév.） Rehder et Hand. -Mazz. ●

332508　Rhus celastroides Sond. = Rhus undulata Jacq. ●☆

332509　Rhus chinensis Mill. ；盐肤木（百虫仓，百药煎，棓子，倍树，倍子柴，犕，鯆林盐，耳大蜈蚣，肤木，肤杨树，麸杨，敷烟树，芙连树，红麸杨，红盐果，红叶桃，猴盐柴，假五味子，角倍，枯盐萁，老公担盐，木椿树，木附子，木五倍子，木盐，女木子，叛奴盐，泡木树，破凉伞，蒲连盐，漆树，青麸杨，山杜仲，山梧桐，山盐菁，酸酱头，天盐，土椿树，文蛤，乌酸桃，乌桃叶，乌烟桃，乌盐泡，五倍柴，五倍树，五倍子，五倍子苗，五倍子树，盐肤子，盐麸子，盐根树，盐梅子，盐樣子，盐酸白，盐酸木，盐酸树，盐桶鬃，盐子树，油盐果，柘柴，猪枣椿）；Chinese Sumac，Chinese Sumach，Nutgall Tree ●

332510　Rhus chinensis Mill. = Rhus javanica L. var. chinensis （Mill.） T. Yamaz. ●

332511　Rhus chinensis Mill. = Rhus javanica L. ●

332512　Rhus chinensis Mill. var. glabra S. B. Liang；光枝盐肤木；Glabrous Chinese Sumac ●

332513　Rhus chinensis Mill. var. roxburghii （DC.） Rehder；滨盐肤木（罗氏盐肤木，埔盐，山盐菁，盐霜白，盐酸树）；China Sumac，Roxburgh Sumac ●

332514　Rhus chinensis Mill. var. roxburghii （DC.） Rehder = Rhus javanica L. ●

332515　Rhus chirindensis Baker f. ；奇林达盐肤木●☆

332516　Rhus chirindensis Baker f. f. legatii （Schönland） R. Fern. et A. Fern. = Rhus chirindensis Baker f. ●☆

332517　Rhus choriophylla Wooton et Standl. ；新常绿漆；New Mexico Evergreen Sumac ●☆

332518　Rhus ciliata Licht. ex Schult. ；睫毛盐肤木●☆

332519　Rhus ciliata Licht. ex Schult. f. fastigiata Schönland = Rhus ciliata Licht. ex Schult. ●☆

332520　Rhus ciliata Licht. ex Schult. f. lepidota Burtt Davy = Rhus ciliata Licht. ex Schult. ●☆

332521　Rhus cinerea R. Fern. et A. Fern. = Rhus magalismontana Sond. ●☆

332522　Rhus cirrhiflora L. f. = Rhoicissus digitata （L. f.） Gilg et M. Brandt ●☆

332523　Rhus coddii R. Fern. et A. Fern. = Rhus magalismontana Sond. subsp. coddii （R. Fern. et A. Fern.） Moffett ●☆

332524　Rhus collina Engl. ；山丘盐肤木●☆

332525　Rhus commiphoroides Engl. et Gilg = Rhus tenuinervis Engl. ●☆

332526　Rhus concinna Burch. = Rhus ciliata Licht. ex Schult. ●☆

332527　Rhus concolor C. Presl = Ozoroa concolor （C. Presl） De Winter ●☆

332528　Rhus conitus L. = Cotinus coggygria Scop. ●

332529　Rhus copallina L. ；亮叶漆树（亮漆）；Black Sumach，Dwarf Sumac，Dwarf Sumach，Flameleaf Sumac，Flame-leaf Sumac，Mountain Sumach，Shining Sumac，Shining Sumach，Winged Sumac ●☆

332530　Rhus copallina L. var. lanceolata A. Gray；燎原亮漆；Prairie Flame-leaf Sumac ●☆

332531　Rhus copallina L. var. latifolia Engl. = Rhus copallina L. ●☆

332532　Rhus coriacea Engl. = Rhus magalismontana Sond. ●☆

332533　Rhus coriaria L. ；西西里漆树（西西里漆）；Elm-leaved Sumach，European Sumach，Sicilian Sumac，Sicilian Sumach，Sumac，Sumach，Tanner's Sumac，Tanning Sumach ●☆

332534　Rhus cotinoides Nutt. = Cotinus americana Nutt. ●

332535　Rhus cotinoides Nutt. ex Torr. et Gray. = Cotinus americana Nutt. ●

332536　Rhus cotinus L. = Cotinus coggygria Scop. ●

332537　Rhus crenata Thunb. ；圆齿盐肤木●☆

332538　Rhus crenulata A. Rich. ；细圆齿盐肤木●☆

332539　Rhus crispa （Harv. ex Engl.） Schönland = Rhus gueinzii Sond. ●☆

332540　Rhus culminum R. Fern. et A. Fern. = Rhus tumulicola S. Moore ●☆

332541　Rhus cuneifolia L. f. ；楔叶盐肤木●☆

332542　Rhus decipiens Wight et Arn. = Filicium decipiens （Wight et Arn.） Thwaites ●☆

332543　Rhus delavayi Franch. = Toxicodendron delavayi （Franch.） F. A. Barkley ●

332544　Rhus delavayi Franch. var. quinquejugum Rehder et E. H. Wilson = Toxicodendron delavayi （Franch.） F. A. Barkley var. quinquejugum （Rehder et E. H. Wilson） C. Y. Wu et T. L. Ming ●

332545　Rhus dentata Thunb. ；罗得西亚盐肤木；Rhodesian Currant ●☆

332546　Rhus dentata Thunb. f. glabra （Schönland） R. Fern. = Rhus dentata Thunb. ●☆

332547　Rhus dentata Thunb. f. glabra Schönland = Rhus dentata Thunb. ●☆

332548　Rhus dentata Thunb. f. parvifolia （Harv. ex Sond.） Schönland = Rhus dentata Thunb. ●☆

332549　Rhus dentata Thunb. f. pilosissima （Engl.） R. Fern. = Rhus dentata Thunb. ●☆

332550　Rhus dentata Thunb. f. puberula Sond. = Rhus dentata Thunb. ●☆

332551　Rhus dentata Thunb. f. sparsepilosa R. Fern. = Rhus dentata Thunb. ●☆

332552　Rhus dentata Thunb. f. villosissima R. Fern. = Rhus dentata Thunb. ●☆

332553　Rhus dentata Thunb. var. fulvescens （Engl.） Burtt Davy = Rhus divaricata Eckl. et Zeyh. ●☆

332554　Rhus dentata Thunb. var. parvifolia （Harv. ex Sond.） Schönland = Rhus dentata Thunb. ●☆

332555　Rhus dentata Thunb. var. truncata Burtt Davy = Rhus rogersii Schönland ●☆

332556　Rhus digitata L. f. = Rhoicissus digitata （L. f.） Gilg et M. Brandt ●☆

332557　Rhus dimidiata Thunb. = Rhoicissus dimidiata （Thunb.） Gilg et Brandt ●☆

332558　Rhus dinteri Engl. = Rhus pyroides Burch. var. dinteri （Engl.） Moffett ●☆

332559　Rhus discolor E. Mey. ex Sond. ；异色盐肤木●☆

332560　Rhus discolor E. Mey. ex Sond. f. grandifolia （Engl.） Schönland = Rhus discolor E. Mey. ex Sond. ●☆

332561　Rhus discolor E. Mey. ex Sond. f. intermedia Burtt Davy = Rhus discolor E. Mey. ex Sond. ●☆

332562　Rhus discolor E. Mey. ex Sond. f. latifolia Schönland = Rhus discolor E. Mey. ex Sond. ●☆

332563　Rhus discolor E. Mey. ex Sond. f. villosissima （Engl.） Schönland = Rhus discolor E. Mey. ex Sond. ●☆

332564　Rhus dispar C. Presl = Ozoroa dispar （C. Presl） R. Fern. et A. Fern. ●☆

332565　Rhus dissecta Thunb. ；深裂盐肤木●☆

332566　Rhus dissecta Thunb. var. obovata Schönland = Rhus dissecta Thunb. ●☆

332567　Rhus dissecta Thunb. var. pinnatifida Schönland = Rhus dissecta Thunb. ●☆

332568　Rhus divaricata Eckl. et Zeyh. ;叉开盐肤木●☆

332569　Rhus divaricata Eckl. et Zeyh. var. fulvescens Engl. = Rhus divaricata Eckl. et Zeyh. ●☆

332570　Rhus diversiloba Torr. et A. Gray;槲叶毒葛;Californian Poison Oak,Poison Oak,Western Poison Oak,Western Poison-oak ●☆

332571　Rhus djalonensis A. Chev. = Ozoroa pulcherrima (Schweinf.) R. Fern. et A. Fern. ●☆

332572　Rhus dracomontana Moffett;德拉科盐肤木●☆

332573　Rhus dregeana Sond. ;德雷盐肤木●☆

332574　Rhus dumetorum Exell;灌丛盐肤木●☆

332575　Rhus dunensis Gand. = Rhus lucida L. f. elliptica (Sond.) Moffett ●☆

332576　Rhus dura Schönland = Rhus tumulicola S. Moore ●☆

332577　Rhus dyeri R. Fern. et A. Fern. = Rhus rigida Mill. var. dentata (Engl.) Moffett ●☆

332578　Rhus eburnea Schönland = Rhus transvaalensis Engl. ●☆

332579　Rhus echinocarpa H. Lév. = Rhus punjabensis J. L. Stewart ex Brandis var. sinica (Diels) Rehder et E. H. Wilson ●

332580　Rhus echinocarpa H. Lév. = Rhus trichocarpa Miq. ●

332581　Rhus echinocarpa H. Lév. = Toxicodendron trichocarpum (Miq.) Kuntze ●

332582　Rhus eckloniana Sond. = Rhus rigida Mill. ●☆

332583　Rhus engleri Britten;恩格勒盐肤木●☆

332584　Rhus ernestii Schönland = Rhus tumulicola S. Moore. f. meeuseana (R. Fern. et A. Fern.) Moffett ●☆

332585　Rhus erosa Thunb. ;啮蚀状盐肤木●☆

332586　Rhus erosus Radlk. = Allophylus natalensis (Sond.) De Winter ●☆

332587　Rhus esquirolii H. Lév. = Rhus punjabensis J. L. Stewart ex Brandis var. sinica (Diels) Rehder et E. H. Wilson ●

332588　Rhus excisa Thunb. = Rhus undulata Jacq. ●☆

332589　Rhus exelliana Meikle = Rhus blanda Meikle. f. exelliana (Meikle) R. Fern. et A. Fern. ●☆

332590　Rhus eylesii Hutch. = Rhus kirkii Oliv. ●☆

332591　Rhus falcata (Becc. ex Mart.) Penz. = Pistacia falcata Becc. ex Martelli ●☆

332592　Rhus fanshawei R. Fern. et A. Fern. = Rhus magalismontana Sond. subsp. trifoliolata (Baker f.) Moffett ●☆

332593　Rhus fastigiata Eckl. et Zeyh. ;帚状盐肤木●☆

332594　Rhus filiformis Schinz = Rhus gracillima Engl. var. glaberrima Schönland ●☆

332595　Rhus flexicaulis Baker;曲茎盐肤木●☆

332596　Rhus flexuosa Diels = Rhus pyroides Burch. var. gracilis (Engl.) Burtt Davy ●☆

332597　Rhus foliosa A. Rich. = Rhus glutinosa Hochst. ex A. Rich. subsp. abyssinica (Hochst. ex Oliv.) M. G. Gilbert ●☆

332598　Rhus fraseri Schönland = Rhus pyroides Burch. var. integrifolia (Engl.) Moffett ●☆

332599　Rhus fraxinifolia D. Don = Evodia fraxinifolia (D. Don) Hook. f. ●

332600　Rhus fraxinifolia D. Don = Rhus succedanea L. ●

332601　Rhus fulva Craib = Toxicodendron fulvum (Craib) G. Y. Wu et T. L. Ming ●

332602　Rhus fulvescens (Engl.) Diels = Rhus divaricata Eckl. et Zeyh. ●☆

332603　Rhus galpinii Schinz = Rhus grandidens Harv. ex Engl. ●☆

332604　Rhus gerrardii (Harv. ex Engl.) Diels;杰勒德盐肤木●☆

332605　Rhus gerrardii (Harv. ex Engl.) Diels var. basutorum Schönland = Rhus gerrardii (Harv. ex Engl.) Diels ●☆

332606　Rhus gerrardii (Harv. ex Engl.) Diels var. latifolia Schönland = Rhus gerrardii (Harv. ex Engl.) Diels ●☆

332607　Rhus gerrardii (Harv. ex Engl.) Diels var. montana (Diels) Schönland = Rhus montana Diels ●☆

332608　Rhus glabra L. ;光叶盐肤木(光滑漆树,光叶漆,黄栌,无毛火炬树,阳葛);Common Sumac, Pennsylvania Sumach, Scarlet Sumac, Scarlet Sumach, Smooth Sumac, Smooth Sumach, Upland Sumach, Vinegar Tree ●☆

332609　Rhus glabra L. var. borealis Britton = Rhus pulvinata Greene ●☆

332610　Rhus glabra L. var. laciniata Carrière;细裂光叶盐肤木●☆

332611　Rhus glabra L. var. laciniata Carrière = Rhus glabra L. ●☆

332612　Rhus glabra L. var. occidentalis Torr. = Rhus glabra L. ●☆

332613　Rhus glauca Thunb. ;灰绿盐肤木●☆

332614　Rhus glaucescens A. Rich. ;浅绿盐肤木●☆

332615　Rhus glaucescens A. Rich. var. macrocarpa Schweinf. = Rhus natalensis Bernh. ex Krauss var. macrocarpa (Schweinf.) Cufod. ●☆

332616　Rhus glaucescens A. Rich. var. natalensis (Bernh. ex Krauss) Engl. = Rhus natalensis Bernh. ex Krauss ●☆

332617　Rhus glaucovirens Engl. = Rhus zeyheri Sond. ●☆

332618　Rhus glutinosa Hochst. ex A. Rich. ;黏性盐肤木●☆

332619　Rhus glutinosa Hochst. ex A. Rich. subsp. abyssinica (Hochst. ex Oliv.) M. G. Gilbert;阿比西尼亚盐肤木●☆

332620　Rhus glutinosa Hochst. ex A. Rich. subsp. neoglutinosa (M. G. Gilbert) M. G. Gilbert;新阿比尼亚盐肤木●☆

332621　Rhus glutinosa Hochst. ex A. Rich. var. obtusifolia Engl. = Rhus longipes Engl. ●☆

332622　Rhus glutinosa Hochst. ex A. Rich. var. unifoliolata Cufod. = Rhus glutinosa Hochst. ex A. Rich. ●☆

332623　Rhus gossweileri Engl. ;戈斯盐肤木●☆

332624　Rhus gracilipes Exell;细梗盐肤木●☆

332625　Rhus gracilis Engl. = Rhus pyroides Burch. var. gracilis (Engl.) Burtt Davy ●☆

332626　Rhus gracillima Engl. ;细长盐肤木●☆

332627　Rhus gracillima Engl. var. glaberrima Schönland;无毛盐肤木●☆

332628　Rhus grandidens Harv. ex Engl. ;大齿盐肤木●☆

332629　Rhus grandifolia Engl. = Rhus discolor E. Mey. ex Sond. ●☆

332630　Rhus griffithii Hook. f. = Toxicodendron griffithii (Hook. f.) Kuntze ●

332631　Rhus grossireticulata Van der Veken;粗网盐肤木●☆

332632　Rhus gueinzii Sond. ;吉内斯盐肤木●☆

332633　Rhus gueinzii Sond. var. crispa Harv. ex Engl. = Rhus gueinzii Sond. ●☆

332634　Rhus gummifera H. Lév. = Pistacia chinensis Bunge ●

332635　Rhus harveyi Moffett;哈维盐肤木●☆

332636　Rhus henryi Diels = Rhus potaninii Maxim. ●

332637　Rhus heptaphylla Hiern;七叶盐肤木●☆

332638　Rhus herbacea A. Chev. = Ozoroa pulcherrima (Schweinf.) R. Fern. et A. Fern. ●☆

332639　Rhus hirta (L.) Sudw. ;毛叶盐肤木;Staghorn Sumac, Staghorn Sumach, Staghorn Velvet Sumac, Staghorn Velvet Sumach, Stagshorn Sumach, Velvet Sumac, Velvet Sumach ●☆

332640　Rhus hirta (L.) Sudw. = Rhus typhina L. ●

332641　Rhus hirta Harv. ex Engl. = Rhus harveyi Moffett ●☆

332642　Rhus hookeri Sahni et Bahadur = Toxicodendron hookeri (Sahni et Bahadur) C. Y. Wu et T. L. Ming ●

332643　Rhus horrida Eckl. et Zeyh. ;多刺盐肤木●☆

332644　Rhus humpatensis Meikle;洪帕塔盐肤木●☆

332645　Rhus humpatensis Meikle. f. subglabra R. Fern. ;近光盐肤木●☆

332646　Rhus hypoleuca Champ. ex Benth. ;白背麸杨（里白漆）; Hongkong Sumac,Lacquertree,Lacquer-tree,Sumac ●

332647　Rhus hypoleuca Champ. ex Benth. var. barbata Z. X. Yu et Q. G. Zhang;髯毛白背麸杨;Barbate Lacquertree,Barbate Sumac ●

332648　Rhus impermeabilis Dinter = Rhus pyroides Burch. var. dinteri（Engl.）Moffett ●☆

332649　Rhus inamoena Standl. ex Bullock = Rhus longipes Engl. var. elgonensis Kokwaro ●☆

332650　Rhus incana Engl. = Rhus engleri Britten ●☆

332651　Rhus incana Mill. = Rhus laevigata L. var. villosa（L. f.）R. Fern. ●☆

332652　Rhus incisa L. f. ;锐裂盐肤木●☆

332653　Rhus incisa L. f. var. effusa（C. Presl）R. Fern. ;开展锐裂盐肤木☆

332654　Rhus incisa L. f. var. obovata（Sond.）Schönland = Rhus incisa L. f. var. effusa（C. Presl）R. Fern. ●☆

332655　Rhus insignis（Delile）Oliv. var. obovata Oliv. = Ozoroa obovata（Oliv.）R. Fern. et A. Fern. ●☆

332656　Rhus insignis Hook. f. = Toxicodendron hookeri（Sahni et Bahadur）C. Y. Wu et T. L. Ming ●

332657　Rhus integerrima Wall. = Pistacia chinensis Bunge subsp. integerrima（J. L. Stewart ex Brand.）Rech. f. ●☆

332658　Rhus integrifolia Benth. et Hook. f. ex S. Watson;全缘叶盐肤木; Lemonade Berry,Lemonade Sumac,Lemonade Tree,Sourberry ●☆

332659　Rhus integrifolia Engl. = Rhus pyroides Burch. var. integrifolia（Engl.）Moffett ●☆

332660　Rhus intermedia Hayata = Toxicodendron radicans（L.）Kuntze subsp. hispidum（Engl.）Gillis ●

332661　Rhus intermedia Schönland = Rhus pyroides Burch. var. integrifolia（Engl.）Moffett ●☆

332662　Rhus javanica L. ;爪哇盐肤木（五倍树,盐肤木,盐麸子）; China Sumac,Chinese Sumac,Chinese Sumach,Japan Galls ●

332663　Rhus javanica L. = Brucea javanica（L.）Merr. ●

332664　Rhus javanica L. = Rhus chinensis Mill. ●

332665　Rhus javanica L. f. toyohashiensis（Hayashi）Satomi = Rhus javanica L. ●

332666　Rhus javanica L. var. chinensis（Mill.）T. Yamaz. = Rhus javanica L. ●

332667　Rhus javanica L. var. roxburghiana（DC.）Rehder et E. H. Wilson = Rhus chinensis Mill. var. roxburghii（DC.）Rehder ●

332668　Rhus javanica L. var. roxburghii（DC.）Rehder et E. H. Wilson = Rhus chinensis Mill. var. roxburghii（DC.）Rehder ●

332669　Rhus javanica L. var. toyohashiensis Hayashi = Rhus javanica L. ●

332670　Rhus juglandifolia Wall. = Toxicodendron wallichii（Hook. f.）Kuntze ●

332671　Rhus kaempferi Sw. = Rhus verniciflua Stokes ●

332672　Rhus keetii Schönland;克特盐肤木●☆

332673　Rhus kirkii Oliv. ;柯克盐肤木●☆

332674　Rhus kirkii Oliv. var. kwangoensis Van der Veken = Rhus kirkii Oliv. ●☆

332675　Rhus kirkii Oliv. var. polyneura（Engl. et Gilg）R. Fern. et A. Fern. = Rhus kirkii Oliv. ●☆

332676　Rhus knysniaca Schinz;克尼斯纳盐肤木●☆

332677　Rhus krebsiana C. Presl ex Engl. ;克雷布斯盐肤木●☆

332678　Rhus kwangoensis（Van der Veken）Kokwaro;宽果河盐肤木●☆

332679　Rhus kwebensis N. E. Br. Engl. ;奎波盐肤木●☆

332680　Rhus laevigata L. ;平滑盐肤木●☆

332681　Rhus laevigata L. var. atomaria（Jacq.）R. Fern. = Rhus laevigata L. ●☆

332682　Rhus laevigata L. var. latifolia（Schönland）R. Fern. = Rhus laevigata L. var. villosa（L. f.）R. Fern. ●☆

332683　Rhus laevigata L. var. villosa（L. f.）R. Fern. ;长毛盐肤木●☆

332684　Rhus lancea E. Mey. ex Harv. et Sond. = Rhus lancea L. f. ●☆

332685　Rhus lancea L. f. ;披针盐肤木;African Sumac,Karee,South African Sumac,Willow Rhus ●☆

332686　Rhus lanceolata（A. Gray）Britton;披针叶盐肤木;Lanceleaf Sumac,Prairie Flameleaf Sumac,Prairie Sumac,Texan Sumac ●☆

332687　Rhus laurina Nutt. ex Torr. et Gray;月桂盐肤木（桂叶漆）; Laurel Sumac ●☆

332688　Rhus legatii Schönland = Rhus chirindensis Baker f. ●☆

332689　Rhus leptodictya Diels;细脉盐肤木●☆

332690　Rhus leptodictya Diels. f. pilosa R. Fern. et A. Fern. ;疏毛细脉盐肤木●☆

332691　Rhus longifolia（Bernh.）Sond. = Protorhus longifolia（Bernh.）Engl. ●☆

332692　Rhus longipes Engl. ;长梗盐肤木●☆

332693　Rhus longipes Engl. var. elgonensis Kokwaro;伸长盐肤木●☆

332694　Rhus longipes Engl. var. grandifolia（Oliv.）Meikle = Rhus longipes Engl. ●☆

332695　Rhus longipes Engl. var. schinoides R. Fern. ;拟欣兹盐肤木●☆

332696　Rhus longispina Eckl. et Zeyh. ;长刺盐肤木●☆

332697　Rhus lucens Hutch. ;光亮盐肤木●☆

332698　Rhus lucida E. Mey. ex Engl. = Rhus lucida L. ●☆

332699　Rhus lucida L. ;银枝盐肤木●☆

332700　Rhus lucida L. f. elliptica（Sond.）Moffett;椭圆银枝盐肤木●☆

332701　Rhus lucida L. f. scoparia（Eckl. et Zeyh.）Moffett;帚状银枝盐肤木●☆

332702　Rhus lucida L. var. elliptica Sond. = Rhus lucida L. f. elliptica（Sond.）Moffett ●☆

332703　Rhus lucida L. var. outeniquensis（Szyszyl.）Schönland = Rhus lucida L. ●☆

332704　Rhus macowanii Schönland = Rhus rehmanniana Engl. var. glabrata（Sond.）Moffett ●☆

332705　Rhus macowanii Schönland. f. rehmanniana（Engl.）Schönland = Rhus rehmanniana Engl. ●☆

332706　Rhus macrocarpa Engl. = Rhus rosmarinifolia Vahl ●☆

332707　Rhus magalismontana Sond. ;马加利斯盐肤木●☆

332708　Rhus magalismontana Sond. subsp. coddii（R. Fern. et A. Fern.）Moffett;科德盐肤木●☆

332709　Rhus magalismontana Sond. subsp. trifoliolata（Baker f.）Moffett;三小叶盐肤木●☆

332710　Rhus mairei H. Lév. = Rhus punjabensis J. L. Stewart ex Brandis var. sinica（Diels）Rehder et E. H. Wilson ●

332711　Rhus marginata E. Mey. ;具边盐肤木●☆

332712　Rhus marlothii Engl. ;马洛斯盐肤木●☆

332713　Rhus metopia St. -Lag. ;西印度盐肤木（西印度毒漆树）●☆

332714　Rhus michauxii Sarg. ;米氏盐肤木;Michaux's Sumac ●☆

332715　Rhus microcarpa Schönland = Rhus pyroides Burch. var. integrifolia（Engl.）Moffett ●☆

332716　Rhus microphylla Engelm. ;小叶漆;Little-leaf Sumac,Squawberry ●☆

332717　Rhus montana Diels;山地盐肤木●☆

332718　Rhus montana Diels var. basutorum（Schönland）R. Fern. = Rhus gerrardii（Harv. ex Engl.）Diels ●☆

332719　Rhus montana Diels var. gerrardii（Harv. ex Engl.）R. Fern. = Rhus gerrardii（Harv. ex Engl.）Diels ●☆

332720　Rhus montana Diels var. latifolia（Schönland）R. Fern. = Rhus gerrardii（Harv. ex Engl.）Diels ●☆

332721　Rhus monticola Meikle;山生盐肤木●☆

332722　Rhus mucronata Thunb. = Rhus laevigata L. ●☆

332723　Rhus mucronata Thunb. var. atomaria（Jacq.）Schönland = Rhus laevigata L. ●☆

332724　Rhus mucronata Thunb. var. laevigata（L.）Schönland = Rhus laevigata L. ●☆

332725　Rhus mucronata Thunb. var. latifolia Schönland = Rhus laevigata L. var. villosa（L. f.）R. Fern. ●☆

332726　Rhus mucronata Thunb. var. villosa（L. f.）Schönland = Rhus laevigata L. var. villosa（L. f.）R. Fern. ●☆

332727　Rhus myriantha Baker = Rhus somalensis Engl. ●☆

332728　Rhus natalensis Bernh. ex Krauss;纳塔尔盐肤木●☆

332729　Rhus natalensis Bernh. ex Krauss var. macrocarpa（Schweinf.）Cufod.;大果纳塔尔盐肤木●☆

332730　Rhus natalensis Bernh. ex Krauss var. obovatifoliolata（Engl.）Chiov.;倒卵纳塔尔盐肤木●☆

332731　Rhus natalensis Krauss = Rhus natalensis Bernh. ex Krauss ●☆

332732　Rhus nebulosa Schönland;星云盐肤木●☆

332733　Rhus nebulosa Schönland. f. pubescens Moffett;短柔毛星云盐肤木●☆

332734　Rhus neoglutinosa M. G. Gilbert = Rhus glutinosa Hochst. ex A. Rich. subsp. neoglutinosa（M. G. Gilbert）M. G. Gilbert ●☆

332735　Rhus nitida Engl. ;亮盐肤木●☆

332736　Rhus oblanceolata Schinz = Rhus magalismontana Sond. ●☆

332737　Rhus obliqua E. Mey. = Clausena anisata（Willd.）Hook. f. ex Benth. ●☆

332738　Rhus oblongifolia E. Mey. = Deinbollia oblongifolia（E. Mey. ex Arn.）Radlk. ●☆

332739　Rhus oblongifolia E. Mey. ex Arn. = Deinbollia oblongifolia（E. Mey. ex Arn.）Radlk. ●☆

332740　Rhus obovata Sond. = Rhus incisa L. f. var. effusa（C. Presl）R. Fern. ●☆

332741　Rhus obtusata（Engl.）Meikle;钝盐肤木●☆

332742　Rhus ochracea Meikle;淡黄褐盐肤木●☆

332743　Rhus ochracea Meikle var. saxicola R. Fern. et A. Fern. ;岩地淡黄褐盐肤木●☆

332744　Rhus odina Buch. -Ham. ex Wall. = Lannea coromandelica（Houtt.）Merr. ●

332745　Rhus orientalis（Green）C. K. Schneid. = Rhus ambigua H. Lév. ex Dippel ●

332746　Rhus orientalis Schneid. ;台湾藤漆●

332747　Rhus orientalis Schneid. = Toxicodendron radicans（L.）Kuntze subsp. hispidum（Engl.）Gillis ●

332748　Rhus osbeckii Decne. ex Steud. = Rhus chinensis Mill. ●

332749　Rhus osbeckii Steud. = Rhus javanica L. ●

332750　Rhus outeniquensis Szyszyl. = Rhus lucida L. ●☆

332751　Rhus ovata S. Watson;糖盐肤木（糖漆）;Chaparral Sumac, Mountain Laurel,Sugar Bush,Sugar Sumac,Sugar Sumach,Sugarbush ●☆

332752　Rhus oxyacantha Cav. var. albida（Schousb.）Ball = Rhus albida Schousb. ●☆

332753　Rhus oxyacantha Cav. var. ballii Maire = Rhus tripartita（Ucria）Grande ●☆

332754　Rhus oxyacantha Cav. var. zizyphina（Tineo）Ball = Rhus tripartita（Ucria）Grande ●☆

332755　Rhus oxyacantha Schousb. ex Cav. ;尖花盐肤木●☆

332756　Rhus oxyacanthoides Dum. Cours. = Rhus tripartita（Ucria）Grande ●☆

332757　Rhus pallens Eckl. et Zeyh. ;苍白盐肤木●☆

332758　Rhus paniculata Wall. = Terminthia paniculata（Wall. ex G. Don）C. Y. Wu et T. L. Ming ●

332759　Rhus paniculata Wall. ex G. Don = Terminthia paniculata（Wall. ex G. Don）C. Y. Wu et T. L. Ming ●

332760　Rhus paniculosa Sond. = Ozoroa paniculosa（Sond.）R. Fern. et A. Fern. ●☆

332761　Rhus parviflora Roxb. ;小花盐肤木●☆

332762　Rhus parvifolia Harv. ex Sond. = Rhus dentata Thunb. ●☆

332763　Rhus pauciflora L. f. = Hippobromus pauciflorus（L. f.）Radlk. ■☆

332764　Rhus pendulina Jacq. = Rhus pendulina Jacq. ex Willd. ●☆

332765　Rhus pendulina Jacq. ex Willd. ;柳漆●☆

332766　Rhus pentaphylla（Jacq.）Desf. ;五叶漆树;Fiveleaf Sumac, Five-leaf Sumac,Five-leaved Sumach,Tiara ●☆

332767　Rhus pentaphylla Desf. = Rhus pentaphylla（Jacq.）Desf. ●☆

332768　Rhus pentheri Zahlbr. ;彭泰尔盐肤木●☆

332769　Rhus petitiana A. Rich. = Rhus glutinosa Hochst. ex A. Rich. subsp. abyssinica（Hochst. ex Oliv.）M. G. Gilbert ●☆

332770　Rhus plukenetiana Eckl. et Zeyh. = Rhus tomentosa L. ●☆

332771　Rhus polyneura Engl. et Gilg = Rhus kirkii Oliv. ●☆

332772　Rhus polyneura Engl. et Gilg var. hylophila ? = Rhus kirkii Oliv. ●☆

332773　Rhus pondoensis Schönland;庞多盐肤木●☆

332774　Rhus populifolia E. Mey. ex Sond. ;杨叶盐肤木●☆

332775　Rhus potaninii Maxim. ;青麸杨（百虫仓,百药煎,栲子,倍子,倍子树,波氏盐肤木,角倍,木附子,铁倍树,文蛤,五倍子）;Chinese Varnish Tree,Potanin Sumac ●

332776　Rhus problematodes Merxm. et Rössler;普罗盐肤木●☆

332777　Rhus pterota C. Presl;小翅盐肤木●☆

332778　Rhus puberula Eckl. et Zeyh. = Rhus pyroides Burch. ●☆

332779　Rhus pubescens Thunb. ;短柔毛盐肤木●☆

332780　Rhus pubescens Thunb. var. caledonica Eckl. et Zeyh. = Rhus rehmanniana Engl. var. glabrata（Sond.）Moffett ●☆

332781　Rhus pubescens Thunb. var. subglabra Eckl. et Zeyh. = Rhus laevigata L. ●☆

332782　Rhus pubescens Thunb. var. tulbaghica Eckl. et Zeyh. = Rhus laevigata L. ●☆

332783　Rhus pubescens Thunb. var. uitenhagensis Eckl. et Zeyh. = Rhus rehmanniana Engl. var. glabrata（Sond.）Moffett ●☆

332784　Rhus pubigera Blume = Rhus succedanea L. ●

332785　Rhus puccionii Chiov. ;普乔尼盐肤木●☆

332786　Rhus pulcherrima（Schweinf.）Oliv. = Ozoroa pulcherrima（Schweinf.）R. Fern. et A. Fern. ●☆

332787　Rhus pulvinata Greene;垫漆;Hybrid Sumac, Northern Sumac, Pulvinate Sumac ☆

332788　Rhus pumila Michx. ;矮漆藤●☆

332789　Rhus punjabensis J. L. Stewart ex Brandis;旁遮普麸杨;Punjab Sumac ●

332790　Rhus punjabensis J. L. Stewart ex Brandis var. pilosa Engl. ;毛叶麸杨（毛红麸杨）;Hairypunjab Sumac ●

332791　Rhus punjabensis J. L. Stewart ex Brandis var. sinica（Diels）Rehder et E. H. Wilson;红麸杨（百虫仓,百药煎,棓子,倍子树,旱倍子,角倍,木附子,漆倍子,铁麸杨,文蛤）;Redpunjab Sumac ●

332792　Rhus pygmaea Moffett;矮小盐肤木●☆

332793　Rhus pyroides Burch.;梨形盐肤木●☆

332794　Rhus pyroides Burch. var. dinteri（Engl.）Moffett;丁特梨形盐肤木●☆

332795　Rhus pyroides Burch. var. glabrata Sond. = Rhus rehmanniana Engl. var. glabrata（Sond.）Moffett ●☆

332796　Rhus pyroides Burch. var. gracilis（Engl.）Burtt Davy;纤细盐肤木●☆

332797　Rhus pyroides Burch. var. integrifolia（Engl.）Moffett;全叶盐肤木●☆

332798　Rhus pyroides Burch. var. puberula（Eckl. et Zeyh.）Schönland = Rhus pyroides Burch. ●☆

332799　Rhus pyroides Burch. var. transvaalensis Schönland = Rhus pyroides Burch. var. gracilis（Engl.）Burtt Davy ●☆

332800　Rhus quartiniana A. Rich.;夸尔廷盐肤木●☆

332801　Rhus quartiniana A. Rich. var. acutifoliolata（Engl.）Meikle = Rhus quartiniana A. Rich. ●☆

332802　Rhus quartiniana A. Rich. var. zambesiensis R. Fern. et A. Fern. = Rhus quartiniana A. Rich. ●☆

332803　Rhus radicans L.;辐射盐肤木;Poison Ivy Oak, Poison Oak, Poisonivy ●☆

332804　Rhus radicans L. = Rhus toxicodendron L. ●

332805　Rhus radicans L. = Toxicodendron radicans（L.）Kuntze ●

332806　Rhus radicans L. var. rydbergii（Small ex Rydb.）Rehder = Toxicodendron rydbergii（Small ex Rydb.）Greene ●☆

332807　Rhus radicans L. var. vulgaris（Michx.）DC. = Toxicodendron rydbergii（Small ex Rydb.）Greene ●☆

332808　Rhus radicans L. var. vulgaris（Michx.）DC. f. negundo（Greene）Fernald = Toxicodendron radicans（L.）Kuntze ●

332809　Rhus rangeana Engl. = Rhus burchellii Sond. ex Engl. ●☆

332810　Rhus refracta Eckl. et Zeyh.;反折盐肤木●☆

332811　Rhus rehmanniana Engl.;雷曼盐肤木●☆

332812　Rhus rehmanniana Engl. var. glabrata（Sond.）Moffett;光滑雷曼盐肤木●☆

332813　Rhus rehmanniana Engl. var. longecuneata R. Fern. et A. Fern. = Rhus rehmanniana Engl. ●☆

332814　Rhus rhodesiensis R. Fern. et A. Fern. = Rhus magalismontana Sond. subsp. trifoliolata（Baker f.）Moffett ●☆

332815　Rhus rhodesiensis R. Fern. et A. Fern. f. glabra ? = Rhus magalismontana Sond. subsp. trifoliolata（Baker f.）Moffett ●☆

332816　Rhus rhombocarpa R. Fern. et A. Fern. = Rhus leptodictya Diels. ●☆

332817　Rhus rigida Mill.;硬盐肤木●☆

332818　Rhus rigida Mill. var. dentata（Engl.）Moffett;尖齿硬盐肤木●☆

332819　Rhus rimosa Eckl. et Zeyh.;龟裂盐肤木●☆

332820　Rhus rogersii Schönland;罗杰斯盐肤木●☆

332821　Rhus rosmarinifolia Vahl;迷迭香叶盐肤木●☆

332822　Rhus rosmarinifolia Vahl var. brevifolia Schönland = Rhus rosmarinifolia Vahl ●☆

332823　Rhus rosmarinifolia Vahl var. stenophylla（Eckl. et Zeyh.）Schönland = Rhus stenophylla Eckl. et Zeyh. ●☆

332824　Rhus rosmarinifolia Vahl var. swellendamensis Eckl. et Zeyh. = Rhus rosmarinifolia Vahl ●☆

332825　Rhus rosmarinifolia Vahl var. typica Schönland = Rhus rosmarinifolia Vahl ●☆

332826　Rhus roxburghii Decne. ex Steud. = Rhus chinensis Mill. var. roxburghii（DC.）Rehder ●

332827　Rhus rudatisii Engl.;鲁达蒂斯盐肤木●☆

332828　Rhus rupicola J. M. Wood et M. S. Evans = Rhus rigida Mill. var. dentata（Engl.）Moffett ●☆

332829　Rhus ruspolii Engl.;鲁斯波利盐肤木●☆

332830　Rhus ruzizensis Engl. = Rhus longipes Engl. ●☆

332831　Rhus saeneb Forssk. = Debregeasia saeneb（Forssk.）Hepper et J. R. I. Wood ●

332832　Rhus salicina Sond. = Ozoroa paniculosa（Sond.）R. Fern. et A. Fern. var. salicina ? ●☆

332833　Rhus schinoides Hutch. = Rhus longipes Engl. var. schinoides R. Fern. ●☆

332834　Rhus schlechteri Diels = Rhus lucida L. f. scoparia（Eckl. et Zeyh.）Moffett ●☆

332835　Rhus schliebenii R. Fern. et A. Fern. = Rhus magalismontana Sond. subsp. coddii（R. Fern. et A. Fern.）Moffett ●☆

332836　Rhus schoenlandii Engl. = Rhus rigida Mill. ●☆

332837　Rhus scoparia Eckl. et Zeyh. = Rhus lucida L. f. scoparia（Eckl. et Zeyh.）Moffett ●☆

332838　Rhus scytophylla Eckl. et Zeyh.;革叶盐肤木●☆

332839　Rhus scytophylla Eckl. et Zeyh. var. dentata Moffett;尖齿盐肤木●☆

332840　Rhus sekhukhuniensis Moffett;塞库盐肤木●☆

332841　Rhus semialata Brandis = Rhus chinensis Mill. var. roxburghii（DC.）Rehder ●

332842　Rhus semialata Brandis f. exalata Franch. = Rhus chinensis Mill. var. roxburghii（DC.）Rehder ●

332843　Rhus semialata Murray = Rhus chinensis Mill. ●

332844　Rhus semialata Murray = Rhus javanica L. ●

332845　Rhus semialata Murray var. osbeckii DC. = Rhus chinensis Mill. ●

332846　Rhus semialata Murray var. osbeckii DC. = Rhus javanica L. ●

332847　Rhus semialata Murray var. roxburghii DC. = Rhus chinensis Mill. var. roxburghii（DC.）Rehder ●

332848　Rhus semialata Murray var. roxburghii DC. = Rhus javanica L. var. roxburghiana（DC.）Rehder et E. H. Wilson ●

332849　Rhus semialata Murray var. roxburghii DC. = Rhus javanica L. ●

332850　Rhus sempervirens Scheele;常绿漆（常绿毒漆藤）;Evergreen Sumac ●☆

332851　Rhus simii Schönland var. lydenburgensis ? = Rhus gueinzii Sond. ●☆

332852　Rhus sinica Diels = Rhus punjabensis J. L. Stewart ex Brandis var. sinica（Diels）Rehder et E. H. Wilson ●

332853　Rhus sinica Koehne = Rhus potaninii Maxim. ●

332854　Rhus sinuata Eckl. et Zeyh. var. effusa C. Presl = Rhus incisa L. f. var. effusa（C. Presl）R. Fern. ●☆

332855　Rhus somalensis Engl.;索马里盐肤木●☆

332856　Rhus sonderi Engl. var. glaberrima ? = Rhus dentata Thunb. ●☆

332857　Rhus sonderi Engl. var. pilosissima ? = Rhus dentata Thunb. ●☆

332858　Rhus sordida Meikle = Rhus arenaria Engl. ●☆

332859　Rhus spicata Thunb. = Allophylus decipiens（Sond.）Radlk. ●☆

332860　Rhus spinescens Diels = Rhus gueinzii Sond. ●☆

332861　Rhus steingroeveri Engl. = Rhus populifolia E. Mey. ex Sond. ●☆

332862　Rhus stenophylla Eckl. et Zeyh.;狭叶盐肤木●☆

332863　Rhus steudneri Engl.;斯托德盐肤木●☆

332864　Rhus stolzii Engl. = Rhus quartiniana A. Rich. ●☆

332865　Rhus succedanea L.；多汁盐肤木；Japan Wax Tree，Japanese Tallow Tree，Japanese Wax，Japanese Wax Tree ●☆

332866　Rhus succedanea L. ＝ Toxicodendron succedaneum（L.）Kuntze ●

332867　Rhus succedanea L. var. acuminata（DC.）Hook. f. ＝ Rhus succedanea L. ●

332868　Rhus succedanea L. var. acuminata（DC.）Hook. f. ＝ Toxicodendron acuminatum（DC.）C. Y. Wu et T. L. Ming ●

332869　Rhus succedanea L. var. dumontieri（Pierre）Kudo et Matsum.；安南漆树(南南漆)●☆

332870　Rhus succedanea L. var. himalaica Hook. f. ＝ Rhus verniciflua Stokes ●

332871　Rhus succedanea L. var. himalaica Hook. f. ＝ Toxicodendron vernicifluum（Stokes）F. A. Barkley ●

332872　Rhus succedanea L. var. japonica Engl. ＝ Rhus succedanea L. ●

332873　Rhus succedanea L. var. japonica Engl. ＝ Toxicodendron succedaneum（L.）Kuntze ●

332874　Rhus succedanea L. var. longipes Franch. ＝ Toxicodendron grandiflorum C. Y. Wu et T. L. Ming var. longipes（Franch.）C. Y. Wu et T. L. Ming ●

332875　Rhus succedanea L. var. sikkimensis ？＝ Rhus succedanea L. ●☆

332876　Rhus succedanea L. var. silvestrii Pamp. ＝ Toxicodendron vernicifluum（Stokes）F. A. Barkley ●

332877　Rhus suffruticosa Meikle ＝ Rhus anchietae Ficalho ex Hiern. f. suffruticosa（Meikle）R. Fern. et A. Fern. ●☆

332878　Rhus sylvestris Siebold et Zucc. ＝ Toxicodendron sylvestre（Siebold et Zucc.）Kuntze ●

332879　Rhus synstylica R. Fern. et A. Fern. ＝ Rhus tumulicola S. Moore ●☆

332880　Rhus synstylica R. Fern. et A. Fern. var. meeuseana R. Fern. et A. Fern. ＝ Rhus tumulicola S. Moore f. meeuseana（R. Fern. et A. Fern.）Moffett ●☆

332881　Rhus taishanensis S. B. Liang；泰山盐肤木；Taishan Sumac ●

332882　Rhus taitensis Guillaumin；泰特盐肤木●☆

332883　Rhus teniana Hand. -Mazz.；滇麸杨；Yunnan Sumac ●

332884　Rhus tenuinervis Engl.；纤脉盐肤木●☆

332885　Rhus tenuipes R. Fern. et A. Fern.；瘦梗盐肤木●☆

332886　Rhus thunbergii Hook. ＝ Heeria argentea（Thunb.）Meisn. ●☆

332887　Rhus tomentosa L.；绒毛盐肤木●☆

332888　Rhus toxicarium Salisb.；箭毒盐肤木●☆

332889　Rhus toxicodendron L. ＝ Toxicodendron pubescens Mill. ●

332890　Rhus toxicodendron L. ＝ Toxicodendron radicans（L.）Kuntze ●

332891　Rhus toxicodendron L. f. radicans McNair ＝ Toxicodendron radicans（L.）Kuntze ●

332892　Rhus toxicodendron L. subsp. radicans（L.）R. T. Clausen ＝ Toxicodendron radicans（L.）Kuntze ●

332893　Rhus toxicodendron L. var. hispida Engl. ＝ Rhus ambigua H. Lév. ex Dippel ●

332894　Rhus toxicodendron L. var. hispida Engl. ＝ Toxicodendron radicans（L.）Kuntze subsp. hispidum（Engl.）Gillis ●

332895　Rhus toxicodendron L. var. radicans（L.）Torr. ＝ Toxicodendron radicans（L.）Kuntze ●

332896　Rhus toxicodendron L. var. vulgaris Michx. ＝ Toxicodendron rydbergii（Small ex Rydb.）Greene ●☆

332897　Rhus transvaalensis Engl.；德兰士瓦盐肤木●☆

332898　Rhus trichocarpa Miq. ＝ Toxicodendron trichocarpum（Miq.）Kuntze ●

332899　Rhus trichocarpa Miq. f. viridis H. Hara ＝ Rhus trichocarpa Miq. ●

332900　Rhus trichocarpa Miq. var. humilis Honda ＝ Toxicodendron trichocarpum（Miq.）Kuntze ●

332901　Rhus trichocarpa Miq. var. serrata Engl.；齿叶毛漆树●☆

332902　Rhus trichocarpa Miq. var. serrata Engl. ＝ Rhus trichocarpa Miq. ●

332903　Rhus trichocarpa Miq. var. virescens ？＝ Rhus trichocarpa Miq. ●

332904　Rhus trichocarpa Miq. var. viridis ？＝ Rhus trichocarpa Miq. ●

332905　Rhus tridactyla Burch.；三指盐肤木●☆

332906　Rhus tridentata L. f. ＝ Rhoicissus tridentata（L. f.）Wild et R. B. Drumm. ●☆

332907　Rhus trifoliolata Baker f. ＝ Rhus magalismontana Sond. subsp. trifoliolata（Baker f.）Moffett ●☆

332908　Rhus trilobata Nutt.；三裂漆树(三裂漆木)；Aromatic Sumac，Hi-scented Sumach，Ill-scented Sumach，Lemonade Berry，Lemonade Sumach，Limitas，Polecat Bush，Skunk Bush，Skunk Sumash，Skunkbush，Skunkbush Sumac，Squaw Bush，Squaw-berry，Squawbush，Squaw-bush，Squawbush Sumac，Three-leaf Sumach，Three-lobes Sumac ●☆

332909　Rhus trilobata Nutt. sensu Fassett ＝ Rhus aromatica Aiton var. serotina（Greene）Rehder ●☆

332910　Rhus trilobata Nutt. var. serotina（Greene）F. A. Barkley ＝ Rhus aromatica Aiton var. serotina（Greene）Rehder ●☆

332911　Rhus tripartita（Ucria）Grande；三深裂盐肤木●☆

332912　Rhus truncata Schinz ＝ Rhus rigida Mill. var. dentata（Engl.）Moffett ●☆

332913　Rhus tumulicola S. Moore；小山盐肤木●☆

332914　Rhus tumulicola S. Moore. f. meeuseana（R. Fern. et A. Fern.）Moffett；梅斯盐肤木●☆

332915　Rhus tumulicola S. Moore. f. pumila Moffett；矮盐肤木●☆

332916　Rhus typhina L.；火炬树(鹿角漆树，美国漆树)；Buck's Horn Tree，Lemonade Tree，Stag's Horn Sumac，Stag's Horn Sumach，Staghorn Sumac，Staghorn Sumach，Stag's-horn Sumach，Sumac，Velvet Sumac，Velvet Sumach，Vinegar Plant，Vinegar-tree，Virginia Sumach，Virginian Sumac ●

332917　Rhus typhina L. ' Laciniata '；条裂火炬树(细裂火炬树)；Cutleaf Staghorn Sumac，Cut-leaf Staghorn Sumac ●☆

332918　Rhus typhina L. ＝ Rhus hirta（L.）Sudw. ●

332919　Rhus typhina L. f. dissecta Rehder；多裂火炬树(深裂叶火炬树，羽裂火炬树)；Shred-leaf Staghorn Sumac ●☆

332920　Rhus typhina L. f. dissecta Rehder ＝ Rhus hirta（L.）Sudw. ●

332921　Rhus typhina L. f. dissecta Rehder ＝ Rhus typhina L. ●

332922　Rhus typhina L. f. viridiflora Duhamel；绿花火炬树；Green Staghorn Sumac ●☆

332923　Rhus typhina L. var. laciniata A. W. Wood ＝ Rhus hirta（L.）Sudw. ●☆

332924　Rhus typhina L. var. laciniata A. W. Wood ＝ Rhus typhina L. ' Laciniata ' ●☆

332925　Rhus tysonii E. Phillips ＝ Rhus rigida Mill. var. dentata（Engl.）Moffett ●☆

332926　Rhus undulata Jacq.；波状盐肤木●☆

332927　Rhus undulata Jacq. f. contracta（Schönland）R. Fern. ＝ Rhus undulata Jacq. ●☆

332928　Rhus undulata Jacq. f. contracta Schönland ＝ Rhus undulata Jacq. ●☆

332929　Rhus undulata Jacq. f. excisa（Thunb.）R. Fern. ＝ Rhus undulata Jacq. ●☆

332930　Rhus undulata Jacq. f. excisa（Thunb.）Schönland ＝ Rhus undulata Jacq. ●☆

332931　Rhus undulata Jacq. var. burchellii（Sond. ex Engl.）Schönland ＝ Rhus burchellii Sond. ex Engl. ●☆

332932　Rhus undulata Jacq. var. celastroides（Sond.）Schönland ＝ Rhus undulata Jacq. ●☆

332933　Rhus undulata Jacq. var. tricrenata（Engl.）R. Fern. ＝ Rhus burchellii Sond. ex Engl. ●☆

332934　Rhus upingtoniae Dinter ＝ Rhus marlothii Engl. ●☆

332935　Rhus vernicifera DC. ＝ Toxicodendron vernicifluum（Stokes）F. A. Barkley ●

332936　Rhus vernicifera DC. ＝ Toxicodendron wallichii（Hook. f.）Kuntze ●

332937　Rhus vernicifera DC. var. silvestri ？ ＝ Toxicodendron vernicifluum（Stokes）F. A. Barkley ●

332938　Rhus verniciflua Stokes；光泽盐肤木（漆树）；Chinese Lacquer，Japan Lacquer，Japanese Lacquer ●

332939　Rhus verniciflua Stokes ＝ Toxicodendron vernicifluum（Stokes）F. A. Barkley ●

332940　Rhus verniciflua Stokes ＝ Toxicodendron wallichii（Hook. f.）Kuntze ●

332941　Rhus vernix L.；漆盐肤木；Poison Elder ●☆

332942　Rhus vernix L. ＝ Rhus sylvestris Siebold et Zucc. ●

332943　Rhus vernix L. ＝ Toxicodendron vernicifluum（Stokes）F. A. Barkley ●

332944　Rhus vernix L. ＝ Toxicodendron vernix（L.）Kuntze ●

332945　Rhus villosa L. f. ＝ Rhus laevigata L. var. villosa（L. f.）R. Fern. ●☆

332946　Rhus villosa L. f. f. acutifoliolata Engl. ＝ Rhus quartiniana A. Rich. ●☆

332947　Rhus villosa L. f. f. obtusifoliolata Engl. ＝ Rhus natalensis Bernh. ex Krauss var. obovatifoliolata（Engl.）Chiov. ●☆

332948　Rhus villosa L. f. var. gracilis Engl. ＝ Rhus pyroides Burch. var. gracilis（Engl.）Burtt Davy ●☆

332949　Rhus villosa L. f. var. grandifolia Oliv. ＝ Rhus longipes Engl. ●☆

332950　Rhus villosa L. f. var. optusata Engl. ＝ Rhus obtusata（Engl.）Meikle ●☆

332951　Rhus villosa L. f. var. tomentosa R. E. Fr. ＝ Rhus ochracea Meikle ●☆

332952　Rhus villosissima Engl. ＝ Rhus discolor E. Mey. ex Sond. ●☆

332953　Rhus viminalis Aiton ＝ Rhus lancea L. f. ●☆

332954　Rhus viminalis Vahl ＝ Rhus laevigata L. ●☆

332955　Rhus viminalis Vahl var. gerrardii Harv. ex Engl. ＝ Rhus gerrardii（Harv. ex Engl.）Diels ●☆

332956　Rhus virens Lindh. ex A. Gray；绿盐肤木（绿漆）；Evergreen Sumac，Lentisco，Tobacco Sumac ●☆

332957　Rhus virens Lindh. ex A. Gray var. choriophylla（Wooten et Standl.）L. D. Benson；绿叶盐肤木；Evergreen Sumac ●☆

332958　Rhus virgata Hiern；条纹盐肤木 ●☆

332959　Rhus virtix L.；毒盐肤木（毒漆）；Poison Sumac ●☆

332960　Rhus volkii Suess.；福尔克盐肤木 ●☆

332961　Rhus vulgaris Meikle；掌漆树 ●☆

332962　Rhus vulgaris Meikle ＝ Rhus pyroides Burch. ●☆

332963　Rhus wallichii Hook. f. ＝ Toxicodendron wallichii（Hook. f.）Kuntze ●

332964　Rhus wellmanii Engl.；韦尔曼盐肤木 ●☆

332965　Rhus welwitschii Engl. ＝ Rhus kirkii Oliv. ●☆

332966　Rhus welwitschii Engl. var. angustifoliolata Baker f. ＝ Rhus kirkii Oliv. ●☆

332967　Rhus wildii R. Fern. et A. Fern.；韦尔德盐肤木 ●☆

332968　Rhus wilmsii Diels；维尔姆斯盐肤木 ●☆

332969　Rhus wilsonii Hemsl.；川麸杨；E. H. Wilson Sumac, Sichuan Sumac, Wilson Sumac ●

332970　Rhus wilsonii Hemsl. var. glabra Y. T. Wu；无毛川麸杨（无毛叶杨）；Smooth-leaf Sichuan Sumac ●

332971　Rhus zeyheri Sond.；泽赫盐肤木 ●☆

332972　Rhus zeyheri Sond. var. dentata Engl. ＝ Rhus rigida Mill. var. dentata（Engl.）Moffett ●☆

332973　Rhus zeyheri Sond. var. parvifolia Burtt Davy ＝ Rhus zeyheri Sond. ●☆

332974　Rhus ziziphina Tineo ＝ Rhus oxyacantha Cav. var. zizyphina（Tineo）Ball ●☆

332975　Rhuyschiana Adans. ＝ Dracocephalum L.（保留属名）■●

332976　Rhyacophila Hochst. ＝ Quartinia Endl. ■

332977　Rhyacophila Hochst. ＝ Rotala L. ■

332978　Rhyacophila repens Hochst. ＝ Rotala repens（Hochst.）Koehne ■☆

332979　Rhyanspermum C. F. Gaertn. ＝ Notelaea Vent. ■☆

332980　Rhyditospermum Walp. ＝ Matricaria L. ■

332981　Rhyditospermum Walp. ＝ Rhytidospermum Sch. Bip. ■

332982　Rhynchadenia A. Rich. ＝ Macradenia R. Br. ■☆

332983　Rhynchandra Rchb. ＝ Rhynchanthera DC.（保留属名）●☆

332984　Rhynchandra Rchb. f. ＝ Corymborkis Thouars ■

332985　Rhynchanthera Blume ＝ Corymborkis Thouars ■

332986　Rhynchanthera DC.（1828）（保留属名）；喙药野牡丹属（喙花牡丹属）●☆

332987　Rhynchanthera grandiflora（Aubl.）DC.；喙药野牡丹（喙花牡丹）●☆

332988　Rhynchanthus Hook. f.（1886）；喙花姜属；Beakflower ■

332989　Rhynchanthus beesianus W. W. Sm.；喙花姜（滇高良姜，岩姜）；Bees Beakflower ■

332990　Rhyncharrhena F. Muell.（1859）；雄喙萝藦属 ■☆

332991　Rhyncharrhena F. Muell. ＝ Pentatropis R. Br. ex Wight et Arn. ■☆

332992　Rhyncharrhena linearis（Decne.）K. L. Wilson；雄喙萝藦 ■☆

332993　Rhynchelytrum Nees ＝ Melinis P. Beauv. ■

332994　Rhynchelytrum Nees ＝ Rhynchelytrum Nees ■

332995　Rhynchelytrum Nees ＝ Melinis P. Beauv. ■

332996　Rhynchelytrum Nees（1836）；红毛草属；Redhairgrass, Rhynchelytrum ■

332997　Rhynchelytrum amethysteum（Franch.）Chiov. ＝ Melinis amethystea（Franch.）Zizka ■☆

332998　Rhynchelytrum ascendens（Mez）Stapf et C. E. Hubb. ＝ Melinis ascendens Mez ■☆

332999　Rhynchelytrum bellespicatum（Rendle）Stapf et C. E. Hubb. ＝ Melinis longiseta（A. Rich.）Zizka subsp. bellespicata（Rendle）Zizka ■☆

333000　Rhynchelytrum bequaertii Robyns ＝ Melinis ambigua Hack. subsp. longicauda（Mez）Zizka ■☆

333001　Rhynchelytrum brevipilum（Hack.）Chiov. ＝ Melinis repens（Willd.）Zizka subsp. grandiflora（Hochst.）Zizka ■☆

333002　Rhynchelytrum catangense Chiov. ＝ Melinis amethystea（Franch.）Zizka ■☆

333003　Rhynchelytrum costatum Stapf et C. E. Hubb. ＝ Melinis repens（Willd.）Zizka subsp. grandiflora（Hochst.）Zizka ■☆

333004　Rhynchelytrum denudatum（Mez）Stapf et C. E. Hubb. ＝

Melinis subglabra Mez ■☆

333005　Rhynchelytrum drakensbergense C. E. Hubb. et Schweick. = Melinis drakensbergensis（C. E. Hubb. et Schweick.）Clayton ■☆

333006　Rhynchelytrum dregeanum Nees；藤红毛草；Dregea Rhynchelytrum ■

333007　Rhynchelytrum dregeanum Nees = Melinis repens（Willd.）Zizka ■

333008　Rhynchelytrum dregeanum Nees var. annuum Chiov.；一年红毛草■☆

333009　Rhynchelytrum dregeanum Nees var. intermedium Chiov.；间型红毛草■☆

333010　Rhynchelytrum eylesii Stapf et C. E. Hubb. = Melinis subglabra Mez ■☆

333011　Rhynchelytrum filifolium（Franch.）Stapf et C. E. Hubb. = Melinis nerviglumis（Franch.）Zizka ■☆

333012　Rhynchelytrum gossweileri Stapf et C. E. Hubb. = Melinis repens（Willd.）Zizka ■

333013　Rhynchelytrum grandiflorum Hochst. = Melinis repens（Willd.）Zizka subsp. grandiflora（Hochst.）Zizka ■☆

333014　Rhynchelytrum longicaudum（Mez）Chiov. = Melinis ambigua Hack. subsp. longicauda（Mez）Zizka ■☆

333015　Rhynchelytrum longisetum（A. Rich.）Stapf et C. E. Hubb. = Melinis longiseta（A. Rich.）Zizka ☆

333016　Rhynchelytrum merkeri（Mez）Stapf et C. E. Hubb. = Melinis subglabra Mez ■☆

333017　Rhynchelytrum microstachyum Balf. f. = Melinis repens（Willd.）Zizka subsp. grandiflora（Hochst.）Zizka ■☆

333018　Rhynchelytrum minutiflorum（Rendle）Stapf et C. E. Hubb. = Melinis longiseta（A. Rich.）Zizka ■☆

333019　Rhynchelytrum minutiflorum（Rendle）Stapf et C. E. Hubb. var. melinoides（Stent）Stapf et C. E. Hubb. = Melinis longiseta（A. Rich.）Zizka ■☆

333020　Rhynchelytrum nerviglume（Franch.）Chiov. = Melinis nerviglumis（Franch.）Zizka ■☆

333021　Rhynchelytrum nigricans（Mez）Stapf et C. E. Hubb. = Melinis repens（Willd.）Zizka subsp. nigricans（Mez）Zizka ■

333022　Rhynchelytrum nyassanum（Mez）Stapf et C. E. Hubb. = Melinis nerviglumis（Franch.）Zizka ■☆

333023　Rhynchelytrum ramosum Stapf et C. E. Hubb. = Melinis nerviglumis（Franch.）Zizka ■☆

333024　Rhynchelytrum repens（Willd.）C. E. Hubb.；红毛草（红茅草）；Creeping Redhairgrass, Creeping Rhynchelytrum, Natal Grass, Natal-grass, Ruby-grass ■

333025　Rhynchelytrum repens（Willd.）C. E. Hubb. = Melinis repens（Willd.）Zizka ■

333026　Rhynchelytrum reynaudioides C. E. Hubb.；丝形草状红毛草■☆

333027　Rhynchelytrum rhodesianum（Rendle）Stapf et C. E. Hubb. = Melinis nerviglumis（Franch.）Zizka ■☆

333028　Rhynchelytrum roseum（Nees）Stapf et C. E. Hubb. = Melinis repens（Willd.）Zizka ■

333029　Rhynchelytrum ruficomum Hochst. ex Steud.；浅红红毛草■☆

333030　Rhynchelytrum rupicola（Rendle）Stapf = Melinis rupicola（Rendle）Zizka ■☆

333031　Rhynchelytrum scabridum（K. Schum.）Chiov. = Melinis scabrida（K. Schum.）Hack. ■☆

333032　Rhynchelytrum setifolium（Stapf）Chiov. = Melinis nerviglumis（Franch.）Zizka ■☆

333033　Rhynchelytrum shantzii Stapf et C. E. Hubb. = Melinis ambigua Hack. subsp. longicauda（Mez）Zizka ■☆

333034　Rhynchelytrum stolzii（Mez）Stapf et C. E. Hubb. = Melinis repens（Willd.）Zizka ■

333035　Rhynchelytrum stuposum Stapf et C. E. Hubb. = Melinis nerviglumis（Franch.）Zizka ■☆

333036　Rhynchelytrum suberostratum Stapf et C. E. Hubb. = Melinis repens（Willd.）Zizka subsp. grandiflora（Hochst.）Zizka ■☆

333037　Rhynchelytrum subglabrum（Mez）Stapf et C. E. Hubb. = Melinis subglabra Mez ■☆

333038　Rhynchelytrum tanatrichum（Rendle）Stapf et C. E. Hubb. = Melinis tanatricha（Rendle）Zizka ■☆

333039　Rhynchelytrum tomentosum（Rendle）Stapf et C. E. Hubb. = Melinis tomentosa Rendle ■☆

333040　Rhynchelytrum villosum（Parl.）Chiov. = Melinis repens（Willd.）Zizka subsp. grandiflora（Hochst.）Zizka ■☆

333041　Rhynchelytrum villosum（Parl.）Chiov. = Rhynchelytrum repens（Willd.）C. E. Hubb. ■

333042　Rhynchelytrum welwitschii（Rendle）Stapf et C. E. Hubb. = Melinis welwitschii Rendle ■☆

333043　Rhynchelytrum wightii（Nees et Arn. ex Steud.）Duthie = Rhynchelytrum repens（Willd.）C. E. Hubb. ■

333044　Rhynchium Dulac = Vicia L. ■

333045　Rhynchocalycaceae L. A. S. Johnson et B. G. Briggs = Crypteroniaceae A. DC.（保留科名）●☆

333046　Rhynchocalycaceae L. A. S. Johnson et B. G. Briggs = Lythraceae J. St. -Hil.（保留科名）●●

333047　Rhynchocalycaceae L. A. S. Johnson et B. G. Briggs（1985）；喙萼花科●☆

333048　Rhynchocalyx Oliv.（1894）；喙萼花属●☆

333049　Rhynchocalyx lawsonioides Oliv.；喙萼花■☆

333050　Rhynchocarpa Backer ex K. Heyne = Dansara Steenis ●☆

333051　Rhynchocarpa Backer ex K. Heyne = Dialium L. ●☆

333052　Rhynchocarpa Becc. = Burretiokentia Pic. Serm. ●☆

333053　Rhynchocarpa Endl. = Kedrostis Medik. ■☆

333054　Rhynchocarpa Schrad. = Kedrostis Medik. ■☆

333055　Rhynchocarpa Schrad. ex Endl. = Kedrostis Medik. ■☆

333056　Rhynchocarpa africana Asch. = Kedrostis africana（L.）Cogn. ■☆

333057　Rhynchocarpa bainesii Hook. f. = Corallocarpus bainesii（Hook. f.）A. Meeuse ■☆

333058　Rhynchocarpa corallina Naudin = Corallocarpus epigaeus（Rottler）Hook. f. ex C. B. Clarke ■☆

333059　Rhynchocarpa courbonii Naudin = Corallocarpus schimperi（Naudin）Hook. f. ■☆

333060　Rhynchocarpa dissecta Naudin = Kedrostis africana（L.）Cogn. ■☆

333061　Rhynchocarpa ehrenbergii Schweinf. = Corallocarpus schimperi（Naudin）Hook. f. ■☆

333062　Rhynchocarpa epigaea（Rottler）Naudin = Corallocarpus epigaeus（Rottler）Hook. f. ex C. B. Clarke ■☆

333063　Rhynchocarpa erostris Schweinf. = Corallocarpus schimperi（Naudin）Hook. f. ■☆

333064　Rhynchocarpa foetida C. B. Clarke = Kedrostis foetidissima（Jacq.）Cogn. ■☆

333065　Rhynchocarpa foetida Schrad. = Kedrostis foetidissima（Jacq.）Cogn. ■☆

333066　Rhynchocarpa hirtella Naudin = Kedrostis hirtella（Naudin）

Cogn. ■☆

333067　Rhynchocarpa pedunculosa Naudin ＝ Corallocarpus schimperi （Naudin）Hook. f. ■☆

333068　Rhynchocarpa schimperi Naudin ＝ Corallocarpus schimperi （Naudin）Hook. f. ■☆

333069　Rhynchocarpa welwitschii Naudin ＝ Corallocarpus welwitschii （Naudin）Hook. f. ex Welw. ■☆

333070　Rhynchocarpus Less. ＝ Relhania L'Hér.（保留属名）●☆

333071　Rhynchocarpus Reinw. ex Blume ＝ Cyrtandra J. R. Forst. et G. Forst. ●■

333072　Rhynchocladium T. Koyama（1972）；委内瑞拉莎草属■☆

333073　Rhynchocladium steyermarkii（T. Koyama）T. Koyama；委内瑞拉莎草■☆

333074　Rhynchocorys Griseb.（1844）（保留属名）；伊朗参属■☆

333075　Rhynchocorys elephas（L.）Griseb.；伊朗参■☆

333076　Rhynchodia Benth.（1876）；尖子藤属（尖种藤属）；Beakseedvine，Rhynchodia ●■

333077　Rhynchodia Benth. ＝ Chonemorpha G. Don（保留属名）●

333078　Rhynchodia rhynchosperma（Wall.）K. Schum.；尖子藤（尖种藤）；Beak-seed Rhynchodia，Beak-seeded Chonemorpha，Beakseedvine ●

333079　Rhynchodia rhynchosperma（Wall.）K. Schum. ＝ Chonemorpha verrucosa（Blume）D. J. Middleton ●

333080　Rhynchodia verrucosa（Blume）Woods. ＝ Chonemorpha verrucosa（Blume）D. J. Middleton ●

333081　Rhynchodium C. Presl ＝ Bituminaria Heist. ex Fabr. ■☆

333082　Rhynchodium C. Presl ＝ Psoralea L. ●■

333083　Rhynchoglossum Blume（1826）（'Rhinchoglossum'）（保留属名）；尖舌苣苔属（尖舌草属，歪冠苣苔属）；Rhynchoglossum ■

333084　Rhynchoglossum hologlossum Hayata ＝ Rhynchoglossum obliquum Blume var. hologlosum（Hayata）W. T. Wang ■

333085　Rhynchoglossum hologlossum Hayata ＝ Rhynchoglossum obliquum Blume ■

333086　Rhynchoglossum obliquum Blume；尖舌苣苔（半边脸，大脖子药，尖舌草，歪冠苦苣苔，一串珍珠）；Oblique Rhynchoglossum ■

333087　Rhynchoglossum obliquum Blume f. albiflorum Kuntze ＝ Rhynchoglossum obliquum Blume ■

333088　Rhynchoglossum obliquum Blume f. coeruleum Kuntze ＝ Rhynchoglossum obliquum Blume ■

333089　Rhynchoglossum obliquum Blume var. hologlosum（Hayata）W. T. Wang；全唇尖舌苣苔（尖舌草）；Hololip Oblique Rhynchoglossum ■

333090　Rhynchoglossum obliquum Blume var. hologlossum（Hayata）W. T. Wang ＝ Rhynchoglossum obliquum Blume ■

333091　Rhynchoglossum obliquum Blume var. parviflorum C. B. Clarke ＝ Rhynchoglossum obliquum Blume ■

333092　Rhynchoglossum omeiense W. T. Wang；峨眉尖舌苣苔；Emei Rhynchoglossum ■

333093　Rhynchoglossum sasakii Hayata ＝ Whytockia sasakii（Hayata）B. L. Burtt ■

333094　Rhynchoglossum zeylanicum Hook. ＝ Rhynchoglossum obliquum Blume ■

333095　Rhynchogyna Seidenf. et Garay（1973）；雌喙兰属■☆

333096　Rhynchogyna luisifolia（Ridl.）Seidenf. et Garay；雌喙兰■☆

333097　Rhyncholacis Tul.（1849）；空喙苔草属■☆

333098　Rhyncholacis macrocarpa Tul.；大果空喙苔草■☆

333099　Rhyncholacis oligandra Wedd.；寡蕊空喙苔草■☆

333100　Rhyncholacis tenuifolia Spruce ex Wedd.；细叶空喙苔草■☆

333101　Rhyncholaelia Schltr.（1918）；喙果兰属（林可蕾利亚属）●☆

333102　Rhyncholaelia digbyana（Lindl.）Schltr.；喙果兰（巴索拉兰，大猪哥，流苏兰）；Common Brassavola ■☆

333103　Rhyncholaelia digbyana（Lindl.）Schltr. ＝ Brassavola digbyana Lindl. ■☆

333104　Rhyncholaelia digbyana Schltr. ＝ Brassavola digbyana Lindl. ■☆

333105　Rhyncholepis Miq. ＝ Piper L. ●■

333106　Rhyncholepsis C. DC. ＝ Rhyncholepis Miq. ●■

333107　Rhynchopappus Dulac ＝ Crepis L. ■

333108　Rhynchopera Boruer ＝ Carex L. ■

333109　Rhynchopera Klotzsch ＝ Pleurothallis R. Br. ■☆

333110　Rhynchopetalum Fresen. ＝ Lobelia L. ●■

333111　Rhynchopetalum montanum Fresen. ＝ Lobelia rhynchopetalum （Hochst. ex A. Rich.）Hemsl. ■☆

333112　Rhynchophora Arènes（1946）；喙梗木属●☆

333113　Rhynchophora humbertii Arènes；喙梗木●☆

333114　Rhynchophora phillipsonii W. R. Anderson；菲氏喙梗木●☆

333115　Rhynchophorum（Miq.）Small ＝ Peperomia Ruiz et Pav. ■

333116　Rhynchophorum floridanum Small ＝ Peperomia obtusifolia（L.）A. Dietr. ■☆

333117　Rhynchophorum obtusifolium（L.）Small ＝ Peperomia obtusifolia（L.）A. Dietr. ■☆

333118　Rhynchophorum spathulifolium（Small）Small ＝ Peperomia magnoliifolia（Jacq.）A. Dietr. ■☆

333119　Rhynchophreatia Schltr.（1921）；喙馥兰属■☆

333120　Rhynchophreatia angustifolia Schltr.；窄叶喙馥兰■☆

333121　Rhynchophreatia micrantha（A. Rich.）N. Hallé；小花喙馥兰■☆

333122　Rhynchopsidium DC.（1836）；棕苞金绒草属■☆

333123　Rhynchopsidium DC. ＝ Relhania L'Hér.（保留属名）●☆

333124　Rhynchopsidium pedunculatum DC. ＝ Rhynchopsidium pumilum（L. f.）DC. ■☆

333125　Rhynchopsidium pumilum（L. f.）DC.；小棕苞金绒草■☆

333126　Rhynchopsidium sessiliflorum（L. f.）DC.；无梗棕苞金绒草■☆

333127　Rhynchopyle Engl. ＝ Piptospatha N. E. Br. ■

333128　Rhynchoryza Baill.（1893）；多年稻属■☆

333129　Rhynchoryza Baill. ＝ Oryza L. ■

333130　Rhynchoryza subulata（Nees）Baill.；多年稻■☆

333131　Rhynchosia Lour.（1790）（保留属名）；鹿藿属（括根属）；Rhynchosia ●■

333132　Rhynchosia acuminata Eckl. et Zeyh. ＝ Rhynchosia caribaea （Jacq.）DC. ■☆

333133　Rhynchosia acuminatifolia Makino；渐尖叶鹿藿；Acuminateleaf Rhynchosia ■

333134　Rhynchosia acuminatissima Miq.；密果鹿藿；Fruitful Rhynchosia，Most-acuminate Rhynchosia ■

333135　Rhynchosia adenodes Eckl. et Zeyh.；腺鹿藿■☆

333136　Rhynchosia adenodes Eckl. et Zeyh. var. cooperi Baker f. ＝ Rhynchosia cooperi（Baker f.）Burtt Davy ■☆

333137　Rhynchosia affinis De Wild. ＝ Rhynchosia insignis（O. Hoffm.）R. E. Fr. subsp. affinis（De Wild.）L. Pauwels ■☆

333138　Rhynchosia airica Miré et H. Gillet ＝ Rhynchosia totta （Thunb.）DC. var. fenchelii Schinz ■☆

333139　Rhynchosia albiflora（Sims）Alston ＝ Rhynchosia hirta （Andréws）Meikle et Verdc. ■☆

333140　Rhynchosia albissima Gand.；白鹿藿■☆

333141　Rhynchosia albomarginata Chiov. ＝ Rhynchosia albissima Gand. ■☆

333142 Rhynchosia alluaudii Sacleux;阿吕鹿藿■☆
333143 Rhynchosia amatymbica Eckl. et Zeyh. = Rhynchosia adenodes Eckl. et Zeyh. ■☆
333144 Rhynchosia ambacensis (Hiern) K. Schum.;安巴卡鹿藿■☆
333145 Rhynchosia ambacensis (Hiern) K. Schum. subsp. cameroonensis Verdc.;喀麦隆鹿藿■☆
333146 Rhynchosia androyensis Du Puy et Labat;安德罗鹿藿■☆
333147 Rhynchosia angulosa Schinz;棱角鹿藿■☆
333148 Rhynchosia angustifolia (Jacq.) DC.;窄叶鹿藿■☆
333149 Rhynchosia antennulifera Baker = Eminia antennulifera (Baker) Taub. ■☆
333150 Rhynchosia arenaria Blatt. et Hallb. = Rhynchosia schimperi Hochst. ex Boiss. ■☆
333151 Rhynchosia argentea (Thunb.) Harv.;银白鹿藿■☆
333152 Rhynchosia argyi H. Lév. = Glycine soja Siebold et Zucc. ■
333153 Rhynchosia arida C. H. Stirt.;旱生鹿藿■☆
333154 Rhynchosia atropurpurea Germish.;暗紫鹿藿■☆
333155 Rhynchosia aureovillosa Hauman = Rhynchosia ferruginea A. Rich. ■☆
333156 Rhynchosia aureovillosa Hauman var. humbertii ? = Rhynchosia ferruginea A. Rich. ■☆
333157 Rhynchosia axilliflora Hauman;腋花鹿藿■☆
333158 Rhynchosia barbertonensis C. H. Stirt.;巴伯顿鹿藿■☆
333159 Rhynchosia baukea Du Puy et Labat;鲍克鹿藿■☆
333160 Rhynchosia baumii Harms;鲍姆鹿藿■☆
333161 Rhynchosia benguellensis Torre = Pseudeminia benguellensis (Torre) Verdc. ■☆
333162 Rhynchosia borianii Schweinf. = Pseudoeriosema borianii (Schweinf.) Hauman ■☆
333163 Rhynchosia braunii Harms;布劳恩鹿藿■☆
333164 Rhynchosia breviracemosa Hauman = Rhynchosia viscosa (Roth) DC. var. breviracemosa (Hauman) Verdc. ■☆
333165 Rhynchosia brunnea Baker f.;褐鹿藿■☆
333166 Rhynchosia buchananii Harms;布坎南鹿藿■☆
333167 Rhynchosia buettneri Harms;比特纳鹿藿■☆
333168 Rhynchosia bullata Benth. ex Harv.;泡状鹿藿■☆
333169 Rhynchosia buramensis Hutch. et E. A. Bruce = Rhynchosia elegans A. Rich. ■☆
333170 Rhynchosia cajanoides Guillaumin et Perr. = Eriosema psoraleoides (Lam.) G. Don ■☆
333171 Rhynchosia calobotrya Harms;美穗鹿藿■☆
333172 Rhynchosia calvescens Meikle;光秃鹿藿■☆
333173 Rhynchosia calycina Guillaumin et Perr. = Rhynchosia pycnostachya (DC.) Meikle ■☆
333174 Rhynchosia candida (Welw. ex Hiern) Torre;纯白鹿藿■☆
333175 Rhynchosia capensis (Burm. f.) Schinz;好望角鹿藿■☆
333176 Rhynchosia capitata (Heyne ex Roth) DC.;头状鹿藿■☆
333177 Rhynchosia caribaea (Jacq.) DC.;加勒比鹿藿■☆
333178 Rhynchosia chapelieri Baill.;沙普鹿藿■☆
333179 Rhynchosia chapmanii Verdc.;查普曼鹿藿■☆
333180 Rhynchosia chevalieri Harms;舍瓦利耶鹿藿■☆
333181 Rhynchosia chimanimaniensis Verdc.;奇马尼马尼鹿藿■☆
333182 Rhynchosia chinensis Hung T. Chang ex Y. T. Wei et S. K. Lee;中华鹿藿;China Rhynchosia, Chinese Rhynchosia ■
333183 Rhynchosia chrysadenia Taub. = Rhynchosia densiflora (Roth) DC. subsp. chrysadenia (Taub.) Verdc. ■☆
333184 Rhynchosia chrysantha Schltr. ex Zahlbr.;金花鹿藿■☆
333185 Rhynchosia chrysoposta Steud.;金色鹿藿■☆
333186 Rhynchosia chrysoscias Benth. ex Harv.;金伞鹿藿■☆
333187 Rhynchosia ciliata (Thunb.) Druce;缘毛鹿藿■☆
333188 Rhynchosia cinnamomea Schinz = Rhynchosia totta (Thunb.) DC. var. fenchelii Schinz ■☆
333189 Rhynchosia clivorum S. Moore;斜坡鹿藿■●☆
333190 Rhynchosia clivorum S. Moore subsp. pycnantha (Harms) Verdc.;密花斜坡鹿藿■☆
333191 Rhynchosia clivorum S. Moore var. caudata Meikle = Rhynchosia clivorum S. Moore subsp. pycnantha (Harms) Verdc. ■●☆
333192 Rhynchosia clivorum S. Moore var. fulvida Meikle = Rhynchosia clivorum S. Moore subsp. pycnantha (Harms) Verdc. ■☆
333193 Rhynchosia clivorum S. Moore var. pycnantha (Harms) Verdc. = Rhynchosia clivorum S. Moore subsp. pycnantha (Harms) Verdc. ■☆
333194 Rhynchosia comosa Baker = Pseudeminia comosa (Baker) Verdc. ■☆
333195 Rhynchosia confertiflora A. Rich. = Pseudarthria confertiflora (A. Rich.) Baker ●☆
333196 Rhynchosia confusa Burtt Davy;混乱鹿藿■☆
333197 Rhynchosia congensis Baker;刚果鹿藿■☆
333198 Rhynchosia congensis Baker subsp. orientalis Verdc.;东方刚果鹿藿■☆
333199 Rhynchosia congensis Baker subsp. pseudobuettneri Verdc.;假比特纳鹿藿■☆
333200 Rhynchosia congestiflora Schinz = Rhynchosia caribaea (Jacq.) DC. ■☆
333201 Rhynchosia connata Baker f.;合生鹿藿■☆
333202 Rhynchosia cooperi (Baker f.) Burtt Davy;库珀鹿藿■☆
333203 Rhynchosia craibiana Rehder = Rhynchosia himalensis Benth. ex Baker var. craibiana (Rehder) E. Peter ■
333204 Rhynchosia crassifolia Benth. ex Harv.;厚叶鹿藿■☆
333205 Rhynchosia crispa Verdc.;皱波鹿藿■☆
333206 Rhynchosia cyanosperma Baker = Rhynchosia hirta (Andréws) Meikle et Verdc. ■☆
333207 Rhynchosia debilis G. Don = Rhynchosia densiflora (Roth) DC. subsp. debilis (G. Don) Verdc. ■☆
333208 Rhynchosia dekindtii Harms;德金鹿藿■☆
333209 Rhynchosia denisii R. Vig. = Rhynchosia chapelieri Baill. ■☆
333210 Rhynchosia densiflora (Roth) DC.;密花鹿藿■☆
333211 Rhynchosia densiflora (Roth) DC. subsp. chrysadenia (Taub.) Verdc.;金腺密花鹿藿■☆
333212 Rhynchosia densiflora (Roth) DC. subsp. debilis (G. Don) Verdc.;弱小鹿藿■☆
333213 Rhynchosia densiflora (Roth) DC. subsp. stuhlmannii (Harms) Verdc.;斯图尔曼鹿藿■☆
333214 Rhynchosia desertorum Harms = Rhynchosia fleckii Schinz ■☆
333215 Rhynchosia dielsii Harms ex Diels;菱叶鹿藿(苦角藤,螃蟹眼睛,山黄豆藤,小角藤,野黄豆,野莞豆);Diels Rhynchosia ■
333216 Rhynchosia difformis (Elliott) DC.;参差鹿藿;Snout Bean ■☆
333217 Rhynchosia dinteri Schinz = Rhynchosia namaensis Schinz ■☆
333218 Rhynchosia discolor Klotzsch = Rhynchosia velutina Wight et Arn. var. discolor (Baker) Verdc. ■☆
333219 Rhynchosia divaricata Baker;叉开鹿藿■☆
333220 Rhynchosia dregei (E. Mey.) Steud. = Eriosema dregei E. Mey. ■☆
333221 Rhynchosia ecklonei Steud.;艾克鹿藿■☆
333222 Rhynchosia elachistantha Chiov. = Rhynchosia pulverulenta Stocks ■☆

333223　Rhynchosia elegans A. Rich.；雅致鹿藿■☆

333224　Rhynchosia elegans A. Rich. var. grandistipulata Verdc.；大托叶雅致鹿藿■☆

333225　Rhynchosia elegantissima Schinz ＝ Rhynchosia totta（Thunb.）DC. var. elegantissima（Schinz）Verdc.■☆

333226　Rhynchosia emarginata Germish.；微缺鹿藿■☆

333227　Rhynchosia erlangeri Harms；厄兰格鹿藿■☆

333228　Rhynchosia erythraeae Schweinf.；浅红鹿藿■☆

333229　Rhynchosia exellii Torre；埃克塞尔氏鹿藿■☆

333230　Rhynchosia fagelioides Taub. ex Engl. ＝ Rhynchosia pseudoviscosa Harms ■☆

333231　Rhynchosia faginea Guillaumin et Perr. ＝ Flemingia faginea（Guillaumin et Perr.）Baker ●☆

333232　Rhynchosia ferruginea A. Rich.；锈色鹿藿■☆

333233　Rhynchosia ficifolia Benth. ex Harv. ＝ Neorautanenia ficifolia（Benth. ex Harv.）C. A. Sm.■☆

333234　Rhynchosia filicaulis Welw. ex Baker ＝ Rhynchosia totta（Thunb.）DC. var. fenchelii Schinz ■☆

333235　Rhynchosia fischeri Harms ＝ Rhynchosia densiflora（Roth）DC. subsp. chrysadenia（Taub.）Verdc.■☆

333236　Rhynchosia flavissima Baker ＝ Rhynchosia malacophylla（Spreng.）Bojer ■☆

333237　Rhynchosia flavissima Baker var. macrocalyx Chiov. ＝ Rhynchosia minima（L.）DC. var. macrocalyx（Chiov.）Verdc.☆

333238　Rhynchosia fleckii Schinz；弗莱克鹿藿■☆

333239　Rhynchosia floribunda Baker ＝ Rhynchosia procurrens（Hiern）K. Schum. subsp. floribunda（Baker）Verdc.■☆

333240　Rhynchosia foliosa Markötter；多叶鹿藿■☆

333241　Rhynchosia friesiorum Harms ＝ Rhynchosia ferruginea A. Rich.■☆

333242　Rhynchosia gabonensis Jongkind；加蓬鹿藿■☆

333243　Rhynchosia galpinii Baker f.；盖尔鹿藿■☆

333244　Rhynchosia gazensis Baker f. ＝ Rhynchosia caribaea（Jacq.）DC.■☆

333245　Rhynchosia genistoides Burtt Davy；金雀鹿藿■☆

333246　Rhynchosia gibba E. Mey. ＝ Rhynchosia caribaea（Jacq.）DC.☆

333247　Rhynchosia glandulosa（Thunb.）DC.；多腺鹿藿■☆

333248　Rhynchosia glomerata Guillaumin et Perr. ＝ Eriosema glomeratum（Guillaumin et Perr.）Hook. f.■☆

333249　Rhynchosia glomerulans Fiori ＝ Rhynchosia densiflora（Roth）DC. subsp. chrysadenia（Taub.）Verdc.■☆

333250　Rhynchosia glutinosa Harms ＝ Rhynchosia nyasica Baker ■☆

333251　Rhynchosia goetzei Harms；格兹鹿藿■☆

333252　Rhynchosia gorsii Berhaut ＝ Rhynchosia orthobotrya Harms ■☆

333253　Rhynchosia gossweileri Baker f.；戈斯鹿藿■☆

333254　Rhynchosia grandifolia Steud.；大叶鹿藿■☆

333255　Rhynchosia grantii Baker ＝ Pseudoeriosema borianii（Schweinf.）Hauman ■☆

333256　Rhynchosia grevei Drake ＝ Rhynchosia viscosa（Roth）DC.●■

333257　Rhynchosia harmsiana Schltr. ex Zahlbr.；哈姆斯鹿藿■☆

333258　Rhynchosia harmsiana Schltr. ex Zahlbr. var. burchellii Burtt Davy；伯切尔鹿藿■☆

333259　Rhynchosia harveyi Eckl. et Zeyh.；哈维鹿藿■☆

333260　Rhynchosia hermannii Baker f. ＝ Rhynchosia elegans A. Rich.■☆

333261　Rhynchosia heterophylla Hauman；互叶鹿藿■☆

333262　Rhynchosia himalensis Benth. ex Baker；喜马拉雅鹿藿；Himalayan Rhynchosia, Himalayas Rhynchosia ■

333263　Rhynchosia himalensis Benth. ex Baker var. craibiana（Rehder）E. Peter；紫脉花鹿藿；Craib Rhynchosia ■

333264　Rhynchosia hirsuta Eckl. et Zeyh.；粗毛鹿藿■☆

333265　Rhynchosia hirsuta Schinz ＝ Rhynchosia venulosa（Hiern）K. Schum.■☆

333266　Rhynchosia hirta（Andréws）Meikle et Verdc.；多毛鹿藿■☆

333267　Rhynchosia hockii De Wild. ＝ Rhynchosia minima（L.）DC. var. prostrata（Harv.）Meikle ■☆

333268　Rhynchosia hockii De Wild. var. grandifolia Hauman ＝ Rhynchosia minima（L.）DC. var. prostrata（Harv.）Meikle ■☆

333269　Rhynchosia holosericea Schinz；全毛鹿藿■☆

333270　Rhynchosia holtzii Harms；霍尔茨鹿藿■☆

333271　Rhynchosia holubii Hemsl. ＝ Eminia holubii（Hemsl.）Taub.■☆

333272　Rhynchosia hookeri G. Don ＝ Rhynchosia malacophylla（Spreng.）Bojer ■☆

333273　Rhynchosia huillensis（Hiern）K. Schum.；威拉鹿藿■☆

333274　Rhynchosia imbricata Baker ＝ Rhynchosia nyasica Baker ■☆

333275　Rhynchosia inflata Bojer ＝ Rhynchosia sublobata（Schumach.）Meikle ■☆

333276　Rhynchosia insignis（O. Hoffm.）R. E. Fr.；显著鹿藿■☆

333277　Rhynchosia insignis（O. Hoffm.）R. E. Fr. subsp. affinis（De Wild.）L. Pauwels；近缘显著鹿藿■☆

333278　Rhynchosia ischnoclada Harms ＝ Rhynchosia minima（L.）DC. var. prostrata（Harv.）Meikle ■☆

333279　Rhynchosia jacottetii Schinz ＝ Rhynchosia reptabunda N. E. Br.■☆

333280　Rhynchosia karaguensis Harms ＝ Rhynchosia orthobotrya Harms ■☆

333281　Rhynchosia katangensis De Wild. ＝ Rhynchosia sublobata（Schumach.）Meikle ■☆

333282　Rhynchosia kerstingii Harms ＝ Rhynchosia nyasica Baker ■☆

333283　Rhynchosia kilimandscharica Harms；基利鹿藿■☆

333284　Rhynchosia klotzschii Cufod. ＝ Rhynchosia velutina Wight et Arn. var. discolor（Baker）Verdc.■☆

333285　Rhynchosia komatiensis Harms；科马蒂鹿藿■☆

333286　Rhynchosia kunmingensis Y. T. Wei et S. K. Lee；昆明鹿藿；Kunming Rhynchosia ■

333287　Rhynchosia laetissima Welw. ex Baker；愉悦鹿藿■☆

333288　Rhynchosia latifolia Nutt. ex Torr. et A. Gray；宽叶鹿藿；Snout Bean ■☆

333289　Rhynchosia laxiflora Cambess. ＝ Rhynchosia minima（L.）DC.☆

333290　Rhynchosia leandrii Du Puy et Labat；利安鹿藿■☆

333291　Rhynchosia ledermannii Harms；莱德鹿藿■☆

333292　Rhynchosia leucoscias Benth. ex Harv.；白伞鹿藿■☆

333293　Rhynchosia longepedunculata Hochst. ex A. Rich. ＝ Eriosema longepedunculatum（Hochst. ex A. Rich.）Baker ■☆

333294　Rhynchosia longiflora Schinz ＝ Rhynchosia totta（Thunb.）DC.■☆

333295　Rhynchosia longipes Harms ＝ Rhynchosia crassifolia Benth. ex Harv.■☆

333296　Rhynchosia longissima Hauman；极长鹿藿■☆

333297　Rhynchosia lukafuensis Baker f.；卢卡夫鹿藿■☆

333298　Rhynchosia lutea Dunn；黄花鹿藿；Yellow Rhynchosia, Yellowflower Rhynchosia ■

333299　Rhynchosia luteola（Hiern）K. Schum.；淡黄鹿藿■☆

333300　Rhynchosia luteola（Hiern）K. Schum. var. velutina Verdc.；黄短绒毛鹿藿■☆

333301　Rhynchosia luteola（Hiern）K. Schum. var. verdickii（De Wild.）Verdc.；韦尔鹿藿■☆

333302　Rhynchosia macinaca A. Chev. ＝ Rhynchosia sublobata

（Schumach.）Meikle ■☆

333303　Rhynchosia macrantha Hauman;大花鹿藿 ■☆

333304　Rhynchosia madagascariensis R. Vig. ;马岛鹿藿 ●■☆

333305　Rhynchosia maitlandii Baker f. = Rhynchosia minima（L.）DC. var. macrocalyx（Chiov.）Verdc. ■☆

333306　Rhynchosia malacophylla（Spreng.）Bojer;软叶鹿藿 ■☆

333307　Rhynchosia malacotricha Harms;软毛鹿藿 ■☆

333308　Rhynchosia mannii Baker;曼氏鹿藿 ■☆

333309　Rhynchosia manobotrya Harms = Rhynchosia goetzei Harms ■☆

333310　Rhynchosia maxima Bojer ex Drake = Rhynchosia baukea Du Puy et Labat ■☆

333311　Rhynchosia megalocalyx Thulin;大萼鹿藿 ■☆

333312　Rhynchosia melanosperma Klotzsch = Rhynchosia sublobata（Schumach.）Meikle ■☆

333313　Rhynchosia memnomia（Delile）DC. = Rhynchosia minima（L.）DC. var. memnonia（Delile）T. Cooke ■☆

333314　Rhynchosia memnonia（Delile）DC. var. candida（Welw. ex Hiern）Baker f. = Rhynchosia candida（Welw. ex Hiern）Torre ■☆

333315　Rhynchosia memnonia（Delile）DC. var. discolor Baker = Rhynchosia velutina Wight et Arn. var. discolor（Baker）Verdc. ■☆

333316　Rhynchosia memnonia（Delile）DC. var. prostrata Harv. = Rhynchosia minima（L.）DC. var. prostrata（Harv.）Meikle ■☆

333317　Rhynchosia mendoncae Torre = Pseudeminia mendoncae（Torre）Verdc. ■☆

333318　Rhynchosia mensensis Schweinf. ;芒斯鹿藿 ■☆

333319　Rhynchosia meyeri Steud. ;迈尔鹿藿 ■☆

333320　Rhynchosia micrantha Harms;小花鹿藿 ■☆

333321　Rhynchosia microscias Benth. ex Harv. ;小伞鹿藿 ■☆

333322　Rhynchosia mildbraedii Harms = Rhynchosia resinosa（Hochst. ex A. Rich.）Baker ■☆

333323　Rhynchosia minima（L.）DC. ;小鹿藿（小花鹿藿,小叶括根）;Little Rhynchosia,Mini Rhynchosia ■

333324　Rhynchosia minima（L.）DC. f. nuda（DC.）H. Ohashi et Tateishi;裸小鹿藿（小叶括根）■☆

333325　Rhynchosia minima（L.）DC. var. falcata（E. Mey.）Verdc. ;镰形小鹿藿 ■☆

333326　Rhynchosia minima（L.）DC. var. macrocalyx（Chiov.）Verdc. ;大萼小鹿藿 ■☆

333327　Rhynchosia minima（L.）DC. var. memnonia（Delile）T. Cooke;深褐小鹿藿 ■☆

333328　Rhynchosia minima（L.）DC. var. nuda（DC.）Kuntze = Rhynchosia minima（L.）DC. f. nuda（DC.）H. Ohashi et Tateishi ■☆

333329　Rhynchosia minima（L.）DC. var. pedicellata Verdc. ;梗花小鹿藿 ■☆

333330　Rhynchosia minima（L.）DC. var. prostrata（Harv.）Meikle;平卧小鹿藿 ■☆

333331　Rhynchosia mollis Burtt Davy = Rhynchosia totta（Thunb.）DC. var. fenchelii Schinz ■☆

333332　Rhynchosia moninensis Harms = Rhynchosia luteola（Hiern）K. Schum. ■☆

333333　Rhynchosia monophylla Schltr. ;单叶小鹿藿 ■☆

333334　Rhynchosia monophylla Schltr. var. eylesii Baker f. = Rhynchosia monophylla Schltr. ■☆

333335　Rhynchosia myriocarpa Quisumb. et Merr. = Rhynchosia acuminatissima Miq. ■

333336　Rhynchosia namaensis Schinz;纳马鹿藿 ■☆

333337　Rhynchosia nervosa Benth. ex Harv. ;多脉鹿藿 ■☆

333338　Rhynchosia nervosa Benth. ex Harv. var. pauciflora Harv. = Rhynchosia confusa Burtt Davy ■☆

333339　Rhynchosia nervosa Benth. ex Harv. var. petiolata Burtt Davy;柄叶多脉鹿藿 ■☆

333340　Rhynchosia nitens Benth. ex Harv. ;光亮鹿藿 ■☆

333341　Rhynchosia nitida（E. Mey.）Steud. ;澳非亮鹿藿 ■☆

333342　Rhynchosia nuda DC. = Rhynchosia minima（L.）DC. var. nuda（DC.）Kuntze ■☆

333343　Rhynchosia nyasica Baker;尼亚斯鹿藿 ■☆

333344　Rhynchosia nyikensis Baker;尼卡鹿藿 ■☆

333345　Rhynchosia oblatifoliolata Verdc. ;扁球小叶鹿藿 ■☆

333346　Rhynchosia oblongifoliolata Hauman;矩圆叶鹿藿 ■☆

333347　Rhynchosia oligantha Harms = Rhynchosia fleckii Schinz ■☆

333348　Rhynchosia oreophila Harms = Rhynchosia clivorum S. Moore subsp. pycnantha（Harms）Verdc. ■●☆

333349　Rhynchosia orthobotrya Harms;直序鹿藿 ■☆

333350　Rhynchosia orthodanum Benth. ex Harv. et Sond. = Rhynchosia sordida（E. Mey.）Schinz ■☆

333351　Rhynchosia ovata J. M. Wood et M. S. Evans;卵形鹿藿 ■☆

333352　Rhynchosia ovatifoliolata Torre;卵小叶鹿藿 ■☆

333353　Rhynchosia paniculata（E. Mey.）Steud. = Rhynchosia totta（Thunb.）DC. ☆

333354　Rhynchosia pauciflora Bolus;寡花鹿藿 ■☆

333355　Rhynchosia peglerae Baker f. ;佩格拉鹿藿 ■☆

333356　Rhynchosia pentheri Schltr. ex Zahlbr. ;彭泰尔鹿藿 ■☆

333357　Rhynchosia pentheri Schltr. ex Zahlbr. var. hutchinsoniana Burtt Davy;哈钦森鹿藿 ■☆

333358　Rhynchosia pilosa（E. Mey.）Steud. = Rhynchosia totta（Thunb.）DC. ■☆

333359　Rhynchosia pinnata Harv. ;羽状鹿藿 ■☆

333360　Rhynchosia preussii（Harms）Taub. ex Harms;普罗伊斯鹿藿 ■☆

333361　Rhynchosia procurrens（Hiern）K. Schum. ;伸展鹿藿 ■☆

333362　Rhynchosia procurrens（Hiern）K. Schum. subsp. floribunda（Baker）Verdc. ;繁花伸展鹿藿 ■☆

333363　Rhynchosia procurrens（Hiern）K. Schum. subsp. latisepala（Hauman）Verdc. ;宽萼伸展鹿藿 ■☆

333364　Rhynchosia procurrens（Hiern）K. Schum. var. rhodesica Baker f. = Rhynchosia procurrens（Hiern）K. Schum. subsp. floribunda（Baker）Verdc. ■☆

333365　Rhynchosia prostrata Suess. = Rhynchosia monophylla Schltr. ■☆

333366　Rhynchosia pseudoteramnoides Hauman;假软荚豆鹿藿 ■☆

333367　Rhynchosia pseudoviscosa Harms;假黏鹿藿 ■☆

333368　Rhynchosia psoraloides（Lam.）DC. = Eriosema psoraleoides（Lam.）G. Don ■☆

333369　Rhynchosia puberula（Eckl. et Zeyh.）Steud. ;微毛鹿藿 ●☆

333370　Rhynchosia pulchra（Vatke）Harms;美丽鹿藿 ■☆

333371　Rhynchosia pulverulenta Stocks;粉粒鹿藿 ■☆

333372　Rhynchosia pycnantha Harms = Rhynchosia clivorum S. Moore subsp. pycnantha（Harms）Verdc. ■☆

333373　Rhynchosia pycnostachya（DC.）Meikle;密穗鹿藿 ■☆

333374　Rhynchosia pyramidalis（Lam.）Urb. ;角锥鹿藿 ■☆

333375　Rhynchosia ramosa Verdc. ;分枝鹿藿 ☆

333376　Rhynchosia remota Conrath = Rhynchosia totta（Thunb.）DC. var. fenchelii Schinz ■☆

333377　Rhynchosia reniformis DC. ;肾叶鹿藿;Dollarleaf Rhynchosia ■☆

333378　Rhynchosia reptabunda N. E. Br. ;澳非鹿藿 ■☆

333379　Rhynchosia reptans Suess. = Rhynchosia monophylla Schltr. ■☆

333380 Rhynchosia resinosa (Hochst. ex A. Rich.) Baker;胶鹿藿■☆

333381 Rhynchosia resinosa (Hochst. ex A. Rich.) Baker var. latisepala Hauman = Rhynchosia procurrens (Hiern) K. Schum. subsp. latisepala (Hauman) Verdc.■☆

333382 Rhynchosia resinosa (Hochst. ex A. Rich.) Baker var. schliebenii (Harms) Hauman = Rhynchosia procurrens (Hiern) K. Schum.■☆

333383 Rhynchosia rhodesica Baker f. = Rhynchosia goetzei Harms■☆

333384 Rhynchosia rhodophylla Baker = Rhynchosia versicolor Baker■☆

333385 Rhynchosia rhombifolia Blatt. et Hallb. = Rhynchosia pulverulenta Stocks■☆

333386 Rhynchosia rigidula DC. = Rhynchosia totta (Thunb.) DC.■☆

333387 Rhynchosia rivae Schweinf. = Rhynchosia ferruginea A. Rich.■☆

333388 Rhynchosia rogersii Schinz;罗杰斯鹿藿■☆

333389 Rhynchosia rothii Benth. ex Aitch.;绒叶鹿藿（绒叶括根）; Floss Rhynchosia,Roth Rhynchosia●■

333390 Rhynchosia rotundifolia (E. Mey.) Steud.;圆叶鹿藿■☆

333391 Rhynchosia rudolfii Harms;鲁道夫鹿藿■☆

333392 Rhynchosia rufescens (Willd.) DC.;淡红鹿藿;Lightred Rhynchosia,Redish Rhynchosia■

333393 Rhynchosia salicifolia Hauman;柳叶鹿藿■☆

333394 Rhynchosia schimperi Hochst. ex Boiss.;欣珀鹿藿■☆

333395 Rhynchosia schlechteri Baker f.;施莱鹿藿■☆

333396 Rhynchosia schliebenii Harms = Rhynchosia procurrens (Hiern) K. Schum.■☆

333397 Rhynchosia schoelleri Schweinf. = Rhynchosia minima (L.) DC. var. prostrata (Harv.) Meikle■☆

333398 Rhynchosia schweinfurthii Harms = Rhynchosia densiflora (Roth) DC. subsp. chrysadenia (Taub.) Verdc.■☆

333399 Rhynchosia secunda (Thunb.) Eckl. et Zeyh.;单侧鹿藿■☆

333400 Rhynchosia sennaarensis Hochst. ex Schweinf. = Rhynchosia malacophylla (Spreng.) Bojer☆

333401 Rhynchosia sennaarensis Hochst. ex Schweinf. var. flavissima Schweinf. = Rhynchosia malacophylla (Spreng.) Bojer☆

333402 Rhynchosia sennaarensis Schweinf. var. macrocalyx (Chiov.) Cufod. = Rhynchosia minima (L.) DC. var. macrocalyx (Chiov.) Verdc.■☆

333403 Rhynchosia sericea Span. = Rhynchosia rothii Benth. ex Aitch.●■

333404 Rhynchosia sericosemium Harms = Rhynchosia luteola (Hiern) K. Schum.■☆

333405 Rhynchosia sigmoides Benth. ex Harv. = Rhynchosia villosa (Meisn.) Druce■☆

333406 Rhynchosia singulifolia Steud.;单叶鹿藿■☆

333407 Rhynchosia sordida (E. Mey.) Schinz;污浊鹿藿■☆

333408 Rhynchosia speciosa Verdc.;丽鹿藿■☆

333409 Rhynchosia spectabilis Schinz;壮观鹿藿■☆

333410 Rhynchosia sphaerocephala Baker = Dolichos sericeus E. Mey.■☆

333411 Rhynchosia splendens Schweinf.;亮鹿藿■☆

333412 Rhynchosia stenodon Baker f.;狭鹿藿■☆

333413 Rhynchosia stenodon Harms = Rhynchosia elegans A. Rich.■☆

333414 Rhynchosia stipata Meikle;堆积鹿藿■☆

333415 Rhynchosia stipitata Verdc.;具柄鹿藿■☆

333416 Rhynchosia stipulosa A. Rich. = Rhynchosia viscosa (Roth) DC. subsp. stipulosa (A. Rich.) Verdc.■☆

333417 Rhynchosia stuhlmannii Harms = Rhynchosia densiflora (Roth) DC. subsp. stuhlmannii (Harms) Verdc.■☆

333418 Rhynchosia subaphylla Baker f. = Rhynchosia insignis (O. Hoffm.) R. E. Fr.■☆

333419 Rhynchosia sublobata (Schumach. et Thonn.) Meikle = Rhynchosia sublobata (Schumach.) Meikle■☆

333420 Rhynchosia sublobata (Schumach.) Meikle;微裂鹿藿■☆

333421 Rhynchosia swynnertonii Baker f.;斯温纳顿鹿藿■☆

333422 Rhynchosia teixeirae Torre;特谢拉鹿藿■☆

333423 Rhynchosia teramnoides Harms;软荚豆鹿藿■☆

333424 Rhynchosia thorncroftii (Baker f.) Burtt Davy;托恩鹿藿■☆

333425 Rhynchosia tibestica Miré, H. Gillet et Quézel = Rhynchosia totta (Thunb.) DC. var. fenchelii Schinz■☆

333426 Rhynchosia tomentosa (Roxb.) Babu;毛叶鹿藿;Velvetleaf Rhynchosia■☆

333427 Rhynchosia torrei Verdc.;托雷鹿藿■☆

333428 Rhynchosia totta (Thunb.) DC.;南美鹿藿■☆

333429 Rhynchosia totta (Thunb.) DC. var. elegantissima (Schinz) Verdc.;极长南美鹿藿■☆

333430 Rhynchosia totta (Thunb.) DC. var. elongatifolia Verdc.;长叶南美鹿藿■☆

333431 Rhynchosia totta (Thunb.) DC. var. fenchelii Schinz;芬切尔鹿藿■☆

333432 Rhynchosia totta (Thunb.) DC. var. pilosa (E. Mey.) Baker f. = Rhynchosia totta (Thunb.) DC.■☆

333433 Rhynchosia totta (Thunb.) DC. var. venulosa (Hiern) Verdc. = Rhynchosia totta (Thunb.) DC. var. fenchelii Schinz■☆

333434 Rhynchosia transjubensis Chiov. = Rhynchosia sublobata (Schumach.) Meikle■☆

333435 Rhynchosia trichocephala Baker = Rhynchosia versicolor Baker■☆

333436 Rhynchosia tricuspidata Baker f.;三尖鹿藿■☆

333437 Rhynchosia trinervis (E. Mey.) Steud.;三脉鹿藿■☆

333438 Rhynchosia uniflora Harv. = Rhynchosia angustifolia (Jacq.) DC.●☆

333439 Rhynchosia usambarensis Taub.;乌桑巴拉鹿藿■☆

333440 Rhynchosia usambarensis Taub. subsp. inelegans Verdc.;不雅乌桑巴拉鹿藿■☆

333441 Rhynchosia usambarensis Taub. var. obtusifoliola Verdc.;钝叶鹿藿■☆

333442 Rhynchosia velutina Wight et Arn.;短绒毛鹿藿■☆

333443 Rhynchosia velutina Wight et Arn. var. discolor (Baker) Verdc.;异色短绒毛鹿藿■☆

333444 Rhynchosia vendae C. H. Stirt.;文达鹿藿■☆

333445 Rhynchosia venulosa (Hiern) K. Schum. = Rhynchosia totta (Thunb.) DC. var. fenchelii Schinz■☆

333446 Rhynchosia verdcourtii Thulin;韦尔德鹿藿■☆

333447 Rhynchosia verdickii De Wild. = Rhynchosia luteola (Hiern) K. Schum. var. verdickii (De Wild.) Verdc.■☆

333448 Rhynchosia versicolor Baker;变色鹿藿●☆

333449 Rhynchosia villosa (Meisn.) Druce;长柔毛鹿藿■☆

333450 Rhynchosia violacea (Hiern) K. Schum. = Rhynchosia viscosa (Roth) DC. subsp. violacea (Hiern) Verdc.■☆

333451 Rhynchosia viscidula Steud.;黏毛鹿藿■☆

333452 Rhynchosia viscosa (Roth) DC.;黏鹿藿（黏质鹿藿）;Sticky Rhynchosia,Viscous Rhynchosia■

333453 Rhynchosia viscosa (Roth) DC. subsp. stipulosa (A. Rich.) Verdc.;托叶黏鹿藿■☆

333454 Rhynchosia viscosa (Roth) DC. subsp. violacea (Hiern) Verdc.;堇色黏鹿藿■☆

333455 Rhynchosia viscosa (Roth) DC. var. breviracemosa (Hauman)

Verdc.；短总花黏鹿藿■☆

333456　Rhynchosia volubilis Lour.；鹿藿（大叶野绿豆，饿蚂蝗，鸡仔目周仁，假荷兰豆，酒壶藤，藰，老鼠豆，老鼠眼，荖豆，鹿藿，门瘦，拑根，鸟眼睛豆，痰切豆，野黄豆，野绿豆，野毛扁豆，野毛豆）；Rhynchosia，Twining Rhynchosia ■

333457　Rhynchosia volubilis Lour. var. longiracemosa Franch.；长穗鹿藿（鹿藿）■

333458　Rhynchosia wellmaniana Harms；韦尔曼鹿藿■☆

333459　Rhynchosia wildii Verdc.；维尔德鹿藿■☆

333460　Rhynchosia woodii Schinz；伍得鹿藿■☆

333461　Rhynchosia yunnanensis Franch.；云南鹿藿；Yunnan Rhynchosia ■

333462　Rhynchosia zernyi Harms；策尼鹿藿■☆

333463　Rhynchosida Fryxell（1978）；喙稆属●●☆

333464　Rhynchosida physocalyx（Gray）Fryxell；喙稆■●☆

333465　Rhynchosinapis Hayek＝Coincya Rouy ■☆

333466　Rhynchosinapis pseudoerucastrum（Brot.）Franco subsp. orophila Franco＝Coincya monensis（L.）Greuter et Burdet subsp. orophila（Franco）Aedo et al. ■☆

333467　Rhynchospermum A. DC.＝Rhyncospermum A. DC. ●■

333468　Rhynchospermum Lindl.＝Rhynchodia Benth. ●■

333469　Rhynchospermum Lindl.＝Rhyncospermum A. DC. ●■

333470　Rhynchospermum Reinw.＝Rhynchospermum Reinw. ex Blume ■

333471　Rhynchospermum Reinw. ex Blume（1825）；秋分草属；Rhynchospermum ■

333472　Rhynchospermum formosanum Yamam.＝Rhynchospermum verticillatum Reinw. ex Blume ■

333473　Rhynchospermum jasminioides Lindl.＝Trachelospermum jasminoides（Lindl.）Lem. ●

333474　Rhynchospermum verticillatum Reinw.＝Rhynchospermum verticillatum Reinw. ex Blume ■

333475　Rhynchospermum verticillatum Reinw. ex Blume；秋分草（白鱼鳅串，大鱼鳅串，调粪草，林荫菊）；Verticillate Rhynchospermum ■

333476　Rhynchospermum verticillatum Reinw. ex Blume var. subsessilis Oliv. ex Miq.＝Rhynchospermum verticillatum Reinw. ex Blume ■

333477　Rhynchospora Vahl（1805）（'Rynchospora'）（保留属名）；刺子莞属；Beak Rush，Beakrush，Beak-rush，Beak-sedge，Rhynchospore，Whitetop Sedge ■

333478　Rhynchospora × hakkodensis Mochizuki；八甲田刺子莞■☆

333479　Rhynchospora africana Cherm.＝Rhynchospora angolensis Turrill ■☆

333480　Rhynchospora alba（L.）Vahl；白鳞刺子莞（白穗刺子莞，新竹莞）；White Beak-rush，White Beak-sedge，White Scale Beakrush，Whitescale Beakrush ■

333481　Rhynchospora alba（L.）Vahl var. fusca（L.）Vahl＝Rhynchospora fusca（L.）W. T. Aiton ■☆

333482　Rhynchospora alba（L.）Vahl var. macra C. B. Clarke ex Britton＝Rhynchospora macra（C. B. Clarke ex Britton）Small ■☆

333483　Rhynchospora angolensis Turrill；安哥拉刺子莞■☆

333484　Rhynchospora arechavaletae Boeck.＝Rhynchospora holoschoenoides（Rich.）Herter ■☆

333485　Rhynchospora aurea Vahl＝Rhynchospora corymbosa（L.）Britton ■

333486　Rhynchospora axillaris（Lam.）Britton var. microcephala Britton＝Rhynchospora microcephala（Britton）Britton ex Small ■☆

333487　Rhynchospora baldwinii A. Gray；鲍尔温刺子莞■☆

333488　Rhynchospora barteri C. B. Clarke＝Rhynchospora brevirostris Griseb. ■☆

333489　Rhynchospora blauneri Britton＝Rhynchospora brachychaeta C. Wright ■☆

333490　Rhynchospora boninensis Nakai ex Tuyama；小笠原刺子莞■☆

333491　Rhynchospora brachychaeta C. Wright；短绒毛刺子莞■☆

333492　Rhynchospora brevirostris Griseb.；短喙刺子莞■☆

333493　Rhynchospora breviseta（Gale）Channell；短刚毛刺子莞■☆

333494　Rhynchospora brown Roem. et Schult.＝Rhynchospora rugosa（Vahl）Gale subsp. brownii（Roem. et Schult.）T. Koyama ■

333495　Rhynchospora brownii Roem. et Schult.；白喙刺子莞；Brown Beakrush，Whitebeak Beakrush ■

333496　Rhynchospora brownii Roem. et Schult.＝Rhynchospora rugosa（Vahl）Gale var. brownii（Roem. et Schult.）T. Koyama ■

333497　Rhynchospora bulbocaulis Boeck.＝Mariscus amomodorus（K. Schum.）Cufod. var. mollipes（C. B. Clarke）Cufod. ■☆

333498　Rhynchospora caduca Elliott；角茎刺子莞；Angle-stem Beak Sedge ■☆

333499　Rhynchospora californica Gale；加州刺子莞■☆

333500　Rhynchospora candida（Nees）Boeck.；纯白刺子莞■☆

333501　Rhynchospora capillacea Torr.；毛刺子莞（喙刺子莞）；Beaked Rush，Hair Beak-rush，Needle Beak Sedge ■☆

333502　Rhynchospora capillacea Torr. f. leviseta（E. J. Hill）Fernald＝Rhynchospora capillacea Torr. ■☆

333503　Rhynchospora capillacea Torr. var. leviseta E. J. Hill ex A. Gray＝Rhynchospora capillacea Torr. ■☆

333504　Rhynchospora capitellata（Michx.）Vahl；小头状刺子莞；Brownish Beak Sedge，Clustered Beak-rush，False Bog Rush ■☆

333505　Rhynchospora capitellata（Michx.）Vahl f. discutiens（C. B. Clarke）Gale＝Rhynchospora capitellata（Michx.）Vahl ■☆

333506　Rhynchospora capitellata（Michx.）Vahl var. discutiens（C. B. Clarke）S. F. Blake＝Rhynchospora capitellata（Michx.）Vahl ■☆

333507　Rhynchospora capitellata（Michx.）Vahl var. leptocarpa（Chapm. ex Britton）S. F. Blake＝Rhynchospora capitellata（Michx.）Vahl ■☆

333508　Rhynchospora capitellata（Michx.）Vahl var. minor（Britton）S. F. Blake＝Rhynchospora capitellata（Michx.）Vahl ■☆

333509　Rhynchospora careyana Fernald；卡雷刺子莞■☆

333510　Rhynchospora caucasica Palla；高加索刺子莞■☆

333511　Rhynchospora cephalantha A. Gray；头花刺子莞■☆

333512　Rhynchospora cephalantha A. Gray var. attenuata Gale＝Rhynchospora cephalantha A. Gray ■☆

333513　Rhynchospora cephalantha A. Gray var. microcephala（Britton）Kük.＝Rhynchospora microcephala（Britton）Britton ex Small ■☆

333514　Rhynchospora cephalantha A. Gray var. pleiocephala Fernald et Gale＝Rhynchospora cephalantha A. Gray ■☆

333515　Rhynchospora chalarocephala Fernald et Gale；疏头刺子莞■☆

333516　Rhynchospora chapmanii Britton＝Rhynchospora brachychaeta C. Wright ■☆

333517　Rhynchospora chapmanii M. A. Curtis；查普曼刺子莞■☆

333518　Rhynchospora chinensis Nees et Meyen ex Nees；华刺子莞；China Beakrush，Chinese Beakrush，Spiked Beaksedge ■

333519　Rhynchospora chinensis Nees et Meyen subsp. fauriei（Franch.）T. Koyama＝Rhynchospora fauriei Franch. ■☆

333520　Rhynchospora chinensis Nees et Meyen var. curvoaristata（Tuyama）Ohwi＝Rhynchospora rugosa（Vahl）Gale ■☆

333521　Rhynchospora chinensis Nees et Meyen var. fauriei（Franch.）T. Koyama＝Rhynchospora fauriei Franch. ■☆

333522　Rhynchospora ciliaris（Michx.）C. Mohr；缘毛刺子莞■☆

333523　Rhynchospora ciliata Vahl = Rhynchospora ciliaris（Michx.）C. Mohr■☆

333524　Rhynchospora colorata（Hitchc.）H. Pfeiff.；具色刺子莞；Starrush, White-topped Sedge■☆

333525　Rhynchospora compressa J. Carey ex Chapm.；扁刺子莞■☆

333526　Rhynchospora contracta（Nees）J. Raynal；紧缩刺子莞■☆

333527　Rhynchospora corniculata（Lam.）A. Gray；小喙刺子莞；Horned Rush■☆

333528　Rhynchospora corniculata（Lam.）Fernald var. interior Fernald = Rhynchospora corniculata（Lam.）A. Gray■☆

333529　Rhynchospora corniculata（Lam.）Fernald var. patula（Chapm.）Britton = Rhynchospora careyana Fernald■☆

333530　Rhynchospora corymbosa（L.）Britton；伞房刺子莞（三俭草）；Corymbose Beakrush■

333531　Rhynchospora corymbosa（L.）Britton var. grandspiculosa Kük. = Rhynchospora spectabilis Hochst. ex Krauss■

333532　Rhynchospora crinipes Gale；发梗刺子莞■☆

333533　Rhynchospora curtissii Britton；柯蒂斯刺子莞■☆

333534　Rhynchospora curtissii Steud. = Rhynchospora pallida M. A. Curtis■☆

333535　Rhynchospora cymosa Elliott var. compressa（J. Carey ex Chapm.）C. B. Clarke ex Britton = Rhynchospora compressa J. Carey ex Chapm.■☆

333536　Rhynchospora cymosa Elliott var. globularis Chapm. = Rhynchospora globularis（Chapm.）Small■☆

333537　Rhynchospora cymosa Muhl. ex Elliott = Rhynchospora glomerata（L.）Vahl■☆

333538　Rhynchospora cyperoides（Sw.）Mart. = Rhynchospora holoschoenoides（Rich.）Herter■☆

333539　Rhynchospora debilis Gale；柔弱刺子莞■☆

333540　Rhynchospora decurrens Chapm.；下延刺子莞■☆

333541　Rhynchospora deightonii Hutch. = Rhynchospora perrieri Cherm.■☆

333542　Rhynchospora distans（Michx.）Vahl = Rhynchospora fascicularis（Michx.）Vahl■☆

333543　Rhynchospora distans（Michx.）Vahl var. fascicularis（Michx.）Kük. = Rhynchospora fascicularis（Michx.）Vahl■☆

333544　Rhynchospora distans（Michx.）Vahl var. gracillima Kük. = Rhynchospora wrightiana Boeck.■☆

333545　Rhynchospora distans（Michx.）Vahl var. tenuis Britton = Rhynchospora wrightiana Boeck.■☆

333546　Rhynchospora distans Elliott = Rhynchospora grayi Kunth■☆

333547　Rhynchospora divergens Chapm. ex M. A. Curtis；叉刺子莞■☆

333548　Rhynchospora dodecrandra Baldwin ex A. Gray = Rhynchospora megalocarpa A. Gray■☆

333549　Rhynchospora dolichostyla K. Schum.；长柱刺子莞■☆

333550　Rhynchospora dommucensis A. H. Moore = Rhynchospora fascicularis（Michx.）Vahl■☆

333551　Rhynchospora drummondiana Boeck. = Rhynchospora gracilenta A. Gray■☆

333552　Rhynchospora drummondiana Steud. = Rhynchospora colorata（Hitchc.）H. Pfeiff.■☆

333553　Rhynchospora edisoniana Small = Rhynchospora microcarpa Baldwin ex A. Gray■☆

333554　Rhynchospora elliottii A. Dietr.；埃利奥特刺子莞■☆

333555　Rhynchospora elliottii A. Gray = Rhynchospora grayi Kunth■☆

333556　Rhynchospora erinacea（Ridl.）C. B. Clarke = Sphaerocyperus erinaceus（Ridl.）Lye■☆

333557　Rhynchospora etuberculata Steud. = Schoenoplectus etuberculatus（Steud.）Soják■☆

333558　Rhynchospora eximia（Nees）Boeck.；卓越刺子莞■☆

333559　Rhynchospora eximia（Nees）Boeck. var. pleiantha（Cherm.）Raymond = Rhynchospora eximia（Nees）Boeck.■☆

333560　Rhynchospora faberi C. B. Clarke；细叶刺子莞；Farer Beakrush■

333561　Rhynchospora faberi C. B. Clarke f. breviseta（Palla）Ohwi et T. Koyama；短毛细叶刺子莞■☆

333562　Rhynchospora faberi C. B. Clarke f. exigua（Takeda）Ohwi et T. Koyama；弱小细叶■☆

333563　Rhynchospora fascicularis（Michx.）Vahl；簇生刺子莞；Fascicled Beaksedge■☆

333564　Rhynchospora fascicularis（Michx.）Vahl var. debilis（Gale）Kük. = Rhynchospora debilis Gale■☆

333565　Rhynchospora fascicularis（Michx.）Vahl var. distans（Michx.）Chapm. = Rhynchospora fascicularis（Michx.）Vahl■☆

333566　Rhynchospora fascicularis（Michx.）Vahl var. fernaldii（Gale）Kük. = Rhynchospora fernaldii Gale■☆

333567　Rhynchospora fascicularis（Michx.）Vahl var. harperi（Small）Kük. = Rhynchospora harperi Small■☆

333568　Rhynchospora fauriei Franch.；法氏刺子莞■☆

333569　Rhynchospora fernaldii Gale；弗纳尔德刺子莞■☆

333570　Rhynchospora filifolia A. Gray；线叶刺子莞■☆

333571　Rhynchospora filifolia A. Gray var. ellipsoidea Kük. = Rhynchospora curtissii Britton■☆

333572　Rhynchospora filifolia A. Gray var. pleiantha Kük. = Rhynchospora pleiantha（Kük.）Gale■☆

333573　Rhynchospora floridensis（Britton）H. Pfeiff.；佛罗里达刺子莞■☆

333574　Rhynchospora floridensis（Britton）H. Sweet = Rhynchospora floridensis（Britton）H. Pfeiff.■☆

333575　Rhynchospora fujiiana Makino；葛刺子莞■☆

333576　Rhynchospora fusca（L.）W. T. Aiton；褐刺子莞；Brown Beak Sedge, Brown Beak-rush, Brown Beak-sedge, Soóty Beak-rush■☆

333577　Rhynchospora fusca Aiton = Rhynchospora fusca（L.）W. T. Aiton■☆

333578　Rhynchospora glauca C. B. Clarke = Rhynchospora brownii Roem. et Schult.■

333579　Rhynchospora glauca C. B. Clarke var. chinensis C. B. Clarke = Rhynchospora chinensis Nees et Meyen ex Nees■

333580　Rhynchospora glauca Vahl = Rhynchospora rugosa（Vahl）Gale■

333581　Rhynchospora glauca Vahl var. chinensis（Nees et Meyen）C. B. Clarke = Rhynchospora chinensis Nees et Meyen ex Nees■

333582　Rhynchospora glauca Vahl var. condensata Degl. = Rhynchospora rugosa（Vahl）Gale subsp. brownii（Roem. et Schult.）T. Koyama■

333583　Rhynchospora glauca Vahl var. juncea（Willd. ex Kunth）Cherm. = Rhynchospora rugosa（Vahl）Gale■

333584　Rhynchospora glauca Vahl var. pauciseta Turrill = Rhynchospora rugosa（Vahl）Gale■

333585　Rhynchospora globularis（Chapm.）Small；玉簪刺子莞；Beaked Rush■☆

333586　Rhynchospora globularis（Chapm.）Small var. obliterata（Gale）Kük. = Rhynchospora globularis（Chapm.）Small■☆

333587　Rhynchospora globularis（Chapm.）Small var. recognita Gale；平滑玉簪刺子莞■☆

333588 Rhynchospora globularis（Chapm.）Small var. saxicola（Small）Kük.；岩地玉簪刺子莞■☆

333589 Rhynchospora glomerata（L.）Vahl；团集刺子莞；Clustered Beak-rush ■☆

333590 Rhynchospora glomerata（L.）Vahl var. angusta Gale = Rhynchospora glomerata（L.）Vahl ■☆

333591 Rhynchospora glomerata（L.）Vahl var. discutiens C. B. Clarke = Rhynchospora capitellata（Michx.）Vahl ■☆

333592 Rhynchospora glomerata（L.）Vahl var. leptocarpa Chapm. ex Britton = Rhynchospora capitellata（Michx.）Vahl ■☆

333593 Rhynchospora glomerata（L.）Vahl var. minor Britton = Rhynchospora capitellata（Michx.）Vahl ■☆

333594 Rhynchospora glomerata（L.）Vahl var. minor Britton f. discutiens（C. B. Clarke）Fernald = Rhynchospora capitellata（Michx.）Vahl ■☆

333595 Rhynchospora glomerata（L.）Vahl var. paniculata（A. Gray）Chapm. = Rhynchospora glomerata（L.）Vahl ■☆

333596 Rhynchospora glomerata（L.）Vahl var. robustior Kunth = Rhynchospora glomerata（L.）Vahl ■☆

333597 Rhynchospora gracilenta A. Gray；细瘦刺子莞■☆

333598 Rhynchospora gracilenta A. Gray var. diversifolia Fernald = Rhynchospora gracilenta A. Gray ■☆

333599 Rhynchospora gracillima C. Wright = Rhynchospora wrightiana Boeck. ■☆

333600 Rhynchospora gracillima Thwaites；细长刺子莞（柔弱刺子莞）■

333601 Rhynchospora gracillima Thwaites subsp. subquadrata（Cherm.）J. Raynal；四方形刺子莞■☆

333602 Rhynchospora grayi Kunth；格雷刺子莞■☆

333603 Rhynchospora harperi Small；哈珀刺子莞■☆

333604 Rhynchospora harveyi W. Boott；哈维刺子莞；Harvey's Beaked Rush ■☆

333605 Rhynchospora holoschoenoides（Rich.）Herter；全箭莎状刺子莞■☆

333606 Rhynchospora intermedia（Chapm.）Britton = Rhynchospora pineticola C. B. Clarke ■☆

333607 Rhynchospora intermixta C. Wright = Rhynchospora pusilla Chapm. ex M. A. Curtis ■☆

333608 Rhynchospora inundata（Oakes）Fernald；水生刺子莞；Inundated Beakrush ■☆

333609 Rhynchospora jacobi C. E. C. Fisch. = Schoenoplectus senegalensis（Hochst. ex Steud.）Palla ex J. Raynal ■☆

333610 Rhynchospora japonica Makino = Rhynchospora chinensis Nees et Meyen ex Nees ■

333611 Rhynchospora juncea Willd. ex Kunth = Rhynchospora rugosa（Vahl）Gale ■

333612 Rhynchospora kamphoeveneri Boeck. = Rhynchospora gracillima Thwaites ■

333613 Rhynchospora knieskernii J. Carey；克尼刺子莞■☆

333614 Rhynchospora kunthii Nees = Rhynchospora kunthii Nees ex Kunth ■☆

333615 Rhynchospora kunthii Nees ex Kunth；库氏刺子莞；Kunth's Beaksedge ■☆

333616 Rhynchospora latifolia（Baldwin）W. W. Thomas；宽叶刺子莞■☆

333617 Rhynchospora laxa Benth. = Rhynchospora chinensis Nees et Meyen ex Nees ■

333618 Rhynchospora laxa R. Br. = Rhynchospora rugosa（Vahl）Gale ■

333619 Rhynchospora leptocarpa（Chapm. ex Britton）Small =

333620 Rhynchospora leptorhyncha Small = Rhynchospora harperi Small ■☆

333621 Rhynchospora longisetigera Hayata = Rhynchospora chinensis Nees et Meyen ex Nees ■

333622 Rhynchospora luguillensis Britton = Rhynchospora alba（L.）Vahl ■

333623 Rhynchospora macra（C. B. Clarke ex Britton）Small；马克拉刺子莞■☆

333624 Rhynchospora macrocarpa Boeck. = Rhynchospora spectabilis Hochst. ex Krauss ■

333625 Rhynchospora macrostachya Torr. ex A. Gray；大穗刺子莞；Horned Rush ■☆

333626 Rhynchospora macrostachya Torr. ex A. Gray var. colpophylla Fernald et Gale = Rhynchospora macrostachya Torr. ex A. Gray ■☆

333627 Rhynchospora macrostachya Torr. ex A. Gray var. inundata（Oakes）Fernald = Rhynchospora inundata（Oakes）Fernald ■☆

333628 Rhynchospora malasica C. B. Clarke；日本刺子莞（马来刺子莞）；Japan Beakrush,Japanese Beakrush ■

333629 Rhynchospora marginata C. B. Clarke = Rhynchospora submarginata Degl. ■

333630 Rhynchospora mauritii Steud. = Rhynchospora holoschoenoides（Rich.）Herter ■☆

333631 Rhynchospora megalocarpa A. Gray；大果刺子莞■☆

333632 Rhynchospora megaplumosa E. L. Bridges et Orzell；大羽状刺子莞■☆

333633 Rhynchospora micrantha Vahl = Rhynchospora contracta（Nees）J. Raynal ■☆

333634 Rhynchospora microcarpa Baldwin ex A. Gray；小果刺子莞■☆

333635 Rhynchospora microcephala（Britton）Britton ex Small；小头刺子莞■☆

333636 Rhynchospora miliacea（Lam.）A. Gray；巨数刺子莞■☆

333637 Rhynchospora mixta Britton；混乱刺子莞■☆

333638 Rhynchospora multiflora A. Gray = Rhynchospora elliottii A. Dietr. ■☆

333639 Rhynchospora nipponica Makino = Rhynchospora malasica C. B. Clarke ■

333640 Rhynchospora nitens（Vahl）A. Gray；亮刺子莞；Bald-rush ■☆

333641 Rhynchospora nitida Spreng. = Tetraria compar（L.）T. Lestib. ■☆

333642 Rhynchospora nivea Boeck.；雪白刺子莞■☆

333643 Rhynchospora obliterata Gale = Rhynchospora globularis（Chapm.）Small ■☆

333644 Rhynchospora ochrocephala Boeck. = Cyperus rhynchosporoides Kük. ■☆

333645 Rhynchospora odorata C. Wright ex Griseb.；香刺子莞■☆

333646 Rhynchospora oligantha A. Gray；少花刺子莞■☆

333647 Rhynchospora oligantha A. Gray var. breviseta Gale = Rhynchospora breviseta（Gale）Channell ■☆

333648 Rhynchospora oritrephes（P. J. Bergius）Boeck. var. semisetacea Kük. = Rhynchospora perrieri Cherm. ■☆

333649 Rhynchospora oxycephala C. Wright = Rhynchospora eximia（Nees）Boeck. ■☆

333650 Rhynchospora pallida C. B. Clarke = Rhynchospora brachychaeta C. Wright ■☆

333651 Rhynchospora pallida Kük. = Rhynchospora brachychaeta C. Wright ■☆

333652 Rhynchospora pallida M. A. Curtis；苍白刺子莞■☆

333653 Rhynchospora paniculata A. Gray = Rhynchospora glomerata

(L.) Vahl ■☆

333654 Rhynchospora parva (Nees) Steud. var. boninensis (Nakai ex Tuyama) T. Koyama = Rhynchospora boninensis Nakai ex Tuyama ■☆

333655 Rhynchospora penniseta Griseb. = Rhynchospora plumosa Elliott ☆

333656 Rhynchospora perplexa Britton;紊乱刺子莞■☆

333657 Rhynchospora perrieri Cherm. ;佩里耶刺子莞■☆

333658 Rhynchospora pineticola C. B. Clarke;松林刺子莞■☆

333659 Rhynchospora plankii Britton = Rhynchospora harveyi W. Boott ■☆

333660 Rhynchospora pleiantha (Kük.) Gale;多花刺子莞■☆

333661 Rhynchospora plumosa Elliott;羽状刺子莞☆

333662 Rhynchospora plumosa Elliott var. intermedia Chapm. = Rhynchospora pineticola C. B. Clarke ■☆

333663 Rhynchospora prolifera Small = Rhynchospora mixta Britton ■☆

333664 Rhynchospora psilocaroides Griseb. = Rhynchospora eximia (Nees) Boeck. ■☆

333665 Rhynchospora punctata Elliott;斑点刺子莞■☆

333666 Rhynchospora pusilla Chapm. ex M. A. Curtis;微小刺子莞■☆

333667 Rhynchospora pycnocarpa A. Gray = Rhynchospora megalocarpa A. Gray ■☆

333668 Rhynchospora rappiana Small = Rhynchospora ciliaris (Michx.) C. Mohr ■☆

333669 Rhynchospora rariflora (Michx.) Elliott;稀花刺子莞■☆

333670 Rhynchospora recognita (Gale) Král;挪威刺子莞■☆

333671 Rhynchospora recognita (Gale) Král = Rhynchospora globularis (Chapm.) Small ■☆

333672 Rhynchospora rubra (Lour.) Makino;刺子莞(大一箭球,龙须草,绣球草,一包刺,一包金);Red Beakrush ■

333673 Rhynchospora rubra (Lour.) Makino subsp. africana J. Raynal;非洲刺子莞■☆

333674 Rhynchospora rubra (Lour.) Makino subsp. senegalensis J. Raynal;塞内加尔刺子莞■☆

333675 Rhynchospora rugosa (Vahl) Gale;布朗氏刺子莞(布朗氏莞)■

333676 Rhynchospora rugosa (Vahl) Gale = Rhynchospora brownii Roem. et Schult. ■☆

333677 Rhynchospora rugosa (Vahl) Gale subsp. brownii (Roem. et Schult.) T. Koyama = Rhynchospora brownii Roem. et Schult. ■

333678 Rhynchospora rugosa (Vahl) Gale var. brownii (Roem. et Schult.) T. Koyama = Rhynchospora brownii Roem. et Schult. ■

333679 Rhynchospora rugosa (Vahl) Gale var. condensata (Degl.) T. Koyama = Rhynchospora rugosa (Vahl) Gale subsp. brownii (Roem. et Schult.) T. Koyama ■

333680 Rhynchospora rugosa (Vahl) Gale var. condensata (Kük.) T. Koyama = Rhynchospora rugosa (Vahl) Gale var. brownii (Roem. et Schult.) T. Koyama ■

333681 Rhynchospora rugosa (Vahl) Gale var. condensata (Kük.) T. Koyama = Rhynchospora brownii Roem. et Schult. ■

333682 Rhynchospora ruppioides Benth. = Websteria confervoides (Poir.) S. S. Hooper ■☆

333683 Rhynchospora saxicola Small = Rhynchospora globularis (Chapm.) Small var. saxicola (Small) Kük. ■☆

333684 Rhynchospora schaffneri Boeck. = Rhynchospora kunthii Nees ex Kunth ■☆

333685 Rhynchospora schoenoides A. W. Wood = Rhynchospora elliottii A. Dietr. ■☆

333686 Rhynchospora schroederi K. Schum. = Rhynchospora eximia (Nees) Boeck. ■☆

333687 Rhynchospora scirpoides (Torr.) Griseb. ;长裂刺子莞;Long-

beaked Bald Sedge,Long-beaked Bald-rush ■☆

333688 Rhynchospora semiplumosa A. Gray = Rhynchospora plumosa Elliott ■☆

333689 Rhynchospora senegalensis Steud. = Fuirena stricta Steud. ■☆

333690 Rhynchospora setacea (Muhl.) MacMill. = Rhynchospora capillacea Torr. ■☆

333691 Rhynchospora setacea (P. J. Bergius) Boeck. var. africana Gross ex Dinkl. = Rhynchospora perrieri Cherm. ■☆

333692 Rhynchospora smallii Britton = Rhynchospora capillacea Torr. ■☆

333693 Rhynchospora solitaria R. M. Harper;单果刺子莞■☆

333694 Rhynchospora sparsa (Michx.) Vahl = Rhynchospora miliacea (Lam.) A. Gray ■☆

333695 Rhynchospora spectabilis Hochst. ex Krauss = Rhynchospora corymbosa (L.) Britton ■

333696 Rhynchospora stellata (Lam.) Griseb. var. latifolia (Baldwin) Kük. = Rhynchospora latifolia (Baldwin) W. W. Thomas ■☆

333697 Rhynchospora stenophylla Chapm. ;窄叶刺子莞■☆

333698 Rhynchospora stipitata Chapm. = Rhynchospora odorata C. Wright ex Griseb. ■☆

333699 Rhynchospora submarginata Degl. ;类缘刺子莞■

333700 Rhynchospora subquadrata Cherm. = Rhynchospora gracillima Thwaites subsp. subquadrata (Cherm.) J. Raynal ■☆

333701 Rhynchospora sulcata Gale = Rhynchospora microcarpa Baldwin ex A. Gray ■☆

333702 Rhynchospora tenerrima Nees ex Spreng. ;极细刺子莞■☆

333703 Rhynchospora tenerrima Nees ex Spreng. subsp. microcarpa J. Raynal;小果极细刺子莞■☆

333704 Rhynchospora testui Cherm. = Rhynchospora gracillima Thwaites subsp. subquadrata (Cherm.) J. Raynal ■☆

333705 Rhynchospora testui Cherm. var. pleiantha ? = Rhynchospora eximia (Nees) Boeck. ■☆

333706 Rhynchospora thornei Král;托纳刺子莞■☆

333707 Rhynchospora torreyana A. Gray;托里刺子莞■☆

333708 Rhynchospora torreyana A. Gray var. microrhyncha Griseb. = Rhynchospora microcarpa Baldwin ex A. Gray ■☆

333709 Rhynchospora tracyi Britton;特拉西刺子莞;Tracy's Beakrush ■☆

333710 Rhynchospora trichophylla Fernald = Rhynchospora gracilenta A. Gray ■☆

333711 Rhynchospora triflora Vahl;三花刺子莞■☆

333712 Rhynchospora trigyna Hochst. = Coleochloa abyssinica (Hochst. ex A. Rich.) Gilly ■☆

333713 Rhynchospora wallichiana Kunth = Rhynchospora rubra (Lour.) Makino ■

333714 Rhynchospora wrightiana Boeck. ;赖特刺子莞■☆

333715 Rhynchospora yasudana Makino;安田刺子莞■☆

333716 Rhynchospora yasudana Makino subsp. leviseta (C. B. Clarke ex H. Lév.) T. Koyama = Rhynchospora fujiiana Makino ■☆

333717 Rhynchospora yasudana Makino var. leviseta (C. B. Clarke ex H. Lév.) T. Koyama = Rhynchospora fujiiana Makino ■☆

333718 Rhynchostele Rchb. f. = Leochilus Knowles et Westc. ■☆

333719 Rhynchostemon Steetz = Thomasia J. Gay ●☆

333720 Rhynchostigma Benth. = Toxocarpus Wight et Arn. ●

333721 Rhynchostigma brevipes Benth. = Secamone brevipes (Benth.) Klack. ●☆

333722 Rhynchostigma lujae De Wild. et T. Durand = Secamone brevipes (Benth.) Klack. ●☆

333723 Rhynchostigma parviflorum Benth. = Secamone brevipes (Benth.)

Klack. ●☆

333724　Rhynchostigma racemosum Benth. = Secamone racemosa（Benth.）Klack. ■☆

333725　Rhynchostylis Blume（1825）；钻喙兰属（狐狸尾属，喙蕊兰属）；Awlbillorchis, Fox Tail Orchid, Fox-tail Orchid, Rhynchostylis ■

333726　Rhynchostylis Tausch = Chaerophyllum L. ■

333727　Rhynchostylis coelestis（Rchb. f.）Rchb. f. ex Veitch；天蓝钻喙兰；Skyblue Awlbillorchis ■☆

333728　Rhynchostylis densiflora（Lindl.）L. O. Williams = Robiquetia spatulata（Blume）J. J. Sm. ■

333729　Rhynchostylis gigantea（Lindl.）Ridl.；大花钻喙兰（大喙蕊兰，海南钻喙兰，钻喙兰）；Bigflower Awlbillorchis, Hainan Awlbillorchis ■

333730　Rhynchostylis gigantea（Lindl.）Ridl. var. petoliana Seidenf.；白喙蕊兰■☆

333731　Rhynchostylis retusa（L.）Blume；钻喙兰（喙蕊兰）；Awlbillorchis, Retuse Rhynchostylis ■

333732　Rhynchostylis violacea（Lindl.）Ridl.；堇色钻喙兰（堇喙兰）；Violet Awlbillorchis ■☆

333733　Rhynchotechum Blume（1826）；线柱苣苔属；Rhynchotechum ●

333734　Rhynchotechum brevipedunculatum J. C. Wang；短梗线柱苣苔（短梗同蕊草，线柱苣苔）●

333735　Rhynchotechum discolor（Maxim.）B. L. Burtt；异色线柱苣苔（同蕊草）；Differentcolor Rhynchotechum, Discolor Rhynchotechum ●

333736　Rhynchotechum discolor（Maxim.）B. L. Burtt f. incisum（Ohwi）Hatus. ex J. C. Wang = Rhynchotechum discolor（Maxim.）B. L. Burtt var. incisum（Ohwi）E. Walker ●

333737　Rhynchotechum discolor（Maxim.）B. L. Burtt var. austrokiushiuense（Ohwi）Ohwi；日本线柱苣苔●

333738　Rhynchotechum discolor（Maxim.）B. L. Burtt var. incisum（Ohwi）E. Walker；羽裂异色线柱苣苔（羽裂线柱苣苔）●

333739　Rhynchotechum ellipticum（Wall. ex D. Dietr.）A. DC.；椭圆线柱苣苔；Elliptic Rhynchotechum ●

333740　Rhynchotechum ellipticum（Wall. ex D. Dietr.）A. DC. = Rhynchotechum formosanum Hatus. ●

333741　Rhynchotechum ellipticum（Wall. ex D. Dietr.）A. DC. var. saurauifolia（S. S. Ying）S. S. Ying = Rhynchotechum formosanum Hatus. ●

333742　Rhynchotechum formosanum Hatus.；冠萼线柱苣苔（蓬莱同蕊草，台湾线柱苣苔）；Taiwan Rhynchotechum ●

333743　Rhynchotechum latifolium Hook. f. et Thomson ex C. B. Clarke = Rhynchotechum ellipticum（Wall. ex D. Dietr.）A. DC. ●

333744　Rhynchotechum longipes W. T. Wang；长梗线柱苣苔；Long-stalk Rhynchotechum ●

333745　Rhynchotechum obovatum（Griff.）B. L. Burtt；线柱苣苔（矮脚白风，白饭公，白马胎，大叶猪食，横脉，山枇杷）；Obovate Rhynchotechum ●

333746　Rhynchotechum obovatum（Griff.）B. L. Burtt = Rhynchotechum ellipticum（Wall. ex D. Dietr.）A. DC. ●

333747　Rhynchotechum omeiensis W. T. Wang；峨眉线柱苣苔；Emei Rhynchotechum, Omei Rhynchotechum ●

333748　Rhynchotechum saurauifolia（S. S. Ying）S. S. Ying = Rhynchotechum formosanum Hatus. ●

333749　Rhynchotechum vestitum Wall. ex C. B. Clarke；毛线柱苣苔；Hairy Rhynchotechum ●

333750　Rhynchotheca Pers. = Rhynchotheca Ruiz et Pav. ●☆

333751　Rhynchotheca Ruiz et Pav.（1794）；喙果木属（刺灌木属）●☆

333752　Rhynchotheca spinosa Ruiz et Pav.；喙果木●☆

333753　Rhynchothecaceae A. Juss. = Rhynchothecaceae Endl. ●☆

333754　Rhynchothecaceae Endl.（1841）；喙果木科（刺灌木科）●☆

333755　Rhynchothecaceae Endl. = Geraniaceae Juss.（保留科名）■●

333756　Rhynchothecaceae J. Agardh = Ledocarpaceae Meyen ●☆

333757　Rhynchothecaceae J. Agardh = Ricinocarpaceae Hurus. ●

333758　Rhynchotoechum Blume = Rhynchotechum Blume ●

333759　Rhynchotropis Harms（1901）；喙龙骨豆属■☆

333760　Rhynchotropis curtisiae I. M. Johnst. = Microcharis microcharoides（Taub.）Schrire ●☆

333761　Rhynchotropis dekindtii Harms = Rhynchotropis poggei（Taub.）Harms ■☆

333762　Rhynchotropis marginata（N. E. Br.）J. B. Gillett；具边喙龙骨豆■☆

333763　Rhynchotropis poggei（Taub.）Harms；喙龙骨豆■☆

333764　Rhynchotropis praecox Baker f. = Rhynchotropis marginata（N. E. Br.）J. B. Gillett ■☆

333765　Rhyncosia Webb = Rhynchosia Lour.（保留属名）●■

333766　Rhyncospermum A. DC. = Rhynchodia Benth. ●■

333767　Rhyncospermum jasminioides Lindl. = Trachelospermum jasminoides（Lindl.）Lem. ●

333768　Rhyncostylis Steud. = Rhynchostylis Blume ■

333769　Rhyncothecum A. DC. = Rhynchotechum Blume ●

333770　Rhynea DC. = Cassinia R. Br.（保留属名）●☆

333771　Rhynea DC. = Tenrhynea Hilliard et B. L. Burtt ■☆

333772　Rhynea Scop. = Mesua L. ●

333773　Rhynea Scop. = Nagassari Adans. ●

333774　Rhynea phylicifolia DC. = Tenrhynea phylicifolia（DC.）Hilliard et B. L. Burtt ■☆

333775　Rhynospermum Walp. = Rhynchodia Benth. ●■

333776　Rhynospermum Walp. = Rhyncospermum A. DC. ●■

333777　Rhyparia Blume ex Hassk. = Ryparosa Blume ●☆

333778　Rhyparia Hassk. = Ryparosa Blume ●☆

333779　Rhysolepis S. F. Blake（1917）；皱鳞菊属（软肋菊属）■☆

333780　Rhysolepis grandiflora（Gardner）H. Rob. et A. J. Moore；大花皱鳞菊■☆

333781　Rhysolepis linearifolia（Chodat）H. Rob. et A. J. Moore；线叶皱鳞菊■☆

333782　Rhysopteris Blume ex A. Juss. = Ryssopterys Blume ex A. Juss.（保留属名）●

333783　Rhysopterus J. M. Coult. et Rose（1900）；皱翅草属☆

333784　Rhysopterus corrugatus J. M. Coult. et Rose；皱翅草☆

333785　Rhysospermum C. F. Gaertn. = Notelaea Vent. ●☆

333786　Rhysotoechia Radlk.（1879）；皱壁无患子属●☆

333787　Rhysotoechia gracilipes Radlk.；细梗皱壁无患子●☆

333788　Rhysotoechia grandifolia Radlk.；大叶皱壁无患子●☆

333789　Rhyssocarpus Endl. = Billiottia DC. ●☆

333790　Rhyssocarpus Endl. = Melanopsidium Colla ●☆

333791　Rhyssolobium E. Mey.（1838）；皱片萝藦属☆

333792　Rhyssolobium dumosum E. Mey.；皱片萝藦☆

333793　Rhyssopteris（Blume）A. Juss. = Ryssopterys Blume ex A. Juss.（保留属名）●

333794　Rhyssopteris Blume ex. A. Juss. = Ryssopterys Blume ex A. Juss.（保留属名）●

333795　Rhyssopteris Rickctt et Stafieu = Ryssopterys Blume ex A. Juss. ●

333796　Rhyssopterys Blume ex A. Juss. = Ryssopterys Blume ex A.

Juss.（保留属名）●

333797 Rhyssopterys dealbata A. Juss. = Ryssopterys timoriensis（DC.）Blume ex A. Juss. ●

333798 Rhyssopterys timoriensis（DC.）Blume ex A. Juss. = Ryssopterys timoriensis（DC.）Blume ex A. Juss. ●

333799 Rhyssopteryx Dalla Torre et Harms = Rhyssopterys Blume ex A. Juss. ●

333800 Rhyssostelma Decne.（1844）;皱冠萝藦属☆

333801 Rhyssostelma nigricans Decne. ;皱冠萝藦☆

333802 Rhytachne Desv. = Rhytachne Desv. ex Ham. ■

333803 Rhytachne Desv. ex Ham.（1825）;皱颖草属■

333804 Rhytachne anisonodis B. S. Sun;异节皱颖草■

333805 Rhytachne anisonodis B. S. Sun = Phacelurus trichophyllus S. L. Zhong ■

333806 Rhytachne benguellensis Rendle = Rhytachne rottboellioides Desv. ex Ham. ■☆

333807 Rhytachne bovonei（Chiov.）Chiov. = Loxodera bovonei（Chiov.）Launert ■☆

333808 Rhytachne caespitosa（Baker）Bosser = Rhytachne rottboellioides Desv. ex Ham. ■☆

333809 Rhytachne congoensis Hack. = Phacelurus gabonensis（Steud.）Clayton ■☆

333810 Rhytachne furtiva Clayton;隐匿皱颖草■☆

333811 Rhytachne gabonensis（Steud.）Hack. = Phacelurus gabonensis（Steud.）Clayton ■☆

333812 Rhytachne geminatosubulata P. A. Duvign. = Rhytachne rottboellioides Desv. ex Ham. ■☆

333813 Rhytachne gigantea Stapf = Urelytrum giganteum Pilg. ■☆

333814 Rhytachne glabra（Gledhill）Clayton;光皱颖草■☆

333815 Rhytachne gracilis Stapf;纤细皱颖草■☆

333816 Rhytachne latifolia Clayton;宽叶皱颖草■☆

333817 Rhytachne lijiangensis B. S. Sun;丽江皱颖草■

333818 Rhytachne lijiangensis B. S. Sun = Phacelurus trichophyllus S. L. Zhong ■

333819 Rhytachne mannii Stapf = Rhytachne rottboellioides Desv. ex Ham. ■☆

333820 Rhytachne megastachya Jacq. -Fél. ;大穗皱颖草■

333821 Rhytachne minor Pilg. = Rhytachne gracilis Stapf ■☆

333822 Rhytachne perfecta Jacq. -Fél. ;完全皱颖草■

333823 Rhytachne pilosa Ballard et C. E. Hubb. = Loxodera bovonei（Chiov.）Launert ■☆

333824 Rhytachne princeps（A. Rich.）T. Durand et Schinz = Thelepogon elegans Roth ex Roem. et Schult. ■☆

333825 Rhytachne robusta Stapf;粗壮皱颖草■

333826 Rhytachne rottboellioides Desv. = Rhytachne rottboellioides Desv. ex Ham. ■☆

333827 Rhytachne rottboellioides Desv. ex Ham. ;筒轴茅皱颖草■☆

333828 Rhytachne rottboellioides Desv. var. guineensis A. Camus et Schnell = Rhytachne rottboellioides Desv. ex Ham. ■☆

333829 Rhytachne triaristata（Steud.）Stapf;三芒皱颖草■☆

333830 Rhytachne triseta Hack. = Rhytachne triaristata（Steud.）Stapf ■☆

333831 Rhyticalymma Bremek. = Justicia L. ●■

333832 Rhyticarpus Sond. = Anginon Raf. ■☆

333833 Rhyticarpus difformis（L.）Briq. = Anginon difforme（L.）B. L. Burtt ■☆

333834 Rhyticarpus ecklonis Sond. = Anginon swellendamense（Eckl. et Zeyh.）B. L. Burtt ■☆

333835 Rhyticarpus rugosus（Thunb.）Sond. = Anginon rugosum（Thunb.）Raf. ■☆

333836 Rhyticarpus swellendamensis（Eckl. et Zeyh.）Briq. = Anginon swellendamense（Eckl. et Zeyh.）B. L. Burtt ■☆

333837 Rhyticarum Boerl. = Rhyticaryum Becc. ●☆

333838 Rhyticaryum Becc.（1877）;新几内亚棕属●☆

333839 Rhyticaryum oleraceum Becc. ;新几内亚棕（皱果壳果棕）●☆

333840 Rhyticocos Becc.（1887）;皱苞椰属（皱果壳果棕属）●☆

333841 Rhyticocos Becc. = Syagrus Mart. ●

333842 Rhyticocos longifolium Schum. et Lauterb. ;长叶皱苞椰（长叶皱果壳果棕）●☆

333843 Rhytidachne K. Schum. = Rhytachne Desv. ex Ham. ■

333844 Rhytidandra A. Gray = Alangium Lam.（保留属名）●

333845 Rhytidanthe Benth. = Leptorhynchos Less. ●☆

333846 Rhytidanthera（Planch.）Tiegh.（1904）;皱药木属●☆

333847 Rhytidanthera（Planch.）Tiegh. = Rhytidanthera Tiegh. ●☆

333848 Rhytidanthera Tiegh. = Rhytidanthera（Planch.）Tiegh. ●☆

333849 Rhytidanthera splendida Tiegh. ;皱药木●☆

333850 Rhytidocaryum K. Schum. et Lauterb. = Rhyticaryum Becc. ●☆

333851 Rhytidocaulon Nyl. ex Elenkin（废弃属名）= Rhytidocaulon P. R. O. Bally ●☆

333852 Rhytidocaulon P. R. O. Bally（1963）;皱茎萝藦属●☆

333853 Rhytidocaulon paradoxum P. R. O. Bally;奇异皱茎萝藦●☆

333854 Rhytidocaulon piliferum Lavranos;纤毛皱茎萝藦●☆

333855 Rhytidocaulon richardianum Lavranos;理查德皱茎萝藦●☆

333856 Rhytidocaulon scandens P. R. O. Bally;攀缘皱茎萝藦●☆

333857 Rhytidocaulon specksii McCoy;斯佩克斯皱茎萝藦●☆

333858 Rhytidocaulon subscandens P. R. O. Bally;亚攀缘皱茎萝藦●☆

333859 Rhytidomene Rydb. = Orbexilum Raf. ●☆

333860 Rhytidomene Rydb. = Psoralea L. ●■

333861 Rhytidophyllum Mart.（1829）（'Rytidophyllum'）（保留属名）;皱叶苣苔属;Rhytidophyllum ●☆

333862 Rhytidophyllum tomentosum Mart. ;毛皱叶苣苔■☆

333863 Rhytidosolen Tiegh. = Arthrosolen C. A. Mey. ●☆

333864 Rhytidospermum Sch. Bip. = Tripleurospermum Sch. Bip. ■

333865 Rhytidosporum F. Muell. = Marianthus Hügel ex Endl. ●☆

333866 Rhytidosporum F. Muell. ex Hook. f. = Marianthus Hügel ex Endl. ●☆

333867 Rhytidostylis Rchb. = Rytidostylis Hook. et Arn. ■☆

333868 Rhytidotus Hook. f. = Bobea Gaudich. ●☆

333869 Rhytidotus Hook. f. = Rytidotus Hook. f. ●☆

333870 Rhytiglossa Nees ex Lindl. = Isoglossa Oerst.（保留属名）■★

333871 Rhytiglossa Nees（废弃属名）= Dianthera L. ■☆

333872 Rhytiglossa Nees（废弃属名）= Isoglossa Oerst.（保留属名）■★

333873 Rhytiglossa Nees（废弃属名）= Justicia L. ●■

333874 Rhytiglossa ciliata Nees = Isoglossa ciliata（Nees）Lindau ■☆

333875 Rhytiglossa eckloniana Nees = Isoglossa eckloniana（Nees）Lindau ■☆

333876 Rhytiglossa origanoides Nees = Isoglossa origanoides（Nees）Lindau ■☆

333877 Rhytiglossa ovata Nees = Isoglossa ovata（Nees）Lindau ■☆

333878 Rhytiglossa pectoralis（Jacq.）Nees = Justicia pectoralis Jacq. ■☆

333879 Rhytiglossa prolixa Nees = Isoglossa prolixa（Nees）Lindau ■☆

333880 Rhytileucoma F. Muell. = Chilocarpus Blume ●☆

333881 Rhytionanthos aemulus（W. W. Sm.）Garay, Hamer et Siegerist = Bulbophyllum forrestii Seidenf. ■

333882 Rhytionanthos sphaericus（Z. H. Tsi et H. Li）Garay, Hamer et

Siegerist　= Bulbophyllum sphaericum Z. H. Tsi et H. Li ■

333883　Rhytis Lour. = Antidesma L. ●

333884　Rhytispermum Link　= Aegonychon Gray ■

333885　Rhytispermum Link　= Alkanna Tausch(保留属名)●☆

333886　Rhytispermum Link　= Lithospermum L. ■

333887　Rhytispermum arvense (L.) Lind = Lithospermum arvense L. ■

333888　Rhyttiglossa T. Anderson = Justicia L. ●■

333889　Riana Aubl. = Rinorea Aubl. (保留属名) ●

333890　Ribeirea Allemão = Schoepfia Schreb. ●

333891　Ribeirea Arruda ex H. Kost. = Hancornia Gomes ●☆

333892　Ribeirta Willis = Ribeirea Arruda ex H. Kost. ●☆

333893　Ribes L. (1753);茶藨子属;Currant, Gooseberry ●

333894　Ribes aciculare Sm.;阿尔泰茶藨子(阿尔泰醋栗,五刺茶藨,西伯利亚茶藨,西伯利亚醋栗,针状茶藨);Altai Gooseberry, Needleshaped Currant, Needle-shaped Currant ●

333895　Ribes acuminatum Wall. et G. Don = Ribes takare D. Don ●

333896　Ribes acuminatum Wall. et G. Don var. desmacarpum (Hook. f. et Thomson) Jancz. = Ribes takare D. Don var. desmacarpum (Hook. f. et Thomson) L. T. Lu ●

333897　Ribes acuminatum Wall. et G. Don var. maius Jancz. = Ribes takare D. Don ●

333898　Ribes albinervium Michx. = Ribes triste Pall. ●

333899　Ribes alpestre Decne. = Ribes alpestre Wall. ex Decne. ●

333900　Ribes alpestre Wall. ex Decne.;长刺茶藨子(茶茹,刺茶藨,刺李,大刺茶藨,大刺茶藨子,高山醋栗,密刺高山茶藨,亚高山茶藨);Hedge Gooseberry, Sensespine Hedge Gooseberry ●

333901　Ribes alpestre Wall. ex Decne. var. commune Jancz. = Ribes alpestre Wall. ex Decne. ●

333902　Ribes alpestre Wall. ex Decne. var. eglandulosum L. T. Lu;无腺茶藨子(无腺茶藨);Glanduleless Hedge Gooseberry ●

333903　Ribes alpestre Wall. ex Decne. var. giganteum Jancz.;大刺茶藨子(长刺茶藨子,长刺李,大刺茶藨,光果高山茶藨,光果高山醋栗);Smoothfruit Hedge Gooseberry ●

333904　Ribes alpinum L.;高山茶藨子(高山茶藨);Alpine Currant, Mountain Currant ●☆

333905　Ribes alpinum L. 'Aureum';金叶高山茶藨子●☆

333906　Ribes alpinum L. 'Compactum';紧凑高山茶藨子●☆

333907　Ribes alpinum L. 'Green Mound';绿堆高山茶藨子●☆

333908　Ribes alpinum L. 'Laciniatum';深裂高山茶藨子●☆

333909　Ribes alpinum L. 'Pumilum';矮生高山茶藨子●☆

333910　Ribes alpinum L. 'Schmidt';施密特高山茶藨子●☆

333911　Ribes alpinum L. var. japonicum Maxim. = Ribes maximowiczii Batalin ●

333912　Ribes alpinum L. var. mandshuricum Maxim. = Ribes maximowiczii Batalin ●

333913　Ribes altissimum Turcz. ex Ledeb. et Pojark.;高茶藨子(高茶藨);High Currant, Tall Currant ●

333914　Ribes ambiguum Maxim.;四川蔓茶藨子(南醋栗,苟,四川茶藨,四川蔓茶藨);Doubtful Currant, Epiphyte Currant ●

333915　Ribes ambiguum Maxim. var. glabrum Ohwi;光四川蔓茶藨子●☆

333916　Ribes americanum Mill.;美洲茶藨子(黄银茶藨,美国茶藨,美国黑茶藨,美洲茶藨,密苏里茶藨);American Black Currant, American Blackcurrant, American Currant, Eastern Black Currant, Missouri Gooseberry, Wild Black Currant, Wild Blackcurrant, Yellow Silver Currant ●

333917　Ribes americanum Mill. f. paucyglandulosum Fassett = Ribes americanum Mill. ●

333918　Ribes atropurpureum C. A. Mey. ;红花茶藨(紫花茶藨)●

333919　Ribes aureum Pursh;金花茶藨子(金茶藨,金花茶藨);Buffalo Currant, Flowering Currant, Golden Currant, Golden Flowering Currant, Missouri Currant, Yellow-flowered Currant ●☆

333920　Ribes aureum Pursh var. villosum DC. = Ribes odoratum H. L. Wendl. ●

333921　Ribes aureum sensu Lindl. = Ribes odoratum H. L. Wendl. ●

333922　Ribes benthamii Lavaleé;本瑟姆茶藨子 (比通茶藨);Beaton Currant ●☆

333923　Ribes bethmontii Jancz. ;柏斯蒙茶藨;Bethmont Currant ●☆

333924　Ribes biebersteinii Berlin;毕氏茶藨子●☆

333925　Ribes billiardii Carrière = Ribes fasciculatum Siebold et Zucc. var. chinense Maxim. ●

333926　Ribes bolivianum Jancz. ;玻尔维亚茶藨子●☆

333927　Ribes brachybotrys (Wedd.) Jancz. ;短穗茶藨子●☆

333928　Ribes bracteosum Douglas ex Hook.;具苞茶藨子(加州黑茶藨,具苞茶藨);Alaska Currant, California Black Currant, Currant ●☆

333929　Ribes burejense F. Schmidt;刺果茶藨子(刺醋栗,刺儿李,刺果茶藨,刺果蔓茶藨,刺梨,刺李,醋栗,拟茶藨,山梨,酸溜溜);Bureja Gooseberry ●

333930　Ribes burejense F. Schmidt var. inermis D. Z. Lu;光果刺李;Smooth-fruited Bureja Gooseberry ●

333931　Ribes burejense F. Schmidt var. villosum L. T. Lu;长毛茶藨子(长毛茶藨);Villose Bureja Gooseberry ●

333932　Ribes californicum Hook. et Arn. ;加洲茶藨子(加洲茶藨);California Currant ●☆

333933　Ribes campanulatum Moench = Ribes americanum Mill. ●

333934　Ribes carpathicum Kit. ex Schult. = Ribes petraeum Wulfen ●☆

333935　Ribes carrierei C. K. Schneid. ;开锐茶藨子 (开锐茶藨);Carier Currant ●☆

333936　Ribes cereum Douglas;蜡茶藨子(蜡茶藨);Squaw Currant, Wax Currant ●☆

333937　Ribes chilense K. Koch;智利茶藨子●☆

333938　Ribes chinense Hance = Ribes fasciculatum Siebold et Zucc. var. chinense Maxim. ●

333939　Ribes coeleste Jancz. = Ribes tenue Jancz. ●

333940　Ribes coloradense Coville;有色茶藨子(有色茶藨);Colorado Currant, Coloured Currant ●☆

333941　Ribes culverwellii Macfarl. ;醋栗叶茶藨子(库维勒茶藨子);Curverwell Currant ●☆

333942　Ribes cuneatum Kar. et Kir. = Ribes saxatile Pall. ●

333943　Ribes curvatum Small;乔治亚茶藨子(乔治亚茶藨);Georgia Gooseberry, Granite Gooseberry ●☆

333944　Ribes cyathiforme Pojark. = Ribes nigrum L. ●

333945　Ribes cynosbati L. ;牧场茶藨子(牧场茶藨);American Gooseberry, Dogberry, Eastern Prickly Gooseberry, Pasture Gooseberry, Prickly Gooseberry, Prickly Wild Gooseberry, Wild Gooseberry ●☆

333946　Ribes cynosbati L. f. inerme Rehder = Ribes cynosbati L. ●☆

333947　Ribes cynosbati L. var. atrox Fernald = Ribes cynosbati L. ●☆

333948　Ribes cynosbati L. var. glabratum Fernald = Ribes cynosbati L. ●☆

333949　Ribes cynosbati L. var. inerme (Rehder) L. H. Bailey = Ribes cynosbati L. ●☆

333950　Ribes davidii Franch. ;革叶茶藨子(草叶茶藨,大卫茶藨,冬茶藨,革叶茶藨,石夹生,小活血,小石生,岩石榴);David Currant, David Gooseberry ●

333951　Ribes davidii Franch. var. ciliatum L. T. Lu;睫毛茶藨子(睫毛

茶藨）；Ciliate David Gooseberry ●

333952 Ribes davidii Franch. var. lobatum L. T. Lu；浅裂茶藨子（浅裂茶藨）；Lobed David Gooseberry ●

333953 Ribes densiflorum Liou ＝Ribes palczewskii（Jancz.）Pojark. ●

333954 Ribes desmacarpum Hook. f. et Thomson ＝Ribes takare D. Don var. desmacarpum（Hook. f. et Thomson）L. T. Lu ●

333955 Ribes diacantha Pall.；双刺茶藨子（二刺茶藨，双刺茶藨，楔叶茶藨）；Siberia Currant，Siberian Currant ●

333956 Ribes diaseantha Pall. ＝Ribes saxatile Pall. ●

333957 Ribes dikuscha Fisch. ex Turcz.；迪库氏茶藨子（迪库氏茶藨）；Dikuscha Currant ●☆

333958 Ribes distans Jancz. ＝Ribes maximowiczianum Kom. ●

333959 Ribes divaricatum Douglas；极叉分茶藨子（极叉分茶藨）；Straggly Currant，Straggly Gooseberry，Worcesterberry ●☆

333960 Ribes domesticum Jancz. ＝Ribes sativum Syme ●☆

333961 Ribes echinellum（Coville）Rehder；刺茶藨子（刺茶藨）；Spiny Currant ●☆

333962 Ribes emodense Rehder；糖茶藨（埃牟茶藨子，滇藏醋栗，糖茶藨子）；Himalayan Currant，Himalayas Currant ●

333963 Ribes emodense Rehder ＝Ribes himalense Royle ex Decne. ●

333964 Ribes emodense Rehder ＝Ribes odoratum H. L. Wendl. ●

333965 Ribes emodense Rehder var. urceolatum（Jancz.）Rehder；坛花糖茶藨；Urceolate Himalayan Currant ●

333966 Ribes emodense Rehder var. urceolatum（Jancz.）Rehder ＝Ribes emodense Rehder ●

333967 Ribes emodense Rehder var. urceolatum（Jancz.）Rehder ＝Ribes himalense Royle ex Decne. ●

333968 Ribes emodense Rehder var. verruculosum Rehder ＝Ribes himalense Royle ex Decne. var. verruculosum（Rehder）L. T. Lu ●

333969 Ribes fargesii Franch.；花茶藨子（城口茶藨，法氏茶藨，花茶藨）；Farges' Currant ●

333970 Ribes fasciculatum Siebold et Zucc.；簇花茶藨子（茶藨子，簇花茶藨，蔓茶藨，日本茶藨）；Clustered Redcurrant，Japan Currant，Japanese Currant，Winterberry Currant ●

333971 Ribes fasciculatum Siebold et Zucc. ＝Ribes fasciculatum Siebold et Zucc. var. chinense Maxim. ●

333972 Ribes fasciculatum Siebold et Zucc. var. chinense Maxim.；华蔓茶藨子（簇花茶藨，大蔓茶藨，华蔓茶藨，华蔓茶藨，蔓蛇藨，三升米，三升末）；China Winterberry Currant，Chinese Winterberry Currant，Fragrant Currant ●

333973 Ribes fasciculatum Siebold et Zucc. var. guizhouense L. T. Lu；贵州茶藨子（贵州茶藨）；Guizhou Currant ●

333974 Ribes floridum L'Hér. ＝Ribes americanum Mill. ●

333975 Ribes floridum L'Hér. var. grandiflorum Lud.？＝Ribes americanum Mill. ●

333976 Ribes floridum L'Hér. var. parviflorum Lud.？＝Ribes americanum Mill. ●

333977 Ribes fontenayense Jancz.；泉水茶藨子（泉水茶藨）；Waterspring Currant ●☆

333978 Ribes formosanum Hayata；台湾茶藨子（台湾茶藨，台湾醋栗）；Formosan Gooseberry，Taiwan Currant ●

333979 Ribes formosanum Hayata var. sinanense（F. Maek.）Kitam. ＝Ribes sinanense F. Maek. ●☆

333980 Ribes fragrans Lodd. ＝Ribes odoratum H. L. Wendl. ●

333981 Ribes fragrans Pall.；甜茶藨子 ●☆

333982 Ribes franchetii Jancz.；鄂西茶藨子（川鄂茶藨，鄂西茶藨，云南茶藨）；Frachet's Currant ●

333983 Ribes fuscescens Jancz.；褐毛茶藨子（褐毛茶藨）；Brownhair Currant ●☆

333984 Ribes fuyunense T. C. Ku et F. Konta；富蕴茶藨子（富蕴茶藨）；Fuyun Currant ●

333985 Ribes gayanum（Spach）Steud.；佳易茶藨子（智利茶藨）；Chilean Currant ●☆

333986 Ribes giraldii Jancz.；腺毛茶藨子（纪氏茶藨，老铁山茶藨，陕西茶藨，腺毛茶藨）；Girald's Currant ●

333987 Ribes giraldii Jancz. var. cuneatum F. T. Wang et H. L. Li；滨海茶藨子（滨海茶藨）；Cuneate Girald Currant ●

333988 Ribes giraldii Jancz. var. polyanthum Kitag.；旅顺茶藨子（旅顺茶藨）；Manyflower Girald Currant ●

333989 Ribes glabricalycinum L. T. Lu；光萼茶藨子（光萼茶藨）；Glabrous-calyx Currant ●

333990 Ribes glabrilolium L. T. Lu；光叶茶藨子（光叶茶藨）；Glabrous-leaf Currant ●

333991 Ribes glaciale Wall.；冰川茶藨子（冰川茶藨，奶浆子）；Glacial Currant，Nepal Currant ●

333992 Ribes glaciale Wall. ＝Ribes takare D. Don ●

333993 Ribes glaciale Wall. f. hirsutum T. C. Ku；腺毛冰川茶藨（粗毛冰川茶藨）●

333994 Ribes glaciale Wall. f. hirsutum T. C. Ku ＝Ribes lucidum Hook. f. et Thomson ●

333995 Ribes glaciale Wall. var. glandulosum Jancz. ＝Ribes glaciale Wall. ●

333996 Ribes glaciale Wall. var. laciniatum（Hook. f. et Thomson）Clarke ＝Ribes laciniatum Hook. f. et Thomson ●

333997 Ribes glandulosum Grauer ex Weber；具腺茶藨；Skunk Currant ●☆

333998 Ribes glandulosum Ruiz et Pav. ＝Ribes glandulosum Weber ●☆

333999 Ribes glandulosum Weber ＝Ribes glandulosum Grauer ex Weber ●☆

334000 Ribes glandulosum Weber ＝Ribes prostratum L'Hér. ●☆

334001 Ribes glutinosum Benth. ＝Ribes villosum Wall. ●

334002 Ribes gongshanense T. C. Ku ＝Ribes griffithii Hook. f. et Thomson var. gongshanense（T. C. Ku）L. T. Lu ●

334003 Ribes gordonianum Lem.；戈登氏茶藨子（暗红茶藨子，戈登氏茶藨）；Currant，Gordon Currant ●☆

334004 Ribes gracile Pursh ＝Ribes missouriense Nutt. ●☆

334005 Ribes gracillimum K. S. Hao ＝Ribes longiracemosum Franch. var. gracillimum L. T. Lu ●

334006 Ribes graveolens Bunge；臭茶藨；Strong-smelling Currant ●

334007 Ribes griffithii Hook. f. et Thomson；曲萼茶藨子（藏南茶藨，曲萼茶藨）；Griffith Currant ●

334008 Ribes griffithii Hook. f. et Thomson var. gongshanense（T. C. Ku）L. T. Lu；贡山茶藨子（贡山茶藨）；Gongshan Currant，Gongshan Griffith Currant ●

334009 Ribes grossularia L. ＝Grossularia uva-rrispa Mill. ●

334010 Ribes grossularia L. ＝Ribes alpestre Wall. ex Decne. ●

334011 Ribes grossularia L. ＝Ribes reclinatum L. ●

334012 Ribes grossularioides Hemsl. ＝Ribes burejense F. Schmidt ●

334013 Ribes guangxiense C. Z. Gao ＝Ribes hunanense Chang Y. Yang et C. J. Qi ●

334014 Ribes haoi C. Y. Yang et Y. L. Han；细穗茶藨（细穗醋栗）；Hao's Currant ●

334015 Ribes haoi C. Y. Yang et Y. L. Han ＝Ribes longiracemosum Franch. var. gracillimum L. T. Lu ●

334016 Ribes henryi Franch.；华中茶藨子（亨利茶藨，亨利茶藨子，

华中茶藨,睫毛茶藨,岩马桑,钻石风);Henry Currant ●

334017　Ribes heterotrichum C. A. Mey. ;小叶茶藨子(异毛茶藨子,圆叶茶藨,圆叶茶藨子);Small-leaved Currant ●

334018　Ribes heterotrichum Royle = Ribes villosum Wall. ●

334019　Ribes himalayense Royle = Ribes himalayense Royle ex Decne. ●

334020　Ribes himalayense Royle ex Decne. var. appendiculatum ? = Ribes himalense Royle ex Decne. ●

334021　Ribes himalayense Royle ex Decne. var. decaisnei Jancz. = Ribes himalense Royle ex Decne. ●

334022　Ribes himalayense Royle ex Decne. var. uvceolatum Jancz. = Ribes himalense Royle ex Decne. ●

334023　Ribes himalense Decne. = Ribes himalense Royle ex Decne. ●

334024　Ribes himalense Royle = Ribes alpestre Decne. ●

334025　Ribes himalense Royle ex Decne. ;糖茶藨子(埃牟茶藨,糖茶藨,喜马拉雅茶藨);Himalayan Currant ●

334026　Ribes himalense Royle ex Decne. var. glandulosum Jancz. ;疏腺茶藨子(疏腺茶藨)●

334027　Ribes himalense Royle ex Decne. var. pubicalycinum L. T. Lu et J. T. Pan;毛萼茶藨子(毛萼茶藨);Hairy-calyx Himalayan Currant ●

334028　Ribes himalense Royle ex Decne. var. trichophyllum T. C. Ku;异毛茶藨子(异毛茶藨);Hairy Himalayan Currant ●

334029　Ribes himalense Royle ex Decne. var. verruculosum (Rehder) L. T. Lu;瘤糖茶藨子(瘤埃牟茶藨,瘤糖茶藨);Verruculose Himalayan Currant ●

334030　Ribes hirtellum Michx. ;毛茎茶藨子(毛茎茶藨);Hairy Gooseberry, Hairystem Gooseberry, Hairy-stem Gooseberry, Northern Gooseberry, Smooth Gooseberry, Swamp Gooseberry ●☆

334031　Ribes hirtellum Michx. var. calcicola (Fernald) Fernald = Ribes hirtellum Michx. ●☆

334032　Ribes hirtellum Michx. var. saxosum (Hook.) Fernald = Ribes hirtellum Michx. ●☆

334033　Ribes hispidulum Pojark. ;刺毛茶藨(糙茶藨)●☆

334034　Ribes horridum Rupr. ex Maxim. ;密刺茶藨子(黑果茶藨,密刺茶藨)●

334035　Ribes hortense Hedl. = Ribes sativum Syme ●☆

334036　Ribes houghtonianum Jancz. ;河通茶藨子(河通茶藨);Houghton Currant ●☆

334037　Ribes hudsonianum Rich. ;哈得逊茶藨子(哈得逊茶藨);Canadian Black Currant, Hudson Currant, Northern Black Currant ●☆

334038　Ribes humile Jancz. ;矮醋栗(黄果矮茶藨,黄果茶藨);Dwarf Currant ●

334039　Ribes hunanense Chang Y. Yang et C. J. Qi;湖南茶藨子(广西茶藨,湖南茶藨);Guangxi Currant, Hunan Currant ●

334040　Ribes huronense Rydb. = Ribes cynosbati L. ●☆

334041　Ribes incarnatum Wedd. ;肉色茶藨子●☆

334042　Ribes inebrians Lindl. ;醉茶藨子(醉茶藨);Squaw Currant, Wild Currant, Wild Gooseberry ●☆

334043　Ribes inerme Rydb. ;白茎茶藨子(白茎茶藨);White-stem Gooseberry, Whitestem Gooseberry, Whitestemmed Gooseberry ●☆

334044　Ribes innominatum Jancz. ;白叶茶藨子(白叶茶藨);Whiteleaf Currant ●☆

334045　Ribes intermedium Carrière = Ribes americanum Mill. ●

334046　Ribes intermedium Tausch = Ribes americanum Mill. ●

334047　Ribes irriguum Douglas;有水茶藨子(有水茶藨);Watering Currant ●☆

334048　Ribes janczevskii Pojark. ;茶藨子(塔城茶藨子);Janczewsky Currant ●

334049　Ribes japonicum Carrière = Ribes fasciculatum Siebold et Zucc. ●

334050　Ribes japonicum Maxim. = Ribes fasciculatum Siebold et Zucc. ●

334051　Ribes jessoniae Stapf = Ribes maximowiczii Batalin ●

334052　Ribes kansuense K. S. Hao = Ribes himalense Royle ex Decne. var. verruculosum (Rehder) L. T. Lu ●

334053　Ribes kialanum Jancz. ;康边茶藨子(康边茶藨)●

334054　Ribes kolymense Kom. ex Pojark. ;科雷马茶藨子●☆

334055　Ribes komarovii Pojark. ;长白茶藨子(长白茶藨);Komarov Currant ●

334056　Ribes komarovii Pojark. var. cuneifolium Liou;楔叶长白茶藨子(楔叶长白茶藨);Wedge-leaved Komarov Currant ●

334057　Ribes laciniatum Hook. f. et Thomson;裂叶茶藨子(裂叶茶藨,深裂茶藨,狭萼茶藨);Laciniate Currant, Lobed Currant ●

334058　Ribes lacustre (Pers.) Poir. ;湖沼茶藨子(湖沼茶藨);Bristly Black Curramt, Prickly Currant, Spiny Swamp Currant, Swamp Currant, Swamp Gooseberry ●☆

334059　Ribes lacustre (Pers.) Poir. var. horridum (Rupr. ex Maxim.) Jancz. = Ribes horridum Rupr. ex Maxim. ●

334060　Ribes lacustre (Pers.) Poir. var. parvulum A. Gray = Ribes lacustre (Pers.) Poir. ●☆

334061　Ribes latifolium Jancz. ;阔叶茶藨子(阔叶茶藨);Broad-leaf Currant ●

334062　Ribes laurifolium Jancz. ;月桂叶茶藨子(桂叶茶藨);Currant, Laurelleaf Currant, Laurel-leaved Currant ●

334063　Ribes laurifolium Jancz. var. yunnanense L. T. Lu;光果茶藨子(光果茶藨);Yunnan Laureleaf Currant ●

334064　Ribes laxiflorum Pursh;疏花茶藨子(疏花茶藨);Alaska Black Currant, Trailing Currant ●☆

334065　Ribes leptanthum A. Gray;喇叭茶藨子(喇叭茶藨);Trumpet Gooseberry ●☆

334066　Ribes leptostachyum Decne. = Ribes orientale Desf. ●

334067　Ribes leptostachyum Decne. = Ribes villosum Wall. ●

334068　Ribes liouanum Kitag. = Ribes palczewskii (Jancz.) Pojark. ●

334069　Ribes liouii Z. Wang et C. Y. Yang = Ribes palczewskii (Jancz.) Pojark. ●

334070　Ribes lobbii A. Gray;洛布氏茶藨子(洛布氏茶藨);Lobb's Gooseberry ●☆

334071　Ribes longiflorum Nutt. ;长花茶藨子;Buffalo Currant, Clove Currant ●☆

334072　Ribes longiracemosum Franch. ;长序茶藨子(长串茶藨,长穗茶藨,长序茶藨,红花茶藨);Long Racemose Gooseberry, Longraceme Currant, Long-racemed Currant, Wistaria Currant ●

334073　Ribes longiracemosum Franch. var. davidii Jancz. ;腺毛长序茶藨子(腺毛茶藨,腺毛茶藨子,腺毛长串茶藨,腺毛长总序茶藨);David Currant ●

334074　Ribes longiracemosum Franch. var. gracillimum L. T. Lu;纤细茶藨子(纤细茶藨);Slender Wistaria Currant ●

334075　Ribes longiracemosum Franch. var. pilosum T. C. Ku;毛长串茶藨子(毛长串茶藨);Pilose Longraceme Currant ●

334076　Ribes longiracemosum Franch. var. wilsonii Jancz. ;小花长串茶藨(小花茶藨);Littleflower Currant ●

334077　Ribes longiracemosum Franch. var. wilsonii Jancz. = Ribes longiracemosum Franch. ●

334078　Ribes lucidum Hook. f. et Thomson;紫花茶藨子(紫花茶藨);Dirty-brown Currant, Purpleflower Currant ●

334079　Ribes macrobotrys Hort. ex K. Koch;大穗茶藨子●☆

334080　Ribes macrocalyx Hance；大花醋栗；Big-calyx Currant ●

334081　Ribes macrocalyx Hance ＝ Ribes burejense F. Schmidt ●

334082　Ribes malvaceum Sm.；丛林茶藨子；Chaparral Currant ●☆

334083　Ribes malvifolium Pojark.；锦葵叶醋栗●☆

334084　Ribes mandschuricum（Maxim.）Kom.；东北茶藨子（醋栗，灯笼果，东北茶藨，东北醋栗，狗葡萄，满洲茶藨，山麻子，山樱桃）；Manschurian Currant ●

334085　Ribes mandschuricum（Maxim.）Kom. var. subglabrum Kom.；光叶东北茶藨子（光叶东北茶藨，毛山麻子，疏毛东北茶藨）；Hairy Manschurian Currant ●

334086　Ribes mandschuricum（Maxim.）Kom. var. villosum Kom.；内蒙茶藨子（内蒙茶藨）；Villose Manschurian Currant ●

334087　Ribes maximowiczianum Kom.；尖叶茶藨子（尖叶茶藨）；Sharp-leaf Currant ●

334088　Ribes maximowiczianum Kom. var. saxatile Kom. ＝ Ribes komarovii Pojark. ●

334089　Ribes maximowiczii Batalin；华西茶藨子（刺果茶藨，多花刺果茶藨，华西茶藨，马氏茶藨）；Manyflower Maximowicz Currant, Maximowicz Currant ●

334090　Ribes maximowiczii Batalin var. floribundum Jesson ＝ Ribes maximowiczii Batalin ●

334091　Ribes melancholicum Siev. ex Pall. ＝ Ribes triste Pall. ●

334092　Ribes melanocarpum K. Koch ＝ Ribes himalense Royle ●

334093　Ribes menziesii Pursh；孟席氏茶藨子（孟席氏茶藨）；Menzies Gooseberry ●☆

334094　Ribes meyeri Maxim.；天山茶藨子（麦氏茶藨，麦氏醋栗，天山茶藨，五裂茶藨）；Meyer Currant, Meyer's Currant ●

334095　Ribes meyeri Maxim. var. pubescens L. T. Lu；北疆茶藨子（北疆茶藨）；Pubescent Meyer. Currant ●

334096　Ribes meyeri Maxim. var. tanguticum Jancz. ＝ Ribes meyeri Maxim. ●

334097　Ribes meyeri Maxim. var. tanguticum Jancz. ＝ Ribes tanguticum（Jancz.）Pojark. ●

334098　Ribes meyeri Maxim. var. turkestanicum Jancz. ＝ Ribes meyeri Maxim. ●

334099　Ribes missouriense Bean；密苏里茶藨子；Missouri Gooseberry, Wild Gooseberry ●☆

334100　Ribes missouriense Bean ＝ Ribes americanum Mill. ●

334101　Ribes missouriense Bean var. ozarkanum Fassett ＝ Ribes missouriense Bean ●☆

334102　Ribes missouriense Nutt. ＝ Ribes missouriense Bean ●☆

334103　Ribes missouriense Nutt. var. ozarkanum Fassett ＝ Ribes missouriense Nutt. ●☆

334104　Ribes moupinense Franch.；宝兴茶藨子（宝兴茶藨，穆坪茶藨，穆坪醋栗）；Baoxing Currant, Paohsing Currant, Tree Currant ●

334105　Ribes moupinense Franch. f. incisoserratum T. C. Ku；缺裂宝兴茶藨●

334106　Ribes moupinense Franch. f. incisoserratum T. C. Ku ＝ Ribes griffithii Hook. f. et Thomson ●

334107　Ribes moupinense Franch. var. laxiflorum Jancz. ＝ Ribes moupinense Franch. ●

334108　Ribes moupinense Franch. var. lobatum Jancz. ＝ Ribes moupinense Franch. ●

334109　Ribes moupinense Franch. var. muliense S. H. Yu et J. M. Xu；木里茶藨子（木里茶藨）；Baoxing Currant, Muli Tree Currant ●

334110　Ribes moupinense Franch. var. pubicarpum L. T. Lu；毛果茶藨子（毛果茶藨）；Hairy-fruit Tree Currant ●

334111　Ribes moupinense Franch. var. tripartitum（Batalin）Jancz.；三裂茶藨子（三裂茶藨，三裂穆坪茶藨）●

334112　Ribes multiflorum Kit. ＝ Ribes multiflorum Kit. ex Schult. ●

334113　Ribes multiflorum Kit. ex Roem. et Schult. ＝ Ribes mandschuricum（Maxim.）Kom. ●

334114　Ribes multiflorum Kit. ex Schult.；多花茶藨子（多花茶藨）；Russian Currant ●

334115　Ribes multiflorum Kit. ex Schult. var. mandshuricum Maxim. ＝ Ribes mandschuricum（Maxim.）Kom. ●

334116　Ribes nevadense Kellogg；内华达茶藨子；Wild Sierra Currant ●☆

334117　Ribes nigrum L.；黑茶藨子（茶藨子，旱葡萄，黑茶藨，黑豆，黑果茶藨，黑加仑，欧洲黑茶藨，兴安茶藨）；Black Currant, Black Gooseberry, Blackberry, Blackcurrant, Curnberries, European Black Currant, Garden Black Currant, Garnet-berry, Gazel, Gazle, Quinsey-berry, Quinsy Berry, Quinsyberry, Squinancy, Squinancy-berry, Wineberry ●

334118　Ribes nigrum L. 'Apiifolium'；芹叶黑茶藨子●☆

334119　Ribes nigrum L. 'Ben Conna'；本·考南黑茶藨子●☆

334120　Ribes nigrum L. 'Ben Lomond'；本·罗蒙德黑茶藨子●☆

334121　Ribes nigrum L. 'Coloratum'；斑叶黑茶藨子●☆

334122　Ribes nigrum L. 'Jet'；黑玉黑茶藨子●☆

334123　Ribes nigrum L. 'Xanthocarpum'；黄果黑茶藨子●☆

334124　Ribes nigrum L. var. pauciflorum（Turcz. ex Ledeb.）Jancz. ＝ Ribes nigrum L. ●

334125　Ribes nigrum L. var. pennsylvanicum Marshall ＝ Ribes americanum Mill. ●

334126　Ribes niveum Lindl.；雪茶藨子（雪茶藨）；Snow Gooseberry ●☆

334127　Ribes odoratum H. L. Wendl.；香茶藨子（香茶藨，香花茶藨）；Buffalo Currant, Buffalo-currant, Clove Currant, Golden Currant, Missouri Currant ●

334128　Ribes odoratum H. L. Wendl. 'Crandall'；大果香茶藨；Large-fruited Buffalo Currant ●☆

334129　Ribes odoratum H. L. Wendl. 'Xanthocarpum'；黄果香茶藨子●☆

334130　Ribes odoratum H. L. Wendl. ＝ Ribes aureum Pursh ●

334131　Ribes olidum Moench. ＝ Ribes nigrum L. ●

334132　Ribes orientale Desf.；东方茶藨子（东方茶藨，东方醋栗，柱腺茶藨）；Oriental Currant ●

334133　Ribes orientale Desf. ＝ Ribes takare D. Don var. desmacarpum（Hook. f. et Thomson）L. T. Lu ●

334134　Ribes orientale Desf. var. genuinum Jancz. ＝ Ribes orientale Desf. ●

334135　Ribes orientale Desf. var. heterotrichum（Mey.）Jancz. ＝ Ribes heterotrichum C. A. Mey. ●

334136　Ribes orientale Desf. var. resinosum Jancz. ＝ Ribes orientale Desf. ●

334137　Ribes oxyacanthoides L.；加拿大茶藨子（加拿大茶藨）；Canada Gooseberry, Hawthorn Currant Tree, Mountai Gooseberry, Northern Gooseberry, Smooth Gooseberry ●☆

334138　Ribes oxyacanthoides L. var. calcicola Fernald ＝ Ribes hirtellum Michx. ●☆

334139　Ribes oxyacanthoides L. var. hirtellum（Michx.）Scoggan ＝ Ribes hirtellum Michx. ●☆

334140　Ribes oxyacanthoides L. var. lacustre Pers. ＝ Ribes lacustre（Pers.）Poir. ●☆

334141　Ribes oxyacanthoides L. var. saxosum（Hook.）Coville ＝ Ribes hirtellum Michx. ●☆

334142　Ribes pachysandroidea H. Lév. ＝ Ribes laurifolium Jancz. ●

334143　Ribes pachysandroides Oliv. = Ribes davidii Franch. ●

334144　Ribes palczewskii（Jancz.）Pojark.;英吉利茶藨子（密果茶藨,密花茶藨,密穗茶藨,英吉利茶藨）;Dense-flowered, Liou's Currant, Palczewsk Currant, Palczewsk Gooseberry ●

334145　Ribes pallidiflorum Pojark.;苍白花茶藨子●☆

334146　Ribes palmatum Desf. = Ribes odoratum H. L. Wendl. ●

334147　Ribes pauciflorum Turcz. ex Ledeb.;兴安茶藨（疏花茶藨）;Dahurian Currant ●

334148　Ribes pauciflorum Turcz. ex Ledeb. = Ribes nigrum L. ●

334149　Ribes petraeum Wulfen;石生茶藨;Carpathian Current, Rock Currant, Rock Redcurrant, Rockliving Currant, Upright Redcurrant ●☆

334150　Ribes petraeum Wulfen = Ribes mandschuricum（Maxim.）Kom. ●

334151　Ribes petraeum Wulfen var. altissimum（Turcz. ex Pojark.）Jancz. = Ribes altissimum Turcz. ex Ledeb. et Pojark. ●

334152　Ribes petraeum Wulfen var. atlanticum Maire;北非茶藨子●☆

334153　Ribes petraeum Wulfen var. atropurpureum Jancz. = Ribes meyeri Maxim. ●

334154　Ribes petraeum Wulfen var. mongolica Franch. = Ribes mandschuricum（Maxim.）Kom. ●

334155　Ribes pilosum Ledeb.;小葡萄茶藨;Pilose Currant ●

334156　Ribes pinetorum Greene;橘茶藨子（橘茶藨）;Orange Gooseberry ●☆

334157　Ribes procumbens Pall.;水葡萄茶藨子（葡萄茶藨,水葡萄,水葡萄茶藨）;Ducumbent Currant, Procumbent Currant ●

334158　Ribes propinquum Turcz. = Ribes triste Pall. ●

334159　Ribes prostratum L' Hér.;具腺茶藨子（具腺茶藨）;Fetid Currant, Skunk Currant ●☆

334160　Ribes prostratum L' Hér. = Ribes glandulosum Weber ●☆

334161　Ribes prostratum L' Hér. var. wisconsinum Fassett = Ribes prostratum L' Hér. ●☆

334162　Ribes pseudofasciculatum K. S. Hao;青海茶藨子（单花茶藨,短毛茶藨,青海茶藨）;Qinghai Gooseberry, Uniflorous Currant, Winterberry-like Currant ●

334163　Ribes pubescens（Sw. ex Hartm.）Hedl.;毛茶藨子（毛茶藨）;Pubescent Currant ●

334164　Ribes pubescens Hedl. = Ribes pubescens（Sw. ex Hartm.）Hedl. ●

334165　Ribes pubescens Kom. = Ribes palczewskii（Jancz.）Pojark. ●

334166　Ribes pulchellum Turcz.;美丽茶藨子（碟花茶藨子,美丽茶藨,酸麻子,小叶茶藨）;Beautiful Currant, Beautiful Gooseberry ●

334167　Ribes pulchellum Turcz. var. manshuriense F. T. Wang et H. L. Li;东北小叶茶藨子（东北小叶茶藨）;NE. China Beautiful Currant ●

334168　Ribes punctatum Lindl. = Ribes orientale Desf. ●

334169　Ribes quercetorum Greene;栎叶醋栗;Oak-belt Gooseberry ●☆

334170　Ribes reclinatum L.;欧洲醋栗（欧洲茶藨子）●

334171　Ribes recurvatum Michx. = Ribes americanum Mill. ●

334172　Ribes repens A. I. Baranov = Ribes triste Pall. var. repens（A. I. Baranov）L. T. Lu ●

334173　Ribes resinosum Pursh = Ribes prostratum L' Hér. ●☆

334174　Ribes rigens Michx. = Ribes glandulosum Weber ●☆

334175　Ribes roezlii Regel;内华达茶藨（山岭茶藨,山岭茶藨子）;Sierra Gooseberry, Sierra Nevada Gooseberry ●☆

334176　Ribes rosthornii Diels;南川茶藨子（南川茶藨）;Nanchuan Gooseberry, Rosthorn Currant ●

334177　Ribes rotundifolium Michx.;圆叶茶藨子（圆叶茶藨）;Roundleaf Gooseberry ●☆

334178　Ribes rubrisepalum L. T. Lu;红萼茶藨子（红萼茶藨）;Red-seepal Gooseberry, Redsepal Currant ●

334179　Ribes rubrum L.;红茶藨子（北方红醋栗,红茶藨,红醋栗,红果茶藨,欧洲红穗醋栗,普通茶藨子,瑞典茶藨）;Cherry-currant, Common Currant, Cultivated Currant, Currant-berry, Garden Currant, Garden Red Currant, Garnet Berry, Garnetberry, Gazel, Gazle, Gozill, Northern Red Currant, Northern Red-currant, Raisin Tree, Raspberry, Red Currant, Red Garden-currant, Red Gooseberry, Redcurrant, Reps, Rizards, Rizzer-berries, Rizzles, Russles, Sweden Currant, Wild Currant, Wineberry ●

334180　Ribes rubrum L. = Ribes himalense Decne. ●

334181　Ribes rubrum L. = Ribes meyeri Maxim. ●

334182　Ribes rubrum L. var. alaskanum（Berger）B. Boivin = Ribes triste Pall. ●

334183　Ribes rubrum L. var. palczewckii Jancz. = Ribes palczewskii（Jancz.）Pojark. ●

334184　Ribes rubrum L. var. propinquum（Turcz.）Trautv. et C. A. Mey. = Ribes triste Pall. ●

334185　Ribes rubrum L. var. pubescens（Hedl.）Jancz. = Ribes pubescens（Sw. ex Hartm.）Hedl. ●

334186　Ribes rubrum L. var. pubescens Sw. ex C. Harm. = Ribes pubescens Hedl. ●

334187　Ribes rubrum Torr. et Gray = Ribes triste Pall. ●

334188　Ribes sachalinense（F. Schmidt）Nakai;库页醋栗●☆

334189　Ribes sanguineum Pursh;血红茶藨子;American Currant, Blood Currant, Coral Plant, Flowering Currant, Red-flowered Currant, Redflowering Currant, Winter Currant ●☆

334190　Ribes sanguineum Pursh 'Brocklebankii';黄叶血红茶藨子●☆

334191　Ribes sanguineum Pursh 'Claremont';克莱尔蒙特血红茶藨子●☆

334192　Ribes sanguineum Pursh 'Elk River Red';爱尔克河红血红茶藨子;Red-flowered Currant ●☆

334193　Ribes sanguineum Pursh 'Inverness White';白披肩血红茶藨子●☆

334194　Ribes sanguineum Pursh 'King Edward Ⅶ';爱德华七世血红茶藨子（密枝血红茶藨子）;Red-flowered Currant ●☆

334195　Ribes sanguineum Pursh 'Plenum';重瓣血红茶藨子●☆

334196　Ribes sanguineum Pursh 'Pulborough Scarlet';白心血红茶藨子●☆

334197　Ribes sanguineum Pursh 'Spring Showers';春光血红茶藨子●☆

334198　Ribes sanguineum Pursh 'Spring Snow';白花血红茶藨子;White-flowered Currant ●☆

334199　Ribes sanguineum Pursh 'Tydeman's White';提德曼白血红茶藨子●☆

334200　Ribes sanguineum Pursh 'White Icicle';白冰血红茶藨子;White-flowered Currant ●☆

334201　Ribes sativum Syme;普通红茶藨子（茶藨子,醋栗,普通红茶藨）;Common Currant, Common Red Currant, Garden Currant, White Currant ●☆

334202　Ribes sativum Syme = Ribes rubrum L. ●

334203　Ribes saxatile Pall.;石生茶藨子（石茶藨,石生茶藨）;Cliff Currant, Saxatile Currant, Siberian Currant ●

334204　Ribes saxosum Hook. = Ribes hirtellum Michx. ●☆

334205　Ribes scandicum Hedl. = Ribes rubrum L. ●

334206　Ribes setchnense Jancz.;四川茶藨子（四川茶藨）;Sichuan Currant ●

334207　Ribes setosum Lindl.;红枝茶藨子（红枝茶藨）;Redshoot

Gooseberry ●☆

334208　Ribes sibiricum Hort. ex K. Koch;西伯利亚茶藨子●☆

334209　Ribes sinanense F. Maek.;信浓茶藨子●☆

334210　Ribes sinanense F. Maek. f. inerme Karayama;无刺茶藨子●☆

334211　Ribes soulieanum Jancz.;滇中茶藨子（滇中茶藨）;Soulie's Currant ●

334212　Ribes speciosum Pursh;紫红茶藨子（美丽茶藨）;Californian Fuchsia, Fuchsia Gooseberry, Fuchsiaflowered Currant, Fuchsia-flowered Currant, Fuchsia-flowered Gooseberry ●☆

334213　Ribes spethianum Koehne = Ribes inebrians Lindl. ●☆

334214　Ribes spicatum E. Robson;毛茶藨;Downy Currant, Northern Red Currant, Spicate Currant ●

334215　Ribes spicatum E. Robson = Ribes rubrum L. ●

334216　Ribes spicatum E. Robson subsp. palczewskii （ Jancz. ） Malyschev = Ribes palczewskii （Jancz.） Pojark. ●

334217　Ribes spicatum Robson subsp. pubescens （C. Hartm.） Hyl. = Ribes pubescens Hedl. ●

334218　Ribes spicatum Robson = Ribes rubrum L. ●

334219　Ribes spicatum Vis. = Ribes multiflorum Kit. ●

334220　Ribes stenocarpum Maxim.;长果茶藨子（长果茶藨,长果醋栗,狭果茶藨）;Gansu Gooseberry, Narrowfruit Currant, Narrow-fruited Currant ●

334221　Ribes subglabrum Kom.;毛山麻子●

334222　Ribes sylvestre （Lam.） Mert. et W. D. J. Koch = Ribes rubrum L. ●

334223　Ribes sylvestre Syme;亮红果茶藨子;Northern Red Currant, Redcurrant ●☆

334224　Ribes sylvestre Syme 'Macrocarpum';大果亮红果茶藨子●☆

334225　Ribes sylvestre Syme 'Red Lake';红湖亮红果茶藨子●☆

334226　Ribes sylvestre Syme 'White Grape';白葡萄亮红果茶藨子●☆

334227　Ribes sylvestre Syme = Ribes rubrum L. ●

334228　Ribes takare D. Don;渐尖茶藨子（尖叶茶藨,尖叶茶藨子,渐尖茶藨,山麻子）;Sharpleaf Gooseberry, Sharp-leaved Currant, Sharp-leaved Gooseberry ●

334229　Ribes takare D. Don f. desmocarpum （Hook. f. et Thomson） Hara = Ribes takare D. Don var. desmacarpum （Hook. f. et Thomson） L. T. Lu ●

334230　Ribes takare D. Don var. desmacarpum （Hook. f. et Thomson） L. T. Lu;束果茶藨子（束果茶藨,束果醋栗）;Bundlefruit Gooseberry ●

334231　Ribes tanguticum （Jancz.） Pojark.;甘青茶藨;Tangut Currant ●

334232　Ribes tenue Jancz.;细枝茶藨子（红枝茶藨,蓝茶藨,三升米,细醋栗,细梗茶藨子,细枝茶藨,狭萼茶藨子）;Asia Currant, Asiatic Currant, Blue-berry Currant, Blue-fruit Currant ●

334233　Ribes tenue Jancz. var. incisum L. T. Lu;深裂茶藨子（深裂茶藨）;Lobed Asiatic Currant ●

334234　Ribes tenue Jancz. var. viridiflorum W. C. Cheng = Ribes viridiflorum （W. C. Cheng） L. T. Lu et G. Yao ●

334235　Ribes tenuiflorum Lindl. = Ribes odoratum H. L. Wendl. ●

334236　Ribes tianquanense S. H. Yu et J. M. Xu;天全茶藨子（天全茶藨）;Tianquan Currant ●

334237　Ribes tricuspe Nakai = Ribes maximowiczianum Kom. ●

334238　Ribes tricuspe Nakai var. japonicum （Maxim.） Nakai = Ribes maximowiczianum Kom. ●

334239　Ribes tripartitum Batalin = Ribes moupinense Franch. var. tripartitum （Batalin） Jancz. ●

334240　Ribes triste Pall.;矮茶藨子（矮茶藨,沼地红茶藨）;American

Red Currant, American Red-currant, Swamp Red Currant, Swamp Red-currant ●

334241　Ribes triste Pall. = Ribes meyeri Maxim. ●

334242　Ribes triste Pall. f. repens （A. I. Baranov） Y. L. Chou = Ribes triste Pall. var. repens （A. I. Baranov） L. T. Lu ●

334243　Ribes triste Pall. var. alaskanum Berger = Ribes triste Pall. ●

334244　Ribes triste Pall. var. albinervium （Michx.） Fernald = Ribes triste Pall. ●

334245　Ribes triste Pall. var. repens （A. I. Baranov） L. T. Lu;伏生茶藨子（伏生矮茶藨,伏生茶藨）;Creeping Currant, Creeping Swamp Red Currant ●

334246　Ribes triste Turcz. = Ribes altissimum Turcz. ex Ledeb. et Pojark. ●

334247　Ribes turbinatum Pojark.;陀螺茶藨子●☆

334248　Ribes uniflorum T. C. Ku = Ribes pseudofasciculatum K. S. Hao ●

334249　Ribes ussuriense Jancz. ex Vilm. et Bois;乌苏里茶藨子（乌苏里茶藨）;Ussuri Currant ●

334250　Ribes uva-crispa L.;欧洲茶藨子（茶藨子,鹅莓,鹅沫,欧洲醋栗,酸子,须具利,圆醋栗）;Berry-bush, Blab, Blob, Catberry, Dabbery, Dayberry, Deberry, Dewberry, English Gooseberry, Europe Currant, European Currant, European Gooseberry, Fabe, Faberry, Faeberry, Fape, Fayberry, Fea, Feabe, Feaberry, Feap, Feapberry, Feberry, Feverberry, Gaskins, Gew-gog, Goggles, Golfob, Gooseberry, Goosegob, Goosegog, Gorstberry, Gosler, Gozill, Grizzle, Grosart, Grosel, Groser, Grosert, Grosier, Grossberry, Grosset, Grozer, Grozet, Grozzle, Gruzel, Gruzzle, Guzzleberry, Honey Blob, Honey-blob, Round Currant, Thape, Theabberry, Thteabe, Wineberry ●

334251　Ribes uva-crispa L. subsp. septentrionale Maire = Ribes uva-crispa L. ●

334252　Ribes uva-crispa L. var. atlanticum Ball = Ribes uva-crispa L. ●

334253　Ribes uva-crispa L. var. glanduligerum （H. Lindb.） Maire = Ribes uva-crispa L. ●

334254　Ribes uva-crispa L. var. grossularia （L.） Maire = Ribes uva-crispa L. ●

334255　Ribes uva-crispa L. var. reclinatum （L.） Berland. = Ribes reclinatum L. ●

334256　Ribes uva-crispa L. var. reclinatum （L.） Berland. = Ribes uva-crispa L. ●

334257　Ribes uva-crispa L. var. sativum DC.;栽培酸子;European Gooseberry ●☆

334258　Ribes uva-crispa L. var. sativum DC. = Ribes uva-crispa L. ●

334259　Ribes uva-crispa L. var. subatlanticum Maire = Ribes uva-crispa L. ●

334260　Ribes villosum Wall.;黄果茶藨子（黄果茶藨）;Blood Currant, Villose Currant ●☆

334261　Ribes villosum Wall. = Ribes orientale Desf. ●

334262　Ribes vilmorinii Jancz.;小果茶藨子（魏氏茶藨,小果茶藨）;Vilmorin Currant ●

334263　Ribes vilmorinii Jancz. var. pubicarpum L. T. Lu;康定茶藨子（康定茶藨）;Kangding Currant ●

334264　Ribes viridiflorum （W. C. Cheng） L. T. Lu et G. Yao;绿花茶藨子（绿花茶藨,绿花细枝茶藨,绿花细枝茶藨子,三升米,细枝茶藨子）;Green-flower Asiatic Currant, Green-flower Currant ●

334265　Ribes viscosissimum Pursh;黏茶藨子（黏茶藨）;Sticky Currant ●☆

334266　Ribes viscosum Ruiz et Pav. var. brachybotrys Wedd. = Ribes brachybotrys （Wedd.） Jancz. ●☆

334267　Ribes vitifolium Host = Ribes multiflorum Kit. ex Schult. ●

334268　Ribes vulgare Lam. = Ribes rubrum L. ●

334269　Ribes wolfii Rothr.；沃尔夫茶藨子（沃尔夫茶藨）；Rothrock Currant ●☆

334270　Ribes xizangense L. T. Lu；西藏茶藨子（西藏茶藨）；Xizang Currant ●

334271　Ribesiaceae A. Rich. = Grossulariaceae DC.（保留科名）●

334272　Ribesiaceae Marquis = Grossulariaceae DC.（保留科名）●

334273　Ribesiodes Kuntze = Embelia Burm. f.（保留属名）●■

334274　Ribesiodes L. = Embelia Burm. f.（保留属名）●■

334275　Ribesiodes arboreum（A. DC.）Kuntze = Embelia arborea A. DC. ●☆

334276　Ribesiodes concinnum（Baker）Kuntze = Embelia concinna Baker ●☆

334277　Ribesiodes floribundum（Wall.）Kuntze = Embelia floribunda Wall. ●

334278　Ribesiodes floribundum Kuntze = Embelia floribunda Wall. ●

334279　Ribesiodes gamblei（Kurz ex C. B. Clarke）Kuntze = Embelia gamblei Kurz ex C. B. Clarke ●

334280　Ribesiodes gamblei Kuntze = Embelia gamblei Kurz ex C. B. Clarke ●

334281　Ribesiodes jussieui（A. DC.）Kuntze = Embelia pyrifolia（Willd. ex Roem. et Schult.）Mez ●☆

334282　Ribesiodes longifolium（Benth.）Kuntze = Embelia undulata（Wall.）Mez ●

334283　Ribesiodes longifolium Kuntze = Embelia undulata（Wall.）Mez ●

334284　Ribesiodes oblongifolium（Hemsl.）Kuntze = Embelia vestita Roxb. ●

334285　Ribesiodes oblongifolium Kuntze = Embelia vestita Roxb. ●

334286　Ribesiodes obovatum（Benth.）Kuntze = Embelia laeta（L.）Mez ●

334287　Ribesiodes obovatum Kuntze = Embelia laeta（L.）Mez ●

334288　Ribesiodes parviflora Kuntze = Embelia parviflora Wall. et A. DC. ●

334289　Ribesiodes parviflorum（Wall. ex A. DC.）Kuntze = Embelia parviflora Wall. et A. DC. ●

334290　Ribesiodes ribes（Burm. f.）Kuntze = Embelia ribes Burm. f. ●

334291　Ribesiodes ribes Kuntze = Embelia ribes Burm. f. ●

334292　Ribesiodes sarmentosum（Baker）Kuntze = Embelia pyrifolia（Willd. ex Roem. et Schult.）Mez ●☆

334293　Ribesiodes sessiliflorum（Kurz）Kuntze = Embelia sessiliflora Kurz ●

334294　Ribesiodes sessiliflorum Kuntze = Embelia sessiliflora Kurz ●

334295　Ribesiodes vestitum（Roxb.）Kuntze = Embelia vestita Roxb. ●

334296　Ribesiodes vestitum Kuntze = Embelia vestita Roxb. ●

334297　Ribesium Medik. = Ribes L. ●

334298　Ricardia Adans. = Richardia L. ■

334299　Ricaurtea Triana = Doliocarpus Rol. ●☆

334300　Richaeia Thouars（废弃属名）= Cassipourea Aubl. ●☆

334301　Richaeia Thouars（废弃属名）= Weihea Spreng.（保留属名）●☆

334302　Richardella Pierre = Lucuma Molina ●

334303　Richardella Pierre = Pouteria Aubl. ●

334304　Richardella afzelii（Engl.）Baehni = Synsepalum afzelii（Engl.）T. D. Penn. ●☆

334305　Richardella dulcifica（Schumach. et Thonn.）Baehni = Synsepalum dulcificum（Schumach. et Thonn.）Daniell ●☆

334306　Richardella subcordata（De Wild.）Baehni = Synsepalum subcordatum De Wild. ●☆

334307　Richardella superba（Vermoesen）Baehni = Pouteria superba（Vermoesen）L. Gaut. ●☆

334308　Richardia Houst. ex L. = Zantedeschia Spreng.（保留属名）■

334309　Richardia Kunth = Zantedeschia Spreng.（保留属名）■

334310　Richardia L.（1753）；墨苜蓿属（波状吐根属，糙独根属，拟鸭舌癀属）；Calla Lily，Mexican Clover，Trumpet Lily ■

334311　Richardia Lindl. = Picridium Desf. ■☆

334312　Richardia Lindl. = Reichardia Roth ■☆

334313　Richardia aethiopica（L.）Spreng. = Zantedeschia aethiopica（L.）Spreng. ■

334314　Richardia africana Kunth；非洲墨苜蓿；Altar Lily ■☆

334315　Richardia africana Kunth = Zantedeschia aethiopica（L.）Spreng. ■

334316　Richardia albomaculata Hook.；白斑墨苜蓿 ■☆

334317　Richardia albomaculata Hook. = Zantedeschia albomaculata（Hook.）Baill. ■

334318　Richardia angustiloba Schott；狭裂墨苜蓿 ■☆

334319　Richardia angustiloba Schott = Zantedeschia albomaculata（Hook.）Baill. ■

334320　Richardia brasiliensis Gomes；巴西墨苜蓿（巴西波状吐根）■☆

334321　Richardia brasiliensis Gomes = Richardia scabra L. ■

334322　Richardia elliottiana W. Watson = Zantedeschia elliottiana（W. Watson）Engl. ■☆

334323　Richardia grandiflora Britton；大花墨苜蓿；Largeflower Mexican Clover ■☆

334324　Richardia hastata Hook. = Zantedeschia hastata（Hook.）Engl. ■☆

334325　Richardia humistrata（Cham. et Schltdl.）Steud.；南美墨苜蓿；South American Mexican Clover ■☆

334326　Richardia humistrata Steud. = Richardia humistrata（Cham. et Schltdl.）Steud. ■☆

334327　Richardia lutwychei N. E. Br. = Zantedeschia albomaculata（Hook.）Baill. ■

334328　Richardia macrocarpa（Engl.）W. Watson = Zantedeschia albomaculata（Hook.）Baill. subsp. macrocarpa（Engl.）Letty ■☆

334329　Richardia macrocarpa W. Watson；大果墨苜蓿 ■☆

334330　Richardia melanoleuca Hook. f. = Zantedeschia albomaculata（Hook.）Baill. ■

334331　Richardia melanoleuca Hook. f. var. tropicalis N. E. Br. = Zantedeschia albomaculata（Hook.）Baill. ■

334332　Richardia metamoleuca Hook. f. = Zantedeschia melanoleuca（Hook. f.）Engl. ■

334333　Richardia pentlandii Whyte ex W. Watson = Zantedeschia pentlandii（Whyte ex W. Watson）Wittm. ■☆

334334　Richardia pilosa Ruiz. et Pav. = Richardia scabra L. ■

334335　Richardia rehmannii（Engl.）N. E. Br. ex Krelage = Zantedeschia rehmannii Engl. ■

334336　Richardia rehmannii N. E. Br. ex Harrow = Zantedeschia rehmannii Engl. ■

334337　Richardia scabra（L.）A. St. -Hil. = Richardia scabra L. ■

334338　Richardia scabra L.；墨苜蓿（波状吐根，糙独根，拟鸭舌癀，鸭舌癀）；False Ipecacuanha，Mexican Clover ■☆

334339　Richardia sprengeri Comes = Zantedeschia pentlandii（Whyte ex W. Watson）Wittm. ■☆

334340　Richardia stellaris（Cham. et Schltdl.）Steud.；星苜蓿 ■☆

334341　Richardsiella Elffers et Kenn. -O'Byrne（1957）；丝秆草属 ■☆

334342　Richardsiella eruciformis Elffers et Kenn. -O'Byrne；丝秆草 ■☆

334343　Richardsonia Kunth ＝ Richardia L. ■

334344　Richardsonia brasiliensis （Gomes） Hayne ＝ Richardia brasiliensis Gomes ■☆

334345　Richardsonia humistrata Cham. et Schltdl. ＝ Richardia humistrata （Cham. et Schltdl.） Steud. ■☆

334346　Richardsonia scabra （L.） A. St. -Hil. ＝ Richardia scabra L. ■

334347　Richardsonia stellaris Cham. et Schltdl. ＝ Richardia stellaris （Cham. et Schltdl.） Steud. ■☆

334348　Richea Kumze ＝ Richaeia Thouars（废弃属名）●☆

334349　Richea Kuntze ＝ Cassipourea Aubl. ●☆

334350　Richea Labill.（废弃属名） ＝ Richea R. Br.（保留属名）●☆

334351　Richea Lablll.（废弃属名） ＝ Craspedia G. Forst. ■☆

334352　Richea R. Br.（1810）（保留属名）；彩穗木属（利切木属，芦荟石南属）●☆

334353　Richea afzelia Kuntze ＝ Cassipourea afzelii （Oliv.） Alston ●☆

334354　Richea dracophylla R. Br.；丰果彩穗木（丰果利切木）●☆

334355　Richea pandanifolia Hook. f.；露兜树叶彩穗木（露兜树叶利切木）；Pandani，Pandanni，Pandanny，Tree Heath ●☆

334356　Richea plumosa Kuntze ＝ Cassipourea plumosa （Oliv.） Alston ●☆

334357　Richea scoparia Hook. f.；帚状彩穗木（帚状利切木）；Kerosene Bush ●☆

334358　Richella A. Gray（1852）；尖花藤属；Pointedflowervine，Richella ●

334359　Richella albida （Engl.） R. E. Fr. ＝ Friesodielsia gracilipes （Benth.） Steenis ●☆

334360　Richella gracilipes （Benth.） R. E. Fr. ＝ Friesodielsia gracilipes （Benth.） Steenis ●☆

334361　Richella gracilis （Hook. f.） R. E. Fr. ＝ Friesodielsia gracilis （Hook. f.） Steenis ●☆

334362　Richella grandiflora （Boutique） R. E. Fr. ＝ Friesodielsia enghiana （Diels） Verdc. ●☆

334363　Richella hainanensis （Tsiang et P. T. Li） Tsiang et P. T. Li；尖花藤（海南尖花藤）；Hainan Pointedflowervine，Hainan Richella ●

334364　Richella hirsuta （Benth.） R. E. Fr. ＝ Friesodielsia hirsuta （Benth.） Steenis ●☆

334365　Richella longipedicellata （Baker f.） R. E. Fr. ＝ Friesodielsia gracilipes （Benth.） Steenis ●☆

334366　Richella montana （Engl. et Diels） R. E. Fr. ＝ Friesodielsia montana （Engl. et Diels） Steenis ●☆

334367　Richella obanensis （Baker f.） R. E. Fr. ＝ Friesodielsia enghiana （Diels） Verdc. ●☆

334368　Richella soyauxii （Sprague et Hutch.） Steenis ＝ Friesodielsia montana （Engl. et Diels） Steenis ●☆

334369　Richella velutina （Sprague et Hutch.） R. E. Fr. ＝ Friesodielsia velutina （Sprague et Hutch.） Steenis ●☆

334370　Richeopsis Arènes ＝ Scolopia Schreb.（保留属名）●

334371　Richeria Vahl（1797）；里谢大戟属●☆

334372　Richeria grandis Vahl；里谢大戟●☆

334373　Richeriella Pax et K. Hoffm.（1922）；龙胆木属（梨查木属）；Gentian-wood，Gentiawood ●

334374　Richeriella gracilis （Merr.） Pax et K. Hoffm.；龙胆木；Gentian-wood，Gentiawood ●

334375　Richetia Heim ＝ Balanocarpus Bedd. ●☆

334376　Richiaea Benth. et Hook. f. ＝ Cassipourea Aubl. ●☆

334377　Richiaea Benth. et Hook. f. ＝ Richaeia Thouars（废弃属名）●☆

334378　Richiea G. Don ＝ Ritchiea R. Br. ex G. Don ●☆

334379　Richtera Rchb. ＝ Annesiea Wall.（保留属名）●

334380　Richterago Kuntze ＝ Gochnatia Kunth ●

334381　Richterago Kuntze（1891）；小绒菊木属●☆

334382　Richterago amplexifolia （Gardner） Kuntze；小绒菊木●☆

334383　Richterago elegans Roque；雅致小绒菊木●☆

334384　Richterago polyphylla （Baker ex Mart.） Ferreyra；多叶小绒菊木●☆

334385　Richteria Kar. et Kir.（1842）；细裂匹菊属（灰叶菊属）●☆

334386　Richteria Kar. et Kit. ＝ Leucopoa Griseb. ●☆

334387　Richteria Karelin et Kir. ＝ Chrysanthemum L.（保留属名）●■

334388　Richteria pyrethroides Kar. et Kir. ＝ Pyrethrum pyrethroides （Kar. et Kir.） B. Fedtsch. ex Krasch. ■

334389　Richtersveldia Meve et Liede ＝ Trichocaulon N. E. Br. ■☆

334390　Richtersveldia Meve et Liede（2002）；柱亚罗汉属■☆

334391　Richtersveldia columnaris （Nel） Meve et Liede；圆柱细裂匹菊●☆

334392　Richthofenia Hosseus ＝ Sapria Griff. ■

334393　Richthofenia siamensis Hosseus ＝ Sapria himalayana Griff. ■

334394　Ricinaceae Barkley ＝ Euphorbiaceae Juss.（保留科名）●■

334395　Ricinaceae Martinov ＝ Euphorbiaceae Juss.（保留科名）●■

334396　Ricinella Müll. Arg. ＝ Adelia L.（保留属名）●☆

334397　Ricinocarpaceae （Müll. Arg.） Hurus. ＝ Euphorbiaceae Juss.（保留科名）●■

334398　Ricinocarpaceae （Pax） Hurus. ＝ Euphorbiaceae Juss.（保留科名）●■

334399　Ricinocarpaceae Hurus.；蓖麻果木科●

334400　Ricinocarpaceae Hurus. ＝ Euphorbiaceae Juss.（保留科名）●■

334401　Ricinocarpodendron Arum. ex Boehm. ＝ ? Dysoxylum Blume ●

334402　Ricinocarpos A. Juss. ＝ Ricinocarpos Desf. ●☆

334403　Ricinocarpos Desf.（1817）；蓖麻果木属；Wedding Bush ●☆

334404　Ricinocarpos pinifolius Desf.；蓖麻果木；Wedding Bush ●☆

334405　Ricinocarpus A. Juss. ＝ Ricinocarpos Desf. ●☆

334406　Ricinocarpus Burm. ex Kuntze ＝ Acalypha L. ●■

334407　Ricinocarpus Kuntze ＝ Acalypha L. ●■

334408　Ricinocarpus australis Kuntze ＝ Acalypha australis L. ■

334409　Ricinocarpus glabratus Kuntze f. pilosior ？ ＝ Acalypha glabrata Thunb. var. pilosa Pax ■☆

334410　Ricinodendron Müll. Arg.（1864）；蓖麻树属●☆

334411　Ricinodendron africanus Müll. Arg.；非洲蓖麻树；African Wood Oil-nut，Musodo Manketti Nut，Musodo Manketti-nut，Wood-oilnut Tree ●☆

334412　Ricinodendron gracilius Mildbr. ＝ Ricinodendron heudelotii （Baill.） Pierre ex Heckel var. africanum （Müll. Arg.） J. Léonard ●☆

334413　Ricinodendron heudelotii （Baill.） Pierre ex Heckel；蓖麻树●☆

334414　Ricinodendron heudelotii （Baill.） Pierre ex Heckel var. africanum （Müll. Arg.） J. Léonard ＝ Ricinodendron africanus Müll. Arg. ●☆

334415　Ricinodendron heudelotii （Baill.） Pierre ex Heckel var. tomentellum （Hutch. et E. A. Bruce） Radcl. -Sm.；绒毛非洲蓖麻树●☆

334416　Ricinodendron heudelotii （Baill.） Pierre ex Pax et K. Hoffm. ＝ Ricinodendron heudelotii （Baill.） Pierre ex Heckel ●☆

334417　Ricinodendron rautanenii Schinz；劳塔宁蓖麻树；Mangetti Tree，Man-ketti Nut ●☆

334418　Ricinodendron rautanenii Schinz ＝ Schinziophyton rautanenii （Schinz） Radcl. -Sm. ●☆

334419　Ricinodendron schliebenii Mildbr. ＝ Ricinodendron heudelotii （Baill.） Pierre ex Heckel var. tomentellum （Hutch. et E. A. Bruce） Radcl. -Sm. ●☆

334420　Ricinodendron staudtii Pax ＝ Lannea welwitschii （Hiern） Engl. ●

334421　Ricinodendron tomentellum Hutch. et E. A. Bruce ＝

Ricinodendron heudelotii（Baill.）Pierre ex Heckel var. tomentellum（Hutch. et E. A. Bruce）Radcl. -Sm. ●☆

334422 Ricinodendron viticoides Mildbr. = Schinziophyton rautanenii（Schinz）Radcl. -Sm. ●☆

334423 Ricinoides Gagnebin = Croton L. ●

334424 Ricinoides Mill. = Jatropha L.（保留属名）●■

334425 Ricinoides Moench = Chrozophora A. Juss.（保留属名）●

334426 Ricinoides Moench = Tournesol Adans.（废弃属名）●

334427 Ricinoides Tourn. ex Moench = Chrozophora A. Juss.（保留属名）●

334428 Ricinophyllum Pall. ex Ledeb.（1844）;蓖麻叶五加属●☆

334429 Ricinophyllum Pall. ex Ledeb. = Fatsia Decne. et Planch. ●

334430 Ricinophyllum americanum Pall. ex Ledeb.;美洲蓖麻叶五加●☆

334431 Ricinophyllum horridum A. Nelson et J. F. Macbr.;蓖麻叶五加●☆

334432 Ricinus L.（1753）;蓖麻属;Castor Bean, Castorbean, Castorbean, Castorbean-oilplant, Castor-oil Bean, Castor-oil Plant, Palma Christi ●■

334433 Ricinus africanus Willd. = Ricinus communis L. var. africanus（Willd.）Müll. Arg. ●☆

334434 Ricinus apelta Lour. = Mallotus apelta（Lour.）Müll. Arg. ●

334435 Ricinus communis L.;蓖麻（八麻, 贝麻, 草麻, 草麻子, 蓖麻子, 草麻, 大麻, 大麻子, 杜麻, 观赏蓖麻, 红蓖麻, 红大麻, 勒菜子, 牛蓖子草, 脾麻, 蝉麻, 天麻子果, 远近子）;Castor, Castor Bean, Castor Bean Plant, Castor Oil Plant, Castor Plant, Castorbean, Castor-bean, Castor-oil Plant, Eroton Oil Plant, Great Spurge, Man's Motherwort, Palm of Christian, Palma Christi, Palma-christi, Stedfast, Tickseed, Wonder Tree ●■

334436 Ricinus communis L. 'Black Beaty';黑美女蓖麻●■

334437 Ricinus communis L. 'Impala';红芽蓖麻●■

334438 Ricinus communis L. 'Sanguineus';血红蓖麻●■

334439 Ricinus communis L. 'Scarlet Queen';红色皇后蓖麻●■

334440 Ricinus communis L. 'Zanzibarensis';桑吉巴尔蓖麻●■

334441 Ricinus communis L. var. africanus（Willd.）Müll. Arg.;非洲蓖麻●☆

334442 Ricinus communis L. var. lividus（Jacq.）Müll. Arg. = Ricinus communis L. ●■

334443 Ricinus communis L. var. megalospermus（Delile）Müll. Arg.;大籽蓖麻●☆

334444 Ricinus communis L. var. sanguineus ? = Ricinus communis L. 'Sanguineus' ●■

334445 Ricinus japonicus Thunb. = Mallotus japonicus（L. f.）Müll. Arg. ●

334446 Ricinus mappa L. = Macaranga tanarius（L.）Müll. Arg. ●

334447 Ricinus megalospermus Delile = Ricinus communis L. var. megalospermus（Delile）Müll. Arg. ●☆

334448 Ricinus tanarius L. = Macaranga tanarius（L.）Müll. Arg. ●

334449 Ricoila Renealm. ex Raf. = Gentiana L. ■

334450 Ricophora Mill. = Dioscorea L.（保留属名）■

334451 Ricotia L.（1763）（保留属名）;凹瓣芥属■☆

334452 Ricotia aegyptiaca L.;凹瓣芥■☆

334453 Ricotia cantoniensis Lour. = Rorippa cantoniensis（Lour.）Ohwi ■

334454 Ridan Adans.（废弃属名）= Actinomeris Nutt.（保留属名）■☆

334455 Ridan alternifolia（L.）Britton = Verbesina alternifolia（L.）Britton ex Kearney ■☆

334456 Ridania Kuntze = Ridan Adans.（废弃属名）■☆

334457 Riddelia Raf. = Melochia L.（保留属名）●■

334458 Riddelia Raf. = Riddellia Raf. ●■

334459 Riddellia Nutt. = Psilostrophe DC. ■☆

334460 Riddellia Raf. = Melochia L.（保留属名）●■

334461 Riddellia cooperi A. Gray = Psilostrophe cooperi（A. Gray）Greene ■☆

334462 Riddellia tagetina Nutt. = Psilostrophe tagetina（Nutt.）Greene ■☆

334463 Riddellia tagetina Nutt. var. sparsiflora A. Gray = Psilostrophe sparsiflora（A. Gray）A. Nelson ■☆

334464 Ridelia Spach = Lantana L.（保留属名）●

334465 Ridelia Spach = Riedelia Cham.（废弃属名）●■

334466 Ridleia Endl. = Melochia L.（保留属名）●■

334467 Ridleia Endl. = Riedlea Vent. ●■

334468 Ridleya（Hook. f.）Pfitzer = Thrixspermum Lour. ■

334469 Ridleya K. Schum. = Risleya King et Pantl. ■

334470 Ridleyandra A. Weber et B. L. Burtt（1998）;里德利苣苔属■☆

334471 Ridleyandra atrocyanea（Ridl.）A. Weber;深蓝里德利苣苔■☆

334472 Ridleyandra atropurpurea（Ridl.）A. Weber;暗紫里德利苣苔■☆

334473 Ridleyella Schltr.（1913）;里德利兰属■☆

334474 Ridleyella paniculata Schltr.;里德利兰■☆

334475 Ridleyinda Kuntze = Isoptera Scheft. ex Burck ●

334476 Ridolfia Moris = Carum L. ■

334477 Ridolfia Moris（1841）;里多尔菲草属;False Fennel ■☆

334478 Ridolfia segetum（L.）Moris;里多尔菲草;False Fennel ■☆

334479 Riedelia Cham.（废弃属名）= Lantana L.（保留属名）●

334480 Riedelia Cham.（废弃属名）= Riedelia Oliv.（保留属名）■☆

334481 Riedelia Kunth = Arundinella Raddi ■

334482 Riedelia Meisn. = Satyria Klotzsch ●☆

334483 Riedelia Oliv.（1883）（保留属名）;里德尔姜属■☆

334484 Riedelia Trin. ex Kunth = Arundinella Raddi ■

334485 Riedelia angustifolia Valeton;窄叶里德尔姜■☆

334486 Riedelia flava Lauterb. ex Valeton;黄里德尔姜■☆

334487 Riedeliella Harms（1903）;醉畜豆属■☆

334488 Riedeliella graciliflora Harms;醉畜豆■☆

334489 Riedeliella sessiliflora Kuhlm.;无梗花醉畜豆■☆

334490 Riedlea Vent. = Melochia L.（保留属名）●■

334491 Riedlea corchorifolia（L.）DC. = Melochia corchorifolia L. ●■

334492 Riedleia DC. = Riedlea Vent. ●■

334493 Riedleja Hassk. = Riedlea Vent. ●■

334494 Riedlia Dumort. = Riedlea Vent. ●■

334495 Riencourtia Cass.（1818）（'Riencurtia'）. ;双凸菊属■☆

334496 Riencourtia angustifolia Gardner;窄叶双凸菊■☆

334497 Riencourtia glomerata Cass.;双凸菊■☆

334498 Riencourtia latifolia Gardner;宽叶双凸菊■☆

334499 Riencourtia longifolia Baker;长叶双凸菊■☆

334500 Riencourtia oblongifolia Gardner;矩圆叶双凸菊■☆

334501 Riencourtia ovata S. F. Blake;卵叶双凸菊■☆

334502 Riencourtia tenuifolia Gardner;细叶双凸菊■☆

334503 Riencurtia Cass. = Riencourtia Cass. ■☆

334504 Riesenbachia C. Presl = Lopezia Cav. ■☆

334505 Riessia Klotzsch = Begonia L. ●■

334506 Riessia Klotzsch = Steineria Klotzsch ●■

334507 Rigidella Lindl.（1840）;硬鸢尾属■☆

334508 Rigidella Lindl. = Tigridia Juss. ■☆

334509 Rigidella flammea Lindl.;硬鸢尾■☆

334510 Rigidella maculata Walp.;斑点硬鸢尾■☆

334511 Rigiocarya Post et Kuntze = Rhigiocarya Miers ●☆

334512 Rigiolepis Hook. f. = Vaccinium L. ●

334513　Rigiopappus A. Gray（1865）；硬冠菀属■☆

334514　Rigiopappus leptocladus A. Gray；硬冠菀■☆

334515　Rigiophyllum（Less.）Spach ＝Relhania L'Hér.（保留属名）●☆

334516　Rigiophyllum Post et Kuntze ＝Rhigiophyllum Hochst. ●☆

334517　Rigiostachys Planch. ＝Recchia Moc. et Sessé ex DC. ●☆

334518　Rigocarpus Neck. ＝Rytidostylis Hook. et Arn. ●☆

334519　Rigospira Post et Kuntze ＝Rhigospira Miers ●☆

334520　Rigospira Post et Kuntze ＝Tabernaemontana L. ●

334521　Rigozum Post et Kuntze ＝Rhigozum Burch. ●☆

334522　Rikliella J. Raynal ＝Lipocarpha R. Br.（保留属名）■

334523　Rikliella kernii（Raymond）J. Raynal ＝Lipocarpha kernii（Raymond）Goetgh. ■☆

334524　Rikliella rehmannii（Ridl.）J. Raynal ＝Lipocarpha rehmannii（Ridl.）Goetgh. ■☆

334525　Rima Sonn. ＝Artocarpus J. R. Forst. et G. Forst.（保留属名）●

334526　Rimacactus Mottram ＝Eriosyce Phil. ●☆

334527　Rimacactus Mottram（2001）；智利极光球属●☆

334528　Rimacola Rupp（1942）；隙居兰属■☆

334529　Rimacola elliptica（R. Br.）Rupp；隙居兰■☆

334530　Rimaria L. Bolus ＝Vanheerdea L. Bolus ex H. E. K. Hartmann ●☆

334531　Rimaria N. E. Br. ＝Gibbaeum Haw. ex N. E. Br. ●☆

334532　Rimaria angusta L. Bolus ＝Vanheerdea roodiae（N. E. Br.）L. Bolus ex H. E. K. Hartmann ●☆

334533　Rimaria comptonii L. Bolus ＝Gibbaeum heathii（N. E. Br.）L. Bolus ■☆

334534　Rimaria divergens L. Bolus ＝Vanheerdea roodiae（N. E. Br.）L. Bolus ex H. E. K. Hartmann ●☆

334535　Rimaria heathii N. E. Br. var. elevata L. Bolus ＝Gibbaeum heathii（N. E. Br.）L. Bolus ■☆

334536　Rimaria heathii N. E. Br. var. major L. Bolus ＝Gibbaeum heathii（N. E. Br.）L. Bolus ■☆

334537　Rimaria primosii L. Bolus ＝Vanheerdea primosii（L. Bolus）L. Bolus ex H. E. K. Hartmann ●☆

334538　Rimaria roodiae N. E. Br. ＝Vanheerdea roodiae（N. E. Br.）L. Bolus ex H. E. K. Hartmann ●☆

334539　Rinanthus Gilib. ＝Rhinanthus L. ■

334540　Rindera Pall.（1771）；翅果草属（凌德草属，运得草属，紫果紫草属）；Rindera ■

334541　Rindera austroechinata Popov；南方翅果草■☆

334542　Rindera baldshuanica Kusn.；巴尔德翅果草■☆

334543　Rindera cyclodonta Bunge；环齿翅果草■☆

334544　Rindera echinata Regel；刺凌德草■☆

334545　Rindera ferganlca Popov；费尔干翅果草■☆

334546　Rindera glochidiata Wall. ＝Hackelia uncinata（Benth.）C. E. C. Fisch. ■

334547　Rindera gymnandra（Coss.）Gürke ＝Cynoglossum gymnandrum（Coss.）Greuter et Burdet ■☆

334548　Rindera holochiton Popov；全被翅果草■☆

334549　Rindera korshinskyi（Lipsky）Brand；考尔翅果草■☆

334550　Rindera laevigata Roem. et Schult. ＝Rindera tetraspis Pall. ■

334551　Rindera lanata（Lam.）Bunge；绵毛翅果草■☆

334552　Rindera oblongifolia Popov；矩叶翅果草■☆

334553　Rindera ochroleuca Kar. et Kir.；绿白翅果草■☆

334554　Rindera oschensis Popov.；奥什翅果草■☆

334555　Rindera tetraspis Pall.；翅果草（四分果运得草）；Rindera ■

334556　Rindera tianschanica Popov；天山翅果草；Tianshan Rindera ■☆

334557　Rindera turkestanica Kusn.；土耳其斯坦翅果草■☆

334558　Rindera umbellata Bunge；伞形运得草■☆

334559　Ringentiarum Nakai ＝Arisaema Mart. ●■

334560　Ringentiarum ringens（Schott）Nakai ＝Arisaema ringens（Thunb.）Schott ●■

334561　Rinopodium Salisb. ＝Scilla L. ■

334562　Rinorea Aubl.（1775）（保留属名）；三角车属（雷诺木属）；Rinorea ●

334563　Rinorea abbreviata Achound. et Bos；缩短三角车●☆

334564　Rinorea abidjanensis Aubrév. et Pellegr. ＝Decorsella paradoxa A. Chev. ●☆

334565　Rinorea acutidens M. Brandt；尖三角车●☆

334566　Rinorea adnata Chipp；贴生三角车●☆

334567　Rinorea adolfi-friderici M. Brandt；弗里德里西三角车●☆

334568　Rinorea afzelii Engl.；阿芙泽尔三角车●☆

334569　Rinorea afzelii Engl. var. pubescens Taton；毛阿芙泽尔三角车●☆

334570　Rinorea albersii Engl. ＝Rinorea angustifolia（Thouars）Baill. subsp. albersii（Engl.）Grey-Wilson ●☆

334571　Rinorea albidiflora Engl.；白花三角车●☆

334572　Rinorea amaniensis M. Brandt ＝Rinorea subintegrifolia（P. Beauv.）Kuntze ●☆

334573　Rinorea angolensis Exell ＝Rinorea ilicifolia（Welw. ex Oliv.）Kuntze ●☆

334574　Rinorea angustifolia（Thouars）Baill.；窄叶三角车●☆

334575　Rinorea angustifolia（Thouars）Baill. subsp. albersii（Engl.）Grey-Wilson；阿伯斯三角车●☆

334576　Rinorea angustifolia（Thouars）Baill. subsp. ardisiiflora（Oliv.）Grey-Wilson；紫金牛花三角车●☆

334577　Rinorea angustifolia（Thouars）Baill. subsp. engleriana（De Wild. et T. Durand）Grey-Wilson；恩格勒三角车●☆

334578　Rinorea angustifolia（Thouars）Baill. subsp. myrsinifolia（Dunkley）Grey-Wilson ＝Rinorea myrsinifolia Dunkley ●☆

334579　Rinorea angustifolia（Thouars）Baill. subsp. natalensis（Engl.）Grey-Wilson；纳塔尔三角车●☆

334580　Rinorea arborea（Thouars）Baill.；树状三角车●☆

334581　Rinorea ardisiiflora（Welw. ex Oliv.）Kuntze ＝Rinorea angustifolia（Thouars）Baill. subsp. ardisiiflora（Oliv.）Grey-Wilson ●☆

334582　Rinorea ardisiiflora（Welw. ex Oliv.）Kuntze var. salicifolia Taton；柳叶三角车●☆

334583　Rinorea arenicola M. Brandt ＝Rinorea welwitschii（Oliv.）Kuntze ●☆

334584　Rinorea aucuparia（Oliv.）Kuntze ＝Rinorea brachypetala（Turcz.）Kuntze ●☆

334585　Rinorea aylmeri Chipp；艾梅三角车●☆

334586　Rinorea banguensis Engl. ＝Rinorea welwitschii（Oliv.）Kuntze ●☆

334587　Rinorea batangae Engl. ＝Allexis batangae（Engl.）Melch. ●☆

334588　Rinorea bengalensis（Wall.）Gagnep. ＝Rinorea bengalensis（Wall.）Kuntze ●

334589　Rinorea bengalensis（Wall.）Kuntze；三角车（雷诺木）；Bengal Rinorea ●

334590　Rinorea beniensis Engl.；贝尼三角车●☆

334591　Rinorea bondjorum A. Chev. ＝Rinorea oblongifolia（C. H. Wright）C. Marquand ex Chipp ●☆

334592　Rinorea botryoides Achound. et Bos；葡萄三角车●☆

334593　Rinorea brachypetala（Turcz.）Kuntze；短瓣三角车●☆

334594　Rinorea brachypetala（Turcz.）Kuntze var. velutina Taton；绒毛短瓣三角车●☆

334595 Rinorea breteleri Achound.；布勒泰尔三角车●☆

334596 Rinorea breviracemosa Chipp；短枝三角车●☆

334597 Rinorea bussei M. Brandt；布瑟三角车●☆

334598 Rinorea cafassii Chiov. = Casearia battiscombei R. E. Fr. ●☆

334599 Rinorea campoensis Brandt ex Engl.；平原三角车●☆

334600 Rinorea castaneoides (Oliv.) Kuntze；栗色三角车●☆

334601 Rinorea caudata (Oliv.) Kuntze；尾状三角车●☆

334602 Rinorea cauliflora (Oliv.) Kuntze = Allexis cauliflora (Oliv.) Pierre ●☆

334603 Rinorea chevalieri Exell；舍瓦利耶三角车●☆

334604 Rinorea claessensii De Wild.；克莱森斯三角车●☆

334605 Rinorea comperei Taton；孔佩尔三角车●☆

334606 Rinorea convallariiflora M. Brandt = Rinorea convallarioides (Baker f.) Eyles subsp. occidentalis Grey-Wilson ●☆

334607 Rinorea convallarioides (Baker f.) Eyles；环绕三角车●☆

334608 Rinorea convallarioides (Baker f.) Eyles subsp. marsabitensis Grey-Wilson；马萨比特三角车●☆

334609 Rinorea convallarioides (Baker f.) Eyles subsp. occidentalis Grey-Wilson；西方三角车●☆

334610 Rinorea crassifolia (Baker f.) De Wild.；厚叶三角车●☆

334611 Rinorea dentata (P. Beauv.) Kuntze；具齿三角车●☆

334612 Rinorea dichroa Mildbr. et Melch. = Rinorea oblongifolia (C. H. Wright) Marquand ex Chipp ●☆

334613 Rinorea djalonensis A. Chev. ex Hutch. et Dalziel；贾隆三角车●☆

334614 Rinorea dubia De Wild.；可疑三角车●☆

334615 Rinorea ebolowensis M. Brandt；埃博洛瓦三角车●☆

334616 Rinorea elliotii Engl. = Rinorea welwitschii (Oliv.) Kuntze ●☆

334617 Rinorea elliptica (Oliv.) Kuntze；椭圆三角车●☆

334618 Rinorea erianthera C. Y. Wu et Chu Ho；毛蕊三角车（毛蕊三角草）；Fairenther Rinorea, Woolly-flowered Rinorea ●

334619 Rinorea exappendiculata Engl.；附属物三角车●☆

334620 Rinorea fausteana Achound.；福斯特三角车●☆

334621 Rinorea ferruginea Engl.；锈色三角车●☆

334622 Rinorea friisii M. G. Gilbert；弗里斯三角车●☆

334623 Rinorea gabunensis Engl.；加蓬三角车（加本三角车）●☆

334624 Rinorea gazensis (Baker f.) M. Brandt = Rinorea ferruginea Engl. ●☆

334625 Rinorea gilletii De Wild.；吉勒特三角车●☆

334626 Rinorea glandulosa Merr. = Rinorea bengalensis (Wall.) Kuntze ●

334627 Rinorea gossweileri Exell；戈斯三角车●☆

334628 Rinorea gracilipes Engl. = Rinorea angustifolia (Thouars) Baill. subsp. engleriana (De Wild. et T. Durand) Grey-Wilson ●☆

334629 Rinorea holtzii Engl. = Rinorea angustifolia (Thouars) Baill. subsp. ardisiiflora (Oliv.) Grey-Wilson ●☆

334630 Rinorea ilicifolia (Welw. ex Oliv.) Kuntze；冬青叶三角车●☆

334631 Rinorea ilicifolia (Welw. ex Oliv.) Kuntze var. amplexicaulis Grey-Wilson；抱茎三角车●☆

334632 Rinorea ilicifolia Engl. var. khutuensis (Engl.) Tennant = Rinorea ilicifolia (Welw. ex Oliv.) Kuntze ●☆

334633 Rinorea insularis Engl.；海岛三角车■☆

334634 Rinorea ituriensis M. Brandt；伊图里三角车●☆

334635 Rinorea johnstonii (Stapf) M. Brandt；约翰斯顿三角车●☆

334636 Rinorea kamerunensis Engl.；喀麦隆三角车●☆

334637 Rinorea kassneri Engl. = Rinorea squamosa (Boivin ex Tul.) Baill. subsp. kassneri (Engl.) Grey-Wilson ●☆

334638 Rinorea keayi Brenan；凯伊三角车●☆

334639 Rinorea kemoensis A. Chev.；凯莫三角车●☆

334640 Rinorea khutuensis Engl. = Rinorea ilicifolia (Welw. ex Oliv.) Kuntze ●☆

334641 Rinorea kibbiensis Chipp；基比三角车●☆

334642 Rinorea kimiloloensis Taton = Rinorea angustifolia (Thouars) Baill. subsp. ardisiiflora (Oliv.) Grey-Wilson ●☆

334643 Rinorea latibracteata M. Brandt；宽苞三角车●☆

334644 Rinorea laurentii De Wild.；洛朗三角车●☆

334645 Rinorea laurentii De Wild. var. velutina Taton；短绒毛三角车●☆

334646 Rinorea ledermannii Engl.；莱德三角车●☆

334647 Rinorea leiophylla M. Brandt；光叶三角车●☆

334648 Rinorea liberica Engl.；利比里亚三角车●☆

334649 Rinorea longicuspis Engl. = Rinorea welwitschii (Oliv.) Kuntze ●☆

334650 Rinorea longifolia De Wild.；长叶三角车●☆

334651 Rinorea longiracemosa (Kurz) Craib；长穗三角车（短柄三角车）●

334652 Rinorea longiracemosa Kurz = Rinorea sessilis (Lour.) Kuntze ●

334653 Rinorea longisepala Engl.；长萼三角车●☆

334654 Rinorea malembaensis Taton；马伦巴三角车●☆

334655 Rinorea mayumbensis Exell；马永巴三角车●☆

334656 Rinorea microdon M. Brandt；小齿三角车●☆

334657 Rinorea microglossa Engl.；小舌三角车●☆

334658 Rinorea mildbraedii M. Brandt；米尔德三角车●☆

334659 Rinorea molleri M. Brandt；默勒三角车●☆

334660 Rinorea multinervis M. Brandt；多脉三角车●☆

334661 Rinorea myrsinifolia Dunkley；铁仔叶三角车●☆

334662 Rinorea natalensis Engl. = Rinorea angustifolia (Thouars) Baill. subsp. natalensis (Engl.) Grey-Wilson ●☆

334663 Rinorea obanensis (Baker f.) Chipp = Allexis obanensis (Baker f.) Melch. ●☆

334664 Rinorea oblanceolata Chipp；倒披针形三角车●☆

334665 Rinorea oblongifolia (C. H. Wright) C. Marquand = Rinorea oblongifolia (C. H. Wright) C. Marquand ex Chipp ●☆

334666 Rinorea oblongifolia (C. H. Wright) C. Marquand ex Chipp；矩圆叶三角车●☆

334667 Rinorea oliveri T. Durand et Schinz；奥里弗三角车●☆

334668 Rinorea oppositifolia Exell；对叶三角车●☆

334669 Rinorea oubanguiensis Tisser.；乌班吉三角车●☆

334670 Rinorea oxycarpa Exell；尖果三角车●☆

334671 Rinorea parviflora Chipp；小花三角车●☆

334672 Rinorea physiphora Kuntze；巴西三角车●☆

334673 Rinorea pierrei (H. Boissieu) Melch. = Scyphellandra pierrei H. Boissieu ●

334674 Rinorea pilosa Chipp；疏毛三角车●☆

334675 Rinorea poggei Engl. = Rinorea brachypetala (Turcz.) Kuntze ●☆

334676 Rinorea prasina (Stapf) Chipp；草绿三角车●☆

334677 Rinorea preusii Engl.；普雷乌斯三角车●☆

334678 Rinorea raymondiana Taton；雷蒙三角车●☆

334679 Rinorea rubrotincta Chipp；红三角车●☆

334680 Rinorea rudolphiana Taton；鲁道夫三角车●☆

334681 Rinorea sapinii De Wild.；萨潘三角车●☆

334682 Rinorea scheffleri Engl.；谢夫勒三角车●☆

334683 Rinorea seleensis De Wild.；塞莱三角车●☆

334684 Rinorea sessilis (Lour.) Kuntze；短柄三角车（短柄雷诺木）；Shortstalk Rinorea, Stipeless Rinorea ●

334685 Rinorea sinuata Chipp；深波三角车●☆

334686 Rinorea somalensis Chiov. = Rinorea elliptica (Oliv.) Kuntze ●☆

334687　Rinorea soyauxii M. Brandt;索亚三角车●☆

334688　Rinorea squamosa（Boivin ex Tul.）Baill.;多鳞三角车●☆

334689　Rinorea squamosa（Boivin ex Tul.）Baill. subsp. kassneri（Engl.）Grey-Wilson;卡斯纳三角车●☆

334690　Rinorea stipulata Exell;托叶三角车●☆

334691　Rinorea strictiflora（Oliv.）Exell et Mendonça;直花三角车●☆

334692　Rinorea subauriculata Chipp;耳形三角车●☆

334693　Rinorea subglandulosa De Wild.;亚腺三角车●☆

334694　Rinorea subintegrifolia（P. Beauv.）Kuntze;近全叶鳞隔堇●☆

334695　Rinorea subintegrifolia（P. Beauv.）Kuntze f. parvifolia Roberty;小叶鳞隔堇●

334696　Rinorea subsessilis M. Brandt;近无柄三角车●☆

334697　Rinorea talbotii（Baker f.）De Wild.;塔尔博特三角车●☆

334698　Rinorea tessmannii M. Brandt;泰斯曼三角车●☆

334699　Rinorea thomasii Achound.;托马斯三角车●☆

334700　Rinorea thomensis Exell;托马三角车●☆

334701　Rinorea tortuosa Taton;扭曲三角车●☆

334702　Rinorea umbricola Engl.;荫地三角车●☆

334703　Rinorea varia Chipp;易变三角车●☆

334704　Rinorea verrucosa Chipp;多疣三角车●☆

334705　Rinorea virgata（Thwaites）Kuntze;鳞隔堇; Pierre Scyphellandra ●

334706　Rinorea wagemansii Taton;瓦格曼斯三角车●☆

334707　Rinorea wallichiana（Hook. f. et Thomson）Kuntze ＝ Rinorea bengalensis（Wall.）Kuntze ●

334708　Rinorea welwitschii（Oliv.）Kuntze;韦尔三角车●☆

334709　Rinorea welwitschii（Oliv.）Kuntze subsp. tanzanica Grey-Wilson;坦桑尼亚三角车●☆

334710　Rinorea whytei（Stapf）M. Brandt;怀特三角车●☆

334711　Rinorea woermanniana（Büttner）Engl.;韦尔曼三角车●☆

334712　Rinorea yaundensis Engl.;雅温德三角车●☆

334713　Rinorea youngii Exell et Mendonça;扬氏三角车●☆

334714　Rinorea zanagensis Achound. et Bos;扎纳加三角车●☆

334715　Rinorea zenkeri Engl.;岑克尔三角车●☆

334716　Rinoreocarpus Ducke（1925）;尖隔堇属■☆

334717　Rinoreocarpus salmoneus Ducke;尖隔堇■☆

334718　Rinxostylis Raf. ＝ Cissus L. ●

334719　Rinzia Schauer ＝ Baeckea L. ●

334720　Rinzia Schauer（1843）;林茨桃金娘属●☆

334721　Rinzia affinis Trudgen;近缘林茨桃金娘●☆

334722　Rinzia communis Trudgen;普通林茨桃金娘●☆

334723　Rinzia crassifolia Turcz.;厚叶林茨桃金娘●☆

334724　Rinzia fumana Schauer;林茨桃金娘●☆

334725　Rinzia rubra Trudgen;红林茨桃金娘●☆

334726　Rinzia sessilis Trudgen;无梗林茨桃金娘●☆

334727　Riocreuxia Decne.（1844）;里奥萝藦属■☆

334728　Riocreuxia aberrans R. A. Dyer;异常里奥萝藦■☆

334729　Riocreuxia alexandrina（H. Huber）R. A. Dyer ＝ Riocreuxia flanaganii Schltr. subsp. alexandrina（H. Huber）R. A. Dyer ■☆

334730　Riocreuxia alexandrina（H. Huber）R. A. Dyer ＝ Riocreuxia flanaganii Schltr. var. alexandrina（H. Huber）Masinde ■☆

334731　Riocreuxia bolusii N. E. Br. ＝ Riocreuxia torulosa（E. Mey.）Decne. var. bolusii（N. E. Br.）Masinde ■☆

334732　Riocreuxia burchellii K. Schum. ＝ Riocreuxia polyantha Schltr. ■☆

334733　Riocreuxia chrysochroma（H. Huber）A. R. Sm.;金色里奥萝藦■☆

334734　Riocreuxia flanaganii Schltr.;弗拉纳根里奥萝藦■☆

334735　Riocreuxia flanaganii Schltr. subsp. alexandrina（H. Huber）R. A. Dyer ＝ Riocreuxia alexandrina（H. Huber）R. A. Dyer ■☆

334736　Riocreuxia flanaganii Schltr. subsp. segregata R. A. Dyer ＝ Riocreuxia polyantha Schltr. ■☆

334737　Riocreuxia flanaganii Schltr. subsp. woodii（N. E. Br.）R. A. Dyer ＝ Riocreuxia woodii N. E. Br. ■☆

334738　Riocreuxia flanaganii Schltr. var. alexandrina（H. Huber）Masinde;亚历山大里奥萝藦■☆

334739　Riocreuxia longiflora K. Schum. ＝ Ceropegia stenantha K. Schum. ■☆

334740　Riocreuxia picta Schltr.;着色里奥萝藦■☆

334741　Riocreuxia polyantha Schltr.;多花里奥萝藦■☆

334742　Riocreuxia profusa N. E. Br. ＝ Riocreuxia polyantha Schltr. ■☆

334743　Riocreuxia splendida K. Schum.;闪光里奥萝藦■☆

334744　Riocreuxia torulosa（E. Mey.）Decne.;结节里奥萝藦■☆

334745　Riocreuxia torulosa（E. Mey.）Decne. var. bolusii（N. E. Br.）Masinde;博卢斯里奥萝藦■☆

334746　Riocreuxia torulosa Decne. var. longidens N. E. Br. ＝ Riocreuxia torulosa（E. Mey.）Decne. ■☆

334747　Riocreuxia torulosa Decne. var. obsoleta N. E. Br. ＝ Riocreuxia torulosa（E. Mey.）Decne. ■☆

334748　Riocreuxia torulosa Decne. var. tomentosa N. E. Br. ＝ Riocreuxia torulosa（E. Mey.）Decne. ■☆

334749　Riocreuxia torulosa Schltr. ＝ Riocreuxia polyantha Schltr. ■☆

334750　Riocreuxia woodii N. E. Br.;伍得里奥萝藦■☆

334751　Riodocea Delprete（1999）;里奥茜属■☆

334752　Riparia Raf. ＝ Baptisia Vent. ■☆

334753　Riparia Raf. ＝ Ripasia Raf. ●☆

334754　Ripartia（Gand.）Gand. ＝ Rosa L. ●

334755　Ripartia Gand. ＝ Rosa L. ●

334756　Ripasia Raf. ＝ Baptisia Vent. ■☆

334757　Ripidium Trin. ＝ Erianthus Michx. ■

334758　Ripidium Trin. ＝ Saccharum L. ■

334759　Ripidium arun-dinaceum（Retz.）Grassl ＝ Saccharum arundinaceum Retz. ■

334760　Ripidium elephantinum（Hook. f.）Grassl ＝ Saccharum ravennae（L.）Murray ■

334761　Ripidium japonicus Trin. ＝ Miscanthus sinensis Andersson ■

334762　Ripidium procerum（Roxb.）Grassl ＝ Saccharum procerum Roxb. ■

334763　Ripidium ravennae（L.）Trin. ＝ Saccharum ravennae（L.）Murray ■

334764　Ripidodendrum Post et Kuntze ＝ Aloe L. ●■

334765　Ripidodendrum Post et Kuntze ＝ Rhipidodendrum Willd. ●■

334766　Ripidostigma Post et Kuntze ＝ Diospyros L. ●

334767　Ripidostigma Post et Kuntze ＝ Rhipidostigma Hassk. ●

334768　Ripogonaceae Conran. et Clifford ＝ Smilacaceae Vent.（保留科名）●

334769　Ripogonaceae Conran. et Clifford（1985）;无须藤科●☆

334770　Ripogonaceae Huber ex Takht. ＝ Ripogonaceae Conran. et Clifford ●☆

334771　Ripogonum J. R. Forst. et G. Forst.（1775）;无须藤属（菝葜藤属,无须菝葜属）●☆

334772　Ripogonum J. R. Forst. et G. Forst. ＝ Rhipogonum J. R. Forst. et G. Forst. ●☆

334773　Ripogonum scandens J. R. Forst. et G. Forst.;无须藤（菝葜藤,无须菝葜）●☆

334774　Ripsalis Post et Kuntze = Rhipsalis Gaertn.（保留属名）●

334775　Ripselaxis Raf. = Salix L.（保留属名）●

334776　Ripsoctis Raf. = Salix L.（保留属名）●

334777　Riqueria Pers. = Riqueuria Ruiz et Pav. ☆334778　Riqueuria Ruiz et Pav.（1794）；里克尔茜属☆334779　Riqueuria avenia Ruiz et Pav.；里克尔茜☆334780　Riseleya Hemsl. = Drypetes Vahl ●

334781　Risleya King et Pantl.（1898）；紫茎兰属；Risleya ■

334782　Risleya atropurpurea King et Pantl.；紫茎兰；Darkpurple Risleya，Risleya ■

334783　Rissoa Arn. = Atalantia Corrêa（保留属名）●

334784　Ristantia Peter G. Wilson et J. T. Waterh.（1982）；昆士兰桃金娘属●☆

334785　Ristantia pachysperma（Bailey）Peter G. Wilson et J. T. Waterh.；昆士兰桃金娘●☆

334786　Ritaia King et Pantl. = Ceratostylis Blume ■

334787　Ritaia himalaica（Hook. f.）King et Pantl. = Ceratostylis himalaica Hook. f. ■

334788　Ritchiea R. Br. = Ritchiea R. Br. ex G. Don ●☆

334789　Ritchiea R. Br. ex G. Don（1831）；里奇山柑属●☆

334790　Ritchiea afzelii Gilg；阿氏里奇山柑●☆

334791　Ritchiea albersii Gilg；阿伯斯山柑●☆

334792　Ritchiea apiculata Gilg et Gilg-Ben. = Ritchiea capparoides（Andréws）Britten ●☆

334793　Ritchiea balbi Chiov. = Ritchiea albersii Gilg ●☆

334794　Ritchiea boukokoensis Tisser. et Sillans = Ritchiea capparoides（Andréws）Britten ●☆

334795　Ritchiea brachypoda Gilg = Ritchiea erecta Hook. f. ●☆

334796　Ritchiea bussei Gilg = Ritchiea capparoides（Andréws）Britten ●☆

334797　Ritchiea capparoides（Andréws）Britten；假山柑●☆

334798　Ritchiea capparoides（Andréws）Britten var. longipedicellata（Gilg）DeWolf；长柄假山柑●☆

334799　Ritchiea carrissoi Exell et Mendonça；卡里索山柑●☆

334800　Ritchiea chlorantha Gilg = Ritchiea albersii Gilg ●☆

334801　Ritchiea dolichocarpa Gilg et Gilg-Ben. = Euadenia eminens Hook. f. ●☆

334802　Ritchiea duchesnei（De Wild.）Keay = Maerua duchesnei（De Wild.）F. White ●☆

334803　Ritchiea ealaensis De Wild. = Ritchiea capparoides（Andréws）Britten ●☆

334804　Ritchiea engleriana Buscal. et Muschl. = Ritchiea capparoides（Andréws）Britten ●☆

334805　Ritchiea erecta Hook. f.；直立假山柑●☆

334806　Ritchiea fragariodora Gilg = Ritchiea capparoides（Andréws）Britten ●☆

334807　Ritchiea fragrans（Sims）R. Br. = Ritchiea capparoides（Andréws）Britten ●☆

334808　Ritchiea fragrans（Sims）R. Br. ex G. Don = Ritchiea capparoides（Andréws）Britten ●☆

334809　Ritchiea gigantocarpa Gilg et Gilg-Ben. = Cladostemon kirkii（Oliv.）Pax et Gilg ●☆

334810　Ritchiea glossopetala Gilg = Ritchiea erecta Hook. f. ●☆

334811　Ritchiea gossweileri Exell et Mendonça；戈斯山柑●☆

334812　Ritchiea grandiflora（Pax）Gilg = Ritchiea reflexa（Thonn.）Gilg et Gilg-Ben. ●☆

334813　Ritchiea heterophylla Gilg = Ritchiea erecta Hook. f. ●☆

334814　Ritchiea immersa De Wild. = Ritchiea capparoides（Andréws）Britten ●☆

334815　Ritchiea insculpta Gilg et Gilg-Ben. = Ritchiea capparoides（Andréws）Britten ●☆

334816　Ritchiea insignis（Pax）Gilg = Ritchiea capparoides（Andréws）Britten ●☆

334817　Ritchiea jansii R. Wilczek；简斯山柑●☆

334818　Ritchiea laurentii De Wild. = Ritchiea capparoides（Andréws）Britten ●☆

334819　Ritchiea leucantha Gilg et Gilg-Ben. = Ritchiea capparoides（Andréws）Britten ●☆

334820　Ritchiea littoralis R. Wilczek；滨海山柑●☆

334821　Ritchiea longipedicellata Gilg = Ritchiea capparoides（Andréws）Britten var. longipedicellata（Gilg）DeWolf ●☆

334822　Ritchiea macrantha Gilg et Gilg-Ben.；大花里奇山柑●☆

334823　Ritchiea macrocarpa Gilg = Ritchiea albersii Gilg ●☆

334824　Ritchiea mayumbensis Exell；马永巴山柑●☆

334825　Ritchiea mildbraedii Gilg = Ritchiea albersii Gilg ●☆

334826　Ritchiea noldeae Exell et Mendonça；诺尔德山柑●☆

334827　Ritchiea obanensis Hutch. et Dalziel = Ritchiea erecta Hook. f. ●☆

334828　Ritchiea oreophila Gilg et Gilg-Ben. = Ritchiea erecta Hook. f. ●☆

334829　Ritchiea ovata R. Wilczek；卵形山柑●☆

334830　Ritchiea pentaphylla Gilg et Gilg-Ben. = Ritchiea erecta Hook. f. ●☆

334831　Ritchiea persicifolia Schinz ex A. Chev. = Maerua duchesnei（De Wild.）F. White ●☆

334832　Ritchiea polypetala Hook. f. = Ritchiea erecta Hook. f. ●☆

334833　Ritchiea pygmaea（Gilg）DeWolf；矮小里奇山柑●☆

334834　Ritchiea pynaertii De Wild. = Ritchiea capparoides（Andréws）Britten ●☆

334835　Ritchiea quarrei R. Wilczek；卡雷山柑●☆

334836　Ritchiea reflexa（Thonn.）Gilg et Gilg-Ben.；反折里奇山柑●☆

334837　Ritchiea simplicifolia Oliv.；单叶里奇山柑●☆

334838　Ritchiea stella-aethiopica Pax = Ritchiea albersii Gilg ●☆

334839　Ritchiea steudneri Gilg = Ritchiea albersii Gilg ●☆

334840　Ritchiea tessmannii Gilg ex Engl. = Ritchiea simplicifolia Oliv. ●☆

334841　Ritchiea thouretiae Gilg et Gilg-Ben. = Ritchiea capparoides（Andréws）Britten ●☆

334842　Ritchiea werthiana Gilg = Ritchiea capparoides（Andréws）Britten ●☆

334843　Ritchiea wilczekiana Bamps；维尔切克山柑●☆

334844　Ritchiea wittei R. Wilczek；维特山柑●☆

334845　Ritchiea youngii Exell；扬氏山柑●☆

334846　Ritchieophyton Pax = Givotia Griff. ●☆

334847　Rithrophyllum Post et Kuntze = Aeschynanthus Jack（保留属名）●■

334848　Rithrophyllum Post et Kuntze = Rheithrophyllum Hassk. ●■

334849　Ritinophora Neck. = Amyris P. Browne ●☆

334850　Ritonia Benoist（1962）；里顿爵床属■☆

334851　Ritonia barbigera Benoist；毛里顿爵床■☆

334852　Ritonia humbertii Benoist；亨伯特里顿爵床■☆

334853　Ritonia poissonii Benoist；里顿爵床■☆

334854　Ritonia rosea Benoist；粉红里顿爵床■☆

334855　Rittcnasia Raf. = Menispermum L. ●■

334856　Rittera Raf. = Centranthus Lam. et DC. ■

334857　Rittera Raf. = Monastes Raf. ■

334858　Rittera Schreb. = Possira Aubl.（废弃属名）●☆

334859　Rittera Schreb. = Swartzia Schreb.（保留属名）●☆

334860　Ritterocactus Doweld = Echinocactus Link et Otto ●

334861 Ritterocactus Doweld(1999);巴西仙人球属●☆

334862 Ritterocereus Backeb. = Lemaireocereus Britton et Rose ●☆

334863 Ritterocereus Backeb. = Stenocereus（A. Berger）Riccob.（保留名）●☆

334864 Ritterocereus griseus（Haw.）Backeb.；象牙仙人柱●☆

334865 Rivasgodaya Esteve = Genista L. ●

334866 Rivea Choisy(1833);里夫藤属(赖维亚属,力夫藤属)●☆

334867 Rivea adenioides（Schinz）Hallier f. = Ipomoea adenioides Schinz ■☆

334868 Rivea argyrophylla（Vatke）Hallier f. = Ipomoea argyrophylla Vatke ■☆

334869 Rivea corymbosa（L.）Hallier f.；伞花力夫藤（赖维亚,墨西哥旋花）；Christmas Gambol, Morning Glory ●☆

334870 Rivea corymbosa（L.）Hallier f. var. mollissima（Webb et Berthel.）Hallier f. = Turbina corymbosa（L.）Raf. ■☆

334871 Rivea corymbosa Hallier f. = Rivea corymbosa（L.）Hallier f. ●☆

334872 Rivea cuneata Wight;截叶力夫藤(截叶旋花)●☆

334873 Rivea decora（Vatke et Hildebrandt）Hallier f. = Ipomoea hildebrandtii Vatke ■☆

334874 Rivea hartmannii（Vatke）Hallier f. = Ipomoea hartmannii Vatke ■☆

334875 Rivea holubii（Baker）Hallier f. = Turbina holubii（Baker）A. Meeuse ■☆

334876 Rivea hypocrateriformis（Desr.）Choisy;高脚杯状力夫藤●☆

334877 Rivea kituiensis（Vatke）Hallier f. = Ipomoea kituiensis Vatke ■☆

334878 Rivea nana Hallier f. = Ipomoea jaegeri Pilg. ■☆

334879 Rivea nervosa（Burm. f.）Hall. f. = Argyreia nervosa（Burm. f.）Bojer ●

334880 Rivea obtecta（Wall.）Choisy = Argyreia mollis（Burm. f.）Choisy ●

334881 Rivea oenotheroides（L. f.）Hallier f. = Ipomoea oenotheroides（L. f.）Raf. ex Hallier f. ■☆

334882 Rivea pringsheimiana Dammer = Stictocardia laxiflora（Baker）Hallier f. ●☆

334883 Rivea pyramidalis（Hallier f.）Hallier f. = Ipomoea pyramidalis Hallier f. ■☆

334884 Rivea shirensis（Oliv.）Hallier f. = Paralepistemon shirensis（Oliv.）Lejoly et Lisowski ●☆

334885 Rivea stenosiphon（Hallier f.）Hallier f. = Ipomoea stenosiphon Hallier f. ■☆

334886 Rivea suffruticosa（Burch.）Hallier f. = Ipomoea suffruticosa Burch. ●☆

334887 Rivea tiliifolia（Desr.）Choisy = Stictocardia tiliifolia（Desr.）Hallier f. ●■

334888 Rivea tiliifolia Choisy = Stictocardia tiliifolia（Choisy）Hallier f. ●■

334889 Rivea tiliifolia Choisy = Stictocardia tiliifolia（Desr.）Hallier f. ●■

334890 Rivea urbaniana Dammer = Ipomoea urbaniana（Dammer）Hallier f. ●☆

334891 Riveria Kunth = Swartzia Schreb.（保留属名）●☆

334892 Rivina L.(1753);数珠珊瑚属(蕾芬属);Rivin, Rouge Plant, Rougeplant ●

334893 Rivina Plum. = Rivina L. ●

334894 Rivina Plum. ex L. = Rivina L. ●

334895 Rivina apetala Schumach. et Thonn. = Hilleria latifolia（Lam.）H. Walter ●☆

334896 Rivina brasiliensis Nocca = Rivina humilis L. ●☆

334897 Rivina humilis L.；数珠珊瑚（蕾芬,小商陆）；Baby Pepper, Blood Berry, Bloodberry, Pigeon Berry, Pigeonberry, Pigeon-berry, Rivin, Rouge Plant, Rougeplant ●

334898 Rivina laevis L. = Rivina humilis L. ●

334899 Rivina latifolia Lam. = Hilleria latifolia（Lam.）H. Walter ●☆

334900 Rivina octandra L. = Trichostigma octandrum（L.）H. Walter ●☆

334901 Rivina portulaccoides Nutt. = Rivina humilis L. ●

334902 Rivina tinctaria Ham. ex G. Don;着色数珠珊瑚(着色商陆)●☆

334903 Rivinaceae C. Agardh = Petiveriaceae C. Agardh ●☆

334904 Rivinaceae C. Agardh = Phytolaccaceae R. Br.（保留科名）●■

334905 Rivinaceae C. Agardh;数珠珊瑚科●

334906 Rivinia L. = Rivina L. ●

334907 Rivinia Mill. = Trichostigma A. Rich. ●☆

334908 Riviniaceae C. Agardh = Phytolaccaceae R. Br.（保留科名）●■

334909 Rivinoides Afzel. ex Prain = Erythrococca Benth. ●☆

334910 Rixea C. Morren = Tropaeolum L. ●

334911 Rixia Lindl. = Rixea C. Morren ●

334912 Rizoa Cav. = Gardoquia Ruiz et Pav. ●■

334913 Rizoa Cav. = Satureja L. ●■

334914 Robbairea Boiss. = Polycarpaea Lam.（保留属名）■●

334915 Robbairea confusa Maire = Polycarpon robbaireum Kuntze ■☆

334916 Robbairea delileana Milne-Redh. = Polycarpon robbaireum Kuntze ■☆

334917 Robbairea delileana Milne-Redh. var. major（Asch. et Schweinf.）Täckh. = Polycarpon robbaireum Kuntze ■☆

334918 Robbairea major（Asch. et Schweinf.）Botsch. = Polycarpon robbaireum Kuntze ■☆

334919 Robbairea prostrata（Asch. et Schweinf.）Botsch. = Polycarpaea robbairea（Kuntze）Greuter et Burdet ☆

334920 Robbairea prostrata Boiss. var. major Asch. et Schweinf. = Polycarpon robbaireum Kuntze ■☆

334921 Robbia A. DC. = Malouetia A. DC. ●☆

334922 Robbrechtia De Block(2003);罗伯茜属●☆

334923 Robbrechtia grandifolia De Block;罗伯茜●☆

334924 Robbrechtia milleri De Block;马岛罗伯茜●☆

334925 Robergia Roxb. = Pegia Colebr. ●

334926 Robergia Schreb. = Rourea Aubl.（保留属名）●

334927 Robergia hirsuta Roxb. = Pegia nitida Colebr. ●

334928 Roberta St.-Lag. = Robertia DC. ■

334929 Robertia DC. = Hypochaeris L. ●

334930 Robertia Merat = Eranthis Salisb.（保留属名）■

334931 Robertia Rich. ex Carrière = Phyllocladus Rich. ex Mirb.（保留属名）●☆

334932 Robertia Rich. ex DC. = Hypochaeris L. ■

334933 Robertia Scop.（废弃属名）= Bumelia Sw.（保留属名）●☆

334934 Robertia Scop.（废弃属名）= Sideroxylon L. ●☆

334935 Robertiella Hanks = Geranium L. ■●

334936 Robertiella Hanks = Robertium Picard ■●

334937 Robertiella robertiana（L.）Hanks = Geranium robertianum L. ■

334938 Robertium Picard = Geranium L. ■●

334939 Robertsia Endl. = Sideroxylon L. ●☆

334940 Robertsonia Haw. = Saxifraga L. ■

334941 Robeschia Hochst. ex E. Fourn. = Descurainia Webb et Berthel.（保留属名）■

334942 Robeschia Hochst. ex O. E. Schulz(1924);中东芥属■☆

334943 Robeschia schimperi（Boiss.）O. E. Schulz;中东芥■☆

334944 Robeschia sinaica Hochst. ex E. Fourn. = Robeschia schimperi

4780

（Boiss.）O. E. Schulz ■☆

334945　Robina Aubl. = Robinia L. ●

334946　Robinia L.（1753）；刺槐属（洋槐属）；Black Locust，False Acacia，Locust，Locust Bean，Robinia ●

334947　Robinia altagana Poir. var. fruticosa Pall. = Caragana fruticosa （Pall.）Besser ●

334948　Robinia ambigua Poir.；安比刺槐（黏枝刺槐）；Purple Robe Locust ●☆

334949　Robinia ambigua Poir. 'Decne. ana'；德氏黏枝刺槐●☆

334950　Robinia ambigua Poir. 'Idahensis'；伊达赫安比刺槐●☆

334951　Robinia boyntonii Ashe；波伊图刺槐●☆

334952　Robinia candida（DC.）Roxb. = Tephrosia candida DC. ●

334953　Robinia candida Roxb. = Tephrosia candida（Roxb.）DC. ●

334954　Robinia caragana L. = Caragana arborescens Lam. ●

334955　Robinia chamlagu L'Hér. = Caragana sinica（Buc'hoz）Rehder ●

334956　Robinia cyanescens Schumach. et Thonn. = Philenoptera cyanescens（Schumach. et Thonn.）Roberty ●☆

334957　Robinia ferruginea Roxb. = Derris ferruginea（Roxb.）Benth. ●

334958　Robinia fertilis Ashe = Robinia hispida L. ●

334959　Robinia flava Lour. = Sophora flavescens Aiton ●■

334960　Robinia frutex L. = Caragana frutex（L.）C. Koch ●

334961　Robinia grandiflora Ashe = Robinia hispida L. ●

334962　Robinia grandiflora L. = Sesbania grandiflora（L.）Pers. ●

334963　Robinia guineensis Willd. = Ormocarpum sennoides（Willd.）DC. subsp. hispidum（Willd.）Brenan et J. Léonard ●☆

334964　Robinia halodendron Pall. = Halimodendron halodendron（Pall.）C. K. Schneid. ●

334965　Robinia hispida L.；毛刺槐（粉花洋槐，江南槐，毛洋槐）；Bristly Locust，Hardhair Locust，Hispid Locust，Moss Locust，Pink Acacia，Rose Acacia，Rose-acacia ●

334966　Robinia hispida L. var. kelseyi（Hutch.）Isely；腺毛刺槐●☆

334967　Robinia inermis Dum. Cours. = Robinia pseudoacacia L. var. umbraculifera DC. ●

334968　Robinia jubata Pall. = Caragana jubata（Pall.）Poir. ●

334969　Robinia kelseyi Hutch.；阿莱干尼刺槐；Allegheny Moss ●☆

334970　Robinia longiloba Ashe = Robinia hispida L. ●

334971　Robinia macrophylla Roxb. = Millettia extensa Benth. ●☆

334972　Robinia mitis L. = Pongamia pinnata（L.）Pierre ●

334973　Robinia multiflora Schumach. et Thonn. = Millettia irvinei Hutch. et Dalziel ●☆

334974　Robinia neomexicana A. Gray；新墨西哥刺槐；Cat's Claws，New Mexican Locust，New Mexico Locust，Southwestern Locust ●☆

334975　Robinia pallida Ashe = Robinia hispida L. ●

334976　Robinia pseudoacacia L.；刺槐（德国槐，海岸玛都那木，胡藤，浆果鹃木，蒙多那木，太平洋玛都那木，太平洋玛都那木树，熊果树，洋槐）；Acacia，Arbuti Tree，Bastard Acacia，Black Acacia，Black Locust，Black Locust Tree，Blackacacia，Coast Madrone，Common Acacia，Common Locust，False Acacia，Locust，Locust Tree，Pacific Madrone，Shipmast Locust，Silver Chain，White Watch-and-chain，Whya-tree，Yellow Locust ●

334977　Robinia pseudoacacia L. 'Appalachia'；阿巴拉契亚刺槐●☆

334978　Robinia pseudoacacia L. 'Aurea'；金叶刺槐●☆

334979　Robinia pseudoacacia L. 'Bessoniana'；拜苏尼纳刺槐●☆

334980　Robinia pseudoacacia L. 'Coluteoides'；科鲁特奥刺槐●☆

334981　Robinia pseudoacacia L. 'Frisia'；福利斯刺槐（金叶刺槐）●☆

334982　Robinia pseudoacacia L. 'Inermis'；无刺槐（无刺洋槐）；Mop-headed Acacia，Spineless Black Locust，Spineless Yellow Locust ●

334983　Robinia pseudoacacia L. 'Semperflorens'；常花刺槐●☆

334984　Robinia pseudoacacia L. 'Tortuosa'；扭枝刺槐●☆

334985　Robinia pseudoacacia L. 'Umbraculifera' = Robinia pseudoacacia L. var. umbraculifera DC. ●

334986　Robinia pseudoacacia L. 'Unifoliola'；单叶刺槐；Single-leaf Black Locust ●☆

334987　Robinia pseudoacacia L. f. decaisneana（Carrière）Voss；红花刺槐（红花洋槐）；Red-flower Black Locust ●

334988　Robinia pseudoacacia L. f. inermis（Mirb.）Rehder = Robinia pseudoacacia L. 'Inermis' ●

334989　Robinia pseudoacacia L. f. umbraculifera（DC.）Rehder = Robinia pseudoacacia L. var. umbraculifera DC. ●

334990　Robinia pseudoacacia L. var. inermis DC. = Robinia pseudoacacia L. ●

334991　Robinia pseudoacacia L. var. inermis DC. = Robinia pseudoacacia L. 'Inermis' ●

334992　Robinia pseudoacacia L. var. pyramidalis（Pepin）C. K. Schneid.；塔形洋槐；Umbrella Locust ●

334993　Robinia pseudoacacia L. var. pyramidalis（Pepin）C. K. Schneid. = Robinia pseudoacacia L. ●

334994　Robinia pseudoacacia L. var. rectissima（L.）Raber = Robinia pseudoacacia L. ●

334995　Robinia pseudoacacia L. var. umbraculifera DC.；伞刺槐（朝鲜槐，球冠无刺槐，伞槐，伞形洋槐，伞洋槐，无刺洋槐，圆冠刺槐）；Mophead Acacia ●

334996　Robinia pseudoacacia L. var. umbraculifera DC. = Robinia pseudoacacia L. ●

334997　Robinia pygmaea L. = Caragana pygmaea（L.）DC. ●

334998　Robinia pyramidalis Pepin = Robinia pseudoacacia L. var. pyramidalis（Pepin）C. K. Schneid. ●

334999　Robinia pyramidalis Pepin = Robinia pseudoacacia L. ●

335000　Robinia sepium Jacq. = Gliricidia sepium（Jacq.）Walp. ●☆

335001　Robinia sericea Poir. = Lonchocarpus sericeus（Poir.）Kunth ex DC. ●☆

335002　Robinia sinica Buc'hoz = Caragana sinica（Buc'hoz）Rehder ●

335003　Robinia slavinii Rehder；斯拉维刺槐●☆

335004　Robinia speciosa Ashe = Robinia hispida L. ●

335005　Robinia spinosa L. = Caragana spinosa（L.）DC. ●

335006　Robinia spinosa L. = Caragana spinosa（L.）Hornem. ●

335007　Robinia subdecandra L'Hér. = Calpurnia aurea（Aiton）Benth. ■☆

335008　Robinia thonningii Schumach. et Thonn. = Millettia thonningii（Schumach. et Thonn.）Baker ●☆

335009　Robinia tragacanthoides Pall. = Caragana tragacanthoides（Pall.）Poir. ●

335010　Robinia uliginosa Roxb. ex Willd. = Derris trifoliata Lour. ●

335011　Robinia uliginosa Willd. = Derris trifoliata Lour. ●

335012　Robinia umbraculifera DC. = Robinia pseudoacacia L. var. umbraculifera DC. ●

335013　Robinia viscosa Vent.；小刺槐（红花刺槐，许枝洋槐，黏刺槐，黏毛刺槐）；Clammy Locust，Rose-acacia ●☆

335014　Robiniaceae Vest = Robiniaceae Welw. ●

335015　Robiniaceae Welw.；刺槐科●

335016　Robiniaceae Welw. = Fabaceae Lindl.（保留科名）●■

335017　Robiniaceae Welw. = Leguminosae Juss.（保留科名）●■

335018　Robinsonecio T. M. Barkley et Janovec（1996）；外苞狗舌草属■☆

335019　Robinsonecio gerberifolius（Sch. Bip.）T. M. Barkley et Janovec；外苞狗舌草■☆

335020 Robinsonella Rose et Baker f. (1897);罗氏锦葵属●☆

335021 Robinsonella brevituba Fryxell;短管罗氏锦葵●☆

335022 Robinsonella cordata Rose et Baker f.;心形罗氏锦葵●☆

335023 Robinsonella densiflora Fryxell;密花罗氏锦葵●☆

335024 Robinsonella discolor Rose et Baker f.;异色罗氏锦葵●☆

335025 Robinsonella glabrifolia Fryxell;光叶罗氏锦葵●☆

335026 Robinsonia DC.(1833)(保留属名);顶叶菊属●☆

335027 Robinsonia Scop.(废弃属名) = Quiina Aubl.●☆

335028 Robinsonia Scop.(废弃属名) = Robinsonia DC.(保留属名)●☆

335029 Robinsonia Scop.(废弃属名) = Touroulia Aubl.☆

335030 Robinsonia gracilis Decne.;细顶叶菊●☆

335031 Robinsonia longifolia Phil.;长叶顶叶菊●☆

335032 Robinsonia macrocephala Decne.;大头顶叶菊●☆

335033 Robinsonia micrantha Phil. ex Hemsl.;小花顶叶菊●☆

335034 Robinsoniodendron Merr. = Maoutia Wedd.●

335035 Robiquetia Gaudich.(1829);寄树兰属;Robiquetia■

335036 Robiquetia hamata Schltr.;钩状寄树兰;Hooked Robiquetia■☆

335037 Robiquetia mooreana (Rolfe) J. J. Sm.;莫里寄树兰;Moore Robiquetia■☆

335038 Robiquetia paniculata (Lindl.) J. J. Sm. = Robiquetia succisa (Lindl.) Seidenf. et Garay

335039 Robiquetia spatulata (Blume) J. J. Sm.;大叶寄树兰(匙唇陆宾兰,匙状寄树兰);Largeleaf Robiquetia,Spatulate Robiquetia■

335040 Robiquetia succisa (Lindl.) Seidenf. et Garay;寄树兰(截叶陆宾兰,小叶寄树兰);Littleleaf Robiquetia■

335041 Roborowskia Batalin = Corydalis DC.(保留属名)■

335042 Roborowskia Batalin(1893);疆芹属(疆堇属);Xinjiang Violet■

335043 Roborowskia mira Batalin;疆芹(疆堇);Xinjiang Violet■

335044 Roborowskia mira Batalin = Corydalis mira (Batalin) C. Y. Wu et H. Chuang■

335045 Robsonia (Berland.) Rchb. = Ribes L.●

335046 Robsonia Rchb. = Ribes L.●

335047 Robsonodendron R. H. Archer(1997);罗氏卫矛属●☆

335048 Robsonodendron eucleiforme (Eckl. et Zeyh.) R. H. Archer;罗氏卫矛●☆

335049 Robsonodendron maritimum (Bolus) R. H. Archer;非洲罗氏卫矛●☆

335050 Robynsia Drapiez(废弃属名) = Bravoa Lex.■☆

335051 Robynsia Drapiez(废弃属名) = Polianthes L.■

335052 Robynsia Drapiez(废弃属名) = Robynsia Hutch.(保留属名)■☆

335053 Robynsia Hutch.(1931)(保留属名);罗宾茜属■☆

335054 Robynsia Hutch. = ? Rytigynia Blume■☆

335055 Robynsia M. Martens et Galeotti = Pachyrhizus Rich. ex DC.(保留属名)■

335056 Robynsia glabrata Hutch.;罗宾茜■☆

335057 Robynsiella Suess. = Centemopsis Schinz■☆

335058 Robynsiella fastigiata Suess. = Centemopsis fastigiata (Suess.) C. C. Towns.■☆

335059 Robynsiochloa Jacq.-Fél. = Rottboellia L. f.(保留属名)■

335060 Robynsiochloa purpurascens (Robyns) Jacq.-Fél. = Chasmopodium purpurascens (Robyns) Clayton■☆

335061 Robynsiophyton R. Wilczek(1953);罗宾豆属●☆

335062 Robynsiophyton vanderystii R. Wilczek;罗宾豆●☆

335063 Rocama Forssk. = Trianthema L.■

335064 Rocama prostrata Forssk. = Zaleya pentandra (L.) C. Jeffrey■☆

335065 Roccardia Neck. = Helipterum DC. ex Lindl.■☆

335066 Roccardia Neck. ex Raf. = Staehelina L.●■☆

335067 Roccardia Neck. ex Voss = Rhodanthe Lindl.●■☆

335068 Roccardia Neck. ex Voss = Syncarpha DC.■☆

335069 Roccardia Raf. = Staehelina L.●☆

335070 Rochea DC.(1802)(保留属名);罗景天属(罗齐阿属)●■☆

335071 Rochea DC.(保留属名) = Crassula L.●■☆

335072 Rochea Salisb. = Geissorhiza Ker Gawl.■☆

335073 Rochea Scop.(废弃属名) = Aeschynomene L.●■

335074 Rochea Scop.(废弃属名) = Rochea DC.(保留属名)●■☆

335075 Rochea albiflora (Sims) DC. = Crassula dejecta Jacq.■☆

335076 Rochea bicolor (Haw.) Steud. = Crassula fascicularis Lam.■☆

335077 Rochea biconvexa (Haw.) DC. = Crassula fascicularis Lam.■☆

335078 Rochea coccinea (L.) DC. = Crassula coccinea L.●☆

335079 Rochea coccinea DC.;绯红罗景天■☆

335080 Rochea cymosa (P. J. Bergius) DC. = Crassula cymosa P. J. Bergius■☆

335081 Rochea falcata (J. C. Wendl.) DC. var. minor (Haw.) G. D. Rowley = Crassula perfoliata L. var. minor (Haw.) G. D. Rowley●☆

335082 Rochea falcata (J. C. Wendl.) DC. var. acuminata Eckl. et Zeyh. = Crassula perfoliata L.●☆

335083 Rochea falcata (J. C. Wendl.) DC. var. minor (Haw.) DC. = Crassula perfoliata L. var. minor (Haw.) G. D. Rowley●☆

335084 Rochea fascicularis (Lam.) DC. = Crassula fascicularis Lam.■☆

335085 Rochea flava (L.) DC. = Crassula flava L.●☆

335086 Rochea jasminea (Haw. ex Sims) DC. = Crassula obtusa Haw.■☆

335087 Rochea media (Haw.) DC. = Crassula fascicularis Lam.■☆

335088 Rochea microphylla E. Mey. ex Drège = Crassula obtusa Haw.■☆

335089 Rochea odoratissima (Andréws) Link = Crassula fascicularis Lam.■☆

335090 Rochea odoratissima (Andréws) Link var. alba DC. = Crassula fascicularis Lam.■☆

335091 Rochea odoratissima (Andréws) Link var. bicolor (Haw.) DC. = Crassula fascicularis Lam.■☆

335092 Rochea perfoliata (L.) DC. = Crassula perfoliata L.●☆

335093 Rochea perfoliata (L.) DC. var. alba (Haw.) Sweet = Crassula perfoliata L.●☆

335094 Rochea perfoliata (L.) DC. var. coccinea Sweet = Crassula perfoliata L. var. coccinea (Sweet) G. D. Rowley●☆

335095 Rochea perfoliata (L.) DC. var. glaberrima E. Mey. ex Drège = Crassula macowaniana Schönland et Baker f.■☆

335096 Rochea undosa (Haw.) Steud. = Crassula dejecta Jacq.■☆

335097 Rochea venusta Salisb. = Geissorhiza radians (Thunb.) Goldblatt■☆

335098 Rochea versicolor (Burch. ex Ker Gawl.) Link = Crassula coccinea L.●☆

335099 Rochefortia Sw.(1788);罗什紫草属☆

335100 Rochefortia cuneata Sw.;楔形罗什紫草☆

335101 Rochefortia fasciculata Gürke;簇生罗什紫草☆

335102 Rochefortia ovata Sw.;卵形罗什紫草☆

335103 Rochelia Rchb.(1824)(保留属名);孪果鹤虱属(双果鹤虱属,弯果鹤虱属,旋果草属);Rochelia■

335104 Rochelia Roem. et Schult.(废弃属名) = Echinospermum Sw.■

335105 Rochelia Roem. et Schult.(废弃属名) = Lappula Wolf■

335106 Rochelia Roem. et Schult.(废弃属名) = Rochelia Rchb.(保留属名)■

335107 Rochelia bungei Trautv.;孪果鹤虱;Flex Rochelia■☆

335108 Rochelia campanulata Popov et Zakirov;风铃草状孪果鹤虱☆

335109 Rochelia cardiosepala Bunge;心萼孪果鹤虱■

335110 Rochelia disperma (L.) Hochr.;二籽孪果鹤虱(二籽旋果

草）■

335111　Rochelia disperma（L.）Koch = Rochelia bungei Trautv. ■

335112　Rochelia disperma Hochr. = Rochelia bungei Trautv. ■

335113　Rochelia karsensis Popov;卡尔斯鹤虱☆

335114　Rochelia leiocarpa Ledeb. ;光果李果鹤虱;Smoothfruit Rochelia ■

335115　Rochelia macrocalyx Bunge;大萼鹤虱■

335116　Rochelia macrocalyx Bunge = Rochelia rectipes Stocks ■

335117　Rochelia macrocalyx Bunge = Rochelia retorta（Pall.）Lipsky ■

335118　Rochelia peluncularis Boiss. ;总梗李果鹤虱(悬垂鹤虱)■

335119　Rochelia persica Bunge ex Boiss. ;波斯李果鹤虱(波斯鹤虱)■☆

335120　Rochelia rectipes Stocks;直柄李果鹤虱■

335121　Rochelia retorta（Pall.）Lipsky = Rochelia bungei Trautv. ■

335122　Rochelia stellulata Rchb. = Rochelia bungei Trautv. ■

335123　Rochelia stellulata Rchb. = Rochelia disperma（L. f.）Hochr. ■

335124　Rochetia Delile = Trichilia P. Browne(保留属名)●

335125　Rochonia DC.（1836）;绒菀木属●☆

335126　Rochonia antandroy Humbert;绒菀木●☆

335127　Rochonia aspera Humbert;粗糙绒菀木●☆

335128　Rochonia cinerarioides DC. ;灰色绒菀木●☆

335129　Rochonia cuneata DC;楔形绒菀木●☆

335130　Rochonia senecionoides Baker = Madagaster senecionoides（Baker）G. L. Nesom ●☆

335131　Rockia Heimerl = Pisonia L. ●

335132　Rockinghamia Airy Shaw(1966);罗金大戟属●☆

335133　Rockinghamia angustifolia（Benth.）Airy Shaw;窄叶罗金大戟●☆

335134　Rockinghamia brevipes Airy Shaw;短梗罗金大戟●☆

335135　Rodatia Raf. = Beloperone Nees ●☆

335136　Rodatia Raf. = Justicia L. ●■

335137　Rodentiophila F. Ritter ex Backeb. = Eriosyce Phil. ●☆

335138　Rodentiophila Ritt. et Y. Itô = Eriosyce Phil. ●☆

335139　Rodetia Moq. = Bosea L. ●☆

335140　Rodetia amherstiana Moq. = Bosea amherstiana（Moq.）Hook. f. ●☆

335141　Rodgersia A. Gray（1858）;鬼灯檠属;Ghost lampstand, Rodgersflower, Rodgersia ■

335142　Rodgersia aesculifolia Batalin;七叶鬼灯檠(宝剑叶,秤杆七,辫合山,鬼灯檠,红骡子,红苔七,红药子,厚朴七,黄药子,金毛狗,老汉求,老蛇莲,老蛇盘,麻鹞子,毛荷叶,慕荷,牛角七,山藕,水五龙,索骨丹,天蓬伞,猪屎七,作合山);Fingerleaf Ghost Lampstand, Fingerleaf Rodgersflower ■

335143　Rodgersia aesculifolia Batalin var. henricii（Franch.）C. Y. Wu ex J. T. Pan;滇西鬼灯檠,W. Yunnan Ghost lampstand, W. Yunnan Rodgersflower ■

335144　Rodgersia henricii（Franch.）Franch. = Rodgersia aesculifolia Batalin var. henricii（Franch.）C. Y. Wu ex J. T. Pan ■

335145　Rodgersia henricii Franch. = Rodgersia aesculifolia Batalin var. henricii（Franch.）C. Y. Wu ex J. T. Pan ■

335146　Rodgersia japonica A. Gray ex Regel = Rodgersia podophylla A. Gray ■

335147　Rodgersia pinnata Franch. ;羽叶鬼灯檠(半边伞,大红袍,红姜,红楼,红升嘛,九叶岩陀,毛头寒,蛇疙瘩,岩七,岩陀,野黄姜,羽叶岩陀,羽状鬼灯檠,紫姜);Featherleaf Ghost lampstand, Featherleaf Rodgersflower ■

335148　Rodgersia pinnata Franch. 'Superba';华丽羽叶鬼灯檠■☆

335149　Rodgersia pinnata Franch. var. estrigosa J. T. Pan;伏毛鬼灯檠(红楼,羽叶鬼灯檠,羽叶岩陀,羽状鬼灯檠);Estrigose Featherleaf Ghost Lampstand, Estrigose Featherleaf Rodgersflower ■

335150　Rodgersia platyphylla Pax et K. Hoffm. = Rodgersia aesculifolia Batalin ■

335151　Rodgersia podophylla A. Gray;鬼灯檠(独脚莲,日本鬼灯檠,矢车草）; Bronzeleaf Rodgersflower, Ghost lampstand, Rodger's Bronze-leaf ■

335152　Rodgersia podophylla A. Gray. = Rodgersia aesculifolia Batalin ■

335153　Rodgersia sambucifolia Hemsl. ;西南鬼灯檠(半边伞,参麻,大红袍,红姜,红升麻,九月岩陀,毛七,毛青杠,毛青红,毛头寒,毛头三七,蛇疙瘩,血三七,岩七,岩陀,岩陀陀,野黄姜,紫姜); Elderleaf Ghost lampstand, Elderleaf Rodgersflower, Rodgersia ■

335154　Rodgersia sambucifolia Hemsl. var. estrigosa J. T. Pan;光腹鬼灯檠;Smooth Elderleaf Rodgersflower ■

335155　Rodgersia tabularis（Hemsl.）Kom. = Astilboides tabularis（Hemsl.）Engl. ■

335156　Rodgersia tabularis Kom. = Astilboides tabularis（Hemsl.）Engl. ■

335157　Rodigia Spreng. = Crepis L. ●

335158　Rodora Adans. = Rhododendron L. ●

335159　Rodora Adans. = Rhodora L. ●

335160　Rodrigoa Braas = Masdevallia Ruiz et Pav. ■☆

335161　Rodriguezia Ruiz et Pav.（1794）;凹萼兰属（茹氏兰属）; Rodriguezia ■☆

335162　Rodriguezia batemanii Lindl. ;巴特曼凹萼兰;Bateman Rodriguezia ■☆

335163　Rodriguezia candida（Lindl.）Christenson;白花凹萼兰; Whiteflower Rodriguezia ■☆

335164　Rodriguezia decora Rchb. f. ;秀丽凹萼兰;Beautiful Rodriguezia ■☆

335165　Rodriguezia granadensis Rchb. f. ;哥伦比亚凹萼兰;Columbian Rodriguezia ■☆

335166　Rodriguezia maculata Rchb. f. ;斑点凹萼兰;Spotted Rodriguezia ■☆

335167　Rodriguezia pubescens Rchb. f. ;软毛凹萼兰■☆

335168　Rodriguezia secunda Kunth;凹萼兰;Secund Rodriguezia ■☆

335169　Rodriguezia suaveolens Lindl. ;芬芳凹萼兰;Fragrant Rodriguezia ■☆

335170　Rodriguezia venusta Rchb. f. ;美丽凹萼兰;Beautiful Rodriguezia ■☆

335171　Rodrigueziella Kuntze(1891);小凹萼兰属■☆

335172　Rodrigueziella petropolitana Pabst;小凹萼兰■☆

335173　Rodrigueziopsis Schltr.（1920）;类凹萼兰属■☆

335174　Rodrigueziopsis eleutherosepala（Barb. Rodr.）Schltr. ;类凹萼兰■☆

335175　Rodschiedia G. Gaertn. , B. Mey. et Scberb. = Capsella Medik. (保留属名)■

335176　Rodschiedia Miq. = Securidaca L. (保留属名)●

335177　Rodwaya F. Muell. = Thismia Griff. ■

335178　Roea Hueg. ex Benth. = Sphaerolobium Sm. ■☆

335179　Roebelia Bngel = Calyptrogyne H. Wendl. ●☆

335180　Roebelia Engel = Geonoma Willd. ●☆

335181　Roegneria C. Koch = Roegneria K. Koch ■

335182　Roegneria K. Koch(1848);鹅观草属;Goosecomb, Roegneria ■

335183　Roegneria K. Koch = Elymus L. ■

335184　Roegneria abolinii（Drobow）Nevski;异芒鹅观草(大穗鹅观草);Abolin Goosecomb, Abolin Roegneria ■

335185　Roegneria abolinii（Drobow）Nevski = Elymus abolinii（Drobow）Tzvelev ■

335186 Roegneria abolinii (Drobow) Nevski f. divaricans Nevski；长芒大穗鹅观草■

335187 Roegneria abolinii (Drobow) Nevski f. divaricans Nevski ＝Elymus abolinii (Drobow) Tzvelev var. divaricans (Nevski) Tzvelev ■

335188 Roegneria abolinii (Drobow) Nevski var. divaricans Nevski ＝ Elymus abolinii (Drobow) Tzvelev var. divaricans (Nevski) Tzvelev ■

335189 Roegneria abolinii (Drobow) Nevski var. pluriflora (D. F. Cui) L. B. Cai ＝Elymus abolinii (Drobow) Tzvelev var. pluriflorus D. F. Cui ■

335190 Roegneria abolinii (Drobow) Nevski var. pluriflorus (D. F. Cui) L. B. Cai；多花大穗鹅观草(多花大穗披碱草)；Manyflower Abolin Goosecomb, Manyflower Roegneria, Manyflower Wildryegrass ■

335191 Roegneria alashanica Keng ＝Elymus alashanicus (Keng) S. L. Chen ■

335192 Roegneria alashanica Keng ex Keng et S. L. Chen ＝ Elymus alashanicus (Keng) S. L. Chen ■

335193 Roegneria alashanica Keng ex Keng et S. L. Chen var. elytrigioides (C. Yen et J. L. Yang) L. B. Cai；昌都鹅观草；Changdu Goosecomb, Changdu Roegneria ■

335194 Roegneria alashanica Keng ex Keng et S. L. Chen var. jufinshanica C. P. Wang et X. L. Yang ＝Roegneria jufinshanica (C. P. Wang et H. L. Yang) L. B. Cai ■

335195 Roegneria alashanica Keng var. elytrigioides (C. Yen et J. L. Yang) L. B. Cai ＝Elymus elytrigioides (C. Yen et J. L. Yang) S. L. Chen ■

335196 Roegneria alashanica Keng var. jufinshanica C. P. Wang et H. L. Yang ＝Elymus jufinshanicus (C. P. Wang et H. L. Yang) S. L. Chen ■

335197 Roegneria aliena Keng ＝Elymus alienus (Keng) S. L. Chen ■

335198 Roegneria aliena Keng ex Keng et S. L. Chen；涞源鹅观草；Laiyuan Goosecomb, Laiyuan Roegneria ■

335199 Roegneria altaica L. B. Cai ＝Elymus pseudocaninus G. Zhu et S. L. Chen ■

335200 Roegneria altaicus (D. F. Cui) L. B. Cai；阿尔泰鹅观草(阿尔泰披碱草)；Altai Goosecomb, Altai Lymegrass, Altai Roegneria, Altai Wildryegrass ■

335201 Roegneria altissima Keng ＝Elymus altissimus (Keng) Á. Löve ex B. Rong Lu ■

335202 Roegneria altissima Keng ex Keng et S. L. Chen；高株鹅观草；Tall Roegneria, Tallest Goosecomb ■

335203 Roegneria altissima Keng ex Keng et S. L. Chen ＝ Elymus altissimus (Keng) Á. Löve ex B. Rong Lu ■

335204 Roegneria amurensis (Drobow) Nevski；毛叶鹅观草；Amur Goosecomb, Amur Roegneria ■

335205 Roegneria amurensis (Drobow) Nevski ＝Elymus ciliaris (Trin. ex Bunge) Tzvelev var. amurensis (Drobow) S. L. Chen ■

335206 Roegneria amurensis (Drobow) Nevski var. hirtiflora (C. P. Wang et H. L. Yang) L. B. Cai ＝Elymus ciliaris (Trin. ex Bunge) Tzvelev var. hirtiflorus (C. P. Wang et H. L. Yang) S. L. Chen ■

335207 Roegneria angusta L. B. Cai；狭穗鹅观草；Narrowspike Goosecomb, Narrowspike Roegneria ■

335208 Roegneria angusta L. B. Cai ＝Elymus angustispiculatus S. L. Chen et G. Zhu ■

335209 Roegneria angustiglumis (Nevski) Nevski ＝Elymus mutabilis (Drobow) Tzvelev ■

335210 Roegneria anthosachnoides Keng ＝Elymus anthosachnoides (Keng) Á. Löve ex B. Rong Lu ■

335211 Roegneria anthosachnoides Keng var. scabrilemmata L. B. Cai ＝ Elymus anthosachnoides (Keng) Á. Löve ex B. Rong Lu var. scabrilemmatus (L. B. Cai) S. L. Chen ■

335212 Roegneria antiqua (Nevski) B. S. Sun ＝Elymus antiquus (Nevski) Tzvelev ■

335213 Roegneria aristiglumis Keng et S. L. Chen ＝Elymus aristiglumis (Keng et S. L. Chen) S. L. Chen ■

335214 Roegneria aristiglumis Keng et S. L. Chen var. hirsuta H. L. Yang；毛芒颖鹅观草；Hairy Awnedglume Roegneria ■

335215 Roegneria aristiglumis Keng et S. L. Chen var. hirsuta H. L. Yang ＝Elymus aristiglumis (Keng et S. L. Chen) S. L. Chen var. hirsutus (H. L. Yang) S. L. Chen ■

335216 Roegneria aristiglumis Keng et S. L. Chen var. hirsuta H. L. Yang ＝ Roegneria aristiglumis Keng et S. L. Chen var. leiantha H. L. Yang ■

335217 Roegneria aristiglumis Keng et S. L. Chen var. leiantha H. L. Yang；光花芒颖鹅观草；Smoothflower Awnedglume Roegneria ■

335218 Roegneria aristiglumis Keng et S. L. Chen var. leiantha H. L. Yang ＝Elymus aristiglumis (Keng et S. L. Chen) S. L. Chen var. leianthus (H. L. Yang) S. L. Chen ■

335219 Roegneria barbicalla (Ohwi) Keng et S. L. Chen ＝Elymus barbicallus (Ohwi) S. L. Chen ■

335220 Roegneria barbicalla (Ohwi) Keng et S. L. Chen var. breviseta Keng ＝Elymus barbicallus (Ohwi) S. L. Chen ■

335221 Roegneria barbicalla (Ohwi) Keng et S. L. Chen var. foliosa (Keng) L. B. Cai ＝Elymus alienus (Keng) S. L. Chen ■

335222 Roegneria barbicalla (Ohwi) Keng et S. L. Chen var. pubifolia Keng ＝ Elymus barbicallus (Ohwi) S. L. Chen var. pubifolius (Keng) S. L. Chen ■

335223 Roegneria barbicalla (Ohwi) Keng et S. L. Chen var. pubinodis Keng ＝ Elymus barbicallus (Ohwi) S. L. Chen var. pubinodis (Keng) S. L. Chen ■

335224 Roegneria barbicalla Ohwi；毛盘鹅观草；Hairy-callus Roegneria, Hairydish Goosecomb ■

335225 Roegneria barbicalla Ohwi var. breviseta Keng；短芒毛盘草；Shortawn Hairy-callus Roegneria ■

335226 Roegneria barbicalla Ohwi var. foliosa (Keng et S. L. Chen) L. B. Cai；多叶鹅观草；Leafy Hairy-callus Roegneria, Manyleaf Goosecomb, Multileaved Roegneria ■

335227 Roegneria barbicalla Ohwi var. pubifolia Keng；毛叶毛盘鹅观草(毛叶鹅观草,毛叶毛盘草)；Hairleaf Hairy-callus Roegneria ■

335228 Roegneria barbicalla Ohwi var. pubifolia Keng ＝Roegneria barbicalla Ohwi ■

335229 Roegneria barbicalla Ohwi var. pubinodis Keng；毛节毛盘鹅观草(毛节鹅观草,毛节毛盘草毛,毛盘鹅观草,叶毛盘草)；Hairynode Hairy-callus Roegneria ■

335230 Roegneria borealis (Turcz.) Nevski；北方鹅观草(北鹅观草)；North Roegneria ■☆

335231 Roegneria borealis (Turcz.) Nevski ＝Elymus borealis (Turcz.) D. F. Cui ■

335232 Roegneria borealis (Turcz.) Nevski ＝Elymus kronokensis (Kom.) Tzvelev ■

335233 Roegneria brachypodioides Nevski；短足鹅观草■☆

335234 Roegneria breviarista (D. F. Cui) L. B. Cai ＝Roegneria breviarista (S. L. Chen ex D. F. Cui) L. B. Cai ■

335235 Roegneria breviarista (S. L. Chen ex D. F. Cui) L. B. Cai；短芒鹅观草(短芒光穗披碱草)；Shortawn Glabrous-spiked Lymegrass, Shortawn Roegneria ■

335236 Roegneria breviarista (S. L. Chen ex D. F. Cui) L. B. Cai ＝

Elymus glaberrimus（Keng et S. L. Chen）S. L. Chen var. breviarista S. L. Chen ex D. F. Cui ■

335237　Roegneria breviglumis Keng ＝ Elymus burchan-buddae（Nevski）Tzvelev ■

335238　Roegneria breviglumis Keng ex Keng et S. L. Chen；短颖鹅观草；Shortglume Goosecomb，Shortglume Roegneria ■

335239　Roegneria breviglumis Keng ex Keng et S. L. Chen ＝ Elymus breviglumis（Keng ex Keng et S. L. Chen）Loeve ■

335240　Roegneria breviglumis Keng ex Keng et S. L. Chen var. brevipes（Keng et S. L. Chen）L. B. Cai；短柄鹅观草；Shortstalk Roegneria ■

335241　Roegneria breviglumis Keng var. brevipes（Keng）L. B. Cai ＝ Elymus brevipes（Keng）S. L. Chen ■

335242　Roegneria brevipes Keng ＝ Elymus brevipes（Keng）S. L. Chen ■

335243　Roegneria brevipes Keng ＝ Roegneria breviglumis Keng ex Keng et S. L. Chen var. brevipes（Keng et S. L. Chen）L. B. Cai ■

335244　Roegneria brevipes Keng et S. L. Chen ＝ Roegneria breviglumis Keng ex Keng et S. L. Chen var. brevipes（Keng et S. L. Chen）L. B. Cai ■

335245　Roegneria burchan-buddae（Nevski）B. S. Sun ＝ Elymus burchan-buddae（Nevski）Tzvelev ■

335246　Roegneria buschiana（Roshev.）Nevski；布氏鹅观草■☆

335247　Roegneria cacumina（B. Rong Lu et Salomon）L. B. Cai；峰峦鹅观草■

335248　Roegneria cacuminis（B. Rong Lu et B. Salomon）L. B. Cai ＝ Elymus cacuminis B. Rong Lu et B. Salomon ■

335249　Roegneria calcicola Keng ＝ Elymus calcicola（Keng）S. L. Chen ■

335250　Roegneria calcicola Keng ＝ Roegneria calcicola Keng ex Keng et S. L. Chen ■

335251　Roegneria calcicola Keng ex Keng et S. L. Chen；钙生鹅观草（草灵芝，茅草箭）；Calciumsoil Goosecomb，Calciumsoil Roegneria ■

335252　Roegneria canaliculata（Nevski）Ohwi ＝ Elymus canaliculatus（Nevski）Tzvelev ■

335253　Roegneria canina（L.）Nevski ＝ Elymus caninus（L.）L. ■

335254　Roegneria carinata Ovcz. et Sidorenko；龙骨状鹅观草■☆

335255　Roegneria caucasica C. Koch；高加索鹅观草（鹅观草）；Caucasus Roegneria ■

335256　Roegneria cheniae L. B. Cai；陈氏鹅观草；Chen Goosecomb，Chen Roegneria ■

335257　Roegneria cheniae L. B. Cai ＝ Elymus cheniae（L. B. Cai）G. Zhu ■

335258　Roegneria ciliaris（Trin. ex Bunge）Nevski；纤毛鹅观草（缘毛披碱草）；Ciliate Goosecomb，Ciliate Roegneria ■

335259　Roegneria ciliaris（Trin. ex Bunge）Nevski ＝ Elymus ciliaris（Trin. ex Bunge）Tzvelev ■

335260　Roegneria ciliaris（Trin. ex Bunge）Nevski f. eriocaulis Kitag.；毛节鹅观草；Woollystem-ciliate Roegneria ■

335261　Roegneria ciliaris（Trin. ex Bunge）Nevski f. eriocaulis Kitag. ＝ Elymus ciliaris（Trin. ex Bunge）Tzvelev ■

335262　Roegneria ciliaris（Trin. ex Bunge）Nevski var. hackeliana（Honda）L. B. Cai ＝ Elymus ciliaris（Trin. ex Bunge）Tzvelev var. hackelianus（Honda）G. Zhu et S. L. Chen ■

335263　Roegneria ciliaris（Trin. ex Bunge）Nevski var. hackeliana（Honda）L. B. Cai；细叶毛节鹅观草（细叶鹅观草）■

335264　Roegneria ciliaris（Trin. ex Bunge）Nevski var. japonensis（Honda）C. Yen et al. ＝ Roegneria ciliaris（Trin.）Nevski var. hackeliana（Honda）L. B. Cai ■

335265　Roegneria ciliaris（Trin. ex Bunge）Nevski var. japonensis（Honda）C. Yen et al. ＝ Elymus ciliaris（Trin. ex Bunge）Tzvelev

var. hackelianus（Honda）G. Zhu et S. L. Chen ■

335266　Roegneria ciliaris（Trin. ex Bunge）Nevski var. lasiophylla（Kitag.）Kitag. ＝ Elymus ciliaris（Trin. ex Bunge）Tzvelev var. lasiophyllus（Kitag.）S. L. Chen ■

335267　Roegneria ciliaris（Trin. ex Bunge）Nevski var. lasiophylla（Kitag.）Kitag.；毛叶纤毛鹅观草（粗毛纤毛草，毛叶纤毛草）；Hairleaf Goosecomb，Woolyleaved Ciliate Roegneria ■

335268　Roegneria ciliaris（Trin. ex Bunge）Nevski var. pilosa（Korsh.）Ohwi ＝ Elymus ciliaris（Trin. ex Bunge）Tzvelev var. amurensis（Drobow）S. L. Chen ■

335269　Roegneria ciliaris（Trin. ex Bunge）Nevski var. submutica（Honda）Keng ＝ Elymus ciliaris（Trin. ex Bunge）Tzvelev var. submuticus（Honda）S. L. Chen ■

335270　Roegneria ciliaris（Trin. ex Bunge）Nevski var. submutica（Honda）Keng；短芒纤毛草；Short-awned Ciliate Roegneria ■

335271　Roegneria ciliaris（Trin.）Nevski ＝ Elymus ciliaris（Trin. ex Bunge）Tzvelev ■

335272　Roegneria ciliaris（Trin.）Nevski ＝ Roegneria ciliaris（Trin. ex Bunge）Nevski ■

335273　Roegneria ciliaris（Trin.）Nevski f. eriocaulis Kitag. ＝ Roegneria ciliaris（Trin. ex Bunge）Nevski f. eriocaulis Kitag. ■

335274　Roegneria ciliaris（Trin.）Nevski var. hackeliana（Honda）L. B. Cai ＝ Roegneria ciliaris（Trin. ex Bunge）Nevski var. hackeliana（Honda）L. B. Cai ■

335275　Roegneria ciliaris（Trin.）Nevski var. japonensis（Honda）C. Yen，J. L. Yang et B. Rong Lu ＝ Roegneria ciliaris（Trin.）Nevski var. hackeliana（Honda）L. B. Cai ■

335276　Roegneria ciliaris（Trin.）Nevski var. lasiophylla（Kitag.）Kitag. ＝ Roegneria ciliaris（Trin. ex Bunge）Nevski var. lasiophylla（Kitag.）Kitag. ■

335277　Roegneria ciliaris（Trin.）Nevski var. submutica（Honda）Keng ＝ Roegneria ciliaris（Trin. ex Bunge）Nevski var. submutica（Honda）Keng ■

335278　Roegneria confusa（Roshev.）Nevski；紊草；Confuse Goosecomb，Confuse Roegneria ■

335279　Roegneria confusa（Roshev.）Nevski subsp. breviaristata（Keng）N. R. Cui ＝ Elymus confusus（Roshev.）Tzvelev var. breviaristatus（Keng）S. L. Chen ■

335280　Roegneria confusa（Roshev.）Nevski var. breviaristata Keng ＝ Elymus confusus（Roshev.）Tzvelev var. breviaristatus（Keng）S. L. Chen ■

335281　Roegneria crassa L. B. Cai；粗壮鹅观草；Robust Roegneria ■

335282　Roegneria crassa L. B. Cai ＝ Elymus strictus（Keng ex Keng et S. L. Chen）S. L. Chen var. crassus（L. B. Cai）S. L. Chen ■

335283　Roegneria curtiarista L. B. Cai ＝ Elymus curtiaristatus（L. B. Cai）S. L. Chen et G. Zhu ■

335284　Roegneria curvata（Nevski）Nevski；弯穗草（曲弯穗草）；Bent Spike Roegneria ■

335285　Roegneria curvata（Nevski）Nevski ＝ Elymus curvatus（Nevski）D. F. Cui ■

335286　Roegneria curvata（Nevski）Nevski ＝ Elymus fedtschenkoi Tzvelev ■

335287　Roegneria czimganica（Drobow）Nevski；捷姆鹅观草；Czimganic Roegneria ■☆

335288　Roegneria debilis L. B. Cai；柔弱鹅观草；Weak Roegneria ■

335289　Roegneria debilis L. B. Cai ＝ Elymus debilis（L. B. Cai）S. L. Chen et G. Zhu ■

335290　Roegneria dolichathera Keng ＝ Elymus dolichatherus（Keng）S. L. Chen ■

335291　Roegneria dolichathera Keng ex Keng et S. L. Chen；长芒鹅观草；Longawn Goosecomb，Longawn Roegneria ■

335292　Roegneria dolichathera Keng ex Keng et S. L. Chen var. glabrifolia Keng ＝ Roegneria dolichathera Keng ex Keng et S. L. Chen ■

335293　Roegneria dolichathera Keng ex Keng et S. L. Chen var. glabrifolia Keng；光叶鹅观草；Smoothleaf Longawn Roegneria ■

335294　Roegneria dolichathera var. glabrifolia Keng ＝ Elymus dolichatherus（Keng）S. L. Chen ■

335295　Roegneria drobovii Nevski；德劳鹅观草■☆

335296　Roegneria dura（Keng）Keng ＝ Elymus durus（Keng）S. L. Chen ■

335297　Roegneria dura（Keng）Keng ex Keng et S. L. Chen ＝ Elymus sclerus Á. Löve ■

335298　Roegneria dura（Keng）Keng ex Keng et S. L. Chen ＝ Roegneria tschimganica（Drobow）Nevski var. varriglumis（Keng ex Keng et S. L. Chen）L. B. Cai ■

335299　Roegneria dura（Keng）Keng ex Keng et S. L. Chen var. variiglumis Keng ＝ Elymus sclerus Á. Löve ■

335300　Roegneria dura（Keng）Keng ex Keng et S. L. Chen var. variiglumis Keng ＝ Roegneria tschimganica（Drobow）Nevski var. varriglumis（Keng ex Keng et S. L. Chen）L. B. Cai ■

335301　Roegneria dura（Keng）Keng ex Keng et S. L. Chen var. variiglumis Keng ＝ Roegneria dura（Keng）Keng ex Keng et S. L. Chen ■

335302　Roegneria dura（Keng）Keng var. variiglumis Keng ＝ Elymus durus（Keng）S. L. Chen ■

335303　Roegneria elytrigioides C. Yen et J. L. Yang ＝ Elymus elytrigioides（C. Yen et J. L. Yang）S. L. Chen ■

335304　Roegneria elytrigioides C. Yen et J. L. Yang ＝ Roegneria alashanica Keng ex Keng et S. L. Chen var. elytrigioides（C. Yen et J. L. Yang）L. B. Cai ■

335305　Roegneria fedtschenkoi（Tzvelev）N. R. Cui ＝ Elymus fedtschenkoi Tzvelev ■

335306　Roegneria festucoides（Maire）Dobignard；羊茅状鹅观草■

335307　Roegneria fibrosa（Schrenk）Nevski；纤维鹅观草（须草）；Fibre Roegneria ■☆

335308　Roegneria flexuosa L. B. Cai；弯曲鹅观草■

335309　Roegneria flexuosa L. B. Cai ＝ Elymus sinoflexuosus（L. B. Cai）S. L. Chen et G. Zhu ■

335310　Roegneria foliosa Keng ＝ Elymus alienus（Keng）S. L. Chen ■

335311　Roegneria foliosa Keng ex Keng et S. L. Chen ＝ Roegneria barbicalla Ohwi var. foliosa（Keng et S. L. Chen）L. B. Cai ■

335312　Roegneria formosana（Honda）Ohwi；台湾鹅观草；Taiwan Goosecomb，Taiwan Roegneria ■

335313　Roegneria formosana（Honda）Ohwi ＝ Elymus formosanus（Honda）Á. Löve ■

335314　Roegneria formosana（Honda）Ohwi var. longearistata Keng；长芒台湾鹅观草（台湾鹅观草）；Long Awned Taiwan Roegneria ■

335315　Roegneria formosana（Honda）Ohwi var. longearistata Keng ＝ Elymus formosanus（Honda）Á. Löve ■

335316　Roegneria formosana（Honda）Ohwi var. pubigera Keng；毛鞘台湾鹅观草；Hairy Taiwan Roegneria ■

335317　Roegneria formosana（Honda）Ohwi var. pubigera Keng ＝ Elymus formosanus（Honda）Á. Löve var. pubigerus（Keng）S. L.

Chen ■

335318　Roegneria geminata Keng et S. L. Chen；孪生鹅观草；Geminate Roegneria，Twin Goosecomb ■

335319　Roegneria geminata Keng et S. L. Chen ＝ Kengyilia geminata（Keng et S. L. Chen）S. L. Chen ■

335320　Roegneria glaberrima Keng et S. L. Chen ＝ Elymus glaberrimus（Keng et S. L. Chen）S. L. Chen ■

335321　Roegneria glaberrima Keng et S. L. Chen var. breviarista（D. F. Cui）L. B. Cai ＝ Roegneria breviarista（D. F. Cui）L. B. Cai ■

335322　Roegneria glaberrima Keng et S. L. Chen var. breviarista（S. L. Chen ex D. F. Cui）L. B. Cai ＝ Elymus glaberrimus（Keng et S. L. Chen）S. L. Chen var. breviarista S. L. Chen ex D. F. Cui ■

335323　Roegneria glabrispicula（D. F. Cui）L. B. Cai；曲芒鹅观草；Bentawned Roegneria ■

335324　Roegneria glabrispicula（D. F. Cui）L. B. Cai ＝ Elymus tschimuganicus（Drobow）Tzvelev var. glabrispiculus D. F. Cui ■

335325　Roegneria glaucifolia Keng ＝ Elymus caesifolius Á. Löve ex S. L. Chen ■

335326　Roegneria glaucifolia Keng ex Keng et S. L. Chen；蓝灰鹅观草（马格草）；Glaucousleaf Roegneria ■

335327　Roegneria gmelinii（Griseb.）Kitag. ＝ Roegneria turczaninovii（Drobow）Nevski ■

335328　Roegneria gmelinii（Ledeb.）Kitag. ＝ Elymus gmelinii（Ledeb.）Tzvelev ■

335329　Roegneria gmelinii（Ledeb.）Kitag. var. macra-thera（Ohwi）Kitag. ＝ Elymus gmelinii（Ledeb.）Tzvelev var. macratherus（Ohwi）S. L. Chen et G. Zhu ■

335330　Roegneria gracilis L. B. Cai；纤瘦鹅观草；Thin Roegneria ■

335331　Roegneria gracilis L. B. Cai ＝ Elymus caianus S. L. Chen et G. Zhu ■

335332　Roegneria grandiglumis Keng ＝ Kengyilia grandiglumis（Keng）J. L. Yang，C. Yen et B. R. Baum ■

335333　Roegneria grandiglumis Keng ex Y. L. Keng et S. L. Chen；大颖草；Largeglume Goosecomb，Largeglume Roegneria ■

335334　Roegneria grandiglumis Keng ex Y. L. Keng et S. L. Chen ＝ Kengyilia grandiglumis（Keng ex Keng et S. L. Chen）J. L. Yang，C. Yen et B. R. Baum ■

335335　Roegneria grandis Keng ＝ Elymus grandis（Keng）S. L. Chen ■

335336　Roegneria grandis Keng ex Keng et S. L. Chen；大鹅观草；Large Roegneria ■

335337　Roegneria hackeliana（Honda）Nakai ＝ Elymus ciliaris（Trin. ex Bunge）Tzvelev var. hackelianus（Honda）G. Zhu et S. L. Chen ■

335338　Roegneria himalayana Nevski；喜马拉雅鹅观草■☆

335339　Roegneria himalayana Nevski ＝ Elymus himalayanus（Nevski）Tzvelev ■☆

335340　Roegneria hirsuta Keng ＝ Kengyilia hirsuta（Keng）J. L. Yang，C. Yen et B. R. Baum ■

335341　Roegneria hirsuta Keng ex Keng et S. L. Chen；糙毛鹅观草；Hirsute Goosecomb，Hirsute Roegneria ■

335342　Roegneria hirsuta Keng ex Keng et S. L. Chen ＝ Elymus kengii（Tzvelev）D. F. Cui ■

335343　Roegneria hirsuta Keng ex Y. L. Keng et S. L. Chen ＝ Kengyilia hirsuta（Keng et S. L. Chen）J. L. Yang，C. Yen et B. R. Baum ■

335344　Roegneria hirsuta Keng ex Y. L. Keng et S. L. Chen var. leiophylla Keng et S. L. Chen；光叶糙毛草；Glabrousleaf Hirsute Roegneria ■

335345　Roegneria hirsuta Keng ex Y. L. Keng et S. L. Chen var.

variabilis Keng;善变糙毛草;Variable Hirsute Roegneria ■

335346　Roegneria hirsuta Keng ex Y. L. Keng et S. L. Chen var. varibilis Keng et S. L. Chen = Kengyilia hirsuta (Keng ex Keng et S. L. Chen) J. L. Yang,C. Yen et B. R. Baum var. varibilis (Keng ex Keng et S. L. Chen) L. B. Cai ■

335347　Roegneria hirsuta Keng var. leiophylla Keng et S. L. Chen = Kengyilia hirsuta (Keng) J. L. Yang,C. Yen et B. R. Baum ■

335348　Roegneria hirsuta Keng var. variabilis Keng = Kengyilia hirsuta (Keng) J. L. Yang,C. Yen et B. R. Baum ■

335349　Roegneria hirtiflora C. P. Wang et H. L. Yang = Elymus ciliaris (Trin. ex Bunge) Tzvelev var. hirtiflorus (C. P. Wang et H. L. Yang) S. L. Chen ■

335350　Roegneria hondae Kitag.;五龙山鹅观草(本田鹅观草,蛊草,河北鹅观草);Fascinate Honda Roegneria, Honda Roegneria, Wulongshan Goosecomb ■

335351　Roegneria hondae Kitag. = Elymus hondae (Kitag.) S. L. Chen ■

335352　Roegneria hondae Kitag. var. fascinata Keng = Elymus hondae (Kitag.) S. L. Chen ■

335353　Roegneria hondae Kitag. var. fascinata Keng = Roegneria hondae Kitag. ■

335354　Roegneria hongyuanensis L. B. Cai;红原鹅观草;Hongyuan Honda Roegneria ■

335355　Roegneria hongyuanensis L. B. Cai = Elymus hongyuanensis (L. B. Cai) S. L. Chen et G. Zhu ■

335356　Roegneria humilis Keng et S. L. Chen = Elymus humilis (Keng et S. L. Chen) S. L. Chen ■

335357　Roegneria hybrida Keng = Elymus hybridus (Keng) S. L. Chen ■

335358　Roegneria hybrida Keng ex Keng et S. L. Chen = Roegneria tsukushiensis (Drobow) Nevski var. hybrida (Keng) L. B. Cai ■

335359　Roegneria interrupta Nevski;间断鹅观草■☆

335360　Roegneria intramongolica Shan Chen et W. Gao;内蒙古鹅观草(短芒鹅观草);Innermongol Goosecomb,Intromongolian Roegneria ■

335361　Roegneria intramongolica Shan Chen et W. Gao = Elymus intramongolicus (Shan Chen et W. Gao) S. L. Chen ■

335362　Roegneria jacquemontii (Hook. f.) Nevski = Elymus jacquemontii (Hook. f.) Tzvelev ■

335363　Roegneria jacquemontii (Hook. f.) Nevski var. pulanen-sis (H. L. Yang) L. B. Cai = Elymus pulanensis (H. L. Yang) S. L. Chen ■

335364　Roegneria jacquemontii (Hook. f.) Ovcz. et Sidorenko;低株鹅观草;Jacquemont. Roegneria ■

335365　Roegneria jacquemontii (Hook. f.) Ovcz. et Sidorenko var. pulanensis (H. L. Yang) L. B. Cai;普兰鹅观草;Pulan Goosecomb, Pulan Roegneria ■

335366　Roegneria jacutensis (Drobow) Nevski;亚库特鹅观草■☆

335367　Roegneria japonensis (Honda) Keng = Agropyron ciliare (Trin.) Franch. f. japonense (Honda) Ohwi ■

335368　Roegneria japonensis (Honda) Keng = Elymus ciliaris (Trin. ex Bunge) Tzvelev var. hackelianus (Honda) G. Zhu et S. L. Chen ■

335369　Roegneria japonensis (Honda) Keng = Roegneria japonensis (Honda) Keng ex Keng et S. L. Chen ■

335370　Roegneria japonensis (Honda) Keng ex Keng et S. L. Chen;竖立鹅观草;Japan Goosecomb,Japanese Roegneria ■

335371　Roegneria japonensis (Honda) Keng ex Keng et S. L. Chen var. hackeliana (Honda) Keng;细叶竖立鹅观草(竖立鹅观草,细叶鹅观草);Hackel Roegneria ■

335372　Roegneria japonensis (Honda) Keng ex Keng et S. L. Chen var. hackeliana (Honda) Keng = Roegneria ciliaris (Trin.) Nevski var.

hackeliana (Honda) L. B. Cai ■

335373　Roegneria japonica B. S. Sun = Elymus ciliaris (Trin. ex Bunge) Tzvelev var. hackelianus (Honda) G. Zhu et S. L. Chen ■

335374　Roegneria jufinshanica (C. P. Wang et H. L. Yang) L. B. Cai;九峰山鹅观草;Jiufengshan Roegneria ■

335375　Roegneria jufinshanica (C. P. Wang et H. L. Yang) L. B. Cai = Elymus jufinshanicus (C. P. Wang et H. L. Yang) S. L. Chen ■

335376　Roegneria kamoji (Ohwi) Keng et S. L. Chen = Elymus kamoji (Ohwi) S. L. Chen ■

335377　Roegneria kamoji (Ohwi) Keng et S. L. Chen subsp. macerrima (Keng) N. R. Cui = Elymus kamoji (Ohwi) S. L. Chen var. macerrimus (Keng) G. Zhu ■

335378　Roegneria kamoji (Ohwi) Keng et S. L. Chen var. macerrima Keng = Elymus kamoji (Ohwi) S. L. Chen var. macerrimus (Keng) G. Zhu ■

335379　Roegneria kamoji (Ohwi) Ohwi ex Keng = Elymus kamojus (Ohwi) S. L. Chen ■

335380　Roegneria kamoji (Ohwi) Ohwi ex Keng = Elymus tsukushiensis Honda var. transiens (Hack.) Osada ■

335381　Roegneria kamoji (Ohwi) Ohwi ex Keng subsp. macerrima (Keng) N. R. Cui = Roegneria macerrima (Keng ex Keng et S. L. Chen) L. B. Cai ■

335382　Roegneria kamoji (Ohwi) Ohwi ex Keng var. macerrima Keng = Roegneria macerrima (Keng ex Keng et S. L. Chen) L. B. Cai ■

335383　Roegneria kaschgarica (D. F. Cui) Y. H. Wu = Kengyilia kaschgarica (D. F. Cui) L. B. Cai ■

335384　Roegneria kengyilia ? var. stenachyra Keng et S. L. Chen = Kengyilia stenachyra (Keng et S. L. Chen) J. L. Yang,C. Yen et B. R. Baum ■

335385　Roegneria kokonorica Keng = Kengyilia kokonorica (Keng) J. L. Yang,C. Yen et B. R. Baum ■

335386　Roegneria kokonorica Keng ex Keng et S. L. Chen;青海鹅观草;Qinghai Goosecomb,Qinghai Roegneria ■

335387　Roegneria kokonorica Keng ex Keng et S. L. Chen = Elymus kokonoricus (Keng ex Keng et S. L. Chen) Á. Löve ■

335388　Roegneria kokonorica Keng ex Keng et S. L. Chen = Kengyilia kokonorica (Keng ex Keng et S. L. Chen) J. L. Yang,C. Yen et B. R. Baum ■

335389　Roegneria komarovii (Nevski) Nevski;偏穗鹅观草;Komalov Goosecomb,Komalov Roegneria ■

335390　Roegneria komarovii (Nevski) Nevski = Elymus komarovii (Nevski) Tzvelev ■

335391　Roegneria kronokensis (Kom.) Tzvelev = Elymus kronokensis (Kom.) Tzvelev ■

335392　Roegneria lachnophylla Ovcz. et Sidorenko;羊毛叶鹅观草■☆

335393　Roegneria latiglumis (Scribn. et Sm.) Nevski;宽颖草;Broadglume Roegneria ■

335394　Roegneria laxiflora Keng = Kengyilia laxiflora (Keng) J. L. Yang,C. Yen et B. R. Baum ■

335395　Roegneria laxiflora Keng ex Keng et S. L. Chen;疏花鹅观草;Looseflower Goosecomb,Looseflower Roegneria ■

335396　Roegneria laxiflora Keng ex Keng et S. L. Chen = Kengyilia laxiflora (Keng ex Keng et S. L. Chen) J. L. Yang,C. Yen et B. R. Baum ■

335397　Roegneria laxinodis L. B. Cai = Elymus laxinodis (L. B. Cai) S. L. Chen et G. Zhu ■

335398　Roegneria leiantha Keng = Elymus leianthus (Keng) S. L. Chen ■

335399 Roegneria leiantha Keng ex Keng et S. L. Chen；光花鹅观草；Brightflower Goosecomb，Glabrousflower Roegneria ■

335400 Roegneria leiantha Keng ex Keng et S. L. Chen ＝Kengyilia leiantha (Keng ex Keng et S. L. Chen) L. B. Cai ■

335401 Roegneria leiotropis Keng ＝Elymus leiotropis (Keng) S. L. Chen ■

335402 Roegneria leiotropis Keng ex Keng et S. L. Chen；光脊鹅观草；Brightridge Goosecomb，Glabrous Keel Roegneria ■

335403 Roegneria leptoura Nevski；细尾鹅观草；Delicate Roegneria ■

335404 Roegneria longearistata (Boiss.) C. Y. Wu；长尖鹅观草；Longgearistate Roegneria ■

335405 Roegneria longearistata (Boiss.) C. Y. Wu var. caniculatus (Nevski) L. B. Cai；沟槽鹅观草；Canaliculate Roegneria ■

335406 Roegneria longearistata (Boiss.) Drobow var. canaliculata (Nevski) L. B. Cai ＝Elymus canaliculatus (Nevski) Tzvelev ■

335407 Roegneria longiaristata (Boiss.) Sidorenko；欧洲长芒鹅观草■☆

335408 Roegneria longiglumis Keng ＝Kengyilia alatavica (Drobow) J. L. Yang，C. Yen et B. R. Baum var. longiglumis (Keng et S. L. Chen) C. Yen，J. L. Yang et B. R. Baum ■

335409 Roegneria longiglumis Keng ex Keng et S. L. Chen；长颖鹅观草；Longglume Goosecomb，Longglume Roegneria ■

335410 Roegneria longiglumis Keng ex Keng et S. L. Chen ＝Kengyilia longiglumis (Keng ex Keng et S. L. Chen) J. L. Yang ■

335411 Roegneria macerrima (Keng ex Keng et S. L. Chen) L. B. Cai；细瘦鹅观草；Steep Roegneria ■

335412 Roegneria macerrima (Keng) L. B. Cai ＝Elymus kamoji (Ohwi) S. L. Chen var. macerrimus (Keng) G. Zhu ■

335413 Roegneria macranthera (Ohwi) L. B. Cai；大药鹅观草；Largeawn Roegneria ■

335414 Roegneria macranthera (Ohwi) L. B. Cai ＝Elymus gmelinii (Ledeb.) Tzvelev var. macrantherus (Ohwi) S. L. Chen et G. Zhu ■

335415 Roegneria macrochaeta Nevski；大毛鹅观草■☆

335416 Roegneria macroura (Turcz.) Nevski；大尾鹅观草■☆

335417 Roegneria magnicaespes (D. F. Cui) L. B. Cai；大丛鹅观草(大丛披碱草)；Bigtuft Roegneria ■

335418 Roegneria magnicaespes (D. F. Cui) L. B. Cai ＝Elymus magnicaespes D. F. Cui ■

335419 Roegneria magnipoda L. B. Cai；大柄鹅观草；Bigstalk Roegneria ■

335420 Roegneria magnipoda L. B. Cai ＝Elymus magnipodus (L. B. Cai) S. L. Chen et G. Zhu ■

335421 Roegneria marginata (H. Lindb.) Dobignard；具边鹅观草■☆

335422 Roegneria marginata (H. Lindb.) Dobignard subsp. kabylica (Maire et Weiller) Dobignard ＝Elymus marginatus (H. Lindb.) Á. Löve subsp. kabylicus (Maire et Weiller) Valdés et H. Scholz ■☆

335423 Roegneria mayebarana (Honda) Ohwi；东瀛鹅观草(前原鹅观草)；Oriental Goosecomb，Oriental Roegneria ■

335424 Roegneria mayebarana (Honda) Ohwi ＝Elymus mayebaranus (Honda) S. L. Chen ■

335425 Roegneria media (Keng ex Keng et S. L. Chen) L. B. Cai；中间鹅观草；Medium Roegneria ■

335426 Roegneria media (Keng ex Keng et S. L. Chen) L. B. Cai ＝Elymus sinicus (Keng) S. L. Chen var. medius (Keng) S. L. Chen et G. Zhu ■

335427 Roegneria media (Keng) L. B. Cai ＝Elymus sinicus (Keng) S. L. Chen var. medius (Keng) S. L. Chen et G. Zhu ■

335428 Roegneria melanthera (Keng) Keng ＝Kengyilia melanthera (Keng) J. L. Yang，C. Yen et B. R. Baum ■

335429 Roegneria melanthera (Keng) Keng ex Keng et S. L. Chen；黑药鹅观草■

335430 Roegneria melanthera (Keng) Keng ex Keng et S. L. Chen ＝Kengyilia melanthera (Keng) J. L. Yang，C. Yen et B. R. Baum ■

335431 Roegneria melanthera (Keng) Keng ex Keng et S. L. Chen var. tahopaica Keng et S. L. Chen ＝Kengyilia melanthera (Keng) J. L. Yang，C. Yen et B. R. Baum var. tahopaica (Keng) S. L. Chen ■

335432 Roegneria melanthera (Keng) Keng ex Keng et S. L. Chen var. tahopaica Keng；大河坝黑药草；Daheba Roegneria，Dahoba Roegneria ■

335433 Roegneria melanthera (Keng) Keng var. tahopaica Keng ＝Kengyilia melanthera (Keng) J. L. Yang，C. Yen et B. R. Baum var. tahopaica (Keng) S. L. Chen ■

335434 Roegneria minor Keng ＝Elymus zhui S. L. Chen ■

335435 Roegneria minor Keng ex Keng et S. L. Chen；小株鹅观草；Small Goosecomb，Small Roegneria ■

335436 Roegneria multiculmis Kitag.；多秆鹅观草；Manyculm Goosecomb，Manystalk Roegneria ■

335437 Roegneria multiculmis Kitag. ＝Elymus pendulinus (Nevski) Tzvelev subsp. multiculmis (Kitag.) Á. Löve ■

335438 Roegneria multiculmis Kitag. var. pubiflora Keng ＝Elymus pendulinus (Nevski) Tzvelev subsp. multiculmis (Kitag.) Á. Löve ■

335439 Roegneria multiculmis Kitag. var. pubiflora Keng ＝Roegneria multiculmis Kitag. ■

335440 Roegneria mutabilis (Drobow) Hyl.；狭颖鹅观草(假颖鹅观草，细鳞鹅观草，狭颖披碱草)；Narrowawn Roegneria，Narrowglume Goosecomb，Narrowglume Roegneria ■

335441 Roegneria mutabilis (Drobow) Hyl. ＝Elymus mutabilis (Drobow) Tzvelev ■

335442 Roegneria mutabilis (Drobow) Hyl. var. nemoralis (D. F. Cui) L. B. Cai ＝Elymus mutabilis (Drobow) Tzvelev var. nemoralis (S. L. Chen ex D. F. Cui) L. B. Cai ■

335443 Roegneria mutica Keng ＝Kengyilia mutica (Keng) J. L. Yang，C. Yen et B. R. Baum ■

335444 Roegneria mutica Keng ex Keng et S. L. Chen；无芒鹅观草；Awnless Roegneria ■

335445 Roegneria mutica Keng ex Keng et S. L. Chen ＝Kengyilia mutica (Keng ex Keng et S. L. Chen) J. L. Yang，C. Yen et B. R. Baum ■

335446 Roegneria nakaii Kitag.；吉林鹅观草(中井鹅观草)；Nakai Goosecomb，Nakai Roegneria ■

335447 Roegneria nakaii Kitag. ＝Elymus nakaii (Kitag.) S. L. Chen ■

335448 Roegneria nudiuscula L. B. Cai；裸穗鹅观草；Nakedspike Goosecomb，Nakedspike Roegneria ■

335449 Roegneria nudiuscula L. B. Cai ＝Elymus abolinii (Drobow) Tzvelev var. nudiusculus (L. B. Cai) S. L. Chen et G. Zhu ■

335450 Roegneria nutans (Keng) Keng ＝Elymus burchan-buddae (Nevski) Tzvelev ■

335451 Roegneria nutans (Keng) Keng ex Keng et S. L. Chen；垂穗鹅观草；Nodding Roegneria，Noddingspike Goosecomb ■

335452 Roegneria nutans (Keng) Keng ex Keng et S. L. Chen ＝Elymus burchan-buddae (Nevski) Tzvelev ■

335453 Roegneria oschensis (Roshev.) Nevski；奥什鹅观草■☆

335454 Roegneria panormitana (Bertol.) Nevski；西西里鹅观草■☆

335455 Roegneria parviglume Keng ＝Elymus antiquus (Nevski) Tzvelev ■

335456 Roegneria parvigluma Keng ex Keng et S. L. Chen；小颖鹅观草；Smallglume Goosecomb，Smallglume Roegneria ■

335457 Roegneria pauciflora（Schwein.）Hyl.；贫花鹅观草；Fewflower Goosecomb, Fewflower Roegneria ■

335458 Roegneria pauciflora（Schwein.）Hyl. = Elymus trachycaulus（Link）Gould ex Shinners ■☆

335459 Roegneria pendulina Keng f. pubinodis（Keng）Kitag. = Elymus pendulinus（Nevski）Tzvelev subsp. pubicaulis（Keng）S. L. Chen ■

335460 Roegneria pendulina Keng var. pubinodis Keng = Elymus pendulinus（Nevski）Tzvelev subsp. pubicaulis（Keng）S. L. Chen ■

335461 Roegneria pendulina Nevski；缘毛鹅观草；Pendent Goosecomb, Pendent Roegneria ■

335462 Roegneria pendulina Nevski = Elymus pendulinus（Nevski）Tzvelev ■

335463 Roegneria pendulina Nevski var. pubinodis Keng；毛节缘毛草；Hairynode Roegneria ■

335464 Roegneria platyllus Keng = Elymus platyllus（Keng）Á. Löve ■

335465 Roegneria platyphylla Keng = Elymus platyphyllus（Keng）Á. Löve ex D. F. Cui ■

335466 Roegneria platyphylla Keng ex Keng et S. L. Chen = Elymus platyphyllus（Keng）Á. Löve ex D. F. Cui ■

335467 Roegneria praecaespitosa（Nevski）Nevski = Elymus mutabilis（Drobow）Tzvelev var. praecaespitosus（Nevski）S. L. Chen ■

335468 Roegneria praecaespitosa（Nevski）Nevski = Elymus praecaespitosus（Nevski）Tzvelev ■

335469 Roegneria pseudonutans（Keng）Keng = Elymus pseudonutans Á. Löve ■

335470 Roegneria puberula Keng = Elymus puberulus（Keng）S. L. Chen ■

335471 Roegneria puberula Keng ex Keng et S. L. Chen；微毛鹅观草；Hair Goosecomb, Hairy Roegneria ■

335472 Roegneria pubescens（Trin.）Nevski；短柔毛鹅观草■☆

335473 Roegneria pubicaulis Keng = Elymus pendulinus（Nevski）Tzvelev subsp. pubicaulis（Keng）S. L. Chen ■

335474 Roegneria pubicaulis Keng ex Keng et S. L. Chen；毛秆鹅观草；Hairculm Goosecomb, Hairy Culm Roegneria ■

335475 Roegneria pulanensis H. L. Yang = Elymus pulanensis（H. L. Yang）S. L. Chen ■

335476 Roegneria pulanensis H. L. Yang = Roegneria jacquemontii（Hook. f.）Ovcz. et Sidorenko var. pulanensis（H. L. Yang）L. B. Cai ■

335477 Roegneria purpurascens Keng = Elymus purpurascens（Keng）S. L. Chen ■

335478 Roegneria purpurascens Keng ex Keng et S. L. Chen；紫穗鹅观草；Purple Spike Roegneria, Purplespike Goosecomb ■

335479 Roegneria racemifera（Steud.）Kitag. = Elymus racemifer（Steud.）Tzvelev ■☆

335480 Roegneria retroflexa（B. Rong Lu et B. Salomon）L. B. Cai；反折鹅观草；Reflexed Roegneria ■

335481 Roegneria retroflexa（B. Rong Lu et B. Salomon）L. B. Cai = Elymus retroflexus B. Rong Lu et B. Salomon ■

335482 Roegneria rigidula Keng = Kengyilia rigidula（Keng et S. L. Chen）J. L. Yang, C. Yen et B. R. Baum ■

335483 Roegneria rigidula Keng ex Keng et S. L. Chen；硬秆鹅观草；Hardculm Goosecomb, Rigid Roegneria ■

335484 Roegneria rigidula Keng ex Keng et S. L. Chen = Kengyilia rigidula（Keng ex Keng et S. L. Chen）J. L. Yang, C. Yen et B. R. Baum ■

335485 Roegneria rigidula Keng ex Keng et S. L. Chen var. intermedia

Keng et S. L. Chen = Kengyilia rigidula（Keng ex Keng et S. L. Chen）J. L. Yang, C. Yen et B. R. Baum var. intermedia（Keng ex Keng et S. L. Chen）L. B. Cai ■

335486 Roegneria rigidula Keng var. intermedia Keng = Kengyilia rigidula（Keng et S. L. Chen）J. L. Yang, C. Yen et B. R. Baum ■

335487 Roegneria sajanensis Nevski；萨因鹅观草■☆

335488 Roegneria scabridula（Ohwi）Melderis = Elymus scabridulus（Ohwi）Tzvelev ■

335489 Roegneria scabridula Ohwi；粗糙鹅观草；Scabrous Goosecomb, Scabrous Roegneria ■

335490 Roegneria scandica Nevski；瑞典鹅观草■☆

335491 Roegneria schrenkiana（Fisch. et C. A. Mey.）Nevski；扭轴鹅观草；Schrenk Goosecomb, Schrenk Roegneria ■

335492 Roegneria schrenkiana（Fisch. et C. A. Mey.）Nevski = Elymus schrenkianus（Fisch. et C. A. Mey.）Tzvelev ■

335493 Roegneria schugnanica Nevski；舒格南鹅观草■☆

335494 Roegneria sclerophylla Nevski；硬叶鹅观草■☆

335495 Roegneria semicostata（Nees ex Steud.）Kitag.；半脊鹅观草；Semicostate Roegneria ■

335496 Roegneria serotina Keng = Elymus serotinus（Keng）Á. Löve ex B. Rong Lu ■

335497 Roegneria serotina Keng ex Keng et S. L. Chen；秋鹅观草（茅草箭，茅灵芝）；Autumn Goosecomb, Autumn Roegneria ■

335498 Roegneria serpentina L. B. Cai；蜿轴鹅观草■

335499 Roegneria serpentina L. B. Cai = Elymus serpentinus（L. B. Cai）S. L. Chen et G. Zhu ■

335500 Roegneria shandongensis（B. Salomon）J. L. Yang et al. = Elymus shandongensis B. Salomon ■

335501 Roegneria shandongensis（B. Salomon）L. B. Cai；山东鹅观草；Shandong Roegneria ■

335502 Roegneria shouliangiae L. B. Cai；守良鹅观草；Shouliang Roegneria ■

335503 Roegneria shouliangiae L. B. Cai = Elymus shouliangiae（L. B. Cai）G. Zhu ■

335504 Roegneria sinica Keng = Elymus sinicus（Keng）S. L. Chen ■

335505 Roegneria sinica Keng ex Keng et S. L. Chen；中华鹅观草；China Goosecomb, Chinese Roegneria ■

335506 Roegneria sinica Keng ex Keng et S. L. Chen var. angustifolia C. P. Wang et H. L. Yang；狭叶鹅观草；Narrowleaf Roegneria ■

335507 Roegneria sinica Keng ex Keng et S. L. Chen var. media Keng = Roegneria media（Keng ex Keng et S. L. Chen）L. B. Cai ■

335508 Roegneria sinica Keng var. angustifolia C. P. Wang et H. L. Yang = Elymus sinicus（Keng）S. L. Chen ■

335509 Roegneria sinica Keng var. media Keng = Elymus sinicus（Keng）S. L. Chen var. medius（Keng）S. L. Chen et G. Zhu ■

335510 Roegneria sinkiangensis（D. F. Cui）L. B. Cai；新疆鹅观草（新疆披碱草）；Xinjiang Roegneria ■

335511 Roegneria sinkiangensis（D. F. Cui）L. B. Cai = Elymus sinkiangensis D. F. Cui ■

335512 Roegneria stenachyra Keng = Kengyilia stenachyra（Keng）J. L. Yang, C. Yen et B. R. Baum ■

335513 Roegneria stenachyra Keng ex Keng et S. L. Chen；窄颖鹅观草；Narrowglume Goosecomb, Narrowglume Roegneria ■

335514 Roegneria striata Keng ex Keng et S. L. Chen = Elymus semicostatus（Nees ex Steud.）Melderis ■☆

335515 Roegneria stricta Keng = Elymus strictus（Keng ex Keng et S. L. Chen）S. L. Chen ■

335516 Roegneria stricta Keng ex Keng et S. L. Chen = Elymus strictus (Keng ex Keng et S. L. Chen) S. L. Chen ■

335517 Roegneria stricta Keng ex Keng et S. L. Chen f. major Keng;大蒜草;Large Strict Roegneria ■

335518 Roegneria stricta Keng ex Keng et S. L. Chen f. major Keng = Elymus strictus (Keng ex Keng et S. L. Chen) S. L. Chen ■

335519 Roegneria sylvatica Keng et S. L. Chen = Elymus sylvaticus (Keng et S. L. Chen) S. L. Chen ■

335520 Roegneria thoroldiana (Oliv.) Keng = Kengyilia thoroldiana (Oliv.) J. L. Yang,C. Yen et B. R. Baum ■

335521 Roegneria thoroldiana (Oliv.) Keng ex Keng et S. L. Chen;梭罗草;Thorold Lymegrass, Thorold Roegneria, Thoroldgrass Goosecomb ■

335522 Roegneria thoroldiana (Oliv.) Keng ex Keng et S. L. Chen = Elymus thoroldianus (Oliv.) G. Singh. ■

335523 Roegneria thoroldiana (Oliv.) Keng ex Keng et S. L. Chen = Kengyilia thoroldiana (Oliv.) J. L. Yang,C. Yen et B. R. Baum ■

335524 Roegneria thoroldiana (Oliv.) Keng ex Keng et S. L. Chen var. laxiuscula (Melderis) H. L. Yang = Kengyilia grandiglumis (Keng ex Keng et S. L. Chen) J. L. Yang,C. Yen et B. R. Baum var. laxiuscula (Melderis) L. B. Cai ■

335525 Roegneria thoroldiana (Oliv.) Keng ex Keng et S. L. Chen var. laxiuscula (Melderis) H. L. Yang;疏穗梭罗草;Loose Spike Roegneria ■

335526 Roegneria tianschanica (Drobow) Nevski;天山鹅观草;Tianshan Goosecomb,Tianshan Lymegrass,Tianshan Roegneria ■

335527 Roegneria tianschanica (Drobow) Nevski = Elymus tianschanigenus Czerep. ■

335528 Roegneria tianschanicus (Drobow) Nevski = Elymus czilikensis (Drobow) Tzvelev ■

335529 Roegneria tibetica (Melderis) H. L. Yang;西藏鹅观草;Xizang Goosecomb,Xizang Roegneria ■

335530 Roegneria tibetica (Melderis) H. L. Yang = Elymus tibeticus (Melderis) G. Singh ■

335531 Roegneria trachycaulis (Link) Nevski = Elymus trachycaulus (Link) Gould ex Shinners ■☆

335532 Roegneria trichospicula L. B. Cai;毛穗鹅观草;Hairyspike Roegneria ■

335533 Roegneria trichospicula L. B. Cai = Elymus trichospiculus (L. B. Cai) S. L. Chen et G. Zhu ■

335534 Roegneria tridentata C. Yen et J. L. Yang;三齿鹅观草 ■

335535 Roegneria tridentata C. Yen et J. L. Yang = Elymus tridentatus (C. Yen et J. L. Yang) S. L. Chen ■

335536 Roegneria tschimganica (Drobow) Nevski;高山鹅观草;Alpine Goosecomb,Alpine Roegneria ■

335537 Roegneria tschimganica (Drobow) Nevski = Elymus tschimuganicus (Drobow) Tzvelev ■

335538 Roegneria tschimganica (Drobow) Nevski var. glabrispicula (D. F. Cui) L. B. Cai = Roegneria glabrispicula (D. F. Cui) L. B. Cai ■

335539 Roegneria tschimganica (Drobow) Nevski var. glabrispicula (D. F. Cui) L. B. Cai = Elymus tschimuganicus (Drobow) Tzvelev var. glabrispiculus D. F. Cui ■

335540 Roegneria tschimganica (Drobow) Nevski var. variiglumis (Keng) L. B. Cai = Elymus durus (Keng) S. L. Chen ■

335541 Roegneria tschimganica (Drobow) Nevski var. varriglumis (Keng ex Keng et S. L. Chen) L. B. Cai;变颖鹅观草;Variedawn Roegneria ■

335542 Roegneria tschimuganicus (Drobow) Nevski = Elymus tschimuganicus (Drobow) Tzvelev ■

335543 Roegneria tsukushiensis (Drobow) Nevski var. hybrida (Keng) L. B. Cai;杂交鹅观草;Hybrid Goosecomb,Hybrid Roegneria ■

335544 Roegneria tsukushiensis (Drobow) Nevski var. transiens (Hack.) B. Rong Lu,C. Yen et J. L. Yang;川西鹅观草 ■

335545 Roegneria tsukushiensis (Honda) B. Rong Lu et al. var. hybrida (Keng) L. B. Cai = Elymus hybridus (Keng) S. L. Chen ■

335546 Roegneria tsukushiensis (Honda) B. Rong Lu et al. var. transiens (Hack.) B. Rong Lu et al. = Elymus kamoji (Ohwi) S. L. Chen ■

335547 Roegneria turczaninovii (Drobow) Nevski;直穗鹅观草;Gmelin Lymegrass,Turczaninov Goosecomb,Turczaninov Roegneria ■

335548 Roegneria turczaninovii (Drobow) Nevski = Elymus gmelinii (Ledeb.) Tzvelev ■

335549 Roegneria turczaninovii (Drobow) Nevski var. macranthera Ohwi = Roegneria macranthera (Ohwi) L. B. Cai ■

335550 Roegneria turczaninovii (Drobow) Nevski var. pohuashanensis Keng;百花山鹅观草;Baihuashan Roegneria ■

335551 Roegneria turczaninovii (Drobow) Nevski var. tenuiseta (Ohwi) Ohwi ex Keng = Elymus gmelinii (Ledeb.) Tzvelev var. tenuisetus (Ohwi) Osada ■☆

335552 Roegneria turczaninovii (Drobow) Nevski var. tenuiseta Ohwi;细穗鹅观草;Finespike Roegneria ■

335553 Roegneria turczaninowii (Drobow) Nevski = Elymus gmelinii (Ledeb.) Tzvelev ■

335554 Roegneria turczaninowii (Drobow) Nevski var. macrathera (Ohwi) H. L. Yang et C. P. Wang = Elymus gmelinii (Ledeb.) Tzvelev var. macratherus (Ohwi) S. L. Chen et G. Zhu ■

335555 Roegneria turczaninowii (Drobow) Nevski var. pohuashanensis Keng = Elymus gmelinii (Ledeb.) Tzvelev ■

335556 Roegneria turczaninowii (Drobow) Nevski var. tenuiseta (Ohwi) H. L. Yang et C. P. Wang = Elymus gmelinii (Ledeb.) Tzvelev ■

335557 Roegneria turuchanensis (Reverdin) Nevski;土鲁罕鹅观草 ■☆

335558 Roegneria ugamica (Drobow) Nevski;乌岗姆鹅观草;Ugam Goosecomb,Ugame Roegneria ■

335559 Roegneria ugamica (Drobow) Nevski = Elymus nevskii Tzvelev ■

335560 Roegneria uralensis (Nevski) Nevski;乌拉尔鹅观草;Ural Roegneria ■☆

335561 Roegneria varia Keng = Elymus strictus (Keng) S. L. Chen ■

335562 Roegneria varia Keng ex Keng et S. L. Chen;多变鹅观草;Variable Goosecomb,Variable Roegneria ■

335563 Roegneria viridiglumis Nevski;绿颖鹅观草 ■☆

335564 Roegneria viridula Keng et S. L. Chen = Elymus virgidulus (Keng ex Keng et S. L. Chen) S. L. Chen ■

335565 Roegneria viridula Keng ex Keng et S. L. Chen = Elymus virgidulus (Keng ex Keng et S. L. Chen) S. L. Chen ■

335566 Roegneria yangiae (B. Rong Lu) L. B. Cai;杨氏鹅观草;Yang Goosecomb,Yang Roegneria ■

335567 Roegneria yangiae (B. Rong Lu) L. B. Cai = Elymus yangiae B. Rong Lu ■

335568 Roegneria yushuensis L. B. Cai;玉树鹅观草;Yushu Goosecomb,Yushu Roegneria ■

335569 Roegneria yushuensis L. B. Cai = Elymus yushuensis (L. B. Cai) S. L. Chen et G. Zhu ■

335570 Roehlingia Dennst. = Tetracera L. ●

335571 Roehlingia Roepert = Eranthis Salisb. (保留属名)■

335572 Roela Scop. = Roella L. ●■☆

335573 Roelana Comm. ex DC. = Erythroxylum P. Browne ●

335574 Roella L. (1753);南非桔梗属●■☆

335575 Roella alpina Bond = Prismatocarpus alpinus (Bond) Adamson ●☆

335576 Roella amplexicaulis Wolley-Dod;抱茎南非桔梗●☆

335577 Roella arenaria Schltr. ;沙地南非桔梗●☆

335578 Roella ciliata L. ;缘毛南非桔梗●☆

335579 Roella ciliata L. var. incurva (A. DC.) Sond. = Roella incurva A. DC. ●☆

335580 Roella ciliata L. var. linnaeana Sond. = Roella ciliata L. ●☆

335581 Roella ciliata L. var. triflora R. D. Good = Roella triflora (R. D. Good) Adamson ●☆

335582 Roella compacta Schltr. ;紧密南非桔梗●☆

335583 Roella cuspidata Adamson;骤尖南非桔梗●☆

335584 Roella cuspidata Adamson = Roella compacta Schltr. ●☆

335585 Roella cuspidata Adamson var. hispida ? = Roella compacta Schltr. ●☆

335586 Roella decurrens L'Hér. ;下延南非桔梗●☆

335587 Roella dregeana A. DC. ;德雷南非桔梗●☆

335588 Roella dregeana A. DC. var. nitida (Schltr.) Adamson;光亮南非桔梗●☆

335589 Roella eckloniana H. Buek = Roella spicata L. f. ●☆

335590 Roella elegans Paxton;雅致南非桔梗●☆

335591 Roella ericoides R. D. Good = Roella goodiana Adamson ●☆

335592 Roella filiformis Lam. = Roella squarrosa P. J. Bergius ●☆

335593 Roella glabra Poir. = Roella decurrens L'Hér. ●☆

335594 Roella glomerata A. DC. ;团集南非桔梗●☆

335595 Roella goodiana Adamson;古德南非桔梗●☆

335596 Roella incurva A. DC. ;内折南非桔梗●☆

335597 Roella incurva A. DC. var. rigida Adamson = Roella prostrata E. Mey. ex A. DC. ●☆

335598 Roella latiloba A. DC. ;宽裂南非桔梗●☆

335599 Roella leptosepala Sond. = Roella dregeana A. DC. ●☆

335600 Roella lightfootioides Schltr. = Roella spicata L. f. ●☆

335601 Roella maculata Adamson;斑点南非桔梗●☆

335602 Roella muscosa L. f. ;苔藓南非桔梗●☆

335603 Roella nitida Schltr. = Roella dregeana A. DC. var. nitida (Schltr.) Adamson ●☆

335604 Roella pedunculata P. J. Bergius = Prismatocarpus pedunculatus (P. J. Bergius) A. DC. ●☆

335605 Roella prostrata E. Mey. ex A. DC. ;平卧南非桔梗●☆

335606 Roella psammophila Schltr. = Roella dregeana A. DC. ●☆

335607 Roella recurvata A. DC. ;反曲南非桔梗●☆

335608 Roella reticulata A. DC. ;网状南非桔梗●☆

335609 Roella reticulata A. DC. = Roella prostrata E. Mey. ex A. DC. ●☆

335610 Roella rhodantha Adamson = Roella incurva A. DC. ●☆

335611 Roella secunda H. Buek;单侧南非桔梗●☆

335612 Roella spicata L. f. ;长穗南非桔梗●☆

335613 Roella spicata L. f. var. burchellii Adamson;伯切尔南非桔梗●☆

335614 Roella squarrosa P. J. Bergius;粗鳞南非桔梗●☆

335615 Roella triflora (R. D. Good) Adamson;三花南非桔梗●☆

335616 Roellana Comm. ex Lam. = Erythroxylum P. Browne ●

335617 Roelloides Banks ex A. DC. = Prismatocarpus L' Hér. (保留属名)●■☆

335618 Roelpinia Scop. = Acronychia J. R. Forst. et G. Forst. (保留属名)●

335619 Roelpinia Scop. = Cunto Adans. ●

335620 Roelpinia Scop. = Koelpinia Scop. ●

335621 Roemera Tratt. = Steriphoma Spreng. (保留属名)●☆

335622 Roemera DC. = Roemera Tratt. ●☆

335623 Roemeria Dennst. = Scaevola L. (保留属名)●■

335624 Roemeria Medik. (1792);裂叶罂粟属(红罂粟属,新疆罂粟属);Asia Poppy,Xinjiang opium ■

335625 Roemeria Moench = Amaranthus L. ■

335626 Roemeria Moench = Amblogyna Raf. ■

335627 Roemeria Roem. et Schult. = Diarrhena P. Beauv. (保留属名)■

335628 Roemeria Thunb. = Heeria Meisn. + Myrsine L. + Sideroxylon L. ●☆

335629 Roemeria Thunb. = Heeria Meisn. ●☆

335630 Roemeria Thunb. = Sideroxylon L. ●☆

335631 Roemeria Tratt. ex DC. = Steriphoma Spreng. (保留属名)●☆

335632 Roemeria Zea ex Roem. et Schult. = Diarrhena P. Beauv. (保留属名)■

335633 Roemeria argemonoides Pomel = Roemeria hybrida (L.) DC. subsp. dodecandra (Forssk.) Maire ■☆

335634 Roemeria argentea Thunb. = Heeria argentea (Thunb.) Meisn. ●☆

335635 Roemeria bicolor Regel = Roemeria refracta DC. ■

335636 Roemeria caudata ?;巴西裂叶罂粟■☆

335637 Roemeria dodecandra (Forssk.) Stapf = Roemeria hybrida (L.) DC. subsp. dodecandra (Forssk.) Maire ■☆

335638 Roemeria hybrida (L.) DC. ;紫裂叶罂粟(紫花疆罂粟,紫勒米花);Purple Horned Poppy,Purple Xinjiang opium,Violet Horned Poppy ■

335639 Roemeria hybrida (L.) DC. subsp. dodecandra (Forssk.) Maire;十二蕊裂叶罂粟■☆

335640 Roemeria hybrida (L.) DC. subsp. tenuifolia (Pamp.) Maire = Roemeria hybrida (L.) DC. ■

335641 Roemeria hybrida (L.) DC. var. dodecandra (Forssk.) Durand et Barratte = Roemeria hybrida (L.) DC. subsp. dodecandra (Forssk.) Maire ■☆

335642 Roemeria hybrida (L.) DC. var. eriocarpa DC. = Roemeria hybrida (L.) DC. ■

335643 Roemeria hybrida (L.) DC. var. latiloba Pamp. = Roemeria hybrida (L.) DC. ■

335644 Roemeria hybrida (L.) DC. var. orientalis (Boiss.) Coss. = Roemeria hybrida (L.) DC. subsp. dodecandra (Forssk.) Maire ■☆

335645 Roemeria hybrida (L.) DC. var. simplex (Fedde) Maire = Roemeria hybrida (L.) DC. ■

335646 Roemeria hybrida (L.) DC. var. subsimplex Maire = Roemeria hybrida (L.) DC. ■

335647 Roemeria hybrida (L.) DC. var. velutina DC. = Roemeria hybrida (L.) DC. ■

335648 Roemeria hybrida (L.) DC. var. velutino-eriocarpa Fedde = Roemeria hybrida (L.) DC. ■

335649 Roemeria nivea ?;白裂叶罂粟;China Grass ■☆

335650 Roemeria refracta (Steven) DC. ;红裂叶罂粟(红花疆罂粟,红勒米花,红罂粟,裂叶罂粟);Red Xinjiang Opium,Spotted Asian Poppy ■

335651 Roemeria refracta DC. = Roemeria refracta (Steven) DC. ■

335652 Roemeria rhoeadiflora Boiss. = Roemeria refracta DC. ■

335653 Roemeria simplex Fedde = Roemeria hybrida (L.) DC. subsp. dodecandra (Forssk.) Maire ■☆

335654　Roemeria tenuifolia Pamp. = Roemeria hybrida（L.）DC. ■

335655　Roemeria violacea（Juss.）Medik. = Roemeria hybrida（L.）DC. ■

335656　Roemeria violacea Medik. = Roemeria hybrida（L.）DC. ■

335657　Roentgenia Urb.（1916）;伦琴紫葳属●☆

335658　Roentgenia bracteomana（K. Schum. ex Sprague）Urb.;伦琴紫葳●☆

335659　Roepera A. Juss.（1825）;勒珀蒺藜属●■☆

335660　Roepera A. Juss. = Zygophyllum L. ●■

335661　Roepera cordifolia（L. f.）Beier et Thulin;心叶勒珀蒺藜●☆

335662　Roepera cuneifolia（Eckl. et Zeyh.）Beier et Thulin;楔叶勒珀蒺藜●☆

335663　Roepera debilis（Cham. et Schltdl.）Beier et Thulin;弱小勒珀蒺藜■☆

335664　Roepera divaricata（Eckl. et Zeyh.）Beier et Thulin;叉开勒珀蒺藜●☆

335665　Roepera flexuosa（Eckl. et Zeyh.）Beier et Thulin;曲折勒珀蒺藜●☆

335666　Roepera foetida（Schrad. et J. C. Wendl.）Beier et Thulin;臭勒珀蒺藜●☆

335667　Roepera fulva（L.）Beier et Thulin;黄褐勒珀蒺藜●☆

335668　Roepera fuscata（Van Zyl）Beier et Thulin;勒珀蒺藜●☆

335669　Roepera hirticaulis（Van Zyl）Beier et Thulin;毛茎勒珀蒺藜■☆

335670　Roepera horrida（Cham.）Beier et Thulin;多刺勒珀蒺藜■☆

335671　Roepera incrustata（E. Mey. ex Sond.）Beier et Thulin;硬壳勒珀蒺藜●☆

335672　Roepera leptopetala（E. Mey. ex Sond.）Beier et Thulin;细瓣勒珀蒺藜●☆

335673　Roepera leucoclada（Diels）Beier et Thulin;白枝勒珀蒺藜●☆

335674　Roepera lichtensteiniana（Cham. et Schltdl.）Beier et Thulin;利希滕勒珀蒺藜●☆

335675　Roepera macrocarpa（Retief）Beier et Thulin;大果勒珀蒺藜■☆

335676　Roepera maculata（Aiton）Beier et Thulin;斑点勒珀蒺藜●☆

335677　Roepera maritima（Eckl. et Zeyh.）Beier et Thulin;滨海勒珀蒺藜●☆

335678　Roepera microphylla（L. f.）Beier et Thulin;小叶勒珀蒺藜■☆

335679　Roepera orbiculata（Welw. ex Oliv.）Beier et Thulin;圆勒珀蒺藜●☆

335680　Roepera pubescens（Schinz）Beier et Thulin;短柔毛勒珀蒺藜●■☆

335681　Roepera pygmaea（Eckl. et Zeyh.）Beier et Thulin;矮小勒珀蒺藜●☆

335682　Roepera rogersii（Compton）Beier et Thulin;罗杰斯勒珀蒺藜●☆

335683　Roepera sessilifolia（L.）Beier et Thulin;无柄叶勒珀蒺藜●☆

335684　Roepera sphaerocarpa（Schltr. ex Huysst.）Beier et Thulin;球果勒珀蒺藜●☆

335685　Roepera spinosa（L.）Beier et Thulin;具刺勒珀蒺藜●☆

335686　Roepera teretifolia（Schltr.）Beier et Thulin;四叶勒珀蒺藜●☆

335687　Roeperia F. Muell. = Justago Kuntze ●■

335688　Roeperia Spreng. = Ricinocarpos Desf. ●☆

335689　Roeperocharis Rchb. f.（1881）;勒珀兰属■☆

335690　Roeperocharis alcicornis Kraenzl.;尖角勒珀兰●☆

335691　Roeperocharis bennettiana Rchb. f.;贝内特勒珀兰■☆

335692　Roeperocharis camerunensis Szlach. et Olszewski;喀麦隆勒珀兰■☆

335693　Roeperocharis elata Schltr. = Roeperocharis bennettiana Rchb. f. ■☆

335694　Roeperocharis occidentalis Kraenzl. = Habenaria obovata Summerh. ■☆

335695　Roeperocharis platyanthera（Rchb. f.）Rchb. f.;宽药勒珀兰■☆

335696　Roeperocharis rendlei（Rolfe）Kraenzl. = Habenaria peristyloides A. Rich. ■☆

335697　Roeperocharis ukingensis Schltr. = Habenaria peristyloides A. Rich. ■☆

335698　Roeperocharis urbaniana Kraenzl.;乌尔巴尼亚勒珀兰■☆

335699　Roeperocharis wentzeliana Kraenzl.;文策尔勒珀兰■☆

335700　Roeslinia Moench = Chironia L. ●■☆

335701　Roettlera Post et Kuntze = Mallotus Lour. ●

335702　Roettlera Post et Kuntze = Rottlera Willd. ●

335703　Roettlera Post et Kuntze = Trewia L. ●

335704　Roettlera Vahl = Didymocarpus Wall.（保留属名）●■

335705　Roettlera Vahl = Henckelia Spreng.（废弃属名）●■

335706　Roettlera anachoreta（Hance）Kuntze = Chirita anachoreta Hance ■

335707　Roettlera aurea Franch. = Ancylostemon aureus（Franch.）B. L. Burtt ●

335708　Roettlera brevipes（C. B. Clarke）Kuntze = Chirita speciosa Kurz ■

335709　Roettlera cortusifolia（Hance）Fritsch. = Didymocarpus cortusifolius（Hance）W. T. Wang ■

335710　Roettlera demissa（Hance）Kuntze = Chirita demissa（Hance）W. T. Wang ■

335711　Roettlera dimidiata（Wall. ex C. B. Clarke）Kuntze = Chirita anachoreta Hance ■

335712　Roettlera eburnea（Hance）Kuntze = Chirita eburnea Hance ■

335713　Roettlera fargesii Franch. = Opithandra fargesii（Franch.）B. L. Burtt ■

335714　Roettlera forrestii Diels = Oreocharis forrestii（Diels）V. Naray. ■

335715　Roettlera hamosa（R. Br.）Kuntze = Chirita hamosa R. Br. ■

335716　Roettlera juliae（Hance）Kuntze = Chirita juliae Hance ■

335717　Roettlera kamerunensis（Engl.）Fritsch = Schizoboea kamerunensis（Engl.）B. L. Burtt ■☆

335718　Roettlera kurzii（C. B. Clarke）Kuntze = Briggsia kurzii（C. B. Clarke）W. E. Evans ■

335719　Roettlera macrosiphon（Hance）Kuntze = Didissandra macrosiphon（Hance）W. T. Wang ■

335720　Roettlera mannii（C. B. Clarke）Fritsch = Trachystigma mannii C. B. Clarke ■☆

335721　Roettlera mekongense Franch. = Loxostigma mekongense（Franch.）B. L. Burtt ■

335722　Roettlera oblongifolia（Roxb.）Kuntze = Chirita oblongifolia（Roxb.）J. Sinclair ■

335723　Roettlera obtusa（C. B. Clarke）Kuntze = Didymostigma obtusum（C. B. Clarke）W. T. Wang ■

335724　Roettlera pumila（D. Don）Kuntze = Chirita pumila D. Don ■

335725　Roettlera sinensis（Lindl.）Kuntze = Chirita sinensis Lindl. ■

335726　Roettlera speciosa（Kurz）Kuntze = Chirita speciosa Kurz ■

335727　Roettlera tibetica Franch. = Chirita tibetica（Franch.）B. L. Burtt ■

335728　Roettlera uniflora Franch. = Chirita dielsii（Borza）B. L. Burtt ■

335729　Roettlera urticifolia（Buch. -Ham. ex D. Don）Kuntze = Chirita urticifolia Buch. -Ham. ex D. Don ■

335730　Roettlera villosa（D. Don）Kuntze = Didymocarpus villosus D. Don ■

335731　Roettlera yunnanensis Franch. = Didymocarpus yunnanensis

（Franch.）W. W. Sm.■

335732 Roettlera yunnanensis Franch. f. cleistogama Diels = Didymocarpus yunnanensis（Franch.）W. W. Sm.■

335733 Roezlia Hort. = Furcraea Vent.●☆

335734 Roezlia Regel = Monochaetum（DC.）Naudin（保留属名）●☆

335735 Roezliella Schltr.（1918）;勒茨兰属■☆

335736 Roezliella Schltr. = Sigmatostalix Rchb. f.■☆

335737 Roezliella dilatata Schltr.;勒茨兰☆

335738 Rogeonella A. Chev. = Afrosersalisia A. Chev.●☆

335739 Rogeonella A. Chev. = Synsepalum（A. DC.）Daniell●☆

335740 Rogeonella chevalieri（Engl.）Chesnais ex A. Chev. = Synsepalum cerasiferum（Welw.）T. D. Penn.●☆

335741 Rogeria J. Gay = Rogeria J. Gay ex Delile■☆

335742 Rogeria J. Gay ex Delile（1827）;罗杰麻属;Rogeria■☆

335743 Rogeria adenophylla J. Gay;腺叶罗杰麻■☆

335744 Rogeria bigibbosa Engl.;罗杰麻■☆

335745 Rogeria longiflora（Royen）J. Gay ex DC.;长花罗杰麻■☆

335746 Rogeria petrophila De Winter;喜岩罗杰麻■☆

335747 Rogersonanthus Maguire et B. M. Boom（1989）;罗杰森龙胆属●☆

335748 Rogersonanthus arboreus（Britton）Maguire et B. M. Boom;罗杰森龙胆●☆

335749 Roggeveldia Goldblatt = Moraea Mill.（保留属名）■

335750 Roggeveldia Goldblatt（1980）;罗格鸢尾属■☆

335751 Roggeveldia fistulosa Goldblatt = Moraea fistulosa（Goldblatt）Goldblatt■☆

335752 Roggeveldia montana Goldblatt = Moraea monticola Goldblatt■☆

335753 Rogiera Planch.（1849）;罗吉茜属■☆

335754 Rogiera Planch. = Rondeletia L.●

335755 Rogiera amoena Planch.;罗吉茜☆

335756 Rogiera amoena Planch. = Rondeletia amoena（Planch.）Hemsl.●☆

335757 Rogiera latifolia Decne.;宽叶罗吉茜■☆

335758 Rohdea Roth（1821）;万年青属;Nippon Lily, Nipponlily, Rogeria■

335759 Rohdea esquirolii H. Lév. = Rohdea japonica（Thunb.）Roth et Kunth■

335760 Rohdea japonica（Thunb.）Roth;万年青（白河车,白重楼,冲天七,冬不凋草,冬不雕草,渡边万年青,九节连,开口剑,牛尾七,千年润,青龙胆,山苞谷,铁扁担,乌木莓,屋周,野郁蕉,斩蛇剑,竹根七,状元红）;Lily-of-China, Nipponlily, Omoto, Omoto Nippon Lily, Omoto Nipponlily, Sacred Lily-of-China■

335761 Rohdea japonica（Thunb.）Roth et Kunth = Rohdea japonica（Thunb.）Roth■

335762 Rohdea japonica（Thunb.）Roth var. marginata Hort.;银边日本万年青■☆

335763 Rohdea japonica（Thunb.）Roth var. variegata Hort.;金边万年青（金缘万年青）■

335764 Rohdea japonica（Thunb.）Roth var. watanabei（Hayata）S. S. Ying;渡边万年青■

335765 Rohdea japonica（Thunb.）Roth var. watanabei（Hayata）S. S. Ying = Campylandra chinensis（Baker）M. N. Tamura, S. Yun Liang et N. J. Turland■

335766 Rohdea sinensis H. Lév. = Rohdea japonica（Thunb.）Roth■

335767 Rohdea tui F. T. Wang et Ts. Tang = Campylandra tui（F. T. Wang et Ts. Tang）M. N. Tamura, S. Yun Liang et N. J. Turland■

335768 Rohdea tui F. T. Wang et Ts. Tang = Tupistra tui（F. T. Wang et Ts. Tang）F. T. Wang et S. Yun Liang■

335769 Rohdea urotepala Hand.-Mazz. = Campylandra urotepala（Hand.-Mazz.）M. N. Tamura, S. Yun Liang et N. J. Turland■

335770 Rohdea watanabei（Hayata）Dandy = Rohdea japonica（Thunb.）Roth var. watanabei（Hayata）S. S. Ying■

335771 Rohdea watanabei Hayata = Campylandra chinensis（Baker）M. N. Tamura, S. Yun Liang et N. J. Turland■

335772 Rohdea watanabei Hayata = Rohdea japonica（Thunb.）Roth var. watanabei（Hayata）S. S. Ying■

335773 Rohmooa Farille et Lachard（2002）;罗姆草属■☆

335774 Rohrbachia（Kronf. ex Riedl）Mavrodiev = Typha L.■

335775 Rohrbachia minima（Funck ex Hoppe）Mavrodiev = Typha minima Funck. ex Hoppe■

335776 Rohria Schreb. = Tapura Aubl.●☆

335777 Rohria Vahl = Berkheya Ehrh.（保留属名）●■☆

335778 Rohria armata Vahl = Berkheya armata（Vahl）Druce■☆

335779 Rohria bisulca Thunb. = Cullumia bisulca（Thunb.）Less.■☆

335780 Rohria carlinoides Vahl = Berkheya carlinoides（Vahl）Willd.■☆

335781 Rohria carthamoides Thunb. = Berkheya armata（Vahl）Druce■☆

335782 Rohria cuneata Thunb. = Berkheya cuneata（Thunb.）Willd.■☆

335783 Rohria cynaroides Vahl = Berkheya herbacea（L. f.）Druce☆

335784 Rohria decurrens Thunb. = Berkheya decurrens（Thunb.）Willd.■☆

335785 Rohria grandiflora Thunb. = Berkheya barbata（L. f.）Hutch.■☆

335786 Rohria ilicifolia Vahl = Berkheya barbata（L. f.）Hutch.■☆

335787 Rohria incana Thunb. = Berkheya fruticosa（L.）Ehrh.■☆

335788 Rohria lanceolata Thunb. = Berkheya angustifolia（Houtt.）Merr.■☆

335789 Rohria obovata Thunb. = Berkheya spinosa（L. f.）Druce■☆

335790 Rohria palmata Thunb. = Heterorhachis aculeata（Burm. f.）Rössler●☆

335791 Rohria patula Thunb. = Cullumia patula（Thunb.）Less.■☆

335792 Rohria pectinata Thunb. = Cullumia pectinata（Thunb.）Less.■☆

335793 Rohria pungens Thunb. = Berkheya carlinoides（Vahl）Willd.■☆

335794 Rohria spinosissima Thunb. = Berkheya spinosissima（Thunb.）Willd.■☆

335795 Rohria sulcata Thunb. = Cullumia sulcata（Thunb.）Less.■☆

335796 Roia Scop. = Swietenia Jacq.●

335797 Roifia Verdc.（2009）;网果槿属●☆

335798 Roifia Verdc. = Hibiscus L.（保留属名）●■

335799 Roigella Borhidi et M. Fernández（1982）;罗伊格茜属●☆

335800 Roigella correifolia（Griseb.）Borhidi et M. Fernández;罗伊格茜●☆

335801 Roigia Britton = Phyllanthus L.●■

335802 Rojasia Malme（1905）;罗加草属●■☆

335803 Rojasia gracilis（Morong）Malme;罗加草●■☆

335804 Rojasianthe Standl. et Steyerm.（1940）;乳丝菊属●■☆

335805 Rojasianthe superba Standl. et Steyerm.;乳丝菊●■☆

335806 Rojasimalva Fryxell（1984）;委内瑞拉锦葵属●☆

335807 Rojasimalva tetrahedralis Fryxell;委内瑞拉锦葵●■☆

335808 Rojasiophyton Hassl. = Xylophragma Sprague●☆

335809 Rojasiophytum Hassl. = Xylophragma Sprague●☆

335810 Rojoc Adans. = Morinda L.●■

335811 Rokejeka Forssk. = Gypsophila L.■●

335812 Rolandra Rottb.（1775）;银菊木属●☆

335813 Rolandra argentea Rottb.;银菊木●☆

335814 Roldana La Llave = Senecio L.■●

335815 Roldana La Llave（1825）;伞蟹甲属（罗达纳菊属）●■☆

335816　Roldana hartwegii（Benth.）H. Rob. et Brettell;哈氏伞蟹甲■☆

335817　Roldana petasitis（Sims）H. Rob. et Brettell;罗达纳菊■●■☆

335818　Rolfea Zahlbr. = Palmorchis Barb. Rodr. ■☆

335819　Rolfeella Schltr. = Benthamia A. Rich. ●☆

335820　Rolfeella glaberrima（Ridl.）Schltr. = Benthamia glaberrima（Ridl.）H. Perrier ●☆

335821　Rolfinkia Zenk. = Centratherum Cass. ■☆

335822　Rollandia Gaudich.（1829）;罗兰桔梗属●☆

335823　Rollandia Gaudich. = Cyanea Gaudich. ●☆

335824　Rollandia crispa Gaudich. ;罗兰桔梗●☆

335825　Rollinia A. St. -Hil.（1824）;娄林果属（比丽巴属,卷团属,罗林果属,罗林木属,罗林属）●

335826　Rollinia deliciosa Saff. = Rollinia mucosa（Jacq.）Baill. ●

335827　Rollinia emarginata Schltdl. ;微缺娄林果（微缺卷团,微缺罗林木）●☆

335828　Rollinia membranacea Triana et Planch. ;膜质娄林果（膜质卷团,膜质罗林木）●☆

335829　Rollinia mucosa（Jacq.）Baill. ;米糕娄林果（黏液卷团,黏液罗林木）;Biriba, Biriba Tree ●

335830　Rollinia mucosa Baill. = Rollinia mucosa（Jacq.）Baill. ●

335831　Rollinia orthopetala A. DC. = Rollinia mucosa（Jacq.）Baill. ●

335832　Rollinia pulchrinervia A. DC. = Rollinia mucosa（Jacq.）Baill. ●

335833　Rollinia sieberi A. DC. ;索瓦热娄林果●☆

335834　Rolliniopsis Saff.（1916）;拟娄林果属●☆

335835　Rolliniopsis Saff. = Rollinia A. St. -Hil. ●

335836　Rolliniopsis discreta Saff. ;拟娄林果●☆

335837　Rolliniopsis ferruginea R. E. Fr. ;锈色拟娄林果●☆

335838　Rolliniopsis leptopetala Saff. ;细瓣拟娄林果●☆

335839　Rolliniopsis parviflora Saff. ;小花拟娄林果●☆

335840　Rollinsia Al-Shehbaz（1982）;罗林斯芥属☆

335841　Rollinsia paysonii（Rollins）Al-Shehbaz;罗林斯芥■☆

335842　Rolofa Adans. = Glinus L. ■

335843　Rolpa Zahlbr. = Palmorchis Barb. Rodr. ■☆

335844　Rolsonia Rchb. = Ribes L. ●

335845　Romana Vell. = Buddleja L. ●■

335846　Romanesia Gand. = Antirrhinum L. ●■

335847　Romanoa Trevis.（1848）;罗马诺大戟属●☆

335848　Romanoa Trevis. = Anabaena A. Juss. ●☆

335849　Romanoa tamnoides（A. Juss.）Radcl. -Sm. ;罗马诺大戟●☆

335850　Romanowia Gander ex André = Ptychosperma Labill. ●☆

335851　Romanowia Sander = Ptychosperma Labill. ●☆

335852　Romanschulzia O. E. Schulz（1933）;罗曼芥属●☆

335853　Romanschulzia alpina Standl. et Steyerm. ;高山罗曼芥■☆

335854　Romanschulzia mexicana Iltis et Al-Shehbaz;墨西哥罗曼芥■☆

335855　Romanzoffia Cham.（1820）;罗氏麻属;Mistmaiden ■☆

335856　Romanzoffia altera Cham. ;罗氏麻■☆

335857　Romanzovia Spreng. = Romanzoffia Cham. ■☆

335858　Romanzowia DC. = Romanzoffia Cham. ■☆

335859　Romboda Post et Kuntze = Hetaeria Blume（保留属名）■

335860　Romboda Post et Kuntze = Rhomboda Lindl. ■☆

335861　Rombolobium Post et Kuntze = Clitoria L. ●

335862　Rombolobium Post et Kuntze = Rhombolobium Rich. ex Kunth ●

335863　Rombut Adans. = Cassytha L. ■●

335864　Rombut Rumph. ex Adans. = Cassytha L. ■●

335865　Romeroa Dugand（1952）;罗梅紫葳属●☆

335866　Romeroa verticillata Dugand;罗梅紫葳●☆

335867　Romnalda P. F. Stevens（1978）;总序点柱花属●☆

335868　Romnalda grallata R. J. F. Hend. ;总序点柱花■☆

335869　Romneya Harv.（1845）;灌木罂粟属（灌状罂粟属,裂叶罂粟属,马梯里亚罂粟属）;Californian Poppy, Matilija Poppy, Matilija-poppy, Tree Poppy, White Bush ●■☆

335870　Romneya coulteri Harv. ;加州灌木罂粟（裂叶罂粟）;California Tree Poppy, California Tree-poppy, Californian Poppy, Coulter's Matilija Poppy, Matilija Poppy, Tree Pop, Tree Poppy ●☆

335871　Romneya coulteri Harv. 'White Cloud';白云裂叶罂粟●☆

335872　Romneya coulteri Harv. var. trichocalyx（Eastw.）Jeps. = Romneya trichocalyx Eastw. ●☆

335873　Romneya trichocalyx Eastw. ;毛萼灌木罂粟;Giant White Bush, Poppy of California ●☆

335874　Romovia Müll. Arg. = Omphalea L.（保留属名）■☆

335875　Romovia Müll. Arg. = Ronnowia Buc'hoz ■☆

335876　Rompelia Koso-Pol. = Angelica L. ■☆

335877　Rompelia polymorpha（Maxim.）Koso-Pol. = Angelica polymorpha Maxim. ■☆

335878　Romualdea Triana et Planch. = Cuervea Triana ex Miers ●☆

335879　Romulea Maratti（1772）（保留属名）;乐母丽属;Onion Grass, Sand Crocus ■☆

335880　Romulea albomarginata M. P. de Vos;白边乐母丽■☆

335881　Romulea alpina L. Bolus = Romulea fibrosa M. P. de Vos ■☆

335882　Romulea alpina Rendle = Romulea camerooniana Baker ■☆

335883　Romulea ambigua Bég. var. biflora ? = Romulea biflora（Bég.）M. P. de Vos ■☆

335884　Romulea amoena Schltr. ex Bég. ;秀丽乐母丽■☆

335885　Romulea antiatlantica Maire;安蒂乐母丽■☆

335886　Romulea aquatica G. J. Lewis;水生乐母丽■☆

335887　Romulea arenaria Eckl. = Romulea flava（Lam.）M. P. de Vos var. viridiflora（Bég.）M. P. de Vos ■☆

335888　Romulea atrandra G. J. Lewis;黑蕊乐母丽■☆

335889　Romulea atrandra G. J. Lewis var. lewisiae M. P. de Vos;刘易斯乐母丽■☆

335890　Romulea atrandra G. J. Lewis var. luteoflora M. P. de Vos = Romulea luteoflora（M. P. de Vos）M. P. de Vos ■☆

335891　Romulea attenuata M. P. de Vos = Romulea flexuosa Klatt ■☆

335892　Romulea aurea Klatt = Romulea tortuosa（Licht. ex Roem. et Schult.）Baker subsp. aurea（Klatt）M. P. de Vos ■☆

335893　Romulea autumnalis L. Bolus;秋乐母丽■☆

335894　Romulea bachmannii Bég. = Romulea flava（Lam.）M. P. de Vos ■☆

335895　Romulea barkerae M. P. de Vos;巴尔凯拉乐母丽■☆

335896　Romulea battandieri Bég. ;巴坦乐母丽■☆

335897　Romulea biflora（Bég.）M. P. de Vos;双花叶乐母丽■☆

335898　Romulea bifrons Pau;双叶乐母丽■☆

335899　Romulea bifrons Pau var. rosea Bég. = Romulea bifrons Pau ■☆

335900　Romulea bulbocodioides sensu Eckl. = Romulea tabularis Eckl. ex Bég. ■☆

335901　Romulea bulbocodium（L.）Sebast. et Mauri;乐母丽■☆

335902　Romulea bulbocodium（L.）Sebast. et Mauri subsp. ligustica（Parl.）Maire et Weiller = Romulea ligustica Parl. ■☆

335903　Romulea bulbocodium（L.）Sebast. et Mauri subsp. maroccana（Bég.）Maire et Weiller = Romulea maroccana Bég. ■☆

335904　Romulea bulbocodium（L.）Sebast. et Mauri subsp. rouyana（Batt.）Maire et Weiller;卢伊乐母丽■☆

335905　Romulea bulbocodium（L.）Sebast. et Mauri var. debilis Bég. = Romulea bulbocodium（L.）Sebast. et Mauri ■☆

335906　Romulea bulbocodium （L.） Sebast. et Mauri var. grandiflora Parl. = Romulea bulbocodium （L.） Sebast. et Mauri ■☆

335907　Romulea bulbocodium （L.） Sebast. et Mauri var. heterodoxa Maire = Romulea bulbocodium （L.） Sebast. et Mauri ■☆

335908　Romulea bulbocodium （L.） Sebast. et Mauri var. major （Schousb.） Maire et Weiller = Romulea major （Schousb.） A. Marin ■☆

335909　Romulea bulbocodium （L.） Sebast. et Mauri var. pylium Baker = Romulea major （Schousb.） A. Marin ■☆

335910　Romulea bulbocoides sensu Baker var. minor Bég. = Romulea flava （Lam.） M. P. de Vos var. minor （Bég.） M. P. de Vos ■☆

335911　Romulea bulbocoides sensu Baker var. viridiflora Bég. = Romulea flava （Lam.） M. P. de Vos var. viridiflora （Bég.） M. P. de Vos ■☆

335912　Romulea camerooniana Baker;喀麦隆乐母丽■☆

335913　Romulea campanuloides Cufod. = Romulea fischeri Pax ■☆

335914　Romulea campanuloides Harms;风铃草乐母丽■☆

335915　Romulea campanuloides Harms subsp. camerooniana （Baker） Cufod. = Romulea camerooniana Baker ■☆

335916　Romulea campanuloides Harms var. gigantea M. P. de Vos = Romulea camerooniana Baker ■☆

335917　Romulea caplandica Bég. = Romulea dichotoma （Thunb.） Baker ■☆

335918　Romulea cedarbergensis M. P. de Vos;锡达伯格乐母丽■☆

335919　Romulea celsii （Planch.） Klatt = Romulea rosea （L.） Eckl. ■☆

335920　Romulea chloroleuca （Jacq.） Baker = Romulea rosea （L.） Eckl. ■☆

335921　Romulea chloroleuca （Jacq.） Eckl. = Romulea flava （Lam.） M. P. de Vos ■☆

335922　Romulea citrina Baker;柠檬乐母丽■☆

335923　Romulea clusiana （Lange） Nyman = Romulea major （Schousb.） A. Marin ■☆

335924　Romulea collina J. C. Manning et Goldblatt;山丘乐母丽■☆

335925　Romulea columnae Cufod. = Romulea fischeri Pax ■☆

335926　Romulea columnae Sebast. et Mauri;沙地乐母丽;Sand Crocus ■☆

335927　Romulea columnae Sebast. et Mauri subsp. rollii （Parl.） Marais = Romulea rollii Parl. ■☆

335928　Romulea columnae Sebast. et Mauri var. immaculata Maire = Romulea columnae Sebast. et Mauri ■☆

335929　Romulea columnae Sebast. et Mauri var. occidentalis Bég. = Romulea columnae Sebast. et Mauri ■☆

335930　Romulea congoensis Bég.;刚果乐母丽■☆

335931　Romulea cruciata （Jacq.） Baker;十字形乐母丽■☆

335932　Romulea cruciata （Jacq.） Baker var. intermedia （Bég.） M. P. de Vos;间型乐母丽■☆

335933　Romulea cruciata Bég. var. hirsuta？ = Romulea flava （Lam.） M. P. de Vos var. hirsuta （Bég.） M. P. de Vos ■☆

335934　Romulea crusiata （Jacq.） Baker var. australis Ewart = Romulea rosea （L.） Eckl. var. australis （Ewart） M. P. de Vos ■☆

335935　Romulea cuprea Herb. ex Baker = Romulea hirsuta （Steud. ex Klatt） Baker var. cuprea （Baker） M. P. de Vos ■☆

335936　Romulea cyrenaica Bég.;昔兰尼乐母丽■☆

335937　Romulea daveauana Emb. et Maire = Romulea bifrons Pau ■☆

335938　Romulea dichotoma （Thunb.） Baker;二歧乐母丽■☆

335939　Romulea diversiformis M. P. de Vos;异形乐母丽■☆

335940　Romulea duthieae L. Bolus = Romulea tabularis Eckl. ex Bég. ■☆

335941　Romulea eburnea J. C. Manning et Goldblatt;象牙白乐母丽■☆

335942　Romulea elegans Klatt = Romulea rosea （L.） Eckl. var. elegans （Klatt） Bég. ■☆

335943　Romulea elliptica M. P. de Vos;椭圆乐母丽■☆

335944　Romulea engleri Bég.;恩格勒乐母丽■☆

335945　Romulea eximia M. P. de Vos;优异乐母丽■☆

335946　Romulea fibrosa M. P. de Vos;纤维质乐母丽■☆

335947　Romulea filifolia （F. Delaroche） Eckl. = Romulea triflora （Burm. f.） N. E. Br. ■☆

335948　Romulea fischeri Pax;菲舍尔乐母丽■☆

335949　Romulea flava （Lam.） M. P. de Vos;鲜黄乐母丽■☆

335950　Romulea flava （Lam.） M. P. de Vos var. hirsuta （Bég.） M. P. de Vos;粗毛黄乐母丽■☆

335951　Romulea flava （Lam.） M. P. de Vos var. minor （Bég.） M. P. de Vos;小黄乐母丽■☆

335952　Romulea flava （Lam.） M. P. de Vos var. viridiflora （Bég.） M. P. de Vos;绿花乐母丽■☆

335953　Romulea flexuosa Klatt;曲折乐母丽■☆

335954　Romulea fragrans Eckl. = Romulea flava （Lam.） M. P. de Vos ■☆

335955　Romulea framesii L. Bolus = Romulea hirsuta （Steud. ex Klatt） Baker var. framesii （L. Bolus） M. P. de Vos ■☆

335956　Romulea gaditana Bég. = Romulea bifrons Pau ■☆

335957　Romulea gigantea Bég.;巨大乐母丽■☆

335958　Romulea gracillima Baker;细长乐母丽■☆

335959　Romulea grandiscapa Baker;大花茎乐母丽■☆

335960　Romulea hallii M. P. de Vos;霍尔乐母丽■☆

335961　Romulea hantamensis （Diels） Goldblatt;汉塔姆乐母丽■☆

335962　Romulea hirsuta （Steud. ex Klatt） Baker;粗毛乐母丽■☆

335963　Romulea hirsuta （Steud. ex Klatt） Baker var. cuprea （Baker） M. P. de Vos;铜色乐母丽■☆

335964　Romulea hirsuta （Steud. ex Klatt） Baker var. framesii （L. Bolus） M. P. de Vos;弗雷斯乐母丽■☆

335965　Romulea hirsuta （Steud. ex Klatt） Baker var. zeyheri （Baker） M. P. de Vos;泽赫乐母丽■☆

335966　Romulea hirsuta Eckl. = Romulea hirsuta （Steud. ex Klatt） Baker ■☆

335967　Romulea hirta Schltr.;多毛乐母丽■☆

335968　Romulea intermedia Bég. = Romulea cruciata （Jacq.） Baker var. intermedia （Bég.） M. P. de Vos ■☆

335969　Romulea iroensis A. Chev. = Gladiolus iroensis （A. Chev.） Marais ■☆

335970　Romulea kamisensis M. P. de Vos;卡米斯乐母丽■☆

335971　Romulea keniensis Hedberg = Romulea congoensis Bég. ■☆

335972　Romulea klattii Bég. = Romulea hirsuta （Steud. ex Klatt） Baker var. zeyheri （Baker） M. P. de Vos ■☆

335973　Romulea latifolia Baker = Romulea flava （Lam.） M. P. de Vos ■☆

335974　Romulea leipoldtii Marais;莱波尔德乐母丽■☆

335975　Romulea ligustica Parl.;利古里亚乐母丽■☆

335976　Romulea ligustica Parl. subsp. rouyana （Batt.） Bég. = Romulea bulbocodium （L.） Sebast. et Mauri subsp. rouyana （Batt.） Maire et Weiller ■☆

335977　Romulea ligustica Parl. var. rouyana （Batt.） Pau = Romulea bulbocodium （L.） Sebast. et Mauri subsp. rouyana （Batt.） Maire et Weiller ■☆

335978　Romulea ligustica Parl. var. stenopetala Bég. = Romulea ligustica Parl. ■☆

335979　Romulea lilacina J. C. Manning et Goldblatt;紫丁香色乐母丽■☆

335980　Romulea linaresii Parl. subsp. abyssinica Bég. = Romulea fischeri Pax ■☆

335981　Romulea linaresii Parl. subsp. rouyana （Batt.） Batt. = Romulea bulbocodium （L.） Sebast. et Mauri subsp. rouyana （Batt.） Maire et

Weiller ■☆

335982 Romulea longifolia（Salisb.）Baker ＝ Romulea rosea（L.）Eckl. var. australis（Ewart）M. P. de Vos ■☆

335983 Romulea longipes Schltr. ;长梗乐母丽■☆

335984 Romulea longituba G. J. Lewis ＝ Romulea macowanii Baker var. alticola（B. L. Burtt）M. P. de Vos ■☆

335985 Romulea longituba L. Bolus ＝ Romulea macowanii Baker var. alticola（B. L. Burtt）M. P. de Vos ■☆

335986 Romulea longituba L. Bolus var. alticola B. L. Burtt ＝ Romulea macowanii Baker var. alticola（B. L. Burtt）M. P. de Vos ■☆

335987 Romulea luteoflora（M. P. de Vos）M. P. de Vos;黄花乐母丽■☆

335988 Romulea macowanii Baker;马氏乐母丽■☆

335989 Romulea macowanii Baker var. alticola（B. L. Burtt）M. P. de Vos;长管乐母丽■☆

335990 Romulea macowanii Baker var. oreophila M. P. de Vos;喜沙马氏乐母丽■☆

335991 Romulea maculata J. C. Manning et Goldblatt;斑点乐母丽■☆

335992 Romulea major（Schousb.）A. Marin;大乐母丽■☆

335993 Romulea malenconiana Maire ＝ Romulea ligustica Parl. ■☆

335994 Romulea malenconiana Maire var. gattefossei（Bég.）Maire ＝ Romulea ligustica Parl. ■☆

335995 Romulea malenconiana Maire var. stenotepala（Bég.）Maire ＝ Romulea ligustica Parl. var. stenopetala Bég. ■☆

335996 Romulea maroccana Bég. ;摩洛哥乐母丽■☆

335997 Romulea membranacea M. P. de Vos;膜质乐母丽■☆

335998 Romulea minutiflora Klatt;微花乐母丽■☆

335999 Romulea monadelpha（Sweet）Baker;单体雄蕊乐母丽■☆

336000 Romulea montana Schltr. ex Bég. ;山地乐母丽■☆

336001 Romulea monticola M. P. de Vos;山生乐母丽■☆

336002 Romulea muirii N. E. Br. ＝ Romulea rosea（L.）Eckl. var. reflexa（Eckl.）Bég. ■☆

336003 Romulea multifida M. P. de Vos;多裂乐母丽■☆

336004 Romulea multisulcata M. P. de Vos;多沟乐母丽■☆

336005 Romulea namaquensis M. P. de Vos;纳马夸乐母丽■☆

336006 Romulea namaquensis M. P. de Vos subsp. bolusii M. P. de Vos ＝ Romulea namaquensis M. P. de Vos ■☆

336007 Romulea neglecta（Schult.）M. P. de Vos;忽视乐母丽■☆

336008 Romulea nivalis（Boiss. et Kotschy）Klatt ＝ Romulea bulbocodia（L.）Sebast. et Mauri ■☆

336009 Romulea numidica Jord. et Fourr. ;努米底亚乐母丽■☆

336010 Romulea numidica Jord. et Fourr. var. aurantioviolacea（Faure et Maire）Maire et Weiller ＝ Romulea numidica Jord. et Fourr. ■☆

336011 Romulea obscura Klatt;隐匿乐母丽■☆

336012 Romulea obscura Klatt var. blanda M. P. de Vos;光滑乐母丽■☆

336013 Romulea obscura Klatt var. campestris M. P. de Vos;田野乐母丽■☆

336014 Romulea obscura Klatt var. subtestacea M. P. de Vos;淡褐乐母丽■☆

336015 Romulea oliveri M. P. de Vos ＝ Romulea neglecta（Schult.）M. P. de Vos ■☆

336016 Romulea papyracea Wolley-Dod ＝ Romulea schlechteri Bég. ■☆

336017 Romulea parviflora Eckl. ＝ Romulea obscura Klatt ■☆

336018 Romulea pearsonii M. P. de Vos;皮尔逊乐母丽■☆

336019 Romulea penzigi Bég. ;彭西格乐母丽■☆

336020 Romulea pratensis M. P. de Vos;草原乐母丽■☆

336021 Romulea pylia（Herb.）Klatt ＝ Romulea bulbocodia（L.）Sebast. et Mauri ■☆

336022 Romulea ramiflora Baker ＝ Romulea fischeri Pax ■☆

336023 Romulea ramiflora Ten. ;岐花乐母丽■☆

336024 Romulea ramiflora Ten. var. contorta Bég. ＝ Romulea ramiflora Ten. ■☆

336025 Romulea ramiflora Ten. var. parlatorei（Tod.）Richt. ＝ Romulea ramiflora Ten. ■☆

336026 Romulea recurva（F. Delaroche）Eckl. ＝ Romulea flava（Lam.）M. P. de Vos ■☆

336027 Romulea reflexa Eckl. ＝ Romulea rosea（L.）Eckl. var. reflexa（Eckl.）Bég. ■☆

336028 Romulea rollii Parl. ;罗尔乐母丽■☆

336029 Romulea rollii Parl. var. algerica Bég. ＝ Romulea rollii Parl. ■☆

336030 Romulea rosea（L.）Eckl. ;粉花乐母丽; Onion-grass, Rosy Sandcrocus ■☆

336031 Romulea rosea（L.）Eckl. var. australis（Ewart）M. P. de Vos;澳洲粉花乐母丽; Guildford Grass, Rosy Sandcrocus ■☆

336032 Romulea rosea（L.）Eckl. var. celsii Planch. ＝ Romulea rosea（L.）Eckl. ■☆

336033 Romulea rosea（L.）Eckl. var. communis M. P. de Vos;普通粉花乐母丽■☆

336034 Romulea rosea（L.）Eckl. var. elegans（Klatt）Bég. ;雅致粉花乐母丽■☆

336035 Romulea rosea（L.）Eckl. var. reflexa（Eckl.）Bég. ;反折乐母丽■☆

336036 Romulea rosea（L.）Eckl. var. zeyheri Baker ＝ Romulea hirsuta（Steud. ex Klatt）Baker var. zeyheri（Baker）M. P. de Vos ■☆

336037 Romulea rouyana Batt. ＝ Romulea bulbocodium（L.）Sebast. et Mauri subsp. rouyana（Batt.）Maire et Weiller ■☆

336038 Romulea rubrolutea Baker ＝ Romulea hirsuta（Steud. ex Klatt）Baker ■☆

336039 Romulea rupestris J. C. Manning et Goldblatt;岩生乐母丽■☆

336040 Romulea sabulosa Schltr. ex Bég. ;红花乐母丽■☆

336041 Romulea saldanhensis M. P. de Vos;萨尔达尼亚乐母丽■☆

336042 Romulea sanguinalis M. P. de Vos;血红乐母丽■☆

336043 Romulea saxatilis M. P. de Vos;岩栖乐母丽■☆

336044 Romulea schlechteri Bég. ;施莱乐母丽■☆

336045 Romulea schlechteriana Schinz ＝ Romulea triflora（Burm. f.）N. E. Br. ■☆

336046 Romulea setifolia N. E. Br. ;毛叶乐母丽■☆

336047 Romulea setifolia N. E. Br. var. aggregata M. P. de Vos;聚集乐母丽■☆

336048 Romulea setifolia N. E. Br. var. ceresiana M. P. de Vos;塞里斯乐母丽■☆

336049 Romulea similis Eckl. ex Baker ＝ Romulea flava（Lam.）M. P. de Vos var. viridiflora（Bég.）M. P. de Vos ■☆

336050 Romulea singularis J. C. Manning et Goldblatt;单一乐母丽■☆

336051 Romulea sladenii M. P. de Vos;斯莱登乐母丽■☆

336052 Romulea sphaerocarpa M. P. de Vos;球果乐母丽■☆

336053 Romulea spiralis（Burch.）Baker ＝ Geissorhiza spiralis（Burch.）M. P. de Vos ex Goldblatt ■☆

336054 Romulea stellata M. P. de Vos;星状乐母丽■☆

336055 Romulea stenopetala Bég. ＝ Romulea ligustica Parl. ■☆

336056 Romulea stenopetala Bég. subsp. gattefossei ? ＝ Romulea ligustica Parl. ■☆

336057 Romulea subfistulosa M. P. de Vos;亚管乐母丽■☆

336058 Romulea sublutea（Lam.）Baker ＝ Romulea triflora（Burm. f.）N. E. Br. ■☆

336059　Romulea subpalustris（Herb.）Klatt ＝ Romulea bulbocodia（L.）Sebast. et Mauri ■☆

336060　Romulea sulphurea Bég. ;硫色乐母丽■☆

336061　Romulea syringodeoflora M. P. de Vos;管花鸢尾乐母丽■☆

336062　Romulea tabularis Eckl. ex Bég. ;扁平乐母丽■☆

336063　Romulea tetragona M. P. de Vos;四角乐母丽■☆

336064　Romulea tetragona M. P. de Vos var. flavandra M. P. de Vos;黄蕊乐母丽☆

336065　Romulea thodei Schltr. ＝ Romulea camerooniana Baker ■☆

336066　Romulea thodei Schltr. subsp. gigantea M. P. de Vos ＝ Romulea camerooniana Baker ■☆

336067　Romulea torta Baker ＝ Romulea tortilis Baker ■☆

336068　Romulea tortilis Baker;螺旋状乐母丽■☆

336069　Romulea tortilis Baker var. dissecta M. P. de Vos;深裂乐母丽■☆

336070　Romulea tortuosa（Licht. ex Roem. et Schult.）Baker;扭曲乐母丽■☆

336071　Romulea tortuosa（Licht. ex Roem. et Schult.）Baker subsp. aurea（Klatt）M. P. de Vos;黄扭曲乐母丽■☆

336072　Romulea tortuosa（Licht. ex Roem. et Schult.）Baker subsp. depauperata M. P. de Vos;萎缩扭曲乐母丽■☆

336073　Romulea tridentifera Klatt ＝ Romulea tortuosa（Licht. ex Roem. et Schult.）Baker ☆

336074　Romulea triflora（Burm. f.）N. E. Br. ;三花乐母丽■☆

336075　Romulea uliginosa Kuntze;沼泽乐母丽■☆

336076　Romulea uncinata Klatt ＝ Romulea hirsuta（Steud. ex Klatt）Baker ■☆

336077　Romulea unifolia M. P. de Vos;单叶乐母丽■☆

336078　Romulea vaillantii Quézel;瓦扬乐母丽■☆

336079　Romulea vanzyliae M. P. de Vos ＝ Romulea subfistulosa M. P. de Vos ☆

336080　Romulea versicolor Bég. ＝ Romulea tabularis Eckl. ex Bég. ■☆

336081　Romulea villaretii Dobignard;维拉雷乐母丽■☆

336082　Romulea vinacea M. P. de Vos;葡萄酒色乐母丽■☆

336083　Romulea viridibracteata M. P. de Vos;绿苞乐母丽■☆

336084　Romulea vlokii M. P. de Vos;弗劳克乐母丽■☆

336085　Romulea zeyheri（Baker）Bég. ＝ Romulea hirsuta（Steud. ex Klatt）Baker var. zeyheri（Baker）M. P. de Vos ☆

336086　Ronabea Aubl. ＝ Psychotria L.（保留属名）●

336087　Ronabia St. -Lag. ＝ Psychotria L.（保留属名）●

336088　Ronabia St. -Lag. ＝ Ronabea Aubl. ●

336089　Ronaldella Luer ＝ Pleurothallis R. Br. ■☆

336090　Ronaldella Luer(2006);罗纳兰属■☆

336091　Roncelia Willk. ＝ Campanula L. ＋ Wahlenbergia Schrad. ex Roth(保留属名)■●

336092　Roncelia Willk. ＝ Roucela Dumort. ■●

336093　Ronconia Raf. ＝ Ammannia L. ■

336094　Rondachine Bosc ＝ Brasenia Schreb. ■

336095　Rondachine Bosc ＝ Hydropeltis Michx. ■☆

336096　Rondeletia L.(1753);郎德木属;Rondeletia ●

336097　Rondeletia africana T. Winterb. ＝ Crossopteryx febrifuga（Afzel. ex G. Don）Benth. ●☆

336098　Rondeletia amoena（Planch.）Hemsl. ;可爱郎德木（美丽郎德木）●☆

336099　Rondeletia amoena Hemsl. ＝ Rondeletia amoena（Planch.）Hemsl. ●☆

336100　Rondeletia cordata Benth. ;心形郎德木●☆

336101　Rondeletia febrifuga Afzel. ex G. Don ＝ Crossopteryx febrifuga（Afzel. ex G. Don）Benth. ●☆

336102　Rondeletia floribunda G. Don ＝ Holarrhena floribunda（G. Don）T. Durand et Schinz ●☆

336103　Rondeletia longifolia Wall. ＝ Mycetia longifolia（Wall.）Kuntze ●

336104　Rondeletia odorata Jacq. ; 郎 德 木; Fragrant Rondeletia, Rondeletia ●

336105　Rondeletia panamensis DC. ;巴拿马郎德木●☆

336106　Rondeletia pendula Wall. ＝ Wendlandia pendula（Wall.）DC. ●

336107　Rondeletia repens L. ＝ Geophila repens（L.）I. M. Johnst. ■☆

336108　Rondeletia tinctoria Roxb. ＝ Wendlandia tinctoria（Roxb.）DC. ●

336109　Rondonanthus Herzog ＝ Paepalanthus Kunth(保留属名)■☆

336110　Rondonanthus Herzog(1931);败蕊谷精草属■☆

336111　Rondonanthus micropetalus Moldenke;小瓣败蕊谷精草■☆

336112　Rondonanthus roraimae（Oliv.）Herzog;败蕊谷精草■☆

336113　Ronnbergia E. Morren et André.（1874）;伦内凤梨属■☆

336114　Ronnbergia brasiliensis E. Pereira et I. A. Penna;巴西伦内凤梨■☆

336115　Ronnbergia columbiana E. Morren;哥伦比亚伦内凤梨■☆

336116　Ronnowia Buc'hoz ＝ Omphalea L.（保留属名）■☆

336117　Ronoria Augier ＝ Rinorea Aubl.（保留属名）●

336118　Roodebergia B. Nord.（2002）;对叶紫菀属■☆

336119　Roodebergia kitamurana B. Nord. ;对叶紫菀■☆

336120　Roodia N. E. Br. ＝ Argyroderma N. E. Br. ●☆

336121　Roodia braunsii（Schwantes）Schwantes ＝ Argyroderma fissum（Haw.）L. Bolus ●☆

336122　Roodia brevipes（Schltr.）L. Bolus ＝ Argyroderma fissum（Haw.）L. Bolus ●☆

336123　Roodia digitifolia N. E. Br. ＝ Argyroderma fissum（Haw.）L. Bolus ●☆

336124　Rooksbya（Backeb.）Backeb. ＝ Neobuxbaumia Backeb. ●☆

336125　Rooksbya Backeb. ＝ Neobuxbaumia Backeb. ●☆

336126　Rooseveltia O. F. Cook ＝ Euterpe Mart.（保留属名）●☆

336127　Ropala J. F. Gmel. ＝ Roupala Aubl. ●☆

336128　Ropalocarpus Bojer ＝ Rhopalocarpus Bojer ●☆

336129　Ropalon Raf. ＝ Nuphar Sm.（保留属名）■

336130　Ropalopetalum Griff. ＝ Artabotrys R. Br. ex Ker Gawl. ■

336131　Ropalophora Post et Kuntze ＝ Aneilema R. Br. ■☆

336132　Ropalophora Post et Kuntze ＝ Rhopalephora Hassk. ■

336133　Rophostemon Endl. ＝ Nervilia Comm. ex Gaudich.（保留属名）■

336134　Rophostemon Endl. ＝ Roptrostemon Blume ■

336135　Rophostemum Rchb. ＝ Nervilia Comm. ex Gaudich.（保留属名）■

336136　Rophostemum Rchb. ＝ Rophostemon Endl. ■

336137　Rophostemum Rchb. ＝ Roptrostemon Blume ■

336138　Ropourea Aubl. ＝ Diospyros L. ●

336139　Roptrostemon Blume ＝ Nervilia Comm. ex Gaudich.（保留属名）■

336140　Roptrostemon discolor（Blume）Blume ＝ Nervilia plicata（Andréws）Schltr. ■

336141　Roraimaea Struwe, S. Nilsson et V. A. Albert(2008);巴西龙胆属■☆

336142　Rorairnanthus Gleason ＝ Sauvagesia L. ●

336143　Roram Endl. ＝ Cenchrus L. ■

336144　Roram Endl. ＝ Raram Adans. ■

336145　Rorella Haller ex All. ＝ Drosera L. ■

336146　Rorella Hill ＝ Drosera L. ■

336147　Rorella Raf. ＝ Drosophyllum Link ●☆

336148 Rorida J. F. Gmel. = Cleome L. ●■

336149 Roridula Burm. f. ex L. (1764);捕蝇幌属(捕虫木属)●☆

336150 Roridula Forssk. = Cleome L. ●■

336151 Roridula Forssk. = Rorida J. F. Gmel. ■

336152 Roridula arabica Roem. et Schult. = Cleome droserifolia (Forssk.) Delile ■☆

336153 Roridula brachysepala Gand. = Roridula dentata L. ●☆

336154 Roridula crinita Gand. = Roridula gorgonias Planch. ●☆

336155 Roridula dentata L. ;捕蝇幌●☆

336156 Roridula droserifolia Forssk. = Cleome droserifolia (Forssk.) Delile ■☆

336157 Roridula gorgonias Planch. ;湿地捕蝇幌●☆

336158 Roridula muscicapa Gaertn. = Roridula dentata L. ●☆

336159 Roridula verticillata Pers. = Roridula dentata L. ●☆

336160 Roridulaceae Engl. et Gilg = Roridulaceae Martinov(保留科名)●☆

336161 Roridulaceae Martinov(1820)(保留科名);捕蝇幌科●☆

336162 Roripa Adans. = Rorippa Scop. ■

336163 Roripella (Maire) Greuter et Burdet = Rorippa Scop. ■

336164 Roripella (Maire) Greuter et Burdet(1983);大西洋蔊菜属(天柱蔊菜属)■☆

336165 Roripella atlantica (Ball) Greuter et Burdet;大西洋蔊菜■☆

336166 Rorippa Scop. (1760);蔊菜属(葶苈属);Marshcress, Water Cress, Water-cress, Yellowcress, Yellow-cress ■

336167 Rorippa × brachyceras (Honda) Kitam. ex T. Shimizu;短角蔊菜■☆

336168 Rorippa × erythrocaulis Borbás ex Nyman;红茎蔊菜;Thames Yellow-cress ■☆

336169 Rorippa × sterilis Airy Shaw;不育蔊菜;Hybrid Water-cress ■☆

336170 Rorippa africana (Braun-Blanq.) Maire = Nasturtium africanum Braun-Blanq. ■☆

336171 Rorippa americana (A. Gray) Britton = Armoracia lacustris (A. Gray) Al-Shehbaz et V. M. Bates ■☆

336172 Rorippa amphibia (L.) Besser;两栖蔊菜;Amphibious Marsh Cress, Amphibious Marsh-cress, Amphibious Water-cress, Bellragges, Giant Yellow-cress, Great Water Cress, Great Yellowcress, Great Yellow-cress, Greater Yellow Cress ■☆

336173 Rorippa anceps (Wahlenb.) Rchb. ;两刃蔊菜;Hybrid Yellow-cress ■☆

336174 Rorippa aquatica (Eaton) E. J. Palmer et Steyerm. = Armoracia lacustris (A. Gray) Al-Shehbaz et V. M. Bates ■☆

336175 Rorippa armoracia (L.) Hitchc. = Armoracia lapathifolia Gilib. ■☆

336176 Rorippa armoracia (L.) Hitchc. = Armoracia rusticana (Lam.) Gaertn. , B. Mey. et Scherb. ■

336177 Rorippa aspera (L.) Maire = Sisymbrella aspera (L.) Spach ■☆

336178 Rorippa aspera (L.) Maire subsp. boissieri (Coss.) Maire = Sisymbrella aspera (L.) Spach subsp. boissieri (Coss.) Heywood ■☆

336179 Rorippa aspera (L.) Maire subsp. munbyanum (Boiss. et Reut.) Maire = Sisymbrella aspera (L.) Spach subsp. munbyana (Boiss. et Reut.) Greuter et Burdet ■☆

336180 Rorippa atlantica (Ball) Maire = Roripella atlantica (Ball) Greuter et Burdet ■☆

336181 Rorippa atrovirens (Hornem.) Ohwi et H. Hara = Rorippa indica (L.) Hiern ■

336182 Rorippa austriaca (Crantz) Besser;奥地利蔊菜;Austrian Field Cress, Austrian Fieldcress, Austrian Field-cress, Austrian Yellow Cress, Austrian Yellowcress, Austrian Yellow-cress, Field Yellow-

336183 Rorippa barbareifolia (DC.) Kitag.;山芥叶蔊菜;Barbarealeaf Rorippa, Barbarealeaf Yellowcress ■

336184 Rorippa benghalensis (DC.) H. Hara;孟加拉蔊菜■

336185 Rorippa brachycarpa (C. A. Mey.) Woronow;短果蔊菜■☆

336186 Rorippa caledonica (Sond.) R. A. Dyer = Rorippa fluviatilis (E. Mey. ex Sond.) Thell. var. caledonica (Sond.) Marais ■☆

336187 Rorippa cantoniensis (Lour.) Ohwi;广州蔊菜(广东葶苈,沙地菜,微子蔊菜,细子蔊菜);Caton Rorippa, Chinese Yellowcress, Guangzhou Yellowcress, Yellow-cress ■

336188 Rorippa cryptantha (Hochst. ex A. Rich.) Robyns et Boutique;隐花蔊菜■☆

336189 Rorippa cryptantha (Hochst. ex A. Rich.) Robyns et Boutique var. mildbraedii (O. E. Schulz) Robyns et Boutique = Rorippa cryptantha (Hochst. ex A. Rich.) Robyns et Boutique ■☆

336190 Rorippa curvipes Greene;平截蔊菜■☆

336191 Rorippa curvisiliqua (Hook.) Bessey ex Britton;弯果蔊菜■☆

336192 Rorippa dogadovae Tzvelev;道氏蔊菜■☆

336193 Rorippa dubia (Pers.) H. Hara;无瓣蔊菜(大叶香荠菜,地豇豆,干油菜,蔊菜,鸡肉菜,江剪刀草,南蔊菜,清明菜,塘葛菜,天葛菜,铁菜子,小葶苈,野菜子,野辣菜,野油菜);Petalless Rorippa, Petalless Yellowcress ■

336194 Rorippa dubia (Pers.) H. Hara = Rorippa heterophylla (Blume) R. O. Williams ■

336195 Rorippa dubia (Pers.) H. Hara var. benghalensis (DC.) Mukerjee = Rorippa benghalensis (DC.) H. Hara ■

336196 Rorippa elata (Hook. f. et Thomson) Hand.-Mazz.;高蔊菜(苦菜,葶苈);Tall Rorippa, Tall Yellowcress ■

336197 Rorippa floridana Al-Shehbaz et Rollins;佛罗里达蔊菜;Florida Water Cress ■☆

336198 Rorippa fluviatilis (E. Mey. ex Sond.) Thell. ;河岸蔊菜■☆

336199 Rorippa fluviatilis (E. Mey. ex Sond.) Thell. var. caledonica (Sond.) Marais;卡利登蔊菜■☆

336200 Rorippa globosa (Turcz. ex Fisch. et C. A. Mey.) Hayek;风花菜(球果蔊菜,球果山芥菜,银条菜,圆果蔊菜);Globate Rorippa, Globate Yellowcress ■

336201 Rorippa globosa (Turcz.) Hayek = Rorippa globosa (Turcz. ex Fisch. et C. A. Mey.) Hayek ■

336202 Rorippa globosa (Turcz.) Thell. = Rorippa globosa (Turcz. ex Fisch. et C. A. Mey.) Hayek ■

336203 Rorippa heterophylla (Blume) R. O. Williams = Rorippa dubia (Pers.) H. Hara ■

336204 Rorippa hispida (DC.) Britton;毛蔊菜■☆

336205 Rorippa hispida (DC.) Britton var. barbareifolia (DC.) Hultén = Rorippa barbareifolia (DC.) Kitag. ■

336206 Rorippa hispida (Desv.) Britton = Rorippa palustris (L.) Besser subsp. hispida (Desv.) Jonsell ■☆

336207 Rorippa hispida (Desv.) Britton var. barbareifolia (DC.) Hultén = Rorippa barbareifolia (DC.) Kitag. ■

336208 Rorippa hispida (Desv.) Britton var. glabrata Lunell = Rorippa palustris (L.) Besser subsp. fernaldiana (Butters et Abbe) Jonsell ■☆

336209 Rorippa humifusa (Guillaumin et Perr.) Hiern;平伏蔊菜■☆

336210 Rorippa indica (L.) Bailey = Rorippa indica (L.) Hiern ■

336211 Rorippa indica (L.) Hiern;蔊菜(白骨山芥菜,地豇豆,独根菜,干油菜,鸡肉菜,江剪刀草,金丝菜,辣米菜,青蓝菜,山芥菜,山萝卜,石豇豆,水辣菜,塘葛菜,天菜子,田蔊菜,铁菜子,葶苈,香荠菜,野菜花,野菜子,野葛菜,野芥菜,野雪里蕻,野油菜,印度薄菜）；India Fieldcress, India Yellowcress, Indian Rorippa, Variableleaf Yellowcress ■

336212 Rorippa indica（L.）Hiern f. longicarpa（Koidz.）Kitam.；长果薄菜■☆

336213 Rorippa indica（L.）Hiern f. viridiflora Hiyama = Rorippa indica（L.）Hiern var. apetala Hochr. ■☆

336214 Rorippa indica（L.）Hiern subsp. benghalensis（DC.）Bennet = Rorippa benghalensis（DC.）H. Hara ■

336215 Rorippa indica（L.）Hiern var. apetala（DC.）Hochr.；变叶薄菜；Variableleaf Yellowcress ■☆

336216 Rorippa indica（L.）Hiern var. apetala（DC.）Hochr. = Rorippa dubia（Pers.）H. Hara ■

336217 Rorippa indica（L.）Hiern var. benghalensis（DC.）Debeaux = Rorippa benghalensis（DC.）H. Hara ■

336218 Rorippa integrifolia Boulos = Nasturtiopsis integrifolia（Boulos）Abdel Kahlik et F. T. Bakker ■☆

336219 Rorippa islandica（Oeder ex Murray）Borbás = Rorippa palustris（L.）Besser ■

336220 Rorippa islandica（Oeder ex Murray）Borbás var. fernaldiana Butters et Abbe = Rorippa palustris（L.）Besser ■

336221 Rorippa islandica（Oeder）Borbás = Rorippa palustris（L.）Besser ■

336222 Rorippa islandica（Oeder）Borbás var. fernaldiana Butters et Abbe = Rorippa palustris（L.）Besser ■

336223 Rorippa islandica（Oeder）Borbás var. nikkoensis（H. Hara）Kitam. = Rorippa nikkoensis H. Hara ■☆

336224 Rorippa islandica Borbás = Rorippa palustris（L.）Besser ■

336225 Rorippa islandica Schinz et Thell. = Rorippa islandica（Oeder）Borbás ■

336226 Rorippa liaotungensis H. T. Tsui et Y. L. Chang；辽东薄菜；Liaodong Rorippa, Liaodong Yellowcress, Liaotung Rorippa ■

336227 Rorippa liaotungensis X. D. Cui et Y. L. Chang = Rorippa sylvestris（L.）Besser ■

336228 Rorippa madagascariensis（DC.）Hara；马岛薄菜■☆

336229 Rorippa micrantha（Roth）Jonsell；小花薄菜■☆

336230 Rorippa microcapsa（Engl. et Gilg）Robyns et Boutique = Rorippa cryptantha（Hochst. ex A. Rich.）Robyns et Boutique ■☆

336231 Rorippa microphylla（Boenn. ex Rchb.）Hyl. = Rorippa microphylla（Boenn. ex Rchb.）Hyl. ex Á. Löve et D. Löve ■☆

336232 Rorippa microphylla（Boenn. ex Rchb.）Hyl. ex Á. Löve et D. Löve；小叶薄菜（小叶豆瓣菜）；Brown-leaved Watercress, Leko, Narrow-fruited Water-cress, Onerow Yellowcress, One-row Yellowcress, One-row Yellow-cress, One-rowed Watercress, Watercress ■☆

336233 Rorippa microphylla（Boenn. ex Rchb.）Hyl. ex Á. Löve et D. Löve = Nasturtium microphyllum Boenn. ex Rchb. ■☆

336234 Rorippa microphyllum（Boenn. ex Rchb.）Hyl. ex Á. Löve et D. Löve = Nasturtium microphyllum Boenn. ex Rchb. ■☆

336235 Rorippa microsperma（DC.）Hand.-Mazz. = Rorippa cantoniensis（Lour.）Ohwi ■

336236 Rorippa microsperma（DC.）L. H. Bailey = Rorippa cantoniensis（Lour.）Ohwi ■

336237 Rorippa microsperma（DC.）Vassilcz.；小籽薄菜■☆

336238 Rorippa microsperma（DC.）Vassilcz. = Rorippa cantoniensis（Lour.）Ohwi ■

336239 Rorippa montana（Wall. ex Hook. f. et Thomson）Small = Rorippa indica（L.）Hiern ■

336240 Rorippa montana（Wall.）Small = Rorippa dubia（Pers.）H. Hara ■

Hara ■

336241 Rorippa munbyana（Boiss. et Reut.）Maire = Sisymbrella aspera（L.）Spach subsp. munbyana（Boiss. et Reut.）Greuter et Burdet ■☆

336242 Rorippa nasturtium-aquaticum（L.）Hayek；水生薄菜；Water Cress, Watercress ■☆

336243 Rorippa nasturtium-aquaticum（L.）Hayek = Nasturtium officinale R. Br. ■

336244 Rorippa nasturtium-aquaticum（L.）Hayek var. longisiliqua（Irmisch）B. Boivin = Nasturtium microphyllum Boenn. ex Rchb. ■☆

336245 Rorippa nasturtium-aquaticum（L.）Hayek var. sterilis（Airy Shaw）B. Boivin = Nasturtium sterile（Airy Shaw）Oefelein ■☆

336246 Rorippa nikkoensis H. Hara；日光薄菜■☆

336247 Rorippa nudiuscula（E. Mey. ex Sond.）Thell.；稍裸薄菜■☆

336248 Rorippa nudiuscula Thell. subsp. serrata（Burtt Davy）Exell = Rorippa nudiuscula（E. Mey. ex Sond.）Thell. ■☆

336249 Rorippa obtusa（Nutt.）Britton = Rorippa teres（Michx.）Stuckey ■☆

336250 Rorippa palustris（L.）Besser；沼生薄菜（风花菜，薄菜，湿生葶苈，水萝卜，水前菜，葶苈，叶香，沼泽薄菜）；Bog Marsh Cress, Bog Marshcress, Bog Marsh-cress, Bog Rorippa, Bog Yellow Cress, Bog Yellow-cress, Common Marsh Water-cress, Common Yellow-cress, Iceland Watercress, Marsh Yellow Cress, Marsh Yellow-cress, Northern Yellow-cress, Yellow Water Cress ■

336251 Rorippa palustris（L.）Besser subsp. fernaldiana（Butters et Abbe）Jonsell；费氏沼生薄菜；Fernald's Yellow-cress, Marsh-cress ■☆

336252 Rorippa palustris（L.）Besser subsp. glabra（O. E. Schulz）Stuckey var. fernaldiana（Butters et Abbe）Stuckey = Rorippa palustris（L.）Besser subsp. fernaldiana（Butters et Abbe）Jonsell ■☆

336253 Rorippa palustris（L.）Besser subsp. glabra（O. E. Schulz）Stuckey = Rorippa palustris（L.）Besser subsp. fernaldiana（Butters et Abbe）Jonsell ■☆

336254 Rorippa palustris（L.）Besser subsp. hispida（Desv.）Jonsell；硬毛沼生薄菜；Hispid Marsh-cress, Hispid Yellow-cress ■☆

336255 Rorippa palustris（L.）Besser var. africana（Braun-Blanq.）Pau = Nasturtium africanum Braun-Blanq. ■☆

336256 Rorippa palustris（L.）Besser var. cernua（Nutt.）Stuckey = Rorippa palustris（L.）Besser subsp. fernaldiana（Butters et Abbe）Jonsell ■☆

336257 Rorippa palustris（L.）Besser var. dictyota（Greene）Stuckey = Rorippa palustris（L.）Besser subsp. fernaldiana（Butters et Abbe）Jonsell ■☆

336258 Rorippa palustris（L.）Besser var. elongata Stuckey = Rorippa palustris（L.）Besser subsp. fernaldiana（Butters et Abbe）Jonsell ■☆

336259 Rorippa palustris（L.）Besser var. fernaldiana（Butters et Abbe）Stuckey = Rorippa palustris（L.）Besser subsp. fernaldiana（Butters et Abbe）Jonsell ■☆

336260 Rorippa palustris（L.）Besser var. glabra（O. E. Schulz）Stuckey = Rorippa palustris（L.）Besser subsp. fernaldiana（Butters et Abbe）Jonsell ■☆

336261 Rorippa palustris（L.）Besser var. glabrata（Lunell）Stuckey = Rorippa palustris（L.）Besser subsp. fernaldiana（Butters et Abbe）Jonsell ■☆

336262 Rorippa palustris（L.）Besser var. hispida（Desv.）Rydb. = Rorippa palustris（L.）Besser subsp. hispida（Desv.）Jonsell ■☆

336263 Rorippa palustris（L.）Moench = Rorippa islandica（Oeder）

Borbás ■

336264　Rorippa palustris（Leyss.）Besser = Rorippa islandica（Oeder）Borbás ■

336265　Rorippa prostrata Schinz et Thell.；平卧薄菜；Prostrate Yellowcress ■☆

336266　Rorippa rusticana（Gaertn.，B. Mey. et Scherb.）Godr. = Armoracia rusticana（Lam.）Gaertn.，B. Mey. et Scherb. ■

336267　Rorippa sarmentosa Klinkova；蔓茎薄菜■☆

336268　Rorippa sessiliflora（Nutt. ex Torr. et A. Gray）Hitchc.；无柄花薄菜；Marsh Cress，Sessile-flowered Cress，Southern Yellow-cress，Stalkless Yellow-cress，Yellow Cress ■☆

336269　Rorippa sessiliflora（Nutt.）Hitchc. = Rorippa sessiliflora（Nutt. ex Torr. et A. Gray）Hitchc. ■☆

336270　Rorippa sinapis（Burm. f.）Keay；芥薄菜■☆

336271　Rorippa sinapis（Burm. f.）Ohwi et H. Hara = Rorippa indica（L.）Hiern ■

336272　Rorippa sinuata（Nutt. ex Torr. et A. Gray）Hitchc.；西部薄菜；Spreading Yellow Cress，Spreading Yellow-cress，Western Yellow-cress ■☆

336273　Rorippa sinuata（Nutt.）Hitchc. = Rorippa sinuata（Nutt. ex Torr. et A. Gray）Hitchc. ■☆

336274　Rorippa sterilis Airy Shaw = Nasturtium sterile（Airy Shaw）Oefelein ☆

336275　Rorippa sublyrata（Franch. et Sav.）T. Y. Cheo = Rorippa indica（L.）Hiern ■

336276　Rorippa sublyrata（Miq.）H. Hara = Rorippa dubia（Pers.）H. Hara ■

336277　Rorippa sylvestris（L.）Besser；欧亚薄菜；Creeping Water Rocket，Creeping Yellow Cress，Creeping Yellowcress，Creeping Yellow-cress，Cress Rocket，Cress-rocket，Forest Yellowcress，Water Rocket，Wild Nasturtium，Wood Water-cress，Woods Rorippa，Yellow Field Cress，Yellow Fieldcress，Yellow Field-cress ■

336278　Rorippa tenerrima Greene；柔弱薄菜；■☆

336279　Rorippa teres（Michx.）Stuckey；北美薄菜；Obtuse Water-cress ■☆

336280　Rorippa truncata（Jeps.）Stuckey = Rorippa curvipes Greene ■☆

336281　Rorippa villosa R. F. Huang；柔毛薄菜；Villose Rorippa，Villose Yellowcress ■

336282　Rorippa villosa R. F. Huang = Yinshania acutangula（O. E. Schulz）Y. H. Zhang ■

336283　Rosa L.（1753）；蔷薇属；Rose ●

336284　Rosa 'Alabama'；阿拉巴马蔷薇；Alabama ●☆

336285　Rosa 'Angel Face'；天使面蔷薇；Angel Face Rose ●☆

336286　Rosa 'Annuad Delband'；苹果蔷薇；Apple Rose ●☆

336287　Rosa 'Aroplumi'；香李蔷薇；Fragrant Plum Rose ●☆

336288　Rosa 'Aroyquel'；金勋章蔷薇；Gold Medal Rose，Golden Medal Rose ●☆

336289　Rosa 'Excelsa'；高大蔷薇；Excelsa Rose ●☆

336290　Rosa 'Gypsy'；喜钙蔷薇；Gypsy Rose ●☆

336291　Rosa 'Harrisonii'；哈里森蔷薇；Harrison Yellow Rose ●☆

336292　Rosa 'Kiboh'；基博蔷薇；Gipsy Rose，Kibo Rose，Kiboh Rose ●☆

336293　Rosa 'New Dawn'；新黎明蔷薇；New Dawn Rose ●☆

336294　Rosa 'New Day'；新日蔷薇；New Day Rose ●☆

336295　Rosa 'Ophelia'；奥菲利亚蔷薇；Ophelia Rose ●☆

336296　Rosa 'Orange Honey'；蜜橙蔷薇；Orange Honey Rose ●☆

336297　Rosa 'Orange Triumph'；橙胜利蔷薇；Orange Triumph Rose ●☆

336298　Rosa 'Pallida'；苍白蔷薇；Pallida Rose ●☆

336299　Rosa 'Pam's Pink'；帕姆粉蔷薇；Pam's Pink Rose ●☆

336300　Rosa 'Parson's Pink'；帕森粉蔷薇；Parson's Pink Rose ●☆

336301　Rosa 'Peace'；和平蔷薇；Peace Rose ●☆

336302　Rosa 'Penelope'；珀涅罗珀蔷薇；Penelope Rose ●☆

336303　Rosa 'Petite Pink Scotch'；苏格兰小粉蔷薇；Petite Pink Scotch Rose ●☆

336304　Rosa 'Pilgrim'；朝圣蔷薇；Pilgrim Rose ●☆

336305　Rosa 'Pink Prosperity'；绯荣蔷薇；Pink Prosperity Rose ●☆

336306　Rosa 'Playboy'；花花公子蔷薇；Playboy Rose ●☆

336307　Rosa 'Pol M Lillie'；极星蔷薇；Baby Faurax Rose ●☆

336308　Rosa 'Rose d'Amour'；重瓣弗吉尼亚蔷薇；Blue Virginia Rose，Double Virginia Rose，St. Mark's Rose ●☆

336309　Rosa 'Sparte'；鹰爪豆蔷薇；Sparte Rose ●☆

336310　Rosa 'Sweet Vivien'；香维维安蔷薇；Sweet Vivian Rose ●☆

336311　Rosa 'Tanolg'；俄勒冈蔷薇；Miss Harp Rose，Oregold Rose ●☆

336312　Rosa 'Thelma'；温泉蔷薇；Thelma Rose ●☆

336313　Rosa × makinoana H. Ohba；牧野氏蔷薇●☆

336314　Rosa × palustriformis Rydb.；拟沼泽蔷薇●☆

336315　Rosa × pteragonis M. Krause ex Kordes；暗红刺蔷薇●☆

336316　Rosa × pulcherrima Koidz.；美丽蔷薇●☆

336317　Rosa abietina H. Christ；冷杉蔷薇●☆

336318　Rosa abyssinica Lindl.；阿比西尼亚蔷薇●☆

336319　Rosa abyssinica Lindl. var. microphylla Crép.；小叶阿比西尼亚蔷薇●☆

336320　Rosa acicularis Lindl.；刺蔷薇（长果蔷薇，大叶蔷薇，多刺大叶蔷薇，柔毛大叶蔷薇，少刺大叶蔷薇，细刺蔷薇）；Bristly Rose，Prickly Rose ●

336321　Rosa acicularis Lindl. = Rosa suavis Willd. ●

336322　Rosa acicularis Lindl. subsp. sayi（Schwein.）W. H. Lewis；赛氏刺蔷薇；Bristly Rose，Prickly Rose ●

336323　Rosa acicularis Lindl. var. albiflora X. Lin et Y. L. Lin = Rosa acicularis Lindl. ●

336324　Rosa acicularis Lindl. var. albifloris X. Lin et Y. L. Lin；白花刺蔷薇；White-flower Prickly Rose ●

336325　Rosa acicularis Lindl. var. bourgeauiana（Crép.）Crép. = Rosa acicularis Lindl. subsp. sayi（Schwein.）W. H. Lewis ●

336326　Rosa acicularis Lindl. var. glandulifolia Y. B. Chang；腺叶大叶蔷薇；Glandularleaf Prickly Rose ●

336327　Rosa acicularis Lindl. var. glandulifolia Y. B. Chang = Rosa acicularis Lindl. ●

336328　Rosa acicularis Lindl. var. glandulosa Liou；腺果刺蔷薇（腺果大叶蔷薇）；Glandularfruit Prickly Rose ●

336329　Rosa acicularis Lindl. var. glandulosa Liou = Rosa acicularis Lindl. ●

336330　Rosa acicularis Lindl. var. gmelinii（Bunge）C. A. Mey. = Rosa acicularis Lindl. ●

336331　Rosa acicularis Lindl. var. gmelinii C. A. Mey. = Rosa acicularis Lindl. ●

336332　Rosa acicularis Lindl. var. nipponensis（Crép.）Koehne = Rosa nipponensis Crép. ●☆

336333　Rosa acicularis Lindl. var. pubescens Liou = Rosa acicularis Lindl. ●

336334　Rosa acicularis Lindl. var. sayana Erlanson = Rosa acicularis Lindl. subsp. sayi（Schwein.）W. H. Lewis ●

336335　Rosa acicularis Lindl. var. setacea Liou = Rosa acicularis Lindl. ●

336336　Rosa acicularis Lindl. var. taquetii（H. Lév.）Nakai = Rosa luciae Franch. et Rochebr. ex Crép. ●

336337　Rosa acicularis Lindl. var. taquetii（H. Lév.）Nakai ＝ Rosa wichuraiana Crép. ex Desegl. ●

336338　Rosa acicularis Lindl. var. taquetii Nakai ＝ Rosa acicularis Lindl. ●

336339　Rosa agrestis Savi；野生蔷薇●☆

336340　Rosa agrestis Savi ＝ Rosa inodora Hook. ●☆

336341　Rosa agrestis Savi var. denudata Keller ＝ Rosa agrestis Savi ●☆

336342　Rosa agrestis Savi var. pubescens（Rapin）Keller ＝ Rosa agrestis Savi ●☆

336343　Rosa alba L.；白蔷薇；Brandy Bottles，Cottage Rose，Incarnation Rose，Jacobite Rose，White Rose，White Rose of York，White Rose-of-York ●

336344　Rosa alba L.'Semi-plena'；半重瓣白蔷薇；Alba Semi-plena Rose ●☆

336345　Rosa albertii Regel；腺齿蔷薇（落萼蔷薇）；Albert Rose ●

336346　Rosa alcea Greene ＝ Rosa arkansana Porter var. suffulta（Greene）Cockerell ●☆

336347　Rosa alexeenkoi Crép.；阿莱蔷薇●☆

336348　Rosa alpina L.；高山蔷薇（垂蔷薇）；Alpine Rose，Drophip Rose，Penduline Rose ●☆

336349　Rosa alpina L. var. macrophylla（Lindl.）Boulenger ＝ Rosa macrophylla Lindl. ●

336350　Rosa alpina L. var. macrophylla Boulenger ＝ Rosa macrophylla Lindl. ●

336351　Rosa altaica Willd. ＝ Rosa spinosissima L. var. altaica（Willd.）Rehder ●

336352　Rosa alticola Boulenger；高原蔷薇●☆

336353　Rosa amblyotis C. A. Mey.；钝耳蔷薇●☆

336354　Rosa amygdalifolia Ser. ＝ Rosa laevigata Michx. ●

336355　Rosa andegavensis Bastard；昂德加维蔷薇●☆

336356　Rosa andrzejowskii Steven ex M. Bieb.；安德氏蔷薇；Andrzejowsk Rose ●☆

336357　Rosa anemoniflora Fortune ex Lindl.；银粉蔷薇（红枝蔷薇，银莲花蔷薇）；Anemoneflower Rose，Anemone-flowered Rose，Anemonyflower Rose ●

336358　Rosa anemonoides Rehder ＝ Rosa lucidissima H. Lév. ●

336359　Rosa anioyensis Hance ＝ Rosa cymosa Tratt. ●

336360　Rosa argyi H. Lév. ＝ Rosa laevigata Michx. ●

336361　Rosa arkansana Porter；阿肯色蔷薇；Arkansas Rose，Dwarf Prairie Rose，Prairie Rose，Sunshine Rose，Wild Prairie Rose ●☆

336362　Rosa arkansana Porter et Coult. ＝ Rosa arkansana Porter ●☆

336363　Rosa arkansana Porter var. suffulta（Greene）Cockerell；支柱蔷薇（草原蔷薇）；Arkansas Rose，Prairie Rose ●☆

336364　Rosa arvensis Huds.；田野蔷薇（野生蔷薇）；Ayrshire Rose，Brid Breer，Bush Briar，Cat Rose，Corn Rose，Field Rose，Gypsy Rose，Musk Rose，Trailing Rose ●☆

336365　Rosa baiyushanensis Q. L. Wang；白玉山蔷薇；Baiyushan Rose ●

336366　Rosa banksiae W. T. Aiton；木香花蔷薇（班克蔷薇，木香，木香花，七里香）；Banks Rose，Banks' Rose，Banksia Rose，Banksian Rose，Lady Banks Rose，Lady Banks' Rose ●

336367　Rosa banksiae W. T. Aiton 'Alba Plena'；重瓣白木香；Banks Rose，Banksian Rose，Double White Banksia Rose，Doublewhite Banks Rose，Lady Banks' Rose，White Lady Banks Rose ●

336368　Rosa banksiae W. T. Aiton 'Lutea'；重瓣黄木香（黄花木香，黄木香花）；White Lady Banks Rose，Yellow Banks Rose，Yellow Banksian Rose，Yellow Lady Banks Rose ●

336369　Rosa banksiae W. T. Aiton f. lutea（Lindl.）Rehder ＝ Rosa banksiae W. T. Aiton 'Lutea' ●

336370　Rosa banksiae W. T. Aiton f. luteiflora H. Lév. ＝ Rosa banksiae W. T. Aiton f. lutea（Lindl.）Rehder ●

336371　Rosa banksiae W. T. Aiton f. luteiflora H. Lév. ＝ Rosa banksiae W. T. Aiton 'Lutea' ●

336372　Rosa banksiae W. T. Aiton f. lutescens Voss；单瓣黄木香（黄木香）；Yellowish Banks Rose ●

336373　Rosa banksiae W. T. Aiton var. albo-plena Rehder ＝ Rosa banksiae W. T. Aiton ●

336374　Rosa banksiae W. T. Aiton var. albo-plena Rehder ＝ Rosa banksiae W. T. Aiton 'Alba Plena' ●

336375　Rosa banksiae W. T. Aiton var. lutea Lindl. ＝ Rosa banksiae W. T. Aiton f. lutea（Lindl.）Rehder ●

336376　Rosa banksiae W. T. Aiton var. lutea Lindl. ＝ Rosa banksiae W. T. Aiton 'Lutea' ●

336377　Rosa banksiae W. T. Aiton var. lutescens Voss ＝ Rosa banksiae W. T. Aiton f. lutescens Voss ●

336378　Rosa banksiae W. T. Aiton var. normalis Regel；白木香（单瓣白木香，单瓣木香花，木香花，七里香，七里香蔷薇，香花刺，香水花）；Wild Banks Rose ●

336379　Rosa banksiopsis Baker；拟木香蔷薇（假木香蔷薇，拟木香，小尾萼蔷薇）；Banksia-like Rose，False Banks Rose ●

336380　Rosa beggeriana Schrenk；弯刺蔷薇（多花蔷薇，落蔷薇，弯刺木香）；Begger Rose，Curved-prickle Rose，Curved-prickled Rose ●

336381　Rosa beggeriana Schrenk var. liouii（Te T. Yu et H. T. Tsai）Te T. Yu et T. C. Ku；毛叶弯刺蔷薇（刘氏弯刺蔷薇，毛叶弯刺木香）；Liou Cuevedprickle Rose ●

336382　Rosa bella Rehder et E. H. Wilson；美蔷薇（美丽蔷薇，山刺玫，油瓶瓶，油瓶子）；Solitary Rose ●

336383　Rosa bella Rehder et E. H. Wilson f. pallens Rehder et E. H. Wilson ＝ Rosa bella Rehder et E. H. Wilson ●

336384　Rosa bella Rehder et E. H. Wilson var. nuda Te T. Yu et H. T. Tsai；光叶美蔷薇；Nakedleaf Solitary Rose ●

336385　Rosa berberifolia Pall.；小檗叶蔷薇●

336386　Rosa berberifolia Pall. ＝ Rosa persica Michx. ●

336387　Rosa bernarditae Sennen et Mauricio ＝ Rosa pouzinii Tratt. ●☆

336388　Rosa biflora T. C. Ku；双花蔷薇；Biflous Rose，Biflowered Rose ●

336389　Rosa blanda Aiton；哈德逊蔷薇；Early Wild Rose，Hudson Bay Rose，Mesdow Rose，Northern Prairie Rose，Smooth Rose，Wild Rose ●☆

336390　Rosa blanda Aiton f. alba（Schuette ex Erlanson）Fernald ＝ Rosa blanda Aiton ●☆

336391　Rosa blanda Aiton f. carpohispida（Schuette）W. H. Lewis ＝ Rosa blanda Aiton ●☆

336392　Rosa blanda Aiton var. alba Schuette ＝ Rosa blanda Aiton ●☆

336393　Rosa blanda Aiton var. carpohispida Schuette ＝ Rosa blanda Aiton ●☆

336394　Rosa blanda Aiton var. glandulosa Schuette ＝ Rosa blanda Aiton ●☆

336395　Rosa blanda Aiton var. hispida Farw. ＝ Rosa blanda Aiton ●☆

336396　Rosa blanda Aiton var. nuda Schuette ＝ Rosa blanda Aiton ●☆

336397　Rosa blanda Aiton var. subgeminata（Schuette）Erlanson ＝ Rosa blanda Aiton ●☆

336398　Rosa blanda Aiton var. subgeminata Schuette ＝ Rosa blanda Aiton ●☆

336399　Rosa blinii H. Lév. ＝ Rosa multiflora Thunb. var. carnea Thory ●

336400　Rosa bodinieri H. Lév. et Vaniot ＝ Rosa cymosa Tratt. ●

336401　Rosa boisii Cardot ＝ Rosa lucidissima H. Lév. ●

336402　Rosa boissieri Crép.；布瓦西耶蔷薇；Boissier Rose ●☆

336403   Rosa borboniana Desp. ;波旁蔷薇(波旁月季);Bourbon Rose ●

336404   Rosa bourgeauiana Crép. = Rosa acicularis Lindl. subsp. sayi (Schwein.) W. H. Lewis ●

336405   Rosa bracteata J. C. Wendl. ;硕苞蔷薇(苞蔷薇,刺柿,大红袍,猴局,猴柿,金柿,琉球野蔷薇,毛刺头,七姊妹,山麻栗子,糖钵,糖球子硬苞蔷薇,野毛栗,野毛粟,圆刺菱);Macariney Rose ●

336406   Rosa bracteata J. C. Wendl. var. scabriacaulis Lindl. ex Koidz. ;密刺硕苞蔷薇(滨野蔷薇);Prickly Macartney Rose ●

336407   Rosa bracteata J. C. Wendl. var. typica Lindl. ex Koidz. = Rosa bracteata J. C. Wendl. ●

336408   Rosa brunonii Lindl. ;复伞房蔷薇(白刺梅,勃朗蔷薇,复伞序蔷薇,麝香蔷薇,麝香月季,万朵刺,喜马拉雅麝香蔷薇);Brunon Rose, Himalayan Musk Rose, Himalayan Musk-rose, Musk Rose, Musk-rose ●

336409   Rosa bungeana Boiss. et Buhse;布氏蔷薇;Bunge Rose ●☆

336410   Rosa caesia Sm. ;杜玛蔷薇;Hairy Dog-rose ●☆

336411   Rosa caesia Sm. subsp. glauca (Nyman) G. G. Graham et Primavesi;阿夫氏蔷薇;Afzel Rose, Glaucous Dog-rose ●☆

336412   Rosa californica Cham. et Schltdl. = Rosa californica E. Willm. ●☆

336413   Rosa californica E. Willm. ;加州蔷薇(粗刺蔷薇)●☆

336414   Rosa californica E. Willm. f. plena Rehder;重瓣加州蔷薇;Double California Rose ●☆

336415   Rosa calva (Franch. et Sav.) Boulenger var. cathayensis (Rehder et E. H. Wilson) Boulenger = Rosa multiflora Thunb. var. cathayensis Rehder et E. H. Wilson ●

336416   Rosa calva (Franch. et Sav.) Boulenger var. cathayensis Boulenger = Rosa multiflora Thunb. var. cathayensis Rehder et E. H. Wilson ●

336417   Rosa calva (Franch. et Sav.) Boulenger var. formosana (Cardot) Boulenger = Rosa multiflora Thunb. ●

336418   Rosa calyptopoda Cardot;短脚蔷薇(短角蔷薇,帽状蔷薇,美人脱衣);Capshaped Rose, Shortfoot Rose, Short-footed Rose ●

336419   Rosa canina L. ;犬蔷薇(钝叶蔷薇,狗蔷薇,狗牙蔷薇,灌丛蔷薇,欧洲蔷薇,野蔷薇);Bachelor's Pear, Bacon, Bird Briar, Bird Pears, Booag, Brandy Bottle, Bread-and-cheese, Bread-and-cheese Tree, Breer, Brere, Briar, Briar Balls, Briar Rose, Briar-bush, Briar-rose, Briar-tree, Brier Bush, Brimmle, Buck Breer, Buckie-lice, Buckle, Bucky, Budde Lice, Bull Beef, Bumbleberry, Canker, Canker Rose, Canker-balls, Canker-berry, Canker-flower, Cat Choops, Cat Hip, Cat Hips, Cat Jug, Cat Rose, Cat Whin, Cat's Hips, Cat's Jugs, Cattijugs, Cheese, Choop Rose, Choop-rose, Choops, Choop-tree, Choups, Chowps, Cock Bramble, Cock-battler, Cock-fighter, Common Briar, Conkerberry, Conkers, Corumb Rose, Cow Itch, Daily Bread, Dike Rose, Dike-rose, Dog Berries, Dog Breer, Dog Briar, Dog Choops, Dog Hip, Dog Hippan, Dog Job, Dog Jump, Dog Rose, Dog's Briar, Dog's Thorn, English Rose, Fogs Cat, Gypsy Nut, Gypsy Nuts, Hagisses, Haps, Haves, Haw, Hawp, Hedge Speak, Hedgespeaks, Hedgy-pedgy, Hep Briar, Hep Tree, Hip, Hip Bramble, Hip Briar, Hip Rose, Hip Tree, Hippan, Hipson, Hiptypips, Horse Bramble, Huggan, Humack, Itching Berry, Itchy-backs, Klonger, Klunger, Lawyers, Locks-and-keys, Lops-and-lice, Muckies, Neddy Grinnel, Nippernail, Old Man's Beard, Pig Rose, Pig's Rose, Pixy Pear, Pixy-pear, Red Berry, Robin Redbreast's Cushion, Robin's Pillow, Robin's Pincush, Robin's Pincushion, Roe Briar, Roe-briar, Round-leaved Dog-rose, Save-whallop, Shoups, Soldiers, Sugar Candy, Tiekling Tommies, Titty Bottle, Titty-bottles, Yew Brimmle, Yew-brimmle, Yoe Brimmle, Yoe-brimmle ●

336420   Rosa canina L. 'Abbotswood';阿波兹伍德犬蔷薇(阿波兹伍德狗蔷薇)●☆

336421   Rosa canina L. 'Canina Abbotswood' = Rosa canina L. 'Abbotswood' ●☆

336422   Rosa canina L. subsp. abietina (Gren.) Rouy = Rosa abietina H. Christ ●☆

336423   Rosa canina L. subsp. andegavensis (Bastard) Batt. = Rosa andegavensis Bastard ●☆

336424   Rosa canina L. subsp. dumetorum (Thuill.) Batt. = Rosa corymbifera Borkh. ●

336425   Rosa canina L. subsp. obtusifolia (Desv.) Maire = Rosa obtusifolia Desv. ●

336426   Rosa canina L. subsp. pouzinii (Tratt.) Crép. = Rosa pouzinii Tratt. ●☆

336427   Rosa canina L. var. andegavensis (Bastard) = Rosa andegavensis Bastard ●☆

336428   Rosa canina L. var. avelloi Sennen et Mauricio = Rosa canina L. ●

336429   Rosa canina L. var. beatricis (Burnat et Gremli) Keller = Rosa canina L. ●

336430   Rosa canina L. var. bellezmensis Keller = Rosa canina L. ●

336431   Rosa canina L. var. bernardita Sennen et Mauricio = Rosa canina L. ●

336432   Rosa canina L. var. bisserata (Mérat) Chevall. = Rosa agrestis Savi ●☆

336433   Rosa canina L. var. blondaeana (Ripart) Crép. = Rosa canina L. ●

336434   Rosa canina L. var. dumetorum Baker = Rosa canina L. ●

336435   Rosa canina L. var. flexibilis (Déségl.) Keller = Rosa canina L. ●

336436   Rosa canina L. var. glaberrima (Dumort.) Christ = Rosa canina L. ●

336437   Rosa canina L. var. hirtella Christ = Rosa canina L. ●

336438   Rosa canina L. var. hispidula (Ripart) Roux = Rosa canina L. ●

336439   Rosa canina L. var. hispidula Christ = Rosa canina L. ●

336440   Rosa canina L. var. lutetiana (Léman) Baker = Rosa canina L. ●

336441   Rosa canina L. var. mairei Keller = Rosa canina L. ●

336442   Rosa canina L. var. nitens (Desv.) Debeaux = Rosa agrestis Savi ●☆

336443   Rosa canina L. var. obtusifolioides Keller = Rosa canina L. ●

336444   Rosa canina L. var. parvipetala Pau et Font Quer = Rosa canina L. ●

336445   Rosa canina L. var. platyphylla Pau = Rosa canina L. ●

336446   Rosa canina L. var. pseudocollina Christ = Rosa canina L. ●

336447   Rosa canina L. var. pseudohispidula Maire = Rosa canina L. ●

336448   Rosa canina L. var. pubescens Crép. = Rosa canina L. ●

336449   Rosa canina L. var. recognita Rouy = Rosa canina L. ●

336450   Rosa canina L. var. scabrata Crép. = Rosa canina L. ●

336451   Rosa canina L. var. serrata Keller = Rosa canina L. ●

336452   Rosa canina L. var. subinermis Ball = Rosa canina L. ●

336453   Rosa canina L. var. substylosa Keller = Rosa canina L. ●

336454   Rosa canina L. var. tenuissima Rouy = Rosa canina L. ●

336455   Rosa canina L. var. thuillieri Christ = Rosa canina L. ●

336456   Rosa canina L. var. tomentella (Léman) Crép. = Rosa canina L. ●

336457   Rosa canina L. var. transitoria Keller = Rosa canina L. ●

336458   Rosa canina L. var. uncinella (Besser) Keller = Rosa canina L. ●

336459   Rosa canina L. var. verticillacantha (Mérat) Baker = Rosa canina L. ●

336460　Rosa canina L. var. vinealis（Ripart）Keller ＝Rosa canina L. ●

336461　Rosa canina L. var. yebalica Pau et Font Quer ＝Rosa canina L. ●

336462　Rosa carolina L.；卡罗来纳蔷薇（北美蔷薇）；Carolina Rose, Pasture Rose ●☆

336463　Rosa carolina L. f. glandulosa（Crép.）Fernald ＝Rosa carolina L. ●☆

336464　Rosa carolina L. var. grandiflora（Baker）Rehder var. villosa（Best）Rehder ＝Rosa carolina L. ●☆

336465　Rosa carolina L. var. inermis Regel ＝Rosa palustris Marshall ●☆

336466　Rosa caryophyllaeea Besser ＝Rosa inodora Hook. ●☆

336467　Rosa cathayensis（Rehder et E. H. Wilson）L. H. Bailey ＝Rosa multiflora Thunb. ex Murray ●

336468　Rosa cathayensis（Rehder et E. H. Wilson）L. H. Bailey ＝Rosa multiflora Thunb. var. cathayensis Rehder et E. H. Wilson ●

336469　Rosa cathayensis（Rehder）L. H. Bailey ＝Rosa multiflora Thunb. var. cathayensis Rehder et E. H. Wilson ●

336470　Rosa cathayensis（Rehder）L. H. Bailey var. platyphylla ?；宽叶红刺玫；Seven Sisters Rose ●☆

336471　Rosa caudata Baker；尾叶蔷薇（尾萼蔷薇）；Caudate Rose, Whiplash Rose ●

336472　Rosa caudata Baker var. maxima Te T. Yu et T. C. Ku；大花尾叶蔷薇（大花尾萼蔷薇）；Largeflower Caudate Rose ●

336473　Rosa cavaleriei H. Lév. ＝Rosa cymosa Tratt. ●

336474　Rosa centifolia L.；洋蔷薇（百叶蔷薇，西洋蔷薇）；Cabbage Rose, Cabbage-rose, Provevce Rose ●

336475　Rosa centifolia L. 'Cristata'；鸡冠洋蔷薇；Crested Cabbage-rose, Crested Moss-rose ●☆

336476　Rosa centifolia L. 'Muscosa'；毛萼洋蔷薇；Moss Cabbage Rose, Moss Cabbage-rose, Moss Rose ●

336477　Rosa centifolia L. 'Parvifolia'；小叶洋蔷薇；Burgundian Rose, Burgundy Cabbage Rose, Burgundy Cabbage-rose, Parvifolia Rose ●☆

336478　Rosa centifolia L. f. albo-muscosa（J. Willm.）Rehder；白花洋蔷薇；White-moss Cabbage Rose ●

336479　Rosa centifolia L. f. muscosa（Aiton）C. K. Schneid. ＝Rosa centifolia L. 'Muscosa' ●

336480　Rosa centifolia L. var. parvifolia ? ＝Rosa centifolia L. 'Parvifolia' ●☆

336481　Rosa centifolia L. var. pomponia Lindl.；红花洋蔷薇（婆波西洋蔷薇）；Pompon Cabbage Rose, Pompon Cabbage-rose, Red Cabbage Rose, Rose-de-Meuse ●☆

336482　Rosa chaffanjoni H. Lév. et Vaniot ＝Rosa cymosa Tratt. ●

336483　Rosa charbonneaui H. Lév. ＝Rosa longicuspis Bertol. ●

336484　Rosa chengkouensis Te T. Yu et T. C. Ku；城口蔷薇；Chengkou Rose ●

336485　Rosa chinensis Jacq.；月季花（长春花，绸春花，斗雪红，勒泡，胜春，瘦容，四季春，四季花，铜棰子，艳雪红，印度蔷薇，月光花，月贵红，月贵花，月桂花，月记，月季，月季红，月月红，月月花，月月开）；Bengal Rose, China Rose, Chinese Rose, Cyme Rose, Everflowering Rose, Fairy Rose, Monthly Rosebush ●

336486　Rosa chinensis Jacq. 'Viridiflora'；绿月季花（绿花蔷薇，绿花月月红）；Green Rose, Greenflower China Rose, Greenflower Chinese Rose ●

336487　Rosa chinensis Jacq. f. spontanea Rehder et E. H. Wilson ＝Rosa chinensis Jacq. var. spontanea（Rehder et E. H. Wilson）Te T. Yu et T. C. Ku ●

336488　Rosa chinensis Jacq. subsp. indica Koehne；春仔花（绸春花，月

季花）●

336489　Rosa chinensis Jacq. var. fragrans Thory；香月季花●

336490　Rosa chinensis Jacq. var. minima Voss；小月季花（小月季）；Fairy Rose, Fairy Rose Bush, Roulett China Rose, Roulett Rose, Roulette Chinese Rose ●

336491　Rosa chinensis Jacq. var. mutabilis ?；易变月季花；Butterfly Rose, Mutabilis Rose, Tipo Ideale Rose ●☆

336492　Rosa chinensis Jacq. var. pseudindica E. Willm. ＝Rosa odorata（Andréws）Sweet 'Pseudindica' ●

336493　Rosa chinensis Jacq. var. pseudindica E. Willm. ＝Rosa odorata（Andréws）Sweet var. pseudoindica（Lindl.）Rehder ●

336494　Rosa chinensis Jacq. var. pseudoindica（Lindl.）E. Willm. ＝Rosa odorata（Andréws）Sweet var. pseudoindica（Lindl.）Rehder ●

336495　Rosa chinensis Jacq. var. semperflorens Koehne；紫月季花（紫花月月红）；Crimson China Rose, Everblooming China Rose, Everblooming Chinese Rose, Purple Rose ●

336496　Rosa chinensis Jacq. var. spontanea（Rehder et E. H. Wilson）Te T. Yu et T. C. Ku；单瓣月季花（单瓣月月红）；Semperflorens Rose, Slater's Crimson China Rose, Spontaneous China Rose ●

336497　Rosa chinensis Jacq. var. viridiflora Dippel ＝Rosa chinensis Jacq. 'Viridiflora' ●

336498　Rosa cinnamomea L.；肉桂蔷薇（褐刺蔷薇，褐色野蔷薇，樟味蔷薇）；Cinnamon Rose, May-rose ●☆

336499　Rosa clavigera H. Lév. ＝Rosa brunonii Lindl. ●

336500　Rosa collaris Rydb. ＝Rosa acicularis Lindl. subsp. sayi（Schwein.）W. H. Lewis ●

336501　Rosa collina Jacq. ＝Rosa alba L. ●

336502　Rosa conjuncta Rydb. ＝Rosa arkansana Porter var. suffulta（Greene）Cockerell ●☆

336503　Rosa cordotii Masam. ＝Rosa bracteata J. C. Wendl. ●

336504　Rosa coriifolia Fr.；革叶犬蔷薇；Cori-leaf Rose, Leather-leaved Rose ●☆

336505　Rosa corymbifera Borkh. ＝Rosa canina L. ●

336506　Rosa corymbulosa Rolfe；伞房蔷薇（伞花蔷薇）；White-eye Rose ●

336507　Rosa cucumerina Tratt. ＝Rosa laevigata Michx. ●

336508　Rosa cuspidata M. Bieb.；骤尖蔷薇●☆

336509　Rosa cymosa Tratt.；小果蔷薇（白花七叶树，倒钩簕，倒钩蓊，狗屎刺，红茨藤，红根，红荆藤，鸡公子，结茧，荆刺甲，苙刺甲，明目茶，七姐妹，青刺，雀梅，山木香，五甲莲，小刺花，小和尚藤，小金樱，鱼杆子）；Smallfruit Rose, Small-fruited Rose ●

336510　Rosa cymosa Tratt. f. plena Z. X. Yu et G. Z. Liu；重瓣小果蔷薇；Doubleflower Smallfruit Rose ●

336511　Rosa cymosa Tratt. var. puberula Te T. Yu et T. C. Ku；毛叶山木香；Hairy-leaf Smallfruit Rose ●

336512　Rosa dahurica Pall. ＝Rosa davurica Pall. ●

336513　Rosa daishanensis T. C. Ku；岱山蔷薇；Daishan Rose ●

336514　Rosa damascena Mill.；突厥蔷薇（大马士革蔷薇）；Autumn Damask, Damascus Rose, Damask Rose, Monthly Rose, Painted Damask, Painted Damask Rose, Rose of Melaxo ●

336515　Rosa damascena Mill. 'Bifera'；两次花突厥蔷薇；Autumn Damask, Autumn Damask Rose, Four Seasons Rose, Rose of Castile ●☆

336516　Rosa damascena Mill. 'Leda'；丽达突厥蔷薇；Painted Damask ●☆

336517　Rosa damascena Mill. 'Versicolor'；彩色突厥蔷薇；York And Lancaster Rose, York-and-lancaster Rose ●☆

336518　Rosa dasistema Raf. ＝Rosa palustris Marshall ●☆

336519　Rosa davidii Crép.；西北蔷薇（大卫蔷薇，花别刺，金樱子，山

刺玫,山刺玫蔷薇,万朵刺);David Rose ●

336520 Rosa davidii Crép. var. ellipsoidea Nakai;长果刺玫蔷薇;Bigfruit David Rose ●

336521 Rosa davidii Crép. var. elongata Rehder et E. H. Wilson;长果西北蔷薇;Longfruit David Rose ●

336522 Rosa davurica Pall.;山刺玫(刺玫,刺玫果,刺玫蔷薇,刺莓果,达呼尔蔷薇,达乌里蔷薇,红根,蔷薇,蔷薇果,柔毛刺玫蔷薇,野刺玫);Dahurian Rose ●

336523 Rosa davurica Pall. var. alpestris (Nakai) Kitag. = Rosa amblyotis C. A. Mey. ●☆

336524 Rosa davurica Pall. var. glabra Y. H. Liu;光叶山刺玫(光叶刺玫蔷薇);Glabrous Dahurian Rose ●

336525 Rosa davurica Pall. var. glabra Y. H. Liu = Rosa marretii H. Lév. ●

336526 Rosa davurica Pall. var. setacea Liou;多刺山刺玫(多刺刺玫蔷薇);Bristle Dahurian Rose ●

336527 Rosa dawoensis Pax et K. Hoffm.;道浮蔷薇;Dawo Rose ●

336528 Rosa deqenensis T. C. Ku;德钦蔷薇;Deqin Rose ●

336529 Rosa derongensis T. C. Ku;得荣蔷薇;Derong Rose ●

336530 Rosa duclouxii H. Lév. = Rosa odorata (Andréws) Sweet var. gigantea (Crép.) Rehder et E. H. Wilson ●

336531 Rosa dumalis Bechst. = Rosa caesia Sm. ●☆

336532 Rosa dumetorum (Thuill.) Debeaux = Rosa canina L. ●

336533 Rosa dumetorum Thuill. = Rosa canina L. ●

336534 Rosa dumetorum Thuill. = Rosa corymbifera Borkh. ●

336535 Rosa duplicata Te T. Yu et T. C. Ku;重齿蔷薇;Doubletoothed Rose, Double-toothed Rose, Duplicate-tooth Rose ●

336536 Rosa ecae Aiton;蕨叶蔷薇●☆

336537 Rosa ecae Aiton = Rosa primula Boulenger ●

336538 Rosa eglanteria L.;香甜蔷薇;Eglantine, Eglantine Rose, Sweetbrier, Sweetbrier Rose ●☆

336539 Rosa eglanteria L. = Rosa rubiginosa L. ●☆

336540 Rosa eglanteria L. = Rosa xanthina Lindl. f. normalis Rehder et E. H. Wilson ●

336541 Rosa elasmacantha Trautv.;薄刺蔷薇●☆

336542 Rosa elegantula Rolfe;秀丽蔷薇(小蔷薇);Fairy Rose, Threepenny-bit Rose ●

336543 Rosa elegantula Rolfe 'Persetosa';多刚毛小蔷薇●☆

336544 Rosa elegantula Rolfe = Rosa persetosa Rolfe ●

336545 Rosa elliptica Tausch ex Tratt. = Rosa inodora Hook. ●☆

336546 Rosa elymaitica Boiss. et Hausskn.;埃利迈特蔷薇●☆

336547 Rosa engelmannii S. Watson = Rosa acicularis Lindl. subsp. sayi (Schwein.) W. H. Lewis ●

336548 Rosa ernestii Stapf ex Bean = Rosa rubus H. Lév. et Vaniot ●

336549 Rosa esquirolii H. Lév. et Vaniot = Rosa cymosa Tratt. ●

336550 Rosa fargesiana Boulenger;川东蔷薇(光梗蔷薇)●

336551 Rosa farreri Stapf ex Cox;刺毛蔷薇;Farrer Rose, Threepenny-bit Rose ●

336552 Rosa farreri Stapf ex Cox f. persetosa Stapf = Rosa farreri Stapf ex Cox ●

336553 Rosa fauriei H. Lév. = Rosa acicularis Lindl. ●

336554 Rosa fedtschenkoana Regel;腺果蔷薇(腺毛蔷薇);Fedtschenk Rose, Fedtschenko Rose ●

336555 Rosa ferox Aiton = Rosa rugosa Thunb. ●◇

336556 Rosa ferox Lawrance = Rosa rugosa Thunb. ●◇

336557 Rosa ferruginea Vill.;锈红蔷薇;Redleaf Rose ●☆

336558 Rosa filipes Rehder et E. H. Wilson;腺梗蔷薇(白桂花);

Glandstalk Rose, Glandular-stalked Rose, Threadstalk Rose ●

336559 Rosa fimbriata Gremli;流苏蔷薇;Casnation Rose ●☆

336560 Rosa floribunda Baker = Rosa helenae Rehder et E. H. Wilson ●

336561 Rosa floribunda Steven ex Besser = Rosa henryi Boulenger ●

336562 Rosa floridana Rydb. = Rosa palustris Marshall ●☆

336563 Rosa foetida Herrm.;异味蔷薇(奥太利蔷薇,臭蔷薇,西亚蔷薇);Austria Rose, Austrian Briar, Austrian Briar Rose, Austrian Brier Rose, Austrian Bush Briar, Austrian Copper Briar, Austrian Rose, Austrian Yellow Rose, Capucine Rose, Levant Rose, Persian Briar ●

336564 Rosa foetida Herrm. 'Austrian Copper Briar' = Rosa foetida Herrm. 'Bicolor' ●☆

336565 Rosa foetida Herrm. 'Bicolor';二色异味蔷薇;Austrian Copper Briar, Austrian Copper Rose ●☆

336566 Rosa foetida Herrm. 'Persian Yellow' = Rosa foetida Herrm. 'Persiana' ●☆

336567 Rosa foetida Herrm. 'Persiana';波斯黄异味蔷薇;Persian Yellow Rose ●☆

336568 Rosa foetida Herrm. f. persiana (Lem.) Rehder;重瓣异味蔷薇;Persia Austrian Rose, Persian Yellow Rose ●

336569 Rosa foetida Herrm. var. bicolor E. Willm. = Rosa foetida Herrm. 'Bicolor' ●☆

336570 Rosa foliosa Leman;狭叶蔷薇●☆

336571 Rosa fontanesii Pomel = Rosa pouzinii Tratt. ●☆

336572 Rosa formosana (Cardot) Koidz. = Rosa multiflora Thunb. ●

336573 Rosa forrestiana Boulenger;滇边蔷薇(和氏蔷薇);Forrest Rose ●

336574 Rosa forrestiana Boulenger f. glanddulosa T. C. Ku;小叶滇边蔷薇;Glandulose Forrest Rose ●

336575 Rosa forrestii Focke = Rosa roxburghii Tratt. f. normalis Rehder et E. H. Wilson ●

336576 Rosa forrestii Focke = Rosa roxburghii Tratt. ●

336577 Rosa fortuneana Lem.;大花白木香;Fortune Rose ●

336578 Rosa fujisanensis (Makino) Makino;富士山蔷薇●☆

336579 Rosa fujisanensis Makino = Rosa fujisanensis (Makino) Makino ●☆

336580 Rosa fukienensis F. P. Metcalf;福建蔷薇●

336581 Rosa gallica L.;法国蔷薇;Apothecaries' Rose, Apotheearies' Rose, Damask Rose, French Rose, Provins Rose, Red Rose, Rose of Provins ●

336582 Rosa gallica L. 'Apothecary's Rose';大花法国蔷薇●☆

336583 Rosa gallica L. 'Rosa Mundi' = Rosa gallica L. 'Versicolor' ●☆

336584 Rosa gallica L. 'Rose of Provins' = Rosa gallica L. 'Apothecary's Rose' ●☆

336585 Rosa gallica L. 'Versicolor';多色法国蔷薇;Mundi Rose, Rosa Mundi, Versicolor Rose ●☆

336586 Rosa gallica L. duplex;二重法国蔷薇●☆

336587 Rosa gallica L. f. pseudolivescens (H. Br.) Vukic. = Rosa gallica L. 'Apothecary's Rose' ●☆

336588 Rosa gallica L. florepleno ?;重瓣法国蔷薇●☆

336589 Rosa gallica L. parvifolia ?;小叶法国蔷薇●☆

336590 Rosa gallica L. var. centifolla Regel = Rosa centifolia L. ●

336591 Rosa gallica L. var. damascena Voss = Rosa damascena Mill. ●

336592 Rosa gallica L. var. eriostyla Keller = Rosa gallica L. ●

336593 Rosa gallica L. var. officinalis Thory;药用蔷薇;Apothecary's Rose, Red Rose-of-lancaster ●

336594 Rosa gallica L. var. officinalis Thory = Rosa gallica L. ●

336595 Rosa gallica L. var. pumila (Jacq.) Regel;矮法国蔷薇;Dwarf French Rose ●☆

336596 Rosa gebleriana Schrenk = Rosa laxa Retz. ●

336597 Rosa gechouitangensis H. Lév. = Rosa odorata (Andréws) Sweet ●◇

336598 Rosa gentiliana H. Lév. et Vaniot = Rosa henryi Boulenger ●

336599 Rosa gentiliana H. Lév. et Vaniot = Rosa multiflora Thunb. var. cathayensis Rehder et E. H. Wilson ●

336600 Rosa gentiliana H. Lév. et Vaniot f. puberula Hand. -Mazz. = Rosa rubus H. Lév. et Vaniot ●

336601 Rosa gentiliana Rehder et E. H. Wilson = Rosa henryi Boulenger ●

336602 Rosa gigantea Collett ex Crép. ;巨大蔷薇●☆

336603 Rosa gigantea Collett ex Crép. = Rosa odorata (Andréws) Sweet var. gigantea (Crép.) Rehder et E. H. Wilson ●

336604 Rosa gigantea Collett ex Crép. = Rosa odorata Sweet ●◇

336605 Rosa gigantea Collett ex Crép. f. erubescens Focke = Rosa odorata (Andréws) Sweet var. erubescens (Focke) Te T. Yu et T. C. Ku ●

336606 Rosa gigantea Collett ex Crépin = Rosa odorata (Andréws) Sweet var. gigantea (Crép.) Rehder et E. H. Wilson ●

336607 Rosa giraldii Crép. ;陕西蔷薇；Girald Rose，Red Rose ●

336608 Rosa giraldii Crép. var. bidendata Te T. Yu et T. C. Ku;重齿陕西蔷薇；Doubletoothed Girald Rose ●

336609 Rosa giraldii Crép. var. venulosa Rehder et E. H. Wilson;毛叶陕西蔷薇；Hairyleaf Girald Rose ●

336610 Rosa glabrifolia C. A. Mey. ex Rupr. ;光叶蔷薇●☆

336611 Rosa glauca Pourr. ;红叶蔷薇；Redleaf Rose，Red-leaf Rose，Red-leaved Rose ●☆

336612 Rosa glomerata Rehder et E. H. Wilson;绣球蔷薇；Glustered Rose ●

336613 Rosa glutinosa Sm. ;黏性蔷薇●☆

336614 Rosa gmelinii Bunge = Rosa acicularis Lindl. ●

336615 Rosa graciliflora Rehder et E. H. Wilson;细梗蔷薇（刺栗子，野人头）；Daintypetal Rose，Slender-flower Rose，Slender-flowered Rose ●

336616 Rosa graciliflora Rehder et E. H. Wilson = Rosa farreri Stapf ex Cox ●

336617 Rosa granulosa Keller = Rosa acicularis Lindl. ●

336618 Rosa haemisphaerica Herrm. ;半球蔷薇●☆

336619 Rosa hakonensis (Franch. et Sav.) Koidz. = Rosa onoei Makino var. hakonensis (Franch. et Sav.) H. Ohba ●☆

336620 Rosa helenae Rehder et E. H. Wilson;卵果蔷薇（巴东蔷薇，牛黄树刺，野牯牛刺）；Helen Rose，Ovate-fruit Rose ●

336621 Rosa helenae Rehder et E. H. Wilson f. duplicata T. C. Ku;重齿卵果蔷薇；Duplicate Ovate-fruit Rose ●

336622 Rosa helenae Rehder et E. H. Wilson f. glandulifera T. C. Ku;腺叶卵果蔷薇；Glandulate Ovate-fruit Rose ●

336623 Rosa hemisphaerica Herrm. ;淡黄月季；Sulphur Rose，Yellow Provence Rose ●☆

336624 Rosa hemsleyana Tackh. = Rosa setipoda Hemsl. et E. H. Wilson ●

336625 Rosa henryi Boulenger;软条七蔷薇（亨氏蔷薇，湖北蔷薇，小金樱，秀蔷薇）；Henry Rose ●

336626 Rosa henryi Boulenger var. glandulosa Ze M. Wu et Z. L. Cheng;密腺湖北蔷薇；Glandulose Henry Rose ●

336627 Rosa henryi Boulenger var. glandulosa Ze M. Wu et Z. L. Cheng = Rosa henryi Boulenger ●

336628 Rosa henryi Boulenger var. puberula (Hand. -Mazz.) F. P. Metcalf = Rosa rubus H. Lév. et Vaniot ●

336629 Rosa hezhangensis T. L. Xu;赫章蔷薇；Hezhang Rose ●☆

336630 Rosa hirtissima Lonacz. ;多毛蔷薇●☆

336631 Rosa hirtula (Regel) Nakai;短毛蔷薇●☆

336632 Rosa hispida Sims;硬毛蔷薇；Hispid Rose ●☆

336633 Rosa hissarica Sloboda;希萨尔蔷薇●☆

336634 Rosa horrida Fisch. = Rosa rugosa Thunb. ●◇

336635 Rosa housei Erlanson;豪斯蔷薇；House's Rose ●☆

336636 Rosa hugonis Hemsl. ;黄蔷薇（大马茄子，红眼刺，鸡蛋黄花）；China Golden Rose，Father Hugo Rose，Father Hugo's Rose，Golden Rose，Golden Rose of China，Hugo Rose，Hugonis Rose ●

336637 Rosa humilis Besser;牧场；Pasture Rose，Prairie Rose ●☆

336638 Rosa hwangata Michx. ;黄山蔷薇；Huangshan Rose ●

336639 Rosa hwangshanensis P. S. Hsu = Rosa sertata Rolfe ●

336640 Rosa hybrida Hort. ;现代蔷薇(现代月季)；Modern Rose ●

336641 Rosa hybrida Hort. 'Floribunda';丰花月季；Floribunda Rose ●

336642 Rosa iberica Stev. ;伊比利亚蔷薇●☆

336643 Rosa iliensis Chrshan. ;中亚蔷薇●☆

336644 Rosa illinoensis Baker = Rosa pimpinellifolia L. ●

336645 Rosa indica L. = Rosa chinensis Jacq. ●

336646 Rosa indica L. var. fragrans Thory = Rosa odorata (Andréws) Sweet ●◇

336647 Rosa indica L. var. odorata Andréws = Rosa odorata (Andréws) Sweet ●◇

336648 Rosa indica L. var. vulgaris ? = Rosa chinensis Jacq. ●

336649 Rosa inermis Bertol. ;无刺蔷薇；Spineless Rose ●☆

336650 Rosa inodora Hook. ;椭圆蔷薇（田蔷薇，爪瓣蔷薇）；Elliptic Rose，Grassland Rose，Narrow-leaved Sweet Briar，Scentless Ruse，Small-leaved Sweet-briar ●☆

336651 Rosa iochanensis H. Lév. = Rosa sertata Rolfe ●

336652 Rosa iwara Siebold ex Regel;虾夷蔷薇●☆

336653 Rosa jacutica Juz. ;屋久岛蔷薇(雅库蔷薇)；Jacu Rose ●☆

336654 Rosa jasminoides Koidz. = Rosa onoei Makino var. hakonensis (Franch. et Sav.) H. Ohba ●☆

336655 Rosa jundzillii Besser = Rosa marginata Wallr. ●☆

336656 Rosa kamtschatica Vent. ;勘察加蔷薇●☆

336657 Rosa kanzanensis Masam. = Rosa pricei Hayata ●

336658 Rosa klukii Besser;克陆氏蔷薇；Kluk Rose ●☆

336659 Rosa kokanica (Regel) Regel ex Juz. ;腺叶蔷薇；Glandleaf Rose，Glandular Rose，Glandularleaf Rose，Kokand Rose ●

336660 Rosa koreana Kom. ;长白蔷薇；Korea Rose，Korean Rose ●

336661 Rosa koreana Kom. var. glandulosa Te T. Yu et T. C. Ku;腺叶长白蔷薇；Glandular Korean Rose ●

336662 Rosa korsakoviensis H. Lév. = Rosa acicularis Lindl. ●

336663 Rosa korshinskiana Boulenger;柯辛氏蔷薇；Kosinsk's Rose ●☆

336664 Rosa kuhitangi Nevski ;库希塘蔷薇●☆

336665 Rosa kunmingensis T. C. Ku;昆明蔷薇；Kunming Rose ●

336666 Rosa kwangsiensis H. L. Li = Rosa multiflora Thunb. var. cathayensis Rehder et E. H. Wilson ●

336667 Rosa kwangtungensis Te T. Yu et H. T. Tsai;广东蔷薇（野蔷薇）；Guangdong Rose，Kwangtung Rose ●

336668 Rosa kwangtungensis Te T. Yu et H. T. Tsai f. roseoliflora Y. B. Chang et Y. L. Xu;粉花广东蔷薇●

336669 Rosa kwangtungensis Te T. Yu et H. T. Tsai var. mollis F. P. Metcalf;毛叶广东蔷薇（柔毛广东蔷薇）；Hairyleaf Guangdong Rose ●

336670 Rosa kwangtungensis Te T. Yu et H. T. Tsai var. plena Te T. Yu et T. C. Ku;重瓣广东蔷薇；Doubleflower Guangdong Rose ●

336671 Rosa kweichowensis Te T. Yu et T. C. Ku;贵州蔷薇(贵州缫丝花)；Guizhou Rose ●

336672 Rosa lacerans Boiss. et Buhse;撕裂蔷薇●☆

336673　Rosa laevigata F. Michx. rosea Makino et Nemoto;紫红金樱子●☆

336674　Rosa laevigata F. Michx. semiplena Te T. Yu et T. C. Ku;重瓣金樱子;Doubleflower Cherokee Rose ●

336675　Rosa laevigata Michx.;金樱子(白玉带子,槟榔果,刺橄榄,刺郎子,刺梨子,刺藤棘子,刺头,刺榆子,刺子,菊兰棵子,大金英,大金樱,倒挂金钩,灯笼果,蜂糖罐,和尚头,黄茶瓶,黄刺果,鸡母卵子,金茶瓶,金刺角树,金壶瓶,金因子,金英,金婴子,金罂子,金樱果,金樱子蔷薇,三叶筋子,山鸡头子,山石榴,棠球,塘莺蔃,糖钵,糖刺果,糖罐,糖罐头,糖罐子,糖果,糖橘子,糖球,糖莺子,糖樱筋,螳螂果,藤勾子,脱骨丹,下山虎子,小石榴,野石榴,油饼果子,月月红花);Cherokee Rose ●

336676　Rosa laevigata Michx. 'Anemone';银莲花蔷薇;Pink Cherokee Rose ●☆

336677　Rosa laevigata Michx. var. kaiscianensis Pamp. = Rosa laevigata Michx. ●

336678　Rosa laevigata Michx. var. leiocarpa Y. Q. Wang et P. Y. Chen;光果金樱子;Smoothfruit Cherokee Rose ●

336679　Rosa laevigata Michx. var. leiocarpa Y. Q. Wang et P. Y. Chen = Rosa laevigata Michx. ●

336680　Rosa lancifolia Small = Rosa palustris Marshall ●☆

336681　Rosa langyashanica D. C. Zhang et J. Z. Shao;琅琊山蔷薇(琅琊蔷薇);Langyashan Rose ●

336682　Rosa lasiosepala F. P. Metcalf;毛萼蔷薇;Hairy-calyx Rose,Hairysepal Rose ●

336683　Rosa latibracteata Boulenger = Rosa multibracteata Hemsl. et E. H. Wilson ●

336684　Rosa laurrenceana ? = Rosa chinensis Jacq. var. minima Voss ●

336685　Rosa laxa Retz.;疏花蔷薇(野蔷薇);Laxiflowered Rose,Looseflower Rose,Turkestan Rose ●

336686　Rosa laxa Retz. var. kaschgarica (Rupr.) Y. L. Han;喀什疏花蔷薇●

336687　Rosa laxa Retz. var. mollis Te T. Yu et T. C. Ku;毛叶疏花蔷薇;Hairyleaf Looseflower Rose ●

336688　Rosa lebrunei H. Lév. = Rosa multiflora Thunb. var. carnea Thory ●

336689　Rosa leucantha M. Bieb.;白花蔷薇●☆

336690　Rosa lichiangensis Te T. Yu et T. C. Ku;丽江蔷薇;Lijiang Rose ●

336691　Rosa lioui Te T. Yu et H. T. Tsai = Rosa beggeriana Schrenk var. liouii (Te T. Yu et H. T. Tsai) Te T. Yu et T. C. Ku ●

336692　Rosa longicuspis Bertol.;长尖叶蔷薇(长光叶蔷薇,长叶尖蔷薇,粉棠果,粉糖果);Longtooth Rose,Long-toothed Rose,Rose ●

336693　Rosa longicuspis Bertol. var. sinowilsonii (Hemsl.) Te T. Yu et T. C. Ku;多花长尖叶蔷薇(多花长光叶蔷薇,多花长叶尖蔷薇);Sino-wilson Longtooth Rose ●

336694　Rosa lucens Paul et Son = Rosa longicuspis Bertol. ●

336695　Rosa lucens Rolfe = Rosa longicuspis Bertol. ●

336696　Rosa luciae Franch. et Rochebr. = Rosa luciae Franch. et Rochebr. ex Crép. ●

336697　Rosa luciae Franch. et Rochebr. ex Crép.;亮叶蔷薇●

336698　Rosa luciae Franch. et Rochebr. ex Crép. = Rosa fujisanensis (Makino) Makino ●☆

336699　Rosa luciae Franch. et Rochebr. ex Crép. = Rosa wichuraiana Crép. ex Desegl. ●

336700　Rosa luciae Franch. et Rochebr. ex Crép. f. glandulifera (Koidz.) H. Ohba;腺点光叶蔷薇●☆

336701　Rosa luciae Franch. et Rochebr. ex Crép. f. ohwii (Oishi) Yonek.;大井氏蔷薇●☆

336702　Rosa luciae Franch. et Rochebr. ex Crép. f. rosea Hayashi = Rosa onoei Makino var. oligantha (Franch. et Sav.) H. Ohba f. rosea (Hayashi) Yonek. ●☆

336703　Rosa luciae Franch. et Rochebr. ex Crép. subsp. hakonensis (Franch. et Sav.) Kitam. = Rosa onoei Makino var. hakonensis (Franch. et Sav.) H. Ohba ●☆

336704　Rosa luciae Franch. et Rochebr. ex Crép. subsp. onoei (Makino) Kitam. = Rosa onoei Makino ●☆

336705　Rosa luciae Franch. et Rochebr. ex Crép. var. formosana Cardot;台湾滨蔷薇;Formosan Seashore Rose ●

336706　Rosa luciae Franch. et Rochebr. ex Crép. var. fujisanensis Makino = Rosa fujisanensis (Makino) Makino ●☆

336707　Rosa luciae Franch. et Rochebr. ex Crép. var. onoei (Makino) Momiy. ex Ohwi = Rosa onoei Makino ●☆

336708　Rosa luciae Franch. et Rochebr. ex Crép. var. paniculigera (Koidz.) Momiy. ex Ohwi = Rosa paniculigera (Koidz.) Makino ex Momiy. ●☆

336709　Rosa luciae Franch. et Rochebr. ex Crép. var. rosea H. L. Li;粉花光叶蔷薇(台湾光叶蔷薇) ●

336710　Rosa luciae Franch. et Rochebr. ex Crép. var. wichurana (Crépin) Koidz. = Rosa luciae Franch. et Rochebr. ex Crép. ●

336711　Rosa luciae Pranch. et Rochebr. = Rosa luciae Franch. et Rochebr. ex Crép. ●

336712　Rosa lucida Ehrh. = Rosa bracteata J. C. Wendl. ●

336713　Rosa lucida Ehrh. = Rosa virginiana Mill. ●☆

336714　Rosa lucidissima H. Lév.;亮叶月季;Lucid-leaved Rose,Shiningleaf Rose,Shinyleaf Rose ●

336715　Rosa lucidissima H. Lév. f. setosa Cardot = Rosa lucidissima H. Lév. ●

336716　Rosa ludingensis T. C. Ku;泸定蔷薇;Luding Rose ●

336717　Rosa lunellii Greene = Rosa arkansana Porter ●☆

336718　Rosa lutea Brot.;黄色蔷薇;Yellow Rose ●☆

336719　Rosa lutea Mill. = Rosa foetida Herrm. ●

336720　Rosa lutea Mill. var. persiana Lem. = Rosa foetida Herrm. f. persiana (Lem.) Rehder ●

336721　Rosa luzoniensis Merr. = Rosa transmorrisonensis Hayata ●

336722　Rosa macartnea Dumont = Rosa bracteata J. C. Wendl. ●

336723　Rosa macrocarpa Watt ex Crépin = Rosa odorata (Andréws) Sweet var. gigantea (Crép.) Rehder et E. H. Wilson ●

336724　Rosa macrocarpa Watt = Rosa odorata (Andréws) Sweet var. gigantea (Crép.) Rehder et E. H. Wilson ●

336725　Rosa macrophylla Lindl.;大叶蔷薇(大叶月季);Bigleaf Rose,Big-leaved Rose,Largeleaf Rose ●

336726　Rosa macrophylla Lindl. = Rosa setipoda Hemsl. et E. H. Wilson ●

336727　Rosa macrophylla Lindl. var. crasseaculeata Vilm. = Rosa setipoda Hemsl. et E. H. Wilson ●

336728　Rosa macrophylla Lindl. var. glandulifera Te T. Yu et T. C. Ku;腺果大叶蔷薇;Glandular Largeleaf Rose ●

336729　Rosa macrophylla Lindl. var. hypoleuca H. Lév. = Rosa multiflora Thunb. var. cathayensis Rehder et E. H. Wilson ●

336730　Rosa maikwai H. Hara;重瓣紫玫瑰(紫花重瓣玫瑰);Double Rugose Rose ●

336731　Rosa mairei H. Lév.;毛叶蔷薇(昭通山不榴);Hairyleaf Rose,Maire Rose ●

336732　Rosa mairei Sennen = Rosa micrantha Borrer ex Sm. ●☆

336733　Rosa majalis Herrm.;月桂蔷薇(五月蔷薇);Cinnamon Rose,Double Cinnamon Rose,May Rose ●☆

336734　Rosa maracandica Bunge；马拉坎达蔷薇●☆

336735　Rosa marginata Wallr.；朱氏蔷薇；Jundzill's Rose，Margined Rose ●☆

336736　Rosa marretii H. Lév.；深山蔷薇（多刺刺玫蔷薇，光叶刺玫蔷薇，库页蔷薇，马氏蔷薇，玛雷迪蔷薇）●☆

336737　Rosa marretii H. Lév. = Rosa amblyotis C. A. Mey. ●☆

336738　Rosa maximowicziana Regel；伞花蔷薇（刺玫果，钩脚藤子，蔓野蔷薇，酸溜溜，牙门太）；Maximowicz Rose ●

336739　Rosa maximowicziana Regel f. adenocalyx Nakai；腺萼伞花蔷薇；Glandcalyx Maximowicz Rose ●

336740　Rosa michiganensis Erlanson = Rosa palustriformis Rydb. ●☆

336741　Rosa micrantha Borrer ex Sm.；小花蔷薇；Eglantine，Littleflower Rose，Small-flower Rose，Smallflower Sweetbrier，Small-flowered Sweetbrier，Smallleaf Sweet Briar，Small-leaf Sweet Briar，Sweet Briar，Sweetbrier ●☆

336742　Rosa micrantha Sm. = Rosa micrantha Borrer ex Sm. ●☆

336743　Rosa micrantha Sm. var. atlantica Ball = Rosa micrantha Borrer ex Sm. ●☆

336744　Rosa micrantha Sm. var. avelloi Sennen et Mauricio = Rosa micrantha Borrer ex Sm. ●☆

336745　Rosa micrantha Sm. var. diminuta（Boreau）= Rosa micrantha Borrer ex Sm. ●☆

336746　Rosa micrantha Sm. var. heterostyla Keller = Rosa micrantha Borrer ex Sm. ●☆

336747　Rosa micrantha Sm. var. mairei（Sennen）Maire = Rosa micrantha Borrer ex Sm. ●☆

336748　Rosa micrantha Sm. var. normalis Rouy = Rosa micrantha Borrer ex Sm. ●☆

336749　Rosa micrantha Sm. var. ovata Rouy = Rosa micrantha Borrer ex Sm. ●☆

336750　Rosa micrantha Sm. var. perglandulosa Keller = Rosa micrantha Borrer ex Sm. ●☆

336751　Rosa micrantha Sm. var. pseudopommaretii Rouy = Rosa micrantha Borrer ex Sm. ●☆

336752　Rosa micrantha Sm. var. septicola Gren. = Rosa micrantha Borrer ex Sm. ●☆

336753　Rosa micrantha Sm. var. septicoloides（Crép.）Rouy = Rosa micrantha Borrer ex Sm. ●☆

336754　Rosa micrantha Sm. var. trichostyla Keller = Rosa micrantha Borrer ex Sm. ●☆

336755　Rosa microcarpa Lindl. = Rosa cymosa Tratt. ●

336756　Rosa microonoei Nakai = Rosa onoei Makino ●☆

336757　Rosa microphylla Desf. var. glabra Regel = Rosa roxburghii Tratt. ●

336758　Rosa microphylla Roxb. = Platyrhodon microphyllum（Roxb.）Hurst ●

336759　Rosa microphylla Roxb. = Rosa roxburghii Tratt. ●

336760　Rosa miyiensis T. C. Ku；米易蔷薇；Miyi Rose ●

336761　Rosa mohavensis Parish；莫哈维蔷薇；Mohave Rose ●☆

336762　Rosa mollis Sm.；软蔷薇；Bloomy-stem Rose，Soft Downy-rose，Soft Rose ●☆

336763　Rosa montana Vill.；山地蔷薇●☆

336764　Rosa morrisonensis Hayata；玉山蔷薇●

336765　Rosa morrisonensis Hayata = Rosa sericea Lindl. var. morrisonensis（Hayata）Masam. ●

336766　Rosa moschata Herrm.；麝香蔷薇；Musk Rose ●☆

336767　Rosa moschata Herrm. = Rosa brunonii Lindl. ●

336768　Rosa moschata Herrm. var. densa Vilm. = Rosa henryi Boulenger ●

336769　Rosa moschata Herrm. var. hupehensis Pampanin = Rosa rubus H. Lév. et Vaniot ●

336770　Rosa moschata Herrm. var. yunnanensis Crépin = Rosa longicuspis Bertol. ●

336771　Rosa moschata Mill. = Rosa brunonii Lindl. ●

336772　Rosa moschata Mill. var. densa Vilmorim = Rosa henryi Boulenger ●

336773　Rosa moschata Mill. var. hupehensis Pamp. = Rosa rubus H. Lév. et Vaniot ●

336774　Rosa moschata Mill. var. nepalensis Lindl. = Rosa brunonii Lindl. ●

336775　Rosa moschata Mill. var. yunnanensis Crép. = Rosa longicuspis Bertol. ●

336776　Rosa moschata Mill. var. yunnanensis Focke = Rosa soulieana Crép. var. yunnanensis C. K. Schneid. ●

336777　Rosa moyesii Hemsl. et E. H. Wilson；华西蔷薇（红花蔷薇，穆氏蔷薇）；Moyes Rose，W. China Rose ●

336778　Rosa moyesii Hemsl. et E. H. Wilson 'Eddie's Jewel'；艾迪的钻石华西蔷薇●☆

336779　Rosa moyesii Hemsl. et E. H. Wilson 'Geranium'；鲜红华西蔷薇●☆

336780　Rosa moyesii Hemsl. et E. H. Wilson var. pubescens Te T. Yu et H. T. Tsai；毛叶华西蔷薇（华西蔷薇）；Hairyleaf Moyes Rose ●

336781　Rosa multibracteata Hemsl. et E. H. Wilson；多苞蔷薇；Manybract Rose，Manybracted Rose，Multi-bracted Rose ●

336782　Rosa multiflora Thunb. = Rosa multiflora Thunb. ex Murray ●

336783　Rosa multiflora Thunb. ex Murray；多花蔷薇（白残花，白玉棠，柴米米花，刺红花，刺花，刺梅花，倒钩刺花，佛见笑，和尚头花，荷花蔷薇，棘子花，牛棘花，牛勒花，七姐妹，七星梅，七姊妹花，墙麻花，墙薇花，蔷靡花，蔷薇，蔷薇刺花，山棘花，山枣花，石珊瑚，台湾野蔷薇，小果蔷薇，小金英，小金樱，药王子，野蔷薇，营实，营实花，营实蘼蘼，玉鸡苗花，枣蘼花）；Baby Rose，Japan Rose，Japanese Rose，Manyflowered Rose Many-flowered Rose，Multiflora Rose，Polyantha Rose，Rambler Rose，Wild Rose ●

336784　Rosa multiflora Thunb. ex Murray var. cathayensis Rehder et E. H. Wilson = Rosa multiflora Thunb. ex Murray ●

336785　Rosa multiflora Thunb. f. rosipetala（Honda）Yonek.；粉瓣多花蔷薇●☆

336786　Rosa multiflora Thunb. var. adenochaeta（Koidz.）Ohwi；腺毛多花蔷薇●☆

336787　Rosa multiflora Thunb. var. albo-plena Te T. Yu et T. C. Ku；白玉堂；White Doubleflowered Manyflowered Rose ●

336788　Rosa multiflora Thunb. var. brachacantha（Focke）Rehder et E. H. Wilson；红刺藤（短刺野蔷薇）●

336789　Rosa multiflora Thunb. var. carnea Thory；七姊妹（荷花蔷薇，阔叶十姊妹，十姊妹，姊妹花）；Flesh-coloured Manyflowered Rose，Seven Sisters Rose ●

336790　Rosa multiflora Thunb. var. carnea Thory f. cathayensis（Rehder et E. H. Wilson）Kitam. = Rosa multiflora Thunb. var. adenochaeta（Koidz.）Ohwi ☆

336791　Rosa multiflora Thunb. var. carnea Thory f. platyphylla（Thory）Rehder et E. H. Wilson；阔叶七姊妹●☆

336792　Rosa multiflora Thunb. var. cathayensis Rehder et E. H. Wilson；红刺玫（白残花，粉团蔷薇，牛刺，蔷薇，野蔷薇，营实蘼蘼）；Cathaya Japan Rose，Cathaya Japanese Rose，Cathaya Manyflowered Rose ●

336793　Rosa multiflora Thunb. var. formosana Cardot；台湾蔷薇（台湾

野蔷薇）；Formosan Rose ●

336794　Rosa multiflora Thunb. var. gentiliana (H. Lév. et E. H. Wilson) Te T. Yu et H. T. Tsai = Rosa multiflora Thunb. var. cathayensis Rehder et E. H. Wilson ●

336795　Rosa multiflora Thunb. var. nanningensis Y. Wan et Z. R. Huang；南宁蔷薇；Nanning Rose ●

336796　Rosa multiflora Thunb. var. nanningensis Y. Wan et Z. R. Huang = Rosa kwangtungensis Te T. Yu et H. T. Tsai var. mollis F. P. Metcalf ●

336797　Rosa multiflora Thunb. var. platyphylla ? = Rosa multiflora Thunb. var. carnea Thory ●

336798　Rosa multiflora Thunb. var. taoyuanensis Z. M. Wu；桃源蔷薇；Taoyuan Manyflowered Rose ●

336799　Rosa multiflora Thunb. var. villosula Metcalf；毛野蔷薇 ●

336800　Rosa murielae Rehder et E. H. Wilson；西南蔷薇(缪雷蔷薇)；Muriel Rose, Southwestern Rose ●

336801　Rosa muscosa Mill. = Rosa centifolia L. 'Muscosa' ●☆

336802　Rosa myriacantha DC. ；多刺蔷薇；Densehairy Rose ●☆

336803　Rosa nankinensis Lour. = Rosa chinensis Jacq. ●

336804　Rosa nanothamnus Boulenger；矮蔷薇 ●

336805　Rosa nipponensis Crép. ；日本小花蔷薇 ●☆

336806　Rosa nipponensis Crép. = Rosa acicularis Lindl. var. nipponensis (Crép.) Koehne ●☆

336807　Rosa nitida Willd. ；亮蔷薇；Nitida Rose, Shining Rose ●☆

336808　Rosa nivea DC. = Rosa laevigata Michx. ●

336809　Rosa noisettiana Thory；繁花蔷薇；Noisette Rose ●☆

336810　Rosa nutkana C. Presl；努特卡蔷薇；Nootka Rose, Nutka Rose ●☆

336811　Rosa nutkana C. Presl 'Plena'；重瓣努特卡蔷薇 ●☆

336812　Rosa obtusifolia Desv. = Rosa canina L. ●

336813　Rosa odorata (Andréws) Sweet；香水月季(芳香月季)；Tea Rose ●◇

336814　Rosa odorata (Andréws) Sweet 'Pseudindica'；橙黄香水月季(橘黄香水月季)；False-Indian Tea Rose, Fortune's Double Yellow Tea Rose, Fortune's Double Yellow, Gold Rose, Orange Tea Rose ●

336815　Rosa odorata (Andréws) Sweet f. erubescens (Focke) Rehder et E. H. Wilson = Rosa odorata (Andréws) Sweet var. erubescens (Focke) Te T. Yu et T. C. Ku ●

336816　Rosa odorata (Andréws) Sweet var. erubescens (Focke) Te T. Yu et T. C. Ku；粉红香水月季(红花香水月季,紫花香水月季)；Blush Tea Rose, Red Tea Rose ●

336817　Rosa odorata (Andréws) Sweet var. gigantea (Crép.) Rehder et E. H. Wilson；大花香水月季(打破碗,大花香蔷薇)；Giant Tea Rose ●

336818　Rosa odorata (Andréws) Sweet var. gigantea (Crép.) Rehder et E. H. Wilson f. erubescens (Focke) Rehder et E. H. Wilson = Rosa odorata (Andréws) Sweet var. erubescens (Focke) Te T. Yu et T. C. Ku ●

336819　Rosa odorata (Andréws) Sweet var. ochroleuca Rehder；黄花香水月季；Amber Tea Rose ●

336820　Rosa odorata (Andréws) Sweet var. pseudoindica (Lindl.) Rehder = Rosa odorata (Andréws) Sweet 'Pseudindica' ●

336821　Rosa odorata Sweet = Rosa odorata (Andréws) Sweet ●◇

336822　Rosa odoratissima Sweet ex Lindl. = Rosa odorata (Andréws) Sweet ●◇

336823　Rosa oligantha (Franch. et Sav.) Koidz. = Rosa onoei Makino var. oligantha (Franch. et Sav.) H. Ohba ●☆

336824　Rosa oligantha (Franch. et Sav.) Koidz. var. fujisanensis (Makino) Koidz. = Rosa fujisanensis (Makino) Makino ●☆

336825　Rosa omeiensis Rolfe；峨眉蔷薇(刺石榴,山石榴)；Emei Rose, Omei Mountain Rose, Omei Rose ●

336826　Rosa omeiensis Rolfe f. glandulosa Te T. Yu et T. C. Ku；腺叶峨眉蔷薇；Glandular Emei Rose ●

336827　Rosa omeiensis Rolfe f. paucijuga Te T. Yu et T. C. Ku；少对峨眉蔷薇；Fewpair Emei Rose ●

336828　Rosa omeiensis Rolfe f. pteracantha Rehder et E. H. Wilson；扁刺峨眉蔷薇(翅刺峨眉蔷薇)；Winged Spine Emei Rose ●

336829　Rosa onoei Makino；斧氏蔷薇 ●☆

336830　Rosa onoei Makino var. hakonensis (Franch. et Sav.) H. Ohba；箱根蔷薇 ●☆

336831　Rosa onoei Makino var. oligantha (Franch. et Sav.) H. Ohba；疏花斧氏蔷薇 ●☆

336832　Rosa onoei Makino var. oligantha (Franch. et Sav.) H. Ohba f. rosea (Hayashi) Yonek. ；粉疏花斧氏蔷薇 ●☆

336833　Rosa orbicularis Baker = Rosa multibracteata Hemsl. et E. H. Wilson ●

336834　Rosa oulengensis H. Lév. = Rosa odorata (Andréws) Sweet ●◇

336835　Rosa oxyacantha M. Bieb. ；尖刺蔷薇；Oxyacanthous Rose, Sharpspine Rose, Sharpspiny Rose ●

336836　Rosa oxyodon Boiss. ；尖齿蔷薇 ●☆

336837　Rosa palustriformis Rydb. ；密歇根蔷薇 ●☆

336838　Rosa palustris Buch. -Ham. ex Lindl. = Rosa palustris Marshall ●☆

336839　Rosa palustris Marshall；沼泽蔷薇；Swamp Rose, Swamp Wild Rose ●☆

336840　Rosa palustris Marshall f. inermis (Regel) W. H. Lewis = Rosa palustris Marshall ●☆

336841　Rosa palustris Marshall var. dasistema (Raf.) E. J. Palmer et Steyerm. = Rosa palustris Marshall ●☆

336842　Rosa paniculigera (Koidz.) Makino ex Momiy. ；圆锥蔷薇 ●☆

336843　Rosa paniculigera (Koidz.) Makino ex Momiy. f. rosiflora Horino；粉花圆锥蔷薇 ●☆

336844　Rosa parmentieri H. Lév. = Rosa davidii Crép. var. elongata Rehder et E. H. Wilson ●

336845　Rosa parvifolia L. ；茅玫 ●

336846　Rosa paucispinosa H. L. Li = Rosa henryi Boulenger ●

336847　Rosa pendulina L. = Rosa alpina L. ●☆

336848　Rosa persetosa Rolfe；全针蔷薇；Acicular Rose, Bristly Rose, Persetose Rose ●

336849　Rosa persica Michx. ；波斯蔷薇(单叶蔷薇,小檗叶蔷薇)；Barberryleaf Rose, Barberry-leaved Rose, Persian Rose ●

336850　Rosa phoenicia Boiss. ；紫红蔷薇；Phoenician Rose ●☆

336851　Rosa pimpinellifolia L. ；芹叶蔷薇(多刺蔷薇,密刺蔷薇,苏格兰蔷薇)；Barrow Rose, Barrow-rose, Brid Breer, Brid Rose, Burnet Rose, Burnett Rose, Burrow Rose, Burrow-rose, Cant-robin, Cat Hep, Cat Rose, Cat Whin, Fox Rose, Pimpernel Rose, Scotch Briar, Scotch Rose, Scots Rose, Soldier Buttons, Soldier's Buttons, St. David's Rose ●

336852　Rosa pimpinellifolia L. = Rosa spinosissima L. ●

336853　Rosa pimpinellifolia L. var. subalpina Bunge ex M. Bieb. = Rosa oxyacantha M. Bieb. ●

336854　Rosa pinnatisepala T. C. Ku；羽萼蔷薇；Burnet Rose, Pinnatesepal Rose ●

336855　Rosa pinnatisepala T. C. Ku f. glandulosa T. C. Ku；多腺羽萼蔷薇；Glandulose Pinnatesepal Rose ●

336856　Rosa pisocarpa A. Gray；簇生蔷薇；Cluster Rose ●☆

336857　Rosa platyacantha Schrenk；宽刺蔷薇（黄蔷薇，宽叶蔷薇，密刺蔷薇）；Broad Spiny Rose，Broadspine Rose，Broad-spined Rose ●

336858　Rosa platyacantha Schrenk var. variabilis Regel = Rosa kokanica（Regel）Regel ex Juz. ●

336859　Rosa polyantha Schrenk 'Grandiflora'；大花宽刺蔷薇；Polyatha Grandiflora Rose ●☆

336860　Rosa polyantha Siebold et Zucc. = Rosa multiflora Thunb. ●

336861　Rosa pomifera Herrm. 'Duplex'；重瓣苹果蔷薇；Double Apple Rose ●☆

336862　Rosa pomifera Herrm. = Rosa villosa M. Bieb. ●☆

336863　Rosa pomifera Herrm. f. duplex（West.）Rehder = Rosa pomifera Herrm. 'Duplex' ●☆

336864　Rosa pouzinii Tratt.；普宁蔷薇●☆

336865　Rosa pouzinii Tratt. var. mauritii Sennen = Rosa pouzinii Tratt. ●☆

336866　Rosa pouzinii Tratt. var. parvipetala Pau et Font Quer = Rosa pouzinii Tratt. ●☆

336867　Rosa pouzinii Tratt. var. yebalica Pau et Font Quer = Rosa micrantha Borrer ex Sm. ●☆

336868　Rosa praelucens Bijh.；中甸刺玫（中甸蔷薇）；Dazzling Rose，Zhongdian Rose ●

336869　Rosa pratincola Greene = Rosa arkansana Porter var. suffulta（Greene）Cockerell ●☆

336870　Rosa prattii Hemsl.；铁杆蔷薇（勃拉蔷薇）；Ironpole Rose，Pratt Rose，Rose ●

336871　Rosa pricei Hayata；太鲁阁蔷薇（能高蔷薇）；Nengkaoshan Rose ●

336872　Rosa primula Boulenger；樱草蔷薇（报春刺玫，报春蔷薇，大马茄子）；Afghan Yellow Rose，Incense Rose，Primrose Rose，Primule Rose ●

336873　Rosa pseudindica Lindl. = Rosa odorata（Andréws）Sweet 'Pseudindica' ●

336874　Rosa pseudindica Lindl. = Rosa odorata（Andréws）Sweet var. pseudoindica（Lindl.）Rehder ●

336875　Rosa pseudobanksiae Te T. Yu et T. C. Ku；粉蕾蔷薇（粉蕾木香）；False Banks Rose，Falsebanksia Rose，Pseudobanks Rose ●

336876　Rosa pseudoindica Lindl. = Rosa odorata（Andréws）Sweet var. pseudoindica（Lindl.）Rehder ●

336877　Rosa pubescens Baker = Rosa rugosa Thunb. ●◇

336878　Rosa pubescens Roxb. = Rosa brunonii Lindl. ●

336879　Rosa pumila Scop.；低矮蔷薇；Dwarf Rose ●☆

336880　Rosa punicea Mill. = Rosa foetida Herrm. 'Bicolor' ●☆

336881　Rosa reducta Baker = Rosa multibracteata Hemsl. et E. H. Wilson ●

336882　Rosa relicta Erlanson = Rosa arkansana Porter var. suffulta（Greene）Cockerell ●☆

336883　Rosa richardii Rehder；理查德蔷薇；Richardii Rose，St. John Rose，St. John's Rose ●☆

336884　Rosa rotundibracteata Cardot = Rosa multibracteata Hemsl. et E. H. Wilson ●

336885　Rosa roxburghii Tratt.；缫丝花（茨梨，刺槟榔，刺梨，单瓣缫丝花，木梨，送春归，团棠二，团糖二，文光果，文先果，野蔷薇）；Bur Rose，Burr Rose，Chestnut Rose，Roxburgh Rose ●

336886　Rosa roxburghii Tratt. 'Plena' = Rosa roxburghii Tratt. var. plena Rehder ●☆

336887　Rosa roxburghii Tratt. f. esetosa T. C. Ku；无刺缫丝花；Esetose Rose ●

336888　Rosa roxburghii Tratt. f. normalis Rehder et E. H. Wilson；单瓣缫丝花（刺石榴，野石榴）；Normal Roxburgh Rose，Single Roxburgh Rose ●

336889　Rosa roxburghii Tratt. var. hirtula（Regel）Rehder et E. H. Wilson = Rosa hirtula（Regel）Nakai ●☆

336890　Rosa roxburghii Tratt. var. normalis Rehder et E. H. Wilson = Rosa roxburghii Tratt. f. normalis Rehder et E. H. Wilson ●

336891　Rosa roxburghii Tratt. var. plena Rehder；重瓣缫丝花；Double Roxburgh Rose ●

336892　Rosa rubiginosa L.；荆蔷薇（多刺蔷薇）；Briar Rose，Eglantine，Eglantine Rose，Sweet Briar，Sweet Briar，Sweet brier，Sweet-briar，Sweetbrier Rose，Wild Rose ●☆

336893　Rosa rubiginosa L. = Rosa eglanteria L. ●☆

336894　Rosa rubra Lam. = Rosa gallica L. ●

336895　Rosa rubrifolia Rupr. ex Boiss. = Rosa glauca Pourr. ●☆

336896　Rosa rubus H. Lév. et Vaniot；悬钩子蔷薇（茶藨花，茶藨花蔷薇，倒挂刺）；Blackberry Rose ●

336897　Rosa rubus H. Lév. et Vaniot f. glandulifera Te T. Yu et T. C. Ku；腺叶悬钩子蔷薇（腺叶茶藨花）；Glandular Blackberry Rose ●

336898　Rosa rubus H. Lév. et Vaniot var. pubescens Hayata = Rosa sambucina Koidz. ●

336899　Rosa rubus H. Lév. et Vaniot var. yunnanensis H. Lév. = Rosa rubus H. Lév. et Vaniot ●

336900　Rosa rudiuscula Greene；粗糙蔷薇；Rough Rose ●☆

336901　Rosa rugosa Thunb.；玫瑰（蓓蕾花，笔头花，赤玫瑰，刺玫花，红玫瑰，湖花，玫瑰花，梅桂，梅槐，猛蔷薇，徘徊花）；Beach Rose，Hedgehog Rose，Hedgerow Rose，Hedge-row Rose，Japanese Rose，Ramanas Rose，Rose，Rough Rose，Rugose Rose，Sea Tomato，Turkestan Rose ●◇

336902　Rosa rugosa Thunb. 'Alba'；白玫瑰（白花单瓣玫瑰，白花玫瑰）；White Rugose Rose ●

336903　Rosa rugosa Thunb. f. alba（Robins）F. Seym. = Rosa rugosa Thunb. 'Alba' ●

336904　Rosa rugosa Thunb. f. alba（Ware）Rehder = Rosa rugosa Thunb. 'Alba' ●

336905　Rosa rugosa Thunb. f. alboplena Rehder；重瓣白玫瑰（白花重瓣玫瑰）；Double-petalous White Rugose Rose ●

336906　Rosa rugosa Thunb. f. magnifica ?；华丽玫瑰；Magnifica Rose ●☆

336907　Rosa rugosa Thunb. f. plena（Regel）Bijh. = Rosa maikwai H. Hara ●

336908　Rosa rugosa Thunb. f. rosea Rehder；红玫瑰（粉红单瓣玫瑰）；Pink Rugose Rose ●

336909　Rosa rugosa Thunb. f. rubra ?；红皱蔷薇；Red Rugose Rose ●☆

336910　Rosa rugosa Thunb. f. typica Regel；紫玫瑰；Purple Rugose Rose ●

336911　Rosa rugosa Thunb. var. alba Rob. = Rosa rugosa Thunb. 'Alba' ●

336912　Rosa rugosa Thunb. var. plena Regel = Rosa maikwai H. Hara ●

336913　Rosa rydbergii Greene = Rosa arkansana Porter ●☆

336914　Rosa sambucina Koidz.；山蔷薇（台湾山蔷薇）；Mountain Rose ●

336915　Rosa sambucina Koidz. var. pubescens Koidz.；毛山蔷薇（山蔷薇）●☆

336916　Rosa sambucina Koidz. var. pubescens Koidz. = Rosa sambucina Koidz. ●

336917　Rosa sancta A. Rich.；神蔷薇；Abyssinian Rose，Holy Rose ●☆

336918　Rosa saturata Baker；大红蔷薇；Coral Rose，Pure Red Rose，Saturate Rose ●

336919　Rosa saturata Baker var. glandulosa Te T. Yu et T. C. Ku；腺叶大红蔷薇；Glandular Saturate Rose ●

336920　Rosa sayi Schwein. = Rosa acicularis Lindl. subsp. sayi

（Schwein.）W. H. Lewis ●

336921　Rosa scabriuscula Winch ex Sm. = Rosa tomentosa Sm. ●☆

336922　Rosa schrenkiana Crép. ;施雷蔷薇●☆

336923　Rosa schuetteana Erlanson = Rosa palustriformis Rydb. ●☆

336924　Rosa semperflorens Curtis = Rosa chinensis Jacq. var. semperflorens Koehne ●

336925　Rosa sempervirens L. ;常绿蔷薇;Evergreen Rose ●☆

336926　Rosa sempervirens L. var. anemoniflora（Fortune ex Lindl.）Regel = Rosa anemoniflora Fortune ex Lindl. ●

336927　Rosa sepium Thuill. = Rosa agrestis Savi ●☆

336928　Rosa sepium Thuill. subsp. micrantha（Sm.）Batt. = Rosa micrantha Borrer ex Sm. ●☆

336929　Rosa seraphinii Viv. = Rosa sicula Tratt. ●☆

336930　Rosa sericea Lindl. ;绢毛蔷薇（刺梨,峨眉蔷薇,色瓦,山刺梨）;Fourpetal Rose,Four-petaled,Maltese Cross Rose,Mount Omei Rose,Silky Rose ●

336931　Rosa sericea Lindl. = Rosa omeiensis Rolfe ●

336932　Rosa sericea Lindl. f. aculeato-eglandulosa Focke = Rosa omeiensis Rolfe ●

336933　Rosa sericea Lindl. f. glabrescens Franch. ;光叶绢毛蔷薇;Glabrous Silky Rose ●

336934　Rosa sericea Lindl. f. glandulosa Te T. Yu et T. C. Ku;腺叶绢毛蔷薇;Glandular Fourpetal Rose ●

336935　Rosa sericea Lindl. f. inermi-eglandulosa Focke = Rosa omeiensis Rolfe ●

336936　Rosa sericea Lindl. f. pteracantha Franch. ;宽刺绢毛蔷薇;Wingedspine Silky Rose ●

336937　Rosa sericea Lindl. subsp. omeiensis（Rolfe）A. V. Roberts = Rosa omeiensis Rolfe ●

336938　Rosa sericea Lindl. var. morrisonensis（Hayata）Masam. ;玉山野蔷薇（高山蔷薇,玉山蔷薇）;Taiwan Rose,Yushan Rose ●

336939　Rosa sericea Lindl. var. morrisonensis（Hayata）Masam. = Rosa morrisonensis Hayata ●

336940　Rosa sertata Rolfe;钝叶蔷薇（美丽蔷薇）;Gerland Rose,Obtuseleaf Rose,Obtuse-leaved Rose ●

336941　Rosa sertata Rolfe var. multijuga Te T. Yu et T. C. Ku;多对钝叶蔷薇;Manypaired Obtuseleaf Rose ●

336942　Rosa setigera Michx. ;草原蔷薇;Climbing Prairie Rose,Climbing Rose,Illinois Rose,Prairie Rose ●☆

336943　Rosa setigera Michx. var. tomentosa Torr. et A. Gray;毛草原蔷薇;Climbing Prairie Rose,Illinois Rose ●☆

336944　Rosa setipoda Hemsl. et E. H. Wilson;刺梗蔷薇（刺柄蔷薇,刺毛蔷薇,黄花蔷薇,色清）;Nodfruit Rose,Nod-fruited Rose,Rose ●

336945　Rosa shangchengensis T. C. Ku;商城蔷薇;Shangcheng Rose ●

336946　Rosa sherardii Davies;希拉尔迪蔷薇;Sherard's Downy-rose ●☆

336947　Rosa sicula Tratt. ;西西里蔷薇;Mediterranean Rose ●☆

336948　Rosa sicula Tratt. var. leiopoda Sennen = Rosa sicula Tratt. ●☆

336949　Rosa sicula Tratt. var. maroccana Pau et Font Quer = Rosa sicula Tratt. ●☆

336950　Rosa sicula Tratt. var. thuretii（Burnat et Gremli）Crép. = Rosa sicula Tratt. ●☆

336951　Rosa sicula Tratt. var. veridicata Burnat et Gremli = Rosa sicula Tratt. ●☆

336952　Rosa sikangensis Te T. Yu et T. C. Ku;川西蔷薇（康藏蔷薇,西康蔷薇）;West Sichuan Rose,Xikang Rose ●

336953　Rosa sikangensis Te T. Yu et T. C. Ku f. pilosa T. C. Ku;疏毛西康蔷薇（疏毛川西蔷薇）;Pilose Xikang Rose ●

336954　Rosa sikokiana Koidz. ;四国蔷薇;Sikoku Rose ●☆

336955　Rosa silenidiflora Nakai;丝花蔷薇●☆

336956　Rosa silverhjelmii Schrenk;伊犁蔷薇;Yili Rose ●

336957　Rosa sinica L. = Rosa chinensis Jacq. ●

336958　Rosa sinica Lindl. = Rosa chinensis Jacq. ●

336959　Rosa sinica Lindl. = Rosa laevigata Michx. ●

336960　Rosa sinobiflora T. C. Ku = Rosa biflora T. C. Ku ●

336961　Rosa sinowilsonii Hemsl. = Rosa longicuspis Bertol. var. sinowilsonii（Hemsl.）Te T. Yu et T. C. Ku ●

336962　Rosa soongarica Bunge = Rosa laxa Retz. ●

336963　Rosa sorbiflora Focke = Rosa cymosa Tratt. ●

336964　Rosa sorbus H. Lév. = Rosa omeiensis Rolfe ●

336965　Rosa soulieana Crép. ;川滇蔷薇（苏刺蔷薇）;Soulie Rose ●

336966　Rosa soulieana Crép. var. microphylla Te T. Yu et T. C. Ku;小叶川滇蔷薇;Small-leaf Soulie Rose ●

336967　Rosa soulieana Crép. var. sungpanensis Rehder;大叶川滇蔷薇（毛叶川滇蔷薇）;Sungpanensis Soulie Rose ●

336968　Rosa soulieana Crép. var. yunnanensis C. K. Schneid. ;毛叶川滇蔷薇;Yunnan Soulie Rose ●

336969　Rosa spinosissima L. ;密刺蔷薇;Burner Rose,Hornet Rose,Scotch Rose ●

336970　Rosa spinosissima L. = Rosa pimpinellifolia L. ●

336971　Rosa spinosissima L. var. altaica（Willd.）Rehder;大花密刺蔷薇;Altai Scotch Rose ●

336972　Rosa squarrosa（Rau）Boreau = Rosa canina L. ●

336973　Rosa stellata Wooton = Rosa stylosa Desv. ●☆

336974　Rosa stylosa Desv. ;沙漠柱蔷薇;Desert Rose,Short-styled Field-rose ●☆

336975　Rosa suavis Willd. = Rosa acicularis Lindl. ●

336976　Rosa suffulta Greene = Rosa arkansana Porte ●☆

336977　Rosa suffulta Greene = Rosa arkansana Porter var. suffulta（Greene）Cockerell ●☆

336978　Rosa suffulta Greene var. relicta（Erlanson）Deam = Rosa arkansana Porter var. suffulta（Greene）Cockerell ●☆

336979　Rosa sulphurea Aiton = Rosa hemisphaerica Herrm. ●☆

336980　Rosa svanetica Crép. ;斯万涅特蔷薇●☆

336981　Rosa sweginzowii Koehne;扁刺蔷薇（裂萼蔷薇）;Flatspine Rose,Sweginzow Rose ●

336982　Rosa sweginzowii Koehne var. glandulosa Cardot;腺叶扁刺蔷薇;Glandular Sweginzow Rose ●

336983　Rosa taiwanensis Nakai;小金樱（小金樱子）;Taiwan Rose ●

336984　Rosa taquetii H. Lév. = Rosa luciae Franch. et Rochebr. ex Crép. ●

336985　Rosa taronensis Te T. Yu et T. C. Ku;俅江蔷薇;Qiujiang Rose ●

336986　Rosa tatsienlouensis Cardot;打箭炉蔷薇 ●

336987　Rosa ternata Poir. = Rosa laevigata Michx. ●

336988　Rosa tetrapetala Royle = Rosa sericea Lindl. ●

336989　Rosa thea Savi = Rosa odorata（Andréws）Sweet ●◇

336990　Rosa tibetica Te T. Yu et T. C. Ku;西藏蔷薇;Tibet Rose,Xizang Rose ●

336991　Rosa tomentella Léman;短绒毛蔷薇;Short-hairy Rose ●☆

336992　Rosa tomentella Léman = Rosa canina L. ●

336993　Rosa tomentosa Sm. ;小绒毛蔷薇（多毛蔷薇）;Downy Rose,Harsh Downy-rose,Tomentose Rose,Whitewoolly Rose ●☆

336994　Rosa tongtchouanensis H. Lév. = Rosa odorata（Andréws）Sweet ●◇

336995　Rosa transmorrisonensis Hayata;高山腺蔷薇;Alpine Rose,

Glandular Rose ●

336996 Rosa transmorrisonensis Hayata var. taiwanensis（Nakai）S. S. Ying;单花蔷薇●

336997 Rosa triphylla Roxb. = Rosa laevigata Michx. ●

336998 Rosa triphylla Roxb. ex Hemsl. = Rosa anemoniflora Fortune ex Lindl. ●

336999 Rosa triphylla Roxb. ex Lindl. = Rosa laevigata Michx. ●

337000 Rosa tsinlingensis Pax et K. Hoffm. ;秦岭蔷薇;Qinling Rose ●◇

337001 Rosa turbinata Aiton;陀螺形蔷薇●☆

337002 Rosa turkestanica Regel;土耳其斯坦蔷薇;Turkestan Rose ●☆

337003 Rosa uncinella Besser;短钩刺蔷薇●☆

337004 Rosa uniflora Te T. Yu et T. C. Ku;单花合柱蔷薇;Oneflower Rose,Single-flowered Rose,Uniflorous Rose ●

337005 Rosa ussuriensis Juz. ;乌苏里蔷薇;Ussuri Rose ●☆

337006 Rosa villosa M. Bieb. ;苹果毛蔷薇（苹果蔷薇）;Apple Rose, Apple-fruited Rose ●☆

337007 Rosa virginiana Mill. ;弗吉尼亚蔷薇（维吉尼亚蔷薇）; Dorsay's Buttonhole Rose,St. Mark's Rose,Virginia Rose,Virginian Rose ●☆

337008 Rosa virginiana Mill. 'Plena' = Rosa 'Rose d'Amour' ●☆

337009 Rosa virginiana Mill. var. plena Rehder =Rosa 'Rose d'Amour' ●☆

337010 Rosa viridiflora ?;绿花蔷薇;Green Rose ●☆

337011 Rosa wallichii Tratt. = Rosa sericea Lindl. ●

337012 Rosa webbiana Wall. ex Royle;藏边蔷薇（大果蔷薇,韦伯氏蔷薇）;Webb Rose ●

337013 Rosa weixiensis Te T. Yu et T. C. Ku;维西蔷薇;Weixi Rose ●

337014 Rosa wichuraiana Crép. = Rosa luciae Franch. et Rochebr. ex Crép. ●

337015 Rosa wichuraiana Crép. ex Desegl. ;维氏蔷薇（光叶蔷薇）; Memorial Rose,Wichura Rose ●

337016 Rosa wichuraiana Crép. ex Desegl. f. simpliciflora T. C. Ku;单花光叶蔷薇;One-flower Wichura Rose ●

337017 Rosa wichuraiana Crép. ex Desegl. var. plena Honda;重瓣光叶蔷薇●☆

337018 Rosa wichuraiana Crép. f. ohwii Oishi = Rosa luciae Franch. et Rochebr. ex Crép. f. ohwii（Oishi）Yonek. ●☆

337019 Rosa wichuraiana Crép. var. glandulifera（Koidz.）Honda = Rosa luciae Franch. et Rochebr. ex Crép. f. glandulifera（Koidz.）H. Ohba ●☆

337020 Rosa wichurana Crépin = Rosa luciae Franch. et Rochebr. ex Crép. ●

337021 Rosa willdenowii Spreng. = Rosa davurica Pall. ●

337022 Rosa willmottiae Hemsl. ; 小叶蔷薇; Miss Willmott's Rose, Willmott Rose ●

337023 Rosa willmottiae Hemsl. var. glandulifera Te T. Yu et T. C. Ku; 多腺小叶蔷薇;Glandular Willmott Rose ●

337024 Rosa willmottiae Hemsl. var. glandulosa Te T. Yu et T. C. Ku = Rosa willmottiae Hemsl. var. glandulifera Te T. Yu et T. C. Ku ●

337025 Rosa willmottiana H. Lév. = Rosa longicuspis Bertol. ●

337026 Rosa woodsii A. Gray;西部野蔷薇;Mountain Rose,Western Rose,Western Wild Rose,Woods' Rose ●☆

337027 Rosa woodsii A. Gray subsp. ultramontana（S. Watson）R. L. Taylor et Macbryde;蒙大拿西部野蔷薇●☆

337028 Rosa woronowii Lonacz. ;高加索蔷薇●☆

337029 Rosa xanthina Lindl. ;黄刺玫（黄刺玫蔷薇,黄刺莓,黄刺梅, 金黄蔷薇,重瓣黄刺玫）;Manchu Rose,Manchurian Rose,Yellow Rose ●

337030 Rosa xanthina Lindl. 'Canary Bird';金丝雀黄刺玫●☆

337031 Rosa xanthina Lindl. 'Cantabrigiensis';坎塔桥黄刺玫●☆

337032 Rosa xanthina Lindl. = Rosa hugonis Hemsl. ●

337033 Rosa xanthina Lindl. f. hugonis（Hemsl.）A. V. Roberts = Rosa hugonis Hemsl. ●

337034 Rosa xanthina Lindl. f. normalis Rehder et E. H. Wilson;野生黄刺玫(单瓣黄刺玫,马茹茹,马茹子);Normal Manchurian Rose, Singlepetal Yellow Rose ●

337035 Rosa xanthina Lindl. f. normalis Rehder et E. H. Wilson = Rosa primula Boulenger ●

337036 Rosa xanthina Lindl. f. spontanea Rehder;单瓣黄刺玫●☆

337037 Rosa xanthina Lindl. f. spontanea Rehder = Rosa primula Boulenger ●

337038 Rosa xanthina Lindl. var. kokanica（Regel）Boulenger = Rosa kokanica（Regel）Regel ex Juz. ●

337039 Rosa xanthina Lindl. var. kokanica Boulenger = Rosa kokanica（Regel）Regel ex Juz. ●

337040 Rosa xanthinoides Nakai = Rosa xanthina Lindl. ●

337041 Rosa xanthocarpa Watt = Rosa odorata（Andréws）Sweet var. gigantea（Crép.）Rehder et E. H. Wilson ●

337042 Rosa xanthocarpa Watt ex E. Willm. = Rosa odorata（Andréws）Sweet var. gigantea（Crép.）Rehder et E. H. Wilson ●

337043 Rosa yakualpina Nakai et Momiy. = Rosa onoei Makino ●☆

337044 Rosa yesoensis（Franch. et Sav.）Makino = Rosa iwara Siebold ex Regel ●☆

337045 Rosa yezoensis Makino = Rosa iwara Siebold ex Regel ●☆

337046 Rosa yokoscensis Koidz. ;横须贺蔷薇●☆

337047 Rosa yunnanensis（Crépin）Boulenger = Rosa longicuspis Bertol. ●

337048 Rosa zhongdianensis T. C. Ku;中甸蔷薇;Zhongdian Rose ●

337049 Rosaceae Adans. = Rosaceae Juss. （保留科名）●■

337050 Rosaceae Juss. （1789）（保留科名）;蔷薇科;Rose Family ●■

337051 Rosaceae L. = Rosaceae Juss. （保留科名）●■

337052 Rosalesia La Llave = Brickellia Elliott（保留属名）■●

337053 Rosanovia Benth. et Hook. f. = Rosanowia Regel ●■☆

337054 Rosanovia Benth. et Hook. f. = Sinningia Nees ●■☆

337055 Rosanowia Regel = Sinningia Nees ●■☆

337056 Rosanthus Small = Gaudichaudia Kunth ●☆

337057 Rosaura Noronha = Ardisia Sw. （保留属名）●■

337058 Roscheria H. Wendl. = Roscheria H. Wendl. ex Balf. f. ●☆

337059 Roscheria H. Wendl. ex Balf. f. （1877）;黑毛棕属（罗舍椰属,若瑟尔桐属,塞舌尔双花棕属,双花刺椰属）;Black Bristle Palm ●☆

337060 Roscheria melanochaetes H. Wendl. ex Balf. f. ;黑毛棕;Roscher Palm ●☆

337061 Roscia D. Dietr. = Boscia Thunb. （废弃属名）●☆

337062 Roscia D. Dietr. = Toddalia Juss. （保留属名）●

337063 Roscia D. Dietr. = Vepris Comm. ex A. Juss. ●☆

337064 Roscoea Roxb. = Sphenodesme Jack ●

337065 Roscoea Sm. （1806）;象牙参属;Roscoea,Himalayan Ginger ■

337066 Roscoea alpina Royle;高山象牙参;Alpine Roscoea,Himalaya Roscoea ■

337067 Roscoea auriculata K. Schum. ;耳叶象牙参;Earleaf Roscoea ■

337068 Roscoea blanda K. Schum. = Roscoea debilis Gagnep. ■

337069 Roscoea blanda K. Schum. var. limprichtii Loes. = Roscoea debilis Gagnep. var. limprichtii（Loes.）Cowley ■

337070 Roscoea blanda K. Schum. var. pumila Hand. -Mazz. = Roscoea wardii Cowley ■

337071　Roscoea brevibracteata Z. Y. Zhu；短苞象牙参；Shortbract Roscoea ■

337072　Roscoea brevibracteata Z. Y. Zhu ＝ Roscoea schneideriana（Loes.）Cowley ■

337073　Roscoea cangshanensis M. H. Luo，X. F. Gao et H. H. Lin；苍山象牙参■

337074　Roscoea capitata Sm.；头花象牙参；Headflower Roscoea ■

337075　Roscoea capitata Sm. var. purpurata Gagnep. ＝ Roscoea cautleoides Gagnep. ■

337076　Roscoea capitata Sm. var. scillifolia Gagnep. ＝ Roscoea scillifolia（Gagnep.）Cowley ■

337077　Roscoea cautleoides Gagnep.；早花象牙参（矮狮花，双唇象牙参）；Primrose Roscoea，Twinlipped Roscoea ■

337078　Roscoea cautleoides Gagnep. var. pubescens（Z. Y. Zhu）T. L. Wu；毛早花象牙参（毛象牙参）；Pubescent Roscoea ■

337079　Roscoea cautleoides Gagnep. var. purpurea Stapf ＝ Roscoea cautleoides Gagnep. ■

337080　Roscoea chamaeleon Gagnep. ＝ Roscoea cautleoides Gagnep. ■

337081　Roscoea debilis Gagnep.；长柄象牙参；Longstalk Roscoea ■

337082　Roscoea debilis Gagnep. var. limprichtii（Loes.）Cowley；白象牙参；White Roscoea ■

337083　Roscoea flava Merr. ＝ Caulokaempferia coenobialis（Hance）K. Larsen ■

337084　Roscoea forrestii Cowley；大理象牙参；Dali Roscoea ■

337085　Roscoea gracilis Sm. ＝ Cautleya gracilis（Sm.）Dandy ■

337086　Roscoea humeana Balf. f. et W. W. Sm.；大花象牙参（大象牙参，双唇象牙参，象牙参）；Bigflower Roscoea ■

337087　Roscoea humeana Balf. f. et W. W. Sm. ＝ Roscoea chamaeleon Gagnep. ■

337088　Roscoea intermedia Gagnep. ＝ Roscoea alpina Royle ■

337089　Roscoea intermedia Gagnep. var. anomala Gagnep. ＝ Roscoea praecox K. Schum. ■

337090　Roscoea intermedia Gagnep. var. macrorrhiza Gagnep. ＝ Roscoea praecox K. Schum. ■

337091　Roscoea intermedia Gagnep. var. minuta Gagnep. ＝ Roscoea tibetica Batalin ■

337092　Roscoea intermedia Gagnep. var. plurifolia Loes. ＝ Roscoea tibetica Batalin ■

337093　Roscoea kunmingensis S. Q. Tong；昆明象牙参；Kunming Roscoea ■

337094　Roscoea kunmingensis S. Q. Tong var. elongata-bractea S. Q. Tong；延苞象牙参；Longbract Kunming Roscoea ■

337095　Roscoea lutea Royle ＝ Cautleya gracilis（Sm.）Dandy ■

337096　Roscoea pentandra Roxb. ＝ Sphenodesme pentandra Jack ●☆

337097　Roscoea praecox K. Schum.；先花象牙参（高山象牙参）■

337098　Roscoea pubescens Z. Y. Zhu ＝ Roscoea cautleoides Gagnep. var. pubescens（Z. Y. Zhu）T. L. Wu ■

337099　Roscoea purpurea Sm.；象牙参；Purple Roscoea ■

337100　Roscoea purpurea Sm. var. auriculata（K. Schum.）H. Hara ＝ Roscoea auriculata K. Schum. ■

337101　Roscoea purpurea Sm. var. procea Wall.；大象牙参；Big Purple Roscoea ■

337102　Roscoea schneideriana（Loes.）Cowley；无柄象牙参（滇象牙参）；Stalkless Roscoea ■

337103　Roscoea scillifolia（Gagnep.）Cowley；绵枣象牙参■

337104　Roscoea sichuanensis R. H. Miao；四川象牙参；Sichuan Roscoea ■

337105　Roscoea sichuanensis R. H. Miao ＝ Roscoea humeana Balf. f. et W. W. Sm. ■

337106　Roscoea sinopurpurea Stapf；华象牙参；China Roscoea，Chinese Purple Roscoea ■

337107　Roscoea sinopurpurea Stapf ＝ Roscoea cautleoides Gagnep. ■

337108　Roscoea spicata Sm. ＝ Cautleya spicata（Sm.）Baker ■

337109　Roscoea tibetica Batalin；藏象牙参（鸡脚参，鸡脚玉兰，土中闻，象牙参）；Tibet Roscoea，Xizang Roscoea ■

337110　Roscoea tibetica Batalin f. albiflora X. D. Dong et J. H. Li；白花藏象牙参■

337111　Roscoea tibetica Batalin var. emarginata S. Q. Tong；微凹象牙参；Emarginate Roscoea ■

337112　Roscoea tibetica Batalin var. emarginata S. Q. Tong ＝ Roscoea tibetica Batalin ■

337113　Roscoea toenrntosa Roxb. ＝ Congea tomentosa Roxb. ●

337114　Roscoea wardii Cowley；苍白象牙参（白象牙参）；Ward Roscoea ■

337115　Roscoea yunnanensis Loes.；川滇象牙参（滇象牙参，云南象牙参）；Yunnan Roscoea ■

337116　Roscoea yunnanensis Loes. ＝ Roscoea cautleoides Gagnep. ■

337117　Roscoea yunnanensis Loes. var. diersiana Loes. ＝ Roscoea schneideriana（Loes.）Cowley ■

337118　Roscoea yunnanensis Loes. var. purpurata（Gagnep.）Loes. ＝ Roscoea cautleoides Gagnep. ■

337119　Roscoea yunnanensis Loes. var. purpurea（Gagnep.）Loes. ＝ Roscoea cautleoides Gagnep. ■

337120　Roscoea yunnanensis Loes. var. schneideriana Loes. ＝ Roscoea schneideriana（Loes.）Cowley ■

337121　Roscoea yunnanensis Loes. var. scillifolia（Gagnep.）Loes. ＝ Roscoea scillifolia（Gagnep.）Cowley ■

337122　Roscyna Spach ＝ Hypericum L. ■●

337123　Roscyna gebleri（Ledeb.）Spach ＝ Hypericum ascyron L. subsp. gebleri（Ledeb.）N. Robson ■

337124　Roscyna gmelinii Spach ＝ Hypericum ascyron L. ●■

337125　Roscyna japonica Blume ＝ Hypericum ascyron L. ●■

337126　Rosea Fabr. ＝ Rhodiola L. ■

337127　Rosea Klotzsch ＝ Neorosea N. Hallé ●

337128　Rosea Klotzsch ＝ Tricalysia A. Rich. ex DC. ●

337129　Rosea Mart. ＝ Iresine P. Browne（保留属名）●■

337130　Rosea jasminiflora Klotzsch ＝ Tricalysia jasminiflora（Klotzsch）Benth. et Hook. f. ex Hiern ●☆

337131　Roseanthus Cogn. ＝ Polyclathra Bertol. ■☆

337132　Roseia Frič ＝ Ancistrocactus Britton et Rose ■☆

337133　Roseia Frič ＝ Coryphantha（Engelm.）Lem.（保留属名）●■

337134　Roseia Frič（1925）；肖菠萝属（顶花球属）●☆

337135　Roseia castanedai Frič；肖菠萝球●☆

337136　Rosenbachia Regel ＝ Ajuga L. ■●

337137　Rosenbergia Oerst.（1856）；洛氏花荵属●☆

337138　Rosenbergia Oerst. ＝ Cobaea Cav. ●■

337139　Rosenbergia gracilis Oerst.；细洛氏花荵●☆

337140　Rosenbergia macrostoma House；大口洛氏花荵●☆

337141　Rosenbergia minor House；小洛氏花荵●☆

337142　Rosenbergia scandens（Cav.）House ＝ Cobaea scandens Cav. ●

337143　Rosenbergia triflora House；三花洛氏花荵●☆

337144　Rosenbergiodendron Fagerl. ＝ Randia L. ●

337145　Rosenia Thunb.（1800）；二色鼠麹木属●☆

337146　Rosenia angustifolia Compton ＝ Rosenia oppositifolia（DC.）K. Bremer ●☆

337147 Rosenia glandulosa Thunb. ;多腺二色鼠麹木●☆

337148 Rosenia humilis（Less.）K. Bremer;低矮二色鼠麹木●☆

337149 Rosenia nestleroides Compton = Rosenia humilis（Less.）K. Bremer ●☆

337150 Rosenia oppositifolia（DC.）K. Bremer;对叶二色鼠麹木●

337151 Rosenia spinescens DC. ;刺二色鼠麹木●☆

337152 Roseocactus A. Berger = Ariocarpus Scheidw. ●

337153 Roseocactus A. Berger(1925);龟甲牡丹属（连山属）■☆

337154 Roseocactus fissuratus（Schum.）A. Berger;龟甲牡丹（龟甲仙人球）;Splitting Living-rock Cactus ●

337155 Roseocactus fissuratus A. Berger = Ariocarpus fissuratus（Engelm.）K. Schum. ●☆

337156 Roseocactus kotschoubeyanus A. Berger;黑牡丹■☆

337157 Roseocactus lloydii（Rose）A. Berger;连山■☆

337158 Roseocereus（Backeb.）Backeb. = Trichocereus（A. Berger）Riccob. ●

337159 Roseocereus Backeb. = Harrisia Britton ●

337160 Roseocereus Backeb. = Trichocereus（A. Berger）Riccob. ●

337161 Roseodendron Miranda = Tabebuia Gomes ex DC. ●☆

337162 Roshevitzia Tsvelev = Diandrochloa De Winter ■

337163 Roshevitzia Tsvelev = Eragrostis Wolf ■

337164 Roshevitzia diplachnoides（Steud.）Tzvelev = Eragrostis japonica（Thunb.）Trin. ■

337165 Roshevitzia japonica（Thunb.）Tzvelev = Eragrostis japonica（Thunb.）Trin. ■

337166 Rosifax C. C. Towns. (1991);耳叶苋属■☆

337167 Rosifax sabuletorum C. C. Towns. ;沙地耳叶苋■☆

337168 Rosilla Less. = Dyssodia Cav. ■☆

337169 Roslinia G. Don = Chironia L. ●■☆

337170 Roslinia G. Don = Roeslinia Moench ●■☆

337171 Roslinia Neck. = Justicia L. ●■

337172 Roslinia angustifolia（Sims）G. Don = Orphium frutescens（L.）E. Mey. ●☆

337173 Roslinia frutescens（L.）G. Don = Orphium frutescens（L.）E. Mey. ●☆

337174 Rosmarinus L. (1753);迷迭香属;Rosemary ●■

337175 Rosmarinus angustifolius Mill. = Rosmarinus officinalis L. ●

337176 Rosmarinus eriocalix Jord. et Fourr. ;毛萼迷迭香●☆

337177 Rosmarinus eriocalyx Jord. et Fourr. var. pallescens（Maire）Upson et Jury = Rosmarinus eriocalyx Jord. et Fourr. ●☆

337178 Rosmarinus latifolius Mill. = Rosmarinus officinalis L. ●

337179 Rosmarinus laxiflorus Noë = Rosmarinus officinalis L. ●

337180 Rosmarinus officinalis L. ;迷迭香;Common Rosemary,Compass Plant,Compass-weed,Creeping Rosemary,Girl's Love,Guard-robe,Herb of Memory,Incensier,Old Man,Polar Plant,Rosecampi,Rosemary,Sea Dew,Sea Spray,Speak,Speek ●

337181 Rosmarinus officinalis L. 'Albus';白花迷迭香●☆

337182 Rosmarinus officinalis L. 'Aureus';金斑迷迭香●☆

337183 Rosmarinus officinalis L. 'Fastigiatus';帚状迷迭香;Miss Jessop's Variety Rosemary ●☆

337184 Rosmarinus officinalis L. 'Huntingdon Carpet';亨廷登球迷迭香●☆

337185 Rosmarinus officinalis L. 'Irece';伊莱尼迷迭香;Rosemary ●☆

337186 Rosmarinus officinalis L. 'Joyce de Baggio';乔伊斯·巴奇奥迷迭香●☆

337187 Rosmarinus officinalis L. 'Lockwood de Forest';洛克伍德·福莱斯特迷迭香●☆

337188 Rosmarinus officinalis L. 'Majorca Pink';马略卡粉迷迭香●☆

337189 Rosmarinus officinalis L. 'Miss Jessopp's Upright';直立迷迭香●☆

337190 Rosmarinus officinalis L. 'Prostratus';平卧迷迭香;Prostrate Rosemary ●☆

's337191 Rosmarinus officinalis L. 'Roseus';玫瑰红迷迭香●☆

337192 Rosmarinus officinalis L. 'Severn Sea';浅蓝迷迭香●☆

337193 Rosmarinus officinalis L. 'Tuscan Blue';托斯卡纳蓝迷迭香●☆

337194 Rosmarinus officinalis L. var. genuina Turrill = Rosmarinus officinalis L. ●

337195 Rosmarinus officinalis L. var. lavandulaceus（Noë）Batt. = Rosmarinus officinalis L. ●

337196 Rosmarinus officinalis L. var. laxiflorus（Masson）Schltr. = Rosmarinus officinalis L. ●

337197 Rosmarinus officinalis L. var. prostratus Pasq. = Rosmarinus officinalis L. ●

337198 Rosmarinus officinalis L. var. subtomentosus Maire et Weiller = Rosmarinus officinalis L. ●

337199 Rosmarinus officinalis L. var. troglodytarum Maire et Weiller = Rosmarinus officinalis L. ●

337200 Rosmarinus officinalis L. var. ubescens Pamp. = Rosmarinus officinalis L. ●

337201 Rosmarinus prostratus Fl. Corc. ;匍匐迷迭香;Creeping Rosemary ●☆

337202 Rosmarinus tournefortii Noë = Rosmarinus eriocalyx Jord. et Fourr. ●☆

337203 Rosmarinus tournefortii Noëvar. pubescens（Pamp.）Maire et Weiller = Rosmarinus eriocalyx Jord. et Fourr. ●☆

337204 Rospidios A. DC. = Diospyros L. ●

337205 Rospidios vaccinioides A. DC. = Diospyros vaccinioides Lindl. ●

337206 Rossatis Thouars = Habenaria Willd. ■

337207 Rossatis Thouars = Satyrium Sw. (保留属名)■

337208 Rosselia Forman(1994);新几内亚橄榄属●☆

337209 Rossenia Vell. = Angostura Roem. et Schult. ●☆

337210 Rossina Steud. = Swartzia Schreb. (保留属名)●☆

337211 Rossioglossum（Schltr.）Garay et G. C. Kenn. (1976);罗斯兰属☆

337212 Rossioglossum grande（Lindl.）Garay et G. C. Kenn. ;大花罗斯兰☆

337213 Rossittia Ewart = Hibbertia Andréws ●☆

337214 Rossmaesslera Rchb. = Fenzlia Endl. ●☆

337215 Rossmaesslera Rchb. = Gilia Ruiz et Pav. ■●☆

337216 Rossmannia Klotzsch = Begonia L. ●■

337217 Rossolis Adans. (1763);肖茅膏菜属;Sundew ■☆

337218 Rossolis Adans. = Drosera L. ■

337219 Rossolis intermedia ? = Drosera intermedia Hayne ■☆

337220 Rossolis rotundifolia Moench = Drosera rotundifolia L. ■

337221 Rossolis septemtrionalis Scop. = Drosera rotundifolia L. ■

337222 Rostellaria C. F. Gaertn. = ? Bumelia Sw. (保留属名)●☆

337223 Rostellaria C. F. Gaertn. = Rostellularia Rchb. ■

337224 Rostellaria Nees = Justicia L. ●■

337225 Rostellaria Nees = Rostellularia Rchb. ■

337226 Rostellaria diffusa（Willd.）Nees = Justicia diffusa Willd. ■

337227 Rostellaria mollissima Nees = Justicia procumbens（L.）Nees var. simplex（D. Don）T. Yamaz. ■

337228 Rostellaria mollissima Nees = Rostellularia rotundifolia Nees ■

337229 Rostellaria parviflora Benth. = Justicia tenella（Nees）T. Anderson ■☆

337230　Rostellaria rotundifolia Nees ＝Justicia procumbens（L.）Nees var. simplex（D. Don）T. Yamaz.■

337231　Rostellaria tenella Nees ＝Justicia tenella（Nees）T. Anderson ■☆

337232　Rostellularia Rchb.（1837）；爵床属；Rostellularia ■

337233　Rostellularia Rchb. ＝Justicia L.●■

337234　Rostellularia abyssinica Nees ＝Justicia diffusa Willd.■

337235　Rostellularia diffusa（Willd.）Nees；小叶散爵床；Diffuse Rostellularia ■

337236　Rostellularia diffusa（Willd.）Nees ＝Justicia diffusa Willd.■

337237　Rostellularia diffusa（Willd.）Nees var. hedyotidifolia（Willd.）C. Y. Wu；耳草叶爵床（耳叶散爵床）■

337238　Rostellularia diffusa（Willd.）Nees var. prostrata（Roxb. ex C. B. Clarke）H. S. Lo；匍匐小叶散爵床（平卧爵床，小叶散爵床）■

337239　Rostellularia glandulosa Nees ＝Monothecium glandulosum（Nees）Hochst.■☆

337240　Rostellularia haplostachya Nees ＝Justicia haplostachya（Nees）T. Anderson ■☆

337241　Rostellularia hayatae（Yamam.）S. S. Ying ＝Rostellularia procumbens（L.）Nees var. ciliata（Yamam.）S. S. Ying ■

337242　Rostellularia hedyotidifolia Nees ＝Rostellularia diffusa（Willd.）Nees var. hedyotidifolia（Willd.）C. Y. Wu ■

337243　Rostellularia heterocarpa Hochst. ＝Justicia heterocarpa T. Anderson ■☆

337244　Rostellularia humilis H. S. Lo；矮爵床；Dwarf Rostellularia ■

337245　Rostellularia khasiana（C. B. Clarke）C. Y. Wu ex C. C. Hu；喀西爵床；Khasi Rostellularia ■

337246　Rostellularia khasiana（C. B. Clarke）C. Y. Wu ex C. C. Hu var. latispica（C. B. Clarke）C. Y. Wu ex C. C. Hu；宽穗爵床；Broadspike Rostellularia ■

337247　Rostellularia khasiana（C. B. Clarke）J. L. Ellis ＝Rostellularia khasiana（C. B. Clarke）C. Y. Wu ex C. C. Hu ■

337248　Rostellularia khasiana（C. B. Clarke）J. L. Ellis var. latispica（C. B. Clarke）C. Y. Wu ＝Rostellularia khasiana（C. B. Clarke）C. Y. Wu ex C. C. Hu var. latispica（C. B. Clarke）C. Y. Wu ex C. C. Hu ■

337249　Rostellularia linearifolia Bremek.；线叶爵床■

337250　Rostellularia linearifolia Bremek. subsp. liangkwangensis H. S. Lo；两广线叶爵床；Liangguang Rostellularia ■

337251　Rostellularia mollissima Nees ＝Rostellularia khasiana（C. B. Clarke）C. Y. Wu ex C. C. Hu var. latispica（C. B. Clarke）C. Y. Wu ex C. C. Hu ■

337252　Rostellularia mollissima Nees ＝Rostellularia khasiana（C. B. Clarke）J. L. Ellis var. latispica（C. B. Clarke）C. Y. Wu ■

337253　Rostellularia procumbens（L.）Nees；爵床（苍蝇草，苍蝇翅，赤眼，赤眼老母草，大鸭草，倒花草，肝火草，疳积草，观音草，孩儿草，黑节草，鸡骨草，节寒寒，爵卿，辣椒草，六方疳积草，六角仙草，六角英，麦穗红，麦穗癀，毛泽兰，奶疲草，奶杨草，澎湖爵床，蜻蜓草，屈胶仔，山苏麻，蛇食草，鼠尾红，鼠尾癀，水竹笋，四季青，瓦子草，五累草，香苏，小寒药，小黑节草，小青草，小青叶，心火草，野万年青，阴牛郎，蚱蜢腿）；Creeping Rostellularia ■

337254　Rostellularia procumbens（L.）Nees ＝Justicia procumbens L. var. riukiuensis Yamam.■

337255　Rostellularia procumbens（L.）Nees var. ciliata（Yamam.）S. S. Ying；睫毛爵床（早田氏爵床）■

337256　Rostellularia procumbens（L.）Nees var. hirsuta（Yamam.）S. S. Ying；密毛爵床（俯仰爵床，爵床，密毛澎湖爵床，澎湖爵床）■

337257　Rostellularia procumbens（L.）Nees var. linearifolia

（Yamam.）S. S. Ying；狭叶爵床■

337258　Rostellularia procumbens（L.）Nees var. riukiuensis（Yamam.）S. S. Ying ＝Rostellularia procumbens（L.）Nees ■

337259　Rostellularia quadrifaria（Wall.）S. S. Ying；花莲爵床■

337260　Rostellularia reptans Nees ＝Justicia trifolioides T. Anderson ■☆

337261　Rostellularia rotundifolia Nees；椭苞爵床（单爵床）；Roundleaf Rostellularia ■

337262　Rostellularia tenella Nees ＝Justicia tenella（Nees）T. Anderson ■☆

337263　Rostellularia trichochila Miq. ＝Rostellularia procumbens（L.）Nees ■

337264　Rostkovia Desv.（1809）；罗斯特草属■☆

337265　Rostkovia brevifolia Phil.；短叶罗斯特草■☆

337266　Rostkovia clandestina Phil. ＝Patosia clandestina Buchenau.■☆

337267　Rostkovia sphaerocarpa Desv.；罗斯特草■☆

337268　Rostraceae Dulac ＝Geraniaceae Juss.（保留科名）■●

337269　Rostraria Trin.（1820）；洛氏禾属（一年生洽草属）；Mediterranean Hair-grass ■☆

337270　Rostraria Trin. ＝Trisetum Pers.■

337271　Rostraria altissima Poir. ＝Hemarthria altissima（Poir.）Stapf et C. E. Hubb.■

337272　Rostraria balansae（Coss. et Durieu）Holub；巴兰萨洛氏禾■☆

337273　Rostraria clarkeana（Domin）Holub；克拉洛氏禾■☆

337274　Rostraria cristata（L.）Tzvelev ＝Koeleria phleoides（Vill.）Pers.■☆

337275　Rostraria fasciculata Desf. ＝Hemarthria altissima（Poir.）Stapf et C. E. Hubb.■

337276　Rostraria festucoides（Link）Romero Zarco；羊茅洛氏禾■☆

337277　Rostraria fuscescens（Pomel）Holub；浅棕色洛氏禾■☆

337278　Rostraria hispida（Savi）Dogan；硬毛洛氏禾■☆

337279　Rostraria incurvata L. ＝Parapholis incurva（L.）C. E. Hubb.■

337280　Rostraria litorea（All.）Holub；海滨洛氏禾■☆

337281　Rostraria phleoides（Desf.）Holub；梯牧草洛氏禾■☆

337282　Rostraria pumila（Desf.）Tzvelev；小洛氏禾■☆

337283　Rostraria ramosa Cav. ＝Monerma cylindrica（Willd.）Coss. et Durieu ■☆

337284　Rostraria recurviflora（Braun-Blanq. et Wilczek）Holub ＝Rostraria cristata（L.）Tzvelev ■☆

337285　Rostraria rohlfsii（Asch.）Holub ＝Lophochloa rohlfsii（Asch.）H. Scholz ■☆

337286　Rostraria salzmannii（Boiss. et Reut.）Holub；萨尔洛氏禾■☆

337287　Rostraria salzmannii（Boiss. et Reut.）Holub subsp. maroccana（Domin）H. Scholz；摩洛哥洛氏禾■☆

337288　Rostrinucula Kudo ＝Elsholtzia Willd.●■

337289　Rostrinucula Kudo（1929）；钩子木属；Hooktree，Rostrinucula ●★

37290　Rostrinucula dependens（Rehder）Kudo；钩子木（钩子，火香）；Hooked Rostrinucula，Hooktree，Weeping Buddleja ●

337291　Rostrinucula sinensis（Hemsl.）C. Y. Wu；长叶钩子木；Chinese Hooktree，Chinese Rostrinucula ●

337292　Rosularia（DC.）Stapf（1923）；瓦莲属（叠叶景天属，小长生草属）；Rosularia，Stonecrop ●

337293　Rosularia Stapf ＝Rosularia（DC.）Stapf

337294　Rosularia adenotricha（Wall. ex Edgew.）C. -A. Jansson et Rech. f.；腺毛瓦莲■☆

337295　Rosularia adenotricha（Wall. ex Edgew.）C. -A. Jansson et Rech. f. subsp. viguieri（Raym. -Hamet）C. -A. Jansson et Rech. f. ＝Rosularia viguieri（Raym. -Hamet ex Fröd.）G. R. Sarwar ■☆

337296　Rosularia adenotricha（Wall. ex Edgew.）C. -A. Jansson et

Rech. f. subsp. chitralica G. R. Sarwar;吉德拉尔瓦莲■☆

337297 Rosularia alpestris (Kar. et Kir.) Boriss.;长叶瓦莲(高山石莲);Alpine Rosularia■

337298 Rosularia elymaitica (Boiss. et Hauss005kn. ex Boiss.) A. Berger;埃利迈特瓦莲■☆

337299 Rosularia glabra (Regel et Winkl.) A. Berger;光瓦莲■☆

337300 Rosularia hissarica Boriss.;里普瓦莲■☆

337301 Rosularia kokanica (Regel et Schmalh.) Boriss.;浩罕瓦莲■☆

337302 Rosularia lipskyi Boriss.;利普斯基瓦莲■☆

337303 Rosularia lutea Boriss.;黄瓦莲■☆

337304 Rosularia paniculata (Regel et Schmalh.) A. Berger;圆锥瓦莲■☆

337305 Rosularia persica (Boiss.) A. Berger;波斯瓦莲■☆

337306 Rosularia pilosa (M. Bieb.) Boriss.;疏毛瓦莲■☆

337307 Rosularia platyphylla (Schrenk) A. Berger;卵叶瓦莲(宽叶石莲);Broadleaf Rosularia■

337308 Rosularia radiciflora (Steud.) Boiss.;根花瓦莲■☆

337309 Rosularia rosulata (Edgew.) H. Ohba;莲座瓦莲■☆

337310 Rosularia schischkinii Boriss.;希施瓦莲■☆

337311 Rosularia sedoides (Decne.) H. Ohba;景天瓦莲■☆

337312 Rosularia semiensis (A. Rich.) H. Ohba = Rosularia semiensis (J. Gay ex A. Rich.) H. Ohba■☆

337313 Rosularia semiensis (J. Gay ex A. Rich.) H. Ohba;塞米亚瓦莲■☆

337314 Rosularia semperivivum (M. Bieb.) A. Berger;长生瓦莲■☆

337315 Rosularia sempervivoides (Fisch.) Boriss.;类长生瓦莲■☆

337316 Rosularia subspicata (Freyn et Sint.) Boriss.;穗状瓦莲■☆

337317 Rosularia tadzhikistana Boriss.;塔吉克瓦莲■☆

337318 Rosularia turkestanica (Regel et Winkl.) A. Berger;小花瓦莲(中亚石莲,中亚瓦莲);Littleflower Rosularia■

337319 Rosularia viguieri (Raym.-Hamet ex Fröd.) G. R. Sarwar;维基耶瓦莲■☆

337320 Rotala L. (1771);节节菜属(水松药属,水猪母乳属);Rotala■

337321 Rotala brevistyla Baker f. = Rotala tenella (Guillaumin et Perr.) Hiern■

337322 Rotala capensis (Harv.) A. Fern. et Diniz;好望角节节菜■☆

337323 Rotala catholica (Cham. et Schltdl.) Leeuwen = Rotala ramosior (L.) Koehne■

337324 Rotala congolensis A. Fern. et Diniz = Rotala filiformis (Bellardi) Hiern■☆

337325 Rotala cordata Koehne;异叶节节菜;Diverseleaf Rotala■

337326 Rotala cordipetala R. E. Fr. = Rotala fontinalis Hiern■☆

337327 Rotala debilissima Chiov. = Rotala filiformis (Bellardi) Hiern■☆

337328 Rotala decumbens A. Fern. = Rotala juniperina A. Fern.■☆

337329 Rotala decussata Hiern = Rotala welwitschii Exell■

337330 Rotala densiflora (Roth ex Roem. et Schult.) Koehne;密花节节菜;Denseflower Rotala■

337331 Rotala densiflora (Roth ex Roem. et Schult.) Koehne subsp. uliginosa Koehne = Rotala densiflora (Roth ex Roem. et Schult.) Koehne■

337332 Rotala densiflora (Roth ex Roem. et Schult.) Koehne var. formosana Hayata = Rotala indica (Willd.) Koehne var. uliginosa (Miq.) Koehne■

337333 Rotala densiflora (Roth) Koehne = Rotala densiflora (Roth ex Roem. et Schult.) Koehne■

337334 Rotala densiflora (Roth) Koehne subsp. uliginosa (Roth) Koehne = Rotala densiflora (Roth ex Roem. et Schult.) Koehne■

337335 Rotala densiflora (Roth) Koehne var. formosana Hayata =

Rotala indica (Willd.) Koehne■

337336 Rotala dentifera (A. Gray) Koehne = Rotala ramosior (L.) Koehne■

337337 Rotala dinteri Koehne;丁特节节菜■☆

337338 Rotala diversifolia Koehne = Rotala cordata Koehne■

337339 Rotala elatinoides (DC.) Hiern;高节节菜■☆

337340 Rotala elatinomorpha Makino = Rotala indica (Willd.) Koehne■

337341 Rotala filiformis (Bellardi) Hiern;线形节节菜■☆

337342 Rotala fluitans Pohnert;漂浮节节菜■☆

337343 Rotala fontinalis Hiern;春节节菜■☆

337344 Rotala gerardii Boutique;杰勒德节节菜■☆

337345 Rotala gossweileri Koehne;微花节节菜■☆

337346 Rotala heteropetala Koehne = Rotala filiformis (Bellardi) Hiern■☆

337347 Rotala heterophylla Welw. ex A. Fern. et Diniz = Rotala filiformis (Bellardi) Hiern■☆

337348 Rotala hexandra Wall. ex Koehne;六蕊节节菜■

337349 Rotala hippuris Makino;杉叶藻节节菜■☆

337350 Rotala hutchinsoniana A. Fern. = Rotala myriophylloides Welw. ex Hiern■☆

337351 Rotala indica (Willd.) Koehne;节节菜(节节草,碌耳草,水马兰);Indian Rotala, Indian Toothcup■

337352 Rotala indica (Willd.) Koehne var. koreana Nakai = Rotala indica (Willd.) Koehne■

337353 Rotala indica (Willd.) Koehne var. uliginosa (Miq.) Koehne;印度水猪母乳(台湾水猪母乳)■

337354 Rotala indica (Willd.) Koehne var. uliginosa (Miq.) Koehne = Rotala indica (Willd.) Koehne■

337355 Rotala indica Blatt. et Hallb. = Rotala indica (Willd.) Koehne■

337356 Rotala juniperina A. Fern.;刺柏状节节菜■☆

337357 Rotala kainantensis Masam. = Rotala hexandra Wall. ex Koehne■

337358 Rotala koreana (Nakai) Mori = Rotala indica (Willd.) Koehne■

337359 Rotala leptopetala (Blume) Koehne = Rotala pentandra (Roxb.) Blatt. et Hallb.■

337360 Rotala leptopetala (Blume) Koehne = Rotala rosea (Poir.) C. D. K. Cook ex H. Hara■

337361 Rotala leptopetala (Blume) Koehne var. littorea (Miq.) Koehne = Rotala rosea (Poir.) C. D. K. Cook ex H. Hara■

337362 Rotala letouzeyana Bamps;勒图节节菜■☆

337363 Rotala littorea (Miq.) Nakai = Rotala pentandra (Roxb.) Blatt. et Hallb.■

337364 Rotala littorea (Miq.) Nakai = Rotala rosea (Poir.) C. D. K. Cook ex H. Hara■

337365 Rotala longicaulis A. Fern. et Diniz = Rotala myriophylloides Welw. ex Hiern■☆

337366 Rotala longistyla Gibbs;长柱节节菜■☆

337367 Rotala lucalensis A. Fern. et Diniz;卢卡拉节节菜■☆

337368 Rotala mexicana Cham. et Schltdl.;轮叶节节菜(轮生叶水猪母乳,墨西哥水松叶,水松叶);Whorlleaf Rotala■

337369 Rotala mexicana Cham. et Schltdl. subsp. pusilla (Tul.) Koehne = Rotala mexicana Cham. et Schltdl.■

337370 Rotala mexicana Cham. et Schltdl. var. spruceana (Benth.) Koehne = Rotala mexicana Cham. et Schltdl.■

337371 Rotala minuta A. Fern. et Diniz = Rotala gossweileri Koehne■☆

337372 Rotala myriophylloides Welw. ex Hiern;多叶节节菜■☆

337373 Rotala nashii A. Fern. = Rotala myriophylloides Welw. ex Hiern■☆

337374 Rotala nummularia Welw. ex Hiern;铜钱节节菜■☆

337375 Rotala oblonga Peter = Rotala tenella (Guillaumin et Perr.)

Hiern ■

337376　Rotala pearsoniana A. Fern. et Diniz = Rotala myriophylloides Welw. ex Hiern ■☆

337377　Rotala pedicellata A. Fern. et Diniz = Rotala tenella（Guillaumin et Perr.）Hiern ■

337378　Rotala pentandra（Roxb.）Blatt. et Hallb.；薄瓣节节菜；Thinpetal Rotala ■

337379　Rotala pentandra（Roxb.）Blatt. et Hallb. = Rotala rosea（Poir.）C. D. K. Cook ex H. Hara ■

337380　Rotala perigrina H. Perrier = Rotala tenella（Guillaumin et Perr.）Hiern ■

337381　Rotala pterocalyx A. Raynal；翅萼节节菜■☆

337382　Rotala pusilla Tul.；苏州节节菜■

337383　Rotala pusilla Tul. = Rotala mexicana Cham. et Schltdl. ■

337384　Rotala ramosior（L.）Koehne；美洲节节菜（美洲水猪母乳）；Lowland Rotala，Toothcup，Tooth-cup，Wheel-wort Tooth-cup ■

337385　Rotala ramosior（L.）Koehne var. interior Fernald et Griscom = Rotala ramosior（L.）Koehne ■

337386　Rotala ramosior（L.）Koehne var. typica Fernald et Griscom = Rotala ramosior（L.）Koehne ■

337387　Rotala repens（Hochst.）Koehne；匍匐节节菜■☆

337388　Rotala robynsiana A. Fern. et Diniz = Rotala capensis（Harv.）A. Fern. et Diniz ■☆

337389　Rotala rosea（Poir.）C. D. K. Cook ex H. Hara；五蕊节节菜（薄瓣节节菜，五蕊水猪母乳）■

337390　Rotala rotunda A. Chev. = Rotala pterocalyx A. Raynal ■☆

337391　Rotala rotundifolia（Buch.-Ham. ex Roxb.）Koehne；圆叶节节菜（禾虾菜，红格草，红眼猫，假桑子，肉矮陀陀，水串，水豆瓣，水瓜子菜，水马桑，水泉，水水花，水松叶，水酸草，水苋菜，水指甲，水猪母乳）；Roundleaf Rotala ■

337392　Rotala roxburghiana Wight = Rotala densiflora（Roth ex Roem. et Schult.）Koehne ■

337393　Rotala scabrida（Franch.）S. H. Fu = Rhodiola coccinea（Royle）Boriss. subsp. scabrida（Franch.）H. Ohba ■

337394　Rotala serpiculoides Welw. ex Hiern；仙草节节菜■☆

337395　Rotala smithii A. Fern. et Diniz；史密斯节节菜■☆

337396　Rotala submersa Pohnert = Rotala tenella（Guillaumin et Perr.）Hiern ■

337397　Rotala submersa Pohnert var. angustipetala A. Fern.；窄瓣节节菜■

337398　Rotala taiwaniana Y. C. Liu et F. Y. Lu；台湾节节菜■

337399　Rotala tenella（Guillaumin et Perr.）Hiern；沉水节节菜■

337400　Rotala tetragonocalyx A. Fern. et Diniz = Rotala fluitans Pohnert ☆

337401　Rotala thymoides Exell；百里香节节菜■☆

337402　Rotala uliginosa（Miq.）Nakai = Rotala indica（Willd.）Koehne ■

337403　Rotala urundiensis A. Fern. et Diniz = Rotala gossweileri Koehne ■☆

337404　Rotala verticillaris L. var. spruceana（Benth.）Hiern = Rotala mexicana Cham. et Schltdl. ■

337405　Rotala wallichii（Hook. f.）Koehne；沃利克节节菜（瓦氏节节菜，瓦氏水猪母乳）■

337406　Rotala welwitschii Exell；韦尔节节菜■

337407　Rotang Adans. = Calamus L. ●

337408　Rotang Arians. = Rotanga Boehm. ●

337409　Rotanga Boehm. = Calamus L. ●

337410　Rotanga Boehm. ex Crantz = Calamus L. ●

337411　Rotantha Baker = Lawsonia L. ●

337412　Rotantha Small = Campanula L. ■●

337413　Rotbolla Zumagl. = Rottboellia L. f.（保留属名）■

337414　Roterbe Klatt = Botherbe Steud. ex Klatt ■

337415　Roterbe Klatt = Calydorea Herb. ■☆

337416　Rotheca Raf.（1838）；肖顙桐属●☆

337417　Rotheca Raf. = Clerodendrum L. ●■

337418　Rotheca alata（Gürke）Verdc.；具翅肖顙桐●☆

337419　Rotheca amplifolia（S. Moore）R. Fern.；大叶肖顙桐●☆

337420　Rotheca aurantiaca（Baker）R. Fern.；橙色肖顙桐●☆

337421　Rotheca aurantiaca（Baker）R. Fern. f. faulkneri（R. Fern.）R. Fern.；福克纳节节菜●☆

337422　Rotheca bukobensis（Gürke）Verdc.；布科巴肖顙桐●☆

337423　Rotheca caerulea（N. E. Br.）Herman et Retief；天蓝顙桐●☆

337424　Rotheca commiphoroides（Verdc.）Steane et Mabb.；没药肖顙桐●☆

337425　Rotheca cuneiformis（Moldenke）Herman et Retief；楔形顙桐●☆

337426　Rotheca cyanea（R. Fern.）R. Fern.；蓝色顙桐●☆

337427　Rotheca hirsuta（Hochst.）R. Fern.；粗毛顙桐●☆

337428　Rotheca hirsuta（Hochst.）R. Fern. f. triphylla（Harv.）R. Fern. = Rotheca hirsuta（Hochst.）R. Fern. ●☆

337429　Rotheca incisa（Klotzsch）Steane et Mabb.；锐裂顙桐●☆

337430　Rotheca kissakensis（Gürke）Verdc.；基萨萨科顙桐●☆

337431　Rotheca luembensis（De Wild.）R. Fern.；卢恩贝顙桐●☆

337432　Rotheca luembensis（De Wild.）R. Fern. f. herbacea（Hiern）R. Fern.；草本顙桐■☆

337433　Rotheca luembensis（De Wild.）R. Fern. var. malawiensis（R. Fern.）R. Fern.；马拉维顙桐■☆

337434　Rotheca mendesii（R. Fern.）R. Fern.；门代斯肖顙桐●☆

337435　Rotheca myricoides（Hochst.）Steane et Mabb.；对叶顙桐；Blue Butterfly，Butterfly Bush ●☆

337436　Rotheca myricoides（Hochst.）Steane et Mabb. f. angustilobata（R. Fern.）R. Fern.；细裂对叶顙桐●☆

337437　Rotheca myricoides（Hochst.）Steane et Mabb. f. brevilobata（R. Fern.）R. Fern.；短裂对叶顙桐●☆

337438　Rotheca myricoides（Hochst.）Steane et Mabb. f. cubangensis（R. Fern.）R. Fern.；古巴对叶顙桐●☆

337439　Rotheca myricoides（Hochst.）Steane et Mabb. f. lanceolatilobata（R. Fern.）R. Fern.；披针对叶顙桐●☆

337440　Rotheca myricoides（Hochst.）Steane et Mabb. f. lobulata（R. Fern.）R. Fern.；小裂片对叶顙桐●☆

337441　Rotheca myricoides（Hochst.）Steane et Mabb. f. reflexilobata（R. Fern.）R. Fern.；反折顙桐●☆

337442　Rotheca myricoides（Hochst.）Steane et Mabb. subsp. austromonticola（Verdc.）Verdc.；南方山生顙桐●☆

337443　Rotheca myricoides（Hochst.）Steane et Mabb. subsp. mafiensis（Verdc.）Verdc.；马菲顙桐●☆

337444　Rotheca myricoides（Hochst.）Steane et Mabb. subsp. namibiensis（R. Fern.）R. Fern. = Rotheca myricoides（Hochst.）Steane et Mabb. ●☆

337445　Rotheca myricoides（Hochst.）Steane et Mabb. var. discolor（Klotzsch）Verdc.；异色对叶顙桐●☆

337446　Rotheca myricoides（Hochst.）Steane et Mabb. var. dumalis（Hiern）R. Fern.；矮丛对叶顙桐●☆

337447　Rotheca myricoides（Hochst.）Steane et Mabb. var. kilimandscharensis（Verdc.）Verdc.；基利顙桐●☆

337448　Rotheca myricoides（Hochst.）Steane et Mabb. var. viridiflora（Verdc.）Verdc.；绿花对叶顙桐●☆

337449 Rotheca pilosa（H. Pearson）Herman et Retief；疏毛肖桢桐●☆

337450 Rotheca prittwitzii（B. Thomas）Verdc.；普里特肖桢桐●☆

337451 Rotheca quadrangulata（B. Thomas）R. Fern. = Rotheca sansibarensis（Gürke）Steane et Mabb. ●☆

337452 Rotheca reflexa（H. Pearson）R. Fern.；反折肖桢桐●☆

337453 Rotheca rupicola（Verdc.）Verdc.；岩生肖桢桐●☆

337454 Rotheca sansibarensis（Gürke）Steane et Mabb.；桑给巴尔肖桢桐●☆

337455 Rotheca sansibarensis（Gürke）Steane et Mabb. f. tomentella（R. Br.）R. Fern.；绒毛肖桢桐●☆

337456 Rotheca sansibarensis（Gürke）Steane et Mabb. subsp. caesia（Gürke）Steane et Mabb.；淡蓝肖桢桐●☆

337457 Rotheca sansibarensis（Gürke）Steane et Mabb. subsp. occidentalis（Verdc.）Steane et Mabb.；西方肖桢桐●☆

337458 Rotheca suffruticosa（Gürke）Verdc.；亚灌木大青●☆

337459 Rotheca taborensis（Verdc.）Verdc.；泰伯肖桢桐●☆

337460 Rotheca taborensis（Verdc.）Verdc. var. latifolia（Verdc.）Verdc.；宽叶泰伯肖桢桐●☆

337461 Rotheca tanneri（Verdc.）Verdc.；坦纳肖桢桐●☆

337462 Rotheca teaguei（Hutch.）R. Fern.；蒂格肖桢桐●☆

337463 Rotheca uncinata（Schinz）Herman et Retief；具钩大青●☆

337464 Rotheca verdcourtii（R. Fern.）R. Fern.；韦尔德肖桢桐●☆

337465 Rotheca violacea（Gürke）Verdc.；堇色大青●☆

337466 Rotheca wildii（Moldenke）R. Fern.；威氏大青●☆

337467 Rotheca wildii（Moldenke）R. Fern. f. glabra（R. Fern.）R. Fern.；无毛威氏大青●☆

337468 Rotheria Meyen = Cruckshanksia Hook. et Arn.（保留属名）●☆

337469 Rothia Borkh. = Mibora Adans. ■☆

337470 Rothia Borkh. = Rothia Pers.（保留属名）■

337471 Rothia Lam. = Hymenopappus L' Hér. ■☆

337472 Rothia Lam. = Rothia Pers.（保留属名）■

337473 Rothia Pers.（1807）（保留属名）；落地豆属（罗思豆属）■

337474 Rothia Schreb.（废弃属名）= Andryala L. ■☆

337475 Rothia Schreb.（废弃属名）= Rothia Pers.（保留属名）■

337476 Rothia hirsuta（Guillaumin et Perr.）Baker；毛落地豆■☆

337477 Rothia indica（L.）Druce；落地豆（罗思豆，印度罗思豆）■

337478 Rothia pinnata（Lam.）Kuntze = Schkuhria pinnata（Lam.）Kuntze ex Thell. ■☆

337479 Rothia trifoliata（Roth）Pers. = Rothia indica（L.）Druce ■

337480 Rothmaleria Font Quer（1940）；毛托苣属■☆

337481 Rothmaleria granatensis（Boiss. ex DC.）Font Quer；毛托苣■☆

337482 Rothmannia Thunb.（1776）；大黄栀子属（非洲栀属，罗斯曼木属，野栀子属）●

337483 Rothmannia Thunb. = Hyperacanthus E. Mey. ex Bridson ●☆

337484 Rothmannia annae（E. P. Wright）Keay；安纳大黄栀子●☆

337485 Rothmannia arcuata Bremek. = Rothmannia urcelliformis（Hiern）Robyns ●☆

337486 Rothmannia bowieana（A. Cunn. ex Hook.）Benth. = Euclinia longiflora Salisb. ●☆

337487 Rothmannia breviflora J. G. Garcia = Hyperacanthus microphyllus（K. Schum.）Bridson ●☆

337488 Rothmannia buchananii（Oliv.）Fagerl. = Rothmannia manganjae（Hiern）Keay ●☆

337489 Rothmannia capensis Thunb.；非洲大黄栀子（非洲栀）●☆

337490 Rothmannia coriacea（K. Schum. ex Hutch. et Dalziel）Fagerl. = Rothmannia lujae（De Wild.）Keay ●☆

337491 Rothmannia daweishanensis Y. M. Shui et W. H. Chen；大围山

野栀子●

337492 Rothmannia eetveldiana（De Wild. et T. Durand）Fagerl. = Rothmannia whitfieldii（Lindl.）Dandy ●☆

337493 Rothmannia engleriana（K. Schum.）Keay；恩氏大黄栀子●☆

337494 Rothmannia engleriana（K. Schum.）Keay var. ternifolia（Ficalho et Hiern）Somers = Rothmannia engleriana（K. Schum.）Keay ●☆

337495 Rothmannia fischeri（K. Schum.）Bullock；菲舍尔大黄栀子●☆

337496 Rothmannia fischeri（K. Schum.）Bullock subsp. moramballae（Hiern）Bridson；莫拉大黄栀子●☆

337497 Rothmannia fischeri（K. Schum.）Bullock subsp. verdcourtii Bridson；韦尔德黄栀子●☆

337498 Rothmannia fratrum（K. Krause）Fagerl. = Rothmannia manganjae（Hiern）Keay ●☆

337499 Rothmannia giganthosphaera K. Schum. ex Fagerl. = Rothmannia lujae（De Wild.）Keay ●☆

337500 Rothmannia globosa（Hochst.）Keay；球冠大黄栀子（球冠罗斯曼木）；Bell Gardenia, Cape Jasmine, September Bells, Tree Gardenia ●☆

337501 Rothmannia hispida（K. Schum.）Fagerl.；硬毛大黄栀子●☆

337502 Rothmannia jollyana N. Hallé；若利大黄栀子●☆

337503 Rothmannia kuhniana（F. Hoffm. et K. Schum.）Fagerl. = Rothmannia engleriana（K. Schum.）Keay ●☆

337504 Rothmannia lateriflora（K. Schum.）Keay；侧花大黄栀子●☆

337505 Rothmannia leptactinoides（K. Schum.）Fagerl. = Aulacocalyx caudata（Hiern）Keay ●☆

337506 Rothmannia liebrechtsiana（De Wild. et T. Durand）Keay；利布黄栀子●☆

337507 Rothmannia longiflora Salisb.；长花大黄栀子（长花罗斯曼木）●☆

337508 Rothmannia lujae（De Wild.）Keay；卢亚大黄栀子●☆

337509 Rothmannia macrantha（Schult.）Robyns = Euclinia longiflora Salisb. ●☆

337510 Rothmannia macrocarpa（Hiern）Keay；大果大黄栀子●☆

337511 Rothmannia macrosiphon（K. Schum. ex Engl.）Bridson；大管大黄栀子●☆

337512 Rothmannia maculata（DC.）Fagerl. = Rothmannia longiflora Salisb. ●☆

337513 Rothmannia malleifera（Hook.）Benth. = Rothmannia whitfieldii（Lindl.）Dandy ●☆

337514 Rothmannia manganjae（Hiern）J. G. Garcia = Rothmannia manganjae（Hiern）Keay ●☆

337515 Rothmannia manganjae（Hiern）Keay；大花大黄栀子（大黄栀子）；Scented Bells ●☆

337516 Rothmannia mayumbensis（R. D. Good）Keay；马永巴大黄栀子●☆

337517 Rothmannia megalostigma（Wernham）Keay = Rothmannia munsae（Schweinf. ex Hiern）E. M. Petit subsp. megalostigma（Wernham）Somers ●☆

337518 Rothmannia microphylla（K. Schum.）J. G. Garcia = Hyperacanthus microphyllus（K. Schum.）Bridson ●☆

337519 Rothmannia microphylla（K. Schum.）J. G. Garcia var. major J. G. Garcia = Hyperacanthus microphyllus（K. Schum.）Bridson ●☆

337520 Rothmannia munsae（Schweinf. ex Hiern）E. M. Petit；芒萨大黄栀子●☆

337521 Rothmannia munsae（Schweinf. ex Hiern）E. M. Petit subsp. megalostigma（Wernham）Somers；大柱头大黄栀子●☆

337522 Rothmannia octomera（Hook.）Fagerl.；八数大黄栀子●☆

337523　Rothmannia physophylla（K. Schum.）Fagerl. = Gardenia imperialis K. Schum. subsp. physophylla（K. Schum.）L. Pauwels ●☆

337524　Rothmannia ravae（Chiov.）Bridson；泰北大黄栀子●☆

337525　Rothmannia riparia（K. Schum.）Fagerl. = Rothmannia urcelliformis（Hiern）Robyns ●☆

337526　Rothmannia sootepensis（Craib）Bremek. = Gardenia sootepensis Hutch. ●

337527　Rothmannia sootepensis（Craib）Bremek. = Randia sootepensis Craib ●☆

337528　Rothmannia sootepensis（Craib）T. Yamaz. = Randia sootepensis Craib ●☆

337529　Rothmannia spathicalyx（De Wild.）Fagerl. = Rothmannia urcelliformis（Hiern）Robyns ●☆

337530　Rothmannia stanleyana（Hook. ex Lindl.）Benth. = Rothmannia longiflora Salisb. ●☆

337531　Rothmannia talbotii（Wernham）Keay；塔尔博特黄栀子●☆

337532　Rothmannia urcelliformis（Hiern）Robyns；小壶大黄栀子●☆

337533　Rothmannia whitfieldii（Lindl.）Dandy；惠特黄栀子●☆

337534　Rothrockia A. Gray（1885）；罗思萝藦属●☆

337535　Rothrockia cordifolia A. Gray；罗思萝藦●☆

337536　Rothrockia fruticosa Brandegee；灌木罗思萝藦●☆

337537　Rotmannia Neck. = Eperua Aubl. ●☆

337538　Rottbodia Scop. = Ximenia L. ●

337539　Rottboelia Scop.（废弃属名）= Heymassoli Aubl. ●

337540　Rottboelia Scop.（废弃属名）= Rottboellia L. f.（保留属名）■

337541　Rottboella Murr. = Rottboellia L. f.（保留属名）■

337542　Rottboellia Host = Parapholis C. E. Hubb. + Pholiurus Trin. ■☆

337543　Rottboellia Host = Rottboellia L. f.（保留属名）■

337544　Rottboellia L. f.（1782）（保留属名）；筒轴茅属（高臭草属，罗氏草属）；Itchgrass，Joint-Tail Grass，Joint-tall-grass ■

337545　Rottboellia afraurita Stapf = Coelorachis afraurita（Stapf）Stapf ■☆

337546　Rottboellia afzelii Hack. = Chasmopodium afzelii（Hack.）Stapf ■☆

337547　Rottboellia agropyroides Hack. = Urelytrum agropyroides（Hack.）Hack. ■☆

337548　Rottboellia altissima Poir. = Hemarthria altissima（Poir.）Stapf et C. E. Hubb. ■

337549　Rottboellia angolensis Rendle = Phacelurus gabonensis（Steud.）Clayton ■☆

337550　Rottboellia anthephoroides Steud. = Ischaemum antephoroides（Steud.）Miq. ■

337551　Rottboellia arundinacea Hochst. ex A. Rich. = Rottboellia exaltata L. f. ■

337552　Rottboellia bovonei Chiov. = Loxodera bovonei（Chiov.）Launert ■☆

337553　Rottboellia caespitosa Baker = Rhytachne rottboellioides Desv. ex Ham. ■☆

337554　Rottboellia caudata Hack. = Chasmopodium caudatum（Hack.）Stapf ■☆

337555　Rottboellia cochinchinensis（Lour.）Clayton；筒轴茅（罗氏草，南部俭草，蛇尾草，筒轴草）；Itchgrass，Kelly Grass ■

337556　Rottboellia compressa L. f. = Hemarthria compressa（L. f.）R. Br. ■

337557　Rottboellia compressa L. f. var. fasciculata（Lam.）Hack. = Hemarthria altissima（Poir.）Stapf et C. E. Hubb. ■

337558　Rottboellia compressa L. f. var. fasciculata Hack. = Hemarthria altissima（Poir.）Stapf et C. E. Hubb. ■

337559　Rottboellia compressa L. f. var. japonica Hack. = Hemarthria sibirica（Gand.）Ohwi ■

337560　Rottboellia corymbosa L. f. = Ophiuros exaltatus（L.）Kuntze ■

337561　Rottboellia cylindrica Vanderyst = Chasmopodium caudatum（Hack.）Stapf ■☆

337562　Rottboellia cylindrica Willd. = Monerma cylindrica（Willd.）Coss. et Durieu ■☆

337563　Rottboellia exaltata（L.）L. f. = Rottboellia cochinchinensis（Lour.）Clayton ■

337564　Rottboellia exaltata L. f. = Rottboellia cochinchinensis（Lour.）Clayton ■

337565　Rottboellia fasciculata Desf. = Hemarthria altissima（Poir.）Stapf et C. E. Hubb. ■

337566　Rottboellia fasciculata Lam. = Hemarthria altissima（Poir.）Stapf et C. E. Hubb. ■

337567　Rottboellia gabonensis（Steud.）Roberty = Phacelurus gabonensis（Steud.）Clayton ■☆

337568　Rottboellia glabra Roxb. = Hemarthria compressa（L. f.）R. Br. ■

337569　Rottboellia glauca Hack. = Phacelurus speciosus（Steud.）C. E. Hubb. ■☆

337570　Rottboellia gracillima Baker = Oxyrhachis gracillima（Baker）C. E. Hubb. ■☆

337571　Rottboellia granularis（L.）Roberty = Hackelochloa granularis（L.）Kuntze ■

337572　Rottboellia heterochroa Gand. = Hemarthria altissima（Poir.）Stapf et C. E. Hubb. ■

337573　Rottboellia hirsuta（Forssk.）Vahl = Lasiurus scindicus Henrard ■☆

337574　Rottboellia hordeoides Munro = Urelytrum agropyroides（Hack.）Hack. ■☆

337575　Rottboellia huillensis Rendle = Phacelurus huillensis（Rendle）Clayton ■☆

337576　Rottboellia incurva（L.）Roem. et Schult. = Parapholis incurva（L.）C. E. Hubb. ■

337577　Rottboellia incurvata L. = Parapholis incurva（L.）C. E. Hubb. ■

337578　Rottboellia inermis Peter = Coelorachis lepidura Stapf ■☆

337579　Rottboellia japonica（Hack.）Honda = Hemarthria sibirica（Gand.）Ohwi ■

337580　Rottboellia kalcicola Keng = Roegneria calcicola Keng ex Keng et S. L. Chen ■

337581　Rottboellia kerstingii Pilg. = Chasmopodium caudatum（Hack.）Stapf ■☆

337582　Rottboellia laevis Retz. = Mnesithea laevis（Retz.）Kunth ■

337583　Rottboellia laevispica Keng；光穗筒轴茅（光穗罗氏草，光穗筒轴草）；Nitidspike Itchgrass，Smoothspike Itchgrass ■

337584　Rottboellia latifolia Steud. = Phacelurus latifolius（Steud.）Ohwi ■

337585　Rottboellia latifolia Steud. var. angustifolia Debeaux = Phacelurus latifolius（Steud.）Ohwi var. angustifolius（Debeaux）Keng ■

337586　Rottboellia latifolia Steud. var. angustifolia Debeaux = Phacelurus latifolius（Steud.）Ohwi ■

337587　Rottboellia lepidura（Stapf）Pilg. = Coelorachis lepidura Stapf ■☆

337588　Rottboellia loliacea Bory et Chaub. = Lolium rigidum Gaudin ■

337589　Rottboellia longiflora Hook. f. = Hemarthria longiflora（Hook. f.）A. Camus ■

337590　Rottboellia longiflora Hook. f. var. tonkinensis（A. Camus）A.

Camus　= Hemarthria longiflora（Hook. f.）A. Camus ■

337591　Rottboellia loricata Trin. var. vautieri Roberty ＝ Rhytachne megastachya Jacq. -Fél. ■

337592　Rottboellia maitlandii（Stapf et C. E. Hubb.）Pilg. ＝ Loxodera ledermannii（Pilg.）Clayton ■☆

337593　Rottboellia mollicoma Hance ＝ Mnesithea mollicoma（Hance）A. Camus ■

337594　Rottboellia papillosa（Hochst.）T. Durand et Schinz ＝ Ophiuros papillosus Hochst. ■☆

337595　Rottboellia perforata Roxb. ＝ Mnesithea laevis（Retz.）Kunth ■

337596　Rottboellia protensa（Nees ex Steud.）Hack. ＝ Hemarthria vaginata Büse ■

337597　Rottboellia protensa（Steud.）Hack. ＝ Hemarthria protensa Steud. ■

337598　Rottboellia purpurascens Robyns ＝ Chasmopodium purpurascens（Robyns）Clayton ■☆

337599　Rottboellia ramosa Cav. ＝ Monerma cylindrica（Willd.）Coss. et Durieu ■☆

337600　Rottboellia repens G. Forst. ＝ Lepturus repens（G. Forst.）R. Br. ■

337601　Rottboellia rhytachne Hack. ＝ Rhytachne rottboellioides Desv. ex Ham. ■☆

337602　Rottboellia robusta（Stapf）Keng ＝ Rhytachne robusta Stapf ■☆

337603　Rottboellia sanguinea Retz. ＝ Schizachyrium sanguineum（Retz.）Alston ■

337604　Rottboellia setifolia K. Schum. ＝ Rhytachne rottboellioides Desv. ex Ham. ■☆

337605　Rottboellia sibirica Gand. ＝ Hemarthria sibirica（Gand.）Ohwi ■

337606　Rottboellia speciosa（Steud.）Hack. ＝ Phacelurus speciosus（Steud.）C. E. Hubb. ■☆

337607　Rottboellia striata Nees ex Steud. ＝ Coelorachis striata（Nees ex Steud.）A. Camus ■

337608　Rottboellia striata Nees ex Steud. ＝ Mnesithea striata（Steud.）de Koning et Sosef ■

337609　Rottboellia striata Nees ex Steud. subsp. genuina var. pubescens Hack. ＝ Coelorachis striata（Nees ex Steud.）A. Camus var. pubescens（Hack.）Bor ■

337610　Rottboellia striata Nees ex Steud. subsp. khasiana Hack. ＝ Mnesithea khasiana（Hack.）de Koning et Sosef ■

337611　Rottboellia striata Nees ex Steud. var. pubescens Hack. ＝ Mnesithea striata（Steud.）de Koning et Sosef var. pubescens（Hack.）S. M. Phillips et S. L. Chen ■

337612　Rottboellia sulcata Peter ＝ Phacelurus huillensis（Rendle）Clayton ■☆

337613　Rottboellia thyrsoidea Hack. ＝ Phacelurus zea（C. B. Clarke）Clayton ■

337614　Rottboellia thyrsoidea Hack. ＝ Thyrsia zea（C. B. Clarke）Stapf ■

337615　Rottboellia tonkinensis A. Camus ＝ Hemarthria longiflora（Hook. f.）A. Camus ■

337616　Rottboellia triaristata（Steud.）Roberty ＝ Rhytachne triaristata（Steud.）Stapf ☆

337617　Rottboellia tripsacoides Lam. ＝ Stenotaphrum secundatum（Walter）Kuntze ■

337618　Rottboellia undulatifolia Chiov. ＝ Phacelurus huillensis（Rendle）Clayton ■☆

337619　Rottboellia vaginata（Büse）Backer ＝ Hemarthria vaginata Büse ■

337620　Rottboellia zea C. B. Clarke ＝ Phacelurus zea（C. B. Clarke）Clayton ■

337621　Rottboellia zea C. B. Clarke ＝ Thyrsia zea（C. B. Clarke）Stapf ■

337622　Rottlera Roem. et Schult. ＝ Didymocarpus Wall.（保留属名）●■

337623　Rottlera Roem. et Schult. ＝ Roettlera Vahl ●■

337624　Rottlera Roxb. ＝ Mallotus Lour. ●

337625　Rottlera Willd.（1797）＝ Tetragastris Gaertn. ●☆

337626　Rottlera Willd.（1797）＝ Trewia L. ●

337627　Rottlera Willd.（1804）＝ Mallotus Lour. ●

337628　Rottlera acuminata A. Juss. ＝ Mallotus tiliifolius（Blume）Müll. Arg. ●

337629　Rottlera alba Roxb. ex Jack ＝ Mallotus paniculatus（Lam.）Müll. Arg. ●

337630　Rottlera aurantiaca Hook. et Arn. ＝ Mallotus philippensis（Lam.）Müll. Arg. ●

337631　Rottlera barbata Wall. ＝ Mallotus barbatus（Wall.）Müll. Arg. ●

337632　Rottlera cantoniensis Spreng. ＝ Mallotus apelta（Lour.）Müll. Arg. ●

337633　Rottlera chinensis A. Juss. ＝ Mallotus apelta（Lour.）Müll. Arg. ●

337634　Rottlera cordifolia Benth. ＝ Mallotus repandus（Willd.）Müll. Arg. ●

337635　Rottlera ferruginea Roxb. ＝ Mallotus tetracoccus（Roxb.）Kurz ●

337636　Rottlera glauca Hassk. ＝ Macaranga denticulata（Blume）Müll. Arg. ●

337637　Rottlera japonica（Thunb.）Spreng. ＝ Mallotus japonicus（L. f.）Müll. Arg. ●

337638　Rottlera japonica Spreng. ＝ Mallotus japonicus（L. f.）Müll. Arg. ●

337639　Rottlera multiglandulosa（Reinw. ex Blume）Blume ＝ Melanolepis multiglandulosa（Reinw. ex Blume）Rchb. f. et Zoll. ●

337640　Rottlera oblongifolia Miq. ＝ Mallotus oblongifolius（Miq.）Müll. Arg. ●

337641　Rottlera paniculata（Lam.）A. Juss. ＝ Mallotus paniculatus（Lam.）Müll. Arg. ●

337642　Rottlera peltata Roxb. ＝ Mallotus roxburghianus Müll. Arg. ●

337643　Rottlera philippinensis Scheff. ＝ Mallotus philippensis（Lam.）Müll. Arg. ●

337644　Rottlera scabrifolia A. Juss. ＝ Mallotus repandus（Willd.）Müll. Arg. ●

337645　Rottlera sinensis（Lindl.）Kuntze ＝ Chirita sinensis Lindl. ■

337646　Rottlera tanaria Hassk. ＝ Macaranga tanarius（L.）Müll. Arg. ●

337647　Rottlera tetracocca Roxb. ＝ Mallotus tetracoccus（Roxb.）Kurz ●

337648　Rottlera tiliifolia Blume ＝ Mallotus tiliifolius（Blume）Müll. Arg. ●

337649　Rottlera tinctoria Roxb. ＝ Mallotus philippensis（Lam.）Müll. Arg. ●

337650　Rottlera tomentosa（Blume）Hassk. ＝ Macaranga tanarius（L.）Müll. Arg. ●

337651　Rotula Lour.（1790）；轮冠木属；Rotula ●

337652　Rotula aquatica Lour. ；轮冠木；Water Rotula ●

337653　Roubieva Moq. ＝ Chenopodium L. ■●

337654　Roubieva Moq. ＝ Dysphania R. Br. ■●

337655　Roubieva multifida（L.）Moq. ＝ Dysphania multifidum（L.）Mosyakin et Clemants ■☆

337656　Roucela Dumort. ＝ Campanula L. ＋ Wahlenbergia Schrad. ex Roth（保留属名）■●

337657　Roucela Dumort. ＝ Campanula L. ■●

337658　Rouchera Hallier. f. = Roucheria Planch. ●☆

337659　Roucheria Miq. = Sarcotheca Blume ●☆

337660　Roucheria Planch. (1847);鲁谢麻属●☆

337661　Roucheria angulata Gleason;窄鲁谢麻●☆

337662　Roucheria columbiana Hallier;哥伦比亚鲁谢麻●☆

337663　Roucheria latifolia Spruce ex Benth. et Hook. f.;宽叶鲁谢麻●☆

337664　Roucheria laxiflora H. J. P. Winkl.;疏花鲁谢麻●☆

337665　Roucheria macrophylla Miq.;大叶鲁谢麻●☆

337666　Roucheria punctata (Ducke) Ducke;斑点鲁谢麻●☆

337667　Rouhamon Aubl. = Strychnos L. ●

337668　Roulinia Brongn. = Nolina Michx. ●☆

337669　Roulinia Decne. = Cynanchum L. ●■

337670　Roulinia Decne. = Rouliniella Vail ●■

337671　Rouliniella Vail = Cynanchum L. ●■

337672　Roumea DC. = Rumea Poit. ●

337673　Roumea DC. = Xylosma G. Forst. (保留属名) ●

337674　Roumea Wall. ex Meisn. = Daphne L. ●

337675　Roumea abyssinica A. Rich. = Dovyalis abyssinica (A. Rich.) Warb. ●☆

337676　Roupala Aubl. (1775);洛佩龙眼属(洛佩拉属);Roupala ●☆

337677　Roupala brachybotrys I. M. Johnst.;短穗洛佩龙眼●☆

337678　Roupala macrophylla Pohl;大叶洛佩龙眼(大叶洛佩拉);Largeleaf Roupala ●☆

337679　Roupala obovata Kunth;倒卵洛佩龙眼●☆

337680　Roupalia T. Moores et Ayres = Roupallia Hassk. ●

337681　Roupallia Hassk. = Roupellia Wall. et Hook. ex Benth. ●

337682　Roupallia Hassk. = Strophanthus DC. ●

337683　Roupelina Pichon = Roupellina (Baill.) Pichon ●

337684　Roupelina Pichon = Strophanthus DC. ●

337685　Roupellia Wall. et Hook. = Strophanthus DC. ●

337686　Roupellia Wall. et Hook. ex Benth. = Strophanthus DC. ●

337687　Roupellia grata Wall. et Hook. = Strophanthus gratus (Wall. et Hook. ex Benth.) Baill. ●

337688　Roupellia grata Wall. ex Hook. et Benth. = Strophanthus gratus (Wall. et Hook. ex Benth.) Baill. ●

337689　Roupellina (Baill.) Pichon = Strophanthus DC. ●

337690　Rourea Aubl. (1775)(保留属名);红叶藤属;Rourea ●

337691　Rourea adiantoides Gilg = Rourea obliquifoliolata Gilg ●☆

337692　Rourea afzelii R. Br. ex Planch. = Rourea minor (Gaertn.) Alston ●

337693　Rourea albidoflavescens Gilg = Rourea thomsonii (Baker) Jongkind ●☆

337694　Rourea bamangensis De Wild. et T. Durand = Rourea minor (Gaertn.) Alston ●

337695　Rourea baumannii Gilg = Rourea thomsonii (Baker) Jongkind ●☆

337696　Rourea bipindensis Gilg = Rourea minor (Gaertn.) Alston ●

337697　Rourea boiviniana Baill. = Rourea coccinea (Thonn. ex Schumach.) Benth. subsp. boiviniana (Baill.) Jongkind ●☆

337698　Rourea buchholzii Gilg = Rourea thomsonii (Baker) Jongkind ●☆

337699　Rourea calophylla (Gilg ex G. Schellenb.) Jongkind;美红叶藤●☆

337700　Rourea calophylloides (G. Schellenb.) Jongkind;拟美红叶藤●☆

337701　Rourea cassioides Hiern;决明红叶藤●☆

337702　Rourea caudata Planch.;长尾红叶藤;Caudate Rourea, Longtail Rourea ●

337703　Rourea chiliantha Gilg = Rourea minor (Gaertn.) Alston ●

337704　Rourea claessensii De Wild. = Rourea thomsonii (Baker) Jongkind ●☆

337705　Rourea coccinea (Thonn. ex Schumach.) Benth.;猩红叶藤●☆

337706　Rourea coccinea (Thonn. ex Schumach.) Benth. subsp. boiviniana (Baill.) Jongkind;鲍伊猩红叶藤●☆

337707　Rourea coccinea (Thonn. ex Schumach.) Benth. var. viridis (Gilg) Jongkind;绿花猩红叶藤●☆

337708　Rourea coriacea De Wild. = Rourea coccinea (Thonn. ex Schumach.) Benth. var. viridis (Gilg) Jongkind ●☆

337709　Rourea dinklagei Gilg = Rourea coccinea (Thonn. ex Schumach.) Benth. var. viridis (Gilg) Jongkind ●☆

337710　Rourea ealensis De Wild. = Rourea coccinea (Thonn. ex Schumach.) Benth. var. viridis (Gilg) Jongkind ●☆

337711　Rourea emarginata (Jack) Jongkind = Roureopsis emarginata (Jack) Merr. ●☆

337712　Rourea erecta Merr.;直立红叶藤●☆

337713　Rourea erythrocalyx (Gilg ex G. Schellenb.) Jongkind;红萼红叶藤●☆

337714　Rourea fasciculata Gilg = Rourea obliquifoliolata Gilg ●☆

337715　Rourea fasciculata Gilg var. flagelliflora Welw. ex Hiern = Rourea obliquifoliolata Gilg ●☆

337716　Rourea foenumgraecum De Wild. et T. Durand = Rourea coccinea (Thonn. ex Schumach.) Benth. var. viridis (Gilg) Jongkind ●☆

337717　Rourea goetzei Gilg = Rourea coccinea (Thonn. ex Schumach.) Benth. subsp. boiviniana (Baill.) Jongkind ●☆

337718　Rourea gudjuana Gilg = Rourea minor (Gaertn.) Alston ●

337719　Rourea heterophylla Baker = Rourea solanderi Baker ●☆

337720　Rourea inodora De Wild. et T. Durand = Rourea coccinea (Thonn. ex Schumach.) Benth. ●☆

337721　Rourea ivorensis A. Chev. = Rourea thomsonii (Baker) Jongkind ●☆

337722　Rourea laurentii De Wild. = Rourea coccinea (Thonn. ex Schumach.) Benth. var. viridis (Gilg) Jongkind ●☆

337723　Rourea lescrauwaetii De Wild. = Rourea thomsonii (Baker) Jongkind ●☆

337724　Rourea lescrauwaetii De Wild. var. seretii ? = Rourea thomsonii (Baker) Jongkind ●☆

337725　Rourea lescrauwaetii De Wild. var. tenuifolia ? = Rourea thomsonii (Baker) Jongkind ●☆

337726　Rourea macrantha Gilg = Rourea coccinea (Thonn. ex Schumach.) Benth. var. viridis (Gilg) Jongkind ●☆

337727　Rourea mannii Gilg = Rourea coccinea (Thonn. ex Schumach.) Benth. var. viridis (Gilg) Jongkind ●☆

337728　Rourea maxima (Baker) Gilg = Rourea coccinea (Thonn. ex Schumach.) Benth. subsp. boiviniana (Baill.) Jongkind ●☆

337729　Rourea microphylla (Hook. et Arn.) Planch.;小叶红叶藤(红叶秋树,红叶藤,荔枝藤,牛见愁,牛栓藤,山洋朵,铁藤,小叶牛栓藤);Little Leaf Rourea, Littleleaf Rourea, Little-leaved Rourea ●

337730　Rourea millettii Planch. = Rourea minor (Gaertn.) Leenh. ●

337731　Rourea mimosoides Planch.;含羞草红叶藤●☆

337732　Rourea minor (Gaertn.) Alston = Rourea minor (Gaertn.) Leenh. ●

337733　Rourea minor (Gaertn.) Leenh.;红叶藤(檀香红叶藤);Redleaf Rourea, Red-leaved Rourea, Rourea, Small Rourea ●

337734　Rourea minor (Gaertn.) Leenh. subsp. caudata (Planch.) Y. M. Shui = Rourea caudata Planch. ●

337735　Rourea minor (Gaertn.) Leenh. subsp. microphylla (Hook. et Arn.) Vidal = Rourea microphylla (Hook. et Arn.) Planch. ●

337736 Rourea minor (Gaertn.) Leenh. subsp. microphylla Vidal = Rourea microphylla (Hook. et Arn.) Planch. ●

337737 Rourea minor (Gaertn.) Leenh. subsp. monadelpha (Roxb.) J. E. Vidal;单体红叶藤●

337738 Rourea monticola Gilg = Rourea thomsonii (Baker) Jongkind ●☆

337739 Rourea myriantha Baill.;繁花红叶藤●☆

337740 Rourea nivea Gilg = Rourea thomsonii (Baker) Jongkind ●☆

337741 Rourea obliquifoliolata Gilg;斜小叶红叶藤●☆

337742 Rourea oblongifolia Hook. et Arn.;斜叶红叶藤●☆

337743 Rourea oddonii De Wild. = Rourea thomsonii (Baker) Jongkind ●☆

337744 Rourea orientalis Baill.;东方红叶藤●☆

337745 Rourea ovalifoliolata Gilg = Rourea coccinea (Thonn. ex Schumach.) Benth. var. viridis (Gilg) Jongkind ●☆

337746 Rourea ovatifolia (Baker) Gilg = Rourea coccinea (Thonn. ex Schumach.) Benth. subsp. boiviniana (Baill.) Jongkind ●☆

337747 Rourea palisotii (Planch.) Baill. = Rourea thomsonii (Baker) Jongkind ●☆

337748 Rourea pallens Hiern = Rourea coccinea (Thonn. ex Schumach.) Benth. var. viridis (Gilg) Jongkind ●☆

337749 Rourea parviflora Gilg;小花红叶藤●☆

337750 Rourea pervilleana Baill.;马岛红叶藤●☆

337751 Rourea platysepala Baker = Rourea minor (Gaertn.) Alston ●

337752 Rourea poggeana Gilg = Rourea coccinea (Thonn. ex Schumach.) Benth. var. viridis (Gilg) Jongkind ●☆

337753 Rourea pseudobaccata Gilg = Rourea thomsonii (Baker) Jongkind ●☆

337754 Rourea ptaeroxyloides Gilg = Rourea obliquifoliolata Gilg ●☆

337755 Rourea santaloides (Vahl) Wight et Arn. = Rourea minor (Gaertn.) Leenh. ●

337756 Rourea santaloides Wight et Arn. = Rourea minor (Gaertn.) Leenh. ●

337757 Rourea solanderi Baker;索兰德红叶藤●☆

337758 Rourea soyauxii Gilg = Rourea myriantha Baill. ●☆

337759 Rourea splendida Gilg = Rourea minor (Gaertn.) Alston ●

337760 Rourea striata De Wild. = Rourea minor (Gaertn.) Alston ●

337761 Rourea strigulosa Gilg = Rourea parviflora Gilg ●☆

337762 Rourea thomsonii (Baker) Jongkind;汤姆逊红叶藤●☆

337763 Rourea thonneri De Wild. = Rourea erythrocalyx (Gilg ex G. Schellenb.) Jongkind ●☆

337764 Rourea unifoliolata Gilg = Rourea coccinea (Thonn. ex Schumach.) Benth. var. viridis (Gilg) Jongkind ●☆

337765 Rourea usaramensis Gilg = Rourea coccinea (Thonn. ex Schumach.) Benth. subsp. boiviniana (Baill.) Jongkind ●☆

337766 Rourea venulosa Hiern = Rourea thomsonii (Baker) Jongkind ●☆

337767 Rourea verruculosa De Wild. = Rourea thomsonii (Baker) Jongkind ●☆

337768 Rourea viridis Gilg = Rourea coccinea (Thonn. ex Schumach.) Benth. var. viridis (Gilg) Jongkind ●☆

337769 Rourea volubilis Merr.;缠绕红叶藤●☆

337770 Roureopsis Planch. (1850);牛果藤属;Roureopsis ●

337771 Roureopsis Planch. = Rourea Aubl. (保留属名)●

337772 Roureopsis emarginata (Jack) Merr.;牛果藤(红果牛果藤,微凹牛果藤);Emarginate Roureopsis, Redfruit Roureopsis, Red-fruited Roureopsis, Roureopsis ●

337773 Roureopsis erythrocalyx Gilg ex G. Schellenb. = Rourea erythrocalyx (Gilg ex G. Schellenb.) Jongkind ●☆

337774 Roureopsis javanica Planch. = Roureopsis emarginata (Jack) Merr. ●

337775 Roureopsis obliquifoliolata (Gilg) G. Schellenb. = Rourea obliquifoliolata Gilg ●☆

337776 Roureopsis pubinervis Planch. = Roureopsis emarginata (Jack) Merr. ●

337777 Roureopsis rubicarpa C. Y. Wu = Roureopsis emarginata (Jack) Merr. ●

337778 Roureopsis rubricarpa C. Y. Wu = Roureopsis emarginata (Jack) Merr. ●

337779 Rouseauvia Bojer = Roussea Sm. ●☆

337780 Roussaea DC. = Roussea Sm. ●☆

337781 Roussea L. = Russelia Jacq. ■●

337782 Roussea L. ex B. D. Jacks. = Bistorta (L.) Adans. ■

337783 Roussea L. ex B. D. Jacks. = Russelia L. f. ■

337784 Roussea Sm. (1789);鲁索木属(卢梭木属,毛岛藤灌属)●☆

337785 Roussea simplex Sm.;鲁索木(毛岛藤灌)●☆

337786 Rousseaceae DC. (1839);鲁梭木科(卢梭木科,毛岛藤灌科)●☆

337787 Rousseaceae DC. = Brexiaceae Lindl. ●☆

337788 Rousseauxia DC. (1828);卢梭野牡丹属●☆

337789 Rousseauxia andringitrensis (H. Perrier) Jacq.-Fél.;安德林吉特拉山卢梭野牡丹●☆

337790 Rousseauxia chrysophylla (Desr.) DC.;金叶卢梭野牡丹●☆

337791 Rousseauxia cistoides Jacq.-Fél.;岩蔷薇卢梭野牡丹●☆

337792 Rousseauxia dionychoides (Cogn.) Jacq.-Fél.;泥双距野牡丹●☆

337793 Rousseauxia glauca Jacq.-Fél.;灰绿卢梭野牡丹●☆

337794 Rousseauxia gracilis Jacq.-Fél.;纤细卢梭野牡丹●☆

337795 Rousseauxia humbertii (H. Perrier) Jacq.-Fél.;亨伯特卢梭野牡丹●☆

337796 Rousseauxia madagascariensis (Cogn.) Jacq.-Fél.;马岛卢梭野牡丹●☆

337797 Rousseauxia mandrarensis (H. Perrier) Jacq.-Fél.;曼德拉卢梭野牡丹●☆

337798 Rousseauxia marojejensis Jacq.-Fél.;马鲁杰卢梭野牡丹●☆

337799 Rousseauxia minimifolia (Jum. et H. Perrier) Jacq.-Fél.;小叶卢梭野牡丹●☆

337800 Rousseauxia tamatavensis (H. Perrier) Jacq.-Fél.;塔马塔夫卢梭野牡丹●☆

337801 Rousselia Gaudich. (1830);耀麻属■☆

337802 Rousselia humilis (Sw.) Urb.;耀麻;Shineseed ■☆

337803 Roussinia Gaudich. = Pandanus Parkinson ex Du Roi ●■

337804 Roussoa Roem. et Schult. = Roussea Sm. ●☆

337805 Rouya Coincy(1901);洛伊草属■☆

337806 Rouya polygama (Desf.) Coincy;洛伊草■☆

337807 Rovaeanthus Borhidi = Bouvardia Salisb. ●■☆

337808 Rovillia Bubani = Polycnemum L. ■

337809 Roxburghia Banks = Stemona Lour. ■

337810 Roxburghia Koeniguer ex Roxb. (1795);钦百部属;Roxburghia, Thew Roxburghia ■☆

337811 Roxburghia Koeniguer ex Roxb. = Olax L. ●

337812 Roxburghia Roxb. = Stemona Lour. ■

337813 Roxburghia W. Jones ex Roxb. = Stemona Lour. ■

337814 Roxburghia gloriosa Pers. = Stemona tuberosa Lour. ■

337815 Roxburghia gloriosoides Roxb.;钦百部■☆

337816 Roxburghia japonica Blume = Stemona japonica (Blume) Miq. ■

337817 Roxburghia sessilifolia Miq. = Stemona sessilifolia (Miq.) Miq. ■

337818 Roxburghia stemona Steud. = Stemona tuberosa Lour. ■

337819 Roxburghia viridiflora Sm. = Stemona tuberosa Lour. ■

337820 Roxburghiaceae Wall. = Roxburghiaceae Wall. et Lindl. ∎

337821 Roxburghiaceae Wall. = Stemonaceae Caruel(保留科名)∎

337822 Roxburghiaceae Wall. et Lindl. ;钦百部科(百部科)∎

337823 Roxburghiaceae Wall. et Lindl. = Stemonaceae Caruel(保留科名)∎

337824 Roycea C. A. Gardner(1948);短被澳藜属●☆

337825 Roycea pycnophylloides C. A. Gardner;短被澳藜●☆

337826 Roydsia Roxb. = Stixis Lour. ●

337827 Roydsia fasciculata King = Stixis ovata (Korth.) Hallier f. subsp. fasciculata (King) Jacobs ●

337828 Roydsia suaveolens Roxb. = Stixis suaveolens (Roxb.) Pierre ●

337829 Royena L. = Diospyros L. ●

337830 Royena acocksii De Winter = Diospyros acocksii (De Winter) De Winter ●☆

337831 Royena ambigua Salisb. = Diospyros pallens (Thunb.) F. White ●☆

337832 Royena amnicola B. L. Burtt = Diospyros zombensis (B. L. Burtt) F. White ●☆

337833 Royena cordata E. Mey. ex A. DC. = Diospyros scabrida (Harv. ex Hiern) De Winter var. cordata (E. Mey. ex A. DC.) De Winter ●☆

337834 Royena dichrophylla Gand. = Diospyros dichrophylla (Gand.) De Winter ●☆

337835 Royena fischeri (Gürke) Mildbr. = Diospyros fischeri Gürke ●☆

337836 Royena galpinii Hiern = Diospyros galpinii (Hiern) De Winter ●☆

337837 Royena glabra L. = Diospyros glabra (L.) De Winter ●☆

337838 Royena glandulosa Harv. ex Hiern = Diospyros glandulifera De Winter ●☆

337839 Royena goetzei Gürke = Diospyros whyteana (Hiern) F. White ●☆

337840 Royena guerkei Kuntze = Diospyros lycioides Desf. subsp. guerkei (Kuntze) De Winter ●☆

337841 Royena heterotricha B. L. Burtt = Diospyros heterotricha (B. L. Burtt) F. White ●☆

337842 Royena hirsuta L. = Diospyros austro-africana De Winter ●☆

337843 Royena hirsuta L. var. rubriflora De Winter = Diospyros austro-africana De Winter var. rubriflora (De Winter) De Winter ●☆

337844 Royena lucida L. = Diospyros whyteana (Hiern) F. White ●☆

337845 Royena lucida L. var. whyteana (Hiern) De Winter et Brenan = Diospyros whyteana (Hiern) F. White ●☆

337846 Royena lycioides (Desf.) A. DC. = Diospyros lycioides Desf. ●☆

337847 Royena lycioides (Desf.) A. DC. subsp. guerkei (Kuntze) De Winter = Diospyros lycioides Desf. subsp. guerkei (Kuntze) De Winter ●☆

337848 Royena lycioides (Desf.) A. DC. subsp. nitens (Harv. ex Hiern) De Winter = Diospyros lycioides Desf. subsp. nitens (Harv. ex Hiern) De Winter ●☆

337849 Royena lycioides (Desf.) A. DC. subsp. sericea (Bernh.) De Winter = Diospyros lycioides Desf. subsp. sericea (Bernh.) De Winter ●☆

337850 Royena macrocalyx (Klotzsch) Gürke = Diospyros loureiriana G. Don ●☆

337851 Royena macrophylla E. Mey. ex A. DC. = Euclea schimperi (A. DC.) Dandy ●☆

337852 Royena microphylla Burch. = Diospyros austro-africana De Winter var. microphylla (Burch.) De Winter ●☆

337853 Royena nitens Harv. ex Hiern = Diospyros lycioides Desf. subsp. nitens (Harv. ex Hiern) De Winter ●☆

337854 Royena nyassae Gürke = Diospyros whyteana (Hiern) F. White ●☆

337855 Royena pallens Thunb. = Diospyros pallens (Thunb.) F. White ●☆

337856 Royena pentandra Gürke = Catunaregam pentandra (Gürke) Bridson ●☆

337857 Royena polyandra L. f. = Euclea polyandra (L. f.) E. Mey. ex Hiern ●☆

337858 Royena ramulosa E. Mey. ex A. DC. = Diospyros ramulosa (E. Mey. ex A. DC.) De Winter ●☆

337859 Royena rugosa E. Mey. ex A. DC. = Diospyros austro-africana De Winter var. rugosa (E. Mey. ex A. DC.) De Winter ●☆

337860 Royena scabrida Harv. ex Hiern = Diospyros scabrida (Harv. ex Hiern) De Winter ●☆

337861 Royena sericea Bernh. = Diospyros lycioides Desf. subsp. sericea (Bernh.) De Winter ●☆

337862 Royena simii Kuntze = Diospyros simii (Kuntze) De Winter ●☆

337863 Royena usambarensis Gürke ex Engl. = Diospyros loureiriana G. Don subsp. rufescens (Caveney) Verdc. ●☆

337864 Royena villosa L. = Diospyros villosa (L.) De Winter ●☆

337865 Royena villosa L. var. parvifolia De Winter = Diospyros villosa (L.) De Winter var. parvifolia (De Winter) De Winter ●☆

337866 Royena whyteana Hiern = Diospyros whyteana (Hiern) F. White ●☆

337867 Royena zombensis B. L. Burtt = Diospyros zombensis (B. L. Burtt) F. White ●☆

337868 Roylea Nees ex Steud. = Melanocenchris Nees ∎☆

337869 Roylea Steud. = Melanocenchris Nees ∎☆

337870 Roylea Wall. = Roylea Wall. ex Benth. ●☆

337871 Roylea Wall. ex Benth. (1829);罗氏草属●☆

337872 Roylea elegans Wall. ex Benth. ;罗氏草∎☆

337873 Roystonea O. F. Cook(1900);王棕属(大王椰属,大王椰子属,王椰属);Palm Hearts,Royal Palm,Royalpalm ●

337874 Roystonea altissima (Mill.) H. E. Moore;东拉王棕●☆

337875 Roystonea borinquena O. F. Cook;波多黎各王棕(波多黎各王棕);Palma Real,Puerto Rican Royal Palm,Puerto Rico Royalpalm, Royal Palm-of-porto Rico ●☆

337876 Roystonea caribaea (Spreng.) P. Wilson = Roystonea oleracea (Jacq.) O. F. Cook ●

337877 Roystonea dunlapiana P. H. Allen;东菜王棕●☆

337878 Roystonea elata (W. Bartram) F. Harper = Roystonea regia (Kunth) O. F. Cook ●

337879 Roystonea elata F. Harper;高王棕(佛罗里达王椰,高王椰); Florida Royal Palm ●☆

337880 Roystonea floridana O. F. Cook;古巴王棕;Cuban Royal Palm ●☆

337881 Roystonea floridana O. F. Cook = Roystonea regia (Kunth) O. F. Cook ●

337882 Roystonea hispaniola L. H. Bailey;西班牙王棕●☆

337883 Roystonea oleracea (Jacq.) O. F. Cook;菜王棕(菜王椰,甘蓝椰子,西印度椰子);Abage Palm, Cabage Palm, Cabage Tree, Cabbage Palm, Cabbage Royal Palm, Cabbage Tree, Carib Royal Palm, Caribbean Royal Palm, Caribbee Royal Palm, Greens Royalpalm,Palmiste ●

337884 Roystonea princeps Burret;帝王棕●☆

337885 Roystonea regia (Kunth) O. F. Cook;王棕(大王椰,大王椰子);Cuba Royalpalm, Cuban Royal Palm, Florida Royal Palm, Mountain Glory,Palmier Royal,Royal Palm,Royalpalm ●

337886 Roystonea venezuelana L. H. Bailey;委内瑞拉王椰(委内瑞拉王棕)●☆

337887 Rrynchoryza B. D. Jacks. = Oryza L. ∎

337888 Rrynchoryza B. D. Jacks. = Rhynchoryza Baill. ∎☆

337889 Ruagea H. Karst. (1863);卢栋属●☆

337890　Ruagea hirsuta（C. DC.）Harms；毛卢楝●☆

337891　Ruagea microphylla W. Palacios；小叶卢楝●☆

337892　Ruagea pubescens H. Karst. ；卢楝●☆

337893　Rubacer Rydb.（1903）；肖悬钩子属●■☆

337894　Rubacer Rydb. = Rubus L. ●■

337895　Rubacer odoratum（L.）Rydb. = Rubus odoratus L. ●☆

337896　Rubacer parviflorum（Nutt.）Rydb. = Rubus parvifolius L. ●

337897　Rubachia O. Berg = Marlierea Cambess. ●☆

337898　Rubellia（Luer）Luer = Pleurothallis R. Br. ☆

337899　Rubellia（Luer）Luer(2004)；巴拿马肋枝兰属■☆

337900　Rubentia Bojer ex Steud. = Toddalia Juss.（保留属名）●

337901　Rubentia Comm. ex Juss. = Elaeodendron J. Jacq. ●☆

337902　Rubeola Hill = Sherardia L. ■☆

337903　Rubeola Mill. = Crucianella L. ●■☆

337904　Rubeola angustifolia（L.）Fourr. = Crucianella angustifolia L. ☆

337905　Rubeola linearifolia Moench = Crucianella angustifolia L. ■☆

337906　Rubia L.（1753）；茜草属；Madder，Wild Madder ■

337907　Rubia akane Nakai；红藤仔草（过山龙）■

337908　Rubia akane Nakai = Rubia argyi（H. Lév. et Vaniot）H. Hara ex Lauener et D. K. Uson ■

337909　Rubia akane Nakai var. erecta Masam. ；直立红藤草■

337910　Rubia alata Wall. ；披针叶茜草（长叶茜草，红丝线，接筋草，金剑草，锯锯藤，老麻藤，茜草，四穗竹，沾沾草）；Lanceolate Madder，Longleaf Indian Madder ■

337911　Rubia angustifolia L. = Rubia peregrina L. var. angustifolia（L.）Webb ■☆

337912　Rubia argyi（H. Lév. et Vaniot）H. Hara ex Lauener et D. K. Uson；东南茜草（高原茜草，过山龙，红藤仔草，茜根，主线草）■

337913　Rubia argyi（H. Lév. et Vant）H. Hara ex L. = Rubia argyi（H. Lév. et Vaniot）H. Hara ex Lauener et D. K. Uson ■

337914　Rubia braunii Hochst. = Rubia discolor Turcz. ■☆

337915　Rubia chekiangensis Deb = Rubia argyi（H. Lév. et Vaniot）H. Hara ex Lauener et D. K. Uson ■

337916　Rubia chinensis Regel et Maack；中国茜草（大砧草，地苏木，疗毒草，茜草，小丹参，中华茜草）；China Madder，Chinese Madder ■

337917　Rubia chinensis Regel et Maack f. glabrescens（Nakai）Kitag. = Rubia chinensis Regel et Maack f. mitis（Miq.）Kitag. ■

337918　Rubia chinensis Regel et Maack f. glabrescens（Nakai）Kitag. = Rubia chinensis Regel et Maack var. glabrescens（Nakai）Kitag. ■

337919　Rubia chinensis Regel et Maack f. mitis（Miq.）Kitag. ；光滑大砧草■

337920　Rubia chinensis Regel et Maack var. esquirolii（H. Lév.）H. Lév. = Rubia schumanniana E. Pritz. ■

337921　Rubia chinensis Regel et Maack var. glabrescens（Nakai）Kitag. ；无毛大砧草；Glabrous Chinese Madder ■

337922　Rubia chinensis Regel et Maack var. glabrescens（Nakai）Kitag. = Rubia chinensis Regel et Maack f. mitis（Miq.）Kitag. ■

337923　Rubia chitralensis Ehrend. ；高原茜草；Highland Madder ■

337924　Rubia conotricha Gand. = Rubia cordifolia L. subsp. conotricha（Gand.）Verdc. ■☆

337925　Rubia cordifolia L. ；茜草（地苏木，过山龙，黑果茜草，红根草，红楝子，红龙须，红茜草，红丝线，红藤仔草，红线草，活血丹，九龙根，锯锯藤，锯子草，拉拉，满江红，破血丹，茜草藤，茜草头，染蛋藤，染绯草，茹藤，三爪龙，沙雨秧，四方红，四轮车，土丹参，娃娃拳，五爪龙，小孩拳，小活血丹，小血藤，心叶茜草，血见愁，血藤，粘粘草）；Black Fruit Madder，India Madder，Indian Madder，Munjeet ■

337926　Rubia cordifolia L. = Rubia wallichiana Decne. ■

337927　Rubia cordifolia L. f. pratensis（Maxim.）Kitag. = Rubia cordifolia L. ■

337928　Rubia cordifolia L. subsp. conotricha（Gand.）Verdc. ；束毛茜草■☆

337929　Rubia cordifolia L. subsp. hexaphylla（Makino）Kitam. = Rubia hexaphylla（Makino）Makino ■☆

337930　Rubia cordifolia L. subsp. pratensis（Maxim.）Kitam. = Rubia cordifolia L. ■

337931　Rubia cordifolia L. var. alaschanica G. H. Liu；阿拉善茜草；Alashan Madder ■

337932　Rubia cordifolia L. var. coriacea Z. Ying Zhang；革叶茜草；Coriaceous Indian Madder ■

337933　Rubia cordifolia L. var. discolor（Turcz.）K. Schum. = Rubia cordifolia L. subsp. conotricha（Gand.）Verdc. ■☆

337934　Rubia cordifolia L. var. herbacea F. H. Chen et F. C. How；肉叶茜草（肉质茜草，小血藤）；Herbaceous Indian Madder ■

337935　Rubia cordifolia L. var. hexaphylla Makino = Rubia hexaphylla（Makino）Makino ■☆

337936　Rubia cordifolia L. var. khasiana Watt = Rubia manjith Roxb. ex Fleming ■

337937　Rubia cordifolia L. var. longifolia Hand. -Mazz. = Rubia alata Wall. ■

337938　Rubia cordifolia L. var. maillardii（H. Lév. et Vaniot）H. Lév. = Rubia schumanniana E. Pritz. ■

337939　Rubia cordifolia L. var. manjista（Roxb.）Miq. f. rubra Kitam. = Rubia manjith Roxb. ex Fleming ■

337940　Rubia cordifolia L. var. mollis F. H. Chen et F. C. How；柔毛茜草；Hairy Indian Madder ■

337941　Rubia cordifolia L. var. mungista Miq. = Rubia argyi（H. Lév. et Vaniot）H. Hara ex Lauener et D. K. Uson ■

337942　Rubia cordifolia L. var. pratensis Maxim. = Rubia cordifolia L. ■

337943　Rubia cordifolia L. var. pubescens（Nakai）Nakai ex W. T. Lee = Rubia pubescens Nakai ■☆

337944　Rubia cordifolia L. var. rotundifolia Franch. = Rubia cordifolia L. ■

337945　Rubia cordifolia L. var. stenophylla Franch. ；狭叶茜草（四棱草，四轮草，土茜草，小叶红丝线）；Narrowleaf Indian Madder ■

337946　Rubia cordifolia L. var. sylvatica Maxim. = Rubia sylvatica（Maxim.）Nakai ■

337947　Rubia crassipes Collett et Hemsl. ；厚柄茜草；Thickstalk Madder ■

337948　Rubia cretacea Pojark. ；白垩茜草■☆

337949　Rubia deserticola Pojark. ；沙生茜草；Sandy Madder ■

337950　Rubia discolor Turcz. ；异色茜草■☆

337951　Rubia discolor Turcz. = Rubia cordifolia L. subsp. conotricha（Gand.）Verdc. ■☆

337952　Rubia dolichophylla Schrenk；长叶茜草；Longleaf Madder ■

337953　Rubia edgeworthii Hook. f. ；川滇茜草；Edgeworth Madder ■

337954　Rubia esquirolii H. Lév. = Rubia schumanniana E. Pritz. ■

337955　Rubia falciformis H. S. Lo；镰叶茜草；Falcateleaf Madder ■

337956　Rubia filiformis F. C. How ex H. S. Lo；丝梗茜草；Filistalk Madder ■

337957　Rubia florida Boiss. ；繁花茜草■☆

337958　Rubia fruticosa Aiton；灌丛茜草■☆

337959　Rubia fruticosa Aiton subsp. melanocarpa（Bornm.）Bramwell；黑果灌丛茜草■☆

337960　Rubia fruticosa Aiton var. angustifolia Kuntze = Rubia fruticosa

Aiton ■☆

337961　Rubia fruticosa Aiton var. melanocarpa Bornm. = Rubia fruticosa Aiton subsp. melanocarpa（Bornm.）Bramwell ■☆

337962　Rubia fruticosa Aiton var. pendula Pit. = Rubia fruticosa Aiton ■☆

337963　Rubia garrettii Craib;嘎氏茜草■☆

337964　Rubia haematantha Airy Shaw;红花茜草（血花茜草）; Redflower Madder ■

337965　Rubia hexaphylla（Makino）Makino;六叶茜草☆

337966　Rubia himalayensis Klotzsch;喜马拉雅茜草■☆

337967　Rubia horrida（Thunb.）Puff;多刺茜草■☆

337968　Rubia iberica K. Koch;格鲁吉亚茜草■☆

337969　Rubia javana DC.;爪哇茜草■☆

337970　Rubia jesoensis（Miq.）Miyabe et Miyake;北海道茜草;Yezo Madder ■☆

337971　Rubia komarovii Pojark.;科马罗夫茜草■☆

337972　Rubia krascheninnikovii Pojark.;克拉氏茜草■☆

337973　Rubia laevis Poir.;平滑茜草■☆

337974　Rubia laevis Thunb. = Rubia thunbergii DC. ■☆

337975　Rubia lanceolata Hayata;金剑草■

337976　Rubia lanceolata Hayata = Rubia alata Wall. ■

337977　Rubia latipetala H. S. Lo;阔瓣茜草;Broadpetal Madder ■

337978　Rubia laxiflora Gontsch.;疏花茜草■☆

337979　Rubia leiocaulis（Franch.）Diels = Rubia schumanniana E. Pritz. ■

337980　Rubia leiocaulis Diels = Rubia schumanniana E. Pritz. ■

337981　Rubia linii J. M. Chao;林氏茜草（圆茎茜草）;Lin Madder ■

337982　Rubia longifolia Poir. = Rubia peregrina L. subsp. longifolia（Poir.）O. Bolòs ■☆

337983　Rubia longipetiolata Bullock;长梗茜草■☆

337984　Rubia longipetiolata Bullock = Rubia cordifolia L. subsp. conotricha（Gand.）Verdc. ☆

337985　Rubia lucida L.;光亮茜草■☆

337986　Rubia lucida L. = Rubia peregrina L. ■☆

337987　Rubia magna P. K. Hsiao;峨眉茜草（大茜草,茜草）;Emei Madder ■

337988　Rubia maillardii H. Lév. et Vaniot = Rubia schumanniana E. Pritz. ■

337989　Rubia mandersii Collett et Hemsl.;黑花茜草（大理茜草,代褐茜草,带褐茜草,鸡血生,小茜草）;Dali Madder ■

337990　Rubia manjista Roxb. = Rubia manjith Roxb. ex Fleming ■

337991　Rubia manjith Roxb. ex Desv. = Rubia manjith Roxb. ex Fleming ■

337992　Rubia manjith Roxb. ex Fleming;梵茜草（茜草,青藏茜草）■

337993　Rubia membranacea Diels;膜茜草（大活血丹,金线草,膜叶茜草,小茜草,猪猪藤）;Membranaceus Madder ■

337994　Rubia membranacea Diels var. caudata Z. Ying Zhang;尾叶茜草;Caudate Membranaceus Madder ■

337995　Rubia membranacea Diels var. incurvata Z. Ying Zhang;内弯茜草;Incurvate Membranaceus Madder ■

337996　Rubia mitis Miq. = Rubia chinensis Regel et Maack ■

337997　Rubia mitis Miq. f. glabrescens Nakai = Rubia chinensis Regel et Maack var. glabrescens（Nakai）Kitag. ■

337998　Rubia mitis Miq. f. glabrescens Nakai = Rubia chinensis Regel et Maack ■

337999　Rubia nankotaizana Masam. = Rubia akane Nakai var. erecta Masam. ■

338000　Rubia nephrophylla Deb = Rubia podantha Diels ■

338001　Rubia olivieri A. Rich. = Rubia tenuifolia d'Urv. ■☆

338002　Rubia oncotricha Hand. -Mazz.;钩毛茜草（红丝线,活血丹,四棱草,小茜草,小血藤）;Hookedhair Madder ■

338003　Rubia ovatifolia Z. Ying Zhang;卵叶茜草（茜草红蛇儿,小红藤）;Ovateleaf Madder ■

338004　Rubia ovatifolia Z. Ying Zhang var. oligantha H. S. Lo;少花卵叶茜草（少花茜草）;Fewflower Madder ■

338005　Rubia pallida Diels;浅色茜草;Pallid Madder ■

338006　Rubia pauciflora Boiss.;少花茜草■☆

338007　Rubia peregrina L.;洋茜草;Evergreen Clivers,Levant Madder, Wild Madder ■☆

338008　Rubia peregrina L. subsp. longifolia（Poir.）O. Bolòs;长叶洋茜草■☆

338009　Rubia peregrina L. var. anglica Huds. = Rubia peregrina L. subsp. longifolia（Poir.）O. Bolòs ■☆

338010　Rubia peregrina L. var. angustifolia（L.）Webb = Rubia peregrina L. subsp. longifolia（Poir.）O. Bolòs ■☆

338011　Rubia peregrina L. var. intermedia Gren. et Godr. = Rubia peregrina L. ■☆

338012　Rubia peregrina L. var. latifolia Gren. et Godr. = Rubia peregrina L. ■☆

338013　Rubia peregrina L. var. rotundifolia Maire = Rubia peregrina L. ■☆

338014　Rubia petiolaris DC.;柄叶茜草■☆

338015　Rubia petiolaris DC. var. heterophylla Sond. = Rubia horrida（Thunb.）Puff ■☆

338016　Rubia petiolaris DC. var. isophylla Sond. = Rubia petiolaris DC. ■☆

338017　Rubia podantha Diels;柄花茜草（大茜草,红花茜草,活血草,逆刺,小血藤）;Redflower Madder ■

338018　Rubia polyphlebia H. S. Lo;多脉茜草;Manyvein Madder ■

338019　Rubia pratensis（Maxim.）Nakai = Rubia cordifolia L. ■

338020　Rubia pterygocaulis H. S. Lo;翅茎茜草;Wingstem Madder ■

338021　Rubia pubescens Nakai;毛茜草■☆

338022　Rubia rechingeri Ehrend.;雷氏茜草■☆

338023　Rubia regelii Pojark.;雷格尔茜草■☆

338024　Rubia rezniczenkoana Litv.;小叶茜草;Littleleaf Madder ■

338025　Rubia ruwenzoriensis Cortesi = Galium ruwenzoriense（Cortesi）Chiov. ■☆

338026　Rubia salicifolia H. S. Lo;柳叶茜草;Willowleaf Madder ■

338027　Rubia schugnanica B. Fedtsch. ex Pojark.;四叶茜草;Fourleaves Madder ■

338028　Rubia schumanniana E. Pritz.;大叶茜草（大茜草,灯儿草,红血儿,女儿红,茜草,茜草藤,沙糖根,四块瓦,四轮筋草,四能草,四片瓦,土茜草,西南茜草,小红参,小红藤,小血散,紫脉茜草）;Smoothstalk Madder ■

338029　Rubia schumanniana E. Pritz. var. maillardii（H. Lév. et Vaniot）Hand. -Mazz. = Rubia schumanniana E. Pritz. ■

338030　Rubia siamensis Craib;对叶茜草;Pairleaf Madder ■

338031　Rubia sikkimensis Kurz;锡金茜草■☆

338032　Rubia sylvatica（Maxim.）Nakai;林生茜草（穿心草,林茜草,茜草）;Forest Madder ■

338033　Rubia tatarica Schmidt;鞑靼茜草;Tatar Madder ■☆

338034　Rubia tenuifolia d'Urv.;细叶茜草■☆

338035　Rubia tenuis H. S. Lo;纤梗茜草;Thin Madder ■

338036　Rubia thunbergii DC.;通贝里茜草■☆

338037　Rubia tibetica Hook. f.;西藏茜草;Tibet Madder,Xizang Madder ■

338038　Rubia tinctoria L.;染色茜草（欧茜草,茜草,染料茜草,西洋茜草,洋茜草）;Alizarin, Common Madder, Dyer's Madder, Dyer's

Root，Madder，Red Madder，Warence ■

338039　Rubia trichocarpa H. S. Lo；毛果茜草；Hairfruit Madder ■

338040　Rubia truppeliana Loes.；山东茜草（披针叶茜草，茜草，狭叶茜草）；Shandong Madder ■

338041　Rubia ustulata Diels；大理茜草（带褐茜草）■

338042　Rubia ustulata Diels = Rubia yunnanensis（Franch.）Diels ■

338043　Rubia wallichiana Decne.；多花茜草（光茎茜草，红丝线，金钱草，三爪龙）；Wallich Madder ■

338044　Rubia yunnanensis（Franch.）Diels；云南茜草（滇茜草，滇紫参，红根，小红参，小红药，小活血，小茜草，小舒筋，紫参）；Yunnan Madder ■

338045　Rubiaceae Juss.（1789）（保留科名）；茜草科；Bedstraw Family，Madder Family ●■

338046　Rubimons B. S. Sun = Miscanthus Andersson ■

338047　Rubimons paniculatus B. S. Sun = Miscanthus paniculatus（B. S. Sun）Renvoize et S. L. Chen ■

338048　Rubina Noronha = Antidesma L. ●

338049　Rubioides Perkins = Opercularia Gaertn. ■☆

338050　Rubioides Sol. ex Gaertn. = Opercularia Gaertn. ■☆

338051　Rubiteucris Kudo = Teucrium L. ●■

338052　Rubiteucris Kudo（1929）；掌叶石蚕属（野藿香属）；Palmgermander ■

338053　Rubiteucris palmata（Benth. ex Hook. f.）Kudo；掌叶石蚕（裂叶苦草，掌叶野藿香，掌状香科，掌状野藿香）；Palmate Germander，Palmgermander ■

338054　Rubovietnamia Tirveng.（1998）；越南茜属●

338055　Rubovietnamia aristata Tirveng.；越南茜（长管越南茜）●

338056　Rubrivena M. Král = Persicaria（L.）Mill. ■

338057　Rubrivena polystachya（Wall. ex Meisn.）Král = Persicaria wallichii Greuter et Burdet ■☆

338058　Rubrivena polystachya（Wall. ex Meisn.）Král = Polygonum polystachyum Wall. ex Meisn. ●■

338059　Rubus L.（1753）；悬钩子属；Blackberry，Bramble，Brambles，Dewberry，Hildaberry，Marionberry，Nectarberry，Phenomenal Berry，Raspberry，Silvaberry，Sun Berry，Taybercy，Tummelberry，Veitchberry，Youngberry ●■

338060　Rubus × babae Naruh.；马场悬钩子●☆

338061　Rubus × calopalmatus Naruh. et Masaki；美掌悬钩子●☆

338062　Rubus × hiraseanus Makino；平世悬钩子●☆

338063　Rubus × kenoensis Koidz.；毛野悬钩子●☆

338064　Rubus × medius Kuntze；中间悬钩子●☆

338065　Rubus × nikaii Ohwi；二阶氏悬钩子●☆

338066　Rubus × ohmineanus Koidz.；大峰悬钩子●☆

338067　Rubus × omogoensis Koidz. = Rubus × ribifolius Siebold et Zucc. ●☆

338068　Rubus × paxii Focke；帕克斯悬钩子●☆

338069　Rubus × pseudohakonensis Sugim.；假箱根悬钩子●☆

338070　Rubus × pseudoyoshinoi Naruh. et Masaki；假吉野氏悬钩子●☆

338071　Rubus × ribifolius Siebold et Zucc.；茶藨叶悬钩子●☆

338072　Rubus × vinaceus T. Yamanaka ex Momiy.；葡萄酒色悬钩子●☆

338073　Rubus abbrevians Blanch. = Rubus vermontanus Blanch. ☆

338074　Rubus abnormis Sudre；异常悬钩子●☆

338075　Rubus aboriginum Rydb.；土著悬钩子●☆

338076　Rubus acadiensis L. H. Bailey = Rubus glandicaulis Blanch. ●☆

338077　Rubus acaenocalyx H. Hara = Rubus alexeterius Focke var. acaenocalyx（H. Hara）Te T. Yu et L. T. Lu ●

338078　Rubus acaulis Michx.；无茎悬钩子；Stemless Arctic Bramble ●☆

338079　Rubus aculeatiflorus Hayata；刺花悬钩子（刺萼悬钩子，皮刺叶悬钩子）；Prickly-calyx Raspberry，Spinyflower Raspberry，Spiny-flowered Raspberry ●

338080　Rubus aculeatiflorus Hayata = Rubus taitoensis Hayata var. aculeatiflorus（Hayata）H. Ohashi et C. F. Hsieh ●

338081　Rubus aculeatiflorus Hayata var. taitoensis（Hayata）Yang S. Liu et T. Y. Yang = Rubus taitoensis Hayata ●

338082　Rubus acuminatus Sm.；尖叶悬钩子（光叶悬钩子）；Acuminate Raspberry ●

338083　Rubus acuminatus Sm. var. puberulus Te T. Yu et L. T. Lu；柔毛尖叶悬钩子（柔毛光叶悬钩子）；Hairy Acuminate Raspberry ●

338084　Rubus adenanthus Finet et Franch. = Rubus swinhoei Hance ●

338085　Rubus adenochlamys（Focke）Focke = Rubus parvifolius L. var. adenochlamys（Focke）Migo ●

338086　Rubus adenochlamys（Focke）Focke var. orientalis F. P. Metcalf = Rubus parvifolius L. var. adenochlamys（Focke）Migo ●

338087　Rubus adenocomus（Focke）Gust.；热非腺毛莓●☆

338088　Rubus adenophorus Rolfe；腺毛莓（红毛草，红毛牛刺，雀不站）；Glandular-haired Raspberry，Glandularhairy Raspberry ●

338089　Rubus adenothyrsus Cardot = Rubus lambertianus Ser. var. glandulosus Cardot ●

338090　Rubus adenotrichopodus Hayata = Rubus swinhoei Hance ●

338091　Rubus adjacens Fernald；泥煤悬钩子；Peaty Dewberry ●☆

338092　Rubus adolfi-friedericii Engl. = Rubus apetalus Poir. ●☆

338093　Rubus aethiopicus R. A. Graham；埃塞俄比亚悬钩子●☆

338094　Rubus affinis Wight et Arn.；近缘悬钩子●☆

338095　Rubus alacer L. H. Bailey = Rubus flagellaris Willd. ●☆

338096　Rubus albescens Roxb. = Rubus niveus Thunb. ●

338097　Rubus alceifolius Poir.；粗叶悬钩子（八月泡，大笋坛，大破皮刺，大叶蛇泡筋，狗头泡，海南悬钩子，虎掌筋，九月泡，老虎泡，牛尾泡，羽萼悬钩子）；Roughleaf Raspberry，Rough-leaved Raspberry ●

338098　Rubus alceifolius Poir. var. diversilobatus（Merr. et Chun）Te T. Yu et L. T. Lu = Rubus alceifolius Poir. ●

338099　Rubus alceifolius Poir. var. diversilobatus（Merr. et Chun）Te T. Yu et L. T. Lu；深裂粗叶悬钩子；Variable-lobe Roughleaf Raspberry ●

338100　Rubus alceifolius Poir. var. emigratus sensu Koidz. = Rubus nagasawanus Koidz. ●

338101　Rubus alexeterius Focke；刺萼悬钩子（黄琐梅）；Spinycalyx Raspberry，Spiny-calyx Raspberry ●

338102　Rubus alexeterius Focke var. acaenocalyx（H. Hara）Te T. Yu et L. T. Lu；腺毛刺萼悬钩子；Glandular Spinycalyx Raspberry ●

338103　Rubus alleghaniensis L. Bailey = Rubus alleghaniensis Porter ex L. H. Bailey ●☆

338104　Rubus alleghaniensis Porter = Rubus alleghaniensis Porter ex L. H. Bailey ●☆

338105　Rubus alleghaniensis Porter ex L. H. Bailey；细枝悬钩子（普通悬钩子）；Aneghany Black-berry Allegheny Blackberry，Blackberry，Common Blackberry，Highbush Blackberry，Sow-teat Blackberry ●☆

338106　Rubus alleghaniensis Porter ex L. H. Bailey f. albinus（L. H. Bailey）Fernald = Rubus alleghaniensis Porter ex L. H. Bailey ●☆

338107　Rubus alleghaniensis Porter ex L. H. Bailey f. calycosus（Fernald）Fernald = Rubus alleghaniensis Porter ex L. H. Bailey ●☆

338108　Rubus alleghaniensis Porter ex L. H. Bailey f. rubrobaccus L. P. Wolfe et Hodgdon = Rubus alleghaniensis Porter ex L. H. Bailey ●☆

338109　Rubus alleghaniensis Porter ex L. H. Bailey var. albinus（L. H.

Bailey) L. H. Bailey ＝ Rubus allegheniensis Porter ex L. H. Bailey ●☆

338110　Rubus allegheniensis Porter ex L. H. Bailey var. calycosus (Fernald) Fernald ＝ Rubus allegheniensis Porter ex L. H. Bailey ●☆

338111　Rubus allegheniensis Porter ex L. H. Bailey var. gravesii Fernald；格氏悬钩子；Allegheny Blackberry，Graves' Blackberry ●☆

338112　Rubus allegheniensis Porter ex L. H. Bailey var. nigrobaccus (L. H. Bailey) Farw. ＝ Rubus allegheniensis Porter ex L. H. Bailey ●☆

338113　Rubus allegheniensis Porter ex L. H. Bailey var. plausus L. H. Bailey ＝ Rubus allegheniensis Porter ex L. H. Bailey ●☆

338114　Rubus allegheniensis Porter ex L. H. Bailey var. populifolius Fernald ＝ Rubus allegheniensis Porter ex L. H. Bailey ●☆

338115　Rubus allegheniensis Porter var. plausus L. H. Bailey ＝ Rubus allegheniensis Porter ●☆

338116　Rubus alnifoliolatus H. Lév. et Vaniot；椆叶悬钩子；Alderleaf Raspberry，Alder-leaved Raspberry ●

338117　Rubus alnifoliolatus H. Lév. et Vaniot var. kotoensis (Hayata) H. L. Li ＝ Rubus fraxinifolius Poir. ●

338118　Rubus althaeoides Hance ＝ Rubus corchorifolius L. f. ●

338119　Rubus alumnus L. H. Bailey；田野莓；Old-field Blackberry ●☆

338120　Rubus amabilis Blanch. ＝ Rubus elegantulus Blanch. ●☆

338121　Rubus amabilis Focke；秀丽莓（美丽悬钩子，秀丽悬钩子）；Elegant Raspberry ●

338122　Rubus amabilis Focke var. aculeatissimus Te T. Yu et L. T. Lu；刺萼秀丽莓；Spinecalyx Elegant Raspberry ●

338123　Rubus amabilis Focke var. microcarpus Te T. Yu et L. T. Lu；小果秀丽莓；Littlefruit Elegant Raspberry，Littlefruit Raspberry ●

338124　Rubus amamianus Hatus. et Ohwi；奄美秀丽莓●☆

338125　Rubus amamianus Hatus. et Ohwi var. minor Hatus.；小奄美秀丽莓●☆

338126　Rubus americanus Britton ＝ Rubus pubescens Raf. ●☆

338127　Rubus amicalis Blanch. ＝ Rubus elegantulus Blanch. ●☆

338128　Rubus ampelinus Focke ＝ Rubus lambertianus Ser. var. glaber Hemsl. ●

338129　Rubus ampelophyllus H. Lév. ＝ Rubus crataegifolius Bunge ●

338130　Rubus amphidasys Focke；周毛悬钩子（红毛猫耳扭，老虎泡藤，全毛泡藤，周毛莓）；Big-leaved Bramble，Hairy Raspberry ●

338131　Rubus ampliflorus H. Lév. et Vaniot ＝ Rubus tephrodes Hance var. ampliflorus (H. Lév. et Vaniot) Hand. -Mazz. ●

338132　Rubus andropogon H. Lév. ＝ Rubus multibracteatus H. Lév. et Vaniot ●

338133　Rubus andropogon H. Lév. ＝ Rubus pluribracteatus L. T. Lu et Boufford ●

338134　Rubus angustibracteatus Te T. Yu et L. T. Lu；狭苞悬钩子；Angustibracteate Raspberry，Narrowbract Raspberry ●

338135　Rubus apetalus Poir.；无瓣悬钩子●☆

338136　Rubus apetalus Poir. f. pyramidalis Gust. ＝ Rubus apetalus Poir. ●☆

338137　Rubus apetalus Poir. var. grossoserratus Hauman；粗齿无瓣悬钩子●☆

338138　Rubus apianus L. H. Bailey ＝ Rubus alumnus L. H. Bailey ●☆

338139　Rubus arachnoideus Y. C. Liu et F. Y. Lu；灰叶悬钩子●

338140　Rubus aralioides Hance ＝ Rubus innominatus S. Moore var. aralioides (Hance) Te T. Yu et L. T. Lu ●

338141　Rubus arbor H. Lév. et Vaniot ＝ Rubus malifolius Focke ●

338142　Rubus arcticus L.；北悬钩子（北极悬钩子）；Arctic Bramble，Arctic Raspberry，Dwarf Raspberry ■

338143　Rubus arcticus L. var. fragarioides (Bertol.) Focke ＝ Rubus fragarioides Bertol. ●■

338144　Rubus arcuatus Kuntze ＝ Rubus treutleri Hook. f. ●

338145　Rubus argutus Link ＝ Rubus ostryifolius Rydb. ●☆

338146　Rubus argutus Link var. randii (L. H. Bailey) L. H. Bailey ＝ Rubus canadensis L. ●☆

338147　Rubus argyi H. Lév. ＝ Rubus hirsutus Thunb. ●

338148　Rubus arisanensis Hayata ＝ Rubus corchorifolius L. f. ●

338149　Rubus arisanensis Hayata var. horishanensis Hayata ＝ Rubus corchorifolius L. f. ●

338150　Rubus arizonicus Rydb.；亚利桑那悬钩子；Arizona Bramble，Arizona Red Raspberry ●☆

338151　Rubus armeniacus Focke；亚美尼亚悬钩子；Himalayan Blackberry ●☆

338152　Rubus asper Wall. ex G. Don ＝ Rubus croceacanthus H. Lév. ●

338153　Rubus asper Wall. ex G. Don ＝ Rubus sumatranus Miq. ●

338154　Rubus asper Wall. ex G. Don var. glaber (Koidz.) C. F. Hsieh ＝ Rubus croceacanthus H. Lév. var. glaber Koidz. ●

338155　Rubus asper Wall. ex G. Don var. myriadenus (H. Lév. et Vaniot) Focke subvar. grandifoliolatus (H. Lév.) Focke ＝ Rubus sumatranus Miq. ●

338156　Rubus asper Wall. ex G. Don var. pekanins Focke ＝ Rubus croceacanthus H. Lév. ●

338157　Rubus asper Wall. ex G. Don var. pekanius Focke ＝ Rubus sumatranus Miq. ●

338158　Rubus assamensis Focke；西南悬钩子；Assam Raspberry，Southwestern Raspberry ●

338159　Rubus assaortinus Chiov. ＝ Rubus apetalus Poir. ●☆

338160　Rubus assaortinus Chiov. var. erythraeus Gust. ＝ Rubus apetalus Poir. ●☆

338161　Rubus associus Hanes ＝ Rubus uvidus L. H. Bailey ●☆

338162　Rubus atlanticus Pomel ＝ Rubus incanescens (DC.) Bertol. ●☆

338163　Rubus atrocoeruleus Gust. ＝ Rubus rigidus Sm. ●☆

338164　Rubus attractus L. H. Bailey ＝ Rubus allegheniensis Porter ex L. H. Bailey ●☆

338165　Rubus atwoodii L. H. Bailey ＝ Rubus glandicaulis Blanch. ●☆

338166　Rubus aurantiacus Focke；橘红悬钩子；Orange Raspberry，Orangedred Raspberry ●

338167　Rubus aurantiacus Focke var. obtusifolius Te T. Yu et L. T. Lu；钝叶橘红悬钩子；Obtuseleaf Orangedred Raspberry ●

338168　Rubus auroralis L. H. Bailey ＝ Rubus allegheniensis Porter ex L. H. Bailey ●☆

338169　Rubus australis G. Forst.；澳洲悬钩子；Bush Lawyer ●☆

338170　Rubus austrinus L. H. Bailey ＝ Rubus steelei L. H. Bailey ●☆

338171　Rubus austro-tibetanus Te T. Yu et L. T. Lu；藏南悬钩子；S. Xizang Raspberry，South Tibet Raspberry，South Xizang Raspberry ●

338172　Rubus avipes L. H. Bailey ＝ Rubus allegheniensis Porter ex L. H. Bailey ●☆

338173　Rubus axilliflorens Cardot ＝ Rubus reflexus Ker Gawl. var. hui (Diels) F. P. Metcalf ●

338174　Rubus bahanensis Hand. -Mazz. ＝ Rubus assamensis Focke ●

338175　Rubus baileyanus Britton；巴氏悬钩子；Bailey's Dewberry ●☆

338176　Rubus baileyanus Britton ＝ Rubus flagellaris Willd. ●☆

338177　Rubus bambusarum Focke；竹叶鸡爪茶（短柄鸡爪茶，观音茶藤，老林茶）；Bambooleaf Raspberry，Bamboo-leaved Raspberry，Bamboosifolious Clawtea ●

338178　Rubus bellobatus L. H. Bailey ＝ Rubus alumnus L. H. Bailey

●☆

338179　Rubus benneri L. H. Bailey　= Rubus semisetosus Blanch. ●☆

338180　Rubus bergii Eckl. et Zeyh. = Rubus fruticosus L. ●

338181　Rubus besseyi L. H. Bailey = Rubus canadensis L. ●☆

338182　Rubus betulinus D. Don = Rubus acuminatus Sm. ●

338183　Rubus biflorus Buch. -Ham. ex Sm. ;二花莓（二花悬钩子，粉枝莓）;Biflorous Raspberry,Twoflower Raspberry ●

338184　Rubus biflorus Buch. -Ham. ex Sm. f. parceglanduliger Focke = Rubus biflorus Buch. -Ham. ex Sm. var. adenophorus Franch. ●

338185　Rubus biflorus Buch. -Ham. ex Sm. var. adenophorus Franch. ;腺毛粉枝莓;Glandular Twoflower Raspberry ●

338186　Rubus biflorus Buch. -Ham. ex Sm. var. pubescens Te T. Yu et L. T. Lu ;柔毛粉枝莓;Softhair Twoflower Raspberry ●

338187　Rubus biflorus Buch. -Ham. ex Sm. var. quinqueflorus Focke = Rubus biflorus Buch. -Ham. ex Sm. ●

338188　Rubus biflorus Buch. -Ham. ex Sm. var. spinocalycinus Y. Gu et W. L. Li ;刺萼粉枝莓;Spiny-calyx Twoflower Raspberry ●

338189　Rubus bifrons Vest ex Tratt. ;喜马拉雅双花悬钩子;Blackberry,Himalayan Berry,Himalayan-berry ●☆

338190　Rubus bispidus ? ;湿地悬钩子;Swamp Blackberry ●☆

338191　Rubus blinii H. Lév. = Rubus lasiotrichos Focke var. blinnii（H. Lév.）L. T. Lu ●

338192　Rubus bodinieri H. Lév. et Vaniot = Rubus buergeri Miq. ●

338193　Rubus bollei Focke;博勒悬钩子●☆

338194　Rubus bonatianus Focke;滇北悬钩子;N. Yunnan Raspberry,North Yunnan Raspberry ●

338195　Rubus bonatii H. Lév. = Rubus niveus Thunb. ●

338196　Rubus boninensis Koidz. ;小笠原悬钩子●☆

338197　Rubus bonus L. H. Bailey = Rubus enslenii Tratt. ●☆

338198　Rubus botruosus L. H. Bailey = Rubus plicatifolius Blanch. ●☆

338199　Rubus boudieri H. Lév. = Rubus niveus Thunb. ●

338200　Rubus bractealis L. H. Bailey = Rubus alleghaniensis Porter ex L. H. Bailey ●☆

338201　Rubus bracteoliferus Fernald = Rubus glandicaulis Blanch. ●☆

338202　Rubus brainerdii Fernald = Rubus frondosus Bigelow ●☆

338203　Rubus brevipetalus Elmer = Rubus pirifolius Sm. ●

338204　Rubus brevipetiolatus Te T. Yu et L. T. Lu ;短柄悬钩子;Shortpetiole Raspberry,Short-petioled Raspberry,Shortstalk Dewberry ●

338205　Rubus buergeri Miq. ;寒莓（大号刺波，大叶寒莓，大叶漂，地莓，冬扎公，肺形草，过江龙，寒刺泡，虎脚扭，假寒莓，咯咯红，聋朵公，猫儿扭，岂陈晃，乔果，山火莓，水漂沙，踏地杨梅）;Buerger's Raspberry ●

338206　Rubus buergeri Miq. var. pseudobuergeri（Sasaki）Y. C. Liu et Yang = Rubus buergeri Miq. ●

338207　Rubus buergeri Miq. var. viridifolius Hand. -Mazz. = Rubus hunanensis Hand. -Mazz. ●

338208　Rubus bullatifolius Merr. = Rubus alceifolius Poir. ●

338209　Rubus caesius L. ;欧洲木莓（欧洲悬钩子）;Blackberry Token,Blue Bramble,Blueberry,Caesian Raspberry,Dewberry,Eggs-and-bacon,European Dewberry,European Raspberry,Heath Bramble,Jubilee Hunter,Mulberry,Token Blackberry ●

338210　Rubus caesius L. subsp. leucosepalus Focke = Rubus caesius L. ●

338211　Rubus caesius L. subsp. turkestanicus Focke = Rubus caesius L. ●

338212　Rubus caesius L. var. turkesianicus Regel = Rubus caesius L. ●

338213　Rubus calycacanthus H. Lév. ;猬莓●

338214　Rubus calycacanthus H. Lév. var. buergerifolius H. Lév. = Rubus pinnatisepalus Hemsl. ●

338215　Rubus calycanthus H. Lév. = Rubus pinnatisepalus Hemsl. ●

338216　Rubus calycanthus H. Lév. var. buergerifolia H. Lév. = Rubus pinnatisepalus Hemsl. ●

338217　Rubus calycinoides Hayata ex Koidz. ;玉山悬钩子●

338218　Rubus calycinoides Hayata ex Koidz. = Rubus rolfei Vidal ●

338219　Rubus calycinoides Hayata ex Koidz. var. macrophyllus H. L. Li ;大叶玉山悬钩子●

338220　Rubus calycinoides Hayata ex Koidz. var. macrophyllus H. L. Li = Rubus rolfei Vidal ●

338221　Rubus calycinoides Hayata var. macrophyllus H. L. Li = Rubus rolfei Vidal ●

338222　Rubus calycinus Wall. = Rubus calycinus Wall. ex D. Don ●■

338223　Rubus calycinus Wall. ex D. Don ;齿萼悬钩子;Toothedcalyx Raspberry ●■

338224　Rubus campester L. H. Bailey = Rubus alleghaniensis Porter ex L. H. Bailey ●☆

338225　Rubus canadensis L. ;加拿大悬钩子（北美悬钩子）;American Dewberry,Canada Blackberry,Canada Dewberry,Smooth Blackberry,Thornless Blackberry ●☆

338226　Rubus canadensis L. var. elegantulus（Blanch.）Farw. = Rubus elegantulus Blanch. ●☆

338227　Rubus canadensis L. var. imus L. H. Bailey = Rubus canadensis L. ●☆

338228　Rubus canadensis L. var. pergratus（Blanch.）L. H. Bailey = Rubus alleghaniensis Porter ex L. H. Bailey ●☆

338229　Rubus canadensis L. var. roribaccus L. H. Bailey = Rubus roribaccus（L. H. Bailey）Rydb. ●☆

338230　Rubus canadensis Torr. et A. Gray = Rubus flagellaris Willd. ●☆

338231　Rubus canariensis Focke;加那利悬钩子●☆

338232　Rubus candicans Weihe ex Rchb. ;白光悬钩子;Candicant Raspberry ●

338233　Rubus cardotii Koidz. ;卡托悬钩子●

338234　Rubus cardotii Koidz. = Rubus croceacanthus H. Lév. ●

338235　Rubus carolinianus Rydb. = Rubus idaeus L. var. strigosus（Michx.）Maxim. ●☆

338236　Rubus caucasicus Focke;高加索悬钩子●

338237　Rubus caucasigenus（Sudre）Juz. ;高加索生悬钩子●☆

338238　Rubus caudifolius Wuzhi;尾叶悬钩子;Caudateleaf Raspberry,Caudate-leaved Raspberry,Tailleaf Raspberry ●

338239　Rubus cauliflorus L. H. Bailey ;茎花悬钩子●☆

338240　Rubus cavaleriei H. Lév. et Vaniot = Rubus setchuenensis Bureau et Franch. ●

338241　Rubus celer L. H. Bailey ;马尔特悬钩子;Bolting Dewberry ●☆

338242　Rubus chaffanjoni H. Lév. et Vaniot = Rubus amphidasys Focke ●

338243　Rubus chamaemorus L. ;兴安悬钩子;Aiverin,Averin Averen,Avrons,Bakeapples,Baked-apple Berry,Cloudberry,Cnoutberry,Dewberry,Everocks,Evron,Ground Mulberry,Knoop,Knotberry,Knoutberry,Moltebeere,Mountain Bramble,Noops,Nub-berry Nub,Roebuck-berry ■

338244　Rubus chapmanianus Kupicha;查普曼悬钩子●☆

338245　Rubus chiesae Chiov. = Rubus volkensii Engl. ●☆

338246　Rubus chiliadenus Focke;长序莓（钩实泡藤）;Longinflorescence Raspberry,Long-racemed Raspberry ●

338247　Rubus chinensis Franch. = Rubus stimulans Focke ●

338248　Rubus chinensis Franch. var. concolor Cardot = Rubus stimulans Focke ●

338249　Rubus chingianus Hand. -Mazz. = Rubus dolichophyllus Hand. -

Mazz. ●

338250　Rubus chingii Hu;掌叶覆盆子(毕楞加,大号角公,覆盆,覆盆子,茎,华东覆盆子,华覆盆子,艻薦子,莓子,牛奶母,盆子,秦氏悬钩子,缺盆,托盘,乌薦子,西国草,小托盘);Palmleaf Raspberry,Palm-leaved Raspberry ●

338251　Rubus chingii Hu = Rubus dolichophyllus Hand. -Mazz. ●

338252　Rubus chingii Hu var. suavissimus (S. K. Lee) L. T. Lu;甜茶(甜搅莓);Fragrant Raspberry,Sweettea Raspberry ●

338253　Rubus chiovendae Gust. = Rubus apetalus Poir. ●☆

338254　Rubus chrolosepalus Focke;毛萼莓(毛萼悬钩子,紫萼莓,紫萼悬钩子);Hairysepal Raspberry,Hairy-sepaled Raspberry ●

338255　Rubus chroosepalus Hand. -Mazz. var. omeiensis Matsuda = Rubus chrolosepalus Focke ●

338256　Rubus chrysobotrys Hand. -Mazz.;黄穗悬钩子;Yellowraceme Raspberry,Yellow-racemed Raspberry ●

338257　Rubus chrysobotrys Hand. -Mazz. var. araneosus Q. H. Chen et T. L. Xu;蛛丝毛萼莓●

338258　Rubus chrysobotrys Hand. -Mazz. var. lobophyllus Hand. -Mazz.;裂叶黄穗悬钩子;Lobedleaf Raspberry ●

338259　Rubus chrysocarpus Mundt = Rubus rigidus Sm. ●☆

338260　Rubus cinclidodictyus Cardot;网纹悬钩子;Netty Raspberry ●

338261　Rubus clemens Focke = Rubus setchuenensis Bureau et Franch. ●

338262　Rubus clinocephalus Focke = Rubus multibracteatus H. Lév. et Vaniot ●

338263　Rubus clinocephalus Focke = Rubus pluribracteatus L. T. Lu et Boufford ●

338264　Rubus clivicola Walker;矮生悬钩子;Dwarf Raspberry ■

338265　Rubus cochinchinensis Tratt.;越南悬钩子(鸡足刺,猫枚筋,蛇泡筋,蛇泡藤,五叶泡,小猛虎);Blackberry,Bramble,Cochinchina Raspberry,Cochin-China Raspberry,Common Bramble,Snakebubble Raspberry ●

338266　Rubus cochinchinensis Tratt. var. stenophyllus Franch. = Rubus playfairianus Hemsl. ex Focke ●

338267　Rubus cockburnianus Hemsl.;华中悬钩子(粉花悬钩子,郭氏悬钩子);Cockburn Raspberry,White-stemmed Bramble ●

338268　Rubus coloniatus L. H. Bailey = Rubus plicatifolius Blanch. ●☆

338269　Rubus columellaris Tutcher;小柱悬钩子(三叶吊杆泡);Collumella Raspberry,Collumellaelike Raspberry,Smallpiiar Raspberry ●

338270　Rubus columellaris Tutcher var. etropicus (Hand. -Mazz.) F. P. Metcalf = Rubus columellaris Tutcher ●

338271　Rubus columellaris Tutcher var. villosus Te T. Yu et L. T. Lu;柔毛小柱悬钩子;Villose Smallpiiar Raspberry ●

338272　Rubus comintanus Blanco = Rubus rosifolius Sm. ●

338273　Rubus commersonii Poir. = Rubus rosifolius Sm. ●☆

338274　Rubus compos L. H. Bailey = Rubus wheeleri (L. H. Bailey) L. H. Bailey ●☆

338275　Rubus concolor Lowe;同色悬钩子●☆

338276　Rubus condignus L. H. Bailey = Rubus setosus Bigelow ●☆

338277　Rubus conduplicatus Duthie ex Hayata = Rubus trianthus Focke ●

338278　Rubus congruus L. H. Bailey = Rubus alleghaniensis Porter ex L. H. Bailey ●☆

338279　Rubus connixus L. H. Bailey = Rubus steelei L. H. Bailey ●☆

338280　Rubus coptophyllus A. Gray = Rubus palmatus Thunb. var. coptophyllus (A. Gray) Kuntze ex Koidz. ●☆

338281　Rubus corchorifolius L. f.;山莓(变叶悬钩子,薦子,刺葫芦,刺泡儿,大麦包,大麦泡,吊杆泡,对口薦,对嘴薦,对嘴泡,高脚泡,薅秧薦,黄莓,莳,龙船泡,馒头菠,木暗桐,木莓,牛奶泡,泡儿刺,撒秧泡,三月薦,三月脬,三月泡,山抛子,树莓,四月泡,五月泡,悬钩,悬钩子,沿钩子,秧苗脬,莜,猪母泡);American Blackberry,Blackberry,Broombeere,Fingerberry,Juteleaf Raspberry,Jute-leaved Raspberry ●

338282　Rubus corchorifolius L. f. f. roseolus Z. X. Yu;粉红山莓;Rose Juteleaf Raspberry ●

338283　Rubus corchorifolius L. f. f. roseolus Z. X. Yu = Rubus corchorifolius L. f. ●

338284　Rubus corchorifolius L. f. f. semiplenus Z. X. Yu;重瓣山莓;Doubleflower Juteleaf Raspberry ●

338285　Rubus corchorifolius L. f. subsp. faberi Focke;光山莓;Faber Juteleaf Raspberry ●

338286　Rubus corchorifolius L. f. subsp. faberi Focke = Rubus glabricarpus W. C. Cheng ●

338287　Rubus corchorifolius L. f. var. glaber Matsum. = Rubus corchorifolius L. f. ●

338288　Rubus corchorifolius L. f. var. neillioides Focke = Rubus glabricarpus W. C. Cheng ●

338289　Rubus corchorifolius L. f. var. oliveri (Miq.) Focke = Rubus corchorifolius L. f. ●

338290　Rubus cordialis L. H. Bailey = Rubus steelei L. H. Bailey ●☆

338291　Rubus cordifrons L. H. Bailey = Rubus curtipes L. H. Bailey ●☆

338292　Rubus coreanus Miq.;插田泡(白龙须,菜子泡,插田薦,朝설悬钩子,刺桑椹,大麦莓,大乌泡,倒触伞,倒生根,端阳莓,复盆子,覆盆子,覆蓝子,高丽悬钩子,茎,过江龙,回头龙,两头草,两头忙,马瘿,荞麦抛子,蕨藃,乌龙毛,乌龙须,乌泡,乌泡倒触伞,乌沙莓);Korea Raspberry,Korean Dewberry,Korean Raspberry ●

338293　Rubus coreanus Miq. var. kouytchenis (H. Lév.) H. Lév. = Rubus coreanus Miq. ●

338294　Rubus coreanus Miq. var. nakaianus H. Lév. = Rubus coreanus Miq. ●

338295　Rubus coreanus Miq. var. tomentosus Cardot;毛叶插田泡(白绒覆盆子,白绒悬钩子);Hairyleaf Korean Raspberry ●

338296　Rubus corei L. H. Bailey = Rubus alumnus L. H. Bailey ●☆

338297　Rubus coronarius Sims = Rubus rosifolius Sm. 'Coronarius' ●

338298　Rubus coronarius Sims = Rubus rosifolius Sm. var. coronarius Sims ●

338299　Rubus crassifolius Te T. Yu et L. T. Lu;厚叶悬钩子;Thickleaf Raspberry,Thick-leaved Raspberry ●

338300　Rubus crataegifolius Bunge;牛叠肚(驴叠肚,马林果,牛迭肚,蓬蘽,蓬蘽悬钩子,婆婆头,沙窝,山楂叶悬钩子,树莓,树梅,托盘,悬钩子);Hawthornleaf Raspberry,Hornthorn-leaf Raspberry,Hornthorn-leaved Raspberry,Korean Raspberry ●

338301　Rubus crataegifolius Bunge f. flavescens Skvortsov = Rubus crataegifolius Bunge ●

338302　Rubus crataegifolius Bunge f. inermis (Honda) Sugim.;无刺牛叠肚●☆

338303　Rubus crataegifolius Bunge f. xanthocarpus Naruh.;黄果牛叠肚●☆

338304　Rubus crataegifolius Bunge var. inermis Honda = Rubus crataegifolius Bunge f. inermis (Honda) Sugim. ●☆

338305　Rubus croceacanthus H. Lév.;薄瓣悬钩子(大玫瑰悬钩子,虎不刺,虎梅刺,虎婆刺);Hupoci Raspberry,Thinpetal Raspberry ●

338306　Rubus croceacanthus H. Lév. var. glaber Koidz. = Rubus okinawensis Koidz. ●

338307　Rubus cuneifolius Pursh;楔叶悬钩子(沙地悬钩子);Biuarberry, Blackberry, Brier-berry, Sand Blackberry, Wedge-leaved Raspberry ●☆

338308 Rubus currulis L. H. Bailey =Rubus steelei L. H. Bailey ●☆

338309 Rubus curtipes L. H. Bailey;短梗悬钩子;Short-stalk Dewberry ●☆

338310 Rubus cyrenaicae Hruby =Rubus ulmifolius Schott ●☆

338311 Rubus cyri Juz.;西尔悬钩子●☆

338312 Rubus darrisii H. Lév. =Rubus pinnatisepalus Hemsl. ●

338313 Rubus davidianus Kuntze =Rubus crataegifolius Bunge ●

338314 Rubus deaneanus L. H. Bailey =Rubus vermontanus Blanch. ●☆

338315 Rubus debilis Ball =Rubus ulmifolius Schott ●☆

338316 Rubus decor L. H. Bailey =Rubus meracus L. H. Bailey ●☆

338317 Rubus delavayi Franch.;三叶悬钩子(绊脚刺,刺茶,刺黄连,倒钩刺,三月藨,伞三叶藨,散血草,小倒钩刺,小黄泡刺,小乌泡);Delavay Raspberry ●

338318 Rubus deliciosus Torr.;落基山悬钩子;Rocky Mountain Raspberry ●☆

338319 Rubus deliciosus Torr. var. neomexicanus Kearney;新墨西哥悬钩子●☆

338320 Rubus densipubens L. H. Bailey =Rubus satis L. H. Bailey ●☆

338321 Rubus dictyophyllus Oliv. =Rubus steudneri Schweinf. var. dictyophyllus (Oliv.) R. A. Graham ●☆

338322 Rubus dictyophyllus Oliv. var. adenocomus Focke =Rubus adenocomus (Focke) Gust. ●☆

338323 Rubus dielsianus Focke =Rubus xanthoneurus Focke ●

338324 Rubus difformis L. H. Bailey =Rubus recurvans Blanch. ●☆

338325 Rubus discolor E. Mey. =Rubus rigidus Sm. ●☆

338326 Rubus discolor Weihe et Nees;异色悬钩子;Himalaya Berry, Himalayan Blackberry ●☆

338327 Rubus discretus L. H. Bailey =Rubus groutianus Blanch. ●☆

338328 Rubus dissitiflorus Fernald =Rubus flagellaris Willd. ●☆

338329 Rubus distans D. Don =Rubus niveus Thunb. ●

338330 Rubus distentus Focke =Rubus poliophyllus Kuntze ●

338331 Rubus distinctus L. H. Bailey =Rubus permixtus Blanch. ●☆

338332 Rubus dives L. H. Bailey =Rubus satis L. H. Bailey ●☆

338333 Rubus dolichocarpus Juz.;长果悬钩子●☆

338334 Rubus dolichocephalus Hayata;长头悬钩子(长果悬钩子);Longfruit Raspberry, Long-fruited Raspberry, Longhead Raspberry ●

338335 Rubus dolichocephalus Hayata =Rubus sumatranus Miq. ●

338336 Rubus dolichophyllus Hand.-Mazz.;长叶悬钩子;Longleaf Raspberry, Long-leaved Raspberry ●

338337 Rubus dolichophyllus Hand.-Mazz. var. pubescens Te T. Yu et L. T. Lu;毛梗长叶悬钩子;Pubescent Longleaf Raspberry ●

338338 Rubus doyonensis Hand.-Mazz.;白蕌(贡山莓);Bairu Raspberry, Doyonon Raspberry ●

338339 Rubus duclouxii H. Lév. =Rubus delavayi Franch. ●

338340 Rubus dunnii F. P. Metcalf;闽粤悬钩子;Dunn Raspberry ●

338341 Rubus dunnii F. P. Metcalf var. glabrescens Te T. Yu et L. T. Lu;光叶闽粤悬钩子;Glabrous Dunn Raspberry ●

338342 Rubus echinoides F. P. Metcalf;猥莓(凌云悬钩子);Hedgehog Raspberry, Hedgehog-like Raspberry ●

338343 Rubus echinoides F. P. Metcalf =Rubus calycacanthus H. Lév. ●

338344 Rubus ecklonii Focke =Rubus apetalus Poir. ●☆

338345 Rubus elegans Hayata =Rubus taiwanicola Koidz. et Ohwi ●

338346 Rubus elegantulus Blanch.;艳丽悬钩子;Showy Blackberry ●☆

338347 Rubus ellipticus Sm.;椭圆悬钩子(黄藨,黄泡,黄喜马莓,老虎泡,切头悬钩子);Elliptic Raspberry, Elliptical Raspberry, Golden Evergreen Raspberry, Yellow Himalayan Raspberry ●

338348 Rubus ellipticus Sm. f. obcordatus Franch. =Rubus ellipticus Sm. var. obcordatus (Franch.) Focke ●

338349 Rubus ellipticus Sm. subsp. fasciculatus (Duthie) Focke =Rubus wallichianus Wight et Arn. ●

338350 Rubus ellipticus Sm. subsp. fasciculatus Focke =Rubus wallichianus Wight et Arn. ●

338351 Rubus ellipticus Sm. var. fasciculatus (Duthie) Masam. =Rubus wallichianus Wight et Arn. ●

338352 Rubus ellipticus Sm. var. fasciculatus Masam. ex Kudo et Masam.;鬼悬钩子●

338353 Rubus ellipticus Sm. var. fasciculatus Masam. ex Kudo et Masam. =Rubus pinnatisepalus Hemsl. ●

338354 Rubus ellipticus Sm. var. obcordatus (Franch.) Focke;栽秧泡(大红黄泡,倒竹伞,红锁梅,黄藨,黄茨果,黄刺果,黄泡,黄泡刺,黄锁梅,雀不站,三月泡,锁地风,锁梅,乌泡,钻地风);Obcordate Golden Evergreen Raspberry, Yellow Himalayan Raspberry ●

338355 Rubus elmeri Focke =Rubus rolfei Vidal ●

338356 Rubus elongatus Brainerd et Peitersen =Rubus permixtus Blanch. ●☆

338357 Rubus enslenii Tratt.;单花悬钩子;One-flowered Dewberry, Southern Dewberry ●☆

338358 Rubus eous Focke =Rubus mesogaeus Focke ex Diels ●

338359 Rubus eriensis L. H. Bailey =Rubus rosa L. H. Bailey ●☆

338360 Rubus erlangeri Engl.;厄兰格悬钩子●☆

338361 Rubus erythrocarpus Te T. Yu et L. T. Lu;红果悬钩子;Redfruit Raspberry, Red-fruited Raspberry ●

338362 Rubus erythrocarpus Te T. Yu et L. T. Lu var. weixiensis Te T. Yu et L. T. Lu;腺萼红果悬钩子;Glandularcalyx Redfruit Raspberry ●

338363 Rubus erythrolasius Focke =Rubus wallichianus Wight et Arn. ●

338364 Rubus esquirolii H. Lév. =Rubus reflexus Ker Gawl. ●

338365 Rubus etropicus (Hand.-Mazz.) Thuan =Rubus columellaris Tutcher ●

338366 Rubus eucalyptus Focke;桉叶悬钩子(六月泡);Eucalyptus Raspberry ●

338367 Rubus eucalyptus Focke var. etomentosus Te T. Yu et L. T. Lu;脱毛桉叶悬钩子;Hairless Eucalyptus Raspberry ●

338368 Rubus eucalyptus Focke var. trillisatus (Focke) Te T. Yu et L. T. Lu;无腺桉叶悬钩子;Glandless Eucalyptus Raspberry ●

338369 Rubus eucalyptus Focke var. yunnanensis Te T. Yu et L. T. Lu;云南桉叶悬钩子;Yunnan Raspberry ●

338370 Rubus eugenius Focke =Rubus ichangensis Hemsl. et Kuntze ●

338371 Rubus euleucus Focke ex Hand.-Mazz. =Rubus mesogaeus Focke ex Diels ●

338372 Rubus euphleobophyllus Hayata =Rubus croceacanthus H. Lév. ●

338373 Rubus euphleobophyllus Hayata =Rubus piptopetalus Hayata ex Koidz. ●

338374 Rubus eustephanus Focke;大红泡(大红袍);Bigred Raspberry, Big-red Raspberry ●

338375 Rubus eustephanus Focke var. coronarius Koidz.;栽培大红泡;Cultivated Raspberry ●

338376 Rubus eustephanus Focke var. glanduliger Te T. Yu et L. T. Lu;腺毛大红泡;Glandulate Bigred Raspberry ●

338377 Rubus evadens Focke =Rubus neoviburnifolius L. T. Lu et Boufford ●

338378 Rubus evadens Focke =Rubus viburnifolius Focke ●

338379 Rubus exemptus L. H. Bailey =Rubus flagellaris Willd. ●☆

338380 Rubus exsuccus Steud. ex A. Rich. =Rubus apetalus Poir. ●☆

338381   Rubus exsularis L. H. Bailey；栅栏悬钩子；Fenceline Dewberry ●☆

338382   Rubus exter L. H. Bailey = Rubus fulleri L. H. Bailey ●☆

338383   Rubus exutus L. H. Bailey = Rubus plicatifolius Blanch. ●☆

338384   Rubus faberi Focke；峨眉悬钩子；Faber Raspberry ●

338385   Rubus facetus L. H. Bailey = Rubus alumnus L. H. Bailey ●☆

338386   Rubus fandus L. H. Bailey = Rubus frondosus Bigelow ●☆

338387   Rubus fanjingshanensis L. T. Lu ex Boufford et al. ；梵净山悬钩子●

338388   Rubus fargesii Franch. = Rubus henryi Hemsl. et Kuntze var. sozostylus（Focke）Rehder ●

338389   Rubus farinaceus Cardot = Rubus tephrodes Hance var. setosissimus Hand. -Mazz. ●

338390   Rubus fasciculatus Duthie = Rubus pinfaensis H. Lév. et Vaniot ●

338391   Rubus fasciculatus Duthie = Rubus wallichianus Wight et Arn. ●

338392   Rubus fassettii L. H. Bailey = Rubus wheeleri（L. H. Bailey）L. H. Bailey ●☆

338393   Rubus feddei H. Lév. et Vaniot；黔桂悬钩子；Fedde Raspberry ●

338394   Rubus felix L. H. Bailey = Rubus flagellaris Willd. ●☆

338395   Rubus fernaldianus L. H. Bailey = Rubus alumnus L. H. Bailey ●☆

338396   Rubus fimbriferus Focke var. diversilobatus Merr. et Chun = Rubus alceifolius Poir. var. diversilobatus（Merr. et Chun）Te T. Yu et L. T. Lu ●

338397   Rubus fimbriiferus Focke = Rubus alceifolius Poir. ●

338398   Rubus fimbriiferus Focke var. diversilobatus Merr. et Chun = Rubus alceifolius Poir. ●

338399   Rubus flagellaris Willd. ；北方悬钩子（美国悬钩子）；American Dewberry, American Northern Dewberry, Common Dewberry, Dewberry, Flagellate Dewberry, Northern Dewberry, Prickly Dewberry ●☆

338400   Rubus flagellaris Willd. var. almus L. H. Bailey = Rubus aboriginum Rydb. ●☆

338401   Rubus flagellaris Willd. var. humifusus（Torr. et A. Gray）B. Boivin = Rubus baileyanus Britton ●☆

338402   Rubus flagellaris Willd. var. michiganensis（L. H. Bailey）L. H. Bailey = Rubus roribaccus（L. H. Bailey）Rydb. ●☆

338403   Rubus flagellaris Willd. var. occidualis L. H. Bailey = Rubus flagellaris Willd. ●☆

338404   Rubus flagellaris Willd. var. occidualis L. H. Bailey = Rubus roribaccus（L. H. Bailey）Rydb. ●☆

338405   Rubus flagellaris Willd. var. roribaccus（L. H. Bailey）L. H. Bailey = Rubus roribaccus（L. H. Bailey）Rydb. ●☆

338406   Rubus flagellaris Willd. var. rosea-plenus E. J. Palmer et Steyerm. = Rubus flagellaris Willd. ●☆

338407   Rubus flagelliflorus Focke ex Diels；攀枝莓（对嘴泡, 老鸦泡, 裂缘苞悬钩子, 少花乌泡）；Cilmbing Raspberry, Flagellate-flowered Raspberry ●

338408   Rubus flagelliformis Hort. = Rubus flagelliflorus Focke ex Diels ●

338409   Rubus florenceae L. H. Bailey = Rubus ithacanus L. H. Bailey ●☆

338410   Rubus floribundopaniculatus Hayata = Rubus pirifolius Sm. ●

338411   Rubus floricomus Blanch. = Rubus alleghemiensis Porter ex L. H. Bailey ●☆

338412   Rubus floridus Tratt. ；繁花悬钩子；Blackberry ●

338413   Rubus flosculosus Focke；弓茎悬钩子（弓茎莓, 山挂牌条, 小花莓）；Littleflower Raspberry, Little-flowered Raspberry ●

338414   Rubus flosculosus Focke f. laxiflorus Focke = Rubus flosculosus Focke ●

338415   Rubus flosculosus Focke f. parvifolius Focke = Rubus flosculosus Focke ●

338416   Rubus flosculosus Focke var. etomentosus Te T. Yu et L. T. Lu；脱毛弓茎悬钩子；Hairless Littleflower Raspberry ●

338417   Rubus flosculosus Focke var. mairei Focke；白花弓茎悬钩子●

338418   Rubus fockeanus Kurz；凉山悬钩子（凉山莓, 匍状悬钩子）；Focke Raspberry ■

338419   Rubus foliaceistipulatus Te T. Yu et L. T. Lu；托叶悬钩子；Stipulate Raspberry, Stipule Raspberry ●

338420   Rubus folioflorus L. H. Bailey = Rubus frondosus Bigelow ●☆

338421   Rubus foliolosus D. Don = Rubus niveus Thunb. ●

338422   Rubus fordii Hance = Rubus hanceanus Kuntze ●

338423   Rubus forestalis L. H. Bailey = Rubus canadensis L. ●☆

338424   Rubus formosanus Maxim. ex Focke = Rubus formosensis Kuntze ●

338425   Rubus formosensis Kuntze；台湾悬钩子（南投悬钩子）；Formosan Bramble, Formosan Raspberry, Taiwan Raspberry ●

338426   Rubus formosensis Matsum. = Rubus nagasawanus Koidz. ●

338427   Rubus formosensis sensu Matsum. = Rubus nagasawanus Koidz. ●

338428   Rubus forrestianus Bertol. ；无腺柄悬钩子（贡山蓬蘽）●

338429   Rubus fragarioides Bertol. ；莓叶悬钩子；Strawberry Raspberry, Strawberryleaf Raspberry ●■

338430   Rubus fragarioides Bertol. var. adenophorus Franch. ；腺毛莓叶悬钩子；Glandular Strawberryleaf Raspberry ●■

338431   Rubus fragarioides Bertol. var. pubescens Franch. ；柔毛莓叶悬钩子；Pubescent Strawberryleaf Raspberry ■

338432   Rubus franchetianus H. Lév. = Rubus fragarioides Bertol. var. adenophorus Franch. ●■

338433   Rubus fraxinifoliolus Hayata；梣叶悬钩子（枪木悬钩子, 台北悬钩子）；Ash-leaved Raspberry, Suzuki Raspberry, Taibei Raspberry ●

338434   Rubus fraxinifolius Matsum. et Hayata = Rubus fraxinifolius Poir. ●

338435   Rubus fraxinifolius Poir. ；兰屿梣叶悬钩子（梣叶悬钩子, 兰屿杞叶悬钩子, 兰屿悬钩子, 紫萼悬钩子）；Ash-leaf Raspberry, Koto Alderleaf Raspberry, Lanyu Alderleaf Raspberry, Lanyu Raspberry ●

338436   Rubus fraxinifolius Poir. sensu Matsum. et Hayata = Rubus alnifoliolatus H. Lév. et Vaniot ●

338437   Rubus fraxinifolius Poir. var. kotoensis（Hayata）Koidz. = Rubus alnifoliolatus H. Lév. et Vaniot var. kotoensis（Hayata）H. L. Li ●

338438   Rubus fraxinifolius Poir. var. kotoensis（Hayata）Koidz. = Rubus fraxinifolius Poir. ●

338439   Rubus fraxinifolius Poir. var. yushuni Suzuki et Yamam. = Rubus ritozanensis Sasaki ●

338440   Rubus fraxinifolius Poir. var. yushunii Suzuki et Yamam. = Rubus inopertus（Diels）Focke ●

338441   Rubus friesiorum Gust. ；弗里斯悬钩子●☆

338442   Rubus friesiorum Gust. subsp. elgonensis（Gust. ）R. A. Graham；埃尔贡悬钩子●☆

338443   Rubus friesiorum Gust. var. elgonensis Gust. = Rubus friesiorum Gust. subsp. elgonensis（Gust. ）R. A. Graham ●☆

338444   Rubus frondosus Bigelow；美国悬钩子；Highbush Blackberry, Yankee Blackberry ●☆

338445   Rubus frustratus L. H. Bailey = Rubus flagellaris Willd. ●☆

338446   Rubus fruticosus Eckl. et Zeyh. = Rubus fruticosus L. ●

338447   Rubus fruticosus L. ；灌木悬钩子（树莓）；Blacebergan, Black Bowour, Black Bowwower, Black Boyd, Black Bums, Black Kite, Black Spice, Blackbern, Blackberry, Blackberry Mouchers,

Blackbleg, Blackbowwowers, Blackite, Blackleg, Blagg, Bly, Boyd, Bramble, Bramble Breer, Brambleberries, Bramblekite, Brammelkite, Brammle, Brannel Kite, Brier, Brimbel, Brimble, Brimmle, Brumble Kite, Brumleyberry Bush, Brummel, Brummel Kite, Brummelkite, Bull Beef, Bumbleberry, Bumblekite, Bumlykite, Bummel, Bummelberries, Bummelkite, Bum-meltykite, Bummull, Cat's Claws, Cock Bramble, Cock Brumble, Cock-battler, Cock-fighter, Country Lawyers, Doctor's Medicine, Drisag, European Blackberry, Ewe Bramble, Ewe-bramble, Followers, Gaitberry, Gaiterberry, Garter-berry, Gatter-berry, Gatter-tree, Hawk's Bill Bramble, He-brimmel, Lady Garten Berries, Lady's Garten-berry, Lady's Garter-berry, Lady's Garters, Lawyers, Mouchers, Mulberry, Mush, Penny Mouchers, Rhubarb, Scaldberry, Scald-head, Thief, Winterpicks, Yne-brimmel, Yoe-brimmle ●

338448 Rubus fruticosus Lour. = Rubus cochinchinensis Tratt. ●

338449 Rubus fujianensis Te T. Yu et L. T. Lu; 福建悬钩子; Fujian Raspberry ●

338450 Rubus fulleri L. H. Bailey; 富勒悬钩子●☆

338451 Rubus fuscifolius Te T. Yu et L. T. Lu; 锈叶悬钩子; Duskyleaf Raspberry, Dusky-leaved Raspberry, Fuscifolius Raspberry ●

338452 Rubus fusco-rubens Focke; 黄毛悬钩子; Duskyred Raspberry, Dusky-red Raspberry ●

338453 Rubus gansuensis X. G. Sun; 甘青悬钩子; Gansu Raspberry ●

338454 Rubus gelatinosus Sasaki = Rubus lambertianus Ser. var. glandulosus Cardot ●

338455 Rubus gelatinosus Sasaki = Rubus morii Hayata ●

338456 Rubus geniculatus Focke; 假喜马悬钩子; False Himalayan Berry ●☆

338457 Rubus gentilianus H. Lév. et Vaniot = Rubus xanthoneurus Focke ●

338458 Rubus geophilus Blanch. = Rubus flagellaris Willd. ●☆

338459 Rubus georgicus Focke; 乔治悬钩子●☆

338460 Rubus gigantiflorus H. Hara = Rubus wardii Merr. ●

338461 Rubus gigantiflorus H. Hara var. chiliocanthus Hand. -Mazz. = Rubus wardii Merr. ●

338462 Rubus gigantiflorus H. Hara var. pluvialis Hand. -Mazz. = Rubus wardii Merr. ●

338463 Rubus gilvus Focke = Rubus alceifolius Poir. ●

338464 Rubus gilvus Focke = Rubus reflexus Ker Gawl. var. hui (Diels) F. P. Metcalf ●

338465 Rubus giraldianus Focke = Rubus cockburnianus Hemsl. ●

338466 Rubus glaberrimus Champ. = Rubus leucanthus Hance ●■

338467 Rubus glabricarpus W. C. Cheng; 光果悬钩子; Glabrousfruit Raspberry, Glabrous-fruited Raspberry ●

338468 Rubus glabricarpus W. C. Cheng var. eglandulosus Y. Gu et W. L. Li; 无腺光果悬钩子; Glanduleless Glabrousfruit Raspberry ●

338469 Rubus glabricarpus W. C. Cheng var. glabratus C. Z. Zheng et Y. Y. Fang; 无毛光果悬钩子●

338470 Rubus glandicaulis Blanch.; 腺茎悬钩子; Gland-stem Blackberry ●☆

338471 Rubus glandulosocalycinus Hayata; 腺萼悬钩子; Glandcalyx Raspberry, Glandular Calyx Raspberry ●

338472 Rubus glandulosocalycinus Hayata = Rubus sumatranus Miq. ●

338473 Rubus glandulosocarpus M. X. Nie; 腺果悬钩子; Glandularfruit Raspberry ●

338474 Rubus glandulosopunctatus Hayata = Rubus rosifolius Sm. ●

338475 Rubus glaucus Benth.; 安第斯山悬钩子; Andes Berry, Andes Berry Bush ●☆

338476 Rubus godongensis Y. Gu et W. L. Li; 果东悬钩子; Guodong Raspberry ●

338477 Rubus gongshanensis Te T. Yu et L. T. Lu; 贡山悬钩子; Gongshan Raspberry ●

338478 Rubus gongshanensis Te T. Yu et L. T. Lu var. eglandulosus Y. Gu et W. L. Li; 无腺贡山悬钩子; Glanduleless Gongshan Raspberry ●

338479 Rubus gongshanensis Te T. Yu et L. T. Lu var. qiujiangensis Te T. Yu et L. T. Lu; 无刺贡山悬钩子; Spineless Gongshan Raspberry ●

338480 Rubus gortanii Chiov. = Rubus apetalus Poir. ●☆

338481 Rubus gracilis Roxb. = Rubus pedunculosus D. Don ●

338482 Rubus gracilis Roxb. var. chiliacanthus Hand. -Mazz. = Rubus pedunculosus D. Don ●

338483 Rubus gracilis Roxb. var. hypargyrus (Edgew.) Focke = Rubus hypargyrus Edgew. ●

338484 Rubus gracilis sensu Roxb. = Rubus hypargyrus Edgew. ●

338485 Rubus grandidens L. H. Bailey = Rubus glandicaulis Blanch. ●☆

338486 Rubus grandipaniculatus Te T. Yu et L. T. Lu; 大序悬钩子; Largepanicle Raspberry, Large-panicled Raspberry, Largepaniculate Raspberry ●

338487 Rubus grayanus Maxim.; 中南悬钩子 (山莓); Central-south Raspberry, Gray Raspberry, Midsouth China Raspberry ●

338488 Rubus grayanus Maxim. var. chaetophorus Koidz.; 毛山莓●☆

338489 Rubus grayanus Maxim. var. trilobatus Te T. Yu et L. T. Lu; 三裂中南悬钩子; Threelobed Central-south Raspberry ●

338490 Rubus gressittii F. P. Metcalf; 江西悬钩子; Jiangxi Raspberry ●

338491 Rubus groutianus Blanch.; 间断悬钩子; Discretus L. H. Bailey ●☆

338492 Rubus gyamdaensis L. T. Lu et Boufford; 柔毛悬钩子; Pubigenous Raspberry, Softhairleaf Raspberry, Softhairy Raspberry ●

338493 Rubus gyamdaensis L. T. Lu et Boufford var. glabriusculus (Te T. Yu et L. T. Lu) L. T. Lu et Boufford; 川西柔毛悬钩子; West Sichuan Softhairleaf Raspberry ●

338494 Rubus hachijoensis Nakai = Rubus nishimuranus Koidz. ●☆

338495 Rubus hainanensis Focke = Rubus alceifolius Poir. ●

338496 Rubus hakonensis Franch. et Sav. = Rubus lambertianus Ser. var. glaber Hemsl. ●

338497 Rubus hakonensis Franch. et Sav. = Rubus lambertianus Ser. ●

338498 Rubus hanceanus Kuntze; 华南悬钩子; Hance Raspberry ●

338499 Rubus hastifolius H. Lév. et Vaniot; 戟叶悬钩子 (红绵藤); Hastateleaf Raspberry, Hastate-leaved Raspberry ●

338500 Rubus hatsushimae Koidz.; 初岛悬钩子●☆

338501 Rubus hayatai Nemoto = Rubus pungens Cambess. var. oldhamii (Miq.) Maxim. ●

338502 Rubus hayatai Nemoto ex Makino = Rubus pungens Cambess. var. oldhamii (Miq.) Maxim. ●

338503 Rubus hayatai Nemoto ex Makino et Nemoto = Rubus pungens Cambess. var. oldhamii (Miq.) Maxim. ●

338504 Rubus hayata-koidzumii Naruh. = Rubus rolfei Vidal ●

338505 Rubus hayatanus Koidz. = Rubus glandulosocalycinus Hayata ●

338506 Rubus hederifolius (Cardot) Thuan = Rubus lasiotrichos Focke ●

338507 Rubus hemithyrsus Hand. -Mazz.; 半锥莓●

338508 Rubus henryi Hemsl. et Kuntze; 鸡爪茶 (亨利莓, 老林茶); Henry Raspberry ●

338509 Rubus henryi Hemsl. et Kuntze var. bambusarum (Focke) Rehder = Rubus bambusarum Focke ●

338510 Rubus henryi Hemsl. et Kuntze var. sozostylus (Focke) Rehder; 大叶鸡爪茶; Bigleaf Henry Raspberry, Largeleaf Henry Raspberry ●

338511 Rubus heterogeneus L. H. Bailey = Rubus recurvans Blanch. ●☆

338512　Rubus hexagynus sensu Merr. = Rubus pirifolius Sm. ●

338513　Rubus hirsutopungens Hayata;高山悬钩子（毛刺悬钩子）;Stiff-hair Raspberry ●

338514　Rubus hirsutopungens Hayata = Rubus pungens Cambess. var. oldhamii（Miq.）Maxim. ●

338515　Rubus hirsutus Hayata = Rubus pungens Cambess. var. oldhamii（Miq.）Maxim. ●

338516　Rubus hirsutus Thunb. ;蓬蘽（波盘，刺菠，地苗，饭包菠，饭消扭，割田藨，空腹莲，空腹妙，篷蘽，泼盘，企晃刺，三月泡，竖藤火梅刺，田角公，田母，托盘，雅旱，野杜荆，野杜利）;Hirsute Raspberry ●

338517　Rubus hirsutus Thunb. f. harae（Makino）Ohwi;原氏蓬蘽●☆

338518　Rubus hirsutus Thunb. f. simplicifolius（Makino）Ohwi;单叶蓬蘽●☆

338519　Rubus hirsutus Thunb. f. xanthocarpus（Nakai）Sugim. ;黄果蓬蘽●☆

338520　Rubus hirsutus Thunb. var. argyi（H. Lév.）Nakai = Rubus hirsutus Thunb. ●

338521　Rubus hirsutus Thunb. var. brevipedicellus Z. M. Wu;短梗蓬蘽●

338522　Rubus hirsutus Thunb. var. glabellus（Focke）Wuzhi;无毛蓬蘽（空心泡）●

338523　Rubus hirsutus Thunb. var. glabellus（Focke）Wuzhi = Rubus rosifolius Sm. ●

338524　Rubus hirsutus Thunb. var. simplicifolius（Makino）Nakai = Rubus hirsutus Thunb. f. simplicifolius（Makino）Ohwi ●☆

338525　Rubus hirsutus Thunb. var. xanthocarpus Nakai = Rubus hirsutus Thunb. f. xanthocarpus（Nakai）Sugim. ●☆

338526　Rubus hirtiflorus Cardot = Rubus hanceanus Kuntze ●

338527　Rubus hirtus Waldst. et Kit. ;硬毛悬钩子●☆

338528　Rubus hispidus L. ;沼地悬钩子;Bristly Dewberry, Hispid Blackberry, Swamp Blackberry, Swamp Dewberry ●☆

338529　Rubus hispidus L. f. pleniflorus Nieuwl. = Rubus plus L. H. Bailey ●☆

338530　Rubus hispidus L. var. cupulifer L. H. Bailey = Rubus hispidus L. ●☆

338531　Rubus hispidus L. var. obovalis（Michx.）Fernald = Rubus hispidus L. ●☆

338532　Rubus hispidus L. var. suberectus Peck = Rubus setosus Bigelow ●☆

338533　Rubus holadenus H. Lév. = Rubus tephrodes Hance var. holadenus（H. Lév.）L. T. Li ●

338534　Rubus hookeri Focke = Rubus wardii Merr. ●

338535　Rubus hopingensis Y. C. Liu et F. Y. Lu = Rubus rosifolius Sm. ●

338536　Rubus housei L. H. Bailey = Rubus baileyanus Britton ●☆

338537　Rubus howii Merr. et Chun;裂叶悬钩子;How Raspberry ●

338538　Rubus huangpingensis Te T. Yu et L. T. Lu;黄平悬钩子;Huangping Raspberry ●

338539　Rubus hui Diels ex Hu = Rubus reflexus Ker Gawl. var. hui（Diels）F. P. Metcalf ●

338540　Rubus hui Diels Hu = Rubus reflexus Ker Gawl. var. hui（Diels）F. P. Metcalf ●

338541　Rubus humifusus Schltdl. = Rubus flagellaris Willd. ●☆

338542　Rubus humilior L. H. Bailey = Rubus uvidus L. H. Bailey ●☆

338543　Rubus humulifolius C. A. Mey. ;葎草叶悬钩子■

338544　Rubus hunanensis Hand. -Mazz. ;湖南悬钩子;Hunan Raspberry ●

338545　Rubus hupehensis Oliv. = Rubus swinhoei Hance ●

338546　Rubus hypargyrus Edgew. ;纤细悬钩子;Slender Raspberry ●

338547　Rubus hypargyrus Edgew. = Rubus pedunculosus D. Don ●

338548　Rubus hypargyrus Edgew. var. niveus H. Hara = Rubus hypargyrus Edgew. var. niveus（Wall. ex G. Don）H. Hara ●

338549　Rubus hypargyrus Edgew. var. niveus H. Hara = Rubus pedunculosus D. Don ●

338550　Rubus hypopitys Focke;滇藏悬钩子;Yunnan-Xizang Raspberry ●

338551　Rubus hypopitys Focke var. hanmiensis Te T. Yu et L. T. Lu;汉密悬钩子;Hanmi Raspberry ●

338552　Rubus hyrcanus Juz. ;希尔康悬钩子●☆

338553　Rubus ibericus Juz. ;伊比利亚悬钩子●☆

338554　Rubus ichangensis Hemsl. et Kuntze;宜昌悬钩子（红五泡，黄蘽子，黄泡叶，黄泡子，拘匍粘藤，牛尾泡，山泡刺藤，小黄泡子）;Ichang Bramble, Yichang Raspberry ●

338555　Rubus ichangensis Hemsl. et Kuntze var. latifolius Cardot = Rubus ichangensis Hemsl. et Kuntze ●

338556　Rubus idaeopsis Focke;拟覆盆子;Idaeuslike Raspberry ●

338557　Rubus idaeus L. ;覆盆子（覆葐子，红树莓，马莓，绒毛悬钩子，树莓）;American Raspberry, Arnberry, Common Raspberry, Dewberry, European Raspberry, European Red Raspberry, Framboise, Framboise Bush, Framboys, Hainberry, Hindberry, Hineberry, Rasp, Raspberry, Red Raspberry, Red-and-yellow Garden Raspberry, Respis, Sivven, Wild Raspberry, Wild Red Raspberry, Wood Rasp ●

338558　Rubus idaeus L. 'Aureus';黄果覆盆子●☆

338559　Rubus idaeus L. 'Fallgold';法尔格德覆盆子●☆

338560　Rubus idaeus L. 'Glen Moy';古伦莫覆盆子●☆

338561　Rubus idaeus L. 'Heritage';遗产覆盆子●☆

338562　Rubus idaeus L. f. concolor（Kom.）Ohwi = Rubus komarovii Nakai ●

338563　Rubus idaeus L. f. inermis Kaufm. = Rubus idaeus L. var. strigosus（Michx.）Maxim. ●☆

338564　Rubus idaeus L. subsp. kanayamensis（H. Lév. et Vaniot）Koidz. = Rubus idaeus L. subsp. melanolasius Focke f. concolor（Kom.）Ohwi ●☆

338565　Rubus idaeus L. subsp. komarovii（Nakai）Vorosch. = Rubus komarovii Nakai ●

338566　Rubus idaeus L. subsp. melanolasius（Dieck）Focke;西伯利亚红树莓●☆

338567　Rubus idaeus L. subsp. melanolasius Focke = Rubus idaeus L. var. strigosus（Michx.）Maxim. ●☆

338568　Rubus idaeus L. subsp. melanolasius Focke f. concolor（Kom.）Ohwi;同色西伯利亚红树莓●☆

338569　Rubus idaeus L. subsp. melanolasius Focke f. concolor Kom. = Rubus komarovii Nakai ●

338570　Rubus idaeus L. subsp. melanolasius Focke var. concolor（Kom.）Nakai = Rubus idaeus L. subsp. melanolasius Focke f. concolor（Kom.）Ohwi ●☆

338571　Rubus idaeus L. subsp. melanolasius Focke var. matsumuranus（H. Lév. et Vaniot）Koidz. = Rubus sachalinensis H. Lév. ●

338572　Rubus idaeus L. subsp. nipponicus Focke;深山悬钩子●☆

338573　Rubus idaeus L. subsp. nipponicus Focke f. marmoratus（H. Lév. et Vaniot）Kitam. ;八岳莓●☆

338574　Rubus idaeus L. subsp. nipponicus Focke var. hondoensis Koidz. ;深山里白莓●☆

338575　Rubus idaeus L. subsp. nipponicus Focke var. marmoratus（H. Lév. et Vaniot）H. Hara = Rubus idaeus L. subsp. nipponicus Focke f. marmoratus（H. Lév. et Vaniot）Kitam. ●☆

338576　Rubus idaeus L. subsp. nipponicus Focke var. shikokianus

（Ohwi et Inobe）Kitam. et Naruh. ;四国悬钩子●☆

338577　Rubus idaeus L. subsp. sachalinensis（A. Lév.）Focke ＝ Rubus strigosus Michx. ex Koidz. ●

338578　Rubus idaeus L. subsp. sachalinensis（H. Lév.）Focke ＝ Rubus idaeus L. subsp. melanolasius Focke ●☆

338579　Rubus idaeus L. subsp. sachalinensis（H. Lév.）Focke ＝ Rubus idaeus L. var. strigosus（Michx.）Maxim. ●☆

338580　Rubus idaeus L. subsp. sachalinensis（H. Lév.）Focke ＝ Rubus sachalinensis H. Lév. ●

338581　Rubus idaeus L. subsp. strigosus（Michx.）Focke ＝ Rubus idaeus L. var. strigosus（Michx.）Maxim. ●☆

338582　Rubus idaeus L. subsp. vulgaris Arrh. ＝ Rubus idaeus L. ●

338583　Rubus idaeus L. var. aculeatissimus C. A. Mey. ;北海道悬钩子;Yezo Raspberry ●☆

338584　Rubus idaeus L. var. aculeatissimus C. A. Mey. ＝ Rubus idaeus L. subsp. melanolasius Focke ●☆

338585　Rubus idaeus L. var. aculeatissimus C. A. Mey. f. concolor（Kom.）Ohwi ＝ Rubus komarovii Nakai ●

338586　Rubus idaeus L. var. aculeatissimus Regel et Tiling ＝ Rubus idaeus L. var. strigosus（Michx.）Maxim. ●☆

338587　Rubus idaeus L. var. aculeatissimus Regel et Tiling ＝ Rubus sachalinensis H. Lév. ●

338588　Rubus idaeus L. var. borealisinensis Te T. Yu et L. T. Lu;华北覆盆子（东北覆盆子）;North China Red Raspberry ●

338589　Rubus idaeus L. var. canadensis Richardson ＝ Rubus idaeus L. var. strigosus（Michx.）Maxim. ●☆

338590　Rubus idaeus L. var. concolor（Kom.）Nakai ＝ Rubus komarovii Nakai ●

338591　Rubus idaeus L. var. concolor Nakai ＝ Rubus komarovii Nakai ●

338592　Rubus idaeus L. var. exsuccus Franch. et Sav. ＝ Rubus mesogaeus Focke ex Diels ●

338593　Rubus idaeus L. var. glabratus Te T. Yu et L. T. Lu;无毛覆盆子;Glabrous Red Raspberry ●

338594　Rubus idaeus L. var. gracilipes M. E. Jones ＝ Rubus idaeus L. var. strigosus（Michx.）Maxim. ●☆

338595　Rubus idaeus L. var. matsumuranus（H. Lév. et Vaniot）Koidz. ＝ Rubus idaeus L. subsp. melanolasius Focke ●☆

338596　Rubus idaeus L. var. matsumuranus（H. Lév. et Vaniot）Koidz. ＝ Rubus sachalinensis H. Lév. ●

338597　Rubus idaeus L. var. melanotrachys（Focke）Fernald ＝ Rubus idaeus L. var. strigosus（Michx.）Maxim. ●☆

338598　Rubus idaeus L. var. microphyllus Turcz. ＝ Rubus sachalinensis H. Lév. ●

338599　Rubus idaeus L. var. sachalinensis（A. Lév.）Focke ＝ Rubus idaeus L. var. strigosus（Michx.）Maxim. ●☆

338600　Rubus idaeus L. var. strigosus（Michx.）Maxim. ;糙毛覆盆子（美国覆盆子）;American Red Raspberry, Red Raspberry, Wild Red Raspberry ●☆

338601　Rubus idaeus L. var. strigosus（Michx.）Maxim. ＝ Rubus strigosus Michx. ex Koidz. ●

338602　Rubus idaeus L. var. strigosus Maxim. ＝ Rubus idaeus L. var. strigosus（Michx.）Maxim. ●☆

338603　Rubus idaeus L. var. strigosus Maxim. ＝ Rubus sachalinensis H. Lév. ●

338604　Rubus idaeus L. var. yabei（H. Lév. et Vaniot）Koidz. ＝ Rubus idaeus L. subsp. nipponicus Focke ●☆

338605　Rubus ignarus L. H. Bailey ＝ Rubus meracus L. H. Bailey ●☆

338606　Rubus ikenoensis H. Lév. et Vaniot;五叶莓（五叶悬钩子）■☆

338607　Rubus ilanensis Y. C. Liu et F. Y. Lu ＝ Rubus liuii Y. P. Yang et S. Y. Lu ●

338608　Rubus illecebrosus Focke;软枝悬钩子;Bramble, Strawberry Raspberry, Strawberry-raspberry ●☆

338609　Rubus illecebrosus Focke f. tokinibara H. Hara ＝ Rubus tokinibara（H. Hara）Naruh. ●☆

338610　Rubus illecebrosus Focke var. yakusimensis（Masam.）Hatus. ;屋久岛悬钩子●☆

338611　Rubus illudens H. Lév. ＝ Rubus mesogaeus Focke ex Diels ●

338612　Rubus illustris L. H. Bailey ＝ Rubus canadensis L. ●☆

338613　Rubus imperiorum Fernald ＝ Rubus roribaccus（L. H. Bailey）Rydb. ●☆

338614　Rubus impos L. H. Bailey ＝ Rubus alumnus L. H. Bailey ●☆

338615　Rubus impressinervius F. P. Metcalf;陷脉悬钩子;Concavenerve Raspberry, Impressednerve Raspberry ■

338616　Rubus incanescens（DC.）Bertol. ;灰毛悬钩子●☆

338617　Rubus incanus Sasaki ex Tang S. Liu et T. Y. Yang;白毛悬钩子（白绒悬钩子）;White-hair Raspberry ●

338618　Rubus incanus Sasaki ex Tang S. Liu et T. Y. Yang subsp. koehneanus var. formosanus Masam. ex Kudo et Masam. ＝ Rubus trianthus Focke ●

338619　Rubus incanus Sasaki ex Tang S. Liu et T. Y. Yang var. conduplicams（Duthie ex Hayata）Koidz. ＝ Rubus trianthus Focke ●

338620　Rubus incanus Sasaki ex Y. C. Liu et T. Y. Yang ＝ Rubus niveus Thunb. ●

338621　Rubus incisus Thunb. subsp. koehneanus（Focke）Koidz. ＝ Rubus trianthus Focke ●

338622　Rubus incisus Thunb. var. conduplicatus（Duthie ex Hayata）Koidz. ＝ Rubus trianthus Focke ●

338623　Rubus incisus Thunb. var. formosanus（Cardot）Masam. ＝ Rubus trianthus Focke ●

338624　Rubus indotibetanus Koidz. ＝ Rubus sumatranus Miq. ●

338625　Rubus inedulis Rolfe. f. umbrosus Gust. ＝ Rubus runssorensis Engl. var. umbrosus（Gust.）Hauman ●☆

338626　Rubus inedulis Rolfe. f. velutinus Hauman;短绒毛悬钩子●☆

338627　Rubus inermis Pourr. ＝ Rubus ulmifolius Schott ●☆

338628　Rubus innominatus S. Moore;白叶莓（白背叶悬钩子,白叶悬钩子,刺泡,酸母子,天青白扭,天青地白扭,无腺白叶莓,早谷蘸）;Whiteleaf Raspberry, White-leaved Raspberry ●

338629　Rubus innominatus S. Moore subsp. plebejus Focke ＝ Rubus spinulosoides F. P. Metcalf ●

338630　Rubus innominatus S. Moore var. aralioides（Hance）Te T. Yu et L. T. Lu;蜜腺白叶莓;Nectarean Whiteleaf Raspberry ●

338631　Rubus innominatus S. Moore var. kuntzeanus（Hemsl.）Bailey;无腺白叶莓（光腺白叶莓,红梅梢,酸母子,天青白扭,早谷蘸,早谷抛子）;Kuntz Whiteleaf Raspberry ●

338632　Rubus innominatus S. Moore var. macrosepalus F. P. Metcalf;宽萼白叶莓;Largesepal Whiteleaf Raspberry ●

338633　Rubus innominatus S. Moore var. quinatus Bailey;五叶白叶莓;Fiveleaved Whiteleaf Raspberry ●

338634　Rubus innoxius Focke ＝ Rubus irenaeus Focke var. innoxius（Focke ex Diels）Te T. Yu et L. T. Lu ●

338635　Rubus innoxius Focke ex Diels ＝ Rubus irenaeus Focke var. innoxius（Focke ex Diels）Te T. Yu et L. T. Lu ●

338636　Rubus inobvius L. H. Bailey ＝ Rubus curtipes L. H. Bailey ●☆

338637　Rubus inopertus（Diels）Focke;红花悬钩子（脉翅悬钩子）;

Redflower Raspberry, Red-flowered Raspberry ●

338638　Rubus inopertus (Diels) Focke var. echinocalyx Cardot；刺萼红花悬钩子；Spinycalyx Redflower Raspberry ●

338639　Rubus interjungens Gust. = Rubus apetalus Poir. ●☆

338640　Rubus involucratus Focke = Rubus corchorifolius L. f. ●

338641　Rubus irenaeus Focke；灰毛泡（地五泡藤，家正牛）；Greyhair Raspberry, Grey-haired Raspberry ●

338642　Rubus irenaeus Focke var. innoxius (Focke ex Diels) Te T. Yu et L. T. Lu；尖裂灰毛泡；Unharmed Greyhair Raspberry ●

338643　Rubus irenaeus Focke var. orogenes (Hand.-Mazz.) F. P. Metcalf = Rubus reflexus Ker Gawl. var. orogenes Hand.-Mazz. ●

338644　Rubus iringanus Gust.；伊林加悬钩子●☆

338645　Rubus irregularis L. H. Bailey = Rubus canadensis L. ●☆

338646　Rubus irritans Focke；紫色悬钩子；Purple Raspberry ●

338647　Rubus ithacanus L. H. Bailey；伊萨卡悬钩子；Ithaca Blackberry ●☆

338648　Rubus jacens Blanch.；蔓延悬钩子；Trailing Dewberry ●☆

338649　Rubus jamaicensis Blanco = Rubus rosifolius Sm. ●

338650　Rubus jambosoides Hance；蒲桃叶悬钩子；Jambosleaf Raspberry, Jamboslike Raspberry, Syzigium-leaved Raspberry ●

338651　Rubus jamini H. Lév. et Vaniot = Rubus irenaeus Focke ●

338652　Rubus japonicus L. = Kerria japonica (L.) DC. ●

338653　Rubus japonicus Maxim. ex Kuntze = Rubus ikenoensis H. Lév. et Vaniot ■☆

338654　Rubus jaysmithii L. H. Bailey var. angustior L. H. Bailey = Rubus flagellaris Willd. ●☆

338655　Rubus jejunus L. H. Bailey = Rubus uvidus L. H. Bailey ●☆

338656　Rubus jianensis L. T. Lu et Boufford；常绿悬钩子；Evergreen Raspberry ●

338657　Rubus jiangxiensis Z. X. Yu, W. T. Ji et H. Zheng；武夷悬钩子；Jiangxi Raspberry ●

338658　Rubus jiangxiensis Z. X. Yu, W. T. Ji et H. Zheng = Rubus glabricarpus W. C. Cheng var. glabratus C. Z. Zheng et Y. Y. Fang ●

338659　Rubus jinfoshanensis Te T. Yu et L. T. Lu；金佛山悬钩子；Jinfoshan Raspberry ●

338660　Rubus junceus Blanch.；灯心草悬钩子；Herbaceous Blackberry ■☆

338661　Rubus junior L. H. Bailey = Rubus setosus Bigelow ●☆

338662　Rubus kalamazoensis L. H. Bailey = Rubus plus L. H. Bailey ●☆

338663　Rubus kanayamensis H. Lév. et Vaniot = Rubus komarovii Nakai ●

338664　Rubus kawakamii Hayata；桑叶悬钩子（川上悬钩子）；Kawakami's Raspberry, Mulberryleaf Raspberry, Mulberry-leaved Raspberry ●

338665　Rubus kelloggii L. H. Bailey = Rubus meracus L. H. Bailey ●☆

338666　Rubus keniensis Standl.；肯尼亚悬钩子●☆

338667　Rubus kennedyanus Fernald；肯尼迪悬钩子；Kennedy's Blackberry ●☆

338668　Rubus kerrifolius H. Lév. et Vaniot = Rubus corchorifolius L. f. ●

338669　Rubus kinashii H. Lév. et Vaniot = Rubus mesogaeus Focke ex Diels ●

338670　Rubus kinashii H. Lév. et Vaniot f. microphyllus Cardot = Rubus mesogaeus Focke ex Diels ●

338671　Rubus kinginsis Engl. = Rubus pinnatus Willd. ●☆

338672　Rubus kisoensis Nakai = Rubus palmatus Thunb. var. kisoensis (Nakai) Ohwi ●☆

338673　Rubus kiwuensis Focke = Rubus runssorensis Engl. ●☆

338674　Rubus koehneanus Focke = Rubus subcrataegifolius (H. Lév. et Vaniot) H. Lév. ●☆

338675　Rubus koehneanus Focke = Rubus trianthus Focke ●

338676　Rubus koehneanus Focke var. formisanus Cardot = Rubus trianthus Focke ●

338677　Rubus koehneanus Focke var. formosanus Cardot = Rubus trianthus Focke ●

338678　Rubus kokonoricus K. S. Hao；青海悬钩子；Qinghai Raspberry ●

338679　Rubus komarovii Nakai；绿叶悬钩子（金山梅）；Komarov Raspberry ●

338680　Rubus komarovii Nakai = Rubus idaeus L. subsp. melanolasius Focke f. concolor (Kom.) Ohwi ●☆

338681　Rubus kotoensis Hayata = Rubus alnifoliolatus H. Lév. et Vaniot var. kotoensis (Hayata) H. L. Li ●

338682　Rubus kotoensis Hayata = Rubus fraxinifolius Poir. ●

338683　Rubus kulinganus L. H. Bailey；牯岭悬钩子；Guling Raspberry, Kuling Raspberry ●

338684　Rubus kuntzeanus Hemsl. = Rubus innominatus S. Moore var. kuntzeanus (Hemsl.) Bailey ●

338685　Rubus kuntzeanus Hemsl. var. glandulosus Cardot = Rubus innominatus S. Moore ●

338686　Rubus kuntzeanus Hemsl. var. xanthacantha (H. Lév.) H. Lév. = Rubus innominatus S. Moore ●

338687　Rubus kwangsiensis H. L. Li；广西悬钩子；Guangxi Raspberry, Kwangsi Raspberry ●

338688　Rubus kwangtungensis H. L. Li = Rubus tsangii Merr. ●

338689　Rubus labbei H. Lév. et Vaniot = Rubus calycacanthus H. Lév. ●

338690　Rubus labber H. Lév. et Vaniot = Rubus pinnatisepalus Hemsl. ●

338691　Rubus lachnocarpus Focke = Rubus piluliferus Focke ●

338692　Rubus lachnocarpus Focke ex Diels = Rubus piluliferus Focke ●

338693　Rubus laciniatostipulatus Hayata ex Koidz. = Rubus alceifolius Poir. ●

338694　Rubus laciniatostipulatus Hayata ex Koidz. = Rubus pinnatisepalus Hemsl. ●

338695　Rubus laciniatus Willd.；深裂悬钩子；Cutleaf Blackberry, Cut-leaf Blackberry, Cut-leaved Blackberry, Evergreen Blackberry, Oregon Evergreen Blackberry ●☆

338696　Rubus laetabilis L. H. Bailey = Rubus canadensis L. ●☆

338697　Rubus laevior (L. H. Bailey) Fernald = Rubus permixtus Blanch. ●☆

338698　Rubus lambertianus Ser.；高粱泡（刺五泡藤，倒盘龙，冬菠，冬牛，蓬蘽）；Lambert's Raspberry ●

338699　Rubus lambertianus Ser. = Rubus xanthoneurus Focke ●

338700　Rubus lambertianus Ser. ex DC. = Rubus lambertianus Ser. ●

338701　Rubus lambertianus Ser. subsp. hakonensis (Franch. et Sav.) Focke = Rubus lambertianus Ser. var. glaber Hemsl. ●

338702　Rubus lambertianus Ser. subsp. xanthoneurus (Focke) Focke = Rubus xanthoneurus Focke ●

338703　Rubus lambertianus Ser. var. glaber Hemsl.；光滑高粱泡（倒盘龙，光蓬蘽，光叶高粱泡，红娘藤，黄花蘽，黄莓刺，黄水蘽，黄水泡，三月泡，山泡刺藤，深山寒莓，酸蘽，小米蘽，小米泡）；Glabrous Lambert Raspberry ●

338704　Rubus lambertianus Ser. var. glandulosus Cardot；腺毛高粱泡（腺叶蓬蘽）；Glandular Lambert Raspberry ●

338705　Rubus lambertianus Ser. var. hakonensis (Franch. et Sav.) Rehder = Rubus lambertianus Ser. var. glaber Hemsl. ●

338706　Rubus lambertianus Ser. var. mekiongensis Hand.-Mazz. = Rubus lambertianus Ser. var. glandulosus Cardot ●

338707　Rubus lambertianus Ser. var. minimiflorus (H. Lév.) Cardot =

Rubus lambertianus Ser. var. glandulosus Cardot ●

338708　Rubus lambertianus Ser. var. morii（Hayata）S. S. Ying ＝ Rubus lambertianus Ser. var. glandulosus Cardot ●

338709　Rubus lambertianus Ser. var. morii（Hayata）S. S. Ying ＝ Rubus morii Hayata ●

338710　Rubus lambertianus Ser. var. paykouangensis（H. Lév.）Hand. - Mazz.；毛叶高粱泡；Hairy-leaf Lambert Raspberry ●

338711　Rubus lambertianus Ser. var. xanthoneurus Focke ＝ Rubus lambertianus Ser. ●

338712　Rubus lambertianus Ser. var. xanthoneurus Focke ＝ Rubus xanthoneurus Focke ●

338713　Rubus lanuginosus Staven ex Ser.；密毛悬钩子●☆

338714　Rubus lanyuensis C. E. Chang；兰屿悬钩子●

338715　Rubus lasiocarpus Sm. ＝ Rubus niveus Thunb. ●

338716　Rubus lasiocarpus Sm. var. ectenothyrsus Cardot ＝ Rubus niveus Thunb. ●

338717　Rubus lasiocarpus Sm. var. micranthus（D. Don）Hook. f. ＝ Rubus niveus Thunb. ●

338718　Rubus lasiococcus Sm.；绵毛果悬钩子；Dwarf Bramble ●☆

338719　Rubus lasiostylus Focke；绵果悬钩子（瞻帽蘸，刺泡花，毛柱莓，毛柱悬钩子）；Woolly Raspberry，Woollystyle Raspberry，Woolly-styled Raspberry ●

338720　Rubus lasiostylus Focke f. glabratus Focke ＝ Rubus lasiostylus Focke ●

338721　Rubus lasiostylus Focke f. glandulosus Focke ＝ Rubus eucalyptus Focke ●

338722　Rubus lasiostylus Focke var. dizygos Focke；五叶绵果悬钩子；Fiveleaved Woollystyle Raspberry ●

338723　Rubus lasiostylus Focke var. eglandulosus Focke；腺梗绵果悬钩子●

338724　Rubus lasiostylus Focke var. hubeiensis Te T. Yu，Spongberg et A. M. Lu；鄂西绵果悬钩子；Hubei Woollystyle Raspberry ●

338725　Rubus lasiostylus Focke var. tomentosus Focke；绒毛绵果悬钩子●

338726　Rubus lasiostylus Focke var. villosus Cardot ＝ Rubus eucalyptus Focke var. trillisatus（Focke）Te T. Yu et L. T. Lu ●

338727　Rubus lasiotrichos Focke；多毛悬钩子；Shaggy Raspberry ●

338728　Rubus lasiotrichos Focke var. blinnii（H. Lév.）L. T. Lu；狭萼多毛悬钩子（五叶悬钩子，五爪风，五爪藤）；Blinn's Raspberry，Blinn's Shaggy Raspberry ●

338729　Rubus latens L. H. Bailey ＝ Rubus alleghENIensis Porter ex L. H. Bailey ●☆

338730　Rubus latifoliolus L. H. Bailey ＝ Rubus wisconsinensis L. H. Bailey ●☆

338731　Rubus latoauriculatus F. P. Metcalf；耳叶悬钩子；Broadauriculate Raspberry，Broad-auriculated Raspberry，Broadleaf Raspberry ●

338732　Rubus laxus Focke；疏毛悬钩子（疏松悬钩子）；Loose Raspberry ●

338733　Rubus ledermannii Engl. ＝ Rubus pinnatus Willd. ●☆

338734　Rubus ledermannii Engl. var. serrulatus Gust. ＝ Rubus pinnatus Willd. ●☆

338735　Rubus lepidulus（Sudre）Juz.；小鳞悬钩子●☆

338736　Rubus leptostemon Juz.；细冠悬钩子●☆

338737　Rubus leucanthus Hance；白钩悬钩子（白钩箣藤，南蛇箣）；Whiteflower Raspberry，White-flowered Raspberry ●■

338738　Rubus leucanthus Hance var. etropicus Hand. -Mazz. ＝ Rubus columellaris Tutcher ●

338739　Rubus leucanthus Hance var. paradoxus（S. Moore）F. P. Metcalf ＝ Rubus leucanthus Hance ●■

338740　Rubus leucanthus Hance var. villosulus Cardot ＝ Rubus leucanthus Hance ●■

338741　Rubus leucodermis Douglas ex Hook.；白皮悬钩子；Black Raspberry，Blackcap，Western Raspberry ●☆

338742　Rubus liboensis T. L. Xu；荔波悬钩子；Libo Raspberry ●

338743　Rubus licens L. H. Bailey ＝ Rubus uvidus L. H. Bailey ●☆

338744　Rubus lichuanensis Te T. Yu et L. T. Lu；黎川悬钩子；Lichuan Raspberry ●

338745　Rubus licitus L. H. Bailey ＝ Rubus alumnus L. H. Bailey ●☆

338746　Rubus limprichtii Pax et K. Hoffm. ＝ Rubus malifolius Focke ●

338747　Rubus limulus L. H. Bailey ＝ Rubus cauliflorus L. H. Bailey ●☆

338748　Rubus linearifoliolus Hayata；雾社悬钩子（细叶悬钩子）；Linear-leaf Raspberry ●

338749　Rubus linearifoliolus Hayata ＝ Rubus tsangii Merr. var. linearifoliolus（Hayata）Te T. Yu et L. T. Lu ●

338750　Rubus linearifoliolus Hayata ＝ Rubus tsangii Merr. ●

338751　Rubus lineatus Reinw. ex Blume；绢毛悬钩子；Lineate Raspberry，Silky Raspberry，Silky-leaved Berry ●

338752　Rubus lineatus Reinw. ex Blume var. angustifolius Hook. f.；狭叶绢毛悬钩子；Narrowleaf Silky Raspberry ●

338753　Rubus lineatus Reinw. ex Blume var. glabrescens Te T. Yu et L. T. Lu；光秃绢毛悬钩子；Glabrous Silky Raspberry ●

338754　Rubus linkianus Ser.；林氏悬钩子；Link's Blackberry ●

338755　Rubus lishuiensis Te T. Yu et L. T. Lu；丽水悬钩子；Lishui Raspberry ●

338756　Rubus liuii Y. P. Yang et S. Y. Lu；柳氏悬钩子（宜兰悬钩子）；Liu Raspberry，Yilan Raspberry ●

338757　Rubus lloydianus Genev.；娄氏悬钩子；Lloyd's Raspberry ●☆

338758　Rubus lobatus Te T. Yu et L. T. Lu；五裂悬钩子；Fivelobed Raspberry，Five-lobed Raspberry ●

338759　Rubus lobophyllus C. Shih ex L. T. Lu；角裂悬钩子；Lobedleaf Raspberry，Lobulate-leaved Raspberry ●

338760　Rubus localis L. H. Bailey ＝ Rubus uvidus L. H. Bailey ●☆

338761　Rubus loganobaccus L. H. Bailey；罗甘莓；Boysenberry，Loganberry ●☆

338762　Rubus lohfauensis Te T. Yu et L. T. Lu；罗浮山悬钩子；Luofushan Raspberry ●

338763　Rubus longepedicellatus（Gust.）C. H. Stirt.；长梗悬钩子●☆

338764　Rubus longipes Fernald ＝ Rubus flagellaris Willd. ●☆

338765　Rubus longissimus L. H. Bailey ＝ Rubus alleghENIensis Porter ex L. H. Bailey ●☆

338766　Rubus longistylus H. Lév. ＝ Rubus niveus Thunb. ●

338767　Rubus loropetalum Franch. ＝ Rubus fockeanus Kurz ■

338768　Rubus lucens Focke；光亮悬钩子；Shining Raspberry ●

338769　Rubus luchunensis Te T. Yu et L. T. Lu；禄春悬钩子（绿春悬钩子）；Luchun Raspberry ●

338770　Rubus luchunensis Te T. Yu et L. T. Lu var. coriseceus Te T. Yu et L. T. Lu；硬叶禄春悬钩子（硬叶绿春悬钩子）；Leatherleaf Luchun Raspberry ●

338771　Rubus ludwigii Eckl. et Zeyh.；路德维格悬钩子●☆

338772　Rubus ludwigii Eckl. et Zeyh. subsp. spatiosus C. H. Stirt.；阔路德维格悬钩子●☆

338773　Rubus lutescens Franch.；黄色悬钩子；Yellow Raspberry ●

338774　Rubus lutescens Franch. f. glabrescens Cardot ＝ Rubus lutescens Franch. ●

338775 Rubus lyi H. Lév. = Rubus setchuenensis Bureau et Franch. ●

338776 Rubus macilentus Cambess. ;细瘦悬钩子;Thin Raspberry ●

338777 Rubus macilentus Cambess. var. angualtus Delavay;棱枝细瘦悬钩子;Angularstem Thin Raspberry ●

338778 Rubus macrocarpus King ex C. B. Clarke = Rubus wardii Merr. ●

338779 Rubus macrophyllus Weihe et Nees;大叶悬钩子;Largeleaf Blackberry ●☆

338780 Rubus mairei H. Lév. = Rubus niveus Thunb. ●

338781 Rubus mairei H. Lév. = Rubus preptanthus Focke var. mairei (H. Lév.) Te T. Yu et L. T. Lu ●

338782 Rubus major Focke = Rubus multibracteatus H. Lév. et Vaniot ●

338783 Rubus major Focke = Rubus pluribracteatus L. T. Lu et Boufford ●

338784 Rubus malifolius Focke;棠叶悬钩子(海棠叶莓,老林茶,羊尿泡);Appleleaf Raspberry, Apple-leaved Raspberry ●

338785 Rubus malifolius Focke var. longisepalus Te T. Yu et L. T. Lu;长萼棠叶悬钩子(长萼悬钩子);Longsepal Raspberry ●

338786 Rubus maliodes Focke = Rubus multibracteatus H. Lév. et Vaniot ●

338787 Rubus malipoensis Te T. Yu et L. T. Lu;麻栗坡悬钩子;Malipo Raspberry ●

338788 Rubus mallodes Focke = Rubus multibracteatus H. Lév. et Vaniot ●

338789 Rubus mallodes Focke = Rubus pluribracteatus L. T. Lu et Boufford ●

338790 Rubus mallotifolius Y. C. Wu ex Te T. Yu et L. T. Lu;楸叶悬钩子;Mallotusleaf Raspberry, Mallotus-leaved Raspberry ●

338791 Rubus maltei L. H. Bailey = Rubus celer L. H. Bailey ●☆

338792 Rubus malus L. H. Bailey = Rubus vermontanus Blanch. ●☆

338793 Rubus marilandicus L. H. Bailey = Rubus allegheniensis Porter ex L. H. Bailey var. gravesii Fernald ●☆

338794 Rubus maruyamae Naruh. = Rubus × ohmineanus Koidz. ●☆

338795 Rubus matsumuranus H. Lév. et Vaniot = Rubus idaeus L. subsp. melanolasius Focke ●☆

338796 Rubus matsumuranus H. Lév. et Vaniot = Rubus sachalinensis H. Lév. ●

338797 Rubus matsumuranus H. Lév. et Vaniot var. concolor (Kom.) Kitag. = Rubus idaeus L. subsp. melanolasius Focke f. concolor (Kom.) Ohwi ●☆

338798 Rubus matsumuranus H. Lév. et Vaniot var. eglanduratus Y. B. Chang = Rubus sachalinensis H. Lév. var. eglanduratus (Y. B. Chang) L. T. Lu ●

338799 Rubus mediocris L. H. Bailey = Rubus missouricus L. H. Bailey ●☆

338800 Rubus megalothyrsus Cardot = Rubus tephrodes Hance var. ampliflorus (H. Lév. et Vaniot) Hand. -Mazz. ●

338801 Rubus melanolasius Focke = Rubus idaeus L. var. strigosus (Michx.) Maxim. ●☆

338802 Rubus melanolasius Focke = Rubus komarovii Nakai ●

338803 Rubus melanolasius Focke var. concolor Kom. = Rubus komarovii Nakai ●

338804 Rubus melanolasius Focke var. discolor Kom. = Rubus sachalinensis H. Lév. ●

338805 Rubus melanolasius var. concolor Kom. = Rubus komarovii Nakai ●

338806 Rubus menglaensis Te T. Yu et L. T. Lu;勐腊悬钩子;Mengla Raspberry ●

338807 Rubus meracus L. H. Bailey;倾斜悬钩子;Dryslope Dewberry ●☆

338808 Rubus mesogaeus Focke ex Diels;喜荫悬钩子(短样泡刺藤,黑帽莓,加拿大悬钩子,里白悬钩子,莓子,深山悬钩子);Blackcap Raspberry, Common Blackcap, Shadily Raspberry, Shady Raspberry, Thimbleberry ●

338809 Rubus mesogaeus Focke ex Diels f. floribus-roseis Focke = Rubus mesogaeus Focke ex Diels ●

338810 Rubus mesogaeus Focke ex Diels f. roseus Koji Ito;红花喜阴悬钩子●☆

338811 Rubus mesogaeus Focke ex Diels var. adenothrix Momiy. ex Hatus. ;日本腺毛喜阴悬钩子●☆

338812 Rubus mesogaeus Focke ex Diels var. glabrescens Te T. Yu et L. T. Lu;脱毛喜阴悬钩子;Glabrous Shady Raspberry ●

338813 Rubus mesogaeus Focke ex Diels var. incisus Cardot = Rubus mesogaeus Focke ex Diels ●

338814 Rubus mesogaeus Focke ex Diels var. oxycomus Focke;腺毛喜阴悬钩子;Glandular Shady Raspberry ●

338815 Rubus mesogaeus Focke f. floribus-roseis Focke = Rubus mesogaeus Focke ex Diels ●

338816 Rubus mesogaeus Focke f. roseus Koji Ito = Rubus mesogaeus Focke ex Diels f. roseus Koji Ito ●☆

338817 Rubus mesogaeus Focke var. adenothrix Momiy. ex Hatus. = Rubus mesogaeus Focke ex Diels var. adenothrix Momiy. ex Hatus. ●☆

338818 Rubus metoensis Te T. Yu et L. T. Lu;墨脱悬钩子;Medog Raspberry, Motuo Raspberry ●

338819 Rubus mgosissimus Hayata = Rubus formosensis Kuntze ●

338820 Rubus michiganensis (L. H. Bailey) L. H. Bailey = Rubus roribaccus (L. H. Bailey) Rydb. ●☆

338821 Rubus micranthus D. Don = Rubus niveus Thunb. ●

338822 Rubus microphyllus L. f. ;苦莓●☆

338823 Rubus microphyllus L. f. = Rubus niveus Thunb. ●

338824 Rubus microphyllus L. f. f. plenus T. Yamanaka;重瓣苦莓●☆

338825 Rubus microphyllus L. subsp. koehneanus (Focke) Sugim. = Rubus subcrataegifolius (H. Lév. et Vaniot) H. Lév. ●☆

338826 Rubus microphyllus L. var. subcrataegifolius (H. Lév. et Vaniot) Ohwi = Rubus subcrataegifolius (H. Lév. et Vaniot) H. Lév. ●☆

338827 Rubus millspaughii Britton = Rubus canadensis L. ●☆

338828 Rubus minensis Pax et K. Hoffm. = Rubus macilentus Cambess. ●

338829 Rubus mingetsensis Hayata = Rubus aculeatiflorus Hayata ●

338830 Rubus mingetsensis Hayata = Rubus taitoensis Hayata var. aculeatiflorus (Hayata) H. Ohashi et C. F. Hsieh ●

338831 Rubus mingetsensis Hayata var. taitoensis (Hayata) Y. C. Liu et Yang = Rubus aculeatiflorus Hayata ●

338832 Rubus minimiflorus H. Lév. = Rubus lambertianus Ser. var. glandulosus Cardot ●

338833 Rubus minnesotanus L. H. Bailey = Rubus wisconsinensis L. H. Bailey ●☆

338834 Rubus minusculus H. Lév. = Rubus rosifolius Sm. ●

338835 Rubus minusculus H. Lév. et Vaniot = Rubus rosifolius Sm. ●

338836 Rubus miriflorus L. H. Bailey = Rubus alumnus L. H. Bailey ●☆

338837 Rubus mirus L. H. Bailey;奇异悬钩子;Youngberry ●☆

338838 Rubus miscix L. H. Bailey;不凡悬钩子●☆

338839 Rubus missouricus L. H. Bailey;密苏里悬钩子;Missouri Dewberry, Prickly Groundberry ●☆

338840 Rubus miszczenkoi Juz. ;米氏悬钩子●☆

338841 Rubus modestus Focke = Rubus pentagonus Wall. ex Focke var. modestus (Focke) Te T. Yu et L. T. Lu ●

338842　Rubus modicus Focke ＝Rubus pentagonus Wall. ex Focke var. modestus（Focke）Te T. Yu et L. T. Lu ●

338843　Rubus mollior L. H. Bailey;宾州悬钩子;Highbush Blackberry, Pennsylvania Blackberry, Yankee Blackberry ●☆

338844　Rubus moluccanus sensu Matsum. et Hayata ＝Rubus pinnatisepalus Hemsl. ●

338845　Rubus mongouilloni H. Lév. et Vaniot ＝Rubus alceifolius Poir. ●

338846　Rubus monguillonii H. Lév. et Vaniot ＝Rubus alceifolius Poir. ●

338847　Rubus montanus（Porter）Porter ＝Rubus allegheniensis Porter ex L. H. Bailey ●☆

338848　Rubus montpelierensis Blanch. ＝Rubus glandicaulis Blanch. ●☆

338849　Rubus morifolius Siebold ex Franch. et Sav. ＝Rubus crataegifolius Bunge ●

338850　Rubus morii Hayata;鳞萼悬钩子（台湾悬钩子,尾叶悬钩子）;Mori's Raspberry ●

338851　Rubus morii Hayata ＝Rubus lambertianus Ser. var. glandulosus Cardot ●

338852　Rubus mouyousensis H. Lév. ＝Rubus chrolosepalus Focke ●

338853　Rubus multibracteatus H. Lév. et Vaniot ＝Rubus pluribracteatus L. T. Lu et Boufford ●

338854　Rubus multibracteatus H. Lév. et Vaniot var. demangei H. Lév. ＝Rubus alceifolius Poir. ●

338855　Rubus multibracteatus H. Lév. et Vaniot var. lobatisepalus Te T. Yu et L. T. Lu ＝Rubus pluribracteatus L. T. Lu et Boufford var. lobatisepalus（Te T. Yu et L. T. Lu）L. T. Lu et Boufford ●

338856　Rubus multifer L. H. Bailey;多育悬钩子;Kinnickinnick Dewberry ●☆

338857　Rubus multiformis Blanch.;多变悬钩子;Variable Blackberry ●☆

338858　Rubus multiformis Blanch. var. delicatior Blanch. ＝Rubus multiformis Blanch. ●☆

338859　Rubus multisetosus Te T. Yu et L. T. Lu;刺毛悬钩子;Bristly Raspberry ●

338860　Rubus mundtii Cham. et Schltdl. ＝Rubus rigidus Sm. ●☆

338861　Rubus myriadenus H. Lév. et Vaniot ＝Rubus sumatranus Miq. ●

338862　Rubus myriadenus H. Lév. et Vaniot var. grandifoliolatus H. Lév. ＝Rubus sumatranus Miq. ●

338863　Rubus mysorensis F. Heyne ＝Rubus niveus Thunb. ●

338864　Rubus nagasawanus Koidz.;高砂悬钩子（粗毛悬钩子）;Nagasawa Raspberry, Rough-hair Raspberry ●

338865　Rubus nagasawanus Koidz. var. arachnoideus（Y. C. Liu et F. Y. Lu）S. S. Ying ＝Rubus arachnoideus Y. C. Liu et F. Y. Lu ●

338866　Rubus nakaianus H. Lév. ＝Rubus coreanus Miq. ●

338867　Rubus nakaii Tuyama;中井氏悬钩子●☆

338868　Rubus nanopetalus Cardot;矮小悬钩子●

338869　Rubus nanopetalus Cardot ＝Rubus neoviburnifolius L. T. Lu et Boufford ●

338870　Rubus nantoensis Hayata ＝Rubus formosensis Kuntze ●

338871　Rubus navus L. H. Bailey;舟莓;Grand Lake Blackberry ●☆

338872　Rubus nefrens L. H. Bailey ＝Rubus enslenii Tratt. ●☆

338873　Rubus neglectus Peck;忽视悬钩子;Purple Raspberry ●☆

338874　Rubus neillioides（Focke）Migo ＝Rubus glabricarpus W. C. Cheng ●

338875　Rubus neonefrens L. H. Bailey ＝Rubus flagellaris Willd. ●☆

338876　Rubus neoviburnifolius L. T. Lu et Boufford;新荚蒾叶悬钩子●

338877　Rubus nepalensis Hort.;尼泊尔悬钩子;Himalayan Carpet Bramble ●☆

338878　Rubus nessensis Hall;半立悬钩子●☆

338879　Rubus nigricans Rydb. ＝Rubus setosus Bigelow ●☆

338880　Rubus nigrobaccus L. H. Bailey;黑果莓;Black-fruit Raspberry ●☆

338881　Rubus nigrobaccus L. H. Bailey ＝Rubus allegheniensis Porter ex L. H. Bailey ●☆

338882　Rubus nigrobaccus L. H. Bailey var. albinus（L. H. Bailey）L. H. Bailey ＝Rubus allegheniensis Porter ex L. H. Bailey ●☆

338883　Rubus nigrobaccus L. H. Bailey var. calycosus Fernald ＝Rubus allegheniensis Porter ex L. H. Bailey ●☆

338884　Rubus nigrobaccus L. H. Bailey var. sativus（L. H. Bailey）L. H. Bailey ＝Rubus allegheniensis Porter ex L. H. Bailey ●☆

338885　Rubus nishimuranus Koidz.;西村悬钩子●☆

338886　Rubus niveus Thunb.;红泡刺藤（白刺泡,白绒悬钩子,白枝泡,倒生根,多叶蔫,覆盆子,钩丝刺,钩撕刺,狗屎,薅秧泡,黑黄袍,黑锁莓,黑锁莓,红刺泡,灰毛果莓,日本小叶悬钩子,疏风草,锁梅,小花悬钩子,硬枝黑锁莓,栽秧苗,紫茵,钻地风）;Littleflower Raspberry, Littleleaf Raspberry, Mysore Raspberry, Snow-peak Raspberry, Snowpeaks Raspberry ●

338887　Rubus niveus Thunb. subsp. inopertus Diels ＝Rubus inopertus（Diels）Focke ●

338888　Rubus niveus Thunb. subsp. inopertus Focke ＝Rubus inopertus（Diels）Focke ●

338889　Rubus niveus Thunb. var. hypargyrus（Edgew.）Hook. f. ＝Rubus hypargyrus Edgew. ●

338890　Rubus niveus Thunb. var. micranthus（D. Don）H. Hara;小花红泡刺藤（小花毛果覆盆子）●

338891　Rubus niveus Thunb. var. micranthus（D. Don）H. Hara ＝Rubus niveus Thunb. ●

338892　Rubus niveus Thunb. var. niveus（Wall.）Hook. f. ＝Rubus hypargyrus Edgew. var. niveus（Wall. ex G. Don）H. Hara ●

338893　Rubus niveus Wall. ＝Rubus hypargyrus Edgew. var. niveus（Wall. ex G. Don）H. Hara ●

338894　Rubus niveus Wall. ex G. Don ＝Rubus pedunculosus D. Don ●

338895　Rubus norikurensis Ohwi;乘鞍悬钩子●☆

338896　Rubus numidicus Focke ＝Rubus incanescens（DC.）Bertol. ●☆

338897　Rubus nuperus L. H. Bailey ＝Rubus allegheniensis Porter ex L. H. Bailey ●☆

338898　Rubus nutans Wall. var. fockeanus（Kuntze）Kuntze ＝Rubus fockeanus Kurz ■

338899　Rubus nyalamensis Te T. Yu et L. T. Lu;聂拉木悬钩子;Nielamu Raspberry ■

338900　Rubus obcordatus（Franch.）Thuan ＝Rubus ellipticus Sm. var. obcordatus（Franch.）Focke ●

338901　Rubus oblongus Te T. Yu et L. T. Lu;长圆悬钩子;Oblong Raspberry ●

338902　Rubus obocordatus（Franch.）Thuan ＝Rubus ellipticus Sm. var. obcordatus（Franch.）Focke ●

338903　Rubus obovalis Michx. ＝Rubus hispidus L. ●☆

338904　Rubus obovatus Pers. ＝Rubus hispidus L. ●☆

338905　Rubus obsessus L. H. Bailey var. unilaris L. H. Bailey ＝Rubus curtipes L. H. Bailey ●☆

338906　Rubus occidentalis H. Lév. ＝Rubus mesogaeus Focke ex Diels ●

338907　Rubus occidentalis H. Lév. var. exsuccus（Franch. et Sav.）Makino ＝Rubus mesogaeus Focke ex Diels ●

338908　Rubus occidentalis H. Lév. var. japonicus Miyabe ＝Rubus mesogaeus Focke ex Diels ●

338909　Rubus occidentalis L.;西方悬钩子;Black Cap, Black Raspberry, Black Rasp-berry, Blackcap, Black-cap, Blackcap

Raspberry, Thimbleberry, Thimble-berry, Virginian Raspberry ●

338910 Rubus occidentalis L. f. pallidus (L. H. Bailey) B. L. Rob. = Rubus occidentalis L. ●

338911 Rubus occidentalis L. var. exsuccus (Franch. et Sav.) Makino = Rubus mesogaeus Focke ex Diels ●

338912 Rubus occidentalis L. var. japonicus Miyabe = Rubus mesogaeus Focke ex Diels ●

338913 Rubus occidentalis L. var. pallidus L. H. Bailey = Rubus occidentalis L. ●

338914 Rubus occidualis (L. H. Bailey) L. H. Bailey = Rubus roribaccus (L. H. Bailey) Rydb. ●☆

338915 Rubus occultus L. H. Bailey = Rubus flagellaris Willd. ●☆

338916 Rubus ochlanthus Hance = Rubus lambertianus Ser. ●

338917 Rubus odoratus L.; 芳香悬钩子 (糙莓, 香莓); American Bramble, Flowering Raspberry, Fragrant Thimbleberry, Ornamental Raspberry, Purple Flowering Raspberry, Purple-flowered Raspberry, Purple-flowering Raspberry, Scented Bramble, Thimbleberry, Thimble-berry ●☆

338918 Rubus officinalis Koidz. = Rubus chingii Hu ●

338919 Rubus ohmatiensis Nakai = Rubus hirsutus Thunb. ●

338920 Rubus ohwianus Koidz. = Rubus inopertus (Diels) Focke ●

338921 Rubus ohwianus Koidz. = Rubus ritozanensis Sasaki ●

338922 Rubus okinawensis Koidz.; 秃悬钩子 (红狭叶悬钩子, 能高悬钩子, 腺毛悬钩子); Glabrous Saffron-coloured Flower Raspberry, Red-narrowleaf Raspberry ●

338923 Rubus okinawensis Koidz. var. formosana Koidz. = Rubus croceacanthus H. Lév. ●

338924 Rubus oldhamii Miq. = Rubus pungens Cambess. var. oldhamii (Miq.) Maxim. ●

338925 Rubus oliveri Miq. = Rubus corchorifolius L. f. ●

338926 Rubus omeiensis Rolfe = Rubus setchuenensis Bureau et Franch. ●

338927 Rubus onustus L. H. Bailey = Rubus satis L. H. Bailey ●☆

338928 Rubus orarius Blanch.; 大灌木悬钩子; Highbush Blackberry ■☆

338929 Rubus oriens L. H. Bailey; 东方悬钩子; Eastern Dewberry ●☆

338930 Rubus oriens L. H. Bailey = Rubus spectatus L. H. Bailey ●☆

338931 Rubus ossicus Juz.; 奥斯悬钩子 ●☆

338932 Rubus ostryifolius L. H. Bailey = Rubus alumnus L. H. Bailey ●☆

338933 Rubus ostryifolius Rydb.; 锐齿悬钩子; Highbush Blackberry, Sawtooth Blackberry ●☆

338934 Rubus ostryifolius Rydb. = Rubus argutus Link ●☆

338935 Rubus otophorus Franch.; 成凤山悬钩子 ●

338936 Rubus otophorus Franch. = Rubus corchorifolius L. f. ●

338937 Rubus ouensanensis H. Lév. et Vaniot = Rubus crataegifolius Bunge ●

338938 Rubus ourosepalus Cardot; 宝兴悬钩子; Baoxing Raspberry, Paosing Raspberry ●

338939 Rubus oxyphyllus Wall. = Rubus acuminatus Sm. ●

338940 Rubus pacatus Focke = Rubus setchuenensis Bureau et Franch. ●

338941 Rubus pacatus Focke var. alypus Focke = Rubus setchuenensis Bureau et Franch. ●

338942 Rubus pacificus Hance; 太平莓 (大叶莓, 老虎扭, 太平洋莓); Pacific Raspberry ●

338943 Rubus pacificus Nakai = Rubus boninensis Koidz. ●☆

338944 Rubus palmatus Hemsl. = Rubus chingii Hu ●

338945 Rubus palmatus Thunb.; 悬钩子 (钩藤子, 拘朴子, 槭叶莓, 掌叶悬钩子); Bramble, Palmate Raspberry ●

338946 Rubus palmatus Thunb. var. coptophyllus (A. Gray) Kuntze ex Koidz.; 东方裂叶悬钩子 ●☆

338947 Rubus palmatus Thunb. var. coptophyllus (A. Gray) Kuntze f. coronarius H. Ohba; 冠饰悬钩子 ●☆

338948 Rubus palmatus Thunb. var. coptophyllus (A. Gray) Kuntze f. inermis Ide; 无刺悬钩子 ●☆

338949 Rubus palmatus Thunb. var. kisoensis (Nakai) Ohwi; 木曽悬钩子 ●☆

338950 Rubus palmatus Thunb. var. yakumontanus (Masam.) Hatus. ex H. Ohba; 屋久岛山地悬钩子 ●☆

338951 Rubus palmensis Hansen; 帕尔马悬钩子 ●☆

338952 Rubus panduratus Hand.-Mazz.; 琴叶悬钩子; Fiddle-shaped Leaf Raspberry, Fiddle-shape-leaved Raspberry, Pandurate Raspberry ●

338953 Rubus panduratus Hand.-Mazz. var. etomentosus Hand.-Mazz.; 脱毛琴叶悬钩子; Hairless Fiddle-shaped Leaf Raspberry ●

338954 Rubus paniculatus Sm.; 圆锥悬钩子; Panicled Raspberry, Paniculate Raspberry ●

338955 Rubus paniculatus Sm. f. tiliaceus (Sm.) H. Hara = Rubus paniculatus Sm. ●

338956 Rubus paniculatus Sm. var. brevifolius Kuntze = Rubus tephrodes Hance ●

338957 Rubus paniculatus Sm. var. glabrescens Te T. Yu et L. T. Lu; 脱毛圆锥悬钩子; Hairless Panicled Raspberry ●

338958 Rubus pappei Eckl. et Zeyh. = Rubus pinnatus Willd. ●☆

338959 Rubus paptrus H. Lév. = Rubus ichangensis Hemsl. et Kuntze ●

338960 Rubus par L. H. Bailey = Rubus alleghentiensis Porter ex L. H. Bailey ●☆

338961 Rubus paradoxus S. Moore = Rubus leucanthus Hance ●■

338962 Rubus pararosifolius F. P. Metcalf; 矮空心泡 (矮生空心泡); Dwarf Raspberry ●

338963 Rubus parcifrondifer L. H. Bailey = Rubus alumnus L. H. Bailey ●☆

338964 Rubus parkeri Hance; 乌泡子 (乌胞, 乌蔗子, 乌泡, 小乌泡); Parker Raspberry ●

338965 Rubus parkeri Hance var. brevisetosus Focke = Rubus parkeri Hance ●

338966 Rubus parkeri Hance var. longisetosus Focke = Rubus parkeri Hance ●

338967 Rubus parviaraliifolius Hayata; 楤叶悬钩子 (疏花小楤叶悬钩子, 小桵叶悬钩子, 小楤叶悬钩子, 爆叶悬钩子); Aralia-leaf Raspberry, Tappasha Raspberry ●

338968 Rubus parviaraliifolius Hayata var. laxiflorus Y. C. Liu et F. Y. Lu = Rubus parviaraliifolius Hayata ●

338969 Rubus parviflorus H. Walter = Rubus cuneifolius Pursh ●☆

338970 Rubus parviflorus Nutt.; 小花悬钩子; Purple Cane Raspberry, Thimbleberry, Thimble-berry ●

338971 Rubus parviflorus Nutt. 'Double'; 重瓣小花悬钩子; Double-white Thimbleberry ●

338972 Rubus parviflorus Nutt. 'Double-flowered Selection'; 双小花悬钩子; Thimbleberry ●☆

338973 Rubus parviflorus Nutt. f. adenius Fassett = Rubus parvifolius L. ●

338974 Rubus parviflorus Nutt. f. bifarius (Fernald) Fassett = Rubus parvifolius L. ●

338975 Rubus parviflorus Nutt. f. glabrifolius Fassett = Rubus parvifolius L. ●

338976 Rubus parviflorus Nutt. f. hypomalacus (Fernald) Fassett = Rubus parvifolius L. ●

338977 Rubus parviflorus Nutt. f. micradenius Fassett = Rubus parvifolius L. ●

338978　Rubus parviflorus Nutt. f. trichophorus Fassett ＝ Rubus parvifolius L. ●

338979　Rubus parviflorus Nutt. var. bifarius Fernald ＝ Rubus parvifolius L. ●

338980　Rubus parviflorus Nutt. var. genuinus Fernald ＝ Rubus parvifolius L. ●

338981　Rubus parviflorus Nutt. var. grandiflorus Farw. ＝ Rubus parvifolius L. ●

338982　Rubus parviflorus Nutt. var. heteradenius Fernald ＝ Rubus parvifolius L. ●

338983　Rubus parviflorus Nutt. var. hypomalacus Fernald ＝ Rubus parvifolius L. ●

338984　Rubus parviflorus Nutt. var. parvifolius（A. Gray）Fernald ＝ Rubus parvifolius L. ●

338985　Rubus parvifolius L. ；茅莓（薦，布田菠草，草杨梅，草杨梅子，陈刺波，过江龙，蒿央胞，薅田藨，薅秧藨，薅秧泡，黑梅，红花孚苈，红梅消，红琐梅，红锁梅，虎波草，黄胞，监监婆，两头粘，龙船藨，毛叶仙桥，茅莓悬钩子，茅抛子，婆婆头，乳痈泡，三月泡，蛇泡苈，蛇泡簕，天青地白，天青地白草，五月红，五月红藨刺，细蛇迯，仙人搭桥，小还魂，小叶悬钩子，牙鹰簕，鹰爪苈，鹰爪簕，栽秧抛，种田蒲）；Japan Raspberry, Japanese Raspberry, Salmon Berry, Small-leaved Raspberry, Thimbleberry, Threeleaf Blackberry, Trailing Raspberry ●

338986　Rubus parvifolius L. ＝ Rubus subornatus Focke ●

338987　Rubus parvifolius L. f. concolor（Koidz.）Sugim. ；同色茅莓●☆

338988　Rubus parvifolius L. f. flavus Akasawa；黄茅莓●☆

338989　Rubus parvifolius L. f. leucanthus Sugim. ；白花茅莓●☆

338990　Rubus parvifolius L. f. parce Focke ＝ Rubus subornatus Focke ●

338991　Rubus parvifolius L. f. yoshinagae（Makino）Sugim. ex Naruh. ；吉永茅莓●☆

338992　Rubus parvifolius L. subvar. subconcolor（Cardot）Masam. ＝ Rubus parvifolius L. ●

338993　Rubus parvifolius L. var. adenochlamys（Focke）Migo；腺花茅莓（倒莓子）；Glandular Flower Japanese Raspberry ●

338994　Rubus parvifolius L. var. concolor（Koidz.）Makino et Nemoto ＝ Rubus parvifolius L. f. concolor（Koidz.）Sugim. ●☆

338995　Rubus parvifolius L. var. purpureus Y. Gu et W. L. Li；紫果茅莓；Purple-fruit Japanese Raspberry ●

338996　Rubus parvifolius L. var. subconcolor（Cardot）Makino et Nemoto ＝ Rubus parvifolius L. ●

338997　Rubus parvifolius L. var. toapiensis（Yamam.）Hosok. ；五叶茅莓（台东红梅消，五叶红梅消）；Fiveleaved Japanese Raspberry ●

338998　Rubus parvifolius L. var. triphyllus（Thunb.）Nakai ＝ Rubus parvifolius L. ●

338999　Rubus parvifraxinifolius Hayata；花莲悬钩子（小梣叶悬钩子）；Small Ash-leaf Raspberry ●

339000　Rubus parvifraxinifolius Hayata ＝ Rubus fraxinifoliolus Hayata ●

339001　Rubus parvifraxinus Hayata ＝ Rubus fraxinifolius Matsum. et Hayata ●

339002　Rubus parvipetalus Odash. ＝ Rubus pirifolius Sm. ●

339003　Rubus parvipungens Hayata ＝ Rubus pungens Cambess. var. oldhamii（Miq.）Maxim. ●

339004　Rubus parvirosifolius Hayata ＝ Rubus rosifolius Sm. ●

339005　Rubus paucidentatus Te T. Yu et L. T. Lu；少齿悬钩子；Fewtooth Raspberry ●

339006　Rubus paucidentatus Te T. Yu et L. T. Lu var. guangxiensis Te

T. Yu et L. T. Lu；广西少齿悬钩子；Guangxi Fewtooth Raspberry, Kwangsi Raspberry ●

339007　Rubus pauciflorus Baker ＝ Rubus parvifolius L. ●

339008　Rubus paulus L. H. Bailey ＝ Rubus alleghaniensis Porter ex L. H. Bailey ●☆

339009　Rubus pauper L. H. Bailey ＝ Rubus ithacanus L. H. Bailey ●☆

339010　Rubus pauperrimus L. H. Bailey ＝ Rubus roribaccus（L. H. Bailey）Rydb. ●☆

339011　Rubus paykouangensis H. Lév. ＝ Rubus lambertianus Ser. var. paykouangensis（H. Lév.）Hand. -Mazz. ●

339012　Rubus pectinarioides H. Hara；匍枝悬钩子；Creeping Raspberry ●

339013　Rubus pectinaris Focke；梳齿悬钩子；Pectinate Raspberry, Pectinatetooth Raspberry ■

339014　Rubus pectinellus Maxim. ；黄泡（刺萼寒莓，小黄泡）；Pectinate Raspberry, Yellowbubble Raspberry ●■

339015　Rubus pectinellus Maxim. var. trilobus Koidz. ；刺萼寒莓；Prickly-calyx Raspberry ●

339016　Rubus pectinellus Maxim. var. trilobus Koidz. ＝ Rubus pectinellus Maxim. ●■

339017　Rubus peculiaris Blanch. ＝ Rubus miscix L. H. Bailey ●☆

339018　Rubus pedatus Sm. ；鸟足悬钩子；Strawberry Bramble ●☆

339019　Rubus pedunculosus D. Don；密毛纤细悬钩子；Snow-white Slender Raspberry ●

339020　Rubus pedunculosus D. Don ＝ Rubus hypargyrus Edgew. var. niveus（Wall. ex G. Don）H. Hara ●

339021　Rubus pedunculosus D. Don var. hypargyrus（Edgew.）Kitag. ＝ Rubus hypargyrus Edgew. ●

339022　Rubus pedunculosus D. Don var. hypargyrus（Edgew.）Kitam. ＝ Rubus pedunculosus D. Don ●

339023　Rubus peii R. H. Miao；阿沛悬钩子●

339024　Rubus peltatus Maxim. ；盾叶莓（大叶覆盆子，对月刺草，牛奶藨，天青地白扭）；Peltateleaf Raspberry, Peltate-leaved Raspberry ●

339025　Rubus penduliflorus Y. C. Wu ex Te T. Yu et L. T. Lu；河口悬钩子；Droopingflower Raspberry, Hekou Raspberry, Penduliflory Raspberry ●

339026　Rubus pensilvanicus Poir. ＝ Rubus frondosus Bigelow ●☆

339027　Rubus pensilvanicus Poir. ＝ Rubus mollior L. H. Bailey ●☆

339028　Rubus pensilvanicus Poir. var. frondosus（Bigelow）B. Boivin ＝ Rubus frondosus Bigelow ●☆

339029　Rubus pentagonus Wall. ex Focke；掌叶悬钩子（里泡刺）；Five-angle Raspberry, Five-angled Raspberry, Palmleaf Raspberry ●

339030　Rubus pentagonus Wall. ex Focke var. eglandulosua Te T. Yu et L. T. Lu；无腺掌叶悬钩子；Glandless Five-angle Raspberry ●

339031　Rubus pentagonus Wall. ex Focke var. longisepalus Te T. Yu et L. T. Lu；长萼掌叶悬钩子；Longsepal Five-angle Raspberry ●

339032　Rubus pentagonus Wall. ex Focke var. modestus（Focke）Te T. Yu et L. T. Lu；无刺掌叶悬钩子；Spineless Five-angle Raspberry ●

339033　Rubus pentalobus Hayata ＝ Rubus calycinoides Hayata var. macrophyllus H. L. Li ●

339034　Rubus pentalobus Hayata ＝ Rubus rolfei Vidal ●

339035　Rubus peracer L. H. Bailey ＝ Rubus multiformis Blanch. ●☆

339036　Rubus pergratus Blanch. ＝ Rubus alleghaniensis Porter ex L. H. Bailey ●☆

339037　Rubus permixtus Blanch. ；灌丛悬钩子；Thicket Dewberry ●☆

339038　Rubus permixtus Blanch. var. laevior L. H. Bailey ＝ Rubus permixtus Blanch. ●☆

339039　Rubus persicus Boiss. ；波斯悬钩子●☆

339040　Rubus perspicuus L. H. Bailey；五大湖悬钩子；Great Lakes Dewberry ●☆

339041　Rubus peruncinatus（Sudre）Juz.；具钩悬钩子●☆

339042　Rubus petalabigens Gust. = Rubus apetalus Poir. ●☆

339043　Rubus petaloideus H. Lév. = Rubus chrolosepalus Focke ●

339044　Rubus petitianus A. Rich. = Rubus apetalus Poir. ●☆

339045　Rubus philippinensis Focke = Rubus pirifolius Sm. ●

339046　Rubus philippinensis Focke ex Elmer = Rubus pirifolius Sm. ●

339047　Rubus phoenicolasius Maxim.；多腺悬钩子（红毛巾，空筒泡，雀不站，树莓，悬钩木，钻地风）；Japanese Wineberry，Wine Raspberry，Wineberry ●

339048　Rubus phoenicolasius Maxim. f. aureiceps（Honda）Honda；黄柄多腺悬钩子；Japanese Wineberry ●☆

339049　Rubus piceetorum Juz.；沥青悬钩子●☆

339050　Rubus pileatus Focke；菰帽悬钩子；Capform Raspberry，Pileate Raspberry ●

339051　Rubus pileatus Focke var. canotomentosus Focke = Rubus lasiostylus Focke var. dizygos Focke ●

339052　Rubus piluliferus Focke；陕西悬钩子；Shaanxi Raspberry ●

339053　Rubus pinfaensis H. Lév. et Vaniot = Rubus wallichianus Wight et Arn. ●

339054　Rubus pinnatiformis Gust. = Rubus apetalus Poir. ●☆

339055　Rubus pinnatisepalus Hemsl.；羽萼悬钩子（爬地泡，新店悬钩子）；Feathery-sepaled Raspberry，Moluccan Bramble，Pinnate-sepal Raspberry，Pinnate-sepaled Raspberry ●

339056　Rubus pinnatisepalus Hemsl. = Rubus alceifolius Poir. ●

339057　Rubus pinnatisepalus Hemsl. var. glandulous Te T. Yu et L. T. Lu；密腺羽萼悬钩子；Glandular Pinnatesepal Raspberry ●

339058　Rubus pinnatus D. Don；南非悬钩子；South African Blackberry ●☆

339059　Rubus pinnatus D. Don = Rubus niveus Thunb. ●

339060　Rubus pinnatus Willd.；羽状悬钩子●☆

339061　Rubus pinnatus Willd. f. subglandulosus Gust. = Rubus pinnatus Willd. ●☆

339062　Rubus pinnatus Willd. subsp. afrotropicus Engl. = Rubus pinnatus Willd. ●☆

339063　Rubus pinnatus Willd. var. afrotropicus（Engl.）Gust. = Rubus pinnatus Willd. ●☆

339064　Rubus pinnatus Willd. var. defensus Gust. = Rubus pinnatus Willd. ●☆

339065　Rubus pinnatus Willd. var. subglandulosus（Gust.）R. A. Graham = Rubus pinnatus Willd. ●☆

339066　Rubus piptopetalus Hayata ex Koidz. = Rubus croceacanthus H. Lév. ●

339067　Rubus pirifolius Sm.；梨叶悬钩子（红筋菜，红筋钩，蛇泡，太平悬钩子，锥花悬钩子）；Paniculateflower Raspberry，Pearleaf Raspberry，Pear-leaved Raspberry，Tai-ping Raspberry ●

339068　Rubus pirifolius Sm. var. cordatus Te T. Yu et L. T. Lu；心状梨叶悬钩子；Cordate Pearleaf Raspberry ●

339069　Rubus pirifolius Sm. var. permollis Merr.；柔毛梨叶悬钩子；Hairy Pearleaf Raspberry ●

339070　Rubus pirifolius Sm. var. tomentosus Kuntze；绒毛梨叶悬钩子；Tomentose Pearleaf Raspberry ●

339071　Rubus pityophilus S. J. Sm. = Rubus ithacanus L. H. Bailey ●☆

339072　Rubus platyphyllus K. Koch；宽叶悬钩子●☆

339073　Rubus platysepalus Hand. -Mazz.；武岗悬钩子●

339074　Rubus platysepalus Hand. -Mazz. var. gracilior Hand. -Mazz. = Rubus platysepalus Hand. -Mazz. ●

339075　Rubus playfairianus Hemsl. ex Focke；五叶鸡爪茶（普莱肥莓，五加皮）；Fiveleaved Raspberry，Five-leaved Raspberry ●

339076　Rubus playfairianus Hemsl. ex Focke var. stenophyllus（Franch.）Cardot = Rubus playfairianus Hemsl. ex Focke ●

339077　Rubus playfairii Hemsl. = Rubus cochinchinensis Tratt. ●

339078　Rubus plicatifolius Blanch.；褶叶悬钩子；Plait-leaf Dewberry ●☆

339079　Rubus plicatus Weihe et Nees；折叠悬钩子●☆

339080　Rubus pluribracteatus L. T. Lu et Boufford；大乌泡（大红黄泡，倒生根，红铁泡刺，黄水泡，马莓，马莓叶，糖泡，糖泡叶，乌龙须，乌泡，无刺乌炮）；Multibract Raspberry，Multibracted Raspberry ●

339081　Rubus pluribracteatus L. T. Lu et Boufford var. lobatisepalus（Te T. Yu et L. T. Lu）L. T. Lu et Boufford；裂萼大乌泡；Lobedsepal Raspberry ●

339082　Rubus plus L. H. Bailey；毛叶悬钩子；Hairy-leaved Dewberry ●☆

339083　Rubus pohlii L. H. Bailey = Rubus ithacanus L. H. Bailey ●☆

339084　Rubus poliophyllus Kuntze；灰白叶悬钩子（毛叶悬钩子）；Hairyleaf Raspberry，Hairy-leaved Raspberry ●

339085　Rubus poliophyllus Kuntze var. ximengensis Y. Y. Qian；西盟悬钩子；Ximeng Raspberry ●

339086　Rubus polyanthus H. L. Li = Rubus nagasawanus Koidz. ●

339087　Rubus polybotrys L. H. Bailey = Rubus multifer L. H. Bailey ●☆

339088　Rubus polyodontus Hand. -Mazz.；多齿悬钩子；Manytooth Raspberry，Polydentate Raspberry ●

339089　Rubus polytrichus Franch. = Rubus multisetosus Te T. Yu et L. T. Lu ●

339090　Rubus ponticus Juz.；蓬特悬钩子●☆

339091　Rubus porotoensis R. A. Graham；波罗托悬钩子●☆

339092　Rubus potentilloides W. E. Evans；委陵悬钩子；Cinquefoil-like Raspberry ●■

339093　Rubus potis L. H. Bailey = Rubus wheeleri（L. H. Bailey）L. H. Bailey ●☆

339094　Rubus prandianus Hand. -Mazz. = Rubus hanceanus Kuntze ●

339095　Rubus pratensis L. H. Bailey = Rubus frondosus Bigelow ●☆

339096　Rubus preptanthus Focke；早花悬钩子；Earlyflower Raspberry，Primrose Raspberry，Wothflower Raspberry ●

339097　Rubus preptanthus Focke var. mairei（H. Lév.）Te T. Yu et L. T. Lu；狭叶早花悬钩子；Maire Wothflower Raspberry ●

339098　Rubus prior L. H. Bailey = Rubus plicatifolius Blanch. ●☆

339099　Rubus problematicus L. H. Bailey = Rubus plicatifolius Blanch. ●☆

339100　Rubus probus L. H. Bailey；昆士兰悬钩子；Queensland Raspberry ●☆

339101　Rubus procerus P. J. Müll；喜马拉雅悬钩子；Himalayan Blackberry ●☆

339102　Rubus procerus P. J. Müll. = Rubus discolor Weihe et Nees ●☆

339103　Rubus procumbens Muhl.；平铺悬钩子；Low Running Blackberry ●☆

339104　Rubus procumbens Muhl. = Rubus flagellaris Willd. ●☆

339105　Rubus procumbens Muhl. subsp. subuniflorus Focke = Rubus flagellaris Willd. ●☆

339106　Rubus procumbens Muhl. var. roribaccus（L. H. Bailey）L. H. Bailey = Rubus roribaccus（L. H. Bailey）Rydb. ●☆

339107　Rubus proprius L. H. Bailey = Rubus elegantulus Blanch. ●☆

339108　Rubus prosper L. H. Bailey var. cordifrons L. H. Bailey = Rubus curtipes L. H. Bailey ●☆

339109　Rubus proteus C. H. Stirt.；易变悬钩子●☆

339110　Rubus przewalskii Prochanov = Rubus sachalinensis H. Lév.

var. przewalskii（Prokh.）L. T. Lu ●

339111   Rubus pseudoacer Makino;深山红叶悬钩子●☆

339112   Rubus pseudoacer Makino subsp. flexuosus（Y. C. Liu et F. Y. Lu）H. Ohashi;清水悬钩子●

339113   Rubus pseudoacer Makino var. flexuosus Y. C. Liu et F. Y. Lu = Rubus pseudoacer Makino subsp. flexuosus（Y. C. Liu et F. Y. Lu）H. Ohashi ●

339114   Rubus pseudobuergeri Sasaki = Rubus buergeri Miq. ●

339115   Rubus pseudojaponicus Koidz. ;假日本悬钩子●☆

339116   Rubus pseudopileatus Cardot;假帽莓;Falsecapform Raspberry, False-capform Raspberry ●

339117   Rubus pseudopileatus Cardot var. glabratus Te T. Yu et L. T. Lu;光梗假帽莓;Glabrous Falsecapform Raspberry ●

339118   Rubus pseudopileatus Cardot var. kangdingensis Te T. Yu et L. T. Lu;康定假帽莓;Kangding Falsecapform Raspberry ●

339119   Rubus pseudosaxatilis H. Lév. = Rubus coreanus Miq. ●

339120   Rubus pseudosaxatilis H. Lév. var. kouytchensis H. Lév. = Rubus coreanus Miq. ●

339121   Rubus psilophyllus Nevski = Rubus caesius L. ●

339122   Rubus ptilocarpus Te T. Yu et L. T. Lu;毛果悬钩子;Hairyfruit Raspberry, Hairy-fruited Raspberry ●

339123   Rubus ptilocarpus Te T. Yu et L. T. Lu var. degensis Te T. Yu et L. T. Lu;长萼毛果悬钩子;Longcalyx Hairyfruit Raspberry ●

339124   Rubus pubescens Raf. ;小红悬钩子;Dwarf Raspberry, Dwarf Red Raspberry, Eyeberry, Plumboy ●☆

339125   Rubus pubescens Raf. var. pilosifolius A. F. Hill = Rubus pubescens Raf. ●☆

339126   Rubus pubifolius L. H. Bailey = Rubus alumnus L. H. Bailey ●☆

339127   Rubus pubifolius Te T. Yu et L. T. Lu = Rubus gyamdaensis L. T. Lu et Boufford ●

339128   Rubus pubifolius Te T. Yu et L. T. Lu var. glabriusculus Te T. Yu et L. T. Lu = Rubus gyamdaensis L. T. Lu et Boufford var. glabriusculus（Te T. Yu et L. T. Lu）L. T. Lu et Boufford ●

339129   Rubus pulcherrimus Hook. = Rubus lineatus Reinw. ex Blume ●

339130   Rubus pungens Cambess. ;针刺悬钩子（刺悬钩子,单花悬钩子,倒毒散,倒扎龙,九里香,葡萄杖,香莓）;Pungent Raspberry, Spiny Raspberry ●

339131   Rubus pungens Cambess. var. discolor Prochanov = Rubus pungens Cambess. ●

339132   Rubus pungens Cambess. var. fargesii Cardot = Rubus pungens Cambess. ●

339133   Rubus pungens Cambess. var. indefensus Focke = Rubus pungens Cambess. var. oldhamii（Miq.）Maxim. ●

339134   Rubus pungens Cambess. var. linearisepalus Te T. Yu et L. T. Lu;线萼针刺悬钩子;Linearsepal Spiny Raspberry ●

339135   Rubus pungens Cambess. var. oldhamii（Miq.）Maxim. ;疏刺悬钩子（九里香,九头饭消扭,落地甬公,毛刺悬钩子,香莓,针刺悬钩子）;Oldham Spring Raspberry ●

339136   Rubus pungens Cambess. var. oldhamii（Miq.）Maxim. f. roseus Nakai;粉花疏刺悬钩子（粉花悬钩子,粉疏刺悬钩子）●☆

339137   Rubus pungens Cambess. var. ternatus Cardot;三叶针刺悬钩子;Ternate Oldham Spring Raspberry ●

339138   Rubus pungens Cambess. var. villosus Cardot;柔毛针刺悬钩子;Villose Oldham Spring Raspberry ●

339139   Rubus purpureus Bunge ex Hook. f. = Rubus irritans Focke ●

339140   Rubus pycnanthus Focke = Rubus lambertianus Ser. ●

339141   Rubus pyi H. Lév. = Rubus niveus Thunb. ●

339142   Rubus pyrifolius Sm. = Rubus pirifolius Sm. ●

339143   Rubus qinglongensis Q. H. Chen et T. L. Xu;晴隆悬钩子;Qinglong Raspberry ●

339144   Rubus qinglongensis Q. H. Chen et T. L. Xu = Rubus assamensis Focke ●

339145   Rubus quaesitus L. H. Bailey;精致悬钩子;Prince Edward Island Blackberry ●☆

339146   Rubus quartinianus A. Rich. = Rubus apetalus Poir. ●☆

339147   Rubus quartinianus A. Rich. var. pappianus Gust. = Rubus apetalus Poir. ●☆

339148   Rubus quelpacrtensis H. Lév. = Rubus coreanus Miq. ●

339149   Rubus quinquefoliolatus Te T. Yu et L. T. Lu;五叶悬钩子;Fiveleaf Raspberry, Fiveleaflets Oldham Spring Raspberry, Five-leaved Raspberry, Quinquefoliate Raspberry ●

339150   Rubus raddeanus Focke;拉德悬钩子●☆

339151   Rubus radicans Focke = Rubus fockeanus Kurz ■

339152   Rubus randaiensis Hayata = Rubus formosensis Kuntze ●

339153   Rubus randii（L. H. Bailey）Rydb. = Rubus canadensis L. ●☆

339154   Rubus raopingensis Te T. Yu et L. T. Lu;饶平悬钩子;Raoping Raspberry ●

339155   Rubus raopingensis Te T. Yu et L. T. Lu var. obtusidentatus Te T. Yu et L. T. Lu;钝齿饶平悬钩子（钝齿悬钩子）;Obtusedentate Raoping Raspberry ●

339156   Rubus rappii L. H. Bailey = Rubus alleghniensis Porter ex L. H. Bailey ●☆

339157   Rubus rarissimus Hayata = Rubus mesogaeus Focke ex Diels ●

339158   Rubus recurvans Blanch. ;反折悬钩子;Recurved Blackberry ●☆

339159   Rubus recurvans Blanch. var. subrecurvans Blanch. = Rubus recurvans Blanch. ●☆

339160   Rubus recurvicaulis Blanch. ;布兰悬钩子;Arching Dewberry, Blanchard's Dewberry ●☆

339161   Rubus recurvicaulis Blanch. var. inarmatus Blanch. = Rubus recurvicaulis Blanch. ●☆

339162   Rubus reflexus Ker Gawl. ;锈毛莓（大叶蛇簕,七指风,七爪风,山烟筒子,蛇包簕,蛇泡簕,蛇泡荔,万枝莓,细锁梅）;Rustyhair Raspberry, Rusty-haired Raspberry ●

339163   Rubus reflexus Ker Gawl. var. hui（Diels）F. P. Metcalf;浅裂锈毛莓（胡氏悬钩子,山佛手,小桔公,锈毛莓）;Hu Rustyhair Raspberry, Hu's Raspberry ●

339164   Rubus reflexus Ker Gawl. var. lanceolobus F. P. Metcalf;深裂锈毛莓（红泡刺,七爪风）;Lanceolobed Rustyhair Raspberry ●

339165   Rubus reflexus Ker Gawl. var. macrophyllus Te T. Yu et L. T. Lu;大叶锈毛莓;Bigleaf Rustyhair Raspberry ●

339166   Rubus reflexus Ker Gawl. var. orogenes Hand. -Mazz. ;长叶锈毛莓;Longleaf Rustyhair Raspberry ●

339167   Rubus refractus H. Lév. ;曲萼悬钩子;Refracted Raspberry ●

339168   Rubus refractus H. Lév. var. latifolius Cardot = Rubus refractus H. Lév. ●

339169   Rubus regionalis L. H. Bailey;绿花悬钩子●☆

339170   Rubus regosissimus Hayata = Rubus formosensis Kuntze ●

339171   Rubus repens（L.）Kuntze;匐匐悬钩子;Dalibarda, Dewdrop ●☆

339172   Rubus reticulatus Wall. ex Hook. f. ;网脉悬钩子;Reticulate Raspberry ●

339173   Rubus retusipetalus Hayata = Rubus trianthus Focke ●

339174   Rubus rhodacantha E. Mey. = Rubus ludwigii Eckl. et Zeyh. ●☆

339175   Rubus rhodinsulanus L. H. Bailey = Rubus plicatifolius Blanch. ●☆

339176   Rubus ribes L. H. Bailey = Rubus setosus Bigelow ●☆

339177　Rubus rigidus Sm. ;硬悬钩子●☆

339178　Rubus rigidus Sm. var. buchananii Focke ＝Rubus rigidus Sm. ●☆

339179　Rubus rigidus Sm. var. camerunensis Letouzey ＝ Rubus rigidus Sm. ●☆

339180　Rubus rigidus Sm. var. chrysocarpus（Mundt）Focke ＝Rubus rigidus Sm. ●☆

339181　Rubus rigidus Sm. var. discolor Hauman ＝Rubus rigidus Sm. ●☆

339182　Rubus rigidus Sm. var. incisus Gust. ＝Rubus rigidus Sm. ●☆

339183　Rubus rigidus Sm. var. longepedicellatus Gust. ＝ Rubus longepedicellatus（Gust.）C. H. Stirt. ●☆

339184　Rubus rigidus Sm. var. mundtii（Cham. et Schltdl.）Focke ＝ Rubus rigidus Sm. ●☆

339185　Rubus ritozanensis Sasaki;李栋山悬钩子; Lidongshan Raspberry ●

339186　Rubus ritozanensis Sasaki ＝Rubus inopertus（Diels）Focke ●

339187　Rubus rocheri H. Lév. ＝Rubus refractus H. Lév. ●

339188　Rubus rolfei Vidal;罗氏悬钩子（大叶玉山悬钩子,高山悬钩子,无刺悬钩子,玉山悬钩子）;Bigleaf Yushan Dewberry, Creeping Bramble, Largeleaf Yushan Dewberry, Rolfe's Raspberry, Taiwanese Creeping Rubus, Yushan Dewberry, Yushan Raspberry ●

339189　Rubus rolfei Vidal var. lanatus Hayata ＝Rubus rolfei Vidal ●

339190　Rubus roribaccus（L. H. Bailey）Rydb. ;贞节悬钩子; Lucretia Dewberry ●☆

339191　Rubus roribaccus（L. H. Bailey）Rydb. ＝ Rubus flagellaris Willd. ●☆

339192　Rubus rosa L. H. Bailey;玫瑰悬钩子; Rose Blackberry ●☆

339193　Rubus rosendahlii L. H. Bailey ＝Rubus plicatifolius Blanch. ●☆

339194　Rubus rosifolius Sm. ;空心泡（刺莓,倒触伞,和平悬钩子,划船泡,黄牛泡,空心藨,龙船泡,毛悬钩子,七时饭消扭,蔷薇莓,三月泡,腺斑悬钩子,洋金银藤）;Brier Rose, Cape Bramble, Glandulary-dotted Raspberry, Heping Raspberry, Mauritius Raspberry, Roseleaf Raspberry, Rose-leaved Raspberry, Salem-rose, Thimbleberry, West Indian Raspberry ●

339195　Rubus rosifolius Sm. 'Coronarius';重瓣空心泡（佛见笑,蔷薇叶悬钩子,荼蘼花,酴醾,重瓣蔷薇）;Brier Rose, Double Cream Blackberry, Doublepetal Roseleaf Raspberry, Roseleaf Raspberry, Salem-rose ●

339196　Rubus rosifolius Sm. f. coronarius（Sims）Kuntze ＝Rubus rosifolius Sm. var. coronarius Sims ●

339197　Rubus rosifolius Sm. subsp. maximowiczii Focke ＝Rubus croceacanthus H. Lév. ●

339198　Rubus rosifolius Sm. subsp. sumatranus（Miq.）Focke ＝Rubus sumatranus Miq. ●

339199　Rubus rosifolius Sm. var. coronarius Sims ＝ Rubus rosifolius Sm. 'Coronarius' ●

339200　Rubus rosifolius Sm. var. formosanus Cardot ＝ Rubus croceacanthus H. Lév. ●

339201　Rubus rosifolius Sm. var. hirsutus Hayata ＝ Rubus pungens Cambess. var. oldhamii（Miq.）Maxim. ●

339202　Rubus rosifolius Sm. var. inermis Z. X. Yu;无刺空心泡; Spineless Salem-rose ●

339203　Rubus rosifolius Sm. var. linearifoliotus（Hayata）H. L. Li ＝ Rubus tsangii Merr. ●

339204　Rubus rosifolius Sm. var. linearifolius（Hayata）H. L. Li ＝Rubus tsangii Merr. var. linearifoliolus（Hayata）Te T. Yu et L. T. Lu ●

339205　Rubus rosifolius Sm. var. polyphyllarius Cardot;红叶悬钩子; Redleaf Salem-rose ●

339206　Rubus rosifolius Sm. var. wuyishanensis Z. X. Yu;武夷山空心泡;Wuyishan Salem-rose ●

339207　Rubus rosifolius Sm. var. wuyishanensis Z. X. Yu ＝ Rubus rosifolius Sm. var. coronarius Sims ●

339208　Rubus rosiformis Sm. var. formosanus Cardot;阿里山悬钩子●

339209　Rubus rosulans Kuntze ＝Rubus treutleri Hook. f. ●

339210　Rubus rotundifolius Reinw. ex Miq. ＝ Rubus pirifolius Sm. ●

339211　Rubus rotundior（L. H. Bailey）L. H. Bailey ＝Rubus fulleri L. H. Bailey ●☆

339212　Rubus rowleei L. H. Bailey ＝Rubus wheeleri（L. H. Bailey）L. H. Bailey ●☆

339213　Rubus rubribracteatus F. P. Metcalf ＝Rubus formosensis Kuntze ●

339214　Rubus rubrisetulosus Cardot;红刺悬钩子;Redspine Raspberry ■

339215　Rubus rubroangustifolius Sasaki;红狭叶悬钩子■

339216　Rubus rubroangustifolius Sasaki ＝Rubus croceacanthus H. Lév. var. glaber Koidz. ●

339217　Rubus rufolanatus H. T. Chang ＝Rubus hastifolius H. Lév. et Vaniot ●

339218　Rubus rufus Focke;棕红悬钩子;Reddish Raspberry ●

339219　Rubus rufus Focke var. hederifolius Cardot ＝Rubus lasiotrichos Focke ●

339220　Rubus rufus Focke var. longipedicellatus Te T. Yu et L. T. Lu;长梗棕红悬钩子;Longpedicel Raspberry ●

339221　Rubus rufus Focke var. palmatifidus Cardot;掌裂棕红悬钩子（掌叶棕红悬钩子）;Palmateleaf Reddish Raspberry ●

339222　Rubus rugosissimus Hayata ＝Rubus formosensis Kuntze ●

339223　Rubus runssorensis Engl. ;伦索悬钩子●☆

339224　Rubus runssorensis Engl. f. umbrosus Gust. ＝ Rubus runssorensis Engl. var. umbrosus（Gust.）Hauman ●☆

339225　Rubus runssorensis Engl. var. kiwuensis（Focke）Engl. ＝ Rubus runssorensis Engl. ●☆

339226　Rubus runssorensis Engl. var. umbrosus（Gust.）Hauman;耐荫伦索悬钩子●☆

339227　Rubus sachalinensis H. Lév. ;库页悬钩子（沙窝窝）;Kuye Raspberry, Sachalin Raspberry ●

339228　Rubus sachalinensis H. Lév. ＝ Rubus idaeus L. subsp. melanolasius Focke ●☆

339229　Rubus sachalinensis H. Lév. var. concolor（Kom.）Lauener et Ferguson ＝Rubus komarovii Nakai ●

339230　Rubus sachalinensis H. Lév. var. eglanduratus（Y. B. Chang）L. T. Lu;无腺里白悬钩子;Glanduleless Sachalin Raspberry ●

339231　Rubus sachalinensis H. Lév. var. przewalskii（Prokh.）L. T. Lu;甘肃悬钩子;Gansu Raspberry ●

339232　Rubus sagatus Focke ＝Rubus adenophorus Rolfe ●

339233　Rubus salwinensis Hand. -Mazz. ;怒江悬钩子;Nujiang Raspberry ●

339234　Rubus sanctus Kuntze;神悬钩子;Palestine Bramble ●☆

339235　Rubus sanctus Schreb. ＝Rubus ulmifolius Schott ●☆

339236　Rubus sanguineus Friv. ;血刺悬钩子（红色悬钩子）●☆

339237　Rubus satis L. H. Bailey;美国毛悬钩子●☆

339238　Rubus sativus（L. H. Bailey）Brainerd ＝Rubus allegheniensis Porter ex L. H. Bailey ●☆

339239　Rubus sativus Brainerd ＝Rubus frondosus Bigelow ●☆

339240　Rubus saxatilis L. ;石生悬钩子（地豆豆,天山悬钩子,悬钩木）; Bummelkite, Bunch-berry, Bunch-of-keys, Rasp, Rocky Raspberry, Roebuck, Roebuck Berry, Roebuck-berry, Saxatile Raspberry, Stone Bramble, Stoneberry ●■

339241 Rubus saxatilis L. var. canadensis Michx. = Rubus pubescens Raf. ●☆

339242 Rubus scheffleri Engl. ;谢夫勒悬钩子●☆

339243 Rubus schindleri Focke = Rubus tephrodes Hance var. ampliflorus (H. Lév. et Vaniot) Hand. -Mazz. ●

339244 Rubus schneideri L. H. Bailey =Rubus missouricus L. H. Bailey ●☆

339245 Rubus schoolcraftianus L. H. Bailey = Rubus ithacanus L. H. Bailey ●☆

339246 Rubus scioanus Chiov. = Rubus steudneri Schweinf. ●☆

339247 Rubus semierectus Blanch. = Rubus plicatifolius Blanch. ●☆

339248 Rubus semisetosus Blanch. ;沼泽悬钩子;Swamp Blackberry ●☆

339249 Rubus semisetosus Blanch. var. wheeleri L. H. Bailey = Rubus wheeleri (L. H. Bailey) L. H. Bailey ●☆

339250 Rubus sempervirens Bigelow = Rubus hispidus L. ●☆

339251 Rubus sempervirens Te T. Yu et L. T. Lu = Rubus jianensis L. T. Lu et Boufford ●

339252 Rubus sepalanthus Focke = Rubus assamensis Focke ●

339253 Rubus separ L. H. Bailey = Rubus allegheniensis Porter ex L. H. Bailey ●☆

339254 Rubus serenus L. H. Bailey = Rubus enslenii Tratt. ●☆

339255 Rubus serpens Weihe;蛇形悬钩子●☆

339256 Rubus serratifolius Te T. Yu et L. T. Lu = Rubus wuzhianus L. T. Lu et Boufford ●

339257 Rubus serratifolius Wuzhi = Rubus serratifolius Te T. Yu et L. T. Lu ●

339258 Rubus serrulatus Wuzhi = Rubus wuzhianus L. T. Lu et Boufford ●

339259 Rubus setchuenensis Bureau et Franch. ;川莓(大乌泡,倒生根,黄水泡,马莓叶,糖泡刺,乌泡,无刺乌泡);Sichuan Raspberry,Szechuan Bramble,Szechwan Raspberry ●

339260 Rubus setchuenensis Bureau et Franch. var. omeiensis (Rolfe) Hand. -Mazz. = Rubus setchuenensis Bureau et Franch. ●

339261 Rubus setospinosus L. H. Bailey = Rubus wisconsinensis L. H. Bailey ●☆

339262 Rubus setosus Bigelow;刚毛悬钩子;Bristly Blackberry,Setose Blackberry ●☆

339263 Rubus setosus Bigelow var. groutianus (Blanch.) L. H. Bailey = Rubus groutianus Blanch. ●☆

339264 Rubus setosus Bigelow var. rotundior L. H. Bailey = Rubus fulleri L. H. Bailey ●☆

339265 Rubus sharpii L. H. Bailey = Rubus permixtus Blanch. ●☆

339266 Rubus shihae F. P. Metcalf;桂滇悬钩子;C. Shih Raspberry, Shiha Raspberry ●

339267 Rubus shimadae Hayata = Rubus buergeri Miq. ●

339268 Rubus shinkoensis Hayata;毛萼悬钩子(变叶悬钩子,双叶悬钩子);Sinko Raspberry ●

339269 Rubus shinkoensis Hayata = Rubus corchorifolius L. f. ●

339270 Rubus sieboldii Blume;炮烙莓(八月泡,包泡天,狗屎泡,过江龙,龙船乌泡,乌莓,乌泡,羊鸡树,羊鸟树);Palmleaf Dewberry,Siebold's Raspberry ●

339271 Rubus significans L. H. Bailey = Rubus setosus Bigelow ●☆

339272 Rubus sikkimensis Hook. f. ;锡金悬钩子;Sikkim Raspberry ●

339273 Rubus simplex Focke;单茎悬钩子(单生莓,单叶悬钩子); Simple Raspberry ●

339274 Rubus singulifolius Focke = Rubus setchuenensis Bureau et Franch. ●

339275 Rubus singulus L. H. Bailey =Rubus vermontanus Blanch. ●☆

339276 Rubus sitiens Focke = Rubus xanthocarpus Bureau et Franch. ●

339277 Rubus somae Hayata = Rubus croceacanthus H. Lév. ●

339278 Rubus somae Hayata = Rubus dolichocephalus Hayata ●

339279 Rubus somae Hayata = Rubus sumatranus Miq. ●

339280 Rubus sorbifolius Maxim. = Rubus sumatranus Miq. ●

339281 Rubus soulieanus Cardot = Rubus stans Focke var. soulianus (Cardot) Te T. Yu et L. T. Lu ●

339282 Rubus sozostylus Focke = Rubus henryi Hemsl. et Kuntze var. sozostylus (Focke) Rehder ●

339283 Rubus sozostylus Focke var. fargesii (Franch.) Cardot = Rubus henryi Hemsl. et Kuntze var. sozostylus (Focke) Rehder ●

339284 Rubus spananthus Z. M. Wu et Z. L. Cheng;少花悬钩子;Few-flower Raspberry ●

339285 Rubus spectabilis Pursh;美莓;Salmonberry,Salmon-berry ●☆

339286 Rubus spectabilis Pursh 'Flore Pleno';重瓣美莓;Double-flowered Salmonberry ●☆

339287 Rubus spectabilis Pursh subsp. vernus (Focke) Focke = Rubus vernus Focke ●☆

339288 Rubus spectabilis Pursh var. franciscanus (Rydb.) J. T. Howell;多毛美莓●☆

339289 Rubus spectatus L. H. Bailey;藓丛悬钩子;Sphagnum Blackberry ●☆

339290 Rubus sphaerocephalus Hayata;圆果悬钩子;Globose Raspberry ●

339291 Rubus sphaerocephalus Hayata = Rubus croceacanthus H. Lév. ●

339292 Rubus sphaerocephalus Hayata = Rubus piptopetalus Hayata ex Koidz. ●

339293 Rubus spinipes Hemsl. = Rubus xanthoneurus Focke ●

339294 Rubus spinulosoides F. P. Metcalf;刺毛白叶莓(刺毛悬钩子,拟刺悬钩子);Spinelike Raspberry,Spine-like Raspberry,Spinulose Raspberry ●

339295 Rubus stans Focke;直立悬钩子(直茎莓);Erect Raspberry ●

339296 Rubus stans Focke var. soulianus (Cardot) Te T. Yu et L. T. Lu;多刺直立悬钩子;Manyspiny Erect Raspberry ●

339297 Rubus steelei L. H. Bailey;斯蒂尔悬钩子;Steele's Dewberry ●☆

339298 Rubus stellatus Sm. ;星形悬钩子●☆

339299 Rubus stephanandra H. Lév. = Rubus hirsutus Thunb. ●

339300 Rubus steudneri Schweinf. ;斯托德悬钩子●☆

339301 Rubus steudneri Schweinf. var. aberensis Gust. = Rubus steudneri Schweinf. ●☆

339302 Rubus steudneri Schweinf. var. dictyophyllus (Oliv.) R. A. Graham;网叶悬钩子●☆

339303 Rubus steudneri Schweinf. var. sidamensis Engl. = Rubus steudneri Schweinf. ●☆

339304 Rubus stimulans Focke;华西悬钩子;W. China Raspberry,West China Raspberry ●

339305 Rubus stimulans Focke var. concolor (Cardot) C. Y. Wu = Rubus stimulans Focke ●

339306 Rubus stipulatus L. H. Bailey;马蹄悬钩子;Big Horseshoe Lake Dewberry ●☆

339307 Rubus stipulosus Te T. Yu et L. T. Lu;巨托悬钩子;Big Horseshoe Lake Dewberry,Stipule Raspberry,Stipuled Raspberry ●

339308 Rubus strigosus Michx. = Rubus idaeus L. var. strigosus (Michx.) Maxim. ●☆

339309 Rubus strigosus Michx. = Rubus strigosus Michx. ex Koidz. ●

339310 Rubus strigosus Michx. ex Koidz. ;野藨莓;American Red Raspberry,Red Raspberry,Strigose Raspberry ●

339311 Rubus strigosus Michx. ex Koidz. = Rubus sachalinensis H. Lév. ●

339312 Rubus strigosus Michx. var. acalyphacea (Greene) L. H. Bailey

= Rubus idaeus L. var. strigosus（Michx.）Maxim. ●☆

339313　Rubus strigosus Michx. var. arizonicus（Greene）Kearney et Peebles　= Rubus idaeus L. var. strigosus（Michx.）Maxim. ●☆

339314　Rubus strigosus Michx. var. canadensis（Richardson）House　= Rubus idaeus L. var. strigosus（Michx.）Maxim. ●☆

339315　Rubus stuhlmannii Engl. ；斯图尔曼悬钩子●☆

339316　Rubus suavissimus S. K. Lee　= Rubus chingii Hu var. suavissimus（S. K. Lee）L. T. Lu ●

339317　Rubus subcoreanus Te T. Yu et L. T. Lu；柱序悬钩子；Big-subkorean Raspberry，Subkorean Raspberry ●

339318　Rubus subcrataegifolius（H. Lév. et Vaniot）H. Lév. ；假牛叠肚●☆

339319　Rubus subiniflorus Rydb.　= Rubus flagellaris Willd. ●☆

339320　Rubus subinopertus Te T. Yu et L. T. Lu；紫红悬钩子；Purplered Raspberry，Purple-red Raspberry ●

339321　Rubus subornatus Focke；美饰悬钩子；Ornamental Raspberry ●

339322　Rubus subornatus Focke var. concolor Cardot　= Rubus subornatus Focke var. melanadenus Focke ●

339323　Rubus subornatus Focke var. fockei H. Lév.　= Rubus subornatus Focke var. melanadenus Focke ●

339324　Rubus subornatus Focke var. fockei H. Lév.　= Rubus subornatus Focke ●

339325　Rubus subornatus Focke var. melanadenus Focke；黑腺美饰悬钩子；Black Ornamental Raspberry ●

339326　Rubus subspicatus Hauman；穗状悬钩子●☆

339327　Rubus subtentus L. H. Bailey　= Rubus curtipes L. H. Bailey ●☆

339328　Rubus subtibetanus Hand. -Mazz. ；密刺悬钩子；Denselyspine Raspberry，Densespine Raspberry，Dense-spiny Raspberry ●

339329　Rubus subtibetanus Hand. -Mazz. var. glandulosus Te T. Yu et L. T. Lu；腺毛密刺悬钩子；Glandular Denselyspine Raspberry ●

339330　Rubus subumbellatus Cardot；拟伞悬钩子●

339331　Rubus subumbellatus Cardot　= Rubus hypopitys Focke ●

339332　Rubus subuniflorus Rydb.　= Rubus flagellaris Willd. ●☆

339333　Rubus succedaneus Nakai et Koidz.　= Rubus minusculus H. Lév. et Vaniot ●

339334　Rubus suishaensis Hayata　= Rubus corchorifolius L. f. ●☆

339335　Rubus sumatranus Miq. ；红腺悬钩子（红刺苔，花楸叶悬钩子，龙泡，马泡，牛奶莓，牛奶藤）；Redglandular Raspberry，Red-glandular Raspberry ●

339336　Rubus suzukianus Y. C. Liu et T. Y. Yang　= Rubus fraxinifoliolus Hayata ●

339337　Rubus swinhoei Dunn et Tutcher　= Rubus swinhoei Hance ●

339338　Rubus swinhoei Hance；木莓悬钩子（高脚老虎扭，里白悬钩子，木莓，斯氏悬钩子）；Awinhoe Raspberry，Swinhoe's Raspberry ●

339339　Rubus swinhoei Hance var. hupehensis（Oliv.）F. P. Metcalf　= Rubus swinhoei Hance ●

339340　Rubus tagallus Cham. et Schltdl.　= Rubus rosifolius Sm. ●

339341　Rubus tagallus Cham. et Schltdl. var. lanyuensis（C. E. Chang）S. S. Ying　= Rubus lanyuensis C. E. Chang ●

339342　Rubus tagalus sensu Forbes et Hemsl.　= Rubus croceacanthus H. Lév. ●

339343　Rubus tagalus sensu Forbes et Hemsl.　= Rubus piptopetalus Hayata ex Koidz. ●

339344　Rubus taitoensis Hayata；台东悬钩子（台东刺花悬钩子）；Taidong Springflower Raspberry，Taitung Raspberry ●

339345　Rubus taitoensis Hayata　= Rubus aculeatiflorus Hayata var. taitoensis（Hayata）Tang S. Liu et T. Y. Yang ●

339346　Rubus taitoensis Hayata var. aculeatiflorus（Hayata）H. Ohashi

et C. F. Hsieh　= Rubus aculeatiflorus Hayata ●

339347　Rubus taiwanianus Matsum. ；刺莓（虎刺莓，台湾悬钩子）；Taiwan Raspberry ●

339348　Rubus taiwanianus Matsum.　= Rubus rosifolius Sm. ●

339349　Rubus taiwanicola Koidz. et Ohwi；小叶悬钩子（美悬钩子，台湾莓）；Small-leaf Raspberry，Taiwan Raspberry ●

339350　Rubus takasagoensis Koidz.　= Rubus sumatranus Miq. ●

339351　Rubus takasagomontanus Hatus.　= Rubus taiwanicola Koidz. et Ohwi ●

339352　Rubus talaikiaensis H. Lév.　= Rubus hirsutus Thunb. ●

339353　Rubus taquitii H. Lév.　= Rubus parvifolius L. ●

339354　Rubus tardatus Blanch. ；晚花悬钩子；Late-flowering Dewberry ●☆

339355　Rubus taronensis Y. C. Wu ex Te T. Yu et L. T. Lu；独龙悬钩子；Dulong Raspberry ●

339356　Rubus tectus L. H. Bailey　= Rubus groutianus Blanch. ●☆

339357　Rubus teledapos Focke　= Rubus spinulosoides F. P. Metcalf ●

339358　Rubus tennesseanus L. H. Bailey　= Rubus alumnus L. H. Bailey ●☆

339359　Rubus tenuicaulis L. H. Bailey　= Rubus enslenii Tratt. ●☆

339360　Rubus tephrodes Hance；灰白毛莓（大勒潭，倒生根，倒水莲，割田藨，过江龙，寒藨，寒莓，黑乌苞，红泡勒，灰绿悬钩子，灰莓，灰山泡，陵藨，蓬藟，蛇乌苞，蛇乌泡，乌龙摆尾，乌泡，乌泡刺，阴藟）；Grey-white Raspberry，Greywhitehair Raspberry ●

339361　Rubus tephrodes Hance var. ampliflorus（H. Lév. et Vaniot）Hand. -Mazz. ；无腺灰白毛莓（大勒潭，倒水莲，黑乌苞，红泡勒，灰绿悬钩子，灰山泡，蛇乌苞，乌龙摆尾，乌泡）；Glandless Greywhitehair Raspberry ●

339362　Rubus tephrodes Hance var. eglandulosus W. C. Cheng　= Rubus tephrodes Hance var. ampliflorus（H. Lév. et Vaniot）Hand. -Mazz. ●

339363　Rubus tephrodes Hance var. holadenus（H. Lév.）L. T. Li；硬腺灰白毛莓●

339364　Rubus tephrodes Hance var. schindleri（Focke）Hand. -Mazz. = Rubus tephrodes Hance var. ampliflorus（H. Lév. et Vaniot）Hand. -Mazz. ●

339365　Rubus tephrodes Hance var. setosissimus Hand. -Mazz. ；长腺灰白毛莓；Longgland Greywhitehair Raspberry ●

339366　Rubus tephrodes Hance var. setosissimus Koidz.　= Rubus nagasawanus Koidz. ●

339367　Rubus tephrodes Hance var. setosissimus senus Koidz.　= Rubus nagasawanus Koidz. ●

339368　Rubus testaceus C. K. Schneid.　= Rubus stans Focke ●

339369　Rubus tetricus L. H. Bailey　= Rubus flagellaris Willd. ●☆

339370　Rubus thibetanus Franch. ；西藏悬钩子（藏莓，红莓子，莓儿刺）；Tibet Dewberry，Tibet Raspberry，Xizang Raspberry ●

339371　Rubus thunbergii Siebold et Zucc.　= Rubus hirsutus Thunb. ●

339372　Rubus thunbergii Siebold et Zucc. var. argyi（H. Lév.）Focke　= Rubus hirsutus Thunb. ●

339373　Rubus thunbergii Siebold et Zucc. var. glabellus Focke　= Rubus rosifolius Sm. ●

339374　Rubus thunbergii Siebold et Zucc. var. talaikiensis（H. Lév.）Focke　= Rubus hirsutus Thunb. ●

339375　Rubus thyrsoideus Wimm. ；英国悬钩子；Great Britain Blackberry ●☆

339376　Rubus tibetanus Focke　= Rubus xanthocarpus Bureau et Franch. ●

339377　Rubus tiliaceus Sm.　= Rubus paniculatus Sm. ●

339378　Rubus tinifolius Y. C. Wu ex Te T. Yu et L. T. Lu；截叶悬钩子；Loppedleaf Raspberry，Tinifoliate Raspberry，Truncateleaf Raspberry ●

339379　Rubus tiponensis Hosok.　= Rubus lambertianus Ser. var.

glandulosus Cardot ●

339380　Rubus tiponensis Hosok. = Rubus morii Hayata ●

339381　Rubus tokinibara（H. Hara）Naruh. ;蔷薇莓●☆

339382　Rubus tokkura Siebold = Rubus coreanus Miq. ●

339383　Rubus tomentosus Borkh. ;毡绒悬钩子;Tomentose Raspberry, Woolly Blackberry ●☆

339384　Rubus tomentosus Borkh. var. canescens ?;灰毡绒悬钩子; Woolly Blackberry ●☆

339385　Rubus tongchouanensis H. Lév. = Rubus niveus Thunb. ●

339386　Rubus tonglooensis Kuntze = Rubus treutleri Hook. f. ●

339387　Rubus tracyi L. H. Bailey = Rubus flagellaris Willd. ●☆

339388　Rubus transvaaliensis Gust. ;德兰士瓦悬钩子●☆

339389　Rubus treutleri Hook. f. ;滇西北悬钩子（滇西悬钩子）; Northwest Yunnan Raspberry, Treutler Raspberry ●

339390　Rubus trianthus Focke;三花悬钩子（苦悬钩子,三花莓,玉花莓）;Bitter Raspberry, Threeflower Raspberry, Triflorous Raspberry ●

339391　Rubus trichopetalus Hand. -Mazz. = Rubus macilentus Cambess. ●

339392　Rubus tricolor Focke;三色莓;Chinese Bramble, Himalayan Bramble, Tricolor Raspberry ●

339393　Rubus tridactylus Focke = Rubus pentagonus Wall. ex Focke ●

339394　Rubus trifidus Thunb. ;三裂悬钩子●☆

339395　Rubus trifidus Thunb. f. semiplenus（Makino）Sugim. ;重瓣三裂悬钩子●☆

339396　Rubus triflorus Richardson = Rubus pubescens Raf. ●☆

339397　Rubus trifoliolatus Suess. ;三小叶悬钩子●☆

339398　Rubus trijugus Focke;三对叶悬钩子;Threepairleaf Raspberry, Trijugate Raspberry ●

339399　Rubus triphyllus Focke = Rubus spinulosoides F. P. Metcalf ●

339400　Rubus triphyllus Thunb. = Rubus parvifolius L. ●

339401　Rubus triphyllus Thunb. var. adenochlamys Focke = Rubus parvifolius L. var. adenochlamys（Focke）Migo ●

339402　Rubus triphyllus Thunb. var. concolor subvar. subconcolor Masam. ex Kudo et Masam. = Rubus parvifolius L. ●

339403　Rubus triphyllus Thunb. var. eglandulosus L. H. Bailey = Rubus parvifolius L. ●

339404　Rubus triphyllus Thunb. var. internuntius Hance = Rubus spinulosoides F. P. Mctcalf ●

339405　Rubus triphyllus Thunb. var. subconcolor Cardot = Rubus parvifolius L. ●

339406　Rubus triphyllus Thunb. var. toapiensis Yamam. = Rubus parvifolius L. var. toapiensis（Yamam.）Hosok. ●

339407　Rubus trivialis Michx. ;南方悬钩子;Dewberry, Southern Dewberry ●☆

339408　Rubus trullisatus Focke = Rubus eucalyptus Focke var. trillisatus（Focke）Te T. Yu et L. T. Lu ●

339409　Rubus tsangii Merr. ;光滑悬钩子;Tsang Raspberry ●

339410　Rubus tsangii Merr. var. linearifoliolus（Hayata）Te T. Yu et L. T. Lu;无腺光滑悬钩子;Glandless Tsang Raspberry ●

339411　Rubus tsangii Merr. var. linearifoliolus（Hayata）Te T. Yu et L. T. Lu = Rubus tsangii Merr. ●

339412　Rubus tsangii Merr. var. yanshanensis（Z. X. Yu et W. T. Ji）L. T. Lu;铅山悬钩子;Qianshan Tsang Raspberry ●

339413　Rubus tsangorum Hand. -Mazz. ;东南悬钩子（浙闽悬钩子）; SE. China Raspberry, Southeast Raspberry ●

339414　Rubus tsangsihsinensis K. S. Hao = Rubus parkeri Hance ●

339415　Rubus tumularis L. H. Bailey = Rubus allegheniensis Porter ex L. H. Bailey var. gravesii Fernald ●☆

339416　Rubus turkestanicus Pavlov = Rubus caesius L. ●

339417　Rubus uber L. H. Bailey = Rubus allegheniensis Porter ex L. H. Bailey var. gravesii Fernald ●☆

339418　Rubus udus L. H. Bailey = Rubus setosus Bigelow ●☆

339419　Rubus ulmifolius Schott; 榆叶悬钩子（榆叶黑莓）; Bellidiflorus, Bramble, Elmleaf Blackberry, Elm-leaf Blackberry, Elmleaf Bramble, Elm-leaved Blackberry ●☆

339420　Rubus ulmifolius Schott ' Bellidiflorus';雏菊花榆叶悬钩子（雏菊花榆叶黑莓）●☆

339421　Rubus ulmifolius Schott subsp. bollei（Focke）Pit. et Proust = Rubus bollei Focke ●☆

339422　Rubus ulmifolius Schott subsp. rusticanus（Mercier）Pit. et Proust = Rubus ulmifolius Schott ●☆

339423　Rubus ulmifolius Schott var. albidiflorus（Sudre）O. Bolòs et Vigo;白花榆叶悬钩子●☆

339424　Rubus ulmifolius Schott var. chilensis H. Lév. ;智利榆叶悬钩子●☆

339425　Rubus ulmifolius Schott var. contractifolius Sudre = Rubus ulmifolius Schott ●☆

339426　Rubus ulmifolius Schott var. cruentiflorus Sudre = Rubus ulmifolius Schott ●☆

339427　Rubus ulmifolius Schott var. inermis ?;无刺榆叶悬钩子; Elmleaf Blackberry ●☆

339428　Rubus ulugurensis Engl. = Rubus steudneri Schweinf. ●☆

339429　Rubus unanimus L. H. Bailey = Rubus vermontanus Blanch. ●☆

339430　Rubus uncatus Wall. = Rubus macilentus Cambess. ●

339431　Rubus uniflorifer L. H. Bailey = Rubus baileyanus Britton ●☆

339432　Rubus uniformis L. H. Bailey;单型悬钩子;Thornless Dewberry, Uniform Bramble ●☆

339433　Rubus univocus L. H. Bailey = Rubus wheeleri（L. H. Bailey）L. H. Bailey ●☆

339434　Rubus urbanianus L. H. Bailey = Rubus flagellaris Willd. ●☆

339435　Rubus urophyllus Y. C. Liu et F. Y. Lu;台湾尾叶悬钩子（尾叶悬钩子）●

339436　Rubus ursinus Cham. et Schltdl. = Rubus vitifolius Cham. et Schltdl. ●

339437　Rubus uvidus L. H. Bailey;卡拉马祖悬钩子;Kalamazoo Dewberry ●☆

339438　Rubus vagus L. H. Bailey;铺散悬钩子;Rambling Dewberry ●☆

339439　Rubus vaniotii H. Lév. et Vaniot = Rubus corchorifolius L. f. ●

339440　Rubus variispinus L. H. Bailey;维克斯堡悬钩子;Vicksburg Blackberry ●☆

339441　Rubus vegrandis L. H. Bailey = Rubus permixtus Blanch. ●☆

339442　Rubus veitchii Rolfe = Rubus thibetanus Franch. ●

339443　Rubus vermontanus Blanch. ;佛特蒙悬钩子;Vermont Blackberry ●☆

339444　Rubus vermontanus Blanch. var. viridiflorus Blanch. = Rubus regionalis L. H. Bailey ●☆

339445　Rubus vernus Focke;春悬钩子●☆

339446　Rubus vestitus Weihe et Nees;包被悬钩子;European Blackberry ●☆

339447　Rubus viburnifolius Focke;荚蒾叶悬钩子;Viburnum Leaf Raspberry, Viburnumleaf Raspberry, Viburnum-leaved Raspberry ●

339448　Rubus viburnifolius Focke = Rubus neoviburnifolius L. T. Lu et Boufford ●

339449　Rubus viburnifolius Focke var. apetalus Y. Gu et W. L. Li;无瓣荚蒾叶悬钩子;Petalless Viburnum Leaf Raspberry ●

339450　Rubus vicarius Focke = Rubus subornatus Focke var. melanadenus Focke ●

339451　Rubus victorinii L. H. Bailey = Rubus plicatifolius Blanch. ●☆

339452　Rubus villosus Aiton var. albinus L. H. Bailey = Rubus allegheniensis Porter ex L. H. Bailey ●☆

339453　Rubus villosus Aiton var. engelmannii Focke = Rubus allegheniensis Porter ex L. H. Bailey ●☆

339454　Rubus villosus Aiton var. enslenii (Tratt.) W. Stone = Rubus enslenii Tratt. ●☆

339455　Rubus villosus Aiton var. frondosus (Bigelow) Torr. = Rubus frondosus Bigelow ●☆

339456　Rubus villosus Aiton var. michiganensis L. H. Bailey = Rubus roribaccus (L. H. Bailey) Rydb. ●☆

339457　Rubus villosus Aiton var. montanus Porter = Rubus allegheniensis Porter ex L. H. Bailey ●☆

339458　Rubus villosus Aiton var. randii L. H. Bailey = Rubus canadensis L. ●☆

339459　Rubus villosus Aiton var. roribaccus (L. H. Bailey) L. H. Bailey = Rubus roribaccus (L. H. Bailey) Rydb. ●☆

339460　Rubus villosus Aiton var. sativus L. H. Bailey = Rubus allegheniensis Porter ex L. H. Bailey ●☆

339461　Rubus villosus Aiton var. villigerus Focke = Rubus allegheniensis Porter ex L. H. Bailey ●☆

339462　Rubus villosus Thunb. = Rubus corchorifolius L. f. ●

339463　Rubus virginianus L. H. Bailey = Rubus allegheniensis Porter ex L. H. Bailey var. gravesii Fernald ●☆

339464　Rubus viridifrons L. H. Bailey = Rubus regionalis L. H. Bailey ●☆

339465　Rubus viscidus Focke = Rubus lambertianus Ser. var. paykouangensis (H. Lév.) Hand. -Mazz. ●

339466　Rubus vitifolius Cham. et Schltdl. ; 加州悬钩子; California Blackberry, Dewberry ●

339467　Rubus volkensii Engl. ; 福尔悬钩子●☆

339468　Rubus wallichianus Wight et Arn. ; 红毛悬钩子(川黔悬钩子, 鬼悬钩子, 红毛巾, 黄刺泡, 空洞泡, 老虎泡, 老熊泡, 牛毛大王); Pinfa Raspberry ●

339469　Rubus wangii F. P. Metcalf; 大苞悬钩子; Wang Raspberry ●

339470　Rubus wardii Merr. ; 大花悬钩子; Ward Raspberry ●

339471　Rubus wawushanensis Te T. Yu et L. T. Lu; 瓦屋山悬钩子; Wawushan Raspberry ●

339472　Rubus wheeleri (L. H. Bailey) L. H. Bailey; 惠勒悬钩子; Wheeler's Blackberry ●☆

339473　Rubus wiegandii L. H. Bailey = Rubus recurvans Blanch. ●☆

339474　Rubus willii ?; 瓦里悬钩子●☆

339475　Rubus wilsonii Duthie; 湖北悬钩子(倒扎龙, 倒扎泡, 过江龙, 两广悬钩子, 十月泡, 威尔氏莓, 威氏莓, 乌龙摆尾); E. H. Wilson Raspberry ●

339476　Rubus wisconsinensis L. H. Bailey; 威斯康星悬钩子; Wisconsin Blackberry ●☆

339477　Rubus woronowii Sudre; 沃氏悬钩子●☆

339478　Rubus wrightii A. Gray = Rubus crataegifolius Bunge ●

339479　Rubus wuchuanensis S. Z. Ho; 务川悬钩子●

339480　Rubus wushanensis Te T. Yu et L. T. Lu; 巫山悬钩子; Wushan Raspberry ●

339481　Rubus wuzhianus L. T. Lu et Boufford; 锯叶悬钩子; Serrateleaf Raspberry, Serrate-leaved Raspberry ●

339482　Rubus xanthacanthus H. Lév. = Rubus innominatus S. Moore ●

339483　Rubus xanthocarpus Bureau et Franch. ; 黄果悬钩子(地莓子, 黄刺儿根, 黄帽子, 黄莓子, 莓子刺, 泡儿刺); Yellowfruit Raspberry ●

339484　Rubus xanthocarpus Bureau et Franch. var. tibetanus (Focke) Cardot = Rubus xanthocarpus Bureau et Franch. ●

339485　Rubus xanthoneurus Focke; 黄脉莓(刺五泡藤, 黄脉泡); Yellownerve Raspberry, Yellow-nerved Raspberry ●

339486　Rubus xanthoneurus Focke ex Diels var. brevipetiolatus Focke; 短柄黄脉莓; Shortpetiole Raspberry, Shortstalk Raspberry ●

339487　Rubus xanthoneurus Focke ex Diels var. glandulosus Te T. Yu et L. T. Lu; 腺毛黄脉莓; Glandular Yellownerve Raspberry ●

339488　Rubus xichouensis Te T. Yu et L. T. Lu; 西畴悬钩子; Xichou Raspberry ●

339489　Rubus yabei H. Lév. et Vaniot = Rubus idaeus L. subsp. nipponicus Focke ●☆

339490　Rubus yabei H. Lév. et Vaniot f. eglandulosus (Ohwi) Ohwi = Rubus idaeus L. subsp. nipponicus Focke ●☆

339491　Rubus yabei H. Lév. et Vaniot f. marmoratus (H. Lév. et Vaniot) Ohwi = Rubus idaeus L. subsp. nipponicus Focke f. marmoratus (H. Lév. et Vaniot) Kitam. ●☆

339492　Rubus yabei H. Lév. et Vaniot var. marmoratus (H. Lév. et Vaniot) Ohwi = Rubus idaeus L. subsp. nipponicus Focke f. marmoratus (H. Lév. et Vaniot) Kitam. ●☆

339493　Rubus yabei H. Lév. et Vaniot var. sikokianus Ohwi et Inobe = Vincetoxicum calcareum (H. Ohashi) Akasawa ■☆

339494　Rubus yakumontanus Masam. = Rubus palmatus Thunb. var. yakumontanus (Masam.) Hatus. ex H. Ohba ●☆

339495　Rubus yamamotoanus H. L. Li = Rubus inopertus (Diels) Focke ●

339496　Rubus yanshanensis Z. X. Yu et W. T. Ji = Rubus tsangii Merr. var. yanshanensis (Z. X. Yu et W. T. Ji) L. T. Lu ●

339497　Rubus yanyunii Y. T. Chang et L. Y. Chen; 九仙莓; Jiuxianshan Raspberry ●

339498　Rubus yiwuanus W. P. Fang; 奕武悬钩子; Yiwu Raspberry ●

339499　Rubus yoshinoi Koidz. ; 吉野氏悬钩子●☆

339500　Rubus yui E. Walker = Rubus fragarioides Bertol. var. adenophorus Franch. ●■

339501　Rubus yuliensis Y. C. Liu et F. Y. Lu; 玉里悬钩子; Yuli Raspberry ●

339502　Rubus yunnanicus Kuntze; 云南悬钩子(滇悬钩子); Yunnan Raspberry ●

339503　Rubus yushunii (Suzuki et Yamam.) Suzuki et Yamam. = Rubus inopertus (Diels) Focke ●

339504　Rubus zhaogoshanensis Te T. Yu et L. T. Lu; 草果山悬钩子; Caoguoshan Raspberry ●

339505　Ruckeria DC. = Euryops (Cass.) Cass. ●■☆

339506　Ruckeria euryopoides Drège = Euryops othonnoides (DC.) B. Nord. ●☆

339507　Ruckeria euryopsidis DC. = Euryops othonnoides (DC.) B. Nord. ●☆

339508　Ruckeria othonnoides DC. = Euryops othonnoides (DC.) B. Nord. ●☆

339509　Ruckeria tagetoides DC. = Euryops tagetoides (DC.) B. Nord. ●☆

339510　Ruckia Regel = Rhodostachys Phil. ■☆

339511　Rudbeckia Adans. = Conocarpus L. ●☆

339512　Rudbeckia L. (1753); 金光菊属; Black-eyed Dais, Black-eyed Susan, Cone Flower, Coneflower, Gloriosa Daisy, Mexican Hat, Rudbeckia ■

339513　Rudbeckia acuminata C. L. Boynton et Beadle = Rudbeckia fulgida Aiton ■

339514　Rudbeckia alismifolia Torr. et A. Gray ＝ Rudbeckia grandiflora（Sweet）C. C. Gmel. ex DC. var. alismifolia（Torr. et A. Gray）Cronquist ■☆

339515　Rudbeckia alpicola Piper；高山金光菊；Showy Coneflower，Washington Coneflower，Wenatchee Mountain Coneflower ■☆

339516　Rudbeckia ampla A. Nelson ＝ Rudbeckia laciniata L. var. ampla（A. Nelson）Cronquist ■☆

339517　Rudbeckia amplectens T. V. Moore ＝ Rudbeckia hirta L. ■

339518　Rudbeckia amplexicaulis Vahl；抱茎金光菊；Clasping Coneflower，Clasping-leaf Coneflower，Coneflower ■

339519　Rudbeckia atrorubens Nutt. ＝ Echinacea atrorubens（Nutt.）Nutt. ■☆

339520　Rudbeckia auriculata（Perdue）Král；耳形金光菊；Alabama Coneflower，Eared Coneflower ■☆

339521　Rudbeckia beadlei Small ＝ Rudbeckia triloba L. ■☆

339522　Rudbeckia bicolor Nutt. ；二色金光菊；Pinewoods Coneflower ■

339523　Rudbeckia bicolor Nutt. ＝ Rudbeckia hirta L. var. pulcherrima Farw. ■☆

339524　Rudbeckia brittonii Small ＝ Rudbeckia hirta L. ■

339525　Rudbeckia californica A. Gray；加州金光菊；California Coneflower ■☆

339526　Rudbeckia californica A. Gray var. glauca S. F. Blake ＝ Rudbeckia glaucescens Eastw. ■☆

339527　Rudbeckia californica A. Gray var. intermedia Perdue ＝ Rudbeckia klamathensis P. B. Cox et Urbatsch ■☆

339528　Rudbeckia chapmanii C. L. Boynton et Beadle ＝ Rudbeckia fulgida Aiton var. umbrosa（C. L. Boynton et Beadle）Cronquist ■☆

339529　Rudbeckia columnaris Pursh ＝ Ratibida columnifera（Nutt.）Wooton et Standl. ■☆

339530　Rudbeckia columnaris Sims ＝ Ratibida columnifera（Nutt.）Wooton et Standl. ■☆

339531　Rudbeckia columnifera Nutt. ＝ Ratibida columnifera（Nutt.）Wooton et Standl. ■☆

339532　Rudbeckia deamii S. F. Blake ＝ Rudbeckia fulgida Aiton var. deamii（S. F. Blake）Perdue ■☆

339533　Rudbeckia decumbens Sm. ＝ Spilanthes decumbens（Sm.）A. H. Moore ■☆

339534　Rudbeckia digitata Mill. ＝ Rudbeckia laciniata L. var. humilis A. Gray ■☆

339535　Rudbeckia divergens T. V. Moore ＝ Rudbeckia hirta L. var. angustifolia（T. V. Moore）Perdue ■☆

339536　Rudbeckia floridana T. V. Moore ＝ Rudbeckia hirta L. var. floridana（T. V. Moore）Perdue ■☆

339537　Rudbeckia floridana T. V. Moore var. angustifolia T. V. Moore ＝ Rudbeckia hirta L. var. angustifolia（T. V. Moore）Perdue ■☆

339538　Rudbeckia foliosa C. L. Boynton et Beadle ＝ Rudbeckia fulgida Aiton ■

339539　Rudbeckia fulgida Aiton；全缘金光菊（全缘叶金光菊）；Black-eyed Susan，Eastern Coneflower，Golden Coneflower，Orange Coneflower，Showy Coneflower ■

339540　Rudbeckia fulgida Aiton var. auriculata Perdue ＝ Rudbeckia auriculata（Perdue）Král ■☆

339541　Rudbeckia fulgida Aiton var. deamii（S. F. Blake）Perdue；第姻金光菊；Deam's Coneflower ■☆

339542　Rudbeckia fulgida Aiton var. missouriensis（Engelm. ex C. L. Boynton et Beadle）Cronquist ＝ Rudbeckia missouriensis Engelm. ex C. L. Boynton et Beadle ■☆

339543　Rudbeckia fulgida Aiton var. palustris（Eggert ex C. L. Boynton et Beadle）Perdue，Rhodora；沼泽金光菊；Marsh Coneflower，Prairie Coneflower ■☆

339544　Rudbeckia fulgida Aiton var. spathulata（Michx.）Perdue；匙状金光菊；Orange Coneflower ■☆

339545　Rudbeckia fulgida Aiton var. speciosa（Wender.）Perdue ＝ Rudbeckia fulgida Aiton ■

339546　Rudbeckia fulgida Aiton var. speciosa（Wender.）Perdue ＝ Rudbeckia speciosa Wender. ■

339547　Rudbeckia fulgida Aiton var. sullivantii（C. L. Boynton et Beadle）Cronquist；沙利文特金光菊；Sullivant's Coneflower ■☆

339548　Rudbeckia fulgida Aiton var. umbrosa（C. L. Boynton et Beadle）Cronquist；查普曼金光菊；Shady Coneflower ■☆

339549　Rudbeckia glabra DC. ＝ Rudbeckia nitida Nutt. ■☆

339550　Rudbeckia glaucescens Eastw. ；蜡金光菊；Waxy Coneflower ■☆

339551　Rudbeckia graminifolia（Torr. et A. Gray）C. L. Boynton et Beadle；禾叶金光菊；Grassleaf Coneflower ■☆

339552　Rudbeckia grandiflora（Sweet）C. C. Gmel. ex DC. ；大花金光菊；Largeflower Coneflower，Rough Coneflower ■☆

339553　Rudbeckia grandiflora（Sweet）C. C. Gmel. ex DC. var. alismifolia（Torr. et A. Gray）Cronquist；泽泻金光菊 ■☆

339554　Rudbeckia heliopsidis Torr. et A. Gray；湿地金光菊；Little River Black-eyed Susan，Sunfacing Coneflower ■☆

339555　Rudbeckia heterophylla Torr. et A. Gray ＝ Rudbeckia laciniata L. var. heterophylla（Torr. et A. Gray）Fernald et B. G. Schub. ■☆

339556　Rudbeckia hirta L. ；黑心金光菊（黑心菊）；Black-eyed Susan，Brown-eyed Susan，Gloriosa Daisy，Roughhair Coneflower，Roughhairy Coneflower，Yellow Daisy ■

339557　Rudbeckia hirta L. 'Goldilocks'；金锁黑心金光菊（金锁黑心菊）■☆

339558　Rudbeckia hirta L. 'Irish Eyes'；绿眼黑心金光菊（绿眼黑心菊）■☆

339559　Rudbeckia hirta L. 'Marmalade'；大花黑心金光菊（大花黑心菊）■☆

339560　Rudbeckia hirta L. 'Rustic Dwarf'；粗矮黑心金光菊（粗矮黑心菊）■☆

339561　Rudbeckia hirta L. var. angustifolia（T. V. Moore）Perdue；窄叶黑心金光菊 ■☆

339562　Rudbeckia hirta L. var. brittonii（Small）Fernald ＝ Rudbeckia hirta L. ■

339563　Rudbeckia hirta L. var. corymbifera Fernald ＝ Rudbeckia hirta L. var. pulcherrima Farw. ■☆

339564　Rudbeckia hirta L. var. floridana（T. V. Moore）Perdue；佛罗里达金光菊 ■☆

339565　Rudbeckia hirta L. var. lanceolata（Bisch.）Core ＝ Rudbeckia hirta L. var. pulcherrima Farw. ■☆

339566　Rudbeckia hirta L. var. monticola（Small）Fernald ＝ Rudbeckia hirta L. ■

339567　Rudbeckia hirta L. var. pulcherrima Farw. ；美丽黑心金光菊；Black-eyed Susan ■☆

339568　Rudbeckia hirta L. var. pulcherrima Farw. ＝ Rudbeckia hirta L. var. sericea（T. V. Moore）Fernald ■☆

339569　Rudbeckia hirta L. var. sericea（T. V. Moore）Fernald；绢毛黑心金光菊 ■☆

339570　Rudbeckia hirta L. var. sericea（T. V. Moore）Fernald ＝ Rudbeckia hirta L. var. pulcherrima Farw. ■☆

339571 Rudbeckia hirta L. var. serotina（Nutt.）Core ＝ Rudbeckia hirta L. var. pulcherrima Farw. ■☆

339572 Rudbeckia hybrida Hort.；杂种金光菊（毛叶金光菊，杂种黑心菊）；Hybrid Coneflower ■☆

339573 Rudbeckia klamathensis P. B. Cox et Urbatsch；克拉马斯金光菊；Klamath Coneflower ■☆

339574 Rudbeckia laciniata L.；裂叶金光菊（黑眼菊，裂叶铳菊，太阳菊）；Coneflower，Cutleaf Coneflower，Cut-leaf Coneflower，Cut-leaved Coneflower，Golden Glow，Golden-glow，Green-headed Coneflower，Tall Coneflower，Thimbleweed，Wild Golden-glow ■

339575 Rudbeckia laciniata L. 'Golden Glow'；重瓣金光菊■☆

339576 Rudbeckia laciniata L. 'Hortensis' ＝ Rudbeckia laciniata L. var. hortensis Bailey ■

339577 Rudbeckia laciniata L. var. ampla（A. Nelson）Cronquist；落基山金光菊；Rocky Mountain Cutleaf Coneflower ■☆

339578 Rudbeckia laciniata L. var. bipinnata Perdue；北方裂叶金光菊；Northeastern Cutleaf Coneflower ■☆

339579 Rudbeckia laciniata L. var. digitata（Mill.）Fiori ＝ Rudbeckia laciniata L. var. humilis A. Gray ■☆

339580 Rudbeckia laciniata L. var. heterophylla（Torr. et A. Gray）Fernald et B. G. Schub.；佛罗里达裂叶金光菊；Florida Coneflower ■☆

339581 Rudbeckia laciniata L. var. hortensis Bailey；庭院金光菊（金光菊，太阳菊，庭园铳菊，庭院铳菊，园金光菊，重瓣金光菊）；Garden Coneflower ■

339582 Rudbeckia laciniata L. var. humilis A. Gray；矮小金光菊（南方裂叶金光菊）；Cutleaf Coneflower，Green-headed Coneflower，Southeastern Cutleaf Coneflower ■☆

339583 Rudbeckia lanceolata Bisch. ＝ Rudbeckia hirta L. var. pulcherrima Farw. ■☆

339584 Rudbeckia longipes T. V. Moore ＝ Rudbeckia hirta L. var. pulcherrima Farw. ■☆

339585 Rudbeckia maxima Nutt.；大金光菊；Cabbage Coneflower，Cabbage-leaf Coneflower，Coneflower，Giant Coneflower，Great Coneflower，Large Coneflower，Swamp Coneflower，Tall Coneflower ■☆

339586 Rudbeckia missouriensis Engelm. ＝ Rudbeckia missouriensis Engelm. ex C. L. Boynton et Beadle ■☆

339587 Rudbeckia missouriensis Engelm. ex C. L. Boynton et Beadle；密苏里金光菊；Missouri Coneflower，Missouri Orange Coneflower ■☆

339588 Rudbeckia mohrii A. Gray；莫尔金光菊；Grassy Coneflower，Mohr's Coneflower ■☆

339589 Rudbeckia mollis Elliott；柔毛金光菊；Softhair Coneflower ■☆

339590 Rudbeckia montana A. Gray；山地金光菊；Montane Coneflower ■☆

339591 Rudbeckia monticola Small ＝ Rudbeckia hirta L. ■

339592 Rudbeckia neurmanii Steud. ＝ Rudbeckia fulgida Aiton ■

339593 Rudbeckia nitida Nutt.；光亮金光菊；Black-eyed Susan，Shiny Coneflower，St. John's Susan ■☆

339594 Rudbeckia nitida Nutt. var. texana Perdue ＝ Rudbeckia texana（Perdue）P. B. Cox et Urbatsch ■☆

339595 Rudbeckia occidentalis Nutt.；西方金光菊；Western Coneflower ■☆

339596 Rudbeckia occidentalis Nutt. var. alpicola（Piper）Cronquist ＝ Rudbeckia alpicola Piper ■☆

339597 Rudbeckia occidentalis Nutt. var. montana（A. Gray）Perdue ＝ Rudbeckia montana A. Gray ■☆

339598 Rudbeckia pallida Nutt. ＝ Echinacea pallida（Nutt.）Nutt. ■☆

339599 Rudbeckia palustris Eggert ex C. L. Boynton et Beadle ＝ Rudbeckia fulgida Aiton var. palustris（Eggert ex C. L. Boynton et Beadle）Perdue，Rhodora ■☆

339600 Rudbeckia pinnata Vent. ＝ Ratibida pinnata（Vent.）Barnhart ■☆

339601 Rudbeckia pinnatiloba（Torr. et A. Gray）Beadle ＝ Rudbeckia triloba L. var. pinnatiloba Torr. et A. Gray ■☆

339602 Rudbeckia porteri A. Gray ＝ Helianthus porteri（A. Gray）Pruski ■☆

339603 Rudbeckia purpurea L. ＝ Echinacea purpurea（L.）Moench ■☆

339604 Rudbeckia radula Pursh ＝ Helianthus radula（Pursh）Torr. et A. Gray ■☆

339605 Rudbeckia rupestris Chick. ＝ Rudbeckia triloba L. var. rupestris（Chick.）A. Gray ■☆

339606 Rudbeckia scabrifolia L. E. Br.；圆叶金光菊；Roughleaf Coneflower ■☆

339607 Rudbeckia sericea T. V. Moore ＝ Rudbeckia hirta L. var. pulcherrima Farw. ■☆

339608 Rudbeckia serotina Nutt. ＝ Rudbeckia hirta L. var. pulcherrima Farw. ■☆

339609 Rudbeckia serotina Nutt. ＝ Rudbeckia hirta L. var. sericea（T. V. Moore）Fernald ■☆

339610 Rudbeckia serotina Nutt. ＝ Rudbeckia hirta L. ■

339611 Rudbeckia serotina Nutt. var. corymbifera（Fernald）Fernald et B. G. Schub. ＝ Rudbeckia hirta L. var. pulcherrima Farw. ■☆

339612 Rudbeckia serotina Nutt. var. lanceolata（Bisch.）Fernald et B. G. Schub. ＝ Rudbeckia hirta L. var. pulcherrima Farw. ■☆

339613 Rudbeckia spathulata Michx. ＝ Rudbeckia fulgida Aiton var. spathulata（Michx.）Perdue ■☆

339614 Rudbeckia speciosa Wender.；齿叶金光菊（美丽金光菊）；Eastern Coneflower，Orange Coneflower，Showy Coneflower ■

339615 Rudbeckia speciosa Wender. ＝ Rudbeckia fulgida Aiton ■

339616 Rudbeckia speciosa Wender. var. sullivantii（C. L. Boynton et Beadle）B. L. Rob. ＝ Rudbeckia fulgida Aiton var. sullivantii（C. L. Boynton et Beadle）Cronquist ■☆

339617 Rudbeckia subtomentosa Pursh；香金光菊；Sweet Black-eyed Susan，Sweet Coneflower ■

339618 Rudbeckia sullivantii C. L. Boynton et Beadle ＝ Rudbeckia fulgida Aiton var. sullivantii（C. L. Boynton et Beadle）Cronquist ■☆

339619 Rudbeckia tagetes E. James ＝ Ratibida tagetes（E. James）Barnhart ■☆

339620 Rudbeckia tenax C. L. Boynton et Beadle ＝ Rudbeckia fulgida Aiton ■

339621 Rudbeckia texana（Perdue）P. B. Cox et Urbatsch；得州金光菊；Texas Coneflower ■☆

339622 Rudbeckia triloba L.；三裂叶金光菊；Brown-eyed Susan，Thin-leaved Coneflower，Three-lobed Coneflower ■☆

339623 Rudbeckia triloba L. var. beadlei（Small）Fernald ＝ Rudbeckia triloba L. ■☆

339624 Rudbeckia triloba L. var. pinnatiloba Torr. et A. Gray；羽裂金光菊■☆

339625 Rudbeckia triloba L. var. rupestris（Chick.）A. Gray；湿地三裂叶金光菊■☆

339626 Rudbeckia truncata Small ＝ Rudbeckia fulgida Aiton ■

339627 Rudbeckia umbrosa C. L. Boynton et Beadle ＝ Rudbeckia fulgida Aiton var. umbrosa（C. L. Boynton et Beadle）Cronquist ■☆

339628 Ruddia Yakovlev ＝ Ormosia Jacks.（保留属名）●

339629 Ruddia Yakovlev（1971）；薄皮红豆属●☆

339630 Ruddia fordiana（Oliv.）Yakovlev；薄皮红豆●☆

339631 Ruddia fordiana（Oliv.）Yakovlev ＝ Ormosia fordiana Oliv. ●

339632 Rudelia B. D. Jacks. ＝ Riedelia Oliv.（保留属名）■☆

339633　Rudelia Oliv. = Riedelia Oliv.（保留属名）■☆

339634　Rudella Loes. = Riedelia Oliv.（保留属名）■☆

339635　Rudella Loes. = Rudelia B. D. Jacks. ■☆

339636　Rudgea Salisb.（1807）;鲁奇茜属■☆

339637　Rudgea chiriquiensis Dwyer = Psychotria nebulosa K. Krause ●☆

339638　Rudgea viburnioides（Cham.）Benth.;荚蒾鲁奇茜■☆

339639　Rudicularia Moc. et Sessé ex Ramfrez = Semeiandra Hook. et Arn. ■☆

339640　Rudliola Baill. = Brillantaisia P. Beauv. ●■☆

339641　Rudolfiella Hoehne（1944）;鲁道兰属■☆

339642　Rudolfiella aurantiaca（Lindl.）Hoehne;橘黄鲁道兰■☆

339643　Rudolfiella bicornaria（Rchb. f.）Hoehne;双角鲁道兰■☆

339644　Rudolfiella floribunda（Schltr.）Hoehne;多花鲁道兰■☆

339645　Rudolfiella saxicola（Schltr.）Hoehne;岩地鲁道兰■☆

339646　Rudolphia Medik. = Malpighia L. ●

339647　Rudolphia Willd. = Neorudolphia Britton ■☆

339648　Rudolphia Willd. = Rhodopis Urb.（保留属名）■☆

339649　Rudolpho-Roemeria Steud. ex Hochst. = Kniphofia Moench（保留属名）■☆

339650　Rudua F. Maek. = Phaseolus L. ■

339651　Rudua aurea（Roxb.）Maek. = Vigna radiata（L.）R. Wilczek ■

339652　Ruehssia H. Karst. = Marsdenia R. Br.（保留属名）●

339653　Ruehssia H. Karst. ex Schltdl. = Marsdenia R. Br.（保留属名）●

339654　Ruelingia Ehrh.（废弃属名）= Anacampseros L.（保留属名）■☆

339655　Ruelingia Ehrh.（废弃属名）= Rulingia R. Br.（保留属名）●■☆

339656　Ruelingia F. Muell. = Rulingia R. Br.（保留属名）●■☆

339657　Ruelingia varians Haw. = Anacampseros telephiastrum DC. ■☆

339658　Ruellia L.（1753）;芦莉草属;Manyroot,Ruellia,Wild Petunia ■●

339659　Ruellia Nees = Hemigraphis Nees ■

339660　Ruellia adhaerens Forssk. = Priva adhaerens（Forssk.）Chiov. ■☆

339661　Ruellia alata Nees = Pteracanthus alatus（Wall.）Bremek. ●■

339662　Ruellia alata Wall. ex Nees = Strobilanthes wallichii Nees ●■

339663　Ruellia albopurpurea Benoist;暗紫芦莉草●☆

339664　Ruellia alopecuroidea Vahl = Lepidagathis alopecuroides（Vahl）R. Br. ex Griseb. ■☆

339665　Ruellia amabilis S. Moore;秀丽芦莉草■☆

339666　Ruellia amoena Nees = Ruellia graecizans Baker ■☆

339667　Ruellia anagallis Burm. f. = Lindernia anagallis（Burm. f.）Pennell ■

339668　Ruellia antipoda L. = Lindernia antipoda（L.）Alston ■

339669　Ruellia arcuta Lingelsh. et Borza = Pararuellia delavayana（Baill.）E. Hossain ■

339670　Ruellia ardeicollis Benoist = Ruellia togoensis（Lindau）Heine ■☆

339671　Ruellia aristata Vahl = Lepidagathis scariosa Nees ■☆

339672　Ruellia aspera（Schinz）E. Phillips;粗糙芦莉草■☆

339673　Ruellia atropurpurea Wall. = Strobilanthes atropurpurea Nees ■☆

339674　Ruellia barlerioides Roth = Petalidium barlerioides（Roth）Nees ■☆

339675　Ruellia batangana J. Braun et K. Schum. = Physacanthus batanganus（J. Braun et K. Schum.）Lindau ■☆

339676　Ruellia baurii C. B. Clarke;巴利芦莉草■☆

339677　Ruellia bignoniiflora S. Moore;紫葳花芦莉草■☆

339678　Ruellia blechum L. = Blechum pyramidatum（Lam.）Urb. ■

339679　Ruellia boranica Ensermu;博兰芦莉草■☆

339680　Ruellia bracteata Roxb. = Petalidium barlerioides（Roth）Nees ■☆

339681　Ruellia bracteophylla Chiov. = Ruellia patula Jacq. ■☆

339682　Ruellia brandbergensis Kers;布兰德山芦莉草■☆

339683　Ruellia brevifolia（Pohl）C. Ezcurra;热带芦莉草;Tropical Wild Petunia ■☆

339684　Ruellia brittoniana Léonard ex Fernald;芦莉草（翠芦莉,红楠草,蓝花草）;Britton Ruellia,Britton's Wild Petunia,Desert Petunia,Mexican Petunia,Ruellia,Ruellia Pink ■

339685　Ruellia californica I. M. Johnst.;加州芦莉草■☆

339686　Ruellia capuronii Benoist;凯普伦芦莉草■☆

339687　Ruellia carolinensis Steud.;卡罗里纳芦莉草;Carolina Wild Petunia,Hairy Ruellia ■☆

339688　Ruellia cavaleriei H. Lév. = Pararuellia cavaleriei（H. Lév.）E. Hossain ■

339689　Ruellia chartacea（T. Anderson）Wassh.;秘鲁芦莉草;Peruvian Wild Petunia ■☆

339690　Ruellia chinensis Nees = Sericocalyx chinensis（Nees）Bremek. ●■

339691　Ruellia chiovendae Fiori = Ruellia discifolia Oliv. ■☆

339692　Ruellia ciliaris L. = Blepharis ciliaris（L.）B. L. Burtt ■☆

339693　Ruellia ciliatiflora Hook.;毛花芦莉草;Hairyflower Wild Petunia ■☆

339694　Ruellia ciliosa Pursh;缘毛芦莉草（缘毛叶芦莉草）;Pringleaf Ruellia ■☆

339695　Ruellia ciliosa Pursh var. longiflora A. Gray = Ruellia humilis Nutt. ■☆

339696　Ruellia cirsioides Nees = Crabbea hirsuta Harv. ■☆

339697　Ruellia coerulea Morong;布里顿芦莉草;Britton's Wild Petunia ■☆

339698　Ruellia congoensis Benoist;刚果芦莉草■☆

339699　Ruellia cordata Thunb. ;心形芦莉草■☆

339700　Ruellia coromandeliana Nees = Asystasia gangetica（L.）T. Anderson ●

339701　Ruellia cumingiana Nees = Hemigraphis cumingiana（Nees）Fern. -Vill. ■

339702　Ruellia currorii T. Anderson;库洛里芦莉草●☆

339703　Ruellia cuspidata Wall. = Lepidagathis cuspidata Nees ■☆

339704　Ruellia cyanea（Nees）T. Anderson;蓝色芦莉草■☆

339705　Ruellia decaryi Benoist;德卡里芦莉草●☆

339706　Ruellia delavayana Baill. = Pararuellia delavayana（Baill.）E. Hossain ■

339707　Ruellia depressa L. = Dyschoriste depressa（L.）Nees ■

339708　Ruellia depressa L. f. = Aptosimum procumbens（Lehm.）Steud. ■☆

339709　Ruellia devosiana Makoy ex E. Morren;锦芦莉草（礼布芦莉草）;Brazilian Wild Petunia,Ruellia ■☆

339710　Ruellia discifolia Oliv. ;盘叶芦莉草■☆

339711　Ruellia divaricata Wall. = Diflugossa divaricata（Nees）Bremek. ■

339712　Ruellia diversifolia S. Moore;异叶芦莉草■☆

339713　Ruellia dorsiflora Retz. = Phaulopsis oppositifolius（J. C. Wendl.）Lindau ■

339714　Ruellia drymophila（Diels）Hand. -Mazz.;喜栎小苞爵床（刀口药,地皮胶,地皮消,红头翁,芦莉草,蛆药,岩威灵仙）■

339715　Ruellia drymophila（Diels）Hand. -Mazz. = Pararuellia delavayana（Baill.）E. Hossain ■

339716　Ruellia drymophila（Diels）Hand. -Mazz. = Pararuellia hainanensis C. Y. Wu et H. S. Lo ■

339717　Ruellia elegans Poir. ;美丽芦莉草;Red Ruellia ■☆

339718　Ruellia elongata P. Beauv. = Whitfieldia elongata（P. Beauv.）De Wild. et T. Durand ■☆

339719　Ruellia erecta Burm. = Dyschoriste erecta（Burm.）Kuntze ■

339720　Ruellia erecta Burm. f. = Hygrophila erecta（Burm. f.）Hochr. ■

339721　Ruellia esquirolii H. Lév. = Pararuellia delavayana（Baill.）E. Hossain ■

339722　Ruellia fasciculata Retz. = Lepidagathis fasciculata（Retz.）Nees ■☆

339723　Ruellia fasciculata Vahl = Lepidagathis fasciculata Nees ■

339724　Ruellia flagelliformis Roxb. = Pararuellia alata H. P. Tsui ■

339725　Ruellia floribunda Hook.；多花芦莉草■☆

339726　Ruellia galactophylla Chiov.；乳白叶芦莉草■☆

339727　Ruellia geniculata（Nees）Benoist；膝曲芦莉草■☆

339728　Ruellia glutinosa Nees = Pseudaechmanthera glutinosa（Nees）Bremek. ■

339729　Ruellia glutinosa Wall. = Strobilanthes glutinosa Nees ■☆

339730　Ruellia gossypina Wall. = Aechmanthera gossypina（Nees）Nees ●

339731　Ruellia graecizans Baker；广布芦莉草（长叶芦莉草，可爱芦莉草）；Cosmopolite Ruellia，Redspray Ruellia ■☆

339732　Ruellia guttata Forssk. = Asystasia guttata（Forssk.）Brummitt ●☆

339733　Ruellia heterosepala Benoist；异萼芦莉草●☆

339734　Ruellia hirta D. Don = Pseudaechmanthera glutinosa（Nees）Bremek. ●

339735　Ruellia hirta D. Don = Strobilanthes glutinosa Nees ■☆

339736　Ruellia humilis Nutt.；矮芦莉草；Dwarf Ruellia，Fringeleaf Ruellia，Fringe-leaf Ruellia，Hairy Ruellia，Hairy Wild Petunia，Wild Petunia ■☆

339737　Ruellia humilis Nutt. var. calvescens Fernald = Ruellia humilis Nutt. ■☆

339738　Ruellia humilis Nutt. var. depauperata Tharp et F. A. Barkley = Ruellia humilis Nutt. ■☆

339739　Ruellia humilis Nutt. var. expansa Fernald = Ruellia humilis Nutt. ■☆

339740　Ruellia humilis Nutt. var. frondosa Fernald = Ruellia humilis Nutt. ■☆

339741　Ruellia humilis Nutt. var. longiflora（A. Gray）Fernald = Ruellia humilis Nutt. ■☆

339742　Ruellia huttonii T. Anderson = Ruellia patula Jacq. ■☆

339743　Ruellia hyssopifolia Hochst. = Dyschoriste hyssopifolia（Nees）Kuntze ■☆

339744　Ruellia ibbensis Lindau = Ruellia sudanica（Schweinf.）Lindau ■☆

339745　Ruellia imbricata Forssk. = Phaulopsis imbricata（Forssk.）Sweet ■

339746　Ruellia indecora Benoist；装饰芦莉草●☆

339747　Ruellia indigofera Griff. = Baphicacanthus cusia（Ness）Bremek. ●

339748　Ruellia indigotica Fortune = Baphicacanthus cusia（Ness）Bremek. ●

339749　Ruellia infundibuliformis（L.）Andréws = Crossandra infundibuliformis（L.）Nees ●☆

339750　Ruellia jacquemontiana Nees = Pseudaechmanthera glutinosa（Nees）Bremek. ■

339751　Ruellia jacquemontiana Nees = Strobilanthes glutinosa Nees ■☆

339752　Ruellia japonica Thunb. = Championella japonica（Thunb.）Bremek. ●■

339753　Ruellia latebrosa Roth = Hemigraphis latebrosa（Roth）Nees ■☆

339754　Ruellia latisepala Benoist；宽萼芦莉草●☆

339755　Ruellia leucoderma Lindau；白皮芦莉草●☆

339756　Ruellia linearibracteolata Lindau；线苞芦莉草●☆

339757　Ruellia linifolia Benoist；亚麻叶芦莉草●☆

339758　Ruellia lithophila Lindau；喜石芦莉草●☆

339759　Ruellia longicalyx Deflers = Duosperma longicalyx（Deflers）Vollesen ■☆

339760　Ruellia lyi H. Lév. = Paragutzlaffia lyi（H. Lév.）H. P. Tsui ■

339761　Ruellia macrantha Mart. ex Nees；大花芦莉草；Christmas Pride ●☆

339762　Ruellia makoyana Hort.；马可芦莉草；Monkey Plant，Trailing Velvet Plant，Vining Velvet ■☆

339763　Ruellia malacophylla C. B. Clarke；软叶芦莉草●☆

339764　Ruellia marlothii Engl. = Ruellia diversifolia S. Moore ■☆

339765　Ruellia megachlamys S. Moore；大被芦莉草■☆

339766　Ruellia misera Benoist；贫弱芦莉草■☆

339767　Ruellia mollissima Vahl = Hypoestes mollissima（Vahl）Nees ●☆

339768　Ruellia monanthos（Nees）Bojer ex T. Anderson；单花芦莉草■☆

339769　Ruellia mysurensis Roth = Asystasia mysurensis（Roth）T. Anderson ●☆

339770　Ruellia nana Nees = Crabbea nana Nees ■☆

339771　Ruellia neesiana Wall. = Asystasiella neesiana（Wall.）Lindau ●

339772　Ruellia nemoralis Vahl；森林芦莉草■☆

339773　Ruellia nocturna Hedrén；夜花芦莉草■☆

339774　Ruellia nummularia Benoist；铜钱芦莉草●☆

339775　Ruellia otaviensis P. G. Mey.；奥塔维芦莉草■☆

339776　Ruellia ovata C. B. Clarke = Ruellia cordata Thunb ■☆

339777　Ruellia oxysepala C. B. Clarke；尖萼芦莉草■☆

339778　Ruellia pallida Vahl = Ruellia patula Jacq. ■☆

339779　Ruellia paniculata Heyne ex C. B. Clarke；圣诞芦莉草；Christmas Pride ■☆

339780　Ruellia paradoxa Lindau = Satanocrater paradoxus（Lindau）Lindau ■☆

339781　Ruellia patula Jacq. = Ruellia cyanea（Nees）T. Anderson ■☆

339782　Ruellia patula Jacq. var. dumicola Chiov. = Ruellia patula Jacq. ■☆

339783　Ruellia pedunculata Torr. ex A. Gray；具梗芦莉草；Wild Petunia ■☆

339784　Ruellia peninsularis I. M. Johnst.；沙地芦莉草；Baja Ruellia，Desert Petunia，Desert Ruellia ■☆

339785　Ruellia persica Burm. f. = Blepharis ciliaris（L.）B. L. Burtt ■☆

339786　Ruellia pilosa L. f.；疏毛芦莉草■☆

339787　Ruellia poissonii Benoist；普瓦松芦莉草●☆

339788　Ruellia praetermissa Schweinf. ex Lindau；疏忽芦莉草■☆

339789　Ruellia primulifolia Nees = Hemigraphis primulifolia（Nees）Fern. -Vill. ■

339790　Ruellia primuloides（T. Anderson ex Benth.）Heine；报春芦莉草■☆

339791　Ruellia primuloides（T. Anderson ex Benth.）Heine var. hirsuta Benoist ex A. Chev. = Ruellia primuloides（T. Anderson ex Benth.）Heine ■☆

339792　Ruellia prostrata Poir.；平卧芦莉草（匍芦利草）；Prostrate Ruellia，Prostrate Wild Petunia ■☆

339793　Ruellia prostrata Poir. = Ruellia rivularis（Benoist）Boivin ex Benoist ■☆

339794　Ruellia prostrata Poir. var. dejecta（Nees）Clarke = Ruellia prostrata Poir. ■☆

339795　Ruellia pseudopatula Ensermu；假张开芦莉草■☆

339796　Ruellia quadrisepala Benoist；四萼芦莉草●☆

339797　Ruellia quadrivalvis Buch. -Ham. = Hygrophila quadrivalvis（Buch. -Ham.）Nees ■☆

339798　Ruellia radicans Hochst. = Dyschoriste radicans Nees ■☆

339799　Ruellia repens L. = Dipteracanthus repens（L.）Hassk. ■

339800　Ruellia repens L. var. kouytcheensis H. Lév. = Calophanoides kouytcheensis（H. Lév.）H. S. Lo ●■

339801　Ruellia reptans G. Forst. = Hemigraphis reptans（G. Forst.）T. Anderson ex Hemsl. ■

339802　Ruellia rivularis（Benoist）Boivin ex Benoist；溪边芦莉草■☆

339803　Ruellia rosea（Nees）Hemsl.；红芦莉草；Rosered Ruellia ■☆

339804　Ruellia ruspolii Lindau = Satanocrater fellatensis Schweinf. ■☆

339805　Ruellia saccifera Benoist；囊状芦莉草■☆

339806　Ruellia salicifolia Vahl = Hygrophila salicifolia（Vahl）Nees ■

339807　Ruellia sclerochiton S. Moore = Eremomastax speciosa（Hochst.）Cufod. ■☆

339808　Ruellia seclusa S. Moore = Leptosiphonium venustum（Hance）E. Hossain ■

339809　Ruellia secunda Blanco = Lepidagathis secunda（Blanco）Nees ■

339810　Ruellia setigera Pers. = Chaetacanthus setiger（Pers.）Lindl. ■☆

339811　Ruellia singularis Benoist；单一芦莉草■☆

339812　Ruellia sokotrana Vierh. = Ruellia patula Jacq. ■☆

339813　Ruellia somalensis Lindau = Satanocrater somalensis（Lindau）Lindau ■☆

339814　Ruellia spatulifolia Benoist；匙叶芦莉草■☆

339815　Ruellia spinescens Thunb. = Aptosimum spinescens（Thunb.）F. E. Weber ■☆

339816　Ruellia squarrosa（Fenzl）Cufod.；糙芦莉草■☆

339817　Ruellia stelligera Benoist；星状芦莉草■☆

339818　Ruellia stenophylla C. B. Clarke；窄叶芦莉草■☆

339819　Ruellia strepens L.；石灰岩芦莉草；Limestone Ruellia, Smooth Ruellia, Wild Petunia ■☆

339820　Ruellia strepens L. f. cleistantha（A. Gray）S. McCoy = Ruellia strepens L. ■☆

339821　Ruellia sudanica（Schweinf.）Lindau；苏丹芦莉草■☆

339822　Ruellia tetrasperma Champ. ex Benth. = Acanthopale tetrasperma（Champ. ex Benth.）Hand. -Mazz. ●

339823　Ruellia tetrasperma Champ. ex Benth. = Championella tetrasperma（Champ. ex Benth.）Bremek. ●

339824　Ruellia thunbergiaeflora T. Anderson = Dischistocalyx thunbergiiflorus（T. Anderson）Benth. ex C. B. Clarke ●☆

339825　Ruellia togoensis（Lindau）Heine；多哥芦莉草■☆

339826　Ruellia tomentosa Wall. = Aechmanthera tomentosa（Wall.）Nees ●

339827　Ruellia transitoria Benoist；中间芦莉草■☆

339828　Ruellia tuberosa L.；块茎芦莉草；Minnieroot Ruellia ■☆

339829　Ruellia urticifolia Wall. = Strobilanthes urticifolia Wall. ex Kuntze ●

339830　Ruellia velutina（C. B. Clarke）E. Phillips = Ruellia diversifolia S. Moore ■☆

339831　Ruellia venusta Hance = Leptosiphonium venustum（Hance）E. Hossain ■

339832　Ruellia woodii C. B. Clarke；伍得芦莉草■☆

339833　Ruellia zeyheri（Sond.）T. Anderson = Ruellia pilosa L. f. ■☆

339834　Ruelliola Baill. = Brillantaisia P. Beauv. ●■☆

339835　Ruelliola grevei Baill. = Brillantaisia pubescens T. Anderson ex Oliv. ■☆

339836　Ruelliopsis C. B. Clarke（1899）；类芦莉草属■☆

339837　Ruelliopsis damarensis S. Moore；类芦莉草■☆

339838　Ruelliopsis mutica C. B. Clarke；非洲类芦莉草■☆

339839　Ruelliopsis setosa（Nees）C. B. Clarke；刚毛类芦莉草■☆

339840　Rueppelia A. Rich. = Aeschynomene L. ●☆

339841　Rueppelia abyssinica A. Rich. = Aeschynomene abyssinica（A.

Rich.）Vatke ●☆

339842　Rufacer Small = Acer L. ●

339843　Rufacer rubrum（L.）Small = Acer rubrum L. ●

339844　Rufodorsia Wiehler(1975)；红背苣苔(鲁福苣苔属)■●☆

339845　Rufodorsia congestiflora（Donn. Sm.）Wiehler；鲁福苣苔●☆

339846　Rufodorsia intermedia Wiehler；间型鲁福苣苔●☆

339847　Rufodorsia major Wiehler；大鲁福苣苔●☆

339848　Rufodorsia minor Wiehler；小鲁福苣苔●☆

339849　Rugelia Shuttlew. ex Chapm.（1860）；冬泉菊属；Rugel's Ragwort, Winter Well ■☆

339850　Rugelia Shuttlew. ex Chapm. = Senecio L. ■●

339851　Rugelia nudicaulis Shuttlew. ex Chapm.；冬泉菊■☆

339852　Rugendasia Schiede ex Schlechid. = Weldenia Schult. f. ■☆

339853　Rugenia Neck. = Eugenia L. ●

339854　Ruhamon Post et Kuntze = Rouhamon Aubl. ●

339855　Ruhamon Post et Kuntze = Strychnos L. ●

339856　Ruilopezia Cuatrec. = Espeletia Mutis ex Humb. et Bonpl. ●☆

339857　Ruizia Cav.（1786）；鲁伊斯桐属；Ruizia ●☆

339858　Ruizia Cav. = Helmiopsiella Arènes ●☆

339859　Ruizia Mutis = Ruizia Cav. ●☆

339860　Ruizia Pav. = Peumus Molina(保留属名)●☆

339861　Ruizia Ruiz et Pay. = Peumus Molina(保留属名)●☆

339862　Ruizia cordata Cav.；鲁伊斯梧桐●☆

339863　Ruizodendron R. E. Fr.（1936）；鲁伊斯木属(鲁泽木属)●☆

339864　Ruizodendrun ovale（Ruiz et Pav.）R. E. Fr.；鲁伊斯木■☆

339865　Ruizterania Marc. -Berti = Qualea Aubl. ●☆

339866　Rulac Adans. = Acer L. ●

339867　Rulac Adans. = Negundo Boehm. ●

339868　Rulingia R. Br.（1820）(保留属名)；龙鳞树属●☆

339869　Rulingia hermaniifolia（DC.）Endl.；龙鳞树●☆

339870　Rulingia macrantha Baill. = Keraudrenia macrantha（Baill.）Arènes ●☆

339871　Rulingia madagascariensis Baker；马岛龙鳞树●☆

339872　Rulingia madagascariensis Baker subsp. andringitrensis（Hochr.）Arènes = Rulingia madagascariensis Baker ●☆

339873　Rulingia madagascariensis Baker var. andringitrensis Hochr. = Rulingia madagascariensis Baker ●☆

339874　Rulingia madagascariensis Baker var. hildebrandtii Baill. ex Arènes = Rulingia madagascariensis Baker ●☆

339875　Rulingia madagascariensis Baker var. luteo-hirta Arènes = Rulingia madagascariensis Baker ●☆

339876　Rulingia madagascariensis Baker var. perrieri Arènes = Rulingia madagascariensis Baker ●☆

339877　Rumania Parl. = Leucojum L. ■●

339878　Rumea Poit. = Xylosma G. Forst.（保留属名）●

339879　Rumea hebecarpa Gardner = Dovyalis hebecarpa（Gardner）Warb. ●

339880　Rumex L.（1753）；酸模属(羊蹄属)；Dock, Doken, Sorrel ■●

339881　Rumex × abortiva Ruhmer；败育酸模■☆

339882　Rumex × halacsyi Rech. = Rumex dentata L. ■

339883　Rumex × oryzetora Rech. f.；稻田酸模■☆

339884　Rumex × pratensis Mert. et W. D. J. Koch；草原酸模■☆

339885　Rumex abyssinica Desf. = Rumex abyssinica Jacq. ■☆

339886　Rumex abyssinica Jacq.；阿比西尼亚酸模；Spanish Rhubarb Dock ■☆

339887　Rumex acetosa L.；酸模(大山七, 当归, 当药, 遏兰菜, 黄根根, 活血莲, 鸡爪黄连, 莫, 莫菜, 牛耳大黄, 牛舌头, 牛舌头棵, 山菠菜,山大黄,山羊蹄,水牛舌头,水乔菜,酸不溜,酸姜,酸浆,酸溜溜,酸迷,酸母,酸木通,酸汤菜,酸汤草,蓣芜,田鸡脚,须,癣草,野菠菜);Bread-and-cheese, Brown Sugar, Cheese, Cock Sorrel, Cock-sorrel, Coffee, Common Sorrel, Cuckoo Sorrel, Cuckoo Sorrow, Cuckoo-meat, Cuckoo's Meat, Cuckoo-sorrow, Dock, Dock Sorrel, Docken, Dockseed, Donkey's Oats, English Sorrel, Garden Sorrel, Garden-sorrel, Gobblety-guts, Gowk Meat, Gowk's Meat, Green Sauce, Green Snob, Green Sorrel, Green Souce, Greensauce Dock, Gypsy's Bacca, Gypsy's Tobacco, Hundreds-and-thousands, Hunters, Lammie Sou Rock, Laramie Sourock, London Green Sauce, Meadow Sorrel, Poor Man's Herb, Red Sour Leek, Redshanks, Sallet, Salry, Sarock, Sauce Sour, Sharp Dock, Sheep Sorrel Dock, Sheep Souce, Sheep's Sorrell, Sheep's Sourack, Sheep-souce, Soldiers, Sollop, Soór Dockin, Soót Dockin, Sorrel, Sorrel Dock, Sorrow, Sour Dock, Sour Duck, Sour Gog, Sour Gogs, Sour Grabs, Sour Grass, Sour Leaves, Sour Leek, Sour Sab, Sour Sabs, Sour Salves, Sour Sauce, Sour Sodge, Sour Sog, Sour Sogs, Sour Sops, Sour Suds, Sourack, Sourdock, Sourock, Sow Sorrel, Tea, Tom Thumb's Thousand Fingers, Wild Sorrel, Wood Dock ■

339888 Rumex acetosa L. subsp. auriculata (Wallr.) A. Blytt et O. C. Dahl = Rumex thyrsiflora Fingerh. ■

339889 Rumex acetosa L. subsp. lapponica Hiitonen = Rumex lapponica (Hiitonen) Czernov ■☆

339890 Rumex acetosa L. subsp. pratensis (Mill.) A. Blytt et O. C. Dahl = Rumex acetosa L. ■

339891 Rumex acetosa L. subsp. thyrsiflora (Fingerh.) Celak. = Rumex thyrsiflora Fingerh. ■

339892 Rumex acetosa L. var. atlantis Maire = Rumex acetosa L. ■

339893 Rumex acetosa L. var. auriculata Wallr. = Rumex thyrsiflora Fingerh. ■

339894 Rumex acetosa L. var. crispa (Roth) Celak. = Rumex thyrsiflora Fingerh. ■

339895 Rumex acetosa L. var. haplorhiza (Czern. ex Turcz.) Trautv. = Rumex thyrsiflora Fingerh. ■

339896 Rumex acetosa L. var. hortensis Dierb. = Rumex acetosa L. ■

339897 Rumex acetosella L. ;小酸模(莙);Common Sheep Sorrel, Cow Serl, Cuckoo Sorrel, Cuckoo's Meat, Dwarf Dock, Field Sorrel, Green Sauce, Laramie Sourock, Red Sorrel, Sheep Sorrel, Sheep Sorrel Dock, Sheep's Sorrel, Sheep's Sourock, Sheep-sorrel Dock, Sorrel, Sour Dock, Sour Grass, Sour Leek, Sourack, Sourweed ■

339898 Rumex acetosella L. subsp. angiocarpa (Murb.) Murb. = Rumex acetosella L. ■

339899 Rumex acetosella L. subsp. pyrenaica (Pourr. et Lapeyr.) Akeroyd;比利牛斯酸模■☆

339900 Rumex acetosella L. var. graminifolia (Rudolph ex Lamb.) Schrenk = Rumex graminifolia Rudolph ex Lamb. ■☆

339901 Rumex acetosella L. var. pyrenaea (Pourr.) Timb.-Lagr. = Rumex acetosella L. ■

339902 Rumex acetosella L. var. tenuifolia Wallr. = Rumex acetosella L. ■

339903 Rumex acetosella L. var. vulgaris W. D. J. Koch = Rumex acetosella L. ■

339904 Rumex acetoselloides Baill. ;拟小酸模■☆

339905 Rumex acuta L. ;尖酸模;Acute Dock ■☆

339906 Rumex acuta Sm. = Rumex conglomerata Murray ■☆

339907 Rumex acutata L. ;锐酸模■☆

339908 Rumex aegyptiaca L. ;埃及酸模■☆

339909 Rumex afromontana T. C. E. Fr. = Rumex ruwenzoriensis Chiov. ■☆

339910 Rumex algeriensis Barratte et Murb. ;阿尔及利亚酸模■☆

339911 Rumex algeriensis Barratte et Murb. var. hipporegiana Batt. = Rumex algeriensis Barratte et Murb. ■☆

339912 Rumex alluvius F. C. Gates et McGregor = Rumex stenophylla Ledeb. ■

339913 Rumex alpestris Jacq. subsp. lapponica (Hiitonen) Jalas = Rumex lapponica (Hiitonen) Czernov ■☆

339914 Rumex alpina L. ;高山酸模(僧大黄);Alpine Dock, Butter Dock, Butter Docken, Butter Leaves, Garden Patience, Monk's Rhubarb, Monks' Rhubarb, Monk's-rhubarb, Rhubarb ■☆

339915 Rumex alpina L. var. subcalligerus Boiss. = Rumex conferta Willd. ■

339916 Rumex altissima A. W. Wood;极高酸模;Pale Dock, Peach-leaved Dock, Smooth Dock, Tall Dock, Water Dock ■☆

339917 Rumex altissima A. Wood subsp. elliptica (Greene) Á. Löve = Rumex elliptica Greene ■☆

339918 Rumex amurensis F. Schmidt ex Maxim. ;黑龙江酸模(黑龙酸模,黑水酸模);Heilongjiang Dock ■

339919 Rumex andreaeana Makino = Rumex nepalensis Spreng. subsp. andreaeana (Makino) Yonek. ■☆

339920 Rumex angiocarpa Murb. = Rumex acetosella L. subsp. angiocarpa (Murb.) Murb. ■

339921 Rumex angiocarpa Murb. = Rumex acetosella L. ■

339922 Rumex angulata Rech. f. ;紫茎酸模;Purplestem Dock ■

339923 Rumex angustifolia Campd. ;窄叶酸模■☆

339924 Rumex angustissima Ledeb. = Rumex graminifolia Rudolph ex Lamb. ■☆

339925 Rumex aquatica L. ;水生酸模(金不换,水酸模,土大黄); Eldin-docken, Elgins, Pond Dock, Scottish Dock, Trossachs Dock, Water Dock ■☆

339926 Rumex aquatica L. = Rumex gmelinii Turcz. ex Ledeb. ■

339927 Rumex aquatica L. subsp. arctica (Trautv.) Hiitonen = Rumex arctica Trautv. ■☆

339928 Rumex aquatica L. subsp. fenestrata (Greene) Hultén = Rumex occidentalis S. Watson ■☆

339929 Rumex aquatica L. subsp. lipschitzii Rech. f. = Rumex popovii Pachom. ■

339930 Rumex aquatica L. subsp. occidentalis (S. Watson) Hultén = Rumex occidentalis S. Watson ■☆

339931 Rumex aquatica L. subsp. protracta (Rech. f.) Rech. f. = Rumex aquatica L. ■

339932 Rumex aquatica L. subsp. schischkinii (Losinsk.) Rech. f. ;紫茎水生酸模■

339933 Rumex aquatica L. subsp. tomentella (Rech. f.) Á. Löve = Rumex tomentella Rech. f. ■☆

339934 Rumex aquatica L. var. japonica Meisn. ;日本水酸模■☆

339935 Rumex arctica L. var. kamtschadala (Kom.) Rech. f. ex Tolm. = Rumex arctica Trautv. ■☆

339936 Rumex arctica L. var. latifolia Tolm. = Rumex arctica Trautv. ■☆

339937 Rumex arctica Trautv. ;北极酸模;Arctic Dock ■☆

339938 Rumex arctica Trautv. var. kamtschadala (Kom.) Rech. f. ex Tolm. = Rumex arctica Trautv. ■☆

339939 Rumex arctica Trautv. var. latifolia Tolm. = Rumex arctica Trautv. ■☆

339940 Rumex arifolia All. ;芋叶酸模■☆

339941 Rumex aristidis Coss. ;三芒草酸模■☆

339942 Rumex arizonica Britton = Rumex hymenosepala Torr. ■

339943　Rumex armena K. Koch；亚美尼亚酸模■☆

339944　Rumex aschabadensis Losinsk.；阿沙巴得酸模■☆

339945　Rumex atlantica Batt.；大西洋酸模■☆

339946　Rumex aurea Mill. = Rumex maritima L.■

339947　Rumex aureostigmatica Kom.；黄柱头酸模■☆

339948　Rumex auriculata（Wallr.）Murb. = Rumex thyrsiflora Fingerh.■

339949　Rumex bakeri Greene = Rumex occidentalis S. Watson■☆

339950　Rumex bequaertii De Wild. = Rumex nepalensis Spreng.■

339951　Rumex bequaertii De Wild. var. quarrei（De Wild.）Robyns = Rumex nepalensis Spreng.■

339952　Rumex beringensis Jurtzev et V. V. Petrovsky；白令海酸模；Bering Sea Sorrel，Beringia Sorrel■☆

339953　Rumex berlandieri Meisn. = Rumex chrysocarpa Moris■☆

339954　Rumex bipinnata L. f.；双羽酸模■☆

339955　Rumex bolosii Stübing et Peris et Romo = Rumex vesicaria L.■☆

339956　Rumex brachypoda Rchb. f.；短足酸模■☆

339957　Rumex britannica L.；黄根水酸模；British Dock，Great Water Dock，Great Water-dock，Greater Water Dock，Yellow-rooted Water Dock■☆

339958　Rumex britannica L. var. borealis Rech. f. = Rumex britannica L.■☆

339959　Rumex britannicus Meisn. = Rumex altissima A. W. Wood■☆

339960　Rumex brownei Campd.；布氏酸模；Browne's Dock，Hooked Dock■☆

339961　Rumex bucephalophora L.；红酸模；Horned Dock，Red Dock，Ruby Dock■☆

339962　Rumex bucephalophora L. subsp. aegaea Rchb. f. = Rumex bucephalophora L. subsp. gallica（Steinh.）Rech. f.■☆

339963　Rumex bucephalophora L. subsp. canariensis（Steinh.）Rech. f.；加那利酸模■☆

339964　Rumex bucephalophora L. subsp. fruticescens Bornm.；灌木状红酸模■☆

339965　Rumex bucephalophora L. subsp. gallica（Steinh.）Rech. f.；法国酸模■☆

339966　Rumex bucephalophora L. subsp. hipporegii（Steinh.）Rech. f. = Rumex bucephalophora L.■☆

339967　Rumex bucephalophora L. var. aegea（Rech. f.）Maire = Rumex bucephalophora L. subsp. gallica（Steinh.）Rech. f.■☆

339968　Rumex bucephalophora L. var. crassissima Maire = Rumex bucephalophora L. subsp. gallica（Steinh.）Rech. f.■☆

339969　Rumex bucephalophora L. var. fruticescens（Bornm.）Press = Rumex bucephalophora L. subsp. canariensis（Steinh.）Rech. f.■☆

339970　Rumex bucephalophora L. var. gallica Steinh. = Rumex bucephalophora L. subsp. gallica（Steinh.）Rech. f.■☆

339971　Rumex bucephalophora L. var. hipporegii（Steinh.）Murb. = Rumex bucephalophora L.■☆

339972　Rumex bucephalophora L. var. platycarpa Batt. = Rumex bucephalophora L. subsp. gallica（Steinh.）Rech. f.■☆

339973　Rumex bucephalophora L. var. stenocarpa（Beck）Press = Rumex bucephalophora L. subsp. gallica（Steinh.）Rech. f.■☆

339974　Rumex bucephalophora L. var. subaegaea Maire = Rumex bucephalophora L. subsp. gallica（Steinh.）Rech. f.■☆

339975　Rumex burchellii Campd. = Rumex sagittata Thunb.■☆

339976　Rumex cacaliifolia H. Lév. = Rheum kialense Franch.■

339977　Rumex californica Rech. f.；加州酸模；California Willow Dock■☆

339978　Rumex callosa（F. Schmidt ex Maxim.）Rech. f. = Rumex patientia L.■

339979　Rumex callosa（F. Schmidt）Rech. f.；硬质酸模■☆

339980　Rumex camptodon Rech. f. = Rumex bequaertii De Wild.■

339981　Rumex camptodon Rech. f. = Rumex nepalensis Spreng.■

339982　Rumex cardiocarpa Pamp. = Rumex japonica Houtt.■

339983　Rumex chalepensis Mill.；网果酸模（红丝酸模，化血莲，金不换，牛西西，乳突酸模，乳突叶酸模，土大黄，血当归，血三七，止血草，中亚酸模）；Netfruit Dock■

339984　Rumex chinensis Campd. = Rumex trisetifer Stokes■

339985　Rumex chrysocarpa Moris；金果酸模；Amamastla Dock■☆

339986　Rumex condylodes M. Bieb. = Rumex sanguinea L.■☆

339987　Rumex conferta Willd.；密生酸模（糙叶酸模，密酸模）；Russian Dock■

339988　Rumex confertoides Bihari = Rumex kerneri Borbás■☆

339989　Rumex conglomerata Murray；紧球酸模；Clustered Dock，Clustered Green Dock，Dock，Green Dock，Red Dock，Sharp Dock■☆

339990　Rumex conglomerata Murray var. borreri Trimen = Rumex conglomerata Murray■☆

339991　Rumex cordata Desf. = Rumex cordata Poir.■☆

339992　Rumex cordata Poir.；心形酸模■☆

339993　Rumex crassa Rech. f.；柳林酸模；Fleshy Willow Dock■☆

339994　Rumex crispa L.；皱叶酸模（荜菝，火风棠，壳菜，牛耳大黄，牛舌，牛舌头，四季菜根，土大黄，蓄，羊蹄，羊蹄草，羊蹄根，皱叶羊蹄）；Crispate Dock，Crisped Dock，Curled Dock，Curly Dock，Dock，Narrow Dock，Narrow-leaved Dock，Sour Dock，Wrinkleleaf Dock，Yellow Dock■

339995　Rumex crispa L. subsp. fauriei（Rech. f.）Mosyakin et W. L. Wagner；法氏皱叶酸模；Curly Dock■☆

339996　Rumex crispa L. subsp. japonica（Houtt.）Kitam. = Rumex japonica Houtt.■

339997　Rumex crispa L. subsp. littorea（Hardy）Akeroyd；海岸皱叶酸模■☆

339998　Rumex crispa L. var. dentata Schur = Rumex stenophylla Ledeb.■

339999　Rumex crispa L. var. elongata Coss. = Rumex crispa L.■

340000　Rumex crispa L. var. japonica（Houtt.）Makino；羊蹄■

340001　Rumex crispa L. var. japonica（Houtt.）Makino = Rumex japonica Houtt.■

340002　Rumex crispa L. var. trigranulata Syme = Rumex crispa L. subsp. littorea（Hardy）Akeroyd■☆

340003　Rumex crispa L. var. unicallosa Peterm.；单瘤皱叶酸模■☆

340004　Rumex crispatula Michx. = Rumex obtusifolia L.■

340005　Rumex crispobtusifolia Meisn. = Rumex acuta L.■☆

340006　Rumex cristata DC.；希腊酸模；Crested Dock，Greek Dock■☆

340007　Rumex cristata DC. subsp. kerneri（Borbás）Akeroyd et D. A. Webb = Rumex kerneri Borbás■☆

340008　Rumex cuneifolia Campd.；阿根廷酸模；Argentine Dock，Wedgeleaf Dock■☆

340009　Rumex cyprius Murb.；铜色酸模■☆

340010　Rumex cyprius Murb. subsp. conjungens Sam.；康朱加酸模■☆

340011　Rumex cyprius Murb. subsp. disciformis Sam. = Rumex cyprius Murb.■☆

340012　Rumex cyprius Murb. subsp. subinteger Sam.；近全缘酸模■☆

340013　Rumex cyprius Murb. subsp. vesceritensis（Murb.）Sam.；韦斯塞里德酸模■☆

340014　Rumex daiwoo Makino = Rumex madaio Makino■

340015　Rumex daiwoo Makino var. andreaeanus（Makino）Makino = Rumex nepalensis Spreng. subsp. andreaeana（Makino）Yonek.■☆

340016　Rumex densiflora Osterh.；北美密花酸模；Dense-flowered Dock■☆

340017　Rumex densiflora Osterh. subsp. orthoneura（Rech. f.）Á. Löve = Rumex orthoneura Rech. f. ■☆

340018　Rumex densiflora Osterh. subsp. pycnantha（Rech. f.）Á. Löve = Rumex pycnantha Rech. f. ■☆

340019　Rumex dentata L. ；齿果酸模（大黄，牛耳大黄，牛舌草，羊蹄大黄）；Aegean Dock，Dentate Dock，Dock，Greek Dock，Indian Dock，Toothed Dock，Toothedfruit Dock，Toothfruit Dock ■

340020　Rumex dentata L. subsp. callosissima（Meisn.）Rchb. f. ；硬皮齿果酸模■☆

340021　Rumex dentata L. subsp. halacsyi（Rech. f.）Rech. f. = Rumex dentata L. ■

340022　Rumex dentata L. subsp. halacsyi（Rech.）Rech. f. = Rumex dentata L. subsp. callosissima（Meisn.）Rchb. f. ■☆

340023　Rumex dentata L. subsp. klotzschiana（Meisn.）Rech. f. ；克氏齿果酸模（刺果羊蹄，小羊蹄）■

340024　Rumex dentata L. subsp. klotzschiana（Meisn.）Rech. f. = Rumex dentata L. ■

340025　Rumex dentata L. subsp. nipponica（Franch. et Sav.）Rech. f. = Rumex dentata L. subsp. klotzschiana（Meisn.）Rech. f. ■

340026　Rumex dictyocarpa Boiss. et Buhse = Rumex chalepensis Mill. ■

340027　Rumex digyna L. = Oxyria digyna（L.）Hill ■

340028　Rumex dissecta H. Lév. = Rumex hastata D. Don ●■

340029　Rumex divaricata L. = Rumex pulcher L. subsp. divaricata（L.）Murb. ☆

340030　Rumex domestica Hartm. = Rumex longifolia DC. ■

340031　Rumex domestica Hartm. var. nana Hook. = Rumex arctica Trautv. ■☆

340032　Rumex domestica Hartm. var. pseudonatronata Borbás = Rumex pseudonatronata（Borbás）Borbás ex Murb. ■

340033　Rumex dregeana Meisn. ；德雷酸模■☆

340034　Rumex dregeana Meisn. subsp. montana B. L. Burtt；山地德雷酸模■☆

340035　Rumex dregei Meisn. = Rumex dregeana Meisn. ■☆

340036　Rumex drobovii Korovin = Rumex chalepensis Mill. ■

340037　Rumex eckloniana Meisn. = Rumex lanceolata Thunb. ■☆

340038　Rumex ecklonii Meisn. = Rumex lanceolata Thunb. ■☆

340039　Rumex ellenbeckii Dammer；埃伦酸模■☆

340040　Rumex elliptica Greene；椭圆酸模；Elliptic Tall Dock ■☆

340041　Rumex elliptica Greene = Rumex altissima A. W. Wood ■☆

340042　Rumex elongata Guss. = Rumex crispa L. ■

340043　Rumex engelmannii Meisn. = Rumex hastatula Baldwin ex Elliott ●■

340044　Rumex engelmannii Meisn. var. geyeri Meisn. ；盖耶氏酸模■☆

340045　Rumex engelmannii Meisn. var. geyeri Meisn. = Rumex paucifolia Nutt. ■☆

340046　Rumex esquirolii H. Lév. = Rumex nepalensis Spreng. ■

340047　Rumex fascicularis Small；簇生酸模■☆

340048　Rumex fascicularis Small = Rumex verticillata L. ■☆

340049　Rumex fauriei Rech. f. ；法氏酸模■☆

340050　Rumex fausse-persicaire？= Rumex persicarioides L. ■☆

340051　Rumex fenestrata Greene = Rumex occidentalis S. Watson ■☆

340052　Rumex fenestrata Greene var. labradorica Rech. f. = Rumex occidentalis S. Watson ■☆

340053　Rumex fennica（Murb.）Murb. = Rumex pseudonatronata（Borbás）Borbás ex Murb. ■

340054　Rumex fimbriata R. Br. = Rumex brownei Campd. ■☆

340055　Rumex fischeri Rchb. ；菲舍尔酸模■☆

340056　Rumex floridana Meisn. ；佛罗里达酸模；Florida Dock ■☆

340057　Rumex floridana Meisn. = Rumex verticillata L. ■☆

340058　Rumex foveolata Losinsk. ；蜂窝酸模■☆

340059　Rumex frutescens Thouars；灌木酸模；Argentine Dock，Wedgeleaf Dock ■☆

340060　Rumex fuegina Phil. ；美洲金酸模；American Golden Dock，Golden Dock ■☆

340061　Rumex garipensis Meisn. ；加里普酸模■☆

340062　Rumex geyeri（Meisn.）Trel. = Rumex paucifolia Nutt. ■☆

340063　Rumex glomerata Schreb. = Rumex conglomerata Murray ■☆

340064　Rumex gmelinii Turcz. ex Ledeb. ；毛脉酸模；Hairyvein Dock ■

340065　Rumex gracilipes Greene = Rumex occidentalis S. Watson ■☆

340066　Rumex graecus Boiss. et Heldreich = Rumex cristata DC. ■☆

340067　Rumex graminifolia Lamb. = Rumex graminifolia Rudolph ex Lamb. ■☆

340068　Rumex graminifolia Rudolph ex Lamb. ；禾叶酸模；Grassleaf Sorrel，Grass-leaved Sorrel ■☆

340069　Rumex hadroocarpa Rech. f. ；锐齿酸模■

340070　Rumex hadroocarpa Rech. f. = Rumex japonica Houtt. ■

340071　Rumex halacsyi Rech. ；矮酸模■

340072　Rumex halacsyi Rech. = Rumex dentata L. ■

340073　Rumex haplorhiza Czern. ex Turcz. = Rumex thyrsiflora Fingerh. ■

340074　Rumex hararensis Dammer；哈拉雷酸模■☆

340075　Rumex hastata D. Don；戟叶酸模（川滇土大黄，浆草，酸浆草，太阳草，土麻黄）；Halbertleaf Dock，Hastate-leaved Dock，Heartwing Sorrel，Sorrel，Sour Dock，Spear-leaved Dock，Wild Sorrel ●■

340076　Rumex hastatula Baldwin ex Elliott = Rumex hastata D. Don ●■

340077　Rumex hastatula Muhl. = Rumex hastata D. Don ●■

340078　Rumex hastulata Sm. = Muehlenbeckia hastulata（Sm.）I. M. Johnst. ●☆

340079　Rumex hesperia Greene；西方柳叶酸模；Western Willow Dock ■☆

340080　Rumex hippolapathum Fr. = Rumex longifolia DC. ■

340081　Rumex hydrolapathum C. H. Wright = Rumex lanceolata Thunb. ■☆

340082　Rumex hydrolapathum Huds. ；大酸模（大水酸模）；Bloodwort，Darele，Great Water Dock，Great Water-dock，Horse Sorrel，Huds，Red Dock，Water Dock，Water Sorrel ■☆

340083　Rumex hydrolapathum Huds. var. americana A. Gray = Rumex britannica L. ■☆

340084　Rumex hymenosepala Torr. ；膜萼酸模（托里酸模）；Arizona Dock，Canaigre，Canaigre Dock，Dock，Ganagra，Sorrel，Tanner's Dock，Wild Rhubarb，Wild-rhubarb ■

340085　Rumex hymenosepala Torr. var. salina（A. Nelson）Rech. f. = Rumex hymenosepala Torr. ■

340086　Rumex indurata Boiss. et Reut. ；坚硬酸模■☆

340087　Rumex intermedia DC. ；间型酸模■☆

340088　Rumex interrupta Rech. f. = Rumex patientia L. ■

340089　Rumex japonica Houtt. ；羊蹄酸模（败毒菜，东方宿，鬼目，鸡脚大黄，金荞麦，连虫陆，牛利菜，牛舌菜，牛舌大黄，牛舌根，牛舌头，土大黄，癣草，羊舌头，羊蹄，羊蹄大黄，野菠菜）；Japan Dock，Japanese Curely Dock，Japanese Dock ■

340090　Rumex japonica Houtt. var. sachalinensis（Regel）H. Hara = Rumex japonica Houtt. ■

340091　Rumex japonica Houtt. var. yezoensis（H. Hara）Ohwi = Rumex japonica Houtt. ■

340092　Rumex japonica Meisn. = Rumex japonica Houtt. ■

340093　Rumex kamtschadalus Kom. ；勘察加酸模■☆

340094　Rumex kamtschadalus Kom. = Rumex arctica Trautv. ■☆

340095　Rumex kaschgarica Chang Y. Yang;喀什酸模■

340096　Rumex kerneri Borbás;克纳酸模■;Kerner's Dock ■☆

340097　Rumex klotzschiana Meisn. = Rumex dentata L. ■

340098　Rumex komarovii Schischk. et Serg.;科马罗夫酸模■☆

340099　Rumex krausei Jurtzev et V. V. Petrovsky;克劳斯酸模;Cape Krause Sorrel,Krause Sorrel ■☆

340100　Rumex lacustris Greene;湖畔酸模;Lake Willow Dock ■☆

340101　Rumex lanceolata Thunb.;剑叶酸模■☆

340102　Rumex langloisii Small = Rumex chrysocarpa Moris ■☆

340103　Rumex lapponica (Hiitonen) Czernov;山酸模;Lapland Mountain Sorrel,Lapland Sorrel ■☆

340104　Rumex lativalvis Meisn.;宽分果片酸模■☆

340105　Rumex lativalvis Meisn. var. acetosoides ? = Rumex lativalvis Meisn. ■☆

340106　Rumex lativalvis Meisn. var. decipiens ? = Rumex lativalvis Meisn. ■☆

340107　Rumex linearis Campd. = Rumex lanceolata Thunb. ■☆

340108　Rumex lonaczevskii Klokov = Rumex patientia L. ■☆

340109　Rumex longifolia DC.;长叶酸模(红牛耳酸模,家酸模,直穗酸模);Butter Dock,Dooryard Dock,Door-yard Dock,Longleaf Dock,Long-leaved Dock,Northern Dock,Yard Dock ■

340110　Rumex longifolia DC. var. nana (Hook.) Meisn. = Rumex arctica Trautv. ■☆

340111　Rumex longiseta A. I. Baranov et B. Skvortsov = Rumex maritima L. ■

340112　Rumex lunaria L.;新月酸模■☆

340113　Rumex luxurians L. f. = Rumex sagittata Thunb. ■☆

340114　Rumex madaio Makino;大黄酸模(土大黄)■

340115　Rumex maderensis Lowe;梅德酸模■☆

340116　Rumex maritima L.;刺酸模(长刺酸模,俄罗斯酸模,海滨酸模,海滨羊蹄,假菠菜,壳菜,连明子,牛舌菜,牛舌草,牛屎草,土大黄,癣药草,血大黄,野菠菜,皱叶羊蹄);Golden Cuvely Dock,Golden Dock,Maritime Dock,Russia Dock,Small Water Dock,Thorn Dock ■

340117　Rumex maritima L. = Rumex trisetifer Stokes ■

340118　Rumex maritima L. subsp. fuegina (Phil.) Hultén = Rumex fuegina Phil. ■☆

340119　Rumex maritima L. subsp. rossica (Murb.) Krylov = Rumex maritima L. ■

340120　Rumex maritima L. var. athrix H. St. John = Rumex fuegina Phil. ■☆

340121　Rumex maritima L. var. fuegina (Phil.) Dusen = Rumex fuegina Phil. ■☆

340122　Rumex maritima L. var. persicarioides (L.) R. S. Mitch. = Rumex persicarioides L. ■☆

340123　Rumex marschalliana Rchb.;单瘤酸模(马氏酸模,盐生酸模);Marschall Dock ■

340124　Rumex marschalliana Rchb. var. brevidens Bong. et C. A. Mey.;短齿单瘤酸模■

340125　Rumex marschalliana Rchb. var. brevidens Bong. et C. A. Mey. = Rumex similans Rech. f. ■

340126　Rumex martima L.;连明子■

340127　Rumex membranosa Poir. = Rumex bucephalophora L. ■☆

340128　Rumex mexicana Meisn.;墨西哥酸模;Mexican Dock,Mexican Willow,Willow-leaved Dock ■☆

340129　Rumex mexicana Meisn. = Rumex salicifolia Weinm. var. mexicana (Meisn.) C. L. Hitchc. ■☆

340130　Rumex mexicana Meisn. = Rumex triangulivalvis (Danser) Rech. f. ■☆

340131　Rumex mexicana Meisn. var. angustifolia (Meisn.) B. Boivin = Rumex salicifolia Weinm. var. mexicana (Meisn.) C. L. Hitchc. ■☆

340132　Rumex mexicana Meisn. var. triangulivalvis (Danser) Lepage = Rumex salicifolia Weinm. var. mexicana (Meisn.) C. L. Hitchc. ■☆

340133　Rumex meyeri Meisn. = Rumex lanceolata Thunb. ■☆

340134　Rumex meyeriana Meisn. = Rumex lanceolata Thunb. ■☆

340135　Rumex microcarpa Campd.;小果酸模(越南酸模);Narrow Dock,Pale Dock,Smallfruit Dock,Willow-leaved Dock ■☆

340136　Rumex montana Desf.;山地酸模;Maiden Sorrel ■☆

340137　Rumex natalensis Dammer ex J. M. Wood = Rumex woodii N. E. Br. ■☆

340138　Rumex nemolapathum Ehrh. = Rumex conglomerata Murray ■☆

340139　Rumex nemorosa Schrad. ex Willd = Rumex sanguinea L. ■☆

340140　Rumex nepalensis C. H. Wright = Rumex steudelii Hochst. ex A. Rich. ■☆

340141　Rumex nepalensis Spreng.;尼泊尔酸模(包金莲,大晕药,广角,红筋大黄,化血莲,箭头草,金不换,救命王,萝卜奇,尼泊尔羊蹄,牛儿黄草,铁蒲扇,土大黄,土三七,吐血草,癣药,血当归,野蒿荬,止血草);Nepal Dock ■

340142　Rumex nepalensis Spreng. subsp. andreaeana (Makino) Yonek.;安氏酸模■☆

340143　Rumex nepalensis Spreng. var. andreaeana (Makino) Kitam. = Rumex nepalensis Spreng. subsp. andreaeana (Makino) Yonek. ■☆

340144　Rumex nepalensis Spreng. var. remotiflora (Sam.) A. J. Li;疏花酸模■

340145　Rumex nervosa Vahl;密脉酸模■

340146　Rumex nervosa Vahl var. usambarensis Dammer = Rumex usambarensis (Dammer) Dammer ■☆

340147　Rumex nipponica Franch. et Sav.;小羊蹄■

340148　Rumex nipponica Franch. et Sav. = Rumex dentata L. subsp. klotzschiana (Meisn.) Rech. f. ■

340149　Rumex nipponica Franch. et Sav. = Rumex dentata L. ■

340150　Rumex obovata Danser;热带酸模;Obovate-leaf Dock,Obovate-leaved Dock,Tropical Dock ■☆

340151　Rumex obtusifolia L.;钝叶酸模(大羊蹄,大晕草,化血莲,金不换,苦酸模,绿当归,土大黄,血三七);Batter Dock,Bitter Dock,Bluntleaf Dock,Blunt-leaved Dock,Broadleaf Dock,Broad-leaved Dock,Bulwand,Butter Dock,Carrag,Celery-seed,Doctor's Medicine,Donkey's Oats,Good King Henry,Kettle Dock,Kettle-dock,Monk's Rhubarb,Rantytanty,Redshanks,Red-veined Dock,Smair Dock,Sour Dock,Wild Rhubarb ■

340152　Rumex obtusifolia L. subsp. agrestis (Fr.) Danser = Rumex obtusifolia L. ■

340153　Rumex obtusifolia L. subsp. sylvestris (Wallr.) Rech. f. = Rumex obtusifolia L. ■

340154　Rumex obtusifolia L. var. agrestis Fr. = Rumex obtusifolia L. ■

340155　Rumex obtusifolia L. var. cristata Neilr. = Rumex stenophylla Ledeb. ■

340156　Rumex obtusifolia L. var. sylvestris (Wallr.) Koch = Rumex obtusifolia L. ■

340157　Rumex occidentalis S. Watson;西方酸模;Western Dock ■☆

340158　Rumex occidentalis S. Watson subsp. fenestrata (Greene) Hultén = Rumex occidentalis S. Watson ■☆

340159　Rumex occidentalis S. Watson var. labradorica (Rech. f.)

Lepage = Rumex occidentalis S. Watson ■☆

340160 Rumex odontocarpa Sandor ex Borbás = Rumex stenophylla Ledeb. ■

340161 Rumex orbiculata A. Gray;圆叶大酸模;Great Water Dock, Great Water-dock ■☆

340162 Rumex orbiculata A. Gray = Rumex britannica L. ■☆

340163 Rumex orbiculata A. Gray var. borealis Rech. f. = Rumex orbiculata A. Gray ■☆

340164 Rumex orientalis Burnham ex Schult. f.;东方酸模;Oriental Dock ■☆

340165 Rumex orispus L.;牛耳大黄■☆

340166 Rumex orthoneura Rech. f.;奇里卡华酸模;Blumer's Dock, Chiricahua Dock ■☆

340167 Rumex pallida Bigelow;苍白酸模;Pale Willow Dock, Seabeach Dock, Seaside Dock, White Dock ☆

340168 Rumex pallida Bigelow subsp. subarctica (Lepage) Á. Löve = Rumex subarctica Lepage ■☆

340169 Rumex paludosus With. = Rumex conglomerata Murray ■☆

340170 Rumex palustris Sm.;沼泽酸模;Marsh Dock ■☆

340171 Rumex pamirica Rech. f.;帕米尔酸模■

340172 Rumex pamirica Rech. f. = Rumex patientia L. ■

340173 Rumex papilio Coss. et Balansa;蝶形酸模■☆

340174 Rumex papilio Coss. et Balansa var. glabrivalvis Maire = Rumex papilio Coss. et Balansa ■☆

340175 Rumex papilio Coss. et Balansa var. hirtivalvis Maire = Rumex papilio Coss. et Balansa ■☆

340176 Rumex paraguayensis D. Parodi;巴拉圭酸模;Paraguay Dock, Paraguayan Dock ■☆

340177 Rumex paraguayensis D. Parodi et Danser = Rumex paraguayensis D. Parodi ■☆

340178 Rumex patientia L.;巴天酸模(菠菜酸模,金不换,牛耳大黄,牛耳酸模,牛西西,山荞麦,土大黄,羊蹄叶,羊铁酸模,洋铁酸模);Bastard Rhubarb, Dock, Garden Dock, Garden Patience, Herb Patience, Monk's Rhubarb, Passion, Passion Dock, Patience, Patience Dock, Patient Dock, Payshun Dock, Payshun-dock, Poor Man's Cabbage, Spinach Dock ■

340179 Rumex patientia L. subsp. callosa (F. Schmidt ex Maxim.) Rech. f. = Rumex patientia L. ■

340180 Rumex patientia L. subsp. graecus (Boiss. et Heldreich) Lindberg = Rumex cristata DC. ■☆

340181 Rumex patientia L. subsp. interrupta Rech. f. = Rumex patientia L. ■

340182 Rumex patientia L. subsp. orientalis Danser = Rumex patientia L. ■

340183 Rumex patientia L. subsp. pamirica (Rech. f.) Rech. f. = Rumex patientia L. ■

340184 Rumex patientia L. subsp. pamirica Rech. f. = Rumex patientia L. ■

340185 Rumex patientia L. subsp. tibetica (Rech. f.) Rech. f. = Rumex patientia L. ■

340186 Rumex patientia L. subsp. tibetica Rech. f.;西藏酸模■

340187 Rumex patientia L. subsp. tibetica Rech. f. = Rumex patientia L. ■

340188 Rumex patientia L. var. callosa F. Schmidt ex Maxim.;洋铁酸模(羊铁酸模)■

340189 Rumex patientia L. var. callosa F. Schmidt ex Maxim. = Rumex patientia L. ■

340190 Rumex patientia L. var. kurdica Boiss. = Rumex patientia L. ■

340191 Rumex patientia L. var. tibetica Rech. f. = Rumex patientia L. ■

340192 Rumex paucifolia Nutt.;高山红花酸模;Alpine Sheep Sorrel ■☆

340193 Rumex paucifolia Nutt. var. gracilescens Rech. = Rumex paucifolia Nutt. ■☆

340194 Rumex paulseniana Rech. f.;保尔森酸模■☆

340195 Rumex persicarioides L.;蔊蓄酸模■☆

340196 Rumex persicarioides L. = Rumex maritima L. ■

340197 Rumex picta Forssk.;着色酸模■☆

340198 Rumex picta Forssk. subsp. bipinnata (L. f.) Maire = Rumex bipinnata L. f. ■☆

340199 Rumex planivalvis Murb. = Rumex simpliciflora Murb. ■☆

340200 Rumex planivalvis Murb. var. hirtivalvis Maire = Rumex simpliciflora Murb. var. hirtivalvis (Maire) Sam. ■☆

340201 Rumex polyrrhiza Greene = Rumex densiflora Osterh. ■☆

340202 Rumex popovii Pachom.;中亚酸模;Central Asia Dock ■

340203 Rumex praecox Rydb.;早熟酸模;Early Dock ■☆

340204 Rumex pratensis Mert. et W. D. J. Koch = Rumex acuta L. ■☆

340205 Rumex protracta Rech. f. = Rumex aquatica L. ■

340206 Rumex pseudoalpina Höfft = Rumex alpina L. ■☆

340207 Rumex pseudonatronata (Borbás) Borbás ex Murb. = Rumex pseudonatronata (Borbás) Murb. ■

340208 Rumex pseudonatronata (Borbás) Borbás ex Murb. subsp. fennica Murb. = Rumex pseudonatronata (Borbás) Borbás ex Murb. ■

340209 Rumex pseudonatronata (Borbás) Murb.;披针叶酸模(欧酸模);Field Dock, Finnish Dock, Lanceolateleaf Dock ■

340210 Rumex pseudonatronata (Borbás) Murb. = Rumex pseudonatronata (Borbás) Borbás ex Murb. ■

340211 Rumex pseudonatronata (Borbás) Murb. subsp. fennica Murb. = Rumex pseudonatronata (Borbás) Murb. ■

340212 Rumex pseudonatronata Borbás ex Rech. = Rumex pseudonatronata (Borbás) Borbás ex Murb. ■

340213 Rumex pseudoscutata Dinter;长圆盾酸模■☆

340214 Rumex pulcher L.;美酸模(美丽酸模);Bloody Dock, Fiddle Dock, Fiddle-leaved Flock ■☆

340215 Rumex pulcher L. subsp. anodonta (Hausskn.) Rech. f.;无齿美酸模■☆

340216 Rumex pulcher L. subsp. divaricata (L.) Murb.;叉开酸模■☆

340217 Rumex pulcher L. subsp. woodsii (De Not.) Arcang.;伍兹酸模■☆

340218 Rumex pycnantha Rech. f.;密花酸模■☆

340219 Rumex quadrangulivalvis (Danser) Rech. f. = Rumex salicifolia Weinm. var. mexicana (Meisn.) C. L. Hitchc. ■☆

340220 Rumex quarrei De Wild. = Rumex nepalensis Spreng. ■

340221 Rumex rechingeriana Losinsk.;红干酸模(新疆酸模)■

340222 Rumex rechingeriana Losinsk. = Rumex patientia L. ■

340223 Rumex regelii F. Schmidt = Rumex japonica Houtt. ■

340224 Rumex remotiflora Sam. = Rumex nepalensis Spreng. var. remotiflora (Sam.) A. J. Li

340225 Rumex reticulata Besser;网酸模■☆

340226 Rumex rhodesia Rech. f.;罗得西亚酸模■☆

340227 Rumex rosea L. = Rumex cyprius Murb. subsp. subinteger Sam. ■☆

340228 Rumex rosea L. var. integra Maire = Rumex rosea L. ■☆

340229 Rumex rossica Murb. = Rumex maritima L. ■

340230 Rumex rothschildiana Aarons.;罗思酸模■☆

340231 Rumex rugelii Meisn. = Rumex obtusifolia L. ■

340232 Rumex rupestris Le Gall;岩生酸模;Shore Dock ■☆

340233　Rumex ruwenzoriensis Chiov. ;鲁文佐里酸模■☆

340234　Rumex sagittata Thunb. ;箭形酸模■☆

340235　Rumex sagittata Thunb. var. latilobus Meisn. = Rumex sagittata Thunb. ■☆

340236　Rumex sagittata Thunb. var. megalotys Meisn. = Rumex sagittata Thunb. ■☆

340237　Rumex salicifolia Weinm. ; 柳叶酸模;Willow Dock, Willow-leaved Dock ■☆

340238　Rumex salicifolia Weinm. subsp. triangulivalvis Danser = Rumex salicifolia Weinm. var. mexicana（Meisn.）C. L. Hitchc. ■☆

340239　Rumex salicifolia Weinm. subsp. triangulivalvis Danser = Rumex triangulivalvis（Danser）Rech. f. ■☆

340240　Rumex salicifolia Weinm. var. angustifolia Meisn. = Rumex salicifolia Weinm. var. mexicana（Meisn.）C. L. Hitchc. ■☆

340241　Rumex salicifolia Weinm. var. angustifolia Meisn. = Rumex sibirica Hultén ■☆

340242　Rumex salicifolia Weinm. var. denticulata Torr. = Rumex californica Rech. f. ■☆

340243　Rumex salicifolia Weinm. var. lacustris（Greene）J. C. Hickman = Rumex lacustris Greene ■☆

340244　Rumex salicifolia Weinm. var. mexicana（Meisn.）C. L. Hitchc. ;墨西哥柳叶酸模;Triangular-valved Dock, White Dock, White Willow Dock, Willow Dock, Willow-leaved Dock ■☆

340245　Rumex salicifolia Weinm. var. transitoria（Rchb. f.）J. C. Hickman = Rumex transitoria Rech. f. ■☆

340246　Rumex salicifolia Weinm. var. triangulivalvis（Danser）Hickman = Rumex salicifolia Weinm. var. mexicana（Meisn.）C. L. Hitchc. ■☆

340247　Rumex salicifolia Weinm. var. triangulivalvis（Danser）J. C. Hickman = Rumex triangulivalvis（Danser）Rech. f. ■☆

340248　Rumex salina A. Nelson = Rumex hymenosepala Torr. ■

340249　Rumex sanguinea L. ;血红酸模;Blethard, Blood-veined Dock, Bloodwort, Bloodwort Dock, Bloody Dock, French Sorrel, Red Dock, Redvein Dock, Red-veined Dock, Red-vien Dock, Wood Dock ■☆

340250　Rumex sarcorhiza Link = Rumex cordata Poir. ■☆

340251　Rumex saxei Kellogg = Rumex hymenosepala Torr. ■

340252　Rumex scandens Burch. = Rumex sagittata Thunb. ■☆

340253　Rumex schischkinii Losinsk. ;希施酸模■☆

340254　Rumex scutata L. ;盾状酸模;Buckler-shaped Sorrel, French Sorrel, French Spinach, Rubble Dock, Shield Dock, Sorrel ■

340255　Rumex scutata L. subsp. indurata（Boiss. et Reut.）Maire et Weiller = Rumex indurata Boiss. et Reut. ■☆

340256　Rumex sibirica Hultén;西伯利亚酸模;Siberian Dock, Siberian Willow Dock ■☆

340257　Rumex similans Rech. f. ;蒙新酸模■

340258　Rumex similans Rech. f. = Rumex marschalliana Rchb. var. brevidens Bong. et C. A. Mey. ■

340259　Rumex simpliciflora Murb. ;单花酸模■☆

340260　Rumex simpliciflora Murb. subsp. libyca（Murb.）Sam. ;利比亚酸模■

340261　Rumex simpliciflora Murb. subsp. murbeckii Maire et Weiller = Rumex simpliciflora Murb. ■☆

340262　Rumex simpliciflora Murb. var. hirtivalvis（Maire）Sam. ;刺果片单花酸模■☆

340263　Rumex simpliciflora Murb. var. maderensis ? = Rumex simpliciflora Murb. ■☆

340264　Rumex simpliciflora Murb. var. planivalvis（Murb.）Sam. = Rumex simpliciflora Murb. ■☆

340265　Rumex simpliciflora Murb. var. subdentata Maire = Rumex simpliciflora Murb. ■☆

340266　Rumex spathulata Thunb. ;匙形酸模■☆

340267　Rumex spinosa L. = Emex spinosa（L.）Campd. ■☆

340268　Rumex spiralis Small;螺旋酸模;Spiral Tall Dock ■☆

340269　Rumex stenophylla Ledeb. ;狭叶酸模（牛西西,窄叶酸模）;Dock, Narrowleaf Dock, Narrow-leaved Dock ■

340270　Rumex stenophylla Ledeb. var. ussuriensis（Losinsk.）Kitag. ;乌苏里酸模;Ussuri Dock ■

340271　Rumex stenophylla Ledeb. var. ussuriensis（Losinsk.）Kitag. = Rumex stenophylla Ledeb. ■

340272　Rumex steudeliana Meisn. = Rumex steudelii Hochst. ex A. Rich. ■☆

340273　Rumex steudelii Hochst. ex A. Rich. ;斯托酸模■☆

340274　Rumex subalpina M. E. Jones = Rumex pycnantha Rech. f. ■☆

340275　Rumex subarctica Lepage; 亚北极酸模;Subarctic Dock, Subarctic Willow Dock ■☆

340276　Rumex subarctica Lepage = Rumex salicifolia Weinm. var. mexicana（Meisn.）C. L. Hitchc. ■☆

340277　Rumex syriaca Meisn. ;叙利亚酸模■☆

340278　Rumex tenuifolia（Wallr.）Á. Löve;细叶酸模;Slender Sheep's Sorrel ■☆

340279　Rumex thyrsiflora Fingerh. ;直根酸模（长根酸模,圆锥花酸模）;Garden Sorrel, Narrow-leaved Sorrel, Taproot Dock ■☆

340280　Rumex thyrsiflora Fingerh. var. mandshurica A. I. Baranov et Skvortsov = Rumex thyrsiflora Fingerh. ■☆

340281　Rumex thyrsiflora Fingerh. var. mandshurica A. I. Baranov et Skvortsov;东北酸模■

340282　Rumex thyrsoides Desf. ;聚伞酸模■☆

340283　Rumex thyrsoides Desf. subsp. intermedia（DC.）Maire et Weiller = Rumex intermedia DC. ■☆

340284　Rumex thyrsoides Desf. var. sagittata Batt. = Rumex thyrsoides Desf. ■☆

340285　Rumex tianschanica Losinsk. ;天山酸模;Tianshan Dock ■

340286　Rumex tingitana L. = Rumex rosea L. ■☆

340287　Rumex tingitana L. var. lacerus Boiss. = Rumex rosea L. ■☆

340288　Rumex tomentella Rech. f. ;毛酸模■☆

340289　Rumex transitoria Rech. f. ;太平洋酸模;Pacific Willow Dock ■☆

340290　Rumex triangulivalvis（Danser）Rech. f. = Rumex salicifolia Weinm. ■☆

340291　Rumex triangulivalvis（Danser）Rech. f. = Rumex salicifolia Weinm. var. mexicana（Meisn.）C. L. Hitchc. ■☆

340292　Rumex triangulivalvis（Danser）Rech. f. var. oreolapathum Rech. f. = Rumex salicifolia Weinm. var. mexicana（Meisn.）C. L. Hitchc. ■☆

340293　Rumex trisetifer Stokes;长刺酸模;Longthorn Dock ■

340294　Rumex tuberosa L. ;块根酸模■☆

340295　Rumex tuberosa Thunb. = Rumex cordata Poir. ■☆

340296　Rumex tunetana Barratte et Murb. ;图内特酸模■☆

340297　Rumex ucranica Fisch. = Rumex urcanica Fisch. ex Spreng. ■

340298　Rumex ucranica Fisch. ex Spreng. ;乌克兰酸模;Ukraine Dock ■

340299　Rumex ursina M. M. Maximova = Rumex arctica Trautv. ■☆

340300　Rumex usambarensis（Dammer）Dammer;乌桑巴拉酸模☆

340301　Rumex ussuriensis Losinsk. = Rumex stenophylla Ledeb. var. ussuriensis（Losinsk.）Kitag. ■

340302　Rumex ussuriensis Losinsk. = Rumex stenophylla Ledeb. ■

340303　Rumex utahensis Rech. f. ;犹他州酸模;Utah Willow Dock ■☆

340304 Rumex venosa Pursh;多脉酸模;Sour Greens,Veined Dock,Veiny Dock,Wild Begonia,Wild Hydrangea,Wild-begonia,Winged Dock ■☆

340305 Rumex verticillata L. ;湿地酸模;Swamp Dock,Water Dock ■☆

340306 Rumex verticillata L. subsp. fascicularis (Small) Á. Löve = Rumex fascicularis Small ■☆

340307 Rumex verticillata L. subsp. floridana (Meisn.) Á. Löve = Rumex floridana Meisn. ■☆

340308 Rumex vesceritensis Murb. = Rumex cyprius Murb. subsp. vesceritensis (Murb.) Sam. ■☆

340309 Rumex vesceritensis Murb. var. papillosa Maire = Rumex cyprius Murb. subsp. vesceritensis (Murb.) Sam. ■☆

340310 Rumex vesicaria L. ;泡囊酸模■☆

340311 Rumex vesicaria L. subsp. simpliciflora (Murb.) Maire = Rumex simpliciflora Murb. ■☆

340312 Rumex vesicaria L. var. rhodophysa Ball = Rumex vesicaria L. ■☆

340313 Rumex vesicaria L. var. subdentata Maire = Rumex vesicaria L. ■☆

340314 Rumex violascens Rech. f. ;堇色酸模;Violet Dock ■☆

340315 Rumex wallichiana Meisn. = Rumex microcarpa Campd. ■

340316 Rumex wallichii Meisn. = Rumex microcarpa Campd. ■

340317 Rumex woodii N. E. Br. ;伍得酸模■☆

340318 Rumex yezoensis H. Hara = Rumex japonica Houtt. ■

340319 Rumex yungningensis Sam. ;永宁酸模;Yongning Dock ■

340320 Rumfordia DC. (1836);翼柄菊属■●☆

340321 Rumfordia floribunda DC. ;翼柄菊■●☆

340322 Rumia Hoffm. (1816);鲁米草属■☆

340323 Rumia crithmifolia (Willd.) Koso-Pol. ;鲁米草■☆

340324 Rumia seseloides Hoffm. = Saposhnikovia divaricata (Turcz.) Schischk. ■

340325 Rumicaceace Durand = Polygonaceae Juss. (保留科名)●■

340326 Rumicaceae Martinov = Polygonaceae Juss. (保留科名)●■

340327 Rumicastrum Ulbr. (1934);离柱马齿苋属■☆

340328 Rumicastrum chamaecladum (Diels) Ulbr. ;离柱马齿苋■☆

340329 Rumicicarpus Chiov. = Triumfetta Plum. ex L. ●■

340330 Rumicicarpus ramosissimus Chiov. = Triumfetta trigona Sprague et Hutch. ■☆

340331 Ruminia Parl. = Leucojum L. ■●

340332 Rumpfia L. = Rumphia L. ●

340333 Rumphia L. = Croton L. ●

340334 Rumputris Raf. = Cassytha L. ■●

340335 Runcina Allem. = Cenchrus L. ●

340336 Rungia Nees (1832);孩儿草属(明萼草属);Childgrass, Rungia ■

340337 Rungia axiliflora H. S. Lo;腋花孩儿草(孩儿草); Axillaryflower Childgrass,Axillaryflower Rungia ■

340338 Rungia baumannii Lindau =Justicia tenella (Nees) T. Anderson ■☆

340339 Rungia bisaccata D. Fang et H. S. Lo;囊花孩儿草■

340340 Rungia buettneri Lindau;比特纳孩儿草■

340341 Rungia caespitosa Lindau;丛生孩儿草■

340342 Rungia camerunensis Champl. ;喀麦隆孩儿草■

340343 Rungia cantonensis F. C. How;广州孩儿草■

340344 Rungia chinensis Benth. ; 中华孩儿草(明萼草);China Childgrass,Chinese Rungia ■

340345 Rungia congoensis C. B. Clarke;刚果孩儿草■☆

340346 Rungia densiflora H. S. Lo;密花孩儿草;Denseflower Childgrass,Denseflower Rungia ■

340347 Rungia dimorpha S. Moore;二型孩儿草■

340348 Rungia eriostachya Hua;毛穗孩儿草■☆

340349 Rungia grandis T. Anderson;大孩儿草■☆

340350 Rungia guangxiensis H. S. Lo et D. Fang;广西孩儿草■

340351 Rungia guineensis Heine;几内亚孩儿草■☆

340352 Rungia henry C. B. Clarke;南鼠尾黄;Henry Rungia ■

340353 Rungia hirpex Benoist;金沙鼠尾黄;Jinshajiang Rungia ■

340354 Rungia letestui Benoist = Adhatoda letestui (Benoist) Heine ■☆

340355 Rungia longipes D. Fang et H. S. Lo;长柄孩儿草■

340356 Rungia mina H. S. Lo;矮孩儿草;Mini Childgrass,Mini Rungia ■

340357 Rungia napoensis D. Fang et H. S. Lo;那坡孩儿草■

340358 Rungia obcordata Lindau = Ascotheca paucinervia (T. Anderson ex C. B. Clarke) Heine ■☆

340359 Rungia obcordata Lindau var. obtusa Benoist = Ascotheca paucinervia (T. Anderson ex C. B. Clarke) Heine ■☆

340360 Rungia parviflora Nees = Rungia pectinata (L.) Nees ■

340361 Rungia parviflora Nees sensu Hemsl. = Rungia pectinata (L.) Nees ■

340362 Rungia parviflora Nees sensu Masam. = Rungia taiwanensis T. Yamaz. ■

340363 Rungia parviflora Nees var. pectinata (L.) C. B. Clarke = Rungia pectinata (L.) Nees ■

340364 Rungia parviflora Nees var. pectinata sensu Matsum. et Hayata = Rungia taiwanensis T. Yamaz. ■

340365 Rungia paucinervia (T. Anderson ex C. B. Clarke) Heine = Ascotheca paucinervia (T. Anderson ex C. B. Clarke) Heine ■☆

340366 Rungia paxiana (Lindau) C. B. Clarke;帕克斯孩儿草■

340367 Rungia pectinata (L.) Nees;孩儿草(痒积草,黄蜂草,火炭草,积药草,节节红,蓝色草,明萼草,鼠尾黄,土夏枯草,由甲草,由甲草);Childgrass, Pectinate Rungia ■

340368 Rungia pectinata (L.) Nees var. clarkeana Hand. -Mazz. = Rungia pectinata (L.) Nees ■

340369 Rungia pinpienensis H. S. Lo;屏边孩儿草; Pingbian Childgrass,Pingbian Rungia ■

340370 Rungia pobeguinii Hutch. et Dalziel = Rungia eriostachya Hua ■☆

340371 Rungia pubinervia T. Anderson = Metarungia pubinervia (T. Anderson) Baden ●☆

340372 Rungia pungens D. Fang et H. S. Lo;尖苞孩儿草■

340373 Rungia robusta C. B. Clarke;粗壮鼠尾黄;Robust Rungia ■

340374 Rungia rosacea Lindau = Justicia trifolioides T. Anderson ■☆

340375 Rungia schliebenii Mildbr. ;施利本孩儿草■☆

340376 Rungia stolonifera C. B. Clarke;匍匐鼠尾黄(毛叶孩儿草); Stolonifer Rungia ■

340377 Rungia taiwanensis T. Yamaz. ;台湾明萼草■

340378 Rungia yunnanensis H. S. Lo;云南孩儿草(云南鼠尾黄); Yunnan Childgrass,Yunnan Rungia ■

340379 Runyonia Rose = Agave L. ■

340380 Runyonia longiflora Rose = Manfreda longiflora (Rose) Verh. -Will. ■☆

340381 Rupala Vahl = Roupala Aubl. ●☆

340382 Rupalleya Morière = Dichelostemma Kunth ■☆

340383 Rupalleya Moriere = Stropholirion Torr. ■☆

340384 Rupalleya volubilis (Kellogg) Morière = Dichelostemma volubile (Kellogg) A. Heller ■☆

340385 Rupertia J. W. Grimes = Psoralea L. ●■

340386 Rupertia J. W. Grimes(1990);北美补骨脂属●■☆

340387 Rupestrina Prov. = Trisetum Pers. ■

340388 Rupicapnos Pomel(1860);岩堇属■☆

340389　Rupicapnos africana（Lam.）Pomel；非洲岩堇■☆

340390　Rupicapnos africana（Lam.）Pomel subsp. ambigua（Pugsley）Emb. et Maire ＝ Rupicapnos africana（Lam.）Pomel subsp. mairei（Pugsley）Maire ■☆

340391　Rupicapnos africana（Lam.）Pomel subsp. argentea Pugsley ＝ Rupicapnos africana（Lam.）Pomel subsp. faurei（Pugsley）Maire ■☆

340392　Rupicapnos africana（Lam.）Pomel subsp. cerefolia（Pomel）Maire；蜡叶岩堇■☆

340393　Rupicapnos africana（Lam.）Pomel subsp. corymbosa（Desf.）Maire ＝ Rupicapnos africana（Lam.）Pomel ■☆

340394　Rupicapnos africana（Lam.）Pomel subsp. decipiens（Pugsley）Maire；迷惑岩堇■☆

340395　Rupicapnos africana（Lam.）Pomel subsp. elegans（Pugsley）Emb. et Maire ＝ Rupicapnos africana（Lam.）Pomel subsp. faurei（Pugsley）Maire ■☆

340396　Rupicapnos africana（Lam.）Pomel subsp. faurei（Pugsley）Maire；福雷岩堇■☆

340397　Rupicapnos africana（Lam.）Pomel subsp. fraterna（Pugsley）Maire ＝ Rupicapnos africana（Lam.）Pomel subsp. gaetula（Maire）Maire ■☆

340398　Rupicapnos africana（Lam.）Pomel subsp. gaetula（Maire）Maire；盖图拉岩堇■☆

340399　Rupicapnos africana（Lam.）Pomel subsp. gracilescens Maire et Weiller ＝ Rupicapnos africana（Lam.）Pomel subsp. gaetula（Maire）Maire ■☆

340400　Rupicapnos africana（Lam.）Pomel subsp. graciliflora（Pomel）Maire ＝ Rupicapnos africana（Lam.）Pomel ■☆

340401　Rupicapnos africana（Lam.）Pomel subsp. mairei（Pugsley）Maire；迈氏岩堇■☆

340402　Rupicapnos africana（Lam.）Pomel subsp. ochracea（Pomel）Maire ＝ Rupicapnos ochracea Pomel ■☆

340403　Rupicapnos africana（Lam.）Pomel subsp. oranensis（Pugsley）Maire；奥兰岩堇■☆

340404　Rupicapnos africana（Lam.）Pomel subsp. parvicalcarata（Pugsley）Maire ＝ Rupicapnos africana（Lam.）Pomel subsp. gaetula（Maire）Maire ■☆

340405　Rupicapnos africana（Lam.）Pomel subsp. platycentra（Pomel）Maire ＝ Rupicapnos africana（Lam.）Pomel subsp. cerefolia（Pomel）Maire ■☆

340406　Rupicapnos africana（Lam.）Pomel subsp. pomeliana（Pugsley）Maire；波梅尔岩堇■☆

340407　Rupicapnos africana（Lam.）Pomel subsp. rifana（Pugsley）Maire ＝ Rupicapnos africana（Lam.）Pomel subsp. decipiens（Pugsley）Maire ■☆

340408　Rupicapnos africana（Lam.）Pomel subsp. speciosa（Pomel）Maire ＝ Rupicapnos africana（Lam.）Pomel subsp. mairei（Pugsley）Maire ■☆

340409　Rupicapnos africana（Lam.）Pomel subsp. splendens（Pugsley）Emb. et Maire ＝ Rupicapnos africana（Lam.）Pomel subsp. mairei（Pugsley）Maire ■☆

340410　Rupicapnos africana（Lam.）Pomel var. battandieri Pugsley ＝ Rupicapnos ochracea Pomel ■☆

340411　Rupicapnos africana（Lam.）Pomel var. colorata Pugsley ＝ Rupicapnos africana（Lam.）Pomel subsp. decipiens（Pugsley）Maire ■☆

340412　Rupicapnos africana（Lam.）Pomel var. dubuisii Maire ＝ Rupicapnos africana（Lam.）Pomel subsp. cerefolia（Pomel）Maire ■☆

340413　Rupicapnos africana（Lam.）Pomel var. gracilis Pugsley ＝ Rupicapnos africana（Lam.）Pomel ■☆

340414　Rupicapnos africana（Lam.）Pomel var. mauritanica Pugsley ＝ Rupicapnos africana（Lam.）Pomel ■☆

340415　Rupicapnos africana（Lam.）Pomel var. mesoceras Maire ＝ Rupicapnos africana（Lam.）Pomel ■☆

340416　Rupicapnos africana（Lam.）Pomel var. minor Pugsley ＝ Rupicapnos africana（Lam.）Pomel ■☆

340417　Rupicapnos africana（Lam.）Pomel var. nedromensis Pugsley ＝ Rupicapnos africana（Lam.）Pomel subsp. faurei（Pugsley）Maire ■☆

340418　Rupicapnos ambigua Pugsley ＝ Rupicapnos africana（Lam.）Pomel subsp. mairei（Pugsley）Maire ■☆

340419　Rupicapnos calcarata Lidén；距岩堇■☆

340420　Rupicapnos cerefolia Pomel ＝ Rupicapnos africana（Lam.）Pomel subsp. cerefolia（Pomel）Maire ■☆

340421　Rupicapnos corymbosa（Desf.）Pomel ＝ Rupicapnos africana（Lam.）Pomel ■☆

340422　Rupicapnos erosa Pomel ＝ Rupicapnos ochracea Pomel ■☆

340423　Rupicapnos gaetula（Maire）Pugsley ＝ Rupicapnos africana（Lam.）Pomel subsp. gaetula（Maire）Maire ■☆

340424　Rupicapnos graciliflora Pomel ＝ Rupicapnos africana（Lam.）Pomel ■☆

340425　Rupicapnos longipes（Coss. et Durieu）Pomel；长梗岩堇■☆

340426　Rupicapnos longipes（Coss. et Durieu）Pomel subsp. aurasiaca Lidén；奥拉斯岩堇■☆

340427　Rupicapnos longipes（Coss. et Durieu）Pomel subsp. reboudiana（Pomel）Lidén；雷博岩堇■☆

340428　Rupicapnos muricaria Pomel；短尖岩堇■☆

340429　Rupicapnos numidica（Coss. et Durieu）Pomel subsp. aurasiaca Maire ＝ Rupicapnos longipes（Coss. et Durieu）Pomel subsp. aurasiaca Lidén ■☆

340430　Rupicapnos numidica（Coss. et Durieu）Pomel subsp. caput-plataleae（Pomel）Maire ＝ Rupicapnos numidica（Coss. et Durieu）Pomel subsp. delicatula（Pomel）Maire ■☆

340431　Rupicapnos numidica（Coss. et Durieu）Pomel subsp. cossonii Pugsley ＝ Rupicapnos numidica（Coss. et Durieu）Pomel ■☆

340432　Rupicapnos numidica（Coss. et Durieu）Pomel subsp. delicatula（Pomel）Maire；姣美岩堇■☆

340433　Rupicapnos numidica（Coss. et Durieu）Pomel subsp. erosa（Pomel）Maire ＝ Rupicapnos ochracea Pomel ■☆

340434　Rupicapnos numidica（Coss. et Durieu）Pomel subsp. longipes（Coss. et Durieu）Maire et Weiller ＝ Rupicapnos longipes（Coss. et Durieu）Pomel ■☆

340435　Rupicapnos numidica（Coss. et Durieu）Pomel subsp. praetermissa Pugsley ＝ Rupicapnos longipes（Coss. et Durieu）Pomel ■☆

340436　Rupicapnos numidica（Coss. et Durieu）Pomel subsp. reboudiana Pomel ＝ Rupicapnos longipes（Coss. et Durieu）Pomel subsp. reboudiana（Pomel）Lidén ■☆

340437　Rupicapnos numidica（Coss. et Durieu）Pomel subsp. sarcocapnoides（Coss. et Durieu）Maire ＝ Rupicapnos sarcocapnoides（Coss. et Durieu）Pomel ■☆

340438　Rupicapnos numidica（Coss. et Durieu）Pomel subsp. sublaevis Pugsley ＝ Rupicapnos longipes（Coss. et Durieu）Pomel ■☆

340439　Rupicapnos numidica（Coss. et Durieu）Pomel subsp. tenuifolia Pomel ＝ Rupicapnos numidica（Coss. et Durieu）Pomel subsp. delicatula（Pomel）Maire ■☆

340440　Rupicapnos numidica（Coss. et Durieu）Pomel var. elkantarica Pugsley = Rupicapnos longipes（Coss. et Durieu）Pomel subsp. aurasiaca Lidén■☆

340441　Rupicapnos numidica（Coss. et Durieu）Pomel var. major Pugsley = Rupicapnos numidica（Coss. et Durieu）Pomel ■☆

340442　Rupicapnos numidica（Coss. et Durieu）Pomel var. pulchella Pugsley = Rupicapnos longipes（Coss. et Durieu）Pomel subsp. reboudiana（Pomel）Lidén■☆

340443　Rupicapnos ochracea Pomel；淡黄褐岩堇■☆

340444　Rupicapnos parvicalcarata Pugsley = Rupicapnos africana（Lam.）Pomel subsp. gaetula（Maire）Maire■☆

340445　Rupicapnos platycentra Pomel = Rupicapnos africana（Lam.）Pomel subsp. cerefolia（Pomel）Maire■☆

340446　Rupicapnos reboudiana Pomel = Rupicapnos longipes（Coss. et Durieu）Pomel subsp. reboudiana（Pomel）Lidén■☆

340447　Rupicapnos sarcocapnoides（Coss. et Durieu）Pomel；烟岩堇■☆

340448　Rupicapnos speciosa Pomel = Rupicapnos africana（Lam.）Pomel ■☆

340449　Rupicapnos splendens Pugsley = Rupicapnos africana（Lam.）Pomel subsp. mairei（Pugsley）Maire■☆

340450　Rupicapnos tenuifolia Pomel = Rupicapnos numidica（Coss. et Durieu）Pomel subsp. delicatula（Pomel）Maire■☆

340451　Rupichloa Salariato et Morrone(2009)；巴西岩禾属■☆

340452　Rupichloa acuminata（Renvoize）Salariato et Morrone = Streptostachys acuminata Renvoize■☆

340453　Rupichloa decidua（Morrone et Zuloaga）Salariato et Morrone；巴西岩禾■☆

340454　Rupichloa decidua（Morrone et Zuloaga）Salariato et Morrone = Urochloa decidua Morrone et Zuloaga■☆

340455　Rupicola Maiden = Rupicola Maiden et Betche●☆

340456　Rupicola Maiden et Betche(1899)；竹柏石南属●☆

340457　Rupicola ciliata I. Telford；缘毛竹柏石南●☆

340458　Rupicola sprengelioides Maiden et Betche；竹柏石南●☆

340459　Rupifraga Raf. = Sekika Medik.

340460　Rupifraga sarmentosa（L. f.）Raf. = Saxifraga stolonifera Curtis

340461　Rupiphila Pimenov et Lavrova = Ligusticum L. ■

340462　Rupiphila Pimenov et Lavrova(1986)；岩茴香属■

340463　Rupiphila tachiroei（Franch. et Sav.）Pimenov et Lavrova；岩茴香（柏子三七,桂花三七,火藥本,细叶藁本）；Rock Ligusticum■

340464　Rupiphila tachiroei（Franch. et Sav.）Pimenov et Lavrova = Ligusticum tachiroei（Franch. et Sav.）M. Hiroe et Constance■

340465　Ruppalleya Krause = Dichelostemma Kunth☆

340466　Ruppalleya Krause = Rupalleya Morière☆

340467　Ruppalleya Krause = Stropholirion Torr.☆

340468　Ruppelia Baker = Aeschynomene L. ●■

340469　Ruppelia Baker = Rueppelia A. Rich. ●■

340470　Ruppia L.（1753）；川蔓藻属（流苏菜属）；Ditch-grass, Ruppia, Tasselweed, Widgeon Grass, Widgeonweed■

340471　Ruppia brachypus Gay = Ruppia maritima L. ■

340472　Ruppia cirrhosa（Petagna）Grande；螺旋川蔓藻；Coiled Pondweed, Spiral Ditch-grass, Spiral Tasselweed■☆

340473　Ruppia cirrhosa（Petagna）Grande var. truncatifolia（Miki）H. Hara = Ruppia occidentalis S. Watson☆

340474　Ruppia drepanensis Guss. ；德雷帕农川蔓藻■☆

340475　Ruppia lacustris Macoun = Ruppia occidentalis S. Watson☆

340476　Ruppia maritima L. ；川蔓藻（海岸川蔓藻,流苏菜）；Beaked Tasselweed, Ditch Grass, Ditchgrass, Maritine Widgeonweed, Sea-grass, Tassel Pondweed, Widgeon Weed, Widgeongrass, Widgeonweed■

340477　Ruppia maritima L. subsp. brevirostris C. Agardh = Ruppia maritima L. ■

340478　Ruppia maritima L. subsp. drepanensis（Guss.）Maire et Weiller = Ruppia drepanensis Guss. ■☆

340479　Ruppia maritima L. subsp. rostellata（Koch ex Rchb.）Asch. et Graebn. = Ruppia maritima L. ■

340480　Ruppia maritima L. subsp. rostellata Koch = Ruppia maritima L. ■

340481　Ruppia maritima L. subsp. spiralis（L. ex Dumort.）Asch. et Graebn. = Ruppia cirrhosa（Petagna）Grande ■☆

340482　Ruppia maritima L. var. brevirostris Agardh = Ruppia maritima L. ■

340483　Ruppia maritima L. var. exigua Fernald et Wiegand = Ruppia maritima L. ■

340484　Ruppia maritima L. var. intermedia（Thed.）Asch. et Graebn. = Ruppia maritima L. ■

340485　Ruppia maritima L. var. longipes Hagstr. = Ruppia maritima L. ■

340486　Ruppia maritima L. var. obliqua（Schur）Asch. et Graebn. = Ruppia maritima L. ■

340487　Ruppia maritima L. var. occidentalis（S. Watson）Graebn. = Ruppia cirrhosa（Petagna）Grande ■☆

340488　Ruppia maritima L. var. rostrata J. Agardh = Ruppia maritima L. ■

340489　Ruppia maritima L. var. spiralis Morris = Ruppia cirrhosa（Petagna）Grande ■☆

340490　Ruppia maritima L. var. subcapitata Fernald et Wiegand = Ruppia maritima L. ■

340491　Ruppia occidentalis S. Watson；西方川蔓藻■☆

340492　Ruppia occidentalis S. Watson = Ruppia cirrhosa（Petagna）Grande ■☆

340493　Ruppia rostellata（J. Agardh）K. Koch = Ruppia maritima L. ■

340494　Ruppia rostellata K. Koch = Ruppia maritima L. ■

340495　Ruppia rostellata K. Koch ex Rchb. = Ruppia maritima L. ■

340496　Ruppia rostrata（C. Agardh）Pamp. = Ruppia maritima L. ■

340497　Ruppia spiralis L. = Ruppia cirrhosa（Petagna）Grande ■☆

340498　Ruppia spiralis L. ex Dumort. = Ruppia cirrhosa（Petagna）Grande ■☆

340499　Ruppia trichoides Durieu = Ruppia drepanensis Guss. ■☆

340500　Ruppia truncatifolia Miki = Ruppia occidentalis S. Watson ■☆

340501　Ruppia zosteroides Lojac. = Ruppia maritima L. ■

340502　Ruppiaceae Horan.（1834）（保留科名）；川蔓藻科（流苏菜科,蔓藻科）；Ditch-grass Family, Widgeonweed Family ■

340503　Ruppiaceae Horan.（保留科名）= Polygonaceae Juss.（保留科名）●■

340504　Ruppiaceae Horan.（保留科名）= Potamogetonaceae Bercht. et J. Presl(保留科名)■

340505　Ruppiaceae Horan. ex Hutch. = Ruppiaceae Horan.（保留科名）■

340506　Ruppiaceae Hutch. = Ruppiaceae Horan.（保留科名）■

340507　Ruprechtia C. A. Mey.（1840）；多花蓼树属（鲁氏蓼属）；Viraru●☆

340508　Ruprechtia Opiz = Thalictrum L. ■

340509　Ruprechtia Reichb. = Plinthus Fenzl●☆

340510　Ruprechtia albida Pendry；白多花蓼●☆

340511　Ruprechtia boliviensis Herzog；玻利维亚多花蓼●☆

340512　Ruprechtia brachysepala Meisn.；短萼多花蓼●☆

340513　Ruprechtia brachystachya Benth. ；短穗多花蓼●☆

340514　Ruprechtia laevigata Pendry；平滑多花蓼●☆

340515　Ruprechtia latifolia Huber;宽叶多花蓼●☆

340516　Ruprechtia laxiflora Meisn.;疏花多花蓼●☆

340517　Ruprechtia macrocalyx Huber;大萼多花蓼●☆

340518　Ruprechtia nitida Brandbyge;光亮多花蓼●☆

340519　Ruprechtia obovata Pendry;倒卵多花蓼●☆

340520　Ruprechtia occidentalis Standl.;西方多花蓼●☆

340521　Ruprechtia oxyphylla S. F. Blake;尖叶多花蓼●☆

340522　Ruprechtia pallida Standl.;苍白多花蓼●☆

340523　Ruprechtia polystachya Griseb.;多穗多花蓼●☆

340524　Ruprechtia salicifolia C. A. Mey.;柳叶多花蓼●☆

340525　Ruprechtia splendens Standl. ex Reko;纤细多花蓼●☆

340526　Ruprechtia triflora Griseb.;三花多花蓼●☆

340527　Ruptfraga (Stemb.) Raf. = Saxifraga L. ■

340528　Ruptiliocarpon Hammel et N. Zamora(1993);无梗鳞球穗属●☆

340529　Ruptiliocarpon caracolito Hammel et N. A. Zamora;无梗鳞球穗●☆

340530　Rurea Post et Kuntze = Rourea Aubl.(保留属名)●

340531　Rureopsis Post et Kuntze = Rourea Aubl.(保留属名)●

340532　Rureopsis Post et Kuntze = Roureopsis Planch. ●

340533　Rusaea J. F. Gruel. = Roussea Sm. ●☆

340534　Rusbya Britton(1893);杉叶莓属●☆

340535　Rusbya boliviana Britton;杉叶莓●☆

340536　Rusbyanthus Gilg = Macrocarpaea (Griseb.) Gilg ●☆

340537　Rusbyanthus cinchonifolius Gilg = Macrocarpaea cinchonifolia (Gilg) Weaver ●☆

340538　Rusbyella Rolfe = Rusbyella Rolfe ex Rusby ■☆

340539　Rusbyella Rolfe ex Rusby(1896);鲁斯兰属■☆

340540　Rusbyella caespitosa Rolfe;鲁斯兰■☆

340541　Ruscaceae M. Roem. (1840)(保留科名);假叶树科;Butchersbroom Family ●

340542　Ruscaceae M. Roem.(保留科名) = Asparagaceae Juss.(保留科名)■●

340543　Ruscaceae M. Roem.(保留科名) = Rutaceae Juss.(保留科名)●■

340544　Ruscaceae Spreng. = Asparagaceae Juss.(保留科名)■●

340545　Ruscaceae Spreng. = Ruscaceae M. Roem.(保留科名)●

340546　Ruscaceae Spreng. ex Hutch. = Ruscaceae M. Roem.(保留科名)●

340547　Ruschia Schwantes = Mesembryanthemum L. emend. N. E. Br. ●☆

340548　Ruschia Schwantes(1926);舟叶花属;Purple Dew-plant ●■☆

340549　Ruschia abbreviata L. Bolus;缩短舟叶花●☆

340550　Ruschia abrupta (A. Berger) G. D. Rowley = Octopoma abruptum (A. Berger) N. E. Br. ●☆

340551　Ruschia acocksii L. Bolus;阿氏舟叶花●☆

340552　Ruschia acuminata L. Bolus;渐尖舟叶花●☆

340553　Ruschia acutangula (Haw.) Schwantes;棱角舟叶花●☆

340554　Ruschia aggregata L. Bolus;团集舟叶花●☆

340555　Ruschia alata L. Bolus;翅舟叶花●☆

340556　Ruschia albertensis L. Bolus = Ruschia spinosa (L.) Dehn ●☆

340557　Ruschia albiflora L. Bolus = Polymita albiflora (L. Bolus) L. Bolus ■☆

340558　Ruschia alborubra L. Bolus = Antimima alborubra (L. Bolus) Dehn ■☆

340559　Ruschia amicorum (L. Bolus) Schwantes;可爱舟叶花●☆

340560　Ruschia amoena Schwantes = Antimima amoena (Schwantes) H. E. K. Hartmann ■☆

340561　Ruschia amphibolia G. D. Rowley = Phiambolia hallii (L. Bolus) Klak ■☆

340562　Ruschia ampliata L. Bolus;膨大舟叶花●☆

340563　Ruschia androsacea Marloth et Schwantes = Antimima androsacea (Marloth et Schwantes) H. E. K. Hartmann ■☆

340564　Ruschia approximata (L. Bolus) Schwantes;相似舟叶花●☆

340565　Ruschia archeri L. Bolus;阿谢尔舟叶花●☆

340566　Ruschia archeri L. Bolus var. sexpartita ? = Ruschia archeri L. Bolus ●☆

340567　Ruschia arenosa L. Bolus = Lampranthus arenarius H. E. K. Hartmann ■☆

340568　Ruschia argentea L. Bolus = Antimima argentea (L. Bolus) H. E. K. Hartmann ■☆

340569　Ruschia aristata L. Bolus = Erepsia aristata (L. Bolus) Liede et H. E. K. Hartmann ●☆

340570　Ruschia aristulata (Sond.) Schwantes = Antimima aristulata (Sond.) Chess. et G. F. Sm. ■☆

340571　Ruschia armata L. Bolus = Arenifera stylosa (L. Bolus) H. E. K. Hartmann ■☆

340572　Ruschia aspera L. Bolus;粗糙舟叶花●☆

340573　Ruschia atrata L. Bolus;黑舟叶花●☆

340574　Ruschia barnardii L. Bolus;巴纳德舟叶花■☆

340575　Ruschia beaufortensis L. Bolus;博福特舟叶花●☆

340576　Ruschia bicolorata L. Bolus = Stoeberia beetzii (Dinter) Dinter et Schwantes ●☆

340577　Ruschia biformis (N. E. Br.) Schwantes = Antimima biformis (N. E. Br.) H. E. K. Hartmann ■☆

340578　Ruschia bina L. Bolus = Antimima viatorum (L. Bolus) Klak ■☆

340579　Ruschia bipapillata L. Bolus;双乳突舟叶花●☆

340580　Ruschia bolusiae Schwantes;博卢斯舟叶花●☆

340581　Ruschia bracteata L. Bolus = Antimima bracteata (L. Bolus) H. E. K. Hartmann ■☆

340582　Ruschia brakdamensis (L. Bolus) L. Bolus;布拉克达姆舟叶花■☆

340583　Ruschia brevibracteata L. Bolus;短苞舟叶花●☆

340584　Ruschia brevicarpa L. Bolus = Antimima brevicarpa (L. Bolus) H. E. K. Hartmann ■☆

340585　Ruschia brevicollis (N. E. Br.) Schwantes = Antimima brevicollis (N. E. Br.) H. E. K. Hartmann ■☆

340586　Ruschia brevicyma L. Bolus;短芽舟叶花●☆

340587　Ruschia brevifolia L. Bolus;短叶舟叶花●☆

340588　Ruschia brevipes L. Bolus;短梗舟叶花●☆

340589　Ruschia brevipes L. Bolus var. gracilis ? = Ruschia brevipes L. Bolus ●☆

340590　Ruschia britteniae L. Bolus;布里滕舟叶花■☆

340591　Ruschia buchubergensis Dinter = Antimima buchubergensis (Dinter) H. E. K. Hartmann ■☆

340592　Ruschia burtoniae L. Bolus;伯顿舟叶花●☆

340593　Ruschia calcarea L. Bolus;距舟叶花●☆

340594　Ruschia calcicola (L. Bolus) L. Bolus;钙生舟叶花●☆

340595　Ruschia callifera L. Bolus;花敷菊●☆

340596　Ruschia calycina L. Bolus = Zeuktophyllum calycinum (L. Bolus) H. E. K. Hartmann ●☆

340597　Ruschia campestris (Burch.) Schwantes;原野舟叶花●☆

340598　Ruschia canonotata (L. Bolus) Schwantes;线舟叶花●☆

340599　Ruschia capornii (L. Bolus) L. Bolus;凯波恩舟叶花●☆

340600　Ruschia capulata (L. Bolus) Schwantes = Ruschia cupulata (L. Bolus) Schwantes ●☆

340601　Ruschia caroli (L. Bolus) Schwantes;紫舟叶花;Purple Dew-plant ●☆

340602　Ruschia caudata L. Bolus;尾状舟叶花●☆

340603　Ruschia cedarbergensis L. Bolus;锡达伯格舟叶花●☆

340604　Ruschia ceresiana L. Bolus;塞里斯舟叶花●☆

340605　Ruschia ceresiana Schwantes = Ruschia ceresiana L. Bolus ●☆

340606　Ruschia cincta (L. Bolus) L. Bolus;围绕舟叶花■☆

340607　Ruschia clavata L. Bolus;棍棒舟叶花●☆

340608　Ruschia cleista L. Bolus = Eberlanzia sedoides (Dinter et A. Berger) Schwantes ●☆

340609　Ruschia compacta L. Bolus = Antimima compacta (L. Bolus) H. E. K. Hartmann ■☆

340610　Ruschia complanata L. Bolus;扁平舟叶花●☆

340611　Ruschia compressa L. Bolus = Antimima compressa (L. Bolus) H. E. K. Hartmann ■☆

340612　Ruschia comptonii L. Bolus = Drosanthemum pulverulentum (Haw.) Schwantes ●☆

340613　Ruschia concava L. Bolus = Antimima dasyphylla (Schltr.) H. E. K. Hartmann ■☆

340614　Ruschia concinna L. Bolus = Antimima concinna (L. Bolus) H. E. K. Hartmann ■☆

340615　Ruschia condensa (N. E. Br.) Schwantes = Antimima condensa (N. E. Br.) H. E. K. Hartmann ■☆

340616　Ruschia congesta (Salm-Dyck) L. Bolus;密集舟叶花●☆

340617　Ruschia congesta L. Bolus = Ruschia ceresiana L. Bolus ●☆

340618　Ruschia conjuncta L. Bolus = Octopoma connatum (L. Bolus) L. Bolus ●☆

340619　Ruschia connata L. Bolus = Octopoma connatum (L. Bolus) L. Bolus ●☆

340620　Ruschia constricta L. Bolus = Acrodon bellidiflorus (L.) N. E. Br. ■☆

340621　Ruschia copiosa L. Bolus;丰富舟叶花●☆

340622　Ruschia coriaria (Burch. ex N. E. Br.) Schwantes = Psilocaulon coriarium (Burch. ex N. E. Br.) N. E. Br. ■☆

340623　Ruschia costata L. Bolus;单脉舟叶花●☆

340624　Ruschia cradockensis (Kuntze) H. E. K. Hartmann et Stüber;克拉多克舟叶花●☆

340625　Ruschia cradockensis (Kuntze) H. E. K. Hartmann et Stüber subsp. triticiformis (L. Bolus) H. E. K. Hartmann et Stüber;小麦舟叶花●☆

340626　Ruschia crassa (L. Bolus) Schwantes;龙骨舟叶花(银孔雀)●☆

340627　Ruschia crassa Schwantes = Ruschia crassa (L. Bolus) Schwantes ●☆

340628　Ruschia crassifolia L. Bolus = Antimima crassifolia (L. Bolus) H. E. K. Hartmann ■☆

340629　Ruschia crassisepala L. Bolus;厚瓣舟叶花●☆

340630　Ruschia crassisepala L. Bolus var. major ? = Ruschia crassisepala L. Bolus ●☆

340631　Ruschia crassuloides L. Bolus = Eberlanzia sedoides (Dinter et A. Berger) Schwantes ●☆

340632　Ruschia cupulata (L. Bolus) Schwantes;杯状舟叶花●☆

340633　Ruschia curta (Haw.) Schwantes;短小舟叶花●☆

340634　Ruschia cyathiformis L. Bolus = Eberlanzia cyathiformis (L. Bolus) H. E. K. Hartmann ●☆

340635　Ruschia cymbifolia (Haw.) L. Bolus;舟叶花●☆

340636　Ruschia cymosa (L. Bolus) Schwantes;聚伞舟叶花●☆

340637　Ruschia dasyphylla (Schltr.) Schwantes = Antimima dasyphylla (Schltr.) H. E. K. Hartmann ■☆

340638　Ruschia decumbens L. Bolus;外倾舟叶花●☆

340639　Ruschia decurrens L. Bolus;下延舟叶花●☆

340640　Ruschia dekenahii (N. E. Br.) Schwantes = Antimima dekenahi (N. E. Br.) H. E. K. Hartmann ■☆

340641　Ruschia deminuta L. Bolus;缩小舟叶花●☆

340642　Ruschia densiflora L. Bolus;密花舟叶花●☆

340643　Ruschia depressa L. Bolus;凹陷舟叶花●☆

340644　Ruschia derenbergiana (Dinter) C. Weber;光芒舟叶花●☆

340645　Ruschia derenbergiana (Dinter) C. Weber = Ebracteola derenbergiana (Dinter) Dinter et Schwantes ■☆

340646　Ruschia dichotoma L. Bolus = Eberlanzia dichotoma (L. Bolus) H. E. K. Hartmann ●☆

340647　Ruschia dichroa (Rolfe) L. Bolus;二色舟叶花●☆

340648　Ruschia dichroa (Rolfe) L. Bolus var. alba L. Bolus = Ruschia dichroa (Rolfe) L. Bolus ●☆

340649　Ruschia dilatata L. Bolus;膨肿舟叶花●☆

340650　Ruschia disarticulata L. Bolus = Antimima hantamensis (Engl.) H. E. K. Hartmann et Stüber ■☆

340651　Ruschia dissimilis G. D. Rowley = Phiambolia stayneri (L. Bolus ex Toelken et Jessop) Klak ●☆

340652　Ruschia distans (L. Bolus) L. Bolus = Antimima distans (L. Bolus) H. E. K. Hartmann ■☆

340653　Ruschia diutina L. Bolus = Polymita albiflora (L. Bolus) L. Bolus ■☆

340654　Ruschia divaricata L. Bolus;叉开舟叶花●☆

340655　Ruschia diversifolia L. Bolus;片敷菊●☆

340656　Ruschia dolomitica (Dinter) Dinter et Schwantes = Antimima dolomitica (Dinter) H. E. K. Hartmann ■☆

340657　Ruschia drepanophylla (Schltr. et A. Berger) L. Bolus = Esterhuysenia drepanophylla (Schltr. et A. Berger) H. E. K. Hartmann ■☆

340658　Ruschia drepanophylla (Schltr. et A. Berger) L. Bolus var. sneeubergensis L. Bolus = Esterhuysenia drepanophylla (Schltr. et A. Berger) H. E. K. Hartmann ■☆

340659　Ruschia dualis L. Bolus;翡翠鉾●☆

340660　Ruschia dubitans (L. Bolus) L. Bolus = Phiambolia unca (L. Bolus) Klak ■☆

340661　Ruschia duplessiae L. Bolus = Acrodon bellidiflorus (L.) N. E. Br. ■☆

340662　Ruschia dyeri L. Bolus = Rhombophyllum albanense (L. Bolus) H. E. K. Hartmann ●☆

340663　Ruschia ebracteata L. Bolus = Eberlanzia ebracteata (L. Bolus) H. E. K. Hartmann ●☆

340664　Ruschia edentula (Haw.) L. Bolus;无齿舟叶花●☆

340665　Ruschia elevata L. Bolus = Antimima elevata (L. Bolus) H. E. K. Hartmann ■☆

340666　Ruschia elineata L. Bolus;无纹舟叶花●☆

340667　Ruschia emarcescens L. Bolus = Antimima emarcescens (L. Bolus) H. E. K. Hartmann ■☆

340668　Ruschia emarcidens L. Bolus ex H. Jacobsen = Antimima emarcescens (L. Bolus) H. E. K. Hartmann ■☆

340669　Ruschia erecta (L. Bolus) Schwantes;直立舟叶花●☆

340670　Ruschia erosa L. Bolus = Antimima erosa (L. Bolus) H. E. K. Hartmann ■☆

340671　Ruschia esterhuyseniae L. Bolus;埃斯特舟叶花●☆

340672　Ruschia evoluta (N. E. Br.) L. Bolus = Antimima evoluta (N. E. Br.) H. E. K. Hartmann ■☆

340673　Ruschia evoluta L. Bolus;铃笼●☆

340674   Ruschia exigua L. Bolus;小舟叶花●☆

340675   Ruschia exsurgens L. Bolus = Antimima exsurgens（L. Bolus）H. E. K. Hartmann ■☆

340676   Ruschia extensa L. Bolus;伸展舟叶花●☆

340677   Ruschia fenestrata L. Bolus = Antimima fenestrata（L. Bolus）H. E. K. Hartmann ■☆

340678   Ruschia fergusoniae L. Bolus = Antimima fergusoniae（L. Bolus）H. E. K. Hartmann ■☆

340679   Ruschia ferox L. Bolus = Ruschia intricata（N. E. Br.）H. E. K. Hartmann et Stüber ●☆

340680   Ruschia festiva（N. E. Br.）Schwantes;华美舟叶花●☆

340681   Ruschia filamentosa（DC.）L. Bolus = Erepsia forficata（L.）Schwantes ●☆

340682   Ruschia filipetala L. Bolus;线瓣舟叶花●☆

340683   Ruschia firma L. Bolus;坚硬舟叶花●☆

340684   Ruschia floribunda L. Bolus;繁花舟叶花●☆

340685   Ruschia foliosa（Haw.）Schwantes;多叶舟叶花●☆

340686   Ruschia forficata（L.）L. Bolus = Erepsia forficata（L.）Schwantes ●☆

340687   Ruschia forficata L. Bolus;彩龙●☆

340688   Ruschia fourcadei L. Bolus;富尔卡德舟叶花■☆

340689   Ruschia framesii L. Bolus;弗雷斯舟叶花■☆

340690   Ruschia fredericii（L. Bolus）L. Bolus;弗雷德里克舟叶花■☆

340691   Ruschia frutescens（L. Bolus）L. Bolus = Stoeberia frutescens（L. Bolus）Van Jaarsv. ●☆

340692   Ruschia fulleri L. Bolus = Ebracteola fulleri（L. Bolus）Glen ■☆

340693   Ruschia gemina L. Bolus = Cerochlamys gemina（L. Bolus）H. E. K. Hartmann ■☆

340694   Ruschia geminiflora（Haw.）Schwantes;对花舟叶花■☆

340695   Ruschia gibbosa L. Bolus = Leipoldtia compacta L. Bolus ●☆

340696   Ruschia glauca L. Bolus;灰绿舟叶花●☆

340697   Ruschia globularis L. Bolus = Ruschia spinosa（L.）Dehn ●☆

340698   Ruschia gracilipes L. Bolus;细梗舟叶花●☆

340699   Ruschia gracilis L. Bolus;纤细舟叶花●☆

340700   Ruschia gracillima L. Bolus = Antimima gracillima（L. Bolus）H. E. K. Hartmann ■☆

340701   Ruschia graminea H. Jacobsen = Acrodon bellidiflorus（L.）N. E. Br. ■☆

340702   Ruschia granitica（L. Bolus）L. Bolus = Antimima granitica（L. Bolus）H. E. K. Hartmann ■☆

340703   Ruschia griquensis（L. Bolus）Schwantes;格里夸舟叶花●☆

340704   Ruschia grisea（L. Bolus）Schwantes;灰舟叶花●☆

340705   Ruschia gydouwensis（L. Bolus）G. D. Rowley = Phiambolia incumbens（L. Bolus）Klak ■☆

340706   Ruschia hallii L. Bolus = Antimima hallii（L. Bolus）H. E. K. Hartmann ■☆

340707   Ruschia hallowayana L. Bolus = Eberlanzia schneideriana（A. Berger）H. E. K. Hartmann ●☆

340708   Ruschia hamata（L. Bolus）Schwantes;顶钩舟叶花●☆

340709   Ruschia haworthii Jacobsen et G. D. Rowley;霍沃斯舟叶花■☆

340710   Ruschia herrei Schwantes = Antimima herrei（Schwantes）H. E. K. Hartmann ■☆

340711   Ruschia heteropetala L. Bolus;异瓣舟叶花●☆

340712   Ruschia hexamera L. Bolus;六数舟叶花●☆

340713   Ruschia hexamera L. Bolus var. longipetala ? = Ruschia hexamera L. Bolus ●☆

340714   Ruschia holensis L. Bolus;侯拉舟叶花●☆

340715   Ruschia hollowayana L. Bolus = Eberlanzia schneideriana（A. Berger）H. E. K. Hartmann ●☆

340716   Ruschia horrescens L. Bolus = Ruschia cradockensis（Kuntze）H. E. K. Hartmann et Stüber ●☆

340717   Ruschia horrescens L. Bolus var. densa ? = Ruschia cradockensis（Kuntze）H. E. K. Hartmann et Stüber ●☆

340718   Ruschia horrida L. Bolus = Ruschia cradockensis（Kuntze）H. E. K. Hartmann et Stüber ●☆

340719   Ruschia hutchinsonii L. Bolus = Amphibolia laevis（Aiton）H. E. K. Hartmann ●☆

340720   Ruschia imbricata（Haw.）Schwantes;覆瓦舟叶花■☆

340721   Ruschia impressa L. Bolus;凹舟叶花■☆

340722   Ruschia inclaudens L. Bolus = Esterhuysenia inclaudens（L. Bolus）H. E. K. Hartmann ●☆

340723   Ruschia inconspicua L. Bolus;显著舟叶花●☆

340724   Ruschia incumbens L. Bolus = Phiambolia incumbens（L. Bolus）Klak ■☆

340725   Ruschia incurvata L. Bolus;内折舟叶花●☆

340726   Ruschia indecora（L. Bolus）Schwantes;装饰舟叶花●☆

340727   Ruschia indurata（L. Bolus）Schwantes = Smicrostigma viride（Haw.）N. E. Br. ●☆

340728   Ruschia integra Schwantes = Smicrostigma viride（Haw.）N. E. Br. ●☆

340729   Ruschia intermedia L. Bolus;间型舟叶花●☆

340730   Ruschia intervallaris L. Bolus = Antimima intervallaris（L. Bolus）H. E. K. Hartmann ■☆

340731   Ruschia intricata（N. E. Br.）H. E. K. Hartmann et Stüber;缠结舟叶花●☆

340732   Ruschia intrusa（Kensit）L. Bolus = Brianhuntleya intrusa（Kensit）Chess.,S. A. Hammer et I. M. Oliv. ●☆

340733   Ruschia ivori（N. E. Br.）Schwantes = Antimima ivori（N. E. Br.）H. E. K. Hartmann ■☆

340734   Ruschia jacobseniana L. Bolus = Astridia longifolia（L. Bolus）L. Bolus ■☆

340735   Ruschia kakamasensis L. Bolus = Ruschia barnardii L. Bolus ■☆

340736   Ruschia karroidea L. Bolus = Antimima karroidea（L. Bolus）H. E. K. Hartmann ■☆

340737   Ruschia karrooica（L. Bolus）L. Bolus;卡卢舟叶花■☆

340738   Ruschia kenhardtensis L. Bolus;肯哈特舟叶花●☆

340739   Ruschia klaverensis（L. Bolus）Schwantes = Antimima klaverensis（L. Bolus）H. E. K. Hartmann ■☆

340740   Ruschia klipbergensis L. Bolus;克勒舟叶花●☆

340741   Ruschia knysnana（L. Bolus）L. Bolus;克尼斯纳舟叶花●☆

340742   Ruschia knysnana（L. Bolus）L. Bolus var. angustifolia L. Bolus = Ruschia knysnana（L. Bolus）L. Bolus ●☆

340743   Ruschia koekenaapensis L. Bolus = Antimima koekenaapensis（L. Bolus）H. E. K. Hartmann ■☆

340744   Ruschia komkansica L. Bolus = Antimima komkansica（L. Bolus）H. E. K. Hartmann ■☆

340745   Ruschia lapidicola L. Bolus;石砾舟叶花●☆

340746   Ruschia lavisii L. Bolus;拉维斯舟叶花●☆

340747   Ruschia lawsonii（L. Bolus）L. Bolus = Antimima lawsonii（L. Bolus）H. E. K. Hartmann ■☆

340748   Ruschia laxa（Willd.）Schwantes;疏松舟叶花■☆

340749   Ruschia laxiflora L. Bolus;疏花舟叶花●☆

340750   Ruschia laxipetala L. Bolus;疏瓣舟叶花●☆

340751   Ruschia leightoniae L. Bolus = Phiambolia unca（L. Bolus）

Klak ■☆

340752　Ruschia leipoldtii L. Bolus = Antimima leipoldtii（L. Bolus）H. E. K. Hartmann ■☆

340753　Ruschia leptocalyx L. Bolus;细萼舟叶花●☆

340754　Ruschia leptophylla L. Bolus = Acrodon subulatus（Mill.）N. E. Br. ■☆

340755　Ruschia lerouxiae（L. Bolus）L. Bolus;勒鲁丹舟叶花■☆

340756　Ruschia leucanthera（L. Bolus）L. Bolus = Antimima leucanthera（L. Bolus）H. E. K. Hartmann ■☆

340757　Ruschia leucanthera L. Bolus;群剑●☆

340758　Ruschia leucosperma L. Bolus;白籽舟叶花●☆

340759　Ruschia levynsiae（L. Bolus）Schwantes = Antimima pumila（L. Bolus ex Fedde et C. Schust.）H. E. K. Hartmann ■☆

340760　Ruschia limbata（N. E. Br.）Schwantes = Antimima limbata（N. E. Br.）H. E. K. Hartmann ■☆

340761　Ruschia lineolata（Haw.）Schwantes;细线舟叶花■☆

340762　Ruschia littlewoodii L. Bolus;利特尔伍德舟叶花●☆

340763　Ruschia lodewykii L. Bolus = Antimima lodewykii（L. Bolus）H. E. K. Hartmann ■☆

340764　Ruschia loganii L. Bolus = Antimima loganii（L. Bolus）H. E. K. Hartmann ■☆

340765　Ruschia lokenbergensis L. Bolus = Antimima lokenbergensis（L. Bolus）H. E. K. Hartmann ■☆

340766　Ruschia longifolia（L. Bolus）L. Bolus = Astridia longifolia（L. Bolus）L. Bolus ■☆

340767　Ruschia longifolia L. Bolus = Acrodon bellidiflorus（L.）N. E. Br. ■☆

340768　Ruschia longipes L. Bolus = Antimima longipes（L. Bolus）Dehn ■☆

340769　Ruschia luckhoffii L. Bolus = Antimima luckhoffii（L. Bolus）H. E. K. Hartmann ■☆

340770　Ruschia lutea L. Bolus = Rhinephyllum luteum（L. Bolus）L. Bolus ●☆

340771　Ruschia macowanii（L. Bolus）Schwantes;紫瓣舟叶花●☆

340772　Ruschia macrophylla L. Bolus = Acrodon bellidiflorus（L.）N. E. Br. ■☆

340773　Ruschia macroura L. Bolus = Ruschia spinosa（L.）Dehn ●☆

340774　Ruschia maleolens L. Bolus = Antimima maleolens（L. Bolus）H. E. K. Hartmann ■☆

340775　Ruschia mallesoniae（L. Bolus）L. Bolus = Esterhuysenia drepanophylla（Schltr. et A. Berger）H. E. K. Hartmann ■☆

340776　Ruschia marginata L. Bolus = Phiambolia unca（L. Bolus）Klak ■☆

340777　Ruschia mariae L. Bolus;玛利亚舟叶花●☆

340778　Ruschia marianae（L. Bolus）Schwantes;玛氏舟叶花■☆

340779　Ruschia maritima（L. Bolus ex Toelken et Jessop）G. D. Rowley = Amphibolia laevis（Aiton）H. E. K. Hartmann ●☆

340780　Ruschia mathewsii L. Bolus = Antimima mucronata（Haw.）H. E. K. Hartmann ■☆

340781　Ruschia maxima（Haw.）L. Bolus;群鉾●☆

340782　Ruschia mesklipensis L. Bolus = Antimima mesklipensis（L. Bolus）H. E. K. Hartmann ■☆

340783　Ruschia meyerae Schwantes = Antimima meyerae（Schwantes）H. E. K. Hartmann ■☆

340784　Ruschia meyeri Schwantes = Antimima papillata（L. Bolus）H. E. K. Hartmann ■☆

340785　Ruschia micropetala L. Bolus = Stoeberia beetzii（Dinter）Dinter et Schwantes ●☆

340786　Ruschia microphylla（Haw.）Schwantes;司宝●☆

340787　Ruschia microphylla（Haw.）Schwantes = Antimima microphylla（Haw.）Dehn ■☆

340788　Ruschia middlemostii L. Bolus;米德尔舟叶花●☆

340789　Ruschia milleflora L. Bolus = Eberlanzia dichotoma（L. Bolus）H. E. K. Hartmann ●☆

340790　Ruschia minutifolia L. Bolus = Antimima minutifolia（L. Bolus）H. E. K. Hartmann ●☆

340791　Ruschia misera（L. Bolus）L. Bolus;贫弱舟叶花■☆

340792　Ruschia modesta L. Bolus. f. glabrescens ? = Antimima modesta（L. Bolus）H. E. K. Hartmann ■☆

340793　Ruschia modesta L. Bolus. f. modesta ? = Antimima modesta（L. Bolus）H. E. K. Hartmann ■☆

340794　Ruschia mollis Schwantes;柔软舟叶花■☆

340795　Ruschia montaguensis L. Bolus;蒙塔古舟叶花●☆

340796　Ruschia monticola（Sond.）G. D. Rowley = Ottosonderia monticola（Sond.）L. Bolus ●☆

340797　Ruschia mucronata（Haw.）Schwantes = Antimima mucronata（Haw.）H. E. K. Hartmann ■☆

340798　Ruschia muelleri（L. Bolus）Schwantes;米勒舟叶花■☆

340799　Ruschia muiriana（L. Bolus）Schwantes;缪里舟叶花●☆

340800　Ruschia multiflora（Haw.）Schwantes;丰花舟叶花■☆

340801　Ruschia muricata L. Bolus;糙舟叶花■☆

340802　Ruschia mutata G. D. Rowley = Lampranthus mutatus（G. D. Rowley）H. E. K. Hartmann ■☆

340803　Ruschia mutica L. Bolus = Antimima mutica（L. Bolus）H. E. K. Hartmann ■☆

340804　Ruschia namaquana L. Bolus = Amphibolia rupis-arcuatae（Dinter）H. E. K. Hartmann ●☆

340805　Ruschia namaquana L. Bolus var. quinqueflora ? = Eberlanzia ebracteata（L. Bolus）H. E. K. Hartmann ●☆

340806　Ruschia nana L. Bolus;矮小舟叶花●☆

340807　Ruschia neilii L. Bolus = Stayneria neilii（L. Bolus）L. Bolus ●☆

340808　Ruschia nelii Schwantes;尼尔舟叶花■☆

340809　Ruschia neovirens Schwantes;新绿舟叶花■☆

340810　Ruschia nivea L. Bolus = Ruschia rupicola（Engl.）Schwantes ●☆

340811　Ruschia nobilis Schwantes = Antimima nobilis（Schwantes）H. E. K. Hartmann ■☆

340812　Ruschia nordenstamii L. Bolus = Antimima nordenstamii（L. Bolus）H. E. K. Hartmann ■☆

340813　Ruschia obtusa L. Bolus;钝舟叶花■☆

340814　Ruschia obtusifolia L. Bolus = Antimima watermeyeri（L. Bolus）H. E. K. Hartmann ■☆

340815　Ruschia octojugis（L. Bolus）L. Bolus = Octopoma octojuge（L. Bolus）N. E. Br. ●☆

340816　Ruschia octonaria（L. Bolus）G. D. Rowley = Enarganthe octonaria（L. Bolus）N. E. Br. ●☆

340817　Ruschia odontocalyx（Schltr. et Diels）Schwantes;齿萼舟叶花■☆

340818　Ruschia orientalis L. Bolus;东方舟叶花■☆

340819　Ruschia ottosonderi G. D. Rowley = Ottosonderia monticola（Sond.）L. Bolus ●☆

340820　Ruschia pakhuisensis L. Bolus = Lampranthus pakpassensis H. E. K. Hartmann ■☆

340821　Ruschia pallens L. Bolus;变苍白舟叶花■☆

340822　Ruschia papillata L. Bolus = Antimima papillata（L. Bolus）H. E. K. Hartmann ■☆

340823　Ruschia paripetala（L. Bolus）L. Bolus;等瓣舟叶花■☆

340824　Ruschia paripetala（L. Bolus）L. Bolus var. occultans L. Bolus = Antimima perforata（L. Bolus）H. E. K. Hartmann ■☆

340825　Ruschia parvibracteata L. Bolus = Eberlanzia parvibracteata（L. Bolus）H. E. K. Hartmann ●☆

340826　Ruschia parviflora Schwantes；小花舟叶花■☆

340827　Ruschia parvifolia L. Bolus；小叶舟叶花■☆

340828　Ruschia patens L. Bolus；铺展舟叶花■☆

340829　Ruschia patulifolia L. Bolus；叉叶舟叶花●☆

340830　Ruschia pauciflora L. Bolus；少花舟叶花●☆

340831　Ruschia paucifolia L. Bolus = Antimima paucifolia（L. Bolus）H. E. K. Hartmann ■☆

340832　Ruschia paucipetala L. Bolus；少瓣舟叶花●☆

340833　Ruschia pauper L. Bolus = Antimima pauper（L. Bolus）H. E. K. Hartmann ■☆

340834　Ruschia peersii L. Bolus = Antimima peersii（L. Bolus）H. E. K. Hartmann ■☆

340835　Ruschia perfoliata（Mill.）Schwantes；讴春玉●☆

340836　Ruschia perfoliata Schwantes = Ruschia perfoliata（Mill.）Schwantes ●☆

340837　Ruschia perforata L. Bolus = Antimima perforata（L. Bolus）H. E. K. Hartmann ■☆

340838　Ruschia persistens L. Bolus = Antimima persistens H. E. K. Hartmann ■☆

340839　Ruschia persistens L. Bolus = Ruschia intricata（N. E. Br.）H. E. K. Hartmann et Stüber ●☆

340840　Ruschia phylicoides L. Bolus；菲利木舟叶花■☆

340841　Ruschia pillansii L. Bolus = Eberlanzia schneideriana（A. Berger）H. E. K. Hartmann ●☆

340842　Ruschia pilosula L. Bolus = Antimima pilosula（L. Bolus）H. E. K. Hartmann ■☆

340843　Ruschia pinguis L. Bolus；肥厚舟叶花●☆

340844　Ruschia piscodora L. Bolus；琴柱舟叶花（琴柱菊）●☆

340845　Ruschia polita L. Bolus = Braunsia geminata（Haw.）L. Bolus ●☆

340846　Ruschia prolongata L. Bolus = Antimima prolongata（L. Bolus）H. E. K. Hartmann ■☆

340847　Ruschia promontorii L. Bolus；普罗蒙特里舟叶花●☆

340848　Ruschia propinqua（N. E. Br.）Schwantes = Antimima propinqua（N. E. Br.）H. E. K. Hartmann ■☆

340849　Ruschia prostrata L. Bolus = Antimima prostrata（L. Bolus）H. E. K. Hartmann ■☆

340850　Ruschia psammophila（Dinter）Dinter et Schwantes = Ruschia ruschiana（Dinter）Dinter et Schwantes ●☆

340851　Ruschia pulchella（Haw.）Schwantes；美丽舟叶花●☆

340852　Ruschia pulchella Schwantes = Ruschia pulchella（Haw.）Schwantes ●☆

340853　Ruschia pulvinaris L. Bolus；垫状舟叶花●☆

340854　Ruschia pumila（L. Bolus ex Fedde et C. Schust.）L. Bolus = Antimima pumila（L. Bolus ex Fedde et C. Schust.）H. E. K. Hartmann ■☆

340855　Ruschia punctulata（L. Bolus）L. Bolus ex H. E. K. Hartmann；小斑舟叶花●☆

340856　Ruschia pungens（A. Berger）H. Jacobsen；锐尖舟叶花●☆

340857　Ruschia puniens L. Bolus = Ruschia intricata（N. E. Br.）H. E. K. Hartmann et Stüber ●☆

340858　Ruschia purpureostyla（L. Bolus）Bruyns = Acrodon purpureostylus（L. Bolus）Burgoyne ■☆

340859　Ruschia pusilla Schwantes = Antimima pusilla（Schwantes）H.

340860　Ruschia pygmaea（Haw.）Schwantes；白天子●☆

340861　Ruschia pygmaea（Haw.）Schwantes = Antimima pygmaea（Haw.）H. E. K. Hartmann ■☆

340862　Ruschia pygmaea Schwantes = Ruschia pygmaea（Haw.）Schwantes ●☆

340863　Ruschia quadrisepala L. Bolus = Octopoma quadrisepalum（L. Bolus）H. E. K. Hartmann ●☆

340864　Ruschia quarzitica（Dinter）Dinter et Schwantes = Antimima quarzitica（Dinter）H. E. K. Hartmann ■☆

340865　Ruschia radicans L. Bolus；具根舟叶花■☆

340866　Ruschia rariflora L. Bolus；稀花舟叶花●☆

340867　Ruschia recurva（Moench）H. E. K. Hartmann；反曲舟叶花●☆

340868　Ruschia renniei（L. Bolus）Schwantes = Ebracteola montis-moltkei（Dinter）Dinter et Schwantes ■☆

340869　Ruschia restituta G. D. Rowley = Octopoma inclusum（L. Bolus）N. E. Br. ●☆

340870　Ruschia rigens L. Bolus；稍坚挺舟叶花■☆

340871　Ruschia rigida（Haw.）Schwantes；坚挺舟叶花●☆

340872　Ruschia rigida L. Bolus = Ruschia bolusiae Schwantes ●☆

340873　Ruschia rigidicaulis（Haw.）Schwantes；硬茎舟叶花●☆

340874　Ruschia robusta L. Bolus；粗壮舟叶花●☆

340875　Ruschia robusta Schwantes = Ruschia valida Schwantes ●☆

340876　Ruschia roseola（N. E. Br.）Schwantes = Antimima roseola（N. E. Br.）H. E. K. Hartmann ■☆

340877　Ruschia rostella（Haw.）Schwantes；喙状舟叶花■●☆

340878　Ruschia rubra（L. Bolus）L. Bolus = Astridia rubra（L. Bolus）L. Bolus ●☆

340879　Ruschia rubricaulis（Haw.）L. Bolus；红茎舟叶花●☆

340880　Ruschia rupicola（Engl.）Schwantes；岩生舟叶花●☆

340881　Ruschia rupicola L. Bolus = Octopoma rupigenum（L. Bolus）L. Bolus ●☆

340882　Ruschia rupigena L. Bolus = Octopoma rupigenum（L. Bolus）L. Bolus ●☆

340883　Ruschia rupis-arcuatae（Dinter）Friedrich = Amphibolia rupis-arcuatae（Dinter）H. E. K. Hartmann ●☆

340884　Ruschia ruralis（N. E. Br.）Schwantes；田野舟叶花■☆

340885　Ruschia ruschiana（Dinter）Dinter et Schwantes；鲁施舟叶花●☆

340886　Ruschia sabulicola Dinter；砂地舟叶花■☆

340887　Ruschia saginata L. Bolus = Amphibolia saginata（L. Bolus）H. E. K. Hartmann ●☆

340888　Ruschia salteri L. Bolus = Hammeria meleagris（L. Bolus）Klak ■☆

340889　Ruschia sandbergensis L. Bolus；桑德舟叶花●☆

340890　Ruschia sarmentosa（Haw.）Schwantes；蔓茎舟叶花■☆

340891　Ruschia sarmentosa（Haw.）Schwantes var. rigida（Salm-Dyck）Schwantes = Ruschia sarmentosa（Haw.）Schwantes ■☆

340892　Ruschia saturata L. Bolus = Antimima saturata（L. Bolus）H. E. K. Hartmann ■☆

340893　Ruschia saxicola L. Bolus = Antimima saxicola（L. Bolus）H. E. K. Hartmann ■☆

340894　Ruschia scabra H. E. K. Hartmann；微糙舟叶花●☆

340895　Ruschia schneideriana（A. Berger）L. Bolus = Eberlanzia schneideriana（A. Berger）H. E. K. Hartmann ●☆

340896　Ruschia schollii（Salm-Dyck）Schwantes；肖尔舟叶花●☆

340897　Ruschia schollii（Salm-Dyck）Schwantes var. caledonica（L. Bolus）Schwantes = Ruschia schollii（Salm-Dyck）Schwantes ●☆

340898　Ruschia scopelogena G. D. Rowley = Scopelogena verruculata

(L.) L. Bolus ●☆

340899 Ruschia sedoides (Dinter et A. Berger) Friedrich = Eberlanzia sedoides (Dinter et A. Berger) Schwantes ●☆

340900 Ruschia semidentata (Haw.) Schwantes;半齿舟叶花●☆

340901 Ruschia semiglobosa L. Bolus;半球形舟叶花●☆

340902 Ruschia serrulata (Haw.) Schwantes;细齿舟叶花●☆

340903 Ruschia sessilis (Thunb.) H. E. K. Hartmann;无柄舟叶花●☆

340904 Ruschia simulans L. Bolus = Antimima simulans (L. Bolus) H. E. K. Hartmann ■☆

340905 Ruschia singula L. Bolus;单一舟叶花●☆

340906 Ruschia socia (N. E. Br.) Schwantes = Argyroderma fissum (Haw.) L. Bolus ●☆

340907 Ruschia solida (L. Bolus) L. Bolus = Antimima solida (L. Bolus) H. E. K. Hartmann ■☆

340908 Ruschia solida (L. Bolus) L. Bolus var. stigmatosa L. Bolus = Antimima solida (L. Bolus) H. E. K. Hartmann ■☆

340909 Ruschia spathulata L. Bolus = Eberlanzia schneideriana (A. Berger) H. E. K. Hartmann ●☆

340910 Ruschia sphaerophylla Dinter = Eberlanzia schneideriana (A. Berger) H. E. K. Hartmann ●☆

340911 Ruschia spinescens L. Bolus = Arenifera spinescens (L. Bolus) H. E. K. Hartmann ■☆

340912 Ruschia spinosa (L.) Dehn;具刺舟叶花●☆

340913 Ruschia stayneri L. Bolus = Antimima stayneri (L. Bolus) H. E. K. Hartmann ■☆

340914 Ruschia steingröveri (Pax) Schwantes = Ruschia rupicola (Engl.) Schwantes ●☆

340915 Ruschia stellata L. Bolus = Antimima hantamensis (Engl.) H. E. K. Hartmann et Stüber ■☆

340916 Ruschia stenopetala L. Bolus = Antimima watermeyeri (L. Bolus) H. E. K. Hartmann ■☆

340917 Ruschia stenophylla (L. Bolus) L. Bolus = Marlothistella stenophylla (L. Bolus) S. A. Hammer ●☆

340918 Ruschia stenophylla (L. Bolus) L. Bolus ex Fourc.;浅茅菊●☆

340919 Ruschia stokoei L. Bolus = Antimima stokoei (L. Bolus) H. E. K. Hartmann ■☆

340920 Ruschia stricta L. Bolus;刚直舟叶花●☆

340921 Ruschia stricta L. Bolus var. turgida ? = Amphibolia saginata (L. Bolus) H. E. K. Hartmann ●☆

340922 Ruschia strubeniae (L. Bolus) Schwantes;剑龙●☆

340923 Ruschia stylosa L. Bolus = Arenifera stylosa (L. Bolus) H. E. K. Hartmann ■☆

340924 Ruschia suaveolens L. Bolus;芳香舟叶花●☆

340925 Ruschia subaphylla Friedrich = Ruschia abbreviata L. Bolus ●☆

340926 Ruschia subglobosa L. Bolus = Octopoma subglobosum (L. Bolus) L. Bolus ●☆

340927 Ruschia subpaniculata L. Bolus;圆锥舟叶花●☆

340928 Ruschia subsphaerica L. Bolus;亚球形舟叶花●☆

340929 Ruschia subteres L. Bolus;圆柱舟叶花●☆

340930 Ruschia subtruncata L. Bolus;平截舟叶花●☆

340931 Ruschia subtruncata L. Bolus = Antimima subtruncata (L. Bolus) H. E. K. Hartmann ■☆

340932 Ruschia subtruncata L. Bolus var. minor ? = Antimima subtruncata (L. Bolus) H. E. K. Hartmann ■☆

340933 Ruschia subtruncata L. Bolus var. minor ? = Antimima subtruncata (L. Bolus) H. E. K. Hartmann ■☆

340934 Ruschia succulenta L. Bolus = Amphibolia succulenta (L.

Bolus) H. E. K. Hartmann ●☆

340935 Ruschia tardissima L. Bolus;迟舟叶花●☆

340936 Ruschia tecta L. Bolus;屋顶舟叶花●☆

340937 Ruschia tenella (Haw.) Schwantes;小屋顶舟叶花■☆

340938 Ruschia testacea L. Bolus;淡褐舟叶花●☆

340939 Ruschia tetrasepala L. Bolus = Octopoma tetrasepalum (L. Bolus) H. E. K. Hartmann ●☆

340940 Ruschia thomae L. Bolus = Esterhuysenia stokoei (L. Bolus) H. E. K. Hartmann ●☆

340941 Ruschia thomae L. Bolus var. microstigma ? = Esterhuysenia stokoei (L. Bolus) H. E. K. Hartmann ●☆

340942 Ruschia translucens L. Bolus = Stoeberia beetzii (Dinter) Dinter et Schwantes ●☆

340943 Ruschia tribracteata L. Bolus;三苞舟叶花●☆

340944 Ruschia triflora L. Bolus;三花舟叶花●☆

340945 Ruschia triquetra L. Bolus = Antimima triquetra (L. Bolus) H. E. K. Hartmann ■☆

340946 Ruschia truteri L. Bolus;特鲁特尔舟叶花●☆

340947 Ruschia tuberculosa L. Bolus = Antimima tuberculosa (L. Bolus) H. E. K. Hartmann ■☆

340948 Ruschia tumidula (Haw.) Schwantes;肿胀舟叶花●☆

340949 Ruschia turneriana L. Bolus = Antimima turneriana (L. Bolus) H. E. K. Hartmann ■☆

340950 Ruschia uitenhagensis (L. Bolus) Schwantes;埃滕哈赫舟叶花●☆

340951 Ruschia umbellata (L.) Schwantes;小伞舟叶花●☆

340952 Ruschia unca (L. Bolus) L. Bolus = Phiambolia unca (L. Bolus) Klak ■☆

340953 Ruschia unca (L. Bolus) L. Bolus var. punctulata ? = Ruschia punctulata (L. Bolus) L. Bolus ex H. E. K. Hartmann ●☆

340954 Ruschia uncinata (L.) Schwantes;登龙舟叶花(登龙)●☆

340955 Ruschia uncinata Schwantes = Ruschia uncinata (L.) Schwantes ●☆

340956 Ruschia uncinella (Haw.) Schwantes = Ruschia uncinata (L.) Schwantes ●☆

340957 Ruschia unidens (Haw.) Schwantes;单齿舟叶花●☆

340958 Ruschia utilis (L. Bolus) L. Bolus;有用舟叶花●☆

340959 Ruschia utilis (L. Bolus) L. Bolus var. giftbergensis L. Bolus = Stoeberia utilis (L. Bolus) Van Jaarsv. ●☆

340960 Ruschia vaginata Schwantes;具鞘舟叶花●☆

340961 Ruschia valida Schwantes;强健舟叶花●☆

340962 Ruschia vanderbergiae L. Bolus;范德舟叶花●☆

340963 Ruschia vanheerdei L. Bolus;黑尔德舟叶花●☆

340964 Ruschia vanzylii L. Bolus = Antimima vanzylii (L. Bolus) H. E. K. Hartmann ■☆

340965 Ruschia varians L. Bolus = Antimima varians (L. Bolus) H. E. K. Hartmann ■☆

340966 Ruschia velutina L. Bolus = Eberlanzia schneideriana (A. Berger) H. E. K. Hartmann ●☆

340967 Ruschia ventricosa (L. Bolus) Schwantes = Antimima ventricosa (L. Bolus) H. E. K. Hartmann ■☆

340968 Ruschia verruculata (L.) G. D. Rowley = Scopelogena verruculata (L.) L. Bolus ●☆

340969 Ruschia verruculosa L. Bolus = Antimima verruculosa (L. Bolus) H. E. K. Hartmann ■☆

340970 Ruschia versicolor L. Bolus;变色舟叶花●☆

340971 Ruschia victoris (L. Bolus) L. Bolus;维多利亚舟叶花■☆

340972 Ruschia villetii L. Bolus = Antimima dualis (N. E. Br.) N. E.

Br. ■☆

340973 Ruschia virens L. Bolus;绿舟叶花●☆

340974 Ruschia virgata (Haw.) L. Bolus;条纹舟叶花●☆

340975 Ruschia viridifolia L. Bolus;绿叶舟叶花●☆

340976 Ruschia viridis (Haw.) G. D. Rowley = Smicrostigma viride (Haw.) N. E. Br. ●☆

340977 Ruschia vulnerans L. Bolus = Ruschia divaricata L. Bolus ●☆

340978 Ruschia watermeyeri L. Bolus = Antimima watermeyeri (L. Bolus) H. E. K. Hartmann ■☆

340979 Ruschia willdenowii Schwantes;威尔舟叶花●☆

340980 Ruschia wilmaniae (L. Bolus) L. Bolus = Ebracteola wilmaniae (L. Bolus) Glen ■☆

340981 Ruschia wilmaniae (L. Bolus) L. Bolus var. angustifolia L. Bolus = Ebracteola wilmaniae (L. Bolus) Glen ■☆

340982 Ruschia wilmaniae (L. Bolus) L. Bolus var. vermeuleniae ? = Ebracteola wilmaniae (L. Bolus) Glen ■☆

340983 Ruschia wittebergensis (L. Bolus) Schwantes = Antimima wittebergensis (L. Bolus) H. E. K. Hartmann ■☆

340984 Ruschianthemum Friedrich(1960);棒玉树属●☆

340985 Ruschianthemum gigas (Dinter) Friedrich = Stoeberia gigas (Dinter) Dinter et Schwantes ●☆

340986 Ruschianthus L. Bolus(1960);镰刀玉属■☆

340987 Ruschianthus falcatus L. Bolus;镰刀玉■☆

340988 Ruschiella Klak(2005);小舟叶花属●☆

340989 Ruscus L. (1753);肖假叶树属;Butcher's Broom, Butchers Broom, Butchersbroom, Butcher's-broom ●

340990 Ruscus aculeatus L.;假叶树;Box Holly, Butcher's Broom, Butchersbroom, Butcher's-broom, Crow Leek, Jew's Myrtle, Knee Holly, Knee Holme, Knee Hulver, Knee Hulyer, Kneed Holly, Kneeholy, Periwinkle, Petigree, Petigrue, Petti Grue, Prickly Pet Tigrue, Prickly Pettigree, Prickly Pettigrue, Shepherd's Myrtle, Sweet Broom, Wild Myrtle ●

340991 Ruscus aculeatus L. var. angustifolia Boiss.;细叶假叶树●☆

340992 Ruscus androgynus L. = Semele androgyna (L.) Kunth ■☆

340993 Ruscus androgynus L. var. gayae (Webb et Berthel.) Christ = Semele gayae (Webb) Svent. et G. Kunkel ■☆

340994 Ruscus hypoglossus L.;长叶假叶树(舌苞假叶树,舌假叶树,下舌假叶树);Butcher's Broom, Double Tongue, Laurel Crown of Caesar, Longleaf Butchersbroom, Spineless Butcher's-broom ●

340995 Ruscus hypoglossus L. = Ruscus hypophyllus L. ●☆

340996 Ruscus hypophyllus L.;里白假叶树●☆

340997 Ruscus hyrcanus Woronow;波斯假叶树●☆

340998 Ruscus ponticus Woronow;黑海假叶树●☆

340999 Ruscus racemosus L. = Danae laurus Medik. ☆

341000 Ruscus racemosus L. = Danae racemosa Moench ●☆

341001 Ruscus reticulatus Thunb. = Behnia reticulata (Thunb.) Didr. ●☆

341002 Ruscus streptophyllus Yeo;旋扭假叶树●☆

341003 Ruspolia Lindau(1895);鲁斯木属●☆

341004 Ruspolia decurrens (Hochst. ex Nees) Milne-Redh.;下延鲁斯木●☆

341005 Ruspolia humbertii Benoist;亨伯特鲁斯木●☆

341006 Ruspolia hypocrateriformis (Vahl) Milne-Redh.;钩子花鲁斯木●☆

341007 Ruspolia hypocrateriformis (Vahl) Milne-Redh. var. australis Milne-Redh.;南方钩子花鲁斯木●☆

341008 Ruspolia paniculata Benoist;圆锥鲁斯木●☆

341009 Ruspolia pseuderanthemoides Lindau;鲁斯木(鲁斯鲍木)●☆

341010 Ruspolia seticalyx (C. B. Clarke) Milne-Redh.;毛鲁斯木(毛鲁斯鲍木)●☆

341011 Russea L. ex B. D. Jacks. = Roussea Sm. ●☆

341012 Russegera Endl. = Lepidagathis Willd. ●■

341013 Russegera Endl. et Fenzl = Lepidagathis Willd. ●■

341014 Russeggera collina Endl. = Lepidagathis hamiltoniana Wall. subsp. collina (Endl.) J. K. Morton ■☆

341015 Russelia J. König ex Roxb. = Ormocarpum P. Beauv. (保留属名)●

341016 Russelia Jacq. (1760);炮仗竹属(爆仗竹属);Coralblow, Coral-blow, Firecracker Bamboo ■●

341017 Russelia L. f. = Bistorta (L.) Adans. ■

341018 Russelia L. f. = Vahlia Dahl ●☆

341019 Russelia capensis L. f. = Vahlia capensis (L. f.) Thunb. ■☆

341020 Russelia equisetiformis Schltdl. et Champ.;炮仗竹(爆仗花,爆仗竹,吉祥草,鲁士拉草,墨西哥炮仗竹,炮竹红,炮竹花,一串红);Coral Bush, Coral Fountain, Coral Plant, Fire Cracker Plant, Firecracker Bamboo, Firecracker Plant, Fountain Plant, Fountainbush ●

341021 Russelia glutinosa Libosch.;怀庆地黄;Glutinous Coral-blow ●

341022 Russelia juncea Zucc. = Russelia equisetiformis Schltdl. et Champ. ●

341023 Russelia sarmentosa Jacq.;卵叶炮仗竹(匍茎炮仗竹)●☆

341024 Russellodendron Britton et Rose = Caesalpinia L. ●

341025 Russeria H. Buek = Bursera Jacq. ex L. (保留属名)●☆

341026 Russowia C. Winkl. (1996);纹苞菊属;Russowia ■

341027 Russowia crupinoides C. Winkl. = Russowia sogdiana (Bunge) B. Fedtsch. ■

341028 Russowia sogdiana (Bunge) B. Fedtsch.;纹苞菊;Sogd Russowia ■

341029 Rustia Klotzsch(1846);鲁斯特茜属●☆

341030 Rustia alba Delprete;白鲁斯特茜●☆

341031 Rustia angustifolia K. Schum.;窄叶鲁斯特茜●☆

341032 Rustia ferruginea Standl.;锈色鲁斯特茜●☆

341033 Rustia gracilis K. Schum.;细鲁斯特茜●☆

341034 Rustia longifolia Standl.;长叶鲁斯特茜●☆

341035 Rustia occidentalis (Benth.) Hemsl.;西方鲁斯特茜●☆

341036 Rustia pauciflora Solereder;少花鲁斯特茜●☆

341037 Rustia rosea K. Schum.;粉红鲁斯特茜●☆

341038 Rustia splendens Standl.;纤细鲁斯特茜●☆

341039 Ruta L. (1753);芸香属;Rue ■●

341040 Ruta acutifolia DC. = Haplophyllum acutifolium (DC.) G. Don ■

341041 Ruta acutifolia DC. = Haplophyllum perforatum (M. Bieb.) Kar. et Kir. ■

341042 Ruta albiflora Hook. = Boenninghausenia albiflora (Hook.) Rchb. ex Meisn. ■

341043 Ruta angustifolia Pers.;狭叶芸香●☆

341044 Ruta arbuscula (Franch.) Cufod.;小乔木芸香●☆

341045 Ruta bracteosa DC. = Ruta chalepensis L. var. bracteosa (DC.) Halácsy ●☆

341046 Ruta bracteosa DC. = Ruta chalepensis L. ●☆

341047 Ruta chalepensis L.;流苏芸香(叙利亚芸香);Fringed Rue ●☆

341048 Ruta chalepensis L. subsp. angustifolia (Pers.) Cout. = Ruta angustifolia Pers. ●☆

341049 Ruta chalepensis L. subsp. bracteosa (DC.) Batt. = Ruta chalepensis L. ●☆

341050 Ruta chalepensis L. subsp. latifolia (Salisb.) H. Lindb.;宽叶流苏芸香●☆

341051　Ruta chalepensis L. var. bracteosa（DC.）Boiss. ＝ Ruta chalepensis L. ●☆

341052　Ruta chalepensis L. var. bracteosa（DC.）Halácsy；多苞芸香（多苞片芸香,芸香）●☆

341053　Ruta chalepensis L. var. intermedia Rouy ＝ Ruta chalepensis L. ●☆

341054　Ruta chalepensis L. var. jacobaea Maire ＝ Ruta chalepensis L. ●☆

341055　Ruta chalepensis L. var. pellucido-punctata Sennen ＝ Ruta chalepensis L. ●☆

341056　Ruta chalepensis L. var. tenuifolia d'Urv. ＝ Ruta chalepensis L. ●☆

341057　Ruta crenulata（Boiss.）Burkill ＝ Haplophyllum crenulatum Boiss. ■☆

341058　Ruta dahurica（L.）DC. ＝ Haplophyllum dauricum（L.）G. Don ■

341059　Ruta divaricata Ten.；欧芸香（叉枝芸香）●☆

341060　Ruta erythraea（Boiss.）Aitch. et Hemsl. ＝ Haplophyllum erythraeum Boiss ■☆

341061　Ruta flexuosa（Boiss.）Engl. ＝ Haplophyllum acutifolium（DC.）G. Don ■

341062　Ruta gilesii Hemsl. ＝ Haplophyllum gilesii（Hemsl.）C. C. Towns. ■☆

341063　Ruta glabra DC. ＝ Haplophyllum tuberculatum（Forssk.）A. Juss. ●☆

341064　Ruta graveolens L.；芸香（百应草,臭艾,臭草,臭芙蓉,干臭草,狗屎灵香,猴仔草,荆芥七,净臭草,全臭草,香草,小香草,小叶香,熊胆草,芸香草）；Ave Grace, Common Rue, Countryman's Treacle, Garden Rue, Herb Grace, Herb of Grace, Herb of Repentance, Herb-a-grass, Herbgrass, Herb-of-grace, Herbygrass, Herby-grass, Rue ●

341065　Ruta graveolens L. 'Jakman's Blue'；蓝粉芸香●☆

341066　Ruta hortensis Mill.；园芸香●☆

341067　Ruta japonica Siebold ex Hook. f. ＝ Boenninghausenia albiflora（Hook.）Rchb. ex Heynh. ■

341068　Ruta japonica Siebold ex Hook. f. ＝ Boenninghausenia albiflora Rchb. ex Meisn. var. japonica（Nakai ex Makino et Nemoto）H. Ohba ■☆

341069　Ruta linifolia L. ＝ Haplophyllum linifolium（L.）G. Don ■☆

341070　Ruta microcarpa Svent.；小果芸香●☆

341071　Ruta montana L.；山地芸香●☆

341072　Ruta obovata（Hochst. ex Boiss.）O. Schwartz ＝ Haplophyllum tuberculatum（Forssk.）A. Juss. ●☆

341073　Ruta oreojasme Webb ＝ Ruta oreojasme Webb et Berthel. ●☆

341074　Ruta oreojasme Webb et Berthel.；山茉莉芸香●☆

341075　Ruta patavina L.；矮芸香；Dwarf Rue ●☆

341076　Ruta pedicellata（Bunge ex Boiss.）Aitch. et Hemsl. ＝ Haplophyllum pedicellatum Bunge ex Boiss. ■☆

341077　Ruta perforata M. Bieb. ＝ Haplophyllum acutifolium（DC.）G. Don ■

341078　Ruta perforata M. Bieb. ＝ Haplophyllum perforatum（M. Bieb.）Kar. et Kir. ■

341079　Ruta pinnata L. f.；羽状芸香●☆

341080　Ruta propinqua（Spach）O. Schwartz ＝ Haplophyllum tuberculatum（Forssk.）A. Juss. ●☆

341081　Ruta sieversii（Fisch. et C. A. Mey.）B. Fedtsch. ＝ Haplophyllum acutifolium（DC.）G. Don ■

341082　Ruta tenuifolia Desf. ＝ Ruta montana L. ●☆

341083　Ruta tuberculata Forssk. ＝ Haplophyllum tuberculatum（Forssk.）A. Juss. ●☆

341084　Ruta tuberculata Forssk. var. forskahlii DC. ＝ Haplophyllum tuberculatum（Forssk.）A. Juss. ●☆

341085　Ruta tuberculata Forssk. var. montbretti DC. ＝ Haplophyllum tuberculatum（Forssk.）A. Juss. ●☆

341086　Rutaceae Juss. (1789)（保留科名）；芸香科；Rue Family ●■

341087　Rutaea M. Roem. ＝ Turraea L. ●

341088　Rutaneblina Steyerm. et Luteyn(1984)；委内瑞拉芸香属●☆

341089　Rutaneblina pusilla Steyerm. et Luteyn；委内瑞拉芸香●☆

341090　Rutaria Webb ex Benth. et Hook. f. ＝ Ruta L. ●■

341091　Rutea M. Roem. ＝ Turraea L. ●

341092　Ruteria Medik. ＝ Psoralea L. ●■

341093　Ruthalicia C. Jeffrey(1962)；卢萨瓜属■☆

341094　Ruthalicia eglandulosa（Hook. f.）C. Jeffrey；卢萨瓜■☆

341095　Ruthalicia longipes（Hook. f.）C. Jeffrey；长梗卢萨瓜■☆

341096　Ruthea Bolle(1862)；露特草属■☆

341097　Ruthea Bolle ＝ Lichtensteinia Cham. et Schltdl.（保留属名）■☆

341098　Ruthea Bolle ＝ Rutheopsis A. Hansen et G. Kunkel ■☆

341099　Ruthea herbanica Bolle ＝ Rutheopsis herbanica（Bolle）A. Hansen et Sunding ■☆

341100　Ruthea interrupta（Thunb.）Druce ＝ Lichtensteinia interrupta（Thunb.）Sond. ■☆

341101　Rutheopsis A. Hansen et G. Kunkel(1976)；拟露特草属■☆

341102　Rutheopsis herbanica（Bolle）A. Hansen et Sunding；拟露特草■☆

341103　Ruthiella Steenis(1965)；露特桔梗属■☆

341104　Ruthiella oblongifolia（Diels）Steenis；矩圆露特桔梗■☆

341105　Ruthiella saxicola（P. Royen）Steenis；岩生露特桔梗■☆

341106　Ruthiella schlechteri（Diels）Steenis；露特桔梗■☆

341107　Ruthrum Hill ＝ Echinops L. ■

341108　Rutica Neck. ＝ Urtica L. ■

341109　Rutidanthera Tiegh. (1904)；皱药金莲木属●☆

341110　Rutidea DC. (1807)；皱茜属●☆

341111　Rutidea albiflora K. Schum. ＝ Rutidea smithii Hiern ●☆

341112　Rutidea atrata Mildbr. ex Hutch. et Dalziel ＝ Nichallea soyauxii（Hiern）Bridson ●☆

341113　Rutidea brachyantha K. Schum. ＝ Rutidea smithii Hiern ●☆

341114　Rutidea breviflora De Wild. ＝ Rutidea smithii Hiern ●☆

341115　Rutidea decorticata Hiern；脱皮皱茜●☆

341116　Rutidea degemensis Wernham ＝ Tarenna eketensis Wernham ●☆

341117　Rutidea dupuisii De Wild. subsp. occidentalis Bridson；西方皱茜●☆

341118　Rutidea ferruginea Hiern；锈色皱茜●☆

341119　Rutidea fuscescens Hiern；浅棕色皱茜●☆

341120　Rutidea fuscescens Hiern subsp. bracteata Bridson；具苞皱茜●☆

341121　Rutidea gabonensis Bridson；加蓬皱茜●☆

341122　Rutidea glabra Hiern；光滑皱茜●☆

341123　Rutidea gracilis Bridson；纤细皱茜●☆

341124　Rutidea hirsuta Hiern；多毛皱茜●☆

341125　Rutidea hispida Hiern；硬毛皱茜●☆

341126　Rutidea insculpta Mildbr. ex Bridson；雕刻皱茜●☆

341127　Rutidea kerstingii K. Krause ＝ Rutidea parviflora DC. ●☆

341128　Rutidea kimuenzae Mildbr.；基姆扎皱茜●☆

341129　Rutidea landolphidoides Wernham ＝ Rutidea decorticata Hiern ●☆

341130　Rutidea lasiosiphon K. Schum. ＝ Rutidea decorticata Hiern ●☆

341131　Rutidea laxiflora K. Schum. ＝ Rutidea decorticata Hiern ●☆

341132　Rutidea leucantha K. Schum. ex De Wild. ＝ Rutidea smithii Hiern ●☆

341133　Rutidea leucotricha T. Durand et H. Durand ＝ Rutidea smithii Hiern ●☆

341134　Rutidea loesneriana K. Schum. = Rutidea hispida Hiern ●☆

341135　Rutidea lomaniensis Bremek. = Rutidea smithii Hiern ●☆

341136　Rutidea lujae De Wild. ;卢亚皱茜●☆

341137　Rutidea melanophylla（K. Schum.）Mildbr. = Nichallea soyauxii（Hiern）Bridson ●☆

341138　Rutidea membranacea Hiern;膜质皱茜●☆

341139　Rutidea nigerica Bridson;尼日利亚皱茜●☆

341140　Rutidea obtusa K. Krause = Rutidea membranacea Hiern ●☆

341141　Rutidea odorata K. Krause = Rutidea fuscescens Hiern ●☆

341142　Rutidea orientalis Bridson;东方皱茜●☆

341143　Rutidea parviflora DC. ;小花皱茜●☆

341144　Rutidea pavettoides Wernham = Rutidea hispida Hiern ●☆

341145　Rutidea rufipilis Hiern;红毛皱茜●☆

341146　Rutidea schlechteri K. Schum. = Rutidea membranacea Hiern ●☆

341147　Rutidea seretii De Wild. ;赛雷皱茜●☆

341148　Rutidea smithii Hiern;史密斯皱茜●☆

341149　Rutidea smithii Hiern subsp. submontana（K. Krause）Bridson;亚山地史密斯皱茜●☆

341150　Rutidea smithii Hiern var. subcordata Scott-Elliot = Rutidea smithii Hiern ●☆

341151　Rutidea smithii Hiern var. welwitschii Scott-Elliot = Rutidea smithii Hiern ●☆

341152　Rutidea striatulata Pellegr. = Rutidea dupuisii De Wild. ●☆

341153　Rutidea syringoides（Webb）Bremek. = Rutidea parviflora DC. ●☆

341154　Rutidea talbotiorum Wernham = Tarenna eketensis Wernham ●☆

341155　Rutidea tarennoides Wernham = Rutidea glabra Hiern ●☆

341156　Rutidea tenuicaulis K. Krause;细茎皱茜●☆

341157　Rutidea tomentosa K. Schum. = Rutidea smithii Hiern ●☆

341158　Rutidea vanderystii Wernham;范德皱茜●☆

341159　Rutidea zombana（K. Schum.）Bremek. = Rutidea fuscescens Hiern ●☆

341160　Rutidochlamys Sond. = Podolepis Labill.（保留属名）■☆

341161　Rutidosis DC.（1838）;锥托棕鼠麹属■☆

341162　Rutidosis leucantha F. Muell. ;白花锥托棕鼠麹■☆

341163　Rutosma A. Gray = Thamnosma Torr. et Frém. ●

341164　Ruttya Harv.（1842）;拉梯爵床属（鲁特亚木属）●☆

341165　Ruttya fruticosa Lindau;灌木拉梯爵床（橙红鲁特亚木）;Jammy-mouth ●☆

341166　Ruttya ovata Harv. ;拉梯爵床（拉梯木,鲁特亚木）●☆

341167　Ruttya speciosa（Hochst.）Engl. ;非洲拉梯爵床（非洲鲁特亚木）●☆

341168　Ruttya tricolor Benoist;三色拉梯爵床●☆

341169　Ruuellodendron Britton et Rose = Caesalpinia L. ●

341170　Ruyschia Fabr. = Dracocephalum L.（保留属名）■●

341171　Ruyschia Fabr. = Ruyschiana Mill. ●

341172　Ruyschia Jacq.（1760）;勒伊斯藤属;Ruyschia ●☆

341173　Ruyschia bicolor Benth. ;二色勒伊斯藤●☆

341174　Ruyschia fragrans Mor. ex Wittm. ;脆勒伊斯藤●☆

341175　Ruyschia longistylis Standl. et Steyerm. ;长柱勒伊斯藤●☆

341176　Ruyschia mexicana Baill. ;墨西哥勒伊斯藤●☆

341177　Ruyschiana Boehr. ex Mill. = Dracocephalum L.（保留属名）■●

341178　Ruyschiana Mill. = Dracocephalum L.（保留属名）■●

341179　Ruyschiana spicata Mill. = Dracocephalum ruyschiana L. ●

341180　Ryacophila Post et Kuntze = Dianella Lam. ex Juss. ●■

341181　Ryacophila Post et Kuntze = Rhuacophila Blume ■☆

341182　Ryanaea DC. = Ryania Vahl（保留属名）●☆

341183　Ryania Vahl（1796）（保留属名）;瑞安木属●☆

341184　Ryania angustifolia（Turcz.）Monach;窄叶瑞安木●☆

341185　Ryania speciosa Vahl;美丽瑞安木●☆

341186　Ryckia Ball f. = Pandanus Parkinson ex Du Roi ●■

341187　Ryckia Ball f. = Rykia de Vriese ●■

341188　Rydbergia Greene = Actinea Juss. ■

341189　Rydbergia Greene = Actinella Pers. ■

341190　Rydbergiella Fedde = Astragalus L. ●■

341191　Rydbergiella Fedde et Syd. ex Rydb. = Astragalus L. ●■

341192　Rydingia Scheen et V. A. Albert（2007）;吕丁草属■☆

341193　Rydingia integrifolia（Benth.）Scheen et V. A. Albert;吕丁草■☆

341194　Ryditophyllum Walp. = Rhytidophyllum Mart.（保留属名）●☆

341195　Ryditophyllum Walp. = Rytidophyllum Mart. ●☆

341196　Rydtostylis Walp. = Rytidostylis Hook. et Arn. ●☆

341197　Rykia de Vriese = Pandanus Parkinson ex Du Roi ●■

341198　Rylstonea R. T. Baker = Homoranthus A. Cunn. ex Schauer ●☆

341199　Rymandra Salisb. = Knightia R. Br.（保留属名）●☆

341200　Rymandra Salisb. ex Knight（废弃属名）= Knightia R. Br.（保留属名）●☆

341201　Rymia Endl. = Euclea L. ●☆

341202　Rynchanthera Blume（废弃属名）= Corymborkis Thouars ■

341203　Rynchanthera Blume（废弃属名）= Rhynchanthera DC.（保留属名）●☆

341204　Ryncholeucaena Britton et Rose = Leucaena Benth.（保留属名）●

341205　Rynchosia Macfad. = Rhynchosia Lour.（保留属名）●■

341206　Rynchospermum Post et Kuntze = Rhynchodia Benth. ●■

341207　Rynchospermum Post et Kuntze = Rhyncospermum A. DC. ●■

341208　Rynchospora Vahl = Rhynchospora Vahl（保留属名）●■

341209　Rynchostylis Blume = Rhynchostylis Blume ■

341210　Rynchostylis Post et Kuntze = Cissus L. ●

341211　Rynchostylis Post et Kuntze = Rinxostylis Raf. ●

341212　Ryparia Blume = Ryparosa Blume ●☆

341213　Ryparosa Blume（1826）;污木属（利帕木属）●☆

341214　Ryparosa cauliflora Merr. ;茎花利帕木●☆

341215　Ryparosa fasciculata King;簇生利帕木●☆

341216　Ryparosa hirsuta J. J. Sm. ;粗毛利帕木●☆

341217　Ryparosa kingii King;金氏利帕木●☆

341218　Ryparosa maculata B. L. Webber;斑点利帕木●☆

341219　Ryparosa minor Ridl. ;小利帕木●☆

341220　Ryparosa multinervosa Slooten;多脉利帕木●☆

341221　Rysodium Steven = Astragalus L. ●■

341222　Ryssopteris Hassk. = Ryssopterys Blume ex A. Juss.（保留属名）●

341223　Ryssopterys Blume ex A. Juss.（1838）（保留属名）;翅实藤属（黎棱翼属,狭翅果属,皱翅果属）;Ryssopterys ●

341224　Ryssopterys cumingiana A. Juss. ;肯氏翅实藤●

341225　Ryssopterys dealbata A. Juss. = Ryssopterys timoriensis（DC.）Blume ex A. Juss. ●

341226　Ryssopterys dealbata Juss. ;皱翅果;Common Ryssopterys ■☆

341227　Ryssopterys timoriensis（DC.）Blume ex A. Juss. ;翅实藤（皱翅果）;Common Ryssopterys, Ryssopterys, Timor Ryssopterys ●

341228　Ryssosciadium Kuntze = Rhysopterus J. M. Coult. et Rose ☆

341229　Ryssotoechia Kuntze = Rhysotoechia Radlk. ●☆

341230　Rytachne Endl. = Rhytachne Desv. ex Ham. ■

341231　Ryticaryum Becc. = Rhyticaryum Becc. ●☆

341232　Rytidea Spreng. = Rutidea DC. ●☆

341233　Rytidocarpus Coss.（1889）;皱果芥属■☆

341234 Rytidocarpus maroccanus (O. E. Schulz) Maire;非洲皱果芥■☆

341235 Rytidocarpus moricandioides Coss.;皱果芥■☆

341236 Rytidocarpus moricandioides Coss. var. maroccanus (O. E. Schulz) Maire = Rytidocarpus moricandioides Coss.■☆

341237 Rytidocarpus moricandioides Coss. var. stenocarpus Emb. et Maire = Rytidocarpus moricandioides Coss.■☆

341238 Rytidochlamys Post et Kuntze = Podolepis Labill.(保留属名)■☆

341239 Rytidochlamys Post et Kuntze = Rutidochlamys Sond.■☆

341240 Rytidolobus Dulac = Hyacinthus L.■☆

341241 Rytidoloma Turcz. = Dictyanthus Decne.●

341242 Rytidophyllum Mart. = Rhytidophyllum Mart.(保留属名)●☆

341243 Rytidosperma Steud.(1854);皱籽草属;Wallaby-grass■☆

341244 Rytidosperma Steud. = Danthonia DC.(保留属名)■

341245 Rytidosperma Steud. = Deschampsia P. Beauv.■

341246 Rytidosperma Steud. = Notodanthonia Zotov■

341247 Rytidosperma davyi (C. E. Hubb.) Cope;戴维皱籽草■☆

341248 Rytidosperma distichum (Nees) Cope;二列皱籽草■☆

341249 Rytidosperma grandiflorum (Hochst. ex A. Rich.) S. M. Phillips;大花皱籽草■☆

341250 Rytidosperma pilosum (R. Br.) Connor et Edgar;毛皱籽草;Hairy Wallaby Grass,Wallaby Grass■☆

341251 Rytidosperma racemosum (R. Br.) Connor et Edgar;总花皱籽草;Wallaby-grass■☆

341252 Rytidosperma semiannulare (Labill.) Connor et Edgar;塔斯马尼亚皱籽草;Tasmanian Wallaby Grass■☆

341253 Rytidosperma subulatum (A. Rich.) Cope;钻形皱籽草■☆

341254 Rytidosperma tenuius (Steud.) A. Hansen et Sunding;细皱籽草■☆

341255 Rytidostylis Hook. et Arn.(1840);纹柱瓜属■☆

341256 Rytidostylis brevisetosa Steyerm.;短刚毛纹柱瓜■☆

341257 Rytidostylis ciliata Kuntze;缘毛纹柱瓜■☆

341258 Rytidostylis glabra Kuntze;光纹柱瓜■☆

341259 Rytidostylis gracilis Hook. et Arn.;细纹柱瓜■☆

341260 Rytidostylis longiflora Kuntze;长花纹柱瓜■☆

341261 Rytidotus Hook. f. = Bobea Gaudich.●☆

341262 Rytiglossa Steud. = Dianthera L.■☆

341263 Rytiglossa Steud. = Rhytiglossa Nees(废弃属名)●■

341264 Rytigynia Blume(1850);纹蕊茜属■☆

341265 Rytigynia acuminatissima (K. Schum.) Robyns;渐尖纹蕊茜■☆

341266 Rytigynia acuminatissima (K. Schum.) Robyns subsp. pedunculata Verdc.;梗花渐尖纹蕊茜■☆

341267 Rytigynia adenodonta (K. Schum.) Robyns;腺齿纹蕊茜■☆

341268 Rytigynia adenodonta (K. Schum.) Robyns var. reticulata (Robyns) Verdc.;网状纹蕊茜■☆

341269 Rytigynia affinis (Robyns) Hepper = Pyrostria affinis (Robyns) Bridson●☆

341270 Rytigynia amaniensis (K. Krause) Bullock = Rytigynia celastroides (Baill.) Verdc.●☆

341271 Rytigynia argentea (Wernham) Robyns;银白纹蕊茜■☆

341272 Rytigynia bagshawei (S. Moore) Robyns;巴格肖纹蕊茜■☆

341273 Rytigynia bagshawei (S. Moore) Robyns var. lebrunii (Robyns) Verdc.;勒布伦银白纹蕊茜■☆

341274 Rytigynia beniensis (De Wild.) Robyns;贝尼纹蕊茜■☆

341275 Rytigynia biflora Robyns = Rytigynia eickii (K. Schum. et K. Krause) Bullock■☆

341276 Rytigynia binata (K. Schum.) Robyns;双纹蕊茜●☆

341277 Rytigynia bomiliensis (De Wild.) Robyns;博米利纹蕊茜■☆

341278 Rytigynia bugoyensis (K. Krause) Verdc.;布戈纹蕊茜■☆

341279 Rytigynia bugoyensis (K. Krause) Verdc. subsp. glabriflora Verdc.;光花纹蕊茜■☆

341280 Rytigynia butaguensis (De Wild.) Robyns = Rytigynia bugoyensis (K. Krause) Verdc.■☆

341281 Rytigynia canthioides (Benth.) Robyns;鱼骨木纹蕊茜■☆

341282 Rytigynia castanea Lebrun et Taton et L. Touss. = Rytigynia monantha (K. Schum.) Robyns■☆

341283 Rytigynia caudatissima Verdc.;极长尾纹蕊茜■☆

341284 Rytigynia celastroides (Baill.) Verdc.;南蛇藤纹蕊茜●☆

341285 Rytigynia celastroides (Baill.) Verdc. var. australis Verdc.;澳洲南蛇藤纹蕊茜●☆

341286 Rytigynia celastroides (Baill.) Verdc. var. nuda Verdc.;裸露纹蕊茜■☆

341287 Rytigynia claessensii (De Wild.) Robyns;克莱森斯纹蕊茜■☆

341288 Rytigynia claviflora Robyns = Rytigynia canthioides (Benth.) Robyns■☆

341289 Rytigynia concolor (Hiern) Robyns = Rytigynia umbellulata (Hiern) Robyns■☆

341290 Rytigynia congesta (K. Krause) Robyns;密集纹蕊茜■☆

341291 Rytigynia constricta Robyns;缢缩纹蕊茜■☆

341292 Rytigynia dasyothamnus (K. Schum.) Robyns;毛枝纹蕊茜■☆

341293 Rytigynia decussata (K. Schum.) Robyns;对生纹蕊茜■☆

341294 Rytigynia demeusei (De Wild.) Robyns;迪米纹蕊茜■☆

341295 Rytigynia dewevrei (De Wild. et T. Durand) Robyns;德韦纹蕊茜●☆

341296 Rytigynia dubiosa (De Wild.) Robyns;可疑纹蕊茜■☆

341297 Rytigynia eickii (K. Schum. et K. Krause) Bullock;艾克纹蕊茜☆

341298 Rytigynia euclioides Robyns = Rytigynia uhligii (K. Schum. et K. Krause) Verdc.■☆

341299 Rytigynia ferruginea Robyns;锈色纹蕊茜■☆

341300 Rytigynia flavida Robyns;浅黄纹蕊茜■☆

341301 Rytigynia friesiorum Robyns = Canthium oligocarpum Hiern subsp. friesiorum (Robyns) Bridson●☆

341302 Rytigynia fuscosetulosa Verdc.;褐毛纹蕊茜■☆

341303 Rytigynia gillettii Tennant = Pachystigma gillettii (Tennant) Verdc.●☆

341304 Rytigynia glabra (K. Schum.) Robyns = Rytigynia celastroides (Baill.) Verdc.●☆

341305 Rytigynia glabrifolia (De Wild.) Robyns;光叶纹蕊茜■☆

341306 Rytigynia gossweileri Robyns;戈斯纹蕊茜■☆

341307 Rytigynia gracilipetiolata (De Wild.) Robyns;细柄纹蕊茜■☆

341308 Rytigynia griseovelutina Verdc.;短绒毛纹蕊茜■☆

341309 Rytigynia hirsutiflora Verdc.;粗毛花纹蕊茜■☆

341310 Rytigynia humbertii Cavaco;亨伯特纹蕊茜■☆

341311 Rytigynia junodii (Schinz) Robyns = Rytigynia umbellulata (Hiern) Robyns■☆

341312 Rytigynia kidaris (K. Schum. et K. Krause) Bullock = Rytigynia uhligii (K. Schum. et K. Krause) Verdc.■☆

341313 Rytigynia kigeziensis Verdc.;基盖济纹蕊茜■☆

341314 Rytigynia kiwuensis (K. Krause) Robyns;基武纹蕊茜■☆

341315 Rytigynia krauseana Robyns;克劳斯纹蕊茜■☆

341316 Rytigynia laurentii (De Wild.) Robyns;洛朗纹蕊茜■☆

341317 Rytigynia lebrunii Robyns = Rytigynia bagshawei (S. Moore) Robyns var. lebrunii (Robyns) Verdc.■☆

341318 Rytigynia lecomtei Robyns;勒孔特纹蕊茜■☆

341319　Rytigynia lenticellata Robyns = Rytigynia uhligii (K. Schum. et K. Krause) Verdc. ■☆

341320　Rytigynia leonensis (K. Schum.) Robyns;莱昂纹蕊茜■☆

341321　Rytigynia lewisii Tennant;刘易斯纹蕊茜■☆

341322　Rytigynia liberica Robyns;利比里亚纹蕊茜■☆

341323　Rytigynia lichenoxenos (K. Schum.) Robyns subsp. glabrituba Verdc. ;光管纹蕊茜■☆

341324　Rytigynia longicaudata Verdc. ;长尾纹蕊茜■☆

341325　Rytigynia longipedicellata Verdc. ;长梗纹蕊茜■☆

341326　Rytigynia longituba Verdc. ;长管纹蕊茜■☆

341327　Rytigynia loranthifolia (K. Schum.) Robyns = Pachystigma loranthifolium (K. Schum.) Verdc. ●☆

341328　Rytigynia macrostipulata Robyns;大托叶纹蕊茜■☆

341329　Rytigynia macrura Verdc. ;大尾纹蕊茜■☆

341330　Rytigynia madagascariensis Homolle ex Cavaco;马岛纹蕊茜■☆

341331　Rytigynia mayumbensis Robyns;马永巴纹蕊茜■☆

341332　Rytigynia membranacea (Hiern) Robyns;膜质纹蕊茜■☆

341333　Rytigynia microphylla (K. Schum.) Robyns = Rytigynia celastroides (Baill.) Verdc. ●☆

341334　Rytigynia monantha (K. Schum.) Robyns;单花纹蕊茜■☆

341335　Rytigynia murifolia Gilli = Rytigynia uhligii (K. Schum. et K. Krause) Verdc. ■☆

341336　Rytigynia mutabilis Robyns;易变纹蕊茜■☆

341337　Rytigynia neglecta (Hiern) Robyns;忽视纹蕊茜■☆

341338　Rytigynia neglecta (Hiern) Robyns var. vatkeana (Hiern) Verdc.;瓦特凯纹蕊茜●☆

341339　Rytigynia nigerica (S. Moore) Robyns;尼日利亚纹蕊茜■☆

341340　Rytigynia nodulosa (K. Schum.) Robyns;多节纹蕊茜●☆

341341　Rytigynia obscura Robyns;隐匿纹蕊茜■☆

341342　Rytigynia oligacantha (K. Schum.) Robyns = Rytigynia celastroides (Baill.) Verdc. ●☆

341343　Rytigynia orbicularis (K. Schum.) Robyns;圆形纹蕊茜■☆

341344　Rytigynia parvifolia Verdc. ;小叶纹蕊茜■☆

341345　Rytigynia pauciflora (Schweinf. ex Hiern) Robyns;少花纹蕊茜■☆

341346　Rytigynia pawekiae Verdc. ;帕维基纹蕊茜■☆

341347　Rytigynia perlucidula Robyns = Rytigynia umbellulata (Hiern) Robyns ■☆

341348　Rytigynia phyllanthoidea (Baill.) Bullock = Pyrostria phyllanthoidea (Baill.) Bridson ●☆

341349　Rytigynia pseudolongicaudata Verdc. ;假长尾纹蕊茜■☆

341350　Rytigynia pubescens Verdc. ;短柔毛纹蕊茜■☆

341351　Rytigynia reticulata Robyns = Rytigynia adenodonta (K. Schum.) Robyns var. reticulata (Robyns) Verdc. ■☆

341352　Rytigynia rhamnoides Robyns;鼠李纹蕊茜■☆

341353　Rytigynia rubiginosa (K. Schum.) Robyns;锈红纹蕊茜■☆

341354　Rytigynia rubra Robyns;红纹蕊茜■☆

341355　Rytigynia ruwenzoriensis (De Wild.) Robyns;鲁文佐里纹蕊茜■☆

341356　Rytigynia ruwenzoriensis (De Wild.) Robyns var. breviflora ? = Rytigynia ruwenzoriensis (De Wild.) Robyns ■☆

341357　Rytigynia saliensis Verdc. ;萨利纹蕊茜■☆

341358　Rytigynia sambavensis Cavaco;桑巴纹蕊茜■☆

341359　Rytigynia schumannii Robyns = Rytigynia uhligii (K. Schum. et K. Krause) Verdc. ■☆

341360　Rytigynia schumannii Robyns var. puberula (K. Schum.) Robyns;微毛纹蕊茜■☆

341361　Rytigynia schumannii Robyns var. uhligii (K. Schum. et K. Krause) Robyns = Rytigynia uhligii (K. Schum. et K. Krause) Verdc. ■☆

341362　Rytigynia senegalensis Blume;塞内加尔纹蕊茜■●☆

341363　Rytigynia senegalensis Blume var. ledermannii Robyns;莱德纹蕊茜■☆

341364　Rytigynia sessilifolia Robyns = Rytigynia decussata (K. Schum.) Robyns ■☆

341365　Rytigynia setosa Robyns;刚毛纹蕊茜■☆

341366　Rytigynia seyrigii Cavaco;塞里格尔纹蕊茜■☆

341367　Rytigynia sparsifolia (S. Moore) Robyns = Rytigynia umbellulata (Hiern) Robyns ■☆

341368　Rytigynia squamata (De Wild.) Robyns;鳞纹蕊茜■☆

341369　Rytigynia stolzii Robyns;斯托尔兹纹蕊茜■●☆

341370　Rytigynia subbiflora (Mildbr.) Robyns;亚双花纹蕊茜■☆

341371　Rytigynia tomentosa (K. Schum.) Robyns = Pachystigma schumannianum (Robyns) Bridson et Verdc. subsp. mucronulatum (Robyns) Bridson et Verdc. ●☆

341372　Rytigynia torrei Verdc. ;托雷纹蕊茜■☆

341373　Rytigynia uhligii (K. Schum. et K. Krause) Verdc. ;乌里希纹蕊茜■☆

341374　Rytigynia umbellulata (Hiern) Robyns;小伞纹蕊茜■☆

341375　Rytigynia undulata Robyns = Rytigynia uhligii (K. Schum. et K. Krause) Verdc. ■☆

341376　Rytigynia verruculosa (K. Krause) Robyns;小疣纹蕊茜■☆

341377　Rytigynia vilhenae Cavaco = Rytigynia rubiginosa (K. Schum.) Robyns ■☆

341378　Rytigynia viridissima (Wernham) Robyns = Rytigynia membranacea (Hiern) Robyns ■☆

341379　Rytigynia welwitschii Robyns = Rytigynia umbellulata (Hiern) Robyns ■☆

341380　Rytigynia xanthotricha (K. Schum.) Verdc. ;黄毛纹蕊茜●☆

341381　Rytilix Hitchc. = Hackelochloa Kuntze ■

341382　Rytilix Raf. = Hackelochloa Kuntze ■

341383　Rytilix Raf. ex Hitchc. = Hackelochloa Kuntze ■

341384　Rytilix granularis (L.) Skeels = Hackelochloa granularis (L.) Kuntze ■

341385　Rzedowskia Cham. et Schltdl. = Smelowskia C. A. Mey. ex Ledebour(保留属名)■

341386　Rzedowskia Medrano(1981);任多卫矛属●☆

341387　Rzedowskia tolantonguensis Medrano;任多卫矛●☆

341388　Saba (Pichon) Pichon(1953);萨巴木属●☆

341389　Saba Pichon = Saba (Pichon) Pichon ●☆

341390　Saba comorensis (Bojer ex A. DC.) Pichon;萨巴木●☆

341391　Saba comorensis (Bojer ex A. DC.) Pichon var. florida (Benth.) Pichon = Saba comorensis (Bojer ex A. DC.) Pichon ●☆

341392　Saba florida (Benth.) Bullock = Saba comorensis (Bojer ex A. DC.) Pichon ●☆

341393　Saba senegalensis (A. DC.) Pichon;塞内加尔萨巴木●☆

341394　Saba senegalensis (A. DC.) Pichon var. glabriflora (Hua) Pichon = Saba senegalensis (A. DC.) Pichon ●☆

341395　Saba thompsonii (A. Chev.) Pichon;汤普森萨巴木●☆

341396　Sabadilla Brandt et Ratzeb. = Schoenocaulon A. Gray ■☆

341397　Sabadilla Raf. = Sabadilla Brandt et Ratzeb. ■☆

341398　Sabadilla drummondii Kuntze = Schoenocaulon ghiesbreghtii Greenm. ■☆

341399　Sabadilla dubia (Michx.) Kuntze = Schoenocaulon dubium (Michx.) Small ■☆

341400 Sabal Adans. = Sabal Adans. ex Guers. ●

341401 Sabal Adans. ex Guers. (1763);菜棕属(蓝棕属,南美棕属,箸棕属,萨巴尔桐属,萨巴尔椰子属,萨巴棕属,沙巴尔桐属,沙巴桐属);Palmetto, Vegetablepalm ●

341402 Sabal adansonii Guers. = Sabal minor (Jacq.) Pers. ●

341403 Sabal adansonii Guers. var. megacarpa Chapm. = Sabal etonia Swingle ex Nash ●☆

341404 Sabal adiantinum Raf. = Sabal minor (Jacq.) Pers. ●

341405 Sabal bermudana L. H. Bailey;百慕大箸棕(百慕大萨巴尔桐);Bermuda Palm, Bermuda Palmetto ●

341406 Sabal blackburniana Glaz. ;大叶箸棕(百慕大萨巴尔桐);Blackburn, Blackburn Palmetto, Fan Palm, Hispaniolan Palmetto ●

341407 Sabal causiarum (Cook) Becc. ;海地菜棕(巨箸棕);Puerto Rico Hat Palm, Purtorican Palmetto ●☆

341408 Sabal causiarum Becc. = Sabal causiarum (Cook) Becc. ●☆

341409 Sabal deeringiana Small = Sabal minor (Jacq.) Pers. ●

341410 Sabal domingensis Becc. ;大果箸棕(海地菜棕);Cana, Dominican Palm, Dominican Palmetto, Hispaniolan Palmetto Palm, Latanier ●☆

341411 Sabal etonia Swingle ex Nash;小箸棕;Dwarf Palmetto, Scrub Palmetto ●☆

341412 Sabal exul (O. F. Cook) L. H. Bailey = Sabal mexicana Mart. ●☆

341413 Sabal jamesiana Small = Sabal palmetto (Walter) Lodd. ex Roem. et Schult. f. ●

341414 Sabal louisiana (Darby) Bomhard = Sabal minor (Jacq.) Pers. ●

341415 Sabal maritima Burret;牙买加箸棕●☆

341416 Sabal mauritiiformis Griseb. et H. Wendl. ;灰绿箸棕;Savannah Palm ●☆

341417 Sabal megacarpa (Chapm.) Small = Sabal etonia Swingle ex Nash ●☆

341418 Sabal mexicana Mart. ;墨西哥菜棕(墨西哥菜棕);Mexican Palmetto, Oaxaca Palmetto, Rio Grande Palmetto ●☆

341419 Sabal miamiensis Zona;迈阿密箸棕;Miami Palmetto ●☆

341420 Sabal mininum Nutt. = Sabal minor (Jacq.) Pers. ●

341421 Sabal minor (Jacq.) Pers. ;矮菜棕(矮茎萨尔桐,矮茎沙巴桐,露沙箸棕,小箸棕,侏儒桐);Bush Palmetto, Dcrub Palmetto, Dwarf Palmetto, Dwarf Vegetablepalm, Little Blue Stem, Palmetto, Swamp Palmetto ●

341422 Sabal palmetto (Walter) Lodd. ex Roem. et Schult. f. ;菜棕(巴尔麦棕桐,甘蓝棕,箸棕);American Palmetto, Cabbage Palm, Cabbage Palmetto, Cabbage Palmetto Palm, Cabbage-palm, Carolina Palmetto, Florida Palmetto, Palmetto, Palmetto Royal, Pond Thatch, Pond Top, Sabal Palm, Thatch Palm, Vegetablepalm ●

341423 Sabal parviflora Becc. ;小花菜棕●☆

341424 Sabal princeps Becc. ;王箸棕●☆

341425 Sabal pumila (Walter) Elliott = Sabal minor (Jacq.) Pers. ●

341426 Sabal rosei Becc. ;粉红箸棕●☆

341427 Sabal serrulata (Michx.) Nutt. ex Schult. et Schult. f. = Serenoa repens (Bartram) Small ●☆

341428 Sabal serrulatum Schult. f. = Serenoa repens (Bartram) Small ●☆

341429 Sabal texana (O. F. Cook) Becc. = Sabal mexicana Mart. ●☆

341430 Sabal texana Becc. ;墨西哥箸棕●☆

341431 Sabal uresana Trel. ;伍勒萨箸棕;Sonoran Palmetto ●☆

341432 Sabalaceae Schultz Sch. ;菜棕科●

341433 Sabalaceae Schultz Sch. = Arecaceae Bercht. et J. Presl(保留科名)●

341434 Sabalaceae Schultz Sch. = Palmae Juss. (保留科名)●

341435 Sabatia Adans. (1763);萨巴特龙胆属;Marsh-pink, Rose-gentian ■☆

341436 Sabatia Adans. = Sabbatia Adans. ■☆

341437 Sabatia angularis (L.) Pursh = Sabbatia angularis (L.) Pursh ■☆

341438 Sabatia australis Cham. et Schltdl. = Zygostigma australe (Cham. et Schltdl.) Griseb. ■☆

341439 Sabatia brachiata Elliott;沼泽萨巴特龙胆;Marsh Pink ■☆

341440 Sabatia campestris Nutt. ;平原萨巴特龙胆;Prairie Rose Gentian ■☆

341441 Sabatia dodecandra (L.) Britton, Sterns et Poggenb. ;十二雄蕊龙胆(萨巴夫龙胆)■☆

341442 Sabatia dodecandra Britton, Sterns et Poggenb. = Sabatia dodecandra (L.) Britton, Sterns et Poggenb. ■☆

341443 Sabaudia Buscal. et Muschl. (1913);萨包草属■●☆

341444 Sabaudia Buscal. et Muschl. = Lavandula L. ●■

341445 Sabaudia helenae Buscal. et Muschl. ;萨包草■☆

341446 Sabaudiella Chiov. (1929);萨包花属●☆

341447 Sabaudiella aloysii Chiov. ;萨包花●☆

341448 Sabaudiella aloysii Chiov. = Hildebrandtia aloysii (Chiov.) Sebsebe ●☆

341449 Sabazia Cass. (1827);粉白菊属■●☆

341450 Sabazia annua (S. F. Blake) B. L. Turner;一年粉白菊■☆

341451 Sabazia elata (Canne) B. L. Turner;高大粉白菊■☆

341452 Sabazia glabra S. Watson;无毛粉白菊■☆

341453 Sabazia microglossa DC. ;小舌粉白菊■☆

341454 Sabazia multiradiata (Seaton) Longpre;多线粉白菊■☆

341455 Sabazia obtusata S. F. Blake;钝粉白菊■☆

341456 Sabazia triangularis S. F. Blake;三角粉白菊■☆

341457 Sabazia trifida J. J. Fay;三裂粉白菊■☆

341458 Sabbata Vell. (1829);萨巴菊属■☆

341459 Sabbata romana Vell. ;萨巴菊■☆

341460 Sabbatia Adans. = Micromeria Benth. (保留属名)■●

341461 Sabbatia Adans. = Sabatia Adans. ■☆

341462 Sabbatia Moench = Micromeria Benth. (保留属名)■●

341463 Sabbatia Moench (1794);美苦草属;Marsh-pink, Rose Gentian, Rose Pink, Rose-gentian ■☆

341464 Sabbatia Post et Kuntze = Sabbata Vell. ☆

341465 Sabbatia Salisb. = Sabatia Adans. ■☆

341466 Sabbatia angularis (L.) Pursh;美苦草;Common Marsh-pink, Rose Gentian, Rose Pink, Rose-gentian, Rose-pink, Sabbatia ■☆

341467 Sabbatia campanulata ?;细美苦草;Slender Marsh-pink ■☆

341468 Sabbatia corymbosa Moench = Micromeria juliana (L.) Rchb. ■☆

341469 Sabbatia dodecandra ?;十二蕊美苦草;Large Marsh-pink ■☆

341470 Sabbatia elliottii Steud. ;翼枝美苦草■☆

341471 Sabbatia stellaris ?;星美苦草(北美美苦草);Sea Pink ■☆

341472 Sabdariffa (DC.) Kostel. = Hibiscus L. (保留属名)●■

341473 Sabdariffa Kostel. = Hibiscus L. (保留属名)●■

341474 Sabdariffa rubra Kostel. = Hibiscus sabdariffa L. ●■

341475 Sabia Colebr. (1819);清风藤属;Sabia ●

341476 Sabia acuminata H. Y. Chen = Sabia purpurea Hook. f. et Thomson subsp. dumicala (W. W. Sm.) Water ●

341477 Sabia angustifolia H. Y. Chen;剑川清风藤●

341478 Sabia angustifolia H. Y. Chen = Sabia yunnanensis Franch. ●

341479 Sabia bicolor H. Y. Chen;两色清风藤;Bicolor Sabia ●

341480 Sabia bicolor H. Y. Chen = Sabia schumanniana Diels subsp. pluriflora (Rehder et E. H. Wilson) Y. F. Wu ●

341481 Sabia bicolor H. Y. Chen = Sabia schumanniana Diels subsp.

pluriflora（Rehder et E. H. Wilson）Y. F. Wu var. bicolor（H. Y. Chen）Y. F. Wu ●

341482　Sabia brevipetiolata H. Y. Chen ＝Sabia dielsii H. Lév. ●

341483　Sabia bullockii Hance ＝Sabia japonica Maxim. ●

341484　Sabia burmanica H. Y. Chen；厚叶清风藤；Burma Sabia ●

341485　Sabia calcicola C. Y. Wu ex S. K. Chen；文山清风藤；Calcicole Sabia，Wenshan Sabia ●

341486　Sabia calcicola C. Y. Wu ex S. K. Chen ＝Sabia fasciculata Lecomte ex H. Y. Chen ●

341487　Sabia callosa H. Y. Chen；纸叶清风藤；Callous Sabia ●

341488　Sabia callosa H. Y. Chen ＝Sabia yunnanensis Franch. ●

341489　Sabia campanulata Wall. ＝Sabia campanulata Wall. ex Roxb. ●

341490　Sabia campanulata Wall. ex Roxb.；钟花清风藤；Bellflower Sabia，Campanulate Sabia ●

341491　Sabia campanulata Wall. ex Roxb. subsp. metcalfana（H. Y. Chen）Y. F. Wu；隆陵清风藤；Metcalf Sabia ●

341492　Sabia campanulata Wall. ex Roxb. subsp. ritchieae（Rehder et E. H. Wilson）Y. F. Wu；鄂西清风藤；West Hubei Sabia ●

341493　Sabia campanulata Wall. subsp. metcalfiana（H. Y. Chen）Y. F. Wu ＝Sabia campanulata Wall. ex Roxb. ●

341494　Sabia cavaleriei H. Lév. ＝Orixa japonica Thunb. ●

341495　Sabia coriacea Rehder et E. H. Wilson；革叶清风藤（厚叶清风藤）；Coriaceousleaf Sabia，Coriaceous-leaved Sabia ●

341496　Sabia croizatiana H. Y. Chen ＝Sabia yunnanensis Franch. ●

341497　Sabia dielsii H. Lév.；平伐清风藤（云雾清风藤）；Diels Sabia ●

341498　Sabia discolor Dunn；灰背清风藤（白背清风藤，风藤，叶上果）；Diversecolor Sabia，Diverse-colored Sabia ●

341499　Sabia dumicala W. W. Sm. ＝Sabia purpurea Hook. f. et Thomson subsp. dumicala（W. W. Sm.）Water ●

341500　Sabia dunnii H. Lév. ＝Sabia swinhoei Hemsl. ex Forbes et Hemsl. ●

341501　Sabia emarginata Lecomte；凹萼清风藤（凹叶清风藤）；Emarginate Sabia ●

341502　Sabia esquirolii H. Lév. ＝Gardneria multiflora Makino ●

341503　Sabia fasciculata Lecomte ex Anon. ＝Sabia fasciculata Lecomte ex Koidz. ●

341504　Sabia fasciculata Lecomte ex H. Y. Chen；簇花清风藤（小发散）；Fascicled-flower Sabia，Fascicled-flowered Sabia ●

341505　Sabia fasciculata Lecomte ex Koidz. ＝Sabia fasciculata Lecomte ex H. Y. Chen ●

341506　Sabia feddei H. Lév. ＝Orixa japonica Thunb. ●

341507　Sabia gaultheriifolia Stapf ex H. Y. Chen ＝Sabia campanulata Wall. ex Roxb. subsp. ritchieae（Rehder et E. H. Wilson）Y. F. Wu ●

341508　Sabia glandulosa H. Y. Chen；中甸清风藤 ●

341509　Sabia glandulosa H. Y. Chen ＝Sabia yunnanensis Franch. ●

341510　Sabia gracilis Hemsl. ＝Sabia swinhoei Hemsl. ex Forbes et Hemsl. ●

341511　Sabia harmandiana Pierre ＝Sabia parviflora Wall. ex Roxb. ●

341512　Sabia heterosepala H. Y. Chen ＝Sabia emarginata Lecomte ●

341513　Sabia japonica Maxim.；清风藤（寻风藤）；Japan Sabia，Japanese Sabia ●

341514　Sabia japonica Maxim. var. sinensis（Stapf）H. Y. Chen；中华清风藤；Chinese Japanese Sabia ●

341515　Sabia japonica Maxim. var. spinosa Lecomte ＝Sabia japonica Maxim. ●

341516　Sabia kachinica H. Y. Chen ＝Sabia lanceolata Colebr. ●

341517　Sabia lanceolata Colebr.；披针清风藤 ●

341518　Sabia latifolia Rehder et E. H. Wilson ＝Sabia yunnanensis Franch. subsp. latifolia（Rehder et E. H. Wilson）Y. F. Wu ●

341519　Sabia latifolia Rehder et E. H. Wilson var. omeiensis（Stapf ex H. Y. Chen）S. K. Chen ＝Sabia yunnanensis Franch. subsp. latifolia（Rehder et E. H. Wilson）Y. F. Wu ●

341520　Sabia leptandra Hook. f. et Thomson；细蕊清风藤；Slenderandrus Sabia ●

341521　Sabia leptandra Hook. f. et Thomson ＝Sabia yunnanensis Franch. ●

341522　Sabia limoniacea Wall. ＝Sabia limoniacea Wall. ex Hook. f. et Thomson ●

341523　Sabia limoniacea Wall. ex Hook. f. et Thomson；柠檬清风藤；Lemon Sabia ●

341524　Sabia limoniacea Wall. var. ardisioides（Hook. et Arn.）H. Y. Chen；毛萼清风藤 ●

341525　Sabia limoniacea Wall. var. ardisioides H. Y. Chen ＝Sabia limoniacea Wall. ex Hook. f. et Thomson ●

341526　Sabia longruiensis X. X. Chen et D. R. Liang；陇瑞清风藤；Longrui Sabia ■

341527　Sabia longruiensis X. X. Chen et D. R. Liang ＝Sabia swinhoei Hemsl. ex Forbes et Hemsl. ●

341528　Sabia metcalfiana H. Y. Chen ＝Sabia campanulata Wall. ex Roxb. subsp. metcalfana（H. Y. Chen）Y. F. Wu ●

341529　Sabia metcalfiana H. Y. Chen ＝Sabia campanulata Wall. ex Roxb. ●

341530　Sabia nervosa Chun ex Y. F. Wu；长脉清风藤；Longiveined Sabia，Longvein Sabia ●

341531　Sabia obovatifolia Y. W. Law et Y. F. Wu；倒卵叶清风藤（倒卵清风藤）；Obovateleaf Sabia，Obovate-leaved Sabia ●

341532　Sabia obovatifolia Y. W. Law et Y. F. Wu ＝Sabia yunnanensis Franch. subsp. latifolia（Rehder et E. H. Wilson）Y. F. Wu ●

341533　Sabia olacifolia Stapf et H. Y. Chen ＝Sabia dielsii H. Lév. ●

341534　Sabia omeiensis Stapf ex H. Y. Chen ＝Sabia yunnanensis Franch. subsp. latifolia（Rehder et E. H. Wilson）Y. F. Wu ●

341535　Sabia ovalifolia S. Y. Liu ＝Sabia swinhoei Hemsl. ex Forbes et Hemsl. ●

341536　Sabia pallida Stapf ex H. Y. Chen ＝Sabia yunnanensis Franch. ●

341537　Sabia paniculata Edgew. ex Hook. f. et Thomson；锥序清风藤（圆锥花清风藤）；Panicle Sabia，Paniculate Sabia ●

341538　Sabia parviflora Wall. ＝Sabia parviflora Wall. ex Roxb. ●

341539　Sabia parviflora Wall. ex Roxb.；小花清风藤；Smallflower Sabia，Small-flowered Sabia ●

341540　Sabia parviflora Wall. ex Roxb. var. harmandiana（Pierre）Lecomte ＝Sabia parviflora Wall. ex Roxb. ●

341541　Sabia parviflora Wall. ex Roxb. var. nitidissima H. Lév. ＝Sabia parviflora Wall. ex Roxb. ●

341542　Sabia parvifolia H. Y. Chen ＝Sabia purpurea Hook. f. et Thomson subsp. dumicala（W. W. Sm.）Water ●

341543　Sabia pentadenia H. Y. Chen；五腺清风藤；Pentagland Sabia ●

341544　Sabia pentadenia H. Y. Chen ＝Sabia yunnanensis Franch. ●

341545　Sabia polyantha Hand. -Mazz. ＝Sabia parviflora Wall. ex Roxb. ●

341546　Sabia puberula Rehder et E. H. Wilson；兴山清风藤（柔毛清风藤）●

341547　Sabia puberula Rehder et E. H. Wilson ＝Sabia yunnanensis Franch. ●

341548　Sabia puberula Rehder et E. H. Wilson var. hupehensis H. Y. Chen ＝Sabia yunnanensis Franch. ●

341549　Sabia pubescens H. Y. Chen = Sabia yunnanensis Franch. ●

341550　Sabia purpurea Hook. f. et Thomson;紫花清风藤;Purple-flower Sabia, Purple-flowered Sabia ●

341551　Sabia purpurea Hook. f. et Thomson subsp. dumicala（W. W. Sm.）Water;灌丛清风藤;Shrubby Sabia ●

341552　Sabia ritchieae Rehder et E. H. Wilson = Sabia campanulata Wall. ex Roxb. subsp. ritchieae（Rehder et E. H. Wilson）Y. F. Wu ●

341553　Sabia rockii H. Y. Chen;丽江清风藤●

341554　Sabia rockii H. Y. Chen = Sabia yunnanensis Franch. ●

341555　Sabia rotundata Stapf ex H. Y. Chen = Sabia yunnanensis Franch. ●

341556　Sabia schumanniana Diels;四川清风藤（女儿藤,青木香,清风藤,清木香,石钻子,铁牛钻石,钻石风）;Sichuan Sabia, Szechwan Sabia ●

341557　Sabia schumanniana Diels subsp. longipes（Rehder et E. H. Wilson）C. Y. Chang = Sabia schumanniana Diels var. longipes Rehder et E. H. Wilson ●

341558　Sabia schumanniana Diels subsp. longipes（Rehder et E. H. Wilson）C. Y. Chang = Sabia schumanniana Diels ●

341559　Sabia schumanniana Diels subsp. pluriflora（Rehder et E. H. Wilson）Y. F. Wu var. bicolor（H. Y. Chen）Y. F. Wu = Sabia bicolor H. Y. Chen ●

341560　Sabia schumanniana Diels subsp. pluriflora（Rehder et E. H. Wilson）Y. F. Wu;多花清风藤;Many-flower Sichuan Sabia ●

341561　Sabia schumanniana Diels var. bicolor（H. Y. Chen）Y. F. Wu = Sabia schumanniana Diels subsp. pluriflora（Rehder et E. H. Wilson）Y. F. Wu ●

341562　Sabia schumanniana Diels var. longipes Rehder et E. H. Wilson;长柄四川清风藤;Longistalked Sichuan Sabia ●

341563　Sabia schumanniana Diels var. longipes Rehder et E. H. Wilson = Sabia schumanniana Diels ●

341564　Sabia schumanniana Diels var. pluriflora Rehder et E. H. Wilson = Sabia schumanniana Diels subsp. pluriflora（Rehder et E. H. Wilson）Y. F. Wu ●

341565　Sabia shensiensis H. Y. Chen;陕西清风藤;Shaanxi Sabia, Shaaxi Sabia ●

341566　Sabia shensiensis H. Y. Chen = Sabia campanulata Wall. ex Roxb. subsp. ritchieae（Rehder et E. H. Wilson）Y. F. Wu ●

341567　Sabia sinensis Stapf ex Anon = Sabia japonica Maxim. var. sinensis（Stapf）H. Y. Chen ●

341568　Sabia spinosa Stapf ex Anon. = Sabia japonica Maxim. ●

341569　Sabia swinhoei Hemsl. = Sabia swinhoei Hemsl. ex Forbes et Hemsl. ●

341570　Sabia swinhoei Hemsl. ex Forbes et Hemsl.;尖叶清风藤（台湾清风藤）;Sharpleaf Sabia, Sharp-leaved Sabia, Swinhoe Sabia, Taiwan Sabia ●

341571　Sabia swinhoei Hemsl. ex Forbes et Hemsl. var. hainanensis H. Y. Chen = Sabia swinhoei Hemsl. ex Forbes et Hemsl. ●

341572　Sabia swinhoei Hemsl. var. hainanensis H. Y. Chen = Sabia swinhoei Hemsl. ex Forbes et Hemsl. ●

341573　Sabia swinhoei Hemsl. var. parvifolia Y. H. Xiang et Q. H. Chen = Sabia swinhoei Hemsl. ex Forbes et Hemsl. ●

341574　Sabia swinhoei Hemsl. var. subcorymbosa H. Y. Chen = Sabia swinhoei Hemsl. ex Forbes et Hemsl. ●

341575　Sabia transarisanensis Hayata;阿里山清风藤;Alishan Sabia, Transarisan Sabia ●

341576　Sabia uropetala Gagnep. = Sabia swinhoei Hemsl. ex Forbes et Hemsl. ●

341577　Sabia wangii H. Y. Chen = Sabia dielsii H. Lév. ●

341578　Sabia yuii H. Y. Chen = Sabia yunnanensis Franch. ●

341579　Sabia yunnanensis Franch.;云南清风藤（滇东清风藤,风藤草,鸡舌头叶,老鼠吹箫,羊饥藤）;East Yunnan Sabia, Yunnan Sabia ●

341580　Sabia yunnanensis Franch. subsp. latifolia（Rehder et E. H. Wilson）Y. F. Wu;宽叶清风藤（峨眉宽叶清风藤,宽叶滇东清风藤,阔叶清风藤）;Broadleaf East Yunnan Sabia, Broad-leaved Sabia, Emei Broad-leaved Sabia ●

341581　Sabia yunnanensis Franch. var. mairei（H. Lév.）H. Y. Chen = Sabia yunnanensis Franch. ●

341582　Sabiaceae Blume(1851)（保留科名）;清风藤科;Sabia Family ●

341583　Sabicea Aubl.（1775）;萨比斯茜属●☆

341584　Sabicea acuminata Baker;渐尖萨比斯茜●☆

341585　Sabicea adamsii Hepper = Bertiera adamsii（Hepper）N. Hallé ■☆

341586　Sabicea affinis De Wild. = Sabicea venosa Benth. ●☆

341587　Sabicea africana（P. Beauv.）Hepper = Stipularia africana P. Beauv. ●☆

341588　Sabicea amomi Wernham;豆蔻萨比斯茜●☆

341589　Sabicea angolensis Wernham = Sabicea venosa Benth. ●☆

341590　Sabicea angustifolia Boivin ex Wernham;窄叶萨比斯茜●☆

341591　Sabicea anomala K. Schum. = Bertiera letouzeyi Hallé ■☆

341592　Sabicea arachnoidea Hutch. et Dalziel;蛛毛萨比斯茜●☆

341593　Sabicea arborea K. Schum. = Pseudosabicea arborea（K. Schum.）N. Hallé ●☆

341594　Sabicea barteri Wernham = Sabicea calycina Benth. ●☆

341595　Sabicea batesii Wernham = Pseudosabicea batesii（Wernham）N. Hallé ●☆

341596　Sabicea bequaertii De Wild. = Pseudosabicea arborea（K. Schum.）N. Hallé subsp. bequaertii（De Wild.）Verdc. ●☆

341597　Sabicea bicarpellata K. Schum. = Bertiera bicarpellata（K. Schum.）N. Hallé ■☆

341598　Sabicea brachiata Wernham;短萨比斯茜●☆

341599　Sabicea bracteolata Wernham;小苞萨比斯茜●☆

341600　Sabicea brevipes Wernham;短梗萨比斯茜●☆

341601　Sabicea brunnea Wernham = Sabicea capitellata Benth. ●☆

341602　Sabicea calycina Benth.;苍白萨比斯茜●☆

341603　Sabicea calycina Benth. var. hirsutiflora Wernham = Sabicea calycina Benth. ●☆

341604　Sabicea cameroonensis Wernham;喀麦隆萨比斯茜●☆

341605　Sabicea capitellata Benth.;头状萨比斯茜●☆

341606　Sabicea capitellata Benth. var. insularis Wernham = Sabicea ingrata K. Schum. var. insularis（Wernham）Joffroy ●☆

341607　Sabicea carminata N. Hallé;洋红萨比斯茜●☆

341608　Sabicea cauliflora Hiern = Ecpoma cauliflorum（Hiern）N. Hallé ■☆

341609　Sabicea composita Wernham;复合萨比斯茜●☆

341610　Sabicea congensis Wernham;刚果萨比斯茜●☆

341611　Sabicea cordata Hutch. et Dalziel;心形萨比斯茜●☆

341612　Sabicea cruciata Wernham;十字形萨比斯茜●☆

341613　Sabicea dewevrei De Wild.;德韦萨比斯茜●☆

341614　Sabicea dewevrei De Wild. et T. Durand var. latifolia De Wild. = Sabicea dewevrei De Wild. ●☆

341615　Sabicea dewevrei De Wild. var. glabra Wernham;光滑萨比斯茜●☆

341616　Sabicea dewildemaniana Wernham;德氏曼萨比斯茜●☆

341617　Sabicea dinklagei K. Schum.;丁氏萨比斯茜●☆

341618 Sabicea discolor Stapf;异色萨比斯茜●☆

341619 Sabicea diversifolia Pers. ;异叶萨比斯茜●☆

341620 Sabicea dubia Wernham;可疑萨比斯茜●☆

341621 Sabicea duparquetiana Baill. ex Wernham;迪帕萨比斯茜●☆

341622 Sabicea efulenensis (Hutch.) Hepper = Sabicea gabonica (Hiern) Hepper ●☆

341623 Sabicea elliptica (Schweinf. ex Hiern) Hepper = Stipularia elliptica Schweinf. ex Hiern ■☆

341624 Sabicea entebbensis Wernham;恩德培萨比斯茜●☆

341625 Sabicea exellii G. Taylor;埃克塞尔萨比斯茜●☆

341626 Sabicea ferruginea (G. Don) Benth. ;锈色萨比斯茜●☆

341627 Sabicea ferruginea (G. Don) Benth. var. lasiocalyx (Stapf) Wernham = Sabicea ferruginea (G. Don) Benth. ●☆

341628 Sabicea floribunda K. Schum. = Pseudosabicea floribunda (K. Schum.) N. Hallé ●☆

341629 Sabicea fulva Wernham;黄褐萨比斯茜●☆

341630 Sabicea fulvovenosa R. D. Good;黄褐脉萨比斯茜●☆

341631 Sabicea gabonica (Hiern) Hepper;加蓬萨比斯茜●☆

341632 Sabicea geophiloides Wernham;爱地草萨比斯茜●☆

341633 Sabicea gigantea Wernham;巨大萨比斯茜●☆

341634 Sabicea gigantostipula K. Schum. = Ecpoma gigantostipulum (K. Schum.) N. Hallé ■☆

341635 Sabicea gilletii De Wild. ;吉勒特萨比斯茜●☆

341636 Sabicea globifera Hutch. et Dalziel = Sabicea vogelii Benth. ●☆

341637 Sabicea goossensii De Wild. ;古斯萨比斯茜●☆

341638 Sabicea gracilis Wernham;纤细萨比斯茜●☆

341639 Sabicea henningsiana Büttner = Pseudosabicea segregata (Hiern) N. Hallé ●☆

341640 Sabicea hierniana Wernham = Ecpoma hiernianum (Wernham) N. Hallé et F. Hallé ■☆

341641 Sabicea homblei De Wild. = Sabicea dinklagei K. Schum. ●☆

341642 Sabicea ingrata K. Schum. var. insularis (Wernham) Joffroy;海岛萨比斯茜●☆

341643 Sabicea insularis (Wernham) G. Taylor = Sabicea ingrata K. Schum. var. insularis (Wernham) Joffroy ●☆

341644 Sabicea johnstonii K. Schum. ex Wernham;约翰斯顿萨比斯茜●☆

341645 Sabicea kolbeana Büttner = Sabicea venosa Benth. ●☆

341646 Sabicea lanata Hepper;绵毛萨比斯茜●☆

341647 Sabicea lanuginosa Wernham;多毛萨比斯茜●☆

341648 Sabicea lasiocalyx Stapf = Sabicea ferruginea (G. Don) Benth. ●☆

341649 Sabicea laurentii De Wild. = Sabicea dinklagei K. Schum. ●☆

341650 Sabicea laurentii De Wild. var. velutina ? = Sabicea dinklagei K. Schum. ●☆

341651 Sabicea laxa Wernham;疏松萨比斯茜●☆

341652 Sabicea laxothyrsus K. Schum. et Dinkl. ex Stapf = Sabicea discolor Stapf ●☆

341653 Sabicea leucocarpa (K. Krause) Mildbr. ;白果萨比斯茜●☆

341654 Sabicea liberica Hepper;利比里亚萨比斯茜●☆

341655 Sabicea linderi (Hutch. et Dalziel) Bremek. = Schizocolea linderi (Hutch. et Dalziel) Bremek. ●☆

341656 Sabicea longepetiolata De Wild. ;长梗萨比斯茜●☆

341657 Sabicea loxothyrsus K. Schum. et Dinkl. ex Stapf;斜序萨比斯茜●☆

341658 Sabicea marojejyensis Razafim. et J. S. Mill. ;马罗萨比斯茜●☆

341659 Sabicea mildbraedii Wernham = Pseudosabicea mildbraedii (Wernham) N. Hallé ●☆

341660 Sabicea mollis K. Schum. ex Wernham;柔软萨比斯茜●☆

341661 Sabicea mortehani De Wild. ;莫特汉萨比斯茜●☆

341662 Sabicea multibracteata J. B. Hall;多苞萨比斯茜●☆

341663 Sabicea neglecta Hepper;疏忽萨比斯茜●☆

341664 Sabicea nobilis R. D. Good = Pseudosabicea nobilis (R. D. Good) N. Hallé ●☆

341665 Sabicea orientalis Wernham;东方萨比斯茜●☆

341666 Sabicea parviflora K. Schum. ex Wernham = Bertiera orthopetala (Hiern) N. Hallé ■☆

341667 Sabicea pedicellata Wernham = Pseudosabicea pedicellata (Wernham) N. Hallé ●☆

341668 Sabicea pilosa Hiern;疏毛萨比斯茜●☆

341669 Sabicea pseudocapitellata Wernham;假头萨比斯茜●☆

341670 Sabicea robbii Wernham = Sabicea duparquetiana Baill. ex Wernham ●☆

341671 Sabicea rosea Hoyle;粉红萨比斯茜●☆

341672 Sabicea rufa Wernham;浅红萨比斯茜●☆

341673 Sabicea salmonea A. Chev. = Sabicea ferruginea (G. Don) Benth. ●☆

341674 Sabicea schumanniana Büttner;舒曼萨比斯茜●☆

341675 Sabicea segregata Hiern = Pseudosabicea segregata (Hiern) N. Hallé ●☆

341676 Sabicea smithii Wernham;史密斯萨比斯茜●☆

341677 Sabicea speciosa K. Schum. ;美丽萨比斯茜●☆

341678 Sabicea speciosissima K. Schum. ;极美萨比斯茜●☆

341679 Sabicea stipularioides Wernham;托叶状萨比斯茜●☆

341680 Sabicea talbotii Wernham;塔尔博特萨比斯茜●☆

341681 Sabicea tchapensis (Hutch.) Hepper var. glabrescens Wernham = Sabicea efulenensis (Hutch.) Hepper ●☆

341682 Sabicea thomensis Joffroy;爱岛萨比斯茜●☆

341683 Sabicea trichochlamys K. Schum. = Sabicea capitellata Benth. ●☆

341684 Sabicea trigemina K. Schum. ;三对萨比斯茜●☆

341685 Sabicea urbaniana Wernham;乌尔班萨比斯茜●☆

341686 Sabicea urceolata Hepper;坛状萨比斯茜●☆

341687 Sabicea venosa Benth. ;多脉萨比斯茜●☆

341688 Sabicea venosa Benth. var. anomala Wernham = Sabicea venosa Benth. ●☆

341689 Sabicea venosa Benth. var. villosa K. Schum. = Sabicea mollis K. Schum. ex Wernham ●☆

341690 Sabicea verticillata Wernham;轮生萨比斯茜●☆

341691 Sabicea vogelii Benth. ;沃格尔萨比斯茜●☆

341692 Sabicea xanthotricha Wernham;黄毛萨比斯茜●☆

341693 Sabiceaceae Blume(保留科名) = Rubiaceae Juss. (保留科名)●■

341694 Sabiceaceae Martinov = Rubiaceae Juss. (保留科名)●■

341695 Sabiceaceae Martinov = Sabiaceae Blume(保留科名)●

341696 Sabina Mill. (1754);圆柏属;Juniper,Sabina,Savin ●

341697 Sabina Mill. = Juniperus L. ●

341698 Sabina aquatica Antoine = Glyptostrobus pensilis (Staunton ex D. Don) K. Koch ●◇

341699 Sabina arenaria (E. H. Wilson) W. C. Cheng et W. T. Wang = Juniperus sabina L. ●

341700 Sabina arenaria (E. H. Wilson) W. C. Cheng et W. T. Wang = Sabina vulgaris Antoine ●

341701 Sabina californica (Carrière) Antoine = Juniperus californica Carrière ●☆

341702 Sabina centrasiatica (Kom.) W. C. Cheng et L. K. Fu = Juniperus centrasiatica Kom. ●

341703 Sabina centrasiatica Kom. = Juniperus tibetica Kom. ●

341704 Sabina chinensis (L.) Antoine 'Aurea' = Juniperus chinensis

L. ' Aurea ' ●

341705　Sabina chinensis (L.) Antoine ' Aureoglobosa ' = Juniperus chinensis L. ' Aureoglobosa ' ●

341706　Sabina chinensis (L.) Antoine ' Globosa ' = Juniperus chinensis L. ' Globosa ' ●

341707　Sabina chinensis (L.) Antoine ' Kaizuca Procumbens ' = Juniperus chinensis L. ' Kaizuca Procumbens ' ●

341708　Sabina chinensis (L.) Antoine ' Kaizuca ' = Juniperus chinensis L. ' Kaizuca ' ●

341709　Sabina chinensis (L.) Antoine ' Pfitzeriana ' = Juniperus chinensis L. ' Pfitzeriana ' ●

341710　Sabina chinensis (L.) Antoine ' Pyramidalis ' = Juniperus chinensis L. ' Pyramidalis ' ●

341711　Sabina chinensis (L.) Antoine = Juniperus chinensis L. ●

341712　Sabina chinensis (L.) Antoine f. aurea (Young) Beissn. = Juniperus chinensis L. ' Aurea ' ●

341713　Sabina chinensis (L.) Antoine f. aureoglobosa (Nash) W. C. Cheng et W. T. Wang = Sabina chinensis (L.) Antoine ' Aureoglobosa ' ●

341714　Sabina chinensis (L.) Antoine f. globosa (Hornibr.) W. C. Cheng et W. T. Wang = Juniperus chinensis L. ' Globosa ' ●

341715　Sabina chinensis (L.) Antoine f. pendula (Franch.) W. C. Cheng et L. K. Fu = Juniperus chinensis L. f. pendula (Franch.) W. C. Cheng et W. T. Wang ●

341716　Sabina chinensis (L.) Antoine f. pyramidalis (Carrière) W. C. Cheng et W. T. Wang = Juniperus chinensis L. ' Pyramidalis ' ●

341717　Sabina chinensis (L.) Antoine var. globosa (Hornibr.) Iwata et Kusaka = Juniperus chinensis L. ' Globosa ' ●

341718　Sabina chinensis (L.) Antoine var. kaizuca W. C. Cheng et W. T. Wang = Juniperus chinensis L. ' Kaizuca ' ●

341719　Sabina chinensis (L.) Antoine var. nana Hochst. ;矮生圆柏; Awarf Chinese Juniper ●

341720　Sabina chinensis (L.) Antoine var. pfitzeriana (Spüth) Moldenke = Juniperus chinensis L. ' Pfitzeriana ' ●

341721　Sabina chinensis (L.) Antoine var. pfitzeriana (Spüth) Moldenke = Sabina chinensis (L.) Antoine ' Pfitzeriana ' ●

341722　Sabina chinensis (L.) Antoine var. procumbens (Siebold ex Endl.) Honda = Juniperus chinensis L. var. procumbens Siebold ex Endl. ●

341723　Sabina chinensis (L.) Antoine var. sargentii (A. Henry) W. C. Cheng et L. K. Fu = Juniperus chinensis L. var. sargentii A. Henry ●

341724　Sabina convallium (Rehder et E. H. Wilson) W. C. Cheng et W. T. Wang;密枝圆柏(密条柏,深山柏,细枝桧,小籽密枝圆柏); Densebranchlet Juniper, Densebranchlet Savin, Littleseed Densebranchlet Savin ●

341725　Sabina convallium (Rehder et E. H. Wilson) W. C. Cheng et W. T. Wang var. microsperma W. C. Cheng et L. K. Fu = Sabina convallium (Rehder et E. H. Wilson) W. C. Cheng et W. T. Wang ●

341726　Sabina convallium (Rehder. et E. H. Wilson) W. C. Cheng et L. K. Fu var. microsperma W. C. Cheng et L. K. Fu = Juniperus convallium Rehder et E. H. Wilson var. microsperma (W. C. Cheng et L. K. Fu) Silba ●

341727　Sabina convallium (Rehder. et E. H. Wilson) W. C. Cheng et L. K. Fu = Juniperus convallium Rehder et E. H. Wilson ●

341728　Sabina davurica (Pall.) Antoine = Juniperus davurica Pall. ●

341729　Sabina fischeri Antoine = Juniperus pseudosabina Fisch. et C. A. Mey. ●

341730　Sabina gaussenii (W. C. Cheng) W. C. Cheng et W. T. Wang = Juniperus gaussenii W. C. Cheng ●

341731　Sabina horizontalis (Moench) Antoine = Juniperus horizontalis Moench ●☆

341732　Sabina komarovii (Florin) W. C. Cheng et W. T. Wang = Juniperus komarovii Florin ●

341733　Sabina lemeeana (H. Lév. et Vaniot) W. C. Cheng et W. T. Wang = Sabina squamata (Buch. -Ham. ex D. Don) Antoine ●

341734　Sabina lemeeana (H. Lév. et Vaniot) W. C. Cheng et W. T. Wang var. meyer (Rehder) W. C. Cheng et W. T. Wang = Sabina squamata (Buch. -Ham. ex D. Don) Antoine ' Meyer ' ●

341735　Sabina lemeeana (H. Lév. et Vaniot) W. C. Cheng et W. T. Wang var. meyer (Rehder) W. C. Cheng et W. T. Wang = Juniperus squamata Buch. -Ham. ex D. Don ' Meyer ' ●

341736　Sabina mekongensis Kom. = Juniperus convallium Rehder et E. H. Wilson ●

341737　Sabina microsperma(W. C. Cheng et L. K. Fu) W. C. Cheng et L. K. Fu = Juniperus convallium Rehder et E. H. Wilson var. microsperma(W. C. Cheng et L. K. Fu) Silba ●

341738　Sabina occidentalis (Hook.) Antoine = Juniperus occidentalis Hook. ●☆

341739　Sabina officinalis Garcke = Juniperus sabina L. ●

341740　Sabina officinalis Garcke = Sabina vulgaris Antoine ●

341741　Sabina osteosperma (Torr.) Antoine = Juniperus osteosperma (Torr.) Little ●

341742　Sabina pingii (W. C. Cheng ex Ferre) W. C. Cheng et W. T. Wang = Juniperus pingii W. C. Cheng ex Ferre ●

341743　Sabina pingii (W. C. Cheng ex Ferre) W. C. Cheng et W. T. Wang var. wilsonii (Rehder) W. C. Cheng et L. K. Fu = Sabina sino-alpina W. C. Cheng et W. T. Wang ●

341744　Sabina pingii (W. C. Cheng ex Ferre) W. C. Cheng et W. T. Wang var. wilsonii (Rehder) W. C. Cheng et L. K. Fu = Juniperus pingii W. C. Cheng ex Ferre var. wilsonii Rehder ●

341745　Sabina potaninii Kom. = Juniperus tibetica Kom. ●

341746　Sabina procumbens (Siebold ex Endl.) Iwata et Kusaka = Juniperus chinensis L. var. procumbens Siebold ex Endl. ●

341747　Sabina procumbens (Siebold ex Endl.) Iwata et Kusaka = Juniperus procumbens (Siebold ex Endl.) Miq. ●

341748　Sabina prostrata (Pers.) Antoine = Juniperus horizontalis Moench ●☆

341749　Sabina przewalskii (Engl.) Iwata et Kusaka = Juniperus przewalskii Kom. ●

341750　Sabina przewalskii Kom. = Juniperus przewalskii Kom. ●

341751　Sabina przewalskii Kom. f. pendula W. C. Cheng et L. K. Fu;垂枝祁连圆柏;Nutantsa Japgarden Vin,Sweepina Przewalsk Juniper ●

341752　Sabina przewalskii Kom. f. pendula W. C. Cheng et L. K. Fu = Juniperus przewalskii Kom. ●

341753　Sabina pseudosabina (Fisch. et C. A. Mey.) W. C. Cheng et W. T. Wang var. turkestanica (Kom.) Chang Y. Yang = Sabina centrasiatica (Kom.) W. C. Cheng et L. K. Fu ●

341754　Sabina pseudosabina (Fisch. et C. A. Mey.) W. C. Cheng et W. T. Wang var. turkestanica (Kom.) Chang Y. Yang = Juniperus pseudosabina Fisch. et C. A. Mey. var. turkestanica (Kom.) Silba ●

341755　Sabina pseudosabina (Fisch. et C. A. Mey.) W. C. Cheng et W. T. Wang = Juniperus pseudosabina Fisch. et C. A. Mey. ●

341756　Sabina recurva (Buch. -Ham. ex D. Don) Antoine = Juniperus recurva Buch. -Ham. ●

341757　Sabina recurva (Buch. -Ham.) Antoine ' Nana ';稠密垂枝柏●☆

341758　Sabina recurva（Buch.-Ham.）Antoine = Juniperus recurva Buch.-Ham. ●

341759　Sabina recurva（Buch.-Ham.）Antoine var. coxii（A. B. Jacks.）W. C. Cheng et L. K. Fu = Juniperus recurva Buch.-Ham. var. coxii（A. B. Jacks.）Melville ●

341760　Sabina saltuaria（Rehder et E. H. Wilson）W. C. Cheng et W. T. Wang；方枝柏（方香柏，方枝桧，木香）；Blackseed Juniper, Blackseed Savin, Blackseeded Juniper ●

341761　Sabina saltuaria（Rehder et E. H. Wilson）W. C. Cheng et W. T. Wang = Juniperus saltuaria Rehder et E. H. Wilson ●

341762　Sabina sargentii（A. Henry）Miyabe et Tatew. = Juniperus chinensis L. var. sargentii A. Henry ●

341763　Sabina sargentii（A. Henry）Miyabe et Tatew. = Sabina chinensis（L.）Antoine var. sargentii（A. Henry）W. C. Cheng et L. K. Fu ●

341764　Sabina scopulorum（Sarg.）Rydb. = Juniperus scopulorum Sarg. ●☆

341765　Sabina semiglobosa（Regel）W. C. Cheng et W. T. Wang；半球圆柏；Half-globe Juniper, Half-globe Savin ●

341766　Sabina silicicola Small = Juniperus virginiana L. var. silicicola（Small）E. Murray ●☆

341767　Sabina sino-alpina W. C. Cheng et W. T. Wang = Juniperus pingii W. C. Cheng ex Ferre var. wilsonii Rehder ●

341768　Sabina squamata（Buch.-Ham. ex D. Don）Antoine 'Meyer' = Juniperus squamata Buch.-Ham. ex D. Don 'Meyer' ●

341769　Sabina squamata（Buch.-Ham. ex D. Don）Antoine = Juniperus squamata Buch.-Ham. ex D. Don ●

341770　Sabina squamata（Buch.-Ham. ex D. Don）Antoine var. fargesii（Rehder et E. H. Wilson）W. C. Cheng et L. K. Fu = Sabina squamata（Buch.-Ham. ex D. Don）Antoine ●

341771　Sabina squamata（Buch.-Ham. ex D. Don）Antoine var. fargesii（Rehder et E. H. Wilson）S. H. Cheng et L. K. Fu = Juniperus squamata Buch.-Ham. ex D. Don var. fargesii Rehder et E. H. Wilson ●

341772　Sabina squamata（Buch.-Ham. ex D. Don）Antoine var. wilsonii（Rehder）W. C. Cheng et L. K. Fu = Juniperus pingii W. C. Cheng ex Ferre var. wilsonii Rehder ●

341773　Sabina squamata（Buch.-Ham.）Antoine = Sabina squamata（Buch.-Ham. ex D. Don）Antoine ●

341774　Sabina tibetica（Kom.）W. C. Cheng et L. K. Fu = Juniperus tibetica Kom. ●

341775　Sabina turkestanica Kom. = Juniperus pseudosabina Fisch. et C. A. Mey. var. turkestanica（Kom.）Silba ●

341776　Sabina turkestanica Kom. = Sabina pseudosabina（Fisch. et C. A. Mey.）W. C. Cheng et W. T. Wang ●

341777　Sabina utahensis（Engelm.）Rydb. = Juniperus osteosperma（Torr.）Little ●

341778　Sabina virginiana（L.）Antoine 'Pendula'；垂枝铅笔柏；Pendulous Red Cedar ●

341779　Sabina virginiana（L.）Antoine 'Pyramidalis'；塔冠铅笔柏；Pyramidal Juniper ●

341780　Sabina virginiana（L.）Antoine = Juniperus virginiana L. ●

341781　Sabina virginiana（L.）Antoine var. crebra Fernald；东北刺柏；Northeastern Juniper ●

341782　Sabina virginiana（L.）Antoine var. scopulorum（Sarg.）Lemmon = Juniperus scopulorum Sarg. ●☆

341783　Sabina vulgaris Antoine = Juniperus sabina L. ●

341784　Sabina vulgaris Antoine var. erectopatens W. C. Cheng et L. K. Fu = Juniperus sabina L. var. erectopatens（W. C. Cheng et L. K. Fu）Y. F. Yu et L. K. Fu ●

341785　Sabina vulgaris Antoine var. jarkendensis（Kom.）Chang Y. Yang = Juniperus semiglobosa Regel ●

341786　Sabina vulgaris Antoine var. yulinensis T. C. Chang et C. G. Chen = Juniperus sabina L. var. yulinensis（T. C. Chang et C. G. Chen）Y. F. Yu et L. K. Fu ●

341787　Sabina wallichiana（Hook. f. et Thomson）Kom. = Juniperus indica Bertol. ●

341788　Sabina wallichiana（Hook. f. et Thomson）Kom. var. meinocarpa（Hand.-Mazz.）W. C. Cheng et L. K. Fu = Juniperus indica Bertol. ●

341789　Sabina zaidamensis Kom. = Juniperus przewalskii Kom. ●

341790　Sabinea DC. = Poitea Vent. ●☆

341791　Sabinella Nakal = Juniperus L. ●

341792　Sabouraea Léandri = Talinella Baill. ●☆

341793　Sabsab Adans. = Paspalum L. ■

341794　Sabularia Small = Sabulina Rchb. ■

341795　Sabulina Rchb. = Minuartia L. ■

341796　Sabulina caroliniana（Walter）Small = Minuartia californica（A. Gray）Mattf. ■☆

341797　Sabulina dawsonensis（Britton）Rydb. = Arenaria stricta Michx. subsp. dawsonensis（Britton）Maguire ■☆

341798　Sabulina dawsonensis（Britton）Rydb. = Minuartia dawsonensis（Britton）House ■☆

341799　Sabulina glabra（Michx.）Small = Minuartia glabra（Michx.）Mattf. ■☆

341800　Sabulina groenlandica（Retz.）Small = Minuartia groenlandica（Retz.）Ostenf. ■☆

341801　Sabulina patula（Michx.）Small = Arenaria patula Michx. ■☆

341802　Sabulina patula（Michx.）Small ex Rydb. = Minuartia patula（Michx.）Mattf. ■☆

341803　Sabulina stricta（Michx.）Small = Arenaria stricta Michx. ■☆

341804　Sabulina stricta（Michx.）Small = Minuartia michauxii（Fenzl）Farw. ■☆

341805　Sabulina uniflora（Walter）Small = Minuartia uniflora（Walter）Mattf. ■☆

341806　Sabulinaceae Döll = Caryophyllaceae Juss.（保留科名）■●

341807　Sacaglottis G. Don = Sacoglottis Mart. ●☆

341808　Saccaceae Dulac = Nymphaeaceae Salisb.（保留科名）■

341809　Saccanthus Herzog = Basistemon Turcz. ●☆

341810　Saccapetalum Benn. = Miliusa Lesch. ex A. DC. ●

341811　Saccardophytum Speg. = Benthamiella Speg. ●☆

341812　Saccarum Sanguin. = Saccharum L. ■

341813　Saccellium Bonpl.（1806）；热美紫草属●☆

341814　Saccellium Humb. et Bonpl. = Saccellium Bonpl. ●☆

341815　Saccellium brasiliense I. M. Johnst.；巴西热美紫草●☆

341816　Saccellium lanceolatum Humb. et Bonpl.；热美紫草●☆

341817　Saccellium oliverii Britton；奥氏热美紫草●☆

341818　Saccharaceae Bercht. et J. Presl = Gramineae Juss.（保留科名）■●

341819　Saccharaceae Bercht. et J. Presl = Poaceae Barnhart（保留科名）■●

341820　Saccharaceae Burnett = Gramineae Juss.（保留科名）■●

341821　Saccharaceae Burnett = Poaceae Barnhart（保留科名）■●

341822　Saccharaceae Martinov = Gramineae Juss.（保留科名）■●

341823　Saccharaceae Martinov = Poaceae Barnhart（保留科名）■●

341824　Saccharifera Stokes = Saccharum L. ■

341825　Saccharodendron Nieuwl. = Acer L. ●

341826　Saccharodendron nigrum (F. Michx.) Small = Acer nigrum F. Michx. ●

341827　Saccharodendron saccharum (Marshall) Moldenke = Acer saccharum Marshall ●

341828　Saccharophorum Neck. = Saccharum L. ■

341829　Saccharum L. (1753) ; 甘蔗属 ; Plume Grass, Sugarcane, Sweet Cane, Sweetcane ■

341830　Saccharum × kanashiroi (Ohwi) Ohwi ; 金城甘蔗 ■☆

341831　Saccharum aegyptiacum L. = Saccharum spontaneum L. subsp. aegyptiacum (Willd.) Hack. ■☆

341832　Saccharum aegyptiacum Willd. = Saccharum spontaneum L. subsp. aegyptiacum (Willd.) Hack. ■☆

341833　Saccharum aegyptiacum Willd. = Saccharum spontaneum L. ■

341834　Saccharum alopecuroideum (L.) Nutt. = Erianthus alopecuroides (L.) Elliott ■☆

341835　Saccharum arenicola Ohwi = Saccharum spontaneum L. var. arenicola (Ohwi) Ohwi ■

341836　Saccharum arundinaceum Hook. f. = Saccharum bengalense Retz. ■☆

341837　Saccharum arundinaceum Retz. ; 斑茅（芭茅, 大密, 管精）; Reedlike Plumegrass, Reedlike Sugarcane, Reedlike Sweetcane ■

341838　Saccharum arundinaceum Retz. var. trichophyllum (Hand.-Mazz.) S. M. Phillips et S. L. Chen ; 毛颖斑茅 ■

341839　Saccharum baldwinii Spreng. = Erianthus strictus Baldwin ■

341840　Saccharum barberi Jeswiet ; 细秆甘蔗 ; Thinculm Sugarcane ■

341841　Saccharum barbicostatum Ohwi = Saccharum arundinaceum Retz. ■

341842　Saccharum bengalense Retz. ; 孟加拉甘蔗 ; Ekar Ekra, Munj Sweetcane ■☆

341843　Saccharum biflorum Forssk. = Saccharum spontaneum L. subsp. aegyptiacum (Willd.) Hack. ■☆

341844　Saccharum brachypogon Stapf = Eriochrysis brachypogon (Stapf) Stapf ■☆

341845　Saccharum canaliculatum Roxb. = Saccharum spontaneum L. ■

341846　Saccharum chinense Osbeck = Saccharum sinense Roxb. ■

341847　Saccharum ciliare Andersson = Saccharum bengalense Retz. ■☆

341848　Saccharum ciliare Andersson var. griffithii (Munro ex Boiss.) Hack. = Saccharum griffithii Munro ex Boiss. ■☆

341849　Saccharum cotuliferum (Thunb.) Roberty = Spodiopogon cotulifer (Thunb.) Hack. ■

341850　Saccharum cylindricum (L.) Lam. = Imperata cylindrica (L.) P. Beauv. ■

341851　Saccharum edule Hassk. ; 可食斑茅 ■☆

341852　Saccharum elephantinum (Hook. f.) V. Naray. = Saccharum ravennae (L.) Murray ■

341853　Saccharum fallax Balansa = Narenga fallax (Balansa) Bor ■

341854　Saccharum fallax Balansa var. aristatum Balansa = Narenga fallax (Balansa) Bor var. aristata (Balansa) L. Liou ■

341855　Saccharum fallax Balansa var. aristatum Balansa = Saccharum fallax Balansa ■

341856　Saccharum filifolium Nees ex Steud. ; 线甘蔗 ■☆

341857　Saccharum floridulum Labill. = Miscanthus floridulus (Labill.) Warb. ex K. Schum. et Lauterb. ■

341858　Saccharum formosanum (Stapf) Ohwi ; 台蔗茅（台湾蔗草, 台湾蔗茅）; Taiwan Plumegrass ■

341859　Saccharum formosanum (Stapf) Ohwi = Erianthus formosanus Stapf ■

341860　Saccharum formosanum (Stapf) Ohwi var. pollinioides (Rendle) Ohwi = Erianthus formosanus Stapf var. pollinioides (Rendle) Ohwi ■

341861　Saccharum formosanum (Stapf) Ohwi var. pollinioides (Rendle) Ohwi = Saccharum formosanum (Stapf) Ohwi ■

341862　Saccharum fuscum Roxb. = Sclerostachya fusca (Roxb.) A. Camus ■☆

341863　Saccharum giganteum (Walter) Pursh = Erianthus giganteus (Walter) Muhl. ■

341864　Saccharum griffithii Munro ex Boiss. ; 格氏甘蔗 ■☆

341865　Saccharum hirsutum Forssk. = Lasiurus scindicus Henrard ■☆

341866　Saccharum hookeri (Hack.) V. Naray. = Saccharum longisetosum (Andersson) V. Naray. ex Bor ■

341867　Saccharum hookeri (Hack.) V. Naray. ex Bor = Erianthus hookeri Hack. ■

341868　Saccharum kajkaiense (Melderis) Melderis ; 卡伊甘蔗 ■☆

341869　Saccharum koenigii Retz. = Imperata cylindrica (L.) P. Beauv. var. major (Nees) C. E. Hubb. et Vaughan ■

341870　Saccharum koenigii Retz. = Imperata cylindrica (L.) Raeusch. ■

341871　Saccharum koenigii Retz. = Imperata koenigii (Retz.) P. Beauv. ■

341872　Saccharum laguroides Pourr. = Imperata cylindrica (L.) P. Beauv. ■

341873　Saccharum longesetosum (Andersson) V. Naray. = Saccharum longisetosum (Andersson) V. Naray. ex Bor ■

341874　Saccharum longesetosum (Andersson) V. Naray. var. hookeri (Hack.) Shukla = Saccharum longisetosum (Andersson) V. Naray. ex Bor ■

341875　Saccharum longifolium Munro ex Benth. = Narenga fallax (Balansa) Bor ■

341876　Saccharum longisetosum (Andersson) V. Naray. ex Bor. ; 长齿蔗茅 ; Longtooth Plumegrass ■

341877　Saccharum longisetosum (Andersson) V. Naray. ex Bor. = Erianthus longisetosus T. Anderson ■

341878　Saccharum longisetum (A. Rich.) Walp. = Melinis longiseta (A. Rich.) Zizka ■☆

341879　Saccharum macratherum (Pilg.) S. Ahmad et R. R. Stewart = Saccharum filifolium Nees ex Steud. ■☆

341880　Saccharum munja Roxb. = Saccharum bengalense Retz. ■☆

341881　Saccharum munroanum Hack. = Eriochrysis pallida Munro ■☆

341882　Saccharum narenga (Nees ex Steud.) Hack. = Narenga porphyrocoma (Hance) Bor ■

341883　Saccharum narenga (Nees ex Steud.) Wall. ex Hack. = Narenga porphyrocoma (Hance) Bor ■

341884　Saccharum narenga (Nees ex Steud.) Wall. ex Hack. var. khasianum Hack. = Saccharum narenga (Nees ex Steud.) Hack. ■

341885　Saccharum narenga Wall. = Narenga porphyrocoma (Hance) Bor ■

341886　Saccharum narenga Wall. = Saccharum narenga (Nees ex Steud.) Hack. ■

341887　Saccharum officinarum L. ; 甘蔗（干蔗, 竿蔗, 果蔗, 红甘蔗, 红蔗, 蚋蔗, 薯蔗, 秀贵甘蔗, 印度甘蔗, 蔗, 竹蔗, 竹楂）; Eulalia Grass, Sugar Cane, Sugarcane, Uba Cane ■

341888　Saccharum officinarum L. subsp. barberi (Jeswiet) Burkill = Saccharum barberi Jeswiet ■

341889　Saccharum officinarum L. subsp. sinense (Roxb.) Burkill = Saccharum sinense Roxb. ■

341890 Saccharum pallidum（Munro）Benth. = Eriochrysis pallida Munro ■☆

341891 Saccharum paniceum Lam. = Pogonatherum paniceum（Lam.）Hack. ■

341892 Saccharum porphyrocomum（Hance ex Trin.）Hack. = Saccharum narenga（Nees ex Steud.）Hack. ■

341893 Saccharum porphyrocomum（Hance）Hack. = Narenga porphyrocoma（Hance）Bor ■

341894 Saccharum procerum Roxb. ;狭叶斑茅■

341895 Saccharum propinquum Steud. = Saccharum spontaneum L. ■

341896 Saccharum purpuratum Rendle = Eriochrysis purpurata（Rendle）Stapf ■☆

341897 Saccharum rarum（R. Br.）Poir. = Perotis rara R. Br. ■

341898 Saccharum ravennae（L.）L. = Erianthus ravennae（L.）P. Beauv. ■

341899 Saccharum ravennae（L.）L. subsp. parviflorum（Pilg.）Scholz = Tripidium ravennae（L.）H. Scholz subsp. parviflorum（Pilg.）H. Scholz ■

341900 Saccharum ravennae（L.）L. var. parviflorum（Pilg.）Maire = Saccharum ravennae（L.）L. subsp. parviflorum（Pilg.）Scholz ■

341901 Saccharum ravennae（L.）Murray = Erianthus ravennae（L.）P. Beauv. ■

341902 Saccharum ravennae（L.）P. Beauv. = Erianthus ravennae（L.）P. Beauv. ■

341903 Saccharum repens Willd. = Melinis repens（Willd.）Zizka ■

341904 Saccharum repens Willd. = Rhynchelytrum repens（Willd.）C. E. Hubb. ■

341905 Saccharum rufipilum Steud. = Erianthus rufipilus（Steud.）Griseb. ■

341906 Saccharum sara Roxb. = Saccharum bengalense Retz. ■☆

341907 Saccharum semidecumbens Roxb. = Saccharum spontaneum L. ■

341908 Saccharum sibiricum（Trin.）Roberty = Spodiopogon sibiricus Trin. ■

341909 Saccharum sinense Roxb. ;竹蔗（草甘蔗,干蔗,甘蔗,竿蔗,接肠草,芦蔗,薯蔗,糖梗）;Bamboo Sugarcane,Chinese Sugarcane ■

341910 Saccharum sphacelatum（Benth.）Walp. = Melinis repens（Willd.）Zizka ■

341911 Saccharum spicatum L. = Perotis indica（L.）Kuntze ■

341912 Saccharum spontaneum L. ;甜根子草（割手密,黑猴蔗,甜茅）;Wild Sugar Cane,Wild Sugarcane,Wild Sweetcane ■

341913 Saccharum spontaneum L. subsp. aegyptiacum（Willd.）Hack. ;埃及甜根子草■☆

341914 Saccharum spontaneum L. var. aegyptiacum（Willd.）Hack. = Saccharum spontaneum L. subsp. aegyptiacum（Willd.）Hack. ■☆

341915 Saccharum spontaneum L. var. arenicola（Ohwi）Ohwi;砂地甜根子草■

341916 Saccharum spontaneum L. var. juncifolium Hack. ;灯芯草叶甜根子草■

341917 Saccharum spontaneum L. var. roxburghii Honda;罗氏甜根子草■

341918 Saccharum spontaneum L. var. roxburghii Honda = Saccharum spontaneum L. ■

341919 Saccharum spontaneum L. var. sinense（Roxb.）Andersson = Saccharum sinense Roxb. ■

341920 Saccharum strictum（Baldwin）Nutt. = Erianthus strictus Baldwin ■

341921 Saccharum teneriffae L. f. = Tricholaena teneriffae（L. f.）Link ■☆

341922 Saccharum thunbergii Retz. = Imperata cylindrica（L.）Raeusch. ■

341923 Saccharum tristachyum Steud. = Eulalia trispicata（Schult.）Henrard ■

341924 Saccharum versicolor Nees ex Steud. = Saccharum filifolium Nees ex Steud. ■☆

341925 Sacchrosphendamnus Nieuwl. = Acer L. ●

341926 Sacchrosphendamnus Nieuwl. = Sachrosphendamnus Nieuwl. ●

341927 Saccia Naudin(1889);囊旋花属☆

341928 Saccia elegans Naudin;囊旋花☆

341929 Saccidium Lindl. (废弃属名) = Holothrix Rich. ex Lindl. (保留属名)■☆

341930 Saccidium pilosum Lindl. = Holothrix pilosa（Lindl.）Rchb. f. ■☆

341931 Saccifoliaceae Maguire et Pires = Gentianaceae Juss. (保留科名)●■

341932 Saccifoliaceae Maguire et Pires = Salazariaceae F. A. Barkley ●■

341933 Saccifoliaceae Maguire et Pires(1978);勺叶木科（袋叶科,囊叶木科）●☆

341934 Saccifolium Maguire et J. M. Pires(1978);勺叶木属●☆

341935 Saccifolium bandeirae Maguire et J. M. Pires;勺叶木●☆

341936 Saccilabium Rottb. = Nepeta L. ■●

341937 Sacciolepis Nash(1901);囊颖草属（滑草属）;Cupscale ■

341938 Sacciolepis africana C. E. Hubb. et Snowden;非洲囊颖草■☆

341939 Sacciolepis albida Stapf = Sacciolepis seslerioides（Rendle）Stapf ■☆

341940 Sacciolepis angusta（Trin.）Stapf = Sacciolepis indica（L.）Chase ■

341941 Sacciolepis antsirabensis A. Camus = Sacciolepis chevalieri Stapf ■☆

341942 Sacciolepis arenaria Mimeur;沙地囊颖草■☆

341943 Sacciolepis auriculata Stapf = Sacciolepis indica（L.）Chase ■

341944 Sacciolepis barbiglandularis Mez = Sacciolepis transbarbata Stapf ■☆

341945 Sacciolepis brevifolia Stapf = Sacciolepis chevalieri Stapf ■☆

341946 Sacciolepis catumbensis（Rendle）Stapf;卡通贝囊颖草■☆

341947 Sacciolepis chevalieri Stapf;舍瓦利耶囊颖草■☆

341948 Sacciolepis ciliocincta（Pilg.）Stapf;毛带囊颖草■☆

341949 Sacciolepis cinereo-vestita（Pilg.）C. E. Hubb. = Sacciolepis typhura（Stapf）Stapf ■☆

341950 Sacciolepis circumciliata Mez = Sacciolepis transbarbata Stapf ■☆

341951 Sacciolepis claviformis B. K. Simon = Sacciolepis indica（L.）Chase ■

341952 Sacciolepis curvata（L.）Chase;内折囊颖草■☆

341953 Sacciolepis cymbiandra Stapf;舟蕊囊颖草■☆

341954 Sacciolepis geniculata B. K. Simon = Sacciolepis spiciformis（Hochst. ex A. Rich.）Stapf ■

341955 Sacciolepis glaucescens Stapf = Sacciolepis typhura（Stapf）Stapf ■☆

341956 Sacciolepis gracilis Stent et J. M. Rattray = Sacciolepis indica（L.）Chase ■

341957 Sacciolepis huillensis（Rendle）Stapf = Sacciolepis spiciformis（Hochst. ex A. Rich.）Stapf ■

341958 Sacciolepis incana Mez = Sacciolepis transbarbata Stapf ■☆

341959 Sacciolepis incurva Stapf = Sacciolepis chevalieri Stapf ■☆

341960 Sacciolepis indica（L.）Chase;囊颖草（长穗稗,狗尾草,滑草,鼠尾黍,水囊颖草,印度稷,英雄草）;Glenwoodgrass,India Cupscale,Indian Cupscale ■

341961 Sacciolepis indica（L.）Chase subsp. oryzetorum（Makino）T. Koyama = Sacciolepis indica（L.）Chase ■

341962 Sacciolepis indica（L.）Chase var. angusta（Trin.）Keng;窄穗

囊颖草;Narrow Spike Cupscale ■

341963 Sacciolepis indica (L.) Chase var. angusta (Trin.) Keng = Sacciolepis indica (L.) Chase ■

341964 Sacciolepis indica (L.) Chase var. oryzetorum (Makino) Ohwi = Sacciolepis indica (L.) Chase ■

341965 Sacciolepis insulicola (Steud.) Ohwi = Hymenachne insulicola (Steud.) L. Liou ■

341966 Sacciolepis insulicola (Steud.) Ohwi = Panicum auritum J. Presl ex Nees ■

341967 Sacciolepis interrupta (Willd.) Stapf;间序囊颖草;Interspike Cupscale ■

341968 Sacciolepis interrupta Chase = Sacciolepis interrupta (Willd.) Stapf ■

341969 Sacciolepis johnstonii C. E. Hubb. et Snowden = Sacciolepis rigens (Mez) A. Chev. ■☆

341970 Sacciolepis kimayalaensis Vanderyst = Sacciolepis rigens (Mez) A. Chev. ■☆

341971 Sacciolepis kimpasaensis Vanderyst = Sacciolepis typhura (Stapf) Stapf ■☆

341972 Sacciolepis lebrunii Robyns = Sacciolepis spiciformis (Hochst. ex A. Rich.) Stapf ■

341973 Sacciolepis leptorhachis Stapf = Sacciolepis rigens (Mez) A. Chev. ■☆

341974 Sacciolepis luciae B. K. Simon = Sacciolepis spiciformis (Hochst. ex A. Rich.) Stapf ■

341975 Sacciolepis micrococca Mez;小果囊颖草■☆

341976 Sacciolepis mukuku Vanderyst = Sacciolepis africana C. E. Hubb. et Snowden ■☆

341977 Sacciolepis myosuroides (R. Br.) Chase ex E. G. Camus et A. Camus;鼠尾囊颖草;Mousetail Cupscale ■

341978 Sacciolepis myosuroides (R. Br.) Chase ex E. G. Camus et A. Camus var. spici-formis (Hochst. ex A. Rich.) Engl. = Sacciolepis myosuroides (R. Br.) Chase ex E. G. Camus et A. Camus ■

341979 Sacciolepis myosuroides (R. Br.) Chase ex E. G. Camus et A. Camus var. nana S. L. Chen et T. D. Zhuang;矮小囊颖草;Dwarf Mousetail Cupscale ■

341980 Sacciolepis nana Stapf = Sacciolepis micrococca Mez ■☆

341981 Sacciolepis palustris Napper = Sacciolepis chevalieri Stapf ■☆

341982 Sacciolepis pergracilis Chiov. = Sacciolepis indica (L.) Chase ■

341983 Sacciolepis rigens (Mez) A. Chev.;硬囊颖草■☆

341984 Sacciolepis scirpoides Stapf = Sacciolepis typhura (Stapf) Stapf ■☆

341985 Sacciolepis semienensis Chiov. = Panicum hymeniochilum Nees ■☆

341986 Sacciolepis seslerioides (Rendle) Stapf;苍白囊颖草■☆

341987 Sacciolepis simaoensis Y. Y. Qian;思茅囊颖草;Simao Cupscale ■

341988 Sacciolepis simaoensis Y. Y. Qian = Sacciolepis interrupta (Willd.) Stapf ■

341989 Sacciolepis spicata (L.) Honda = Sacciolepis indica (L.) Chase ■

341990 Sacciolepis spicata Honda ex Masam. = Sacciolepis indica (L.) Chase ■

341991 Sacciolepis spiciformis (Hochst. ex A. Rich.) Stapf = Sacciolepis myosuroides (R. Br.) Chase ex E. G. Camus et A. Camus ■

341992 Sacciolepis squamigera (Pilg.) C. E. Hubb. = Sacciolepis catumbensis (Rendle) Stapf ■☆

341993 Sacciolepis striata (L.) Nash;美洲囊颖草(显脉囊颖草);American Cupscale, American Cupscale-grass ■☆

341994 Sacciolepis strictula Pilg. = Sacciolepis chevalieri Stapf ■☆

341995 Sacciolepis transbarbata Stapf;直毛囊颖草■☆

341996 Sacciolepis trollii Pilg. = Sacciolepis typhura (Stapf) Stapf ■☆

341997 Sacciolepis typhura (Stapf) Stapf;南非囊颖草■☆

341998 Sacciolepis velutina Napper = Sacciolepis typhura (Stapf) Stapf ■☆

341999 Sacciolepis viguieri A. Camus;维基耶囊颖草■☆

342000 Sacciolepis wittei Robyns = Sacciolepis typhura (Stapf) Stapf ■☆

342001 Sacciolepis wombaliensis Vanderyst = Sacciolepis typhura (Stapf) Stapf ■☆

342002 Saccocalyx Coss. et Durieu(1853)(保留属名);囊萼属●☆

342003 Saccocalyx Steven = Astragalus L. ●■

342004 Saccocalyx satureioides Coss. et Durieu;非洲囊萼■☆

342005 Saccochilus Blume = Saccolabium Blume(保留属名)■

342006 Saccochilus Blume = Thrixspermum Lour. ■

342007 Saccoglossum Schltr. (1912);囊舌兰属■☆

342008 Saccoglossum maculatum Schltr.;斑点囊舌兰■☆

342009 Saccoglossum papuanum Schltr.;囊舌兰■☆

342010 Saccoglottis Endl. = Sacoglottis Mart. ●☆

342011 Saccolabiopsis J. J. Sm. (1918);拟囊唇兰属(小囊唇兰属)■☆

342012 Saccolabiopsis bakhuizenii J. J. Sm.;巴氏拟囊唇兰■☆

342013 Saccolabiopsis taiwaniana S. W. Chung et T. C. Hsu;台湾拟囊唇兰■

342014 Saccolabiopsis wulaokenensis W. M. Lin, Kuo Huang et T. P. Lin;拟囊唇兰■

342015 Saccolabium Blume (1825) (保留属名);囊唇兰属;Dishspurorchis, Gastrochilus, Saccolabium ■

342016 Saccolabium Post et Kuntze = Nepeta L. ■●

342017 Saccolabium Post et Kuntze = Saccilabium Rottb. ■

342018 Saccolabium acuminatum (Lindl.) Rchb. f. = Uncifera acuminata Lindl. ■

342019 Saccolabium bellinum Rchb. f. = Gastrochilus bellinus (Rchb. f.) Kuntze ■

342020 Saccolabium buccosum Rchb. f. = Robiquetia succisa (Lindl.) Seidenf. et Garay ■

342021 Saccolabium calceolare (Buch. -Ham. ex Sm.) Lindl. = Gastrochilus calceolaris (Buch. -Ham. ex Sm.) D. Don ■

342022 Saccolabium ciliare (F. Maek.) Ohwi;囊唇兰■☆

342023 Saccolabium densiflorum Lindl. = Robiquetia spatulata (Blume) J. J. Sm. ■

342024 Saccolabium distichum Lindl. = Gastrochilus distichus (Lindl.) Kuntze ■

342025 Saccolabium eberhardtii Finet = Cleisostomopsis eberhardtii (Finet) Seidenf. et Smitinand ■

342026 Saccolabium fargesii Kraenzl. = Gastrochilus fargesii (Kraenzl.) Schltr. ■

342027 Saccolabium formosanum Hayata = Gastrochilus formosanus (Hayata) Hayata ■

342028 Saccolabium fuscopunctatum Hayata = Gastrochilus fuscopunctatus (Hayata) Hayata ■

342029 Saccolabium gemmatum Lindl. = Schoenorchis gemmata (Lindl.) J. J. Sm. ■

342030 Saccolabium giganteum Lindl. ;大囊唇兰■☆

342031 Saccolabium giganteum Lindl. = Rhynchostylis gigantea (Lindl.) Ridl. ■

342032 Saccolabium hainanense Rolfe = Schoenorchis gemmata (Lindl.) J. J. Sm. ■

342033 Saccolabium himalaicum Deb, Sengupta et Malick =

Ascocentrum himalaicum（Deb et Malick）Christensen ■

342034　Saccolabium hoyopse Rolfe ex Downie ＝ Gastrochilus pseudodistichus（King et Pantl.）Schltr. ■

342035　Saccolabium intermedium Griff. ex Lindl. ＝ Gastrochilus acinacifolius Z. H. Tsi ■

342036　Saccolabium intermedium Griff. ex Lindl. ＝ Gastrochilus intermedius（Griff. ex Lindl.）Kuntze ■

342037　Saccolabium japonicum Makino ＝ Gastrochilus japonicus （Makino）Schltr. ■

342038　Saccolabium kotoense（Yamam.）Yamam. ＝ Tuberolabium kotoense Yamam. ■

342039　Saccolabium matsudae（Hayata）Makino et Nemoto ＝ Gastrochilus matsudae Hayata ■

342040　Saccolabium matsudae Makino et Nemoto ＝ Gastrochilus matsudae Hayata ■

342041　Saccolabium micranthum Lindl. ＝ Smitinandia micrantha （Lindl.）Holttum ■

342042　Saccolabium microphyton Frapp. ＝ Angraecum tenellum （Ridl.）Schltr. ■☆

342043　Saccolabium mombasense（Rendle）Rolfe ＝ Acampe pachyglossa Rchb. f. ■☆

342044　Saccolabium monticola Rolfe ex Downie ＝ Gastrochilus yunnanensis Schltr. ■

342045　Saccolabium nebulosum（Fukuy.）S. Y. Hu ＝ Gastrochilus formosanus（Hayata）Hayata ■

342046　Saccolabium obliquum Lindl. ＝ Gastrochilus hainanensis Z. H. Tsi ■

342047　Saccolabium obliquum Lindl. ＝ Gastrochilus obliquus（Lindl.） Kuntze ■

342048　Saccolabium occidentale Kraenzl. ＝ Angraecopsis tridens （Lindl.）Schltr. ■☆

342049　Saccolabium ochraceum Lindl. ＝ Acampe ochracea（Lindl.） Hochr. ■

342050　Saccolabium odoratum（Kudo）Makino ＝ Haraella retrocalla （Hayata）Kudo ■

342051　Saccolabium odoratum（Kudo）Makino et Nemoto ＝ Haraella retrocalla（Hayata）Kudo ■

342052　Saccolabium oeonioides Kraenzl. ＝ Solenangis clavata（Rolfe） Schltr. ■☆

342053　Saccolabium pachyglossum（Rchb. f.）Rolfe ＝ Acampe pachyglossa Rchb. f. ■☆

342054　Saccolabium papillosum Lindl. ＝ Acampe papillosa（Lindl.） Lindl. ■

342055　Saccolabium parviflorum（Thouars）Cordem. ＝ Angraecopsis parviflora（Thouars）Schltr. ■☆

342056　Saccolabium platycalcaratum Rolfe ＝ Gastrochilus platycalcaratus（Rolfe）Schltr. ■

342057　Saccolabium pseudodistichum King et Pantl. ＝ Gastrochilus pseudodistichus（King et Pantl.）Schltr. ■

342058　Saccolabium pumilum Hayata ＝ Ascocentrum pumilum （Hayata）Schltr. ■

342059　Saccolabium quasipinifolium Hayata ＝ Holcoglossum quasipinifolium（Hayata）Schltr. ■

342060　Saccolabium quetcetorum（Fukuy.）S. Y. Hu ＝ Gastrochilus formosanus（Hayata）Hayata ■

342061　Saccolabium racemiferum Lindl. ＝ Cleisostoma racemiferum （Lindl.）Garay ■

342062　Saccolabium radicosum A. Rich. ＝ Microcoelia globulosa （Ridl.）L. Jonss. ■☆

342063　Saccolabium raraensis（Fukuy.）S. Y. Hu ＝ Gastrochilus raraensis Fukuy. ■

342064　Saccolabium retrocallum Hayata ＝ Haraella retrocalla（Hayata） Kudo ■

342065　Saccolabium retusum（L.）Voigt ＝ Rhynchostylis retusa（L.） Blume ■

342066　Saccolabium rupestre（Fukuy.）S. Y. Hu ＝ Gastrochilus formosanus（Hayata）Hayata ■

342067　Saccolabium scolopendrifolium Makino ＝ Cleisostoma scolopendrifolium（Makino）Garay ■

342068　Saccolabium shaoyaoii S. S. Ying ＝ Gastrochilus formosanus （Hayata）Hayata ■

342069　Saccolabium somai Hayata ＝ Gastrochilus japonicus（Makino） Schltr. ■

342070　Saccolabium taiwanianum（Hayata）Ts. Tang et F. T. Wang ＝ Sarcophyton taiwanianum（Hayata）Garay ■

342071　Saccolabium taiwanianum S. S. Ying ＝ Gastrochilus japonicus （Makino）Schltr. ■

342072　Saccolabium tixieri Guillaumin ＝ Schoenorchis tixieri （Guillaumin）Seidenf. ■

342073　Saccolabium toramanum Makino ＝ Gastrochilus toramanus （Makino）Schltr. ■

342074　Saccolabium triflorum Guillaumin ＝ Trichoglottis triflora （Guillaumin）Garay et Seidenf. ■

342075　Saccolabium uteriferum（Hook. f.）Ridl. ＝ Pomatocalpa spicatum Breda ■

342076　Saccolabium wendlandorum（Rchb. f.）Kraenzl. ＝ Pomatocalpa spicatum Breda ■

342077　Saccolabium yunnanense（Schltr.）S. Y. Hu ＝ Gastrochilus yunnanensis Schltr. ■

342078　Saccolabium yunpeense Ts. Tang et F. T. Wang ＝ Holcoglossum flavescens（Schltr.）Z. H. Tsi ■

342079　Saccolaria Kuhlmann ＝ Biovularia Karnienski ■

342080　Saccolaria Kuhlmann ＝ Utricularia L. ■

342081　Saccolena Gleason ＝ Salpinga Mart. ex DC. ●☆

342082　Saccolepis Nash ＝ Sacciolepis Nash ■

342083　Sacconia Endl. ＝ Chione DC. ■☆

342084　Saccopetalum Benn.（1840）；囊瓣木属（囊瓣花属）； Bagpetaltree, Bag-petal-tree ●

342085　Saccopetalum Benn. ＝ Miliusa Lesch. ex A. DC. ●

342086　Saccopetalum prolificum（Chun et F. C. How）Tsiang；囊瓣木 （多子阿芳）；China Bagpetaltree, Chinese Bagpetaltree, Chinese Bag-petal-tree, Manyseed Alphonsea ●◇

342087　Saccopetalum prolificum（Chun et F. C. How）Tsiang ＝ Miliusa prolifica（Chun et F. C. How）P. T. Li ●◇

342088　Saccoplectus Oerst. ＝ Alloplectus Mart.（保留属名）●■☆

342089　Saccoplectus Oerst. ＝ Drymonia Mart. ●☆

342090　Saccostoma Wall. ex Voigt ＝ Coleus Lour. ●■

342091　Saccostoma Wall. ex Voigt ＝ Plectranthus L'Hér.（保留属名）●■

342092　Saccularia Kellogg ＝ Gambelia Nutt. ●☆

342093　Sacculina Bosser ＝ Utricularia L. ■

342094　Sacculina madecassa Bosser ＝ Utricularia cymbantha Oliv. ■☆

342095　Saccus Kuntze ＝ Artocarpus J. R. Forst. et G. Forst.（保留属 名）●

342096　Saccus Rumph. ＝ Artocarpus J. R. Forst. et G. Forst.（保留属

342097   Saccus Rumph. ex Kuntze = Artocarpus J. R. Forst. et G. Forst.（保留属名）●

342098   Sacellium Spreng. = Saccellium Humb. et Bonpl. ●☆

342099   Saceoglottis Walp. = Sacoglottis Mart. ●☆

342100   Sacharum Scop. = Saccharum L. ■

342101   Sacharum repens Willd. = Rhynchelytrum repens（Willd.）C. E. Hubb. ■

342102   Sachokiella Kolak.（1985）;梭根桔梗属■☆

342103   Sachokiella Kolak. = Campanula L. ■●

342104   Sachokiella macrochlamys（Boiss. et Huet.）Kolak.;梭根桔梗■☆

342105   Sachrosphendamnus Nieuwl. = Acer L. ●

342106   Sachsia Griseb.（1866）;银蓬属■☆

342107   Sachsia bahamensis Urb. = Sachsia polycephala Griseb. ■☆

342108   Sachsia divaricata Griseb. = Sachsia polycephala Griseb. ■☆

342109   Sachsia polycephala Griseb.;哈马银蓬;Bahama Sachsia ■☆

342110   Sacleuxia Baill.（1890）;萨克萝摩属■☆

342111   Sacleuxia salicina Baill.;柳叶萨克萝摩■☆

342112   Sacleuxia tuberosa（E. A. Bruce）Bullock;块根萨克萝摩■☆

342113   Sacodon Raf. = Cypripedium L. ■

342114   Sacodon macranthos（Sw.）Raf. = Cypripedium macranthum Sw. ■

342115   Sacoglottis Mart.（1827）;盾舌核果树属●☆

342116   Sacoglottis gabonensis（Baill.）Urb.;盾舌核果树;Liberian Cherry ●☆

342117   Sacoila Raf. = Spiranthes Rich.（保留属名）■

342118   Sacoila Raf. = Stenorrhynchos Rich. ex Spreng. ■☆

342119   Sacoila lanceolata（Aubl.）Garay = Limodorum lanceolatum Aubl. ■☆

342120   Sacoila lanceolata（Aubl.）Garay = Stenorrhynchos lanceolatum（Willd.）Rich. ■☆

342121   Sacoila lanceolata（Aubl.）Garay var. paludicola（Luer）Sauleda, Wunderlin et B. F. Hansen = Spiranthes lanceolata（Aubl.）León var. paludicola Luer ■☆

342122   Sacoila lanceolata（Aubl.）Garay var. paludicola（Luer）Sauleda, Wunderlin et B. F. Hansen = Stenorrhynchos cinnabarinum Lindl. var. paludicola（Luer）W. J. Schrenk ■☆

342123   Sacoila lanceolata（Aubl.）Garay var. squamulosa（Kunth）Szlach. = Sacoila squamulosa（Kunth）Garay ■☆

342124   Sacoila squamulosa（Kunth）Garay = Stenorrhynchos squamulosum（Kunth）Spreng. ■☆

342125   Sacosperma G. Taylor（1944）;盾籽茜属●☆

342126   Sacosperma paniculatum（Benth.）G. Taylor;圆锥盾籽茜●☆

342127   Sacosperma parviflorum（Benth.）G. Taylor;小花盾籽茜●☆

342128   Sacranthus Endl. = Nicotiana L. ●■

342129   Sacranthus Endl. = Sairanthus G. Don ●■☆

342130   Sacropteryx Radlk. = Sarcopteryx Radlk. ●☆

342131   Sacrosphendamus Willis = Acer L. ●

342132   Sacrosphendamus Willis = Sachrosphendamnus Nieuwl. ●

342133   Sadiria Mez（1902）;印度紫金牛属●☆

342134   Sadleria cyatheoides Kaulf.;印度紫金牛;Pulu ●☆

342135   Sadokum D. Tiu et Cootes = Cymbidium Sw. ■

342136   Sadokum D. Tiu et Cootes（2007）;爪哇蕙兰属■☆

342137   Sadrum Sol. ex Baill. = Pyrenacantha Wight（保留属名）●

342138   Sadymia Griseb. = Samyda Jacq.（保留属名）●☆

342139   Saelanthus Forssk. = Cissus L. ●

342140   Saelanthus digitatus Forssk. = Cyphostemma digitatum（Forssk.）Desc. ●☆

342141   Saelanthus rotundifolius Forssk. = Cissus rotundifolia（Forssk.）Vahl ●☆

342142   Saelanthus ternatus Forssk. = Cyphostemma ternatum（Forssk.）Desc. ●☆

342143   Saeranthus Post et Kuntze = Nicotiana L. ●■

342144   Saeranthus Post et Kuntze = Sairanthus G. Don ●■☆

342145   Saerocarpus Post et Kuntze = Antirrhinum L. ●■

342146   Saerocarpus Post et Kuntze = Sairocarpus Nutt. ex A. DC. ●■

342147   Saffordiella Merr. = Myrtella F. Muell. ●☆

342148   Safran Medik. = Crocus L. ■

342149   Sagaceae Schultz Sch. = Arecaceae Bercht. et J. Presl（保留科名）●

342150   Sagaceae Schultz Sch. = Palmae Juss.（保留科名）●

342151   Sagapenon Raf. = Danaa All. ■☆

342152   Sagapenon Raf. = Physospermum Cusson ex Juss. ■☆

342153   Sageraea Dalzell（1851）;陷药玉盘属●☆

342154   Sageraea laurina Dalzell;陷药玉盘●☆

342155   Sageretia Brongn.（1826）;雀梅藤属（对节刺属）;Sageretia ●☆

342156   Sageretia affinis（Blume）G. Don = Sageretia hamosa（Wall. ex Roxb.）Brongn. ●

342157   Sageretia apiculata C. K. Schneid. = Sageretia gracilis J. R. Drumm. et Sprague ●

342158   Sageretia apiculata Schneid. = Sageretia gracilis J. R. Drumm. et Sprague ●

342159   Sageretia bodinieri H. Lév. = Rhamnus esquirolii H. Lév. ●

342160   Sageretia brandrethiana Aitch.;窄叶雀梅藤;Narrowleaf Sageretia, Narrow-leaved Sageretia ●

342161   Sageretia brandrethiana Aitch. = Sageretia thea（Osbeck）M. C. Johnst. var. brandrethiana（Aitch.）Qaiser et Nazim. ●

342162   Sageretia camellifolia Y. L. Chen et P. K. Chou;茶叶雀梅藤;Tealeaf Sageretia, Tea-leaved Sageretia ●

342163   Sageretia cavaleriei（H. Lév.）C. K. Schneid. = Sageretia henryi J. R. Drumm. et Sprague ●

342164   Sageretia chanetii（H. Lév.）C. K. Schneid. = Sageretia thea（Osbeck）M. C. Johnst. ●

342165   Sageretia compacta Drumm. et Sprague = Sageretia gracilis J. R. Drumm. et Sprague ●

342166   Sageretia costata Miq. = Sageretia hamosa（Wall. ex Roxb.）Brongn. ●

342167   Sageretia ferruginea Oliv. = Sageretia rugosa Hance ●

342168   Sageretia filiformis（Roth）G. Don;线叶雀梅藤●☆

342169   Sageretia gracilis J. R. Drumm. et Sprague;纤细雀梅藤（筛子簸箕果,铁藤）;Thin Sageretia ●

342170   Sageretia hamosa（Wall. ex Roxb.）Brongn.;钩枝雀梅藤（钩刺梅藤,钩刺雀梅藤,钩状雀梅藤,猴栗）;Hooked Sageretia ●

342171   Sageretia hamosa（Wall. ex Roxb.）Brongn. = Sageretia randaiensis Hayata ●

342172   Sageretia hamosa（Wall. ex Roxb.）Brongn. var. trichoclada C. Y. Wu ex Y. L. Chen;毛枝雀梅藤;Hairybranch Sageretia ●

342173   Sageretia hamosa（Wall.）Brongn. = Sageretia hamosa（Wall. ex Roxb.）Brongn. ●

342174   Sageretia hamosa（Wall.）Brongn. = Sageretia randaiensis Hayata ●

342175   Sageretia hayatae Kaneh. = Sageretia thea（Osbeck）M. C. Johnst. ●

342176   Sageretia henryi J. R. Drumm. et Sprague;梗花雀梅藤（柄花雀

梅藤,红雀梅藤,红藤,皱锦,皱锦藤);Henry Sageretia ●

342177 Sageretia horrida Pax et K. Hoffm.;凹叶雀梅藤;Emarginate Sageretia ●

342178 Sageretia laxiflora Hand.-Mazz.;疏花雀梅藤;Sparseflower Sageretia,Sparse-flowered Sageretia ●

342179 Sageretia lucida Merr.;亮叶雀梅藤(倒钩茶,钩状雀梅藤);Lucidleaf Sageretia,Lucid-leaved Sageretia ●

342180 Sageretia melliana Hand.-Mazz.;刺藤子;Mell Sageretia,Spinevine Sageretia ●

342181 Sageretia minutiflora (Michx.) Trel.;小花雀梅藤;Gulfcost Sageretia ●☆

342182 Sageretia omeiensis C. K. Schneid.;峨眉雀梅藤;Emei Sageretia,Omei Sageretia ●

342183 Sageretia oppositifolia Brongn. = Sageretia filiformis (Roth) G. Don ●☆

342184 Sageretia parviflora G. Don = Sageretia filiformis (Roth) G. Don ●☆

342185 Sageretia paucicostata Maxim.;少脉雀梅藤(对节刺,对节木,对结刺,对结子);Fewvein Sageretia,Sparse-veined Sageretia ●

342186 Sageretia pauciflora H. T. Tsai;弯花雀梅藤;Few-flowered Sageretia ●

342187 Sageretia pedicellata C. Z. Gao;南丹雀梅藤;Pedicellate Sageretia ●

342188 Sageretia perpusilla C. K. Schneid. = Sageretia pycnophylla C. K. Schneid. ●

342189 Sageretia pycnophylla C. K. Schneid.;密叶雀梅藤(对节刺,对节木,木对节刺,砂糖果,铁箭鞭棵棵);Denseleaf Sageretia,Densileaved Sageretia,Woollytwig Sageretia,Woolly-twig Sageretia ●

342190 Sageretia pycnophylla Schneid. = Sageretia paucicostata Maxim. ●

342191 Sageretia randaiensis Hayata;峦大雀梅藤;Luanda Sageretia,Luanta Sageretia,Randa Sageretia ●

342192 Sageretia rugosa Hance;皱叶雀梅藤(九把伞,锈毛雀梅藤);Wrinkledleaf Sageretia,Wrinkle-leaved Sageretia ●

342193 Sageretia sikayoensis Masam.;高山雀梅藤;Sikayo Sageretia ●

342194 Sageretia spiciflora (A. Rich.) Chiov. = Sageretia thea (Osbeck) M. C. Johnst. ●

342195 Sageretia subcaudata C. K. Schneid.;尾叶雀梅藤;Acutelcaf Sageretia,Acute-leaved Sageretia ●

342196 Sageretia taiwaniana Hosok. ex Masam. = Sageretia thea (Osbeck) M. C. Johnst. var. taiwaniana (Hosok. ex Masam.) Y. C. Liu et C. M. Wang ●

342197 Sageretia taiwaniana Hosok. ex Masam. = Sageretia thea (Osbeck) M. C. Johnst. ●

342198 Sageretia thea (Osbeck) M. C. Johnst.;雀梅藤(柄花雀梅藤,刺冻绿,刺藤子,对角刺,对节刺,梗花雀梅藤,马沙刺,米碎木,牛鼻刺,沙穷勒,酸刺,酸梅箹,酸铜子,酸味,碎米子,台湾雀梅藤);Hedge Sageretia,Pauper's-tea ●

342199 Sageretia thea (Osbeck) M. C. Johnst. subsp. brandrethiana (Aitch.) Zeilinski = Sageretia thea (Osbeck) M. C. Johnst. var. brandrethiana (Aitch.) Qaiser et Nazim. ●

342200 Sageretia thea (Osbeck) M. C. Johnst. var. bilocularis S. Y. Liu;二室雀梅藤●

342201 Sageretia thea (Osbeck) M. C. Johnst. var. bilocularis S. Y. Liu = Sageretia thea (Osbeck) M. C. Johnst. ●

342202 Sageretia thea (Osbeck) M. C. Johnst. var. brandrethiana (Aitch.) Qaiser et Nazim. = Sageretia brandrethiana Aitch. ●

342203 Sageretia thea (Osbeck) M. C. Johnst. var. cordiformis Y. L. Chen et P. K. Chou;心叶雀梅藤;Cordateleaf Hedge Sageretia ●

342204 Sageretia thea (Osbeck) M. C. Johnst. var. taiwaniana (Hosok. ex Masam.) Y. C. Liu et C. M. Wang = Sageretia thea (Osbeck) M. C. Johnst. ●

342205 Sageretia thea (Osbeck) M. C. Johnst. var. taiwaniana (Hosok. ex Masam.) Y. C. Liu et C. M. Wang;台湾雀梅藤●

342206 Sageretia thea (Osbeck) M. C. Johnst. var. tomentosa (C. K. Schneid.) Y. L. Chen et P. K. Chou;毛叶雀梅藤;Hairyleaf Sageretia ●

342207 Sageretia theezanas (L.) Brongn. var. brandrethiana Parker = Sageretia thea (Osbeck) M. C. Johnst. var. brandrethiana (Aitch.) Qaiser et Nazim. ●

342208 Sageretia theezans (L.) Brongn. = Sageretia thea (Osbeck) M. C. Johnst. ●

342209 Sageretia theezans (L.) Brongn. f. tomentosa (C. K. Schneid.) H. Hara = Sageretia thea (Osbeck) M. C. Johnst. var. tomentosa (C. K. Schneid.) Y. L. Chen et P. K. Chou ●

342210 Sageretia theezans (L.) Brongn. var. glabrescens Kitam. = Sageretia thea (Osbeck) M. C. Johnst. ●

342211 Sageretia theezans (L.) Brongn. var. hildebrandii (Engl.) Chiov. = Sageretia thea (Osbeck) M. C. Johnst. ●

342212 Sageretia theezans (L.) Brongn. var. schweinfurthii Chiov. = Sageretia thea (Osbeck) M. C. Johnst. ●

342213 Sageretia theezans (L.) Brongn. var. spiciflora (A. Rich.) Chiov. = Sageretia thea (Osbeck) M. C. Johnst. ●

342214 Sageretia theezans Brongn. = Sageretia thea (Osbeck) M. C. Johnst. ●

342215 Sageretia theezans Brongn. var. tomentosa C. K. Schneid. = Sageretia thea (Osbeck) M. C. Johnst. var. tomentosa (C. K. Schneid.) Y. L. Chen et P. K. Chou ●

342216 Sageretia tibetica Pax et K. Hoffm. = Sageretia paucicostata Maxim. ●

342217 Sageretia wrightii S. Watson;赖特雀梅藤;Wright Sageretia ●☆

342218 Sageretia yemensis (Deflers) Suess. = Sageretia thea (Osbeck) M. C. Johnst. ●

342219 Sageretia yilinii G. S. Fan et S. K. Chen;脱毛雀梅藤●

342220 Sagina Druce = Minuartia L. ■

342221 Sagina L. (1753);漆姑草属(瓜槌草属);Pearlweed,Pearlwort ■

342222 Sagina abyssinica Hochst. ex A. Rich.;阿比西尼亚漆姑草■☆

342223 Sagina abyssinica Hochst. ex A. Rich. f. apetala Hauman = Sagina afroalpina Hedberg ■☆

342224 Sagina abyssinica Hochst. ex A. Rich. subsp. aequinoctialis Hedberg;昼夜漆姑草■☆

342225 Sagina afroalpina Hedberg;非洲山地漆姑草■☆

342226 Sagina apetala Ard.;无瓣漆姑草;Annual Pearlwort,Apetalous Pearlwort ■☆

342227 Sagina apetala Ard. subsp. ciliata (Fr.) Hook. f. = Sagina apetala Ard. ■☆

342228 Sagina apetala Ard. subsp. erecta (Hornem.) F. Herm. = Sagina apetala Ard. ■☆

342229 Sagina apetala Ard. subsp. filicaulis (Jord.) Berher = Sagina apetala Ard. ■☆

342230 Sagina apetala Ard. var. barbata Fenzl = Sagina apetala Ard. ■☆

342231 Sagina apetala Ard. var. barbata Fenzl ex Ledeb. = Sagina apetala Ard. ■☆

342232 Sagina apetala Ard. var. echinosperma Thell. = Sagina apetala Ard. ■☆

342233 Sagina apetala Ard. var. imberbis Fenzl = Sagina apetala Ard. ■☆

342234　Sagina apetala Ard. var. leiosperma Thell. = Sagina apetala Ard. ■☆

342235　Sagina apetala Ard. var. obtusisepala (Faure et al.) Maire = Sagina apetala Ard. ■☆

342236　Sagina boydii F. B. White;曲叶漆姑草;Boyd's Pearlwort ■☆

342237　Sagina brachysepala Chiov. ;短瓣漆姑草■☆

342238　Sagina caespitosa (J. Vahl) Lange;簇生漆姑草■☆

342239　Sagina ciliata Fr. ;缘毛漆姑草;Fringed Pearlwort ■☆

342240　Sagina ciliata Fr. = Sagina apetala Ard. ■☆

342241　Sagina condensata Sennen = Sagina apetala Ard. ■☆

342242　Sagina crassicaulis S. Watson = Sagina maxima A. Gray subsp. crassicaulis (S. Watson) G. E. Crow ■☆

342243　Sagina crassicaulis S. Watson var. litoralis (Hultén) Hultén = Sagina maxima A. Gray ■

342244　Sagina crassicaulis S. Watson var. littorea (Makino) Hara = Sagina maxima A. Gray ■

342245　Sagina decumbens (Elliott) Torr. et A. Gray;铺散漆姑草;Trailing Pearlwort ■☆

342246　Sagina decumbens (Elliott) Torr. et A. Gray var. smithii (A. Gray) S. Watson = Sagina decumbens (Elliott) Torr. et A. Gray ■☆

342247　Sagina decumbens (Elliott) Torry et A. Gray subsp. occidentalis (S. Watson) G. E. Crow;西方漆姑草■☆

342248　Sagina erecta L. = Moenchia erecta (L.) P. Gaertn. ,B. Mey. et Scherb. ■☆

342249　Sagina fontinalis Short et R. Peter = Stellaria fontinalis (Short et R. Peter) B. L. Rob. ■☆

342250　Sagina intermedia Fenzl ex Ledeb. = Sagina nivalis (Lindblom) Fr. ■☆

342251　Sagina japonica (Sw. ex Steud.) Ohwi;漆姑草(大龙叶,地兰,地松,瓜槌草,瓜槌草,牛毛毡,漆姑,日本漆姑草,踏地草,腺漆姑草,星宿草,星秀草,羊儿草,针包草,珍珠草);Japanese Pearlwort,Pearlwort ■

342252　Sagina japonica (Sw. ex Steud.) Ohwi var. parviflora C. Y. Wu;小花漆姑草(珍珠草);Littleflower Pearlwort ■

342253　Sagina japonica (Sw.) Ohwi = Sagina japonica (Sw. ex Steud.) Ohwi ■

342254　Sagina karakorensis (Em. Schmid) Kozhevn. ;克拉克漆姑草■

342255　Sagina karakorensis (Em. Schmid) Kozhevn. = Arenaria karakorensis Em. Schmid ■

342256　Sagina linnaei C. Presl = Sagina saginoides (L.) H. Karst. ■

342257　Sagina litoralis Hultén = Sagina maxima A. Gray ■

342258　Sagina maritima D. Don;海滨漆姑草;Sea Pearlwort ■☆

342259　Sagina maxima A. Gray;根叶漆姑草(大瓜槌草,腺毛漆姑草);Miximoewicz Pearlwort ■

342260　Sagina maxima A. Gray = Sagina japonica (Sw. ex Steud.) Ohwi ■

342261　Sagina maxima A. Gray f. crassicaulis (S. Watson) M. Mizush. ;粗茎根叶漆姑草■☆

342262　Sagina maxima A. Gray f. littorea Makino = Sagina maxima A. Gray ■

342263　Sagina maxima A. Gray subsp. crassicaulis (S. Watson) G. E. Crow = Sagina maxima A. Gray f. crassicaulis (S. Watson) M. Mizush. ■☆

342264　Sagina maxima A. Gray var. crassicaulis (S. Watson) H. Hara = Sagina maxima A. Gray f. crassicaulis (S. Watson) M. Mizush. ■☆

342265　Sagina maxima A. Gray var. littorea (Makino) H. Hara = Sagina maxima A. Gray ■

342266　Sagina micrantha (Bunge) Fernald = Sagina saginoides (L.) H. Karst. ■

342267　Sagina nevadensis Boiss. et Reut. = Sagina saginoides (L.) H. Karst. subsp. nevadensis (Boiss. et Reut.) Greuter et Burdet ■☆

342268　Sagina nivalis (Lindblom) Fr. ;雪线漆姑草(全叶漆姑草);Lesser Alpine Pearlwort,Snow Pearlwort ■☆

342269　Sagina nivalis (Lindblom) Fr. var. caespitosa (J. Vahl) B. Boivin = Sagina caespitosa (J. Vahl) Lange ■☆

342270　Sagina nivalis Fr. = Sagina nivalis (Lindblom) Fr. ■☆

342271　Sagina nodosa (L.) Fenzl;节漆姑草;Knotted Pearlwort,Knotted Spurry ■☆

342272　Sagina nodosa (L.) Fenzl subsp. borealis G. E. Crow;北方漆姑草■☆

342273　Sagina nodosa (L.) Fenzl var. borealis (G. E. Crow) Cronquist = Sagina nodosa (L.) Fenzl subsp. borealis G. E. Crow ■☆

342274　Sagina nodosa (L.) Fenzl var. pubescens (Besser) Koch = Sagina nodosa (L.) Fenzl ■☆

342275　Sagina normaniana Lagerh. ;诺尔曼漆姑草;Druee's Pearlwort, Scottish Pearlwort ■☆

342276　Sagina occidentalis S. Watson = Sagina decumbens (Elliott) Torry et A. Gray subsp. occidentalis (S. Watson) G. E. Crow ■☆

342277　Sagina oxysepala Boiss. ;尖瓣漆姑草■☆

342278　Sagina pilifera Fenzl = Sagina subulata C. Presl ■

342279　Sagina procumbens L. ;仰卧漆姑草(平铺漆姑草,匍匐漆姑草);Biddy's Eyes, Birdeye Pearlwort, Bird's Eye, Bird's-eye Pearlwort, Breakstone, Little Chickweed, Make-beggar, Makebeggars, Matted Pearlwort, Mossy Pearl Wort, Mothan, Pearlweed, Pearlwort, Poverty,Procumbent Pearlwort,Trailing Pearlwort ■

342280　Sagina procumbens L. = Sagina japonica (Sw. ex Steud.) Ohwi ■

342281　Sagina procumbens L. subsp. atlasica Dobignard;小花仰卧漆姑草■☆

342282　Sagina procumbens L. var. apetala Fenzl = Sagina procumbens L. ■

342283　Sagina procumbens L. var. compacta Lange = Sagina procumbens L. ■

342284　Sagina procumbens L. var. glaberrima Nielr. ? = Sagina procumbens L. ■

342285　Sagina procumbens L. var. parviflora Ball = Sagina procumbens L. subsp. atlasica Dobignard ■☆

342286　Sagina sabuletorum Lange;沙地漆姑草■☆

342287　Sagina sabuletorum Lange var. atlantica Litard. et Maire = Sagina sabuletorum Lange ■☆

342288　Sagina sabuletorum Lange var. longifolia Pau et Font Quer = Sagina sabuletorum Lange ■☆

342289　Sagina sabuletorum Lange var. loscosii (Boiss.) Pau = Sagina sabuletorum Lange ■☆

342290　Sagina saginoides (L.) H. Karst. ;无毛漆姑草(凉风草,漆姑草,藓状漆姑草,珍珠草);Alpine Pearlwort, Arctic Peariwort, Hairless Pearlwort ■

342291　Sagina saginoides (L.) H. Karst. subsp. nevadensis (Boiss. et Reut.) Greuter et Burdet;内华达漆姑草■☆

342292　Sagina saginoides (L.) H. Karst. subsp. parviflora Litard. et Maire;小花无毛漆姑草■☆

342293　Sagina saginoides (L.) H. Karst. var. hesperia Fernald = Sagina saginoides (L.) H. Karst. ■

342294　Sagina saginoides (L.) H. Karst. var. nevadensis (Boiss. et Reut.) Briq. = Sagina saginoides (L.) H. Karst. subsp. nevadensis

（Boiss. et Reut.）Greuter et Burdet ■☆

342295 Sagina saginoides（L.）H. Karst. var. stenophylla Pau = Sagina saginoides（L.）H. Karst. ■

342296 Sagina sinensis Hance = Sagina japonica（Sw. ex Steud.）Ohwi ■

342297 Sagina stricta Fr. = Sagina maritima G. Don ■☆

342298 Sagina subulata（Sw.）C. Presl；钻叶漆姑草；Corsican Pearlwort，Heath Pearlwort，Scottish Moss ■☆

342299 Sagina subulata（Sw.）C. Presl = Arenaria verna L. ■

342300 Sagina subulata C. Presl = Arenaria verna L. ■

342301 Sagina subulata C. Presl = Sagina subulata（Sw.）C. Presl ■☆

342302 Sagina taquetii H. Lév. = Sagina japonica（Sw. ex Steud.）Ohwi ■

342303 Sagina taquetii H. Lév. = Sagina maxima A. Gray ■

342304 Sagina virginica L. = Bartonia virginica（L.）Britton，Sterns et Poggenb. ■☆

342305 Saginaceae Bercht. et J. Presl = Caryophyllaceae Juss.（保留科名）■●

342306 Sagitta Adans. = Sagitta Guett. ■

342307 Sagitta Guett. = Sagittaria L. ■

342308 Sagittaria L.（1753）；慈姑属（慈菇属）；Arrow Head，Arrowhead，Wapato ■

342309 Sagittaria Rupp. ex L. = Sagittaria L. ■

342310 Sagittaria aginashi Makino；长叶慈姑（长叶泽泻，姑妈菜，剪刀草，水慈姑）；Longleaf Arrowhead ■

342311 Sagittaria altigena Hand.-Mazz.；高原慈姑；Plateau Arrowhead ■

342312 Sagittaria altigena Hand.-Mazz. ex Sam. = Sagittaria pygmaea Miq. ■

342313 Sagittaria ambigua J. G. Sm.；堪萨斯慈姑；Kansas Arrowhead，Plains Sagittaria ■☆

342314 Sagittaria angustifolia Lindl. = Sagittaria lancifolia L. ■☆

342315 Sagittaria arifolia Nutt. ex J. G. Sm.；海芋叶慈姑；Arum-leaved Arrowhead，Tuleroot，Wild Potato ■☆

342316 Sagittaria arifolia Nutt. ex J. G. Sm. = Sagittaria cuneata E. Sheld. ■☆

342317 Sagittaria australis（J. G. Sm.）Small；南方慈姑；Appalachian Arrowhead ■☆

342318 Sagittaria blumei Kunth = Sagittaria guyanensis Kunth subsp. lappula（D. Don）Bogin ■

342319 Sagittaria brevirostra Mack. et Bush；短喙慈姑；Midwestern Arrowhead，Short-beaked Arrowhead ■☆

342320 Sagittaria calycina Engelm. = Sagittaria montevidensis Cham. et Schltdl. subsp. calycina（Engelm.）Bogin ■☆

342321 Sagittaria chapmanii（J. G. Sm.）C. Mohr = Sagittaria graminea Michx. subsp. chapmanii（J. G. Sm.）R. R. Haynes et Hellq. ■☆

342322 Sagittaria chinensis Pursh = Sagittaria latifolia Willd. ■

342323 Sagittaria cordifolia Roxb. = Sagittaria guyanensis Kunth subsp. lappula（D. Don）Bogin ■

342324 Sagittaria cristata Engelm.；冠状慈姑；Crested Arrowhead ■☆

342325 Sagittaria cuneata E. Sheld.；楔叶慈姑；Arum-leaved Arrowhead，Cuneate-leaved Arrowleaf，Northern Arrowhead，Wapato ■☆

342326 Sagittaria cycloptera（J. G. Sm.）C. Mohr = Sagittaria graminea Michx. ■☆

342327 Sagittaria doniana J. G. Sm. = Sagittaria trifolia L. ■

342328 Sagittaria eatonii J. G. Sm. = Sagittaria graminea Michx. ■☆

342329 Sagittaria edulis Siebold ex Schltdl. = Sagittaria trifolia L. subsp. sinensis（Sims）Q. F. Wang ■

342330 Sagittaria engelmanniana J. G. Sm.；酸水慈姑；Acid-water Arrowhead ■☆

342331 Sagittaria engelmanniana J. G. Sm. subsp. brevirostra（Mack. et Bush）Bogin = Sagittaria brevirostra Mack. et Bush ■☆

342332 Sagittaria engelmanniana J. G. Sm. subsp. longirostra（Micheli）Bogin = Sagittaria australis（J. G. Sm.）Small ■

342333 Sagittaria esculenta Howell = Sagittaria latifolia Willd. ■

342334 Sagittaria falcata Pursh = Sagittaria lancifolia L. subsp. media（Micheli）Bogin ■☆

342335 Sagittaria fasciculata E. O. Beal；簇生慈姑；Duck Potato ■☆

342336 Sagittaria filiformis J. G. Sm.；线形慈姑■☆

342337 Sagittaria formosana Hayata = Sagittaria guyanensis Kunth subsp. lappula（D. Don）Bogin ■

342338 Sagittaria graminea Michx.；禾叶慈姑（禾状慈姑）；Arrowhead，Grass-leaved Arrowhead，Grass-leaved Sagittaria，Grass-like Arrowleaf，Grassy Arrowhead ■☆

342339 Sagittaria graminea Michx. subsp. chapmanii（J. G. Sm.）R. R. Haynes et Hellq.；查普曼慈姑■☆

342340 Sagittaria graminea Michx. subsp. weatherbiana（Fernald）R. R. Haynes，Hellq. et Novon；韦瑟斯比慈姑■☆

342341 Sagittaria graminea Michx. var. chapmanii J. G. Sm. = Sagittaria graminea Michx. subsp. chapmanii（J. G. Sm.）R. R. Haynes et Hellq. ■☆

342342 Sagittaria graminea Michx. var. cristata（Engelm.）Bogin = Sagittaria cristata Engelm. ■☆

342343 Sagittaria graminea Michx. var. platyphylla Engelm. = Sagittaria platyphylla（Engelm.）J. G. Sm. ■☆

342344 Sagittaria graminea Michx. var. weatherbiana（Fernald）Bogin = Sagittaria graminea Michx. subsp. weatherbiana（Fernald）R. R. Haynes，Hellq. et Novon ■☆

342345 Sagittaria greggii J. G. Sm. = Sagittaria longiloba Engelm. ex J. G. Sm. ■☆

342346 Sagittaria guayanensis Kunth；圭亚那慈姑；Guyanese Arrowhead ■☆

342347 Sagittaria guayanensis Kunth subsp. lappula（D. Don）Bogin；冠果草（鹤虱慈姑，假菱角，水菱角，台湾冠果草，田连藕，土紫蔻）；Guiana Arrowhead ■

342348 Sagittaria hastata Pursh = Sagittaria trifolia L. ■

342349 Sagittaria heterophylla Pursh；互叶慈姑；Heterophyllous Arrowleaf ■☆

342350 Sagittaria heterophylla Pursh = Sagittaria rigida Pursh ■☆

342351 Sagittaria heterophylla Pursh f. elliptica（Engelm.）S. F. Blake = Sagittaria rigida Pursh ■☆

342352 Sagittaria heterophylla Pursh f. fluitans（Engelm.）S. F. Blake = Sagittaria rigida Pursh ■☆

342353 Sagittaria heterophylla Pursh f. rigida（Pursh）S. F. Blake = Sagittaria rigida Pursh ■☆

342354 Sagittaria hirundinacea Blume = Sagittaria trifolia L. ■

342355 Sagittaria humilis（Rich.）Kuntze = Ranalisma humile（Rich.）Hutch. ■☆

342356 Sagittaria japonica Vilm. = Sagittaria sagittifolia L. ■

342357 Sagittaria japonica Vilm. = Sagittaria trifolia L. ■

342358 Sagittaria kurziana Glück；带叶慈姑；Strap-leaf Sagittaria ■☆

342359 Sagittaria lancifolia L.；矛叶慈姑；Duck Potato，Lanceleaved Arrowhead ■☆

342360 Sagittaria lancifolia L. subsp. media（Micheli）Bogin；间型矛叶慈姑■☆

342361 Sagittaria lancifolia L. var. media Micheli = Sagittaria lancifolia L. subsp. media（Micheli）Bogin ■☆

342362 Sagittaria lappula D. Don = Sagittaria guyanensis Kunth subsp.

lappula（D. Don）Bogin ■

342363 Sagittaria latifolia Willd. ；弯喙慈姑（宽叶慈姑）；American Arrowhead，Arrowhead，Arrow-leaf，Broadleaf Arrowhead，Broad-leaved Arrowhead，Common Arrowhead，Duck Potato，Duck-potato，Tule Potato，Tulepotato，Wapato，Wapatoo ■

342364 Sagittaria latifolia Willd. = Sagittaria trifolia L. ■

342365 Sagittaria latifolia Willd. f. gracilis（Pursh）B. L. Rob. = Sagittaria latifolia Willd. ■

342366 Sagittaria latifolia Willd. f. hastata（Pursh）B. L. Rob. = Sagittaria latifolia Willd. ■

342367 Sagittaria latifolia Willd. var. obtusa（Muhl. ex Willd.）Wiegand = Sagittaria latifolia Willd. ■

342368 Sagittaria latifolia Willd. var. obtusa（Muhl.）Wiegand = Sagittaria latifolia Willd. ■

342369 Sagittaria latifolia Willd. var. pubescens（Engelm.）J. G. Sm. = Sagittaria latifolia Willd. ■

342370 Sagittaria latifolia Willd. var. pubescens（Muhl. ex Nutt.）J. G. Sm. = Sagittaria latifolia Willd. ■

342371 Sagittaria leucopetala（Miq.）Bergman = Sagittaria trifolia L. ■

342372 Sagittaria lichuanensis J. K. Chen，X. Z. Sun et H. Q. Wang；利川慈姑；Lichuan Arrowhead ■

342373 Sagittaria longiloba Engelm. = Sagittaria longiloba Engelm. ex J. G. Sm. ■☆

342374 Sagittaria longiloba Engelm. ex J. G. Sm. ；长裂慈姑；Longbarb Arrowhead ■☆

342375 Sagittaria longirostra（Micheli）J. G. Sm. ；长喙慈姑；Longbeak Arrowhead ■☆

342376 Sagittaria longirostra（Micheli）J. G. Sm. var. australis J. G. Sm. = Sagittaria australis（J. G. Sm.）Small ■☆

342377 Sagittaria longirostra J. G. Sm. = Sagittaria longirostra（Micheli）J. G. Sm. ■☆

342378 Sagittaria lorata（Chapm.）Small = Sagittaria subulata（L.）Buchenau ■☆

342379 Sagittaria macrocarpa J. G. Sm. = Sagittaria graminea Michx. ■☆

342380 Sagittaria mohrii J. G. Sm. = Sagittaria platyphylla（Engelm.）J. G. Sm. ■☆

342381 Sagittaria montevidensis Cham. et Schltdl. ；大慈姑；Giant Arrowhead ■☆

342382 Sagittaria montevidensis Cham. et Schltdl. subsp. calycina（Engelm.）Bogin = Sagittaria calycina Engelm. ■☆

342383 Sagittaria montevidensis Cham. et Schltdl. subsp. calycina（Engelm.）Bogin；密西西比慈姑（大萼慈姑）；Giant Arrowhead，Hooded Arrowhead，Long-lobed Arrowhead，Mississippi Arrowhead ■☆

342384 Sagittaria montevidensis Cham. et Schltdl. subsp. spongiosa（Engelm.）Bogin；海绵慈姑；Tidal Arrowhead ■☆

342385 Sagittaria natans Pall. ；浮叶慈姑（小慈姑）；Floatleaf Arrowhead ■

342386 Sagittaria nymphaeifolia Hochst. = Sagittaria guayanensis Kunth subsp. lappula（D. Don）Bogin ■

342387 Sagittaria obtusa Muhl. ex Willd. = Sagittaria latifolia Willd. ■

342388 Sagittaria obtusifolia L. = Limnophyton obtusifolium（L.）Miq. ■☆

342389 Sagittaria obtusissima Hassk. = Sagittaria guyanensis Kunth subsp. lappula（D. Don）Bogin

342390 Sagittaria ornithorhyncha Small = Sagittaria latifolia Willd. ■

342391 Sagittaria papillosa Buchenau；乳突慈姑■☆

342392 Sagittaria planipes Fernald = Sagittaria latifolia Willd. ■

342393 Sagittaria platyphylla（Engelm.）J. G. Sm. ；宽叶慈姑（油叶慈姑）；Arrowhead，Delta Arrowhead ■☆

342394 Sagittaria potamogetifolia Merr. ；小慈姑；Small Arrowhead ■

342395 Sagittaria pubescens Muhl. = Sagittaria latifolia Willd. ■

342396 Sagittaria pygmaea Miq. ；矮慈姑（扁簪草，瓜皮草，水充草，水蒜，线慈姑，鸭舌草，鸭舌头，鸭舌子）；Dwarf Arrowhead ■

342397 Sagittaria rigida Pursh；美洲慈姑；American Arrowhead，Arrowhead，Canadian Arrowhead，Sessile-fruited Arrowhead，Stiff Arrowhead ■☆

342398 Sagittaria rigida Pursh f. elliptica（Engelm.）Fernald = Sagittaria rigida Pursh ■☆

342399 Sagittaria rigida Pursh f. fluitans（Engelm.）Fernald = Sagittaria rigida Pursh ■☆

342400 Sagittaria sagittifolia L. ；欧洲慈姑（白地栗，槎牙，芘菇，芘菰，茨姑，茨菇，慈姑，慈菇，慈菰，河凫茨，华夏慈姑，剪搭草，剪刀草，箭搭草，藕剪，梨头草，水慈姑，水慈菇，水慈菰，水萍，乌芋，燕尾草，野茨菰，野慈姑，泽泻，张口草）；Adder's Tongue，Arrowhead，Common Arrowhead，Japanese Arrowhead，Kyor，Moses-in-the-bulrushes，Old World Arrowhead，Oldworld Arrowhead，Old-world Arrowhead，Swamp Potato，Wapatoo，Water Archer ■

342401 Sagittaria sagittifolia L. 'Flore Pleno'；重瓣欧洲慈姑；Japanese Arrowhead ■☆

342402 Sagittaria sagittifolia L. = Sagittaria trifolia L. ■

342403 Sagittaria sagittifolia L. f. sinensis（Sims）Makino = Sagittaria trifolia L. ■

342404 Sagittaria sagittifolia L. sensu Hohen. = Sagittaria trifolia L. subsp. sinensis（Sims）Q. F. Wang ■

342405 Sagittaria sagittifolia L. subsp. leucopetala（Miq.）Hartog = Sagittaria trifolia L. subsp. sinensis（Sims）Q. F. Wang ■

342406 Sagittaria sagittifolia L. var. angustifolia Siebold = Sagittaria trifolia L. subsp. sinensis（Sims）Q. F. Wang ■

342407 Sagittaria sagittifolia L. var. edulis Siebold ex Miq. = Sagittaria trifolia L. ■

342408 Sagittaria sagittifolia L. var. leucopetala Miq. = Sagittaria trifolia L. subsp. sinensis（Sims）Q. F. Wang ■

342409 Sagittaria sagittifolia L. var. longiloba Turcz. ；长瓣欧洲慈姑（长瓣慈姑，长裂慈姑）■

342410 Sagittaria sagittifolia L. var. longiloba Turcz. = Sagittaria trifolia L. subsp. sinensis（Sims）Q. F. Wang ■

342411 Sagittaria sagittifolia L. var. pygmaea（Miq.）Makino = Sagittaria pygmaea Miq. ■

342412 Sagittaria sagittifolia L. var. sinensis Sims = Sagittaria trifolia L. subsp. sinensis（Sims）Q. F. Wang ■

342413 Sagittaria sanfordii Greene；桑福德慈姑■☆

342414 Sagittaria secundifolia Král；侧叶慈姑■☆

342415 Sagittaria sinensis Sims = Sagittaria trifolia L. subsp. sinensis（Sims）Q. F. Wang ■

342416 Sagittaria stagnorum Small = Sagittaria filiformis J. G. Sm. ■☆

342417 Sagittaria subulata（L.）Buchenau；钻叶慈姑；Awlleaf Arrowhead，Narrow-leaved Arrowhead ■☆

342418 Sagittaria subulata（L.）Buchenau var. natans（Michx.）J. G. Sm. = Sagittaria subulata（L.）Buchenau ■☆

342419 Sagittaria tengtsungensis H. Li；腾冲慈姑；Tengchong Arrowhead ■

342420 Sagittaria teres S. Watson；纤细慈姑；Quill-leaved Sagittaria，Slender Arrowhead ■☆

342421 Sagittaria trifolia L. ；野慈姑（长瓣慈姑，长裂叶慈姑，慈姑草，慈姑苗，慈果子，剪刀草，三角剪，三脚剪，水慈姑，水箭草，水马慈，水芋，狭叶慈姑，燕尾草）；Japanese Arrowhead，Longlobe

Oldworld Arrowhead, Narrowleaf Wild Arrowhead, Old World Arrowhead, Old-world Arrowhead, Wild Arrowhead ■

342422　Sagittaria trifolia L. 'Caerulea' = Sagittaria trifolia L. 'Sinensis' ■

342423　Sagittaria trifolia L. 'Flore Pleno' = Sagittaria sagittifolia L. 'Flore Pleno' ■☆

342424　Sagittaria trifolia L. 'Sinensis';慈姑(白地栗,茨菇,华夏慈姑);Arrowhead, Chinese Arrowhead, Old-world Arrowhead ■

342425　Sagittaria trifolia L. f. longiloba (Turcz.) Makino = Sagittaria sagittifolia L. var. longiloba Turcz. ■

342426　Sagittaria trifolia L. f. longiloba (Turcz.) Makino = Sagittaria trifolia L. ■

342427　Sagittaria trifolia L. subsp. leucopetala (Miq.) Hartog = Sagittaria trifolia L. ■

342428　Sagittaria trifolia L. subsp. sinensis (Sims) Q. F. Wang;华夏慈姑■

342429　Sagittaria trifolia L. var. alismifolia (Makino) Makino;泽泻慈姑■☆

342430　Sagittaria trifolia L. var. angustifolia (Siebold) Kitag. = Sagittaria trifolia L. ■

342431　Sagittaria trifolia L. var. edulis (Siebold et Zucc.) Ohwi = Sagittaria trifolia L. 'Sinensis' ■

342432　Sagittaria trifolia L. var. edulis (Siebold ex Miq.) Ohwi = Sagittaria trifolia L. subsp. sinensis (Sims) Q. F. Wang ■

342433　Sagittaria trifolia L. var. longiloba (Turcz.) Kitag. = Sagittaria trifolia L. ■

342434　Sagittaria trifolia L. var. retusa J. K. Chen, S. C. Sun et H. Q. Wang = Sagittaria trifolia L. ■

342435　Sagittaria trifolia L. var. retusa J. K. Chen, X. Z. Sun et H. Q. Wang;微凹慈姑;Retuse Wild Arrowhead ■

342436　Sagittaria trifolia L. var. retusa J. K. Chen, X. Z. Sun et H. Q. Wang = Sagittaria trifolia L. ■

342437　Sagittaria trifolia L. var. retusa J. K. Chen, X. Z. Sun et H. Q. Wang = Sagittaria trifolia L. f. longiloba (Turcz.) Makino ■

342438　Sagittaria trifolia L. var. sinensis (Sims) Makino = Sagittaria trifolia L. 'Sinensis' ■

342439　Sagittaria trifolia L. var. sinensis (Sims) Makino = Sagittaria trifolia L. subsp. sinensis (Sims) Q. F. Wang ■

342440　Sagittaria trifolia L. var. typica Makino = Sagittaria trifolia L. ■

342441　Sagittaria variabilis Engelm. var. obtusa (Muhl. ex Willd.) Engelm. = Sagittaria latifolia Willd. ■

342442　Sagittaria viscosa C. Mohr = Sagittaria latifolia Willd. ■

342443　Sagittaria weatherbiana Fernald = Sagittaria graminea Michx. subsp. weatherbiana (Fernald) R. R. Haynes, Hellq. et Novon ■☆

342444　Sagittaria wuyiensis J. K. Chen = Sagittaria lichuanensis J. K. Chen, X. Z. Sun et H. Q. Wang ■

342445　Sagittaria wuyiensis J. K. Chen, X. Z. Sun et H. Q. Wang = Sagittaria lichuanensis J. K. Chen, X. Z. Sun et H. Q. Wang ■

342446　Sagittipetalum Merr. = Carallia Roxb. (保留属名) ●

342447　Saglorithys Rizzini = Justicia L. ●■

342448　Saglorithys Rizzini (1949);巴西鸭嘴花属●

342449　Sagmen Hill = Centaurea L. (保留属名) ●■

342450　Sagoaceae Schultz Sch. = Arecaceae Bercht. et J. Presl(保留科名) ●

342451　Sagoaceae Schultz Sch. = Palmae Juss. (保留科名) ●

342452　Sagonea Aubl. = Hydrolea L. (保留属名) ■

342453　Sagonea palustris Aubl. = Hydrolea palustris (Aubl.) Raeusch. ■☆

342454　Sagoneaceae Martinov = Boraginaceae Juss. (保留科名) ■●

342455　Sagotanthus Tiegh. = Chaunochiton Benth. ●☆

342456　Sagotia Baill. (1860) (保留属名);萨戈大戟属●☆

342457　Sagotia Duchass. et Walp. (废弃属名) = Desmodium Desv. (保留属名) ●■

342458　Sagotia Duchass. et Walp. (废弃属名) = Sagotia Baill. (保留属名) ●☆

342459　Sagotia racemosa Baill. ;萨戈大戟●☆

342460　Sagraea DC. = Clidemia D. Don ●☆

342461　Saguaster Kuntze = Drymophloeus Zipp. ●☆

342462　Saguerus Steck(废弃属名) = Arenga Labill. (保留属名) ●

342463　Saguerus pinnata Wurmb. = Arenga pinnata (Wurmb) Merr. ●

342464　Sagus Gaertn. = Raphia P. Beauv. ●

342465　Sagus Rumph. ex Gaertn. = Raphia P. Beauv. ●

342466　Sagus Steck(废弃属名) = Metroxylon Rottb. (保留属名) ●

342467　Sagus farinifera Gaertn. = Raphia farinifera (Gaertn.) Hyl. ●

342468　Sagus gomutus Perr. = Arenga pinnata (Wurmb) Merr. ●

342469　Sagus ruffia Jacq. = Raphia farinifera (Gaertn.) Hyl. ●

342470　Sagus taedigera Mart. = Raphia taedigera (Mart.) Mart. ●☆

342471　Sagus vinifera Poir. = Raphia vinifera P. Beauv. ●

342472　Sahagunia Liebm. = Clarisia Ruiz et Pav. (保留属名) ●☆

342473　Saharanthus M. B. Crespo et Lledó(2000);撒哈拉属●☆

342474　Saharanthus ifniensis (Caball.) M. B. Crespo et Lledó;萨哈花●☆

342475　Saheria Fenzl ex Durand = Maerua Forssk. ●☆

342476　Sahlbergia Neck. = Gardenia Ellis(保留属名) ●

342477　Sahlbergia Rchb. = Gardenia Ellis(保留属名) ●

342478　Sahlbergia Rchb. = Salhbergia Neck. ●

342479　Saintlegeria Cordem. = Chloranthus Sw. ■●

342480　Saintmorysia Endl. = Athanasia L. ●☆

342481　Saintmorysia Endl. = Morysia Cass. ●☆

342482　Saintpaulia H. Wendl. (1893);非洲紫罗兰属(非洲堇属,非洲紫苣苔属);African Violet, Common African Violet, Saintpaulia ■☆

342483　Saintpaulia alba E. A. Bruce = Streptocarpus albus (E. A. Bruce) I. Darbysh ■☆

342484　Saintpaulia amaniensis F. Roberts = Saintpaulia ionantha H. Wendl. subsp. grotei (Engl.) I. Darbysh. ■☆

342485　Saintpaulia brevipilosa B. L. Burtt = Saintpaulia ionantha H. Wendl. subsp. velutina (B. L. Burtt) I. Darbysh. ■☆

342486　Saintpaulia confusa B. L. Burtt = Saintpaulia ionantha H. Wendl. subsp. grotei (Engl.) I. Darbysh. ■☆

342487　Saintpaulia diplotricha B. L. Burtt = Saintpaulia ionantha H. Wendl. var. diplotricha (B. L. Burtt) I. Darbysh. ■☆

342488　Saintpaulia goetzeana Engl. ;格兹紫罗兰■☆

342489　Saintpaulia grandifolia B. L. Burtt = Saintpaulia ionantha H. Wendl. subsp. grandifolia (B. L. Burtt) I. Darbysh. ■☆

342490　Saintpaulia grotei Engl. = Saintpaulia ionantha H. Wendl. subsp. grotei (Engl.) I. Darbysh. ■☆

342491　Saintpaulia inconspicua B. L. Burtt;显著非洲紫罗兰■☆

342492　Saintpaulia intermedia B. L. Burtt = Saintpaulia ionantha H. Wendl. subsp. pendula (B. L. Burtt) I. Darbysh. ■☆

342493　Saintpaulia ionantha H. Wendl. ;非洲紫罗兰(非洲堇,非洲紫苣苔);African Violet, Common African Violet, Usambara Violet ■☆

342494　Saintpaulia ionantha H. Wendl. subsp. grandifolia (B. L. Burtt) I. Darbysh. ;大叶非洲紫罗兰■☆

342495　Saintpaulia ionantha H. Wendl. subsp. grotei (Engl.) I. Darbysh. ;格氏非洲紫罗兰■☆

342496　Saintpaulia ionantha H. Wendl. subsp. mafiensis I. Darbysh. et

Pócs;马菲紫罗兰■☆

342497　Saintpaulia ionantha H. Wendl. subsp. nitida（B. L. Burtt）I. Darbysh.;光亮非洲紫罗兰■☆

342498　Saintpaulia ionantha H. Wendl. subsp. occidentalis（B. L. Burtt）I. Darbysh. ;西部非洲紫罗兰■☆

342499　Saintpaulia ionantha H. Wendl. subsp. orbicularis（B. L. Burtt）I. Darbysh. ;圆形非洲紫罗兰■☆

342500　Saintpaulia ionantha H. Wendl. subsp. pendula（B. L. Burtt）I. Darbysh. ;下垂非洲紫罗兰■☆

342501　Saintpaulia ionantha H. Wendl. subsp. rupicola（B. L. Burtt）I. Darbysh. ;岩地非洲紫罗兰■☆

342502　Saintpaulia ionantha H. Wendl. subsp. velutina（B. L. Burtt）I. Darbysh. ;绒毛非洲紫罗兰■☆

342503　Saintpaulia ionantha H. Wendl. var. diplotricha（B. L. Burtt）I. Darbysh. ;双毛非洲紫罗兰■☆

342504　Saintpaulia kewensis C. B. Clarke = Saintpaulia ionantha H. Wendl.■☆

342505　Saintpaulia magungensis E. P. Roberts = Saintpaulia ionantha H. Wendl. subsp. grotei（Engl.）I. Darbysh.■☆

342506　Saintpaulia magungensis E. P. Roberts var. minima B. L. Burtt = Saintpaulia ionantha H. Wendl. subsp. grotei（Engl.）I. Darbysh.■☆

342507　Saintpaulia magungensis E. P. Roberts var. occidentalis B. L. Burtt = Saintpaulia ionantha H. Wendl. subsp. occidentalis（B. L. Burtt）I. Darbysh.■☆

342508　Saintpaulia nitida B. L. Burtt = Saintpaulia ionantha H. Wendl. subsp. nitida（B. L. Burtt）I. Darbysh.■☆

342509　Saintpaulia orbicularis B. L. Burtt = Saintpaulia ionantha H. Wendl. subsp. orbicularis（B. L. Burtt）I. Darbysh.■☆

342510　Saintpaulia orbicularis B. L. Burtt var. purpurea ? = Saintpaulia ionantha H. Wendl. subsp. orbicularis（B. L. Burtt）I. Darbysh.■☆

342511　Saintpaulia pendula B. L. Burtt = Saintpaulia ionantha H. Wendl. subsp. pendula（B. L. Burtt）I. Darbysh.■☆

342512　Saintpaulia pendula B. L. Burtt var. kizarae ? = Saintpaulia ionantha H. Wendl. subsp. pendula（B. L. Burtt）I. Darbysh.■☆

342513　Saintpaulia pusilla Engl. ;微小非洲紫罗兰■☆

342514　Saintpaulia rupicola B. L. Burtt = Saintpaulia ionantha H. Wendl. subsp. rupicola（B. L. Burtt）I. Darbysh.■☆

342515　Saintpaulia shumensis B. L. Burtt;舒梅非洲紫罗兰■☆

342516　Saintpaulia teitensis B. L. Burtt;泰塔非洲紫罗兰■☆

342517　Saintpaulia tongwensis B. L. Burtt;坦噶尼喀非洲紫罗兰■☆

342518　Saintpaulia velutina B. L. Burtt = Saintpaulia ionantha H. Wendl. subsp. velutina（B. L. Burtt）I. Darbysh.■☆

342519　Saintpauliopsis Staner(1934);类非洲紫罗兰属■☆

342520　Saintpauliopsis Starter = Staurogyne Wall.■

342521　Saintpauliopsis lebrunii Staner;类非洲紫罗兰■☆

342522　Saintpauliopsis lebrunii Staner var. obtusa ? = Saintpauliopsis lebrunii Staner■☆

342523　Saintpierrea Germ. = Rosa L.●

342524　Saionia Hatus.（1976）;日本三蕊杯属■☆

342525　Saionia Hatus. = Oxygyne Schltr.■☆

342526　Saionia shinzatoi Hatus. = Oxygyne shinzatoi（Hatus.）C. Abe et Akasawa■☆

342527　Saiothra Raf. = Hypericum L.■●

342528　Saiothra Raf. = Sarothra L.■●

342529　Saipania Hosok. = Croton L.●

342530　Sairanthus G. Don = Nicotiana L.●■

342531　Sairanthus G. Don = Tabacus Moench●■

342532　Sairanthus G. Don(1838);洁花茄属●■☆

342533　Sairanthus glutinosus G. Don;洁花茄●■☆

342534　Sairocarpus D. A. Sutton(1988);净果婆婆纳属(净果玄参属)■☆

342535　Sairocarpus Nutt. ex A. DC. = Antirrhinum L.●■

342536　Sairocarpus breweri（A. Gray）D. A. Sutton;净果婆婆纳■☆

342537　Sairocarpus kingii（S. Watson）D. A. Sutton;金氏净果婆婆纳■☆

342538　Sairocarpus multiflorus（Pennell）D. A. Sutton;多花净果婆婆纳■☆

342539　Saivala Jones = Blyxa Noronha ex Thouars■

342540　Sajanella Soják(1980);西伯利亚萝卜属■☆

342541　Sajanella monstrosa（Willd.）Soják;西伯利亚萝卜■☆

342542　Sajania Pimenov = Sajanella Soják■☆

342543　Sajorium Endl. = Plukenetia L.●☆

342544　Sajorium Endl. = Pterococcus Hassk.（保留属名）●☆

342545　Sajorium africanum（Sond.）Baill. = Plukenetia africana Sond.●☆

342546　Sakakia Nakai = Cleyera Thunb.（保留属名）●

342547　Sakakia Nakai = Eurya Thunb.●

342548　Sakakia hayatae Masam. et Yamam. = Cleyera japonica Thunb. var. hayatae（Masam. et Yamam.）Kobuski●

342549　Sakakia hayatae Masam. et Yamam. = Cleyera japonica Thunb.●

342550　Sakakia longicarpa Yamam. = Cleyera japonica Thunb. var. longicarpa（Gogelein）L. K. Ling et C. F. Hsieh●

342551　Sakakia longicarpa Yamam. = Cleyera longicarpa（Yamam.）L. K. Ling●

342552　Sakakia matsudae（Hayata）Masam. = Eurya loquaiana Dunn et Kobuski●

342553　Sakakia morii（Yamam.）Yamam. et Masam. = Cleyera japonica Thunb. var. morii（Yamam.）Masam.●

342554　Sakakia morii（Yamam.）Yamam. et Masam. = Cleyera japonica Thunb.●

342555　Sakakia ochnacea（DC.）Nakai = Cleyera japonica Thunb.●

342556　Sakersia Hook. f. = Dichaetanthera Endl.●☆

342557　Sakersia africana Hook. f. = Dichaetanthera africana（Hook. f.）Jacq. -Fél.●☆

342558　Sakersia calodendron Gilg et Ledermann ex Engl. = Dichaetanthera corymbosa（Cogn.）Jacq. -Fél.●☆

342559　Sakersia corymbosa（Cogn.）Jacq. -Fél. = Dichaetanthera corymbosa（Cogn.）Jacq. -Fél.●☆

342560　Sakersia echinulata Hook. f. = Dichaetanthera echinulata（Hook. f.）Jacq. -Fél.●☆

342561　Sakersia laurentii Cogn. = Dichaetanthera corymbosa（Cogn.）Jacq. -Fél.●☆

342562　Sakersia mirabilis A. Chev. = Dichaetanthera echinulata（Hook. f.）Jacq. -Fél.●☆

342563　Sakersia strigosa Cogn. = Dichaetanthera strigosa（Cogn.）Jacq. -Fél.●☆

342564　Sakoanala R. Vig.（1951）;海滨森林豆属●☆

342565　Sakoanala capuroniana Yakovlev = Sakoanala madagascariensis R. Vig.●☆

342566　Sakoanala madagascariensis R. Vig. ;海滨森林豆●☆

342567　Sakoanala villosa R. Vig. ;柔毛海滨森林豆●☆

342568　Salabertia Neck. = Tapiria Juss.●☆

342569　Salabertia Neck. = Tapirira Aubl.●☆

342570　Salacca Reinw.（1825）;萨拉卡棕属（鳞果椰属,沙拉卡椰子属,蛇皮果属）;Salacca,Salak,Salak Palm,Snakefruit●

342571　Salacca affinis Griff. ;近亲萨拉卡棕;Affined Salacca●☆

342572　Salacca conferta Griff. ;密集萨拉卡棕;Dense Salacca,Kelubi

Palm ●☆

342573　Salacca edulis Reinw. = Salacca zalacca（Gaertn.）Voss ex Vilm. ●☆

342574　Salacca flabellata Furtado;扇形萨拉卡棕;Fanshaped Salacca ●☆

342575　Salacca glabroscens Griff.;光滑萨拉卡棕;Smoothish Salacca ●☆

342576　Salacca rumphii Wall.;容佛萨拉卡棕;Rumph Salacca ●☆

342577　Salacca scortechinii Becc.;斯考氏萨拉卡棕;Scortechin Salacca ●☆

342578　Salacca secunda Griff.;滇西蛇皮果（蛇皮果）;W. Yunnan Snakefruit ●◇

342579　Salacca wallichiana Mart.;沃利克萨拉卡棕;Rakum ●☆

342580　Salacca zalacca（Gaertn.）Voss ex Vilm.;蛇皮果（可食萨拉卡棕）;Buak Salak,Edible Salacca,Salak ●☆

342581　Salacia L.（1771）（保留属名）;五层龙属（柃椤木属,五层楼属）;Salacia ●

342582　Salacia adolfi-friderici Loes. ex Harms;弗里德里西五层龙 ●☆

342583　Salacia africana（Willd.）DC. = Loeseneriella africana（Willd.）N. Hallé ●☆

342584　Salacia alata De Wild.;翅五层龙 ●☆

342585　Salacia alata De Wild. var. superba N. Hallé;华美五层龙 ●☆

342586　Salacia alpestris A. Chev. = Salacia erecta（G. Don）Walp. ●☆

342587　Salacia alternifolia Hochst. = Salacia kraussii（Harv.）Harv. ●☆

342588　Salacia alveolata Louis ex R. Wilczek;蜂窝五层龙 ●☆

342589　Salacia amplifolia Merr. ex Chun et F. C. How;阔叶五层龙（阔叶柃拉木）;Broadleaf Salacia,Broad-leaved Salacia ●

342590　Salacia angustifolia Scott-Elliot = Salacia chlorantha Oliv. ●☆

342591　Salacia aurantiaca C. Y. Wu;橙果五层龙;Orange Salacia,Orangefruit Salacia,Orange-fruited Salacia ●

342592　Salacia bangalensis Vermoesen ex R. Wilczek;班加拉五层龙 ●☆

342593　Salacia baumannii Loes.;鲍曼五层龙 ●☆

342594　Salacia baumii（Loes.）Exell et Mendonça = Salacia bussei Loes. ●☆

342595　Salacia bayakensis Pellegr. = Salacia letestui Pellegr. ●☆

342596　Salacia bequaerti De Wild. = Salacia pyriformis（Sabine）Steud. ●☆

342597　Salacia bipindensis Loes. = Salacia nitida（Benth.）N. E. Br. var. bipindensis（Loes.）N. Hallé ●☆

342598　Salacia brachypoda Peyr.;短柄五层龙 ●☆

342599　Salacia bussei Loes.;布瑟五层龙 ●☆

342600　Salacia callensii R. Wilczek;卡伦斯五层龙 ●☆

342601　Salacia camerunensis Loes. = Salacia longipes（Oliv.）N. Hallé var. camerunensis（Loes.）N. Hallé ●☆

342602　Salacia camerunensis Loes. var. longipetiolata？= Salacia longipes（Oliv.）N. Hallé var. longipetiolata（Loes.）N. Hallé ●☆

342603　Salacia capitulata N. Hallé;头状五层龙 ●☆

342604　Salacia cerasifera Welw. ex Oliv.;角五层龙 ●☆

342605　Salacia cerasifera Welw. ex Oliv. var. wendjiensis（R. Wilczek）N. Hallé;文德五层龙 ●☆

342606　Salacia chlorantha Oliv.;热非绿花五层龙 ●☆

342607　Salacia chlorantha Oliv. subsp. dalziellii（Hutch. et M. B. Moss）N. Hallé = Salacia chlorantha Oliv. ●☆

342608　Salacia chlorantha Oliv. subsp. demeusei（De Wild. et T. Durand）N. Hallé = Salacia chlorantha Oliv. ●☆

342609　Salacia cochinchinensis Lour.;柳叶五层龙;Willowleaf Salacia,Willow-leaved Salacia ●

342610　Salacia columna N. Hallé;圆柱五层龙 ●☆

342611　Salacia confertiflora Merr.;密花五层龙;Crowded-flower

Salacia,Crowded-flowered Salacia ●

342612　Salacia congolensis De Wild. et T. Durand;刚果五层龙 ●☆

342613　Salacia conraui Loes.;康氏五层龙 ●☆

342614　Salacia cornifolia Hook. f.;角叶五层龙 ●☆

342615　Salacia coronata N. Hallé;花冠五层龙 ●☆

342616　Salacia crampeli A. Chev. = Salacia laurentii De Wild. ●☆

342617　Salacia cuspidicoma Loes. = Salacia mannii Oliv. ●☆

342618　Salacia dalzielii Hutch. et M. B. Moss = Salacia chlorantha Oliv. ●☆

342619　Salacia debilis（G. Don）Walp.;弱小五层龙 ●☆

342620　Salacia demeusei De Wild. et T. Durand = Salacia chlorantha Oliv. ●☆

342621　Salacia denudata A. Chev. = Salacighia letestuana（Pellegr.）Blakelock ●☆

342622　Salacia devredii R. Wilczek;德夫雷五层龙 ●☆

342623　Salacia dewevrei De Wild. et T. Durand;德韦五层龙 ●☆

342624　Salacia dewildemaniana R. Wilczek = Salacia erecta（G. Don）Walp. var. dewildemaniana（R. Wilczek）N. Hallé ●☆

342625　Salacia doeringii Loes. = Salacia chlorantha Oliv. ●☆

342626　Salacia dusenii Loes.;杜森五层龙 ●☆

342627　Salacia echinulata Louis ex R. Wilczek = Salacia adolfi-friderici Loes. ex Harms ●☆

342628　Salacia elegans Welw. ex Oliv.;雅致五层龙 ●☆

342629　Salacia elegans Welw. ex Oliv. var. pynaertii（De Wild.）R. Wilczek = Salacia pynaertii De Wild. ●☆

342630　Salacia elliotii Loes. = Salacia cerasifera Welw. ex Oliv. ●☆

342631　Salacia elongata Hook. f. = Salacia pyriformis（Sabine）Steud. ●☆

342632　Salacia erecta（G. Don）Walp.;直立五层龙 ●☆

342633　Salacia erecta（G. Don）Walp. var. dewildemaniana（R. Wilczek）N. Hallé;德怀尔德曼五层龙 ●☆

342634　Salacia erecta（G. Don）Walp. var. kabweensis（R. Wilczek）N. Hallé;卡布韦五层龙 ●☆

342635　Salacia erecta（G. Don）Walp. var. leonardii（R. Wilczek）N. Hallé;莱奥五层龙 ●☆

342636　Salacia euryoides Hutch. et M. B. Moss = Salacia chlorantha Oliv. ●☆

342637　Salacia eurypetala Loes.;宽瓣五层龙 ●☆

342638　Salacia fimbrisepala Loes.;线萼五层龙 ●☆

342639　Salacia fredericqii R. Wilczek = Salacia laurentii De Wild. ●☆

342640　Salacia gabunensis Loes.;加蓬五层龙（加本五层龙）●☆

342641　Salacia gabunensis O. Loes. volkensiana（Loes. ex Harms）N. Hallé;福尔五层龙 ●☆

342642　Salacia germainii R. Wilczek;杰曼五层龙 ●☆

342643　Salacia germainii R. Wilczek var. cordata？;心形五层龙 ●☆

342644　Salacia gerrardii Harv. ex Sprague;杰勒德五层龙 ●☆

342645　Salacia gilgiana Loes. = Salacia stuhlmanniana Loes. ●☆

342646　Salacia glaucifolia C. Y. Wu;粉叶五层龙;Glaucousleaf Salacia,Glaucous-leaved Salacia,Paleleaf Salacia ●

342647　Salacia hainanensis Chun et F. C. How;海南五层龙（海南沙拉木,海南柃拉木）;Hainan Salacia ●

342648　Salacia hallei Jongkind;哈勒五层龙 ●☆

342649　Salacia hispida Blakelock;硬毛五层龙 ●☆

342650　Salacia howesii Hutch. et M. B. Moss;豪斯五层龙 ●☆

342651　Salacia ituriensis Loes.;伊图里五层龙 ●☆

342652　Salacia johannis-albrechti Loes. et Winkl. = Salacia oliveriana Loes. ●☆

342653　Salacia kabweensis R. Wilczek = Salacia erecta（G. Don）Walp. var. kabweensis（R. Wilczek）N. Hallé ●☆

342654　Salacia kivuensis R. Wilczek;基伍五层龙●☆

342655　Salacia klainei Pierre ex R. Wilczek;克莱恩五层龙●☆

342656　Salacia kraussii（Harv.）Harv.;克劳斯五层龙●☆

342657　Salacia laurentii De Wild.;洛朗五层龙●☆

342658　Salacia lebrunii R. Wilczek;勒布伦五层龙●☆

342659　Salacia lehmbachii Loes.;莱姆巴赫五层龙●☆

342660　Salacia lehmbachii Loes. var. aurantiaca（N. Hallé）N. Hallé;橘色五层龙●☆

342661　Salacia lehmbachii Loes. var. cucumerella（N. Hallé）N. Hallé;瓜叶莱姆五层龙●☆

342662　Salacia lehmbachii Loes. var. leonensis（Hutch. et M. B. Moss）N. Hallé;莱昂五层龙●☆

342663　Salacia lehmbachii Loes. var. manus-lacertae N. Hallé;撕裂五层龙●☆

342664　Salacia lehmbachii Loes. var. uregaensis（R. Wilczek）N. Hallé;乌雷加五层龙●☆

342665　Salacia leonardii R. Wilczek ＝Salacia erecta（G. Don）Walp. var. leonardii（R. Wilczek）N. Hallé ●☆

342666　Salacia leonardii R. Wilczek var. kivuensis ? ＝Salacia erecta（G. Don）Walp. var. leonardii（R. Wilczek）N. Hallé ●☆

342667　Salacia leonensis Hutch. et M. B. Moss ＝Salacia lehmbachii Loes. leonensis（Hutch. et M. B. Moss）N. Hallé ●☆

342668　Salacia leonensis Hutch. et M. B. Moss var. cucumerella N. Hallé ＝ Salacia lehmbachii Loes. cucumerella（N. Hallé）N. Hallé ●☆

342669　Salacia leptoclada Tul.;细枝五层龙●☆

342670　Salacia letestuana Pellegr. ＝Salacighia letestuana（Pellegr.）Blakelock ●☆

342671　Salacia letestui Pellegr.;莱泰斯图五层龙●☆

342672　Salacia letouzeyana N. Hallé;勒图五层龙●☆

342673　Salacia linderi Loes. ex Harms ＝Salacighia linderi（Loes. ex Harms）Blakelock ●☆

342674　Salacia livingstonii Loes. ＝Salacia stuhlmanniana Loes. ●☆

342675　Salacia loloensis Loes.;洛洛五层龙●☆

342676　Salacia loloensis Loes. var. sibangana N. Hallé;西邦五层龙●☆

342677　Salacia lomensis Loes. ＝Salacia stuhlmanniana Loes. ●☆

342678　Salacia longipes（Oliv.）N. Hallé;长梗五层龙●☆

342679　Salacia longipes（Oliv.）N. Hallé var. camerunensis（Loes.）N. Hallé;喀麦隆五层龙●☆

342680　Salacia longipes（Oliv.）N. Hallé var. longipetiolata（Loes.）N. Hallé;长柄五层龙●☆

342681　Salacia louisii R. Wilczek ＝Salacia preussii Loes. var. louisii（R. Wilczek）N. Hallé ●☆

342682　Salacia lovettii N. Hallé et B. Mathew;洛维特五层龙●☆

342683　Salacia lucida Oliv.;光亮五层龙●☆

342684　Salacia luebbertii Loes.;吕贝特五层龙●☆

342685　Salacia macrocarpa Welw. ex Oliv. ＝Salacia chlorantha Oliv. ●☆

342686　Salacia macrosperma Wight;大籽五层龙（大子五层龙）;Bigseed Salacia ●☆

342687　Salacia madagascariensis（Lam.）DC.;马达加斯加五层龙●☆

342688　Salacia madagascariensis DC. ＝ Salacia madagascariensis（Lam.）DC. ●☆

342689　Salacia mannii Oliv.;曼氏五层龙●☆

342690　Salacia mayumbensis Exell et Mendonça;马永巴五层龙●☆

342691　Salacia mildbraediana Loes. ＝Salacia pynaertii De Wild. ●☆

342692　Salacia mortehanii R. Wilczek ＝Salacia erecta（G. Don）Walp. var. dewildemaniana（R. Wilczek）N. Hallé ●☆

342693　Salacia ngaziensis Louis ex R. Wilczek ＝Salacia tuberculata Blakelock var. ngaziensis（Louis et R. Wilczek）N. Hallé ●☆

342694　Salacia nitida（Benth.）N. E. Br.;亮五层龙●☆

342695　Salacia nitida（Benth.）N. E. Br. var. bipindensis（Loes.）N. Hallé;比平迪五层龙●☆

342696　Salacia oblongifolia Oliv. ＝Salacia oliveriana Loes. ●☆

342697　Salacia obovatilimba S. Y. Pao;河口五层龙;Hekou Salacia, Obovatelimb Salacia, Obovate-limb Salacia ●

342698　Salacia oliveriana Loes.;奥里弗五层龙●☆

342699　Salacia orientalis N. Robson;东方五层龙●☆

342700　Salacia owabiensis Hoyle;奥瓦比五层龙●☆

342701　Salacia pallescens Oliv.;变苍白五层龙●☆

342702　Salacia pengheensis De Wild. ＝Salacia pyriformis（Sabine）Steud. ●☆

342703　Salacia pierlotii R. Wilczek ＝Salacia bangalensis Vermoesen ex R. Wilczek ●☆

342704　Salacia pierrei N. Hallé;皮埃尔五层龙●☆

342705　Salacia polysperma Hu;多籽五层龙（沙拉藤）;Manyseeds Salacia, Polyspermous Salacia, Seedy Salacia ●

342706　Salacia polysperma Hu subsp. verrucosorugosa H. W. Li;皱果五层龙●

342707　Salacia preussii Loes.;普罗伊斯五层龙●

342708　Salacia preussii Loes. var. louisii（R. Wilczek）N. Hallé;路易斯五层龙●☆

342709　Salacia prinoides（Willd.）DC.;五层龙（沙拉木）;Prinos-like Salacia, Prinus Salacia, Salacia ●

342710　Salacia pynaertii De Wild.;皮那五层龙●☆

342711　Salacia pyriformioides Loes.;拟梨形五层龙●☆

342712　Salacia pyriformis（Sabine）Steud.;梨形五层龙●☆

342713　Salacia quadrangulata R. Wilczek;四棱五层龙●☆

342714　Salacia racemosa Loes. ex Harms ＝ Thyrsosalacia racemosa（Loes. ex Harms）N. Hallé ●☆

342715　Salacia rehmannii Schinz;拉赫曼五层龙●☆

342716　Salacia rehmannii Schinz var. baumii Loes. ＝ Salacia bussei Loes. ●☆

342717　Salacia reticulata Wight;网状五层龙●☆

342718　Salacia rhodesiaca Blakelock;罗得西亚五层龙●☆

342719　Salacia rivularis Louis ex R. Wilczek;溪边五层龙●☆

342720　Salacia rufescens Hook. f.;浅红五层龙●☆

342721　Salacia schlechteri Loes. ＝Salacia loloensis Loes. ●☆

342722　Salacia semlikiensis De Wild. ＝Salacia elegans Welw. ex Oliv. ●☆

342723　Salacia senegalensis（Lam.）DC.;塞内加尔五层龙●☆

342724　Salacia sessiliflora Hand.-Mazz.;无柄五层龙（狗卵子，鸡卵黄，棱子藤，野柑子，野黄果）;Sessileflower Salacia, Sessile-flowered Salacia ●

342725　Salacia simtata Loes. ＝Salacia madagascariensis（Lam.）DC.●☆

342726　Salacia somalensis Chiov. ＝Salacia stuhlmanniana Loes.●☆

342727　Salacia soyauxii Loes. ＝Salacia mannii Oliv.●☆

342728　Salacia staudtiana Loes.;索马里五层龙●☆

342729　Salacia staudtiana Loes. var. cerasiocarpa（R. Wilczek）N. Hallé;角果五层龙●☆

342730　Salacia staudtiana Loes. var. tshopoensis（De Wild.）N. Hallé;乔波五层龙●☆

342731　Salacia stuhlmanniana Loes.;斯图尔曼五层龙●☆

342732　Salacia sulphur Loes. et H. J. P. Winkl.;硫色五层龙●☆

342733　Salacia talbotii Baker f.;塔尔博特五层龙●☆

342734　Salacia tessmannii Loes.;泰斯曼五层龙●☆

342735　Salacia togoica Loes.;多哥五层龙●☆

342736　Salacia tomiensis A. Chev. ;富江五层龙●☆

342737　Salacia toussaintii R. Wilczek = Salacia whytei Loes. var. toussaintii（R. Wilczek）N. Hallé ●☆

342738　Salacia transvaalensis Burtt Davy = Elaeodendron transvaalense（Burtt Davy）R. H. Archer ●☆

342739　Salacia tshopoensis De Wild. = Salacia staudtiana Loes. var. tshopoensis（De Wild.）N. Hallé ●☆

342740　Salacia tshopoensis De Wild. var. cerasiocarpa R. Wilczek = Salacia staudtiana Loes. var. cerasiocarpa（R. Wilczek）N. Hallé ●☆

342741　Salacia tuberculata Blakelock;多疣五层龙●☆

342742　Salacia tuberculata Blakelock var. ngaziensis（Louis et R. Wilczek）N. Hallé;恩加齐多疣五层龙●☆

342743　Salacia unguiculata De Wild. et T. Durand = Helictonema velutinum（Afzel.）Pierre ex N. Hallé ●☆

342744　Salacia uregaensis R. Wilczek = Salacia lehmbachii Loes. var. uregaensis（R. Wilczek）N. Hallé ●☆

342745　Salacia uregaensis R. Wilczek var. aurantiaca N. Hallé = Salacia lehmbachii Loes. var. aurantiaca（N. Hallé）N. Hallé ●☆

342746　Salacia vermoeseniana R. Wilczek = Salacia whytei Loes. var. vermoeseniana（R. Wilczek）N. Hallé ●☆

342747　Salacia villiersii N. Hallé;维利尔斯五层龙●☆

342748　Salacia viridiflora R. Wilczek;刚果绿花五层龙●☆

342749　Salacia volkensiana Loes. ex Harms = Salacia gabunensis O. Loes. volkensiana（Loes. ex Harms）N. Hallé ●☆

342750　Salacia volubilis Loes. et H. J. P. Winkl.;缠绕五层龙●☆

342751　Salacia wardii I. Verd. = Salacia leptoclada Tul. ●☆

342752　Salacia wendjiensis R. Wilczek = Salacia cerasifera Welw. ex Oliv. var. wendjiensis（R. Wilczek）N. Hallé ●☆

342753　Salacia whytei Loes. ;怀特五层龙●☆

342754　Salacia whytei Loes. var. ophiurella N. Hallé = Salacia whytei Loes. ●☆

342755　Salacia whytei Loes. var. toussaintii（R. Wilczek）N. Hallé;图森特五层龙●☆

342756　Salacia whytei Loes. var. vermoeseniana（R. Wilczek）N. Hallé;韦尔蒙森五层龙●☆

342757　Salacia zenkeri Loes. ;泽赫五层龙●☆

342758　Salacia zeyheri Planch. ex Harv. = Elaeodendron zeyheri Spreng. ex Turcz. ●☆

342759　Salaciaceae Raf. ;五层龙科●

342760　Salaciaceae Raf. = Celastraceae R. Br.（保留科名）●

342761　Salacicratea Loes. = Salacia L.（保留属名）●

342762　Salacighia Loes.（1940）;萨拉卫矛属●☆

342763　Salacighia letestuana（Pellegr.）Blakelock;萨拉卫矛●☆

342764　Salacighia linderi（Loes. ex Harms）Blakelock;非洲萨拉卫矛●☆

342765　Salacighia malpighioides Loes. = Salacighia letestuana（Pellegr.）Blakelock ●☆

342766　Salaciopsis Baker f.（1921）;拟萨拉卫矛属●☆

342767　Salaciopsis longistyla I. H. Müller;长柱拟萨拉卫矛●☆

342768　Salaciopsis neocaledonica Baker f.;拟萨拉卫矛●☆

342769　Salacistis Rchb. f. = Goodyera R. Br. ■

342770　Salacistis Rchb. f. = Hetaeria Blume（保留属名）■

342771　Salakka Reinw. ex Blume = Salacca Reinw. ●

342772　Salaxidaceae J. Agardh = Ericaceae Juss.（保留科名）●

342773　Salaxis Salisb.（1802）;筛子杜鹃属●☆

342774　Salaxis Salisb. = Erica L. ●☆

342775　Salaxis Salisb. et E. Phillips = Salaxis Salisb. ●☆

342776　Salaxis aristata（Benth.）D. Dietr. = Erica bojeri Dorr et E. G. H. Oliv. ●☆

342777　Salaxis artemisioides Klotzsch = Erica artemisioides（Klotzsch）E. G. H. Oliv. ●☆

342778　Salaxis axillaris（Thunb.）G. Don = Erica axillaris Thunb. ●☆

342779　Salaxis benguelensis Welw. ex Engl. = Erica benguelensis（Welw. ex Engl.）E. G. H. Oliv. ●☆

342780　Salaxis ciliata（Benth.）D. Dietr. = Erica boutonii Dorr et E. G. H. Oliv. ●☆

342781　Salaxis ciliata Benth. = Erica serrata Thunb. ●☆

342782　Salaxis densa（Benth.）D. Dietr. = Erica densata Dorr et E. G. H. Oliv. ●☆

342783　Salaxis flexuosa Klotzsch = Erica axillaris Thunb. ●☆

342784　Salaxis flexuosa Klotzsch var. cognata N. E. Br. = Erica axillaris Thunb. ●☆

342785　Salaxis floribunda（Benth.）D. Dietr. = Erica baroniana Dorr et E. G. H. Oliv. ●☆

342786　Salaxis goudotiana（Klotzsch）D. Dietr. = Erica goudotiana（Klotzsch）Dorr et E. G. H. Oliv. ●☆

342787　Salaxis gracilis（Benth.）D. Dietr. = Erica rakotozafyana Dorr et E. G. H. Oliv. ●☆

342788　Salaxis hexandra Klotzsch = Erica axillaris Thunb. ●☆

342789　Salaxis major N. E. Br. = Erica axillaris Thunb. ●☆

342790　Salaxis micrantha Benth. = Erica artemisioides（Klotzsch）E. G. H. Oliv. ●☆

342791　Salaxis octandra Klotzsch = Erica axillaris Thunb. ●☆

342792　Salaxis octandra Klotzsch var. artemisioides（Klotzsch）N. E. Br. = Erica artemisioides（Klotzsch）E. G. H. Oliv. ●☆

342793　Salaxis parviflora（Benth.）D. Dietr. = Erica lyallii Dorr et E. G. H. Oliv. ●☆

342794　Salaxis puberula Klotzsch = Erica axillaris Thunb. ●☆

342795　Salaxis rugosa（Klotzsch）Benth. = Erica rugata E. G. H. Oliv. ●☆

342796　Salaxis sieberi Benth. = Erica axillaris Thunb. ●☆

342797　Salaxis tenuifolia（Benth.）D. Dietr. = Erica goudotiana（Klotzsch）Dorr et E. G. H. Oliv. ●☆

342798　Salaxis tenuissima（Klotzsch）D. Dietr. = Erica rakotozafyana Dorr et E. G. H. Oliv. ●☆

342799　Salaxis triflora Compton = Erica terniflora E. G. H. Oliv. ●☆

342800　Salazaria Torr. = Scutellaria L. ●■

342801　Salazaria mexicana Torr. = Scutellaria mexicana（Torr.）A. J. Paton ●☆

342802　Salazariaceae F. A. Barkley = Labiatae Juss.（保留科名）●■

342803　Salazariaceae F. A. Barkley = Lamiaceae Martinov（保留科名）●■

342804　Salazia T. Durand et Jacks. = Sabazia Cass. ■●☆

342805　Salceda Blanco = Camellia L. ●

342806　Salcedoa Jiménez Rodr. et Katinas（2004）;簇花红菊木属●☆

342807　Salcedoa mirabaliarum Jiménez Rodr. et Katinas;簇花红菊木●☆

342808　Saldanha Vell. = Hillia Jacq. ●☆

342809　Saldanhaea Bureau = Cuspidaria DC.（保留属名）●☆

342810　Saldanhaea Kuntze = Hillia Jacq. ●☆

342811　Saldanhaea Kuntze = Saldanha Vell. ●☆

342812　Saldanhaea Post et Kuntze = Hillia Jacq. ●☆

342813　Saldanhaea Post et Kuntze = Saldanha Vell. ●☆

342814　Saldania Sim = Ormocarpum P. Beauv.（保留属名）●

342815　Saldania acanthocarpa Sim = Ormocarpum trichocarpum（Taub.）Engl. ●☆

342816　Saldinia A. Rich. = Saldinia A. Rich. ex DC. ■☆

342817　Saldinia A. Rich. ex DC.（1830）;马岛茜草属■☆

342818　Saldinia acuminata Bremek. ;尖马岛茜草■☆

342819　Saldinia axillaris (Lam. ex Poir.) Bremek. ;腋花马岛茜草■☆

342820　Saldinia bullata Bremek. ;泡状马岛茜草■☆

342821　Saldinia coursiana Bremek. ;库尔斯马岛茜草■☆

342822　Saldinia dasyclada Bremek. ;毛枝马岛茜草■☆

342823　Saldinia hirsuta Bremek. ;粗毛马岛茜草■☆

342824　Saldinia littoralis Bremek. ;滨海马岛茜草■☆

342825　Saldinia longistipulata Bremek. ;长托叶马岛茜草■☆

342826　Saldinia mandracensis Bremek. ;曼德拉斯马岛茜草■☆

342827　Saldinia oblongifolia Bremek. ;矩圆叶马岛茜草■☆

342828　Saldinia obovatifolia Bremek. ;倒卵叶马岛茜草■☆

342829　Saldinia obtusata Bremek. ;钝马岛茜草☆

342830　Saldinia pallida Bremek. ;苍白马岛茜草■☆

342831　Saldinia phlebophylla Bremek. ;脉叶马岛茜草■☆

342832　Saldinia platyclada Bremek. ;粗枝马岛茜草■☆

342833　Saldinia proboscidea Hochr. ;长角马岛茜草■☆

342834　Saldinia pycnophylla Bremek. ;密叶马岛茜草■☆

342835　Saldinia stenophylla Bremek. ;窄叶马岛茜草■☆

342836　Saldinia subacuminata Bremek. ;渐尖马岛茜草■☆

342837　Salgada Blanco ＝ Cryptocarya R. Br. (保留属名)●

342838　Salhbergia Neck. ＝ Gardenia Ellis(保留属名)●

342839　Salica Hill ＝ Lythastrum Hill ●■

342840　Salica Hill ＝ Lythrum L. ●■

342841　Salicaceae Mirb. (1815)(保留科名);杨柳科;Willow Family ●

342842　Salicaria Adans. ＝ Lythrum L. ●■

342843　Salicaria Mill. ＝ Lythrum L. ●■

342844　Salicaria Moench ＝ Nesaea Comm. ex Kunth(保留属名)■●☆

342845　Salicaria Tourn. ex Mill. ＝ Lythrum L. ●■

342846　Salicariaceae Juss. ＝ Lythraceae J. St. -Hil. (保留科名)■●

342847　Salicornia L. (1753);盐角草属(海蓬子属,盐角属);Glasswort,Marsh Samphire,Salicornia,Saltwort ■●

342848　Salicornia alpini Lag. ＝ Sarcocornia perennis (Mill.) A. J. Scott subsp. alpini (Lag.) Castrov. ●■☆

342849　Salicornia ambigua Micheli ＝ Sarcocornia perennis (Mill.) A. J. Scott ■☆

342850　Salicornia amplexicaulis Vahl ＝ Halopeplis amplexicaulis (Vahl) Ung. -Sternb. ex Ces. ,Pass. et Gibelli ■☆

342851　Salicornia arabica L. var. perennis (Mill.) Fiori ＝ Sarcocornia perennis (Mill.) A. J. Scott ■☆

342852　Salicornia bigelovii Torr. ;毕氏盐角草■☆

342853　Salicornia borealis S. L. Wolff et Jefferies ＝ Salicornia rubra A. Nelson ■☆

342854　Salicornia caspica L. ＝ Kalidium caspicum (L.) Ung. -Sternb. ●

342855　Salicornia caspica Pall. ＝ Halostachys caspica C. A. Mey. ex Schrenk ●

342856　Salicornia depressa Standl. ;凹陷盐角草■☆

342857　Salicornia deserticola A. Chev. ;沙地盐角草■☆

342858　Salicornia dolichostachya Moss;长穗盐角草;Bushy Glasswort,Long-spiked Glasswort ■☆

342859　Salicornia europaea L. ;盐角草(草盐角,海蒿,海蓬子,胖蒿子);Chicken-claws, Common Glasswort, Crab Grass, English Sea Grape,Frog Grass,Glass Saltwort,Glasswort,Jointed Glasswort,Marsh Samphire, Marshfire Glasswort, Pigeon-foot, Pigeon's Foot, Rock Samphire,Saltwort,Samphire,Sampion,Sea Grape,Seagrass,Semper, Slender Glasswort ■

342860　Salicornia europaea L. subsp. duvalii (A. Chev.) Maire ＝ Salicornia patula Duval-Jouve ■☆

342861　Salicornia europaea L. var. fruticosa L. ＝ Arthrocnemum fruticosum (L.) Moq. ●☆

342862　Salicornia europaea L. var. herbacea L. ＝Salicornia europaea L. ■

342863　Salicornia foliata Pall. ＝ Kalidium foliatum (Pall.) Moq. ●

342864　Salicornia fragilis P. W. Ball et Tutin;纤细盐角草;Yellow Glasswort ■☆

342865　Salicornia fruticosa (L.) L. ;欧洲海蓬子■☆

342866　Salicornia fruticosa (L.) L. ＝ Arthrocnemum fruticosum (L.) Moq. ●☆

342867　Salicornia herbacea L. ＝ Salicornia europaea L. ■

342868　Salicornia indica Willd. ;印度盐角草■☆

342869　Salicornia indica Willd. ＝ Arthrocnemum indicum (Willd.) Moq. ●☆

342870　Salicornia lignosa Woods ＝ Salicornia perennis Mill. var. lignosa (Woods) Moss ■☆

342871　Salicornia longispicata A. Chev. ＝ Sarcocornia perennis (Mill.) A. J. Scott ■☆

342872　Salicornia macrostachya Moric. ＝ Arthrocnemum macrostachyum (Moric.) K. Koch ●☆

342873　Salicornia maritima S. L. Wolff et Jefferies;沼泽盐角草■☆

342874　Salicornia media ?;中间盐角草;Common Chickweed ■☆

342875　Salicornia meyeriana Moss;迈尔盐角草■☆

342876　Salicornia natalensis Bunge ex Ung. -Sternb. ＝ Arthrocnemum natalense (Bunge ex Ung. -Sternb.) Moss ●☆

342877　Salicornia neglecta ?;忽视盐角草;Greater Chickweed ■☆

342878　Salicornia nitens;光亮盐角草;Shiny Glasswort ■☆

342879　Salicornia obscura P. W. Ball et Tutin;灰绿盐角草;Glaucous Glasswort ■☆

342880　Salicornia pachystachya Bunge ex Ung. -Sternb. ;粗穗盐角草■☆

342881　Salicornia pacifica Standl. ＝ Sarcocornia pacifica (Standl.) A. J. Scott ●☆

342882　Salicornia patula Duval-Jouve;张开盐角草■☆

342883　Salicornia perennis Mill. ＝ Sarcocornia perennis (Mill.) A. J. Scott ■☆

342884　Salicornia perennis Mill. var. lignosa (Woods) Moss ＝ Sarcocornia perennis (Mill.) A. J. Scott ■☆

342885　Salicornia perfoliata Forssk. ＝ Halopeplis perfoliata (Forssk.) Bunge ex Asch. et Schweinf. ☆

342886　Salicornia perrieri A. Chev. ;佩里耶盐角草■☆

342887　Salicornia pojarkovae N. Semenova;波氏盐角草■☆

342888　Salicornia praecox A. Chev. ＝Salicornia senegalensis A. Chev. ■☆

342889　Salicornia prostrata Pall. ;平卧盐角草;Seablite Glasswort ■☆

342890　Salicornia pusilla Woods;单花盐角草;Fragile Glasswort,One-flowered Glasswort ■☆

342891　Salicornia pygmaea Pall. ＝ Halopeplis pygmaea (Pall.) Bunge ex Ung. -Sternb. ■

342892　Salicornia radicans Sm. ＝ Sarcocornia perennis (Mill.) A. J. Scott ■☆

342893　Salicornia ramosissima Woods;紫盐角草;Purple Glasswort, Twiggy Glasswort ■☆

342894　Salicornia rubra A. Nelson;红盐角草;Western Glasswort ■☆

342895　Salicornia senegalensis A. Chev. ;塞内加尔盐角草■☆

342896　Salicornia strobilacea Pall. ＝ Halocnemum strobilaceum (Pall.) M. Bieb. ●

342897　Salicornia subterminalis Parish ＝ Arthrocnemum subterminale (Parish) Standl. ●☆

342898　Salicornia uniflora Toelken;独花盐角草■☆

342899　Salicornia utahensis Tidestr. = Sarcocornia utahensis (Tidestr.) A. J. Scott ●☆

342900　Salicornia virginica L. ;弗吉尼亚盐角草(维尔吉亚盐角草); Perennial Glasswort ■☆

342901　Salicorniaceae J. Agardh = Amaranthaceae Juss. (保留科名)●■

342902　Salicorniaceae J. Agardh = Chenopodiaceae Vent. (保留科名)●■

342903　Salicorniaceae J. Agardh;盐角草科●■

342904　Salicorniaceae Martinov = Chenopodiaceae Vent. (保留科名)●■

342905　Salimori Adans. = Cordia L. (保留属名)●

342906　Salisburia Sm. = Ginkgo L. ●★

342907　Salisburia adiantifolia Sm. = Ginkgo biloba L. ●

342908　Salisburiaceae Link = Ginkgoaceae Engl. (保留科名)●

342909　Salisburiana Wood = Ginkgo L. ●★

342910　Salisburiana Wood = Salisburia Sm. ●★

342911　Salisburya Hoffmanns. = Salisburiana Wood ●

342912　Salisburya biloba Hoffm. = Ginkgo biloba L. ●

342913　Salisburyaceae Kuntze = Ginkgoaceae Engl. (保留科名)●

342914　Salisburyaceae Kuntze = Salisburiaceae Link ●

342915　Salisburyodendron A. V. Bobrov et Melikyan = Agathis Salisb. (保留属名)●

342916　Salisburyodendron A. V. Bobrov et Melikyan = Dammara Lam. ●

342917　Salisburyodendron A. V. Bobrov et Melikyan(2006);澳洲南洋杉属●☆

342918　Salisia Brongn. et Gris = Xanthostemon F. Muell. (保留属名)●☆

342919　Salisia Lindl. = Kunzea Rchb. (保留属名)●☆

342920　Salisia Panch. ex Brongn. et Gris = Xanthostemon F. Muell. (保留属名)●☆

342921　Salisia Regel = Gloxinia L'Hér. ■☆

342922　Saliunea Raf. = Valerianella Mill. ■

342923　Salix L. (1753)(保留属名);柳属;Osier,Sallow,Willow ●

342924　Salix × arakiana Koidz. ;荒木柳●☆

342925　Salix × boulayi F. Gerard;鲍迪柳●☆

342926　Salix × chrysocoma Dode;金荑柳;Golden Weeping Willow ●☆

342927　Salix × conifera Wangenh. ;球果柳;Conifer Willow ●☆

342928　Salix × gracilistyloides Kimura;拟细柱柳●☆

342929　Salix × hatusimae Kimura;初岛柳●☆

342930　Salix × hayatana Kimura;早田氏柳●☆

342931　Salix × hisauchiana Koidz. ;久内柳●☆

342932　Salix × ishikawae Kimura = Salix × yamatoensis Koidz. ●☆

342933　Salix × kawamurana Kimura;川邑柳●☆

342934　Salix × koidzumii Kimura;小泉柳●☆

342935　Salix × lasiogyne Seemen;毛蕊柳●☆

342936　Salix × lasiogyne Seemen nothosubsp. yuhkii (Kimura) H. Ohashi = Salix × yuhkii Kimura ●☆

342937　Salix × laurina Sm. ;月桂柳;Laurel-leaved Willow ●☆

342938　Salix × matsumurae Seemen;松村氏柳●☆

342939　Salix × meyeriana Rostl. ;迈尔柳;Shiny-leaved Willow ●☆

342940　Salix × microstemon Kimura;小茎柳●☆

342941　Salix × nasuensis Kimura;那须柳●☆

342942　Salix × pendulina Wender. ;下垂柳;Weeping Crack-willow,Weeping Willow,Wisconsin Weeping Willow ●☆

342943　Salix × sendaica Kimura nothosubsp. ultima (Koidz.) H. Ohashi et Yonek. ;远仙台柳●☆

342944　Salix × sepulchralis Simonk. ;裂皮柳;Weeping Willow ●☆

342945　Salix × sepulchralis Simonk. var. chrysocoma (Dode) Meikle;金垂柳●☆

342946　Salix × sericans Tausch ex Kem. ;绢毛柳;Broad-leaved Osier ●☆

342947　Salix × sirakawensis Kimura;白河柳●☆

342948　Salix × stipularis Kimura;托叶柳;Eared Osier ●☆

342949　Salix × sumiyosensis Kimura;住吉柳●☆

342950　Salix × tsugaluensis Koidz. ;津轻柳●☆

342951　Salix × ultima Koidz. = Salix × sendaica Kimura nothosubsp. ultima (Koidz.) H. Ohashi et Yonek. ●☆

342952　Salix × yamatoensis Koidz. ;矢本柳●☆

342953　Salix × yoitiana Kimura = Salix leucopithecia Kimura ●

342954　Salix × yuhkii Kimura;结城柳●☆

342955　Salix abscondita Laksch. ;隐匿柳●☆

342956　Salix acuminata Lange = Salix smithiana Willd. ●☆

342957　Salix acuminata Mill. = Salix cinerea L. ●☆

342958　Salix acuminata Schleich. ex Ser. = Salix grandifolia Ser. ●☆

342959　Salix acuminatomicrophylla K. S. Hao ex C. F. Fang et A. K. Skvortsov = Salix ovatomicrophylla K. S. Hao ●

342960　Salix acutidens Rydb. = Salix eriocephala Michx. ●☆

342961　Salix acutifolia Willd. ;尖叶柳(锐叶柳);Caspian Willow,Sharp-leaf Willow,Sharp-leafed Willow,Siberian Violet-willow ●☆

342962　Salix acutifolia Willd. 'Blue Streak';蓝带尖叶柳●☆

342963　Salix acutifolia Willd. 'Pendulifolia';垂叶尖叶柳●☆

342964　Salix adamauensis Seemen = Salix mucronata Thunb. ●☆

342965　Salix adenophylla Hook. ;腺叶柳;Furry Willow ●☆

342966　Salix adenophylla Hook. = Salix cordata Michx. ●☆

342967　Salix aegyptiaca L. ;埃及柳(麝香柳);Calaf of Persia Willow,Egyptian Willow,Musk Willow ●☆

342968　Salix aegyptiaca L. = Salix mucronata Thunb. ●☆

342969　Salix alatavica Kar. et Kir. ex Stschegl. ;阿拉套柳;Alatao Willow,Alataw Mountain Willow ●

342970　Salix alaxensis (Andersson ex DC.) Coville;北美毡柳;Feltleaf Willow ●☆

342971　Salix alaxensis (Andersson) Coville = Salix alaxensis (Andersson ex DC.) Coville ●☆

342972　Salix alaxensis Coville = Salix alaxensis (Andersson ex DC.) Coville ●☆

342973　Salix alba L. ;白柳(新疆长叶柳);Cricket-bat Willow,Duck Willow,European White Willow,European Willow,Great Willow,Hoary Willow,Huntingdon Willow,Popple,Sauch Saugh,Saugh-tree,Swallow-tailed Willow,White Salix,White Willow ●

342974　Salix alba L. 'Aurea';黄叶白柳●☆

342975　Salix alba L. 'Britzensis';猩红柳;Coral-bark Willow,Golden Willow,Scarlet Willow ●☆

342976　Salix alba L. 'Chermesina' = Salix alba L. 'Britzensis'●☆

342977　Salix alba L. 'Coerulea';天蓝柳;Cricket-bat Willow ●☆

342978　Salix alba L. 'Tristis';黯淡白柳;Golden Weeping Willow,Weeping Willow ●☆

342979　Salix alba L. f. argentea ?;银白叶柳;Silver Willow,White Willow ●☆

342980　Salix alba L. subsp. coerulea (Sm.) Rchb. f. ;蓝白柳;Cricket Bat Willow,Cricket-bat Willow ●☆

342981　Salix alba L. subsp. vitellina (L.) Arcang. = Salix alba L. subsp. vitellina (L.) Schübl. et M. Martens ●☆

342982　Salix alba L. subsp. vitellina (L.) Schübl. et M. Martens;金色柳(黄枝白柳);Babylon Willow,Coral-bark Willow,Dutch Osier,Golden Osier,Golden Weeping Willow,Golden Willow,Seale Tree,Weeping Willow,Yellow Willow,Yellow-stem White Willow,Yolk-of-egg Willow ●☆

342983　Salix alba L. var. caerulea (Sm.) Koch = Salix alba L. subsp.

coerulea (Sm.) Rchb. f. ●☆

342984   Salix alba L. var. cardinalis；绯红柳；Belgian Red Willow ●☆

342985   Salix alba L. var. pendula ?；垂白柳；Weeping Willow ●☆

342986   Salix alba L. var. sericea ?；绢毛白柳；Silky Willow, Silver Willow, White Willow ●☆

342987   Salix alba L. var. splendens (Bray) Andersson ＝Salix alba L. ●

342988   Salix alba L. var. subintegra (Z. Wang et P. Y. Fu) N. Chao ＝ Salix paraplesia C. K. Schneid. var. subintegra Z. Wang et P. Y. Fu ●

342989   Salix alba L. var. tristis ? ＝ Salix alba L. 'Tristis' ●☆

342990   Salix alba L. var. typica Andersson ＝Salix alba L. ●

342991   Salix alba L. var. vitellina (L.) Ser. ＝ Salix alba L. subsp. vitellina (L.) Schübl. et M. Martens ●☆

342992   Salix alba L. var. vitellina (L.) Stokes 'Britzensis' ＝ Salix alba L. 'Britzensis' ●☆

342993   Salix alba L. var. vitellina (L.) Stokes 'Pendula'；垂枝金色柳；Babylon Willow, Drooping Willow ●☆

342994   Salix alba L. var. vitellina (L.) Stokes ＝ Salix alba L. subsp. vitellina (L.) Schübl. et M. Martens ●☆

342995   Salix albertii Regel；二色柳；Albert Willow, Bicolor Willow ●

342996   Salix albertii Regel ＝ Salix tenuijulis Ledeb. ●

342997   Salix alfredii Goerz；秦岭柳；Alfred Willow ●

342998   Salix alfredii Goerz var. fengxianica (N. Chao) G. Zhu；凤县柳；Fengxian Willow ●

342999   Salix alfredii Goerz var. fengxianica (N. Chao) G. Zhu ＝ Salix fengxianica N. Chao ●

343000   Salix allochroa C. K. Schneid. ＝Salix plocotricha C. K. Schneid. ●

343001   Salix alopochroa Kimura ＝ Salix vulpina Andersson subsp. alopochroa (Kimura) H. Ohashi et Yonek. ●☆

343002   Salix alopochroa Kimura var. psilostachys (Kimura) Kimura ＝ Salix vulpina Andersson subsp. alopochroa (Kimura) H. Ohashi et Yonek. f. psilostachys (Kimura) H. Ohashi et Yonek. ●☆

343003   Salix alpina Scop. ；高山柳●☆

343004   Salix alpina Walter ＝ Salix humilis Marshall var. tristis (Aiton) Griggs ●☆

343005   Salix ambigua Pursh ＝ Salix nigra Marshall ●☆

343006   Salix amnematchinensis K. S. Hao ＝ Salix oritrepha C. K. Schneid. var. amnematchinensis (K. S. Hao) Z. Wang et C. F. Fang ●

343007   Salix amnematchinensis K. S. Hao ex C. F. Fang et A. K. Skvortsov ＝ Salix oritrepha C. K. Schneid. var. amnematchinensis (K. S. Hao) Z. Wang et C. F. Fang ●

343008   Salix amphibola C. K. Schneid. ；九鼎柳；Amphibola Willow, Jiuding Mountain Willow ●

343009   Salix amygdalina L. ＝ Salix triandra L. subsp. amygdalina (L.) Schübl. et M. Martens ●☆

343010   Salix amygdalina L. ＝ Salix triandra L. ●

343011   Salix amygdalina L. var. nipponica (Franch. et Sav.) C. K. Schneid. ＝ Salix triandra L. var. nipponica (Franch. et Sav.) Seemen ●

343012   Salix amygdalina L. var. nipponica (Franch. et Sav.) C. K. Schneid. ＝ Salix nipponica Franch. et Sav. ●

343013   Salix amygdaloides Andersson；毛柳；Almond Willow, Almondleaved Willow, Peach Willow, Peachleaf Willow, Peach-leaved Willow ●

343014   Salix amygdaloides Andersson var. wrightii (Andersson) C. K. Schneid. ＝ Salix amygdaloides Andersson ●

343015   Salix anadyrensis Flod. ；阿纳代尔柳●☆

343016   Salix ancorifera Fernald ＝ Salix discolor Muhl. ●☆

343017   Salix andersoniana Sm. ；安氏柳；Green Mountain Sallow ●☆

343018   Salix andersoniana Sm. ＝ Salix nigricans (Sm.) Enander ●☆

343019   Salix andropogon H. Lév. ＝ Salix variegata Franch. ●

343020   Salix andropogon H. Lév. et Vaniot ＝Salix variegata Franch. ●

343021   Salix angiolepis H. Lév. ＝ Salix rosthornii Seemen ex Diels ●

343022   Salix angiolepis H. Lév. et Vaniot ＝Salix rosthornii Seemen ex Diels ●

343023   Salix angustata Pursh ＝Salix eriocephala Michx. ●☆

343024   Salix angustifolia Willd. ＝ Salix linearistipularis (Franch.) K. S. Hao ●

343025   Salix angustifolia Willd. ＝ Salix wilhelmsiana M. Bieb. ●

343026   Salix anisandra H. Lév. et Vaniot ＝ Salix rosthornii Seemen ex Diels ●

343027   Salix annularis Forbes；环叶柳；Curled-leaved Weeping Willow, Ring-leaved Willow ●☆

343028   Salix annulifera C. Marquand et Airy Shaw；环纹矮柳；Annular Willow, Annularline Willow ●

343029   Salix annulifera C. Marquand et Airy Shaw var. dentata S. D. Zhao；齿苞矮柳；Toothed-bract Willow ●

343030   Salix annulifera C. Marquand et Airy Shaw var. glabra P. Y. Mao et W. Z. Li；无毛矮柳；Smooth Annular Willow ●

343031   Salix annulifera C. Marquand et Airy Shaw var. macriula C. Marquand et Airy Shaw；匙叶矮柳；Spoonleaf Annular Willow ●

343032   Salix anomala E. L. Wolf；异常柳●☆

343033   Salix ansoniana Forbes ＝ Salix nigricans (Sm.) Enander ●☆

343034   Salix antiatlantica Maire et Wilczek ＝ Salix pedicellata Desf. subsp. antiatlantica (Maire et Wilczek) Maire ●☆

343035   Salix anticecrenata Kimura；圆齿垫柳；Antecrenate Willow, Crenate Willow ●

343036   Salix apatela C. K. Schneid. ＝ Salix magnifica Hemsl. var. apetala (C. K. Schneid.) K. S. Hao ●

343037   Salix apoda Trautv. ；灰皮柳(无柄柳)●☆

343038   Salix aquatica Ser. ＝ Salix aurita L. ●

343039   Salix aquatica Sm. ＝ Salix cinerea L. ●

343040   Salix araeostachya C. K. Schneid. ；纤序柳(纤穗柳)；Araeostachya Willow, Slender Catkin Willow, Slender-spiked Willow ●

343041   Salix arbuscula L. ；小树柳(灌木柳, 密枝柳)；Mountain Willow, Plum-leaved Willow, Small Tree Willow ●☆

343042   Salix arbusculoides Andersson；拟小树柳；Littletree Willow, Little-tree Willow ●☆

343043   Salix arbutifolia Pall. ＝ Chosenia arbutifolia (Pall.) A. K. Skvortsov ●◇

343044   Salix arbutifolia Pall. f. adenantha (Kimura) Kimura ＝ Chosenia arbutifolia (Pall.) A. K. Skvortsov ●◇

343045   Salix arctica Pall. ；北极柳；Arctic Willow ●

343046   Salix arenaria L. ；沙地柳；Creeping Willow, Sand Willow ●☆

343047   Salix areostachya C. K. Schneid. ＝ Salix araeostachya C. K. Schneid ●

343048   Salix arguta Andersson var. alpigena Andersson ＝ Salix serissima (L. H. Bailey) Fernald ●☆

343049   Salix arguta Andersson var. hirtisquama (Andersson) Andersson ＝ Salix lucida Muhl. ●☆

343050   Salix arguta Andersson var. pallescens Andersson ＝ Salix serissima (L. H. Bailey) Fernald ●☆

343051   Salix argyi H. Lév. ＝ Salix rosthornii Seemen ex Diels ●

343052   Salix argyracea E. L. Wolf；银柳；Sivery Willow ●

343053   Salix argyracea E. L. Wolf f. obovata Girz ＝ Salix argyracea E.

L. Wolf ●

343054   Salix argyri H. Lév. = Salix rosthornii Seemen ex Diels ●

343055   Salix argyrophegga C. K. Schneid. ; 银光柳;Silverlight Willow, Silver-lighted Willow ●

343056   Salix argyrotrichocarpa C. F. Fang; 银毛果柳;Silver-hairyfruit Willow,Silver-hairy-fruited Willow ●

343057   Salix atopantha C. K. Schneid. ;奇花柳;Sessileflower Willow, Wonder Willow,Wonderflower Willow ●

343058   Salix atopantha C. K. Schneid. var. glabra K. S. Hao = Salix chienii W. C. Cheng ●

343059   Salix atopantha C. K. Schneid. var. pedicellata C. F. Fang et J. Q. Wang;长柄奇花柳;Long-pedicel Wonderflower Willow ●

343060   Salix atrocinerea Brot. ; 深灰柳; Gray Willow, Grey Sallow, Large Gray Willow ● ☆

343061   Salix atrocinerea Brot. subsp. catalaunica Goerz = Salix atrocinerea Brot. ● ☆

343062   Salix atrocinerea Brot. subsp. jahandiezii（Chass.）Ibn Tattou; 贾汉深灰柳●☆

343063   Salix aurita L. ; 耳柳; Auriculate Willow, Ear Willow, Eared Sallow,Eared Willow,Ear-leaved Willow,Round-ear Willow,Round-eared Sallow,Trailing Sallow,Withe-tree ●

343064   Salix australior Andersson;南方柳;Southern Sallow ● ☆

343065   Salix australis Forbes = Salix cinerea L. ●

343066   Salix australis Hilsenb. et Bojer ex Fr. = Salix madagascariensis Bojer ex Andersson ●☆

343067   Salix australis Schleich. ex Spreng. = Salix nigricans（Sm.）Enander ●☆

343068   Salix austro-tibetica N. Chao;藏南柳; S. Xizang Willow, South Tibet Willow,South Xizang Willow ●

343069   Salix axillaris A. Rich. = Salix mucronata Thunb. ● ☆

343070   Salix babylonica L. ;垂柳(垂丝柳,垂杨柳,倒垂柳,倒柳,倒栽柳,吊柳,河柳,柳,柳树,青龙须,青丝柳,清明柳,水柳,水杨柳,仙柳,线柳,杨柳);Babylon Weeping Willow, Green Weeping Willow,Weeping Willow ●

343071   Salix babylonica L. ‘Crispa’;皱叶垂柳;Ramshorn Willow ● ☆

343072   Salix babylonica L. ‘Tortuosa’ = Salix babylonica L. f. tortuosa Y. L. Chou ●

343073   Salix babylonica L. f. pekinensis（A. Henry）Geerinck ‘Tortuosa’ = Salix babylonica L. f. tortuosa Y. L. Chou ●

343074   Salix babylonica L. f. tortuosa Y. L. Chou;曲枝垂柳(龙爪柳); Tortuous Weeping Willow,Twisted Babylon Weeping Willow ●

343075   Salix babylonica L. f. tortuosa Y. L. Chou = Salix matsudana Koidz. f. tortuosa（Vilm.）Rehder ●

343076   Salix babylonica L. f. villosa C. F. Fang;长毛苞垂柳;Villose Babylon Weeping Willow ●

343077   Salix babylonica L. f. villosa C. F. Fang =Salix babylonica L. ●

343078   Salix babylonica L. var. glandulipilosa P. Y. Mao et W. Z. Li;腺毛垂柳;Glandulipilose Babylon Weeping Willow ●

343079   Salix babylonica L. var. hunanensis N. Chao;湘柳; Hunan Babylon Weeping Willow ●

343080   Salix babylonica L. var. lavallei Dode = Salix babylonica L. ●

343081   Salix babylonica L. var. szechuanica Goerz = Salix babylonica L. ●

343082   Salix babylonica L. var. tortuosa Y. L. Chou = Salix babylonica L. f. tortuosa Y. L. Chou ●

343083   Salix baileyi C. K. Schneid. ;百里柳;Bailey Willow ●

343084   Salix bakko Kimura = Salix caprea L. ●

343085   Salix balansaei Seemen;中越柳●

343086   Salix balansaei Seemen var. szechuanica Goerz = Salix radinostachya C. K. Schneid. ●

343087   Salix balfouriana C. K. Schneid. ;白背柳;Balfour Willow ●

343088   Salix balsamifera Barratt ex Andersson = Salix pyrifolia Anderson ●☆

343089   Salix balsamifera Barratt ex Andersson var. alpestris Bebb = Salix pyrifolia Anderson ●☆

343090   Salix balsamifera Barratt ex Andersson var. lanceolata Bebb = Salix pyrifolia Anderson ●☆

343091   Salix balsamifera Barratt ex Andersson var. vegeta Bebb = Salix pyrifolia Anderson ●☆

343092   Salix bangongensis Z. Wang et C. F. Fang;班公柳; Bangong Willow ●

343093   Salix barclayi Andersaon;巴尔柳●☆

343094   Salix basfordiana Scaling ex J. Salter;巴斯柳;Basford Willow ●☆

343095   Salix bebbiana Sarg. ; 长喙柳（喙柳）; Beak Willow, Beaked Willow,Bebb Willow,Bebb's Willow,Diamond,Diamond Willow ●

343096   Salix bebbiana Sarg. var. capreifolia（Fernald）Fernald = Salix bebbiana Sarg. ●

343097   Salix bebbiana Sarg. var. depilis Raup = Salix bebbiana Sarg. ●

343098   Salix bebbiana Sarg. var. luxurians（Fernald）Fernald = Salix bebbiana Sarg. ●

343099   Salix bebbiana Sarg. var. perrostrata（Rydb.）C. K. Schneid. = Salix bebbiana Sarg. ●

343100   Salix bebbiana Sarg. var. projecta（Fernald）C. K. Schneid. = Salix bebbiana Sarg. ●

343101   Salix berberifolia Pall. ;刺叶柳;Barberryleaf Willow,Barberry-leaved Willow ●

343102   Salix bhutanensis Flod. ;不丹柳 ●

343103   Salix bhutanensis Flod. = Salix karelinii Turcz. ex Stschegl. ●

343104   Salix bhutanensis Flod. var. lasiopes（Z. Wang et P. Y. Fu）N. Chao = Salix lasiopes Z. Wang et P. Y. Fu ●

343105   Salix bhutanensis Flod. var. yadongensis（N. Chao）N. Chao = Salix yadongensis N. Chao ●

343106   Salix bicarpa Nakai;双果柳●☆

343107   Salix bikouensis Y. L. Chou;碧口柳;Bikou Willow ●

343108   Salix bikouensis Y. L. Chou var. villosa Y. L. Chou;毛碧口柳; Villose Bikou Willow ●

343109   Salix biondiana Seemen;华柳(庙王柳);Biond Willow ●

343110   Salix bistyla Hand. -Mazz. ;双柱柳;Bistyle Willow ●

343111   Salix blakii Goerz;黄线柳; Linearleaf Willow, Linear-leaved Willow,Willow,Yellowlinear Willow ●

343112   Salix blanda Andersson;威斯康星柳; Wisconsin Weeping Willow ●☆

343113   Salix bockii Seemen ex Diels;宝克柳●☆

343114   Salix bockii Seemen ex Diels = Salix variegata Franch. ●

343115   Salix bordensis Nakai = Salix microstachya Turcz. var. bordensis（Nakai）C. F. Fang ●

343116   Salix borealis Fr. ;北方柳;Nortern Willow ●☆

343117   Salix boseensis N. Chao;桂柳;Baise Willow,Bose Willow ●

343118   Salix boydii E. F. Linton;波氏柳●☆

343119   Salix brachista C. K. Schneid. ;小垫柳(垫柳,沼柳);Cushion Willow,Shortbranch Willow ●

343120   Salix brachista C. K. Schneid. = Salix ovatomicrophylla K. S. Hao ●

343121   Salix brachista C. K. Schneid. var. integra Z. Wang et C. F. Fang;全缘小垫柳;Entire Shortbranch Willow ●

343122　Salix brachista C. K. Schneid. var. multiflora Z. Wang et P. Y. Fu = Salix serpyllum Andersson ●

343123　Salix brachista C. K. Schneid. var. pilifera N. Chao;毛果小垫柳;Hairyfruit Shortbranch Willow ●

343124　Salix brachypoda (Trautv. et C. A. Mey.) Kom. = Salix brachista C. K. Schneid. ●

343125　Salix brachypoda (Trautv. et C. A. Mey.) Kom. = Salix rosmarinifolia L. var. brachypoda (Trautv. et C. A. Mey.) Y. L. Chou ●

343126　Salix bracteosa Turcz. ex Trautv. = Chosenia arbutifolia (Pall.) A. K. Skvortsov ●◇

343127　Salix brayi Ledeb. = Salix berberifolia Pall. ●

343128　Salix brevijulis Turcz.;短序柳●☆

343129　Salix buergeriana Miq. = Salix sieboldiana Blume ●☆

343130　Salix burkingensis Chang Y. Yang;布尔津柳(布尔青柳); Buljing Willow, Bulking Willow ●

343131　Salix caenomeloides Kimura f. obtusa (Z. Wang et C. Y. Yu) C. F. Fang;钝头腺柳●

343132　Salix caerulea Sm. = Salix alba L. subsp. coerulea (Sm.) Rchb. f. ●☆

343133　Salix caesia Vill.;欧杞柳;Bluegrey Willow, Lavender Blue Willow, Lightgrey Willow, Light-grey Willow ●

343134　Salix calodendron;美柳;Winuner's Osier ●☆

343135　Salix caloneura C. K. Schneid. = Salix radinostachya C. K. Schneid. ●

343136　Salix calyculata Hook. f.;长柄垫柳;Calycular Willow, Longstalk Willow ●

343137　Salix calyculata Hook. f. var. glabrifolia Hand.-Mazz. = Salix calyculata Hook. f. ●

343138　Salix calyculata Hook. f. var. gongshanica Z. Wang et C. F. Fang;贡山长柄垫柳;Gongshan Longstalk Willow ●

343139　Salix camusii H. Lév. = Salix etosia C. K. Schneid. ●

343140　Salix canariensis Link;加那利柳●☆

343141　Salix candida Flügge = Salix candida Flügge ex Willd. ●☆

343142　Salix candida Flügge ex Willd.;北美灰毛柳;Hoary Willow, Sage Willow, Sage-leaved Willow ●☆

343143　Salix candida Flügge ex Willd. f. denudata (Andersson) Rouleau = Salix candida Flügge ex Willd. ●☆

343144　Salix candida Flügge ex Willd. var. denudata Andersson = Salix candida Flügge ex Willd. ●☆

343145　Salix candidula Nieuwl. = Salix candida Flügge ex Willd. ●☆

343146　Salix cantoniensis Hance = Salix babylonica L. ●

343147　Salix capensis Thunb. = Salix mucronata Thunb. subsp. capensis (Thunb.) Immelman ●☆

343148　Salix capensis Thunb. var. gariepina (Burch.) Anderson = Salix mucronata Thunb. ●☆

343149　Salix capensis Thunb. var. mucronata (Thunb.) Anderson = Salix mucronata Thunb. ●☆

343150　Salix capitata Y. L. Chou et Skvortsov;圆头柳;Capitate Willow ●

343151　Salix caprea L.;黄花柳(黄花儿柳,山毛柳);Black Sally, English Palm, European Pussywillow, Fluffy Buttons, French Pussy Willow, Fussy Cats, Geese-and-gullies, Gezlins, Gibs, Goat Willow, Goat's Willow, Golden Pussies, Goose Chicks, Goose Withy, Goose-and-goslings, Goose-and-gubblies, Gosling Tree, Goslings, Great Round-leaved Willow, Great Sallow, Great Sallow Willow, Great Willow, Gulls, Kilmarnock Willow, Lamb's Tails, May Goslings, Palm Willow, Palmer, Pink Pussy Willow, Pussy Willow, Sallow, Sally, Sally

Withy, Sally-wood, Sauch Saugh, Sauf, Saugh-tree, Saute, Sawg, Seale Tree, Selly, Sollar, Tassel-rags, Watersal, Wilf, Yellowflower Willow ●

343152　Salix caprea L. 'Kilmarnock';吉尔马诺克黄花柳;Kilmarnock Willow, Kilmarnock's Willow ●☆

343153　Salix caprea L. 'Pendula';垂枝黄花柳;Kilmarnock Willow, Weeping Pussy Willow ●☆

343154　Salix caprea L. 'Pendula' = Salix caprea L. 'Kilmarnock' ●☆

343155　Salix caprea L. = Salix raddeana Laksch. ex Nasarow ●

343156　Salix caprea L. = Salix sinica (K. S. Hao ex C. F. Fang et A. K. Skvortsov) Z. Wang et C. F. Fang ●

343157　Salix caprea L. = Salix wallichiana Andersson ●

343158　Salix caprea L. f. elongata (Nakai) Kitag. = Salix raddeana Laksch. ex Nasarow ●

343159　Salix caprea L. f. pendula (T. Lang) Geerinck = Salix caprea L. 'Pendula' ●☆

343160　Salix caprea L. f. subglabra (Y. L. Chang et Skvortsov) Kitag. = Salix raddeana Laksch. ex Nasarow var. subglabra Y. L. Chang et Skvortsov ●

343161　Salix caprea L. subsp. hultenii (Flod.) Kom. = Salix caprea L. ●

343162　Salix caprea L. var. dentata K. S. Hao = Salix sinica (K. S. Hao ex C. F. Fang et A. K. Skvortsov) Z. Wang et C. F. Fang var. dentata (K. S. Hao ex C. F. Fang et A. K. Skvortsov) Z. Wang et C. F. Fang ●

343163　Salix caprea L. var. dentata K. S. Hao ex C. F. Fang et A. K. Skvortsov = Salix sinica (K. S. Hao ex C. F. Fang et A. K. Skvortsov) Z. Wang et C. F. Fang var. dentata (K. S. Hao ex C. F. Fang et A. K. Skvortsov) Z. Wang et C. F. Fang ●

343164　Salix caprea L. var. sinica K. S. Hao = Salix sinica (K. S. Hao ex C. F. Fang et A. K. Skvortsov) Z. Wang et C. F. Fang ●

343165　Salix caprea L. var. sinica K. S. Hao ex C. F. Fang et A. K. Skvortsov = Salix sinica (K. S. Hao ex C. F. Fang et A. K. Skvortsov) Z. Wang et C. F. Fang ●

343166　Salix caprea L. var. subsessilis K. S. Hao ex C. F. Fang et A. K. Skvortsov = Salix sinica (K. S. Hao ex C. F. Fang et A. K. Skvortsov) Z. Wang et C. F. Fang var. subsessilis (K. S. Hao ex C. F. Fang et A. K. Skvortsov) G. Zhu ●

343167　Salix capusii Franch.;蓝叶柳;Capus Willow, Cricket-bat Willow ●

343168　Salix cardiophylla Trautv. et C. A. Mey.;心叶柳●☆

343169　Salix cardiophylla Trautv. et C. A. Mey. subsp. urbaniana (Seemen) A. K. Skvortsov = Salix cardiophylla Trautv. et C. A. Mey. var. urbaniana (Seemen) Kudo ●☆

343170　Salix cardiophylla Trautv. et C. A. Mey. var. urbaniana (Seemen) Kudo;乌尔班心叶柳●☆

343171　Salix carmanica Roem.;黄皮柳;Carman Willow, Yellowbark Willow ●

343172　Salix caroliniana Michx.;卡罗来纳柳(卡罗林柳);Carolina Willow, Coastal Plain Willow, Soutern Willow, Swamp Willow, Ward Willow, Ward's Willow ●☆

343173　Salix carpinifolia Schleich. ex Spreng.;鹅耳枥叶柳;Hornbeam-leaved Sallow ●☆

343174　Salix caspica Pall.;油柴柳(里海柳);Caspian Sea Willow, Caspiansea Willow ●

343175　Salix caspica Pall. = Salix gracilior (Siuzew) Nakai ●

343176　Salix caspica Pall. var. michelsonii (Goerz ex Nasarow) Poljakov = Salix michelsonii Goerz ex Nasarow ●

343177　Salix caspica Pall. var. michelsonii (Goerz) Poljakov = Salix michelsonii Goerz ex Nasarow ●

43178 Salix catalaunica Sennen = Salix atrocinerea Brot. ●☆

43179 Salix cathayana Diels;中华柳（华柳，山柳，中国柳）; China Willow, Chinese Willow ●

43180 Salix cathayana Diels var. denticulata Andersson;齿叶柳（山杨柳）; Denticulate Willow, Serrateleaf Willow ●

43181 Salix caucasica Andersson;高加索柳●☆

43182 Salix cavaleriei H. Lév.;云南柳; Cavalerie Willow, Yunnan Willow ●

43183 Salix cereifolia Goerz = Salix dissa C. K. Schneid. var. cerifolia (Goerz) C. F. Fang ●

343184 Salix cereifolia Goerz ex Rehder et Kobuski = Salix dissa C. K. Schneid. var. cerifolia (Goerz) C. F. Fang ●

343185 Salix chaenomeloides Kimura;腺柳（河柳，红心柳，苦柳，魁柳,紫心柳）; Asian Pussy Willow, Flowering-quince Willow, Giant Pussy Willow, Japanese Pussywillow ●

343186 Salix chaenomeloides Kimura ‘ Mt. Asama’;红花腺柳; Red-flowered Pussy Willow ●☆

343187 Salix chaenomeloides Kimura f. obtusa (Z. Wang et C. Y. Yu) C. F. Fang;钝叶腺柳; Obtuseleaf Flowering-quince Willow ●

343188 Salix chaenomeloides Kimura var. glandulifolia (Z. Wang et C. Y. Fu) C. F. Fang;腺叶腺柳; Glandule Flowering-quince Willow ●

343189 Salix chaenomeloides Kimura var. pilosa Kimura = Salix chaenomeloides Kimura ●

343190 Salix chamissonis Andersson;沙米逊柳●☆

343191 Salix changchouwensis F. P. Metcalf = Salix dunnii C. K. Schneid. ●

343192 Salix characta C. K. Schneid.;密齿柳（陇山柳）; Densetoothed Willow, Dense-toothed Willow ●

343193 Salix cheilophila C. K. Schneid.;乌柳（降马，筐柳，毛柳，沙柳,沙柳根）; Sand Willow ●

343194 Salix cheilophila C. K. Schneid. var. acuminata Z. Wang et Y. L. Chou;宽叶乌柳（宽叶柳）; Acuminate Sand Willow ●

343195 Salix cheilophila C. K. Schneid. var. cyanolimnea (Hance) Chang Y. Yang;光叶乌柳●

343196 Salix cheilophila C. K. Schneid. var. cyanolimnea (Hance) Chang Y. Yang = Salix cyanolimnea Hance ●

343197 Salix cheilophila C. K. Schneid. var. microstachyoides (Z. Wang et P. Y. Fu) C. F. Fang = Salix cheilophila C. K. Schneid. var. microstachyoides (Z. Wang et P. Y. Fu) Z. Wang et C. F. Fang ●

343198 Salix cheilophila C. K. Schneid. var. microstachyoides (Z. Wang et P. Y. Fu) Z. Wang et C. F. Fang;大红柳; Like Littlespike Willow, Smallspike Sand Willow ●

343199 Salix cheilophila C. K. Schneid. var. villosa G. H. Wang;毛苞乌柳; Villose Sand Willow ●

343200 Salix chekiangensis W. C. Cheng;浙江柳; Chekiang Willow, Zhejiang Willow ●

343201 Salix chekiangensis W. C. Cheng =Salix dunnii C. K. Schneid. ●

343202 Salix chevalieri Seemen;舍瓦利耶柳●☆

343203 Salix chienii W. C. Cheng;银叶柳（钱氏柳）; Chien's Willow, Silverleaf Willow, Silver-leaved Willow ●

343204 Salix chienii W. C. Cheng var. glabra K. S. Hao = Salix chienii W. C. Cheng ●

343205 Salix chienii W. C. Cheng var. pubigera N. Chao;常宁柳; Changning Chien's Willow ●

343206 Salix chikungensis C. K. Schneid.;鸡公山柳（鸡公柳）; Jigong Mountain Willow, Jigong Willow, Jigongshan Willow ●

343207 Salix chilensis Molina;智利柳; Chilean Willow, Humboldt Willow ●☆

343208 Salix chinensis Burm. f. = Salix babylonica L. ●

343209 Salix chingiana K. S. Hao ex C. F. Fang et A. K. Skvortsov;秦氏柳（秦柳）; Ching Willow ●

343210 Salix chingshuishanensis S. S. Ying = Salix taiwanalpina Kimura ●

343211 Salix chlorophylla Andersson = Salix planifolia Pursh ●☆

343212 Salix chlorophylla Andersson var. monica (Bebb) Flod. = Salix planifolia Pursh ●☆

343213 Salix chlorophylla Andersson var. nelsonii (C. R. Ball) Flod. = Salix planifolia Pursh ●☆

343214 Salix chlorophylla Andersson var. pellita Andersson = Salix pellita (Andersson) Andersson ex C. K. Schneid. ●☆

343215 Salix chlorophylla Andersson var. pychnocarpa (Andersson) Andersson = Salix planifolia Pursh ●☆

343216 Salix chlorostachya Turcz.;绿穗柳●☆

343217 Salix chrysanthis;金花柳;Golden-flowered Norway Willow ●☆

343218 Salix chumulamanica Z. Wang et P. Y. Fu;珠穆垫柳; Zhumu Willow ●

343219 Salix chumulamanica Z. Wang et P. Y. Fu = Salix serpyllum Andersson ●

343220 Salix chuniana C. F. Fang = Salix hylonoma C. K. Schneid. ●

343221 Salix cinerea L.;灰柳（灰毛柳）; Gray Willow, Grey Sallow, Grey Willow, Large Gray Willow ●

343222 Salix cinerea L. ‘ Variegata’;三色灰柳; Tricolor Willow ●☆

343223 Salix cinerea L. = Salix gracilistyla Miq. ●

343224 Salix cinerea L. = Salix raddeana Laksch. ex Nasarow ●

343225 Salix cinerea L. subsp. atrocinerea (Brot.) Guinier = Salix atrocinerea Brot. ●☆

343226 Salix cinerea L. subsp. catalaunica (Goerz) Maire et Weiller = Salix atrocinerea Brot. ●☆

343227 Salix cinerea L. subsp. jahandiezii (Chass.) Maire et Weiller = Salix atrocinerea Brot. subsp. jahandiezii (Chass.) Ibn Tattou ●☆

343228 Salix clathrata Hand. -Mazz.;栅枝垫柳; Clathrate Willow, Sieve-like Willow ●

343229 Salix clathrata Hand. -Mazz. var. rockiana Hand. -Mazz. = Salix clathrata Hand. -Mazz. ●

343230 Salix coactilis Fernald = Salix sericea Marshall ●☆

343231 Salix coaetanea Flod.;一时柳●☆

343232 Salix coerulea E. L. Wolf = Salix capusii Franch. ●

343233 Salix coggygria Hand. -Mazz.;怒江矮柳;Nujiang Willow ●

343234 Salix coluteoides Mirb.;膀胱豆柳●☆

343235 Salix conformis Forbes = Salix discolor Muhl. ●☆

343236 Salix conifera Wangenh.;针叶柳;Conifer Willow ●☆

343237 Salix contortiapiculata P. Y. Mao et W. Z. Li;扭尖柳●

343238 Salix cordata Michx. = Salix eriocephala Michx. ●☆

343239 Salix cordata Muhl. = Salix eriocephala Michx. ●☆

343240 Salix cordata Muhl. subsp. rigida (Muhl.) Andersson = Salix eriocephala Michx. ●☆

343241 Salix cordata Muhl. var. abrasa Fernald = Salix eriocephala Michx. ●☆

343242 Salix cordata Muhl. var. angustata (Pursh) Andersson = Salix eriocephala Michx. ●☆

343243 Salix cordata Muhl. var. balsamifera (Barratt ex Andersson) Hook. = Salix pyrifolia Anderson ●☆

343244 Salix cordata Muhl. var. glaucophylla Bebb = Salix myricoides Muhl. ●☆

343245 Salix cordata Muhl. var. missouriensis (Bebb) Mack. et Bush =

Salix eriocephala Michx. ● ☆

343246 Salix cordata Muhl. var. myricoides (Muhl.) J. Carey = Salix myricoides Muhl. ● ☆

343247 Salix cordata Muhl. var. rigida (Muhl.) Andersson = Salix eriocephala Michx. ● ☆

343248 Salix cordata Muhl. var. rigida (Muhl.) B. Boivin = Salix eriocephala Michx. ● ☆

343249 Salix cordata Muhl. var. rigida (Muhl.) J. Carey = Salix eriocephala Michx. ● ☆

343250 Salix cordata Muhl. var. vestita (Andersson) Bebb = Salix eriocephala Michx. ● ☆

343251 Salix crassa Barratt = Salix discolor Muhl. ● ☆

343252 Salix crassifolia Schleich. ex Ser. ；厚叶柳；Thick-leaved Sallow ● ☆

343253 Salix crassifolia Schleich. ex Ser. = Salix nigricans (Sm.) Enander ● ☆

343254 Salix crataegifolia Bertol. ；山楂叶柳；Apuan Willow ● ☆

343255 Salix crenata K. S. Hao ex C. F. Fang et A. K. Skvortsov；锯齿叶垫柳；Crenate Willow ●

343256 Salix cuneata Turcz. ；楔叶柳● ☆

343257 Salix cuprea ?；紫铜柳；Tassel Rags ● ☆

343258 Salix cupularis Rehder；怀腺柳（高山柳）；Cupular Willow ●

343259 Salix cupularis Rehder var. acutifolia S. Q. Zhou；尖叶怀腺柳；Acuminate-leaf Cupular Willow ●

343260 Salix cupularis Rehder var. lasiogyne Rehder；毛蕊怀腺柳●

343261 Salix cupularis Rehder var. lasiogyne Rehder = Salix oritrepha C. K. Schneid. ●

343262 Salix cyanolimnea Hance；光果乌柳（光叶乌柳）；Cyanolimne Sand Willow ●

343263 Salix cyathipoda Andersson ex A. Rich. = Salix mucronata Thunb. ● ☆

343264 Salix dabeshanensis B. C. Ding et T. B. Chao；大别山柳；Dabieshan Willow ●

343265 Salix daddeana Laksch. = Salix raddeana Laksch. ex Nasarow ●

343266 Salix daghestanica Goerz；达赫斯坦柳●☆

343267 Salix daguanensis P. Y. Mao et P. X. He；大关柳；Daguan Willow ●

343268 Salix dahurica Turcz. ；达呼尔柳●☆

343269 Salix dailingensis Y. L. Chou et C. Y. King；岱岭柳；Dailing Willow ●

343270 Salix dailingensis Y. L. Chou et C. Y. King = Salix schwerinii E. L. Wolf ●

343271 Salix dailingensis Y. L. Chou et C. Y. King = Salix viminalis L. ●

343272 Salix daiseniensis Seemen = Salix sieboldiana Blume ● ☆

343273 Salix daliensis C. F. Fang et S. D. Zhao；大理柳；Dali Willow ●

343274 Salix daliensis C. F. Fang et S. D. Zhao f. longispica C. F. Fang；长穗大理柳；Long-ament Dali Willow ●

343275 Salix daltoniana Andersson；褐背柳；Brownback Willow, Dalton Willow ●

343276 Salix daltoniana Andersson var. franchetiana Burkill = Salix ernestii C. K. Schneid. ●

343277 Salix dalungensis Z. Wang et P. Y. Fu；节枝柳；Dalung Willow ●

343278 Salix damascena Forbes；李叶柳；Damson-leaved Sallow ●

343279 Salix damascena Forbes = Salix nigricans (Sm.) Enander ● ☆

343280 Salix daphnoides Vill. ；瑞香柳（集穗柳, 紫罗兰柳, 紫毛柳）；Daphne Willow, Daphne-like Willow, European Violet-willow, Violet Willow ● ☆

343281 Salix daphnoides Vill. = Salix kangensis Nakai var. leiocarpa Kitag. ●

343282 Salix daphnoides Vill. = Salix rorida Laksch. ●

343283 Salix dasyclados Wimm. ；毛枝柳；Hairytwig Willow Thickbranch Willow, Thick-branched Willow, Thickhair Willow Woolly Willow, Woolly-twigged Willow ●

343284 Salix daviesii Boiss. ；戴维斯柳●☆

343285 Salix decipiens Hoffm. = Salix fragilis L. ●

343286 Salix delavayana Hand. -Mazz. ；腹毛柳；Delavay Willow ●

343287 Salix delavayana Hand. -Mazz. f. glabra C. F. Fang；光苞腹毛柳●

343288 Salix delavayana Hand. -Mazz. var. pilosoputuralis T. Y. Chou et W. P. Fang f. glabra C. F. Fang = Salix delavayana Hand. -Mazz. var. pilosoputuralis T. Y. Chou et W. P. Fang ●

343289 Salix delavayana Hand. -Mazz. var. pilosoputuralis T. Y. Chou et W. P. Fang f. glabra C. F. Fang = Salix delavayana Hand. -Mazz. f. glabra C. F. Fang ●

343290 Salix delavayana Hand. -Mazz. var. pilosoputuralis T. Y. Chou et W. P. Fang；毛缝腹毛柳；Pilose-sutural Delavay Willow ●

343291 Salix densifoliata Seemen = Salix variegata Franch. ●

343292 Salix denticulata Andersson = Salix cathayana Diels var. denticulata Andersson ●

343293 Salix denudata Raf. = Salix nigra Marshall ● ☆

343294 Salix dependens Nakai；龟白柳●

343295 Salix depressa L. subsp. rostrata (Richardson) Hiitonen = Salix bebbiana Sarg. ●

343296 Salix depressa Poljak. = Salix iliensis Regel ●

343297 Salix dibapha C. K. Schneid. ；异色柳；Dichromatic Willow ●

343298 Salix dibapha C. K. Schneid. var. biglandulosa C. F. Fang；二腺异色柳；Triglandulose Dichromatic Willow ●

343299 Salix dictyoneura Seemen = Salix rosthornii Seemen ex Diels ●

343300 Salix dictyoneura Seemen ex Diels = Salix rosthornii Seemen ex Diels ●

343301 Salix dieckiana Suksd. = Salix pedicellaris Pursh ● ☆

343302 Salix discolor Muhl. ；褪色柳（柔毛柳）；American Black Willow, Discolor Willow, Holy Thorn of Christmas, Pussy Willow, Pussywillow ● ☆

343303 Salix discolor Muhl. var. eriocephala (Michx.) Andersson = Salix eriocephala Michx. ● ☆

343304 Salix discolor Muhl. var. latifolia Andersson = Salix conifera Wangenh. ● ☆

343305 Salix discolor Muhl. var. overi C. R. Ball = Salix discolor Muhl. ● ☆

343306 Salix discolor Muhl. var. prinoides (Pursh) Andersson = Salix discolor Muhl. ● ☆

343307 Salix discolor Muhl. var. rigidior (Andersson) C. K. Schneid. = Salix discolor Muhl. ● ☆

343308 Salix disperma Don = Salix tetrasperma Roxb. ●

343309 Salix disperma Roxb. ex D. Don = Salix tetrasperma Roxb. ●

343310 Salix dissa C. K. Schneid. ；异型柳；Different Willow, Dissimilar Willow ●

343311 Salix dissa C. K. Schneid. f. angustifolia C. F. Fang；狭叶异型柳；Narrowleaf Dissimilar Willow ●

343312 Salix dissa C. K. Schneid. f. angustifolia C. F. Fang = Salix dissa C. K. Schneid. ●

343313 Salix dissa C. K. Schneid. var. cerifolia (Goerz) C. F. Fang；单腺异型柳●

343314 Salix divaricata Pall. et Flod. ；叉枝柳；Divaricate Willow ●

343315 Salix divaricata Pall. et Flod. var. meta-formosa (Nakai) Kitag. ；长圆叶柳●

43316　Salix divaricata Pall. et Flod. var. orthostemma（Nakai）Kitag. = Salix divaricata Pall. et Flod. var. meta-formosa（Nakai）Kitag. ●

43317　Salix divaricata var. orthostemma（Nakai）Kitag. = Salix divaricata Pall. et Flod. var. meta-formosa（Nakai）Kitag. ●

343318　Salix divergentistyla C. F. Fang；叉柱柳；Divergentstyle Willow, Divergent-styled Willow, Forkstyle Willow ●

343319　Salix dodecandra H. Lév. = Salix rosthornii Seemen ex Diels ●

343320　Salix dodecandra H. Lév. et Vaniot = Salix rosthornii Seemen ex Diels ●

343321　Salix doii Hayata；台湾柳（薄叶柳，森氏柳，台东柳）；Mori Willow, Taiwan Willow ●

343322　Salix doii Hayata = Salix fulvopubescens Hayata var. doii（Hayata）K. C. Yang et T. C. Huang ●

343323　Salix doii Hayata = Salix kusanoi（Hayata）C. K. Schneid. et Kimura ●

343324　Salix dolia C. K. Schneid. = Salix rehderiana C. K. Schneid. var. dolia（C. K. Schneid.）N. Chao ●

343325　Salix dolia C. K. Schneid. var. lineariloba N. Chao = Salix eriostachya Wall. ex Andersson var. lineariloba（N. Chao）G. Zhu ●

343326　Salix dolichostyla Seemen = Salix eriocarpa Franch. et Sav. ●

343327　Salix donggouxianica C. F. Fang；东沟柳；Donggou Willow ●

343328　Salix driophila C. K. Schneid.；林柳；Forest Willow ●

343329　Salix dubia Trautv. = Salix nigra Marshall ●☆

343330　Salix duclouxii H. Lév. = Salix variegata Franch. ●

343331　Salix duclouxii H. Lév. var. kouytchensis H. Lév. = Salix kouytchensis C. K. Schneid. ●

343332　Salix dunnii C. K. Schneid.；长柄柳（长梗柳）；Dunn Willow, Longgynophore Willow ●

343333　Salix dunnii C. K. Schneid. var. tsoongii（W. C. Cheng）C. Y. Yu et S. D. Zhao；钟氏柳；Tsoong Willow ●

343334　Salix dyscrita C. K. Schneid. = Salix luctuosa H. Lév. ●

343335　Salix elaeagnoides Schleich. ex Ser.；灰枝柳；Hoary Willow, Rosemary Willow ●☆

343336　Salix elaeagnos Scop.；胡颓子柳；Elaeagnus Willow, Hoary Willow, Olive Willow, Olive-leaved Sallow ●☆

343337　Salix elaeagnos Scop. subsp. angustifolia（Cariot）Rech. f.；窄叶胡颓子柳●☆

343338　Salix elegans Wall. = Salix denticulata Andersson ●

343339　Salix elegans Wall. ex Andersson = Salix denticulata Andersson ●

343340　Salix elegans Wall. var. himalensis ？ = Salix denticulata Andersson ●

343341　Salix elegantissima K. Koch；秀丽柳；Thurlow Weeping Willow, Thurlow's Weeping Willow ●☆

343342　Salix eriocarpa Franch. et Sav.；长柱柳（洪台柳）；Eriocarpous Willow, Longstyle Willow ●

343343　Salix eriocephala Michx.；密苏里柳（绵毛柳，心叶柳）；Diamond Willow, Dune Willow, Heart-leaf Willow, Heart-leaved Willow, Missouri River Willow, Sand Dune Willow, Willow ●☆

343344　Salix erioclada H. Lév. et Vaniot；绵毛柳（毛枝柳）；Eriocladous Willow, Woollybranch Willow ●

343345　Salix eriophylla Andersson = Salix psilostigma Andersson ●

343346　Salix eriophylla Burkill = Salix daliensis C. F. Fang et S. D. Zhao ●

343347　Salix eriostachya Wall. ex Andersson；绵穗柳●

343348　Salix eriostachya Wall. ex Andersson var. lineariloba（N. Chao）G. Zhu；线裂绵穗柳（线裂棉穗柳）●

343349　Salix eriostroma Hayata = Salix doii Hayata ●

343350　Salix eriostroma Hayata = Salix fulvopubescens Hayata var. doii（Hayata）K. C. Yang et T. C. Huang ●

343351　Salix ernestii C. K. Schneid.；银背柳；Ernest Willow ●

343352　Salix ernestii C. K. Schneid. f. glabrescens Y. L. Chou et C. F. Fang；脱毛银背柳；Glabrescent Ernest Willow ●

343353　Salix ernestii C. K. Schneid. f. glabrescens Y. L. Chou et C. F. Fang = Salix ernestii C. K. Schneid. ●

343354　Salix ernestii C. K. Schneid. var. wangii（Goerz）N. Chao = Salix ernestii C. K. Schneid. ●

343355　Salix erythrocarpa Kom.；红果柳●☆

343356　Salix etosia C. K. Schneid.；巴柳（巴东柳）；Badong Willow ●

343357　Salix etosia C. K. Schneid. f. longipes N. Chao et C. F. Fang；长柄巴柳；Long-stalk Badong Willow ●

343358　Salix euapiculata Nasarow；细尖柳●☆

343359　Salix eucalyptoides Mey. ex Schneid. = Chosenia arbutifolia（Pall.）A. K. Skvortsov ●◇

343360　Salix exigua Nutt.；沙皮柳（曲枝柳，小柳）；Coyote Willow, Green-leaved Willow, Narrowleaf Willow, Red Willow, Sandbar Willow, Slender Willow ●☆

343361　Salix exigua Nutt. subsp. interior（Rowlee）Cronquist；沙洲柳；Sandbar Willow ●☆

343362　Salix exigua Nutt. subsp. interior（Rowlee）Cronquist var. pedicellata（Andersson）Cronquist = Salix exigua Nutt. subsp. interior（Rowlee）Cronquist ●☆

343363　Salix exigua Nutt. var. exterior（Fernald）C. F. Reed = Salix exigua Nutt. subsp. interior（Rowlee）Cronquist ●☆

343364　Salix exigua Nutt. var. luteosericea（Rydb.）C. K. Schneid. = Salix exigua Nutt. subsp. interior（Rowlee）Cronquist ●☆

343365　Salix falcata Pursh = Salix nigra Marshall ●☆

343366　Salix fargesii Burkill；川鄂柳（巫山柳）；Chinese Willow, Farges' Willow ●

343367　Salix fargesii Burkill var. hypotricha N. Chao；湘鄂柳；Hairy Farges' Willow ●

343368　Salix fargesii Burkill var. kansuensis（K. S. Hao ex C. F. Fang et A. K. Skvortsov）G. H. Zhu；甘肃柳；Gansu Willow, Kansu Willow ●

343369　Salix fargesii Burkill var. kansuensis（K. S. Hao）N. Chao = Salix fargesii Burkill var. kansuensis（K. S. Hao ex C. F. Fang et A. K. Skvortsov）G. H. Zhu ●

343370　Salix faxoniana C. K. Schneid. = Salix oreinoma C. K. Schneid. ●

343371　Salix faxonianoides Z. Wang et P. Y. Fu；藏匐柳；Faxonian-like Willow, Tibet Willow, Xizang Willow ●

343372　Salix faxonianoides Z. Wang et P. Y. Fu var. villosa S. D. Zhao；毛轴藏匐柳（毛轴藏柳）；Villose Xizang Willow ●

343373　Salix fedtschenkoi Goerz；山羊柳（菲氏柳）；Fedtschenko Willow ●

343374　Salix fenghuangschanica Y. L. Chou et Skvortsov = Salix kangensis Nakai var. leiocarpa Kitag. ●

343375　Salix fengiana C. F. Fang et Chang Y. Yang；贡山柳；Feng Willow, Gongshan Willow ●

343376　Salix fengiana C. F. Fang et Chang Y. Yang var. gymnocarpa P. I. Mao et W. Z. Li；裸果贡山柳；Nakedfruit Gongshan Willow ●

343377　Salix fengxianica N. Chao = Salix alfredii Goerz var. fengxianica（N. Chao）G. Zhu ●

343378　Salix ferganensis Nasarow；费尔干柳●☆

343379　Salix filistyla Z. Wang et P. Y. Fu = Salix bhutanensis Flod. ●

343380　Salix filistyla Z. Wang et P. Y. Fu = Salix himalayensis（Andersson）Flod. var. filistyla（Z. Wang et P. Y. Fu）C. F. Fang ●

343381　Salix flabellaris Andersson；扇叶垫柳（扇叶柳）；Fanleaf Willow，Flabellate Willow，Flabellateleaf Willow ●

343382　Salix flavicans (Andersson) K. S. Hao = Salix rosmarinifolia L. var. brachypoda (Trautv. et C. A. Mey.) Y. L. Chou ●

343383　Salix flavida Y. L. Chang et Skvortsov = Salix gordejevii Y. L. Chang et Skvortsov ●

343384　Salix flavovirens Hornem. = Salix nigra Marshall ●☆

343385　Salix floccosa Burkill；丛毛矮柳；Floccose Willow ●

343386　Salix floccosa Burkill var. leiogyna P. Y. Mao et W. Z. Li；光果丛毛矮柳；Smoothfruit Floccose Willow ●

343387　Salix floccosa Burkill var. leiogyna P. Y. Mao et W. Z. Li = Salix resectoides Hand. -Mazz. ●

343388　Salix floderusii Nakai；崖柳（狐柳，山柳，王八柳）；Floderus Willow ●

343389　Salix floderusii Nakai f. glabra Nakai = Salix taraikensis Kimura ●

343390　Salix floderusii Nakai f. manshurica Nakai = Salix floderusii Nakai ●

343391　Salix floridana Chapm.；佛罗里达柳；Florida Willow ●☆

343392　Salix fluviatilis Nutt.；西方河柳；River Willow，Sandbar Willow ●☆

343393　Salix fluviatilis Nutt. var. sericans (Nees) B. Boivin = Salix exigua Nutt. subsp. interior (Rowlee) Cronquist ●☆

343394　Salix forrestii K. S. Hao = Salix balfouriana C. K. Schneid. ●

343395　Salix forrestii K. S. Hao ex C. F. Fang et A. K. Skvortsov = Salix balfouriana C. K. Schneid. ●

343396　Salix fragilis L.；爆竹柳（脆枝柳）；Brittle Willow，Cat's Tails，Cat's Tall，Crack Willow，Fragile Willow，Lamb's Tails，Red-wood Willow，Snap Willow，Snapwillow，Widow's Willow ●

343397　Salix franchetiana (Burkill) Hand. -Mazz. = Salix ernestii C. K. Schneid. ●

343398　Salix fruticosa Döll；灌丛柳；Shrubby Osier ●☆

343399　Salix fruticulosa Kern = Salix furcata Andersson ●☆

343400　Salix fulvopubescens Hayata；褐毛柳（阿里山柳）；Brownhair Willow，Brown-hairy Willow，Mountain Willow ●

343401　Salix fulvopubescens Hayata var. doii (Hayata) K. C. Yang et T. C. Huang = Salix doii Hayata ●

343402　Salix fulvopubescens Hayata var. tagawana (Koidz.) K. C. Yang et T. C. Huang = Salix tagawana Koidz. ●

343403　Salix fumosa Turcz.；烟柳 ●☆

343404　Salix funebris H. Lév. = Salix wallichiana Andersson ●

343405　Salix furcata Andersson；分叉柳 ●☆

343406　Salix furcata Andersson = Salix lindleyana Wall. ●

343407　Salix fuscata Pursh = Salix discolor Muhl. ●☆

343408　Salix fuscescens Andersson var. hebecarpa Fernald = Salix pedicellaris Pursh ●☆

343409　Salix fustescens Andersson；阿拉斯加柳；Alaska Bog Willow，Alaska Bog-willow ●☆

343410　Salix futura Seemen f. rufa (Kimura) H. Ohashi；浅红柳 ●☆

343411　Salix gariepina Burch. = Salix mucronata Thunb. ●☆

343412　Salix geminata Y. L. Chang et Skvortsov = Salix hsinganica Y. L. Chang et Skvortsov ●

343413　Salix gilashanica Z. Wang et P. Y. Fu；吉拉柳；Gilashan Willow ●

343414　Salix gilgiana Seemen；日本河柳（河柳）；Willow ●☆

343415　Salix gilgiana Seemen = Salix miyabeana Seemen subsp. gymnolepis (H. Lév. et Vaniot) H. Ohashi et Yonek. ●☆

343416　Salix glabra Scop.；光柳；Alps Willow，Hairless Willow ●☆

343417　Salix glandulifera Flod.；西方腺柳；Glandule Willow ●☆

343418　Salix glandulosa Seemen = Salix chaenomeloides Kimura ●

343419　Salix glandulosa Seemen f. obtusa Z. Wang et C. Y. Yu = Salix caenomeloides Kimura f. obtusa (Z. Wang et C. Y. Yu) C. F. Fang ●

343420　Salix glandulosa Seemen var. glandulifolia Z. Wang et C. Y. Yu = Salix chaenomeloides Kimura var. glandulifolia (Z. Wang et C. Y. Yu) C. F. Fang ●

343421　Salix glandulosa Seemen var. stenophylla Z. Wang et C. Y. Yu = Salix rosthornii Seemen ex Diels ●

343422　Salix glandulosa Seemen var. warburgii (Seemen) Koidz. = Salix warburgii Seemen ●

343423　Salix glareorum P. Y. Mao et W. Z. Li；石流垫柳 ●

343424　Salix glatfelteri C. K. Schneid.；格拉特柳；Glatfelter Willow Glatfelter's Willow，Hybrid Black Willow ●☆

343425　Salix glauca L.；灰蓝柳（灰叶柳）；Arctic Gray Willow Glaucous Willow，Gray-leaved Willow ●

343426　Salix glaucophylla (Bebb) Bebb = Salix myricoides Muhl. ●☆

343427　Salix glaucophylla (Bebb) Bebb var. albovestita C. R. Ball = Salix myricoides Muhl. var. albovestita (C. R. Ball) Dorn ●☆

343428　Salix glaucophylla (Bebb) Bebb var. angustifolia Bebb = Salix myricoides Muhl. ●☆

343429　Salix glaucophylla (Bebb) Bebb var. brevifolia Bebb = Salix myricoides Muhl. ●☆

343430　Salix glaucophylla (Bebb) Bebb var. latifolia Bebb = Salix myricoides Muhl. ●☆

343431　Salix glaucophylloides Fernald = Salix myricoides Muhl. ●☆

343432　Salix glaucophylloides Fernald var. albovestita (C. R. Ball) Fernald = Salix myricoides Muhl. var. albovestita (C. R. Ball) Dorn ●☆

343433　Salix glaucophylloides Fernald var. brevifolia (Bebb) C. R. Ball = Salix myricoides Muhl. ●☆

343434　Salix glaucophylloides Fernald var. glaucophylla (Bebb) C. K. Schneid. = Salix myricoides Muhl. ●☆

343435　Salix gmelinii Pall. = Salix schwerinii E. L. Wolf ●

343436　Salix gonggashanica C. F. Fang et A. K. Skvortsov；贡嘎山柳 ●

343437　Salix gooddingi C. R. Ball；古丁柳；Goodding Willow ●☆

343438　Salix gordejevii Y. L. Chang et Skvortsov；黄柳（戈尔柳，砂柳，小黄柳）；Gordejev Willow，Yellow Willow ●

343439　Salix graciliglans Nakai = Salix gracilistyla Miq. var. graciliglans (Nakai) H. Ohashi ●☆

343440　Salix gracilior (Siuzew) Nakai；细枝蒙古柳（蒙古柳，细柳）；Gracile Willow，Slender-branched Willow，Thinner Willow ●

343441　Salix gracilis Andersson = Salix petiolaris Sm. ●☆

343442　Salix gracilis Andersson var. rosmarinoides Andersson = Salix petiolaris Sm. ●☆

343443　Salix gracilis Andersson var. textoris Fernald = Salix petiolaris Sm. ●☆

343444　Salix gracilistyla Miq.；细柱柳（红毛柳，猫柳，银芽柳）；Bigcatkin Willow，Big-catkined Willow，Chinese Pussy Willow，Gracile-style Willow，Rosegold Pussy Willow，Slenderstyle Willow ●

343445　Salix gracilistyla Miq. ' Melanostachys'；黑穗细柱柳；Black Willow ●☆

343446　Salix gracilistyla Miq. f. adscendens (Kimura) H. Ohashi = Salix gracilistyla Miq. ●

343447　Salix gracilistyla Miq. f. melanostachys (Makino) H. Ohashi = Salix gracilistyla Miq. ' Melanostachys' ●☆

343448　Salix gracilistyla Miq. f. pendula (Kimura) H. Ohashi；垂枝细柱柳 ●☆

343449　Salix gracilistyla Miq. f. variegata (Kimura) Kimura；斑点细柱柳 ●☆

343450　Salix gracilistyla Miq. subsp. melanostachys（Makino）Makino = Salix gracilistyla Miq. f. melanostachys（Makino）H. Ohashi ●☆

343451　Salix gracilistyla Miq. var. acuminata Skvortsov = Salix gracilior（Siuzew）Nakai ●

343452　Salix gracilistyla Miq. var. adscendens Kimura = Salix gracilistyla Miq. ●

343453　Salix gracilistyla Miq. var. graciliglans（Nakai）H. Ohashi；细腺细柱柳 ●☆

343454　Salix gracilistyla Miq. var. latifolia Skvortsov = Salix gracilior（Siuzew）Nakai ●

343455　Salix gracilistyla Miq. var. latifolia Skvortsov = Salix gracilistyla Miq. ●

343456　Salix gracilistyla Miq. var. melanostachys（Makino）C. K. Schneid. = Salix gracilistyla Miq. f. melanostachys（Makino）H. Ohashi ●☆

343457　Salix gracilistyla Miq. var. pendula Kimura = Salix gracilistyla Miq. f. pendula（Kimura）H. Ohashi ●☆

343458　Salix grandifolia Ser.；欧洲大叶柳；Bigleaf Willow，Big-leaf Willow ●☆

343459　Salix grisea Willd. = Salix sericea Marshall ●☆

343460　Salix grisonensis Forbes；格里森柳；Grisons Sallow ●☆

343461　Salix guebrianthiana C. K. Schneid.；细序柳（黑杨柳，山黑柳，山杨柳）；Guebrianth Willow，Smallcatkin Willow ●

343462　Salix gyamdaensis C. F. Fang；江达柳；Jiangda Willow ●

343463　Salix gyirongensis S. D. Zhao et C. F. Fang；吉隆垫柳；Jilong Willow ●

343464　Salix hainanica A. K. Skvortsov；海南柳 ●

343465　Salix haoana W. P. Fang；四川红柳（川红柳）；Hao Willow ●

343466　Salix harmsiana Seemen = Salix sieboldiana Blume ●☆

343467　Salix hastata L.；戟叶柳（戟柳，戟形柳）；Halberd Willow，Halberd-leaved Willow，Hastate Willow，Spear-leaved Willow ●

343468　Salix hastata L. 'Wehrhahnii'；韦氏戟叶柳 ●☆

343469　Salix hastata L. = Salix karelinii Turcz. ex Stschegl. ●

343470　Salix hastata L. var. himalayensis Andersson = Salix karelinii Turcz. ex Stschegl. ●

343471　Salix hastata Poljak. = Salix fedtschenkoi Goerz ●

343472　Salix hebecarpa（Fernald）Fernald = Salix pedicellaris Pursh ●☆

343473　Salix heishuiensis N. Chao；黑水柳；Heishui Willow ●

343474　Salix helvetica Forbes；瑞士柳；Swiss Willow ●☆

343475　Salix henryi Burkill = Salix heterochroma Seemen ●

343476　Salix herbacea L.；草柳（矮柳，刺叶柳）；Dwarf Willow，Least Willow，Pygmy Willow ●☆

343477　Salix heterochroma Seemen；紫枝柳；Heterochromia Willow，Purplebranch Willow，Purple-branched Willow ●

343478　Salix heterochroma Seemen var. concolor Goerz = Salix heterochroma Seemen ●

343479　Salix heterochroma Seemen var. glabra C. Y. Wu et C. F. Fang；无毛紫枝柳；Glabrous Purplebranch Willow ●

343480　Salix heteromera Hand. -Mazz.；异蕊柳；Diverse Willow，Heteromerous Willow，Hetero-stamenned Willow ●

343481　Salix heteromera Hand. -Mazz. var. villosior Hand. -Mazz. = Salix heteromera Hand. -Mazz. ●

343482　Salix heterostemon Flod.；异雄柳；Diverse Male Willow，Heterostamen Willow ●

343483　Salix heterostemon Flod. = Salix resectoides Hand. -Mazz. ●

343484　Salix hibernica Rech. f.；茶叶柳；Tea-leaved Willow ●☆

343485　Salix hidakamontana H. Hara = Salix nakamurana Koidz. subsp. kurilensis（Koidz.）H. Ohashi ●☆

343486　Salix hidewoi Koidz. = Salix reinii Franch. et Sav. ex Seemen f. eriocarpa（Kimura）T. Shimizu ●☆

343487　Salix himalayensis（Andersson）Flod.；喜马拉雅柳；Himalayan Willow，Himalayas Willow ●

343488　Salix himalayensis（Andersson）Flod. = Salix karelinii Turcz. ex Stschegl. ●

343489　Salix himalayensis（Andersson）Flod. var. filistyla（Z. Wang et P. Y. Fu）C. F. Fang；丝柱柳；Filiformstyle Willow，Filistyle Willow ●

343490　Salix himalayensis（Andersson）Flod. var. filistyla（Z. Wang et P. Y. Fu）C. F. Fang = Salix bhutanensis Flod. ●

343491　Salix hindsiana Benth.；海因兹柳；Hinds Willow，Sandbar Willow，Valley Willow ●☆

343492　Salix hirsuta Thunb. = Salix mucronata Thunb. subsp. hirsuta（Thunb.）Immelman ●☆

343493　Salix hirticaulis Hand. -Mazz.；毛枝垫柳；Hairy-branched Willow ●

343494　Salix hirtula Andersson；多毛柳；Hairy-branched Sallow ●☆

343495　Salix holargyrea Bornm. et Goerz；全银柳 ●☆

343496　Salix hookeriana Barratt；胡克柳；Bigleaf Willow，Coast Willow，Coastal Willow，Hooker Willow，Hooker's Willow ●☆

343497　Salix hsinganica Y. L. Chang et Skvortsov；兴安柳；Hsingan Willow，Xing'an Willow ●

343498　Salix hsinhsuaniana W. P. Fang = Salix cathayana Diels ●

343499　Salix huiana Goerz = Salix dissa C. K. Schneid. ●

343500　Salix huiana Goerz = Salix luctuosa H. Lév. ●

343501　Salix huiana Goerz var. tricholepis Goerz = Salix luctuosa H. Lév. ●

343502　Salix hultenii Flod. = Salix caprea L. ●

343503　Salix hultenii Flod. var. angustifolia Kimura = Salix caprea L. ●

343504　Salix humaensis Y. L. Chou et R. C. Chou；呼玛柳；Huma Willow ●

343505　Salix humboldtiana Willd.；南美柳；Humboldt's Willow，Pencil Willow，South American Willow ●☆

343506　Salix humilis Marshall；矮柳；Brown Willow，Dwarf Gray Willow，Dwarf Willow，Prairie Willow，Sage Willow，Upland Willow ●☆

343507　Salix humilis Marshall var. microphylla（Andersson）Fernald = Salix humilis Marshall var. tristis（Aiton）Griggs ●☆

343508　Salix humilis Marshall var. tristis（Aiton）Griggs；灰矮柳；Prairie Willow ●☆

343509　Salix hupehensis K. S. Hao ex C. F. Fang et A. K. Skvortsov；湖北柳；Hubei Willow ●

343510　Salix hutchinsii V. Naray. = Salix mucronata Thunb. ●☆

343511　Salix hylematica C. K. Schneid. = Salix furcata Andersson ●☆

343512　Salix hylonoma C. K. Schneid.；川柳；Sichuan Willow，Szechwan Willow ●

343513　Salix hylonoma C. K. Schneid. f. liocarpa Goerz；无毛川柳（光果川柳）；Glabrous Sichuan Willow ●

343514　Salix hylonoma C. K. Schneid. var. isochroma（Schneid.）Schneid. = Salix hylonoma C. K. Schneid. ●

343515　Salix hypoleuca C. K. Schneid. f. trichorachis C. F. Fang；毛轴小叶柳；Hairy-rachilla Littleleaf Willow ●

343516　Salix hypoleuca Seemen；小叶柳（翻白柳，红蜡梅，红梅蜡，山杨柳）；Littleleaf Willow，Small-leaved Willow ●

343517　Salix hypoleuca Seemen f. trichorachis C. F. Fang = Salix hypoleuca C. K. Schneid. f. trichorachis C. F. Fang ●

343518　Salix hypoleuca Seemen var. kansuensis Goerz ex Rehder et

Kobuski ＝Salix hypoleuca Seemen ●

343519 Salix hypoleuca Seemen var. platyphylla C. K. Schneid. ;宽叶翻白柳；Broadleaf Small-leaved Willow ●

343520 Salix ilectica var. integristyla Y. L. Chou ＝Salix hsinganica Y. L. Chang et Skvortsov ●

343521 Salix ilectica Y. L. Chou ＝Salix hsinganica Y. L. Chang et Skvortsov ●

343522 Salix ilectica Y. L. Chou var. integristyla Y. L. Chou ＝Salix hsinganica Y. L. Chang et Skvortsov ●

343523 Salix iliensis Regel;伊犁柳；Ili Willow, Yili Willow ●

343524 Salix inamoena Hand. -Mazz. ;丑柳(山白蜡条)；Ugly Willow, Unlovely Willow ●

343525 Salix inamoena Hand. -Mazz. var. glabra C. F. Fang;无毛丑柳；Glabrous Unlovely Willow ●

343526 Salix incana Schrank ＝Salix elaeagnos Scop. ●☆

343527 Salix incanescens Forbes;灰毛柳；White-leaved Willow ●☆

343528 Salix insignis Andersson;藏西柳；Insignis Willow, Marked Willow, West Tibet Willow, West Xizang Willow ●

343529 Salix integra Thunb. ;杞柳(白箕柳, 白杞柳)；Entire Willow, White Tip Willow ●

343530 Salix integra Thunb. ' Hakuro Nishiki'；杂色杞柳；Japanese Dappled Willow, Nishiki Willow, Variegated Willow ●☆

343531 Salix integra Thunb. f. albovariegata Kimura;白斑杞柳●☆

343532 Salix integra Thunb. f. pendula Kimura;垂枝杞柳●☆

343533 Salix interior Rowlee ＝Salix exigua Nutt. subsp. interior (Rowlee) Cronquist ●☆

343534 Salix interior Rowlee ＝Salix exigua Nutt. ●☆

343535 Salix interior Rowlee ＝Salix longifolia Muhl. ●☆

343536 Salix interior Rowlee f. wheeleri (Rowlee) Rouleau ＝Salix exigua Nutt. subsp. interior (Rowlee) Cronquist ●☆

343537 Salix interior Rowlee var. exterior Fernald ＝Salix exigua Nutt. subsp. interior (Rowlee) Cronquist ●☆

343538 Salix interior Rowlee var. luteosericea (Rydb.) C. K. Schneid. ＝Salix exigua Nutt. subsp. interior (Rowlee) Cronquist ●☆

343539 Salix interior Rowlee var. pedicellata (Andersson) C. R. Ball ＝Salix exigua Nutt. subsp. interior (Rowlee) Cronquist ●☆

343540 Salix interior Rowlee var. wheeleri Rowlee ＝Salix exigua Nutt. subsp. interior (Rowlee) Cronquist ●☆

343541 Salix irrorata Andersson;亚利桑那柳(露珠柳)；Arizona Willow ●☆

343542 Salix isochroma C. K. Schneid. ＝Salix hylonoma C. K. Schneid. ●

343543 Salix issykiensis Goerz;伊塞克柳●☆

343544 Salix jacutica Nasarow;雅库特柳●☆

343545 Salix jahandiezii Chass. ＝Salix atrocinerea Brot. subsp. jahandiezii (Chass.) Ibn Tattou ●☆

343546 Salix japonica Thunb. ;日本柳；Japanese Willow ●☆

343547 Salix jeholensis Nakai ＝Salix matsudana Koidz. ●

343548 Salix jessoensis K. S. Hao ＝Salix koreensis Andersson var. shandongensis C. F. Fang ●

343549 Salix jessoensis Seemen;虾夷柳●

343550 Salix jessoensis Seemen ＝Salix koreensis Andersson var. shandongensis C. F. Fang ●

343551 Salix jessoensis Seemen subsp. serissifolia (Kimura) H. Ohashi ＝Salix serissifolia Kimura ●☆

343552 Salix jinchuanica N. Chao;金川柳；Jinchuan Willow ●

343553 Salix jingdongensis C. F. Fang;景东矮柳；Jingdong Willow ●

343554 Salix jishiensis C. F. Fang et J. Q. Wang;积石柳；Jishi Willow ●

343555 Salix juparica Goerz;贵南柳；Guinan Willow, Juburishan

Willow, S. Guizhou Willow ●

343556 Salix juparica Goerz var. tibetica (Goerz) C. F. Fang;光果贵南柳；Smoothfruit Guinan Willow ●

343557 Salix kamanica Z. Wang et P. Y. Fu;卡马垫柳(毛枝垫柳)；Kama Willow ●

343558 Salix kamerunensis Seemen ＝Salix mucronata Thunb. ●☆

343559 Salix kangdingensis S. D. Zhao et C. F. Fang;康定垫柳；Kangding Willow ●

343560 Salix kangensis Nakai;江界柳(凤凰柳)；Jiangjie Willow ●

343561 Salix kangensis Nakai f. leiocarpa (Kitag.) Kitag. ＝Salix kangensis Nakai var. leiocarpa Kitag. ●

343562 Salix kangensis Nakai var. leiocarpa Kitag. ;光果江界柳(凤凰柳)；Glabrousfruit Willow ●

343563 Salix kansuensis K. S. Hao ＝Salix fargesii Burkill var. kansuensis (K. S. Hao ex C. F. Fang et A. K. Skvortsov) G. H. Zhu ●

343564 Salix kansuensis K. S. Hao ex C. F. Fang et A. K. Skvortsov ＝Salix fargesii Burkill var. kansuensis (K. S. Hao ex C. F. Fang et A. K. Skvortsov) G. H. Zhu ●

343565 Salix karelinii Turcz. ex Stschegl. ;枸子叶柳；Karelin Willow ●

343566 Salix kenoensis Koidz. ＝Salix shiraii Seemen var. kenoensis (Koidz.) Sugim. ●☆

343567 Salix kinuyanagii Kimura;日本绢柳●☆

343568 Salix kirilowiana Stschegl. ;天山筐柳●

343569 Salix kochiana Trautv. ;沙杞柳(砂杞柳)；Koch Willow ●

343570 Salix kolaensis Schljakov;科拉柳●☆

343571 Salix kolymensis Seemen;科雷马柳●☆

343572 Salix komarovii E. L. Wolf;科马罗夫柳●☆

343573 Salix kongbanica Z. Wang et P. Y. Fu;康巴柳；Kangba Willow, Kongba Willow ●

343574 Salix koreensis Andersson;朝鲜柳(白皮柳, 韩国柳)；Korea Willow, Korean Willow ●

343575 Salix koreensis Andersson ＝Salix pierotii Miq. ●

343576 Salix koreensis Andersson var. brevistyla Y. L. Chou et Skvortsov;短柱朝鲜柳；Shortstyle Korean Willow ●

343577 Salix koreensis Andersson var. pedunculata Y. L. Chou;长梗朝鲜柳；Longpedicel Korean Willow ●

343578 Salix koreensis Andersson var. shandongensis C. F. Fang;山东柳(山东朝鲜柳, 虾夷柳)；Shandong Willow ●

343579 Salix koriyanagi Kimura ex Goerz;尖叶紫柳；Koriyanagi Willow, Sharpleaf Purple Willow ●

343580 Salix korshinskyi Goerz;考尔柳●☆

343581 Salix kouytchensis C. K. Schneid. ;贵州柳；Guizhou Willow ●

343582 Salix krylovii E. L. Wolf;克氏柳●

343583 Salix kulashanensis Z. Wang et P. Y. Fu;古拉柳；Gulashan Willow ●

343584 Salix kulashanensis Z. Wang et P. Y. Fu ＝Salix gilashanica Z. Wang et P. Y. Fu ●

343585 Salix kungmuensis P. Y. Mao et W. Z. Li;孔目矮柳；Kongmu Willow ●

343586 Salix kurilensis Koidz. ＝Salix nakamurana Koidz. subsp. kurilensis (Koidz.) H. Ohashi ●☆

343587 Salix kusanoi (Hayata) C. K. Schneid. et Kimura;水社柳(草野氏柳)；Kusano Willow ●

343588 Salix kusanoi (Hayata) Hayata ＝Salix kusanoi (Hayata) C. K. Schneid. et Kimura ●

343589 Salix kusnetzowii Laksch. ;库兹柳●☆

343590 Salix laevigata Bebb;光滑红柳(平滑柳)；Lake Sallow,

Polished Willow, Red Willow ●☆

343591 Salix lamashanensis K. S. Hao ex C. F. Fang et A. K. Skvortsov; 拉马山柳; Lamashan Willow ●

343592 Salix lanata L.; 北密毛柳 (绵毛柳, 羊毛柳); Arctic Willow, Woolly Willow ●☆

343593 Salix lanata L. 'Stuartii'; 斯图尔特柳 ●☆

343594 Salix lanifera C. F. Fang and S. D. Zhao; 白毛柳; Lanose Willow, Woolly Willow ●

343595 Salix lapponum L.; 拉普兰柳; Downy Willow, Lapland Willow ●☆

343596 Salix lasiandra Benth.; 太平洋柳 (毛柳柳); Pacific Willow, Western Black Willow, Western Willow, Yellow Willow ●☆

343597 Salix lasiolepis Benth.; 粉叶柳; Arroyo Willow ●☆

343598 Salix lasiopes Z. Wang et P. Y. Fu; 毛柄柳; Hairpetiole Willow, Lasiopetiole Willow, Woollypetiole Willow, Woolly-petioled Willow ●

343599 Salix ledermannii Seemen; 莱德柳 ●☆

343600 Salix lepidostachys Seemen; 鳞穗柳 ●☆

343601 Salix leptoclados Andersson; 细枝柳 ●☆

343602 Salix leucopithecia Kimura; 棉花柳 (银柳); Cotton Willow ●

343603 Salix leveilleana C. K. Schneid.; 井冈柳; Leveille Willow ●

343604 Salix liangshuiensis Y. L. Chou et C. Y. King = Salix raddeana Laksch. ex Nasarow ●

343605 Salix ligustrina F. Michx. = Salix nigra Marshall ●☆

343606 Salix limprichtii Pax et K. Hoffm.; 黑皮柳; Blackbark Willow, Limpricht's Willow ●

343607 Salix lindleyana Wall.; 青藏垫柳; Lindley Willow ●

343608 Salix linearifolia E. L. Wolf = Salix blakii Goerz ●

343609 Salix linearifolia Rydb. = Salix exigua Nutt. subsp. interior (Rowlee) Cronquist ●☆

343610 Salix linearistipularis (Franch.) K. S. Hao; 筐柳 (白箕柳, 蒙古柳, 棉花柳); Linear-stypulate Willow, Linearstypule Willow ●

343611 Salix liouana Z. Wang et Chang Y. Yang; 黄龙柳; Huanglong Willow, Liou's Willow ●

343612 Salix lipskyi (Goerz) Nasarow; 利普斯基柳 ●☆

343613 Salix litwinowii Goerz; 利特氏柳 ●☆

343614 Salix livida Wahlenb.; 铅色柳 (谷柳) ●☆

343615 Salix livida Wahlenb. = Salix taraikensis Kimura ●

343616 Salix livida Wahlenb. var. occidentalis (Andersson) A. Gray = Salix bebbiana Sarg. ●

343617 Salix livida Wahlenb. var. rostrata (Richardson) Dippel = Salix bebbiana Sarg. ●

343618 Salix longiflora Burkill = Salix cathayana Diels ●

343619 Salix longiflora Wall. ex Andersson; 长花柳; Longflower Willow, Long-flowered Willow ●

343620 Salix longiflora Wall. ex Andersson var. albescens Burkill; 小叶长花柳 (小花长花柳); Smallleaf Longflower Willow ●

343621 Salix longiflora Wall. ex Andersson var. psilolepis Hand.-Mazz. = Salix tenela C. K. Schneid. ●

343622 Salix longifolia Muhl.; 长叶柳 (美洲长叶柳, 沙洲柳); Sandbar Willow, Sand-bar Willow ●☆

343623 Salix longifolia Muhl. = Salix exigua Nutt. subsp. interior (Rowlee) Cronquist ●☆

343624 Salix longifolia Muhl. var. interior (Rowlee) M. E. Jones = Salix exigua Nutt. subsp. interior (Rowlee) Cronquist ●☆

343625 Salix longifolia Muhl. var. pedicellata Andersson = Salix exigua Nutt. subsp. interior (Rowlee) Cronquist ●☆

343626 Salix longifolia Muhl. var. sericans Nees = Salix exigua Nutt. subsp. interior (Rowlee) Cronquist ●☆

343627 Salix longifolia Muhl. var. wheeleri (Rowlee) C. K. Schneid. = Salix exigua Nutt. subsp. interior (Rowlee) Cronquist ●☆

343628 Salix longipes Hook. f. et Thomson ex Hook. f.; 长梗柳; Coastal-plain Willow ●☆

343629 Salix longirostris Michx. = Salix humilis Marshall var. tristis (Aiton) Griggs ●☆

343630 Salix longissimipedicellaris N. Chao ex P. Y. Mao; 苍山长梗柳; Long-pedicel Willow ●

343631 Salix longistamina Z. Wang et P. Y. Fu; 长蕊柳; Longstamen Willow, Long-stamened Willow ●

343632 Salix longistamina Z. Wang et P. Y. Fu var. glabra Y. L. Chou; 无毛长蕊柳; Glabrous Longstamen Willow ●

343633 Salix lucida Muhl.; 亮叶柳; Shining Willow, Shiny Willow ●☆

343634 Salix lucida Muhl. var. angustifolia (Andersson) Andersson = Salix lucida Muhl. ●☆

343635 Salix lucida Muhl. var. intonsa Fernald = Salix lucida Muhl. ●☆

343636 Salix lucida Muhl. var. latifolia (Andersson) Andersson = Salix lucida Muhl. ●☆

343637 Salix lucida Muhl. var. ovatifolia Andersson = Salix lucida Muhl. ●☆

343638 Salix lucida Muhl. var. serissima L. H. Bailey = Salix serissima (L. H. Bailey) Fernald ●☆

343639 Salix luctuosa H. Lév.; 丝毛柳; Silkhairy Willow, Silky-hairy Willow ●

343640 Salix luctuosa H. Lév. var. pubescens Z. Wang et P. Y. Fu; 毛果丝毛柳; Hairy-fruit Silkhairy Willow ●

343641 Salix luctuosa H. Lév. var. pubescens Z. Wang et P. Y. Fu = Salix rehderiana C. K. Schneid. ●

343642 Salix ludingensis T. Y. Ding et C. F. Fang; 泸定垫柳; Luding Willow ●

343643 Salix ludoviciana Raf. = Salix nigra Marshall ●☆

343644 Salix luteosericea Rydb. = Salix exigua Nutt. subsp. interior (Rowlee) Cronquist ●☆

343645 Salix luzhongensis X. W. Li et Y. Q. Zhu; 鲁中柳; Central Shandong Willow ●

343646 Salix maboulasensis S. S. Ying = Salix taiwanalpina Kimura var. takasagoalpina (Koidz.) S. S. Ying ●

343647 Salix maboulasensis S. S. Ying = Salix takasagoalpina Koidz. ●

343648 Salix macilenta Anders.; 贫弱柳 ●☆

343649 Salix mackenzieana (Hook.) Barratt ex Andersson; 马氏柳; Diamond Willow, Mackenzie Willow ●☆

343650 Salix macroblasta C. K. Schneid.; 灌西柳; Longshoot Willow, Long-shooted Willow ●

343651 Salix macrolepis Turcz. = Chosenia arbutifolia (Pall.) A. K. Skvortsov ●◇

343652 Salix macropoda Stschegl.; 大梗柳 ●☆

343653 Salix macrostachya E. L. Wolf; 大穗柳 ●☆

343654 Salix madagascariensis Bojer ex Andersson; 马岛柳 ●☆

343655 Salix maerkangensis N. Chao; 簇毛柳; Maerkaang Willow, Markaang Willow ●

343656 Salix magnifica Hemsl.; 大叶柳; Footcatkin Willow, Foot-catkin Willow, Magnificent Willow ●

343657 Salix magnifica Hemsl. var. apetala (C. K. Schneid.) K. S. Hao; 倒卵叶大叶柳; Apetalus Footcatkin Willow ●

343658 Salix magnifica Hemsl. var. microphylla P. Y. Mao; 小叶大叶柳; Small Footcatkin Willow ●

343659 Salix magnifica Hemsl. var. microphylla P. Y. Mao = Salix zangica N. Chao ●

343660　Salix magnifica Hemsl. var. ulotricha（C. K. Schneid.）N. Chao；卷毛大叶柳；Ulotrichy Footcatkin Willow ●

343661　Salix mairei H. Lév. = Salix wallichiana Andersson ●

343662　Salix maizhokunggarensis N. Chao；墨竹柳（墨竹工卡柳）；Mozhugongka Willow，Mozugongka Willow ●

343663　Salix malifolia Besser = Salix livida Wahlenb. ●☆

343664　Salix malifolia J. Walker；苹果叶柳；Apple-leaved Willow ●☆

343665　Salix malifolia Schleich. ex Ser. = Salix nigricans（Sm.）Enander ●☆

343666　Salix malifolia Sm. = Salix hastata L. ●

343667　Salix margaritifera E. L. Wolf；珍珠柳●☆

343668　Salix matsudana Koidz.；旱柳（白柳，白皮柳，长叶柳，汉宫柳，河柳，山杨柳，小叶柳，言叶柳，羊角柳，杨柳，窄叶柳）；Corkscrew Willow，Dryland Willow，Hankow Willow，Peking Willow，Tortured Willow ●

343669　Salix matsudana Koidz.‘Scarlet Curls’；猩红旱柳；Hankow Willow，Pekin Willow，Scarlet Curls Willow ●☆

343670　Salix matsudana Koidz.‘Tortuosa’；龙爪柳；Contorted Willow，Corkscrew Willow，Dragon's-claw Willow，Drangon-claw Hankow，Permanent Wave Tree，Tortuous Hankow Willow，Zag Willow ●

343671　Salix matsudana Koidz.‘Umbraculifera’ = Salix matsudana Koidz. f. umbraculifera Rehder ●

343672　Salix matsudana Koidz. f. lobato-glandulosa C. F. Fang et W. D. Liu；漳河旱柳；Zhanghe Hankow Willow ●

343673　Salix matsudana Koidz. f. pendula C. K. Schneid.；绦柳（倒栽柳，条柳）；Pendulous Hankow Willow ●

343674　Salix matsudana Koidz. f. rubriflora C. F. Fang；红花龙须柳；Red-flower Hankow Willow ●

343675　Salix matsudana Koidz. f. tortuosa（Vilm.）Rehder = Salix matsudana Koidz.‘Tortuosa’ ●

343676　Salix matsudana Koidz. f. umbraculifera Rehder；馒头柳；Globe Willow，Umbelliform Hankow Willow ●

343677　Salix matsudana Koidz. var. anshanensis Z. Wang et J. Z. Yan；鞍山柳（旱快柳）；Anshan Willow ●

343678　Salix matsudana Koidz. var. pseudomatsudana（Y. L. Chou et Skvortsov）Y. L. Chou；旱垂柳；Fake Hankow Willow ●

343679　Salix matsudana Koidz. var. tortuosa Vilm. = Salix matsudana Koidz.‘Tortuosa’ ●

343680　Salix maximowiczii Kom.；大白柳；Maximowicz Willow ●

343681　Salix medogensis Y. L. Chou；墨脱柳；Motuo Willow ●

343682　Salix medwedewii Dode；迈德柳●☆

343683　Salix melanostachys（Makino）Makino；黑穗柳；Black Pussy Willow ●☆

343684　Salix melea C. K. Schneid. = Salix rehderiana C. K. Schneid. ●

343685　Salix mesinyi Hance；粤柳；Guangdong Willow，Kwangtung Willow，Mesny Willow ●

343686　Salix meta-formosa Nakai = Salix divaricata Pall. et Flod. var. meta-formosa（Nakai）Kitag. ●

343687　Salix metaglauca Chang Y. Yang；绿叶柳；Greenleaf Willow，Green-leaved Willow ●

343688　Salix micans Andersson = Salix alba L. ●

343689　Salix michelsonii Goerz ex Nasarow；米黄柳；Michelson Willow ●

343690　Salix microphyta Franch.；宝兴矮柳；Baoxing Little Willow，Small Willow，Small-spiked Willow ●

343691　Salix microstachya Turcz.；小穗柳（少柳，乌柳，小红柳）；Little Spike Willow，Littlespike Willow，Micro-spicated Willow ●

343692　Salix microstachya Turcz. var. bordensis（Nakai）C. F. Fang；小红柳；Borden Littlespike Willow ●

343693　Salix microstachyoides Z. Wang et P. Y. Fu = Salix cheilophila C. K. Schneid. var. microstachyoides（Z. Wang et P. Y. Fu）Z. Wang et C. F. Fang ●

343694　Salix microstachyoides Z. Wang et P. Y. Fu = Salix cheilophila C. K. Schneid. ●

343695　Salix microtricha C. K. Schneid.；兴山柳；Xingshan Willow ●

343696　Salix minjiangensis N. Chao；岷江柳；Minjiang Willow ●

343697　Salix minutiflora E. L. Wolf var. pubescnes E. L. Wolf = Salix rosmarinifolia L. ●

343698　Salix minutiflora Turcz. ex E. L. Wolf = Salix caesia Vill. ●

343699　Salix minutifolia E. L. Wolf = Salix rosmarinifolia L. ●

343700　Salix missouriensis Bebb = Salix eriocephala Michx. ●☆

343701　Salix mixta Korsh. = Salix eriocarpa Franch. et Sav. ●

343702　Salix miyabeana Seemen；宫部氏柳；Miyabe Willow ●☆

343703　Salix miyabeana Seemen subsp. gilgiana（Seemen）H. Ohashi = Salix miyabeana Seemen subsp. gymnolepis（H. Lév. et Vaniot）H. Ohashi et Yonek. ●☆

343704　Salix miyabeana Seemen subsp. gilgiana（Seemen）H. Ohashi f. pendula（Kimura）H. Ohashi；垂枝宫部氏柳●☆

343705　Salix miyabeana Seemen subsp. gymnolepis（H. Lév. et Vaniot）H. Ohashi et Yonek.；裸鳞宫部氏柳●☆

343706　Salix mongolica（Franch.）Siuzew；蒙古柳；Mongolian Willow ●

343707　Salix mongolica（Franch.）Siuzew = Salix linearifolia E. L. Wolf ●

343708　Salix mongolica（Franch.）Siuzew = Salix linearistipularis（Franch.）K. S. Hao ●

343709　Salix mongolica（Franch.）Siuzew f. bicolor Y. L. Chang et Skvortsov；杂色苞蒙古柳；Bicolor Mongolian Willow ●

343710　Salix mongolica（Franch.）Siuzew f. bicolor Y. L. Chang et Skvortsov = Salix linearistipularis（Franch.）K. S. Hao ●

343711　Salix mongolica（Franch.）Siuzew f. bicolor Y. L. Chang et Skvortsov = Salix linearifolia E. L. Wolf ●

343712　Salix mongolica（Franch.）Siuzew f. gracilior Siuzew = Salix gracilior（Siuzew）Nakai ●

343713　Salix mongolica（Franch.）Siuzew f. latifolia Nasarow = Salix linearistipularis（Franch.）K. S. Hao ●

343714　Salix mongolica（Franch.）Siuzew f. sericea Y. L. Chang et Skvortsov；毛蒙古柳；Sericeous Mongolian Willow ●

343715　Salix mongolica（Franch.）Siuzew f. sericea Y. L. Chang et Skvortsov = Salix linearistipularis（Franch.）K. S. Hao ●

343716　Salix mongolica（Franch.）Siuzew f. sericea Y. L. Chang et Skvortsov = Salix linearifolia E. L. Wolf ●

343717　Salix mongolica（Franch.）Siuzew var. yanbianica（C. F. Fang et Chang Y. Yang）Y. L. Chou = Salix yanbianica C. F. Fang et Chang Y. Yang ●

343718　Salix monica Bebb = Salix planifolia Pursh ●☆

343719　Salix morii Hayata = Salix doii Hayata ●

343720　Salix morii Hayata = Salix fulvopubescens Hayata var. doii（Hayata）K. C. Yang et T. C. Huang ●

343721　Salix morrisonicola Kimura；玉山柳（磨里山柳）；Molishan Willow，Yushan Willow ●

343722　Salix morrisonicola Kimura = Salix taiwanalpina Kimura var. morrisonicola（Kimura）K. C. Yang et T. C. Huang ●

343723　Salix moupinensis Franch.；宝兴柳（穆坪柳）；Baoxing Willow，Mupin Willow，Muping Willow ●

343724　Salix moupinensis Franch. f. elliptica Goerz = Salix moupinensis

Franch. ●

343725 Salix moupinensis Franch. f. obovata Goerz = Salix moupinensis Franch. ●

343726 Salix mucronata Thunb. ;短尖柳;African Willow, Cape Willow, Native Willow, Palestine Willow, River Willow, Wild Willow ●☆

343727 Salix mucronata Thunb. subsp. capensis（Thunb.）Immelman;好望角短尖柳●☆

343728 Salix mucronata Thunb. subsp. hirsuta（Thunb.）Immelman;粗毛短尖柳●☆

343729 Salix mucronata Thunb. subsp. wilmsii（Seemen）Immelman = Salix mucronata Thunb. subsp. woodii（Seemen）Immelman ●

343730 Salix mucronata Thunb. subsp. woodii（Seemen）Immelman = Salix woodii Seemen ●

343731 Salix mucronata Thunb. var. caffra Burtt Davy = Salix mucronata Thunb. ●☆

343732 Salix mucronata Thunb. var. integra Burtt Davy = Salix mucronata Thunb. ●☆

343733 Salix muhlenbergiana Willd. = Salix humilis Marshall var. tristis（Aiton）Griggs ●☆

343734 Salix muliensis Goerz;木里柳;Muli Willow ●

343735 Salix multinervis Franch. et Sav. = Salix integra Thunb. ●

343736 Salix myricifolia Andersson = Salix caesia Vill. ●

343737 Salix myricoides（Muhl.）J. Carey = Salix myricoides Muhl. ●☆

343738 Salix myricoides Muhl. ;野梅柳;Bayberry Willow, Blue-leaf Willow ●☆

343739 Salix myricoides Muhl. var. albovestita（C. R. Ball）Dorn;白鳞野梅柳;Bayberry Willow, Blue-leaf Willow ●☆

343740 Salix myricoides Muhl. var. angustata（Pursh）Dippel = Salix eriocephala Michx. ●☆

343741 Salix myricoides Muhl. var. cordata（Muhl.）Dippel = Salix eriocephala Michx. ●☆

343742 Salix myricoides Muhl. var. rigida（Muhl.）Dippel = Salix eriocephala Michx. ●☆

343743 Salix myrsinites L. ;铁仔柳;Black-twig Willow, Myrsine-like Willow, Myrtle Willow, Whortle-leaved Willow ●☆

343744 Salix myrtillacea Andersson;坡柳;Mountain Willow, Slope Willow ●

343745 Salix myrtilloides L. ;越橘柳（越桔柳）;Blueberry-like Willow, Whortleberry Willow, Whortleberry-like Willow ●

343746 Salix myrtilloides L. var. hypoglauca（Fernald）C. R. Ball = Salix pedicellaris Pursh ●☆

343747 Salix myrtilloides L. var. mandshurica Nakai;东北越橘柳（东北越桔柳）;Manchurian Whortleberry-like Willow ●

343748 Salix myrtilloides L. var. pedicellaris（Pursh）Andersson = Salix pedicellaris Pursh ●☆

343749 Salix nakamurana Koidz. ;日本银毛柳●☆

343750 Salix nakamurana Koidz. f. eriocarpa（Kimura）T. Shimizu;毛果日本银毛柳●☆

343751 Salix nakamurana Koidz. f. stenophylla（Kimura）Kimura = Salix nakamurana Koidz. ●☆

343752 Salix nakamurana Koidz. subsp. kurilensis（Koidz.）H. Ohashi;库里尔柳●☆

343753 Salix nakamurana Koidz. subsp. yezoalpina（Koidz.）H. Ohashi;虾夷山柳●☆

343754 Salix nakamurana Koidz. subsp. yezoalpina（Koidz.）H. Ohashi f. noreticulata（Nakai）H. Ohashi;网虾夷山柳●☆

343755 Salix nakamurana Koidz. var. eriocarpa Kimura = Salix

nakamurana Koidz. f. eriocarpa（Kimura）T. Shimizu ●☆

343756 Salix nakamurana Koidz. var. yezoalpina（Koidz.）Kimura = Salix nakamurana Koidz. subsp. yezoalpina（Koidz.）H. Ohashi ●☆

343757 Salix nankingensis Z. Wang et S. L. Tung;南京柳;Nanjing Willow ●

343758 Salix nelamunensis Z. Wang et P. Y. Fu = Salix serpyllum Andersson ●

343759 Salix nelsonii C. R. Ball = Salix planifolia Pursh ●☆

343760 Salix neoamnematchinensis T. Y. Ding et C. F. Fang;新生山柳●

343761 Salix neoforbesii Toepff. = Salix petiolaris Sm. ●☆

343762 Salix neolapponum Chang Y. Yang;绢柳;Silky Willow ●

343763 Salix neowilsonii W. P. Fang;新紫柳;Neowilson Willow, New Purple Willow, New Wilson Willow ●

343764 Salix niedzwieckii Goerz = Salix capusii Franch. ●

343765 Salix nielamunensis Z. Wang et P. Y. Fu;聂拉木柳;Nielamu Willow ●

343766 Salix nigerica V. Naray. = Salix mucronata Thunb. ●☆

343767 Salix nigra Marshall;黑柳（北美黑柳）;Black Willow, Dudley Willow, Goodding Willow, Swamp Willow ●☆

343768 Salix nigra Marshall var. altissima Sarg. = Salix nigra Marshall ●☆

343769 Salix nigra Marshall var. amygdaloides（Andersson）Andersson = Salix amygdaloides Andersson ●

343770 Salix nigra Marshall var. brevifolia Andersson = Salix nigra Marshall ●☆

343771 Salix nigra Marshall var. falcata（Pursh）Torr. = Salix nigra Marshall ●☆

343772 Salix nigra Marshall var. lindheimeri C. K. Schneid. = Salix nigra Marshall ●☆

343773 Salix nigra Marshall var. longifolia Andersson = Salix nigra Marshall ●☆

343774 Salix nigra Marshall var. marginata Wimm. ex Andersson = Salix nigra Marshall ●☆

343775 Salix nigra Marshall var. wrightii（Andersson）Andersson = Salix amygdaloides Andersson ●

343776 Salix nigricans（Sm.）Enander;铁仔叶柳（暗叶柳）;Anson's Sallow, Blueberry Willow, Dark-leaved Willow ●☆

343777 Salix nipponica Franch. et Sav. = Salix triandra L. var. nipponica（Franch. et Sav.）Seemen ●

343778 Salix nipponica Franch. et Sav. var. mengshanensis（S. B. Liang）G. Zhu;蒙山柳;Mengshan Willow ●

343779 Salix nujiangensis N. Chao;怒江柳;Nujiang Willow ●

343780 Salix nummularia Andersson;扁圆柳（长白柳, 多腺柳）;Changbaishan Willow, Manygland Willow, Poluglandular Willow ●

343781 Salix nummularia Andersson f. hebecarpa（Kimura）Kimura = Salix nummularia Andersson subsp. pauciflora Kimura ●☆

343782 Salix nummularia Andersson subsp. pauciflora（Koidz.）Kimura = Salix nummularia Andersson ●

343783 Salix nummularia Andersson subsp. pauciflora（Koidz.）Kimura var. hebecarpa（Kimura）Kimura ex Ohwi et Kitag. = Salix nummularia Andersson f. hebecarpa（Kimura）Kimura ●☆

343784 Salix nummularia Andersson subsp. pauciflora Kimura;小花扁圆柳●☆

343785 Salix nyiwensis Kimura;内韦柳●☆

343786 Salix oblongifolia Liou = Salix divaricata Pall. et Flod. var. metaformosa（Nakai）Kitag. ●

343787 Salix oblongifolia Trautv. et C. A. Mey. ;矩圆叶柳●☆

343788 Salix obovata Pursh = Salix pellita（Andersson）Andersson ex

C. K. Schneid. ●☆

343789　Salix obscura Andersson；毛坡柳；Obscure Willow ●

343790　Salix obscura Andersson var. lanifera（C. F. Fang et S. D. Zhao）N. Chao = Salix lanifera C. F. Fang et S. D. Zhao ●

343791　Salix occidentalis Walter = Salix humilis Marshall var. tristis（Aiton）Griggs ●☆

343792　Salix occidentali-sinensis N. Chao；华西柳；W. China Willow，West China Willow ●

343793　Salix ochetophylla Goerz；汶川柳；Wenchuan Willow ●

343794　Salix octandra A. Rich. = Salix mucronata Thunb. ●☆

343795　Salix ohsidara Kimura；青皮垂柳●

343796　Salix ohsidare Kimura = Salix babylonica L. ●

343797　Salix okamotoana Koidz. ；关山岭柳（冈本氏柳，高雄柳，台矮柳，台湾矮柳）；Okamoto Willow ●

343798　Salix oldhamiana Miq. = Salix warburgii Seemen ●

343799　Salix oleifolia Host ex Wimm. = Salix purpurea L. ●

343800　Salix oleifolia Ser. = Salix incana Schrank ●☆

343801　Salix oleifolia Sm. = Salix cinerea L. ●

343802　Salix oleninii Nasarow；奥列氏柳●☆

343803　Salix olgae Regel；奥氏柳●☆

343804　Salix omeiensis C. K. Schneid. ；峨眉柳；Emei Willow，Omei Willow ●

343805　Salix opaca Andersson = Salix sachalinensis F. Schmidt ●

343806　Salix opaca Andersson ex Seemen = Salix sachalinensis F. Schmidt ●

343807　Salix opsimantha C. K. Schneid. ；迟花柳；Lateflowered Willow，Late-flowered Willow ●

343808　Salix opsimantha C. K. Schneid. var. wawashanica（P. Y. Mao et P. X. He）G. Zhu；娃娃山矮柳（娃娃山柳）；Wawashan Lateflowered Dwarf Willow ●

343809　Salix orbicularis Andersson；圆柳；Round Netleaf Willow，Round Netleaf-willow ●☆

343810　Salix oreinoma C. K. Schneid. ；迟花矮柳；Lateflowered Dwarf Willow，Small Lateflowered Willow ●

343811　Salix oreinoma C. K. Schneid. var. wawashanica P. Y. Mao et P. X. He = Salix opsimantha C. K. Schneid. var. wawashanica（P. Y. Mao et P. X. He）G. Zhu ●

343812　Salix oreophila Hook. f. ex Anderson；尖齿叶垫柳；Mount Willow，Sharp-toothed Willow，Sharptoothleaf Willow ●

343813　Salix oreophila Hook. f. ex Anderson var. secta（Andersson）Andersson；五齿叶垫柳；Fivetooth Mount Willow ●

343814　Salix oritrepha C. K. Schneid. ；山生柳；Montane Willow，Mountain Willow，Mountaineer Willow ●

343815　Salix oritrepha C. K. Schneid. var. amnematchinensis（K. S. Hao）Z. Wang et C. F. Fang；青山生柳●

343816　Salix oritrepha C. K. Schneid. var. tibetica Goerz = Salix sclerophylla Andersson var. tibetica（Goerz）C. F. Fang ●

343817　Salix oritrepha C. K. Schneid. var. tibetica Goerz ex Rehder et Kobuski = Salix sclerophylla Andersson var. tibetica（Goerz）C. F. Fang ●

343818　Salix oritrepha C. K. Schneid. var. tibetica Goerz f. uniglandulosa Goerz ex K. S. Hao = Salix sclerophylla Andersson var. tibetica（Goerz）C. F. Fang ●

343819　Salix orthostemma Nakai = Salix divaricata Pall. et Flod. var. meta-formosa（Nakai）Kitag. ●

343820　Salix ovalifoliar Trautv. ；卵叶柳；Oval-leaved Willow ●☆

343821　Salix ovatomicrophylla K. S. Hao；卵小叶垫柳（卵状小叶垫

柳）；Ovate Littleleaf Willow，Ovate-small-leaved Willow ●

343822　Salix oxycarpa Andersson；尖果柳●☆

343823　Salix oxycarpa Andersson = Salix pycnostachya Andersson var. oxycarpa（Andersson）Y. L. Chou et C. F. Fang ●

343824　Salix pachyclada H. Lév. et Vaniot = Salix wallichiana Andersson var. pachyclada（H. Lév. et Vaniot）Z. Wang et C. F. Fang ●

343825　Salix palibinii Goerz；帕里宾柳●☆

343826　Salix pallasii Andersson；帕拉斯柳●☆

343827　Salix paludicola Koidz. = Salix fustescens Andersson ●☆

343828　Salix panchetiana（Burkill）Hand. -Mazz. = Salix ernestii C. K. Schneid. ●

343829　Salix pannosa Schleich. ；毡柳；Cloth-leaved Sallow ●☆

343830　Salix paraflabellaris S. D. Zhao；类扇叶垫柳；Fake Flabellateleaf Willow，False Fan-shaped Willow，Fanleaflike Willow ●

343831　Salix paraheterochroma Z. Wang et P. Y. Fu；藏紫枝柳；Tibet Purple-branched Willow，Xizang Purplebranch Willow ●

343832　Salix parallelinervis Flod. ；平行侧脉柳●☆

343833　Salix paraphylicifolia Chang Y. Yang；光叶柳；Brightleaf Willow，Glabrousleaf Willow，Glabrous-leaved Willow ●

343834　Salix paraplesia C. K. Schneid. ；康定柳（拟五蕊柳）；False Bayleaf Willow，Kangding Willow ●

343835　Salix paraplesia C. K. Schneid. f. lanceolata Z. Wang et C. Y. Yu；狭叶康定柳；Lanceoleaf Kangding Willow ●

343836　Salix paraplesia C. K. Schneid. var. pubescens Z. Wang et C. F. Fang；毛枝康定柳；Pubescent Kangding Willow ●

343837　Salix paraplesia C. K. Schneid. var. subintegra Z. Wang et P. Y. Fu；左旋柳●

343838　Salix paratetradenia Z. Wang et P. Y. Fu；类四腺柳；Fourglandlike Willow，Paratetragland Willow ●

343839　Salix paratetradenia Z. Wang et P. Y. Fu var. yatungensis Z. Wang et P. Y. Fu；亚东柳（亚东类四腺柳）；Yadong Paratetradgland Willow ●

343840　Salix parvidenticulata C. F. Fang；小齿叶柳；Little-serrated Willow，Small-denticulated Willow，Smallserrated Willow ●

343841　Salix pauciflora Koidz. = Salix nummularia Andersson ●

343842　Salix pauciflora Koidz. f. cyclophylla（Kimura）Kimura = Salix nummularia Andersson ●

343843　Salix pauciflora Koidz. f. hebecarpa（Kimura）Kimura = Salix nummularia Andersson f. hebecarpa（Kimura）Kimura ●☆

343844　Salix pauciflora Koidz. f. stenophylla（Kimura）Kimura = Salix nummularia Andersson ●

343845　Salix pedicellaris Pursh；梗花柳；Bog Willow ●☆

343846　Salix pedicellaris Pursh var. hypoglauca Fernald = Salix pedicellaris Pursh ●☆

343847　Salix pedicellaris Pursh var. tenuescens Fernald = Salix pedicellaris Pursh ●☆

343848　Salix pedicellata Desf. ；短花梗柳●☆

343849　Salix pedicellata Desf. subsp. antiatlantica（Maire et Wilczek）Maire；安蒂柳●☆

343850　Salix pedicellata Desf. subsp. hesperia Maire et Weiller；西方柳●☆

343851　Salix pedicellata Desf. var. longidentata Maire = Salix pedicellata Desf. ●☆

343852　Salix pedicellata Desf. var. santamariae Goerz et Sennen = Salix pedicellata Desf. ●☆

343853　Salix pedicellata Desf. var. villipes Maire et Weiller = Salix pedicellata Desf. ●☆

343854 Salix pella C. K. Schneid. ;黑枝柳;Blackbranch Willow,Black-branched Willow ●

343855 Salix pellita（Andersson）Andersson ex C. K. Schneid. ;光滑柳;Satiny Willow ●☆

343856 Salix pennata C. R. Ball = Salix planifolia Pursh ●☆

343857 Salix pensylvanica Forbes = Salix sericea Marshall ●☆

343858 Salix pentandra L. ;五蕊柳（柳条,柳芽子）;Bay Willow,Bay-leaf Willow, Bay-leaved Willow, Black Willow, French Sally, Goslings, Laurel Willow, Laurel-leaf Willow, Laurel-leaved Willow, May-goslings,Sweet Bay Willow, Sweet Willie, Sweet Willow, Sweet Willy,Willow Bay,Willow-bay ●

343859 Salix pentandra L. var. intermedia Nakai;白背五蕊柳;Whiteback Bay-leaf Willow ●

343860 Salix pentandra L. var. intermedia Nakai = Salix pseudopentandra（Flod.）Flod. ●

343861 Salix pentandra L. var. lucida（Muhl.）Kuntze = Salix lucida Muhl. ●☆

343862 Salix pentandra L. var. obovata C. Y. Yu;卵苞五蕊柳;Ovate Bay-leaf Willow ●

343863 Salix pentandra L. var. pseudopentandra（Flod.）Kitag. = Salix pentandra L. var. intermedia Nakai ●

343864 Salix pentandra L. var. pseudopentandra（Flod.）Kitag. = Salix pseudopentandra（Flod.）Flod. ●

343865 Salix permollis Z. Wang et C. Y. Yu;山毛柳;Mountain Hairy Willow,Pubescent Willow ●

343866 Salix perrostrata Rydb. = Salix bebbiana Sarg. ●

343867 Salix persica Boiss. ;波斯柳●☆

343868 Salix petiolaris Sm. ;具柄草地柳（牧场柳）;Meadow Willow, Slender Willow ●☆

343869 Salix petiolaris Sm. var. angustifolia Andersson = Salix petiolaris Sm. ●☆

343870 Salix petiolaris Sm. var. gracilis（Andersson）Andersson = Salix petiolaris Sm. ●☆

343871 Salix petiolaris Sm. var. grisea（Willd.）Torr. = Salix sericea Marshall ●☆

343872 Salix petiolaris Sm. var. rosmarinoides（Andersson）C. K. Schneid. = Salix petiolaris Sm. ●☆

343873 Salix petiolaris Sm. var. sericea（Marshall）Andersson = Salix sericea Marshall ●☆

343874 Salix petiolaris Sm. var. subsericea Andersson = Salix petiolaris Sm. ●☆

343875 Salix petiolata ? = Salix phylicifolia L. ●☆

343876 Salix petraea G. Anders. ;岩生柳;Rock Sallow ●☆

343877 Salix petraea G. Anders. = Salix nigricans（Sm.）Enander ●☆

343878 Salix pet-susu Kimura = Salix schwerinii E. L. Wolf ●

343879 Salix pet-susu Kimura f. abbreviata（Kimura）Kimura = Salix schwerinii E. L. Wolf ●

343880 Salix phaidima C. K. Schneid. ;纤柳;Linearleaf Willow, Thin Willow,Weak Willow ●

343881 Salix phanera C. K. Schneid. ;显叶柳（长叶柳）;Longleaf Willow,Long-leaved Willow ●

343882 Salix phanera C. K. Schneid. var. weixiensis C. F. Fang;维西长叶柳（维西显叶柳）;Weixi Longleaf Willow ●

343883 Salix phaneroides Goerz = Salix phanera C. K. Schneid. ●

343884 Salix phlebophylla Andersson;脉叶柳●☆

343885 Salix phylicifolia Buikill ? = Salix taishanensis Z. Wang et C. F. Fang var. hobeinica C. F. Fang ●

343886 Salix phylicifolia L. ;东陵山柳（深山柳菲利木柳）;Tealeaf Willow,Tea-leaved Willow ●☆

343887 Salix phylicifolia L. subsp. planifolia（Pursh）Hiitonen = Salix planifolia Pursh ●☆

343888 Salix phylicifolia L. var. monica（Bebb）Jeps. = Salix planifolia Pursh ●☆

343889 Salix phylicifolia L. var. pennata（C. R. Ball）Cronquist = Salix planifolia Pursh ●☆

343890 Salix pierotii Miq. ;白皮柳;Pierot Willow, Tea-leaved Willow, Whitebark Willow ●

343891 Salix pilosomicrophylla Z. Wang et P. Y. Fu;毛小叶垫柳;Hairy Little-leaved Willow, Hairy Smallleaf Willow, Hairy-microphyllous Willow, Pilose-little-leaf Willow ●

343892 Salix pingliensis Y. L. Chou;平利柳;Pingli Willow ●

343893 Salix piptotricha Hand. -Mazz. ; 毛果垫柳; Downy-capsule Willow, Hairfruit Willow, Hairy-fruited Cushion Willow ●

343894 Salix planifolia Pursh;菱叶柳;Diamond-leaf Willow, Flat-leaved Willow, Tea-leaved Willow ●☆

343895 Salix planifolia Pursh var. monica（Bebb）C. K. Schneid. = Salix planifolia Pursh ●☆

343896 Salix planifolia Pursh var. nelsonii（C. R. Ball）C. R. Ball ex E. C. Sm. = Salix planifolia Pursh ●☆

343897 Salix planifolia Pursh var. pennata（C. R. Ball）C. R. Ball ex Dutilly, Lepage et Duman = Salix planifolia Pursh ●☆

343898 Salix plocotricha C. K. Schneid. ;曲毛柳;Retrose-hairy Willow, Villose Willow ●

343899 Salix podophylla Andersson;梗叶柳●☆

343900 Salix polaris Wahlenb. ;极柳;Polar Willow ●☆

343901 Salix polyadenia Hand. -Mazz. ;多腺柳●

343902 Salix polyadenia Hand. -Mazz. = Salix nummularia Andersson ●

343903 Salix polyadenia Hand. -Mazz. var. tschanbaischanica（Y. L. Chou et Y. L. Chang）Y. L. Chang = Salix nummularia Andersson ●

343904 Salix polyadenia Hand. -Mazz. var. tschanbaischanica（Y. L. Chou et Y. L. Chang）Y. L. Chang;长白柳●

343905 Salix polyandra H. Lév. = Salix cavaleriei H. Lév. ●

343906 Salix polyclona C. K. Schneid. ;多枝柳;Manybranch Willow, Multibranched Willow,Polybrached Willow ●

343907 Salix pominica Z. Wang et P. Y. Fu;波密柳;Bomi Willow ●

343908 Salix pominica Z. Wang et P. Y. Fu = Salix brachista C. K. Schneid. ●

343909 Salix praticola Hand. -Mazz. ex Enander;草地柳;Meadow Willow,Pastural Willow,Prarie Willow ●

343910 Salix prinoides Pursh = Salix discolor Muhl. ●☆

343911 Salix propitia Koidz. ;日本深山柳●☆

343912 Salix propitia Koidz. = Salix sieboldiana Blume ☆

343913 Salix przewalskii E. L. Wolf;普尔柳●☆

343914 Salix psammophila Z. Wang et Chang Y. Yang;北沙柳（沙柳,西北沙柳）;Psammophil Willow, Sand Willow, Sandlive Willow, Sand-loving Willow ●

343915 Salix pseudalba E. L. Wolf;假白柳●☆

343916 Salix pseudoernestii Goerz = Salix ernestii C. K. Schneid. ●

343917 Salix pseudokoreensis Koidz. ;假朝鲜柳●☆

343918 Salix pseudolasiogyne H. Lév. ;朝鲜垂柳（韩国垂柳）;Fake Woollypistil Willow, False Harpetiole Willow, Korean Weeping Willow ●

343919 Salix pseudolasiogyne H. Lév. var. bilofolia J. Q. Wang et Dian M. Li;垦绥垂柳;Kensui False Harpetiole Willow ●

343920　Salix pseudolasiogyne H. Lév. var. erythrantha C. F. Fang；红花朝鲜垂柳；Red-flower Fake Woollypistil Willow ●

343921　Salix pseudolinearis Nasarow = Salix viminalis L. var. angustifolia Turcz. ●

343922　Salix pseudomatsudana Y. L. Chou et Skvortsov = Salix matsudana Koidz. var. pseudomatsudana（Y. L. Chou et Skvortsov）Y. L. Chou ●

343923　Salix pseudopaludicola Kimura；假大雪山柳●☆

343924　Salix pseudopentandra（Flod.）Flod. = Salix pentandra L. var. intermedia Nakai ●

343925　Salix pseudopentandra Flod. = Salix pentandra L. var. intermedia Nakai ●

343926　Salix pseudopermollis C. Y. Yu et Chang Y. Yang；小叶山毛柳；Fake Pubescent Willow, False Mountain Hairy Willow, Smallleaf Pubescent Willow ●

343927　Salix pseudospissa Goerz；大苞柳；Bigbract Willow, Big-bracted Willow, Largebract Willow ●

343928　Salix pseudotangii Z. Wang et C. Y. Yu；山柳；Fake Tang Willow, False Tang Willow ●

343929　Salix pseudowallichiana Goerz；青皂柳；Fake Wallich Willow, False Soap Willow, False Wallich Willow ●

343930　Salix pseudowolohoensis K. S. Hao ex C. F. Fang et A. K. Skvortsov；西柳；Fake Woloho Willow, False Woloho Willow, West Willow ●

343931　Salix psiloides（Flod.）Kom.；平滑柳●☆

343932　Salix psilostigma Andersson；裸柱头柳；Bare-stigma Willow, Nakedstigma Willow, Nakestigma Willow ●

343933　Salix psilostigma Andersson = Salix daliensis C. F. Fang et S. D. Zhao f. longispica C. F. Fang ●

343934　Salix psilostigma Andersson = Salix daliensis C. F. Fang et S. D. Zhao ●

343935　Salix pubescens（E. L. Wolf）K. S. Hao = Salix rosmarinifolia L. ●

343936　Salix pulchra Cham.；美丽柳●☆

343937　Salix purpurea L.；欧洲杞柳（地中海紫柳，红皮柳，萑苻，欧紫柳，蒲柳根，青杨根，水杨根，紫柳）；Alaska Blue Willow, Arctic Willow, Basket Willow, Bitter Willow, Europe Willow, European Osier, Purple Osier, Purple Osier-willow, Purple Willow, Purpleosier, Purpleosier Willow, Red Osier, Stone Osier, Willow ●

343938　Salix purpurea L. 'Dicky Meadows'；紫色欧洲杞柳；Purple Osier Willow ●☆

343939　Salix purpurea L. 'Eugene'；欧根欧洲杞柳；Purple Osier Willow ●☆

343940　Salix purpurea L. 'Gracilis' = Salix purpurea L. 'Nana' ●☆

343941　Salix purpurea L. 'Nana'；矮欧洲杞柳（矮生地中海紫柳）；Dwarf Purple Willow ●☆

343942　Salix purpurea L. = Salix sinopurpurea Z. Wang et Chang Y. Yang ●

343943　Salix purpurea L. subsp. amplexicaulis Boiss. var. multinervis C. K. Schneid. = Salix integra Thunb. ●

343944　Salix purpurea L. subsp. angustifolia Koidz. = Salix purpurea L. var. angustifolia Koidz. ●☆

343945　Salix purpurea L. subsp. embergeri Chass. = Salix purpurea L. ●

343946　Salix purpurea L. var. amplexicaulis Boiss. = Salix integra Thunb. ●

343947　Salix purpurea L. var. angustifolia Koidz.；狭叶红皮柳（蒲柳）；Narrowleaf Purple Willow ●☆

343948　Salix purpurea L. var. embergeri（Chass.）Maire = Salix purpurea L. ●

343949　Salix purpurea L. var. gracilis Gren. et Gordon；细红皮柳（狭叶紫柳）；Thin Purple Willow ●

343950　Salix purpurea L. var. helix Koch = Salix purpurea L. ●

343951　Salix purpurea L. var. hispanica Goerz = Salix purpurea L. ●

343952　Salix purpurea L. var. japonica Nakai = Salix koriyanagi Kimura ex Goerz ●

343953　Salix purpurea L. var. longipetiolata C. Y. Yu；长柄红皮柳；Long-pedicel Purple Willow ●

343954　Salix purpurea L. var. longipetiolata C. Y. Yu = Salix sinopurpurea Z. Wang et Chang Y. Yang ●

343955　Salix purpurea L. var. multinervis Matsum.；多脉红皮柳（杞柳）；Manynerv Purple Willow ●

343956　Salix purpurea L. var. smithiana Trautv.；史氏红皮柳（红皮柳）；Smith Purple Willow ●

343957　Salix purpurea L. var. smithiana Trautv. = Salix gracilior（Siuzew）Nakai ●

343958　Salix purpurea L. var. stipularis Franch. = Salix linearistipularis（Franch.）K. S. Hao ●

343959　Salix purshiana Spreng. = Salix nigra Marshall ●☆

343960　Salix pychnocarpa Andersson = Salix planifolia Pursh ●☆

343961　Salix pycnostachya Andersson；密穗柳；Dense-spikated Willow, Densispike Willow, Densispikeed Willow ●

343962　Salix pycnostachya Andersson var. oxycarpa（Andersson）Y. L. Chou et C. F. Fang；尖果密穗柳（尖果柳）●

343963　Salix pyi H. Lév. = Salix cavaleriei H. Lév. ●

343964　Salix pyrifolia Anderson；梨叶柳；Balsam Willow, Bog Willow ●☆

343965　Salix pyrifolia Andersson var. lanceolata（Bebb）Fernald = Salix pyrifolia Anderson ●☆

343966　Salix pyrina K. S. Hao = Salix tetrasperma Roxb. ●

343967　Salix pyrina Wall.；梨形柳●☆

343968　Salix pyrolifolia Ledeb.；鹿蹄柳（鹿蹄叶柳）；Balsam Willow, Pyrolaleaf Willow, Pyrola-leaved Willow, Shinyleaf Willow ●

343969　Salix pyrolifolia Ledeb. var. cordata Trautv. = Salix pyrolifolia Ledeb. ●

343970　Salix pyrolifolia Ledeb. var. ovata Trautv. = Salix pyrolifolia Ledeb. ●

343971　Salix pyrolifolia Ledeb. var. pubescens Nasarow = Salix pyrolifolia Ledeb. ●

343972　Salix qinghaiensis Y. L. Chou；青海柳；Qinghai Willow ●

343973　Salix qinghaiensis Y. L. Chou var. microphylla Y. L. Chou；小叶青海柳；Littleleaf Qinghai Willow ●

343974　Salix qinlingica Z. Wang et N. Chao；秦岭山柳（秦岭柳，陕西柳）；Qinling Willow ●

343975　Salix raddeana Laksch. ex Nasarow；大黄柳（凉水柳）；Liangshui Willow, Radde's Willow ●

343976　Salix raddeana Laksch. ex Nasarow = Salix abscondita Laksch. ●☆

343977　Salix raddeana Laksch. ex Nasarow var. liangshuiensis（Y. L. Chou et C. Y. King）Y. L. Chou = Salix raddeana Laksch. ex Nasarow ●

343978　Salix raddeana Laksch. ex Nasarow var. subglabra Y. L. Chang et Skvortsov；稀毛大黄柳；Few-hairs Radde's Willow ●

343979　Salix radinostachya C. K. Schneid.；长穗柳；Longspike Willow, Long-spiked Willow, Long-stachys Willow ●

343980　Salix radinostachya C. K. Schneid. var. pseudophanera C. F. Fang；绒毛长穗柳；Tomentose Longspike Willow ●

343981　Salix radinostachya C. K. Schneid. var. szechuanica（Goerz）N. Chao = Salix radinostachya C. K. Schneid. ●

343982　Salix rectijulis Ledeb. ex Trautv.；欧越橘柳（直穗柳）；European Blueberry Willow, Wineberry-leaved Willow ●

343983　Salix recurvata Pursh = Salix humilis Marshall var. tristis（Aiton）Griggs ●☆

343984　Salix rehderiana C. K. Schneid.；川滇柳；Rehder Willow ●

343985　Salix rehderiana C. K. Schneid. var. brevisericea C. K. Schneid. = Salix rehderiana C. K. Schneid. ●

343986　Salix rehderiana C. K. Schneid. var. dolia（C. K. Schneid.）N. Chao；灌柳；Dolium Rehder Willow ●

343987　Salix rehderiana C. K. Schneid. var. lasiogyna Z. Wang et P. Y. Fu；穿鱼柳 ●

343988　Salix rehderiana C. K. Schneid. var. lasiogyna Z. Wang et P. Y. Fu = Salix sericocarpa Andersson ●

343989　Salix reinii Franch. et Sav. ex Seemen；雷恩柳 ●☆

343990　Salix reinii Franch. et Sav. ex Seemen f. cyclophylloides（Koidz.）Kimura ex Ohwi et Kitag. = Salix reinii Franch. et Sav. ex Seemen ●☆

343991　Salix reinii Franch. et Sav. ex Seemen f. eriocarpa（Kimura）T. Shimizu；毛果雷恩柳 ●☆

343992　Salix reinii Franch. et Sav. ex Seemen f. pendula Kimura；垂枝雷恩柳 ●☆

343993　Salix reinii Franch. et Sav. ex Seemen var. eriocarpa Kimura = Salix reinii Franch. et Sav. ex Seemen f. eriocarpa（Kimura）T. Shimizu ●☆

343994　Salix repens L.；匍匐柳；Cran-commer, Creeping Willow, Dwarf Silky Willow, Dwarf Willow ●☆

343995　Salix repens L. ‘Boyds Pendulous’；矮匍匐柳；Dwarf Willow ●☆

343996　Salix repens L. var. argentea G. Camus et A. Camus；银白匍匐柳 ●☆

343997　Salix repens L. var. brachypoda Trautv. et C. A. Mey. = Salix rosmarinifolia L. var. brachypoda（Trautv. et C. A. Mey.）Y. L. Chou ●

343998　Salix repens L. var. flavicans Andersson = Salix rosmarinifolia L. var. brachypoda（Trautv. et C. A. Mey.）Y. L. Chou ●

343999　Salix repens L. var. rosmarinifolia（L.）Wimm. et Grab. = Salix rosmarinifolia L. ●

344000　Salix reptans Rupr.；北极匍匐柳；Arctic Creeping Willow ●☆

344001　Salix resecta Diels；截苞柳；Truncatebract Willow, Truncate-bracted Willow ●

344002　Salix resectoides Hand. -Mazz.；藏截苞矮柳；Xizang Truncate-bract Willow, Xizang Truncate-bracted Willow ●

344003　Salix reticulata L.；网脉柳（网脉叶柳, 网皱柳）；Netleaf Willow, Net-leaf Willow, Net-leafed Willow, Netted Willow, Net-veined Willow, Reticulate Willow, Wrinkled-leaved Willow, Wrinkle-leaved Willow ●☆

344004　Salix retusa L.；欧洲小叶柳（小叶柳）；Blunt-leaved Willow, Notchleaf Willow ●☆

344005　Salix rhamnifolia Pall.；鼠李叶柳 ●☆

344006　Salix rhododendrifolia Z. Wang et P. Y. Fu；杜鹃叶柳；Rhododendron-leaved Willow ●

344007　Salix rhododendroides Z. Wang et C. Y. Yu = Salix wangiana K. S. Hao ex C. F. Fang et A. K. Skvortsov ●

344008　Salix rhoophila C. K. Schneid.；房县柳（房山柳）；Fangxian Willow ●

344009　Salix richardsonii Hook.；理氏柳；Richardson's Willow ●☆

344010　Salix rigida Muhl.；坚柳；Missouri River Willow, Richardson

344011　Salix rigida Muhl. = Salix eriocephala Michx. ●☆

344012　Salix rigida Muhl. f. subintegra（E. J. Palmer et Steyerm.）Steyerm. = Salix eriocephala Michx. ●☆

344013　Salix rigida Muhl. var. angustata（Pursh）Fernald = Salix eriocephala Michx. ●☆

344014　Salix rigida Muhl. var. rigida（E. J. Palmer et Steyerm.）Fernald = Salix eriocephala Michx. ●☆

344015　Salix rigida Muhl. var. vestita（Andersson）C. R. Ball = Salix eriocephala Michx. ●☆

344016　Salix rivularis Schleich.；溪边柳；River Sallow ●☆

344017　Salix rivulicola P. Y. Mao et W. Z. Li；溪旁矮柳；Riverside Willow ●

344018　Salix rivulicola P. Y. Mao et W. Z. Li = Salix annulifera C. Marquand et Airy Shaw var. dentata S. D. Zhao ●

344019　Salix rnacarolepis Turcz. = Chosenia arbutifolia（Pall.）A. K. Skvortsov ●◇

344020　Salix rockii Goerz；拉加柳（山柳）；Rock Willow ●

344021　Salix rockii Goerz f. biglandulosa C. F. Fang；二腺拉加柳；Bigland Rock Willow ●

344022　Salix rorida Laksch.；粉枝柳；Pruina-shoot Willow, Pruinous-shooted Willow ●

344023　Salix rorida Laksch. f. pendula Kimura；粉垂柳 ●☆

344024　Salix rorida Laksch. f. roridaeformis（Nakai）Kimura ex H. Ohashi = Salix rorida Laksch. var. roridaeformis（Nakai）Ohwi ●

344025　Salix rorida Laksch. var. oblanceolata T. Y. Chou et Skvortsov；倒披针叶粉枝柳；Oblanceolate Pruinashoot Willow ●

344026　Salix rorida Laksch. var. oblanceolata T. Y. Chou et Skvortsov = Salix rorida Laksch. ●

344027　Salix rorida Laksch. var. pendula Skvortsov = Salix rorida Laksch. ●

344028　Salix rorida Laksch. var. roridaeformis（Nakai）Ohwi；伪粉枝柳；Fake Pruinashoot Willow ●

344029　Salix roridaeformis Nakai = Salix rorida Laksch. var. roridaeformis（Nakai）Ohwi ●

344030　Salix roridiformis Nakai = Salix rorida Laksch. var. roridaeformis（Nakai）Ohwi ●

344031　Salix rosmarinifolia L.；细叶沼柳（迷迭香叶柳, 西伯利亚沼柳）；Rosemaryleaf Willow, Rosemary-leaved Willow ●

344032　Salix rosmarinifolia L. = Salix elaeagnos Scop. ●☆

344033　Salix rosmarinifolia L. var. brachypoda（Trautv. et C. A. Mey.）Y. L. Chou；沼柳；Short-stalk Rosemaryleaf Willow ●

344034　Salix rosmarinifolia L. var. brachypoda（Trautv. et C. A. Mey.）Y. L. Chou = Salix brachypoda（Trautv. et C. A. Mey.）Kom. ●

344035　Salix rosmarinifolia L. var. gannanensis C. F. Fang；甘南沼柳（甘肃沼柳）；Gannan Willow ●

344036　Salix rosmarinifolia L. var. tungbeiana Y. L. Chou et Skvortsov；东北细叶沼柳；Northeastern Rosemaryleaf Willow ●

344037　Salix rossica Nasarow；俄罗斯柳；Russian Willow ●☆

344038　Salix rossica Nasarow = Salix schwerinii E. L. Wolf ●

344039　Salix rosthornii Seemen ex Diels；南川柳；Nanchuan Willow, Rosthorn Willow ●

344040　Salix rostrata Richardson = Salix bebbiana Sarg. ●

344041　Salix rostrata Richardson var. capreifolia Fernald = Salix bebbiana Sarg. ●

344042　Salix rostrata Richardson var. luxurians Fernald = Salix bebbiana Sarg. ●

344043 Salix rostrata Richardson var. perrostrata（Rydb.）Fernald = Salix bebbiana Sarg. ●

344044 Salix rostrata Richardson var. projecta Fernald = Salix bebbiana Sarg. ●

344045 Salix rotundifolia Trautv. ；圆叶柳；Least Willow, Round-leaf Willow, Round-leaved Willow ●

344046 Salix rubens Schrank；红枝柳（橙枝柳，淡红柳）；Hybrid Crack Willow, Hybrid Crack-willow, Willow ●☆

344047 Salix rubens Schrank 'Basfordiana'；巴斯福特红枝柳（巴斯福特橙枝柳）●☆

344048 Salix rubra Richardson = Salix exigua Nutt. ●☆

344049 Salix rufescens Turcz. ；岩地柳；Silky Rock Sallow ●☆

344050 Salix rupifraga Koidz. ；石隙柳；●☆

344051 Salix rupifraga Koidz. f. eriocarpa（Kimura）Kimura ex H. Ohashi；毛果石隙(绵毛柳)●☆

344052 Salix rupifraga Koidz. var. eriocarpa Kimura = Salix rupifraga Koidz. f. eriocarpa（Kimura）Kimura ex H. Ohashi ☆

344053 Salix sachalinensis F. Schmidt；龙江柳●

344054 Salix sachalinensis F. Schmidt 'Sekka' = Salix udensis Trautv. 'Sekka'●☆

344055 Salix sachalinensis F. Schmidt = Salix udensis Trautv. et C. A. Mey. ●☆

344056 Salix sadleri Syme；扎德尔柳；Sadler's Willow ●☆

344057 Salix safsaf Forssk. = Salix mucronata Thunb. ●☆

344058 Salix safsaf Forssk. ex Trautv. = Salix mucronata Thunb. ●☆

344059 Salix saidaeana Seemen = Salix sieboldiana Blume ●☆

344060 Salix sajanensis Nasarow；萨彦柳●

344061 Salix salviifoliae Brot. var. affinis Bunge；近缘柳●☆

344062 Salix salwinensis Hand. -Mazz. ；对叶柳●

344063 Salix salwinensis Hand. -Mazz. var. longiamentifera C. F. Fang；长穗对叶柳●

344064 Salix salwinensis Hand. -Mazz. var. radinostachya Hand. -Mazz. = Salix salwinensis Hand. -Mazz. ●

344065 Salix saposhnikovii A. K. Skvortsov；灌木柳；Saposhnikov Willow ●

344066 Salix saxatilis Turcz. ex Ledeb. ；岩栖柳●☆

344067 Salix schneideriana K. S. Hao = Salix kouytchensis C. K. Schneid. ●

344068 Salix schneideriana K. S. Hao ex C. F. Fang et A. K. Skvortsov = Salix kouytchensis C. K. Schneid. ●

344069 Salix schugnanica Goerz；阿克苏柳●

344070 Salix schugnanica Goerz = Salix rosmarinifolia L. ●

344071 Salix schwerinii E. L. Wolf；伪蒿柳(蒿柳)；Fake Basket Willow ●☆

344072 Salix sclerophylla Andersson；硬叶柳；Hardleaf Willow, Hard-leaved Willow, Sclerophyllous Willow ●

344073 Salix sclerophylla Andersson var. obtusa（Z. Wang et P. Y. Fu）C. F. Fang；宽苞金背柳；Obtuse Greyleaf Willow, Obtuse Hardleaf Willow ●

344074 Salix sclerophylla Andersson var. tibetica（Goerz）C. F. Fang；小叶硬叶柳；Tibet Hardleaf Willow ●

344075 Salix sclerophylloides Y. L. Chou；近硬叶柳；Hardleaf-like Willow, Sclerophyllous-like Willow ●

344076 Salix scopulicola P. Y. Mao et W. Z. Li；岩壁垫柳●

344077 Salix scouleriana Barra；斯考氏柳；Black Willow, Fire Willow, Nuttall Willow, Scouler Willow ●☆

344078 Salix secta Andersson = Salix oreophila Hook. f. ex Anderson var. secta（Andersson）Andersson ●

344079 Salix secta Hook. f. ex Andersson = Salix oreophila Hook. f. ex Anderson var. secta（Andersson）Andersson ●

344080 Salix seemafinii Rydb. ；西玛柳●☆

344081 Salix semiviminalis E. L. Wolf；细条柳●☆

344082 Salix sendaica Kimura；仙台柳●☆

344083 Salix sensitiva Barratt = Salix discolor Muhl. ●☆

344084 Salix sepulcralis Simonk. ；埋生柳；Weeping Willow ●☆

344085 Salix sericea Marshall；丝柳；Fire Willow, Mountain Willow, Silky Willow ●☆

344086 Salix sericea Marshall f. glabra E. J. Palmer et Steyerm. = Salix sericea Marshall ●☆

344087 Salix sericea Marshall var. subsericea（Andersson）Rydb. = Salix petiolaris Sm. ●☆

344088 Salix sericocarpa Andersson；绢果柳；Sericeousfruit Willow, Sericeousfruited Willow, Silkfruit Willow, Silky-fruited Willow ●

344089 Salix seriocarpa Buser = Salix pellita（Andersson）Andersson ex C. K. Schneid. ●☆

344090 Salix serissifolia Kimura；晚叶柳●☆

344091 Salix serissifolia Kimura = Salix jessoensis Seemen subsp. serissifolia（Kimura）H. Ohashi ●☆

344092 Salix serissifolia Kimura f. pendula Okuhara；垂枝晚叶柳●☆

344093 Salix serissima（L. H. Bailey）Fernald；秋柳；Autumn Willow ●☆

344094 Salix serpyllifolia Scop. ；百里香叶柳；Thyme-leafed Willow ●☆

344095 Salix serpyllum Andersson；多花小垫柳(聂拉木垫柳，珠穆垫柳)；Flowery Willow, Manyflower Willow, Multiflorous Cushion Willow ●

344096 Salix serrulatifolia E. L. Wolf；锯齿柳(齿叶柳)；Serrate-leaved Willow, Serrulateleaf Willow, Serrulate-leaved Willow ●

344097 Salix serrulatifolia E. L. Wolf = Salix tenuijulis Ledeb. ●

344098 Salix serrulatifolia E. L. Wolf var. subintegrifolia Chang Y. Yang；疏齿柳(疏锯齿柳)；Subentire Serralateleaf Willow ●

344099 Salix sessilifolia Nutt. ；西北柳；Northwest Willow, Sandbar Willow, Soft-leaf Willow, Velvet Willow ●☆

344100 Salix sessilifolia Nutt. var. hindsiana（Benth.）Bebb = hindsiana Benth. ●☆

344101 Salix shandanensis C. F. Fang；山丹柳；Shandan Willow ●

344102 Salix shangchengensis B. C. Ding et T. B. Chao；商城柳；Shangcheng Willow ●

344103 Salix shansiensis K. S. Hao = Salix linearistipularis（Franch.）K. S. Hao ●

344104 Salix shihtsuanensis Z. Wang et C. Y. Yu；石泉柳；Shihtsuan Willow, Shiquan Willow ●

344105 Salix shihtsuanensis Z. Wang et C. Y. Yu var. glabrata C. F. Fang et J. Q. Wang；光果石泉柳；Glabrous-fruit Shiquan Willow ●

344106 Salix shihtsuanensis Z. Wang et C. Y. Yu var. globosa C. Y. Yu；球果石泉柳；Globosefruit Shiquan Willow ●

344107 Salix shihtsuanensis Z. Wang et C. Y. Yu var. sessilis C. Y. Yu；无柄石泉柳；Sessile Shiquan Willow ●

344108 Salix shiraii Seemen；日光柳●☆

344109 Salix shiraii Seemen var. kenoensis（Koidz.）Sugim. ；毛野柳●☆

344110 Salix sibirica Pall. = Salix rosmarinifolia L. ●

344111 Salix sibirica Seemen var. brachypoda（Trautv. et C. A. Mey.）Nakai = Salix rosmarinifolia L. var. brachypoda（Trautv. et C. A. Mey.）Y. L. Chou ●

344112 Salix sieboldiana Blume；西氏柳(席氏柳)；Siebold Willow ●☆

344113 Salix sikkimensis Andersson；锡金柳；Sikkim Willow ●

344114 Salix sikkimensis Burkill = Salix delavayana Hand. -Mazz. ●

344115 Salix silesiaca Willd. ；塞利西亚柳；Silesian Willow ●☆

344116 *Salix simulatrix* B. White；无茎柳●☆

344117 *Salix sinica*（K. S. Hao ex C. F. Fang et A. K. Skvortsov）Z. Wang et C. F. Fang；中国黄花柳（黄花儿柳，黄华柳）；China Yellowflower Willow，Chinese Goat Willow ●

344118 *Salix sinica*（K. S. Hao ex C. F. Fang et A. K. Skvortsov）Z. Wang et C. F. Fang var. dentata（K. S. Hao ex C. F. Fang et A. K. Skvortsov）Z. Wang et C. F. Fang；齿叶中国黄花柳（齿叶黄花柳）；Dentate Chinese Goat Willow ●

344119 *Salix sinica*（K. S. Hao ex C. F. Fang et A. K. Skvortsov）Z. Wang et C. F. Fang var. semicannexa（K. S. Hao ex C. F. Fang et A. K. Skvortsov）C. F. Fang et W. D. Liu；关帝柳；Guandishan Chinese Goat Willow ●

344120 *Salix sinica*（K. S. Hao ex C. F. Fang et A. K. Skvortsov）Z. Wang et C. F. Fang var. subsessilis（K. S. Hao ex C. F. Fang et A. K. Skvortsov）G. Zhu；无柄黄花柳●

344121 *Salix sinopurpurea* Z. Wang et Chang Y. Yang；红皮柳（簸箕柳，蒲柳，水杨）；China Purple Willow，Chinese Purple Willow，Red-barked Willow ●

344122 *Salix sitchensis* Sanson ex Bong.；锡特卡柳；Coulter Willow，Silky Willow，Sitka Willow，Velvet Willow ●☆

344123 *Salix sitchensis* Sanson ex Bong. var. pellita（Andersson）Jeps. = Salix pellita（Andersson）Andersson ex C. K. Schneid. ●☆

344124 *Salix siuzevii* Seemen；卷边柳；Coilmargin Willow，Siuzev Willow ●

344125 *Salix skvortzovii* Y. L. Chang et Y. L. Chou；司氏柳；Skvortsov Willow ●

344126 *Salix smithiana* Willd.；史密斯柳（渐尖叶柳）；Long-leaved Sallow，Silky-leaf Osier，Smith Willow，Smith's Willow ●☆

344127 *Salix songarica* Andersson；准噶尔柳；Dzungar Willow，Songar Willow ●

344128 *Salix souliei* Seemen；黄花垫柳；Soulie Willow，Yellowflower Willow ●

344129 *Salix spathulifolia* Seemen；匙叶柳；Spatulate-leaved Willow，Spoonleaf Willow ●

344130 *Salix spathulifolia* Seemen f. lobata C. F. Fang et J. Q. Wang；浅裂柱匙叶柳；Lobed Spatulateleaf Willow ●

344131 *Salix spathulifolia* Seemen f. lobata C. F. Fang et J. Q. Wang = Salix spathulifolia Seemen ●

344132 *Salix speciosa* Hook. et Arn.；丽柳●☆

344133 *Salix sphaeronymphe* Goerz；巴郎柳；Balang Willow ●

344134 *Salix sphaeronymphoides* Y. L. Chou；光果巴郎柳；Balang-like Willow，Brightfruit Balang Willow，Glabrousfruit Willow ●

344135 *Salix spinidens* E. L. Wolf；刺齿柳●☆

344136 *Salix spissa* Andersson = Salix alatavica Kar. et Kir. ex Stschegl. ●

344137 *Salix splendens* Bray = Salix alba L. var. splendens（Bray）Andersson ●

344138 *Salix splendens* Turcz.；光亮柳●☆

344139 *Salix splendida* Nakai = Chosenia arbutifolia（Pall.）A. K. Skvortsov ●◇

344140 *Salix spodiophylla* Hand.-Mazz.；灰叶柳；Greyleaf Willow，Grey-leaved Willow ●

344141 *Salix spodiophylla* Hand.-Mazz. f. liocarpa K. S. Hao ex C. F. Fang et A. K. Skvortsov = Salix spodiophylla Hand.-Mazz. var. leiocarpa（K. S. Hao ex C. F. Fang et A. K. Skvortsov）G. Zhu ●

344142 *Salix spodiophylla* Hand.-Mazz. var. angustifolia C. F. Fang；狭叶灰叶柳；Narrowleaf Greyleaf Willow ●

344143 *Salix spodiophylla* Hand.-Mazz. var. leiocarpa（K. S. Hao ex C. F. Fang et A. K. Skvortsov）G. Zhu；光果灰叶柳（无毛灰叶柳）；

Glabrousfruit Greyleaf Willow ●

344144 *Salix spodiophylla* Hand.-Mazz. var. leiocarpa K. S. Hao = Salix spodiophylla Hand.-Mazz. var. leiocarpa（K. S. Hao ex C. F. Fang et A. K. Skvortsov）G. Zhu ●

344145 *Salix spodiophylla* Hand.-Mazz. var. obtusa Z. Wang et P. Y. Fu = Salix sclerophylla Andersson var. obtusa（Z. Wang et P. Y. Fu）C. F. Fang ●

344146 *Salix squamata* Rydb. = Salix discolor Muhl. ●☆

344147 *Salix squarrosa* C. K. Schneid. = Salix myrtillacea Andersson ●

344148 *Salix starkeana* Willd.；波纹柳；Starke Willow ●

344149 *Salix starkeana* Willd. = Salix taraikensis Kimura ●

344150 *Salix starkeana* Willd. subsp. bebbiana（Sarg.）Youngberg = Salix bebbiana Sarg. ●

344151 *Salix stipulifera* Flod. ex Hayren；长托叶柳；Stipule Willow ●☆

344152 *Salix strobilacea*（E. L. Wolf）Nasarow；球穗柳●☆

344153 *Salix subfragilis* Andersson = Salix nipponica Franch. et Sav. ●

344154 *Salix subfragilis* Andersson = Salix triandra L. subsp. nipponica（Franch. et Sav.）A. K. Skvortsov ●

344155 *Salix subopposita* Miq.；近对生柳●☆

344156 *Salix subpycnostachya* Burkill = Salix myrtillacea Andersson ●

344157 *Salix subpyroliformis* Y. L. Chang et A. K. Skvortsov = Salix pyrolifolia Ledeb. ●

344158 *Salix subsericea*（Andersson）C. K. Schneid. = Salix petiolaris Sm. ●☆

344159 *Salix subserrata* Willd.；小齿柳；Wild Willow ●☆

344160 *Salix subserrata* Willd. var. cyathipoda（Andersson ex A. Rich.）Cufod. = Salix mucronata Thunb. ●☆

344161 *Salix suchowensis* W. C. Cheng ex G. H. Zhu；簸箕柳；Dustpan Willow，Suchow Willow ●

344162 *Salix suishaensis* Hayata = Salix kusanoi（Hayata）C. K. Schneid. et Kimura ●

344163 *Salix sungkianica* Y. L. Chou et Skvortsov；松江柳；Songjiang Willow，Sungjiang Willow ●

344164 *Salix sungkianica* Y. L. Chou et Skvortsov f. brevistachys Y. L. Chou et S. L. Tung；短序松江柳；Short-stachys Songjiang Willow ●

344165 *Salix sushanensis* Hayata = Salix kusanoi（Hayata）C. K. Schneid. et Kimura ●

344166 *Salix syrticola* Fernald = Salix cordata Michx. ●☆

344167 *Salix tagawana* Koidz.；花莲柳（白毛柳，田川氏柳）；Hualian Willow，Tagawa Willow ●

344168 *Salix tagawana* Koidz. = Salix fulvopubescens Hayata var. tagawana（Koidz.）K. C. Yang et T. C. Huang ●

344169 *Salix taipaiensis* Chang Y. Yang；太白柳；Taibai Willow，Taibaishan Willow ●

344170 *Salix taishanensis* Z. Wang et C. F. Fang；泰山柳（太山柳）；Taishan Procumbent Willow，Taishan Willow ●

344171 *Salix taishanensis* Z. Wang et C. F. Fang var. glabra C. F. Fang et W. D. Liu；光子房泰山柳；Glabrous Taishan Willow ●

344172 *Salix taishanensis* Z. Wang et C. F. Fang var. hobeinica C. F. Fang；河北柳；Hebei Willow ●

344173 *Salix taiwanalpina* Kimura；台湾山柳（高山柳，马勃拉斯柳，南湖山柳，台高山柳，台湾高山柳）；Nanhushan Willow，Taiwan Alpine Willow ●

344174 *Salix taiwanalpina* Kimura var. morrisonicola（Kimura）K. C. Yang et T. C. Huang = Salix morrisonicola Kimura ●

344175 *Salix taiwanalpina* Kimura var. takasagoalpina（Koidz.）S. S. Ying = Salix takasagoalpina Koidz. ●

344176 *Salix takasagoalpina* Koidz.；台湾匐柳（高山柳,台湾山柳）；Alpine Stolon Willow,Taiwan Stolon Willow,Taiwan Willow ●

344177 *Salix takasagoalpina* Koidz. = *Salix taiwanalpina* Kimura var. takasagoalpina（Koidz.）S. S. Ying ●

344178 *Salix tangii* K. S. Hao ex C. F. Fang et A. K. Skvortsov；周至柳；Tang Willow,Zhouzhi Willow ●

344179 *Salix tangii* K. S. Hao f. villosa C. F. Fang et J. Q. Wang；缘毛周至柳；Villose Tang Willow ●

344180 *Salix tangii* K. S. Hao var. angustifolia C. Y. Yu；细叶周至柳；Narrowleaf Tang Willow ●

344181 *Salix taoensis* Golz；洮河柳；Taohe Willow ●

344182 *Salix taoensis* Golz var. leiocarpa T. Y. Ding et C. F. Fang；光果洮河柳；Grabrous-fruit Taohe Willow ●

344183 *Salix taoensis* Golz var. pedicellata C. F. Fang et J. Q. Wang；柄果洮河柳（柄洮河柳）；Pedicellaate Taohe Willow ●

344184 *Salix taraikensis* Kimura；谷柳；Taraike Willow ●

344185 *Salix taraikensis* Kimura f. latifolia（Kimura）Kimura = *Salix taraikensis* Kimura ●

344186 *Salix taraikensis* Kimura var. latifolia Kimura；宽叶谷柳；Broadleaf Taraike Willow ●

344187 *Salix taraikensis* Kimura var. latifolia Kimura = *Salix taraikensis* Kimura ●

344188 *Salix taraikensis* Kimura var. oblanceolata Z. Wang et C. F. Fang；倒披针叶谷柳；Oblanceolate Willow ●

344189 *Salix tarbagataica* Chang Y. Yang；塔城柳；Tacheng Willow ●

344190 *Salix taxifolia* Kunth；红豆杉叶柳（紫杉叶柳）；Yew Willow,Yewleaf Willow,Yew-leaf Willow ●☆

344191 *Salix tenela* C. K. Schneid.；光苞柳；Glabrousbract Willow,Glabrous-bracted Willow,Smooth-bracted Willow ●

344192 *Salix tenela* C. K. Schneid. var. trichadenia Hand.-Mazz.；基毛光苞柳●

344193 *Salix tengchongensis* C. F. Fang；腾冲柳；Tengchong Willow ●

344194 *Salix tenuifolia* Turcz. ex Laksch.；细叶柳●☆

344195 *Salix tenuijulis* Ledeb.；细穗柳；Smallkatkin Willow,Tenui-inflorescence Willow,Thin-spiked Willow ●

344196 *Salix tenuijulis* Ledeb. var. alberti（Regel）Poljak. = *Salix tenuijulis* Ledeb. ●

344197 *Salix tetradenia* Hand.-Mazz. = *Salix guebrianthiana* C. K. Schneid. ●

344198 *Salix tetrasperma* Burkill ex Forbes et Hemsl. = *Salix kusanoi*（Hayata）C. K. Schneid. et Kimura ●

344199 *Salix tetrasperma* Roxb.；四子柳；Fourseed Willow,Four-seed Willow ●

344200 *Salix tetrasperma* Roxb. var. kusanoi Hayata = *Salix kusanoi*（Hayata）C. K. Schneid. et Kimura ●

344201 *Salix tetrasperrna* Roxb. = *Salix cavaleriei* H. Lév. ●

344202 *Salix thunbergiana* Blume ex Andersaon = *Salix gracilistyla* Miq. ●

344203 *Salix tianschanica* Regel；天山柳；Tianshan Willow ●

344204 *Salix tibetica* Goerz = *Salix juparica* Goerz var. tibetica（Goerz）C. F. Fang ●

344205 *Salix tibetica* Goerz ex Rehder et Kobuski = *Salix juparica* Goerz var. tibetica（Goerz）C. F. Fang ●

344206 *Salix tontomussirensis* Koidz.；托恩拖姆西尔柳●☆

344207 *Salix torreyana* Barratt = *Salix eriocephala* Michx. ●☆

344208 *Salix torulosa* Trautv.；结节柳●☆

344209 *Salix transarisanensis* Hayata = *Salix fulvopubescens* Hayata var. doii（Hayata）K. C. Yang et T. C. Huang ●

344210 *Salix transarisanensis* Hayata = *Salix fulvopubescens* Hayata ●

344211 *Salix triandra* L.；三蕊柳（白浆柳,剑叶柳,毛柳）；Almond Willow,Almond-leaved Willow,French Willow,Kit Willow,Snake's Skin Willow,Snake's-skin Willow,Threestamen Willow,Three-threaded Willow,Triandrous Willow ●

344212 *Salix triandra* L. subsp. amygdalina（L.）Schübl. et M. Martens；膀胱三蕊柳●☆

344213 *Salix triandra* L. subsp. nipponica（Franch. et Sav.）A. K. Skvortsov；日本三蕊柳（赛三蕊柳）；Japanese Threestamen Willow ●

344214 *Salix triandra* L. subsp. nipponica（Franch. et Sav.）A. K. Skvortsov = *Salix nipponica* Franch. et Sav. ●

344215 *Salix triandra* L. var. concolor Wimm. et Grab. = *Salix triandra* L. ●

344216 *Salix triandra* L. var. glaucophylla Ser. = *Salix triandra* L. subsp. amygdalina（L.）Schübl. et M. Martens ●☆

344217 *Salix triandra* L. var. mengshanensis S. B. Liang = *Salix nipponica* Franch. et Sav. var. mengshanensis（S. B. Liang）G. Zhu

344218 *Salix triandra* L. var. nipponica（Franch. et Sav.）Seemen = *Salix nipponica* Franch. et Sav. ●

344219 *Salix triandra* L. var. nipponica（Franch. et Sav.）Seemen = *Salix triandra* L. subsp. nipponica（Franch. et Sav.）A. K. Skvortsov ●

344220 *Salix triandroides* C. F. Fang；川三蕊柳；Threestamen-like Willow,Triandrous-like Willow,Tristamen Willow ●

344221 *Salix trichocarpa* C. F. Fang；毛果柳；Hairyfruit Willow,Hairy-fruited Willow ●

344222 *Salix trichomicrophylla* Z. Wang et P. Y. Fu = *Salix pilosomicrophylla* Z. Wang et P. Y. Fu ●

344223 *Salix tristis* Aiton = *Salix humilis* Marshall var. tristis（Aiton）Griggs ●☆

344224 *Salix tristis* Aiton var. longifolia Andersson = *Salix humilis* Marshall var. tristis（Aiton）Griggs ●☆

344225 *Salix tristis* Aiton var. microphylla Andersson = *Salix humilis* Marshall var. tristis（Aiton）Griggs ●☆

344226 *Salix tschanbaischanica* Y. L. Chou et Y. L. Chang = *Salix nummularia* Andersson ●

344227 *Salix tschanbaischanica* Y. L. Chou et Y. L. Chang = *Salix polyadenia* Hand.-Mazz. var. tschanbaischanica（Y. L. Chou et Y. L. Chang）Y. L. Chang ●

344228 *Salix tsoongii* W. C. Cheng = *Salix dunnii* C. K. Schneid. var. tsoongii（W. C. Cheng）C. Y. Yu et S. D. Zhao ●

344229 *Salix tundricola* Schljakov；冻原柳●☆

344230 *Salix turanica* Nasarow；吐兰柳（土兰柳,土伦柳）；Tulan Willow ●

344231 *Salix turczaninowii* Laksch. ex Printz；蔓柳；Turczaninov Willow ●

344232 *Salix udensis* Trautv. 'Sekka'；扇尾于登柳（扇尾龙江柳）；Fantail Willow ●☆

344233 *Salix udensis* Trautv. et C. A. Mey.；于登柳●☆

344234 *Salix ulotricha* C. K. Schneid. = *Salix magnifica* Hemsl. var. ulotricha（C. K. Schneid.）N. Chao ●

344235 *Salix urbaniana* Seemen = *Salix cardiophylla* Trautv. et C. A. Mey. var. urbaniana（Seemen）Kudo ●☆

344236 *Salix uva-ursi* Pursh；熊果柳；Bearberry Willow ●☆

344237 *Salix vaccinioides* Hand.-Mazz.；乌饭叶矮柳；Blueberryleaf Willow,Blueberry-leaved Willow ●

344238 *Salix vagans* Hook. f. var. occidentalis Andersson = *Salix bebbiana* Sarg. ●

344239 *Salix vagans* Hook. f. var. rostrata（Richardson）Andersson =

Salix bebbiana Sarg. ●

344240　Salix variegata Franch. ;秋华柳（变色柳，布克柳，秋花柳）；Variegated Willow, Vesture Willow ●

344241　Salix vaudensis Forbes;沃杜柳;Vaudois Sallow ●☆

344242　Salix veroviminalis Nasarow;真青冈柳●☆

344243　Salix vestita Pursh;皱纹柳;Clothed Willow, Rugose Willow, Vesture Willow ●

344244　Salix viminalis L. ;蒿柳（绢柳，柳茅子，柳树蒿，柳树毫，清钢柳）；Augers, Basket Willow, Common Osier, Common Willow, Osier, Osier Willow, Silky Osier, Twig Withy, Twig-withy, Welgers, Widdy, Wilgers, Wilgers Welgers, Withwine ●

344245　Salix viminalis L. var. angustifolia Turcz. ;细叶蒿柳;Narrowleaf Basket Willow ●

344246　Salix viminalis L. var. angustifolia Turcz. = Salix schwerinii E. L. Wolf ●

344247　Salix viminalis L. var. gmelinii（Pall.）Andersson = Salix schwerinii E. L. Wolf ●

344248　Salix viminalis L. var. gmelinii Turcz. = Salix schwerinii E. L. Wolf ●

344249　Salix viminalis L. var. songarica Andersson = Salix turanica Nasarow ●

344250　Salix viridis;绿柳;Bedford Willow ●☆

344251　Salix viridula Andersson ex A. Gray;浅绿柳●☆

344252　Salix vitellina L. ‘Pendula’ = Salix alba L. ‘Tristis’●☆

344253　Salix vitellina L. = Salix alba L. subsp. vitellina（L.）Schübl. et M. Martens ●☆

344254　Salix vitellina L. = Salix alba L. var. vitellina（L.）Stokes ●☆

344255　Salix vulpina Andersson;孤柳●☆

344256　Salix vulpina Andersson subsp. alopochroa（Kimura）H. Ohashi et Yonek. ;东亚孤柳●☆

344257　Salix vulpina Andersson subsp. alopochroa（Kimura）H. Ohashi et Yonek. f. psilostachys（Kimura）H. Ohashi et Yonek. ;光果孤柳●☆

344258　Salix vulpina Andersson var. subalpina Koidz. = Salix vulpina Andersson ●☆

344259　Salix vulpina Andersson var. tomentosa Koidz. ;大毛孤柳●☆

344260　Salix vupinoides Koidz. ;拟孤柳●☆

344261　Salix wallichiana Andersson;皂柳（红心柳，毛狗条，山柳，山杨柳，杨柳树）；Aoap Willow, Wallich Willow ●

344262　Salix wallichiana Andersson f. longistyla C. F. Fang;长柱皂柳；Longstyle Wallich Willow ●

344263　Salix wallichiana Andersson var. grisea Andersson = Salix wallichiana Andersson ●

344264　Salix wallichiana Andersson var. pachyclada（H. Lév. et Vaniot）Z. Wang et C. F. Fang;绒毛皂柳；Thickbranch Willow ●

344265　Salix wangiana K. S. Hao ex C. F. Fang et A. K. Skvortsov;眉柳（杜鹃柳）；Azalealeaf Willow, Brow Willow, Rhododendron-like Willow, Wang Willow ●

344266　Salix wangiana K. S. Hao var. tibetica Z. Wang et C. F. Fang;红柄柳；Tibet Wang Willow ●

344267　Salix wangii Goerz = Salix ernestii C. K. Schneid. ●

344268　Salix warburgii Seemen;水柳（河柳）；Warburg Willow, Water Willow ●

344269　Salix weixiensis Y. L. Chou;维西柳;Weixi Willow ●

344270　Salix weixiensis Y. L. Chou var. eucalyptifolia S. S. Ying = Salix kusanoi（Hayata）C. K. Schneid. et Kimura ●

344271　Salix wenchuanica Goerz = Salix argyrophegga C. K. Schneid. ●

344272　Salix wheeleri（Rowlee）Rydb. = Salix exigua Nutt. subsp. interior（Rowlee）Cronquist ●☆

344273　Salix wilhelmsiana M. Bieb. ;线叶柳（威汉氏柳）；Linearleaf Willow, Wilhelms Willow ●

344274　Salix wilhelmsiana M. Bieb. var. latifolia Chang Y. Yang;宽线叶柳；Broad-leaf Wilhelms Willow ●

344275　Salix wilhelmsiana M. Bieb. var. leiocarpa Chang Y. Yang;光果线叶柳；Smooth-fruit Wilhelms Willow ●

344276　Salix wilmsii Seemen = Salix mucronata Thunb. subsp. woodii（Seemen）Immelman ●

344277　Salix wilsonii Seemen;紫柳（红柳，山杨柳，威氏柳，野杨柳）；E. H. Wilson Willow, Purple Willow, Wilson Willow ●

344278　Salix wolohoensis C. K. Schneid. ;川南柳；S. Sichuan Willow, South Sichuan Willow, South Szechwan Willow ●

344279　Salix woodii Seemen;伍得柳;Wild Willow, Wood's Willow ●

344280　Salix woodii Seemen = Salix mucronata Thunb. subsp. woodii（Seemen）Immelman ●

344281　Salix woodii Seemen var. wilmsii（Seemen）V. Naray. = Salix mucronata Thunb. subsp. woodii（Seemen）Immelman ●

344282　Salix wrightii Andersson = Salix amygdaloides Andersson ●

344283　Salix wuiana K. S. Hao = Salix alfredii Goerz ●

344284　Salix wuiana K. S. Hao ex C. F. Fang et A. K. Skvortsov = Salix alfredii Goerz ●

344285　Salix wuxuhaiensis N. Chao;伍须柳;Wuxuhai Willow ●

344286　Salix xerophila Flod. ;燥柳（崖柳）；Xerophilous Willow ●

344287　Salix xerophila Flod. = Salix bebbiana Sarg. ●

344288　Salix xerophila Flod. f. glabra（Nakai）Kitag. = Salix taraikensis Kimura ●

344289　Salix xerophila Flod. f. ilectica（Y. L. Chou）Y. L. Chou = Salix hsinganica Y. L. Chang et Skvortsov ●

344290　Salix xerophila Flod. f. manshurica（Nakai）Kitag. = Salix floderusii Nakai ●

344291　Salix xerophila Flod. var. ilectica（Y. L. Chou）Y. L. Chou = Salix hsinganica Y. L. Chang et Skvortsov ●

344292　Salix xerophila Nasarow = Salix floderusii Nakai ●

344293　Salix xiaoguangshanica Y. L. Chou et N. Chao;小光山柳；Xiaoguangshan Willow, Xiaoguongshan Willow ●

344294　Salix xizangensis Y. L. Chou;西藏柳;Tibet Willow, Xizang Willow ●

344295　Salix yadongensis N. Chao;亚东毛柳（亚东柳）；Yadong Willow ●

344296　Salix yanbianica C. F. Fang et Chang Y. Yang;延边柳（白河柳）；Baihe Willow, Yanbian Willow ●

344297　Salix yezoalpina Koidz. = Salix nakamurana Koidz. var. yezoalpina（Koidz.）Kimura ●☆

344298　Salix yezoalpina Koidz. f. grandiflora（Nakai）Kimura = Salix nakamurana Koidz. var. yezoalpina（Koidz.）Kimura ●☆

344299　Salix yezoensis（C. K. Schneid.）Kimura = Salix schwerinii E. L. Wolf ●

344300　Salix yoshinoi Koidz. ;吉野氏柳●☆

344301　Salix yuhuangshanensis Z. Wang et C. Y. Yu;玉皇柳；Yuhuangshan Willow ●

344302　Salix yuhuangshanensis Z. Wang et C. Y. Yu var. weiheensis N. Chao;渭柳;Weihe Yuhuangshan Willow ●

344303　Salix yuhuangshanensis Z. Wang et C. Y. Yu var. weiheensis N. Chao = Salix yuhuangshanensis Z. Wang et C. Y. Yu ●

344304　Salix yumenensis H. L. Yang;玉门柳;Yumen Willow ●

344305　Salix yunnanensis H. Lév. = Salix cavaleriei H. Lév. ●

344306　Salix zangica N. Chao;藏柳;Tibet Willow, Zang Willow ●

344307 Salix zayulica Z. Wang et C. F. Fang;察隅矮柳;Chayu Willow, Zayu Willow ●

344308 Salix zhegushanica N. Chao;鹧鸪柳;Zhegushan Willow ●

344309 Salix zhouquensis X. G. Sun;舟曲柳●

344310 Salizaria A. Gray = Salazaria Torr. ●■

344311 Salizaria A. Gray = Scutellaria L. ●■

344312 Salkea Steud. = Derris Lour. (保留属名)●

344313 Salken Adans. (废弃属名) = Derris Lour. (保留属名)●

344314 Salloa Walp. = Cocculus DC. (保留属名)●

344315 Salloa Walp. = Galloa Hassk. ●

344316 Salmalia Schott et Endl. = Bombax L. (保留属名)●

344317 Salmalia malabarica (DC.) Schott et Endl. = Bombax ceiba L. ●

344318 Salmalia malabarica (DC.) Schott et Endl. = Bombax malabaricum DC. ●

344319 Salmasia Bubani = Aira L. (保留属名)■

344320 Salmasia Rchb. = Salmalia Schott et Endl. ●

344321 Salmasia Schreb. = Hirtella L. ●☆

344322 Salmasia Schreb. = Tachibota Aubl. ●☆

344323 Salmea DC. (1813) (保留属名);银钮扣属■☆

344324 Salmea angustifolia Benth. ;窄叶银钮扣■☆

344325 Salmea curviflora R. Br. ;弯花银钮扣■☆

344326 Salmea montana (Britton et S. F. Blake) M. R. Bolick et R. K. Jansen;山地银钮扣■☆

344327 Salmea nitida Sch. Bip. ;光亮银钮扣■☆

344328 Salmea oligocephala Hemsl. ;寡头银钮扣■☆

344329 Salmea pubescens (S. F. Blake) Standl. et Steyerm. ;毛银钮扣■☆

344330 Salmea salicifolia Brongn. ;柳叶银钮扣■☆

344331 Salmeopsis Benth. (1873);拟银钮扣属■☆

344332 Salmeopsis claussenii Benth. ;拟银钮扣■☆

344333 Salmia Cav. (废弃属名) = Salmea DC. (保留属名)■☆

344334 Salmia Cav. (废弃属名) = Sansevieria Thunb. (保留属名)■

344335 Salmia Willd. = Carludovica Ruiz et Pav. ●■

344336 Salmiopuntia Fri ? = Opuntia Mill. ●

344337 Salmonea Vahl = Salomonia Lour. (保留属名)■

344338 Salmonia Scop. = Vochysia Aubl. (保留属名)●☆

344339 Salmonopuntia P. V. Heath = Opuntia Mill. ●

344340 Salmonopuntia P. V. Heath(1999);萨尔蒙掌属●☆

344341 Saloa Stuntz = Blumenbachia Schrad. (保留属名)■☆

344342 Salomonia Fabr. = Polygonatum Mill. ■

344343 Salomonia Heist. ex Fabr. (废弃属名) = Polygonatum Mill. ■

344344 Salomonia Heist. ex Fabr. (废弃属名) = Salomonia Lour. (保留属名)■

344345 Salomonia Lour. (1790) (保留属名);齿果草属(莎萝莽属);Salomonia ■

344346 Salomonia aphylla Griff. = Salomonia elongata (Blume) Kurz ex Koord. ■

344347 Salomonia cantoniensis Lour. ;齿果草(川风,吹云草,公儿草,过路蛇,过山龙,过山蛇,莎萝莽,细黄药,一碗泡,斩蛇剑);Canton Salomonia,Guangzhou Salomonia ■

344348 Salomonia cantoniensis Lour. var. edentula (DC.) C. Y. Wu = Salomonia cantoniensis Lour. var. edentula (DC.) Gagnep. ■

344349 Salomonia cantoniensis Lour. var. edentula (DC.) Gagnep. ;小果齿果草(无齿齿果草,小齿果草,小腻药);Littlefruit Canton Salomonia,Toothless Guangzhou Salomonia ■

344350 Salomonia cavaleriei H. Lév. = Salomonia oblongifolia DC. ■

344351 Salomonia ciliata (L.) DC. ;睫毛齿果草(齿果草,睫毛莎萝莽,椭圆齿果草,椭圆叶齿果草,缘毛齿果草);Ciliate Salomonia,

Oblongleaf Salomonia ■

344352 Salomonia ciliata (L.) DC. = Salomonia oblongifolia DC. ■

344353 Salomonia edentula DC. = Salomonia cantoniensis Lour. var. edentula (DC.) Gagnep. ■

344354 Salomonia elongata (Blume) Kurz ex Koord. = Epirixanthes elongata Blume ■

344355 Salomonia elongata (Blume) S. K. Chen = Salomonia elongata (Blume) Kurz ex Koord. ■

344356 Salomonia martinii H. Lév. = Polygala tatarinowii Regel ■

344357 Salomonia oblongifolia DC. = Salomonia ciliata (L.) DC. ■

344358 Salomonia obovata Wight = Salomonia ciliata (L.) DC. ■

344359 Salomonia parasitica Griff. = Salomonia elongata (Blume) Kurz ex Koord. ■

344360 Salomonia seguinii H. Lév. = Polygala furcata Royle ■

344361 Salomonia sessiliflora D. Don = Salomonia oblongifolia DC. ■

344362 Salomonia stricta Siebold et Zucc. = Salomonia ciliata (L.) DC. ■

344363 Salomonia stricta Siebold et Zucc. = Salomonia oblongifolia DC. ■

344364 Salomonia tenella Hook. f. = Salomonia elongata (Blume) Kurz ex Koord. ■

344365 Salpianthus Bonpl. (1807);沙茉莉属●☆

344366 Salpianthus Humb. et Bonpl. = Salpianthus Bonpl. ●☆

344367 Salpianthus arenarius Bonpl. = Salpianthus arenarius Humb. et Bonpl. ●☆

344368 Salpianthus arenarius Humb. et Bonpl. ;沙茉莉●☆

344369 Salpichroa Miers(1845);鸡蛋茄属;Cock's-eggs ●☆

344370 Salpichroa diffusa Miers;铺散鸡蛋茄●☆

344371 Salpichroa microloba Keel;小裂片鸡蛋茄●☆

344372 Salpichroa origanifolia (Lam.) Baill. ;鸡蛋茄;Cock's-eggs, Lily of the Valley Vine ●☆

344373 Salpichroa rhomboidea (Gillies et Hook.) Miers = Salpichroa origanifolia (Lam.) Baill. ●☆

344374 Salpichroa rhomboidea Miers = Salpichroa origanifolia (Lam.) Baill. ●☆

344375 Salpichroa tristis Miers;暗淡鸡蛋茄●☆

344376 Salpichroa weberbauerii Dammer. ;韦伯鲍尔鸡蛋茄●☆

344377 Salpichroma Miers = Salpichroa Miers ●☆

344378 Salpiglossidaceae (Benth.) Hutch. = Solanaceae Juss. (保留科名)●■

344379 Salpiglossidaceae Hutch. ;智利喇叭花科(美人襟科)●■☆

344380 Salpiglossidaceae Hutch. = Solanaceae Juss. (保留科名)●■

344381 Salpiglossis Ruiz et Pav. (1794);智利喇叭花属(蛾蝶花属,猴面花属,美人襟属);Salpiglossis,Tube-Tongue ■☆

344382 Salpiglossis atropurpurea Graham;紫色智利喇叭花●☆

344383 Salpiglossis sinuata Hook. et Arn. ex Miers;智利喇叭花(朝颜烟草,猴面花,美人襟);Painted Tongue, Painted-tongue, Scalloped Salpiglossis, Silky Flossy, Velvet Flower, Velvet Trumpet Flower ■☆

344384 Salpiglossis sinuata Hook. et Arn. ex Miers 'Bolero';包列罗舞●☆

344385 Salpiglossis sinuata Hook. et Arn. ex Miers 'Splash';炫耀●☆

344386 Salpiglottis Hort. ex C. Koch = Salpiglossis Ruiz et Pav. ●☆

344387 Salpiglottis Hort. ex K. Koch = Salpiglossis Ruiz et Pav. ■☆

344388 Salpinctes Woodson(1931);委内瑞拉夹竹桃属●☆

344389 Salpinctium T. J. Edwards = Asystasia Blume ●■

344390 Salpinctium hirsutum T. J. Edwards;毛委内瑞拉夹竹桃●☆

344391 Salpinctium natalense (C. B. Clarke) T. J. Edwards;纳塔尔夹竹桃●☆

344392 Salpinctium stenosiphon (C. B. Clarke) T. J. Edwards;细管委

内瑞拉夹竹桃●☆

344393 Salpinga Mart. = Salpinga Mart. ex DC. ●☆

344394 Salpinga Mart. ex DC. (1828);号角野牡丹属●☆

344395 Salpinga ciliata Pilg. ;睫毛号角野牡丹●☆

344396 Salpinga glandulosa (Gleason) Wurdack;多腺号角野牡丹●☆

344397 Salpinga longifolia Triana;长叶号角野牡丹●☆

344398 Salpinga monostachya Pittier;单穗号角野牡丹●☆

344399 Salpingacanthus S. Moore = Ruellia L. ■●

344400 Salpingantha Hort. ex Lem. = Salpixantha Hook. ■☆

344401 Salpingantha Lem. = Salpixantha Hook. ■☆

344402 Salpingia (Torr. et A. Gray) Raim. = Calylophus Spach ■☆

344403 Salpingia (Torr. et A. Gray) Raim. = Galpinsia Britton ■☆

344404 Salpingia (Torr. et A. Gray) Raim. = Oenothera L. ●■

344405 Salpingia Raim. = Calylophus Spach ■☆

344406 Salpingia Raim. = Oenothera L. ●■

344407 Salpingoglottis C. Koch = Salpiglossis Ruiz et Pav. ■☆

344408 Salpingoglottis K. Koch = Salpiglossis Ruiz et Pav. ■☆

344409 Salpingolobivia Y. Ito = Echinopsis Zucc. ●

344410 Salpingostylis Small = Calydorea Herb. ■☆

344411 Salpingostylis Small = Ixia L. (保留属名)■☆

344412 Salpingostylis coelestina (W. Bartram) Small = Calydorea coelestina (W. Bartram) Goldblatt et Henrich ■☆

344413 Salpinxantha Hook. (1845);日本爵床属■☆

344414 Salpinxantha Hook. = Salpingantha Hort. ex Lem. ■☆

344415 Salpinxantha Urb. = Salpixantha Hook. ■☆

344416 Salpistele Dressier(1979);鱼柱兰属■☆

344417 Salpistele brunnea Dressler;鱼柱兰■☆

344418 Salpistele lutea Dressler;黄鱼柱兰■☆

344419 Salpixantha Hook. (1845);黄鱼爵床属■☆

344420 Salpixantha Hook. = Salpingantha Lem. ■☆

344421 Salpixantha coccinea Hook. ;黄鱼爵床■☆

344422 Salpixanthus Lindl. = Salpingantha Lem. ■☆

344423 Salpixanthus Lindl. = Salpinxantha Hook. ■☆

344424 Salsa Feuillee ex Ruiz et Pav. = Herreria Ruiz et Pav. ■☆

344425 Salsola L. (1753);猪毛菜属;Barilla, Glasswort, Russian Thistle, Russianthistle, Russian-thistle, Saltwort, Salt-wort ●■

344426 Salsola abrotanoides Bunge;蒿叶猪毛菜(灰叶猪毛菜,蒿叶猪毛菜);Sagebrushleaf Russianthistle ●

344427 Salsola acanthoclada Botsch. ;枝刺猪毛菜●☆

344428 Salsola acocksii Botsch. ;阿氏猪毛菜■☆

344429 Salsola adisca Botsch. ;无盘猪毛菜■☆

344430 Salsola adversariifolia Botsch. ;对叶猪毛菜■☆

344431 Salsola aellenii Botsch. ;埃伦猪毛菜■☆

344432 Salsola aethiopica Botsch. = Salsola spinescens Moq. ■☆

344433 Salsola affinis C. A. Mey. = Salsola affinis C. A. Mey. ex Schrenk ■

344434 Salsola affinis C. A. Mey. ex Schrenk;紫翅猪毛菜;Purplewinged Russianthistle ■

344435 Salsola africana (Brenan) Botsch. ;非洲猪毛菜■☆

344436 Salsola albida Botsch. ;白猪毛菜■☆

344437 Salsola albisepala Aellen;白瓣猪毛菜■☆

344438 Salsola algeriensis Botsch. ;阿尔及利亚猪毛菜■☆

344439 Salsola alopecuroides Delile = Agathophora alopecuroides (Delile) Bunge ■☆

344440 Salsola altissima L. = Suaeda altissima (L.) Pall. ■

344441 Salsola androssovii Litv. ;安氏猪毛菜■☆

344442 Salsola angolensis Botsch. ;安哥拉猪毛菜■☆

344443 Salsola aperta Paulsen;露果猪毛菜;Wingless Russianthistle ■

344444 Salsola aphylla L. f. ;无叶猪毛菜●■☆

344445 Salsola aphylla L. f. var. canescens Fenzl ex Drège = Salsola aphylla L. f. ●■☆

344446 Salsola aphylla L. f. var. virescens Fenzl ex Drège = Salsola rabieana I. Verd. ■☆

344447 Salsola aptera Hand. -Mazz. = Halogeton arachnoides Moq. ■

344448 Salsola aralensis Iljin;阿拉尔猪毛菜■☆

344449 Salsola araneosa Botsch. ;纳米比亚猪毛菜■☆

344450 Salsola arborea C. A. Sm. ex Aellen;北方猪毛菜■☆

344451 Salsola arborescens L. f. = Salsola arbuscula Pall. ●

344452 Salsola arbuscula Pall. ;木本猪毛菜(白木猪毛菜,灌木猪毛菜,灌木状猪毛菜);Woody Russianthistle ●

344453 Salsola arbusculaeformis Drobow;白枝猪毛菜;Whitebranch Russianthistle, White-branched Russianthistle ●

344454 Salsola arenaria Maerkl. = Kochia laniflora (S. G. Gmel.) Borbás ■

344455 Salsola armata C. A. Sm. ex Aellen;微花猪毛菜■☆

344456 Salsola articulata Cav. = Hammada scoparia (Pomel) Iljin ●☆

344457 Salsola articulata Forssk. = Anabasis articulata (Forssk.) Moq. ●☆

344458 Salsola asparagoides Miq. = Suaeda glauca (Bunge) Bunge ■

344459 Salsola atrata Botsch. ;黑猪毛菜■☆

344460 Salsola atriplicifolia Spreng. = Cycloloma atriplicifolium (Spreng.) J. M. Coult. ☆

344461 Salsola aucheri Bunge;奥氏猪毛菜■☆

344462 Salsola auriculata C. A. Sm. = Salsola namibica Botsch. ■☆

344463 Salsola australis R. Br. ;俄罗斯南方猪毛菜;Russian Thistle ■☆

344464 Salsola australis R. Br. = Salsola kali L. ■☆

344465 Salsola australis R. Br. = Salsola tragus L. ■

344466 Salsola baranovii Iljin;巴拉猪毛菜■☆

344467 Salsola barbata Aellen;髯毛猪毛菜■☆

344468 Salsola baryosma (Schult. ex Roem. et Schult.) Dandy = Salsola imbricata Forssk. ■

344469 Salsola baryosma (Schult. ex Roem. et Schult.) Dandy subsp. gaetula (Maire) Freitag = Salsola imbricata Forssk. subsp. gaetula (Maire) Boulos ■☆

344470 Salsola beticolor Iljin;赭紫猪毛菜;Beticolor Russianthistle ■

344471 Salsola bottae (Jaub. et Spach) Boiss. = Halothamnus bottae Jaub. et Spach ●☆

344472 Salsola bottae (Jaub. et Spach) Boiss. var. farinulenta Chiov. = Halothamnus somalensis (N. E. Br.) Botsch. ●☆

344473 Salsola bottae (Jaub. et Spach) Boiss. var. faurotii Franch. = Halothamnus somalensis (N. E. Br.) Botsch. ●☆

344474 Salsola brachiata (Pall.) Botsch. ;散枝猪毛菜(散枝梯翅蓬);Scatterbranch Russianthistle ■

344475 Salsola brevifolia Desf. ;短叶猪毛菜■☆

344476 Salsola bucharica Iljin;布哈尔猪毛菜■☆

344477 Salsola bullata Fenzl ex Drège = Salsola luederitzensis Botsch. ■☆

344478 Salsola caffra Sparrm. = Salsola aphylla L. f. ●■☆

344479 Salsola calluna Drège ex C. H. Wright;黑色猪毛菜■☆

344480 Salsola camphorosmoides Desf. = Noaea mucronata (Forssk.) Asch. et Schweinf. ■☆

344481 Salsola campyloptera Botsch. ;弯翅猪毛菜■☆

344482 Salsola cana K. Koch;灰色猪毛菜■☆

344483 Salsola candida Fenzl ex Drège = Salsola araneosa Botsch. ■☆

344484 Salsola canescens Boiss. ;灰白猪毛菜■☆

344485 Salsola capensis Botsch. ;好望角猪毛菜■☆

344486 Salsola carinata C. A. Mey. ;龙骨状猪毛菜■☆

344487　Salsola caroxylon Moq. = Salsola aphylla L. f. ●■☆

344488　Salsola cauliflora Botsch. ;茎花猪毛菜■☆

344489　Salsola centralasiatica Iljin = Salsola pestifer A. Nelson ■

344490　Salsola ceresica Botsch. ;塞里斯猪毛菜■☆

344491　Salsola chellalensis Botsch. ;舍拉勒猪毛菜■☆

344492　Salsola chinensis Gand. = Salsola collina Pall. ■

344493　Salsola chinghaiensis A. J. Li;青海猪毛菜;Chinghai Russianthistle,Qinghai Russianthistle ■

344494　Salsola chiwensis Popov;奇瓦猪毛菜■☆

344495　Salsola claviflora Pall. = Salsola foliosa (L.) Schrad. ex Schult. ■

344496　Salsola collina Pall. ;猪毛菜(刺蓬,牛尾巴,三叉明棵,沙蓬,山叉明棵,扎蓬蒿,扎蓬棵,猪毛草,猪毛蒿,猪毛缨);Common Russianthistle, Katune, Russian Thistle, Slender Russian Thistle, Slender Russian-thistle,Tumbleweed ■

344497　Salsola columnaris Botsch. ;圆柱猪毛菜■☆

344498　Salsola congesta N. E. Br. = Salsola spinescens Moq. ☆

344499　Salsola crassa Popov;粗枝猪毛菜■

344500　Salsola crassa Popov subsp. turcomanica (Litv.) Freitag;土库曼粗枝猪毛菜■☆

344501　Salsola cruciata Batt. et Trab. ;十字形猪毛菜■☆

344502　Salsola cryptoptera Aellen;隐翅猪毛菜■☆

344503　Salsola cyclophylla Baker;圆叶猪毛菜■☆

344504　Salsola cycloptera Stapf = Lagenantha cycloptera (Stapf) M. G. Gilbert et Friis ■☆

344505　Salsola cyrenaica (Maire et Weiller) Brullo;昔兰尼猪毛菜■☆

344506　Salsola daghestanica (Turcz.) Lipsky;达赫斯坦猪毛菜■☆

344507　Salsola dasyantha Pall. = Kochia laniflora (S. G. Gmel.) Borbás ■

344508　Salsola decussata C. A. Sm. ex Botsch. ;对生猪毛菜■

344509　Salsola delileana Botsch. = Salsola vermiculata L. ●☆

344510　Salsola dendroides Pall. ;树状猪毛菜■☆

344511　Salsola dendroides Pall. var. africana Brenan = Salsola africana (Brenan) Botsch. ■☆

344512　Salsola denudata Botsch. ;裸露猪毛菜■☆

344513　Salsola deschaseauxiana Litard. et Maire = Salsola longifolia Forssk. ■☆

344514　Salsola deserticola Iljin;荒漠猪毛菜■☆

344515　Salsola dichracantha Kitag. = Salsola ruthenica Iljin ■

344516　Salsola dichracantha Kitag. = Salsola tragus L. ■

344517　Salsola diffusa Thunb. = Bassia diffusa (Thunb.) Kuntze ■☆

344518　Salsola dinteri Botsch. ;丁特猪毛菜■☆

344519　Salsola dioica (Nutt.) Spreng. = Atriplex suckleyi (Torr.) Rydb. ■☆

344520　Salsola diplantha Botsch. ;平花猪毛菜■☆

344521　Salsola divaricata (Moq.) Moq. = Salsola capensis Botsch. ■☆

344522　Salsola dolichostigma Botsch. ;长柱头猪毛菜■☆

344523　Salsola drummondii Ulbr. ;德拉蒙德猪毛菜■☆

344524　Salsola dshungarica Iljin;准噶尔猪毛菜;Dzungar Russianthistle ●■

344525　Salsola engleri Ulbr. = Salsola armata C. A. Sm. ex Aellen ■☆

344526　Salsola ericoides M. Bieb. ;石南状猪毛菜●☆

344527　Salsola eriosepala C. A. Sm. = Salsola scopiformis Botsch. ■☆

344528　Salsola esterhuyseniae Botsch. ;埃斯特猪毛菜■☆

344529　Salsola etoshensis Botsch. ;埃托沙猪毛菜■☆

344530　Salsola exalata Botsch. ;南非猪毛菜■☆

344531　Salsola ferganica Drobow;费尔干猪毛菜;Fergan Russianthistle ■●

344532　Salsola flavescens Cav. ;浅黄猪毛菜■☆

344533　Salsola flexuosa C. A. Sm. = Salsola angolensis Botsch. ■☆

344534　Salsola foetida Delile ex Spreng. = Salsola imbricata Forssk. ■

344535　Salsola foetida Delile ex Spreng. var. gaetula Maire = Salsola imbricata Forssk. subsp. gaetula (Maire) Boulos ■☆

344536　Salsola foetida Delile var. glabrescens Maire = Salsola flavescens Cav. ■☆

344537　Salsola foetida Delile var. scopiformis Maire = Salsola gaetula (Maire) Botsch. ■☆

344538　Salsola foliosa (L.) Schrad. ex Schult. ;浆果猪毛菜;Berry Russianthistle ■

344539　Salsola forcipitata Iljin;钳猪毛菜■☆

344540　Salsola forskaolii Schweinf. = Salsola spinescens Moq. ■☆

344541　Salsola frankenioides (Caball.) Botsch. ;瓣鳞花猪毛菜■☆

344542　Salsola fuliginosa C. A. Sm. = Salsola gemmifera Botsch. ■☆

344543　Salsola gaetula (Maire) Botsch. = Salsola imbricata Forssk. subsp. gaetula (Maire) Boulos ■☆

344544　Salsola garubica Botsch. ;加鲁布猪毛菜■☆

344545　Salsola geminiflora Fenzl ex C. H. Wright;对花猪毛菜■☆

344546　Salsola gemmascens Pall. ;芽猪毛菜■☆

344547　Salsola gemmascens Pall. subsp. maroccana Botsch. ;摩洛哥猪毛菜■☆

344548　Salsola gemmascens Pall. subsp. passerina (Bunge) Botsch. = Salsola passerina Bunge ●

344549　Salsola gemmifera Botsch. ;具芽猪毛菜■☆

344550　Salsola genistoides Poir. ;金雀猪毛菜■☆

344551　Salsola giessii Botsch. ;吉斯猪毛菜■☆

344552　Salsola glabra Botsch. ;光滑猪毛菜■☆

344553　Salsola glabrescens Burtt Davy;渐光猪毛菜■☆

344554　Salsola glauca M. Bieb. = Aellenia glauca (M. Bieb.) Aellen ●

344555　Salsola glauca M. Bieb. = Halothamnus glaucus (M. Bieb.) Botsch. ●

344556　Salsola globulifera Fenzl = Salsola atrata Botsch. ■☆

344557　Salsola glomerata (Maire) Brullo;团集猪毛菜■☆

344558　Salsola gobicola Iljin = Salsola pellucida Litv. ■

344559　Salsola gobicola Litv. = Salsola pellucida Litv. ■

344560　Salsola griffithii (Bunge) Freitag et Khani;格氏猪毛菜■☆

344561　Salsola gymnomaschala Maire;裸窝猪毛菜■☆

344562　Salsola gypsacea Botsch. ;喜钙猪毛菜■☆

344563　Salsola henriciae I. Verd. ;昂里克猪毛菜■☆

344564　Salsola heptapotamica Iljin;钝叶猪毛菜;Obtuseleaf Russianthistle ■

344565　Salsola hispanica Botsch. = Salsola vermiculata L. ●☆

344566　Salsola hispidula Bunge;细毛猪毛菜■☆

344567　Salsola hottentottica Botsch. ;霍屯督猪毛菜■☆

344568　Salsola humifusa A. Brückn. ;平伏猪毛菜■☆

344569　Salsola hyssopifolia Pall. = Bassia hyssopifolia (Pall.) Kuntze ■

344570　Salsola iberica (Sennen et Pau) Botsch. ex De Moor = Salsola tragus L. ■

344571　Salsola iberica Sennen et Pau;伊比利亚猪毛菜;Russian Thistle,Saltwort,Tumbleweed ■☆

344572　Salsola iberica Sennen et Pau = Salsola kali L. ■☆

344573　Salsola ikonnikovii Iljin;蒙古猪毛菜(展苞猪毛菜);Mongol Russianthistle, Mongolian Russianthistle ■

344574　Salsola iliensis Lipsky;伊犁猪毛菜■☆

344575　Salsola imbricata Forssk. ; 密枝猪毛菜; Densebranched Russianthistle ■

344576　Salsola imbricata Forssk. subsp. gaetula (Maire) Boulos;盖图拉猪毛菜■☆

344577　Salsola imbricata Forssk. var. hirtisepala Freitag;毛萼猪毛菜■☆

344578　Salsola incanescens C. A. Mey. ;灰毛猪毛菜■☆

344579　Salsola inermis Forssk. ;无刺猪毛菜■☆

344580　Salsola intramongolica H. C. Fu et Z. Y. Chu;红翅猪毛菜;Redwing Russianthistle ■

344581　Salsola intricata Iljin;缠结猪毛菜■☆

344582　Salsola jacquemontii Moq. ;雅克蒙猪毛菜■☆

344583　Salsola junatovii Botsch. ;天山猪毛菜;Tianshan Russianthistle ●

344584　Salsola kalaharica Botsch. ;卡拉哈利猪毛菜■☆

344585　Salsola kali L. ;普通猪毛菜;Barilla Plant, Buck Bush, Buckbush, Common Russian Thistle, Common Saltwort, Glasswort, Kali, Kelp, Prickly Glasswort, Prickly Saltwort, Russian Thistle, Russian Tumbleweed, Saltwort, Sea Grape, Sea Thongs, Sea Wrack, Soapweed, Soda Plant, Sowd-wort, Tumble Weed, Tumbleweed ■☆

344586　Salsola kali L. = Salsola iberica Sennen et Pau ■☆

344587　Salsola kali L. = Salsola ruthenica Iljin ■

344588　Salsola kali L. subsp. austroafricana Aellen = Salsola kali L. ■☆

344589　Salsola kali L. subsp. pontica (Pall.) S. L. Mosyakin;俄罗斯猪毛菜;Russian Thistle ■☆

344590　Salsola kali L. subsp. ruthenica (Iljin) Soó = Salsola ruthenica Iljin ■

344591　Salsola kali L. subsp. ruthenica Soó = Salsola tragus L. ■

344592　Salsola kali L. subsp. tragus (L.) Celak. = Salsola tragus L. ■

344593　Salsola kali L. var. angustifolia Fenzl = Salsola kali L. ■☆

344594　Salsola kali L. var. angustifolia Fenzl = Salsola tragus L. ■

344595　Salsola kali L. var. glabra Forssk. = Salsola kali L. ■☆

344596　Salsola kali L. var. hirta Ten. = Salsola kali L. ■☆

344597　Salsola kali L. var. pontica Pall. = Salsola kali L. subsp. pontica (Pall.) S. L. Mosyakin ■☆

344598　Salsola kali L. var. praecox Litv. = Salsola praecox (Litv.) Iljin ■

344599　Salsola kali L. var. pseudotragus Beck = Salsola tragus L. ■

344600　Salsola kali L. var. tenuifolia Tausch = Salsola iberica Sennen et Pau ■☆

344601　Salsola kali L. var. tenuifolia Tausch = Salsola tragus L. ■

344602　Salsola kali L. var. tenuifolia Tausch ex Moq. = Salsola tragus L. ■

344603　Salsola kali L. var. tragus (L.) Moq. = Salsola kali L. ■☆

344604　Salsola kali L. var. tragus (L.) Moq. = Salsola tragus L. ■

344605　Salsola kochii Guss. ex Tod. = Suaeda pruinosa Lange var. kochii (Tod.) Maire et Weiller ■☆

344606　Salsola komarovii Iljin;无翅猪毛菜;Komalov Russianthistle ■

344607　Salsola korshinskyi Drobow;褐翅猪毛菜;Korshinsky Russianthistle ■

344608　Salsola lachnophylla Iljin;白叶猪毛菜■☆

344609　Salsola lanata Pall. ;短柱猪毛菜(梯翅蓬);Lanate Russianthistle ■

344610　Salsola laniflora S. G. Gmel. = Kochia laniflora (S. G. Gmel.) Borbás ■

344611　Salsola laricifolia Turcz. ex Litv. ;落叶松叶猪毛菜(松叶猪毛菜);Larchleaf Russianthistle, Larch-leaved Russianthistle ●

344612　Salsola laricina Pall. ;落叶松猪毛菜■☆

344613　Salsola libyca Botsch. ;利比亚猪毛菜■☆

344614　Salsola linearis Elliott = Suaeda linearis (Elliott) Moq. ■☆

344615　Salsola longifolia Forssk. ;长叶猪毛菜■☆

344616　Salsola longifolia Forssk. = Darniella longifolia (Forssk.) Brullo ■☆

344617　Salsola longifolia Forssk. var. verticillata (Schousb.) Ball = Salsola verticillata Schousb. ■☆

344618　Salsola longistylosa Iljin;土耳其斯坦猪毛菜■☆

344619　Salsola lueдеritzensis Botsch. ;吕德里茨猪毛菜■☆

344620　Salsola macera Litv. ;瘦弱猪毛菜■☆

344621　Salsola mairei Botsch. ;迈雷猪毛菜■☆

344622　Salsola makranica Freitag;莫克兰猪毛菜■☆

344623　Salsola maracandica Iljin;马拉坎达猪毛菜■☆

344624　Salsola marginata Botsch. ;具边猪毛菜■☆

344625　Salsola melanantha Botsch. ;黑花猪毛菜■☆

344626　Salsola merxmuelleri Aellen;梅尔猪毛菜■☆

344627　Salsola micranthera Botsch. ;小药猪毛菜;Littleanther Russianthistle ■

344628　Salsola microphylla Cav. = Salsola vermiculata L. ●☆

344629　Salsola microtricha Botsch. ;小毛猪毛菜■☆

344630　Salsola minkvitziae Korovin;敏克猪毛菜■☆

344631　Salsola minutiflora C. A. Sm. = Salsola armata C. A. Sm. ex Aellen ■☆

344632　Salsola minutifolia Botsch. ;微叶猪毛菜■☆

344633　Salsola mirabilis Botsch. ;奇异猪毛菜■☆

344634　Salsola mollis Desf. = Suaeda vermiculata Forssk. ex J. F. Gmel. ■☆

344635　Salsola monoptera Bunge;单翅猪毛菜(刺蓬,沙蓬);Monopterous Russianthistle, Monowing Russianthistle ■

344636　Salsola montana Litv. ;山地猪毛菜■☆

344637　Salsola mucronata Forssk. = Noaea mucronata (Forssk.) Asch. et Schweinf. ■☆

344638　Salsola muricata L. = Bassia muricata (L.) Asch. ■☆

344639　Salsola mutica C. A. Mey. ;无尖猪毛菜■☆

344640　Salsola namaqualandica Botsch. ;纳马夸兰猪毛菜■☆

344641　Salsola namibica Botsch. ;纳米布猪毛菜■☆

344642　Salsola nepalensis Grubov;尼泊尔猪毛菜;Nepal Russianthistle ■

344643　Salsola nigrescens I. Verd. = Salsola calluna Drège ex C. H. Wright ■☆

344644　Salsola nitraria Pall. ;喜硝猪毛菜;Sodium Russianthistle ■

344645　Salsola nodulosa (Moq.) Iljin;多节猪毛菜■☆

344646　Salsola okaukuejensis Botsch. ;奥考奎约猪毛菜■☆

344647　Salsola olgae Iljin;奥尔嘎猪毛菜■☆

344648　Salsola omaruruensis Botsch. ;奥马鲁鲁猪毛菜■☆

344649　Salsola oppositiflora Pall. = Girgensohnia oppositiflora (Pall.) Fenzl ●■

344650　Salsola oppositifolia Desf. = Salsola longifolia Forssk. ■☆

344651　Salsola oppositifolia Desf. var. verticillata (Schousb.) Moq. = Salsola verticillata Schousb. ■☆

344652　Salsola orientalis S. G. Gmel. ;东方猪毛菜(直立猪毛菜);Oriental Russianthistle ●

344653　Salsola pachyphylla Botsch. ;延叶猪毛菜●

344654　Salsola paletzkiana Litv. ;帕来猪毛菜■☆

344655　Salsola parviflora Botsch. ;小花猪毛菜■☆

344656　Salsola passerina Bunge;珍珠猪毛菜(雀猪毛菜,珍珠柴);Pearl Russianthistle ●

344657　Salsola patentipilosa Botsch. ;展毛猪毛菜■☆

344658　Salsola paulsenii Litv. ;长刺猪毛菜(蒙古沙蓬,蒙古猪毛菜);Barbwire Russian Thistle, Barbwire Russian-thistle, Longspine Russian Thistle, Paulse Russian Thistle ■

344659　Salsola paulsenii Litv. subsp. praecox (Litv.) Rilke = Salsola praecox (Litv.) Iljin ■

344660　Salsola paulsenii Litv. var. potaninii (Iljin) Grubov = Salsola potaninii Iljin ■

344661　Salsola pearsonii Botsch. ;皮尔逊猪毛菜■☆

344662　Salsola pellucida Litv. ;薄翅猪毛菜(戈壁沙蓬,戈壁猪毛菜);Thinwinged Russianthistle ■

344663　Salsola pentandra Botsch. = Salsola tetrandra Forssk. ■☆

344664　Salsola pestifer A. Nelson = Salsola kali L. subsp. ruthenica (Iljin) So6 ■

344665　Salsola pestifer A. Nelson = Salsola ruthenica Iljin ■

344666　Salsola pestifer A. Nelson = Salsola tragus L. ■

344667　Salsola phillipsii Botsch. ;菲利猪毛菜■☆

344668　Salsola physophora (Pall.) Schrad. = Suaeda physophora Pall. ●

344669　Salsola physophora Schrad. = Suaeda physophora Pall. ●

344670　Salsola pillansii Botsch. ;皮朗斯猪毛菜■☆

344671　Salsola platyphylla Michx. = Cycloloma atriplicifolium (Spreng.) J. M. Coult. ■☆

344672　Salsola postii Eig = Agathophora postii (Eig) Botsch. ■☆

344673　Salsola potaninii Iljin = Salsola paulsenii Litv. ■

344674　Salsola praecox (Litv.) Iljin;早熟猪毛菜;Early Russianthistle ■

344675　Salsola praecox Litv. = Salsola praecox (Litv.) Iljin ■

344676　Salsola procera Botsch. ;高大猪毛菜●■

344677　Salsola prostrata L. = Kochia prostrata (L.) Schrad. ●

344678　Salsola ptiloptera Botsch. ;毛翅猪毛菜■☆

344679　Salsola pycnophylla Brenan;密叶猪毛菜■☆

344680　Salsola rabieana I. Verd. ;密拉比猪毛菜■☆

344681　Salsola regelii (Bunge) Litv. ex Popov = Iljinia regelii (Bunge) Korovin ●

344682　Salsola richteri Kar. ex Moq. ;鹿尾草(猪毛菜)■☆

344683　Salsola rigida Pall. = Salsola orientalis S. G. Gmel. ●

344684　Salsola robinsonii Botsch. ;鲁滨逊猪毛菜■☆

344685　Salsola roborowskii Iljin = Salsola affinis C. A. Mey. ex Schrenk ■

344686　Salsola rosacea L. ;蔷薇猪毛菜;Rose Russianthistle ■

344687　Salsola roshevitzii Iljin;罗塞猪毛菜■☆

344688　Salsola rubescens Franch. ;变红猪毛菜■☆

344689　Salsola ruschii Aellen;鲁施猪毛菜■☆

344690　Salsola ruthenica Iljin = Salsola kali L. subsp. ruthenica (Iljin) So6 ■

344691　Salsola ruthenica Iljin = Salsola kali L. ■☆

344692　Salsola ruthenica Iljin = Salsola pestifer A. Nelson ■

344693　Salsola ruthenica Iljin = Salsola tragus L. ■

344694　Salsola ruthenica Iljin var. filifolia A. J. Li;细叶猪毛菜;Filiformis-leaved Russianthistle ■

344695　Salsola ruthenica Iljin var. filifolia A. J. Li = Salsola tragus L. ■

344696　Salsola sabaphylla C. A. Mey. ;近无叶猪毛菜■☆

344697　Salsola salsa (L.) L. = Suaeda salsa (L.) Pall. ●

344698　Salsola sativa L. = Halogeton sativus (L.) Moq. ●

344699　Salsola schweinfurthii Solms;施韦猪毛菜■☆

344700　Salsola schweinfurthii Solms = Darniella schweinfurthii (Solms) Brullo ■☆

344701　Salsola scopiformis Botsch. ;帚状猪毛菜■☆

344702　Salsola sedoides Pall. = Bassia sedoides (Pall.) Asch. ■

344703　Salsola semhahensis Vierh. = Lagenantha cycloptera (Stapf) M. G. Gilbert et Friis ■☆

344704　Salsola seminuda Botsch. ;半裸猪毛菜■☆

344705　Salsola sericata Botsch. ;绢毛猪毛菜■☆

344706　Salsola sericea Aiton = Bassia diffusa (Thunb.) Kuntze ■☆

344707　Salsola sieberi C. Presl subsp. cyrenaica (Maire et Weiller) Brullo et Furnari = Salsola cyrenaica (Maire et Weiller) Brullo ■☆

344708　Salsola sieberi C. Presl subsp. deschaseauxiana (Litard. et Maire) Sauvage = Salsola verticillata Schousb. ■☆

344709　Salsola sieberi C. Presl var. cyrenaica Maire et Weiller = Salsola cyrenaica (Maire et Weiller) Brullo ■☆

344710　Salsola sieberi C. Presl var. deschaseauxiana (Litard. et Maire) Maire = Salsola verticillata Schousb. ■☆

344711　Salsola sieberi C. Presl var. glomerata Maire = Salsola glomerata (Maire) Brullo ■☆

344712　Salsola sieberi C. Presl var. gymnomaschala (Maire) Maire = Salsola gymnomaschala Maire ■☆

344713　Salsola sieberi C. Presl var. gymnomaschala (Maire) Sauvage = Salsola gymnomaschala Maire ■☆

344714　Salsola sieberi C. Presl var. vesceritensis Chevall. = Salsola zygophylla Batt. ■☆

344715　Salsola sieberi C. Presl var. zygophylla (Batt.) Maire = Salsola zygophylla Batt. ■☆

344716　Salsola sinkiangensis A. J. Li;新疆猪毛菜;Sinkiang Russianthistle, Xinjiang Russianthistle ■

344717　Salsola smithii Botsch. ;史密斯猪毛菜■☆

344718　Salsola soda L. ;苏打猪毛菜;Barilla-plant, Oppositeleaf Russian Thistle, Saltwort ■

344719　Salsola soda L. = Salsola komarovii Iljin ■

344720　Salsola somalensis N. E. Br. = Halothamnus somalensis (N. E. Br.) Botsch. ●☆

344721　Salsola spicata Pall. = Halothamnus glaucus (M. Bieb.) Botsch. ●

344722　Salsola spicata Willd. = Suaeda spicata (Willd.) Moq. ■☆

344723　Salsola spinescens Moq. ;小刺猪毛菜■☆

344724　Salsola splendens Pourr. = Suaeda splendens (Pourr.) Gren. et Godr. ■☆

344725　Salsola squarrosula Botsch. ;粗鳞猪毛菜■☆

344726　Salsola stellulata Korovin;星状猪毛菜■☆

344727　Salsola subcrassa Popov = Salsola subcrassa Popov ex Iljin ■

344728　Salsola subcrassa Popov ex Iljin;亚粗枝猪毛菜;Stoutbranch Russianthistle ■

344729　Salsola subglabra Botsch. ;近光猪毛菜■☆

344730　Salsola subsericea C. A. Sm. = Salsola dinteri Botsch. ■☆

344731　Salsola sukaczevii (Botsch.) A. J. Li;长柱猪毛菜;Longstyle Russianthistle ■

344732　Salsola swakopmundi Botsch. ;斯瓦科普蒙德猪毛菜■☆

344733　Salsola takhtadshjanii Iljin;塔赫猪毛菜■☆

344734　Salsola tamamschjanae Iljin;塔麻姆猪毛菜■☆

344735　Salsola tamariscina Pall. ;柽柳叶猪毛菜;Tamarisk-leaf Russianthistle ■

344736　Salsola tetragona Delile = Salsola tetrandra Forssk. ■☆

344737　Salsola tetramera Botsch. ;四数猪毛菜■☆

344738　Salsola tetrandra Forssk. ;四蕊猪毛菜■☆

344739　Salsola tetrandra Forssk. subsp. occidentalis Botsch. ;西方猪毛菜■☆

344740　Salsola tetrandra Forssk. var. glabrescens Sauvage = Salsola tetrandra Forssk. ■☆

344741　Salsola tetrandra Forssk. var. puberula Le Houér. = Salsola tetrandra Forssk. ■☆

344742　Salsola tetrandra Forssk. var. pubescens Le Houér. = Salsola tetrandra Forssk. ■☆

344743　Salsola tragus L. ;刺沙蓬(刺蓬,刺沙蓬猪毛菜,大翅猪毛菜,俄罗斯多刺猪毛菜,风滚草,沙蓬,苏联猪毛菜,细叶猪毛菜,扎蓬棵,猪毛菜);Prickly Russian Thistle, Russian Thistle, Russian Tumbleweed, Russianthistle, Spineless Saltwort, Tumbleweed ■

344744　Salsola tragus L. = Salsola kali L. ■☆

344745　Salsola tragus L. subsp. iberica Sennen et Pau = Salsola tragus

L. ■

344746　Salsola tragus L. subsp. pontica（Pall.）Rilke ＝ Salsola kali L. subsp. pontica（Pall.）S. L. Mosyakin ■☆

344747　Salsola transhyrcanica Iljin;外吉尔康猪毛菜■☆

344748　Salsola transoxana Iljin;外阿穆达尔猪毛菜■☆

344749　Salsola tuberculata（Fenzl ex Moq.）Schinz;多疣猪毛菜■☆

344750　Salsola tuberculata（Fenzl ex Moq.）Schinz var. flavovirens Fenzl ＝ Salsola gemmifera Botsch. ■☆

344751　Salsola tuberculata（Fenzl ex Moq.）Schinz var. tomentosa Aellen ＝ Salsola tuberculatiformis Botsch. ■☆

344752　Salsola tuberculatiformis Botsch. ;瘤状猪毛菜■☆

344753　Salsola tunetana Brullo;图内特猪毛菜■☆

344754　Salsola turcomanica Litv. ;土库曼猪毛菜■☆

344755　Salsola turgaica Iljin;图尔嘎猪毛菜■☆

344756　Salsola turkestanica Litv. ;突厥斯坦猪毛菜■☆

344757　Salsola verdoorniae Toelken;韦尔猪毛菜■☆

344758　Salsola vermiculata L. ;灌木猪毛菜; Damascus Saltwort, Mediterranean Saltwort, Shrubby Russian Thistle ●☆

344759　Salsola vermiculata L. subsp. frankenioides Caball. ＝ Salsola frankenioides（Caball.）Botsch. ■☆

344760　Salsola vermiculata L. var. albescens Maire ＝ Salsola vermiculata L. ●☆

344761　Salsola vermiculata L. var. brevifolia（Desf.）Maire et Weiller ＝ Salsola brevifolia Desf. ■☆

344762　Salsola vermiculata L. var. flavescens（Cav.）Moq. ＝ Salsola flavescens Cav. ■☆

344763　Salsola vermiculata L. var. frankenioides（Caball.）Maire ＝ Salsola frankenioides（Caball.）Botsch. ■☆

344764　Salsola vermiculata L. var. glabrescens Moq. ＝ Salsola vermiculata L. ●☆

344765　Salsola vermiculata L. var. microphylla（Cav.）Moq. ＝ Salsola vermiculata L. ●☆

344766　Salsola vermiculata L. var. pseudopapillosa Caball. ＝ Salsola vermiculata L. ●☆

344767　Salsola vermiculata L. var. pubescens Moq. ＝ Salsola flavescens Cav. ■☆

344768　Salsola vermiculata L. var. spinescens（Moq.）Maire et Weiller ＝ Salsola acanthoclada Botsch. ●☆

344769　Salsola vermiculata L. var. villosa（Delile）Moq. ＝ Salsola vermiculata L. ●☆

344770　Salsola vermiformis C. A. Sm. ＝ Salsola gemmifera Botsch. ■☆

344771　Salsola verticillata Schousb. ;轮生猪毛菜■☆

344772　Salsola villosa Delile ＝ Salsola vermiculata L. var. villosa（Delile）Moq. ●☆

344773　Salsola villosa Schult. ;长柔毛猪毛菜■☆

344774　Salsola volkensii Schweinf. et Asch. ;福尔猪毛菜■☆

344775　Salsola vvedenskyi Iljin et Popov;韦德猪毛菜■☆

344776　Salsola webbii Moq. ;韦布猪毛菜■☆

344777　Salsola zaidamica Iljin;柴达木猪毛菜; Chaidamu Russianthistle,Zaidam Russianthistle ■

344778　Salsola zeyheri（Moq.）Bunge;泽赫猪毛菜■☆

344779　Salsola zygophylla Batt. ;对称叶猪毛菜■☆

344780　Salsola zygophylla Batt. var. vesceritensis L. Chevall. ＝ Salsola cruciata Batt. et Trab. ■☆

344781　Salsolaceae Menge ＝ Salsolaceae Moq. ●■

344782　Salsolaceae Moq. ;猪毛菜科●■

344783　Salsolaceae Moq. ＝ Amaranthaceae Juss.（保留科名）●■

344784　Salsolaceae Moq. ＝ Chenopodiaceae Vent.（保留科名）●■

344785　Saltera Bullock（1958）;长丝管萼木属●☆

344786　Saltera sarcocolla（L.）Bullock;长丝管萼木;Sarcocolla Tree ●☆

344787　Saltia R. Br. ＝ Cometes L. ■☆

344788　Saltia R. Br. ex Moq.（1849）;毛苋木属●☆

344789　Saltia papposa（Forssk.）Moq. ;毛苋木●☆

344790　Saltugilia（V. E. Grant）L. A. Johnson ＝ Gilia Ruiz et Pav. ■●☆

344791　Saltugilia（V. E. Grant）L. A. Johnson（2000）;美澳吉莉花属■●☆

344792　Saltzwedelia P. Gaertn. , B. Mey. et Scherb. ＝ Chamaespartium Adans. ●

344793　Saltzwedelia P. Gaertn. ,B. Mey. et Scherb. ＝ Genistella Moench ●

344794　Salutiaea Colla ＝ Achimenes Pers.（保留属名）■☆

344795　Salutiea Griseb. ＝ Salutiaea Colla ■☆

344796　Salutiea Griseb. ex Pfeiff. ＝ Salutiaea Colla ■☆

344797　Salvadora Garcin ex L. ＝ Salvadora L. ●

344798　Salvadora L.（1753）;牙刷树属（萨瓦杜属）;Salvadora ●

344799　Salvadora angustifolia Turrill;狭叶牙刷树（狭叶萨瓦杜）●☆

344800　Salvadora angustifolia Turrill var. australis（Schweick.）I. Verd. ＝ Salvadora australis Schweick. ●☆

344801　Salvadora australis Schweick. ;澳洲牙刷树●☆

344802　Salvadora cyclophylla Chiov. ＝ Salvadora persica L. var. cyclophylla（Chiov.）Cufod. ●☆

344803　Salvadora oleoides Decne. ;南亚牙刷树（南亚萨瓦杜）●☆

344804　Salvadora persica L. ;牙刷树;Mustard Tree, Peelu Extract, Salt Bush,Toothbrush Tree ●☆

344805　Salvadora persica L. var. angustifolia Verdc. ;窄叶牙刷树●☆

344806　Salvadora persica L. var. crassifolia Verdc. ;厚叶牙刷树●☆

344807　Salvadora persica L. var. cyclophylla（Chiov.）Cufod. ;圆叶牙刷树●☆

344808　Salvadora persica L. var. parviflora Verdc. ;小花牙刷树●☆

344809　Salvadora persica L. var. pubescens Brenan;毛牙刷树●☆

344810　Salvadora persica L. var. wightiana（Planch. ex Thwaites）Verdc. ＝ Salvadora persica L. ●☆

344811　Salvadora persica Wall. ＝ Cansjera rheedii J. F. Gmel. ●

344812　Salvadora persica Wall. ＝ Salvadora persica L. ●☆

344813　Salvadora persica Wall. var. wightiana（Planch. ex Thwaites）Verdc. ＝ Salvadora persica L. ●☆

344814　Salvadora stocksii Wight ＝ Salvadora oleoides Decne. ●☆

344815　Salvadora wightiana Planch. ex Thwaites ＝ Salvadora persica L. ●☆

344816　Salvadoraceae Lindl.（1836）（保留科名）;牙刷树科（刺茉莉科）;Salvadora Family ●

344817　Salvadoropsis H. Perrier（1944）;拟牙刷树属●☆

344818　Salvadoropsis arenicola H. Perrier;拟牙刷树●☆

344819　Salvertia A. St. -Hil.（1820）;巴西囊萼花属●☆

344820　Salvertia convallariodora A. St. -Hil. ;巴西囊萼花●☆

344821　Salvia L.（1753）;鼠尾草属;Chia Seeds, Clary, Sage, Salvia ●■

344822　Salvia × jamensis J. Compton;墨西哥秋鼠尾草;Autumn Sage ■☆

344823　Salvia × sakuensis Naruh. et Hihara;佐久鼠尾草■☆

344824　Salvia × superba Vilm. ;华美鼠尾草;Hybrid Sage, Salvia ■☆

344825　Salvia abyssinica Jacq. ＝ Salvia nilotica Juss. ex Jacq. ■☆

344826　Salvia abyssinica L. f. ＝ Salvia merjamie Forssk. ●☆

344827　Salvia abyssinica R. Br. ＝ Meriandra bengalensis（J. König ex Roxb.）Benth. ■☆

344828　Salvia adenostachya Juz. ;腺穗鼠尾草■☆

344829　Salvia adiantifolia E. Peter;铁线蕨叶鼠尾草■☆

344830　Salvia adiantifolia Stibal ＝ Salvia adiantifolia E. Peter ■☆

344831　Salvia adoxoides C. Y. Wu;五福花鼠尾草;Adoxalike Sage,

Muskroot-like Sage ■

344832　Salvia aegyptiaca L.；埃及鼠尾草；Aegypt Sage ■☆

344833　Salvia aequidens Borsch.；等齿鼠尾草■☆

344834　Salvia aerea H. Lév.；橙色鼠尾草（大叶丹参，红丹参，红秦艽，马蹄叶红仙茅，铜色鼠尾，紫丹参）；Orange Sage, Orangecolour Sage ■

344835　Salvia aerea H. Lév. = Salvia brevilabra Franch. ■

344836　Salvia aethiopis L.；埃塞俄比亚鼠尾草（非洲鼠尾草）；Aethiopia Sage, Meditteranean Sage, Woolly Clary ■☆

344837　Salvia africana L. = Salvia africana-caerulea L. ●☆

344838　Salvia africana-caerulea L.；非洲鼠尾草；Aromatic Sage, Purple Sage, Wild Sage ●☆

344839　Salvia africana-lutea L.；非洲黄鼠尾草；Beach Sage ■☆

344840　Salvia alatipetiolata Y. Z. Sun；翅柄鼠尾草；Wingedpetiole Sage, Wingstalk Sage ■

344841　Salvia albicaulis Benth.；白茎鼠尾草■☆

344842　Salvia albicaulis Benth. var. dregeana（Benth.）V. Naray. = Salvia albicaulis Benth. ☆

344843　Salvia alexandri Pobed.；阿赖鼠尾草■☆

344844　Salvia algeriensis Desf.；阿尔及利亚鼠尾草■☆

344845　Salvia algeriensis Desf. var. mariae（Sennen）Maire et Sennen = Salvia algeriensis Desf. ■☆

344846　Salvia amasiaca Freyn et Bornm.；阿马斯鼠尾草■☆

344847　Salvia ambigua Salisb. = Salvia merjamie Forssk. ☆

344848　Salvia andiantifolia E. Peter；铁线鼠尾草；Maidenhairleaf Sage, Wire Sage ■

344849　Salvia andreji Pobed.；安氏鼠尾草■☆

344850　Salvia anhweiensis Migo = Salvia chienii Stibal ■

344851　Salvia anomala Vaniot = Salvia miltiorrhiza Bunge f. alba C. Y. Wu et H. W. Li ■

344852　Salvia anomala Vaniot = Salvia miltiorrhiza Bunge ■

344853　Salvia apiana Jeps.；加州白鼠尾草（阿平鼠尾草）；Bee Sage, California White Sage ●☆

344854　Salvia appendiculata Stibal；附片鼠尾草；Appendiculate Sage ■

344855　Salvia argentea L.；银叶鼠尾草；Selver Clary, Silver Sage ■☆

344856　Salvia argentea L. subsp. patula（Desf.）Maire；开展银叶鼠尾草■☆

344857　Salvia argentea L. var. aurasiaca（Pomel）Maire = Salvia argentea L. ■☆

344858　Salvia argentea L. var. fontanesiana Maire = Salvia argentea L. subsp. patula（Desf.）Maire ■☆

344859　Salvia argentea L. var. mesatlantica Maire = Salvia argentea L. ■☆

344860　Salvia argentea L. var. patula（Desf.）Maire = Salvia argentea L. ■☆

344861　Salvia argentea L. var. pomelii Maire = Salvia argentea L. ■☆

344862　Salvia arisanensis Hayata = Salvia hayatana Makino ex Hayata ■

344863　Salvia armeniaca（Bordz.）Grossh.；亚美尼亚鼠尾草■☆

344864　Salvia aspera M. Martens et Galeotti；粗糙鼠尾草■☆

344865　Salvia asperata Falc. ex Benth.；微糙鼠尾草；Rough Sage ■☆

344866　Salvia atropurpurea C. Y. Wu；暗紫鼠尾草（暗紫鼠尾）；Darkpurple Sage ■

344867　Salvia atrorubra C. Y. Wu；暗红鼠尾草（红花鼠尾）；Darkred Sage ■

344868　Salvia aucheri Boiss. = Scutellaria orientalis L. subsp. demnatensis Batt. ■☆

344869　Salvia aucheri Boiss. subsp. blancoana Webb et Heldr. = Salvia lavandulifolia Vahl subsp. blancoana（Webb et Heldr.）Rosua et Blanca ●☆

344870　Salvia aucheri Boiss. var. amethystea Emb. et Maire = Salvia lavandulifolia Vahl subsp. amethystea（Emb. et Maire）Rosua et Blanca ■☆

344871　Salvia aucheri Boiss. var. aurasiaca Maire = Salvia lavandulifolia Vahl ■☆

344872　Salvia aucheri Boiss. var. claryi Faure et Maire = Salvia lavandulifolia Vahl ■☆

344873　Salvia aucheri Boiss. var. maurorum（Ball）Maire = Salvia lavandulifolia Vahl subsp. maurorum（Ball）Rosua et Blanca ●☆

344874　Salvia aucheri Boiss. var. oranensis Maire = Salvia lavandulifolia Vahl ■☆

344875　Salvia aucheri Boiss. var. reboudiana Maire = Salvia lavandulifolia Vahl ■☆

344876　Salvia aucheri Boiss. var. tananica Maire = Salvia lavandulifolia Vahl ■☆

344877　Salvia aurasiaca Pomel = Salvia argentea L. subsp. patula（Desf.）Maire ■☆

344878　Salvia aurea L.；海滨鼠尾草；Beach Sage, Beach Salvia, Brown Salvia ●☆

344879　Salvia aurea L. ‘Kirstenbosch’；凯尔斯特波奇海滨鼠尾草●☆

344880　Salvia aurea L. = Salvia africana-lutea L. ■☆

344881　Salvia aurita L. f.；耳状鼠尾草■☆

344882　Salvia aurita L. f. var. galpinii（V. Naray.）Hedge；盖尔鼠尾草■☆

344883　Salvia austriaca Jacq.；南方鼠尾草■☆

344884　Salvia azurea Lam. = Salvia azurea Michx. ex Lam. ■☆

344885　Salvia azurea Michx. ex Lam.；蓝花鼠尾草；Azure Blue Sage, Blue Sage, Blue Salvia, Texas Sage ■☆

344886　Salvia azurea Michx. ex Lam. subsp. pitcheri（Torr. ex Benth.）Epling = Salvia azurea Michx. ex Lam. var. grandiflora Benth. ■☆

344887　Salvia azurea Michx. ex Lam. var. grandiflora Benth.；大蓝花鼠尾草；Azure Blue Sage, Blue Sage, Pitcher Sage ■☆

344888　Salvia baimaensis S. W. Su et Z. A. Shen；白马鼠尾草；Baima Sage ■

344889　Salvia balansae Noë；巴兰萨鼠尾草■☆

344890　Salvia baldshuanica Lipsky；巴尔德鼠尾草■☆

344891　Salvia ballotiflora Benth.；宽萼苏鼠尾草■☆

344892　Salvia barbata Lam. = Salvia africana-caerulea L. ●☆

344893　Salvia bariensis Thulin；巴里鼠尾草●☆

344894　Salvia barrelieri Etl.；巴雷鼠尾草■☆

344895　Salvia barrelieri Etl. subsp. pseudobicolor（Batt. et Pit.）Maire；假二色鼠尾草●☆

344896　Salvia barrelieri Etl. var. dichroa（Hook. f.）Maire = Salvia barrelieri Etl. ☆

344897　Salvia barrelieri Etl. var. pallida Maire = Salvia barrelieri Etl. ■☆

344898　Salvia barrelieri Etl. var. pluripartita（Pau）Maire = Salvia barrelieri Etl. ■☆

344899　Salvia beckeri Trautv.；巴凯尔鼠尾草■☆

344900　Salvia benecincta W. W. Sm. = Salvia kiaometiensis H. Lév. ■

344901　Salvia bengalensis J. König ex Roxb. = Meriandra bengalensis（J. König ex Roxb.）Benth. ■☆

344902　Salvia betonicoides H. Lév. = Salvia cavaleriei H. Lév. ■

344903　Salvia bicolor Lam. = Salvia barrelieri Etl. ■☆

344904　Salvia bifidocalyx C. Y. Wu et Y. C. Huang；开萼鼠尾草（开萼鼠尾）；Bifidcalyx Sage, Opencalyx Sage ■

344905　Salvia blancoana Webb et Heldr. = Salvia lavandulifolia Vahl subsp. blancoana（Webb et Heldr.）Rosua et Blanca ●☆

344906　Salvia blancoana Webb et Heldr. ex Walp. ;直布罗陀鼠尾草●☆

344907　Salvia blepharophylla Brandegee ex Epling;睫叶鼠尾草; English Sage ■☆

344908　Salvia blinii H. Lév. = Salvia brevilabra Franch. ■

344909　Salvia bodinieri Vaniot = Salvia yunnanensis C. H. Wright ■

344910　Salvia bowleyana Dunn;南丹参(八莲麻,奔马草,赤参,赤丹参,丹参,红根,红萝卜,木羊乳,七里蕉,七里麻,紫丹参,紫根); Bowley Sage ■

344911　Salvia bowleyana Dunn var. subbipinnata C. Y. Wu;二回羽裂南丹参(近二回羽裂南丹参,羽裂南丹参);Pinnatifid Sage ■

344912　Salvia brachiata Roxb. = Salvia plebeia R. Br. ■

344913　Salvia brachyantha (Bordz.) Pobed. ;短花鼠尾草■☆

344914　Salvia brachyloma Stibal;短冠鼠尾草(短冠鼠尾);Shortpappo Sage ■

344915　Salvia breviconnectivata Y. Z. Sun ex C. Y. Wu;短隔鼠尾草(短隔鼠尾);Shortconnective Sage ■

344916　Salvia brevilabra Franch. ;短唇鼠尾草;Shortlip Sage ■

344917　Salvia broussonetii Benth. ;布鲁索内鼠尾草●☆

344918　Salvia bulleyana Diels;戟叶鼠尾草(糙叶鼠尾草,戟叶鼠尾);Hastateleaf Sage ■

344919　Salvia burchellii N. E. Br. ;伯切尔鼠尾草●☆

344920　Salvia burchellii N. E. Br. = Salvia namaensis Schinz ■☆

344921　Salvia burchellii N. E. Br. var. hispidula V. Naray. = Salvia namaensis Schinz ■☆

344922　Salvia cabiosifolia Lam. ;山萝卜叶鼠尾草■☆

344923　Salvia cacaliifolia Benth. ;蓝鼠尾草;Blue Vine Sage ■☆

344924　Salvia calthifolla H. Lév. = Salvia mairei H. Lév. ■

344925　Salvia calycina Sibth. et Sm. ;大萼鼠尾草■☆

344926　Salvia campanulata Wall. ex Benth. ;钟萼鼠尾草;Campanulate Sage ■

344927　Salvia campanulata Wall. ex Benth. = Salvia hylocharis Diels ■

344928　Salvia campanulata Wall. ex Benth. var. codonantha (E. Peter) E. Peter;平截钟萼鼠尾草(黄花紫丹参,截萼鼠尾草); Truncatesepal Bellsepal Sage ■

344929　Salvia campanulata Wall. ex Benth. var. fissa Stibal;裂钟萼鼠尾草(裂萼鼠尾草);Lobedsepal Bellsepal Sage ■

344930　Salvia campanulata Wall. ex Benth. var. fissa Stibal = Salvia campanulata Wall. ex Benth. ■

344931　Salvia campanulata Wall. ex Benth. var. hirtella Stibal;微硬毛鼠尾草;Hairy Bellsepal Sage ■

344932　Salvia campanulata Wall. ex Benth. var. hirtella Stibal = Salvia campanulata Wall. ex Benth. ■

344933　Salvia campanulata Wall. ex Benth. var. nepalensis ? = Salvia campanulata Wall. ex Benth. ■

344934　Salvia campanulata Wall. ex Benth. var. piliniphylla ? = Salvia campanulata Wall. ex Benth. ■

344935　Salvia campylodonta Botsch. ;弯齿鼠尾草■☆

344936　Salvia canariensis L. ;加那利鼠尾草;Canary Island Sage ●☆

344937　Salvia candelabrum Boiss. subsp. maurorum Ball. = Salvia lavandulifolia Vahl subsp. maurorum (Ball) Rosua et Blanca ●☆

344938　Salvia canescens C. A. Mey. ;灰鼠尾草■☆

344939　Salvia carduacea Benth. ;飞廉鼠尾草;Thistle Sage ■☆

344940　Salvia carnosa Douglas ex Benth. ;紫鼠尾草■☆

344941　Salvia castanea Diels;栗色鼠尾草(栗色鼠尾);Chestnut Sage, Nutbrwon Sage ■

344942　Salvia castanea Diels f. glabrescens Stibal;光叶栗色鼠尾草; Glabrous Chestnut Sage ■

344943　Salvia castanea Diels f. pubescens Stibal;柔毛栗色鼠尾草; Pubescens Chestnut Sage ■

344944　Salvia castanea Diels f. tomentosa Stibal;绒毛栗色鼠尾草; Tomentose Chestnut Sage ■

344945　Salvia cavaleriei H. Lév. ;贵州鼠尾草(反背红,贵州鼠尾,红青菜,红青草,气草,血盆草,叶下红,朱砂草);Guizhou Sage, Kweichow Sage ■

344946　Salvia cavaleriei H. Lév. var. erythrophylla (Hemsl.) Stibal;紫背贵州鼠尾草(女菀,紫背鼠尾草);Purpleback Guizhou Sage ■

344947　Salvia cavaleriei H. Lév. var. simplicifolia E. Peter;血盆草(单叶波罗子,单叶鼠尾草,单叶血盆草,翻背红,反背红,红薄洛,红肺筋,红青菜,红青叶,红五达,红五匹,罗汉草,破罗子,破锣子,破落子,铺地虎,气喘药,鼠雀菜,退节草,小丹参,野丹参,叶下红,朱砂草,紫背丹参);Bloodbasin Sage, Unileaf Sage ■

344948　Salvia cavaleriei H. Lév. var. simplicifolia Stibal = Salvia cavaleriei H. Lév. var. simplicifolia E. Peter ■

344949　Salvia ceratophylla L. ;角叶鼠尾草■☆

344950　Salvia chamelaeagnea P. J. Bergius;南非鼠尾草;Germander Sage, Mexican Blue Sage ●☆

344951　Salvia charbonnelii H. Lév. = Salvia miltiorhiza Bunge var. charbonnelii (H. Lév.) C. Y. Wu ■

344952　Salvia chiapensis Brandegee;恰帕斯鼠尾草;Chiapas Sage ■☆

344953　Salvia chienii E. Peter;黄山鼠尾草;Huangshan Sage ■

344954　Salvia chienii E. Peter wuyuania Y. Z. Sun;婺源鼠尾草; Wuyuan Sage ■

344955　Salvia chienii Stibal = Salvia chienii E. Peter ■

344956　Salvia chienii Stibal var. wuyuania Y. Z. Sun = Salvia chienii E. Peter wuyuania Y. Z. Sun ■

344957　Salvia chinensis Benth. ;华鼠尾草(半支莲,大发汗,黑面风,华鼠尾,活血草,石打穿,石大川,石见穿,乌沙草,小丹参,小红参,野沙参,月下红,紫参,紫丹花);China Sage, Chinese Sage ■

344958　Salvia chinensis Benth. f. alatopinnata Matsum. et Kudo = Salvia japonica Thunb. ■

344959　Salvia chingii C. Y. Wu;秦氏鼠尾草;Ching's Sage ■

344960　Salvia chingii C. Y. Wu = Salvia flava Forrest ex Diels ■

344961　Salvia chloroleuca Rech. f. et Aellen;绿白鼠尾草■☆

344962　Salvia chlorophylla Briq. = Salvia stenophylla Burch. ex Benth. ■☆

344963　Salvia chudaei Batt. et Trab. ;朱丹鼠尾草■☆

344964　Salvia chudaei Batt. et Trab. var. lanuginosa Maire = Salvia chudaei Batt. et Trab. ■☆

344965　Salvia chudaei Batt. et Trab. var. tefedestica Maire = Salvia chudaei Batt. et Trab. ■☆

344966　Salvia chudaei Batt. et Trab. var. tibestiensis (A. Chev.) Maire = Salvia chudaei Batt. et Trab. ■☆

344967　Salvia chunganensis C. Y. Wu et Y. C. Huang;崇安鼠尾草; Chong' an Sage ■

344968　Salvia cinnabarina Mart. et Aellen;朱红鼠尾草■☆

344969　Salvia clandestina L. ;隐匿鼠尾草■☆

344970　Salvia clandestina L. = Salvia verbenaca L. ■☆

344971　Salvia clandestina L. var. angustifolia Benth. = Salvia verbenaca L. ■☆

344972　Salvia cleistogama de Bary et Paul = Salvia verbenaca L. ■☆

344973　Salvia clevelandii Greene;克利夫兰鼠尾草;Blue Ball Sage, California Blue Sage, Cleveland Sage, Jim Sage ●☆

344974　Salvia clevelandii Greene 'Winifred Gilman';威尼夫尔德・吉尔曼克利夫兰鼠尾草●☆

344975　Salvia coahuilensis Fernald;瓦湖岛鼠尾草;Coahuila Sage ■☆

344976　Salvia coccinea Buc'hoz ex Etl. = Salvia coccinea L. ■

344977　Salvia coccinea Etl. = Salvia coccinea L. ■

344978　Salvia coccinea Etl. var. pseudococcinea（Jacq.）A. Gray ＝ Salvia coccinea Etl. ■

344979　Salvia coccinea Juss. ex Murray = Salvia coccinea L. ■

344980　Salvia coccinea L.；红花鼠尾草（红花紫参，红鼠尾草，小红花，一串红，朱唇鼠尾草）；Bloody Sage, Cherry Red Sage, Herb Robert, Indian Fire, Red Sage, Redlip Sage, Scarlet Sage, South American Sage, Texas Sage, Tropical Sage ■

344981　Salvia coccinea L. var. pseudococcinea（Jacq.）A. Gray ＝ Salvia coccinea L. ■

344982　Salvia codonantha E. Peter = Salvia campanulata Wall. ex Benth. var. codonantha（E. Peter）E. Peter ■

344983　Salvia columbariae Benth.；哥伦比亚鼠尾草（美国鼠尾草）；Chia, Chia Sage, Chia Seeds ■☆

344984　Salvia compar Trautv.；伴侣鼠尾草■☆

344985　Salvia confertiflora Pohl；密花鼠尾草；Sabra Spike Sage ●☆

344986　Salvia congesta A. Rich. = Salvia merjamie Forssk. ●☆

344987　Salvia cooperi V. Naray. = Salvia repens Burch. ex Benth. ■☆

344988　Salvia crispula Benth. = Salvia dentata Aiton ■☆

344989　Salvia cryptantha Montbret et Aucher ex Benth.；隐花鼠尾草■☆

344990　Salvia cryptoclada Baker；隐枝鼠尾草●☆

344991　Salvia cyclostegia Stibal；圆苞鼠尾草；Roundbract Sage ■

344992　Salvia cyclostegia Stibal var. purpurascens C. Y. Wu；紫花圆苞鼠尾草；Purpleflower Roundbract Sage ■

344993　Salvia cynica Dunn；犬形鼠尾草（山藿香）；Doglike Sage ■

344994　Salvia cypria Unger et Kotschy；塞浦路斯鼠尾草■☆

344995　Salvia dabieshanensis J. Q. He；大别山鼠尾草（大别山丹参）；Dabieshan Sage ■

344996　Salvia daghestanica Sosn.；达赫斯坦鼠尾草■☆

344997　Salvia delavayi H. Lév. = Salvia cavaleriei H. Lév. var. simplicifolia Stibal ■

344998　Salvia dentata Aiton；尖齿鼠尾草■☆

344999　Salvia deserta Schangin；新疆鼠尾草；Desert Sage ■

345000　Salvia deserti Decne.；荒漠鼠尾草■☆

345001　Salvia dianthera Roth = Meriandra bengalensis（J. König ex Roxb.）Benth. ■☆

345002　Salvia digitaloides Diels；毛地黄鼠尾草（白背丹参，白元参，白云参，丹参，毛地黄鼠尾，银紫丹参，玉名喇叭）；Foxglove-like Sage ■

345003　Salvia digitaloides Diels var. glabrescens Stibal；无毛毛地黄鼠尾草（无毛鼠尾草）；Glabrous Foxglove-like Sage ■

345004　Salvia dinteri Briq. = Salvia garipensis E. Mey. ex Benth. ●☆

345005　Salvia discolor Sessé et Moc.；西方异色鼠尾草；Andean Sage, Andean Silver Leaf, Peruvian Black Salvia ☆

345006　Salvia disermas L.；波叶鼠尾草■☆

345007　Salvia divinorum Epling et Jativa；神鼠尾草（神地榆）■☆

345008　Salvia dolichantha Stibal；长花鼠尾草；Longflower Sage ■

345009　Salvia dolichorhiza Caball. = Salvia viridis L. ■☆

345010　Salvia dolomitica Codd；多罗米蒂鼠尾草●☆

345011　Salvia dorrii（Kellogg）Abrams；沙漠鼠尾草；Desert Sage, Great Basin Blue Sage, Purple Sage ●☆

345012　Salvia dregeana Benth. = Salvia albicaulis Benth. ■☆

345013　Salvia drobovii Botsch.；德罗博夫鼠尾草■☆

345014　Salvia dumetorum Andrz.；灌丛鼠尾草；Thicket Sage ■☆

345015　Salvia eckloniana Benth. = Salvia africana-lutea L. ■☆

345016　Salvia elegans Vahl；雅致鼠尾草；Pineapple Sage, Pineapple Scented Sage, Pineapple-scented Sage ●☆

345017　Salvia elegans Vahl 'Scarlet Pineapple'；猩红凤梨美丽鼠尾草●☆

345018　Salvia esquirolii H. Lév. = Salvia yunnanensis C. H. Wright ■

345019　Salvia evansiana Hand. -Mazz. et Stibal；雪山鼠尾草（埃望鼠尾草，雪山鼠尾，紫花丹参）；Snowmountain Sage ■

345020　Salvia evansiana Hand. -Mazz. et Stibal var. scaposa E. Peter ＝ Salvia evansiana Hand. -Mazz. et Stibal var. scaposa Stibal ■

345021　Salvia evansiana Hand. -Mazz. et Stibal var. scaposa Stibal；葶花雪山鼠尾草（黄花雪山鼠尾草，葶花鼠尾草）■

345022　Salvia fargesii H. Lév. = Salvia maximowicziana Hemsl. ■

345023　Salvia farinacea Benth.；蓝花鼠尾（萼松鼠尾草，粉萼鼠尾草，一串蓝，一串蓝草）；Blue Sage, Blue Victoria Salvia, Mealy Cup Sage, Mealy Sage, Mealycup Sage, Mealy-cup Sage, Starchcontaining Sage, Texas Violet ■☆

345024　Salvia farinacea Benth. 'Strata'；白萼一串蓝■☆

345025　Salvia farinacea Benth. 'Victoria'；维多利亚一串蓝■☆

345026　Salvia farinosa Mart. et Aellen；被粉鼠尾草■☆

345027　Salvia feddei H. Lév. = Salvia przewalskii Maxim. var. mandarinorum（Diels）Stibal ■

345028　Salvia filicifolia Merr. = Salvia japonica Thunb. f. filicifolia（Merr.）T. C. Huang et J. T. Wu ■

345029　Salvia flava Forrest ex Diels；黄花鼠尾草（黄花丹参，黄花鼠尾，黄鼠狼花）；Yellowflower Sage ■

345030　Salvia flava Forrest ex Diels var. megalantha Diels；大黄花鼠尾草；Big Yellowflower Sage ■

345031　Salvia foetida Lam. = Salvia argentea L. ■☆

345032　Salvia fominii Grossh.；福明鼠尾草■☆

345033　Salvia formosana（Murata）T. Yamaz. = Salvia nipponica Miq. var. formosana（Hayata）Kudo ■

345034　Salvia formosana（Murata）T. Yamaz. var. matsudae（Kudo）T. C. Huang et J. T. Wu = Salvia matsudae Kudo ■

345035　Salvia formosana Hayata = Salvia nipponica Miq. var. formosana（Hayata）Kudo ■

345036　Salvia forrestii Diels = Salvia hylocharis Diels ■

345037　Salvia forskaehlei L.；福尔鼠尾草■☆

345038　Salvia fortunei Benth. = Salvia japonica Thunb. ■

345039　Salvia fragarioides C. Y. Wu；草莓状鼠尾草（草莓鼠尾草）；Strawberrylike Sage ■

345040　Salvia fruticosa（L.）Mill.；灌木鼠尾草；Chahomiiia ●☆

345041　Salvia fruticosa Mill. = Salvia fruticosa（L.）Mill. ●☆

345042　Salvia fugax Pobed.；早萎鼠尾草■☆

345043　Salvia fulgens Cav.；亮花鼠尾草（辉红鼠尾草）；Cardinal Sage, Mexican Red Sage ●☆

345044　Salvia funerea M. E. Jones；墓地鼠尾草；Death Valley Sage ■☆

345045　Salvia galeottii M. Martens = Salvia coccinea Buc'hoz ex Etl. ■

345046　Salvia galpinii V. Naray. = Salvia aurita L. f. var. galpinii（V. Naray.）Hedge ■☆

345047　Salvia garipensis E. Mey. ex Benth.；加里普鼠尾草●☆

345048　Salvia gattefossei Emb.；加特福塞鼠尾草●☆

345049　Salvia gesneriifolia Lindl. ex Lem.；金鱼花鼠尾草；Sage ●☆

345050　Salvia glabrescens（Franch. et Sav.）Makino；光鼠尾草■☆

345051　Salvia glabrescens（Franch. et Sav.）Makino f. purpureomaculata（Makino）Honda = Salvia glabrescens（Franch. et Sav.）Makino var. purpureomaculata（Makino）K. Inoue ☆

345052　Salvia glabrescens（Franch. et Sav.）Makino f. robusta（Koidz.）Murata；粗壮光鼠尾草☆

345053　Salvia glabrescens（Franch. et Sav.）Makino var.

purpureomaculata（Makino）K. Inoue；紫斑光鼠尾草☆

345054 Salvia glabrescens（Franch. et Sav.）Makino var. repens（Koidz.）Kurosaki；匍匐光鼠尾草■☆

345055 Salvia glabricaulis Pobed.；光茎鼠尾草■☆

345056 Salvia glaucescens Pohl = Salvia coccinea Buc'hoz ex Etl. ■

345057 Salvia glutinosa L.；胶质鼠尾草（胶黏鼠尾，黏地榆，黏毛地榆，黏毛鼠尾草，黏质鼠尾草，黏质鼠尾）；Glutinous Sage，Jupiter's Distaff，Rubbery Sage，Sticky Sage ■

345058 Salvia glutinosa L. = Salvia przewalskii Maxim. ■

345059 Salvia glutinosa L. subsp. nubicola ？ = Salvia nubicola Wall. ex Sweet ■

345060 Salvia glutinosa L. var. nubicola ？ = Salvia nubicola Wall. ex Sweet ■

345061 Salvia gontscharovii Kudr.；高恩恰洛夫鼠尾草■☆

345062 Salvia goudotii Benth. = Salvia sessilifolia Baker ●☆

345063 Salvia grahami Benth. = Salvia microphylla Torr. ●☆

345064 Salvia grandiflora Sessé et Moc.；大花鼠尾草；Big-flower Sage ●☆

345065 Salvia grandifolia W. W. Sm.；大叶鼠尾草；Bigleaf Sage ■

345066 Salvia granitica Hochst.；花岗岩鼠尾草●☆

345067 Salvia greggii A. Gray；多色鼠尾草（可氏鼠尾草，秋鼠尾草）；Autumn Sage，Cherry Sage ●☆

345068 Salvia greggii A. Gray 'Alba'；白花多色鼠尾草●☆

345069 Salvia greggii A. Gray 'Peach'；桃红多色鼠尾草●☆

345070 Salvia greggii A. Gray 'Raspberry Royal'；悬沟子多色鼠尾草●☆

345071 Salvia grossheimii Sosn.；格罗鼠尾草■☆

345072 Salvia guaranitica A. St. -Hil. ex Benth.；茴芹鼠尾草；Anise-scented sage ☆

345073 Salvia hajastanica Pobed.；哈贾斯坦鼠尾草■☆

345074 Salvia handelii Stibal；木里鼠尾草；Muli Sage ■

345075 Salvia hastifolia Benth. = Salvia lanceolata Lam. ■☆

345076 Salvia hayatae Makino ex Hayata；阿里山鼠尾草（阿里山紫参，阿里山紫花鼠尾草，阿里山紫缘花鼠尾草，白花鼠尾草，柔叶紫参，羽叶紫参，早田氏鼠尾草）；Alishan Sage，Arishan Sage ■

345077 Salvia hayatae Makino ex Hayata var. pinnata（Hayata）C. Y. Wu；羽叶阿里山鼠尾草（溪头紫参，隐囊鼠尾草，隐药鼠尾草）；Pinnate Alishan Sage ■

345078 Salvia hayatana Makino ex Hayata = Salvia hayatae Makino ex Hayata ■

345079 Salvia heterochroa Stibal；异色鼠尾草；Differentcolor Sage，Diverscolour Sage ■

345080 Salvia hians Royle ex Benth.；裂鼠尾草（裂地榆）；Sage ■☆

345081 Salvia hildebrandtii Briq. = Salvia sessilifolia Baker ●☆

345082 Salvia himmelbaurii Stibal；瓦山鼠尾草；Washan Sage ■

345083 Salvia hispanica Garsault；西班牙鼠尾草（野鼠尾草）；Chia，Chia Seeds ■☆

345084 Salvia hochstetteri Baker = Salvia nilotica Juss. ex Jacq. ■☆

345085 Salvia holwayi S. F. Blake；霍氏鼠尾草；Holway's Sage ■☆

345086 Salvia honania L. H. Bailey；河南鼠尾草（丹参）；Henan Sage，Honan Sage ■

345087 Salvia horminioides Pourr. = Salvia verbenaca L. ■☆

345088 Salvia horminum L. = Salvia viridis L. ■☆

345089 Salvia horminum L. var. intermedia Briq. = Salvia viridis L. ■☆

345090 Salvia horminum L. var. viridis（L.）Caruel = Salvia viridis L. ■☆

345091 Salvia hupehensis Stibal；湖北鼠尾草；Hubei Sage，Hupeh Sage ■☆

345092 Salvia hybrid Hort.；杂种鼠尾草；Ecuador Sage ■☆

345093 Salvia hylocharis Diels；林华鼠尾草；Forestloving Sage，Linhua Sage ■

345094 Salvia hylocharis Diels var. subsimplex C. Y. Wu；单序林华鼠尾草；Subsimplex Sage ■

345095 Salvia hylocharis Diels var. subsimplex C. Y. Wu = Salvia hylocharis Diels ■

345096 Salvia hypargeia Fisch. et C. A. Mey.；黄背鼠尾草■☆

345097 Salvia hypoleuca Hochst. = Salvia schimperi Benth. ■☆

345098 Salvia incisa Benth. = Salvia repens Burch. ex Benth. ■☆

345099 Salvia insignis Kudr.；显著鼠尾草■☆

345100 Salvia integerrima Mill. = Salvia africana-caerulea L. ●☆

345101 Salvia integrifolia Hardw. = Salvia lanata Roxb. ■☆

345102 Salvia intercedens Pobed.；中间鼠尾草■☆

345103 Salvia interrupta Schousb.；间断鼠尾草■☆

345104 Salvia interrupta Schousb. subsp. paui（Maire）Maire；波氏鼠尾草■☆

345105 Salvia involucrata Cav.；总苞鼠尾草；Roseleaf Sage ■☆

345106 Salvia involucrata Cav. 'Bethellii'；伯舍尔总苞鼠尾草■☆

345107 Salvia isensis Nakai ex H. Hara；伊势鼠尾草■☆

345108 Salvia japonica Thunb.；鼠尾草（霸王鞭，劲枝丹参，南丹参，秋丹参，日本紫花鼠尾草，山陵翘，水青，乌草，消炎草，紫参，紫花鼠尾草）；Japanese Sage，Sage ■

345109 Salvia japonica Thunb. f. alatopinnata（Matsum. et Kudo）Kudo = Salvia japonica Thunb. ■

345110 Salvia japonica Thunb. f. albiflora Hiyama；白花鼠尾草■☆

345111 Salvia japonica Thunb. f. erythrophylla（Hemsl.）Kudo = Salvia cavaleriei H. Lév. var. erythrophylla（Hemsl.）Stibal ■

345112 Salvia japonica Thunb. f. filicifolia（Merr.）T. C. Huang et J. T. Wu；蕨叶鼠尾草（蕨叶太平南丹参）；Fernleaf Sage ■

345113 Salvia japonica Thunb. f. lanuginosa（Franch.）E. Peter；多毛鼠尾草■☆

345114 Salvia japonica Thunb. f. polakioides（Honda）T. Yamaz. = Polakiastrum longipes Nakai ■☆

345115 Salvia japonica Thunb. subsp. taipingshanensis T. C. Huang et J. T. Wu = Salvia japonica Thunb. var. taipingshanensis（T. C. Huang et J. T. Wu）T. C. Huang et J. T. Wu ■

345116 Salvia japonica Thunb. var. chinensis（Benth.）E. Peter = Salvia chinensis Benth. ■

345117 Salvia japonica Thunb. var. chinensis（Benth.）Stibal = Salvia chinensis Benth. ■

345118 Salvia japonica Thunb. var. erythrophylla Hemsl. = Salvia cavaleriei H. Lév. var. erythrophylla（Hemsl.）Stibal ■

345119 Salvia japonica Thunb. var. filicifolia（Merr.）Metcalf et E. Peter = Salvia filicifolia Merr. ■

345120 Salvia japonica Thunb. var. filicifolia（Merr.）Metcalf et Stibal = Salvia filicifolia Merr. ■

345121 Salvia japonica Thunb. var. formosana Murata = Salvia formosana（Murata）T. Yamaz. var. matsudae（Kudo）T. C. Huang et J. T. Wu ■

345122 Salvia japonica Thunb. var. fortunei（Benth.）Kudo = Salvia japonica Thunb. ■

345123 Salvia japonica Thunb. var. fortunei（Benth.）Kudo f. pinnata（Diels）Kudo = Salvia nanchuanensis H. Z. Sun ■

345124 Salvia japonica Thunb. var. gracillima Diels = Salvia plectranthoides Griff. et Stibal ■

345125 Salvia japonica Thunb. var. integrifolia Franch. et Sav. = Salvia chinensis Benth. ■

345126 Salvia japonica Thunb. var. kaiscianensis Pamp. = Salvia plectranthoides Griff. et Stibal ■

345127 Salvia japonica Thunb. var. lanuginosa（Franch.）E. Peter =

Salvia japonica Thunb. ■

345128　Salvia japonica Thunb. var. lanuginosa Franch. = Salvia japonica Thunb. ■

345129　Salvia japonica Thunb. var. multifoliolata E. Peter = Salvia japonica Thunb. var. multifoliolata Stibal ■

345130　Salvia japonica Thunb. var. multifoliolata Stibal;多小叶鼠尾草（朱砂草）;Multifoliolate Sage ■

345131　Salvia japonica Thunb. var. parvifoliola Hemsl. = Salvia nanchuanensis H. Z. Sun ■

345132　Salvia japonica Thunb. var. parvifoliola Hemsl. = Salvia plectranthoides Griff. et Stibal ■

345133　Salvia japonica Thunb. var. pinnata Diels = Salvia nanchuanensis H. Z. Sun ■

345134　Salvia japonica Thunb. var. prionitis (Hance) Kudo = Salvia prionitis Hance ■

345135　Salvia japonica Thunb. var. taipingshanensis (T. C. Huang et J. T. Wu) T. C. Huang et J. T. Wu;太平山鼠尾草（太平南丹参,太平紫花鼠尾草）■

345136　Salvia japonica Thunb. var. ternata Franch. = Salvia japonica Thunb. ■

345137　Salvia judaica Boiss.;朱达鼠尾草;Judean Sage ■☆

345138　Salvia jurisicii Kosanin ex Jurisic;西方毛唇鼠尾草■☆

345139　Salvia karabachensis Pobed.;卡拉巴丹参■☆

345140　Salvia keitaoensis Hayata;隐药鼠尾草（溪头紫参）■

345141　Salvia keitaoensis Hayata = Salvia hayatae Makino ex Hayata var. pinnata (Hayata) C. Y. Wu ■

345142　Salvia kiangsiensis C. Y. Wu;关公须（根下红,关爷须,关羽须,江西鼠尾,落地红,扑地红,扑地消,土活血,小活血,叶下红）;Jiangxi Sage,Kiangxi Sage ■

345143　Salvia kiaometiensis H. Lév.;荞麦地鼠尾草（丹参,红根,荞麦地鼠尾,土丹参）;Kiaometi Sage,Qiaomaidi Sage ■

345144　Salvia kiaometiensis H. Lév. f. pubescens Stibal;柔毛荞麦地鼠尾草（柔毛鼠尾草,土丹参）;Pubescent Qiaomaidi Sage ■

345145　Salvia kiaometiensis H. Lév. f. tomentella Stibal;绒毛荞麦地鼠尾草（绒毛鼠尾草）;Tomentose Qiaomaidi Sage ■

345146　Salvia komarovii Pobed.;科马罗夫鼠尾草■☆

345147　Salvia kopetdaghensis Kudr.;科佩特鼠尾草■☆

345148　Salvia koyamae Makino;小山氏鼠尾草■☆

345149　Salvia kuznetzovii Sosn.;库兹鼠尾草■☆

345150　Salvia labellifera H. Lév. = Salvia przewalskii Maxim. var. mandarinorum (Diels) Stibal ■

345151　Salvia lanata Roxb.;绵毛鼠尾草;Cottony Sage ■☆

345152　Salvia lanceifolia Poir. = Salvia reflexa Hornem. ■☆

345153　Salvia lanceolata Lam.;剑叶鼠尾草;Lance-leaf Sage ■☆

345154　Salvia lanigera Poir.;旱鼠尾草■☆

345155　Salvia lankongensis C. Y. Wu;洱源鼠尾草（红须须）;Eryuan Sage,Lankong Sage ■

345156　Salvia lanuginosa Burm. f. = Salvia africana-caerulea L. ●☆

345157　Salvia lasiostachys Benth. = Salvia aurita L. f. ■☆

345158　Salvia lavandulifolia Vahl;薰衣草叶鼠尾草（西班牙鼠尾草）■☆

345159　Salvia lavandulifolia Vahl subsp. amethystea (Emb. et Maire) Rosua et Blanca;紫晶鼠尾草■☆

345160　Salvia lavandulifolia Vahl subsp. blancoana (Webb et Heldr.) Rosua et Blanca;布兰科鼠尾草●☆

345161　Salvia lavandulifolia Vahl subsp. maurorum (Ball) Rosua et Blanca;模糊鼠尾草●☆

345162　Salvia lavandulifolia Vahl var. aurasiaca (Maire) Rosua et Blanca = Salvia lavandulifolia Vahl ■☆

345163　Salvia lavandulifolia Vahl var. blancoana (Webb et Heldr.) Rosua et Blanca = Salvia lavandulifolia Vahl ■☆

345164　Salvia lavanduloides Benth.;假薰衣草叶鼠尾草（薰衣鼠尾草）■☆

345165　Salvia leclerei H. Lév. = Salvia mairei H. Lév. ■

345166　Salvia leonuroides Gloxin;光亮鼠尾草;Shining Sage ■☆

345167　Salvia leucantha Cav.;厚毛鼠尾草（白花鼠尾草）;Mexican Bush Sage,Slenderleaf Sage,Velvet Sage ●☆

345168　Salvia leucodermis Baker;白皮鼠尾草●☆

345169　Salvia leucophylla Greene;白叶鼠尾草;Chaparral Sage,Gray Sage,Purple Sage ●☆

345170　Salvia leveillana Fedde = Salvia kiachiangensis H. Lév. ■

345171　Salvia lichiangensis W. W. Sm. = Salvia aerea H. Lév. ■

345172　Salvia liguliloba Y. Z. Sun et Migo;舌瓣鼠尾草（长叶丹参,毛瓣鼠尾草,舌瓣鼠尾）;Ligulelobe Sage ■

345173　Salvia lilacinocoerulea Nevski;堇蓝鼠尾草■☆

345174　Salvia limbata C. A. Mey.;具边鼠尾草■☆

345175　Salvia linczevskii Kudr.;林克柴夫斯基鼠尾草■☆

345176　Salvia lipskyi Pobed.;利普斯基鼠尾草■☆

345177　Salvia longistyla Benth.;长柱鼠尾草;Mexican Sage ■☆

345178　Salvia lutescens (Koidz.) Koidz.;淡黄花鼠尾草■☆

345179　Salvia lutescens (Koidz.) Koidz. f. crenata (Makino) G. Nakai = Salvia lutescens (Koidz.) Koidz. var. crenata (Makino) Murata ■☆

345180　Salvia lutescens (Koidz.) Koidz. f. lobatocrenata (Makino) G. Nakai = Salvia lutescens (Koidz.) Koidz. var. intermedia (Makino) Murata ■☆

345181　Salvia lutescens (Koidz.) Koidz. var. crenata (Makino) Murata;圆齿鼠尾草■☆

345182　Salvia lutescens (Koidz.) Koidz. var. crenata (Makino) Murata f. leucantha Murata;白花圆齿鼠尾草■☆

345183　Salvia lutescens (Koidz.) Koidz. var. intermedia (Makino) Murata;居间淡黄花鼠尾草■☆

345184　Salvia lutescens (Koidz.) Koidz. var. intermedia (Makino) Murata f. albiflora Murata;白花居间鼠尾草■☆

345185　Salvia lutescens (Koidz.) Koidz. var. stolonifera G. Nakai;匍匐淡黄花鼠尾草■☆

345186　Salvia lycioides A. Gray;枸杞状鼠尾草;Blue Canyon Sage ■☆

345187　Salvia lyrata L.;癌草;Cancerweed,Lyre-leaf Sage,Lyre-leaved Sage ■☆

345188　Salvia macilenta Boiss.;细弱鼠尾草■☆

345189　Salvia macrorrhiza Chiov. = Salvia nilotica Juss. ex Jacq. ■☆

345190　Salvia macrosiphon Boiss.;大管鼠尾草■☆

345191　Salvia madrensis Seemen;马德雷鼠尾草;Forsythia Sage ■☆

345192　Salvia mairei H. Lév.;东川鼠尾草;E. Sichuan Sage,Maire Sage ■☆

345193　Salvia mairei H. Lév. = Salvia kiaometiensis H. Lév. ■

345194　Salvia mandarinorum Diels = Salvia przewalskii Maxim. var. mandarinorum (Diels) Stibal ■

345195　Salvia mans Royle = Salvia prattii Hemsl. ■

345196　Salvia marchandii H. Lév. = Salvia cavaleriei H. Lév. ■

345197　Salvia margaritae Botsch.;马尔鼠尾草■☆

345198　Salvia marginata Benth. = Salvia obtusata Thunb. ■☆

345199　Salvia mariae Sennen = Salvia algeriensis Desf. ■☆

345200　Salvia maroccana Batt. et Pit. = Salvia mouretii Batt. et Pit. ■☆

345201　Salvia marretii H. Lév. = Salvia tricuspis Franch. ■

345202　Salvia matsudae Kudo;蕨叶紫花鼠尾草■

345203　Salvia matsudae Kudo = Salvia formosana (Murata) T. Yamaz.

var. matsudae（Kudo）T. C. Huang et J. T. Wu ■

345204　Salvia maurorum Ball ＝ Salvia lavandulifolia Vahl subsp. maurorum（Ball）Rosua et Blanca ●☆

345205　Salvia maximowicziana Hemsl.；鄂西鼠尾草（红秦艽）；Maximowicz Sage ■

345206　Salvia maximowicziana Hemsl. var. floribunda Stibal；多花鄂西鼠尾草；Manyflower Maximowicz Sage ■

345207　Salvia meiliensis S. W. Su；美丽鼠尾草（梅里鼠尾草）；Beautiful Sage ■

345208　Salvia mekongensis Stibal；湄公鼠尾草（沧江鼠尾）；Meigong Sage，Mekong Sage ■

345209　Salvia melissodora Lag.；葡萄香鼠尾草；Grape-scented Sage ●☆

345210　Salvia mellifera Greene；蜜鼠尾草；Black Sage，Honey Sage ●☆

345211　Salvia merjamie Forssk.；迈尔鼠尾草●☆

345212　Salvia mexicana L.；墨西哥鼠尾草；Mexican Sage ■☆

345213　Salvia microphylla Torr.；小叶鼠尾草；Baby Sage，Little-leafed Sage，Red Canyon Sage ●☆

345214　Salvia miltiorhiza Bunge；丹参（白花丹参，奔马草，长鼠尾草，炒丹参，赤参，赤丹参，大红袍，大叶活血丹，滇丹参，朵朵花，蜂糖罐，褐毛丹参，红参，红丹参，红根，红根赤参，红根红参，红暖药，活血根，靠山红，蜜罐头，木羊乳，却蝉草，壬参，山参，山红萝卜，山苏子，烧酒壶，水羊耳，五凤花，郏蝉草，夏丹参，小丹参，血参，血参根，血山根，血生根，野苏子，阴行草，郁蝉草，逐马，逐乌，紫参，紫丹参，紫党参）；Dan-shen ■

345215　Salvia miltiorhiza Bunge f. alba C. Y. Wu et H. W. Li；白花丹参■

345216　Salvia miltiorhiza Bunge f. alba C. Y. Wu et H. W. Li ＝ Salvia miltiorhiza Bunge ■

345217　Salvia miltiorhiza Bunge var. australis Stibal ＝ Salvia bowleyana Dunn ■

345218　Salvia miltiorhiza Bunge var. australis Stibal ＝Salvia sinica Migo ■

345219　Salvia miltiorhiza Bunge var. charbonnelii（H. Lév.）C. Y. Wu；单叶丹参（血参）■

345220　Salvia miltiorhiza Bunge var. hupehensis E. Peter ＝ Salvia paramiltiorrhiza H. W. Li et X. L. Wang ■

345221　Salvia miltiorhiza Bunge var. hupehensis Stibal ＝ Salvia sinica Migo ■

345222　Salvia miltiorrhiza Bunge ＝ Salvia miltiorhiza Bunge ■

345223　Salvia miltiorhiza Bunge f. alba C. Y. Wu et H. W. Li ＝Salvia miltiorhiza Bunge ■

345224　Salvia miltiorhiza Bunge var. australis E. Peter ＝ Salvia bowleyana Dunn ■

345225　Salvia miltiorhiza Bunge var. hupehensis E. Peter ＝ Salvia paramiltiorrhiza H. W. Li et X. L. Wang ■

345226　Salvia minutiflora Bunge ＝ Salvia plebeia R. Br. ■

345227　Salvia mohavensis Greene；莫哈鼠尾草；Mohave Sage ■☆

345228　Salvia monticola Benth. ＝ Salvia runcinata L. f. ■☆

345229　Salvia monticola Benth. var. angustiloba V. Naray. ＝ Salvia schlechteri Briq. ■☆

345230　Salvia moorcroftiana Wall. ex Benth.；穆尔鼠尾草（木尔克洛夫鼠尾草，穆尔地榆）；Moorcroft Sage ■☆

345231　Salvia mouretii Batt. et Pit.；穆雷鼠尾草■☆

345232　Salvia mouretii Batt. et Pit. var. maroccana（Batt. et Pit.）Maire ＝ Salvia mouretii Batt. et Pit. ☆

345233　Salvia mouretii Batt. et Pit. var. violacea Maire，Weiller et Wilczek ＝ Salvia mouretii Batt. et Pit. ■☆

345234　Salvia muirii L. Bolus；缪里鼠尾草■☆

345235　Salvia muirii L. Bolus var. grandiflora？＝ Salvia muirii L. Bolus ■☆

345236　Salvia multicaulis Vahl；多茎鼠尾草■☆

345237　Salvia multifida Sm. ＝ Salvia verbenaca L. ■☆

345238　Salvia multiorrhiza Crevost et Petelot；多根丹参■☆

345239　Salvia namaensis Schinz；纳马鼠尾草■☆

345240　Salvia nanchuanensis H. Z. Sun；南川鼠尾草；Nanchuan Sage ■

345241　Salvia nanchuanensis H. Z. Sun f. intermedia H. Z. Sun；居间南川鼠尾草；Intermediate Nanchuan Sage ■

345242　Salvia nanchuanensis H. Z. Sun f. intermedia H. Z. Sun ＝ Salvia nanchuanensis H. Z. Sun ■

345243　Salvia nanchuanensis H. Z. Sun var. pteridifolia H. Z. Sun；蕨叶南川鼠尾草；Fernleaf Nanchuan Sage ■

345244　Salvia napifolia Jacq.；西亚鼠尾草■☆

345245　Salvia natalensis Briq. et Schinz ＝ Salvia repens Burch. ex Benth. ☆

345246　Salvia nemorosa L.；林生鼠尾草（林鼠尾草，森林鼠尾草）；Balkan Clary，May Night Sage，Meadow Sage，Violet Sage，Wild Sage，Woodland Sage ■☆

345247　Salvia nemorosa L. 'East Friesland' ＝ Salvia nemorosa L. 'Ostfriesland' ■☆

345248　Salvia nemorosa L. 'Lubecca'；卢贝卡森林鼠尾草■☆

345249　Salvia nemorosa L. 'Ostfriesland'；东弗里斯兰森林鼠尾草（西弗里斯兰林地鼠尾草）■☆

345250　Salvia nemorosa L. ＝ Salvia sylvestris L. ■☆

345251　Salvia nilotica Juss. ex Jacq.；尼罗河鼠尾草■☆

345252　Salvia nipponica Miq.；琴柱草；Japanese Sage，Lyrestyle Sage ■

345253　Salvia nipponica Miq. f. repens Koidz. ＝ Salvia glabrescens（Franch. et Sav.）Makino var. repens（Koidz.）Kurosaki ■☆

345254　Salvia nipponica Miq. var. formosana（Hayata）Kudo；台湾鼠尾草（黄花鼠尾草，台湾琴柱草，台湾日紫参，台湾紫花鼠尾草）；Taiwan Sage ■

345255　Salvia nipponica Miq. var. glabrescens Franch. et Sav. ＝ Salvia glabrescens（Franch. et Sav.）Makino ■☆

345256　Salvia nipponica Miq. var. kisoensis K. Imai；木曾鼠尾草■☆

345257　Salvia nipponica Miq. var. repens（Koidz.）Kudo ＝ Salvia glabrescens（Franch. et Sav.）Makino var. repens（Koidz.）Kurosaki ■☆

345258　Salvia nipponica Miq. var. trisecta（Matsum. ex Kudo）Honda；三深裂鼠尾草■☆

345259　Salvia nivea Thunb. ＝ Salvia lanceolata Lam. ■☆

345260　Salvia nubia Juss. ex Murray ＝ Salvia merjamie Forssk. ●☆

345261　Salvia nubicola Wall. ex Sweet；云南丹参■

345262　Salvia nudicaulis Vahl ＝ Salvia merjamie Forssk. ●☆

345263　Salvia nudicaulis Vahl var. congesta（A. Rich.）Engl. ＝ Salvia merjamie Forssk. ●☆

345264　Salvia nudicaulis Vahl var. nubia Baker ＝ Salvia merjamie Forssk. ●☆

345265　Salvia nudicaulis Vahl var. pubescens Benth. ＝ Salvia merjamie Forssk. ●☆

345266　Salvia nutans L.；俯垂鼠尾草；Nodding Sage ■☆

345267　Salvia obtusa M. Martens et Galeotti；钝叶鼠尾草■☆

345268　Salvia obtusata Thunb.；钝鼠尾草■☆

345269　Salvia occidentalis Sw.；西方鼠尾草；Hopweed ■☆

345270　Salvia ochreuleuca Coss. et Balansa ＝ Salvia verbenaca L. ■☆

345271　Salvia officinalis L.；药用地榆（欧鼠尾草，撒尔维亚，洋苏草，药鼠尾草，药用丹参，药用鼠尾草）；Berggarten Sage，Common Sage，Garden Sage，Kitchen Sage，Medicinal Sage，Officinal Sage，Red Sage，Sage，Save ●

345272　Salvia officinalis L. 'Icterina';黄斑药用鼠尾草■☆

345273　Salvia officinalis L. 'Purpurascens';紫芽药用鼠尾草;Purple-leaved Sage ■☆

345274　Salvia omeiana Stibal;峨眉鼠尾草(白气草,白生麻,南茄草,野苏麻);Emei Sage, Omei Sage ■

345275　Salvia omeiana Stibal var. grandibracteata Stibal;宽苞峨眉鼠尾草;Bradbract Emei Sage ■

345276　Salvia omerocalyx Hayata var. crenata (Makino) F. Maek. = Salvia lutescens (Koidz.) Koidz. var. crenata (Makino) Murata ■☆

345277　Salvia omerocalyx Hayata var. intermedia (Makino) F. Maek. = Salvia lutescens (Koidz.) Koidz. var. intermedia (Makino) Murata ■☆

345278　Salvia omerocalyx Hayata var. intermedia (Makino) F. Maek. f. lutescens (Koidz.) Ohwi = Salvia lutescens (Koidz.) Koidz. ■☆

345279　Salvia omerocalyx Hayata var. isensis (Nakai ex H. Hara) Okuyama = Salvia isensis Nakai ex H. Hara ■☆

345280　Salvia omerocalyx Hayata var. prostrata Satake;平卧鼠尾草■☆

345281　Salvia pachyphylla Munz;厚叶鼠尾草;Rose Sage ■☆

345282　Salvia pachystachya Trautv.;粗穗鼠尾草●☆

345283　Salvia palaestina Benth.;燥地鼠尾草●☆

345284　Salvia pallida Benth.;苍白鼠尾草;Pale Sage ■☆

345285　Salvia pallida Dinter ex Engl. = Salvia stenophylla Burch. ex Benth. ■☆

345286　Salvia pallidiflora V. Naray. = Salvia aurita L. f. ■☆

345287　Salvia paniculata L. = Salvia chamelaeagnea P. J. Bergius ●☆

345288　Salvia paohsingensis C. Y. Wu;宝兴鼠尾草(宝兴鼠尾,拟丹参);Baoxing Sage, Paohsing Sage ■

345289　Salvia paramiltiorrhiza H. W. Li et X. L. Wang;皖鄂丹参(鄂皖丹参,拟丹参,紫花拟丹参,紫花皖鄂丹参);Anhui-Hubei Sage, Purple Anhui-Hubei Sage ■

345290　Salvia paramiltiorrhiza H. W. Li et X. L. Wang f. purpureorubra H. W. Li = Salvia paramiltiorrhiza H. W. Li et X. L. Wang ■

345291　Salvia parviflora Salisb. = Salvia nilotica Juss. ex Jacq. ■☆

345292　Salvia patens Cav.;长蕊鼠尾草(蓝花鼠尾草);Blue Salvia, Gentian Sage ■☆

345293　Salvia patens Cav. 'Cambridge Blue';剑桥蓝长蕊鼠尾草■☆

345294　Salvia patula Desf. = Salvia argentea L. subsp. patula (Desf.) Maire ■☆

345295　Salvia pauciflora Stibal;少花鼠尾草;Fewflower Sage ■

345296　Salvia paui Maire = Salvia interrupta Schousb. subsp. paui (Maire) Maire ■☆

345297　Salvia peglerae V. Naray. = Salvia aurita L. f. ■☆

345298　Salvia pendula Vahl;下垂鼠尾草■☆

345299　Salvia pentstemonoides Kunth et Baucher;钓钟柳鼠尾草;Big Red Sage ■☆

345300　Salvia perrieri Hedge;佩里耶鼠尾草●☆

345301　Salvia phlomoides Asso;糙苏鼠尾草●☆

345302　Salvia phlomoides Asso subsp. africana (Maire) Greuter et Burdet;非洲糙苏鼠尾草●☆

345303　Salvia piasezkii Maxim.;秦岭鼠尾草;Qinling Sage ■

345304　Salvia pinetorum Hand.-Mazz. = Salvia aerea H. Lév. ■

345305　Salvia pinguifolia Wooton et Standl.;肥叶鼠尾草;Rock Sage ■☆

345306　Salvia pinnata Pavol. = Salvia plectranthoides Griff. et Stibal ■

345307　Salvia pitcheri Torr. = Salvia pitcheri Torr. ex Benth. ■☆

345308　Salvia pitcheri Torr. ex Benth.;皮彻鼠尾草;Salvia ■☆

345309　Salvia pitcheri Torr. ex Benth. = Salvia azurea Michx. ex Lam. var. grandiflora Benth. ■☆

345310　Salvia plebeia R. Br.;荔枝草(波罗子,臭草,大塔花,地榆,方骨苦草,凤眼草,蛤蟆草,隔冬青,根下红,沟香薷,鼓胀草,关公须,过冬青,黑紫苏,猴臂草,假苏,薑芥,节毛鼠尾草,蚧肚宗,荆芥,赖断头草,赖师草,赖团草,赖子草,癞肚皮棵,癞肚子苗,癞疙包草,癞疙宝草,癞蛤蟆草,癞客蚂草,癞头草,癞虾蟆,癞疒草,落地红,麻鸡婆,麻鸡婆草,麻麻草,毛苦菜,内红消,女菀,膨胀草,朴地消,荠宁,荠苎,荠苎蛤蟆草,青蛙草,山苘苔,鼠蓑,鼠尾草,水羊耳,天明精,土荆芥,土犀角,虾蟆草,小花地榆,小花鼠尾草,小活血,旋涛草,雪见草,雪里青,野薄荷,野芥菜,野茄子,野苏麻,野芝麻,野猪菜,野紫苏,鱼味草,泽泻,皱皮草,皱皮葱,皱皮大菜,猪婆草,紫丹参);Common Sage, Litchi Sage ■

345311　Salvia plebeia R. Br. f. leucantha Kawas.;白花荔枝草■☆

345312　Salvia plebeia R. Br. var. kiangsiensis C. Y. Wu = Salvia kiangsiensis C. Y. Wu ■

345313　Salvia plebeia R. Br. var. latifolia E. Peter = Salvia plebeia R. Br. ■

345314　Salvia plebeia R. Br. var. latifolia Stibal = Salvia plebeia R. Br. ■

345315　Salvia plectranthoides Griff. et Stibal;长冠鼠尾草(长冠鼠尾,串皮猫药,丹参,红骨参,活血草,劲枝丹参,毛丹参,散血草,山胡椒,四花菜叶丹参,小丹参,野藿香,紫参);Longcoronate Sage ■

345316　Salvia pogonocalyx Hance = Salvia miltiorrhiza Bunge ■

345317　Salvia pogonochila Diels;毛唇鼠尾草;Hairylip Sage ■

345318　Salvia polystachya Ortega;多穗鼠尾草■☆

345319　Salvia porphyrocalyx Baker;紫萼鼠尾草●☆

345320　Salvia potaninii Krylov;洪桥鼠尾草;Hongqiao Sage, Potanin Sage ■

345321　Salvia pratensis L.;草原鼠尾草(草甸鼠尾草,草原地榆);Introduced Sage, Meadow Clary, Meadow Sage, Wild Clary ■☆

345322　Salvia prattii Hemsl.;康定鼠尾草(康定鼠尾);Kangding Sage ■

345323　Salvia prattii Hemsl. var. souliei (H. Lév.) Kudo = Salvia prattii Hemsl. ■

345324　Salvia prionitis Hance;黄埔鼠尾草(根下红,关公须,红地胆,红根草,红根子,黄埔鼠尾,落地红,小丹参);Hispid Sage, Redroot Sage ■

345325　Salvia procurrens Benth.;伸展鼠尾草■☆

345326　Salvia prunelloides Kunth;夏枯草状鼠尾草■☆

345327　Salvia przewalskii Maxim.;甘西鼠尾草(丹参,甘青丹参,甘肃丹参,甘西鼠尾,高原丹参,红秦艽,华西丹参,华西党参,紫丹参);Przewalsk Sage ■

345328　Salvia przewalskii Maxim. var. alba X. L. Huang et H. W. Li;白花甘西鼠尾草;Whieflower Przewalsk Sage ■

345329　Salvia przewalskii Maxim. var. glabrescens Stibal;少毛甘西鼠尾草(灵兰香);Glabrous Przewalsk Sage ■

345330　Salvia przewalskii Maxim. var. mandarinorum (Diels) E. Peter = Salvia przewalskii Maxim. var. mandarinorum (Diels) Stibal ■

345331　Salvia przewalskii Maxim. var. mandarinorum (Diels) Stibal;褐毛甘西鼠尾草(褐毛丹参,褐毛甘西鼠尾,褐毛鼠尾草,黄毛华西丹参);Brownhair Przewalsk Sage ■

345332　Salvia przewalskii Maxim. var. rubrobrunnea C. Y. Wu;红褐甘西鼠尾草(红褐毛鼠尾草);Redbrown Przewalsk Sage ■

345333　Salvia pseudococcinea Jacq. = Salvia coccinea Buc'hoz ex Etl. ■

345334　Salvia pygmaea Matsum.;矮小鼠尾草■☆

345335　Salvia pygmaea Matsum. var. simplicior Hatus. ex T. Yamaz.;简单鼠尾草■☆

345336　Salvia qimenensis S. W. Su et J. Q. He;祁门鼠尾草;Qimen Sage ■

345337　Salvia radula Benth.;刮刀鼠尾草■☆

345338　Salvia ranzaniana Makino;小野鼠尾草■☆

345339 Salvia raphanifolia Benth. = Salvia repens Burch. ex Benth. ■☆

345340 Salvia reflexa Hornem.；岩山鼠尾草（毒苏草）；Annuul Sage，Lance-leaf Sage，Lance-leaved Sage，Mintweed，Rocky Mountain Sage ■☆

345341 Salvia regla Cav.；山鼠尾草；Mountain Sage ●☆

345342 Salvia repens Benth. = Salvia repens Burch. ex Benth. ■☆

345343 Salvia repens Burch. ex Benth.；匍匐鼠尾草■☆

345344 Salvia repens Burch. ex Benth. var. keiensis Hedge；凯亚鼠尾草■☆

345345 Salvia repens Burch. ex Benth. var. transvaalensis Hedge；德兰士瓦鼠尾草■☆

345346 Salvia ringens Sibth. et Sm.；坚挺鼠尾草■☆

345347 Salvia roborowskii Maxim.；黏毛鼠尾草（黄花地榆，黄花鼠尾草，野芝麻，粘毛鼠尾）；Robomwsk Sage，Stickyhair Sage ■

345348 Salvia rockiana E. Peter = Salvia evansiana Hand. -Mazz. et Stibal var. scaposa Stibal ■

345349 Salvia rockiana Stibal = Salvia evansiana Hand. -Mazz. et Stibal var. scaposa Stibal ■

345350 Salvia roemeriana Scheele；勒默尔鼠尾草；Cedar Sage ■☆

345351 Salvia rosifolia Sm.；粉叶鼠尾草■☆

345352 Salvia rotundifolia Salisb. = Salvia africana-caerulea L. ●☆

345353 Salvia rudis Benth. = Salvia repens Burch. ex Benth. ■☆

345354 Salvia rugosa Aiton = Salvia disermas L. ■☆

345355 Salvia rugosa Thunb.；皱褶鼠尾草■☆

345356 Salvia rugosa Thunb. var. angustifolia Benth. = Salvia disermas L. ■☆

345357 Salvia runcinata L. f.；倒齿形鼠尾草■☆

345358 Salvia runcinata L. f. var. grandiflora V. Naray. = Salvia runcinata L. f. ■☆

345359 Salvia runcinata L. f. var. nana V. Naray. = Salvia runcinata L. f. ■☆

345360 Salvia rutilans Carrière；橙红鼠尾草；Pineapple Sage ■☆

345361 Salvia sabulicola Pomel = Salvia verbenaca L. ■☆

345362 Salvia saxicola Wall. ex Benth.；岩生鼠尾草；Stoneliving Sage ■☆

345363 Salvia scabiosifolia Lam.；糙叶鼠尾草■☆

345364 Salvia scabra L. f.；糙鼠尾草；Coastal Blue Sage ■☆

345365 Salvia scapiformis Hance；地埂鼠尾草（白补药，地埂鼠尾，花茎状丹参，菌柱紫参，卵叶鼠尾草，山字止，田芹菜，紫花丹参）；Scape-like Sage ■

345366 Salvia scapiformis Hance f. keitaoensis（Hayata）Kudo = Salvia hayatae Makino ex Hayata var. pinnata（Hayata）C. Y. Wu ■

345367 Salvia scapiformis Hance var. arisanensis（Hayata）Kudo = Salvia hayatae Makino ex Hayata ■

345368 Salvia scapiformis Hance var. carphocalyx Stibal；钟萼地埂鼠尾草；Bellcalyx Scape-like Sage ■

345369 Salvia scapiformis Hance var. hirsuta Stibal；硬毛地埂鼠尾草（白补药，硬毛鼠尾草）；Hisute Scape-like Sage ■

345370 Salvia scapiformis Hance var. pinnata Hayata = Salvia hayatae Makino ex Hayata var. pinnata（Hayata）C. Y. Wu ■

345371 Salvia scapiformis Hance var. pinnata Hayata f. gracilis Hayata = Salvia scapiformis Hance ■

345372 Salvia scapiformis Hance var. pinnata Hayata f. gracilis Hayata = Salvia hayatae Makino ex Hayata ■

345373 Salvia schenckii Briq. = Salvia repens Burch. ex Benth. ■☆

345374 Salvia schimperi Benth.；欣珀鼠尾草■☆

345375 Salvia schizocalyx Stibal；裂萼鼠尾草；Splitcalyx Sage，Splittedcalyx Sage ■

345376 Salvia schizochila Stibal；裂瓣鼠尾草；Splitpetal Sage，Splittedpetal Sage ■

345377 Salvia schlechteri Briq.；施莱鼠尾草■☆

345378 Salvia schmalhausenii Regel；史马尔鼠尾草■☆

345379 Salvia sclarea L.；南欧丹参（土耳其粉红鼠尾草，薰衣苏草）；Clary，Clary Sage，Clear-eye，Clear-eyes，Clereye，Clury Sage，Common Clary，Europe Sage，Fetid Clary Sage，God's Eye，God's Eyes，Goody's Eye，Goody's Eyes，Muscatel Sage，Seebright ■☆

345380 Salvia scutellarioides Kunth；黄芩鼠尾草■☆

345381 Salvia semiatrata Zucc.；二色鼠尾草；Bicolor Sage，Wo-tone Sage ●☆

345382 Salvia semilanata Czerniak.；半毛鼠尾草■☆

345383 Salvia sennenii Font Quer = Salvia verbenaca L. ■☆

345384 Salvia seravschanica Regel et Schmalh.；塞拉夫鼠尾草■☆

345385 Salvia serrata Benth.；锯齿鼠尾草■☆

345386 Salvia sessei Benth.；塞斯鼠尾草■☆

345387 Salvia sessilifolia Baker；无柄鼠尾草●☆

345388 Salvia sessilifolia Baker var. auriculata Hedge = Salvia sessilifolia Baker ●☆

345389 Salvia sibthorpii Sibth. et Sm.；西氏鼠尾草；Sibthorp Sage，Sibthorp's Sage ■☆

345390 Salvia sikkimensis Stibal；锡金鼠尾草；Sikkim Sage ■

345391 Salvia sikkimensis Stibal var. chaenocalyx Stibal；张萼锡金鼠尾草（张萼鼠尾草）■

345392 Salvia sinaloensis Fernald；锡那罗亚鼠尾草；Sinaloa Sage ■☆

345393 Salvia sinica Migo；拟丹参（浙皖丹参，紫花浙皖丹参）；Chinese Sage，Purpleflower Chinese Sage ■

345394 Salvia sinica Migo f. purpurea H. W. Li = Salvia sinica Migo ■

345395 Salvia sisymbrifolia V. Naray. = Salvia runcinata L. f. ■☆

345396 Salvia smithii Stibal；橙香鼠尾草；Orange Sage，Smith Sage ■

345397 Salvia somalensis Vatke；索马里鼠尾草■☆

345398 Salvia sonchifolia C. Y. Wu；苣叶鼠尾草；Sowthistleleaf Sage ■

345399 Salvia sonomensis Greene；索诺门鼠尾草（虎尾兰）；Sonomen Sage ■☆

345400 Salvia souliei Duthie ex J. H. Veitch = Salvia brevilabra Franch. ■

345401 Salvia souliei H. Lév. = Salvia prattii Hemsl. ■

345402 Salvia spathecifolia ?；苞叶鼠尾草；Tarahumara Sage ■☆

345403 Salvia spinosa L.；刺鼠尾草■☆

345404 Salvia spinosa L. subsp. maroccana Dobignard；摩洛哥鼠尾草■☆

345405 Salvia splendens Ker Gawl.；一串红（炮仔花，墙下红，鼠尾草，西洋红，象牙海棠，象牙红，象洋红）；Red Salvia，Redstring，Salvia，Scarlet Sage ■

345406 Salvia splendens Sellow ex Roem. et Schult. = Salvia splendens Ker Gawl. ■

345407 Salvia stachyoides Kunth；水苏鼠尾草■☆

345408 Salvia steingroeveri Briq. = Salvia garipensis E. Mey. ex Benth. ●☆

345409 Salvia stenodonta Briq. = Salvia sessilifolia Baker ●☆

345410 Salvia stenophylla Burch. ex Benth.；窄叶鼠尾草■☆

345411 Salvia stenophylla Burch. ex Benth. var. subintegra V. Naray. = Salvia stenophylla Burch. ex Benth. ■☆

345412 Salvia stepposa Des. -Shost.；欧洲草原鼠尾草■☆

345413 Salvia suaveolens Pomel = Salvia argentea L. ■☆

345414 Salvia submutica Botsch. et Vved.；无尖鼠尾草■☆

345415 Salvia subpalmatinervis Stibal；近掌脉鼠尾草；Similarpalmvein Sage，Subpalatenerve Sage ■

345416 Salvia subsessilis Benth. = Salvia repens Burch. ex Benth. ■☆

345417 Salvia subspathulata Lehm. = Salvia repens Burch. ex Benth. ■☆

345418 Salvia substolonifera Stibal；佛光草（湖广草，荔枝肾，蔓茎鼠尾，蔓茎鼠尾草，乌痧草，小灯台，小灯台草，小铜钱草，小退火草，盐咳草，盐咳药，走茎丹参）；Buddhahalo Sage，Creeping Sage ■

345419　Salvia summa A. Nelson；苏马鼠尾草；Summa Sage ●☆

345420　Salvia sylvestris L. = Salvia nemorosa L. ■☆

345421　Salvia sylvicola Burch. ex Benth. = Salvia aurita L. f. ■☆

345422　Salvia syriaca L. ；叙利亚鼠尾草■☆

345423　Salvia szechuanica Yamaz. = Salvia japonica Thunb. var. multifoliolata Stibal ■

345424　Salvia tananarivensis Briq. = Salvia sessilifolia Baker ●☆

345425　Salvia taraxacifolia Hook. f. ；蒲公英叶鼠尾草■☆

345426　Salvia tashiroi Hayata；田代氏鼠尾草■

345427　Salvia tashiroi Hayata = Salvia chinensis Benth. ■

345428　Salvia tatsiensis Franch. = Salvia przewalskii Maxim. ■

345429　Salvia thermara Van Jaarsv. ；温泉鼠尾草●☆

345430　Salvia thibetica H. Lév. = Salvia przewalskii Maxim. var. mandarinorum（Diels）Stibal ■

345431　Salvia tibestiensis A. Chev. = Salvia chudaei Batt. et Trab. ■☆

345432　Salvia tiliifolia Vahl；椴叶鼠尾草；Lindenleaf Sage ●☆

345433　Salvia tingitana Etl. ；丹吉尔鼠尾草■☆

345434　Salvia tomentosa Mill. ；茸毛鼠尾草（茸毛地榆）■☆

345435　Salvia trautvetteri Regel；特劳特鼠尾草■☆

345436　Salvia triangularis Thunb. ；三角鼠尾草■☆

345437　Salvia tricuspis Franch. ；黄鼠尾草（黄鼠狼花，三角鼠尾）；Weasel Flower ■

345438　Salvia trigonocalyx Woronow；三棱萼鼠尾草■☆

345439　Salvia trijuga Diels；三叶鼠尾草（红根根药，三对叶丹参，三对叶鼠参，三角叶鼠尾草，三叶鼠尾，小红参，小红丹参，紫丹参）；Threecusp Sage，Threeleaf Sage ■

345440　Salvia triloba L. f. ；希腊鼠尾草（希腊地榆，希腊苏草）；Three-lobed Sage ■☆

345441　Salvia triloba L. f. = Salvia fruticosa（L.）Mill. ■☆

345442　Salvia tsaiana E. Peter = Salvia cavaleriei H. Lév. var. simplicifolia E. Peter ■

345443　Salvia tsaiana Stibal = Salvia cavaleriei H. Lév. var. simplicifolia Stibal ■

345444　Salvia tuberifera H. Lév. = Salvia plectranthoides Griff. et Stibal ■

345445　Salvia turcomanica Pobed. ；土库曼鼠尾草■☆

345446　Salvia turdi A. Rich. = Otostegia tomentosa A. Rich. ■☆

345447　Salvia turkestanica Hort. ex Mottet；土耳其斯坦鼠尾草；Clary ■☆

345448　Salvia tysonii V. Naray. ；泰森鼠尾草■☆

345449　Salvia uliginosa Benth. ；沼泽鼠尾草；Bog Sage，Sky-blue Sage ■☆

345450　Salvia umbratica Hance；荫生鼠尾草（山椒子，山苏子）；Shady Sage ■

345451　Salvia undulata Benth. = Salvia africana-caerulea L. ●☆

345452　Salvia vaseyi Parish；瓦齐鼠尾草；Wand Sage ●☆

345453　Salvia vasta H. W. Li；野丹参；Wild Sage ■

345454　Salvia vasta H. W. Li f. fimbriata H. W. Li；齿唇野丹参（齿唇丹参）；Fimbriate Wild Sage ■

345455　Salvia vasta H. W. Li f. purpurea H. W. Li；紫花野丹参；Purple Wild Sage ■

345456　Salvia vasta H. W. Li f. purpurea H. W. Li = Salvia vasta H. W. Li ■

345457　Salvia verbascifolia M. Bieb. ；毛蕊花叶鼠尾草；Mullein-leaved Sage ■

345458　Salvia verbenaca L. ；马鞭鼠尾草（野鼠尾草）；Christ's Eye，Christ's Eyes，Clary，Clear-eye，Clear-eyes，Eyeseed，Verbena Sage，Vervain Sage，Wild Clary，Wild Clear-eye，Wild Sage ■☆

345459　Salvia verbenaca L. subsp. battandieri Maire = Salvia verbenaca L. ■☆

345460　Salvia verbenaca L. subsp. controversa（Ten.）Batt. = Salvia verbenaca L. ■☆

345461　Salvia verbenaca L. subsp. foetens Maire = Salvia verbenaca L. ■☆

345462　Salvia verbenaca L. subsp. horminoides（Pourr.）Pugsley = Salvia verbenaca L. ■☆

345463　Salvia verbenaca L. subsp. lanigera（Poir.）Batt. = Salvia lanigera Poir. ■☆

345464　Salvia verbenaca L. subsp. multifida（Sibth. et Sm.）Murb. = Salvia verbenaca L. ■☆

345465　Salvia verbenaca L. subsp. ochroleuca（Coss. et Balansa）Maire = Salvia verbenaca L. ■☆

345466　Salvia verbenaca L. var. bicolor Maire = Salvia verbenaca L. ■☆

345467　Salvia verbenaca L. var. oblongata（Vahl）Briq. = Salvia verbenaca L. ■☆

345468　Salvia verbenaca L. var. praecox Lange = Salvia lanigera Poir. ■☆

345469　Salvia verbenaca L. var. sabulicola（Pomel）Batt. = Salvia verbenaca L. ■☆

345470　Salvia verbenaca L. var. typica Maire = Salvia verbenaca L. ■☆

345471　Salvia verticillata L. ；轮生鼠尾草；Lilac Sage，Whorled Clary，Whorled Sage，Wild Clary ■☆

345472　Salvia verticillata L. ‘Purple Rain’；紫色轮生鼠尾草；Hardy Purple Sage ■☆

345473　Salvia virgata Jacq. ；条纹鼠尾草；Wand Sage ■☆

345474　Salvia viridis L. ；紫苞鼠尾草（彩包花，彩包鼠尾草，紫顶鼠尾草，紫叶草）；Annual Clary，Blue Beard，Joseph Sage，Red-topped Sage ■☆

345475　Salvia viridis L. ‘Bouquet’；布凯彩包花■☆

345476　Salvia viridis L. ‘Monarch Bouquet’= Salvia viridis L. ‘Bouquet’■☆

345477　Salvia viridis L. ‘Oxford Blue’；牛津蓝彩包花■☆

345478　Salvia viridis L. var. horminum（L.）Batt. et Trab. ；彩包鼠尾草■☆

345479　Salvia viridis L. var. horminum（L.）Batt. et Trab. = Salvia viridis L. ■☆

345480　Salvia wardii Stibal；西藏鼠尾草；Xizang Sage ■

345481　Salvia weihaiensis C. Y. Wu et H. W. Li；威海鼠尾草；Weihai Sage ■

345482　Salvia woodii Gürke = Salvia repens Burch. ex Benth. ■☆

345483　Salvia xanthocheila Boiss. ；黄舌鼠尾草■☆

345484　Salvia xerobia Briq. = Salvia stenophylla Burch. ex Benth. ■☆

345485　Salvia yunnanensis C. H. Wright；云南鼠尾草（奔马草，丹参，滇丹参，山槟榔，小丹参，小红参，小红草乌，小红党参，云南丹参，朱砂理肺散，紫参，紫丹参）；Yunnan Sage ■

345486　Salviacanthus Lindau = Justicia L. ●■

345487　Salviacanthus preussii Lindau = Justicia preussii（Lindau）C. B. Clarke ■☆

345488　Salviaceae Bercht. et J. Presl = Salviaceae Raf. ■

345489　Salviaceae Raf. ；鼠尾草科■

345490　Salviaceae Raf. = Labiatae Juss. （保留科名）●■

345491　Salviaceae Raf. = Lamiaceae Martinov（保留科名）●■

345492　Salviastrum Fabr. = Tarchonanthus L. ●☆

345493　Salviastrum Heist. ex Fabr. = Tarchonanthus L. ●☆

345494　Salviastrum Scheele = Salvia L. ●■

345495　Salweenia Baker f. （1935）；冬麻豆属；Salweenia ●★

345496　Salweenia wardii Baker f. ；冬麻豆（冬麻）；Ward Salweenia ●◇

345497　Salzmannia DC. （1830）；扎尔茜属☆

345498　Salzmannia nitida DC. ；扎尔茜☆

345499 Salzwedelia O. F. Lang = Genista L. ●

345500 Salzwedelia O. F. Lang = Saltzwedelia P. Gaertn. , B. Mey. et Scherb. ●

345501 Samadera Gaertn. (1791) (保留属名) ;黄楝树属(干果樗属)●☆

345502 Samadera Gaertn. (保留属名) = Quassia L. ●☆

345503 Samadera indica Gaertn. ;印度黄楝树●☆

345504 Samadera indica Gaertn. = Quassia indica (Gaertn.) Noot. ●☆

345505 Samadera lucida (Wall.) Voigt;光亮黄楝树●☆

345506 Samadera madagascariensis A. Juss. ;马岛黄楝树●☆

345507 Samaipaticereus Cárdenas (1952) ;棍棒花柱属●☆

345508 Samaipaticereus corroanus Cárdenas;棍棒花柱●☆

345509 Samama Kuntze = Anthocephalus A. Rich. ●☆

345510 Samama Rumph. = Anthocephalus A. Rich. ●☆

345511 Samama Rumph. ex Kuntze = Anthocephalus A. Rich. ●☆

345512 Samandura Baill. = Quassia L. ●☆

345513 Samandura Baill. = Samadera Gaertn. (保留属名) ●☆

345514 Samandura L. (1886) ;萨曼苦木属●☆

345515 Samandura indica Baill. ;萨曼苦木;Niepa-bark Tree ●☆

345516 Samandura madagascariensis Gaertn. ;马岛萨曼苦木●☆

345517 Samanea (Benth.) Merr. (1916) ;雨树属;Rain Tree, Raintree, Rain-tree, Saman ●

345518 Samanea (DC.) Merr. = Albizia Durazz. ●

345519 Samanea (DC.) Merr. = Samanea (Benth.) Merr. ●

345520 Samanea Merr. = Albizia Durazz. ●

345521 Samanea Merr. = Samanea (Benth.) Merr. ●

345522 Samanea dinklagei (Harms) Keay = Albizia dinklagei (Harms) Harms ●☆

345523 Samanea guineensis (G. C. C. Gilbert et Boutique) Brenan et Brummitt = Samanea leptophylla (Harms) Brenan et Brummitt ●☆

345524 Samanea leptophylla (Harms) Brenan et Brummitt;细叶雨树●☆

345525 Samanea saman (Jacq.) Merr. ;雨树(雨豆树) ;Cow Tamarind, Monkeypod, Rain Tree, Raintree, Rain-tree, Saman, Saman Rain Tree, Zaman ●

345526 Samara L. = Embelia Burm. f. (保留属名) ●■

345527 Samara Sw. = Myrsine L. ●

345528 Samara floribunda (Wall.) Kurz = Embelia floribunda Wall. ●

345529 Samara floribunda Kurz = Embelia floribunda Wall. ●

345530 Samara laeta L. = Embelia laeta (L.) Mez ●

345531 Samara laeta L. var. papilligera Nakai = Embelia laeta (L.) Mez subsp. papilligera (Nakai) Pipoly et C. Chen ●

345532 Samara longifolia Benth. = Embelia undulata (Wall.) Mez ●

345533 Samara obovata Benth. = Embelia laeta (L.) Mez ●

345534 Samara parviflora (Wall. ex A. DC.) Kurz = Embelia parviflora Wall. et A. DC. ●

345535 Samara parviflora Kurz = Embelia parviflora Wall. et A. DC. ●

345536 Samara ribes (Burm. f.) Kurz = Embelia ribes Burm. f. ●

345537 Samara ribes Kurz = Embelia ribes Burm. f. ●

345538 Samara sessiliflora (Kurz) Kurz = Embelia sessiliflora Kurz ●

345539 Samara sessiliflora Kurz = Embelia sessiliflora Kurz ●

345540 Samara undulata (Wall.) Arn. = Embelia undulata (Wall.) Mez ●

345541 Samara undulata Arn. = Embelia undulata (Wall.) Mez ●

345542 Samara vestila Kurz = Embelia vestita Roxb. ●

345543 Samara vestita (Roxb.) Kurz = Embelia vestita Roxb. ●

345544 Samaraceae Dulac = Ulmaceae Mirb. (保留科名) ●

345545 Samaroceltis J. Poiss. = Phyllostylon Capan. ex Benth. et Hook. f. ●☆

345546 Samaropyxis Miq. = Hymenocardia Wall. ex Lindl. ●☆

345547 Samarorchis Ormerod (2008) ;菲律宾兰属●☆

345548 Samarorchis sulitiana Ormerod;菲律宾兰■☆

345549 Samarpsea Raf. = Fraxinus L. ●

345550 Samba Roberty = Triplochiton Alef. (废弃属名) ●☆

345551 Samba Roberty = Triplochiton K. Schum. (保留属名) ●☆

345552 Sambucaceae Batsch ex Borkh. (1797) ;接骨木科●■

345553 Sambucaceae Batsch ex Borkh. = Caprifoliaceae Juss. (保留科名) ●■

345554 Sambucaceae Link = Adoxaceae E. Mey. (保留科名) ●■

345555 Sambucaceae Link = Caprifoliaceae Juss. (保留科名) ●■

345556 Sambucaceae Link = Sambucaceae Batsch ex Borkh. ●■

345557 Sambucaceae Link = Samolaceae Dumort. ■

345558 Sambucus L. (1753) ;接骨木属(蒴藋属) ;Elder, Elderberry ●■

345559 Sambucus adnata DC. var. puberula Schwer. = Sambucus ebulus L. subsp. africana (Engl.) Bolli ●☆

345560 Sambucus adnata Wall. = Sambucus adnata Wall. ex DC. ●

345561 Sambucus adnata Wall. ex DC. ;贴生接骨木(大血草,红山花,接骨草,接骨丹,接骨木,接骨药,苛草,血管草,血满草,血荞草,珍珠麻) ;Adnate Elder ●

345562 Sambucus africana Standl. = Sambucus ebulus L. subsp. africana (Engl.) Bolli ●☆

345563 Sambucus alba Raf. ;白接骨木;Whithe Adnate Elder ●☆

345564 Sambucus albulus L. ;稍白接骨木;Whitish Adnate Elder ●☆

345565 Sambucus arborescens Koehne;乔木接骨木;Tree Scarlet Elder, Tree Scarlet-elder ●☆

345566 Sambucus argyi H. Lév. = Sambucus chinensis Lindl. ●■

345567 Sambucus argyi H. Lév. = Sambucus javanica Blume ■●

345568 Sambucus aurea Carrière = Sambucus nigra L. ' Aurea' ●☆

345569 Sambucus australasica (Lindl.) Fritsch;澳洲接骨木; Australian Elder ●☆

345570 Sambucus barbinerivis Nakai = Sambucus williamsii Hance ●

345571 Sambucus buergeriana (Nakai) Blume ex Nakai = Sambucus williamsii Hance var. miquelii (Nakai ex Kom. et Aliss.) Y. C. Tang ex J. Q. Hu ●

345572 Sambucus buergeriana (Nakai) Blume ex Nakai = Sambucus williamsii Hance ●

345573 Sambucus buergeriana (Nakai) Blume ex Nakai var. miquelii (Nakai) Nakai = Sambucus sibirica Nakai ●

345574 Sambucus buergeriana (Nakai) Blume f. cordifoliata Skvortsov et W. Wang = Sambucus williamsii Hance var. miquelii (Nakai ex Kom. et Aliss.) Y. C. Tang ex J. Q. Hu ●

345575 Sambucus buergeriana Blume = Sambucus williamsii Hance var. miquelii (Nakai ex Kom. et Aliss.) Y. C. Tang ex J. Q. Hu ●

345576 Sambucus buergeriana Blume f. cordifoliata Skvortsov et W. Wang = Sambucus williamsii Hance ●

345577 Sambucus buergeriana Blume var. miquelii Nakai = Sambucus williamsii Hance var. miquelii (Nakai ex Kom. et Aliss.) Y. C. Tang ex J. Q. Hu ●

345578 Sambucus caerulea Raf. ;西方蓝果接骨木;Blue Berry Elder, Blue Elder, Blue Elderberry ●☆

345579 Sambucus callicarpa Greene;太平洋红果接骨木;Blue Elder, Blueberry Elder, Coast Red Elder, Coast Red Elderberry, Pacific Coast Red Elder, Pacific Red Elder, Red Elderberry, Red-berry Elder ●☆

345580 Sambucus canadensis L. ;美洲接骨木(加拿大接骨木) ; America Elder, American Elder, American Elderberry, Autumn-flowering Elder, Canadian Elder, Common Elder, Common

Elderberry, Elder, Elderberry, Sweet Elder, Sweet Elderberry, White Elder ●☆

345581　Sambucus canadensis L. 'Adams'；丰果美洲接骨木；Fruiting Elderberry ●☆

345582　Sambucus canadensis L. 'Aurea'；金叶美洲接骨木(黄叶加拿大接骨木)；Golden-leaved Elderberry ●☆

345583　Sambucus canadensis L. 'Johns'；约翰美洲接骨木；Fruiting Elderberry ●☆

345584　Sambucus canadensis L. var. laciniata A. Gray ＝ Sambucus canadensis L. ●☆

345585　Sambucus canadensis L. var. submollis Rehder ＝ Sambucus canadensis L. ●☆

345586　Sambucus chinensis Lindl.；接骨草(八棱麻,八里麻,赤苓叶,臭草,臭根草,臭黄金,戳戳苗,大臭草,大血草,地马桑,藋,赶山虎,葛辣,公道老,茇,鸡䔖风,尖尾青,接骨木,堇草,苛草,龙州三七,陆英,落得打,马鞭三七,马龙符,冇骨消,排风草,排风藤,七叶金,七叶麻,扫地风,珊瑚花,水马桑,蒴藋,蒴藋苗,蒴翟,台湾接骨木,铁篱笆,乌鸡腿,五甲皮,小臭牡丹,小接骨丹,血满草,秧心草,英雄草,游民草,珍珠花,珍珠连,珍珠麻,真珠花,走马风,走马箭)；China Elder, Chinese Elder, Taiwan Elder ●■

345587　Sambucus chinensis Lindl. ＝ Sambucus javanica Blume ■●

345588　Sambucus chinensis Lindl. var. formosana (Nakai) H. Hara；台湾接骨草●■

345589　Sambucus chinensis Lindl. var. formosana (Nakai) H. Hara ＝ Sambucus javanica Blume ■●

345590　Sambucus chinensis Lindl. var. pinnatilobatus G. W. Hu ＝ Sambucus javanica Blume ■●

345591　Sambucus coerulea Raf.；蓝色接骨木(蓝果接骨木)；Blue Elder, Blue Elderberry, Blueberry Elder ●☆

345592　Sambucus coreana (Nakai) Kom. et Aliss.；朝鲜接骨木(马尿烧)；Korean Elder ●

345593　Sambucus coreana (Nakai) Kom. et Aliss. ＝ Sambucus williamsii Hance ●

345594　Sambucus ebulus L.；矮接骨木；Bloodwort, Borr-tree, Daneball, Dane's Blood, Dane's Elder, Dane's Wood, Daneweed, Danewort, Deadwort, Deathwort, Dwarf Elder, Dwarf Elderberry, Eble, Ground Elder, Herbaceous Elder, Lithewort, Lithwort, Mediterranean Herb Elder, Mediterranean Herb-elder, Red Elder, Walewort, Wallwort, Water Elder, Wild Elder ●☆

345595　Sambucus ebulus L. subsp. africana (Engl.) Bolli；非洲接骨木●☆

345596　Sambucus ebulus L. var. africana Engl. ＝ Sambucus ebulus L. subsp. africana (Engl.) Bolli ●☆

345597　Sambucus foetidissima Nakai ＝ Sambucus williamsii Hance ●

345598　Sambucus foetidissima Nakai f. flava Skvortsov et W. Wang ＝ Sambucus williamsii Hance ●

345599　Sambucus formosana Nakai ＝ Sambucus chinensis Lindl. var. formosana (Nakai) H. Hara ●■

345600　Sambucus formosana Nakai ＝ Sambucus chinensis Lindl. ●■

345601　Sambucus formosana Nakai ＝ Sambucus javanica Blume ■●

345602　Sambucus formosana Nakai var. arborescens Kaneh. et Sasaki ＝ Sambucus javanica Blume ■●

345603　Sambucus gautschii Wettst. ＝ Sambucus adnata Wall. ●

345604　Sambucus glauca Nutt. ex Torr. et Gray；蓝果接骨木；Blue Elder, Blueberry Elder, Bluefruit Elder, California Elder, Californian Elder ●

345605　Sambucus henriana Samutina ＝ Sambucus javanica Blume ■●

345606　Sambucus hookeri Rehder ＝ Sambucus chinensis Lindl. ●■

345607　Sambucus hookeri Rehder ＝ Sambucus javanica Blume ■●

345608　Sambucus japonica Thunb. ＝ Euscaphis japonica (Thunb.) Kanitz ●

345609　Sambucus javanica Blume；爪哇接骨草■●

345610　Sambucus javanica Blume subsp. chinensis (Lindl.) Fukuoka ＝ Sambucus chinensis Lindl. ●■

345611　Sambucus javanica Blume subsp. chinensis (Lindl.) Fukuoka ＝ Sambucus javanica Blume ■●

345612　Sambucus javanica Blume var. argyi (H. Lév.) Rehder ＝ Sambucus javanica Blume ■●

345613　Sambucus javanica Blume var. formosana (Nakai) Schwer. ＝ Sambucus chinensis Lindl. var. formosana (Nakai) H. Hara ●■

345614　Sambucus javanica Reinw. ex Blume ＝ Sambucus chinensis Lindl. ●■

345615　Sambucus javanica Thunb. var. argyi (H. Lév.) Rehder ＝ Sambucus chinensis Lindl. ●■

345616　Sambucus junnanica J. J. Vassil. ＝ Sambucus williamsii Hance ●

345617　Sambucus kamtschatica E. L. Wolf ＝ Sambucus racemosa L. subsp. kamtschatica (E. L. Wolf) Hultén ●☆

345618　Sambucus lanceolata R. Br.；披针接骨木●☆

345619　Sambucus latipinata Nakai var. pendula Skvortsov ＝ Sambucus williamsii Hance ●

345620　Sambucus latipinna Nakai；宽叶接骨木；Arrow-bearing Tree, Broadleaf Elder ●

345621　Sambucus latipinna Nakai ＝ Sambucus williamsii Hance ●

345622　Sambucus latipinna Nakai var. coreana Nakai ＝ Sambucus coreana (Nakai) Kom. et Aliss. ●

345623　Sambucus leucocarpa Hort. ex K. Koch；白果接骨木●☆

345624　Sambucus manshurica Kitag.；东北接骨木(马尿烧)；Manchrian Elder ●

345625　Sambucus manshurica Kitag. ＝ Sambucus williamsii Hance ●

345626　Sambucus melanocarpa A. Gray；黑果接骨木；Black Elderberry, Blackbead Elder, Black-fruit Elder ●☆

345627　Sambucus mexicana C. Presl ex DC.；墨西哥接骨木；Arizina Elder, Blue Elderberry, Desert Elderberry, Mexican Elder ●☆

345628　Sambucus microbotrys Rydb.；山地红接骨木；Bunchberry Elder, Mountain Red Elderberry ●☆

345629　Sambucus microsperma Nakai；小籽接骨木(小花接骨木)●☆

345630　Sambucus miquelii Nakai ex Kom. et Aliss ＝ Sambucus williamsii Hance var. miquelii (Nakai ex Kom. et Aliss.) Y. C. Tang ex J. Q. Hu ●

345631　Sambucus nigra L.；西洋接骨木(黑果接骨木,黑接骨木,欧洲接骨木,洋接骨木)；Aldern, Alderne, Arntree, Baw-tree, Bertery, Bitter Medicine, Bitter-flower, Bitter-medicine, Black Beauty Elderberry, Black Elder, Boon Tree, Boon-tree, Bootry, Bore-tree, Borral, Bortree, Bor-tree, Bothery-tree, Bottery, Bottry, Bountree, Boun-tree, Bour Tree, Bourtree, Boutrey, Bown-tree, Bulltree, Bur Tree, Buttery, Cauliflower, Common Elder, Devil's Wood, Dog Tree, Edder, Elder, Elderberry, Eldern, Eldon, Ellarne, Ellen, Ellen-tree, Eller, Ellern, Ellern Aul, Ellet, Ell-shinders, Ellum, Elren, European Black Elderberry, European Elder, God's Stinking Tree, Golden Elder, Gypsy's Treacle, Hertfordshire Weed, Judas Tree, Purple-berry, Scaw, Scow, Skaw, Skew, Stinking Elder, Tea Flower, Tea Tree, Tramman, Welsh Vine, Whit Aller, Witch Tree, Woody Elder, Yellow Shinders Ell ●

345632　Sambucus nigra L. 'Albopunctata'；白斑西洋接骨木；Black Elderberry ●☆

345633　Sambucus nigra L. 'Aurea';金叶西洋接骨木（黄叶西洋接骨木）;Golden American Elder, Golden Elder, Golden Elderberry, Yellow-leaved Elderberry ●☆

345634　Sambucus nigra L. 'Aureomarginata';黄边西洋接骨木;Variegated Black Elder ●☆

345635　Sambucus nigra L. 'Guincho Purple';紫叶西洋接骨木;Black Elderberry, Purple-leaved Elderberry ●☆

345636　Sambucus nigra L. 'Laciniata';深裂西洋接骨木;Black Elder, Cut-leaf Elderberry, Fern-leaved Black Elder, Fern-leaved Elder ●☆

345637　Sambucus nigra L. 'Linearis';线叶西洋接骨木;Fern-leaved Elder, Thread-leaved Elderberry ●☆

345638　Sambucus nigra L. 'Madonna';金斑西洋接骨木;Golden-variegata Black Elder, Variegated Black Elderberry ●☆

345639　Sambucus nigra L. 'Marginata';金边西洋接骨木;Variegated Black Elder ●☆

345640　Sambucus nigra L. 'Nana';矮生西洋接骨木;Dwarf Black Elder ●☆

345641　Sambucus nigra L. 'Plena';重瓣西洋接骨木;Black Elderberry ●☆

345642　Sambucus nigra L. 'Viridis';绿果西洋接骨木●☆

345643　Sambucus nigra L. f. porphyrifolia E. C. Nelson;英国紫叶接骨木;Purple-leaved Black Elderberry ●☆

345644　Sambucus nigra L. subsp. canadensis（L.）Bolli = Sambucus canadensis L. ●☆

345645　Sambucus nigra L. subsp. canadensis L. 'Acutiloba';蕨叶美洲接骨木;Fernleaf American Elder ●☆

345646　Sambucus nigra L. subsp. canadensis L. 'Maxima';大美洲接骨木;American Elderberry ●☆

345647　Sambucus nigra L. subsp. canadensis L. 'Variegata';彩色美洲接骨木;American Elderberry ●☆

345648　Sambucus nigra L. subsp. cerulea（Raf.）Bolli;天蓝美洲接骨木;Blue Elderberry, Blueberry Elder ●☆

345649　Sambucus nigra L. var. aurea ? = Sambucus nigra L. 'Aurea' ●☆

345650　Sambucus nigra L. var. rotundifolia ?;圆叶西洋接骨木;Round-leaved Elder ●☆

345651　Sambucus palmensis Link;帕尔马接骨木●☆

345652　Sambucus pendula Nakai;悬垂接骨木●☆

345653　Sambucus peninsularis Kitag.;长尾叶接骨木●

345654　Sambucus peninsularis Kitag. = Sambucus williamsii Hance ●

345655　Sambucus peruviana Kunth;秘鲁接骨木;Peruvian Elder ●☆

345656　Sambucus potaninii J. J. Vassil. = Sambucus williamsii Hance ●

345657　Sambucus pubens F. Michx. dissecta（Britton）Schwer. = Sambucus racemosa L. subsp. pubens（Michx.）House ●☆

345658　Sambucus pubens Michx. = Sambucus racemosa L. subsp. pubens（Michx.）House ●☆

345659　Sambucus pubens Michx. = Sambucus racemosa L. ●☆

345660　Sambucus racemosa L.;欧洲接骨木（戳树,公道老树,红果接骨木,接骨木,欧接骨木,欧洲红果接骨木,柔毛接骨木,蒴树,西洋赤接骨木,也卜涛,野黄杨,总花接骨木）;Alpine Elder, American Red Elder, Eastern Red Elderberry, Elderberry, European Red Elder, European Red Elderberry, Mountain Elder, Red Elder, Red Elderberry, Red-berried Elder, Red-berrier Elder, Red-berriered Elder, Scarlet Elder, Scarlet Elderberry, Scarlet-berried Elder, Stinking Elder, Stinking Elderberry ●☆

345661　Sambucus racemosa L. 'Aurea';金叶欧洲接骨木（金叶红果接骨木）●☆

345662　Sambucus racemosa L. 'Laciniata';深裂欧洲接骨木（深裂红果接骨木）●☆

345663　Sambucus racemosa L. 'Plumosa Aurea';金羽欧洲接骨木（欧洲接骨木）●☆

345664　Sambucus racemosa L. 'Plumosa';羽毛欧洲接骨木●☆

345665　Sambucus racemosa L. 'Tenuifolia';薄叶欧洲接骨木（薄叶红果接骨木）;Fernleaf Red Elderberry ●☆

345666　Sambucus racemosa L. = Sambucus williamsii Hance ●

345667　Sambucus racemosa L. f. microphylla Honda;小叶欧洲接骨木（小叶红果接骨木,小叶兰筛朴）●☆

345668　Sambucus racemosa L. f. purpurascens Nakai;紫花欧洲接骨木（紫花红果接骨木,紫花兰筛朴）●☆

345669　Sambucus racemosa L. subsp. kamtschatica（E. L. Wolf）Hultén;勘察加接骨木●☆

345670　Sambucus racemosa L. subsp. kamtschatica（E. L. Wolf）Hultén f. aureocarpa（H. Hara）H. Hara;黄果勘察加接骨木●☆

345671　Sambucus racemosa L. subsp. kamtschatica（E. L. Wolf）Hultén f. hiraii（Miyabe et Tatew.）H. Hara;平井氏接骨木●☆

345672　Sambucus racemosa L. subsp. kamtschatica（E. L. Wolf）Hultén f. lacera（Nakai）H. Hara;撕裂接骨木●☆

345673　Sambucus racemosa L. subsp. kamtschatica（E. L. Wolf）Hultén f. nakayamae（Yanagita）H. Hara;中山氏接骨木●☆

345674　Sambucus racemosa L. subsp. kamtschatica（E. L. Wolf）Hultén f. nuda H. Hara;裸勘察加接骨木●☆

345675　Sambucus racemosa L. subsp. kamtschatica（E. L. Wolf）Hultén f. platyphylla（Miyabe et Tatew.）H. Hara;阔叶勘察加接骨木●☆

345676　Sambucus racemosa L. subsp. manshurica（Kitag.）Vorosh. = Sambucus williamsii Hance ●

345677　Sambucus racemosa L. subsp. pubens（Michx.）House;柔毛欧洲接骨木;American Red Elderberry, Red Elderberry, Red-berried Elder, Scarlet Elderberry, Stinking Elder ●☆

345678　Sambucus racemosa L. subsp. sibirica H. Hara = Sambucus sibirica Nakai ●

345679　Sambucus racemosa L. subsp. sieboldiana（Miq.）H. Hara;无梗接骨木（接骨木,蓝筛朴,小接骨丹）;Chinese Elderberry, Siebold Elder ●

345680　Sambucus racemosa L. subsp. sieboldiana（Miq.）H. Hara f. aurantiaca（Nakai）Murata;橘色无梗接骨木●☆

345681　Sambucus racemosa L. subsp. sieboldiana（Miq.）H. Hara f. dissecta Murata;深裂接骨木●☆

345682　Sambucus racemosa L. subsp. sieboldiana（Miq.）H. Hara f. glaberrima（Nakai）Murata;光无梗接骨木;Bigleaf Siebold Elder ●☆

345683　Sambucus racemosa L. subsp. sieboldiana（Miq.）H. Hara f. macrophylla H. Hara;大叶无梗接骨木●☆

345684　Sambucus racemosa L. subsp. sieboldiana（Miq.）H. Hara f. nakaiana Murata;中井氏接骨木●☆

345685　Sambucus racemosa L. subsp. sieboldiana（Miq.）H. Hara f. ovatifolia H. Hara;卵叶无梗接骨木●☆

345686　Sambucus racemosa L. subsp. sieboldiana（Miq.）H. Hara f. stenophylla（Nakai）H. Hara;狭叶无梗接骨木●☆

345687　Sambucus racemosa L. subsp. sieboldiana（Miq.）H. Hara var. major（Nakai）Murata;大无梗接骨木●☆

345688　Sambucus racemosa L. var. laciniata Koch;裂叶欧洲接骨木（红果接骨木）●☆

345689　Sambucus racemosa L. var. pubens（Michx.）Koehne = Sambucus racemosa L. subsp. pubens（Michx.）House ●☆

345690　Sambucus sachalinensis Pojark.;萨哈林接骨木（库页接骨木）;Sachalin Elder ●☆

345691　Sambucus schweriniana Rehder = Sambucus adnata Wall. ex

DC. ●

345692　Sambucus sibirica Nakai；西伯利亚接骨木；Siberia Elder, Siberian Elder ●

345693　Sambucus sieboldiana (Miq.) Blume ex Graebn. = Sambucus racemosa L. subsp. sieboldiana (Miq.) H. Hara ●

345694　Sambucus sieboldiana (Miq.) Blume ex Graebn. f. glaberrima (Nakai) H. Hara = Sambucus racemosa L. subsp. sieboldiana (Miq.) H. Hara f. glaberrima (Nakai) Murata ●☆

345695　Sambucus sieboldiana (Miq.) Blume ex Graebn. f. microphylla Honda；小叶无梗接骨木；Smallleaf Siebold Elder ●☆

345696　Sambucus sieboldiana (Miq.) Blume ex Graebn. f. purpurascens Nakai；紫花无梗接骨木；Purpleflower Siebold Elder ●☆

345697　Sambucus sieboldiana (Miq.) Blume ex Graebn. f. xanthocarpa Rehder；黄果无梗接骨木；Yellowfruit Siebold Elder ●☆

345698　Sambucus sieboldiana (Miq.) Blume ex Graebn. var. miquelii (Nakai ex Kom. et Aliss) H. Hara = Sambucus williamsii Hance var. miquelii (Nakai ex Kom. et Aliss.) Y. C. Tang ex J. Q. Hu ●

345699　Sambucus sieboldiana (Miq.) Blume ex Graebn. var. miquelii (Nakai) H. Hara = Sambucus racemosa L. subsp. kamtschatica (E. L. Wolf) Hultén ●☆

345700　Sambucus sieboldiana (Miq.) Blume ex Graebn. var. miquelii (Nakai) H. Hara = Sambucus williamsii Hance ●

345701　Sambucus sieboldiana (Miq.) Blume ex Graebn. var. pinnatisecta G. Y. Luo et P. H. Huang；羽裂接骨木；Pinnateleaf Siebold Elder ●

345702　Sambucus sieboldiana Blume = Sambucus williamsii Hance ●

345703　Sambucus simpsonii Rehder；西印度接骨木(辛普森接骨木)；West Indian Elder ●☆

345704　Sambucus simpsonii Rehder ex Sarg. = Sambucus canadensis L. ●☆

345705　Sambucus tiliifolia Wall. = Torricellia tiliifolia (Wall.) DC. ●

345706　Sambucus wightiana Wall. = Sambucus adnata Wall. ●

345707　Sambucus wightiana Wall. ex Wight et Arn. = Sambucus adnata Wall. ●

345708　Sambucus williamsii Hance；接骨木(戳树,大接骨丹,大叶接骨木,大叶蒴藋,放棍行,公道老,公道老树,钩齿接骨木,钩齿叶接骨木,接骨草,接骨丹,接骨风,九节风,马尿骚,马尿梢,马尿烧,木蒴藋,木英,七叶黄荆,七叶金,扦扦活,芊芊活,珊瑚配,舒筋树,蒴树,铁箍散,铁骨散,透骨草,小油木,续骨草,续骨木,野黄杨,野杨树,章漆木,樟木树)；Chinese Elder, Foetid Elder, Williams Elder ●

345709　Sambucus williamsii Hance var. miquelii (Nakai ex Kom. et Aliss.) Y. C. Tang ex J. Q. Hu；毛接骨木(本巴木,本把木,公道老,接骨木,马尿梢,马尿烧,米氏接骨木,扦扦活,兴安接骨木)；Buerger's Adnate Elder, Buerger's Elder, Miquel Elder ●

345710　Sambucus williamsii Hance var. miquelii (Nakai) Y. C. Tang ex J. Q. Hu = Sambucus sibirica Nakai ●

345711　Sameraria Desv. (1815)；翅果菘蓝属■☆

345712　Sameraria Desv. = Isatis L. ■

345713　Sameraria armena (L.) Desv.；亚美尼亚翅果菘蓝■☆

345714　Sameraria armena Desv. = Sameraria armena (L.) Desv. ■☆

345715　Sameraria bullata (Arch. et Hemsl.) B. Fedtsch.；泡状翅果菘蓝■☆

345716　Sameraria canaliculata Vassilcz.；具沟翅果菘蓝■☆

345717　Sameraria deserti N. Busch；荒漠翅果菘蓝■☆

345718　Sameraria glastifolia (Fisch. et C. A. Mey.) Boiss.；菘蓝叶翅果菘蓝■☆

345719　Sameraria litvinovii N. Busch；利特维诺夫翅果菘蓝■☆

345720　Sameraria sclerocarpa Bordz.；硬果翅果菘蓝■☆

345721　Sameraria turcomanica (Korsh.) B. Fedtsch.；土库曼翅果菘蓝■☆

345722　Samodia Baudo = Samolus L. ■

345723　Samolaceae Dumort. = Primulaceae Batsch ex Borkh. (保留科名)●■

345724　Samolaceae Dumort. = Theophrastaceae D. Don(保留科名)●■

345725　Samolaceae Raf. (1820)；水茴草科■

345726　Samolaceae Raf. = Primulaceae Batsch ex Borkh. (保留科名)●■

345727　Samolus Ehrh. = Samolus L. ■

345728　Samolus L. (1753)；水茴草属(水繁缕属)；Brookweed ■

345729　Samolus floribundus Kunth；多花水茴草；Brookweed, Water Pimpernel ■☆

345730　Samolus floribundus Kunth = Samolus valerandi L. subsp. parviflorus (Raf.) Hultén ■☆

345731　Samolus parviflorus Raf.；小花水茴草；Brookweed, Seaside Brookweed, Small-flowered Brookweed, Water Pimpernel ■☆

345732　Samolus valerandi L. subsp. parviflorus (Raf.) Hultén = Samolus parviflorus Raf. ■☆

345733　Samolus valerandii L.；水茴草；Broohweed, Brookweed, Round Pimpernel, Round-leaved Water Pimpernel, Seaside Brookweed, Water Pimpernel, Worldwise ■

345734　Samolus valerandii L. = Samolus parviflorus Raf. ■☆

345735　Samolus valerandii L. subsp. parviflorus (Raf.) Hultén = Samolus parviflorus Raf. ■☆

345736　Sampaca Raf. = Michelia L. ●

345737　Sampacca Kuntze = Michelia L. ●

345738　Sampacca Kuntze = Sampaea Raf. ●

345739　Sampacca excelsa (Wall.) Kuntze = Michelia doltsopa Buch. -Ham. ex DC. ●

345740　Sampacca kisopa (Buch. -Ham. ex DC.) Kuntze = Michelia kisopa Buch. -Ham. ex DC. ●◇

345741　Sampacca lanuginosa (Wall.) Kuntze = Michelia velutina DC. ●

345742　Sampacca longifolia (Blume) Kuntze = Michelia alba DC. ●

345743　Sampacca parviflora Kuntze = Michelia figo (Lour.) Spreng. ●

345744　Sampacca suaveolens (Pers.) Kuntze = Michelia champaca L. ●

345745　Sampaea Raf. = Champaca Adans. ●

345746　Sampaiella J. C. Gomes = Arrabidaea DC. ●☆

345747　Sampantaea Airy Shaw(1972)；东南亚大戟属●☆

345748　Sampantaea amentiflora (Airy Shaw) Airy Shaw；东南亚大戟●☆

345749　Samudra Raf. = Argyreia Lour. ●

345750　Samuela Trel. (1902)；南美龙舌兰属●☆

345751　Samuela Trel. = Yucca L. ●■

345752　Samuela carnerosana Trel.；南美龙舌兰；Palma Fibre ●☆

345753　Samuela faxoniana Trel. = Yucca faxoniana Sarg. ●☆

345754　Samuelssonia Urb. et Ekman(1929)；萨姆爵床属☆

345755　Samuelssonia verrucosa Urb. et Ekman；萨姆爵床☆

345756　Samyda Jacq. (1760)(保留属名)；美天料木属●☆

345757　Samyda L. (废弃属名) = Guarea F. Allam. (保留属名)●☆

345758　Samyda L. (废弃属名) = Samyda Jacq. (保留属名)●☆

345759　Samyda P. Br. = Casearia Jacq. ●

345760　Samydaceae Vent. (1808)(保留科名)；天料木科；Samyda Family ●

345761　Samydaceae Vent. (保留科名) = Flacourtiaceae Rich. ex DC. (保留科名)●

345762　Samydaceae Vent. (保留科名) = Salicaceae Mirb. (保留科名)●

345763　Sanamunda Adans. = Passerina L. ●☆

345764　Sanamunda Neck. = Daphne L. ●

345765　Sanango G. S. Bunting et J. A. Duke(1961);萨那木属●☆

345766　Sanango durum Bunting et Duke;萨那木●☆

345767　Sanblasia L. Andersson(1984);小卷苞竹芋属■☆

345768　Sanblasia dressleri L. Andersson;小卷苞竹芋■☆

345769　Sanchezia Ruiz et Pav. (1794);黄脉爵床属(金鸡蜡属,金脉木属,金叶木属);Sanchezia●■

345770　Sanchezia nobilis Hook. f. ;黄脉爵床(金鸡蜡,金叶木,山邱氏爵床);Edle Sanchezia,Noble Sanchezia●■

345771　Sanchezia parvibracteata Sprague et Hutch. ;小苞黄脉爵床■

345772　Sanchezia speciosa Léonard;美叶黄脉爵床(黄脉爵床,金鸡蜡,金脉爵床,金脉木,金叶木);Shrubby Whitevein●☆

345773　Sanctambrosia Skottsb. (1962);圣仙木属●☆

345774　Sanctambrosia Skottsb. = Sanctambrosia Skottsb. ex Kuschel ●☆

345775　Sanctambrosia Skottsb. ex Kuschel = Sanctambrosia Skottsb. ●☆

345776　Sanctambrosia manicata (Skottsb.) Skottsb. ;圣仙木●☆

345777　Sandbergia Greene = Halimolobos Tausch ■☆

345778　Sandemania Gleason(1939);桑氏野牡丹属●☆

345779　Sandemania hoehnei (Cogn.) Wurdack. ;桑氏野牡丹●☆

345780　Sanderella Kuntze(1891);桑德兰属■☆

345781　Sanderella discolor Cogn. ;桑德兰■☆

345782　Sandersonia Hook. (1853);灯笼百合属(圣诞钟属,提灯花属);Chinese Lantern,Christmas Bells,Sandersonia ■☆

345783　Sandersonia aurantiaca Hook. ;灯笼百合(圣诞钟,提灯花,中国灯);Chinese-lantern Lily,Christmas Bells ■☆

345784　Sandersonia littonioides Welw. ex Baker = Littonia littonioides (Welw. ex Baker) K. Krause ■☆

345785　Sandoricum Cav. (1789);山道楝属;Sandoricum,Santol ●☆

345786　Sandoricum indicum Cav. ;印度山道楝(山陀儿);Indian Katon,Indian Krathon,Santol,Sentul ●☆

345787　Sandoricum indicum Cav. = Sandoricum koetjape (Burm. f.) Merr. ●☆

345788　Sandoricum koetjape (Burm. f.) Merr. ;考特山道楝(山陀儿,酸多果);Katon,Santol,Sentul ●☆

345789　Sandoricum koetjape (Burm. f.) Merr. = Sandoricum indicum Cav. ●☆

345790　Sandoricum nervosum Blume;网脉山道楝●☆

345791　Sandwithia Lanj. (1932);桑德大戟属☆

345792　Sandwithia guyanensis Lanj. ;桑德大戟☆

345793　Sandwithia heterocalyx Secco;异萼桑德大戟☆

345794　Sandwithiodendron Aubrév. et Pellegr. = Pouteria Aubl. ●

345795　Sandwithiodendron Aubrév. et Pellegr. = Sandwithiodoxa Aubrév. et Pellegr. ●

345796　Sandwithiodoxa Aubrév. et Pellegr. = Pouteria Aubl. ●

345797　Sanfordia J. Drumm. ex Harv. = Geleznovia Turcz. ●☆

345798　Sanguilluma Plowes = Boucerosia Wight et Arn. ■☆

345799　Sanguilluma Plowes = Caralluma R. Br. ■☆

345800　Sanguilluma Plowes(1995);血红水牛角属■☆

345801　Sanguilluma socotrana (Balf. f.) Plowes = Caralluma socotrana (Balf. f.) N. E. Br. ■☆

345802　Sanguinaria Bubani = Digitaria Haller(保留属名)■

345803　Sanguinaria L. (1753);血根草属(血根属);Bloodroot,Blood-root,Puccoon,Sanguinaria ■☆

345804　Sanguinaria australis Greene = Sanguinaria canadensis L. ■☆

345805　Sanguinaria canadensis L. ;血根草(美洲血根草,血红根,血罂粟);Blood Root,Bloodroot,Blood-root,Bloodwort,Coon Root,Indian Paint,Puccoon,Red Puccoon,Redroot,Sanguinaria,Snakebite,Sweet Slumber,Tetterwort ■☆

345806　Sanguinaria canadensis L. 'Flore Pleno' = Sanguinaria canadensis L. 'Plena' ■☆

345807　Sanguinaria canadensis L. 'Multiplex' = Sanguinaria canadensis L. 'Plena' ■☆

345808　Sanguinaria canadensis L. 'Plena';重瓣血根草■☆

345809　Sanguinaria canadensis L. var. rotundifolia (Greene) Fedde = Sanguinaria canadensis L. ■☆

345810　Sanguinaria dilleniana Greene = Sanguinaria canadensis L. ■☆

345811　Sanguinaria rotundifolia Greene = Sanguinaria canadensis L. ■☆

345812　Sanguinaria vaginata (Sw.) Bubani = Paspalum vaginatum Sw. ■

345813　Sanguinella Gleichen = Digitaria Haller(保留属名)■

345814　Sanguinella Gleichen ex Steud. = Digitaria Haller(保留属名)■

345815　Sanguinella P. Beauv. = Manisuris L. (废弃属名)■

345816　Sanguisorba L. (1753);地榆属;Burnet ■

345817　Sanguisorba Rupp. ex L. = Sanguisorba L. ■

345818　Sanguisorba × poroshirensis S. Watan. ;幌尻地榆■☆

345819　Sanguisorba affinis C. A. Mey. ex Regel et Tiling = Sanguisorba tenuifolia Fisch. ex Link ■

345820　Sanguisorba albiflora (Makino) Makino;白花地榆■☆

345821　Sanguisorba alpina Bunge;高山地榆;Alpine Burnet ■

345822　Sanguisorba ancistroides (Desf.) A. Br. ;非洲地榆■☆

345823　Sanguisorba ancistroides (Desf.) A. Br. var. ballii Maire = Sanguisorba ancistroides (Desf.) A. Br. ■☆

345824　Sanguisorba ancistroides (Desf.) A. Br. var. battandieri Maire = Sanguisorba ancistroides (Desf.) A. Br. ■☆

345825　Sanguisorba ancistroides (Desf.) A. Br. var. castellorum Maire = Sanguisorba ancistroides (Desf.) A. Br. ■☆

345826　Sanguisorba ancistroides (Desf.) A. Br. var. diania Maire = Sanguisorba ancistroides (Desf.) A. Br. ■☆

345827　Sanguisorba ancistroides (Desf.) A. Br. var. dyris Maire = Sanguisorba ancistroides (Desf.) A. Br. ■☆

345828　Sanguisorba ancistroides (Desf.) A. Br. var. fontqueri Maire = Sanguisorba ancistroides (Desf.) A. Br. ■☆

345829　Sanguisorba ancistroides (Desf.) A. Br. var. glaberrima Maire = Sanguisorba ancistroides (Desf.) A. Br. ■☆

345830　Sanguisorba ancistroides (Desf.) A. Br. var. humbertii Maire = Sanguisorba ancistroides (Desf.) A. Br. ■☆

345831　Sanguisorba ancistroides (Desf.) A. Br. var. parviflora Pomel = Sanguisorba ancistroides (Desf.) A. Br. ■☆

345832　Sanguisorba applanata Te T. Yu et C. L. Li;宽蕊地榆;Broadstamen Burnet ■

345833　Sanguisorba applanata Te T. Yu et C. L. Li var. villosa Te T. Yu et C. L. Li;柔毛宽蕊地榆;Hairy Broadstamen Burnet,Villose Broadstamen Burnet ■

345834　Sanguisorba canadensis L. ;美洲地榆(加拿大地榆);American Burnet,American Great Burnet,American Great-burnet,Canadian Burnet,White Bumet ■☆

345835　Sanguisorba canadensis L. subsp. latifolia (Hook.) Calder et R. L. Taylor var. riishirensis (Makino) T. Shimizu;利尻地榆■☆

345836　Sanguisorba canadensis L. subsp. latifolia (Hook.) Calder et R. L. Taylor var. pilosa (H. Hara) T. Shimizu;毛宽叶美洲地榆■☆

345837　Sanguisorba canadensis L. subsp. latifolia (Hook.) Calder et R. L. Taylor;宽叶美洲地榆■☆

345838　Sanguisorba canadensis L. var. latifolia Hook. = Sanguisorba stipulata Raf. ■

345839　Sanguisorba canadensis L. var. sitchensis (C. A. Mey.) Koidz. = Sanguisorba stipulata Raf. ■

345840 Sanguisorba canadensis Torr. et A. Gray subsp. latifolia（Hook.）Calder = Sanguisorba stipulata Raf. ■

345841 Sanguisorba carnea Fisch. ex Link = Sanguisorba officinalis L. var. carnea（Fisch. ex Link）Regel ex Maxim. ■

345842 Sanguisorba diandra（Hook. f.）Nordborg；疏花地榆；Looseflower Burnet ■

345843 Sanguisorba diandra Wall. ex Nordborg = Sanguisorba diandra（Hook. f.）Nordborg ■

345844 Sanguisorba dissita Te T. Yu et C. L. Li = Sanguisorba diandra（Hook. f.）Nordborg ■

345845 Sanguisorba filiformis（Hook. f.）Hand.-Mazz.；矮地榆（虫莲,海参,五母那包,线叶地榆）；Dwarf Burnet ■

345846 Sanguisorba formosana Hayata；台湾地榆■

345847 Sanguisorba formosana Hayata = Sanguisorba officinalis L. var. longifolia（Bertol.）Te T. Yu et C. L. Li ■

345848 Sanguisorba glandulosa Kom. = Sanguisorba officinalis L. var. glandulosa（Kom.）Vorosch. ■

345849 Sanguisorba grandiflora（Maxim.）Makino = Sanguisorba tenuifolia Fisch. ex Link var. grandiflora Maxim. ■☆

345850 Sanguisorba hakusanensis Makino；白山地榆■☆

345851 Sanguisorba hakusanensis Makino var. coreana H. Hara；朝鲜地榆■☆

345852 Sanguisorba hakusanensis Makino var. japonensis（Makino）Ohwi = Sanguisorba japonensis（Makino）Kudo ■☆

345853 Sanguisorba japonensis（Makino）Kudo；日本地榆■☆

345854 Sanguisorba latifolia（Hook.）Coville = Sanguisorba stipulata Raf. ■

345855 Sanguisorba linostemon Hand.-Mazz. = Sanguisorba alpina Bunge ■

345856 Sanguisorba longifolia Bertol. = Sanguisorba officinalis L. var. longifolia（Bertol.）Te T. Yu et C. L. Li ■

345857 Sanguisorba longifolia Bertol. var. longifila（Kitag.）Kitag. = Sanguisorba officinalis L. var. longifila（Kitag.）Te T. Yu et C. L. Li ■

345858 Sanguisorba magnifica Schischk. et Kom.；大蕊地榆■☆

345859 Sanguisorba mauritanica Desf.；毛里塔尼亚地榆■☆

345860 Sanguisorba minor Scop.；小地榆；Bloodwort, Burnet, Lesser Burnet, Salad Burnet, Small Burnet ■☆

345861 Sanguisorba minor Scop. subsp. alveolosa（Spach）Maire；蜂窝小地榆■☆

345862 Sanguisorba minor Scop. subsp. anceps（Ball）Maire = Sanguisorba minor Scop. subsp. maroccana（Coss.）Maire ■☆

345863 Sanguisorba minor Scop. subsp. balearica（Nyman）Munoz Garm. et C. Navarro；巴莱小地榆；Small Burnet ■☆

345864 Sanguisorba minor Scop. subsp. magnolii（Spach）Briq. = Sanguisorba verrucosa（Don）Ces. ■☆

345865 Sanguisorba minor Scop. subsp. maroccana（Coss.）Maire；摩洛哥小地榆■☆

345866 Sanguisorba minor Scop. subsp. muricata（Bonnier et Layens）Briq.；粗糙小地榆；Fodder Burnet, Pimpinell, Salad Burnet, Small Burnet ■☆

345867 Sanguisorba minor Scop. subsp. muricata（Spach）Briq. = Sanguisorba minor Scop. subsp. balearica（Nyman）Munoz Garm. et C. Navarro ■☆

345868 Sanguisorba minor Scop. subsp. rupicola（Boiss. et Reut.）Nordborg = Sanguisorba rupicola（Boiss. et Reut.）A. Braun et Bouché ■☆

345869 Sanguisorba minor Scop. subsp. spachiana（Coss.）Munoz Garm. et Pedrol = Sanguisorba verrucosa（Don）Ces. ■☆

345870 Sanguisorba minor Scop. subsp. verrucosa（G. Don）Cout. = Sanguisorba verrucosa（Don）Ces. ■☆

345871 Sanguisorba minor Scop. subsp. vestita（Pomel）Maire；包被小地榆■☆

345872 Sanguisorba minor Scop. var. atlantica Litard. et Maire = Sanguisorba minor Scop. subsp. maroccana（Coss.）Maire ■☆

345873 Sanguisorba minor Scop. var. megacarpa（Lowe）Briq. = Sanguisorba verrucosa（Don）Ces. ■☆

345874 Sanguisorba minor Scop. var. microcarpa（Boiss.）Briq. = Sanguisorba verrucosa（Don）Ces. ■☆

345875 Sanguisorba minor Scop. var. stenolopha（Spach）Maire = Sanguisorba minor Scop. ■☆

345876 Sanguisorba minor Scop. var. transiens Maire = Sanguisorba minor Scop. ■☆

345877 Sanguisorba montana Jord. = Sanguisorba officinalis L. ■

345878 Sanguisorba multicaulis（Boiss. et Reut.）Asch. et Graebn. = Sanguisorba rupicola（Boiss. et Reut.）A. Braun et Bouché ■☆

345879 Sanguisorba muricata（Spach）Gremli = Sanguisorba minor Scop. subsp. muricata（Spach）Briq. ■☆

345880 Sanguisorba obtusa Maxim.；钝地榆■☆

345881 Sanguisorba officinalis L.；地榆（白地榆,鞭枣胡子,赤地榆,地榆炭,菜菜,红地榆,红绣球,花椒地榆,黄根子,黄瓜香,蕨菜参,马边鞍薯,马猴枣,马连鞍薯,蒙古枣,涩地榆,山红枣,山枣参,山枣仁,山枣子,鼠尾地榆,水槟榔,水橄榄,酸赭,台湾地榆,土儿红,豚榆系,无名印,线形地榆,小紫草,血箭草,岩地芨,岩地莈,野升麻,野生地榆,一枝箭,玉豉,玉札,杂花地榆,枣儿红,赭酢枣,紫地榆,紫朵苗子）；Bloodwort, Burnet, Burnet Bloodwort, Drumsticks, Evergreen Burnet, Garden Burner, Garden Burnet, Great Burnet, Greater Salad Burnet, Hardhead, Lamb's Lettuce, Maiden's Head, Maiden's Heads, Old Man's Pepper, Parasol, Pimpernel, Poor Man's Pepper, Prieldy Salad Burnet, Red Head, Red Knob, Salad Burnet, Toper's Plant, Upland Burnet ■

345882 Sanguisorba officinalis L. f. dulutiflora Kitag.；浅花地榆■

345883 Sanguisorba officinalis L. f. pallescens Asai = Sanguisorba longifolia Bertol. ■

345884 Sanguisorba officinalis L. f. pilosella（Ohwi）H. Hara；毛地榆■☆

345885 Sanguisorba officinalis L. subsp. longifolia（Bertol.）K. M. Purohit et Panigrahi = Sanguisorba officinalis L. var. longifolia（Bertol.）Te T. Yu et C. L. Li ■

345886 Sanguisorba officinalis L. subsp. microcephala（C. Presl）Calder = Sanguisorba officinalis L. ■

345887 Sanguisorba officinalis L. var. carnea（Fisch. ex Link）Regel ex Maxim. = Sanguisorba officinalis L. ■

345888 Sanguisorba officinalis L. var. carnea（Fisch. ex Link）Regel ex Maxim.；粉花地榆（肉色地榆）；Pinkflower Burnet ■

345889 Sanguisorba officinalis L. var. carnea（Fisch.）Regel ex Maxim. = Sanguisorba officinalis L. var. carnea（Fisch. ex Link）Regel ex Maxim. ■

345890 Sanguisorba officinalis L. var. glandulosa（Kom.）Vorosch.；腺地榆；Glandular Burnet ■

345891 Sanguisorba officinalis L. var. glandulosa（Kom.）Vorosch. = Sanguisorba officinalis L. f. pilosella（Ohwi）H. Hara ■☆

345892 Sanguisorba officinalis L. var. latifolia Liou et C. Y. Li；宽叶地榆；Broadleaf Garden Burnet ■

345893 Sanguisorba officinalis L. var. longa Kitag. = Sanguisorba officinalis L. ■

345894 Sanguisorba officinalis L. var. longifila（Kitag.）Te T. Yu et C. L. Li；长蕊地榆（白地榆，长地榆，长穗地榆，长叶地榆，赤地榆，大花地榆，地榆，红地榆，花椒地榆，黄根子，黄瓜香，蕨苗参，马边鞍薯，马猴枣，马连鞍薯，绵地榆，涩地榆，山红枣，山枣参，山枣子，鼠尾地榆，水槟榔，水橄榄，台湾地榆，线形地榆，岩地芰，枣儿红，直穗粉花地榆，紫地榆）；Longleaf Burnet, Longstamen Burnet ■

345895 Sanguisorba officinalis L. var. longifolia（Bertol.）Te T. Yu et C. L. Li；长叶地榆 ■

345896 Sanguisorba officinalis L. var. microcephala Kitag. = Sanguisorba officinalis L. ■

345897 Sanguisorba officinalis L. var. montana（Jord.）Focke = Sanguisorba officinalis L. ■

345898 Sanguisorba officinalis L. var. pilosella Ohwi = Sanguisorba officinalis L. f. pilosella（Ohwi）H. Hara ■☆

345899 Sanguisorba officinalis L. var. polygama（Nyl.）Serg. = Sanguisorba officinalis L. ■

345900 Sanguisorba parviflora（Maxim.）Takeda = Sanguisorba tenuifolia Fisch. ex Link var. alba Trautv. ■

345901 Sanguisorba parviflora（Maxim.）Takeda = Sanguisorba tenuifolia Fisch. ex Link var. parviflora Maxim. ■☆

345902 Sanguisorba polygama Nyl. = Sanguisorba officinalis L. ■

345903 Sanguisorba rectispicata Kitag. = Sanguisorba officinalis L. var. longifolia（Bertol.）Te T. Yu et C. L. Li ■

345904 Sanguisorba rectispicata Kitag. var. longifila Kitag. = Sanguisorba officinalis L. var. longifila（Kitag.）Te T. Yu et C. L. Li ■

345905 Sanguisorba rhodopaea（Velen.）Hayek = Sanguisorba minor Scop. subsp. muricata（Spach）Briq. ■☆

345906 Sanguisorba riparia Juz.；河岸地榆 ■☆

345907 Sanguisorba rupicola（Boiss. et Reut.）A. Braun et Bouché；岩生地榆 ■☆

345908 Sanguisorba sitchensis C. A. Mey. = Sanguisorba canadensis L. subsp. latifolia（Hook.）Calder et R. L. Taylor ■

345909 Sanguisorba sitchensis C. A. Mey. = Sanguisorba stipulata Raf. ■

345910 Sanguisorba spinosa（L.）Bertol. = Sarcopoterium spinosum（L.）Spach ●☆

345911 Sanguisorba stipulata Raf.；大白花地榆（白花地榆，大白地榆）；Sikta Burnet ■

345912 Sanguisorba stipulata Raf. = Sanguisorba canadensis L. subsp. latifolia（Hook.）Calder et R. L. Taylor ■☆

345913 Sanguisorba stipulata Raf. var. latifolia（Hook.）H. Hara = Sanguisorba stipulata Raf. ■

345914 Sanguisorba stipulata Raf. var. latifolia H. Hara = Sanguisorba sitchensis C. A. Mey. ■

345915 Sanguisorba stipulata Raf. var. pilosa（H. Hara）H. Hara = Sanguisorba canadensis L. subsp. latifolia（Hook.）Calder et R. L. Taylor var. pilosa（H. Hara）T. Shimizu ■☆

345916 Sanguisorba stipulata Raf. var. riishirensis（Makino）H. Hara = Sanguisorba canadensis L. subsp. latifolia（Hook.）Calder et R. L. Taylor var. riishirensis（Makino）T. Shimizu ■☆

345917 Sanguisorba tenuifolia Fisch. ex Link；细叶地榆（垂穗粉花地榆，地榆，黄攸香，狭叶地榆，野鸡冠花）；Burnet, Siberia Burnet, Siberian Burnet ■

345918 Sanguisorba tenuifolia Fisch. ex Link f. alba（Trautv. et C. A. Mey.）Kitam. = Sanguisorba tenuifolia Fisch. ex Link ■

345919 Sanguisorba tenuifolia Fisch. ex Link f. purpurea（Trautv. et C. A. Mey.）W. T. Lee = Sanguisorba tenuifolia Fisch. ex Link ■

345920 Sanguisorba tenuifolia Fisch. ex Link subsp. grandiflora（Maxim.）Toyok. = Sanguisorba tenuifolia Fisch. ex Link var. grandiflora Maxim. ■☆

345921 Sanguisorba tenuifolia Fisch. ex Link var. alba Trautv.；小白花地榆（地榆，小花地榆）；Smallflower Burnet, Whiteflower Siberia Burnet ■

345922 Sanguisorba tenuifolia Fisch. ex Link var. alba Trautv. et C. A. Mey. = Sanguisorba tenuifolia Fisch. ex Link ■

345923 Sanguisorba tenuifolia Fisch. ex Link var. grandiflora Maxim.；大花细叶地榆 ■☆

345924 Sanguisorba tenuifolia Fisch. ex Link var. parviflora Maxim.；小花细叶地榆 ■☆

345925 Sanguisorba tenuifolia Fisch. ex Link var. parviflora Maxim. = Sanguisorba tenuifolia Fisch. ex Link ■

345926 Sanguisorba tenuifolia Fisch. ex Link var. parviflora Maxim. f. coccinea（Koidz.）Sugim. = Sanguisorba tenuifolia Fisch. ex Link var. parviflora Maxim. ■☆

345927 Sanguisorba tenuifolia Fisch. ex Link var. parviflora Maxim. f. pilosa H. Hara；毛小花细叶地榆 ■☆

345928 Sanguisorba tenuifolia Fisch. ex Link var. purpurea Trautv. et C. A. Mey. = Sanguisorba tenuifolia Fisch. ex Link ■

345929 Sanguisorba tenuifolia Korsh. = Sanguisorba tenuifolia Fisch. ex Link ■

345930 Sanguisorba tenuifolia var. parviflora Maxim. = Sanguisorba tenuifolia Fisch. ex Link var. alba Trautv. ■

345931 Sanguisorba tenuifolia var. purpurea Trautv. et C. A. Mey. = Sanguisorba tenuifolia Fisch. ex Link ■

345932 Sanguisorba verrucosa（Don）Ces.；多疣地榆 ■☆

345933 Sanguisorba vestita（Pomel）Nordborg = Sanguisorba minor Scop. subsp. vestita（Pomel）Maire ■☆

345934 Sanguisorbaceae Bercht. et J. Presl = Sanguisorbaceae Loisel. ■

345935 Sanguisorbaceae Loisel.；地榆科 ■

345936 Sanguisorbaceae Loisel. = Rosaceae Juss.（保留科名）●■

345937 Sanguisorbaceae Marquis = Rosaceae Juss.（保留科名）●■

345938 Sanhilaria Baill. = Pappobolus S. F. Blake ■●☆

345939 Sanhilaria Leandro ex DC. = Stifftia J. C. Mikan（保留属名）●☆

345940 Sanicula L.（1753）；变豆菜属（山蕲菜属，山芹菜属）；Black Snakeroot, Sanicle ■

345941 Sanicula astrantiifolia H. Wolff ex Krecz.；川滇变豆菜（草本三角枫，昆明变豆菜，三角枫，三台草，铜脚威灵仙，五角枫，五匹风，小黑药，叶三七）；Chuan-Dian Sanicle, Szechwan-Yunnan Sanicle ■

345942 Sanicula bipinnata Hook. et Arn.；二回羽状变豆菜；Poison Sanicle ■☆

345943 Sanicula bipinnatifida Douglas；紫变豆菜；Purple Sanicle ■☆

345944 Sanicula caerulescens Franch.；天蓝变豆菜（蓝山蕲菜，散血草，山五爪龙，心肺草）；Azure Sanicle ■

345945 Sanicula canadensis L.；加拿大变豆菜；Black Snakeroot, Canadian Black Snakeroot, Canadian Sanicle, Short-styled Snakeroot ■☆

345946 Sanicula canadensis L. var. grandis Fernald = Sanicula canadensis L. ■☆

345947 Sanicula capensis（Cham. et Schltdl.）Eckl. et Zeyh. = Sanicula elata Buch.-Ham. ex D. Don ■

345948 Sanicula chinensis Bunge；变豆菜（蓝布正，山芹菜，鸭巴芹，鸭脚板，鸭掌芹）；China Sanicle, Chinese Sanicle ■

345949 Sanicula chinensis Bunge f. paupera（Nakai）Kitag.；贫乏变豆菜 ■☆

345950　Sanicula costata H. Wolff ＝Sanicula orthacantha S. Moore ■

345951　Sanicula dielsiana H. Wolff ＝Sanicula caerulescens Franch. ■

345952　Sanicula elata Buch. -Ham. ex D. Don；软雀花（高变豆菜，三叶七，水茯苓）；High Sanicle ■

345953　Sanicula elata Buch. -Ham. ex D. Don var. japonica Koidz. ＝ Sanicula chinensis Bunge ■

345954　Sanicula elata Buch. -Ham. ex D. Don var. partita Kuntze ＝ Sanicula elata Buch. -Ham. ex D. Don ■

345955　Sanicula elongata K. T. Fu；长序变豆菜；Longcluster Sanicle ■

345956　Sanicula erythrophylla Bobrov ＝Sanicula caerulescens Franch. ■

345957　Sanicula europaea L.；欧洲变豆菜（变豆菜，欧变豆菜，山芥菜，鸭脚斑）；Butterwort, European Sanicle, Leechwort, Poolroot, Sanicle, Saniker, Self-heal, Wild London Pride, Wood Elder, Wood Marche, Wood Sanical, Wood Sanicle ■☆

345958　Sanicula europaea L. ＝Sanicula elata Buch. -Ham. ex D. Don ■

345959　Sanicula europaea L. sensu Hemsl. et Forbes ＝ Sanicula chinensis Bunge ■

345960　Sanicula europaea L. subsp. chinensis（Bunge）Hultén ＝ Sanicula chinensis Bunge ■

345961　Sanicula europaea L. subsp. elata（Buch. -Ham. ex D. Don）H. Boissieu ＝Sanicula elata Buch. -Ham. ex D. Don ■

345962　Sanicula europaea L. var. capensis Cham. et Schltdl. ＝Sanicula elata Buch. -Ham. ex D. Don ■

345963　Sanicula europaea L. var. chinensis（Bunge）Diels ＝Sanicula chinensis Bunge ■

345964　Sanicula europaea L. var. chinensis Diels ＝Sanicula chinensis Bunge ■

345965　Sanicula europaea L. var. elata（Buch. -Ham. ex D. Don）Boiss. ＝Sanicula elata Buch. -Ham. ex D. Don ■

345966　Sanicula europaea L. var. elata Boiss. ＝Sanicula elata Buch. -Ham. ex D. Don ■

345967　Sanicula europaea L. var. javanica H. Wolff ＝Sanicula elata Buch. -Ham. ex D. Don ■

345968　Sanicula europaea L. var. partita（Kuntze）M. Hiroe ＝Sanicula elata Buch. -Ham. ex D. Don ■

345969　Sanicula giraldii H. Wolff；首阳变豆菜（秦岭变豆菜，小黑药）；Clustered Black Snakeroot, Girald Sanicle, Shouyang Sanicle ■

345970　Sanicula giraldii H. Wolff var. ovicalycina R. H. Shan et S. L. Liou；卵萼变豆菜；Ovatecalyx Sanicle ■

345971　Sanicula gregaria E. P. Bicknell；丛生变豆菜；Clustered Snakeroot, Yellow Sanide ■☆

345972　Sanicula gregaria E. P. Bicknell ＝Sanicula odorata（Raf.）Pryer et Phillippe ■☆

345973　Sanicula hacquetioides Franch.；鳞果变豆菜（肾叶变豆菜）；Scalefruit Sanicle ■

345974　Sanicula henryi H. Wolff ＝Sanicula orthacantha S. Moore ■

345975　Sanicula hermaphrodita Buch. -Ham. ex D. Don ＝Sanicula elata Buch. -Ham. ex D. Don ■

345976　Sanicula ichangensis H. Wolff ＝Sanicula lamelligera Hance ■

345977　Sanicula javanica Blume；南山蕲菜；Java Sanicle ■

345978　Sanicula lamelligera Hance；薄片变豆菜（半边钱，大肺筋草，大肺经草，滇北山蕲菜，鹅掌脚草，反背红，肺筋草，肺经草，脐风草，三叶山芹菜，散血草，山果菜，松叶防风，乌兜，乌豆草，血经草，野芹菜，一枝箭）；Laminated Sanicle, Yunnan Sanicle ■

345979　Sanicula lamelligera Hance var. wakayamensis（Masam.）Murata；和歌山变豆菜●☆

345980　Sanicula marilandica L.；马里兰德变豆菜（黑变豆菜）；Black Sanicle, Black Snakeroot, Maryland Sanicle, Sanicle ■☆

345981　Sanicula marilandica L. var. petiolulata Fernald ＝Sanicula marilandica L. ■☆

345982　Sanicula menziesii Hook. et Arn.；门氏变豆菜；Gamble-weed ■☆

345983　Sanicula montana Reinw. ex Blume ＝Sanicula elata Buch. -Ham. ex D. Don ■

345984　Sanicula nanchuanensis R. H. Shan ＝Sanicula orthacantha S. Moore ■

345985　Sanicula natalensis Gand. ＝Sanicula elata Buch. -Ham. ex D. Don ■

345986　Sanicula odorata（Raf.）Pryer et Phillippe；香变豆菜；Bblack Snakeroot ■☆

345987　Sanicula odorata（Raf.）Pryer et Phillippe ＝Sanicula gregaria E. P. Bicknell ■☆

345988　Sanicula orthacantha S. Moore；直刺变豆菜（地黄连，黑鹅脚板，蒙自山蕲菜，蒙自山芹菜，水虎掌草，小紫花菜，野鹅脚板，直刺山芹菜）；Erectspine Sanicle, Henry Sanicle ■

345989　Sanicula orthacantha S. Moore var. brevispina H. Boissieu；短刺变豆菜（短刺鹅脚板，鸭脚七）；Shortspine Sanicle ■

345990　Sanicula orthacantha S. Moore var. costata（H. Wolff）K. T. Fu ＝Sanicula orthacantha S. Moore ■

345991　Sanicula orthacantha S. Moore var. longispina H. Wolff ＝Sanicula lamelligera Hance ■

345992　Sanicula orthacantha S. Moore var. pumila H. Boissieu ＝Sanicula orthacantha S. Moore ■

345993　Sanicula orthacantha S. Moore var. stolonifera R. H. Shan et S. L. Liou；走茎变豆菜（走茎鹅脚板）；Creeping Sanicle, Stoloniferous Sanicle ■

345994　Sanicula oviformis X. T. Liu et Z. Y. Liu；卵叶变豆菜；Ovateleaf Sanicle ■

345995　Sanicula pengshuiensis M. L. Sheh et Z. Y. Liu；彭水变豆菜；Pengshui Sanicle ■

345996　Sanicula petagnioides Hayata；台湾变豆菜（五叶山芹菜）；Taiwan Sanicle ■

345997　Sanicula potaninii Bobrov. ＝Sanicula astrantiifolia H. Wolff ex Krecz. ■

345998　Sanicula rubriflora F. Schmidt ex Maxim.；红花变豆菜（鸡爪芹，紫花变豆菜，紫花芹）；Redflower Sanicle ■

345999　Sanicula rugulosa Diels；皱叶变豆菜；Wrinkleleaf Sanicle ■

346000　Sanicula satsumana Maxim. ＝Sanicula lamelligera Hance ■

346001　Sanicula serrata H. Wolff；锯叶变豆菜（锯齿山蕲菜）；Serrateleaf Sanicle ■

346002　Sanicula smallii E. P. Bicknell；斯莫尔变豆菜；Black Snakeroot ■☆

346003　Sanicula stapfiana H. Wolff ＝Sanicula caerulescens Franch. ■

346004　Sanicula subgiraldii R. H. Shan ＝Sanicula giraldii H. Wolff var. ovicalycina R. H. Shan et S. L. Liou ■

346005　Sanicula tienmusis R. H. Shan et Constance；天目变豆菜；Tianmu Sanicle, Tienmu Sanicle ■

346006　Sanicula tienmusis R. H. Shan et Constance var. pauciflora R. H. Shan et F. T. Pu；疏花变豆菜；Looseflower Tianmu Sanicle ■

346007　Sanicula trifoliata E. P. Bicknell；三小叶变豆菜；Beaked Sanicle, Large-fruited Black Snakeroot, Long-fruited Sanide, Long-fruited Snakeroot ■☆

346008　Sanicula tuberculata Maxim.；瘤果变豆菜；Tubercled-fruit Sanicle ■

346009　Sanicula tuberosa Torr.；土耳其变豆菜；Turkey Pea ■☆

346010　Sanicula yunnanensis Franch. ＝Sanicula lamelligera Hance ■

346011　Saniculaceae（Drude）A. Löve et D. Löve ＝ Apiaceae Lindl.（保留科名）●■

346012　Saniculaceae（Drude）A. Löve et D. Löve ＝ Sansevieriaceae Nakai ■

346013　Saniculaceae（Drude）A. Löve et D. Löve ＝ Umbelliferae Juss.（保留科名）■●

346014　Saniculaceae A. Löve et D. Löve ＝ Apiaceae Lindl.（保留科名）●■

346015　Saniculaceae A. Löve et D. Löve ＝ Sansevieriaceae Nakai ■

346016　Saniculaceae A. Löve et D. Löve ＝ Umbelliferae Juss.（保留科名）■●

346017　Saniculaceae A. Löve et D. Löve；变豆菜科■

346018　Saniculaceae Bercht. et J. Presl ＝ Saniculaceae A. Löve et D. Löve ■

346019　Saniculiphyllum C. Y. Wu et T. C. Ku（1992）；变豆菜叶属（变豆叶草属）■★

346020　Saniculiphyllum guangxiense C. Y. Wu et T. C. Ku；变豆菜叶（变豆叶草）■

346021　Sanidophyllum Small ＝ Hypericum L. ■●

346022　Saniella Hilliard et B. L. Burtt（1978）；萨尼仙茅属■☆

346023　Saniella occidentalis（Nel）B. L. Burtt；西部萨尼仙茅■☆

346024　Saniella verna Hilliard et B. L. Burtt；萨尼仙茅■☆

346025　Sanilum Raf. ＝ Hewittia Wight et Arn. ■

346026　Sankowskya P. I. Forst.（1995）；昆士兰大戟属☆

346027　Sanmartina Traub ＝ Castellanoa Traub ■☆

346028　Sanmartinia M. Buchinger ＝ Eriogonum Michx. ●■☆

346029　Sannantha Peter G. Wilson ＝ Leptospermum J. R. Forst. et G. Forst.（保留属名）●☆

346030　Sannantha Peter G. Wilson（2007）；萨恩薄子木属●☆

346031　Sanopodium Hort. ex Rchb. ＝ Epigeneium Gagnep. ■

346032　Sanopodium Hort. ex Rchb. ＝ Sarcopodium Lindl. ■

346033　Sanrafaelia Verdc.（1996）；坦桑尼亚番荔枝属●☆

346034　Sansevera Stokes ＝ Sansevieria Thunb.（保留属名）■

346035　Sanseveria Raf. ＝ Sansevieria Thunb.（保留属名）■

346036　Sanseverina Thunb. ＝ Sansevieria Thunb.（保留属名）■

346037　Sanseverinia Petagna（废弃属名）＝ Sansevieria Thunb.（保留属名）■

346038　Sanseverinia rorida Lanza ＝ Sansevieria rorida（Lanza）N. E. Br. ■☆

346039　Sanseviella Rchb. ＝ Reineckea H. Karst. ●☆

346040　Sanseviella Rchb. ＝ Synechanthus H. Wendl. ●☆

346041　Sanseviera Willd. ＝ Sansevieria Thunb.（保留属名）■

346042　Sanseviera carnea Andr. ＝ Reineckea carnea（Andréws）Kunth ■

346043　Sanseviera guineensis Willd. ＝ Sansevieria guineensis Willd. ■☆

346044　Sanseviera metallica Gerome et Labroy ＝ Sansevieria metallica Gerome et Labroy ■☆

346045　Sanseviera sessiliflora Ker Gawl. ＝ Reineckea carnea（Andréws）Kunth ■

346046　Sanseviera trifasciata Prain ＝ Sansevieria trifasciata Prain ■

346047　Sansevieria Thunb.（1794）（保留属名）；虎尾兰属；African Bowstring Hemp，Bowstring Hemp，Bowstringhemp，Bowstring-hemp，Mother-in-law's Tongue，Sansevieria，Tigertaillily ■

346048　Sansevieria abyssinica N. E. Br. ＝ Sansevieria forskaliana（Schult. f.）Hepper et J. R. I. Wood ■☆

346049　Sansevieria abyssinica N. E. Br. var. angustior（Engl.）Cufod. ＝ Sansevieria forskaliana（Schult. f.）Hepper et J. R. I. Wood ■☆

346050　Sansevieria abyssinica N. E. Br. var. sublaevigata（Chiov.）Cufod. ＝ Sansevieria forskaliana（Schult. f.）Hepper et J. R. I. Wood ■☆

346051　Sansevieria aethiopica Thunb.；埃塞俄比亚虎尾兰■☆

346052　Sansevieria andradae God. -Leb. ＝ Sansevieria stuckyi God. -Leb. ■☆

346053　Sansevieria angolensis Welw. ex Baker ＝ Sansevieria cylindrica Bojer ex Hook. ■☆

346054　Sansevieria angustiflora Lindb. ＝ Sansevieria hyacinthoides（L.）Druce ■☆

346055　Sansevieria arborescens Cornu ex Servett. et Labroy；灌木状虎尾兰■●

346056　Sansevieria bagamoyensis N. E. Br.；巴加莫约虎尾兰■☆

346057　Sansevieria ballyi L. E. Newton；博利虎尾兰■☆

346058　Sansevieria bella L. E. Newton；雅致虎尾兰■☆

346059　Sansevieria bequaertii De Wild. ＝ Sansevieria parva N. E. Br. ■☆

346060　Sansevieria bracteata Baker；具苞虎尾兰■☆

346061　Sansevieria braunii Engl. et K. Krause；布劳恩虎尾兰■☆

346062　Sansevieria caespitosa Dinter ＝ Sansevieria aethiopica Thunb. ■☆

346063　Sansevieria canaliculata Carrière；柱叶虎尾兰（羊角虎尾兰）；Canaliculate Sansevieria ■☆

346064　Sansevieria carnea Andréws ＝ Reineckea carnea（Andréws）Kunth ■

346065　Sansevieria caulescens N. E. Br.；无茎虎尾兰■☆

346066　Sansevieria concinna N. E. Br.；整洁虎尾兰■☆

346067　Sansevieria conspicua N. E. Br.；显著虎尾兰■☆

346068　Sansevieria cornui Servett. et Labroy ＝ Sansevieria senegambica Baker ■☆

346069　Sansevieria cylindrica Bojer ＝ Sansevieria cylindrica Bojer ex Hook. ■☆

346070　Sansevieria cylindrica Bojer ex Hook.；圆叶虎尾兰（棒叶虎尾兰，鹿角掌，筒叶虎尾兰，羊角兰，圆柱虎尾兰）；Cylindrical Sansevieria ■☆

346071　Sansevieria cylindrica Bojer ex Hook. var. patula N. E. Br.；张开虎尾兰■☆

346072　Sansevieria dawei Stapf；道氏虎尾兰■☆

346073　Sansevieria desertii N. E. Br. ＝ Sansevieria pearsonii N. E. Br. ■☆

346074　Sansevieria ehrenbergii Schweinf. ex Baker；爱伦堡虎尾兰■☆

346075　Sansevieria eilensis Chahin.；埃勒虎尾兰■☆

346076　Sansevieria elliptica（Chiov.）Cufod. ＝ Sansevieria forskaliana（Schult. f.）Hepper et J. R. I. Wood ■☆

346077　Sansevieria ensifolia Haw. ＝ Sansevieria grandicuspis Haw. ■☆

346078　Sansevieria erythraeae Mattei；浅红虎尾兰■☆

346079　Sansevieria fasciata Cornu ex Servett. et Labroy；带状虎尾兰■☆

346080　Sansevieria fischeri（Baker）Marais；菲舍尔虎尾兰■☆

346081　Sansevieria forskaliana（Schult. f.）Hepper et J. R. I. Wood；福斯虎尾兰■☆

346082　Sansevieria fragrans Jacq.；芳香虎尾兰■☆

346083　Sansevieria francisii Chahin.；弗朗西斯虎尾兰■☆

346084　Sansevieria frequens Chahin.；常见虎尾兰■☆

346085　Sansevieria glauca Servett. et Labroy ＝ Sansevieria aethiopica Thunb. ■☆

346086　Sansevieria gracilis N. E. Br.；细虎尾兰■☆

346087　Sansevieria gracillima Chahin.；细长虎尾兰■☆

346088　Sansevieria grandicuspis Haw.；大尖虎尾兰■☆

346089　Sansevieria grandis Hook. f.；大花虎尾兰（扇叶虎尾兰）；Largeflower Sansevieria ■☆

346090　Sansevieria grandis Hook. f. ＝ Sansevieria hyacinthoides（L.）Druce ■☆

346091　Sansevieria grandis Hook. f. var. zuluensis N. E. Br. = Sansevieria hyacinthoides (L.) Druce ■☆

346092　Sansevieria guineensis (L.) Willd. = Sansevieria hyacinthoides (L.) Druce ■☆

346093　Sansevieria guineensis (L.) Willd. var. angustior Engl. = Sansevieria forskaoliana (Schult. f.) Hepper et J. R. I. Wood ■☆

346094　Sansevieria guineensis Willd.;非洲虎尾兰;African Hemp, Bowstring Hemp ■☆

346095　Sansevieria guineensis Willd. = Sansevieria thyrsiflora Thunb. ■☆

346096　Sansevieria hahnii Hort.;小虎尾(金边小虎尾)■☆

346097　Sansevieria hallii Chahin.;霍尔虎尾兰 ■☆

346098　Sansevieria hargeisana Chahin. = Sansevieria phillipsiae N. E. Br. ■☆

346099　Sansevieria humbertiana Guillaumin = Sansevieria volkensii Gürke ■☆

346100　Sansevieria hyacinthoides (L.) Druce;大叶虎尾兰;African Bowstring Hemp, Bowstring Hemp, Iguanatail ■☆

346101　Sansevieria intermedia N. E. Br.;间型虎尾兰 ■☆

346102　Sansevieria intermedia N. E. Br. = Sansevieria volkensii Gürke ■☆

346103　Sansevieria jacquinii N. E. Br. = Sansevieria trifasciata Prain ■☆

346104　Sansevieria kirkii Baker;柯克虎尾兰 ■☆

346105　Sansevieria kirkii Baker var. pulchra N. E. Br.;星柯克虎尾兰;Star Sansevieria ■☆

346106　Sansevieria latifolia Bojer = Sansevieria hyacinthoides (L.) Druce ■☆

346107　Sansevieria laurentii De Wild. = Sansevieria trifasciata Prain var. laurentii (De Wild.) N. E. Br. ■

346108　Sansevieria liberica Gentil = Sansevieria liberica Servett. et Labroy ■☆

346109　Sansevieria liberica Servett. et Labroy;利比里亚虎尾兰;Liberia Sansevieria ■☆

346110　Sansevieria liberiensis Cornu ex A. Chev. = Sansevieria senegambica Baker ■☆

346111　Sansevieria livingstoniae Rendle = Sansevieria cylindrica Bojer ex Hook. ■☆

346112　Sansevieria longiflora Sims;长花虎尾兰 ■☆

346113　Sansevieria longiflora Sims var. fernandopoensis N. E. Br.;费尔南虎尾兰 ■☆

346114　Sansevieria longistyla la Croix;长柱虎尾兰 ■☆

346115　Sansevieria masoniana Chahin.;梅森虎尾兰 ■☆

346116　Sansevieria massae (Chiov.) Cufod. = Sansevieria nilotica Baker ■☆

346117　Sansevieria metallica Servett. et Labroy;亮叶虎尾兰;Bowstring Hemp ■☆

346118　Sansevieria metallica Servett. et Labroy var. longituba N. E. Br.;长管虎尾兰 ■☆

346119　Sansevieria metallica Servett. et Labroy var. nyasica N. E. Br.;尼亚斯虎尾兰 ■☆

346120　Sansevieria nilotica Baker;尼罗河虎尾兰(尼罗虎尾兰,千岁兰)■☆

346121　Sansevieria nilotica Baker var. obscura N. E. Br.;隐匿虎尾兰 ■☆

346122　Sansevieria nitida Chahin.;光亮虎尾兰 ■☆

346123　Sansevieria paniculata Schinz = Dracaena aletriformis (Haw.) Bos ●☆

346124　Sansevieria parva N. E. Br.;肯尼亚虎尾兰;Kenya Hyacinth ■☆

346125　Sansevieria patens N. E. Br.;铺展虎尾兰 ■☆

346126　Sansevieria pearsonii N. E. Br.;皮尔逊虎尾兰 ■☆

346127　Sansevieria pedicellata la Croix;梗花虎尾兰 ■☆

346128　Sansevieria perrotii Warb.;佩罗虎尾兰 ■☆

346129　Sansevieria phillipsiae N. E. Br.;菲利虎尾兰 ■☆

346130　Sansevieria pinguicula Bally;肥厚虎尾兰 ■☆

346131　Sansevieria polyrhitis (Chiov.) Cufod. = Sansevieria volkensii Gürke ■☆

346132　Sansevieria powellii N. E. Br.;鲍威尔虎尾兰 ■☆

346133　Sansevieria pumila Haw. = Sansevieria grandicuspis Haw. ■☆

346134　Sansevieria quarrei De Wild.;卡雷虎尾兰 ■☆

346135　Sansevieria raffillii N. E. Br.;拉菲尔虎尾兰 ■☆

346136　Sansevieria raffillii N. E. Br. var. glauca ?;灰绿虎尾兰 ■☆

346137　Sansevieria rhodesiana N. E. Br.;罗得西亚虎尾兰 ■☆

346138　Sansevieria robusta N. E. Br.;粗壮虎尾兰 ■☆

346139　Sansevieria rorida (Lanza) N. E. Br.;雾状虎尾兰 ■☆

346140　Sansevieria roxburghiana Schult. f.;罗氏虎尾兰 ■☆

346141　Sansevieria scabrifolia Dinter = Sansevieria aethiopica Thunb. ■☆

346142　Sansevieria schimperi Baker = Sansevieria canaliculata Carrière ■☆

346143　Sansevieria schweinfurthii Täckh. et Drar = Sansevieria erythraeae Mattei ■☆

346144　Sansevieria senegambica Baker;塞内虎尾兰 ■☆

346145　Sansevieria sessiliflora Ker Gawl. = Reineckea carnea (Andréws) Kunth ■

346146　Sansevieria singularis N. E. Br. = Sansevieria fischeri (Baker) Marais ■☆

346147　Sansevieria sordida N. E. Br.;暗色虎尾兰 ■☆

346148　Sansevieria stuckyi God. -Leb.;石笔虎尾兰 ■☆

346149　Sansevieria subspicata Baker;穗状虎尾兰 ■☆

346150　Sansevieria subtilis N. E. Br.;纤细虎尾兰 ■☆

346151　Sansevieria suffruticosa N. E. Br.;亚灌木虎尾兰 ●☆

346152　Sansevieria suffruticosa N. E. Br. var. longituba Pfennig;长管亚灌木虎尾兰 ●☆

346153　Sansevieria sulcata Bojer ex Baker = Sansevieria canaliculata Carrière ■☆

346154　Sansevieria thunbergii Mattei =Sansevieria aethiopica Thunb. ■☆

346155　Sansevieria thyrsiflora Thunb.;伞花虎尾兰(圆锥花虎尾兰);Sweet Sansevieria, Thyrsus Sansevieria ■☆

346156　Sansevieria thyrsiflora Thunb. = Sansevieria hyacinthoides (L.) Druce ■☆

346157　Sansevieria trifasciata Hort. = Sansevieria zeylanica (L.) Willd. ■☆

346158　Sansevieria trifasciata Prain;虎尾兰(弓弦麻,花蛇草,老虎尾,龙舌兰,千岁兰);Bayonet Plant, Congo Snake, Leopard Lily, Mother-in-law's Tongue, Snake Plant, Snake Sansevieria, Sweet Sansevierin, Tigertaillily, Viper's Bowstring Hemp ■

346159　Sansevieria trifasciata Prain 'Argentea-striata';银纹虎尾兰 ■

346160　Sansevieria trifasciata Prain 'Bantels Sensation';白斑虎尾兰 ■

346161　Sansevieria trifasciata Prain 'Craigii';黄斑虎尾兰 ■

346162　Sansevieria trifasciata Prain 'Golden Hahnii';黄纹短叶虎尾兰(黄纹短叶虎尾兰,金边短叶虎尾兰)■

346163　Sansevieria trifasciata Prain 'Hahnii';短叶虎尾兰;Shortleaf Sansevieria ■

346164　Sansevieria trifasciata Prain 'Laurentii Compacta';密叶金边虎尾兰 ■

346165　Sansevieria trifasciata Prain 'Laurentii';金边虎尾兰(虎尾兰,金边短叶虎尾兰);Golden Margin Sansevieria ■

346166　Sansevieria trifasciata Prain 'Silver Hahnii Marginata';镶边银短叶虎尾兰 ■

346167　Sansevieria trifasciata Prain 'Silver Hahnii';银短叶虎尾兰■

346168　Sansevieria trifasciata Prain var. hanhnii Hort. = Sansevieria trifasciata Prain 'Hahnii'■

346169　Sansevieria trifasciata Prain var. laurentii（De Wild.）N. E. Br. = Sansevieria trifasciata Prain 'Laurentii'■

346170　Sansevieria trifasciata Prain var. macrophylla Hort.;栽培大叶虎尾兰;Largeleaf Sansevieria ■☆

346171　Sansevieria volkensii Gürke;福尔虎尾兰■☆

346172　Sansevieria zanzibarica Servett. et Labroy;桑给巴尔虎尾兰■☆

346173　Sansevieria zeylanica（L.）Willd.;锡兰虎尾兰（虎耳兰麻,虎尾兰）;Bowstring Hemp,Ceylon Sansevieria,Ceylon Tigertaillily ■☆

346174　Sansevieria zeylanica Willd. = Sansevieria zeylanica（L.）Willd.■☆

346175　Sansevieriaceae Nakai = Agavaceae Dumort.（保留科名）●■

346176　Sansevieriaceae Nakai = Dracaenaceae Salisb.（保留科名）●

346177　Sansevieriaceae Nakai = Ruscaceae M. Roem.（保留科名）●

346178　Sansevieriaceae Nakai;虎尾兰科■

346179　Sansevieroa Post et Kuntze = Sansevieria Thunb.（保留属名）■

346180　Sansovinia Scop. = Leea D. Royen ex L.（保留属名）●■

346181　Santalaceae R. Br.（1810）（保留科名）;檀香科;Bastard-toadflax Family,Sandalwood Family ●■

346182　Santalina Baill.（1890）;檀茜草属●☆

346183　Santalina Baill. = Enterospermum Hiern ●

346184　Santalina Baill. = Tarenna Gaertn. ●

346185　Santalina madagascariensis Baill.;马岛檀茜草●☆

346186　Santalodes Kuntze（废弃属名）= Rourea Aubl.（保留属名）●

346187　Santalodes Kuntze（废弃属名）= Santaloides G. Schellenb.（保留属名）●☆

346188　Santalodes L. ex Kuntze = Rourea Aubl.（保留属名）●

346189　Santalodes bakeri Kuntze = Rourea solanderi Baker ●☆

346190　Santalodes hermanniana Kuntze = Rourea minor（Gaertn.）Leenh. ●

346191　Santaloidella G. Schellenb. = Rourea Aubl.（保留属名）●

346192　Santaloidella G. Schellenb. = Santaloides G. Schellenb.（保留属名）●☆

346193　Santaloidella gilletii G. Schellenb. = Rourea parviflora Gilg ●☆

346194　Santaloides G. Schellenb.（1910）（保留属名）;肖红叶藤属●☆

346195　Santaloides G. Schellenb.（保留属名）= Rourea Aubl.（保留属名）●

346196　Santaloides L. = Rourea Aubl.（保留属名）●

346197　Santaloides afzelii（R. Br. ex Planch.）G. Schellenb. = Rourea minor（Gaertn.）Alston ●

346198　Santaloides brachyandra（F. Muell.）G. Schellenb.;肖红叶藤●☆

346199　Santaloides caudata Kuntze = Rourea caudata Planch. ●

346200　Santaloides gossweileri Exell et Mendonça = Rourea minor（Gaertn.）Alston ●

346201　Santaloides gudjuana（Gilg）G. Schellenb. = Rourea minor（Gaertn.）Alston ●

346202　Santaloides microphylla（Hook. et Arn.）G. Schellenb. = Rourea microphylla（Hook. et Arn.）Planch. ●

346203　Santaloides microphylla Schellenb. = Rourea microphylla（Hook. et Arn.）Planch. ●

346204　Santaloides minor（Gaertn.）G. Schellenb. = Rourea minor（Gaertn.）Leenh. ●

346205　Santaloides roxburghii Kuntze = Rourea minor（Gaertn.）Leenh. ●

346206　Santaloides splendida（Gilg）G. Schellenb. = Rourea minor（Gaertn.）Alston ●

346207　Santaloides urophylla G. Schellenb. = Rourea minor（Gaertn.）Alston ●

346208　Santalum L.（1753）;檀香属;Sandal,Sandal Wood,Sandalwood ●

346209　Santalum acuminatum（R. Br.）A. DC.;渐尖檀香（框档树）;Dong, Dong Nut, Indian Sandalwood, Northern Sandalwood, Quandong,Sweet Quandong,Whie Sandalwood ●☆

346210　Santalum album L.;檀香（白檀,白檀木,白檀香,白银香,白英古,白英石,白栴檀,驳马,黄檀香,黄英香,老山檀,六驳,檀香树,雪梨檀,浴檀,浴香,栴,栴檀,栴檀娜,真檀,真檀香,真香）;Sandal,Sandal Wood,Sandaltree,Sandal-tree,Sandalwood,Sanders,White Sandal Wood,White Sandalwood,Yellow Sandalwood ●

346211　Santalum boninense（Nakai）Tuyama;小笠原檀香●☆

346212　Santalum lanceolatum R. Br.;披针叶檀香（垂枝檀香）;Australian Sandalwood, Lance-leaf SandalwoodPlum Wood, Northern Sandalwood,Plumwood ●☆

346213　Santalum longifolium Meurisse;长叶檀香●☆

346214　Santalum multiflorum J. W. Moore;多花檀香●☆

346215　Santalum murrayanum C. A. Gardner;穆氏檀香●☆

346216　Santalum myrtifolium L. = Santalum album L. ●

346217　Santalum obtusifolium R. Br.;钝叶檀香●☆

346218　Santalum papuanum Summerh.;巴布亚檀香;Papua Sandalwood ●

346219　Santalum spicatum（R. Br.）A. DC.;穗花檀香;Fragrant Sandalwood ●☆

346220　Santanderia Cespedes ex Triana et Planch. = Talauma Juss. ●

346221　Santapaua N. P. Balakr. et Subram. = Hygrophila R. Br. ●■

346222　Santia Savi = Polypogon Desf. ■

346223　Santia Wight et Arn. = Lasianthus Jack（保留属名）●

346224　Santiera Span. = Sautiera Decne. ☆

346225　Santiria Blume（1850）;山地榄属（滇榄属）;Santiria ●

346226　Santiria balsamifera Oliv. = Santiria trimera（Oliv.）Aubrév. ●☆

346227　Santiria glaberrima（Engl.）H. J. Lam;无毛山地榄●☆

346228　Santiria kamerunensis（Engl.）H. J. Lam;喀麦隆山地榄●☆

346229　Santiria obovata（Pierre）H. J. Lam;倒卵山地榄●☆

346230　Santiria tessmannii（K. Krause）H. J. Lam;泰斯曼山地榄●☆

346231　Santiria trimera（Oliv.）Aubrév.;三数山地榄●☆

346232　Santiria yunnanensis Hu;山地榄（滇榄）;Yunnan Santiria ●

346233　Santiria yunnanensis Hu = Protium yunnanense（Hu）Kalkman ●◇

346234　Santiridium Pierre = Dacryodes Vahl ●☆

346235　Santiridium Pierre = Pachylobus G. Don ●☆

346236　Santiriopsis Engl. = Santiria Blume ●

346237　Santiriopsis balsamifera（Oliv.）Engl. = Santiria trimera（Oliv.）Aubrév. ●☆

346238　Santiriopsis glaberrima Engl. = Santiria glaberrima（Engl.）H. J. Lam ●☆

346239　Santiriopsis klaineana Pierre = Dacryodes klaineana（Pierre）H. J. Lam ●☆

346240　Santiriopsis mayumbensis（Exell）Exell et Mendonça = Santiria trimera（Oliv.）Aubrév. ●☆

346241　Santisukia Brummitt（1992）;桑蒂紫葳属●☆

346242　Santisukia kerrii（Barnett et Sandwith）Brummitt;桑蒂紫葳●☆

346243　Santolina L.（1753）;银香菊属（绵杉菊属,棉杉菊属,神麻菊属,神圣亚麻属,圣麻属）;Cotton Lavender, Holy Flax, Lavender Cotton, Lavender-cotton, Santolina ●☆

346244　Santolina Tourn. = Santolina L. ●☆

346245　Santolina africana Jord. et Fourr.;非洲银香菊●☆

346246　Santolina alpina Bertol. = Lasiospermum pedunculare Lag. ■☆

346247 Santolina ascensionis Sennen ＝Santolina africana Jord. et Fourr. ●☆

346248 Santolina canescens Lag. ＝Santolina pectinata Lag. ●☆

346249 Santolina capitata L. ＝Athanasia capitata（L.）L. ●☆

346250 Santolina chamaecyparissus L. ；白神麻菊（白神圣麻，棉杉菊，圣麻，薰衣草棉）；Common Lavender-cotton，Cotton Lavender，Cypress Lavender Cotton，Cypress Lavender-cotton，French Lavender，Garden Cypress，Ground Cypress，Lavender Corn，Lavender Cotton，Lavender-cotton，Santolina，Silver Lavender ●☆

346251 Santolina chamaecyparissus L. 'Nana'；矮神麻菊（矮白神圣亚麻）；Dwarf Silver Santolina ●☆

346252 Santolina crenata L. ＝Athanasia crenata（L.）L. ●☆

346253 Santolina crithmifolia L. ＝Athanasia crithmifolia（L.）L. ●☆

346254 Santolina dentata L. ＝Athanasia dentata（L.）L. ●☆

346255 Santolina elegans Boiss. ex DC. ；雅致神麻菊（雅致神圣亚麻）●☆

346256 Santolina erecta Lam. ＝Lasiospermum pedunculare Lag. ■☆

346257 Santolina eriosperma Pers. ＝Lasiospermum pedunculare Lag. ■☆

346258 Santolina fragrantissima Forssk. ＝Achillea fragrantissima（Forssk.）Sch. Bip. ■☆

346259 Santolina incana Lam. ＝Santolina chamaecyparissus L. ●☆

346260 Santolina laevigata L. ＝Athanasia dentata（L.）L. ●☆

346261 Santolina neapolitana Jord. et Fourr. ＝Santolina pinnata Viv. subsp. neapolitana（Jord. et Fourr.）Guinea ex C. Jeffrey ●☆

346262 Santolina oblongifolia Boiss. ；长圆叶神麻菊●☆

346263 Santolina oppositifolia L. ＝Isocarpha oppositifolia（L.）Cass. ■☆

346264 Santolina pectinata Lag. ；篦状神麻菊●☆

346265 Santolina pectinata Lag. subsp. subclausa H. Lindb. ＝Santolina pectinata Lag. ●☆

346266 Santolina pinnata Donn ＝Lasiospermum pedunculare Lag. ■☆

346267 Santolina pinnata Viv. ；羽叶神麻菊（羽裂圣麻，羽叶圣亚麻）；Green Santolina ●☆

346268 Santolina pinnata Viv. subsp. neapolitana（Jord. et Fourr.）Guinea ex C. Jeffrey；柠檬神麻菊（柠檬羽裂圣麻）●☆

346269 Santolina pinnata Viv. subsp. neapolitana（Jord. et Fourr.）Guinea ex C. Jeffrey 'Sulphurea'；硫黄神麻菊（硫黄柠檬羽裂圣麻）●☆

346270 Santolina pubescens L. ＝Athanasia pubescens（L.）L. ●☆

346271 Santolina rosmarinifolia L. ；绿神麻菊（绿神圣亚麻，迷迭香叶圣麻）；Green Lavender Cotton，Green Santolina，Holy Flax ●☆

346272 Santolina rosmarinifolia L. 'Primrose Gem'；黄宝石神麻菊（淡黄宝石迷迭香叶圣麻）●☆

346273 Santolina rosmarinifolia L. subsp. canescens（Lag.）Nyman ＝Santolina pectinata Lag. ●☆

346274 Santolina rosmarinifolia L. subsp. pectinata（Lag.）Maire ＝Santolina pectinata Lag. ●☆

346275 Santolina rosmarinifolia L. var. fruticosa Maire ＝Santolina pectinata Lag. ●☆

346276 Santolina rosmarinifolia L. var. leptocephala（Webb）Pau et Font Quer ＝Santolina pectinata Lag. ●☆

346277 Santolina rosmarinifolia L. var. pharaonis Maire ＝Santolina pectinata Lag. ●☆

346278 Santolina rosmarinifolia L. var. subclausa（H. Lindb.）Jahand. et Maire ＝Santolina pectinata Lag. ●☆

346279 Santolina scariosa Ball ＝Cladanthus scariosus（Ball）Oberpr. et Vogt ●☆

346280 Santolina squarrosa L. ＝Oedera squarrosa（L.）Anderb. et K. Bremer ●☆

346281 Santolina suaveolens Pursh；香神麻菊（香母菊，香神圣亚麻）●☆

346282 Santolina suaveolens Pursh ＝Matricaria discoidea DC. ■

346283 Santolina suaveolens Pursh ＝Matricaria matricarioides（Less.）Ced. Porter ex Britton ■

346284 Santolina trifurcata L. ＝Athanasia trifurcata（L.）L. ●☆

346285 Santolina virens Mill. ＝Santolina rosmarinifolia L. ●☆

346286 Santolinaceae Augier ex Martinov ＝Asteraceae Bercht. et J. Presl(保留科名)●■

346287 Santolinaceae Augier ex Martinov ＝Compositae Giseke(保留科名)●■

346288 Santolinaceae Augier ex Martinov；银香菊科（神麻菊科，圣麻科）●■

346289 Santolinaceae Martinov ＝Asteraceae Bercht. et J. Presl(保留科名)●■

346290 Santolinaceae Martinov ＝Compositae Giseke(保留科名)●■

346291 Santomasia N. Robson(1981)；南美藤黄属●☆

346292 Santomasia N. Robson ＝Hypericum L. ■●

346293 Santomasia steyermarkii（Standl.）N. Robson；南美藤黄●☆

346294 Santonica Griff.（1848）；阿富汗菊属☆

346295 Santosia R. M. King et H. Rob.（1980）；藤本亮泽兰属●☆

346296 Santosia talmonii R. M. King et H. Rob. ；藤本亮泽兰●☆

346297 Santotomasia Ormerod(2008)；瓦尔德兰属■☆

346298 Sanvitalia Gualt. ＝Sanvitalia Lam. ■●

346299 Sanvitalia Gualt. ex Lam. ＝Sanvitalia Lam. ■●

346300 Sanvitalia Lam.（1792）；蛇目菊属（蛇纹菊属）；Creeping Zinnia，Sanvitalia ■●

346301 Sanvitalia aberti A. Gray；艾伯蛇目菊■☆

346302 Sanvitalia mexicana Vilm. ；墨西哥蛇目菊■☆

346303 Sanvitalia ocymoides DC. ；罗勒蛇目菊■☆

346304 Sanvitalia procumbens Lam. ；蛇目菊（凉菊，匍匐蛇目菊，蛇纹菊，卧茎蛇目菊）；Common Sanvitalia，Creeping Sanvitalia，Creeping Zinnia，Mexican Creeping Zinnia，Monarch of the Veldt，Monarch-of-the-veldt，Namaqua Daisy，Namaqualand Daisy ■

346305 Sanvitalia procumbens Lam. 'Mandarin Orange'；橙红蛇目菊（橙红蛇纹菊）■☆

346306 Sanvitalia versicolor Griseb. ；变色蛇目菊■☆

346307 Sanvitalia villosa Cav. ；毛蛇目菊■☆

346308 Sanvitaliopsis Sch. Bip. ex Benth. et Hook. f. ＝Zinnia L. (保留属名)●■

346309 Sanvitaliopsis Sch. Bip. ex Greenm.（1905）；类蛇目菊属■☆

346310 Sanvitaliopsis liebmannii（Klatt）K. H. Schultz ex Greenm. ；类蛇目菊■☆

346311 Sanvitaliopsis liebmannii Sch. Bip. ＝Sanvitaliopsis liebmannii（Klatt）K. H. Schultz ex Greenm. ■☆

346312 Sanvitaliopsis liebmannii Sch. Bip. ex Klatt ＝Sanvitaliopsis liebmannii（Klatt）K. H. Schultz ex Greenm. ■☆

346313 Saouari Aubl. ＝Caryocar F. Allam. ex L. ●☆

346314 Saphesia N. E. Br.（1932）；白环花属■☆

346315 Saphesia flaccida（Jacq.）N. E. Br. ；白环花●☆

346316 Sapindaceae Juss.（1789）(保留科名)；无患子科；Pride-of-India Family，Soapberry Family ●■

346317 Sapindaceae Juss. (保留科名)＝Turpinia Vent. (保留属名)●

346318 Sapindopsis F. C. How et C. N. Ho ＝Aphania Blume ●

346319 Sapindopsis F. C. How et C. N. Ho ＝Howethoa Rauschert ●

346320 Sapindopsis F. C. How et C. N. Ho ＝Lepisanthes Blume ●

346321 Sapindopsis oligophylla（Merr. et Chun）F. C. How et C. N. Ho ＝Aphania oligophylla（Merr. et Chun）H. S. Lo ●

346322 Sapindopsis oligophylla（Merr. et Chun）F. C. How et C. N. Ho

= Lepisanthes oligophylla (Merr. et Chun) N. H. Xia et Gadek ●
346323 Sapindus L. (1753)(保留属名);无患子属;Soap Nut, Soap Nuts, Soapberry, Soaptree ●
346324 Sapindus Tourn. ex L. = Sapindus L. (保留属名) ●
346325 Sapindus abruptus Lour. = Sapindus mukorossi Gaertn. ●
346326 Sapindus abruptus Lour. = Sapindus saponaria Lam. ●
346327 Sapindus acuminatus Wall. = Sapindus mukorossi Gaertn. ●
346328 Sapindus attenuatus Wall. = Lepisanthes senegalensis (Juss. ex Poir.) Leenh. ●
346329 Sapindus boninensis Tuyama = Sapindus mukorossi Gaertn. ●
346330 Sapindus capensis Sond. = Smelophyllum capense (Sond.) Radlk. ●☆
346331 Sapindus chinensis Lour. = Koelreuteria paniculata Laxm. ●
346332 Sapindus chinensis Murray = Koelreuteria paniculata Laxm. ●
346333 Sapindus delavayi (Franch.) Radlk.;川滇无患子(打冷冷,黑苦楝,皮哨子,菩提味,菩提子,无患子,油患心,油患子);Chuandian Soapberry, Delavay Soapberry ●
346334 Sapindus detergens Roxb. = Sapindus mukorossi Gaertn. ●
346335 Sapindus detergens Wall. = Sapindus mukorossi Gaertn. ●
346336 Sapindus drummondii Hook. et Arn.;德拉蒙德无患子(西方无患子);Chinaberry, Drummond's Soapberry, Jaboncillo, Soapberry, Western Soapberry, Wild China Tree, Wild Chinatree ●☆
346337 Sapindus laurifolius Vahl = Sapindus trifoliatus L. ●☆
346338 Sapindus mukorossi Gaertn.;无患子(肥皂树,肥珠子,桂圆肥皂,猴儿皂,桓,槵,黄木子,黄目树,黄目子,噤娄,苦患树,苦患子,苦枝子,卢鬼木,栌木,栾树,木桓,木患,木患树,木患子,木槵子,目浪,目浪树,槵,菩提子,棙娄,拾栌鬼木,拾栌木,无患树,无槵,洗衫,洗手果,油患子,油罗树,油珠子,圆肥皂);China Soapberry, Chinese Soapberry, Soap Nut Tree, Soapnut Tree, Soap-nut Tree, Soap-nut-tree ●
346339 Sapindus mukorossi Gaertn. = Sapindus saponaria Lam. ●
346340 Sapindus oblongifolius (E. Mey. ex Arn.) Sond. = Deinbollia oblongifolia (E. Mey. ex Arn.) Radlk. ●☆
346341 Sapindus oligophyllus Merr. et Chun = Aphania oligophylla (Merr. et Chun) H. S. Lo ●
346342 Sapindus oligophyllus Merr. et Chun = Lepisanthes oligophylla (Merr. et Chun) N. H. Xia et Gadek ●
346343 Sapindus pappea Sond. = Pappea capensis Eckl. et Zeyh. ●☆
346344 Sapindus rarak DC.;毛瓣无患子(买马萨,毛瓣,皮哨子);Hairypetal Soapberry, Hairy-petaled Soapberry ●
346345 Sapindus rarak DC. var. velutinus C. Y. Wu;石屏无患子(黄绒毛瓣无患子);Velutinous Soapberry ●
346346 Sapindus ruber (Roxb.) Kurz = Aphania rubra (Roxb.) Radlk. ●
346347 Sapindus ruber (Roxb.) Kurz = Lepisanthes senegalensis (Juss. ex Poir.) Leenh. ●
346348 Sapindus rubiginosus Roxb. = Erioglossum rubiginosum (Roxb.) Blume ●
346349 Sapindus rubiginosus Roxb. = Lepisanthes rubiginosa (Roxb.) Leenh. ●
346350 Sapindus saponaria L. var. drummondii (Hook. et Arn.) L. D. Benson = Sapindus drummondii Hook. et Arn. ●☆
346351 Sapindus saponaria Lam.;南无患子(菲律宾无患子);Soapberry, Soap-nut Tree, Southern Soapberry ●
346352 Sapindus senegalensis Juss. ex Poir. = Lepisanthes senegalensis (Juss. ex Poir.) Leenh. ●
346353 Sapindus senegalensis Poir. = Lepisanthes senegalensis (Juss. ex Poir.) Leenh. ●
346354 Sapindus tomentosus Kurz;绒毛无患子(茸毛无患子);Tomentose Soapberry ●
346355 Sapindus tomentosus Kurz. = Sapindus delavayi (Franch.) Radlk. ●
346356 Sapindus trifoliatus L.;三叶无患子;Soap-nut Tree, Threeleaf Soapberry ●☆
346357 Sapindus xanthocarpus Klotzsch = Deinbollia xanthocarpa (Klotzsch) Radlk. ●☆
346358 Sapiopsis Müll. Arg. (1863);类乌桕属●☆
346359 Sapiopsis Müll. Arg. = Sapium Jacq. (保留属名)●
346360 Sapiopsis cremostachys Müll. Arg.;类乌桕;Bolivian Rubber ●☆
346361 Sapium Jacq. (1760)(保留属名);乌桕属(乌臼属);Sapium, Tallow Tree, Tallowtree, Tallow-tree ●
346362 Sapium P. Browne(废弃属名) = Gymnanthes Sw. ●☆
346363 Sapium P. Browne(废弃属名) = Sapium Jacq. (保留属名)●
346364 Sapium abyssinicum (Müll. Arg.) Pax = Shirakiopsis elliptica (Hochst.) Esser ●☆
346365 Sapium acetosella Milne-Redh. = Microstachys acetosella (Milne-Redh.) Esser ●☆
346366 Sapium acetosella Milne-Redh. var. elatius Radcl.-Sm. = Microstachys acetosella (Milne-Redh.) Esser ●☆
346367 Sapium acetosella Milne-Redh. var. lineare J. Léonard = Microstachys acetosella (Milne-Redh.) Esser var. linearis (J. Léonard) Radcl.-Sm. ●☆
346368 Sapium africanum (Sond.) Kuntze = Spirostachys africana Sond. ■☆
346369 Sapium armatum Pax et K. Hoffm. = Sclerocroton integerrimus Hochst. ●☆
346370 Sapium atrobadiomaculatum F. P. Metcalf;斑籽乌桕;Spotted-seed Tallow-tree ●
346371 Sapium aubrevillei Léandri = Shirakiopsis aubrevillei (Léandri) Esser ●☆
346372 Sapium baccatum Roxb.;浆果乌桕(山柏木);Baccate Tallow-tree, Berry-like Sapium ●
346373 Sapium biglandulosum Müll. Arg.;双腺乌桕●☆
346374 Sapium bussei Pax = Excoecaria bussei (Pax) Pax ●☆
346375 Sapium carterianum J. Léonard = Sclerocroton carterianus (J. Léonard) Kruijt et Roebers ●☆
346376 Sapium chihsinianum S. K. Lee;桂林乌桕(济新乌桕,文县乌桕);Chinsin Tallow-tree, Guilin Sapium ●
346377 Sapium cornutum Pax = Sclerocroton cornutus (Pax) Kruijt et Roebers ●☆
346378 Sapium dalzielii Hutch. = Microstachys dalzielii (Hutch.) Esser ●☆
346379 Sapium discolor (Champ. ex Benth.) Müll. Arg.;山乌桕(白臼,白柏,红乌桕,红心乌桕,红叶乌桕,冇拱,琼仔,山柳乌桕,山桕);Mountain Tallow Tree, Mountain Tallow-tree, Taiwan Sapium, Taiwan Tallowtree, Taiwan Tallow-tree, Wild Tallowtree ●
346380 Sapium discolor (Champ. ex Benth.) Müll. Arg. var. wenhsienensis S. B. Ho = Sapium chihsinianum S. K. Lee ●
346381 Sapium ellipticum (Hochst.) Pax = Shirakiopsis elliptica (Hochst.) Esser ●☆
346382 Sapium eugeniifolium Buch.-Ham. ex Wall.;云南乌桕;Yunnan Sapium, Yunnan Tallow-tree ●
346383 Sapium faradiananse (Beille) Pax = Microstachys faradianensis (Beille) Esser ■☆
346384 Sapium grahamii (Stapf) Prain = Excoecaria grahamii Stapf ●☆

346385 Sapium guineense Benth. = Excoecaria guineensis (Benth.) Müll. Arg. ●☆

346386 Sapium indicum Willd. ;木果乌桕●☆

346387 Sapium indicum Willd. = Shirakiopsis indica (Willd.) Esser ●

346388 Sapium insigne (Royle) Benth. ;异序乌桕;Standing Tallowtree ●

346389 Sapium integerrimum (Hochst.) J. Léonard = Sclerocroton integerrimus Hochst. ●☆

346390 Sapium japonicum (Siebold et Zucc.) Pax et K. Hoffm. ;白木乌桕(白木,白乳木,猛树,日本乌桕,野笸麻);Japan Tallowtree, Japanese Sapium, Japanese Soaptree, Japanese Tallow Tree, Japanese Tallow-tree, Japanese Wax ●

346391 Sapium japonicum (Siebold et Zucc.) Pax et K. Hoffm. = Neoshirakia japonica (Siebold et Zucc.) Esser ●☆

346392 Sapium japonicum (Siebold et Zucc.) Pax et K. Hoffm. var. ryukyuense Masam. = Neoshirakia japonica (Siebold et Zucc.) Esser ●☆

346393 Sapium jenmanni Hemsl. ;詹曼乌桕;Esmeralda Rubber ●☆

346394 Sapium kerstingii Pax = Shirakiopsis elliptica (Hochst.) Esser ●☆

346395 Sapium laui Croizat = Sapium discolor (Champ. ex Benth.) Müll. Arg. ●

346396 Sapium leonardii-crispi J. Léonard = Duvigneaudia leonardii-crispi (J. Léonard) Kruijt et Roebers ●☆

346397 Sapium leonardii-crispi J. Léonard var. pubescentifolium ? = Duvigneaudia leonardii-crispi (J. Léonard) Kruijt et Roebers ●☆

346398 Sapium madagascariense (Baill.) Prain = Excoecaria madagascariensis (Baill.) Müll. Arg. ●☆

346399 Sapium mannianum (Müll. Arg.) Hiern = Shirakiopsis elliptica (Hochst.) Esser ●☆

346400 Sapium oblongifolium (Müll. Arg.) Pax = Sclerocroton oblongifolius (Müll. Arg.) Kruijt et Roebers ●☆

346401 Sapium pleiocarpum Y. C. Tseng;多果乌桕(糠柏,米柏);Manyfruit Sapium, Multifruited Tallow-tree ●

346402 Sapium poggei Pax = Sclerocroton cornutus (Pax) Kruijt et Roebers ●☆

346403 Sapium reticulatum (Hochst. ex C. Krauss) Pax = Sclerocroton integerrimus Hochst. ●☆

346404 Sapium rotundifolium Hemsl. ;圆叶乌桕(红叶树,妹妗,雁来红,圆叶柏木);Roundleaf Sapium, Roundleaf Tallowtree, Round-leaved Tallow-tree ●

346405 Sapium rotundifolium Hemsl. var. obcordatum S. K. Lee = Sapium rotundifolium Hemsl. ●

346406 Sapium salicifolium Kunth;柳叶乌桕●☆

346407 Sapium schmitzii J. Léonard = Sclerocroton schmitzii (J. Léonard) Kruijt et Roebers ●☆

346408 Sapium sebiferum (L.) Roxb. ;乌桕(虹树,柏,柏安树,柏树,柏子树,蜡烛树,蜡子树,丁白,木蜡树,木油树,木子树,木梓树,椰,琼仔,琼子,桊子,桊子树,乌茶子,乌臼,乌白木,乌桕树,鸦臼);Candleberry Tree, China Tallowtree, Chinese Sapium, Chinese Tallow, Chinese Tallow Tree, Chinese Tallowtree, Chinese Tallow-tree, Chinese Vegetable Tallow, Popcorn Tree, Tallowtree, Tallow-tree, Vegetable Tallow, Vegetable Tallow Tree ●

346409 Sapium sebiferum (L.) Roxb. = Triadica sebifera (L.) Small ●

346410 Sapium simii Kuntze = Excoecaria simii (Kuntze) Pax ●☆

346411 Sapium suffruticosum Pax = Sclerocroton oblongifolius (Müll. Arg.) Kruijt et Roebers ●☆

346412 Sapium sylvaticum Torr. ;草乌桕(林生假乌桕)■☆

346413 Sapium triloculare Pax et K. Hoffm. = Shirakiopsis trilocularis (Pax et K. Hoffm.) Esser ●☆

346414 Sapium xylocarpum Pax = Sclerocroton cornutus (Pax) Kruijt et Roebers ●☆

346415 Sapium xylocarpum Pax var. genuinum ? = Sclerocroton cornutus (Pax) Kruijt et Roebers ●☆

346416 Sapium xylocarpum Pax var. lineolatum Pax et K. Hoffm. = Sclerocroton cornutus (Pax) Kruijt et Roebers ●☆

346417 Saponaceae Vent. = Sapindaceae Juss. (保留科名)●■

346418 Saponaria L. (1753);肥皂草属;Soapwort ■

346419 Saponaria bellidifolia Sm. ;匙叶肥皂草;Spoon-leaved Soapwort ■☆

346420 Saponaria caespitosa DC. ;丛生肥皂草;Caespitose, Pyrenees Soapwort, Soapwort ■☆

346421 Saponaria calabrica Guss. ;喀拉布利亚肥皂草;Calabrian Soapwort, Manyflowersd Soapwort ■☆

346422 Saponaria cerastoides Fisch. ;角形肥皂草■☆

346423 Saponaria depressa Biv. = Saponaria sicula Raf. ■☆

346424 Saponaria depressa Biv. var. djudjurae Chabert = Saponaria sicula Raf. ■☆

346425 Saponaria glutinosa M. Bieb. ;黏肥皂草■☆

346426 Saponaria griffithiana Boiss. ;格氏肥皂草■☆

346427 Saponaria hispanica Mill. = Vaccaria hispanica (Mill.) Raeusch. ■

346428 Saponaria levantica ?;西亚肥皂草■☆

346429 Saponaria lutea L. ;黄肥皂草;Yellow Soapwort, Yellowflowered Soapwort ■☆

346430 Saponaria montana (Balf. f.) Barkoudah;山地肥皂草■☆

346431 Saponaria montana (Balf. f.) Barkoudah subsp. somalensis (Franch.) Barkoudah = Gypsophila montana Balf. f. subsp. somalensis (Franch.) M. G. Gilbert ■☆

346432 Saponaria montana (Balf. f.) Barkoudah var. diffusa Barkoudah;铺散山地肥皂草■☆

346433 Saponaria ocymoides L. ;岩生肥皂草(罗勒石碱花);Basillike Bruise-wort, Basil-like Bruise-wort, Basil-like Soapwort, Bouncing Bet, Rock Soapwort, Tumbling Ted, Tumbling-ted ■☆

346434 Saponaria officinalis L. ;肥皂草(肥皂花,石碱花,皂质草);Bladder Soapwort, Bouncing Bet, Bouncing Betty, Bouncingbet, Bouncing-bet, Bruisewort, China Cockle, Cockle, Common Soapwort, Cow Basil, Cow Cockle, Cow Foot, Cow Soapwort, Cowherb, Crow Soap, Crowther Soap, Farewell Summer, Farewell-summer, Foam Dock, Fuller's Grass, Fuller's Herb, Gill-run-by-the-street, Ground Soap, Hedge Pink, Latherwort, Mock Gilliflower, My Lady's Washing Bowl, Scourwort, Soapwort, Soapwort Gentian, Suds, Sweet Betty, Wild Sweet William ■

346435 Saponaria officinalis L. 'Rubra Plena';红重瓣肥皂草;Double Soapwort ■☆

346436 Saponaria orientalis L. ;东方肥皂草;Bouncing Bet, Common Soapwort ■☆

346437 Saponaria parvula Bunge;较小肥皂草■☆

346438 Saponaria prostrata Willd. ;平卧肥皂草■☆

346439 Saponaria pumilio Fenzl ex A. Braun;小肥皂草;Pygmy Pink ■☆

346440 Saponaria pungens Bunge = Acanthophyllum pungens (Bunge) Boiss. ■

346441 Saponaria segetalis Neck. = Vaccaria hispanica (Mill.) Raeusch. ■

346442 Saponaria sewerzowii Regel et Schmalh. ;塞氏肥皂草■☆

346443 Saponaria sicula Raf. ;西西里肥皂草■☆

346444 Saponaria vaccaria L. = Vaccaria hispanica (Mill.) Raeusch. ■

346445 Saponaria vaccaria L. var. grandiflora Fisch. ex Ser. = Vaccaria hispanica（Mill.）Rauschert ■

346446 Saponaria viscosa C. A. Mey.；黏性肥皂草■☆

346447 Saposhnikovia Schischk.（1951）；防风属（北防风属）；Fangfeng，Saposhnikovia ■

346448 Saposhnikovia divaricata（Turcz.）Schischk.；防风（白毛草，白毛花，百蕐，百枝，北防风，北风，炒防风，东防风，防丰，防风炭，风肉，公防风，关防风，黄防风，黄风，回草，回辛，回芸，茴草，茴芸，蕌根，蕌根，口防风，旁风，屏风，青防风，曲方氏，软防风，山防风，山花菜，山芹菜，松叶防风，苏风，铜芸，西防风，细叶防风，新疆防风，绫弦胶，云防风）；Divaricate Saposhnikovia，Fangfeng ■

346449 Saposhnikovia divaricata（Turcz.）Schischk. = Saposhnikovia seseloides（Hoffm.）Kitag. ■

346450 Saposhnikovia seseloides（Hoffm.）Kitag. = Saposhnikovia divaricata（Turcz.）Schischk. ■

346451 Sapota Mill. = Achras L.（废弃属名）●

346452 Sapota Mill. = Manilkara Adans.（保留属名）●

346453 Sapota Plum. ex Mill. = Achras L.（废弃属名）●

346454 Sapota Plum. ex Mill. = Manilkara Adans.（保留属名）●

346455 Sapota achras Mill. = Manilkara zapota（L.）P. Royen ●

346456 Sapota cerasifera Welw. = Synsepalum cerasiferum（Welw.）T. D. Penn. ●☆

346457 Sapota mammosa Gaertn. = Chrysophyllum albidum G. Don ●☆

346458 Sapota marginata Decne. = Sideroxylon marginatum（Decne.）Cout. ●☆

346459 Sapota sericea（Schumach.）A. DC. = Chrysophyllum albidum G. Don ●☆

346460 Sapotaceae Juss.（1789）（保留科名）；山榄科；Sapdilla Family，Sapodilla Family，Sapote Family ●

346461 Sapphoa Urb.（1922）；硬叶爵床属■☆

346462 Sapphoa rigidifolia Urb.；硬叶爵床■☆

346463 Sapranthus Seem.（1866）；腐花木属●☆

346464 Sapranthus borealis R. E. Fr.；北方腐花木●☆

346465 Sapranthus foetidus Saff.；烈味腐花木●☆

346466 Sapranthus microcarpus R. E. Fr.；小果腐花木●☆

346467 Sapranthus stenopetalus Saff. ex Standl.；窄瓣腐花木●☆

346468 Sapranthus viridiflorus G. E. Schatz；绿花腐花木●☆

346469 Sapria Griff.（1844）；寄生花属（崖藤寄生属）；Sapria ■

346470 Sapria himalayana Griff.；寄生花；Himalaya Sapria，Himalayas Sapria ■

346471 Saprosma Blume（1827）；染木树属（染料木属，染料树属）；Dyeing Tree，Dyeingtree，Dyeing-tree ●

346472 Saprosma anisophylla Merr.；婆罗洲染木树●☆

346473 Saprosma annamensis Pierre ex Pit.；越南染木树●☆

346474 Saprosma arborea Blume；北爪哇染木树●☆

346475 Saprosma borneensis Wernham；爪哇染木树●☆

346476 Saprosma brassii Merr. et L. M. Perry；所罗门染木树●☆

346477 Saprosma brevipes（Craib）H. Zhu；文山粗叶木（卵叶粗叶木，文山鸡屎树，小毛鸡屎树）；Wenshan Lasianthus，Wenshan Roughleaf ●

346478 Saprosma brunnea Craib；褐色染木树●☆

346479 Saprosma cochinchinensis Pierre ex Pit.；印度支那染木树●☆

346480 Saprosma crassipes H. S. Lo；厚梗染木树；Crassipediceled Dyeing-tree，Thickstalk Dyeingtree ●

346481 Saprosma hainanensis Merr.；海南染木树；Hainan Dyeingtree，Hainan Dyeing-tree ●

346482 Saprosma henryi Hutch.；云南染木树（滇南尸臭树）；Henry Dyeingtree，Henry Dyeing-tree ●

346483 Saprosma hirsuta Korth.；苏门答腊染木树●☆

346484 Saprosma indica Dalzell；印度染木树●☆

346485 Saprosma latifolia Craib；宽叶染木树●☆

346486 Saprosma longicalyx Craib；长萼染木树●☆

346487 Saprosma longifolia Pit.；长叶染木树●☆

346488 Saprosma lowiana（King et Gamble）H. Zhu；娄氏染木树●☆

346489 Saprosma lowiana（King）H. Zhu = Saprosma lowiana（King et Gamble）H. Zhu ●☆

346490 Saprosma membranacea Merr.；膜质染木树●☆

346491 Saprosma merrillii H. S. Lo；琼岛染木树（海南粗叶木）；Hainan Lasianthus，Island Dyeing-tree，Merrill Dyeingtree ●

346492 Saprosma parvifolia Craib；小叶染木树●☆

346493 Saprosma philippinensis Elmer；菲律宾染木树●☆

346494 Saprosma spathulata Valeton；匙叶染木树●☆

346495 Saprosma ternata Benth. et Hook. f.；染木树；Ternate Dyeingtree，Ternate Dyeing-tree ●

346496 Sapucaya R. Knuth = Lecythis Loefl. ●☆

346497 Saraca L.（1767）；无忧花属（无忧树属）；Saraca ●

346498 Saraca asoca（Roxb.）J. J. de Wilde = Saraca indica L. ●

346499 Saraca cauliflora Baker；茎花树（干花树）●☆

346500 Saraca chinensis Merr. et Chun = Saraca dives Pierre ●

346501 Saraca declinata Miq.；垂枝无忧树●☆

346502 Saraca dives Pierre；中国无忧花(黄莺花，火焰花，火焰木，马树，马叶树，四方木，无忧树，云南无忧花，中国无忧树)；China Saraca，Chinese Saraca ●

346503 Saraca griffithiana Prain；云南无忧花（马树）；Griffith Saraca，Yunnan Saraca ●

346504 Saraca indica L.；无忧花(火焰花，四方木，无忧树)；Asoca Tree，Asoka，Common Saraca，Indian Saraca，Saraca，Sorrowless Tree ●

346505 Saraca indica L. = Saraca dives Pierre ●

346506 Saraca thaipingensis Cantley ex King；厚叶无忧花(台屏无忧树)●☆

346507 Saracha Ruiz et Pav.（1794）；萨拉茄属●☆

346508 Saracha acutifolia Miers；尖叶萨拉茄●☆

346509 Saracha biflora Ruiz et Pav.；双花萨拉茄●☆

346510 Saracha contorta Ruiz et Pav.；扭转萨拉茄●☆

346511 Saracha microsperma Bitter；小籽萨拉茄●☆

346512 Saracha punctata Ruiz et Pav.；斑点萨拉茄●☆

346513 Saracha quitensis（Hook.）Miers.；基塔萨拉茄●☆

346514 Saragodra Hort. ex Steud. = Suregada Roxb. ex Rottler ●

346515 Sarana Fisch. ex Baker = Fritillaria L. ■

346516 Saranthe（Regel et Körn.）Eichler（1884）；肉花竹竽属■☆

346517 Saranthe leptostachya Eichl.；肉花竹竽■☆

346518 Sararanga Hemsl.（1894）；四列属●☆

346519 Sararanga philippinensis Merr.；菲律宾四列叶●☆

346520 Sararanga sinuosa Hemsl.；四列叶●☆

346521 Sararenia Spreng. = Sarracenia L. ■☆

346522 Sarawakodendron Ding Hou（1967）；沙捞越卫矛属（婆罗洲卫矛属）●☆

346523 Sarawakodendron filamentosum Ding Hou；沙捞越卫矛●☆

346524 Sarazina Raf. = Sarracenia L. ■☆

346525 Sarcandra Gardner（1845）；草珊瑚属（接骨木属）；Sarcandra，Herbcoral ●

346526 Sarcandra chloranthoides Gardner = Sarcandra glabra（Thunb.）Nakai ●

346527   Sarcandra glabra (Thunb.) Nakai;草珊瑚(驳骨兰,驳节茶,草珠兰,大威灵仙,隔年红,骨风消,观音茶,红果金粟兰,鸡骨香,鸡膝风,接骨茶,接骨金粟兰,接骨兰,接骨莲,接骨木,节骨茶,金粟兰草,九节茶,九节风,九节红,九节花,九节兰,九节蒲,满山香,嫩头子,牛膝头,千两,青甲子,山胡椒,山鸡茶,山石兰,珊瑚,十月红,仙蓼,仙灵草,鱼子兰,肿节风,竹节草,竹节茶);Glabrous Herbcoral,Glabrous Sarcandra ●

346528   Sarcandra glabra (Thunb.) Nakai f. flava (Makino) Okuyama;黄草珊瑚●☆

346529   Sarcandra glabra (Thunb.) Nakai subsp. brachystachys (Blume) Verdc.;短穗草珊瑚(海南草珊瑚)●

346530   Sarcandra glabra Nakai = Sarcandra glabra (Thunb.) Nakai ●

346531   Sarcandra glabra Nakai subsp. brachystachys (Blume) Verdc. = Sarcandra glabra (Thunb.) Nakai subsp. brachystachys (Blume) Verdc. ●

346532   Sarcandra hainanensis (C. P'ei) Swamy et I. W. Bailey;海南草珊瑚(驳节莲树,九节风,山牛耳青,山泽兰); Hainan Herbcoral, Hainan Sarcandra ●

346533   Sarcandra hainanensis (C. P'ei) Swamy et I. W. Bailey = Sarcandra glabra Nakai subsp. brachystachys (Blume) Verdc. ●

346534   Sarcandra hainanensis (C. P'ei) Swamy et I. W. Bailey var. lingshuiensis C. Z. Qiao et Q. H. Zhang;陵水草珊瑚;Lingshui Herbcoral ●

346535   Sarcandra hainanensis (C. P'ei) Swamy et I. W. Bailey var. pingbianensis C. Z. Qiao et Q. H. Zhang;屏边草珊瑚;Pingbian Herbcoral ●

346536   Sarcanthemum Cass. (1818);黏菀木属●☆

346537   Sarcanthemum Cass. = Psiadia Jacq. ●☆

346538   Sarcanthemum coronopus Cass.;黏菀木●☆

346539   Sarcanthera Raf. (1838);肉药爵床属■☆

346540   Sarcanthera Raf. = Cryptophragmium Nees ■☆

346541   Sarcanthera Raf. = Gymnostachyum Nees ■

346542   Sarcanthera venusta Raf.;肉药爵床■☆

346543   Sarcanthidion Baill. = Citronella D. Don ●☆

346544   Sarcanthidium Endl. et Prantl = Sarcanthidion Baill. ●☆

346545   Sarcanthopsis Garay(1972);肉花兰属■☆

346546   Sarcanthopsis curvata (J. J. Sm.) Garay;弯肉花兰■☆

346547   Sarcanthopsis nagarensis (Rchb. f.) Garay;肉花兰■☆

346548   Sarcanthus Andersson = Heliotropium L. ●■

346549   Sarcanthus Lindl. (废弃属名) = Acampe Lindl. (保留属名) ■

346550   Sarcanthus Lindl. (废弃属名) = Cleisostoma Blume ■

346551   Sarcanthus bicuspidatus Rolfe ex Downie = Pelatantheria bicuspidata (Rolfe ex Downie) Ts. Tang et F. T. Wang ■

346552   Sarcanthus birmanicus (Schltr.) Seidenf. et Smitinand = Cleisostoma birmanicum (Schltr.) Garay ■

346553   Sarcanthus brevipes (Hook. f.) J. J. Sm. = Cleisostoma striatum (Rchb. f.) Garay ■

346554   Sarcanthus cerinus (Hance) Rolfe = Cleisostoma paniculatum (Ker Gawl.) Garay ■

346555   Sarcanthus dalatensis Guillaumin = Stereochilus dalatensis (Guillaumin) Garay ■

346556   Sarcanthus densiflorus (Lindl.) E. C. Parish et Rchb. f. = Robiquetia spatulata (Blume) J. J. Sm. ■

346557   Sarcanthus eberhardtii (Finet) Ts. Tang et F. T. Wang = Cleisostomopsis eberhardtii (Finet) Seidenf. et Smitinand ■

346558   Sarcanthus elongatus Rolfe = Cleisostoma williamsonii (Rchb. f.) Garay ■

346559   Sarcanthus filiformis Lindl. = Cleisostoma filiforme (Lindl.) Garay ■

346560   Sarcanthus flagellaris Schltr. = Cleisostoma fuerstenbergianum Kraenzl. ■

346561   Sarcanthus flagellaris Schltr. = Cleisostoma williamsonii (Rchb. f.) Garay ■

346562   Sarcanthus flagelliformis Rolfe ex Downie = Cleisostoma fuerstenbergianum Kraenzl. ■

346563   Sarcanthus fordii (Hance) Rolfe = Cleisostoma rostratum (Lodd.) Seidenf. ex Aver. ■

346564   Sarcanthus formosanus (Hance) Rolfe = Cleisostoma paniculatum (Ker Gawl.) Garay ■

346565   Sarcanthus fuerstenbergianus (Kraenzl.) J. J. Sm. = Cleisostoma fuerstenbergianum Kraenzl. ■

346566   Sarcanthus fuscomaculatus Hayata = Cleisostoma paniculatum (Ker Gawl.) Garay ■

346567   Sarcanthus henryi Schltr. = Robiquetia succisa (Lindl.) Seidenf. et Garay ■

346568   Sarcanthus hongkongensis Rolfe = Cleisostoma williamsonii (Rchb. f.) Garay ■

346569   Sarcanthus micranthus Ames = Cleisostoma uraiense (Hayata) Garay et H. R. Sweet ■

346570   Sarcanthus micranthus Ames = Rhynchostylis retusa (L.) Blume ■

346571   Sarcanthus ophioglossa Guillaumin = Cleisostoma birmanicum (Schltr.) Garay ■

346572   Sarcanthus pallidus Lindl. = Cleisostoma racemiferum (Lindl.) Garay ■

346573   Sarcanthus paniculatus (Ker Gawl.) Lindl. = Cleisostoma paniculatum (Ker Gawl.) Garay ■

346574   Sarcanthus parishii Hook. f. = Cleisostoma parishii (Hook. f.) Garay ■

346575   Sarcanthus poilanei Guillaumin = Micropera poilanei (Guillaumin) Garay ■

346576   Sarcanthus racemifer (Lindl.) Rchb. f. = Cleisostoma racemiferum (Lindl.) Garay ■

346577   Sarcanthus rivesii Guillaumin = Pelatantheria rivesii (Guillaumin) Ts. Tang et F. T. Wang ■

346578   Sarcanthus rostratus Lindl. = Cleisostoma rostratum (Lodd.) Seidenf. ex Aver. ■

346579   Sarcanthus rostratus Lodd. ex Lindl. = Cleisostoma rostratum (Lodd.) Seidenf. ex Aver. ■

346580   Sarcanthus scolopendrifolius Makino = Cleisostoma scolopendrifolium (Makino) Garay ■

346581   Sarcanthus scolopendrifolius Makino = Pelatantheria scolopendrifolia (Makino) Aver. ■

346582   Sarcanthus smithianus Kerr = Sarcoglyphis smithianus (Kerr) Seidenf. ■

346583   Sarcanthus striatus (Rchb. f.) J. J. Sm. = Cleisostoma striatum (Rchb. f.) Garay ■

346584   Sarcanthus succisus Lindl. = Robiquetia succisa (Lindl.) Seidenf. et Garay ■

346585   Sarcanthus taiwanianus Hayata = Sarcophyton taiwanianum (Hayata) Garay ■

346586   Sarcanthus teretifolius (Lindl.) Lindl. = Cleisostoma simondii (Gagnep.) Seidenf. var. guangdongense Z. H. Tsi ■

346587   Sarcanthus teretifolius (Lindl.) Lindl. = Cleisostoma simondii

（Gagnep.）Seidenf. ■

346588 Sarcanthus teretifolius（Lindl.）Lindl. = Vanda watsonii Rolfe ■☆

346589 Sarcanthus uncifer Schltr. = Cleisostoma paniculatum（Ker Gawl.）Garay ■

346590 Sarcanthus uraiensis Hayata = Cleisostoma uraiense（Hayata）Garay et H. R. Sweet ■

346591 Sarcanthus viridescens Fukuy. = Cleisostoma uraiense（Hayata）Garay et H. R. Sweet ■

346592 Sarcanthus williamsonii Rchb. f. = Cleisostoma williamsonii（Rchb. f.）Garay ■

346593 Sarcanthus yunnanensis Schltr. = Cleisostoma racemiferum（Lindl.）Garay ■

346594 Sarcathria Raf. = Halocnemum M. Bieb. ●

346595 Sarcathria Raf. = Salicornia L. ■●

346596 Sarcaulis B. D. Jacks. = Sarcaulus Radlk. ●☆

346597 Sarcaulis Radlk. = Sarcaulus Radlk. ●☆

346598 Sarcaulus Radlk.（1882）;肉山榄属●☆

346599 Sarcaulus brasiliensis Eyma;巴西肉山榄●☆

346600 Sarcauthemum Cass. = Psiadia Jacq. ●☆

346601 Sarchochilus S. Vidal = Sarcochilus R. Br. ■☆

346602 Sarcinanthus Oerst.（废弃属名）= Asplundia Harling（保留属名）■☆

346603 Sarcinanthus Oerst.（废弃属名）= Carludovica Ruiz et Pav. ●■

346604 Sarcobataceae Behnke = Chenopodiaceae Vent.（保留科名）●■

346605 Sarcobataceae Behnke（1997）;肉叶刺藜科（夷藜科）●☆

346606 Sarcobatus K. Schum. = Sarcolobus R. Br. ●☆

346607 Sarcobatus Nees（1839）;肉叶刺藜（肉刺藜属,肉叶刺茎藜属,夷藜属）;Greasewood ●☆

346608 Sarcobatus baileyi Coville;肉叶刺藜;Bailey Greasewood ●☆

346609 Sarcobatus maximiliani Nees = Sarcobatus vermiculatus（Hook.）Torr. ●☆

346610 Sarcobatus vermiculatus（Hook.）Torr.;黑肉叶刺藜;Black Greasewood,Caterpillar Greasewood,Greasewood,Saltbush,Seepwood ●☆

346611 Sarcobatus vermiculatus（Hook.）Torr. var. baileyi（Coville）Jeps. = Sarcobatus baileyi Coville ●☆

346612 Sarcobodium Beer = Bulbophyllum Thouars（保留属名）■

346613 Sarcobotrya R. Vig. = Kotschya Endl. ex Endl. et Fenzl ●☆

346614 Sarcobotrya strigosa（Benth.）R. Vig. = Kotschya strigosa（Benth.）Dewit et P. A. Duvign. ■☆

346615 Sarcoca Raf. = Phytolacca L. ●■

346616 Sarcocaceae Raf. = Phytolaccaceae R. Br.（保留科名）●■

346617 Sarcocadetia（Schltr.）M. A. Clem. et D. L. Jones = Cadetia Gaudich. ■☆

346618 Sarcocadetia（Schltr.）M. A. Clem. et D. L. Jones（2002）;萨尔兰属■☆

346619 Sarcocadetia wariana（Schltr.）M. A. Clem. et D. L. Jones;萨尔兰■☆

346620 Sarcocadetia wariana（Schltr.）M. A. Clem. et D. L. Jones = Cadetia wariana Schltr. ■☆

346621 Sarcocalyx Walp. = Aspalathus L. ●☆

346622 Sarcocalyx Zipp. = Exocarpos Labill.（保留属名）●☆

346623 Sarcocalyx capensis Walp. = Aspalathus capensis（Walp.）R. Dahlgren ●☆

346624 Sarcocampsa Miers = Peritassa Miers ●☆

346625 Sarcocapnos DC.（1821）;肉烟堇属（假紫堇属）■☆

346626 Sarcocapnos crassifolia（Desf.）DC.;厚叶肉烟堇■☆

346627 Sarcocapnos crassifolia（Desf.）DC. var. atlantica Braun-Blanq.

et Wilczek = Sarcocapnos crassifolia（Desf.）DC. ■☆

346628 Sarcocapnos crassifolia（Desf.）DC. var. balneora Jahand. et Weiller = Sarcocapnos crassifolia（Desf.）DC. ■☆

346629 Sarcocapnos crassifolia（Desf.）DC. var. bekritensis Maire = Sarcocapnos crassifolia（Desf.）DC. ■☆

346630 Sarcocapnos crassifolia（Desf.）DC. var. brachyceras Maire = Sarcocapnos crassifolia（Desf.）DC. ■☆

346631 Sarcocapnos crassifolia（Desf.）DC. var. brevicalcarata Emb. = Sarcocapnos crassifolia（Desf.）DC. ■☆

346632 Sarcocapnos crassifolia（Desf.）DC. var. dissidens Maire = Sarcocapnos crassifolia（Desf.）DC. ■☆

346633 Sarcocapnos crassifolia（Desf.）DC. var. fallax Maire = Sarcocapnos crassifolia（Desf.）DC. ■☆

346634 Sarcocapnos crassifolia（Desf.）DC. var. micrantha Jahand. et Maire = Sarcocapnos crassifolia（Desf.）DC. ■☆

346635 Sarcocapnos crassifolia（Desf.）DC. var. purpurascens Maire = Sarcocapnos crassifolia（Desf.）DC. ■☆

346636 Sarcocapnos crassifolia（Desf.）DC. var. stenoceras Maire = Sarcocapnos crassifolia（Desf.）DC. ■☆

346637 Sarcocapnos crassifolia（Desf.）DC. var. subspeciosa Maire = Sarcocapnos crassifolia（Desf.）DC. ■☆

346638 Sarcocapnos enneaphylla（L.）DC.;九叶假紫堇（九叶假黄堇）■☆

346639 Sarcocapnos enneaphylla（L.）DC. var. pubescens Lange = Sarcocapnos enneaphylla（L.）DC. ■☆

346640 Sarcocapnos enneaphylla DC. = Sarcocapnos enneaphylla（L.）DC. ■☆

346641 Sarcocarpon Blume = Kadsura Kaempf. ex Juss. ●

346642 Sarcocaulon（DC.）Sweet（1827）;肉茎犄牛儿苗属（龙骨葵属）;Sarcocaulon ●■☆

346643 Sarcocaulon Sweet = Sarcocaulon（DC.）Sweet ●■☆

346644 Sarcocaulon burmannii（DC.）Sweet;布尔曼犄牛儿苗;Bushman's Candle,Bushman's Candles ●☆

346645 Sarcocaulon burmannii Sweet = Sarcocaulon burmannii（DC.）Sweet ●☆

346646 Sarcocaulon camdeboense Moffett;坎博犄牛儿苗●☆

346647 Sarcocaulon ciliatum Moffett;缘毛肉茎犄牛儿苗●☆

346648 Sarcocaulon crassicaule Rehm;粗茎犄牛儿苗;Bushman Candle,Groot Kersbos ●☆

346649 Sarcocaulon currali Heckel = Kalanchoe grandidieri Baill. ■☆

346650 Sarcocaulon flavescens Rehm;浅黄肉茎犄牛儿苗●☆

346651 Sarcocaulon herrei L. Bolus;赫勒肉茎犄牛儿苗●☆

346652 Sarcocaulon inerme Rehm;无刺肉茎犄牛儿苗●☆

346653 Sarcocaulon marlothii Engl. ;马洛斯犄牛儿苗●☆

346654 Sarcocaulon mossamedense（Welw. ex Oliv.）Hiern;穆萨犄牛儿苗●☆

346655 Sarcocaulon multifidum E. Mey. ex R. Knuth;细裂叶肉茎犄牛儿苗（多裂肉茎犄牛儿苗）●■☆

346656 Sarcocaulon multifidum R. Knuth = Sarcocaulon multifidum E. Mey. ex R. Knuth ●■☆

346657 Sarcocaulon patersonii（DC.）G. Don;帕特森犄牛儿苗●☆

346658 Sarcocaulon rigidum Schinz;硬刺肉茎犄牛儿苗●■☆

346659 Sarcocaulon rigidum Schinz = Sarcocaulon patersonii（DC.）G. Don ●☆

346660 Sarcocaulon salmoniflorum Moffett;鲑花犄牛儿苗■☆

346661 Sarcocaulon vanderietiae L. Bolus;范德犄牛儿苗●☆

346662 Sarcoccaceae Dulac = Coriariaceae DC.（保留科名）●

346663　Sarcocephalus Afzel. ex R. Br. = Nauclea L. ●

346664　Sarcocephalus Afzel. ex Sabine（1824）；肉序茜属（肉序茜草属，肉序属）●☆

346665　Sarcocephalus badi Aubrév. = Nauclea diderrichii（De Wild. et T. Durand）Merr. ●☆

346666　Sarcocephalus cadamba（Roxb.）Kurz = Neolamarckia cadamba（Roxb.）Bosser ●

346667　Sarcocephalus diderrichii De Wild.；迪氏肉序茜草（迪氏肉序）●☆

346668　Sarcocephalus diderrichii De Wild. et T. Durand = Nauclea diderrichii（De Wild. et T. Durand）Merr. ●☆

346669　Sarcocephalus esculentus Afzel. ex Sabine；可食肉序茜草（可食肉序）●☆

346670　Sarcocephalus esculentus Afzel. ex Sabine = Sarcocephalus latifolius（Sm.）E. A. Bruce ●☆

346671　Sarcocephalus esculentus Sabine var. amarissima A. Chev. = Sarcocephalus latifolius（Sm.）E. A. Bruce ●☆

346672　Sarcocephalus esculentus Sabine var. velutina A. Chev. = Sarcocephalus latifolius（Sm.）E. A. Bruce ●☆

346673　Sarcocephalus gilletii De Wild. = Nauclea gilletii（De Wild.）Merr. ●☆

346674　Sarcocephalus latifolius（Sm.）E. A. Bruce；宽叶肉序茜草；African Peach, Guinea Peach ●☆

346675　Sarcocephalus nervosus Hutch. et Dalziel = Nauclea vanderguchtii（De Wild.）E. M. Petit ●☆

346676　Sarcocephalus nervosus Hutch. et Dalziel var. cordifolia A. Chev. = Nauclea vanderguchtii（De Wild.）E. M. Petit ●☆

346677　Sarcocephalus officinalis Pierre ex Pit. = Nauclea officinalis（Pierre ex Pit.）Merr. et Chun ●

346678　Sarcocephalus pobeguinii Pobeg.；波别肉序茜●☆

346679　Sarcocephalus richardianus Baill. = Breonia chinensis（Lam.）Capuron ●☆

346680　Sarcocephalus richardii Drake = Breonia chinensis（Lam.）Capuron ●☆

346681　Sarcocephalus russeggeri Schweinf. = Sarcocephalus latifolius（Sm.）E. A. Bruce ●☆

346682　Sarcocephalus trillesii Pierre ex A. Chev. var. lancifolia A. Chev. = Nauclea gilletii（De Wild.）Merr. var. lancifolia（A. Chev.）N. Hallé ●☆

346683　Sarcocephalus trillesii Pierre ex De Wild. = Nauclea diderrichii（De Wild. et T. Durand）Merr. ●☆

346684　Sarcocephalus vanderguchtii De Wild. = Nauclea vanderguchtii（De Wild.）E. M. Petit ●☆

346685　Sarcocephalus xanthoxylon A. Chev. = Nauclea xanthoxylon（A. Chev.）Aubrév. ●☆

346686　Sarcochilus R. Br.（1810）；肉唇兰属（狭唇兰属）；Sarcochilus ■☆

346687　Sarcochilus aphyllus Makino = Taeniophyllum glandulosum Blume ■

346688　Sarcochilus arachnites（Blume）Rchb. f. = Thrixspermum centipeda Lour. ■

346689　Sarcochilus asperatus Schltr. = Pteroceras asperatum（Schltr.）P. F. Hunt ■

346690　Sarcochilus aurifer（Lindl.）Rchb. f. = Thrixspermum centipeda Lour. ■

346691　Sarcochilus australis（Lindl.）Rchb. f.；澳洲狭唇兰；Australian Sarcochilus ■☆

346692　Sarcochilus berkeleyi Rchb. f.；伯氏狭唇兰■☆

346693　Sarcochilus centipeda（Lour.）Náves = Thrixspermum centipeda Lour. ■

346694　Sarcochilus difformis（Wall. ex Lindl.）Ts. Tang et F. T. Wang = Ornithochilus difformis（Lindl.）Schltr. ■

346695　Sarcochilus difformis（Lindl.）Ts. Tang et F. T. Wang = Ornithochilus difformis（Lindl.）Schltr. ■

346696　Sarcochilus difformis（Wall. ex Lindl.）Ts. Tang et F. T. Wang = Ornithochilus difformis（Lindl.）Schltr. ■

346697　Sarcochilus falcatus R. Br.；镰叶狭唇兰；Falcateleaf Sarcochilus ■☆

346698　Sarcochilus fitzgeraldii F. Muell.；费氏狭唇兰；Fitzgerald Sarcochilus ■☆

346699　Sarcochilus formosanus Hayata = Thrixspermum formosanum（Hayata）Schltr. ■

346700　Sarcochilus hainanensis Rolfe = Thrixspermum centipeda Lour. ■

346701　Sarcochilus hartmannii F. Muell.；哈氏狭唇兰；Hartmann Sarcochilus ■☆

346702　Sarcochilus hirtulus Hook. f. = Grosourdya appendiculatum（Blume）Rchb. f. ■

346703　Sarcochilus japonicus（Rchb. f.）Miq.；日本狭唇兰■☆

346704　Sarcochilus japonicus（Rchb. f.）Miq. = Thrixspermum japonicum（Miq.）Rchb. f. ■

346705　Sarcochilus japonicus Miq. = Thrixspermum japonicum（Miq.）Rchb. f. ■

346706　Sarcochilus kusukusensis Hayata = Thrixspermum merguense（Hook. f.）Kuntze ■

346707　Sarcochilus laurisilvaticus Fukuy. = Thrixspermum laurisilvaticum（Fukuy.）Garay ■

346708　Sarcochilus laurisilvaticus Fukuy. = Thrixspermum saruwatarii（Hayata）Schltr. ■

346709　Sarcochilus leopardinus（E. C. Parish et Rchb. f.）Hook. f. = Pteroceras leopardinum（Parl. et Rchb. f.）Seidenf. et Sm. ■

346710　Sarcochilus merguensis Hook. f. = Thrixspermum merguense（Hook. f.）Kuntze ■

346711　Sarcochilus olivacenus Lindl.；绿花狭唇兰；Greenflower Sarcochilus ■☆

346712　Sarcochilus pallidus（Blume）Rchb. f.；白花狭唇兰；Whiteflower Sarcochilus ■☆

346713　Sarcochilus parviflorus Lindl.；小花狭唇兰；Smallflower Sarcochilus ■☆

346714　Sarcochilus purpureus Benth. et Hook. f.；紫花狭唇兰■☆

346715　Sarcochilus saruwatarii Hayata = Thrixspermum saruwatarii（Hayata）Schltr. ■

346716　Sarcochilus segawae Masam. = Chiloschista segawai（Masam.）Masam. et Fukuy. ■

346717　Sarcochilus subulatus（Blume）Rchb. f. = Thrixspermum subulatum Rchb. f. ■

346718　Sarcochilus trichoglottis Hook. f. = Thrixspermum trichoglottis（Hook. f.）Kuntze ■

346719　Sarcochlaena Spreng. = Sarcolaena Thouars ●☆

346720　Sarcochlamys Gaudich.（1844）；肉被麻属（球隔麻属）；Fleshtepal, Sarcochlamys ●

346721　Sarcochlamys pulcherrima Gaudich.；肉被麻（球隔麻）；Fleshtepal, Sarcochlamys ●

346722　Sarcoclinium Wight = Agrostistachys Dalzell ■☆

346723　Sarcococca Lindl.（1826）；野扇花属（清香桂属）；Cristmas Box, Sarcococca, Sweet Box ●

346724　Sarcococca balansae Gagnep. = Sarcococca vagans Stapf ●

346725　Sarcococca confertiflora Sealy；聚花野扇花（聚花清香桂）；Confertedflower Sarcococca，Densiflowered Sarcococca ●

346726　Sarcococca confertiflora Sealy ＝ Sarcococca hookeriana Baill. var. digyna Franch. ●

346727　Sarcococca confusa Sealy；黑果野扇花（美丽野扇，相似野扇花）；Sarcococca ●☆

346728　Sarcococca coriacea（Hook.）Sweet；革质野扇花●☆

346729　Sarcococca coriacea Müll. Arg. ＝ Sarcococca wallichii Stapf ●

346730　Sarcococca euphlebia Merr. ＝ Sarcococca vagans Stapf ●

346731　Sarcococca hookeriana Baill.；羽脉野扇花（百年青，黑果青香桂，厚叶子树，千年青，西藏野扇花，云南野扇花，紫果野扇花）；Himalaya Sarcococca，Hooker Sarcococca ●

346732　Sarcococca hookeriana Baill. var. digyna Franch.；双蕊野扇花（矮树八爪龙，八爪金龙，山豆根，树八爪龙，小叶野扇花，岩花子）；Twopistil Sarcococca ●

346733　Sarcococca hookeriana Baill. var. humilis Rehder et E. H. Wilson；矮树八爪龙（矮生野扇花）；Dwarf Hooker Sarcococca，Dwarf Sarcococca，Sarcococca，Sweetbox ●

346734　Sarcococca hookeriana Baill. var. humilis Rehder et E. H. Wilson ＝ Sarcococca hookeriana Baill. var. digyna Franch. ●

346735　Sarcococca humilis Stapf；小叶野扇花；Sweet Box ●

346736　Sarcococca longifolia M. Cheng et K. F. Wu；长叶野扇花；Longileaved Sarcococca，Longleaf Sarcococca ●

346737　Sarcococca longipetiolata M. Cheng et K. F. Wu；长叶柄野扇花（长柄野扇花，柑子风，千年青）；Longipetioled Sarcococca ●

346738　Sarcococca orientalis C. Y. Wu ex M. Cheng；东方野扇花（大风消，三两根，土丹皮，象天雷）；Oriental Sarcococca ●

346739　Sarcococca pauciflora C. Y. Wu；少花清香桂；Fewflower Sarcococca，Few-flowered Sarcococca ●

346740　Sarcococca pruniformis Lindl. var. angustifolia Lindl. ＝ Sarcococca saligna（D. Don）Müll. Arg. ●

346741　Sarcococca pruniformis Lindl. var. dioica Hayata ＝ Sarcococca saligna（D. Don）Müll. Arg. ●

346742　Sarcococca pruniformis Lindl. var. hookeriana Hook. f. ＝ Sarcococca hookeriana Baill. ●

346743　Sarcococca ruscifolia Stapf；野扇花（矮陀，大风消，观音柴，桂花矮陀陀，花子藤，棉草木，千年崖，清香桂，全青，土丹皮，万年春，万年青，胃友，香野扇花，野樱桃，叶上花）；Fragrant Sarcococca，Sweet Box ●

346744　Sarcococca ruscifolia Stapf var. chinensis（Franch.）Rehder et E. H. Wilson ＝ Sarcococca ruscifolia Stapf ●

346745　Sarcococca ruscifolia Stapf var. chinensis Rehder et E. H. Wilson；窄叶野扇花（大叶野扇花，狭叶清香桂）；Chinese Fragrant Sarcococca ●

346746　Sarcococca salicifolia Baill. ＝ Sarcococca saligna（D. Don）Müll. Arg. ●

346747　Sarcococca saligna（D. Don）Müll. Arg.；柳叶野扇花（柳状野扇花，台湾野扇花，喜马拉雅清香桂，喜马拉雅野扇花）；Willowleaf Sarcococca，Willow-leaved Sarcococca ●

346748　Sarcococca saligna（D. Don）Müll. Arg. var. chinensis Franch. ＝ Sarcococca ruscifolia Stapf ●

346749　Sarcococca vagans Stapf；海南野扇花（大叶清香桂）；Hainan Sarcococca ●

346750　Sarcococca wallichii Stapf；云南野扇花（厚叶清香桂，厚叶野扇花）；Yunnan Sarcococca ●

346751　Sarcocodon N. E. Br. ＝ Caralluma R. Br. ■

346752　Sarcocodon speciosus N. E. Br. ＝ Caralluma speciosa（N. E. Br.）N. E. Br. ■☆

346753　Sarcocolla Boehm. ＝ Penaea L. ●☆

346754　Sarcocolla Kunth ＝ Saltera Bullock ●☆

346755　Sarcocolla minor A. DC. ＝ Saltera sarcocolla（L.）Bullock ●☆

346756　Sarcocolla squamosa（L.）Endl. ＝ Saltera sarcocolla（L.）Bullock ●☆

346757　Sarcocolla tetragona（P. J. Bergius）Salter ＝ Saltera sarcocolla（L.）Bullock ●☆

346758　Sarcocordylis Wall. ＝ Balanophora J. R. Forst. et G. Forst. ■

346759　Sarcocordylis indica Wall. ex Steud. ＝ Balanophora indica（Arn.）Griff. ■

346760　Sarcocornia A. J. Scott ＝ Salicornia L. ■●

346761　Sarcocornia A. J. Scott（1978）；肉角藜属（冈羊栖菜属）；Glasswort，Perennial Glasswort，Saltwort，Samphire ●☆

346762　Sarcocornia capensis（Moss）A. J. Scott；好望角肉角藜●☆

346763　Sarcocornia decumbens（Toelken）A. J. Scott ＝ Arthrocnemum decumbens Toelken ●☆

346764　Sarcocornia fruticosa（L.）A. J. Scott ＝ Arthrocnemum fruticosum（L.）Moq. ●☆

346765　Sarcocornia littorea（Moss）A. J. Scott；滨海肉角藜●☆

346766　Sarcocornia mossambicensis Brenan；莫桑比克肉角藜●☆

346767　Sarcocornia mossiana（Toelken）A. J. Scott；莫西肉角藜●☆

346768　Sarcocornia natalensis（Bunge ex Ung. -Sternb.）A. J. Scott；纳塔尔肉角藜●☆

346769　Sarcocornia natalensis（Bunge ex Ung. -Sternb.）A. J. Scott var. affinis（Moss）O'Call.；近缘纳塔尔肉角藜●☆

346770　Sarcocornia pacifica（Standl.）A. J. Scott；太平洋肉角藜●☆

346771　Sarcocornia perennis（Mill.）A. J. Scott；多年肉角藜（多年盐角草）；Creeping Glasswort，Perennial Glasswort ●☆

346772　Sarcocornia perennis（Mill.）A. J. Scott ＝ Arthrocnemum perenne（Mill.）Moss ex Fourc. ●☆

346773　Sarcocornia perennis（Mill.）A. J. Scott subsp. alpini（Lag.）Castrov.；高山太平洋肉角藜●■☆

346774　Sarcocornia perennis（Mill.）A. J. Scott var. lignosa（Woods）O'Call.；木质肉角藜●☆

346775　Sarcocornia pillansii（Moss）A. J. Scott；皮氏肉角藜●☆

346776　Sarcocornia pillansii（Moss）A. J. Scott var. dunensis（Moss）O'Call.；砂丘皮氏肉角藜●☆

346777　Sarcocornia terminalis（Toelken）A. J. Scott；顶生肉角藜●☆

346778　Sarcocornia utahensis（Tidestr.）A. J. Scott；犹他肉角藜●☆

346779　Sarcocornia xerophila（Toelken）A. J. Scott；旱生肉角藜●☆

346780　Sarcocyphula Harv. ＝ Cynanchum L. ●■

346781　Sarcocyphula gerrardii Harv. ＝ Cynanchum gerrardii（Harv.）Liede ●☆

346782　Sarcodactilis C. F. Gaertn. ＝ Citrus L. ●

346783　Sarcodactilis helicteroides C. F. Gaertn. ＝ Citrus medica L. ●

346784　Sarcodes Torr.（1851）；血晶兰属；Snow Plant ●☆

346785　Sarcodes sanguinea Torr.；血晶兰；Snow Plant ●☆

346786　Sarcodiscaceae Dulac ＝ Rutaceae Juss.（保留科名）●■

346787　Sarcodiscus Griff. ＝ Kibara Endl. ●☆

346788　Sarcodiscus Mart. ex Miq. ＝ Sorocea A. St. -Hil. ●☆

346789　Sarcodium Pers. ＝ Clianthus Sol. ex Lindl.（保留属名）●

346790　Sarcodraba Gilg et Muschl.（1909）；肉葶苈属■☆

346791　Sarcodraba karraikensis Gilg et Muschl.；肉葶苈■☆

346792　Sarcodrimys（Baill.）Baum. -Bod. ＝ Zygogynum Baill. ●☆

346793　Sarcodum Lour.（废弃属名）＝ Clianthus Sol. ex Lindl.（保留属名）●

346794 Sarcodum scandens Lour. = Clianthus scandens (Lour.) Merr. ●■

346795 Sarcoglossum Beer = Cirrhaea Lindl. ■☆

346796 Sarcoglossum suaveolens Beer = Cirrhaea dependens (Lodd.) Rchb. f. ■☆

346797 Sarcoglottis C. Presl(1827);肉舌兰属■☆

346798 Sarcoglottis acaulis Schltr.;无茎肉舌兰■☆

346799 Sarcoglottis acutata (Rchb. f. et Warm.) Garay;尖肉舌兰■☆

346800 Sarcoglottis albiflos Schltr. ex Hoehne;白花肉舌兰■☆

346801 Sarcoglottis biflora Schltr.;双花肉舌兰■☆

346802 Sarcoglottis grandiflora Klotzsch;大花肉舌兰■☆

346803 Sarcoglottis latifolia Schltr.;宽叶肉舌兰■☆

346804 Sarcoglottis micrantha Christenson;小花肉舌兰■☆

346805 Sarcoglottis pauciflora (A. Rich. et Galeotti) Schltr.;少花肉舌兰■☆

346806 Sarcoglottis tenuiflora (Greenm.) Conz.;细花肉舌兰■☆

346807 Sarcoglyphis Garay(1972);大喙兰属;Bigbillorchis ■

346808 Sarcoglyphis magnirostris Z. H. Tsi;短帽大喙兰;Bigbillorchis ■

346809 Sarcoglyphis smithianus (Kerr) Seidenf.;大喙兰;Smith Bigbillorchis ■

346810 Sarcoglyphis yunnanensis Z. H. Tsi = Sarcoglyphis smithianus (Kerr) Seidenf. ■

346811 Sarcogonum G. Don = Muehlenbeckia Meisn.(保留属名)●☆

346812 Sarcolaena Thouars(1805);苞杯花属(旋花树属)●☆

346813 Sarcolaena bojeriana Baill. = Leptolaena bojeriana (Baill.) Cavaco ●☆

346814 Sarcolaena delphinensis Cavaco;德尔菲苞杯花●☆

346815 Sarcolaena diospyroidea Baill. = Leptolaena diospyroidea (Baill.) Cavaco ●☆

346816 Sarcolaena eriophora Thouars;绵毛苞杯花●☆

346817 Sarcolaena grandidieri Baill. = Leptolaena bojeriana (Baill.) Cavaco ●☆

346818 Sarcolaena grandiflora Thouars;大花苞杯花●☆

346819 Sarcolaena humbertiana Cavaco;亨伯特苞杯花●☆

346820 Sarcolaena isaloensis Randrian. et J. S. Mill.;伊萨卢苞杯花●☆

346821 Sarcolaena multiflora Thouars;多花苞杯花●☆

346822 Sarcolaena oblongifolia F. Gérard;矩圆叶苞杯花●☆

346823 Sarcolaenaceae Caruel(1881)(保留科名);苞杯花科(旋花树科)●☆

346824 Sarcolemma Griseb. ex Lorentz = Sarcostemma R. Br. ■

346825 Sarcolipes Eckl. et Zeyh. = Crassula L. ●■☆

346826 Sarcolipes pubescens Eckl. et Zeyh. = Crassula strigosa L. ■☆

346827 Sarcolobus R. Br. (1810);肉片萝藦属●☆

346828 Sarcolobus minor Schltr.;小肉片萝藦●☆

346829 Sarcolobus multiflorus K. Schum. et Lauterb.;多花肉片萝藦●☆

346830 Sarcolobus oblongus R. E. Rintz;倒卵肉片萝藦●☆

346831 Sarcolobus sulphureus Schltr.;硫色肉片萝藦●☆

346832 Sarcolophium Troupin(1960);肉脊藤属●☆

346833 Sarcolophium tuberosum (Diels) Troupin;肉脊藤●☆

346834 Sarcomelicope Engl. (1896);肉稷芸香属●☆

346835 Sarcomelicope argyrophylla Guillaumin;银叶肉稷芸香●☆

346836 Sarcomelicope glauca T. G. Hartley;灰绿肉稷芸香●☆

346837 Sarcomelicope leiocarpa (P. S. Green) T. G. Hartley;光果肉稷芸香●☆

346838 Sarcomelicope pubescens C. T. White;毛肉稷芸香●☆

346839 Sarcomeris Naudin = Pachyanthus A. Rich. ●☆

346840 Sarcomorphis Bojer ex Moq. = Salsola L. ●■

346841 Sarcomphalium Dulac = Brimeura Salisb. ■☆

346842 Sarcomphalium Dulac = Hyacinthus L. ■☆

346843 Sarcomphalodes (DC.) Kuntze = Noltea Rchb. ●☆

346844 Sarcomphalus P. Browne = Ziziphus Mill. ●

346845 Sarcopera Bedell(1997);穗状附生藤属●

346846 Sarcopera anomala (Kunth) Bedell;穗状附生藤●

346847 Sarcoperis Raf. = Campelia Rich. ■

346848 Sarcopetalum F. Muell. (1862);肉瓣藤属●☆

346849 Sarcopetalum harveyanum F. Muell. ;肉瓣藤●☆

346850 Sarcophagophilus Dinter = Quaqua N. E. Br. ■☆

346851 Sarcophagophilus armatus (N. E. Br.) Dinter = Quaqua armata (N. E. Br.) Bruyns ■☆

346852 Sarcophagophilus armatus Dinter = Quaqua mammillaris (L.) Bruyns ■☆

346853 Sarcophagophilus winklerianus Dinter = Quaqua mammillaris (L.) Bruyns ■☆

346854 Sarcopharyngia (Stapf) Boiteau = Tabernaemontana L. ●

346855 Sarcopharyngia Boiteau = Tabernaemontana L. ●

346856 Sarcopharyngia angolensis (Stapf) L. Allorge = Tabernaemontana pachysiphon Stapf ●☆

346857 Sarcopharyngia contorta (Stapf) Boiteau = Tabernaemontana contorta Stapf ●☆

346858 Sarcopharyngia crassa (Benth.) Boiteau = Tabernaemontana crassa Benth. ●☆

346859 Sarcopharyngia gentilii (De Wild.) Boiteau = Tabernaemontana crassa Benth. ●☆

346860 Sarcopharyngia stapfiana (Britten) Boiteau = Tabernaemontana stapfiana Britten ●☆

346861 Sarcopharyngia ventricosa (Hochst. ex A. DC.) Boiteau = Tabernaemontana ventricosa Hochst. ex A. DC. ●☆

346862 Sarcophrynium K. Schum. (1902);肉柊叶属■☆

346863 Sarcophrynium adenocarpum (K. Schum.) K. Schum. = Megaphrynium macrostachyum (Benth.) Milne-Redh. ■☆

346864 Sarcophrynium arnoldianum De Wild. = Megaphrynium macrostachyum (Benth.) Milne-Redh. ■☆

346865 Sarcophrynium baccatum (K. Schum.) K. Schum. = Sarcophrynium schweinfurthianum (Kuntze) Milne-Redh. ■☆

346866 Sarcophrynium bisubulatum (K. Schum.) K. Schum. ;双秃肉柊叶■☆

346867 Sarcophrynium brachystachyum (Benth.) K. Schum. ;短穗肉柊叶■☆

346868 Sarcophrynium brachystachyum (Benth.) K. Schum. var. puberulifolium Koechlin;微毛短穗肉柊叶■☆

346869 Sarcophrynium leiogonium (K. Schum.) K. Schum. = Sarcophrynium prionogonium (K. Schum.) K. Schum. ■☆

346870 Sarcophrynium macrostachyum (Benth.) K. Schum. = Megaphrynium macrostachyum (Benth.) Milne-Redh. ■☆

346871 Sarcophrynium oxycarpum (K. Schum.) K. Schum. = Megaphrynium macrostachyum (Benth.) Milne-Redh. ■☆

346872 Sarcophrynium prionogonium (K. Schum.) K. Schum. ;膝齿肉柊叶■☆

346873 Sarcophrynium prionogonium (K. Schum.) K. Schum. var. ivorense Schnell;伊沃里肉柊叶■☆

346874 Sarcophrynium prionogonium (K. Schum.) K. Schum. var. puberulifolium Schnell;微毛叶肉柊叶■☆

346875 Sarcophrynium schweinfurthianum (Kuntze) Milne-Redh. ;施韦肉柊叶■☆

346876 Sarcophrynium schweinfurthianum (Kuntze) Milne-Redh. var.

puberulifolium Koechlin;微毛施韦肉枨叶■☆

346877 Sarcophrynium spicatum K. Schum. ;穗状肉枨叶■☆

346878 Sarcophrynium strictifolium Schnell = Sarcophrynium brachystachyum (Benth.) K. Schum. ■☆

346879 Sarcophrynium velutinum (Baker) K. Schum. = Megaphrynium velutinum (Baker) Koechlin ■☆

346880 Sarcophrynium villosum (Benth.) K. Schum. ;长柔毛肉枨叶■☆

346881 Sarcophyllum E. Mey. = Lebeckia Thunb. ●☆

346882 Sarcophyllum Willd. = Sarcophyllus Thunb. ●☆

346883 Sarcophyllum carnosum E. Mey. = Lebeckia carnosa (E. Mey.) Druce ■☆

346884 Sarcophyllum gnidioides Walp. = Amphithalea tortilis (E. Mey.) Steud. ●☆

346885 Sarcophyllus Thunb. = Aspalathus L. ●☆

346886 Sarcophyllus carnosus Thunb. = Aspalathus capensis (Walp.) R. Dahlgren ●☆

346887 Sarcophyllus grandiflorus E. Mey. = Lebeckia grandiflora (E. Mey.) Benth. ●☆

346888 Sarcophysa Miers = Juanulloa Ruiz et Pav. ●☆

346889 Sarcophytaceae (Engl.) Tiegh. = Balanophoraceae Rich. (保留科名)●■

346890 Sarcophytaceae A. Kern. (1891);肉蛇菰科(肉草科)■☆

346891 Sarcophytaceae A. Kern. = Balanophoraceae Rich. (保留科名)●■

346892 Sarcophytaceae Tiegh. = Balanophoraceae Rich. (保留科名)●■

346893 Sarcophytaceae Tiegh. = Sarcophytaceae Tiegh. ex Takht. ■☆

346894 Sarcophytaceae Tiegh. ex Takht. = Balanophoraceae Rich. (保留科名)●■

346895 Sarcophyte Sparrm. (1776);肉蛇菰属(肉草属)■☆

346896 Sarcophyte piriei Hutch. = Sarcophyte sanguinea Sparrm. subsp. piriei (Hutch.) Hansen ■☆

346897 Sarcophyte sanguinea Sparrm. ;肉蛇菰(肉草)■☆

346898 Sarcophyte sanguinea Sparrm. subsp. piriei (Hutch.) Hansen;皮里肉蛇菰■☆

346899 Sarcophyton Garay(1972);肉兰属;Fleshyorchis ■

346900 Sarcophyton taiwanianum (Hayata) Garay;肉兰(厚唇兰,台湾肉兰,台湾山兰);Taiwan Fleshyorchis ■

346901 Sarcopilea Urb. (1912);绿珠草属■☆

346902 Sarcopilea domingensis Urb. ;绿珠草■☆

346903 Sarcopodaceae Gagnep. ;肉足兰科

346904 Sarcopodaceae Gagnep. = Santalaceae R. Br. (保留科名)●■

346905 Sarcopodium Lindl. = Epigeneium Gagnep. ■

346906 Sarcopodium Lindl. et Paxton = Epigeneium Gagnep. ■

346907 Sarcopodium Lindl. et Paxton(1850);肉足兰属(厚唇兰属)■

346908 Sarcopodium affine (Lindl.) Lindl. et Paxton = Bulbophyllum affine Lindl. ■

346909 Sarcopodium amplum (Lindl.) Lindl. = Epigeneium amplum (Lindl. ex Wall.) Summerh. ■

346910 Sarcopodium amplum (Lindl.) Lindl. et Paxton = Epigeneium amplum (Lindl. ex Wall.) Summerh. ■

346911 Sarcopodium clemensiae (Gagnep.) Ts. Tang et F. T. Wang = Epigeneium clemensiae Gagnep. ■

346912 Sarcopodium coelogyne (Rchb. f.) Rolfe = Epigeneium amplum (Lindl. ex Wall.) Summerh. ■

346913 Sarcopodium dearei Hort. = Bulbophyllum dearei Rchb. f. ■☆

346914 Sarcopodium fargesii (Finet) Ts. Tang et F. T. Wang = Epigeneium fargesii (Finet) Gagnep. ■

346915 Sarcopodium fuscescens (Griff.) Lindl. = Epigeneium

fuscescens (Griff.) Summerh. ■

346916 Sarcopodium griffithii Lindl. = Bulbophyllum griffithii (Lindl.) Rchb. f. ■

346917 Sarcopodium leopardinum (Wall.) Lindl. = Bulbophyllum leopardinum (Wall.) Lindl. ■

346918 Sarcopodium psittacoglossum (Rchb. f.) Hook. = Bulbophyllum psittacoglossum Rchb. f. ■

346919 Sarcopodium romndamm Lindl. = Epigeneium rotundatum (Lindl.) Summerh. ■

346920 Sarcopodium striatum (Griff.) Lindl. = Bulbophyllum striatum (Griff.) Rchb. f. ■

346921 Sarcopodium uniflorum Lindl. = Bulbophyllum pteroglossum Schltr. ■

346922 Sarcopoterium Spach(1846);肉棘蔷薇属(肉棘属)●☆

346923 Sarcopoterium spinosum (L.) Spach;肉棘蔷薇●☆

346924 Sarcopteryx Radlk. (1879);肉翼无患子属●☆

346925 Sarcopteryx acuminata S. T. Reynolds;渐尖肉翼无患子●☆

346926 Sarcopteryx brachyphylla Radlk. ;短叶肉翼无患子●☆

346927 Sarcopteryx caudata Welzen;尾状肉翼无患子●☆

346928 Sarcopteryx montana S. T. Reynolds;山地肉翼无患子●☆

346929 Sarcopteryx reticulata S. T. Reynolds;网脉肉翼无患子●☆

346930 Sarcopteryx rigida Radlk. ;坚挺肉翼无患子●☆

346931 Sarcopus Gagnep. = Exocarpos Labill. (保留属名)●☆

346932 Sarcopygme Setch. et Christoph. (1935);矮肉茜属●☆

346933 Sarcopygme intermedia Setch. et Christoph. ;间型矮肉茜●☆

346934 Sarcopygme multinervis Setch. et Christoph. ;多脉矮肉茜●☆

346935 Sarcopyramis Wall. (1824);肉穗草属(肉穗野牡丹属);Fleshspike ■

346936 Sarcopyramis bodinieri H. Lév. et Vaniot;肉穗草(肉穗野牡丹,小肉穗草);Bodinier Fleshspike, Fleshspike ■

346937 Sarcopyramis bodinieri H. Lév. et Vaniot var. delicata (C. B. Rob.) C. Chen = Sarcopyramis bodinieri H. Lév. et Vaniot ■

346938 Sarcopyramis bodinieri H. Lév. et Vaniot var. delicata (C. B. Rob.) C. Chen;东方肉穗草(东方肉穗野牡丹,肉穗草);Delicate Bodinier Fleshspike ■

346939 Sarcopyramis crenata H. L. Li;圆齿肉穗草;Crenate Fleshspike, Roundtooth Fleshspike ■

346940 Sarcopyramis crenata H. L. Li = Sarcopyramis bodinieri H. Lév. et Vaniot ■

346941 Sarcopyramis delicata C. B. Rob. = Sarcopyramis bodinieri H. Lév. et Vaniot var. delicata (C. B. Rob.) C. Chen ■

346942 Sarcopyramis delicata C. B. Rob. = Sarcopyramis bodinieri H. Lév. et Vaniot ■

346943 Sarcopyramis dielsii Hu = Sarcopyramis nepalensis Wall. ■

346944 Sarcopyramis grandiflora Griff. = Sarcopyramis nepalensis Wall. ■

346945 Sarcopyramis lanceolata Wall. = Sarcopyramis nepalensis Wall. ■

346946 Sarcopyramis lanceolata Wall. ex Benn. = Sarcopyramis nepalensis Wall. ■

346947 Sarcopyramis napalensis var. delicata (C. B. Rob.) S. F. Huang et T. C. Huang = Sarcopyramis bodinieri H. Lév. et Vaniot ■

346948 Sarcopyramis napalensis Wall. var. bodinieri (H. Lév. et Vaniot) H. Lév. = Sarcopyramis bodinieri H. Lév. et Vaniot ■

346949 Sarcopyramis napalensis Wall. var. maculata C. Y. Wu = Sarcopyramis nepalensis Wall. ■

346950 Sarcopyramis nepalensis Wall. ;尼泊尔肉穗草(楮头红,满江红,肉穗野牡丹);Nepal Fleshspike ■

346951 Sarcopyramis nepalensis Wall. = Sarcopyramis bodinieri H. Lév.

et Vaniot ■

346952 Sarcopyramis nepalensis Wall. = Sarcopyramis bodinieri H. Lév. et Vaniot var. delicata (C. B. Rob.) C. Chen ■

346953 Sarcopyramis nepalensis Wall. var. bodinieri H. Lév. et Vaniot = Sarcopyramis bodinieri H. Lév. et Vaniot ■

346954 Sarcopyramis nepalensis Wall. var. delicata (C. B. Rob.) S. F. Huang et T. C. Huang = Sarcopyramis bodinieri H. Lév. et Vaniot var. delicata (C. B. Rob.) C. Chen ■

346955 Sarcopyramis nepalensis Wall. var. delicata (C. B. Rob.) S. F. Huang et T. C. Huang;东方肉穗野牡丹■

346956 Sarcopyramis nepalensis Wall. var. maculata C. Y. Wu et C. Chen;斑点尼泊尔肉穗草(斑点楮头红);Spotted Nepal Fleshspike ■

346957 Sarcopyramis parvifolia Merr. ex H. L. Li;小叶肉穗草;Smallleaf Fleshspike ■

346958 Sarcopyramis parvifolia Merr. ex H. L. Li = Sarcopyramis bodinieri H. Lév. et Vaniot ■

346959 Sarcorhachis Trel. (1927);肉胡椒属●☆

346960 Sarcorhachis anomala Trel.;肉胡椒●☆

346961 Sarcorhachis obtusata Trel.;钝肉胡椒●☆

346962 Sarcorhyna C. Presl = Bumelia Sw. (保留属名)●☆

346963 Sarcorhyna C. Presl = Rostellaria C. F. Gaertn.●☆

346964 Sarcorhyna C. Presl = Rostellularia Rchb. ■

346965 Sarcorhynchus Schltr. (1918);肉喙兰属■☆

346966 Sarcorhynchus Schltr. = Diaphananthe Schltr. ■☆

346967 Sarcorhynchus bilobatus Summerh. = Rhipidoglossum bilobatum (Summerh.) Szlach. et Olszewski ■☆

346968 Sarcorhynchus orientalis Mansf. = Rhipidoglossum orientalis (Mansf.) Szlach. et Olszewski ■☆

346969 Sarcorhynchus polyanthus (Kraenzl.) Schltr. = Rhipidoglossum polyanthum (Kraenzl.) Szlach. et Olszewski ■☆

346970 Sarcorhynchus saccolabioides Schltr. = Rhipidoglossum polyanthum (Kraenzl.) Szlach. et Olszewski ■☆

346971 Sarcorrhiza Bullock(1962);肉根萝藦属☆

346972 Sarcorrhiza epiphytica Bullock;肉根萝藦☆

346973 Sarcoryna Post et Kuntze = Bumelia Sw. (保留属名)●☆

346974 Sarcoryna Post et Kuntze = Rostellaria C. F. Gaertn. ■

346975 Sarcoryna Post et Kuntze = Rostellularia Rchb. ■

346976 Sarcoryna Post et Kuntze = Sarcorhyna C. Presl ●☆

346977 Sarcosiphon Blume = Thismia Griff. ■

346978 Sarcosiphon Reinw. ex Blume = Thismia Griff. ■

346979 Sarcosperma Hook. f. (1876);肉实树属(肉实属);Fleshseed Tree,Flesh-seed Tree,Fleshspike Tree ●

346980 Sarcosperma arboreum Hook. f.;大肉实树(肉实树);Arboreous Fleshseed Tree, Treelike Fleshseed Tree, Treelike Fleshspike Tree ●

346981 Sarcosperma cheliense Hu ex Lam. et Royen = Sarcosperma griffithii Hook. f. ●

346982 Sarcosperma griffithii Hook. f.;小叶肉实树;Griffith Fleshseed Tree, Griffith Fleshspike Tree ●

346983 Sarcosperma kachinense (King et Prain) Exell;绒毛肉实树(毛序肉实树, 枚辣叶肉实);Hairy Fleshseed Tree, Hairy Fleshspike Tree, Kachin Fleshseed Tree ●

346984 Sarcosperma kachinense (King et Prain) Exell var. simondii (Gagnep.) Lam. et Royen;光序肉实树;Glabrous Fleshseed Tree, Glabrous Fleshspike Tree ●

346985 Sarcosperma laurinum (Benth.) Hook. f.;肉实树(山苦瓜,水石梓);Laurell Fleshseed Tree, Laurelike Fleshseed Tree, Laurelike

Fleshspike Tree ●

346986 Sarcosperma pedunculata Hemsl. = Sinosideroxylon pedunculatum (Hemsl.) H. Chuang ●

346987 Sarcosperma simondii Gagnep. = Sarcosperma kachinense (King et Prain) Exell var. simondii (Gagnep.) Lam. et Royen ●

346988 Sarcospermaceae H. J. Lam = Sarcospermataceae H. J. Lam(保留科名)●

346989 Sarcospermataceae H. J. Lam(1925)(保留科名);肉实树科●

346990 Sarcospermataceae H. J. Lam(保留科名) = Sapotaceae Juss. (保留科名)●

346991 Sarcospermum Reinw. ex de Vriese = Gunnera L. ■☆

346992 Sarcostachys Juss. = Stachytarpheta Vahl(保留属名)■●

346993 Sarcostemma R. Br. (1810);肉珊瑚属(无叶藤属);Fleshcoral ■

346994 Sarcostemma R. Br. = Cynanchum L. ●■

346995 Sarcostemma acidum (Roxb.) Voigt;肉珊瑚(珊瑚,铁珊,无叶藤);Acid Fleshcoral, Moon Plant, Rapunzel Plant, Somabutta, Sour Creeper ■

346996 Sarcostemma acidum (Roxb.) Voigt = Sarcostemma viminale (L.) R. Br. ■

346997 Sarcostemma andongense Hiern = Sarcostemma viminale (L.) R. Br. ■

346998 Sarcostemma annular Roth = Holostemma annularium (Roxb.) K. Schum. ●■

346999 Sarcostemma antsiranense Meve et Liede = Cynanchum antsiranense (Meve et Liede) Liede et Meve ■☆

347000 Sarcostemma aphyllum (Thunb.) R. Br. = Sarcostemma viminale (L.) R. Br. ■

347001 Sarcostemma aphyllum Thunb. = Sarcostemma viminale (L.) R. Br. ■

347002 Sarcostemma australe R. Br.;澳洲肉珊瑚■☆

347003 Sarcostemma bicolor Decne.;二色肉珊瑚■☆

347004 Sarcostemma brevistigma Wight et Arn. = Sarcostemma acidum (Roxb.) Voigt ■

347005 Sarcostemma brevistigma Wight et Arn. = Sarcostemma viminale (L.) R. Br. ■

347006 Sarcostemma clausum Schult.;白肉珊瑚;White Twinevine ■☆

347007 Sarcostemma crassifolium Decne.;厚叶肉珊瑚■☆

347008 Sarcostemma cynanchoides Decne.;北美肉珊瑚;Climbing Milkweed, Twining Milkweed ■☆

347009 Sarcostemma daltonii Decne.;多尔顿肉珊瑚■☆

347010 Sarcostemma decorsei Costantin et Gallaud = Cynanchum decorsei (Costantin et Gallaud) Liede et Meve ■☆

347011 Sarcostemma elachistemmoides Liede et Meve = Cynanchum elachistemmoides (Liede et Meve) Liede et Meve ■☆

347012 Sarcostemma esculentum (L. f.) R. W. Holm = Oxystelma esculentum (L. f.) Sm. ■

347013 Sarcostemma hirtellum (Vail) R. W. Holm;毛肉珊瑚;Rambling Milkweed ■☆

347014 Sarcostemma implicatum Jum. et H. Perrier = Cynanchum implicatum (Jum. et H. Perrier) Jum. et H. Perrier ■☆

347015 Sarcostemma insigne (N. E. Br.) Desc. = Cynanchum insigne (N. E. Br.) Liede et Meve ■☆

347016 Sarcostemma intermedium Decne.;间型肉珊瑚■☆

347017 Sarcostemma membranaceum Liede et Meve = Cynanchum membranaceum (Liede et Meve) Liede et Meve ●☆

347018 Sarcostemma mulanjense Liede et Meve;姆兰杰肉珊瑚■☆

347019 Sarcostemma nudum C. Sm. ex Decne. = Sarcostemma daltonii

Decne. ■☆

347020　Sarcostemma odontolepis Balf. f. = Sarcostemma viminale（L.）R. Br. subsp. odontolepis（Balf. f.）Meve et Liede ■☆

347021　Sarcostemma pearsonii N. E. Br. ;皮尔逊肉珊瑚■☆

347022　Sarcostemma pyrotechnicum（Forssk.）Schult. = Leptadenia pyrotechnica（Forssk.）Decne. ■☆

347023　Sarcostemma resiliens B. R. Adams et R. W. K. Holland;雷西里肉珊瑚■☆

347024　Sarcostemma stipitaceum Forssk. = Sarcostemma viminale（L.）R. Br. subsp. stipitaceum（Forssk.）Meve et Liede ■☆

347025　Sarcostemma stocksii Hook. f. = Sarcostemma viminale（L.）R. Br. subsp. stocksii（Hook. f.）Ali ■☆

347026　Sarcostemma stoloniferum B. R. Adams et R. W. K. Holland;匍匐肉珊瑚■☆

347027　Sarcostemma tetrapterum Turcz. = Sarcostemma viminale（L.）R. Br. ■

347028　Sarcostemma thunbergii Don = Sarcostemma viminale（L.）R. Br. subsp. thunbergii（Don）Liede et Meve ■☆

347029　Sarcostemma viminale（L.）R. Br. = Sarcostemma acidum（Roxb.）Voigt ■

347030　Sarcostemma viminale（L.）R. Br. subsp. odontolepis（Balf. f.）Meve et Liede;齿鳞肉珊瑚■☆

347031　Sarcostemma viminale（L.）R. Br. subsp. stipitaceum（Forssk.）Meve et Liede;托叶状肉珊瑚■☆

347032　Sarcostemma viminale（L.）R. Br. subsp. stocksii（Hook. f.）Ali;斯托克斯肉珊瑚■☆

347033　Sarcostemma viminale（L.）R. Br. subsp. suberosum Meve et Liede;木栓质肉珊瑚■☆

347034　Sarcostemma viminale（L.）R. Br. subsp. thunbergii（Don）Liede et Meve;通贝里肉珊瑚■☆

347035　Sarcostemma welwitschii Hiern;韦尔肉珊瑚■☆

347036　Sarcostigma Wight et Arn.（1833）;肉柱铁青树属●

347037　Sarcostigma brevipes（Engl.）Engl. = Desmostachys brevipes（Engl.）Sleumer ■☆

347038　Sarcostigma kleimi Wight et Arn. ;肉柱铁青树●☆

347039　Sarcostigma vogelii Miers = Desmostachys vogelii（Miers）Stapf ■☆

347040　Sarcostigmataceae Tiegh. = Icacinaceae Miers（保留科名）●■

347041　Sarcostigmataceae Tiegh. = Sarcostigmataceae Tiegh. ex Bullock ●

347042　Sarcostigmataceae Tiegh. ex Bullock = Icacinaceae Miers（保留科名）●■

347043　Sarcostigmataceae Tiegh. ex Bullock;肉柱铁青树科●

347044　Sarcostoma Blume（1825）;肉口兰属■☆

347045　Sarcostoma brevipes J. J. Sm. ;短梗肉口兰■☆

347046　Sarcostoma javanica Blume;爪哇肉口兰■☆

347047　Sarcostyles C. Presl ex DC. = Hydrangea L. ●

347048　Sarcotheca Blume（1851）;肉囊木属（肉囊酢浆草属）●☆

347049　Sarcotheca Kuntze = Justicia L. ●■

347050　Sarcotheca Kuntze = Sarotheca Nees ●■

347051　Sarcotheca Turcz. = Schinus L. ●

347052　Sarcotheca celebica Veldkamp;肉囊木●☆

347053　Sarcotoechia Radlk.（1879）;肉壁无患子属●☆

347054　Sarcotoechia heterophylla S. T. Reynolds;异叶肉壁无患子●☆

347055　Sarcotoechia lanceolata（C. T. White）S. T. Reynolds;披针叶肉壁无患子●☆

347056　Sarcotoechia villosa S. T. Reynolds;毛肉壁无患子●☆

347057　Sarcotoxicum Cornejo et Iltis = Capparis L. ●

347058　Sarcotoxicum Cornejo et Iltis（2008）;柳叶山柑属●☆

347059　Sarcoyucca（Engelm.）Linding. = Yucca L. ●■

347060　Sarcoyucca（Trel.）Linding. = Yucca L. ●■

347061　Sarcozona J. M. Black = Carpobrotus N. E. Br. ●☆

347062　Sarcozygium Bunge = Zygophyllum L. ●■

347063　Sarcozygium Bunge（1843）;霸王属（肉蒺藜属）●

347064　Sarcozygium kaschgaricum（Boiss.）Y. X. Liou = Zygophyllum kaschgaricum Boriss. ●

347065　Sarcozygium xanthoxylon Bunge = Sarcozygium xanthoxylum Bunge ●

347066　Sarcozygium xanthoxylum Bunge;霸王（霸王树,木霸王）;Common Beancaper,Overlord ●

347067　Sarcozygium xanthoxylum Bunge = Zygophyllum xanthoxylum（Bunge）Maxim. ●

347068　Sardinia Vell. = Guettarda L. ●

347069　Sardonula Raf. = Ranunculus L. ■

347070　Sarga Ewart = Chrysopogon Trin.（保留属名）■

347071　Sarga Ewart = Sorghum Moench（保留属名）■

347072　Sargentia H. Wendl. et Drude ex Salomon（废弃属名）= Pseudophoenix H. Wendl. ex Sarg.（废弃属名）●☆

347073　Sargentia H. Wendl. et Drude ex Salomon（废弃属名）= Sargentia S. Watson（保留属名）●☆

347074　Sargentia S. Watson（1890）（保留属名）;肖刺葵属（葫芦椰子属,假海枣属,樱桃椰属）;Cherry Palm ●☆

347075　Sargentia S. Watson（保留属名）= Casimiroa La Lave et Lex. ●

347076　Sargentia aricocca H. Wendl. et Drude ex Salomon;肖刺葵（樱桃椰）;Buccaneer Palm,Sargent's Cherry Palm ●☆

347077　Sargentodoxa Rehder et E. H. Wilson（1913）;大血藤属;Bloodvine,Sargent's Glory Vine,Sargentglorivine,Sargent-glory-vine ●★

347078　Sargentodoxa cuneata（Oliv.）Rehder et E. H. Wilson;大血藤（八卦藤,半血莲,槟榔钻,赤沙藤,穿尖龙,大活血,大血通,单叶血藤,过血莲,过血藤,红菊花心,红牛鼻陈,红皮藤,红藤,红血藤,花血藤,黄梗藤,黄鸡藤,黄省藤,活血藤,鸡血藤,蕨心藤,千年健,山红藤,山血藤,省藤,五花七,五花血藤,五花血通,五血藤,血陈根,血风藤,血灌肠,血木通,血藤,血通,鱼藤）;Bloodvine,Sargent's Glory Vine,Sargentglorivine,Sargent-glory-vine,Simpleleaf Sargentglorivine,Simple-leaved Sargent-glory-vine ●

347079　Sargentodoxa cuneata（Oliv.）Rehder et E. H. Wilson var. setosa S. C. Li et Z. M. Wu;刺毛大血藤●

347080　Sargentodoxa simplicifolia S. Z. Qu et C. L. Min = Sargentodoxa cuneata（Oliv.）Rehder et E. H. Wilson ●

347081　Sargentodoxaceae Stapf = Lardizabalaceae R. Br.（保留科名）●

347082　Sargentodoxaceae Stapf = Sargentodoxaceae Stapf ex Hutch.（保留科名）●

347083　Sargentodoxaceae Stapf ex Hutch.（1926）（保留科名）;大血藤科;Sargent-glory-vine Family,Sargentodoxa Family ●

347084　Sargentodoxaceae Stapf ex Hutch.（保留科名）= Lardizabalaceae R. Br.（保留科名）●

347085　Sariava Reinw. = Symplocos Jacq. ●

347086　Saribus Blume = Livistona R. Br. ●

347087　Saribus chinensis Blume = Livistona chinensis（Jacq.）R. Br. ex Mart. ●

347088　Saribus rotundifolia（Lam.）Blume = Livistona rotundifolia（Lam.）Mart. ●

347089　Sarinia O. F. Cook = Attalea Kunth ●☆

347090　Sarissus Gaertn. = Hydrophylax L. f. ■☆

347091　Saritaea Dugand（1945）;蒜香藤属●☆

347092　Saritaea magnifica（Sprague ex Steenis）Dugand;蒜香藤（紫玲

藤）；Glow Vine，Glowvine ●

347093 Saritaea magnifica（W. Bull）Dugand ＝ Bignonia magnifica W. Bull ●

347094 Sarlina Guillaumin ＝？Linociera Sw. ex Schreb.（保留属名）●

347095 Sarlina Guillaumin（1952）；柱果木犀榄属●☆

347096 Sarlina cylindrocarpa Guillaumin；柱果木犀榄●☆

347097 Sarmasikia Bubani ＝ Cynanchum L. ●■

347098 Sarmentaceae Schultz Sch. ＝ Dioscoreaceae R. Br.（保留科名）＋ Liliaceae Juss.（保留科名）●■

347099 Sarmentaceae Vent. ＝ Vitaceae Juss.（保留科名）●■

347100 Sarmentaria Naudin ＝ Adelobotrys DC. ●☆

347101 Sarmenticola Senghas et Garay（1996）；秘鲁翅冠兰属■☆

347102 Sarmenticola calceolaris（Garay）Senghas et Garay；秘鲁翅冠兰■☆

347103 Sarmienta Ruiz et Pav.（1794）（保留属名）；吊钟苣苔属（萨民托苣苔属）；Sarmienta ■●☆

347104 Sarmienta Siebold ex Baill. ＝ Cnestis Juss. ●

347105 Sarmienta repens Ruiz et Pav.；吊钟苣苔（攀缘萨民托苣苔）；Climbing Sarmienta ■☆

347106 Sarmienta scandens Pers. ＝ Sarmienta repens Ruiz et Pav. ■☆

347107 Sarna H. Karst. ＝ Pilostyles Guill. ■☆

347108 Sarocalamus Stapleton ＝ Arundinaria Michx. ●

347109 Sarocalamus racemosus（Munro）Stapleton ＝ Arundinaria racemosa Munro ●

347110 Sarocalamus spanostachyus（T. P. Yi）Stapleton. ＝ Arundinaria spanostachya（T. P. Yi）D. Z. Li ●◇

347111 Sarocolla Kunth ＝ Saltera Bullock ●☆

347112 Sarojusticia Bremek. ＝ Justicia L. ●■

347113 Saropsis B. G. Briggs et L. A. S. Johnson（1998）；密鞘帚灯草属■☆

347114 Saropsis fastigiata（R. Br.）B. G. Briggs et L. A. S. Johnson；密鞘帚灯草■☆

347115 Sarosanthera Korth. ＝ Adinandra Jack ●

347116 Sarotes Lindl. ＝ Guichenotia J. Gay ●☆

347117 Sarothamnus Wimm.（1832）（保留属名）；帚灌豆属●☆

347118 Sarothamnus Wimm.（保留属名）＝ Cytisus Desf.（保留属名）●

347119 Sarothamnus arboresceus（Desf.）Webb ＝ Cytisus arboreus（Desf.）DC. ●☆

347120 Sarothamnus arboresceus（Desf.）Webb var. barbarus Jahand. et Maire ＝ Cytisus arboreus（Desf.）DC. ●☆

347121 Sarothamnus baeticus Webb ＝ Cytisus arboreus（Desf.）DC. subsp. baeticus（Webb）Maire ●☆

347122 Sarothamnus baeticus Webb var. tetuanensis Pau ＝ Cytisus arboreus（Desf.）DC. subsp. baeticus（Webb）Maire ●☆

347123 Sarothamnus grandiflorus（DC.）Webb ＝ Cytisus grandiflorus（Brot.）DC. ●☆

347124 Sarothamnus maurus（Humbert et Maire）Raynaud ＝ Cytisus maurus Humbert et Maire ●☆

347125 Sarothamnus megalanthus Pau et Font Quer ＝ Cytisus megalanthus（Pau et Font Quer）Font Quer ●☆

347126 Sarothamnus scoparius（L.）Wimm. ex K. Koch ＝ Cytisus scoparius（L.）Link ●

347127 Sarothamnus virgatus Webb；帚灌豆●☆

347128 Sarotheca Nees ＝ Justicia L. ●■

347129 Sarothra L. ＝ Hypericum L. ■●

347130 Sarothra gentianoides L. ＝ Hypericum gentianoides（L.）Britton，Sterns et Poggenb. ■☆

347131 Sarothra graminea（G. Forst.）Y. Kimura ＝ Hypericum gramineum G. Forst. ■

347132 Sarothra japonica（Thunb. ex Murray）Y. Kimura ＝ Hypericum japonicum Thunb. ■

347133 Sarothra japonica（Thunb.）Y. Kimura ＝ Hypericum japonicum Thunb. ■

347134 Sarothra laxa（Blume）Y. Kimura ＝ Hypericum japonicum Thunb. ■

347135 Sarothra major（A. Gray）Y. Kimura ＝ Hypericum majus（A. Gray）Britton ●☆

347136 Sarothra saginoides Y. Kimura ＝ Hypericum gramineum G. Forst. ■

347137 Sarothrochilus Schltr. ＝ Staurochilus Ridl. ex Pfitzer ■

347138 Sarothrochilus dawsonianus（Rchb. f.）Schltr. ＝ Staurochilus dawsonianus（Rchb. f.）Schltr. ■

347139 Sarothroehilus Schltr. ＝ Trichoglottis Blume ■

347140 Sarothrostachys Klotzsch ＝ Sebastiania Spreng. ●

347141 Sarpedonia Raf. ＝ Ranunculus L. ■

347142 Sarracena L. ＝ Sarracenia L. ■☆

347143 Sarracenella Luer ＝ Pleurothallis R. Br. ■☆

347144 Sarracenia L.（1753）；瓶子草属（管叶草属，管子草属，肖瓶子草属）；Devil's Boot，Indian Cup，North American Pitcherplant，North American Pitcher-plant，Pitcher Plant，Pitcher's Plants，Sarracenia，Side-saddle Flower，Trumpet ■☆

347145 Sarracenia L. ＝ Sarraceniaceae Dumort.（保留科名）■☆

347146 Sarracenia × excellens G. Nicholson；优雅瓶子草■☆

347147 Sarracenia alata（A. W. Wood）A. W. Wood；翼瓶子草；Sweet Pitcher Plant，Trumpet Pitcherplant，Yellow Trumpets ■☆

347148 Sarracenia areolata Macfarl.；斑纹瓶子草■☆

347149 Sarracenia drummondii Croom；德拉蒙德瓶子草；Crimson Pitcher Plant，Drummond's Side-saddle Flower，Fiddler's Trumpet ■☆

347150 Sarracenia flava L.；黄花瓶子草（黄瓶子草）；Huntsman's Horn，Huntsman's Trumpet，Trumpet，Trumpet Leaf，Yellow Pitcher Plant，Yellow Pitcherplant，Yellow Side-saddle Flower ■☆

347151 Sarracenia heterophylla Eaton ＝ Sarracenia purpurea L. ■☆

347152 Sarracenia leucophylla Raf.；白叶瓶子草；Crimson Pitcherplant ■☆

347153 Sarracenia minor Walter；小瓶子草；Hooded Pitcher Plant，Hooded Pitcherplant，Hooded Pitcher-plant，Varioeal Side-saddle Flower ■☆

347154 Sarracenia oreophila Wherry；山地瓶子草■☆

347155 Sarracenia psittacina Michx.；鹦鹉瓶子草；Parrot Head，Parrot Pitcher Plant，Parrot Pitcherplant，Parrot Pitcher-plant ■☆

347156 Sarracenia purpurea L.；瓶子草；Boots，Common Pitcher Plant，Common Pitcherplant，Common Pitcher-plant，Devil's Boots，Eve's Cup，Eve's Cups，Flycatcher，Flytrap，Forefather's Cup，Huntsman's Cup，Huntsman's Horn，Indian Cup，Northern Pitcher Plant，Northern Pitcherplant，Owl's Socks，Owpa Socks，Pitcher Plant，Pitcherplant，Purple Pitcher Plant，Purple Pitcher-plant，Purple Side-saddle Flower，Sidesaddle Flower，Side-saddle Plant，Soldier's Drinking Cup，Trumpet，Trumpet Leaf，Watches，Water Cup ■☆

347157 Sarracenia purpurea L. f. heterophylla（Eaton）Fernald ＝ Sarracenia purpurea L. ■☆

347158 Sarracenia purpurea L. subsp. gibbosa（Raf.）Wherry；囊瓶子草；Purple Pitcher-plant ■☆

347159 Sarracenia purpurea L. subsp. heterophylla（Eaton）Torr. ＝ Sarracenia purpurea L. ■☆

347160 Sarracenia purpurea L. var. ripicola B. Boivin ＝ Sarracenia purpurea L. ■☆

347161 Sarracenia purpurea L. var. stolonifera Macfarl. et Steckbeck ＝

Sarracenia purpurea L. ■☆

347162　Sarracenia purpurea L. var. terrae-novae Bach. Pyl. = Sarracenia purpurea L. ■☆

347163　Sarracenia rubra Walter；红花瓶子草；Miniature Huntsman Horn，Red Pitcher Plant，Red Side-saddle Flower，Sweet Pitcher Plant，Sweet Pitcher-plant ■☆

347164　Sarracenia sledgei Macfarl.；斯氏瓶子草；Sledge's Pitcher-plant，Yellow Trumpet ☆

347165　Sarracenia trifoliata ?；三小叶瓶子草；Pitcherplant ■☆

347166　Sarraceniaceae Dumort. (1829)（保留科名）；瓶子草科（管叶草科，管子草科）；American Pitcherplant Family，American Pitcher-Plant Family，Pitcherplant Family ■☆

347167　Sarracha Rchb. = Saracha Ruiz et Pav. ●☆

347168　Sarratia Moq. = Amaranthus L. ■

347169　Sarratia berlandieri Moq. var. emarginata Torr. = Amaranthus torreyi (A. Gray) S. Watson ■☆

347170　Sarratia berlandieri Moq. var. fimbriata Torr. = Amaranthus fimbriatus (Torr.) Benth. ex S. Watson ■☆

347171　Sarrazinia Hoffmanns. = Sarracenia L. ■☆

347172　Sarsaparilla Kuntze = Smilax L. ●

347173　Sartidia De Winter(1963)；健三芒草属■☆

347174　Sartidia angolensis (C. E. Hubb.) De Winter；安哥拉健三芒草■☆

347175　Sartidia jucunda (Schweick.) De Winter；愉悦健三芒草■☆

347176　Sartidia perrieri (A. Camus) Bourreil；佩里耶健三芒草■☆

347177　Sartidia vanderystii (De Wild.) De Winter；范德三芒草■☆

347178　Sartoria Boiss. = Onobrychis Mill. ■

347179　Sartoria Boiss. et Heldr. (1849)；岩黄耆状驴豆属■☆

347180　Sartoria hedysaroides Boiss. et Heldr.；岩黄耆状驴豆■☆

347181　Sartorina R. M. King et H. Rob. (1974)；毛柱泽兰属■☆

347182　Sartorina schultzii R. M. King et H. Rob.；毛柱泽兰■☆

347183　Sartwellia A. Gray(1852)；黄光菊属；Glowwort ■☆

347184　Sartwellia flaveriae A. Gray；黄光菊■☆

347185　Saruma Oliv. (1889)；马蹄香属；Saruma ■★

347186　Saruma henryi Oliv.；马蹄香（高脚细辛，高足细辛，狗肉香，冷水丹）；Henry Saruma ■

347187　Sarumaceae (O. C. Schmidt) Nakai = Aristolochiaceae Juss.（保留科名）■●

347188　Sarumaceae Nakai = Aristolochiaceae Juss.（保留科名）■●

347189　Sarumaceae Nakai；马蹄香科■

347190　Sarx H. St. John = Sicyos L. ■

347191　Sasa Makino et Shibata (1901)；赤竹属（箬竹属）；Sasa，Bamboo ●

347192　Sasa akiuensis (Sad. Suzuki) Sad. Suzuki；秋保赤竹●☆

347193　Sasa albomarginata (Miq.) Makino et Shibata = Sasa veitchii (Carrière) Rehder ●

347194　Sasa albosericea W. T. Lin et J. Y. Lin；银环赤竹；Silver-sericeous Sasa ●

347195　Sasa argenteostriata (Regel) E. G. Camus；铺地竹；Silverstripe-leaved Bamboo，Silverstripe-leaved Japanese Bamboo，Silvery White-striped Sasa ●

347196　Sasa asagishiana Makino et Uchida = Neosasamorpha asagishiana (Makino et Uchida) Tatew. ●

347197　Sasa auricoma E. G. Camus；菲黄竹；Aureo-coma Sasa ●

347198　Sasa bashanensis C. D. Chu et C. S. Chao = Indocalamus bashanensis (C. D. Chu et C. S. Chao) H. R. Zhao et Y. L. Yang ●

347199　Sasa bitchuensis Makino；备中赤竹●☆

347200　Sasa borealis (Hack.) Makino et Shibata；北方华箬竹；North

Sasa ●

347201　Sasa cernua Makino = Sasa spiculosa (F. Schmidt) Makino ●☆

347202　Sasa cernua Makino f. nebulosa (Makino et Shibata) Tatew.；星云华箬竹●☆

347203　Sasa chartacea (Makino) Makino et Shibata；纸质赤竹●☆

347204　Sasa chartacea (Makino) Makino et Shibata f. hattoriana (Koidz.) Sad. Suzuki = Cerbera manghas L. ●

347205　Sasa chartacea (Makino) Makino var. mollis (Nakai) Sad. Suzuki；柔软赤竹●☆

347206　Sasa chartacea (Makino) Makino var. nana (Makino) Sad. Suzuki；矮赤竹●☆

347207　Sasa chartacea (Makino) Makino var. simotsukensis Sad. Suzuki；下野赤竹●☆

347208　Sasa depauperata (Takeda) Nakai = Sasa yahikoensis Makino var. depauperata (Takeda) Sad. Suzuki ●☆

347209　Sasa disticha (Mitford) E. G. Camus = Pleioblastus distichus (Mitford) Nakai ●

347210　Sasa disticha (Mitford) E. G. Camus = Sasa pygmaea (Miq.) E. G. Camus var. disticha (Mitford) C. S. Chao et G. G. Tang ●

347211　Sasa duplicata W. T. Lin et Z. J. Feng；孖竹仔；Duplicate Sasa ●

347212　Sasa duplicata W. T. Lin et Z. J. Feng = Sasa rubrovaginata C. H. Hu ●

347213　Sasa elegantissima Koidz.；雅致赤竹●☆

347214　Sasa fortunei (Van Houtte) Fiori；福氏竹（菲白竹，缟竹，花叶苦竹，稚子竹）；Dwarf Whitestripe Bamboo，Fortune Bamboo，Fortune Bitterbamboo，Fortune Sasa ●

347215　Sasa fortunei (Van Houtte) Fiori = Pleioblastus fortunei (Van Houtte) Nakai ●

347216　Sasa fugeshiensis Koidz.；风至赤竹●☆

347217　Sasa gracillima Nakai；细长赤竹●☆

347218　Sasa guangdongensis W. T. Lin et X. B. Ye；广东赤竹；Guangdong Sasa ●

347219　Sasa guangxiensis C. D. Chu et C. S. Chao；广西赤竹；Guangxi Sasa ●

347220　Sasa hainanensis C. D. Chu et C. S. Chao；海南赤竹；Hainan Sasa ●

347221　Sasa hayatae Makino；早田氏赤竹●☆

347222　Sasa hayatae Makino var. hirtella (Nakai) Sad. Suzuki；多毛早田氏赤竹（多毛赤竹）●☆

347223　Sasa heterotricha Koidz.；异毛赤竹●☆

347224　Sasa heterotricha Koidz. var. nagatoensis Sad. Suzuki；长门赤竹●☆

347225　Sasa hirta (Koidz.) Tzvelev = Sasa kurilensis (Rupr.) Makino et Shibata var. hirta (Koidz.) Sad. Suzuki ●☆

347226　Sasa hubeiensis (C. H. Hu) C. H. Hu；湖北华箬竹；Hubei Sasa ●

347227　Sasa japonica (Siebold et Zucc. ex Steud.) Makino = Pseudosasa japonica (Siebold et Zucc. ex Steud.) Makino ex Nakai ●

347228　Sasa jotanii (Kenji Inoue et Tanim.) M. Kobay.；约坦赤竹●☆

347229　Sasa kagamiana Makino et Uchida；镜赤竹●☆

347230　Sasa kagamiana Makino et Uchida subsp. yoshinoi (Koidz.) Sad. Suzuki；吉野氏赤竹●☆

347231　Sasa kesuzu Muroi et H. Okamura = Sasa sikokiana Koidz. ●☆

347232　Sasa kogasensis Nakai = Sasaella kogasensis (Nakai) Nakai ex Koidz. ●☆

347233　Sasa kogasensis Nakai var. nasuensis (Kimura et Sad. Suzuki) Sad. Suzuki；那须赤竹●☆

347234　Sasa kurilensis (Rupr.) Makino et Shibata；千岛赤竹●☆

347235　Sasa kurilensis (Rupr.) Makino et Shibata f. pseudokurilensis

(Nakai) Sad. Suzuki ＝Sasa kurilensis (Rupr.) Makino et Shibata ●☆

347236 Sasa kurilensis (Rupr.) Makino et Shibata f. uchidae (Makino) Sad. Suzuki ＝Sasa kurilensis (Rupr.) Makino et Shibata var. uchidae (Makino) Makino ●☆

347237 Sasa kurilensis (Rupr.) Makino et Shibata var. gigantea Tatew. ＝Sasa spiculosa (F. Schmidt) Makino ●☆

347238 Sasa kurilensis (Rupr.) Makino et Shibata var. hirta (Koidz.) Sad. Suzuki;多毛千岛赤竹(多毛赤竹)●☆

347239 Sasa kurilensis (Rupr.) Makino et Shibata var. jotanii Kenji Inoue et Tanim. ＝Sasa jotanii (Kenji Inoue et Tanim.) M. Kobay. ●☆

347240 Sasa kurilensis (Rupr.) Makino et Shibata var. uchidae (Makino) Makino;内多赤竹●☆

347241 Sasa longiligulata McClure;赤竹;Longtongue Sasa,Longtongued Bamboo,Longtongued Sasa,Long-tongued Sasa ●

347242 Sasa maculata Nakai;斑点赤竹●☆

347243 Sasa maculata Nakai var. abei Sad. Suzuki;阿拜赤竹●☆

347244 Sasa magnifica (Nakai) Sad. Suzuki;华丽赤竹●☆

347245 Sasa magnifica (Nakai) Sad. Suzuki subsp. fujitae Sad. Suzuki;藤田赤竹●☆

347246 Sasa magnifica (Nakai) Sad. Suzuki var. igaensis (Nakai) Sad. Suzuki ＝Sasa magnifica (Nakai) Sad. Suzuki ●☆

347247 Sasa magnonoda T. H. Wen et J. C. Liao;大节赤竹●

347248 Sasa masamuneana (Makino) C. S. Chao et Renvoize ＝Sasaella masamuneana (Makino) Hatus. et Muroi ●☆

347249 Sasa matsudae Nakai ＝Sasa spiculosa (F. Schmidt) Makino ●☆

347250 Sasa megalophylla Makino et Uchida;大叶赤竹●☆

347251 Sasa megalophylla Makino et Uchida f. aureovariegata Sad. Suzuki;黄斑大叶赤竹●☆

347252 Sasa miakeana Sad. Suzuki;三宅赤竹●☆

347253 Sasa minensis Sad. Suzuki;峰山赤竹●☆

347254 Sasa minensis Sad. Suzuki var. awaensis Sad. Suzuki;安和赤竹●☆

347255 Sasa nana (Makino) Makino ＝Sasa chartacea (Makino) Makino var. nana (Makino) Sad. Suzuki ●☆

347256 Sasa niijimae Tatew. ex Nakai ＝Sasa palmata (Lat. -Marl. ex Burb.) E. G. Camus var. niijimae (Tatew. ex Nakai) Sad. Suzuki ●☆

347257 Sasa niitakayamensis (Hayata) E. G. Camus ＝Yushania niitakayamensis (Hayata) P. C. Keng ●

347258 Sasa niitakayamensis (Hayata) E. G. Camus var. microcarpa E. G. Camus ＝Yushania niitakayamensis (Hayata) P. C. Keng ●

347259 Sasa nipponica (Makino) Makino et Shibata;日本赤竹●☆

347260 Sasa nubigena P. C. Keng ＝Indocalamus wilsonii (Rendle) C. S. Chao et C. D. Chu ●

347261 Sasa oblongula C. H. Hu;矩叶赤竹;Oblong-foliate Sasa, Oblong-leaf Sasa ●

347262 Sasa occidentalis Sad. Suzuki;西方赤竹●☆

347263 Sasa oseana (Makino) Uchida ＝Sasa yahikoensis Makino var. oseana (Makino) Sad. Suzuki ●☆

347264 Sasa oshidensis Makino et Uchida;押田赤竹●☆

347265 Sasa oshidensis Makino et Uchida subsp. glabra (Koidz.) Sad. Suzuki;光滑赤竹●☆

347266 Sasa oshidensis Makino et Uchida var. shigaensis (Koidz.) Sad. Suzuki ＝Sasa oshidensis Makino et Uchida ●☆

347267 Sasa owatarii (Makino) Makino ＝Pseudosasa owatarii (Makino) Makino ex Nakai ●☆

347268 Sasa palmata (Lat. -Marl. ex Burb.) E. G. Camus;黄脉赤竹(紫纹赤竹);Bamboo, Broadleaf Bamboo, Broad-leaved Bamboo, Palmate Bamboo ●☆

347269 Sasa palmata (Lat. -Marl. ex Burb.) E. G. Camus f. australis (Makino) Sad. Suzuki ＝Sasa palmata (Lat. -Marl. ex Burb.) E. G. Camus ●☆

347270 Sasa palmata (Lat. -Marl. ex Burb.) E. G. Camus f. nebulosa (Makino) Sad. Suzuki;星云黄脉赤竹●☆

347271 Sasa palmata (Lat. -Marl. ex Burb.) E. G. Camus var. niijimae (Tatew. et Nakai) Sad. Suzuki f. linearifolia (Koidz.) Sad. Suzuki ＝Sasa palmata (Lat. -Marl. ex Burb.) E. G. Camus ●☆

347272 Sasa palmata (Lat. -Marl. ex Burb.) E. G. Camus var. niijimae (Tatew. ex Nakai) Sad. Suzuki;似岛赤竹●☆

347273 Sasa palmata (Lat. -Marl. ex Burb.) E. G. Camus var. yosaensis (Koidz.) Sad. Suzuki;与谢赤竹●☆

347274 Sasa paniculata (F. Schmidt) Makino et Shibata ＝Sasa senanensis (Franch. et Sav.) Rehder ●☆

347275 Sasa pubens Nakai;短柔毛赤竹●☆

347276 Sasa pubiculmis Makino;短毛秆赤竹●☆

347277 Sasa pubiculmis Makino subsp. sugimotoi (Nakai) Sad. Suzuki;杉本赤竹●☆

347278 Sasa pulcherrima Koidz. ;美丽赤竹●☆

347279 Sasa pygmaea (Miq.) E. G. Camus;翠竹(矮苦竹);Dwarfich Sasa,Jadegreen Sasa,Pygmy Bamboo ●

347280 Sasa pygmaea (Miq.) E. G. Camus ＝Pleioblastus argenteostriatus (Regel) Nakai 'Distichus' ●

347281 Sasa pygmaea (Miq.) E. G. Camus var. disticha (Mitford) C. S. Chao et G. G. Tang ＝Pleioblastus argenteostriatus (Regel) Nakai 'Distichus' ●

347282 Sasa pygmaea (Miq.) Rehder ＝Pleioblastus fortunei (Van Houtte) Nakai ●

347283 Sasa pygmaea (Miq.) Rehder var. disticha (Mitford) C. S. Chao et G. G. Tang ＝Pleioblastus distichus (Mitford) Nakai ●

347284 Sasa qingyuanensis (C. H. Hu) C. H. Hu;庆元华箬竹;Qingyuan Sasa ●◇

347285 Sasa ramosa (Makino) Makino et Shibata ＝Sasaella ramosa (Makino) Makino ●

347286 Sasa rubrovaginata C. H. Hu;红壳赤竹;Red-sheath Sasa, Red-shelled Sasa ●

347287 Sasa samaniana Nakai;萨马尼赤竹●☆

347288 Sasa samaniana Nakai var. villosa (Makino et Nakai) Sad. Suzuki ＝Sasa samaniana Nakai ●☆

347289 Sasa samaniana Nakai var. yoshinoi (Koidz.) Sad. Suzuki ＝Sasa kagamiana Makino et Uchida subsp. yoshinoi (Koidz.) Sad. Suzuki ●☆

347290 Sasa samaniana Nakai var. yoshinoi (Koidz.) Sad. Suzuki f. hidejiroana (Koidz.) Sad. Suzuki ＝Sasa samaniana Nakai var. yoshinoi (Koidz.) Sad. Suzuki ●☆

347291 Sasa scytophylla Koidz. ;革叶赤竹●☆

347292 Sasa senanensis (Franch. et Sav.) Rehder;信浓赤竹(锥花赤竹)●☆

347293 Sasa senanensis (Franch. et Sav.) Rehder f. nobilis (Makino et Uchida) Sad. Suzuki;名贵信浓赤竹●☆

347294 Sasa senanensis (Franch. et Sav.) Rehder var. harae (Nakai) Sad. Suzuki f. argillacea (Koidz.) Sad. Suzuki ＝Sasa senanensis (Franch. et Sav.) Rehder var. harae (Nakai) Sad. Suzuki ●☆

347295 Sasa senanensis (Franch. et Sav.) Rehder var. harae (Nakai) Sad. Suzuki;原氏赤竹●☆

347296 Sasa sendaica Makino ＝Sasa chartacea (Makino) Makino et Shibata ●☆

347297　Sasa septentrionalis Makino；北方赤竹●☆

347298　Sasa septentrionalis Makino f. kuzakaina（Koidz.）Sad. Suzuki = Sasa septentrionalis Makino ●☆

347299　Sasa septentrionalis Makino var. membranacea（Makino et Uchida）Sad. Suzuki；膜质赤竹●☆

347300　Sasa shimidzuana Makino；清水峠竹●☆

347301　Sasa shimidzuana Makino subsp. kashidensis（Makino ex Koidz.）Sad. Suzuki；樫田赤竹●☆

347302　Sasa shimidzuana Makino var. asagishiana（Makino et Uchida）Sad. Suzuki = Sasa shimidzuana Makino subsp. kashidensis（Makino ex Koidz.）Sad. Suzuki ●☆

347303　Sasa sikokiana Koidz. ；四国赤竹●☆

347304　Sasa sinica Keng = Sasamorpha sinica（Keng）Koidz. ●

347305　Sasa spiculosa（F. Schmidt）Makino；细刺赤竹●☆

347306　Sasa stenophylla Koidz. ；狭叶赤竹●☆

347307　Sasa subglabra McClure；香港赤竹（光笹竹，光屉竹）；Hongkong Sasa ●

347308　Sasa sulcata W. T. Lin；沟槽赤竹；Culcate Sasa ●

347309　Sasa suzukii Nakai；铃木赤竹●☆

347310　Sasa takizawana Makino et Uchida subsp. nakashimana（Koidz.）Sad. Suzuki；长岛赤竹●☆

347311　Sasa takizawana Makino et Uchida var. kumagaiana（Uchida）Murata = Sasa takizawana Makino et Uchida subsp. nakashimana（Koidz.）Sad. Suzuki ●☆

347312　Sasa tatewakiana Makino；馆肋赤竹●☆

347313　Sasa tesselata（Munro）Makino et Shibata = Indocalamus tessellatus（Munro）P. C. Keng ●

347314　Sasa tessellata（Munro）Makino et Shibata；方斑赤竹；Chequer-shaped Indocalamus ●☆

347315　Sasa tessellata（Munro）Makino et Shibata = Indocalamus tessellatus（Munro）P. C. Keng ●

347316　Sasa tomentosa C. D. Chu et C. S. Chao；绒毛赤竹；Tomentosa Sasa ●

347317　Sasa tsukubensis Nakai；筑波赤竹●☆

347318　Sasa tsukubensis Nakai subsp. pubifolia（Koidz.）Sad. Suzuki；短毛叶赤竹●☆

347319　Sasa tsukubensis Nakai subsp. pubifolia（Koidz.）Sad. Suzuki var. ashikagaensis Sad. Suzuki = Sasa tsukubensis Nakai subsp. pubifolia（Koidz.）Sad. Suzuki ●☆

347320　Sasa tyugokensis Makino = Sasa veitchii（Carrière）Rehder var. tyugokensis（Makino）Sad. Suzuki ●☆

347321　Sasa variegata（Siebold ex Miq.）E. G. Camus = Bambusa variegata Siebold ex Miq. ●

347322　Sasa variegata（Siebold ex Miq.）E. G. Camus = Pleioblastus fortunei（Van Houtte）Nakai ●

347323　Sasa variegata（Siebold ex Miq.）E. G. Camus = Sasa fortunei（Van Houtte）Fiori ●

347324　Sasa veitchii（Carrière）Rehder；山白竹（白边赤竹，箬，箬竹，维奇赤竹，熊竹）；Kuma Sasa, Kuma Zasa, Nagasa Bamboo, Veitch Sasa, Veitch's Bamboo, White-margin Sasa ●

347325　Sasa veitchii（Carrière）Rehder var. grandiflora（Koidz.）Sad. Suzuki f. myojinensis（Koidz.）Sad. Suzuki = Sasa veitchii（Carrière）Rehder var. grandifolia（Koidz.）Sad. Suzuki ●☆

347326　Sasa veitchii（Carrière）Rehder var. grandifolia（Koidz.）Sad. Suzuki；大花山白竹●☆

347327　Sasa veitchii（Carrière）Rehder var. tyugokensis（Makino）Sad. Suzuki；中乡赤竹●☆

347328　Sasa yahikoensis Makino；弥彦赤竹●☆

347329　Sasa yahikoensis Makino var. depauperata（Takeda）Sad. Suzuki；萎缩赤竹●☆

347330　Sasa yahikoensis Makino var. oseana（Makino）Sad. Suzuki；逢濑赤竹●☆

347331　Sasa yahikoensis Makino var. oseana（Makino）Sad. Suzuki f. mogamensis（Nakai）Sad. Suzuki = Sasa yahikoensis Makino var. oseana（Makino）Sad. Suzuki ●☆

347332　Sasa yahikoensis Makino var. rotundissima（Makino et Uchida）Sad. Suzuki；圆弥彦赤竹●☆

347333　Sasaella Makino = Sasa Makino et Shibata ●

347334　Sasaella Makino（1929）；小赤竹属（东竹属，小箸竹属）；Hairy Bamboo, Bamboo ●☆

347335　Sasaella atamiana（Nakai）Sad. Suzuki = Elaeagnus arakiana Koidz. ●☆

347336　Sasaella atamiana（Nakai）Sad. Suzuki f. leucorhoda（Koidz.）Sad. Suzuki = Elaeagnus arakiana Koidz. ●☆

347337　Sasaella atamiana（Nakai）Sad. Suzuki var. kanayamensis（Nakai）Sad. Suzuki = Sasaella leucorhoda（Koidz.）Koidz. var. kanayamensis（Nakai）Sad. Suzuki ●☆

347338　Sasaella bitchuensis（Makino）Makino ex Koidz.；松柏小赤竹●☆

347339　Sasaella bitchuensis（Makino）Makino ex Koidz. var. tashirozentaroana（Koidz.）Sad. Suzuki f. praestantissima（Koidz.）Sad. Suzuki = Sasaella bitchuensis（Makino）Makino ex Koidz. var. tashirozentaroana（Koidz.）Sad. Suzuki ●☆

347340　Sasaella bitchuensis（Makino）Makino ex Koidz. var. tashirozentaroana（Koidz.）Sad. Suzuki；田代小赤竹●☆

347341　Sasaella caudiceps（Koidz.）Koidz.；尾头小赤竹●☆

347342　Sasaella hashimotoi（Makino）Makino ex Koidz. ；端元小赤竹●☆

347343　Sasaella hidaensis（Makino）Makino；斐太赤竹●☆

347344　Sasaella hidaensis（Makino）Makino f. kishinoana（Koidz.）Sad. Suzuki = Sasaella hidaensis（Makino）Makino ●☆

347345　Sasaella hidaensis（Makino）Makino f. yenaensis（Koidz.）Sad. Suzuki = Sasaella hidaensis（Makino）Makino var. muraii（Makino et Uchida）Sad. Suzuki ●☆

347346　Sasaella hidaensis（Makino）Makino var. iwatekensis（Makino et Uchida）Sad. Suzuki = Sasaella hidaensis（Makino）Makino var. muraii（Makino et Uchida）Sad. Suzuki ●☆

347347　Sasaella hidaensis（Makino）Makino var. muraii（Makino et Uchida）Sad. Suzuki；村井赤竹●☆

347348　Sasaella hisauchii（Makino）Makino = Arundinaria hakonensis Nakai ●☆

347349　Sasaella ikegamii（Nakai）Sad. Suzuki；池上小赤竹●☆

347350　Sasaella kogasensis（Nakai）Nakai ex Koidz. ；古河小赤竹●☆

347351　Sasaella kogasensis（Nakai）Nakai ex Koidz. f. uchidai（Makino ex Uchida）Sad. Suzuki = Sasaella kogasensis（Nakai）Nakai ex Koidz. ●☆

347352　Sasaella kogasensis（Nakai）Nakai ex Koidz. var. gracillima Sad. Suzuki；细长小赤竹●☆

347353　Sasaella kogasensis（Nakai）Nakai ex Koidz. var. yoshinoi（Koidz.）Sad. Suzuki；吉野氏小赤竹●☆

347354　Sasaella leucorhoda（Koidz.）Koidz.；白红小赤竹●☆

347355　Sasaella leucorhoda（Koidz.）Koidz. var. kanayamensis（Nakai）Sad. Suzuki；金山小赤竹●☆

347356　Sasaella masamuneana（Makino）Hatus. et Muroi；正宗小赤竹●☆

347357　Sasaella masamuneana（Makino）Hatus. et Muroi f. hashimotoi（Makino）Sad. Suzuki = Sasaella hashimotoi（Makino）Makino ex

Koidz. ●☆

347358 Sasaella masamuneana（Makino）Hatus. et Muroi var. amoena
（Nakai）Sad. Suzuki f. muramatsuana（Koidz.）Sad. Suzuki =
Sasaella hashimotoi（Makino）Makino ex Koidz. ●☆

347359 Sasaella masamuneana（Makino）Hatus. et Muroi var. amoena
（Nakai）Sad. Suzuki;秀丽小赤竹●☆

347360 Sasaella midoensis Hatak.;御堂小赤竹●☆

347361 Sasaella ramosa（Makino）Makino;东竹（狭叶青苦竹）;Hairy
Bamboo ●

347362 Sasaella ramosa（Makino）Makino = Arundinaria ramosa
Makino ●

347363 Sasaella ramosa（Makino）Makino = Pleioblastus chino
（Franch. et Sav.）Makino var. hisauchii Makino ●

347364 Sasaella ramosa（Makino）Makino = Sasa ramosa（Makino）
Makino et Shibata ●

347365 Sasaella ramosa（Makino）Makino f. tomikusensis（Nakai）
Sad. Suzuki;富草小赤竹●☆

347366 Sasaella ramosa（Makino）Makino var. latifolia（Nakai）Sad.
Suzuki;宽叶东竹●☆

347367 Sasaella ramosa（Makino）Makino var. latifolia（Nakai）Sad.
Suzuki f. trichophila（Koidz.）Sad. Suzuki;毛宽叶东竹●☆

347368 Sasaella sadoensis（Nakai）Sad. Suzuki;佐渡东竹●☆

347369 Sasaella sasakiana Makino et Uchida;佐佐木小赤竹●☆

347370 Sasaella sawadae（Makino）Makino ex Koidz.;泽田东竹●☆

347371 Sasaella sawadae（Makino）Makino ex Koidz. var. aobayamana
Sad. Suzuki = Sasaella ramosa（Makino）Makino var. latifolia
（Nakai）Sad. Suzuki ●☆

347372 Sasaella shiobarensis（Nakai）Nakai ex Koidz.;盐原小赤竹●☆

347373 Sasaella shiobarensis（Nakai）Nakai ex Koidz. var. yessaensis
（Koidz.）Sad. Suzuki;越佐小赤竹●☆

347374 Sasaella takinagawaensis Hatak.;龟波川小赤竹●☆

347375 Sasali Adans. = Microcos Burm. ex L. ●

347376 Sasamorpha Nakai = Sasa Makino et Shibata ●

347377 Sasamorpha Nakai（1931）;华箬竹属;Sasamorpha ●

347378 Sasamorpha amabilis Nakai = Sasa borealis（Hack.）Makino et
Shibata ●

347379 Sasamorpha borealis（Hack.）Nakai = Sasa borealis（Hack.）
Makino et Shibata ●

347380 Sasamorpha borealis（Hack.）Nakai var. angustior（Makino）
Hiyama;狭北方华箬竹●☆

347381 Sasamorpha borealis（Hack.）Nakai var. pilosa（Uchida）Sad.
Suzuki;疏毛北方华箬竹●☆

347382 Sasamorpha borealis（Hack.）Nakai var. purpurascens
（Hack.）Hiyama = Sasa borealis（Hack.）Makino et Shibata ●

347383 Sasamorpha borealis（Hack.）Nakai var. viridescens（Nakai）
Sad. Suzuki;绿北方华箬竹●

347384 Sasamorpha hubeiensis C. H. Hu = Sasa hubeiensis（C. H. Hu）
C. H. Hu ●

347385 Sasamorpha latifolia（Keng）Nakai ex Migo = Indocalamus
latifolius（Keng）McClure ●

347386 Sasamorpha migoi Nakai = Indocalamus latifolius（Keng）
McClure ●

347387 Sasamorpha migoi Nakai ex Migo = Indocalamus latifolius
（Keng）McClure ●

347388 Sasamorpha mollis Nakai = Sasa sikokiana Koidz. ●☆

347389 Sasamorpha purpurascens（Hack.）Nakai = Sasa borealis
（Hack.）Makino et Shibata ●

347390 Sasamorpha qingyuanensis C. H. Hu = Sasa qingyuanensis（C.
H. Hu）C. H. Hu ●◇

347391 Sasamorpha sinica（Keng）Koidz.;华箬竹（华赤竹）;China
Sasa,Chinese Bamboo,Chinese Sasa,Chinese Sasamorpha ●

347392 Sasamorpha sinica（Keng）Koidz. = Sasa sinica Keng ●

347393 Sasamorpha sinica（Keng）Koidz. f. glabra C. H. Hu;光叶华箬
竹;Glabrous Chinese Sasa ●

347394 Sasamorpha sinica（Keng）Koidz. f. glabra C. H. Hu = Sasa
sinica Keng ●

347395 Sasamorpha sinica（Keng）Koidz. f. glabra C. H. Hu =
Sasamorpha sinica（Keng）Koidz. ●

347396 Sasamorpha tesselatus（Munro）Koidz. = Indocalamus
tessellatus（Munro）P. C. Keng ●

347397 Sasanqua Nees = Camellia L. ●

347398 Sasanqua Nees ex Esenbeck = Camellia L. ●

347399 Sassa Bruce ex J. F. Gmel. = Acacia Mill.（保留属名）●■

347400 Sassa gummifera J. F. Gmel. = Albizia gummifera（J. F. Gmel.）
C. A. Sm. ●☆

347401 Sassafras Bercht. et J. Presl = Lindera Thunb.（保留属名）●

347402 Sassafras J. Presl（1825）;檫木属（檫树属）;Sassafras ●

347403 Sassafras Nees = Sassafras J. Presl ●

347404 Sassafras Nees et Eberm. = Sassafras J. Presl ●

347405 Sassafras Trew = Sassafras Nees ●

347406 Sassafras albidum（Nutt.）Nees var. molle（Raf.）Fernald =
Sassafras albidum（Nutt.）Ness ●☆

347407 Sassafras albidum（Nutt.）Ness;美国檫木（红檫木,黄樟,灰
白檫木,美国黄樟,美洲檫木,肉桂木,洋檫木）;Ague-tree,
American Sassafras, Cinnamon Wood, Common Sassafras, Red
Sassafras,Saffron Tree,Sassafras,Sassafras Tree,White Sassafras ●☆

347408 Sassafras officinale Nees et Eberm. = Sassafras albidum（Nutt.）
Ness ●☆

347409 Sassafras officinale T. Nees ex C. H. Eberm. = Sassafras albidum
（Nutt.）Ness ●☆

347410 Sassafras parthenoxylon（Jack）Nees = Cinnamomum
parthenoxylum（Jack）Meisn. ●

347411 Sassafras randaiense（Hayata）Rehder;台湾檫木（台湾檫树）;
Taiwan Sassafras ●

347412 Sassafras sassafras（L.）H. Karst. = Sassafras albidum（Nutt.）
Ness ●☆

347413 Sassafras tsumu（Hemsl.）Hemsl.;檫木（半枫樟,蔡木,檫树,
独脚樟,鹅脚板,枫荷桂,花楸树,黄楸树,梨火哄,南树,青檫,山
檫,刷木,天鹅枫,铜梓树,洋檫木,梓木,梓木树）;Chinese
Sassafras,Common Sassafras,Sassafras ●

347414 Sassafras variifolium（Salisb.）Kuntze = Sassafras albidum
（Nutt.）Ness ●☆

347415 Sassafras variifolium Kuntze = Sassafras albidum（Nutt.）Ness ●☆

347416 Sassafridium Meisn. = Cinnamomum Schaeff.（保留属名）●

347417 Sassafridium Meisn. = Ocotea Aubl. ●☆

347418 Sassea Klotzsch = Begonia L. ●■

347419 Sassia Molina = Oxalis L. ■●

347420 Satakentia H. E. Moore.（1969）;琉球椰属●☆

347421 Satakentia liukiuensis（Hatus.）H. E. Moore;琉球椰●☆

347422 Satania Noronha = Flacourtia Comm. ex L' Hér. ●

347423 Satanocrater Schweinf.（1868）;魔王杯属■☆

347424 Satanocrater berhautii Benoist = Satanocrater fellatensis
Schweinf. ■☆

347425 Satanocrater coccineus（S. Moore）Lindau = Satanocrater somalensis（Lindau）Lindau ■☆

347426　Satanocrater fellatensis Schweinf.；魔王杯■☆

347427　Satanocrater fruticulosus （Rolfe） Lindau = Dyschoriste hildebrandtii（S. Moore）Lindau ■☆

347428　Satanocrater paradoxus（Lindau）Lindau；奇异魔王杯■☆

347429　Satanocrater ruspolii（Lindau）Lindau = Satanocrater fellatensis Schweinf. ■☆

347430　Satanocrater somalensis（Lindau）Lindau；索马里魔王杯■☆

347431　Sataria Raf. = Oxypolis Raf. ■☆

347432　Saterna Noronha = Rauvolfia L. + Urceola Roxb.（保留属名）●

347433　Satirium Neck. = Satyrium Sw.（保留属名）■

347434　Satorchis Thouars = Satorkis Thouars ■

347435　Satorkis Thouars = Coeloglossum Hartm. ■

347436　Satorkis Thouars = Habenaria Willd. + Cynorkis Thouars + Benthamia A. Rich. ●☆

347437　Satorkis Thouars = Satyrium L.（废弃属名）■

347438　Satranala J. Dransf. et Beentje（1995）；翅果棕属●☆

347439　Satranala decussilvae Beentje et J. Dransf.；翅果棕●☆

347440　Sattadia E. Fourn.（1885）；肖异冠藤属●☆

347441　Sattadia Fourn. = Metastelma R. Br. ●☆

347442　Sattadia burchellii E. Fourn.；肖异冠藤●☆

347443　Saturegia Leers = Satureja L. ●■

347444　Satureia L. = Satureja L. ●■

347445　Satureia nepeta Scheele = Satureja nepeta Scheele ■☆

347446　Satureja L.（1753）；香草属（冬薄荷属，塔花属，夏薄荷属）；Savory, Winter Savory ●■

347447　Satureja abyssinica（Hochst. ex Benth.）Briq. = Clinopodium abyssinicum（Hochst. ex Benth.）Kuntze ■☆

347448　Satureja abyssinica （Hochst. ex Benth.） Briq. subsp. condensata （Hedberg） Seybold = Clinopodium abyssinicum （Hochst. ex Benth.）Kuntze var. condensata（Hedberg）Ryding ■☆

347449　Satureja abyssinica （Hochst. ex Benth.） Briq. var. condensata Hedberg = Clinopodium abyssinicum （Hochst. ex Benth.） Kuntze var. condensata（Hedberg）Ryding ■☆

347450　Satureja acinos（L.）Scheele = Acinos arvense（Lam.）Dandy ■☆

347451　Satureja acinos（L.）Scheele = Calamintha arvensis Lam. ■☆

347452　Satureja acinos（L.）Scheele = Clinopodium acinos Kuntze ■☆

347453　Satureja alpina（L.）Scheele = Acinos alpinus（L.）Moench ■☆

347454　Satureja alpina （L.） Scheele subsp. granatensis （Boiss. et Reut.） Maire = Acinos alpinus （L.） Moench subsp. meridionalis （Nyman）P. W. Ball ■☆

347455　Satureja alpina （L.） Scheele subsp. meridionalis （Nyman） Greuter et Burdet = Acinos alpinus （L.） Moench subsp. meridionalis（Nyman）P. W. Ball ■☆

347456　Satureja alpina （L.） Scheele var. amplifoliata （Pau） Maire = Acinos alpinus（L.）Moench ■☆

347457　Satureja alpina （L.） Scheele var. aurasiaca Maire = Acinos alpinus（L.）Moench ■☆

347458　Satureja alpina （L.） Scheele var. baumgarteni （Simonk.） Briq. = Acinos alpinus（L.）Moench ■☆

347459　Satureja alpina （L.） Scheele var. chabertii Maire = Acinos alpinus（L.）Moench ■☆

347460　Satureja alpina （L.） Scheele var. erecta （Lange） Maire = Acinos alpinus（L.）Moench ■☆

347461　Satureja alpina （L.） Scheele var. granatensis （Boiss. et Reut.） H. Lindb. = Acinos alpinus （L.） Moench subsp. meridionalis （Nyman）P. W. Ball ■☆

347462　Satureja alpina （L.） Scheele var. kestica Maire et Weiller = Acinos alpinus（L.）Moench ■☆

347463　Satureja alpina （L.） Scheele var. latior （Schott） Briq. = Acinos alpinus（L.）Moench ■☆

347464　Satureja alpina （L.） Scheele var. macrantha （H. Lindb.） Maire = Acinos alpinus（L.）Moench ■☆

347465　Satureja alpina （L.） Scheele var. parviflora （Ball） Maire = Acinos alpinus（L.）Moench ■☆

347466　Satureja alpina （L.） Scheele var. purpurascens （Pers.） Pau et Font Quer = Acinos alpinus（L.）Moench ■☆

347467　Satureja alpina （L.） Scheele var. subinodora Maire = Acinos alpinus（L.）Moench ■☆

347468　Satureja alpina Scheele = Acinos alpinus（L.）Moench ■☆

347469　Satureja altaica Boriss.；阿尔泰香草■☆

347470　Satureja amplifoliata Pau = Acinos alpinus （L.） Moench subsp. meridionalis（Nyman）P. W. Ball ■☆

347471　Satureja anagae R. H. Willemse = Micromeria glomerata Pérez ■☆

347472　Satureja annua （Schrenk） B. Fedtsch. = Calamintha debilis （Bunge）Benth. ■

347473　Satureja arkansana （Nutt.） Briq. = Calamintha arkansana （Nutt.）Shinners ■☆

347474　Satureja arkansana （Nutt.） Briq. f. alba Steyerm. = Calamintha arkansana（Nutt.）Shinners ■☆

347475　Satureja ascendens （Jord.） K. Malý = Calamintha sylvatica Bromf. subsp. ascendens（Jord.）P. W. Ball ■☆

347476　Satureja ascendens （Jord.） K. Malý = Clinopodium ascendens Samp. ■☆

347477　Satureja atlantica （Ball） H. Lindb. = Clinopodium atlanticum （Ball）N. Galland ■☆

347478　Satureja baborensis （Batt.） Briq. = Calamintha grandiflora （L.）Moench subsp. baborensis（Batt.）N. Galland ■☆

347479　Satureja baborensis （Batt.） Briq. var. occidentalis Pau = Calamintha grandiflora （L.） Moench subsp. baborensis （Batt.） N. Galland ■☆

347480　Satureja baetica （Boiss. et Heldr.） Pau = Calamintha sylvatica Bromf. subsp. ascendens（Jord.）P. W. Ball ■☆

347481　Satureja barceloi （Willk.） Pau = Micromeria inodora （Desf.） Benth. ■☆

347482　Satureja barosma （W. W. Sm.） Kudo = Micromeria barosma （W. W. Sm.）Hand. -Mazz. ■

347483　Satureja battandieri Briq. = Micromeria fontanesii Pomel ■☆

347484　Satureja battandieri Briq. var. villosa （Benth.） Briq. = Micromeria fontanesii Pomel ■☆

347485　Satureja benthamii （Webb et Berthel.） Briq. = Micromeria benthamii Webb et Berthel. ■☆

347486　Satureja biflora （Buch. -Ham. ex D. Don） Briq. = Micromeria imbricata（Forssk.）C. Chr. ■☆

347487　Satureja biflora （Buch. -Ham. ex D. Don） Briq. f. discolor Maire = Micromeria imbricata（Forssk.）C. Chr. ■☆

347488　Satureja biflora （Buch. -Ham. ex D. Don） Briq. f. nana Maire = Micromeria imbricata（Forssk.）C. Chr. ■☆

347489　Satureja biflora （Buch. -Ham. ex D. Don） Briq. var. cinereotomentonsa （A. Rich.） Cufod. = Micromeria imbricata （Forssk.）C. Chr. ■☆

347490　Satureja biflora Briq. = Micromeria biflora （Buch. -Ham. ex D. Don）Benth. ●■

347491　Satureja briquetii Maire = Micromeria debilis Pomel ■☆

347492　Satureja briquetii Maire var. villosissima ? = Micromeria debilis Pomel ■☆

347493　Satureja brivesii (Batt.) Murb. = Micromeria brivesii Batt. ■☆

347494　Satureja cacondensis (G. Taylor) Brenan = Clinopodium myrianthum (Baker) Ryding ■☆

347495　Satureja calamintha (L.) Scheele = Calamintha nepeta (L.) Savi ■☆

347496　Satureja calamintha (L.) Scheele = Calamintha officinalis Moench ■☆

347497　Satureja calamintha (L.) Scheele = Clinopodium calamintha Kuntze ■☆

347498　Satureja calamintha (L.) Scheele subsp. ascendens (Jord.) Briq. = Calamintha nepeta (L.) Savi ■☆

347499　Satureja calamintha (L.) Scheele subsp. nepeta Briq. = Calamintha nepeta (L.) Savi ■☆

347500　Satureja calamintha (L.) Scheele subsp. sylvatica Briq. = Calamintha sylvatica Bromf. ■☆

347501　Satureja calamintha (L.) Scheele var. calaminthoides (Rchb.) Briq. = Calamintha nepeta (L.) Savi ■☆

347502　Satureja calamintha (L.) Scheele var. heterotricha (Boiss. et Reut.) Briq. = Calamintha heterotricha Boiss. et Reut. ■☆

347503　Satureja candidissima (Munby) Briq. = Calamintha candidissima (Munby) Benth. ■☆

347504　Satureja candidissima (Munby) Briq. var. laxiflora Faure et Maire = Calamintha candidissima (Munby) Benth. ■☆

347505　Satureja capitata L. = Coridothymus capitatus (L.) Rchb. f. ■☆

347506　Satureja caroliniana (Michx.) Briq. ;卡罗林香草（加州香草）;Bean Herb,Savonette,Summer Savory,Sweet Herb ■☆

347507　Satureja chinensis (Benth.) Briq. = Clinopodium chinense (Benth.) Kuntze subsp. grandiflorum (Maxim.) H. Hara var. urticifolium (Hance) Koidz. ■

347508　Satureja chinensis (Benth.) Briq. = Clinopodium chinense (Benth.) Kuntze ■

347509　Satureja chinensis (Benth.) Briq. = Clinopodium urticifolium (Hance) C. Y. Wu et S. J. Hsuan ex H. W. Li ■

347510　Satureja chinensis (Benth.) Briq. var. discolor (Diels) Kudo = Clinopodium discolor (Diels) C. Y. Wu et S. J. Hsuan ex H. W. Li ■

347511　Satureja chinensis Briq. = Clinopodium chinense (Benth.) Kuntze ■

347512　Satureja chinensis Briq. var. discolor (Diels) Kudo = Clinopodium discolor (Diels) C. Y. Wu et S. J. Hsuan ex H. W. Li ■

347513　Satureja chinensis Briq. var. megalantha (Diels) Kudo = Clinopodium megalanthum (Diels) C. Y. Wu et S. J. Hsuan ex H. W. Li ■

347514　Satureja chinensis Briq. var. megalantha Kudo = Clinopodium megalanthum (Diels) C. Y. Wu et S. J. Hsuan ex H. W. Li ■

347515　Satureja chinensis Briq. var. parviflora Kudo = Clinopodium polycephalum (Vaniot) C. Y. Wu et S. J. Hsuan ex H. W. Li ■

347516　Satureja chinensis Briq. var. parviflora Kudo = Clinopodium repens (D. Don) Vell. ■

347517　Satureja chinensis Briq. var. repens (Buch. -Ham. ex D. Don) Kudo = Clinopodium repens (D. Don) Vell. ■

347518　Satureja chinensis Briq. var. repens (D. Don) Kudo = Clinopodium repens (D. Don) Vell. ■

347519　Satureja clinopodium L. = Clinopodium vulgare L. ■☆

347520　Satureja compacta Killick;紧密塔花■☆

347521　Satureja conferta (Coss.) Bég. et Vacc. = Micromeria conferta (Coss.) Stefani ■☆

347522　Satureja conferta (Coss.) Bég. et Vacc. var. elongata Maire et Weiller = Micromeria conferta (Coss.) Stefani ■☆

347523　Satureja confinis (Hance) Kudo = Clinopodium confine (Hance) Kuntze ■

347524　Satureja confinis Boriss. = Clinopodium confine (Hance) Kuntze ■

347525　Satureja contardoi Pic. Serm. = Micromeria imbricata (Forssk.) C. Chr. ■☆

347526　Satureja debilis (Bunge) Briq. = Calamintha debilis (Bunge) Benth. ■

347527　Satureja debilis (Pomel) Briq. = Micromeria debilis Pomel ■☆

347528　Satureja douglasii (Benth.) Briq. ;道氏塔花■☆

347529　Satureja eugenioides Loes. ex R. Fr. ;南美塔花■☆

347530　Satureja euosma (W. W. Sm.) Kudo = Micromeria euosma (W. W. Sm.) C. Y. Wu ■

347531　Satureja filiformis Desf. = Micromeria fontanesii Pomel ■☆

347532　Satureja fontanesii Briq. = Micromeria inodora (Desf.) Benth. ■☆

347533　Satureja fontanesii Briq. var. barceloi (Willk.) Maire = Micromeria inodora (Desf.) Benth. ■☆

347534　Satureja fontanesii Briq. var. elata Maire = Micromeria inodora (Desf.) Benth. ■☆

347535　Satureja forbesii (Benth.) Briq. ;福布斯塔花■☆

347536　Satureja forbesii (Benth.) Briq. var. altitudinum (Bolle) R. H. Willemse = Satureja forbesii (Benth.) Briq. ■☆

347537　Satureja forbesii (Benth.) Briq. var. inodora (J. A. Schmidt) R. H. Willemse = Satureja forbesii (Benth.) Briq. ■☆

347538　Satureja fortii Pamp. ;福特塔花■☆

347539　Satureja fruticosa (L.) Briq. ;灌丛塔花■☆

347540　Satureja glabella (Michx.) Briq. var. angustifolia (Torr.) Svenson = Calamintha arkansana (Nutt.) Shinners ■☆

347541　Satureja glabra (Nutt.) Fernald = Calamintha arkansana (Nutt.) Shinners ■☆

347542　Satureja gracilis (Benth.) Briq. = Clinopodium gracile (Benth.) Matsum. ■

347543　Satureja graeca L. ;希腊塔花■☆

347544　Satureja graeca L. = Micromeria graeca (L.) Rchb. ■☆

347545　Satureja graeca L. subsp. micrantha (Brot.) Greuter et Burdet = Micromeria graeca (L.) Rchb. ■☆

347546　Satureja granatensis (Boiss. et Reut.) Sennen et Mauricio = Acinos alpinus (L.) Moench subsp. meridionalis (Nyman) P. W. Ball ■☆

347547　Satureja grandibracteata Killick;大苞塔花■☆

347548　Satureja grandiflora (L.) Scheele = Calamintha grandiflora (L.) Moench ■☆

347549　Satureja grandiflora (L.) Scheele subsp. baborensis (Batt.) Maire = Calamintha grandiflora (L.) Moench subsp. baborensis (Batt.) N. Galland ■☆

347550　Satureja grandiflora (L.) Scheele var. baborensis (Batt.) Pau et Font Quer = Calamintha grandiflora (L.) Moench subsp. baborensis (Batt.) N. Galland ■☆

347551　Satureja grandiflora Scheele = Calamintha grandiflora (L.) Moench ■☆

347552　Satureja guichardii Quézel et Zaffran = Micromeria guichardii (Quézel et Zaffran) Brullo et Furnari ■☆

347553　Satureja helianthemifolia (Webb et Berthel.) Briq. = Micromeria helianthemifolia Webb et Berthel. ■☆

347554　Satureja helianthemifolia（Webb et Berthel.）Briq. var. mary-annae（Pérez et G. Kunkel）R. H. Willemse = Micromeria helianthemifolia Webb et Berthel. ■☆

347555　Satureja heterotricha（Boiss. et Reut.）Greuter et Burdet = Calamintha heterotricha Boiss. et Reut. ■☆

347556　Satureja hispidula（Boiss. et Reut.）Briq. = Calamintha hispidula Boiss. et Reut. ■☆

347557　Satureja hochreutineri Briq. = Micromeria hochreutineri（Briq.）Maire ■☆

347558　Satureja hochreutineri Briq. var. brevidens Maire = Micromeria hochreutineri（Briq.）Maire ■☆

347559　Satureja hochreutineri Briq. var. citriodora Maire, Weiller et Wilczek = Micromeria hochreutineri（Briq.）Maire ■☆

347560　Satureja hortensis L. ;家园香草（新塔花，园圃塔花）;Savory, Summer Savory ■☆

347561　Satureja imbricata（Forssk.）Briq. = Micromeria imbricata（Forssk.）C. Chr. ■☆

347562　Satureja incana（Sm.）Briq. = Calamintha incana（Sm.）Boiss. ■☆

347563　Satureja inodora Benth. = Argantoniella salzmanii（P. W. Ball）G. López et R. Morales ●☆

347564　Satureja insularis Greuter et Burdet = Calamintha incana（Sm.）Boiss. ■☆

347565　Satureja intermedia C. A. Mey. ;间型香草■☆

347566　Satureja japonica（Miq.）Matsum. et Kudo = Mentha japonica（Miq.）Makino ■☆

347567　Satureja juliana L. ;七月香草（七月塔花）■☆

347568　Satureja juliana L. = Micromeria juliana（L.）Rchb. ■☆

347569　Satureja kilimandschari（Gürke）Hedberg = Clinopodium kilimandschari（Gürke）Ryding ■☆

347570　Satureja kudoi Hosok. = Clinopodium repens（D. Don）Vell. ■

347571　Satureja kuegleri（Bornm.）R. H. Willemse;屈格勒香草■☆

347572　Satureja kuegleri（Bornm.）R. H. Willemse var. glabrescens（Webb et Berthel.）R. H. Willemse = Satureja kuegleri（Bornm.）R. H. Willemse ■☆

347573　Satureja kuegleri（Bornm.）R. H. Willemse var. hyssopifolia（Webb et Berthel.）R. H. Willemse = Satureja kuegleri（Bornm.）R. H. Willemse ■☆

347574　Satureja lachnophylla（Webb et Berthel.）Briq. = Micromeria lachnophylla Webb et Berthel. ■☆

347575　Satureja lanata Link = Micromeria lanata（Link）Benth. ■☆

347576　Satureja lasiophylla（Webb et Berthel.）R. H. Willemse = Micromeria lasiophylla Webb et Berthel. ■☆

347577　Satureja lasiophylla（Webb et Berthel.）R. H. Willemse subsp. palmensis（Bolle）R. H. Willemse = Micromeria lasiophylla Webb et Berthel. subsp. palmensis（Bolle）Pérez ☆

347578　Satureja laxiflora（Hayata）Matsum. et Kudo = Clinopodium laxiflorum（Hayata）Mori ■

347579　Satureja laxiflora K. Koch;疏花香草■☆

347580　Satureja lepida（Webb et Berthel.）Briq. = Micromeria lepida Webb et Berthel. ■☆

347581　Satureja lepida（Webb et Berthel.）Briq. subsp. bolleana（Pérez）R. H. Willemse = Micromeria lepida Webb et Berthel. subsp. bolleana Pérez ■☆

347582　Satureja lepida（Webb et Berthel.）Briq. subsp. lepida ? = Micromeria lepida Webb et Berthel. ■☆

347583　Satureja lepida（Webb et Berthel.）Briq. var. argagae（Pérez）

A. Hansen et Sunding = Micromeria lepida Webb et Berthel. ■☆

347584　Satureja lepida（Webb et Berthel.）Briq. var. bolleana（Pérez）R. H. Willemse = Micromeria lepida Webb et Berthel. subsp. bolleana Pérez ■☆

347585　Satureja lepida（Webb et Berthel.）Briq. var. fernandezii（Pérez）A. Hansen et Sunding = Micromeria lepida Webb et Berthel. subsp. bolleana Pérez ■☆

347586　Satureja leucantha（Pérez）R. H. Willemse = Micromeria leucantha Pérez ■☆

347587　Satureja linearifolia（Brullo et Furnari）Greuter;线叶塔花■☆

347588　Satureja macrantha C. A. Mey. ;大花香草■☆

347589　Satureja macrosiphon（Coss.）Maire = Micromeria hochreutineri（Briq.）Maire ■☆

347590　Satureja macrosiphon（Coss.）Maire var. glaucescens Emb. et Maire = Micromeria macrosiphon Coss. ■☆

347591　Satureja masukuensis（Baker）Eyles = Clinopodium myrianthum（Baker）Ryding ■☆

347592　Satureja menthifolia（Host）Fritsch = Calamintha sylvatica Bromf. ■☆

347593　Satureja menthifolia Fritsch = Clinopodium menthifolium（Host）Stace ■☆

347594　Satureja mexicana（Benth.）Briq. ;墨西哥塔花■☆

347595　Satureja microphylla（d'Urv.）Guss. = Micromeria microphylla（d'Urv.）Benth. ■☆

347596　Satureja monantha Font Quer;单花塔花●☆

347597　Satureja montana L. ;冬香草（山塔花，塔花）;Mountain Savory, Winter Savory ●■☆

347598　Satureja montana L. 'Prostrate White';平卧白塔花●■☆

347599　Satureja mutica Fisch. et C. A. Mey. ;无尖塔花■☆

347600　Satureja myriantha（Baker）Brenan = Clinopodium myrianthum（Baker）Ryding ■☆

347601　Satureja myriantha（Baker）Brenan var. brachytricha Brenan = Clinopodium myrianthum（Baker）Ryding ■☆

347602　Satureja myriantha（Baker）Brenan var. wellmanii（C. H. Wright）Brenan = Clinopodium myrianthum（Baker）Ryding ■☆

347603　Satureja nepeta（L.）Scheele = Calamintha nepeta（L.）Savi ☆

347604　Satureja nepeta Scheele;荆芥塔花■☆

347605　Satureja nervosa Desf. = Micromeria nervosa（Desf.）Benth. ■☆

347606　Satureja oacana（Fernald）Standl. ;瓦哈卡塔花■☆

347607　Satureja obovata Lag. ;倒卵塔花（卵圆香草）■☆

347608　Satureja ovata（Benth.）Pic. Serm. = Micromeria imbricata（Forssk.）C. Chr. ■☆

347609　Satureja ovata（Benth.）Pic. Serm. var. cinereotomentosa（A. Rich.）Pic. Serm. = Micromeria imbricata（Forssk.）C. Chr. ■☆

347610　Satureja pachyphylla K. Koch;厚叶塔花■☆

347611　Satureja paradoxa（Vatke）Engl. ex Seybold = Clinopodium paradoxum（Vatke）Ryding ■☆

347612　Satureja peltieri Maire = Micromeria peltieri（Maire）R. Morales ■☆

347613　Satureja pomelii Briq. = Calamintha nervosa Pomel ■☆

347614　Satureja pseudosimensis Brenan = Clinopodium uhligii（Gürke）Ryding var. obtusifolium（Avetta）Ryding ■☆

347615　Satureja pseudosimensis Brenan var. micrantha Cufod. = Clinopodium uhligii（Gürke）Ryding var. obtusifolium（Avetta）Ryding ■☆

347616　Satureja punctata（Benth.）Briq. = Micromeria imbricata（Forssk.）C. Chr. ■☆

347617　Satureja punctata（Benth.）Briq. subsp. ovata（Benth.）

Seybold = Micromeria imbricata（Forssk.）C. Chr. ■☆

347618 Satureja punctata（Benth.）Briq. var. rigida Pic. Serm. = Micromeria imbricata（Forssk.）C. Chr. ■☆

347619 Satureja quartiniana（A. Rich.）Cufod. = Satureja punctata（Benth.）Briq. ■☆

347620 Satureja reptans Killick；匍匐塔花■☆

347621 Satureja rivas-martinezii（Wildpret）R. H. Willemse = Micromeria rivas-martinezii Wildpret ■☆

347622 Satureja robusta（Hook. f.）Brenan = Clinopodium robustum（Hook. f.）Ryding ■☆

347623 Satureja rotundifolia（Pers.）Briq. = Acinos rotundifolius Pers. ■☆

347624 Satureja rotundifolia（Pers.）Briq. var. micrantha Murb. = Acinos rotundifolius Pers. ■☆

347625 Satureja salzmannii P. W. Ball = Argantoniella salzmanii（P. W. Ball）G. López et R. Morales ●☆

347626 Satureja schimperi（Vatke）Cufod. = Micromeria imbricata（Forssk.）C. Chr. ■☆

347627 Satureja simensis（Benth.）Briq.；锡米塔花■☆

347628 Satureja simensis（Benth.）Briq. = Clinopodium simense（Benth.）Kuntze ■☆

347629 Satureja sinaica（Benth.）Briq. = Micromeria sinaica Benth. ■☆

347630 Satureja spicigera（K. Koch）Boiss.；黑塔花■☆

347631 Satureja subdentata Boriss.；小齿香草■☆

347632 Satureja taurica Velen.；克里木香草■☆

347633 Satureja teneriffae（Poir.）Briq. = Micromeria teneriffae（Poir.）Benth. ■☆

347634 Satureja teneriffae（Poir.）Briq. var. cordifolia（Pérez）R. H. Willemse = Micromeria teneriffae（Poir.）Benth. ■☆

347635 Satureja tenuis Link = Micromeria tenuis（Link）Webb et Berthel. ■☆

347636 Satureja tenuis Link subsp. linkii（Webb et Berthel.）R. H. Willemse = Micromeria tenuis（Link）Webb et Berthel. ■☆

347637 Satureja tenuis Link var. soriae（Pérez）Link = Micromeria tenuis（Link）Webb et Berthel. ■☆

347638 Satureja thymbra L.；夏塔花；Summer Savory ■☆

347639 Satureja uhligii Gürke = Clinopodium uhligii（Gürke）Ryding ■☆

347640 Satureja umbrosa（M. Bieb.）Greuter et Burdet var. repens（Buch.-Ham. ex D. Don）Briq. = Clinopodium repens（D. Don）Vell. ■

347641 Satureja umbrosa（M. Bieb.）Greuter et Burdet var. repens（D. Don）Briq. = Clinopodium repens（D. Don）Vell. ■

347642 Satureja unguentaria（Schweinf.）Cufod. = Micromeria unguentaria Schweinf. ■☆

347643 Satureja ussuriensis Kudo = Clinopodium gracile（Benth.）Matsum. ■

347644 Satureja varia（Benth.）Briq. = Micromeria varia Benth. ■☆

347645 Satureja varia（Benth.）Briq. subsp. canariensis（Pérez）A. Hansen et Sunding = Micromeria varia Benth. ■☆

347646 Satureja varia（Benth.）Briq. subsp. hierrensis（Pérez）A. Hansen et Sunding = Micromeria varia Benth. ■☆

347647 Satureja varia（Benth.）Briq. subsp. meridialis（Pérez）A. Hansen et Sunding = Micromeria varia Benth. ■☆

347648 Satureja varia（Benth.）Briq. subsp. rupestris（Webb et Berthel.）A. Hansen et Sunding = Micromeria varia Benth. ■☆

347649 Satureja varia（Benth.）Briq. subsp. thymoides（Lowe）A. Hansen et Sunding = Micromeria varia Benth. ■☆

347650 Satureja varia（Benth.）Briq. var. cacuminicolae（Pérez）A. Hansen et Sunding = Micromeria varia Benth. ■☆

347651 Satureja varia（Benth.）Briq. var. thymoides（Lowe）A. Hansen et Sunding = Micromeria varia Benth. ■☆

347652 Satureja vernayana Brenan = Clinopodium vernayanum（Brenan）Ryding ●☆

347653 Satureja vulgaris（L.）Fritsch = Clinopodium vulgare L. ■☆

347654 Satureja vulgaris（L.）Fritsch subsp. arundana（Boiss.）Greuter et Burdet = Clinopodium vulgare L. subsp. arundanum（Boiss.）Nyman ■☆

347655 Satureja vulgaris（L.）Fritsch subsp. clinopodium？ = Clinopodium vulgare L. ■☆

347656 Satureja vulgaris（L.）Fritsch subsp. villosa（Noë）Maire = Clinopodium vulgare L. subsp. arundanum（Boiss.）Nyman ■☆

347657 Satureja vulgaris（L.）Fritsch var. diminuta（Simon）Fernald et Wiegand = Clinopodium vulgare L. ■☆

347658 Satureja vulgaris（L.）Fritsch var. gattefossei Maire = Clinopodium vulgare L. ■☆

347659 Satureja vulgaris（L.）Fritsch var. neogaea Fernald = Clinopodium vulgare L. ■☆

347660 Satureja vulgaris（L.）Fritsch var. plumosa（Hoffmanns. et Link）Pau = Clinopodium vulgare L. ■☆

347661 Satureja vulgaris（L.）Fritsch var. transiens Maire = Clinopodium vulgare L. ■☆

347662 Satureja weilleri Maire = Micromeria weilleri（Maire）R. Morales ■☆

347663 Saturejaceae Döll = Labiatae Juss.（保留科名）●■

347664 Saturejaceae Döll = Lamiaceae Martinov（保留科名）●■

347665 Saturiastrum Fourr. = Satureja L. ●■

347666 Saturna B. D. Jacks. = Rauvolfia L. + Urceola Roxb.（保留属名）●

347667 Saturna B. D. Jacks. = Saterna Noronha ●

347668 Saturnia Maratti = Allium L. ■

347669 Saturnia littoralis Jord. et Fourr. = Allium chamaemoly L. var. littoralis（Jord. et Fourr.）Maire et Weiller ■☆

347670 Saturnia viridulum Jord. et Fourr. = Allium chamaemoly L. var. viridulum（Jord. et Fourr.）Maire et Weiller ■☆

347671 Satyria Klotzsch（1851）；合丝莓属●☆

347672 Satyria allenii A. C. Sm.；阿伦合丝莓●☆

347673 Satyria arborea A. C. Sm.；乔木合丝莓●☆

347674 Satyria boliviana Luteyn；玻利维亚合丝莓●☆

347675 Satyria breviflora Hoerold；短花合丝莓●☆

347676 Satyria grandifolia Hoerold；大叶合丝莓●☆

347677 Satyria latifolia A. C. Sm.；宽叶合丝莓●☆

347678 Satyria leptantha A. C. Sm.；细花合丝莓●☆

347679 Satyria leucostoma Sleumer；白口合丝莓●☆

347680 Satyria minutiflora A. C. Sm.；小花合丝莓●☆

347681 Satyria nitida A. C. Sm.；光亮合丝莓●☆

347682 Satyria ovata A. C. Sm.；卵形合丝莓●☆

347683 Satyria pilosa A. C. Sm.；毛合丝莓●☆

347684 Satyria polyantha A. C. Sm.；多花合丝莓●☆

347685 Satyridium Lindl.（1838）；醉兰属■☆

347686 Satyridium rostratum Lindl.；醉兰■☆

347687 Satyridium rostratum Lindl. = Satyrium rhynchanthum Bolus ■☆

347688 Satyrium L.（废弃属名）= Coeloglossum Hartm. ■

347689 Satyrium L.（废弃属名）= Satyrium Sw.（保留属名）■

347690 Satyrium Sw.（1800）（保留属名）；鸟足兰属；Birdfootorchis, Satyrium ■

347691　Satyrium aberrans Summerh. ;异常鸟足兰■☆

347692　Satyrium aceras Schltr. = Satyrium ciliatum Lindl. ■

347693　Satyrium aceras Schltr. ex Limpr. = Satyrium nepalense D. Don var. ciliatum（Lindl.）Hook. f.

347694　Satyrium aciculare van der Niet et P. J. Cribb;针形鸟足兰■☆

347695　Satyrium aculeatum L. f. = Eulophia aculeata（L. f.）Spreng. ■☆

347696　Satyrium acuminatum Lindl. ;渐尖鸟足兰■☆

347697　Satyrium acutirostrum Summerh. = Satyrium sceptrum Schltr. ■☆

347698　Satyrium adnatum Sw. = Pelexia adnata（Sw.）Spreng. ■☆

347699　Satyrium aethiopicum Summerh. ;埃塞俄比亚鸟足兰■☆

347700　Satyrium afromontanum la Croix et P. J. Cribb;非洲山生鸟足兰■☆

347701　Satyrium albidum L. = Pseudorchis albida（L.）Á. Löve et D. Löve ■☆

347702　Satyrium albiflorum A. Rich. = Satyrium nepalense D. Don ■

347703　Satyrium amoenum A. Rich. ;秀丽鸟足兰■☆

347704　Satyrium anomalum Schltr. ;畸形鸟足兰■☆

347705　Satyrium aphyllum Schltr. = Satyrium parviflorum Sw. ■☆

347706　Satyrium atherstonei Rchb. f. = Satyrium trinerve Lindl. ■☆

347707　Satyrium aureum Paxton = Satyrium coriifolium Sw. ■☆

347708　Satyrium australis（R. Br.）Lindl. ;盘龙参■☆

347709　Satyrium barbatum（L. f.）Thunb. = Disa barbata（L. f.）Sw. ■☆

347710　Satyrium baronii Schltr. ;巴龙鸟足兰■☆

347711　Satyrium beyrichianum Kraenzl. = Satyrium neglectum Schltr. subsp. woodii（Schltr.）A. V. Hall ■☆

347712　Satyrium bicallosum Thunb. ;二硬皮鸟足兰■☆

347713　Satyrium bicallosum Thunb. var. ocellatum Bolus = Satyrium bicallosum Thunb. ■☆

347714　Satyrium bicallosum Thunb. var. thunbergianum Bolus = Satyrium bicallosum Thunb. ■☆

347715　Satyrium bicorne（L.）Thunb. ;二角鸟足兰■☆

347716　Satyrium bifidum Thunb. = Schizodium bifidum（Thunb.）Rchb. f. ■☆

347717　Satyrium bifolium A. Rich. ;二叶鸟足兰;Twoleaves Satyrium ■☆

347718　Satyrium bifolium A. Rich. = Satyrium aethiopicum Summerh. ■☆

347719　Satyrium bowiei Rolfe = Satyrium bracteatum（L. f.）Thunb. ■☆

347720　Satyrium brachypetalum A. Rich. ;短瓣鸟足兰■☆

347721　Satyrium brachyrhynchum Schltr. = Satyrium macrophyllum Lindl. ■☆

347722　Satyrium bracteatum（L. f.）Thunb. ;具苞鸟足兰■☆

347723　Satyrium bracteatum（L. f.）Thunb. var. glandulosum Sond. = Satyrium bracteatum（L. f.）Thunb. ■☆

347724　Satyrium bracteatum（L. f.）Thunb. var. latebracteatum Sond. = Satyrium bracteatum（L. f.）Thunb. ■☆

347725　Satyrium bracteatum（L. f.）Thunb. var. lineatum（Lindl.）Sond. = Satyrium bracteatum（L. f.）Thunb. ■☆

347726　Satyrium bracteatum（L. f.）Thunb. var. nanum Bolus = Satyrium bracteatum（L. f.）Thunb. ■☆

347727　Satyrium bracteatum（L. f.）Thunb. var. pictum（Lindl.）Schltr. = Satyrium bracteatum（L. f.）Thunb. ■☆

347728　Satyrium bracteatum（L. f.）Thunb. var. saxicola（Bolus）Schltr. = Satyrium bracteatum（L. f.）Thunb. ■☆

347729　Satyrium bracteatum Lindl. = Satyrium retusum Lindl. ■☆

347730　Satyrium breve Rolfe;短鸟足兰■☆

347731　Satyrium breve Rolfe var. minor Summerh. = Satyrium breve Rolfe ■☆

347732　Satyrium buchananii Rolfe = Satyrium macrophyllum Lindl. ■☆

347733　Satyrium buchananii Schltr. ;布坎南鸟足兰■☆

347734　Satyrium calceatum Ridl. = Disa buchenaviana Kraenzl. ■☆

347735　Satyrium candidum Lindl. ;纯白鸟足兰■☆

347736　Satyrium candidum Lindl. var. minus Sond. = Satyrium candidum Lindl. ■☆

347737　Satyrium capense（P. J. Bergius）Houtt. = Acrolophia capensis（P. J. Bergius）Fourc. ■☆

347738　Satyrium carneum（Dryand.）R. Br. ;肉色鸟足兰■☆

347739　Satyrium carsonii Rolfe;卡森鸟足兰■☆

347740　Satyrium cassideum Lindl. = Satyrium parviflorum Sw. ■☆

347741　Satyrium cernuum Thunb. = Disa cernua（Thunb.）Sw. ■☆

347742　Satyrium cheirophorum Rolfe = Satyrium macrophyllum Lindl. ■☆

347743　Satyrium ciliatum Lindl. = Satyrium nepalense D. Don var. ciliatum（Lindl.）Hook. f. ■

347744　Satyrium colliferum Schltr. = Satyrium neglectum Schltr. ■☆

347745　Satyrium compactum Summerh. ;紧密鸟足兰■☆

347746　Satyrium comptum Summerh. ;装饰鸟足兰■☆

347747　Satyrium confusum Summerh. ;混乱鸟足兰■☆

347748　Satyrium cordifolium Lindl. = Satyrium bracteatum（L. f.）Thunb. ■☆

347749　Satyrium coriifolium Sw. ;革叶鸟足兰■☆

347750　Satyrium coriifolium Sw. var. maculatum Hook. f. = Satyrium coriifolium Sw. ■☆

347751　Satyrium coriophoroides A. Rich. ;革梗鸟足兰■☆

347752　Satyrium coriophoroides A. Rich. var. sacculata Rendle = Satyrium coriophoroides A. Rich. ■☆

347753　Satyrium cornutum（L.）Thunb. = Disa cornuta（L.）Sw. ■☆

347754　Satyrium crassicaule Rendle;粗茎鸟足兰■☆

347755　Satyrium cristatum Sond. ;冠状唇鸟足兰;Cristate Satyrium ■☆

347756　Satyrium cristatum Sond. var. longilabiatum A. V. Hall;长唇冠状鸟足兰■☆

347757　Satyrium cucullatum Sw. = Satyrium bicorne（L.）Thunb. ■☆

347758　Satyrium cylindricum Thunb. = Disa cylindrica（Thunb.）Sw. ■☆

347759　Satyrium debile Bolus = Satyrium retusum Lindl. ■☆

347760　Satyrium densiflorum Lindl. = Satyrium parviflorum Sw. ■☆

347761　Satyrium densum Rolfe = Satyrium neglectum Schltr. ■☆

347762　Satyrium dizygoceras Summerh. = Satyrium volkensii Schltr. ■☆

347763　Satyrium djalonis A. Chev. = Satyrium trinerve Lindl. ■☆

347764　Satyrium draconis（L. f.）Thunb. = Disa draconis（L. f.）Sw. ■☆

347765　Satyrium dregei Rolfe = Satyrium bracteatum（L. f.）Thunb. ■☆

347766　Satyrium ecalcaratum Schltr. ;无距鸟足兰■☆

347767　Satyrium elatum Sw. = Cyclopogon elatus（Sw.）Schltr. ■☆

347768　Satyrium elongatum Rolfe;伸长鸟足兰■☆

347769　Satyrium emarcidum Bolus = Satyrium ligulatum Lindl. ■☆

347770　Satyrium epipogium L. = Epipogium aphyllum（F. W. Schmidt）Sw. ■

347771　Satyrium erectum Sw. ;直立鸟足兰■☆

347772　Satyrium eriostomum Lindl. = Satyrium parviflorum Sw. ■☆

347773　Satyrium excelsum Thunb. = Disa tripetaloides（L. f.）N. E. Br. ■☆

347774　Satyrium fanniniae Rolfe = Satyrium bracteatum（L. f.）Thunb. ■☆

347775　Satyrium ferrugineum Thunb. = Disa ferruginea（Thunb.）Sw. ■☆

347776　Satyrium fimbriatum Summerh. ;流苏鸟足兰■☆

347777　Satyrium fischerianum Kraenzl. = Satyrium crassicaule Rendle ■☆

347778　Satyrium flavum la Croix;黄鸟足兰■☆

347779　Satyrium flexuosum（L.）Thunb. = Schizodium flexuosum（L.）Lindl. ■☆

347780　Satyrium foliosum Sw. ;多叶鸟足兰■☆

347781　Satyrium foliosum Sw. var. helonioides Lindl. = Satyrium hallackii Bolus ■☆

347782　Satyrium gigas Ridl. = Satyrium rostratum Lindl. ■☆

347783　Satyrium gilletii De Wild. = Satyrium coriophoroides A. Rich. ■☆

347784　Satyrium goetzenianum Kraenzl. = Satyrium crassicaule Rendle ■☆

347785　Satyrium gracile Lindl. = Satyrium amoenum A. Rich. ■☆

347786　Satyrium gramineum Thouars = Cynorkis graminea (Thouars) Schltr. ■☆

347787　Satyrium grandiflora (L. f.) Thunb. = Disa uniflora P. J. Bergius ■☆

347788　Satyrium guthriei Bolus;格斯里鸟足兰■☆

347789　Satyrium hallackii Bolus;哈拉克鸟足兰■☆

347790　Satyrium hallackii Bolus subsp. ballei (van der Niet et P. J. Cribb) van der Niet et P. J. Cribb;巴勒鸟足兰■☆

347791　Satyrium hallackii Bolus subsp. ocellatum (Bolus) A. V. Hall;单眼鸟足兰■☆

347792　Satyrium henryi Schltr. ;滇南鸟足兰■

347793　Satyrium henryi Schltr. = Satyrium nepalense D. Don ■

347794　Satyrium hians L. f. = Disa hians (L. f.) Spreng. ☆

347795　Satyrium humile Lindl. ;矮小鸟足兰■☆

347796　Satyrium jacottetiae Kraenzl. = Satyrium membranaceum Sw. ■

347797　Satyrium jacottetianum Kraenzl. = Satyrium longicauda Lindl. var. jacottetianum (Kraenzl.) A. V. Hall ■☆

347798　Satyrium johnsonii Rolfe;约翰斯顿鸟足兰■☆

347799　Satyrium kassnerianum Kraenzl. = Satyrium buchananii Schltr. ☆

347800　Satyrium kermesinum Kraenzl. ;克迈斯鸟足兰■☆

347801　Satyrium kirkii Rolfe = Satyrium crassicaule Rendle ■☆

347802　Satyrium lanceolata (Aubl.) León = Sacoila lanceolata (Aubl.) Garay ☆

347803　Satyrium lanceum (Thunb. ex Sw.) Pers. = Herminium lanceum (Thunb. ex Sw.) Vuijk ■

347804　Satyrium leptopetalum Kraenzl. = Satyrium volkensii Schltr. ☆

347805　Satyrium leucanthum Schltr. = Satyrium carsonii Rolfe ■☆

347806　Satyrium leucocomos Rchb. f. ;白簇毛鸟足兰■☆

347807　Satyrium ligulatum Lindl. ;舌状鸟足兰■☆

347808　Satyrium lindleyanum Bolus = Satyrium retusum Lindl. ■☆

347809　Satyrium lineatum Lindl. = Satyrium bracteatum (L. f.) Thunb. ☆

347810　Satyrium longebracteatum Rolfe;长苞片鸟足兰■☆

347811　Satyrium longicauda Lindl. ;长尾鸟足兰■☆

347812　Satyrium longicauda Lindl. var. jacottetianum (Kraenzl.) A. V. Hall;贾科泰鸟足兰■☆

347813　Satyrium longicolle Lindl. ;长颈鸟足兰■☆

347814　Satyrium longissimum Rolfe = Satyrium buchananii Schltr. ■☆

347815　Satyrium lupulinum Lindl. ;狼鸟足兰■☆

347816　Satyrium lydenburgense Rchb. f. = Satyrium parviflorum Sw. ■☆

347817　Satyrium macrophyllum Lindl. ;大叶鸟足兰■☆

347818　Satyrium maculatum Burch. ex Lindl. = Satyrium longicolle Lindl. ■☆

347819　Satyrium maculatum Desf. = Neotinea maculata (Desf.) Stearn ■☆

347820　Satyrium mairei Schltr. = Satyrium ciliatum Lindl. ■

347821　Satyrium mairei Schltr. = Satyrium nepalense D. Don var. ciliatum (Lindl.) Hook. f. ■

347822　Satyrium marginatum Bolus = Satyrium stenopetalum Lindl. subsp. brevicalcaratum (Bolus) A. V. Hall ■☆

347823　Satyrium mechowianum Kraenzl. ;梅休鸟足兰■☆

347824　Satyrium membranaceum Sw. ;膜质鸟足兰■☆

347825　Satyrium microcephalum Kraenzl. = Satyrium yunnanense Rolfe ■☆

347826　Satyrium microcorys Schltr. ;小鸟足兰■☆

347827　Satyrium microrrhynchum Schltr. ;小喙鸟足兰■☆

347828　Satyrium militare Lindl. = Satyrium sphaerocarpum Lindl. ■☆

347829　Satyrium minax Rolfe = Satyrium anomalum Schltr. ■☆

347830　Satyrium mirum Summerh. ;奇异鸟足兰■☆

347831　Satyrium miserum Kraenzl. ;贫弱鸟足兰■☆

347832　Satyrium monadenum Schltr. ;单蕊鸟足兰■☆

347833　Satyrium monopetalum Kraenzl. = Satyrium trinerve Lindl. ■☆

347834　Satyrium monophyllum Kraenzl. ;单叶鸟足兰■☆

347835　Satyrium monorchis (L.) Pers. = Herminium monorchis (L.) R. Br. ■

347836　Satyrium morrumbalaense De Wild. = Satyrium macrophyllum Lindl. ■☆

347837　Satyrium muticum Lindl. ;无尖鸟足兰■☆

347838　Satyrium mystacinum Kraenzl. = Satyrium crassicaule Rendle ■☆

347839　Satyrium neglectum Schltr. ;疏忽鸟足兰■☆

347840　Satyrium neglectum Schltr. subsp. woodii (Schltr.) A. V. Hall;伍得鸟足兰■☆

347841　Satyrium neglectum Schltr. var. brevicalcar Summerh. ;短距鸟足兰■☆

347842　Satyrium nepalense D. Don;鸟足兰(长距鸟足兰,对对参,小鸡腿,壮精丹);Nepal Birdfootorchis, Nepal Rattlesnake Plantain ■

347843　Satyrium nepalense D. Don subsp. yunnanense (Rolfe) Soó = Satyrium yunnanense Rolfe ■

347844　Satyrium nepalense D. Don var. ciliatum (Lindl.) Hook. f. ;缘毛鸟足兰(对对参,鸡肾草,假天麻,蜡烛花,绿毛鸟足兰,四川鸟足兰,云南鸟足兰);Ciliate Birdfootorchis, Ciliate Rattlesnake Plantain ■

347845　Satyrium nepalense D. Don var. ciliatum (Lindl.) Hook. f. = Satyrium ciliatum Lindl. ■

347846　Satyrium nepalense D. Don var. yunnanense (Rolfe) Soó = Satyrium yunnanense Rolfe ■

347847　Satyrium nigericum Hutch. = Satyrium carsonii Rolfe ■☆

347848　Satyrium niloticum Rendle = Satyrium crassicaule Rendle ■☆

347849　Satyrium nutans Kraenzl. = Satyrium hallackii Bolus subsp. ocellatum (Bolus) A. V. Hall ☆

347850　Satyrium nuttii Rolfe = Satyrium trinerve Lindl. ■☆

347851　Satyrium nyassense Kraenzl. = Satyrium buchananii Schltr. ■☆

347852　Satyrium occultum Rolfe = Satyrium trinerve Lindl. ■☆

347853　Satyrium ocellatum Bolus = Satyrium hallackii Bolus subsp. ocellatum (Bolus) A. V. Hall ■☆

347854　Satyrium ochroleucum Bolus = Satyrium humile Lindl. ■☆

347855　Satyrium odorum Sond. ;芳香鸟足兰■☆

347856　Satyrium oliganthum Schltr. ;寡花鸟足兰■☆

347857　Satyrium orbiculare Rolfe;圆形鸟足兰■☆

347858　Satyrium orchioides Sw. = Sacoila lanceolata (Aubl.) Garay ■☆

347859　Satyrium orobanchoides L. f. = Corycium orobanchoides (L. f.) Sw. ■☆

347860　Satyrium pallens S. D. Johnson et Kurzweil;变苍白鸟足兰■☆

347861　Satyrium pallidiflorum Schltr. = Satyrium lupulinum Lindl. ■☆

347862　Satyrium pallidum A. Rich. = Satyrium nepalense D. Don ■

347863　Satyrium paludicola Schltr. = Satyrium bracteatum (L. f.) Thunb. ■☆

347864　Satyrium paludosum Rchb. f. ;沼泽鸟足兰■☆

347865　Satyrium paludosum Rchb. f. var. parvibracteatum Schltr. = Satyrium oliganthum Schltr. ■☆

347866 Satyrium papillosum Lindl. = Satyrium erectum Sw. ■☆

347867 Satyrium papyrrtorum Schltr. = Satyrium neglectum Schltr. var. brevicalcar Summerh. ■☆

347868 Satyrium parviflorum Lindl. = Satyrium stenopetalum Lindl. subsp. brevicalcaratum（Bolus）A. V. Hall ■☆

347869 Satyrium parviflorum Sw. ;小花鸟足兰;Smallflower Satyrium ■☆

347870 Satyrium parviflorum Sw. var. schimperi（A. Rich.）Schltr. = Satyrium schimperi Hochst. ex A. Rich. ■☆

347871 Satyrium pedicellatum L. f. = Eulophia aculeata（L. f.）Spreng. ■☆

347872 Satyrium pentadactylum Kraenzl. = Satyrium cristatum Sond. ☆

347873 Satyrium pentherianum Kraenzl. = Satyrium pygmaeum Sond. ■☆

347874 Satyrium perrieri Schltr. ;佩里耶鸟足兰☆

347875 Satyrium perrottetianum A. Rich. = Satyrium nepalense D. Don ■

347876 Satyrium pictum Lindl. = Satyrium bracteatum（L. f.）Thunb. ■☆

347877 Satyrium platystigma Schltr. = Satyrium longicauda Lindl. ■☆

347878 Satyrium praealtum Thouars = Habenaria praealta（Thouars）Spreng. ■☆

347879 Satyrium princeps Bolus;帝王鸟足兰■☆

347880 Satyrium proschii Briq. = Satyrium trinerve Lindl. ■☆

347881 Satyrium pulchrum S. D. Johnson et Kurzweil;美丽鸟足兰■☆

347882 Satyrium pumilum Thunb. ;矮鸟足兰■☆

347883 Satyrium pustulatum Lindl. = Satyrium erectum Sw. ■☆

347884 Satyrium pycnostachyum Schltr. = Satyrium yunnanense Rolfe ■

347885 Satyrium pygmaeum Sond. ;微小鸟足兰■☆

347886 Satyrium repens L. = Goodyera repens（L.）R. Br. ■

347887 Satyrium retusum Lindl. ;微凹鸟足兰■☆

347888 Satyrium rhodanthum Schltr. ;粉红花鸟足兰■☆

347889 Satyrium rhynchanthum Bolus;喙鸟足兰■☆

347890 Satyrium rhynchantoides Schltr. ;拟喙花鸟足兰■☆

347891 Satyrium riparium Rchb. f. ;河岸鸟足兰■☆

347892 Satyrium robustum Schltr. ;粗壮鸟足兰■☆

347893 Satyrium rosellatum Thouars = Cynorkis rosellata（Thouars）Bosser ■☆

347894 Satyrium rostratum Lindl. ;喙状鸟足兰■☆

347895 Satyrium rufescens Thunb. = Disa rufescens（Thunb.）Sw. ■☆

347896 Satyrium rupestre Schltr. ex Bolus;岩生鸟足兰■☆

347897 Satyrium sacculatum（Rendle）Rolfe = Satyrium coriophoroides A. Rich. ■☆

347898 Satyrium sagittale（L. f.）Thunb. = Disa sagittalis（L. f.）Sw. ☆

347899 Satyrium saxicola Bolus = Satyrium bracteatum（L. f.）Thunb. ■☆

347900 Satyrium sceptrum Schltr. ;王杖鸟足兰■☆

347901 Satyrium schimperi Hochst. ex A. Rich. ;欣珀鸟足兰■☆

347902 Satyrium schinzii T. Durand et Kraenzl. ;欣兹鸟足兰■☆

347903 Satyrium schlechteri Rolfe = Satyrium pygmaeum Sond. ■☆

347904 Satyrium secundum Thunb. = Disa racemosa L. f. ■☆

347905 Satyrium setchuenicum Kraenzl. = Satyrium ciliatum Lindl. ■

347906 Satyrium setchuenicum Kraenzl. = Satyrium nepalense D. Don var. ciliatum（Lindl.）Hook. f. ■

347907 Satyrium setchuenicum Kraenzl. ex Diels;四川鸟足兰■

347908 Satyrium setchuenicum Kraenzl. ex Diels = Satyrium cristatum Sond. ■☆

347909 Satyrium shirense Rolfe;希尔鸟足兰■☆

347910 Satyrium spathulatum（L. f.）Thunb. = Disa spathulata（L. f.）Sw. ■☆

347911 Satyrium speciosum Rolfe;美花鸟足兰;Beautyflower Satyrium ■☆

347912 Satyrium speciosum Rolfe = Satyrium macrophyllum Lindl. ■☆

347913 Satyrium sphaeranthum Schltr. ;球花鸟足兰■☆

347914 Satyrium sphaerocarpum Lindl. ;球果鸟足兰■☆

347915 Satyrium spirale Thouars = Benthamia spiralis（Thouars）A. Rich. ●☆

347916 Satyrium stenopetalum Lindl. ;窄瓣鸟足兰■☆

347917 Satyrium stenopetalum Lindl. subsp. brevicalcaratum（Bolus）A. V. Hall;短距窄瓣鸟足兰■☆

347918 Satyrium stenopetalum Lindl. var. brevicalcaratum Bolus = Satyrium stenopetalum Lindl. subsp. brevicalcaratum（Bolus）A. V. Hall ■☆

347919 Satyrium stenopetalum Lindl. var. parviflorum（Lindl.）Schltr. = Satyrium stenopetalum Lindl. subsp. brevicalcaratum（Bolus）A. V. Hall ■☆

347920 Satyrium stolzianum Kraenzl. = Satyrium buchananii Schltr. ■☆

347921 Satyrium stolzii Kraenzl. = Satyrium coriophoroides A. Rich. ■☆

347922 Satyrium striatum Thunb. ;条纹鸟足兰■☆

347923 Satyrium tabulare L. f. = Eulophia tabularis（L. f.）Bolus ■☆

347924 Satyrium tenellum（L. f.）Thunb. = Disa tenella（L. f.）Sw. ■☆

347925 Satyrium tenii Schltr. = Satyrium ciliatum Lindl. ■

347926 Satyrium tenii Schltr. = Satyrium nepalense D. Don var. ciliatum（Lindl.）Hook. f. ■

347927 Satyrium tenuifolium Kraenzl. = Satyrium parviflorum Sw. ■☆

347928 Satyrium trachypetalum Kraenzl. = Satyrium volkensii Schltr. ■☆

347929 Satyrium trinerve Lindl. ;三脉鸟足兰■☆

347930 Satyrium triphyllum Kraenzl. = Satyrium trinerve Lindl. ■☆

347931 Satyrium triste L. f. = Acrolophia capensis（P. J. Bergius）Fourc. ■☆

347932 Satyrium tschangii Schltr. = Satyrium ciliatum Lindl. ■

347933 Satyrium tschangii Schltr. = Satyrium nepalense D. Don var. ciliatum（Lindl.）Hook. f. ■

347934 Satyrium usambarae Kraenzl. = Satyrium crassicaule Rendle ■☆

347935 Satyrium utriculatum Sond. = Satyrium candidum Lindl. ■☆

347936 Satyrium viride L. = Coeloglossum viride（L.）Hartm. ■

347937 Satyrium viride L. = Dactylorhiza viridis（L.）R. M. Bateman, Pridgeon et M. W. Chase ■

347938 Satyrium volkensii Schltr. ;糙瓣鸟足兰■☆

347939 Satyrium welwitschii Rchb. f. ;韦尔鸟足兰■☆

347940 Satyrium wilmsianum Kraenzl. = Satyrium parviflorum Sw. ■☆

347941 Satyrium woodii Schltr. = Satyrium neglectum Schltr. subsp. woodii（Schltr.）A. V. Hall ■☆

347942 Satyrium yunnanense Rolfe; 云南鸟足兰; Yunnan Birdfootorchis,Yunnan Rattlesnake Plantain ■

347943 Satyrium zombense Rolfe = Satyrium trinerve Lindl. ■☆

347944 Saubinetia J. Rémy = Verbesina L.（保留属名）●■☆

347945 Saueria Klotzsch = Begonia L. ●■

347946 Saugetia Hitchc. et Chase = Enteropogon Nees ■

347947 Saul Roxb. ex Wight et Arn. = Shorea Roxb. ex C. F. Gaertn. ●

347948 Saulcya Michon = Odontospermum Neck. ex Sch. Bip. ■☆

347949 Saulcya hierichuntica Michon = Pallenis hierochuntica（Michon）Greuter ■☆

347950 Saundersia Rchb. f.（1866）;桑德斯兰属☆

347951 Saundersia mirabilis Rchb. f. ;桑德斯兰☆

347952 Saurauia Willd.（1801）（保留属名）;水东哥属（水冬瓜属）;Saurauia ●

347953 Saurauia aequatoriensis Sprague var. boliviana Buscal. = Saurauia glabra（Ruiz et Pav.）Soejarto ●☆

347954 Saurauia brevipes Rusby = Saurauia spectabilis Hook. ●☆

347955　Saurauia cerea Griff. ex Dyer;蜡质水东哥;Waxy Saurauia ●

347956　Saurauia erythrocarpa C. F. Liang et Y. S. Wang;红果水东哥（大木钩,红马耳);Redfruit Saurauia,Red-fruited Saurauia ●

347957　Saurauia erythrocarpa C. F. Liang et Y. S. Wang var. grosseserrata C. F. Liang et Y. S. Wang;粗齿水东哥;Fascite Saurauia ●

347958　Saurauia excelsa Willd.;高大水东哥●☆

347959　Saurauia floribunda Linden et Planch. var. peruviana Buscal. = Saurauia peruviana Buscal. ●☆

347960　Saurauia glabra（Ruiz et Pav.）Soejarto;光滑水东哥●☆

347961　Saurauia griffithii Dyer;绵毛水东哥;Griffith Saurauia,Lanose Saurauia,Tonkin Saurauia ●

347962　Saurauia griffithii Dyer var. annamica Gagnep.;越南水东哥;Griffith Saurauia,Viatnam ●

347963　Saurauia hirsuta C. F. Liang = Saurauia sinohirsuta J. Q. Li et Soejarto ●

347964　Saurauia kawagoeana Hatus. = Saurauia tristyla DC. var. oldhamii（Hemsl.）Finet et Gagnep. ●

347965　Saurauia lantsangensis Hu;澜沧水东哥（鼻涕果,大叶杜仲,大叶仲憋促,马耳子果,密心果,牛嗓管树,野枇杷,粘心果);Lancang Saurauia ●

347966　Saurauia macrotricha Kurz ex Dyer;长毛水东哥;Longhair Saurauia,Long-haired Saurauia ●

347967　Saurauia miniata C. F. Liang et Y. S. Wang;朱毛水东哥（野枇杷,珠毛水东哥);Cinnabarhair Saurauia,Cinnabar-haired Saurauia ●

347968　Saurauia napaulensis DC.;尼泊尔水东哥（鼻涕果,大叶杜仲,马耳子果,蜜心果,牛嗓管树,撒罗夷,野枇杷,粘心果,锥序水东哥);Nepal Saurauia ●

347969　Saurauia napaulensis DC. var. montana C. F. Liang et Y. S. Wang;山地水东哥（大叶杜仲,粉心果,澜沧水东哥,马耳子果,密心果,牛嗓管树,少牢木,野枇杷);Montane Saurauia ●

347970　Saurauia napaulensis DC. var. montana C. F. Liang et Y. S. Wang = Saurauia napaulensis DC. ●

347971　Saurauia napaulensis DC. var. omeiensis C. F. Liang et Y. S. Wang;峨眉水东哥;Emei Nepal Saurauia,Omei Nepal Saurauia ●

347972　Saurauia natalicia Sleumer;纳塔尔水东哥●☆

347973　Saurauia oldhamii Hemsl. = Saurauia tristyla DC. var. oldhamii（Hemsl.）Finet et Gagnep. ●

347974　Saurauia parviflora Triana et Planch.;小花水东哥●☆

347975　Saurauia paucinervis C. F. Liang et Y. S. Wang;少脉水东哥;Few-nerved Saurauia,Fewvein Saurauia ●

347976　Saurauia peruviana Buscal.;秘鲁水东哥●☆

347977　Saurauia polyneura C. F. Liang et Y. S. Wang;多脉水东哥;Manyvein Saurauia,Poly-nerved Saurauia ●

347978　Saurauia polyneura C. F. Liang et Y. S. Wang var. paucinervis J. Q. Li et Soejarto = Saurauia paucinervis C. F. Liang et Y. S. Wang ●

347979　Saurauia pseudoparviflora Buscal. = Saurauia peruviana Buscal. ●☆

347980　Saurauia pseudoparviflora Buscal. var. rusbyana Buscal. = Saurauia natalicia Sleumer ●☆

347981　Saurauia punduana Wall.;大花水东哥;Bigflower Saurauia,Big-flowered Saurauia,Large-flower Saurauia ●

347982　Saurauia pyramidata Sleumer = Saurauia peruviana Buscal. ●☆

347983　Saurauia rubricalyx C. F. Liang et Y. S. Wang;红萼水东哥;Redcalyx Saurauia,Red-sepaled Saurauia ●

347984　Saurauia rusbyi Britton;鲁斯比水东哥●☆

347985　Saurauia rusbyi Britton var. glabrata Buscal. = Saurauia

spectabilis Hook. ●☆

347986　Saurauia rusbyi Britton var. spectabilis Buscal. = Saurauia spectabilis Hook. ●☆

347987　Saurauia scabra Britton var. boliviana Buscal. = Saurauia peruviana Buscal. ●☆

347988　Saurauia serrata DC.;尖齿水东哥●☆

347989　Saurauia sibcordata Korth.;印尼水东哥●☆

347990　Saurauia sinohirsuta J. Q. Li et Soejarto;糙毛水东哥;Hirsute Saurauia,Rough-haired Saurauia ●

347991　Saurauia spectabilis Hook.;壮观水东哥●☆

347992　Saurauia thyrsiflora C. F. Liang et Y. S. Wang;聚锥水东哥（个毛,筋苗,水枇杷,羊桃山枇杷);Thyrse Saurauia ●

347993　Saurauia tristyla DC.;水东哥（白饭果,白饭木,鼻涕果,红毛树,水冬瓜,水牛奶,水枇杷,水自环);Common Saurauia ●

347994　Saurauia tristyla DC. var. hekouensis C. F. Liang et Y. S. Wang;河口水东哥（白饭果,鼻涕果);Hekou Saurauia,Saurauia ●

347995　Saurauia tristyla DC. var. hekouensis C. F. Liang et Y. S. Wang = Saurauia thyrsiflora C. F. Liang et Y. S. Wang ●

347996　Saurauia tristyla DC. var. oldhamii（Hemsl.）Finet et Gagnep.;台湾水东哥（大冇树,火筒冇,水东哥,水冬瓜,水管心);Taiwan Saurauia ●

347997　Saurauia tristyla DC. var. oldhamii（Hemsl.）Finet et Gagnep. = Saurauia thyrsiflora C. F. Liang et Y. S. Wang ●

347998　Saurauia tristyla DC. var. oldhamii Finet et Gagnep. = Saurauia tristyla DC. var. oldhamii（Hemsl.）Finet et Gagnep. ●

347999　Saurauia trolliana Sleumer;特洛尔水东哥●☆

348000　Saurauia vanioti H. Lév. = Celastrus vaniotii（H. Lév.）Rehder ●

348001　Saurauia yunnanensis C. F. Liang et Y. S. Wang;云南水东哥;Yunnan Saurauia ●

348002　Saurauia zetigeri Korth.;泽提水东哥●☆

348003　Saurauiaceae Griseb.（1854）(保留科名);水东哥科（伞罗夷科,水冬瓜科);Saurauia Family ●

348004　Saurauiaceae Griseb.（保留科名） = Actinidiaceae Gilg et Werderm.（保留科名)●

348005　Saurauiaceae J. Agardh = Actinidiaceae Gilg et Werderm.（保留科名)●

348006　Saurauiaceae J. Agardh = Saurauiaceae Griseb.（保留科名)●

348007　Sauravia Spreng. = Saurauia Willd.（保留属名)●

348008　Sauria Bajtenov（1995）;蜥蜴紫草属 ☆

348009　Sauria akkolia Bajtenov;蜥蜴紫草 ☆

348010　Saurobroma Raf. = Celtis L. ●

348011　Saurobroma Raf. = Colletia Scop.（废弃属名)●

348012　Sauroglossum Lindl.（1833）;蜥蜴兰属 ■☆

348013　Sauroglossum candidum Kraenzl.;白蜥蜴兰 ■☆

348014　Sauroglossum cranichoides（Griseb.）Ames = Cyclopogon cranichoides（Griseb.）Schltr. ■☆

348015　Sauroglossum elatum Lindl.;蜥蜴兰 ■☆

348016　Sauroglossum nigricans Schltr.;黑蜥蜴兰 ■☆

348017　Sauroglossum richardii Ames = Cyclopogon elatus（Sw.）Schltr. ■☆

348018　Saurolluma Plowes = Caralluma R. Br. ■

348019　Saurolluma Plowes（1995);蜥蜴角属 ■☆

348020　Saurolluma furta（P. R. O. Bally）Plowes = Caralluma furta P. R. O. Bally ■☆

348021　Saurolophorkis Marg. et Szlach. = Crepidium Blume ■

348022　Sauromatum Schott（1832);斑龙芋属;Lizard Arum,Lizardtaro,Sauromatum ■

348023 Sauromatum abyssinicum Schott = Typhonium venosum (Dryand. ex Aiton) Hett. et P. C. Boyce ■☆

348024 Sauromatum angolense N. E. Br. = Typhonium venosum (Dryand. ex Aiton) Hett. et P. C. Boyce ■☆

348025 Sauromatum brevipes (Hook. f.) N. E. Br.；短柄斑龙芋；Shortstalk Lizardtaro, Shortstalk Sauromatum ■

348026 Sauromatum guttatum (Wall.) Schott = Sauromatum venosum (Aiton) Kunth ■

348027 Sauromatum guttatum (Wall.) Schott = Typhonium venosum (Dryand. ex Aiton) Hett. et P. C. Boyce ■☆

348028 Sauromatum guttatum (Wall.) Schott var. typicum Engl. = Sauromatum venosum (Aiton) Kunth ■

348029 Sauromatum guttatum Schott = Sauromatum venosum (Aiton) Schott ■

348030 Sauromatum guttatum Wall. = Sauromatum venosum (Aiton) Schott ■

348031 Sauromatum nubicum Schott = Typhonium venosum (Dryand. ex Aiton) Hett. et P. C. Boyce ■☆

348032 Sauromatum pedatum (Willd.) Schott = Sauromatum venosum (Aiton) Schott ■

348033 Sauromatum pedatum (Willd.) Schott = Typhonium venosum (Dryand. ex Aiton) Hett. et P. C. Boyce ■☆

348034 Sauromatum venosum (Aiton) Kunth = Sauromatum venosum (Dryand. ex Aiton) Kunth ■

348035 Sauromatum venosum (Aiton) Schott = Sauromatum venosum (Dryand. ex Aiton) Kunth ■

348036 Sauromatum venosum (Dryand. ex Aiton) Kunth；斑龙芋（蛇芋,掌叶斑龙,掌叶斑龙芋）；Monarch of The East, Monarch-of-the-east, Palmleaf Lizardtaro, Veiny Lizardtaro, Veiny Sauromatum, Voodoo Lily ■

348037 Sauromatum venosum (Dryand. ex Aiton) Kunth = Typhonium venosum (Dryand. ex Aiton) Hett. et P. C. Boyce ■☆

348038 Sauropus Blume(1826)；守宫木属（梭罗巴属,越南菜属）；Geckowood, Sauropus ●■

348039 Sauropus albicans Blume = Sauropus androgynus (L.) Merr. ●

348040 Sauropus albicans Blume = Sauropus garrettii Craib ●

348041 Sauropus androgynus (L.) Merr.；守宫木（篱笆菜,树仔菜,甜菜,同序守宫木,越南菜）；Common Sauropus, Geckowood ●

348042 Sauropus androgynus (L.) Merr. = Sauropus garrettii Craib ●

348043 Sauropus bacciformis (L.) Airy Shaw；艾堇（艾堇守宫木,艾茎守宫木,红果草,假叶下珠）；Berry-shaped Sauropus ●

348044 Sauropus bacciformis (L.) Airy Shaw = Synostemon bacciforme (L.) Webster ■

348045 Sauropus bonii Beille；茎花守宫木；Bon Sauropus, Stem-flower Sauropus ●

348046 Sauropus changianus S. Y. Hu = Sauropus spatulifolius Beille ●

348047 Sauropus chorisepalus Merr. et Chun = Sauropus garrettii Craib ●

348048 Sauropus compressus Müll. Arg. = Sauropus quadrangularis (Willd.) Müll. Arg. ●

348049 Sauropus delavayi Croizat；石山守宫木（石山越南菜）；Delavay Sauropus ●

348050 Sauropus garrettii Craib；苍叶守宫木（滇越南菜）；Garrett Sauropus ●

348051 Sauropus grandifolius Pax et K. Hoffm. = Sauropus macranthus Hassk. ●

348052 Sauropus guadrangularis (Willd.) Müll. Arg.；细叶守宫木；Thin-leaf Sauropus ●

348053 Sauropus longipedicellatus Merr. et Chun = Sauropus macranthus Hassk. ●

348054 Sauropus macranthus Hassk.；长梗守宫木；Longipetiole Sauropus ●

348055 Sauropus macrophyllus Hook. f. = Sauropus macranthus Hassk. ●

348056 Sauropus orbicularis Craib = Sauropus delavayi Croizat ●

348057 Sauropus parviflorus Pax et K. Hoffm. = Sauropus androgynus (L.) Merr. ●

348058 Sauropus pierrei (Beille) Croizat；盈江守宫木；Pierre Sauropus ●

348059 Sauropus quadrangularis (Willd.) Müll. Arg.；方枝守宫木（四角越南菜,细叶守宫木）；Fourangular Sauropus, Quadrangular Sauropus ●

348060 Sauropus quadrangularis (Willd.) Müll. Arg. var. compressus (Müll. Arg.) Airy Shaw；扁枝守宫木●

348061 Sauropus ramosissimus (F. Muell.) Airy Shaw；多枝守宫木●☆

348062 Sauropus repandus Müll. Arg.；波萼守宫木；Repand Sauropus, Repandous Sauropus ●

348063 Sauropus reticulatus X. L. Mo ex P. T. Li；网脉守宫木；Netted Sauropus, Net-veined Sauropus ●

348064 Sauropus rhamnoides Blume；鼠李叶守宫木●☆

348065 Sauropus rostratus Miq. = Sauropus spatulifolius Beille ●

348066 Sauropus spatulifolius Beille；龙脷叶（龙利叶,龙舌叶,龙味叶,龙疢叶）；Spatulateleaf Sauropus, Spatulate-leaved Sauropus ●

348067 Sauropus spectabilis Miq. = Sauropus macranthus Hassk. ●

348068 Sauropus sumatramus Miq. = Sauropus androgynus (L.) Merr. ●

348069 Sauropus trinervius (Wall.) Hook. f. et Thomson ex Müll. Arg.；三脉守宫木；Trinervate Sauropus, Trinerves Sauropus ●

348070 Sauropus tsiangii P. T. Li；尾叶守宫木；Tsiang Sauropus ●

348071 Sauropus yanhuianus P. T. Li；多脉守宫木；Yanhui Sauropus ●

348072 Sauropus yunnanensis Pax et K. Hoffm. = Sauropus garrettii Craib ●

348073 Saururaceae A. Rich. = Saururaceae Rich. ex T. Lestib. (保留科名) ■

348074 Saururaceae E. Mey. = Saururaceae Rich. ex T. Lestib. (保留科名) ■

348075 Saururaceae F. Voigt = Saururaceae Rich. ex T. Lestib. (保留科名) ■

348076 Saururaceae Rich. ex E. Mey. = Saururaceae Rich. ex T. Lestib. (保留科名) ■

348077 Saururaceae Rich. ex T. Lestib. (1826)(保留科名)；三白草科；Lizard's-tail Family, Lizardtail Family ■

348078 Saururopsis Turcz. = Saururus L. ■

348079 Saururopsis chinensis (Lour.) Turcz. = Saururus chinensis (Lour.) Baill. ■

348080 Saururopsis chinensis Turcz. = Saururus chinensis (Lour.) Baill. ■

348081 Saururopsis cumingii C. DC. = Saururus chinensis (Lour.) Baill. ■

348082 Saururus L. (1753)；三白草属；Lizard's Tail, Lizards Tail, Lizard's-tail, Lizardtail ■

348083 Saururus Mill. = Piper L. ●■

348084 Saururus cavaleriei H. Lév. = Gymnotheca chinensis Decne. ■

348085 Saururus cernuus L.；美国三白草（垂穗三白草,沼泽三白草）；American Swamp Lily, Lizard's Tail, Lizards Tail, Lizard's-tail, Lizardtail, Swamp Lily, Water Dragon ■☆

348086 Saururus cernuus Thunb. = Saururus chinensis (Lour.) Baill. ■

348087 Saururus chinensis (Lour.) Baill.；三白草(白花莲,白花照水

莲,白黄脚,白节藕,白桔朝,白莲藕,白面姑,白舌骨,白水鸡,白
尾刁,白叶莲,百节藕,地藕,东方三白草,过山龙,过塘莲,过塘
藕,九节藕,洛氏三白草,三白根,三点白,三叶白,三叶白草,水
伴深乌,水边兰,水槟榔,水九节莲,水莲藕,水木通,水牛草,塘
边藕,天性草,田三白,土玉竹,五路白,五叶白,一白二白);
China Lizardtail,Chinese Lizardtail,Oriental Lizardtail ■

348088　Saururus loureiri Decne. = Saururus chinensis (Lour.) Baill. ■

348089　Saussurea DC. (1810)(保留属名);风毛菊属(青木香属,雪
莲属); Alpine Saw-wort, Saussurea, Saw-wort, Snowlotus,
Snowrabbiten, Windhairdaisy ●■

348090　Saussurea Salisb. = Hosta Tratt. (保留属名) ■

348091　Saussurea × iwateyamensis M. Kikuchi;岩手山风毛菊■☆

348092　Saussurea × karuizawensis H. Hara;轻井泽风毛菊■☆

348093　Saussurea × mirabilis Kitam.;奇异风毛菊■☆

348094　Saussurea × subgracilis Sugim.;亚纤风毛菊■☆

348095　Saussurea × tobitae Kitam.;飞田风毛菊■☆

348096　Saussurea abnormis Lipsch.;普兰风毛菊;Pulan Windhairdaisy ■

348097　Saussurea acaulis Klatt = Saussurea thomsonii C. B. Clarke ■

348098　Saussurea acromelaena Hand.-Mazz.;肾叶风毛菊;Kidneyleaf
Windhairdaisy ■

348099　Saussurea acrophila Diels;破血丹(光叶风毛菊);Broken
Blood Windhairdaisy ■

348100　Saussurea acrophila Diels = Saussurea oligocephala (Y. Ling)
Y. Ling ■

348101　Saussurea acrophila Diels var. oligocephala Y. Ling = Saussurea
oligocephala (Y. Ling) Y. Ling ■

348102　Saussurea acropilina Diels = Saussurea populifolia Hemsl. ■

348103　Saussurea acroura Cummins;川甘风毛菊;Chuanigan
Windhairdaisy ■

348104　Saussurea acuminata Turcz. ex Fisch. et C. A. Mey.;渐尖风毛
菊;Acuminate Windhairdaisy ■

348105　Saussurea acuminata Turcz. ex Fisch. et C. A. Mey. subsp.
sachalinensis (F. Schmidt) Kitam. = Saussurea sachalinensis F.
Schmidt ■☆

348106　Saussurea acuminata Turcz. ex Fisch. et C. A. Mey. var.
sachalinensis (F. Schmidt) Herder = Saussurea sachalinensis F.
Schmidt ■☆

348107　Saussurea aegirophylla Diels = Saussurea cordifolia Hemsl. ■

348108　Saussurea aerjingensis K. M. Shen;阿尔金风毛菊;Aerjin
Windhairdaisy ■

348109　Saussurea affinis Spreng. = Hemistepta lyrata (Bunge) Bunge ■

348110　Saussurea affinis Spreng. ex DC. = Hemistepta lyrata (Bunge)
Bunge ■

348111　Saussurea ajanensis (Regel) Lipsch.;阿雅风毛菊■☆

348112　Saussurea alaschanica Maxim.;阿拉善风毛菊;Alashan
Windhairdaisy ■

348113　Saussurea alata DC.;翼茎风毛菊(翅茎风毛菊);Wingstem
Windhairdaisy ■

348114　Saussurea alata DC. var. laciniata Herder = Saussurea laciniata
Ledeb. ■

348115　Saussurea alata DC. var. runcinata Herder = Saussurea runcinata
DC. ■

348116　Saussurea alatipes Hemsl.;翼柄风毛菊(翅风毛菊,岩牛蒡
子);Wingstalk Windhairdaisy ■

348117　Saussurea alatipes Hemsl. var. huashanensis (Y. Ling) Y. Ling
= Saussurea huashanensis (Y. Ling) X. Y. Wu ■

348118　Saussurea albertii Regel et C. Winkl.;新疆风毛菊;Xinjiang
Windhairdaisy ■

348119　Saussurea alginophylla Diels = Saussurea cordifolia Hemsl. ■

348120　Saussurea alpicola Kitam. = Saussurea tomentosa Kom. ■

348121　Saussurea alpina (L.) DC.;高山风毛菊;Alp Windhairdaisy,
Alpine Saussurea, Alpine Sawwort, Alpine Saw-wort, Purple
Hawkweed ■

348122　Saussurea alpina (L.) DC. = Saussurea nuda Ledeb. ■☆

348123　Saussurea alpina (L.) DC. var. ledebourii A. Gray = Saussurea
nuda Ledeb. ■☆

348124　Saussurea alpina (L.) DC. var. leucophylla Hance = Saussurea
chinensis (Maxim.) Lipsch. ■

348125　Saussurea alpina (L.) DC. var. manhanschanensis H. C. Fu =
Saussurea parviflora (Poir.) DC. ■

348126　Saussurea alpincola Kitam. = Saussurea tomentosa Kom. ■

348127　Saussurea amabilis Kitam.;秀丽风毛菊■☆

348128　Saussurea amabilis Kitam. f. pinnatiloba Kitam.;羽裂秀丽风毛
菊■☆

348129　Saussurea amara (L.) DC.;草地风毛菊(狗舌头,狗舌头草,
驴耳朵,驴耳朵草,驴耳风毛菊,羊耳朵);Meadow Saussurea,
Meadow Windhairdaisy, Tall Saw-wort ■

348130　Saussurea amara (L.) DC. f. microcephala Franch. = Saussurea
amara (L.) DC. ■

348131　Saussurea amara (L.) DC. var. exappendiculata H. C. Fu;尖苞
草地风毛菊■

348132　Saussurea amara (L.) DC. var. glomerata (Poir.) Trautv.;苦
风毛菊■☆

348133　Saussurea amara (L.) DC. var. glomerata (Poir.) Trautv. =
Saussurea amara (L.) DC. ■

348134　Saussurea amara (L.) DC. var. glomerata Trautv. = Saussurea
amara (L.) DC. ■

348135　Saussurea amara (L.) DC. var. microcephala (Franch.)
Lipsch.;小花草地风毛菊;Smallflower Meadow Windhairdaisy ■

348136　Saussurea amara (L.) DC. var. microcephala (Franch.)
Lipsch. = Saussurea amara (L.) DC. ■

348137　Saussurea ambigua Krylov ex Serg.;含糊风毛菊■☆

348138　Saussurea amblyophylla C. Winkl. = Saussurea thomsonii C. B.
Clarke ■

348139　Saussurea americana D. C. Eaton;美国风毛菊;American Sawwort ■☆

348140　Saussurea amoena Kar. et Kir. = Saussurea elegans Ledeb. ■

348141　Saussurea amurensis Turcz. ex DC.;龙江风毛菊(齿叶风毛菊,东
北燕尾风毛菊);Amur Saussurea,Heilongjiang Windhairdaisy ■

348142　Saussurea amurensis Turcz. ex DC. subsp. duiensis (F. Schmidt)
Kitam.;杜伊风毛菊■☆

348143　Saussurea amurensis Turcz. ex DC. subsp. stenophylla (Freyn)
Kitam. = Saussurea amurensis Turcz. ex DC. ■

348144　Saussurea andersonii C. B. Clarke;卵苞风毛菊;Andreson
Windhairdaisy ■

348145　Saussurea andryaloides (DC.) Sch. Bip.;吉隆风毛菊;Jilong
Windhairdaisy ■

348146　Saussurea angustifolia (L.) DC.;狭叶风毛菊;Common
Saussurea,Narrow-leaved Saw-wort ■☆

348147　Saussurea angustifolia (L.) DC. var. viscida (Hultén) S. L.
Welsh;黏狭叶风毛菊;Sticky saw-wort ■☆

348148　Saussurea angustifolia (L.) DC. var. yukonensis A. E. Porsild;
育空风毛菊■☆

348149　Saussurea angustifolia (Willd.) DC. = Saussurea angustifolia

（L.）DC. ■☆

348150 Saussurea angustifolia（Willd.）DC. subsp. yukonensis（A. E. Porsild）Cody ＝ Saussurea angustifolia（L.）DC. var. yukonensis A. E. Porsild ■☆

348151 Saussurea anochaete Hand. -Mazz. ＝ Saussurea katochaete Maxim. ■

348152 Saussurea apus Maxim. ；无梗风毛菊；Stalkless Windhairdaisy ■

348153 Saussurea apus Maxim. ＝ Saussurea pumila C. Winkl. ■

348154 Saussurea arenaria Maxim. ；沙生风毛菊；Sandy Windhairdaisy ■

348155 Saussurea aristata Lipsch. ＝ Saussurea stenolepis Nakai ■

348156 Saussurea asbukinii Iljin；阿斯布风毛菊■☆

348157 Saussurea aspera Hand. -Mazz. ＝ Saussurea odontolepis（Herder）Sch. Bip. ex Maxim. ■

348158 Saussurea aster Hemsl. ；云状雪兔子（绵毛雪莲）；Cloudy Windhairdaisy ■

348159 Saussurea atriplicifolia Fisch. ex Herder ＝ Saussurea parviflora（Poir.）DC. ■

348160 Saussurea auriculata（DC.）Sch. Bip. ；白背风毛菊；Whiteback Saussurea，Whiteback Windhairdaisy ■

348161 Saussurea auriculata Franch. ＝ Saussurea neofranchetii Lipsch. ■

348162 Saussurea auriculata Hemsl. ＝ Saussurea macrota Franch. ■

348163 Saussurea baicalensis（Adams）B. L. Rob. ；大头风毛菊；Largehead Windhairdaisy ■

348164 Saussurea baroniana Diels；棕脉风毛菊；Palmvein Windhairdaisy ■

348165 Saussurea bella Lipsch. ＝ Saussurea pulchra Lipsch. ■

348166 Saussurea bella Y. Ling；漂亮风毛菊■

348167 Saussurea blanda Schrenk；绿风毛菊（美丽风毛菊）；Green Windhairdaisy ■

348168 Saussurea bodinieri H. Lév. ＝ Saussurea pachyneura Franch. ■

348169 Saussurea bomiensis Y. L. Chen et S. Yun Liang；波密风毛菊；Bomi Windhairdaisy ■

348170 Saussurea botschantzevii Lipsch. ＝ Saussurea bella Lipsch. ■

348171 Saussurea botschantzevii Lipsch. ＝ Saussurea pulchra Lipsch. ■

348172 Saussurea brachycephala Franch. ；短头风毛菊■☆

348173 Saussurea brachylepis Hand. -Mazz. ；短苞风毛菊；Shortbract Windhairdaisy ■

348174 Saussurea bracteata Decne. ；膜苞雪莲；Filmbract Windhairdaisy ■

348175 Saussurea broussonetifolia F. H. Chen ＝ Saussurea baroniana Diels ■

348176 Saussurea brunneopilosa Hand. -Mazz. ；褐毛风毛菊（绵毛风毛菊，异色风毛菊）；Differcolor Windhairdaisy ■

348177 Saussurea brussonetiifolia Chenin ＝ Saussurea baroniana Diels ■

348178 Saussurea bullata W. W. Sm. ；泡叶风毛菊；Bullateleaf Windhairdaisy ■

348179 Saussurea bullockii Dunn；庐山风毛菊；Bullock's Saussurea，Lushan Windhairdaisy ■

348180 Saussurea caeruleo-violacea H. Lév. ＝ Saussurea chetchozensis Franch. ■

348181 Saussurea caespitosa Iljin；丛生风毛菊■☆

348182 Saussurea calobotrys Diels ＝ Saussurea baicalensis（Adams）B. L. Rob. ■

348183 Saussurea cana Ledeb. ；灰白风毛菊（宽叶西北风毛菊）；Grey Saussurea，Pale Windhairdaisy ■

348184 Saussurea cana Ledeb. var. angustifolia Ledeb. ex Handel-Mazzeti ＝ Saussurea cana Ledeb. ■

348185 Saussurea canescens C. Winkl. ；伊犁风毛菊（伊宁风毛菊）；Yining Windhairdaisy ■

348186 Saussurea carduicephala（Iljin）Iljin；飞廉风毛菊■☆

348187 Saussurea carduiformis Franch. ；蓟状风毛菊；Thistlelike Windhairdaisy ■

348188 Saussurea carduiformis Franch. var. megaphylla X. Y. Wu；大叶蓟状风毛菊；Bigleaf Thistlelike Windhairdaisy ■

348189 Saussurea carthamoides Buch. -Ham. ＝ Hemistepta lyrata（Bunge）Bunge ■

348190 Saussurea carthamoides Buch. -Ham. ex DC. ＝ Hemistepta lyrata（Bunge）Bunge ■

348191 Saussurea caudata Franch. ；尾叶风毛菊；Tailleaf Windhairdaisy ■

348192 Saussurea cauloptera Hand. -Mazz. ；翅茎风毛菊；Wingstem Windhairdaisy ■

348193 Saussurea cavaleriei H. Lév. et Vaniot ＝ Saussurea cordifolia Hemsl. ■

348194 Saussurea centiloba Hand. -Mazz. ；百裂风毛菊；Hunderedlobe Windhairdaisy ■

348195 Saussurea ceterach Hand. -Mazz. ；康定风毛菊；Kangding Windhairdaisy ■

348196 Saussurea chapmannii C. E. C. Fisch. ＝ Saussurea nepalensis Spreng. ■

348197 Saussurea chetchozensis Franch. ；大坪风毛菊（臭威灵，大叶兔耳风，绵毛风毛菊，威灵仙）；Lanate Saussurea ■

348198 Saussurea chetchozensis Franch. var. glabrescens（Hand. -Mazz.）Lipsch. ；光叶风毛菊■

348199 Saussurea chinensis（Maxim.）Lipsch. ；中华风毛菊；China Windhairdaisy，Chinese Windhairdaisy ■

348200 Saussurea chingiana Hand. -Mazz. ；抱茎风毛菊（甘青风毛菊，冷风菊）；Ching Windhairdaisy ■

348201 Saussurea chinnampoensis H. Lév. et Vaniot；京风毛菊；Chinnampo Saussurea，Chinnampo Windhairdaisy ■

348202 Saussurea chinnampoensis H. Lév. et Vaniot var. gracilis H. C. Fu et D. S. Wen；细弱京风毛菊■

348203 Saussurea chionophora Hand. -Mazz. ；显脉雪兔子；Veined Windhairdaisy ■

348204 Saussurea chionophora Hand. -Mazz. ＝ Saussurea medusa Maxim. ■

348205 Saussurea chionophora Hand. -Mazz. ＝ Saussurea wellbyi Hemsl. ■

348206 Saussurea chionophylla Takeda；雪叶风毛菊■☆

348207 Saussurea chionophylla Takeda f. albiflora Sugim. ；白花雪叶风毛菊■☆

348208 Saussurea chondrilloides C. Winkl. ；木质风毛菊；Woody Windhairdaisy ●

348209 Saussurea chowana F. H. Chen；雾灵风毛菊；Chow Windhairdaisy ■

348210 Saussurea chrysanthemoides F. H. Chen ＝ Saussurea compta Franch. ■

348211 Saussurea ciliaris Franch. ；硬叶风毛菊；Hardleaf Windhairdaisy ■

348212 Saussurea ciliaris Franch. var. major Y. Ling ＝ Saussurea ciliaris Franch. ■

348213 Saussurea cinerea Franch. ；昆仑风毛菊；Kunlun Windhairdaisy ■

348214 Saussurea cirsioides Hemsl. ＝ Saussurea likiangensis Franch. ■

348215 Saussurea cirsium H. Lév. ＝ Saussurea crispa Vaniot ■

348216 Saussurea cissioides Hemsl. ＝ Saussurea likiangensis Franch. ■

348217 Saussurea clarkeana H. Lév. ＝ Saussurea subulata C. B. Clarke ■

348218 Saussurea cochlearifolia Y. L. Chen et S. Yun Liang；匙叶风毛

菊;Spoonleaf Windhairdaisy ■

348219 Saussurea coeruleo-violacea H. Lév. = Saussurea chetchozensis Franch. ■

348220 Saussurea colpodes Y. L. Chen et S. Yun Liang;鞘基风毛菊; Sheathingbase Windhairdaisy ■

348221 Saussurea columnaris Hand.-Mazz.;柱茎风毛菊; Pole Windhairdaisy ■

348222 Saussurea compta Franch.;华美风毛菊; Gaudineess Windhairdaisy ■

348223 Saussurea congesta Turcz.;密集风毛菊■☆

348224 Saussurea conica C. B. Clarke;肿柄雪莲; Turgidstipe Windhairdaisy ■

348225 Saussurea conodasys Hand.-Mazz. = Saussurea pinetorum Hand.-Mazz. ■

348226 Saussurea controversa DC.;疑惑风毛菊■☆

348227 Saussurea conyzoides Hemsl.;假蓬风毛菊(假蓬叶风毛菊); Conyzalike Windhairdaisy ■

348228 Saussurea cordifolia Hemsl.;心叶风毛菊(马蹄细辛,山牛蒡, 水葫芦,锈毛风毛菊); Heartleaf Saussurea, Heartleaf Windhairdaisy,Rustyhair Windhairdaisy ■

348229 Saussurea cordifolia Hemsl. var. ombrophilla Hand.-Mazz. = Saussurea cordifolia Hemsl. ■

348230 Saussurea coriacea Y. L. Chen et S. Yun Liang;革苞风毛菊; Leatherbract Windhairdaisy ■

348231 Saussurea coriolepis Hand.-Mazz.;硬苞风毛菊; Hardbract Windhairdaisy ■

348232 Saussurea coronata Schrenk;副冠风毛菊■

348233 Saussurea corymbosa F. H. Chen = Saussurea paleata Maxim. ■

348234 Saussurea costus (Falc.) Lipsch. = Aucklandia lappa Decne. ■

348235 Saussurea crassifolia DC. = Saussurea salsa (Pall.) Spreng. ■

348236 Saussurea crassifolia DC. f. papposa (Turcz.) Ledeb. = Saussurea davurica Adams ■

348237 Saussurea crassifolia DC. var. papposa (Turcz. ex DC.) Ledeb. = Saussurea davurica Adams ■

348238 Saussurea crepidifolia Turcz. = Saussurea runcinata DC. ■

348239 Saussurea crispa Vaniot;小头风毛菊;Smallhead Windhairdaisy ■

348240 Saussurea dainellii Pamp. = Saussurea medusa Maxim. ■

348241 Saussurea davidii Franch. = Saussurea pectinata Bunge ex DC. ■

348242 Saussurea davidii Franch. var. macrocephala Franch. = Saussurea pectinata Bunge ex DC. ■

348243 Saussurea davurica Adams;达乌里风毛菊(毛苞风毛菊); Davur Windhairdaisy ■

348244 Saussurea dealbata Collett et Hemsl. = Saussurea peguensis C. B. Clarke ■

348245 Saussurea decurens Hemsl. = Saussurea hemsleyi Lipsch. ■

348246 Saussurea decurens Hemsl. = Saussurea parviflora (Poir.) DC. ■

348247 Saussurea decurrens Hemsl. = Saussurea hemsleyi Lipsch. ■

348248 Saussurea decurrens Hemsl. = Saussurea parviflora (Poir.) DC. ■

348249 Saussurea delavayi Franch.;大理雪兔子; Delavay Windhairdaisy ■

348250 Saussurea delavayi Franch. f. hirsuta J. Anthony;硬毛雪兔子; Hirsute Delavay Windhairdaisy ■

348251 Saussurea delavayi Hemsl. f. hirsuta J. Anthony = Saussurea delavayi Franch. ■

348252 Saussurea deltoidea (DC.) C. B. Clarke = Saussurea deltoidea (DC.) Sch. Bip. ■

348253 Saussurea deltoidea (DC.) C. B. Clarke ex Hand.-Mazz. = Saussurea deltoidea (DC.) Sch. Bip. ■

348254 Saussurea deltoidea (DC.) C. B. Clarke var. nivea Hook. f. = Saussurea crispa Vaniot

348255 Saussurea deltoidea (DC.) C. B. Clarke var. peguensis (C. B. Clarke) Hook. f. = Saussurea peguensis C. B. Clarke

348256 Saussurea deltoidea (DC.) Sch. Bip.;三角叶风毛菊(白牛 蒡,大叶防风,翻白叶,海肥干,毛叶威灵仙,台湾青木香,娃儿 草,野烟); Deltoidleaf Saussurea, Taiwan Windhairdaisy, Triangle-leaf Windhairdaisy ■

348257 Saussurea deltoidea (DC.) Sch. Bip. var. peguensis (C. B. Clarke) Hook. f. = Saussurea peguensis C. B. Clarke

348258 Saussurea deltoidea (DC.) Sch. Bip. var. polycephala C. B. Clarke = Saussurea crispa Vaniot ■

348259 Saussurea densa (Hook.) Rydb. = Saussurea nuda Ledeb. ■☆

348260 Saussurea denticulata Ledeb. = Saussurea amurensis Turcz. ex DC. ■

348261 Saussurea denticulata Ledeb. = Saussurea chinensis (Maxim.) Lipsch. ■

348262 Saussurea denticulata Ledeb. var. chinensis Y. Ling = Saussurea chinensis (Maxim.) Lipsch. ■

348263 Saussurea denticulata Wall. ex C. B. Clarke = Saussurea fastuosa (Decne.) Sch. Bip. ■

348264 Saussurea depsangensis Pamp.;昆仑雪兔子; Kunlun Windhairdaisy ■

348265 Saussurea deserticola H. C. Fu;荒漠风毛菊■

348266 Saussurea dielsiana Koidz.;狭头风毛菊; Narrowhead Windhairdaisy ■

348267 Saussurea dimorphaea Franch.;东川风毛菊; E. Sichuan Windhairdaisy ■

348268 Saussurea discolor (Willd.) DC.;异色风毛菊■☆

348269 Saussurea discolor DC. var. eriolepis Bunge ex Maxim. = Saussurea nivea Turcz. ■

348270 Saussurea discolor Diels;北风毛菊(北地风毛菊)■

348271 Saussurea discolor Diels var. nana F. H. Chen = Saussurea acromelaena Hand.-Mazz. ■

348272 Saussurea dolichopoda Diels;长梗风毛菊;Longstalk Saussurea, Longstalk Windhairdaisy ■

348273 Saussurea dorogostaiskii Palib.;多罗高风毛菊■☆

348274 Saussurea dschungdienensis Hand.-Mazz.;中甸风毛菊; Zhongdian Windhairdaisy ■

348275 Saussurea dubia Freyn;可疑风毛菊■☆

348276 Saussurea duiensis F. Schmidt = Saussurea amurensis Turcz. ex DC. subsp. duiensis (F. Schmidt) Kitam. ■☆

348277 Saussurea dumetorum J. Anthony = Saussurea uliginosa Hand.-Mazz. ■

348278 Saussurea dutaillyana Franch. = Saussurea cordifolia Hemsl. ■

348279 Saussurea dutaillyana Franch. var. shensiensis Pai ? = Saussurea sutchuenensis Franch. ■

348280 Saussurea dzeurensis Franch.;川西风毛菊; W. Sichuan Windhairdaisy,West Szechuan Saussurea ■

348281 Saussurea edulis Franch. = Dolomiaea edulis (Franch.) C. Shih ■

348282 Saussurea edulis Franch. var. berardioidea Franch. = Dolomiaea berardioides (Franch.) C. Shih ■

348283 Saussurea elata Ledeb.;高风毛菊■

348284 Saussurea elegans Ledeb.;优雅风毛菊;Elegant Windhairdaisy ■

348285 Saussurea elegans Ledeb. var. latifolia Kar. et Kir. = Saussurea elegans Ledeb. ■

348286　Saussurea elegans Ledeb. var. nivea Lipsch. = Saussurea elegans Ledeb. ■

348287　Saussurea elliptica C. B. Clarke ex Hook. f. = Saussurea ovata Benth. ■

348288　Saussurea elongata DC. var. recurvata Maxim. = Saussurea recurvata（Maxim.）Lipsch. ■

348289　Saussurea elongate DC.；伸长风毛菊■☆

348290　Saussurea enostemon Wall. = Saussurea nepalensis Spreng. ■

348291　Saussurea eopygmaea Hand. -Mazz.；矮丛风毛菊■

348292　Saussurea epilobioides Maxim.；柳叶菜风毛菊（灰毛柳叶菜风毛菊，柳兰叶风毛菊）；Valley Windhairdaisy, Willowweedleaf Saussurea, Willowweedleaf Windhairdaisy ■

348293　Saussurea epilobioides Maxim. var. cana Hand. -Mazz. = Saussurea epilobioides Maxim. ■

348294　Saussurea eriocephala Franch.；绵头风毛菊（大拇花，大木花，毛头雪莲花，棉头风毛菊，雪荷花，雪莲，雪莲花，雪兔子）；Cottonhead Windhairdaisy ■

348295　Saussurea eriocepis Bunge var. caudata Herder = Saussurea subtriangulata Kom. ■

348296　Saussurea eriolepis Bunge ex DC. = Saussurea crispa Vaniot ■

348297　Saussurea eriolepis Bunge ex DC. = Saussurea nivea Turcz. ■

348298　Saussurea eriolepis Bunge ex DC. f. paleata（Maxim.）Herder = Saussurea paleata Maxim. ■

348299　Saussurea eriolepis Bunge ex DC. var. huashanensis Y. Ling- = Saussurea huashanensis（Y. Ling）X. Y. Wu ■

348300　Saussurea eriolepis Bunge var. huashanensis Y. Ling = Saussurea huashanensis（Y. Ling）X. Y. Wu ■

348301　Saussurea eriolepis Bunge var. paleata（Maxim.）Herder = Saussurea paleata Maxim. ■

348302　Saussurea eriophylla Nakai f. alpina Nakai = Saussurea tomentosa Kom. ■

348303　Saussurea eriostemon Wall. ex C. B. Clarke = Saussurea nepalensis Spreng. ■

348304　Saussurea eriphylla Nakai var. alpina Nakai = Saussurea tomentosa Kom. ■

348305　Saussurea erubescens Lipsch.；红柄雪莲（变红风毛菊）；Redstipe Windhairdaisy ■

348306　Saussurea euodonta Diels；锐齿风毛菊；Tinetooth Windhairdaisy ■

348307　Saussurea falconeri Hook. f. = Saussurea andryaloides（DC.）Sch. Bip. ■

348308　Saussurea famintziniana Krasn.；中新风毛菊■☆

348309　Saussurea fargesii Franch.；川东风毛菊；E. Sichuan Windhairdaisy, Farges Windhairdaisy ■

348310　Saussurea fastuosa（Decne.）Benth. et Hook. f. ex Klatt = Saussurea fastuosa（Decne.）Sch. Bip. ■

348311　Saussurea fastuosa（Decne.）Hand. -Mazz. = Saussurea fastuosa（Decne.）Sch. Bip. ■

348312　Saussurea fastuosa（Decne.）Sch. Bip.；奇形风毛菊；Queer Windhairdaisy, Wonderful Saussurea ■

348313　Saussurea fauriei Franch.；法氏风毛菊■☆

348314　Saussurea filifolia Regel et Schmalh. = Pilostemon filifolia（C. Winkl.）Iljin ■

348315　Saussurea firma（Kitag.）Kitag. = Saussurea ussuriensis Maxim. var. firma Kitag. ■

348316　Saussurea firma Kitam. = Saussurea ussuriensis Maxim. var. firma Kitag. ■

348317　Saussurea fistulosa J. Anthony；管茎雪兔子；Tube-stem Windhairdaisy ■

348318　Saussurea flaccida Y. Ling；萎软风毛菊（纤细风毛菊）；Weak Windhairdaisy ■

348319　Saussurea flavo-virens Y. L. Chen et S. Yun Liang；黄绿苞风毛菊（黄绿风毛菊）；Yellowgreenbract Windhairdaisy ■

348320　Saussurea flexuosa Franch.；城口风毛菊；Chengkou Windhairdaisy ■

348321　Saussurea flexuosa Franch. var. penicillata Franch. = Saussurea tsinlingensis Hand. -Mazz. ■

348322　Saussurea foliosa Ledeb.；繁叶风毛菊■☆

348323　Saussurea formosana Hayata；台湾青木香■

348324　Saussurea formosana Hayata = Saussurea deltoidea（DC.）Sch. Bip. ■

348325　Saussurea forrestii Diels = Saussurea fastuosa（Decne.）Sch. Bip. ■

348326　Saussurea franchetiana H. Lév. = Saussurea leucoma Diels ■

348327　Saussurea franchetii Koidz.；弗氏风毛菊■☆

348328　Saussurea frolovii Ledeb.；大序风毛菊（福罗风毛菊）；Frolov Windhairdaisy ■

348329　Saussurea frondosa Hand. -Mazz.；狭翼风毛菊（窄翼风毛菊）；Narrowwing Saussurea, Narrowwing Windhairdaisy ■

348330　Saussurea fruticosa Kar. et Kir. = Saussurea cana Ledeb. ■

348331　Saussurea georgei J. Anthony；川滇雪兔子；Chuandian Windhairdaisy ■

348332　Saussurea giraldii Diels = Saussurea likiangensis Franch. ■

348333　Saussurea glabrata（DC.）C. Shih；无毛叶风毛菊；Glabrous Windhairdaisy ■

348334　Saussurea glacialis Herder；冰川雪兔子（冰河雪兔子）；Glacial Windhairdaisy ■

348335　Saussurea glanduligera Sch. Bip. ex Hook. f.；腺毛风毛菊；Glandhair Windhairdaisy ■

348336　Saussurea glandulosa Kitam.；腺点风毛菊（高山青木香，腺毛青木香）；Glandular Windhairdaisy ■

348337　Saussurea globosa F. H. Chen；球花雪莲（球花风毛菊）；Ballflower Windhairdaisy, Globularflower Saussurea ■

348338　Saussurea glomerata Poir. = Saussurea amara（L.）DC. var. glomerata（Poir.）Trautv. ■☆

348339　Saussurea glomerata Poir. = Saussurea amara（L.）DC. ■

348340　Saussurea glomerata Poir. var. chinensis F. H. Chen = Saussurea amara（L.）DC. ■

348341　Saussurea glomerata Poir. var. chinensis F. H. Chen f. alata F. H. Chen = Saussurea amara（L.）DC. ■

348342　Saussurea gnaphaloides（Royle）Sch. Bip.；鼠曲雪兔子（鼠曲风毛菊）；Cudweed-like Saussurea, Cudweed-like Windhairdaisy ■

348343　Saussurea gossypina Wall. = Saussurea gossypiphora D. Don ■

348344　Saussurea gossypiphora D. Don；雪兔子（白絮风毛菊，棉青木香）；Cotton Saussurea, Snowrabbiten ■

348345　Saussurea gossypiphora D. Don = Saussurea laniceps Hand. -Mazz. ■

348346　Saussurea gossypiphora D. Don = Saussurea leucoma Diels ■

348347　Saussurea gossypiphora D. Don var. conaensis S. W. Liu；错那雪兔子；Cuona Windhairdaisy ■

348348　Saussurea graciliformis Lipsch.；拟纤细风毛菊；Fine Windhairdaisy ■

348349　Saussurea gracilis Maxim.；纤细风毛菊■☆

348350　Saussurea graminea Dunn；禾叶风毛菊（线叶风毛菊）；Grassleaf Saussurea, Grassleaf Windhairdaisy ■

348351 Saussurea graminea Dunn var. ortholepis Hand. -Mazz. = Saussurea graminea Dunn ■

348352 Saussurea graminicola F. H. Chen = Saussurea wardii J. Anthony ■

348353 Saussurea graminifolia Wall. ex DC. ;密毛风毛菊;Densehair Windhairdaisy ■

348354 Saussurea grandiceps S. W. Liu;硕首雪兔子;Bighead Windhairdaisy ■

348355 Saussurea grandifolia Maxim. ;大花风毛菊(大叶风毛菊); Largeleaf Windhairdaisy ■

348356 Saussurea grandifolia Maxim. var. asperifolia Herder = Saussurea grandifolia Maxim. ■

348357 Saussurea grandifolia Maxim. var. caudata (Herder) Kom. = Saussurea subtriangulata Kom. ■

348358 Saussurea grandifolia Maxim. var. coarctata Herder = Saussurea grandifolia Maxim. ■

348359 Saussurea grandifolia Maxim. var. tenuior Herder = Saussurea grandifolia Maxim. ■

348360 Saussurea grosseserrata Franch. ;粗裂风毛菊;Large-serrete Windhairdaisy ■

348361 Saussurea grum-grshimailoi C. Winkl. ex Bretschneider = Saussurea sylvatica Maxim. ■

348362 Saussurea gyacaensis S. W. Liu;加查雪兔子;Jiacha Windhairdaisy ■

348363 Saussurea hakoensis Franch. et Sav. = Saussurea maximowiczii Herder ■

348364 Saussurea handeliana Y. Ling = Saussurea leptolepis Hand. - Mazz. ■

348365 Saussurea haoi Y. Ling ex Y. L. Chen et S. Yun Liang;青藏风毛菊;Hao Windhairdaisy ■

348366 Saussurea hegueensis C. B. Clarke;鹰爪莲■

348367 Saussurea hemsleyana Hand. -Mazz. = Saussurea macrota Franch. ■

348368 Saussurea hemsleyi Lipsch. ;湖北风毛菊;Hubei Windhairdaisy ■

348369 Saussurea henryi Hemsl. ;巴东风毛菊;Badong Windhairdaisy ■

348370 Saussurea hieracioides Hook. f. ;长毛风毛菊(华丽风毛菊,漏子多吾,美丽风毛菊);Hawkweed-like Saussurea, Longhair Windhairdaisy ■

348371 Saussurea higomontana Honda = Saussurea nipponica Miq. subsp. savatieri (Franch.) Kitam. var. higomontana (Honda) H. Koyama ■☆

348372 Saussurea hirsuta (Anthony) Hand. -Mazz. = Saussurea delavayi Franch. f. hirsuta J. Anthony ■

348373 Saussurea hisauchii Nakai;久内氏风毛菊■☆

348374 Saussurea hisauchii Nakai = Saussurea triptera Maxim. var. hisauchii (Nakai) Kitam. ■☆

348375 Saussurea hsiaowutaishanensis F. H. Chen = Saussurea sylvatica Maxim. var. hsiaowutaishanensis (F. H. Chen) Lipsch. ■

348376 Saussurea huashanensis (Y. Ling) X. Y. Wu;华山风毛菊;Huashan Windhairdaisy ■

348377 Saussurea hultenii Lipsch. ;雅龙江风毛菊;Yalongjiang Windhairdaisy ■

348378 Saussurea humilis Ostenf. = Saussurea apus Maxim. ■

348379 Saussurea hwangshanensis Y. Ling;黄山风毛菊;Huangshan Windhairdaisy ■

348380 Saussurea hylophila Hand. -Mazz. = Saussurea stricta Franch. ■

348381 Saussurea hypargyrea Lipsch. et Vved. ;下银风毛菊■☆

348382 Saussurea hyperiophora Hand. -Mazz. = Saussurea wellbyi Hemsl. ■

348383 Saussurea hypoleuca Spreng. ex DC. = Saussurea auriculata (DC.) Sch. Bip. ■

348384 Saussurea hypsipeta Diels;黑毛雪兔子;Blackhair Windhairdaisy ■

348385 Saussurea inaensis Kitam. ;伊那风毛菊■☆

348386 Saussurea incisa F. H. Chen;锐裂风毛菊;Tinelobe Windhairdaisy ■

348387 Saussurea inconspicua Hand. -Mazz. = Saussurea leptolepis Hand. -Mazz. ■

348388 Saussurea insularis Kitam. ;海岛风毛菊■☆

348389 Saussurea integrifolia Hand. -Mazz. ;全缘叶风毛菊(全缘风毛菊);Entireleaf Windhairdaisy ■

348390 Saussurea intermedia Turcz. = Saussurea japonica (Thunb.) DC. ■

348391 Saussurea involucrata (Kar. et Kir.) Maxim. = Saussurea involucrata (Kar. et Kir.) Sch. Bip. ■

348392 Saussurea involucrata (Kar. et Kir.) Sch. Bip. ;雪莲花(大苞雪莲花,大拇花,大木花,荷莲,天山雪莲花,新疆雪莲,新疆雪莲花,雪荷,雪荷花,雪莲);Snow Lotus,Snowlotus ■

348393 Saussurea involucrata (Kar. et Kir.) Sch. Bip. var. axillicalathina J. S. Li;腋序雪莲■

348394 Saussurea iodoleuca Hand. -Mazz. ;滇川风毛菊;Dian-chuan Windhairdaisy ■

348395 Saussurea iodostegia Hance;紫苞雪莲(紫苞风毛菊); Purplebract Saussurea,Purplebract Windhairdaisy ■

348396 Saussurea iodostegia Hance var. ferruginipes Drumm. ex Hand. -Mazz. ;锈色雪莲;Rust Purplebract Saussurea ■

348397 Saussurea iodostegia Hance var. glandulifera X. Y. Wu;腺毛紫苞风毛菊(腺毛风毛菊);Glanduiferous Purplebract Saussurea ■

348398 Saussurea ionodasys Hand. -Mazz. = Saussurea pinetorum Hand. -Mazz. ■

348399 Saussurea irregularis Y. L. Chen et S. Yun Liang;异裂风毛菊;Differsplit Windhairdaisy ■

348400 Saussurea ischnoides J. S. Li;狭苞雪莲;Naroowbract Windhairdaisy ■

348401 Saussurea ispajensis Iljin;伊斯帕伊风毛菊■☆

348402 Saussurea jadrinzevii Krylov;雅氏风毛菊■☆

348403 Saussurea japonica (Thunb.) DC. ;风毛菊(八棱麻,八楞麻,八楞木,八面风,风毛菊花,风毛菊,青竹标,日本风毛菊,三草,山苦子);Japanese Saussurea,Japanese Windhairdaisy ■

348404 Saussurea japonica (Thunb.) DC. = Saussurea pulchella (Fisch. ex Hornem.) Fisch. ■

348405 Saussurea japonica (Thunb.) DC. f. leucocephala (Nakai et Kitag.) Nakai et Kitag. = Saussurea amara (L.) DC. ■

348406 Saussurea japonica (Thunb.) DC. f. leucocephala (Nakai et Kitag.) Nakai et Kitag. ;白头风毛菊■☆

348407 Saussurea japonica (Thunb.) DC. var. alata (F. H. Chen) Nakai et Kitag. = Saussurea amara (L.) DC. ■

348408 Saussurea japonica (Thunb.) DC. var. alata Regel = Saussurea amara (L.) DC. ■

348409 Saussurea japonica (Thunb.) DC. var. dentata Kom. ;齿叶风毛菊;Toothedleaf Japanese Windhairdaisy ■

348410 Saussurea japonica (Thunb.) DC. var. lineariloba Nakai;细叶风毛菊;Linearleaf Japanese Windhairdaisy ■

348411 Saussurea japonica (Thunb.) DC. var. longicephala Hayata;长头风毛菊;Longhead Japanese Windhairdaisy ■

348412　Saussurea japonica（Thunb.）DC. var. longicephala Hayata ＝ Saussurea japonica（Thunb.）DC. ■

348413　Saussurea japonica（Thunb.）DC. var. maritima Kitag. ＝ Saussurea chingiana Hand. -Mazz. ■

348414　Saussurea japonica（Thunb.）DC. var. maritima Kitag. ＝ Saussurea japonica（Thunb.）DC. ■

348415　Saussurea japonica（Thunb.）DC. var. subintegra（Regel）Kom. ；全叶风毛菊(绿叶风毛菊)■

348416　Saussurea jurineioides H. C. Fu;阿右风毛菊■

348417　Saussurea kaialpina Nakai ＝ Saussurea triptera Maxim. var. minor（Takeda）Kitam. ■☆

348418　Saussurea kansuensis Hand. -Mazz. ；甘肃风毛菊；Gansu Windhairdaisy, Kansu Saussurea ■

348419　Saussurea kanzanensis Kitam. ；台湾风毛菊(关山青木香)；Taiwan Windhairdaisy ■

348420　Saussurea karaartscha Saposhn. ；卡拉阿尔恰风毛菊■☆

348421　Saussurea karateginii Lipsky ＝ Pilostemon karateginii（Lipsky）Iljin ■

348422　Saussurea karelinii Stschegl. ＝ Saussurea involucrata（Kar. et Kir.）Sch. Bip. ■

348423　Saussurea karlongensis Hand. -Mazz. ＝ Saussurea epilobioides Maxim. ■

348424　Saussurea kaschgarica Rupr. ；喀什风毛菊；Kaschgar Windhairdaisy ■

348425　Saussurea katochaete Maxim. ；重齿风毛菊(大通风毛菊，蓝紫风毛菊，细齿风毛菊)；Double-serrate Windhairdaisy ■

348426　Saussurea katochaetoides Hand. -Mazz. ；类风毛菊(风毛菊)■

348427　Saussurea katochaetoides Hand. -Mazz. ＝ Saussurea katochaete Maxim. ■

348428　Saussurea kingii J. R. Drumm. ex C. E. C. Fisch. ；拉萨雪兔子(拉萨风毛菊)；King's Saussurea, Lasa Windhairdaisy ■

348429　Saussurea kingii J. R. Drumm. ex Hand. -Mazz. ＝ Saussurea kingii J. R. Drumm. ex C. E. C. Fisch. ■

348430　Saussurea kiraisanensis Masam. ；台岛风毛菊(奇莱青木香)；Kirais Windhairdaisy ■

348431　Saussurea kiraisanensis Masam. ex Kitam. ＝ Saussurea kiraisanensis Masam. ■

348432　Saussurea kirigaminensis Kitam. ＝Saussurea modesta Kitam. ■☆

348433　Saussurea kitamurae S. Y. Hu ＝Saussurea macrota Franch. ■

348434　Saussurea kitamurana Miyabe et Tatew. ；北村氏风毛菊■☆

348435　Saussurea kiusiana Franch. ＝Saussurea nipponica Miq. ■☆

348436　Saussurea kokonorensis Y. Ling ＝ Saussurea subulisquama Hand. -Mazz. ■

348437　Saussurea komamitzkii Lipsch. ；腋头风毛菊；Komarnitzk Windhairdaisy ■

348438　Saussurea konuroba Saposhn. ＝ Saussurea blanda Schrenk ■

348439　Saussurea koslowii C. Winkl. ＝ Saussurea apus Maxim. ■

348440　Saussurea kouytcheensis H. Lév. ＝ Saussurea deltoidea（DC.）Sch. Bip. ■

348441　Saussurea krylovii Schischk. et Serg. ；阿尔泰风毛菊■

348442　Saussurea kudoana Tatew. et Kitam. ；久藤风毛菊■☆

348443　Saussurea kudoana Tatew. et Kitam. var. uryuensis Kadota;雨龙风毛菊■☆

348444　Saussurea kungii Y. Ling ；洋县风毛菊；Yangxian Windhairdaisy ■

348445　Saussurea kunthiana C. B. Clarke ＝ Saussurea leontodontoides（DC.）Sch. Bip. ■

348446　Saussurea kunthiana C. B. Clarke ＝ Saussurea polypodioides J. Anthony ■

348447　Saussurea kunthiana C. B. Clarke var. caulescens Kitam. ＝ Saussurea pachyneura Franch. ■

348448　Saussurea kunthiana C. B. Clarke var. filicifolia Hook. f. ＝ Saussurea leontodontoides（DC.）Sch. Bip. ■

348449　Saussurea kunthiana C. B. Clarke var. major Hook. f. ＝ Saussurea pachyneura Franch. ■

348450　Saussurea kunthiana C. B. Clarke Wall. ex C. B. Clarke ＝ Saussurea pachyneura Franch. ■

348451　Saussurea kunthiana Wall. ex C. B. Clarke var. caulescens Kitam. ＝ Saussurea pachyneura Franch. ■

348452　Saussurea kurilensis Tatew. ；库里尔风毛菊■☆

348453　Saussurea kurosawae Kitam. ＝ Saussurea nipponica Miq. subsp. savatieri（Franch.）Kitam. ■☆

348454　Saussurea kuschakeviczii C. Winkl. ；库夏风毛菊■☆

348455　Saussurea kwangtungensis F. H. Chen ＝ Saussurea bullockii Dunn ■

348456　Saussurea laciniata Ledeb. ；裂叶风毛菊（翼茎风毛菊）；Wingstem Windhairdaisy ■

348457　Saussurea laciniata Ledeb. var. pygmaea Lipsch. ＝ Saussurea laciniata Ledeb. ■

348458　Saussurea lacostei Danguy；高盐地风毛菊；Lacoste Windhairdaisy ■

348459　Saussurea ladyginii Lipsch. ；拉氏风毛菊；Ladygin Windhairdaisy ■

348460　Saussurea lamprocarpa Hemsl. ＝ Saussurea deltoidea（DC.）Sch. Bip. ■

348461　Saussurea lampsanifolia Franch. ；鹤庆风毛菊；Heqing Windhairdaisy ■

348462　Saussurea lanata Y. L. Chen et S. Yun Liang；白毛风毛菊；White Windhairdaisy ■

348463　Saussurea lancifolia Hand. -Mazz. ＝Saussurea minuta C. Winkl. ■

348464　Saussurea lanicaulis Hand. -Mazz. ＝Saussurea graminea Dunn ■

348465　Saussurea laniceps Hand. -Mazz. ；绵头雪兔子(大拇花，大木花，大雪兔子，锦头雪莲花，绵毛雪莲，绵头雪莲花，雪荷花，雪莲，雪莲花，雪兔子)；Cottonhead Windhairdaisy, Lanatehead Saussurea ■

348466　Saussurea lanuginosa Vaniot;绵毛风毛菊■

348467　Saussurea lanuginosa Vaniot ＝Saussurea chetchozensis Franch. ■

348468　Saussurea lanuginosa Vaniot var. glabrescens Hand. -Mazz. ＝ Saussurea chetchozensis Franch. var. glabrescens（Hand. -Mazz.）Lipsch. ■

348469　Saussurea lappa（Decne.）C. B. Clarke ＝ Aucklandia lappa Decne. ■

348470　Saussurea lappa（Decne.）C. B. Clarke ＝ Saussurea costus（Falc.）Lipsch. ■

348471　Saussurea lappa（Decne.）Sch. Bip. ＝ Saussurea costus（Falc.）Lipsch. ■

348472　Saussurea lappa（Falc.）Sch. Bip. ＝ Aucklandia lappa Decne. ■

348473　Saussurea lappa（Falc.）Sch. Bip. ＝ Saussurea costus（Falc.）Lipsch. ■

348474　Saussurea larionowii C. Winkl. ；天山风毛菊；Tianshan Windhairdaisy ■

348475　Saussurea latifolia Ledeb. ；宽叶风毛菊■

348476　Saussurea lavrenkoana Lipsch. ；双齿风毛菊；Lavrencko Windhairdaisy ■

348477　Saussurea leclerei H. Lév. ；利马川风毛菊(利马风毛菊)；

Leclere Windhairdaisy ■

348478　Saussurea leiocarpa Hand. -Mazz. = Saussurea stoliczkae C. B. Clarke ■

348479　Saussurea lenensis Popov. ex Lipsch.；勒拿风毛菊■☆

348480　Saussurea leontodon Dunn = Saussurea scabrida Franch. ■

348481　Saussurea leontodontoides（DC.）Sch. Bip.；狮牙草状风毛菊（松潘风毛菊）；Edelweisslike Windhairdaisy ■

348482　Saussurea leontodontoides（DC.）Sch. Bip. var. filicifolia（Hook. f.）Hand. -Mazz. = Saussurea leontodontoides（DC.）Sch. Bip. ■

348483　Saussurea leontodontoies（DC.）Hand. -Mazz. = Saussurea leontodontoides（DC.）Sch. Bip. ■

348484　Saussurea leontopodium H. Lév. = Saussurea peguensis C. B. Clarke ■

348485　Saussurea leontopodium H. Lév. et Vaniot = Saussurea peguensis C. B. Clarke ■

348486　Saussurea leptolepis Hand. -Mazz.；薄苞风毛菊（毛苞雪莲）；Thinbract Windhairdaisy ■

348487　Saussurea leucoma Diels；羽裂雪兔子（白毛雪兔子，瞎果羔贝，雪兔，雪兔子，羽裂雪莲）；Pinnate Windhairdaisy ■

348488　Saussurea leucophylla Schrenk；白叶风毛菊；Whiteleaf Windhairdaisy ■

348489　Saussurea leucota Hand. -Mazz. = Saussurea salicifolia（L.）DC. ■

348490　Saussurea levdillei F. H. Chen = Saussurea centiloba Hand. -Mazz. ■

348491　Saussurea licentiana Hand. -Mazz.；川陕风毛菊；Chuanshan Windhairdaisy，Licent's Saussurea ■

348492　Saussurea likiangensis Franch.；丽江风毛菊；Lijiang Windhairdaisy，Likiang Saussurea ■

348493　Saussurea likiangensis Franch. var. integrifolia Hand. -Mazz. = Saussurea likiangensis Franch. ■

348494　Saussurea likiangensis Franch. var. siningensis Hand. -Mazz. = Saussurea likiangensis Franch. ■

348495　Saussurea limprichtii Diels；巴塘风毛菊；Limpricht Windhairdaisy ■

348496　Saussurea linearis Champ. et Benth. = Saussurea japonica（Thunb.）DC. ■

348497　Saussurea lingulata Franch.；小舌风毛菊；Lingulate Windhairdaisy ■

348498　Saussurea lioui Y. Ling = Saussurea involucrata（Kar. et Kir.）Sch. Bip. ■

348499　Saussurea lomatolepis Lipsch.；纹苞风毛菊（苞鳞风毛菊）；Tasselscale Windhairdaisy ■

348500　Saussurea longifolia Franch.；长叶雪莲（长叶风毛菊）；Longleaf Saussurea，Longleaf Windhairdaisy ■

348501　Saussurea loriformis W. W. Sm.；带叶风毛菊；Beltleaf Windhairdaisy ■

348502　Saussurea lunzhubensis Y. L. Chen et S. Yun Liang；林周风毛菊；Linzhou Windhairdaisy ■

348503　Saussurea lyrata（Bunge）Franch. = Hemistepta lyrata（Bunge）Bunge ■

348504　Saussurea lyratifolia Y. L. Chen et S. Yun Liang；大头羽裂风毛菊；Lyrateleaf Windhairdaisy ■

348505　Saussurea macrolepis（Nakai）Kitam.；大鳞风毛菊■☆

348506　Saussurea macrota Franch.；大耳叶风毛菊；Largeear Windhairdaisy ■

348507　Saussurea mai H. C. Fu；毓泉风毛菊；Ma's Windhairdaisy ■

348508　Saussurea mairei H. Lév. = Saussurea yunnanensis Franch. ■

348509　Saussurea malitiosa Maxim.；尖头风毛菊（柴达木风毛菊）；Evil Windhairdaisy ■

348510　Saussurea manshurica Kom.；东北风毛菊；Manchurian Saussurea，NE. China Windhairdaisy ■

348511　Saussurea manshurica Kom. = Saussurea triangulata Trautv. et C. A. Mey. ■☆

348512　Saussurea manshurica Kom. var. pinnatifida Nakai = Saussurea manshurica Kom. ■

348513　Saussurea masarica Lipsky；马萨尔风毛菊■☆

348514　Saussurea matsumurae Nakai = Saussurea mongolica（Franch.）Franch. ■

348515　Saussurea maximowiczii Herder；羽叶风毛菊；Maximowicz Windhairdaisy ■

348516　Saussurea maximowiczii Herder f. serrata（Nakai）Kitam.；具齿羽叶风毛菊■☆

348517　Saussurea maximowiczii Herder f. serrata（Nakai）Kitam. = Saussurea maximowiczii Herder ■

348518　Saussurea maximowiczii Herder var. platyphylla Makino = Saussurea maximowiczii Herder f. serrata（Nakai）Kitam. ■☆

348519　Saussurea maximowiczii Herder var. serrata Nakai = Saussurea maximowiczii Herder ■

348520　Saussurea maximowiczii Herder var. triceps（H. Lév. et Vaniot）Kitam. = Saussurea maximowiczii Herder ■

348521　Saussurea medusa Maxim.；水母雪兔子（甘青雪莲花，水母莲，水田雪莲花，雪莲，雪莲花）；Medusa Saussurea，Medusa Windhairdaisy ■

348522　Saussurea melanotricha Hand. -Mazz.；黑苞风毛菊；Blackbract Windhairdaisy ■

348523　Saussurea merinoi H. Lév.；截叶风毛菊；Cuneateleaf Windhairdaisy ■

348524　Saussurea micradenia Hand. -Mazz.；滇风毛菊；Yunnan Windhairdaisy ■

348525　Saussurea microcephala C. A. Mey. = Saussurea cana Ledeb. ■

348526　Saussurea microcephala Diels = Saussurea dielsiana Koidz. ■

348527　Saussurea microcephala Franch. = Saussurea amara（L.）DC. ■

348528　Saussurea microcephala Franch. ex Hemsl. = Saussurea amara（L.）DC. ■

348529　Saussurea microcephala Franch. var. aptera Nakai et Kitag. = Saussurea amara（L.）DC. ■

348530　Saussurea microcephala Franch. var. aptera Nakai et Kitag. f. leucocephala Nakai et Kitag. = Saussurea amara（L.）DC. ■

348531　Saussurea microcephala Franch. var. pteroclada Nakai et Kitag. = Saussurea amara（L.）DC. ■

348532　Saussurea microdeltoidea Kitam. = Saussurea crispa Vaniot ■

348533　Saussurea mikeschinii Iljin；米凯风毛菊■☆

348534　Saussurea minuta C. Winkl.；小风毛菊（披针叶风毛菊）；Small Windhairdaisy ■

348535　Saussurea modesta Hand. -Mazz. = Saussurea leptolepis Hand. -Mazz. ■

348536　Saussurea modesta Kitam.；适度风毛菊■☆

348537　Saussurea mollis Franch. = Saussurea leclerei H. Lév. ■

348538　Saussurea mongolica（Franch.）Franch.；蒙古风毛菊（华北风毛菊）；Mongol Windhairdaisy，Mongolian Saussurea ■

348539　Saussurea mongolica（Franch.）Franch. f. shansiensis（F. H. Chen）Y. Ling = Saussurea mongolica（Franch.）Franch. ■

348540　Saussurea mongolica（Franch.）Franch. var. viridior Hand. -Mazz. = Saussurea mongolica（Franch.）Franch. ■

348541　Saussurea montana J. Anthony；山地风毛菊；Montane Windhairdaisy ■

348542　Saussurea morifolia F. H. Chen；桑叶风毛菊；Mulberryleaf Windhairdaisy ■

348543　Saussurea mucronulata Lipsch.；小尖风毛菊（尖苞风毛菊）；Mucronulate Windhairdaisy ■

348544　Saussurea muliensis Hand. -Mazz.；木里雪莲；Muli Windhairdaisy ■

348545　Saussurea multiflora（L.）DC. = Saussurea salicifolia（L.）DC. ■

348546　Saussurea muramatsui Kitam. = Saussurea sugimurae Honda ■☆

348547　Saussurea mutabilis Diels；变叶风毛菊；Changeable Windhairdaisy ■

348548　Saussurea mutabilis Diels var. diplochaeta Ling = Saussurea mutabilis Diels ■

348549　Saussurea namikawae Kitam. = Saussurea medusa Maxim. ■

348550　Saussurea neglecta Ludlow = Saussurea abnormis Lipsch. ■

348551　Saussurea nematolepis Y. Ling；钻状风毛菊；Awlshape Windhairdaisy ■

348552　Saussurea neofranchetii Lipsch.；耳叶风毛菊；Earleaf Saussurea, Earleaf Windhairdaisy ■

348553　Saussurea neopulchella Lipsch.；新美丽风毛菊■☆

348554　Saussurea neoserrata Nakai；新齿叶风毛菊；Serrateleaf Windhairdaisy ■

348555　Saussurea nepalensis Spreng.；尼泊尔风毛菊；Nepal Windhairdaisy ■

348556　Saussurea nidulans Hand. -Mazz.；鸟巢状雪莲；Nest Windhairdaisy ■

348557　Saussurea nigrescens Maxim.；钝苞雪莲（黑紫风毛菊，瑞苓草，紫苞风毛菊）；Blacken Saussurea, Obtusebract Windhairdaisy ■

348558　Saussurea nigrescens Maxim. var. acutisquama Y. Ling = Saussurea polycolea Hand. -Mazz. var. acutisquama（Y. Ling）Lipsch. ■

348559　Saussurea nikoensis Franch. et Sav.；日光雪莲■☆

348560　Saussurea nikoensis Franch. et Sav. var. involucrata（Matsum. et Koidz.）Kitam.；总苞日光雪莲■☆

348561　Saussurea nikoensis Franch. et Sav. var. sessiliflora（Koidz.）Kitam.；无柄花日光雪莲■☆

348562　Saussurea nimborum W. W. Sm.；倒披针叶风毛菊；Oblanceotaeleaf Windhairdaisy ■

348563　Saussurea ninae Iljin；尼娜风毛菊■☆

348564　Saussurea nipponica Miq.；本州风毛菊■☆

348565　Saussurea nipponica Miq. subsp. higomontana（Honda）H. T. Im = Saussurea nipponica Miq. subsp. savatieri（Franch.）Kitam. var. higomontana（Honda）H. Koyama ■☆

348566　Saussurea nipponica Miq. subsp. hokurokuensis Kitam. = Lecanorchis hokurikuensis Masam. ■☆

348567　Saussurea nipponica Miq. subsp. kurosawae（Kitam.）Kitam. = Saussurea nipponica Miq. subsp. savatieri（Franch.）Kitam. ■☆

348568　Saussurea nipponica Miq. subsp. muramatsui（Kitam.）Kitam. = Saussurea sugimurae Honda ■☆

348569　Saussurea nipponica Miq. subsp. savatieri（Franch.）Kitam.；萨瓦捷风毛菊■☆

348570　Saussurea nipponica Miq. subsp. savatieri（Franch.）Kitam. var. higomontana（Honda）H. Koyama；肥厚山风毛菊■☆

348571　Saussurea nipponica Miq. subsp. savatieri（Franch.）Kitam. var. robusta（Makino）Ohwi ex Lipsch.；粗壮萨瓦捷风毛菊■☆

348572　Saussurea nipponica Miq. subsp. savatieri（Franch.）Kitam. var. yakusimensis（Masam.）H. Koyama；屋久岛风毛菊■☆

348573　Saussurea nipponica Miq. subsp. sendaica（Franch.）= Saussurea nipponica Miq. subsp. savatieri（Franch.）Kitam. ■☆

348574　Saussurea nipponica Miq. subsp. sikokiana（Makino）Kitam. = Saussurea nipponica Miq. subsp. savatieri（Franch.）Kitam. var. robusta（Makino）Ohwi ex Lipsch. ■☆

348575　Saussurea nipponica Miq. subsp. yakusimensis（Masam.）H. T. Im = Saussurea nipponica Miq. subsp. savatieri（Franch.）Kitam. var. yakusimensis（Masam.）H. Koyama ■☆

348576　Saussurea nipponica Miq. subsp. yoshinagae（Kitam.）Kitam. = Saussurea nipponica Miq. var. yoshinagae（Kitam.）H. Koyama ■☆

348577　Saussurea nipponica Miq. var. glabrescens（Nakai）Kitam. ex Lipsch. = Saussurea sinuatoides Nakai var. glabrescens Nakai ■☆

348578　Saussurea nipponica Miq. var. hokurokuensis（Kitam.）Ohwi；北陆风毛菊■☆

348579　Saussurea nipponica Miq. var. kurosawae（Kitam.）Ohwi = Saussurea nipponica Miq. subsp. savatieri（Franch.）Kitam. ■☆

348580　Saussurea nipponica Miq. var. muramatsui（Kitam.）Ohwi = Saussurea sugimurae Honda ■☆

348581　Saussurea nipponica Miq. var. sendaica（Franch.）Ohwi = Saussurea nipponica Miq. subsp. savatieri（Franch.）Kitam. ■☆

348582　Saussurea nipponica Miq. var. sikokiana（Makino）Ohwi = Saussurea nipponica Miq. subsp. savatieri（Franch.）Kitam. var. robusta（Makino）Ohwi ex Lipsch. ■☆

348583　Saussurea nipponica Miq. var. yoshinagae（Kitam.）H. Koyama；吉永风毛菊■☆

348584　Saussurea nivea（DC.）Sch. Bip. = Saussurea crispa Vaniot ■

348585　Saussurea nivea Turcz.；银背风毛菊（华北风毛菊，羊耳白背）；Silverback Saussurea, Silverback Windhairdaisy ■

348586　Saussurea nivea Turcz. var. huashanensis（Y. Ling）S. Y. Hu = Saussurea huashanensis（Y. Ling）X. Y. Wu ■

348587　Saussurea nivea Turcz. var. nana（F. H. Chen）Hand. -Mazz. = Saussurea acromelaena Hand. -Mazz. ■

348588　Saussurea nobilis Franch. = Saussurea woodiana Hemsl. ■

348589　Saussurea nodesta Hand. -Mazz. = Saussurea leptolepis Hand. -Mazz. ■

348590　Saussurea nuda Ledeb.；裸风毛菊；Chaffless Saw-wort, Dwarf Saw-wort ■☆

348591　Saussurea nuda Ledeb. subsp. densa（Hook.）G. W. Douglas = Saussurea nuda Ledeb. ■☆

348592　Saussurea nuda Ledeb. var. densa（Hook.）Hultén = Saussurea nuda Ledeb. ■☆

348593　Saussurea nupuripoensis Miyabe et Miyake；努普里普风毛菊■☆

348594　Saussurea nyalamensis Y. L. Chen et S. Yun Liang；聂拉木风毛菊；Nielamu Windhairdaisy ■

348595　Saussurea oblongifolia F. H. Chen；长圆叶风毛菊；Oblongleaf Windhairdaisy ■

348596　Saussurea obvallata（DC.）Edgew.；苞叶雪莲（苞叶风毛菊，苞叶风毛菊，苞叶雪莲花，紫苞风毛菊）；Bractleaf Saussurea, Bractleaf Windhairdaisy ■

348597　Saussurea obvallata（DC.）Edgew. var. gymnocephala Y. Ling = Saussurea erubescens Lipsch. ■

348598　Saussurea obvallata（DC.）Edgew. var. orientalis Diels = Saussurea tangutica Maxim. ■

348599　Saussurea obvallata (DC.) Sch. Bip. = Saussurea obvallata (DC.) Edgew. ■

348600　Saussurea obvallata Nakai var. gymnocephala Y. Ling = Saussurea erubescens Lipsch. ■

348601　Saussurea obvallata Nakai var. orientalis Diels = Saussurea tangutica Maxim. ■

348602　Saussurea obvallata Wall. = Saussurea obvallata (DC.) Edgew. ■

348603　Saussurea ochrochlaena Hand.-Mazz.；褐黄色风毛菊；Brown-yellow Windhairdaisy ■

348604　Saussurea odontolepis (Herder) Sch. Bip. ex Maxim.；齿苞风毛菊；Toothscale Windhairdaisy ■

348605　Saussurea odontolepis Sch. Bip. ex Herder = Saussurea odontolepis (Herder) Sch. Bip. ex Maxim. ■

348606　Saussurea oligantha Franch.；少花风毛菊；Fewflower Saussurea, Fewflower Windhairdaisy ■

348607　Saussurea oligantha Franch. var. oligolepis (Y. Ling) X. Y. Wu = Saussurea oligantha Franch. ■

348608　Saussurea oligantha Franch. var. parvifolia Y. Ling = Saussurea oligantha Franch. ■

348609　Saussurea oligocephala (Y. Ling) Y. Ling；少头风毛菊；Poorhead Windhairdaisy ■

348610　Saussurea oligolepis Y. Ling = Saussurea oligantha Franch. ■

348611　Saussurea oppositicolor H. Lév. et Vaniot = Saussurea conyzoides Hemsl. ■

348612　Saussurea otophylla Diels = Saussurea macrota Franch. ■

348613　Saussurea otophylla Diels var. cinerea Y. Ling = Saussurea macrota Franch. ■

348614　Saussurea ovata Benth.；乌恰风毛菊（卵圆叶风毛菊）；Ovate Windhairdaisy ■

348615　Saussurea ovatifolia Y. L. Chen et S. Yun Liang；卵叶风毛菊；Ovateleaf Windhairdaisy ■

348616　Saussurea oxyodonta Hultén；尖齿风毛菊■☆

348617　Saussurea pachyneura Franch.；东俄洛风毛菊（八面风，滇西风毛菊，天蓬草，羽裂风毛菊）；Bodinier's Saussurea, Tongol Windhairdaisy ■

348618　Saussurea paleacea Y. L. Chen et S. Yun Liang；糠秕毛风毛菊；Chaff Windhairdaisy ■

348619　Saussurea paleata Maxim.；膜片风毛菊；Scaly Windhairdaisy ■

348620　Saussurea pallidiceps Hand.-Mazz. = Saussurea eriocephala Franch. ■

348621　Saussurea pamirica C. Winkl. = Saussurea glacialis Herder ■

348622　Saussurea papposa Turcz. = Saussurea davurica Adams ■

348623　Saussurea papposa Turcz. ex DC. = Saussurea davurica Adams ■

348624　Saussurea paradoxa Lipsch.；俄罗斯奇异风毛菊■☆

348625　Saussurea parasclerolepis A. I. Baranov et Skvortsov = Saussurea recurvata (Maxim.) Lipsch. ■

348626　Saussurea parviflora (Poir.) DC.；小花风毛菊（蛮汗风毛菊，燕北风毛菊，燕尾风毛菊）；Smallflower Saussurea, Smallflower Windhairdaisy ■

348627　Saussurea parviflora (Poir.) DC. var. amurensis (Herder) S. Yhu = Saussurea neoserrata Nakai ■

348628　Saussurea parviflora (Poir.) DC. var. atriplicifolia (Fisch. ex Herder) Hand.-Mazz. = Saussurea parviflora (Poir.) DC. ■

348629　Saussurea parviflora (Poir.) DC. var. atriplicifolia (Fisch.) Hand.-Mazz. = Saussurea hemsleyi Lipsch. ■

348630　Saussurea parviflora (Poir.) DC. var. cinerascens Hand.-Mazz. = Saussurea parviflora (Poir.) DC. ■

348631　Saussurea parviflora (Poir.) DC. var. cuspidata Hand.-Mazz. = Saussurea parviflora (Poir.) DC. ■

348632　Saussurea paucijuga Y. Ling；深裂风毛菊；Deepsplit Windhairdaisy ■

348633　Saussurea paxiana Diels = Saussurea paxiana Diels ex H. Limpr. ■

348634　Saussurea paxiana Diels ex H. Limpr.；红叶雪兔子；Redleaf Windhairdaisy ■

348635　Saussurea pectinata Bunge ex DC.；篦苞风毛菊（羽苞风毛菊）；Pectinatebract Windhairdaisy, Pectinatebracted Saussurea ■

348636　Saussurea pectinata Bunge ex DC. var. amurensis Maxim. = Saussurea odontolepis Sch. Bip. ex Herder ■

348637　Saussurea pectinata Bunge ex DC. var. macrocephala (Franch.) Hand.-Mazz. = Saussurea pectinata Bunge ex DC. ■

348638　Saussurea pectinata Bunge ex DC. var. pekinensis Maxim. = Saussurea pectinata Bunge ex DC. ■

348639　Saussurea peduncularis Franch.；显梗风毛菊；Pedunculate Windhairdaisy ■

348640　Saussurea peduncularis Franch. var. diversifolia Franch. = Saussurea peduncularis Franch. ■

348641　Saussurea peduncularis Franch. var. lobata (Franch.) Lipsch. = Saussurea peduncularis Franch. ■

348642　Saussurea peduncularis Franch. var. lobata Franch. = Saussurea peduncularis Franch. ■

348643　Saussurea peguensis C. B. Clarke；叶头风毛菊（蛇咬药，鹰爪莲）；Leafyinvolucre Saussurea, Pegu Windhairdaisy ■

348644　Saussurea peipingensis F. H. Chen = Saussurea chinnampoensis H. Lév. et Vaniot ■

348645　Saussurea pennata Koidz.；羽状风毛菊■☆

348646　Saussurea petrovii Lipsch.；西北风毛菊；Petrov Windhairdaisy ■

348647　Saussurea petrovii Lipsch. var. latifolia H. C. Fu = Saussurea cana Ledeb. ■

348648　Saussurea petrovii Lipsch. var. latifolia H. C. Fu = Saussurea petrovii Lipsch. ■

348649　Saussurea phaeantha Maxim.；褐花雪莲（褐花风毛菊，褐毛风毛菊）；Brownflower Saussurea, Brownflower Windhairdaisy ■

348650　Saussurea phyllocephala Collett et Hemsl. = Saussurea peguensis C. B. Clarke ■

348651　Saussurea pinetorum Hand.-Mazz.；松林风毛菊；Pineforest Saussurea, Pineforest Windhairdaisy ■

348652　Saussurea pinnatidentata Lipsch.；羽裂风毛菊；Pinnate Windhairdaisy ■

348653　Saussurea platypoda Hand.-Mazz.；川南风毛菊；Broadstalk Windhairdaisy ■

348654　Saussurea poljakovii Glehn；波尔风毛菊■☆

348655　Saussurea polycephala Hand.-Mazz.；多头风毛菊；Manyhead Saussurea, Manyhead Windhairdaisy ■

348656　Saussurea polycolea Hand.-Mazz.；多鞘雪莲；Manysheath Windhairdaisy ■

348657　Saussurea polycolea Hand.-Mazz. var. acutisquama (Y. Ling) Lipsch.；尖苞雪莲■

348658　Saussurea polygonifolia F. H. Chen；蓼叶风毛菊；Knowtweedleaf Windhairdaisy ■

348659　Saussurea polylada J. S. Li；簇枝雪莲■

348660　Saussurea polypodifolia Turcz. var. angustifolia Turcz. ex Herder = Saussurea runcinata DC. ■

348661　Saussurea polypodioides J. Anthony；水龙骨风毛菊；Wallfern Windhairdaisy ■

348662　Saussurea polystichoides Hook. f. = Saussurea kansuensis Hand. -Mazz. ■

348663　Saussurea poochlamys Hand. -Mazz. ;草叶风毛菊(草苞蛇眼草,草叶风毛菊,一枝箭);Leatherleaf Windhairdaisy ■

348664　Saussurea poophylla Diels = Saussurea graminea Dunn ■

348665　Saussurea popovii Lipsch. ;寡头风毛菊(乌鲁木齐风毛菊);Popov Windhairdaisy ■

348666　Saussurea populifolia Hemsl. ;杨叶风毛菊;Poplarleaf Saussurea,Poplarleaf Windhairdaisy ■

348667　Saussurea porcii Degen;鲍尔风毛菊■☆

348668　Saussurea porphyroleuca Hand. -Mazz. ;紫白风毛菊;Lilac Windhairdaisy ■

348669　Saussurea pratensis J. Anthony;草原雪莲;Grassland Windhairdaisy ■

348670　Saussurea prexima Diels;红毛雪兔子■

348671　Saussurea pricei Simpson;普里风毛菊■☆

348672　Saussurea propinqua Iljin;亲缘风毛菊■☆

348673　Saussurea propinqua Iljin = Saussurea controversa DC. ■☆

348674　Saussurea prostrata C. Winkl. ;展序风毛菊(平卧风毛菊)■☆

348675　Saussurea przewalskii Maxim. ;弯齿风毛菊(祁连风毛菊);Przewalsk Windhairdaisy ■

348676　Saussurea pseudoalpina N. D. Simpson;假高山风毛菊;False Alpine Windhairdaisy ■

348677　Saussurea pseudoangustifolia Lipsch. ;假狭叶风毛菊■☆

348678　Saussurea pseudobullockii Lipsch. ;洮河风毛菊;Taohe Windhairdaisy ■

348679　Saussurea pseudocolorata Danguy = Saussurea ovata Benth. ■

348680　Saussurea pseudoleontodon F. H. Chen = Saussurea scabrida Franch. ■

348681　Saussurea pseudomalitiosa Lipsch. ;类尖头风毛菊;Similar Evil Windhairdaisy ■

348682　Saussurea pseudosalsa Lipsch. ;假盐地风毛菊(喀什风毛菊)■☆

348683　Saussurea pseudosquarrosa Popov. et Lipsch. ;假粗糙风毛菊■☆

348684　Saussurea pseudotilesii Lipsch. ;假索氏风毛菊■☆

348685　Saussurea pteridophylla Hand. -Mazz. ;蕨叶风毛菊(延翅风毛菊);Wingleaf Windhairdaisy ■

348686　Saussurea pubescens Y. L. Chen et S. Yun Liang;毛果风毛菊;Hairfruit Windhairdaisy ■

348687　Saussurea pubifolia S. W. Liu;毛背雪莲;Hairback Windhairdaisy ■

348688　Saussurea pubifolia S. W. Liu var. acutisquama (Y. Ling) Lipsch. = Saussurea polycolea Hand. -Mazz. var. acutisquama (Y. Ling) Lipsch. ■

348689　Saussurea pubifolia S. W. Liu var. lhasaensis S. W. Liu;小苞雪莲■

348690　Saussurea pulchella (Fisch. ex Hornem.) Fisch. ;美花风毛菊(球花风毛菊);Beautiful Windhairdaisy,Beautifulflower Saussurea ■

348691　Saussurea pulchella (Fisch. ex Hornem.) Fisch. f. albiflora (Kitam.) Kitam. ;白色美花风毛菊■☆

348692　Saussurea pulchella (Fisch. ex Hornem.) Fisch. f. latifolia (Maxim.) Kitag. = Saussurea pulchella (Fisch. ex Hornem.) Fisch. ■

348693　Saussurea pulchella (Fisch. ex Hornem.) Fisch. f. subintegra (Regel) Kitag. = Saussurea pulchella (Fisch. ex Hornem.) Fisch. ■

348694　Saussurea pulchella (Fisch.) Fisch. = Saussurea pulchella (Fisch. ex Hornem.) Fisch. ■

348695　Saussurea pulchra Lipsch. ;美丽风毛菊(漂亮风毛菊);Finery Windhairdaisy,Spiffy Windhairdaisy ■

348696　Saussurea pulvinata Maxim. ;垫风毛菊;Pulvinate Windhairdaisy ■

348697　Saussurea pulviniformis C. Winkl. ;垫状风毛菊■

348698　Saussurea pumila C. Winkl. ;矮小风毛菊;Dwarf Windhairdaisy ■

348699　Saussurea purpurascens Y. L. Chen et S. Yun Liang;紫苞风毛菊(紫红苞风毛菊);Purplebract Windhairdaisy ■

348700　Saussurea pycnocephala Ledeb. var. sordida (Kar. et Kir.) Herder = Saussurea sordida Kar. et Kir. ■

348701　Saussurea pycnocephala Y. L. Chen et S. Yun Liang var. sordida (Kar. et Kir.) Herder = Saussurea sordida Kar. et Kir. ■

348702　Saussurea qinghaiensis S. W. Liu et T. N. Ho;青海雪兔子;Qinghai Saussurea ■

348703　Saussurea quercifolia W. W. Sm. ;槲叶雪兔子(川西雪莲,黑毛雪兔子,槲叶雪莲花,玄果搜花);Oakleaf Saussurea, Oakleaf Windhairdaisy ■

348704　Saussurea quercifolia W. W. Sm. var. major Anthony = Saussurea quercifolia W. W. Sm. ■

348705　Saussurea radiata Franch. = Saussurea deltoidea (DC.) Sch. Bip. ■

348706　Saussurea recurvata (Maxim.) Lipsch. ;折苞风毛菊(长叶风毛菊,全叶折苞风毛菊,弯苞风毛菊);Reflexbract Windhairdaisy ■

348707　Saussurea recurvata (Maxim.) Lipsch. var. angustata H. C. Fu = Saussurea recurvata (Maxim.) Lipsch. ■

348708　Saussurea reniforme Y. Ling = Saussurea acromelaena Hand. -Mazz. ■

348709　Saussurea retroserrata Y. L. Chen et S. Yun Liang;倒齿风毛菊;Reversetooth Windhairdaisy ■

348710　Saussurea rhytidocarpa Hand. -Mazz. ;皱果风毛菊;Wrinklefruit Windhairdaisy ■

348711　Saussurea riederi Herder;里德尔风毛菊■☆

348712　Saussurea riederi Herder f. insularis (Tatew. et Kitam.) Ohwi = Saussurea riederi Herder subsp. yezoensis (Maxim.) Kitam. var. insularis Tatew. et Kitam. ■☆

348713　Saussurea riederi Herder f. japonica (Koidz.) Ohwi = Saussurea riederi Herder subsp. yezoensis (Maxim.) Kitam. var. japonica Koidz. ■☆

348714　Saussurea riederi Herder subsp. kudoana (Tatew. et Kitam.) Kitam. = Saussurea kudoana Tatew. et Kitam. ■☆

348715　Saussurea riederi Herder subsp. yezoensis (Maxim.) Kitam. ;北海道风毛菊■☆

348716　Saussurea riederi Herder subsp. yezoensis (Maxim.) Kitam. f. albiflora (Koidz.) H. Hara;白花北海道风毛菊■☆

348717　Saussurea riederi Herder subsp. yezoensis (Maxim.) Kitam. var. daisetsuenis (Nakai) Kitam. ;大雪风毛菊■☆

348718　Saussurea riederi Herder subsp. yezoensis (Maxim.) Kitam. var. elongata Kitam. ;伸长里德尔风毛菊■☆

348719　Saussurea riederi Herder subsp. yezoensis (Maxim.) Kitam. var. insularis Tatew. et Kitam. ;岛生北海道风毛菊■☆

348720　Saussurea riederi Herder subsp. yezoensis (Maxim.) Kitam. var. japonica Koidz. ;日本北海道风毛菊■☆

348721　Saussurea riederi Herder var. kudoana (Tatew. et Kitam.) T. Shimizu = Saussurea kudoana Tatew. et Kitam. ■☆

348722　Saussurea riederi Herder var. yezoensis Maxim. f. daisetsuensis (Nakai) Ohwi = Saussurea riederi Herder subsp. yezoensis (Maxim.) Kitam. var. daisetsuensis (Nakai) Kitam. ■☆

348723　Saussurea riederi Herder var. yezoensis Maxim. f. elongata

（Kitam.）Ohwi ＝ Saussurea riederi Herder subsp. yezoensis （Maxim.）Kitam. var. elongata Kitam. ■☆

348724 Saussurea riederi Herder var. yezoensis Maxim. f. insularis （Tatew. et Kitam.）Ohwi ＝ Saussurea riederi Herder subsp. yezoensis（Maxim.）Kitam. var. insularis Tatew. et Kitam. ■☆

348725 Saussurea riederi Herder var. yezoensis Maxim. f. japonica （Koidz.）Ohwi ＝ Saussurea riederi Herder subsp. yezoensis （Maxim.）Kitam. var. japonica Koidz. ■☆

348726 Saussurea riederi Herder var. yezoensis Maxim. f. kudoana （Tatew. et Kitam.）Ohwi ＝ Saussurea kudoana Tatew. et Kitam. ■☆

348727 Saussurea rigida Ledeb.；硬风毛菊■☆

348728 Saussurea robusta Ledeb.；强壮风毛菊；Strong Windhairdaisy ■

348729 Saussurea rockii J. Anthony；显鞘风毛菊；Rock Windhairdaisy ■

348730 Saussurea rohmooana C. Marquand et Airy Shaw ＝ Saussurea katochaete Maxim. ■

348731 Saussurea romuleifolia Franch.；鸢尾叶风毛菊（大麻草,粉草,钱叶风毛菊,蛇箭,蛇眼草,线叶风毛菊,雨过天青,雨过天晴,鸢尾叶风尾菊）；Irisleaf Saussurea,Irisleaf Windhairdaisy）■

348732 Saussurea romuleifolia Franch. var. ortholepis （Hand. -Mazz.）Handel-Mazzetii ex S. Y. Hu ＝ Saussurea graminea Dunn ■

348733 Saussurea romuleifolia Franch. var. ortholepis Hand. -Mazz. ＝ Saussurea graminea Dunn ■

348734 Saussurea rosthornii Diels ＝ Saussurea conyzoides Hemsl. ■

348735 Saussurea rosthornii Diels var. oppositicolor （H. Lév. et Vaniot）F. H. Chen ex Hand. -Mazz. ＝ Saussurea conyzoides Hemsl. ■

348736 Saussurea rosthornii Diels var. sessilifolia Diels ＝ Saussurea cauloptera Hand. -Mazz. ■

348737 Saussurea rosthornii Diels var. sessilifolia Diels ＝ Saussurea leclerei H. Lév. ■

348738 Saussurea rotundifolia F. H. Chen；圆叶风毛菊；Roundleaf Windhairdaisy ■

348739 Saussurea rufostrigillosa Y. Ling ＝ Saussurea sutchuenensis Franch. ■

348740 Saussurea rufostrigillosa Y. Ling var. macrocephala Y. Ling ＝ Saussurea sutchuenensis Franch. ■

348741 Saussurea rufotricha Y. Ling ＝ Saussurea baroniana Diels ■

348742 Saussurea runcinata DC.；倒羽叶风毛菊（碱地风毛菊）；Runcinate Saussurea,Runcinate Windhairdaisy ■

348743 Saussurea runcinata DC. var. dentata Ledeb. ＝ Saussurea alata DC. ■

348744 Saussurea runcinata DC. var. integrifolia H. C. Fu et D. S. Wen；全叶碱地风毛菊；Entireleaf Runcinate Saussurea ■

348745 Saussurea runcinata DC. var. integrifolia H. C. Fu et D. S. Wen ＝ Saussurea runcinata DC. ■

348746 Saussurea runcinata DC. var. pinnatidentata （Lipsch.）H. C. Fu et D. S. Wen ＝ Saussurea pinnatidentata Lipsch. ■

348747 Saussurea ruoqiangensis K. M. Shen；若羌风毛菊；Ruoqiang Windhairdaisy ■

348748 Saussurea russowi C. Winkl. ＝ Saussurea sordida Kar. et Kir. ■

348749 Saussurea sachalinensis F. Schmidt；库页风毛菊■☆

348750 Saussurea sachalinensis F. Schmidt ＝ Saussurea acuminata Turcz. ex Fisch. et C. A. Mey. var. sachalinensis （F. Schmidt）Herder ■☆

348751 Saussurea sacra Edgew. ＝ Saussurea simpsoniana （Fielding et Gardner）Lipsch. ■

348752 Saussurea sagitta Franch.；箭状风毛菊■☆

348753 Saussurea sagitta Franch. f. alba Kitam.；白花箭状风毛菊■☆

348754 Saussurea sagitta Franch. var. yoshizawae Kitam.；义泽风毛菊■☆

348755 Saussurea sajanensis Gudoschn.；萨因风毛菊■☆

348756 Saussurea salemannii C. Winkl.；倒卵叶风毛菊（赛里木风毛菊）；Obovateleaf Windhairdaisy ■

348757 Saussurea salicifolia （L.）DC.；柳叶风毛菊；Willowleaf Windhairdaisy ■

348758 Saussurea salicifolia （L.）DC. var. chinensis Maxim. ＝ Saussurea chinensis （Maxim.）Lipsch. ■

348759 Saussurea salicifolia （L.）DC. var. elegans （Ledeb.）Trautv. ＝ Saussurea elegans Ledeb. ■

348760 Saussurea salicifolia （L.）DC. var. shensiensis Y. Ling ＝ Saussurea chinensis （Maxim.）Lipsch. ■

348761 Saussurea saliginiformis Hand. -Mazz. ＝ Saussurea dolichopoda Diels ■

348762 Saussurea saligna Franch.；尾尖风毛菊（尖尾风毛菊）；Tail Windhairdaisy ■

348763 Saussurea salsa （Pall.）Spreng.；盐地风毛菊；Saline Saussurea,Saline Windhairdaisy ■

348764 Saussurea salsa （Pall.）Spreng. var. papposa （Turcz.）Ledeb. ＝ Saussurea davurica Adams ■

348765 Saussurea salwinensis J. Anthony；怒江风毛菊；Salwin Windhairdaisy ■

348766 Saussurea sawadae Kitam. ＝ Saussurea sinuatoides Nakai var. glabrescens Nakai ■☆

348767 Saussurea saxosa Lipsch.；岩栖风毛菊■☆

348768 Saussurea scabrida Franch.；糙毛风毛菊；Hispid Windhairdaisy ■

348769 Saussurea scaposa Franch. et Sav.；花茎风毛菊■☆

348770 Saussurea schanginiana （Wydler）Fisch. ex Herder；暗苞风毛菊■

348771 Saussurea schultzii Hook. f. ＝ Saussurea bracteata Decne. ■

348772 Saussurea sclerolepis Nakai et Kitag.；卷苞风毛菊；Coilbract Windhairdaisy ■

348773 Saussurea scrratuloides Turcz.；麻花头风毛菊■☆

348774 Saussurea semiamplexicaulis Lipsch.；半抱茎风毛菊；Semi-amplectant Windhairdaisy ■

348775 Saussurea semifasciata Hand. -Mazz.；锯叶风毛菊；Serrateleaf Windhairdaisy ■

348776 Saussurea semilyrata Bureau et Franch.；半琴叶风毛菊；Semi-lyrate Windhairdaisy ■

348777 Saussurea sericea Y. L. Chen et S. Yun Liang；绢毛风毛菊；Sericeous Windhairdaisy ■

348778 Saussurea serrata DC.；锯齿风毛菊■☆

348779 Saussurea serrata DC. ＝ Saussurea parviflora （Poir.）DC. ■

348780 Saussurea serrata DC. var. amurensis Herder ＝ Saussurea neoserrata Nakai ■

348781 Saussurea setifolia Klatt ＝ Saussurea subulata C. B. Clarke ■

348782 Saussurea shiretokoensis Susaw.；知床风毛菊■☆

348783 Saussurea sikangensis F. H. Chen ＝ Saussurea pachyneura Franch. ■

348784 Saussurea sikokiana Makino ＝ Saussurea nipponica Miq. subsp. savatieri （Franch.）Kitam. var. robusta （Makino）Ohwi ex Lipsch. ■☆

348785 Saussurea silvestri Pamp. ＝ Saussurea conyzoides Hemsl. ■

348786 Saussurea simpsoniana （Fielding et Gardner）Lipsch.；小果雪兔子；Smallfruit Windhairdaisy ■

348787 Saussurea sinensis Sch. Bip. ＝ Saussurea japonica （Thunb.）DC. ■

348788 Saussurea sinuata Kom.；林风毛菊；Sinuate Windhairdaisy ■

348789 Saussurea sinuata Kom. f. japonica Nakai ＝ Saussurea sinuata

Kom. ■

348790　Saussurea sinuata Kom. var. cordata F. H. Chen　＝ Saussurea hwangshanensis Y. Ling ■

348791　Saussurea sinuata Kom. var. shansiensis F. H. Chen　＝ Saussurea mongolica（Franch.）Franch. ■

348792　Saussurea sinuatoides Nakai；拟林风毛菊■☆

348793　Saussurea sinuatoides Nakai　＝ Saussurea sinuata Kom. ■

348794　Saussurea sinuatoides Nakai var. bosopeninsularis Ohba et Yoko Kimura；海岛拟林风毛菊■☆

348795　Saussurea sinuatoides Nakai var. glabrescens Nakai；光拟林风毛菊■☆

348796　Saussurea sinuatoides Nakai var. serrata Nakai　＝ Saussurea nipponica Miq. subsp. savatieri（Franch.）Kitam. ■☆

348797　Saussurea smithiana Hand. -Mazz.　＝ Saussurea iodoleuca Hand. -Mazz. ■

348798　Saussurea sobarocephala Diels；昂头风毛菊；Elevatehead Saussurea，Look up Windhairdaisy ■

348799　Saussurea soczavae Lipsch.；索氏风毛菊■☆

348800　Saussurea sordida Kar. et Kir.；污花风毛菊；Dirtflower Windhairdaisy ■

348801　Saussurea sordida Kar. et Kir.　＝ Saussurea globosa F. H. Chen ■

348802　Saussurea sordida Kar. et Kir. var. oligocephala C. Winkl.　＝ Saussurea sordida Kar. et Kir. ■

348803　Saussurea sordida Kar. et Kir. var. oligocephala C. Winkl. ex Lipsch.　＝ Saussurea sordida Kar. et Kir. ■

348804　Saussurea sorocephala（Schrenk）Hook. f. et Thomson ex C. B. Clarke　＝ Saussurea gnaphaloides（Royle）Sch. Bip. ■

348805　Saussurea sorocephala（Schrenk）Sch. Bip.　＝ Saussurea gnaphaloides（Royle）Sch. Bip. ■

348806　Saussurea sorocephala（Schrenk）Schrenk　＝ Saussurea gnaphaloides（Royle）Sch. Bip. ■

348807　Saussurea souliei Franch.；披针叶风毛菊；Lanceolateleaf Windhairdaisy ■

348808　Saussurea sovietica Kom.；索维特风毛菊■☆

348809　Saussurea spathulifolia Franch.；维西风毛菊；Weixi Windhairdaisy ■

348810　Saussurea splendida Kom.；节毛风毛菊；Nodehair Windhairdaisy ■

348811　Saussurea squarrosa Turcz.；粗糙风毛菊■☆

348812　Saussurea stella Maxim.；星状雪兔子（匍地风毛菊，星状风毛菊，紫星菊）；Starry Saussurea，Starry Windhairdaisy ■

348813　Saussurea stenocephala Y. Ling　＝ Saussurea dielsiana Koidz. ■

348814　Saussurea stenolepis Nakai；窄苞风毛菊；Narrowbract Windhairdaisy ■

348815　Saussurea stenophylla Freyn　＝ Saussurea amurensis Turcz. ex DC. ■

348816　Saussurea stoetzneriana Diels　＝ Saussurea semilyrata Bureau et Franch. ■

348817　Saussurea stoliczkae C. B. Clarke；川藏风毛菊（藏西风毛菊，光果风毛菊）；Stoliczka Windhairdaisy ■

348818　Saussurea stricta Franch.；喜林风毛菊；Strict Windhairdaisy，Woodland Windhairdaisy ■

348819　Saussurea stricta S. Y. Hu　＝ Saussurea stricta Franch. ■

348820　Saussurea stricta Spreng. ex DC.　＝ Hemistepta lyrata（Bunge）Bunge ■

348821　Saussurea stubendorfii Herder；斯图风毛菊■☆

348822　Saussurea subcordata F. H. Chen　＝ Saussurea stricta Franch. ■

348823　Saussurea subtriangulata Kom.；吉林风毛菊；Jilin Windhairdaisy ■

348824　Saussurea subulata C. B. Clarke；钻叶风毛菊；Awl-leaf Windhairdaisy ■

348825　Saussurea subulata C. B. Clarke　＝ Saussurea columnaris Hand. -Mazz. ■

348826　Saussurea subulata C. B. Clarke　＝ Saussurea wernerioides Sch. Bip. ex Hook. f. ■

348827　Saussurea subulisquama Hand. -Mazz.；尖苞风毛菊；Tinebract Windhairdaisy ■

348828　Saussurea sughoo C. B. Clarke　＝ Saussurea nimborum W. W. Sm. ■

348829　Saussurea sugimurae Honda；杉村风毛菊■☆

348830　Saussurea sukaczevii Lipsch.；苏卡风毛菊■☆

348831　Saussurea sulcata Iljin；纵沟风毛菊■☆

348832　Saussurea sumneviczii Serg.；苏木奈风毛菊■☆

348833　Saussurea sungpanensis Hand. -Mazz.　＝ Saussurea leontodontoides（DC.）Sch. Bip. ■

348834　Saussurea superba J. Anthony　＝ Saussurea hieracioides Hook. f. ■

348835　Saussurea sutchuenensis Franch.；四川风毛菊；Sichuan Windhairdaisy ■

348836　Saussurea sylvatica Maxim.；林生风毛菊；Woodland Windhairdaisy ■

348837　Saussurea sylvatica Maxim. var. hsiaowutaishanensis（F. H. Chen）Lipsch.；小五台山风毛菊■

348838　Saussurea tadschikorum Iljin；塔什克风毛菊■☆

348839　Saussurea taipaiensis Y. Ling　＝ Saussurea globosa F. H. Chen ■

348840　Saussurea takhtadganii Lipsch.　＝ Saussurea larionowii C. Winkl. ■

348841　Saussurea tanakae Franch. et Sav. ex Maxim.；田中氏风毛菊■☆

348842　Saussurea tanakae Franch. et Sav. ex Maxim.　＝ Saussurea bullockii Dunn ■

348843　Saussurea tanakae Franch. et Sav. ex Maxim. f. albiflora M. Tash.；白花田中氏风毛菊■☆

348844　Saussurea tanguensis J. R. Drumm.　＝ Cavea tanguensis（Drumm.）W. W. Sm. ■

348845　Saussurea tangutica Maxim.；唐古特雪莲（东方风毛菊，东方雪莲花，唐古特风毛菊，血莲）；Tangut Windhairdaisy ■

348846　Saussurea taquetii H. Lév.　＝ Saussurea japonica（Thunb.）DC. ■

348847　Saussurea taquetii H. Lév. et Vaniot　＝ Saussurea japonica（Thunb.）DC. ■

348848　Saussurea taquetii H. Lév. et Vaniot var. paniculata H. Lév. et Vaniot　＝ Saussurea japonica（Thunb.）DC. ■

348849　Saussurea taraxacifolia Wall. ex DC.；蒲公英叶风毛菊（公英叶风毛菊）；Dandelionleaf Windhairdaisy ■

348850　Saussurea tatsienensis Franch.；打箭风毛菊；Dajian Windhairdaisy ■

348851　Saussurea tatsienensis Franch. var. monocephala Diels　＝ Saussurea tatsienensis Franch. ■

348852　Saussurea tenella Y. Ling　＝ Saussurea rotundifolia F. H. Chen ■

348853　Saussurea tenuicaulis Y. Ling　＝ Saussurea amara（L.）DC. ■

348854　Saussurea tenuifolia Kitag.；长白山风毛菊；Changbaishan Windhairdaisy ■

348855　Saussurea tenuis Ledeb.　＝ Saussurea elegans Ledeb. ■

348856　Saussurea thibetica Franch.　＝ Saussurea polycephala Hand. -Mazz. ■

348857　Saussurea thomsonii C. B. Clarke；肉叶雪兔子；Succulentleaf Windhairdaisy ■

348858 Saussurea thoroldii Hemsl. ; 草甸雪兔子（草甸风毛菊）; Meadow Windhairdaisy ■

348859 Saussurea tianshuiensis X. Y. Wu; 天水风毛菊; Tianshui Windhairdaisy ■

348860 Saussurea tianshuiensis X. Y. Wu var. huxianensis X. Y. Wu; 户县风毛菊; Huxian Windhairdaisy ■

348861 Saussurea tibetica C. Winkl. ; 西藏风毛菊; Tibet Windhairdaisy, Xizang Windhairdaisy ■

348862 Saussurea tienmoshanensis F. H. Chen = Saussurea bullockii Dunn ■

348863 Saussurea tilesii Ledeb. ; 梯来风毛菊 ■☆

348864 Saussurea tomentosa Kom. ; 高岭风毛菊; Tomentose Windhairdaisy ■

348865 Saussurea triangulata Trautv. et C. A. Mey. ; 三角风毛菊 ■☆

348866 Saussurea triangulata Trautv. et C. A. Mey. subsp. manshurica （Kom.) Kitam. = Saussurea manshurica Kom. ■

348867 Saussurea triangulata Trautv. et C. A. Mey. var. pinnatifida （Nakai) Kitam. = Saussurea manshurica Kom. ■

348868 Saussurea triceps H. Lév. = Saussurea maximowiczii Herder ■

348869 Saussurea triceps H. Lév. et Vaniot = Saussurea maximowiczii Herder ■

348870 Saussurea tridactyla Sch. Bip. ex Hook. f. ; 三指雪兔子（三指雪莲，西藏雪莲花）; Threefinger Windhairdaisy ■

348871 Saussurea tridactyla Sch. Bip. ex Hook. f. var. maiduoganla S. W. Liu; 丛株雪兔子 ■

348872 Saussurea triptera Maxim. ; 三翼风毛菊 ■☆

348873 Saussurea triptera Maxim. f. hisauchii （Nakai) Ohwi = Saussurea hisauchii Nakai ■☆

348874 Saussurea triptera Maxim. f. kaialpina （Nakai) Ohwi = Saussurea triptera Maxim. var. kaialpina （Nakai) Kitam. ■☆

348875 Saussurea triptera Maxim. f. major （Takeda) Ohwi = Saussurea triptera Maxim. ■☆

348876 Saussurea triptera Maxim. var. hisauchii （Nakai) Kitam. = Saussurea hisauchii Nakai ■☆

348877 Saussurea triptera Maxim. var. kaialpina （Nakai) Kitam. = Saussurea triptera Maxim. var. minor （Takeda) Kitam. ■☆

348878 Saussurea triptera Maxim. var. major （Takeda) Kitam. ; 大三翼风毛菊 ■☆

348879 Saussurea triptera Maxim. var. major （Takeda) Kitam. = Saussurea triptera Maxim. var. minor （Takeda) Kitam. ■☆

348880 Saussurea triptera Maxim. var. major （Takeda) Kitam. = Saussurea triptera Maxim. ■☆

348881 Saussurea triptera Maxim. var. mikurasimensis Kitam. ; 御藏风毛菊 ■☆

348882 Saussurea triptera Maxim. var. minor （Takeda) Kitam. ; 小三翼风毛菊 ■☆

348883 Saussurea triptera Maxim. var. minor （Takeda) Kitam. = Saussurea triptera Maxim. ■☆

348884 Saussurea trullifolia W. W. Sm. = Saussurea medusa Maxim. ■

348885 Saussurea trullifolia W. W. Sm. var. pinnatibracteata Anthony = Saussurea medusa Maxim. ■

348886 Saussurea tsarongensis Anthony = Saussurea phaeantha Maxim. ■

348887 Saussurea tsinlingensis Hand. -Mazz. ; 秦岭风毛菊; Qinling Windhairdaisy ■

348888 Saussurea tuoliensis G. M. Shen; 托里风毛菊; Tuoli Windhairdaisy ■

348889 Saussurea turgaiensis B. Fedtsch. ; 图尔盖风毛菊（太加风毛菊）■

348890 Saussurea uliginosa Hand. -Mazz. ; 湿地雪兔子; Marsh Windhairdaisy ■

348891 Saussurea uliginosa Hand. -Mazz. var. vittifolia （Anthony) Hand. -Mazz. = Saussurea uliginosa Hand. -Mazz. ■

348892 Saussurea umbrosa Kom. ; 湿地风毛菊; Everglade Windhairdaisy ■

348893 Saussurea undulata Hand. -Mazz. ; 波缘风毛菊; Undulate Windhairdaisy ■

348894 Saussurea uniflora （DC.) Wall. ex Sch. Bip. = Saussurea uniflora （Wall. ex DC.) Sch. Bip. ■

348895 Saussurea uniflora （Wall. ex DC.) Sch. Bip. ; 单花雪莲; Oneflower Windhairdaisy ■

348896 Saussurea uniflora （Wall.) C. B. Clarke = Saussurea uniflora （Wall. ex DC.) Sch. Bip. ■

348897 Saussurea uniflora （Wall.) C. B. Clarke var. conica （C. B. Clarke) Hook. f. = Saussurea conica C. B. Clarke ■

348898 Saussurea uniflora C. B. Clarke = Saussurea uniflora （Wall. ex DC.) Sch. Bip. ■

348899 Saussurea uniflora Wall. = Saussurea uniflora （Wall. ex DC.) Sch. Bip. ■

348900 Saussurea uralensis Lipsch. ; 乌拉尔风毛菊 ■☆

348901 Saussurea ussiriensis Lipsch. var. pinnatifida Maxim. = Saussurea ussuriensis Maxim. ■

348902 Saussurea ussuriensis Maxim. ; 乌苏里风毛菊（山牛蒡，乌苏里苦麻，乌苏里野苦麻）; Ussuri Saussurea, Wusuli Windhairdaisy ■

348903 Saussurea ussuriensis Maxim. var. firma Kitag. ; 硬叶乌苏里风毛菊（硬叶风毛菊）; Rigidleaf Windhairdaisy ■

348904 Saussurea ussuriensis Maxim. var. firma Kitag. = Saussurea firma Kitam. ■

348905 Saussurea ussuriensis Maxim. var. incisa Maxim. = Saussurea ussuriensis Maxim. ■

348906 Saussurea ussuriensis Maxim. var. laxiodontolepis Q. Zh. Han et Shu Y. Wang; 疏齿苞乌苏里风毛菊 ■

348907 Saussurea ussuriensis Maxim. var. mongolica Franch. = Saussurea mongolica （Franch.) Franch. ■

348908 Saussurea ussuriensis Maxim. var. nivea Kitam. ; 雪白乌苏里风毛菊 ■☆

348909 Saussurea ussuriensis Maxim. var. odontolepis Herder = Saussurea odontolepis Sch. Bip. ex Herder ■

348910 Saussurea ussuriensis Maxim. var. pinnatifida Maxim. = Saussurea ussuriensis Maxim. ■

348911 Saussurea ussuriensis Maxim. var. triangulata H. C. Fu; 长叶乌苏里风毛菊 ■

348912 Saussurea vaginata Dunn = Saussurea yunnanensis Franch. ■

348913 Saussurea vaniotii H. Lév. = Saussurea centiloba Hand. -Mazz. ■

348914 Saussurea variilobs Y. Ling; 变裂风毛菊; Changelobe Windhairdaisy ■

348915 Saussurea veitchiana J. R. Drumm. et Hutch. ; 华中雪莲; Veitch Windhairdaisy ■

348916 Saussurea velutina W. W. Sm. ; 毡毛雪莲（黄绒风毛菊，毡毛风毛菊，黏毛雪莲）; Velvety Saussurea, Velvety Windhairdaisy ■

348917 Saussurea velutina W. W. Sm. = Saussurea muliensis Hand. -Mazz. ■

348918 Saussurea venosa Kerr = Saussurea peguensis C. B. Clarke ■

348919 Saussurea vestita Franch. ; 绒背风毛菊; Flossback Windhairdaisy ■

348920　Saussurea vestitiformis Hand. -Mazz. ; 河谷风毛菊; Valley Windhairdaisy ■

348921　Saussurea villosa Franch. = Saussurea hieracioides Hook. f. ■

348922　Saussurea violacea Pamp. = Saussurea glacialis Herder ■

348923　Saussurea virgata Franch. ; 帚状风毛菊; Fastigiate Windhairdaisy ■

348924　Saussurea viridibracteata F. H. Chen = Saussurea semilyrata Bureau et Franch. ■

348925　Saussurea viscida Hultén; 黏风毛菊■☆

348926　Saussurea viscida Hultén = Saussurea angustifolia（L.）DC. var. viscida（Hultén）S. L. Welsh ■☆

348927　Saussurea viscida Hultén var. yukonensis（A. E. Porsild）Hultén = Saussurea angustifolia（L.）DC. var. yukonensis A. E. Porsild ■☆

348928　Saussurea vittifolia J. Anthony = Saussurea uliginosa Hand. -Mazz. ■

348929　Saussurea vvedenskyi Lipsch. ; 韦氏风毛菊■☆

348930　Saussurea wakasugiana Kadota; 若杉风毛菊■☆

348931　Saussurea wallichi Sch. Bip. = Saussurea fastuosa（Decne.）Sch. Bip. ■

348932　Saussurea wardii J. Anthony; 川滇风毛菊; Ward Windhairdaisy ■

348933　Saussurea weberi Hultén; 韦伯风毛菊■☆

348934　Saussurea wellbyi Hemsl. ; 羌塘雪兔子; Wellby Windhairdaisy ■

348935　Saussurea wernerioides Sch. Bip. ex Hook. f. ; 锥叶风毛菊; Awlleaf Windhairdaisy ■

348936　Saussurea wettsteiniana Hand. -Mazz. ; 垂头雪莲; Nutant Windhairdaisy ■

348937　Saussurea wilsoniana Hand. -Mazz. = Saussurea dolichopoda Diels ■

348938　Saussurea woodiana Hemsl. ; 牛耳风毛菊; Oxear Windhairdaisy ■

348939　Saussurea woodiana Hemsl. f. caulescens Lipsch. = Saussurea woodiana Hemsl. ■

348940　Saussurea xanthotricha Hand. -Mazz. = Saussurea melanotricha Hand. -Mazz. ■

348941　Saussurea yabulaiensis Y. Y. Yao; 雅布赖风毛菊; Yabulai Windhairdaisy ■

348942　Saussurea yakusimensis Masam. = Saussurea nipponica Miq. subsp. savatieri（Franch.）Kitam. var. yakusimensis（Masam.）H. Koyama ■☆

348943　Saussurea yamatensis Honda = Saussurea nipponica Miq. ■☆

348944　Saussurea yanagisawae Honda var. elegans（Koidz.）Nakai = Saussurea yanagisawae Takeda ■☆

348945　Saussurea yanagisawae Takeda; 柳泽风毛菊■☆

348946　Saussurea yanagisawae Takeda f. angustifolia（Nakai）Ohwi = Saussurea yanagisawae Takeda ■☆

348947　Saussurea yanagisawae Takeda f. elegans（Koidz.）Ohwi = Saussurea yanagisawae Takeda ■☆

348948　Saussurea yanagisawae Takeda f. imperialis（Koidz.）Ohwi = Saussurea yanagisawae Takeda ■☆

348949　Saussurea yanagisawae Takeda f. nivea（Koidz.）Toyok. = Saussurea yanagisawae Takeda var. nivea（Koidz.）Nakai ■☆

348950　Saussurea yanagisawae Takeda f. vestita（Kitam.）Ohwi = Saussurea yanagisawae Takeda var. nivea（Koidz.）Nakai ■☆

348951　Saussurea yanagisawae Takeda var. angustifolia（Nakai）Kitam. = Saussurea yanagisawae Takeda ■☆

348952　Saussurea yanagisawae Takeda var. elegans（Koidz.）Nakai = Saussurea yanagisawae Takeda ■☆

348953　Saussurea yanagisawae Takeda var. imperialis（Koidz.）Nakai

= Saussurea yanagisawae Takeda ■☆

348954　Saussurea yanagisawae Takeda var. nivea（Koidz.）Nakai; 雪白柳泽风毛菊■☆

348955　Saussurea yanagisawae Takeda var. vestita Kitam. = Saussurea yanagisawae Takeda var. nivea（Koidz.）Nakai ■☆

348956　Saussurea yatagaiana Mori = Saussurea glandulosa Kitam. ■

348957　Saussurea yoshinagae Kitam. = Saussurea nipponica Miq. var. yoshinagae（Kitam.）H. Koyama ■☆

348958　Saussurea yosthornii Diels var. sessilifolia Diels = Saussurea cauloptera Hand. -Mazz. ■

348959　Saussurea yunnanensis Franch. ; 云南风毛菊; Yunnan Windhairdaisy ■

348960　Saussurea yunnanensis Franch. var. integrifolia Franch. = Saussurea yunnanensis Franch. ■

348961　Saussurea yunnanensis Franch. var. runcinata Franch. = Saussurea yunnanensis Franch. ■

348962　Saussurea yunnanensis Franch. var. sessilifolia Anthony = Saussurea yunnanensis Franch. ■

348963　Saussurea yushuensis S. W. Liu et T. N. Ho; 玉树雪兔子; Yushu Windhairdaisy ■

348964　Saussuria Moench（废弃属名）= Nepeta L. ■●

348965　Saussuria Moench（废弃属名）= Saussurea DC.（保留属名）●■

348966　Saussuria Moench（废弃属名）= Schizonepeta（Benth.）Briq. ■

348967　Saussuria St. -Lag. = Saussurea DC.（保留属名）●■

348968　Sautiera Decne.（1834）; 帝汶爵床属☆

348969　Sautiera tinctorum Decne. ; 帝汶爵床☆

348970　Sauvagea Adans. = Sauvagesia L. ●

348971　Sauvagea L. = Sauvagesia L. ●

348972　Sauvagesia L.（1753）; 旱金莲木属（合柱金莲木属）●

348973　Sauvagesia africana（Baill.）Bamps; 非洲旱金莲木●☆

348974　Sauvagesia congensis Engl. = Sauvagesia africana（Baill.）Bamps ●☆

348975　Sauvagesia erecta L. ; 旱金莲木; Creole Tea, St. Martin's Herb ●☆

348976　Sauvagesia rhodoleuca（Diels）M. C. E. Amaral; 合柱金莲木●

348977　Sauvagesiaceae（DC）Dumort. = Ochnaceae DC.（保留科名）●■

348978　Sauvagesiaceae Dumort.（1829）; 旱金莲木科（辛木科）●■

348979　Sauvagesiaceae Dumort. = Ochnaceae DC.（保留科名）●■

348980　Sauvagia St. -Lag. = Sauvagesia L. ●

348981　Sauvalella Willis = Sauvallella Rydb. ●☆

348982　Sauvallea C. Wright（1871）; 独焰草属■☆

348983　Sauvallea blainii C. Wright; 独焰草■☆

348984　Sauvallella Rydb.（1924）; 独焰豆属●☆

348985　Sauvallella Rydb. = Poitea Vent. ●☆

348986　Sauvallella immarginata（Wright et Sauvalle）Rydberg; 独焰豆■☆

348987　Sauvetrea Szlach.（2006）; 美洲鳃兰属■☆

348988　Sauvetrea Szlach. = Maxillaria Ruiz et Pav. ■☆

348989　Sava Adans. = Onosma L. ■

348990　Savannosiphon Goldblatt et W. Marais（1980）; 草原鸢尾属■☆

348991　Savannosiphon euryphyllus（Harms）Goldblatt et W. Marais; 草原鸢尾■☆

348992　Savastana Raf. = Savastania Scop. ●■☆

348993　Savastana Raf. = Tibouchina Aubl. ●■☆

348994　Savastana Schrank（废弃属名）= Hierochloe R. Br.（保留属名）■

348995　Savastania Scop. = Tibouchina Aubl. ●■☆

348996　Savastonia Neck. ex Steud. = Savastania Scop. ●■☆

348997　Savia Raf. = Amphicarpaea Elliott ex Nutt.（保留属名）■

348998　Savia Willd.（1806）;萨维大戟属●☆

348999　Savia elegans（Baill.）Müll. Arg. = Wielandia elegans Baill. ●☆

349000　Savia fadenii Radcl. -Sm. = Petalodiscus fadenii（Radcl. -Sm.）Radcl. -Sm. ●☆

349001　Savia laurifolia Griseb. ;桂叶萨维大戟●☆

349002　Savia maculata Urb. ;斑点萨维大戟●☆

349003　Savia ovalis（E. Mey. ex Sond.）Pax et K. Hoffm. = Andrachne ovalis（E. Mey. ex Sond.）Müll. Arg. ●☆

349004　Savignya DC.（1821）;肉叶长柄芥属■☆

349005　Savignya aegyptiaca DC. = Savignya parviflora（Delile）Webb ■☆

349006　Savignya longistyla Boiss. et Reut. = Savignya parviflora（Delile）Webb subsp. longistyla（Boiss. et Reut.）Maire ■☆

349007　Savignya parviflora（Delile）Webb;小花肉叶长柄芥☆

349008　Savignya parviflora（Delile）Webb subsp. aegyptiaca（DC.）Maire = Savignya parviflora（Delile）Webb ■☆

349009　Savignya parviflora（Delile）Webb subsp. globosa Jafri;球形小花肉叶长柄芥■☆

349010　Savignya parviflora（Delile）Webb subsp. longistyla（Boiss. et Reut.）Maire;长柱小花肉叶长柄芥■☆

349011　Savignya parviflora（Delile）Webb var. intermedia Maire = Savignya parviflora（Delile）Webb ■☆

349012　Saviniona Webb et Berthel. = Lavatera L. ■●

349013　Saviniona acerifolia（Cav.）Webb et Berthel. = Lavatera acerifolia Cav. ■☆

349014　Saviona Pritz. = Lavatera L. ■●

349015　Saviona Pritz. = Saviniona Webb et Berthel. ■●

349016　Saxegothaea Lindl.（1851）（保留属名）;智利杉属（艾伯特王子松属,智利紫杉属）;Prince Albert Yew,Prince Albert's Yew ●☆

349017　Saxe-Gothaea Lindl. = Saxegothaea Lindl.（保留属名）●☆

349018　Saxegothaea conspicua Lindl. ;智利杉（艾伯特王子松）;Prince Albert Yew,Prince Albert's Yew,Prince Alberts Yew ●☆

349019　Saxegothaeaceae Doweld et Reveal = Podocarpaceae Endl.（保留科名）●

349020　Saxegothea Benth. = Saxegothaea Lindl.（保留属名）●☆

349021　Saxegothea Benth. et Hook. f. = Saxegothaea Lindl.（保留属名）●☆

349022　Saxe-Gothea Gay = Saxegothaea Lindl.（保留属名）●☆

349023　Saxegotheaceae Doweld et Reveal = Podocarpaceae Endl.（保留科名）●

349024　Saxicolella Engl.（1926）;石苔草属■☆

349025　Saxicolella amicorum J. B. Hall;可爱石苔草■☆

349026　Saxicolella flabellata（G. Taylor）C. Cusset;扇状石苔草■☆

349027　Saxicolella laciniata（Engl.）C. Cusset;撕裂石苔草■☆

349028　Saxicolella marginalis（G. Taylor）C. Cusset ex Cheek;尼日利亚苔草■☆

349029　Saxicolella marginalis（G. Taylor）C. Cusset ex Cheek = Butumia marginalis G. Taylor ■☆

349030　Saxicolella nana Engl. ;小石苔草■☆

349031　Saxicolella submersa（J. B. Hall）C. D. K. Cook et Rutish. ;沉水石苔草■☆

349032　Saxifraga L.（1753）;虎耳草属;Rockfoil,Saxifraga,Saxifrage,Strawberry Begonia ■☆

349033　Saxifraga Tourn. ex L. = Saxifraga L. ■

349034　Saxifraga × anglica Horny,Soják et Webr;英国虎耳草■☆

349035　Saxifraga × anglica Horny,Soják et Webr 'Cranbourne';克兰伯恩英国虎耳草■☆

349036　Saxifraga × apiculata Engl. ;短尖虎耳草■☆

349037　Saxifraga × apiculata Engl. 'Gregor Mendel';格雷戈里·门德尔短尖虎耳草■☆

349038　Saxifraga × boydi Dewar;波氏虎耳草■☆

349039　Saxifraga × boydi Dewar 'Hindhead Seedling';藏头苗波氏虎耳草■☆

349040　Saxifraga × irvingii A. S. Thomps.;欧文虎耳草■☆

349041　Saxifraga × irvingii A. S. Thomps. 'Jenkinsiae';詹金斯·欧文虎耳草■☆

349042　Saxifraga × irvingii A. S. Thomps. 'Walter Irving';瓦尔特·欧文虎耳草■☆

349043　Saxifraga × stribrnyi（Velen.）Podp.;紫序虎耳草■☆

349044　Saxifraga × urbium D. A. Webb;伦敦虎耳草;Biddy's Eyes, Chickens, Children-of-israel, Curds-and-cream, Dots-and-dashes, Edging, Gleaming Star, Goddard, Hen-and-chickens, Hundreds-and-thousands, Kiss-behind-the-garden-gate, Kiss-me, Kiss-me-love, Kiss-me-love-at-the-garden-gate, Kiss-me-love-behind-the-gardengate, Kiss-me-not, Kiss-me-quick, Ladies-in-white, Little Pink, Little-and-pretty, London Pretty, London Pride, London Tuft, Look-vp-and-kiss-me, Lorelon Pride, Meet-me-love, Mother-of-thousands, Nancy Pretty, Nancy-none-so-pretty, Nancy-pretty, None-so-pretty, Painted Lady, Peace-and-plenty, Pheasant's Feathers, Pins-and-needles, Prattling Parnell, Pretty Betsy, Pretty Betty, Pretty Lady, Pretty Nancy, Pride-o'-London, Prince's Feathers, Queen of the Mist, Queen's Feather, Queen's Feathers, St. Anne's Needlework, Sweet William ■☆

349045　Saxifraga abchasica Oett. ;阿伯哈斯■☆

349046　Saxifraga acerifolia Wakab. et Satomi;尖叶虎耳草■☆

349047　Saxifraga aculeata Balf. f. = Saxifraga mengtzeana Engl. et Irmsch. ■

349048　Saxifraga adenophora K. Koch;具腺虎耳草■☆

349049　Saxifraga adscenends L. ;上升虎耳草■☆

349050　Saxifraga aestivalis Fisch. et C. A. Mey. = Saxifraga nelsoniana D. Don ■

349051　Saxifraga afghanica Aitch. et Hemsl. ;具梗虎耳草;Afghan Rockfoil, Stalked Rockfoil ■

349052　Saxifraga aizoides L. ;番杏叶虎耳草（艾宗状虎耳草,番杏虎耳草）;Aizoon Rockfoil, Yellow Mountain Saxifrage, Yellow Saxifrage ■☆

349053　Saxifraga aizoides L. var. autumnalis（L.）Engl. et Irmsch. = Saxifraga hirculus L. ■

349054　Saxifraga aizoon Jacq. ;长绿虎耳草;Aizoon Saxifrage, Evergreen Saxifrage, Livelong Saxifrage, Live-long Saxifrage ■☆

349055　Saxifraga albertii Regel et Schmalh. ;阿氏虎耳草■☆

349056　Saxifraga altissima J. Kern. ;高虎耳草;Tall Saxifrage ■☆

349057　Saxifraga anadena Harry Sm. = Saxifraga heterotricha C. Marquand et Airy Shaw var. anadena（Harry Sm.）J. T. Pan et Gornall ■

349058　Saxifraga anadyrensis Losinsk. ;阿纳代尔虎耳草■☆

349059　Saxifraga andersonii Engl. ;短瓣虎耳草;Anderson Rockfoil ■

349060　Saxifraga androsacea L. ;点地梅虎耳草;Scree Saxifrage ■☆

349061　Saxifraga androsacea L. var. tridentata;三齿点地梅虎耳草■☆

349062　Saxifraga androsacea L. var. uniflora ?;单花点地梅虎耳草■☆

349063　Saxifraga angustata Harry Sm. ;狭叶虎耳草（川北虎耳草）;Narrow Rockfoil ■

349064　Saxifraga angustata Harry Sm. = Saxifraga microgyna Engl. et Irmsch. ■

349065　Saxifraga aphylla Sternb. ;无叶虎耳草;Tinyleaf Saxifrage ■☆

349066　Saxifraga aquatica Lapeyr. ;水生虎耳草;Water Saxifrage ■☆

349067　Saxifraga arco-vafleyi Sund. 'Arco';阿可虎耳草■☆

349068　Saxifraga aretioides Lapeyr. ;黄虎耳草;Yellow Saxifrage ■☆

349069 Saxifraga aristulata Hook. f. et Thomson；小芒虎耳草（大柱头虎耳草）；Aristulate Rockfoil ■

349070 Saxifraga aristulata Hook. f. et Thomson f. longipila（Engl. et Irmsch.）J. T. Pan ex T. C. Ku ＝Saxifraga aristulata Hook. f. et Thomson var. longipila（Engl. et Irmsch.）J. T. Pan ■

349071 Saxifraga aristulata Hook. f. et Thomson var. earistulata T. C. Ku；无毛小芒虎耳草；Hairless Aristulate Rockfoil ■

349072 Saxifraga aristulata Hook. f. et Thomson var. earistulata T. C. Ku ＝Saxifraga sublinearifolia J. T. Pan ■

349073 Saxifraga aristulata Hook. f. et Thomson var. longipila（Engl. et Irmsch.）J. T. Pan；长毛小芒虎耳草（长毛虎耳草）；Longhair Aristulate Rockfoil ■

349074 Saxifraga aristulata Hook. f. et Thomson var. microcephala Engl. et Irmsch. ＝Saxifraga aristulata Hook. f. et Thomson ■

349075 Saxifraga arundana Boiss. ＝Saxifraga dichotoma Willd. ■☆

349076 Saxifraga ascendens L.；直立虎耳草；Ascendent Saxifrage ■☆

349077 Saxifraga asiatica Hayek ＝Saxifraga oppositifolia L. ■

349078 Saxifraga aspera L.；粗糙虎耳草；Rough Saxifrage, Rough stone-break ■☆

349079 Saxifraga astilboides Losinsk.；落新妇虎耳草■☆

349080 Saxifraga atlantica Boiss. et Reut. ＝Saxifraga dichotoma Willd. ■☆

349081 Saxifraga atlantica Boiss. et Reut. subsp. carpetana（Boiss. et Reut.）Batt. ＝Saxifraga dichotoma Willd. ■☆

349082 Saxifraga atrata Engl.；黑虎耳草（阿仲茶保, 黑化虎耳草）；Black Rockfoil, Blackened Rockfoil ■

349083 Saxifraga atrata Engl. var. subcorymbosa Engl. ＝Saxifraga melanocentra Franch. ■

349084 Saxifraga atrosanguinea J. Anthony ＝Saxifraga pardanthina Hand.-Mazz. ■

349085 Saxifraga atuntsinensis W. W. Sm.；阿墩子虎耳草；Adunz Rockfoil ■

349086 Saxifraga aurantiaca Franch.；橙黄虎耳草；Orange Rockfoi ■

349087 Saxifraga aurantiaca Franch. f. lanceolata T. C. Ku；披针瓣橙黄虎耳草；Lanceolate Orange Rockfoil ■

349088 Saxifraga aurantiaca Franch. f. lanceolata T. C. Ku ＝Saxifraga unguiculata Engl. ■

349089 Saxifraga auriculata Engl. et Irmsch.；耳状虎耳草；Auriculate Rockfoil, Rarshape Rockfoil ■

349090 Saxifraga auriculata Engl. et Irmsch. var. conaensis J. T. Pan；错那虎耳草；Cuona Rockfoil ■

349091 Saxifraga autumnalis L. ＝Saxifraga hirculus L. ■

349092 Saxifraga baborensis Batt. ＝Saxifraga cymbalaria L. ■☆

349093 Saxifraga baimashanensis C. Y. Wu；白马山虎耳草；Baimashan Rockfoil ■

349094 Saxifraga balfourii Engl. et Irmsch.；马耳山虎耳草；Balfour Rockfoil ■

349095 Saxifraga balongshanensis T. C. Ku；巴郎山虎耳草；Balangshan Rockfoil ■

349096 Saxifraga balongshanensis T. C. Ku ＝Saxifraga pseudohirculus Engl. ■

349097 Saxifraga benzilanensis H. Chuang；奔子栏虎耳草；Benzilan Rockfoil ■

349098 Saxifraga bergenioides C. Marquand；紫花虎耳草；Purpleflower Rockfoil ■

349099 Saxifraga bicuspidata Hook. ＝Saxifragella bicuspidata（Hook.）Engl. ■☆

349100 Saxifraga biflora All.；二花虎耳草；Twoflower Saxifrage ■

349101 Saxifraga biflora T. C. Ku ＝Saxifraga aurantiaca Franch. ■

349102 Saxifraga bijiangensis H. Chuang；碧江虎耳草；Bijiang Rockfoil ■

349103 Saxifraga birostris Engl. et Irmsch. ＝Saxifraga davidii Franch. ■

349104 Saxifraga blinii H. Lév. ＝Saxifraga clavistaminea Engl. et Irmsch. ■

349105 Saxifraga bonatiana Engl. et Irmsch. ＝Saxifraga candelabrum Franch. ■

349106 Saxifraga brachyphylla Franch.；短叶虎耳草；Shortleaf Rockfoil ■

349107 Saxifraga brachypoda D. Don；短柄虎耳草；Shortstalk Rockfoil ■

349108 Saxifraga brachypoda D. Don var. eglandulosa（Harry Sm.）Akiyama et al. ＝Saxifraga gouldii C. E. C. Fisch. var. eglandulosa Harry Sm. ■

349109 Saxifraga brachypoda D. Don var. eglandulosa（Harry Sm.）Akiyama et al. ＝Saxifraga wardii W. W. Sm. ■

349110 Saxifraga brachypoda D. Don var. fimbriata（Ser.）Engl. et Irmsch. ＝Saxifraga wallichiana Sternb. ■

349111 Saxifraga brachypoda D. Don var. fimbriata（Wall. ex Ser.）Engl. et Irmsch. ＝Saxifraga wallichiana Sternb. ■

349112 Saxifraga brachypoda D. Don var. gouldii（C. E. C. Fisch.）Akiyama et al. ＝Saxifraga gouldii C. E. C. Fisch. ■

349113 Saxifraga bracteata D. Don；苞片虎耳草■☆

349114 Saxifraga brechypodoidea J. T. Pan；光花梗虎耳草；Nakestalk Rockfoil ■

349115 Saxifraga brevicaulis Harry Sm.；短茎虎耳草■

349116 Saxifraga brevicaulis Harry Sm. ＝Saxifraga sessiliflora Harry Sm. ■

349117 Saxifraga bronchialis L.；刺虎耳草（芒虎耳草）；Thorn Rockfoil ■

349118 Saxifraga bronchialis L. subsp. funstonii（Small）Hultén var. cherlerioides（D. Don）Engl.；米努草状虎耳草■☆

349119 Saxifraga bronchialis L. subsp. funstonii（Small）Hultén var. rebunshirensis（Engl. et Irmsch.）H. Hara；礼文虎耳草■☆

349120 Saxifraga bronchialis L. subsp. funstonii（Small）Hultén var. rebunshirensis（Engl. et Irmsch.）H. Hara f. laxa（H. Hara）T. Shimizu；疏松刺虎耳草■☆

349121 Saxifraga bronchialis L. subsp. funstonii（Small）Hultén var. rebunshirensis（Engl. et Irmsch.）H. Hara f. togakushiensis（H. Hara）Toyok. et Nosaka；户隐山虎耳草■☆

349122 Saxifraga bronchialis L. subsp. funstonii（Small）Hultén var. yuparensis（Nosaka）T. Shimizu；北海道虎耳草■☆

349123 Saxifraga bronchialis L. subsp. spinulosa Hultén ＝Saxifraga bronchialis L. ■

349124 Saxifraga brunneopunctata Harry Sm.；褐斑虎耳草■

349125 Saxifraga brunneopunctata Harry Sm. ＝Saxifraga signatella C. Marquand ■

349126 Saxifraga brunoniana Sternb. ＝Saxifraga brunonis Wall. ex Ser. ■

349127 Saxifraga brunoniana Sternb. var. majuscula Engl. et Irmsch. ＝Saxifraga brunonis Wall. ex Ser. ■

349128 Saxifraga brunoniana Sternb. var. majuscula Engl. et Irmsch. subvar. exunguiculata Engl. et Irmsch. ＝Saxifraga brunonis Wall. ex Ser. ■

349129 Saxifraga brunoniana Sternb. var. majuscula Engl. et Irmsch. subvar. unguiculata Engl. et Irmsch. ＝Saxifraga brunonis Wall. ex Ser. ■

349130 Saxifraga brunoniana Wall. ＝Saxifraga brunonis Wall. ex Ser. ■

349131 Saxifraga brunoniana Wall. ex Sternb. var. majuscula Engl. et Irmsch. ＝Saxifraga brunonis Wall. ex Ser. ■

349132 Saxifraga brunoniana Wall. ex Sternb. var. majuscula subvar.

exunguiculata Engl. et Irmsch. = Saxifraga brunonis Wall. ex Ser. ■

349133 Saxifraga brunoniana Wall. ex Sternb. var. majuscula subvar. unguiculata Engl. et Irmsch. = Saxifraga brunonis Wall. ex Ser. ■

349134 Saxifraga brunonis Wall. ex Ser. ；喜马拉雅虎耳草（布鲁诺虎耳草，褐虎耳草，须弥虎耳草）；Himalayan Rockfoil, Himalayas Rockfoil ■

349135 Saxifraga brunonis Wall. ex Ser. var. majuscula Engl. et Irmsch. ；大喜马拉雅虎耳草（大褐虎耳草）■

349136 Saxifraga buceras Harry Sm. = Saxifraga elliotii Harry Sm. ■

349137 Saxifraga bulbifera L. ；鳞茎虎耳草；Bulb-bearing Rockfoil ■☆

349138 Saxifraga bulleyana Engl. et Irmsch. ；小泡虎耳草；Bulley Rockfoil ■

349139 Saxifraga burseriana L. ；伯舌虎耳草（伯寒氏虎耳草）；Burser's Saxifrage, One-flowered Cushion Saxifrage ■☆

349140 Saxifraga burseriana L. 'Brookside'；布鲁克塞伯舌虎耳草■☆

349141 Saxifraga burseriana L. 'Crenata'；圆齿伯舌虎耳草■☆

349142 Saxifraga burseriana L. 'Gloria'；华丽伯舌虎耳草■☆

349143 Saxifraga cacuminum Harry Sm. ；顶峰虎耳草；Pinnacle Rockfoil ■

349144 Saxifraga caesia L. ；青灰虎耳草；Glaucous Saxifrage, Grey Rockfoil ■☆

349145 Saxifraga caespitosa L. ；簇生虎耳草；California Saxifrage, Tufted Saxifrage ■☆

349146 Saxifraga calcicola J. Anthony = Saxifraga likiangensis Franch. ■

349147 Saxifraga callosa Sm. ；硬叶虎耳草（舌状虎耳草，硬虎耳草）；Hard Rockfoil, Thick-leaved Saxifrage, Tongue Saxifrage ■☆

349148 Saxifraga calycina Sternb. ；萼状虎耳草■☆

349149 Saxifraga canaliculata Boiss. et Reut. ；沟茎虎耳草■☆

349150 Saxifraga candelabrum Franch. ；灯架虎耳草（灯台虎耳草，虎耳草）；Lampstand Rockfoil ■

349151 Saxifraga candelabrum Franch. var. patentiramea Engl. et Irmsch. = Saxifraga candelabrum Franch. ■

349152 Saxifraga cardiophylla Franch. ；心叶虎耳草；Cordateleaf Rockfoil ■

349153 Saxifraga carinata Oett. ；龙骨状虎耳草■☆

349154 Saxifraga carnosula Mattf. ；肉质虎耳草（单脉虎耳草）；Carnose Rockfoil ■

349155 Saxifraga carpetana Boiss. et Reut. ；卡尔佩特虎耳草■☆

349156 Saxifraga carpetana Boiss. et Reut. subsp. atlantica (Boiss. et Reut.) Romo = Saxifraga carpetana Boiss. et Reut. var. atlantica (Boiss. et Reut.) Engl. et Irmsch. ■☆

349157 Saxifraga carpetana Boiss. et Reut. var. atlantica (Boiss. et Reut.) Engl. et Irmsch. = Saxifraga dichotoma Willd. ■☆

349158 Saxifraga cartilaginea Willd. ex Sternb. ；软骨虎耳草；Gristleleaf Rockfoil, Gristleleaf Saxifrage ■☆

349159 Saxifraga caucasica Sommier et H. Lév. ；高加索虎耳草■☆

349160 Saxifraga caveana W. W. Sm. ；近岩梅虎耳草；Cavea Rockfoil ■

349161 Saxifraga caveana W. W. Sm. var. lanceolata J. T. Pan；狭萼虎耳草；Lanceolate Cavea Rockfoil ■

349162 Saxifraga cernua L. ；零余虎耳草（点头虎耳草，分枝虎耳草，零余子虎耳草，珠芽虎耳草）；Branched Rockfoil, Drooping Saxifrage, Nodding Rockfoil ■

349163 Saxifraga cernua L. f. bulbillosa Engl. et Irmsch. = Saxifraga cernua L. ■

349164 Saxifraga cernua L. f. ramosa J. G. Gmel. = Saxifraga cernua L. ■

349165 Saxifraga cernua L. f. simplicissima Ledeb. = Saxifraga cernua L. ■

349166 Saxifraga cernua L. var. bulbillosa Engl. et Irmsch. ；鳞茎点头虎耳草（鳞茎虎耳草）；Bulbiferous Rockfoil ■☆

349167 Saxifraga cernua L. var. linnaeana Ser. = Saxifraga cernua L.

349168 Saxifraga chaffanjoni H. Lév. = Saxifraga stolonifera Curtis ■

349169 Saxifraga changputungensis H. Chuang；菖蒲桶虎耳草；Changputong Rockfoil ■

349170 Saxifraga cherlerioides D. Don = Saxifraga bronchialis L. subsp. funstonii (Small) Hultén var. cherlerioides (D. Don) Engl. ■☆

349171 Saxifraga cherlerioides D. Don var. rebunshirensis (Engl. et Irmsch.) H. Hara = Saxifraga bronchialis L. subsp. funstonii (Small) Hultén var. rebunshirensis (Engl. et Irmsch.) H. Hara ■☆

349172 Saxifraga cherlerioides D. Don var. rebunshirensis (Engl. et Irmsch.) H. Hara f. togakushiensis (H. Hara) Ohwi = Saxifraga bronchialis L. subsp. funstonii (Small) Hultén var. rebunshirensis (Engl. et Irmsch.) H. Hara f. togakushiensis (H. Hara) Toyok. et Nosaka ■☆

349173 Saxifraga cherlerioides D. Don var. togakushiensis H. Hara = Saxifraga bronchialis L. subsp. funstonii (Small) Hultén var. rebunshirensis (Engl. et Irmsch.) H. Hara f. togakushiensis (H. Hara) Toyok. et Nosaka ■☆

349174 Saxifraga chinensis Lour. = Saxifraga stolonifera Curtis ■

349175 Saxifraga chionophila Franch. ；雪地虎耳草；Snowloving Rockfoil ■

349176 Saxifraga chrysantha Franch. = Saxifraga chrysanthoides Engl. et Irmsch. ■

349177 Saxifraga chrysanthoides Engl. et Irmsch. ；拟黄花虎耳草；Yellowflower Rockfoil ■

349178 Saxifraga chrysosplenifolia Boiss. ；金腰叶虎耳草（金腰虎耳草）；Gold-saxifrage-leaf Rockfoil ■☆

349179 Saxifraga chumbiensis Engl. et Irmsch. ；春丕虎耳草；Chunpi Rockfoil ■

349180 Saxifraga ciliata Royle = Bergenia ciliata (Haw.) Sternb. ■☆

349181 Saxifraga ciliatopetala (Engl. et Irmsch.) J. T. Pan；毛瓣虎耳草；Hairypetal Rockfoil ■

349182 Saxifraga ciliatopetala (Engl. et Irmsch.) J. T. Pan var. ciliata J. T. Pan = Saxifraga ciliatopetala (Engl. et Irmsch.) J. T. Pan ■

349183 Saxifraga ciliatopetala (Engl. et Irmsch.) J. T. Pan var. ciliata J. T. Pan；毛缘虎耳草■

349184 Saxifraga cinerascens Engl. et Irmsch. ；灰虎耳草；Grey Rockfoil ■

349185 Saxifraga cinerascens Engl. et Irmsch. f. major Engl. et Irmsch. = Saxifraga cinerascens Engl. et Irmsch. ■

349186 Saxifraga clavistaminea Engl. et Irmsch. ；棒蕊虎耳草；Clavatestamen Rockfoil ■

349187 Saxifraga clavistamineoides T. C. Ku = Saxifraga pallida Wall. ex Ser. ■

349188 Saxifraga clivorum Harry Sm. ；截叶虎耳草；Truncateleaf Rockfoil ■

349189 Saxifraga coarctata W. W. Sm. ；矮虎耳草；Dwarf Rockfoil ■

349190 Saxifraga cochlearis Rchb. ；蜗牛虎耳草（匙叶虎耳草）；Snail Saxifrage ■☆

349191 Saxifraga cochlearis Rchb. 'Minor'；小型匙叶虎耳草■☆

349192 Saxifraga colchica Albov；黑海虎耳草■☆

349193 Saxifraga columnaris Schmalh. ；圆柱虎耳草■☆

349194 Saxifraga confertifolia Engl. et Irmsch. ；聚叶虎耳草；Fascicleleaf Rockfoil ■

349195 Saxifraga confertifolia Engl. et Irmsch. = Saxifraga aurantiaca Franch. ■

349196　Saxifraga confertifolia Engl. et Irmsch. var. glabrifolia Engl. et Irmsch. = Saxifraga unguiculata Engl. ■

349197　Saxifraga congestiflora Engl. et Irmsch. ;密花虎耳草■

349198　Saxifraga conifera Coss. et Durieu；松柏虎耳草；Conifer Saxifrage ■☆

349199　Saxifraga consanguinea W. W. Sm. ;棒腺虎耳草；Clavate-glandular Rockfoil ■

349200　Saxifraga contrarea Harry Sm. ;对叶虎耳草；Opposite Rockfoil ■

349201　Saxifraga cordifolia Haw. = Bergenia crassifolia（L.）Fritsch ■

349202　Saxifraga cordigera Hook. f. et Thomson；心虎耳草；Heart Rockfoil ■

349203　Saxifraga coriifolia（Sommier et H. Lév.）Grossh. ;革叶虎耳草■☆

349204　Saxifraga cortusifolia Siebold et Zucc. ; 大字草；Cortuseleaf Rockfoil ■☆

349205　Saxifraga cortusifolia Siebold et Zucc. = Saxifraga fortunei Hook. f. ■

349206　Saxifraga cortusifolia Siebold et Zucc. = Saxifraga rufescens Balf. f. ■

349207　Saxifraga cortusifolia Siebold et Zucc. f. atropurpurea（Makino）Nemoto；暗紫大字草■☆

349208　Saxifraga cortusifolia Siebold et Zucc. f. breviloba H. Hara；浅裂大字草■☆

349209　Saxifraga cortusifolia Siebold et Zucc. f. incisa（Takeda）Nemoto；锐裂大字草■☆

349210　Saxifraga cortusifolia Siebold et Zucc. f. partita（Makino）Ohwi = Saxifraga acerifolia Wakab. et Satomi ■☆

349211　Saxifraga cortusifolia Siebold et Zucc. f. variegata（Nakai）H. Hara；斑点大字草■☆

349212　Saxifraga cortusifolia Siebold et Zucc. var. fortunei（Hook. f.）Maxim. = Saxifraga fortunei Hook. f. ■

349213　Saxifraga cortusifolia Siebold et Zucc. var. partita Makino = Saxifraga acerifolia Wakab. et Satomi ■☆

349214　Saxifraga cortusifolia Siebold et Zucc. var. stolonifera（Makino）Koidz. ;匍匐大字草■☆

349215　Saxifraga corymbosa Hook. f. et Thomson = Saxifraga hookeri Engl. et Irmsch. ■

349216　Saxifraga cotyledon L. ;少妇虎耳草（圣塔虎耳草）；Jungfrau Saxifrage, Mountain Saxifrage, Pyramidal Saxifrage, Thick-leaved Saxifrage ■☆

349217　Saxifraga crassicarpa A. M. Johnson = Saxifraga pensylvanica L. ■

349218　Saxifraga crassifolia（L.）Fritsch var. elliptica Ledeb. = Bergenia crassifolia（L.）Fritsch ■

349219　Saxifraga crassifolia（L.）Fritsch var. obovata Ser. = Bergenia crassifolia（L.）Fritsch ■

349220　Saxifraga crassifolia L. = Bergenia crassifolia（L.）Fritsch ■

349221　Saxifraga crassifolia L. var. elliptica Ledeb. = Bergenia crassifolia（L.）Fritsch ■

349222　Saxifraga crassifolia L. var. obovata Ser. = Bergenia crassifolia（L.）Fritsch ■

349223　Saxifraga crassulifolia Engl. = Saxifraga atuntsinensis W. W. Sm. ■

349224　Saxifraga crinalis Franch. = Saxifraga tsangchanensis Franch. ■

349225　Saxifraga crustata Vest;萨尔斯堡虎耳草；Salzburg Saxifrage ■☆

349226　Saxifraga culcitosa Mattf. ;枕状虎耳草；Pillowlike Rockfoil ■

349227　Saxifraga cuneifolia L. ; 楔叶虎耳草；Lesser Londonpride, Spoon-leaved Saxifrage, Wedgeleaf Saxifrage, Wedge-leaved Saxifrage, Wood Saxifrage ■☆

349228　Saxifraga cuscutiformis Lodd. = Saxifraga stolonifera Curtis ■

349229　Saxifraga cymbalaria L. ;船状虎耳草（常春藤叶虎耳草）；Celandine Saxifrage, Ivyleaf Saxifrage, Persian Saxifrage ■☆

349230　Saxifraga cymbalaria L. var. atlantica Batt. = Saxifraga cymbalaria L. ■☆

349231　Saxifraga cymiformis T. C. Ku;假聚伞虎耳草■

349232　Saxifraga cymiformis T. C. Ku = Saxifraga glaucophylla Franch. ■

349233　Saxifraga dahaiensis H. Chuang；大海虎耳草；Dahai Rockfoil ■

349234　Saxifraga dahurica Willd. ;达呼尔虎耳草■☆

349235　Saxifraga daochengensis J. T. Pan；稻城虎耳草；Daocheng Rockfoil ■

349236　Saxifraga davidii Franch. ;双喙虎耳草；Twinbill Rockfoil ■

349237　Saxifraga debeauxii Pomel = Saxifraga dichotoma Willd. ■☆

349238　Saxifraga decipiens Ehrh. = Saxifraga rosacea Moench ■☆

349239　Saxifraga decora Harry Sm. ;滇藏虎耳草■

349240　Saxifraga decussata J. Anthony；十字虎耳草（矮生虎耳草）；Deccusate Rockfoil ■

349241　Saxifraga delavayi Franch. = Bergenia purpurascens（Hook. f. et Thomson）Engl. ■

349242　Saxifraga demnatensis Batt. ;德姆纳特虎耳草■☆

349243　Saxifraga densifoliata Engl. et Irmsch. ;密叶虎耳草；Denseleaf Rockfoil ■

349244　Saxifraga densifoliata Engl. et Irmsch. var. nedongensis J. T. Pan;乃东虎耳草；Naidong Rockfoil ■

349245　Saxifraga deqenensis C. Y. Wu;德钦虎耳草；Deqin Rockfoil ■

349246　Saxifraga desoulavyi Oett. ;代苏虎耳草■

349247　Saxifraga dianxibeiensis J. T. Pan;滇西北虎耳草（高山异叶虎耳草）；Dianxibei Rockfoil ■

349248　Saxifraga diapensia Harry Sm. var. glabrisepala J. T. Pan = Saxifraga elliptica Engl. et Irmsch. ■

349249　Saxifraga diapensis Harry Sm. ;岩梅虎耳草；Diapensia Rockfoil ■

349250　Saxifraga dichotoma Willd. ;二歧虎耳草■☆

349251　Saxifraga dichotoma Willd. var. hervieri（Debeaux）Engl. et Irmsch. = Saxifraga dichotoma Willd. ■☆

349252　Saxifraga dielsiana Engl. et Irmsch. ;川西虎耳草；Diels Rockfoil ■

349253　Saxifraga diffusicallosa C. Y. Wu;散痂虎耳草；Diffusicallose Rockfoil ■

349254　Saxifraga dinnikii Schmalh. ;迪尼虎耳草■☆

349255　Saxifraga divaricata Engl. et Irmsch. ;叉枝虎耳草；Divaricate Rockfoil ■

349256　Saxifraga diversifolia Wall. ex Ser. ;异叶虎耳草（山羊参）；Diversefolious Rockfoil ■

349257　Saxifraga diversifolia Wall. ex Ser. = Saxifraga erectisepala J. T. Pan ■

349258　Saxifraga diversifolia Wall. ex Ser. f. alpina Engl. et Irmsch. = Saxifraga dianxibeiensis J. T. Pan ■

349259　Saxifraga diversifolia Wall. ex Ser. f. amplexifolia Irmsch. ;抱茎虎耳草；Amplexicaul Rockfoil ■☆

349260　Saxifraga diversifolia Wall. ex Ser. f. amplexifolia Irmsch. = Saxifraga diversifolia Wall. ex Ser. ■

349261　Saxifraga diversifolia Wall. ex Ser. f. angustibracteata Engl. et Irmsch. = Saxifraga diversifolia Wall. ex Ser. var. angustibracteata（Engl. et Irmsch.）J. T. Pan ■

349262　Saxifraga diversifolia Wall. ex Ser. f. foliata Engl. et Irmsch. = Saxifraga diversifolia Wall. ex Ser. ■

349263　Saxifraga diversifolia Wall. ex Ser. f. haematophylla（Franch.）Engl. et Irmsch. = Saxifraga diversifolia Wall. ex Ser. ■

349264　Saxifraga diversifolia Wall. ex Ser. f. parviflora（Franch.）Engl.

et Irmsch. = Saxifraga glaucophylla Franch. ■

349265 Saxifraga diversifolia Wall. ex Ser. var. angustibracteata (Engl. et Irmsch.) J. T. Pan; 狭苞异叶虎耳草; Narrowbract Diversefolious Rockfoil ■

349266 Saxifraga diversifolia Wall. ex Ser. var. haematophylla Franch. = Saxifraga diversifolia Wall. ex Ser. ■

349267 Saxifraga diversifolia Wall. ex Ser. var. lanceolata Ser. = Saxifraga diversifolia Wall. ex Ser. ■

349268 Saxifraga diversifolia Wall. ex Ser. var. moorcroftiana Ser. = Saxifraga moorcroftiana Wall. ex Sternb. ■

349269 Saxifraga diversifolia Wall. ex Ser. var. parnassifolia (D. Don) Engl.; 梅花草叶虎耳草(梅花异叶虎耳草); Parnassia Rockfoil, Parnassialeaf Rockfoil ■☆

349270 Saxifraga diversifolia Wall. ex Ser. var. parnassifolia a (D. Don) Ser. = Saxifraga parnassifolia D. Don ■

349271 Saxifraga diversifolia Wall. ex Ser. var. parviflora Franch. = Saxifraga glaucophylla Franch. ■

349272 Saxifraga diversifolia Wall. ex Ser. var. parviflora Franch. = Saxifraga diversifolia Wall. ex Ser. ■

349273 Saxifraga diversifolia Wall. ex Ser. var. soulieana Engl. et Irmsch. = Saxifraga egregia Engl. ■

349274 Saxifraga dongchuanensis H. Chuang; 东川虎耳草; Dongchuan Rockfoil ■

349275 Saxifraga dongwanensis H. Chuang; 东旺虎耳草; Dongwang Rockfoil ■

349276 Saxifraga doyalana Harry Sm.; 白瓣虎耳草; Whitepetal Rockfoil ■

349277 Saxifraga drabiformis Franch.; 莛苈虎耳草; Drablike Rockfoil ■

349278 Saxifraga draboides C. Y. Wu; 中甸虎耳草; Zhongdian Rockfoil ■

349279 Saxifraga dshagalensis Engl.; 无爪虎耳草; Clawless Rockfoil ■

349280 Saxifraga dumetorum Balf. f. = Saxifraga stolonifera Curtis ■

349281 Saxifraga dungbooi Engl. et Irmsch.; 邓波虎耳草 ■

349282 Saxifraga dunniana H. Lév. = Saxifraga diversifolia Wall. ex Ser. ■

349283 Saxifraga dunniana H. Lév. = Saxifraga glaucophylla Franch. ■

349284 Saxifraga echinophora H. Lév. = Saxifraga strigosa Wall. ex Ser. ■

349285 Saxifraga eglandulosa Engl.; 长毛梗虎耳草; Glandless Rockfoil ■

349286 Saxifraga egregia Engl.; 优越虎耳草; Superior Rockfoil ■

349287 Saxifraga egregia Engl. var. eciliata J. T. Pan; 无睫毛虎耳草; Eciliate Superior Rockfoil ■

349288 Saxifraga egregia Engl. var. xiaojinensis J. T. Pan; 小金虎耳草; Xiaojin Superior Rockfoil ■

349289 Saxifraga egregioides J. T. Pan; 矮优越虎耳草; Dwarf Superior Rockfoil ■

349290 Saxifraga elatinoides Hand.-Mazz.; 沟繁缕虎耳草; Waterwortlike Rockfoil ■

349291 Saxifraga elliotii Harry Sm.; 索白拉虎耳草; Suobaila Rockfoil ■

349292 Saxifraga elliptica Engl. et Irmsch.; 光萼虎耳草; Elliptic Rockfoil, Glabrous-calyx Rockfoil ■

349293 Saxifraga embergeri Maire; 恩贝格尔虎耳草 ■☆

349294 Saxifraga engleriana Harry Sm.; 藏南虎耳草; S. Xizang Rockfoil, Zangnan Rockfoil ■

349295 Saxifraga epiphylla Gornall et O. Ohba; 卵心叶虎耳草 ■

349296 Saxifraga erectisepala J. T. Pan; 直萼虎耳草; Erectsepal Rockfoil ■

349297 Saxifraga erinacea Harry Sm.; 猥状虎耳草(猬状虎耳草); Erinuslike Rockfoil ■

349298 Saxifraga erosa Pursh; 蚀状虎耳草; Eroded Rockfoil, Eroded Saxifrage ■☆

349299 Saxifraga eryuan J. T. Pan = Saxifraga peplidifolia Franch. ■

349300 Saxifraga eschscholtzii Cham. ex DC.; 塞地虎耳草; Eschscholtz Saxifrage ■☆

349301 Saxifraga eschscholtzii Sternb. = Saxifraga hemisphaerica Hook. f. et Thomson ■

349302 Saxifraga evolvuloides Wall. ex Selinge = Saxifraga hispidula D. Don ■

349303 Saxifraga exarata Vill.; 具沟虎耳草; Furrowed Saxifrage, White Musky Saxifrage ■☆

349304 Saxifraga exarata Vill. subsp. moschata (Wulfen) Cavill.; 香虎耳草 ■☆

349305 Saxifraga exilis Steph.; 细弱虎耳草 ■☆

349306 Saxifraga ferdinandi-coburgi Keller et Sund.; 菲迪南氏虎耳草 (巴尔干虎耳草); Ferdinand-coburg Saxifrage ■☆

349307 Saxifraga fiagrans Harry Sm. = Saxifraga tangutica Engl. ■

349308 Saxifraga filicaulis Wall. ex Ser.; 线茎虎耳草; Filicauline Rockfoil ■

349309 Saxifraga filifolia Anthony; 细叶虎耳草 ■

349310 Saxifraga filifolia Anthony = Saxifraga llonakhensis W. W. Sm. ■

349311 Saxifraga filifolia Anthony var. rosettifolia C. Y. Wu ex H. Chuang; 小线叶虎耳草 ■

349312 Saxifraga fimbriata Wall. ex Ser. = Saxifraga wallichiana Sternb. ■

349313 Saxifraga fimbriatoides J. T. Pan = Saxifraga wallichiana Sternb. ■

349314 Saxifraga finitima W. W. Sm.; 区限虎耳草; Finite Rockfoil ■

349315 Saxifraga firma Litv.; 硬虎耳草 ■☆

349316 Saxifraga flabellifolia Franch. = Saxifraga rufescens Balf. f. var. flabellifolia C. Y. Wu et J. T. Pan ■

349317 Saxifraga flaccida J. T. Pan; 柔弱虎耳草; Flaccid Rockfoil ■

349318 Saxifraga flagellarioides Engl. = Saxifraga mucronulata Royle ■

349319 Saxifraga flagellaris Sternb. et Willd.; 鞭状虎耳草(匍匐鞭状虎耳草, 匍匐虎耳草); Flagellate Rockfoil ■

349320 Saxifraga flagellaris Willd. ex Sternb. = Saxifraga stenophylla Royle ■

349321 Saxifraga flagellaris Willd. ex Sternb. subsp. euflagellaris Engl. et Irmsch. = Saxifraga stenophylla Royle ■

349322 Saxifraga flagellaris Willd. ex Sternb. subsp. komarovii ? = Saxifraga komarovii Losinsk. ■☆

349323 Saxifraga flagellaris Willd. ex Sternb. subsp. megistantha Hand.-Mazz. = Saxifraga stenophylla Royle ■

349324 Saxifraga flagellaris Willd. ex Sternb. subsp. mucronulata (Royle) Engl. et Irmsch. = Saxifraga mucronulata Royle ■

349325 Saxifraga flagellaris Willd. ex Sternb. subsp. sikkimensis Hultén = Saxifraga mucronulatoides J. T. Pan ■

349326 Saxifraga flagellaris Willd. ex Sternb. subsp. stenophylla (Royle) Hultén = Saxifraga stenophylla Royle ■

349327 Saxifraga flagellaris Willd. ex Sternb. subsp. stenopylla Hultén = Saxifraga stenophylla Royle ■

349328 Saxifraga flagellaris Willd. ex Sternb. var. komarovii ? = Saxifraga komarovii Losinsk. ■☆

349329 Saxifraga flagellaris Willd. ex Sternb. var. mucronulata (Royle) C. B. Clarke = Saxifraga mucronulata Royle ■

349330 Saxifraga flagellaris Willd. ex Sternb. var. stenophylla ? = Saxifraga stenophylla Royle ■

349331 Saxifraga flagellaris Willd. ex Sternb. var. stenosepala Trautv. = Saxifraga stenophylla Royle ■

349332 Saxifraga flagellaris Willd. ex Sternb. var. stenosepala Trautv. f. alta Engl. et Irmsch. = Saxifraga stenophylla Royle ■

349333　Saxifraga flagellaris Willd. ex Sternb. var. stenosepala Trautv. f. humilis Engl. et Irmsch. = Saxifraga stenophylla Royle ■

349334　Saxifraga flagellaris Willd. ex Sternb. var. stenosepala Trautv. f. pauciflora Engl. et Irmsch. = Saxifraga stenophylla Royle ■

349335　Saxifraga flagrans Harry Sm. = Saxifraga tangutica Engl. ■

349336　Saxifraga flagrans Harry Sm. var. platyphylla Harry Sm. = Saxifraga tangutica Engl. var. platyphylla（Harry Sm.）J. T. Pan ■

349337　Saxifraga flexilis W. W. Sm. ;曲茎虎耳草;Flexuose Rockfoil ■

349338　Saxifraga florulenta Moretti;欧洲小花虎耳草■☆

349339　Saxifraga foliolosa R. Br. ;多小叶虎耳草■☆

349340　Saxifraga foliosa Orsini ex Ten. ;茂叶虎耳草■☆

349341　Saxifraga forbesii Vasey = Saxifraga pensylvanica L. ■

349342　Saxifraga forbesii Vasey f. bracteosa G. W. Burns = Saxifraga pensylvanica L. ■

349343　Saxifraga forrestii Engl. et Irmsch. ;玉龙虎耳草;Forrest Rockfoil ■

349344　Saxifraga fortunei Hook. f. ;齿瓣虎耳草（福氏虎耳草,华中虎耳草）;Fortune Rockfoil ■

349345　Saxifraga fortunei Hook. f. 'Rubrifolia';红叶福氏虎耳草■☆

349346　Saxifraga fortunei Hook. f. var. alpina（Matsum. et Nakai）Nakai;高山齿瓣虎耳草■☆

349347　Saxifraga fortunei Hook. f. var. alpina（Matsum. et Nakai）Nakai f. rubrifolia Honda;红花高山齿瓣虎耳草■☆

349348　Saxifraga fortunei Hook. f. var. crassa Nakai = Saxifraga fortunei Hook. f. var. alpina（Matsum. et Nakai）Nakai ■☆

349349　Saxifraga fortunei Hook. f. var. incisolobata（Engl. et Irmsch.）Nakai = Saxifraga fortunei Hook. f. var. alpina（Matsum. et Nakai）Nakai ■☆

349350　Saxifraga fortunei Hook. f. var. jotanii（Honda）Wakab. ;太白虎耳草（纤细虎耳草）;Joseph Rockfoil,Taibai Rockfoil ■

349351　Saxifraga fortunei Hook. f. var. koraiensis Nakai;镜叶虎耳草;Korean Rockfoil ■

349352　Saxifraga fortunei Hook. f. var. minima Nakai = Saxifraga fortunei Hook. f. var. obtusocuneata（Makino）Nakai f. minima（Nakai）Masam. ■☆

349353　Saxifraga fortunei Hook. f. var. mutabilis（Koidz.）H. Nakai et H. Ohashi = Saxifraga fortunei Hook. f. var. alpina（Matsum. et Nakai）Nakai ■☆

349354　Saxifraga fortunei Hook. f. var. obtusocuneata（Makino）Nakai;钝楔齿瓣虎耳草■☆

349355　Saxifraga fortunei Hook. f. var. obtusocuneata（Makino）Nakai f. minima（Nakai）Masam. ;小钝楔齿瓣虎耳草■☆

349356　Saxifraga fortunei Hook. f. var. partita（Makino）Nakai = Saxifraga acerifolia Wakab. et Satomi ☆

349357　Saxifraga fortunei Hook. f. var. pilosissima Nakai;多毛齿瓣虎耳草■☆

349358　Saxifraga fortunei Hook. f. var. suwoensis Nakai;诹访虎耳草■☆

349359　Saxifraga fortunei Hook. f. var. tricolor Lem. = Saxifraga stolonifera Curtis ■

349360　Saxifraga frederici-augusti Biasol. ;石莲虎耳草■☆

349361　Saxifraga fusca Maxim. ;棕色虎耳草■☆

349362　Saxifraga fusca Maxim. f. intermedia H. Hara = Saxifraga fusca Maxim. ■☆

349363　Saxifraga fusca Maxim. f. kurilensis（Ohwi）Ohwi = Saxifraga fusca Maxim. var. kurilensis Ohwi ☆

349364　Saxifraga fusca Maxim. var. divaricata Franch. et Sav. =

Saxifraga fusca Maxim. ■☆

349365　Saxifraga fusca Maxim. var. kiusiana H. Hara;九州虎耳草■☆

349366　Saxifraga fusca Maxim. var. kurilensis Ohwi;千岛虎耳草■☆

349367　Saxifraga gageana Engl. et Irmsch. = Saxifraga kingiana Engl. et Irmsch.

349368　Saxifraga gageana W. W. Sm. = Saxifraga melanocentra Franch. ■

349369　Saxifraga gasterostens H. Lév. = Saxifraga gemmipara Franch. ■

349370　Saxifraga gatogombensis Engl. = Saxifraga aurantiaca Franch. ■

349371　Saxifraga gatogombensis Engl. = Saxifraga unguiculata Engl. ■

349372　Saxifraga gedangensis J. T. Pan;格当虎耳草;Gedang Rockfoil ■

349373　Saxifraga geifolia Balf. f. = Saxifraga mengtzeana Engl. et Irmsch.

349374　Saxifraga gemmigera Engl. ex Diels;芽虎耳草（小芽虎耳草）;Budbearing Rockfoil ■

349375　Saxifraga gemmigera Engl. ex Diels var. gemmuligera（Engl.）J. T. Pan et Gornall;小芽虎耳草（小伞虎耳草）;Smallbud Rockfoil ■

349376　Saxifraga gemmipara Franch. ;芽生虎耳草;Bud-growing Rockfoil,Bud-producing Rockfoil ■

349377　Saxifraga gemmuligera（Engl.）Engl. = Saxifraga gemmigera Engl. ex Diels var. gemmuligera（Engl.）J. T. Pan et Gornall ■

349378　Saxifraga gemmuligera Engl. = Saxifraga gemmigera Engl. ex Diels var. gemmuligera（Engl.）J. T. Pan et Gornall ■

349379　Saxifraga georgei J. Anthony;对生叶虎耳草;Oppositeleaf Rockfoil ■

349380　Saxifraga geranioides L. ;牻牛儿苗虎耳草;Geranium Saxifrage ■☆

349381　Saxifraga geum L. = Saxifraga hirsuta L. ■☆

349382　Saxifraga giraldiana Engl. ex Diels;秦岭虎耳草（太白虎耳草）;Qinling Rockfoil ■

349383　Saxifraga giraldiana Engl. ex Diels var. biondiana Engl. = Saxifraga giraldiana Engl. ex Diels ■

349384　Saxifraga giraldiana Engl. ex Diels var. hupehensis Engl. = Saxifraga giraldiana Engl. ex Diels ■

349385　Saxifraga glabricaulis Harry Sm. ;光茎虎耳草;Smoothstem Rockfoil ■

349386　Saxifraga glacialis Harry Sm. ;冰雪虎耳草;Glacial Rockfoil,Ice Rockfoil ■

349387　Saxifraga glacialis Harry Sm. var. rubra J. Anthony = Saxifraga glacialis Harry Sm. ■

349388　Saxifraga glandulosa Wall. = Saxifraga brachypoda D. Don ■

349389　Saxifraga glaucophylla Franch. ;灰叶虎耳草;Greyleaf Rockfoil ■

349390　Saxifraga globulifera Desf. ;球花虎耳草;Globularflower Rockfoil ■☆

349391　Saxifraga globulifera Desf. subsp. oranensis（Munby）Batt. = Saxifraga globulifera Desf. ■☆

349392　Saxifraga globulifera Desf. subsp. spathulata（Desf.）Engl. et Irmsch. = Saxifraga globulifera Desf. ■☆

349393　Saxifraga globulifera Desf. subsp. trabutiana（Engl. et Irmsch.）Maire = Saxifraga trabutiana Engl. et Irmsch. ■☆

349394　Saxifraga globulifera Desf. var. acutisepala Maire = Saxifraga trabutiana Engl. et Irmsch. ■☆

349395　Saxifraga globulifera Desf. var. coronata Emb. et Maire = Saxifraga globulifera Desf. ■☆

349396　Saxifraga globulifera Desf. var. divaricata Pau = Saxifraga globulifera Desf. ■☆

349397　Saxifraga globulifera Desf. var. gibraltarica Ser. = Saxifraga globulifera Desf. ■☆

349398　Saxifraga globulifera Desf. var. glandulosissima Maire et Wilczek

= Saxifraga globulifera Desf. ■☆

349399　Saxifraga globulifera Desf. var. granatensis（Boiss. et Reut.）Engl. = Saxifraga globulifera Desf. ■☆

349400　Saxifraga globulifera Desf. var. guruguensis Sennen = Saxifraga globulifera Desf. ■☆

349401　Saxifraga globulifera Desf. var. integrifolia Pons et Quézel = Saxifraga globulifera Desf. ■☆

349402　Saxifraga globulifera Desf. var. major Batt. = Saxifraga trabutiana Engl. et Irmsch. ■☆

349403　Saxifraga globulifera Desf. var. minuscula（Pau et Font Quer）Emb. et Maire = Saxifraga globulifera Desf. ■☆

349404　Saxifraga globulifera Desf. var. oranensis（Munby）Engl. = Saxifraga globulifera Desf. ■☆

349405　Saxifraga globulifera Desf. var. oscillans（Pau）Maire = Saxifraga globulifera Desf. ■☆

349406　Saxifraga globulifera Desf. var. pseudogranatensis Humbert et Maire = Saxifraga globulifera Desf. ■☆

349407　Saxifraga globulifera Desf. var. pseudomaweana Font Quer = Saxifraga globulifera Desf. ■☆

349408　Saxifraga globulifera Desf. var. spathulata（Desf.）Engl. et Irmsch. = Saxifraga globulifera Desf. ■☆

349409　Saxifraga globulifera Desf. var. villigemma Maire = Saxifraga globulifera Desf. ■☆

349410　Saxifraga gonggashanensis J. T. Pan；贡嘎山虎耳草；Gonggashan Rockfoil ■

349411　Saxifraga gongshanensis T. C. Ku；小刚毛虎耳草（贡山虎耳草）；Gongshan Rockfoil, Setulose Rockfoil ■

349412　Saxifraga gouldii C. E. C. Fisch.；顶腺虎耳草 ■

349413　Saxifraga gouldii C. E. C. Fisch. = Saxifraga wardii W. W. Sm. ■

349414　Saxifraga gouldii C. E. C. Fisch. var. eglandulosa Harry Sm.；无顶腺虎耳草 ■

349415　Saxifraga gouldii C. E. C. Fisch. var. eglandulosa Harry Sm. = Saxifraga wardii W. W. Sm. ■

349416　Saxifraga granatensis Boiss. et Reut. = Saxifraga globulifera Desf. ■☆

349417　Saxifraga granatensis Boiss. et Reut. var. minuscula Pau et Font Quer = Saxifraga globulifera Desf. ■☆

349418　Saxifraga grandipetala（Engl. et Irmsch.）Losinsk.；大瓣虎耳草 ☆

349419　Saxifraga granulata L.；细粒虎耳草（草地虎耳草，颗粒虎耳草）；Billy Buttons, Bulbous Saxifrage, Darmell Goddard, Dry Cuckoo, Dryland Cuckoo, Fair Maid of France, Fair Maids, Fair Maids of France, First of May, Lady's Featherbeds, Meadow Rockfoil, Meadow Saxifrage, Milkmaids, Pretty Maids, Sassifrax, Sengreen, Snow-on-the-mountain, Stonebreak, Stone-break, Strawberry Geranium, Sundcorns, White Saxifrage, White Stoncbreak, White Stone-break ■☆

349420　Saxifraga granulata L. 'Plena'；重瓣颗粒虎耳草 ■☆

349421　Saxifraga granulata L. var. glaucescens（Boiss. et Reut.）Engl. = Saxifraga granulata L. ■☆

349422　Saxifraga granulata L. var. mauiritii Sennen = Saxifraga granulata L. ■☆

349423　Saxifraga granulifera Harry Sm.；珠芽虎耳草 ■

349424　Saxifraga granulifera Harry Sm. = Saxifraga cernua L. ■

349425　Saxifraga grisebachii Degen et Dörfl.；格里泽巴赫虎耳草；Grisebach Saxifrage ■☆

349426　Saxifraga gyacaensis J. T. Pan = Saxifraga sessiliflora Harry Sm. ■

349427　Saxifraga gyalana C. Marquand et Airy Shaw；加拉虎耳草；Jiala Rockfoil ■

349428　Saxifraga haagi Steud.；哈格氏虎耳草；Haag Saxifrage ■☆

349429　Saxifraga habaensis C. Y. Wu ex H. Chuang；拟繁缕虎耳草；Haba Rockfoil ■

349430　Saxifraga haematochroa Harry Sm. = Saxifraga bergenioides C. Marquand ■

349431　Saxifraga haplophylloides Engl. et Irmsch.；六痂虎耳草（轮叶虎耳草）；Six-scab Rockfoil, Verticillateleaf Rockfoil ■

349432　Saxifraga hastigera H. Lév. = Saxifraga giraldiana Engl. ex Diels ■

349433　Saxifraga hecherifolia Griseb. et Schenk；矾根叶虎耳草；Heckrialeaf Rockfoil ■☆

349434　Saxifraga hederacea L.；常春藤状虎耳草 ■☆

349435　Saxifraga hederifolia Hochst. ex A. Rich.；常春藤叶虎耳草 ■☆

349436　Saxifraga heleonastes Harry Sm.；沼地虎耳草；Bog Rockfoil ■

349437　Saxifraga hemisphaerica Hook. f. et Thomson；半球虎耳草；Hemispherical Rockfoil ■

349438　Saxifraga hengduanensis H. Chuang；横断山虎耳草；Hengduanshan Rockfoil ■

349439　Saxifraga henryi Balf. f. = Saxifraga mengtzeana Engl. et Irmsch. ■

349440　Saxifraga heteroclada Harry Sm. var. aurantia Harry Sm.；异枝虎耳草；Aureate Rockfoil ■

349441　Saxifraga heterocladoides J. T. Pan；近异枝虎耳草；Aureatelike Rockfoil ■

349442　Saxifraga heterotricha C. Marquand et Airy Shaw；异毛虎耳草；Differenthair Rockfoil ■

349443　Saxifraga heterotricha C. Marquand et Airy Shaw var. anadena（Harry Sm.）J. T. Pan et Gornall；波密虎耳草；Bomi Rockfoil ■

349444　Saxifraga hieracifolia Waldst. et Kit.；大叶虎耳草；Hawkweed Saxifrage ■☆

349445　Saxifraga himalaica N. P. Balak. = Saxifraga pallida Wall. ex Ser. ■

349446　Saxifraga hirculoides Decne.；唐古拉虎耳草；Tangula Rockfoil ■

349447　Saxifraga hirculoides Engl. = Saxifraga pseudohirculus Engl. ■

349448　Saxifraga hirculoides Engl. f. abbreviata Engl. = Saxifraga pseudohirculus Engl. ■

349449　Saxifraga hirculus（L.）Small f. intermedia Engl. et Irmsch. = Saxifraga hirculus L. ■

349450　Saxifraga hirculus（L.）Small f. major Engl. et Irmsch. = Saxifraga hirculus L. ■

349451　Saxifraga hirculus（L.）Small var. major（Engl. et Irmsch.）J. T. Pan = Saxifraga hirculus L. ■

349452　Saxifraga hirculus L.；山羊臭虎耳草（较大膛虎耳草）；Goatsmell Rockfoil, Marsh Saxifrage, Yellow Marsh Saxifrage ■

349453　Saxifraga hirculus L. f. intermedia Engl. et Irmsch. = Saxifraga hirculus L. ■

349454　Saxifraga hirculus L. f. minor Engl. et Irmsch. = Saxifraga hirculus L. var. alpina Engl. ■

349455　Saxifraga hirculus L. f. vestita Engl. = Saxifraga sinomontana J. T. Pan et Gornall ■

349456　Saxifraga hirculus L. subsp. alpina（Engl.）Podlech = Saxifraga hirculus L. var. alpina Engl. ■

349457　Saxifraga hirculus L. subsp. compacta Hedberg = Saxifraga hirculus L. var. alpina Engl. ■

349458　Saxifraga hirculus L. var. alpina Engl.；高山虎耳草；Alpine Goatsmell Rockfoil ■

349459　Saxifraga hirculus L. var. alpina Engl. f. ciliatopetala Engl. et Irmsch. = Saxifraga ciliatopetala（Engl. et Irmsch.）J. T. Pan ■

349460　Saxifraga hirculus L. var. alpina Engl. f. elata Engl. et Irmsch. = Saxifraga hirculus L. var. alpina Engl. ■

349461　Saxifraga hirculus L. var. alpina Engl. f. humilis Engl. et Irmsch. = Saxifraga hirculus L. var. alpina Engl. ■

349462　Saxifraga hirculus L. var. hirculoides（Decne.）C. B. Clarke = Saxifraga hirculoides Decne. ■

349463　Saxifraga hirculus L. var. indica C. B. Clarke = Saxifraga hirculus L. var. alpina Engl. ■

349464　Saxifraga hirculus L. var. indica C. B. Clarke = Saxifraga sinomontana J. T. Pan et Gornall ■

349465　Saxifraga hirculus L. var. kansuensis Kanitz = Saxifraga sinomontana J. T. Pan et Gornall ■

349466　Saxifraga hirculus L. var. major（Engl. et Irmsch.）J. T. Pan = Saxifraga hirculus L. ■

349467　Saxifraga hirculus L. var. platypetala Franch. = Saxifraga nigroglandulosa Engl. et Irmsch. ■

349468　Saxifraga hirculus L. var. subdioica C. B. Clarke = Saxifraga tangutica Engl. ■

349469　Saxifraga hirculus L. var. tafeliana Engl. et Irmsch. = Saxifraga parva Hemsl. ■

349470　Saxifraga hirculus L. var. typica C. B. Clarke = Saxifraga hirculus L. ■

349471　Saxifraga hirculus L. var. typica C. B. Clarke f. intermedia Engl. Irmsch. = Saxifraga hirculus L. ■

349472　Saxifraga hirculus L. var. typica C. B. Clarke f. major Engl. et Irmsch. = Saxifraga hirculus L. ■

349473　Saxifraga hirsuta L.；硬毛虎耳草（繁星虎耳草，肾叶虎耳草）；Kidney Saxifrage，Scarce Londonpride ■☆

349474　Saxifraga hispidula D. Don；齿叶虎耳草；Toothleaf Rockfoil ■

349475　Saxifraga hispidula D. Don var. dentata Franch. = Saxifraga hispidula D. Don ■

349476　Saxifraga hispidula D. Don var. doniana Engl. = Saxifraga hispidula D. Don ■

349477　Saxifraga hookeri Engl. et Irmsch.；近优越虎耳草；Hooker Rockfoil ■

349478　Saxifraga hookeri Engl. et Irmsch. var. aequifolia Marquand et Airy Shaw = Saxifraga hookeri Engl. et Irmsch. ■

349479　Saxifraga hookeri Engl. et Irmsch. var. glabrisepala Engl. et Irmsch. = Saxifraga hookeri Engl. et Irmsch. ■

349480　Saxifraga hookeri Engl. et Irmsch. var. smithii Engl. et Irmsch. = Saxifraga hookeri Engl. et Irmsch. ■

349481　Saxifraga hostii Tausch；霍斯特氏虎耳草；Hosts Saxifrage ■☆

349482　Saxifraga humilis Engl. et Irmsch.；短虎耳草■

349483　Saxifraga humilis Engl. et Irmsch. = Saxifraga coarctata W. W. Sm. ■

349484　Saxifraga hypericoides Franch.；金丝桃虎耳草（多花虎耳草）；Hipericumlike Rockfoil ■

349485　Saxifraga hypericoides Franch. f. latifolia（Engl. et Irmsch.）J. T. Pan ex T. C. Ku = Saxifraga peplidifolia Franch. ■

349486　Saxifraga hypericoides Franch. f. longipetala T. C. Ku；长瓣多花虎耳草；Longpetal Hipericumlike Rockfoil ■

349487　Saxifraga hypericoides Franch. f. longipetala T. C. Ku = Saxifraga hypericoides Franch. ■

349488　Saxifraga hypericoides Franch. f. longistyla（Franch.）J. T. Pan et T. C. Ku = Saxifraga hypericoides Franch. ■

349489　Saxifraga hypericoides Franch. var. aurantiascens（Engl. et Irmsch.）J. T. Pan et Gornall；橙瓣虎耳草（三芒虎耳草）；Trinervose Rockfoil ■

349490　Saxifraga hypericoides Franch. var. glabrescens T. C. Ku；无毛多花虎耳草；Hairless Hipericumlike Rockfoil ■

349491　Saxifraga hypericoides Franch. var. likiangensis（Engl. et Irmsch.）J. T. Pan = Saxifraga peplidifolia Franch. ■

349492　Saxifraga hypericoides Franch. var. likiangensis（Engl. et Irmsch.）J. T. Pan；丽江岩虎耳草■

349493　Saxifraga hypericoides Franch. var. longistyla（Franch.）J. T. Pan；长花柱虎耳草；Longstyle Hipericumlike Rockfoil ■

349494　Saxifraga hypericoides Franch. var. longistyla（Franch.）J. T. Pan = Saxifraga hypericoides Franch. ■

349495　Saxifraga hypericoides Franch. var. rockii（Mattf.）J. T. Pan；贡嘎虎耳草■

349496　Saxifraga hypnoides L.；藓状虎耳草；Dovedale Moss, Eve's Cushion, Indian Moss, Lady's Cushion, Moss Saxifrage, Mossy Saxifrage, Queen's Cushion ■☆

349497　Saxifraga idsuroei Franch. et Sav. = Saxifraga merkii Fisch. var. idsuroei（Franch. et Sav.）Engl. ex Matsum. ■☆

349498　Saxifraga imbricata Royle = Saxifraga pulvinaria Harry Sm. ■

349499　Saxifraga imparilis Balf. f.；大字虎耳草（滇大萼虎耳草，滇大字草）；Bigword-shaped Rockfoil ■

349500　Saxifraga implicans Harry Sm.；藏东虎耳草；E. Xizang Rockfoil ■

349501　Saxifraga implicans Harry Sm. var. weixiensis C. Y. Wu；维西虎耳草；Weixi E. Xizang Rockfoil ■

349502　Saxifraga implicans Harry Sm. var. weixiensis C. Y. Wu = Saxifraga implicans Harry Sm. ■

349503　Saxifraga insolens Irmsch.；贡山虎耳草；Gongshan Rockfoil ■

349504　Saxifraga iochanensis H. Lév. = Saxifraga stolonifera Curtis ■

349505　Saxifraga irrigua M. Bieb.；克里木虎耳草；Kelimu Rockfoil ■☆

349506　Saxifraga isophylla Harry Sm.；林芝虎耳草；Linzhi Rockfoil ■

349507　Saxifraga jacquemontiana Decne.；隐茎虎耳草；Hiddenstem Rockfoil ■

349508　Saxifraga jacquemontiana Decne. var. stella-aurea（Hook. f. et Thomson）Clarek = Saxifraga stella-aurea Hook. f. et Thomson ■

349509　Saxifraga jainzhuglaensis J. T. Pan；金珠拉虎耳草；Jinzhula Rockfoil ■

349510　Saxifraga japonica H. Boissieu；日本虎耳草■☆

349511　Saxifraga jingdongensis H. Chuang；景东虎耳草；Jingdong Rockfoil ■

349512　Saxifraga josephii Engl.；约瑟夫虎耳草■☆

349513　Saxifraga jotanii Honda = Saxifraga fortunei Hook. f. var. jotanii（Honda）Wakab. ■

349514　Saxifraga juniperifolia Adams；柏叶虎耳草；Juniper Saxifrage ■☆

349515　Saxifraga kangdingensis T. C. Ku = Saxifraga culcitosa Mattf. ■

349516　Saxifraga kansuensis Mattf. = Saxifraga unguipetala Engl. et Irmsch. ■

349517　Saxifraga kingdonii C. Marquand；金冬虎耳草■

349518　Saxifraga kingdonii C. Marquand = Saxifraga eglandulosa Engl. ■

349519　Saxifraga kingiana Engl. et Irmsch.；毛叶虎耳草；Hairyleaf Rockfoil ■

349520　Saxifraga komarovii Losinsk.；科马罗夫虎耳草；Japanese Saxifrage ■☆

349521　Saxifraga komarovii Losinsk. = Saxifraga flagellaris Willd. ex Sternb. subsp. komarovii ? ■☆

349522　Saxifraga kongboensis Harry Sm.；九窝虎耳草；Kongbo Rockfoil ■

349523　Saxifraga korshinskyi Kom.；考尔虎耳草■☆

349524　Saxifraga kotschyi Boiss.；考奇虎耳草■☆

349525 Saxifraga kuana Zhmylev = Saxifraga moorcroftiana Wall. ex Sternb. ■

349526 Saxifraga kusnezowiana Oett. ;库兹涅佐夫虎耳草■☆

349527 Saxifraga kwangsiensis Chun et F. C. How ex C. Z. Gao et G. Z. Li;广西虎耳草(龙胜虎耳草);Guangxi Rockfoil,Kwangsi Rockfoil ■

349528 Saxifraga laciniata (Nakai ex H. Hara) Akasawa = Saxifraga sendaica Maxim. f. laciniata (Nakai ex H. Hara) Ohwi ■☆

349529 Saxifraga laciniata Nakai et Takeda;长白虎耳草(条裂虎耳草);Changbai Rockfoil,Laciniate Rockfoil ■

349530 Saxifraga laciniata Nakai et Takeda f. takedana (Nakai) Toyok. ;武田虎耳草■☆

349531 Saxifraga laciniata Nakai et Takeda var. takedana (Nakai) H. Hara = Saxifraga laciniata Nakai et Takeda f. takedana (Nakai) Toyok. ■☆

349532 Saxifraga lactea Turcz. ;乳白虎耳草■☆

349533 Saxifraga laevis M. Bieb. ;平滑虎耳草■☆

349534 Saxifraga lamarum Harry Sm. = Saxifraga decora Harry Sm. ■

349535 Saxifraga lamarum Harry Sm. = Saxifraga meeboldii Engl. et Irmsch. ■☆

349536 Saxifraga lancangensis Y. Y. Qian = Saxifraga mengtzeana Engl. et Irmsch. ■

349537 Saxifraga latipetala T. C. Ku;宽瓣虎耳草;Broadpetal Rockfoil ■

349538 Saxifraga latipetala T. C. Ku = Saxifraga montanella Harry Sm. ■

349539 Saxifraga latipetala T. C. Ku var. speciosa (J. Anthony) T. C. Ku = Saxifraga montanella Harry Sm. ■

349540 Saxifraga lepidostolonosa Harry Sm. ;异条叶虎耳草;Lepidotestoloniform Rockfoil ■

349541 Saxifraga leptarrhenifolia Engl. et Irmsch. = Saxifraga davidii Franch. ■

349542 Saxifraga lhasana Harry Sm. = Saxifraga umbellulata Hook. f. et Thomson var. muricola (C. Marquand et Airy Shaw) J. T. Pan ■

349543 Saxifraga lhasana Harry Sm. var. decapitula Harry Sm. = Saxifraga umbellulata Hook. f. et Thomson var. muricola (C. Marquand et Airy Shaw) J. T. Pan ■

349544 Saxifraga lhasana Harry Sm. var. decapitulata Harry Sm. = Saxifraga umbellulata Hook. f. et Thomson var. muricola (C. Marquand et Airy Shaw) J. T. Pan ■

349545 Saxifraga ligulata Murray = Saxifraga stolonifera Curtis ■

349546 Saxifraga ligulata Murray var. densiflora Ser. = Bergenia pacumbis (Buch. -Ham. ex D. Don) C. Y. Wu et J. T. Pan ■

349547 Saxifraga ligulata Murray var. minor Wall. ex DC. = Bergenia pacumbis (Buch. -Ham. ex D. Don) C. Y. Wu et J. T. Pan ■

349548 Saxifraga ligulata Wall. = Bergenia ciliata (Haw.) Sternb. f. ligulata (Engl.) P. F. Yeo ■

349549 Saxifraga ligulata Wall. = Bergenia pacumbis (Buch. -Ham. ex D. Don) C. Y. Wu et J. T. Pan ■

349550 Saxifraga ligulata Wall. var. densiflora Ser. = Bergenia pacumbis (Buch. -Ham. ex D. Don) C. Y. Wu et J. T. Pan ■

349551 Saxifraga ligulata Wall. var. minor Wall. ex DC. = Bergenia pacumbis (Buch. -Ham. ex D. Don) C. Y. Wu et J. T. Pan ■

349552 Saxifraga likiangensis Franch. ;丽江虎耳草;Lijiang Rockfoil ■

349553 Saxifraga lilacina Duthie;紫丁香色虎耳草■☆

349554 Saxifraga limprichtii Engl. et Irmsch. = Saxifraga unguiculata Engl. var. limprichtii (Engl. et Irmsch.) J. T. Pan ex S. Y. He ■

349555 Saxifraga linearifolia Engl. et Irmsch. ;条叶虎耳草;Linearleaf Rockfoil ■

349556 Saxifraga lingulata Bellardi = Saxifraga callosa Sm. ■☆

349557 Saxifraga litangensis Engl. ;理塘虎耳草;Litang Rockfoil ■

349558 Saxifraga litangensis Engl. f. minor Engl. = Saxifraga litangensis Engl. ■

349559 Saxifraga lixianensis T. C. Ku;倒卵瓣虎耳草(理县虎耳草);Lixian Rockfoil ■

349560 Saxifraga llonakhensis W. W. Sm. ;近加拉虎耳草;Similar Jiala Rockfoil ■

349561 Saxifraga lolaensis Harry Sm. = Saxifraga subsessiliflora Engl. et Irmsch. ■

349562 Saxifraga longifolia Lapeyr. ;长叶虎耳草;Longleaf Saxifrage,Pyrenean Saxifrage ■☆

349563 Saxifraga longifolia Lapeyr. var. ghatica Quézel = Saxifraga longifolia Lapeyr. ■☆

349564 Saxifraga longifolia Lapeyr. var. orientalis Quézel = Saxifraga longifolia Lapeyr. ■☆

349565 Saxifraga longifolia Lapeyr. var. pyrenaica Emb. = Saxifraga longifolia Lapeyr. ■☆

349566 Saxifraga longipetala T. C. Ku;长瓣虎耳草;Longpetal Rockfoil ■

349567 Saxifraga longipetala T. C. Ku = Saxifraga pseudohirculus Engl. ■

349568 Saxifraga longistyla Franch. = Saxifraga hypericoides Franch. var. longistyla (Franch.) J. T. Pan ■

349569 Saxifraga longistyla Franch. = Saxifraga hypericoides Franch. ■

349570 Saxifraga longshengensis J. T. Pan;龙胜虎耳草;Longsheng Rockfoil ■

349571 Saxifraga longshengensis J. T. Pan = Saxifraga kwangsiensis Chun et F. C. How ex C. Z. Gao et G. Z. Li ■

349572 Saxifraga loripes Anthony;鞭枝虎耳草■

349573 Saxifraga loripes Anthony = Saxifraga brunonis Wall. ex Ser. ■

349574 Saxifraga ludlowii Harry Sm. ;红瓣虎耳草;Redpetal Rockfoil ■

349575 Saxifraga lumpuensis Engl. ;道孚虎耳草;Daofu Rockfoil ■

349576 Saxifraga lushuiensis H. Chuang;泸水虎耳草;Lushui Rockfoil ■

349577 Saxifraga lychnitis Hook. f. et Thomson;燃灯虎耳草(灯架虎耳草,虎耳草);Lightting Rockfoil ■

349578 Saxifraga lysimachioides Klotzsch. = Saxifraga moorcroftiana Wall. ex Sternb. ■

349579 Saxifraga macrocalyx Tolm. ;大萼虎耳草■

349580 Saxifraga macrostigma Franch. ;大柱头虎耳草;Bigstigma Rockfoil ■

349581 Saxifraga macrostigma Franch. = Saxifraga aristulata Hook. f. et Thomson ■

349582 Saxifraga macrostigma Franch. = Saxifraga peplidifolia Franch. ■

349583 Saxifraga macrostigma Franch. f. hastifolia Engl. et Irmsch. = Saxifraga aristulata Hook. f. et Thomson ■

349584 Saxifraga macrostigma Franch. var. aurantiascens Engl. et Irmsch. ;宽叶大柱头虎耳草(宽叶虎耳草);Wideleaf Rockfoil ■

349585 Saxifraga macrostigma Franch. var. aurantiascens Engl. et Irmsch. = Saxifraga peplidifolia Franch. ■

349586 Saxifraga macrostigma Franch. var. aurantiascens Engl. et Irmsch. = Saxifraga hypericoides Franch. var. aurantiascens (Engl. et Irmsch.) J. T. Pan et Gornall ■

349587 Saxifraga macrostigma Franch. var. cordifolia W. W. Sm. = Saxifraga peplidifolia Franch. ■

349588 Saxifraga macrostigma Franch. var. georgeana Engl. et Irmsch. ;腺芒虎耳草;George Rockfoil ■

349589 Saxifraga macrostigma Franch. var. georgeana Engl. et Irmsch. = Saxifraga aristulata Hook. f. et Thomson ■

349590 Saxifraga macrostigma Franch. var. georgeana Engl. et Irmsch. f.

longipila Engl. et Irmsch. = Saxifraga aristulata Hook. f. et Thomson var. longipila（Engl. et Irmsch.）J. T. Pan ■

349591 Saxifraga macrostigma Franch. var. gracillima Engl. et Irmsch. = Saxifraga aristulata Hook. f. et Thomson ■

349592 Saxifraga macrostigma Franch. var. hypericoides（Franch.）Engl. et Irmsch. = Saxifraga hypericoides Franch. ■

349593 Saxifraga macrostigma Franch. var. hypericoides（Franch.）Engl. et Irmsch. f. latifolia Engl. et Irmsch. = Saxifraga peplidifolia Franch. ■

349594 Saxifraga macrostigma Franch. var. hypericoides（Franch.）Engl. et Irmsch. subvar. longistyla（Franch.）Engl. et Irmsch. = Saxifraga hypericoides Franch. ■

349595 Saxifraga macrostigma Franch. var. hypericoides（Franch.）Engl. et Irmsch. subvar. longistyla Engl. et Irmsch. = Saxifraga hypericoides Franch. ■

349596 Saxifraga macrostigma Franch. var. hypericoides（Franch.）Engl. et Irmsch. subvar. macrantha Engl. et Irmsch. = Saxifraga peplidifolia Franch. ■

349597 Saxifraga macrostigma Franch. var. hypericoides（Franch.）Engl. et Irmsch. ;多花虎耳草;Manyflower Rockfoil ■

349598 Saxifraga macrostigma Franch. var. typica Engl. et Irmsch. = Saxifraga aristulata Hook. f. et Thomson ■

349599 Saxifraga macrostigmatoides Engl. ;假大柱头虎耳草; False Bigstigma Rockfoil ■

349600 Saxifraga macrostigmatoides Engl. var. habaensis H. Chuang;哈巴虎耳草;Haba False Bigstigma Rockfoil ■

349601 Saxifraga maderensis D. Don;梅德虎耳草■☆

349602 Saxifraga maderensis D. Don var. pickeringii（Simon）D. A. Webb et Press = Saxifraga maderensis D. Don ■☆

349603 Saxifraga madida（Maxim.）Makino = Saxifraga cortusifolia Siebold et Zucc. ■☆

349604 Saxifraga madida Makino;人字草;Moist Rockfoil ■☆

349605 Saxifraga maireana Luizet;迈雷虎耳草■☆

349606 Saxifraga mairei H. Lév. = Saxifraga filicaulis Wall. ex Ser. ■

349607 Saxifraga manshuriensis（Engl.）Kom. ;腺毛虎耳草（东北虎耳草）;Manchur Rockfoil,Manchurian Rockfoil ■

349608 Saxifraga marginata Sternb. ;角状边虎耳草;Hornrim Saxifrage ■☆

349609 Saxifraga marginata Sternb. var. rockeliana（Sternb.）Engl. et Irmsch. ;洛氏角状边虎耳草;Rochel Hornrim Saxifrage ■☆

349610 Saxifraga martinii H. Lév. et Vaniot = Saxifraga imparilis Balf. f. ■

349611 Saxifraga matta-florida Harry Sm. = Saxifraga subsessiliflora Engl. et Irmsch. ■

349612 Saxifraga maweana Baker;马韦虎耳草■☆

349613 Saxifraga maweana Baker var. oscilans Pau = Saxifraga maweana Baker ■☆

349614 Saxifraga maximowiczii Losinsk. = Saxifraga nigroglandulosa Engl. et Irmsch. ■

349615 Saxifraga maxionggouensis J. T. Pan;马熊沟虎耳草;Maxionggou Rockfoil ■

349616 Saxifraga media Gouan;中间虎耳草;Middle Rockfoil ■☆

349617 Saxifraga medogensis J. T. Pan;墨脱虎耳草;Motuo Rockfoil ■

349618 Saxifraga meeboldii Engl. et Irmsch. ;米波虎耳草（滇藏虎耳草）;Meebold Rockfoil ■☆

349619 Saxifraga meeboldii Engl. et Irmsch. = Saxifraga decora Harry Sm. ■

349620 Saxifraga megacordia C. Y. Wu ex H. Chuang;大心虎耳草;Bigheart Rockfoil ■

349621 Saxifraga megalantha C. Marquand = Saxifraga wardii W. W. Sm. ■

349622 Saxifraga melaleuca Fisch. ;黑白虎耳草■

349623 Saxifraga melanocentra Franch. ;黑蕊虎耳草（大柱头虎耳草,黑心虎耳草）;Blackcentral Rockfoil ■

349624 Saxifraga melanocentra Franch. f. angustispathulata Engl. = Saxifraga melanocentra Franch. ■

349625 Saxifraga melanocentra Franch. f. franchetiana Engl. et Irmsch. = Saxifraga melanocentra Franch. ■

349626 Saxifraga melanocentra Franch. f. pluriflora Engl. et Irmsch. = Saxifraga melanocentra Franch. ■

349627 Saxifraga mengtzeana Engl. et Irmsch. ;蒙自虎耳草（大虎耳草,反背红,反面红,红岩草,红岩耳,卵心叶虎耳草,马莲花,心叶虎耳草,心叶蒙自虎耳草,岩巴草,云南虎耳草）;Aculeate Rockfoil,Ovatecordateleaf Rockfoil,Yunnan Rockfoil ■

349628 Saxifraga mengtzeana Engl. et Irmsch. var. cordatifolia Engl. et Irmsch. = Saxifraga mengtzeana Engl. et Irmsch. ■

349629 Saxifraga mengtzeana Engl. et Irmsch. var. foliolata H. Chuang;具小叶虎耳草■

349630 Saxifraga mengtzeana Engl. et Irmsch. var. peltifolia Engl. et Irmsch. = Saxifraga mengtzeana Engl. et Irmsch. ■

349631 Saxifraga merkii Fisch. ;迈尔克虎耳草■☆

349632 Saxifraga merkii Fisch. var. idsuroei（Franch. et Sav.）Engl. ex Matsum. ;伊藤谦虎耳草■☆

349633 Saxifraga mertensiana Bong. ; 梅尔滕斯虎耳草;Merten's Saxifrag ■☆

349634 Saxifraga michauxii Edgew. ;米氏虎耳草;Michaux's Saxifrage ■☆

349635 Saxifraga micrantha Edgew. = Saxifraga pallida Wall. ex Ser. ■

349636 Saxifraga micrantha Edgew. f. corymbiflora Engl. et Irmsch. = Saxifraga pallida Wall. ex Ser. ■

349637 Saxifraga micrantha Edgew. f. foliosa Engl. et Irmsch. = Saxifraga pallida Wall. ex Ser. ■

349638 Saxifraga micrantha Edgew. f. minor Engl. et Irmsch. = Saxifraga pallida Wall. ex Ser. ■

349639 Saxifraga micrantha Edgew. var. monbeigii Engl. et Irmsch. = Saxifraga pallida Wall. ex Ser. ■

349640 Saxifraga micrantha Edgew. var. yunnanensis Franch. = Saxifraga pallida Wall. ex Ser. ■

349641 Saxifraga micranthoides Engl. = Saxifraga pallida Wall. ex Ser. ■

349642 Saxifraga microgyna Engl. et Irmsch. ;小果虎耳草;Littlefruit Rockfoil,Smallfruit Rockfoil ■

349643 Saxifraga microgyna Engl. et Irmsch. f. uniflora T. C. Ku;单花小果虎耳草;Uniflower Littlefruit Rockfoil ■

349644 Saxifraga microgyna Engl. et Irmsch. f. uniflora T. C. Ku = Saxifraga microgyna Engl. et Irmsch. ■

349645 Saxifraga microgyna Engl. et Irmsch. var. ramosior Engl. et Irmsch. = Saxifraga microgyna Engl. et Irmsch. ■

349646 Saxifraga microphylla Royle ex Hook. f. et Thomson = Saxifraga microviridis H. Hara ■☆

349647 Saxifraga microviridis H. Hara;小绿虎耳草■☆

349648 Saxifraga milesii Baker = Bergenia stracheyi（Hook. f. et Thomson）Engl. ■

349649 Saxifraga minlingensis J. T. Pan = Saxifraga tigrina Harry Sm. ■

349650 Saxifraga minor Wall. ex DC. = Bergenia pacumbis（Buch.-Ham. ex D. Don）C. Y. Wu et J. T. Pan ■

349651 Saxifraga minutifolia Pau = Saxifraga tricrenata Pau et Font Quer ■☆

349652 Saxifraga minutifoliosa C. Y. Wu = Saxifraga minutifoliosa C. Y.

Wu ex H. Chuang ■

349653　Saxifraga minutifoliosa C. Y. Wu ex H. Chuang；小叶虎耳草；Small-leaved Rockfoil ■

349654　Saxifraga miralana Harry Sm.；白毛茎虎耳草；Whitehairy Rockfoil ■

349655　Saxifraga mollis Sm.；柔软虎耳草■☆

349656　Saxifraga mollis Sm. = Saxifraga sibirica L. ■

349657　Saxifraga monantha Harry Sm.；四数花虎耳草；Fourpetal Rockfoil ■

349658　Saxifraga montana Harry Sm. = Saxifraga sinomontana J. T. Pan et Gornall ■

349659　Saxifraga montana Harry Sm. f. densifolia T. C. Ku = Saxifraga sinomontana J. T. Pan et Gornall ■

349660　Saxifraga montana Harry Sm. f. humilis Harry Sm.；矮山地虎耳草；Dwarf Montane Rockfoil ■

349661　Saxifraga montana Harry Sm. f. humilis Harry Sm. = Saxifraga sinomontana J. T. Pan et Gornall ■

349662　Saxifraga montana Harry Sm. f. oblongipetala T. C. Ku；密叶山地虎耳草■

349663　Saxifraga montana Harry Sm. f. oblongipetala T. C. Ku = Saxifraga hirculus L. ■

349664　Saxifraga montana Harry Sm. f. platypetala C. Y. Wu；宽瓣山地虎耳草■

349665　Saxifraga montana Harry Sm. var. speciosa J. Anthony = Saxifraga montanella Harry Sm. ■

349666　Saxifraga montana Harry Sm. var. splendens Harry Sm. = Saxifraga sinomontana J. T. Pan et Gornall ■

349667　Saxifraga montana Harry Sm. var. subdioica (C. B. Clarke) C. Marquand = Saxifraga tangutica Engl. ■

349668　Saxifraga montanella Harry Sm.；类毛瓣虎耳草；Hairypetallike Rockfoil ■

349669　Saxifraga montanella Harry Sm. var. retusa J. T. Pan；凹瓣虎耳草；Concavepetal Rockfoil ■

349670　Saxifraga moorcroftiana Wall. ex Sternb.；聂拉木虎耳草；Nielamu Rockfoil，Nyanang Rockfoil ■

349671　Saxifraga moschata Wulfen；麝香虎耳草；Musk Saxifrage ■☆

349672　Saxifraga mucronulata Royle；小短尖虎耳草；Pointletted Rockfoil ■

349673　Saxifraga mucronulata Royle subsp. sikkimensis (Hultén) H. Hara = Saxifraga mucronulatoides J. T. Pan ■

349674　Saxifraga mucronulatoides J. T. Pan；痂虎耳草；Scab Rockfoil ■

349675　Saxifraga muliensis Hand. -Mazz. = Saxifraga consanguinea W. W. Sm. ■

349676　Saxifraga mundula Harry Sm. = Saxifraga likiangensis Franch. ■

349677　Saxifraga muricola C. Marquand et Airy Shaw = Saxifraga umbellulata Hook. f. et Thomson var. muricola (C. Marquand et Airy Shaw) J. T. Pan ■

349678　Saxifraga muricola C. Marquand et Airy Shaw var. brachypetala Marquand et Airy Shaw = Saxifraga umbellulata Hook. f. et Thomson var. muricola (C. Marquand et Airy Shaw) J. T. Pan ■

349679　Saxifraga muricola C. Marquand et Airy Shaw var. quinquenervis C. Marquand et Airy Shaw = Saxifraga umbellulata Hook. f. et Thomson var. muricola (C. Marquand et Airy Shaw) J. T. Pan ■

349680　Saxifraga muscoides All.；薛虎耳草；Musky Saxifrage ■☆

349681　Saxifraga mutabilis Koidz. = Saxifraga fortunei Hook. f. var. alpina (Matsum. et Nakai) Nakai ■☆

349682　Saxifraga mutata L.；粗根虎耳草；Thickroot Saxifrage ■☆

349683　Saxifraga nakaoides J. T. Pan；平脉腺虎耳草；Nakao Shaped Rockfoil ■

349684　Saxifraga nambulana Harry Sm.；南布拉虎耳草；Nanbula Rockfoil ■

349685　Saxifraga nana Engl.；矮生虎耳草（青海虎耳草）；Low Rockfoil ■

349686　Saxifraga nanella Engl. et Irmsch.；光缘虎耳草；Glabrate Rockfoil，Glabrous Margin Rockfoil ■

349687　Saxifraga nanella Engl. et Irmsch. var. glabrisepala J. T. Pan；秃萼虎耳草；Glabroussepal Glabrate Rockfoil ■

349688　Saxifraga nanelloides C. Y. Wu；拟光缘虎耳草；Similar Glabrate Rockfoil ■

349689　Saxifraga nangqenica J. T. Pan；囊谦虎耳草；Nangqian Rockfoil ■

349690　Saxifraga nangxianensis J. T. Pan；朗县虎耳草；Langxian Rockfoil，Nangxian Rockfoil ■

349691　Saxifraga nelsoniana D. Don；斑点虎耳草（涅尔虎耳草）；Dotted Rockfoil，Punctate Rockfoil ■

349692　Saxifraga nelsoniana D. Don subsp. reniformis (Ohwi) Hultén = Saxifraga reniformis Ohwi ■☆

349693　Saxifraga nelsoniana D. Don var. reniformis (Ohwi) H. Ohba = Saxifraga reniformis Ohwi ■☆

349694　Saxifraga nelsoniana D. Don var. tateyamensis H. Ohba；馆山虎耳草■☆

349695　Saxifraga nigroglandulifera N. P. Balakr.；垂头虎耳草；Drooping Rockfoil ■

349696　Saxifraga nigroglandulosa Engl. et Irmsch.；黑腺虎耳草；Blackgland Rockfoil ■

349697　Saxifraga nipponica Makino；本州虎耳草■☆

349698　Saxifraga nipponica Makino f. rosea Togashi et Satomi；红色本州虎耳草■☆

349699　Saxifraga nishidae Miyabe et Kudo；西田虎耳草■☆

349700　Saxifraga nivalis L.；雪线虎耳草（细长虎耳草，雪球虎耳草）；Alpine Saxifrage，Arctic Saxifrage，Snowball Rockfoil，Snowball Saxifrage ■☆

349701　Saxifraga nudicaulis D. Don；裸茎虎耳草■☆

349702　Saxifraga numidica Maire；努米底亚虎耳草■☆

349703　Saxifraga nutans Adams = Saxifraga hirculus L. ■

349704　Saxifraga nutans Hook. f. et Thomson = Saxifraga nigroglandulifera N. P. Balakr. ■

349705　Saxifraga nutans Hook. f. et Thomson f. swertioides Engl. = Saxifraga nigroglandulifera N. P. Balakr. ■

349706　Saxifraga nyanangensis J. T. Pan = Saxifraga moorcroftiana Wall. ex Sternb. ■

349707　Saxifraga oblongifolia Nakai；矩圆叶虎耳草■☆

349708　Saxifraga obovatipetala T. C. Ku = Saxifraga umbellulata Hook. f. et Thomson var. pectinata (C. Marquand et Airy Shaw) J. T. Pan ■

349709　Saxifraga octandra Harry Sm. = Saxifraga nana Engl. ■

349710　Saxifraga oligantha Zhmylev = Saxifraga wallichiana Sternb. ■

349711　Saxifraga oligophylla T. C. Ku；少叶虎耳草■

349712　Saxifraga oligophylla T. C. Ku = Saxifraga aristulata Hook. f. et Thomson var. longipila (Engl. et Irmsch.) J. T. Pan ■

349713　Saxifraga omphalodifolia Hand. -Mazz.；无斑虎耳草；Spotless Rockfoil ■

349714　Saxifraga omphalodifolia Hand. -Mazz. var. callosa C. Y. Wu；具痂虎耳草■

349715　Saxifraga omphalodifolia Hand. -Mazz. var. callosa C. Y. Wu = Saxifraga omphalodifolia Hand. -Mazz. ■

349716　Saxifraga omphalodifolia Hand. -Mazz. var. callosa C. Y. Wu = Saxifraga omphalodifolia Hand. -Mazz. var. retusipetala J. T. Pan ■

349717　Saxifraga omphalodifolia Hand. -Mazz. var. retusipetala J. T. Pan;微凹无斑虎耳草(微凹虎耳草)■

349718　Saxifraga omphalodifolia Hand. -Mazz. var. retusipetala J. T. Pan = Saxifraga omphalodifolia Hand. -Mazz. ■

349719　Saxifraga oppositifolia L. ;挪威虎耳草(对生叶虎耳草,对叶虎耳草); Norway Rockfoil, Purple Mountain Saxifrage, Purple Saxifrage,Twinleaf Rockfoil,Twinleaf Saxifrage ■

349720　Saxifraga oppositifolia L. ' Ruth Draper';鲁斯·德雷伯对叶虎耳草■☆

349721　Saxifraga oppositifolia L. subsp. asiatica (Hayek) Engl. et Irmsch. = Saxifraga oppositifolia L. ■

349722　Saxifraga oranensis Munby = Saxifraga numidica Maire ■☆

349723　Saxifraga oranensis Munby var. major Alleiz. = Saxifraga numidica Maire ■☆

349724　Saxifraga oregana Howell;俄勒冈虎耳草;Oregan Rockfoil ■☆

349725　Saxifraga oreophila Franch. ;刚毛虎耳草(毛叶虎耳草,山生虎耳草);Bristly Rockfoil ■

349726　Saxifraga oreophila Franch. var. dapaoshanensis J. T. Pan;大炮山虎耳草;Dapaoshan Rockfoil ■

349727　Saxifraga oreophila Franch. var. dapaoshanensis J. T. Pan = Saxifraga macrostigmatoides Engl. ■

349728　Saxifraga oresbia J. Anthony;山生虎耳草;Montane Rockfoil ■

349729　Saxifraga ovatipetala T. C. Ku;卵瓣虎耳草;Ovatepetal Rockfoil ■

349730　Saxifraga ovatipetala T. C. Ku = Saxifraga ciliatopetala (Engl. et Irmsch.) J. T. Pan ■

349731　Saxifraga ovatocordata Hand. -Mazz. = Saxifraga aculeata Balf. f. ■

349732　Saxifraga ovatocordata Hand. -Mazz. = Saxifraga mengtzeana Engl. et Irmsch. ■

349733　Saxifraga pacumbis Buch. -Ham. = Bergenia pacumbis (Buch. -Ham. ex D. Don) C. Y. Wu et J. T. Pan ■

349734　Saxifraga pacumbis Buch. -Ham. ex D. Don = Bergenia pacumbis (Buch. -Ham. ex D. Don) C. Y. Wu et J. T. Pan ■

349735　Saxifraga paiquensis J. T. Pan;派区虎耳草;Paiqu Rockfoil ■

349736　Saxifraga pallida Wall. ex Ser. ;多叶虎耳草(淡白虎耳草,小花虎耳草);Leafy Rockfoil,Manyleaf Rockfoil,Pallid Rockfoil ■

349737　Saxifraga pallida Wall. ex Ser. f. folliosa Engl. et Irmsch. = Saxifraga pallida Wall. ex Ser. ■

349738　Saxifraga pallida Wall. ex Ser. var. manbeigii Engl. et Irmsch. = Saxifraga pallida Wall. ex Ser. ■

349739　Saxifraga pallida Wall. ex Ser. var. typica ? f. bracteosa Engl. et Irmsch. = Saxifraga pallida Wall. ex Ser. ■

349740　Saxifraga pallida Wall. ex Ser. var. typica ? f. corymbiflora Engl. et Irmsch. = Saxifraga pallida Wall. ex Ser. ■

349741　Saxifraga pallida Wall. ex Ser. var. typica ? f. foliosa Engl. et Irmsch. = Saxifraga pallida Wall. ex Ser. ■

349742　Saxifraga pallida Wall. ex Ser. var. typica ? f. geoides Anthony = Saxifraga pallida Wall. ex Ser. ■

349743　Saxifraga pallidiformis Engl. = Saxifraga pallida Wall. ex Ser. ■

349744　Saxifraga palpebrata Hook. f. et Thomson var. elliptica W. W. Sm. = Saxifraga glabricaulis Harry Sm. ■

349745　Saxifraga palpebrata Hook. f. et Thomson var. parceciliata Engl. et Irmsch. = Saxifraga glabricaulis Harry Sm. ■

349746　Saxifraga paludosa J. Anthony = Saxifraga melanocentra Franch. ■

349747　Saxifraga paniculata Cav. ;圆锥虎耳草■☆

349748　Saxifraga paradoxa Sternb. ;奇异虎耳草■☆

349749　Saxifraga pardanthina Hand. -Mazz. ;豹纹虎耳草;Leopard Rockfoil ■

349750　Saxifraga parisii Pomel = Saxifraga carpetana Boiss. et Reut. subsp. atlantica (Boiss. et Reut.) Romo ■☆

349751　Saxifraga parkaensis J. T. Pan;巴格虎耳草;Parka Rockfoil ■

349752　Saxifraga parnassifolia D. Don = Saxifraga diversifolia Wall. ex Ser. var. parnassifolia (D. Don) Engl. ■

349753　Saxifraga parnassifolia D. Don var. obscuricallosa J. T. Pan;隐痂虎耳草;Obscuricallose Rockfoil ■

349754　Saxifraga parnassifolia D. Don var. obscuricallosa J. T. Pan = Saxifraga parnassifolia D. Don ■

349755　Saxifraga parnassioides Regel et Schmalh. ;假梅花草叶虎耳草■☆

349756　Saxifraga parva Hemsl. ;弱小虎耳草;Little Rockfoil ■

349757　Saxifraga parvula Engl. et Irmsch. ;微虎耳草;Mini Rockfoil ■

349758　Saxifraga pasumensis C. Marquand et Airy Shaw;伞梗虎耳草■

349759　Saxifraga pasumensis C. Marquand et Airy Shaw = Saxifraga umbellulata Hook. f. et Thomson var. pectinata (C. Marquand et Airy Shaw) J. T. Pan ■

349760　Saxifraga pasumensis C. Marquand et Airy Shaw f. gracilis C. Marquand et Airy Shaw = Saxifraga umbellulata Hook. f. et Thomson var. pectinata (C. Marquand et Airy Shaw) J. T. Pan ■

349761　Saxifraga pauciflora T. C. Ku;少花虎耳草■

349762　Saxifraga pauciflora T. C. Ku = Saxifraga wallichiana Sternb. ■

349763　Saxifraga pedemontana All. ;皮埃蒙特虎耳草;Piedmont Saxifrage ■☆

349764　Saxifraga pedemontana All. subsp. demnatensis (Batt.) Quézel = Saxifraga demnatensis Batt. ■☆

349765　Saxifraga pedemontana All. var. ayachica Quézel = Saxifraga demnatensis Batt. ■☆

349766　Saxifraga pedemontana All. var. demnatensis (Batt.) Emb. = Saxifraga demnatensis Batt. ■☆

349767　Saxifraga pekinensis Maxim. = Saxifraga sibirica L. ■

349768　Saxifraga pellucida C. Y. Wu;透明虎耳草;Pellucid Rockfoil ■

349769　Saxifraga peltata Torr. = Saxifraga peltata Torr. ex Benth. ■☆

349770　Saxifraga peltata Torr. ex Benth. ;印度虎耳草;Indian Rhubarb, Umbrella Plant ■☆

349771　Saxifraga pensylvanica L. ;宾州虎耳草;Eastern Swamp Saxifrage, Forbes ' Saxifrage, Pennsylvania Saxifrage, Swamp Saxifrage,Wild Beet ■

349772　Saxifraga pensylvanica L. f. fultior Fernald = Saxifraga pensylvanica L. ■

349773　Saxifraga pensylvanica L. f. purpuripetala (A. M. Johnson) House = Saxifraga pensylvanica L. ■

349774　Saxifraga pensylvanica L. subsp. eupensylvanica G. W. Burns = Saxifraga pensylvanica L. ■

349775　Saxifraga pensylvanica L. subsp. eupensylvanica G. W. Burns f. fultior (Fernald) G. W. Burns = Saxifraga pensylvanica L. ■

349776　Saxifraga pensylvanica L. subsp. interior G. W. Burns = Saxifraga pensylvanica L. ■

349777　Saxifraga pensylvanica L. subsp. interior G. W. Burns var. congesta G. W. Burns = Saxifraga pensylvanica L. ■

349778　Saxifraga pensylvanica L. subsp. interior G. W. Burns var. crassicarpa (A. M. Johnson) G. W. Burns = Saxifraga pensylvanica L. ■

349779　Saxifraga pensylvanica L. subsp. interior G. W. Burns var. crassicarpa (A. M. Johnson) G. W. Burns f. bracteata G. W. Burns = Saxifraga pensylvanica L. ■

349780 Saxifraga pensylvanica L. subsp. tenuirostrata G. W. Burns = Saxifraga pensylvanica L. ■

349781 Saxifraga pensylvanica L. var. crassicarpa (A. M. Johnson) Bush = Saxifraga pensylvanica L. ■

349782 Saxifraga pensylvanica L. var. forbesii (Vasey) Engl. et Irmsch. = Saxifraga pensylvanica L. ■

349783 Saxifraga pensylvanica L. var. fultior (Fernald) Bush = Saxifraga pensylvanica L. ■

349784 Saxifraga pensylvanica L. var. purpuripetala (A. M. Johnson) Bush = Saxifraga pensylvanica L. ■

349785 Saxifraga peplidifolia Franch. ;洱源虎耳草(丽江虎耳草); Eryuan Rockfoil ■

349786 Saxifraga peplidifolia Franch. var. angustipetala T. C. Ku;窄瓣洱源虎耳草(窄瓣丽江虎耳草)■

349787 Saxifraga peplidifolia Franch. var. angustipetala T. C. Ku = Saxifraga hypericoides Franch. ■

349788 Saxifraga peplidifolia Franch. var. foliata Franch. = Saxifraga peplidifolia Franch. ■

349789 Saxifraga peraristulata Mattf. ;川滇虎耳草(三芒虎耳草)■

349790 Saxifraga peraristulata Mattf. = Saxifraga hypericoides Franch. var. aurantiscens (Engl. et Irmsch.) J. T. Pan et Gornall ■

349791 Saxifraga perpusilla Hook. f. et Thomson;矮小虎耳草; Small Rockfoil ■

349792 Saxifraga petrophylla Franch. = Saxifraga peplidifolia Franch. ■

349793 Saxifraga petrophylla Franch. var. likiangensis Engl. et Irmsch. = Saxifraga peplidifolia Franch. ■

349794 Saxifraga petrophylla Franch. var. likiangensis Engl. et Irmsch. = Saxifraga hypericoides Franch. var. likiangensis (Engl. et Irmsch.) J. T. Pan ■

349795 Saxifraga phaenophylla Franch. = Saxifraga wallichiana Sternb. ■

349796 Saxifraga pickeringii Simon = Saxifraga maderensis D. Don ☆

349797 Saxifraga polita Link;假伦敦虎耳草; False Londonpride ■☆

349798 Saxifraga polita Ser. = Saxifraga polita Link ■☆

349799 Saxifraga pontica Albov;蓬特虎耳草■☆

349800 Saxifraga potentilliflora H. Lév. = Saxifraga hispidula D. Don ■

349801 Saxifraga pratensis Engl. et Irmsch. ; 草地虎耳草; Meadow Rockfoil ■

349802 Saxifraga prattii Engl. et Irmsch. ;康定虎耳草(无爪虎耳草); Pratt Rockfoil ■

349803 Saxifraga prattii Engl. et Irmsch. var. obtusata Engl. ;毛茎虎耳草; Hairystem Rockfoil ■

349804 Saxifraga prattii Engl. et Irmsch. var. obtusata Engl. = Saxifraga nangqenica J. T. Pan ■

349805 Saxifraga prattii Engl. et Irmsch. var. trinervia Engl. = Saxifraga dshagalensis Engl. ■

349806 Saxifraga propagulifera Harry Sm. ;匍茎虎耳草■

349807 Saxifraga propagulifera Harry Sm. = Saxifraga consanguinea W. W. Sm. ■

349808 Saxifraga przewalskii Engl. ;青藏虎耳草(大通虎耳草,大同虎耳草); Przewalsk Rockfoil ■

349809 Saxifraga pseudohirculus Engl. ;狭瓣虎耳草(长瓣虎耳草); Narrowpetal Rockfoil ■

349810 Saxifraga pseudohirculus Engl. var. shensinensis Engl. et Irmsch. = Saxifraga pseudohirculus Engl. ■

349811 Saxifraga pseudohirculus Engl. var. tenuiflora Harry Sm. = Saxifraga pseudohirculus Engl. ■

349812 Saxifraga pseudolaevis Oett. ;假平滑虎耳草■☆

349813 Saxifraga pseudopallida Engl. et Irmsch. = Saxifraga melanocentra Franch. ■

349814 Saxifraga pseudopallida Engl. et Irmsch. f. bracteata Engl. et Irmsch. = Saxifraga melanocentra Franch. ■

349815 Saxifraga pseudopallida Engl. et Irmsch. f. foliosa Engl. et Irmsch. = Saxifraga melanocentra Franch. ■

349816 Saxifraga pseudoparvula H. Chuang;细虎耳草;Thin Rockfoil ■

349817 Saxifraga pulchra Engl. et Irmsch. ;美丽虎耳草;Beautiful Rockfoil ■

349818 Saxifraga pulchra Engl. et Irmsch. = Saxifraga meeboldii Engl. et Irmsch. ■☆

349819 Saxifraga pulvinaria Harry Sm. ;垫状虎耳草;Cushion Rockfoil ■

349820 Saxifraga pumila Harry Sm. = Saxifraga stella-aurea Hook. f. et Thomson ■

349821 Saxifraga punctata L. = Saxifraga nelsoniana D. Don ■

349822 Saxifraga punctata L. subsp. nelsoniana (D. Don) Hultén = Saxifraga nelsoniana D. Don ■

349823 Saxifraga punctata L. subsp. reniformis (Ohwi) H. Hara = Saxifraga nelsoniana D. Don var. reniformis (Ohwi) H. Ohba ■☆

349824 Saxifraga punctata L. subsp. reniformis (Ohwi) H. Hara = Saxifraga reniformis Ohwi ■☆

349825 Saxifraga punctata L. var. manschuriensis Engl. = Saxifraga manshuriensis (Engl.) Kom. ■

349826 Saxifraga punctata L. var. nelsoniana (D. Don) Engl. = Saxifraga nelsoniana D. Don ■

349827 Saxifraga punctulata Engl. ;小斑虎耳草;Punctulate Rockfoil ■

349828 Saxifraga punctulata Engl. var. minuta J. T. Pan;矮小斑虎耳草;Dwarf Punctulate Rockfoil ■

349829 Saxifraga punctulatoides J. T. Pan;拟小斑虎耳草;Pseudopunctulate Rockfoil ■

349830 Saxifraga purpurascens Hook. f. et Thomson = Bergenia purpurascens (Hook. f. et Thomson) Engl. ■

349831 Saxifraga purpurascens Hook. f. et Thomson var. macrantha Franch. = Bergenia purpurascens (Hook. f. et Thomson) Engl. ■

349832 Saxifraga purpurascens Kom. ;紫色虎耳草■☆

349833 Saxifraga purpurascens Kom. var. macrantha Franch. = Bergenia purpurascens (Hook. f. et Thomson) Engl. ■

349834 Saxifraga purpuripetala A. M. Johnson = Saxifraga pensylvanica L. ■

349835 Saxifraga qinghaiensis J. T. Pan;青海虎耳草;Qinghai Rockfoil ■

349836 Saxifraga qinghaiensis J. T. Pan = Saxifraga nana Engl. ■

349837 Saxifraga quadricallosa Hand. -Mazz. = Saxifraga tsangchanensis Franch. ■

349838 Saxifraga radiata Small;辐射虎耳草■☆

349839 Saxifraga redowskiana Sternb. ;列多夫斯基虎耳草■☆

349840 Saxifraga redowskii Adams;列氏虎耳草■☆

349841 Saxifraga reflexa T. C. Ku;反萼虎耳草;Reflexed Saxifrage ■

349842 Saxifraga reflexa T. C. Ku = Saxifraga moorcroftiana Wall. ex Sternb. ■

349843 Saxifraga reniformis Ohwi;日本肾叶虎耳草■☆

349844 Saxifraga reniformis Ohwi = Saxifraga nelsoniana D. Don var. reniformis (Ohwi) H. Ohba ■☆

349845 Saxifraga repanda Willd. ex Sternb. ;浅波状虎耳草;Repand Saxifrage ■☆

349846 Saxifraga retusa Gouan;微凹虎耳草;Tricorner Saxifrage ■☆

349847 Saxifraga reuteriana Boiss. ;路透虎耳草■☆

349848 Saxifraga reuteriana Boiss. var. riphaea Pau = Saxifraga globulifera Desf. ■☆

349849　Saxifraga rigdomensis T. C. Ku　= Saxifraga kingdonii C. Marquand ■

349850　Saxifraga rigoi Porta subsp. maroccana Luizet et Maire;摩洛哥虎耳草■☆

349851　Saxifraga rivularia L. ;河边虎耳草(溪虎耳草);Alpine Brook Saxifrage,Brook Saxifrage,Brookside Rockfoil,Highland Saxifrage ■☆

349852　Saxifraga rizhaoshanenris J. T. Pan;日照山虎耳草;Rizhaoshan Rockfoil ■

349853　Saxifraga rockii Irmsch. = Saxifraga eglandulosa Engl. ■

349854　Saxifraga rockii Matff. = Saxifraga peplidifolia Franch. ■

349855　Saxifraga rockii Mattf. = Saxifraga hypericoides Franch. var. rockii(Mattf.)J. T. Pan ■

349856　Saxifraga rosacea Moench;蔷薇虎耳草(假虎耳草);False Saxifrage,Irish Saxifrage,Mossy Saxifrage,Rose Rockfoil ■☆

349857　Saxifraga rossii Oliv. = Aceriphyllum rossii(Oliv.)Engl. ■

349858　Saxifraga rossii Oliv. = Mukdenia rossii(Oliv.)Koidz. ■

349859　Saxifraga rotundifolia L. ;圆叶虎耳草;Broadleaf Saxifrage,Broad-leaf Saxifrage,Round-leaved Saxifrage ■☆

349860　Saxifraga rotundipetala J. T. Pan;圆瓣虎耳草;Roundpetal Rockfoil ■

349861　Saxifraga rufescens Balf. f. ;红毛虎耳草(红毛大字草,扇叶虎耳草);Redhair Rockfoil ■

349862　Saxifraga rufescens Balf. f. var. flabellifolia C. Y. Wu et J. T. Pan;扇叶虎耳草;Fanleaf Redhair Rockfoil,Flabellateleaf Rockfoil ■

349863　Saxifraga rufescens Balf. f. var. uninervata J. T. Pan;单脉红毛虎耳草;Univerve Redhair Rockfoil ■

349864　Saxifraga rupestris T. C. Ku;长毛虎耳草;Longhairy Rockfoil ■

349865　Saxifraga rupestris T. C. Ku = Saxifraga gongshanensis T. C. Ku ■

349866　Saxifraga rupicola Franch. ;石生虎耳草(崖生虎耳草)■

349867　Saxifraga rupinarum J. Anthony = Saxifraga flexilis W. W. Sm. ■

349868　Saxifraga sabulicola Pomel = Saxifraga dichotoma Willd. ■☆

349869　Saxifraga sachalinensis F. Schmidt;库页虎耳草■☆

349870　Saxifraga saginoides Hook. f. et Thomson;漆姑虎耳草;Pearlweed Rockfoil ■

349871　Saxifraga sancta Griseb. ;神圣虎耳草■☆

349872　Saxifraga sanguinea Franch. ;红虎耳草;Red Rockfoil ■

349873　Saxifraga sarmentosa L. = Saxifraga stolonifera Curtis ■

349874　Saxifraga sarmentosa L. var. cuscutiformis(Lodd.)Ser. = Saxifraga stolonifera Curtis ■

349875　Saxifraga sarmentosa L. var. immaculata Diels = Saxifraga stolonifera Curtis ■

349876　Saxifraga sarmentosa L. var. tricolor(Lem.)Maxim. = Saxifraga stolonifera Curtis ■

349877　Saxifraga saxatilis Harry Sm. ;灰岩虎耳草■

349878　Saxifraga saxatilis Harry Sm. = Saxifraga unguipetala Engl. et Irmsch. ■

349879　Saxifraga saxicola Harry Sm. ;岩生虎耳草■

349880　Saxifraga scardica Griseb. ;沙尔山虎耳草■☆

349881　Saxifraga schneideri Engl. = Saxifraga chionophila Franch. ■

349882　Saxifraga scleropoda Sommier et H. Lév. ;硬梗虎耳草■☆

349883　Saxifraga sediformis Engl. et Irmsch. ;景天虎耳草;Sediformis Rockfoil ■

349884　Saxifraga selgenensis K. S. Hao = Saxifraga pseudohirculus Engl. ■

349885　Saxifraga sempervivum K. Koch;长生虎耳草■☆

349886　Saxifraga sendaica Maxim. ;苔藓虎耳草■☆

349887　Saxifraga sendaica Maxim. f. laciniata(Nakai ex H. Hara)Ohwi;条裂苔藓虎耳草■☆

349888　Saxifraga serpyllifolia Pursh;百里香叶虎耳草■☆

349889　Saxifraga serpyllifolia Pursh var. pallasiana Hance = Saxifraga unguiculata Engl. ■

349890　Saxifraga sessiliflora Harry Sm. ;加查虎耳草;Gyaca Rockfoil,Jiacha Rockfoil ■

349891　Saxifraga setigera Pursh;具刺虎耳草■☆

349892　Saxifraga setulosa C. Y. Wu = Saxifraga gongshanensis T. C. Ku ■

349893　Saxifraga setulosa C. Y. Wu var. gombalana C. Y. Wu et H. Chuang;石山虎耳草■

349894　Saxifraga sheqilaensis J. T. Pan;舍季拉虎耳草;Shejila Rockfoil ■

349895　Saxifraga sibirica L. ;球茎虎耳草(北京虎耳草,西伯利亚虎耳草,楔基虎耳草);Beijing Rockfoil,Bock Rockfoil,Siberia Rockfoil,Siberian Rockfoil ■☆

349896　Saxifraga sibirica L. = Saxifraga cernua L. ■

349897　Saxifraga sibirica L. var. bockiana Engl. = Saxifraga sibirica L. ■

349898　Saxifraga sibirica L. var. bulbillifera Harry Sm. = Saxifraga cernua L. ■

349899　Saxifraga sibirica L. var. bulbillifera Harry Sm. = Saxifraga granulifera Harry Sm. ■

349900　Saxifraga sibirica L. var. eusibirica Engl. et Irmsch. = Saxifraga sibirica L. ■

349901　Saxifraga sibirica L. var. pekingensis(Maxim.)Engl. et Irmsch. = Saxifraga sibirica L. ■

349902　Saxifraga sibirica L. var. pycnoloba Franch. = Saxifraga sibirica L. ■

349903　Saxifraga sibirica L. var. schindleri Engl. et Irmsch. = Saxifraga sibirica L. ■

349904　Saxifraga sibthorpii Boiss. ;西布索普虎耳草;Yellow Saxifrage ■☆

349905　Saxifraga sieversiana Sternb. ;西维尔虎耳草■☆

349906　Saxifraga signata Engl. et Irmsch. ;西南虎耳草(标记虎耳草,箭头虎耳草);Southwest Rockfoil,SW. China Rockfoil ■

349907　Saxifraga signata Engl. et Irmsch. var. lancipetala Hand. -Mazz. ;二痂虎耳草;Lanceolate-petal Rockfoil ■

349908　Saxifraga signata Engl. et Irmsch. var. lancipetala Hand. -Mazz. = Saxifraga signata Engl. et Irmsch. ■

349909　Saxifraga signatella C. Marquand;藏中虎耳草(红虎耳草,松吉斗);C. Xizang Rockfoil ■

349910　Saxifraga sileniflora Sternb. ;蝇子草花虎耳草■☆

349911　Saxifraga sinensis Engl. et Irmsch. = Saxifraga rufescens Balf. f. ■

349912　Saxifraga sinensis Engl. et Irmsch. var. discolor Engl. et Irmsch. = Saxifraga rufescens Balf. f. ■

349913　Saxifraga sinomontana J. T. Pan et Gornall;山地虎耳草(寒仁交木);Montane Rockfoil ■

349914　Saxifraga sinomontana J. T. Pan et Gornall var. amabilis Harry Sm. ;可观山地虎耳草■

349915　Saxifraga smensis Lour. = Saxifraga stolonifera Curtis ■

349916　Saxifraga smensis Lour. var. discolor Engl. et Irmsch. = Saxifraga rufescens Balf. f. ■

349917　Saxifraga smithiana Irmsch. ;剑川虎耳草;Smith Rockfoil ■

349918　Saxifraga spathularia Desf. ;鸟眼虎耳草;Bird's Eye, Pretty Betsy,Saint Patrick's Cabbage, St. Patrick's Cabbage, St. Patrick's-cabbage ■☆

349919　Saxifraga spathulata Desf. = Saxifraga globulifera Desf. ■☆

349920　Saxifraga spathulifolia T. C. Ku;匙叶虎耳草;Spoonleaf Rockfoil ■

349921　Saxifraga spathulifolia T. C. Ku = Saxifraga pseudohirculus

Engl. ■

349922　Saxifraga sphaeradena Harry Sm.；禿叶虎耳草；Bareleaf Rockfoil ■

349923　Saxifraga sphaeradena Harry Sm. subsp. dhwojii Harry Sm.；隆痂虎耳草■

349924　Saxifraga spinulosa Adams ＝Saxifraga bronchialis L. ■

349925　Saxifraga spinulosa Royle ＝Saxifraga mucronulata Royle ■

349926　Saxifraga stella-aurea Hook. f. et Thomson；金星虎耳草；Goldenstar Rockfoil ■

349927　Saxifraga stella-aurea Hook. f. et Thomson var. ciliata C. Marquand et Airy Shaw ＝Saxifraga llonakhensis W. W. Sm. ■

349928　Saxifraga stella-aurea Hook. f. et Thomson var. ciliata C. Marquand et Airy Shaw ＝Saxifraga stella-aurea Hook. f. et Thomson ■

349929　Saxifraga stella-aurea Hook. f. et Thomson var. macrostellata H. Chuang；大金星虎耳草■

349930　Saxifraga stella-aurea Hook. f. et Thomson var. polyadena Harry Sm. ＝Saxifraga stella-aurea Hook. f. et Thomson ■

349931　Saxifraga stellariifolia Franch.；繁缕虎耳草；Starwort-leaf Rockfoil ■

349932　Saxifraga stellaris L.；星状虎耳草（星虎耳草）；Kidneywort, Star Rockfoil, Star Saxifrage, Starry Saxifrage ■☆

349933　Saxifraga stenensis Engl. et Irmsch. ＝Saxifraga rufescens Balf. f. ■

349934　Saxifraga stenophylla Royle；大花虎耳草（鞭枝虎耳草）；Bigflower Rockfoil, Bigflower Saxifrage ■

349935　Saxifraga stenophylla Royle ＝Saxifraga flagellaris Willd. ex Sternb. subsp. stenopylla Hultén ■

349936　Saxifraga stolonifera Curtis；虎耳草（草莓虎耳草，搽耳草，澄耳草，短虎耳草，耳朵草，耳朵红，耳聋草，佛耳草，红丝络，红线草，红线绳，虎耳，金钱荷叶，金丝草，金丝芙蓉，金丝荷叶，金丝叶，金线吊芙蓉，金线荷叶，金线莲，老虎草，老虎耳，猫耳朵，猫耳朵草，巧家虎耳草，狮耳草，狮子草，狮子耳，狮子耳草，石丹药，石荷叶，水耳朵，丝绵吊梅，丝丝草，疼耳草，天荷叶，天青地红，通耳草，铜钱草，系系草，系系叶，小虎耳草，蟹壳草，月下红，猪耳草）；Aaron's Beard, Aaron's-beard, Bear's Ear, Bear's Ears, Creeping Rockfoil, Creeping Sailor, Creeping Saxifrage, Hanging Geranium, Hen-and-chickens, Ice-plant, Mother of Thousands, Mother-of-thousands, Old Man's Beard, Pedlar's Basket, Poor Man's Geranium, Rambling Sailor, Roving Jenny, Spider Plant, Strawberry Begonia, Strawberry Geranium, Strawberry Plant, Strawberry Saxifrage, Strawberry-leaved Geranium, Strawberry-stone-break, Thread-of-life, Wandering Jew, Wandering Sailor ■

349937　Saxifraga stolonifera Curtis 'Tricolor'；三色虎耳草■☆

349938　Saxifraga stolonifera Curtis f. aptera (Makino) H. Hara；无翼虎耳草■☆

349939　Saxifraga stolonifera Curtis f. cuscutiformis (Lodd.) Tebbitt ＝Saxifraga stolonifera Curtis ■

349940　Saxifraga stolonifera Curtis f. leuconeura (Makino) H. Hara；白花虎耳草■☆

349941　Saxifraga stolonifera Curtis f. viridifolia (Makino) H. Hara；绿叶虎耳草■☆

349942　Saxifraga stolonifera Curtis var. immaculata (Diels) Hand. -Mazz. ＝Saxifraga stolonifera Curtis ■

349943　Saxifraga stolonifera Meerb. ＝Saxifraga stolonifera Curtis ■

349944　Saxifraga stolonifera Meerb. f. cuscutiformis (Lodd.) Tebbitt ＝Saxifraga stolonifera Curtis ■

349945　Saxifraga stolonifera Meerb. var. immaculata (Diels) Hand. -Mazz. ＝Saxifraga stolonifera Curtis ■

349946　Saxifraga stracheyi Hook. f. et Thomson ＝Bergenia stracheyi (Hook. f. et Thomson) Engl. ■

349947　Saxifraga strigosa Wall. ex Ser.；伏毛虎耳草；Strigose Rockfoil ■

349948　Saxifraga strigosa Wall. ex Ser. f. ramosa Engl. et Irmsch. ＝Saxifraga strigosa Wall. ex Ser. ■

349949　Saxifraga strigosa Wall. ex Ser. f. simplex Engl.；草茎虎耳草；Simplestem Rockfoil ■

349950　Saxifraga strigosa Wall. ex Ser. f. simplex Engl. ＝Saxifraga strigosa Wall. ex Ser. ■

349951　Saxifraga strigosa Wall. ex Ser. f. subasexualis Engl. et Irmsch. ＝Saxifraga strigosa Wall. ex Ser. ■

349952　Saxifraga subaequifoliata Irmsch.；近等叶虎耳草；Subaequifoliate Rockfoil ■

349953　Saxifraga subaequifoliata Irmsch. var. striata H. Chuang；横纹虎耳草■

349954　Saxifraga subamplexicaulis Engl. et Irmsch.；近抱茎虎耳草；Subamplexicauline Rockfoil ■

349955　Saxifraga subdioica (C. B. Clarke) Engl. ex W. W. Sm. et Cave ＝Saxifraga tangutica Engl. ■

349956　Saxifraga sublinearifolia J. T. Pan；四川虎耳草；Sichuan Rockfoil ■

349957　Saxifraga subomphalodifolia J. T. Pan；川西南虎耳草；SW. Sichuan Rockfoil ■

349958　Saxifraga subrhombifolia Irmsch. ＝Saxifraga pratensis Engl. et Irmsch. ■

349959　Saxifraga subsediformis J. T. Pan ＝Saxifraga lixianensis T. C. Ku ■

349960　Saxifraga subsessiliflora Engl. et Irmsch.；单窝虎耳草；Subsessiliflorous Rockfoil ■

349961　Saxifraga subspathulata Engl. et Irmsch.；近匙叶虎耳草■

349962　Saxifraga subspathulata Engl. et Irmsch. var. kumaunensis Engl. et Irmsch. ＝Saxifraga subspathulata Engl. et Irmsch. ■

349963　Saxifraga substrigosa J. T. Pan；疏叶虎耳草；Laxhair Rockfoil ■

349964　Saxifraga substrigosa J. T. Pan var. gemmifera J. T. Pan；展萼虎耳草■

349965　Saxifraga substrigosa J. T. Pan var. gemmifera J. T. Pan ＝Saxifraga substrigosa J. T. Pan ■

349966　Saxifraga subternata Harry Sm.；对轮叶虎耳草（对轮虎耳草）；Subternate Rockfoil ■

349967　Saxifraga subtsangchanensis J. T. Pan；藏东南虎耳草；SE. Xizang Rockfoil ■

349968　Saxifraga sullivantii Torr. et A. Gray ＝Sullivantia sullivantii (Torr. et A. Gray) Britton ■☆

349969　Saxifraga sulphurascens Hand. -Mazz. ＝Saxifraga melanocentra Franch. ■

349970　Saxifraga swertiiflora H. Lév. ＝Saxifraga brachyphylla Franch. ■

349971　Saxifraga tabularis Hemsl. ＝Astilboides tabularis (Hemsl.) Engl. ■

349972　Saxifraga takedana Nakai；斑瓣虎耳草■

349973　Saxifraga takedana Nakai ＝Saxifraga laciniata Nakai et Takeda f. takedana (Nakai) Toyok. ■☆

349974　Saxifraga takedana Nakai ＝Saxifraga laciniata Nakai et Takeda ■

349975　Saxifraga tangulaensis J. T. Pan ＝Saxifraga hirculoides Decne. ■

349976　Saxifraga tangutica Engl.；唐古特虎耳草（大通虎耳草，甘青虎耳草，甘肃虎耳草）；Tangut Rockfoil ■

349977　Saxifraga tangutica Engl. var. minutiflora Engl.；小花虎耳草；Littleflower Rockfoil ■

349978　Saxifraga tangutica Engl. var. minutiflora Engl. ＝Saxifraga tangutica Engl. ■

349979　Saxifraga tangutica Engl. var. platyphylla （Harry Sm.） J. T. Pan；宽叶小花虎耳草；Broadleaf Rockfoil ■

349980　Saxifraga taraktophylla C. Marquand et Airy Shaw；线叶虎耳草（条叶虎耳草）；Linearifolious Rockfoil ■

349981　Saxifraga tatsienluensis Engl.；打箭炉虎耳草；Dajianlu Rockfoil ■

349982　Saxifraga taygetea Boiss. et Heldr.；台日土斯虎耳草；Taygetus Saxifrage ■☆

349983　Saxifraga taylorii Harry Sm. = Saxifraga diffusicallosa C. Y. Wu ■

349984　Saxifraga tellimoides Maxim. = Peltoboykinia tellimoides （Maxim.） H. Hara ■

349985　Saxifraga tenella Wulfen；纤细虎耳草■☆

349986　Saxifraga tentaculata C. E. C. Fisch.；秃茎虎耳草；Bladstem Rockfoil ■

349987　Saxifraga tenuis Harry Sm. = Saxifraga nivalis L. ■☆

349988　Saxifraga terektensis Bunge；西阿虎耳草■☆

349989　Saxifraga texana Buckley；得州虎耳草；Saxifrage ■☆

349990　Saxifraga tibetica Losinsk.；西藏虎耳草；Tibet Rockfoil, Xizang Rockfoil ■

349991　Saxifraga tigrina Harry Sm.；米林虎耳草；Milin Rockfoil, Miling Rockfoil, Tiger-like Rockfoil ■

349992　Saxifraga tilingiana Regel et Tiling.；梯林虎耳草■☆

349993　Saxifraga trabutiana Engl. et Irmsch.；特拉布虎耳草■☆

349994　Saxifraga triaristulata Hand.-Mazz. = Saxifraga hypericoides Franch. var. aurantiscens （Engl. et Irmsch.） J. T. Pan et Gornall ■

349995　Saxifraga triastulata Hand.-Mazz.；三芒虎耳草■

349996　Saxifraga tricrenata Pau et Font Quer；微叶虎耳草■☆

349997　Saxifraga tricrenata Pau et Font Quer var. minutifolia （Pau） Font Quer = Saxifraga tricrenata Pau et Font Quer ■☆

349998　Saxifraga tricrenata Pau et Font Quer var. villosissima Maire = Saxifraga tricrenata Pau et Font Quer ■☆

349999　Saxifraga tridactylites L.；三指虎耳草；Fingered Saxifrage, Nailwort, Rue-leaved Saxifrage, Three-fingered Jack, White Blow, Whitlow Grass ■☆

350000　Saxifraga tridactylites L. var. submeridionalis Engl. et Irmsch. = Saxifraga tridactylites L. ■☆

350001　Saxifraga tridens （Jan ex Engl.） Jan ex Engl. et Irmsch.；三齿虎耳草；Threeteeth Rockfoil ■☆

350002　Saxifraga trifurcata Schrad.；三叉虎耳草；Threefork Saxifrage ■☆

350003　Saxifraga trinervia Franch. = Saxifraga hypericoides Franch. var. aurantiscens （Engl. et Irmsch.） J. T. Pan et Gornall ■

350004　Saxifraga tsangchanensis Franch.；苍山虎耳草；Cangshan Rockfoil ■

350005　Saxifraga tsarongensis Anthony = Saxifraga stella-aurea Hook. f. et Thomson ■

350006　Saxifraga turfosa Engl. et Irmsch. = Saxifraga haplophylloides Engl. et Irmsch. ■

350007　Saxifraga umbellulata Hook. f. et Thomson；小伞虎耳草；Umbellulate Rockfoil ■

350008　Saxifraga umbellulata Hook. f. et Thomson f. pectinata C. Marquand et Airy Shaw = Saxifraga umbellulata Hook. f. et Thomson var. pectinata （C. Marquand et Airy Shaw） J. T. Pan ■

350009　Saxifraga umbellulata Hook. f. et Thomson var. muricola （C. Marquand et Airy Shaw） J. T. Pan；白小伞虎耳草 ■

350010　Saxifraga umbellulata Hook. f. et Thomson var. pectinata （C. Marquand et Airy Shaw） J. T. Pan；篦齿虎耳草（伞梗虎耳草，栉齿虎耳草）；Pectinate Rockfoil ■

350011　Saxifraga umbrosa L.；耐荫虎耳草；French Mignonette, King's Feather, King's Feathers, London Pride, London Pride Saxifrage, Londonpride, Londonpride Saxifrage, None-so-pretty, Pyrenean Saxifrage, St. Patrick's Cabbage, Wood Saxifrage ■☆

350012　Saxifraga unguiculata Engl.；爪瓣虎耳草（赛滴，爪虎耳草）；Clawy Rockfoil ■

350013　Saxifraga unguiculata Engl. f. auctiflora （Engl.） Engl. et Irmsch. = Saxifraga unguiculata Engl. ■

350014　Saxifraga unguiculata Engl. subvar. aurea Engl. = Saxifraga unguiculata Engl. ■

350015　Saxifraga unguiculata Engl. var. auctiflora Engl. = Saxifraga unguiculata Engl. ■

350016　Saxifraga unguiculata Engl. var. auctiflora Engl. subvar. aurea Engl. = Saxifraga unguiculata Engl. ■

350017　Saxifraga unguiculata Engl. var. gemmuligera Engl. = Saxifraga gemmigera Engl. ex Diels var. gemmuligera （Engl.） J. T. Pan et Gornall ■

350018　Saxifraga unguiculata Engl. var. limprichtii （Engl. et Irmsch.） J. T. Pan ex S. Y. He；五台虎耳草；Limpricht Rockfoil ■

350019　Saxifraga unguiculata Engl. var. subglabra Engl. = Saxifraga unguiculata Engl. ■

350020　Saxifraga unguipetala Engl. et Irmsch.；鄂西虎耳草（爪瓣虎耳草，爪虎耳草）；Clawpetal Rockfoil, W. Hubei Rockfoil ■

350021　Saxifraga uninervia Anthony；单脉虎耳草■

350022　Saxifraga uninervia Anthony = Saxifraga carnosula Mattf. ■

350023　Saxifraga untans Hook. f. et Thomson = Saxifraga nigroglandulifera N. P. Balakr. ■

350024　Saxifraga valleculosa H. Chuang；山箐虎耳草■

350025　Saxifraga veitchiana Balf. f. = Saxifraga stolonifera Curtis ■

350026　Saxifraga versicallosa C. Y. Wu ex H. Chuang；多痂虎耳草■

350027　Saxifraga verticillata Losinsk.；轮叶虎耳草■☆

350028　Saxifraga vilmoriniana Engl. et Irmsch.；长圆叶虎耳草；Vilmorin Rockfoil ■

350029　Saxifraga vilmoriniana Engl. et Irmsch. = Saxifraga unguiculata Engl. ■

350030　Saxifraga vilmoriniana Engl. et Irmsch. var. yungningensis Hand.-Mazz. = Saxifraga flexilis W. W. Sm. ■

350031　Saxifraga vilmoriniana Engl. et Irmsch. var. yungningensis Hand.-Mazz. = Saxifraga glacialis Harry Sm. ■

350032　Saxifraga virginiensis Michx.；弗吉尼亚虎耳草；Early Saxifrage, Michaux's Saxifrage, Rose Pennywort, St. Peter's Cabbage, Virginia Saxifrage ■☆

350033　Saxifraga wallichiana Sternb.；流苏虎耳草；Tassel Rockfoil ■

350034　Saxifraga wangiana Zhmylev = Saxifraga aurantiaca Franch. ■

350035　Saxifraga wardii W. W. Sm.；腺瓣虎耳草；Glandpetal Rockfoil ■

350036　Saxifraga wardii W. W. Sm. var. glabripedicellata J. T. Pan；光梗虎耳草；Smoothpedicel Glandpetal Rockfoil ■

350037　Saxifraga wayredana Luizet；瓦里虎耳草■☆

350038　Saxifraga wenchuanensis T. C. Ku；汶川虎耳草；Wenchuan Saxifrage ■

350039　Saxifraga werneri Pau et Font Quer；维尔纳虎耳草■☆

350040　Saxifraga xiaojinensis T. C. Ku = Saxifraga egregia Engl. var. xiaojinensis J. T. Pan ■

350041　Saxifraga xiaojinensis T. C. Ku = Saxifraga nigroglandulifera N. P. Balakr. ■

350042　Saxifraga yaluzangbuensis J. T. Pan；雅鲁藏布虎耳草；Yaluzangbu Rockfoil, Yarlungzangbo Rockfoil ■

350043　Saxifraga yezhiensis C. Y. Wu；叶枝虎耳草；Yezhi Rockfoil ■

350044　Saxifraga yuana Zhmylev = Saxifraga gongshanensis T. C. Ku ■

350045　Saxifraga yunlingensis C. Y. Wu；云岭虎耳草；Yunling Rockfoil ■

350046　Saxifraga yunlingensis C. Y. Wu = Saxifraga zayuensis T. C. Ku ■

350047　Saxifraga yuparensis Nosaka = Saxifraga bronchialis L. subsp. funstonii（Small）Hultén var. yuparensis（Nosaka）T. Shimizu ■☆

350048　Saxifraga yushuensis J. T. Pan；玉树虎耳草；Yushu Rockfoil ■

350049　Saxifraga zangnanensis J. T. Pan = Saxifraga engleriana Harry Sm. ■

350050　Saxifraga zayuensis T. C. Ku；察隅虎耳草（云岭虎耳草）；Chayu Rockfoil ■

350051　Saxifraga zayuensis T. C. Ku f. angustipetala T. C. Ku；窄瓣察隅虎耳草；Narrowpetal Chayu Rockfoil ■

350052　Saxifraga zayuensis T. C. Ku f. angustipetala T. C. Ku = Saxifraga yunlingensis C. Y. Wu ■

350053　Saxifraga zayuensis T. C. Ku f. angustipetala T. C. Ku = Saxifraga zayuensis T. C. Ku ■

350054　Saxifraga zekoensis J. T. Pan；泽库虎耳草；Zeko Rockfoil, Zeku Rockfoil ■

350055　Saxifraga zhejiangensis Z. Wei et Y. B. Chang；浙江虎耳草；Zhejiang Rockfoil ■

350056　Saxifraga zhejiangensis Z. Wei et Y. B. Chang = Saxifraga rufescens Balf. f. var. flabellifolia C. Y. Wu et J. T. Pan ■

350057　Saxifraga zhidoensis J. T. Pan；治多虎耳草；Zhiduo Rockfoil ■

350058　Saxifraga zogangensis T. C. Ku = Saxifraga egregia Engl. ■

350059　Saxifraga zogongensis T. C. Ku；左贡虎耳草；Zuogong Rockfoil ■

350060　Saxifraga zogongensis T. C. Ku = Saxifraga egregia Engl. ■

350061　Saxifraga zogongensis T. C. Ku var. pilosa T. C. Ku；少毛左贡虎耳草；Fewhair Zuogong Rockfoil ■

350062　Saxifraga zogongensis T. C. Ku var. pilosa T. C. Ku = Saxifraga egregia Engl. var. eciliata J. T. Pan ■

350063　Saxifragaceae Juss.（1789）（保留科名）；虎耳草科；Saxifrage Family ●■

350064　Saxifragella Engl.（1891）；小虎耳草属■☆

350065　Saxifragella Engl. = Saxifraga L. ■

350066　Saxifragella albowiana Engl.；小虎耳草☆

350067　Saxifragella bicuspidata（Hook.）Engl.；骤尖小虎耳草■☆

350068　Saxifragella bicuspidata（Hook.）Engl. = Saxifraga bicuspidata Hook. ■☆

350069　Saxifragella bicuspidata Engl. = Saxifragella bicuspidata（Hook.）Engl. ■☆

350070　Saxifragites Gagnep. = Distylium Siebold et Zucc. ●

350071　Saxifragodes D. M. Moore（1969）；类虎耳草属■☆

350072　Saxifragodes albowiana（Kurtz）D. M. Moore；类虎耳草■☆

350073　Saxifragopsis Small（1896）；拟虎耳草属■☆

350074　Saxifragopsis fragarioides Small；拟虎耳草■☆

350075　Saxipoa Soreng, L. J. Gillespie et S. W. L. Jacobs = Poa L. ■

350076　Saxipoa Soreng, L. J. Gillespie et S. W. L. Jacobs（2009）；岩禾属■☆

350077　Saxipoa saxicola（R. Br.）Soreng, L. J. Gillespie et S. W. L. Jacobs；岩禾■☆

350078　Saxipoa saxicola（R. Br.）Soreng, L. J. Gillespie et S. W. L. Jacobs = Poa saxicola R. Br. ■☆

350079　Saxofridericia R. H. Schomb.（1845）；弗里石草属■☆

350080　Saxo-fridericia R. H. Schomb. = Saxofridericia R. H. Schomb. ■☆

350081　Saxofridericia inermis Ducke；弗里石草■☆

350082　Saxogothaea Dalla Torre et Harms = Saxegothaea Lindl.（保留属名）●☆

350083　Sayera Post et Kuntze = Dendrobium Sw.（保留属名）■

350084　Sayera Post et Kuntze = Sayeria Kraenzl. ■

350085　Sayeria Kraenzl. = Dendrobium Sw.（保留属名）■

350086　Scabiosa L.（1753）；蓝盆花属（山萝卜属，松虫草属）；Bluebasin, Cmpet Pink Pincushion, Mourning Bride, Pincushion Flower, Piscushion-flower, Scabious, Sweet Scabious ●■

350087　Scabiosa afghanica Aitch. et Hemsl. = Pterocephalus afghanica（Aitch. et Hemsl.）Boiss. ■☆

350088　Scabiosa africana L.；非洲蓝盆花■☆

350089　Scabiosa albanensis R. A. Dyer；奥尔本蓝盆花■☆

350090　Scabiosa alpestris Kar. et Kir.；高山蓝盆花；Alpine Bluebasin, Alpine Scabious ■

350091　Scabiosa altissima Jacq. = Scabiosa africana L. ■☆

350092　Scabiosa amoena J. Jacq.；秀丽蓝盆花■☆

350093　Scabiosa angustiloba（Sond.）Hutch.；狭裂蓝盆花■☆

350094　Scabiosa anthemifolia Eckl. et Zeyh. = Scabiosa columbaria L. ■

350095　Scabiosa arenaria Forssk. = Sixalix arenaria（Forssk.）Greuter et Burdet ■☆

350096　Scabiosa argentea L.；银色蓝盆花■☆

350097　Scabiosa argentea L. = Lomelosia argentea（L.）Greuter et Burdet ■☆

350098　Scabiosa arvensis L. = Knautia arvensis（L.）Coult. ■☆

350099　Scabiosa atropurpurea L.；紫盆花（蓝松虫草，轮锋菊，轮蜂菊，松虫草，洋冠笄花）；Blacky-Moor's Beauty, Egyptian Rose, Gypsy Rose, Indian Scabious, Lady's Needlework, Lady's Pincushion, Morningbride, Mournful Widow, Mourning Bride, Mourning Widow, Pincushion, Pincushion Flower, Poor Widow, Scabiosa, Sweet Bluebasin, Sweet Kabious, Sweet Scabious, Widow's Flower ■

350100　Scabiosa atropurpurea L. = Sixalix atropurpurea（L.）Greuter et Burdet ■

350101　Scabiosa atropurpurea L. subsp. maritima（L.）Maire = Sixalix atropurpurea（L.）Greuter et Burdet subsp. maritima（L.）Greuter et Burdet ■☆

350102　Scabiosa atropurpurea L. subsp. thysdrusiana（Le Houér.）Pott.-Alap. = Sixalix thysdrusiana（Le Houér.）Greuter et Burdet ■☆

350103　Scabiosa atropurpurea L. var. adenocalyx（Batt.）Maire = Sixalix atropurpurea（L.）Greuter et Burdet ■

350104　Scabiosa atropurpurea L. var. grandiflora（Scop.）Fiori et = Sixalix atropurpurea（L.）Greuter et Burdet ■

350105　Scabiosa atropurpurea L. var. mauritii（Sennen）Maire = Sixalix atropurpurea（L.）Greuter et Burdet ■

350106　Scabiosa atropurpurea L. var. minor Ball = Sixalix atropurpurea（L.）Greuter et Burdet ■

350107　Scabiosa atropurpurea L. var. paui（Sennen）Maire = Sixalix atropurpurea（L.）Greuter et Burdet ■

350108　Scabiosa attenuata L. f. = Cephalaria attenuata（L. f.）Roem. et Schult. ■☆

350109　Scabiosa australis Wulfen = Succisa australis Borbás ■☆

350110　Scabiosa austro-africana Heine = Scabiosa columbaria L. ■☆

350111　Scabiosa austro-altaica Bobrov；阿尔泰蓝盆花；Altai Bluebasin, Altai Scabious ●

350112　Scabiosa austromongolica Hurus. = Scabiosa comosa Fisch. ex Roem. et Schult. ■

350113　Scabiosa bipinnata K. Koch；双羽蓝盆花■☆

350114　Scabiosa bretschneideri Batalin = Pterocephalus bretschneideri（Batalin）Pritz. ex Diels ●■

50115　Scabiosa buekiana Eckl. et Zeyh. var. angustiloba Sond. = Scabiosa angustiloba（Sond.）Hutch. ■☆

50116　Scabiosa camelorum Coss. et Durieu = Sixalix cartenniana（Pons et Quézel）Greuter et Burdet ■☆

50117　Scabiosa canescens Taur. ex DC.；灰蓝盆花；Grey Scabious ☆

50118　Scabiosa cartenniana Pons et Quézel = Sixalix cartenniana（Pons et Quézel）Greuter et Burdet ■☆

50119　Scabiosa caucasica M. Bieb.；高加索蓝盆花（高加索轮锋菊，高加索山萝卜，高卡萨山萝卜）；Butterfly Pincushion，Caucasia Bluebasin，Pincushion Flower ■☆

350120　Scabiosa caucasica M. Bieb.‘Clive Greaves’；克莱夫胚甲高加索蓝盆花■☆

350121　Scabiosa caucasica M. Bieb.‘Floral Queen’；花后高加索蓝盆花■☆

350122　Scabiosa caucasica M. Bieb.‘Miss Willmott’；威尔莫特小姐高加索蓝盆花■☆

350123　Scabiosa cochinchinensis Lour. = Elephantopus scaber L. ■

350124　Scabiosa colchica Stev.；黑海蓝盆花■☆

350125　Scabiosa columbaria L.；鸽子山萝卜（灰蓝盆花）；Bachelor's Buttons，Blue Buttons，Dove Pincushions，Dove Scabious，Pincushion Flower，Rice Flower，Small Scabious ■☆

350126　Scabiosa columbaria L. subsp. ochroleuca（L.）= Scabiosa ochroleuca L. ■

350127　Scabiosa columbaria L. var. longebracteolata Chiov. = Scabiosa columbaria L. ■☆

350128　Scabiosa columbaria L. var. webbiana（G. Don）Matthews = Scabiosa webbiana D. Don ■☆

350129　Scabiosa comosa Fisch. ex Roem. et Schult.；窄叶蓝盆花（蓝盆花，蒙古山萝卜，山萝卜，细叶山萝卜）；Narrowleaf Bluebasin，Narrowleaf Scabious ■☆

350130　Scabiosa comosa Fisch. ex Roem. et Schult. var. japonica（Miq.）Tatew. = Scabiosa japonica Miq. ■

350131　Scabiosa comosa Fisch. ex Roem. et Schult. var. lachnophylla（Kitag.）Kitag. = Scabiosa comosa Fisch. ex Roem. et Schult. ■

350132　Scabiosa comosa Fisch. ex Roem. et Schult. var. lachnophylla（Kitag.）Kitag.；毛叶蓝盆花（山萝卜）；Hairyleaf Scabious ■

350133　Scabiosa comosa Roem. et Schult. = Pterocephalus bretschneideri（Batalin）Pritz. ●■

350134　Scabiosa crenata Cirillo = Lomelosia crenata（Cirillo）Greuter et Burdet ■☆

350135　Scabiosa crenata Cirillo subsp. robertii（Bonnet et Barratte）Pott. -Alap. = Lomelosia robertii（Barratte）Greuter et Burdet ■☆

350136　Scabiosa crenata Cirillo var. breviseta Maire = Lomelosia crenata（Cirillo）Greuter et Burdet ■☆

350137　Scabiosa crenata Cirillo var. hirsuta Guss. = Lomelosia crenata（Cirillo）Greuter et Burdet ■☆

350138　Scabiosa cretica L. = Zygostemma creticum（L.）Tiegh. ■☆

350139　Scabiosa daucoides Desf. = Sixalix daucoides（Desf.）Raf. ■☆

350140　Scabiosa decurrens Thunb. = Cephalaria decurrens（Thunb.）Roem. et Schult. ■☆

350141　Scabiosa diandra Lag. = Pterocephalidium diandrum（Lag）G. López ■☆

350142　Scabiosa dichotoma Ucria = Scabiosa parviflora Desf. ■☆

350143　Scabiosa djurdjurae Chabert = Scabiosa columbaria L. ■☆

350144　Scabiosa djurdjurae Chabert var. fulva ? = Scabiosa columbaria L. ■☆

350145　Scabiosa drakensbergensis B. L. Burtt；德拉肯斯蓝盆花■☆

350146　Scabiosa eremophila Boiss. = Sixalix eremophila（Boiss.）Greuter et Burdet ■☆

350147　Scabiosa farinosa Coss. = Sixalix farinosa（Coss.）Greuter et Burdet ■☆

350148　Scabiosa fenestrata Pomel = Sixalix arenaria（Forssk.）Greuter et Burdet ■☆

350149　Scabiosa fischeri DC. = Pterocephalus bretschneideri（Batalin）Pritz. ●■

350150　Scabiosa fischeri DC. = Scabiosa comosa Fisch. ex Roem. et Schult. ■

350151　Scabiosa fischeri DC. f. breviseta Hand. -Mazz. = Pterocephalus bretschneideri（Batalin）Pritz. ●■

350152　Scabiosa fischeri DC. f. breviseta Hand. -Mazz. = Scabiosa comosa Fisch. ex Roem. et Schult. ■

350153　Scabiosa fischeri DC. var. japonica（Miq.）Nakai = Scabiosa japonica Miq. ■

350154　Scabiosa fischeri DC. var. longiseta H. Hara = Scabiosa japonica Miq. var. alpina Takeda ■☆

350155　Scabiosa georgica Sulak.；乔治亚蓝盆花■☆

350156　Scabiosa graminifolia L.；禾叶紫盆花；Grass-leaved Scabious ■☆

350157　Scabiosa graminifolia L. = Lomelosia graminifolia（L.）Greuter et Burdet ■☆

350158　Scabiosa graminifolia L. var. condensata Emb. et Maire = Lomelosia graminifolia（L.）Greuter et Burdet ■☆

350159　Scabiosa gramuntia L.；分叉山萝卜■☆

350160　Scabiosa gramuntia L. = Scabiosa triandra L. ■☆

350161　Scabiosa grandiflora Desf. = Sixalix atropurpurea（L.）Greuter et Burdet subsp. maritima（L.）Greuter et Burdet ■☆

350162　Scabiosa hairalensis Nakai = Scabiosa comosa Fisch. ex Roem. et Schult. ■

350163　Scabiosa henanensis Y. K. Yang et J. K. Wu = Scabiosa comosa Fisch. ex Roem. et Schult. ■

350164　Scabiosa hookeri C. B. Clarke = Pterocephalus hookeri（C. B. Clarke）Höck ●■

350165　Scabiosa hopeiensis Nakai = Pterocephalus bretschneideri（Batalin）Pritz. ●■

350166　Scabiosa hopeiensis Nakai = Scabiosa comosa Fisch. ex Roem. et Schult. ■

350167　Scabiosa humilis Thunb. = Cephalaria humilis（Thunb.）Roem. et Schult. ■☆

350168　Scabiosa hybrida Bouch.；杂种山萝卜；Hybrid Scabious ■☆

350169　Scabiosa imeretica Sulak.；伊梅里特蓝盆花■☆

350170　Scabiosa incisa Mill.；锐裂紫盆花■☆

350171　Scabiosa indurata L. = Scabiosa africana L. ■☆

350172　Scabiosa inflexa Kluk = Succisella inflexa（Kluk）Beck ■☆

350173　Scabiosa isetensis L.；伊赛特山萝卜（细叶山萝卜）■☆

350174　Scabiosa isetensis L. = Scabiosa comosa Fisch. ex Roem. et Schult. ■

350175　Scabiosa japonica Miq.；日本蓝盆花（日本轮锋菊，山萝卜，山萝卜蓝盆花，玉球花）；Japan Bluebasin，Japanese Scabious ■

350176　Scabiosa japonica Miq. f. albiflora（Honda）H. Hara；白花日本蓝盆花■☆

350177　Scabiosa japonica Miq. f. littoralis Nakai；海滨日本蓝盆花■☆

350178　Scabiosa japonica Miq. subsp. tschiliensis（Grüning）Hurus. = Scabiosa tschiliensis Grüning ■

350179　Scabiosa japonica Miq. subsp. tschiliensis Hurus. = Scabiosa comosa Fisch. ex Roem. et Schult. ■

350180　Scabiosa japonica Miq. var. acutiloba H. Hara;尖裂日本蓝盆花■☆

350181　Scabiosa japonica Miq. var. acutiloba H. Hara = Scabiosa comosa Fisch. ex Roem. et Schult. ■

350182　Scabiosa japonica Miq. var. acutiloba H. Hara = Scabiosa tschiliensis Grüning ■

350183　Scabiosa japonica Miq. var. alpina Takeda;山地日本蓝盆花（高山日本蓝盆花）■☆

350184　Scabiosa japonica Miq. var. alpina Takeda f. alba Sugim.;白花高山日本蓝盆花■☆

350185　Scabiosa lacerifolia Hayata;台湾蓝盆花（玉山山萝卜）;Taiwan Bluebasin, Taiwan Scabious ■

350186　Scabiosa lacerifolia Hayata f. leucanta Masam.;白花玉山山萝卜■

350187　Scabiosa lachnophylla Kitag. = Scabiosa comosa Fisch. ex Roem. et Schult. ■

350188　Scabiosa lachnophylla Kitag. = Scabiosa comosa Fisch. ex Roem. et Schult. var. lachnophylla（Kitag.）Kitag. ■

350189　Scabiosa liaoningensis Y. K. Yang et J. K. Wu = Scabiosa comosa Fisch. ex Roem. et Schult. ■

350190　Scabiosa libyca Alavi = Sixalix libyca（Alavi）Greuter et Burdet ☆

350191　Scabiosa lucida Vill. ;亮叶紫盆花;Shining Scabious ■☆

350192　Scabiosa mairei H. Lév. = Myriactis wallichii Less. ■

350193　Scabiosa mansenensis Nakai = Scabiosa comosa Fisch. ex Roem. et Schult. ■

350194　Scabiosa mansenensis Nakai = Scabiosa tschiliensis Grüning ■

350195　Scabiosa maritima L. = Sixalix atropurpurea（L.）Greuter et Burdet subsp. maritima（L.）Greuter et Burdet ■☆

350196　Scabiosa maritima L. subsp. mauritii Sennen = Sixalix atropurpurea（L.）Greuter et Burdet subsp. maritima（L.）Greuter et Burdet ■☆

350197　Scabiosa maritima L. var. ateridoi Pau = Sixalix atropurpurea（L.）Greuter et Burdet subsp. maritima（L.）Greuter et Burdet ■☆

350198　Scabiosa maritima L. var. cephalarioides Alleiz. = Sixalix atropurpurea（L.）Greuter et Burdet subsp. maritima（L.）Greuter et Burdet ■☆

350199　Scabiosa maritima L. var. mauritii（Sennen）Maire = Sixalix atropurpurea（L.）Greuter et Burdet subsp. maritima（L.）Greuter et Burdet ■☆

350200　Scabiosa maritima L. var. minor Ball = Sixalix atropurpurea（L.）Greuter et Burdet subsp. maritima（L.）Greuter et Burdet ■☆

350201　Scabiosa maritima L. var. paui（Sennen）Maire = Sixalix atropurpurea（L.）Greuter et Burdet subsp. maritima（L.）Greuter et Burdet ■☆

350202　Scabiosa maslakhensis Y. J. Nasir;马斯拉克蓝盆花■☆

350203　Scabiosa micrantha Desf. ;小花山萝卜■☆

350204　Scabiosa monspeliensis Jacq. = Lomelosia stellata（L.）Raf. ■☆

350205　Scabiosa montana（Pomel）Batt. = Pycnocomon rutifolium（Vahl）Hoffmanns. et Link ■☆

350206　Scabiosa oberti-manettii Pamp. = Lomelosia oberti-manettii（Pamp.）Greuter et Burdet ■☆

350207　Scabiosa ochroleuca L. ;黄盆花（黄色蓝盆花,黄山萝卜）;Cream Pincushions, Creamy Scabious, Yellow Pincushion Flower, Yellow Scabious ■

350208　Scabiosa olgae Alb. ;奥尔加蓝盆花■☆

350209　Scabiosa olivieri Coult. ;小花蓝盆花;Oliver Scabious, Small Bluebasin ■

350210　Scabiosa olivieri Coult. var. longinvolucra Y. K. Yang, N. R. Cu et Hazit = Scabiosa olivieri Coult. ■

350211　Scabiosa olivieri Coult. var. longinvolucra Y. K. Yang, N. R. Cu et Hazit;长总苞小花蓝盆花■

350212　Scabiosa olivieri Coult. var. pinnatisecta Boiss. ;羽裂小花蓝盆花■☆

350213　Scabiosa orientalis Lag. ;东方山萝卜■☆

350214　Scabiosa overinii Boiss. ;奥氏蓝盆花■☆

350215　Scabiosa palaestina L. = Lomelosia palaestina（L.）Raf. ■☆

350216　Scabiosa palaestrina L. ;巴尔干蓝盆花;Balkan Pincushions ■☆

350217　Scabiosa parielii Maire = Sixalix parielii（Maire）Greuter et Burdet ■☆

350218　Scabiosa parielii Maire var. intermedia Emb. = Sixalix parielii（Maire）Greuter et Burdet ■☆

350219　Scabiosa parielii Maire var. pinnatisecta Litard. et Maire = Sixalix parielii（Maire）Greuter et Burdet ☆

350220　Scabiosa parviflora Desf. ;热带小花蓝盆花■☆

350221　Scabiosa paui Sennen = Sixalix atropurpurea（L.）Greuter et Burdet subsp. maritima（L.）Greuter et Burdet ■☆

350222　Scabiosa persica Boiss. ;波斯蓝盆花■☆

350223　Scabiosa procera Salisb. = Scabiosa africana L. ■☆

350224　Scabiosa prolifera L. = Lomelosia prolifera（L.）Greuter et Burdet ■☆

350225　Scabiosa pterocephala L. ;希腊蓝盆花;Grecian Scabious ■☆

350226　Scabiosa rhizantha Viv. = Sixalix arenaria（Forssk.）Greuter et Burdet ■☆

350227　Scabiosa rhizantha Viv. var. ochroleuca Batt. = Sixalix arenaria（Forssk.）Greuter et Burdet ■☆

350228　Scabiosa rigida L. = Cephalaria rigida（L.）Roem. et Schult. ■☆

350229　Scabiosa robertii Barratte = Lomelosia robertii（Barratte）Greuter et Burdet ■☆

350230　Scabiosa rotata M. Bieb. ;轮状山萝卜■☆

350231　Scabiosa rutifolia Vahl = Pycnocomon rutifolium（Vahl）Hoffmanns. et Link ☆

350232　Scabiosa saxatilis Cav. var. africana Font Quer = Pseudoscabiosa africana（Font Quer）Romo et al. ■☆

350233　Scabiosa semipapposa Salzm. = Sixalix semipapposa（DC.）Greuter et Burdet ■☆

350234　Scabiosa semipapposa Salzm. var. integriloba Batt. = Sixalix semipapposa（DC.）Greuter et Burdet ■☆

350235　Scabiosa simplex Desf. = Lomelosia simplex（Desf.）Raf. ■☆

350236　Scabiosa simplex Desf. subsp. dentata（Jord. et Fourr.）Devesa = Lomelosia simplex（Desf.）Raf. subsp. dentata（Jord. et Fourr.）Greuter et Burdet ■☆

350237　Scabiosa soongorica Schrenk;准噶尔蓝盆花■

350238　Scabiosa sosnowskyi Sulak. ;锁斯诺夫斯基蓝盆花■☆

350239　Scabiosa speciosa Royle;美丽蓝盆花■☆

350240　Scabiosa stellata L. ;星花蓝盆花;Starflower Pincushions ■☆

350241　Scabiosa stellata L. = Lomelosia stellata（L.）Raf. ■☆

350242　Scabiosa stellata L. subsp. monspeliensis（Jacq.）Rouy = Lomelosia stellata（L.）Raf. ■☆

350243　Scabiosa stellata L. var. monspeliensis（Jacq.）Ball = Lomelosia stellata（L.）Raf. ■☆

350244　Scabiosa stellata L. var. simplex（Desf.）Coult. = Lomelosia simplex（Desf.）Raf. ■☆

350245　Scabiosa succisa L. = Succisa pratensis Moench ■☆

350246　Scabiosa succisa L. var. gigantea Maire = Succisa pratensis

Moench ■☆

350247　Scabiosa superba Grüning ＝ Scabiosa comosa Fisch. ex Roem. et Schult. ■

350248　Scabiosa superba Grüning ＝ Scabiosa tschiliensis Grüning var. superba（Grüning）S. Y. He ■

350249　Scabiosa superba Grüning f. elatior Grüning ＝ Scabiosa comosa Fisch. ex Roem. et Schult. ■

350250　Scabiosa superba Grüning f. elatior Grüning ＝ Scabiosa tschiliensis Grüning ■

350251　Scabiosa superba Grüning f. nana Grüning ＝ Scabiosa comosa Fisch. ex Roem. et Schult. ■

350252　Scabiosa superba Grüning f. nana Grüning ＝ Scabiosa tschiliensis Grüning var. superba（Grüning）S. Y. He ■

350253　Scabiosa syriaca L. ＝ Cephalaria syriaca（L.）Roem. et Schult. ■☆

350254　Scabiosa tenuis Boiss. ;细蓝盆花■☆

350255　Scabiosa thysdrusiana Le Houér. ＝ Sixalix thysdrusiana（Le Houér.）Greuter et Burdet ■☆

350256　Scabiosa togashiana Hurus. ＝ Scabiosa comosa Fisch. ex Roem. et Schult. ■

350257　Scabiosa togashiana Hurus. ＝ Scabiosa tschiliensis Grüning var. superba（Grüning）S. Y. He ■

350258　Scabiosa tomentosa Cav. ＝ Scabiosa turolensis Pau ■☆

350259　Scabiosa tomentosa Cav. subsp. grosii（Pau）Font Quer ＝ Scabiosa turolensis Pau subsp. grosii（Pau）Devesa ■☆

350260　Scabiosa tomentosa Cav. var. fallax Font Quer ＝ Scabiosa turolensis Pau subsp. grosii（Pau）Devesa ■☆

350261　Scabiosa tomentosa Cav. var. maroccana Pau et Font Quer ＝ Scabiosa turolensis Pau subsp. grosii（Pau）Devesa ■☆

350262　Scabiosa tomentosa Cav. var. weyleri（Pau）Maire ＝ Scabiosa turolensis Pau subsp. grosii（Pau）Devesa ■☆

350263　Scabiosa transvaalensis S. Moore;德兰士瓦蓝盆花■☆

350264　Scabiosa triandra L. ;三蕊蓝盆花■☆

350265　Scabiosa tschiliensis Grüning;华北蓝盆花（大花蓝盆花，河北山萝卜，山萝卜）; Huapei Scabious, N. China Bluebasin, North China Scabious ■

350266　Scabiosa tschiliensis Grüning ＝ Scabiosa comosa Fisch. ex Roem. et Schult. ■

350267　Scabiosa tschiliensis Grüning var. brevisecta Hurus. ＝ Scabiosa comosa Fisch. ex Roem. et Schult. ■

350268　Scabiosa tschiliensis Grüning var. brevisecta Hurus. ＝ Scabiosa tschiliensis Grüning ■

350269　Scabiosa tschiliensis Grüning var. japonica（Miq.）Hurus. ＝ Scabiosa japonica Miq. ■

350270　Scabiosa tschiliensis Grüning var. longiseta Hurus. ＝ Scabiosa comosa Fisch. ex Roem. et Schult. ■

350271　Scabiosa tschiliensis Grüning var. longiseta Hurus. ＝ Scabiosa tschiliensis Grüning ■

350272　Scabiosa tschiliensis Grüning var. superba（Grüning）S. Y. He; 大花蓝盆花;Bigflower Scabious, Largeflower Scabious ■

350273　Scabiosa tschiliensis Grüning var. superba（Grüning）S. Y. He ＝ Scabiosa comosa Fisch. ex Roem. et Schult. ■

350274　Scabiosa turolensis Pau;毛蓝盆花■☆

350275　Scabiosa turolensis Pau subsp. grosii（Pau）Devesa;格罗斯蓝盆花■☆

350276　Scabiosa turolensis Pau subsp. maroccana（Pau et Font Quer）Romo ＝ Scabiosa turolensis Pau subsp. grosii（Pau）Devesa ■☆

350277　Scabiosa turolensis Pau subsp. weyleri（Pau）Romo ＝ Scabiosa

turolensis Pau subsp. grosii（Pau）Devesa ■☆

350278　Scabiosa tysonii L. Bolus;泰森蓝盆花■☆

350279　Scabiosa ucranica L. ;乌克兰山萝卜;Ukraine Scabious ■☆

350280　Scabiosa ucranica L. ＝ Lomelosia argentea（L.）Greuter et Burdet ■☆

350281　Scabiosa ulugbekii Zakirov;乌氏蓝盆花■☆

350282　Scabiosa urceolata Desf. ＝ Pycnocomon rutifolium（Vahl）Hoffmanns. et Link ■☆

350283　Scabiosa ustulata Thunb. ＝ Cephalaria decurrens（Thunb.）Roem. et Schult. ■☆

350284　Scabiosa velenovskiana Bobrov;韦莱诺夫斯基蓝盆花■☆

350285　Scabiosa webbiana D. Don;韦布蓝盆花■☆

350286　Scabiosa weyleri Pau ＝ Scabiosa turolensis Pau subsp. weyleri（Pau）Romo ■☆

350287　Scabiosa xinjiangensis Y. K. Yang ＝ Scabiosa austro-altaica Bobrov ●

350288　Scabiosa xinjiangensis Y. K. Yang, G. J. Liu et J. K. Wu;新疆蓝盆花;Xinjiang Bluebasin ■

350289　Scabiosaceae Adans. ex T. Post et Kuntze ＝ Dipsacaceae Juss. （保留科名）■●

350290　Scabiosaceae Adans. ex T. Post et Kuntze;蓝盆花科■

350291　Scabiosaceae Martinov ＝ Scabiosaceae Adans. ex T. Post et Kuntze ■

350292　Scabiosella Tiegh. ＝ Scabiosa L. ●■

350293　Scabiosiopsis Rech. f.（1989）;类蓝盆花属■☆

350294　Scabiosiopsis Rech. f. ＝ Scabiosa L. ●■

350295　Scabiosiopsis enigmatica Rech. f. ;类蓝盆花■☆

350296　Scabrethia W. A. Weber ＝ Wyethia Nutt. ■☆

350297　Scabrethia W. A. Weber（1998）;糙韦斯菊属■☆

350298　Scabrethia scabra（Hook.）W. A. Weber;糙韦斯菊■☆

350299　Scabrethia scabra（Hook.）W. A. Weber subsp. attenuata（W. A. Weber）W. A. Weber;渐尖糙韦斯菊■☆

350300　Scabrethia scabra（Hook.）W. A. Weber subsp. canescens（W. A. Weber）W. A. Weber;灰糙韦斯菊■☆

350301　Scabrita L. ＝ Nyctanthes L. ●

350302　Scabrita scabra L. ＝ Nyctanthes arbor-tristis L. ●

350303　Scabrita triflora L. ＝ Nyctanthes arbor-tristis L. ●

350304　Scadianus Raf. ＝ Crinum L. ●

350305　Scadiasis Raf. ＝ Angelica L. ●

350306　Scadoxus Raf.（1838）;虎耳兰属（血百合属）■

350307　Scadoxus Raf. ＝ Haemanthus L. ●

350308　Scadoxus cinnabarinus（Decne.）Friis et Nordal;朱红虎耳兰（朱红网球花）;Fire-ball Lily ■☆

350309　Scadoxus cyrtanthiflorus（C. H. Wright）Friis et Nordal;曲花虎耳兰■☆

350310　Scadoxus longifolius（De Wild. et T. Durand）Friis et Nordal;长叶虎耳兰■☆

350311　Scadoxus membranaceus（Baker）Friis et Nordal;膜状虎耳兰■☆

350312　Scadoxus multiflorus（Martyn）Raf. ;虎耳兰（火球花，网球花，血百合）;African Blood Lily, Blood Flower, Blood Lily, Common Bloodlily, Powderpuff Lily, Salmon Blood-lily, Torch Lily ■

350313　Scadoxus multiflorus（Martyn）Raf. subsp. katharinae（Baker）Friis et Nordal;凯氏虎耳兰（绣球百合）;Blood Lily, Bloodflower, Blood-flower, Katharine Blood Lily, Katherine Bloodlily, Katherine Blood-lily ■☆

350314　Scadoxus multiflorus（Martyn）Raf. subsp. longitubus（C. H. Wright）I. Björnstad et Friis;长管虎耳兰■☆

350315　Scadoxus multiflorus Raf. = Scadoxus multiflorus (Martyn) Raf. ■

350316　Scadoxus multiflorus Raf. subsp. katherinae (Baker) Friis et Nordal = Scadoxus multiflorus (Martyn) Raf. subsp. katharinae (Baker) Friis et Nordal ■☆

350317　Scadoxus multiflorus Raf. subsp. longitubus (C. H. Wright) Friis et Nordal;长筒虎耳兰■☆

350318　Scadoxus nutans (Friis et I. Björnstad) Friis et Nordal;俯垂虎耳兰■☆

350319　Scadoxus pole-evansii (Oberm.) Friis et Nordal;埃文斯虎耳兰■☆

350320　Scadoxus pseudocaulus (I. Björnstad et Friis) Friis et Nordal;假茎虎耳兰●☆

350321　Scadoxus puniceus (L.) Friis et Nordal;石榴花虎耳兰(鲜红网球花);Blood-flower,Blood-lily,Royal Paintbrush,Salmon Bloodlily ■☆

350322　Scaduakintos Raf. = Brodiaea Sm.(保留属名)■☆

350323　Scaevola L.(1771)(保留属名);草海桐属;Fan Flower,Fanflower,Herb Seatung,Scaevola,White Fan Flower ●■

350324　Scaevola aemula R. Br.;紫草海桐;Fairy Fanflower,Fan Flower,Purple Fan☆

350325　Scaevola calendulacea (Andréws) Druce;沙丘草海桐;Dune Fan Flower☆

350326　Scaevola frutescens (Mill.) K. Krause = Scaevola taccada (Gaertn.) Roxb. ●■

350327　Scaevola frutescens K. Krause = Scaevola sericea Vahl ●■

350328　Scaevola frutescens K. Krause = Scaevola taccada (Gaertn.) Roxb. ●■

350329　Scaevola frutescens K. Krause var. sericea (G. Forst.) Merr. = Scaevola sericea Vahl ●■

350330　Scaevola frutescens K. Krause var. sericea (G. Forst.) Merr. = Scaevola taccada (Gaertn.) Roxb. ●■

350331　Scaevola hainanensis Hance;小叶草海桐(草海桐,海南草海桐,水草,小草海桐);Hainan Scaevola,Littleleaf Herb seatung,Littleleaf Scaevola,Little-leaved Scaevola ●■

350332　Scaevola humilis R. Br.;矮生草海桐;Fairy Fan Flower,Fan Flower ■☆

350333　Scaevola koenigii Vahl = Scaevola sericea Vahl ●■

350334　Scaevola koenigii Vahl = Scaevola taccada (Gaertn.) Roxb. ●■

350335　Scaevola koenigii Vahl var. glabra ? = Scaevola taccada (Gaertn.) Roxb. ●■

350336　Scaevola lobelia Murray = Scaevola plumieri (L.) Vahl ■☆

350337　Scaevola microcarpa Cav.;小果草海桐;Smallfruit Herb Seatung ●☆

350338　Scaevola plumieri (L.) Vahl;普氏草海桐■☆

350339　Scaevola ramosissima (Sm.) K. Krause;多枝草海桐;Fan Flowe ■☆

350340　Scaevola ramosissima K. Krause = Scaevola ramosissima (Sm.) K. Krause ■☆

350341　Scaevola sericea Vahl = Scaevola taccada (Gaertn.) Roxb. f. moomomiana (O. Deg. et Greenwell) T. Yamaz. ■☆

350342　Scaevola sericea Vahl = Scaevola taccada (Gaertn.) Roxb. ●■

350343　Scaevola sericea Vahl f. moomomiana (O. Deg. et Greenwell) O. Deg. et I. Deg. = Scaevola taccada (Gaertn.) Roxb. f. moomomiana (O. Deg. et Greenwell) T. Yamaz. ■☆

350344　Scaevola taccada (Gaertn.) Roxb.;草海桐(大网梢,海南草海桐,水草);Beach Naupaka,Half-flower,Herb Seatung,Malayan Rice Paper,Scaevola,Sea Lettuce,Sea Scaevola ●■

350345　Scaevola taccada (Gaertn.) Roxb. = Scaevola sericea Vahl ●■

350346　Scaevola taccada (Gaertn.) Roxb. f. moomomiana (O. Deg. et Greenwell) T. Yamaz.;穆蒙草海桐■☆

350347　Scaevola taccada (Gaertn.) Roxb. var. sericea (Vahl) St. John = Scaevola taccada (Gaertn.) Roxb. f. moomomiana (O. Deg. et Greenwell) T. Yamaz. ■☆

350348　Scaevola thunbergii Eckl. et Zeyh. = Scaevola plumieri (L.) Vahl ■☆

350349　Scaevola uniflora Stocks = Scaevola plumieri (L.) Vahl ■☆

350350　Scaevolaceae Lindl. = Goodeniaceae R. Br.(保留科名)●■

350351　Scagea McPherson(1986);卡莱大戟属●☆

350352　Scagea depauperata (Baill.) McPherson;卡莱大戟●☆

350353　Scalesia Arn.(1836);木雏菊属(歧伞葵属)●☆

350354　Scalesia affinis Hook. f.;近缘木雏菊●☆

350355　Scalesia microcephala B. L. Rob.;小头木雏菊●☆

350356　Scalia Sieber ex Sims = Podolepis Labill.(保留属名)●☆

350357　Scalia Sims = Podolepis Labill.(保留属名)●☆

350358　Scaligera Adans.(废弃属名)= Aspalathus L. ●☆

350359　Scaligera Adans.(废弃属名)= Scaligera DC.(保留属名)■

350360　Scaligeria DC.(1829)(保留属名);丝叶芹属;Scaligeria ■

350361　Scaligeria alaica (Lipsky) Korovin;阿拉丝叶芹■☆

350362　Scaligeria allioides (Regel et Schmalh.) Boiss.;葱状丝叶芹■☆

350363　Scaligeria bucharica Korovin;布赫丝叶芹■☆

350364　Scaligeria cretica (Mill.) Boiss.;克里特丝叶芹■☆

350365　Scaligeria ferganensis Lipsky;费尔干丝叶芹■☆

350366　Scaligeria glaucescens (DC.) Boiss.;白霜丝叶芹■☆

350367　Scaligeria hirtula (Regel et Schmalh.) Lipsky;毛丝叶芹■☆

350368　Scaligeria knorringiana Korovin;克诺丝叶芹■☆

350369　Scaligeria kopetdaghensis (Korovin) Schischk.;科佩特丝叶芹■☆

350370　Scaligeria korovinii Bobrov;科罗温丝叶芹■☆

350371　Scaligeria korshinskyi (Lipsky) Korovin;科尔丝叶芹■☆

350372　Scaligeria lipskyi Korovin;利普斯基丝叶芹■☆

350373　Scaligeria microcarpa DC.;小叶丝叶芹■☆

350374　Scaligeria platyphylla Korovin;阔叶丝叶芹;Broadleaf Scaligeria ■☆

350375　Scaligeria polycarpa Korovin;多果丝叶芹;Manyfruit Scaligeria ■☆

350376　Scaligeria samarcandica Korovin;萨马丝叶芹■☆

350377　Scaligeria setacea (Schrenk) Korovin;丝叶芹;Scaligeria,Setaceous Scaligeria ■

350378　Scaligeria transcaspica Korovin;里海丝叶芹■☆

350379　Scaligeria tschimganica Korovin;契穆干丝叶芹■☆

350380　Scaligeria ugamica Korovin;乌噶姆丝叶芹■☆

350381　Scaliopsis Walp. = Podolepis Labill.(保留属名)■☆

350382　Scambopus O. E. Schulz(1924);澳洲曲足芥属■☆

350383　Scambopus curvipes (F. Muell.) O. E. Schulz;澳洲曲足芥■☆

350384　Scammonea Raf. = Convolvulus L. ■●

350385　Scandalida Adans.(废弃属名)= Tetragonolobus Scop.(保留属名)■☆

350386　Scandederis Thouars = Bulbophyllum Thouars(保留属名)■

350387　Scandederis Thouars = Neottia Guett.(保留属名)■

350388　Scandia J. W. Dawson(1967);攀缘草属■☆

350389　Scandia geniculata (G. Forst.) J. W. Dawson;攀缘草■☆

350390　Scandia rosifolia (Hook. f.) J. W. Dawson;蔷薇攀缘草■☆

350391　Scandicaceae Bercht. et J. Presl = Labiatae Juss.(保留科名)●■

350392　Scandicaceae Bercht. et J. Presl = Lamiaceae Martinov(保留科名)●■

350393　Scandicium (K. Koch) Thell. = Birostula Raf. ■

350394　Scandicium (K. Koch) Thell. = Scandix L. ■

350395　Scandicium Thell. = Scandix L. ■

350396　Scandicium stellatum (Banks et Sol.) Thell. = Scandix stellata

Banks et Sol. ■

50397 Scandicium stellatum (Sol.) Thell. = Scandix stellata Banks et Sol. ■

50398 Scandicium stellatum (Sol.) Thell. var. decipiens (Bornm.) Thell. = Scandix stellata Banks et Sol. ■

50399 Scandicium stellatum (Sol.) Thell. var. hirsutum (Koch) Thell. = Scandix stellata Banks et Sol. ■

50400 Scandivepres Loes. (1910);攀缘卫矛属●☆

50401 Scandivepres mexicanus Loes. ;攀缘卫矛☆

350402 Scandix L. (1753);针果芹属(鸡冠芹属);Shepherd's-needle ■

350403 Scandix Molina = Erodium L' Hér. ex Aiton ■●

350404 Scandix amurensis Koso-Pol. = Osmorhiza aristata (Thunb.) Makino et Y. Yabe ■

350405 Scandix aristata (Thunb.) Makino = Osmorhiza aristata (Thunb.) Makino et Y. Yabe ■

350406 Scandix aristata Koso-Pol. = Osmorhiza aristata (Thunb.) Makino et Y. Yabe ■

350407 Scandix aucheri Boiss. ;奥氏针果芹■☆

350408 Scandix australis L. ;南方针果芹■☆

350409 Scandix australis L. subsp. curvirostris (Murb.) Jahand. et Maire;弯喙南方针果芹■☆

350410 Scandix australis L. subsp. cyrenaica Maire et Weiller;昔兰尼针果芹■☆

350411 Scandix australis L. subsp. gallica Vierh. = Scandix australis L. ■☆

350412 Scandix australis L. subsp. occidentalis Vierh. = Scandix australis L. ■☆

350413 Scandix australis L. var. glabricaulis Faure et Maire = Scandix australis L. ■☆

350414 Scandix australis L. var. hirticaulis Faure et Maire = Scandix australis L. ■☆

350415 Scandix australis L. var. leiobasis Maire et Weiller = Scandix australis L. ■☆

350416 Scandix australis L. var. villicaulis Maire et Weiller = Scandix australis L. ■☆

350417 Scandix brachycarpa Guss. = Scandix pecten-veneris L. subsp. brachycarpa (Guss.) Thell. ■☆

350418 Scandix cerefolius L. = Anthriscus cerefolius (L.) Hoffm. ■☆

350419 Scandix claytonii (Michx.) Koso-Pol. = Osmorhiza aristata (Thunb.) Makino et Y. Yabe ■

350420 Scandix curvirostris Murb. = Scandix australis L. subsp. curvirostris (Murb.) Jahand. et Maire ■☆

350421 Scandix falcata Londes;镰形针果芹■☆

350422 Scandix glaberrima Desf. = Conopodium glaberrimum (Desf.) Engstrand ■☆

350423 Scandix iberica M. Bieb. ;伊比利亚针果芹■☆

350424 Scandix infesta L. = Torilis arvensis (Huds.) Link ■☆

350425 Scandix pecten-veneris L. ;针果芹;Adam's-needle, Beggar's Needle, Beggar's-needle, Cammock, Clock Needle, Clock-needle, Comb, Coombs, Corn Chervil, Crake Needle, Crow Needle, Crow Peck, Darning Old Wife's Needle, Devil's Darning Needle, Devil's Darning-needle, Devil's Elshins, Grandmother's Darning Needle, Grandmother's Darning Needles, Ground Enell, Hedgehogs, Lady's Comb, Lady's Needle, Lady's-comb, Long Beaks, Mock Chervil, Nadelkerbel, Needle Chervil, Needle Points, Needles, Needleweed, Old Wife's Darning-needle, Old Woman's Needle, Pink-needle, Pins-and-needles, Poke Needle, Poke-needle, Pook-needle, Pound Needle, Prick Devil, Prick-devil, Puck Needle, Pucker Needle, Punk Needle,

Shepherd's Needle, Shepherd's Bodkin, Shepherd's Comb, Shepherd's-needle, Tailders, Tailor's Needle, Venuskamm, Venus' Comb, Venus' Needle, Venus'-comb, Wild Chervil, Wild Parsley, Witches' Needle ■☆

350426 Scandix pecten-veneris L. subsp. brachycarpa (Guss.) Thell. ;短果针果芹■☆

350427 Scandix pecten-veneris L. var. brevirostris Boiss. = Scandix pecten-veneris L. ■☆

350428 Scandix persica Mart. ;波斯针果芹■☆

350429 Scandix persica Mart. = Scandix pecten-veneris L. ■☆

350430 Scandix pinnatifida Vent. = Scandix stellata Banks et Sol. ■

350431 Scandix stellata Banks et Sol. ;星形针果芹(针果芹)■

350432 Scandix stellata Banks et Sol. var. velutina (Coss.) Charpin et Fern. Casas = Scandix stellata Banks et Sol. ■

350433 Scandix stellata Sol. = Scandix stellata Banks et Sol. ■

350434 Scapha Noronha = Saurauia Willd. (保留属名)●

350435 Scaphespermum Edgew. = Eriocycla Lindl. ■

350436 Scaphiophora Schltr. (1921);舟梗玉簪属■☆

350437 Scaphiophora Schltr. = Thismia Griff. ■

350438 Scaphiophora appendiculata Schltr. ;舟梗玉簪■☆

350439 Scaphispatha Brongn. ex Schott(1860);舟苞南星属■☆

350440 Scaphispatha gracilis Brongn. ex Schott;舟苞南星■☆

350441 Scaphium Endl. = Scaphium Schott et Endl. ●☆

350442 Scaphium Post et Kuntze = Skaphium Miq. ●

350443 Scaphium Post et Kuntze = Xanthophyllum Roxb. (保留属名)●

350444 Scaphium Schott et Endl. (1832);舟梧桐属(胖大海属)●☆

350445 Scaphium lychnophorum Pierre = Sterculia lychnophora Hance ●☆

350446 Scaphium macropodum Beumee ex K. Heyne;大柄舟梧桐●☆

350447 Scaphium wallichii Schott et Endl. ;沃利克舟梧桐●☆

350448 Scaphocalyx Ridl. (1920);舟萼木属●☆

350449 Scaphocalyx parviflora Ridl. ;舟萼木●☆

350450 Scaphochlamys Baker(1892);舟被姜属■☆

350451 Scaphochlamys gracilipes (K. Schum.) S. Sakai et Nagam. ;细梗舟被姜■☆

350452 Scaphoglottis T. Durand et Jacks. = Scaphyglottis Poepp. et Endl. (保留属名)■☆

350453 Scaphopetalum Mast. (1867);舟瓣梧桐属●☆

350454 Scaphopetalum acuminatum Engl. et K. Krause;渐尖舟瓣梧桐●☆

350455 Scaphopetalum amoenum A. Chev. ;秀丽舟瓣梧桐●☆

350456 Scaphopetalum blackii Mast. ;布莱克舟瓣梧桐●☆

350457 Scaphopetalum blackii Mast. var. letestui (Pellegr.) N. Hallé;莱泰斯图舟瓣梧桐●☆

350458 Scaphopetalum brunneo-purpureum Engl. et K. Krause;褐紫舟瓣梧桐●☆

350459 Scaphopetalum dewevrei De Wild. et T. Durand;德韦舟瓣梧桐●☆

350460 Scaphopetalum dewevrei De Wild. et T. Durand var. suborophila R. Germ. ;喜山舟瓣梧桐●☆

350461 Scaphopetalum discolor Engl. et K. Krause;异色舟瓣梧桐●☆

350462 Scaphopetalum letestui Pellegr. = Scaphopetalum blackii Mast. var. letestui (Pellegr.) N. Hallé ●☆

350463 Scaphopetalum longipedunculatum Mast. ;长梗舟瓣梧桐●☆

350464 Scaphopetalum macranthum K. Schum. ;大花舟瓣梧桐●☆

350465 Scaphopetalum mannii Mast. ;曼氏舟瓣梧桐●☆

350466 Scaphopetalum monophysca K. Schum. = Scaphopetalum thonneri De Wild. et T. Durand ●☆

350467 Scaphopetalum ngouniense Pellegr. ;恩古涅舟瓣梧桐●☆

350468 Scaphopetalum obiangianum M. E. Leal;奥比昂舟瓣梧桐●☆

350469　Scaphopetalum pallidinerve Engl. et K. Krause;白脉舟瓣梧桐●☆
350470　Scaphopetalum parvifolium Baker f. ;小叶舟瓣梧桐●☆
350471　Scaphopetalum paxii H. Winkl. ;帕克斯舟瓣梧桐●☆
350472　Scaphopetalum riparium Engl. et K. Krause;河岸舟瓣梧桐●☆
350473　Scaphopetalum stipulosum K. Schum. ;托叶舟瓣梧桐●☆
350474　Scaphopetalum talbotii Baker f. ;塔尔博特舟瓣梧桐●☆
350475　Scaphopetalum thonneri De Wild. et T. Durand;托内舟瓣梧桐●☆
350476　Scaphopetalum thonneri De Wild. et T. Durand var. klainei Pierre ex N. Hallé;克莱恩舟瓣梧桐●☆
350477　Scaphopetalum vanderystii R. Germ. ;范德舟瓣梧桐●☆
350478　Scaphopetalum zenkeri K. Schum. ;岑克尔舟瓣梧桐●☆
350479　Scaphosepalum Pfitzer(1888);舟萼兰属(碗萼兰属)■☆
350480　Scaphosepalum bicolor Luer et R. Escobar;二色舟萼兰■☆
350481　Scaphosepalum breve Rolfe;短舟萼兰■☆
350482　Scaphosepalum platypetalum Schltr. ;宽瓣舟萼兰■☆
350483　Scaphosepalum punctatum Rolfe;斑点舟萼兰■☆
350484　Scaphospatha Post et Kuntze = Scaphispatha Brongn. ex Schott ■☆
350485　Scaphospermum Korovin = Parasilaus Leute ■
350486　Scaphospermum Korovin ex Schischk. (1951);舟籽芹属■☆
350487　Scaphospermum Korovin ex Schischk. = Parasilaus Leute ■
350488　Scaphospermum Post et Kuntze = Eriocycla Lindl. ■
350489　Scaphospermum Post et Kuntze = Scaphespermum Edgew. ■
350490　Scaphospermum asiaticum Korovin;舟籽芹■☆
350491　Scaphula R. Parker = Anisoptera Korth. ●☆
350492　Scaphula R. Parker(1932);缅甸龙脑香属●☆
350493　Scaphula glabra (Kurz) R. Parker;缅甸龙脑香●☆
350494　Scaphyglottis Poepp. et Endl. (1836)(保留属名);碗唇兰属■☆
350495　Scaphyglottis parviflora Poepp. et Endl. = Maxillaria parviflora (Poepp. et Endl.) Garay ■☆
350496　Scapicephalus Ovcz. et Czukav. (1974);箭头草属■☆
350497　Scapicephalus Ovcz. et Czukav. = Pseudomertensia Riedl ■
350498　Scapicephalus rosulatus Ovcz. et Czukav. ;箭头草■☆
350499　Scaraboides Magee et B. -E. van Wyk(2009);南非草属☆
350500　Scaredederis Thouars = Dendrobium Sw. (保留属名)■
350501　Scaredederis Thouars = Hederorkis Thouars ■☆
350502　Scaredederis Thouars = Scandederis Thouars ■
350503　Scariola F. W. Schmidt = Lactuca L. ■
350504　Scariola F. W. Schmidt(1795);雀苣属;Scariola ■●
350505　Scariola acanthifolia (Willd.) Soják;刺叶雀苣■☆
350506　Scariola albertoregelia (C. Winkl.) Kirp. ;阿氏雀苣■☆
350507　Scariola orientalis (Boiss.) Soják;东方雀苣;Oriental Scariola ■☆
350508　Scariola orientalis (Boiss.) Soják = Lactuca orientalis (Boiss.) Boiss. ■☆
350509　Scariola viminea F. W. Schmidt;雀苣■☆
350510　Scassellatia Chiov. = Lannea A. Rich. (保留属名)●
350511　Scassellatia heterophylla Chiov. = Lannea schweinfurthii (Engl.) Engl. ●☆
350512　Scatohyacinthus Post et Kuntze = Brodiaea Sm. (保留属名)■☆
350513　Scatohyacinthus Post et Kuntze = Scaduakintos Raf. ■☆
350514　Scatoxis Post et Kuntze = Haemanthus L. ■
350515　Scatoxis Post et Kuntze = Scadoxus Raf. ■
350516　Sceletium N. E. Br. (1925);肖凤卵草属●☆
350517　Sceletium N. E. Br. = Phyllobolus N. E. Br. ●☆
350518　Sceletium albanense L. Bolus = Sceletium crassicaule (Haw.) L. Bolus ●☆
350519　Sceletium anatomicum (Haw.) L. Bolus = Sceletium emarcidum (Thunb.) L. Bolus ex H. Jacobsen ●☆

350520　Sceletium archeri L. Bolus = Sceletium rigidum L. Bolus ●☆
350521　Sceletium boreale L. Bolus = Sceletium tortuosum (L.) N. E. Br. ●☆
350522　Sceletium compactum L. Bolus = Sceletium tortuosum (L.) N. E. Br. ●☆
350523　Sceletium concavum (Haw.) Schwantes = Sceletium tortuosum (L.) N. E. Br. ●☆
350524　Sceletium crassicaule (Haw.) L. Bolus;粗茎肖凤卵草●☆
350525　Sceletium dejagerae L. Bolus = Sceletium emarcidum (Thunb.) L. Bolus ex H. Jacobsen ●☆
350526　Sceletium emarcidum (Thunb.) L. Bolus ex H. Jacobsen;萎蔫肖凤卵草●☆
350527　Sceletium exalatum Gerbaulet;无翅肖凤卵草●☆
350528　Sceletium expansum (L.) L. Bolus;扩展肖凤卵草●☆
350529　Sceletium framesii L. Bolus = Sceletium tortuosum (L.) N. E. Br. ●☆
350530　Sceletium gracile L. Bolus = Sceletium tortuosum (L.) N. E. Br. ●☆
350531　Sceletium joubertii L. Bolus = Sceletium tortuosum (L.) N. E. Br. ●☆
350532　Sceletium namaquense L. Bolus = Sceletium tortuosum (L.) N. E. Br. ●☆
350533　Sceletium namaquense L. Bolus var. subglobosum ? = Sceletium tortuosum (L.) N. E. Br. ●☆
350534　Sceletium ovatum L. Bolus = Sceletium tortuosum (L.) N. E. Br. ●☆
350535　Sceletium regium L. Bolus = Sceletium expansum (L.) L. Bolus ●☆
350536　Sceletium rigidum L. Bolus;硬肖凤卵草●☆
350537　Sceletium strictum L. Bolus;刚直肖凤卵草●☆
350538　Sceletium subvelutinum L. Bolus. = Sceletium varians (Haw.) Gerbaulet ●☆
350539　Sceletium subvelutinum L. Bolus. f. luxurians ? = Sceletium varians (Haw.) Gerbaulet ●☆
350540　Sceletium tortuosum (L.) N. E. Br. ;扭曲肖凤卵草●☆
350541　Sceletium tugwelliae L. Bolus = Sceletium tortuosum (L.) N. E. Br. ●☆
350542　Sceletium varians (Haw.) Gerbaulet;易变肖凤卵草●☆
350543　Scelochiloides Dodson et M. W. Chase(1989);拟肋唇兰属■☆
350544　Scelochiloides vasquezii Dodson et M. W. Chase;拟肋唇兰■☆
350545　Scelochilus Klotzsch(1841);肋唇兰属■☆
350546　Scelochilus aureus Schltr. ;黄肋唇兰■☆
350547　Scelochilus heterophyllus Rchb. f. ;异叶肋唇兰■☆
350548　Scepa Lindl. = Aporusa Blume ●
350549　Scepa chinensis Champ. ex Benth. = Aporusa dioica (Roxb.) Airy Shaw ●
350550　Scepa villosa Lindl. = Aporusa villosa (Lindl.) Baill. ●
350551　Scepaceae Lindl. = Euphorbiaceae Juss. (保留科名)●■
350552　Scepaceae Lindl. = Phyllanthaceae J. Agardh ●■
350553　Scepanium Ehrh. = Pedicularis L. ■
350554　Scepasma Blume = Phyllanthus L. ●■
350555　Scepinia Neck. = Pteronia L. (保留属名)●☆
350556　Scepinia Neck. ex Cass. = Pteronia L. (保留属名)●☆
350557　Scepocarpus Wedd. = Urera Gaudich. ●☆
350558　Scepocarpus mannii Wedd. = Urera mannii (Wedd.) Benth. et Hook. f. ex Rendle ■☆
350559　Scepseotharanus Cham. = Alibertia A. Rich. ex DC. ●☆

350560　Scepsothamnus Steud. = Scepseothamnus Cham. ●☆

350561　Sceptranthes R. Graham = Cooperia Herb. ■☆

350562　Sceptranthes R. Graham = Zephyranthes Herb. (保留属名)■

350563　Sceptranthus Benth. et Hook. f. = Sceptranthes R. Graham ■☆

350564　Sceptrocnide Maxim. (1877);杖麻属●■

350565　Sceptrocnide Maxim. = Laportea Gaudich. (保留属名)●■

350566　Sceptrocnide macarostachya Maxim. = Laportea cuspidata (Wedd.) Friis ■

350567　Sceura Forssk. = Avicennia L. ●

350568　Sceura marina Forssk. = Avicennia marina (Forssk.) Vierh. ●

350569　Scevola Raf. = Scaevola L. (保留属名)●■

350570　Schachtia H. Karst. = Duroia L. f. (保留属名)●☆

350571　Schaeffera Schreb. = Schaefferia Jacq. ●☆

350572　Schaefferia Jacq. (1760);沙氏木属●☆

350573　Schaefferia cuneifolia A. Gray.;肾叶沙氏木;Capul, Desert Yaupon, Panalero ●☆

350574　Schaefferia frutescens Jacq.;沙氏木;Florida Box, Florida Boxwood ●☆

350575　Schaeffnera Benth. et Hook. f. = Dicoma Cass. ●☆

350576　Schaeffnera Benth. et Hook. f. = Schaffnera Sch. Bip. ●☆

350577　Schaenfeldia Edgew. = Schoenefeldia Kunth ■☆

350578　Schaenolaena Lindl. = Schoenolaena Bunge ■☆

350579　Schaenolaena Lindl. = Xanthosia Rudge ■☆

350580　Schaenomorphus Thorei ex Gagnep. = Tropidia Lindl. ■

350581　Schaenoprasum Franch. et Sav. = Allium L. ■

350582　Schaenoprasum Franch. et Sav. = Schoenoprasum Kunth ■

350583　Schaenus Gouan = Schoenus L. ■

350584　Schaetzelia Sch. Bip. = Schaetzellia Sch. Bip. ■●☆

350585　Schaetzellia Klotzsch = Onoseris Willd. ●■☆

350586　Schaetzellia Sch. Bip. (1849) = Hinterhubera Sch. Bip. ex Wedd. ●☆

350587　Schaetzellia Sch. Bip. (1850) = Macvaughiella R. M. King et H. Rob. ■●☆

350588　Schaffnera Benth. = Schaffnerella Nash ■☆

350589　Schaffnera Sch. Bip. = Dicoma Cass. ●☆

350590　Schaffnerella Nash(1912);沙夫草属■☆

350591　Schaffnerella gracilis Nash;沙夫草■☆

350592　Schanginia C. A. Mey. = Suaeda Forssk. ex J. F. Gmel. (保留属名)●■

350593　Schanginia Sievers ex Pall. = Hololachna Ehrenb. ●

350594　Schanginia aegyptiaca (Hasselq.) Aellen = Suaeda aegyptiaca (Hasselq.) Zohary ■☆

350595　Schanginia baccata (Forssk. ex J. F. Gmel.) Moq. = Suaeda aegyptiaca (Hasselq.) Zohary ■☆

350596　Schanginia hortensis (Forssk. ex J. F. Gmel.) Moq. = Suaeda aegyptiaca (Hasselq.) Zohary ■☆

350597　Schanginia linifolia (Pall.) C. A. Mey. = Suaeda linifolia Pall. ■

350598　Schanginia linifolia C. A. Mey. = Suaeda linifolia Pall. ■

350599　Schaphespermum Edgew. = Scaphespermum Edgew. ■

350600　Schaphespermum Edgew. = Seseli L. ■

350601　Schauera Nees (废弃属名) = Endlicheria Nees (保留属名)●☆

350602　Schauera Nees (废弃属名) = Schaueria Nees (保留属名)■☆

350603　Schauera graveolens Hassk. = Hyptis suaveolens (L.) Poit. ●■

350604　Schauera suaveolens (L.) Hassk. = Hyptis suaveolens (L.) Poit. ●■

350605　Schaueria Hassk. = Hyptis Jacq. (保留属名)●■

350606　Schaueria Meisn. = Endlicheria Nees (保留属名)●☆

350607　Schaueria Nees(1838)(保留属名);绍尔爵床属■☆

350608　Schaueria populifolia C. B. Clarke;杨叶绍尔爵床■☆

350609　Schauerla flavicoma (Lindl.) N. E. Br.;黄花绍尔爵床■☆

350610　Scheadendron G. Bertol = Combretum Loefl. (保留属名)●

350611　Scheadendron G. Bertol = Sheadendron G. Bertol ●

350612　Schedonnardus Steud. (1854);异留草属■☆

350613　Schedonnardus paniculatus (Nutt.) Trel.;异留草;Tumble Grass, Tumblegrass, Tumble-grass ■☆

350614　Schedonorus P. Beauv. (1812);杂雀麦属■☆

350615　Schedonorus P. Beauv. = Bromus L. + Festuca L. ■

350616　Schedonorus P. Beauv. = Festuca L. ■

350617　Schedonorus arundinaceus (Schreb.) Dumort.;苇状杂雀麦■

350618　Schedonorus arundinaceus (Schreb.) Dumort. subsp. cirtensis (St.-Yves) Scholz et Valdés;锡尔塔杂雀麦■☆

350619　Schedonorus arundinaceus (Schreb.) Dumort. subsp. mediterraneus (Hack.) Scholz et Valdés;地中海杂雀麦■☆

350620　Schedonorus arundinaceus Roem. et Schult. = Scolochloa festucacea (Willd.) Link ■

350621　Schedonorus benekenii Lange = Bromus benekenii (Lange) Trimen ■

350622　Schedonorus elatior (L.) P. Beauv. = Festuca arundinacea Schreb. ■

350623　Schedonorus erectus (Huds.) Gaudin ex Roem. et Schult. = Bromus erectus Huds. ■

350624　Schedonorus fontqueri (St.-Yves) Scholz et Valdés;丰特杂雀麦■☆

350625　Schedonorus giganteus (L.) Holub;巨大杂雀麦;Giant Fescue ■☆

350626　Schedonorus inermis (Leyss.) P. Beauv. = Bromus inermis Leyss. ■

350627　Schedonorus interruptus (Desf.) Tzvelev;间断杂雀麦■☆

350628　Schedonorus littoreus (Retz.) Tzvelev = Schedonorus arundinaceus (Schreb.) Dumort. ■

350629　Schedonorus phoenix (Scop.) Holub;高杂雀麦;Tall Fescue ■☆

350630　Schedonorus pratensis (Huds.) P. Beauv. = Festuca pratensis Huds. ■

350631　Schedonorus pratensis P. Beauv.;草原杂雀麦;Meadow Fescue ■☆

350632　Schedonorus sterilis (L.) Fr. = Bromus sterilis L. ■

350633　Schedonorus tectorum (L.) Fr. = Bromus tectorum L. ■

350634　Schedonorus unioloides (Kunth) Roem. et Schult. = Bromus catharticus Vahl ■

350635　Scheelea H. Karst. (1857);希乐棕属(休雷氏椰子属,迤逦椰子属,迤逦棕属);Scheelea ●☆

350636　Scheelea H. Karst. = Attalea Kunth ●☆

350637　Scheelea curvifrons L. H. Bailey;曲叶希乐棕●☆

350638　Scheelea liebmannii Becc.;迤逦棕●☆

350639　Scheeria Seem. = Achimenes Pers. (保留属名)■☆

350640　Schefferella Pierre = Burckella Pierre ●☆

350641　Schefferella Pierre = Payena A. DC. ●☆

350642　Schefferomitra Diels(1912);赛帽花属●☆

350643　Schefferomitra subaequalis (Scheff.) Diels;赛帽花●☆

350644　Scheffieldia Scop. = Samolus L. ■

350645　Scheffieldia Scop. = Sheffieldia J. R. Forst. et G. Forst. ■

350646　Schefflera J. R. Forst. et G. Forst. (1775)(保留属名);鹅掌柴属(鸭脚木属,鸭母树属);Schefflera ●

350647　Schefflera abyssinica (Hochst. ex A. Rich.) Harms;阿比西尼亚鹅掌柴●☆

350648　Schefflera actinophylla (Endl.) Harms;辐叶鹅掌柴(澳洲鹅

掌柴,澳洲鸭脚木,伞树,星叶罗伞);Actinoleaf Schefflera, Actino-leaved Schefflera, Australian Ivy Plam, Australian Umbrella Tree, Octopus Tree, Queensland Umbrella Tree, Rubber Tree, Schefflera, Umbrella Tree ●

350649　Schefflera acutifoliolata De Wild. = Schefflera abyssinica (Hochst. ex A. Rich.) Harms ●☆

350650　Schefflera adolfi-fridericii Harms =Schefflera goetzenii Harms ●☆

350651　Schefflera angiensis De Wild. = Schefflera myriantha (Baker) Drake ●☆

350652　Schefflera angustifoliolata C. N. Ho;狭叶鹅掌藤(狭叶鸭脚木)●

350653　Schefflera angustifoliolata C. N. Ho = Schefflera minutistellata Merr. ex H. L. Li ●

350654　Schefflera arboricola (Hayata) Merr.;鹅掌藤(鹅掌蘗,狗脚蹄,汉桃叶,七加皮,七叶烂,七叶莲,七叶藤,手树,小叶鸭脚木,鸭脚木);Dwarf Umbrella Tree, Hawallan Elf Schefflera, Parasol Plant,Parasol Tree,Scandent Schefflera ●

350655　Schefflera arboricola Hayata 'Hong Kong Variegata';香港斑叶鸭脚木●☆

350656　Schefflera arboricola Hayata 'Hong Kong';香港鸭脚木●

350657　Schefflera arboricola Hayata 'Renata White Variegata';斑叶耳突鸭脚木●☆

350658　Schefflera arboricola Hayata 'Renata';裂叶鹅掌柴●☆

350659　Schefflera arboricola Hayata 'Trinette';黄金鸭脚木●☆

350660　Schefflera atrifoliata R. H. Miao;黑叶鹅掌柴(黑叶鹅掌藤);Blak-leaf Schefflera ●

350661　Schefflera atrifoliata R. H. Miao = Schefflera heptaphylla (L.) Frodin ●

350662　Schefflera baikiei (Seem.) Harms = Schefflera barteri (Seem.) Harms ●☆

350663　Schefflera barteri (Seem.) Harms;巴氏鹅掌柴;Barter Schefflera ●☆

350664　Schefflera barteri (Seem.) Harms var. urostachya (Harms) Tennant =Schefflera urostachya Harms ●☆

350665　Schefflera bengalensis Gamble;孟加拉鹅掌柴●☆

350666　Schefflera bequaertii De Wild. = Schefflera myriantha (Baker) Drake ●☆

350667　Schefflera bodinieri (H. Lév.) Rehder;短序鹅掌柴(川黔鸭脚木);Bodinier Schefflera ●

350668　Schefflera bojeri (Seem.) R. Vig.;博耶尔鹅掌柴●☆

350669　Schefflera bracteolifera Frodin;苞片鹅掌柴●☆

350670　Schefflera brevipedicellata Harms;多核鹅掌柴(鸡脚,泡笼桐,泡通杆,鸭脚木);Manypit Schefflera, Polystone Schefflera, Short-pediceled Schefflera ●

350671　Schefflera capitata Harms;头状鹅掌柴;Capitate Schefflera ●☆

350672　Schefflera capuroniana (Bernardi) Bernardi;凯普伦鹅掌柴●☆

350673　Schefflera cephalotes (C. B. Clarke) Harms;印度鹅掌柴;India Schefflera,Indian Schefflera ●

350674　Schefflera chapana Harms;异叶鹅掌柴(鸭脚木);Differentleaf Schefflera, Diverseleaf Schefflera, Diversifoliolate Schefflera, Diversifolious Schefflera ●

350675　Schefflera chinensis (Dunn) H. L. Li;中华鹅掌柴;China Schefflera,Chinese Schefflera ●

350676　Schefflera chinpingensis C. J. Tseng et G. Hoo = Schefflera petelotii Merr. ●

350677　Schefflera cinnamomifoliolata C. B. Shang = Schefflera pesavis R. Vig. ●

350678　Schefflera compacta Frodin ex Lauener = Schefflera bodinieri

(H. Lév.) Rehder ●

350679　Schefflera congesta De Wild. = Schefflera myriantha (Baker Drake ●☆

350680　Schefflera delavayi (Franch.) Harms;穗序鹅掌柴(柴厚朴,广加皮,大泡通,大通草,大通塔,大五加皮,德氏鸭脚木,隔子通假通脱木,绒毛鸭脚木,山加皮,野巴戟);Delavay Schefflera ●

350681　Schefflera delavayi (Franch.) Harms var. ochrascens Hand. Mazz. = Schefflera delavayi (Franch.) Harms ●

350682　Schefflera digitata J. R. Forst. et G. Forst.;新西兰鹅掌柴(掌叶鹅掌柴,指状鹅掌柴);New Zealand Schefflera,Pate ●☆

350683　Schefflera discolor Merr. =Schefflera delavayi (Franch.) Harms ●

350684　Schefflera diversifoliolata H. L. Li = Schefflera chapana Harms ●

350685　Schefflera dumicola W. W. Sm. = Schefflera hoi (Dunn) R. Vig. ●

350686　Schefflera dumicola W. W. Sm. f. acuta (C. J. Tseng et G. Hoo Frodin = Schefflera hoi (Dunn) R. Vig. ●

350687　Schefflera elata (C. B. Clarke) Harms;高鹅掌柴;Higl Schefflera ●

350688　Schefflera elegantissima (Veitch ex Mast.) Lowry et Frodin;极美鹅掌柴(孔雀木,雅致鹅掌柴);Aralia, False Aralia ●☆

350689　Schefflera elliptica (Blume) Harms;密脉鹅掌柴●

350690　Schefflera elliptica sensu R. R. Stewart = Schefflera bengalensis Gamble ●☆

350691　Schefflera evrardii Bamps;埃夫拉尔鹅掌柴●

350692　Schefflera favargeri Bernardi;法瓦尔热鹅掌柴●☆

350693　Schefflera fengii C. J. Tseng et G. Hoo;文山鹅掌柴(国楣鹅掌柴);Feng Schefflera, Wenshan Schefflera ●

350694　Schefflera fosbergiana (Bernardi) Bernardi;福斯伯格鹅掌柴●☆

350695　Schefflera fukienensis Merr.;福建鹅掌柴;Fujian Schefflera ●

350696　Schefflera fukienensis Merr. = Schefflera elliptica (Blume) Harms ●

350697　Schefflera fukienensis Merr. = Schefflera khasiana (C. B. Clarke) R. Vig. ●

350698　Schefflera glabrescens (C. J. Tseng et G. Hoo) Frodin;光叶鹅掌柴●

350699　Schefflera glomerulata H. L. Li;球序鹅掌柴(团花鸭脚木,五加皮);Glomerule Schefflera, Glomerulose Schefflera ●

350700　Schefflera glomerulata H. L. Li = Schefflera pauciflora R. Vig. ●

350701　Schefflera goetzenii Harms;格兹鹅掌柴●☆

350702　Schefflera guizhouensis C. B. Shang;贵州鹅掌柴●

350703　Schefflera hainanensis Merr. et Chun;海南鹅掌柴;Hainan Schefflera ●

350704　Schefflera halleana Bernardi;哈勒鹅掌柴●☆

350705　Schefflera henriquesiana Harms ex Henriq. = Schefflera barteri (Seem.) Harms ●☆

350706　Schefflera heptaphylla (L.) Frodin;鹅掌柴(公母,公母树,七叶鹅掌柴,七叶莲,伞托树,五指通,西加皮,小叶鸭脚木,鸭达木,鸭脚板,鸭脚木,鸭母树);Common Schefflera, Ivy Tree, Octopus Tree, Schefflera ●

350707　Schefflera hierniana Harms =Schefflera barteri (Seem.) Harms ●☆

350708　Schefflera hoi (Dunn) R. Vig.;红河鹅掌柴(何氏鹅掌柴,通草);Ho's Schefflera, Honghe Schefflera ●

350709　Schefflera hoi (Dunn) R. Vig. f. acuta C. J. Tseng et G. Hoo = Schefflera hoi (Dunn) R. Vig. ●

350710　Schefflera hoi (Dunn) R. Vig. var. acuta C. J. Tseng et G. Hoo;急尖叶红河鹅掌柴;Acute Ho Schefflera ●

350711　Schefflera hoi (Dunn) R. Vig. var. macrophylla H. L. Li;大叶

红河鹅掌柴;Bigleaf Ho Schefflera ●

350712　Schefflera hoi（Dunn）R. Vig. var. macrophylla H. L. Li ＝ Schefflera hoi（Dunn）R. Vig. ●

350713　Schefflera hookeriana Harms ＝ Schefflera abyssinica（Hochst. ex A. Rich.）Harms ●☆

350714　Schefflera humblotiana Drake;洪布鹅掌柴●☆

350715　Schefflera humblotii Harms ＝ Schefflera myriantha（Baker）Drake ●☆

350716　Schefflera hypoleuca（Kurz）Harms;白背鹅掌柴（白背叶鹅掌柴,大豆豉叶,大木通,大泡通,大通塔,豆豉叶,饭包叶,隔子通,牛噪管,泡通,三叉木,伞把木）;Whiteback Schefflera, Whitebacked Schefflera ●

350717　Schefflera hypoleuca（Kurz）Harms var. hypochlorum Dunn ex K. M. Feng et Y. R. Li;绿背叶鹅掌柴;Greenback Schefflera ●

350718　Schefflera hypoleucoides Harms;离柱鹅掌柴（拟白背叶鹅掌柴）;Dispersalstyle Schefflera, Separatestyle Schefflera, Separate-styled Schefflera ●

350719　Schefflera hypoleucoides Harms var. tomentosa Grushv. et Skvortsova ＝ Schefflera hypoleucoides Harms ●

350720　Schefflera hypoleucoides Harms var. truncata C. B. Shang;截叶离柱鹅掌柴;Truncate Separatestyle Schefflera ●

350721　Schefflera hypoleucoides Harms var. truncata C. B. Shang ＝ Schefflera hypoleucoides Harms ●

350722　Schefflera impressa（C. B. Clarke）Harms ＝ Schefflera rhododendrifolia（Griff.）Frodin ●

350723　Schefflera impressa（C. B. Clarke）Harms var. glabrescens C. J. Tseng et G. Hoo ＝ Schefflera glabrescens（C. J. Tseng et G. Hoo）Frodin ●

350724　Schefflera impressa（C. B. Clarke）Harms var. glabrescens C. J. Tseng et G. Hoo;光叶凹脉鹅掌柴;Glabrousleaf Sunkenvein Schefflera ●

350725　Schefflera insignis C. N. Ho;粉背鹅掌柴（粉背叶鸭脚木）;Distinctia Schefflera, Distinctive Schefflera, Farinoseback Schefflera ●

350726　Schefflera khasiana（C. B. Clarke）R. Vig.;扁盘鹅掌柴（扁鹅掌柴,粗芽鹅掌柴,季川鹅掌柴,龙爪树,七多,七叶加,七叶莲,西南鹅掌柴,鸭脚木,印度鹅掌柴）;Fujian Schefflera, Khas Schefflera, Khasia Schefflera, Ruggedbud Schefflera ●

350727　Schefflera kivuensis Bamps;基伍鹅掌柴●☆

350728　Schefflera kwangsiensis Merr. ex H. L. Li;广西鹅掌柴（肥达汗,费丁必,勾虽,广西鸭脚木,计进占,棵肥档,苗留堆,七多,七加风,七叶莲,鸭脚木）;Guangxi Schefflera ●

350729　Schefflera kwangsiensis Merr. ex H. L. Li ＝ Schefflera leucantha R. Vig. ●

350730　Schefflera ledermannii Harms ＝ Schefflera barteri（Seem.）Harms ●☆

350731　Schefflera leucantha R. Vig.;白花鹅掌柴（广西鹅掌柴,广西鸭脚木）;Whiteflower Schefflera, White-flowered Schefflera ●

350732　Schefflera lociana R. Vig. var. megaphylla C. B. Shang;龙州鹅掌柴（大叶鹅掌柴,琼山鹅掌柴）;Longzhou Schefflera ●

350733　Schefflera lociana var. megaphylla C. B. Shang ＝ Schefflera lociana R. Vig. var. megaphylla C. B. Shang ●

350734　Schefflera longipedicellata（Lecomte）Bernardi;长梗鹅掌柴●☆

350735　Schefflera lukwangulensis（Tennant）Bernardi;卢夸古尔鹅掌柴●☆

350736　Schefflera macerosa Bernardi;瘦弱鹅掌柴●☆

350737　Schefflera macrophylla（Lour.）Harms;大叶鹅掌柴;Bigleaf Schefflera, Big-leaved Schefflera ●

350738　Schefflera mannii（Hook. f.）Harms;曼氏鹅掌柴●☆

350739　Schefflera marlipoensis C. J. Tseng et G. Hoo;麻栗坡鹅掌柴;Malipo Schefflera ●

350740　Schefflera megalobotrya Harms ＝ Schefflera delavayi（Franch.）Harms ●

350741　Schefflera megalobotrya Harms ex Diels ＝ Schefflera delavayi（Franch.）Harms ●

350742　Schefflera menglaensis H. Chu et H. Wang;勐腊鹅掌柴;Mengla Schefflera ●

350743　Schefflera menglaensis H. Chu et H. Wang ＝ Schefflera brevipedicellata Harms ●

350744　Schefflera metcalfiana Merr. ex H. L. Li;多叶鹅掌柴;Manyleaf Schefflera, Multifolious Schefflera, Multileaf Schefflera ●

350745　Schefflera microphylla Merr.;吕宋鹅掌柴;Littleleaf Schefflera, Luzon Schefflera, Microphyllous Schefflera ●

350746　Schefflera mildbraedii Harms ＝ Schefflera goetzenii Harms ●☆

350747　Schefflera minutistellata Merr. ex H. L. Li;星毛鹅掌柴（卡汤,七加皮,通脱木,微星毛鸭母树,狭叶鹅掌柴,狭叶鸭脚木,小泡通,小齿通树,小星鸭脚木,星毛鸭脚木,鸭麻木）;Narrowleaf Schefflera, Star-haired Schefflera, Staryhair Schefflera ●

350748　Schefflera monophylla（Baker）Bernardi;单叶鹅掌柴●☆

350749　Schefflera multinervia H. L. Li;多脉鹅掌柴;Manyvein Schefflera, Multinerved Schefflera, Multivein Schefflera ●

350750　Schefflera myriantha（Baker）Drake;繁花鹅掌柴●☆

350751　Schefflera myriantha（Baker）Drake var. attenuata Bernardi ＝ Schefflera myriantha（Baker）Drake ●☆

350752　Schefflera napuoensis C. B. Shang;矩圆叶鹅掌柴（那坡鹅掌柴）;Napo Schefflera, Oblongleaf Schefflera ●

350753　Schefflera nyasensis De Wild. ＝ Schefflera myriantha（Baker）Drake ●☆

350754　Schefflera oblonga C. B. Shang ＝ Schefflera napuoensis C. B. Shang ●

350755　Schefflera octophylla（Lour.）Harms ＝ Schefflera heptaphylla（L.）Frodin ●

350756　Schefflera odorata（Blanco）Merr. et Rolfe;香鹅掌柴（鹅掌藤,香港鸭脚木）;Odorate Schefflera ●

350757　Schefflera parvifoliolata C. J. Tseng et G. Hoo;小叶鹅掌柴（七叶莲,伞把木,小豆豉杆）;Littleleaf Schefflera, Smallleaf Schefflera ●

350758　Schefflera pauciflora R. Vig.;贫花鹅掌柴（球序鹅掌柴）●

350759　Schefflera pentagyra C. J. Tseng et G. Hoo;五柱鹅掌柴;Fivestyle Schefflera ●

350760　Schefflera pentagyra C. J. Tseng et G. Hoo ＝ Schefflera chinensis（Dunn）H. L. Li ●

350761　Schefflera pesavis R. Vig.;樟叶鹅掌柴（火柴木）;Cinnamon-leaved Schefflera ●

350762　Schefflera petelotii Merr.;金平鹅掌柴;Jinping Schefflera ●

350763　Schefflera pingpienensis C. J. Tseng et G. Hoo;屏边鹅掌柴;Pingbian Schefflera ●

350764　Schefflera pingpienensis C. J. Tseng et G. Hoo ＝ Schefflera chapana Harms ●

350765　Schefflera pingpienensis C. J. Tseng et G. Hoo ＝ Schefflera diversifoliolata H. L. Li ●

350766　Schefflera polypyrena C. J. Tseng et G. Hoo ＝ Schefflera brevipedicellata Harms ●

350767　Schefflera polysciada Harms ＝ Schefflera myriantha（Baker）Drake ●☆

350768　Schefflera producta（Dunn）R. Vig.;尾叶鹅掌柴;Tailleaf

Schefflera,Tailshapeleaf Schefflera ●

350769　Schefflera producta（Dunn）R. Vig. = Brassaiopsis producta（Dunn）C. B. Shang ●

350770　Schefflera pubigera（Brongn. et Planch.）Frodin = Schefflera elliptica（Blume）Harms ●

350771　Schefflera pueckleri（K. Koch）Frodin = Tupidanthus calyptratus Hook. f. et Thomson ●

350772　Schefflera racemosa（Wight）Harms；总序鹅掌柴；Raceme Schefflera ●☆

350773　Schefflera racemosa Harms = Schefflera taiwaniana（Nakai）Kaneh. ●

350774　Schefflera rhododendrifolia（Griff.）Frodin；凹脉鹅掌柴；Sunkenvein Schefflera,Sunken-veined Schefflera ●

350775　Schefflera roxburghii Gamble；顶叶鹅掌柴●☆

350776　Schefflera rubriflora C. J. Tseng et G. Hoo；红花鹅掌柴（当遁,七叶莲,五加风,五加皮）；Redflower Schefflera, Red-flowered Schefflera ●

350777　Schefflera rubriflora C. J. Tseng et G. Hoo = Schefflera heptaphylla（L.）Frodin ●

350778　Schefflera salweenensis W. W. Sm. = Schefflera hoi（Dunn）R. Vig. ●

350779　Schefflera salweenensis W. W. Sm. var. macrophylla（H. L. Li）Frodin = Schefflera hoi（Dunn）R. Vig. ●

350780　Schefflera saveenensis W. W. Sm. = Schefflera hoi（Dunn）R. Vig. ●

350781　Schefflera schweliensis W. W. Sm.；瑞丽鹅掌柴；Ruili Schefflera ●

350782　Schefflera singalangensis R. Vig.；新加兰鹅掌藤；Singalang Schefflera ●☆

350783　Schefflera stenomera Hand. -Mazz. = Schefflera hoi（Dunn）R. Vig. ●

350784　Schefflera stolzii Harms；斯托尔兹鹅掌柴●☆

350785　Schefflera stuhlmannii Harms = Schefflera goetzenii Harms ●☆

350786　Schefflera taiwaniana（Nakai）Kaneh.；台湾鹅掌柴（台湾鸭脚木）；Taiwan Schefflera ●

350787　Schefflera tenuis H. L. Li；细序鹅掌柴；Tenuousspanicle Schefflera,Thin Schefflera,Thinpanicle Schefflera ●

350788　Schefflera tenuis H. L. Li = Schefflera leucantha R. Vig. ●

350789　Schefflera tessmannii Harms；泰斯曼鹅掌柴●☆

350790　Schefflera thorelii Vig. = Brassaiopsis glomerulata（Blume）Regel ●

350791　Schefflera trevesioides Harms =Schefflera hypoleucoides Harms ●

350792　Schefflera trevesioides Harms var. tomentosa（Grushv. et Skvortsova）Frodin = Schefflera hypoleucoides Harms ●

350793　Schefflera tridentata De Wild. = Schefflera urostachya Harms ●☆

350794　Schefflera uinnanensis H. L. Li = Schefflera leucantha R. Vig. ●

350795　Schefflera umbellifera（Sond.）Baill.；长柄鹅掌柴；Bastard Cabbage Tree,Forest Cabbage Tree ●☆

350796　Schefflera umbellifera（Sond.）Baill. var. buchananii（Harms）Bernardi = Schefflera umbellifera（Sond.）Baill. ●☆

350797　Schefflera umbellifera Baill. = Schefflera umbellifera（Sond.）Baill. ●☆

350798　Schefflera urostachya Harms；尾穗鹅掌柴●☆

350799　Schefflera venulosa（Wight et Arn.）Harms；七叶藤（狗脚蹄,汉桃叶,密脉鹅掌柴,七叶莲,屋加风,五加皮）；Densevein Schefflera, Dense-veined Schefflera ●

350800　Schefflera volkensii（Harms）Harms；福尔鹅掌柴●■

350801　Schefflera wangii H. L. Li = Schefflera chinensis（Dunn）H. L. Li ●

350802　Schefflera wangii H. L. Li = Schefflera pentagyra C. J. Tseng et G. Hoo ●

350803　Schefflera wardii C. Marquand et Airy Shaw；西藏鹅掌柴；Ward Schefflera ●

350804　Schefflera yui C. J. Tseng et G. Hoo；独龙鹅掌柴（粗芽鹅掌柴）；Dulong Schefflera ●

350805　Schefflera yui C. J. Tseng et G. Hoo = Schefflera khasiana（C. B. Clarke）R. Vig. ●

350806　Schefflera yunnanensis H. L. Li；云南鹅掌柴；Yunnan Schefflera ●

350807　Schefflera yunnanensis H. L. Li = Schefflera leucantha R. Vig. ●

350808　Schefflera zhuana Lowry et C. B. Shang；光华鹅掌柴●

350809　Schefflerodendron Harms ex Engl. = Schefflerodendron Harms ●☆

350810　Schefflerodendron Harms（1901）；舍夫豆属●☆

350811　Schefflerodendron adenopetalum（Taub.）Harms；腺瓣舍夫豆●☆

350812　Schefflerodendron gabonense Pellegr.；加蓬舍夫豆●☆

350813　Schefflerodendron gazense Baker f. = Craibia brevicaudata（Vatke）Dunn subsp. baptistarum（Büttner）J. B. Gillett ●☆

350814　Schefflerodendron gilbertianum J. Léonard et Latour；吉尔伯特舍夫豆●☆

350815　Schefflerodendron usambarense Harms ex Engl. ；乌桑巴拉舍夫豆●☆

350816　Schefflerodendron usambarense Harms ex Engl. var. macrophyllum Pellegr. ；大叶舍夫豆●☆

350817　Scheffleropsis Ridl.（1922）；类鹅掌柴属●☆

350818　Scheffleropsis Ridl. = Schefflera J. R. Forst. et G. Forst.（保留属名）●

350819　Scheffleropsis polyandra（Ridl.）Ridl.；类鹅掌柴●☆

350820　Scheidweileria Klotzsch = Begonia L. ●■

350821　Schelhameria Fabr. = Matthiola W. T. Aiton（保留属名）■●

350822　Schelhameria Heist. ex Fabr.（废弃属名）= Schelhammera R. Br.（保留属名）■☆

350823　Schelhammera R. Br.（1810）（保留属名）；谢勒水仙属■☆

350824　Schelhammera multiflora R. Br.；多花谢勒水仙■☆

350825　Schelhammera undulata R. Br.；谢勒水仙■☆

350826　Schelhammeria Moench = Carex L. ■

350827　Schellanderia Francisci = Physoplexis（Endl.）Schur ■☆

350828　Schellanderia Francisci = Synotoma（G. Don）R. Schulz ■☆

350829　Schellenbergia C. E. Parkinson = Vismianthus Mildbr. ●☆

350830　Schellhammeria capitata Moench = Carex bohemica Schreb. ■

350831　Schellingia Steud. = Aegopogon Humb. et Bonpl. ex Willd. ■☆

350832　Schellingia Steud. = Aegopogon P. Beauv. ■☆

350833　Schellingia Steud. = Amphipogon R. Br. ■☆

350834　Schellolepis J. Sm. = Goniorrhachis Taub. ●☆

350835　Schelveria Nees = Angelonia Bonpl. ■●☆

350836　Schelveria Nees et Mart. = Angelonia Bonpl. ■●☆

350837　Schema Seem. = Achimenes Pers.（保留属名）■☆

350838　Schenckia K. Schum. = Deppea Cham. et Schltdl. ●☆

350839　Schenckochloa J. J. Ortíz（1991）；申克禾属■☆

350840　Schenckochloa barbata（Hack.）J. J. Ortíz；申克禾■☆

350841　Schenkia Griseb.（1853）；申克龙胆属■☆

350842　Schenkia Griseb. = Centaurium Hill ■

350843　Schenkia Griseb. = Erythraea Borkh. ■

350844　Schenkia japonica（Maxim.）G. Mans. = Centaurium japonicum（Maxim.）Druce ■☆

350845　Schenkia sebaeoides Griseb. ；申克龙胆■☆

350846 Schenkia spicata（L.）G. Mans. = Centaurium spicatum（L.）Fritsch ■

350847 Schenodorus P. Beauv. = Festuca L. + Bromus L. ■

350848 Schenodorus P. Beauv. = Festuca L. ■

350849 Schenodorus P. Beauv. = Schedonorus P. Beauv. ☆

350850 Scheperia Raf. = Cadaba Forssk. ●☆

350851 Scheperia Raf. = Schepperia Neck. ●☆

350852 Schepperia Neck. = Cadaba Forssk. ●☆

350853 Schepperia juncea（Sparrm.）DC. = Cadaba aphylla（Thunb.）Wild ●☆

350854 Scherardia Neck. = Sherardia L. ■☆

350855 Scherya R. M. King et H. Rob.（1977）;彩片菊属■☆

350856 Scherya bahiensis R. M. King et H. Rob. ;彩片菊☆

350857 Schetti Adans. = Ixora L. ●

350858 Scheuchzera St. -Lag. = Scheuchzeria L. ■

350859 Scheuchzeria L.（1753）;芝菜属（冰沼草属）;Iceboggrass, Rannoch-rush, Scheuchzeria ■

350860 Scheuchzeria americana（Fernald）G. N. Jones = Scheuchzeria palustris L. ■

350861 Scheuchzeria palustris L. ;芝菜（冰沼草）;Arrow-grass, Iceboggrass, Marsh Scheuchzeria, Pod-grass, Rannoch-rush, Scheuchzeria, Swampy Scheuchzeria ■

350862 Scheuchzeria palustris L. subsp. americana（Fernald）Hultén = Scheuchzeria palustris L. ■

350863 Scheuchzeria palustris L. var. americana Fernald = Scheuchzeria palustris L. ■

350864 Scheuchzeriaceae F. Rudolphi(1830)（保留科名）;芝菜科（冰沼草科）;Iceboggrass Family, Rannoch-rush Family, Scheuchzeria Family ■

350865 Scheuchzeriaceae F. Rudolphi（保留科名）= Juncaginaceae Rich.（保留科名）■

350866 Scheutzia（Gand.）Gand. = Rosa L. ●

350867 Scheutzia Gand. = Rosa L. ●

350868 Schewykerta S. G. Gmel. = Nymphoides Ség. ■

350869 Schickendantzia Pax(1889);希肯花属■☆

350870 Schickendantzia Speg. = Schickendantziella（Speg.）Speg. ■☆

350871 Schickendantzia hieronymi Pax;希肯花■☆

350872 Schickendantziella（Speg.）Speg.(1903);希肯葱属■☆

350873 Schickendantziella Speg. = Schickendantziella（Speg.）Speg. ■☆

350874 Schickendantziella trichosepala Speg. ;希肯葱■☆

350875 Schickia Tischer = Mesembryanthemum L.（保留属名）■●

350876 Schidiomyrtus Schauer = Baeckea L. ●

350877 Schidorhynchos Szlach.（1993）;裂喙绶草属■☆

350878 Schidorhynchos Szlach. = Spiranthes Rich.（保留属名）■

350879 Schidospermum Griseb. = Fosterella L. B. Sm. ■☆

350880 Schidospermum Griseb. ex Lechl. = Fosterella L. B. Sm. ■☆

350881 Schieckia H. Karst. = Celastrus L.（保留属名）●

350882 Schieckia Benth. et Hook. f. = Schiekia Meisn. ■☆

350883 Schieckia H. Karst. = Celastrus L.（保留属名）●

350884 Schiedea A. Rich. = Machaonia Humb. et Bonpl. ■☆

350885 Schiedea A. Rich. = Tertrea DC. ■☆

350886 Schiedea Bartl. = Richardia L. ■

350887 Schiedea Bartl. = Richardsonia Kunth ■

350888 Schiedea Cham. et Schltdl.（1826）;合丝繁缕木属■●☆

350889 Schiedea ligustrina Cham. et Schltdl. ;合丝繁缕木●☆

350890 Schiedeella Schltr.（1920）;希德兰属■☆

350891 Schiedeella arizonica P. M. Br. ;亚利桑那希德兰■☆

350892 Schiedeella confusa（Garay）Espejo et López-Ferr. = Deiregyne confusa Garay ■☆

350893 Schiedophytum H. Wolff = Donnellsmithia J. M. Coult. et Rose ■☆

350894 Schiedta Willis = Schiedea Cham. et Schltdl. ■●☆

350895 Schiekea Walp. = Celastrus L.（保留属名）●

350896 Schiekea Walp. = Schieckia H. Karst. ●

350897 Schiekia Meisn.（1842）;热美血草属■☆

350898 Schiekia orinocensis（Kunth）Meisn. ;热美血草■☆

350899 Schievereckia Nyman = Schivereckia Andrz. ex DC. ■☆

350900 Schillera Rchb. = Eriolaena DC. ●

350901 Schilleria Kunth = Oxodium Raf. ●■

350902 Schilleria Kunth = Piper L. ●■

350903 Schima Reinw. = Schima Reinw. ex Blume ●

350904 Schima Reinw. ex Blume（1823）;木荷属（桐树属）;Guger Tree, Gugertree, Guger-tree ●

350905 Schima argentea Pritz. ;银木荷;Silvery Gugertree ●

350906 Schima bambusifolia Hu;竹叶木荷;Bambooleaf Gugertree, Bamboo-leaved Gugertree ●

350907 Schima bambusifolia Hu = Schima argentea Pritz. ●

350908 Schima boninensis Nakai = Schima wallichii（DC.）Korth. subsp. noronhae（Reinw. ex Blume）Bloemb. ●

350909 Schima brevipedicellata Hung T. Chang;短梗木荷;Shortpedicel Gugertree, Shortstlak Gugertree ●

350910 Schima brevipes Craib = Schima wallichii（DC.）Korth. ●

350911 Schima confertiflora Merr. = Schima superba Gardner et Champ. ●

350912 Schima crenata Korth. ;钝齿木荷;Crenate Gugertree ●

350913 Schima dulungensis Hung T. Chang et C. X. Ye = Schima sericans（Hand. -Mazz.）T. L. Ming var. paracrenata（Hung T. Chang）T. L. Ming ●

350914 Schima forrestii Airy Shaw;大花木荷（腾冲木荷）;Bigflower Gugertree, Forrest Gugertree ●

350915 Schima forrestii Airy Shaw = Schima khasiana Dyer ●

350916 Schima grandiperulata Hung T. Chang;大苞木荷;Bigbract Gugertree, Big-bracted Gugertree ●

350917 Schima grandiperulata Hung T. Chang = Schima sinensis（Hemsl. et E. H. Wilson）Airy Shaw ●

350918 Schima kanhaoensis Hayata = Schima superba Gardner et Champ. var. kanhoensis（Hayata）H. Keng ●

350919 Schima kankaoensis Hayata = Schima superba Gardner et Champ. ●

350920 Schima khasiana Dyer;尖齿木荷（印度木荷）;Indian Gugertree, Khas Gugertree, Khasia Gugertree ●

350921 Schima khasiana Dyer var. sericans Hand. -Mazz. ;尖齿毛木荷（齿叶荷树,绢毛印度木荷,毛木树）;Sericeous Indian Gugertree ●

350922 Schima khasiana Dyer var. sericans Hand. -Mazz. = Schima sericans（Hand. -Mazz.）T. L. Ming ●

350923 Schima kwangtungensis Hung T. Chang;广东木荷;Guangdong Gugertree, Kwangtung Gugertree ●

350924 Schima kwangtungensis Hung T. Chang = Schima remotiserrata Hung T. Chang ●

350925 Schima liukiuensis Nakai = Schima superba Gardner et Champ. ●

350926 Schima liukiuensis Nakai = Schima wallichii（DC.）Korth. subsp. noronhae（Reinw. ex Blume）Bloemb. ●

350927 Schima macrosepala Hung T. Chang;大萼木荷;Bigcalyx Gugertree, Bigsepal Gugertree, Macrosepal Gugertree ●

350928 Schima macrosepala Hung T. Chang = Schima villosa Hu ●

350929 Schima mairei Hochr. = Schima argentea Pritz. ●

350930 Schima mertensiana (Siebold et Zucc.) Koidz. = Schima wallichii (DC.) Korth. subsp. noronhae (Reinw. ex Blume) Bloemb.

350931 Schima multibracteata Hung T. Chang;多苞木荷;Bracteose Gugertree,Manybract Gugertree,Multibracted Gugertree ●

350932 Schima noronhae Reinw. ex Blume;南洋木荷(成凤山木荷,滇木荷,杆仔皮);Burma Gugertree ●

350933 Schima noronhae Reinw. ex Blume = Schima wallichii (DC.) Korth. subsp. noronhae (Reinw. ex Blume) Bloemb. ●

350934 Schima noronhae sensu Matsum. = Schima superba Gordon et Champ. ●

350935 Schima paracrenata Hung T. Chang;拟钝齿木荷(钝叶木荷);Obtuse-leaved Gugertree, Paracrenate Gugertree, Subcrenate Gugertree ●

350936 Schima paracrenata Hung T. Chang = Schima sericans (Hand.-Mazz.) T. L. Ming var. paracrenata (Hung T. Chang) T. L. Ming ●

350937 Schima parviflora Hung T. Chang et W. C. Cheng ex Hung T. Chang;小花木荷;Littleflower Gugertree,Little-flowered Gugertree ●

350938 Schima polyneura Hung T. Chang;多脉木荷;Manyvein Gugertree,Neurose Gugertree ●

350939 Schima polyneura Hung T. Chang = Schima brevipedicellata Hung T. Chang ●

350940 Schima remotiserrata Hung T. Chang;疏齿木荷;Laxtooth Gugertree,Remote-serrated Gugertree,Scattredtooth Gugertree ●

350941 Schima sericans (Hand.-Mazz.) T. L. Ming;贡山木荷●

350942 Schima sericans (Hand.-Mazz.) T. L. Ming var. paracrenata (Hung T. Chang) T. L. Ming;独龙木荷;Dulong Gugertree ●

350943 Schima sinensis (Hemsl. et E. H. Wilson) Airy Shaw;中华木荷(华木荷);China Gugertree,Chinese Gugertree ●

350944 Schima stellata Pierre = Craibiodendron stellatum (Pierre) W. W. Sm. ●

350945 Schima superba Gardner et Champ.;木荷(椿木,果槁,何树,荷木,荷树,横柴,回树,木艾树,木荷柴);Chinese Gugertree,Chinese Guger-tree,Guger Tree,Gugertree,Schima ●

350946 Schima superba Gardner et Champ. var. kankoensis (Hayata) H. Keng;甘木荷(港口木荷,恒春木荷);Gangkou Gugertree,Taiwan Guger-tree ●

350947 Schima superba Gordon et Champ. = Schima wallichii (DC.) Korth. subsp. noronhae (Reinw. ex Blume) Bloemb. ●

350948 Schima villosa Hu;毛木荷(毛毛树,箐毛木,柔毛木荷);Softhair Gugertree,Villose Gugertree,Villous Gugertree ●

350949 Schima wallichii (DC.) Korth.;西南木荷(峨眉木荷,红木荷,麻木树,毛木树,乌叶木荷,乌叶子);Chilauni,Darjeeling Guger Tree,Darjeeling Gugertree,Darjeeling Guger-tree,Wallich Gugertree ●

350950 Schima wallichii (DC.) Korth. subsp. liukiuensis (Nakai) Bloemb. = Schima wallichii (DC.) Korth. subsp. noronhae (Reinw. ex Blume) Bloemb. ●

350951 Schima wallichii (DC.) Korth. subsp. mertensiana (Siebold et Zucc.) Bloemb. = Schima wallichii (DC.) Korth. subsp. noronhae (Reinw. ex Blume) Bloemb. ●

350952 Schima wallichii (DC.) Korth. subsp. noronhae (Reinw. ex Blume) Bloemb. = Schima noronhae Reinw. ex Blume ●

350953 Schima wallichii Choisy = Schima wallichii (DC.) Korth. ●

350954 Schima xinyiensis Hung T. Chang et Z. Y. Su = Schima superba Gardner et Champ. ●

350955 Schima xinyiensis Hung T. Chang et Z. Y. Su ex Hung T. Chang et S. X. Ren;信宜木荷;Xinyi Gugertree ●

350956 Schimmelia Holmes = Amyris P. Browne ●☆

350957 Schimpera Hochst. et Steud. ex Endl. = Schimpera Steud. et Hochst. ex Endl. ■☆

350958 Schimpera Hochst. ex Steud. = Schimpera Steud. et Hochst. ex Endl. ■☆

350959 Schimpera Steud. et Hochst. ex Endl. (1839);厚喙荠属■☆

350960 Schimpera arabica Hochst. et Steud.;厚喙荠■☆

350961 Schimperella H. Wolff = Oreoschimperella Rauschert ■☆

350962 Schimperella H. Wolff(1927);小山厚喙荠属■☆

350963 Schimperella aberdarensis C. Norman = Oreoschimperella aberdarensis (C. Norman) Rauschert ■☆

350964 Schimperella verrucosa (M. J. Gay ex A. Rich.) H. Wolff = Oreoschimperella verrucosa (J. Gay ex A. Rich.) Rauschert ■☆

350965 Schimperina Tiegh. = Agelanthus Tiegh. ●☆

350966 Schimperina Tiegh. = Tapinanthus (Blume) Rchb. (保留属名)●☆

350967 Schimperina platyphylla (Hochst. ex A. Rich.) Tiegh. = Agelanthus platyphyllus (Hochst. ex A. Rich.) Balle ●☆

350968 Schinaceae Raf. = Anacardiaceae R. Br. (保留科名)●

350969 Schindleria H. Walter(1906);胞果珊瑚木属(胞果珊瑚属)●☆

350970 Schindleria glabra H. Walter;胞果珊瑚木●☆

350971 Schinnongia Schrank(1822);睫毛鸢尾属■☆

350972 Schinocarpus K. Schum. = Selinocarpus A. Gray ●☆

350973 Schinopsis Engl. (1876);拟破斧木属（破斧木属）;Quebracho,Red Quebracho ●☆

350974 Schinopsis balansae Engl.;红破斧木;Willow-leaf Red Quebracho ●☆

350975 Schinopsis balansae Engl. var. pendula Tortorelli;垂枝红破斧木●☆

350976 Schinopsis brasiliensis Engl.;巴西拟破斧木●☆

350977 Schinopsis brasiliensis Engl. var. glabra Engl.;光巴西拟破斧木●☆

350978 Schinopsis lorentzii Engl. = Quebrachia lorentzii Griseb. ●☆

350979 Schinopsis quebrachocolorado (Schltdl.) F. A. Barkley et T. Mey.;白破斧木●☆

350980 Schinus L. (1753);肖乳香属(胡椒木属);Brazil Pepper, Pepper Tree,Peppertree,Pepper-tree,Weeping Willow ●

350981 Schinus areira L. = Schinus molle L. var. areira (L.) DC. ●☆

350982 Schinus dependens Ortega;垂枝肖乳香●☆

350983 Schinus indicus Burm. = Rhus chinensis Mill. ●

350984 Schinus indicus Burm. f. = Rhus chinensis Mill. ●

350985 Schinus latifolius Engl.;宽叶肖乳香(阔叶肖乳香);Broadleaf Peppertree,Molle ●☆

350986 Schinus lentiscifolius Marchand;紫红花肖乳香;Coroba,Molle ●☆

350987 Schinus limonia L. = Feronia limonia (L.) Swingle ●

350988 Schinus limonia L. = Limonia acidissima L. ●

350989 Schinus longifolius Speg.;长叶肖乳香;Longleaf Peppertree ●☆

350990 Schinus molle L.;加州肖乳香(加州胡椒树,柔毛肖乳香);American Mastic, California Pepper, California Pepper Tree, California Peppertree, California Pepper-tree, Pepper Tree, Peru Peppertree,Peruvian Mastic, Peruvian Mastic Tree, Peruvian Mastic-tree, Peruvian Pepper Tree, Peruvian Peppertree, Peruvian Pepper-tree,Weeping Willow ●☆

350991 Schinus molle L. var. areira (L.) DC.;秘鲁肖乳香;Peppercom,Peppertree ●☆

350992 Schinus molleoides Vell. = Lithraea molleoides (Vell.) Engl. ●☆

350993　Schinus myricoides L. = Cuscuta africana Willd. ■☆

350994　Schinus polygamus（Cav.）Cabrera et I. M. Johnst.；杂性肖乳香；Hardee Peppertree，Peruvian Pepper Tree ●☆

350995　Schinus terebinthifolius Raddi；肖乳香（巴西胡椒木，巴西乳香，胡椒木）；Brazil Pepper，Brazil Peppertree，Brazil Pepper-tree，Brazilian Pepper，Brazilian Pepper Tree，Brazilian Peppertree，Brazilian Pepper-tree，Christmas Berry，Christmas Peppertree，Christmas-berry-tree ●

350996　Schinus terebinthifolius Raddi var. acutifolius Engl. = Schinus terebinthifolius Raddi ●

350997　Schinus terebinthifolius Raddi var. raddianus Engl.；巴西肖乳香；Brazilian Peppertree ●☆

350998　Schinzafra Kuntze = Thamnea Sol. ex Brongn.（保留属名）●☆

350999　Schinzia Dennst.（1818）；欣兹堇属■☆

351000　Schinzia inconspicua Dennst.；欣兹堇■☆

351001　Schinziella Gilg（1895）；欣兹龙胆属■☆

351002　Schinziella tetragona（Schinz）Gilg；欣兹龙胆■☆

351003　Schinziella tetragona（Schinz）Gilg var. parviflora Schinz ex De Wild. = Schinziella tetragona（Schinz）Gilg ■☆

351004　Schinziophyton Hutch. ex Radcl. -Sm.（1990）；欣兹大戟属●☆

351005　Schinziophyton rautanenii（Schinz）Radcl. -Sm.；欣兹大戟；Mongonogo，Mugongo ●☆

351006　Schippia Burret（1933）；单雌棕属（洪都拉斯棕属，康科罗棕属，撕裂柄棕属）●☆

351007　Schippia concolor Burret；单雌棕（康科罗棕）●☆

351008　Schirostachyum de Vriese = Schizostachyum Nees ●

351009　Schisachyrium Munro = Andropogon L.（保留属名）■

351010　Schisachyrium Munro = Schizachyrium Nees ■

351011　Schisandra Michx.（1803）（保留属名）；五味子属；Magnolia Vine，Magnoliavine，Magnolia-vine，Star-vine ●

351012　Schisandra arisanensis Hayata；阿里山五味子（阿里山北五味子，台湾五味子）；Alishan Magnolia Vine，Alishan Magnoliavine，Taiwan Magnoliavine，Taiwan Magnolia-vine ●

351013　Schisandra arisanensis Hayata = Schisandra elongata Hook. f. et Thomson ●

351014　Schisandra arisanensis Hayata subsp. viridis（A. C. Sm.）R. M. K. Saunders；绿叶五味子（凤沙藤，绿五味子）；Green Magnoliavine，Greenleaf Magnoliavine ●

351015　Schisandra axillaris Hook. f. et Thomson；腋生五味子●

351016　Schisandra bicolor W. C. Cheng；二色五味子（北五味子，二色内风消，两色五味子，香苏子）；Bicolor Magnoliavine，Bicolor Magnolia-vine ●

351017　Schisandra bicolor W. C. Cheng = Schisandra repanda（Siebold et Zucc.）Radlk. ●

351018　Schisandra bicolor W. C. Cheng var. tuberculata（Y. W. Law）Y. W. Law；瘤枝五味子（龙藤，罗裙子）；Tuberculate Bicolor Magnoliavine ●

351019　Schisandra bicolor W. C. Cheng var. tuberculata（Y. W. Law）Y. W. Law = Schisandra bicolor W. C. Cheng ●

351020　Schisandra bicolor W. C. Cheng var. tuberculata（Y. W. Law）Y. W. Law = Schisandra repanda（Siebold et Zucc.）Radlk. ●

351021　Schisandra chinensis（Turcz.）Baill.；五味子（北五味，北五味子，荎蕏，红血藤，华中五味子，会及，金铃子，酒五味子，辽五味，辽五味子，六亭剂，面藤，内风消，南五味子，山花椒，山花椒秧，嗽神，菋，乌梅子，五梅子，五味，五子，西五味子，玄及，血藤，壮味）；China Magnoliavine，Chinese Kadsura，Chinese Magnolia Vine，Chinese Magnoliavine，Chinese Magnolia-vine，Magnolia Vine ●

351022　Schisandra chinensis（Turcz.）Baill. var. leucocarpa P. H. Huang et L. H. Zhuo = Schisandra chinensis（Turcz.）Baill. ●

351023　Schisandra chinensis（Turcz.）Baill. var. leucocarpa P. H. Huang et L. H. Zhuo；白果五味子；White-fruit Chinese Magnoliavine ●

351024　Schisandra chinensis（Turcz.）Baill. var. rubriflora Franch. = Schisandra elongata Hook. f. et Thomson ●

351025　Schisandra chinensis（Turcz.）Baill. var. rubriflora Franch. = Schisandra rubriflora（Franch.）Rehder et E. H. Wilson ●

351026　Schisandra chinensis（Turcz.）Baill. var. rubriflora Franch. = Schisandra chinensis（Turcz.）Baill. ●

351027　Schisandra chinensis（Turcz.）Baill. var. typica Nakai = Schisandra chinensis（Turcz.）Baill. ●

351028　Schisandra chinensis Baill. = Schisandra sphenanthera Rehder et E. H. Wilson ●

351029　Schisandra coccinea Michx.；美洲五味子；Carolina Magnoliavine ●☆

351030　Schisandra coccinea Michx. = Schisandra glabra（Brickell）Rehder ●☆

351031　Schisandra discolor Nakai = Schisandra repanda（Siebold et Zucc.）Radlk. f. hypoleuca（Makino）Ohwi ●

351032　Schisandra discolor Nakai = Schisandra repanda（Siebold et Zucc.）Radlk. ●

351033　Schisandra elongata Baill. = Schisandra glaucescens Diels ●

351034　Schisandra elongata Hook. f. et Thomson；长蕊五味子（东亚五味子，面藤，五味子）；Elangate Magnoliavine ●

351035　Schisandra elongata Hook. f. et Thomson var. dentata Finet et Gagnep. = Schisandra elongata Hook. f. et Thomson ●

351036　Schisandra elongata Hook. f. et Thomson var. dentata Finet et Gagnep. = Schisandra micrantha A. C. Sm. ●

351037　Schisandra elongata Hook. f. et Thomson var. longissima Dunn = Schisandra henryi C. B. Clarke ●

351038　Schisandra flaccidiramosa C. R. Sun = Schisandra chinensis（Turcz.）Baill. ●

351039　Schisandra flaccidiramosa C. R. Sun = Schisandra grandiflora（Wall.）Hook. f. et Thomson ●

351040　Schisandra glabra（Brickell）Rehder；光五味子；Star-vine ●☆

351041　Schisandra glaucescens Diels；金山五味子（饭巴团，粉绿五味子，粉叶内风消，粉叶内消散，花血藤，灰五味子，冷饭团，南五味）；Donald Magnoliavine，Palegreen Magnoliavine ●

351042　Schisandra glaucescens Diels = Schisandra grandiflora（Wall.）Hook. f. et Thomson ●

351043　Schisandra gracilis A. C. Sm. = Schisandra elongata Hook. f. et Thomson ●

351044　Schisandra grandiflora（Wall.）Hook. f. et Thomson；大花五味子（五味藤）；Bigbloom Magnoliavine，Bigflower Magnoliavine，Big-flowered Magnolia-vine ●

351045　Schisandra grandiflora（Wall.）Hook. f. et Thomson var. cathayensis C. K. Schneid. = Schisandra grandiflora（Wall.）Hook. f. et Thomson ●

351046　Schisandra grandiflora（Wall.）Hook. f. et Thomson var. cathayensis C. K. Schneid. = Schisandra sphaerandra Stapf f. pallida A. C. Sm. ●

351047　Schisandra grandiflora（Wall.）Hook. f. et Thomson var. cathayensis C. K. Schneid. = Schisandra sphaerandra Stapf ●

351048　Schisandra grandiflora（Wall.）Hook. f. et Thomson var. cathayensis C. K. Schneid. = Schisandra rubriflora（Franch.）Rehder et E. H. Wilson ●

351049　Schisandra grandiflora （Wall.） Hook. f. et Thomson var. cathayensis C. K. Schneid. ; 小叶兴山五味子; Littleleaf Bigbloom Magnoliavine ●

351050　Schisandra grandiflora （Wall.） Hook. f. et Thomson var. rubriflora （Franch.） C. K. Schneid. = Schisandra chinensis （Turcz.） Baill. ●

351051　Schisandra grandiflora （Wall.） Hook. f. et Thomson var. rubriflora Rehder et E. H. Wilson = Schisandra rubriflora （Franch.） Rehder et E. H. Wilson ●

351052　Schisandra grandiflora （Wall.） Hook. f. et Thomson var. rubriflora Rehder et E. H. Wilson; 大红花五味子●

351053　Schisandra hanceana Baill. = Kadsura coccinea （Lem.） A. C. Sm. ●

351054　Schisandra henryi C. B. Clarke; 翼梗五味子(白背铁箍散, 边生五味子, 翅茎五味子, 大伸筋, 大血藤, 峨眉五味子, 红九股牛, 红香血藤, 黄皮血藤, 活血藤, 棱枝五味子, 毛香藤, 气藤, 铁骨散, 西五味, 香石藤, 小血藤, 血藤, 药五味, 紫金血藤); Henry Magnoliavine, Henry Magnolia-vine, Margined Henry Magnoliavine, Underbloue Magnoliavine ●

351055　Schisandra henryi C. B. Clarke subsp. marginalis （A. C. Sm.） R. M. K. Saunders; 东南五味子●

351056　Schisandra henryi C. B. Clarke subsp. yunnanensis （A. C. Sm.） R. M. K. Saunders; 云南五味子(白马五味子, 臭八角, 川茴香, 滇翅梗五味子, 滇五味子, 吊果, 吊山花椒, 断肠草, 山八角, 上八角, 铁骨散, 土大香, 香石藤, 小血藤, 云南八角, 云南茴香); Yunnan Magnoliavine ●

351057　Schisandra henryi C. B. Clarke var. longipes （Merr. et Chun） A. C. Sm. = Schisandra longipes （Merr. et Chun） R. M. K. Saunders ●

351058　Schisandra henryi C. B. Clarke var. marginalis A. C. Sm. = Schisandra henryi C. B. Clarke subsp. marginalis （A. C. Sm.） R. M. K. Saunders ●

351059　Schisandra henryi C. B. Clarke var. marginalis A. C. Sm. = Schisandra henryi C. B. Clarke ●

351060　Schisandra henryi C. B. Clarke var. yunnanensis A. C. Sm. = Schisandra henryi C. B. Clarke subsp. yunnanensis （A. C. Sm.） R. M. K. Saunders ●

351061　Schisandra henryi C. B. Clarke var. yunnanensis A. C. Sm. = Schisandra henryi C. B. Clarke ●

351062　Schisandra hypoglauca H. Lév. = Schisandra henryi C. B. Clarke ●

351063　Schisandra incarnata Stapf; 兴山五味子(北五味子); North Magnoliavine, Xingshan Magnoliavine ●

351064　Schisandra incarnata Stapf = Schisandra grandiflora （Wall.） Hook. f. et Thomson ●

351065　Schisandra japonica Hance = Schisandra chinensis （Turcz.） Baill. ●

351066　Schisandra lancifolia （Rehder et E. H. Wilson） A. C. Sm.; 长叶五味子(黄袍, 满山香, 披针叶五味子, 狭叶五味子, 香石藤, 小密细藤, 小血藤, 小钻地风); Long-leaved Magnolia-vine, Narrowleaf Magnoliavine, Whipash Orange Magnoliavine ●

351067　Schisandra lancifolia （Rehder et E. H. Wilson） A. C. Sm. var. polycarpa Z. Ho = Schisandra elongata Hook. f. et Thomson ●

351068　Schisandra longipes （Merr. et Chun） R. M. K. Saunders; 长柄五味子●

351069　Schisandra micrantha A. C. Sm.; 小花五味子(大伸筋, 过山龙, 黄袍小血藤, 接筋藤, 满山香, 铁骨散, 香石藤, 小红藤, 小花五味子藤, 小密细藤, 血木通); Little-flower Magnoliavine, Parviflorous Magnolia-vine, Smallflower Magnoliavine ●

351070　Schisandra micrantha A. C. Sm. = Schisandra elongata Hook. f. et Thomson ●

351071　Schisandra neglecta A. C. Sm.; 滇藏五味子(小血藤); Neglect Magnoliavine, Neglected Magnoliavine, Neglectus Magnolia-vine ●

351072　Schisandra neglecta A. C. Sm. = Schisandra elongata Hook. f. et Thomson ●

351073　Schisandra nigra Maxim.; 内风消五味子（松藤）; Black Magnoliavine ●

351074　Schisandra nigra Maxim. = Schisandra repanda （Siebold et Zucc.） Radlk. ●

351075　Schisandra nigra Maxim. var. hypoleuca Makino = Schisandra repanda （Siebold et Zucc.） Radlk. f. hypoleuca （Makino） Ohwi ●☆

351076　Schisandra perulata Gagnep. = Schisandra henryi C. B. Clarke ●

351077　Schisandra plena A. C. Sm.; 重瓣五味子(复瓣黄龙藤); Abundant Magnoliavine, Double Magnolia-vine, Doubleflower Magnoliavine ●

351078　Schisandra propinqua （Wall.） Baill.; 合蕊五味子(黄龙藤, 近缘五味子, 满山香, 蒙自五味子, 中间五味子); Angledtwig Magnoliavine, Sib Magnoliavine, Yellow Twigged Magnolia-vine ●

351079　Schisandra propinqua （Wall.） Baill. subsp. intermedia （A. C. Sm.） R. M. K. Saunders = Schisandra propinqua （Wall.） Baill. ●

351080　Schisandra propinqua （Wall.） Baill. subsp. intermedia A. C. Sm.; 中间五味子(拔毒散, 黄龙藤, 蛇毒药, 铁骨散, 通气香, 五香藤, 五香血藤, 小红袍, 小血藤, 岩青叶, 中间黄龙藤, 中间近缘五味子, 中间型黄龙藤); Intermediate Angledtwig Magnoliavine ●

351081　Schisandra propinqua （Wall.） Baill. subsp. sinensis Oliv.; 铁箍散(八仙草, 秤砣根, 滑藤, 黄龙藤, 老蛇盘, 满山香, 爬岩香, 蛇毒药, 天青地红, 五香血藤, 狭叶五味子, 香巴戟, 香血藤, 小血藤, 血糊藤, 野五味, 中华五味子, 钻骨风, 钻石风); Yellow Angledtwig Magnoliavine ●

351082　Schisandra propinqua （Wall.） Baill. var. intermedia A. C. Sm. = Schisandra propinqua （Wall.） Baill. subsp. intermedia A. C. Sm. ●

351083　Schisandra propinqua （Wall.） Baill. var. intermedia A. C. Sm. = Schisandra propinqua （Wall.） Baill. ●

351084　Schisandra propinqua （Wall.） Baill. var. linearis Finet et Gagnep. = Schisandra propinqua （Wall.） Baill. ●

351085　Schisandra propinqua （Wall.） Baill. var. lineraris Finet et Gagnep. = Schisandra propinqua （Wall.） Baill. var. sinensis Oliv. ●

351086　Schisandra propinqua （Wall.） Baill. var. sinensis Oliv. = Schisandra propinqua （Wall.） Baill. subsp. sinensis Oliv. ●

351087　Schisandra propinqua （Wall.） Baill. var. sinensis Oliv. = Schisandra propinqua （Wall.） Baill. ●

351088　Schisandra propinqua Baill. = Schisandra neglecta A. C. Sm. ●

351089　Schisandra pubescens Hemsl. et E. H. Wilson; 毛叶五味子(北五味子, 毛脉五味子, 柔毛五味子, 五味子, 西五味子); Downy Magnolia-vine, Fuzzy Magnoliavine, Hairleaf Magnoliavine ●

351090　Schisandra pubescens Hemsl. et E. H. Wilson var. pubinervis （Rehder et E. H. Wilson） A. C. Sm. = Schisandra pubinervis （Rehder et E. H. Wilson） R. M. K. Saunders ●

351091　Schisandra pubinervis （Rehder et E. H. Wilson） R. M. K. Saunders; 毛脉五味子; Hairy-nerve Fuzzy Magnoliavine, Hairynerve Magnoliavine ●

351092　Schisandra pubinervis （Rehder et E. H. Wilson） R. M. K. Saunders = Schisandra pubescens Hemsl. et E. H. Wilson ●

351093　Schisandra repanda （Siebold et Zucc.） Radlk.; 浅波叶五味子(瘤枝五味子); Repand Magnoliavine ●

351094　Schisandra repanda （Siebold et Zucc.） Radlk. f. hypoleuca

（Makino）Ohwi；里白浅波叶五味子●☆

351095　Schisandra rubriflora（Franch.）Rehder et E. H. Wilson；红花五味子（滇五味，滇五味子，过山龙，五味子，香血藤，猩红五味子）；Red Bigbloom Magnoliavine, Redflower Magnoliavine, Red-flowered Magnolia-vine ●

351096　Schisandra sphaerandra Rehder et E. H. Wilson ＝ Schisandra sphaerandra Stapf ●

351097　Schisandra sphaerandra Rehder et E. H. Wilson f. pallida A. C. Sm. ＝ Schisandra sphaerandra Stapf f. pallida A. C. Sm. ●

351098　Schisandra sphaerandra Stapf；球蕊五味子（过山龙，满山香，山包谷，香石藤）；Globose-stamen Magnoliavine ●

351099　Schisandra sphaerandra Stapf ＝ Schisandra grandiflora（Wall.）Hook. f. et Thomson ●

351100　Schisandra sphaerandra Stapf f. pallida A. C. Sm.；白花球蕊五味子（白蕊五味子，滇藏五味子，球蕊五味子，山包谷，山花椒，五味子，翼梗五味子，圆药五味子）；Pale Globose-stamen Magnoliavine ●

351101　Schisandra sphaerandra Stapf f. pallida A. C. Sm. ＝ Schisandra grandiflora（Wall.）Hook. f. et Thomson ●

351102　Schisandra sphaerandra Stapf f. pallida A. C. Sm. ＝ Schisandra sphaerandra Stapf ●

351103　Schisandra sphenanthera Rehder et E. H. Wilson；华中五味子（大血藤，红铃子，活血藤，南五味子，山包谷，五味子，五香血藤）；Orange Magnoliavine, Orange Magnolia-vine ●

351104　Schisandra sphenanthera Rehder et E. H. Wilson ＝ Schisandra lancifolia（Rehder et E. H. Wilson）A. C. Sm. ●

351105　Schisandra sphenanthera Rehder et E. H. Wilson ＝ Schisandra viridis A. C. Sm. ●

351106　Schisandra sphenanthera Rehder et E. H. Wilson var. lancifolia Rehder et E. H. Wilson ＝ Schisandra lancifolia（Rehder et E. H. Wilson）A. C. Sm. ●

351107　Schisandra sphenanthera Rehder et E. H. Wilson var. longipes Merr. et Chun ＝ Schisandra longipes（Merr. et Chun）R. M. K. Saunders ●

351108　Schisandra sphenanthera Rehder et E. H. Wilson var. pubinervis Rehder et E. H. Wilson ＝ Schisandra pubinervis（Rehder et E. H. Wilson）R. M. K. Saunders ●

351109　Schisandra sphenanthera Rehder et E. H. Wilson var. pubinervis Rehder et E. H. Wilson ＝ Schisandra pubescens Hemsl. et E. H. Wilson var. pubinervis（Rehder et E. H. Wilson）A. C. Sm. ●

351110　Schisandra tomentella A. C. Sm.；柔毛五味子；Softhair Magnoliavine, Tomentose Magnoliavine ●

351111　Schisandra tomentella A. C. Sm. ＝ Schisandra pubescens Hemsl. et E. H. Wilson ●

351112　Schisandra tuberculata Y. W. Law ＝ Schisandra bicolor W. C. Cheng var. tuberculata（Y. W. Law）Y. W. Law ●

351113　Schisandra tuberculata Y. W. Law ＝ Schisandra bicolor W. C. Cheng ●

351114　Schisandra tuberculata Y. W. Law ＝ Schisandra repanda（Siebold et Zucc.）Radlk. ●

351115　Schisandra vestita Pax et K. Hoffm. ＝ Schisandra pubescens Hemsl. et E. H. Wilson ●

351116　Schisandra viridis A. C. Sm. ＝ Schisandra arisanensis Hayata subsp. viridis（A. C. Sm.）R. M. K. Saunders ●

351117　Schisandra viridis A. C. Sm. ＝ Schisandra elongata Hook. f. et Thomson ●

351118　Schisandra wilsoniana A. C. Sm.；鹤庆五味子（马耳山五味子）；E. H. Wilson Magnoliavine, Heqing Magnoliavine, Wilson Magnolia-vine ●

351119　Schisandra wilsoniana A. C. Sm. ＝ Schisandra bicolor W. C. Cheng ●

351120　Schisandra wilsoniana A. C. Sm. ＝ Schisandra henryi C. B. Clarke ●

351121　Schisandraceae Blume（1830）（保留科名）；五味子科；Schisandra Family, Star-vine Family ●

351122　Schisandraceae Blume（保留科名）＝ Illiciaceae A. C. Sm.（保留科名）●

351123　Schisanthes Haw. ＝ Narcissus L. ■

351124　Schischkinia Iljin（1935）；白刺菊属；Schischkinia, Whitespinedaisy ■

351125　Schischkinia albispina（Bunge）Iljin；白刺菊；Whitespine Schischkinia, Whitespinedaisy ■

351126　Schischkiniella Steenis ＝ Silene L.（保留属名）■

351127　Schismaceras Post et Kuntze ＝ Dendrobium Sw.（保留属名）■

351128　Schismaceras Post et Kuntze ＝ Schismoceras C. Presl ■

351129　Schismatoclada Baker ＝ Coursiana Homolle ■☆

351130　Schismatoclada Baker（1883）；裂枝茜属 ■☆

351131　Schismatoclada aurantiaca Homolle；耳裂枝茜 ■☆

351132　Schismatoclada aurea Homolle；黄裂枝茜 ■☆

351133　Schismatoclada bracteata Homolle ex Cavaco；具苞裂枝茜 ■☆

351134　Schismatoclada citrifolia（Lam. ex Poir.）Homolle；橘叶裂枝茜 ■☆

351135　Schismatoclada concinna Baker；整洁裂枝茜 ■☆

351136　Schismatoclada coursiana Cavaco；库尔斯裂枝茜 ■☆

351137　Schismatoclada humbertiana Homolle；亨伯特裂枝茜 ■☆

351138　Schismatoclada longistipula Cavaco；长托叶裂枝茜 ■☆

351139　Schismatoclada lutea Homolle；黄色裂枝茜 ■☆

351140　Schismatoclada marojejyensis Humbert；马罗裂枝茜 ■☆

351141　Schismatoclada psychotrioides Baker；九节裂枝茜 ■☆

351142　Schismatoclada pubescens Homolle；短柔毛裂枝茜 ■☆

351143　Schismatoclada purpurea Homolle；紫裂枝茜 ■☆

351144　Schismatoclada rubra Homolle；红裂枝茜 ■☆

351145　Schismatoclaea Willis ＝ Schismatoclada Baker ■☆

351146　Schismatoglottis Zoll. et Moritzi（1846）；落檐属（电光芋属，落舌蕉属）；Dropptogue, Falleneaves, Schismatoglottis ■

351147　Schismatoglottis asperata Engl.；粗糙落檐（粗糙落舌蕉）；Rough Schismatoglottis ■☆

351148　Schismatoglottis calyptrata（Roxb.）Zoll. et Moritzi；广西落檐（过山龙，落檐）；Guangxi Falleneaves, Kwangsi Falleneaves, Overhilldregon ■

351149　Schismatoglottis concinna Schott；优雅落檐（优雅落舌蕉）；Elegant Schismatoglottis ■☆

351150　Schismatoglottis hainanensis H. Li；落檐（万年青草）；Falleneaves ■

351151　Schismatoglottis kotoensis（Hayata）T. C. Huang, J. L. Hsiao et H. Y. Yeh；兰屿落檐（兰屿芋）■

351152　Schismatoglottis longipes Miq. ＝ Schismatoglottis calyptrata（Roxb.）Zoll. et Moritzi ■

351153　Schismatoglottis novo-guineensis（André）N. E. Br.；巴布亚落檐（过山龙，新几内亚落舌蕉）；New Guinea Falleneaves, New Guinea Schismatoglottis ■

351154　Schismatoglottis picta Schott；花叶落檐（花叶电光芋，花叶落舌蕉）；Painted-leaf Schismatoglottis ■☆

351155　Schismatoglottis pulchra N. E. Br.；美丽落檐（美丽落舌蕉）；Pretty Schismatoglottis ■☆

351156　Schismatoglottis riparia Schott = Schismatoglottis calyptrata （Roxb.）Zoll. et Moritzi ■

351157　Schismatopera Klotzsch = Pera Mutis ●☆

351158　Schismaxon Steud. = Xyris L. ■

351159　Schismocarpus S. F. Blake（1918）；裂果刺莲花属●☆

351160　Schismocarpus pachypus S. F. Blake；裂果刺莲花●☆

351161　Schismoceras C. Presl = Dendrobium Sw.（保留属名）■

351162　Schismus P. Beauv.（1812）；齿稃草属（双齿稃草属）；Kelch-grass, Schismus ■

351163　Schismus arabicus Nees；齿稃草；Arabia Schismus, Arabian Schismus ■

351164　Schismus aristulatus Stapf = Schismus scaberrimus Nees ■☆

351165　Schismus barbatus（L.）Thell. = Schismus barbatus（Loefl. ex L.）Thell. ■

351166　Schismus barbatus（L.）Thell. subsp. arabicus（Nees）Maire et Will. = Schismus arabicus Nees ■

351167　Schismus barbatus（Loefl. ex L.）Thell.；髯毛齿稃草；Bearded Schismus, Common Mediterranean Grass, Kelch-grass, Meditteranean Grass ■

351168　Schismus barbatus（Loefl. ex L.）Thell. subsp. arabicus（Nees）Maire et Weiller = Schismus arabicus Nees ■

351169　Schismus barbatus（Loefl. ex L.）Thell. subsp. calycinus（Loefl.）Maire et Weiller = Schismus barbatus（Loefl. ex L.）Thell. ■

351170　Schismus barbatus（Loefl. ex L.）Thell. var. minutus（Roem. et Schult.）Maire et Weiller = Schismus barbatus（Loefl. ex L.）Thell. ■

351171　Schismus brevifolius Nees = Schismus barbatus（Loefl. ex L.）Thell. ■

351172　Schismus calycinus（L.）Duval-Jouve；萼状齿稃草■☆

351173　Schismus calycinus（Loefl. ex L.）K. Koch = Schismus barbatus（Loefl. ex L.）Thell. ■

351174　Schismus calycinus（Loefl. ex L.）K. Koch var. arabicus（Nees）Batt. et Trab. = Schismus arabicus Nees ■

351175　Schismus calycinus（Loefl.）K. Koch = Schismus barbatus（L.）Thell. ■

351176　Schismus inermis（Stapf）C. E. Hubb.；无刺齿稃草■☆

351177　Schismus koelerioides Stapf = Schismus inermis（Stapf）C. E. Hubb. ■☆

351178　Schismus marginatus Hook. f. = Schismus arabicus Nees ■

351179　Schismus marginatus P. Beauv. = Schismus barbatus（L.）Thell. ■

351180　Schismus minutus（Hoffm.）Roem. et Schult. = Schismus barbatus（L.）Thell. ■

351181　Schismus ovalis Nees = Schismus barbatus（Loefl. ex L.）Thell. ■

351182　Schismus perennis Ducell. et Maire = Schismus barbatus（Loefl. ex L.）Thell. ■

351183　Schismus pleuropogon Stapf；侧毛齿稃草■☆

351184　Schismus scaberrimus Nees；粗糙齿稃草■☆

351185　Schismus spectabilis Fig. et De Not. = Schismus arabicus Nees ■

351186　Schismus tenuis Steud. = Schismus barbatus（Loefl. ex L.）Thell. ■

351187　Schistachne Fig. et De Not. = Aristida L. ■

351188　Schistachne Fig. et De Not. = Stipagrostis Ness ■

351189　Schistanthe Kunze = Alonsoa Ruiz et Pav. ■☆

351190　Schistanthe peduncularis Kunze = Alonsoa peduncularis

（Kunze）Wettst. ●☆

351191　Schistocarpaea F. Muell.（1891）；裂果鼠李属●☆

351192　Schistocarpaea johnsonii F. Muell.；裂果鼠李●☆

351193　Schistocarpha Less.（1831）；裂托菊属■●☆

351194　Schistocarpha sinforosii Cuatrec.；裂托菊●☆

351195　Schistocarpia Pritz. = Schistocarpha Less. ■●☆

351196　Schistocaryum Franch. = Microula Benth. ■

351197　Schistocaryum cilare Bureau et Franch. = Microula ciliaris（Bureau et Franch.）I. M. Johnst. ■

351198　Schistocaryum myosotideum Franch. = Microula myosotidea（Franch.）I. M. Johnst. ■

351199　Schistocaryum ovalifolium Bureau et Franch. = Microula ovalifolia（Bureau et Franch.）I. M. Johnst. ■

351200　Schistocodon Schauer = Toxocarpus Wight et Arn. ●

351201　Schistocodon meyenii Schauer = Toxocarpus wightianus Hook. et Arn. ●

351202　Schistogyne Hook. et Arn.（1834）；裂蕊萝藦属☆

351203　Schistogyne boliviensis Schltr.；玻利维亚裂蕊萝藦☆

351204　Schistogyne sylvestris Hook. et Arn.；森林裂蕊萝藦☆

351205　Schistolobos W. T. Wang = Opithandra B. L. Burtt ■

351206　Schistolobos pumilus W. T. Wang = Opithandra pumila（W. T. Wang）W. T. Wang ■

351207　Schistonema Schltr.（1906）；裂丝萝藦属☆

351208　Schistonema weberbaueri Schltr.；裂丝萝藦☆

351209　Schistophragma Benth.（1839）；裂隔玄参属■☆

351210　Schistophragma Benth. ex Endl. = Schistophragma Benth. ■☆

351211　Schistophragma pusilla Benth.；裂隔玄参■☆

351212　Schistophyllidium（Juz. ex Fed.）Ikonn. = Potentilla L. ■●

351213　Schistophyllidium bifurcum（L.）Ikonn. = Sibbaldianthe bifurca（L.）Kurtto et T. Erikss. ■

351214　Schistostemon（Urb.）Cuatrec.（1961）；裂蕊核果树属●☆

351215　Schistostemon densiflorum（Benth.）Cuatrec.；密花裂蕊核果树●☆

351216　Schistostemon macrophyllum（Benth.）Cuatrec.；大叶裂蕊核果树●☆

351217　Schistostemon oblongifolium（Benth.）Cuatrec.；矩圆裂蕊核果树●☆

351218　Schistostemon sylvaticum D. Sabatier；林地裂蕊核果树●☆

351219　Schistostephium Less.（1832）；平菊木属●☆

351220　Schistostephium argyreum（DC.）Fenzl ex Harv. = Schistostephium flabelliforme Less. ●☆

351221　Schistostephium artemisiifolium Baker；蒿叶平菊木●☆

351222　Schistostephium artemisiifolium Baker subsp. marungensis Lisowski；马龙加平菊木●☆

351223　Schistostephium crataegifolium（DC.）Fenzl ex Harv.；山楂叶平菊木●☆

351224　Schistostephium dactyliferum Hutch.；指状平菊木●☆

351225　Schistostephium flabelliforme Less.；扇状平菊木●☆

351226　Schistostephium griseum（Harv.）Hutch.；灰平菊木●☆

351227　Schistostephium heptalobum（DC.）Oliv. et Hiern；七裂片平菊木●☆

351228　Schistostephium heptalobum Oliv. et Hiern = Schistostephium dactyliferum Hutch. ●☆

351229　Schistostephium heptalobum S. Moore = Schistostephium mollissimum Hutch. ●☆

351230　Schistostephium hippiifolium（DC.）Hutch.；平果菊状平菊木●☆

351231　Schistostephium homblei De Wild. = Schistostephium

artemisiifolium Baker ●☆

351232 Schistostephium microcephalum Baker = Schistostephium crataegifolium（DC.）Fenzl ex Harv. ●☆

351233 Schistostephium mollissimum Hutch. ；柔软平菊木●☆

351234 Schistostephium oxylobum S. Moore；尖裂平菊木●☆

351235 Schistostephium radicale Killick et C. Claassen = Cotula radicalis（Killick et C. Claassen）Hilliard et B. L. Burtt ■☆

351236 Schistostephium rogersii Hutch. ；罗杰斯平菊木●☆

351237 Schistostephium rotundifolium（DC.）Fenzl ex Harv. ；圆叶平菊木●☆

351238 Schistostephium saxicola Hutch. = Schistostephium heptalobum（DC.）Oliv. et Hiern ●☆

351239 Schistostephium scandens Hutch. ；攀缘平菊木●☆

351240 Schistostephium umbellatum（L. f.）Bremer et Humphries；伞形平菊木●☆

351241 Schistostephium villosum Hutch. = Schistostephium crataegifolium（DC.）Fenzl ex Harv. ●☆

351242 Schistostigma Lauterb. = Cleistanthus Hook. f. ex Planch. ●

351243 Schistotylus Dockrill(1967)；裂头兰属■☆

351244 Schistotylus purpuratus（Rupp）Dockrill；裂头兰■☆

351245 Schivereckia Andrz. ex DC. (1821)；席氏葶苈属■☆

351246 Schivereckia Rchb. = Schivereckia Andrz. ex DC. ■☆

351247 Schivereckia podolica Andrz. ；席氏葶苈■☆

351248 Schiverekia Rchb. = Schivereckia Andrz. ex DC. ■☆

351249 Schiwereckia Andrz. ex DC. = Schivereckia Andrz. ex DC. ■☆

351250 Schizachne Hack. (1909)；裂稃茅属■

351251 Schizachne callosa（Turcz. ex Griseb.）Ohwi；裂稃茅；False Melic Grass ■

351252 Schizachne fauriei Hack. = Schizachne callosa（Turcz. ex Griseb.）Ohwi ■

351253 Schizachne komarovii Roshev. ；科马罗夫裂稃茅■☆

351254 Schizachne purpurascens（Torr.）Swallen；紫裂稃茅；False Melic Grass，Purple Schizachne ■☆

351255 Schizachne purpurascens（Torr.）Swallen subsp. callosa（Turcz. ex Griseb.）T. Koyama et Kawano = Schizachne callosa（Turcz. ex Griseb.）Ohwi ■

351256 Schizachne purpurascens（Torr.）Swallen var. callosa（Turcz. ex Griseb.）Kitag. = Schizachne callosa（Turcz. ex Griseb.）Ohwi ■

351257 Schizachne purpurascens（Torr.）Swallen var. pubescens Dore = Schizachne purpurascens（Torr.）Swallen ■☆

351258 Schizachne purpurascens Kitag. = Schizachne callosa（Turcz. ex Griseb.）Ohwi ■

351259 Schizachne purpurascens Kitag. subsp. callosa（Turcz. ex Griseb.）T. Koyama et Kawano = Schizachne callosa（Turcz. ex Griseb.）Ohwi ■

351260 Schizachne striata（Michx.）Hultén = Schizachne purpurascens（Torr.）Swallen ■☆

351261 Schizachyrium Nees（1829）；裂稃草属；Schizachyrium，Splitlemma ■

351262 Schizachyrium bootanense（Hook. f.）A. Camus = Eremopogon delavayi（Hack.）A. Camus ■

351263 Schizachyrium bootanense（Hook. f.）A. Camus = Schizachyrium delavayi（Hack.）Bor ■

351264 Schizachyrium brevifolium（Sw.）Büse var. flaccidum（A. Rich.）Stapf = Schizachyrium brevifolium（Sw.）Nees ex Büse ■

351265 Schizachyrium brevifolium（Sw.）Büse var. maclaudii Jacq.-Fél. = Schizachyrium maclaudii（Jacq.-Fél.）S. T. Blake ■☆

351266 Schizachyrium brevifolium（Sw.）Nees ex Büse；裂稃草(白露红，短叶裂稃草，短叶蜀黍，金字草，牛草，晚碎红)；Shoredt Schizachyrium，Shortleaf Splitlemma ■

351267 Schizachyrium claudopus（Chiov.）Chiov. ；刚果裂稃草■☆

351268 Schizachyrium compressum（Stapf）Stapf = Schizachyrium rupestre（K. Schum.）Stapf ■☆

351269 Schizachyrium condensatum（Kunth）Nees；密裂稃草；Colombian Bluestem，Condensed Schizachyrium ■☆

351270 Schizachyrium delavayi（Hack.）Bor = Eremopogon delavayi（Hack.）A. Camus ■

351271 Schizachyrium delicatum Stapf；姣美裂稃草■☆

351272 Schizachyrium djalonicum Jacq.-Fél. ；贾隆裂稃草■☆

351273 Schizachyrium domingense（Spreng. ex Schult.）Nash = Schizachyrium sanguineum（Retz.）Alston ■

351274 Schizachyrium engleri Pilg. ；恩格勒裂稃草■☆

351275 Schizachyrium exile（Hochst.）Pilg. ；瘦小裂稃草■☆

351276 Schizachyrium fasciculatum Jacq.-Fél. = Schizachyrium brevifolium（Sw.）Nees ex Büse ■

351277 Schizachyrium fragile（R. Br.）A. Camus；斜须裂稃草；Oblique-beard Schizachyrium，Slantingawn Splitlemma ■

351278 Schizachyrium fragile（R. Br.）A. Camus var. shimadae（Ohwi）C. Hsu；尖叶裂稃草■

351279 Schizachyrium fragile（R. Br.）A. Camus var. sinense（Rendle）Jansen = Schizachyrium fragile（R. Br.）A. Camus ■

351280 Schizachyrium glabrescens（Rendle）Stapf = Schizachyrium exile（Hochst.）Pilg. ■☆

351281 Schizachyrium griseum Stapf = Schizachyrium sanguineum（Retz.）Alston ■

351282 Schizachyrium hirtiflorum Nees = Schizachyrium sanguineum（Retz.）Alston ■

351283 Schizachyrium impressum（Hack.）A. Camus；陷脉裂稃草■☆

351284 Schizachyrium inclusum Stent = Schizachyrium exile（Hochst.）Pilg. ■☆

351285 Schizachyrium inspersum Pilg. = Schizachyrium sanguineum（Retz.）Alston ■

351286 Schizachyrium iringense Pilg. = Andropogon schirensis Hochst. ex A. Rich. ■☆

351287 Schizachyrium jeffreysii（Hack.）Stapf；杰弗里斯裂稃草■☆

351288 Schizachyrium kelleri（Hack.）Stapf = Andropogon kelleri Hack. ■☆

351289 Schizachyrium kwiluense Vanderyst；奎卢裂稃草■☆

351290 Schizachyrium lindiense Pilg. = Schizachyrium sanguineum（Retz.）Alston ■

351291 Schizachyrium lomaense A. Camus = Anadelphia lomaense（A. Camus）Jacq.-Fél. ■☆

351292 Schizachyrium lopollense（Rendle）Sales；洛波尔裂稃草■☆

351293 Schizachyrium maclaudii（Jacq.-Fél.）S. T. Blake；马克洛裂稃草■☆

351294 Schizachyrium microstachyum（Desv.）Roseng. ，B. R. Arill. et Izag. ；小穗裂稃草；Small-spike Schizachyrium ■☆

351295 Schizachyrium minutum Gledhill = Schizachyrium brevifolium（Sw.）Nees ex Büse ■

351296 Schizachyrium monostachyon P. A. Duvign. = Schizachyrium thollonii（Franch.）Stapf ■☆

351297 Schizachyrium mukuluense（Vanderyst）Vanderyst；穆库卢裂稃草■☆

351298 Schizachyrium nodulosum（Hack.）Stapf；多节裂稃草■☆

351299 Schizachyrium obliquiberbe （ Hack. ） A. Camus ＝ Schizachyrium fragile （ R. Br. ） A. Camus ■

351300 Schizachyrium platyphyllum （ Franch. ） Stapf;宽叶裂秆草■☆

351301 Schizachyrium praematurum （ Fernald ） C. F. Reed ＝ Schizachyrium scoparium （ Michx. ） Nash ■☆

351302 Schizachyrium pratorum C. E. Hubb. ＝ Schizachyrium rupestre （ K. Schum. ） Stapf ■☆

351303 Schizachyrium pulchellum （ D. Don ex Benth. ） Stapf;美丽裂秆草■☆

351304 Schizachyrium radicosum Jacq. -Fél.;多根裂秆草■☆

351305 Schizachyrium ruderale Clayton;荒地裂秆草■☆

351306 Schizachyrium rupestre （ K. Schum. ） Stapf;岩生裂秆草■☆

351307 Schizachyrium sanguineum （ Retz. ） Alston;血红裂秆草（红裂秆草）;Red Splitlemma,Sanguine Schizachyrium ■☆

351308 Schizachyrium schweinfurthii （ Hack. ） Stapf;施韦裂秆草■☆

351309 Schizachyrium scoparium （ Michx. ） Nash;帚状裂秆草(扫帚状须芒草,蓑衣草,帚状须芒草）;Blue Stem, Bluestem, Broom Beard Grass, Broomlike Bluestem, Bunch Grass, Bunch-grass, Little Bluestem, Little Blue-stem, Prairie Beard Grass ■☆

351310 Schizachyrium scoparium （ Michx. ） Nash subsp. divergens （ Hack. ） Gandhi et Smeins ＝ Schizachyrium scoparium （ Michx. ） Nash var. divergens （ Hack. ） Gould ■☆

351311 Schizachyrium scoparium （ Michx. ） Nash subsp. neomexicanum （ Nash ） Gandhi et Smeins ＝ Schizachyrium scoparium （ Michx. ） Nash ■☆

351312 Schizachyrium scoparium （ Michx. ） Nash var. divergens （ Hack. ） Gould;叉帚状裂秆草;Little Blue-stem, Pinehill Blue-stem ■☆

351313 Schizachyrium scoparium （ Michx. ） Nash var. frequens （ F. T. Hubb. ） Gould ＝ Schizachyrium scoparium （ Michx. ） Nash ■☆

351314 Schizachyrium scoparium （ Michx. ） Nash var. neomexicanum （ Nash ） Gould ＝ Schizachyrium scoparium （ Michx. ） Nash ■☆

351315 Schizachyrium scoparium （ Michx. ） Nash var. polycladum （ Scribn. et Ball ） C. F. Reed ＝ Schizachyrium scoparium （ Michx. ） Nash ■☆

351316 Schizachyrium scoparium （ Michx. ） Nash var. virile （ Shinners ） Gould ＝ Schizachyrium scoparium （ Michx. ） Nash var. divergens （ Hack. ） Gould ■☆

351317 Schizachyrium semiberbe Nees ＝ Schizachyrium sanguineum （ Retz. ） Alston ■

351318 Schizachyrium semiberbe Nees var. flocculiferum Stapf ＝ Schizachyrium sanguineum （ Retz. ） Alston ■

351319 Schizachyrium semiberbe Nees var. hemileium Stapf ＝ Schizachyrium sanguineum （ Retz. ） Alston ■

351320 Schizachyrium semiberbe Nees var. pilosa Gilli ＝ Schizachyrium sanguineum （ Retz. ） Alston ■

351321 Schizachyrium tenuispicatum Pilg. ＝ Schizachyrium sanguineum （ Retz. ） Alston ■

351322 Schizachyrium thollonii （ Franch. ） Stapf;托伦裂秆草■☆

351323 Schizachyrium thollonii （ Franch. ） Stapf var. compressum （ Stapf ） Jacq. -Fél. ＝ Schizachyrium rupestre （ K. Schum. ） Stapf ■☆

351324 Schizachyrium urceolatum （ Hack. ） Stapf;坛状裂秆草■☆

351325 Schizachyrium yangambiense R. Germ.;扬甘比裂秆草■☆

351326 Schizacme Dunlop ＝ Mitrasacme Labill. ■

351327 Schizacme Dunlop(1996);裂缘尖帽草属■☆

351328 Schizacme archeri （ Hook. f. ） Dunlop;裂缘尖帽草■☆

351329 Schizacme montana （ Hook. f. ex Benth. ） Dunlop;山地裂缘尖帽草■☆

351330 Schizandra DC. ＝ Schisandra Michx. （保留属名）●

351331 Schizandra Michx. ＝ Schisandra Michx. （保留属名）●

351332 Schizandra chinensis （ Turcz. ） Baill. ＝ Schisandra chinensis （ Turcz. ） Baill. ●

351333 Schizangium Bartl. ex DC. ＝ Mitracarpus Zucc. ■

351334 Schizanthera Turcz. (1862);裂药野牡丹属●☆

351335 Schizanthera Turcz. ＝ Miconia Ruiz et Pav. （保留属名）●☆

351336 Schizanthera bullata Turcz.;裂药野牡丹●☆

351337 Schizanthera bullata Turcz. ＝ Miconia bullata Triana ■☆

351338 Schizanthes Endl. ＝ Narcissus L. ■

351339 Schizanthes Endl. ＝ Schisanthes Haw. ■

351340 Schizanthes Endl. et Pav. ＝ Narcissus L. ■

351341 Schizanthes Endl. et Pav. ＝ Schisanthes Haw. ■

351342 Schizanthoseddera （ Roberty ） Roberty ＝ Seddera Hochst. ●☆

351343 Schizanthus Ruiz et Pav. (1794);蛾蝶花属;Butterfly Flower, Butterfly-flower, Fringe Flower, Fringe-flower, Poor Man's Orchid, Poorman's Orchid ■☆

351344 Schizanthus × wisetonensis Hort.;怀斯顿蛾蝶花;Poor Man's Orchid ■☆

351345 Schizanthus pinnatus Ruiz et Pav.;蛾蝶花(蝴蝶草);Butterfly Flower, Poor Man's Orchid, Poorman's Orchid, Wingleaf Butterfly Flower ■☆

351346 Schizanthus pinnatus Ruiz et Pav. 'Giant';大蛾蝶花■☆

351347 Schizeilema （ Hook. f. ） Domin ＝ Pozoa Lag. ■☆

351348 Schizeilema （ Hook. f. ） Domin(1908);裂壳草属■☆

351349 Schizeilema Domin ＝ Schizeilema （ Hook. f. ） Domin ■☆

351350 Schizeilema cuneifolium （ F. Muell. ） M. Hiroe;楔叶裂壳草■☆

351351 Schizeilema cyanopetalum Domin;裂壳草■☆

351352 Schizeilema pallidum Domin;苍白裂壳草■☆

351353 Schizeilema reniforme Domin;肾形裂壳草■☆

351354 Schizeilema trifoliolatum Domin;三小叶裂壳草■☆

351355 Schizenterospermum Homolle ex Arènes(1960);星裂籽属■☆

351356 Schizenterospermum grevei Homolle ex Arènes;格雷弗星裂籽■☆

351357 Schizenterospermum majungense Homolle ex Arènes;星裂籽■☆

351358 Schizenterospermum rotundifolium Homolle ex Arènes;圆叶星裂籽■☆

351359 Schizmaxon Steud. ＝ Xyris L. ■

351360 Schizobasis Baker(1873);基裂风信子属■☆

351361 Schizobasis angolensis Baker;安哥拉基裂风信子■☆

351362 Schizobasis cuscutoides （ Burch. ex Baker ） Benth. et Hook. ＝ Drimia cuscutoides （ Burch. ex Baker ） J. C. Manning et Goldblatt ■☆

351363 Schizobasis flagelliformis （ Baker ） Baker ＝ Eriospermum flagelliforme （ Baker ） J. C. Manning ■☆

351364 Schizobasis gracilis R. E. Fr.;纤细基裂风信子■☆

351365 Schizobasis intricata （ Baker ） Baker ＝ Drimia intricata （ Baker ） J. C. Manning et Goldblatt ■☆

351366 Schizobasis macowanii Baker ＝ Drimia intricata （ Baker ） J. C. Manning et Goldblatt ■☆

351367 Schizobasopsis J. F. Macbr. ＝ Bowiea Harv. ex Hook. f. （保留属名）●☆

351368 Schizobasopsis volubilis （ Harv. ex Hook. f. ） J. F. Macbr. ＝ Bowiea volubilis Harv. ex Hook. f. ■☆

351369 Schizoboea （ Fritsch ） B. L. Burtt(1974);单腺苣苔属■☆

351370 Schizoboea kamerunensis （ Engl. ） B. L. Burtt;单腺苣苔■☆

351371 Schizocalomyrtus Kausel ＝ Calycorectes O. Berg ●☆

351372 Schizocalyx Hochst. ＝ Dobera Juss. ●☆

351373　Schizocalyx O. Berg ＝ Calycorectes O. Berg ●☆

351374　Schizocalyx O. Berg ＝ Schizocalomyrtus Kausel ●☆

351375　Schizocalyx Scheele（废弃属名）＝ Origanum L. ●■

351376　Schizocalyx Scheele（废弃属名）＝ Schizocalyx Wedd.（保留属名）■☆

351377　Schizocalyx Wedd.（1854）（保留属名）；裂萼茜属■☆

351378　Schizocalyx bracteosa Wedd. ；裂萼茜■☆

351379　Schizocapsa Hance ＝ Tacca J. R. Forst. et G. Forst.（保留属名）■

351380　Schizocapsa Hance（1881）；裂果薯属（水田七属）；Gapfruityam，Schizocapsa ■

351381　Schizocapsa guangxiensis P. P. Ling et C. T. Ting；广西裂果薯；Guangxi Gapfruityam，Guangxi Schizocapsa ■

351382　Schizocapsa itagakii Yamam. ＝ Tacca chantrieri André ■

351383　Schizocapsa plantaginea Hance；裂果薯（长须果，冬叶七，黑冬叶，凼头鸡，箭根薯，蒟蒻薯，马老头，米荷瓦，屈头鸡，山大黄，水槟榔，水狗仔，水鸡头，水鸡仔，水鸡子，水萝卜，水三七，水田七，水虾公）；Common Gapfruityam，Common Schizocapsa，Lobedfruit Tacca ■

351384　Schizocardia A. C. Sm. et Standl. ＝ Purdiaea Planch. ●☆

351385　Schizocarphus Van der Merwe ＝ Scilla L. ■

351386　Schizocarphus Van der Merwe（1943）；裂果绵枣儿属■☆

351387　Schizocarphus acerosus Van der Merwe ＝ Schizocarphus nervosus（Burch.）Van der Merwe ■☆

351388　Schizocarphus gerrardii（Baker）Van der Merwe ＝ Schizocarphus nervosus（Burch.）Van der Merwe ■☆

351389　Schizocarphus nervosus（Burch.）Van der Merwe；多脉裂果绵枣儿■☆

351390　Schizocarphus rigidifolius（Kunth）Van der Merwe ＝ Schizocarphus nervosus（Burch.）Van der Merwe ■☆

351391　Schizocarpum Schrad.（1830）；裂果葫芦属■☆

351392　Schizocarpum filiforme Schrad. ；线形裂果葫芦■☆

351393　Schizocarpum longisepalum C. Jeffrey；长萼裂果葫芦■☆

351394　Schizocarpum parviflorum B. L. Rob. et Greenm. ；小花裂果葫芦■☆

351395　Schizocarya Spach ＝ Gaura L. ■

351396　Schizocasia Schott ＝ Alocasia（Schott）G. Don（保留属名）■

351397　Schizocasia Schott ＝ Xenophya Schott ■

351398　Schizocasia Schott ex Engl. ＝ Alocasia（Schott）G. Don（保留属名）■

351399　Schizocentron Meisn.（1843）；西班牙野牡丹属●☆

351400　Schizocentron Meisn. ＝ Heterocentron Hook. et Arn. ●■☆

351401　Schizocentron elegans Meisn. ；西班牙野牡丹；Spanish Shawl ●☆

351402　Schizochilus Sond.（1846）；裂唇兰属■☆

351403　Schizochilus albiflos Schltr. ＝ Schizochilus angustifolius Rolfe ■☆

351404　Schizochilus angustifolius Rolfe；窄叶裂唇兰■☆

351405　Schizochilus baurii Schltr. ＝ Schizochilus flexuosus Harv. ex Rolfe ■☆

351406　Schizochilus bolusii Schltr. ＝ Schizochilus zeyheri Sond. ■☆

351407　Schizochilus bulbinellus（Rchb. f.）Bolus；小鳞茎裂唇兰■☆

351408　Schizochilus burchellii Bolus ＝ Schizochilus bulbinellus（Rchb. f.）Bolus ■☆

351409　Schizochilus caffrus Schltr. ＝ Schizochilus zeyheri Sond. ■☆

351410　Schizochilus cecilii Rolfe；塞西尔裂唇兰■☆

351411　Schizochilus cecilii Rolfe subsp. culveri（Schltr.）H. P. Linder；卡尔弗裂唇兰■☆

351412　Schizochilus cecilii Rolfe subsp. transvaalensis（Rolfe）H. P. Linder；德兰士瓦裂唇兰■☆

351413　Schizochilus clavatus Schltr. ＝ Schizochilus zeyheri Sond. ■☆

351414　Schizochilus crenulatus H. P. Linder；细圆齿裂唇兰■☆

351415　Schizochilus culveri Schltr. ＝ Schizochilus cecilii Rolfe subsp. culveri（Schltr.）H. P. Linder ■☆

351416　Schizochilus flexuosus Harv. ex Rolfe；曲折裂唇兰■☆

351417　Schizochilus galpinii Schltr. ＝ Schizochilus cecilii Rolfe subsp. culveri（Schltr.）H. P. Linder ■☆

351418　Schizochilus gerrardii（Rchb. f.）Bolus；杰勒德裂唇兰■☆

351419　Schizochilus grandiflorus Schltr. ＝ Schizochilus zeyheri Sond. ■☆

351420　Schizochilus grandiflorus Schltr. var. crenulatus ？ ＝ Schizochilus zeyheri Sond. ■☆

351421　Schizochilus huttonae Schltr. ＝ Schizochilus zeyheri Sond. ■☆

351422　Schizochilus lepidus Summerh. ；小鳞裂唇兰■☆

351423　Schizochilus lilacinus Schelpe ex H. P. Linder；紫丁香色裂唇兰■☆

351424　Schizochilus pulchellus Schltr. ＝ Schizochilus flexuosus Harv. ex Rolfe ■☆

351425　Schizochilus rehmannii Rolfe ＝ Schizochilus zeyheri Sond. ■☆

351426　Schizochilus rudatisii Schltr. ＝ Schizochilus zeyheri Sond. ■☆

351427　Schizochilus sandersonii Harv. ex Rolfe ＝ Schizochilus zeyheri Sond. ■☆

351428　Schizochilus strictus Rolfe ＝ Schizochilus zeyheri Sond. ■☆

351429　Schizochilus sulphureus Schltr. ；硫色裂唇兰■☆

351430　Schizochilus tenellus Schltr. ＝ Schizochilus cecilii Rolfe subsp. transvaalensis（Rolfe）H. P. Linder ■☆

351431　Schizochilus transvaalensis Rolfe ＝ Schizochilus cecilii Rolfe subsp. transvaalensis（Rolfe）H. P. Linder ■☆

351432　Schizochilus trilobus Rolfe ＝ Schizochilus zeyheri Sond. ■☆

351433　Schizochilus woodii Schltr. ＝ Schizochilus zeyheri Sond. ■☆

351434　Schizochilus zeyheri Sond. ；泽耶尔裂唇兰■☆

351435　Schizochiton Spreng. ＝ Chisocheton Blume ●

351436　Schizochlaena Spreng. ＝ Schizolaena Thouars ●☆

351437　Schizochlaenaceae Wettst. ＝ Sarcolaenaceae Caruel（保留科名）●☆

351438　Schizococcus Eastw. ＝ Arctostaphylos Adans.（保留属名）●☆

351439　Schizocodon Siebold et Zucc.（1843）；小裂缘花属（岩镜属）；Fringe Bell，Hinge Bell ■☆

351440　Schizocodon Siebold et Zucc. ＝ Shortia Torr. et A. Gray（保留属名）■

351441　Schizocodon ilicifolius Maxim. ；冬青叶小裂缘花■☆

351442　Schizocodon ilicifolius Maxim. ＝ Shortia ilicifolia（Maxim.）L. H. Li ■☆

351443　Schizocodon ilicifolius Maxim. f. akaishialpinus T. Yamaz. ；明石山小裂缘花■☆

351444　Schizocodon ilicifolius Maxim. f. purpureiflorus（Makino）Takeda ＝ Schizocodon ilicifolius Maxim. var. australis T. Yamaz. ■☆

351445　Schizocodon ilicifolius Maxim. var. australis T. Yamaz. ；南方冬青叶小裂缘花■☆

351446　Schizocodon ilicifolius Maxim. var. intercedens（Ohwi）T. Yamaz. ；中间小裂缘花■☆

351447　Schizocodon ilicifolius Maxim. var. minimus（Makino）T. Yamaz. ；小冬青叶小裂缘花■☆

351448　Schizocodon ilicifolius Maxim. var. nankaiensis T. Yamaz. ；南海道冬青叶小裂缘花■☆

351449　Schizocodon intercedens（Ohwi）T. Yamaz. ＝ Schizocodon ilicifolius Maxim. var. intercedens（Ohwi）T. Yamaz. ■☆

351450　Schizocodon soldanelloides Siebold et Zucc. ；须边岩扉（流苏岩

扇)■☆

351451 Schizocodon soldanelloides Siebold et Zucc. f. alpinus Maxim. ;山地须边岩扇■☆

351452 Schizocodon soldanelloides Siebold et Zucc. f. leucanthus H. Hara;白花须边岩扇■☆

351453 Schizocodon soldanelloides Siebold et Zucc. var. illicifolius （Maxim.）Makino = Schizocodon ilicifolius Maxim.■☆

351454 Schizocodon soldanelloides Siebold et Zucc. var. longifolius（T. Yamaz.）T. Shimizu f. albiflorus T. Shimizu;白花长叶须边岩扇■☆

351455 Schizocodon soldanelloides Siebold et Zucc. var. longifolius（T. Yamaz.）T. Shimizu;长叶须边岩扇■☆

351456 Schizocodon soldanelloides Siebold et Zucc. var. magnus （Makino）H. Hara f. longifolius T. Yamaz. = Schizocodon soldanelloides Siebold et Zucc. var. longifolius（T. Yamaz.）T. Shimizu ■☆

351457 Schizocodon soldanelloides Siebold et Zucc. var. magnus （Makino）H. Hara;大须边岩扇■☆

351458 Schizocodon soldanelloides Siebold et Zucc. var. magnus （Makino）H. Hara f. niveus H. Hara;雪白大须边岩扇■☆

351459 Schizocodon soldanelloides Siebold et Zucc. var. minimus （Makino）H. Hara = Schizocodon ilicifolius Maxim. var. minimus （Makino）T. Yamaz.■☆

351460 Schizocolea Bremek.（1950）;裂鞘茜属●☆

351461 Schizocolea linderi（Hutch. et Dalziel）Bremek.;裂鞘茜●☆

351462 Schizocolea ochreata E. M. Petit;非洲裂鞘茜●☆

351463 Schizocorona F. Muell.（1853）;裂冠鹅绒藤属●☆

351464 Schizocorona F. Muell. = Cynanchum L.●■

351465 Schizodium Lindl.（1838）;小裂兰属■☆

351466 Schizodium antenniferum Schltr. = Schizodium longipetalum Lindl.■☆

351467 Schizodium arcuatum Lindl. = Schizodium satyrioides（L.）Garay ■☆

351468 Schizodium bifidum（Thunb.）Rchb. f. ;二裂小裂兰■☆

351469 Schizodium bifidum （Thunb.）Rchb. f. var. clavigerum （Lindl.）Schltr. = Schizodium obliquum Lindl. subsp. clavigerum （Lindl.）H. P. Linder ■☆

351470 Schizodium biflorum（L.）T. Durand et Schinz = Schizodium satyrioides（L.）Garay ■☆

351471 Schizodium clavigerum Lindl. = Schizodium obliquum Lindl. subsp. clavigerum（Lindl.）H. P. Linder ■☆

351472 Schizodium cornutum （L.）Schltr. = Schizodium satyrioides （L.）Garay ■☆

351473 Schizodium flexuosum（L.）Lindl.;曲折小裂兰■☆

351474 Schizodium gueinzii Rchb. f. = Schizodium obliquum Lindl. subsp. clavigerum（Lindl.）H. P. Linder ■☆

351475 Schizodium inflexum Lindl.;内折小裂兰■☆

351476 Schizodium longipetalum Lindl.;长瓣小裂兰■☆

351477 Schizodium maculatum（L. f.）Lindl. = Disa maculata L. f. ■☆

351478 Schizodium modestum L. Bolus = Schizodium obliquum Lindl. subsp. clavigerum（Lindl.）H. P. Linder ■☆

351479 Schizodium obliquum Lindl.;偏斜小裂兰■☆

351480 Schizodium obliquum Lindl. subsp. clavigerum（Lindl.）H. P. Linder;棍棒偏斜小裂兰■☆

351481 Schizodium obtusatum Lindl. = Schizodium obliquum Lindl. subsp. clavigerum（Lindl.）H. P. Linder ■☆

351482 Schizodium rigidum Lindl. = Schizodium bifidum（Thunb.）Rchb. f. ■☆

351483 Schizodium satyrioides（L.）Garay;鸟足兰状小裂兰■☆

351484 Schizoglossum E. Mey.（1838）;裂舌萝藦属■☆

351485 Schizoglossum addoense N. E. Br. = Aspidoglossum gracile （E. Mey.）Kupicha ■☆

351486 Schizoglossum aemulum Schltr. = Schizoglossum cordifolium E. Mey.■☆

351487 Schizoglossum alpestre K. Schum.;高山裂舌萝藦■☆

351488 Schizoglossum altissimum Schltr. = Aspidoglossum interruptum （E. Mey.）Bullock ■☆

351489 Schizoglossum altum N. E. Br. = Aspidoglossum masaicum（N. E. Br.）Kupicha ■☆

351490 Schizoglossum angolense Schltr. et Rendle;安哥拉裂舌萝藦■☆

351491 Schizoglossum angustissimum K. Schum. = Aspidoglossum angustissimum（K. Schum.）Bullock ■☆

351492 Schizoglossum anomalum N. E. Br. = Miraglossum anomalum （N. E. Br.）Kupicha ■☆

351493 Schizoglossum aschersonianum Schltr.;阿舌森裂舌萝藦■☆

351494 Schizoglossum aschersonianum Schltr. var. longipes N. E. Br. ;长梗阿舌森裂舌萝藦■☆

351495 Schizoglossum aschersonianum Schltr. var. pygmaeum（Schltr.）N. E. Br. ;矮小阿舌森裂舌萝藦■☆

351496 Schizoglossum aschersonianum Schltr. var. radiatum N. E. Br. ;辐射阿舌森裂舌萝藦■☆

351497 Schizoglossum atropurpureum E. Mey. ;暗紫裂舌萝藦■☆

351498 Schizoglossum atropurpureum E. Mey. subsp. tridentatum （Schltr.）Kupicha;三齿暗紫裂舌萝藦■☆

351499 Schizoglossum atropurpureum E. Mey. subsp. virens（E. Mey.）Kupicha;绿裂舌萝藦■☆

351500 Schizoglossum atropurpureum E. Mey. var. lineatum Schltr. = Schizoglossum cordifolium E. Mey. ■☆

351501 Schizoglossum atrorubens Schltr. = Schizoglossum bidens E. Mey. subsp. atrorubens（Schltr.）Kupicha ■☆

351502 Schizoglossum auriculatum N. E. Br. = Aspidoglossum glanduliferum（Schltr.）Kupicha ■☆

351503 Schizoglossum barbatum Britten et Rendle;髯毛裂舌萝藦■☆

351504 Schizoglossum barbatum Schltr. = Aspidoglossum glabrescens （Schltr.）Kupicha ■☆

351505 Schizoglossum barberae Schltr. = Aspidoglossum interruptum （E. Mey.）Bullock ■☆

351506 Schizoglossum baumii Schltr. ex N. E. Br. = Aspidoglossum masaicum（N. E. Br.）Kupicha ■☆

351507 Schizoglossum bidens E. Mey. ;双齿裂舌萝藦■☆

351508 Schizoglossum bidens E. Mey. subsp. atrorubens（Schltr.）Kupicha;暗红双齿裂舌萝藦■☆

351509 Schizoglossum bidens E. Mey. subsp. galpinii（Schltr.）Kupicha;盖尔双齿裂舌萝藦■☆

351510 Schizoglossum bidens E. Mey. subsp. gracile Kupicha;纤细双齿裂舌萝藦■☆

351511 Schizoglossum bidens E. Mey. subsp. hirtum Kupicha;多毛双齿裂舌萝藦■☆

351512 Schizoglossum bidens E. Mey. subsp. pachyglossum（Schltr.）Kupicha;粗舌双齿裂舌萝藦■☆

351513 Schizoglossum bidens E. Mey. subsp. productum（N. E. Br.）Kupicha;伸展双齿裂舌萝藦■☆

351514 Schizoglossum biflorum（E. Mey.）Schltr. = Aspidoglossum biflorum E. Mey. ■☆

351515 Schizoglossum biflorum （E. Mey.）Schltr. var. concinnum

（ Schltr. ） N. E. Br. = Aspidoglossum biflorum E. Mey. ■☆

351516 Schizoglossum biflorum （ E. Mey. ） Schltr. var. integrum N. E. Br. = Aspidoglossum biflorum E. Mey. ■☆

351517 Schizoglossum bilamellatum Schltr. = Aspidoglossum lamellatum （ Schltr. ） Kupicha ■☆

351518 Schizoglossum bilamellatum Schltr. var. cordylogynoides ？ = Aspidoglossum lamellatum （ Schltr. ） Kupicha ■☆

351519 Schizoglossum bolusii Schltr. = Aspidoglossum gracile （ E. Mey. ） Kupicha ■☆

351520 Schizoglossum bowkerae N. E. Br. = Aspidoglossum gracile （ E. Mey. ） Kupicha ■☆

351521 Schizoglossum buchananii N. E. Br. = Aspidoglossum gracile （ E. Mey. ） Kupicha ■☆

351522 Schizoglossum burchellii N. E. Br. = Aspidoglossum gracile （ E. Mey. ） Kupicha ■☆

351523 Schizoglossum cabrae De Wild. ;卡布拉裂舌萝藦■☆

351524 Schizoglossum carinatum Schltr. = Aspidoglossum carinatum （ Schltr. ） Kupicha ■☆

351525 Schizoglossum carsonii （ N. E. Br. ） N. E. Br. = Glossostelma carsonii （ N. E. Br. ） Bullock ■☆

351526 Schizoglossum chirindense S. Moore = Pachycarpus chirindensis （ S. Moore ） Goyder ■☆

351527 Schizoglossum chlorojodinum （ K. Schum. ） N. E. Br. = Glossostelma carsonii （ N. E. Br. ） Bullock ■☆

351528 Schizoglossum ciliatum Schltr. = Aspidoglossum fasciculare E. Mey. ■☆

351529 Schizoglossum commixtum N. E. Br. = Aspidoglossum glanduliferum （ Schltr. ） Kupicha ■☆

351530 Schizoglossum connatum N. E. Br. = Aspidoglossum connatum （ N. E. Br. ） Bullock ■☆

351531 Schizoglossum conrathii Schltr. = Aspidoglossum biflorum E. Mey. ■☆

351532 Schizoglossum consimile N. E. Br. = Aspidoglossum heterophyllum E. Mey. ■☆

351533 Schizoglossum contracurvum N. E. Br. = Aspidoglossum ovalifolium （ Schltr. ） Kupicha ■☆

351534 Schizoglossum cordatum （ S. Moore ） S. Moore;心形裂舌萝藦■☆

351535 Schizoglossum cordifolium E. Mey. ;心叶裂舌萝藦■☆

351536 Schizoglossum cordifolium E. Mey. var. centralis N. E. Br. = Schizoglossum cordifolium E. Mey. ■☆

351537 Schizoglossum crassipes S. Moore;粗梗裂舌萝藦■☆

351538 Schizoglossum davyi N. E. Br. = Miraglossum davyi （ N. E. Br. ） Kupicha ■☆

351539 Schizoglossum debile Schltr. = Aspidoglossum elliotii （ Schltr. ） Kupicha ■☆

351540 Schizoglossum decipiens N. E. Br. = Schizoglossum stenoglossum Schltr. subsp. flavum （ N. E. Br. ） Kupicha ■☆

351541 Schizoglossum decipiens N. E. Br. var. flavum ？ = Schizoglossum stenoglossum Schltr. subsp. flavum （ N. E. Br. ） Kupicha ■☆

351542 Schizoglossum delagoense Schltr. = Aspidoglossum delagoense （ Schltr. ） Kupicha ■☆

351543 Schizoglossum dissimile N. E. Br. ;不似裂舌萝藦■☆

351544 Schizoglossum distinctum （ N. E. Br. ） N. E. Br. = Pachycarpus distinctus （ N. E. Br. ） Bullock ■☆

351545 Schizoglossum divaricatum N. E. Br. = Schizoglossum cordifolium E. Mey. ■☆

351546 Schizoglossum diversum N. E. Br. = Schizoglossum bidens E.

Mey. subsp. gracile Kupicha ■☆

351547 Schizoglossum dolichoglossum （ K. Schum. ） N. E. Br. = Pachycarpus spurius （ N. E. Br. ） Bullock ■☆

351548 Schizoglossum dregei N. E. Br. = Aspidoglossum gracile （ E. Mey. ） Kupicha ■☆

351549 Schizoglossum elatum K. Schum. = Aspidoglossum angustissimum （ K. Schum. ） Bullock ■☆

351550 Schizoglossum elingue N. E. Br. ;无舌裂舌萝藦■☆

351551 Schizoglossum elingue N. E. Br. subsp. purpureum Kupicha;紫无舌裂舌萝藦■☆

351552 Schizoglossum elliotii Schltr. = Aspidoglossum elliotii （ Schltr. ） Kupicha ■☆

351553 Schizoglossum erubescens Schltr. = Aspidoglossum erubescens （ Schltr. ） Bullock ■☆

351554 Schizoglossum euphorbioides E. Mey. = Schizoglossum atropurpureum E. Mey. subsp. virens （ E. Mey. ） Kupicha ■☆

351555 Schizoglossum eustegioides （ E. Mey. ） Druce;良盖裂舌萝藦■☆

351556 Schizoglossum excisum Schltr. = Aspidoglossum biflorum E. Mey. ■☆

351557 Schizoglossum exile （ Decne. ） Schltr. = Aspidoglossum gracile （ E. Mey. ） Kupicha ■☆

351558 Schizoglossum eximium （ Schltr. ） N. E. Br. = Pachycarpus eximius （ Schltr. ） Bullock ■☆

351559 Schizoglossum eylesii S. Moore = Aspidoglossum eylesii （ S. Moore ） Kupicha ■☆

351560 Schizoglossum fasciculare （ E. Mey. ） Schltr. = Aspidoglossum fasciculare E. Mey. ■☆

351561 Schizoglossum filifolium Schltr. = Aspidoglossum gracile （ E. Mey. ） Kupicha ■☆

351562 Schizoglossum filiforme （ L. f. ） Druce;丝状裂舌萝藦■☆

351563 Schizoglossum filipes Schltr. = Aspidoglossum gracile （ E. Mey. ） Kupicha ■☆

351564 Schizoglossum flanaganii Schltr. = Aspidoglossum flanaganii （ Schltr. ） Kupicha ■☆

351565 Schizoglossum flavum Schltr. ;黄裂舌萝藦■☆

351566 Schizoglossum flavum Schltr. var. lineare N. E. Br. = Schizoglossum flavum Schltr. ■☆

351567 Schizoglossum fusco-purpureum Schltr. et Rendle = Aspidoglossum masaicum （ N. E. Br. ） Kupicha ■☆

351568 Schizoglossum galpinii Schltr. = Schizoglossum bidens E. Mey. subsp. galpinii （ Schltr. ） Kupicha ■☆

351569 Schizoglossum garcianum Schltr. ;加西亚裂舌萝藦■☆

351570 Schizoglossum garuanum Schltr. = Aspidoglossum interruptum （ E. Mey. ） Bullock ■☆

351571 Schizoglossum gerrardii （ Harv. ） Benth. et Hook. f. = Pachycarpus campanulatus （ Harv. ） N. E. Br. var. sutherlandii N. E. Br. ■☆

351572 Schizoglossum gigantiglossum Weim. ;巨舌裂舌萝藦■☆

351573 Schizoglossum glabrescens Schltr. = Aspidoglossum glabrescens （ Schltr. ） Kupicha ■☆

351574 Schizoglossum glabrescens Schltr. var. longirostre （ Schltr. ） N. E. Br. = Aspidoglossum glabrescens （ Schltr. ） Kupicha ■☆

351575 Schizoglossum glanduliferum Schltr. = Aspidoglossum glanduliferum （ Schltr. ） Kupicha ■☆

351576 Schizoglossum glanvillei Hutch. et Dalziel = Tylophora congolana （ Baill. ） Bullock ●☆

351577 Schizoglossum goetzei K. Schum. = Pachycarpus goetzei （ K.

351578　Schizoglossum gossweileri S. Moore；戈斯裂舌萝藦■☆
351579　Schizoglossum graminifolium C. Norman；禾叶裂舌萝藦■☆
351580　Schizoglossum grandiflorum Schltr. = Aspidoglossum grandiflorum（Schltr.）Kupicha ■☆
351581　Schizoglossum grantii Oliv. = Pachycarpus grantii（Oliv.）Bullock ●☆
351582　Schizoglossum guthriei Schltr. = Aspidoglossum gracile（E. Mey.）Kupicha ■☆
351583　Schizoglossum gwelense N. E. Br. = Aspidoglossum biflorum E. Mey. ■☆
351584　Schizoglossum hamatum E. Mey.；顶钩裂舌萝藦■☆
351585　Schizoglossum hamatum E. Mey. var. elegans N. E. Br. = Schizoglossum hamatum E. Mey. ■☆
351586　Schizoglossum hamatum E. Mey. var. pallidum N. E. Br. = Schizoglossum hamatum E. Mey. ■☆
351587　Schizoglossum harveyi N. E. Br. = Aspidoglossum heterophyllum E. Mey. ■☆
351588　Schizoglossum heterophyllum（E. Mey.）Schltr.；互叶裂舌萝藦■☆
351589　Schizoglossum heterophyllum（E. Mey.）Schltr. var. majus N. E. Br. = Aspidoglossum heterophyllum E. Mey. ■☆
351590　Schizoglossum heterophyllum（E. Mey.）Schltr. var. schinzianum（Schltr.）N. E. Br. = Aspidoglossum heterophyllum E. Mey. ■☆
351591　Schizoglossum heudelotianum（Decne.）Roberty = Xysmalobium heudelotianum Decne. ■☆
351592　Schizoglossum hilliardiae Kupicha；希利亚德裂舌萝藦■☆
351593　Schizoglossum hirsutum Turcz. = Schizoglossum cordifolium E. Mey. ■☆
351594　Schizoglossum hirtiflorum N. E. Br. = Aspidoglossum glabrescens（Schltr.）Kupicha ■☆
351595　Schizoglossum huttoniae S. Moore = Sisyranthus huttoniae（S. Moore）S. Moore ■☆
351596　Schizoglossum ingomense N. E. Br.；热非裂舌萝藦■☆
351597　Schizoglossum interruptum（E. Mey.）Schltr. = Aspidoglossum interruptum（E. Mey.）Bullock ■☆
351598　Schizoglossum kassneri S. Moore = Glossostelma carsonii（N. E. Br.）Bullock ■☆
351599　Schizoglossum lamellatum Schltr. = Aspidoglossum lamellatum（Schltr.）Kupicha ■☆
351600　Schizoglossum lanatum Weim. = Aspidoglossum lanatum（Weim.）Kupicha ■☆
351601　Schizoglossum lasiopetalum Schltr. = Aspidoglossum interruptum（E. Mey.）Bullock ■☆
351602　Schizoglossum leptoglossum Weim. = Aspidoglossum nyasae（Britten et Rendle）Kupicha ■☆
351603　Schizoglossum linifolium Schltr.；麻叶裂舌萝藦■☆
351604　Schizoglossum lividiflorum K. Schum. = Glossostelma carsonii（N. E. Br.）Bullock ■☆
351605　Schizoglossum longirostre Schltr. = Aspidoglossum glabrescens（Schltr.）Kupicha ■☆
351606　Schizoglossum loreum S. Moore = Aspidoglossum glabrescens（Schltr.）Kupicha ■☆
351607　Schizoglossum lunatum Schltr. = Aspidoglossum gracile（E. Mey.）Kupicha ■☆
351608　Schizoglossum macowanii N. E. Br.；麦克欧文裂舌萝藦■☆

351609　Schizoglossum macowanii N. E. Br. = Aspidoglossum grandiflorum（Schltr.）Kupicha ■☆
351610　Schizoglossum macowanii N. E. Br. var. tugelense ? = Aspidoglossum grandiflorum（Schltr.）Kupicha ■☆
351611　Schizoglossum macroglossum K. Schum. = Glossostelma lisianthoides（Decne.）Bullock ■☆
351612　Schizoglossum masaicum N. E. Br. = Aspidoglossum masaicum（N. E. Br.）Kupicha ■☆
351613　Schizoglossum montanum R. A. Dyer；山地裂舌萝藦■☆
351614　Schizoglossum monticola Schltr. = Aspidoglossum gracile（E. Mey.）Kupicha ■☆
351615　Schizoglossum multifolium N. E. Br. = Aspidoglossum nyasae（Britten et Rendle）Kupicha ■☆
351616　Schizoglossum nitidum Schltr.；光亮裂舌萝藦■☆
351617　Schizoglossum nyassae Britten et Rendle = Aspidoglossum nyasae（Britten et Rendle）Kupicha ■☆
351618　Schizoglossum oblongum Schltr. = Schizoglossum atropurpureum E. Mey. subsp. virens（E. Mey.）Kupicha ■☆
351619　Schizoglossum orbiculare Schltr.；圆形裂舌萝藦■☆
351620　Schizoglossum ovalifolium Schltr. = Aspidoglossum ovalifolium（Schltr.）Kupicha ■☆
351621　Schizoglossum pachyglossum Schltr.；厚裂舌萝藦■☆
351622　Schizoglossum pachyglossum Schltr. = Schizoglossum bidens E. Mey. subsp. pachyglossum（Schltr.）Kupicha ■☆
351623　Schizoglossum pachyglossum Schltr. var. abbreviatum N. E. Br. = Schizoglossum bidens E. Mey. subsp. pachyglossum（Schltr.）Kupicha ■☆
351624　Schizoglossum pachyglossum Schltr. var. productum N. E. Br. = Schizoglossum bidens E. Mey. subsp. productum（N. E. Br.）Kupicha ■☆
351625　Schizoglossum pallidum Schltr. = Aspidoglossum restioides（Schltr.）Kupicha ■☆
351626　Schizoglossum parcum N. E. Br. = Aspidoglossum gracile（E. Mey.）Kupicha ■☆
351627　Schizoglossum parile N. E. Br. = Aspidoglossum glanduliferum（Schltr.）Kupicha ■☆
351628　Schizoglossum parvulum Schltr.；较小裂舌萝藦■☆
351629　Schizoglossum parvulum Schltr. = Aspidoglossum gracile（E. Mey.）Kupicha ■☆
351630　Schizoglossum parvulum Schltr. var. sessile N. E. Br. = Aspidoglossum gracile（E. Mey.）Kupicha ■☆
351631　Schizoglossum pedunculatum Schltr. = Asclepias aurea（Schltr.）Schltr. ■☆
351632　Schizoglossum peglerae N. E. Br.；佩格拉裂舌萝藦■☆
351633　Schizoglossum pentheri Schltr. = Aspidoglossum erubescens（Schltr.）Bullock ■☆
351634　Schizoglossum periglossoides Schltr.；舌萝藦状裂舌萝藦■☆
351635　Schizoglossum petherickianum Oliv. = Pachycarpus petherickianus（Oliv.）Goyder ■☆
351636　Schizoglossum petherickianum Oliv. var. cordata S. Moore = Schizoglossum cordatum（S. Moore）S. Moore ■☆
351637　Schizoglossum pilosum Schltr. = Miraglossum pilosum（Schltr.）Kupicha ■☆
351638　Schizoglossum propinquum S. Moore = Aspidoglossum lamellatum（Schltr.）Kupicha ■☆
351639　Schizoglossum pulchellum Schltr. = Miraglossum pulchellum（Schltr.）Kupicha ■☆

351640 Schizoglossum pumilum Schltr. = Aspidoglossum ovalifolium (Schltr.) Kupicha ■☆

351641 Schizoglossum quadridens N. E. Br.;四齿裂舌萝藦■☆

351642 Schizoglossum randii S. Moore = Aspidoglossum restioides (Schltr.) Kupicha ■☆

351643 Schizoglossum restioides Schltr. = Aspidoglossum restioides (Schltr.) Kupicha ■☆

351644 Schizoglossum rhodesicum Weim. = Aspidoglossum rhodesicum (Weim.) Kupicha ■☆

351645 Schizoglossum robustum Schltr.;粗壮裂舌萝藦■☆

351646 Schizoglossum robustum Schltr. var. inandense N. E. Br. = Aspidoglossum ovalifolium (Schltr.) Kupicha ■☆

351647 Schizoglossum robustum Schltr. var. pubiflorum N. E. Br. = Aspidoglossum ovalifolium (Schltr.) Kupicha ■☆

351648 Schizoglossum robustum Schltr. var. robustum = Aspidoglossum ovalifolium (Schltr.) Kupicha ■☆

351649 Schizoglossum rubiginosum Hilliard;锈红裂舌萝藦■☆

351650 Schizoglossum saccatum Bruyns;囊状裂舌萝藦■☆

351651 Schizoglossum schinzianum Schltr. = Aspidoglossum heterophyllum E. Mey. ■☆

351652 Schizoglossum schlechteri N. E. Br. = Aspidoglossum glabrescens (Schltr.) Kupicha ■☆

351653 Schizoglossum scyphostigma K. Schum.;杯柱头裂舌萝藦■☆

351654 Schizoglossum semlikense S. Moore = Aspidoglossum masaicum (N. E. Br.) Kupicha ■☆

351655 Schizoglossum shirense N. E. Br.;希雷裂舌萝藦■☆

351656 Schizoglossum shirense N. E. Br. = Aspidoglossum biflorum E. Mey. ■☆

351657 Schizoglossum simulans N. E. Br. = Pachycarpus goetzei (K. Schum.) Bullock ■☆

351658 Schizoglossum singulare Kupicha;单一裂舌萝藦■☆

351659 Schizoglossum spathulatum K. Schum. = Glossostelma spathulatum (K. Schum.) Bullock ■☆

351660 Schizoglossum spurium (N. E. Br.) N. E. Br. = Pachycarpus spurius (N. E. Br.) Bullock ■☆

351661 Schizoglossum stenoglossum Schltr.;窄裂舌萝藦■☆

351662 Schizoglossum stenoglossum Schltr. subsp. flavum (N. E. Br.) Kupicha;黄窄裂舌萝藦■☆

351663 Schizoglossum stenoglossum Schltr. subsp. latifolium Kupicha;宽叶裂舌萝藦■☆

351664 Schizoglossum stenoglossum Schltr. var. longipes ? = Schizoglossum stenoglossum Schltr. ■☆

351665 Schizoglossum striatum Schltr. = Aspidoglossum ovalifolium (Schltr.) Kupicha ☆

351666 Schizoglossum strictissimum S. Moore = Aspidoglossum erubescens (Schltr.) Bullock ☆

351667 Schizoglossum strictum Schltr. = Aspidoglossum biflorum E. Mey. ■☆

351668 Schizoglossum tenellum (Turcz.) Druce;柔弱裂舌萝藦■☆

351669 Schizoglossum tenue (Arn.) Druce = Aspidoglossum gracile (E. Mey.) Kupicha ■☆

351670 Schizoglossum tenuissimum Schltr. = Aspidoglossum glabrescens (Schltr.) Kupicha ■☆

351671 Schizoglossum theileri S. Moore;泰勒裂舌萝藦■☆

351672 Schizoglossum thorbeckii Schltr. = Pachycarpus petherickianus (Oliv.) Goyder ■☆

351673 Schizoglossum togoense Schltr.;多哥裂舌萝藦■☆

351674 Schizoglossum tomentosum Schltr. = Aspidoglossum gracile (E. Mey.) Kupicha ■☆

351675 Schizoglossum tricorniculatum K. Schum. = Xysmalobium andongense Hiern ■☆

351676 Schizoglossum tricuspidatum Schltr. = Aspidoglossum carinatum (Schltr.) Kupicha ■☆

351677 Schizoglossum tridens N. E. Br. = Aspidoglossum glabrescens (Schltr.) Kupicha ■☆

351678 Schizoglossum tridentatum Schltr. = Schizoglossum atropurpureum E. Mey. subsp. tridentatum (Schltr.) Kupicha ■☆

351679 Schizoglossum truncatulum K. Schum. = Xysmalobium reticulatum N. E. Br. ■☆

351680 Schizoglossum truncatum Schltr. = Schizoglossum bidens E. Mey. ■☆

351681 Schizoglossum tubulosum Schltr. = Aspidoglossum biflorum E. Mey. ■☆

351682 Schizoglossum umbellatum Schltr. = Schizoglossum bidens E. Mey. subsp. gracile Kupicha ■☆

351683 Schizoglossum umbelluliferum Schltr. = Stenostelma umbelluliferum (Schltr.) S. P. Bester et Nicholas ■☆

351684 Schizoglossum uncinatum N. E. Br. = Aspidoglossum uncinatum (N. E. Br.) Kupicha ■☆

351685 Schizoglossum unicum N. E. Br. = Aspidoglossum glabrescens (Schltr.) Kupicha ■☆

351686 Schizoglossum venustum Schltr. = Aspidoglossum biflorum E. Mey. ■☆

351687 Schizoglossum venustum Schltr. var. concinnum ? = Aspidoglossum biflorum E. Mey. ■☆

351688 Schizoglossum verticillare Schltr. = Miraglossum verticillare (Schltr.) Kupicha ■☆

351689 Schizoglossum villosum Schltr. = Aspidoglossum heterophyllum E. Mey. ■☆

351690 Schizoglossum violaceum K. Schum. = Glossostelma lisianthoides (Decne.) Bullock ■☆

351691 Schizoglossum virens E. Mey. = Schizoglossum atropurpureum E. Mey. subsp. virens (E. Mey.) Kupicha ■☆

351692 Schizoglossum virgatum (E. Mey.) Schltr. = Aspidoglossum virgatum (E. Mey.) Kupicha ■☆

351693 Schizoglossum viridulum K. Schum.;浅绿裂舌萝藦■☆

351694 Schizoglossum vulcanorum J.-P. Lebrun et Taton = Aspidoglossum connatum (N. E. Br.) Bullock ■☆

351695 Schizoglossum wallacei Schltr. = Schizoglossum nitidum Schltr. ■☆

351696 Schizoglossum welwitschii (Rendle) N. E. Br.;韦尔裂舌萝藦■☆

351697 Schizoglossum whytei N. E. Br. = Aspidoglossum angustissimum (K. Schum.) Bullock ☆

351698 Schizoglossum woodii Schltr. = Aspidoglossum woodii (Schltr.) Kupicha ☆

351699 Schizoglossum xyphostigma K. Schum. = Pachycarpus distinctus (N. E. Br.) Bullock ■☆

351700 Schizogyna Willis = Schizogyne Cass. ●☆

351701 Schizogyne Cass. (1828);分蕊菊属●☆

351702 Schizogyne Cass. = Inula L. ●■

351703 Schizogyne Ehrenb. ex Pax = Acalypha L. ●■

351704 Schizogyne glaberrima DC.;非洲分蕊菊●☆

351705 Schizogyne sericea (L. f.) DC.;分蕊菊●☆

351706 Schizogyne sericea (L. f.) DC. var. glaberrima (DC.) Webb et Berthel. = Schizogyne glaberrima DC. ●☆

351707 Schizoica Alef. = Napaea L. ■☆

351708 Schizojacquemontia (Roberty) Roberty = Jacquemontia Choisy ☆

351709 Schizolaena Thouars(1805);马岛外套花属●☆

351710 Schizolaena capuronii Lowry et al.;凯普伦马岛外套花●☆

351711 Schizolaena cauliflora Thouars;茎花马岛外套花●☆

351712 Schizolaena elongata Thouars;伸长马岛外套花●☆

351713 Schizolaena exinvolucrata Baker;外曲马岛外套花☆

351714 Schizolaena hystrix Capuron;豪猪马岛外套花☆

351715 Schizolaena manomboensis Lowry et al.;马农布马岛外套花●☆

351716 Schizolaena masoalensis Lowry et al.;马苏阿拉马岛外套花●☆

351717 Schizolaena microphylla H. Perrier;小叶马岛外套花☆

351718 Schizolaena milleri Lowry et al.;米勒马岛外套花●☆

351719 Schizolaena parviflora (F. Gérard) H. Perrier;小花马岛外套花●☆

351720 Schizolaena pectinata Capuron;篦状马岛外套花☆

351721 Schizolaena raymondii Lowry et Rabeh.;雷蒙马岛外套花●☆

351722 Schizolaena rosea Thouars;粉红马岛外套花●☆

351723 Schizolaena tampoketsana Lowry et al.;唐波凯茨马岛外套花●☆

351724 Schizolaena turkii Lowry et al.;图尔克马岛外套花☆

351725 Schizolaena viscosa F. Gérard;黏马岛外套花●☆

351726 Schizolaenaceae Barnhart = Sarcolaenaceae Caruel(保留科名)●☆

351727 Schizolaenaceae Barnhart;马岛外套花科●

351728 Schizolepis Schrad. ex Nees = Scleria P. J. Bergius ■

351729 Schizolobium Vogel(1837);塔木属(裂瓣苏木属);Tower Tree ●☆

351730 Schizolobium excelsum Vogel;塔木(黏叶豆)●☆

351731 Schizolobium excelsum Vogel = Schizolobium parahybum (Vell.) Blake ●☆

351732 Schizolobium parahybum (Vell.) Blake;巴西塔木;Brazilian Firetree,Tower Tree,Tree-fern Tree ●☆

351733 Schizolobium parahybum Blake = Schizolobium parahybum (Vell.) Blake ●☆

351734 Schizomeria D. Don(1830);裂苞火把树属●☆

351735 Schizomeria adenophylla L. M. Perry;腺叶裂苞火把树●☆

351736 Schizomeria floribunda Schltr.;多花裂苞火把树☆

351737 Schizomeria parvifolia L. M. Perry;小叶裂苞火把树●☆

351738 Schizomeryta R. Vig. = Meryta J. R. Forst. et G. Forst. ●☆

351739 Schizomussaenda H. L. Li = Mussaenda L. ●■

351740 Schizomussaenda H. L. Li (1943);裂果金花属;Schizomussaenda ●

351741 Schizomussaenda dehiscens (Craib) H. L. Li;裂果金花(白头将军,大树甘草,根辣,木辣,树甘草);Common Schizomussaenda ●

351742 Schizonepeta (Benth.) Briq. (1896);北荆芥属(裂叶荆芥属)■

351743 Schizonepeta Briq. = Schizonepeta (Benth.) Briq. ■

351744 Schizonepeta annua (Pall.) Schischk. = Nepeta annua Pall. ■

351745 Schizonepeta botryoides (Sol.) Briq. = Nepeta annua Pall. ■

351746 Schizonepeta deserticola H. C. Fu et Ninbu = Nepeta annua Pall. ■

351747 Schizonepeta deserticola H. C. Fu et Ninbu = Schizonepeta annua (Pall.) Schischk. ■

351748 Schizonepeta multifida (L.) Briq. = Nepeta multifida L. ■

351749 Schizonepeta tenuifolia (Benth.) Briq. = Nepeta tenuifolia Benth. ■

351750 Schizonepeta tenuifolia (Benth.) Briq. var. japonica (Maxim.) Kitag. = Nepeta japonica Maxim. ■

351751 Schizonephos Griff. = Piper L. ●■

351752 Schizonotus A. Gray = Asclepias L. ■

351753 Schizonotus A. Gray = Solanoa Greene ■

351754 Schizonotus Lindl. (废弃属名) = Sorbaria (Ser. ex DC.) A. Braun(保留属名)●

351755 Schizonotus Lindl. ex Wall. = Sorbaria (Ser. ex DC.) A. Braun (保留属名)●

351756 Schizonotus Raf. = Holodiscus (C. Koch) Maxim.(保留属名)●☆

351757 Schizonotus sorbifolius (L.) Lindl. = Sorbaria sorbifolia (L.) A. Braun ●

351758 Schizopedium Salisb. = Cypripedium L. ■

351759 Schizopepon Maxim. (1859);裂瓜属;Splitmelon ■

351760 Schizopepon bicirrhosus (C. B. Clarke) C. Jeffrey;新裂瓜;New Splitmelon ■

351761 Schizopepon bomiensis A. M. Lu et Zhi Y. Zhang;喙裂瓜(波密裂瓜);Beak Splitmelon,Bill Splitmelon ■

351762 Schizopepon bryoniifolius Maxim.;裂瓜;Common Splitmelon,Splitmelon ■

351763 Schizopepon bryoniifolius Maxim. var. japonicus Cogn. = Schizopepon bryoniifolius Maxim. ■

351764 Schizopepon bryoniifolius Maxim. var. paniculatus Kom. = Schizopepon bryoniifolius Maxim. ■

351765 Schizopepon dioicus Cogn.;湖北裂瓜(毛瓜,毛水瓜蔓);Hubei Splitmelon,Hupeh Splitmelon ■

351766 Schizopepon dioicus Cogn. var. trichogynus Hand.-Mazz.;毛蕊裂瓜;Hairypistil Splitmelon ■

351767 Schizopepon dioicus Cogn. var. wilsonii (Gagnep.) A. M. Lu et Zhi Y. Zhang;四川裂瓜;Sichuan Splitmelon ■

351768 Schizopepon fargesii Gagnep. = Bolbostemma paniculatum (Maxim.) Franquet ■

351769 Schizopepon longipes Gagnep.;长柄裂瓜;Longstalk Splitmelon ■

351770 Schizopepon macranthus Hand.-Mazz.;大花裂瓜;Bigflower Splitmelon ■

351771 Schizopepon monoicus A. M. Lu et Zhi Y. Zhang;峨眉裂瓜;Emei Splitmelon,Omei Splitmelon ■

351772 Schizopepon wardii Chakr. = Schizopepon bicirrhosus (C. B. Clarke) C. Jeffrey ■

351773 Schizopepon wilsonii Gagnep. = Schizopepon dioicus Cogn. var. wilsonii (Gagnep.) A. M. Lu et Zhi Y. Zhang ■

351774 Schizopepon xizangensis A. M. Lu et Zhi Y. Zhang;西藏裂瓜;Xizang Splitmelon ■

351775 Schizopetaceae A. Juss. = Brassicaceae Burnett(保留科名)●■

351776 Schizopetaceae A. Juss. = Cruciferae Juss. (保留科名)●■

351777 Schizopetalon Sims(1823);裂瓣芥属■☆

351778 Schizopetalon walkeri Sims;杏香裂瓣芥;Good Luck Flower ■☆

351779 Schizopetalum DC. = Schizopetalon Sims ■☆

351780 Schizophragma Siebold et Zucc. (1838);钻地风属;Hydrangea Vine,Hydrangeavine,Hydrangea-vine ●

351781 Schizophragma amplum Chun = Schizophragma integrifolium (Franch.) Oliv. ●

351782 Schizophragma choufenianum Chun;临桂钻地风(畴芬钻地风);Lingui Hydrangeavine ●

351783 Schizophragma corylifolium Chun;秦榛钻地风;Corylus-leaf Hydrangeavine ●

351784 Schizophragma crassum Hand.-Mazz.;厚叶钻地风;Thick-leaf Hydrangeavine,Thick-leaved Hydrangea-vine ●

351785 Schizophragma crassum Hand.-Mazz. var. hsitaoiana (Chun) C. F. Wei;维西钻地风(希陶钻地风);Hsitao Thick-leaf Hydrangeavine,Weixi Hydrangea-vine ●

351786 Schizophragma ellipsophyllum C. F. Wei;椭圆钻地风;Ellipticleaf Hydrangeavine ●

351787 Schizophragma fauriei Hayata ex Matsum. et Hayata；圆叶钻地风（台湾钻地风）；Round-leaf Hydrangeavine ●

351788 Schizophragma fauriei Hayata ex Matsum. et Hayata = Schizophragma integrifolium（Franch.）Oliv. var. fauriei（Hayata ex Matsum. et Hayata）Hayata ●

351789 Schizophragma glaucescens（Rehder）Chun = Schizophragma hypoglaucum Rehder ●

351790 Schizophragma glaucescens（Rehder）Chun = Schizophragma integrifolium（Franch.）Oliv. var. glaucescens Rehder ●

351791 Schizophragma glaucescens（Rehder）Chun f. minus（Rehder）Chun = Schizophragma integrifolium（Franch.）Oliv. var. glaucescens Rehder ●

351792 Schizophragma hsitaoanum Chun = Schizophragma crassum Hand.-Mazz. var. hsitaoiana（Chun）C. F. Wei ●

351793 Schizophragma hydrangeoides Siebold et Zucc.；绣球钻地风；Japanese Hydrangea Vine ■☆

351794 Schizophragma hydrangeoides Siebold et Zucc. f. concolor（Hatus.）H. Ohba = Schizophragma hydrangeoides Siebold et Zucc. var. concolor Hatus. ●☆

351795 Schizophragma hydrangeoides Siebold et Zucc. f. formosum（Nakai）Sugim.；台湾钻地风●

351796 Schizophragma hydrangeoides Siebold et Zucc. f. molle（Honda）H. Hara ex H. Ohba；毛钻地风●☆

351797 Schizophragma hydrangeoides Siebold et Zucc. f. sinicum C. C. Yang；华东钻地风；Chinese Hydrangeavine, East China Hydrangeavine ●

351798 Schizophragma hydrangeoides Siebold et Zucc. var. concolor Hatus.；同色钻地风●☆

351799 Schizophragma hydrangeoides Siebold et Zucc. var. fauriei（Hayata）Hayata = Schizophragma fauriei Hayata ex Matsum. et Hayata ●

351800 Schizophragma hydrangeoides Siebold et Zucc. var. integrifolium Franch. = Schizophragma integrifolium（Franch.）Oliv. ●

351801 Schizophragma hypoglaucum Rehder；白背钻地风（散血藤，钻地风）；Whiteback Hydrangeavine, White-backed Hydrangea-vine ●

351802 Schizophragma integrifolium（Franch.）Oliv.；钻地风（齿缘钻地风，地枫，阔瓣钻地风，利筋藤，全叶钻地风，桐叶藤，小齿钻地风，追地枫）；China Hydrangeavine, Chinese Hydrangeavine, Chinese Hydrangea-vine, Littletooth Hydrangeavine ●

351803 Schizophragma integrifolium（Franch.）Oliv. f. denticulatum（Rehder）Chun = Schizophragma integrifolium（Franch.）Oliv. ●

351804 Schizophragma integrifolium（Franch.）Oliv. var. denticutataum Rehder = Schizophragma integrifolium（Franch.）Oliv. ●

351805 Schizophragma integrifolium（Franch.）Oliv. var. fauriei（Hayata ex Matsum. et Hayata）Hayata = Schizophragma fauriei Hayata ex Matsum. et Hayata ●

351806 Schizophragma integrifolium（Franch.）Oliv. var. glaucescens Rehder；粉绿钻地风（粉背钻地风，小粉绿钻地风）；Glaucescent Chinese Hydrangeavine ●

351807 Schizophragma integrifolium（Franch.）Oliv. var. minus Rehder = Schizophragma integrifolium（Franch.）Oliv. var. glaucescens Rehder ●

351808 Schizophragma integrifolium（Franch.）Oliv. var. molle Rehder = Schizophragma molle（Rehder）Chun ●

351809 Schizophragma integrifolium Oliv. var. denticutatum Rehder = Schizophragma integrifolium（Franch.）Oliv. ●

351810 Schizophragma integrifolium Oliv. var. fauriei（Hayata）Hayata = Schizophragma fauriei Hayata ex Matsum. et Hayata ●

351811 Schizophragma integrifolium Oliv. var. molle Rehder = Schizophragma molle（Rehder）Chun ●

351812 Schizophragma macrosepalum Hu = Schizomussaenda dehiscens（Craib）H. L. Li ●

351813 Schizophragma macrosepalum Hu = Schizophragma integrifolium（Franch.）Oliv. ●

351814 Schizophragma megalocarpum Chun；大果钻地风；Bigfruit Hydrangeavine ●

351815 Schizophragma molle（Rehder）Chun；柔毛钻地风；Villose Hydrangeavine ●

351816 Schizophragma molle（Rehder）Chun var. grande Chun = Schizophragma molle（Rehder）Chun ●

351817 Schizophragma molle（Rehder）Chun var. grandis Chun；长叶柔毛钻地风；Long-leaf Villose Hydrangeavine ●

351818 Schizophragma molle（Rehder）Chun var. grandis Chun = Schizophragma molle（Rehder）Chun ●

351819 Schizophragma molle（Rehder）Chun var. rubidum N. Chao et C. C. Yang；红毛钻地风；Red-villose Hydrangeavine ●

351820 Schizophragma molle（Rehder）Chun var. rubidum N. Chao et C. C. Yang = Schizophragma molle（Rehder）Chun ●

351821 Schizophragma obtusifolium Hu = Decumaria sinensis Oliv. ●

351822 Schizophragma viburnoides Stapf = Pileostegia viburnoides Hook. f. et Thomson ●

351823 Schizophyllum Nutt. = Aphanopappus Endl. ■☆

351824 Schizophyllum Nutt. = Lipochaeta DC. ■☆

351825 Schizopleura（Lindl.）Endl. = Beaufortia R. Br. ●☆

351826 Schizopleura Endl. = Beaufortia R. Br. ●☆

351827 Schizopogon Rchb. = Andropogon L.（保留属名）■

351828 Schizopogon Rchb. ex Spreng. = Schizachyrium Nees ■

351829 Schizopremna Baill. = Faradaya F. Muell. ●☆

351830 Schizopsera Turcz. = Schizoptera Turcz. ■☆

351831 Schizopsis Bureau = Tynanthus Miers ●☆

351832 Schizopsis Bureau ex Baill. = Tynanthus Miers ●☆

351833 Schizoptera Turcz.（1851）；裂翅菊属■☆

351834 Schizoptera peduncularis S. F. Blake；裂翅菊■☆

351835 Schizoscyphus K. Schum. ex Taub. = Maniltoa Scheff. ■☆

351836 Schizoscyphus K. Schum. ex Taub. = Schizosiphon K. Schum. ■☆

351837 Schizoscyphus Taub. = Maniltoa Scheff. ■☆

351838 Schizoseddera Roberty = Seddera Hochst. ●☆

351839 Schizosepala G. M. Barroso（1955）；裂萼玄参属■☆

351840 Schizosepala glandulosa G. M. Barroso；裂萼玄参■☆

351841 Schizosiphon K. Schum. = Maniltoa Scheff. ■☆

351842 Schizospatha Furtado = Calamus L. ●

351843 Schizospermum Boiv. ex Baill. = Cremaspora Benth. ●☆

351844 Schizostachyum Nees（1829）；思劳竹属（裂穗草属，沙箪竹属，沙勒竹属，莎莉竹属，思笏竹属）；Schizostachyum ●

351845 Schizostachyum annula-tum J. R. Xue et W. P. Zhang = Bambusa distegia（Keng et P. C. Keng）L. C. Chia et H. L. Fung ●

351846 Schizostachyum annulatum J. R. Xue et W. P. Zhang = Bambusa yunnanensis N. H. Xia ●

351847 Schizostachyum auriculatum Q. H. Dai et D. Y. Huang；耳垂竹●

351848 Schizostachyum blumei Nees et McClure；爪哇思劳竹；Blume Schizostachyum ●☆

351849 Schizostachyum brachycladum（Kurz）Kurz；短枝黄金竹（短枝莎簕竹，黄金沙勒竹，黄金莎勒竹）；Short-branch Sasa, Short-branched Schizostachyum,Shorttwig Schizostachyum ●

351850 Schizostachyum chinense Rendle;薄竹(华薄竹,华思劳竹);China Schizostachyum,Chinese Leptocanna,Chinese Schizostachyum ●

351851 Schizostachyum coradatum（T. H. Wen et Q. H. Dai）N. H. Xia;糯米竹;Heart-shaped Neohouzeaua ●

351852 Schizostachyum diffusum（Blanco）Merr.;沙簕竹(簕竹,沙勒竹,莎莉竹);Climbing Bamboo, Diffuse Schizostachyum, Sprawling Schizostachyum ●

351853 Schizostachyum dumetorum（Hance）Munro;苗竹仔(簕竹,细叶苗竹);Bushy Schizostachyum,Shrub Schizostachyum ●

351854 Schizostachyum dumetorum var. xinwuense（T. H. Wen et J. Y. Chin）N. H. Xia = Schizostachyum xinwuense T. H. Wen et J. Y. Chin ●

351855 Schizostachyum funghomii McClure;沙罗单竹（罗竹）;Funghom Schizostachyum ●

351856 Schizostachyum glaucifolium Munro;灰绿叶思劳竹;Polynesian Ohe ●☆

351857 Schizostachyum hainanense Merr. ex McClure;山骨罗竹（藤竹）;Hainan Schizostachyum ●

351858 Schizostachyum jaculans Holttum;岭南思劳竹(岭南竹);Lingnan Schizostachyum ●

351859 Schizostachyum leviculme McClure = Pseudostachyum polymorphum Munro ●

351860 Schizostachyum lima Merr. = Schizostachyum pseudolima McClure ●

351861 Schizostachyum perrieri A. Camus;佩里耶思劳竹 ●☆

351862 Schizostachyum pseudolima McClure;思劳竹(沙簕竹,沙园竹,山铁罗竹,山竹,苏罗单竹);Common Schizostachyum ●

351863 Schizostachyum sanguineum W. P. Zhang;红毛竹 ●

351864 Schizostachyum subvexorum Q. H. Dai et D. Y. Huang = Schizostachyum funghomii McClure ●

351865 Schizostachyum xinwuense T. H. Wen et J. Y. Chin;火筒竹;Xinhu Schizostachyum ●

351866 Schizostachyum xinwuense T. H. Wen et J. Y. Chin = Schizostachyum dumetorum var. xinwuense（T. H. Wen et J. Y. Chin）N. H. Xia ●

351867 Schizostemma Decne.（1838）;肖尖瓣花属 ●☆

351868 Schizostemma Decne. = Oxypetalum R. Br.（保留属名）●■☆

351869 Schizostemma longifolium Decne.;长叶肖尖瓣花 ●☆

351870 Schizostemma microphyllum Decne.;小花肖尖瓣花 ●☆

351871 Schizostemma parviflorum Decne.;小花肖尖瓣花 ●☆

351872 Schizostephanus Hochst. ex Benth. et Hook. f.（1876）;裂冠萝藦属 ●☆

351873 Schizostephanus Hochst. ex Benth. et Hook. f. = Vincetoxicum Wolf ●■

351874 Schizostephanus Hochst. ex K. Schum. = Cynanchum L. ●■

351875 Schizostephanus Hochst. ex K. Schum. = Schizostephanus Hochst. ex Benth. et Hook. f. ●☆

351876 Schizostephanus alatus Hochst. ex K. Schum. = Cynanchum validum N. E. Br. ●☆

351877 Schizostephanus gossweileri（S. Moore）Liede;裂冠萝藦 ●☆

351878 Schizostephanus somaliensis N. E. Br. = Pentarrhinum somaliense（N. E. Br.）Liede ●☆

351879 Schizostigma Arn.（1840）= Cucurbitella Walp. ■☆

351880 Schizostigma Arn. = Schizostigma Arn. ex Meisn. ■☆

351881 Schizostigma Arn. ex Meisn.（1838）;裂柱葫芦属 ■☆

351882 Schizostigma asperata（Gillies ex Hook. et Arn.）Arn.;裂柱葫芦 ■☆

351883 Schizostylis Backh. et Harv.（1864）;裂柱莲属(切柱花属);Crimson Flag, Cutstyle Flower, Kaffir Lily, Kaffir-Lily ■☆

351884 Schizostylis coccinea Backh. et Harv.;红花裂柱莲(红旗花,细根红旗花);Crimson Flag, Fewflower Crimson Flag, Kaffir Lily, Kaffir-lily,Winter Gladiolus ■☆

351885 Schizostylis coccinea Backh. et Harv.‘Grandiflora’= Schizostylis coccinea Backh. et Harv.‘Major’■☆

351886 Schizostylis coccinea Backh. et Harv.‘Major’;大花裂柱莲 ■☆

351887 Schizostylis coccinea Backh. et Harv.‘Mrs Hegarty’;赫佳提裂柱莲 ■☆

351888 Schizostylis coccinea Backh. et Harv.‘Sunrise’;旭日裂柱莲 ■☆

351889 Schizostylis coccinea Backh. et Harv.‘Viscountess Byng’;拜恩裂柱莲 ■☆

351890 Schizostylis coccinea Backh. et Harv. = Hesperantha coccinea（Backh. et Harv.）Goldblatt et J. C. Manning ■☆

351891 Schizostylis ixioides Harv. ex Baker = Hesperantha coccinea（Backh. et Harv.）Goldblatt et J. C. Manning ■☆

351892 Schizostylis pauciflora Klatt = Hesperantha coccinea（Backh. et Harv.）Goldblatt et J. C. Manning ■☆

351893 Schizotechium（Fenzl）Rchb. = Stellaria L. ■

351894 Schizotechium Rchb. = Stellaria L. ■

351895 Schizotheca Ehrenb. = Thalassia Banks ex K. D. König ■

351896 Schizotheca Ehrenb. ex Solras = Thalassia Banks ex K. D. König ■

351897 Schizotheca Lindl. = Atriplex L. ■●

351898 Schizotheca hemprichii Ehrenb. ex Solms = Thalassia hemprichii（Ehrenb. ex Solms）Asch. ■

351899 Schizothrinax H. Wendl. = ? Thrinax L. f. ex Sw. ●☆

351900 Schizotorenia T. Yamaz.（1978）;肖母草属 ■☆

351901 Schizotorenia T. Yamaz. = Lindernia All. ■

351902 Schizotorenia atropurpurea（Ridl.）T. Yamaz.;深紫肖母草 ■☆

351903 Schizotorenia finetiana（Boneti）T. Yamaz.;肖母草 ■☆

351904 Schizotrichia Benth.（1873）;裂毛菊属 ●☆

351905 Schizotrichia eupatorioides Benth.;裂毛菊 ●☆

351906 Schizozygia Baill.（1888）;裂轭夹竹桃属 ●☆

351907 Schizozygia coffaeoides Baill.;裂轭夹竹桃 ●☆

351908 Schkuhria Moench = Sigesbeckia L. ■

351909 Schkuhria Roth（1797）（保留属名）;假丝叶菊属（史库属）;Dwarf Marigold, False Threadleaf ■☆

351910 Schkuhria abrotanoides Roth = Schkuhria pinnata（Lam.）Kuntze ex Thell. ■☆

351911 Schkuhria anthemoidea（DC.）J. M. Coult. = Schkuhria pinnata（Lam.）Kuntze ex Thell. ■☆

351912 Schkuhria anthemoidea（DC.）J. M. Coult. var. wislizeni（A. Gray）Heiser = Schkuhria pinnata（Lam.）Kuntze ex Thell. ■☆

351913 Schkuhria bonariensis Hook. et Arn. = Schkuhria pinnata（Lam.）Kuntze ex Thell. ■☆

351914 Schkuhria integrifolia A. Gray = Platyschkuhria integrifolia（A. Gray）Rydb. ■☆

351915 Schkuhria multiflora Hook. et Arn.;多花假丝叶菊(多花史库菊)■☆

351916 Schkuhria neomexicana A. Gray = Schkuhria multiflora Hook. et Arn. ■☆

351917 Schkuhria pinnata（Lam.）Kuntze ex Thell.;假丝叶菊(史库菊);Dwarf Marigold, Pinnate False Threadleaf ■☆

351918 Schkuhria pinnata（Lam.）Kuntze ex Thell. var. abrotanoides（Roth）Cabrera = Schkuhria pinnata（Lam.）Kuntze ex Thell. ■☆

351919　Schkuhria pinnata（Lam.）Thell. = Schkuhria pinnata（Lam.）Kuntze ex Thell. ■☆

351920　Schkuhria virgata（La Llave et Lex.）DC. = Schkuhria pinnata（Lam.）Kuntze ex Thell. ■☆

351921　Schkuhria wislizeni A. Gray = Schkuhria pinnata（Lam.）Kuntze ex Thell. ■☆

351922　Schkuhria wislizeni A. Gray var. frustrata S. F. Blake = Schkuhria pinnata（Lam.）Kuntze ex Thell. ■☆

351923　Schkuhria wislizeni A. Gray var. wrightii（A. Gray）S. F. Blake = Schkuhria pinnata（Lam.）Kuntze ex Thell. ■☆

351924　Schlagintweitia Griseb. = Hieracium L. ■

351925　Schlagintweitiella Ulbr. = Thalictrum L. ■

351926　Schlagintweitiella fumarioides Ulbr. = Thalictrum squamiferum Lecoy. ■

351927　Schlagintweitiella glareosa（Hand.-Mazz.）Ulbr. = Thalictrum squamiferum Lecoy. ■

351928　Schlechtendahlia Benth. et Hook. f. = Adenophyllum Pers. ■●☆

351929　Schlechtendahlia Benth. et Hook. f. = Schlechtendalia Less.（保留属名）■☆

351930　Schlechtendalia Less.（1830）（保留属名）；长叶钝菊属■☆

351931　Schlechtendalia Spreng. = Mollia Mart.（保留属名）●☆

351932　Schlechtendalia Willd.（废弃属名）= Adenophyllum Pers. ■●☆

351933　Schlechtendalia Willd.（废弃属名）= Schlechtendalia Less.（保留属名）■☆

351934　Schlechtendalia luzulifolia Less.；长叶钝菊■☆

351935　Schlechteranthus Schwantes（1929）；合叶玉属●☆

351936　Schlechteranthus hallii L. Bolus；哈尔合叶玉■☆

351937　Schlechteranthus maximiliani Schwantes；合叶玉■☆

351938　Schlechterella Hoehne = Lindleyella Schltr. ■☆

351939　Schlechterella Hoehne = Rudolfiella Hoehne ■☆

351940　Schlechterella K. Schum.（1899）；施莱杠柳属（施莱萝藦属）■●☆

351941　Schlechterella K. Schum. = Pleurostelma Baill. ■☆

351942　Schlechterella abyssinica（Chiov.）Venter et R. L. Verh.；阿比西尼亚施莱萝藦■☆

351943　Schlechterella africana（Schltr.）K. Schum.；非洲施莱萝藦●☆

351944　Schlechteria Bolus = Schlechteria Bolus ex Schltr. ■☆

351945　Schlechteria Bolus ex Schltr.（1897）；好望角芥属■☆

351946　Schlechteria Mast. = Phyllocomos Mast. ■☆

351947　Schlechteria capensis Bolus = Heliophila monosperma Al-Shehbaz et Mummenhoff ■☆

351948　Schlechteria capensis Bolus ex Schltr.；好望角芥■☆

351949　Schlechterianthus Quisumb. = Quisumbingia Merr. ●☆

351950　Schlechterina Harms（1902）；施莱莲属■☆

351951　Schlechterina mitostemmatoides Harms；施莱莲■☆

351952　Schlechterorchis Szlach.（2003）；施莱兰属■☆

351953　Schlechterorchis Szlach. = Amphorchis Thouars ■☆

351954　Schlechterorchis Szlach. = Cynorkis Thouars ■☆

351955　Schlechterosciadium H. Wolff = Chamarea Eckl. et Zeyh. ■☆

351956　SchlechterStaehelina Raf. = Helipterum DC. ex Lindl. ■☆

351957　Schlegelia Miq.（1844）；夷地黄属●☆

351958　Schlegelia albiflora Kuhlm.；白花夷地黄●☆

351959　Schlegelia aurea Ducke；黄夷地黄●☆

351960　Schlegelia axillaris Griseb.；腋生夷地黄●☆

351961　Schlegelia brachyantha Griseb.；短花夷地黄●☆

351962　Schlegelia cauliflora A. H. Gentry；茎花夷地黄●☆

351963　Schlegelia macrocarpa Lundell；大果夷地黄●☆

351964　Schlegelia macrophylla Ducke；大叶夷地黄●☆

351965　Schlegelia parviflora（Oerst.）Monach.；小花夷地黄●☆

351966　Schlegeliaceae Reveal（1996）；夷地黄科●☆

351967　Schleichera Willd.（1806）（保留属名）；印度无患子属（枕树属）●☆

351968　Schleichera oleosa（Lour.）Merr.；印度无患子（多油无患子，枕树，锡兰栎树）；Ceylon Oak, Gum-lac Tree, Honey Tree, Kosum Oil, Kosumba, Kusam, Lac Tree, Macasaar Oil ●☆

351969　Schleichera pentapetala Roxb. = Mischocarpus pentapetalus（Roxb.）Radlk. ●

351970　Schleichera trijuga Willd. = Schleichera oleosa（Lour.）Merr. ●☆

351971　Schleidenia Endl. = Heliotropium L. ●■

351972　Schleidenia Endl. = Preslaea Mart. ●■

351973　Schleinitzia Warb. = Schleinitzia Warb. ex Nevling et Niezgoda ■☆

351974　Schleinitzia Warb. ex Nevling et Niezgoda（1978）；异牧豆树属■☆

351975　Schleinitzia microphylla Warb.；异牧豆树■☆

351976　Schleranthus L. = Scleranthus L. ■☆

351977　Schlerochloa Parl. = Sclerochloa P. Beauv. ■

351978　Schleropelta Buckley = Hilaria Kunth ■☆

351979　Schliebenia Mildbr. = Isoglossa Oerst.（保留属名）■★

351980　Schliebenia salviiflora Mildbr. = Isoglossa salviiflora（Mildbr.）Brummitt ■☆

351981　Schliebenia secunda Mildbr. = Isoglossa salviiflora（Mildbr.）Brummitt ■☆

351982　Schlimia Regel = Lisianthus P. Browne ■☆

351983　Schlimia Regel（1875）；施利龙胆属■☆

351984　Schlimia princeps Regel；施利龙胆☆

351985　Schlimmia Planch. et Linden（1852）；施利兰属■☆

351986　Schlimmia alpina Rchb. et Warsz.；高山施利兰■☆

351987　Schlimmia jasminodora Planch. et Linden；施利兰■☆

351988　Schlosseria Ellis = Styrax L. ●

351989　Schlosseria Mill. ex Steud. = Coccoloba P. Browne（保留属名）●

351990　Schlosseria Vuk. = Peucedanum L. ■

351991　Schluckebieria Braem = Cattleya Lindl. ■

351992　Schlumbergera E. Morren = Guzmania Ruiz et Pav. ■☆

351993　Schlumbergera Lem.（1858）；仙人指属（蟹爪属，蟹足霸王树属）；Christmas Cactus, Crab Cactus ●

351994　Schlumbergera 'Nicole'；圣诞仙人指；Christmas Cactus, Holiday Cactus ■☆

351995　Schlumbergera 'Thoralise'；节日仙人指；Christmas Cactus, Holiday Cactus ■☆

351996　Schlumbergera × buckleyi（T. Moore）Tjaden = Schlumbergera bridgesii（Lem.）Loefgr. ■

351997　Schlumbergera bridgesii（Lem.）Loefgr.；仙人指（绿蟹爪，螃蟹兰，圣诞节仙人掌，圆齿蟹爪）；Christmas Cactus ■

351998　Schlumbergera bridgesii（Lem.）Loefgr. = Schlumbergera truncata（Haw.）Moran ■

351999　Schlumbergera gaertneri（Regel）Britton et Rose；盖特纳仙人指；Easter Cactus, Spring Cactus ■☆

352000　Schlumbergera russelliana（Hook.）Britton et Rose；倒吊莲（钝齿仙人指，蟹足霸王树）■

352001　Schlumbergera truncata（Haw.）Moran = Zygocactus truncatus（Haw.）K. Schum. ■

352002　Schlumbergeria E. Morren = Guzmania Ruiz et Pav. ■☆

352003　Schlumbergia Lem. = Schlumbergera Lem. ●

352004　Schmalhausenia C. Winkl.（1892）；虎头蓟属；Tigerheadthistle Schmalhausenia ■★

352005　Schmalhausenia eriophora C. Winkl. = Schmalhausenia nidulans

（Regel）Petr. ■

352006　Schmalhausenia nidulans（Regel）Petr.；虎头蓟；Common Tigerheadthistle Schmalhausenia ■

352007　Schmalhausenia oriophora C. Winkl. = Schmalhausenia nidulans（Regel）Petr. ■

352008　Schmaltzia Desv. = Rhus L. ●

352009　Schmaltzia Desv. ex Small = Lobadium Raf. ●

352010　Schmaltzia Desv. ex Small = Rhus L. ●

352011　Schmaltzia Desv. ex Steud. = Rhus L. ●

352012　Schmaltzia Steud. = Schmaltzia Desv. ●

352013　Schmaltzia crenata（Mill.）Greene = Rhus aromatica Aiton ●☆

352014　Schmaltzia illinoensis Greene = Rhus aromatica Aiton ●☆

352015　Schmardaea H. Karst.（1861）；施马楝属●☆

352016　Schmardaea micrephyna（Hook.）C. Müll.；施马楝●☆

352017　Schmidelia Boehm. = Calophyllum L. ●

352018　Schmidelia Boehm. = Ehretia P. Browne ●

352019　Schmidelia L. = Allophylus L. ●

352020　Schmidelia abyssinica Hochst. = Allophylus abyssinicus（Hochst.）Radlk. ●☆

352021　Schmidelia alnifolia Baker = Allophylus rubifolius（Hochst. ex A. Rich.）Engl. var. alnifolius（Baker）Friis et Vollesen ●☆

352022　Schmidelia chartacea Kurz = Allophylus chartaceus（Kurz）Radlk. ●

352023　Schmidelia decipiens Sond. = Allophylus decipiens（Sond.）Radlk. ●☆

352024　Schmidelia dregeana Sond. = Allophylus dregeanus（Sond.）De Winter ●☆

352025　Schmidelia grandifolia Baker = Allophylus grandifolius（Baker）Radlk. ●☆

352026　Schmidelia hirtella Hook. f. = Allophylus hirtellus（Hook. f.）Radlk. ●☆

352027　Schmidelia leucocarpa Arn. ex Sond. = Allophylus africanus P. Beauv. ●☆

352028　Schmidelia melanocarpa Arn. ex Sond. = Allophylus africanus P. Beauv. ●☆

352029　Schmidelia minutiflora Mattei = Allophylus rubifolius（Hochst. ex A. Rich.）Engl. var. alnifolius（Baker）Friis et Vollesen ●☆

352030　Schmidelia monophylla C. Presl = Allophylus dregeanus（Sond.）De Winter ●☆

352031　Schmidelia natalensis Sond. = Allophylus natalensis（Sond.）De Winter ●☆

352032　Schmidelia repanda Baker = Allophylus rubifolius（Hochst. ex A. Rich.）Engl. var. alnifolius（Baker）Friis et Vollesen ●☆

352033　Schmidelia rubifolia Hochst. ex A. Rich. = Allophylus rubifolius（Hochst. ex A. Rich.）Engl. ●☆

352034　Schmidelia timorensis DC. = Allophylus timorensis（DC.）Blume ●

352035　Schmidia Wight = Thunbergia Retz.（保留属名）●■

352036　Schmidtia Moench（废弃属名）= Schmidtia Steud. ex J. A. Schmidt（保留属名）■☆

352037　Schmidtia Moench（废弃属名）= Tolpis Adans. ●■☆

352038　Schmidtia Sieber = Schmidtia Steud. ex J. A. Schmidt（保留属名）■☆

352039　Schmidtia Steud. = Schmidtia Steud. ex J. A. Schmidt（保留属名）■☆

352040　Schmidtia Steud. ex J. A. Schmidt（1852）（保留属名）；丛林草属■☆

352041　Schmidtia Tratt. = Coleanthus Seidel（保留属名）■

352042　Schmidtia bulbosa Stapf = Schmidtia pappophoroides Steud. ■☆

352043　Schmidtia farinulosa Webb = Tolpis farinulosa（Webb）Walp. ■

352044　Schmidtia glabra Pilg. = Schmidtia pappophoroides Steud. ■

352045　Schmidtia kalahariensis Stent；卡拉哈利丛林草■☆

352046　Schmidtia pappophoroides Steud.；冠毛丛林草■☆

352047　Schmidtia quinqueseta Benth. ex Ficalho et Hiern = Schmidtia pappophoroides Steud. ■☆

352048　Schmidtia subtilis Tratt. = Coleanthus subtilis（Tratt.）Seidl ■

352049　Schmidtottia Urb.（1923）；施米茜属●☆

352050　Schmidtottia cubensis Urb.；古巴施米茜●☆

352051　Schmidtottia elliptica Urb.；椭圆施米茜●☆

352052　Schmidtottia monticola Borhidi；山地施米茜●☆

352053　Schmidtottia multiflora Urb.；多花施米茜●☆

352054　Schmidtottia parvifolia Alain；小叶施米茜●☆

352055　Schmidtottia sessilifolia Urb.；无柄施米茜●☆

352056　Schmiedelia Murr. = Schmidelia L. ●

352057　Schmiedtia Raf. = Coleanthus Seidel（保留属名）■

352058　Schmiedtia Raf. = Schmidtia Tratt. ■

352059　Schnabelia Hand. -Mazz.（1924）；四棱草属；Schnabelia ■●★

352060　Schnabelia oligophylla Hand. -Mazz.；四棱草（长叶四棱草，假马鞭草，箭羽草，箭羽筋骨草，箭羽舒筋草，筋骨草，筋骨连，舒筋箭羽草，四方草，四棱筋骨草，四楞筋骨草）；Fewleaf Schnabelia，Oblongleaf Schnabelia，Schnabelia

352061　Schnabelia oligophylla Hand. -Mazz. var. oblongifolia C. Y. Wu et C. Chen = Schnabelia oligophylla Hand. -Mazz. ■

352062　Schnabelia oligophylla Hand. -Mazz. var. oblongifolia C. Y. Wu et C. Chen；长叶四棱草■

352063　Schnabelia tetrodonta（Y. Z. Sun）C. Y. Wu et C. Chen；四齿四棱草（四齿筋骨草）；Fourteeth Schnabelia ■

352064　Schnarfia Speta（1998）；施纳风信子属■☆

352065　Schnarfia albanica（Turrill）Speta；阿尔班施纳风信子■☆

352066　Schnarfia messeniaca（Boiss.）Speta；施纳风信子■☆

352067　Schnella Raddi = Bauhinia L. ●

352068　Schnittspahnia Rchb. = Mitrella Miq. ●☆

352069　Schnittspahnia Rchb. = Polyalthia Blume ●

352070　Schnittspahnia Sch. Bip. = Landtia Less. ■☆

352071　Schnittspahnia rueppellii Sch. Bip. = Haplocarpha rueppellii（Sch. Bip.）Beauverd ■☆

352072　Schnitzleinia Steud. ex Walp. = Schnizleinia Steud. ex Hochst. ■☆

352073　Schnitzleinia Steud. ex Walp. = Vellozia Vand. ■☆

352074　Schnitzleinia Walp. = Schnizleinia Steud. ex Hochst. ■☆

352075　Schnitzleinia Walp. = Vellozia Vand. ■☆

352076　Schnizleinia Mart. ex Engl. = Emmotum Desv. ex Ham. ●☆

352077　Schnizleinia Steud.（1840）= Boissiera Hochst. ex Steud. ■

352078　Schnizleinia Steud.（1841）= Trochiscanthes W. D. J. Koch ■☆

352079　Schnizleinia Steud. ex Hochst. = Vellozia Vand. ■☆

352080　Schnizleinia amica Steud. = Xerophyta schnizleinia（Hochst.）Baker ■☆

352081　Schobera Scop. = Heliotropium L. ●■

352082　Schoberia G. A. Mey. = Suaeda Forssk. ex J. F. Gmel.（保留属名）●■

352083　Schoberia acuminata C. A. Mey. = Suaeda acuminata（C. A. Mey.）Moq. ■

352084　Schoberia corniculata C. A. Mey. = Suaeda corniculata（C. A. Mey.）Bunge ■

352085　Schoberia dendroides C. A. Mey. = Suaeda dendroides（C. A.

Mey.）Moq. ●

352086　Schoberia glauca Bunge ＝ Suaeda glauca（Bunge）Bunge ■

352087　Schoberia heterophylla Kar. et Kir. ＝ Suaeda heterophylla（Kar. et Kir.）Bunge ■

352088　Schoberia leiosperma C. A. Mey. ＝ Suaeda altissima（L.）Pall. ■

352089　Schoberia maritima C. A. Mey. ＝ Suaeda prostrata Pall. ■

352090　Schoberia microphylla（Pall.）C. A. Mey. ＝ Suaeda microphylla（C. A. Mey.）Pall. ●

352091　Schoberia microphylla C. A. Mey. ＝ Suaeda microphylla（C. A. Mey.）Pall. ●

352092　Schoberia obtusifolia Bunge ＝ Suaeda crassifolia Pall. ■

352093　Schoberia occidentalis S. Watson ＝ Suaeda occidentalis（S. Watson）S. Watson ■☆

352094　Schoberia pterantha Kar. et Kir. ＝ Suaeda pterantha（Kar. et Kir.）Bunge ■

352095　Schoberia pygmaea Kar. et Kir. ＝ Suaeda pterantha（Kar. et Kir.）Bunge ■

352096　Schoberia salsa（L.）C. A. Mey. ＝ Suaeda salsa（L.）Pall. ■

352097　Schoberia stanntonii Moq. ＝ Suaeda glauca（Bunge）Bunge ■

352098　Schoebera Neck. ＝ Heliotropium L. ●■

352099　Schoebera Neck. ＝ Schobera Scop. ●■

352100　Schoedonardus Scribn. ＝ Schedonnardus Steud. ■☆

352101　Schoenanthus Adans. ＝ Ischaemum L. ■

352102　Schoenefeldia Kunth（1829）；苇禾属 ■☆

352103　Schoenefeldia gracilis Kunth；纤细苇禾 ■☆

352104　Schoenefeldia nutans Steud. ＝ Schoenefeldia gracilis Kunth ■☆

352105　Schoenefeldia pallida Edgew. ＝ Schoenefeldia gracilis Kunth ■☆

352106　Schoenefeldia transiens（Pilg.）Chiov.；苇禾 ■☆

352107　Schoenia Steetz（1845）；粉苞鼠麹草属 ■☆

352108　Schoenia cassiniana Steetz；粉苞鼠麹草 ■☆

352109　Schoenia filifolia（Turcz.）Paul G. Wilson；线叶粉苞鼠麹草 ■☆

352110　Schoenidium Nees ＝ Ficinia Schrad.（保留属名）■☆

352111　Schoenissa Salisb. ＝ Allium L. ■

352112　Schoenlandia Cornu ＝ Cyanastrum Oliv. ■☆

352113　Schoenlandia gabonensis Cornu ＝ Cyanastrum cordifolium Oliv. ■☆

352114　Schoenleinia Klotzsch ＝ Bathysa C. Presl ■☆

352115　Schoenleinia Klotzsch ex Lindl. ＝ Ponthieva R. Br. ■☆

352116　Schoenobiblos Endl. ＝ Schoenobiblus Mart. et Zucc. ●☆

352117　Schoenobiblus Mart. ＝ Schoenobiblus Mart. et Zucc. ●☆

352118　Schoenobiblus Mart. et Zucc.（1824）；热美瑞香属 ●☆

352119　Schoenobiblus peruvianus Standl.；热美瑞香 ●☆

352120　Schoenocaulon A. Gray（1837）；苇茎百合属（沙巴草属）；Sabadilla ■☆

352121　Schoenocaulon dubium（Michx.）Small；可疑苇茎百合 ■☆

352122　Schoenocaulon ghiesbreghtii Greenm.；吉氏苇茎百合；Greenlily ■☆

352123　Schoenocaulon gracile A. Gray ＝ Schoenocaulon dubium（Michx.）Small ■☆

352124　Schoenocaulon officinale A. Gray；苇茎百合（沙巴草，种子藜芦）；Cevadilla, Sevadilla, Veratrin ■☆

352125　Schoenocaulon texanum Scheele；得州苇茎百合 ■☆

352126　Schoenocaulon yucatanese Brinker ＝ Schoenocaulon ghiesbreghtii Greenm. ■☆

352127　Schoenocephalium Seub.（1847）；芦头草属 ■☆

352128　Schoenocephalium arthrophyllum Seub.；节叶芦头草 ■☆

352129　Schoenocephalium martianum Seub.；芦头草 ■☆

352130　Schoenochlaena Post et Kuntze ＝ Schoenolaena Bunge ■☆

352131　Schoenocrambe Greene ＝ Sisymbrium L. ■

352132　Schoenocrambe Greene（1896）；苇节荠属（灯芯草两节荠属）■☆

352133　Schoenocrambe linearifolia（A. Gray）Rollins；线叶苇节荠 ■☆

352134　Schoenocrambe linifolia Greene；亚麻叶苇节荠 ■☆

352135　Schoenocrambe pinnata Greene；羽状苇节荠 ■☆

352136　Schoenodendron Engl. ＝ Microdracoides Hua ■☆

352137　Schoenodendron buecherei Engl. ＝ Microdracoides squamosus Hua ■☆

352138　Schoenodorus Roem. et Schult. ＝ Festuca L. ■

352139　Schoenodorus Roem. et Schult. ＝ Schedonorus P. Beauv. ■☆

352140　Schoenodum Labill.（废弃属名）＝ Leptocarpus R. Br.（保留属名）■

352141　Schoenodum Labill.（废弃属名）＝ Lyginia R. Br. + Leptocarpus R. Br. ■

352142　Schoenoides Seberg ＝ Oreobolus R. Br. ■☆

352143　Schoenolaena Bunge（1845）；苇被草属 ■☆

352144　Schoenolaena juncea Bunge；苇被草 ■☆

352145　Schoenolaena tenuior Bunge；细苇被草 ■☆

352146　Schoenolirion Durand ＝ Schoenolirion Torr.（保留属名）■☆

352147　Schoenolirion Torr.（1855）（保留属名）；舒安莲属（灯芯草百合属）；Rush-lily ■☆

352148　Schoenolirion Torr. ex Durand ＝ Schoenolirion Torr.（保留属名）■☆

352149　Schoenolirion albiflorum（Raf.）R. R. Gates；白花舒安莲（白花灯芯草百合）■☆

352150　Schoenolirion album Durand ＝ Hastingsia alba（Durand）S. Watson ■☆

352151　Schoenolirion croceum（Michx.）A. W. Wood；镉黄舒安莲（镉黄灯芯草百合）；Sunnybells, Swamp Candle ■☆

352152　Schoenolirion texanum（Scheele）A. Gray ＝ Camassia scilloides（Raf.）Cory ■☆

352153　Schoenolirion wrightii Sherman；赖特舒安莲 ■☆

352154　Schoenomorphus Thorel ex Gagnep. ＝ Tropidia Lindl. ■

352155　Schoenomorphus capitatus Thorel ex Gagnep. ＝ Tropidia pedunculata Blume ■☆

352156　Schoenoplectiella Lye（2003）；类莞属 ■☆

352157　Schoenoplectiella articulata（L.）Lye；类莞 ■☆

352158　Schoenoplectus（Rchb.）Palla（1888）（保留属名）；拟莞属（湖边藨草属，水葱属，萤蔺属）；Club-rush, Naked-stemmed Bulrushes, Scirpes, Schoenoplecte ■

352159　Schoenoplectus Palla ＝ Schoenoplectus（Rchb.）Palla（保留属名）■

352160　Schoenoplectus × oguraensis（T. Koyama）Hayas. et H. Ohashi；御座拟莞 ■☆

352161　Schoenoplectus × osoreyamensis（M. Kikuchi）Hayas.；恐山拟莞 ■☆

352162　Schoenoplectus × trapezoideus（Koidz.）Hayas. et H. Ohashi；五棱藨草；Fiveangular Bulrush ■

352163　Schoenoplectus × uzenensis（Ohwi ex T. Koyama）Hayas. et H. Ohashi；羽前拟莞 ■☆

352164　Schoenoplectus acutus（Muhl. ex J. M. Bigelow）Á. Löve et D. Löve；硬杆拟莞；Great Bulrush, Hardstem Bulrush, Hard-stem Bulrush, Pointed Bulrush ■☆

352165　Schoenoplectus acutus（Muhl. ex J. M. Bigelow）Á. Löve et D. Löve var. occidentalis（S. Watson）S. G. Sm.；西部硬杆拟莞；Common Ttule ■☆

352166　Schoenoplectus americanus（Pers.）Volkart ex Schinz et R.

Keller；盐沼拟莞；Olney Three-square，Olney's Three-square Bulrush，Saltmarsh Bulrush ■☆

352167 Schoenoplectus articulatus（L.）Palla = Schoenoplectiella articulata（L.）Lye ■☆

352168 Schoenoplectus brachyceras（Hochst. ex A. Rich.）Lye；短角拟莞■☆

352169 Schoenoplectus californicus（C. A. Mey.）Soják；加州拟莞；California Bulrush，California Clubrush，Giant Bulrush，Southern Bulrush，Tule ■☆

352170 Schoenoplectus chenmoui（Ts. Tang et F. T. Wang）C. Y. Wu；陈谋藨草；Chen Mou Bulrush，Chen-mo Bulrush ■

352171 Schoenoplectus chuanus（Ts. Tang et F. T. Wang）S. Yun；曲氏水葱■

352172 Schoenoplectus confusus（N. E. Br.）Lye；混乱拟莞■☆

352173 Schoenoplectus confusus（N. E. Br.）Lye subsp. natalitius Browning；蝙蝠拟莞■☆

352174 Schoenoplectus confusus（N. E. Br.）Lye var. rogersii（N. E. Br.）Lye；罗杰斯拟莞■☆

352175 Schoenoplectus corymbosus（Roth ex Roem. et Schult.）J. Raynal；伞序拟莞■☆

352176 Schoenoplectus corymbosus（Roth ex Roem. et Schult.）J. Raynal var. brachyceras（Hochst. ex A. Rich.）Lye = Schoenoplectus brachyceras（Hochst. ex A. Rich.）Lye ■☆

352177 Schoenoplectus decipiens（Nees）J. Raynal；迷惑拟莞■☆

352178 Schoenoplectus deltarum（Schuyler）Soják；三角州拟莞；Delta Bulrush ■☆

352179 Schoenoplectus ehrenbergii（Boeck.）Y. C. Yang et M. Zhan；剑苞水葱■

352180 Schoenoplectus erectus（Poir.）Palla ex J. Raynal；直立拟莞■☆

352181 Schoenoplectus erectus（Poir.）Palla ex J. Raynal subsp. raynalii（Schuyler）Lye；雷纳尔直立拟莞■☆

352182 Schoenoplectus erectus（Poir.）Palla ex J. Raynal subsp. sinuatus（Schuyler）Lye；深波拟莞■☆

352183 Schoenoplectus etuberculatus（Steud.）Soják；康比拟莞；Canby's Bulrush ■☆

352184 Schoenoplectus fluviatilis（Torr.）M. T. Strong = Bolboschoenus fluviatilis（Torr.）Soják ■☆

352185 Schoenoplectus fohaiensis（Ts. Tang et F. T. Wang）C. Y. Wu；佛海拟莞（佛海藨草，佛海水葱，勐海藨草）；Fohai Bulrush ■

352186 Schoenoplectus grossus（L. f.）C. Y. Wu = Schoenoplectus grossus（L. f.）Palla ■

352187 Schoenoplectus grossus（L. f.）Palla；台南水葱（硕大藨草，猪毛草）；Giant Bulrush，Gross Bulrush，Tuberous Bulrush ■

352188 Schoenoplectus grossus（L. f.）Palla = Actinoscirpus grossus（L. f.）Goetgh. et D. A. Simpson ■

352189 Schoenoplectus grossus（L. f.）Palla = Actinoscripus（Ohwi）R. Haines et Lyo ■

352190 Schoenoplectus grossus（L. f.）Palla = Scirpus wallichii Nees ■

352191 Schoenoplectus hallii（A. Gray）S. G. Sm.；霍尔拟莞；Hall's Bulrush ■☆

352192 Schoenoplectus heterochaetus（Chase）Soják；异毛拟莞；Great Bulrush，Slender Bulrush ■☆

352193 Schoenoplectus hondoensis（Ohwi）Soják；本州拟莞■

352194 Schoenoplectus hotarui（Ohwi）Holub = Schoenoplectus juncoides（Roxb.）Palla subsp. hotarui（Ohwi）Soják ■

352195 Schoenoplectus igaensis（T. Koyama）Hayas. et H. Ohashi；伊贺拟莞■☆

352196 Schoenoplectus inclinatus（Delile ex Barbey）Lye = Schoenoplectus corymbosus（Roth ex Roem. et Schult.）J. Raynal ■☆

352197 Schoenoplectus jacobi（C. E. C. Fisch.）Lye = Schoenoplectus senegalensis（Hochst. ex Steud.）Palla ex J. Raynal ■☆

352198 Schoenoplectus jingmenensis（Ts. Tang et F. T. Wang）S. Yun Liang et S. R. Zhang；荆门水葱■

352199 Schoenoplectus junceus（Willd.）J. Raynal；灯芯草拟莞■☆

352200 Schoenoplectus juncoides（Roxb.）Palla；萤蔺（大井氏水莞，灯芯草，滑关草，假碱草，假马蹄，牛毛草，水灯心，细秆萤蔺，野马蹄草，直立席草）；Rush-like Bulrush，Slender-culm Rush-like Bulrush ■

352201 Schoenoplectus juncoides（Roxb.）Palla = Scirpus juncoides Roxb. ■

352202 Schoenoplectus juncoides（Roxb.）Palla subsp. hotarui（Ohwi）Soják；细秆萤蔺■

352203 Schoenoplectus juncoides（Roxb.）Palla subsp. hotarui（Ohwi）Soják = Schoenoplectus hotarui（Ohwi）Holub ■

352204 Schoenoplectus juncoides（Roxb.）Palla subsp. ohwianus（T. Koyama）Soják = Schoenoplectus juncoides（Roxb.）Palla ■

352205 Schoenoplectus komarovii（Roshev.）Soják；吉林藨草（吉林水葱，头藨草）；Jilin Bulrush，Komalov Bulrush ■

352206 Schoenoplectus lacustris（L.）Palla；湖边拟莞（沼生水葱）；Bulrush，Clubrush，Club-rush，Common Club-rush，Great Bulrush，Groat Bulrush，Lakeshore Bulrush，Zebra Rush ■☆

352207 Schoenoplectus lacustris（L.）Palla subsp. tabernaemontani（C. C. Gmel.）Á. Löve et D. Löve = Schoenoplectus tabernaemontani（C. C. Gmel.）Palla ■

352208 Schoenoplectus lacustris（L.）Palla subsp. validus（Vahl）T. Koyama = Schoenoplectus tabernaemontani（C. C. Gmel.）Palla ■

352209 Schoenoplectus lateriflorus（J. F. Gmel.）Lye；侧花拟莞■☆

352210 Schoenoplectus lateriflorus（J. F. Gmel.）Lye = Schoenoplectus supinus（L.）Palla subsp. lateriflorus（J. F. Gmel.）Soják ■

352211 Schoenoplectus leucanthus（Boeck.）J. Raynal；白花拟莞■☆

352212 Schoenoplectus lineolatus（Franch. et Sav.）T. Koyama；兰屿莞（匍匐莞草，细匍匐茎水葱）■

352213 Schoenoplectus litoralis（Schrad.）Palla；羽状刚毛水葱■

352214 Schoenoplectus litoralis（Schrad.）Palla subsp. thermalis（Trab.）Hooper = Schoenoplectus subulatus（Vahl）Lye ■

352215 Schoenoplectus litoralis（Schrad.）Palla var. pterolepis（Nees）C. C. Towns. = Schoenoplectus subulatus（Vahl）Lye ■

352216 Schoenoplectus littoralis（Schrad.）Palla；匍匐莞草■

352217 Schoenoplectus littoralis（Schrad.）Palla = Scirpus littoralis Schrad. ■

352218 Schoenoplectus littoralis（Schrad.）Palla subsp. subulatus（Vahl）Soják = Scirpus subulatus Vahl ■

352219 Schoenoplectus lupulinus（Nees）Krecz. = Schoenoplectus roylei（Nees）Ovcz. et Czukav. ■☆

352220 Schoenoplectus maritimus（L.）Lye = Bolboschoenus maritimus（L.）Palla ■☆

352221 Schoenoplectus microglumis Lye；小颖拟莞■☆

352222 Schoenoplectus monocephalus（J. Q. He）S. Yun. Liang et S. R. Zhang；单穗水葱■

352223 Schoenoplectus mucronatus（L.）Palla；北水毛花（有葱，水毛花，蔗草）；Bog Bulrush，Bulrush，Mucronate Bulrush，Ricefield Bulrush，Rough-seed Bulrush ■

352224 Schoenoplectus mucronatus（L.）Palla = Scirpus mucronatus L. ■

352225　Schoenoplectus mucronatus（L.）Palla f. tataranus（Honda）T. Koyama = Schoenoplectus mucronatus（L.）Palla var. tataranus（Honda）Kohno, Iokawa et Daigobo ■☆

352226　Schoenoplectus mucronatus（L.）Palla subsp. robustus（Miq.）T. Koyama = Schoenoplectus mucronatus（L.）Palla var. robustus（Miq.）T. Koyama ■

352227　Schoenoplectus mucronatus（L.）Palla subsp. robustus（Miq.）T. Koyama = Schoenoplectus triangulatus（Roxb.）Soják ■☆

352228　Schoenoplectus mucronatus（L.）Palla subsp. robustus（Miq.）T. Koyama f. hosoiri Hayas. et H. Ohashi；细井拟莞■☆

352229　Schoenoplectus mucronatus（L.）Palla var. robustus（Miq.）T. Koyama；粗壮水毛花(水毛花)■

352230　Schoenoplectus mucronatus（L.）Palla var. sanguineus（Ts. Tang et F. T. Wang）P. C. Li；红鳞水毛花；Redscale Triangular Bulrush ■

352231　Schoenoplectus mucronatus（L.）Palla var. tataranus（Honda）Kohno, Iokawa et Daigobo；鞑靼拟莞■☆

352232　Schoenoplectus mucronatus（L.）Palla var. tataranus（Honda）Kohno, Iokawa et Daigobo f. brevisetaceus Horiuchi；短毛鞑靼拟莞■☆

352233　Schoenoplectus mucronatus（L.）Palla var. trialatus（Ts. Tang et F. T. Wang）C. Y. Wu；三翅水毛花；Three-wing Triangular Bulrush ■

352234　Schoenoplectus mucronatus Palla = Scirpus mucronatus L. ■

352235　Schoenoplectus multisetus Hayas. et C. Sato；多毛水毛花■☆

352236　Schoenoplectus muricinux（C. B. Clarke）J. Raynal；短尖拟莞■☆

352237　Schoenoplectus muriculatus（Kük.）Browning；粗糙拟莞■☆

352238　Schoenoplectus nipponicus（Makino）Soják；日本水毛花■☆

352239　Schoenoplectus nobilis（Ridl.）Lye = Bolboschoenus nobilis（Ridl.）Goetgh. et D. A. Simpson ■☆

352240　Schoenoplectus novae-angliae（Britton）M. T. Strong = Bolboschoenus novae-angliae（Britton）S. G. Sm. ■☆

352241　Schoenoplectus ohwianus（T. Koyama）Holub = Schoenoplectus juncoides（Roxb.）Palla ■

352242　Schoenoplectus orthorhizomatus（Arai et Miyam.）Hayas. et H. Ohashi；直根茎拟莞■☆

352243　Schoenoplectus oxyjulos（S. S. Hooper）J. Raynal；尖序拟莞■☆

352244　Schoenoplectus paludicola（Kunth）J. Raynal；沼泽拟莞■☆

352245　Schoenoplectus patentiglumis Hayas.；展颖拟莞■☆

352246　Schoenoplectus proximus（Steud.）J. Raynal；近基拟莞■☆

352247　Schoenoplectus pseudoarticulatus（L. K. Dai et S. M. Huang）S. Yun. Liang et S. R. Zhang；节苞水葱■

352248　Schoenoplectus pulchellus（Kunth）J. Raynal；美丽拟莞■☆

352249　Schoenoplectus pungens（Vahl）Palla；美国蔗草；American Bulrush, Chairmaker's Rush, Chair-Maker's Rush, Common Three-square, Common Three-square Bulrush, Sharp Club-rush, Three-square Bulrush ■☆

352250　Schoenoplectus pungens（Vahl）Palla var. badius（J. Presl et C. Presl）S. G. Sm. = Schoenoplectus pungens（Vahl）Palla ■☆

352251　Schoenoplectus pungens（Vahl）Palla var. longispicatus（Britton）S. G. Sm. = Schoenoplectus pungens（Vahl）Palla ■☆

352252　Schoenoplectus purshianus（Fernald）M. T. Strong；珀什拟莞；Bluntscale Bulrush, Pursh's Bulrush, Weak Bulrush, Weak-stalk Bulrush, Weak-stalk Club-rush ■☆

352253　Schoenoplectus purshianus（Fernald）M. T. Strong var. williamsii（Fernald）S. G. Sm.；威廉斯拟莞；Pursh's Bulrush, Weak Bulrush, Weak-stalk Bulrush ■☆

352254　Schoenoplectus raynalianus Scholz；雷纳尔拟莞■☆

352255　Schoenoplectus rechingeri；雷钦格尔拟莞■☆

352256　Schoenoplectus rhodesicus（Podlech）Lye；罗得西亚拟莞■☆

352257　Schoenoplectus robustus（Pursh）M. T. Strong = Bolboschoenus robustus（Pursh）Soják ■☆

352258　Schoenoplectus rogersii（N. E. Br.）Lye = Schoenoplectus confusus（N. E. Br.）Lye var. rogersii（N. E. Br.）Lye ■☆

352259　Schoenoplectus roylei（Nees）Ovcz. et Czukav.；罗伊尔拟莞■☆

352260　Schoenoplectus saximontanus（Fernald）J. Raynal；落基山拟莞；Rocky Mountain Bulrush ■☆

352261　Schoenoplectus schoofii（Beetle）C. Y. Wu；滇水葱■

352262　Schoenoplectus scirpoideus（Schrad.）Browning；蔗草拟莞■☆

352263　Schoenoplectus senegalensis（Hochst. ex Steud.）Palla ex J. Raynal；塞内加尔拟莞■☆

352264　Schoenoplectus setaceus（L.）Palla = Isolepis setacea（L.）R. Br. ■

352265　Schoenoplectus setaceus（L.）Palla = Scirpus setaceus L. ■

352266　Schoenoplectus smithii（A. Gray）Soják；史密斯拟莞；Bluntscale Bulrush, Perianth Bristles Absent, Perianth Rudimentary, Smith's Bulrush ■☆

352267　Schoenoplectus smithii（A. Gray）Soják subsp. leiocarpus（Kom.）Soják = Schoenoplectus komarovii（Roshev.）Soják ■

352268　Schoenoplectus smithii（A. Gray）Soják var. levisetus（Fernald）S. G. Sm.；微刚毛史密斯拟莞■☆

352269　Schoenoplectus smithii（A. Gray）Soják var. setosus（Fernald）S. G. Sm.；刚毛史密斯拟莞；Blunt-scale Bulrush, Smith's Bulrush ■☆

352270　Schoenoplectus steinmetzii（Fernald）S. G. Sm.；斯坦梅茨拟莞；Steinmetz's Bulrush ■☆

352271　Schoenoplectus subterminalis（Torr.）Soják；水拟莞；Swaying Bulrush, Water Bulrush, Water Club-rush ■☆

352272　Schoenoplectus subulatus（Vahl）Lye = Schoenoplectus littoralis（Schrad.）Palla subsp. subulatus（Vahl）Soják ■

352273　Schoenoplectus subulatus（Vahl）S. Yun. Liang et S. R. Zhang；钻苞水葱■

352274　Schoenoplectus supinus（L.）Palla；小水莞(仰卧秆水葱)■

352275　Schoenoplectus supinus（L.）Palla subsp. densicorrugatus（Ts. Tang et F. T. Wang）S. Yun. Liang et S. R. Zhang；多纹小水莞(多皱纹果仰卧秆水葱)■

352276　Schoenoplectus supinus（L.）Palla subsp. lateriflorus（J. F. Gmel.）Soják；稻田仰卧秆水葱（侧花水拟莞,稻田仰卧秆蔗草,水灯草,小水莞）；Paddyfield Prostrate Bulrush ■

352277　Schoenoplectus supinus（L.）Palla subsp. lateriflorus（J. F. Gmel.）T. Koyama = Scirpus supinus L. var. lateriflorus（S. G. Gmel.）T. Koyama ■

352278　Schoenoplectus tabernaemontani（C. C. Gmel.）Palla；水葱（冲天草,翠管草,欧水葱,软茎蔗草,三棱蔗草,水丈葱,莞,莞蒲,小蔗草）；Banded Bulrush, Glaucous Bulrush, Great Bulrush, Grey Club-rash, Softstem Bulrush, Soft-stem Bulrush, Strong Bulrush, Tabernaemontanus Bulrush, Water Chive ■

352279　Schoenoplectus tabernaemontani（C. C. Gmel.）Palla 'Zebrinus'；斑马水葱；Striped Bulrush, Zebra Rush ■☆

352280　Schoenoplectus tenerimus Peter = Schoenoplectus microglumis Lye ■☆

352281　Schoenoplectus torreyi（Olney）Palla；托里拟莞；Torrey Three-square, Torrey's Bulrush, Torrey's Three-square Bulrush ■☆

352282　Schoenoplectus trapezoideus（Koidz.）Hagasaka et H. Ohashi；五棱水葱■

352283　Schoenoplectus triangulatus（Roxb.）Soják；三棱拟莞■☆

352284 Schoenoplectus triqueter (L.) Palla;三棱水葱(淼蔗,藁莩,蔗草,大甲草,光棍草,光棍子,蒲,三棱茎草,芀,茶,席草,蓆草,野荸荠,野三棱);Common Bulrush,Streambank Bulrush,Triangular Clubrush,Triangular Club-rush,Triangular Scirpus ■

352285 Schoenoplectus triqueter (L.) Palla = Scirpus triqueter L. ■

352286 Schoenoplectus validus (C. C. Gmel.) Á. Löve et D. Löve = Schoenoplectus tabernaemontani (C. C. Gmel.) Palla ■

352287 Schoenoplectus validus (Vahl) Á. Löve et D. Löve = Schoenoplectus tabernaemontani (C. C. Gmel.) Palla ■

352288 Schoenoplectus validus (Vahl) Á. Löve et D. Löve f. luxurians (Miq.) T. Koyama = Schoenoplectus tabernaemontani (C. C. Gmel.) Palla ■

352289 Schoenoplectus validus (Vahl) T. Koyama;莞(冲天草,翠管草,大水莞,水葱,水丈葱,莞蒲,沼生蔗草)■

352290 Schoenoplectus validus (Vahl) T. Koyama var. laeviglumis (Ts. Tang et F. T. Wang) C. Y. Wu;南水葱■

352291 Schoenoplectus validus Vahl = Schoenoplectus tabernaemontani (C. C. Gmel.) Palla ■

352292 Schoenoplectus wallichii (Nees) Ohwi = Schoenoplectus wallichii (Nees) T. Koyama ■

352293 Schoenoplectus wallichii (Nees) T. Koyama;猪毛草(台湾水莞);Wallich Bulrush ■

352294 Schoenoprasum Kunth = Allium L. ■

352295 Schoenopsis P. Beauv. = Tetraria P. Beauv. ■☆

352296 Schoenopsis burmanni (Vahl) Nees = Tetraria burmannii (Vahl) C. B. Clarke ■☆

352297 Schoenopsis flexuosa (Thunb.) Nees = Tetraria flexuosa (Thunb.) C. B. Clarke ■☆

352298 Schoenorchis Blume(1825);匙唇兰属(芦兰属,莞兰属);Schoenorchis ■

352299 Schoenorchis Reinw. = Schoenorchis Blume ■

352300 Schoenorchis eberhardtii (Finet) Aver. = Cleisostomopsis eberhardtii (Finet) Seidenf. et Smitinand ■

352301 Schoenorchis fragrans (Parish et Rchb. f.) Seidenf. = Schoenorchis tixieri (Guillaumin) Seidenf. ■

352302 Schoenorchis fragrans (Parish et Rchb. f.) Seidenf. et Smitinand;芳香匙唇兰;Fragrant Schoenorchis ■

352303 Schoenorchis gemmata (Lindl.) J. J. Sm.;匙唇兰(海南匙唇兰);Budding Schoenorchis,Common Schoenorchis ■

352304 Schoenorchis hainanensis (Rolfe) Schltr.;海南匙唇兰;Hainan Schoenorchis ■

352305 Schoenorchis hainanensis (Rolfe) Schltr. = Schoenorchis gemmata (Lindl.) J. J. Sm. ■

352306 Schoenorchis juncifolia Reinw.;灯芯草叶匙唇兰;Rushleaf Schoenorchis ■☆

352307 Schoenorchis micrantha Reinw. ex Blume;小花匙唇兰;Smallflower Schoenorchis ■☆

352308 Schoenorchis paniculata Blume;羞花兰■

352309 Schoenorchis paniculata Blume = Schoenorchis venoverberghii Ames ■

352310 Schoenorchis paniculata Blume var. vanoverberghii (Ames) S. S. Ying = Schoenorchis venoverberghii Ames ■

352311 Schoenorchis tixieri (Guillaumin) Seidenf.;圆叶匙唇兰;Roundleaf Schoenorchis ■

352312 Schoenorchis venoverberghii Ames;台湾匙唇兰(芦兰,密花芦兰,羞花兰);Taiwan Schoenorchis ■

352313 Schoenoxiphium Nees(1832);剑苇莎属■☆

352314 Schoenoxiphium basutorum Turrill;巴苏托剑苇莎■☆

352315 Schoenoxiphium bracteosum Kukkonen;多苞片剑苇莎■☆

352316 Schoenoxiphium buchananii C. B. Clarke = Schoenoxiphium rufum Nees ■☆

352317 Schoenoxiphium burkei C. B. Clarke = Schoenoxiphium rufum Nees var. dregeanum (Kunth) Kük. ■☆

352318 Schoenoxiphium burttii Kukkonen;伯特非洲莎草■☆

352319 Schoenoxiphium capense Nees = Schoenoxiphium lanceum (Thunb.) Kük. ■☆

352320 Schoenoxiphium caricoides C. B. Clarke = Schoenoxiphium sparteum (Wahlenb.) C. B. Clarke ■☆

352321 Schoenoxiphium clarkeanum Kük. = Kobresia clarkeana (Kük.) Kük. ■

352322 Schoenoxiphium clarkeanum Kük. = Kobresia fragilis C. B. Clarke ■

352323 Schoenoxiphium dregeanum Kunth = Schoenoxiphium rufum Nees var. dregeanum (Kunth) Kük. ■☆

352324 Schoenoxiphium ecklonii Nees;埃氏剑苇莎■☆

352325 Schoenoxiphium ecklonii Nees var. unisexuale Kük.;单性剑苇莎■☆

352326 Schoenoxiphium filiforme Kük.;丝状剑苇莎■☆

352327 Schoenoxiphium kuekenthaliana (Hand. -Mazz.) N. A. Ivanova = Kobresia kuekenthaliana Hand. -Mazz. ■

352328 Schoenoxiphium kunthianum Kük. = Schoenoxiphium sparteum (Wahlenb.) C. B. Clarke ■☆

352329 Schoenoxiphium lanceum (Thunb.) Kük.;披针状剑苇莎■☆

352330 Schoenoxiphium laxum (Nees) N. A. Ivanova = Kobresia laxa Nees ■

352331 Schoenoxiphium lehmannii (Nees) Steud.;莱曼剑苇莎■☆

352332 Schoenoxiphium ludwigii Hochst. = Schoenoxiphium rufum Nees ■☆

352333 Schoenoxiphium madagascariense Cherm.;马岛剑苇莎■☆

352334 Schoenoxiphium meyerianum Kunth = Schoenoxiphium lanceum (Thunb.) Kük. ■☆

352335 Schoenoxiphium molle Kukkonen;柔软剑苇莎■☆

352336 Schoenoxiphium rufum Nees;浅红剑苇莎■☆

352337 Schoenoxiphium rufum Nees var. dregeanum (Kunth) Kük.;德雷剑苇莎■☆

352338 Schoenoxiphium rufum Nees var. pondoense Kük.;庞多剑苇莎■☆

352339 Schoenoxiphium schimperianum (Boeck.) C. B. Clarke = Schoenoxiphium sparteum (Wahlenb.) C. B. Clarke ■☆

352340 Schoenoxiphium schweickerdtii Merxm. et Podlech;施韦剑苇莎■☆

352341 Schoenoxiphium sickmannianum Kunth = Schoenoxiphium lanceum (Thunb.) Kük. ■☆

352342 Schoenoxiphium sparteum (Wahlenb.) C. B. Clarke;鹰爪豆剑苇莎■☆

352343 Schoenoxiphium sparteum (Wahlenb.) C. B. Clarke var. lehmannii (Nees) Kük. = Schoenoxiphium lehmannii (Nees) Steud. ■☆

352344 Schoenoxiphium sparteum (Wahlenb.) C. B. Clarke var. schimperianum (Boeck.) Kük. = Schoenoxiphium sparteum (Wahlenb.) C. B. Clarke ■☆

352345 Schoenoxiphium strictum Kukkonen;刚直剑苇莎■☆

352346 Schoenoxiphium thunbergii Nees = Schoenoxiphium ecklonii Nees ■☆

352347 Schoenoxiphium thunbergii Nees f. elongata Boeck. = Schoenoxiphium ecklonii Nees ■☆

352348 Schoenus L. (1753);赤箭莎属(签草属,舒安属);Blackhead

Sedge，Bog Rush，Bogrush，Bog-rush ■

352349　Schoenus aculeatus L. = Crypsis aculeata（L.）Aiton ■

352350　Schoenus albus L. = Rhynchospora alba（L.）Vahl ■

352351　Schoenus apogon Roem. et Schult. ;矮赤箭莎;Smooth Bogrush ■

352352　Schoenus arenarius Schrad. = Tetraria compar（L.）T. Lestib. ■☆

352353　Schoenus brevifolius R. Br. ;短叶赤箭莎■☆

352354　Schoenus bromoides Lam. = Tetraria bromoides（Lam.）Pfeiff. ■☆

352355　Schoenus bulbosus L. = Ficinia bulbosa（L.）Nees ■☆

352356　Schoenus burmannii Vahl = Tetraria burmannii（Vahl）C. B. Clarke ■☆

352357　Schoenus calostachyus（R. Br.）Poir. ;长穗赤箭莎;Longspike Bogrush ■

352358　Schoenus capensis L. = Ischyrolepis capensis（L.）H. P. Linder ■☆

352359　Schoenus capillaceus Thunb. = Tetraria capillacea（Thunb.）C. B. Clarke ■☆

352360　Schoenus capitellatus Michx. = Rhynchospora capitellata（Michx.）Vahl ■☆

352361　Schoenus capitellum Thunb. = Ficinia capitella（Thunb.）Nees ■☆

352362　Schoenus ciliaris Michx. = Rhynchospora ciliaris（Michx.）C. Mohr ■☆

352363　Schoenus circinalis Schrad. = Tetraria microstachys（Vahl）Pfeiff. ■☆

352364　Schoenus cladium Sw. = Cladium jamaicense Crantz ■

352365　Schoenus coloratus L. = Kyllinga brevifolia Rottb. ■

352366　Schoenus coloratus L. = Rhynchospora colorata（Hitchc.）H. Pfeiff. ■☆

352367　Schoenus coloratus Lour. = Kyllinga monocephala Rottb. ■

352368　Schoenus coloratus Lour. = Kyllinga nemoralis（J. R. Forst. et G. Forst.）Dandy ex Hutch. ■

352369　Schoenus compar L. = Tetraria compar（L.）T. Lestib. ■☆

352370　Schoenus compressus L. = Blysmus compressus（L.）Panz. ■

352371　Schoenus corniculatus Lam. = Rhynchospora corniculata（Lam.）A. Gray ■☆

352372　Schoenus cuspidatus Rottb. = Tetraria cuspidata（Rottb.）C. B. Clarke ■☆

352373　Schoenus cyperoides Sw. = Rhynchospora holoschoenoides（Rich.）Herter ■☆

352374　Schoenus dactyloides Vahl = Carpha glomerata（Thunb.）Nees ■☆

352375　Schoenus deustus P. J. Bergius = Ficinia deusta（P. J. Bergius）Levyns ■☆

352376　Schoenus dispar Spreng. = Ficinia secunda（Vahl）Kunth ■☆

352377　Schoenus distans Michx. = Rhynchospora fascicularis（Michx.）Vahl ■☆

352378　Schoenus distans Muhl. = Rhynchospora grayi Kunth ■☆

352379　Schoenus dregeanus（Boeck.）Kuntze = Epischoenus dregeanus（Boeck.）Levyns ■☆

352380　Schoenus erinaceus Ridl. = Sphaerocyperus erinaceus（Ridl.）Lye ■☆

352381　Schoenus erraticus Hook. f. = Abildgaardia erratica（Hook. f.）Lye ■☆

352382　Schoenus falcatus R. Br. ;赤箭莎;Common Bogrush ■

352383　Schoenus fasciatus Rottb. = Tetraria fasciata（Rottb.）C. B. Clarke ■☆

352384　Schoenus fascicularis Michx. = Rhynchospora fascicularis（Michx.）Vahl ■☆

352385　Schoenus ferrugineus L. ;锈色赤箭莎;Brown Bog-rush，Rusty Bog Rush，Rusty Bog-rush ■☆

352386　Schoenus filiformis Lam. = Ficinia filiformis（Lam.）Schrad. ■☆

352387　Schoenus flexuosus Thunb. = Tetraria flexuosa（Thunb.）C. B. Clarke ■☆

352388　Schoenus fuscus L. = Rhynchospora fusca（L.）W. T. Aiton ■☆

352389　Schoenus fuscus Muhl. = Rhynchospora grayi Kunth ■☆

352390　Schoenus glomeratus L. = Rhynchospora glomerata（L.）Vahl ■☆

352391　Schoenus glomeratus Thunb. = Carpha glomerata（Thunb.）Nees ■☆

352392　Schoenus hattorianus Nakai = Schoenus brevifolius R. Br. ■☆

352393　Schoenus holoschoenoides Rich. = Rhynchospora holoschoenoides（Rich.）Herter ■☆

352394　Schoenus inanis Thunb. = Pseudoschoenus inanis（Thunb.）Oteng-Yeb. ■☆

352395　Schoenus indicus Lam. = Ficinia indica（Lam.）Pfeiff. ■☆

352396　Schoenus involucratus Rottb. = Tetraria involucrata（Rottb.）C. B. Clarke ■☆

352397　Schoenus junceus Willd. = Schoenoplectus junceus（Willd.）J. Raynal ■☆

352398　Schoenus juncoides Vahl = Bulbostylis juncoides（Vahl）Kük. ex Osten ■☆

352399　Schoenus laevis Thunb. ;平滑赤箭莎■☆

352400　Schoenus lanceus Thunb. = Schoenoxiphium lanceum（Thunb.）Kük. ■☆

352401　Schoenus lateralis Vahl = Ficinia lateralis（Vahl）Kunth ■☆

352402　Schoenus longirostris Michx. = Rhynchospora corniculata（Lam.）A. Gray ■☆

352403　Schoenus macrocephalus Boeck. ;大头赤箭莎■☆

352404　Schoenus mariscoides Muhl. = Cladium mariscoides（Muhl.）Torr. ■☆

352405　Schoenus microstachys Vahl = Tetraria microstachys（Vahl）Pfeiff. ■☆

352406　Schoenus miliaceus Lam. = Rhynchospora miliacea（Lam.）A. Gray ■☆

352407　Schoenus nemorum Vahl = Hypolytrum nemorum（Vahl）Spreng. ■

352408　Schoenus nigricans L. ;黑赤箭莎;Black Bog Rush，Black Bog-rush，Black Sedge，Bog-rush ■☆

352409　Schoenus nudifructus C. Chen;无刚毛赤箭莎（昆明赤箭莎）;Bristleless Bogrush ■

352410　Schoenus pallens Schrad. = Ficinia pallens（Schrad.）Nees ■☆

352411　Schoenus pilosus Willd. = Abildgaardia pilosa（Willd.）Nees ■☆

352412　Schoenus punctorius Vahl = Neesenbeckia punctoria（Vahl）Levyns ■☆

352413　Schoenus quadrangularis Boeck. = Epischoenus quadrangularis（Boeck.）C. B. Clarke ■☆

352414　Schoenus radiatus L. f. = Ficinia radiata（L. f.）Kunth ■☆

352415　Schoenus rariflorus Michx. = Rhynchospora rariflora（Michx.）Elliott ■☆

352416　Schoenus ruber Lour. = Rhynchospora rubra（Lour.）Makino ■

352417　Schoenus rufus Huds. = Blysmus rufus（Huds.）Link ■

352418　Schoenus rugosus Vahl = Rhynchospora brownii Roem. et Schult. ■☆

352419　Schoenus rugosus Vahl = Rhynchospora rugosa（Vahl）Gale subsp. brownii（Roem. et Schult.）T. Koyama ■

352420　Schoenus secundus Vahl = Ficinia secunda（Vahl）Kunth ■☆

352421　Schoenus sinensis Hand. -Mazz. = Schoenus falcatus R. Br. ■

352422　Schoenus sparsus Michx. = Rhynchospora miliacea（Lam.）A.

Gray ■☆

352423　Schoenus spicatus Burm. f. = Tetraria burmannii（Vahl）C. B. Clarke ■☆

352424　Schoenus striatus Thunb. = Ficinia indica（Lam.）Pfeiff. ■☆

352425　Schoenus thermalis L. = Tetraria thermalis（L.）C. B. Clarke ■☆

352426　Schoenus thermalis Willd. = Tetraria bromoides（Lam.）Pfeiff. ■☆

352427　Schoenus triceps Vahl = Rhynchospora tracyi Britton ■☆

352428　Schoenus ustulatus L. = Tetraria ustulata（L.）C. B. Clarke ■☆

352429　Schoenus viscosus Schrad. = Tetraria compar（L.）T. Lestib. ■☆

352430　Schoepfia Schreb.（1789）；青皮木属（香芙木属）；Greentwig, Grey Twig, Greytwig, Grey-twig, Schoepfia ●

352431　Schoepfia acuminata Wall. ex DC. = Schoepfia fragrans Wall. ●

352432　Schoepfia chinensis Gardner et Champ.；华南青皮木（管花青皮木，红旦木，华青皮木，青皮木，碎骨仔树，退骨王，香芙木）；China Greentwig, Chinese Greytwig, Chinese Grey-twig ●

352433　Schoepfia chrysophylloides Planch. = Diplocalyx chrysophylloides A. Rich. ●☆

352434　Schoepfia emeiensis Z. Y. Zhu；峨眉香芙木●

352435　Schoepfia fragrans Wall.；香芙木；Fragrant Greentwig, Fragrant Greytwig, Fragrant Grey-twig, Fragrant Schoepfia ●

352436　Schoepfia griffithiana Valeton = Schoepfia fragrans Wall. ●

352437　Schoepfia griffithii Tiegh. ex Steenis；小果青皮木●

352438　Schoepfia jasminodora Siebold et Zucc.；青皮木（脆骨风，幌幌木，素馨地锦树，羊脆骨）；Blue-bark Schoepfia, Common Greentwig, Common Grey Twig, Common Greytwig, Common Grey-twig, Jasminodora Greytwig, Jasminodora Grey-twig ●

352439　Schoepfia jasminodora Siebold et Zucc. = Schoepfia chinensis Gardner et Champ. ●

352440　Schoepfia jasminodora Siebold et Zucc. var. malipoensis Y. R. Ling；麻栗坡青皮木（大果青皮木）；Malipo Common Greentwig, Malipo Common Greytwig, Malipo Grey-twig ●

352441　Schoepfia mierisii Pierre = Schoepfia fragrans Wall. ●

352442　Schoepfia schreberi J. F. Gmel.；秘鲁青皮木；Peru Greentwig ●☆

352443　Schoepfiaceae Blume = Olacaceae R. Br.（保留科名）●

352444　Schoepfiaceae Blume（1850）；青皮木科（香芙木科）●

352445　Schoepfianthus Engl. ex De Wild. = Ongokea Pierre ●☆

352446　Schoepfianthus zenkeri Engl. ex De Wild. = Ongokea gore（Hua）Pierre ●☆

352447　Schoepfiopsis Miers = Schoepfia Schreb. ●

352448　Schoepfiopsis acuminata（DC.）Miers = Schoepfia fragrans Wall. ●

352449　Schoepfiopsis acuminata（Wall. ex DC.）Miers = Schoepfia fragrans Wall. ●

352450　Schoepfiopsis chineisis（Gardner et Champ.）Miers = Schoepfia chinensis Gardner et Champ. ●

352451　Schoepfiopsis chinensis（Gardner et Champ.）Miers = Schoepfia chinensis Gardner et Champ. ●

352452　Schoepfiopsis fragrans（Wall. ex DC.）Miers = Schoepfia fragrans Wall. ●

352453　Schoepfiopsis fragrans（Wall.）Miers = Schoepfia fragrans Wall. ●

352454　Schoepfiopsis jasminodora（Siebold et Zucc.）Miers = Schoepfia jasminodora Siebold et Zucc. ●

352455　Schoepfiopsis miersii Pierre = Schoepfia fragrans Wall. ●

352456　Scholera Hook. f. = Hoya R. Br. ●

352457　Scholera Hook. f. = Schollia J. Jacq. ●

352458　Schollera Rohr = Microtea Sw. ■☆

352459　Schollera Roth = Oxycoccus Hill ●

352460　Schollera Schreb.（1791）；斯霍勒花属■☆

352461　Schollera Schreb. = Heteranthera Ruiz et Pav.（保留属名）■☆

352462　Schollera graminea（Michx.）Raf. = Heteranthera dubia（Jacq.）MacMill. ■☆

352463　Schollera graminea Raf.；斯霍勒花■☆

352464　Scholleropsis H. Perrier（1936）；四裂雨久花属■☆

352465　Scholleropsis lutea H. Perrier；四裂雨久花■☆

352466　Schollia J. Jacq. = Hoya R. Br. ●

352467　Scholtzia Schauer（1843）；肖尔桃金娘属（澳洲桃金娘属）●☆

352468　Scholtzia capitata F. Muell. ex Benth.；头状肖尔桃金娘●☆

352469　Scholtzia laxiflora Benth.；疏花肖尔桃金娘●☆

352470　Scholtzia leptantha Benth.；细花肖尔桃金娘●☆

352471　Scholtzia obovata（DC.）Schauer；倒卵肖尔桃金娘●☆

352472　Scholtzia parviflora F. Muell.；小花肖尔桃金娘●☆

352473　Schomburghia DC. = Geissopappus Benth. ●■☆

352474　Schomburgkia Benth. et Hook. f. = Calea L. ●■☆

352475　Schomburgkia Benth. et Hook. f. = Geissopappus Benth. ●■☆

352476　Schomburgkia Benth. et Hook. f. = Schomburghia DC. ●■☆

352477　Schomburgkia Lindl.（1838）；熊保兰属■☆

352478　Schomburgkia albopurpurea（I. Strachan ex Fawc.）Withner；紫白熊保兰■☆

352479　Schomburgkia rosea Linden ex Lindl.；粉红熊保兰■☆

352480　Schomburgkia undulata Lindl.；波缘熊保兰■☆

352481　Schonlandia L. Bolus = Corpuscularia Schwantes ●☆

352482　Schonlandia lehmannii（Eckl. et Zeyh.）L. Bolus = Corpuscularia lehmannii（Eckl. et Zeyh.）Schwantes ●☆

352483　Schorigeram Adans. = Tragia L. ●☆

352484　Schortia E. Vilm. = Actinolepis DC. ●■☆

352485　Schotia Jacq.（1787）（保留属名）；豆木属；Boer Bean, Boer Beans, Kaffir Bean Tree ●☆

352486　Schotia afra（L.）Thunb.；非洲豆木●☆

352487　Schotia afra（L.）Thunb. var. angustifolia（E. Mey.）Harv.；窄叶非洲豆木●☆

352488　Schotia africana（Baill.）Keay = Leonardoxa africana（Baill.）Aubrév. ●☆

352489　Schotia angustifolia E. Mey. = Schotia afra（L.）Thunb. var. angustifolia（E. Mey.）Harv. ●☆

352490　Schotia bequaertii（De Wild.）De Wild. = Normandiodendron bequaertii（De Wild.）J. Léonard ■☆

352491　Schotia bequaertii（De Wild.）De Wild. var. rubriflora（De Wild.）J. Léonard = Normandiodendron bequaertii（De Wild.）J. Léonard ■☆

352492　Schotia bergeri De Wild. = Normandiodendron bequaertii（De Wild.）J. Léonard ■☆

352493　Schotia brachypetala Sond.；短瓣豆木；African Walnut, Fuchsia Tree, Hottentot Bean, Tree Fuchsia, Weeping Boer-bean ●☆

352494　Schotia brachypetala Sond. var. pubescens Burtt Davy = Schotia brachypetala Sond. ●☆

352495　Schotia capitata Bolle；头状豆木●☆

352496　Schotia claessensii（De Wild.）J. -P. Lebrun = Normandiodendron bequaertii（De Wild.）J. Léonard ■☆

352497　Schotia cuneifolia Gand. = Schotia latifolia Jacq. ●☆

352498　Schotia diversifolia Walp. = Schotia latifolia Jacq. ●☆

352499　Schotia humboldtioides Oliv. = Leonardoxa africana（Baill.）Aubrév. ■●☆

352500　Schotia latifolia Jacq.；阔叶豆木；Bean Tree, Elephant Hedge ●☆

352501 Schotia parvifolia Jacq. = Schotia afra（L.）Thunb. ●☆

352502 Schotia rogersii Burtt Davy = Schotia brachypetala Sond. ●☆

352503 Schotia romii De Wild. = Normandiodendron romii（De Wild.）J. Léonard ■☆

352504 Schotia rubriflora（De Wild.）De Wild. = Normandiodendron bequaertii（De Wild.）J. Léonard ■☆

352505 Schotia semireducta Merxm. = Schotia brachypetala Sond. ●☆

352506 Schotia simplicifolia Vahl ex DC. = Griffonia simplicifolia（Vahl ex DC.）Baill. ■☆

352507 Schotia speciosa Jacq. = Schotia afra（L.）Thunb. ●☆

352508 Schotia speciosa Jacq. var. tamarindifolia（Afzel. ex Sims）Harv. = Schotia afra（L.）Thunb. ●☆

352509 Schotia tamarindifolia Afzel. ex Sims = Schotia afra（L.）Thunb. ●☆

352510 Schotia tamarindifolia Afzel. ex Sims var. forbesiana Baill. = Schotia capitata Bolle ●☆

352511 Schotia transvaalensis Rolfe = Schotia capitata Bolle ●☆

352512 Schotiaria（DC.）Kuntze = Griffonia Baill. ■☆

352513 Schottariella P. C. Boyce et S. Y. Wong（2009）；小舍特尔南星属■☆

352514 Schottarum P. C. Boyce et S. Y. Wong = Hottarum Bogner et Nicolson ■☆

352515 Schottarum P. C. Boyce et S. Y. Wong（2008）；舍特尔南星属■☆

352516 Schousbea Raf. = Cacoucia Aubl. ●

352517 Schousbea Raf. = Combretum Loefl.（保留属名）●

352518 Schousbea Raf. = Schousboea Willd. ●

352519 Schousboea Nicotra = Stipa L. ■

352520 Schousboea Schumach. = Alchornea Sw. ●

352521 Schousboea Schumach. et Thonn. = Alchornea Sw. ●

352522 Schousboea Willd. = Cacoucia Aubl. ●

352523 Schousboea Willd. = Combretum Loefl.（保留属名）●

352524 Schousboea cordifolia Schumach. et Thonn. = Alchornea cordifolia（Schumach. et Thonn.）Müll. Arg. ●☆

352525 Schoutenia Korth.（1848）；蒚椴属●☆

352526 Schoutenia excelsa Pierre = Marquesia excelsa（Pierre）R. E. Fr. ●☆

352527 Schoutenia ovata Korth.；卵蒚椴●☆

352528 Schoutensia Endl. = Pittosporum Banks ex Gaertn.（保留属名）●

352529 Schouwia DC.（1821）（保留属名）；沙蝗芥属■☆

352530 Schouwia Schrad. = Goethea Nees ●☆

352531 Schouwia arabica DC. = Schouwia purpurea（Forssk.）Schweinf. ■☆

352532 Schouwia arabica DC. var. schimperi（Jaub. et Spach）Coss. = Schouwia purpurea（Forssk.）Schweinf. ■☆

352533 Schouwia purpurea（Forssk.）Schweinf.；紫沙蝗芥■☆

352534 Schouwia purpurea（Forssk.）Schweinf. subsp. schimperi（Jaub. et Spach）Maire = Schouwia purpurea（Forssk.）Schweinf. ■☆

352535 Schouwia schimperi Jaub. et Spach = Schouwia purpurea（Forssk.）Schweinf. ■☆

352536 Schouwia thebaica Webb；沙蝗芥■☆

352537 Schowia Sweet = Goethea Nees ●☆

352538 Schradera Vahl = Salvia L. ●■

352539 Schradera Vahl（1796）（保留属名）；施拉茜属■☆

352540 Schradera Willd. = Croton L. ●

352541 Schradera acuminata Standl.；渐尖施拉茜■☆

352542 Schradera brevipes Steyerm.；短梗施拉茜■☆

352543 Schradera capitata Vahl；头状施拉茜■☆

352544 Schradera cubensis Steyerm.；古巴施拉茜■☆

352545 Schradera glabriflora Steyerm.；光花施拉茜■☆

352546 Schradera longifolia Spruce ex Hook. f.；长叶施拉茜■☆

352547 Schradera polycephala DC.；多头施拉茜■☆

352548 Schradera polysperma（Jack）Puff, R. Buchner et Greimler；多籽施拉茜■☆

352549 Schradera rotundata Standl. ex Steyerm.；圆施拉茜■☆

352550 Schraderanthus Averett = Jaltomata Schltdl. ●☆

352551 Schraderanthus Averett = Saracha Ruiz et Pav. ●☆

352552 Schraderanthus Averett（2009）；施拉花属●☆

352553 Schraderia Fabr. ex Medik. = Arischrada Pobed.（废弃属名）●■

352554 Schraderia Fabr. ex Medik. = Salvia L. ●■

352555 Schraderia Heist. ex Medik.（废弃属名）= Schradera Vahl（保留属名）■☆

352556 Schraderia Medik.（1791）；肖鼠尾草属■☆

352557 Schraderia Medik. = Salvia L. ●■

352558 Schraderia acetabulosa（Vahl）Pobed.；碟状肖鼠尾草■☆

352559 Schraderia bucharica（Popov）Nevski；布赫肖鼠尾草■☆

352560 Schraderia dracocephaloides（Boiss.）Pobed.；龙头肖鼠尾草■☆

352561 Schraderia korotkovii（Regel et Schmalh.）Pobed.；科罗肖鼠尾草■☆

352562 Schrameckia Danguy = Tambourissa Sonn. ●☆

352563 Schrammia Britwn et Rose = Hoffmannseggia Cav.（保留属名）■☆

352564 Schranckia J. F. Gmel.（废弃属名）= Schrankia Willd.（保留属名）●☆

352565 Schranckia Scop. ex J. F. Gmel. = Goupia Aubl. ●☆

352566 Schranckia Scop. ex J. F. Gmel. = Schrankia Willd.（保留属名）●☆

352567 Schranckiastrum Hassl. = Mimosa L. ●■

352568 Schrankia Medik. = Neslia Desv.（保留属名）■

352569 Schrankia Medik. = Rapistrum Haller f. ■

352570 Schrankia Willd.（1806）（保留属名）；施兰木属（施兰克亚木属）●☆

352571 Schrankia leptocarpa DC.；施兰木●☆

352572 Schrankia nuttallii（DC. ex Britton et Rose）Standl. = Mimosa quadrivalvis L. ■☆

352573 Schrankia nuttallii（DC.）Standl. = Mimosa nuttallii（DC.）B. L. Turner ●☆

352574 Schrankia uncinata Willd. = Mimosa quadrivalvis L. ■☆

352575 Schrankiastrum Willis = Mimosa L. ●■

352576 Schrankiastrum Willis = Schranckiastrum Hassl. ●■

352577 Schrebera L.（废弃属名）= Cuscuta L. ●■

352578 Schrebera L.（废弃属名）= Schrebera Roxb.（保留属名）●☆

352579 Schrebera L. ex Schreb. = Myrica L. + Cuscuta L. ■

352580 Schrebera Retz. = Elaeodendron J. Jacq. ●☆

352581 Schrebera Retz. = Loureira Meisn. ●

352582 Schrebera Retz. = Schrebera Roxb.（保留属名）●☆

352583 Schrebera Roxb.（1799）（保留属名）；施莱木犀属●☆

352584 Schrebera Schreb. = Myrica L. + Cuscuta L. ■

352585 Schrebera Schreb. = Schrebera Roxb.（保留属名）●☆

352586 Schrebera Thunb. = Cassinopsis Sond. ●☆

352587 Schrebera Thunb. = Hartogia Hochst. ●☆

352588 Schrebera Thunb. = Hartogiella Codd ●☆

352589 Schrebera Thunb. = Schrebera Roxb.（保留属名）●☆

352590 Schrebera affinis Lingelsh. = Schrebera trichoclada Welw. ●☆

352591 Schrebera alata（Hochst.）Welw.；翼施莱木犀●☆

352592 Schrebera alata（Hochst.）Welw. var. tomentella Welw. =

Schrebera alata (Hochst.) Welw. ●☆

352593　Schrebera arborea A. Chev.；树状施莱木犀●☆

352594　Schrebera argyrotricha Gilg ＝Schrebera alata (Hochst.) Welw. ●☆

352595　Schrebera buchananii Baker ＝Schrebera trichoclada Welw. ●☆

352596　Schrebera chevalieri Hutch. et Dalziel ＝Schrebera arborea A. Chev. ●☆

352597　Schrebera excelsa Lingelsh.；高大木犀●☆

352598　Schrebera gilgiana Lingelsh. ＝Schrebera alata (Hochst.) Welw. ●☆

352599　Schrebera goetzeana Gilg ＝Schrebera alata (Hochst.) Welw. ●☆

352600　Schrebera greenwayi Turrill ＝Schrebera alata (Hochst.) Welw. ●☆

352601　Schrebera holstii (Engl. et Gilg) Gilg ＝Schrebera alata (Hochst.) Welw. ●☆

352602　Schrebera koiloneura Gilg ＝Schrebera trichoclada Welw. ●☆

352603　Schrebera koiloneura Gilg var. delevoyi De Wild. ＝Schrebera trichoclada Welw. ●☆

352604　Schrebera koiloneura Gilg var. kakomensis Lingelsh. ＝Schrebera trichoclada Welw. ●☆

352605　Schrebera latialata Gilg ＝Schrebera alata (Hochst.) Welw. ●☆

352606　Schrebera macrantha Gilg et G. Schellenb. ＝Schrebera arborea A. Chev. ●☆

352607　Schrebera macrocarpa Gilg et G. Schellenb. ＝Schrebera arborea A. Chev. ●☆

352608　Schrebera mazoensis S. Moore ＝Schrebera alata (Hochst.) Welw. ●☆

352609　Schrebera merkeri Lingelsh. ＝Schrebera alata (Hochst.) Welw. ●☆

352610　Schrebera nyassae Lingelsh. ＝Schrebera alata (Hochst.) Welw. ●☆

352611　Schrebera obliquifoliolata Gilg ＝Schrebera alata (Hochst.) Welw. ●☆

352612　Schrebera oligantha Gilg ＝Schrebera trichoclada Welw. ●☆

352613　Schrebera platyphylla Gilg；宽叶施莱木犀●☆

352614　Schrebera saundersiae Harv. ＝Schrebera alata (Hochst.) Welw. ●☆

352615　Schrebera schellenbergii Lingelsh. ＝Schrebera trichoclada Welw. ●☆

352616　Schrebera schinoides Thunb. ＝Cassine schinoides (Spreng.) R. H. Archer ●☆

352617　Schrebera tomentella (Welw.) Gilg ＝Schrebera alata (Hochst.) Welw. ●☆

352618　Schrebera trichoclada Welw.；毛枝施莱木犀●☆

352619　Schrebera welwitschii Gilg ＝Schrebera alata (Hochst.) Welw. ●☆

352620　Schreberaceae (R. Wight) Schnizl. ＝Olacaceae R. Br. (保留科名)●

352621　Schreberaceae Schnizl. ＝Olacaceae R. Br. (保留科名)●

352622　Schreiberia Steud. ＝Augusta Pohl(保留属名)■☆

352623　Schreiberia Steud. ＝Schreibersia Pohl ■☆

352624　Schreibersia Pohl ＝Augusta Pohl(保留属名)■☆

352625　Schreibersia Pohl ex Endl. ＝Augusta Pohl(保留属名)■☆

352626　Schreiteria Carolin(1985)；长荫苋属■☆

352627　Schreiteria macrocarpa (Speg.) Carolin.；长荫苋■☆

352628　Schrenkia Benth. et Hook. f. ＝Schrenkia Fisch. et C. A. Mey. ■

352629　Schrenkia Fisch. et C. A. Mey. (1841)；双球芹属；Sensitivebrier ■

352630　Schrenkia Regel et Schmalh. ＝Schrenkia Fisch. et C. A. Mey. ■

352631　Schrenkia golickeana (Regel et Schmalh.) B. Fedtsch.；高里双球芹■☆

352632　Schrenkia insignis Lipsky；显著双球芹■☆

352633　Schrenkia involucrata Regel et Schmalh.；总苞双球芹；Involucrate Sensitivebrier ■☆

352634　Schrenkia kultiassovii Korovin；库尔双球芹■☆

352635　Schrenkia papillaris Regel et Schmalh.；乳突双球芹■☆

352636　Schrenkia pungens Regel et Schmalh.；锐利双球芹；Pungent Sensitivebrier ■☆

352637　Schrenkia vaginata (Ledeb.) Fisch. et C. A. Mey.；双球芹；Common Sensitivebrier ■

352638　Schroeterella Briq. ＝Larrea Cav. (保留属名)●☆

352639　Schroeterella Briq. ＝Neoschroetera Briq. ●☆

352640　Schrophularia Medik. ＝Scrophularia L. ■ ●

352641　Schstocaryum cilliare Bureau et Franch. ＝Microula ciliaris (Bureau et Franch.) I. M. Johnst. ■

352642　Schstschurowskia Willis ＝Schtschurowskia Regel et Schmalh. ■☆

352643　Schtschurowskia Regel et Schmalh. (1882)；希茨草属■☆

352644　Schtsschurowskia meifolia Regel et Schmalh.；希茨草■☆

352645　Schtsschurovskia pentaceros (Korovin) Schischk.；五角希茨草■☆

352646　Schubea Pax ＝Cola Schott et Endl. (保留属名)＋Trichoscypha Hook. f. ●☆

352647　Schubertia St. -Lag. ＝Schubertia Blume ex DC. ●☆

352648　Schubertia Blume ＝Horsfieldia Blume ●☆

352649　Schubertia Blume ex DC. ＝Harmsiopanax Warb. ●☆

352650　Schubertia Mart. (1824)(保留属名)；舒巴特萝藦属●■☆

352651　Schubertia Mart. et Zucc. ＝Schubertia Mart. (保留属名)●■☆

352652　Schubertia Mirb. (废弃属名)＝Schubertia Mart. (保留属名)●■☆

352653　Schubertia Mirb. (废弃属名)＝Taxodium Rich. ●

352654　Schubertia grandiflora Mart. et Zucc.；大花舒巴特萝藦●☆

352655　Schubertia longiflora (Jacq.) Mart.；长花舒巴特萝藦■☆

352656　Schubertia multiflora Mart.；多花舒巴特萝藦■☆

352657　Schubertia schreberi S. G. Gmel. ＝Schoepfia schreberi J. F. Gmel. ●☆

352658　Schubertia sempervirens (Lamb.) Spach ＝Sequoia sempervirens (D. Don ex Lamb.) Endl. ●

352659　Schudia Molina ex Gay ＝Osmorhiza Raf. (保留属名)■

352660　Schuebleria Mart. ＝Curtia Cham. et Schltdl. ■☆

352661　Schuechia Endl. ＝Qualea Aubl. ●☆

352662　Schuenkia Raf. ＝Schwenckia L. ■●☆

352663　Schuermannia F. Muell. ＝Darwinia Rudge ●☆

352664　Schufia Spach ＝Fuchsia L. ●■

352665　Schuitemania Ormerod ＝Herpysma Lindl. ■

352666　Schuitemania Ormerod(2002)；菲律宾爬兰属■☆

352667　Schultesia Mart. (1827)(保留属名)；舒尔龙胆属■☆

352668　Schultesia Roth ＝Wahlenbergia Schrad. ex Roth(保留属名)■●

352669　Schultesia Schrad. ＝Gomphrena L. ●■

352670　Schultesia Spreng. (废弃属名)＝Chloris Sw. ●■

352671　Schultesia Spreng. (废弃属名)＝Eustachys Desv. ■

352672　Schultesia Spreng. (废弃属名)＝Schultesia Mart. (保留属名)■☆

352673　Schultesia senegalensis Baker ＝Schultesia stenophylla Mart. var. latifolia Mart. ex Progel ■☆

352674　Schultesia stenophylla Mart.；舒尔龙胆■☆

352675　Schultesia stenophylla Mart. var. latifolia Mart. ex Progel；宽叶舒尔龙胆■☆

352676　Schultesianthus Hunz. (1977)；舒尔花属●☆

352677　Schultesianthus leucanthus (Donn. Sm.) Hunz.；白舒尔花●☆

352678 Schultesianthus uniflorus（Lundell）S. Knapp;单花舒尔花●☆

352679 Schultesiophytum Harling(1958);舒尔草属■☆

352680 Schultesiophytum chorianthum Harling;舒尔草■☆

352681 Schultzia Nees ＝ Herpetacanthus Nees ■☆

352682 Schultzia Raf. ＝ Obolaria L. ■☆

352683 Schultzia Spreng. ＝ Schulzia Spreng.（保留属名）■

352684 Schultzia Wall. ＝ Cortia DC. ■

352685 Schultzia albiflora（Kar. et Kir.）Popov ＝ Schulzia albiflora（Kar. et Kir.）Popov ■

352686 Schultzia crinita（Pall.）Spreng. ＝ Schulzia crinita（Pall.）Spreng. ■

352687 Schultzia lindlei Wall. ＝ Cortia depressa（D. Don）C. Norman ■

352688 Schulzia Spreng.（1813）（保留属名）;裂苞芹属;Schultzia ■

352689 Schulzia albiflora（Kar. et Kir.）Popov;白花裂苞芹;Whiteflower Schultzia ■

352690 Schulzia crinita（Pall.）Spreng. ;长毛裂苞芹;Longhair Schultzia ■

352691 Schulzia dissecta（C. B. Clarke）C. Norman;裂苞芹■

352692 Schulzia hookeri（C. B. Clarke）M. Hiroe ＝ Cortiella hookeri（C. B. Clarke）C. Norman ■

352693 Schulzia nepalensis（C. Norman）M. Hiroe ＝ Cortia depressa（D. Don）C. Norman ■

352694 Schulzia prostrata Pimenov et Kljuykov;天山苞裂芹■

352695 Schumacheria Spreng. ＝ Tricliceras Thonn. ex DC. ■☆

352696 Schumacheria Spreng. ＝ Wormskioldia Schumach. et Thonn. ■☆

352697 Schumacheria Vahl(1810);舒马草属■☆

352698 Schumacheria alnifolia Hook. f. et Thomson;桤叶舒马草■☆

352699 Schumacheria angustifolia Hook. f. et Thomson;窄叶舒马草■☆

352700 Schumacheria castaneifolia Vahl;舒马草■☆

352701 Schumannia Kuntze ＝ Ferula L. ■

352702 Schumannia Kuntze(1887);球根阿魏属;Schumannia ■

352703 Schumannia karelinii（Bunge）Korovin;球根阿魏;Common Schumannia ■

352704 Schumannia turcomanica Kuntze ＝ Schumannia karelinii（Bunge）Korovin ■

352705 Schumannianthus Gagnep.（1904）;双岐柊叶属■☆

352706 Schumannianthus virgatus Rolfe;双岐柊叶■☆

352707 Schumanniophyton Harms ＝ Tetrastigma（Miq.）Planch. ●■

352708 Schumanniophyton Harms et R. D. Good ＝ Schumanniophyton Harms ●☆

352709 Schumanniophyton Harms(1897);舒曼木属●☆

352710 Schumanniophyton arboreum A. Chev.;树状舒曼木●☆

352711 Schumanniophyton hirsutum（Hiern）R. D. Good;粗毛舒曼木●☆

352712 Schumanniophyton klaineanum Pierre ex A. Chev. ＝ Schumanniophyton magnificum（K. Schum.）Harms var. klaineanum（Pierre ex A. Chev.）N. Hallé ●☆

352713 Schumanniophyton magnificum（K. Schum.）Harms;大舒曼木●☆

352714 Schumanniophyton magnificum（K. Schum.）Harms var. klaineanum（Pierre ex A. Chev.）N. Hallé;克莱恩舒曼木●☆

352715 Schumanniophyton magnificum（K. Schum.）Harms var. trimerum（R. D. Good）N. Hallé;三数舒曼木●☆

352716 Schumanniophyton magnificum（K. Schum.）Harms. f. letestuanum N. Hallé;莱泰斯图舒曼木●☆

352717 Schumanniophyton magnificum（K. Schum.）Harms. f. umbraticola（G. Taylor）N. Hallé;荫蔽大舒曼木●☆

352718 Schumanniophyton problematicum（A. Chev.）Aubrév. ;普罗舒曼木●☆

352719 Schumanniophyton trimerum R. D. Good ＝ Schumanniophyton magnificum（K. Schum.）Harms var. trimerum（R. D. Good）N. Hallé ●☆

352720 Schumanniophyton umbraticola G. Taylor ＝ Schumanniophyton magnificum（K. Schum.）Harms. f. umbraticola（G. Taylor）N. Hallé ●☆

352721 Schumeria Iljin ＝ Serratula L. ■

352722 Schumeria Iljin(1960);舒默菊属■☆

352723 Schumeria latifolia（Boiss.）Iljin;宽叶舒默菊■☆

352724 Schumeria litwinowii（Iljin）Iljin;舒默菊■☆

352725 Schunda-Pana Adans. ＝ Caryota L. ●

352726 Schunkea Senghas(1994);申克兰属■☆

352727 Schuurmansia Blume(1850);斯胡木属●☆

352728 Schuurmansia angustifolia Hook. f.;窄叶斯胡木●☆

352729 Schuurmansia crassinervia Gilg;粗脉斯胡木●☆

352730 Schuurmansia elegans Blume;雅致斯胡木●☆

352731 Schuurmansia grandiflora A. C. Sm. ;大花斯胡木●☆

352732 Schuurmansia longifolia Gilg;长叶斯胡木●☆

352733 Schuurmansia microcarpa Capit. ;小果斯胡木●☆

352734 Schuurmansia parviflora Ridl. ;小花斯胡木●☆

352735 Schuurmansiella Hallier f. (1913);小斯胡木属●☆

352736 Schuurmansiella angustifolia（Hook. f.）Hallier. ;窄叶小斯胡木●☆

352737 Schwabea Endl. ＝ Monechma Hochst. ■●☆

352738 Schwabea anisacanthus（Schweinf.）Lindau ＝ Megalochlamys violacea（Vahl）Vollesen ●☆

352739 Schwabea ciliaris Nees ＝ Monechma ciliatum（Jacq.）Milne-Redh. ●☆

352740 Schwabea ecbolioides Lindau ＝ Megalochlamys revoluta（Lindau）Vollesen ●☆

352741 Schwabea ecbolioides Lindau var. tomentosa ？ ＝ Megalochlamys trinervia（C. B. Clarke）Vollesen ●☆

352742 Schwabea revoluta Lindau ＝ Megalochlamys revoluta（Lindau）Vollesen ●☆

352743 Schwabea spicigera Nees ＝ Monechma ciliatum（Jacq.）Milne-Redh. ●☆

352744 Schwackaea Cogn.（1891）;施瓦野牡丹属●☆

352745 Schwackaea cupheoides（Benth.）Cogn. ;施瓦野牡丹●☆

352746 Schwaegerichenia Steud. ＝ Anigozanthos Labill. ■☆

352747 Schwaegrichenia Rchb. ＝ Hedwigia Sw. ■☆

352748 Schwaegrichenia Rchb. ＝ Tetragastris Gaertn. ■☆

352749 Schwaegrichenia Spreng. ＝ Anigozanthos Labill. ■☆

352750 Schwalbea L.（1753）;施瓦尔列当属■☆

352751 Schwalbea americana L. ;施瓦尔列当;Chaffseed ■☆

352752 Schwannia Endl. ＝ Janusia A. Juss. ex Endl. ●☆

352753 Schwantesia Dinter(1927);施旺花属■☆

352754 Schwantesia L. Bolus ＝ Mitrophyllum Schwantes ●☆

352755 Schwantesia L. Bolus ＝ Monilaria Schw. ＋ Conophytum N. E. Br. ■☆

352756 Schwantesia acutipetala L. Bolus;尖瓣施旺花■☆

352757 Schwantesia australis L. Bolus;南非施旺花■☆

352758 Schwantesia herrei L. Bolus;融香玉■☆

352759 Schwantesia herrei L. Bolus f. major G. D. Rowley ＝ Schwantesia herrei L. Bolus ■☆

352760 Schwantesia herrei L. Bolus var. minor ＝ Schwantesia herrei L. Bolus ■☆

352761 Schwantesia loeschiana Tischer;凝香玉■☆

352762　Schwantesia marlothii L. Bolus；马洛斯施旺花■☆

352763　Schwantesia pillansii L. Bolus；皮朗斯施旺花■☆

352764　Schwantesia ramulosa L. Bolus ＝ Dicrocaulon ramulosum（L. Bolus）Ihlenf. ■☆

352765　Schwantesia ruedebuschii Dinter；灰绿叶施旺花（龙玉）■☆

352766　Schwantesia ruedebuschii Dinter ＝ Schwantesia australis L. Bolus ■☆

352767　Schwantesia speciosa L. Bolus；美丽施旺花■☆

352768　Schwantesia succumbens（Dinter）Dinter；俯卧施旺花■☆

352769　Schwantesia triebneri L. Bolus；鸟舟■☆

352770　Schwantesia watermeyeri L. Bolus ＝ Dicrocaulon microstigma（L. Bolus）Ihlenf. ■☆

352771　Schwartzia Vell.（1829）；施瓦茨藤属●☆

352772　Schwartzia Vell. ＝ Norantea Aubl. ●☆

352773　Schwartzia glabra Vell.；施瓦茨藤●☆

352774　Schwartzkopffia Kraenzl.（1900）；施瓦兰属■☆

352775　Schwartzkopffia Kraenzl. ＝ Brachycorythis Lindl. ■

352776　Schwartzkopffia angolensis Schltr. ＝ Schwartzkopffia lastii（Rolfe）Schltr. ■☆

352777　Schwartzkopffia buettneriana Kraenzl. ＝ Schwartzkopffia pumilio（Lindl.）Schltr. ■☆

352778　Schwartzkopffia lastii（Rolfe）Schltr.；安哥拉施瓦兰■☆

352779　Schwartzkopffia pumilio（Lindl.）Schltr.；矮施瓦兰■☆

352780　Schwarzia Vell. ＝ Norantea Aubl. ●☆

352781　Schwarzia Vell. ＝ Schwartzia Vell. ●☆

352782　Schweiggera E. Mey. ex Baker ＝ Gladiolus L. ■

352783　Schweiggera Mart. ＝ Renggeria Meisn. ●☆

352784　Schweiggera nemorosa E. Mey. ＝ Tritoniopsis nemorosa（Klatt）G. J. Lewis ■☆

352785　Schweiggeria Spreng.（1820）；异萼堇属■☆

352786　Schweiggeria floribunda A. St. -Hil.；多花异萼堇■☆

352787　Schweiggeria fruticosa Spreng.；异萼堇■☆

352788　Schweiggeria mexicana Schltdl.；墨西哥异萼堇■☆

352789　Schweiggeria pauciflora Lindl.；少花异萼堇■☆

352790　Schweinfurthafra Kuntze ＝ Glyphaea Hook. f. ex Planch. ●☆

352791　Schweinfurthia A. Braun（1996）；施氏婆婆纳属●☆

352792　Schweinfurthia papilionacea（Burm. f.）Boiss.；蝶花施氏婆婆纳■☆

352793　Schweinfurthia pedicellata（T. Anderson）Balf. f.；具梗施氏婆婆纳■☆

352794　Schweinfurthia pterosperma（A. Rich.）A. Braun；翅果施氏婆婆纳■☆

352795　Schweinitzia Elliott ＝ Monotropsis Schwein. ex Elliott ●☆

352796　Schweinitzia Elliott ex Nutt. ＝ Monotropsis Schwein. ex Elliott ●☆

352797　Schwenckea Post et Kuntze ＝ Schwenckia L. ■●☆

352798　Schwenckia L.（1764）；施文克茄属■●☆

352799　Schwenckia Royen ex L. ＝ Schwenckia L. ■●☆

352800　Schwenckia americana L.；施文克茄☆

352801　Schwenckiopsis Dammer ＝ Protoschwenkia Soler. ■☆

352802　Schwendenera K. Schum.（1886）；施文茜属☆

352803　Schwendenera tetrapyxis K. Schum.；施文茜☆

352804　Schwenkfelda Schreb. ＝ Sabicea Aubl. ●☆

352805　Schwenkia L. ＝ Schwenckia L. ■●☆

352806　Schwenkiopsis Dammer ＝ Schwenckiopsis Dammer ■☆

352807　Schwerinia H. Karst. ＝ Meriania Sw.（保留属名）●☆

352808　Schweyckerta C. C. Gmel. ＝ Nymphoides Ség. ■

352809　Schweykerta Griseb. ＝ Schweyckerta C. C. Gmel. ■

352810　Schychowskia Endl. ＝ Urtica L. ■

352811　Schychowskia Wedd. ＝ Urtica L. ■

352812　Schychowskya Endl. ＝ Laportea Gaudich.（保留属名）●■

352813　Schyzogyne Cass. ＝ Inula L. ●■

352814　Sciacassia Britton ＝ Cassia L.（保留属名）●■

352815　Sciacassia Britton ＝ Senna Mill. ●■

352816　Sciadiara Raf. ＝ Convolvulus L. ■●

352817　Sciadicarpus Hassk.（1842）；伞果树属●☆

352818　Sciadicarpus Hassk. ＝ Kibara Endl. ●☆

352819　Sciadicarpus brongniartii Hassk.；伞果树●☆

352820　Sciadicarpus brongniartii Hassk. ＝ Kibara blumei Steud. ●☆

352821　Sciadiodaphne Rchb.（废弃属名）＝ Umbellularia（Nees）Nutt.（保留属名）●☆

352822　Sciadioseris Kuntze ＝ Senecio L. ■●

352823　Sciadiphyllum Hassk. ＝ Sciadophyllum P. Browne ●■

352824　Sciadocalyx Regel ＝ Isoloma Decne. ●■☆

352825　Sciadocalyx Regel ＝ Kohleria Regel ●■☆

352826　Sciadocalyx Regel（1853）；伞萼苣苔属■☆

352827　Sciadocalyx warszewiczii Regel；伞萼苣苔■☆

352828　Sciadocarpus Post et Kuntze ＝ Kibara Endl. ●☆

352829　Sciadocarpus Post et Kuntze ＝ Sciadicarpus Hassk. ●☆

352830　Sciadocephala Mattf.（1938）；伞头菊属■☆

352831　Sciadocephala schultze-rhonhofiae Mattf.；伞头菊■☆

352832　Sciadodendron Griseb.（1858）；伞状木属（伞木属）●☆

352833　Sciadodendron excelsum Griseb.；伞状木（高伞木）●☆

352834　Sciadodendron excelsum Griseb. ＝ Aralia excelsa（Griseb.）J. Wen ■☆

352835　Sciadonardus Steud. ＝ Gymnopogon P. Beauv. ■☆

352836　Sciadopanax Seem.（1865）；伞参属●☆

352837　Sciadopanax Seem. ＝ Polyscias J. R. Forst. et G. Forst. ●

352838　Sciadopanax aubrevillei Bernardi ＝ Polyscias aubrevillei（Bernardi）Bernardi ●☆

352839　Sciadopanax boivinii Seem. ＝ Polyscias boivinii（Seem.）Bernardi ●☆

352840　Sciadopanax floccosa（Drake）R. Vig. ＝ Polyscias floccosa（Drake）Bernardi ●☆

352841　Sciadopanax grevei（Drake）Vig. ＝ Polyscias boivinii（Seem.）Bernardi ●☆

352842　Sciadophila Phil. ＝ Rhamnus L. ●

352843　Sciadophyllum P. Browne ＝ Schefflera J. R. Forst. et G. Forst.（保留属名）●

352844　Sciadophyllum P. Browne ＝ Sciadophyllum Rchb. ●■

352845　Sciadophyllum P. Browne（1756）；伞叶五加属●■

352846　Sciadophyllum Rchb. ＝ Schefflera J. R. Forst. et G. Forst.（保留属名）●

352847　Sciadophyllum Rchb. ＝ Sciadophyllum P. Browne ●■

352848　Sciadophyllum umbellatum N. E. Br.；伞叶五加●☆

352849　Sciadopityaceae（Pilg.）J. Doyle ＝ Taxodiaceae Saporta（保留科名）●

352850　Sciadopityaceae J. Doyle ＝ Taxodiaceae Saporta（保留科名）●

352851　Sciadopityaceae Luerss.（1877）；金松科●

352852　Sciadopityaceae Luerss. ＝ Cupressaceae Gray（保留科名）●

352853　Sciadopityaceae Luerss. ＝ Taxodiaceae Saporta（保留科名）●

352854　Sciadopitys Siebold et Zucc.（1842）；金松属（日本金松属）；Japan Umbrella Pine，Japanese Umbrella Pine，Umbrella Pine ●

352855　Sciadopitys verticillata（Thunb. ex A. Murray）Siebold et Zucc. ＝ Sciadopitys verticillata（Thunb.）Siebold et Zucc. ●

352856 Sciadopitys verticillata（Thunb.）Siebold et Zucc.；金松（日本金松）；Japan Umbrella Pine，Japanese Umbrella Pine，Parasol Pine，Umbrella Pine ●

352857 Sciadoseris C. Muell. = Sciadioseris Kuntze ■●

352858 Sciadoseris C. Muell. = Senecio L. ■●

352859 Sciadostima Nied. = Sonneratia L. f.（保留属名）●

352860 Sciadotaenia Benth. = Sciadotenia Miers ●☆

352861 Sciadotenia Miers（1851）；阴毒藤属●☆

352862 Sciadotenia brachypoda Diels；阴毒藤●☆

352863 Sciaphila Blume（1826）；喜荫草属（霉草属，喜阴草属）；Sciaphila，Shadegrass

352864 Sciaphila arfakiana Beccari；兰屿霉草■

352865 Sciaphila boninensis Tuyama = Sciaphila nana Blume ■☆

352866 Sciaphila japonica Makino = Sciaphila nana Blume ■☆

352867 Sciaphila maculata Miers；斑点霉草■

352868 Sciaphila megastyla Fukuy. et T. Suzuki = Sciaphila secundiflora Thwaites ex Benth. ■

352869 Sciaphila nana Blume；日本喜荫草（小喜荫草）■☆

352870 Sciaphila okabeana Tuyama = Sciaphila ramosa Fukuy. et T. Suzuki ■

352871 Sciaphila ramosa Fukuy. et T. Suzuki；多枝霉草（兰屿霉草）；Branchy Shadegrass ■

352872 Sciaphila secundiflora Benth. = Sciaphila secundiflora Thwaites ex Benth. ■

352873 Sciaphila secundiflora Thwaites ex Benth.；大柱霉草（锡兰霉草）；Bigstyle Shadegrass ■

352874 Sciaphila takakumensis Ohwi；高隈喜荫草■☆

352875 Sciaphila tenella Blume；喜荫草（霉草）；Common Sciaphila，Shadegrass ■

352876 Sciaphila tosaensis Makino = Sciaphila secundiflora Thwaites ex Benth. ■

352877 Sciaphyllum Bremek.（1940）；亭叶爵床属☆

352878 Sciaphyllum amoenum Bremek.；亭叶爵床☆

352879 Sciaplea Rauschert = Dialium L. ●☆

352880 Scilla L.（1753）；绵枣儿属（绵枣属）；Bluebell，Scilla，Squill，Wild Hyacinth ■

352881 Scilla alboviridis Hand.-Mazz. = Barnardia japonica（Thunb.）Schult. et Schult. f. ●

352882 Scilla angusta Engelm. et A. Gray = Camassia angusta（Engelm. et A. Gray）Blank. ■●

352883 Scilla armena Grossh.；星花绵枣儿；Star Hyacinth Squill，Star Squill，Star-hyacinth，Star-hyacinth Squill ■☆

352884 Scilla atropatana Grossh.；阿特罗绵枣儿■☆

352885 Scilla autumnalis L.；秋绵枣儿；Autumn Squill，Autumnal Squill，Autumnal Star Hyacinth，Winter Hyacinth ■☆

352886 Scilla bifolia L.；二叶绵枣儿（双叶绵枣儿）；Alpine Squill，Squill，Squll Onion，Twinleaf Squill，Twin-leaved Squill，Twoleaf Squill ■☆

352887 Scilla bispatha Hand.-Mazz. = Barnardia japonica（Thunb.）Schult. et Schult. f. ●

352888 Scilla borealijaponica M. Kikuchi = Barnardia japonica（Thunb.）Schult. et Schult. f. ●

352889 Scilla bucharica Des.-Shost.；布哈拉绵枣儿■☆

352890 Scilla caucasica Miscz.；高加索绵枣儿■☆

352891 Scilla chinensis Benth. = Barnardia japonica（Thunb.）Schult. et Schult. f. ●

352892 Scilla chinensis Benth. = Scilla scilloides（Lindl.）Druce ■

352893 Scilla chinensis Benth. var. mounsei H. Lév. = Barnardia japonica（Thunb.）Schult. et Schult. f. ●

352894 Scilla cooperi Hook. f.；库珀绵枣儿；Wild Squill ■☆

352895 Scilla diziensis Grossh.；迪扎绵枣儿■☆

352896 Scilla engleri T. Durand et Schinz；恩格勒绵枣儿■☆

352897 Scilla genadendalensis Poelln. = Ledebouria ovalifolia（Schrad.）Jessop ■☆

352898 Scilla gerrardii Baker = Schizocarphus nervosus（Burch.）Van der Merwe ■☆

352899 Scilla glandulosa（Chiov.）Chiov. = Ledebouria cordifolia（Baker）Stedje et Thulin ■☆

352900 Scilla glandulosa（Chiov.）Chiov. f. major Chiov. = Ledebouria cordifolia（Baker）Stedje et Thulin ■☆

352901 Scilla glaucescens Van der Merwe = Ledebouria cooperi（Hook. f.）Jessop ■☆

352902 Scilla globosa Baker = Ledebouria cooperi（Hook. f.）Jessop ■☆

352903 Scilla gracillima Engl.；细长绵枣儿■☆

352904 Scilla graminifolia Baker = Ledebouria graminifolia（Baker）Jessop ■☆

352905 Scilla guttata C. A. Sm. = Ledebouria ovatifolia（Baker）Jessop ■☆

352906 Scilla hemisphaerica Boiss. = Oncostema peruviana（L.）Speta ■☆

352907 Scilla hildebrandtii Baker = Ledebouria kirkii（Baker）Stedje et Thulin ■☆

352908 Scilla hispanica Mill. = Hyacinthoides hispanica（Mill.）Rothm. ■☆

352909 Scilla hispanica Mill. var. algeriensis（Batt.）Maire = Hyacinthoides cedretorum（Pomel）Dobignard ■☆

352910 Scilla hispanica Mill. var. cedretorum（Pomel）Maire = Hyacinthoides cedretorum（Pomel）Dobignard ■☆

352911 Scilla hispidula Baker = Scilla nervosa（Burch.）Jessop ■☆

352912 Scilla hohenackeri Fisch. et C. A. Mey.；豪氏绵枣儿■☆

352913 Scilla humifusa Baker = Resnova humifusa（Baker）U. Müll.-Doblies et D. Müll.-Doblies ■☆

352914 Scilla hyacinthina（Roth）J. F. Macbr. = Ledebouria revoluta（L. f.）Jessop ■☆

352915 Scilla hyacinthoides L.；风信子状绵枣儿；Hyacinth Squill ■☆

352916 Scilla hyacinthoides L. = Nectaroscilla hyacinthoides（L.）Parl. ■☆

352917 Scilla hypoxidioides Schönland = Ledebouria hypoxidioides（Schönland）Jessop ■☆

352918 Scilla inandensis Baker = Ledebouria cooperi（Hook. f.）Jessop ■☆

352919 Scilla indica Baker = Ledebouria revoluta（L. f.）Jessop ■☆

352920 Scilla indica Roxb.；印度绵枣儿；India Squill ■☆

352921 Scilla indica Roxb. = Drimia indica（Roxb.）Jessop ■☆

352922 Scilla inquinata C. A. Sm. = Ledebouria inquinata（C. A. Sm.）Jessop ■☆

352923 Scilla italica L. = Hyacinthoides italica（L.）Rothm. ■☆

352924 Scilla jaegeri K. Krause；耶格绵枣儿■☆

352925 Scilla japonica Baker = Barnardia japonica（Thunb.）Schult. et Schult. f. ●

352926 Scilla japonica Thunb. = Barnardia japonica（Thunb.）Schult. et Schult. f. ●

352927 Scilla japonica Thunb. = Scilla scilloides（Lindl.）Druce ■

352928 Scilla japponica Baker = Barnardia japonica（Thunb.）Schult. et Schult. f. ●

352929 Scilla johnstonii Baker = Ledebouria kirkii（Baker）Stedje et Thulin ■☆

352930 Scilla kabylica Chabert = Hyacinthoides hispanica（Mill.）

Rothm. subsp. algeriensis（Batt.）Förther et Podlech ■☆

352931　Scilla katendensis De Wild.；卡滕代绵枣儿■☆

352932　Scilla kirkii Baker ＝Ledebouria kirkii（Baker）Stedje et Thulin ■☆

352933　Scilla kraussii Baker ＝Merwilla plumbea（Lindl.）Speta ■☆

352934　Scilla lachenalioides Baker ＝Resnova lachenalioides（Baker）Van der Merwe ■☆

352935　Scilla lanceifolia（Jacq.）Baker ＝Ledebouria revoluta（L. f.）Jessop ■☆

352936　Scilla lanceolata（Schrad.）Baker ＝Ledebouria ovalifolia（Schrad.）Jessop ■☆

352937　Scilla latifolia Willd. ＝Autonoë latifolia（Willd.）Speta ■☆

352938　Scilla lauta N. E. Br. ＝Ledebouria floribunda（Baker）Jessop ■☆

352939　Scilla ledienii Engl.；莱丁绵枣儿■☆

352940　Scilla leichtlinii Baker ＝Ledebouria cooperi（Hook. f.）Jessop ■☆

352941　Scilla lepida N. E. Br. ＝Ledebouria cooperi（Hook. f.）Jessop ■☆

352942　Scilla leptophylla Baker ＝Ledebouria cooperi（Hook. f.）Jessop ■☆

352943　Scilla lilacina（Fenzl ex Kunth）Baker ＝Ledebouria revoluta（L. f.）Jessop ■☆

352944　Scilla liliohyacinthus L.；核绵枣儿；Pyrenean Squill ■☆

352945　Scilla linearifolia Baker ＝Ledebouria ensifolia（Eckl.）S. Venter et T. J. Edwards ■☆

352946　Scilla lingulata Poir. ＝Hyacinthoides lingulata（Poir.）Rothm. ■☆

352947　Scilla lingulata Poir. var. ciliolata（Pomel）Batt. ＝Hyacinthoides lingulata（Poir.）Rothm. ■☆

352948　Scilla litardierei Breistr.；南斯拉夫绵枣儿■☆

352949　Scilla livida Baker ＝Ledebouria floribunda（Baker）Jessop ■☆

352950　Scilla londonensis Baker ＝Ledebouria cooperi（Hook. f.）Jessop ■☆

352951　Scilla lorata Baker ＝Ledebouria ensifolia（Eckl.）S. Venter et T. J. Edwards ■☆

352952　Scilla luciliae（Boiss.）Speta ＝Chionodoxa luciliae Boiss. ■☆

352953　Scilla ludwigii（Miq.）Baker；卢氏绵枣儿■☆

352954　Scilla macowanii Baker ＝Ledebouria cooperi（Hook. f.）Jessop ■☆

352955　Scilla maculata Schrank ＝Ledebouria revoluta（L. f.）Jessop ■☆

352956　Scilla madeirensis Menezes ＝Autonoe madeirensis（Menezes）Speta ●☆

352957　Scilla madeirensis Menezes var. melliodora Svent. ＝Autonoe madeirensis（Menezes）Speta ●☆

352958　Scilla mankonensis A. Chev. ＝Scilla sudanica A. Chev. ■☆

352959　Scilla marginata Baker ＝Ledebouria marginata（Baker）Jessop ■☆

352960　Scilla maritima（L.）Baker；海葱（滨海葱，海滨海葱，海天蒜，绵枣儿海葱，仙葱）；Coastal Squill, Crusader's Spear, Crusaders' Spears, French Onion, Gladene, Officinal Squill, Red Squill, Sea Leek, Sea Onion, Sea Squill, Sea-onion, Shore Drugs Quill, Shore Sea Onion, Shore Sea-onion, Squill, Squill Onion, White Squill ■☆

352961　Scilla maritima（L.）Baker ＝Urginea maritima（L.）Baker ■☆

352962　Scilla maritima（L.）Baker var. alba Baker；白海葱■☆

352963　Scilla maritima L. ＝Charybdis maritima（L.）Speta ■☆

352964　Scilla mauritanica Schousb. ＝Hyacinthoides mauritanica（Schousb.）Speta ■☆

352965　Scilla megaphylla Baker ＝Ledebouria floribunda（Baker）Jessop ■☆

352966　Scilla micrantha A. Rich. ＝Drimia altissima（L. f.）Ker Gawl. ■☆

352967　Scilla microscypha Baker ＝Ledebouria floribunda（Baker）Jessop ■☆

352968　Scilla minima Baker ＝Ledebouria cooperi（Hook. f.）Jessop ■☆

352969　Scilla mischtschenkoana Grossh.；伊朗绵枣儿（白花绵枣儿）；White Squill ■☆

352970　Scilla misczenkoana Grossh.；米谢绵枣儿■☆

352971　Scilla modesta Baker ＝Scilla engleri T. Durand et Schinz ■☆

352972　Scilla monanthos K. Koch；单花绵枣儿■☆

352973　Scilla monophyllos Link var. tingitana（Schousb.）Pau ＝Tractema tingitana（Schousb.）Speta ■☆

352974　Scilla moschata Schönland ＝Ledebouria floribunda（Baker）Jessop ■☆

352975　Scilla natalensis Planch.；纳塔尔绵枣儿（南非绵枣儿）；Wild Squill ■☆

352976　Scilla natalensis Planch. ＝Merwilla plumbea（Lindl.）Speta ■☆

352977　Scilla neglecta Van der Merwe ＝Ledebouria marginata（Baker）Jessop ■☆

352978　Scilla nelsonii Baker ＝Ledebouria undulata（Jacq.）Jessop ■☆

352979　Scilla nervosa（Burch.）Jessop ＝Schizocarphus nervosus（Burch.）Van der Merwe ■☆

352980　Scilla neumannii Engl. ＝Ledebouria revoluta（L. f.）Jessop ■☆

352981　Scilla nivalis Boiss.；雪白绵枣儿■☆

352982　Scilla non-scripta（L.）Hoffmanns. et Link ＝Hyacinthoides non-scripta（L.）Chouard ex Rothm. ■☆

352983　Scilla nonscripta Hoffmanns. et Link ＝Hyacinthoides non-scripta（L.）Chouard ex Rothm. ■☆

352984　Scilla numidica Poir.；努米底亚绵枣儿■☆

352985　Scilla nutans Sm. ＝Hyacinthoides non-scripta（L.）Chouard ex Rothm. ■☆

352986　Scilla obtusifolia Poir. ＝Prospero obtusifolium（Poir.）Speta ■☆

352987　Scilla obtusifolia Poir. var. fallax（Steinh.）Baker ＝Prospero fallax（Steinh.）Speta ■☆

352988　Scilla obtusifolia Poir. var. glauca Gatt. et Weiller ＝Prospero obtusifolium（Poir.）Speta ■☆

352989　Scilla obtusifolia Poir. var. intermedia（Guss.）Baker ＝Prospero obtusifolium（Poir.）Speta ■☆

352990　Scilla ondongensis Schinz；翁东白绵枣儿■☆

352991　Scilla oostachys Baker ＝Ledebouria cooperi（Hook. f.）Jessop ■☆

352992　Scilla oubanguiensis Hua；乌班吉绵枣儿■☆

352993　Scilla ovalifolia（Schrad.）C. A. Sm. ＝Ledebouria ovalifolia（Schrad.）Jessop ■☆

352994　Scilla ovatifolia Baker ＝Ledebouria ovatifolia（Baker）Jessop ■☆

352995　Scilla pallidiflora Baker ＝Scilla nervosa（Burch.）Jessop ■☆

352996　Scilla palustris Wood et Evans ＝Ledebouria cooperi（Hook. f.）Jessop ■☆

352997　Scilla parviflora Desf. ＝Barnardia numidica（Poir.）Speta ■☆

352998　Scilla paucifolia Baker ＝Ledebouria socialis（Baker）Jessop ■☆

352999　Scilla pearsonii P. E. Glover ＝Lachenalia pearsonii（P. E. Glover）W. F. Barker ■☆

353000　Scilla pendula Baker ＝Ledebouria floribunda（Baker）Jessop ■☆

353001　Scilla peruviana L.；秘鲁绵枣儿（地中海蓝钟花，地中海绵枣儿，聚铃花，野风信子，锥序绵枣儿）；Corymbose Squill, Cuban Lily, Cubanlily, Cuban-lily, Hyacinth of Peru, Hyacinth-of-Peru, Peruvian Lily, Peruvian Squill, Portuguese Squill ■☆

353002　Scilla peruviana L. ＝Oncostema peruviana（L.）Speta ■☆

353003　Scilla peruviana L. subsp. elongata（Parl.）Maire ＝Oncostema elongata（Parl.）Speta ■☆

353004　Scilla peruviana L. subsp. peruviana ＝Oncostema peruviana（L.）Speta ■☆

353005　Scilla peruviana L. var. diluta Maire ＝Oncostema peruviana

(L.) Speta ■☆

353006　Scilla peruviana L. var. elegans（Jord. et Fourr.）Maire et Weiller = Oncostema elongata（Parl.）Speta ■☆

353007　Scilla peruviana L. var. flaveola（Jord. et Fourr.）Maire et Weiller = Oncostema elongata（Parl.）Speta ■☆

353008　Scilla peruviana L. var. gattefossei Maire = Oncostema elongata（Parl.）Speta ■☆

353009　Scilla peruviana L. var. grandiflora（Jord. et Fourr.）Maire et Weiller = Oncostema elongata（Parl.）Speta ■☆

353010　Scilla peruviana L. var. hipponensis（Jord. et Fourr.）Maire et Weiller = Oncostema africana（Borzí et Mattei）Speta ■☆

353011　Scilla peruviana L. var. ifniensis Font Quer = Oncostema elongata（Parl.）Speta ■☆

353012　Scilla peruviana L. var. ifniensis Font Quer = Oncostema peruviana（L.）Speta ■☆

353013　Scilla peruviana L. var. killiani Maire = Oncostema elongata（Parl.）Speta ■☆

353014　Scilla peruviana L. var. livida（Jord. et Fourr.）Maire et Weiller = Oncostema elongata（Parl.）Speta ■☆

353015　Scilla peruviana L. var. minor Maire et Weiller = Oncostema africana（Borzí et Mattei）Speta ■☆

353016　Scilla peruviana L. var. pallidiflora（Jord. et Fourr.）Maire et Weiller = Oncostema elongata（Parl.）Speta ■☆

353017　Scilla peruviana L. var. stenopetala Maire = Oncostema elongata（Parl.）Speta ■☆

353018　Scilla peruviana L. var. subalbida（Jord. et Fourr.）Maire et Weiller = Oncostema elongata（Parl.）Speta ■☆

353019　Scilla peruviana L. var. sublivida Maire = Oncostema peruviana（L.）Speta ■☆

353020　Scilla peruviana L. var. venusta（Jord. et Fourr.）Maire et Weiller = Oncostema elongata（Parl.）Speta ■☆

353021　Scilla peruviana L. var. zaborskiana Emb. = Oncostema elongata（Parl.）Speta ■☆

353022　Scilla petersii Engl. ;彼得斯绵枣儿■☆

353023　Scilla petiolata Van der Merwe = Ledebouria cooperi（Hook. f.）Jessop ■☆

353024　Scilla petitiana A. Rich. = Albuca abyssinica Jacq. ■☆

353025　Scilla picta A. Chev. ex Hutch. = Scilla sudanica A. Chev. ■☆

353026　Scilla platyphylla Baker;阔叶绵枣儿■☆

353027　Scilla plumbea Lindl. = Merwilla plumbea（Lindl.）Speta ■☆

353028　Scilla polyantha Baker = Ledebouria floribunda（Baker）Jessop ■☆

353029　Scilla polyphylla Baker = Scilla welwitschii Poelln. ■☆

353030　Scilla pomeridiana DC. = Chlorogalum pomeridianum Kunth ■☆

353031　Scilla prasina Baker = Ledebouria undulata（Jacq.）Jessop ■☆

353032　Scilla pratensis K. Koch = Scilla litardierei Breistr. ■☆

353033　Scilla princeps Baker = Ledebouria floribunda（Baker）Jessop ■☆

353034　Scilla pubescens Baker = Scilla nervosa（Burch.）Jessop ■☆

353035　Scilla pulchella Munby = Prospero autumnale（L.）Speta ■☆

353036　Scilla puschkinioides Regel;普什绵枣儿■☆

353037　Scilla pusilla Baker = Ledebouria cooperi（Hook. f.）Jessop ■☆

353038　Scilla quartiniana A. Rich. = Albuca abyssinica Jacq. ■☆

353039　Scilla raevskiana Regel;赖氏绵枣儿■☆

353040　Scilla ramburei Boiss. = Tractema ramburei（Boiss.）Speta ■☆

353041　Scilla rautanenii Schinz = Ledebouria undulata（Jacq.）Jessop ■☆

353042　Scilla rehmannii Baker = Ledebouria cooperi（Hook. f.）Jessop ■☆

353043　Scilla revoluta（L. f.）Baker = Ledebouria revoluta（L. f.）Jessop ■☆

353044　Scilla revoluta Baker = Ledebouria revoluta（L. f.）Jessop ■☆

353045　Scilla richardiana Buchinger ex Baker = Ledebouria revoluta（L. f.）Jessop ■☆

353046　Scilla rigidifolia Baker = Schizocarphus nervosus（Burch.）Van der Merwe ■☆

353047　Scilla rigidifolia Baker var. acerosa Van der Merwe = Schizocarphus nervosus（Burch.）Van der Merwe ■☆

353048　Scilla rigidifolia Baker var. gerrardi（Baker）Baker = Schizocarphus nervosus（Burch.）Van der Merwe ■☆

353049　Scilla rigidifolia Baker var. nervosa ? = Schizocarphus nervosus（Burch.）Van der Merwe ■☆

353050　Scilla rogersii Baker = Lachenalia mediana Jacq. var. rogersii（Baker）W. F. Barker ■☆

353051　Scilla rosenii K. Koch;罗森绵枣儿■☆

353052　Scilla rupestris Van der Merwe = Ledebouria cooperi（Hook. f.）Jessop ■☆

353053　Scilla sandersonii Baker = Ledebouria cooperi（Hook. f.）Jessop ■☆

353054　Scilla saturata Baker = Ledebouria cooperi（Hook. f.）Jessop ■☆

353055　Scilla schlechteri Baker = Resnova humifusa（Baker）U. Müll. -Doblies et D. Müll. -Doblies ■☆

353056　Scilla schweinfurthii Engl. ;施韦绵枣儿■☆

353057　Scilla scilloides（Lindl.）Druce;绵枣儿（催生草,地兰,地枣,地枣儿,独叶芹,独叶一枝枪,老鸦葱,老鸦蒜,散血草,山大蒜,石枣儿,双芽,天蒜,鲜白头,鲜蕹白,药狗蒜,野扁九）;Common Squill ■

353058　Scilla scilloides（Lindl.）Druce = Barnardia japonica（Thunb.）Schult. et Schult. f. ●

353059　Scilla scilloides（Lindl.）Druce f. albida Y. N. Lee = Barnardia japonica（Thunb.）Schult. et Schult. f. ●

353060　Scilla scilloides（Lindl.）Druce var. alboviridis（Hand. -Mazz.）F. T. Wang et Y. C. Tang = Barnardia japonica（Thunb.）Schult. et Schult. f. ●

353061　Scilla scilloides（Lindl.）Druce var. alboviridis（Hand. -Mazz.）F. T. Wang et Y. C. Tang;白绿绵枣儿;Whitegreen Squill ■

353062　Scilla scilloides（Lindl.）Druce var. mounsei（H. Lév.）McKean = Barnardia japonica（Thunb.）Schult. et Schult. f. ●

353063　Scilla scilloides（Lindl.）Druce var. pulchella（Kitag.）Kitag. = Barnardia japonica（Thunb.）Schult. et Schult. f. ●

353064　Scilla setifera Baker = Scilla nervosa（Burch.）Jessop ■☆

353065　Scilla siberica Haw. ;西伯利亚绵枣儿;Siberian Scilla, Siberian Squill, Squill ■☆

353066　Scilla sibirica Andréws 'Atrocoerulea';深蓝西伯利亚绵枣儿■☆

353067　Scilla sibirica Andréws = Scilla siberica Haw. ■☆

353068　Scilla sibirica Haw. = Scilla siberica Haw. ■☆

353069　Scilla sicula Tineo = Puschkinia scilloides Adams ■☆

353070　Scilla simensis Hochst. ex A. Rich. = Drimia simensis（Hochst. ex A. Rich.）Stedje ☆

353071　Scilla sinensis（Lour.）Merr. = Barnardia japonica（Thunb.）Schult. et Schult. f. ●

353072　Scilla sinensis（Lour.）Merr. = Scilla scilloides（Lindl.）Druce ■

353073　Scilla sinensis（Lour.）Merr. var. pulchella（Kitag.）Kitag. = Barnardia japonica（Thunb.）Schult. et Schult. f. ●

353074　Scilla sinensis Merr. = Barnardia japonica（Thunb.）Schult. f. ●

353075　Scilla sinensis Merr. var. pulchella（Kitag.）Kitag. = Barnardia japonica（Thunb.）Schult. et Schult. f. ●

353076　Scilla socialis Baker = Ledebouria socialis（Baker）Jessop ■☆

353077　Scilla somaliensis Baker = Ledebouria somaliensis（Baker）Stedje et Thulin ■☆

353078　Scilla spathulata Baker = Ledebouria floribunda（Baker）Jessop ■☆

353079　Scilla sphaerocephala Baker = Ledebouria cooperi（Hook. f.）Jessop ■☆

353080　Scilla spicata Baker;穗状绵枣儿■☆

353081　Scilla stenophylla Van der Merwe = Ledebouria graminifolia（Baker）Jessop ■☆

353082　Scilla subglauca Baker = Ledebouria cooperi（Hook. f.）Jessop ■☆

353083　Scilla subsecunda Baker = Ledebouria floribunda（Baker）Jessop ■☆

353084　Scilla sudanica A. Chev.;苏丹绵枣儿■☆

353085　Scilla tayloriana Rendle = Ledebouria kirkii（Baker）Stedje et Thulin ■☆

353086　Scilla tetraphylla L. f. = Chlorophytum tetraphyllum（L. f.）Baker ■☆

353087　Scilla textilis Rendle;编织绵枣儿■☆

353088　Scilla thunbergii Miyabe et Kudo = Barnardia japonica（Thunb.）Schult. et Schult. f. ●

353089　Scilla thunbergii Miyabe et Kudo = Scilla scilloides（Lindl.）Druce ■

353090　Scilla thunbergii Miyabe et Kudo var. pulchella Kitag. = Barnardia japonica（Thunb.）Schult. et Schult. f. ●

353091　Scilla tingitana Schousb. = Tractema tingitana（Schousb.）Speta ■☆

353092　Scilla tricolor Baker = Ledebouria floribunda（Baker）Jessop ■☆

353093　Scilla tristachya Baker = Ledebouria cooperi（Hook. f.）Jessop ■☆

353094　Scilla tysonii Baker = Ledebouria cooperi（Hook. f.）Jessop ■☆

353095　Scilla undulata（Jacq.）Baker = Ledebouria undulata（Jacq.）Jessop ■☆

353096　Scilla undulata Desf. = Charybdis undulata（Desf.）Speta ■☆

353097　Scilla uyuiensis Rendle;乌尤伊绵枣儿■☆

353098　Scilla verdickii De Wild.;韦尔迪克绵枣儿■☆

353099　Scilla verna Huds.;春花绵枣儿;Sea Onion,Spring Squill,Star Hyacinth,Swine's Beads,Swine's Murrill,Swine's Murrills ■☆

353100　Scilla verna Huds. subsp. ramburei（Boiss.）Maire = Tractema ramburei（Boiss.）Speta ■☆

353101　Scilla verna Huds. var. iberica Maire = Tractema ramburei（Boiss.）Speta ■☆

353102　Scilla verna Huds. var. major Boiss. = Tractema ramburei（Boiss.）Speta ■☆

353103　Scilla verna Huds. var. maroccana Maire = Tractema ramburei（Boiss.）Speta ■☆

353104　Scilla versicolor Baker = Scilla nervosa（Burch.）Jessop ■☆

353105　Scilla villosa Desf. = Oncostema villosa（Desf.）Raf. ■☆

353106　Scilla villosa Desf. var. barba-caprae（Asch. et Barbey）Maire et Weiller = Oncostema barba-caprae（Asch. et Barbey）Speta ■☆

353107　Scilla villosa Desf. var. genuina Maire et Weiller = Oncostema villosa（Desf.）Raf. ■☆

353108　Scilla violacea Hutch. = Ledebouria socialis（Baker）Jessop ■☆

353109　Scilla viridiflora（Kunze）Baker;绿花绵枣儿■☆

353110　Scilla volkensii Baker = Drimiopsis botryoides Baker ●☆

353111　Scilla volkensii Engl. = Drimiopsis botryoides Baker ●☆

353112　Scilla welwitschii Poelln.;韦尔绵枣儿■☆

353113　Scilla werneri De Wild.;维尔纳绵枣儿■☆

353114　Scilla winogradowii Sosn.;维诺绵枣儿■☆

353115　Scilla xanthobotrya Poelln.;黄穗绵枣儿☆

353116　Scilla zambesiaca Baker;赞比西绵枣儿☆

353117　Scilla zebrina Baker = Ledebouria floribunda（Baker）Jessop ■☆

353118　Scillaceae Vest = Hyacinthaceae Batsch ex Borkh. ■

353119　Scillaceae Vest;绵枣儿科■●

353120　Scillopsis Lem. = Lachenalia J. Jacq. ex Murray ■☆

353121　Scindapsus Schott（1832）;藤芋属;Ivy Arum, Ivyarum, Ivy-Arum ■

353122　Scindapsus aureus（Linden et André）Engl. = Epipremnum pinnatum（L.）Engl. 'Aureum' ●■

353123　Scindapsus aureus（Linden et André）Engl. = Epipremnum pinnatum（L.）Engl. ●■

353124　Scindapsus aureus（Lindl. ex André）Engl. et Krause = Epipremnum aureum（Linden ex André）Bunting ●■

353125　Scindapsus decursivus Schott = Rhaphidophora decursiva（Roxb.）Schott ●■

353126　Scindapsus hedraceus（Zoll. et Moritzi）Miq. ex Schott;星藤芋（星点藤）■☆

353127　Scindapsus maclurei（Merr.）Merr. et F. P. Metcalf;海南藤芋（大叶绿萝,大叶藤芋,吊东根藤,吊头藤）;Hainan Ivyarum, Maclure Ivyarum ■

353128　Scindapsus megalophylla Merr. = Scindapsus maclurei（Merr.）Merr. et F. P. Metcalf ■

353129　Scindapsus montanus Kunth = Anadendrum montanum（Blume）Schott ●

353130　Scindapsus officinalis Schott;青竹标;Officinale Ivyarum ■

353131　Scindapsus peepla Schott = Rhaphidophora peepla（Roxb.）Schott ●■

353132　Scindapsus pictus Hassk.;彩叶绿萝;Pothos, Silver Vine ■☆

353133　Scindapsus pictus Hassk. 'Argyraeus' = Epipremnum pictum（L.）Engl. 'Argyraeus' ■☆

353134　Scindapsus pinnatus Schott = Epipremnum pinnatum（L.）Engl. ●■

353135　Scindapsus sinensis Engl. = Amydrium sinense（Engl.）H. Li ■

353136　Sciobia Rchb. = Procris Comm. ex Juss. ●

353137　Sciobia Rchb. = Sciophila Gaudich. ●

353138　Sciodaphyllum P. Browne（废弃属名）= Schefflera J. R. Forst. et G. Forst.（保留属名）●

353139　Sciodaphyllum P. Browne（废弃属名）= Sciadophyllum P. Browne ●■

353140　Sciodaphyllum ellipticum Blume = Schefflera elliptica（Blume）Harms ●

353141　Sciophila Gaudich. = Procris Comm. ex Juss. ●

353142　Sciophila Gaudich. = Sciobia Rchb. ●

353143　Sciophila Post et Kuntze = Columnea L. ●■☆

353144　Sciophila Post et Kuntze = Skiophila Hanst. ■☆

353145　Sciophila Wibel = Maianthemum F. H. Wigg.（保留属名）■

353146　Sciophylla F. Heller = Sciophila Wibel ■

353147　Sciothamnus Endl.（1839）;阴灌草属●☆

353148　Sciothamnus Endl. = Dregea E. Mey.（保留属名）●

353149　Sciothamnus Endl. = Dregea Eckl. et Zeyh.（废弃属名）●

353150　Sciothamnus Endl. = Peucedanum L. ●

353151　Sciothamnus montanus Endl.;阴灌草●☆

353152　Scirpaceae Burnett = Cyperaceae Juss.（保留科名）■

353153　Scirpaceae Burnett = Scirpaceae Burnett ex Borkh. ■

353154　Scirpaceae Burnett ex Borkh.;蔗草科■

353155　Scirpaceae Burnett ex Borkh. = Cyperaceae Juss.（保留科名）■

353156　Scirpidiella Rauschert = Eleogiton Link ■

353157　Scirpidiella Rauschert = Isolepis R. Br. ■

353158　Scirpidiella Rauschert(1983);小针蔺属;Floating Club-rush ■☆

353159　Scirpidiella beccarii（Boeck.）Rauschert;肖细莞■☆

353160　Scirpidium Nees = Eleocharis R. Br. ■

353161　Scirpidium nigrescens Nees = Eleocharis nigrescens（Nees）Steud. ■☆

353162　Scirpobambus（A. Rich.）Post et Kuntze = Oxytenanthera Munro ●☆

353163　Scirpobambus Kuntze = Oxytenanthera Munro ●☆

353164　Scirpocyperus Ség. = Scirpus L.（保留属名）■

353165　Scirpo-cyperus Ség. = Scirpus L.（保留属名）■

353166　Scirpodendron Engl. = Microdracoides Hua ■☆

353167　Scirpodendron Engl. = Schoenodendron Engl. ■☆

353168　Scirpodendron Zipp. ex Kurz(1869);皱果莎草属■☆

353169　Scirpodendron costatum Kurz;皱果莎草■☆

353170　Scirpoides Ség.（1754）;拟藨草属;Round-headed Club-rush ■☆

353171　Scirpoides Ség. = Scirpus L.（保留属名）■

353172　Scirpoides burkei（C. B. Clarke）Goetgh., Muasya et D. A. Simpson;伯克拟藨草■☆

353173　Scirpoides globifera（L. f.）Soják = Scirpoides holoschoenus（L.）Soják ■☆

353174　Scirpoides holoschoenus（L.）Soják;圆头拟藨草;Roundhead Bulrush, Round-headed Club-rush ■☆

353175　Scirpoides holoschoenus（L.）Soják subsp. australis（Murray）Soják;南方圆头拟藨草■☆

353176　Scirpoides nodosus（Rottb.）Soják = Ficinia nodosa（Rottb.）Goetgh., Muasya et D. A. Simpson ■☆

353177　Scirpoides romanus（L.）Soják;罗马拟藨草■☆

353178　Scirpoides thunbergii（Schrad.）Soják;通贝里拟藨草■☆

353179　Scirpus L.（1753）（保留属名）;藨草属（莞草属,莞属）;Bulrush, Club-rush, Floating Club-rush, Scirpus, Wood Club-rush, Wool Grass ■

353180　Scirpus L. = Schoenoplectus（Rchb.）Palla（保留属名）■

353181　Scirpus × igaensis T. Koyama = Schoenoplectus igaensis（T. Koyama）Hayas. et H. Ohashi ■☆

353182　Scirpus × intermedius;中间藨草;Intermediate Bulrush ■

353183　Scirpus × oguraensis T. Koyama = Schoenoplectus × oguraensis（T. Koyama）Hayas. et H. Ohashi ■☆

353184　Scirpus × osoreyamensis M. Kikuchi = Schoenoplectus × osoreyamensis（M. Kikuchi）Hayas. ■☆

353185　Scirpus × trapezoideus Koidz. = Schoenoplectus × trapezoideus（Koidz.）Hayas. et H. Ohashi ■

353186　Scirpus × trapezoideus Koidz. nothovar. triangularis（Honda）Sugim. ;三角藨草■☆

353187　Scirpus × uzenensis Ohwi ex T. Koyama = Schoenoplectus × uzenensis（Ohwi ex T. Koyama）Hayas. et H. Ohashi ■☆

353188　Scirpus abactus Ohwi = Schoenoplectus mucronatus（L.）Palla ■

353189　Scirpus abnormalis（C. B. Clarke）T. Koyama = Fuirena abnormalis C. B. Clarke ■☆

353190　Scirpus acicularis L. = Eleocharis acicularis（L.）Roem. et Schult. ■

353191　Scirpus aciformis B. Nord. = Isolepis hemiuncialis（C. B. Clarke）J. Raynal ■☆

353192　Scirpus acrostachys Steud. = Ficinia acrostachys（Steud.）C. B. Clarke ■☆

353193　Scirpus acutangulus Roxb. = Eleocharis acutangula（Roxb.）

353194　Scirpus acutus Muhl. ex Bigelow = Schoenoplectus acutus（Muhl. ex J. M. Bigelow）Á. Löve et D. Löve ■☆

353195　Scirpus acutus Muhl. ex J. M. Bigelow = Schoenoplectus acutus（Muhl. ex J. M. Bigelow）Á. Löve et D. Löve ■☆

353196　Scirpus aegyptiacus Decne. = Schoenoplectus subulatus（Vahl）Lye ■

353197　Scirpus aestivalis Retz. = Fimbristylis aestivalis（Retz.）Vahl ■

353198　Scirpus affinis Roth = Bolboschoenus strobilinus（Roxb.）V. I. Krecz. ■

353199　Scirpus affinis Roth = Scirpus strobilinus Roxb. ■

353200　Scirpus afflatus（Steud.）Benth. = Eleocharis pellucida J. Presl et C. Presl ■

353201　Scirpus afflatus（Steud.）Benth. = Heleocharis pellucida J. Presl et C. Presl ■

353202　Scirpus albomarginatus Schult. = Eleocharis quadrangulata（Michx.）Roem. et Schult. ■☆

353203　Scirpus alpinus（L.）Dalla = Trichophorum alpinum（L.）Pers. ■☆

353204　Scirpus alpinus（L.）Dalla, Torre et Sarnth. = Trichophorum alpinum（L.）Pers. ■☆

353205　Scirpus americanus Pers. = Schoenoplectus americanus（Pers.）Volkart ex Schinz et R. Keller ■☆

353206　Scirpus americanus Pers. = Schoenoplectus pungens（Vahl）Palla ■☆

353207　Scirpus americanus Pers. var. longispicatus Britton = Schoenoplectus pungens（Vahl）Palla ■☆

353208　Scirpus americanus Pers. var. monophyllus（J. Presl et C. Presl）T. Koyama = Schoenoplectus pungens（Vahl）Palla ■☆

353209　Scirpus angolensis C. B. Clarke = Nemum spadiceum（Lam.）Desv. ex Ham. ■☆

353210　Scirpus angolensis C. B. Clarke var. briziformis（Hutch.）S. S. Hooper = Nemum spadiceum（Lam.）Desv. ex Ham. ■☆

353211　Scirpus angolensis C. B. Clarke var. megastachyum Cherm. = Nemum megastachyum（Cherm.）J. Raynal ■☆

353212　Scirpus angustifolius（Honck.）T. Koyama subsp. latifolius（Hoppe）T. Koyama = Eriophorum latifolium Hoppe ■

353213　Scirpus annamicus Raymond = Schoenoplectus erectus（Poir.）Palla ex J. Raynal ■☆

353214　Scirpus annuus All. = Fimbristylis annua（All.）Roem. et Schult. ■☆

353215　Scirpus annuus All. = Fimbristylis dichotoma（L.）Vahl ■

353216　Scirpus antarcticus L. = Isolepis antarctica（L.）Roem. et Schult. ■☆

353217　Scirpus apus A. Gray = Fimbristylis vahlii（Lam.）Link ■☆

353218　Scirpus ardea T. Koyama = Eriophorum gracile K. Koch ■

353219　Scirpus ardea T. Koyama var. coreanus（Palla）T. Koyama = Eriophorum gracile K. Koch ■

353220　Scirpus argea T. Koyama var. coreanus（Palla）T. Koyama = Eriophorum gracile K. Koch ■

353221　Scirpus argyrolepis Meisn. = Eleocharis argyrolepis Kierulff ex Bunge ■

353222　Scirpus argyrolepis Meisn. = Heleocharis argyrolepis Kierulff ex Bunge ■

353223　Scirpus articulatus L. = Schoenoplectiella articulata（L.）Lye ■☆

353224　Scirpus articulatus L. var. major Boeck. = Schoenoplectiella articulata（L.）Lye ■☆

353225　Scirpus articulatus L. var. stramineus Engl. = Schoenoplectus senegalensis（Hochst. ex Steud.）Palla ex J. Raynal ■☆

353226　Scirpus asiaticus Beetle；茸球藨草■

353227　Scirpus asiaticus Beetle = Scirpus lushanensis Ohwi ■

353228　Scirpus atrocinctus Fernald；北美藨草；Black-girdled Wool-grass ■☆

353229　Scirpus atropurpureus Retz. = Eleocharis atropurpurea（Retz.）J. Presl et C. Presl ■

353230　Scirpus atrosanguineus Boeck. = Abildgaardia setifolia（Hochst. ex A. Rich.）Lye ■☆

353231　Scirpus atrovirens Willd.；暗绿藨草；Black Bulrush，Common Bulrush，Dark-green Bulrush ■☆

353232　Scirpus atrovirens Willd. f. proliferus F. J. Herm. = Scirpus atrovirens Willd. ■☆

353233　Scirpus atrovirens Willd. f. sychnocephalus（Cowles）S. F. Blake = Scirpus atrovirens Willd. ■☆

353234　Scirpus atrovirens Willd. subvar. viviparus Farw. = Scirpus hattorianus Makino ■☆

353235　Scirpus atrovirens Willd. var. flaccidifolius Fernald = Scirpus flaccidifolius（Fernald）Schuyler ■☆

353236　Scirpus atrovirens Willd. var. georgianus（R. M. Harper）Fernald = Scirpus georgianus R. M. Harper ■☆

353237　Scirpus atrovirens Willd. var. pallidus Britton = Scirpus pallidus（Britton）Fernald ■☆

353238　Scirpus atrovirens Willd. var. pycnocephalus Fernald = Scirpus atrovirens Willd. ■☆

353239　Scirpus atrovirens Willd. var. viviparus Farw. = Scirpus hattorianus Makino ■☆

353240　Scirpus attenuatus Franch. et Sav. = Eleocharis attenuata（Franch. et Sav.）Palla ■

353241　Scirpus attenuatus Franch. et Sav. = Heleocharis attenuata（Franch. et Sav.）Palla ■

353242　Scirpus aureiglumis Hooper；金颖藨草■☆

353243　Scirpus aureiglumis S. S. Hooper = Schoenoplectus junceus（Willd.）J. Raynal ■☆

353244　Scirpus autumnalis L. = Fimbristylis autumnalis（L.）Roem. et Schult. ■

353245　Scirpus avatschensis Kom.；勘察加藨草■☆

353246　Scirpus baldwinianus Schult. = Fimbristylis annua（All.）Roem. et Schult. ■☆

353247　Scirpus barbatus Rottb. = Abildgaardia wallichiana（Schult.）Lye ■☆

353248　Scirpus barbatus Rottb. = Bulbostylis barbata（Rottb.）C. B. Clarke ■

353249　Scirpus bengalensis Pers. = Fimbristylis miliacea（L.）Vahl ■

353250　Scirpus bergianus Spreng. = Isolepis antarctica（L.）Roem. et Schult. ■☆

353251　Scirpus bergsonii Schuyler = Schoenoplectus saximontanus（Fernald）J. Raynal ■☆

353252　Scirpus bernardinus Munz et I. M. Johnst. = Eleocharis bernardina（Munz et I. M. Johnst.）Munz ■☆

353253　Scirpus biconcavus Ohwi = Bolboschoenus koshevnikovii（Litv.）A. E. Kozhevn. ■

353254　Scirpus biconcavus Ohwi = Scirpus planiculmis F. Schmidt ■

353255　Scirpus bisumbellatus Forssk. = Fimbristylis bisumbellata（Forssk.）Bubani ■

353256　Scirpus bivalvis Lam. = Fimbristylis bivalvis（Lam.）Lye ■☆

353257　Scirpus boeckelerianus Schweinf. = Abildgaardia boeckeleriana（Schweinf.）Lye ■☆

353258　Scirpus borealis（Ohwi）T. Koyama = Scirpus asiaticus Beetle ■

353259　Scirpus brachyceras Hochst. ex A. Rich. = Abildgaardia boeckeleriana（Schweinf.）Lye ■☆

353260　Scirpus brachyiphyllus Link = Bulbostylis capillaris（L.）Kunth ex C. B. Clarke ■☆

353261　Scirpus bracteatus Bigelow = Trichophorum cespitosum（L.）Schur ■☆

353262　Scirpus brevicaulis Levyns = Isolepis brevicaulis（Levyns）J. Raynal ■☆

353263　Scirpus breviculmis Boeck. = Bulbostylis humilis（Kunth）C. B. Clarke ■☆

353264　Scirpus briziformis Hutch. = Nemum spadiceum（Lam.）Desv. ex Ham. ■☆

353265　Scirpus bucharicus Roshev.；布哈尔藨草■☆

353266　Scirpus buettnerianus Boeck.；比特纳藨草■☆

353267　Scirpus bulbiferus Boeck. = Isolepis bulbifera（Boeck.）Muasya ■☆

353268　Scirpus bulbostyloides S. S. Hooper = Nemum bulbostyloides（S. S. Hooper）J. Raynal ■☆

353269　Scirpus burchellii C. B. Clarke = Isolepis brevicaulis（Levyns）J. Raynal ■☆

353270　Scirpus burkei C. B. Clarke = Scirpoides burkei（C. B. Clarke）Goetgh.，Muasya and D. A. Simpson ■☆

353271　Scirpus caducus Delile = Eleocharis caduca（Delile）Schult. ■☆

353272　Scirpus caespitosus L.；丛藨草；Deer-hair，Deer-hair Brush，Deerhair Bulrush，Deer's Hair，Tufted Club-rush ■☆

353273　Scirpus caespitosus L. var. austriacus（Palla）Asch. et Graebn. = Scirpus caespitosus L. ■☆

353274　Scirpus californicus（C. A. Mey.）Steud.；加州藨草；Southern Bulrush，Totora ■☆

353275　Scirpus californicus（C. A. Mey.）Steud. = Schoenoplectus californicus（C. A. Mey.）Soják ■☆

353276　Scirpus campestris L.；平原藨草；Marsh Bulrush ■☆

353277　Scirpus capillaris L. = Abildgaardia capillaris（L.）Lye ■☆

353278　Scirpus capillaris L. = Bulbostylis capillaris（L.）Kunth ex C. B. Clarke ■☆

353279　Scirpus capillifolius Parl. = Isolepis striata（Nees）Kunth ■☆

353280　Scirpus capitatus Willd. = Eleocharis braviseta Kurz et Skvortsov ■☆

353281　Scirpus capitatus Willd. = Heleocharis caribaea（Rottb.）Blake ■☆

353282　Scirpus caribaeus Rottb. = Eleocharis geniculata（L.）Roem. et Schult. ■

353283　Scirpus caribaeus Rottb. = Heleocharis caribaea（Rottb.）Blake ■☆

353284　Scirpus caricis C. B. Clarke = Blysmus sinocompressus Ts. Tang et F. T. Wang ■

353285　Scirpus carinatus（Hook. et Arn. ex Torr.）A. Gray = Isolepis carinata Hook. et Arn. ex Torr. ■☆

353286　Scirpus carolinianus Lam. = Fimbristylis caroliniana（Lam.）Fernald ■☆

353287　Scirpus cartilagineus（R. Br.）Poir. = Isolepis marginata（Thunb.）A. Dietr. ■☆

353288　Scirpus castaneus Michx. = Fimbristylis castanea（Michx.）Vahl ■☆

353289　Scirpus cephalotes Walter = Rhynchospora colorata（Hitchc.）H. Pfeiff. ■☆

353290　Scirpus cernuus Vahl；虎须草；Fiber Optics Plant，Slender Club-rush，Weeping Bulrush ■☆

353291　Scirpus cernuus Vahl = Isolepis cernua（Vahl）Roem. et

Schult. ■☆

53292　Scirpus cernuus Vahl subsp. californicus（Torr.）Thorne ＝ Isolepis cernua（Vahl）Roem. et Schult. ■☆

53293　Scirpus cernuus Vahl var. californicus（Torr.）Beetle ＝ Isolepis cernua（Vahl）Roem. et Schult. ■☆

53294　Scirpus cernuus Vahl var. subtilis（Kunth）C. B. Clarke ＝ Isolepis sepulcralis Steud. ■☆

53295　Scirpus cespitosus L. ＝ Baeothryon cespitosum（L.）A. Dietr. ■☆

353296　Scirpus cespitosus L. ＝ Trichophorum cespitosum（L.）Schur ■☆

353297　Scirpus cespitosus L. var. austriacus（Pall.）Asch. et Graebn. ＝ Trichophorum cespitosum（L.）Schur ■☆

353298　Scirpus cespitosus L. var. callosus Bigelow ＝ Baeothryon cespitosum（L.）A. Dietr. ■☆

353299　Scirpus cespitosus L. var. callosus Bigelow ＝ Trichophorum cespitosum（L.）Schur ■☆

353300　Scirpus cespitosus L. var. delicatulus Fernald ＝ Trichophorum cespitosum（L.）Schur ■☆

353301　Scirpus chen-moui Ts. Tang et F. T. Wang ＝ Schoenoplectus chen-moui（Ts. Tang et F. T. Wang）C. Y. Wu ■

353302　Scirpus chilensis Nees et Meyen ex Kunth ＝ Schoenoplectus pungens（Vahl）Palla ☆

353303　Scirpus chinensis Diels ＝ Scirpus rosthornii Diels ■

353304　Scirpus chinensis Munro ＝ Scirpus ternatanus Reinw. ex Miq. ■

353305　Scirpus chinensis Osbeck ＝ Lipocarpha chinensis（Osbeck）J. Kern. ■

353306　Scirpus chinensis Raymond ＝ Scirpus neochinensis Ts. Tang et F. T. Wang ■

353307　Scirpus chlorostachyus Levyns ＝ Isolepis sepulcralis Steud. ■☆

353308　Scirpus chuanus Ts. Tang et F. T. Wang；曲氏藨草；Chu Bulrush,Qu Bulrush ■

353309　Scirpus chuanus Ts. Tang et F. T. Wang ＝ Schoenoplectus chuanus（Ts. Tang et F. T. Wang）S. Yun ■

353310　Scirpus chunianus Ts. Tang et F. T. Wang；保亭藨草（陈氏藨草）；Chen Bulrush,Chun Bulrush ■

353311　Scirpus ciliaris L. ＝ Fuirena ciliaris（L.）Roxb. ■

353312　Scirpus ciliatifolius Elliott ＝ Bulbostylis ciliatifolia（Elliott）Fernald ■☆

353313　Scirpus cinnamometorus Vahl ＝ Fimbristylis cinnamometorum（Vahl）Kunth ■

353314　Scirpus cinnamomeus Boeck. ＝ Bulbostylis cinnamomea（Boeck.）C. B. Clarke ■☆

353315　Scirpus cinnamomeus Boeck. var. buchananii C. B. Clarke ＝ Abildgaardia buchananii（C. B. Clarke）Lye ■☆

353316　Scirpus clarkei Stapf ＝ Scirpus subcapitatus Thwaites et Hook. ■

353317　Scirpus clementis M. E. Jones ＝ Trichophorum clementis（M. E. Jones）S. G. Sm. ■☆

353318　Scirpus clintonii A. Gray；克林顿藨草；Clinton's Club-rush ■☆

353319　Scirpus clintonii A. Gray ＝ Trichophorum clintonii（A. Gray）S. G. Sm. ■☆

353320　Scirpus coarctatus Elliott ＝ Bulbostylis ciliatifolia（Elliott）Torr. var. coarctata（Elliott）Král ■☆

353321　Scirpus coerulescens Kuntze ＝ Fuirena coerulescens Steud. ■☆

353322　Scirpus cognatus Hance ＝ Scirpus triangulatus Roxb. ■

353323　Scirpus coleotrichus（Hochst. ex A. Rich.）Boeck. ＝ Abildgaardia coleotricha（Hochst. ex A. Rich.）Lye ■☆

353324　Scirpus collinus（Kunth）Boeck. var. boeckelerianus（Schweinf.）Schweinf. ＝ Abildgaardia boeckeleriana（Schweinf.）Lye ■☆

353325　Scirpus coloradoensis Britton ＝ Eleocharis coloradoensis（Britton）Gilly ■☆

353326　Scirpus complanatus Retz. ＝ Fimbristylis complanata（Retz.）Link ■

353327　Scirpus compressus（L.）Pers. ＝ Blysmus compressus（L.）Panz. ■

353328　Scirpus compressus Kük. ＝ Blysmus sinocompressus Ts. Tang et F. T. Wang ■

353329　Scirpus confervoides Poir. ＝ Websteria confervoides（Poir.）S. S. Hooper ■☆

353330　Scirpus confusus N. E. Br. ＝ Schoenoplectus confusus（N. E. Br.）Lye ■☆

353331　Scirpus congdonii Britton；康登藨草 ■☆

353332　Scirpus corymbosus（Roem. et Schult.）Roth ＝ Schoenoplectus corymbosus（Roth ex Roem. et Schult.）J. Raynal ■☆

353333　Scirpus corymbosus L. ＝ Rhynchospora corymbosa（L.）Britton ■

353334　Scirpus costatus（Hochst. ex A. Rich.）Boeck. ＝ Isolepis costata Hochst. ex A. Rich. ■☆

353335　Scirpus costatus（Hochst. ex A. Rich.）Boeck. var. macer（Boeck.）Cherm. ＝ Isolepis costata Hochst. ex A. Rich. ■☆

353336　Scirpus crassiusculus（Hook. f.）Benth. ＝ Isolepis crassiuscula Hook. f. ■☆

353337　Scirpus criniger A. Gray ＝ Eriophorum crinigerum（A. Gray）Beetle ■☆

353338　Scirpus cubensis Poepp. et Kunth ＝ Oxycaryum cubense（Poepp. et Kunth）Lye ■☆

353339　Scirpus cuspidatus Roem. et Schult. ＝ Tetraria cuspidata（Rottb.）C. B. Clarke ■☆

353340　Scirpus cymosus Lam. ＝ Fimbristylis cymosa（L.）R. Br. ■☆

353341　Scirpus cyperiformis Muhl. ＝ Cyperus lupulinus（Spreng.）Marcks ■☆

353342　Scirpus cyperinus（L.）Kunth；莎藨草（狼尾草）；Wool Grass, Wool-grass, Woolgrass Bulrush ■☆

353343　Scirpus cyperinus（L.）Kunth var. andrewsii（Fernald）Fernald ＝ Scirpus cyperinus（L.）Kunth ■☆

353344　Scirpus cyperinus（L.）Kunth var. brachypodus（Fernald）Gilly ＝ Scirpus atrocinctus Fernald ■☆

353345　Scirpus cyperinus（L.）Kunth var. concolor（Maxim.）Makino f. cylindricus Makino ＝ Scirpus wichurae Boeck. f. cylindricus（Makino）Nemoto ■☆

353346　Scirpus cyperinus（L.）Kunth var. condensatus Fernald ＝ Scirpus cyperinus（L.）Kunth ■☆

353347　Scirpus cyperinus（L.）Kunth var. eriophorum（Michx.）Kuntze ＝ Scirpus cyperinus（L.）Kunth ■☆

353348　Scirpus cyperinus（L.）Kunth var. laxus（A. Gray）Beetle ＝ Scirpus cyperinus（L.）Kunth ■☆

353349　Scirpus cyperinus（L.）Kunth var. pedicellatus（Fernald）Schuyler ＝ Scirpus pedicellatus Fernald ■☆

353350　Scirpus cyperinus（L.）Kunth var. pelius Fernald ＝ Scirpus cyperinus（L.）Kunth ■☆

353351　Scirpus cyperinus（L.）Kunth var. pelius Fernald f. condensatus（Fernald）S. F. Blake ＝ Scirpus cyperinus（L.）Kunth ■☆

353352　Scirpus cyperinus（L.）Kunth var. rubricosus（Fernald）Gilly ＝ Scirpus cyperinus（L.）Kunth ■☆

353353　Scirpus cyperinus（Vahl）Suringar ＝ Scirpus cyperinus（L.）Kunth ■☆

353354　Scirpus cyperoides L. = Cyperus cyperoides（L.）Kuntze ■☆

353355　Scirpus cyperoides L. = Mariscus sumatrensis（Retz.）J. Raynal ■

353356　Scirpus cyperoides L. = Mariscus umbellatus Vahl ■

353357　Scirpus cyperus Kük. = Scirpus asiaticus Beetle ■

353358　Scirpus debilis Pursh = Schoenoplectus purshianus（Fernald）M. T. Strong ■☆

353359　Scirpus debilis Pursh var. williamsii Fernald = Schoenoplectus purshianus（Fernald）M. T. Strong var. williamsii（Fernald）S. G. Sm. ■☆

353360　Scirpus delicatulus Levyns = Isolepis bulbifera（Boeck.）Muasya ■☆

353361　Scirpus deltarum Schuyler = Schoenoplectus deltarum（Schuyler）Soják ■☆

353362　Scirpus densus Wall. = Abildgaardia densa（Wall.）Lye ■☆

353363　Scirpus densus Wall. = Bulbostylis densa（Wall.）Hand. -Mazz. ■

353364　Scirpus densus Wall. ex Roxb. = Bulbostylis dense（Wall. ex Roxb.）Hand. -Mazz. ■

353365　Scirpus depauperatus Kom. = Schoenoplectus nipponicus（Makino）Soják ■☆

353366　Scirpus diabolicus Steud. = Isolepis diabolica（Steud.）Schrad. ■☆

353367　Scirpus dichotomus L. = Fimbristylis dichotoma（L.）Vahl ■

353368　Scirpus diffusus Schuyler；铺散藨草 ■☆

353369　Scirpus digitatus（Schrad.）Boeck. = Isolepis digitata Schrad. ■☆

353370　Scirpus diphyllus Retz. = Fimbristylis dichotoma（L.）Vahl ■

353371　Scirpus dipsaceus Rottb. = Fimbristylis dipsacea（Rottb.）C. B. Clarke ■

353372　Scirpus distichus Peterm. = Blysmus compressus（L.）Link ■

353373　Scirpus distigmaticus（Degl.）Ts. Tang et F. T. Wang = Trichophorum distigmaticum（Degl.）Egorova ■

353374　Scirpus distigmaticus（Kük.）Ts. Tang et F. T. Wang；双柱头藨草；Bitigma Bulrush，Double-stigma Bulrush ■

353375　Scirpus divaricatus Elliott；叉枝藨草；Bulrush ■☆

353376　Scirpus dregeanus C. B. Clarke = Isolepis capensis Muasya ■☆

353377　Scirpus dussii Boeck. = Bulbostylis barbata（Rottb.）C. B. Clarke ■

353378　Scirpus echinatus L. = Cyperus echinatus（L.）A. W. Wood ■☆

353379　Scirpus echrenbergii Boeck. = Schoenoplectus ehrenbergii（Boeck.）Y. C. Yang et M. Zhan ■

353380　Scirpus eckloneus Steud. = Ficinia ecklonea（Steud.）Nees ■☆

353381　Scirpus ehrenbergii Boeck. ；剑苞藨草（伊氏藨草）；Ehrenberg Bulrush ■

353382　Scirpus emergens（Norman）Fernald = Scirpus pumilus Vahl ■

353383　Scirpus emergens（Norman）Fernald = Trichophorum pumilum（Vahl）Schinz et Thell. ■

353384　Scirpus enodis（C. B. Clarke）T. Koyama = Fuirena coerulescens Steud. ■☆

353385　Scirpus equisetoides Elliott = Eleocharis equisetoides（Elliott）Torr. ■☆

353386　Scirpus equitans Kük. = Nemum equitans（Kük.）J. Raynal ■☆

353387　Scirpus erectogracilis Hayata = Schoenoplectus lateriflorus（J. F. Gmel.）Lye ■☆

353388　Scirpus erectogracilis Hayata = Schoenoplectus supinus（L.）Palla subsp. lateriflorus（J. F. Gmel.）Soják ■

353389　Scirpus erectogracilis Hayata = Scirpus supinus L. var. lateriflorus（S. G. Gmel.）T. Koyama ■

353390　Scirpus erectus Diels = Schoenoplectus juncoides（Roxb.）Palla ■

353391　Scirpus erectus Diels = Scirpus juncoides Roxb. ■

353392　Scirpus erectus Poir. = Schoenoplectus erectus（Poir.）Palla ex J. Raynal ■☆

353393　Scirpus erectus Poir. = Scirpus supinus L. var. lateriflorus（S. G. Gmel.）T. Koyama ■

353394　Scirpus erectus Poir. var. triangularis Honda = Schoenoplectus trapezoideus（Koidz.）Hagasaka et H. Ohashi ■

353395　Scirpus erectus Poir. var. wallichii（Nees）Beetle = Schoenoplectus wallichii（Nees）T. Koyama ■

353396　Scirpus eriophorum C. B. Clarke = Scirpus asiaticus Beetle ■

353397　Scirpus eriophorum Michx. = Scirpus cyperinus（L.）Kunth ■☆

353398　Scirpus erismanae Schuyler = Schoenoplectus erectus（Poir.）Palla ex J. Raynal subsp. raynalii（Schuyler）Lye ■☆

353399　Scirpus etuberculatus（Steud.）Kuntze = Schoenoplectus etuberculatus（Steud.）Soják ■☆

353400　Scirpus etuberculatus（Steud.）Kuntze subsp. nipponicus（Makino）T. Koyama = Schoenoplectus nipponicus（Makino）Soják ■☆

353401　Scirpus eupaluster H. Lindb. = Eleocharis palustris（L.）Roem. et Schult. ■

353402　Scirpus expallescens（Kunth）Boeck. = Isolepis expallescens Kunth ■☆

353403　Scirpus expansus Fernald；扩展藨草 ■☆

353404　Scirpus falsus C. B. Clarke；假藨草 ■☆

353405　Scirpus fastigiatus Thunb. = Ficinia fastigiata（Thunb.）Nees ■☆

353406　Scirpus fauriei（E. G. Camus）T. Koyama = Eriophorum vaginatum L. ■

353407　Scirpus fernaldii E. P. Bicknell = Bolboschoenus maritimus（L.）Palla ■☆

353408　Scirpus ferrugineus L. = Fimbristylis ferruginea（L.）Vahl ■

353409　Scirpus ficinioides Kunth；无花果藨草 ■☆

353410　Scirpus filamentosus Vahl = Bulbostylis filamentosa（Vahl）C. B. Clarke ■☆

353411　Scirpus filipes C. B. Clarke；细辐射枝藨草（细枝藨草）；Threadstalk Bulrush ■

353412　Scirpus filipes C. B. Clarke var. paucispiculatus Ts. Tang et F. T. Wang；少花细枝藨草 ■

353413　Scirpus filipes C. B. Clarke var. paurhyncispiculatus Ts. Tang et F. T. Wang；少穗细柄藨草；Few-spiked et Slender-stem Bulrush ■

353414　Scirpus fistulosus Forssk. = Schoenoplectiella articulata（L.）Lye ■☆

353415　Scirpus flaccidifolius（Fernald）Schuyler；软叶藨草 ■☆

353416　Scirpus flaccifolius Steud. = Isolepis digitata Schrad. ■☆

353417　Scirpus flavescens Poir. = Eleocharis flavescens（Poir.）Urb. ■☆

353418　Scirpus fluitans L. = Isolepis fluitans（L.）R. Br. ■☆

353419　Scirpus fluitans L. subsp. pseudofluitans（Makino）T. Koyama = Isolepis crassiuscula Hook. f. ■☆

353420　Scirpus fluitans L. var. fasciicularis（Nees）Boeck. = Isolepis fluitans（L.）R. Br. ■☆

353421　Scirpus fluitans L. var. robustus Boeck. = Isolepis striata（Nees）Kunth ■☆

353422　Scirpus fluviatilis（Torr.）A. Gray；河岸藨草；River Bulrush ■☆

353423　Scirpus fluviatilis（Torr.）A. Gray = Bolboschoenus fluviatilis（Torr.）Soják ■☆

353424　Scirpus fluviatilis（Torr.）A. Gray var. yagara（Ohwi）T. Koyama = Bolboschoenus fluviatilis（Torr.）Soják subsp. yagara（Ohwi）T. Koyama ■☆

353425　Scirpus fohaiensis Ts. Tang et F. T. Wang = Schoenoplectus fohaiensis（Ts. Tang et F. T. Wang）C. Y. Wu ■

53426　Scirpus fontinalis R. M. Harper = Scirpus lineatus Michx. ■☆

53427　Scirpus fuirena T. Koyama = Fuirena umbellata Rottb. ■

53428　Scirpus fuirenoides Courtois = Scirpus karuizawensis Makino ■

53429　Scirpus fuirenoides Maxim. ；芙兰蔗草■☆

53430　Scirpus fuirenoides Maxim. subsp. jaluanus（Kom.）T. Koyama = Scirpus karuizawensis Makino ■

53431　Scirpus fuirenoides Maxim. var. jaluanus Kom. = Scirpus karuizawensis Makino ■

353432　Scirpus fuirenoides Maxim. var. karuisawensis（Makino）H. Hara = Scirpus karuizawensis Makino ■

353433　Scirpus geniculatus L. = Eleocharis geniculata（L.）Roem. et Schult. ■

353434　Scirpus georgianus R. M. Harper；乔治亚蔗草；Bristleless Dark-green Bulrush, Common Bulrush, Georgia Bulrush ■☆

353435　Scirpus glaucus Lam. = Bolboschoenus glaucus（Lam.）S. G. Sm. ■☆

353436　Scirpus glaucus Sm. = Schoenoplectus lacustris（L.）Palla subsp. tabernaemontanii（C. C. Gmel.）Á. Löve et D. Löve ■

353437　Scirpus glaucus Torr. = Eleocharis erythropoda Steud. ■☆

353438　Scirpus globiceps C. B. Clarke = Isolepis rubicunda（Nees）Kunth ■☆

353439　Scirpus globulosus Retz. = Fimbristylis globulosa（Retz.）Kunth ■☆

353440　Scirpus glomeratus L. = Kyllinga brevifolia Rottb. ■

353441　Scirpus gracillimus Boeck. = Abildgaardia densa（Wall.）Lye ■☆

353442　Scirpus graminoides R. W. Haines et Lye = Isolepis graminoides（R. W. Haines et Lye）Lye ■☆

353443　Scirpus grandispicus（Steud.）Berhaut = Bolboschoenus grandispicus（Steud.）Lewej. et Lobin ■☆

353444　Scirpus griquensium C. B. Clarke = Isolepis sepulcralis Steud. ■☆

353445　Scirpus grossus L. f. = Actinoscirpus grossus（L. f.）Goetgh. et D. A. Simpson ■

353446　Scirpus grossus L. f. = Schoenoplectus grossus（L. f.）Palla ■☆

353447　Scirpus guaraniticus Pedersen = Schoenoplectus erectus（Poir.）Palla ex J. Raynal ■☆

353448　Scirpus hainanensis S. M. Hwang；海南蔗草；Hainan Bulrush ■

353449　Scirpus hallii A. Gray = Schoenoplectus hallii（A. Gray）S. G. Sm. ■☆

353450　Scirpus hamulosus（M. Bieb.）Steven = Mariscus hamulosus（M. Bieb.）S. S. Hooper ■☆

353451　Scirpus hattorianus Makino；早蔗草；Early Dark-green Bulrush, Mosquito Bulrush ■☆

353452　Scirpus heleocharidioides F. T. Wang et Ts. Tang；羊胡子蔗草■☆

353453　Scirpus heleocharidioides F. T. Wang et Ts. Tang = Scirpus schansiensis（Hand.-Mazz.）Ts. Tang et F. T. Wang ■

353454　Scirpus heleocharidioides F. T. Wang et Ts. Tang = Trichophorum schansiense Hand.-Mazz. ■

353455　Scirpus hemisphaericus Roth = Lipocarpha hemisphaerica（Roth）Goetgh. ■☆

353456　Scirpus hemiuncialis C. B. Clarke = Isolepis hemiuncialis（C. B. Clarke）J. Raynal ■☆

353457　Scirpus heterochaetus Chase；细蔗草；Heterochetous Bulrush, Slender Bulrush ■☆

353458　Scirpus heterochaetus Chase = Schoenoplectus heterochaetus（Chase）Soják ■☆

353459　Scirpus hildebrandtii Boeck. = Abildgaardia hispidula（Vahl）Lye ■☆

353460　Scirpus hispidulus Vahl = Abildgaardia hispidula（Vahl）Lye ■☆

353461　Scirpus hochstetteri Boeck. = Abildgaardia pusilla（Hochst. ex A. Rich.）Lye ■☆

353462　Scirpus holoschoenus L. ；西班牙蔗草■☆

353463　Scirpus holoschoenus L. = Scirpoides holoschoenus（L.）Soják ■☆

353464　Scirpus holoschoenus L. subsp. globiferus（L. f.）Husn. = Scirpoides holoschoenus（L.）Soják ■☆

353465　Scirpus holoschoenus L. var. australis（Murray）Koch = Scirpoides holoschoenus（L.）Soják subsp. australis（Murray）Soják ■☆

353466　Scirpus holoschoenus L. var. globiferus（L. f.）Parl. = Scirpoides holoschoenus（L.）Soják ■☆

353467　Scirpus holoschoenus L. var. hayekii Maire = Scirpoides holoschoenus（L.）Soják ■☆

353468　Scirpus holoschoenus L. var. macrostachyus Husn. = Scirpoides holoschoenus（L.）Soják ■☆

353469　Scirpus holoschoenus L. var. romanus（L.）Koch = Scirpoides romanus（L.）Soják ■☆

353470　Scirpus holoschoenus L. var. thunbergii（Schrad.）C. B. Clarke = Scirpoides thunbergii（Schrad.）Soják ■☆

353471　Scirpus holoschoenus L. var. vulgaris Koch = Scirpoides holoschoenus（L.）Soják ■☆

353472　Scirpus hondoensis Ohwi = Schoenoplectus hondoensis（Ohwi）Soják ■

353473　Scirpus hondoensis Ohwi var. leiocarpus（Kom.）Ohwi = Schoenoplectus komarovii（Roshev.）Soják ■

353474　Scirpus hotarui Ohwi = Schoenoplectus hotarui（Ohwi）Holub ■

353475　Scirpus hotarui Ohwi = Schoenoplectus juncoides（Roxb.）Palla subsp. hotarui（Ohwi）Soják ■

353476　Scirpus hotarui Ohwi = Scirpus juncoides Roxb. subsp. hotarui（Ohwi）T. Koyama ■

353477　Scirpus hottentotus L. = Fuirena hirsuta（P. J. Bergius）P. L. Forbes ■☆

353478　Scirpus hudsonianus（Michx.）Fernald = Trichophorum alpinum（L.）Pers. ■☆

353479　Scirpus hystricoides B. Nord. = Lipocarpha rehmannii（Ridl.）Goetgh. ■☆

353480　Scirpus hystrix Thunb. = Isolepis hystrix（Thunb.）Nees ■☆

353481　Scirpus inanis（Thunb.）Steud. = Pseudoschoenus inanis（Thunb.）Oteng-Yeb. ■☆

353482　Scirpus inclinatus（Delile ex Barbey）Asch. et Schweinf. = Schoenoplectus corymbosus（Roth ex Roem. et Schult.）J. Raynal ■☆

353483　Scirpus inclinatus（Delile ex Barbey）Boiss. = Schoenoplectus corymbosus（Roth ex Roem. et Schult.）J. Raynal ■☆

353484　Scirpus incomtulus Boeck. = Isolepis incomtula Nees ■☆

353485　Scirpus inconspicuus Levyns = Isolepis inconspicua（Levyns）J. Raynal ■☆

353486　Scirpus iniliacea L. ；龙须草■

353487　Scirpus intermedius F. T. Wang et Ts. Tang；荆门蔗草；Jingmen Bulrush ■

353488　Scirpus intermedius Muhl. = Eleocharis intermedia（Muhl.）Schult. ■☆

353489　Scirpus interstinctus Vahl = Eleocharis interstincta（Vahl）Roem. et Schult. ■☆

353490　Scirpus intricatus L. = Mariscus intricatus（L.）Cufod. ■☆

353491　Scirpus iseensis T. Koyama et T. Shimizu = Scirpus planiculmis F. Schmidt ■

353492　Scirpus isolepis（Nees）Boeck. = Lipocarpha hemisphaerica

（Roth）Goetgh. ■☆

353493 Scirpus jacobi（C. E. C. Fisch.）Lye = Schoenoplectus senegalensis（Hochst. ex Steud.）Palla ex J. Raynal ■☆

353494 Scirpus jaluanus（Kom.）Nakai ex Mori = Scirpus karuizawensis Makino ■

353495 Scirpus japonicus Fernald = Eriophorum japonicum Maxim. ■

353496 Scirpus jingmenensis Ts. Tang et F. T Wang = Schoenoplectus jingmenensis（Ts. Tang et F. T. Wang）S. Yun Liang et S. R. Zhang ■

353497 Scirpus jingmenensis Ts. Tang et F. T. Wang = Scirpus intermedius F. T. Wang et Ts. Tang ■

353498 Scirpus juncoides Roxb. = Schoenoplectus juncoides（Roxb.）Palla ■

353499 Scirpus juncoides Roxb. subsp. hotarui（Ohwi）T. Koyama = Schoenoplectus hotarui（Ohwi）Holub ■

353500 Scirpus juncoides Roxb. var. digynus（Boeck.）T. Koyama = Schoenoplectus purshianus（Fernald）M. T. Strong ■☆

353501 Scirpus juncoides Roxb. var. hotarui（Ohwi）Ohwi = Schoenoplectus hotarui（Ohwi）Holub ■

353502 Scirpus juncoides Roxb. var. hotarui（Ohwi）Ohwi = Schoenoplectus juncoides（Roxb.）Palla ■

353503 Scirpus juncoides Roxb. var. hotarui（Owhi）Ohwi = Schoenoplectus juncoides（Roxb.）Palla subsp. hotarui（Ohwi）Soják ■

353504 Scirpus juncoides Roxb. var. ohwianus（T. Koyama）T. Koyama = Schoenoplectus juncoides（Roxb.）Palla ■

353505 Scirpus juncoides Roxb. var. triangularis（Honda）Ohwi = Schoenoplectus trapezoideus（Koidz.）Hagasaka et H. Ohashi ■

353506 Scirpus juncoides Roxb. var. williamsii（Fernald）T. Koyama = Schoenoplectus purshianus（Fernald）M. T. Strong var. williamsii（Fernald）S. G. Sm. ■☆

353507 Scirpus kalli Forssk. = Cyperus capitatus Vand. ■☆

353508 Scirpus kamtschaticus C. A. Mey. = Heleocharis kamtschatica（C. A. Mey.）Kom. ■

353509 Scirpus karroicus C. B. Clarke = Isolepis karroica（C. B. Clarke）J. Raynal ■☆

353510 Scirpus karuizawensis Makino；华东藨草；Oriental Bulrush ■

353511 Scirpus kernii Raymond = Lipocarpha kernii（Raymond）Goetgh. ■☆

353512 Scirpus kirkii（C. B. Clarke）K. Schum. = Bulbostylis contexta（Nees）M. Bodard ■☆

353513 Scirpus kiushuensis Ohwi = Scirpus rosthornii Diels var. kiushuensis（Ohwi）Ohwi ■☆

353514 Scirpus koilolepis（Steud.）Gleason = Isolepis carinata Hook. et Arn. ex Torr. ■☆

353515 Scirpus komarovii Roshev. = Schoenoplectus komarovii（Roshev.）Soják ■

353516 Scirpus komarovii Roshev. f. laevis Hiyama；平滑藨草■☆

353517 Scirpus kyllingioides（A. Rich.）Boeck. = Kyllingiella microcephala（Steud.）R. W. Haines et Lye ■☆

353518 Scirpus kysoor Roxb. ；印度藨草■☆

353519 Scirpus laciniatus Thunb. = Ficinia laciniata（Thunb.）Nees ■☆

353520 Scirpus lacustris Bunge = Scirpus validus Vahl ■

353521 Scirpus lacustris L. = Schoenoplectus lacustris（L.）Palla ■☆

353522 Scirpus lacustris L. = Schoenoplectus validus（Vahl）T. Koyama ■

353523 Scirpus lacustris L. = Scirpus asiaticus Beetle ■

353524 Scirpus lacustris L. = Scirpus validus Vahl ■

353525 Scirpus lacustris L. subsp. creber（Fernald）T. Koyama =

Schoenoplectus tabernaemontani（C. C. Gmel.）Palla ■

353526 Scirpus lacustris L. subsp. glaucus（Rchb.）Hartm. = Schoenoplectus tabernaemontani（C. C. Gmel.）Palla ■

353527 Scirpus lacustris L. subsp. glaucus（Sm.）Hartm. = Schoenoplectus lacustris（L.）Palla subsp. tabernaemontanii（C. C. Gmel.）Á. Löve et D. Löve ■

353528 Scirpus lacustris L. subsp. tabernaemontani（C. C. Gmel.）Syme = Schoenoplectus lacustris（L.）Palla subsp. tabernaemontanii（C. C. Gmel.）Á. Löve et D. Löve ■

353529 Scirpus lacustris L. subsp. tabernaemontani（C. C. Gmel.）Syme = Schoenoplectus tabernaemontani（C. C. Gmel.）Palla ■

353530 Scirpus lacustris L. subsp. validus（Vahl）T. Koyama = Schoenoplectus tabernaemontani（C. C. Gmel.）Palla ■

353531 Scirpus lacustris L. var. tabernaemontani（C. C. Gmel.）Döll = Schoenoplectus tabernaemontani（C. C. Gmel.）Palla ■

353532 Scirpus lacustris L. var. tabernaemontani（C. C. Gmel.）Palla = Schoenoplectus tabernaemontani（C. C. Gmel.）Palla ■

353533 Scirpus lacustris L. var. tabernaemontani（C. C. Gmel.）Trab. = Schoenoplectus lacustris（L.）Palla subsp. tabernaemontanii（C. C. Gmel.）Á. Löve et D. Löve ■

353534 Scirpus lacustris L. var. tabernaemontani（C. C. Gmel.）Trautv. ex Regel f. alboviridis Makino = Schoenoplectus tabernaemontani（C. C. Gmel.）Palla 'Zebrinus' ■☆

353535 Scirpus lacustris L. var. tabernaernontani C. P'ei et Shan = Scirpus validus Vahl ■

353536 Scirpus lacustris L. var. tenuiculmis E. Sheld. = Schoenoplectus heterochaetus（Chase）Soják ■☆

353537 Scirpus lacustris L. var. validus（Vahl）Kük. = Scirpus validus Vahl ■

353538 Scirpus laeteflorens C. B. Clarke = Bolboschoenus nobilis（Ridl.）Goetgh. et D. A. Simpson ■☆

353539 Scirpus laniferus Boeck. = Abildgaardia lanifera（Boeck.）Lye ■☆

353540 Scirpus lateriflorus J. F. Gmel. = Schoenoplectus lateriflorus（J. F. Gmel.）Lye ■☆

353541 Scirpus lateriflorus J. F. Gmel. = Schoenoplectus supinus（L.）Palla subsp. lateriflorus（J. F. Gmel.）Soják ■

353542 Scirpus lateriflorus J. F. Gmel. = Scirpus supinus L. var. lateriflorus（S. G. Gmel.）T. Koyama ■

353543 Scirpus laxiflorus Thwaites = Eleocharis laxiflora（Thwaites）H. Pfeiff. ■☆

353544 Scirpus laxiflorus Thwaites = Eleocharis ochrostachys Steud. ■

353545 Scirpus laxiflorus Thwaites = Heleocharis ochrostachys Steud. ■

353546 Scirpus leptostachyus（Kunth）Boeck. = Isolepis leptostachya Kunth ■☆

353547 Scirpus leptus（C. B. Clarke）Levyns = Eleocharis lepta C. B. Clarke ■☆

353548 Scirpus leucanthus Boeck. = Schoenoplectus leucanthus（Boeck.）J. Raynal ■☆

353549 Scirpus leucocoleus K. Schum. = Ficinia stolonifera Boeck. ■☆

353550 Scirpus limosus Schrad. = Eleocharis limosa（Schrad.）Schult. ■☆

353551 Scirpus lineatus Michx. ；条纹藨草■☆

353552 Scirpus lineatus Michx. subsp. wichurae（Boeck.）T. Koyama = Scirpus asiaticus Beetle ■

353553 Scirpus lineatus Michx. subsp. wichurae（Boeck.）T. Koyama = Scirpus wichurae Boeck. ■

353554 Scirpus lineatus Michx. subsp. wichurae（Boeck.）T. Koyama var. lushanensis（Ohwi）T. Koyama = Scirpus wichurae Boeck. var.

lushanensis（Ohwi）T. Koyama ■

353555 Scirpus lineatus Michx. subsp. wichurae（Boeck.）T. Koyama var. lushanensis（Ohwi）T. Koyama ＝ Scirpus asiaticus Beetle ■

353556 Scirpus lineatus Michx. subsp. wichurai（Bockeler）T. Koyama var. lushanensis（Ohwi）T. Koyama ＝ Scirpus lushanensis Ohwi ■

353557 Scirpus lineolatus（Franch. et Sav.）T. Koyama；线状匍匐茎藨草（匍匐莞草，线状藨草）；Linear Stole Bulrush ■

353558 Scirpus lineolatus Franch. et Sav. ＝ Schoenoplectus lineolatus（Franch. et Sav.）T. Koyama ■

353559 Scirpus lineolatus Franch. et Sav. ＝ Scirpus lineolatus（Franch. et Sav.）T. Koyama ■

353560 Scirpus lithospermus L. ＝ Scleria lithosperma（L.）Sw. ■

353561 Scirpus litoralis Schrad. ＝ Schoenoplectus litoralis（Schrad.）Palla ■

353562 Scirpus litoralis Schrad. subsp. thermalis（Trab.）Murb. ＝ Schoenoplectus subulatus（Vahl）Lye ■

353563 Scirpus litoralis Schrad. var. foliosus Jahand. et Weiller ＝ Schoenoplectus litoralis（Schrad.）Palla ■

353564 Scirpus litoralis Schrad. var. pterolepis（Nees）C. C. Towns. ＝ Schoenoplectus subulatus（Vahl）Lye ■

353565 Scirpus litoralis Schrad. var. subulatus（Vahl）Chiov. ＝ Schoenoplectus subulatus（Vahl）Lye ■

353566 Scirpus litoralis Schrad. var. subulatus（Vahl）T. Koyama ＝ Schoenoplectus littoralis（Schrad.）Palla subsp. subulatus（Vahl）Soják ■

353567 Scirpus litoralis Schrad. var. thermalis（Trab.）Maire ＝ Schoenoplectus subulatus（Vahl）Lye ■

353568 Scirpus litoralis Schrad. var. thermalis Trab. ＝ Schoenoplectus subulatus（Vahl）Lye ■

353569 Scirpus littoralis Schrad.；钻苞藨草（羽状刚毛藨草）；Coastal Bulrush ■

353570 Scirpus longii Fernald；朗氏藨草■☆

353571 Scirpus lorentzii Boeck. ＝ Bulbostylis juncoides（Vahl）Kük. ex Osten ☆

353572 Scirpus ludwigii（Steud.）Boeck. ＝ Isolepis ludwigii（Steud.）Kunth ☆

353573 Scirpus lugardii C. B. Clarke ＝ Cyperus hamulosus M. Bieb. ■☆

353574 Scirpus lupulinus（Nees）Roshev.；狼草■☆

353575 Scirpus lupulinus Spreng. ＝ Cyperus lupulinus（Spreng.）Marcks ■☆

353576 Scirpus lushanensis Ohwi；庐山藨草（茸球藨草，茸球荆三棱，三棱草，山交染）；Asian Bulrush，Lushan Bulrush ■

353577 Scirpus lushanensis Ohwi ＝ Scirpus asiaticus Beetle ■

353578 Scirpus lushanensis Ohwi ＝ Scirpus wichurae Boeck. var. lushanensis（Ohwi）T. Koyama ■

353579 Scirpus macer Boeck. ＝ Isolepis costata Hochst. ex A. Rich. ■☆

353580 Scirpus macrostachys Willd. ＝ Bolboschoenus glaucus（Lam.）S. G. Sm. ■☆

353581 Scirpus mamillatus H. Lindb. ＝ Eleocharis mamillata（H. Lindb.）H. Lindb. ■

353582 Scirpus marginatus Muhl. ＝ Eleocharis quadrangulata（Michx.）Roem. et Schult. ■☆

353583 Scirpus marginatus Thunb. ＝ Isolepis marginata（Thunb.）A. Dietr. ☆

353584 Scirpus mariqueter Ts. Tang et F. T. Wang；海三棱藨草■

353585 Scirpus maritimus Bunge ＝ Scirpus planiculmis F. Schmidt ■

353586 Scirpus maritimus C. B. Clarke ＝ Scirpus yagara Ohwi ■

353587 Scirpus maritimus L.；荆三棱（滨海藨草，三棱草，薹，砖子苗）；Maritime Bulrush，Saltmarsh Rush，Salt-Marsh Rush，Sea Bulrush，Sea Club-rush，Sea Scirpus ■

353588 Scirpus maritimus L. ＝ Bolboschoenus maritimus（L.）Palla ■☆

353589 Scirpus maritimus L. ＝ Scirpus planiculmis F. Schmidt ■

353590 Scirpus maritimus L. var. affinis C. B. Clarke ＝ Bolboshoenus strobilinus（Roxb.）V. I. Krecz. ■

353591 Scirpus maritimus L. var. affinis C. B. Clarke ＝ Scirpus planiculmis F. Schmidt ■

353592 Scirpus maritimus L. var. affinis C. B. Clarke ＝ Scirpus strobilinus Roxb. ■

353593 Scirpus maritimus L. var. compactus（Hoffm.）＝ Bolboschoenus maritimus（L.）Palla ■☆

353594 Scirpus maritimus L. var. digynus Godr. ＝ Bolboschoenus maritimus（L.）Palla ■☆

353595 Scirpus maritimus L. var. distigmaticus Maxim. ＝ Scirpus planiculmis F. Schmidt ■

353596 Scirpus maritimus L. var. fernaldii（E. P. Bicknell）Beetle ＝ Bolboschoenus maritimus（L.）Palla ■☆

353597 Scirpus maritimus L. var. fluviatilis Torr. ＝ Bolboschoenus fluviatilis（Torr.）Soják ■☆

353598 Scirpus maritimus L. var. laeteflorens（C. B. Clarke）Kük. ＝ Bolboschoenus nobilis（Ridl.）Goetgh. et D. A. Simpson ■☆

353599 Scirpus maritimus L. var. macrostachyus（Willd.）Vis. ＝ Bolboschoenus glaucus（Lam.）S. G. Sm. ■☆

353600 Scirpus maritimus L. var. macrostachyus Michx. ＝ Bolboschoenus robustus（Pursh）Soják ■☆

353601 Scirpus maritimus L. var. monostachyus E. Mey. ＝ Bolboschoenus maritimus（L.）Palla ■☆

353602 Scirpus maritimus L. var. nobilis（Ridl.）C. B. Clarke ＝ Bolboschoenus nobilis（Ridl.）Goetgh. et D. A. Simpson ■☆

353603 Scirpus maritimus L. var. paludosus（A. Nelson）Kük. ＝ Bolboschoenus maritimus（L.）Palla ■☆

353604 Scirpus mattfeldianus Degl. ＝ Trichophorum mattfeldianum（Kük.）S. Yun Liang ■

353605 Scirpus mattfeldianus Kük.；三棱秆藨草；Trianglestalk Bulrush，Triangulastalk Bulrush ■

353606 Scirpus maximowiczii C. B. Clarke；马氏藨草（佛焰苞藨草，日本羊胡子草）；Japan Cottonsedge ■

353607 Scirpus maximowiczii C. B. Clarke ＝ Eriophorum japonicum Maxim. ■

353608 Scirpus melanospermus C. A. Mey.；黑籽藨草■☆

353609 Scirpus membranaceus Thunb. ＝ Hellmuthia membranacea（Thunb.）R. W. Haines et Lye ■☆

353610 Scirpus michauxii Pers. ＝ Fimbristylis autumnalis（L.）Roem. et Schult. ■

353611 Scirpus michelianus L.；米歇尔藨草■

353612 Scirpus michelianus L. ＝ Cyperus michelianus（L.）Link ■

353613 Scirpus micranthus Vahl ＝ Lipocarpha maculata（Michx.）Torr. ■☆

353614 Scirpus micranthus Vahl ＝ Lipocarpha micrantha（Vahl）G. C. Tucker ■☆

353615 Scirpus micranthus Vahl var. drummondii（Nees）Mohlenbr. ＝ Lipocarpha drummondii（Nees）G. C. Tucker ■☆

353616 Scirpus microcarpus J. Presl et C. Presl var. longispicatus M. Peck ＝ Scirpus microcarpus J. R. Presl et G. Presl ■☆

353617 Scirpus microcarpus J. Presl et C. Presl var. rubrotinctus（Fernald）

M. E. Jones ＝Scirpus microcarpus J. R. Presl et G. Presl ■☆

353618　Scirpus microcarpus J. R. Presl et G. Presl；小果藨草；Panicled Bulrush ■☆

353619　Scirpus microcephalus (Steud.) Dandy ＝ Kyllingiella microcephala (Steud.) R. W. Haines et Lye ■☆

353620　Scirpus miliaceus L. ＝Fimbristylis miliacea (L.) Vahl ■

353621　Scirpus minutus Turrill ＝Isolepis minuta (Turrill) J. Raynal ■☆

353622　Scirpus mitratus Franch. et Sav. ＝ Eleocharis kamtschatica (C. A. Mey.) Kom. f. reducta Ohwi ■

353623　Scirpus mitratus Franch. et Sav. ＝ Heleocharis kamtschatica (C. A. Mey.) Kom. f. reducta Ohwi ■

353624　Scirpus mitsukurianus Makino；箕作藨草■☆

353625　Scirpus molestus M. C. Johnst. ＝ Isolepis pseudosetacea (Daveau) Gand. ■☆

353626　Scirpus monocephalus J. Q. He；单头藨草；Monospke Bulrush ■

353627　Scirpus monocephalus J. Q. He ＝Schoenoplectus monocephalus (J. Q. He) S. Yun. Liang et S. R. Zhang ■

353628　Scirpus monophyllus J. Presl et C. Presl ＝ Schoenoplectus pungens (Vahl) Palla ■☆

353629　Scirpus montanus Kunth ＝Eleocharis montana (Kunth) Roem. et Schult. ■☆

353630　Scirpus morrisonensis Hayata ＝Scirpus subcapitatus Thwaites et Hook. ■

353631　Scirpus morrisonensis Hayata ＝Scirpus subcapitatus Thwaites et Hook. var. morrisonensis (Hayata) Ohwi ■

353632　Scirpus morrisonensis Hayata ＝ Trichophorum subcapitatum (Thwaites et Hook.) D. A. Simpson ■

353633　Scirpus mucronatus Diels ＝Scirpus triangulatus Roxb. ■

353634　Scirpus mucronatus L. ＝Schoenoplectus mucronatus (L.) Palla ■

353635　Scirpus mucronatus L. f. tataranus (Honda) T. Koyama ＝ Schoenoplectus mucronatus (L.) Palla var. tataranus (Honda) Kohno , Iokawa et Daigobo ■☆

353636　Scirpus mucronatus L. robustus Miq. ＝ Schoenoplectus mucronatus (L.) Palla var. robustus (Miq.) T. Koyama ■

353637　Scirpus mucronatus L. subsp. robustus (Miq.) T. Koyama ＝ Schoenoplectus mucronatus (L.) Palla subsp. robustus (Miq.) T. Koyama ■

353638　Scirpus mucronatus L. subsp. robustus (Miq.) T. Koyama ＝ Schoenoplectus triangulatus (Roxb.) Soják ■☆

353639　Scirpus mucronatus L. var. robustus Miq. ＝ Schoenoplectus mucronatus (L.) Palla subsp. robustus (Miq.) T. Koyama ■

353640　Scirpus mucronatus L. var. robustus Miq. ＝ Schoenoplectus triangulatus (Roxb.) Soják ■☆

353641　Scirpus mucronatus L. var. subieiocarpa Franch. et Savater ＝ Schoenoplectus mucronatus (L.) Palla subsp. robustus (Miq.) T. Koyama ■

353642　Scirpus mucronatus L. var. tataranus Honda ＝ Schoenoplectus mucronatus (L.) Palla var. tataranus (Honda) Kohno , Iokawa et Daigobo ■☆

353643　Scirpus mucronulatus Michx. ＝ Fimbristylis autumnalis (L.) Roem. et Schult. ■

353644　Scirpus muhlenbergii Spreng. ＝ Bulbostylis capillaris (L.) Kunth ex C. B. Clarke ■☆

353645　Scirpus multicaulis Sm. ＝Eleocharis multicaulis (Sm.) Desv. ■☆

353646　Scirpus multicostatus Baker ＝Isolepis costata Hochst. ex A. Rich. ■☆

353647　Scirpus muricinux C. B. Clarke ＝Schoenoplectus muricinux (C. B. Clarke) J. Raynal ■☆

353648　Scirpus muriculatus Kük. ＝ Schoenoplectus muriculatus (Kük.) Browning ■☆

353649　Scirpus mutatus L. ＝Eleocharis mutata (L.) Roem. et Schult. ■☆

353650　Scirpus nanodes Levyns ＝Isolepis pusilla Kunth ■☆

353651　Scirpus nanus Spreng. ＝Eleocharis parvula (Roem. et Schult.) Link ex Bluff , Nees et Schauer ■☆

353652　Scirpus nanus Spreng. ＝ Heleocharis parvula (Roem. et Schult.) Link ex Bluff , Nees et Schauer ■☆

353653　Scirpus natans Thunb. ＝Isolepis natans (Thunb.) A. Dietr. ■☆

353654　Scirpus neesii Boeck. ；尼斯藨草■☆

353655　Scirpus neochinensis Ts. Tang et F. T. Wang；新华藨草；New China Bulrush , Xinhua Bulrush ■

353656　Scirpus neochinensis Ts. Tang et F. T. Wang ＝ Isolepis squarrosa (L.) Roem. et Schult. ■

353657　Scirpus nervosus (Hochst. ex A. Rich.) Boeck. ＝ Isolepis fluitans (L.) R. Br. var. nervosa (Hochst. ex A. Rich.) Lye ■☆

353658　Scirpus nevadensis S. Watson ＝ Amphiscirpus nevadensis (S. Watson) Oteng-Yeb. ■☆

353659　Scirpus nindensis Ficalho et Hiern ＝ Abildgaardia sphaerocarpa (Boeck.) Lye ■☆

353660　Scirpus nipponicus Makino；三江藨草（日本藨草）■

353661　Scirpus nipponicus Makino ＝ Schoenoplectus nipponicus (Makino) Soják ■☆

353662　Scirpus nitens Vahl ＝ Rhynchospora nitens (Vahl) A. Gray ■☆

353663　Scirpus nobilis Ridl. ＝ Bolboschoenus nobilis (Ridl.) Goetgh. et D. A. Simpson ■☆

353664　Scirpus nodosus Rottb. ＝ Ficinia nodosa (Rottb.) Goetgh. , Muasya et D. A. Simpson ■☆

353665　Scirpus novae-angliae Britton ＝ Bolboschoenus novae-angliae (Britton) S. G. Sm. ■☆

353666　Scirpus nutans Retz. ＝ Fimbristylis nutans (Retz.) Vahl ■

353667　Scirpus obtusifolius Lam. ＝ Fimbristylis cymosa (L.) R. Br. ■

353668　Scirpus obtusifolius Lam. ＝ Fimbristylis obtusifolia (Lam.) Kunth ■

353669　Scirpus obtusus Willd. ＝ Eleocharis nitida Fernald ■☆

353670　Scirpus obtusus Willd. ＝ Eleocharis obtusa (Willd.) Schult. ■☆

353671　Scirpus occultus C. B. Clarke ＝ Cyperus michelianus (L.) Link subsp. pygmaeus (Rottb.) Asch. et Graebn. ■

353672　Scirpus ohwianus T. Koyama ＝ Schoenoplectus juncoides (Roxb.) Palla ■

353673　Scirpus okuyamae Ohwi ＝Schoenoplectus komarovii (Roshev.) Soják ■

353674　Scirpus oliganthus Steud. ＝ Ficinia oligantha (Steud.) J. Raynal ■☆

353675　Scirpus olneyi A. Gray ＝ Schoenoplectus americanus (Pers.) Volkart ex Schinz et R. Keller ■☆

353676　Scirpus orientalis Ohwi；东方藨草（朔北林生藨草）；Maximowicz Woodland Bulrush ■

353677　Scirpus orientalis Ohwi ＝Scirpus sylvaticus L. var. maximowiczii Regel ■

353678　Scirpus orientalis Ohwi ＝Scirpus sylvaticus L. ■☆

353679　Scirpus orthorhizomatus Arai et Miyam. ＝ Schoenoplectus orthorhizomatus (Arai et Miyam.) Hayas. et H. Ohashi ■☆

353680　Scirpus oryzetorum (Staud.) Ohwi ＝Scirpus supinus L. var. lateriflorus (S. G. Gmel.) T. Koyama ■

353681　Scirpus oryzetorum (Steud.) Ohwi ＝Schoenoplectus lateriflorus

（J. F. Gmel.）Lye ■☆

353682　Scirpus oryzetorum（Steud.）Ohwi = Schoenoplectus supinus（L.）Palla subsp. lateriflorus（J. F. Gmel.）Soják ■

353683　Scirpus ovatus Roth = Eleocharis ovata（Roth）Reom. et Schult. ■

353684　Scirpus ovatus Roth. = Heleocharis soloniensis（Dubois）H. Hara ■

353685　Scirpus oxyjulos S. S. Hooper = Schoenoplectus oxyjulos（S. S. Hooper）J. Raynal ■☆

353686　Scirpus pallidus（Britton）Fernald；苍白藨草；Common Bulrush，Pale Bulrush ■☆

353687　Scirpus paludicola Kunth = Schoenoplectus paludicola（Kunth）J. Raynal ■☆

353688　Scirpus paludicola Kunth f. decipiens（Nees）C. B. Clarke = Schoenoplectus decipiens（Nees）J. Raynal ■☆

353689　Scirpus paludosus A. Nelson；碱土藨草（硷土藨草）；Alkali Bulrush，Bayonet Grass，Prairie Bulrush ■☆

353690　Scirpus paludosus A. Nelson = Bolboschoenus maritimus（L.）Palla ■☆

353691　Scirpus palustris Franch. = Heleocharis valleculosa Ohwi f. setosa（Ohwi）Kitag. ■

353692　Scirpus palustris L. = Eleocharis palustris（L.）Roem. et Schult. ■

353693　Scirpus paniculato-corymbosus Kük.；高山藨草；Alpine Bulrush ■

353694　Scirpus parvinux（C. B. Clarke）K. Schum. = Bulbostylis parvinux C. B. Clarke ■☆

353695　Scirpus parvulus Roem. et Schult. = Eleocharis parvula（Roem. et Schult.）Link ex Bluff，Nees et Schauer ■☆

353696　Scirpus pauciflorus Lightf.；少花藨草；Fewflower Bulrush ■☆

353697　Scirpus pauciflorus Lightf. = Eleocharis quinqueflora（Hartm.）O. Schwarz ■

353698　Scirpus pauciflorus Lightf. var. campester（Roth）Asch. et Graebn. = Eleocharis quinqueflora（Hartm.）O. Schwarz ■

353699　Scirpus pedicellatus Fernald；具梗藨草；Bulrush，Pedicillate Wool-grass，Stalked Wool-grass ■☆

353700　Scirpus pedicellatus Fernald var. pullus Fernald = Scirpus pedicellatus Fernald ■☆

353701　Scirpus pendulus Muhl.；匍匐藨草；Bulrush，Rufous Bulrush ■☆

353702　Scirpus petasatus Maxim. = Heleocharis wichurai Boeck. ■

353703　Scirpus pictus Boeck. = Isolepis cernua（Vahl）Roem. et Schult. ■☆

353704　Scirpus pilosus Poir. = Fimbristylis pilosa（Poir.）Vahl ■☆

353705　Scirpus pinguiculus C. B. Clarke；肥厚藨草 ■☆

353706　Scirpus pitardii Trab. ex Pit. = Mariscus hamulosus（M. Bieb.）S. S. Hooper ■☆

353707　Scirpus planiculmis F. Schmidt；扁秆藨草（扁秆荆三棱，扁茎藨草，三棱，水莎草，云林莞草）；Flatstalk Bulrush ■

353708　Scirpus planiculmis F. Schmidt = Bolboschoenus planiculmis（F. Schmidt）T. V. Egorova ■

353709　Scirpus planifolius Muhl. = Trichophorum planifolium（Spreng.）Palla ■☆

353710　Scirpus planifolius Muhl. var. brevifolius Torr. = Trichophorum clintonii（A. Gray）S. G. Sm. ■☆

353711　Scirpus plantagineus Retz. = Heleocharis dulcis（Burm. f.）Trin. ex Hensch. ■

353712　Scirpus pollichi Gren. et Godr. = Scirpus triqueter L. ■

353713　Scirpus polyphyllus Vahl；繁叶藨草；Bulrush ■☆

353714　Scirpus polytrichoides Retz. = Fimbristylis polytrichoides（Retz.）R. Br. ■

353715　Scirpus preslii A. Dietr. = Schoenoplectus triangulatus（Roxb.）Soják ■☆

353716　Scirpus prestii A. Dietr. = Schoenoplectus mucronatus（L.）Palla subsp. robustus（Miq.）T. Koyama ■

353717　Scirpus pringlei Britton = Bulbostylis schaffneri（Boeck.）C. B. Clarke ■☆

353718　Scirpus prolifer Rottb. = Isolepis prolifera（Rottb.）R. Br. ■☆

353719　Scirpus pseudoarticulatus L. K. Dai et S. M. Huang = Schoenoplectus pseudoarticulatus（L. K. Dai et S. M. Huang）S. Yun. Liang et S. R. Zhang ■

353720　Scirpus pseudoarticulatus L. K. Dai et S. M. Hwang；节苞藨草 ■

353721　Scirpus pseudofluitans Makino = Isolepis crassiuscula Hook. f. ■☆

353722　Scirpus pseudosetaceus Daveau = Isolepis pseudosetacea（Daveau）Gand. ■☆

353723　Scirpus pterolepis（Nees）Kunth = Schoenoplectus subulatus（Vahl）Lye ■

353724　Scirpus puberula Poir. = Bulbostylis puberula（Poir.）C. B. Clarke ■

353725　Scirpus puberulus Michx. = Fimbristylis puberula（Michx.）Vahl ■☆

353726　Scirpus puberulus Poir. = Bulbostylis puberula（Poir.）C. B. Clarke ■

353727　Scirpus pubescens（Poir.）Lam. = Fuirena pubescens（Poir.）Kunth ■

353728　Scirpus pulchellus（Kunth）Boeck. = Schoenoplectus pulchellus（Kunth）J. Raynal ■☆

353729　Scirpus pumilus Vahl = Trichophorum pumilum（Vahl）Schinz et Thell. ■

353730　Scirpus pumilus Vahl subsp. distigmaticus Degl. = Trichophorum distigmaticum（Degl.）Egorova ■

353731　Scirpus pumilus Vahl subsp. distigmaticus Kük. = Scirpus distigmaticus（Kük.）Ts. Tang et F. T. Wang ■

353732　Scirpus pumilus Vahl subsp. rollandii（Fernald）Raymond = Scirpus pumilus Vahl ■

353733　Scirpus pumilus Vahl subsp. rollandii（Fernald）Raymond = Trichophorum pumilum（Vahl）Schinz et Thell. ■

353734　Scirpus pumilus Vahl var. rollandii（Fernald）Beetle = Scirpus pumilus Vahl ■

353735　Scirpus pumilus Vahl var. rollandii（Fernald）Beetle = Trichophorum pumilum（Vahl）Schinz et Thell. ■

353736　Scirpus pungens Vahl = Schoenoplectus pungens（Vahl）Palla ■☆

353737　Scirpus pungens Vahl subsp. monophyllus（J. Presl et C. Presl）R. L. Taylor et MacBryde = Schoenoplectus pungens（Vahl）Palla ■☆

353738　Scirpus pungens Vahl var. longispicatus（Britton）R. L. Taylor et MacBryde = Schoenoplectus pungens（Vahl）Palla ■☆

353739　Scirpus purpureoater Boeck. = Abildgaardia oligostachys（Hochst. ex A. Rich.）Lye ■☆

353740　Scirpus purshianus Fernald = Schoenoplectus purshianus（Fernald）M. T. Strong ■☆

353741　Scirpus purshianus Fernald f. williamsii（Fernald）Fernald = Schoenoplectus purshianus（Fernald）M. T. Strong var. williamsii（Fernald）S. G. Sm. ■☆

353742　Scirpus purshianus Fernald var. williamsii Fernald = Schoenoplectus purshianus（Fernald）M. T. Strong var. williamsii（Fernald）S. G. Sm. ■☆

353743 Scirpus quadrangulatus Michx. = Eleocharis quadrangulata（Michx.）Roem. et Schult. ■☆

353744 Scirpus quinquangularis Vahl = Fimbristylis miliacea（L.）Vahl ■

353745 Scirpus quinquangularis Vahl = Fimbristylis quinquangularis（Vahl）Kunth ■

353746 Scirpus quinquefarius Buch. -Ham. ex Boeck. = Schoenoplectus roylei（Nees）Ovcz. et Czukav. ■☆

353747 Scirpus quinqueflorus Hartm. = Eleocharis quinqueflora（Hartm.）O. Schwarz ■

353748 Scirpus radicans Franch. = Scirpus sylvaticus L. var. maximowiczii Regel ■

353749 Scirpus radicans Poir. = Eleocharis radicans（Poir.）Kunth ■☆

353750 Scirpus radicans Schkuhr;单穗藨草（东北藨草,根藨草）;NE. China Bulrush,Northeast Bulrush ■

353751 Scirpus ramosus Boeck. = Isolepis fluitans（L.）R. Br. ■☆

353752 Scirpus raynalii Schuyler = Schoenoplectus erectus（Poir.）Palla ex J. Raynal subsp. raynalii（Schuyler）Lye ■☆

353753 Scirpus raynalii Schuyler = Schoenoplectus erectus（Poir.）Palla ex J. Raynal ■☆

353754 Scirpus rehmannianus Boeck. = Schoenoplectiella articulata（L.）Lye ■☆

353755 Scirpus rehmannii Ridl. = Lipocarpha rehmannii（Ridl.）Goetgh. ■☆

353756 Scirpus retroflexus Poir. = Eleocharis retroflexa（Poir.）Urb. ■☆

353757 Scirpus rhodesicus Podlech = Schoenoplectus rhodesicus（Podlech）Lye ■☆

353758 Scirpus rivularis（Schrad.）Boeck. = Isolepis natans（Thunb.）A. Dietr. ■☆

353759 Scirpus robustus Pursh;粗壮藨草;Salt-marsh Bulrush ■☆

353760 Scirpus robustus Pursh = Bolboschoenus robustus（Pursh）Soják ■☆

353761 Scirpus robustus Pursh var. novae-angliae（Britton）Beetle = Bolboschoenus novae-angliae（Britton）S. G. Sm. ■☆

353762 Scirpus rogersii N. E. Br. = Schoenoplectus confusus（N. E. Br.）Lye var. rogersii（N. E. Br.）Lye ■☆

353763 Scirpus rollandii Fernald = Scirpus pumilus Vahl ■

353764 Scirpus rollandii Fernald = Trichophorum pumilum（Vahl）Schinz et Thell. ■

353765 Scirpus rongchengensis F. Z. Li;荣城藨草;Rongcheng Bulrush ■

353766 Scirpus rostellatus Torr. = Eleocharis rostellata（Torr.）Torr. ■☆

353767 Scirpus rosthornii Diels;百球藨草（百球荆三棱）;Rosthorn Bulrush ■

353768 Scirpus rosthornii Diels var. kiushuensis（Ohwi）Ohwi;基舒藨草 ■☆

353769 Scirpus rougchenensis F. Z. Li = Scirpus lushanensis Ohwi ■

353770 Scirpus roylei（Nees）R. Parker = Schoenoplectus roylei（Nees）Ovcz. et Czukav. ■☆

353771 Scirpus rubicundus（Nees）Parl. = Isolepis rubicunda（Nees）Kunth ■☆

353772 Scirpus rubricosus Fernald = Scirpus cyperinus（L.）Kunth ■☆

353773 Scirpus rubrotinctus Fernald;红藨草;Red-sheathed Bulrush ■☆

353774 Scirpus rubrotinctus Fernald = Scirpus microcarpus J. R. Presl et G. Presl ■☆

353775 Scirpus rubrotinctus Fernald f. confertus（Fernald）Weath. = Scirpus microcarpus J. R. Presl et G. Presl ■☆

353776 Scirpus rubrotinctus Fernald var. confertus Fernald = Scirpus microcarpus J. R. Presl et G. Presl ■☆

353777 Scirpus rufus（Huds.）Schrad. = Blysmus rufus（Huds.）Link ■

353778 Scirpus rufus（Huds.）Schrad. var. neogaeus Fernald = Blysmus rufus（Huds.）Link ■

353779 Scirpus russeolum（Fries）T. Koyama var. major（Sommier）T. Koyama = Eriophorum russeolum Fr. ■

353780 Scirpus russeolum（Fries）T. Koyama var. major T. Koyama = Eriophorum russeolum Fr. var. majus Sommier ■

353781 Scirpus sachcdinensis Meisn. = Eleocharis kamtschatica（C. A. Mey.）Kom. ■

353782 Scirpus sachcdinensis Meisn. = Heleocharis kamtschatica（C. A. Mey.）Kom. ■

353783 Scirpus sasakii Hayata = Schoenoplectus wallichii（Nees）T. Koyama ■

353784 Scirpus sasakii Hayata var. leiocarpus（Kom.）Kitag. = Schoenoplectus komarovii（Roshev.）Soják ■

353785 Scirpus savii Sebast. et Mauri = Isolepis cernua（Vahl）Roem. et Schult. ■☆

353786 Scirpus saximontanus Fernald;落基山藨草;Rocky Mountain Bulrush ■☆

353787 Scirpus saximontanus Fernald = Schoenoplectus saximontanus（Fernald）J. Raynal ■☆

353788 Scirpus schaffneri Boeck. = Bulbostylis schaffneri（Boeck.）C. B. Clarke ■☆

353789 Scirpus schansiensis（Hand. -Mazz.）Ts. Tang et F. T. Wang;太行山藨草;Taihang Bulrush,Taihang Mountain Bulrush ■

353790 Scirpus schansiensis（Hand. -Mazz.）Ts. Tang et F. T. Wang = Trichophorum schansiense Hand. -Mazz. ■

353791 Scirpus schimperianus（A. Rich.）Boeck. = Abildgaardia schimperiana（A. Rich.）Lye ■☆

353792 Scirpus schoenoides Elliott = Rhynchospora elliottii A. Dietr. ■☆

353793 Scirpus schoenoides Retz. = Fimbristylis schoenoides（Retz.）Vahl ■

353794 Scirpus schoofii Beetle;滇藨草（东川萤蔺）;Yunnan Bulrush ■

353795 Scirpus schoofii Beetle = Schoenoplectus schoofii（Beetle）C. Y. Wu ■

353796 Scirpus schweinfurthianus Boeck. = Abildgaardia abortiva（Steud.）Lye ■☆

353797 Scirpus scirpoideus（Michx.）T. Koyama = Fuirena scirpoidea Michx. ■☆

353798 Scirpus scleropus（C. B. Clarke）K. Schum. = Bulbostylis scleropus C. B. Clarke ■☆

353799 Scirpus semiuncialis C. B. Clarke;半钩藨草 ■☆

353800 Scirpus senegalensis Lam. = Lipocarpha chinensis（Osbeck）J. Kern. ■

353801 Scirpus senegalensis Lam. = Lipocarpha senegalensis（Lam.）Dandy ■

353802 Scirpus sericeus Poir. = Fimbristylis sericea（Poir.）R. Br. ■

353803 Scirpus setaceus L. = Isolepis setacea（L.）R. Br. ■

353804 Scirpus silvaticus L. var. subradcans Degl. ex Ts. Tang = Scirpus orientalis Ohwi ■

353805 Scirpus sinuatus Schuyler = Schoenoplectus erectus（Poir.）Palla ex J. Raynal ■☆

353806 Scirpus smithii A. Gray;史密斯藨草;Smith's Club-rush ■☆

353807 Scirpus smithii A. Gray = Schoenoplectus smithii（A. Gray）Soják ■☆

353808 Scirpus smithii A. Gray f. setosus（Fernald）Fernald = Schoenoplectus smithii（A. Gray）Soják var. setosus（Fernald）S. G. Sm. ■☆

353809　Scirpus smithii A. Gray subsp. leiocarpus（Kom.）T. Koyama ＝Schoenoplectus komarovii（Roshev.）Soják ■

353810　Scirpus smithii A. Gray var. leiocarpus（Kom.）T. Koyama ＝ Schoenoplectus komarovii（Roshev.）Soják ■

353811　Scirpus smithii A. Gray var. setosus Fernald ＝ Schoenoplectus smithii（A. Gray）Soják var. setosus（Fernald）S. G. Sm. ■☆

353812　Scirpus smithii A. Gray var. williamsii（Fernald）Beetle ＝ Schoenoplectus purshianus（Fernald）M. T. Strong var. williamsii（Fernald）S. G. Sm. ■☆

353813　Scirpus soloniensis Dubois ＝ Heleocharis soloniensis（Dubois）H. Hara ■

353814　Scirpus sororius（Kunth）C. B. Clarke ＝Isolepis sororia Kunth ■☆

353815　Scirpus spadiceus（Lam.）Boeck. ＝ Nemum spadiceum（Lam.）Desv. ex Ham. ■☆

353816　Scirpus spathaceus Hochst. ＝ Pseudoschoenus inanis（Thunb.）Oteng-Yeb. ■☆

353817　Scirpus sphaerocarpus Boeck. ＝ Abildgaardia sphaerocarpa（Boeck.）Lye ■☆

353818　Scirpus spiralis Rottb. ＝ Eleocharis spiralis（Rottb.）Roem. et Schult. ■

353819　Scirpus spiralis Rottb. ＝ Heleocharis spiralis（Rottb.）R. Br. ■

353820　Scirpus squafrosus L. ＝ Lipocarpha chinensis（Osbeck）J. Kern. ■

353821　Scirpus squarrosus L. ＝ Isolepis squarrosa（L.）Roem. et Schult. ■

353822　Scirpus squarrosus L. ＝ Lipocarpha microcephala（R. Br.）Kunth ■

353823　Scirpus squarrosus Poir. ＝Fimbristylis squarrosa（Poir.）Vahl ■

353824　Scirpus steinmetzii Fernald ＝ Schoenoplectus steinmetzii（Fernald）S. G. Sm. ■☆

353825　Scirpus stenophyllus Elliott ＝ Bulbostylis stenophylla（Elliott）C. B. Clarke ■☆

353826　Scirpus steudneri Boeck. ＝ Kyllingiella polyphylla（A. Rich.）Lye ■☆

353827　Scirpus striatus（Nees）Fourc. ＝Isolepis striata（Nees）Kunth ■☆

353828　Scirpus strobilinus Roxb.；球穗蔗草；Strobiloides Bulrush ■

353829　Scirpus strobilinus Roxb. ＝ Bolboshoenus strobilinus（Roxb.）V. I. Krecz. ■

353830　Scirpus subcapitatus Thwaites et Hook. ＝ Trichophorum subcapitatum（Thwaites et Hook.）D. A. Simpson ■

353831　Scirpus subcapitatus Thwaites et Hook. var. morrisonensis（Hayata）Ohwi ＝Scirpus subcapitatus Thwaites et Hook. ■

353832　Scirpus subcapitatus Thwaites et Hook. var. morrisonensis（Hayata）Ohwi ＝Trichophorum subcapitatum（Thwaites et Hook.）D. A. Simpson ■

353833　Scirpus subcapitatus Thwaites et Hook. var. morrisonensis（Hayata）Ohwi；台湾蔗草 ■

353834　Scirpus submersus C. Wright ＝ Websteria confervoides（Poir.）S. S. Hooper ■☆

353835　Scirpus submersus Sauvalle ＝ Websteria confervoides（Poir.）S. S. Hooper ■☆

353836　Scirpus subsquarrosus Muhl. ＝ Lipocarpha micrantha（Vahl）G. C. Tucker ■☆

353837　Scirpus subterminalis Torr.；亚顶生蔗草；Subterminal Club-rush ■☆

353838　Scirpus subterminalis Torr. ＝ Schoenoplectus subterminalis（Torr.）Soják ■☆

353839　Scirpus subterminalis Torr. subsp. nipponicus（Makino）T. Koyama ＝Schoenoplectus nipponicus（Makino）Soják ■☆

353840　Scirpus subulatus Vahl；羽状刚毛蔗草；Awlshaped Bulrush ■

353841　Scirpus subulatus Vahl ＝ Schoenoplectus littoralis（Schrad.）Palla subsp. subulatus（Vahl）Soják ■

353842　Scirpus subulatus Vahl ＝Schoenoplectus subulatus（Vahl）Lye ■

353843　Scirpus subulatus Vahl ＝Schoenoplectus subulatus（Vahl）S. Yun. Liang et S. R. Zhang ■

353844　Scirpus supinus Hell. subsp. uninodis（Delile）Trab. ＝ Schoenoplectus erectus（Poir.）Palla ex J. Raynal ■☆

353845　Scirpus supinus L.；仰卧秆蔗草；Prostrate Bulrush, Supine Bulrush ■

353846　Scirpus supinus L. ＝ Schoenoplectus supinus（L.）Palla ■

353847　Scirpus supinus L. subsp. uninodis（Delile）Trab. ＝ Schoenoplectus erectus（Poir.）Palla ex J. Raynal ■☆

353848　Scirpus supinus L. var. densicorrugatus Ts. Tang et F. T. Wang；多皱纹果仰卧秆蔗草；Densicorrugate Prostrate Bulrush ■

353849　Scirpus supinus L. var. digynus Boiss. ＝ Schoenoplectus supinus（L.）Palla ■

353850　Scirpus supinus L. var. hallii（A. Gray）A. Gray ＝ Schoenoplectus hallii（A. Gray）S. G. Sm. ■☆

353851　Scirpus supinus L. var. lateriflorus（S. G. Gmel.）T. Koyama ＝ Schoenoplectus supinus（L.）Palla subsp. lateriflorus（J. F. Gmel.）Soják ■

353852　Scirpus supinus L. var. leiocarpus Kom. ＝ Schoenoplectus komarovii（Roshev.）Soják ■

353853　Scirpus supinus L. var. saximontanus（Fernald）T. Koyama ＝ Schoenoplectus saximontanus（Fernald）J. Raynal ■☆

353854　Scirpus supinus W. C. Boeck var. lateriflorus（J. F. Gmel.）Koyama ＝ Schoenoplectus lateriflorus（J. F. Gmel.）Lye ■☆

353855　Scirpus supinus W. C. Boeck var. uninodis（Delile）Asch. et Schweinf. ＝ Schoenoplectus erectus（Poir.）Palla ex J. Raynal ■☆

353856　Scirpus sylvaticus L.；林生蔗草；Wood Club-rush, Woodland Bulrush ■

353857　Scirpus sylvaticus L. subsp. maximowiczii（Regel）N. R. Cui ＝ Scirpus orientalis Ohwi ■

353858　Scirpus sylvaticus L. subsp. maximowiczii（Regel）T. Koyama ＝ Scirpus orientalis Ohwi ■

353859　Scirpus sylvaticus L. var. digynus Boeck. ＝ Scirpus microcarpus J. Presl et C. Presl ■☆

353860　Scirpus sylvaticus L. var. maximowiczii Regel ＝ Scirpus orientalis Ohwi ■

353861　Scirpus sylvaticus L. var. radicans（Schkuhr）Willd. ＝ Scirpus radicans Schkuhr ■

353862　Scirpus sylvaticus L. var. subradicans Kük. ex Ts. Tang ＝ Scirpus sylvaticus L. var. maximowiczii Regel ■

353863　Scirpus sylvaticus Maxim. ＝ Scirpus sylvaticus L. var. maximowiczii Regel ■

353864　Scirpus tabernaemontani（C. C. Gmel.）Palla ＝Schoenoplectus tabernaemontani（C. C. Gmel.）Palla ■

353865　Scirpus tabernaemontani C. C. Gmel. 'Zebrinus' ＝Schoenoplectus tabernaemontani（C. C. Gmel.）Palla 'Zebrinus' ■☆

353866　Scirpus tabernaemontani C. C. Gmel. 'Zebrinus' ＝ Scirpus tabernaemontani C. C. Gmel. ■

353867　Scirpus tabernaemontani C. C. Gmel. ＝ Schoenoplectus tabernaemontani（C. C. Gmel.）Palla ■

353868　Scirpus tabernaemontani C. C. Gmel. f. australis Ohwi ＝ Schoenoplectus tabernaemontani（C. C. Gmel.）Palla ■

353869 Scirpus tabernaemontani C. C. Gmel. f. luxurians Miq. = Schoenoplectus tabernaemontani (C. C. Gmel.) Palla ■

353870 Scirpus tabernaemontani C. C. Gmel. f. luxuriaus Miq. = Schoenoplectus tabernaemontani (C. C. Gmel.) Palla ■

353871 Scirpus tabernaemontani C. C. Gmel. f. zebrinus (André) Asch. et Graebn. = Schoenoplectus tabernaemontani (C. C. Gmel.) Palla 'Zebrinus' ■☆

353872 Scirpus tabernaemontani Maxim. = Scirpus validus Vahl ■

353873 Scirpus tenerrimus Peter = Schoenoplectus microglumis Lye ■☆

353874 Scirpus tenuis Spreng. = Isolepis pusilla Kunth ■☆

353875 Scirpus tenuis Willd. = Eleocharis tenuis (Willd.) Schult. ■☆

353876 Scirpus tenuissimus (Nees) Boeck. = Isolepis tenuissima (Nees) Kunth ■☆

353877 Scirpus ternatanus C. B. Clarke = Scirpus ternatanus Reinw. ex Miq. ■

353878 Scirpus ternatanus Reinw. ex Miq. ; 百穗藨草 (大莞草) ; Manyspike Bulrush ■

353879 Scirpus ternatanus Reinw. ex Miq. subsp. kiushuensis (Ohwi) T. Koyama = Scirpus rosthornii Diels var. kiushuensis (Ohwi) Ohwi ■☆

353880 Scirpus ternatanus Reinw. ex Miq. var. kiushuensis (Ohwi) T. Koyama = Scirpus rosthornii Diels var. kiushuensis (Ohwi) Ohwi ■☆

353881 Scirpus thermalis Trab. = Schoenoplectus subulatus (Vahl) Lye ■

353882 Scirpus thunbergianus (Nees) Levyns = Scirpoides thunbergii (Schrad.) Soják ■☆

353883 Scirpus thunbergii A. Spreng. = Schoenoplectus leucanthus (Boeck.) J. Raynal ■☆

353884 Scirpus torreyi Olney = Schoenoplectus torreyi (Olney) Palla ■☆

353885 Scirpus tortilis Link = Eleocharis tortilis (Link) Schult. ■☆

353886 Scirpus trachyspermus (Nees) C. B. Clarke = Isolepis trachysperma Nees ■☆

353887 Scirpus trapezoideus Koidz. = Schoenoplectus trapezoideus (Koidz.) Hagasaka et H. Ohashi ■

353888 Scirpus trialatus Boeck. = Mariscus trialatus (Boeck.) Ts. Tang et F. T. Wang ■

353889 Scirpus triangulatus Roxb. ; 水毛花 (茫草, 蒲草, 三角草, 三棱观, 水三棱, 水三棱草, 丝毛草, 席草, 蓆草) ; Triangular Bulrush ■

353890 Scirpus triangulatus Roxb. = Schoenoplectus mucronatus (L.) Palla subsp. robustus (Miq.) T. Koyama ■

353891 Scirpus triangulatus Roxb. = Schoenoplectus triangulatus (Roxb.) Soják ■☆

353892 Scirpus triangulatus Roxb. var. brevibracteatus T. Koyama = Schoenoplectus triangulatus (Roxb.) Soják ■☆

353893 Scirpus triangulatus Roxb. var. sanguineus Ts. Tang et F. T. Wang = Schoenoplectus mucronatus (L.) Palla subsp. robustus (Miq.) T. Koyama ■

353894 Scirpus triangulatus Roxb. var. sanguineus Ts. Tang et F. T. Wang = Schoenoplectus mucronatus (L.) Palla var. sanguineus (Ts. Tang et F. T. Wang) P. C. Li ■

353895 Scirpus triangulatus Roxb. var. trialatus Ts. Tang et F. T. Wang = Schoenoplectus mucronatus (L.) Palla subsp. robustus (Miq.) T. Koyama ■

353896 Scirpus triangulatus Roxb. var. tripteris Ts. Tang et F. T. Wang; 台水毛花 ; Taiwan Bulrush ■

353897 Scirpus triangulatus Roxb. var. tripteris Ts. Tang et F. T. Wang = Schoenoplectus mucronatus (L.) Palla subsp. robustus (Miq.) T. Koyama ■

353898 Scirpus trigynus L. = Ficinia deusta (P. J. Bergius) Levyns ■☆

353899 Scirpus triqueter L. ; 藨草 (三棱藨草) ; Triangular Bulrush, Triangular Scirpus ■

353900 Scirpus triqueter L. = Schoenoplectus triqueter (L.) Palla ■

353901 Scirpus trisetosus Ts. Tang et F. T. Wang ; 青岛藨草 ; Chingdao Bulrush, Qingdao Bulrush, Threesect Bulrush ■

353902 Scirpus trisetosus Ts. Tang et F. T. Wang = Schoenoplectus triqueter (L.) Palla ■

353903 Scirpus trispicatus L. f. = Ficinia trispicata (L. f.) Druce ■☆

353904 Scirpus tristachyos Rottb. = Ficinia tristachya (Rottb.) Nees ■☆

353905 Scirpus trollii Kük. = Ficinia trollii (Kük.) Muasya et D. A. Simpson ■☆

353906 Scirpus truncatus Thunb. = Ficinia truncata (Thunb.) Schrad. ■☆

353907 Scirpus tuberculosa Michx. var. pubnicoensis Fernald = Eleocharis tuberculosa (Michx.) Roem. et Schult. ■☆

353908 Scirpus tuberculosus Michx. = Eleocharis tuberculosa (Michx.) Roem. et Schult. ■☆

353909 Scirpus tuberosus Desf. = Bolboschoenus glaucus (Lam.) S. G. Sm. ■☆

353910 Scirpus tuberosus Roxb. = Heleocharis dulcis (Burm. f.) Trin. ex Hensch. ■

353911 Scirpus umbellaris Lam. = Fimbristylis umbellaris (Lam.) Vahl ■

353912 Scirpus uniglumis Link = Eleocharis uniglumis (Link) Schult. et Zinserl. ■

353913 Scirpus uniglumis Link = Heleocharis uniglumis (Link) Schult. ■

353914 Scirpus uninodis (Delile) Coss. = Schoenoplectus erectus (Poir.) Palla ex J. Raynal ■☆

353915 Scirpus uninodis Delile = Schoenoplectus erectus (Poir.) Palla ex J. Raynal ■☆

353916 Scirpus ustulatus Podlech = Nemum spadiceum (Lam.) Desv. ex Ham. ■☆

353917 Scirpus vahlii Lam. = Fimbristylis vahlii (Lam.) Link ■☆

353918 Scirpus validus Vahl = Schoenoplectus tabernaemontani (C. C. Gmel.) Palla ■

353919 Scirpus validus Vahl = Schoenoplectus validus (Vahl) T. Koyama ■

353920 Scirpus validus Vahl = Scirpus tabernaemontani C. C. Gmel. ■

353921 Scirpus validus Vahl var. creber Fernald = Schoenoplectus tabernaemontani (C. C. Gmel.) Palla ■

353922 Scirpus validus Vahl var. creber Fernald f. megastachyus Fernald = Schoenoplectus tabernaemontani (C. C. Gmel.) Palla ■

353923 Scirpus validus Vahl var. laeviglumis Ts. Tang et F. T. Wang = Schoenoplectus tabernaemontani (Gmel.) Palla ■

353924 Scirpus validus Vahl var. laeviglumis Ts. Tang et F. T. Wang = Schoenoplectus validus (Vahl) T. Koyama var. laeviglumis (Ts. Tang et F. T. Wang) C. Y. Wu ■

353925 Scirpus variegatus Poir. = Eleocharis variegata (Poir.) C. Presl ■☆

353926 Scirpus varius C. B. Clarke ; 多态藨草 ■☆

353927 Scirpus venustulus (Kunth) Boeck. = Isolepis venustula Kunth ■☆

353928 Scirpus verecundus Fernald = Trichophorum planifolium (Spreng.) Palla ■☆

353929 Scirpus verruciferus (Maxim.) Meinsh. = Fimbristylis verrucifera (Maxim.) Makino ■

353930 Scirpus verrucosulus Steud. = Isolepis cernua (Vahl) Roem. et Schult. var. setiformis (Benth.) Muasya ■☆

353931 Scirpus verrucosulus Steud. var. pterocaryon C. B. Clarke = Isolepis cernua (Vahl) Roem. et Schult. var. setiformis (Benth.) Muasya ■☆

353932　Scirpus wallichianus Spreng. = Bulbostylis barbata（Rottb.）Kunth ■

353933　Scirpus wallichii Nees = Schoenoplectus wallichii（Nees）T. Koyama ■

353934　Scirpus wichurae Boeck. = Scirpus asiaticus Beetle ■

353935　Scirpus wichurae Boeck. f. concolor（Maxim.）Ohwi;同色藨草■☆

353936　Scirpus wichurae Boeck. f. cylindricus（Makino）Nemoto;柱形藨草■☆

353937　Scirpus wichurae Boeck. subsp. asiaticus（Beetle）T. Koyama = Scirpus asiaticus Beetle ■

353938　Scirpus wichurae Boeck. subsp. lushanensis（Ohwi）T. Koyama = Scirpus wichurae Boeck. var. lushanensis（Ohwi）T. Koyama ■

353939　Scirpus wichurae Boeck. var. asiaticus（Beetle）T. Koyama ex Ohwi = Scirpus asiaticus Beetle ■

353940　Scirpus wichurae Boeck. var. borealis Ohwi = Scirpus asiaticus Beetle ■

353941　Scirpus wichurae Boeck. var. lushanensis（Ohwi）T. Koyama = Scirpus asiaticus Beetle ■

353942　Scirpus wichurae Kom. = Scirpus asiaticus Beetle ■

353943　Scirpus wightianus Boeck. = Ficinia gracilis Schrad. ■☆

353944　Scirpus wilkensii Schuyler = Schoenoplectus erectus（Poir.）Palla ex J. Raynal ■☆

353945　Scirpus wilkensii Schuyler = Schoenoplectus erectus（Poir.）Palla ex J. Raynal subsp. raynalii（Schuyler）Lye ■☆

353946　Scirpus wolfii A. Gray = Eleocharis wolfii（A. Gray）A. Gray ex Britton ■☆

353947　Scirpus yagara Ohwi = Bolboschoenus fluviatilis（Torr.）Soják subsp. yagara（Ohwi）T. Koyama ■☆

353948　Scirpus yagara Ohwi = Bolboschoenus yagara（Ohwi）A. E. Kozhevn. ■

353949　Scirpus yagara Ohwi = Bolboschoenus yagara（Ohwi）Y. C. Yang et M. Zhan ■

353950　Scirpus yokoscensis（Franch. et Sav.）Ts. Tang et F. T. Wang = Eleocharis acicularis（L.）Roem. et Schult. subsp. yokoscensis（Franch. et Sav.）T. V. Egorova ■

353951　Scirpus yokoscensis Franch. et Sav. = Eleocharis acicularis（L.）Roem. et Schult. subsp. yokoscensis（Franch. et Sav.）T. V. Egorova ■

353952　Scirpus yokoscensis Franch. et Sav. = Heleocharis yokoscensis（Franch. et Sav.）Ts. Tang et F. T. Wang ■

353953　Scirpus yosemitanus Smiley = Trichophorum clementis（M. E. Jones）S. G. Sm. ■☆

353954　Scirpus zebrinus？ = Schoenoplectus tabernaemontani（C. C. Gmel.）Palla 'Zebrinus' ■☆

353955　Scirpus zeyheri Boeck. = Bulbostylis contexta（Nees）M. Bodard ■☆

353956　Scirrhophorus Turcz. = Angianthus J. C. Wendl.（保留属名）■●☆

353957　Sciuris Nees et Mart. = Galipea Aubl. ●☆

353958　Sciuris Schreb. = Raputia Aubl. ●☆

353959　Sciurus D. Dietr. = Sciuris Schreb. ●☆

353960　Scizanthus Pers. = Schizanthus Ruiz et Pav. ■☆

353961　Sckuhria Moench（废弃属名）= Schkuhria Roth（保留属名）■☆

353962　Sclaeranthus Thunb. = Scleranthus L. ■☆

353963　Sclaraea Steud. = Sclarea Mill. ●■

353964　Sclarea Mill. = Salvia L. ●■

353965　Sclarea Tourn. ex Mill. = Salvia L. ●■

353966　Sclarea aethiopis（L.）Mill. = Salvia aethiopis L. ■☆

353967　Sclareastrum Fabr. = Salvia L. ●■

353968　Sclepsion Raf. ex Wedd. = Laportea Gaudich.（保留属名）●■

353969　Sclerachne R. Br.（1838）;斑点葫芦草属（硬颖草属）;Sclerachne ■☆

353970　Sclerachne R. Br. = Chionachne R. Br. ■

353971　Sclerachne Torr. ex Trin. = Limnodea L. H. Dewey ■☆

353972　Sclerachne Torr. ex Trin. = Thurberia Benth. ■☆

353973　Sclerachne Trin. = Limnodea L. H. Dewey ■☆

353974　Sclerachne punctata R. Br.;斑点葫芦草;Dotted Sclerachne ■☆

353975　Sclerandrium Stapf et C. E. Hubb. = Germainia Balansa et Poitr. ■

353976　Sclerandrium intermedium（A. Camus）C. E. Hubb. = Apocopis intermedia（A. Camus）Chai-Anan ■

353977　Scleranthaceae Bartl. = Caryophyllaceae Juss.（保留科名）■●

353978　Scleranthaceae Bartl. = Illecebraceae R. Br.（保留科名）●■

353979　Scleranthaceae Bartl. et J. Presl = Caryophyllaceae Juss.（保留科名）■●

353980　Scleranthaceae Bartl. et J. Presl = Illecebraceae R. Br.（保留科名）■●

353981　Scleranthaceae J. Presl et C. Presl = Caryophyllaceae Juss.（保留科名）■●

353982　Scleranthaceae J. Presl et C. Presl = Illecebraceae R. Br.（保留科名）●■

353983　Scleranthera Pichon = Wrightia R. Br. ●

353984　Scleranthopsis Rech. f.（1967）;多刺线球草属●☆

353985　Scleranthopsis aphanantha（Rech. f.）Rech. f.;多刺线球草●☆

353986　Scleranthus L.（1753）;硬萼花属（线球草属）;Knawel ■☆

353987　Scleranthus annuus L.;一年硬萼花;Annual Knawel, German Knotgrass, Green Eyes, Knawel, Parsley Piert ■☆

353988　Scleranthus annuus L. subsp. delortii（Gren.）Meikle = Scleranthus delortii Gren. ■☆

353989　Scleranthus annuus L. subsp. polycarpos（L.）Bonnier et Layens = Scleranthus polycarpos L. ■☆

353990　Scleranthus annuus L. var. micranthus Maire = Scleranthus annuus L. ■☆

353991　Scleranthus delortii Gren.;德洛尔硬萼花■☆

353992　Scleranthus hamatus Chiov. = Scleranthus orientalis Rössler ■☆

353993　Scleranthus orientalis Rössler;东方硬萼花■☆

353994　Scleranthus perennis L.;多年硬萼花;Perennial Knawel ■☆

353995　Scleranthus perennis L. subsp. atlanticus（Maire）Maire;北非多年硬萼花■☆

353996　Scleranthus polycarpos L.;多果硬萼花■☆

353997　Scleranthus uncinatus Schur.;具钩硬萼花■☆

353998　Scleria P. J. Bergius（1765）;珍珠茅属;Pearlsedge, Razorsedge, Razor-sedge ■

353999　Scleria achtenii De Wild.;阿赫顿珍珠茅■☆

354000　Scleria achtenii De Wild. var. subintegriloba（De Wild.）Piérart = Scleria achtenii De Wild. ■☆

354001　Scleria acriulus C. B. Clarke = Scleria griegiifolia（Ridl.）C. B. Clarke ■☆

354002　Scleria acriulus C. B. Clarke f. leopoldiana De Wild. ex C. B. Clarke = Scleria griegiifolia（Ridl.）C. B. Clarke ■☆

354003　Scleria adpresso-hirta（Kük.）E. A. Rob.;伏毛珍珠茅■☆

354004　Scleria africana Benth. = Diplacrum africanum（Benth.）C. B. Clarke ■☆

354005　Scleria afroreflexa Lye;非洲反折珍珠茅■☆

354006　Scleria amphigaea Raymond = Diplacrum longifolium（Griseb.）C. B. Clarke ■☆

354007　Scleria angolensis Turrill = Scleria induta Turrill ■☆

354008　Scleria angusta Nees ex Kunth;狭珍珠茅■☆

354009　Scleria angustifolia E. A. Rob. ;窄叶珍珠茅■☆

354010　Scleria aquatica Cherm. = Scleria lacustris C. Wright ■☆

354011　Scleria arcuata E. A. Rob. ;拱珍珠茅■☆

354012　Scleria aterrima (Ridl.) Napper;黑色珍珠茅■☆

354013　Scleria atrosanguinea Hochst. ex Steud. = Scleria bulbifera Hochst. ex A. Rich. ■☆

354014　Scleria baldwinii (Torr.) Steud. ;鲍尔温珍珠茅■☆

354015　Scleria bambariensis Cherm. ;班巴里珍珠茅■☆

354016　Scleria baroni-clarkei De Wild. ;巴龙珍珠茅■☆

354017　Scleria barteri Boeck. = Scleria boivinii Steud. ☆

354018　Scleria bequaertii De Wild. ;贝卡尔珍珠茅■☆

354019　Scleria bequaertii De Wild. var. laevis Piérart = Scleria bequaertii De Wild. ■☆

354020　Scleria bicolor Nelmes;二色珍珠茅■☆

354021　Scleria biflora C. B. Clarke = Scleria tessellata Willd. ■

354022　Scleria biflora Roxb. ; 二花珍珠茅; Twoflower Pearlsedge, Twoflower Razorsedge ■

354023　Scleria biflora Roxb. subsp. ferruginea (Ohwi) J. Kern = Scleria biflora Roxb. ■

354024　Scleria boivinii Steud. ;博伊文珍珠茅■☆

354025　Scleria bojeri C. B. Clarke = Scleria distans Poir. ■☆

354026　Scleria bourgeaui Boeck. ;鲍氏珍珠茅;Bourgeau's Nutrush ■☆

354027　Scleria brittonii Core = Scleria ciliata Michx. var. glabra (Chapm.) Fairey ☆

354028　Scleria buchananii Boeck. = Scleria bulbifera Hochst. ex A. Rich. ■☆

354029　Scleria buchananii Boeck. var. laevinux Gross = Scleria bulbifera Hochst. ex A. Rich. var. mechowiana (Boeck.) Kük. ■☆

354030　Scleria buchananii Boeck. var. latifolia De Wild. = Scleria bulbifera Hochst. ex A. Rich. var. latifolia (De Wild.) Piérart ■☆

354031　Scleria buchananii Boeck. var. typica Gross = Scleria bulbifera Hochst. ex A. Rich. ■☆

354032　Scleria buettneri Boeck. ;比特纳珍珠茅■☆

354033　Scleria bulbifera Hochst. ex A. Rich. ;球根珍珠茅■☆

354034　Scleria bulbifera Hochst. ex A. Rich. var. latifolia (De Wild.) Piérart;宽叶球根珍珠茅■☆

354035　Scleria bulbifera Hochst. ex A. Rich. var. mechowiana (Boeck.) Kük. ;梅休珍珠茅■☆

354036　Scleria bulbifera Hochst. ex A. Rich. var. schweinfurthiana (Boeck.) Piérart = Scleria bulbifera Hochst. ex A. Rich. ■☆

354037　Scleria caespitosa Welw. ex Ridl. = Scleria dregeana Kunth ■☆

354038　Scleria calcicola E. A. Rob. ;钙生珍珠茅■☆

354039　Scleria canaliculato-triquetra Boeck. = Scleria lagoensis Boeck. ■☆

354040　Scleria canaliculato-triquetra Boeck. var. adpresso-hirta Kük. = Scleria adpresso-hirta (Kük.) E. A. Rob. ☆

354041　Scleria canaliculato-triquetra Boeck. var. clarkeana Piérart = Scleria lagoensis Boeck. ■☆

354042　Scleria caricina (R. Br.) Benth. = Diplacrum caricinum R. Br. ■

354043　Scleria catophylla C. B. Clarke = Scleria aterrima (Ridl.) Napper ■☆

354044　Scleria cenchroides Kunth = Scleria distans Poir. ■☆

354045　Scleria centralis Cherm. = Scleria melanomphala Kunth ☆

354046　Scleria cervina Ridl. = Scleria lagoensis Boeck. ■☆

354047　Scleria chevalieri J. Raynal;舍瓦利耶珍珠茅■☆

354048　Scleria chinensis Kunth = Scleria ciliaris Nees ■

354049　Scleria chlorocalyx E. A. Rob. ;绿萼珍珠茅■☆

354050　Scleria ciliaris Nees;华珍珠茅;Chinese Razorsedge ■

354051　Scleria ciliaris Nees = Scleria terrestris (L.) Fassett ■

354052　Scleria ciliata Michx. ;纤毛珍珠茅(华珍珠茅,缘毛珍珠茅); Ciliate Pearlsedge, Ciliate Razorsedge, Hairy Nut Grass, Nut Rush ■

354053　Scleria ciliata Michx. var. curtissii (Britton) J. W. Kessler = Scleria curtissii Britton ■☆

354054　Scleria ciliata Michx. var. elliottii (Britton) Fernald = Scleria ciliata Michx. ■

354055　Scleria ciliata Michx. var. elliottii (Chapm.) Fernald;埃利奥特珍珠茅■☆

354056　Scleria ciliata Michx. var. glabra (Chapm.) Fairey;光珍珠茅■☆

354057　Scleria ciliata Michx. var. pauciflora (Muhl. ex Willd.) Kük. = Scleria pauciflora Muhl. ex Willd. ■☆

354058　Scleria ciliata Nees = Scleria ciliata Michx. ■

354059　Scleria ciliolata Boeck. = Scleria racemosa Poir. ■☆

354060　Scleria clarkei De Wild. = Scleria baroni-clarkei De Wild. ■☆

354061　Scleria clathrata Hochst. ex A. Rich. ;格子珍珠茅■☆

354062　Scleria complanata Boeck. ;扁平珍珠茅■☆

354063　Scleria congolensis De Wild. ;刚果珍珠茅■☆

354064　Scleria corymbosa Roxb. ;伞房珍珠茅■

354065　Scleria costata (Britton) Small = Scleria baldwinii (Torr.) Steud. ■☆

354066　Scleria curtissii Britton;柯氏珍珠茅■☆

354067　Scleria delicatula Nelmes;姣美珍珠茅■☆

354068　Scleria depressa (C. B. Clarke) Nelmes;凹陷珍珠茅■☆

354069　Scleria distans Poir. ;远离珍珠茅■☆

354070　Scleria distans Poir. var. glomerulata (Oliv.) Lye;团集俯垂珍珠茅■☆

354071　Scleria distans Poir. var. interrupta (Rich.) Kük. = Scleria interrupta Rich. ■☆

354072　Scleria diurensis Boeck. = Scleria lagoensis Boeck. ■☆

354073　Scleria doederleiniana Boeck. = Scleria elata Thwaites ■

354074　Scleria doederleiniana Boeck. = Scleria terrestris (L.) Fassett ■

354075　Scleria dregeana Kunth;德雷珍珠茅■☆

354076　Scleria dulongensis P. C. Li;独龙珍珠茅;Dulong Pearlsedge, Dulong Razorsedge ■

354077　Scleria dumicola Ridl. = Scleria foliosa Hochst. ex A. Rich. ■☆

354078　Scleria duvigneaudii Piérart = Scleria induta Turrill ■☆

354079　Scleria elata Thwaites = Scleria terrestris (L.) Fassett ■

354080　Scleria elata Thwaites var. latior C. B. Clarke = Scleria terrestris (L.) Fassett ■

354081　Scleria elegantissima Piérart;雅致珍珠茅■☆

354082　Scleria elliottii Chapm. = Scleria ciliata Michx. var. elliottii (Chapm.) Fernald ■☆

354083　Scleria elongata Piérart = Scleria lithosperma (L.) Sw. ■

354084　Scleria erythrorrhiza Ridl. ;红根珍珠茅■☆

354085　Scleria fauriei Ohwi = Scleria sumatrensis Retz. ■

354086　Scleria fenestrata Franch. et Sav. = Scleria biflora Roxb. ■

354087　Scleria fenestrata Franch. et Sav. = Scleria parvula Steud. ■

354088　Scleria fenestrata Franch. et Sav. var. pubigera (Makino) Ohwi = Scleria rugosa R. Br. ■

354089　Scleria fenestrata Franch. et Sav. var. pubigera Ohwi = Scleria onoei Franch. et Sav. var. pubigera Ohwi ■

354090　Scleria ferruginea Ohwi = Scleria biflora Roxb. ■

354091　Scleria ferruginea Ohwi = Scleria tessellata Willd. ■

354092　Scleria ferruginea Peter;锈色珍珠茅■☆

354093　Scleria flaccida Steud. = Scleria triglomerata Michx. ■☆

354094　Scleria flexuosa Boeck. ;曲折珍珠茅■☆

354095　Scleria foliosa Hochst. ex A. Rich. ;多叶珍珠茅■☆

354096　Scleria friesii Kük. =Scleria griegiifolia（Ridl.）C. B. Clarke ■☆

354097　Scleria fulvipilosa E. A. Rob. ;黄褐珍珠茅■☆

354098　Scleria georgiana Core;乔治亚珍珠茅■☆

354099　Scleria glabra Boeck. ;光滑珍珠茅■☆

354100　Scleria glabra Boeck. var. pallidior J. Raynal;苍白光滑珍珠茅■☆

354101　Scleria glabroreticulata De Wild. =Scleria mikawana Makino ■☆

354102　Scleria glandiformis Boeck. =Scleria tessellata Willd. ■

354103　Scleria glomerulata Oliv. =Scleria distans Poir. var. glomerulata（Oliv.）Lye ■☆

354104　Scleria goossensii De Wild. ;古森斯珍珠茅■☆

354105　Scleria gracilis Elliott =Scleria georgiana Core ■☆

354106　Scleria gracillima Boeck. ;细长珍珠茅■☆

354107　Scleria grata Nelmes =Scleria melanotricha Hochst. ex A. Rich. var. grata（Nelmes）Lye ■☆

354108　Scleria griegiifolia（Ridl.）C. B. Clarke;热非珍珠茅■☆

354109　Scleria guineensis J. Raynal;几内亚珍珠茅■☆

354110　Scleria harlandii Hance;圆秆珍珠茅（紫珍珠茅）;Harland Pearlsedge,Harland Razorsedge ■

354111　Scleria hebecarpa Nees =Scleria levis Retz. ■

354112　Scleria hebecarpa Nees var. pubescens（Steud.）C. B. Clarke =Scleria levis Retz. ■

354113　Scleria hildebrandtii Boeck. ;希尔德珍珠茅■☆

354114　Scleria hirtella Sw. var. aterrima Ridl. =Scleria aterrima（Ridl.）Napper ■☆

354115　Scleria hispidior（C. B. Clarke）Nelmes;硬毛珍珠茅■☆

354116　Scleria hispidula Hochst. ex A. Rich. ;细毛珍珠茅■☆

354117　Scleria hispidula Hochst. ex A. Rich. var. hispidior C. B. Clarke =Scleria hispidior（C. B. Clarke）Nelmes ■☆

354118　Scleria holcoides Kunth =Scleria dregeana Kunth ■☆

354119　Scleria hookeriana Boeck. ;黑鳞珍珠茅;Hooker Pearlsedge,Hooker Razorsedge ■

354120　Scleria hypoxis Schweinf. ex Boeck. =Scleria schimperiana Boeck. ■☆

354121　Scleria induta Turrill;印度珍珠茅■☆

354122　Scleria interrupta Rich. ;间断珍珠茅■☆

354123　Scleria jiangchengensis Y. Y. Qian;江城珍珠茅;Jiangcheng Pearlsedge ■

354124　Scleria junciformis Welw. ex Ridl. =Scleria welwitschii C. B. Clarke ■☆

354125　Scleria junghuhniana Boeck. =Scleria lacustris C. Wright ■☆

354126　Scleria kwangtungensis Chun et F. C. How;广东珍珠茅;Guangdong Pearlsedge,Kwangtung Razorsedge ■

354127　Scleria kwangtungensis Chun et F. C. How =Scleria radula Hance ■

354128　Scleria kwangtungensis Chun et F. C. How =Scleria tonkinensis C. B. Clarke ■

354129　Scleria lacustris C. Wright;湖滨珍珠茅;Lakeshore Nutrush,Wright's Nutrush,Wright's Nut-rush ■☆

354130　Scleria laeviformis Ts. Tang et F. T. Wang =Scleria radula Hance ■

354131　Scleria laevis Retz. =Scleria levis Retz. ■

354132　Scleria laevis Retz. var. pubescens（Steud.）C. Y. Wu =Scleria levis Retz. ■

354133　Scleria laevis Retz. var. pubescens（Steud.）C. Y. Wu =Scleria laevis Retz. ■

354134　Scleria laevis Retz. var. scabberima Benth. =Scleria tonkinensis C. B. Clarke ■

354135　Scleria laevis Retz. var. scaberrima Benth. =Scleria radula Hance ■

354136　Scleria lagoensis Boeck. ;拉戈珍珠茅■☆

354137　Scleria lagoensis Boeck. subsp. canaliculato-triquetra（Boeck.）Lye =Scleria lagoensis Boeck. ■☆

354138　Scleria lateritica Nelmes =Scleria flexuosa Boeck. ■☆

354139　Scleria laxiflora Gross;疏花珍珠茅■☆

354140　Scleria lelyi Hutch. et Dalziel =Scleria woodii C. B. Clarke ■☆

354141　Scleria levis Retz. ;毛果珍珠茅(割鸡刀,三角草,三稔草,微毛果珍珠茅,微毛珍珠茅,蜈蚣七,珍珠茅);Hairfruit Pearlsedge,Hairyfruit Razorsedge, Pubescentfruit Pearlsedge, Pubescentfruit Razorsedge ■

354142　Scleria levis Retz. var. pubescens（Steud.）C. Z. Zheng =Scleria levis Retz. ■

354143　Scleria lithosperma（L.）Sw. ;石果珍珠茅;Stonefruit Pearlsedge,Stonefruit Razorsedge ■

354144　Scleria longifolia（Griseb.）Roberty =Diplacrum longifolium（Griseb.）C. B. Clarke ■☆

354145　Scleria longifolia Boeck. =Scleria melaleuca Rchb. ex Schltr. et Cham. ■☆

354146　Scleria longigluma Kük. =Scleria melanomphala Kunth ■☆

354147　Scleria longispiculata Nelmes;长细刺珍珠茅■☆

354148　Scleria macrantha Boeck. =Scleria melanomphala Kunth ■☆

354149　Scleria mayottensis C. B. Clarke =Scleria lagoensis Boeck. ■☆

354150　Scleria mechowiana Boeck. =Scleria bulbifera Hochst. ex A. Rich. var. mechowiana（Boeck.）Kük. ■☆

354151　Scleria melaleuca Rchb. ex Schltr. et Cham. ;黑白珍珠茅■☆

354152　Scleria melanomphala Kunth;黑脐珍珠茅■☆

354153　Scleria melanomphala Kunth f. oculo-albo C. B. Clarke =Scleria melanomphala Kunth ■☆

354154　Scleria melanomphala Kunth var. macrantha（Boeck.）C. B. Clarke =Scleria melanomphala Kunth ■☆

354155　Scleria melanotricha Hochst. ex A. Rich. ;黑毛珍珠茅■☆

354156　Scleria melanotricha Hochst. ex A. Rich. var. glabrior C. B. Clarke =Scleria interrupta Rich. ■☆

354157　Scleria melanotricha Hochst. ex A. Rich. var. grata（Nelmes）Lye;可爱珍珠茅■☆

354158　Scleria meyeriana Kunth =Scleria dregeana Kunth ■☆

354159　Scleria mikawana Makino;三河国珍珠茅■☆

354160　Scleria mildbraedii Graebn. ;米尔德珍珠茅■☆

354161　Scleria minor（Britton）W. Stone;小珍珠茅■☆

354162　Scleria mollis Kunth =Scleria distans Poir. ■☆

354163　Scleria monticola Nelmes ex Napper;山地珍珠茅■☆

354164　Scleria moritziana Boeck. =Scleria lagoensis Boeck. ■☆

354165　Scleria muehlenbergii Steud. ;米伦贝格珍珠茅■☆

354166　Scleria muhlenbergii Steud. =Scleria reticularis Michx. ■☆

354167　Scleria nankingensis Ts. Tang et F. T. Wang;南京珍珠茅;Nanjing Pearlsedge,Nanjing Razorsedge ■

354168　Scleria natalensis C. B. Clarke;纳塔尔珍珠茅■☆

354169　Scleria naumanniana Boeck. ;瑙曼珍珠茅■☆

354170　Scleria nitida Willd. =Scleria triglomerata Michx. ■☆

354171　Scleria nutans Willd. ex Kunth =Scleria distans Poir. ■☆

354172　Scleria nyasensis C. B. Clarke;尼亚斯珍珠茅■☆

354173　Scleria oligantha Michx. ;少花珍珠茅;Few-flowered Nut Grass,

Nut Rush ■☆

354174　Scleria oligochondra Nelmes;寡粒珍珠茅■☆

354175　Scleria onoei Franch. et Sav.;垂序珍珠茅(皱果珍珠茅);Drooping Razorsedge, Droopingspike Pearlsedge ■

354176　Scleria onoei Franch. et Sav. var. pubigera Ohwi;毛垂序珍珠茅;Pubescent Drooping Razorsedge ■

354177　Scleria onoei Makino = Diplacrum caricinum R. Br. ■

354178　Scleria onoei Makino = Scleria rugosa R. Br. ■

354179　Scleria onoei Makino var. pubigera (Makino) Ohwi = Scleria rugosa R. Br. ■

354180　Scleria oryzoides C. Presl = Scleria poaeformis Retz. ■

354181　Scleria oryzoides J. Presl et C. Presl = Scleria poaeformis Retz. ■

354182　Scleria ovuligera Nees = Scleria induta Turrill ■☆

354183　Scleria pachyrrhyncha Nelmes;粗喙珍珠茅■☆

354184　Scleria palmifolia Hoffmanns. = Scleria racemosa Poir. ■☆

354185　Scleria parvula Steud.;小型珍珠茅;Small Razorsedge ■

354186　Scleria parvula Steud. subsp. ferruginea (Ohwi) J. Kern.;锈红色珍珠茅■

354187　Scleria parvula Steud. subsp. ferruginea (Ohwi) J. Kern. = Scleria biflora Roxb. subsp. ferruginea (Ohwi) J. Kern. ■

354188　Scleria patula E. A. Rob.;张开珍珠茅■☆

354189　Scleria pauciflora Muhl. ex Willd.;贫花珍珠茅;Carolina Nut Grass, Nut Rush ■☆

354190　Scleria pauciflora Muhl. ex Willd. var. caroliniana (Willd.) A. W. Wood = Scleria pauciflora Muhl. ex Willd. ■☆

354191　Scleria pauciflora Muhl. ex Willd. var. caroliniana A. W. Wood;加州疏花珍珠茅■☆

354192　Scleria pauciflora Muhl. ex Willd. var. curtissii (Britton) Fairey = Scleria curtissii Britton ■☆

354193　Scleria pauciflora Muhl. ex Willd. var. effusa C. B. Clarke = Scleria pauciflora Muhl. ex Willd. ■☆

354194　Scleria pauciflora Muhl. ex Willd. var. glabra Chapm. = Scleria ciliata Michx. var. glabra (Chapm.) Fairey ■☆

354195　Scleria pauciflora Muhl. ex Willd. var. kansana Fernald = Scleria pauciflora Muhl. ex Willd. var. caroliniana A. W. Wood ■☆

354196　Scleria paupercula E. A. Rob.;贫乏珍珠茅■☆

354197　Scleria pergracilis (Nees) Kunth;纤秆珍珠茅;Finestalk Pearlsedge, Finestalk Razorsedge ■

354198　Scleria pergracilis (Nees) Kunth var. brachystachys Nelmes = Scleria pergracilis (Nees) Kunth ■

354199　Scleria poaeformis Retz.;稻形珍珠茅;Rice-like Razorsedge, Riceshape Pearlsedge ■

354200　Scleria polyrrhiza E. A. Rob.;多根珍珠茅■☆

354201　Scleria porphyrocarpa E. A. Rob.;紫果珍珠茅■☆

354202　Scleria procumbens E. A. Rob.;平铺珍珠茅■☆

354203　Scleria psilorrhiza C. B. Clarke ex Hook. f.;细根茎珍珠茅(细根珍珠茅);Slenderrhizome Pearlsedge ■

354204　Scleria pterota C. Presl = Scleria melaleuca Rchb. ex Schltr. et Cham. ■☆

354205　Scleria pubescens Steud. = Scleria hebecarpa Nees var. pubescens (Steud.) C. B. Clarke ■

354206　Scleria pubescens Steud. = Scleria levis Retz. ■

354207　Scleria pubigera Makino = Scleria onoei Franch. et Sav. var. pubigera Ohwi ■

354208　Scleria pubigera Makino = Scleria rugosa R. Br. ■

354209　Scleria pulchella Ridl.;美丽珍珠茅■☆

354210　Scleria purpurascens Benth. = Scleria harlandii Hance ■

354211　Scleria purpurascens Steud.;紫珍珠茅■

354212　Scleria purpurascens Steud. = Scleria harlandii Hance ■

354213　Scleria puzzoleana K. Schum. = Scleria lithosperma (L.) Sw. ■

354214　Scleria racemosa Benth. = Scleria vogelii C. B. Clarke ■☆

354215　Scleria racemosa Poir.;多枝珍珠茅■☆

354216　Scleria racemosa Poir. subsp. depressa (C. B. Clarke) J. Raynal = Scleria depressa (C. B. Clarke) Nelmes ■☆

354217　Scleria racemosa Poir. var. depressa C. B. Clarke = Scleria depressa (C. B. Clarke) Nelmes ■☆

354218　Scleria radula Chun et F. C. How = Scleria laeviformis Ts. Tang et F. T. Wang ■

354219　Scleria radula Hance;光果珍珠茅(香港珍珠茅);Hongkong Pearlsedge, Hongkong Razorsedge, Smoothfruit Pearlsedge, Smoothfruit Razorsedge ■

354220　Scleria rehmannii C. B. Clarke;拉赫曼珍珠茅■☆

354221　Scleria rehmannii C. B. Clarke var. ornata Berhaut = Scleria rehmannii C. B. Clarke ■☆

354222　Scleria rehmannii C. B. Clarke var. ornata Cherm. = Scleria woodii C. B. Clarke ■☆

354223　Scleria remota Ridl. = Scleria flexuosa Boeck. ■☆

354224　Scleria remota Ridl. var. hispida ? = Scleria bicolor Nelmes ■☆

354225　Scleria reticularis Michx.;网状珍珠茅;Muhlenberg's Nut Grass, Netted Nut-rush, Nut Rush, Reticulated Nut-rush ■☆

354226　Scleria reticularis Michx. var. pumila Britton = Scleria reticularis Michx. ■☆

354227　Scleria reticulata (Holttum) Kern = Diplacrum reticulatum Holttum ■

354228　Scleria retroserrata Kük. = Scleria gracillima Boeck. ■☆

354229　Scleria richardsiae E. A. Rob.;理查兹珍珠茅■☆

354230　Scleria ridleyi C. B. Clarke = Scleria corymbosa Roxb. ■

354231　Scleria robinsoniana J. Raynal;鲁滨逊珍珠茅■☆

354232　Scleria robinsoniana J. Raynal var. acanthocarpa ?;刺果珍珠茅■☆

354233　Scleria rugosa R. Br. = Scleria onoei Franch. et Sav. ■

354234　Scleria rugosa R. Br. f. glabrescens (Koidz.) T. Koyama = Scleria rugosa R. Br. ■

354235　Scleria rugosa R. Br. var. glabrescens (Koidz.) Ohwi et T. Koyama = Scleria rugosa R. Br. ■

354236　Scleria schimperiana Boeck.;欣珀珍珠茅■☆

354237　Scleria schimperiana Boeck. var. hypoxis (Schweinf. ex Boeck.) C. B. Clarke = Scleria schimperiana Boeck. ■☆

354238　Scleria schliebenii Gross = Scleria bulbifera Hochst. ex A. Rich. var. latifolia (De Wild.) Piérart ■☆

354239　Scleria schmitzii Piérart;施密茨珍珠茅■☆

354240　Scleria schweinfurthiana Boeck. = Scleria bulbifera Hochst. ex A. Rich. ■☆

354241　Scleria scrobiculata C. B. Clarke = Scleria elata Thwaites var. latior C. B. Clarke ■

354242　Scleria scrobiculata Nees et Meyen ex Nees;轮叶珍珠茅■

354243　Scleria scrobiculata Nees et Meyen ex Nees = Scleria elata Thwaites var. latior C. B. Clarke ■

354244　Scleria setulosa Boeck. = Scleria dregeana Kunth ■☆

354245　Scleria sphaerocarpa (E. A. Rob.) Napper;球果珍珠茅■☆

354246　Scleria spiciformis Benth.;矛珍珠茅■☆

354247　Scleria spinulosa Boeck. = Scleria verrucosa Willd. ■☆

354248　Scleria spondylogona Nelmes = Scleria delicatula Nelmes ■☆

354249　Scleria striatonux De Wild. = Scleria woodii C. B. Clarke ■☆

354250　Scleria suaveolens Nelmes;芳香珍珠茅■☆

354251 Scleria subintegriloba De Wild. = Scleria achtenii De Wild. ■☆

354252 Scleria substriato-alveolata De Wild. =Scleria achtenii De Wild. ■☆

354253 Scleria sumatrensis Retz. ;印尼珍珠茅(印度尼西亚珍珠茅); Sumatra Pearlsedge, Sumatran Pearlsedge ■

354254 Scleria terrestris (L.) Fassett;高秆珍珠茅(陆生珍珠茅,三棱筋骨草,三楞筋骨草);Highstalk Pearlsedge, Highstalk Razorsedge, Highstem Pearlsedge, Highstem Razorsedge ■

354255 Scleria terrestris (L.) Fassett var. latior (C. B. Clarke) Fassett;宽叶珍珠茅(高秆珍珠茅,三楞筋骨草);Broadleaf Pearlsedge, Wideleaf Razorsedge ■

354256 Scleria terrestris How = Scleria elata Thwaites var. latior C. B. Clarke ■

354257 Scleria tessellata C. B. Clarke = Scleria biflora Roxb. ■

354258 Scleria tessellata C. B. Clarke = Scleria parvula Steud. ■

354259 Scleria tessellata Ts. Tang et F. T. Wang =Scleria biflora Roxb. ■

354260 Scleria tessellata Willd.；网果珍珠茅；Chequer-fruit Razorsedge, Netfruit Pearlsedge ■

354261 Scleria tessellata Willd. var. sphaerocarpa E. A. Rob. = Scleria sphaerocarpa (E. A. Rob.) Napper ■☆

354262 Scleria thomasii Piérart = Scleria bulbifera Hochst. ex A. Rich. ■☆

354263 Scleria tisserantii Cherm. ;蒂斯朗特珍珠茅■☆

354264 Scleria tonkinensis C. B. Clarke;越南珍珠茅■

354265 Scleria transvaalensis E. F. Franklin;德兰士瓦珍珠茅■☆

354266 Scleria tricholepis Nelmes = Scleria interrupta Rich. ■☆

354267 Scleria triglomerata Michx. ;三头珍珠茅;Stone-rush, Tall Nut Grass, Tall Nut-rush, Whip Nut-rush, Whipgrass, Whip-grass ■☆

354268 Scleria triglomerata Michx. var. minor Britton = Scleria minor (Britton) W. Stone ■☆

354269 Scleria unguiculata E. A. Rob. ;爪状珍珠茅■☆

354270 Scleria ustulata Ridl. ;泡状珍珠茅■☆

354271 Scleria vanderystii De Wild. = Scleria lagoensis Boeck. ■☆

354272 Scleria verdickii De Wild. = Scleria bulbifera Hochst. ex A. Rich. ■☆

354273 Scleria verrucosa Willd. ;多疣珍珠茅■☆

354274 Scleria verticillata Muhl. ex Willd. ;轮生珍珠茅;Low Nut Grass, Low Nut-rush, Nut Rush, Whorled Nut Grass ■☆

354275 Scleria vogelii C. B. Clarke ;沃格尔珍珠茅■☆

354276 Scleria welwitschii C. B. Clarke;韦尔珍珠茅■☆

354277 Scleria woodii C. B. Clarke;伍德珍珠茅■☆

354278 Scleria woodii C. B. Clarke var. ornata (Cherm.) W. Schultze-Motel = Scleria woodii C. B. Clarke ■☆

354279 Scleria xerophila E. A. Rob. ;旱生珍珠茅■☆

354280 Scleria zambesica E. A. Rob. ;赞比西珍珠茅■☆

354281 Scleriaceae Bercht. et J. Presl = Cyperaceae Juss. (保留科名)■

354282 Sclerobasis Cass. = Senecio L. ■●

354283 Sclerobassia Ulbr. = Bassia All. ■●

354284 Scleroblitum Ulbr. (1934);莲座藜属■☆●

354285 Scleroblitum Ulbr. = Chenopodium L. ■●

354286 Scleroblitum atriplicinum (F. Muell.) Ulbr. ;莲座藜■☆

354287 Sclerocactus Britton et Rose(1922);鲶鲶玉属(白红山属,白虹山属,琥球属);Eagle-claw Cactus, Fishhook Cactus ●☆

354288 Sclerocactus blainei S. L. Welsh et K. H. Thorne;布莱恩鲶玉; Blaine Fishhook Cactus ●☆

354289 Sclerocactus brevihamatus (Engelm.) D. R. Hunt;短钩鲶玉; Shorthook Fishhook Cactus, Short-spine Fishhook, Tobusch Fishhook Cactus ●☆

354290 Sclerocactus brevihamatus (Engelm.) D. R. Hunt subsp. tobuschii (T. Marshall) N. P. Taylor = Sclerocactus brevihamatus (Engelm.) D. R. Hunt ●☆

354291 Sclerocactus brevispinus K. D. Heil et J. M. Porter;短刺鲶玉; Parriette Fishhook Cactus, Shortspine Fishhook Cactus ●☆

354292 Sclerocactus cloverae K. D. Heil et J. M. Porter;克洛弗鲶玉; Clover Eagle-claw Cactus, Clover's Fishhook Ccactus ●☆

354293 Sclerocactus contortus K. D. Heil = Sclerocactus parviflorus Clover et Jotter ●☆

354294 Sclerocactus erectocentrus (J. M. Coult.) N. P. Taylor = Echinomastus erectocentrus (J. M. Coult.) Britton et Rose ■☆

354295 Sclerocactus glaucus (K. Schum.) L. D. Benson;苍白玉;Uinta Basin Hookless Cactus ●☆

354296 Sclerocactus havasupaiensis Clover var. roseus Clover = Sclerocactus parviflorus Clover et Jotter ●☆

354297 Sclerocactus intermedius Peebles = Sclerocactus parviflorus Clover et Jotter ●☆

354298 Sclerocactus intertextus (Engelm.) N. P. Taylor;络合鲶玉; White Fishook Cactus ●☆

354299 Sclerocactus intertextus (Engelm.) N. P. Taylor = Echinomastus intertextus (Engelm.) Britton et Rose ■☆

354300 Sclerocactus intertextus (Engelm.) N. P. Taylor var. dasyacanthus (Engelm.) N. P. Taylor = Echinomastus intertextus (Engelm.) Britton et Rose var. dasyacanthus (Engelm.) Backeb. ■☆

354301 Sclerocactus johnsonii (Parry ex Engelm.) N. P. Taylor = Echinomastus johnsonii (Parry ex Engelm.) E. M. Baxter ●☆

354302 Sclerocactus mariposensis (Hester) N. P. Taylor;藤荣球(南美美刺球);Lloyd's Mariposa ●☆

354303 Sclerocactus mariposensis (Hester) N. P. Taylor = Echinomastus mariposensis Hester ■☆

354304 Sclerocactus mesae-verdae (Boissev. et C. Davidson) L. D. Benson;月想曲;Mesa Verde Fishhook Cactus ●☆

354305 Sclerocactus nyensis Hochstätter;尼氏鲶玉;Nye Fishhook Cactus ●☆

354306 Sclerocactus papyracanthus (Engelm.) N. P. Taylor;月童(月之童子);Grama-grass Cactus, Gramma Grass Cactus, Paper-spined Cactus ●☆

354307 Sclerocactus parviflorus Clover et Jotter;彩虹山;Devil's-claw Cactus, Small Flower Fishhook Cactus ●☆

354308 Sclerocactus parviflorus Clover et Jotter subsp. havasupaiensis (Clover) Hochstätter = Sclerocactus parviflorus Clover et Jotter ●☆

354309 Sclerocactus parviflorus Clover et Jotter subsp. terrae-canyonae (K. D. Heil) K. D. Heil et J. M. Porter = Sclerocactus parviflorus Clover et Jotter ●☆

354310 Sclerocactus parviflorus Clover et Jotter var. intermedius (Peebles) D. Woodruff et L. D. Benson = Sclerocactus parviflorus Clover et Jotter ●☆

354311 Sclerocactus polyancistrus (Engelm. et J. M. Bigelow) Britton et Rose;白虹山;Hermit Cactus, Mojave Eagle-claw Cactus, Pineapple Cactus, Red-spined Fishhook Cactus ●☆

354312 Sclerocactus polyancistrus Britton et Rose = Sclerocactus polyancistrus (Engelm. et J. M. Bigelow) Britton et Rose ●☆

354313 Sclerocactus pubispinus (Engelm.) D. Woodruff et L. D. Benson var. sileri L. D. Benson = Sclerocactus sileri (L. D. Benson) K. D. Heil et J. M. Porter ●☆

354314 Sclerocactus pubispinus (Engelm.) L. D. Benson;毛刺球; Great Basin Fishhook Cactus ●☆

354315 Sclerocactus pubispinus (Engelm.) L. D. Benson var. spinosior

（Engelm.）S. L. Welsh ＝ Sclerocactus spinosior（Engelm.）D. Woodruff et L. D. Benson ●☆

354316 Sclerocactus scheeri（Salm-Dyck）N. P. Taylor；鹰爪球（黑罗纱，天宝球）●☆

354317 Sclerocactus schlesseri K. D. Heil et S. L. Welsh ＝Sclerocactus blainei S. L. Welsh et K. H. Thorne ●☆

354318 Sclerocactus sileri（L. D. Benson）K. D. Heil et J. M. Porter；赛勒鲵玉；Gypsum Cactus，Siler Fishhook Cactus ●☆

354319 Sclerocactus spinosior（Engelm.）D. Woodruff et L. D. Benson，恩氏鲵玉；Desert Valley Fishhook Cactus，Engelmann Fishhook Cactus ●☆

354320 Sclerocactus spinosior（Engelm.）D. Woodruff et L. D. Benson subsp. blainei（S. L. Welsh et K. H. Thorne）Hochstätter ＝ Sclerocactus blainei S. L. Welsh et K. H. Thorne ●☆

354321 Sclerocactus terrae-canyonae K. D. Heil ＝ Sclerocactus parviflorus Clover et Jotter ●☆

354322 Sclerocactus uncinatus（Galeotti）N. P. Taylor ＝Glandulicactus uncinatus（Galeotti）Backeb. ■☆

354323 Sclerocactus uncinatus（Sweet et Otto）N. P. Taylor subsp. wrightii（Engelm.）N. P. Taylor ＝ Glandulicactus uncinatus（Galeotti）Backeb. var. wrightii（Engelm.）Backeb. ■☆

354324 Sclerocactus warnockii（L. D. Benson）N. P. Taylor ＝ Echinomastus warnockii（L. D. Benson）Glass et R. A. Foster ■☆

354325 Sclerocactus wetlandicus Hochstätter；喜湿鲵玉；Parriette Hookless Cactus ●☆

354326 Sclerocactus wetlandicus Hochstätter var. ilseae Hochstätter ＝ Sclerocactus brevispinus K. D. Heil et J. M. Porter ●☆

354327 Sclerocactus whipplei（Engelm. et J. M. Bigelow）Britton et Rose；黑虹山（白虹，惠普尔鲵玉）；Whipple Fishhook Cactus ●☆

354328 Sclerocactus whipplei（Engelm. et J. M. Bigelow）Britton et Rose var. heilii Castetter, P. Pierce et K. H. Sehwer. ＝ Sclerocactus cloverae K. D. Heil et J. M. Porter ●☆

354329 Sclerocactus whipplei（Engelm. et J. M. Bigelow）Britton et Rose var. intermedius（Peebles）L. D. Benson ＝ Sclerocactus parviflorus Clover et Jotter ●☆

354330 Sclerocactus whipplei（Engelm. et J. M. Bigelow）Britton et Rose var. roseus（Clover）L. D. Benson ＝ Sclerocactus parviflorus Clover et Jotter ●☆

354331 Sclerocactus whipplei（Engelm. et J. M. Bigelow）Britton et Rose var. glaucus（K. Schum.）S. L. Welsh ＝ Sclerocactus glaucus（K. Schum.）L. D. Benson ●☆

354332 Sclerocactus whipplei（Engelm. et J. M. Bigelow）Britton et Rose var. pygmaeus Peebles ＝ Sclerocactus whipplei（Engelm. et J. M. Bigelow）Britton et Rose ●☆

354333 Sclerocactus wrightiae L. D. Benson；怀氏虹山（倒吊白虹山，黑虹山）；Wright's Cactus，Wright's Fishhook Cactus ●☆

354334 Sclerocalyx Nees ＝ Gymnacanthus Nees ■☆

354335 Sclerocarpa Sond. ＝ Sclerocarya Hochst. ●☆

354336 Sclerocarpus Jacq.（1781）；硬果菊属（骨苞菊属）；Bone-bract，Mexican Bone-bract ■

354337 Sclerocarpus africanus Jacq. ＝ Sclerocarpus africanus Jacq. ex Murray ■

354338 Sclerocarpus africanus Jacq. ex Murray；硬果菊；African Bonebract ■

354339 Sclerocarpus discoideus Vatke ＝ Micractis discoidea（Vatke）D. L. Schulz ■☆

354340 Sclerocarpus exigua Sm. ＝ Madia exigua（Sm.）A. Gray ■☆

354341 Sclerocarpus gracilis Sm. ＝ Madia gracilis（Sm.）D. D. Keck ■☆

354342 Sclerocarpus uniserialis（Hook.）Benth. et Hook. f. ex Hemsl.；单列骨苞菊■☆

354343 Sclerocarpus uniserialis（Hook.）Benth. et Hook. f. var. austrotexanus B. L. Turner ＝ Sclerocarpus uniserialis（Hook.）Benth. et Hook. f. ex Hemsl. ■☆

354344 Sclerocarya Hochst.（1844）；硬果漆属（玛鲁拉木属）●☆

354345 Sclerocarya birrea（A. Rich.）Hochst.；硬果漆（玛鲁拉木）；Cat-thorn，Danya，Maroola Plum，Marula ●☆

354346 Sclerocarya birrea（A. Rich.）Hochst. subsp. caffra（Sond.）Kokwaro；马鲁拉硬果漆；Maroola Plum，Marula ●☆

354347 Sclerocarya birrea（A. Rich.）Hochst. subsp. multifoliolata（Engl.）Kokwaro；多小叶硬果漆●☆

354348 Sclerocarya birrea（A. Rich.）Hochst. var. multifoliolata Engl. ＝ Sclerocarya birrea（A. Rich.）Hochst. subsp. multifoliolata（Engl.）Kokwaro ●☆

354349 Sclerocarya birrea Hochst. ＝ Sclerocarya birrea（A. Rich.）Hochst. ●☆

354350 Sclerocarya caffra Sond. ＝ Sclerocarya birrea（A. Rich.）Hochst. subsp. caffra（Sond.）Kokwaro ●☆

354351 Sclerocarya caffra Sond. var. dentata Engl. ＝ Sclerocarya birrea（A. Rich.）Hochst. subsp. caffra（Sond.）Kokwaro ●☆

354352 Sclerocarya caffra Sond. var. oblongifoliolata Engl. ＝ Sclerocarya birrea（A. Rich.）Hochst. subsp. caffra（Sond.）Kokwaro ●☆

354353 Sclerocarya gillettii Kokwaro；吉莱硬果漆●☆

354354 Sclerocarya schweinfurthiana Schinz ＝ Sclerocarya birrea（A. Rich.）Hochst. subsp. caffra（Sond.）Kokwaro ●☆

354355 Sclerocaryopsis Brand ＝ Lappula Moench ■

354356 Sclerocaryopsis spinocarpos（Forssk.）Brand ＝ Lappula spinocarpa（Forssk.）Asch. ex Kuntze ■

354357 Sclerocatyx Nees ＝ Ruellia L. ■●

354358 Sclerocephalus Boiss.（1843）；硬头草属■☆

354359 Sclerocephalus arabicus Boiss.；硬头花●☆

354360 Sclerocephalus arabicus Boiss. var. leianthus（Murb.）Maire ＝ Gymnocarpos sclerocephalus（Decne.）Ahlgren et Thulin ●☆

354361 Sclerochaetium Nees ＝ Tetraria P. Beauv. ■☆

354362 Sclerochaetium angustifolium Hochst. ＝ Tetraria bromoides（Lam.）Pfeiff. ■☆

354363 Sclerochaetium involucratum（Rottb.）Nees ＝ Tetraria involucrata（Rottb.）C. B. Clarke ■☆

354364 Sclerochaetium koenigii Hochst. ＝ Tetraria bromoides（Lam.）Pfeiff. ■☆

354365 Sclerochaetium rottboelli Nees ＝ Tetraria bromoides（Lam.）Pfeiff. ■☆

354366 Sclerochaetium spirale Hochst. ＝ Tetraria involucrata（Rottb.）C. B. Clarke ■☆

354367 Sclerochaetium thermale（L.）Nees ＝ Tetraria bromoides（Lam.）Pfeiff. ■☆

354368 Sclerochiton Harv.（1842）；硬衣爵床属●☆

354369 Sclerochiton albus De Wild. ＝ Sclerochiton vogelii（Nees）T. Anderson subsp. congolanus（De Wild.）Vollesen ●☆

354370 Sclerochiton apiculatus Vollesen；细尖硬衣爵床●☆

354371 Sclerochiton bequaertii De Wild.；贝卡尔硬衣爵床●☆

354372 Sclerochiton boivinii（Baill.）C. B. Clarke；博伊文硬衣爵床●☆

354373 Sclerochiton coeruleus（Lindau）S. Moore；天蓝硬衣爵床●☆

354374 Sclerochiton cyaneus De Wild. ＝ Sclerochiton vogelii（Nees）T. Anderson subsp. congolanus（De Wild.）Vollesen ●☆

354375 Sclerochiton gilletii De Wild. ＝ Sclerochiton nitidus（S. Moore）

C. B. Clarke ●☆

354376 Sclerochiton glandulosissimus Vollesen;多腺硬衣爵床●☆

354377 Sclerochiton harveyanus Nees;哈维硬衣爵床●☆

354378 Sclerochiton hirsutus Vollesen;粗毛硬衣爵床●☆

354379 Sclerochiton holstii (Lindau) C. B. Clarke = Sclerochiton vogelii (Nees) T. Anderson subsp. holstii (Lindau) Napper ●☆

354380 Sclerochiton ilicifolius A. Meeuse;冬青叶硬衣爵床●☆

354381 Sclerochiton insignis (Mildbr.) Vollesen;显著硬衣爵床●☆

354382 Sclerochiton kirkii (T. Anderson) C. B. Clarke;柯克硬衣爵床●☆

354383 Sclerochiton kirkii (T. Anderson) C. B. Clarke var. insignis (Mildbr.) Napper = Sclerochiton insignis (Mildbr.) Vollesen ●☆

354384 Sclerochiton nitidus (S. Moore) C. B. Clarke;光亮硬衣爵床●☆

354385 Sclerochiton obtusisepalus C. B. Clarke;钝萼硬衣爵床●☆

354386 Sclerochiton odoratissimus Hilliard;极香硬衣爵床●☆

354387 Sclerochiton preussii (Lindau) C. B. Clarke;普罗伊斯硬衣爵床●☆

354388 Sclerochiton scissisepalus C. B. Clarke = Sclerochiton vogelii (Nees) T. Anderson subsp. holstii (Lindau) Napper ●☆

354389 Sclerochiton sousai Benoist = Sclerochiton vogelii (Nees) T. Anderson subsp. congolanus (De Wild.) Vollesen ●☆

354390 Sclerochiton stenostachyus Lindau = Crossandra stenostachya (Lindau) C. B. Clarke ●☆

354391 Sclerochiton tanzaniensis Vollesen;坦桑尼亚硬衣爵床●☆

354392 Sclerochiton triacanthus A. Meeuse;三刺硬衣爵床●☆

354393 Sclerochiton uluguruensis Vollesen;乌卢古尔硬衣爵床●☆

354394 Sclerochiton vogelii (Nees) T. Anderson;沃格尔硬衣爵床●☆

354395 Sclerochiton vogelii (Nees) T. Anderson subsp. congolanus (De Wild.) Vollesen;刚果硬衣爵床●☆

354396 Sclerochiton vogelii (Nees) T. Anderson subsp. holstii (Lindau) Napper;霍尔硬衣爵床●☆

354397 Sclerochlaena T. Post et Kuntze = Sclerolaena R. Br. ●☆

354398 Sclerochlamys F. Muell. = Sclerolaena R. Br. ●☆

354399 Sclerochlamys Morrone et Zuloaga = Sclerolaena R. Br. ●☆

354400 Sclerochloa P. Beauv. (1812);硬草属(粗茅属,硬茅属);Hard Grass,Hardgrass,Hard-grass,Stiffgrass ■

354401 Sclerochloa Rchb. = Festuca L. ■

354402 Sclerochloa angusta Nees = Puccinellia angusta (Nees) C. A. Sm. et C. E. Hubb. ■☆

354403 Sclerochloa dura (L.) P. Beauv.;硬草(硬茅);Common Hardgrass,Hard Grass,Hardgrass ■

354404 Sclerochloa kengiana (Ohwi) Tzvelev;耿氏硬草(耿氏碱茅,硬草);Keng Alkaligrass,Keng Stiffgrass ■

354405 Sclerochloa kengiana (Ohwi) Tzvelev = Pseudosclerochloa kengiana (Ohwi) Tzvelev ■

354406 Sclerochorton Boiss. (1872);硬芹属☆

354407 Sclerochorton haussknechtii Boiss.;硬芹☆

354408 Sclerocladus Raf. = Bumelia Sw. (保留属名)●☆

354409 Sclerocladus Raf. = Sideroxylon L. ●☆

354410 Sclerococcus Bartl. = Hedyotis L. (保留属名)●■

354411 Sclerocroton Hochst. (1845);肖乌桕属●☆

354412 Sclerocroton Hochst. = Excoecaria L. ●

354413 Sclerocroton Hochst. = Sapium Jacq. (保留属名)●

354414 Sclerocroton carterianus (J. Léonard) Kruijt et Roebers;利比里亚肖乌桕●☆

354415 Sclerocroton cornutus (Pax) Kruijt et Roebers;角状肖乌桕●☆

354416 Sclerocroton cornutus (Pax) Kruijt et Roebers var. africanum Pax ex Engl. = Sclerocroton cornutus (Pax) Kruijt et Roebers ●☆

354417 Sclerocroton cornutus (Pax) Kruijt et Roebers var. coriaceum Pax = Sclerocroton cornutus (Pax) Kruijt et Roebers ●☆

354418 Sclerocroton cornutus (Pax) Kruijt et Roebers var. poggei (Pax) Pax et K. Hoffm. = Sclerocroton cornutus (Pax) Kruijt et Roebers ●☆

354419 Sclerocroton ellipticus Hochst. = Shirakiopsis elliptica (Hochst.) Esser ●☆

354420 Sclerocroton integerrimus Hochst.;全缘肖乌桕●☆

354421 Sclerocroton oblongifolius (Müll. Arg.) Kruijt et Roebers;矩圆叶肖乌桕●☆

354422 Sclerocroton reticulatus Hochst. = Sclerocroton integerrimus Hochst. ●☆

354423 Sclerocroton schmitzii (J. Léonard) Kruijt et Roebers;施密茨肖乌桕●☆

354424 Sclerocyathium Prokh. = Euphorbia L. ●■

354425 Sclerodactylon Stapf(1911);假龙爪茅属■☆

354426 Sclerodactylon macrostachyum (Benth.) A. Camus;大穗假龙爪茅■☆

354427 Sclerodactylon micrandrum P. C. Keng et L. Liou = Acrachne racemosa (K. Heyne ex Roth et Schult.) Ohwi ■

354428 Sclerodeyeuxia (Stapf) Pilg. = Calamagrostis Adans. ■

354429 Sclerodeyeuxia Pilg. = Calamagrostis Adans. ■

354430 Sclerodictyon Pierre = Dictyophleba Pierre ●☆

354431 Scleroehlamys F. Muell. = Kochia Roth ●■

354432 Sclerolaena A. Camus = Cyphochlaena Hack. ■☆

354433 Sclerolaena Boivin ex A. Camus = Boivinella A. Camus ■☆

354434 Sclerolaena Boivin ex A. Camus = Cyphochlaena Hack. ■☆

354435 Sclerolaena R. Br. (1810);澳藜属;Copper-buff ●☆

354436 Sclerolaena paradoxa R. Br.;澳藜●☆

354437 Scleroleima Hook. f. = Abrotanella Cass. ■☆

354438 Sclerolepis Cass. (1816);硬鳞菊属;Bogbutton ■☆

354439 Sclerolepis Monn. = Crepis L. ■

354440 Sclerolepis Monn. = Pachylepis Less. ■

354441 Sclerolepis Monn. = Rodigia Spreng. ■

354442 Sclerolepis uniflora (Walter) Britton,Sterns et Poggenb.;硬鳞菊;One-flowered Sclerolepis,Pink Bogbutton ■☆

354443 Sclerolinon C. M. Rogers(1966);糙果亚麻属■☆

354444 Sclerolinon digynum (A. Gray) C. M. Rogers;糙果亚麻■☆

354445 Sclerolobium Vogel(1837);硬瓣苏木属■☆

354446 Sclerolobium albiflorum Benoist;白花硬瓣苏木■☆

354447 Sclerolobium chrysophyllum Poepp.;金叶硬瓣苏木■☆

354448 Sclerolobium leiocalyx Ducke;光萼硬瓣苏木■☆

354449 Sclerolobium macropetalum Ducke;大瓣硬瓣苏木■☆

354450 Sclerolobium melanocarpum Ducke;黑果硬瓣苏木■☆

354451 Sclerolobium micranthum L. O. Williams;小花硬瓣苏木■☆

354452 Sclerolobium micropetalum Ducke;小瓣硬瓣苏木■☆

354453 Scleromelum K. Schum. et Lauterb. = Scleropyrum Arn. (保留属名)●

354454 Scleromitrion (Wight et Arn.) Meisn. = Hedyotis L. (保留属名)●■

354455 Scleromitrion Wight et Arn. = Hedyotis L. (保留属名)●■

354456 Scleromitrion coronarium Kurz = Hedyotis coronaria (Kurz) Craib ■

354457 Scleromitrion sinense Miq. = Borreria stricta (L. f.) G. Mey. ●

354458 Scleromitrion sinense Miq. = Spermacoce stricta L. f. ■

354459 Scleromphalos Griff. = Withania Pauquy(保留属名)●■

354460 Scleronema Benth. (1862);硬丝木棉属●☆

354461 Scleronema Brongn. et Gris = Xeronema Brongn. et Gris ■☆

354462 Scleronema grandiflorum Huber;大花硬丝木棉●☆

354463 Scleronema micranthum (Ducke) Ducke;小花硬丝木棉●☆

354464 Scleronema spruceanum Benth. ;硬丝木棉●☆

354465 Scleroolaena Baill. = Xyloolaena Baill. ●☆

354466 Scleroon Benth. = Petitia Jacq. ●☆

354467 Sclerophylacaceae Miers = Solanaceae Juss. (保留科名)●■

354468 Sclerophylacaceae Miers(1848);盐生茄科(南美茄科)●☆

354469 Sclerophylax Miers(1848);盐生茄属■☆

354470 Sclerophylax spinescens Miers;盐生茄■☆

354471 Sclerophyllum Gaudin = Crepis L. ■

354472 Sclerophyllum Gaudin = Phaecasium Cass. ■

354473 Sclerophyllum Griff. = Oryza L. ■

354474 Sclerophyllum Griff. = Porteresia Tateoka ■☆

354475 Sclerophyllum coarctatum (Roxb.) Griff. = Oryza coarctata Roxb. ☆

354476 Sclerophyrum Hieron. = Scleropyrum Arn. (保留属名)●

354477 Scleropoa Griseb. (1846);硬蕨禾属;Hard Meadow Grass, Stiff Grass ■☆

354478 Scleropoa Griseb. = Catapodium Link ■☆

354479 Scleropoa Griseb. = Festuca L. ■

354480 Scleropoa divaricata (Desf.) Batt. et Trab. = Cutandia divaricata (Desf.) Benth. ■☆

354481 Scleropoa divaricata (Desf.) Batt. et Trab. var. laxiflora (Hack.) Cavara = Cutandia divaricata (Desf.) Benth. ■☆

354482 Scleropoa hemipoa (Spreng.) Parl. = Catapodium hemipoa (Spreng.) Lainz ■☆

354483 Scleropoa maritima (L.) Parl. = Cutandia maritima (L.) Barbey ■☆

354484 Scleropoa memphitica (Spreng.) Batt. et Trab. var. dichotoma (Forssk.) Pamp. = Cutandia dichotoma (Forssk.) Trab. ■☆

354485 Scleropoa rigescens (Trin.) Grossh. ;渐硬蕨禾■☆

354486 Scleropoa rigescens (Trin.) Grossh. = Fstuca rigescens Trin. ■☆

354487 Scleropoa rigida (L.) Griseb. ;硬蕨禾(硬绳柄草);Fern Grass, Ferngrass, Fern-grass, Stiff Catapodium ■☆

354488 Scleropoa rigida (L.) Griseb. = Catapodium rigidum (L.) C. E. Hubb. ex Dony ■☆

354489 Scleropoa rigida (L.) Griseb. = Desmazeria rigida (L.) Tutin ■☆

354490 Scleropoa rigida (L.) Griseb. var. patens (C. Presl) Coss. et Durieu = Desmazeria rigida (L.) Tutin ■☆

354491 Scleropoa rigida (L.) Griseb. var. robusta Duval-Jouve = Desmazeria rigida (L.) Tutin ■☆

354492 Scleropoa villaris Sennen et Mauricio = Catapodium rigidum (L.) C. E. Hubb. ■☆

354493 Scleropoa woronowii Hack. ;沃氏硬蕨禾■☆

354494 Scleropogon Phil. (1870);短花硬芒草属;Burrograss ■☆

354495 Scleropogon brevifolius Phil. ;短花硬芒草;Burrograss ■☆

354496 Scleropteris Scheidw. = Cirrhaea Lindl. ☆

354497 Scleropterys Scheidw. = Cirrhaea Lindl. ☆

354498 Scleropus Schrad. = Amaranthus L. ■

354499 Scleropyron Endl. = Scleropyrum Arn. (保留属名)●

354500 Scleropyrum Arn. (1838)(保留属名);硬核属;Scleropyrum, Stiffdrupe ●

354501 Scleropyrum mekongense Gagnep. = Scleropyrum wallichianum (Wight et Arn.) Arn. var. mekongense (Gagnep.) Lecomte ●

354502 Scleropyrum wallichianum (Wight et Arn.) Arn. ; 硬核; Wallich Scleropyrum, Wallich Stiffdrupe ●

354503 Scleropyrum wallichianum (Wight et Arn.) Arn. var.

mekongense (Gagnep.) Lecomte;无刺硬核(野葫芦);Mekong Wallich Scleropyrum ●

354504 Sclerorhachis (Rech. f.) Rech. f. (1969);宿轴菊属■☆

354505 Sclerosciadium Koch = Capnophyllum Gaertn. ☆

354506 Sclerosciadium Koch ex DC. = Capnophyllum Gaertn. ■☆

354507 Sclerosciadium nodiflorum (Schousb.) Ball;宿轴菊■☆

354508 Sclerosdadium W. D. J. Koch ex DC. = Capnophyllum Gaertn. ■☆

354509 Sclerosia Klotzsch(1849);硬莲木属●☆

354510 Sclerosiphon Nevski = Iris L. ☆

354511 Sclerosperma G. Mann et H. Wendl. (1864);西非椰属(石籽榈属,硬籽椰属)●☆

354512 Sclerosperma dubium Becc. ;西非椰●☆

354513 Sclerosperma mannii H. Wendl. ;曼氏西非椰●☆

354514 Sclerosperma walkeri A. Chev. ;瓦尔西非椰●☆

354515 Sclerostachya (Andersson ex Hack.) A. Camus = Miscanthus Andersson ■

354516 Sclerostachya (Andersson ex Hack.) A. Camus(1922);硬穗草属■☆

354517 Sclerostachya (Hack.) A. Camus = Miscanthus Andersson ■

354518 Sclerostachya (Hack.) A. Camus = Sclerostachya (Andersson ex Hack.) A. Camus ■☆

354519 Sclerostachya A. Camus = Miscanthus Andersson ■

354520 Sclerostachya A. Camus = Sclerostachya (Andersson ex Hack.) A. Camus ■☆

354521 Sclerostachya fallax (Balansa) Grassl = Saccharum fallax Balansa ■

354522 Sclerostachya fusca (Roxb.) A. Camus;硬穗草■☆

354523 Sclerostachya narenga (Nees ex Steud.) Grassl = Narenga porphyrocoma (Hance) Bor ■

354524 Sclerostachya narenga (Nees ex Steud.) Grassl = Saccharum narenga (Nees ex Steud.) Hack. ■

354525 Sclerostachyum Stapf ex Ridl. = Sclerostachya (Andersson ex Hack.) A. Camus ■☆

354526 Sclerostegia Paul G. Wilson(1980);小叶盐角木属●☆

354527 Sclerostemma Schott ex Roem. et Schult. = Scabiosa L. ●■

354528 Sclerostemma altissimum Schott = Scabiosa africana L. ■☆

354529 Sclerostephane Chiov. (1929);硬冠菊属■●☆

354530 Sclerostephane Chiov. = Pulicaria Gaertn. ■●

354531 Sclerostephane adenophora (Franch.) Chiov. = Pulicaria hildebrandtii Vatke ■☆

354532 Sclerostephane discoidea Chiov. = Pulicaria discoidea (Chiov.) N. Kilian ■☆

354533 Sclerostephane hildebrandtii (Vatke) E. Gamal-Eldin = Pulicaria hildebrandtii Vatke ■☆

354534 Sclerostephane longifolia Wagenitz et E. Gamal-Eldin = Pulicaria uniseriata N. Kilian ■☆

354535 Sclerostylis Blume = Atalantia Corrêa(保留属名)●

354536 Sclerostylis hindsii Champ. ex Benth. = Citrus japonica Thunb. ●

354537 Sclerostylis hindsii Champ. ex Benth. = Fortunella hindsii (Champ. ex Benth.) Swingle ●

354538 Sclerostylis venosa Champ. ex Benth. = Citrus japonica Thunb. ●

354539 Sclerostylis venosa Champ. ex Benth. = Fortunella venosa (Champ. ex Benth.) C. C. Huang ●

354540 Sclerothamnus Fedde = Hesperothamnus Brandegee ●■

354541 Sclerothamnus Fedde = Millettia Wight et Arn. (保留属名)●■

354542 Sclerothamnus Fedde = Selerothamnus Harms ●■

354543 Sclerothamnus Harms = Hesperothamnus Brandegee ●■

354544 Sclerothamnus Harms ＝ Millettia Wight et Arn.（保留属名）●■

354545 Sclerothamnus Harms ＝ Selerothamnus Harms ●■

354546 Sclerothamnus R. Br. ＝ Eutaxia R. Br. ●☆

354547 Sclerothamnus R. Br. ex W. T. Aiton ＝ Eutaxia R. Br. ●☆

354548 Scledotheca DC.（1839）；硬囊桔梗属●☆

354549 Sclerotheca DC. ＝ Lobelia L. ●

354550 Sclerotheca arborea DC.；硬囊桔梗●☆

354551 Sclerothrix C. Presl ＝ Klaprothia Kunth ■☆

354552 Sclerothrix C. Presl（1834）；硬毛刺莲花属■☆

354553 Sclerothrix fasciculata C. Presl；硬毛刺莲花■☆

354554 Sclerotiaria Korovin（1962）；硬巾草属■☆

354555 Sclerotiaria pentaceros（Korovin）Korovin；硬巾草■☆

354556 Sclerotriaria Czerep. ＝ Sclerotiaria Korovin ■☆

354557 Scleroxylon Bertol. ＝ Chrysophyllum L. ●

354558 Scleroxylon Steud. ＝ Scleroxylum Willd. ●

354559 Scleroxylum Willd. ＝ Myrsine L. ●

354560 Sclerozus Raf. ＝ Bumelia Sw.（保留属名）●☆

354561 Sclerozus Raf. ＝ Sclerocladus Raf. ●☆

354562 Sclerozus Raf. ＝ Sideroxylon L. ●☆

354563 Scobia Noronha ＝ Lagerstroemia L. ●

354564 Scobinaria Seibert ＝ Arrabidaea DC. ☆

354565 Scobinaria Seibert（1940）；锉紫葳属☆

354566 Scobinaria verrucosa（Standl.）Seibert；锉紫葳●☆

354567 Scoliaxon Payson（1924）；曲轴芥属■☆

354568 Scoliaxon mexicanus Payson；曲轴芥■☆

354569 Scoliochilus Rchb. f. ＝ Appendicula Blume ■

354570 Scoliopaceae Takht.；紫脉花科（伏地草科）■

354571 Scoliopaceae Takht. ＝ Calochortaceae Dumort. ■

354572 Scoliopaceae Takht. ＝ Liliaceae Juss.（保留科名）■●

354573 Scoliopus Torr.（1857）；紫脉花属；Slink Pod, Fetid Adder's-tongue, Slink-lily ■☆

354574 Scoliopus bigelovii Torr.；紫脉花；Brownies, California Fetid Adder's-tongue ■☆

354575 Scoliopus hallii S. Watson；霍尔紫脉花；Oregon Fetid Adder's-tongue ■☆

354576 Scoliotheca Baill. ＝ Monopyle Moritz ex Benth. et Hook. f. ■☆

354577 Scolleropsis Hutch. ＝ Scholleropsis H. Perrier ■☆

354578 Scolobus Raf. ＝ Thermopsis R. Br. ex W. T. Aiton ■

354579 Scolochloa Link（1827）（保留属名）；水茅属（河茅属）；Hardgrass, River Grass, Scolochloa ■

354580 Scolochloa Mert. et W. D. J. Koch（废弃属名）＝ Arundo L. ●

354581 Scolochloa Mert. et W. D. J. Koch（废弃属名）＝ Scolochloa Link（保留属名）■

354582 Scolochloa donax（L.）Gaudin. ＝ Arundo donax L. ●

354583 Scolochloa festucacea（Willd.）Link；水茅；Fescue Like Scolochloa, Sprangletup ■

354584 Scolochloa festucacea Link ＝ Scolochloa festucacea（Willd.）Link ■

354585 Scolochloa spiculosa F. Schmidt ＝ Glyceria spiculosa（F. Schmidt）Roshev. ex B. Fedtsch. ■

354586 Scolodia Raf. ＝ Cassia L.（保留属名）●■

354587 Scolodrys Raf. ＝ Quercus L. ●

354588 Scolopacium Eckl. et Zeyh. ＝ Erodium L' Hér. ex Aiton ■●

354589 Scolopendrogyne Szlach. et Mytnik ＝ Quekettia Lindl. ■☆

354590 Scolopendrogyne Szlach. et Mytnik（2009）；苏里南兰属■☆

354591 Scolopendrogyne vermeuleniana（Determann）Szlach. et Mytnik；苏里南兰■☆

354592 Scolophyllum T. Yamaz.（1978）；针叶母草属■☆

354593 Scolophyllum T. Yamaz. ＝ Lindernia All. ■

354594 Scolophyllum ilicifolium（Bonati）T. Yamaz.；东南亚母草■☆

354595 Scolophyllum longitubum T. Yamaz. et W. Chuakul；长管东南亚母草■☆

354596 Scolopia Schreb.（1789）（保留属名）；箣柊属（刺柊属，莿冬属，鲁花树属）；Scolopia ●

354597 Scolopia buxifolia Gagnep.；黄杨叶箣柊（海南箣柊，黄杨叶刺柊，鲁花）；Boxleaf Scolopia, Box-leaved Scolopia ●

354598 Scolopia chinensis（Lour.）Clos；箣柊（刺柊,土乌药）；China Scolopia, Chinese Scolopia ●

354599 Scolopia cinnamomifolia Gagnep. ＝ Scolopia saeva（Hance）Hance ●

354600 Scolopia crenata（Wight）Clos ＝ Scolopia oldhamii Hance ●

354601 Scolopia crenata Clos ＝ Scolopia chinensis（Lour.）Clos ●

354602 Scolopia cuneata Warb. ＝ Scolopia zeyheri（Nees）Harv. ●☆

354603 Scolopia dekindtiana Gilg ＝ Scolopia zeyheri（Nees）Harv. ●☆

354604 Scolopia ecklonii（Nees）Harv. ＝ Scolopia zeyheri（Nees）Harv. ●☆

354605 Scolopia ecklonii Szyszyl.；埃氏箣柊；Thorn Pear ●☆

354606 Scolopia engleri Gilg ＝ Scolopia zeyheri（Nees）Harv. ●☆

354607 Scolopia flanaganii（Bolus）Sim；弗拉纳根箣柊●☆

354608 Scolopia flanaganii（Bolus）Sim var. oreophila Sleumer ＝ Scolopia oreophila（Sleumer）Killick ●☆

354609 Scolopia gerardii Harv. ＝ Scolopia zeyheri（Nees）Harv. ●☆

354610 Scolopia gossweileri Sleumer ＝ Scolopia zeyheri（Nees）Harv. ●☆

354611 Scolopia guerkeana Volkens ＝ Scolopia rhamniphylla Gilg ●☆

354612 Scolopia hainanensis Sleumer ＝ Scolopia buxifolia Gagnep. ●

354613 Scolopia henryi Sleumer；珍珠箣柊；Henry Scolopia ●

354614 Scolopia henryi Sleumer ＝ Scolopia saeva（Hance）Hance ●

354615 Scolopia ledermannii Gilg ＝ Scolopia rhamniphylla Gilg ●☆

354616 Scolopia lucida Wall. ex Kurz；光亮箣柊（光亮刺柊）；Lucid Scolopia ●

354617 Scolopia minutiflora Sleumer ＝ Ludia mauritiana J. F. Gmel. ●☆

354618 Scolopia mundtii（Eckl. et Zeyh.）Warb.；红箣柊（湿生箣柊）；Red Pear, Red Pear-wood ●☆

354619 Scolopia mundtii Warb. ＝ Scolopia mundtii（Eckl. et Zeyh.）Warb. ●☆

354620 Scolopia nana Gagnep. ＝ Scolopia buxifolia Gagnep. ●

354621 Scolopia oldhamii Hance；台湾箣柊（俄氏箣柊，俄氏莿冬，鲁花，鲁花树，台湾刺柊，有刺赤兰）；Oldham Scolopia, Taiwan Scolopia ●

354622 Scolopia oreophila（Sleumer）Killick；喜山箣柊●☆

354623 Scolopia platyphylla Chiov. ＝ Maytenus undata（Thunb.）Blakelock ●☆

354624 Scolopia rhamniphylla Gilg；鼠李叶箣柊●☆

354625 Scolopia rigida R. E. Fr. ＝ Scolopia zeyheri（Nees）Harv. ●☆

354626 Scolopia riparia Mildbr. et Sleumer ＝ Scolopia stolzii Gilg ex Sleumer var. riparia（Mildbr. et Sleumer）Sleumer ●☆

354627 Scolopia saeva（Hance）Hance；广东箣柊（白皮,箣咸,箣血,箣仔,广东刺柊,红箣）；Guangdong Scolopia, Kwangtung Scolopia ●

354628 Scolopia siamensis Warb. ＝ Scolopia chinensis（Lour.）Clos ●

354629 Scolopia stolzii Gilg ex Sleumer；斯托尔兹箣柊●☆

354630 Scolopia stolzii Gilg ex Sleumer var. riparia（Mildbr. et Sleumer）Sleumer；河岸箣柊●☆

354631 Scolopia stuhlmannii Warb. et Gilg ＝ Scolopia zeyheri（Nees）Harv. ●☆

354632 Scolopia theifolia Gilg;热非簕柊●☆

354633 Scolopia thorncroftii E. Phillips = Scolopia zeyheri (Nees) Harv. ●☆

354634 Scolopia zavattarii Chiov. = Scolopia theifolia Gilg ●☆

354635 Scolopia zeyheri (Nees) Harv.;泽赫簕柊●☆

354636 Scolopospermum Hemsl. = Baltimora L. (保留属名)■☆

354637 Scolopospermum Hemsl. = Scolospermum Less. ■☆

354638 Scolosanthes Willis = Scolosanthus Vahl ☆

354639 Scolosanthus Vahl(1796);针花茜属 ☆

354640 Scolosanthus densiflorus Urb.;密花针花茜 ☆

354641 Scolosanthus lucidus Britton;亮针花茜 ☆

354642 Scolosanthus multiflorus Krug et Urb.;多花针花茜 ☆

354643 Scolosanthus pycnophyllus Borhidi;密叶针花茜 ☆

354644 Scolosperma Raf. = Cleome L. ●■

354645 Scolospermum Less. = Baltimora L. (保留属名)■☆

354646 Scolymanthus Willd. ex DC. = Perezia Lag. ■☆

354647 Scolymocephalus Kuntze = Protea L. (保留属名)●☆

354648 Scolymus L. (1753);金黄蓟属(刺苣属);Golden Thistle, Goldenthistle, Oyster Plant ■☆

354649 Scolymus Tourn. ex L. = Scolymus L. ■☆

354650 Scolymus grandiflorus Desf.;大花金黄蓟(大花刺苣)■☆

354651 Scolymus hispanicus L.;金黄蓟(刺苣,通常金黄蓟);Common Goldenthistle, Golden Thistle, Spanish Oyster, Spanish Oyster Plant, Spanish Oysterplant, Spanish Salsify ■☆

354652 Scolymus hispanicus L. subsp. occidentalis R. Vásquez;西方金黄蓟■☆

354653 Scolymus hispanicus L. var. aggregatus (Ruch.) R. Vásquez = Scolymus hispanicus L. ■☆

354654 Scolymus hispanicus L. var. aurantiacus Maire = Scolymus hispanicus L. ■☆

354655 Scolymus maculatus L.;斑点金黄蓟(斑叶刺苣);Golden Thistle, Spotted Goldenthistle, Spotted Spanish Oysterplant ☆

354656 Scoparebutia Frič et Kreuz. = Lobivia Britton et Rose ■

354657 Scoparebutia Fričet Kreuz. ex Buining = Echinopsis Zucc. ●

354658 Scoparia L. (1753);野甘草属;Broomwort ■

354659 Scoparia arborea L. f. = Buddleja saligna Willd. ●☆

354660 Scoparia dulcis L.;野甘草(冰糖草,假甘草,假枸杞,节节珠,金荔枝,米碎草,钮吊金英,热痱草,四时茶,通草草,土甘草,香仪,珠仔草,珠子草);Beet Broomwort, Sweet Broomwort ■

354661 Scoparia flava Cham. et Schltdl.;黄野甘草;Yellow Licorice Weed ■☆

354662 Scoparia montevidensis R. E. Fr.;蒙得维的亚野甘草;Broomwort ■☆

354663 Scoparia ternata Forssk. = Scoparia dulcis L. ■

354664 Scopariaceae Link = Plantaginaceae Juss. (保留科名)■

354665 Scopella W. J. De Wilde et Duyfjes = Scopellaria W. J. De Wilde et Duyfjes ■

354666 Scopella marginata (Blume) W. J. de Wilde et Duyfjes = Scopellaria marginata (Blume) W. J. de Wilde et Duyfjes ■

354667 Scopellaria W. J. De Wilde et Duyfjes(2006);肖马㼎儿属 ■

354668 Scopellaria marginata (Blume) W. J. de Wilde et Duyfjes;云南马㼎儿;Yunnan Zehneria ■

354669 Scopellaria marginata (Blume) W. J. de Wilde et Duyfjes = Zehneria marginata (Blume) Rabenant. ■

354670 Scopelogena L. Bolus(1971);群黄玉属 ●☆

354671 Scopelogena bruynsii Klak;布氏群黄玉 ●☆

354672 Scopelogena gracilis L. Bolus = Scopelogena verruculata (L.) L. Bolus ●☆

354673 Scopelogena verruculata (L.) L. Bolus;群黄玉 ●☆

354674 Scopola Jacq. = Scopolia Jacq. (保留属名)■

354675 Scopolia Adans. (废弃属名) = Ricotia L. (保留属名)■☆

354676 Scopolia Adans. (废弃属名) = Scopolia Jacq. (保留属名)■

354677 Scopolia J. R. Forst. et G. Forst. = Griselinia J. R. Forst. et G. Forst. ●☆

354678 Scopolia Jacq. (1764)('Scopola')(保留属名);赛莨菪属(东莨菪属,莨菪属,七厘散属,搜山虎属,新莨菪属);Scopolia ■

354679 Scopolia L. f. = Daphne L. ●

354680 Scopolia L. f. = Eriosolena Blume ●

354681 Scopolia Lam. = Scolopia Schreb. (保留属名)●

354682 Scopolia Sm. = Toddalia Juss. (保留属名)●

354683 Scopolia aculeata Sm. = Toddalia asiatica (L.) Lam. ●

354684 Scopolia anomala (Link et Otto) Airy Shaw = Anisodus luridus Link et Otto ■

354685 Scopolia atropoides Bercht. et C. Presl;黑莨菪■☆

354686 Scopolia cariniolica Jacq.;欧莨菪;European Scopolia ■☆

354687 Scopolia carniolicoides C. Y. Wu et C. Chen = Anisodus carniolicoides (C. Y. Wu et C. Chen) D'Arcy et Zhi Y. Zhang ■

354688 Scopolia carniolicoides C. Y. Wu et C. Chen var. dentata C. Y. Wu et C. Chen ex C. Chen et Chun L. Chen = Anisodus carniolicoides (C. Y. Wu et C. Chen) D'Arcy et Zhi Y. Zhang ■

354689 Scopolia composita L. f. = Eriosolena composita (L. f.) Tiegh. ●

354690 Scopolia datora (Forssk.) Dunal = Hyoscyamus muticus L. ■

354691 Scopolia japonica Maxim.;东莨菪(日本颠茄,日莨菪,唐充);Japanese Scopolia ■☆

354692 Scopolia japonica Maxim. f. lutescens Sugim.;黄东莨菪■☆

354693 Scopolia likiangensis C. Y. Wu et C. Chen;丽江莨菪■

354694 Scopolia lurida (Link) Dunal = Anisodus luridus Link et Otto ■

354695 Scopolia mairei H. Lév. = Anisodus mairei (H. Lév.) C. Y. Wu et C. Chen ■

354696 Scopolia mutica (L.) Dunal = Hyoscyamus muticus L. ■

354697 Scopolia physaloides (L.) Dunal = Physochlaina physaloides (L.) G. Don ■

354698 Scopolia physaloides Dunal = Physochlaina physaloides (L.) G. Don ■

354699 Scopolia praealta (Decne.) Dunal. = Physochlaina praealta (Decne.) Miers ■

354700 Scopolia praealta Dunal = Physochlaina praealta (Decne.) Miers ■

354701 Scopolia sinensis Hemsl. = Atropanthe sinensis (Hemsl.) Pascher ●■

354702 Scopolia stramonifolia Sem. = Anisodus luridus Link et Otto ■

354703 Scopolia tangutica Maxim. = Anisodus tanguticus (Maxim.) Pascher ■

354704 Scopolina Schult. = Scopolia Jacq. (保留属名)■

354705 Scopularia Lindl. (废弃属名) = Holothrix Rich. ex Lindl. (保留属名)■☆

354706 Scopularia burchellii Lindl. = Holothrix burchellii (Lindl.) Rchb. f. ■☆

354707 Scopularia grandiflora Sond. = Holothrix grandiflora (Sond.) Rchb. f. ■☆

354708 Scopularia secunda Lindl. = Holothrix scopularia Rchb. f. ■☆

354709 Scopulophila M. E. Jones(1908);岩生指甲草属■☆

354710 Scopulophila rixfordii (Brandegee) Munz et I. M. Johnst.;岩生指甲草;Rixford's Rockwort ■☆

354711 Scorbion Raf. = Scordium Mill. ●■

354712 Scorbion Raf. = Teucrium L. ●■

354713 Scordium Mill. = Teucrium L. ●■

354714 Scordium inerme Cav. = Teucrium resupinatum Desf. ■☆

354715 Scoria Raf. = Carya Nutt. (保留属名) ●

354716 Scoria Raf. = Hicoria Raf. ●

354717 Scorias Raf. = Carya Nutt. (保留属名) ●

354718 Scorias Raf. = Scoria Raf. ●

354719 Scorias Raf. ex Endl. = Carya Nutt. (保留属名) ●

354720 Scorias Raf. ex Endl. = Scoria Raf. ●

354721 Scorodendron Blume = Lepisanthes Blume ●

354722 Scorodendron Blume = Scorododendron Blume ●

354723 Scorodendron Pierre = Lepisanthes Blume ●

354724 Scorodendron Pierre = Scorododendron Blume ●

354725 Scorodocarpaceae Tiegh. = Erythropalaceae Planch. ex Miq. (保留科名) ●

354726 Scorodocarpaceae Tiegh. = Olacaceae R. Br. (保留科名) ●

354727 Scorodocarpus Becc. (1877); 蒜果木属 ●☆

354728 Scorodocarpus borneensis (Baill.) Becc.; 婆罗洲蒜果木 ●☆

354729 Scorododendron Blume = Lepisanthes Blume ●

354730 Scorodon Fourr. = Allium L. ■

354731 Scorodonia Hill = Teucrium L. ●■

354732 Scorodophloeus Harms(1901); 蒜皮苏木属 ●☆

354733 Scorodophloeus fischeri (Taub.) J. Léonard; 菲舍尔蒜皮苏木 ●☆

354734 Scorodophloeus torrei Lock; 非洲蒜皮苏木 ●☆

354735 Scorodophloeus zenkeri Harms; 蒜皮苏木; Garlic Tree ●☆

354736 Scorodosma Bunge = Ferula L. ■

354737 Scorodoxylum Nees = Ruellia L. ■●

354738 Scorpia Ewart et A. H. K. Petrie = Corchorus L. ■●

354739 Scorpiaceae Dulac = Boraginaceae Juss. (保留科名) ■●

354740 Scorpianthes Raf. = Heliotropium L. ●■

354741 Scorpioides Gilib. = Myosotis L. ■

354742 Scorpioides Hill = Scorpiurus L. ■☆

354743 Scorpiothyrsus H. L. Li(1944); 卷花丹属; Scorpiothyrsus ●★

354744 Scorpiothyrsus erythrotrichus (Merr. et Chun) H. L. Li; 红毛卷花丹; Red Hairy Scorpiothyrsus, Redhair Scorpiothyrsus ●

354745 Scorpiothyrsus glabrifolius H. L. Li; 光叶卷花丹; Glabrous-leaf Scorpiothyrsus, Smoothleaf Scorpiothyrsus ●

354746 Scorpiothyrsus glabrifolius H. L. Li = Scorpiothyrsus xanthostictus (Merr. et Chun) H. L. Li ●◇

354747 Scorpiothyrsus oligotrichus H. L. Li; 疏毛卷花丹; Fewhair Scorpiothyrsus, Scatteredhair Scorpiothyrsus ●

354748 Scorpiothyrsus oligotrichus H. L. Li = Scorpiothyrsus erythrotrichus (Merr. et Chun) H. L. Li ●

354749 Scorpiothyrsus shangszeensis C. Chen; 上思卷花丹; Shangsi Scorpiothyrsus ●

354750 Scorpiothyrsus tetrandrus Nayar = Kerriothyrsus tetrandrus (Nayar) C. Hansen ●☆

354751 Scorpiothyrsus xanthostictus (Merr. et Chun) H. L. Li; 卷花丹 (黄毛卷花丹); Yellowspot Scorpiothyrsus ●◇

354752 Scorpiothyrsus xanthotrichus (Merr. et Chun) H. L. Li = Scorpiothyrsus erythrotrichus (Merr. et Chun) H. L. Li ●

354753 Scorpiurus Fabr. = Heliotropium L. ●■

354754 Scorpiurus Haller = Myosotis L. ■

354755 Scorpiurus L. (1753); 蝎尾豆属 (蝎荚草属); Caterpillar Plant, Caterpillar-plant, Scorpion's-tail ●☆

354756 Scorpiurus acutifolius Viv. = Scorpiurus muricatus L. subsp.

354757 Scorpiurus laevigatus Sibth. et Sm. = Scorpiurus muricatus L. ■☆

354758 Scorpiurus laevigatus Sm. var. glaberrimus Maire = Scorpiurus muricatus L. ■☆

354759 Scorpiurus laevigatus Sm. var. papillosus Maire = Scorpiurus muricatus L. ■☆

354760 Scorpiurus minima Losinsk.; 小蝎尾豆 ■☆

354761 Scorpiurus muricatus L.; 瘤荚蝎尾豆 (地中海蝎尾豆); Caterpilar Plant, Caterpillar-plant, Prickly Scorpion's-tail ■☆

354762 Scorpiurus muricatus L. subsp. laevigatus (Sibth. et Sm.) Thell. = Scorpiurus muricatus L. ■☆

354763 Scorpiurus muricatus L. subsp. subvillosus (L.) Thell. = Scorpiurus muricatus L. ■☆

354764 Scorpiurus muricatus L. subsp. sulcatus (L.) Thell. = Scorpiurus sulcatus L. ■☆

354765 Scorpiurus muricatus L. var. laevigatus (Sibth. et Sm.) Boiss. = Scorpiurus muricatus L. ■☆

354766 Scorpiurus muricatus L. var. margaritae (Paiau Ferrer) E. Domínguez et Galuno = Scorpiurus muricatus L. ■☆

354767 Scorpiurus muricatus L. var. subvillosus (L.) Fiori = Scorpiurus muricatus L. ■☆

354768 Scorpiurus muricatus L. var. sulcatus (L.) Fiori = Scorpiurus muricatus L. ■☆

354769 Scorpiurus purpureus Desf. = Scorpiurus vermiculatus L. ■☆

354770 Scorpiurus subvillosus L. = Scorpiurus muricatus L. subsp. subvillosus (L.) Thell. ■☆

354771 Scorpiurus subvillosus L. = Scorpiurus muricatus L. ■☆

354772 Scorpiurus subvillosus L. var. acutifolius (Viv.) Burnat = Scorpiurus muricatus L. subsp. subvillosus (L.) Thell. ■☆

354773 Scorpiurus subvillosus L. var. breviaculeatus Batt. = Scorpiurus muricatus L. subsp. subvillosus (L.) Thell. ■☆

354774 Scorpiurus sulcatus L.; 具沟蝎尾豆; Furrowed Caterpilar ■☆

354775 Scorpiurus sulcatus L. = Scorpiurus muricatus L. subsp. sulcatus (L.) Thell. ■☆

354776 Scorpiurus sulcatus L. subsp. laevigatus (Sibth. et Sm.) Batt. = Scorpiurus muricatus L. subsp. subvillosus (L.) Thell. ■☆

354777 Scorpiurus sulvatus L. = Scorpiurus muricatus L. ■☆

354778 Scorpiurus vermiculatus L.; 柔毛蝎尾豆; Common Caterpilar ☆

354779 Scorpius Loisel. = Scorpiurus L. ■☆

354780 Scorpius Medik. = Coronilla L. (保留属名) ●■

354781 Scorpius Moench = Genista L. ●

354782 Scorpius Moench = Voglera P. Gaertn., B. Mey. et Scherb. ●

354783 Scortechinia Hook. f. (1887); 斯科大戟属 ☆

354784 Scortechinia Hook. f. = Neoscortechinia Pax ☆

354785 Scortechinia forbesii Hook. f.; 斯科大戟 ☆

354786 Scortechinia parvifolia Merr.; 小叶斯科大戟 ☆

354787 Scorzonella Nutt. = Microseris D. Don ■☆

354788 Scorzonella borealis (Bong.) Greene = Microseris borealis (Bong.) Sch. Bip. ■☆

354789 Scorzonella howellii (A. Gray) Greene = Microseris howellii A. Gray ■☆

354790 Scorzonella laciniata (Hook.) Nutt. = Microseris laciniata (Hook.) Sch. Bip. ■☆

354791 Scorzonella leptosepala Nutt. = Microseris laciniata (Hook.) Sch. Bip. subsp. leptosepala (Nutt.) K. L. Chambers ■☆

354792 Scorzonella nutans Hook. = Microseris nutans (Hook.) Sch. Bip. ■☆

354793 Scorzonella paludosa Greene = Microseris paludosa（Greene）J. T. Howell ■☆

354794 Scorzonella procera（A. Gray）Greene = Microseris laciniata（Hook.）Sch. Bip. ■☆

354795 Scorzonella sylvatica Benth. = Microseris sylvatica（Benth.）Sch. Bip. ■☆

354796 Scorzonella troximoides（A. Gray）Jeps. = Nothocalais troximoides（A. Gray）Greene ■☆

354797 Scorzonera L.（1753）；鸦葱属（雅葱属）；Serpent Root，Serpentroot，Serpent-root，Viper's-grass

354798 Scorzonera acanthoclada Franch.；刺枝鸦葱；Teke-saghyz Rubber ■☆

354799 Scorzonera acrolasia Bunge = Epilasia acrolasia（Bunge）C. B. Clarke ■

354800 Scorzonera albertoregelia C. Winkl.；阿氏鸦葱■☆

354801 Scorzonera albicaulis Bunge；华北鸦葱（白茎雅葱，笔管草，倒扎草，倒扎花，独角茅草，黄好花，箭头草，毛车七，茅草细辛，水防风，水风，丝茅七，条参，细叶鸦葱，仙茅参，雅葱，羊奶子，猪尾巴）；Whitestem Serpentroot ■

354802 Scorzonera albicaulis Bunge f. flavescens Nakai = Scorzonera albicaulis Bunge ■

354803 Scorzonera albicaulis Bunge f. rosea Nakai = Scorzonera albicaulis Bunge ■

354804 Scorzonera albicaulis Bunge var. macrosperma（C. A. Mey. ex Turcz.）Kitag. = Scorzonera albicaulis Bunge ■

354805 Scorzonera albicaulis Bunge var. macrosperma（Turcz.）Kitag. = Scorzonera albicaulis Bunge ■

354806 Scorzonera alexandrina Boiss. = Scorzonera undulata Vahl ■☆

354807 Scorzonera ammophila Bunge = Epilasia acrolasia（Bunge）C. B. Clarke ■

354808 Scorzonera angustifolia L.；窄叶鸦葱■☆

354809 Scorzonera angustifolia Thomson = Scorzonera curvata（Popl.）Lipsch. ■

354810 Scorzonera armeniaca Boiss.；亚美尼亚鸦葱■☆

354811 Scorzonera astrachiana DC. = Scorzonera pusilla Pall. ■

354812 Scorzonera aubertii Braun-Blanq. et Maire = Scorzonera baetica（DC.）Boiss. ■☆

354813 Scorzonera austriaca Willd.；鸦葱（奥地利鸦葱，奥国鸦葱，白茎鸦葱，黄花地丁，菊花参，老观笔，老鹳嘴子，罗谷葱，罗罗葱，人头发，土参，兔儿奶，雅葱）；Common Serpentroot，Serpentroot ■

354814 Scorzonera austriaca Willd. subsp. sinensis Lipsch. et Krasch. = Scorzonera sinensis Lipsch. et Krasch. ■

354815 Scorzonera austriaca Willd. var. curvata Popl. = Scorzonera curvata（Popl.）Lipsch. ■

354816 Scorzonera austriaca Willd. var. intermedia Regel = Scorzonera subacaulis（Regel）Lipsch. ■

354817 Scorzonera austriaca Willd. var. intermedia Regel et Herder = Scorzonera subacaulis（Regel）Lipsch. ■

354818 Scorzonera austriaca Willd. var. plantaginifolla Kitag. = Scorzonera austriaca Willd. ■

354819 Scorzonera austriaca Willd. var. subacaulis Regel = Scorzonera subacaulis（Regel）Lipsch. ■

354820 Scorzonera austriaca Willd. var. typica Trautv. ex Kom. = Scorzonera austriaca Willd. ■

354821 Scorzonera baetica（DC.）Boiss.；伯蒂卡鸦葱■☆

354822 Scorzonera baetica（DC.）Boiss. var. aubertii（Braun-Blanq. et Maire）Maire = Scorzonera baetica（DC.）Boiss. ■☆

354823 Scorzonera baldschuanica Lipsch.；巴尔德鸦葱■☆

354824 Scorzonera bicolor Freyn et Sint.；二色鸦葱■

354825 Scorzonera biebersteinii Lipsch.；毕氏鸦葱■☆

354826 Scorzonera bracteosaC. Winkl.；多苞片鸦葱■☆

354827 Scorzonera brevicaulis Vahl = Scorzonera coronopifolia Desf. ■☆

354828 Scorzonera bungei Krasch. et Lipsch.；邦奇鸦葱■☆

354829 Scorzonera caespitosa Pomel；丛生鸦葱■☆

354830 Scorzonera caespitosa Pomel subsp. longifolia（Emb. et Maire）Dobignard；矩圆叶丛生鸦葱■☆

354831 Scorzonera capito Maxim.；棉毛鸦葱；Lanate Serpentroot ■

354832 Scorzonera caricifolia Pall. = Scorzonera parviflora Jacq. ■

354833 Scorzonera cenopleura Bunge = Epilasia hemilasia（Bunge）C. B. Clarke ■

354834 Scorzonera chantavica Pavlov；哈恩塔夫鸦葱■☆

354835 Scorzonera circumflexa Krasch. et Lipsch.；皱波球根鸦葱（波皱球根鸦葱）；Wrinkl Serpentroot ■

354836 Scorzonera coronopifolia Desf.；鸟足叶鸦葱■☆

354837 Scorzonera crassifolia Krasch. et Lipsch.；厚叶鸦葱■☆

354838 Scorzonera crispa M. Bieb.；皱波鸦葱；Wrinkl ■☆

354839 Scorzonera crispatula（DC.）Boiss. = Scorzonera hispanica L. ■☆

354840 Scorzonera crispatula Boiss.；细卷鸦葱■☆

354841 Scorzonera curvata（Popl.）Lipsch.；丝叶鸦葱（山鸦葱，丝叶雅葱）；Silkleaf Serpentroot，Threadleaf Serpentroot ■

354842 Scorzonera decumbens Guss. = Scorzonera laciniata L. ■☆

354843 Scorzonera deliciosa Guss. = Scorzonera undulata Vahl subsp. deliciosa（Guss.）Maire ■☆

354844 Scorzonera deliciosa Guss. var. tetuanensis Ball = Scorzonera undulata Vahl subsp. deliciosa（Guss.）Maire ■☆

354845 Scorzonera dianthoides（Lipsch. et Krasch.）Lipsch.；石竹鸦葱■☆

354846 Scorzonera divaricata Turcz.；拐轴鸦葱（叉枝鸦葱，分枝鸦葱，及及，极叉枝鸦葱，苦葵鸦葱，苦屈，女苦奶，散枝鸦葱，羊奶，羊奶及及，紫花拐轴鸦葱）；Divaricate Serpentroot，Purpleflower Divaricate Serpentroot ■

354847 Scorzonera divaricata Turcz. = Hexinia polydichotoma（Ostenf.）H. L. Yang ■

354848 Scorzonera divaricata Turcz. var. foliosa Maxim. = Scorzonera pseudodivaricata Lipsch. ■

354849 Scorzonera divaricata Turcz. var. intricatissima Maxim. = Scorzonera divaricata Turcz. ■

354850 Scorzonera divaricata Turcz. var. sublilacina Maxim.；紫花拐轴鸦葱；Purpleflower Divaricate Serpentroot ■

354851 Scorzonera divaricata Turcz. var. sublilacina Maxim. = Scorzonera divaricata Turcz. ■

354852 Scorzonera divaricata Turcz. var. virgata Maxim. = Scorzonera pseudodivaricata Lipsch. ■

354853 Scorzonera elongata C. H. An et X. L. He；长茎鸦葱；Elongate Serpentroot ■

354854 Scorzonera ensifolia M. Bieb.；剑叶鸦葱■☆

354855 Scorzonera eriosperma M. Bieb.；毛子鸦葱；Hairfruit Serpentroot ■☆

354856 Scorzonera fasciata Pomel = Scorzonera coronopifolia Desf. ■☆

354857 Scorzonera fengtienensis Nakai = Scorzonera mongolica Maxim. ■

354858 Scorzonera ferganica Krasch.；费尔干鸦葱■☆

354859 Scorzonera franchetii Lipsch.；弗氏鸦葱■☆

354860 Scorzonera glabra Rupr. = Scorzonera austriaca Willd. ■

354861 Scorzonera glabra Rupr. var. manshurica（Nakai）Kitag. = Scorzonera manshurica Nakai ■

354862　Scorzonera gracilis Lipsch. ;纤细鸦葱■☆

354863　Scorzonera grigoraschvilii (Sosn.) Lipsch. ;格里鸦葱■☆

354864　Scorzonera grossheimii Lipsch. et Vassilcz. ;格劳鸦葱■☆

354865　Scorzonera halophila Fisch. et C. A. Mey. = Scorzonera parviflora Jacq. ■

354866　Scorzonera halophila Fisch. et C. A. Mey. ex DC. = Scorzonera parviflora Jacq. ■

354867　Scorzonera hebecarpa C. H. An et X. L. He;毛果鸦葱■

354868　Scorzonera hemilasia Bunge = Epilasia hemilasia (Bunge) C. B. Clarke ■

354869　Scorzonera heterophylla Pomel;互叶鸦葱■☆

354870　Scorzonera hispanica L. ;西班牙鸦葱(菊牛蒡);Black Salsify, Scorsonera, Scorzonera, Spanish Saisify, Viper's-grass ■☆

354871　Scorzonera hissarica C. Winkl. ;希萨尔鸦葱;Violet Grass ■☆

354872　Scorzonera hotanica C. H. An;和田鸦葱;Hetian Serpentroot ■

354873　Scorzonera humilis L. ;矮小鸦葱;Bohemian Serpent Root, Bohemian Serpent-root, Dwarf Scorzonera, Scorzoner, Viper's Grass, Viper's-grass ■☆

354874　Scorzonera humilis L. var. linearifolia DC. = Scorzonera curvata (Popl.) Lipsch. ■

354875　Scorzonera idae (Sosn.) Lipsch. ;伊达鸦葱■☆

354876　Scorzonera ikonnikovii Lipsch. et Krasch. ex Lipsch. ;伊氏鸦葱(毛果鸦葱,毛果雅葱);Hairfruit Serpentroot, Hairyfruit Serpentroot, Ikonnikov Serpentroot ■

354877　Scorzonera iliensis Krasch. ;北疆鸦葱(伊犁鸦葱);Yili Serpentroot ■

354878　Scorzonera inconspicua Lipsch. ex Pavlov;皱叶鸦葱(不显鸦葱,鸦葱,皱叶雅葱);Undulate Serpentroot, Wrinkleleaf Serpentroot ■

354879　Scorzonera intermedia Bunge = Epilasia hemilasia (Bunge) C. B. Clarke ■

354880　Scorzonera ketzkhovelii Sosn. ;凯氏鸦葱■☆

354881　Scorzonera laciniata L. ;裂叶鸦葱;Cutleaf Vipergrass, Mediterranean Serpent Root, Mediterranean Serpent-root ■☆

354882　Scorzonera laciniata L. subsp. calcitrapifolia (Vahl) Maire = Scorzonera laciniata L. subsp. decumbens (Guss.) Greuter ■☆

354883　Scorzonera laciniata L. subsp. decumbens (Guss.) Greuter;外倾裂叶鸦葱■☆

354884　Scorzonera laciniata L. var. intermedia (Guss.) DC. = Scorzonera laciniata L. subsp. decumbens (Guss.) Greuter ■☆

354885　Scorzonera laciniata L. var. octangularis (Willd.) Maire = Scorzonera laciniata L. subsp. decumbens (Guss.) Greuter ■☆

354886　Scorzonera lanata (L.) Hoffm. ;绵毛鸦葱■☆

354887　Scorzonera latifolia (Fisch. et C. A. Mey.) DC. ;土耳其宽叶鸦葱(宽叶鸦葱)■☆

354888　Scorzonera leptophylla (DC.) Krasch. et Lipsch. ;狭叶鸦葱■☆

354889　Scorzonera lipskyi Lipsch. ;利普斯基鸦葱■☆

354890　Scorzonera litwinowi Krasch. et Lipsch. ;利特氏鸦葱■☆

354891　Scorzonera longifolia (Emb. et Maire) Greuter = Scorzonera caespitosa Pomel subsp. longifolia (Emb. et Maire) Dobignard ■☆

354892　Scorzonera luntaiensis C. Shih;轮台鸦葱;Luntai Serpentroot ■

354893　Scorzonera macrosperma Turcz. = Scorzonera albicaulis Bunge ■

354894　Scorzonera macrosperma Turcz. ex DC. = Scorzonera albicaulis Bunge ■

354895　Scorzonera macrosperma Turcz. f. angustifolia Debeaux = Scorzonera albicaulis Bunge ■

354896　Scorzonera manshunca Nakai = Scorzonera ikonnikovii Lipsch. et Krasch. ex Lipsch. ■

354897　Scorzonera manshurica Nakai;东北鸦葱(东北雅葱);Dongbei Serpentroot, North-eastern Serpentroot ■

354898　Scorzonera marschalliana C. A. Mey. = Scorzonera pubescens DC. ■

354899　Scorzonera marschalliana C. A. Mey. var. latifolia Rupr. = Scorzonera inconspicua Lipsch. ex Pavlov ■

354900　Scorzonera marschalliana C. A. Mey. var. oblongifolia Trautv. = Scorzonera inconspicua Lipsch. ex Pavlov ■

354901　Scorzonera meanspika Lipsch. ;白尖雅葱■

354902　Scorzonera meyeri (K. Koch) Lipsch. ;梅氏鸦葱■☆

354903　Scorzonera mollis M. Bieb. ;柔软鸦葱■☆

354904　Scorzonera mongolica Maxim. ;蒙古鸦葱(滨鸦葱,蒙古雅葱,蒙古野葱,普蒙古野葱,羊角菜);Mongol Serpentroot, Mongolian Serpentroot ■

354905　Scorzonera mongolica Maxim. var. putjatae C. Winkl. = Scorzonera mongolica Maxim. ■

354906　Scorzonera muriculata C. C. Chang;叉枝鸦葱(叉枝雅葱);Suffurutescent Serpentroot ■

354907　Scorzonera muriculata C. C. Chang = Hexinia polydichotoma (Ostenf.) H. L. Yang ■

354908　Scorzonera nana Boiss. et Buhse = Epilasia hemilasia (Bunge) C. B. Clarke ■

354909　Scorzonera orientalis L. = Reichardia tingitana (L.) Roth ■☆

354910　Scorzonera ovata Trautv. ;卵形鸦葱■☆

354911　Scorzonera oxiana Popov;阿穆达尔鸦葱■☆

354912　Scorzonera pamirica C. Shih;帕米尔鸦葱(帕米尔雅葱);Pamir Serpentroot ■

354913　Scorzonera parviflora Jacq. ;光鸦葱(光雅葱,小花雅葱);Glabrous Serpentroot, Smallhead Serpentroot ■☆

354914　Scorzonera petrovii Lipsch. ;排氏鸦葱■☆

354915　Scorzonera pinnatifida Lour. = Launaea intybacea (Jacq.) Beauverd ■☆

354916　Scorzonera popovii Lipsch. = Scorzonera pusilla Pall. ■

354917　Scorzonera pratorum (Krasch.) Stank. ;草甸鸦葱;Meadow Serpentroot ■☆

354918　Scorzonera pseudodivaricata Lipsch. ;帚状鸦葱(假叉枝鸦葱,岐枝雅葱,帚枝鸦葱,帚状雅葱);Broomlike Serpentroot, Virgate Serpentroot ■

354919　Scorzonera pseudodivaricata Lipsch. var. leiocarpa C. H. An;光果鸦葱■

354920　Scorzonera pseudolanata Grossh. ;假绵毛鸦葱■☆

354921　Scorzonera pseudopygmaea Lipsch. = Scorzonera caespitosa Pomel ■☆

354922　Scorzonera pubescens DC. ;基枝鸦葱(基枝雅葱,柔毛雅葱);Hairy Serpentroot, Shortvelvet Serpentroot ■

354923　Scorzonera pulchra Lomakin;美丽雅葱■☆

354924　Scorzonera pulvinata Lipsch. ;叶枕雅葱■☆

354925　Scorzonera purpurea L. ;紫花雅葱(深红鸦葱);Purple Viper's Grass, Purplehead Serpentroot ■☆

354926　Scorzonera pusilla Pall. ;细叶鸦葱(细丝雅葱,小鸦葱);Fineleaf Serpentroot, Slender Serpentroot ■

354927　Scorzonera pusilla Pall. var. latifolia Lipsch. ;宽叶鸦葱;Broadleaf Slender Serpentroot ■

354928　Scorzonera pygmaea Sibth. et Sm. subsp. longifolia Emb. et Maire = Scorzonera caespitosa Pomel subsp. longifolia (Emb. et Maire) Dobignard ■☆

354929　Scorzonera raddeana C. Winkl. ;拉得鸦葱■☆

354930 Scorzonera radiata Fisch. ex DC. ;毛梗鸦葱（草防风，毛梗雅葱，狭叶鸦葱，狭叶野葱）; Hairstalk Serpentroot, Hairystalk Serpentroot ■

354931 Scorzonera radiata Fisch. ex DC. var. linearifolia H. Lév. = Scorzonera albicaulis Bunge ■

354932 Scorzonera radiata Fisch. ex DC. var. rebuensis (Tatew. et Kitam. ex Kitam.) Nakai = Scorzonera radiata Fisch. ex DC. ■

354933 Scorzonera radiata Fisch. ex Ledeb. = Scorzonera radiata Fisch. ex DC. ■

354934 Scorzonera radiata Fisch. var. rebunese (Tatew. et Kitam.) Nakai = Scorzonera radiata Fisch. ex Ledeb. ■

354935 Scorzonera radiata Fisch. var. subacaulis Lipsch. et Krasch. ;矮鸦葱; Low Hairstalk Serpentroot ■

354936 Scorzonera ramosissima DC. ;多枝鸦葱■☆

354937 Scorzonera rebunensis Tatew. et Kitam. = Scorzonera radiata Fisch. ex DC. ■

354938 Scorzonera rebunensis Tatew. et Kitam. ex Kitam. = Scorzonera radiata Fisch. ex DC. ■

354939 Scorzonera resedifolia L. = Scorzonera laciniata L. ■☆

354940 Scorzonera rigida Aucher ex DC. ;硬鸦葱■☆

354941 Scorzonera rosea Waldst. et Kit. ;粉鸦葱■☆

354942 Scorzonera rugulosa C. C. Chang;皱果鸦葱（黑果鸦葱）■

354943 Scorzonera ruprechtiana Lipsch. et Krasch. = Scorzonera austriaca Willd. ■

354944 Scorzonera ruprechtiana Lipsch. et Krasch. ex Lipsch. = Scorzonera austriaca Willd. ■

354945 Scorzonera safievii Grossh. ;萨非鸦葱■☆

354946 Scorzonera schischkinii Lipsch. et Vassilcz. ;希施鸦葱■☆

354947 Scorzonera schweinfurthii Boiss. ;施韦鸦葱■☆

354948 Scorzonera scopolii Hoppe;斯氏鸦葱;Scopol Serpentroot ■☆

354949 Scorzonera seidlitzii Boiss. ;塞氏鸦葱■☆

354950 Scorzonera sericeolanata (Bunge) Krasch. et Lipsch. ;灰枝鸦葱（长毛雅葱，灰枝雅葱）; Greybranched Serpentroot, Greyshoot Serpentroot ■

354951 Scorzonera sericeolanata (Bunge) Krasch. et Lipsch. = Scorzonera circumflexa Krasch. et Lipsch. ■

354952 Scorzonera sinensis Lipsch. et Krasch. ;桃叶鸦葱（老虎嘴,桃叶雅葱）;China Serpentroot,Chinese Serpentroot ■

354953 Scorzonera sinensis Lipsch. et Krasch. f. plantaginifolia (Kitag.) Nakai = Scorzonera austriaca Willd. ■

354954 Scorzonera songorica (Kar. et Kir.) Lipsch. et Vassilcz. ;准噶尔鸦葱（羽裂鸦葱，羽裂雅葱）; Dzungar Serpentroot, Pinnate Serpentroot ■

354955 Scorzonera stricta M. Bieb. ;直鸦葱（直茎鸦葱）■☆

354956 Scorzonera subacaulis (Regel) Lipsch. ;小鸦葱（矮鸦葱,矮雅葱,短雅葱）;Low Serpentroot ■

354957 Scorzonera suberosa K. Koch;粉白鸦葱■☆

354958 Scorzonera tadshikorum Krasch. et Lipsch. ;塔什克鸦葱■☆

354959 Scorzonera taurica M. Bieb. ;克里木鸦葱;Taur Serpentroot ■☆

354960 Scorzonera taurica M. Bieb. = Scorzonera hispanica L. ■☆

354961 Scorzonera tau-saghyz Lipsch. et Bosse;黄花鸦葱（含胶雅葱）; Tau-saghyz, Tau-saghyz Rubber ■

354962 Scorzonera tianshanensis C. H. An;天山鸦葱; Tianshan Serpentroot ■

354963 Scorzonera tingitana L. = Reichardia tingitana (L.) Roth ■☆

354964 Scorzonera tomentosa L. ;绒毛雅葱■☆

354965 Scorzonera tragopogonoides Regel et Schmalh. ;禾兰鸦葱■☆

354966 Scorzonera transiliensis Popov;橙黄鸦葱（横伊雅葱，天山鸦葱，天山雅葱）;Tianshan Serpentroot ■

354967 Scorzonera transiliensis Popov = Scorzonera albicaulis Bunge ■

354968 Scorzonera tuberosa Pall. ;块茎鸦葱（块根鸦葱,球茎鸦葱）; Tuberose Serpentroot ■☆

354969 Scorzonera tuberosa Pall. var. sericeolanata Bunge = Scorzonera sericeolanata (Bunge) Krasch. et Lipsch. ■

354970 Scorzonera turcomanica Krasch. et Lipsch. ;土库曼鸦葱■☆

354971 Scorzonera turkestanica Franch. ;土耳其鸦葱■☆

354972 Scorzonera turkeviczii Krasch. et Lipsch. ;图尔凯维契鸦葱■☆

354973 Scorzonera undulata Vahl;波状鸦葱■☆

354974 Scorzonera undulata Vahl subsp. alexandrina (Boiss.) Maire = Scorzonera undulata Vahl ■☆

354975 Scorzonera undulata Vahl subsp. deliciosa (Guss.) Maire;姣美鸦葱■☆

354976 Scorzonera undulata Vahl var. alexandrina (Boiss.) Barratte = Scorzonera undulata Vahl ■☆

354977 Scorzonera undulata Vahl var. filifolia Batt. = Scorzonera undulata Vahl ■☆

354978 Scorzonera undulata Vahl var. lancifolia Pomel = Scorzonera undulata Vahl ■☆

354979 Scorzonera undulata Vahl var. tetuanensis (Webb) Pau = Scorzonera undulata Vahl ■☆

354980 Scorzoneroides Moench = Leontodon L. (保留属名)■☆

354981 Scorzoneroides Moench(1794);拟鸦葱属■☆

354982 Scorzoneroides atlantica (Ball) Holub;大西洋拟鸦葱■☆

354983 Scorzoneroides cichoriacea (Ten.) Greuter;菊苣拟鸦葱■☆

354984 Scorzoneroides garnironii (Emb. et Maire) Greuter et Talavera;加尔尼龙拟鸦葱■☆

354985 Scorzoneroides hispidula (Delile) Greuter et Talavera;硬毛拟鸦葱■☆

354986 Scorzoneroides kralikii (Pomel) Greuter et Talavera;克拉利克拟鸦葱■☆

354987 Scorzoneroides laciniata (Bertol.) Greuter;撕裂拟鸦葱■☆

354988 Scorzoneroides muelleri (Sch. Bip.) Greuter et Talavera;米勒拟鸦葱■☆

354989 Scorzoneroides muelleri (Sch. Bip.) Greuter et Talavera subsp. reboudiana (Pomel) Greuter;雷布德拟鸦葱■☆

354990 Scorzoneroides oraria (Maire) Greuter et Talavera;海崖拟鸦葱■☆

354991 Scorzoneroides palisiae (Izuzq.) Greuter et Talavera;帕利西拟鸦葱■☆

354992 Scorzoneroides salzmanii (Sch. Bip.) Greuter et Talavera;萨尔兹曼拟鸦葱■☆

354993 Scorzoneroides simplex (Viv.) Greuter et Talavera;简单拟鸦葱■☆

354994 Scotanthus Naudin = Gymnopetalum Arn. ■

354995 Scotanthus Naudin = Tripodanthera M. Roem. ■

354996 Scotanthus tubiflorus Naudin = Gymnopetalum chinense (Lour.) Merr. ■

354997 Scotanum Adans. = Ficaria Guett. ■☆

354998 Scotanum Adans. = Ranunculus L. ■☆

354999 Scotia Thtmb. = Schotia Jacq. (保留属名)●☆

355000 Scottea DC. = Bossiaea Vent. ●☆

355001 Scottea DC. = Scottia R. Br. ●☆

355002 Scottellia Oliv. (1893);热非大风子属■☆

355003 Scottellia chevalieri Chipp = Scottellia klaineana Pierre ●☆

355004 Scottellia coriacea A. Chev. ;革质热非大风子;Odoko ●☆

355005 Scottellia coriacea A. Chev. ex Hutch. et Dalziel = Scottellia klaineana Pierre ●☆

355006 Scottellia gabonensis Pierre ex A. Chev. = Scottellia klaineana Pierre ●☆

355007 Scottellia gossweileri Exell = Scottellia klaineana Pierre ●☆

355008 Scottellia kamerunensis Gilg = Scottellia klaineana Pierre ●☆

355009 Scottellia klaineana Pierre；喀麦隆热非大风子●☆

355010 Scottellia klaineana Pierre var. kamerunensis（Gilg）Pellegr. = Scottellia klaineana Pierre ●☆

355011 Scottellia klaineana Pierre var. mimfiensis（Gilg）Pellegr. = Scottellia klaineana Pierre ●☆

355012 Scottellia leonensis Oliv.；莱昂热非大风子●☆

355013 Scottellia macrocarpa Tisser. et Sillans = Scottellia orientalis Gilg ●☆

355014 Scottellia macrocarpus Gilg et Dinkl. = Scottellia leonensis Oliv. ●☆

355015 Scottellia mimfiensis Gilg = Scottellia klaineana Pierre ●☆

355016 Scottellia montana Gilg = Gymnosporia buchananii Loes. ●☆

355017 Scottellia orientalis Gilg；东方热非大风子●☆

355018 Scottellia polyantha Gilg = Scottellia klaineana Pierre ●☆

355019 Scottellia schweinfurthii Gilg ex Tisser. et Sillans = Scottellia orientalis Gilg ●☆

355020 Scottellia thyrsiflora Gilg = Scottellia klaineana Pierre ●☆

355021 Scottia R. Br. = Bossiaea Vent. ●☆

355022 Scottia Thunb. = Schotia Jacq.（保留属名）●☆

355023 Scovitzia Walp. = Szovitsia Fisch. et C. A. Mey. ■☆

355024 Scribaea Borkh. = Cucubalus L. ■

355025 Scribneria Hack.（1886）；红泥丝草属■☆

355026 Scribneria bolanderi（Thurb.）Hack.；红泥丝草■☆

355027 Scrithacola Alava（1980）；陡坡草属■☆

355028 Scrithacola kurramensis（Kitam.）Alava，陡坡草■☆

355029 Scrobicaria Cass.（1827）；多榔木属●☆

355030 Scrobicaria Cass. = Gynoxys Cass. ●☆

355031 Scrobicaria ilicifolia Cass.；多榔木●☆

355032 Scrobicularia Mansf. = Poikilogyne Baker f. ●☆

355033 Scrofella Maxim.（1888）；细穗玄参属；Scrofella ■★

355034 Scrofella chinensis Maxim.；细穗玄参■

355035 Scrofularia Spreng. = Scrophularia L. ■●

355036 Scrophucephalus A. P. Khokhr.（1993）；胀头玄参属■☆

355037 Scrophucephalus A. P. Khokhr. = Scrophularia L. ■●

355038 Scrophucephalus minimus（M. Bieb.）A. P. Khokhr.；胀头玄参■☆

355039 Scrophularia L.（1753）；玄参属；Figwort，Variegated Figwort ■●

355040 Scrophularia aequilabris P. C. Tsoong；等唇玄参；Equallip Figwort ■

355041 Scrophularia alaschanica Batalin；贺兰玄参；Alashan Figwort ■

355042 Scrophularia alata A. Gray；日本玄参■☆

355043 Scrophularia alata Gilib. = Scrophularia umbrosa Dum. Cours. ■

355044 Scrophularia alpestris J. Gay ex Benth.；山生玄参■☆

355045 Scrophularia altaica Murray；阿尔泰玄参■☆

355046 Scrophularia amgunensis F. Schmidt；岩玄参■

355047 Scrophularia amplexicaulis Benth.；单茎玄参■☆

355048 Scrophularia aquatica L. ‘Variegata’ = Scrophularia auriculata L. ‘Variegata’■☆

355049 Scrophularia aquatica L. = Scrophularia auriculata L. ■☆

355050 Scrophularia aquatica L. subsp. auriculata（L.）Quézel et Santa = Scrophularia auriculata L. ■☆

355051 Scrophularia aquatica L. var. durandii（Boiss. et Reut.）Pau =

355052 Scrophularia aquatica L. var. glabra Cout. = Scrophularia auriculata L. ■☆

355053 Scrophularia aquatica L. var. laxa Maire = Scrophularia auriculata L. ■☆

355054 Scrophularia aquatica L. var. parviflora Ball = Scrophularia auriculata L. ■☆

355055 Scrophularia aquatica L. var. pubescens Cout. = Scrophularia auriculata L. ■☆

355056 Scrophularia arguta Sol.；亮玄参■☆

355057 Scrophularia armeniaca Bordz.；亚美尼亚玄参■☆

355058 Scrophularia atropatana Grossh.；阿特罗驴喜豆■☆

355059 Scrophularia auriculata Brot. = Scrophularia trifoliata L. ■☆

355060 Scrophularia auriculata Heldr. ex Boiss. = Scrophularia cretica L. ■☆

355061 Scrophularia auriculata L.；水玄参（欧洲水玄参）；Angler's Flower, Babe-in-the-cradle, Bishop's Leaf, Black Doctor, Brook Betony, Brownet, Brown-net, Brownwort, Brunnet, Bullwort, Cressel, Cresset, Cressil, Fiddle, Fiddlesticks, Fiddlestrings, Fiddlewood, Figwort, Huntsman's Cap, Marsh Figwort, Poor Man's Salve, Rose Noble, Shoreline Figwort, Stinking Christopher, Stinking Roger, Venus-in-her-car, Water Betony, Water Bitny, Water Figwort ■☆

355062 Scrophularia auriculata L. ‘Variegata’；黄斑水玄参；Water Figwort ☆

355063 Scrophularia auriculata L. subsp. pseudoauriculata（Sennen）O. Bolòs et Vigo = Scrophularia pseudoauriculata Sennen ■☆

355064 Scrophularia auriculata L. subsp. pseudoauriculata（Sennen）O. Bolòs et Vigo；西班牙水玄参（西班牙地黄）■☆

355065 Scrophularia auriculata L. var. parviflora Ball = Scrophularia auriculata L. ■☆

355066 Scrophularia auriculata Scop. = Scrophularia scopolii Hoppe ex Pers. ■☆

355067 Scrophularia buergeriana Miq.；北玄参（玄参）；Buerger Figwort, North Figwort ■

355068 Scrophularia buergeriana Miq. var. tsinglingensis P. C. Tsoong；秦岭玄参（秦岭北玄参）；Chinling Mountains Burger Figwort, Qinling Figwort ■

355069 Scrophularia californica Cham. et Schltdl.；加州玄参■☆

355070 Scrophularia calliantha Webb et Berthel.；美花玄参■☆

355071 Scrophularia calycina Benth.；萼状玄参；Calyxform Figwort ■☆

355072 Scrophularia campanulata H. L. Li = Scrophularia delavayi Franch. ■

355073 Scrophularia canescens Bong.；灰玄参■☆

355074 Scrophularia canescens Bong. var. glabrata Franch. = Scrophularia incisa Weinm. ■

355075 Scrophularia canina L.；欧玄参（犬玄参）；Dog Figwort, French Figwort ■☆

355076 Scrophularia canina L. subsp. frutescens（L.）O. Bolòs et Vigo = Scrophularia frutescens L. ●☆

355077 Scrophularia canina L. subsp. hoppii ?；豪氏玄参；Hopp's Figwort ■☆

355078 Scrophularia canina L. subsp. ramosissima（Loisel.）P. Fourn. = Scrophularia ramosissima Loisel. ■☆

355079 Scrophularia canina L. var. baetica Boiss. = Scrophularia canina L. ■☆

355080 Scrophularia canina L. var. dissecta Rouy = Scrophularia canina L. ■☆

355081　Scrophularia canina L. var. frutescens （L.） Boiss. = Scrophularia frutescens L. ●☆

355082　Scrophularia canina L. var. humifusa （Timb.-Lagr. et Gaut.） Gaut. = Scrophularia canina L. ■☆

355083　Scrophularia canina L. var. pinnatifida （Brot.） Boiss. = Scrophularia canina L. ■☆

355084　Scrophularia charadzei Kem.-Nath. ;哈拉玄参■☆

355085　Scrophularia chasmophila W. W. Sm. ;岩隙玄参; Chasm Figwort, Rockgap Figwort ■

355086　Scrophularia chasmophila W. W. Sm. subsp. xizangensis D. Y. Hong;西藏岩隙玄参■

355087　Scrophularia chinensis L. = Verbascum chinense （L.） Santapau ■

355088　Scrophularia chlorantha Kotschy et Boiss. ;绿花玄参■☆

355089　Scrophularia chrysantha Jaub. et Spach;金花玄参■☆

355090　Scrophularia crenatosepala H. L. Li = Scrophularia diplodonta Franch. ■

355091　Scrophularia cretacea Fisch. ;白垩玄参■☆

355092　Scrophularia cretacea Fisch. ex Spreng. var. glabrata （Franch.） Stiefelh. = Scrophularia incisa Weinm. ■

355093　Scrophularia cretacea Fisch. var. glabrata （Franch.） Stiefelh. = Scrophularia incisa Weinm. ■

355094　Scrophularia cretica L. ;克里特玄参■☆

355095　Scrophularia czernjakowskiana B. Fedtsch. ;契尔玄参■☆

355096　Scrophularia decipiens Boiss. et Kotschy ;迷惑玄参■☆

355097　Scrophularia delavayi Franch. ;大花玄参; Bigflower Figwort, Delavay Figwort ■

355098　Scrophularia dentata Royle ex Benth. ;齿叶玄参; Dentate Figwort ■

355099　Scrophularia deserti Delile;荒漠玄参■☆

355100　Scrophularia diplodonta Franch. ;重齿玄参（苦玄参,小黑药）; Doubleteeth Figwort, Duplicatetooth Figwort ■

355101　Scrophularia diplodonta Franch. var. tsanchanensis Franch. = Scrophularia diplodonta Franch. ■

355102　Scrophularia dissecta （B. Fedtsch.） Gorschk. ;深裂玄参■☆

355103　Scrophularia divaricata Ledeb. ;双叉玄参■☆

355104　Scrophularia duclouxii Stiefelh. = Scrophularia mandarinorum Franch. ■

355105　Scrophularia duclouxii Stiefelh. et Bonati = Scrophularia mandarinorum Franch. ■

355106　Scrophularia duplicatoserrata （Miq.） Makino;日本重齿玄参■☆

355107　Scrophularia duplicatoserrata （Miq.） Makino var. surugensis （Honda） Honda ex H. Hara;骏河湾玄参■☆

355108　Scrophularia durandii Boiss. et Reut. = Scrophularia auriculata L. ■☆

355109　Scrophularia edgeworthii Benth. ;埃奇沃斯玄参; Edgeworth Figwort ■☆

355110　Scrophularia elatior Benth. ;高玄参; High Figwort ■

355111　Scrophularia eriocalyx Emb. et Maire;毛萼玄参■☆

355112　Scrophularia eriocalyx Emb. et Maire var. defoliata Maire = Scrophularia eriocalyx Emb. et Maire ■☆

355113　Scrophularia eriocalyx Emb. et Maire var. foliosa Font Quer et Maire = Scrophularia eriocalyx Emb. et Maire ■☆

355114　Scrophularia exilis Popl. ;柔弱玄参■☆

355115　Scrophularia ezapandaghii B. Fedtsch. ;恰氏玄参■☆

355116　Scrophularia fargesii Franch. ;长梗玄参（鄂玄参）; Farges Figwort, Longstalk Figwort ■

355117　Scrophularia fedtschenkoi Gorschk. ;费氏玄参■☆

355118　Scrophularia foliosa Pomel = Scrophularia laevigata Vahl ■☆

355119　Scrophularia fontqueri Ortega Oliv. et Devesa;丰特玄参■☆

355120　Scrophularia formosana H. L. Li;楔叶玄参(台湾玄参)■

355121　Scrophularia forrestii Diels = Scrophularia urticifolia Wall. ■

355122　Scrophularia franchetiana P. C. Tsoong;傅氏玄参; Franch et Figwort ■

355123　Scrophularia franchetiana P. C. Tsoong = Scrophularia fargesii Franch. ■

355124　Scrophularia frigida Boiss. ;耐寒玄参■☆

355125　Scrophularia frutescens L. ;灌木玄参●☆

355126　Scrophularia glabrata Aiton;光滑玄参■☆

355127　Scrophularia goldeana Juz. ;戈尔德玄参■☆

355128　Scrophularia gontscharovii Gorschk. ;贡氏玄参■☆

355129　Scrophularia grayana Maxim. ex Kom. ;北海道玄参; Yezo Figwort ■☆

355130　Scrophularia grayana Maxim. ex Kom. = Scrophularia alata A. Gray ■☆

355131　Scrophularia grayana Maxim. ex Kom. var. grayanoides （M. Kikuchi） T. Yamaz. = Scrophularia grayanoides M. Kikuchi ■☆

355132　Scrophularia grayanoides M. Kikuchi;假北海道玄参■☆

355133　Scrophularia grossheimii Schischk. ;格罗玄参■☆

355134　Scrophularia haematantha Boiss. et Heldr. ;血花玄参■☆

355135　Scrophularia henryi Hemsl. ;鄂西玄参（黑参,玄台）; Henry Figwort ■

355136　Scrophularia henryi Hemsl. var. glabrescens Hemsl. = Scrophularia henryi Hemsl. ■

355137　Scrophularia heucheriiflora Schrenk;新疆玄参; Heucher Figwort, Xinjiang Figwort ■

355138　Scrophularia himalensis Royle ex Benth. ;喜马拉雅玄参; Himalayan Figwort ■☆

355139　Scrophularia hirta Lowe;毛玄参■☆

355140　Scrophularia hispida Desf. = Scrophularia laevigata Vahl subsp. hispida （Desf.） Maire ■☆

355141　Scrophularia hispida Desf. var. subcrispa （Pomel） Batt. = Scrophularia hirta Lowe ■☆

355142　Scrophularia hypsophila Hand.-Mazz. ;高山玄参; Alpine Figwort ■

355143　Scrophularia hyrcana Grossh. ;希尔康玄参■☆

355144　Scrophularia ilvensis K. Koch;伊尔瓦玄参■☆

355145　Scrophularia incisa Weinm. ;砾玄参(野辛巴);Incised Figwort ■

355146　Scrophularia integrifolia Pavlov;全叶玄参■☆

355147　Scrophularia kabadianensis B. Fedtsch. ;卡巴迪亚玄参■☆

355148　Scrophularia kakudensis Franch. ;丹东玄参（安东玄参,川玄参,大山玄参,卡库玄参,马氏玄参,土玄参,腋花玄参）;Dandong Figwort, Kakuda Figwort, Maximowicz Figwort ■

355149　Scrophularia kakudensis Franch. var. latisepala （Kitag.） Kitag. = Scrophularia kakudensis Franch. ■

355150　Scrophularia kakudensis Franch. var. toyamae （Hatus. ex T. Yamaz.） T. Yamaz. ;富山玄参■☆

355151　Scrophularia kansuensis Batalin;甘肃玄参; Gansu Figwort, Kansu Figwort ■

355152　Scrophularia kiriloviana Schischk. ;羽裂玄参;Kirilow Figwort ■

355153　Scrophularia koelzii Pennell;凯氏玄参;Koelz Figwort ■☆

355154　Scrophularia kotschyana Benth. ;考奇玄参■☆

355155　Scrophularia laevigata Vahl;平滑玄参■☆

355156　Scrophularia laevigata Vahl subsp. hispida （Desf.） Maire;长毛玄参■☆

355157　Scrophularia laevigata Vahl subsp. pellucida（Pomel）Murb. = Scrophularia laevigata Vahl ■☆

355158　Scrophularia laevigata Vahl subsp. simplicifolia（Batt.）Maire；单叶光滑玄参■☆

355159　Scrophularia laevigata Vahl var. decomposita Maire = Scrophularia laevigata Vahl ■☆

355160　Scrophularia laevigata Vahl var. dissecta Faure et Maire = Scrophularia hirta Lowe ■☆

355161　Scrophularia laevigata Vahl var. glabrescens Batt. = Scrophularia hirta Lowe ■☆

355162　Scrophularia laevigata Vahl var. pubescens Maire = Scrophularia laevigata Vahl ■☆

355163　Scrophularia laevigata Vahl var. simplicifolia Batt. = Scrophularia laevigata Vahl subsp. simplicifolia（Batt.）Maire ■☆

355164　Scrophularia laevigata Vahl var. wallii Maire = Scrophularia laevigata Vahl ■☆

355165　Scrophularia lanceolata Pursh；剑叶玄参；American Figwort, Early Figwort, Figwort, Hare Figwort, Lance-leaf Figwort, Lanceolate Figwort ☆

355166　Scrophularia lateriflora Trautv.；宽花玄参■☆

355167　Scrophularia latisepala Kitag. = Scrophularia kakudensis Franch. ■

355168　Scrophularia laxiflora Lange；疏花玄参■☆

355169　Scrophularia leporella E. P. Bicknell = Scrophularia lanceolata Pursh ■☆

355170　Scrophularia leucoclada Bunge；白枝玄参■☆

355171　Scrophularia lhasaensis D. Y. Hong；拉萨玄参■

355172　Scrophularia libanotica Boiss. var. nevshehirensis R. R. Mill；内尖玄参■☆

355173　Scrophularia litwinowii B. Fedtsch.；利特氏玄参■☆

355174　Scrophularia lowei Dalgaard；洛氏玄参■☆

355175　Scrophularia lucida L.；光泽玄参；Shining Figwort ■☆

355176　Scrophularia lunariifolia Boiss. et Balansa；新月玄参■☆

355177　Scrophularia lyrata Willd. = Scrophularia auriculata L. ■☆

355178　Scrophularia macrobotrys Ledeb.；大穗玄参■☆

355179　Scrophularia macrocarpa P. C. Tsoong；大果玄参（一扫光）；Bigfruit Figwort ■

355180　Scrophularia macrorrhyncha（Humbert et al.）Ibn Tattou = Scrophularia ramosissima Loisel. subsp. macrorrhyncha Humbert et Litard. et Maire ■☆

355181　Scrophularia mandarinorum Franch.；单齿玄参（黑玄参，玄参）；China Figwort，Chinese Figwort ■

355182　Scrophularia mandshurica Maxim.；东北玄参■

355183　Scrophularia mapienensis P. C. Tsoong；马边玄参；Mabian Figwort，Mapian Figwort ■

355184　Scrophularia marilandica L.；北美玄参；Carpenter's Square, Carpenter's-square, Eastern Figwort, Figwort, Late Figwort, Maryland Figwort ■☆

355185　Scrophularia marilandica L. f. neglecta（Rydb. ex Small）Pennell = Scrophularia marilandica L. ■☆

355186　Scrophularia maximowiczii Gorschk. = Scrophularia kakudensis Franch. ■

355187　Scrophularia mellifera Aiton = Scrophularia sambucifolia L. ■☆

355188　Scrophularia microdonta Franch. = Scrophularia ningpoensis Hemsl. ■

355189　Scrophularia minima M. Bieb.；小玄参■☆

355190　Scrophularia modesta Kitag.；山西玄参；Shansi Figwort, Shanxi

Figwort ■☆

355191　Scrophularia moellendorffii Maxim.；华北玄参；Moellendorff Figwort，N. China Figwort ■

355192　Scrophularia mollis Sommier et H. Lév.；柔软玄参■☆

355193　Scrophularia muliensis H. L. Li = Scrophularia delavayi Franch. ■

355194　Scrophularia multicaulis Turcz.；多茎玄参■☆

355195　Scrophularia musashiensis Bonati；武藏玄参■☆

355196　Scrophularia nana H. L. Li = Scrophularia chasmophila W. W. Sm. ■

355197　Scrophularia nankinensis P. C. Tsoong；南京玄参；Nanjing Figwort，Nanking Figwort ■

355198　Scrophularia neglecta Rydb. ex Small = Scrophularia marilandica L. ■☆

355199　Scrophularia nervosa Benth.；多脉玄参■☆

355200　Scrophularia nikitinii Gorschk.；尼氏玄参■☆

355201　Scrophularia ningpoensis Hemsl.；玄参（八秽麻，北玄参，大玄参，端，馥草，鬼藏，黑参，黑玄参，角玄，凌消草，鹿肠，鹿阳生，能消草，润玄参，山麻，水萝卜，乌元参，咸，咸端，玄台，玄武精，野芝麻，野脂麻，元参，浙玄参，正马，重台，逐马）；Figwort, Ningbo Figwort, Ningpo Figwort ■

355202　Scrophularia nodosa L.；林生玄参（具节玄参）；Arym, Bore-tree, Brownet, Brown-net, Brownwort, Brunnet, Carpenter's Square, Common Figwort, Cut-finger, Fairy's Bed, Fairy's Beds, Fiddle, Fiddlesticks, Figwort, Great Pilewort, Hasty Roger, Kernelwort, Knotted Figwort, Murrain-grass, Pokeweed, Poor Man's Salve, Rose Noble, Scaribeus, Scrofula-plant, Slinking Christopher, Squarrib, Stinking Christopher, Stinking Roger, Throatwort, Wood Figwort ■☆

355203　Scrophularia obtusa Edgew. ex Hook. f.；钝形玄参；Obtuse Figwort ■☆

355204　Scrophularia oldhamii Oliv.；奥尔德玄参■☆

355205　Scrophularia oldhamii Oliv. = Scrophularia buergeriana Miq. ■

355206　Scrophularia olgae Grossh.；奥氏玄参■☆

355207　Scrophularia olympica Boiss.；奥林匹克玄参■☆

355208　Scrophularia orientalis L.；东方玄参；Oriental Figwort ■☆

355209　Scrophularia pamirooalaica Gorschk.；帕米尔玄参■☆

355210　Scrophularia papillaris Boiss. et Reut. = Scrophularia scorodonia L. ■☆

355211　Scrophularia patriniana Wydler；帕氏玄参（山玄参）；Patrin Figwort ■☆

355212　Scrophularia patriniana Wydler = Scrophularia incisa Weinm. ■

355213　Scrophularia pauciflora Benth.；轮花玄参（疏花玄参）；Fewflower Figwort ■

355214　Scrophularia pectinata Raf. = Scrophularia lanceolata Pursh ■☆

355215　Scrophularia pellucida Pomel = Scrophularia laevigata Vahl ■☆

355216　Scrophularia peregrina L.；外来玄参；Mediterranean Figwort, Nettle-leaved Figwort ■☆

355217　Scrophularia petitmenginii Bonati = Scrophularia elatior Benth. ■

355218　Scrophularia polyantha Royle ex Benth.；多花玄参；Manyflower Figwort ■☆

355219　Scrophularia pruinosa Boiss.；粉玄参■☆

355220　Scrophularia przewalskii Batalin；青海玄参■

355221　Scrophularia pseudoauriculata Sennen = Scrophularia auriculata L. subsp. pseudoauriculata（Sennen）O. Bolòs et Vigo ■☆

355222　Scrophularia racemosa Lowe；总花玄参■☆

355223　Scrophularia ramosissima Loisel.；多枝玄参■☆

355224　Scrophularia ramosissima Loisel. subsp. macrorrhyncha Humbert et Litard. et Maire；大喙多枝玄参■☆

355225 Scrophularia rockii H. L. Li = Scrophularia chasmophila W. W. Sm. ■

355226 Scrophularia rostrata Boiss. et Buhse;喙状玄参■☆

355227 Scrophularia rupestris M. Bieb. ex Willd.;岩地玄参■☆

355228 Scrophularia ruprechtii Boiss.;卢普玄参■☆

355229 Scrophularia rutifolia Boiss.;芸香叶玄参■☆

355230 Scrophularia saharae Batt. = Scrophularia syriaca A. DC.■☆

355231 Scrophularia sambucifolia L.;接骨木玄参■☆

355232 Scrophularia sambucifolia L. subsp. mellifera (Aiton) Maire;蜜接骨木玄参■☆

355233 Scrophularia sambucifolia L. var. vidalii Pau = Scrophularia sambucifolia L. ■☆

355234 Scrophularia sangtodensis B. Fedtsch.;桑托登玄参■☆

355235 Scrophularia scabiosifolia Benth.;蓝盆花叶玄参;Scabiousflower Figwort■☆

355236 Scrophularia scopolii Hoppe ex Pers.;斯氏玄参(司卡氏玄参,斯克波玄参);Scopol Figwort■☆

355237 Scrophularia scorodonia L.;蒜叶玄参;Balm-leaved Figwort■☆

355238 Scrophularia scorodonia L. var. papillaris (Boiss. et Reut.) Ball = Scrophularia scorodonia L. ■☆

355239 Scrophularia shikokiana Kitam. = Scrophularia kakudensis Franch. ■

355240 Scrophularia silvestrii Bonati = Scrophularia ningpoensis Hemsl. ■

355241 Scrophularia silvestrii Bonati et Pamp. = Scrophularia ningpoensis Hemsl. ■

355242 Scrophularia sinaica Benth.;西奈玄参■☆

355243 Scrophularia smithii Hornem.;史密斯小花玄参■☆

355244 Scrophularia smithii Hornem. subsp. hierrensis Dalgaard;耶洛玄参■☆

355245 Scrophularia souliei Franch.;小花玄参;Littleflower Figwort■

355246 Scrophularia spicata Franch.;穗花玄参;Spikate Figwort,Spikeflower Figwort■

355247 Scrophularia sprengeriana Sommier et H. Lév.;斯普玄参■☆

355248 Scrophularia stiefelhagenii Bonati = Scrophularia mandarinorum Franch. ■

355249 Scrophularia striata Boiss.;直玄参■☆

355250 Scrophularia stylosa P. C. Tsoong;长柱玄参;Longstyle Figwort■

355251 Scrophularia subcrispa Pomel = Scrophularia laevigata Vahl subsp. hispida (Desf.) Maire■☆

355252 Scrophularia syriaca A. DC.;叙利亚玄参■☆

355253 Scrophularia tadshicorum Gontsch.;塔什克玄参■☆

355254 Scrophularia taihangshanensis C. S. Zhu et H. W. Yang;太行玄参(太行山玄参);Taihangshan Figwort■

355255 Scrophularia tenuipes Coss. et Durieu;细梗玄参■☆

355256 Scrophularia thesioides Boiss. et Buhse;百蕊草玄参■☆

355257 Scrophularia toyamae Hatus. ex T. Yamaz. = Scrophularia kakudensis Franch. var. toyamae (Hatus. ex T. Yamaz.) T. Yamaz. ■☆

355258 Scrophularia trifoliata L.;三小叶玄参■☆

355259 Scrophularia trisecta Pau = Scrophularia sambucifolia L. ■☆

355260 Scrophularia turcomanica Bornm. et Sint. ex Rech. f.;土库曼玄参■☆

355261 Scrophularia umbrosa Dum. Cours.;翅茎玄参(翼玄参);Green Figwort, Scarce Water-figwort, Water Figwort, Western Figwort, Wingstem Figwort■

355262 Scrophularia uquatica ?;亨氏玄参;Huntsman's Cap■☆

355263 Scrophularia urticifolia Wall.;荨麻叶玄参;Nettleleaf Figwort■

355264 Scrophularia variegata M. Bieb.;变色玄参;Variegated Figwort■☆

355265 Scrophularia vernalis L.;春玄参;Spring Figwort,Yellow Figwort■☆

355266 Scrophularia verticillata Gontsch. et Grig.;轮叶玄参■☆

355267 Scrophularia viridiflora Poir. = Scrophularia sambucifolia L.

355268 Scrophularia wilsonii Bonati = Scrophularia fargesii Franch. ■

355269 Scrophularia xanthoglossa Boiss.;黄舌玄参■☆

355270 Scrophularia xanthoglossa Boiss. var. decipiens (Boiss. et Kotschy) Boiss. = Scrophularia xanthoglossa Boiss. ■☆

355271 Scrophularia xanthoglossa Boiss. var. deserticola Eig = Scrophularia xanthoglossa Boiss. ■☆

355272 Scrophularia yoshimurae T. Yamaz.;双锯齿玄参(双锯叶玄参,台湾玄参);Yoshimura Figwort■

355273 Scrophularia yunnanensis Franch.;云南玄参;Yunnan Figwort■

355274 Scrophularia zaravschanica Gorschk. et Zakirov;萨拉坦玄参■☆

355275 Scrophularia zuvandica Grossh.;祖万德玄参■☆

355276 Scrophulariaceae Juss. (1789)(保留科名);玄参科;Figwort Family■●

355277 Scrophularioides G. Forst. = Premna L. (保留属名)●■

355278 Scrotalaria Scr. ex Pfeiff. = Teucrium L. ●■

355279 Scrotochloa Judz. = Leptaspis R. Br. ■

355280 Scubalia Noronha = Lasianthus Jack(保留属名)●

355281 Scubulon Raf. = Lycopersicon Mill. ■

355282 Scubulus Raf. = Solanum L. ●■

355283 Sculeria Raf. = Vancouveria C. Moore et Decne. ■☆

355284 Sculertia K. Schum. = Brodiaea Sm. (保留属名)■☆

355285 Sculertia K. Schum. = Seubertia Kunth ■☆

355286 Scuria Raf. = Carex L. ■

355287 Scurrula L. (1753)(废弃属名);梨果寄生属(大叶枫寄生属,梨果桑寄生属);Scurrula ●

355288 Scurrula L. (1753)(废弃属名)= Loranthus Jacq. (保留属名)●

355289 Scurrula atropurpurea (Blume) Danser;梨果寄生(菲律宾桑寄生);Philippine Scurrula ●

355290 Scurrula buddleioides (Danser) G. Don;滇藏梨果寄生●

355291 Scurrula buddleioides (Danser) G. Don = Loranthus scurrula L. ●

355292 Scurrula buddleioides (Danser) G. Don var. heynei (DC.) H. S. Kiu;藏南梨果寄生●

355293 Scurrula canescens (Burch.) G. Don = Septulina glauca (Thunb.) Tiegh. ●☆

355294 Scurrula chinensis (DC.) G. Don = Taxillus chinensis (DC.) Danser ●

355295 Scurrula chinensis (DC.) G. Don = Taxillus pseudochinensis (Yamam.) Danser ●

355296 Scurrula chinensis (DC.) G. Don var. formosana Hosok. = Taxillus pseudochinensis (Yamam.) Danser ●

355297 Scurrula chingii (W. C. Cheng) H. S. Kiu;卵叶梨果寄生(卵果梨果寄生,卵叶寄生,水东哥寄生);Ching's Scurrula ●

355298 Scurrula chingii (W. C. Cheng) H. S. Kiu = Loranthus chingii W. C. Cheng ●

355299 Scurrula chingii (W. C. Cheng) H. S. Kiu var. yunnanensis H. S. Kiu;短柄梨果寄生(澜沧江寄生);Yunnan Ching Scurrula ●

355300 Scurrula elata (Edgew.) Danser;高山寄生;Alp Scurrula, Tall Scurrula ●

355301 Scurrula elata (Edgew.) Danser = Loranthus elatus Edgew. ●

355302 Scurrula ferruginea (Jack) Danser;锈毛梨果寄生(滇南寄生);Rustcoloured Scurrula, Rusthair Scurrula, Rusty Scurrula ●

355303 Scurrula ferruginea (Jack) Danser = Loranthus ferrugineus Jack ●

355304 Scurrula ferruginea (Jack) Danser = Scurrula sootepensis (Craib) Danser ●

355305 Scurrula glauca （Thunb.） G. Don ＝Septulina glauca （Thunb.） Tiegh. ●☆

355306 Scurrula gongshanensis H. S. Kiu；贡山梨果寄生；Gongshan Scurrula ●

355307 Scurrula graciliflora （Roxb. ex Schult. f.） Danser ＝Scurrula parasitica L. var. graciliflora （Wall. ex DC.） H. S. Kiu ●

355308 Scurrula gracilifolia （Schult. f.） Danser ＝Scurrula parasitica L. var. graciliflora （Wall. ex DC.） H. S. Kiu ●

355309 Scurrula levinei （Merr.） Danser ＝Taxillus levinei （Merr.） H. S. Kiu ●

355310 Scurrula liquidambaricola （Hayata） Danser ＝Taxillus limprichtii （Grüning） H. S. Kiu var. liquidambaricola （Hayata） H. S. Kiu ●

355311 Scurrula liquidambaricola （Hayata） Danser ＝Taxillus liquidambaricola （Hayata） Hosok. ●

355312 Scurrula lonicerifolia （Hayata） Danser ＝Scurrula parasitica L. var. lonicerifolius （Hayata） Y. C. Liu ●

355313 Scurrula lonicerifolia （Hayata） Danser ＝Taxillus nigrans （Hance） Danser ●

355314 Scurrula lonicerifolia J. M. Chao ＝Scurrula phoebe-formosanae （Hayata） Danser ●

355315 Scurrula notothixoides （Hance） Danser；小叶梨果寄生（蓝木桑寄生）；Notothixos-like Scurrula, Small-leaf Scurrula, Small-leaved Scurrula ●

355316 Scurrula oleifolia （J. C. Wendl.） G. Don ＝Tapinanthus oleifolius （J. C. Wendl.） Danser ●☆

355317 Scurrula parasitica L.；红花寄生（柏寄生，冰粉树，炒寄生，冬青，冻青，杜寄生，蠱心宝，广寄生，华南桑寄生，混沌螟蛉，寄居花童，寄生，寄生草，寄生树，寄童，寄屑，窠童，窠屑，毛叶桑寄生，莴，莴木，柠檬寄生，枇杷寄生，桑寄，桑寄生，桑络，桑上寄生，桑上羊儿藤，桑树上羊儿藤，沙梨寄生，柿寄生，双寄生，娑罗双桑寄生，桃木寄生，宛童，油茶寄生，寓木，樟寄生）；China Scurrula, Parasitic Scurrula, Redflower Scurrula, Red-flowered Scurrula, Sal Mistletoe, Scurrula ●

355318 Scurrula parasitica L. var. graciliflora （Wall. ex DC.） H. S. Kiu；小红花寄生（小叶寄生）；Little Redflower Scurrula ●

355319 Scurrula parasitica L. var. lonicerifolius （Hayata） Y. C. Liu；忍冬叶桑寄生；Honeysuckle-leaf Scurrula ●

355320 Scurrula pentagonia （DC.） G. Don ＝Tapinanthus pentagonia （DC.） Tiegh. ●☆

355321 Scurrula philippensis （Cham. et Schltdl.） G. Don ＝Scurrula atropurpurea （Blume） Danser ●

355322 Scurrula phoebe-formosanae （Hayata） Danser；楠树梨果寄生；Nanmu Scurrula, Phoebe Scurrula ●

355323 Scurrula pseudochinensis （Yamam.） Y. C. Liu et K. L. Chen ＝Taxillus pseudochinensis （Yamam.） Danser ●

355324 Scurrula pulverulenta （Wall.） G. Don；白梨果寄生（白花寄生）；Pawdered Scurrula, White Flower Scurrula ●

355325 Scurrula pulverulenta （Wall.） G. Don ＝Loranthus pulverulentus Wall. ●

355326 Scurrula rhododendricola （Hayata） Danser ＝Taxillus nigrans （Hance） Danser ●

355327 Scurrula ritozanensis （Hayata） Danser ＝Taxillus limprichtii （Grüning） H. S. Kiu ●

355328 Scurrula ritozanensis （Hayata） Danser ＝Taxillus ritozanensis （Hayata） S. T. Chiu ●

355329 Scurrula rufescens （DC.） G. Don ＝Phragmanthera rufescens （DC.） Balle ●☆

355330 Scurrula seraggodostemon （Hayata） Danser ＝Taxillus nigrans （Hance） Danser ●

355331 Scurrula sessilifolia （P. Beauv.） G. Don ＝Tapinanthus sessilifolius （P. Beauv.） Tiegh. ●☆

355332 Scurrula sootepensis （Craib） Danser；元江梨果寄生（元江寄生）；Soótep Scurrula, Yuanjiang Scurrula ●

355333 Scurrula theifer （Hayata） Danser ＝Taxillus theifer （Hayata） H. S. Kiu ●

355334 Scurrula thonningii （DC.） G. Don ＝Phragmanthera capitata （Spreng.） Balle ●☆

355335 Scurrula tsaii （S. T. Chiu） Yuen P. Yang et S. Y. Lu ＝Taxillus tsaii S. T. Chiu ●

355336 Scurrula umbellifer （Schult. f.） G. Don ＝Taxillus umbellifer （Schult.） Danser ●

355337 Scurrula yadoriki （Siebold ex Maxim.） Danser；毛叶梨果寄生（寄生草，毛叶桑寄生，桑寄生，桑上寄生）；Hairleaf Scurrula, Yadorik Scurrula ●☆

355338 Scutachne Hitchc. et Chase(1911)；岩坡草属■☆

355339 Scutachne amphistemon Hitchc. et Chase；岩坡草■☆

355340 Scutellaria L. （1753）；黄芩属；Helmet Flower, Skull Cap, Skullcap ●■

355341 Scutellaria Riv. ex L. ＝Scutellaria L. ●■

355342 Scutellaria abbreviata H. Hara ＝Scutellaria laeteviolacea Koidz. var. abbreviata （H. Hara） H. Hara ■☆

355343 Scutellaria adenophylla Miq. ＝Scutellaria barbata D. Don ■

355344 Scutellaria adenostegia Briq.；腺盖黄芩■☆

355345 Scutellaria adenotricha X. H. Guo et S. B. Zhou；腺毛黄芩；Glandhai Meadowrue ■

355346 Scutellaria adsurgens Popov；直立黄芩；Erect Skullcap ■☆

355347 Scutellaria africana Hochst. ＝Scutellaria arabica Jaub. et Spach ■☆

355348 Scutellaria ahaicola C. Y. Wu et H. W. Li ＝Scutellaria altaica Fisch. ex Sweet ■

355349 Scutellaria alaschanica Tschern.；阿拉善黄芩；Alashan Meadowrue ■

355350 Scutellaria alaschanica Tschern. ＝Scutellaria rehderiana Diels et Hand. -Mazz. ■

355351 Scutellaria albertii Juz.；微尖苞黄芩（微尖黄芩）；Albert Skullcap ●

355352 Scutellaria albertii Juz. ＝Scutellaria sieversii Bunge ●

355353 Scutellaria albida L.；微白黄芩；Whitish Skullcap ■☆

355354 Scutellaria alexeenkoi Juz.；阿列黄芩■☆

355355 Scutellaria alpina L.；高山黄芩（阿尔卑斯黄芩）■☆

355356 Scutellaria alpina L. var. lupulina ? ＝Scutellaria lupulina L. ■☆

355357 Scutellaria altaica Fisch. ex Sweet；阿尔泰黄芩；Altai Skullcap ■

355358 Scutellaria altaicola C. Y. Wu et H. W. Li ＝Scutellaria altaica Fisch. ex Sweet ■

355359 Scutellaria altissima L.；高黄芩（极高黄芩）；Somerset Skullcap, Tall Skullcap, Tallest Skullcap ■☆

355360 Scutellaria amabilis H. Hara；秀丽黄芩■☆

355361 Scutellaria ambigua Nutt. ＝Scutellaria parvula Michx. var. missouriensis （Torr.） Goodman et C. A. Lawson ■☆

355362 Scutellaria ambigua Nutt. var. missouriensis Torr. ＝Scutellaria parvula Michx. var. missouriensis （Torr.） Goodman et C. A. Lawson ■☆

355363 Scutellaria amoena C. H. Wright；云南黄芩（川黄芩，滇黄芩，黄芩，枯子芩，条子芩，土黄芩，西南黄芩，小黄芩）；Yunnan Skullcap ■

355364　Scutellaria amoena C. H. Wright = Scutellaria rehderiana Diels et Hand. -Mazz. ■

355365　Scutellaria amoena C. H. Wright var. cinerea Hand. -Mazz. ；灰毛滇黄芩(灰毛黄芩)；Greyhair Yunnan Skullcap ■

355366　Scutellaria amphichlora Juz. ；绿黄芩■☆

355367　Scutellaria andrachnoides Vved. ；黑钩叶状黄芩；Andrachnelike Skullcap ■☆

355368　Scutellaria androssovii Juz. ；安得罗索夫黄芩；Androssov Skullcap ■☆

355369　Scutellaria angrenica Juz. et Vved. ；安德伦黄芩■☆

355370　Scutellaria angulosa Benth. = Scutellaria franchetiana H. Lév. ■

355371　Scutellaria angulosa Benth. = Scutellaria scandens D. Don ■

355372　Scutellaria angulosa Benth. var. franchetiana (H. Lév.) Kudo = Scutellaria franchetiana H. Lév. ■

355373　Scutellaria angustifolia (Regel) Kom. = Scutellaria regeliana Nakai ■

355374　Scutellaria angustifolia Pursh；细叶黄芩■☆

355375　Scutellaria anhweiensis C. Y. Wu；安徽黄芩；Anhui Skullcap, Anhwei Skullcap ■

355376　Scutellaria anhweiensis C. Y. Wu var. fanchangnica X. L. Liu, W. C. Ye et W. Wu；繁昌黄芩■

355377　Scutellaria anitae Juz. ；阿妮塔黄芩■☆

355378　Scutellaria aquarrosa Nevski；粗糙黄芩；Rough Skullcap ■☆

355379　Scutellaria arabica Jaub. et Spach；阿拉伯黄芩■☆

355380　Scutellaria araratica Grossh. ；亚拉腊黄芩■☆

355381　Scutellaria araxensis Grossh. ；亚美尼亚黄芩；Armenia Skullcap ■☆

355382　Scutellaria artvinensis Grossh. ；阿尔特温黄芩■☆

355383　Scutellaria australis (Fassett) Epling；南部小黄芩；Small Skullcap, Southern Small Skullcap ■☆

355384　Scutellaria austrotaiwanensis T. H. Hsieh et T. C. Huang；南台湾黄芩；S. Taiwan Skullcap ■

355385　Scutellaria axilliflora Hand. -Mazz. ；腋花黄芩；Axillaryflower Skullcap ■

355386　Scutellaria axilliflora Hand. -Mazz. var. medulifera (Y. Z. Sun ex C. H. Hu) C. Y. Wu et H. W. Li；大腋花黄芩(大花腋花黄芩,髓黄芩)■

355387　Scutellaria baicalensis Georgi；黄芩(瓣芩,北芩,薄叶黄芩,大花黄芩,大条,大枝芩,淡芩,滇芩,东芩,妒妇,腐肠,杠谷草,红胜,虹胜,黄花黄芩,黄金条根,黄芩瓣,黄芩茶,黄芩炭,黄芩条,黄文,尖芩,经芩,空肠,空心草,枯肠,枯黄芩,枯芩,苦督邮,蓝靛花,烂心草,烂心子,里腐草,里朽草,里朽斤草,内虚,片芩,片子芩,片子黄芩,釜,芩,山茶根,鼠尾芩,宿芩,条黄芩,条芩,条子芩,土金茶,土金茶条,豚尾芩,尾芩,西芩,细芩,香水水草,野树豆花,印头,元芩,黏毛黄芩,枝芩,子黄芩,子芩,紫花黄芩)；Baikal Skullcap, Skullcap ■

355388　Scutellaria baicalensis Georgi f. albiflora H. Wei Jen et Y. J. Chang；白花黄芩；Whiteflower Baikal Skullcap ■

355389　Scutellaria baldshuanica Nevski ex Juz. ；巴得栓黄芩；Baldshuan Skullcap ■☆

355390　Scutellaria bambusetorum C. Y. Wu；竹林黄芩；Bambooforest Skullcap ■

355391　Scutellaria barbata D. Don；半枝莲(半面花,半向花,半支莲,并头草,耳挖草,方草儿,赶山鞭,虎咬红,金挖草,金挖耳草,偏头草,乞丐碗,瘦黄芩,水韩信,水黄芩,四方草,四方马兰,天基草,铁棍草,通经草,外来黄芩,溪岸黄芩,溪边黄芩,狭叶韩信草,狭叶向天盏,向天盏,小韩信草,小号向天盏,小挖耳草,牙刷草,洋黄芩,野夏枯草,再生草,紫连草,昨日荷草)；Barbate

Skullcap, Barbed Skullcap, Half Lotus, Riverbank Skullcap ■

355392　Scutellaria brachyspica Nakai et H. Hara；短穗黄芩■☆

355393　Scutellaria brachyspica Nakai et H. Hara f. albiflora Hayashi et Y. Kobay. ；白花短穗黄芩■☆

355394　Scutellaria briquetii A. Chev. = Scutellaria schweinfurthii Briq. subsp. pauciflora (Baker) A. J. Paton ■☆

355395　Scutellaria bucharica Juz. ；布加尔黄芩；Buchar Skullcap ■☆

355396　Scutellaria bushii Britton；布什黄芩；Bush's Skullcap, Skullcap ■☆

355397　Scutellaria bussei Gürke = Scutellaria schweinfurthii Briq. subsp. pauciflora (Baker) A. J. Paton ■☆

355398　Scutellaria calcarata C. Y. Wu et H. W. Li；囊距黄芩；Spur Skullcap ■

355399　Scutellaria calyopteroides Hand. -Mazz. ；莐状黄芩；Bluebeard-like Skullcap ■

355400　Scutellaria canescens Nutt. ；西黄芩■☆

355401　Scutellaria catharinae Juz. ；卡塔氏黄芩■☆

355402　Scutellaria caudifolia Y. Z. Sun ex C. H. Hu；尾叶黄芩；Caudateleaf Skullcap, Tailleaf Skullcap ■

355403　Scutellaria caudifolia Y. Z. Sun ex C. H. Hu var. obliquifolia C. Y. Wu et S. Chow；斜尾叶黄芩(野鸡黄)；Oblique Caudateleaf Skullcap ■

355404　Scutellaria cavaleriei H. Lév. et Vaniot. = Scutellaria barbata D. Don ■

355405　Scutellaria celtidifolia A. Ham. = Scutellaria scandens D. Don ■

355406　Scutellaria chekiangensis C. Y. Wu；浙江黄芩(散血丹)；Chekiang Skullcap, Zhejiang Skullcap ■

355407　Scutellaria chenopodiifolia Juz. ；藜叶黄芩；Goosefootleaf Skullcap ■☆

355408　Scutellaria chihshuiensis C. Y. Wu et H. W. Li；赤水黄芩；Chihshui Skullcap, Chishui Skullcap ■

355409　Scutellaria chimenensis C. Y. Wu；祁门黄芩；Chimen Skullcap, Qimen Skullcap ■

355410　Scutellaria chitrovoi Juz. ；黑氏黄芩■☆

355411　Scutellaria chungtienensis C. Y. Wu；中甸黄芩；Chungtien Skullcap, Zhongdian Skullcap ■

355412　Scutellaria coerulea Moc. et Sessé ex Benth. ；蓝黄芩；Blue Skullcap ■☆

355413　Scutellaria coleifolia H. Lév. = Scutellaria violacea K. Heyne ex Wall. var. sikkimensis Hook. f. ■

355414　Scutellaria colpodea Nevski；鞘状黄芩；Colpodes Skullcap ■☆

355415　Scutellaria columnae All. ；柱黄芩■☆

355416　Scutellaria comosa Juz. ；丛毛黄芩；Comose Skullcap ■☆

355417　Scutellaria cordifolia Muhl. = Scutellaria ovata Hill ■☆

355418　Scutellaria cordifrons Juz. ；心叶黄芩■☆

355419　Scutellaria creticcola Juz. ；克里特黄芩■☆

355420　Scutellaria cristata Popov；鸡冠黄芩；Cristate Skullcap ■☆

355421　Scutellaria cyrtopoda Miq. = Scutellaria coleifolia H. Lév. ■

355422　Scutellaria daghestanica Grossh. ；达赫斯坦黄芩；Daghestan Skullcap ■☆

355423　Scutellaria darriensis Grossh. ；达连黄芩■☆

355424　Scutellaria darvasica Juz. ；达尔瓦斯黄芩■☆

355425　Scutellaria debeerstii Briq. = Scutellaria schweinfurthii Briq. subsp. pauciflora (Baker) A. J. Paton ■☆

355426　Scutellaria delavayi H. Lév. ；方枝黄芩；Delavay Skullcap, Squaretwig Skullcap ■

355427　Scutellaria dentata H. Lév. = Scutellaria pekinensis Maxim. var. ussuriensis (Regel) Hand. -Mazz. ■

355428　Scutellaria dependens Maxim.；纤弱黄芩（小花黄芩）；Dependent Skullcap，Think Skullcap ■

355429　Scutellaria discolor Wall. ex Benth.；异色黄芩（四天红，土黄芩，挖耳草，熊胆草，夜行草，一支箭，一枝蒿，一枝箭，追天花根，紫背草，紫背黄芩）；Discolored Skullcap ■

355430　Scutellaria discolor Wall. ex Benth. var. hirta Hand. -Mazz.；地盆草（结筋草）；Hairy Think Skullcap ■

355431　Scutellaria discolor Wall. ex Benth. var. pubescens C. Y. Wu et H. W. Li；多毛异色黄芩（多毛黄芩）■

355432　Scutellaria dubia Taliev et Sirj.；可疑黄芩（疑黄芩）；Doubtful Skullcap ■☆

355433　Scutellaria elliptica Muhl.；椭圆黄芩；Hairy Skullcap ■☆

355434　Scutellaria engleriana Bornm.；恩格勒黄芩■☆

355435　Scutellaria epilobiifolia A. Ham.；柳叶黄芩；Marsh Skullcap，Willowherb Skullcap ■☆

355436　Scutellaria epilobiifolia A. Ham. = Scutellaria galericulata L. ■

355437　Scutellaria esquirolii H. Lév. et Vaniot = Melampyrum roseum Maxim. var. obtusifolium（Bonati）D. Y. Hong ■

355438　Scutellaria filicaulis Regel；丝茎黄芩；Threadshapedstem Skullcap ■☆

355439　Scutellaria flabellulata Juz.；扇形黄芩；Fanshaped Skullcap ■☆

355440　Scutellaria formosana N. E. Br.；蓝花黄芩；Blue Skullcap，Blueflower Skullcap ■

355441　Scutellaria formosana N. E. Br. = Scutellaria tsinyunensis C. Y. Wu et S. Chow ■

355442　Scutellaria formosana N. E. Br. var. pubescens C. Y. Wu et H. W. Li；多毛蓝花黄芩（多毛台湾黄芩）；Pubescent Blueflower Skullcap ■

355443　Scutellaria forrestii Diels；灰岩黄芩；Forrest Skullcap，Limestone Skullcap ■

355444　Scutellaria forrestii Diels var. intermedia C. Y. Wu et H. W. Li；居间灰岩黄芩；Intermediate Forrest Skullcap ■

355445　Scutellaria forrestii Diels var. intermedia C. Y. Wu et H. W. Li = Scutellaria forrestii Diels ■

355446　Scutellaria forrestii Diels var. muliensis C. Y. Wu；木里灰岩黄芩；Muli Skullcap ■

355447　Scutellaria forrestii Diels var. muliensis C. Y. Wu = Scutellaria forrestii Diels ■

355448　Scutellaria franchetiana H. Lév.；岩霍黄芩（方茎犁头草，犁头草，岩藿香）；Franchet Skullcap，Rock Skullcap ■

355449　Scutellaria galericulata L.；盔状黄芩（并头草，兜冠黄芩，克拉塞夫黄芩）；Blue Skullcap，Common Skullcap，Galericulate Skullcap，Helmet-flower，Hooded Skullcap，Hooded Willow，Hooded Willowherb，Hoodwort，Krasev Skullcap Helmit Skullcap，Marsh Skullcap，Skullcap ■

355450　Scutellaria galericulata L. = Scutellaria scordifolia Fisch. ex Schrank ■

355451　Scutellaria galericulata L. var. angustifolia Regel = Scutellaria regeliana Nakai ■

355452　Scutellaria galericulata L. var. epilobiifolia（A. Ham.）Jordal = Scutellaria galericulata L. ■

355453　Scutellaria galericulata L. var. pubescens Benth. = Scutellaria galericulata L. ■

355454　Scutellaria galericulata L. var. scordifolia Regel = Scutellaria scordifolia Fisch. ex Schrank ■

355455　Scutellaria glabrata Vved.；脱毛黄芩；Glabrate Skullcap ■☆

355456　Scutellaria glandulosa Hook. f.；腺体黄芩；Glandular Skullcap ■☆

355457　Scutellaria glecomoides Migo = Scutellaria tuberifera C. Y. Wu et C. Chen ■

355458　Scutellaria grandiflora Sims = Scutellaria baicalensis Georgi ■

355459　Scutellaria grandiflora Sims = Scutellaria orientalis L. ■☆

355460　Scutellaria granulosa Juz.；颗粒黄芩；Granular Skullcap ■☆

355461　Scutellaria grossa Wall.；大黄芩；Gross Skullcap ■

355462　Scutellaria grossecrenata Merr. et Chun ex C. Y. Wu；粗齿黄芩；Bigtooth Skullcap，Grosstooth Skullcap ■

355463　Scutellaria grossheimiana Juz.；格罗黄芩■☆

355464　Scutellaria guilielmii A. Gray；连钱黄芩；Guilielm Skullcap ■

355465　Scutellaria guttata Nevski ex Juz.；斑点黄芩■☆

355466　Scutellaria haematochlora Juz.；血绿黄芩■☆

355467　Scutellaria hainanenais C. Y. Wu；海南黄芩；Hainan Skullcap ■

355468　Scutellaria hansuensis Hand. -Mazz. = Scutellaria rehderiana Diels et Hand. -Mazz. ■

355469　Scutellaria hastifolia L.；戟叶黄芩；Hastateleaf Skullcap，Norfolk Skullcap ■☆

355470　Scutellaria hebeclada W. W. Sm. = Scutellaria mairei H. Lév. ■

355471　Scutellaria hederacea Kunth et Bouché = Scutellaria guilielmii A. Gray ■

355472　Scutellaria hederacea Kunth et Bouché = Scutellaria tuberifera C. Y. Wu et C. Chen ■

355473　Scutellaria helenae Alb.；海伦黄芩；Helen Skullcap ■☆

355474　Scutellaria heterochroa Juz.；克里曼异色黄芩（色异黄芩）；Diversecolour Skullcap ■☆

355475　Scutellaria heterotricha Juz. et Vved.；异毛黄芩■☆

355476　Scutellaria heydei Hook. f.；海地黄芩；Heyde Skullcap ■☆

355477　Scutellaria hirtella Juz.；多毛黄芩■☆

355478　Scutellaria hissarica B. Fedtsch. et Popov；锡萨尔黄芩；Hissar Skullcap ■☆

355479　Scutellaria holosericea Gontsch. ex Juz.；全毛黄芩■☆

355480　Scutellaria honanensis C. Y. Wu et H. W. Li；河南黄芩；Henan Skullcap，Honan Skullcap ■

355481　Scutellaria huangshanensis X. W. Wang et Z. W. Xue；黄山黄芩；Huangshan Skullcap ■

355482　Scutellaria huangshanensis X. W. Wang et Z. W. Xue = Scutellaria anhweiensis C. Y. Wu ■

355483　Scutellaria hunanensis C. Y. Wu；湖南黄芩；Hunan Skullcap ■

355484　Scutellaria hypericifolia H. Lév.；连翘叶黄芩（草地黄芩，川黄芩，魁芩，连翘黄芩，条芩，条子芩，土大芩，子芩）；St. Johnsedleaf Skullcap ■

355485　Scutellaria hypericifolia H. Lév. var. pilosa C. Y. Wu；多毛连翘叶黄芩；Pilose St. Johnsedleaf Skullcap ■

355486　Scutellaria hypopolia Juz.；灰黄芩；Grey Skullcap ■☆

355487　Scutellaria ikonnikovii Juz. = Scutellaria regeliana Nakai var. ikonnikovii（Juz.）C. Y. Wu et H. W. Li ■

355488　Scutellaria ikonnikovii Juz. = Scutellaria regeliana Nakai ■

355489　Scutellaria immaculata Nevski ex Juz.；无斑点黄芩；Immaculate Skullcap ■☆

355490　Scutellaria incana Biehler；灰毛黄芩；Downy Skullcap，Hoary Skullcap ■☆

355491　Scutellaria incana Spreng. = Scutellaria incana Biehler ■☆

355492　Scutellaria incisa Y. Z. Sun ex C. H. Hu；裂叶黄芩（刻叶黄芩）；Incisedleaf Skullcap ■

355493　Scutellaria incurva Wall.；内弯黄芩；Incurved Skullcap ■☆

355494　Scutellaria indica L.；韩信草（大韩信草，大力草，大叶半支莲，大叶韩信草，地苏麻，调羹草，疔疮草，耳挖草，红头犁头草，红叶犁头尖,虎咬癀,虎咬蟥,金茶匙,金锁匙,蓝花茶匙癀,力浪草,立浪草,疗疮草,木勺草,偏向花,三合香,顺筋草,顺经草,田代氏黄芩,铁灯盏,向天盏,笑花草,烟管草,印度黄芩,油灯盏,紫花地丁);India Skullcap,Indian Skullcap,Tashiro Skullcap ■

355495　Scutellaria indica L. f. amagiensis Sugim.;白花韩信草■☆

355496　Scutellaria indica L. f. japonica Franch. et Sav. = Scutellaria indica L. ■

355497　Scutellaria indica L. f. leucantha T. Yamaz. = Scutellaria L. f. amagiensis Sugim. ■☆

355498　Scutellaria indica L. f. parvifolia Makino = Scutellaria indica L. var. parvifolia (Makino) Makino ■

355499　Scutellaria indica L. f. parvifolia Matsum. et Kudo = Scutellaria indica L. var. parvifolia (Makino) Makino ■

355500　Scutellaria indica L. f. ramosa C. Y. Wu et C. Chen;多枝韩信草;Manybranch Indian Skullcap ■

355501　Scutellaria indica L. f. ramosa C. Y. Wu et C. Chen = Scutellaria indica L. ■

355502　Scutellaria indica L. f. subacaulis Y. Z. Sun ex C. H. Hu = Scutellaria indica L. var. subacaulis (Y. Z. Sun ex C. H. Hu) C. Y. Wu et C. Chen ■

355503　Scutellaria indica L. var. ambigua Hand. -Mazz. = Scutellaria teniana Hand. -Mazz. ■

355504　Scutellaria indica L. var. elliptica Y. Z. Sun ex C. H. Hu;长毛韩信草;Elliptic Indian Skullcap ■

355505　Scutellaria indica L. var. humilis Makino;元元草■☆

355506　Scutellaria indica L. var. indica f. ramosa C. Y. Wu et C. Chen = Scutellaria indica L. ■

355507　Scutellaria indica L. var. japonica (Morren et Decne.) Franch. et Sav. f. humilis Makino = Scutellaria laeteviolacea Koidz. ■

355508　Scutellaria indica L. var. japonica (Morren et Decne.) Franch. et Sav. f. parvifolia Makino = Scutellaria indica L. var. parvifolia (Makino) Makino ■

355509　Scutellaria indica L. var. leucantha T. Yamaz. = Scutellaria indica L. f. leucantha T. Yamaz. ■☆

355510　Scutellaria indica L. var. parvifolia (Makino) Makino;小叶韩信草;Smallleaf Indian Skullcap ■

355511　Scutellaria indica L. var. parvifolia (Makino) Makino f. alba (H. Hara) H. Hara ex Murata et T. Yamaz.;白花小叶韩信草■☆

355512　Scutellaria indica L. var. parvifolia (Makino) Makino f. lilacina (H. Hara) H. Hara ex Murata et T. Yamaz.;紫花小叶韩信草■☆

355513　Scutellaria indica L. var. pekinensis Franch. = Scutellaria pekinensis Maxim. ■

355514　Scutellaria indica L. var. subacaulis (Y. Z. Sun ex C. H. Hu) C. Y. Wu et C. Chen;缩茎韩信草(金耳挖,天田盏,小金疮草);Shortstem Indian Skullcap ■

355515　Scutellaria indica L. var. tsusimensis (H. Hara) Ohwi = Scutellaria tsusimensis H. Hara ■☆

355516　Scutellaria indica L. var. typica Kudo = Scutellaria indica L. var. parvifolia (Makino) Makino ■

355517　Scutellaria inghokensis F. P. Metcalf;永泰黄芩;Yongtai Skullcap ■

355518　Scutellaria insignis Nakai;显著黄芩■☆

355519　Scutellaria integrifolia L.;全叶黄芩(全缘叶黄芩);Hyssop Skullcap,Large Skullcap ■☆

355520　Scutellaria intermedia Popov;中间黄芩■☆

355521　Scutellaria irregularis Juz.;不齐齿黄芩;Irregularteeth Skullcap,Irregulartooth Skullcap ●

355522　Scutellaria irregularis Juz. = Scutellaria supina L. ●

355523　Scutellaria iskanderi Juz.;伊斯黄芩■☆

355524　Scutellaria japonica Burm. f. = Isodon japonicus (Burm. f.) H. Hara ■

355525　Scutellaria japonica Burm. f. var. purpureicaulis Migo = Scutellaria pekinensis Maxim. var. purpureicaulis (Migo) C. Y. Wu et H. W. Li ■

355526　Scutellaria japonica Burm. f. var. usssuriensis f. humilis Matsum. et Kudo = Scutellaria indica L. var. subacaulis (Y. Z. Sun ex C. H. Hu) C. Y. Wu et C. Chen ■

355527　Scutellaria japonica Franch. et Sav. = Scutellaria pekinensis Maxim. var. purpureicaulis (Migo) C. Y. Wu et H. W. Li ■

355528　Scutellaria japonica Morren et Decne. = Scutellaria pekinensis Maxim. var. ussuriensis (Regel) Hand. -Mazz. ■

355529　Scutellaria japonica Morren et Decne. var. purpureicaulis Migo = Scutellaria pekinensis Maxim. var. purpureicaulis (Migo) C. Y. Wu et H. W. Li ■

355530　Scutellaria japonica Morren et Decne. var. ussuriensis Regel = Scutellaria pekinensis Maxim. var. ussuriensis (Regel) Hand. -Mazz. ■

355531　Scutellaria javanica Jungh.;爪哇黄芩;Java Skullcap ■

355532　Scutellaria javanica Jungh. = Scutellaria formosana N. E. Br. ■

355533　Scutellaria javanica Jungh. var. luzonica (Rolfe) Keng;吕宋黄芩(爪哇立浪草)■

355534　Scutellaria javanica Jungh. var. luzonica (Rolfe) Keng = Scutellaria luzonica Rolfe ■

355535　Scutellaria javanica Jungh. var. playfairi (Kudo) Huang et Cheng;台湾力浪草(台湾黄芩,台湾立浪草);Taiwan Java Skullcap ■

355536　Scutellaria juzepczukii Gontsch.;究采坡组克黄芩;Juzepczuk Skullcap ■☆

355537　Scutellaria kansuensis Hand. -Mazz. = Scutellaria rehderiana Diels et Hand. -Mazz. ■

355538　Scutellaria karatavica Juz.;喀拉塔夫黄芩;Karatav Skullcap ■☆

355539　Scutellaria karjaginii Grossh.;卡尔黄芩■☆

355540　Scutellaria karkaralensis Juz.;喀尔喀拉勒黄芩;Karkaral Skullcap ■☆

355541　Scutellaria katangensis Robyns et Lebrun = Scutellaria schweinfurthii Briq. subsp. pauciflora (Baker) A. J. Paton ■☆

355542　Scutellaria khasiana C. B. Clarke ex Hook. f.;克哈锡黄芩;Khasi Skullcap ■☆

355543　Scutellaria kingiana Prain;藏黄芩■

355544　Scutellaria kiusiana H. Hara;九州黄芩■☆

355545　Scutellaria kiusiana H. Hara f. albiflora H. Hara;白花九州黄芩■☆

355546　Scutellaria kiusiana H. Hara var. discolor ? = Scutellaria kiusiana H. Hara ■☆

355547　Scutellaria kiusiana H. Hara var. kuromidakensis Yahara = Scutellaria kuromidakensis (Yahara) T. Yamaz. ■☆

355548　Scutellaria knorringiae Juz.;克诺黄芩■☆

355549　Scutellaria komarovii H. Lév. et Vaniot = Scutellaria barbata D. Don ■

355550　Scutellaria krasevii Kom. et I. Schischk. ex Juz. = Scutellaria galericulata L. ■

355551　Scutellaria krylovii Juz. = Scutellaria sieversii Bunge ●

355552　Scutellaria kuromidakensis (Yahara) T. Yamaz.;黑耳黄芩■☆

355553　Scutellaria kurssanovii Pavlov;库萨纳夫黄芩;Kurssanov Skullcap ■☆

355554　Scutellaria laeteviolacea Koidz.;光紫黄芩(荔枝草,元元草);

Brightviolet Skullcap，Shiningpurple Skullcap ■

355555 Scutellaria laeteviolacea Koidz. f. concolor Honda；同色光紫黄芩■☆

355556 Scutellaria laeteviolacea Koidz. var. abbreviata（H. Hara）H. Hara；缩短光紫黄芩■☆

355557 Scutellaria laeteviolacea Koidz. var. discolor（H. Hara）H. Hara ＝ Scutellaria kiusiana H. Hara ■☆

355558 Scutellaria laeteviolacea Koidz. var. kurokawae（H. Hara）H. Hara；黑川黄芩■☆

355559 Scutellaria laeteviolacea Koidz. var. maekawae（H. Hara）H. Hara；前川黄芩■☆

355560 Scutellaria lanceolaria Miq. ＝ Scutellaria baicalensis Georgi ■

355561 Scutellaria lanipes Juz.；毛梗黄芩■☆

355562 Scutellaria lantienensis Hand. -Mazz. ＝ Scutellaria guilielmii A. Gray ■

355563 Scutellaria lateriflora L.；侧花黄芩（美黄芩）；Blue Skullcap，Hoodwort，Mad Dog Skullcap，Mad-dog Skullcap，Side-flowering Skullcap，Virginian Skullcap ■☆

355564 Scutellaria laxa Dunn；散黄芩；Scatter Skullcap ■

355565 Scutellaria leonardii Epling；莱奥纳尔黄芩；Leonard's Skullcap，Small Skullcap ■☆

355566 Scutellaria leonardii Epling ＝ Scutellaria parvula Michx. var. missouriensis（Torr.）Goodman et C. A. Lawson ■☆

355567 Scutellaria leptosiphon Nevski；细管黄芩；Thintube Skullcap ■☆

355568 Scutellaria leptostegia Juz.；细盖黄芩■☆

355569 Scutellaria leucodasys Miq. ＝ Scutellaria indica L. ■

355570 Scutellaria likiangensis Diels；丽江黄芩（白花黄芩，小黄芩）；Lijiang Skullcap，Likiang Skullcap ■

355571 Scutellaria lilungense S. S. Ying ＝ Scutellaria taiwanensis C. Y. Wu ■

355572 Scutellaria linarioides C. Y. Wu；长叶并头草；Longleaf Skullcap ■

355573 Scutellaria linczewskii Juz.；林契黄芩■☆

355574 Scutellaria linearis Benth.；线形黄芩；Linear Skullcap ■☆

355575 Scutellaria lipskyi Juz.；利普斯基黄芩■☆

355576 Scutellaria litwinowii Bornm. et Sint.；利特氏黄芩；Litwinow Skullcap ■☆

355577 Scutellaria livingstonei Baker ＝ Scutellaria schweinfurthii Briq. subsp. pauciflora（Baker）A. J. Paton ■☆

355578 Scutellaria longituba Koidz.；长管黄芩■☆

355579 Scutellaria lotienensis C. Y. Wu et S. Chow；罗甸黄芩；Lotien Skullcap，Luodian Skullcap ■

355580 Scutellaria lupulina L.；啤酒花黄芩；Yellow Alpine Skullcap ■☆

355581 Scutellaria lupulina L. var. violacea Bunge ＝ Scutellaria altaica Fisch. ex Sweet ■

355582 Scutellaria luteocoerulea Bornm. et Sint.；黄蓝黄芩；Yellow-blue Skullcap ■☆

355583 Scutellaria lutescens C. Y. Wu；淡黄黄芩；Lightyellow Skullcap，Whitish Skullcap ■

355584 Scutellaria luzonica Rolfe；吕宋黄芩；Luzon Skullcap ■

355585 Scutellaria luzonica Rolfe ＝ Scutellaria playfairi Kudo ■

355586 Scutellaria luzonica Rolfe var. lotungensis C. Y. Wu et C. Chen；乐东黄芩（乐东吕宋黄芩）；Ledong Skullcap ■

355587 Scutellaria luzonica Rolfe var. playfairi（Kudo）Yamam. ＝ Scutellaria playfairi Kudo ■

355588 Scutellaria macrantha Fisch. ＝ Scutellaria baicalensis Georgi ■

355589 Scutellaria macrantha Fisch. ex Rchb.；大花黄芩■

355590 Scutellaria macrantha Fisch. ex Rchb. ＝ Scutellaria baicalensis Georgi ■

355591 Scutellaria macrodonta Hand. -Mazz.；大齿黄芩；Bigtooth Skullcap，Largetooth Skullcap ■

355592 Scutellaria macrosiphon C. Y. Wu；大管黄芩（长管黄芩）；Bigtube Skullcap ■

355593 Scutellaria maekawae H. Hara ＝ Scutellaria laeteviolacea Koidz. var. maekawae（H. Hara）H. Hara ■☆

355594 Scutellaria maekawae H. Hara f. abbreviata（H. Hara）Murata ＝ Scutellaria laeteviolacea Koidz. var. abbreviata（H. Hara）H. Hara ■☆

355595 Scutellaria maekawae H. Hara var. kurokawae（H. Hara）Murata ＝ Scutellaria laeteviolacea Koidz. var. kurokawae（H. Hara）H. Hara ■☆

355596 Scutellaria mairei H. Lév.；毛茎黄芩；Hairstem Skullcap，Maire Skullcap ■

355597 Scutellaria medulifera Y. Z. Sun ex C. H. Hu ＝ Scutellaria axilliflora Hand. -Mazz. var. medulifera（Y. Z. Sun ex C. H. Hu）C. Y. Wu et H. W. Li ■

355598 Scutellaria meehanioides C. Y. Wu；龙头黄芩；Dregonheadsage Skullcap，Meehanialike Skullcap ■

355599 Scutellaria meehanioides C. Y. Wu var. paucidentata C. Y. Wu et H. W. Li；少齿龙头黄芩；Fewtooth Skullcap ■

355600 Scutellaria megaphylla C. Y. Wu et H. W. Li；大叶黄芩；Bigleaf Skullcap ■

355601 Scutellaria mesostegia Juz.；中被黄芩；Middlecover Skullcap ■☆

355602 Scutellaria mexicana（Torr.）A. J. Paton；墨西哥黄芩；Bladder Sage，Bladdersage，Paper-bag Bush，Paperbagbush ●☆

355603 Scutellaria mexicana（Torr.）A. J. Paton ＝ Salazaria mexicana Torr. ●☆

355604 Scutellaria microdasys Juz.；小刺黄芩；Smallsping Skullcap ■☆

355605 Scutellaria microflora F. P. Metcalf ＝ Scutellaria indica L. var. parvifolia（Makino）Makino ■

355606 Scutellaria microflora Metcalf ＝ Scutellaria indica L. var. parvifolia（Makino）Makino ■

355607 Scutellaria microphysa Juz.；小囊黄芩■☆

355608 Scutellaria microviolacea C. Y. Wu；小紫黄芩；Smallpurple Skullcap，Smallviolet Skullcap ■

355609 Scutellaria minor Huds.；小黄芩；Hedge Hyssop，Lesser Skullcap，Small Hooded Willowherb ■☆

355610 Scutellaria minor Huds. f. indica Benth. ＝ Scutellaria barbata D. Don ■

355611 Scutellaria minor L. var. indica Benth. ＝ Scutellaria barbata D. Don ■

355612 Scutellaria mississippiensis M. Martens ＝ Scutellaria ovata Hill ■☆

355613 Scutellaria mollifolia C. Y. Wu et H. W. Li；毛叶黄芩；Hairleaf Skullcap，Hairyleaf Skullcap ■

355614 Scutellaria mongolica Sobolevsk.；蒙古黄芩；Mongolian Skullcap ■☆

355615 Scutellaria moniliorrhiza Kom.；念珠根茎黄芩（串珠黄芩，念珠根黄芩，念珠黄芩）；Moniliform Skullcap ■

355616 Scutellaria muramatsui H. Hara；村松黄芩■☆

355617 Scutellaria navicularis Juz.；舟状黄芩；Boat-shaped Skullcap ■☆

355618 Scutellaria nepetoides Popov ex Juz.；荆芥黄芩■☆

355619 Scutellaria nervosa Pursh；多脉黄芩；Veined Skullcap，Veiny Skullcap ■☆

355620 Scutellaria nervosa Pursh var. ambigua（Nutt.）Fernald ＝ Scutellaria parvula Michx. var. missouriensis（Torr.）Goodman et C. A. Lawson ■☆

355621 Scutellaria nevskii Juz. et Vved.；涅氏黄芩■☆

355622 Scutellaria nigricans C. Y. Wu;变黑黄芩;Becoming-black Skullcap,Nigrescent Skullcap ■

355623 Scutellaria nigrocaldia C. Y. Wu et H. W. Li;黑心黄芩; Blackheart Skullcap ■

355624 Scutellaria nipponica Franch. et Sav. = Scutellaria dependens Maxim. ■

355625 Scutellaria novorossica Juz.;新俄黄芩;Newrussia Skullcap ■☆

355626 Scutellaria obbreviata Hara = Scutellaria indica L. var. subacaulis (Y. Z. Sun ex C. H. Hu) C. Y. Wu et C. Chen ■

355627 Scutellaria oblonga Benth.;长圆黄芩;Oblong Skullcap ■☆

355628 Scutellaria obtusifolia Hemsl.;钝叶黄芩(蛇头花);Obtuseleaf Skullcap ■

355629 Scutellaria obtusifolia Hemsl. var. trinervata (Vaniot) C. Y. Wu et H. W. Li;三脉钝叶黄芩(大叶耳挖草);Trinervate Obtuseleaf Skullcap ■

355630 Scutellaria ocellata Juz.;单眼黄芩■☆

355631 Scutellaria oldhamii Miq. = Scutellaria dependens Maxim. ■

355632 Scutellaria oligodonta Juz.;少齿黄芩;Fewtooth Skullcap, Scattertooth Skullcap ●

355633 Scutellaria oligophlebia Merr. et Chun ex C. Y. Wu et S. Chow;少脉黄芩;Fewvein Skullcap ■

355634 Scutellaria omeiensis C. Y. Wu;峨眉黄芩(白藿香);Emei Skullcap,Omei Skullcap ■

355635 Scutellaria omeiensis C. Y. Wu var. serratifolia C. Y. Wu et S. Chow;锯叶峨眉黄芩(钝叶峨眉黄芩);Serrateleaf Emei Skullcap ■

355636 Scutellaria orbicularis Bunge;圆叶黄芩■☆

355637 Scutellaria oreophila Grossh.;山地黄芩;Montane Skullcap ■☆

355638 Scutellaria orientalis L.;东方黄芩;Oriental Skullcap ■☆

355639 Scutellaria orientalis L. subsp. demnatensis Batt.;代姆纳特黄芩■☆

355640 Scutellaria orientalis L. var. porphyrantha Litard. et Maire = Scutellaria orientalis L. ■☆

355641 Scutellaria orthocalyx Hand.-Mazz.;直尊黄芩(半支莲,半枝莲,滇紫花地丁,兰花地丁,蓝花地丁,屏风草,小黄芩,紫花地丁);Erectcalyx Skullcap ■

355642 Scutellaria orthotricha C. Y. Wu et H. W. Li;展毛黄芩; Standhair Skullcap,Uprighthair Skullcap ●

355643 Scutellaria ovata Hill;卵叶黄芩(心叶黄芩);Egg-leaved Skullcap,Forest Skullcap,Heart-leaved Skullcap ■☆

355644 Scutellaria ovata Hill subsp. calcarea Epling = Scutellaria ovata Hill ■☆

355645 Scutellaria ovata Hill subsp. mississippiensis (M. Martens) Epling = Scutellaria ovata Hill ■☆

355646 Scutellaria ovata Hill subsp. rugosa (A. W. Wood) Epling = Scutellaria ovata Hill ■☆

355647 Scutellaria ovata Hill subsp. versicolor (Nutt.) Epling = Scutellaria ovata Hill ■☆

355648 Scutellaria ovata Hill var. calcarea (Epling) Gleason = Scutellaria ovata Hill ■☆

355649 Scutellaria ovata Hill var. versicolor (Nutt.) Fernald = Scutellaria ovata Hill ■☆

355650 Scutellaria ovata Hill var. versicolor (Nutt.) Fernald subsp. rugosa (A. W. Wood) Epling = Scutellaria ovata Hill ■☆

355651 Scutellaria oxyphylla Juz.;尖叶黄芩■☆

355652 Scutellaria oxystegia Juz.;尖附属物黄芩;Sharpappendage Skullcap ■☆

355653 Scutellaria pachyrrhiza Pax et K. Hoffm. = Scutellaria hypericifolia H. Lév. ■

355654 Scutellaria pacifica Juz. = Scutellaria pekinensis Maxim. var. transitra (Makino) H. Hara ex H. W. Li ■

355655 Scutellaria pallida L.;苍白黄芩;Pallid Skullcap ■☆

355656 Scutellaria pamirica Juz.;帕米尔黄芩;Pamir Skullcap ■☆

355657 Scutellaria parpureocoerulea Pax et K. Hoffm. = Scutellaria amoena C. H. Wright

355658 Scutellaria parvifolia (Makino) Koidz. = Scutellaria indica L. var. parvifolia (Makino) Makino ■

355659 Scutellaria parvifolia Koidz. = Scutellaria indica L. var. parvifolia (Makino) Makino ■

355660 Scutellaria parvifolia Koidz. var. vulgaris H. Hara = Scutellaria indica L. var. parvifolia (Makino) Makino ■

355661 Scutellaria parvula Michx.;矮小黄芩;Small Skullcap ■☆

355662 Scutellaria parvula Michx. var. ambigua (Nutt.) Fernald = Scutellaria parvula Michx. var. missouriensis (Torr.) Goodman et C. A. Lawson ■☆

355663 Scutellaria parvula Michx. var. australis Fassett = Scutellaria australis (Fassett) Epling ■☆

355664 Scutellaria parvula Michx. var. leonardii (Epling) Fernald = Scutellaria parvula Michx. var. missouriensis (Torr.) Goodman et C. A. Lawson ■☆

355665 Scutellaria parvula Michx. var. leonardii (Epling) Fernald = Scutellaria leonardii Epling ■☆

355666 Scutellaria parvula Michx. var. missouriensis (Torr.) Goodman et C. A. Lawson = Scutellaria leonardii Epling ■☆

355667 Scutellaria parvula Michx. var. missouriensis (Torr.) Goodman et C. A. Lawson;密苏里矮小黄芩;Leonard's Skullcap,Smooth Small Skullcap ■☆

355668 Scutellaria pauciflora Baker = Scutellaria schweinfurthii Briq. subsp. pauciflora (Baker) A. J. Paton ■☆

355669 Scutellaria pekinensis Maxim.;京黄芩(丹参,黄底芩,筋骨草,乌苏里黄芩);Beijing Skullcap,Peking Skullcap ■

355670 Scutellaria pekinensis Maxim. var. alpina (Nakai) H. Hara = Scutellaria pekinensis Maxim. var. ussuriensis (Regel) Hand.-Mazz. ■

355671 Scutellaria pekinensis Maxim. var. grandiflora C. Y. Wu et H. W. Li;大花京黄芩;Bigflower Beijing Skullcap ■

355672 Scutellaria pekinensis Maxim. var. purpureicaulis (Migo) C. Y. Wu et H. W. Li;紫茎京黄芩;Purplestem Beijing Skullcap ■

355673 Scutellaria pekinensis Maxim. var. transitra (Makino) H. Hara = Scutellaria pekinensis Maxim. var. transitra (Makino) H. Hara ex H. W. Li ■

355674 Scutellaria pekinensis Maxim. var. transitra (Makino) H. Hara ex H. W. Li;短促京黄芩(太平洋黄芩);Pacific Skullcap ■

355675 Scutellaria pekinensis Maxim. var. ussuriensis (Regel) Hand.-Mazz.;黑龙江黄芩(黄底芩,京黄芩,乌苏里黄芩);Ussuri Skullcap ■

355676 Scutellaria peregrina L. = Scutellaria barbata D. Don ■

355677 Scutellaria phyllostachya Juz.;叶花黄芩■☆

355678 Scutellaria physocalyx Regel et Schmalh. ex Regel;囊萼黄芩; Bladderlikecalyx Skullcap ■☆

355679 Scutellaria picta Juz.;有色黄芩;Painted Skullcap ■☆

355680 Scutellaria pingbienensis C. Y. Wu et H. W. Li;屏边黄芩; Pingbian Skullcap ■

355681 Scutellaria planipes Nakai et Kitag. = Scutellaria pekinensis Maxim. ■

355682 Scutellaria platystegia Juz.;宽盖黄芩■☆

355683　Scutellaria playfairi Kudo;伏黄芩（布烈氏黄芩）;Procumbent Skullcap ■

355684　Scutellaria playfairi Kudo var. megalantha S. Suzuki et Hosoki = Scutellaria playfairi Kudo ■

355685　Scutellaria playfairi Kudo var. procumbens（Ohwi）C. Y. Wu et H. L. Li;少毛伏黄芩■

355686　Scutellaria playfairi Kudo var. procumbens（Ohwi）C. Y. Wu et H. L. Li = Scutellaria tashiroi Hayata ■

355687　Scutellaria pobeguinii Hua = Scutellaria schweinfurthii Briq. subsp. pauciflora（Baker）A. J. Paton ■☆

355688　Scutellaria poecilantha Nevski ex Juz.;杂色黄芩■☆

355689　Scutellaria polyadenia Briq.;多腺黄芩■☆

355690　Scutellaria polyodon Juz.;多齿黄芩■☆

355691　Scutellaria polyphylla Juz.;多叶黄芩;Manyleaf Skullcap ■☆

355692　Scutellaria polytricha Juz.;密毛黄芩■☆

355693　Scutellaria pontica K. Koch;黑海黄芩;Black Sea Skullcap ■☆

355694　Scutellaria popovii Vved.;波氏黄芩■☆

355695　Scutellaria procumbens Ohwi = Scutellaria playfairi Kudo var. procumbens（Ohwi）C. Y. Wu et H. L. Li ■

355696　Scutellaria procumbens Ohwi = Scutellaria tashiroi Hayata ■

355697　Scutellaria procumbens Ohwi var. tomentosa Ohwi = Scutellaria playfairi Kudo ■

355698　Scutellaria prostrata Jacq. ex Benth.;平卧黄芩;Prostrate Skullcap ■

355699　Scutellaria przewalskii Juz.;深裂叶黄芩（艾叶黄芩）;Przewalsk Skullcap ●

355700　Scutellaria pseudotenax C. Y. Wu;假韧黄芩;Falsecatch Skullcap,Falsegrip Skullcap ■

355701　Scutellaria pseudotenax C. Y. Wu f. brevipeha C. Y. Wu ex C. Chen = Scutellaria pseudotenax C. Y. Wu ■

355702　Scutellaria purpureocardia C. Y. Wu;紫心黄芩（连钱草）;Purpleheart Skullcap ■

355703　Scutellaria purpureocoerulea Pax et K. Hoffm. = Scutellaria amoena C. H. Wright ■

355704　Scutellaria pusilla Gürke = Scutellaria schweinfurthii Briq. subsp. pauciflora（Baker）A. J. Paton ■☆

355705　Scutellaria pycnoclada Juz.;密枝黄芩■☆

355706　Scutellaria quadrilobulata Y. Z. Sun ex C. H. Hu;四裂花黄芩（四季花,四香花,土薄荷）;Fourlobed Skullcap ■

355707　Scutellaria quadrilobulata Y. Z. Sun ex C. H. Hu var. pilosa C. Y. Wu et S. Chow;硬毛四裂花黄芩（硬毛黄芩）;Pilose Fourlobed Skullcap ■

355708　Scutellaria racemosa Pers.;南美黄芩;South American Skullcap ■☆

355709　Scutellaria raddeana Juz.;拉第黄芩;Raddeana Skullcap ■☆

355710　Scutellaria ramosissima Popov;多分枝黄芩;Ramified Skullcap ■☆

355711　Scutellaria regeliana Nakai;狭叶黄芩（薄叶黄芩,意孔尼可夫黄芩）;Ikonnikov Skullcap,Narrowleaf Skullcap ■

355712　Scutellaria regeliana Nakai var. ikonnikovii（Juz.）C. Y. Wu et H. W. Li;塔头狭叶黄芩（薄叶黄芩,香水水草）■

355713　Scutellaria rehderiana Diels et Hand. -Mazz.;甘肃黄芩;Gansu Skullcap,Rehder Skullcap ■

355714　Scutellaria repens Buch. -Ham. ex D. Don;匍匐黄芩;Repent Skullcap ■☆

355715　Scutellaria reticulata C. Y. Wu et W. T. Wang;显脉黄芩;Reticulate Skullcap,Showvein Skullcap ●

355716　Scutellaria rhomboidalis Grossh.;菱形黄芩;Rhomboid Skullcap ■☆

355717　Scutellaria ringoetii De Wild. = Scutellaria schweinfurthii Briq. subsp. pauciflora（Baker）A. J. Paton ■☆

355718　Scutellaria rivularis Wall. = Scutellaria barbata D. Don ■

355719　Scutellaria rivularis Wall. ex Benth. = Scutellaria barbata D. Don ■

355720　Scutellaria rubicunda Hornem.;稍红黄芩■☆

355721　Scutellaria rubromaculata Juz. et Vved.;中亚红斑黄芩（红斑黄芩）■☆

355722　Scutellaria rubropunctata Hayata;红斑黄芩■☆

355723　Scutellaria rubropunctata Hayata var. minima T. Yamaz. = Scutellaria rubropunctata Hayata var. yakusimensis（Masam.）Yahara ■☆

355724　Scutellaria rubropunctata Hayata var. yakusimensis（Masam.）Yahara;屋久岛黄芩■☆

355725　Scutellaria salvia H. Lév. = Scutellaria discolor Wall. ex Benth. ■

355726　Scutellaria salviifolia Benth.;鼠尾草叶黄芩■☆

355727　Scutellaria scandens D. Don;棱茎黄芩（有棱黄芩）;Angular Skullcap,Scandent Skullcap ■

355728　Scutellaria schmidtii Kudo = Scutellaria strigillosa Hemsl. ■

355729　Scutellaria schugnanica B. Fedtsch.;舒格南黄芩■☆

355730　Scutellaria schweinfurthii Briq.;施韦黄芩■☆

355731　Scutellaria schweinfurthii Briq. subsp. pauciflora（Baker）A. J. Paton;寡花施韦黄芩■☆

355732　Scutellaria sciaphila S. Moore;喜荫黄芩;Shadeloving Skullcap ■☆

355733　Scutellaria scordifolia Fisch. ex Schrank;并头黄芩（并头草,山麻子,头巾草,头中草）;Twinflower Skullcap ■

355734　Scutellaria scordifolia Fisch. ex Schrank f. glabrescens Franch. = Scutellaria scordifolia Fisch. ex Schrank ■

355735　Scutellaria scordifolia Fisch. ex Schrank f. pubescens Diels ex Rehder et Kobuski = Scutellaria sessilifolia Hemsl. var. villosissima C. Y. Wu et W. T. Wang ■

355736　Scutellaria scordifolia Fisch. ex Schrank var. ammophila（Kitag.）C. Y. Wu et W. T. Wang = Scutellaria sessilifolia Hemsl. var. ammophila（Kitag.）C. Y. Wu et W. T. Wang ■

355737　Scutellaria scordifolia Fisch. ex Schrank var. ammophila（Kitag.）C. Y. Wu et W. T. Wang;喜沙黄芩（喜沙并头黄芩）■

355738　Scutellaria scordifolia Fisch. ex Schrank var. hirta F. Schmidt = Scutellaria strigillosa Hemsl. ■

355739　Scutellaria scordifolia Fisch. ex Schrank var. puberula Regel ex Kom.;微柔毛并头黄芩■

355740　Scutellaria scordifolia Fisch. ex Schrank var. subglabra Kom. = Scutellaria scordifolia Fisch. ex Schrank ■

355741　Scutellaria scordifolia Fisch. ex Schrank var. subglabra Kom. f. ammophila Kitag. = Scutellaria scordifolia Fisch. ex Schrank var. ammophila（Kitag.）C. Y. Wu et W. T. Wang ■

355742　Scutellaria scordifolia Fisch. ex Schrank var. villosissima C. Y. Wu et W. T. Wang = Scutellaria sessilifolia Hemsl. var. villosissima C. Y. Wu et W. T. Wang ■

355743　Scutellaria scordifolia Fisch. ex Schrank var. wulingshanensis（Nakai et Kitag.）C. Y. Wu et W. T. Wang = Scutellaria sessilifolia Hemsl. var. wulingshanensis（Nakai et Kitag.）C. Y. Wu et W. T. Wang ■

355744　Scutellaria scordifolia Hemsl. var. subglabra Kom. f. ammophila Kitag. = Scutellaria scordifolia Fisch. ex Schrank var. ammophila（Kitag.）C. Y. Wu et W. T. Wang ■

355745　Scutellaria scordiifolia Fisch. ex Schrank subsp. strigillosa（Hemsl.）H. Hara = Scutellaria strigillosa Hemsl. ■

355746　Scutellaria scordiifolia Fisch. ex Schrank var. pubescens Miq. = Scutellaria strigillosa Hemsl. ■

355747　Scutellaria serrata Spreng. ;具齿黄芩;Showy Skullcap ■☆

355748　Scutellaria sessilifolia Hemsl. ;石蜈蚣草(吊鱼杆,钓鱼杆,胡豆草,毛柄黄芩,无柄黄芩); Sessile Skullcap, Stonecentiped Skullcap ■

355749　Scutellaria sessilifolia Hemsl. f. ramiflora C. Y. Wu et S. Chow;枝花石蜈蚣■

355750　Scutellaria sessilifolia Hemsl. f. ramiflora C. Y. Wu et S. Chow = Scutellaria sessilifolia Hemsl. ■

355751　Scutellaria sessilifolia Hemsl. f. terminalis C. Y. Wu et S. Chow;顶序石蜈蚣■

355752　Scutellaria sessilifolia Hemsl. f. terminalis C. Y. Wu et S. Chow = Scutellaria sessilifolia Hemsl. ■

355753　Scutellaria sessilifolia Hemsl. var. ammophila (Kitag.) C. Y. Wu et W. T. Wang = Scutellaria scordifolia Fisch. ex Schrank var. ammophila (Kitag.) C. Y. Wu et W. T. Wang ■

355754　Scutellaria sessilifolia Hemsl. var. delavayi (H. Lév.) Doan = Scutellaria delavayi H. Lév. ■

355755　Scutellaria sessilifolia Hemsl. var. puberula Regel ex Kom. = Scutellaria scordifolia Fisch. ex Schrank var. puberula Regel ex Kom. ■

355756　Scutellaria sessilifolia Hemsl. var. villosissima C. Y. Wu et W. T. Wang;多毛并头黄芩■

355757　Scutellaria sessilifolia Hemsl. var. wulingshanensis (Nakai et Kitag.) C. Y. Wu et W. T. Wang;雾灵山并头黄芩;Wulingshan Sessile Skullcap ■

355758　Scutellaria sevanensis Sosn. ex Grossh. ;劲直黄芩;Rigid Skullcap ■☆

355759　Scutellaria shansiensis C. Y. Wu et H. W. Li;山西黄芩;Shanxi Skullcap ■

355760　Scutellaria shikokiana Makino;四国黄芩(日本黄芩);Japanese Skullcap ■☆

355761　Scutellaria shikokiana Makino f. pubicaulis Ohwi = Scutellaria shikokiana Makino var. pubicaulis (Ohwi) Kitam. ■☆

355762　Scutellaria shikokiana Makino var. pubicaulis (Ohwi) Kitam. ;毛茎日本黄芩■☆

355763　Scutellaria shweliensis W. W. Sm. ;瑞丽黄芩(挖耳草);Ruili Skullcap ●

355764　Scutellaria sichourensis C. Y. Wu et H. W. Li;西畴黄芩;Xichou Skullcap ■

355765　Scutellaria sieversii Bunge;宽苞黄芩(大花准喀尔黄芩,新疆黄芩,准喀尔黄芩); Bigflower Dzungar Skullcap, Broadbract Skullcap, Dzungar Skullcap, Xinjiang Skullcap ●

355766　Scutellaria simplex Migo = Scutellaria laeteviolacea Koidz. ■

355767　Scutellaria somalensis Thulin;索马里黄芩■☆

355768　Scutellaria soongoria Juz. = Scutellaria sieversii Bunge ●

355769　Scutellaria soongoria Juz. var. grandiflora C. Y. Wu et H. W. Li = Scutellaria sieversii Bunge ●

355770　Scutellaria soongorica Juz. = Scutellaria sieversii Bunge ●

355771　Scutellaria soongorica Juz. var. grandiflora C. Y. Wu et H. W. Li = Scutellaria sieversii Bunge ●

355772　Scutellaria sosnowskyi Takht. ;锁斯诺夫斯基黄芩■☆

355773　Scutellaria spectabilis Pax et K. Hoffm. ;西方白花黄芩;Whiteflower Skullcap ■☆

355774　Scutellaria squarosa Nevski;糙黄芩■

355775　Scutellaria stenosiphon Hemsl. ;狭管黄芩;Narrowtube Skullcap ■

355776　Scutellaria stevenii Juz. ;斯太温黄芩;Steven Skullcap ■☆

355777　Scutellaria striatella Gontsch. ;条纹黄芩;Striate Skullcap ■☆

355778　Scutellaria strigillosa Hemsl. ;沙滩黄芩(瓜子兰,海滨黄芩,塔盖黄芩); Sandybeach Skullcap, Strigose Skullcap, Taquet Skullcap ■

355779　Scutellaria strigillosa Hemsl. f. albiflora Kawano;白花沙滩黄芩■☆

355780　Scutellaria strigillosa Hemsl. f. hirta (F. Schmidt) H. Hara;毛沙滩黄芩■☆

355781　Scutellaria strigillosa Hemsl. var. yezoensis (Kudo) Kitam. = Scutellaria yezoensis Kudo ■☆

355782　Scutellaria subcaespitosa Pavlov;近丛簇黄芩(珊广黄芩);Subcaespitose Skullcap ■☆

355783　Scutellaria subcordata Juz. ;近心形黄芩■☆

355784　Scutellaria subintegra C. Y. Wu et H. W. Li;两广黄芩;Subentire Skullcap, Yue-Gui Skullcap ■

355785　Scutellaria suffrutescens S. Watson;得州黄芩;Red Skullcap, Texas Rose ■☆

355786　Scutellaria supina L. ;仰卧黄芩;Backlie Skullcap, Supine Skullcap ●

355787　Scutellaria taipeiensis T. C. Huang;台北黄芩■

355788　Scutellaria taiwanense S. S. Ying = Scutellaria indica L. ■

355789　Scutellaria taiwanensis C. Y. Wu;台湾黄芩;Taiwan Skullcap ■☆

355790　Scutellaria talassica Juz. ;塔拉斯黄芩■☆

355791　Scutellaria tapintzeensis C. Y. Wu et H. W. Li;大坪子黄芩;Dapingzi Skullcap, Tapingtze Skullcap ■

355792　Scutellaria taquetii H. Lév. et Vaniot = Scutellaria strigillosa Hemsl. ■

355793　Scutellaria taroensis T. Yamaz. = Scutellaria tashiroi Hayata ■

355794　Scutellaria tashiroi Hayata;田代氏黄芩■

355795　Scutellaria tashiroi Hayata = Scutellaria indica L. ■

355796　Scutellaria tashiroi Hayata var. haianshanensis T. Yamaz. = Scutellaria tashiroi Hayata ■

355797　Scutellaria tashiroi Hayata var. playfairi (Kudo) T. Yamaz. = Scutellaria playfairi Kudo ■

355798　Scutellaria tatianae Juz. ;塔天黄芩;Tatian Skullcap ■☆

355799　Scutellaria taurica Juz. ;西伯利亚黄芩;Siberian Skullcap ■☆

355800　Scutellaria tayloriana Dunn;偏花黄芩(土黄芩);Secundflower Skullcap, Taylor Skullcap ■

355801　Scutellaria tayloriana Dunn = Scutellaria discolor Wall. ex Benth. var. hirta Hand. -Mazz. ■

355802　Scutellaria tayloriana Dunn = Scutellaria teniana Hand. -Mazz. ■

355803　Scutellaria tayloriana Dunn var. polytricha Hand. -Mazz. = Scutellaria tayloriana Dunn ■

355804　Scutellaria tenax W. W. Sm. ;韧黄芩(大黄芩);Catch Skullcap, Rigid Skullcap ■

355805　Scutellaria tenax W. W. Sm. var. patentipilosa (Hand. -Mazz.) C. Y. Wu;展毛韧黄芩(大黄芩)■

355806　Scutellaria tenera C. Y. Wu et H. W. Li;柔弱黄芩;Slender Skullcap, Soft Skullcap ■

355807　Scutellaria teniana Hand. -Mazz. ;大姚黄芩;Dayao Skullcap ■

355808　Scutellaria tenuiflora C. Y. Wu;细花黄芩;Smallflower Skullcap, Tenuousflower Skullcap ■

355809　Scutellaria tibetica C. Y. Wu et H. W. Li = Scutellaria kingiana Prain ■

355810　Scutellaria tienchuanensis C. Y. Wu et C. Chen;天全黄芩;Tianquan Skullcap, Tienchuan Skullcap ■

355811　Scutellaria titovii Juz. ;梯托黄芩;Titov Skullcap ■☆

355812　Scutellaria toguztoravensis Juz. ;托古黄芩■☆

355813　Scutellaria tournefortii Benth. ;图氏黄芩(土奈佛特黄芩);Tournefort Skullcap ■☆

355814　Scutellaria transiliensis Juz. = Scutellaria sieversii Bunge ●

355815　Scutellaria transitra Makino ＝ Scutellaria pekinensis Maxim. var. transitra（Makino）H. Hara ex H. W. Li ■

355816　Scutellaria transitra Makino var. ussuriensis（Regel）H. Hara ＝ Scutellaria pekinensis Maxim. var. ussuriensis（Regel）Hand. -Mazz. ■

355817　Scutellaria trinervata Vaniot ＝ Scutellaria obtusifolia Hemsl. var. trinervata（Vaniot）C. Y. Wu et H. W. Li ■

355818　Scutellaria tschimganica Juz. ;琴干黄芩;Tschimgan Skullcap ●

355819　Scutellaria tschimganica Juz. ＝ Scutellaria supina L. ●

355820　Scutellaria tsinyunensis C. Y. Wu et S. Chow;缙云黄芩;Jinyun Skullcap,Tsinyun Skullcap ■

355821　Scutellaria tsusimensis H. Hara;津岛黄芩■☆

355822　Scutellaria tuberifera C. Y. Wu et C. Chen;假活血草（假活血丹）;Tuberous Skullcap ■

355823　Scutellaria tuberosa Benth. ＝ Scutellaria amoena C. H. Wright ■

355824　Scutellaria tuminensis Nakai;图门黄芩;Tumen Skullcap,Tumin Skullcap ■

355825　Scutellaria turgaica Juz. ;图尔盖黄芩;Turgai Skullcap ■☆

355826　Scutellaria tuvensis Juz. ;图文黄芩;Tuven Skullcap ■☆

355827　Scutellaria urticifolia C. Y. Wu et H. W. Li ＝ Scutellaria yangiensis H. W. Li ■

355828　Scutellaria ussuriensis（Regel）Hand. -Mazz. var. typica Nakai ＝ Scutellaria pekinensis Maxim. var. ussuriensis（Regel）Hand. -Mazz. ■

355829　Scutellaria ussuriensis（Regel）Kudo ＝ Scutellaria pekinensis Maxim. var. ussuriensis（Regel）Hand. -Mazz. ■

355830　Scutellaria ussuriensis（Regel）Kudo var. tomentosa ? ＝ Scutellaria pekinensis Maxim. var. transitra（Makino）H. Hara ex H. W. Li ■

355831　Scutellaria ussuriensis（Regel）Kudo var. transitra（Makino）Nakai ＝ Scutellaria pekinensis Maxim. var. transitra（Makino）H. Hara ex H. W. Li ■

355832　Scutellaria ussuriensis（Regel）Kudo var. typica Nakai ＝ Scutellaria pekinensis Maxim. var. ussuriensis（Regel）Hand. -Mazz. ■

355833　Scutellaria ussuriensis（Regel）Kudo var. typica Nakai f. humilis Kudo ＝ Scutellaria laeteviolacea Koidz. ■

355834　Scutellaria ussuriensis Kudo var. typica f. humilis Kudo ＝ Scutellaria laeteviolacea Koidz. ■

355835　Scutellaria vaniotiana H. Lév. ex Dunn ＝ Scutellaria obtusifolia Hemsl. var. trinervata（Vaniot）C. Y. Wu et H. W. Li ■

355836　Scutellaria verna Besser;春天黄芩;Spring Skullcap ■☆

355837　Scutellaria veronicifolia H. Lév. ＝ Scutellaria tenax W. W. Sm. ■

355838　Scutellaria veronicifolia H. Lév. var. patentipilosa Hand. -Mazz. ＝ Scutellaria tenax W. W. Sm. var. patentipilosa（Hand. -Mazz.）C. Y. Wu ■

355839　Scutellaria versicolor Nutt. ＝ Scutellaria ovata Hill ■☆

355840　Scutellaria villosissima Gontsch. ex Juz. ;长柔毛黄芩;Villose Skullcap ■☆

355841　Scutellaria violacea K. Heyne ex Benth. ＝ Scutellaria indica L. ■

355842　Scutellaria violacea K. Heyne ex Wall. ;堇色黄芩;Violet Skullcap ■

355843　Scutellaria violacea K. Heyne ex Wall. var. sikkimensis Hook. f. ;紫苏叶黄芩（反叶红）;Coleusleaf Skullcap ■

355844　Scutellaria violacea K. Heyne ex Wall. var. sikkimensis Hook. f. ＝ Scutellaria violacea K. Heyne ex Wall. ■

355845　Scutellaria violascens Gürke;浅堇色黄芩■☆

355846　Scutellaria viscidula Bunge;黏毛黄芩（黄花黄芩,黄芩,下巴子,腺毛黄芩）;Viscidhair Skullcap ■

355847　Scutellaria weishanensis C. Y. Wu et H. W. Li;魏山黄芩;Weishan Skullcap ■

355848　Scutellaria wenshanensis C. Y. Wu et H. W. Li;文山黄芩;Wenshan Skullcap ■

355849　Scutellaria wongkei Dunn;南粤黄芩;S. Guangdong Skullcap,South Kwangtung Skullcap ■

355850　Scutellaria woronowii Juz. ;沃罗诺黄芩;Woronow Skullcap ■☆

355851　Scutellaria wrightii A. Gray;赖氏黄芩;Blue Skullcap ■☆

355852　Scutellaria wulingshanensis Nakai et Kitag. ＝ Scutellaria scordifolia Fisch. ex Schrank var. wulingshanensis（Nakai et Kitag.）C. Y. Wu et W. T. Wang ■

355853　Scutellaria wulingshanensis Nakai et Kitag. ＝ Scutellaria sessilifolia Hemsl. var. wulingshanensis（Nakai et Kitag.）C. Y. Wu et W. T. Wang ■

355854　Scutellaria xanthosiphon Juz. ;黄管黄芩■☆

355855　Scutellaria yangiensis H. W. Li;荨麻叶黄芩;Nettleleaf Skullcap ■

355856　Scutellaria yezoensis Kudo;北海道黄芩;Yezo Skullcap ■☆

355857　Scutellaria yingtakensis Y. Z. Sun;英德黄芩;Yingde Skullcap,Yingtak Skullcap ■

355858　Scutellaria yunnanensis H. Lév. ;红茎黄芩（多子草）;Redstem Skullcap,Yunnan Skullcap ■

355859　Scutellaria yunnanensis H. Lév. var. cuneata C. Y. Wu et W. T. Wang;楔叶红茎黄芩;Cuneate Redstem Skullcap ■

355860　Scutellaria yunnanensis H. Lév. var. salicifolia Y. Z. Sun ex C. H. Hu;柳叶红茎黄芩（血沟丹）;Willowleaf Redstem Skullcap ■

355861　Scutellaria yunnanensis H. Lév. var. subsessilifolia Y. Z. Sun ex C. H. Hu ＝ Scutellaria tsinyunensis C. Y. Wu et S. Chow ■

355862　Scutellaria yunnanensis H. Lév. var. subsessilifolia Y. Z. Sun ex C. H. Hu;短柄红茎黄芩;Shortstalk Redstem Skullcap ■

355863　Scutellariaceae Caruel ＝ Labiatae Juss. （保留科名）●■

355864　Scutellariaceae Caruel ＝ Lamiaceae Martinov（保留科名）●■

355865　Scutellariaceae Caruel;黄芩科 ■

355866　Scutellariaceae Döll ＝ Labiatae Juss. （保留科名）●■

355867　Scutellariaceae Döll ＝ Lamiaceae Martinov（保留科名）●■

355868　Scutia（Comm. ex DC.）Brongn. （1826）（保留属名）;对刺藤属（双刺藤属）;Scutia ●

355869　Scutia（DC.）Brongn. ＝ Scutia（Comm. ex DC.）Brongn. （保留属名）●

355870　Scutia（DC.）Comm. ex Brongn. ＝ Scutia（Comm. ex DC.）Brongn. （保留属名）●

355871　Scutia Comm. ex Brongn. ＝ Scutia（Comm. ex DC.）Brongn. （保留属名）●

355872　Scutia circumscissa（L. f.）Radlk. ;印度对刺藤;India Scutia,Indian Scutia ●☆

355873　Scutia circumscissa（L. f.）W. Theob. ＝ Scutia myrtina（Burm. f.）Kurz ●

355874　Scutia commersonii Brongn. ＝ Scutia myrtina（Burm. f.）Kurz ●

355875　Scutia discolor Klotzsch ＝ Berchemia discolor（Klotzsch）Hemsl. ●☆

355876　Scutia eberhardtii Tardieu ＝ Scutia myrtina（Burm. f.）Kurz ●

355877　Scutia indica Brongn. ＝ Scutia myrtina（Burm. f.）Kurz ●

355878　Scutia indica Brongn. var. oblongifolia Engl. ＝ Scutia myrtina（Burm. f.）Kurz ●

355879　Scutia myrtina（Burm. f.）Kurz;对刺藤（钩刺藤,双刺藤）;Eberhard Scutia,Eberhard Twinspinevine ●

355880　Scutia myrtina（Burm. f.）Kurz var. oblongifolia（Engl.）Evrard ＝ Scutia myrtina（Burm. f.）Kurz ●

355881　Scutia obcordata Boivin ex Tul. = Scutia myrtina（Burm. f.）Kurz ●

355882　Scuticaria Lindl.（1843）;鞭叶兰属■☆

355883　Scuticaria hadwenii（Lindl.）Hoehne;哈氏鞭叶兰■☆

355884　Scuticaria steelii Lindl.;斯氏鞭叶兰■☆

355885　Scutinanthe Thwaites（1856）;盾花榄属●☆

355886　Scutinanthe brevisepala Leenh.;短萼盾花榄●☆

355887　Scutinanthe brunnea Thwaites;盾花榄●☆

355888　Scutis Endl. = Scutia（Comm. ex DC.）Brongn.（保留属名）●

355889　Scutula Lour. = Memecylon L. ●

355890　Scutula scutellata Lour. = Memecylon scutellatum（Lour.）Hook. et Arn. ●

355891　Scybaliaceae A. Kern.（1891）;膜叶菰科■☆

355892　Scybaliaceae A. Kern. = Balanophoraceae Rich.（保留科名）●■

355893　Scybaliaceae A. Kern. = Scyphostegiaceae Hutch.（保留科名）●☆

355894　Scybalium Schott = Scybalium Schott et Endl. ■☆

355895　Scybalium Schott et Endl.（1832）;膜叶菰属■☆

355896　Scybalium fungiforme Schott et Endl.;膜叶菰■☆

355897　Scynopsole Rchb. = Balanophora J. R. Forst. et G. Forst. ■

355898　Scynopsole Rchb. = Cynopsole Endl. ■

355899　Scyphaea C. Presl = Marila Sw. ☆

355900　Scyphanthus D. Don（1828）;杯莲花属■☆

355901　Scyphanthus Sweet = Scyphanthus D. Don ■☆

355902　Scyphanthus stenocarpus Urb. et Gilg;杯莲花■☆

355903　Scypharia Miers = Scutia（Comm. ex DC.）Brongn.（保留属名）●

355904　Scyphellandra Thwaites = Rinorea Aubl.（保留属名）●

355905　Scyphellandra Thwaites（1858）;鳞隔堇属（茜菲堇属）;Scyphellandra ●

355906　Scyphellandra pierrei H. Boissieu = Scyphellandra virgata Thwaites ●

355907　Scyphellandra virgata Thwaites = Rinorea virgata（Thwaites）Kuntze ●

355908　Scyphiphora C. F. Gaertn.（1806）;瓶花木属;Scyphiphora ●

355909　Scyphiphora hydrophyllacea C. F. Gaertn.;瓶花木（厚皮）;Common Scyphiphora ●

355910　Scyphocephalium Warb.（1897）;杯首木属（杯头木属）●☆

355911　Scyphocephalium Warb. = Ochocoa Pierre ●☆

355912　Scyphocephalium chrysothrix Warb.;杯首木●☆

355913　Scyphocephalium mannii（Benth.）Warb.;曼氏杯首木●☆

355914　Scyphocephalium ochocoa Warb. = Scyphocephalium mannii（Benth.）Warb. ●☆

355915　Scyphochlamys Balf. f.（1879）;杯鞘茜属☆

355916　Scyphochlamys revoluta Balf. f.;杯鞘茜☆

355917　Scyphocoronis A. Gray（1851）;绿苞鼠麴草属■☆

355918　Scyphocoronis viscosa A. Gray;绿苞鼠麴草■☆

355919　Scyphoglottis Pritz. = Scaphyglottis Poepp. et Endl.（保留属名）■☆

355920　Scyphogyne Brongn. = Erica L. ●☆

355921　Scyphogyne Brongn. = Scyphogyne Decne. ●☆

355922　Scyphogyne Brongn. et E. Phillips = Scyphogyne Brongn. ●☆

355923　Scyphogyne Decne.（1829-1834）;杯蕊杜鹃属●☆

355924　Scyphogyne biconvexa N. E. Br. = Erica rigidula（N. E. Br.）E. G. H. Oliv. ●☆

355925　Scyphogyne brevifolia Benth. = Erica phacelanthera E. G. H. Oliv. ●☆

355926　Scyphogyne brownii Compton = Erica rigidula（N. E. Br.）E. G. H. Oliv. ●☆

355927　Scyphogyne burchellii N. E. Br. = Erica urceolata（Klotzsch）E. G. H. Oliv. ●☆

355928　Scyphogyne calcicola E. G. H. Oliv. = Erica calcicola（E. G. H. Oliv.）E. G. H. Oliv. ●☆

355929　Scyphogyne capitata（Klotzsch）Benth.;头状杯蕊杜鹃●☆

355930　Scyphogyne capitata（Klotzsch）Benth. = Erica phacelanthera E. G. H. Oliv. ●☆

355931　Scyphogyne capitata（Klotzsch）Benth. var. brevifolia（Benth.）N. E. Br. = Erica phacelanthera E. G. H. Oliv. ●☆

355932　Scyphogyne divaricata（Klotzsch）Benth. = Erica rigidula（N. E. Br.）E. G. H. Oliv. ●☆

355933　Scyphogyne eglandulosa（Klotzsch）Benth. = Erica eglandulosa（Klotzsch）E. G. H. Oliv. ●☆

355934　Scyphogyne fasciculata Benth. = Erica eglandulosa（Klotzsch）E. G. H. Oliv. ●☆

355935　Scyphogyne glandulifera N. E. Br. = Erica rigidula（N. E. Br.）E. G. H. Oliv. ●☆

355936　Scyphogyne inconspicua Decne.;显著杯蕊杜鹃●☆

355937　Scyphogyne inconspicua Decne. = Erica muscosa（Aiton）E. G. H. Oliv. ●☆

355938　Scyphogyne inconspicua Decne. var. ciliata N. E. Br. = Erica muscosa（Aiton）E. G. H. Oliv. ●☆

355939　Scyphogyne inconspicua Decne. var. glabriflora N. E. Br. = Erica muscosa（Aiton）E. G. H. Oliv. ●☆

355940　Scyphogyne inconspicua Decne. var. pubescens N. E. Br. = Erica muscosa（Aiton）E. G. H. Oliv. ●☆

355941　Scyphogyne inconspicua Decne. var. vestita N. E. Br. = Erica muscosa（Aiton）E. G. H. Oliv. ●☆

355942　Scyphogyne longistyla N. E. Br. = Erica rigidula（N. E. Br.）E. G. H. Oliv. ●☆

355943　Scyphogyne micrantha（Benth.）N. E. Br. = Erica artemisioides（Klotzsch）E. G. H. Oliv. ●☆

355944　Scyphogyne muscosa（Aiton）Druce = Erica muscosa（Aiton）E. G. H. Oliv. ●☆

355945　Scyphogyne orientalis E. G. H. Oliv. = Erica melanomontana E. G. H. Oliv. ●☆

355946　Scyphogyne puberula（Klotzsch）Benth. = Erica urceolata（Klotzsch）E. G. H. Oliv. ●☆

355947　Scyphogyne remota N. E. Br. = Erica remota（N. E. Br.）E. G. H. Oliv. ●☆

355948　Scyphogyne rigidula N. E. Br. = Erica rigidula（N. E. Br.）E. G. H. Oliv. ●☆

355949　Scyphogyne rigidula N. E. Br. var. breviciliata ? = Erica rigidula（N. E. Br.）E. G. H. Oliv. ●☆

355950　Scyphogyne schlechteri N. E. Br. = Erica muscosa（Aiton）E. G. H. Oliv. ●☆

355951　Scyphogyne tenuis（Benth.）E. G. H. Oliv. = Erica miniscula E. G. H. Oliv. ●☆

355952　Scyphogyne trimera N. E. Br. = Erica urceolata（Klotzsch）E. G. H. Oliv. ●☆

355953　Scyphogyne urceolata（Klotzsch）Benth. = Erica urceolata（Klotzsch）E. G. H. Oliv. ●☆

355954　Scyphogyne viscida N. E. Br. = Erica phacelanthera E. G. H. Oliv. ●☆

355955　Scyphonychium Radlk.（1879）;杯距无患子属●☆

355956　Scyphonychium multiflorum（C. Mart.）Radlk.;杯距无患子●☆

355957　Scyphonychium multiflorum Radlk. = Scyphonychium multiflorum (C. Mart.) Radlk. ●☆

355958　Scyphopappus B. Nord. = Argyranthemum Webb ex Sch. Bip. ●

355959　Scyphopetalum Hiern = Paranephelium Miq. ●

355960　Scyphophora Post et Kuntze = Scyphiphora C. F. Gaertn. ●

355961　Scyphostachys Thwaites(1859);杯穗茜属●☆

355962　Scyphostachys coffeoides Thwaites;杯穗茜●☆

355963　Scyphostachys pedunculatus Thwaites;下垂杯穗茜●☆

355964　Scyphostegia Stapf(1894);杯盖属●☆

355965　Scyphostegia borneensis Stapf;杯盖花●☆

355966　Scyphostegiaceae Hutch. (1926)(保留科名);杯盖花科(婆罗州大风子科,肉盘树科)●☆

355967　Scyphostelma Baill. (1890);杯冠萝藦属☆

355968　Scyphostelma granatense Baill. ;杯冠萝藦☆

355969　Scyphostigma M. Roem. (1846);杯柱楝属●☆

355970　Scyphostigma M. Roem. = Turraea L. ●

355971　Scyphostigma bennetii M. Roem. ;杯柱楝●☆

355972　Scyphostigma philippinense M. Roem. ;菲律宾杯柱楝●☆

355973　Scyphostrychnos S. Moore = Strychnos L. ●

355974　Scyphostrychnos psittaconyx P. A. Duvign. = Strychnos camptoneura Gilg et Busse ●☆

355975　Scyphostrychnos talbotii S. Moore = Strychnos camptoneura Gilg et Busse ●☆

355976　Scyphosyce Baill. (1875);杯桑属●☆

355977　Scyphosyce manniana Baill. ;曼氏杯桑●☆

355978　Scyphosyce pandurata Hutch. ;杯桑●☆

355979　Scyrtocarpa Miers = Scyrtocarpus Miers ●

355980　Scyrtocarpus Miers = Barberina Vell. ●

355981　Scyrtocarpus Miers = Symplocos Jacq. ●

355982　Scytala E. Mey. ex DC. = Oldenburgia Less. ●☆

355983　Scytalanthus Schauer = Skytanthus Meyen ●☆

355984　Scytalia Gaertn. = Litchi Sonn. ●

355985　Scytalia Gaertn. = Nephelium L. ●

355986　Scytalia chinensis (Sonna.) Gaertn. = Litchi chinensis Sonn. ●

355987　Scytalia rubra Roxb. = Aphania rubra (Roxb.) Radlk. ●

355988　Scytalia rubra Roxb. = Lepisanthes senegalensis (Juss. ex Poir.) Leenh. ●

355989　Scytalis E. Mey. = Vigna Savi(保留属名)■

355990　Scytalis helicopus E. Mey. = Vigna luteola (Jacq.) Benth. ■

355991　Scytalis hispida E. Mey. = Vigna unguiculata (L.) Walp. subsp. protracta (E. Mey.) B. J. Pienaar ■☆

355992　Scytalis protracta E. Mey. = Vigna unguiculata (L.) Walp. subsp. protracta (E. Mey.) B. J. Pienaar ■☆

355993　Scytalis retusa E. Mey. = Vigna marina (Burm.) Merr. ■

355994　Scytalis tenuis E. Mey. = Vigna unguiculata (L.) Walp. var. tenuis (E. Mey.) Maréchal et Mascherpa et Stainier ■☆

355995　Scytalis tenuis E. Mey. var. oblonga E. Mey. = Vigna unguiculata (L.) Walp. var. tenuis (E. Mey.) Maréchal et Mascherpa et Stainier ■☆

355996　Scytalis tenuis E. Mey. var. ovata E. Mey. = Vigna unguiculata (L.) Walp. var. ovata (E. Mey.) B. J. Pienaar ■☆

355997　Scytanthus Hook. = Hoodia Sweet ex Decne. ●☆

355998　Scytanthus Liebm. = Bdallophyton Eichler ■☆

355999　Scytanthus Liebm. = Bdallophytum Eichler ■☆

356000　Scytanthus Post et Kuntze = Skytanthus Meyen ●☆

356001　Scytanthus T. Anderson ex Benth. = Thomandersia Baill. ●☆

356002　Scytanthus T. Anderson ex Benth. et Hook. f. = Thomandersia Baill. ●☆

356003　Scytanthus currorii Hook. = Hoodia currorii (Hook.) Decne. ■☆

356004　Scytanthus laurifolius T. Anderson ex Benth. = Thomandersia laurifolia (T. Anderson ex Benth.) Baill. ●☆

356005　Scytopetalaceae Engl. (1897)(保留科名);革瓣花科(木果树科)●☆

356006　Scytopetalaceae Engl. (保留科名) = Lecythidaceae A. Rich. (保留科名)●

356007　Scytopetalaceae Engl. (保留科名) = Selaginaceae Choisy(保留科名)●■

356008　Scytopetalum Pierre ex Engl. (1897);革瓣花属(木果树属)●☆

356009　Scytopetalum Pierre ex Tiegh. = Scytopetalum Pierre ex Engl. ●☆

356010　Scytopetalum brevipes Pierre ex Tiegh. ;革瓣花●☆

356011　Scytopetalum duchesnei Engl. = Oubanguia africana Baill. ●☆

356012　Scytopetalum kamerunianum Engl. = Scytopetalum klaineanum Pierre ex Engl. ●☆

356013　Scytopetalum klaineanum Pierre ex Engl. ;克雷革瓣花●☆

356014　Scytopetalum klaineanum Pierre ex Engl. var. kamerunianum (Engl.) Letouzey = Scytopetalum klaineanum Pierre ex Engl. ●☆

356015　Scytopetalum pierreanum (De Wild.) Tiegh. ;皮氏革瓣花●☆

356016　Scytopetalum tieghemii A. Chev. ex Hutch. et Dalziel;梯氏革瓣花●☆

356017　Scytophyllum Eckl. et Zeyh. (废弃属名) = Elaeodendron J. Jacq. ●☆

356018　Scytophyllum Eckl. et Zeyh. (废弃属名) = Gymnosporia (Wight et Arn.) Benth. et Hook. f. (保留属名)●

356019　Scytophyllum angustifolium C. Presl ex Sond. = Robsonodendron eucleiforme (Eckl. et Zeyh.) R. H. Archer ●☆

356020　Scytophyllum apiculatum Sond. = Robsonodendron maritimum (Bolus) R. H. Archer ●☆

356021　Scytophyllum laurinum (Thunb.) Eckl. et Zeyh. = Maytenus oleoides (Lam.) Loes. ●☆

356022　Sczegleewia Turcz. (1858) = Pterospermum Schreb. (保留属名)●

356023　Sczegleewia Turcz. (1863) = Symphorema Roxb. ●

356024　Sczukinia Turcz. = Swertia L. ■

356025　Sczukinia diluta Turcz. = Swertia diluta (Turcz.) Benth. et Hook. f. ■

356026　Sdadiodaphne Rchb. = Umbellularia (Nees) Nutt. (保留属名)●☆

356027　Sddzogyne Ehrenb. ex Pax = Acalypha L. ●■

356028　Seaforthia R. Br. = Ptychosperma Labill. ●☆

356029　Seala Adans. = Pectis L. ■☆

356030　Searsia F. A. Barkley = Rhus L. ●

356031　Searsia F. A. Barkley = Terminthia Bernh. ●

356032　Searsia africana (Mill.) F. A. Barkley = Rhus lucida L. ●☆

356033　Searsia legatii (Schönland) F. A. Barkley = Rhus chirindensis Baker f. ●☆

356034　Sebaea Sol. ex R. Br. (1810);小黄管属;Sebaea ■

356035　Sebaea acuminata Hill = Sebaea longicaulis Schinz ■☆

356036　Sebaea acutiloba Schinz = Sebaea zeyheri Schinz subsp. acutiloba (Schinz) Marais ■☆

356037　Sebaea africana Paiva et I. Nogueira;非洲小黄管■☆

356038　Sebaea alata Paiva et I. Nogueira;具翅小黄管■☆

356039　Sebaea albens (L. f.) Sm. ;白小黄管■☆

356040　Sebaea ambigua Cham. ;可疑小黄管■☆

356041　Sebaea amicorum I. M. Oliv. et Beyers;可爱小黄管■☆

356042　Sebaea aurea (L. f.) Sm. ;金小黄管■☆

356043 Sebaea barbeyana Schinz = Sebaea pentandra E. Mey. var. burchellii (Gilg) Marais ■☆

356044 Sebaea baumiana (Gilg) Boutique;鲍姆小黄管■☆

356045 Sebaea baumii Schinz = Sebaea brachyphylla Griseb. ■☆

356046 Sebaea bequaertii De Wild. = Sebaea leiostyla Gilg ■☆

356047 Sebaea bojeri Griseb. ;博耶尔小黄管■☆

356048 Sebaea brachyphylla Griseb. ;短叶小黄管■☆

356049 Sebaea brehmeri Schinz = Sebaea macrophylla Gilg ■☆

356050 Sebaea brevicaulis Sileshi;短茎小黄管■☆

356051 Sebaea burchellii Gilg = Sebaea pentandra E. Mey. var. burchellii (Gilg) Marais ■☆

356052 Sebaea butaguensis De Wild. = Sebaea brachyphylla Griseb. ■☆

356053 Sebaea capitata Cham. et Schltdl. ;头状小黄管■☆

356054 Sebaea capitata Cham. et Schltdl. var. sclerosepala (Schinz) Marais;硬萼头状小黄管■☆

356055 Sebaea caudata Paiva et I. Nogueira;尾状小黄管■☆

356056 Sebaea clavata Paiva et I. Nogueira;棍棒小黄管■☆

356057 Sebaea cleistantha R. A. Dyer = Sebaea zeyheri Schinz subsp. cleistantha (R. A. Dyer) Marais ■☆

356058 Sebaea compacta A. W. Hill;紧密小黄管■☆

356059 Sebaea condensata Klack. ;密集小黄管■☆

356060 Sebaea confertiflora Schinz = Sebaea sedoides Gilg var. confertiflora (Schinz) Marais ■☆

356061 Sebaea congesta Schrad. = Sebaea aurea (L. f.) Sm. ■☆

356062 Sebaea conrathii Schinz = Sebaea pentandra E. Mey. var. burchellii (Gilg) Marais ■☆

356063 Sebaea conspicua Hill = Sebaea procumbens A. W. Hill ■☆

356064 Sebaea cordata (L. f.) Roem. et Schult. = Sebaea exacoides (L.) Schinz ■☆

356065 Sebaea cordata (L. f.) Roem. et Schult. var. intermedia Cham. et Schltdl. = Sebaea micrantha (Cham. et Schltdl.) Schinz var. intermedia (Cham. et Schltdl.) Marais ■☆

356066 Sebaea cordata (L. f.) Roem. et Schult. var. macrantha Cham. et Schltdl. = Sebaea exacoides (L.) Schinz ■☆

356067 Sebaea cordata (L. f.) Roem. et Schult. var. micrantha Cham. et Schltdl. = Sebaea micrantha (Cham. et Schltdl.) Schinz ■☆

356068 Sebaea crassulifolia Cham. et Schltdl. var. lanceolata Schinz = Sebaea longicaulis Schinz ■☆

356069 Sebaea crassulifolia Cham. et Schltdl. var. stricta E. Mey. = Sebaea stricta (E. Mey.) Gilg ■☆

356070 Sebaea cuspidata Schinz = Sebaea elongata E. Mey. ■☆

356071 Sebaea cymosa Jaroscz = Sebaea aurea (L. f.) Sm. ■☆

356072 Sebaea debilis (Welw.) Schinz;弱小黄管■☆

356073 Sebaea dimidiata Sileshi;半片小黄管■☆

356074 Sebaea dinteri Gilg ex Dinter = Sebaea pentandra E. Mey. var. burchellii (Gilg) Marais ■☆

356075 Sebaea dregei Schinz = Sebaea stricta (E. Mey.) Gilg ■☆

356076 Sebaea elongata E. Mey. ;伸长小黄管■☆

356077 Sebaea erecta Hill = Sebaea longicaulis Schinz ■☆

356078 Sebaea erosa Schinz;啮蚀状小黄管■☆

356079 Sebaea evansii N. E. Br. = Sebaea repens Schinz ■☆

356080 Sebaea exacoides (L.) Schinz;藻百年小黄管■☆

356081 Sebaea exigua (Oliv.) Schinz;澳非小黄管■☆

356082 Sebaea fastigiata Hill = Sebaea macrophylla Gilg ■☆

356083 Sebaea fernandesiana Paiva et I. Nogueira;费尔南小黄管■☆

356084 Sebaea filiformis Schinz;丝形小黄管■☆

356085 Sebaea flanaganii (Schinz) Schinz = Sebaea spathulata (E. Mey.) Steud. ■☆

356086 Sebaea fourcadei Marais;富尔卡德小黄管■☆

356087 Sebaea gariepina Gilg = Sebaea pentandra E. Mey. ■☆

356088 Sebaea gentilii (De Wild.) Boutique;让蒂小黄管■☆

356089 Sebaea gibbosa Wolley-Dod = Sebaea aurea (L. f.) Sm. ■☆

356090 Sebaea gilgii Schinz = Sebaea schlechteri Schinz ■☆

356091 Sebaea glauca Hill = Sebaea aurea (L. f.) Sm. ■☆

356092 Sebaea gracilis (Welw.) Paiva et I. Nogueira;纤细小黄管■☆

356093 Sebaea grandiflora Schinz = Sebaea longicaulis Schinz ■☆

356094 Sebaea grandis (E. Mey.) Steud. ;大株小黄管■☆

356095 Sebaea grisebachiana Schinz;格里泽巴赫小黄管■☆

356096 Sebaea hockii (De Wild.) Boutique;霍克小黄管■☆

356097 Sebaea humilis N. E. Br. = Sebaea grisebachiana Schinz ■☆

356098 Sebaea hymenosepala Gilg;膜萼小黄管■☆

356099 Sebaea imbricata Hill = Sebaea natalensis Schinz ■☆

356100 Sebaea intermedia (Cham. et Schltdl.) Schinz = Sebaea micrantha (Cham. et Schltdl.) Schinz var. intermedia (Cham. et Schltdl.) Marais ■☆

356101 Sebaea involucrata Klotzsch = Faroa involucrata (Klotzsch) Knobl. ■☆

356102 Sebaea jasminiflora Schinz =Sebaea thomasii (S. Moore) Schinz ■☆

356103 Sebaea junodii Schinz;朱诺德小黄管■☆

356104 Sebaea khasiana C. B. Clarke = Sebaea microphylla (Edgew.) Knobl. ■

356105 Sebaea laxa N. E. Br. ;疏松小黄管■☆

356106 Sebaea leiostyla Gilg;光萼小黄管■☆

356107 Sebaea linearifolia Schinz = Sebaea exigua (Oliv.) Schinz ■☆

356108 Sebaea lineariformis Sileshi;线形小黄管■☆

356109 Sebaea longicaulis Schinz;长茎小黄管■☆

356110 Sebaea luteo-alba (A. Chev.) P. Taylor;黄白小黄管■☆

356111 Sebaea macowanii Gilg ex Schinz = Sebaea longicaulis Schinz ■☆

356112 Sebaea macrantha Gilg = Sebaea rehmannii Schinz ■☆

356113 Sebaea macrophylla Gilg;大叶小黄管■☆

356114 Sebaea macroptera Sileshi;大翅小黄管■☆

356115 Sebaea macrosepala Gilg = Sebaea longicaulis Schinz ■☆

356116 Sebaea macrostigma Gilg = Sebaea grisebachiana Schinz ■☆

356117 Sebaea madagascariensis Klack. ;马岛小黄管■☆

356118 Sebaea marlothii Gilg;马洛斯小黄管■☆

356119 Sebaea membranacea A. W. Hill;膜状小黄管■☆

356120 Sebaea micrantha (Cham. et Schltdl.) Schinz;小花小黄管■☆

356121 Sebaea micrantha (Cham. et Schltdl.) Schinz var. intermedia (Cham. et Schltdl.) Marais;间型小花小黄管■☆

356122 Sebaea microphylla (Edgew.) Knobl. ; 小黄管; Littleleaf Sebaea ■

356123 Sebaea mildbraedii Gilg = Sebaea oligantha (Gilg) Schinz ■☆

356124 Sebaea minima Jaroscz = Sebaea aurea (L. f.) Sm. ■☆

356125 Sebaea minuta Paiva et I. Nogueira;微小小黄管■☆

356126 Sebaea minutiflora Schinz;微花小黄管■☆

356127 Sebaea minutissima Hilliard et B. L. Burtt;极微小黄管■☆

356128 Sebaea mirabilis Gilg = Sebaea bojeri Griseb. ■☆

356129 Sebaea monantha Gilg ex Schinz;山地小黄管■☆

356130 Sebaea multiflora Schinz = Sebaea macrophylla Gilg ■☆

356131 Sebaea multinodis N. E. Br. = Sebaea brachyphylla Griseb. ■☆

356132 Sebaea natalensis (Schinz) Schinz = Sebaea grandis (E. Mey.) Steud. ■☆

356133 Sebaea natalensis Schinz;纳塔尔小黄管■☆

356134 Sebaea ochroleuca Wolley-Dod = Sebaea schlechteri Schinz ■☆

356135　Sebaea oldenlandioides S. Moore ＝ Exacum oldenlandioides（S. Moore）Klack. ■☆

356136　Sebaea oligantha（Gilg）Schinz；寡花小黄管■☆

356137　Sebaea oreophila Gilg ＝ Sebaea longicaulis Schinz ■☆

356138　Sebaea paludosa Levyns ＝ Sebaea aurea（L. f.）Sm. ■☆

356139　Sebaea pentandra E. Mey. ；五蕊小黄管■☆

356140　Sebaea pentandra E. Mey. var. burchellii（Gilg）Marais；伯切尔小黄管■☆

356141　Sebaea perpusilla Paiva et I. Nogueira；罗得西亚小黄管■☆

356142　Sebaea platyptera（Baker）Boutique；阔翅小黄管■☆

356143　Sebaea pleurostigmatosa Hilliard et B. L. Burtt；侧柱头小黄管■☆

356144　Sebaea polyantha Gilg ＝ Sebaea leiostyla Gilg ■☆

356145　Sebaea pratensis Gilg ＝ Sebaea bojeri Griseb. ■☆

356146　Sebaea primuliflora（Welw.）Sileshi；报春花小黄管■☆

356147　Sebaea primulina Hill ＝ Sebaea pentandra E. Mey. ■☆

356148　Sebaea procumbens A. W. Hill；平铺小黄管■☆

356149　Sebaea pseudobelmontia Schinz ＝ Sebaea micrantha（Cham. et Schltdl.）Schinz var. intermedia（Cham. et Schltdl.）Marais ■☆

356150　Sebaea pumila（Baker）Schinz；矮小黄管■☆

356151　Sebaea pusilla Eckl. ex Cham. ；瘦小黄管■☆

356152　Sebaea pusilla Eckl. ex Cham. var. major Hill ＝ Sebaea pusilla Eckl. ex Cham. ■☆

356153　Sebaea pygmaea Schinz ＝ Sebaea erosa Schinz ■☆

356154　Sebaea radiata Hilliard et B. L. Burtt；辐射小黄管■☆

356155　Sebaea ramosissima Gilg；多枝小黄管■☆

356156　Sebaea rara Wolley-Dod；稀少小黄管■☆

356157　Sebaea rehmannii Schinz；拉赫曼小黄管■☆

356158　Sebaea repens Schinz；匍匐小黄管■☆

356159　Sebaea rotundifolia Hill ＝ Sebaea procumbens A. W. Hill ■☆

356160　Sebaea rudolfii Schinz ＝ Sebaea natalensis Schinz ■☆

356161　Sebaea rutenbergiana Vatke ＝ Exacum spathulatum Baker ■☆

356162　Sebaea saccata Schinz ＝ Sebaea bojeri Griseb. ■☆

356163　Sebaea scabra Schinz；粗糙小黄管■☆

356164　Sebaea schimperiana Buchinger ex Schweinf. ＝ Sebaea brachyphylla Griseb. ■☆

356165　Sebaea schinziana Gilg ＝ Sebaea macrophylla Gilg ■☆

356166　Sebaea schizostigma Gilg ＝ Sebaea grisebachiana Schinz ■☆

356167　Sebaea schlechteri Schinz；施莱小黄管■☆

356168　Sebaea schoenlandii Schinz ＝ Sebaea sedoides Gilg var. schoenlandii（Schinz）Marais ■☆

356169　Sebaea sclerosepala Gilg ex Schinz ＝ Sebaea capitata Cham. et Schltdl. var. sclerosepala（Schinz）Marais ■☆

356170　Sebaea sedoides Gilg；景天小黄管■☆

356171　Sebaea sedoides Gilg var. confertiflora（Schinz）Marais；密花景天小黄管■☆

356172　Sebaea sedoides Gilg var. schoenlandii（Schinz）Marais；绍氏景天小黄管■☆

356173　Sebaea semialata Gilg ＝ Sebaea hymenosepala Gilg ■☆

356174　Sebaea spathulata（E. Mey.）Steud.；匙形小黄管■☆

356175　Sebaea stricta（E. Mey.）Gilg ＝ Sebaea madagascariensis Klack. ■☆

356176　Sebaea stricta Gilg ＝ Sebaea madagascariensis Klack. ■☆

356177　Sebaea sulphurea Cham. et Schltdl.；硫色小黄管■☆

356178　Sebaea tabularis Eckl. ＝ Sebaea sulphurea Cham. et Schltdl. ■☆

356179　Sebaea teuszii（Schinz）P. Taylor；托兹小黄管■☆

356180　Sebaea thomasii（S. Moore）Schinz；托马斯小黄管■☆

356181　Sebaea transvaalensis Schinz ＝ Sebaea leiostyla Gilg ■☆

356182　Sebaea trinervia（Desr.）Schinz ＝ Ornichia trinervis（Desr.）Klack. ■☆

356183　Sebaea tysonii Schinz ＝ Sebaea zeyheri Schinz subsp. acutiloba（Schinz）Marais ■☆

356184　Sebaea vitellina Schinz ＝ Sebaea natalensis Schinz ■☆

356185　Sebaea welwitschii Schinz ＝ Sebaea microphylla（Edgew.）Knobl. ■

356186　Sebaea wildemaniana Boutique；怀尔德曼小黄管■☆

356187　Sebaea wittebergensis Schinz ＝ Sebaea macrophylla Gilg ■☆

356188　Sebaea woodii Gilg ＝ Sebaea longicaulis Schinz ■☆

356189　Sebaea zeyheri Schinz；泽赫小黄管■☆

356190　Sebaea zeyheri Schinz subsp. acutiloba（Schinz）Marais；尖浅裂泽赫小黄管■☆

356191　Sebaea zeyheri Schinz subsp. cleistantha（R. A. Dyer）Marais；闭花泽赫小黄管■☆

356192　Sebastiana Benth. et Hook. f. ＝ Chrysanthellum Rich. ex Pers. ■☆

356193　Sebastiana Benth. et Hook. f. ＝ Sebastiania Bertol. ■☆

356194　Sebastiana Spreng. ＝ Sebastiania Spreng. ●

356195　Sebastiania Bertol. ＝ Chrysanthellum Rich. ex Pers. ■☆

356196　Sebastiania Spreng.（1820）；地阳桃属（地杨桃属）；Sebastiania ●

356197　Sebastiania acetosella（Milne-Redh.）Kruijt ＝ Microstachys acetosella（Milne-Redh.）Esser ●☆

356198　Sebastiania bidentata（Mart. et Zucc.）Pax；二齿地阳桃●☆

356199　Sebastiania bidentata Pax ＝ Sebastiania bidentata（Mart. et Zucc.）Pax ●☆

356200　Sebastiania bilocularis S. Watson；二室地阳桃（二室地杨桃）●☆

356201　Sebastiania chamaelea（L.）F. Muell.；地阳桃（地杨桃，色巴木）；Creeping Sebastiania ●

356202　Sebastiania chamaelea（L.）F. Muell. var. asperococca（F. Muell.）Pax et K. Hoffm. ＝ Sebastiania chamaelea（L.）F. Muell. ●

356203　Sebastiania chamaelea（L.）Müll. Arg. ＝ Microstachys chamaelea（L.）Müll. Arg. ●☆

356204　Sebastiania faradianensis（Beille）Kruijt ＝ Microstachys faradianensis（Beille）Esser ■☆

356205　Sebastiania inopinata Prain ＝ Gymnanthes inopinata（Prain）Esser ●☆

356206　Sebastiania ligustrina Müll. Arg. ；女贞地阳桃；Candleberry ■☆

356207　Sebastiania multiramea（Klotzsch）Mart. var. luschnathiana Müll. Arg. ＝ Gymnanthes inopinata（Prain）Esser ●☆

356208　Sebastiano-Schaueria Nees(1847)；塞沙爵床属☆

356209　Sebastiano-schaueria oblongata Nees；塞沙爵床☆

356210　Sebeckia Stead. ＝ Sebeokia Neck. ●

356211　Sebeokia Neck. ＝ Gentiana L. ■

356212　Sebertia Pierre ＝ Apostasia Blume ■

356213　Sebertia Pierre ＝ Niemeyera F. Muell.（保留属名）●☆

356214　Sebertia Pierre ex Engl. ＝ Niemeyera F. Muell.（保留属名）●☆

356215　Sebesten Adans. ＝ Sebestena Boehm. ●

356216　Sebestena Boehm. ＝ Cordia L.（保留属名）●

356217　Sebestena Gaertn. ＝ Cordia L.（保留属名）●

356218　Sebestenaceae Vent. ＝ Boraginaceae Juss.（保留科名）■●

356219　Sebestenaceae Vent. ＝ Ehretiaceae Mart.（保留科名）●

356220　Sebicea Pierre ex Diels ＝ Tiliacora Colebr.（保留属名）●☆

356221　Sebifera Lour. ＝ Litsea Lam.（保留属名）●☆

356222　Sebifera glutinosa Lour. ＝ Litsea glutinosa（Lour.）C. B. Rob. ●

356223　Sebipira Mart. ＝ Bowdichia Kunth ●☆

356224　Sebizia Mart. ＝ Mappia Jacq.（保留属名）●☆

356225 Sebizia Mart. ex Meisn. = Mappia Jacq. (保留属名)●☆

356226 Sebophora Neck. = Myristica Gronov. (保留属名)●

356227 Seborium Raf. = Sapium Jacq. (保留属名)●

356228 Seborium chinense Raf. = Sapium sebiferum (L.) Roxb. ●

356229 Seborium sebiferum (L.) Hurus. = Sapium sebiferum (L.) Roxb. ●

356230 Seborium sebiferum (L.) Hurus. = Triadica sebifera (L.) Small ●

356231 Sebschauera Kuntze = Sebastiano-Schaueria Nees ☆

356232 Secale L. (1753);黑麦属;Rye ■

356233 Secale afghanicum (Vavilov) Roshev. = Secale segetale (Zhuk.) Roshev. ■

356234 Secale africanum Stapf = Secale strictum (C. Presl) C. Presl subsp. africanum (Stapf) K. Hammer ■☆

356235 Secale anatolicum Boiss. ;阿纳托里黑麦■☆

356236 Secale ancestrale Zhuk. var. afghanicum (Vavilov) A. P. Ivanov et G. V. Yakovlev = Secale segetale (Zhuk.) Roshev. ■

356237 Secale cereale L. ;黑麦;Cereal Rye,Rye ■

356238 Secale cereale L. subsp. afghanicum (Vavilov) K. Hammer = Secale segetale (Zhuk.) Roshev. ■

356239 Secale cereale L. subsp. segetale Zhuk. = Secale segetale (Zhuk.) Roshev. ■

356240 Secale cereale L. var. afghanicum Vavilov = Secale segetale (Zhuk.) Roshev. ■

356241 Secale cereale L. var. multicaule Metzg. ;多茎黑麦;Multistalk Rye ■☆

356242 Secale cornutum Offic. ex Nees;角状黑麦■☆

356243 Secale fragile M. Bieb. = Secale sylvestre Host ■

356244 Secale kuprijanovii Grossh. ;库泊黑麦■☆

356245 Secale montanum Guss. ;山黑麦;Mounrain Rye ■☆

356246 Secale orientale L. = Eremopyrum orientale (L.) Jaub. et Spach ■

356247 Secale prostratum Pall. = Eremopyrum triticeum (Gaertn.) Nevski ■

356248 Secale segetale (Zhuk.) Roshev. ;脆轴黑麦■

356249 Secale segetale (Zhuk.) Roshev. subsp. afghanicum (Vavilov) Bondar ex Korovina = Secale segetale (Zhuk.) Roshev. ■

356250 Secale segetale (Zhuk.) Roshev. var. afghanicum (Vavilov) Tzvelev = Secale segetale (Zhuk.) Roshev. ■

356251 Secale strictum (C. Presl) C. Presl;刚直黑麦■☆

356252 Secale strictum (C. Presl) C. Presl subsp. africanum (Stapf) K. Hammer;非洲黑麦■☆

356253 Secale sylvestre Host;小黑麦(野黑麦);Wild Rye ■

356254 Secale triflorum P. Beauv. ;三花黑麦■☆

356255 Secale vavilovii Grossh. ;瓦维黑麦■☆

356256 Secale villosum L. = Dasypyrum villosum (L.) P. Candargy ■☆

356257 Secalidium Schur = Agropyron Gaertn. ■

356258 Secalidium Schur = Dasypyrum (Coss. et Durieu) T. Durand ■☆

356259 Secamone R. Br. (1810);鲫鱼藤属;Crucianvine,Secamone ●■

356260 Secamone acutifolia Sond. = Cryptolepis oblongifolia (Meisn.) Schltr. ●☆

356261 Secamone africana (Oliv.) Bullock;非洲鲫鱼藤●☆

356262 Secamone afzelii (Schult.) K. Schum. ;阿芙泽尔鲫鱼藤●☆

356263 Secamone alpini Schult. ;高山鲫鱼藤●☆

356264 Secamone ankarensis (Jum. et H. Perrier) Klack. ;安卡拉鲫鱼藤●☆

356265 Secamone bemarahensis Klack. ;贝马拉哈鲫鱼藤●☆

356266 Secamone bicolor Decne. ;二色鲫鱼藤●☆

356267 Secamone bifida Klack. ;二裂鲫鱼藤●☆

356268 Secamone bonii Costantin;斑皮鲫鱼藤;Bon Crucianvine, Bon Secamone ●■

356269 Secamone brevicoronata Klack. ;短冠鲫鱼藤●☆

356270 Secamone brevipes (Benth.) Klack. ;短梗鲫鱼藤●☆

356271 Secamone clavistyla T. M. Harris et Goyder;棒柱鲫鱼藤●☆

356272 Secamone conostyla S. Moore = Tylophora oblonga N. E. Br. ●☆

356273 Secamone delagoensis Schltr. ;迪拉果鲫鱼藤●☆

356274 Secamone dewevrei De Wild. ;德韦鲫鱼藤●☆

356275 Secamone dewevrei De Wild. subsp. elliptica Goyder;椭圆鲫鱼藤●☆

356276 Secamone discolor K. Schum. et Vatke;异色鲫鱼藤●☆

356277 Secamone drepanoloba Klack. ;镰状鲫鱼藤●☆

356278 Secamone elegans Klack. ;雅致鲫鱼藤●☆

356279 Secamone elliptica R. Br. ;鲫鱼藤(黄花藤,四粉块藤);Common Secamone,Crucianvine ●■

356280 Secamone elliptica R. Br. subsp. minutiflora (Woodson) Klack. = Secamone minutiflora (Woodson) Tsiang ●■

356281 Secamone emetica (Retz.) R. Br. ex Schult. var. glabra K. Schum. = Secamone parvifolia (Oliv.) Bullock ●☆

356282 Secamone erythradenia K. Schum. ;红腺鲫鱼藤●☆

356283 Secamone esquirolii Schltr. ex H. Lév. = Secamone sinica Hand. -Mazz. ●■

356284 Secamone falcata Klack. ;镰叶鲫鱼藤●☆

356285 Secamone ferruginea Pierre ex Costantin;锈毛鲫鱼藤;Rustyhair Crucianvine,Rustyhair Secamone,Rusty-haired Secamone ●■

356286 Secamone filiformis (L. f.) J. H. Ross;丝形鲫鱼藤●☆

356287 Secamone floribunda N. E. Br. = Secamone stuhlmannii K. Schum. ●☆

356288 Secamone frutescens (E. Mey.) Decne. = Secamone filiformis (L. f.) J. H. Ross ●☆

356289 Secamone gabonensis P. T. Li = Secamone dewevrei De Wild. ●☆

356290 Secamone gerrardii Harv. ex Benth. ;杰勒德鲫鱼藤●☆

356291 Secamone gracilis N. E. Br. ;纤细鲫鱼藤■☆

356292 Secamone grandiflora Klack. ;大花鲫鱼藤●☆

356293 Secamone humbertii Choux;亨伯特鲫鱼藤●☆

356294 Secamone jongkindii Klack. ;容金德鲫鱼藤●☆

356295 Secamone kirkii N. E. Br. = Secamone parvifolia (Oliv.) Bullock ●☆

356296 Secamone lanceolata Blume = Secamone elliptica R. Br. ●■

356297 Secamone leonensis (Scott-Elliot) N. E. Br. ;莱昂鲫鱼藤●☆

356298 Secamone letouzeana (H. Huber) Klack. ;喀麦隆鲫鱼藤●☆

356299 Secamone likiangensis Tsiang;丽江鲫鱼藤;Lijiang Crucianvine,Lijiang Secamone,Likiang Secamone ●■

356300 Secamone linearis Klack. ;线形鲫鱼藤●☆

356301 Secamone micrandra K. Schum. = Secamone punctulata Decne. ●☆

356302 Secamone micrantha Decne. = Secamone elliptica R. Br. ●■

356303 Secamone minutiflora (Woodson) Tsiang;催吐鲫鱼藤;Emetic Crucianvine, Minute-flowered Secamone, Sichuan Secamone, Szechuan Secamone ●■

356304 Secamone mombasiana N. E. Br. = Secamone parvifolia (Oliv.) Bullock ●☆

356305 Secamone myrtifolia Benth. = Secamone afzelii (Schult.) K. Schum. ●☆

356306 Secamone obovata Decne. ;卵叶鲫鱼藤●☆

356307 Secamone oleifolia Decne. ;橄榄叶鲫鱼藤●☆

356308 Secamone pachystigma Jum. et H. Perrier;大柱头鲫鱼藤●☆

356309　Secamone parvifolia（Oliv.）Bullock；小叶鲫鱼藤●☆

356310　Secamone pedicellaris Klack.；梗花鲫鱼藤●☆

356311　Secamone phillyreoides S. Moore ＝ Secamone stuhlmannii K. Schum. ●☆

356312　Secamone pinnata Choux；羽状鲫鱼藤●☆

356313　Secamone platystigma K. Schum. ＝ Secamone africana（Oliv.）Bullock ●☆

356314　Secamone punctulata Decne.；小斑鲫鱼藤●☆

356315　Secamone punctulata Decne. var. stenophylla（K. Schum.）N. E. Br. ＝ Secamone punctulata Decne. ●☆

356316　Secamone racemosa（Benth.）Klack.；总花鲫鱼藤●☆

356317　Secamone rariflora S. Moore ＝ Secamone stuhlmannii K. Schum. ●☆

356318　Secamone reticulata Klack.；网脉鲫鱼藤●☆

356319　Secamone retusa N. E. Br.；微凹鲫鱼藤●☆

356320　Secamone rubiginosa K. Schum. ＝ Secamone brevipes（Benth.）Klack. ●☆

356321　Secamone sansibariensis K. Schum. ＝ Secamone punctulata Decne. ●☆

356322　Secamone schatzii Klack.；沙茨鲫鱼藤●☆

356323　Secamone schweinfurthii K. Schum. ＝ Secamone parvifolia（Oliv.）Bullock ●☆

356324　Secamone sinica Hand.-Mazz.；吊山桃（细叶青藤）；China Crucianvine，Chinese Secamone ●■

356325　Secamone stenophylla K. Schum. ＝ Secamone punctulata Decne. ●☆

356326　Secamone stuhlmannii K. Schum.；斯图尔曼鲫鱼藤●☆

356327　Secamone stuhlmannii K. Schum. var. whytei（N. E. Br.）Goyder et T. M. Harris；怀特鲫鱼藤●☆

356328　Secamone sulfurea（Jum. et H. Perrier）Klack.；硫色鲫鱼藤●☆

356329　Secamone szechuanensis Tsiang et P. T. Li ＝ Secamone minutiflora（Woodson）Tsiang ●■

356330　Secamone thunbergii E. Mey. ＝ Secamone alpini Schult. ●☆

356331　Secamone thunbergii E. Mey. var. retusa ？ ＝ Secamone alpini Schult. ●☆

356332　Secamone toxocarpoides Choux；弓果藤鲫鱼藤●☆

356333　Secamone trichostemon Klack.；毛冠鲫鱼藤●☆

356334　Secamone tuberculata Klack.；多疣鲫鱼藤●☆

356335　Secamone uncinata Choux；具钩鲫鱼藤●☆

356336　Secamone uniflora Decne.；单花鲫鱼藤●☆

356337　Secamone usambarica N. E. Br. ＝ Secamone parvifolia（Oliv.）Bullock ●☆

356338　Secamone valvata Klack.；锯合鲫鱼藤●☆

356339　Secamone villosa Blume ＝ Toxocarpus villosus（Blume）Decne. ●

356340　Secamone whytei N. E. Br. ＝ Secamone stuhlmannii K. Schum. var. whytei（N. E. Br.）Goyder et T. M. Harris ●☆

356341　Secamone wightiana（Hook. et Arn.）K. Schum. ＝ Toxocarpus wightianus Hook. et Arn. ●

356342　Secamone wightiana K. Schum. ＝ Toxocarpus wightianus Hook. et Arn. ●

356343　Secamone zambeziaca Schltr. ＝ Secamone parvifolia（Oliv.）Bullock ●☆

356344　Secamone zambeziaca Schltr. var. parvifolia N. E. Br. ＝ Secamone parvifolia（Oliv.）Bullock ●☆

356345　Secamonopsis Jum.（1908）；类鲫鱼藤属●☆

356346　Secamonopsis madagascariensis Jum.；类鲫鱼藤●☆

356347　Secamonopsis microphylla Civeyrel et Klack.；大叶类鲫鱼藤●☆

356348　Sechiopsis Naudin（1866）；类佛手瓜属■☆

356349　Sechiopsis triquetra（Moc. et Sessé ex Ser.）Naudin；类佛手瓜■☆

356350　Sechium P. Browne（1756）（保留属名）；佛手瓜属（洋丝瓜属）；Chayote ■

356351　Sechium americanum Poir. ＝ Sechium edule（Jacq.）Sw. ■

356352　Sechium edule（Jacq.）Sw.；佛手瓜（梨瓜，手瓜，隼人瓜，香木缘瓜，洋丝瓜）；Chayota，Chayote，Cho-Cho，Choco，Choko，Chow Chow，Choyote，Christophine，Madeira Marrow，Pipinola，Vegetable Pear ■

356353　Secondatia A. DC.（1844）；塞考木属●☆

356354　Secondatia arborea Müll. Arg.；乔木塞考木●☆

356355　Secondatia densiflora A. DC.；塞考木●☆

356356　Secondatia floribunda A. DC.；多花塞考木●☆

356357　Secondatia foliosa A. DC.；多叶塞考木●☆

356358　Secretania Müll. Arg. ＝ Minquartia Aubl. ●☆

356359　Secula Small ＝ Aeschynomene L. ●■

356360　Securidaca L.（1753）（废弃属名）＝ Securidaca L.（1759）（保留属名）●

356361　Securidaca L.（1759）（保留属名）；蝉翼藤属；Securidaca，Cicadawingvine ●

356362　Securidaca Mill. ＝ Coronilla L.（保留属名）●■

356363　Securidaca Mill. ＝ Securigera DC.（保留属名）●■

356364　Securidaca Tourn. ex Mill. ＝ Coronilla L.（保留属名）●■

356365　Securidaca Tourn. ex Mill. ＝ Securigera DC.（保留属名）●■

356366　Securidaca diversifolia S. F. Blake；东部蝉翼藤；Easter Flower ●☆

356367　Securidaca inappendiculata Hassk.；蝉翼藤（蝉翼木，刁了棒，丢了棒，五味藤，象皮藤，一摩消）；Cicadawingvine，Common Securidaca ●

356368　Securidaca longepedunculata Fresen.；长梗蝉翼藤；Fibre-tree，Rhodes' Violet，Wild Violet Tree，Wild Wistaria ●☆

356369　Securidaca longipedunculata Fresen. var. parvifolia Oliv.；小叶蝉翼藤；Buaze ●☆

356370　Securidaca scandens Jacq. ＝ Securidaca inappendiculata Hassk. ●

356371　Securidaca tavoyana Wall. ＝ Securidaca inappendiculata Hassk. ●

356372　Securidaca tavoyana Wall. ex A. W. Benn. ＝ Securidaca inappendiculata Hassk. ●

356373　Securidaca virgata Sw.；帚状蝉翼藤●☆

356374　Securidaca welwitschii Oliv.；韦氏蝉翼藤●☆

356375　Securidaca yaoshanensis K. S. Hao；瑶山蝉翼藤；Yaoshan Cicadawingvine，Yaoshan Securidaca ●

356376　Securidaea Turcz. ＝ Securidaca Mill. ●■

356377　Securigera DC.（1805）（保留属名）；斧冠花属；Crown Vetch，Hatchet Vetch ●■

356378　Securigera DC.（保留属名）＝ Coronilla L.（保留属名）●■

356379　Securigera atlantica Boiss. et Reut.；大西洋斧冠花■☆

356380　Securigera bicapsularis（L.）Roxb.；双裂果斧冠花■☆

356381　Securigera coronilla DC. ＝ Securigera securidaca（L.）Degen. et Dorf ■☆

356382　Securigera corymbosa（Lam.）H. S. Irwin et Barneby；伞序斧冠花■☆

356383　Securigera cretica（L.）Lassen；克里特斧冠花；Cretan Crownvetch ●☆

356384　Securigera didymobotrya（Fresen.）H. S. Irwin et Barneby；双穗斧冠花■☆

356385　Securigera globosa（Lam.）Lassen；白斧冠花；White Crownvetch ■☆

356386　Securigera securidaca（L.）Degen ＝ Securigera securidaca（L.）Degen. et Dorf ■☆

356387　Securigera securidaca (L.) Degen. et Dorf;斧冠花;Goat Pea ■☆

356388　Securigera somalensis (Thulin) Lassen;索马里斧冠花●☆

356389　Securigera varia (L.) Lassen = Coronilla varia L. ●

356390　Securilla Gaertn. ex Steud. = Securigera DC. (保留属名)●■

356391　Securina Medik. = Securigera DC. (保留属名)●■

356392　Securinega Comm. ex Juss. (1789)(保留属名);叶底球属(一叶萩属);Securinega ●☆

356393　Securinega acicularis Croizat = Flueggea acicularis (Croizat) G. L. Webster ●

356394　Securinega bailloniana Müll. Arg. = Margaritaria discoidea (Baill.) G. L. Webster var. triplosphaera Radcl.-Sm. ●☆

356395　Securinega durissima J. F. Gmel.;硬叶底球;Otaheite Myrtle ●☆

356396　Securinega fluggeoides Müll. Arg. = Flueggea suffruticosa (Pall.) Baill. ●

356397　Securinega japonica Miq. = Flueggea suffruticosa (Pall.) Baill. ●

356398　Securinega leucopyra (Willd.) Müll. Arg. = Flueggea leucopyrus Willd. ●

356399　Securinega multiflora S. B. Liang = Flueggea virosa (Roxb. ex Willd.) Voigt ●

356400　Securinega obovata (Willd.) Müll. Arg. = Flueggea virosa (Roxb. ex Willd.) Voigt ●

356401　Securinega ramiflora (Aiton) Müll. Arg. = Flueggea suffruticosa (Pall.) Baill. ●

356402　Securinega schlechteri Pax = Cleistanthus schlechteri (Pax) Hutch. ●☆

356403　Securinega schweinfurthii Balf. f. = Andrachne schweinfurthii (Balf. f.) Radcl.-Sm. ●☆

356404　Securinega suffruticosa (Pall.) Rehder = Flueggea suffruticosa (Pall.) Baill. ●

356405　Securinega suffruticosa (Pall.) Rehder f. japonica (Miq.) Hurus. = Flueggea suffruticosa (Pall.) Baill. ●

356406　Securinega suffruticosa (Pall.) Rehder var. amamiensis Hurus. = Flueggea suffruticosa (Pall.) Baill. ●

356407　Securinega suffruticosa (Pall.) Rehder var. amamiensis Hurus. = Flueggea trigonoclada (Ohwi) T. Kuros. ●

356408　Securinega suffruticosa (Pall.) Rehder var. japonica (Miq.) Hurus. = Flueggea suffruticosa (Pall.) Baill. ●

356409　Securinega tinctoria L.;染用一叶萩●☆

356410　Securinega virosa (Roxb. et Willd.) Baill. = Flueggea virosa (Roxb. ex Willd.) Voigt ●

356411　Sedaceae Barkley = Crassulaceae J. St.-Hil. (保留科名)●■

356412　Sedaceae Roussel = Crassulaceae J. St.-Hil. (保留科名)●■

356413　Sedaceae Vest = Crassulaceae J. St.-Hil. (保留科名)●■

356414　Sedastrum Rose = Sedum L. ●■

356415　Seddera Hochst. (1844);赛德旋花属●☆

356416　Seddera Hochst. et Steud. ex Moq. = Saltia R. Br. ex Moq. ●☆

356417　Seddera arabica (Forssk.) Choisy;阿拉伯赛德旋花●☆

356418　Seddera bagshawei Rendle;巴格肖赛德旋花●☆

356419　Seddera bracteata Verdc.;具苞赛德旋花●☆

356420　Seddera capensis (E. Mey. ex Choisy) Hallier f.;好望角赛德旋花●☆

356421　Seddera cinerea Hutch. et E. A. Bruce;灰色赛德旋花●☆

356422　Seddera erlangeriana Engl. et Pilg.;厄兰格赛德旋花●☆

356423　Seddera glomerata (Balf. f.) O. Schwartz;团集赛德旋花●☆

356424　Seddera gracilis Chiov. = Seddera hirsuta Hallier f. ●☆

356425　Seddera hallieri Engl. et Pilg.;哈里赛德旋花●☆

356426　Seddera hirsuta Hallier f.;毛赛德旋花●☆

356427　Seddera hirsuta Hallier f. var. glabrescens Verdc.;渐光赛德旋花●☆

356428　Seddera hirsuta Hallier f. var. gracilis (Chiov.) Verdc. = Seddera hirsuta Hallier f. ●☆

356429　Seddera humilis Hallier f. var. erlangeriana (Engl. et Pilg.) Verdc. = Seddera erlangeriana Engl. et Pilg. ●☆

356430　Seddera intermedia Hochst. et Steud.;间型赛德旋花●☆

356431　Seddera latifolia Hochst. et Steud.;宽叶赛德旋花●☆

356432　Seddera latifolia Hochst. et Steud. var. argentea (A. Terracc.) Capua;银宽叶赛德旋花●☆

356433　Seddera latifolia Hochst. et Steud. var. micrantha (Pilg.) Verdc.;小花宽叶赛德旋花●☆

356434　Seddera micrantha Pilg. = Seddera latifolia Hochst. et Steud. var. micrantha (Pilg.) Verdc. ●☆

356435　Seddera microphylla Engl.;小叶赛德旋花●☆

356436　Seddera pedunculata (Balf. f.) Hallier f.;梗花赛德旋花●☆

356437　Seddera saturejoides Chiov. = Seddera latifolia Hochst. et Steud. ●☆

356438　Seddera schizantha Hallier f.;安哥拉赛德旋花●☆

356439　Seddera simmonsii Verdc.;西蒙斯赛德旋花●☆

356440　Seddera somalensis (Vatke) Hallier f. = Seddera arabica (Forssk.) Choisy ●☆

356441　Seddera spinescens Peter ex Hallier f. = Seddera latifolia Hochst. et Steud. ●☆

356442　Seddera suffruticosa (Schinz) Hallier f.;灌木赛德旋花●☆

356443　Seddera suffruticosa (Schinz) Hallier f. var. hirsutissima Hallier f.;毛灌木赛德旋花●☆

356444　Seddera virgata Hochst. et Steud.;条纹赛德旋花●☆

356445　Seddera welwitschii Hallier f. = Seddera suffruticosa (Schinz) Hallier f. ●☆

356446　Seddera welwitschii Hallier f. subsp. tenuisepala Verdc. = Seddera suffruticosa (Schinz) Hallier f. ●☆

356447　Seddera welwitschii Hallier f. var. bakeri Hiern = Seddera suffruticosa (Schinz) Hallier f. ●☆

356448　Sedderopsis Roberty = Seddera Hochst. ●☆

356449　Sedderopsis capensis (E. Mey. ex Choisy) Roberty = Seddera capensis (E. Mey. ex Choisy) Hallier f. ●☆

356450　Sedella Britton et Rose = Parvisedum R. T. Clausen ■☆

356451　Sedella Britton et Rose = Sedum L. ●■

356452　Sedella Britton et Rose(1903);小景天属■☆

356453　Sedella Fourr. = Sedum L. ●■

356454　Sedella pumila (Benth.) Britton et Rose;偃小景天■☆

356455　Sedgwichia Griff. = Altingia Noronha ●

356456　Sedirea Garay et H. R. Sweet(1974);萼脊兰属;Sedirea ■

356457　Sedirea japonica (Linden et Rchb. f.) Garay et H. R. Sweet;萼脊兰(吊兰,风兰,日本指甲兰,仙人指甲兰);Japan Nailorchis, Japan Sedirea,Japanese Fox-taoil Orchis,Japanese Sedirea ■

356458　Sedirea japonica (Rchb. f.) Garay et H. R. Sweet = Sedirea japonica (Linden et Rchb. f.) Garay et H. R. Sweet ■

356459　Sedirea subparishii (Z. H. Tsi) K. I. Chr.;短茎萼脊兰(指甲兰);Shortstem Sedirea ■

356460　Sedopsis (Engl. ex Legrand) Exell et Mendonça = Portulaca L. ■

356461　Sedopsis (Engl.) Exell et Mendonça = Portulaca L. ■

356462　Sedopsis (Legrand) Exell et Mendonça = Portulaca L. ■

356463　Sedopsis carrissoana Exell et Mendonça = Portulaca hereroensis Schinz ■☆

356464　Sedopsis saxifragoides (Welw. ex Oliv.) Exell et Mendonça = Portulaca saxifragoides Welw. ex Oliv. ■☆

356465　Sedopsis sedoides（Welw.）Exell et Mendonça = Portulaca sedoides Welw. ex Oliv. ■☆

356466　Sedum Adans. = Sempervivum L. ■☆

356467　Sedum L.（1753）；景天属（费菜属）；Buddhanailm, Live Long, L, Live-for-ever, Orpine, Sedum, Stone Crop, Stonecrop ●■

356468　Sedum acre L.；苔景天（金苔，金毡景天，锐叶景天，苔千里光，田景天，田千里光，辛辣景天）；Bird's Bread, Biting Stone Crop, Biting Stonecrop, Bullock's Eye, Bullock's Eyes, Common Stonecrop, Common Yellow Stonecrop, Country Pepper, Creeping Charlie, Creeping Jack, Creeping Jenny, Creeping Sailor, Creeping Tom, Crowdy-kit-o'-the-wall, Ginger, Gold Chain, Gold Dust, Golden Carpet, Golden Dust, Golden Stonecrop, Golden Wire, Golden-carpet, Goldmoss, Gold-moss, Goldmoss Sedum, Goldmoss Stonecrop, Gold-moss Stonecrop, Grapes, Hen-and-chickens, Hundreds-and-thousands, Jack-in-the-buttery, Jack-of-the-buttery, Least Houseleek, Little Houseleek, Little Stonecrop, London Pride, Love Entanggle, Love-entangled, Love-in-a-tangle, Money, Mossy Stonecrop, Mousetail, Paper Crop, Pepper Crop, Pickpocket, Pig's Ear, Plenty, Poor Man's Pepper, Pricket, Prick-madam, Queen's Cushion, Rock Plant, Rock-crop, Sengreen, Star-flower, Stone Crop, Stonecrop, Stonehore, Wall Ginger, Wall Grass, Wall Moss, Wall Paper, Wall Pepper, Wall Stonecrop, Wallpepper, Wall-pepper, Wallwort, Yellow Sedum, Yellow Stonecrop ■

356469　Sedum acre L. 'Aureum'；黄色锐叶景天■☆

356470　Sedum acre L. subsp. neglectum（Ten.）Arcang.；忽视苔景天■☆

356471　Sedum acre L. var. atlanticum Batt. et Trab. = Sedum acre L. ■

356472　Sedum acre L. var. morbifugum Chabert = Sedum acre L. ■

356473　Sedum actinocarpum Yamam.；星果佛甲草；Starfruit Stonecrop ■

356474　Sedum adenotrichum Wall. ex Edgew. = Rosularia adenotricha（Wall. ex Edgew.）C. -A. Jansson et Rech. f. ■☆

356475　Sedum adolphii Raym. -Hamet；名月（明月，铭月）；Golden Sedum ■☆

356476　Sedum aetnense Tineo；埃特纳景天■☆

356477　Sedum affine（Schrenk）Raym. -Hamet = Pseudosedum affine（Schrenk）A. Berger ●■

356478　Sedum aggregeatum（Makino）Makino；群集景天■☆

356479　Sedum aizoon L. = Phedimus aizoon（L.）'t Hart ■

356480　Sedum aizoon L. f. angustifolium Franch. = Phedimus aizoon（L.）'t Hart var. yamatutae（Kitag.）H. Ohba, K. T. Fu et B. M. Barthol. ■

356481　Sedum aizoon L. f. glaberrimum（Kitag.）Kitag. = Phedimus aizoon（L.）'t Hart ■

356482　Sedum aizoon L. f. latifolium（Maxim.）Y. C. Zhu = Sedum aizoon L. var. latifolium（Maxim.）H. Ohba et al. ■

356483　Sedum aizoon L. subsp. kamtschaticum（Fisch.）Fröd. = Phedimus kamtschaticum（Fisch. et C. A. Mey.）'t Hart ■

356484　Sedum aizoon L. subsp. middendorffianum（Maxim.）Fröd. = Phedimus middendorffianum（Maxim.）'t Hart ■

356485　Sedum aizoon L. subsp. selskianum（Regel et Maack）Fröd. = Phedimus selskianum（Regel et Maack）'t Hart ■

356486　Sedum aizoon L. var. aizoon = Phedimus aizoon（L.）'t Hart ■

356487　Sedum aizoon L. var. angustifolium（Franch.）Y. C. Zhu = Phedimus aizoon（L.）'t Hart var. yamatutae（Kitag.）H. Ohba, K. T. Fu et B. M. Barthol. ■

356488　Sedum aizoon L. var. angustifolium（Franch.）Y. C. Zhu = Sedum aizoon L. f. angustifolium Franch. ■

356489　Sedum aizoon L. var. austromanshuricum（Nakai et Kitag.）

Kitag. = Sedum aizoon L. f. latifolium（Maxim.）Y. C. Zhu ■

356490　Sedum aizoon L. var. floribundum Nakai = Phedimus aizoon（L.）'t Hart var. floribundus（Nakai）H. Ohba ■☆

356491　Sedum aizoon L. var. floribundum Nakai = Phedimus aizoon（L.）'t Hart ■

356492　Sedum aizoon L. var. glabrifolium（Kitag.）Kitag. = Phedimus aizoon（L.）'t Hart ■

356493　Sedum aizoon L. var. latifolium（Maxim.）H. Ohba et al. = Phedimus aizoon（L.）'t Hart var. latifolium（Maxim.）H. Ohba, K. T. Fu et B. M. Barthol. ■

356494　Sedum aizoon L. var. latifolium Maxim. = Phedimus aizoon（L.）'t Hart var. latifolium（Maxim.）H. Ohba, K. T. Fu et B. M. Barthol. ■

356495　Sedum aizoon L. var. latifolium Maxim. = Sedum aizoon L. f. latifolium（Maxim.）Y. C. Zhu ■

356496　Sedum aizoon L. var. latifolium Maxim. = Sedum aizoon L. ■

356497　Sedum aizoon L. var. scabrum Maxim. = Phedimus aizoon（L.）'t Hart var. scabrum（Maxim.）H. Ohba, K. T. Fu et B. M. Barthol. ■

356498　Sedum aizoon L. var. yamutae Kitag. = Phedimus aizoon（L.）'t Hart var. yamatutae（Kitag.）H. Ohba, K. T. Fu et B. M. Barthol. ■

356499　Sedum albertii Regel = Pseudosedum affine（Schrenk）A. Berger ●■

356500　Sedum albiflorum（Maxim.）Maxim. ex Kom. et Aliss. = Hylotelephium pallescens（Freyn）H. Ohba ■

356501　Sedum alboroseum Baker = Hylotelephium erythrostictum（Miq.）H. Ohba ■

356502　Sedum alboroseum Baker = Sedum erythrostictum Miq. ■

356503　Sedum album L.；白千里光（白景天，土三七，玉米石）；Prick-madam, Small Houseleek, White Lacey, White Stone Crop, White Stonecrop, Worm-grass ■☆

356504　Sedum album L. subsp. gypsicola（Boiss. et Reut.）Maire = Sedum gypsicola Boiss. et Reut. ■☆

356505　Sedum album L. subsp. micranthum（DC.）Syme；小花白千里光■☆

356506　Sedum album L. subsp. teretifolium（Lam.）Syme = Sedum album L. ■☆

356507　Sedum album L. var. clusianum（Guss.）Maire = Sedum gypsicola Boiss. et Reut. ■☆

356508　Sedum album L. var. glanduliferum Ball = Sedum gypsicola Boiss. et Reut. ■☆

356509　Sedum album L. var. micranthum（DC.）Syme = Sedum album L. subsp. micranthum（DC.）Syme ■☆

356510　Sedum album L. var. micranthum DC. = Sedum album L. ■☆

356511　Sedum album L. var. purpureum Maire = Sedum album L. ■☆

356512　Sedum alfredi Hance；东南景天（变叶景天，东南佛甲草，马屎苋，石板菜，石上瓜子菜，石上老鼠耳，台湾景天，台湾千里光，珠芽景天，珠芽千里光）；Alfred Stonecrop ■

356513　Sedum alfredii Hance var. bulbiferum（Makino）Fröd. = Sedum bulbiferum Makino ■

356514　Sedum alfredii Hance var. makinoi（Maxim.）Fröd. = Sedum makinoi Maxim. ■

356515　Sedum algidum（Ledeb.）Fisch. et C. A. Mey. var. tanguticum Maxim. = Rhodiola tangutica（Maxim.）S. H. Fu ■

356516　Sedum algidum Ledeb. var. tangutica Maxim. = Rhodiola algida（Ledeb.）Fisch. et C. A. Mey. var. tangutica（Maxim.）S. H. Fu ■

356517　Sedum algidum Ledeb. var. tanguticum Maxim. = Rhodiola tangutica（Maxim.）S. H. Fu ■

356518 Sedum aliciae（Raym.-Hamet）A. Berger var. komarovii Raym.-Hamet = Kungia aliciae（Raym.-Hamet）K. T. Fu var. komarovii（Raym.-Hamet）K. T. Fu ■

356519 Sedum aliciae（Raym.-Hamet）Raym.-Hamet = Kungia aliciae（Raym.-Hamet）K. T. Fu ■

356520 Sedum aliciae（Raym.-Hamet）Raym.-Hamet = Orostachys aliciae（Raym.-Hamet）H. Ohba ■

356521 Sedum aliciae（Raym.-Hamet）Raym.-Hamet var. komarovii Raym.-Hamet = Orostachys aliciae（Raym.-Hamet）H. Ohba ■

356522 Sedum allantoides Rose;大叶菩提■☆

356523 Sedum almae Fröd. = Hylotelephium almae（Fröd.）K. T. Fu et G. Y. Rao ■

356524 Sedum almae Fröd. = Hylotelephium tatarinowii（Maxim.）H. Ohba var. integrifolium（Palib.）S. H. Fu ■

356525 Sedum alsium Fröd. = Rhodiola alsia（Fröd.）S. H. Fu ■

356526 Sedum altissimum Poir. = Sedum sediforme（Jacq.）Pau ■☆

356527 Sedum ambiguum Praeger = Sinocrassula ambigua（Praeger）A. Berger ■

356528 Sedum amplexicaule DC.;抱茎景天■☆

356529 Sedum amplexicaule DC. subsp. tenuifolium（Sm.）Greuter;细叶抱茎景天■☆

356530 Sedum amplibracteatum K. T. Fu;大苞景天（苞叶景天,灯笼草,活血草,鸡爪七,亮杆草,山胡豆,水生还阳草,一朵云）;Bigbract Stonecrop,Bractleaf Stonecrop ■

356531 Sedum amplibracteatum K. T. Fu = Sedum oligocarpum Fröd. ■

356532 Sedum amplibracteatum K. T. Fu = Sedum oligospermum Maire ■

356533 Sedum amplibracteatum K. T. Fu var. emarginatum（S. H. Fu）S. H. Fu;凹叶大苞景天;Emarginate Bractleaf Stonecrop ■

356534 Sedum amplibracteatum K. T. Fu var. emarginatum（S. H. Fu）S. H. Fu = Sedum oligocarpum Fröd. ■

356535 Sedum anacampseros L.;洋吊金钱;Love-restoring Stonecrop, Reddish Stonecrop ■☆

356536 Sedum andegavense（DC.）Desv.;昂德加维景天■☆

356537 Sedum anglicum Huds.;英国景天;English Stonecrop, Prick-madam,Stonecrop ■☆

356538 Sedum anglicum Huds. subsp. melanantherum（DC.）Maire = Sedum melanantherum DC. ■☆

356539 Sedum angustifolium Z. B. Hu et X. L. Huang = Sedum sarmentosum Bunge ■

356540 Sedum angustipetalum Fröd. = Sedum dielsii Raym.-Hamet ■

356541 Sedum angustum Maxim. = Hylotelephium angustum（Maxim.）H. Ohba ■

356542 Sedum anhuiense S. H. Fu et X. W. Wang;安徽景天;Anhui Stonecrop ■

356543 Sedum anhuiense S. H. Fu et X. W. Wang = Sedum lineare Thunb. ■

356544 Sedum annuum L.;一年景天;Annual Stonecrop ■☆

356545 Sedum anopetalum DC. = Sedum ochroleucum Chaix ■☆

356546 Sedum anthoxanthum Fröd. = Sedum perrotii Raym.-Hamet ■

356547 Sedum aporonticum Fröd. = Rhodiola aporontica（Fröd.）S. H. Fu ■

356548 Sedum aporonticum Fröd. = Rhodiola atuntsuensis（Praeger）S. H. Fu ■

356549 Sedum arisanense Yamam. = Sedum erythrospermum Hayata ■

356550 Sedum asiaticum Aitch. = Rhodiola heterodonta（Hook. f. et Thomson）Boriss. ■

356551 Sedum asiaticum Aitch. = Sedum heterodontum Hook. f. et Thomson ■

356552 Sedum asiaticum Spreng.;亚洲景天■☆

356553 Sedum atlanticum（Ball）Maire = Monanthes atlantica Ball ■☆

356554 Sedum atlanticum（Ball）Maire var. fuscum Emb. = Monanthes atlantica Ball ■☆

356555 Sedum atlanticum（Ball）Maire var. luteum Emb. = Monanthes atlantica Ball ■☆

356556 Sedum atratum L.;暗景天■☆

356557 Sedum atropurpureum Turcz. = Rhodiola rosea L. ■

356558 Sedum atsaense Fröd. = Rhodiola atsaensis（Fröd.）H. Ohba ■

356559 Sedum atuntsuense Praeger = Rhodiola atuntsuensis（Praeger）S. H. Fu ■

356560 Sedum austromanshuricum Nakai et Kitag. = Phedimus aizoon（L.）'t Hart var. latifolium（Maxim.）H. Ohba,K. T. Fu et B. M. Barthol. ■

356561 Sedum azureum Royle = Hylotelephium ewersii（Ledeb.）H. Ohba ■

356562 Sedum baileyi Praeger;拜氏景天(对叶景天);Bailey Stonecrop ■

356563 Sedum baleensis M. G. Gilbert;巴莱景天■☆

356564 Sedum balfouri Raym.-Hamet = Ohbaea balfourii（Raym.-Hamet）Byalt et I. V. Sokolova ■

356565 Sedum banlanense W. Limpr. = Ohbaea balfourii（Raym.-Hamet）Byalt et I. V. Sokolova ■

356566 Sedum barbeyi Raym.-Hamet;离瓣景天;Barbey Stonecrop ■☆

356567 Sedum barcense Maire et Weiller = Sedum littoreum Guss. ■☆

356568 Sedum barnesianum Praeger = Rhodiola humilis（Hook. f. et Thomson）S. H. Fu ■

356569 Sedum beauverdii Raym.-Hamet;短尖景天;Beauverd Stonecrop ■

356570 Sedum bergeri Raym.-Hamet;长丝景天;Berger Stonecrop, Longsilk Stonecrop ■

356571 Sedum bhutanense Praeger = Rhodiola bupleuroides（Wall. ex Hook. f. et Thomson）S. H. Fu ■

356572 Sedum blepharophyllum Fröd.;缝叶景天;Eyelashleaf Stonecrop,Tasselleaf Stonecrop ■

356573 Sedum bodinieri H. Lév. et Vaniot = Sedum stellariifolium Franch. ■

356574 Sedum boehmeri（Makino）Makino = Orostachys malacophylla（Pall.）Fisch. var. boehmeri（Makino）H. Hara ■☆

356575 Sedum boninense Yamam. ex Tuyama = Sedum japonicum Siebold ex Miq. subsp. boninense（Yamam. ex Tuyama）H. Ohba ■☆

356576 Sedum bonnafousii Raym.-Hamet. = Hylotelephium bonnafousii（Raym.-Hamet）H. Ohba ■

356577 Sedum bonnieri Raym.-Hamet;城口景天;Chengkou Stonecrop ■

356578 Sedum brachyrhinchum Yamam. = Sedum erythrospermum Hayata ■

356579 Sedum brachystylum Fröd. = Rhodiola coccinea（Royle）Boriss. subsp. scabrida（Franch.）H. Ohba ■

356580 Sedum bracteatum Diels = Sedum oligocarpum Fröd. ■

356581 Sedum bracteatum Diels = Sedum oligospermum Maire ■

356582 Sedum bracteatum Diels var. emarginatum S. H. Fu = Sedum amplibracteatum K. T. Fu var. emarginatum（S. H. Fu）S. H. Fu ■

356583 Sedum bracteatum Diels var. emarginatum S. H. Fu = Sedum oligocarpum Fröd. ■

356584 Sedum bracteatum Diels var. emarginatum S. H. Fu = Sedum oligospermum Maire ■

356585 Sedum brevifolium DC.;短叶景天■☆

356586 Sedum brevifolium DC. var. induratum Coss. = Sedum

brevifolium DC. ■☆

356587　Sedum brevipetiolatum Fröd. = Rhodiola atuntsuensis（Praeger）S. H. Fu ■

356588　Sedum brevipetiolatum Fröd. = Rhodiola brevipetiolata（Fröd.）S. H. Fu ■

356589　Sedum bucharicum Boriss. ;布哈尔景天■☆

356590　Sedum bulbiferum Makino;珠芽景天(零余子佛甲草,零余子景天,马屎花,马屎花,水三七,小箭草,珠芽半枝,珠芽佛甲草,珠芽石板菜);Bulbiferous Stonecrop ■

356591　Sedum bupleuroides Wall. = Rhodiola bupleuroides（Wall. ex Hook. f. et Thomson）S. H. Fu ■

356592　Sedum bupleuroides Wall. ex Hook. f. et Thomson = Rhodiola bupleuroides（Wall. ex Hook. f. et Thomson）S. H. Fu ■

356593　Sedum bupleuroides Wall. ex Hook. f. et Thomson var. discolor（Franch.）Fröd. = Rhodiola discolor（Franch.）S. H. Fu ■

356594　Sedum bupleuroides Wall. ex Hook. f. et Thomson var. purpureoviride（Praeger）Fröd. = Rhodiola purpureoviridis（Praeger）S. H. Fu ■

356595　Sedum bupleuroides Wall. ex Hook. f. et Thomson var. rotundatum（Hemsl.）Fröd. = Rhodiola crenulata（Hook. f. et Thomson）H. Ohba ■

356596　Sedum caerulans H. Lév. et Vaniot = Rhodiola kirilowii Regel ex Maxim. ■

356597　Sedum caerulans H. Lév. et Vaniot = Rhodiola rosea L. ■

356598　Sedum caeruleum L. ;蓝花景天■☆

356599　Sedum caeruleum L. var. pusillum Maire = Sedum caeruleum L. ■☆

356600　Sedum caeruleum L. var. versicolor Raym. -Hamet = Sedum versicolor（Raym. -Hamet）Maire ■☆

356601　Sedum caespitosum（Cav.）DC. ;丛生景天■☆

356602　Sedum caffrum Kuntze = Crassula sarcocaulis Eckl. et Zeyh. subsp. rupicola Toelken ■☆

356603　Sedum callianthum H. Ohba = Rhodiola calliantha（H. Ohba）H. Ohba ■

356604　Sedum campestre（Eckl. et Zeyh.）Kuntze = Crassula campestris（Eckl. et Zeyh.）Endl. ex Walp. ●☆

356605　Sedum canescens（Schult.）Kuntze var. caulescens Kuntze = Crassula rogersii Schönland ■☆

356606　Sedum caucasicum（Grossh.）Boriss. ;高加索景天■☆

356607　Sedum cauticola Praeger;岩缝景天■☆

356608　Sedum cauticola Praeger = Hylotelephium cauticola（Praeger）H. Ohba ■☆

356609　Sedum cauticola Praeger f. montanum H. Hara = Hylotelephium cauticola（Praeger）H. Ohba ■☆

356610　Sedum cavaleriei H. Lév. = Sinocrassula indica（Decne.）A. Berger ■

356611　Sedum cavaleriense H. Lév. = Sinocrassula indica（Decne.）A. Berger ■

356612　Sedum celatum Fröd. ;隐匿景天(小景天,小千里光,隐匿山景天);Hidden Stonecrop ■

356613　Sedum celatum Fröd. f. calcaratum K. T. Fu;距萼隐匿景天■

356614　Sedum celatum Fröd. f. calcaratum K. T. Fu = Sedum celatum Fröd. ■

356615　Sedum celiae Raym. -Hamet;镰座景天(莲座山景天);Celia Stonecrop ■

356616　Sedum chanetii H. Lév. = Orostachys chanetii（H. Lév.）A. Berger ■

356617　Sedum chauveaudii Raym. -Hamet;轮叶景天(轮叶山景天);Threeleaf Stonecrop ■

356618　Sedum chauveaudii Raym. -Hamet var. margaritae（Raym. -Hamet）Fröd. ;互生叶景天■

356619　Sedum chingtungense K. T. Fu;景东景天;Chingtung Stonecrop ,Jingdong Stonecrop ■

356620　Sedum chrysanthemifolium H. Lév. = Rhodiola chrysanthemifolia（H. Lév.）S. H. Fu ■

356621　Sedum chrysastrum Hance = Sedum japonicum Siebold ex Miq. ■

356622　Sedum chuhsingense K. T. Fu;楚雄景天;Chuxiong Stonecrop ■

356623　Sedum chumbicum Prain ex Raym. -Hamet = Rhodiola smithii（Raym. -Hamet）S. H. Fu ■

356624　Sedum ciliatum（L.）Kuntze = Crassula ciliata L. ■☆

356625　Sedum coccineum Royle = Rhodiola coccinea（Royle）Boriss. ■

356626　Sedum coccineum Royle = Rhodiola quadrifida（Pall.）Fisch. et C. A. Mey. ■

356627　Sedum cogmansense Kuntze = Crassula subaphylla（Eckl. et Zeyh.）Harv. ■☆

356628　Sedum compressum Rose;叠叶景天(红化妆)■☆

356629　Sedum concarpum Fröd. ;合果景天(合果佛甲草,湖北合果景天);Connectedfruit Stonecrop ■

356630　Sedum concarpum Fröd. var. hupehense S. H. Fu = Sedum concarpum Fröd. ■

356631　Sedum concinnum Praeger = Rhodiola atuntsuensis（Praeger）S. H. Fu ■

356632　Sedum concinnum Praeger = Rhodiola concinna（Praeger）S. H. Fu ■

356633　Sedum confusum Hemsl. ;混乱景天;Lesser Mexican Stonecrop, Stonecrop ■☆

356634　Sedum cooperi Praeger = Rhodiola bupleuroides（Wall. ex Hook. f. et Thomson）S. H. Fu ■

356635　Sedum correptum Fröd. ;单花景天;Singleflower Stonecrop ■

356636　Sedum corybulosum（Link et Otto）Kuntze = Crassula capitella Thunb. subsp. thyrsiflora（Thunb.）Toelken ■☆

356637　Sedum corymbosum Grossh. ;伞序景天■☆

356638　Sedum costantinii Raym. -Hamet;三裂距景天(山裂距景天);Costantin Stonecrop ■

356639　Sedum crassiflorum Kuntze = Crassula vaginata Eckl. et Zeyh. ■☆

356640　Sedum crassipes Hook. f. et Thomson var. cholaense Praeger = Rhodiola wallichiana（Hook.）S. H. Fu var. cholaensis（Praeger）S. H. Fu ■

356641　Sedum crassipes Wall. = Rhodiola wallichiana（Hook.）S. H. Fu ■

356642　Sedum crassipes Wall. = Sedum asiaticum Spreng. ■☆

356643　Sedum crassipes Wall. ex Hook. f. et Thomson = Rhodiola wallichiana（Hook.）S. H. Fu ■

356644　Sedum crassipes Wall. ex Hook. f. et Thomson var. cretinii（Raym. -Hamet）Fröd. = Rhodiola cretinii（Raym. -Hamet）H. Ohba ■

356645　Sedum crassipes Wall. ex Hook. f. et Thomson var. stephanii（Cham.）Fröd. = Rhodiola stephanii（Cham.）Trautv. et C. A. Mey. ■

356646　Sedum crassipes Wall. , Hook. f. et Thomson var. cholaensis Praeger = Rhodiola wallichiana（Hook.）S. H. Fu var. cholaensis（Praeger）S. H. Fu ■

356647　Sedum crassipes Wall. , Hook. f. et Thomson var. cretinii（Raym. -Hamet）Fröd. = Rhodiola cretinii（Raym. -Hamet）H. Ohba ■

356648　Sedum crassipes Wall. , Hook. f. et Thomson var. stephanii（Cham.）Fröd. = Rhodiola stephanii（Cham.）Trautv. et C. A. Mey. ■

356649　Sedum cremnophila（Rose）R. T. Clausen；日本悬崖景天■☆

356650　Sedum crenulatum Hook. f. et Thomson = Rhodiola crenulata（Hook. f. et Thomson）H. Ohba ■

356651　Sedum cretense Maire = Sedum cyrenaicum Brullo et Furnari ■☆

356652　Sedum cretinii Raym. -Hamet = Rhodiola cretinii（Raym. -Hamet）H. Ohba ■

356653　Sedum cryptomerioides Bartlett et Yamam. ；杉叶佛甲草■

356654　Sedum cryptomerioides Bartlett et Yamam. = Sedum morrisonense Hayata ■

356655　Sedum cyaneum Rudolph；蓝色景天■☆

356656　Sedum cymosum（P. J. Bergius）Kuntze = Crassula cymosa P. J. Bergius ■☆

356657　Sedum cyrenaicum Brullo et Furnari；昔兰尼景天■☆

356658　Sedum daigremontianum Raym. -Hamet；啮瓣景天；Erosepetalous Stonecrop ■

356659　Sedum daigremontianum Raym. -Hamet var. macrosepalum Fröd. ；大萼啮瓣景天■

356660　Sedum dasyphyllum L. ；粗毛叶景天（姬星美人）；Thick-leaf Stonecrop, Thick-leaved Stone Crop, Thick-leaved Stonecrop ■☆

356661　Sedum dasyphyllum L. subsp. oblongifolium（Ball）Maire；矩圆叶景天■☆

356662　Sedum dasyphyllum L. var. dyris Maire = Sedum dasyphyllum L. subsp. oblongifolium（Ball）Maire ■☆

356663　Sedum dasyphyllum L. var. glanduliferum（Guss.）Moris = Sedum dasyphyllum L. ■☆

356664　Sedum dasyphyllum L. var. glutinosum Maire = Sedum dasyphyllum L. subsp. oblongifolium（Ball）Maire ■☆

356665　Sedum dasyphyllum L. var. mesatlanticum Litard. et Maire = Sedum dasyphyllum L. ■☆

356666　Sedum dasyphyllum L. var. oblongifolium Ball = Sedum dasyphyllum L. subsp. oblongifolium（Ball）Maire ■☆

356667　Sedum dasyphyllum L. var. pulligerum（Pomel）Batt. = Sedum dasyphyllum L. ■☆

356668　Sedum dasyphyllum L. var. rifanum Maire = Sedum dasyphyllum L. subsp. oblongifolium（Ball）Maire ■☆

356669　Sedum definitum H. Lév. = Hylotelephium tatarinowii（Maxim.）H. Ohba ■

356670　Sedum dendroideum DC. = Sedum praealtum A. DC. ■☆

356671　Sedum didymocalyx Fröd. ；双萼景天；Twocalyx Stonecrop ■

356672　Sedum dielsianum W. Limpr. = Rhodiola chrysanthemifolia（H. Lév. ）S. H. Fu ■

356673　Sedum dielsianum W. Limpr. = Rhodiola dielsiana（W. Limpr.）S. H. Fu ■

356674　Sedum dielsii Raym. -Hamet；乳瓣景天；Diels Stonecrop ■

356675　Sedum dimorphophyllum K. T. Fu et G. Y. Rao；二型叶景天；Dimorphleaf Stonecrop ■

356676　Sedum discolor Franch. = Rhodiola discolor（Franch.）S. H. Fu ■

356677　Sedum dolosum K. T. Fu；惑景天；Crafty Stonecrop ■

356678　Sedum dolosum K. T. Fu = Sedum multicaule Wall. ■

356679　Sedum dongzhiense D. Q. Wang et Y. L. Shi；东至景天；Dongzhi Stonecrop ■

356680　Sedum doratocarpum Fröd. = Rhodiola alsia（Fröd.）S. H. Fu ■

356681　Sedum doratocarpum Fröd. = Rhodiola fastigiata（Hook. f. et Thomson）S. H. Fu ■

356682　Sedum dregeanum（Harv.）Kuntze var. adscendens Kuntze = Crassula setulosa Harv. ■☆

356683　Sedum dregeanum（Harv.）Kuntze var. erectum Kuntze = Crassula setulosa Harv. var. rubra（N. E. Br.）G. D. Rowley = ■☆

356684　Sedum droseum var. atropurpureum（Turcz.）Praeger = Rhodiola rosea L. ■

356685　Sedum drymarioides Hance；大叶火焰草（光板猫叶草，光板毛叶草，荷莲豆景天，火焰草，龙鳞草，毛佛甲，毛佛甲草，毛舌辣草，卧儿菜）；Bigleaf Stonecrop ■

356686　Sedum drymarioides Hance = Sedum stellariifolium Franch. ■

356687　Sedum drymarioides Hance var. genuinum Hamer = Sedum drymarioides Hance ■

356688　Sedum drymarioides Hance var. stellariifolium（Franch.）Raym. -Hamet = Sedum stellariifolium Franch. ■

356689　Sedum drymarioides Hance var. toyamae H. Hara = Sedum drymarioides Hance ■

356690　Sedum dubium Paulsen = Rhodiola gelida Schrenk ■

356691　Sedum ducis-aprutii Cortesi = Sedum ruwenzoriense Baker f. ■☆

356692　Sedum dugneyi Raym. -Hamet；薛茎景天（薛状景天）；Duguey Stonecrop, Mossstem Stonecrop ■

356693　Sedum dumulosum Franch. = Rhodiola dumulosa（Franch.）S. H. Fu ■

356694　Sedum dumulosum Franch. var. rendleri（Raym. -Hamet）Fröd. = Rhodiola dumulosa（Franch.）S. H. Fu ■

356695　Sedum durisii Raym. -Hamet = Rosularia alpestris（Kar. et Kir.）Boriss. ■

356696　Sedum elatinoides Franch. ；细叶景天（半边莲，灯台草，沟繁缕景天，细叶山飘风，小鹅儿肠，崖松，疣果景天）；Slenderleaf Stonecrop ■

356697　Sedum elegans Lej. = Sedum forsterianum Sm. ■☆

356698　Sedum ellacombeanum Praeger = Phedimus aizoon（L.）' t Hart var. latifolium（Maxim.）H. Ohba, K. T. Fu et B. M. Barthol. ■

356699　Sedum elongatum Ledeb. = Rhodiola rosea L. ■

356700　Sedum elongatum Ledeb. = Rhodiola sachalinensis Boriss. ■

356701　Sedum elongatum Wall. = Rhodiola bupleuroides（Wall. ex Hook. f. et Thomson）S. H. Fu ■

356702　Sedum emarginatum Migo；凹叶景天（凹叶佛甲草，凹叶佛甲花，半边莲，打不死，豆瓣菜，豆瓣草，佛甲草，狗牙瓣，旱半支，酱瓣半支，酱瓣草，酱瓣豆草，九月寒，六月雪，马牙半支，马牙半枝莲，马牙苋，山马齿苋，石板菜，石板还阳，石马齿苋，石雀还阳，水辣椒，铁梗半支，铁梗山半支，仙人指甲）；Emarginate Stonecrop ■

356703　Sedum engleri Raym. -Hamet；粗壮景天（长圆佛甲草，滇边景天）；Engler Stonecrop ■

356704　Sedum engleri Raym. -Hamet var. dentatum S. H. Fu；远齿粗壮景天（齿圆佛甲草）；Dentate Engler Stonecrop ■

356705　Sedum engleri Raym. -Hamet var. forrestii Raym. -Hamet = Sedum engleri Raym. -Hamet ■

356706　Sedum epidendrum Hochst. ex A. Rich. ；柱瓣兰景天■☆

356707　Sedum erici-magnusii Fröd. ；大炮山景天（景天）；Dapaoshan Stonecrop ■

356708　Sedum erici-magnusii Fröd. subsp. chilianense K. T. Fu；祁连山景天；Qilianshan Stonecrop ■

356709　Sedum erici-magnusii Fröd. var. subalpinum Fröd. ；亚高山景天；Subalpine Dapaoshan Stonecrop ■

356710　Sedum erici-magnusii Fröd. var. subalpinum Fröd. = Sedum erici-magnusii Fröd. ■

356711　Sedum erubescens（Maxim.）Ohwi = Orostachys cartilagineus

Boriss. ■

356712　Sedum erubescens（Maxim.）Ohwi ＝ Orostachys spinosa（L.）Sweet ■

356713　Sedum erubescens（Maxim.）Ohwi var. japonicum（Maxim.）Ohwi ＝ Orostachys japonica（Maxim.）A. Berger ■

356714　Sedum erubescens（Maxim.）Ohwi var. polycephalum（Makino）Ohwi ＝ Orostachys japonica（Maxim.）A. Berger f. polycephala（Makino）H. Ohba ■☆

356715　Sedum erythrospermum Hayata；红籽佛甲草（红子佛甲草）；Redseed Stonecrop ■

356716　Sedum erythrostictum Miq. ＝ Hylotelephium erythrostictum（Miq.）H. Ohba ■

356717　Sedum erythrostictum sensu Mast. ＝ Hylotelephium erythrostictum（Miq.）H. Ohba ■

356718　Sedum esquirolii H. Lév. ＝ Sedum stellariifolium Franch. ■

356719　Sedum ettuense Tomida ＝ Hylotelephium sieboldii（Sweet ex Hook.）H. Ohba var. ettyuense（Tomida）H. Ohba ■☆

356720　Sedum eupatorioides（Kom.）Kom. ＝ Hylotelephium pallescens（Freyn）H. Ohba ■

356721　Sedum eupatorioides Kom. ＝ Hylotelephium pallescens（Freyn）H. Ohba ■

356722　Sedum eurycarpum Fröd. ＝ Rhodiola eurycarpa（Fröd.）S. H. Fu ■

356723　Sedum euryphyllum Fröd. ＝ Rhodiola crenulata（Hook. f. et Thomson）H. Ohba ■

356724　Sedum ewersii Ledeb.；尤尔斯景天■☆

356725　Sedum ewersii Ledeb. ＝ Hylotelephium ewersii（Ledeb.）H. Ohba ■

356726　Sedum expansum（Dryand.）Kuntze ＝ Crassula expansa Dryand. ■☆

356727　Sedum fabaria Koch var. mongolicum Franch. ＝ Hylotelephium mongolicum（Franch.）S. H. Fu ■

356728　Sedum farinosum Lowe；粉景天■☆

356729　Sedum farreri Raym. -Hamet ＝ Rhodiola dumulosa（Franch.）S. H. Fu ■

356730　Sedum farreri W. W. Sm. ＝ Rhodiola dumulosa（Franch.）S. H. Fu ■

356731　Sedum fastigiatum Hook. f. et Thomson ＝ Rhodiola fastigiata（Hook. f. et Thomson）S. H. Fu ■

356732　Sedum feddei Raym. -Hamet；折多景天；Fedde Stonecrop ■

356733　Sedum fedtschenkoi Raym. -Hamet；尖叶景天；Fedschenko Stonecrop, Sharpleaf Stonecrop, Little Stonecrop ■

356734　Sedum fenzelii Fröd. ＝ Rhodiola angusta Nakai ■

356735　Sedum filipes Hemsl. ＝ Sedum major（Hemsl.）Migo ■

356736　Sedum filipes Hemsl. ex Forbes et Hemsl.；小山飘风（豆瓣还阳）；Little Stonecrop ■

356737　Sedum filipes Hemsl. var. major Hemsl. ＝ Sedum major（Hemsl.）Migo ■

356738　Sedum fimbriatum（Turcz.）Franch. ＝ Orostachys fimbriata（Turcz.）A. Berger ■

356739　Sedum fimbriatum（Turcz.）Franch. var. chanetii（H. Lév.）Fröd. ＝ Orostachys chanetii（H. Lév.）A. Berger ■

356740　Sedum fimbriatum（Turcz.）Franch. var. chanetii H. Lév. ＝ Orostachys chanetii（H. Lév.）A. Berger ■

356741　Sedum fimbriatum（Turcz.）Franch. var. genuinum Fröd. ＝ Orostachys fimbriata（Turcz.）A. Berger ■

356742　Sedum fimbriatum（Turcz.）Franch. var. ramosissimum

（Maxim.）Fuckel ＝ Orostachys fimbriata（Turcz.）A. Berger ■

356743　Sedum fischeri Raym. -Hamet；小景天；Fischer Stonecrop ■

356744　Sedum flavum（L.）Kuntze ＝ Crassula flava L. ●☆

356745　Sedum flavum（L.）Kuntze var. brevifolium Kuntze ＝ Crassula flava L. ●☆

356746　Sedum flavum（L.）Kuntze var. lanceolatum Kuntze ＝ Crassula vaginata Eckl. et Zeyh. ■☆

356747　Sedum flavum（L.）Kuntze var. lorifolium Kuntze ＝ Crassula flava L. ●☆

356748　Sedum flavum（L.）Kuntze var. retroflexifolium Kuntze ＝ Crassula vaginata Eckl. et Zeyh. ■☆

356749　Sedum flavum（L.）Kuntze var. subulatum Kuntze ＝ Crassula vaginata Eckl. et Zeyh. ■☆

356750　Sedum floriferum Praeger ＝ Phedimus floriferum（Praeger）’t Hart ■

356751　Sedum formosanum N. E. Br.；台湾佛甲草■

356752　Sedum formosanum N. E. Br. ＝ Sedum alfredii Hance ■

356753　Sedum forrestii Raym. -Hamet；川滇景天（甘肃景天）；Chuan-Dian Stonecrop, Forrest Stonecrop ■

356754　Sedum forrestii Raym. -Hamet ＝ Rhodiola forrestii（Raym. -Hamet）S. H. Fu ■

356755　Sedum forsterianum Phillips ＝ Sedum forsterianum Sm. ■☆

356756　Sedum forsterianum Sm.；福氏景天；Rock Stonecrop, Small Welsh Stonecrop ■☆

356757　Sedum fostigiatum Hook. f. et Thomson ＝ Rhodiola fastigiata（Hook. f. et Thomson）S. H. Fu ■

356758　Sedum franchetii Grande；细叶山景天（细叶景天）；Franchet Stonecrop, Franchet Wild Stonecrop ■

356759　Sedum frutescens Rose；灌木状景天■☆

356760　Sedum fui Rowley；宽叶景天；Broadleaf Stonecrop, Fu Stonecrop, Fu's Stonecrop ■

356761　Sedum fui Rowley var. longisepalum（K. T. Fu）S. H. Fu；长萼宽叶景天；Longcalyx Stonecrop, Longsepal Stonecrop ■

356762　Sedum fusiforme Lowe；梭景天■☆

356763　Sedum gagei Raym. -Hamet；锡金景天；Sikkim Stonecrop ■

356764　Sedum gattefossei Batt. et Jahand.；加特福塞景天■☆

356765　Sedum gelidum（Schrenk）Kar. et Kir. ＝ Rhodiola gelida Schrenk ■

356766　Sedum giajai Raym. -Hamet；柔毛景天；Pubescent Stonecrop ■

356767　Sedum glaciale Franch. ＝ Sedum sinoglaciale K. T. Fu ■

356768　Sedum glaebosum Fröd.；道孚景天；Daofu Stonecrop, Land Stonecrop ■

356769　Sedum glanduliferum Guss. ＝ Sedum dasyphyllum L. ■☆

356770　Sedum glaucum Sm.；印第安景天；Indian Fog ■☆

356771　Sedum glomerifolium M. G. Gilbert；团叶景天■☆

356772　Sedum gorisii Raym. -Hamet；格林景天（高原景天）；Goris Stonecrop ■

356773　Sedum gorisii Raym. -Hamet ＝ Rhodiola bupleuroides（Wall. ex Hook. f. et Thomson）S. H. Fu ■

356774　Sedum gracile C. A. Mey.；纤细景天■☆

356775　Sedum grammophyllum Fröd.；禾叶景天；Grassleaf Stonecrop ■

356776　Sedum griseum Praeger；灰叶景天■☆

356777　Sedum gypsicola Boiss. et Reut.；钙景天■☆

356778　Sedum hakonense Makino；本州景天■

356779　Sedum hametianum H. Lév. ＝ Phedimus stevenianum（Rouy et Camus）’t Hart ■

356780　Sedum hametianum H. Lév. ＝ Sedum stevenianum Rouy et

Camus ■

356781 Sedum hangzhouense K. T. Fu et G. Y. Rao；杭州景天；
Hangzhou Stonecrop ■

356782 Sedum heckelii Raym. -Hamet；巴塘景天；Batang Stonecrop,
Heckel Stonecrop ■

356783 Sedum hengduanense K. T. Fu；横断山景天；Hengduanshan
Stonecrop ■

356784 Sedum henrici-robertii Raym. -Hamet；山岭景天；Montane
Stonecrop ■

356785 Sedum henryi Diels ＝Rhodiola henryi（Diels）S. H. Fu ■

356786 Sedum henryi Diels ＝Rhodiola yunnanensis（Franch.）S. H.
Pu ■

356787 Sedum heptapetalum Poir. ＝Sedum caeruleum L. ■☆

356788 Sedum heterodontum Hook. f. et Thomson ＝Rhodiola
heterodonta（Hook. f. et Thomson）Boriss. ■

356789 Sedum hidakanum Tatew. ex S. Watan. ＝Hylotelephium
pluricaule（Kudo）H. Ohba ■☆

356790 Sedum himalense D. Don ＝Rhodiola himalensis（D. Don）S.
H. Fu ■

356791 Sedum himalense D. Don subsp. taohoense（S. H. Fu）Kozhevn.
＝Rhodiola himalensis（D. Don）S. H. Fu ■

356792 Sedum himalense subsp. taohoense（S. H. Fu）Kozhevn. ＝
Rhodiola himalensis（D. Don）S. H. Fu var. taohoensis（S. H. Fu）
H. Ohba ■

356793 Sedum hirsutum All. ；粗毛景天■☆

356794 Sedum hirsutum All. subsp. baeticum Rouy；伯蒂卡景天■☆

356795 Sedum hirsutum All. subsp. wilczekianum（Font Quer）Maire ＝
Sedum wilczekianum Font Quer ■☆

356796 Sedum hirsutum All. var. baeticum（Rouy）Willk. ＝Sedum
hirsutum All. subsp. baeticum Rouy ■☆

356797 Sedum hirsutum All. var. gattefossei Maire, Weiller et Wilczek
＝Sedum hirsutum All. ■☆

356798 Sedum hirsutum All. var. jahandiezii Maire ＝Sedum hirsutum
All. ■☆

356799 Sedum hirsutum All. var. maroccanum（Font Quer）Jahand. et
Maire ＝Sedum hirsutum All. ■☆

356800 Sedum hirsutum All. var. tazzekanum Sauvage ＝Sedum hirsutum
All. subsp. baeticum Rouy ■☆

356801 Sedum hirsutum All. var. thermarum Maire, Weiller et Wilczek
＝Sedum hirsutum All. ■☆

356802 Sedum hispanicum L. ；绿玉景天（矾小松，西班牙景天）；
Spanish Stone Crop, Spanish Stonecrop ■

356803 Sedum hispidum Desf. ＝Sedum pubescens Vahl ■☆

356804 Sedum hobsonii Prain ex Raym. -Hamet ＝Rhodiola hobsonii
（Prain ex Raym. -Hamet）S. H. Fu ■

356805 Sedum horridum Praeger ＝Rhodiola nobilis（Franch.）S. H. Fu ■

356806 Sedum hsinganicum Y. C. Chu ex S. H. Fu et Y. H. Huang ＝
Phedimus hsinganicum（Y. C. Chu ex S. H. Fu et Y. H. Huang）H.
Ohba, K. T. Fu et B. M. Barthol. ■

356807 Sedum humilis Hook. f. et Thomson ＝Rhodiola humilis（Hook.
f. et Thomson）S. H. Fu ■

356808 Sedum hybridum L. ＝Phedimus hybridum（L.）'t Hart ■

356809 Sedum imbricatum Walp. ＝Rhodiola imbricata Edgew. ■☆

356810 Sedum indicum（Decne.）Raym. -Hamet ＝Sinocrassula indica
（Decne.）A. Berger ■

356811 Sedum indicum（Decne.）Raym. -Hamet var. ambiguum
（Praeger）Raym. -Hamet ＝Sinocrassula ambigua（Praeger）A.

Berger ■

356812 Sedum indicum（Decne.）Raym. -Hamet var. densirosulatum
Praeger ＝Sinocrassula densirosulata（Praeger）A. Berger ■ ■

356813 Sedum indicum（Decne.）Raym. -Hamet var. forrestii Raym. -
Hamet ＝Sinocrassula indica（Decne.）A. Berger var. forrestii
（Raym. -Hamet）A. Berger ■

356814 Sedum indicum（Decne.）Raym. -Hamet var. longistylum
（Praeger）Fröd. ＝Sinocrassula longistyla（Praeger）S. H. Fu ■

356815 Sedum indicum（Decne.）Raym. -Hamet var. luteorubrum
Praeger ＝Sinocrassula indica（Decne.）A. Berger var. luteorubra
（Praeger）S. H. Fu ■

356816 Sedum indicum（Decne.）Raym. -Hamet var. obtusifolium
Fröd. ＝Sinocrassula indica（Decne.）A. Berger var. obtusifolia
（Fröd.）S. H. Fu ■

356817 Sedum indicum（Decne.）Raym. -Hamet var. serratum Raym. -
Hamet ＝Sinocrassula indica（Decne.）A. Berger var. serrata
（Raym. -Hamet）S. H. Fu ■

356818 Sedum indicum（Decne.）Raym. -Hamet var. yunnanense
（Franch.）Raym. -Hamet ＝Sinocrassula yunnanensis（Franch.）
A. Berger ■

356819 Sedum indicum Raym. -Hamet var. ambiguum（Praeger）
Raym. -Hamet ＝Sinocrassula ambigua（Praeger）A. Berger ■

356820 Sedum indicum Raym. -Hamet var. densirosulatum Praeger ＝
Sinocrassula densirosulata（Praeger）A. Berger ■

356821 Sedum indicum Raym. -Hamet var. forrestii Raym. -Hamet ＝
Sinocrassula indica（Decne.）A. Berger var. forrestii（Raym. -
Hamet）A. Berger ■

356822 Sedum indicum Raym. -Hamet var. genuinum Raym. -Hamet ＝
Sinocrassula indica（Decne.）A. Berger ■

356823 Sedum indicum Raym. -Hamet var. luteorubrum Praeger ＝
Sinocrassula indica（Decne.）A. Berger var. luteorubra（Praeger）
S. H. Fu ■

356824 Sedum indicum Raym. -Hamet var. obtusifolium Fröd. ＝
Sinocrassula indica（Decne.）A. Berger var. obtusifolia（Fröd.）S.
H. Fu ■

356825 Sedum indicum Raym. -Hamet var. serratum Raym. -Hamet ＝
Sinocrassula indica（Decne.）A. Berger var. serrata（Raym. -
Hamet）S. H. Fu ■

356826 Sedum indicum Raym. -Hamet var. silvaticum Fröd. ＝
Sinocrassula indica（Decne.）A. Berger ■

356827 Sedum indicum Raym. -Hamet var. yunnanense（Franch.）
Raym. -Hamet ＝Sinocrassula yunnanensis（Franch.）A. Berger ■

356828 Sedum involueratum M. Bieb. ；总苞景天■☆

356829 Sedum ishidae Miyabe et Kudo ＝Rhodiola ishidae（Miyabe et
Kudo）H. Hara ■☆

356830 Sedum iwarenge（Makino）Makino ＝Orostachys malacophylla
（Pall.）Fisch. var. iwarenge（Makino）H. Ohba ■☆

356831 Sedum iwarenge（Makino）Makino var. aggregeatum（Makino）
Ohwi ＝Orostachys malacophylla（Pall.）Fisch. var. aggregata
（Makino）H. Ohba ■☆

356832 Sedum iwarenge（Makino）Makino var. aggregeatum（Makino）
Ohwi ＝Sedum aggregeatum（Makino）Makino ■☆

356833 Sedum iwarenge（Makino）Makino var. boehmeri（Makino）
Ohwi ＝Orostachys malacophylla（Pall.）Fisch. var. boehmeri
（Makino）H. Hara ■☆

356834 Sedum iwarenge（Makino）Makino var. furusei（Ohwi）Ohwi
＝Orostachys furusei Ohwi ■☆

356835　Sedum iwarenge（Makino）Makino var. genkaiense（Ohwi）Ohwi ＝ Orostachys malacophylla（Pall.）Fisch. ■

356836　Sedum jaeschkei Kurz ＝ Sedum oreades（Decne.）Raym. -Hamet ■

356837　Sedum jahandiezii Batt. ;贾汉景天■☆

356838　Sedum jahandiezii Batt. subsp. persicinum Maire et Sam. ＝ Sedum jahandiezii Batt. ■☆

356839　Sedum jahandiezii Batt. var. battandieri Maire ＝ Sedum jahandiezii Batt. ■☆

356840　Sedum jahandiezii Batt. var. persicinum（Maire et Sam.）Maire ＝ Sedum jahandiezii Batt. ■☆

356841　Sedum japonicum Siebold ex Miq. ;日本景天(佛甲草,黄花方,日本佛甲草,深山佛甲草,石板菜,指甲草);Japan Stonecrop,Japanese Stonecrop ■

356842　Sedum japonicum Siebold ex Miq. f. rugosum Fröd. ＝ Sedum wenchuanense S. H. Fu ■

356843　Sedum japonicum Siebold ex Miq. subsp. boninense（Yamam. ex Tuyama）H. Ohba;小笠原景天■☆

356844　Sedum japonicum Siebold ex Miq. subsp. oryzifolium（Makino）H. Ohba;稻叶日本景天■☆

356845　Sedum japonicum Siebold ex Miq. subsp. oryzifolium（Makino）H. Ohba var. pumilum（H. Ohba）H. Ohba;矮景天■☆

356846　Sedum japonicum Siebold ex Miq. subsp. uniflorum H. Ohba;单花日本景天(疏花佛甲草)■☆

356847　Sedum japonicum Siebold ex Miq. var. senanense（Makino）Makino;信浓景天■☆

356848　Sedum japonicum Siebold ex Miq. var. senanense（Makino）Makino f. leucanthemum Honda;白花信浓景天■☆

356849　Sedum jiaodongense Y. M. Zhang et X. D. Chen;胶东景天;Jiaodong Stonecrop ■

356850　Sedum jinianum X. H. Guo;皖景天;Anhui Stonecrop ■

356851　Sedum jinianum X. H. Guo ＝ Sedum bulbiferum Makino ■

356852　Sedum jiulungshanense Y. C. Ho;九龙山景天;Jiulongshan Stonecrop ■

356853　Sedum juparense Fröd. ＝ Rhodiola coccinea（Royle）Boriss. ■

356854　Sedum juparense Fröd. ＝ Rhodiola juparensis（Fröd.）S. H. Fu ■

356855　Sedum kagamontanum Maxim. ;堪察加景天■

356856　Sedum kamtschaticum Fisch. et C. A. Mey. ＝ Phedimus kamtschaticum（Fisch. et C. A. Mey.）'t Hart ■

356857　Sedum kamtschaticum Fisch. f. angustifolium Kom. ;狭花堪察加景天■☆

356858　Sedum kamtschaticum Fisch. f. viviparum Tak. Hashim. ;胎生景天■☆

356859　Sedum karpelesae Raym. -Hamet ＝ Rhodiola humilis（Hook. f. et Thomson）S. H. Fu ■

356860　Sedum kiangnanense D. Q. Wang et Z. F. Wu;江南景天;Jiangnan Stonecrop ■

356861　Sedum kirilowii Regel ＝ Rhodiola kirilowii Regel ex Maxim. ■

356862　Sedum kirilowii Regel var. altum Fröd. ＝ Rhodiola kirilowii Regel ex Maxim. ■

356863　Sedum kirilowii Regel var. linifolium Regel et Schmalh. ＝ Rhodiola kirilowii Regel ex Maxim. ■

356864　Sedum kirilowii Regel var. rubrum Praeger ＝ Rhodiola kirilowii Regel ex Maxim. ■

356865　Sedum kiusianum Makino ＝ Sedum polytrichoides Hemsl. ex Forbes et Hemsl. ■

356866　Sedum komarovii（A. Berger）Chu ＝ Rhodiola angusta Nakai ■

356867　Sedum kouyangense H. Lév. et Vaniot ＝ Sedum sarmentosum Bunge ■

356868　Sedum labordei H. Lév. et Vaniot ＝ Hylotelephium erythrostictum（Miq.）H. Ohba ■

356869　Sedum lagascae Pau;拉加景天■☆

356870　Sedum latentibulbosum K. T. Fu et G. Y. Rao;潜茎景天■

356871　Sedum leblancae Raym. -Hamet;钝萼景天;Obtusesepal Stonecrop ■

356872　Sedum leblancae Raym. -Hamet var. dielsii（Raym. -Hamet）Fröd. ＝ Sedum dielsii Raym. -Hamet ■

356873　Sedum leblancae Raym. -Hamet var. torquatum Fröd. ＝ Sedum tsiangii Fröd. var. torquatum（Fröd.）K. T. Fu ■

356874　Sedum lenkoranicum Grossh. ;连科兰景天■☆

356875　Sedum lepidopodum Nakai ＝ Sedum polytrichoides Hemsl. ex Forbes et Hemsl. ■

356876　Sedum leptophyllum Fröd. ;薄叶景天;Thinleaf Stonecrop ■

356877　Sedum leucocarpum Franch. ;白果景天(白果佛甲草,堆山花,红花岩松,黄花岩松,山瓦松,石莲花,瓦松,瓦指甲,岩松,玉蝴蝶);Whitefruit Stonecrop ■

356878　Sedum leveilleanum Raym. -Hamet ＝ Meterostachys sikokianus（Makino）Nakai ■☆

356879　Sedum levii Raym. -Hamet ＝ Rhodiola humilis（Hook. f. et Thomson）S. H. Fu ■

356880　Sedum liciae Raym. -Hamet ＝ Rhodiola liciae（Raym. -Hamet）S. H. Fu ■

356881　Sedum lievenii（Ledeb.）Raym. -Hamet ＝ Pseudosedum lievenii（Ledeb.）A. Berger ●■

356882　Sedum likiangense Fröd. ＝ Rhodiola coccinea（Royle）Boriss. subsp. scabrida（Franch.）H. Ohba ■

356883　Sedum limuloides Praeger ＝ Orostachys fimbriata（Turcz.）A. Berger ■

356884　Sedum lineare Thunb. ;佛甲草(半支莲,半枝连,打不死,地蜈蚣,豆瓣草,佛指甲,狗牙半支,狗牙瓣,狗牙菜,瓜只玉,禾雀,禾雀脒,禾雀舌,火烧草,火焰草,尖叶佛指甲,尖叶石指甲,金莉插,金枪药,麻雀舌,晒不死,鼠牙半支,鼠牙半支莲,铁指甲,土三七,万年草,枉开口,午时花,小佛指甲,小叶刀燃草,养鸡草,指甲草,猪牙齿);Buddhanail,Linear Stonecrop,Needle Stonecrop ■

356885　Sedum lineare Thunb. 'Variegatum';斑点佛甲草;Variegated Stonecrop ■☆

356886　Sedum lineare Thunb. f. variegatum Praeger ＝ Sedum lineare Thunb. 'Variegatum' ■☆

356887　Sedum linearifolium Royle var. balfouri（Raym. -Hamet）Raym. -Hamet ＝ Rhodiola chrysanthemifolia（H. Lév.）S. H. Fu ■

356888　Sedum linearifolium Royle var. dielsianum（W. Limpr.）Raym. -Hamet ＝ Rhodiola chrysanthemifolia（H. Lév.）S. H. Fu ■

356889　Sedum linearifolium Royle var. forrestii（Raym. -Hamet）Raym. -Hamet ＝ Rhodiola chrysanthemifolia（H. Lév.）S. H. Fu ■

356890　Sedum linearifolium Royle var. ovatisepelum Raym. -Hamet ＝ Rhodiola ovatisepela（Raym. -Hamet）S. H. Fu ■

356891　Sedum linearifolium Royle var. sacrum（Prain ex Raym. -Hamet）Raym. -Hamet ＝ Rhodiola sacra（Prain ex Raym. -Hamet）S. H. Fu ■

356892　Sedum linearifolium Royle var. sinuatum（Edgew.）Raym. -Hamet ＝ Rhodiola sinuata（Royle ex Edgew.）S. H. Fu ■

356893　Sedum linearifolium Royle var. sinuatum（Royle ex Edgew.）Raym. -Hamet ＝ Rhodiola sinuata（Royle ex Edgew.）S. H. Fu ■

356894　Sedum linearifolium Royle var. tieghemii（Raym. -Hamet）

Raym. -Hamet ＝ Rhodiola tieghemii（Raym. -Hamet）S. H. Fu ■

356895　Sedum listoniae Vis. ;里斯顿顿景天■☆

356896　Sedum litorale Kom. ;滨海景天■☆

356897　Sedum longicaule Praeger ＝ Rhodiola kirilowii Regel ex Maxim. ■

356898　Sedum longifuniculatum K. T. Fu;长珠柄景天（长珠景天）; Longfunicle Stonecrop ■

356899　Sedum longistylum Praeger ＝ Sinocrassula longistyla（Praeger）S. H. Fu ■

356900　Sedum longyanense K. T. Fu;浪岩景天;Langyan Stonecrop ■

356901　Sedum luchuanicum K. T. Fu;禄劝景天;Luchuan Stonecrop, Luquan Stonecrop ■

356902　Sedum lucidum R. T. Clausen;光菩提■☆

356903　Sedum lungtsuanense S. H. Fu;龙泉景天;Longquan Stonecrop, Lungtsuan Stonecrop ■

356904　Sedum lutzii Raym. -Hamet;康定景天;Kangding Stonecrop, Lutz Stonecrop ■

356905　Sedum lutzii Raym. -Hamet var. viridiflavum K. T. Fu;黄绿景天;Greenyellow Stonecrop ■

356906　Sedum lycopodioides（Lam.）Kuntze ＝ Crassula muscosa L. ●☆

356907　Sedum lydium Boiss. ;吕地亚景天;Least Stouecrop ■☆

356908　Sedum macrocarpum Praeger ＝ Rhodiola macrocarpa（Praeger）S. H. Fu ■

356909　Sedum macrolepis Franch. ＝ Rhodiola kirilowii Regel ex Maxim. ■

356910　Sedum macrolepis Franch. ＝ Rhodiola macrolepis（Franch.）S. H. Fu ■

356911　Sedum madagascariense H. Perrier ＝ Perrierosedum madagascariense（H. Perrier）H. Ohba ■☆

356912　Sedum magellense Ten. ;麦哲伦景天■☆

356913　Sedum magniflorum K. T. Fu;大花景天;Bigflower Stonecrop ■

356914　Sedum maireanum Sennen;迈氏景天■☆

356915　Sedum mairei Praeger ＝ Sedum somenii Raym. -Hamet ex H. Lév. ■

356916　Sedum major（Hemsl.）Migo;山飘风（半边莲,豆瓣七,岩豆瓣菜）;Big Stonecrop ■

356917　Sedum makinoi Maxim. ;圆叶景天（圆叶佛甲草）;Makino Stonecrop ■

356918　Sedum makinoi Maxim. ' Ogon';金色圆叶景天;Gold-leaf Sedum ■☆

356919　Sedum makinoi Maxim. var. emarginatum（Migo）S. H. Fu ＝ Sedum emarginatum Migo ■

356920　Sedum malacophyllum（Pall.）Fisch. ＝ Orostachys malacophylla（Pall.）Fisch. ■

356921　Sedum malacophyllum（Pall.）Steud. ＝ Orostachys malacophylla（Pall.）Fisch. ■

356922　Sedum malacophyllum Steud. ;锦葵叶景天■☆

356923　Sedum malladrae Chiov. ＝ Hypagophytum abyssinicum（Hochst. ex A. Rich.）A. Berger ■☆

356924　Sedum marganianum E. Walther;翡翠景天;Malachite Stonecrop ■☆

356925　Sedum margaritae Raym. -Hamet ＝ Sedum chauveaudii Raym. -Hamet var. margaritae（Raym. -Hamet）Fröd. ■

356926　Sedum martinii H. Lév. ＝ Sinocrassula indica（Decne.）A. Berger ■

356927　Sedum maurum Humbert et Maire;模糊景天■☆

356928　Sedum maximum（L.）Suter;大景天（大千里光）;Great Stonecrop ■☆

356929　Sedum megalanthum Fröd. ＝ Rhodiola crenulata（Hook. f. et Thomson）H. Ohba ■

356930　Sedum megalophyllum Fröd. ＝ Rhodiola crenulata（Hook. f. et Thomson）H. Ohba ■

356931　Sedum megalophyllum Fröd. ＝ Rhodiola megalophylla（Fröd.）S. H. Fu ■

356932　Sedum mekongense Praeger ＝ Sedum multicaule Wall. ■

356933　Sedum melananthum DC. ;黑药景天■☆

356934　Sedum mexicanum Britton;松叶佛甲草（松叶景天）;Mexican Stonecrop ■

356935　Sedum meyeri-johannis Engl. ;迈尔约翰景天■☆

356936　Sedum meyeri-johannis Engl. var. keniae Fröd. ＝ Sedum meyeri-johannis Engl. ■☆

356937　Sedum microsepalum Hayata;小萼佛甲草（小萼景天）;Littlesepal Stonecrop ■

356938　Sedum middendorffianum Maxim. ＝ Phedimus middendorffianum（Maxim.）'t Hart ■

356939　Sedum middendorffianum Maxim. var. diffusum Praeger ＝ Phedimus middendorffianum（Maxim.）'t Hart ■

356940　Sedum mingjinianum S. H. Fu ＝ Hylotelephium mingjinianum（S. H. Fu）H. Ohba ■

356941　Sedum mirum Pamp. ＝ Umbilicus mirus（Pamp.）Greuter ■☆

356942　Sedum modestum Ball;适度景天■☆

356943　Sedum mooneyi M. G. Gilbert;穆尼景天■☆

356944　Sedum morganianum E. Walther;串珠草（白菩提, 松鼠尾）;Beavertail, Burro Tail, Burro's Tail, Burro's Tails, Burrow's Tail, Burrow's Tails, Donkey's Tai, Donkey's Tails, Donkey's-taill, Donkey-tail ■

356945　Sedum morotii Raym. -Hamet;倒卵叶景天;Morot Stonecrop ■

356946　Sedum morotii Raym. -Hamet var. pinoyi（Raym. -Hamet）Fröd. ;小倒卵叶景天;Small Morot Stonecrop ■

356947　Sedum morrisonense Hayata;玉山佛甲草（新高山景天, 玉山景天, 玉山千里光）;Morrison Stonecrop, Yushan Stonecrop ■

356948　Sedum mosoynense Franch. ＝ Sedum obtusipetalum Franch. ■

356949　Sedum mossii Raym. -Hamet ＝ Ohbaea balfourii（Raym. -Hamet）Byalt et I. V. Sokolova ■

356950　Sedum mucizonia（Ortega）Raym. -Hamet;西班牙景天■☆

356951　Sedum mucizonia（Ortega）Raym. -Hamet subsp. abylaeum（Font Quer et Maire）Spring. ;阿比拉景天■☆

356952　Sedum multicaule Wall. ＝ Sedum multicaule Wall. ex Lindl. ■

356953　Sedum multicaule Wall. ex Lindl. ;多茎景天（滇瓦花, 滇瓦松, 多茎佛甲草, 佛指甲, 九头狮子草, 石根, 石花, 瓦花, 瓦松, 岩如意, 指甲菜, 指甲草）;Manystem Stonecrop ■

356954　Sedum multicaule Wall. subsp. rugosum K. T. Fu;皱茎景天;Rough Manystem Stonecrop ■

356955　Sedum multiceps Coss. et Durieu;小松绿■☆

356956　Sedum muyaicum K. T. Fu;木雅景天;Muya Stonecrop ■

356957　Sedum nagasakianum（H. Hara）H. Ohba;长崎景天■☆

356958　Sedum nanchuanense K. T. Fu et G. Y. Rao;金佛山景天（南川景天）;Nanchuan Stonecrop ■

356959　Sedum nanum Boiss. ;侏儒景天■☆

356960　Sedum nevadense Coss. ;内华达景天■☆

356961　Sedum nicaeense All. ;苍白景天;Pale Stonecrop ■☆

356962　Sedum nicaeense All. ＝ Sedum sediforme（Jacq.）Pau ■☆

356963　Sedum nobile Franch. ＝ Rhodiola nobilis（Franch.）S. H. Fu ■

356964　Sedum nokoense Yamam. ;能高佛甲草（台岛景天）;Noko Stonecrop ■

356965　Sedum nothodugueyi K. T. Fu;距萼景天（伪藓状景天）;Nothoduguey Stonecrop ■

356966　Sedum nudum Aiton；裸露景天■☆

356967　Sedum nudum Aiton subsp. lancerottense（Murray）A. Hansen et Sunding；兰瑟景天■☆

356968　Sedum nussbaumerianum Bitter；铜景天；Coppertone Sedum ■☆

356969　Sedum nuttallianum Raf.；纳托尔景天；Nuttall's Sedum ■☆

356970　Sedum obcordatum R. T. Clausen；倒心叶景天■☆

356971　Sedum obtrullatum K. T. Fu；铲瓣景天；Gulped Stonecrop, Spadepetal Stonecrop ■

356972　Sedum obtusatum A. Gray；齿叶景天；Sierra Sedum ■☆

356973　Sedum obtusifolium C. A. Mey.；钝叶景天■☆

356974　Sedum obtusipetalum Franch.；钝瓣景天；Obtusepetal Stonecrop ■

356975　Sedum obtusipetalum Franch. subsp. danyanum H. Ohba = Sedum obtusipetalum Franch. ■

356976　Sedum obtusipetalum subsp. danyanum H. Ohba = Sedum obtusipetalum Franch. ■

356977　Sedum obtusolineare Hayata = Sedum lineare Thunb. ■

356978　Sedum obtusolineare Hayata = Sedum yvesii Raym. -Hamet ■

356979　Sedum ochroleucum Chaix；欧洲景天；Creamish Stonecrop, European Stonecrop ■☆

356980　Sedum odontophyllum Fröd. = Phedimus odontophyllum（Fröd.）'t Hart ■

356981　Sedum ohbae J. P. Kozhevn. = Rhodiola angusta Nakai ■

356982　Sedum oishii Ohwi = Hylotelephium sordidum（Maxim.）H. Ohba var. oishii（Ohwi）H. Ohba et M. Amano ■☆

356983　Sedum olgae Regel et Schmalh. ex Regel = Rosularia alpestris（Kar. et Kir.）Boriss. ■

356984　Sedum oligocarpum Fröd.；少果景天；Fewfruit Stonecrop ■

356985　Sedum oligospermum Maire；寡籽景天（大苞景天）■

356986　Sedum onychopetalum Fröd.；爪瓣景天；Clawpetal Stonecrop ■

356987　Sedum oppositifolium Sims = Umbilicus oppositifolius Ledeb. ■☆

356988　Sedum oreades（Decne.）Raym. -Hamet；山景天（高山景天）；Alpine Stonecrop, Wild Stonecrop ■

356989　Sedum oreganum Nutt.；俄勒冈景天；Stonecrop ■☆

356990　Sedum orichalcum W. W. Sm. = Ohbaea balfourii（Raym. -Hamet）Byalt et I. V. Sokolova ■

356991　Sedum oryzifolium Makino = Sedum japonicum Siebold ex Miq. subsp. oryzifolium（Makino）H. Ohba ■☆

356992　Sedum ovatisepelum（Raym. -Hamet）H. Ohba = Rhodiola ovatisepela（Raym. -Hamet）S. H. Fu ■

356993　Sedum oxypetalum Kunth；尖瓣景天■☆

356994　Sedum pachyphyllum Rose；八千代■☆

356995　Sedum pagetodes Fröd.；寒地景天；Coldland Stonecrop ■

356996　Sedum pakistanicum G. R. Sarwar = Hylotelephium pakistanicum（G. R. Sarwar）G. R. Sarwar ■☆

356997　Sedum pallescens Freyn = Hylotelephium pallescens（Freyn）H. Ohba ■

356998　Sedum pallidum M. Bieb.；青景天■☆

356999　Sedum palmeri S. Watson；土耳其景天（薄化妆）■☆

357000　Sedum pamiroalaicum（Boriss.）C. -A. Jansson；帕米尔景天■☆

357001　Sedum pamiroalaicum（Boriss.）C. -A. Jansson = Rhodiola pamiro-alaica Boriss. ■

357002　Sedum pampaninii Raym. -Hamet；秦岭景天；Pampanin Stonecrop, Qinling Stonecrop ■

357003　Sedum paoshingense（S. H. Fu）H. Ohba；宝兴景天；Baoxing Stonecrop, Paohsing Stonecrop ■

357004　Sedum paoshingense（S. H. Fu）H. Ohba = Sinocrassula indica（Decne.）A. Berger var. luteorubra（Praeger）S. H. Fu ■

357005　Sedum paoshingense S. H. Fu = Sinocrassula indica（Decne.）A. Berger var. luteorubra（Praeger）S. H. Fu ■

357006　Sedum paracelatum Fröd.；敏感景天；Sensitive Stonecrop ■

357007　Sedum parvisepalum Yamam.；尖萼佛甲草■

357008　Sedum parvisepalum Yamam. = Sedum microsepalum Hayata ■

357009　Sedum parvistamineum Petr.；小雄蕊景天■☆

357010　Sedum paui Sennen = Sedum lagascae Pau ■☆

357011　Sedum pekinense H. Lév. et Vaniot；北京景天；Beijing Stonecrop ■

357012　Sedum pekinense H. Lév. et Vaniot = Hylotelephium tatarinowii（Maxim.）H. Ohba var. integrifolium（Palib.）S. H. Fu ■

357013　Sedum pentapetalum Boriss.；五瓣景天■☆

357014　Sedum pentapetalum Boriss. = Sedum hispanicum L. ■

357015　Sedum perrotii Raym. -Hamet；甘肃景天；Gansu Stonecrop, Perrot Stonecrop ■

357016　Sedum persicinum Maire et Sam. var. persicinum（Maire et Sam.）Guitt. = Sedum jahandiezii Batt. ■☆

357017　Sedum petiolata Fröd. = Sedum perrotii Raym. -Hamet ■

357018　Sedum petiolatum Fröd. = Rhodiola petiolata（Fröd.）S. H. Fu ■

357019　Sedum petiolatum Fröd. = Rhodiola prainii（Raym. -Hamet）H. Ohba ■

357020　Sedum phariense H. Ohba = Rhodiola phariensis（H. Ohba）S. H. Fu ■

357021　Sedum phariense H. Ohba = Rhodiola purpureoviridis（Praeger）S. H. Fu subsp. phariensis（H. Ohba）H. Ohba ■

357022　Sedum phyllanthum H. Lév. et Vaniot；叶花景天；Leafflower Stonecrop ■

357023　Sedum piloshanense Fröd. = Sedum oreades（Decne.）Raym. -Hamet ■

357024　Sedum pinnatifidum（Boriss.）J. P. Kozhevn. = Rhodiola pinnatifida Boriss. ■

357025　Sedum pinoyi Raym. -Hamet = Sedum morotii Raym. -Hamet var. pinoyi（Raym. -Hamet）Fröd. ■

357026　Sedum planifolium K. T. Fu；平叶景天（狗牙瓣）；Flatleaf Stonecrop ■

357027　Sedum platyphyllum S. H. Fu = Sedum fui Rowley ■

357028　Sedum platyphyllum S. H. Fu var. longisepalum K. T. Fu = Sedum fui Rowley var. longisepalum（K. T. Fu）S. H. Fu ■

357029　Sedum platysepalum Franch.；宽萼景天（宽萼山景天）；Broadsepal Stonecrop ■

357030　Sedum pleumgynanthum Hand. -Mazz. = Rhodiola primuloides（Franch.）S. H. Fu ■

357031　Sedum pluricaule Kudo = Hylotelephium pluricaule（Kudo）H. Ohba ■☆

357032　Sedum pluricaule Kudo subsp. ezawae Nosaka = Hylotelephium pluricaule（Kudo）H. Ohba ■☆

357033　Sedum pluricaule Kudo subsp. hidakanum（Tatew. ex Kawano）Nosaka = Hylotelephium pluricaule（Kudo）H. Ohba ■☆

357034　Sedum pluricaule Kudo var. yezoense（Miyabe et Tatew.）Tatew. et Kawano = Hylotelephium pluricaule（Kudo）H. Ohba ■☆

357035　Sedum polytrichoides Hemsl. = Sedum polytrichoides Hemsl. ex Forbes et Hemsl. ■

357036　Sedum polytrichoides Hemsl. ex Forbes et Hemsl.；藓状景天（柳叶景天, 藓状佛甲草）；Mosslike Stonecrop ■

357037　Sedum polytrichoides Hemsl. ex Forbes et Hemsl. subsp. yabeanum（Makino）H. Ohba var. setouchiense（Murata et Yuasa）H. Ohba；濑户内景天■☆

357038　Sedum polytrichoides Hemsl. ex Forbes et Hemsl. var. yabeanum (Makino) H. Ohba;矢部薛状景天■☆

357039　Sedum populifolium Pall. ;杨叶景天●☆

357040　Sedum praealtum A. DC. ;灌丛景天■☆

357041　Sedum praegerianum W. W. Sm. = Rhodiola hobsonii (Prain ex Raym. -Hamet) S. H. Fu ■

357042　Sedum praeinopetalum Fröd. ;绿瓣景天;Greenpetal Stonecrop ■

357043　Sedum prainii Raym. -Hamet = Rhodiola prainii (Raym. -Hamet) H. Ohba ■

357044　Sedum pratoalpinum Fröd. ;牧山景天;Alpine Meadow Stonecrop, Alpine-meadow Stonecrop ■

357045　Sedum primuloides Franch. = Rhodiola primuloides (Franch.) S. H. Fu ■

357046　Sedum primuloides Franch. var. pleurogynanthum (Hand. -Mazz.) Fröd. = Rhodiola primuloides (Franch.) S. H. Fu ■

357047　Sedum progressum Diels = Rhodiola macrocarpa (Praeger) S. H. Fu ■

357048　Sedum pruinatum Brot. = Sedum forsterianum Sm. ■☆

357049　Sedum przewalskii Maxim. ;高原景天(高原山景天);Plateau Stonecrop, Przewalsk Stonecrop ■

357050　Sedum pseudoaizoon Debeaux = Phedimus aizoon (L.) 't Hart ■

357051　Sedum pseudospectabile Praeger = Hylotelephium pseudospectabile (Praeger) S. H. Fu ■

357052　Sedum pubescens Vahl;短柔毛景天■☆

357053　Sedum pulchellum Michx. ;美丽景天;Rock-moss, Widow's Cross, Widow's-cross ■☆

357054　Sedum purdomii W. W. Sm. ;裂鳞景天;Purdom Stonecrop ■

357055　Sedum purpureoviride Praeger = Rhodiola purpureoviridis (Praeger) S. H. Fu ■

357056　Sedum purpureum (L.) Schult. = Hylotelephium purpureum (L.) Holub ■

357057　Sedum purpureum (L.) Schult. = Hylotelephium triphyllum (Haw.) Holub ■

357058　Sedum purpureum (L.) Schult. = Sedum telephium L. ■☆

357059　Sedum pyramidale Praeger = Orostachys chanetii (H. Lév.) A. Berger ■

357060　Sedum quadrifidum Pall. = Rhodiola quadrifida (Pall.) Fisch. et C. A. Mey. ■

357061　Sedum quadrifidum Pall. var. coccineum (Royle) Hook. f. et Thomson = Rhodiola coccinea (Royle) Boriss. ■

357062　Sedum quadrifidum Pall. var. coccineum (Royle) Hook. f. et Thomson = Rhodiola quadrifida (Pall.) Fisch. et C. A. Mey. ■

357063　Sedum quadrifidum Pall. var. fastigiatum (Hook. f. et Thomson) Fröd. = Rhodiola fastigiata (Hook. f. et Thomson) S. H. Fu ■

357064　Sedum quadrifidum Pall. var. himalense (D. Don) Fröd. = Rhodiola himalensis (D. Don) S. H. Fu ■

357065　Sedum quadrifidum Pall. var. tibeticum (Hook. f. et Thomson) Fröd. = Rhodiola tibetica (Hook. f. et Thomson) S. H. Fu ■

357066　Sedum quaternatum Praeger;四叶景天(四部景天,四叶佛甲草);Fourleaf Stonecrop, Fourleaves Stonecrop ■

357067　Sedum quaternatum Praeger = Sedum phyllanthum H. Lév. et Vaniot ■

357068　Sedum radicans (Harv.) Kuntze = Crassula tetragona L. subsp. acutifolia (Lam.) Toelken ●☆

357069　Sedum ramentaceum K. T. Fu;糠秕景天;Scurfy Stonecrop, Shavings Stonecrop ■

357070　Sedum ramosissimum (Maxim.) Franch. = Orostachys fimbriata (Turcz.) A. Berger ■

357071　Sedum ramuliflorum (Link et Otto) Kuntze = Crassula obovata Haw. ■☆

357072　Sedum ramuliflorum (Link et Otto) Kuntze f. rubriflorum Kuntze = Crassula obovata Haw. ■☆

357073　Sedum ramuliflorum (Link et Otto) Kuntze var. oblongifolia Kuntze = Crassula obovata Haw. ■☆

357074　Sedum rarifiroum N. E. Br. = Rhodiola dumulosa (Franch.) S. H. Fu ■

357075　Sedum raymondii Fröd. ;膨果景天;Inflated Stonecrop, Raymond Stonecrop ■

357076　Sedum recticaule (Boriss.) Wendelbo = Rhodiola recticaulis Boriss. ■

357077　Sedum reflexum L. = Sedum rupestre L. ■☆

357078　Sedum regelii Kuntze = Crassula exilis Harv. subsp. cooperi (Regel) Toelken ●☆

357079　Sedum rendleri Raym. -Hamet = Rhodiola dumulosa (Franch.) S. H. Fu ■

357080　Sedum retusum Hemsl. ;微凹叶景天■☆

357081　Sedum rhodiola DC. = Rhodiola rosea L. ■

357082　Sedum rhodiola DC. var. atropurpureum (Turcz.) Maxim. = Rhodiola rosea L. ■

357083　Sedum rhodiola DC. var. linifolia Regel = Rhodiola kirilowii Regel ex Maxim. ■

357084　Sedum roborowskii Maxim. ;阔叶景天(草原景天,甘青景天,甘青千里光);Broadleaf Stonecrop, Roborowsk Stonecrop ■

357085　Sedum robustum Praeger = Rhodiola kirilowii Regel ex Maxim. ■

357086　Sedum robustum Praeger = Rhodiola robusta (Praeger) S. H. Fu ■

357087　Sedum rosea (L.) Scop. ;粉红景天(玫瑰景天);Rose Root ■☆

357088　Sedum rosei Raym. -Hamet;川西景天;Rose Stonecrop, Roseroot ■

357089　Sedum rosei Raym. -Hamet var. brevistamineum Fröd. = Sedum rosei Raym. -Hamet ■

357090　Sedum rosei Raym. -Hamet var. magniflorum Fröd. ;大花川西景天;Bigflower Rose Stonecrop ■

357091　Sedum roseum (L.) Scop. = Rhodiola rosea L. ■

357092　Sedum roseum (L.) Scop. subsp. sino-alpina Fröd. = Rhodiola cretinii (Raym. -Hamet) H. Ohba subsp. sino-alpina (Fröd.) H. Ohba ■

357093　Sedum roseum (L.) Scop. var. heterodontum (Hook. f. et Thomson) Fedtsch. ex Fröd. = Rhodiola heterodonta (Hook. f. et Thomson) Boriss. ■

357094　Sedum roseum (L.) Scop. var. microphyllum Fröd. = Rhodiola rosea L. var. microphylla (Fröd.) S. H. Fu ■

357095　Sedum roseum (L.) Scop. var. sinoalpinum Fröd. = Rhodiola cretinii (Raym. -Hamet) H. Ohba subsp. sino-alpina (Fröd.) H. Ohba ■

357096　Sedum roseum Stev. = Phedimus stevenianum (Rouy et Camus) 't Hart ■

357097　Sedum roseum Stev. = Sedum stevenianum Rouy et Camus ■

357098　Sedum roseum Stev. var. heterodontum (Hook. f. et Thomson) Fedtsch. ex Fröd. = Rhodiola heterodonta (Hook. f. et Thomson) Boriss. ■

357099　Sedum roseum Stev. var. sino-alpinum Fröd. = Rhodiola cretinii (Raym. -Hamet) H. Ohba subsp. sino-alpina (Fröd.) H. Ohba ■

357100　Sedum roseum Stev. var. tschangbai-shcanicum A. I. Baranov et Skvortsov = Rhodiola angusta Nakai ■

357101　Sedum rosthornianum Diels;南川景天;Nanchuan Stonecrop,

Rosthorn Stonecrop ■

357102 Sedum rosulatobulbosum Koidz. ;莲座景天■☆

357103 Sedum rosulatum Edgew. = Rosularia rosulata (Edgew.) H. Ohba ■☆

357104 Sedum rotundatum Hemsl. = Rhodiola crenulata (Hook. f. et Thomson) H. Ohba ■

357105 Sedum rotundatum Hemsl. var. oblongatum C. Marquand et Airy Shaw = Rhodiola crenulata (Hook. f. et Thomson) H. Ohba ■

357106 Sedum rotundatum Hemsl. var. oblongum Marquand et Shaw = Rhodiola crenulata (Hook. f. et Thomson) H. Ohba ■

357107 Sedum rubens L. ;变红景天;Red Sedum, Red Stonecrop ■☆

357108 Sedum rubens L. var. haouzense Andr. = Sedum rubens L. ■☆

357109 Sedum rubens L. var. mediterraneum (Jord. et Fourr.) Rouy et Camus = Sedum rubens L. ■☆

357110 Sedum rubens L. var. pallidiflorum (Jord. et Fourr.) Rouy et Camus = Sedum rubens L. ■☆

357111 Sedum rubrotinctum R. T. Clausen;耳坠草（虹玉）;Jelly-beans, Pork and Beans ■

357112 Sedum rubrum (L.) Thell. ;簇生景天■☆

357113 Sedum rubrum (L.) Thell. = Sedum caespitosum (Cav.) DC. ■☆

357114 Sedum rupestre L. ;反折景天（石生景天,岩生景天）;Blue Stonecrop, Creeping Jenny, Crooked Yellow Stonecrop, Dwarf Houseleek, Ginger, Indian Fog, Indian Moss, Jenny Stone Crop, Jenny Stonecrop, Jenny's Stonecrop, Large Yellow Stonecrop, Link Moss, Love Links, Love-in-a-chain, Love-links, Madam Trip, Prick Madam, Prick-madam, Reflexed Stonecrop, Rock Stonecrop, Spruce Stonecrop, Spruced-leaved Stonecrop, Stone Orpine, Stonecrop, Stone-hore, Stone-hot, Stonewort, Stonor Stonnord, Thrift, Trick-madam, Trip-madam, Yellow Stonecrop ■☆

357115 Sedum rupestre L. subsp. elegans (Lej.) Hegi et Schmid = Sedum forsterianum Sm. ■☆

357116 Sedum rupifragum Koidz. ;石隙景天■☆

357117 Sedum ruwenzoriense Baker f. ;鲁文佐里景天■☆

357118 Sedum sachalinense (A. Berger) Vorosch. = Rhodiola sachalinensis Boriss. ■

357119 Sedum sachalinense (Boriss.) Vorosch. = Rhodiola sachalinensis Boriss. ■

357120 Sedum sacrum Prain ex Raym. -Hamet = Rhodiola sacra (Prain ex Raym. -Hamet) S. H. Fu ■

357121 Sedum sagittipetalum Fröd. ;箭瓣景天;Arrowpetal Stonecrop, Sagittatepatal Stonecrop ■

357122 Sedum sangpotibetanum Fröd. = Rhodiola smithii (Raym. -Hamet) S. H. Fu ■

357123 Sedum sarmentosum Bunge;垂盆草(白蜈蚣,半支莲,半枝莲,豆瓣菜,豆瓣子菜,佛甲草,佛指甲,匐行景天,肝炎草,狗牙半支,狗牙半支莲,狗牙瓣,狗牙草,狗牙齿,瓜子草,瓜子莲,黄瓜子草,火连草,鸡舌草,金钱挂,爬景天,葡茎佛甲草,三七仔,石头菜,石指甲,鼠牙半支,鼠牙半支莲,水马齿苋,太阳花,土三七,枉开口,卧茎景天,狭叶垂盆草,养鸡草,野马齿苋）;Narrowleaf Stonecrop, Stringy Stonecrop, Yellow Stonecrop ■

357124 Sedum sarmentosum Bunge f. major Diels = Sedum sarmentosum Bunge ■

357125 Sedum sasakii Hayata = Sedum uniflorum Hook. et Arn. ■

357126 Sedum satumense Hatus. ;沙都木景天■☆

357127 Sedum saxifragoides Fröd. = Rhodiola saxifragoides (Fröd.) H. Ohba ■☆

357128 Sedum scabridum Franch. = Rhodiola coccinea (Royle) Boriss.

subsp. scabrida (Franch.) H. Ohba ■

357129 Sedum scabridum Franch. = Rhodiola scabrida (Franch.) S. H. Fu ■

357130 Sedum scallanii Diels = Sinocrassula indica (Decne.) A. Berger ■

357131 Sedum scallanii Diels var. majus Pamp. = Sinocrassula indica (Decne.) A. Berger ■

357132 Sedum schimperi Britten = Sedum epidendrum Hochst. ex A. Rich. ■☆

357133 Sedum schlagintweitii Fröd. = Rosularia alpestris (Kar. et Kir.) Boriss. ■

357134 Sedum schoenlandii Raym. -Hamet = Kungia schoenlandii (Raym. -Hamet) K. T. Fu ■

357135 Sedum schrenkii Fröd. = Pseudosedum affine (Schrenk) A. Berger ●■

357136 Sedum sediforme (Jacq.) Pau;座景天■☆

357137 Sedum sediforme (Jacq.) Pau var. brevirostratum Faure et Maire = Sedum sediforme (Jacq.) Pau ■☆

357138 Sedum sekiteiense Yamam. ;石碇佛甲草（台湾景天）;Sekitei Stonecrop ■

357139 Sedum selskianum Regel et Maack = Phedimus selskianum (Regel et Maack) 't Hart ■

357140 Sedum selskianum Regel et Maack var. glabrifolium Kitag. = Phedimus aizoon (L.) 't Hart ■

357141 Sedum semenovii (Regel et Herder) Mast. = Rhodiola semenovii (Regel et Herder) Boriss. ■

357142 Sedum semenovii (Regel et Herder) Mast. var. kansuense Fröd. = Rhodiola kansuensis (Fröd.) S. H. Fu ■

357143 Sedum semilunatum K. T. Fu;月座景天;Halfmoon Stonecrop ■

357144 Sedum sempervivoides Fisch. ex M. Bieb. ;长生景天;False Houseleek ■☆

357145 Sedum senanense Makino = Sedum japonicum Siebold ex Miq. var. senanense (Makino) Makino ■☆

357146 Sedum serratum (H. Ohba) J. P. Kozhevn. = Rhodiola serrata H. Ohba ■

357147 Sedum sexangulare L. ;六角景天;Insipid Stonecrop, Tasteless Stonecrop, Tasteless Yellow Stonecrop ■☆

357148 Sedum sheareri S. Moore = Sedum sarmentosum Bunge ■

357149 Sedum sherriffii (H. Ohba) J. P. Kozhevn. = Rhodiola sherriffii H. Ohba ■

357150 Sedum sieboldii Sweet ex Hook. = Hylotelephium sieboldii (Sweet ex Hook.) H. Ohba ■

357151 Sedum sikokianum Maxim. = Phedimus sikokianus (Maxim.) 't Hart ■☆

357152 Sedum silvestrii Pamp. = Sedum elatinoides Franch. ■

357153 Sedum sinicum Diels = Rhodiola yunnanensis (Franch.) S. H. Pu ■

357154 Sedum sinoglaciale K. T. Fu;冰川景天;Chinese Glacial Stonecrop, Glacial Stonecrop ■

357155 Sedum sinuatum Royle = Rhodiola sinuata (Royle ex Edgew.) S. H. Fu ■

357156 Sedum sinuatum Royle ex Edgew. = Rhodiola sinuata (Royle ex Edgew.) S. H. Fu ■

357157 Sedum smallii (Britton) H. E. Ahles;斯莫尔景天■☆

357158 Sedum smithii Raym. -Hamet = Rhodiola smithii (Raym. -Hamet) S. H. Fu ■

357159 Sedum smithii Raym. -Hamet ex W. W. Sm. et Cave = Rhodiola smithii (Raym. -Hamet) S. H. Fu ■

357160 Sedum somenii Raym. -Hamet ex H. Lév.；邓川景天；Somen Stonecrop ■

357161 Sedum sordidum Maxim. = Hylotelephium sordidum（Maxim.）H. Ohba ■☆

357162 Sedum spathulatum D. Dietr.；匙叶景天(白菊)■☆

357163 Sedum spathulatum D. Dietr.‘Cape Blanco’；卡普布朗匙叶景天(白菊)■☆

357164 Sedum spathulifolium Hook.；白菊；Colorado Stonecrop ■☆

357165 Sedum spectabile Boreau = Hylotelephium spectabile（Boreau）H. Ohba ■

357166 Sedum spectabile Boreau var. angustifolium Kitag. = Hylotelephium spectabile（Boreau）H. Ohba var. angustifolium（Kitag.）S. H. Fu ■

357167 Sedum spinosum（L.）Thunb. = Orostachys spinosa（L.）Sweet ■

357168 Sedum spinosum（L.）Thunb. var. thyrsiflorum（Fisch.）Fröd. = Orostachys thyrsiflora Fisch. ■

357169 Sedum spurium M. Bieb.；大花费菜；Caucasian Stonecrop，Creeping Stonecrop，Dragon's Blood Sedum，Stonecrop，Tworow Stonecrop，Two-row Stonecrop，Two-rowed Stonecrop ■☆

357170 Sedum stahlii Solms；玉叶景天(玉叶)；Stahl Stone Crop ■☆

357171 Sedum stamineum Paulsen = Rhodiola alsia（Fröd.）S. H. Fu ■

357172 Sedum stamineum Paulsen = Rhodiola staminea（Paulsen）S. H. Fu ■

357173 Sedum stapfii Raym. -Hamet = Rhodiola stapfii（Raym. -Hamet）S. H. Fu ■

357174 Sedum stellariifolium Franch.；繁缕叶景天(繁缕景天，狗牙风，火焰草，火焰山景天，毛狗，石豆瓣，卧儿菜)；Flame Grass，Starwoetleaf Stonecrop ■

357175 Sedum stellatum L.；星状景天■☆

357176 Sedum stenostachyum Fröd. = Kungia schoenlandii（Raym. -Hamet）K. T. Fu var. stenostachya（Fröd.）K. T. Fu ■

357177 Sedum stenostachyum Fröd. = Kungia schoenlandii（Raym. -Hamet）K. T. Fu ■

357178 Sedum stephanii Cham. = Rhodiola stephanii（Cham.）Trautv. et C. A. Mey. ■

357179 Sedum stevenianum Rouy et Camus = Phedimus stevenianum（Rouy et Camus）'t Hart ■

357180 Sedum stimulosum K. T. Fu；刺毛景天；Stimulant Stonecrop，Stimulose Stonecrop ■

357181 Sedum stoloniferum S. G. Gmel.；匍匐景天；Lesser Caucasian Stonecrop，Stolon Stonecrop ■☆

357182 Sedum stracheyi Hook. f. et Thomson = Rhodiola tibetica（Hook. f. et Thomson）S. H. Fu ■

357183 Sedum subcapitatum Hayata = Hylotelephium subcapitatum（Hayata）H. Ohba ■

357184 Sedum subgaleatum K. T. Fu；盔瓣景天；Helmetpetal Stonecrop ■

357185 Sedum subgaleatum K. T. Fu = Sedum susannae Raym. -Hamet ■

357186 Sedum suboppositum Maxim. = Rhodiola subopposita（Maxim.）Jacobsen ■

357187 Sedum suboppositum Maxim. var. telephioides Maxim. = Rhodiola rosea L. ■

357188 Sedum suboppositum Maxim. var. telephioides Maxim. = Rhodiola telephioides（Maxim.）S. H. Fu ■

357189 Sedum subtile Miq.；细小景天(姬莲花，小莲花，小万年草)；Fine Stonecrop ■

357190 Sedum subtile Miq. subsp. chinense H. Ohba = Sedum subtile Miq. ■

357191 Sedum subulatum Boiss.；锥状景天■☆

357192 Sedum surculosum Coss. = Monanthes atlantica Ball ■☆

357193 Sedum surculosum Coss. var. fuscum（Emb.）Maire = Monanthes atlantica Ball ■☆

357194 Sedum susannae Raym. -Hamet；方腺景天；Susanna Stonecrop ■

357195 Sedum susannae Raym. -Hamet var. macrosepalum K. T. Fu；大萼方腺景天；Bigsepal Susanna Stonecrop ■

357196 Sedum taiwanianum S. S. Ying = Sedum nokoense Yamam. ■

357197 Sedum talihsiense Fröd. = Rhodiola fastigiata（Hook. f. et Thomson）S. H. Fu ■

357198 Sedum talihsiense Fröd. = Rhodiola dumulosa（Franch.）S. H. Fu ■

357199 Sedum tatarinowii Maxim. = Hylotelephium tatarinowii（Maxim.）H. Ohba ■

357200 Sedum tatarinowii Maxim. var. integrifolium Palib. = Hylotelephium tatarinowii（Maxim.）H. Ohba var. integrifolium（Palib.）S. H. Fu ■

357201 Sedum techinensis S. H. Fu = Sinocrassula techinensis（S. H. Fu）S. H. Fu ■

357202 Sedum telephioides Michx.；肖紫花景天；Wild Live-forever ■☆

357203 Sedum telephium L.；紫花景天(梢梢金花，紫景天)；Alpine，Arpent，Arpent-weed，Fat Hen，Harping Johnny，Heal-all，Healing Leaf，Jacob's Ladder，Large Stonecrop，Liblong，Life-long，Live-forever，Live-forever Sedum，Live-long，Live-longand-love-long，Livelong-love-long，Long-and-love-long，Love-long，Matrona Sedum，Midsummer Men，Orphan John，Orphyne，Orpies，Orpine，Orpy-leaf，Purple Stonecrop，Solomon' Puzzle，St. Giles' Orpine，Taliesin's Cress，Witch's-moneybags ■☆

357204 Sedum telephium L. f. verticillatum（L.）Fröd. = Hylotelephium verticillatum（L.）H. Ohba ■

357205 Sedum telephium L. subsp. alboroseum（Baker）Fröd. = Hylotelephium erythrostictum（Miq.）H. Ohba ■

357206 Sedum telephium L. subsp. alboroseum（Baker）Fröd. = Sedum erythrostictum Miq. ■

357207 Sedum telephium L. subsp. angustum（Maxim.）Fröd. = Hylotelephium angustum（Maxim.）H. Ohba ■

357208 Sedum telephium L. subsp. purpureum（L.）Schinz et R. Keller = Hylotelephium triphyllum（Haw.）Holub ■

357209 Sedum telephium L. subsp. purpureum（L.）Schinz et R. Keller = Sedum telephium L. ■☆

357210 Sedum telephium L. subsp. purpureum（L.）Schinz et R. Keller var. orientale Fröd. = Hylotelephium triphyllum（Haw.）Holub ■

357211 Sedum telephium L. subsp. verticillatum（L.）Fröd. = Hylotelephium verticillatum（L.）H. Ohba ■

357212 Sedum telephium L. subsp. viviparum（Maxim.）Fröd. = Hylotelephium viviparum（Maxim.）H. Ohba ■

357213 Sedum telephium L. var. albiflorum Maxim. = Hylotelephium pallescens（Freyn）H. Ohba ■

357214 Sedum telephium L. var. kirinense Kom. = Hylotelephium spectabile（Boreau）H. Ohba ■

357215 Sedum telephium L. var. pallescens（Freyn）Kom. = Hylotelephium pallescens（Freyn）H. Ohba ■

357216 Sedum telephium L. var. purpureum L. = Hylotelephium purpureum（L.）Holub ■

357217 Sedum telephium L. var. purpureum L. = Hylotelephium triphyllum（Haw.）Holub ■

357218　Sedum telephium L. var. purpureum L. ＝Sedum telephium L. ■☆

357219　Sedum tenellum M. Bieb.；柔弱景天■☆

357220　Sedum tenuifolium（Sm.）Strobl ＝Sedum amplexicaule DC. subsp. tenuifolium（Sm.）Greuter ■☆

357221　Sedum tenuifolium（Sm.）Strobl var. ciliatum（Lange）Fröd. ＝Sedum amplexicaule DC. subsp. tenuifolium（Sm.）Greuter ■☆

357222　Sedum tenuifolium Franch. ＝Sedum franchetii Grande ■

357223　Sedum tenuifolium Franch. ＝Sedum pampaninii Raym. -Hamet ■

357224　Sedum ternatum Michx.；北美景天；Mountain Stonecrop, Three-leaved Stonecrop, Whorled Stonecrop, Wild Stonecrop ■☆

357225　Sedum tetractinum Fröd.；四芒景天（石上开花，四芒苞景天，四射景天，岩莲花）；Chinese Sedum, Fourarista Stonecrop ■

357226　Sedum tetragonum（L.）Kuntze ＝Crassula tetragona L. ●☆

357227　Sedum tetramerum Trautv.；四数景天■☆

357228　Sedum tianmushanense Y. C. Ho et F. Chai；天目山景天；Tianmushan Stonecrop ■

357229　Sedum tibeticum Hook. f. et Thomson ＝Rhodiola tibetica（Hook. f. et Thomson）S. H. Fu ■

357230　Sedum tibeticum Hook. f. et Thomson var. stracheyi（Hook. f. et Thomson）Clarke ＝Rhodiola tibetica（Hook. f. et Thomson）S. H. Fu ■

357231　Sedum tieghemii Raym. -Hamet ＝Rhodiola tieghemii（Raym. -Hamet）S. H. Fu ■

357232　Sedum tosaense Makino；土佐景天■

357233　Sedum tosaense Makino subsp. sinense K. T. Fu et G. Y. Rao；浙景天；Zhejiang Stonecrop ■

357234　Sedum tosaense Makino subsp. sinense K. T. Fu et G. Y. Rao ＝Sedum tosaense Makino ■

357235　Sedum transvaalense Kuntze ＝Crassula lanceolata（Eckl. et Zeyh.）Endl. ex Walp. subsp. transvaalensis（Kuntze）Toelken ■☆

357236　Sedum triactina Berger；三芒景天；Threearista Stonecrop ■

357237　Sedum triactina Berger subsp. leptum Fröd.；小三芒景天■

357238　Sedum triangulosepalum Tang S. Liu et N. J. Chung；等萼佛甲草；Equalsepal Stonecrop, Triangularsepal Stonecrop ■

357239　Sedum triangulosepalum Tang S. Liu et N. J. Chung ＝Sedum microsepalum Hayata ■

357240　Sedum tricarpum Makino；三果景天■☆

357241　Sedum tricarpum Makino f. viride Hatus.；绿三果景天■☆

357242　Sedum trichospermum K. T. Fu；毛籽景天；Hairyseed Stonecrop ■

357243　Sedum trifidum Hook. f. et Thomson var. balfourii Raym. -Hamet ＝Rhodiola chrysanthemifolia（H. Lév.）S. H. Fu ■

357244　Sedum trifidum Hook. f. et Thomson var. forrestii Raym. -Hamet ＝Rhodiola chrysanthemifolia（H. Lév.）S. H. Fu ■

357245　Sedum triphyllum（Haw.）Gray ＝Sedum telephium L. ■☆

357246　Sedum triphyllum Praeger ＝Sedum chauveaudii Raym. -Hamet ■

357247　Sedum trullipetalum Hook. f. et Thomson；馒瓣景天；Trowelpetal Stonecrop ■

357248　Sedum trullipetalum Hook. f. et Thomson var. ciliatum Fröd.；缘毛景天；Ciliate Trowelpetal Stonecrop ■

357249　Sedum trullipetalum Hook. f. et Thomson var. gagei（Raym. -Hamet）H. Ohba ＝Sedum gagei Raym. -Hamet ■

357250　Sedum truncatistigmum T. S. Liu et N. J. Chung ＝Sedum microsepalum Hayata ■

357251　Sedum truncatistigmum Tang S. Liu et N. J. Chung；截柱佛甲草；Truncatestigma Stonecrop ■

357252　Sedum truncatistigmum Tang S. Liu et N. J. Chung ＝Sedum microsepalum Hayata ■

357253　Sedum tsiangii Fröd.；安龙景天；Anlong Stonecrop, Tsiang Stonecrop ■

357254　Sedum tsiangii Fröd. var. torquatum（Fröd.）K. T. Fu；珠节景天■

357255　Sedum tsinghaicum K. T. Fu；青海景天；Chinghai Stonecrop, Qinghai Stonecrop ■

357256　Sedum tsonanum K. T. Fu；错那景天；Cona Stonecrop, Cuona Stonecrop, Tsona Stonecrop ■

357257　Sedum tsugaruense H. Hara ＝Hylotelephium ussuriense（Kom.）H. Ohba var. tsugaruense（H. Hara）H. Ohba ■☆

357258　Sedum tuberosum Coss. et Letourn.；块状景天■☆

357259　Sedum ulricae Fröd.；甘南景天；S. Gansu Stonecrop, South Kansu Stonecrop ■

357260　Sedum umbilicoides Regel ＝Rosularia alpestris（Kar. et Kir.）Boriss. ■

357261　Sedum uniflorum Hook. et Arn.；疏花佛甲草（梳花佛甲草）■

357262　Sedum uniflorum Hook. et Arn. ＝Sedum japonicum Siebold ex Miq. subsp. uniflorum H. Ohba ■☆

357263　Sedum uniflorum Hook. et Arn. subsp. boninense（Yamam. ex Tuyama）H. Ohba ＝Sedum japonicum Siebold ex Miq. subsp. boninense（Yamam. ex Tuyama）H. Ohba ■☆

357264　Sedum uniflorum Hook. et Arn. subsp. japonicum（Siebold ex Miq.）H. Ohba var. senanense（Makino）H. Ohba ＝Sedum japonicum Siebold ex Miq. var. senanense（Makino）Makino ■☆

357265　Sedum uniflorum Hook. et Arn. subsp. japonicum（Siebold ex Miq.）H. Ohba ＝Sedum japonicum Siebold ex Miq. ■

357266　Sedum uniflorum Hook. et Arn. subsp. oryzifolium（Makino）H. Ohba var. pumilum H. Ohba ＝Sedum japonicum Siebold ex Miq. subsp. oryzifolium（Makino）H. Ohba var. pumilum（H. Ohba）H. Ohba ■☆

357267　Sedum uniflorum Hook. et Arn. subsp. oryzifolium（Makino）H. Ohba ＝Sedum japonicum Siebold ex Miq. subsp. oryzifolium（Makino）H. Ohba ■☆

357268　Sedum uniflorum Hook. et Arn. subsp. rugosum（Fröd.）K. T. Fu ＝Sedum wenchuanense S. H. Fu ■

357269　Sedum uraiense Hayata ＝Sedum drymarioides Hance ■

357270　Sedum uraiense Hayata ＝Sedum stellariifolium Franch. ■

357271　Sedum ussuriense Kom.；乌苏里景天■☆

357272　Sedum vaginatum（Eckl. et Zeyh.）Kuntze ＝Crassula vaginata Eckl. et Zeyh. ■☆

357273　Sedum valerianoides Diels ＝Rhodiola yunnanensis（Franch.）S. H. Pu ■

357274　Sedum variicolor Praeger ＝Sedum leucocarpum Franch. ■

357275　Sedum venustum Praeger ＝Rhodiola atuntsuensis（Praeger）S. H. Fu ■

357276　Sedum venustum Praeger ＝Rhodiola fastigiata（Hook. f. et Thomson）S. H. Fu ■

357277　Sedum versicolor（Raym. -Hamet）Maire；变色景天■☆

357278　Sedum verticillatum（Hook. f. et Thomson）Raym. -Hamet ＝Sedum triactina Berger ■

357279　Sedum verticillatum L. ＝Hylotelephium verticillatum（L.）H. Ohba ■

357280　Sedum verticillatum L. f. bulbiferum N. Yonez. ＝Hylotelephium verticillatum（L.）H. Ohba f. bulbiferum（N. Yonez.）Yonek. ■☆

357281　Sedum viguieri Raym. -Hamet ex Fröd. ＝Rosularia viguieri（Raym. -Hamet ex Fröd.）G. R. Sarwar ■☆

357282　Sedum villosum L.；欧洲柔毛景天；Hairy Stonecrop, Pink

Stonecrop ■☆

357283 Sedum villosum L. subsp. aristatum (Emb. et Maire) Lainz = Sedum maireanum Sennen ■☆

357284 Sedum villosum L. subsp. nevadense (Coss.) Batt. = Sedum nevadense Coss. ■☆

357285 Sedum villosum L. var. aristatum Emb. et Maire = Sedum maireanum Sennen ■☆

357286 Sedum villosum L. var. decarrhenum Maire = Sedum maireanum Sennen ■☆

357287 Sedum villosum L. var. pentarrhenum Maire = Sedum maireanum Sennen ■☆

357288 Sedum viride Makino = Hylotelephium viride (Makino) H. Ohba ■☆

357289 Sedum viscosum Praeger = Sedum stellariifolium Franch. ■

357290 Sedum viviparum Maxim. = Hylotelephium viviparum (Maxim.) H. Ohba ■

357291 Sedum volkensii Engl. = Sedum meyeri-johannis Engl. ■☆

357292 Sedum wallichianum Hook. = Rhodiola wallichiana (Hook.) S. H. Fu ■

357293 Sedum wallichianum Hook. var. cretinii (Raym.-Hamet) Hara = Rhodiola cretinii (Raym.-Hamet) H. Ohba ■

357294 Sedum wangii S. H. Fu; 德钦景天 (启无景天); Deqin Stonecrop, Wang Stonecrop ■

357295 Sedum wannanense X. H. Guo, X. P. Zhang et X. H. Chen; 皖南景天; Wannan Stonecrop ■

357296 Sedum wannanense X. H. Guo, X. P. Zhang et X. H. Chen var. incamatum X. H. Guo, X. P. Zhang et X. H. Chen; 肉红景天■

357297 Sedum wenchuanense S. H. Fu; 汶川景天; Wenchuan Stonecrop ■

357298 Sedum wilczekianum Font Quer; 维尔切克景天■☆

357299 Sedum wilsollii Fröd.; 兴山景天; E. H. Wilson Stonecrop ■

357300 Sedum winkleri (Willk.) Font Quer = Sedum hirsutum All. subsp. baeticum Rouy ■☆

357301 Sedum winkleri (Willk.) Font Quer var. maroccanum Font Quer = Sedum hirsutum All. subsp. baeticum Rouy ■☆

357302 Sedum woronowii Raym.-Hamet; 长萼景天; Longcalyx Stonecrop ■

357303 Sedum wuanum K. S. Hao = Sedum celatum Fröd. ■

357304 Sedum wulingense (Nakai) Kitag. = Rhodiola dumulosa (Franch.) S. H. Fu ■

357305 Sedum yabeanum Makino = Sedum polytrichoides Hemsl. ex Forbes et Hemsl. subsp. yabeanum (Makino) H. Ohba ■☆

357306 Sedum yabeanum Makino var. setouchiense Murata et Yuasa = Sedum polytrichoides Hemsl. ex Forbes et Hemsl. subsp. yabeanum (Makino) H. Ohba var. setouchiense (Murata et Yuasa) H. Ohba ■☆

357307 Sedum yantaiense Debeaux = Phedimus aizoon (L.) 't Hart ■

357308 Sedum yunnanense (Franch.) S. H. Fu var. henryi (Diels) Raym.-Hamet = Rhodiola yunnanensis (Franch.) S. H. Pu ■

357309 Sedum yunnanense (Franch.) S. H. Fu var. oxyphyllum Fröd. = Rhodiola yunnanensis (Franch.) S. H. Pu ■

357310 Sedum yunnanense (Franch.) S. H. Fu var. papillocarpum Fröd. = Rhodiola yunnanensis (Franch.) S. H. Pu ■

357311 Sedum yunnanense (Franch.) S. H. Fu var. rotundifolium Fröd. = Rhodiola yunnanensis (Franch.) S. H. Pu ■

357312 Sedum yunnanense (Franch.) S. H. Fu var. valerianoides (Diels) Raym.-Hamet = Rhodiola yunnanensis (Franch.) S. H. Pu ■

357313 Sedum yunnanense Franch. = Rhodiola yunnanensis (Franch.) S. H. Pu ■

357314 Sedum yunnanense Franch. var. forrestii Raym.-Hamet = Rhodiola forrestii (Raym.-Hamet) S. H. Fu ■

357315 Sedum yunnanense Franch. var. henryi (Diels) Raym.-Hamet = Rhodiola yunnanensis (Franch.) S. H. Pu ■

357316 Sedum yunnanense Franch. var. muliense Fröd. = Rhodiola forrestii (Raym.-Hamet) S. H. Fu ■

357317 Sedum yunnanense Franch. var. oblanceolatum Fröd. = Rhodiola forrestii (Raym.-Hamet) S. H. Fu ■

357318 Sedum yunnanense Franch. var. oxyphyllum Fröd. = Rhodiola yunnanensis (Franch.) S. H. Pu ■

357319 Sedum yunnanense Franch. var. papillocarpum Fröd. = Rhodiola yunnanensis (Franch.) S. H. Pu ■

357320 Sedum yunnanense Franch. var. rotundifolia Fröd. = Rhodiola yunnanensis (Franch.) S. H. Pu ■

357321 Sedum yunnanense Franch. var. strictum Fröd. = Rhodiola forrestii (Raym.-Hamet) S. H. Fu ■

357322 Sedum yunnanense Franch. var. valerianoides (Diels) Raym.-Hamet = Rhodiola yunnanensis (Franch.) S. H. Pu ■

357323 Sedum yunnanensis Franch. = Rhodiola yunnanensis (Franch.) S. H. Pu ■

357324 Sedum yunnanensis Franch. var. papillocarpa Fröd. = Rhodiola papillocarpa (Fröd.) S. H. Fu ■

357325 Sedum yunnanensis Franch. var. rotundifolia Fröd. = Rhodiola rotundifolia (Fröd.) S. H. Fu ■

357326 Sedum yvesii Raym.-Hamet; 短蕊景天 (伙连草); Yves Stonecrop ■

357327 Sedum zentaro-tashiroi Makino; 田代善景天■☆

357328 Seegeriella Senghas(1983); 西格兰属■☆

357329 Seemannantha Alef. = Macronyx Dalzell ●■

357330 Seemannantha Alef. = Tephrosia Pers. (保留属名) ●■

357331 Seemannaralia R. Vig. (1906); 西曼五加属●☆

357332 Seemannaralia R. Vig. et R. A. Dyer = Seemannaralia R. Vig. ●☆

357333 Seemannaralia gerrardii (Seem.) Harms; 西曼五加●☆

357334 Seemannia Hook. = Pentagonia Benth. (保留属名) ■☆

357335 Seemannia Regel(1855) (保留属名); 秘鲁苣苔属 (苦乐花属) ■☆

357336 Seemannia Regel(保留属名) = Gloxinia L'Hér. ■☆

357337 Seetzenia R. Br. = Seetzenia R. Br. ex Decne. ■☆

357338 Seetzenia R. Br. ex Decne. (1835); 西茨蒺藜属■☆

357339 Seetzenia africana R. Br. = Seetzenia lanata (Willd.) Bullock ■☆

357340 Seetzenia lanata (Willd.) Bullock; 西茨蒺藜■☆

357341 Seetzenia orientalis Decne. = Seetzenia lanata (Willd.) Bullock ■☆

357342 Seetzenia prostrata (Thunb.) Eckl. et Zeyh. = Seetzenia lanata (Willd.) Bullock ■☆

357343 Seezenia Nees = Seetzenia R. Br. ex Decne. ■☆

357344 Segeretia G. Don = Sageretia Brongn. ●

357345 Segetella (Pers.) R. Hedw. = Alsine Druce ■

357346 Segetella Desv. = Delia Dumort. ■

357347 Segetella Desv. = Spergularia (Pers.) J. Presl et C. Presl(保留属名) ■

357348 Segregatia A. Wood = Sageretia Brongn. ●

357349 Seguidera Rchb. ex Oliv. = Combretum Loefl. (保留属名) ●

357350 Seguiera Adans. = Seguieria Loefl. ●☆

357351 Seguiera Kuntze = Blackstonia Huds. ■☆

357352 Seguiera Rchb. ex Oliv. = Combretum Loefl. (保留属名) ●

357353 Seguieria Loefl. (1758); 翅果商陆属 (翅果珊瑚属) ●☆

357354 Seguieria americana L.; 翅果商陆●☆

357355　Seguieria asiatica Lour. = Tetracera asiatica (Lour.) Hoogland ●

357356　Seguieria asiatica Lour. = Tetracera sarmentosa (L.) Vahl ●

357357　Seguieriaceae Nakai = Petiveriaceae C. Agardh ■☆

357358　Seguieriaceae Nakai = Phytolaccaceae R. Br. (保留科名)●■

357359　Seguinum Raf. = Dieffenbachia Schott ●■

357360　Segurola Larranaga = ? Aeschynomene L. ●■

357361　Sehima Forssk. (1775);沟颖草属;Furrowglume,Sehima ■

357362　Sehima elegans (Roth) Roberty = Thelepogon elegans Roth ex Roem. et Schult. ■☆

357363　Sehima galpinii Stent;盖尔沟颖草■☆

357364　Sehima ischaemoides Forssk.;鸭嘴草状沟颖草■☆

357365　Sehima nervosa (Rottler) Stapf;沟颖草;Common Sehima, Furrowglume ■

357366　Sehima variegata (Stapf) Roberty = Andropterum stolzii (Pilg.) C. E. Hubb. ■☆

357367　Seidelia Baill. (1858);赛德尔大戟属■☆

357368　Seidelia firmula (Prain) Pax et K. Hoffm.;坚硬赛德尔大戟■☆

357369　Seidelia mercurialis Baill. = Seidelia triandra (E. Mey.) Pax ■☆

357370　Seidelia pumila (Sond.) Baill.;矮赛德尔大戟■☆

357371　Seidelia triandra (E. Mey.) Pax;三蕊赛德尔大戟■☆

357372　Seidenfadenia Garay(1972);塞氏兰属■☆

357373　Seidenfadenia mitrata (Rchb. f.) Garay.;塞氏兰■☆

357374　Seidenfadeniella C. S. Kumar(1994);小塞氏兰属■☆

357375　Seidenfadeniella chrysantha (Alston) C. S. Kumar;金花小塞氏兰■☆

357376　Seidenfadeniella filiformis (Rchb. f.) Christenson et Ormerod;线形小塞氏兰■☆

357377　Seidenfadeniella rosea (Wight) C. S. Kumar;粉红小塞氏兰■☆

357378　Seidenfia Szlach. = Crepidium Blume ■

357379　Seidenforchis Marg. = Crepidium Blume ■

357380　Seidenforchis mackinnonii (Duthie) Marg. = Crepidium mackinnonii (Duthie) Szlach. ■

357381　Seidlia Kostel. = Vatica L. ●

357382　Seidlia Opiz = Scirpus L. (保留属名)■

357383　Seidlia radicans (Schkuhr.) Opiz = Scirpus radicans Schkuhr ■

357384　Seidlitzia Bunge ex Boiss. (1879);裂盘藜属■☆

357385　Seidlitzia florida (M. Bieb.) Boiss. = Seidlitzia florida (M. Bieb.) Bunge ex Boiss. ■☆

357386　Seidlitzia florida (M. Bieb.) Bunge ex Boiss.;裂盘藜■☆

357387　Seidlitzia rosmarinus (Ehrh.) Bunge;迷迭香叶裂盘藜■☆

357388　Seidlitzia rosmarinus Boiss. = Seidlitzia rosmarinus (Ehrh.) Bunge ■☆

357389　Sekanama Speta = Urginea Steinh. ■☆

357390　Sekanama Speta(2001);塞卡风信子属■☆

357391　Sekanama burkei (Baker) Speta = Urginea burkei Baker ■☆

357392　Sekanama delagoensis (Baker) Speta = Urginea delagoensis Baker ■☆

357393　Sekanama sanguinea (Schinz) Speta = Urginea sanguinea Schinz ■☆

357394　Sekika Medik. = Saxifraga L. ■

357395　Sekika sarmentosa (L. f.) Moench = Saxifraga stolonifera Curtis ■

357396　Selaginaceae Choisy(1823)(保留科名);穗花科●■

357397　Selaginaceae Choisy(保留科名) = Scrophulariaceae Juss. (保留科名)●■

357398　Selaginastrum Schinz et Thell. = Antherothamnus N. E. Br. ●☆

357399　Selaginastrum karasmontanum Schinz et Thell. = Antherothamnus pearsonii N. E. Br. ●☆

357400　Selaginastrum rigidum (L. Bolus) Schinz et Thell. = Antherothamnus pearsonii N. E. Br. ●☆

357401　Selaginastrum rigidum (L. Bolus) Schinz et Thell. var. karasmontanum Schinz et Thell. = Antherothamnus pearsonii N. E. Br. ●☆

357402　Selago Adans. = Camphorosma L. + Polycnemum L. ■

357403　Selago Adans. = Camphorosma L. ●■

357404　Selago L. (1753);塞拉玄参属●☆

357405　Selago abietina Burm. f. ;冷杉塞拉玄参●☆

357406　Selago acocksii Hilliard;阿氏塞拉玄参●☆

357407　Selago acutibractea Hilliard;尖苞片塞拉玄参●☆

357408　Selago adenodes Hilliard;腺塞拉玄参●☆

357409　Selago adpressa Choisy = Globulariopsis adpressa (Choisy) Hilliard ●☆

357410　Selago aggregata Rolfe = Tetraselago wilmsii (Rolfe) Hilliard et B. L. Burtt ●☆

357411　Selago albanensis Schltr. = Selago recurva E. Mey. ●☆

357412　Selago albida Choisy;微白塞拉玄参●☆

357413　Selago albomarginata Hilliard;白边塞拉玄参●☆

357414　Selago albomontana Hilliard;山地白塞拉玄参●☆

357415　Selago alliodora Dinter ex Engl. = Selago albomarginata Hilliard ●☆

357416　Selago alopecuroides Rolfe;看麦娘塞拉玄参●☆

357417　Selago amboensis Rolfe;安博塞拉玄参●☆

357418　Selago angolensis Rolfe;安哥拉塞拉玄参●☆

357419　Selago angustibractea Hilliard;狭苞塞拉玄参●☆

357420　Selago angustifolia Thunb. = Microdon dubius (L.) Hilliard ●☆

357421　Selago apiculata E. Mey. = Glumicalyx apiculatus (E. Mey.) Hilliard et B. L. Burtt ■☆

357422　Selago appressa E. Mey. = Selago divaricata L. f. ●☆

357423　Selago arguta E. Mey. = Pseudoselago arguta (E. Mey.) Hilliard ●☆

357424　Selago articulata Thunb.;关节塞拉玄参●☆

357425　Selago ascendens E. Mey. = Pseudoselago ascendens (E. Mey.) Hilliard ●☆

357426　Selago aspera Choisy;粗糙拉塞拉玄参●☆

357427　Selago atherstonei Rolfe;澳非拉塞拉玄参●☆

357428　Selago barabei (Mielcarek) Hilliard;巴尔伯塞拉玄参●☆

357429　Selago barbula Harv. ex Rolfe;髯毛塞拉玄参●☆

357430　Selago baurii (Hiern) Hilliard;巴利塞拉玄参■☆

357431　Selago beaniana Hilliard;比恩塞拉玄参●☆

357432　Selago bilacunosa Hilliard;双腔塞拉玄参●☆

357433　Selago blantyrensis Rolfe;布兰太尔塞拉玄参●☆

357434　Selago bolusii Rolfe;博卢斯塞拉玄参●☆

357435　Selago bracteata Thunb. = Microdon parviflorus (P. J. Bergius) Hilliard ●☆

357436　Selago brevifolia Rolfe;短叶塞拉玄参●☆

357437　Selago buchananii Rolfe = Selago blantyrensis Rolfe ●☆

357438　Selago burchellii Rolfe;伯切尔塞拉玄参●☆

357439　Selago burkei Rolfe;伯克塞拉玄参●☆

357440　Selago burmannii Choisy = Pseudoselago burmannii (Choisy) Hilliard ●☆

357441　Selago caerulea Rolfe;天蓝塞拉玄参●☆

357442　Selago canescens L. f. ;灰塞拉玄参●☆

357443　Selago capitata P. J. Bergius = Microdon capitatus (P. J. Bergius) Levyns ●☆

357444　Selago capitellata Schltr. ;小头塞拉玄参●☆

357445　Selago capituliflora Rolfe;头花塞拉玄参●☆

357446　Selago cephalophora Thunb. = Phyllopodium cephalophorum (Thunb.) Hilliard ■☆

357447　Selago chenopodioides Diels ＝ Chenopodiopsis chenopodioides（Diels）Hilliard ■☆

357448　Selago choisiana E. Mey. ＝ Selago paniculata Thunb. ●☆

357449　Selago chongweensis（Rolfe）Torre et Harms ＝ Selago angolensis Rolfe ●☆

357450　Selago ciliata L. f. ;缘毛塞拉玄参●☆

357451　Selago cinerascens E. Mey. ＝ Selago cinerea L. f. ●☆

357452　Selago cinerascens E. Mey. var. exasperata ? ＝ Selago cinerea L. f. ●☆

357453　Selago cinerea Drège ＝ Selago scabrida Thunb. ●☆

357454　Selago cinerea E. Mey. ＝ Selago polystachya L. ●☆

357455　Selago cinerea L. f. ;灰色塞拉玄参●☆

357456　Selago coccinea L. ＝ Pseudoselago rapunculoides（L.）Hilliard ●☆

357457　Selago coerulea Burch. ex Hochst. ＝ Selago decipiens E. Mey. ●☆

357458　Selago comosa E. Mey. ;簇毛塞拉玄参●☆

357459　Selago compacta Rolfe ;紧密塞拉玄参●☆

357460　Selago comptonii Hilliard ;康普顿塞拉玄参●☆

357461　Selago confusa Hilliard ;混乱塞拉玄参●☆

357462　Selago congesta Rolfe ;密集塞拉玄参●☆

357463　Selago cooperi Rolfe ＝ Selago galpinii Schltr. ●☆

357464　Selago cordata E. Mey. ＝ Cromidon decumbens（Thunb.）Hilliard ■☆

357465　Selago cordata Thunb. ＝ Phyllopodium cordatum（Thunb.）Hilliard ■☆

357466　Selago corymbosa L. ;伞序塞拉玄参●☆

357467　Selago corymbosa L. var. polystachya（L.）E. Mey. ＝ Selago polystachya L. ●☆

357468　Selago crassifolia（Rolfe）Hilliard ;厚叶塞拉玄参●☆

357469　Selago cryptadenia Hilliard ;隐腺塞拉玄参●☆

357470　Selago cucullata Hilliard ;僧帽状塞拉玄参●☆

357471　Selago cupressoides Hilliard ;铜色塞拉玄参●☆

357472　Selago curvifolia Rolfe ;折叶塞拉玄参●☆

357473　Selago cylindrica Levyns ＝ Selago scabrida Thunb. ●☆

357474　Selago decipiens E. Mey. ;迷惑塞拉玄参●☆

357475　Selago decumbens Choisy ＝ Pseudoselago quadrangularis（Choisy）Hilliard ●☆

357476　Selago decumbens Thunb. ＝ Cromidon decumbens（Thunb.）Hilliard ■☆

357477　Selago densiflora Rolfe ;密花塞拉玄参●☆

357478　Selago densifolia Hochst. ＝ Pseudoselago densifolia（Hochst.）Hilliard ●☆

357479　Selago dentata Poir. ＝ Pseudoselago spuria（L.）Hilliard ●☆

357480　Selago diabolica Hilliard ;魔鬼塞拉玄参●☆

357481　Selago diffusa Hochst. ＝ Selago linearis Rolfe ●☆

357482　Selago diffusa Thunb. ;松散塞拉玄参●☆

357483　Selago dinteri Rolfe ;丁特塞拉玄参●☆

357484　Selago dinteri Rolfe subsp. pseudodinteri Hilliard ;异丁特塞拉玄参●☆

357485　Selago diosmoides Rolfe ＝ Selago diffusa Thunb. ●☆

357486　Selago distans E. Mey. ;远离塞拉玄参●☆

357487　Selago distans Lindl. ＝ Selago bolusii Rolfe ●☆

357488　Selago divaricata L. f. ;叉开塞拉玄参●☆

357489　Selago dolichonema Hilliard ;长丝塞拉玄参●☆

357490　Selago dolosa Hilliard ;假塞拉玄参●☆

357491　Selago dregeana Hilliard ;德雷塞拉玄参●☆

357492　Selago dregei Rolfe ＝ Selago glandulosa Choisy ●☆

357493　Selago dubia L. ＝ Microdon dubius（L.）Hilliard ●☆

357494　Selago eckloniana Choisy ;埃氏塞拉玄参●☆

357495　Selago elata Choisy ＝ Selago eckloniana Choisy ●☆

357496　Selago elata Rolfe ＝ Selago procera Hilliard ●☆

357497　Selago elegans Choisy ＝ Phyllopodium elegans（Choisy）Hilliard ■☆

357498　Selago elongata Hilliard ;伸长塞拉玄参●☆

357499　Selago elsiae Hilliard ;埃尔西亚塞拉玄参●☆

357500　Selago ericina Drège ＝ Selago ramosissima Rolfe ●☆

357501　Selago ericina E. Mey. ＝ Selago triquetra L. f. ●☆

357502　Selago ericoides L. ＝ Stilbe ericoides（L.）L. ●☆

357503　Selago esterhuyseniae Hilliard ;埃斯特塞拉玄参●☆

357504　Selago exigua Hilliard ;小塞拉玄参●☆

357505　Selago fasciculata L. ＝ Pseudoselago serrata（P. J. Bergius）Hilliard ●☆

357506　Selago fasciculata L. var. hirta（Choisy）E. Mey. ＝ Pseudoselago quadrangularis（Choisy）Hilliard ●☆

357507　Selago fasciculata L. var. lanceolata Walp. ＝ Pseudoselago rapunculoides（L.）Hilliard ●☆

357508　Selago ferruginea Rolfe ;锈色塞拉玄参●☆

357509　Selago flanaganii Rolfe ;弗拉纳根塞拉玄参●☆

357510　Selago foliosa Rolfe ;密叶塞拉玄参●☆

357511　Selago forbesii Rolfe ＝ Selago canescens L. f. ●☆

357512　Selago fourcadei Hilliard ;富尔克塞拉玄参●☆

357513　Selago fruticosa L. ;灌丛塞拉玄参●☆

357514　Selago fruticulosa Rolfe ＝ Selago fruticosa L. ●☆

357515　Selago fulvomaculata Link ＝ Pseudoselago spuria（L.）Hilliard ●☆

357516　Selago galpinii Schltr. ;盖尔塞拉玄参●☆

357517　Selago geniculata L. f. ;膝曲塞拉玄参●☆

357518　Selago gillii Hook. ＝ Selago myrtifolia Rchb. ●☆

357519　Selago glabrata Choisy ;光滑塞拉玄参●☆

357520　Selago glandulosa Choisy ;具腺塞拉玄参●☆

357521　Selago gloioides Hilliard ;胶质塞拉玄参●☆

357522　Selago glomerata Thunb. ;团聚塞拉玄参●☆

357523　Selago glutinosa E. Mey. ;黏性塞拉玄参●☆

357524　Selago glutinosa E. Mey. ＝ Selago dregeana Hilliard ●☆

357525　Selago glutinosa E. Mey. subsp. cylindriphylla Hilliard ;柱叶塞拉玄参●☆

357526　Selago goetzei Rolfe ;格兹塞拉玄参●☆

357527　Selago goetzei Rolfe subsp. ambigua Hilliard ;可疑塞拉玄参●☆

357528　Selago gracilis（Rolfe）Hilliard ;纤细塞拉玄参●☆

357529　Selago grandiceps Hilliard ;大头塞拉玄参●☆

357530　Selago guttata E. Mey. ＝ Pseudoselago guttata（E. Mey.）Hilliard ●☆

357531　Selago hamulosa E. Mey. ＝ Cromidon hamulosum（E. Mey.）Hilliard ■☆

357532　Selago herbacea Choisy ＝ Phyllopodium cuneifolium（L. f.）Benth. ■☆

357533　Selago hermannioides E. Mey. ;密钟木塞拉玄参●☆

357534　Selago heterophylla Rolfe ＝ Pseudoselago ascendens（E. Mey.）Hilliard ●☆

357535　Selago heterophylla Thunb. ＝ Pseudoselago spuria（L.）Hilliard ●☆

357536　Selago heterotricha Hilliard ;异毛塞拉玄参●☆

357537　Selago hirta Choisy ＝ Pseudoselago quadrangularis（Choisy）Hilliard ●☆

357538　Selago hirta L. f. ＝ Chenopodiopsis hirta（L. f.）Hilliard ■☆

357539　Selago hispida L. f. ;粗毛塞拉玄参●☆

357540　Selago hispida L. f. var. major Choisy ＝ Globulariopsis stricta

（P. J. Bergius）Hilliard ●☆

357541　Selago hispida L. f. var. nana Choisy = Globulariopsis stricta （P. J. Bergius）Hilliard ●☆

357542　Selago hispida Sieber ex Choisy = Dischisma capitatum （Thunb.）Choisy ■☆

357543　Selago hoepfneri Rolfe = Selago welwitschii Rolfe ●☆

357544　Selago holstii Rolfe = Selago thomsonii Rolfe ●☆

357545　Selago holubii Rolfe = Selago welwitschii Rolfe var. holubii （Rolfe）Brenan ●☆

357546　Selago humilis Rolfe = Pseudoselago humilis （Rolfe）Hilliard ●☆

357547　Selago hyssopifolia E. Mey. ;神香草叶塞拉玄参●☆

357548　Selago hyssopifolia E. Mey. subsp. retrotricha Hilliard;折毛塞拉玄参●☆

357549　Selago immersa Rolfe;水中塞拉玄参●☆

357550　Selago impedita Hilliard;累赘塞拉玄参●☆

357551　Selago inaequifolia Hilliard;不等叶塞拉玄参●☆

357552　Selago incisa Hochst. = Pseudoselago ascendens （E. Mey.）Hilliard ●☆

357553　Selago intermedia Hilliard;间型塞拉玄参●☆

357554　Selago johnstonii Rolfe = Selago thomsonii Rolfe ●☆

357555　Selago junodii Rolfe = Selago rehmannii Rolfe ●☆

357556　Selago karooica Hilliard;卡鲁塞拉玄参●☆

357557　Selago lacunosa Klotzsch;具腔塞拉玄参●☆

357558　Selago lamprocarpa Schltr. ex Rolfe;亮果塞拉玄参●☆

357559　Selago lanceolata Choisy = Pseudoselago burmannii （Choisy）Hilliard ●☆

357560　Selago laticorymbosa Gilli = Selago viscosa Rolfe ●☆

357561　Selago laxiflora Choisy = Globulariopsis tephrodes （E. Mey.）Hilliard ●☆

357562　Selago lepida Hilliard;小鳞塞拉玄参●☆

357563　Selago lepidioides Rolfe = Selago peduncularis E. Mey. ●☆

357564　Selago leptostachya E. Mey. = Selago geniculata L. f. ☆

357565　Selago leptostachya E. Mey. var. eckloniana Choisy = Selago ferruginea Rolfe ●☆

357566　Selago leptothrix Hilliard;细毛塞拉玄参●☆

357567　Selago levynsiae Hilliard;勒温斯塞拉玄参●☆

357568　Selago lilacina Hilliard;紫丁香色塞拉玄参●☆

357569　Selago linearifolia Rolfe;线叶塞拉玄参●☆

357570　Selago linearis Rolfe;线状塞拉玄参●☆

357571　Selago lithospermoides Rolfe = Selago flanaganii Rolfe ●☆

357572　Selago longicalyx Hilliard;长萼塞拉玄参●☆

357573　Selago longiflora Rolfe;长花塞拉玄参●☆

357574　Selago longipedicellata Rolfe;长梗塞拉玄参●☆

357575　Selago longithyrsis Gilli = Selago nyasae Rolfe ■☆

357576　Selago longituba Rolfe = Tetraselago longituba （Rolfe）Hilliard et B. L. Burtt ■☆

357577　Selago lucida Vent. = Microdon parviflorus （P. J. Bergius）Hilliard ●☆

357578　Selago luxurians Choisy;茂盛塞拉玄参●☆

357579　Selago lychnidea L. = Lyperia lychnidea （L.）Druce ■☆

357580　Selago lydenburgensis Rolfe;莱登堡塞拉玄参●☆

357581　Selago magnakarooica Hilliard;马格纳塞拉玄参●☆

357582　Selago marlothii Hilliard;马洛斯塞拉玄参●☆

357583　Selago mcclouniei Rolfe = Selago viscosa Rolfe ●☆

357584　Selago mediocris Hilliard;中位塞拉玄参●☆

357585　Selago melleri Rolfe = Selago blantyrensis Rolfe ●☆

357586　Selago melliodora Hilliard;蜜味塞拉玄参●☆

357587　Selago meyeri Choisy = Selago canescens L. f. ●☆

357588　Selago michelliae Hilliard;米歇尔塞拉玄参●☆

357589　Selago micradenia Hilliard;小腺塞拉玄参●☆

357590　Selago micrantha Choisy = Selago paniculata Thunb. ●☆

357591　Selago milanjiensis Rolfe = Selago whyteana Rolfe ●☆

357592　Selago minutissima Choisy = Selago divaricata L. f. ●☆

357593　Selago mixta Hilliard;混杂塞拉玄参●☆

357594　Selago monticola J. M. Wood et M. S. Evans;山生塞拉玄参●☆

357595　Selago montis-shebae Brenan = Tetraselago longituba （Rolfe）Hilliard et B. L. Burtt ■☆

357596　Selago mucronata Hilliard;短尖塞拉玄参●☆

357597　Selago muddii Rolfe = Selago atherstonei Rolfe ●☆

357598　Selago multiflora Hilliard;多花塞拉玄参●☆

357599　Selago multispicata Hilliard;多穗塞拉玄参●☆

357600　Selago mundii Rolfe;蒙德塞拉玄参●☆

357601　Selago muralis Benth. et Hook. f. ;厚壁塞拉玄参●☆

357602　Selago myriophylla Hilliard;多叶塞拉玄参●☆

357603　Selago myrtifolia Rchb. ;香桃木叶塞拉玄参●☆

357604　Selago nachtigalii Rolfe;纳赫蒂加尔塞拉玄参●☆

357605　Selago namaquensis Schltr. ;纳马夸塞拉玄参●☆

357606　Selago natalensis Rolfe = Tetraselago natalensis （Rolfe）Junell ■☆

357607　Selago neglecta Hilliard;疏忽塞拉玄参●☆

357608　Selago nelsonii Rolfe = Tetraselago nelsonii （Rolfe）Hilliard et B. L. Burtt ●☆

357609　Selago nigrescens Rolfe;变黑塞拉玄参●☆

357610　Selago nigromontana Hilliard;山地塞拉玄参●☆

357611　Selago nitida （E. Mey.）Schltr. = Selago myrtifolia Rchb. ●☆

357612　Selago nutans Rolfe = Glumicalyx nutans （Rolfe）Hilliard et B. L. Burtt ■☆

357613　Selago nyasae Rolfe;尼亚萨塞拉玄参■☆

357614　Selago nyikensis Rolfe = Selago thyrsoidea Baker ●☆

357615　Selago ohlendorffiana Lehm. = Selago myrtifolia Rchb. ●☆

357616　Selago oppositifolia Hilliard;对叶塞拉玄参●☆

357617　Selago ovata （L.）Aiton = Microdon capitatus （P. J. Bergius）Levyns ●☆

357618　Selago ovata Rolfe = Selago pinguicula E. Mey. ☆

357619　Selago pachypoda Rolfe;粗足塞拉玄参●☆

357620　Selago pallida Salisb. = Pseudoselago spuria （L.）Hilliard ●☆

357621　Selago paniculata Thunb. ;圆锥塞拉玄参●☆

357622　Selago parvibractea Hilliard;小苞塞拉玄参●☆

357623　Selago peduncularis E. Mey. ;梗花塞拉玄参●☆

357624　Selago pentheri Gand. = Pseudoselago serrata （P. J. Bergius）Hilliard ●☆

357625　Selago perplexa Hilliard;缠结塞拉玄参●☆

357626　Selago persimilis Hilliard;相似塞拉玄参●☆

357627　Selago phyllopodioides Schltr. = Phyllopodium phyllopodioides （Schltr.）Hilliard ■☆

357628　Selago pinastra L. = Stilbe vestita P. J. Bergius ●☆

357629　Selago pinea Drège = Selago eckloniana Choisy ☆

357630　Selago pinea Link;松林塞拉玄参●☆

357631　Selago pinguicula E. Mey. ;肥厚塞拉玄参●☆

357632　Selago polycephala Otto ex Walp. ;多头塞拉玄参●☆

357633　Selago polygala S. Moore;远志塞拉玄参●☆

357634　Selago polygaloides L. f. = Microdon polygaloides （L. f.）Druce ●☆

357635　Selago polystachya L. ;密穗塞拉玄参●☆

357636　Selago praetermissa Hilliard;含糊塞拉玄参●☆

357637　Selago procera Hilliard;高大塞拉玄参●☆

357638　Selago prostrata Hilliard;平卧塞拉玄参●☆

357639　Selago psammophila Hilliard;喜沙塞拉玄参●☆

357640　Selago pterophylla Otto ex Sweet;翅叶塞拉玄参●☆

357641　Selago pubescens Rolfe = Selago burchellii Rolfe ●☆

357642　Selago pulchella Salisb. = Pseudoselago rapunculoides（L.）Hilliard ●☆

357643　Selago pulchra Hilliard;美丽塞拉玄参●☆

357644　Selago punctata Rolfe;斑点塞拉玄参●☆

357645　Selago purpurea Cels;紫塞拉玄参●☆

357646　Selago pustulosa Hilliard;刚毛塞拉玄参●☆

357647　Selago quadrangularis Choisy = Pseudoselago quadrangularis（Choisy）Hilliard ●☆

357648　Selago racemosa Bernh. = Selago trinervia E. Mey. ●☆

357649　Selago ramosissima Rolfe;多枝塞拉玄参●☆

357650　Selago ramulosa E. Mey. = Selago canescens L. f. ●☆

357651　Selago ramulosa Link = Pseudoselago verbenacea（L. f.）Hilliard ●☆

357652　Selago rapunculoides L. = Pseudoselago rapunculoides（L.）Hilliard ●☆

357653　Selago rapunculoides L. var. densifolia（Hochst.）Choisy = Pseudoselago densifolia（Hochst.）Hilliard ●☆

357654　Selago recurva E. Mey.;反折塞拉玄参●☆

357655　Selago rehmannii Rolfe;拉赫曼塞拉玄参●☆

357656　Selago retropilosa Hilliard;卷毛塞拉玄参●☆

357657　Selago rigida Rolfe;硬塞拉玄参●☆

357658　Selago robusta Rolfe = Selago pinguicula E. Mey. ●☆

357659　Selago rotundifolia L. f. ;圆叶塞拉玄参●☆

357660　Selago rubromontana Hilliard;山红塞拉玄参●☆

357661　Selago rudolphii（Hiern）Levyns = Pseudoselago humilis（Rolfe）Hilliard ●☆

357662　Selago rustii Rolfe = Phyllopodium rustii（Rolfe）Hilliard ●☆

357663　Selago sandersonii Rolfe = Selago galpinii Schltr. ●☆

357664　Selago saundersiae Rolfe = Selago capitellata Schltr. ●☆

357665　Selago saxatilis E. Mey. ;岩栖塞拉玄参●☆

357666　Selago scabribractea Hilliard;糙苞塞拉玄参●☆

357667　Selago scabrida Thunb. ;微糙塞拉玄参●☆

357668　Selago schlechteri Rolfe = Selago galpinii Schltr. ●☆

357669　Selago serpentina Hilliard;蛇形塞拉玄参●☆

357670　Selago serrata P. J. Bergius = Pseudoselago serrata（P. J. Bergius）Hilliard ●☆

357671　Selago seticaulis Hilliard;毛茎塞拉玄参●☆

357672　Selago setulosa Rolfe;多刚毛塞拉玄参●☆

357673　Selago singularis Hilliard;单一塞拉玄参●☆

357674　Selago speciosa Rolfe;美花塞拉玄参●☆

357675　Selago spectabilis Hilliard;壮观塞拉玄参●☆

357676　Selago spicata Link = Microdon polygaloides（L. f.）Druce ●☆

357677　Selago spuria L. = Pseudoselago spuria（L.）Hilliard ●☆

357678　Selago squarrosa Choisy = Hebenstretia ramosissima Jaroscz ●☆

357679　Selago stenostachya Hilliard;狭穗塞拉玄参●☆

357680　Selago stewartii S. Moore;斯图尔特塞拉玄参●☆

357681　Selago stricta P. J. Bergius = Globulariopsis stricta（P. J. Bergius）Hilliard ●☆

357682　Selago subspinosa Hilliard;具刺塞拉玄参●☆

357683　Selago swaziensis Rolfe;斯威士塞拉玄参●☆

357684　Selago swynnertonii（S. Moore）Hilliard;斯温纳顿塞拉玄参●☆

357685　Selago swynnertonii（S. Moore）Hilliard var. leiophylla（Brenan）Hilliard;光叶塞拉玄参●☆

357686　Selago tenuicaulis Rolfe = Selago caerulea Rolfe ●☆

357687　Selago tenuifolia（Rolfe）Hilliard;细叶塞拉玄参●☆

357688　Selago tenuis E. Mey.;细塞拉玄参●☆

357689　Selago tephrodes Drège = Selago canescens L. f. ●☆

357690　Selago tephrodes E. Mey. = Globulariopsis tephrodes（E. Mey.）Hilliard ●☆

357691　Selago teretifolia Link = Pseudoselago rapunculoides（L.）Hilliard ●☆

357692　Selago teucriifolia Burm. f. ;香科叶塞拉玄参●☆

357693　Selago thermalis Hilliard;温泉塞拉玄参●☆

357694　Selago thomii Rolfe;汤姆塞拉玄参●☆

357695　Selago thomsonii Rolfe;托马森塞拉玄参●☆

357696　Selago thomsonii Rolfe var. caerulea（Rolfe）Brenan = Selago caerulea Rolfe ●☆

357697　Selago thomsonii Rolfe var. whyteana（Rolfe）Brenan = Selago whyteana Rolfe ●☆

357698　Selago thunbergii Choisy = Selago canescens L. f. ●☆

357699　Selago thyrsoidea Baker;聚伞塞拉玄参●☆

357700　Selago thyrsoidea Baker var. austrorhodesica Brenan;南罗得西亚塞拉玄参●☆

357701　Selago thyrsoidea Baker var. nyikensis（Rolfe）Brenan = Selago thyrsoidea Baker ●☆

357702　Selago tomentosa L. = Manulea tomentosa（L.）L. ■☆

357703　Selago transvaalensis Rolfe = Tetraselago nelsonii（Rolfe）Hilliard et B. L. Burtt ●☆

357704　Selago trichophylla Hilliard;毛叶塞拉玄参●☆

357705　Selago trinervia E. Mey. ;三脉塞拉玄参●☆

357706　Selago triquetra E. Mey. = Selago fruticosa L. ●☆

357707　Selago triquetra L. f. ;三棱塞拉玄参●☆

357708　Selago tysonii Rolfe = Selago trinervia E. Mey. ●☆

357709　Selago variicalyx Hilliard;杂萼塞拉玄参●☆

357710　Selago venosa Hilliard;多脉塞拉玄参●☆

357711　Selago verbenacea L. f. = Pseudoselago verbenacea（L. f.）Hilliard ●☆

357712　Selago verbenacea L. f. var. villosa Choisy = Pseudoselago verbenacea（L. f.）Hilliard ●☆

357713　Selago verna Hilliard;春塞拉玄参●☆

357714　Selago villicalyx Rolfe = Selago pachypoda Rolfe ●☆

357715　Selago villicaulis Rolfe;柔毛茎塞拉玄参●☆

357716　Selago villosa Rolfe;长柔毛塞拉玄参●☆

357717　Selago viscosa Rolfe;黏塞拉玄参●☆

357718　Selago welwitschii Rolfe;韦尔塞拉玄参●☆

357719　Selago welwitschii Rolfe var. holubii（Rolfe）Brenan;霍勒布塞拉玄参●☆

357720　Selago whyteana Rolfe;怀特塞拉玄参●☆

357721　Selago wilmsii Rolfe = Tetraselago wilmsii（Rolfe）Hilliard et B. L. Burtt ●☆

357722　Selago witbergensis E. Mey. ;沃特山塞拉玄参●☆

357723　Selago wittebergensis Compton = Globulariopsis montana Hilliard ●☆

357724　Selago woodii Rolfe = Selago peduncularis E. Mey. ●☆

357725　Selago zeyheri Choisy;泽赫塞拉玄参●☆

357726　Selago zeyheri Rolfe = Selago acocksii Hilliard ●☆

357727　Selago zuluensis Hilliard;祖卢塞拉玄参●☆

357728　Selas Spreng. = Acronychia J. R. Forst. et G. Forst.（保留属名）

357729　Selas Spreng. = Gela Lour. ●

357730　Selatium D. Don ex G. Don = Gentianella Moench（保留属名）■

357731　Selatium G. Don = Gentianella Moench（保留属名）■

357732　Selbya M. Roem. = Aglaia Lour. (保留属名)●

357733　Selenia Nutt. (1825);金月芥属■☆

357734　Selenia aurea Nutt. ;金月芥■☆

357735　Selenicereus(A. Berger) Britton et Rose(1909);蛇鞭柱属(大轮柱属,神堂属,天轮柱属,月光掌属,月光柱属);Moon Cactus, Moon Cereus, Moonlight Cactus, Night-blooming Cereus ●

357736　Selenicereus Britton et Rose = Selenicereus(A. Berger) Britton et Rose ●

357737　Selenicereus anthonyanus(Alexander) D. R. Hunt;角叶孔雀●☆

357738　Selenicereus coniflorus(Weing.) Britton et Rose = Cereus grandiflorus(L.) Mill. ●

357739　Selenicereus coniflorus Britton et Rose;锥花蛇鞭柱●☆

357740　Selenicereus donkelaarii Britton et Rose;栋克蛇鞭柱;Chacuob ●☆

357741　Selenicereus grandiflorus(L.) Britton et Rose;夜皇后(花蛇鞭柱,夜后花)●☆

357742　Selenicereus grandiflorus(L.) Britton et Rose = Cereus grandiflorus Mill. ●

357743　Selenicereus hamatus Britton et Rose;钩刺蛇鞭柱;Queen of the Night ●☆

357744　Selenicereus macdonaldiae(Hook.) Britton et Rose;蛇鞭柱(马氏蛇鞭柱,夜之女王,月光仙人柱);Queen-of-night ●

357745　Selenicereus megalanthus(K. Schum. ex Vaupel) Moran;大花蛇鞭柱;Yellow Pitaya ●☆

357746　Selenicereus nycticalus(Link) T. Marshall = Selenicereus pteranthus(Link ex A. Dietr.) Britton et Rose ●

357747　Selenicereus pteranthus(Link et Otto) Britton et Rose = Selenicereus pteranthus(Link ex A. Dietr.) Britton et Rose ●

357748　Selenicereus pteranthus(Link ex A. Dietr.) Britton et Rose;翼花蛇鞭柱(嫦娥仙人柱,夜公主,夜美人柱,夜之女王);King-of-the-night, Pitahaya Real, Princess of the Night, Princess-of-the-night, Snake Cactus ●

357749　Selenicereus spinulosus(DC.) Britton et Rose;小刺蛇鞭柱;Vinelike moonlight cactus ●☆

357750　Selenipedilum Pfitzer = Selenipedium Rchb. f. ■☆

357751　Selenipedium Rchb. f. (1854);月兰属;Selenipedium ■☆

357752　Selenipedium isabelianum Barb. Rodr. ;深黄月兰;Darkyellow Selenipedium ■☆

357753　Selenipedium palmifolium(Lindl.) Rchb. f. ;棕叶月兰;Palmleaf Selenipedium ■☆

357754　Selenipedium parishii(Rchb. f.) André = Paphiopedilum parishii(Rchb. f.) Pfitzer ■

357755　Selenocarpaea(DC.) Eckl. et Zeyh. = Heliophila Burm. f. ex L. ●■☆

357756　Selenocarpaea Eckl. et Zeyh. = Heliophila Burm. f. ex L. ●■☆

357757　Selenocera Zipp. ex Span. = Mitreola L. ■

357758　Selenocera secundiflora Zipp. ex Span. = Mitreola petiolata(J. F. Gmel.) Torr. et A. Gray ■

357759　Selenogyne DC. = Lagenophora Cass. (保留属名)■●

357760　Selenogyne DC. = Solenogyne Cass. ■●☆

357761　Selenothamnus Melville = Lawrencia Hook. ●☆

357762　Selepsion Raf. = Urtica L. ■

357763　Selera Ulbr. = Gossypium L. ●■

357764　Seleranthus Hill = Scleranthus L. ■☆

357765　Seleria Boeck. = Scleria P. J. Bergius ■

357766　Selerothamnus Harms = Hesperothamnus Brandegee ●■

357767　Selerothamnus Harms = Millettia Wight et Arn. (保留属名)●■

357768　Selinaceae Bercht. et J. Presl = Labiatae Juss. (保留科名)●■

357769　Selinaceae Bercht. et J. Presl = Lamiaceae Martinov(保留科名)●■

357770　Selinocarpus A. Gray(1853);翅果茉莉属●☆

357771　Selinocarpus diffusus A. Gray;松散翅果茉莉■☆

357772　Selinocarpus somalensis Chiov. = Acleisanthes somalensis(Chiov.) R. A. Levin ■☆

357773　Selinon Adans. = Apium L. ■

357774　Selinon Raf. = Selinum L. (保留属名)■

357775　Selinopsis Coss. et Durieu ex Batt. et Trab. (1905);拟亮蛇床属■☆

357776　Selinopsis Coss. et Durieu ex Munby = Carum L. ■

357777　Selinopsis foetida Batt. = Carum foetidum(Batt.) Drude ■☆

357778　Selinopsis montana Batt. = Carum montanum(Batt.) Benth. et Hook. f. ■☆

357779　Selinum L. (1753)(废弃属名) = Selinum L. (1762)(保留属名)■

357780　Selinum L. (1762)(保留属名);亮蛇床属(滇前胡属);Cambridge Milk-parsley, Milk Parsley, Selinum ■

357781　Selinum ammoides E. H. L. Krause = Ammi majus L. ■

357782　Selinum angolense C. Norman = Pseudoselinum angolense(C. Norman) C. Norman ■☆

357783　Selinum anisum(L.) E. H. L. Krause = Pimpinella anisum L. ■

357784　Selinum annesorhizum F. Muell. = Annesorhiza nuda(Aiton) B. L. Burtt ■☆

357785　Selinum baicalense I. Redowsk = Peucedanum baicalense(I. Redowsky ex Willd.) W. D. J. Koch ■

357786　Selinum baicalense I. Redowsk ex Willd. = Peucedanum baicalense(I. Redowsky ex Willd.) W. D. J. Koch ■

357787　Selinum candollei DC. = Selinum wallichianum(DC.) Raizada et H. O. Saxena ■

357788　Selinum candollei Edgew. = Selinum wallichianum(DC.) Raizada et H. O. Saxena ■

357789　Selinum carvifolia L. ;小叶亮蛇床;Cambridge Milk-parsley, Cambridge Parsley, False Milk Parsley, Little-leaf Angelica ■☆

357790　Selinum chinense(L.) Druce = Conioselinum chinense(L.) Britton, Sterns et Poggenb. ■

357791　Selinum coreanum H. Boissieu = Angelica polymorpha Maxim. ■

357792　Selinum coriandrum E. H. L. Krause = Coriandrum sativum L. ■

357793　Selinum cortioides C. Norman;无茎亮蛇床;Stemless Selinum ■

357794　Selinum cryptotaenium H. Boissieu;亮蛇床(滇前胡);Selinum ■

357795　Selinum dubium subsp. salinum(Turcz.) Leute = Cnidium salinum Turcz. ■

357796　Selinum filicifolium(Edgew.) Nasir;喜马拉雅亮蛇床■☆

357797　Selinum foeniculum(L.) E. H. L. Krause = Foeniculum vulgare(L.) Mill. ■

357798　Selinum galbanum(L.) Spreng. = Peucedanum galbanum(L.) Drude ■☆

357799　Selinum graveolens(L.) Krause = Apium graveolens L. ■

357800　Selinum gummiferum(L.) Spreng. = Peucedanum gummiferum(L.) Wijnands ■☆

357801　Selinum japonicum(Miq.) Franch. et Sav. = Cnidium japonicum Miq. ■

357802　Selinum kultiassovii Korovin;库尔亮蛇床■☆

357803　Selinum leptophyllum(Pers.) E. H. L. Krause ex Sturm = Cyclospermum leptophyllum(Pers.) Sprague ex Britton et P. Wilson ■

357804　Selinum levisticum(L.) E. H. L. Krause. = Levisticum officinale W. D. J. Koch ■

357805　Selinum levisticum E. H. L. Krause ＝ Levisticum officinale K. Koch ■

357806　Selinum longicalycium M. L. Sheh；长萼亮蛇床■

357807　Selinum melanotilingia H. Boissieu ＝ Angelica decursiva（Miq.）Franch. et Sav. ■

357808　Selinum monnieri L. ＝ Cnidium monnieri（L.）Cusson ■

357809　Selinum nullivittatum（K. T. Fu）C. C. Yuan et L. B. Li ＝ Ligusticum nullivittatum（K. T. Fu）F. T. Pu et M. F. Watson ■

357810　Selinum oliverianum H. Boissieu ＝ Ligusticum oliverianum（H. Boissieu）R. H. Shan ■

357811　Selinum palustre L. ＝ Peucedanum palustre（L.）Moench ■☆

357812　Selinum papyraceum C. B. Clarke；印巴亮蛇床■☆

357813　Selinum pastinaca（L.）Crantz. ＝ Pastinaca sativa L. ■

357814　Selinum popovii（Korovin）Schischk. ；波氏亮蛇床☆

357815　Selinum salinum（Turcz.）Vodop. ＝ Cnidium salinum Turcz. ■

357816　Selinum sinchianum（K. T. Fu）C. Q. Yuan et L. B. Li ＝ Cnidium sinchianum K. T. Fu ■

357817　Selinum stellatum D. Don ＝ Pleurospermum stellatum（D. Don）C. B. Clarke ■

357818　Selinum stewartii Hiroe ＝ Ligusticum stewartii（Hiroe）Nasir ■☆

357819　Selinum striatum（DC.）Benth. et Hook. f. ＝ Ligusticum striatum Wall. ex DC. ■

357820　Selinum striatum Benth. ex C. B. Clarke ＝ Ligusticum striatum Wall. ex DC. ■

357821　Selinum suffruticosum（P. J. Bergius）Benth. et Hook. f. ＝ Dasispermum suffruticosum（P. J. Bergius）B. L. Burtt ☆

357822　Selinum sylvestre L. ＝ Peucedanum palustre（L.）Moench ■☆

357823　Selinum tenuifolium Wall. ＝ Selinum wallichianum（DC.）Raizada et H. O. Saxena ■

357824　Selinum tenuifolium Wall. ex C. B. Clarke ＝ Selinum candollei DC. ■

357825　Selinum tenuifolium Wall. ex C. B. Clarke ＝ Selinum wallichianum（DC.）Raizada et H. O. Saxena ■

357826　Selinum tenuifolium Wall. ex C. B. Clarke var. filicifolia（Edgew.）C. B. Clarke ＝ Selinum filicifolium（Edgew.）Nasir ■☆

357827　Selinum terebinthaceum Fisch. ex Trevir. ＝ Peucedanum terebinthaceum（Fisch. ex Trevir.）Fisch. ex Turcz. ■

357828　Selinum thianschanicum Korovin；天山亮蛇床；Tianshan Selinum ■☆

357829　Selinum tilingia（Regel）Maxim. ＝ Ligusticum ajanense（Regel et Tiling）Koso-Pol. ■

357830　Selinum tilingia Maxim. ＝ Ligusticum ajanense（Regel et Tiling）Koso-Pol. ■

357831　Selinum vaginatum（Edgew.）C. B. Clarke；具鞘亮蛇床■☆

357832　Selinum visnaga E. H. L. Krause ＝ Ammi visnaga（L.）Lam. ■

357833　Selinum wallichianum（DC.）Nasir ＝ Selinum wallichianum（DC.）Raizada et H. O. Saxena ■

357834　Selinum wallichianum（DC.）Raizada et H. O. Saxena；细叶亮蛇床（细叶滇前胡）；Smallleaf Selinum，Thinleaf Selinum ■

357835　Selinum wallichianum（DC.）Raizada et Saxena ＝ Selinum candollei DC. ■

357836　Selkirkia Hemsl.（1884）；塞尔紫草属☆

357837　Selkirkia berteroi Hemsl. ；塞尔紫草☆

357838　Selleola Urb. ＝ Minuartia L. ■

357839　Selleophytum Urb.（1915）；钩芒菊属●☆

357840　Selleophytum Urb. ＝ Coreopsis L. ●■

357841　Selleophytum buchii Urb. ；钩芒菊■☆

357842　Selliera Cav.（1799）；塞利草海桐属■☆

357843　Selliera radicans Cav. ；塞利草海桐■☆

357844　Selloa Kunth（1818）（保留属名）；车前菊属（鞍菊属）■☆

357845　Selloa Spreng.（废弃属名）＝ Gymnosperma Less. ●☆

357846　Selloa Spreng.（废弃属名）＝ Selloa Kunth（保留属名）■☆

357847　Selloa glutinosa Spreng. ＝ Gymnosperma glutinosum（Spreng.）Less. ■☆

357848　Selloa multiflora Kuntze；多花车前菊■☆

357849　Selloa obtusata（S. F. Blake）Longpre；钝车前菊■☆

357850　Selloa plantaginea Kunth；车前菊■☆

357851　Sellocharis Taub.（1889）；鞍豆属■☆

357852　Sellocharis paradoxa Taub. ；鞍豆■☆

357853　Sellowia Roth ex Roem. et Schult. ＝ Ammannia L. ■

357854　Sellowia Schult. ＝ Ammannia L. ■

357855　Sellowia uliginosa Roth ＝ Rotala densiflora（Roth ex Roem. et Schult.）Koehne ■

357856　Sellulocalamus W. T. Lin ＝ Dendrocalamus Nees ●

357857　Sellulocalamus W. T. Lin（1989）；椅子竹属●

357858　Sellulocalamus bambusoides（J. R. Xue et D. Z. Li）W. T. Lin ＝ Dendrocalamus bambusoides J. R. Xue et D. Z. Li ●

357859　Sellulocalamus tibeticus（J. R. Xue et T. P. Yi）W. T. Lin ＝ Dendrocalamus tibeticus J. R. Xue et T. P. Yi ●

357860　Sellunia Alef. ＝ Vicia L. ■

357861　Selmation T. Durand ＝ Metastelma R. Br. ●☆

357862　Selmation T. Durand ＝ Stelmation E. Fourn. ●☆

357863　Selnorition Raf. ＝ Rubus L. ●■

357864　Selonia Regel ＝ Eremurus M. Bieb. ■

357865　Selwynia F. Muell. ＝ Cocculus DC.（保留属名）●

357866　Selysia Cogn.（1881）；塞利瓜属■☆

357867　Selysia cordata Cogn. ；心形塞利瓜■☆

357868　Selysia prunifera Cogn. ；塞利瓜■☆

357869　Selysia smithii（Standl.）C. Jeffrey；史密斯塞利瓜■☆

357870　Semaphyllanthe L. Andersson ＝ Warscewiczia Klotzsch ■☆

357871　Semaphyllanthe L. Andersson（1995）；巴西芸香属●☆

357872　Semaquilegia Post et Kuntze ＝ Semiaquilegia Makino ■

357873　Semarilla Raf. ＝ Gymnosporia（Wight et Arn.）Benth. et Hook. f.（保留属名）●

357874　Semarillaria Ruiz et Pav. ＝ Paullinia L. ●☆

357875　Sematanthera Pierre ex Harms ＝ Efulensia C. H. Wright ■☆

357876　Semecarpos St. -Lag. ＝ Semecarpus L. f. ●

357877　Semecarpus L. f.（1782）；肉托果属（打印果属，大果漆树属，台东漆属）；Marking Nut，Markingnut，Marking-nut ●

357878　Semecarpus anacardium L. f. ；鸡腰肉托果（打印果，毛肉托果，印度肉托果）；Marking Nut，Oriental Cashew Nut ●

357879　Semecarpus australiensis Engl. ；澳洲肉托果（澳洲大果漆）；Tar Tree ●☆

357880　Semecarpus cuneiformis Blanco；钝叶肉托果（钝叶大果漆）●

357881　Semecarpus gigantifolius Vidal；大叶肉托果（蓄漆树，肉托果，台东漆，台东漆树）；Bigleaf Markingnut，Big-leaved Marking-nut，Giant-leaved Marking-nut，Largeleaf Markingnut ●

357882　Semecarpus microcarpus Wall. ；小果肉托果；Littlefruit Markingnut，Small-fruited Marking-nut ●◇

357883　Semecarpus reticulatus Lecomte；网脉肉托果（网眼肉托果）；Retinerve Marking Nut，Retinerve Markingnut，Retinerved Marking-nut，Retivein Markingnut ●

357884　Semecarpus subracemosus Kurz. ＝ Semecarpus microcarpus Wall. ●◇

357885　Semecarpus verniciferus Hayata et Kawak. = Semecarpus gigantifolius Vidal ●

357886　Semeiandra Hook. et Arn. (1838)；半雄花属；Semeiandra ■☆

357887　Semeiandra Hook. et Arn. = Lopezia Cav. ■☆

357888　Semeiandra grandiflora Hook. et Arn.；大花半雄花；Bigflower Semeiandra ■☆

357889　Semeiocardium Hassk. = Polygala L. ●■

357890　Semeiocardium Zoll. = Impatiens L. ■

357891　Semeionotis Schott = Dalbergia L. f.（保留属名）●

357892　Semeionotis Schott ex Endl. = Dalbergia L. f.（保留属名）●

357893　Semeiostachys Drobov = Agropyron Gaertn. ■

357894　Semeiostachys Drobow = Elymus L. ■

357895　Semeiostachys tianschanica（Drobow）Drobow = Elymus tianschanigenus Czerep. ■

357896　Semeiostachys turczaninowii（Drobow）Drobow = Elymus gmelinii（Ledeb.）Tzvelev ■

357897　Semeiostachys ugamica（Drobow）Drobow = Elymus nevskii Tzvelev ■

357898　Semele Kunth(1844)；仙蔓属；Climbing Butcher's Broom ●☆

357899　Semele androgyna（L.）Kunth；仙蔓；Climbing Butcher's Broom，Climbing Butcher's-broom ■☆

357900　Semele androgyna Kunth = Semele androgyna（L.）Kunth ■☆

357901　Semele gayae（Webb）Svent. et G. Kunkel；盖伊仙蔓■☆

357902　Semele maderensis C. G. Costa；梅德仙蔓■☆

357903　Semele menezesii C. G. Costa；梅内塞斯仙蔓■☆

357904　Semenovia Regel et Herder = Heracleum L. ■

357905　Semenovia Regel et Herder(1866)；大瓣芹属（伊犁独活属）；Bigpetalcelery ■

357906　Semenovia dasycarpa（Regel et Schmalh.）Korovin ex Pimenov et V. N. Tikhom.；毛果大瓣芹；Hairfruit Bigpetalcelery ■

357907　Semenovia komarovii Manden. = Semenovia dasycarpa（Regel et Schmalh.）Korovin ex Pimenov et V. N. Tikhom. ■

357908　Semenovia millefolia（Diels）V. M. Vinogr. et Kamelin = Heracleum millefolium Diels ■

357909　Semenovia montana Kamelin et V. M. Vinogr. = Heracleum millefolium Diels var. longilobum Norman ■

357910　Semenovia pimpinelloides（Nevski）Manden.；密毛大瓣芹；Densehair Bigpetalcelery ■

357911　Semenovia rubtzovii（Schischk.）Manden.；光果大瓣芹；Rubtzov Bigpetalcelery ■

357912　Semenovia transiliensis Regel et Herder；大瓣芹；Bigpetalcelery ■

357913　Semetor Raf. = Derris Lour.（保留属名）●

357914　Semetum Raf. = Lepidium L. ■

357915　Semialarium N. Hallé(1983)；半腋生卫矛属●☆

357916　Semialarium mexicanum（Miers）Mennega；墨西哥半腋生卫矛●☆

357917　Semialarium paniculatum（Schultes）N. Hallé；半腋生卫矛●☆

357918　Semiaquilegia Makino（1902）；天葵属；Semiaquilegia, Skymallow ■

357919　Semiaquilegia adoxoides（DC.）Makino；天葵（旱铜钱草，耗子屎，老鼠屎草，雷丸草，麦无踪，千年耗子屎，千年老鼠屎，蛇不见，天葵草，菟葵，夏无踪，小鸟头，直根天葵，紫背天葵）；Carrot-shaped Semiaquilegia，Muskroot-like Semiaquilegia，Skymallow ■

357920　Semiaquilegia adoxoides（DC.）Makino var. grandis D. Q. Wang；大天葵；Big Skymallow，Muskroot-like Semiaquilegia Big ■

357921　Semiaquilegia adoxoides（DC.）Makino var. grandis D. Q. Wang = Semiaquilegia adoxoides（DC.）Makino ■

357922　Semiaquilegia dauciformis D. Q. Wang；直根天葵■

357923　Semiaquilegia dauciformis D. Q. Wang = Semiaquilegia adoxoides（DC.）Makino ■

357924　Semiaquilegia ecalcarata（Maxim.）Sprague et Hutch. = Aquilegia ecalcarata Maxim. ■

357925　Semiaquilegia ecalcarata（Maxim.）Sprague et Hutch. f. semicalcarata Schipcz. = Aquilegia ecalcarata Maxim. ■

357926　Semiaquilegia henryi（Oliv.）J. R. Drumm. et Hutch. = Urophysa henryi（Oliv.）Ulbr. ■

357927　Semiaquilegia manshurica Kom. = Isopyrum manshuricum（Kom.）Kom. ex W. T. Wang et P. K. Hsiao ■

357928　Semiaquilegia rockii（Ulbr.）Hutch. = Urophysa rockii Ulbr. ■

357929　Semiaquilegia rockii（Ulbr.）J. R. Drumm. et Hutch. = Urophysa rockii Ulbr. ■

357930　Semiaquilegia simulatrix J. R. Drumm. et Hutch. = Aquilegia ecalcarata Maxim. ■

357931　Semiarundinaria Makino = Semiarundinaria Makino ex Nakai ●

357932　Semiarundinaria Makino ex Nakai（1925）；业平竹属；Semiarundinaria, Bamboo ●

357933　Semiarundinaria Nakai = Semiarundinaria Makino ex Nakai ●

357934　Semiarundinaria densiflora（Rendle）T. H. Wen = Brachystachyum densiflorum（Rendle）Keng ●

357935　Semiarundinaria densiflora（Rendle）T. H. Wen f. villosa（S. L. Chen et C. Y. Yao）= Brachystachyum densiflorum（Rendle）Keng var. villosum S. L. Chen et C. Y. Yao ●

357936　Semiarundinaria farinosa McClure = Sinobambusa farinosa（McClure）T. H. Wen ●

357937　Semiarundinaria fastuosa（Lat.-Marl. ex Mitford）Makino ex Nakai = Semiarundinaria fastuosa（Mitford）Makino ex Nakai ●

357938　Semiarundinaria fastuosa（Mitford）Makino = Semiarundinaria fastuosa（Mitford）Makino ex Nakai ●

357939　Semiarundinaria fastuosa（Mitford）Makino ex Nakai；业平竹；Japanese Temple Bamboo，Narihira Bamboo，Narihira Cane，Semiarundinaria，Temple Bamboo ●

357940　Semiarundinaria fastuosa（Mitford）Makino ex Nakai var. kagamiana（Makino）Ohwi = Semiarundinaria kagamiana Makino ●☆

357941　Semiarundinaria fastuosa（Mitford）Makino ex Nakai var. viridis（Makino）Makino ex Sad. Suzuki；绿花业平竹●☆

357942　Semiarundinaria fortis Koidz.；福特业平竹●☆

357943　Semiarundinaria gracilipes McClure = Oligostachyum gracilipes（McClure）G. H. Ye et Z. P. Wang ●

357944　Semiarundinaria henryi McClure = Sinobambusa henryi（McClure）C. D. Chu et C. S. Chao ●

357945　Semiarundinaria kagamiana Makino；镜氏业平竹●☆

357946　Semiarundinaria lima McClure = Oligostachyum nuspiculum（McClure）Z. P. Wang et G. H. Ye ●

357947　Semiarundinaria lubrica T. H. Wen = Oligostachyum lubricum（T. H. Wen）P. C. Keng ●

357948　Semiarundinaria nuspicula McClure = Oligostachyum nuspiculum（McClure）Z. P. Wang et G. H. Ye ●

357949　Semiarundinaria okuboi Makino = Sinobambusa tootsik（Siebold）Makino ex Nakai ●

357950　Semiarundinaria scabriflora McClure = Oligostachyum scabriflorum（McClure）Z. P. Wang et G. H. Ye ●

357951　Semiarundinaria scopula McClure = Oligostachyum scopulum（McClure）Z. P. Wang et G. H. Ye ●

357952　Semiarundinaria sinica T. H. Wen；中华业平竹●

357953　Semiarundinaria tenuifolia Koidz. = Sinobambusa tootsik

（Siebold）Makino ex Nakai var. maeshimana Muroi ex Sugim. ●

357954　Semiarundinaria tenuifolia Koidz. ＝ Sinobambusa tootsik（Siebold）Makino ex Nakai var. tenuifolia（Koidz.）S. Susaki ●

357955　Semiarundinaria tootsik（Makino）Muroi ＝ Sinobambusa tootsik（Siebold）Makino ex Nakai ●

357956　Semiarundinaria tootsik（Siebold）Makino ex Nakai var. laeta（McClure）T. H. Wen ＝ Sinobambusa tootsik（Siebold）Makino ex Nakai var. laeta（McClure）T. H. Wen ●

357957　Semiarundinaria venusta McClure ＝ Acidosasa venusa（McClure）Z. P. Wang et G. H. Ye ex C. S. Chao et C. D. Chu ●

357958　Semiarundinaria viridis（Makino）Makino；绿业平竹●☆

357959　Semiarundinaria yashadake（Makino）Makino；夜叉竹●

357960　Semibegoniella C. DC.（1908）；类小秋海棠属（塞米秋海棠属）■☆

357961　Semibegoniella C. E. C. Fisch. ＝ Begonia L. ●■

357962　Semibegoniella jamesoniana（A. DC.）C. DC.；类小秋海棠■☆

357963　Semicipium Pierre ＝ Mimusops L. ●☆

357964　Semicipium boivinii Pierre ＝ Labramia boivinii（Pierre）Aubrév. ●☆

357965　Semicireulaceae Dulac ＝ Monotropaceae Nutt.（保留科名）■

357966　Semidopsis Zumagl. ＝ Duschekia Opiz ●

357967　Semiliquidambar Hung T. Chang ＝ Liquidambar L. ●

357968　Semiliquidambar Hung T. Chang（1962）；半枫荷属；Banfenghe, Semiliquidambar ●★

357969　Semiliquidambar cathayensis Hung T. Chang；半枫荷（厚叶半枫荷, 金缕半枫荷）；Banfenghe, Cathay Semiliquidambar, Chinese Semiliquidambar, Thickleaf Banfenghe ●◇

357970　Semiliquidambar cathayensis Hung T. Chang var. fukienensis Hung T. Chang ＝ Semiliquidambar cathayensis Hung T. Chang ●◇

357971　Semiliquidambar cathayensis Hung T. Chang var. fukienensis Hung T. Chang；福建半枫荷（闽半枫荷）；Fijian Banfenghe, Fijian Semiliquidambar ●

357972　Semiliquidambar cathayensis Hung T. Chang var. parvifolia（Chun）Hung T. Chang ＝ Semiliquidambar cathayensis Hung T. Chang ●◇

357973　Semiliquidambar cathayensis Hung T. Chang var. parvifolia（Chun）Hung T. Chang；小叶半枫荷；Littleleaf Banfenghe, Littleleaf Semiliquidambar ●

357974　Semiliquidambar caudata Hung T. Chang；长尾半枫荷（长叶半枫荷）；Caudate Semiliquidambar, Tailleaf Banfenghe ●

357975　Semiliquidambar caudata Hung T. Chang var. cuspidata（Hung T. Chang）Hung T. Chang ＝ Semiliquidambar caudata Hung T. Chang ●

357976　Semiliquidambar caudata Hung T. Chang var. cuspidata（Hung T. Chang）Hung T. Chang；尖叶半枫荷；Cuspidate Banfenghe, Cuspidate Semiliquidambar ●

357977　Semiliquidambar chingii（F. P. Metcalf）Hung T. Chang；细柄半枫荷；Ching Banfenghe, Ching Semiliquidambar ●

357978　Semiliquidambar chingii（F. P. Metcalf）Hung T. Chang var. longipes Y. K. Li et X. M. Wang ＝ Semiliquidambar chingii（F. P. Metcalf）Hung T. Chang ●

357979　Semiliquidambar chingii（F. P. Metcalf）Hung T. Chang var. longipes Y. K. Li et X. M. Wang；长梗半枫荷；Longstalk Banfenghe ●

357980　Semiliquidambar coriacea Hung T. Chang ＝ Semiliquidambar cathayensis Hung T. Chang ●◇

357981　Semiliquidambar cuspidata Hung T. Chang ＝ Semiliquidambar caudata Hung T. Chang ●

357982　Semiliquidambar cuspidata Hung T. Chang ＝ Semiliquidamba caudata Hung T. Chang var. cuspidata（Hung T. Chang）Hung T. Chang ●

357983　Semilta Raf. ＝ Croton L. ●

357984　Semiphaius Gagnep. ＝ Eulophia R. Br.（保留属名）■

357985　Semiphaius chevalieri Gagnep. ＝ Eulophia spectabilis（Dennst.）Suresh ■

357986　Semiphaius evrardii Gagnep. ＝ Eulophia yunnanensis Rolfe ■

357987　Semiphajus chevalieri Gagnep. ＝ Eulophia spectabilis（Dennst.）Suresh ■

357988　Semiphajus evrardii Gagnep. ＝ Cymbidium faberi Rolfe ■

357989　Semiramisia Klotzsch（1851）；礼裙莓属（安第斯杜鹃属）●☆

357990　Semiramisia alata Luteyn；翅礼裙莓●☆

357991　Semiramisia fragilis A. C. Sm.；脆礼裙莓●☆

357992　Semiramisia karsteniana Klotzsch；礼裙莓●☆

357993　Semiria D. J. N. Hind（1999）；翼果柄泽兰属●☆

357994　Semiria viscosa D. J. N. Hind；翼果柄泽兰■☆

357995　Semnanthe N. E. Br. ＝ Erepsia N. E. Br. ●☆

357996　Semnanthe lacera（Haw.）N. E. Br. ＝ Erepsia lacera（Haw.）Liede ●☆

357997　Semnanthe lacera（Haw.）N. E. Br. var. densipetala L. Bolus ＝ Erepsia lacera（Haw.）Liede ●☆

357998　Semnos Raf. ＝ Chilianthus Burch. ●☆

357999　Semnostachya Bremek.（1944）；长穗马蓝属（糯米香属）；Semnostachya ●■

358000　Semnostachya Bremek. ＝ Strobilanthes Blume ●■

358001　Semnostachya longispicata（Hayata）C. F. Hsieh et C. C. Huang；长穗马蓝（长穗糯米香, 长穗紫云菜）；Longspicate Semnostachya, Longspike Conehead ●■

358002　Semnostachya menglaensis H. P. Tsui；糯米香；Mengla Semnostachya ■

358003　Semnothyrsus Bremek. ＝ Strobilanthes Blume ●■

358004　Semonvillea J. Gay ＝ Limeum L. ■●☆

358005　Semonvillea fenestrata Fenzl ＝ Limeum fenestratum（Fenzl）Heimerl ■☆

358006　Semonvillea pterocarpa J. Gay ＝ Limeum pterocarpum（J. Gay）Heimerl ■☆

358007　Sempervivaceae Juss.；长生草科■

358008　Sempervivaceae Juss. ＝ Asteraceae Bercht. et J. Presl（保留科名）●■

358009　Sempervivaceae Juss. ＝ Crassulaceae J. St. -Hil.（保留科名）●■

358010　Sempervivella Stapf ＝ Rosularia（DC.）Stapf ■

358011　Sempervivella Stapf ＝ Sedum L. ●■

358012　Sempervivella Stapf（1923）；藏瓦莲属（小长生草属）■☆

358013　Sempervivella acuminata（Decne.）A. Berger ＝ Rosularia alpestris（Kar. et Kir.）Boriss. ■

358014　Sempervivella alba（Edgew.）Stapf ＝ Rosularia sedoides（Decne.）H. Ohba ■☆

358015　Sempervivum L.（1753）；长生草属（卷绢属, 蜘蛛巢万代草属）；House Leek, Houseleek, Liveforver, Sempervivum ■☆

358016　Sempervivum abyssinicum Hochst. ex A. Rich. ＝ Hypagophytum abyssinicum（Hochst. ex A. Rich.）A. Berger ■☆

358017　Sempervivum acuminatum Decne. ＝ Rosularia alpestris（Kar. et Kir.）Boriss. ■

358018　Sempervivum album Edgew. ＝ Rosularia sedoides（Decne.）H. Ohba ■☆

358019　Sempervivum allionii Nyman；阿氏长生草■☆

358020　Sempervivum annuum Link　＝ Aichryson dichotomum Webb et Berthel. ■☆

358021　Sempervivum arachnoideum L. ;蛛网长生草（卷绢，蜘蛛巢万代草，蛛丝长生草，蛛丝长生花）；Cobweb Houseleek, Houseleek, Spiderweb Houseleek, Spider-web Houseleek ■☆

358022　Sempervivum arboreum L. ＝ Aeonium arboreum（L. ）Webb et Berthel. ●☆

358023　Sempervivum atlanticum（Ball）Ball;大西洋长生草■☆

358024　Sempervivum aureum Hornem. ＝ Aeonium aureum（Hornem. ）T. Mes ●☆

358025　Sempervivum caespitosum C. Sm. ＝ Aeonium caespitosum（C. Sm. ）Webb et Berthel. ●☆

358026　Sempervivum calyciforme Haw. ＝ Aeonium aureum（Hornem. ）T. Mes ●☆

358027　Sempervivum canariense L. ＝ Aeonium canariense（L. ）Webb et Berthel. ●☆

358028　Sempervivum caucasicum Rupr. ex Boiss. ;高加索长生草■☆

358029　Sempervivum chrysanthum Britten var. glandulosum Chiov. ＝ Aeonium leucoblepharum Webb ex A. Rich. ●☆

358030　Sempervivum ciliosum Craib;缘毛长生草（芳春）■☆

358031　Sempervivum coccinolepis Steud. ＝ Rosularia semiensis（J. Gay ex A. Rich. ）H. Ohba ■☆

358032　Sempervivum cruentum Webb et Berthel. ＝ Aeonium cruentum（Webb et Berthel. ）Webb et Berthel. ●☆

358033　Sempervivum dodrantale Willd. ＝ Aeonium dodrantale（Willd. ）T. Mes ●☆

358034　Sempervivum dolomiticum Facc. ;白云石长生草■☆

358035　Sempervivum erythraeum Velen. ;红色长生草■☆

358036　Sempervivum giuseppii Wale;西班牙长生草■☆

358037　Sempervivum glabrifolium Boriss. ;光叶长生草■☆

358038　Sempervivum glandulosum Aiton ＝ Aeonium glandulosum（Aiton）Webb et Berthel. ●☆

358039　Sempervivum globiferum L. ;球长生草●☆

358040　Sempervivum glutinosum Aiton ＝ Aeonium glutinosum（Aiton）Webb et Berthel. ●☆

358041　Sempervivum goochiae Webb et Berthel. ＝ Aeonium goochiae（Webb et Berthel. ）Webb et Berthel. ●☆

358042　Sempervivum gorgoneum（J. A. Schmidt）Cout. ＝ Aeonium gorgoneum J. A. Schmidt ●☆

358043　Sempervivum gracilis Christ ＝ Aeonium dodrantale（Willd. ）T. Mes ●☆

358044　Sempervivum grandiflorum Haw. ;大花长生草（大花卷绢）；Bigflower Houseleek ■☆

358045　Sempervivum haworthii Webb et Berthel. ＝ Aeonium haworthii（Webb et Berthel. ）Webb et Berthel. ●☆

358046　Sempervivum heuffelii Schott;赫氏长生草■☆

358047　Sempervivum hirtum L. ＝ Jovibarba hirta Opiz ■☆

358048　Sempervivum holochrysum（Webb et Berthel. ）Christ ＝ Aeonium holochrysum Webb et Berthel. ●☆

358049　Sempervivum kindingeri Adam. ;金长生草■☆

358050　Sempervivum kosaninii Praeger;科氏长生草■☆

358051　Sempervivum leucanthum Pancic. ;白花长生草■☆

358052　Sempervivum montanum L. ;山长生草（小云雀）；Mountain Houseleek ■☆

358053　Sempervivum pittonii Schott, Nyman et Kotschy;皮氏长生草■☆

358054　Sempervivum pumilum M. Bieb. ;小长生草■☆

358055　Sempervivum radiscescens（Webb et Berthel. ）Christ ＝

Aichryson tortuosum（Aiton）Webb et Berthel. ■☆

358056　Sempervivum ruthenicum（Koch）Schnittsp. et Lehm. ;神风■☆

358057　Sempervivum schlehani Schott;圣代长生草（圣代）■☆

358058　Sempervivum sedoides Decne. ＝ Rosularia sedoides（Decne. ）H. Ohba ■☆

358059　Sempervivum smithii Sims ＝ Aeonium smithii（Sims）Webb et Berthel. ●☆

358060　Sempervivum soboliferum Sims;绫樱；Hen and Chicken Houseleek, Hen-and-chickens ■☆

358061　Sempervivum strepsicladum Webb et Berthel. ＝ Aeonium spathulatum（Hornem. ）Praeger ●☆

358062　Sempervivum tectorum L. ;长生草（屋顶长生草，屋顶长生花，屋卷绢）；Aigreen, Avron, Ayegreen, Bullock's Eyes, Chicks-and-hens, Common Houseleek, Cyphel, Devil's Bane, Devil's Beard, Erewort, Fews, Foo, Foose, Fouets, Fouse, Fouze, Healing Blade, Healing Leaf, Hen-and-chickens, Hens and Chicks, Hockerie-topner, Hollick, Homewort, House Leek, House Leeks, Housegreen, Houseleek, Hundreds-and-thousands, Huslock, Imbreke, Jew's Beard, Jobarbe, Jove's Beard, Jubarbe, Jubard, Jupiter's Beard, Jupiter's Eye, Jupiter's Eyes, Mallow-rock, Oldman and Woman, Ollick, Poor Jan's Leaf, Prick-madam, Roof Houseleek, Selgreen, Sell Green, Sendgreen, Silgreen, Silly Green, Sinfull, Singreen, St. Patrick's Cabbage, Sungreen, Syphelt, Thor's Beard, Thunder Plant, Tourpin, Welcome-home-husbandthough-never-so-drunk,　Welcome-home-husbandthough-never-so-late ■☆

358063　Sempervivum tectorum L. subsp. atlanticum Ball　＝ Sempervivum atlanticum（Ball）Ball ■☆

358064　Sempervivum urbicum Hornem. ＝ Aeonium urbicum（Hornem. ）Webb et Berthel. ●☆

358065　Sempervivum viscatum（Bolle）Christ ＝ Aeonium viscatum Bolle ●☆

358066　Sempervivum wulfenii Hoppe ex Mert. et Koch;武氏长生草；Yellow Houseteek ■☆

358067　Senacia Comm. ex DC. ＝ Pittosporum Banks ex Gaertn. （保留属名）+ Maytenus Molina + Celastrus L. （保留属名）●

358068　Senacia Comm. ex Lam. ＝ Pittosporum Banks ex Gaertn. （保留属名）+ Maytenus Molina + Celastrus L. （保留属名）●

358069　Senacia Comm. ex Lam. emend. Thouars ＝ Pittosporum Banks ex Gaertn. （保留属名）●

358070　Senacia Lam. ＝ Pittosporum Banks ex Gaertn. （保留属名）●

358071　Senacia napaulensis DC. ＝ Pittosporum napaulense（DC. ）Rehder et E. H. Wilson ●

358072　Senaea Taub. （1893）;塞纳龙胆属■☆

358073　Senaea coerulea Taub. ;塞纳龙胆■☆

358074　Senckenbergia P. Gaertn. , B. Mey. et Scherb. ＝ Lepidium L. ■

358075　Senckenbergia Post et Kuntze ＝ Cyphomeris Standl. ●☆

358076　Senckenbergia Post et Kuntze ＝ Mendoncia Vell. ex Vand. ●☆

358077　Senckenbergia Post et Kuntze ＝ Senkebergia Neck. ●☆

358078　Senckenbergia Post et Kuntze ＝ Senkenbergia S. Schauer ■☆

358079　Senebiera DC. ＝ Coronopus Zinn（保留属名）■

358080　Senebiera DC. ＝ Lepidium L. ■

358081　Senebiera Post et Kuntze ＝ Ocotea Aubl. ●☆

358082　Senebiera Post et Kuntze ＝ Senneberia Neck. ●☆

358083　Senebiera coronopus（L. ）Poir. ＝ Coronopus squamatus（Forssk. ）Asch. ■☆

358084　Senebiera didyma（L. ）Pers. ＝ Coronopus didymus（L. ）Sm. ■

358085　Senebiera integrifolia DC. ＝ Coronopus integrifolius（DC. ）

Spreng. ■

358086　Senebiera lepidioides Coss. et Durieu = Lepidium lepidioides（Coss. et Durieu）Al-Shehbaz ■☆

358087　Senebiera linoides DC. = Coronopus integrifolius（DC.）Spreng. ■

358088　Senebiera pinnatifida DC. = Coronopus didymus（L.）Sm. ■

358089　Senebiera violacea Munby = Lepidium violaceum（Munby）Al-Shehbaz ■☆

358090　Senecillis Gaertn.（废弃属名）= Ligularia Cass.（保留属名）■

358091　Senecillis Gaertn.（废弃属名）= Senecio L. ■●

358092　Senecillis achyrotricha（Diels）Kitam. = Ligularia achyrotricha（Diels）Y. Ling ■

358093　Senecillis altaica（DC.）Kitam. = Ligularia altaica DC. ■

358094　Senecillis arnicoides（DC.）Kitam. = Cremanthodium arnicoides（DC. ex Royle）R. D. Good ■

358095　Senecillis atroviolacea（Franch.）Kitam. = Ligularia atroviolacea（Franch.）Hand. -Mazz. ■

358096　Senecillis botryodes（C. Winkl.）Kitam. = Ligularia botryodes（C. Winkl.）Hand. -Mazz. ■

358097　Senecillis cahhifolia（Maxim.）Kitam. = Ligularia calthifolia Maxim. ■

358098　Senecillis caloxantha（Diels）Kitam. = Ligularia caloxantha（Diels）Hand. -Mazz. ■

358099　Senecillis calthifolia（Maxim.）Kitam. = Ligularia calthifolia Maxim. ■

358100　Senecillis cymbulifera（W. W. Sm.）Kitam. = Ligularia cymbulifera（W. W. Sm.）Hand. -Mazz. ■

358101　Senecillis dentata（A. Gray）Kitam. = Ligularia dentata（A. Gray）H. Hara ■

358102　Senecillis dictyoneura（Franch.）Kitam. = Ligularia dictyoneura（Franch.）Hand. -Mazz. ■

358103　Senecillis dolichobotrys（Diels）Kitam. = Ligularia dolichobotrys Diels ■

358104　Senecillis duciformis（C. Winkl.）Kitam. = Ligularia duciformis（C. Winkl.）Hand. -Mazz. ■

358105　Senecillis dux（C. B. Clarke）Kitam. = Ligularia dux（C. B. Clarke）R. Mathur ■

358106　Senecillis euryphylla（C. Winkl.）Kitam. = Ligularia euryphylla（C. Winkl.）Hand. -Mazz. ■

358107　Senecillis fargesii（Franch.）Kitam. = Ligularia fargesii（Franch.）Diels ■

358108　Senecillis fischeri（Ledeb.）Kitam. = Ligularia fischeri（Ledeb.）Turcz. ■

358109　Senecillis franchetiana（H. Lév.）Kitam. = Ligularia franchetiana（H. Lév.）Hand. -Mazz. ■

358110　Senecillis glauca Gaertn. = Ligularia altaica DC. ■

358111　Senecillis hodgsonii（Hook.）Kitam. = Ligularia hodgsonii Hook. f. ■

358112　Senecillis hookeri（C. B. Clarke）Kitam. = Ligularia hookeri（C. B. Clarke）Hand. -Mazz. ■

358113　Senecillis interrnedia（Nakai）Kitam. = Ligularia intermedia Nakai ■

358114　Senecillis jaluensis（Kom.）Kitam. = Ligularia jaluensis Kom. ■

358115　Senecillis jamesii（Hemsl.）Kitam. = Ligularia jamesii（Hemsl.）Kom. ■

358116　Senecillis japonica（Thunb.）Kitam. = Ligularia japonica（Thunb.）Less. ■

358117　Senecillis kanaitzensis（Franch.）Kitam. = Ligularia kanaitzensis（Franch.）Hand. -Mazz. ■

358118　Senecillis kojimae（Kitam.）Kitam. = Ligularia kojimae Kitam. ■

358119　Senecillis lankongensis（Franch.）Kitam. = Ligularia lankongensis（Franch.）Hand. -Mazz. ■

358120　Senecillis lapathifolia（Franch.）Kitam. = Ligularia lapathifolia（Franch.）Hand. -Mazz. ■

358121　Senecillis latihastata（W. W. Sm.）Kitam. = Ligularia latihastata（W. W. Sm.）Hand. -Mazz. ■

358122　Senecillis leveillei（Vaniot）Kitam. = Ligularia leveillei（Vaniot）Hand. -Mazz. ■

358123　Senecillis liatroides（C. Winkl.）Kitam. = Ligularia liatroides（C. Winkl.）Hand. -Mazz. ■

358124　Senecillis limprichtii（Diels）Kitam. = Ligularia limprichtii（Diels）Hand. -Mazz. ■

358125　Senecillis macrophylla（Ledeb.）Kitam. = Ligularia macrophylla（Ledeb.）DC. ■

358126　Senecillis melanocephala（Franch.）Kitam. = Ligularia melanocephala（Franch.）Hand. -Mazz. ■

358127　Senecillis mongolica（Turcz.）Kitam. = Ligularia mongolica（Turcz.）DC. ■

358128　Senecillis nelumbifolia（Bureau et Franch.）Kitam. = Ligularia nelumbifolia（Bureau et Franch.）Hand. -Mazz. ■

358129　Senecillis nosoyinensis（Franch.）Kitam. = Ligularia kanaitzensis（Franch.）Hand. -Mazz. ■

358130　Senecillis ovato-oblonga Kitam. = Ligularia sagitta（Maxim.）Mattf. ex Rehder et Kobuski ■

358131　Senecillis phoenicochaeta（Franch.）Kitam. = Ligularia phoenicochaeta（Franch.）S. W. Liu ■

358132　Senecillis platyglossa（Franch.）Kitam. = Ligularia platyglossa（Franch.）Hand. -Mazz. ■

358133　Senecillis potaninii（C. Winkl.）Kitam. = Ligularia potaninii（C. Winkl.）Y. Ling ■

358134　Senecillis przewalskii（Maxim.）Kitam. = Ligularia przewalskii（Maxim.）Diels ■

358135　Senecillis putjatae（C. Winkl.）Kitam. = Ligularia mongolica（Turcz.）DC. ■

358136　Senecillis retusa（DC.）Kitam. = Ligularia retusa DC. ■

358137　Senecillis ruficoma（Franch.）Kitam. = Ligularia ruficoma（Franch.）Hand. -Mazz. ■

358138　Senecillis sagitta（Maxim.）Kitam. = Ligularia sagitta（Maxim.）Mattf. ex Rehder et Kobuski ■

358139　Senecillis schizopetala（W. W. Sm.）Kitam. = Ligularia stenoglossa（Franch.）Hand. -Mazz. ■

358140　Senecillis schmidtii Maxim. = Ligularia schmidtii（Maxim.）Makino ■

358141　Senecillis sibirica（L.）Simonk = Ligularia sibirica（L.）Cass. ■

358142　Senecillis songarica（Fisch.）Kitam. = Ligularia songarica（Fisch.）Y. Ling ■

358143　Senecillis stenocephala（Maxim.）Kitam. = Ligularia stenocephala（Maxim.）Matsum. et Koidz. ■

358144　Senecillis stenoglossa（Franch.）Kitam. = Ligularia stenoglossa（Franch.）Hand. -Mazz. ■

358145　Senecillis subspicata（Bureau et Franch.）Kitam. = Ligularia subspicata（Bureau et Franch.）Hand. -Mazz. ■

358146　Senecillis tangutica（Maxim.）Kitam. ＝ Ligularia tangutica（Maxim.）J. Mattf. ■

358147　Senecillis tangutica（Maxim.）Kitam. ＝ Sinacalia tangutica（Maxim.）B. Nord. ■

358148　Senecillis tenuipes（Franch.）Kitam. ＝ Ligularia tenuipes（Franch.）Diels ■

358149　Senecillis thyrsoidea（Ledeb.）Kitam. ＝ Ligularia thyrsoidea（Ledeb.）DC. ■

358150　Senecillis tongolensis（Franch.）Kitam. ＝ Ligularia tongolensis（Franch.）Hand. -Mazz. ■

358151　Senecillis tsangchanensis（Franch.）Kitam. ＝ Ligularia tsangchanensis（Franch.）Hand. -Mazz. ■

358152　Senecillis veitchiana（Hemsl.）Kitam. ＝ Ligularia veitchiana（Hemsl.）Greenm. ■

358153　Senecillis vellerea（Franch.）Kitam. ＝ Ligularia vellerea（Franch.）Hand. -Mazz. ■

358154　Senecillis virgaurea（Maxim.）Hand. -Mazz. ＝ Ligularia virgaurea（Maxim.）Mattf. ex Rehder et Kobuski ■

358155　Senecillis virgaurea（Maxim.）Kitam. ＝ Ligularia virgaurea（Maxim.）Mattf. ex Rehder et Kobuski ■

358156　Senecillis wilsoniana（Hemsl.）Kitam. ＝ Ligularia wilsoniana（Hemsl.）Greenm. ■

358157　Senecillis xanthotricha（Grüning）Kitam. ＝ Ligularia xanthotricha（Grüning）Y. Ling ■

358158　Senecillis yunnanensis（Franch.）Kitam. ＝ Ligularia yunnanensis（Franch.）C. C. Chang ■

358159　Senecio L.（1753）；千里光属（黄菀属）；Butterweed，Dusty Miller，Groundsel，Ragwort，Senecio，Squaw-weed ■●

358160　Senecio abbreviatus S. Moore；缩短千里光■☆

358161　Senecio aberdaricus R. E. Fr. et T. C. E. Fr. ＝ Dendrosenecio battiscombei（R. E. Fr. et T. C. E. Fr.）E. B. Knox ●☆

358162　Senecio abrotanoides E. Mey. ＝ Senecio pinnatifidus（P. J. Bergius）Less. ■☆

358163　Senecio abruptus Thunb. ；平截千里光■☆

358164　Senecio abyssinicus Sch. Bip. ex A. Rich. ＝ Emilia abyssinica（Sch. Bip. ex A. Rich.）C. Jeffrey ■☆

358165　Senecio acaulis（L. f.）Sch. Bip. ；无茎千里光■☆

358166　Senecio accedens Greene ＝ Senecio bigelovii A. Gray var. hallii A. Gray ■☆

358167　Senecio acerifolius C. Winkl. ＝ Sinosenecio euosmus（Hand. -Mazz.）B. Nord. ■

358168　Senecio acervatus S. Moore ＝ Solanecio mannii（Hook. f.）C. Jeffrey ■☆

358169　Senecio achilleifolius DC. ；蓍草叶千里光■☆

358170　Senecio achilleifolius DC. var. glanduloso-scaber Thell. ＝ Senecio achilleifolius DC. ■☆

358171　Senecio achilleifolius DC. var. glaucescens（DC.）Harv. ＝ Senecio achilleifolius DC. ■☆

358172　Senecio achyrotrichus Diels ＝ Ligularia achyrotricha（Diels）Y. Ling ■

358173　Senecio aconitifolius（Bunge）Turcz. ＝ Syneilesis aconitifolia（Bunge）Maxim. ■

358174　Senecio aconitifolius Turcz. ＝ Syneilesis australis Y. Ling ■

358175　Senecio acromaculus Y. Ling ＝ Senecio thianschanicus Regel et Schmalh. ■

358176　Senecio acromaculus Y. Ling f. elatus Y. Ling ＝ Senecio thianschanicus Regel et Schmalh. ■

358177　Senecio actinotus Hand. -Mazz. ；湖南千里光；Hunan Groundsel ■

358178　Senecio actinotus Hand. -Mazz. f. simplicifolius Y. Ling ＝ Senecio actinotus Hand. -Mazz. ■

358179　Senecio acuminatus Wall. ex DC. ＝ Synotis acuminata（Wall. ex DC.）C. Jeffrey et Y. L. Chen ■

358180　Senecio acuminatus Wall. ex DC. f. breviligulatus Hand. -Mazz. ＝ Synotis triligulata（Buch. -Ham. ex D. Don）C. Jeffrey et Y. L. Chen ■

358181　Senecio acuminatus Wall. ex DC. var. latifolius C. C. Chang ＝ Synotis calocephala C. Jeffrey et Y. L. Chen ■

358182　Senecio acutidens Rydb. ＝ Packera tridenticulata（Rydb.）W. A. Weber et Á. Löve ■☆

358183　Senecio acutifolius DC. ；尖叶千里光■☆

358184　Senecio acutipinnus Hand. -Mazz. ；尖羽千里光（锐羽千里光）；Sharppinnate Groundsel ■

358185　Senecio adamsii Howell ＝ Packera streptanthifolia（Greene）W. A. Weber et ÁLöve ■☆

358186　Senecio addoensis Compton ＝ Senecio scaposus DC. var. addoensis（Compton）G. D. Rowley ■☆

358187　Senecio adenocalyx Dinter ＝ Senecio radicans（L. f.）Sch. Bip. ■

358188　Senecio adenodontus DC. ＝ Hubertia adenodonta（DC.）C. Jeffrey ●☆

358189　Senecio adenostylifolius Humbert；柱腺叶千里光■☆

358190　Senecio adnatus DC. ；贴生狗舌草■☆

358191　Senecio adnatus DC. var. monactis Sch. Bip. ＝ Senecio humidanus C. Jeffrey ■☆

358192　Senecio adnivalis Stapf ＝ Dendrosenecio adnivalis（Stapf）E. B. Knox ●☆

358193　Senecio adnivalis Stapf subsp. refractisquamatus（De Wild.）Hauman ＝ Dendrosenecio adnivalis（Stapf）E. B. Knox ●☆

358194　Senecio adnivalis Stapf var. alticola（T. C. E. Fr.）Hedberg ＝ Dendrosenecio erici-rosenii（R. E. Fr. et T. C. E. Fr.）E. B. Knox subsp. alticola（Mildbr.）E. B. Knox ●☆

358195　Senecio adnivalis Stapf var. erioneuron（Cotton）Hedberg ＝ Dendrosenecio adnivalis（Stapf）E. B. Knox ●☆

358196　Senecio adnivalis Stapf var. intermedia Hauman ＝ Dendrosenecio adnivalis（Stapf）E. B. Knox ●☆

358197　Senecio adnivalis Stapf var. oligochaeta Hauman ＝ Dendrosenecio adnivalis（Stapf）E. B. Knox ●☆

358198　Senecio adnivalis Stapf var. petiolatus Hedberg ＝ Dendrosenecio adnivalis（Stapf）E. B. Knox var. petiolatus（Hedberg）E. B. Knox ●☆

358199　Senecio adnivalis Stapf var. stanleyi（Hauman）Hedberg ＝ Dendrosenecio adnivalis（Stapf）E. B. Knox var. petiolatus（Hedberg）E. B. Knox ●☆

358200　Senecio adolfi-friederici Muschl. ＝ Solanecio nandensis（S. Moore）C. Jeffrey ■☆

358201　Senecio adscendens Bojer ex DC. ；上举千里光■☆

358202　Senecio adustus S. Moore ＝ Senecio pachyrhizus O. Hoffm. ■☆

358203　Senecio aegyptius L. ；埃及千里光■☆

358204　Senecio aegyptius L. var. discoideus Boiss. ；盘状埃及千里光■☆

358205　Senecio aequinoctialis R. E. Fr. ；昼夜千里光■☆

358206　Senecio affinis DC. ；近缘千里光■☆

358207　Senecio afromontanum（R. E. Fr.）Humbert et Staner ＝ Crassocephalum montuosum（S. Moore）Milne-Redh. ■☆

358208　Senecio agapetes C. Jeffrey；澳非千里光■☆

358209　Senecio agathionanthes Muschl. = Gynura scandens O. Hoffm. ■☆

358210　Senecio ainsliiflorus Franch. = Parasenecio ainsliiflorus (Franch.) Y. L. Chen ■

358211　Senecio aizoides (DC.) Sch. Bip. ;番杏千里光■☆

358212　Senecio alaskanus Hultén = Tephroseris yukonensis (A. E. Porsild) Holub ■☆

358213　Senecio alatipes C. C. Chang = Synotis sciatrephes (W. W. Sm.) C. Jeffrey et Y. L. Chen ■

358214　Senecio alatus Wall. ex DC. = Synotis alata (Wall. ex DC.) C. Jeffrey et Y. L. Chen ■

358215　Senecio alatus Wall. ex DC. var. oligocephala Y. L. Chen et K. Y. Pan = Synotis alata (Wall. ex DC.) C. Jeffrey et Y. L. Chen ■

358216　Senecio albanensis DC. ;阿尔邦千里光■☆

358217　Senecio albanensis DC. var. doroniciflorus (DC.) Harv. ;多榔菊花千里光■☆

358218　Senecio albanensis DC. var. leiophyllus Harv. = Senecio affinis DC. ■☆

358219　Senecio albanensis DC. var. pseudodecurrens Thell. = Senecio affinis DC. ■☆

358220　Senecio albescens De Wild. = Dendrosenecio adnivalis (Stapf) E. B. Knox subsp. friesiorum (Mildbr.) E. B. Knox ●☆

358221　Senecio albifolius DC. ;白叶千里光■☆

358222　Senecio albopunctatus Bolus;白斑千里光■☆

358223　Senecio albopurpureus Kitam. ; 白 紫 千 里 光; Whitepurple Groundsel, Whiteviolet Groundsel ■

358224　Senecio alboranicus Maire = Senecio gallicus Chaix ■☆

358225　Senecio alleizettei (Humbert) Humbert subsp. pleianthus Humbert = Senecio pleianthus (Humbert) Humbert ■☆

358226　Senecio aloides DC. ;芦荟状千里光■☆

358227　Senecio alpicola Rydb. = Packera werneriifolia (A. Gray) W. A. Weber et Á. Löve ☆

358228　Senecio altaicus (DC.) Sch. Bip. = Ligularia altaica DC. ■

358229　Senecio altaicus Sch. Bip. = Ligularia altaica DC. ■

358230　Senecio alticola T. C. E. Fr. = Dendrosenecio erici-rosenii (R. E. Fr. et T. C. E. Fr.) E. B. Knox subsp. alticola (Mildbr.) E. B. Knox ●☆

358231　Senecio alticola T. C. E. Fr. var. subcalvescens Hauman = Dendrosenecio adnivalis (Stapf) E. B. Knox ●☆

358232　Senecio altus Rydb. = Senecio sphaerocephalus Greene ■☆

358233　Senecio amabilis DC. = Senecio agapetes C. Jeffrey ■☆

358234　Senecio amaniensis (Engl.) H. Jacobsen;青光木■☆

358235　Senecio amaniensis (Engl.) H. Jacobsen = Kleinia amaniensis (Engl.) A. Berger ■☆

358236　Senecio ambiguus DC. = Jacobaea maritima (L.) Pelser et Meijden ■☆

358237　Senecio amblyphyllus Cotton = Dendrosenecio elgonensis (T. C. E. Fr.) E. B. Knox ●☆

358238　Senecio ambondrombeensis Humbert = Io ambondrombeensis (Humbert) B. Nord. ■☆

358239　Senecio ambositrensis Humbert;安布西特拉千里光■☆

358240　Senecio ambraceus Turcz. ex DC. ;琥珀千里光(大花千里光,东北千里光);Amber Groundsel ■

358241　Senecio ambraceus Turcz. ex DC. var. glaber Kitam. ;东北千里光;Manchur Groundsel, Manchurian Groundsel ■

358242　Senecio ambraceus Turcz. ex DC. var. glaber Kitam. = Senecio ambraceus Turcz. ex DC. ■

358243　Senecio ambrosioides Rydb. = Senecio eremophilus Richardson var. kingii Greenm. ■☆

358244　Senecio ammophilus Greene = Senecio californicus DC. ■☆

358245　Senecio amoenus Sch. Bip. = Senecio nanus Sch. Bip. ex A. Rich. ■☆

358246　Senecio amplectens A. Gray;抱茎千里光■☆

358247　Senecio amplectens A. Gray var. holmii (Greene) H. D. Harr. ;豪氏千里光■☆

358248　Senecio amplectens A. Gray var. taraxacoides A. Gray = Senecio taraxacoides (A. Gray) Greene ■☆

358249　Senecio amplexifolius Humbert = Humbertacalia amplexifolia (Humbert) C. Jeffrey ●☆

358250　Senecio ampullaceus Hook. ;得州千里光;Texas Groundsel ■☆

358251　Senecio ampullaceus Hook. var. floccosus Engelm. et A. Gray = Senecio ampullaceus Hook. ■☆

358252　Senecio ampullaceus Hook. var. glaberrimus Engelm. et A. Gray = Senecio ampullaceus Hook. ■☆

358253　Senecio amurensis Schischk. ;阿穆尔千里光■☆

358254　Senecio amurensis Schischk. = Tephroseris kirilowii (Turcz. ex DC.) Holub ■☆

358255　Senecio anacletus Greene = Senecio wootonii Greene ■☆

358256　Senecio andapensis Humbert;安达帕千里光■☆

358257　Senecio andersonii Clokey = Senecio spartioides Torr. et Gray ■☆

358258　Senecio andinus Nutt. = Senecio serra Hook. ■☆

358259　Senecio andohahelensis Humbert;安杜千里光■☆

358260　Senecio andringitrensis Humbert = Hubertia andringitrensis (Humbert) C. Jeffrey ●☆

358261　Senecio angulatus L. f. ;棱角千里光■☆

358262　Senecio angustifolius (Thunb.) Willd. ;窄叶千里光■☆

358263　Senecio angustifolius Hayata = Senecio morrisonensis Hayata var. dentatus Kitam. ■

358264　Senecio angustifolius Hayata = Senecio nemorensis L. var. dentatus (Kitam.) H. Koyama ■

358265　Senecio anonymus A. W. Wood = Packera anonyma (A. W. Wood) W. A. Weber et Á. Löve ■☆

358266　Senecio antandroi Scott-Elliot;安坦德鲁千里光■☆

358267　Senecio antandroi var. sakamaliensis Humbert = Senecio sakamaliensis (Humbert) Humbert ■☆

358268　Senecio antennariifolius Britton = Packera antennariifolia (Britton) W. A. Weber et Á. Löve ■☆

358269　Senecio anteuphorbium (L.) Sch. Bip. = Kleinia anteuphorbium (L.) Haw. ■☆

358270　Senecio anteuphorbium (L.) Sch. Bip. var. odorus (Forssk.) G. D. Rowley = Kleinia odora (Forssk.) DC. ■☆

358271　Senecio anteuphorbius (L.) Sch. Bip. ;柳叶七宝树(凤美柳,凤美龙)●☆

358272　Senecio anthemifolius Harv. ;春黄菊叶千里光■☆

358273　Senecio apiifolius (DC.) Benth. et Hook. f. ex O. Hoffm. = Mesogramma apiifolium DC. ■☆

358274　Senecio appendiculatus (L. f.) Sch. Bip. var. concolor Bornm. = Pericallis appendiculata (L. f.) B. Nord. ■☆

358275　Senecio appendiculatus (L. f.) Sch. Bip. var. longifolia Bornm. = Pericallis appendiculata (L. f.) B. Nord. ■☆

358276　Senecio appendiculatus (L. f.) Sch. Bip. var. preauxiana Sch. Bip. = Pericallis appendiculata (L. f.) B. Nord. ■☆

358277　Senecio appendiculatus (L. f.) Sch. Bip. var. viridis Pit. =

Pericallis appendiculata (L. f.) B. Nord. ■☆

358278 Senecio appendiculatus Greenm. = Packera neomexicana (A. Gray) W. A. Weber et Á. Löve ■☆

358279 Senecio appendiculatus Poir. = Senecio lyratus Forssk. ■☆

358280 Senecio aquaticus Hill;水生千里光（沼泽千里光）;Golden Ragwort,Marsh Groundsel,Marsh Ragwort,Water Ragwort ■☆

358281 Senecio aquaticus Hill subsp. erraticus (Bertol.) Tourlet = Jacobaea aquatica (Hill) P. Gaertn. et B. Mey. et Schreb. var. erratica (Bertol.) Pelser et Meijden ■☆

358282 Senecio aquilonaris Schischk. ;紫红千里光■☆

358283 Senecio arabicus L. = Senecio aegyptius L. var. discoideus Boiss. ■☆

358284 Senecio arachnanthus Franch. ;长舌千里光（蛛花千里光）;Long-ligulate Groundsel,Spidery Groundsel ■

358285 Senecio araneosus DC. = Cissampelopsis volubilis (Blume) Miq. ●

358286 Senecio araneosus Koyama = Cissampelopsis corifolia C. Jeffrey et Y. L. Chen ■

358287 Senecio arcticus Rupr. ;湿生千里光（北极千里光,密集千里光）;Marsh Flawort ■

358288 Senecio arcticus Rupr. = Senecio palustris (L.) Hook. ■

358289 Senecio arcticus Rupr. = Tephroseris palustris (L.) Fourr. ■

358290 Senecio arenarius M. Bieb. = Senecio grandidentatus Ledeb. ■☆

358291 Senecio arenarius Thunb. ;沙地千里光■☆

358292 Senecio argunensis Turcz. ;额河千里光（阿贡千里光,大蓬蒿,光明草,黄花败酱,黄花一枝蒿,千里光,羽叶千里光,斩龙草）;Argun Groundsel ■

358293 Senecio argunensis Turcz. = Senecio ambraceus Turcz. ex DC. ■

358294 Senecio argunensis Turcz. = Senecio laetus Edgew. ■

358295 Senecio argunensis Turcz. f. antugstifolius Kom. = Senecio argunensis Turcz. ■

358296 Senecio argunensis Turcz. f. latifolius Kom. = Senecio argunensis Turcz. ■

358297 Senecio argunensis Turcz. var. blinii (H. Lév.) Hand. -Mazz. = Senecio argunensis Turcz. ■

358298 Senecio argunensis Turcz. var. tenuisectus Nakai = Senecio laetus Edgew. ■

358299 Senecio arizonicus Greene;亚利桑那千里光■☆

358300 Senecio armerifolius Franch. = Cremanthodium lineare Maxim. ■

358301 Senecio arnicoides (DC.) Wall. ex C. B. Clarke = Cremanthodium arnicoides (DC. ex Royle) R. D. Good ■

358302 Senecio arnicoides (DC.) Wall. ex C. B. Clarke var. frigida Hook. f. = Cremanthodium ellisii (Hook. f.) Kitam. ■

358303 Senecio arnicoides Wall. = Cremanthodium arnicoides (DC. ex Royle) R. D. Good ■

358304 Senecio arnicoides Wall. var. frigida Hook. f. = Cremanthodium ellisii (Hook. f.) Kitam. ■

358305 Senecio aronicoides DC. ;细鳞千里光■☆

358306 Senecio articulatus (L. f.) Sch. Bip. ;七宝菊;Candle Plant ■

358307 Senecio articulatus (L. f.) Sch. Bip. 'Variegatus';花叶仙人笔(七宝锦)■☆

358308 Senecio articulatus (L. f.) Sch. Bip. = Kleinia articulata (L. f.) Haw. ■☆

358309 Senecio aschersonianus Muschl. ex Dinter = Anisopappus pinnatifidus (Klatt) O. Hoffm. ex Hutch. ■☆

358310 Senecio asiaticus Schischk. et Serg. = Tephroseris praticola (Schischk. et Serg.) Holub ■

358311 Senecio asper Aiton = Senecio rosmarinifolius L. f. ■☆

358312 Senecio asperifolius Franch. ;糙叶千里光（毛叶红杆草）;Roughleaf Groundsel ■

358313 Senecio asperulus DC. ;粗糙千里光■☆

358314 Senecio atkinsonii C. B. Clarke = Ligularia atkinsonii (C. B. Clarke) S. W. Liu ■

358315 Senecio atlanticus Boiss. et Reut. = Senecio leucanthemifolius Poir. ■☆

358316 Senecio atractylidifolius Y. Ling = Synotis atractylidifolia (Y. Ling) C. Jeffrey et Y. L. Chen ■

358317 Senecio atratus Greene;黑千里光■☆

358318 Senecio atratus Greene var. milleflorus (Greene) Greenm. = Senecio atratus Greene ■☆

358319 Senecio atriplicifolius (L.) Hook. = Arnoglossum atriplicifolium (L.) H. Rob. ■☆

358320 Senecio atriplicifolius (L.) Hook. var. reniformis Hook. = Arnoglossum reniforme (Hook.) H. Rob. ■☆

358321 Senecio atriplicifolius Hook. var. reniformis Hook. = Arnoglossum reniforme (Hook.) H. Rob. ■☆

358322 Senecio atrofuscus Grierson;黑褐千里光;Blackbrown Groundsel ■

358323 Senecio atropurpureus (Ledeb.) B. Fedtsch. ;暗紫千里光■☆

358324 Senecio atropurpureus (Ledeb.) B. Fedtsch. subsp. frigidus (Richardson) Hultén = Tephroseris frigida (Richardson) Holub ■☆

358325 Senecio atropurpureus (Ledeb.) B. Fedtsch. subsp. tomentosus (Kjellm.) Hultén = Tephroseris kjellmanii (A. E. Porsild) Holub ■☆

358326 Senecio atropurpureus (Ledeb.) B. Fedtsch. var. dentatus A. Gray ex Hultén = Tephroseris kjellmanii (A. E. Porsild) Holub ■☆

358327 Senecio atropurpureus (Ledeb.) B. Fedtsch. var. ulmeri (Steffen) A. E. Porsild = Tephroseris frigida (Richardson) Holub ■☆

358328 Senecio atroviolaceus Franch. = Ligularia atroviolacea (Franch.) Hand. -Mazz. ■

358329 Senecio auleticus Greene = Packera hesperia (Greene) W. A. Weber et Á. Löve ■☆

358330 Senecio aurantiacus (Hoppe ex Willd.) Less. = Tephroseris rufa (Hand. -Mazz.) B. Nord. var. chaetocarpa C. Jeffrey et Y. L. Chen ■

358331 Senecio aurantiacus (Hoppe) Less. = Senecio capitatus Steud. ■☆

358332 Senecio aurantiacus (Hoppe) Less. var. glabratus Ledeb. = Tephroseris rufa (Hand. -Mazz.) B. Nord. var. chaetocarpa C. Jeffrey et Y. L. Chen ■

358333 Senecio aurantiacus (Hoppe) Less. var. leiocarpa Boiss. = Tephroseris phaeantha (Nakai) C. Jeffrey et Y. L. Chen ■

358334 Senecio aurantiacus (Hoppe) Less. var. spathulifolius Miq. = Tephroseris kirilowii (Turcz. ex DC.) Holub ■

358335 Senecio aurantiacus sensu Y. Ling = Tephroseris rufa (Hand. -Mazz.) B. Nord. var. chaetocarpa C. Jeffrey et Y. L. Chen ■

358336 Senecio aureus L. ;金色千里光;False Valerian, Female Regulator,Golden Groundsel,Golden Ragwort,Golden Senecio,Liferoot,Swamp Squaw-weed ■☆

358337 Senecio aureus L. = Packera aurea (L.) Á. Löve et D. Löve ■☆

358338 Senecio aureus L. var. alpinus A. Gray = Packera porteri (Greene) C. Jeffrey ■☆

358339　Senecio aureus L. var. angustifolius Britton = Packera anonyma (A. W. Wood) W. A. Weber et Á. Löve ■☆

358340　Senecio aureus L. var. aquilonius Fernald = Packera aurea (L.) Á. Löve et D. Löve ■☆

358341　Senecio aureus L. var. ashei Greenm. = Packera aurea (L.) Á. Löve et D. Löve ■☆

358342　Senecio aureus L. var. aurantiacus Farw. = Packera aurea (L.) Á. Löve et D. Löve ■☆

358343　Senecio aureus L. var. balsamitae (Muhl. ex Willd.) Torr. et A. Gray = Packera paupercula (Michx.) Á. Löve et D. Löve ■☆

358344　Senecio aureus L. var. borealis Torr. et A. Gray = Packera streptanthifolia (Greene) W. A. Weber et Á. Löve ■☆

358345　Senecio aureus L. var. croceus A. Gray = Packera crocata (Rydb.) W. A. Weber et Á. Löve ■☆

358346　Senecio aureus L. var. discoideus Hook. = Packera pauciflora (Pursh) Á. Löve et D. Löve ■☆

358347　Senecio aureus L. var. gracilis (Pursh) Hook. = Packera aurea (L.) Á. Löve et D. Löve ■☆

358348　Senecio aureus L. var. intercursus Fernald = Packera aurea (L.) Á. Löve et D. Löve ■☆

358349　Senecio aureus L. var. lanceolatus Oakes ex Torr. et A. Gray = Packera schweinitziana (Nutt.) W. A. Weber et Á. Löve ■☆

358350　Senecio aureus L. var. obovatus (Muhl. ex Willd.) Torr. et A. Gray = Packera obovata (Muhl. ex Willd.) W. A. Weber et Á. Löve ■☆

358351　Senecio aureus L. var. semicordatus (Mack. et Bush) Greenm. = Packera pseudaurea (Rydb.) W. A. Weber et Á. Löve var. semicordata (Mack. et Bush) Trock et T. M. Barkley ■☆

358352　Senecio aureus L. var. semicordatus (Mack. et Bush) Greenm. = Packera pseudaurea (Rydb.) W. A. Weber et Á. Löve ■☆

358353　Senecio aureus L. var. subnudus (DC.) A. Gray = Packera subnuda (DC.) Trock et T. M. Barkley ■☆

358354　Senecio aureus L. var. werneriifolius A. Gray = Packera werneriifolia (A. Gray) W. A. Weber et Á. Löve ■☆

358355　Senecio auriculatus Desf. = Senecio lividus L. ■☆

358356　Senecio auriculatus Vahl = Senecio lyratus Forssk. ■☆

358357　Senecio auriculus Coss. ;耳状千里光■☆

358358　Senecio austinae Greene = Packera eurycephala (Torr. et A. Gray) W. A. Weber et Á. Löve ■☆

358359　Senecio austromontanus Hilliard;南方山地千里光■☆

358360　Senecio avasimontanus Dinter = Lopholaena cneorifolia (DC.) S. Moore ■☆

358361　Senecio bagshawei S. Moore = Senecio subsessilis Oliv. et Hiern ■☆

358362　Senecio bakeri Scott-Elliot = Senecio madagascariensis Poir. ■☆

358363　Senecio balensis S. Ortiz et Vivero;巴莱千里光■☆

358364　Senecio ballyi G. D. Rowley = Kleinia schweinfurthii (Oliv. et Hiern) A. Berger ■☆

358365　Senecio balsamitae Muhl. ex Willd. = Packera paupercula (Michx.) Á. Löve et D. Löve ■☆

358366　Senecio balsamitae Muhl. ex Willd. var. firmifolius Greenm. = Packera paupercula (Michx.) Á. Löve et D. Löve ■☆

358367　Senecio balsamitae Muhl. ex Willd. var. thomsoniensis Greenm. = Packera paupercula (Michx.) Á. Löve et D. Löve ■☆

358368　Senecio bambuseti R. E. Fr. = Mikaniopsis bambuseti (R. E. Fr.) C. Jeffrey ■☆

358369　Senecio barbareifolius Krock. ;山芥叶千里光■☆

358370　Senecio barbatipes Hedberg = Dendrosenecio elgonensis (T. C. E. Fr.) E. B. Knox subsp. barbatipes (Hedberg) E. B. Knox ●☆

358371　Senecio barbatus DC. ;刚毛千里光;Bristly Foxtail ■☆

358372　Senecio barbellatus DC. = Senecio retrorsus DC. ■☆

358373　Senecio barbertonicus Klatt;巴伯顿千里光■☆

358374　Senecio baronii Humbert;巴隆千里光■☆

358375　Senecio basipinnatus Baker = Senecio lyratus Forssk. ■☆

358376　Senecio basutensis Thell. = Senecio subcoriaceus Schltr. ■☆

358377　Senecio battiscombei R. E. Fr. et T. C. E. Fr. = Dendrosenecio battiscombei (R. E. Fr. et T. C. E. Fr.) E. B. Knox ●☆

358378　Senecio baumii O. Hoffm. = Emilia baumii (O. Hoffm.) S. Moore ■☆

358379　Senecio baurii Oliv. ;巴利千里光■☆

358380　Senecio beauverdianus (H. Lév.) H. Lév. = Senecio pseudomairei H. Lév. ■

358381　Senecio beauverdianus H. Lév. ;川滇千里光;Sichuan-yunnan Groundsel ■

358382　Senecio beauverdianus H. Lév. = Senecio pseudomairiei H. Lév. ■

358383　Senecio begoniifolius (Franch.) Hand. -Mazz. = Parasenecio begoniifolius (Franch.) Y. L. Chen ■

358384　Senecio begoniifolius Franch. = Parasenecio begoniifolius (Franch.) Y. L. Chen ■

358385　Senecio beguei Humbert;贝格千里光■☆

358386　Senecio behmianus Muschl. = Crassocephalum ducis-aprutii (Chiov.) S. Moore ■☆

358387　Senecio behmianus Muschl. var. variostipulatus De Wild. = Crassocephalum ducis-aprutii (Chiov.) S. Moore ■☆

358388　Senecio bellidifolius A. Rich. = Emilia abyssinica (Sch. Bip. ex A. Rich.) C. Jeffrey ■☆

358389　Senecio bellioides Chiov. = Emilia bellioides (Chiov.) C. Jeffrey ■☆

358390　Senecio bellis Harv. ;雅致千里光■☆

358391　Senecio bernardinus Greene = Packera bernardina (Greene) W. A. Weber et Á. Löve ■☆

358392　Senecio bernardinus Greene var. sparsilobatus (Parish) Greenm. = Packera bernardina (Greene) W. A. Weber et Á. Löve ■☆

358393　Senecio besserianus Minderova;贝瑟千里光■☆

358394　Senecio bicolor (Willd.) Balb. ;银毛千里光(雪叶菊);Silver Ragwort ■☆

358395　Senecio bicolor (Willd.) Tod. subsp. cineraria (DC.) Chater;灰银毛千里光(白艾,白妙菊;瓜叶菊,南欧千里光,雪叶莲,银叶菊);Cineraria, Dusty Bob, Dusty Miller, Jerusalem Star, Sea Ragwort, Silver Cineraria, Silver Groundsel, Silver Ragwort, Silver-frosted Plant, Silver-leaved Cineraria ■☆

358396　Senecio bicolor (Willd.) Tod. subsp. cineraria (DC.) Chater = Jacobaea maritima (L.) Pelser et Meijden ■☆

358397　Senecio bicolor (Willd.) Tod. subsp. cineraria (DC.) Chater = Senecio cineraria DC. ■☆

358398　Senecio bigelovii A. Gray;毕氏千里光■☆

358399　Senecio bigelovii A. Gray var. hallii A. Gray;霍尔千里光■☆

358400　Senecio biligulatus W. W. Sm. ;双舌千里光;Doubleligulate Groundsel ■

358401　Senecio bipinnatus (Thunb.) Less. ;双羽千里光■☆

358402　Senecio birubonensis Kitam. = Tephroseris phaeantha (Nakai) C. Jeffrey et Y. L. Chen ■

358403　Senecio bivestitus Cronquist = Tephroseris lindstroemii (Ostenf.) Á Löve et D. Löve ■☆

358404　Senecio blattariifolius Franch. = Senecio nudicaulis Buch. - Ham. ex D. Don ■

358405　Senecio blattarioides DC. = Senecio oxyodontus DC. ■☆

358406　Senecio blinii H. Lév. = Senecio argunensis Turcz. ■

358407　Senecio blitoides Greene = Senecio fremontii Torr. et A. Gray var. blitoides (Greene) Cronquist ■☆

358408　Senecio blochmanae Greene;布洛赫曼千里光■☆

358409　Senecio blumei DC. = Cissampelopsis volubilis (Blume) Miq. ●

358410　Senecio blumeri Greene = Packera neomexicana (A. Gray) W. A. Weber et Á. Löve var. toumeyi (Greene) Trock et T. M. Barkley ■☆

358411　Senecio bodinieri Vaniot = Sinosenecio bodinieri (Vaniot) B. Nord. ■

358412　Senecio bodinieri Vaniot var. brevior Vaniot = Senecio bodinieri Vaniot ■

358413　Senecio bodinieri Vaniot var. elatior Vaniot = Senecio bodinieri Vaniot ■

358414　Senecio bodinieri Vaniot var. elatissimus Hand. -Mazz. = Senecio bodinieri Vaniot ■

358415　Senecio bodinieri Vaniot var. parcepilosa Vaniot = Sinosenecio palmatilobus (Kitam.) C. Jeffrey et Y. L. Chen ■

358416　Senecio bogoroensis De Wild. = Solanecio mannii (Hook. f.) C. Jeffrey ■☆

358417　Senecio boiteaui Humbert;博特千里光■☆

358418　Senecio bojeri DC. = Solanecio angulatus (Vahl) C. Jeffrey ■☆

358419　Senecio bolanderi A. Gray = Packera bolanderi (A. Gray) W. A. Weber et Á. Löve ■☆

358420　Senecio bolanderi A. Gray var. harfordii (Greenm.) T. M. Barkley = Packera bolanderi (A. Gray) W. A. Weber et Á. Löve var. harfordii (Greenm.) Trock et T. M. Barkley ■☆

358421　Senecio bollei Sunding et G. Kunkel;博勒千里光■☆

358422　Senecio bollei Sunding et G. Kunkel var. flaccidus (Bolle) Sunding et G. Kunkel;柔软博勒千里光■☆

358423　Senecio bolusii Oliv. = Phaneroglossa bolusii (Oliv.) B. Nord. ●☆

358424　Senecio borysthenicus Andrz. ex DC. ;第聂伯千里光■☆

358425　Senecio boscianus Sch. Bip. = Arnoglossum ovatum (Walter) H. Rob. ■☆

358426　Senecio botryodes C. Winkl. = Ligularia botryodes (C. Winkl.) Hand. -Mazz. ■

358427　Senecio brachyantherus (Hiern) S. Moore;短药千里光●☆

358428　Senecio brachycephalus R. E. Fr. = Emilia brachycephala (R. E. Fr.) C. Jeffrey ■☆

358429　Senecio brachyglossus Turcz. = Senecio glutinosus Thunb. ■☆

358430　Senecio brachypodus DC. ;短足千里光■☆

358431　Senecio bracteolatus Hook. f. = Senecio albopurpureus Kitam. ■

358432　Senecio brasiliensis (Spreng.) Less. ;巴西千里光;Brazil Groundsel ■☆

358433　Senecio brassica R. E. Fr. et T. C. E. Fr. = Dendrosenecio keniensis (Baker f.) Mabb. ●☆

358434　Senecio brassica R. E. Fr. et T. C. E. Fr. subsp. brassiciformis (R. E. Fr. et T. C. E. Fr.) Mabb. = Dendrosenecio brassiciformis (R. E. Fr. et T. C. E. Fr.) Mabb. ●☆

358435　Senecio brassiciformis R. E. Fr. et T. C. E. Fr. = Dendrosenecio brassiciformis (R. E. Fr. et T. C. E. Fr.) Mabb. ●☆

358436　Senecio brevidentatus M. D. Hend. ;短齿千里光■☆

358437　Senecio brevilimbus S. Moore = Senecio flavus (Decne.) Sch. Bip. ■☆

358438　Senecio brevilorus Hilliard;短千里光■☆

358439　Senecio breviscapus (DC.) H. Jacobsen = Senecio cicatricosus Sch. Bip. ■☆

358440　Senecio breweri Burtt Davy = Packera breweri (Burtt Davy) W. A. Weber et Á. Löve ■☆

358441　Senecio breyeri S. Moore = Senecio polyodon DC. ■☆

358442　Senecio brittenianus Hiern;布里滕千里光■☆

358443　Senecio bryoniifolius Harv. ;泻根叶千里光■☆

358444　Senecio buchwaldii O. Hoffm. = Solanecio buchwaldii (O. Hoffm.) C. Jeffrey ●☆

358445　Senecio buimalia Buch. -Ham. ex D. Don = Cissampelopsis buimalia (Buch. -Ham. ex D. Don) C. Jeffrey et Y. L. Chen ●

358446　Senecio buimalia Buch. -Ham. ex D. Don = Cissampelopsis spelaeicola (Vaniot) C. Jeffrey et Y. L. Chen ●

358447　Senecio buimalia Buch. -Ham. ex D. Don var. bambusetorum Hand. -Mazz. = Cissampelopsis erythrochaeta C. Jeffrey et Y. L. Chen ●

358448　Senecio buimalia sensu Dunn = Cissampelopsis spelaeicola (Vaniot) C. Jeffrey et Y. L. Chen ●

358449　Senecio bulbiferus Maxim. = Parasenecio hwangshanicus (Y. Ling) Y. L. Chen ■

358450　Senecio bulbinifolius DC. ;球百合千里光■☆

358451　Senecio bulleyanus Diels = Synotis lucorum (Franch.) C. Jeffrey et Y. L. Chen ■

358452　Senecio bungei Franch. = Ligularia thomsonii (C. B. Clarke) Pojark. ■

358453　Senecio bupleuriformis Sch. Bip. = Senecio glaberrimus DC. ■☆

358454　Senecio bupleuroides DC. ;柴胡状千里光■☆

358455　Senecio bupleuroides DC. var. denticulatus ? = Senecio bupleuroides DC. ■☆

358456　Senecio bupleuroides DC. var. falcatus Sch. Bip. = Senecio bupleuroides DC. ■☆

358457　Senecio bupleuroides DC. var. latifolius Harv. = Senecio glaberrimus DC. ■☆

358458　Senecio burchellii DC. ;伯切尔千里光■☆

358459　Senecio burkei Greenm. = Packera indecora (Greene) Á. Löve et D. Löve ■☆

358460　Senecio burtonii Hook. f. ;伯顿千里光■☆

358461　Senecio buschianus Sosn. ex Grossh. ;布什千里光■☆

358462　Senecio bussei Muschl. = Senecio purpureus L. ■☆

358463　Senecio butaguensis Muschl. = Crassocephalum montuosum (S. Moore) Milne-Redh. ■☆

358464　Senecio cacaliifolius Sch. Bip. = Ligularia sibirica (L.) Cass. ■

358465　Senecio cacaliifolius Sch. Bip. var. araneosus (A. DC.) Franch. = Cissampelopsis volubilis (Blume) Miq. ●

358466　Senecio cacaliifolius Sch. Bip. var. atkinsonii (C. B. Clarke) Franch. = Ligularia atkinsonii (C. B. Clarke) S. W. Liu ■

358467　Senecio cacaliifolius Sch. Bip. var. atkinsonii Diels = Ligularia hookeri (C. B. Clarke) Hand. -Mazz. ■

358468　Senecio cacaliifolius Sch. Bip. var. polycephalus (Hemsl.) Franch. = Ligularia wilsoniana (Hemsl.) Greenm. ■

358469　Senecio cacaliifolius Sch. Bip. var. polycephalus Franch. = Ligularia wilsoniana (Hemsl.) Greenm. ■

358470　Senecio cacaliifolius Sch. Bip. var. speciosus (Schrad.) DC. = Ligularia alatipes Hand. -Mazz. ■

358471　Senecio cacaliifolius Sch. Bip. var. stenocephalus (Maxim.) Franch. = Ligularia stenocephala (Maxim.) Matsum. et Koidz. ■

358472　Senecio cacaliiformis Sch. Bip. f. araneosa DC. = Ligularia anoleuca Hand. -Mazz. ■

358473　Senecio cacteaeformis Klatt = Othonna graveolens O. Hoffm. ●☆

358474　Senecio cakilefolius DC. ;海滨芥叶千里光■☆

358475　Senecio cakilefolius DC. var. hispidulus Thell. = Senecio arenarius Thunb. ■☆

358476　Senecio calamifolius Hook. = Senecio scaposus DC. ■

358477　Senecio californicus DC. ;加州千里光■☆

358478　Senecio californicus DC. var. ammophilus Greenm. = Senecio californicus DC. ■☆

358479　Senecio calocephalus C. C. Chang = Synotis calocephala C. Jeffrey et Y. L. Chen ■

358480　Senecio caloxanthus Diels = Ligularia caloxantha (Diels) Hand. -Mazz. ■

358481　Senecio calthifolius (Maxim.) Maxim. = Ligularia calthifolia Maxim. ■

358482　Senecio calthifolius Hook. f. = Ligularia hookeri (C. B. Clarke) Hand. -Mazz. ■

358483　Senecio calthifolius Maxim. = Ligularia calthifolia Maxim. ■

358484　Senecio calvertii Boiss. ;卡尔千里光■☆

358485　Senecio cambrensis Rosser;坎布雷千里光; Welsh Groundsel ■☆

358486　Senecio campanulatus Franch. = Cremanthodium campanulatum (Franch.) Diels ■

358487　Senecio campestris (Retz.) DC. = Tephroseris kirilowii (Turcz. ex DC.) Holub ■

358488　Senecio campestris (Retz.) DC. subsp. kirilowii (Turcz. ex DC.) Kitag. = Tephroseris integrifolia (L.) Holub subsp. kirilowii (Turcz. ex DC.) B. Nord. ■

358489　Senecio campestris (Retz.) DC. subsp. kirilowii (Turcz. ex DC.) Kitag. = Tephroseris kirilowii (Turcz. ex DC.) Holub ■

358490　Senecio campestris (Retz.) DC. var. glabratus DC. = Tephroseris praticola (Schischk. et Serg.) Holub ■

358491　Senecio campestris (Retz.) DC. var. puberula Miq. = Tephroseris kirilowii (Turcz. ex DC.) Holub ■

358492　Senecio campestris (Retz.) DC. var. rupicola (Kitag.) Kitag. ;岩生狗舌草■

358493　Senecio campestris (Retz.) DC. var. subdentatus (Bunge) Maxim. = Tephroseris pierotii (Miq.) Holub ■

358494　Senecio campestris (Retz.) DC. var. tomentosus Franch. = Tephroseris kirilowii (Turcz. ex DC.) Holub ■

358495　Senecio camptodontus Franch. = Senecio wightii (DC. ex Wight) Benth. ex C. B. Clarke ■

358496　Senecio campylodes DC. = Senecio scandens Buch. -Ham. ex D. Don ■

358497　Senecio canaliculatus Bojer ex DC. ;具沟千里光■☆

358498　Senecio canescens Cuatrec. ;灰千里光■☆

358499　Senecio canicida Sessé et Moc. ;墨西哥狗舌草■☆

358500　Senecio cannabifolius Less. ;麻叶千里光(返魂草,还魂草,宽叶还魂草,毛被返魂草);Hempleaf Groundsel, Hempleaf Ragwort ■

358501　Senecio cannabifolius Less. f. integrifolius (Koidz.) Kitag. = Senecio cannabifolius Less. var. integrifolius (Koidz.) Kitam. ■

358502　Senecio cannabifolius Less. f. pubinervis Kitag. = Senecio cannabifolius Less. ■

358503　Senecio cannabifolius Less. var. davuricus (Herder) Kitag. = Senecio cannabifolius Less. ■

358504　Senecio cannabifolius Less. var. integrifolius (Koidz.) Kitam. ;全叶千里光(单叶还魂草,单叶千里光)■

358505　Senecio canovirens Rydb. = Packera fendleri (A. Gray) W. A. Weber et Á. Löve ■☆

358506　Senecio canus Hook. = Packera cana (Hook.) W. A. Weber et Á. Löve ■☆

358507　Senecio canus Hook. var. eradiatus D. C. Eaton = Packera cana (Hook.) W. A. Weber et Á. Löve ■☆

358508　Senecio canus Hook. var. purshianus (Nutt.) A. Nelson = Packera cana (Hook.) W. A. Weber et Á. Löve ■☆

358509　Senecio capitatus (Wahlenb.) Steud. = Tephroseris rufa (Hand. -Mazz.) B. Nord. var. chaetocarpa C. Jeffrey et Y. L. Chen ■

358510　Senecio capitatus Steud. ;头状千里光(橙舌千里光,橙叶千里光);Orange Groundsel ■☆

358511　Senecio cappa Buch. -Ham. ex D. Don = Synotis cappa (Buch. -Ham. ex D. Don) C. Jeffrey et Y. L. Chen ■

358512　Senecio capuronii Humbert;凯普伦千里光■☆

358513　Senecio caranianus Chiov. = Solanecio nandensis (S. Moore) C. Jeffrey ■☆

358514　Senecio cardamine Greene = Packera cardamine (Greene) W. A. Weber et Á. Löve ■☆

358515　Senecio carnosus Thunb. ;肉质千里光■☆

358516　Senecio caroli C. Winkl. = Sinacalia caroli (C. Winkl.) C. Jeffrey et Y. L. Chen ■

358517　Senecio carolinianus Spreng. = Packera glabella (Poir.) C. Jeffrey ■☆

358518　Senecio carpathicus Herb. ;卡尔帕索斯千里光■☆

358519　Senecio carthamoides Greene = Senecio fremontii Torr. et A. Gray var. blitoides (Greene) Cronquist ■☆

358520　Senecio caryophyllus Mattf. = Senecio jacksonii S. Moore ■☆

358521　Senecio castoreus S. L. Welsh = Packera castoreus (S. L. Welsh) Kartesz ■☆

358522　Senecio cathcartensis O. Hoffm. ;卡斯卡特千里光■☆

358523　Senecio caucasigenus Schischk. ;高加索千里光■☆

358524　Senecio caudatus DC. ;尾状千里光■☆

358525　Senecio caulopterus DC. = Senecio inornatus DC. ■☆

358526　Senecio caulopterus Harv. = Senecio scoparius Harv. ■☆

358527　Senecio cavaleriei H. Lév. = Synotis cavaleriei (H. Lév.) C. Jeffrey et Y. L. Chen ■

358528　Senecio cephalophorus (Compton) H. Jacobsen = Kleinia cephalophora Compton ■☆

358529　Senecio cernuus A. Gray = Crassocephalum rubens (Juss. ex Jacq.) S. Moore ■

358530　Senecio cernuus A. Gray = Senecio pudicus Greene ■☆

358531　Senecio cernuus L. f. = Crassocephalum rubens (Juss. ex Jacq.) S. Moore ■

358532　Senecio chalureaui Humbert;沙吕千里光■☆

358533　Senecio chalureaui Humbert var. eradiatus Maire = Senecio chalureaui Humbert ■☆

358534　Senecio chalureaui Humbert var. radiatus ? = Senecio chalureaui Humbert ■☆

358535　Senecio chamaemelifolius DC. = Senecio diffusus L. f. ■☆

358536　Senecio chenopodifolius DC. = Parasenecio chola (W. W. Sm.) Y. L. Chen ■

358537　Senecio chenopodioides Kunth = Pseudogynoxys chenopodioides (Kunth) Cabrera ■☆

358538　Senecio cheranganiensis Cotton et Blakelock = Dendrosenecio cheranganiensis (Cotton et Blakelock) E. B. Knox ●☆

358539 Senecio chienii Hand. -Mazz. = Sinosenecio chienii（Hand. - Mazz.）B. Nord. ■

358540 Senecio chillaloensis Cufod. = Senecio schultzii Hochst. ex A. Rich. subsp. chillaloensis（Cufod.）S. Ortiz et Vivero ■☆

358541 Senecio chinensis（Spreng.）DC. = Senecio scandens Buch. -Ham. ex D. Don ■

358542 Senecio chiovendeanus Muschl. = Emilia chiovendeana（Muschl.）Lisowski ■☆

358543 Senecio chlorocephalus Muschl. = Senecio hochstetteri Sch. Bip. ex A. Rich. ■☆

358544 Senecio chola W. W. Sm. = Parasenecio chola（W. W. Sm.）R. C. Srivast. et C. Jeffrey ■

358545 Senecio chrysanthemifolius Poir. ;菊叶千里光(菊状千里光)■

358546 Senecio chrysanthemoides DC. = Senecio laetus Edgew. ■

358547 Senecio chrysanthemoides DC. var. eustegius Hand. -Mazz. = Senecio laetus Edgew. ■

358548 Senecio chrysanthemoides DC. var. khasianus（C. B. Clarke）Hook. f. = Senecio laetus Edgew. ■

358549 Senecio chrysocoma Meerb. ;金黄千里光■☆

358550 Senecio chungtienensis C. Jeffrey et Y. L. Chen;中甸千里光; Chungtien Groundsel ,Zhongdian Groundsel ■

358551 Senecio cicatricosus Baker = Senecio canaliculatus Bojer ex DC. ■☆

358552 Senecio cicatricosus Sch. Bip. ;疤痕千里光■☆

358553 Senecio cichorifolius H. Lév. = Synotis duclouxii（Dunn）C. Jeffrey et Y. L. Chen ■

358554 Senecio cinarifolius H. Lév. ;瓜叶千里光(耳叶千里光); Ear Leaf Groundsel ,Earleaf Groundsel ■

358555 Senecio cineraria DC. 'Cirrus';卷云银叶菊■☆

358556 Senecio cineraria DC. 'Silver Dust';银粉银叶菊■☆

358557 Senecio cineraria DC. = Senecio bicolor（Willd.）Tod. subsp. cineraria（DC.）Chater ■☆

358558 Senecio cineraria DC. var. bicolor（Willd.）Caruel = Jacobaea maritima（L.）Pelser et Meijden ■☆

358559 Senecio cinerascens Aiton;浅灰千里光■☆

358560 Senecio cinerifolius H. Lév. ;灰叶千里光■☆

358561 Senecio citriformis G. D. Rowley;黄花翡翠珠(白寿乐); Citrusform Groundsel ■

358562 Senecio claessensii（De Wild.）Humbert et Staner = Gynura amplexicaulis Oliv. et Hiern ■☆

358563 Senecio clarenceanus Hook. f. = Senecio purpureus L. ■☆

358564 Senecio clarkeanus Franch. = Cremanthodium nanum（Decne.）W. W. Sm. ■

358565 Senecio clarkianus A. Gray;克拉克千里光■☆

358566 Senecio claviseta Pomel = Senecio flavus（Decne.）Sch. Bip. ■☆

358567 Senecio clematoides Sch. Bip. ex A. Rich. = Mikaniopsis clematoides（Sch. Bip. ex A. Rich.）Milne-Redh. ■☆

358568 Senecio clevelandii Greene = Packera clevelandii（Greene）W. A. Weber et Á. Löve ■☆

358569 Senecio clevelandii Greene var. heterophyllus Hoover = Packera clevelandii（Greene）W. A. Weber et Á. Löve ■☆

358570 Senecio cliffordianus Hutch. = Kleinia cliffordiana（Hutch.）C. D. Adams ■☆

358571 Senecio cliffordii N. D. Atwood et S. L. Welsh = Packera spellenbergii（T. M. Barkley）C. Jeffrey ■☆

358572 Senecio clivorum（Maxim.）Maxim. = Ligularia dentata（A. Gray）H. Hara ■

358573 Senecio clivorum Maxim. = Ligularia dentata（A. Gray）H. Hara ■

358574 Senecio coccineiflorus G. D. Rowley = Kleinia grantii（Oliv. et Hiern）Hook. f. ■☆

358575 Senecio coccineus（Oliv. et Hiern）Muschl. = Kleinia grantii（Oliv. et Hiern）Hook. f. ■☆

358576 Senecio cochlearifolius Bojer ex DC. ;螺状千里光■☆

358577 Senecio coferifer H. Lév. = Sinosenecio bodinieri（Vaniot）B. Nord. ■

358578 Senecio cognatus Greene = Packera streptanthifolia（Greene）W. A. Weber et Á. Löve ■☆

358579 Senecio colensoensis O. Hoffm. ex Kuntze = Senecio scoparius Harv. ■☆

358580 Senecio coleophyllus Turcz. ;鞘叶千里光■☆

358581 Senecio compactiflorum C. C. Chang = Senecio humbertii C. C. Chang ■

358582 Senecio compactus（A. Gray）Rydb. = Packera tridenticulata（Rydb.）W. A. Weber et Á. Löve ■☆

358583 Senecio concinnus Franch. = Nemosenecio concinnus（Franch.）C. Jeffrey et Y. L. Chen ■

358584 Senecio concolor DC. = Senecio speciosus Willd. ■☆

358585 Senecio concolor DC. var. subglaber O. Hoffm. ex Kuntze = Senecio polyodon DC. var. subglaber（O. Hoffm. ex Kuntze）Hilliard et B. L. Burtt ■☆

358586 Senecio confertoides De Wild. = Senecio transmarinus S. Moore ■☆

358587 Senecio confertus Sch. Bip. ex A. Rich. ;密集千里光■☆

358588 Senecio confusus Britten;疑仙年;Mexican Flame Vine ,Mexican Flame-vine ■☆

358589 Senecio confusus Britten = Pseudogynoxys chenopodioides（Kunth）Cabrera ■☆

358590 Senecio congestus（R. Br.）DC. ;北方沼泽千里光;Marsh Fleabane ,Marsh Ragwort ,Northern Swamp Groundsel ■☆

358591 Senecio congestus（R. Br.）DC. = Senecio arcticus Rupr. ■

358592 Senecio congestus（R. Br.）DC. = Tephroseris palustris（L.）Fourr. ■

358593 Senecio congestus（R. Br.）DC. var. laceratus（Ledeb.）Fernald = Tephroseris palustris（L.）Fourr. ■

358594 Senecio congestus（R. Br.）DC. var. palustris（L.）Fernald = Senecio congestus（R. Br.）DC. ■☆

358595 Senecio congestus（R. Br.）DC. var. palustris（L.）Fernald = Tephroseris palustris（L.）Fourr. ■

358596 Senecio congestus（R. Br.）DC. var. palustris L. = Tephroseris palustris（L.）Fourr. ■

358597 Senecio congestus（R. Br.）DC. var. tonsus Fernald = Senecio congestus（R. Br.）DC. ■☆

358598 Senecio congestus（R. Br.）DC. var. tonsus Fernald = Tephroseris palustris（L.）Fourr. ■

358599 Senecio congolensis De Wild. = Solanecio mannii（Hook. f.）C. Jeffrey ■☆

358600 Senecio conrathii N. E. Br. ;康拉特千里光■☆

358601 Senecio consanguineus DC. ;亲缘千里光■☆

358602 Senecio convallium Greenm. = Packera cana（Hook.）W. A. Weber et Á. Löve ■☆

358603 Senecio cordatus Nutt. = Senecio integerrimus Nutt. var. ochroleucus（A. Gray）Cronquist ■☆

358604 Senecio cordifolius L. f. ;心叶千里光■☆

358605 Senecio coreopsoides Chiov. = Senecio transmarinus S. Moore var. sycephalus（S. Moore）Hedberg ■☆

358606 Senecio coriaceisquamus C. C. Chang；革苞千里光；Coriaceous Phyllaries Groundsel, Leatherbract Groundsel ■

358607 Senecio coronatus（Thunb.）Harv.；冠千里光（伪泥胡菜）■☆

358608 Senecio coronatus Harv. = Senecio coronatus（Thunb.）Harv. ■☆

358609 Senecio coronopifolius Desf. = Senecio desfontainei Druce ■

358610 Senecio coronopifolius Desf. = Senecio glaucus L. subsp. coronopifolius（Desf.）Alexander ■☆

358611 Senecio coronopifolius Desf. subsp. massaicus Maire = Senecio massaicus（Maire）Maire ■☆

358612 Senecio coronopifolius Desf. var. calyculatus Emb. et Maire = Senecio glaucus L. subsp. coronopifolius（Desf.）Alexander ■☆

358613 Senecio coronopifolius Desf. var. discoideus C. Winkl. ex Danguy = Senecio dubitabilis C. Jeffrey et Y. L. Chen ■

358614 Senecio coronopifolius Desf. var. oasicola Hochr. = Jacobaea maritima（L.）Pelser et Meijden ■☆

358615 Senecio coronopifolius Desf. var. subdentatus（Ledeb.）Boiss. = Senecio subdentatus（Bunge）Turcz. ■

358616 Senecio coronopifolius Desf. var. subdentatus（Ledeb.）Boiss. = Senecio subdentatus Ledeb. ■

358617 Senecio coronopifolius Desf. var. subdentatus（Ledeb.）Boiss. = Tephroseris subdentata（Bunge）Holub ■

358618 Senecio coronopus Nutt. = Senecio californicus DC. ■☆

358619 Senecio cortesianus Muschl. = Senecio transmarinus S. Moore var. sycephalus（S. Moore）Hedberg ■☆

358620 Senecio cortusifolius Hand. -Mazz. = Sinosenecio cortusifolius（Hand. -Mazz.）B. Nord. ■

358621 Senecio corymbiferus DC.；翠珊瑚树 ■☆

358622 Senecio corymbiferus DC. = Senecio sarcoides C. Jeffrey ■☆

358623 Senecio cottonii Hutch. et G. Taylor = Dendrosenecio kilimanjari（Mildbr.）E. B. Knox subsp. cottonii（Hutch. et G. Taylor）E. B. Knox ●☆

358624 Senecio cotyledonis DC.；琉璃草千里光 ■☆

358625 Senecio coursii Humbert = Humbertacalia coursii（Humbert）C. Jeffrey ●☆

358626 Senecio covillei Greene = Senecio scorzonella Greene ■☆

358627 Senecio covillei Greene var. scorzonella（Greene）Jeps. = Senecio scorzonella Greene ■☆

358628 Senecio crassifolius Willd. var. falcifolius Bolle = Senecio leucanthemifolius Poir. ■☆

358629 Senecio crassifolius Willd. var. giganteus Caball. = Senecio leucanthemifolius Poir. ■☆

358630 Senecio crassifolius Willd. var. latisectus Pau et Font Quer = Senecio leucanthemifolius Poir. ■☆

358631 Senecio crassifolius Willd. var. pusillus Bolle = Senecio leucanthemifolius Poir. ■☆

358632 Senecio crassipes H. Lév. et Vaniot = Gynura pseudochina（L.）DC. ■

358633 Senecio crassissimus Humb.；紫金章（鱼尾冠, 紫龙）；Vertical Leaf ■☆

358634 Senecio crassorhizus De Wild.；粗根千里光 ■☆

358635 Senecio crassulifolius（DC.）Sch. Bip.；厚叶千里光 ■☆

358636 Senecio crassulus A. Gray；厚千里光 ■☆

358637 Senecio crataegifolius Hayata；小蔓黄菀；Crataegifoliate Groundsel ■

358638 Senecio crataegifolius Hayata = Senecio scandens Buch. -Ham. ex D. Don var. crataegifolius（Hayata）Kitam. ■

358639 Senecio crawfordii（Britton）G. W. Douglas et G. R. Douglas = Packera paupercula（Michx.）Á. Löve et D. Löve ■☆

358640 Senecio crawfordii Britton = Packera paupercula（Michx.）Á. Löve et D. Löve ■☆

358641 Senecio crenatus Thunb.；圆齿千里光 ■☆

358642 Senecio crenulatus DC.；小齿叶千里光；Florists Cineraria ■☆

358643 Senecio crepidineus Greene = Senecio elmeri Piper ■☆

358644 Senecio crispatipilosus C. Jeffrey；皱毛千里光 ■☆

358645 Senecio crispus Thunb.；皱波千里光 ■☆

358646 Senecio crocatus Rydb. = Packera crocata（Rydb.）W. A. Weber et Á. Löve ■☆

358647 Senecio cruentus（L'Hér.）DC. = Pericallis cruenta（L'Hér.）Bolle ■☆

358648 Senecio cruentus（L'Hér.）DC. var. bracteatus（Link）Christ = Pericallis cruenta（L'Hér.）Bolle ■☆

358649 Senecio cruentus（Masson ex L'Hér.）DC. = Pericallis hybrida B. Nord. ■

358650 Senecio cryptolanatus Killick；隐毛千里光 ■☆

358651 Senecio cupulatus Volkens et Muschl. = Senecio microglossus DC. ■☆

358652 Senecio curtophyllus Klatt = Senecio tysonii MacOwan ●☆

358653 Senecio curvatus Baker = Humbertacalia pyrifolia（DC.）C. Jeffrey ●☆

358654 Senecio curvisquama（Hand. -Mazz.）C. C. Chang = Ligularia curvisquama Hand. -Mazz. ■

358655 Senecio curvisquama（Hand. -Mazz.）C. C. Chang var. robustus C. C. Chang = Ligularia curvisquama Hand. -Mazz. ■

358656 Senecio cyaneus O. Hoffm.；蓝色千里光 ■☆

358657 Senecio cyclamniifolius Franch. = Sinosenecio cyclamniifolius（Franch.）B. Nord. ■

358658 Senecio cyclocladus Baker = Senecio canaliculatus Bojer ex DC. ■☆

358659 Senecio cyclotus Bureau et Franch. = Parasenecio cyclotus（Bureau et Franch.）Y. L. Chen ■

358660 Senecio cydoniifolius O. Hoffm. = Solanecio cydoniifolius（O. Hoffm.）C. Jeffrey ■☆

358661 Senecio cydoniifolius O. Hoffm. ex Engl. = Solanecio cydoniifolius（O. Hoffm.）C. Jeffrey ■☆

358662 Senecio cymatocrepis Diels = Synotis alata（Wall. ex DC.）C. Jeffrey et Y. L. Chen ■

358663 Senecio cymbalaria Pursh = Packera cymbalaria（Pursh）Á. Löve et D. Löve ■

358664 Senecio cymbalariifolius（Thunb.）Less. = Senecio hastifolius（L. f.）Less. ■☆

358665 Senecio cymbalarioides H. Buek = Packera subnuda（DC.）Trock et T. M. Barkley ■☆

358666 Senecio cymbalarioides H. Buek subsp. moresbiensis Calder et R. L. Taylor = Packera subnuda（DC.）Trock et T. M. Barkley var. moresbiensis（Calder et R. L. Taylor）Trock ■☆

358667 Senecio cymbalarioides H. Buek var. moresbiensis（Calder et R. L. Taylor）C. C. Freeman = Packera subnuda（DC.）Trock et T. M. Barkley var. moresbiensis（Calder et R. L. Taylor）Trock ■☆

358668 Senecio cymbalarioides Nutt. = Packera streptanthifolia（Greene）W. A. Weber et Á. Löve ■☆

358669 Senecio cymbalarioides Nutt. var. borealis（Torr. et A. Gray）

Greenm. = Packera streptanthifolia（Greene）W. A. Weber et Á. Löve ■☆

358670　Senecio cymbalarioides Nutt. var. streptanthifolius（Greene）Greenm. = Packera streptanthifolia（Greene）W. A. Weber et Á. Löve ■☆

358671　Senecio cymbulifer W. W. Sm. = Ligularia cymbulifera（W. W. Sm.）Hand. -Mazz. ■

358672　Senecio cynthioides Greene = Packera cynthioides（Greene）W. A. Weber et Á. Löve ■☆

358673　Senecio cyrenaicus（E. A. Durand et Barratte）Pamp. = Senecio leucanthemifolius Poir. subsp. cyrenaicus（E. A. Durand et Barratte）Greuter ■☆

358674　Senecio dahuricus Fisch. ex Turcz. = Parasenecio auriculatus（DC.）J. R. Grant ■

358675　Senecio dahuricus Sch. Bip. = Parasenecio otopteryx（Hand. -Mazz.）Y. L. Chen ■

358676　Senecio dalei Cotton et Blakelock = Dendrosenecio cheranganiensis（Cotton et Blakelock）E. B. Knox subsp. dalei Cotton et Blakelock ●☆

358677　Senecio daochengensis Y. L. Chen；稻城千里光；Daocheng Groundsel ■

358678　Senecio davidii Franch. = Sinacalia davidii（Franch.）H. Koyama ■

358679　Senecio davuricus Fisch. = Parasenecio auriculatus（DC.）J. R. Grant ■

358680　Senecio davuricus Fisch. var. ochotensis Maxim. = Parasenecio auriculatus（DC.）H. Koyama ■

358681　Senecio deaniensis Muschl. = Senecio meyeri-johannis Engl. ■☆

358682　Senecio debilis Harv. = Senecio infirmus C. Jeffrey ■☆

358683　Senecio debilis Nutt. = Packera debilis（Nutt.）W. A. Weber et Á. Löve ■☆

358684　Senecio decaisnei DC. = Senecio flavus（Decne.）Sch. Bip. ■☆

358685　Senecio decaryi Humbert；德卡里千里光■☆

358686　Senecio decorticans A. Nelson = Senecio lemmonii A. Gray ■☆

358687　Senecio decumbens C. C. Chang = Senecio coriaceisquamus C. C. Chang ■

358688　Senecio decurrens DC. ；下延千里光■☆

358689　Senecio deflersii O. Schwartz；德弗莱尔千里光；Pickle Plant ■☆

358690　Senecio degiensis Pic. Serm. = Senecio farinaceus Sch. Bip. ex A. Rich. ■☆

358691　Senecio dekindtianus Volkens et O. Hoffm. = Senecio pachyrhizus O. Hoffm. ■☆

358692　Senecio delavayi Franch. = Cremanthodium delavayi（Franch.）Diels ex H. Lév. ■

358693　Senecio delavayi Franch. = Senecio nigrocinctus Franch. ■

358694　Senecio delphinifolius Vahl；翠雀花叶千里光■☆

358695　Senecio delphinifolius Vahl var. occidentalis Dobignard et Dutartre = Jacobaea delphiniifolia（Vahl）Pelser et Veldkamp ■☆

358696　Senecio delphiniphyllus H. Lév. = Parasenecio delphiniphyllus（H. Lév.）Y. L. Chen ■

358697　Senecio deltoideus Less. ；三角千里光■☆

358698　Senecio deltophyllus Maxim. = Parasenecio deltophyllus（Maxim.）Y. L. Chen ■

358699　Senecio densiflorus M. Martens = Packera glabella（Poir.）C. Jeffrey ■☆

358700　Senecio densiflorus M. Martens var. mishmiensis Hook. f. = Synotis nagensis（C. B. Clarke）C. Jeffrey et Y. L. Chen ■

358701　Senecio densiflorus Wall. ex A. DC. = Synotis cappa（Buch. -Ham. ex D. Don）C. Jeffrey et Y. L. Chen ■

358702　Senecio densiflorus Wall. ex A. DC. var. fargesii Hand. -Mazz. = Synotis nagensis（C. B. Clarke）C. Jeffrey et Y. L. Chen ■

358703　Senecio densiflorus Wall. ex A. DC. var. lobbi Hook. f. = Synotis cappa（Buch. -Ham. ex D. Don）C. Jeffrey et Y. L. Chen ■

358704　Senecio densiflorus Wall. ex DC. = Synotis cappa（Buch. -Ham. ex D. Don）C. Jeffrey et Y. L. Chen ■

358705　Senecio densiflorus Wall. ex DC. var. fargesii Hand. -Mazz. = Synotis nagensis（C. B. Clarke）C. Jeffrey et Y. L. Chen ■

358706　Senecio densiflorus Wall. ex DC. var. lobbi Hook. f. = Synotis cappa（Buch. -Ham. ex D. Don）C. Jeffrey et Y. L. Chen ■

358707　Senecio densiflorus Wall. ex DC. var. mishmiensis Hook. f. = Synotis nagensis（C. B. Clarke）C. Jeffrey et Y. L. Chen ■

358708　Senecio densiserratus C. C. Chang；密齿千里光；Dese Serrate Groundsel ■

358709　Senecio densiserratus C. C. Chang var. glaber C. C. Chang = Synotis atractylidifolia（Y. Ling）C. Jeffrey et Y. L. Chen ■

358710　Senecio densus Greene = Packera tridenticulata（Rydb.）W. A. Weber et Á. Löve ■☆

358711　Senecio dentato-alatus Mildbr. ex C. Jeffrey；齿翅千里光■☆

358712　Senecio denticulatus Engl. = Senecio subsessilis Oliv. et Hiern ■☆

358713　Senecio depauperatus Mattf. ；萎缩千里光■☆

358714　Senecio dephiniphyllus H. Lév. = Parasenecio delphiniphyllus（H. Lév.）Y. L. Chen ■

358715　Senecio dernburgianus Muschl. = Senecio subsessilis Oliv. et Hiern ■☆

358716　Senecio descoingsii（Humbert）G. D. Rowley = Kleinia descoingsii（Humbert）C. Jeffrey ■☆

358717　Senecio desfontainei Druce；苞叶千里光(荒漠千里光,芥叶千里光,苔叶千里光)；Mossleaf Groundsel, Mustard Leaf Groundsel ■

358718　Senecio desfontainei Druce = Senecio glaucus L. subsp. coronopifolius（Desf.）Alexander ■☆

358719　Senecio dewevrei O. Hoffm. ex T. Durand et De Wild. = Solanecio cydoniifolius（O. Hoffm.）C. Jeffrey ■☆

358720　Senecio dewildemanianus Muschl. = Senecio transmarinus S. Moore var. major C. Jeffrey ■☆

358721　Senecio dianthus Franch. = Synotis erythropappa（Bureau et Franch.）C. Jeffrey et Y. L. Chen ■

358722　Senecio dictyoneurus Franch. = Ligularia dictyoneura（Franch.）Hand. -Mazz. ■

358723　Senecio didymanthus Dunn = Sinacalia davidii（Franch.）H. Koyama ■

358724　Senecio dielsii H. Lév. = Cremanthodium coriaceum S. W. Liu ■

358725　Senecio dieterlenii E. Phillips = Senecio rhomboideus Harv. ■☆

358726　Senecio diffusus L. f. ；松散千里光■☆

358727　Senecio digitalifolius DC. ；指叶千里光■☆

358728　Senecio dilungensis Lisowski；迪龙千里光■☆

358729　Senecio dimorphophyllus Greene = Packera dimorphophylla（Greene）W. A. Weber et Á. Löve ■☆

358730　Senecio dimorphophyllus Greene var. intermedius T. M. Barkley = Packera dimorphophylla（Greene）W. A. Weber et Á. Löve var. intermedia（T. M. Barkley）Trock et T. M. Barkley ■☆

358731　Senecio dimorphophyllus Greene var. paysonii T. M. Barkley =

Packera dimorphophylla（Greene）W. A. Weber et Á. Löve var. paysonii（T. M. Barkley）Trock et T. M. Barkley ■☆

358732　Senecio dinteri Muschl. ex Dinter = Emilia ambifaria（S. Moore）C. Jeffrey ■☆

358733　Senecio diphyllus De Wild. et Muschl.；二叶千里光■☆

358734　Senecio discifolius Oliv. = Emilia discifolia（Oliv.）C. Jeffrey ■☆

358735　Senecio discifolius Oliv. var. scaposus O. Hoffm. = Emilia somalensis（S. Moore）C. Jeffrey ■☆

358736　Senecio discoideus（Hook.）Britton = Packera pauciflora（Pursh）Á. Löve et D. Löve ■☆

358737　Senecio discoideus（Maxim.）Franch. = Cremanthodium discoideum Maxim. ■

358738　Senecio dispar A. Nelson = Senecio integerrimus Nutt. var. exaltatus（Nutt.）Cronquist ■☆

358739　Senecio dissimulans Hilliard；不似千里光■☆

358740　Senecio divaricatus L. = Gynura divaricata（L.）DC. ■

358741　Senecio diversidentatus Muschl. = Senecio inornatus DC. ■☆

358742　Senecio diversifoilus Wall. ex A. DC. = Senecio raphanifolius Wall. ex DC. ■

358743　Senecio diversifolius A. Rich. = Crassocephalum crepidioides（Benth.）S. Moore ■

358744　Senecio diversifolius Harv.；异叶千里光■☆

358745　Senecio diversifolius Wall. ex DC. = Senecio raphanifolius Wall. ex DC. ■

358746　Senecio diversipinnus Y. Ling；异羽千里光（羽裂千里光）；Pinnate Groundsel ■

358747　Senecio diversipinnus Y. Ling var. discoideus C. Jeffrey et Y. L. Chen；无舌异羽千里光（无舌千里光）；Tongueless Groundsel ■

358748　Senecio dodrans C. Winkl.；黑缘千里光；Black Marginate Groundsel, Blackmarginate Groundsel ■

358749　Senecio dolichopappus O. Hoffm. = Lopholaena dolichopappa（O. Hoffm.）S. Moore ■☆

358750　Senecio doria L.；多里亚千里光；Golden Ragwort ■☆

358751　Senecio doria L. var. canescens Willk. = Senecio doria L. ■☆

358752　Senecio doroniciflorus DC. = Senecio albanensis DC. var. doroniciflorus（DC.）Harv. ■☆

358753　Senecio doronicum（L.）L.；多榔千里光；Chamois Ragwort ■☆

358754　Senecio doronicum L. var. hosmariensis Ball = Senecio eriopus Willk. subsp. hosmariensis（Ball）Blanca ■☆

358755　Senecio doryotus Hand. -Mazz. = Sinosenecio euosmus（Hand. -Mazz.）B. Nord. ■

358756　Senecio doryphoroides C. Jeffrey；檫木香千里光■☆

358757　Senecio douglasii DC.；道氏千里光；Butterweed, Creek Senecio, Douglas Groundsel, Douglas Ragwort, Douglas Squaw-weed, Old Man, Sand Wash Groundsel, Threadleaf Groundsel ☆

358758　Senecio douglasii DC. = Senecio flaccidus Less. var. douglasii（DC.）B. L. Turner et T. M. Barkley ■☆

358759　Senecio douglasii DC. var. jamesii（Torr. et A. Gray）Ediger ex Correll et M. C. Johnst. = Senecio flaccidus Less. ■☆

358760　Senecio douglasii DC. var. longilobus（Benth.）L. D. Benson = Senecio flaccidus Less. ■☆

358761　Senecio douglasii DC. var. tularensis Munz = Senecio flaccidus Less. var. douglasii（DC.）B. L. Turner et T. M. Barkley ■☆

358762　Senecio dracunculoides DC. = Senecio burchellii DC. ■☆

358763　Senecio drakensbergis Klatt = Senecio sandersonii Harv. ■☆

358764　Senecio dregeanus DC.；德雷千里光■☆

358765　Senecio drukensis C. Marquand et Airy Shaw；垂头千里光；Pendulous Heads Groundsel, Penduloushead Groundsel ■

358766　Senecio drukensis C. Marquand et Airy Shaw var. nodiflorus（C. C. Chang）Hand. -Mazz. = Senecio nodiflorus C. C. Chang ■

358767　Senecio drummondii Babu et S. N. Biswas = Senecio thianschanicus Regel et Schmalh. ■

358768　Senecio dryas Dunn = Sinosenecio dryas（Dunn）C. Jeffrey et Y. L. Chen ■

358769　Senecio dubitabilis C. Jeffrey et Y. L. Chen；北千里光（可疑千里光，疑千里光）；Nothern Groundsel ■

358770　Senecio dubitabilis C. Jeffrey et Y. L. Chen var. densicapitatus C. H. An = Senecio dubitabilis C. Jeffrey et Y. L. Chen ■

358771　Senecio dubitabilis C. Jeffrey et Y. L. Chen var. densicapitatus C. H. An；密头北千里光（密头千里光）■

358772　Senecio dubitabilis C. Jeffrey et Y. L. Chen var. linearifolius C. H. An et S. L. King = Senecio dubitabilis C. Jeffrey et Y. L. Chen ■

358773　Senecio dubitabilis C. Jeffrey et Y. L. Chen var. linearifolius C. H. An et S. L. Keng；线叶北千里光（线叶千里光）；Linearifolious Groundsel ■

358774　Senecio dubius Ledeb. = Senecio dubitabilis C. Jeffrey et Y. L. Chen ■

358775　Senecio duciformis C. Winkl. = Ligularia duciformis（C. Winkl.）Hand. -Mazz. ■

358776　Senecio ducis-aprutii Chiov. = Crassocephalum ducis-aprutii（Chiov.）S. Moore ■☆

358777　Senecio duclouxii Dunn = Synotis duclouxii（Dunn）C. Jeffrey et Y. L. Chen ■

358778　Senecio ductoris Piper = Senecio fremontii Torr. et A. Gray ■☆

358779　Senecio dumeticola S. Moore；灌生千里光■☆

358780　Senecio dumosus Fourc.；灌丛千里光■☆

358781　Senecio durbanensis Gand. = Senecio deltoideus Less. ■☆

358782　Senecio dux C. B. Clarke = Ligularia dux（C. B. Clarke）R. Mathur ■

358783　Senecio earlei Small = Packera anonyma（A. W. Wood）W. A. Weber et Á. Löve ■☆

358784　Senecio echaetus Y. L. Chen et K. Y. Pan；裸缨千里光（无缨千里光）；Pappusless Groundsel ■

358785　Senecio echinatus（L. f.）DC. = Pericallis echinata（L. f.）B. Nord. ☆

358786　Senecio eenii（S. Moore）Merxm.；埃恩千里光■☆

358787　Senecio effusus Mattf. = Crassocephalum effusum（Mattf.）C. Jeffrey ■☆

358788　Senecio elegans L.；绮丽千里光（雅仙年）；Elegant Groundsel, Jacobaea, Purple Groundsel, Purple Jacobea, Purple Ragwort, Redpurple Ragwort, Wild Cineraria ■☆

358789　Senecio elegans Thunb. = Senecio arenarius Thunb. ■☆

358790　Senecio elegans Willd. = Senecio elegans L. ■☆

358791　Senecio elgonensis Mattf. = Senecio snowdenii Hutch. ■☆

358792　Senecio elgonensis T. C. E. Fr. = Dendrosenecio elgonensis（T. C. E. Fr.）E. B. Knox ●☆

358793　Senecio elivorum？ = Ligularia dentata（A. Gray）H. Hara ■

358794　Senecio ellenbeckii O. Hoffm.；埃伦千里光■☆

358795　Senecio elliotii S. Moore = Senecio syringifolius O. Hoffm. ■☆

358796　Senecio elliottii Torr. et A. Gray = Packera obovata（Muhl. ex Willd.）W. A. Weber et Á. Löve ■☆

358797　Senecio elmeri Piper；埃尔默千里光■☆

358798　Senecio elongatus Pursh = Packera obovata (Muhl. ex Willd.) W. A. Weber et Á. Löve ■☆

358799　Senecio elskensii De Wild. = Solanecio cydoniifolius (O. Hoffm.) C. Jeffrey ■☆

358800　Senecio emilioides Sch. Bip. = Emilia emilioides (Sch. Bip.) C. Jeffrey ■☆

358801　Senecio eminens Compton;显著千里光■☆

358802　Senecio emirnensis DC.;埃米千里光■☆

358803　Senecio emirnensis DC. var. vittarifolius (Bojer ex DC.) Humbert = Senecio vittarifolius Bojer ex DC. ■☆

358804　Senecio encelia Greene = Packera neomexicana (A. Gray) W. A. Weber et Á. Löve ■☆

358805　Senecio engleranus O. Hoffm. ;恩格勒千里光■☆

358806　Senecio epidendricus Mattf. = Solanecio epidendricus (Mattf.) C. Jeffrey ■☆

358807　Senecio eremophilus Richardson;喜沙千里光☆

358808　Senecio eremophilus Richardson var. kingii Greenm. ;金氏千里光☆

358809　Senecio eremophilus Richardson var. macdougalii (A. Heller) Cronquist;马克千里光■☆

358810　Senecio erici-rosenii R. E. Fr. et T. C. E. Fr. = Dendrosenecio erici-rosenii (R. E. Fr. et T. C. E. Fr.) E. B. Knox ●☆

358811　Senecio eriobasis DC. ;毛基千里光■☆

358812　Senecio erioneuron Cotton = Dendrosenecio adnivalis (Stapf) E. B. Knox ●☆

358813　Senecio eriopodus Cummins = Sinosenecio eriopodus (Cummins) C. Jeffrey et Y. L. Chen ■

358814　Senecio eriopus Willk. ;毛足千里光■☆

358815　Senecio eriopus Willk. subsp. hosmariensis (Ball) Blanca;霍斯马里千里光■☆

358816　Senecio eriopus Willk. var. hosmariensis (Ball) Pau = Senecio eriopus Willk. ■☆

358817　Senecio erlangeri O. Hoffm. ;厄兰格千里光■☆

358818　Senecio erosus L. f. ;啮蚀千里光■☆

358819　Senecio erraticus Bertol. = Jacobaea aquatica (Hill) P. Gaertn. et B. Mey. et Schreb. var. erratica (Bertol.) Pelser et Meijden ■☆

358820　Senecio erraticus Bertol. = Senecio aquaticus Hill ☆

358821　Senecio ertterae T. M. Barkley;埃尔千里光■☆

358822　Senecio erubescens Aiton;变红千里光■☆

358823　Senecio erubescens Aiton var. crepidifolius DC. ;还阳参千里光■☆

358824　Senecio erubescens Aiton var. crepidifolius Harv. = Senecio variabilis Sch. Bip. ■☆

358825　Senecio erubescens Aiton var. dichotomus DC. ;二歧变红千里光■☆

358826　Senecio erubescens Aiton var. incisus DC. ;锐裂变红千里光■☆

358827　Senecio erubescens Aiton var. lyratus Harv. = Senecio erubescens Aiton ☆

358828　Senecio erucifolius L. ;芝麻菜叶千里光(羽叶千里光,芸芥叶千里光);Hoary Groundsel, Hoary Ragwort, Narrow-leaved Senecio, Pinnateleaf Groundsel ■

358829　Senecio erucifolius L. = Senecio jacobaea L. ■

358830　Senecio erysimoides DC. ;糖芥千里光■☆

358831　Senecio erythropappus Bureau et Franch. = Synotis erythropappa (Bureau et Franch.) C. Jeffrey et Y. L. Chen ■

358832　Senecio esquirolii H. Lév. = Senecio nudicaulis Buch. -Ham. ex D. Don ■

358833　Senecio euosmus Hand. -Mazz. = Sinosenecio euosmus (Hand. -Mazz.) B. Nord. ■

358834　Senecio eupapposus (Cufod.) G. D. Rowley = Kleinia squarrosa Cufod. ■☆

358835　Senecio eurycephalus Torr. et A. Gray = Packera eurycephala (Torr. et A. Gray) W. A. Weber et Á. Löve ■☆

358836　Senecio eurycephalus Torr. et A. Gray var. lewisrosei (J. T. Howell) T. M. Barkley = Packera eurycephala (Torr. et A. Gray) W. A. Weber et Á. Löve var. lewisrosei (J. T. Howell) J. F. Bain ■☆

358837　Senecio eurycephalus Torr. et A. Gray var. major A. Gray = Senecio integerrimus Nutt. var. major (A. Gray) Cronquist ■☆

358838　Senecio euryopoides DC. ;宽梗千里光■☆

358839　Senecio euryphyllus C. Winkl. = Ligularia euryphylla (C. Winkl.) Hand. -Mazz. ■

358840　Senecio eurypterus Greenm. = Packera neomexicana (A. Gray) W. A. Weber et Á. Löve ■☆

358841　Senecio evansii N. E. Br. = Heteromma decurrens (DC.) O. Hoffm. ■☆

358842　Senecio exaltatus Nutt. = Senecio integerrimus Nutt. var. exaltatus (Nutt.) Cronquist ■☆

358843　Senecio exaltatus Nutt. subsp. ochraceus Piper = Senecio integerrimus Nutt. var. ochroleucus (A. Gray) Cronquist ■☆

358844　Senecio exarachnoideus C. Jeffrey;蛛网千里光■☆

358845　Senecio exsertiflorus Baker = Senecio syringifolius O. Hoffm. ■☆

358846　Senecio exsertus Sch. Bip. = Humbertacalia racemosa (DC.) C. Jeffrey ●☆

358847　Senecio exul Hance;散生千里光;Scattered Groundsel ■

358848　Senecio faberi Hemsl. ex Forbes et Hemsl. ;峨眉千里光(长梗千里光,光梗千里光,密花千里光,密伞千里光,野青菜);Denseumbel Groundsel Groundsel, Faber's Groundsel, Longstalk Groundsel ■

358849　Senecio faberi Hemsl. ex Forbes et Hemsl. var. discoideus Lanener et Ferguson = Senecio liangshanensis C. Jeffrey et Y. L. Chen ■

358850　Senecio faberi Hemsl. var. discoideus Lauener et D. K. Ferguson = Senecio liangshanensis C. Jeffrey et Y. L. Chen ■

358851　Senecio fallax Mattf. = Emilia fallax (Mattf.) C. Jeffrey ■☆

358852　Senecio fargesii Franch. = Ligularia fargesii (Franch.) Diels ■

358853　Senecio farinaceus Sch. Bip. ex A. Rich. ;粉质千里光■☆

358854　Senecio fastigiatus Nutt. = Packera macounii (Greene) W. A. Weber et Á. Löve ■☆

358855　Senecio fastigiatus Nutt. subsp. macounii (Greene) Greenm. = Packera macounii (Greene) W. A. Weber et Á. Löve ■☆

358856　Senecio fastigiatus Nutt. var. layneae (Greene) A. Gray = Packera layneae (Greene) W. A. Weber et Á. Löve ■☆

358857　Senecio faujasioides Baker = Hubertia faujasioides (Baker) C. Jeffrey ●☆

358858　Senecio fauriae Dunn = Ligularia tsangchanensis (Franch.) Hand. -Mazz. ■

358859　Senecio fauriae H. Lév. et Vaniot = Tephroseris phaeantha (Nakai) C. Jeffrey et Y. L. Chen ■

358860　Senecio fauriei H. Lév. = Tephroseris phaeantha (Nakai) C. Jeffrey et Y. L. Chen ■

358861　Senecio feddei H. Lév. = Ligularia hookeri (C. B. Clarke) Hand. -Mazz. ■

358862　Senecio fedifolius Rydb. = Packera debilis (Nutt.) W. A. Weber et Á. Löve ■☆

358863　Senecio fendleri A. Gray = Packera fendleri （A. Gray） W. A. Weber et Á. Löve ■☆

358864　Senecio fendleri A. Gray var. molestus Greenm. = Packera fendleri （A. Gray） W. A. Weber et Á. Löve ■☆

358865　Senecio fendleri A. Gray var. subintegra Greene = Packera cynthioides （Greene） W. A. Weber et Á. Löve ■☆

358866　Senecio ferganensis Schischk. ；费尔干千里光■☆

358867　Senecio fernaldii Greenm. = Packera cymbalaria （Pursh） Á. Löve et D. Löve ■

358868　Senecio fibrillosus Dunn = Ligularia subspicata （Bureau et Franch.） Hand. -Mazz. ■

358869　Senecio fibrosus O. Hoffm. ex Kuntze = Senecio albanensis DC. var. doroniciflorus （DC.） Harv. ■☆

358870　Senecio ficariifolius H. Lév. et Vaniot = Ligularia hookeri （C. B. Clarke） Hand. -Mazz. ■

358871　Senecio ficariifolius H. Lév. Vaniot = Ligularia hookeri （C. B. Clarke） Hand. -Mazz. ■

358872　Senecio ficoides （L.） Sch. Bip. ；大型万宝■☆

358873　Senecio ficoides Sch. Bip. = Senecio ficoides （L.） Sch. Bip. ■☆

358874　Senecio filicifolius Greenm. = Senecio flaccidus Less. var. monoensis （Greene） B. L. Turner et T. M. Barkley ■☆

358875　Senecio filiferus Franch. ；匍枝千里光；Creeping Groundsel, Stoloniferous Groundsel ■

358876　Senecio filiferus Franch. var. dilatams Hand. -Mazz. = Senecio filiferus Franch. ■

358877　Senecio filifolius Nutt. var. fremontii Torr. et A. Gray = Senecio riddellii Torr. et A. Gray ■☆

358878　Senecio fimbrillifer B. L. Rob. = Senecio deltoideus Less. ■☆

358879　Senecio fistulosus De Wild. = Senecio subsessilis Oliv. et Hiern ■☆

358880　Senecio fistulosus De Wild. var. mukuleensis ？ = Senecio subsessilis Oliv. et Hiern ■☆

358881　Senecio flaccidus Less. ；柔软千里光；Gordolobos, Threadleaf Groundsel ■☆

358882　Senecio flaccidus Less. var. douglasii （DC.） B. L. Turner et T. M. Barkley；道格拉斯千里光；Creek Senecio, Threadleaf Groundsel ■☆

358883　Senecio flaccidus Less. var. monoensis （Greene） B. L. Turner et T. M. Barkley；线叶柔软千里光■☆

358884　Senecio flammeus Turcz. ex DC. = Tephroseris flammea （Turcz. ex DC.） Holub ■

358885　Senecio flammeus Turcz. ex DC. f. glabrescens H. Hara = Tephroseris flammea （Turcz. ex DC.） Holub subsp. glabrifolia （Cufod.） B. Nord. ■☆

358886　Senecio flammeus Turcz. ex DC. f. limprichtii Cufod. = Tephroseris flammea （Turcz. ex DC.） Holub ■

358887　Senecio flammeus Turcz. ex DC. f. simplex Y. Ling = Tephroseris flammea （Turcz. ex DC.） Holub ■

358888　Senecio flammeus Turcz. ex DC. subsp. glabrifolius （Cufod.） Kitam. = Tephroseris flammea （Turcz. ex DC.） Holub subsp. glabrifolia （Cufod.） B. Nord. ■☆

358889　Senecio flammeus Turcz. ex DC. var. glabrifolius Cufod. = Tephroseris flammea （Turcz. ex DC.） Holub ■

358890　Senecio flammeus Turcz. ex DC. var. glabrifolius Cufod. = Tephroseris flammea （Turcz. ex DC.） Holub subsp. glabrifolia （Cufod.） B. Nord. ■☆

358891　Senecio flammeus Turcz. ex DC. var. rufus （Hand. -Mazz.） Z. Y. Zhang et Y. H. Guo = Tephroseris rufa （Hand. -Mazz.） B. Nord. ■

358892　Senecio flanaganii E. Phillips = Senecio conrathii N. E. Br. ■☆

358893　Senecio flavovirens Rydb. = Packera paupercula （Michx.） Á. Löve et D. Löve ■☆

358894　Senecio flavulus Greene = Packera pseudaurea （Rydb.） W. A. Weber et Á. Löve var. flavula （Greene） Trock et T. M. Barkley ■☆

358895　Senecio flavus （Decne.） Sch. Bip. ；黄千里光■☆

358896　Senecio flavus （Decne.） Sch. Bip. subsp. brevifolius Kadereit = Senecio mohavensis A. Gray subsp. breviflorus （Kadereit） M. Coleman ■☆

358897　Senecio fletcheri Hemsl. = Cremanthodium ellisii （Hook. f.） Kitam. ■

358898　Senecio flettii Wiegand = Packera flettii （Wiegand） W. A. Weber et Á. Löve ■☆

358899　Senecio flexicaulis Edgew. = Senecio scandens Buch. -Ham. ex D. Don var. incisus Franch. ■

358900　Senecio flocciferus DC. = Malacothrix floccifera （DC.） S. F. Blake ■☆

358901　Senecio floridanus Sch. Bip. = Arnoglossum diversifolium （Torr. et A. Gray） H. Rob. ■☆

358902　Senecio fluviatilis Wallr. ；活水千里光（宽叶千里光）；Broad Leaved Ragwort, Broadleaf Groundsel, Broad-leaved Groundsel, Broad-leaved Ragwort, Saracen's Consound, Saracen's Woundwort, Sarncen's Comfrey, Woundwort ■☆

358903　Senecio foeniculoides Harv. ；茴香千里光■☆

358904　Senecio foetidus Howell = Senecio hydrophiloides Rydb. ■☆

358905　Senecio foetidus Howell var. hydrophiloides （Rydb.） T. M. Barkley ex Cronquist = Senecio hydrophiloides Rydb. ■☆

358906　Senecio foliosus DC. = Jacobaea vulgaris Gaertn. ■

358907　Senecio formosanus （Sasaki） Kitam. = Nemosenecio formosanus （Kitam.） B. Nord. ■

358908　Senecio formosanus Kitam. = Nemosenecio formosanus （Kitam.） B. Nord. ■

358909　Senecio frachetianus H. Lév. = Ligularia franchetiana （H. Lév.） Hand. -Mazz. ■

358910　Senecio fradinii Pomel = Senecio leucanthemifolius Poir. ■☆

358911　Senecio franchetianus H. Lév. = Ligularia franchetiana （H. Lév.） Hand. -Mazz. ■

358912　Senecio franchetii C. Winkl. ；弗氏千里光■☆

358913　Senecio franciscanus Greene = Packera franciscana （Greene） W. A. Weber et Á. Löve ■☆

358914　Senecio francoisii Humbert；弗朗千里光■☆

358915　Senecio fraudulentus E. Phillips et C. A. Sm. ；无饰千里光■☆

358916　Senecio fremontii Torr. et A. Gray；弗雷蒙千里光■☆

358917　Senecio fremontii Torr. et A. Gray var. blitoides （Greene） Cronquist；藜千里光■☆

358918　Senecio fremontii Torr. et A. Gray var. occidentalis A. Gray；西方千里光■☆

358919　Senecio fresenii Sch. Bip. ；弗雷森千里光■☆

358920　Senecio friesiorum Mildbr. = Dendrosenecio adnivalis （Stapf） E. B. Knox subsp. friesiorum （Mildbr.） E. B. Knox ■☆

358921　Senecio frigidus （Richardson） Less. = Tephroseris frigida （Richardson） Holub ■☆

358922　Senecio frigidus （Richardson） Less. var. tomentosus （Kjellm.） Cufod. = Tephroseris kjellmanii （A. E. Porsild） Holub ■☆

358923　Senecio fuchsii C. C. Gmel. = Senecio nemorensis L. ■

358924　Senecio fukienensis Y. Ling ex C. Jeffrey et Y. L. Chen；闽千里光；Fujian Groundsel，Fukien Groundsel ■

358925　Senecio fulgens（Hook. f.）G. Nicholson；翠叶菊（白银龙）；Soóty Groundsel ■

358926　Senecio fulgens（Hook. f.）G. Nicholson = Kleinia fulgens Hook. f. ■☆

358927　Senecio fulgens Rydb. = Packera streptanthifolia（Greene）W. A. Weber et Á. Löve ■☆

358928　Senecio fulvipes Y. Ling = Synotis fulvipes（Y. Ling）C. Jeffrey et Y. L. Chen ■

358929　Senecio furusei Kitam. = Tephroseris furusei（Kitam.）B. Nord. ■☆

358930　Senecio gabonicus Oliv. et Hiern = Solanecio angulatus（Vahl）C. Jeffrey ■☆

358931　Senecio gaffatensis Vatke = Senecio aegyptius L. var. discoideus Boiss. ■☆

358932　Senecio gallicus Chaix；法国千里光 ■☆

358933　Senecio gallicus Chaix subsp. coronopifolius（Desf.）Maire = Senecio glaucus L. subsp. coronopifolius（Desf.）Alexander ■☆

358934　Senecio gallicus Chaix subsp. hesperidum（Jahand. et al.）Maire = Senecio hesperidum Jahand. et Degen et Weiller ■☆

358935　Senecio gallicus Chaix subsp. mauritanicus（Pomel）Maire = Senecio leucanthemifolius Poir. ■☆

358936　Senecio gallicus Chaix var. calyculatus Emb. et Maire = Senecio gallicus Chaix ■☆

358937　Senecio gallicus Chaix var. lanigerus（Batt.）Maire = Senecio leucanthemifolius Poir. ■☆

358938　Senecio gallicus Chaix var. laxiflorus（Viv.）DC. = Senecio glaucus L. subsp. coronopifolius（Desf.）Alexander ■☆

358939　Senecio gallicus Chaix var. laxiflorus DC. = Senecio gallicus Chaix ■☆

358940　Senecio gallicus Chaix var. mauritanicus（Pomel）Pau = Senecio leucanthemifolius Poir. ■☆

358941　Senecio gallicus Chaix var. sonchifolius Ball = Senecio gallicus Chaix ■☆

358942　Senecio galpinii（Hook. f.）H. Jacobsen = Kleinia galpinii Hook. f. ■☆

358943　Senecio galpinii（Hook. f.）Hook. f. = Kleinia galpinii Hook. f. ■☆

358944　Senecio ganderi T. M. Barkley et R. M. Beauch. = Packera ganderi（T. M. Barkley et R. M. Beauch.）W. A. Weber et Á. Löve ■☆

358945　Senecio ganpinensis Vaniot = Senecio nemorensis L. ■

358946　Senecio gardneri Cotton = Dendrosenecio elgonensis（T. C. E. Fr.）E. B. Knox subsp. barbatipes（Hedberg）E. B. Knox ●☆

358947　Senecio gardneri Cotton var. ligulatus Cotton et Blakelock = Dendrosenecio elgonensis（T. C. E. Fr.）E. B. Knox subsp. barbatipes（Hedberg）E. B. Knox ●☆

358948　Senecio gariepiensis Cron；加里普千里光 ■☆

358949　Senecio gaspensis Greenm. = Packera paupercula（Michx.）Á. Löve et D. Löve ■☆

358950　Senecio gaspensis Greenm. var. firmifolius（Greenm.）Fernald = Packera paupercula（Michx.）Á. Löve et D. Löve ■☆

358951　Senecio geniorum Humbert；多育千里光 ■☆

358952　Senecio gentilianus Vaniot = Senecio wightii（DC. ex Wight）Benth. ex C. B. Clarke ■

358953　Senecio gerrardii Harv.；杰勒德千里光 ■☆

358954　Senecio gibbonsii Greene = Senecio triangularis Hook. ■☆

358955　Senecio giessii Merxm.；吉斯千里光 ■☆

358956　Senecio giganteus Desf.；巨大千里光 ■☆

358957　Senecio gigas Vatke = Solanecio gigas（Vatke）C. Jeffrey ■☆

358958　Senecio giorgii De Wild. = Solanecio cydoniifolius（O. Hoffm.）C. Jeffrey ■☆

358959　Senecio glabellus（Turcz.）DC. = Tephroseris praticola（Schischk. et Serg.）Holub ■

358960　Senecio glabellus DC. = Tephroseris pseudosonchus（Vaniot）C. Jeffrey et Y. L. Chen ■

358961　Senecio glabellus Poir. = Packera glabella（Poir.）C. Jeffrey ■☆

358962　Senecio glaberrimus DC.；无毛千里光 ■☆

358963　Senecio glabrifolius DC.；光叶千里光 ■☆

358964　Senecio glanduloso-lanosus Thell.；具腺千里光 ■☆

358965　Senecio glanduloso-pilosus Volkens et Muschl.；腺毛千里光 ■☆

358966　Senecio glastifolius L. f.；菘蓝叶千里光 ■☆

358967　Senecio glaucescens DC. = Senecio achilleifolius DC. ■☆

358968　Senecio glauciifolius Rydb. = Senecio eremophilus Richardson ■☆

358969　Senecio glaucus L.；灰蓝千里光 ■☆

358970　Senecio glaucus L. subsp. coronopifolius（Desf.）Alexander；鸟足叶千里光 ■☆

358971　Senecio glaucus L. subsp. coronopifolius（Desf.）Alexander = Senecio desfontainei Druce ■

358972　Senecio globigerus C. C. Chang = Sinosenecio globigerus（C. C. Chang）B. Nord. ■

358973　Senecio glomeratus Desf. ex Poir. = Erechtites glomeratus（Desf. ex Poir.）DC. ■☆

358974　Senecio glomeratus Jeffrey = Synotis glomerata（Jeffrey）C. Jeffrey et Y. L. Chen ■

358975　Senecio glumaceus Dunn = Synotis erythropappa（Bureau et Franch.）C. Jeffrey et Y. L. Chen ■

358976　Senecio glutinarius DC.；稍黏千里光 ■☆

358977　Senecio glutinosus E. Mey. = Senecio glutinarius DC. ■☆

358978　Senecio glutinosus Thunb.；黏性千里光 ■☆

358979　Senecio goetzei O. Hoffm. = Solanecio goetzei（O. Hoffm.）C. Jeffrey ■☆

358980　Senecio goetzenii O. Hoffm. = Crassocephalum goetzenii（O. Hoffm.）S. Moore ■☆

358981　Senecio goodianus Hand. -Mazz. = Sinosenecio hederifolius（Dümmer）B. Nord. ■

358982　Senecio goodianus Hand. -Mazz. var. angulatifolius Y. Ling = Sinosenecio hederifolius（Dümmer）B. Nord. ■

358983　Senecio goringensis Hemsl. = Cremanthodium ellisii（Hook. f.）Kitam. ■

358984　Senecio gossweileri Torre；戈斯千里光 ■☆

358985　Senecio graciliflorus DC.；纤花千里光（多翼千里光）；Fineflower Groundsel，Manywing Groundsel ■

358986　Senecio graciliflorus DC. var. hookeri C. B. Clarke = Senecio royleanus DC. ■

358987　Senecio graciliflorus DC. var. pleopterus Hand. -Mazz. = Senecio graciliflorus DC. ■

358988　Senecio gracilis Pursh；纤细千里光 ■☆

358989　Senecio gracilis Pursh = Packera aurea（L.）Á. Löve et D. Löve ■☆

358990　Senecio graciliserra Mattf. = Senecio lelyi Hutch. ■☆

358991 Senecio gracillimus C. Winkl. = Tephroseris palustris（L.）Fourr. ■

358992 Senecio gramineus Harv. ;禾状千里光■☆

358993 Senecio graminicola C. A. Sm. = Senecio retrorsus DC. ■☆

358994 Senecio graminifolius Jacq. = Senecio chrysocoma Meerb. ■☆

358995 Senecio grandidentatus Ledeb. ;砂千里光■☆

358996 Senecio grandiflorus P. J. Bergius;大紫花千里光;Purple Ragwort ■☆

358997 Senecio grantii（Hook. f.）Sch. Bip. ;绯冠菊(红鹰)■☆

358998 Senecio greenei A. Gray = Packera greenei（A. Gray）W. A. Weber et Á. Löve ■☆

358999 Senecio greenwayi C. Jeffrey;格林韦千里光■☆

359000 Senecio gregatus Hilliard;聚生千里光■☆

359001 Senecio greggii Rydb. = Packera tampicana（DC.）C. Jeffrey ■☆

359002 Senecio gregorii（S. Moore）H. Jacobsen = Kleinia gregorii（S. Moore）C. Jeffrey ■☆

359003 Senecio gunnisii Baker = Kleinia pendula（Forssk.）DC. ■☆

359004 Senecio gyirongensis Y. L. Chen et K. Y. Pan = Senecio biligulatus W. W. Sm. ■

359005 Senecio gyirongensis Y. Ling ex Y. L. Chen,S. Yun Liang et K. Y. Pan = Senecio biligulatus W. W. Sm. ■

359006 Senecio gynura C. Winkl. = Nannoglottis gynura（C. Winkl.）Y. Ling et Y. L. Chen ■

359007 Senecio gynuroides S. Moore = Crassocephalum ducis-aprutii（Chiov.）S. Moore ■☆

359008 Senecio gynuropsis Muschl. = Crassocephalum ducis-aprutii（Chiov.）S. Moore ■☆

359009 Senecio gyrophyllus Klatt = Senecio macrocephalus DC. ■☆

359010 Senecio hadiensis Forssk. ;哈迪千里光■☆

359011 Senecio hageniae R. E. Fr. = Senecio maranguensis O. Hoffm. ■☆

359012 Senecio hainanensis C. C. Chang et Y. C. Tseng = Sinosenecio hainanensis（C. C. Chang et Y. C. Tseng）C. Jeffrey et Y. L. Chen ■

359013 Senecio halimifolius L. ;滨藜叶千里光■☆

359014 Senecio halleri Dandy;单头千里光■☆

359015 Senecio hallianus G. D. Rowley;哈里千里光■☆

359016 Senecio hallii Britton = Packera cana（Hook.）W. A. Weber et Á. Löve ■☆

359017 Senecio hallii Britton var. discoidea W. A. Weber = Packera cana（Hook.）W. A. Weber et Á. Löve ■☆

359018 Senecio handelianus B. Nord. = Synotis fulvipes（Y. Ling）C. Jeffrey et Y. L. Chen ■

359019 Senecio harbourii Rydb. = Packera cana（Hook.）W. A. Weber et Á. Löve ■☆

359020 Senecio harfordii Greenm. = Packera bolanderi（A. Gray）W. A. Weber et Á. Löve var. harfordii（Greenm.）Trock et T. M. Barkley ■☆

359021 Senecio hartianus A. Heller = Packera hartiana（A. Heller）W. A. Weber et Á. Löve ■☆

359022 Senecio hartwegii Benth. = Roldana hartwegii（Benth.）H. Rob. et Brettell ■☆

359023 Senecio hastatus L. ;戟形千里光■☆

359024 Senecio hastifolius（L. f.）Less. ;戟叶千里光■☆

359025 Senecio hastulatus L. = Senecio hastatus L. ■☆

359026 Senecio haworthii（Sweet）Sch. Bip. ;银锤掌(银月);Cocoon Plant,Haworth Groundsel ■

359027 Senecio haworthii Sch. Bip. = Senecio haworthii（Sweet）Sch. Bip. ■☆

359028 Senecio haygarthii Hilliard;海加斯千里光■☆

359029 Senecio hedbergii C. Jeffrey;赫德千里光■☆

359030 Senecio hederiformis Cron;常春藤千里光■☆

359031 Senecio helianthus Franch. = Cremanthodium helianthus（Franch.）W. W. Sm. ■

359032 Senecio helminthioides（Sch. Bip.）Hilliard;蠕虫千里光■☆

359033 Senecio henrici Vaniot = Senecio asperifolius Franch. ■

359034 Senecio henryi Hemsl. = Sinacalia tangutica（Maxim.）B. Nord. ■

359035 Senecio heritieri DC. = Pericallis lanata（L'Hér.）B. Nord. ■☆

359036 Senecio hermannii B. Nord. ;赫尔曼千里光■☆

359037 Senecio herreianus Dinter;大弦月城■☆

359038 Senecio herreianus Tinter = Senecio herreianus Dinter ■☆

359039 Senecio hesperidum Jahand. et Degen et Weiller;金星千里光■☆

359040 Senecio hesperius Greene = Packera hesperia（Greene）W. A. Weber et Á. Löve ■☆

359041 Senecio heteroclinius DC. = Senecio abruptus Thunb. ■☆

359042 Senecio heterodoxus Greene ex Rydb. = Packera dimorphophylla（Greene）W. A. Weber et Á. Löve ■☆

359043 Senecio heteromorphus Hutch. et B. L. Burtt = Crassocephalum radiatum S. Moore ■☆

359044 Senecio hibernus Makino = Senecio scandens Buch. -Ham. ex D. Don ■

359045 Senecio hieracifolius H. Lév. = Synotis hieracifolia（H. Lév.）C. Jeffrey et Y. L. Chen ■

359046 Senecio hieracifolius L. = Erechtites hieracifolius（L.）Raf. ex DC. ■

359047 Senecio hieracioides DC. ;山柳菊千里光■☆

359048 Senecio hildebrandtii Baker;希尔德千里光■☆

359049 Senecio hillebrandii H. Christ = Pericallis papyracea（DC.）B. Nord. ■☆

359050 Senecio himalayensis Franch. = Cremanthodium pinnatifidum Benth. ■

359051 Senecio hindsii Benth. = Senecio scandens Buch. -Ham. ex D. Don ■

359052 Senecio hirsutilobus Hilliard;毛裂片千里光■☆

359053 Senecio hirtellus DC. ;多毛千里光■☆

359054 Senecio hirtifolius DC. ;毛叶千里光■☆

359055 Senecio hochstetteri Sch. Bip. ex A. Rich. ;霍赫千里光■☆

359056 Senecio hochstetteri Sch. Bip. ex A. Rich. var. radiatus Chiov. = Senecio hochstetteri Sch. Bip. ex A. Rich. ■☆

359057 Senecio hockii De Wild. et Muschl. = Emilia hockii（De Wild. et Muschl.）C. Jeffrey ■☆

359058 Senecio hoffmannianus Muschl. = Emilia discifolia（Oliv.）C. Jeffrey ■☆

359059 Senecio hoggariensis Batt. et Trab. ;霍加里千里光■☆

359060 Senecio hoggariensis Batt. et Trab. var. eradiatus Maire = Senecio hoggariensis Batt. et Trab. ■☆

359061 Senecio hoi Dunn;滇南千里光■

359062 Senecio hoi Dunn = Cissampelopsis volubilis（Blume）Miq. ●

359063 Senecio hollandii Compton;霍兰千里光■☆

359064 Senecio holmii Greene = Senecio amplectens A. Gray var. holmii（Greene）H. D. Harr. ■☆

359065 Senecio holubii Hutch. et Burtt Davy;霍勒布千里光■☆

359066　Senecio homblei De Wild. = Emilia homblei（De Wild.）C. Jeffrey ■☆

359067　Senecio homogyniphyllus Cummins = Sinosenecio homogyniphyllus（Cummins）B. Nord. ■

359068　Senecio homogyniphyllus Cummins var. subumbellatus C. C. Chang = Sinosenecio chienii（Hand. -Mazz.）B. Nord. ■

359069　Senecio hookeri Torr. et A. Gray = Senecio integerrimus Nutt. var. exaltatus（Nutt.）Cronquist ■☆

359070　Senecio hookerianus H. Jacobsen = Kleinia fulgens Hook. f. ■☆

359071　Senecio hornogyniphyllus Cummins var. subumbellatus Chang = Sinosenecio chienii（Hand. -Mazz.）B. Nord. ■

359072　Senecio howellii Greene = Packera cana（Hook.）W. A. Weber et Á. Löve ■☆

359073　Senecio hubertia Pers. ;胡伯千里光■☆

359074　Senecio hugonis S. Moore = Synotis nagensis（C. B. Clarke）C. Jeffrey et Y. L. Chen ■

359075　Senecio hui C. C. Chang = Synotis hieraciifolia（H. Lév.）C. Jeffrey et Y. L. Chen ■

359076　Senecio humbertii C. C. Chang;弥勒千里光;Mile Groundsel ■

359077　Senecio humbertii C. C. Chang = Notonia madagascariensis Humbert ■☆

359078　Senecio humidanus C. Jeffrey;潮湿狗舌草■☆

359079　Senecio humilis Desf. = Senecio leucanthemifolius Poir. ■☆

359080　Senecio humilis Desf. var. mauritii Sennen = Senecio leucanthemifolius Poir. ■☆

359081　Senecio hunanensis Hand. -Mazz. = Sinosenecio hunanensis（Y. Ling）B. Nord. ■

359082　Senecio hunanensis Hand. -Mazz. = Synotis fulvipes（Y. Ling）C. Jeffrey et Y. L. Chen ■

359083　Senecio hunanensis Y. Ling = Sinosenecio hunanensis（Y. Ling）B. Nord. ■

359084　Senecio hunnanensis Hand. -Mazz. = Synotis fulvipes（Y. Ling）C. Jeffrey et Y. L. Chen ■

359085　Senecio hybridus Regel = Pericallis hybrida B. Nord. ■

359086　Senecio hydrophiloides Rydb. ;拟湿生狗舌草■☆

359087　Senecio hydrophilus Nutt. ;喜湿狗舌草■☆

359088　Senecio hydrophilus Nutt. var. pacificus Greene = Senecio hydrophilus Nutt. ■☆

359089　Senecio hygrophilus Cuatrec. = Senecio hydrophilus Nutt. ■☆

359090　Senecio hygrophilus Klatt = Cineraria anampoza（Baker）Baker f. hygrophila（Klatt）Cron ■☆

359091　Senecio hygrophilus R. A. Dyer et C. A. Sm. = Senecio humidanus C. Jeffrey ■☆

359092　Senecio hypargyraeus DC. = Hubertia hypargyrea（DC.）C. Jeffrey ●☆

359093　Senecio hyperborealis Greenm. = Packera hyperborealis（Greenm.）Á. Löve et D. Löve ■☆

359094　Senecio hypoleucus Muschl. = Dendrosenecio adnivalis（Stapf）E. B. Knox subsp. friesiorum（Mildbr.）E. B. Knox ■☆

359095　Senecio ianthinus Mattf. = Senecio erubescens Aiton ■☆

359096　Senecio ianthophyllus Franch. = Parasenecio ianthophyllus（Franch.）Y. L. Chen ■

359097　Senecio idahoensis Rydb. = Packera indecora（Greene）Á. Löve et D. Löve ■☆

359098　Senecio ilicifolius L. ;冬青叶千里光■☆

359099　Senecio iljinii Schischk. ;伊尔金千里光■☆

359100　Senecio imaii Nakai = Tephroseris subdentata（Bunge）Holub ■

359101　Senecio imparipinnatus Klatt = Packera tampicana（DC.）C. Jeffrey ■☆

359102　Senecio implexus P. R. O. Bally = Kleinia implexa（P. R. O. Bally）C. Jeffrey ■☆

359103　Senecio inaequidens DC. ;不等齿千里光;Narrow-leaved Rag' wort ■☆

359104　Senecio inamoenus DC. = Senecio glutinosus Thunb. ■☆

359105　Senecio incanus L. ;灰毛千里光;Grey Alpine Groundsel ■☆

359106　Senecio incertus DC. = Senecio tuberosus（DC.）Harv. ■☆

359107　Senecio incisifolius Jeffrey = Nemosenecio incisifolius（Jeffrey）B. Nord. ■

359108　Senecio incisifolius Jeffrey var. gracilior Y. Ling = Nemosenecio incisifolius（Jeffrey）B. Nord. ■

359109　Senecio incisus Thunb. ;锐裂千里光■☆

359110　Senecio incognitus Cabrera = Senecio madagascariensis Poir. ■☆

359111　Senecio incomptus DC. ;装饰千里光■☆

359112　Senecio inconstans DC. = Senecio repandus Thunb. ■☆

359113　Senecio incrassatus Lowe;粗千里光■☆

359114　Senecio incurvus A. Nelson = Senecio spartioides Torr. et Gray ■☆

359115　Senecio indecorus Greene = Packera indecora（Greene）Á. Löve et D. Löve ■☆

359116　Senecio infirmus C. Jeffrey;柔弱千里光■☆

359117　Senecio inophyllus E. Phillips et C. A. Sm. = Senecio glaberrimus DC. ■☆

359118　Senecio inornatus DC. = Senecio fraudulentus E. Phillips et C. A. Sm. ■☆

359119　Senecio integerrimus Nutt. var. exaltatus（Nutt.）Cronquist;虎克千里光■☆

359120　Senecio integerrimus Nutt. var. major（A. Gray）Cronquist;大全缘千里光■☆

359121　Senecio integerrimus Nutt. var. ochroleucus（A. Gray）Cronquist;浅黄全缘千里光■☆

359122　Senecio integerrimus Nutt. var. scribneri（Rydb.）T. M. Barkley;斯克里布纳千里光■☆

359123　Senecio integerrimus Nutt. var. vaseyi（Greenm.）Cronquist = Senecio integerrimus Nutt. var. exaltatus（Nutt.）Cronquist ■☆

359124　Senecio integrifolius（L.）Clairv. = Tephroseris kirilowii（Turcz. ex DC.）Holub ■

359125　Senecio integrifolius（L.）Clairv. subsp. atropurpureus var. robustus（Herder）Curl = Tephroseris turczaninowii（DC.）Holub ■

359126　Senecio integrifolius（L.）Clairv. subsp. capitatus（Wahlenb.）Cufod. = Tephroseris rufa（Hand. -Mazz.）B. Nord. var. chaetocarpa C. Jeffrey et Y. L. Chen ■

359127　Senecio integrifolius（L.）Clairv. subsp. capitatus（Wahlenb.）Cufod. var. aurantiacus f. pseudocampestris（Roem.）Briq. et Cavill. = Tephroseris rufa（Hand. -Mazz.）B. Nord. ■

359128　Senecio integrifolius（L.）Clairv. subsp. capitatus sensu Hand. -Mazz. = Tephroseris rufa（Hand. -Mazz.）B. Nord. var. chaetocarpa C. Jeffrey et Y. L. Chen ■

359129　Senecio integrifolius（L.）Clairv. subsp. fauriei（H. Lév. et Vaniot）Kitam. = Tephroseris kirilowii（Turcz. ex DC.）Holub ■

359130　Senecio integrifolius（L.）Clairv. subsp. fauriei（H. Lév. et Vaniot）Kitam. = Tephroseris phaeantha（Nakai）C. Jeffrey et Y. L. Chen ■

359131　Senecio integrifolius（L.）Clairv. subsp. fauriei Kitam. = Tephroseris phaeantha（Nakai）C. Jeffrey et Y. L. Chen ■

359132　Senecio integrifolius（L.）Clairv. subsp. fauriei sensu Kitam. = Tephroseris kirilowii（Turcz. ex DC.）Holub ■

359133　Senecio integrifolius（L.）Clairv. subsp. kirilowii（Turcz. ex DC.）Kitag. = Tephroseris kirilowii（Turcz. ex DC.）Holub ■

359134　Senecio integrifolius（L.）Clairv. var. glabratus（A. DC.）Cufod. = Tephroseris praticola（Schischk. et Serg.）Holub ■

359135　Senecio integrifolius（L.）Clairv. var. lindstroemii Ostenf. = Tephroseris lindstroemii（Ostenf.）Á. Löve et D. Löve ■☆

359136　Senecio integrifolius（L.）Clairv. var. spathulatus（Miq.）H. Hara = Tephroseris kirilowii（Turcz. ex DC.）Holub ■

359137　Senecio integrifolius（L.）Clairv. var. spathulifolius（Miq.）Hara = Tephroseris kirilowii（Turcz. ex DC.）Holub ■

359138　Senecio integrimus Nutt. ; 全缘千里光（缘叶千里光）; Entire-leaved Groundsel, Field Groundsel ■☆

359139　Senecio intermedius Hayata = Syneilesis intermedia（Hayata）Kitam. ■

359140　Senecio intermedius Wight = Senecio scandens Buch. -Ham. ex D. Don ■

359141　Senecio intricatus S. Moore; 缠结千里光■☆

359142　Senecio invenustus Greene = Senecio fremontii Torr. et A. Gray var. blitoides（Greene）Cronquist ■☆

359143　Senecio iochaneose H. Lév. = Ligularia wilsoniana（Hemsl.）Greenm. ■

359144　Senecio ionodasys Hand. -Mazz. = Synotis ionodasys（Hand. -Mazz.）C. Jeffrey et Y. L. Chen ■

359145　Senecio ionophyllus Greene = Packera ionophylla（Greene）W. A. Weber et Á. Löve ■☆

359146　Senecio ionophyllus Greene var. bernardinus（Greene）H. M. Hall = Packera bernardina（Greene）W. A. Weber et Á. Löve ■☆

359147　Senecio ionophyllus Greene var. intrepidus Greenm. = Packera ionophylla（Greene）W. A. Weber et Á. Löve ■☆

359148　Senecio ionophyllus Greene var. sparsilobatus（Parish）H. M. Hall = Packera bernardina（Greene）W. A. Weber et Á. Löve ■☆

359149　Senecio iosensis G. D. Rowley = Senecio sulcicalyx Baker ■☆

359150　Senecio irregularibracteatus De Wild. = Emilia irregularibracteata（De Wild.）C. Jeffrey ■☆

359151　Senecio isatideus DC. ; 菘蓝千里光■☆

359152　Senecio isatideus DC. var. macrophyllus Thell. = Senecio isatidioides E. Phillips et C. A. Sm. ■☆

359153　Senecio isatidioides E. Phillips et C. A. Sm. ; 拟菘蓝千里光■☆

359154　Senecio ivohibeensis Humbert = Hubertia ivohibeensis（Humbert）C. Jeffrey ●☆

359155　Senecio jacksonii S. Moore; 杰克逊千里光■☆

359156　Senecio jacksonii S. Moore subsp. caryophyllus（Mattf.）Hedberg = Senecio jacksonii S. Moore ■☆

359157　Senecio jacksonii S. Moore var. sympodialis（R. E. Fr.）Hedberg = Senecio jacksonii S. Moore ■☆

359158　Senecio jacobaea L. ; 草甸千里光（夹可千里光, 美狗舌草, 千里光, 新疆千里光, 牙克贝千里光, 异果千里光, 羽叶千里光）; Agreen, Beaweed, Bennel, Benweed, Bindweed, Binweed, Boholawn, Boliaun, Booin, Bouin, Bowen, Bowlocks, Bucalaun, Bundweed, Bunnel, Bunwede, Cammick, Cammock, Canker Weed, Cankerwort, Cheadle Dock, Common Ragwort, Cowfoot, Cradle Dock, Cradle-dock, Crowfoot, Curley-doddies, Cushag, Daisy, Devildums, Dog Stalk, Dog Standard, Dog Standers, Ell-shinders, Fairy Horse, Farenut, Fellon-weed, Fizz-gigg, Flea Nit, Flea Nut, Fleawort, Flee-dod, Fly-dod, Gandergoose, Grundswaith, Grundswathe, Grunsel, Grunsil, Gypsy, He Bulkishawn, He-bulkishawn, Jacob's Groundsel, Jacoby Fleawort, James' Weed, Kadle Dock, Keddle Dock, Keddle-dock, Kedlock, Keedle Dock, Keedle-dock, Kettle Dock, Kettle-dock, Mare Fart, Muggert, Orchis Morio, Ragged Jack, Ragged Robin, Ragweed, Ragwort, Ragwort Groundsel, Scrape-clean, Seattle Dock, Seggrum, Seggy, Sigrim, Skedlock, Sleepydose, St. James' Ragwort, Staggerwort, Staggwort, Stammerwort, Staverwort, Stink Davie, Stinking Alisander, Stinking Alisanders, Stinking Billy, Stinking Davies, Stinking Elshinder, Stinking Elshinders, Stinking Weed, Stinking Willie, Summer's Farewell, Swine's Cress, Swine's Grass, Tansy, Tansy Ragwort, Weebow, Weeby, Wild Chrysanthemum, Yackrod, Yackyar, Yaller, Yarkrod, Yellow Boys, Yellow Daisy, Yellow Ragwort, Yellow Shinders Ell, Yellow-boy, Yellowtop, Yellow-weed ■

359159　Senecio jacobaea L. subsp. barbareifolius ? = Jacobaea aquatica（Hill）P. Gaertn. et B. Mey. et Schreb. ■☆

359160　Senecio jacobaea L. var. grandiflorus Turcz. ex DC. = Senecio ambraceus Turcz. ex DC. ■

359161　Senecio jacobaea L. var. grandiflorus Turcz. ex DC. = Senecio argunensis Turcz. ■

359162　Senecio jacobsenii G. D. Rowley; 悬垂千里光; Jacobsen Groundsel ■☆

359163　Senecio jacobsenii G. D. Rowley = Kleinia petraea（R. E. Fr.）C. Jeffrey ■☆

359164　Senecio jacquemontianus Benth. ex Hook. f. ; 雅克山千里光■☆

359165　Senecio jacuticus Schischk. ; 雅库特千里光■☆

359166　Senecio jamesii Hemsl. = Ligularia jamesii（Hemsl.）Kom. ■

359167　Senecio japonicus（Thunb.）Sch. Bip. = Ligularia japonica（Thunb.）Less. ■

359168　Senecio japonicus Less. = Farfugium japonicum（L.）Kitam. ■

359169　Senecio japonicus Sch. Bip. = Ligularia japonica（Thunb.）Less. ■

359170　Senecio japonicus Sch. Bip. var. integrifolius Matsum. = Ligularia dentata（A. Gray）H. Hara ■

359171　Senecio japonicus Sch. Bip. var. scaberrimus Hayata = Ligularia japonica（Thunb.）Less. var. scaberrima（Hayata）Y. Ling ■

359172　Senecio japonicus Thunb. = Gynura japonica（Thunb.）Juel ■

359173　Senecio jeffreyanus Diels = Ligularia kanaitzensis（Franch.）Hand. -Mazz. ■

359174　Senecio jeffreyanus Lisowski; 杰弗里千里光■☆

359175　Senecio johannesburgensis S. Moore = Senecio hieracioides DC. ■☆

359176　Senecio johnstonii Oliv. ; 约翰斯顿千里光■☆

359177　Senecio johnstonii Oliv. subsp. barbatipes（Hedberg）Mabb. = Dendrosenecio elgonensis（T. C. E. Fr.）E. B. Knox subsp. barbatipes（Hedberg）E. B. Knox ●☆

359178　Senecio johnstonii Oliv. subsp. cheranganiensis（Cotton et Blakelock）Mabb. = Dendrosenecio cheranganiensis（Cotton et Blakelock）E. B. Knox ●☆

359179　Senecio johnstonii Oliv. subsp. cottonii（Hutch. et G. Taylor）Mabb. = Dendrosenecio kilimanjari（Mildbr.）E. B. Knox subsp. cottonii（Hutch. et G. Taylor）E. B. Knox ●☆

359180　Senecio johnstonii Oliv. subsp. refractisquamatus（De Wild.）Mabb. = Dendrosenecio adnivalis（Stapf）E. B. Knox ●☆

359181　Senecio johnstonii Oliv. var. adnivalis（Stapf）C. Jeffrey = Dendrosenecio adnivalis（Stapf）E. B. Knox ●☆

359182　Senecio johnstonii Oliv. var. alticola（T. C. E. Fr.）C. Jeffrey =

Dendrosenecio erici-rosenii（R. E. Fr. et T. C. E. Fr.）E. B. Knox subsp. alticola（Mildbr.）E. B. Knox ●☆

359183　Senecio johnstonii Oliv. var. battiscombei（R. E. Fr. et T. C. E. Fr.）Mabb. = Dendrosenecio battiscombei（R. E. Fr. et T. C. E. Fr.）E. B. Knox ●☆

359184　Senecio johnstonii Oliv. var. cheranganiensis（Cotton et Blakelock）C. Jeffrey;切兰加尼千里光■☆

359185　Senecio johnstonii Oliv. var. cottonii（Hutch. et G. Taylor）C. Jeffrey = Dendrosenecio kilimanjari（Mildbr.）E. B. Knox subsp. cottonii（Hutch. et G. Taylor）E. B. Knox ●☆

359186　Senecio johnstonii Oliv. var. dalei（Cotton et Blakelock）C. Jeffrey = Dendrosenecio cheranganiensis（Cotton et Blakelock）E. B. Knox subsp. dalei Cotton et Blakelock ●☆

359187　Senecio johnstonii Oliv. var. elgonensis（T. C. E. Fr.）Mabb. = Dendrosenecio elgonensis（T. C. E. Fr.）E. B. Knox ●☆

359188　Senecio johnstonii Oliv. var. erici-rosenii（R. E. Fr. et T. C. E. Fr.）C. Jeffrey = Dendrosenecio erici-rosenii（R. E. Fr. et T. C. E. Fr.）E. B. Knox ■☆

359189　Senecio johnstonii Oliv. var. friesiorum（Mildbr.）Mabb. = Dendrosenecio adnivalis（Stapf）E. B. Knox subsp. friesiorum（Mildbr.）E. B. Knox ■☆

359190　Senecio johnstonii Oliv. var. kilimanjari（Mildbr.）C. Jeffrey = Dendrosenecio kilimanjari（Mildbr.）E. B. Knox ●☆

359191　Senecio johnstonii Oliv. var. ligulatus（Cotton et Blakelock）C. Jeffrey = Dendrosenecio elgonensis（T. C. E. Fr.）E. B. Knox subsp. barbatipes（Hedberg）E. B. Knox ●☆

359192　Senecio johnstonii Oliv. var. meruensis（Cotton et Blakelock）C. Jeffrey = Dendrosenecio meruensis（Cotton et Blakelock）E. B. Knox ●☆

359193　Senecio jonesii Rydb. = Packera streptanthifolia（Greene）W. A. Weber et Á. Löve ■☆

359194　Senecio jugicola S. Moore = Senecio maranguensis O. Hoffm. ■☆

359195　Senecio junceus（DC.）Harv.;灯芯草千里光■☆

359196　Senecio juniperinus L. f.;刺柏状千里光■☆

359197　Senecio junodianus O. Hoffm. = Senecio madagascariensis Poir. ■☆

359198　Senecio junodii Hutch. et Burtt Davy;朱诺德千里光■☆

359199　Senecio kaempferi DC. = Farfugium japonicum（L.）Kitam. ■

359200　Senecio kahuzicus Humb. = Dendrosenecio erici-rosenii（R. E. Fr. et T. C. E. Fr.）E. B. Knox ●☆

359201　Senecio kanaitzensis Franch. = Ligularia kanaitzensis（Franch.）Hand. -Mazz. ■

359202　Senecio kansuensis Franch. = Cremanthodium humile Maxim. ■

359203　Senecio kanzibiensis Humbert et Staner = Solanecio kanzibiensis（Humbert et Staner）C. Jeffrey ■☆

359204　Senecio karaguensis O. Hoffm.;卡拉古千里光■☆

359205　Senecio karelinioides C. Winkl.;花花柴千里光■☆

359206　Senecio karjaginii Sofieva;卡里亚千里光■☆

359207　Senecio kaschkarowi sensu Hand. -Mazz. = Senecio diversipinnus Y. Ling ■

359208　Senecio kaschkarowii C. Winkl. = Senecio faberi Hemsl. ex Forbes et Hemsl. ■

359209　Senecio kassnerianum Muschl. = Crassocephalum kassnerianum（Muschl.）Lisowski ■☆

359210　Senecio katangensis O. Hoffm.;加丹加千里光■☆

359211　Senecio katangensis O. Hoffm. var. latifolia Lisowski;宽叶千里光■☆

359212　Senecio kawaguchii Kitam. = Senecio thianschanicus Regel et Schmalh. ■

359213　Senecio kawakamii Kitam. = Tephroseris phaeantha（Nakai）C. Jeffrey et Y. L. Chen ■

359214　Senecio kawakamii Makino = Tephroseris kawakamii（Makino）Holub ■☆

359215　Senecio kawanguchii Kitam. = Senecio thianschanicus Regel et Schmalh. ■

359216　Senecio kebdanicus Maire et Sennen = Senecio leucanthemifolius Poir. ■☆

359217　Senecio kematogensis Vaniot = Senecio nemorensis L. ■

359218　Senecio keniensis Baker f.;肯尼亚千里光■☆

359219　Senecio keniensis Baker f. subsp. brassiciformis（R. E. Fr. et T. C. E. Fr.）C. Jeffrey;芥状肯尼亚千里光■☆

359220　Senecio keniodendron R. E. Fr. et T. C. E. Fr. = Dendrosenecio keniodendron（R. E. Fr. et T. C. E. Fr.）B. Nord. ●☆

359221　Senecio khasianus N. P. Balakr. = Senecio obtusatus Wall. ex DC. ■

359222　Senecio kialensis Franch. = Cremanthodium potaninii C. Winkl. ■

359223　Senecio kilimanjari Mildbr. = Dendrosenecio kilimanjari（Mildbr.）E. B. Knox ●☆

359224　Senecio kingii Rydb. = Senecio eremophilus Richardson var. kingii Greenm. ■☆

359225　Senecio kirghisicus DC.;吉尔吉斯千里光■☆

359226　Senecio kirilowii Turcz. ex DC. = Tephroseris integrifolia（L.）Holub subsp. kirilowii（Turcz. ex DC.）B. Nord. ■

359227　Senecio kirilowii Turcz. ex DC. = Tephroseris kirilowii（Turcz. ex DC.）Holub ■

359228　Senecio kirschsteineanus Muschl. = Cineraria deltoidea Sond. ■☆

359229　Senecio kivuensis Muschl. = Emilia kivuensis（Muschl.）C. Jeffrey ■☆

359230　Senecio kjellmanii A. E. Porsild = Tephroseris kjellmanii（A. E. Porsild）Holub ☆

359231　Senecio kleinia（L.）Sch. Bip. = Kleinia neriifolia Haw. ■☆

359232　Senecio kleinia Less.;克里尼千里光（天龙）■☆

359233　Senecio kleiniiformis Suess.;仙人笔千里光;Spear Head ■☆

359234　Senecio kleinioides（Sch. Bip.）Oliv. et Hiern = Kleinia kleinioides（Sch. Bip.）M. Taylor ■☆

359235　Senecio klinghardtianus Dinter = Senecio sulcicalyx Baker ■☆

359236　Senecio kolenatianus C. A. Mey.;考来纳蒂千里光■☆

359237　Senecio kongboensis Ludlow;工布千里光（贡波千里光）;Gongbu Groundsel,Kongbo Groundsel ■

359238　Senecio koreanus Kom. = Sinosenecio koreanus（Kom.）B. Nord. ■

359239　Senecio koreanus Kom. = Tephroseris koreana（Kom.）B. Nord. et Pelser ■

359240　Senecio korshinskyi Krasch.;考尔千里光■☆

359241　Senecio koualapensis Franch. = Parasenecio koualapensis（Franch.）Y. L. Chen ■

359242　Senecio krascheninnikovii Schischk.;细梗千里光（卷舌千里光,细裂千里光,序柄千里光）;Kraschenninkov Groundsel ■

359243　Senecio kuanshanensis C. I. Peng et S. W. Chung;关山千里光■

359244　Senecio kubensis Grossh.;库波千里光■☆

359245　Senecio kuluensis S. Moore;库卢千里光■☆

359246　Senecio kumaonensis Duthie ex C. Jeffrey et Y. L. Chen;须弥千里光;Nepal Groundsel ■

359247　Senecio kundelungensis Lisowski；扎伊尔千里光■☆

359248　Senecio kunlunshanicus C. H. An；昆仑千里光；Kunlunshan Groundsel ■

359249　Senecio kuntzei O. Hoffm. = Senecio glaberrimus DC. ■☆

359250　Senecio kyimbilensis Mattf. = Crassocephalum uvens（Hiern）S. Moore ■☆

359251　Senecio labordei Vaniot = Ligularia dentata（A. Gray）H. Hara ■

359252　Senecio lachnorhizus O. Hoffm. = Senecio coronatus（Thunb.）Harv. ■☆

359253　Senecio laetus Edgew.；菊状千里光（大红青菜，菊三七，菊叶千里光，山青菜，天青地红，土三七，野青菜）；Chrysanthemum-like Groundsel, Daisylike Groundsel ■

359254　Senecio laevigatus Thunb.；光滑千里光■☆

359255　Senecio laevigatus Thunb. var. integrifolius Harv.；全叶光滑千里光■☆

359256　Senecio laevis Humbert；平滑千里光■☆

359257　Senecio lagotis W. W. Sm. = Ligularia virgaurea（Maxim.）Mattf. ex Rehder et Kobuski ■

359258　Senecio lamarum Diels = Ligularia lamarum（Diels）C. C. Chang ■

359259　Senecio lamarum Diels ex H. Limpr. = Ligularia lamarum（Diels）C. C. Chang ■

359260　Senecio lampsanifolius Baker = Hubertia lampsanifolia（Baker）C. Jeffrey ●☆

359261　Senecio lanatus Thunb. = Oresbia heterocarpa Cron et B. Nord. ■☆

359262　Senecio lanceolatus Torr. et A. Gray = Senecio serra Hook. ■☆

359263　Senecio lanceus Aiton；披针状千里光■☆

359264　Senecio lancifolius Turcz. = Senecio pellucidus DC. ■☆

359265　Senecio lankongensis Franch. = Ligularia lankongensis（Franch.）Hand. -Mazz. ■

359266　Senecio lankongensis Franch. var. laxus Franch. = Ligularia lankongensis（Franch.）Hand. -Mazz. ■

359267　Senecio lanuriensis De Wild. = Senecio transmarinus S. Moore ■☆

359268　Senecio lapathifolius Franch. = Ligularia lapathifolia（Franch.）Hand. -Mazz. ■

359269　Senecio lapathifolius Greene = Senecio crassulus A. Gray ■☆

359270　Senecio lapsanoides DC.；稻槎菜千里光■☆

359271　Senecio laramiensis A. Nelson = Packera cana（Hook.）W. A. Weber et Á. Löve ■☆

359272　Senecio lasiorhizoides Sch. Bip. = Senecio coronatus（Thunb.）Harv. ■☆

359273　Senecio lasiorhizus DC. = Senecio coronatus（Thunb.）Harv. ■☆

359274　Senecio latealatopetiolatus De Wild. = Senecio subsessilis Oliv. et Hiern ■☆

359275　Senecio lathyroides Greene = Senecio flaccidus Less. var. monoensis（Greene）B. L. Turner et T. M. Barkley ■☆

359276　Senecio latibracteatus Humbert；宽苞千里光■☆

359277　Senecio laticorymbosus Gilli；宽伞序千里光■☆

359278　Senecio latifolius Banks et Sol. ex Hook. f. = Senecio latifolius DC. ■☆

359279　Senecio latifolius DC.；皮克特千里光（宽叶千里光）；Dan's Cabbage , Pictou Disease ■☆

359280　Senecio latifolius DC. var. barbellatus（DC.）Harv. = Senecio retrorsus DC. ■☆

359281　Senecio latifolius DC. var. retrorsus（DC.）Harv. = Senecio retrorsus DC. ■☆

359282　Senecio latihastatus W. W. Sm. = Ligularia latihastata（W. W. Sm.）Hand. -Mazz. ■

359283　Senecio latipes Franch. = Parasenecio latipes（Franch.）Y. L. Chen ■

359284　Senecio latouchei Jeffrey = Sinosenecio latouchei（Jeffrey）B. Nord. ■

359285　Senecio latus Rydb. = Senecio sphaerocephalus Greene ■☆

359286　Senecio launayifolius O. Hoffm. = Senecio paucicalyculatus Klatt ■☆

359287　Senecio lautus Sol. ex G. Forst. = Senecio inaequidens DC. ■☆

359288　Senecio lavandulifolius Wall.；薰衣草叶千里光■☆

359289　Senecio lawalreeanus Lisowski；拉瓦尔千里光■☆

359290　Senecio laxiflorus Viv. = Senecio glaucus L. subsp. coronopifolius（Desf.）Alexander ■☆

359291　Senecio laxus DC.；疏松千里光■☆

359292　Senecio layneae Greene = Packera layneae（Greene）W. A. Weber et Á. Löve ■☆

359293　Senecio leandrii Humbert；利安千里光■☆

359294　Senecio lebrunei H. Lév. = Senecio asperifolius Franch. ■

359295　Senecio lecleri H. Lév. = Parasenecio koualapensis（Franch.）Y. L. Chen ■

359296　Senecio ledebourii Sch. Bip. = Ligularia macrophylla（Ledeb.）DC. ■

359297　Senecio leibergii Greene = Senecio integerrimus Nutt. var. ochroleucus（A. Gray）Cronquist ■☆

359298　Senecio leiocarpus DC. = Senecio albanensis DC. ■☆

359299　Senecio lejolyanus Lisowski；勒若利千里光■☆

359300　Senecio lelyi Hutch.；莱利千里光■☆

359301　Senecio lembertii Greene = Packera pauciflora（Pursh）Á. Löve et D. Löve ■☆

359302　Senecio lemmonii A. Gray；莱蒙千里光■☆

359303　Senecio lenensis Schischk.；列恩千里光■☆

359304　Senecio lentior S. Moore = Senecio eenii（S. Moore）Merxm. ■☆

359305　Senecio leonardii Rydb. = Packera streptanthifolia（Greene）W. A. Weber et Á. Löve ■☆

359306　Senecio leptocephalus Mattf. = Emilia leptocephala（Mattf.）C. Jeffrey ■☆

359307　Senecio leptolepis Greene = Senecio aronicoides DC. ■☆

359308　Senecio leptophyllus DC.；澳非细叶千里光■☆

359309　Senecio leptopterus Mesfin；细翅千里光■☆

359310　Senecio lessingii Harv.；莱辛千里光■☆

359311　Senecio letouzeyanus Lisowski；勒图千里光■☆

359312　Senecio leucanthemifolius Poir.；滨菊叶千里光■☆

359313　Senecio leucanthemifolius Poir. subsp. crassifolius（Willd.）Ball；厚滨菊叶千里光■☆

359314　Senecio leucanthemifolius Poir. subsp. cyrenaicus（E. A. Durand et Barratte）Greuter；昔兰尼千里光■☆

359315　Senecio leucanthemifolius Poir. subsp. humilis（Desf.）Murb.；低矮千里光■☆

359316　Senecio leucanthemifolius Poir. subsp. mauritanicus（Pomel）Greuter；毛里塔尼亚千里光■☆

359317　Senecio leucanthemifolius Poir. subsp. poiretianus Maire = Senecio leucanthemifolius Poir. ■☆

359318　Senecio leucanthemifolius Poir. var. atlanticus（Boiss. et Reut.）Batt. = Senecio leucanthemifolius Poir. ■☆

359319　Senecio leucanthemifolius Poir. var. casablancae Alexander =

Senecio leucanthemifolius Poir. ■☆

359320 Senecio leucanthemifolius Poir. var. cyrenaicus E. A. Durand et Barratte = Senecio leucanthemifolius Poir. subsp. cyrenaicus (E. A. Durand et Barratte) Greuter ● ☆

359321 Senecio leucanthemifolius Poir. var. eradiatus Leredde = Senecio leucanthemifolius Poir. ■☆

359322 Senecio leucanthemifolius Poir. var. falcifolius (Bolle) G. Kunkel = Senecio leucanthemifolius Poir. ■☆

359323 Senecio leucanthemifolius Poir. var. fradinii (Pomel) Batt. = Senecio leucanthemifolius Poir. ■☆

359324 Senecio leucanthemifolius Poir. var. giganteus Caball. = Senecio leucanthemifolius Poir. ■☆

359325 Senecio leucanthemifolius Poir. var. humilis (Desf.) Batt. = Senecio leucanthemifolius Poir. subsp. humilis (Desf.) Murb. ■☆

359326 Senecio leucanthemifolius Poir. var. lanuginosus Batt. = Senecio leucanthemifolius Poir. ■☆

359327 Senecio leucanthemifolius Poir. var. latisectus Pau et Font Quer = Senecio leucanthemifolius Poir. ■☆

359328 Senecio leucanthemifolius Poir. var. major Ball = Senecio leucanthemifolius Poir. ■☆

359329 Senecio leucanthemifolius Poir. var. mauritanicus (Pomel) Batt. = Senecio leucanthemifolius Poir. subsp. mauritanicus (Pomel) Greuter ■☆

359330 Senecio leucanthemifolius Poir. var. paui Maire = Senecio leucanthemifolius Poir. ■☆

359331 Senecio leucanthemifolius Poir. var. salzmannii Rouy = Senecio leucanthemifolius Poir. ■☆

359332 Senecio leucanthemifolius Poir. var. vernus (Biv.) Fiori = Senecio leucanthemifolius Poir. ■☆

359333 Senecio leucanthemus Dunn = Parasenecio ainsliiflorus (Franch.) Y. L. Chen ■

359334 Senecio leucanthothamnus Humbert = Hubertia leucanthothamnus (Humbert) C. Jeffrey ● ☆

359335 Senecio leucocephalus Franch. = Parasenecio leucocephalus (Franch.) Y. L. Chen ■

359336 Senecio leucocrinus Greene = Packera macounii (Greene) W. A. Weber et Á. Löve ■☆

359337 Senecio leucoglossus Sond. ;白舌千里光■☆

359338 Senecio leucopappus (DC.) Bojer ex Humbert = Humbertacalia leucopappa (DC.) C. Jeffrey ● ☆

359339 Senecio leucopappus Franch. var. volutus (Baker) Humbert = Senecio volutus (Baker) Humbert ■☆

359340 Senecio leucophyllus DC. ;白毛千里光■☆

359341 Senecio leucoreus Greenm. = Packera multilobata (Torr. et A. Gray) W. A. Weber et Á. Löve ■☆

359342 Senecio leveillei Vaniot = Ligularia leveillei (Vaniot) Hand.-Mazz. ■

359343 Senecio lewallei Lisowski;勒瓦莱千里光■☆

359344 Senecio lewisrosei J. T. Howell = Packera eurycephala (Torr. et A. Gray) W. A. Weber et Á. Löve var. lewisrosei (J. T. Howell) J. F. Bain ■☆

359345 Senecio lhasaeansis Y. Ling ex C. Jeffrey et Y. L. Chen;拉萨千里光;Lasa Groundsel, Lhasa Groundsel ■

359346 Senecio liangshanensis C. Jeffrey et Y. L. Chen;凉山千里光;Liangshan Groundsel ■

359347 Senecio liatroides C. Winkl. = Ligularia liatroides (C. Winkl.) Hand.-Mazz. ■

359348 Senecio lichtensteinensis Dinter = Senecio burchellii DC. ■☆

359349 Senecio ligularia Hook. f. = Ligularia fischeri (Ledeb.) Turcz. ■

359350 Senecio ligularia Hook. f. var. araneosa (DC.) H. Lév. = Ligularia sibirica (L.) Cass. var. araneosa DC. ■

359351 Senecio ligularia Hook. f. var. araneosus (DC.) H. Lév. = Cissampelopsis volubilis (Blume) Miq. ●

359352 Senecio ligularia Hook. f. var. atkinsonii (C. B. Clarke) Hook. f. = Ligularia atkinsonii (C. B. Clarke) S. W. Liu ■

359353 Senecio ligularia Hook. f. var. polycephalus Hemsl. = Ligularia wilsoniana (Hemsl.) Greenm. ■

359354 Senecio ligularioides C. Jeffrey et Y. L. Chen = Sinosenecio ligularioides (Hand.-Mazz.) B. Nord. ■

359355 Senecio ligularioides Hand.-Mazz. = Sinosenecio ligularioides (Hand.-Mazz.) B. Nord. ■

359356 Senecio ligulifolius Greene = Packera macounii (Greene) W. A. Weber et Á. Löve ■☆

359357 Senecio lijiangensis C. Jeffrey et Y. L. Chen;丽江千里光;Lijiang Groundsel ■

359358 Senecio limosus O. Hoffm. = Emilia limosa (O. Hoffm.) C. Jeffrey ☆

359359 Senecio limprichetii Diels = Ligularia limprichtii (Diels) Hand.-Mazz. ■

359360 Senecio lindstroemii (Ostenf.) A. E. Porsild = Tephroseris lindstroemii (Ostenf.) Á Löve et D. Löve ■☆

359361 Senecio lineatus (L. f.) DC. ;条纹千里光■☆

359362 Senecio lingianus C. Jeffrey et Y. L. Chen;君范千里光;Ling Groundsel ■

359363 Senecio linifolius L. ;亚麻叶千里光■☆

359364 Senecio littoreus Thunb. ;滨海千里光■☆

359365 Senecio littoreus Thunb. var. hispidulus Harv. ;细毛滨海千里光■☆

359366 Senecio litvinovii Schischk. = Senecio cannabifolius Less. var. integrifolius (Koidz.) Kitam. ■

359367 Senecio lividus L. ;铅色千里光■☆

359368 Senecio lividus L. subsp. foeniculaceus (Ten.) Braun-Blanq. et Maire = Senecio lividus L. ■☆

359369 Senecio lividus L. var. biennis Humbert = Senecio lividus L. ■☆

359370 Senecio lividus L. var. major Gren. et Godr. = Senecio lividus L. ■☆

359371 Senecio lividus L. var. pinguis Ball = Senecio lividus L. ■☆

359372 Senecio lobatus Pers. = Packera glabella (Poir.) C. Jeffrey ■☆

359373 Senecio lobelioides DC. ;半边莲千里光■☆

359374 Senecio lonchophyllus Hand.-Mazz. = Synotis hieraciifolia (H. Lév.) C. Jeffrey et Y. L. Chen ■

359375 Senecio longeligulatus De Wild. = Dendrosenecio erici-rosenii (R. E. Fr. et T. C. E. Fr.) E. B. Knox ● ☆

359376 Senecio longeligulatus H. Lév. et Vaniot = Tephroseris flammea (Turcz. ex DC.) Holub ■

359377 Senecio longiflorus (DC.) Sch. Bip. ;长花千里光;Woolly Groundsel ● ☆

359378 Senecio longiflorus (DC.) Sch. Bip. = Kleinia longiflora DC. ■☆

359379 Senecio longiflorus (DC.) Sch. Bip. subsp. madagascariensis (Humbert) G. D. Rowley = Kleinia madagascariensis (Humbert) P. Halliday ■☆

359380 Senecio longifolius L. = Senecio linifolius L. ■☆

359381 Senecio longilobus Benth. ; 长 裂 千 里 光;Thread-leaf

Groundsel, Woolly Groundsel ■☆

359382 Senecio longilobus Benth. = Senecio flaccidus Less. ■☆

359383 Senecio longipedunculatus Dinter = Senecio maydae Merxm. ■☆

359384 Senecio longipes Baker = Kleinia grantii (Oliv. et Hiern) Hook. f. ■☆

359385 Senecio longipetiolatus Rydb. = Packera streptanthifolia (Greene) W. A. Weber et Á. Löve ■☆

359386 Senecio longiscapus Bojer ex DC. ;长花茎千里光■☆

359387 Senecio lorentii Hochst. ;劳氏千里光■☆

359388 Senecio lubumbashianus De Wild. = Senecio purpureus L. ■☆

359389 Senecio lucorum Franch. = Synotis lucorum (Franch.) C. Jeffrey et Y. L. Chen ■

359390 Senecio luembensis De Wild. et Muschl. ;卢恩贝千里光■☆

359391 Senecio lugardae Bullock = Senecio hochstetteri Sch. Bip. ex A. Rich. ■☆

359392 Senecio lugens Richardson var. exaltatus (Nutt.) D. C. Eaton = Senecio integerrimus Nutt. var. exaltatus (Nutt.) Cronquist ■☆

359393 Senecio lugens Richardson var. hookeri D. C. Eaton = Senecio sphaerocephalus Greene ■☆

359394 Senecio lugens Richardson var. megacephalus Jeps. = Senecio integerrimus Nutt. var. major (A. Gray) Cronquist ■☆

359395 Senecio lugens Richardson var. ochroleucus A. Gray = Senecio integerrimus Nutt. var. ochroleucus (A. Gray) Cronquist ■☆

359396 Senecio lunulatus (Chiov.) H. Jacobsen = Kleinia lunulata (Chiov.) Thulin ■☆

359397 Senecio luticola Dunn = Senecio asperifolius Franch. ■

359398 Senecio lyallii Hook. f. ;莱尔千里光;Mountain Marigold ■☆

359399 Senecio lycopodioides Schltr. ;石松千里光■☆

359400 Senecio lydenburgensis Hutch. et Burtt Davy;莱登堡千里光■☆

359401 Senecio lygodes Hiern = Senecio inornatus DC. ■☆

359402 Senecio lynceus Greene = Packera multilobata (Torr. et A. Gray) W. A. Weber et Á. Löve ■☆

359403 Senecio lyonii A. Gray;里氏千里光●☆

359404 Senecio lyratipartitus Sch. Bip. ex A. Rich. = Senecio lyratus Forssk. ■☆

359405 Senecio lyratus Forssk. ;大头羽裂千里光■☆

359406 Senecio lyratus L. f. var. subcanescens (DC.) Harv. = Senecio subcanescens (DC.) Compton ■☆

359407 Senecio lyratus Michx. = Packera glabella (Poir.) C. Jeffrey ■☆

359408 Senecio macdougalii A. Heller = Senecio eremophilus Richardson var. macdougalii (A. Heller) Cronquist ■☆

359409 Senecio macounii Greene = Packera macounii (Greene) W. A. Weber et Á. Löve ■☆

359410 Senecio macowanii Hilliard;麦克欧文千里光■☆

359411 Senecio macranthus C. B. Clarke = Ligularia japonica (Thunb.) Less. ■

359412 Senecio macroalatus M. D. Hend. = Senecio inornatus DC. ■☆

359413 Senecio macrocephalus DC. ;大头千里光■☆

359414 Senecio macroglossoides Hilliard;拟大舌千里光■☆

359415 Senecio macroglossus DC. ;大舌千里光;Cape Ivy, Natal Ivy, Wax Vine ■☆

359416 Senecio macroglossus DC. 'Variegatum';金玉菊(黄斑叶大舌千里光);Variegated Wax Vine ■

359417 Senecio macropappus Sch. Bip. ex A. Rich. = Crassocephalum macropappum (Sch. Bip. ex A. Rich.) S. Moore ■☆

359418 Senecio macrophyllus E. Phillips = Senecio isatidioides E. Phillips et C. A. Sm. ■☆

359419 Senecio macrophyllus M. Bieb. ;大叶千里光;Bigleaf Groundsel ■☆

359420 Senecio macropodus DC. = Senecio angulatus L. f. ■☆

359421 Senecio macropus Greenm. = Packera quercetorum (Greene) C. Jeffrey ■☆

359422 Senecio macrospermus DC. ;大籽千里光■☆

359423 Senecio madagascariensis Poir. ;马达加斯加千里光;Fire Weed, Madagascar Ragwort ■☆

359424 Senecio madagascariensis Poir. var. boutonii (Baker) Humbert = Senecio madagascariensis Poir. ■☆

359425 Senecio madagascariensis Poir. var. crassifolius Humbert = Senecio madagascariensis Poir. ■☆

359426 Senecio madagascariensis Poir. var. crassifolius Humbert = Senecio skirrhodon DC. ■☆

359427 Senecio maderensis DC. = Pericallis aurita (L'Hér.) B. Nord. ■☆

359428 Senecio mairiei H. Lév. = Senecio graciliflorus DC. ■

359429 Senecio maisonii H. Lév. = Ligularia nelumbifolia (Bureau et Franch.) Hand. -Mazz. ■

359430 Senecio malaissei Lisowski;马莱泽千里光■☆

359431 Senecio malmstenii S. F. Blake ex Tidestr. = Packera malmstenii (S. F. Blake ex Tidestr.) Kartesz ■☆

359432 Senecio manchuricus Kitam. = Senecio ambraceus Turcz. ex DC. var. glaber Kitam. ■

359433 Senecio mandrarensis Humbert;曼德拉千里光■☆

359434 Senecio mannii Hook. f. ;曼氏千里光(曼千里光)■☆

359435 Senecio mannii Hook. f. = Solanecio mannii (Hook. f.) C. Jeffrey ■☆

359436 Senecio mannii Hook. f. var. kikuyuensis Chiov. = Solanecio mannii (Hook. f.) C. Jeffrey ■☆

359437 Senecio manshuricus Kitam. = Senecio ambraceus Turcz. ex DC. ■

359438 Senecio maranguensis O. Hoffm. ;马兰古千里光■☆

359439 Senecio margaritae C. Jeffrey;珍珠千里光■☆

359440 Senecio marginalis Hilliard;边生千里光■☆

359441 Senecio maritimus (L.) Koidz. = Cineraria maritima (L.) L. ■☆

359442 Senecio maritimus (L.) Koidz. = Senecio bicolor (Willd.) Tod. subsp. cineraria (DC.) Chater ■☆

359443 Senecio maritimus (L.) Koidz. = Senecio pseudoarnica Less. ■

359444 Senecio maritimus L. f. ;海岸千里光■☆

359445 Senecio marlothianus O. Hoffm. = Emilia marlothiana (O. Hoffm.) C. Jeffrey ■☆

359446 Senecio maroccanus P. H. Davis;摩洛哥千里光■☆

359447 Senecio marojejyensis Humbert;马罗千里光■☆

359448 Senecio martinii Vaniot = Sinosenecio oldhamianus (Maxim.) B. Nord. ■

359449 Senecio masonii De Wild. = Senecio lyratus Forssk. ■☆

359450 Senecio massagetovii Schischk. ;马氏千里光■☆

359451 Senecio massaicus (Maire) Maire;马萨千里光■☆

359452 Senecio massaiensis Muschl. = Senecio schweinfurthii O. Hoffm. ■☆

359453 Senecio mattfeldii R. E. Fr. = Senecio snowdenii Hutch. ■☆

359454 Senecio mattirolii Chiov. ;马蒂奥里千里光■☆

359455 Senecio mauritanicus Pomel = Senecio leucanthemifolius Poir. ■☆

359456 Senecio maximowiczii C. Winkl. = Sinacalia caroli (C. Winkl.) C. Jeffrey et Y. L. Chen ■

359457　Senecio maximowiczii Franch. = Cremanthodium ellisii (Hook. f.) Kitam. ■

359458　Senecio maydae Merxm. ;长梗千里光■☆

359459　Senecio mearnsii De Wild. = Solanecio cydoniifolius (O. Hoffm.) C. Jeffrey ■☆

359460　Senecio medley-woodii Hutch. ;迈伍千里光■☆

359461　Senecio megalanthus Y. L. Chen;大花千里光(大头千里光) ; Bigflower Groundsel,Large Head Groundsel ■

359462　Senecio megamontanus Cufod. = Emilia somalensis (S. Moore) C. Jeffrey ■☆

359463　Senecio melanocephalus Franch. = Ligularia melanocephala (Franch.) Hand. -Mazz. ■

359464　Senecio melanophyllus Muschl. = Senecio schweinfurthii O. Hoffm. ■☆

359465　Senecio melastomifolius Baker var. microphyllus Humbert = Senecio saboureaui Humbert ■☆

359466　Senecio membranifolius DC. = Senecio repandus Thunb. ■☆

359467　Senecio memmingeri Britton ex Small = Packera millefolia (Torr. et A. Gray) W. A. Weber et Á. Löve ■☆

359468　Senecio mendocinensis A. Gray = Senecio integerrimus Nutt. var. major (A. Gray) Cronquist ■☆

359469　Senecio meruensis Cotton et Blakelock = Dendrosenecio meruensis (Cotton et Blakelock) E. B. Knox ●☆

359470　Senecio mesembryanthemoides Bojer ex DC. ;日中花千里光 ■☆

359471　Senecio mesogrammoides O. Hoffm. ;芹状千里光■☆

359472　Senecio metallicorum S. Moore = Senecio polyodon DC. ■☆

359473　Senecio metcalfei Greene ex Wooton et Standl. = Packera neomexicana (A. Gray) W. A. Weber et Á. Löve var. mutabilis (Greene) W. A. Weber et Á. Löve ■☆

359474　Senecio meyeri-johannis Engl. ;迈尔约翰千里光■☆

359475　Senecio meyeri-johannis Engl. subsp. olomotiensis C. Jeffrey = Senecio meyeri-johannis Engl. ■☆

359476　Senecio microalatus C. Jeffrey;小翅千里光■☆

359477　Senecio microdontus Baker = Senecio adscendens Bojer ex DC. ■☆

359478　Senecio microdontus Bureau et Franch. = Ligularia sagitta (Maxim.) Mattf. ex Rehder et Kobuski ■

359479　Senecio microglossus DC. ;小舌千里光■☆

359480　Senecio microspermus DC. ;小籽千里光■☆

359481　Senecio mikaniae DC. = Senecio deltoideus Less. ■☆

359482　Senecio mikaniiformis DC. = Senecio deltoideus Less. ■☆

359483　Senecio mikanioides Otto ex Walp. = Delairea odorata Lem. ●☆

359484　Senecio mikanioides Walp. ;假泽兰千里光■☆

359485　Senecio milanjianus S. Moore ;米兰吉千里光■☆

359486　Senecio milleflorus Greene = Senecio atratus Greene ■☆

359487　Senecio milleflorus H. Lév. = Senecio liangshanensis C. Jeffrey et Y. L. Chen ■

359488　Senecio millefolium Torr. et A. Gray = Packera millefolia (Torr. et A. Gray) W. A. Weber et Á. Löve ■☆

359489　Senecio millikenii Eastw. = Senecio serra Hook. ■☆

359490　Senecio mimetes Hutch. et R. A. Dyer;相似千里光■☆

359491　Senecio miniatus (Welw.) Staner = Gynura pseudochina (L.) DC. ■

359492　Senecio minimus Poir. = Erechtites minima (Poir.) DC. ■☆

359493　Senecio minutus (Cav.) DC. ;微小千里光■☆

359494　Senecio mirabilis Muschl. = Solanecio mirabilis (Muschl.) C. Jeffrey ■☆

359495　Senecio mississipianus DC. = Packera glabella (Poir.) C. Jeffrey ■☆

359496　Senecio mitophyllus C. Jeffrey;线叶千里光■☆

359497　Senecio mohavensis A. Gray = Senecio mohavensis Torr. et A. Gray ■☆

359498　Senecio mohavensis A. Gray subsp. breviflorus (Kadereit) M. Coleman;短花莫哈维千里光■☆

359499　Senecio mohavensis Torr. et A. Gray;莫哈维千里光;Mohave Groundsel ■☆

359500　Senecio moisonii H. Lév. = Ligularia nelumbifolia (Bureau et Franch.) Hand. -Mazz. ■

359501　Senecio molinarius Greenm. = Packera werneriifolia (A. Gray) W. A. Weber et Á. Löve ■☆

359502　Senecio momordicifolius Dinter et Muschl. = Cineraria canescens J. C. Wendl. ex Link ■☆

359503　Senecio monanthus Hayata = Parasenecio roborowskii (Maxim.) Y. L. Chen ■

359504　Senecio monbeigii H. Lév. = Ligularia tongolensis (Franch.) Hand. -Mazz. ■

359505　Senecio mongolicus (Turcz.) Sch. Bip. = Ligularia mongolica (Turcz.) DC. ■

359506　Senecio mongolicus Sch. Bip. = Ligularia mongolica (Turcz.) DC. ■

359507　Senecio monocephalus Baker = Senecio baronii Humbert ■☆

359508　Senecio monoensis Greene = Senecio flaccidus Less. var. monoensis (Greene) B. L. Turner et T. M. Barkley ■☆

359509　Senecio monroi Hook. f. = Brachyglottis monroi (Hook. f.) B. Nord. ●☆

359510　Senecio monticola DC. ;山生千里光■☆

359511　Senecio montuosus S. Moore = Crassocephalum montuosum (S. Moore) Milne-Redh. ■☆

359512　Senecio mooreanus Hutch. et Burtt Davy;穆尔千里光■☆

359513　Senecio moorei R. E. Fr. ;穆氏千里光■☆

359514　Senecio mooreioides C. Jeffrey;拟穆氏千里光■☆

359515　Senecio moresbiensis (Calder et R. L. Taylor) G. W. Douglas et Ruyle-Douglas = Packera subnuda (DC.) Trock et T. M. Barkley var. moresbiensis (Calder et R. L. Taylor) Trock ■☆

359516　Senecio morrisonensis Hayata;玉山千里光(玉山黄菀) ;Jade Groundsel,Yushan Groundsel ■

359517　Senecio morrisonensis Hayata var. dentatus Kitam. = Senecio nemorensis L. var. dentatus (Kitam.) H. Koyama ■

359518　Senecio morrumbalensis De Wild. = Solanecio mannii (Hook. f.) C. Jeffrey ■☆

359519　Senecio mosoyinensis (Franch.) Kitam. = Ligularia kanaitzensis (Franch.) Hand. -Mazz. ■

359520　Senecio mosoyinensis Franch. = Ligularia intermedia Nakai ■

359521　Senecio mosoyinensis Franch. = Ligularia kanaitzensis (Franch.) Hand. -Mazz. ■

359522　Senecio mucronatus (Thunb.) Willd. ;短尖千里光■☆

359523　Senecio mucronulatus Sch. Bip. = Senecio purpureus L. ■☆

359524　Senecio muhlenbergii Sch. Bip. = Arnoglossum reniforme (Hook.) H. Rob. ■☆

359525　Senecio muirii Greenm. = Packera werneriifolia (A. Gray) W. A. Weber et Á. Löve ☆

359526　Senecio muirii L. Bolus;缪里千里光■☆

359527　Senecio muliensis C. Jeffrey et Y. L. Chen;木里千里光;Muli

Groundsel ■

359528 Senecio multibracteatus Baker = Senecio pleistocephalus S. Moore ■☆

359529 Senecio multibracteolatus C. Jeffrey et Y. L. Chen;多苞千里光;Manybract Groundsel, Multibracteolate Groundsel ■

359530 Senecio multicapitatus Greenm. = Senecio spartioides Torr. et Gray ■☆

359531 Senecio multicaulis DC. ;多茎千里光■☆

359532 Senecio multicorymbosus Klatt = Solanecio mannii (Hook. f.) C. Jeffrey ■☆

359533 Senecio multidenticulatus Humbert;多小齿千里光■☆

359534 Senecio multiflorus (L'Hér.) DC. = Pericallis multiflora (L'Hér.) B. Nord. ■☆

359535 Senecio multilobatus Torr. et A. Gray = Packera multilobata (Torr. et A. Gray) W. A. Weber et Á. Löve ■☆

359536 Senecio multilobus C. C. Chang;多裂千里光;Manylobe Groundsel, Multifid Groundsel ■

359537 Senecio multnomensis Greenm. = Packera paupercula (Michx.) Á. Löve et D. Löve ■☆

359538 Senecio muninensis Koidz. = Erechtites hieraciifolius (L.) Raf. ex DC. var. cacalioides (Fisch. ex Spreng.) Griseb. ■☆

359539 Senecio muricatus Thunb. ;糙千里光■☆

359540 Senecio musiniensis S. L. Welsh = Packera musiniensis (S. L. Welsh) Trock ■☆

359541 Senecio mutabilis Greene;易变千里光■☆

359542 Senecio mweroensis Baker = Kleinia mweroensis (Baker) C. Jeffrey ■☆

359543 Senecio mweroensis Baker f. schwartzii (L. E. Newton) G. D. Rowley = Kleinia schwartzii L. E. Newton ■☆

359544 Senecio myricifolius (Bojer ex DC.) Humbert = Hubertia myricifolia (Bojer ex DC.) C. Jeffrey ●☆

359545 Senecio myriocephalus Sch. Bip. ex A. Rich. ;多头千里光;Manyhead Groundsel ■

359546 Senecio myriocephalus Y. Ling ex Y. L. Chen et K. Y. Pan = Senecio lingianus C. Jeffrey et Y. L. Chen ■

359547 Senecio myriocephalus Y. Ling ex Y. L. Chen, S. Yun Liang et K. Y. Pan = Senecio lingianus C. Jeffrey et Y. L. Chen ■

359548 Senecio myrrhifolius Thunb. = Senecio arenarius Thunb. ■☆

359549 Senecio myrtifolius Klatt = Hubertia myrtifolia (Klatt) C. Jeffrey ●☆

359550 Senecio nagensis C. B. Clarke = Synotis nagensis (C. B. Clarke) C. Jeffrey et Y. L. Chen ■

359551 Senecio nagensis C. B. Clarke var. lobbi (Hook. f.) Craib = Synotis cappa (Buch. -Ham. ex D. Don) C. Jeffrey et Y. L. Chen ■

359552 Senecio namaquanus Bolus = Senecio cinerascens Aiton ■☆

359553 Senecio nandensis S. Moore = Solanecio nandensis (S. Moore) C. Jeffrey ■☆

359554 Senecio nanus Sch. Bip. ex A. Rich. ;侏儒千里光■☆

359555 Senecio napifolius MacOwan;芜菁千里光■☆

359556 Senecio narynensis C. Winkl. = Ligularia narynensis (C. Winkl.) O. Fedtsch. et B. Fedtsch. ■

359557 Senecio natalensis Sch. Bip. = Senecio brachypodus DC. ■☆

359558 Senecio natalicola Hilliard;纳塔尔千里光■☆

359559 Senecio navicularis Humbert;船状千里光■☆

359560 Senecio nebrodensis L. var. balansae Boiss. et Reut. = Senecio squalidus L. subsp. aurasiacus (Batt. et Trab.) Alexander ■☆

359561 Senecio nebrodensis L. var. rupestris (Waldst. et Kit.) Fiori =

Senecio squalidus L. subsp. rupestris (Waldst. et Kit.) Greuter ■☆

359562 Senecio nelsonii Rydb. = Packera fendleri (A. Gray) W. A. Weber et Á. Löve ■☆

359563 Senecio nelsonii Rydb. var. uintahensis A. Nelson = Packer multilobata (Torr. et A. Gray) W. A. Weber et Á. Löve ■☆

359564 Senecio nelumbifolius Bureau et Franch. = Ligulari nelumbifolia (Bureau et Franch.) Hand. -Mazz. ■

359565 Senecio nemorensis L. ;林荫千里光(福克氏千里光,傅氏千里光,黄菀,林阴千里光,森林千里光);Alpine Ragwort, Fuchs Groundsel, Shady Groundsel ■

359566 Senecio nemorensis L. = Senecio nemorensis L. var. dentatus (Kitam.) H. Koyama ■

359567 Senecio nemorensis L. subsp. fuchsii (C. C. Gmel.) Durand = Senecio nemorensis L. ■

359568 Senecio nemorensis L. var. dentatus (Kitam.) H. Koyama;齿叶玉山千里光(黄菀,狭叶玉山黄菀);Toothleaf Yushan Groundsel ■

359569 Senecio nemorensis L. var. dentatus (Kitam.) H. Koyama = Senecio morrisonensis Hayata var. dentatus Kitam. ■

359570 Senecio nemorensis L. var. octoglossus (DC.) Koch ex Ledeb. = Senecio nemorensis L. ■

359571 Senecio nemorensis L. var. subinteger Hara = Senecio nemorensis L. ■

359572 Senecio nemorensis L. var. taiwanensis (Hayata) Yamam. = Senecio nemorensis L. ■

359573 Senecio nemorensis L. var. turczaninowii (DC.) Kom. = Senecio nemorensis L. ■

359574 Senecio neobakeri Humbert;新贝克千里光■☆

359575 Senecio neohelogetus Franch. = Cremanthodium thomsonii C. B. Clarke ■

359576 Senecio neomexicanus A. Gray = Packera neomexicana (A. Gray) W. A. Weber et Á. Löve ■☆

359577 Senecio neomexicanus A. Gray var. metcalfei (Greene ex Wooton et Standl.) T. M. Barkley = Packera neomexicana (A. Gray) W. A. Weber et Á. Löve var. mutabilis (Greene) W. A. Weber et Á. Löve ■☆

359578 Senecio neomexicanus A. Gray var. mutabilis (Greene) T. M. Barkley = Packera neomexicana (A. Gray) W. A. Weber et Á. Löve var. mutabilis (Greene) W. A. Weber et Á. Löve ■☆

359579 Senecio neomexicanus A. Gray var. toumeyi (Greene) T. M. Barkley = Packera neomexicana (A. Gray) W. A. Weber et Á. Löve var. toumeyi (Greene) Trock et T. M. Barkley ■☆

359580 Senecio neowebsteri S. F. Blake;韦伯斯特千里光■☆

359581 Senecio nephelagetus Franch. = Cremanthodium thomsonii C. B. Clarke ■

359582 Senecio nephrophyllus Rydb. = Packera debilis (Nutt.) W. A. Weber et Á. Löve ■☆

359583 Senecio nevadensis Boiss. et Reut. ;内华达千里光■☆

359584 Senecio newcombei Greene = Sinosenecio newcombei (Greene) Janovec et T. M. Barkley ■☆

359585 Senecio ngoyanus Hilliard;恩戈伊千里光■☆

359586 Senecio nigrocinctus Franch. ;黑苞千里光; Black Margine Groundsel, Blackmargine Groundsel ■

359587 Senecio nikoensis Miq. = Nemosenecio nikoensis (Miq.) B. Nord. ■☆

359588 Senecio nikoensis Miq. f. albiflorus Sugim. ;白花日本羽叶菊 ■☆

359589 Senecio nikoensis Miq. var. formosanus Sasaki = Nemosenecio formosanus（Kitam.）B. Nord. ■

59590 Senecio nimborus Franch. = Ligularia hookeri（C. B. Clarke）Hand.-Mazz. ■

59591 Senecio nimicola Franch.；肾叶千里光■

59592 Senecio niveus（Thunb.）Willd.；雪白千里光■☆

59593 Senecio nobilis Franch. = Cremanthodium nobile（Franch.）Diels ex H. Lév. ■

59594 Senecio nodiflorus C. C. Chang；节花千里光；Nodal Flower Groundsel, Nodalhead Groundsel ■

59595 Senecio nogalensis Chiov. = Kleinia nogalensis（Chiov.）Thulin ■☆

59596 Senecio nudibasis H. Lév. = Gynura nepalensis DC. ■

59597 Senecio nudibasis H. Lév. et Vaniot = Gynura nepalensis DC. ■

359598 Senecio nudicaulis Buch.-Ham. ex D. Don；裸茎千里光（草本反背红，反背红，反背绿丸，老母猪花头，天青地红，紫背鹿含草，紫背鹿衔草，紫背千里光，紫背天葵草）■

359599 Senecio nuttallii Sch. Bip. = Arnoglossum plantagineum Raf. ■☆

359600 Senecio nyikensis Baker = Kleinia abyssinica（A. Rich.）A. Berger ■☆

359601 Senecio nyikensis Baker = Senecio syringifolius O. Hoffm. ■☆

359602 Senecio nyikensis Baker var. hildebrandtii（Vatke）G. D. Rowley = Kleinia abyssinica（A. Rich.）A. Berger var. hildebrandtii（Vatke）C. Jeffrey ■☆

359603 Senecio nyungwensis Maquet = Senecio rugegensis Muschl. ■☆

359604 Senecio oblanceolatus Rydb. = Packera tridenticulata（Rydb.）W. A. Weber et Á. Löve ■☆

359605 Senecio oblongatus（C. B. Clarke）Franch. = Cremanthodium oblongatum C. B. Clarke ■

359606 Senecio obovatus Muhl. ex Willd.；倒卵千里光；Round-leaved Squaw-weed ■☆

359607 Senecio obovatus Muhl. ex Willd. = Packera obovata（Muhl. ex Willd.）W. A. Weber et Á. Löve ■☆

359608 Senecio obtusatus Wall. ex DC.；钝叶千里光；Obtuseleaf Groundsel ■

359609 Senecio occidentalis（A. Gray）Greene = Senecio fremontii Torr. et A. Gray var. occidentalis A. Gray ■☆

359610 Senecio occidentalis（A. Gray）Greene var. rotundatus Rydb. = Senecio fremontii Torr. et A. Gray ■☆

359611 Senecio ochraceus（Piper）Piper = Senecio integerrimus Nutt. var. ochroleucus（A. Gray）Cronquist ■☆

359612 Senecio ochrocarpoides Cufod. = Senecio subsessilis Oliv. et Hiern ■☆

359613 Senecio ochrocarpus Oliv. et Hiern；赭黄果千里光■☆

359614 Senecio octoglossus DC. = Senecio nemorensis L. ■

359615 Senecio odontopterus DC.；齿翼千里光■☆

359616 Senecio odorus（Forssk.）Sch. Bip. = Kleinia odora（Forssk.）DC. ■☆

359617 Senecio ogotorukensis Packer = Packera ogotorukensis（Packer）Á. Löve et D. Löve ■☆

359618 Senecio oldhamianus Maxim. = Sinosenecio oldhamianus（Maxim.）B. Nord. ■

359619 Senecio olgae Regel et Schmalh. ex Regel；奥尔加千里光■☆

359620 Senecio oliganthus DC. = Senecio pauciflosculosus C. Jeffrey ■☆

359621 Senecio olivaceus Klatt = Hubertia olivacea（Klatt）C. Jeffrey ●☆

359622 Senecio ommanei S. Moore = Senecio inornatus DC. ■☆

359623 Senecio oödes Rydb. = Packera streptanthifolia（Greene）W. A. Weber et Á. Löve ■☆

359624 Senecio orbicularis Sond. ex Harv. = Senecio oxyriifolius DC. ■☆

359625 Senecio oreganus Howell = Senecio hydrophiloides Rydb. ■☆

359626 Senecio oreophilus Greenm. = Packera neomexicana（A. Gray）W. A. Weber et Á. Löve ■☆

359627 Senecio oreophilus Muschl. ex Dinter = Anisopappus pinnatifidus（Klatt）O. Hoffm. ex Hutch. ■☆

359628 Senecio oreopolus Greenm. = Packera cana（Hook.）W. A. Weber et Á. Löve ■☆

359629 Senecio oreotrephas W. W. Sm. = Ligularia atroviolacea（Franch.）Hand.-Mazz. ■

359630 Senecio oresbius Greenm. = Packera neomexicana（A. Gray）W. A. Weber et Á. Löve ■☆

359631 Senecio ornatus S. Moore；装点千里光■☆

359632 Senecio oryzetorus Diels；田野千里光（大白顶草）；Field Groundsel ■

359633 Senecio oryzetorus Diels = Senecio yungningensis Hand.-Mazz. ■

359634 Senecio othonnae M. Bieb.；奥氏千里光■☆

359635 Senecio othonniflorus DC.；厚花千里光■☆

359636 Senecio othonniformis Fourc.；鲁文佐基千里光■☆

359637 Senecio otophorus Maxim. = Senecio cannabifolius Less. var. integrifolius（Koidz.）Kitam. ■

359638 Senecio ovatus（Walter）MacMill. = Senecio nemorensis L. ■

359639 Senecio ovatus MacMill. = Senecio nemorensis L. ■

359640 Senecio ovoideus（Compton）Jacobsen；澳非卵形千里光■☆

359641 Senecio oxyodontus DC.；尖齿千里光■☆

359642 Senecio oxyriifolius DC. subsp. milanjianus（S. Moore）G. D. Rowley = Senecio milanjianus S. Moore ■☆

359643 Senecio paarlensis DC.；帕尔千里光■☆

359644 Senecio paberensis Franch. = Cremanthodium ellisii（Hook. f.）Kitam. ■

359645 Senecio pachyrhizus O. Hoffm.；大根千里光■☆

359646 Senecio pachythelis E. Phillips et C. A. Sm. = Senecio conrathii N. E. Br. ■☆

359647 Senecio pagosanus A. Heller = Senecio amplectens A. Gray ■☆

359648 Senecio pallens Wall. ex DC. = Senecio nudicaulis Buch.-Ham. ex D. Don ■

359649 Senecio palmatifidus（Siebold et Zucc.）Wittr. et Juel = Ligularia japonica（Thunb.）Less. ■

359650 Senecio palmatifidus Wittr. et Juel = Ligularia japonica（Thunb.）Less. ■

359651 Senecio palmatilobus Kitam. = Sinosenecio palmatilobus（Kitam.）C. Jeffrey et Y. L. Chen ■

359652 Senecio palmatisectus Jeffrey = Parasenecio palmatisectus（Jeffrey）Y. L. Chen ■

359653 Senecio palmatisectus Jeffrey var. pubeescens Jeffrey = Parasenecio palmatisectus（Jeffrey）Y. L. Chen var. moupingensis（Franch.）Y. L. Chen ■

359654 Senecio palmatus（Pall.）Ledeb.；掌叶千里光（刘寄奴）；Palmleaf Groundsel ■

359655 Senecio palmatus（Pall.）Ledeb. = Senecio cannabifolius Less. ■

359656 Senecio palmatus（Pall.）Ledeb. f. davuricus Herder = Senecio cannabifolius Less. ■

359657 Senecio palmatus（Pall.）Ledeb. var. integrifolius Koidz. =

Senecio cannabifolius Less. var. integrifolius（Koidz.）Kitam. ■

359658　Senecio palmatus Pall. ex Ledeb. =Senecio cannabifolius Less. ■

359659　Senecio palmensis（Nees）Link = Bethencourtia palmensis（Nees）Link ■☆

359660　Senecio paludosus L.；沼泽千里光（沼生千里光）；Bird's Tongue, Fen Groundsel, Fen Ragwort, Great Fen Ragwort, Marsh Groundsel, Pale Ragwort ■☆

359661　Senecio palustris（L.）Hook. =Senecio congestus（R. Br.）DC. ■☆

359662　Senecio palustris（L.）Hook. =Tephroseris palustris（L.）Fourr. ■

359663　Senecio palustris（L.）Hook. var. congestus（R. Br.）Kom. =Tephroseris palustris（L.）Fourr. ■

359664　Senecio panduratus（Thunb.）Less. =Senecio erosus L. f. ■☆

359665　Senecio panduratus DC. =Senecio hastulatus L. ■☆

359666　Senecio pandurifolius Harv. = Senecio panduriformis Hilliard ■☆

359667　Senecio pandurifolius K. Koch；琴叶千里光■☆

359668　Senecio panduriformis Hilliard；琴形千里光■☆

359669　Senecio paniculatus J. M. Wood = Senecio chrysocoma Meerb. ■☆

359670　Senecio paniculatus P. J. Bergius；滕曲千里光■☆

359671　Senecio paniculatus P. J. Bergius var. reclinatus（L. f.）Harv. =Senecio chrysocoma Meerb. ■☆

359672　Senecio papyraceus DC. =Pericallis papyracea（DC.）B. Nord. ■☆

359673　Senecio parascitus Hilliard；艳丽千里光■☆

359674　Senecio parkeri Baker =Senecio hadiensis Forssk. ■☆

359675　Senecio parkeri Baker =Senecio petitianus A. Rich. ■☆

359676　Senecio parnassifolia a De Wild. et Muschl. = Emilia parnassifolia a（De Wild. et Muschl.）S. Moore ■☆

359677　Senecio parryi A. Gray；帕里千里光■☆

359678　Senecio parvifolius DC.；小叶千里光■☆

359679　Senecio pattersonensis Hoover；加利福尼亚千里光■☆

359680　Senecio paucicalyculatus Klatt；少副萼千里光■☆

359681　Senecio paucicephalus R. A. Dyer =Senecio affinis DC. ■☆

359682　Senecio pauciflorus Pursh；少花千里光；Few-flowered Senecio ■☆

359683　Senecio pauciflorus Pursh =Packera pauciflora（Pursh）Á. Löve et D. Löve ■☆

359684　Senecio pauciflorus Pursh subsp. fallax Greenm. =Packera indecora（Greene）Á. Löve et D. Löve ■☆

359685　Senecio pauciflorus Pursh var. fallax（Greenm.）Greenm. =Packera indecora（Greene）Á. Löve et D. Löve ■☆

359686　Senecio pauciflorus Pursh var. jucundulus Jeps. =Packera pseudaurea（Rydb.）W. A. Weber et Á. Löve ■☆

359687　Senecio pauciflorus Thunb. =Senecio grandiflorus P. J. Bergius ■☆

359688　Senecio pauciflosculosus C. Jeffrey；寡小花千里光■☆

359689　Senecio paucifoliatus C. C. Chang =Sinosenecio subrosulatus（Hand. -Mazz.）B. Nord. ■

359690　Senecio paucifolius DC. =Senecio ruwenzoriensis S. Moore ■☆

359691　Senecio paucifolius S. G. Gmel.；少叶千里光■☆

359692　Senecio pauciligulatus R. A. Dyer et C. A. Sm. = Senecio adnatus DC. ■☆

359693　Senecio paucilobus DC.；少裂萼千里光■☆

359694　Senecio paucinervis Dunn =Synotis erythropappa（Bureau et Franch.）C. Jeffrey et Y. L. Chen ■

359695　Senecio paucinervis Dunn var. brachylepis Marquand et Shaw =Synotis solidaginea（Hand. -Mazz.）C. Jeffrey et Y. L. Chen ■

359696　Senecio paulsenii O. Hoffm. ex Paulsen；保尔森千里光■☆

359697　Senecio pauperculus Michx.；细弱千里光；Balsam Ragwort Depauperate Senecio ■☆

359698　Senecio pauperculus Michx. =Packera paupercula（Michx.）Á. Löve et D. Löve ■☆

359699　Senecio pauperculus Michx. var. balsamitae（Muhl. ex Willd.）Fernald =Packera paupercula（Michx.）Á. Löve et D. Löve ■☆

359700　Senecio pauperculus Michx. var. crawfordii（Britton）T. M. Barkley =Packera paupercula（Michx.）Á. Löve et D. Löve ■☆

359701　Senecio pauperculus Michx. var. firmifolius（Greenm.）Greenm. =Packera paupercula（Michx.）Á. Löve et D. Löve ■☆

359702　Senecio pauperculus Michx. var. neoscoticus Fernald =Packera paupercula（Michx.）Á. Löve et D. Löve ■☆

359703　Senecio pauperculus Michx. var. praelongus（Greenm.）House =Packera paupercula（Michx.）Á. Löve et D. Löve ■☆

359704　Senecio pauperculus Michx. var. thompsoniensis（Greenm.）B. Boivin =Packera paupercula（Michx.）Á. Löve et D. Löve ■☆

359705　Senecio pearsonii Hutch.；皮尔逊千里光■☆

359706　Senecio pectinatus A. Nelson =Senecio flaccidus Less. var. monoensis（Greene）B. L. Turner et T. M. Barkley ■☆

359707　Senecio pectinatus C. C. Chang =Senecio acutipinnus Hand. -Mazz. ■

359708　Senecio peculiaris Dinter =Mesogramma apiifolium DC. ■☆

359709　Senecio pedunculatus Edgew. = Senecio krascheninnikovii Schischk. ■

359710　Senecio pellucidus DC. =Senecio madagascariensis Poir. ■☆

359711　Senecio peltophorus Brenan；盾梗千里光■☆

359712　Senecio pendulus（Forssk.）Sch. Bip.；泥鳅掌（初鹰）；Inch Worm, Inchworm, Tapeworm ■

359713　Senecio pendulus（Forssk.）Sch. Bip. = Kleinia pendula（Forssk.）DC. ■☆

359714　Senecio penicillatus（Cass.）Sch. Bip.；笔状千里光■☆

359715　Senecio penninervius DC.；羽脉千里光■☆

359716　Senecio pentactinus Klatt；五数千里光■☆

359717　Senecio pentanthus Merr. =Synotis triligulata（Buch. -Ham. ex D. Don）C. Jeffrey et Y. L. Chen ■

359718　Senecio perennans A. Nelson =Packera werneriifolia（A. Gray）W. A. Weber et Á. Löve ■☆

359719　Senecio pergamentaceus Baker；羊皮纸千里光■☆

359720　Senecio perplexus A. Nelson =Senecio integerrimus Nutt. var. exaltatus（Nutt.）Cronquist ■☆

359721　Senecio perrieri Humbert；佩里耶千里光■☆

359722　Senecio perrottetii DC.；佩罗千里光■☆

359723　Senecio persicifolius Burm. =Senecio lineatus（L. f.）DC. ■☆

359724　Senecio persicifolius L.；桃叶千里光■☆

359725　Senecio petasites（Sims）DC.；蜂斗菜千里光■☆

359726　Senecio petasitoides H. Lév. =Parasenecio petasitoides（H. Lév.）Y. L. Chen ■

359727　Senecio petiolaris DC.；柄叶千里光■☆

359728　Senecio petiolaris Less. =Senecio lessingii Harv. ■☆

359729　Senecio petiolatus Cotton ex Hauman =Dendrosenecio adnivalis（Stapf）E. B. Knox var. petiolatus（Hedberg）E. B. Knox ●☆

359730　Senecio petitianus A. Rich.；佩蒂蒂千里光■☆

359731　Senecio petitianus A. Rich. =Senecio hadiensis Forssk. ■☆

359732　Senecio petraeus (R. E. Fr.) Muschl. = Kleinia petraea (R. E. Fr.) C. Jeffrey ■☆

359733　Senecio petraeus Boiss. et Reut. ;岩生千里光■☆

359734　Senecio petraeus Klatt = Packera werneriifolia (A. Gray) W. A. Weber et Á. Löve ■☆

359735　Senecio petrocallis Greene = Packera werneriifolia (A. Gray) W. A. Weber et Á. Löve ■☆

359736　Senecio petrophilus Greene = Packera werneriifolia (A. Gray) W. A. Weber et Á. Löve ■☆

359737　Senecio petrophilus Klatt = Senecio canaliculatus Bojer ex DC. ■☆

359738　Senecio phaeanthus Nakai = Tephroseris phaeantha (Nakai) C. Jeffrey et Y. L. Chen ■

359739　Senecio phalacrocarpoides C. C. Chang = Sinosenecio phalacrocarpioides (C. C. Chang) B. Nord. ■

359740　Senecio phalacrocarpus Hance = Sinosenecio phalacrocarpus (Hance) B. Nord. ■

359741　Senecio phalacrocarpus Hance var. globigerus Oliv. = Sinosenecio globigerus (C. C. Chang) B. Nord. ■

359742　Senecio phalacrolaenus DC. = Senecio incisus Thunb. ■☆

359743　Senecio phellorhizus Muschl. = Kleinia grantii (Oliv. et Hiern) Hook. f. ■☆

359744　Senecio phoenicochaetus Franch. = Ligularia phoenicochaeta (Franch.) S. W. Liu ■

359745　Senecio phonolithicus Dinter = Senecio sarcoides C. Jeffrey ■☆

359746　Senecio phyllolepis Franch. = Parasenecio phyllolepis (Franch.) Y. L. Chen ■

359747　Senecio picticaulis P. R. O. Bally = Kleinia picticaulis (P. R. O. Bally) C. Jeffrey ■☆

359748　Senecio picticaulis P. R. O. Bally subsp. patriciae (C. Jeffrey) G. D. Rowley = Kleinia patriciae C. Jeffrey ■☆

359749　Senecio pierotii Miq. = Tephroseris pierotii (Miq.) Holub ■

359750　Senecio pierotii Miq. subsp. taitoensis (Hayata) Kitam. = Tephroseris taitoensis (Hayata) Holub ■

359751　Senecio pierotii Miq. var. taitoensis (Hayata) Kitam. = Tephroseris taitoensis (Hayata) Holub ■

359752　Senecio pilgerianus Diels = Parasenecio pilgerianus (Diels) Y. L. Chen ■

359753　Senecio pillansii Levyns ;皮朗斯千里光■☆

359754　Senecio pinguiculus Pomel = Senecio leucanthemifolius Poir. ■☆

359755　Senecio pinguifolius (DC.) Sch. Bip. ;肥叶千里光■☆

359756　Senecio pinifolius (L.) Lam. ;松叶千里光■☆

359757　Senecio pinnatifidus (P. J. Bergius) Less. ;羽裂千里光■☆

359758　Senecio pinnatipartitus Sch. Bip. ex Oliv. et Hiern;羽状深裂千里光■☆

359759　Senecio pinnatus Sch. Bip. ;羽状千里光■☆

359760　Senecio pinnulatus Thunb. ;小羽千里光■☆

359761　Senecio piptocoma O. Hoffm. ;早落千里光■☆

359762　Senecio pirottae Chiov. ;皮罗特千里光■☆

359763　Senecio plantagineoides C. Jeffrey ;车前状千里光■☆

359764　Senecio plantaginifolius Franch. = Ligularia virgaurea (Maxim.) Mattf. ex Rehder et Kobuski ■

359765　Senecio plattensis Nutt. = Packera plattensis (Nutt.) W. A. Weber et Á. Löve ■☆

359766　Senecio platyglossus Franch. = Ligularia platyglossa (Franch.) Hand. -Mazz. ■

359767　Senecio platylobus Rydb. = Packera streptanthifolia (Greene) W. A. Weber et Á. Löve ■☆

359768　Senecio platyphylloides Sommier et H. Lév. ;耳叶千里光■☆

359769　Senecio platyphyllus DC. ;阔叶千里光(宽叶狗舌草,欧狗舌草) ; Broadleaf Groundsel ■

359770　Senecio platypleurus Cufod. = Senecio aequinoctialis R. E. Fr. ■☆

359771　Senecio plebeius DC. = Senecio laevigatus Thunb. ■☆

359772　Senecio pleianthus (Humbert) Humbert ;多花千里光■☆

359773　Senecio pleistocephalus S. Moore ;繁头千里光■☆

359774　Senecio pleistophyllus C. Jeffrey ;多叶千里光■☆

359775　Senecio pleopteris Diels = Senecio graciliflorus DC. ■

359776　Senecio pleurocaulis Franch. = Ligularia pleurocaulis (Franch.) Hand. -Mazz. ■

359777　Senecio pojarkovae Schischk. ;波亚千里光■☆

359778　Senecio polelensis Hilliard ;博莱尔千里光■☆

359779　Senecio polyadenus Hedberg ;多腺千里光■☆

359780　Senecio polyanthemoides Sch. Bip. ;拟多花千里光■☆

359781　Senecio polycephalus Ledeb. ;密头千里光■☆

359782　Senecio polycotomus (Chiov.) H. Jacobsen = Kleinia odora (Forssk.) DC. ■☆

359783　Senecio polycotomus (Chiov.) H. Jacobsen subsp. squarrosus (Cufod.) G. D. Rowley = Kleinia squarrosa Cufod. ■☆

359784　Senecio polygonoides Muschl. = Senecio hochstetteri Sch. Bip. ex A. Rich. ■☆

359785　Senecio polyodon DC. ;多齿千里光■☆

359786　Senecio polyodon DC. var. subglaber (O. Hoffm. ex Kuntze) Hilliard et B. L. Burtt ;近光多齿千里光■☆

359787　Senecio polyrhizus Baker = Senecio vittarifolius Bojer ex DC. ■☆

359788　Senecio populifolius L. = Senecio halimifolius L. ■☆

359789　Senecio porphyranthus Schischk. ;紫花千里光■☆

359790　Senecio porteri Greene = Packera porteri (Greene) C. Jeffrey ■☆

359791　Senecio potaninii C. Winkl. = Ligularia potaninii (C. Winkl.) Y. Ling ■

359792　Senecio pratensis (Hoppe) DC. var. polycephalus Regel = Tephroseris subdentata (Bunge) Holub ■

359793　Senecio pratensis Phil. var. polycephalus Regel = Tephroseris subdentata (Bunge) Holub ■

359794　Senecio praticola Schischk. et Serg. = Tephroseris praticola (Schischk. et Serg.) Holub ■

359795　Senecio prattii Hemsl. = Cremanthodium prattii (Hemsl.) R. D. Good ■

359796　Senecio primulifolius H. Lév. = Ligularia vellerea (Franch.) Hand. -Mazz. ■

359797　Senecio principis Franch. = Cremanthodium principis (Franch.) R. D. Good ■

359798　Senecio prionites MacOwan var. laxus ? = Senecio macowanii Hilliard ■☆

359799　Senecio prionophyllus Franch. = Synotis nagensis (C. B. Clarke) C. Jeffrey et Y. L. Chen ■

359800　Senecio profundorus Dunn = Parasenecio profundorum (Dunn) Y. L. Chen ■

359801　Senecio prolixus Greenm. = Packera multilobata (Torr. et A. Gray) W. A. Weber et Á. Löve ■☆

359802　Senecio propinquus Schischk. ;邻近千里光■☆

359803　Senecio prostratus Klatt ;平卧千里光■☆

359804　Senecio protractus (S. Moore) Eyles = Emilia protracta S. Moore ■☆

359805 Senecio przewalskii Maxim. = Ligularia przewalskii (Maxim.) Diels ■

359806 Senecio pseudaureus Rydb. = Packera pseudaurea (Rydb.) W. A. Weber et Á. Löve ■☆

359807 Senecio pseudaureus Rydb. subsp. semicordatus (Mack. et Bush) G. W. Douglas et G. R. Douglas = Packera pseudaurea (Rydb.) W. A. Weber et Á. Löve var. semicordata (Mack. et Bush) Trock et T. M. Barkley ■☆

359808 Senecio pseudaureus Rydb. var. semicordatus (Mack. et Bush) T. M. Barkley = Packera pseudaurea (Rydb.) W. A. Weber et Á. Löve ■☆

359809 Senecio pseudaureus Rydb. var. semicordatus (Mack. et Bush) T. M. Barkley = Packera pseudaurea (Rydb.) W. A. Weber et Á. Löve var. semicordata (Mack. et Bush) Trock et T. M. Barkley ■☆

359810 Senecio pseudoalata C. C. Chang = Synotis pseudoalata (C. C. Chang) C. Jeffrey et Y. L. Chen ■

359811 Senecio pseudoarnica Less. ;多肉千里光(滨海山金车,假山金车千里光,肿柄千里光);Inflared Peduncle Groundsel,Seabeach Senecio,Swellingpedicelled Groundsel ■

359812 Senecio pseudoarnica Less. var. rollandii Vict. = Senecio pseudoarnica Less. ■

359813 Senecio pseudoauranticus Kom. ;假耳千里光■☆

359814 Senecio pseudochina L. = Gynura pseudochina (L.) DC. ■

359815 Senecio pseudoelegans Less. = Senecio elegans L. ■☆

359816 Senecio pseudomairiei H. Lév. ;西南千里光;South-western Groundsel,SW. China Groundsel ■

359817 Senecio pseudorhyncholaenus Thell. = Senecio subrubriflorus O. Hoffm. ■☆

359818 Senecio pseudorhyncholaenus Thell. var. auriculatus ? = Senecio rhyncholaenus DC. ■☆

359819 Senecio pseudosceptrum Steud. = Senecio spiraeifolius Thunb. ■☆

359820 Senecio pseudosonchus Vaniot = Tephroseris pseudosonchus (Vaniot) C. Jeffrey et Y. L. Chen ■

359821 Senecio pseudosonchus Vaniot var. borealis (Cufod.) S. Y. Hu = Tephroseris subdentata (Bunge) Holub ■

359822 Senecio pseudosonchus Vaniot var. polycephalus (Regel) Kitam. = Tephroseris subdentata (Bunge) Holub ■

359823 Senecio pseudosubsessilis C. Jeffrey;假近无柄千里光■☆

359824 Senecio pseudotites Griseb. ;秘鲁千里光■☆

359825 Senecio pseudotomentosus Mack. et Bush = Packera plattensis (Nutt.) W. A. Weber et Á. Löve ■☆

359826 Senecio psiadioides O. Hoffm. = Senecio maranguensis O. Hoffm. ■☆

359827 Senecio pteridophyllus Franch. ;蕨叶千里光;Fernleaf Groundsel ■

359828 Senecio pteroneurus (DC.) Ball = Kleinia anteuphorbium (L.) Haw. ☆

359829 Senecio pteroneurus (DC.) Sch. Bip. = Kleinia anteuphorbium (L.) Haw. ■☆

359830 Senecio pterophorus DC. ;具翅千里光;Shoddy Ragwort ■☆

359831 Senecio pterophorus DC. var. apterus Harv. = Senecio pterophorus DC. ■☆

359832 Senecio pteropodus W. W. Sm. = Senecio nigrocinctus Franch. ■

359833 Senecio puberulus DC. ;微毛千里光■☆

359834 Senecio pubigerus L. ;短毛千里光■☆

359835 Senecio pudicus Greene;纯洁千里光■☆

359836 Senecio pulcher Hook. et Arn. ;优雅千里光■☆

359837 Senecio pullus Klatt = Senecio umgeniensis Thell. ■☆

359838 Senecio purdomii Turrill = Ligularia purdomii (Turrill) Chitt. ■

359839 Senecio purpureo-viridis Baker = Senecio resectus Bojer ex DC. ●☆

359840 Senecio purpureus L. ;紫千里光■☆

359841 Senecio purshianus Nutt. = Packera cana (Hook.) W. A. Weber et Á. Löve ■☆

359842 Senecio putjatae C. Winkl. = Ligularia mongolica (Turcz.) DC. ■

359843 Senecio pyramidatus DC. ;塔形千里光■☆

359844 Senecio pyroglossus Kar. et Kir. ;红花千里光■

359845 Senecio pyrrhochrous Greene = Packera crocata (Rydb.) W. A. Weber et Á. Löve ■☆

359846 Senecio pyrropappus Franch. = Ligularia cymbulifera (W. W. Sm.) Hand. -Mazz. ■

359847 Senecio quartinianus Asch. = Emilia abyssinica (Sch. Bip. ex A. Rich.) C. Jeffrey ■☆

359848 Senecio quartziticola Humbert;阔茨千里光●☆

359849 Senecio quaylei T. M. Barkley;奎尔千里光■☆

359850 Senecio quercetorum Greene = Packera quercetorum (Greene) C. Jeffrey ■☆

359851 Senecio quercifolius Thunb. = Senecio ilicifolius L. ■☆

359852 Senecio quinqueflorus DC. = Senecio bipinnatus (Thunb.) Less. ■☆

359853 Senecio quinquelobus (Thunb.) DC. ;五裂千里光■☆

359854 Senecio quinquelobus (Thunb.) DC. var. helminthioides Sch. Bip. = Senecio helminthioides (Sch. Bip.) Hilliard ■☆

359855 Senecio quinquelobus (Wall. ex DC.) Hook. f. et Thomson ex C. B. Clarke = Parasenecio quinquelobus (Wall. ex DC.) Y. L. Chen ■

359856 Senecio quinquelobus (Wall. ex DC.) Hook. f. et Thomson ex C. B. Clarke var. moupingensis Franch. = Parasenecio palmatisectus (Jeffrey) Y. L. Chen var. moupingensis (Franch.) Y. L. Chen ■

359857 Senecio quinquelobus DC. = Senecio quinquelobus (Thunb.) DC. ■☆

359858 Senecio quinquenervius Sond. ex Harv. ;五脉千里光■☆

359859 Senecio racemosus DC. ;总状千里光■☆

359860 Senecio radicans (L. f.) Sch. Bip. = Kleimis radicans L. f. ■

359861 Senecio ramsbottomii Hand. -Mazz. = Senecio biligulatus W. W. Sm. ■

359862 Senecio randii S. Moore ;兰德千里光■☆

359863 Senecio raphanifolius Wall. ex DC. ;莱菔千里光(莱菔叶千里光,异叶千里光);Radishleaf Groundsel ■

359864 Senecio rapifolius Nutt. ;萝卜叶千里光■☆

359865 Senecio rautaneni S. Moore = Senecio eenii (S. Moore) Merxm. ■☆

359866 Senecio ravidus C. Winkl. = Nannoglottis ravida (C. Winkl.) Y. L. Chen ■

359867 Senecio rawsonianus Greene = Senecio aronicoides DC. ■☆

359868 Senecio rectiramus Baker = Austrosynotis rectirama (Baker) C. Jeffrey ■☆

359869 Senecio redivivus Mabb. = Lachanodes arbores (Roxb.) B. Nord. ●☆

359870 Senecio redivivus Mabb. = Mikania arborea Roxb ●☆

359871 Senecio refractisquamatus De Wild. = Dendrosenecio adnivalis (Stapf) E. B. Knox ●☆

359872 Senecio refractisquamatus De Wild. var. intermedia (Hauman) Robyns = Dendrosenecio erici-rosenii (R. E. Fr. et T. C. E. Fr.) E.

B. Knox subsp. alticola（Mildbr.）E. B. Knox ●☆

359873　Senecio rehmannii Bolus；拉赫曼千里光■☆

359874　Senecio remipes W. W. Sm. = Ligularia tsangchanensis（Franch.）Hand. -Mazz. ■

359875　Senecio renatus Franch. = Cremanthodium decaisnei C. B. Clarke ■

359876　Senecio renifolius Porter = Packera porteri（Greene）C. Jeffrey ■☆

359877　Senecio reniformis（Hook.）Macmillan = Arnoglossum reniforme（Hook.）H. Rob. ■☆

359878　Senecio reniformis Wall. = Cremanthodium reniforme（DC.）Benth. ■

359879　Senecio repandus Thunb.；浅波状千里光■☆

359880　Senecio repens（L.）Muschl.；匍匐千里光（万宝）■☆

359881　Senecio resectus Bojer ex DC.；截形千里光●☆

359882　Senecio resedifolius Less. = Packera cymbalaria（Pursh）Á. Löve et D. Löve ■

359883　Senecio resedifolius Less. var. moresbiensis（Calder et R. L. Taylor）B. Boivin = Packera subnuda（DC.）Trock et T. M. Barkley var. moresbiensis（Calder et R. L. Taylor）Trock ■☆

359884　Senecio retortus（DC.）Benth.；旋扭千里光■☆

359885　Senecio retrorsus DC.；倒千里光（下向千里光）■☆

359886　Senecio retrorsus DC. var. subedentulus？ = Senecio retrorsus DC. ■☆

359887　Senecio retusus（DC.）Wall. ex Hook. f. = Ligularia retusa DC. ■

359888　Senecio retusus Wall. = Ligularia retusa DC. ■

359889　Senecio revolutus Hoover = Senecio pattersonensis Hoover ■☆

359890　Senecio rhammatophyllus Mattf.；假糖芥千里光■☆

359891　Senecio rhodanthus Baker = Emilia graminea DC. ■☆

359892　Senecio rhombifolius（Willd.）Sch. Bip.；菱叶千里光■☆

359893　Senecio rhomboideus Harv.；菱形千里光■☆

359894　Senecio rhopalodenia Dinter = Senecio radicans（L. f.）Sch. Bip. ■

359895　Senecio rhopalophyllus（Dinter）Merxm. = Senecio aloides DC. ■☆

359896　Senecio rhyncholaenus DC.；喙被千里光■☆

359897　Senecio riddellii Torr. et A. Gray；里德尔千里光；Riddell's Groundsel，Sand Groundsel ■☆

359898　Senecio riddellii Torr. et A. Gray var. parksii Cory = Senecio riddellii Torr. et A. Gray ■☆

359899　Senecio rigens L. = Othonna parviflora P. J. Bergius ●☆

359900　Senecio rigescens Jacq. = Senecio rosmarinifolius L. f. ■☆

359901　Senecio rigidus L.；硬千里光●☆

359902　Senecio riparius DC.；河岸千里光●☆

359903　Senecio rivularis DC.；溪岸千里光■☆

359904　Senecio robbinsii Oakes ex Rusby；罗氏千里光；Robbins' Senecio ■☆

359905　Senecio robbinsii Oakes ex Rusby = Packera schweinitziana（Nutt.）W. A. Weber et Á. Löve ■☆

359906　Senecio robbinsii Oakes ex Rusby var. subtomentosus Peck = Packera paupercula（Michx.）Á. Löve et D. Löve ■☆

359907　Senecio roberti-friesii K. Afzel. = Senecio schweinfurthii O. Hoffm. ■☆

359908　Senecio roberti-friesii K. Afzel. var. subcanescens？ = Senecio schweinfurthii O. Hoffm. ■☆

359909　Senecio roborowskii Maxim. = Parasenecio roborowskii（Maxim.）Y. L. Chen ■

359910　Senecio robustus（DC.）Sch. Bip. var. kareliniana Trautv. = Ligularia narynensis（C. Winkl.）O. Fedtsch. et B. Fedtsch. ■

359911　Senecio robustus（DC.）Sch. Bip. var. karelinianus Trautv. = Ligularia narynensis（C. Winkl.）O. Fedtsch. et B. Fedtsch. ■

359912　Senecio robustus（DC.）Sch. Bip. var. typica Trautv. = Ligularia narynensis（C. Winkl.）O. Fedtsch. et B. Fedtsch. ■

359913　Senecio roccatii Chiov. = Senecio maranguensis O. Hoffm. ■☆

359914　Senecio rogersii S. Moore = Emilia hockii（De Wild. et Muschl.）C. Jeffrey ■☆

359915　Senecio rollandii（Vict.）Vict. = Senecio pseudoarnica Less. ■

359916　Senecio rooseveltianus De Wild. = Senecio deltoideus Less. ■☆

359917　Senecio rosellatus Bojer ex DC. = Hubertia rosellata（Bojer ex DC.）C. Jeffrey ●☆

359918　Senecio rosmarinifolius L. f.；迷迭香叶千里光■☆

359919　Senecio rosulatus Rydb. = Packera fendleri（A. Gray）W. A. Weber et Á. Löve ■☆

359920　Senecio rosulifer H. Lév. et Vaniot = Senecio nudicaulis Buch. -Ham. ex D. Don ■

359921　Senecio rosuliferus C. C. Chang = Sinosenecio changii（B. Nord.）B. Nord. et Pelser ■

359922　Senecio rosuliferus C. C. Chang = Tephroseris changii B. Nord. ■

359923　Senecio rotundus（Britton）Small = Packera obovata（Muhl. ex Willd.）W. A. Weber et Á. Löve ■☆

359924　Senecio rowleyanus H. Jacobsen；翡翠珠（绿串珠，绿铃，绿玉）；String of Pearls，String-of-beads Plant，String-of-beads Senecio ■☆

359925　Senecio royleanus DC.；珠峰千里光；Jolmolungma Groundsel，Royle's Groundsel ■

359926　Senecio royleanus DC. = Senecio cinerifolius H. Lév. ■☆

359927　Senecio rubens Jacq. = Crassocephalum rubens（Juss. ex Jacq.）S. Moore ■

359928　Senecio rubens Juss. ex Jacq. = Crassocephalum rubens（Juss. ex Jacq.）S. Moore ■

359929　Senecio rubescens S. Moore = Parasenecio rubescens（S. Moore）Y. L. Chen ■

359930　Senecio ruderalis Harv. = Senecio madagascariensis Poir. ■☆

359931　Senecio ruepellii Sch. Bip. = Senecio glaucus L. subsp. coronopifolius（Desf.）Alexander ■☆

359932　Senecio ruficomus Franch. = Ligularia ruficoma（Franch.）Hand. -Mazz. ■

359933　Senecio rufipilus Franch. = Parasenecio rufipilis（Franch.）Y. L. Chen ■

359934　Senecio rufopilosulus De Wild. = Crassocephalum montuosum（S. Moore）Milne-Redh. ■☆

359935　Senecio rufus Hand. -Mazz. = Tephroseris rufa（Hand. -Mazz.）B. Nord. ■

359936　Senecio rugegensis Muschl.；鲁吉千里光■☆

359937　Senecio rumicifolius Drumm. = Ligularia rumicifolia（Drumm.）S. W. Liu ■

359938　Senecio rupestris Waldst. et Kit.；岩地千里光；Rock Ragwort ■☆

359939　Senecio rusbyi Greene = Senecio bigelovii A. Gray ■☆

359940　Senecio rusisiensis R. E. Fr. = Senecio hochstetteri Sch. Bip. ex A. Rich. ■☆

359941　Senecio rutshuruensis De Wild. = Gynura scandens O. Hoffm. ■☆

359942　Senecio ruwenzoriensis S. Moore = Senecio othonniformis Fourc. ■☆

359943 Senecio rydbergii A. Nelson = Packera streptanthifolia (Greene) W. A. Weber et Á. Löve ■☆

359944 Senecio sabinjoensis Muschl. ;萨比尼奥千里光■☆

359945 Senecio saboureaui Humbert;萨博千里光■☆

359946 Senecio sacco-flabellatus H. Lév. = Ligularia hookeri (C. B. Clarke) Hand. -Mazz. ■

359947 Senecio saccoso-flabellatus H. Lév. = Ligularia hookeri (C. B. Clarke) Hand. -Mazz. ■

359948 Senecio sagitta Maxim. = Ligularia sagitta (Maxim.) Mattf. ex Rehder et Kobuski ■

359949 Senecio sagittatus Sch. Bip. = Parasenecio hastatus (L.) H. Koyama ■

359950 Senecio sagittatus Sch. Bip. = Parasenecio lancifolius (Franch.) Y. L. Chen ■

359951 Senecio sagittatus Sch. Bip. var. lancifolia Franch. = Parasenecio lancifolius (Franch.) Y. L. Chen ■

359952 Senecio sagittatus Sch. Bip. var. pubescens (Ledeb.) Maxim. = Parasenecio hastatus (L.) H. Koyama ■

359953 Senecio sagittatus Sch. Bip. var. pubescens Maxim. = Parasenecio hastatus (L.) H. Koyama ■

359954 Senecio sagittatus Waldst. et Kit. var. lancifolia Franch. = Parasenecio lancifolius (Franch.) Y. L. Chen ■

359955 Senecio sakalavorum Humbert;萨卡拉瓦千里光■☆

359956 Senecio sakamaliensis (Humbert) Humbert;萨卡马利千里光■☆

359957 Senecio salicinus Rydb. = Packera fendleri (A. Gray) W. A. Weber et Á. Löve ☆

359958 Senecio saliens Rydb. = Senecio triangularis Hook. ■☆

359959 Senecio salignus DC. ;巴氏千里光;Barkley's Ragwort, Willow Leaf Groundsel ■☆

359960 Senecio salignus DC. = Barkleyanthus salicifolius (Kunth) H. Rob. et Brettell ■☆

359961 Senecio saluenensis Diels = Synotis saluenensis (Diels) C. Jeffrey et Y. L. Chen ■

359962 Senecio saluenensis Diels ex Gagnep. = Synotis saluenensis (Diels) C. Jeffrey et Y. L. Chen ■

359963 Senecio salviifolius Sch. Bip. ;鼠尾草叶千里光■☆

359964 Senecio sandersonii Harv. ;桑德森千里光■☆

359965 Senecio sanguisorboides Rydb. = Packera sanguisorboides (Rydb.) W. A. Weber et Á. Löve ■☆

359966 Senecio saniensis Hilliard et B. L. Burtt;萨尼千里光■☆

359967 Senecio saposhnikovii Krasch. et Schipcz. ;萨波夫千里光■☆

359968 Senecio saracenicus Koch;瓶千里光■☆

359969 Senecio sarcoides C. Jeffrey;拟肉质千里光■☆

359970 Senecio sarmentosus O. Hoffm. = Senecio deltoideus Less. ■☆

359971 Senecio sarracenicus L. var. turczaninowii (DC.) Nakai = Senecio nemorensis L. ■

359972 Senecio sattimae K. Afzel. = Senecio schweinfurthii O. Hoffm. ■☆

359973 Senecio saundersii F. W. H. Sauer et Beck;桑德斯千里光■☆

359974 Senecio saussureoides Hand. -Mazz. ;风毛菊状千里光;Saussurea-like Groundsel, Windhairdaisylike Groundsel ■

359975 Senecio savatieri Franch. = Sinosenecio oldhamianus (Maxim.) B. Nord. ■

359976 Senecio saxatilis Wall. ex DC. = Senecio wightii (DC. ex Wight) Benth. ex C. B. Clarke ■

359977 Senecio saxosus Klatt = Packera werneriifolia (A. Gray) W. A. Weber et Á. Löve ■☆

359978 Senecio scabiosifolius C. C. Chang = Senecio diversipinnus Y. Ling ■

359979 Senecio scabriusculus DC. = Senecio pinnulatus Thunb. ■☆

359980 Senecio scandens Buch. -Ham. ex D. Don;千里光(白苏杆,百花草,粗糠花,黄花草,黄花母,黄花演,黄花枝草,箭草,金钗草,金花草,金素英,九里光,九里明,九岭光,九龙光,九龙明,蔓黄菀,木莲草,七里光,千里及,千里急,千里明,青龙硬,软藤黄花草,天青红,眼明草,野菊花,一扫光);Climbing Groundsel, Groundsel ■

359981 Senecio scandens Buch. -Ham. ex D. Don f. incisus (Franch.) Kitam. = Senecio scandens Buch. -Ham. ex D. Don var. incisus Franch. ■

359982 Senecio scandens Buch. -Ham. ex D. Don var. crataegifolius (Hayata) Kitam. = Senecio crataegifolius Hayata ■

359983 Senecio scandens Buch. -Ham. ex D. Don var. crataegifolius (Hayata) Kitam. ;山楂叶千里光(小蔓黄菀) ■

359984 Senecio scandens Buch. -Ham. ex D. Don var. hulienensis S. S. Ying = Senecio scandens Buch. -Ham. ex D. Don var. incisus Franch. ■

359985 Senecio scandens Buch. -Ham. ex D. Don var. incisus Franch. ;缺裂千里光(花莲蔓黄菀,裂叶蔓黄菀,裂叶千里光,深裂千里光);Incised Groundsel ■

359986 Senecio scandens DC. = Delairea odorata Lem. ●☆

359987 Senecio scapiflorus (L'Hér.) C. A. Sm. ;花茎千里光■☆

359988 Senecio scapiformis Y. L. Chen et K. Y. Pan = Senecio laetus Edgew. ■

359989 Senecio scapiformis Y. Ling ex Y. L. Chen, S. Yun Liang et K. Y. Pan = Senecio laetus Edgew. ■

359990 Senecio scaposus A. Nelson = Packera werneriifolia (A. Gray) W. A. Weber et Á. Löve ■☆

359991 Senecio scaposus DC. ;筒叶菊(新月);Scape Groundsel ■

359992 Senecio scaposus DC. var. addoensis (Compton) G. D. Rowley;阿多千里光■☆

359993 Senecio sceleratus Schweick. ;辣千里光■☆

359994 Senecio sceleratus Schweick. = Senecio latifolius DC. ■☆

359995 Senecio schimidtii (Maxim.) Franch. et Sav. = Ligularia schmidtii (Maxim.) Makino ■

359996 Senecio schimperi Sch. Bip. ex A. Rich. ;欣珀千里光■☆

359997 Senecio schinzii O. Hoffm. = Emilia ambifaria (S. Moore) C. Jeffrey ■☆

359998 Senecio schischkinii Lipsch. ex N. I. Rubtzov = Ligularia schischkinii Lipsch. ex N. I. Rubtzov ■

359999 Senecio schizopetalus W. W. Sm. = Ligularia stenoglossa (Franch.) Hand. -Mazz. ■

360000 Senecio schmidtii (Maxim.) Franch. et Sav. = Ligularia schmidtii (Maxim.) Makino ■

360001 Senecio schubotzianus Muschl. = Cineraria deltoidea Sond. ■☆

360002 Senecio schultzii Hochst. ex A. Rich. ;舒尔茨千里光■☆

360003 Senecio schultzii Hochst. ex A. Rich. subsp. chillaloensis (Cufod.) S. Ortiz et Vivero;埃塞俄比亚千里光■☆

360004 Senecio schultzii Hochst. ex A. Rich. var. lanatus D. F. Otieno et Mesfin;绵毛千里光■☆

360005 Senecio schweinfurthii O. Hoffm. ;施韦千里光■☆

360006 Senecio schweinitzianus Nutt. = Packera schweinitziana (Nutt.) W. A. Weber et Á. Löve ■☆

360007 Senecio schwetzovii Korsh. ;施威氏千里光;Schwetzov Groundsel ■☆

360008 Senecio sciatrephedioides Y. Ling = Synotis cavaleriei（H. Lév.）C. Jeffrey et Y. L. Chen ■

360009 Senecio sciatrephes W. W. Sm. = Synotis sciatrephes（W. W. Sm.）C. Jeffrey et Y. L. Chen ■

360010 Senecio scitus Hutch. et Burtt Davy;雅丽千里光■☆

360011 Senecio scoparius Harv.;帚状千里光■☆

360012 Senecio scorzonella Greene;科维尔千里光■☆

360013 Senecio scottii Balf. f. = Kleinia scottii（Balf. f.）P. Halliday ■☆

360014 Senecio scribneri Rydb. = Senecio integerrimus Nutt. var. scribneri（Rydb.）T. M. Barkley ■☆

360015 Senecio scrophulariifolius O. Hoffm. = Senecio maranguensis O. Hoffm. ■☆

360016 Senecio scytophyllus Diels = Cremanthodium coriaceum S. W. Liu ■

360017 Senecio segmentatus Oliv. = Lopholaena segmentata（Oliv.）S. Moore ■☆

360018 Senecio semiamplexicaulis Rydb. = Senecio crassulus A. Gray ■☆

360019 Senecio semiamplexifolius De Wild.;褶叶千里光■☆

360020 Senecio semicordatus Mack. et Bush = Packera pseudaurea（Rydb.）W. A. Weber et Á. Löve var. semicordata（Mack. et Bush）Trock et T. M. Barkley ■☆

360021 Senecio seminiveus J. M. Wood et M. S. Evans;半雪千里光■☆

360022 Senecio sempervivus（Forssk.）Sch. Bip. subsp. grantii（Oliv. et Hiern）G. D. Rowley = Kleinia grantii（Oliv. et Hiern）Hook. f. ■☆

360023 Senecio septilobus C. C. Chang = Sinosenecio septilobus（C. C. Chang）B. Nord. ■

360024 Senecio seretii De Wild. = Gynura scandens O. Hoffm. ■☆

360025 Senecio seridophyllus Greene = Senecio amplectens A. Gray var. holmii（Greene）H. D. Harr. ■☆

360026 Senecio serpens G. D. Rowley;蛇形千里光■☆

360027 Senecio serpens Rowley;万宝; Blue Chalksticks, Serpent Groundsel,Window Plant ●

360028 Senecio serra Hook. ;披针叶千里光■☆

360029 Senecio serra Hook. var. altior Jeps. = Senecio serra Hook. ■☆

360030 Senecio serra Hook. var. sanctus H. M. Hall = Senecio spartioides Torr. et Gray ■☆

360031 Senecio serra Schweinf. = Senecio schweinfurthii O. Hoffm. ■☆

360032 Senecio serra Sond. = Senecio inornatus DC. ■☆

360033 Senecio serratuloides DC.;齿状千里光■☆

360034 Senecio serratuloides DC. var. dieterlenii Thell. = Senecio gregatus Hilliard ■☆

360035 Senecio serratuloides DC. var. glabratus ? = Senecio serratuloides DC. ■☆

360036 Senecio serratuloides DC. var. gracilis Harv. = Senecio gregatus Hilliard ■☆

360037 Senecio serratuloides DC. var. holubii Thell. = Senecio gregatus Hilliard ■☆

360038 Senecio serratuloides DC. var. rehmannii Thell. = Senecio gregatus Hilliard ■☆

360039 Senecio serratus（Thunb.）Sond. = Senecio glanduloso-pilosus Volkens et Muschl. ■☆

360040 Senecio serratus E. Mey. = Senecio incomptus DC. ■☆

360041 Senecio serrulatus DC.;细齿千里光■☆

360042 Senecio serrurioides Turcz.;色罗山龙眼千里光■☆

360043 Senecio sessilifolius Sch. Bip. = Cremanthodium nanum（Decne.）W. W. Sm. ■

360044 Senecio sessilis Thunb. = Senecio halimifolius L. ■☆

360045 Senecio setchuanensis Franch. = Synotis setchuanensis（Franch.）C. Jeffrey et Y. L. Chen ■

360046 Senecio shabensis Lisowski;沙巴千里光■☆

360047 Senecio sheldonensis A. E. Porsild;谢尔登千里光■☆

360048 Senecio sichotensis Kom. ;西豪特千里光■☆

360049 Senecio sikkimensis Franch. = Ligularia hookeri（C. B. Clarke）Hand. -Mazz. ■

360050 Senecio simplicissimus Bojer ex DC. ;单一千里光■☆

360051 Senecio simulans Chiov. = Solanecio cydoniifolius（O. Hoffm.）C. Jeffrey ■☆

360052 Senecio sinicus（Diels）C. C. Chang = Synotis sinica（Diels）C. Jeffrey et Y. L. Chen ■

360053 Senecio sisymbriifolius DC. ;大蒜芥叶千里光■☆

360054 Senecio sisymbriifolius Dinter = Mesogramma apiifolium DC. ■☆

360055 Senecio skirrhodon DC. = Senecio madagascariensis Poir. ■☆

360056 Senecio smallii Britton = Packera anonyma（A. W. Wood）W. A. Weber et Á. Löve ■☆

360057 Senecio smithii DC. ;革叶千里光;Magellan Ragwort ■☆

360058 Senecio sneeuwbergensis Bolus = Senecio inornatus DC. ■☆

360059 Senecio snowdenii Hutch. ;斯诺登千里光■☆

360060 Senecio sociorum Bolus;群生千里光■☆

360061 Senecio solanifolius Jeffrey = Senecio scandens Buch. -Ham. ex D. Don ■

360062 Senecio solanoides Sch. Bip. ex Asch. = Solanecio tuberosus（Sch. Bip. ex A. Rich.）C. Jeffrey ■☆

360063 Senecio solenoides Dunn = Nemosenecio solenoides（Dunn）B. Nord. ■

360064 Senecio solidaginea Hand. -Mazz. = Synotis solidaginea（Hand. -Mazz.）C. Jeffrey et Y. L. Chen ■

360065 Senecio solidagineus Spreng. = Senecio halimifolius L. ■☆

360066 Senecio solidaginoides P. J. Bergius = Senecio halimifolius L. ■☆

360067 Senecio solidago Rydb. = Senecio serra Hook. ■☆

360068 Senecio somalensis Chiov. = Gynura pseudochina（L.）DC. ■

360069 Senecio sonchifolius（L.）Moench. = Emilia sonchifolia（L.）DC. ex Wight ■

360070 Senecio songaricus Fisch. = Ligularia songarica（Fisch.）Y. Ling ■

360071 Senecio soongaricus Fisch. = Ligularia songarica（Fisch.）Y. Ling ■

360072 Senecio sophioides DC. ;播娘蒿千里光■☆

360073 Senecio sororius C. Jeffrey;堆积千里光■☆

360074 Senecio sosnowskyi Sofieva;索斯千里光■☆

360075 Senecio sotikensis S. Moore;索蒂克千里光■☆

360076 Senecio souliei Franch. = Parasenecio souliei（Franch.）Y. L. Chen ■

360077 Senecio sparsilobatus Parish = Packera bernardina（Greene）W. A. Weber et Á. Löve ■☆

360078 Senecio spartioides Torr. et A. Gray var. fremontii（Torr. et A. Gray）Greenm. = Senecio riddellii Torr. et A. Gray ■☆

360079 Senecio spartioides Torr. et A. Gray var. parksii（Cory）Shinners = Senecio riddellii Torr. et A. Gray ■☆

360080 Senecio spartioides Torr. et A. Gray var. riddellii（Torr. et A. Gray）Greenm. = Senecio riddellii Torr. et A. Gray ■☆

360081   Senecio spartioides Torr. et Gray;西方帚状千里光;Broom Groundsel,Many-headed Groundsel ■☆

360082   Senecio spartioides Torr. et Gray var. granularis Maguire et A. H. Holmgren ex Cronquist = Senecio spartioides Torr. et Gray ■☆

360083   Senecio spartioides Torr. et Gray var. parksii (Cory) Shinners = Senecio riddellii Torr. et A. Gray ■☆

360084   Senecio spartioides Torr. et Gray var. riddellii (Torr. et A. Gray) Greenm. = Senecio riddellii Torr. et A. Gray ■☆

360085   Senecio spathiphyllus Franch.;匙叶千里光;Spathulate Fleawort,Spatulate Groundsel,Spoonleaf Groundsel ■

360086   Senecio spatuliformis A. Heller = Packera macounii (Greene) W. A. Weber et Á. Löve ■☆

360087   Senecio speciosissimus J. C. Manning et Goldblatt;极美千里光 ■☆

360088   Senecio speciosus Willd.;美丽千里光■☆

360089   Senecio spelaeicola (Vaniot) Gagnep.;岩穴千里光■

360090   Senecio spelaeicola (Vaniot) Gagnep. = Cissampelopsis spelaeicola (Vaniot) C. Jeffrey et Y. L. Chen ●

360091   Senecio spelaeicola Vaniot = Cissampelopsis spelaeicola (Vaniot) C. Jeffrey et Y. L. Chen ●

360092   Senecio spellenbergii T. M. Barkley = Packera spellenbergii (T. M. Barkley) C. Jeffrey ■☆

360093   Senecio sphaerocephalus Greene;球头千里光■☆

360094   Senecio spiraeifolius Thunb.;绣线菊叶千里光■☆

360095   Senecio spribillei W. A. Weber;斯普里千里光■☆

360096   Senecio squalidus L.;牛津千里光(混型千里光,污染千里光);Dirty Groundsel, Oxford and Cambridge Bush, Oxford Groundsel,Oxford Ragwort ■☆

360097   Senecio squalidus L. subsp. aurasiacus (Batt. et Trab.) Alexander;奥拉斯千里光■☆

360098   Senecio squalidus L. subsp. rupestris (Waldst. et Kit.) Greuter;岩生牛津千里光■☆

360099   Senecio squamosus Thunb. = Senecio pubigerus L. ■☆

360100   Senecio stanleyi Hauman;斯坦千里光;Tree Groundsel ■☆

360101   Senecio stanleyi Hauman = Dendrosenecio adnivalis (Stapf) E. B. Knox var. petiolatus (Hedberg) E. B. Knox ●☆

360102   Senecio stapeliiformis E. Phillips;豹皮花千里光(混型千里光,铁锡杖,肖犀角);Stapeliaform Groundsel ■☆

360103   Senecio stapeliiformis E. Phillips = Kleinia stapeliiformis (E. Phillips) Stapf ■☆

360104   Senecio stauntonii DC.;闽粤千里光(狭叶千里光);Min-Yue Groundsel ■

360105   Senecio steezii Bolle = Pericallis steetzii (Bolle) B. Nord. ■☆

360106   Senecio stenocephalus Maxim. = Ligularia stenocephala (Maxim.) Matsum. et Koidz. ■

360107   Senecio stenoglossus Franch. = Ligularia stenoglossa (Franch.) Hand. -Mazz. ■

360108   Senecio steudelii Sch. Bip. ex A. Rich.;斯托千里光■☆

360109   Senecio steudelii Sch. Bip. ex A. Rich. var. albido-tomentosa A. Rich. = Senecio ochrocarpus Oliv. et Hiern ■☆

360110   Senecio steudelii Sch. Bip. ex A. Rich. var. rosenianus Pax = Senecio steudelii Sch. Bip. ex A. Rich. ■☆

360111   Senecio steudelioides Sch. Bip.;拟斯托千里光■☆

360112   Senecio stipulatus Wall. ex DC. = Senecio scandens Buch. -Ham. ex D. Don ■

360113   Senecio stolonifer Cufod. = Tephroseris stolonifera (Cufod.) Holub ■

360114   Senecio stolzii Mattf. = Senecio inornatus DC. ■☆

360115   Senecio streptanthifolius Greene = Packera streptanthifolia (Greene) W. A. Weber et Á. Löve ■☆

360116   Senecio streptanthifolius Greene var. borealis (Torr. et A. Gray) J. F. Bain = Packera streptanthifolia (Greene) W. A. Weber et Á. Löve ■☆

360117   Senecio streptanthifolius Greene var. kluanei J. F. Bain = Packera streptanthifolia (Greene) W. A. Weber et Á. Löve ■☆

360118   Senecio streptanthifolius Greene var. laetiflorus (Greene) J. F. Bain = Packera streptanthifolia (Greene) W. A. Weber et Á. Löve ■☆

360119   Senecio streptanthifolius Greene var. oödes (Rydb.) J. F. Bain = Packera streptanthifolia (Greene) W. A. Weber et Á. Löve ■☆

360120   Senecio streptanthifolius Greene var. rubricaulis (Greene) J. F. Bain = Packera streptanthifolia (Greene) W. A. Weber et Á. Löve ■☆

360121   Senecio streptanthifolius Greene var. wallowensis J. F. Bain = Packera streptanthifolia (Greene) W. A. Weber et Á. Löve ■☆

360122   Senecio striatifolius DC.;纹叶千里光■☆

360123   Senecio striatus Thunb. = Osteospermum bidens Thunb. ■☆

360124   Senecio strictifolius Hiern;刚叶千里光■☆

360125   Senecio stuhlmannii Klatt = Solanecio cydoniifolius (O. Hoffm.) C. Jeffrey ■☆

360126   Senecio stygius Greene = Packera multilobata (Torr. et A. Gray) W. A. Weber et Á. Löve ■☆

360127   Senecio suaveolens (L.) Elliott = Hasteola suaveolens (L.) Pojark. ■☆

360128   Senecio subalatipetiolatus De Wild. = Gynura scandens O. Hoffm. ■☆

360129   Senecio subalpinus Koch;亚高山千里光■☆

360130   Senecio subcanescens (DC.) Compton;银灰千里光■☆

360131   Senecio subcarnosulus De Wild. = Senecio maranguensis O. Hoffm. ■☆

360132   Senecio subcoriaceus Schltr.;革质千里光■☆

360133   Senecio subcrassifolius De Wild. = Senecio hadiensis Forssk. ■☆

360134   Senecio subcuneatus Rydb. = Packera streptanthifolia (Greene) W. A. Weber et Á. Löve ■☆

360135   Senecio subdentatus (Bunge) Turcz. = Tephroseris subdentata (Bunge) Holub ■

360136   Senecio subdentatus (Bunge) Turcz. var. taitoensis (Hayata) Cufod. = Tephroseris taitoensis (Hayata) Holub ■

360137   Senecio subdentatus Ledeb.;近全缘千里光(疏齿千里光);Nearly Entire Groundsel,Nearlyentireleaf Groundsel ■

360138   Senecio subdentatus Ledeb. = Senecio glaucus L. subsp. coronopifolius (Desf.) Alexander ■☆

360139   Senecio subdentatus Ledeb. var. borealis (Herder) Cufod. = Tephroseris subdentata (Bunge) Holub ■

360140   Senecio subdentatus Ledeb. var. borealis Cufod. = Tephroseris subdentata (Bunge) Holub ■

360141   Senecio subdentatus Ledeb. var. glabellus (Turcz. ex DC.) Cufod. = Tephroseris praticola (Schischk. et Serg.) Holub ■

360142   Senecio subdentatus Ledeb. var. glabellus Cufod. = Tephroseris praticola (Schischk. et Serg.) Holub ■

360143   Senecio subdentatus Ledeb. var. pierotii (Miq.) Cufod. = Tephroseris pierotii (Miq.) Holub ■

360144   Senecio subdentatus Ledeb. var. polycephalus (Regel) Kitam.

= Tephroseris subdentata（Bunge）Holub ■

360145　Senecio subdentatus Ledeb. var. taitoensis（Hayata）Cufod. = Tephroseris taitoensis（Hayata）Holub ■

360146　Senecio subfloccosus Schischk. ;亚软千里光■☆

360147　Senecio subfrigidus Kom. ;亚硬千里光■☆

360148　Senecio submontanus Hilliard et B. L. Burtt;亚山生千里光■☆

360149　Senecio subnudus DC. = Packera subnuda（DC.）Trock et T. M. Barkley ☆

360150　Senecio subpetitianus Baker = Senecio syringifolius O. Hoffm. ■☆

360151　Senecio subradiatus（DC.）Sch. Bip. = Senecio cicatricosus Sch. Bip. ■☆

360152　Senecio subrosulatus Hand. -Mazz. = Sinosenecio subrosulatus（Hand. -Mazz.）B. Nord. ■

360153　Senecio subrubriflorus O. Hoffm. ;浅红花千里光■☆

360154　Senecio subscandens Hochst. ex A. Rich. = Solanecio angulatus（Vahl）C. Jeffrey ■☆

360155　Senecio subscaposus Kom. ;亚花茎千里光■☆

360156　Senecio subsessilis Oliv. et Hiern;近无柄千里光■☆

360157　Senecio subsinuatus DC. ;深波千里光■☆

360158　Senecio subspicatus Bureau et Franch. = Ligularia subspicata（Bureau et Franch.）Hand. -Mazz. ■

360159　Senecio subulatifolius G. D. Rowley = Kleinia picticaulis（P. R. O. Bally）C. Jeffrey ■☆

360160　Senecio succisifolius Kom. ;肉花千里光■☆

360161　Senecio succulentus DC. = Senecio sarcoides C. Jeffrey ■☆

360162　Senecio sukaczevii Schischk. ;苏卡千里光■☆

360163　Senecio suksdorfii Greenm. = Packera streptanthifolia（Greene）W. A. Weber et Á. Löve ☆

360164　Senecio sulcatus DC. = Senecio crispus Thunb. ☆

360165　Senecio sulcicalyx Baker;沟萼千里光■☆

360166　Senecio sumneviczii Schischk. et Serg. ;苏穆千里光■☆

360167　Senecio sungpanensis Hand. -Mazz. = Sinosenecio sungpanensis（Hand. -Mazz.）B. Nord. ■

360168　Senecio superbus De Wild. et Muschl. = Kleinia abyssinica（A. Rich.）A. Berger ■☆

360169　Senecio surculosus MacOwan;木质千里光■☆

360170　Senecio swaziensis Compton = Senecio glaberrimus DC. ■☆

360171　Senecio sycephalus S. Moore = Senecio transmarinus S. Moore var. sycephalus（S. Moore）Hedberg ■☆

360172　Senecio sylvaticus L. ;林生千里光（林千里光）;Groundsel Heath, Heath Groundsel, Mountain Groundsel, Wood Groundsel, Woodland Groundsel, Woodland Ragwort ■☆

360173　Senecio sympodialis R. E. Fr. = Senecio jacksonii S. Moore ■☆

360174　Senecio syringifolius O. Hoffm. ;丁香叶千里光■☆

360175　Senecio tabulicola Baker;平台千里光■☆

360176　Senecio taitoensis Hayata = Tephroseris taitoensis（Hayata）Holub ■

360177　Senecio taitungensis S. S. Ying;台东黄菀;Taidong Groundsel ■

360178　Senecio taiwanensis Hayata = Senecio nemorensis L. ■

360179　Senecio taliensis Franch. = Parasenecio taliensis（Franch.）Y. L. Chen ■

360180　Senecio talongensis Franch. = Synotis erythropappa（Bureau et Franch.）C. Jeffrey et Y. L. Chen ■

360181　Senecio tamoides DC. ;塔谟千里光（假常春藤）■☆

360182　Senecio tampicanus DC. = Packera tampicana（DC.）C. Jeffrey ■☆

360183　Senecio tanacetoides Kunth et C. D. Bouché = Senecio royleanus DC. ■

360184　Senecio tanacetoides Sond. ex Harv. = Senecio tanacetopsis Hilliard ■☆

360185　Senecio tanacetopsis Hilliard;艾菊千里光■☆

360186　Senecio tanganyikensis R. E. Fr. = Mikaniopsis tanganyikensis（R. E. Fr.）Milne-Redh. ■☆

360187　Senecio tanggutica Maxim. = Sinacalia tangutica（Maxim.）B. Nord. ■

360188　Senecio tanzaniensis Cufod. = Senecio deltoideus Less. ■☆

360189　Senecio taquetii H. Lév. et Vaniot = Ligularia mongolica（Turcz.）DC. ■

360190　Senecio taraxacoides（A. Gray）Greene;蒲公英千里光■☆

360191　Senecio tarokoensis C. I. Peng;太鲁阁千里光;Taroko Groundsel ■

360192　Senecio tashiroi Hayata = Tephroseris kirilowii（Turcz. ex DC.）Holub ■

360193　Senecio tatsienensis Bureau et Franch. = Ligularia pleurocaulis（Franch.）Hand. -Mazz. ■

360194　Senecio tatsienensis Bureau et Franch. = Parasenecio roborowskii（Maxim.）Y. L. Chen ■

360195　Senecio tatsienensis Franch. = Ligularia pleurocaulis（Franch.）Hand. -Mazz. ■

360196　Senecio tedliei Oliv. et Hiern = Mikaniopsis tedliei（Oliv. et Hiern）C. D. Adams ■☆

360197　Senecio teixeirae Torre;特谢拉千里光■☆

360198　Senecio telekii（Schweinf.）O. Hoffm. ;泰莱吉千里光■☆

360199　Senecio telmatophyllus O. Hoffm. = Crassocephalum uvens（Hiern）S. Moore ■☆

360200　Senecio tenellulus S. Moore = Emilia tenellula（S. Moore）C. Jeffrey ■☆

360201　Senecio tenellus DC. ;弱小千里光■☆

360202　Senecio tener O. Hoffm. = Emilia tenera（O. Hoffm.）C. Jeffrey ■☆

360203　Senecio teneriffae Sch. Bip. ;特纳千里光■☆

360204　Senecio tenuicaulis Muschl. ;细茎千里光■☆

360205　Senecio tenuifolius Jacq. ;细叶千里光■☆

360206　Senecio tenuilobus DC. = Senecio umbellatus L. ■☆

360207　Senecio tenuipes Franch. = Ligularia tenuipes（Franch.）Diels ■

360208　Senecio tenuiscapus Bojer ex DC. = Senecio simplicissimus Bojer ex DC. ■☆

360209　Senecio tessmannii Mattf. = Emilia tessmannii（Mattf.）C. Jeffrey ■☆

360210　Senecio tetranthus DC. = Synotis tetrantha（DC.）C. Jeffrey et Y. L. Chen ■

360211　Senecio thamathuensis Hilliard;南非千里光■☆

360212　Senecio theodori K. Afzel. = Senecio schweinfurthii O. Hoffm. ■☆

360213　Senecio thermarum Bolus = Emilia transvaalensis（Bolus）C. Jeffrey ■☆

360214　Senecio thianschanicus Regel et Schmalh. ;天山千里光;Tianschan Mountain Groundsel, Tianshan Groundsel ■

360215　Senecio thomsianus Muschl. = Senecio subsessilis Oliv. et Hiern ■☆

360216　Senecio thomsonii C. B. Clarke = Ligularia thomsonii（C. B. Clarke）Pojark. ■

360217　Senecio thornberi Greenm. = Packera multilobata（Torr. et A.

Gray) W. A. Weber et Á. Löve ■☆

360218　Senecio thunbergii Harv. ;通贝里千里光■☆

360219　Senecio thyrsoideus DC. = Senecio barbatus DC. ■☆

360220　Senecio tibeticus Hook. f. ;西藏千里光;Xizang Groundsel ■

360221　Senecio tibeticus Hook. f. = Senecio albopurpureus Kitam. ■

360222　Senecio tnligulatus Buch. -Ham. ex D. Don = Synotis triligulata (Buch. -Ham. ex D. Don) C. Jeffrey et Y. L. Chen ■

360223　Senecio toiyabensis S. L. Welsh et Goodrich = Senecio spartioides Torr. et Gray ■☆

360224　Senecio tomentosus Michx. = Packera tomentosa (Michx.) C. Jeffrey ■☆

360225　Senecio tomentosus Salisb. = Senecio cinerascens Aiton ■☆

360226　Senecio tongoensis De Wild. = Senecio hochstetteri Sch. Bip. ex A. Rich. ■☆

360227　Senecio tongolensis Franch. = Ligularia tongolensis (Franch.) Hand. -Mazz. ■

360228　Senecio tongtchuanensis H. Lév. = Ligularia lapathifolia (Franch.) Hand. -Mazz. ■

360229　Senecio torticaulis Merxm. ;曲茎千里光■☆

360230　Senecio tortuosus DC. ;扭曲千里光■☆

360231　Senecio toumeyi Greene = Packera neomexicana (A. Gray) W. A. Weber et Á. Löve var. toumeyi (Greene) Trock et T. M. Barkley ■☆

360232　Senecio tozanensis Hayata = Senecio nemorensis L. ■

360233　Senecio trachylaenus Harv. ;糙被千里光■☆

360234　Senecio trachyphyllus Schltr. ;粗叶千里光■☆

360235　Senecio tracyi Rydb. = Packera crocata (Rydb.) W. A. Weber et Á. Löve ■☆

360236　Senecio transmarinus S. Moore;外海千里光■☆

360237　Senecio transmarinus S. Moore var. major C. Jeffrey;大外海千里光■☆

360238　Senecio transmarinus S. Moore var. sycephalus (S. Moore) Hedberg;合头外海千里光■☆

360239　Senecio transvaalensis Bolus = Emilia transvaalensis (Bolus) C. Jeffrey ■☆

360240　Senecio triangularis Hook. ;窄三角千里光;Arrowhead Groundsel, Arrow-leaf Groundsel ■☆

360241　Senecio triangularis Hook. var. angustifolius G. N. Jones = Senecio triangularis Hook. ■☆

360242　Senecio trianthemos O. Hoffm. = Lopholaena trianthema (O. Hoffm.) Burtt ■☆

360243　Senecio trichopterygius Muschl. = Senecio subsessilis Oliv. et Hiern ■☆

360244　Senecio tricuspis Franch. ;三尖千里光(三角千里光);Deltoid-leaf Groundsel, Triangule Groundsel ■

360245　Senecio tridenticulatus Rydb. = Packera tridenticulata (Rydb.) W. A. Weber et Á. Löve ■☆

360246　Senecio trifurcatus Klatt = Senecio triodontiphyllus C. Jeffrey ■☆

360247　Senecio triligulata Buch. -Ham. ex D. Don = Synotis triligulata (Buch. -Ham. ex D. Don) C. Jeffrey et Y. L. Chen ■

360248　Senecio trilobus L. ;三裂千里光■☆

360249　Senecio trinervius C. C. Chang = Sinosenecio trinervius (C. C. Chang) B. Nord. ■

360250　Senecio triodontiphyllus C. Jeffrey;三齿叶千里光■☆

360251　Senecio triplinervius DC. ;三脉千里光■☆

360252　Senecio tropaeolifolius MacOwan ex F. Muell. ;旱金莲叶千里光■☆

360253　Senecio tropaeolifolius O. Hoffm. = Senecio milanjianus S. Moore ■☆

360254　Senecio tsangchanensis Franch. = Ligularia tsangchanensis (Franch.) Hand. -Mazz. ■

360255　Senecio tsaratananensis Humbert;察拉塔纳纳千里光■☆

360256　Senecio tshabaensis De Wild. = Senecio caudatus DC. ■☆

360257　Senecio tshitirungensis De Wild. = Senecio hochstetteri Sch. Bip. ex A. Rich. ■☆

360258　Senecio tsoongianus Y. Ling = Synotis cappa (Buch. -Ham. ex D. Don) C. Jeffrey et Y. L. Chen ■

360259　Senecio tuberivagus W. W. Sm. = Sinacalia davidii (Franch.) H. Koyama ■

360260　Senecio tuberosus (DC.) Harv. ;块状千里光■☆

360261　Senecio tuberosus Sch. Bip. ex A. Rich. = Solanecio tuberosus (Sch. Bip. ex A. Rich.) C. Jeffrey ■☆

360262　Senecio tuberosus Sch. Bip. ex A. Rich. var. pubescens (Mesfin) Rowley = Solanecio tuberosus (Sch. Bip. ex A. Rich.) C. Jeffrey var. pubescens Mesfin ■☆

360263　Senecio tubicaulis Mansf. = Tephroseris palustris (L.) Fourr. ■

360264　Senecio tugelensis J. M. Wood et M. S. Evans;图盖拉千里光■☆

360265　Senecio tundricola Tolm. ;冻原千里光■☆

360266　Senecio tundricola Tolm. = Tephroseris tundricola (Tolm.) Holub ■☆

360267　Senecio tundricola Tolm. subsp. lindstroemii (Ostenf.) Korobkov = Tephroseris lindstroemii (Ostenf.) Á Löve et D. Löve ■☆

360268　Senecio turczaninowii DC. = Senecio nemorensis L. ■

360269　Senecio turczaninowii DC. = Tephroseris turczaninowii (DC.) Holub ■

360270　Senecio turkestanicus C. Winkl. = Ligularia songarica (Fisch.) Y. Ling ■

360271　Senecio tussilagineus (Burm. f.) Kuntze = Farfugium japonicum (L.) Kitam. ■

360272　Senecio tussilaginis (L'Hér.) DC. = Pericallis tussilaginis (L'Hér.) D. Don ■☆

360273　Senecio tweedyi Rydb. = Packera paupercula (Michx.) Á. Löve et D. Löve ■☆

360274　Senecio tysonii MacOwan;泰森千里光●☆

360275　Senecio uhligii Muschl. = Senecio meyeri-johannis Engl. ■☆

360276　Senecio uintahensis (A. Nelson) Greenm. = Packera multilobata (Torr. et A. Gray) W. A. Weber et Á. Löve ■☆

360277　Senecio ukambensis O. Hoffm. = Emilia ukambensis (O. Hoffm.) C. Jeffrey ■☆

360278　Senecio ukingensis O. Hoffm. = Emilia ukingensis (O. Hoffm.) C. Jeffrey ■☆

360279　Senecio ulopterus Thell. ;卷翅千里光■☆

360280　Senecio umbellatus L. ;小伞千里光■☆

360281　Senecio umbrosus Waldst. et Kit. ;阴地千里光■☆

360282　Senecio umgeniensis Thell. ;乌姆加尼千里光■☆

360283　Senecio unionis Sch. Bip. ex A. Rich. ;单千里光■☆

360284　Senecio urophyllus Conrath ;尾叶千里光■☆

360285　Senecio urundensis S. Moore;乌隆迪千里光■☆

360286　Senecio usambarensis Muschl. = Mikaniopsis usambarensis (Muschl.) Milne-Redh. ■☆

360287　Senecio ussanguensis O. Hoffm. = Lopholaena ussanguensis (O. Hoffm.) S. Moore ■☆

360288　Senecio ustulatus DC. = Senecio littoreus Thunb. ■☆

360289　Senecio uvens Hiern = Crassocephalum uvens（Hiern）S. Moore ■☆

360290　Senecio valerianifolius Wolf ex Rchb. = Erechtites valerianifolius（Wolf ex Rchb.）DC. ■

360291　Senecio valeriifolius Wolf. = Erechtites valerianifolius（Wolf ex Rchb.）DC. ■

360292　Senecio vaniotii H. Lév. = Synotis vaniotii（H. Lév.）C. Jeffrey et Y. L. Chen ■

360293　Senecio variabilis Sch. Bip. = Senecio erubescens Aiton ■☆

360294　Senecio variifolius DC. = Senecio lyratus Forssk. ■☆

360295　Senecio variifolius DC. var. subcanescens = Senecio subcanescens（DC.）Compton ■☆

360296　Senecio variostipellatus De Wild. = Gynura scandens O. Hoffm. ■☆

360297　Senecio vaseyi Greenm. = Senecio integerrimus Nutt. var. exaltatus（Nutt.）Cronquist ■☆

360298　Senecio veitchianus Hemsl. = Ligularia veitchiana（Hemsl.）Greenm. ■

360299　Senecio vellereus Franch. = Ligularia vellerea（Franch.）Hand. -Mazz. ■

360300　Senecio velutinus H. Lév. et Vaniot = Blumea lacera（Burm. f.）DC. ■

360301　Senecio venosus Harv. ；凸脉千里光■☆

360302　Senecio venustus Aiton = Senecio grandiflorus P. J. Bergius ■☆

360303　Senecio verbascifolius Burm. f. ；毛蕊花千里光■☆

360304　Senecio verbenifolius Jacq. = Senecio aegyptius L. var. discoideus Boiss. ■☆

360305　Senecio verdoorniae R. A. Dyer = Senecio lydenburgensis Hutch. et Burtt Davy ■☆

360306　Senecio vernalis Waldst. et Kit. ；春千里光；Eastern Groundsel，Spring Groundsel ■☆

360307　Senecio vernicosus Baker = Senecio neobakeri Humbert ■☆

360308　Senecio vernonioides Sch. Bip. = Senecio erubescens Aiton ■☆

360309　Senecio versicolor Hiern = Senecio polyodon DC. ■☆

360310　Senecio vespertilo Franch. = Parasenecio vespertilo（Franch.）Y. L. Chen ■

360311　Senecio vestitus（Thunb.）P. J. Bergius；包被千里光■☆

360312　Senecio vicinus S. Moore；相近千里光■☆

360313　Senecio villiferus Franch. = Sinosenecio villiferus（Franch.）B. Nord. ■

360314　Senecio viminalis Bremek. ；细枝千里光■☆

360315　Senecio virgaureus Maxim. = Ligularia virgaurea（Maxim.）Mattf. ex Rehder et Kobuski ■

360316　Senecio viridiflavus Hand. -Mazz. = Synotis erythropappa（Bureau et Franch.）C. Jeffrey et Y. L. Chen ■

360317　Senecio viridiflorus Hutch. = Emilia marlothiana（O. Hoffm.）C. Jeffrey ■☆

360318　Senecio viridis Phil. ；绿花千里光；Bristly Foxtail ■☆

360319　Senecio viscidulus Compton；微黏千里光■☆

360320　Senecio viscidus N. E. Br. = Senecio subrubriflorus O. Hoffm. ■☆

360321　Senecio viscosus L. ；黏质千里光（黏毛千里光，黏千里光）；Fetid Groundsel，Sticky Groundsel，Sticky Ragwort，Stickyhair Groundsel，Stinking Cotton，Stinking Groundsel ■☆

360322　Senecio vitellinoides Merxm. ；蛋黄色千里光■☆

360323　Senecio vittarifolius Bojer ex DC. ；圈叶千里光■☆

360324　Senecio volkameri Sch. Bip. = Senecio arenarius Thunb. ■☆

360325　Senecio volkensii O. Hoffm. = Senecio telekii（Schweinf.）O. Hoffm. ■☆

360326　Senecio volutus（Baker）Humbert；旋卷千里光■☆

360327　Senecio vulgari-humilis Batt. et Trab. = Senecio vulgaris L. ■

360328　Senecio vulgaris L. ；欧洲千里光（欧洲狗舌草，欧洲黄菀，普通千里光）；Ascension，Birdseed，Canary Food，Canary Seed，Chickenweed，Chinchone，Common Groundsel，Europe Groundsel，Grinning Swallow，Grinsel，Ground Glutton，Groundie Swallow，Groundie-swallow，Groundsel，Groundswell，Groundwill，Grundy Swallow，Grundy-swallow，Grunnishule，Grunnistule，Grunsil，Lady's Fingers，Lie-abed，Little，Little Lie-abed，Old-man-in-the-spring，Ragwort，Sen Cion，Senshon，Sension，Sention，Simpson，Sinslon，Swallow Grundy，Swichen，Wattery Drum，Wattery Drums，Yellow Head ■

360329　Senecio vulgaris L. = Senecio dubitabilis C. Jeffrey et Y. L. Chen ■

360330　Senecio vulgaris L. var. dubius Trautv. = Senecio dubitabilis C. Jeffrey et Y. L. Chen ■

360331　Senecio vulgaris L. var. teneriffae（Sch. Bip.）Ball = Senecio teneriffae Sch. Bip. ■☆

360332　Senecio walkeri sensu Rehder = Cissampelopsis spelaeicola（Vaniot）C. Jeffrey et Y. L. Chen ●

360333　Senecio wallichii DC. = Synotis wallichii（DC.）C. Jeffrey et Y. L. Chen ■

360334　Senecio walteri Sch. Bip. = Arnoglossum ovatum（Walter）H. Rob. ■☆

360335　Senecio wardii Greene = Packera streptanthifolia（Greene）W. A. Weber et Á. Löve ■☆

360336　Senecio warnockii Shinners；瓦氏千里光●☆

360337　Senecio waterbergensis S. Moore；沃特千里光■☆

360338　Senecio webbii（Sch. Bip.）Christ = Pericallis webbii（Sch. Bip.）Bolle ■☆

360339　Senecio websteri Greenm. = Senecio neowebsteri S. F. Blake ■☆

360340　Senecio welwitschii O. Hoffm. = Kleinia fulgens Hook. f. ■☆

360341　Senecio werneriifolius（A. Gray）A. Gray = Packera werneriifolia（A. Gray）W. A. Weber et Á. Löve ■☆

360342　Senecio whippleanus A. Gray = Senecio integerrimus Nutt. var. major（A. Gray）Cronquist ■☆

360343　Senecio wightianus DC. ex Wight = Senecio scandens Buch. -Ham. ex D. Don ■

360344　Senecio wightii（DC. ex Wight）Benth. ex C. B. Clarke；弯齿千里光（水泽兰，岩生千里光）；Wight Groundsel ■

360345　Senecio wilsonianus Hemsl. = Ligularia wilsoniana（Hemsl.）Greenm. ■

360346　Senecio windhoekensis Merxm. ；温得和千里光■☆

360347　Senecio winklerianus Hand. -Mazz. = Sinosenecio euosmus（Hand. -Mazz.）B. Nord. ■

360348　Senecio wittebergensis Compton；维特伯格千里光■☆

360349　Senecio wollastonii S. Moore = Senecio subsessilis Oliv. et Hiern ■☆

360350　Senecio wootonii Greene；伍顿千里光■☆

360351　Senecio wrightii Greenm. = Packera cynthioides（Greene）W. A. Weber et Á. Löve ☆

360352　Senecio xantholeucus Hand. -Mazz. = Synotis xantholeuca（Hand. -Mazz.）C. Jeffrey et Y. L. Chen ■

360353　Senecio xenostylus O. Hoffm. ；花柱千里光■☆

360354　Senecio yakoensis Jeffrey = Synotis yakoensis（Jeffrey ex Diels）

C. Jeffrey et Y. L. Chen ■

360355　Senecio yalungensis Hand. -Mazz. = Cissampelopsis spelaeicola（Vaniot）C. Jeffrey et Y. L. Chen ●

360356　Senecio yesoensis Franch. = Ligularia hodgsonii Hook. f. ■

360357　Senecio yesoensis Franch. var. creniferus Franch. = Ligularia hodgsonii Hook. f. ■

360358　Senecio yesoensis Franch. var. sutchuenensis Franch. = Ligularia hodgsonii Hook. f. ■

360359　Senecio yukonensis A. E. Porsild = Tephroseris yukonensis（A. E. Porsild）Holub ■☆

360360　Senecio yungningensis Hand. -Mazz. ;永宁千里光;Yongning Groundsel,Yungning Groundsel ■

360361　Senecio yunnanensis Franch. = Ligularia yunnanensis（Franch.）C. C. Chang ■

360362　Senecio yunnanensis Franch. = Senecio nudicaulis Buch. -Ham. ex D. Don ■

360363　Senecio zeyheri Turcz. = Senecio oxyodontus DC. ■☆

360364　Senecioides Post et Kuntze = Vernonia Schreb.（保留属名）●■

360365　Senecionaceae Bercht. et J. Presl = Asteraceae Bercht. et J. Presl（保留科名）●■

360366　Senecionaceae Bercht. et J. Presl = Compositae Giseke（保留科名）●■

360367　Senecionaceae Bessey = Asteraceae Bercht. et J. Presl（保留科名）●■

360368　Senecionaceae Bessey = Compositae Giseke（保留科名）●■

360369　Senecionaceae Bessey = Senecionidaceae Bessey ●■

360370　Senecionaceae Spenn. = Asteraceae Bercht. et J. Presl（保留科名）●■

360371　Senecionaceae Spenn. = Compositae Giseke（保留科名）●■

360372　Senecionidaceae Bessey = Asteraceae Bercht. et J. Presl（保留科名）●■

360373　Senecionidaceae Bessey;千里光科●■

360374　Seneciuneulus Opiz = Senecio L. ■●

360375　Senefeldera Mart.（1841）;塞内大戟属■☆

360376　Senefeldera angustifolia Klotzsch;窄叶塞内大戟■☆

360377　Senefeldera latifolia Klotzsch;宽叶塞内大戟■☆

360378　Senefeldera macrophylla Ducke;大叶塞内大戟■☆

360379　Senefeldera multiflora Mart. ;多花塞内大戟■☆

360380　Senefeldera nitida Croizat;亮塞内大戟■☆

360381　Senefeldera verticillata（Vell.）Croizat;轮生塞内大戟■☆

360382　Senefelderopsis Steyerm.（1951）;拟塞内大戟属■☆

360383　Senefelderopsis chiribiquetensis（R. E. Schult. et Croizat）Steyerm. ;拟塞内大戟■☆

360384　Senega（DC.）Spach = Polygala L. ●■

360385　Senega（DC.）Spach = Senegaria Raf. ●■

360386　Senega Spach = Polygala L. ●■

360387　Senega Spach = Senegaria Raf. ●■

360388　Senegalia Raf. = Acacia Mill.（保留属名）●■

360389　Senegaria Raf. = Polygala L. ●■

360390　Seneico Hill = Senecio L. ■●

360391　Senftenbergia Klotzsch et H. Karst. ex Klotzsch = Langsdorffia Mart. ■☆

360392　Senghasia Szlach.（2003）;森哈斯兰属■☆

360393　Senghasiella Szlach. = Habenaria Willd. ■

360394　Senghasiella glaucifolia（Bureau et Franch.）Szlach. = Habenaria glaucifolia Bureau et Franch. ■

360395　Senisetum Honda = Agrostis L.（保留属名）■

360396　Senites Adans. = Zeugites P. Browne ■☆

360397　Senkebergia Neck. = Mendoncia Vell. ex Vand. ●☆

360398　Senkenbergia Rchb. = Lepidium L. ■

360399　Senkenbergia Rchb. = Senckenbergia P. Gaertn. , B. Mey. et Scherb. ■

360400　Senkenbergia S. Schauer = Cyphomeris Standl. ■☆

360401　Senkenbergia coulteri Hook. f. = Boerhavia coulteri（Hook. f.）S. Watson ■☆

360402　Senkenbergia crassifolia Standl. = Cyphomeris crassifolia（Standl.）Standl. ■☆

360403　Senkenbergia gypsophiloides（M. Martens et Galeotti）Benth. et Hook. f. = Cyphomeris gypsophiloides（M. Martens et Galeotti）Standl. ■☆

360404　Senna Mill.（1754）;番泻决明属（番泻属,决明属,山扁豆属,异决明属）;Weedy ●■

360405　Senna Mill. = Cassia L.（保留属名）●■

360406　Senna Tourn. ex Mill. = Cassia L.（保留属名）●■

360407　Senna acutifolia（Delile）Batka = Cassia senna L. ●

360408　Senna alata（L.）Irwin et Barneby = Cassia alata L. ●■

360409　Senna alata（L.）Roxb. ;翅荚决明（翅叶槐,对叶豆,非洲木通,具翅异决明,翼柄决明,翼叶决明,有翅决明）;Candlestick Senna, Crawcraw, Craw-craw, French Guava, Ringworm Cassia, Ringworm Cassia Bark, Ringworm Senna, Ringworm Shrub, Winged Senna, Wingpod Senna, Wing-podded Senna, Yellow Craw ●

360410　Senna alata（L.）Roxb. = Cassia alata L. ●■

360411　Senna alexandrina Mill. = Cassia angustifolia Wahlenb. ●

360412　Senna alexandrina Mill. = Cassia senna L. ●

360413　Senna alexandrina Mill. var. obtusata（Brenan）Lock = Cassia senna L. var. obtusata Brenan ●☆

360414　Senna angustifolia（Vahl）Batka = Senna alexandrina Mill. ●

360415　Senna ankaranensis Du Puy et R. Rabev. ;安卡兰异决明●☆

360416　Senna anthoxantha（Capuron）Du Puy;黄花异决明●☆

360417　Senna apiculata（M. Martens et Galeotti）H. S. Irwin et Barneby;尖头异决明●☆

360418　Senna artemisioides（DC.）Randell = Senna artemisioides（Gaudich. ex DC.）Randell ●☆

360419　Senna artemisioides（Gaudich. ex DC.）Randell;羽毛叶异决明（蒿状决明）;Feathery Cassia, Silver Cassia, Silver Senna, Wormwood Cassia ●☆

360420　Senna artemisioides（Gaudich. ex DC.）Randell subsp. filifolia Randell;狭羽毛叶异决明●☆

360421　Senna artemisioides（Gaudich. ex DC.）Randell subsp. oligophylla（F. Muell.）Randell;灰岩异决明;Limestone Cassia, Oval-leaf Cassia ●☆

360422　Senna artemisioides（Gaudich. ex DC.）Randell subsp. sturtii（R. Br.）Randell;肆图特异决明;Dense Senna, Grey Desert Senna ●☆

360423　Senna auriculata（L.）Roxb. = Cassia auriculata L. ●

360424　Senna baccarinii（Chiov.）Lock = Cassia beccarinii Chiov. ●☆

360425　Senna bauhinioides（A. Gray）H. S. Irwin et Barneby;羊蹄甲异决明;Twinleaf Cassia ●☆

360426　Senna bicapsularis（L.）Roxb. ;双荚决明（腊肠仔树,双荚槐）;Bicapsular Senna, Bipod Senna, Butterfly Bush, Money Bush, Shrub Senna, Winter Cassia ●

360427　Senna bicapsularis（L.）Roxb. = Cassia bicapsularis L. ●

360428　Senna bosseri Du Puy et R. Rabev. ;博瑟异决明●☆

360429　Senna corymbosa（Lam.）H. S. Irwin et Barneby;尖叶黄槐（巴

西丁香树,尖叶槐,尖叶决明);Argentina Senna,Flowering Cassia, Flowery Senna ●☆

360430　Senna covesii（A. Gray）H. S. Irwin et Barneby;沙地异决明; Desert Senna ●☆

360431　Senna didymobotrya（Fresen.）H. S. Irwin et Barneby;直穗异决明(长穗决明);African Senna,Golden Wonder,Longspike Senna, Popcorn Senna ●

360432　Senna dimidiata Roxb. = Cassia hochstetteri Ghesq. ●☆

360433　Senna floribunda（Cav.）Irwin et Barneby = Cassia floribunda Cav. ●

360434　Senna fruticosa（Mill.）H. S. Irwin et Barneby;大叶决明; Bigleaf Senna,Shrubby Senna ●

360435　Senna fruticosa（Mill.）H. S. Irwin et Barneby = Cassia fruticosa Mill. ●

360436　Senna gossweileri（Baker f.）Lock = Cassia gossweileri Baker f. ●☆

360437　Senna hebecarpa（Fernald）H. S. Irwin et Barneby;美国异决明;American Wild Sensitive-plant,Northern Wild Senna,Wild Senna ●☆

360438　Senna hebecarpa（Fernald）H. S. Irwin et Barneby var. longipila（E. L. Braun）C. F. Reed = Senna hebecarpa（Fernald）H. S. Irwin et Barneby ●☆

360439　Senna hirsuta（L.）H. S. Irwin et Barneby;毛荚决明(毛决明);Hairpod Senna,Hirsute Senna ●

360440　Senna hirsuta（L.）H. S. Irwin et Barneby = Cassia hirsuta L. ●

360441　Senna holosericea（Fresen.）Greuter = Cassia holosericea Fresen. ●☆

360442　Senna hookeriana Batka = Cassia adenensis Benth. ●☆

360443　Senna humifusa（Brenan）Lock = Cassia humifusa Brenan ●☆

360444　Senna italica Mill.;蓝绿叶异决明(意大利决明);Dog Senna, Italian Senna,Port Royal Senna,Senna,Spanish Senna ●☆

360445　Senna italica Mill. = Cassia italica（Mill.）F. W. Andréws ●☆

360446　Senna italica Mill. subsp. arachoides（Burch.）Lock;蛛网异决明●☆

360447　Senna italica Mill. subsp. micrantha（Brenan）Lock;小花蓝绿叶异决明(小花意大利决明)●☆

360448　Senna lactea（Vatke）Du Puy;乳白异决明●☆

360449　Senna leandrii（Ghesq.）Du Puy;利安异决明●☆

360450　Senna ligustrina（L.）H. S. Irwin et Barneby;女贞异决明(女贞番泻)●☆

360451　Senna lindheimerana（Scheele）H. S. Irwin et Barneby;林氏异决明;Lindheimer Senna,Velvet Leaf Senna ●☆

360452　Senna longiracemosa（Vatke）Lock = Cassia longiracemosa Vatke ●☆

360453　Senna marilandica（L.）Link;马里兰德异决明(马里兰德决明,马里兰番泻);American Senna,Maryland Senna,Partridge Pea, Partridge-pea,Southern Wild Senna,Wild Senna,Wild Shower ●☆

360454　Senna meridionalis（R. Vig.）Du Puy;南方决明 ●☆

360455　Senna multiglandulosa（Jacq.）H. S. Irwin et Barneby;多腺异决明;Glandular Senna ●☆

360456　Senna multiglandulosa（Jacq.）H. S. Irwin et Barneby = Cassia multiglandulosa Jacq. ●☆

360457　Senna multijuga（Rich.）H. S. Irwin et Barneby;密叶决明(多对铁刀苏木,密叶黄槐);Denseleaf Senna,Dense-leaved Senna, Leaf Cassia ●

360458　Senna multijuga（Rich.）H. S. Irwin et Barneby = Cassia multijuga Rich. ●

360459　Senna nemophila ?;澳洲异决明;Australian Senna,Desert Cassia ●☆

360460　Senna nomame（Siebold）T. Chen = Cassia nomame（Siebold）Kitag. ●

360461　Senna obtusifolia（L.）H. S. Irwin et Barneby;倒卵叶决明(钝叶决明,钝叶山扁豆,钝叶异决明,决明);Coffee Weed,Coffee-weed,Italian Senna,Kawal,Sicklepod,Sickle-pod ●☆

360462　Senna occidentalis（L.）Irwin et Barneby = Cassia occidentalis L. ●

360463　Senna occidentalis（L.）Link;望江南(草决明,大更药,大羊角菜,大夜明,毒扁豆,饭匙倩草,风寒豆,凤凰花草,狗屎豆,喉百草,槐豆,黄豇豆,假槐花,假决明,江南豆,金豆子,金花豹子,金角儿,金角子,决明,黎草,黎茶,山咖啡,山绿豆,蛇灭门草,石决明,石芙明,水瓜豆,头晕菜,望江南决明,羊角豆,野扁豆,野鸡子豆,夜关门,猪骨棉);Coffee Senna,Coffee Weed,Coffeesenna, Coffeeweed,French Wild Guava,Mogdad Coffee,Negro Coffee, Stinking Bush,Stinking Pea,Stinking Weed,Stinkweed,Styptic Weed,Wild Coffee,Wild French Guava ●

360464　Senna occidentalis（L.）Link = Cassia occidentalis L. ●

360465　Senna occidentalis（L.）Roxb. = Cassia occidentalis L. ●

360466　Senna odorata（Morris）Randell;香决明(甜香异决明)●☆

360467　Senna pendula（Humb. et Bonpl. ex Willd.）H. S. Irwin et Barneby;攀缘决明;Climbing Cassia,Valamuerto ●☆

360468　Senna pendula（Willd.）H. S. Irwin et Barneby = Cassia pendula Humb. et Bonpl. ex Willd. ●☆

360469　Senna pendula（Willd.）H. S. Irwin et Barneby var. advena（Vogel）H. S. Irwin et Barneby;外来异决明;Valamuerto ●☆

360470　Senna pendula（Willd.）H. S. Irwin et Barneby var. glabrata（Vogel）H. S. Irwin et Barneby = Cassia pendula Humb. et Bonpl. ex Willd. ●☆

360471　Senna pendula（Willd.）H. S. Irwin et Barneby var. glabrata（Vogel）H. S. Irwin et Barneby;光滑攀缘异决明;Christmas Cassia, Christmas Senna,Climbing Cassia,Valamuerto ●☆

360472　Senna perrieri（R. Vig.）Du Puy;佩里耶异决明●☆

360473　Senna petersiana（Bolle）Lock;彼得斯异决明●☆

360474　Senna phyllodinea（R. Br.）Symon;叶状柄异决明;Silver Cassia,Silver Leaf Cassia,Silvery Cassia ●☆

360475　Senna podocarpa（Guillaumin et Perr.）Lock = Cassia podocarpa Guillaumin et Perr. ●☆

360476　Senna polyantha（Collad.）H. S. Irwin et Barneby;硬枝异决明●☆

360477　Senna roemeriana（Scheele）H. S. Irwin et Barneby;勒默尔异决明;Twoleaf Cassia ●☆

360478　Senna ruspolii（Chiov.）Lock = Cassia ruspolii Chiov. ●☆

360479　Senna sensitiva Roxb. = Cassia mimosoides L. ●

360480　Senna septemtrionalis（Viv.）H. S. Irwin et Barneby = Cassia septemtrionalis Viv. ●

360481　Senna siamea（Lam.）H. S. Irwin et Barneby;铁刀木(挨刀树,黑心树);Bombay Blackwood,Kassod Tree,Minjiri,Rosewood, Siam Senna,Siamese Senna ●

360482　Senna siamea（Lam.）H. S. Irwin et Barneby = Cassia siamea Lam. ●

360483　Senna sophera（L.）Roxb.;槐叶决明(茶花儿,江南槐,茳芒,茳芒决明,决明子,苦参类决明,望江南,野苦参子);Inflatable-fruited Senna,Inflatedfruit Senna ●

360484　Senna sophera（L.）Roxb. = Cassia sophera L. ●

360485　Senna spectabilis（DC.）H. S. Irwin et Barneby;美丽决明(绚

丽决明）；Beautiful Senna, Pretty Senna ●

360486 Senna splendida (Vogel) H. S. Irwin et Barneby；闪光异决明；Golden Wonder Senna ●☆

360487 Senna suarezensis (Capuron) Du Puy；苏亚雷斯异决明●☆

360488 Senna sulfurea (Collad.) H. S. Irwin et Barneby = Senna sulfurea (DC. ex Collad.) H. S. Irwin et Barneby ●

360489 Senna sulfurea (DC. ex Collad.) H. S. Irwin et Barneby；粉叶决明(硫色异决明)；Smooth Senna ●

360490 Senna surattensis (Burm. f.) H. S. Irwin et Barneby；黄槐决明（粉叶决明，黄槐，金凤）；Glaucous Cassia, Glossyshower Senna, Glossy-shower Senna, Largeanther Senna, Large-anthered Senna, Suffruticose Senna, Sunshine-tree, Surat Senna ●

360491 Senna surattensis (Burm. f.) H. S. Irwin et Barneby = Cassia surattensis Burm. f. ●

360492 Senna tora (L.) Irwin et Barneby = Cassia tora L. ●

360493 Senna tora (L.) Roxb.；决明(草决明，大山土豆，狄小豆，独占缸，独占缸子，钝叶决明，狗屎豆，合明草子，槐豆，槐藤，还瞳子，假花生，假咖啡豆，假绿豆，假羊角菜，江南豆，茳，茳芒茳芒决明，金豆儿，芙明，芙明子，马蹄草，马蹄决明，马蹄子，千里光，芹决，细叶猪屎豆，小决明，羊触龙，羊角，羊角豆，羊明，羊尾豆，野花生，野青豆，夜关门，夜合草，夜拉子，猪骨明，猪屎豆，猪屎蓝豆)；Foetld Cassia, Oriental Senna, Sickle Senna, Sickle-pod ●

360494 Senna tora (L.) Roxb. = Cassia tora L. ●

360495 Senna truncata (Brenan) Lock = Cassia truncata Brenan ●☆

360496 Senna viguierella (Ghesq.) Du Puy；维基耶异决明●☆

360497 Senna wislizeni (A. Gray) H. S. Irwin et Barneby；灌木异决明；Shrubby Senna ●☆

360498 Senneberia Neck. = Ocotea Aubl. ●☆

360499 Sennebiera Willd. = Coronopus Zinn(保留属名)■

360500 Sennebiera Willd. = Lepidium L. ■

360501 Sennebiera Willd. = Nasturtiolum Gray ■

360502 Sennebiera Willd. = Senebiera DC. ■

360503 Sennefeldera Endl. = Senefeldera Mart. ■☆

360504 Sennenia Pau ex Sennen = Trisetum Pers. ■

360505 Sennenia Sennen = Trisetaria Forssk. ■☆

360506 Sennia Chiov. = Dialium L. ●☆

360507 Sennia Chiov. = Sciaplea Rauschert ●☆

360508 Sennia sciap-sciaple Chiov. = Dialium orientale Baker f. ●☆

360509 Senniella Aellen = Atriplex L. ■●

360510 Senniella Aellen(1938)；决明藜属■●☆

360511 Senniella spongiosa (F. Muell.) Aellen；决明藜■☆

360512 Senra Cav. (1786)；线托叶锦葵属●☆

360513 Senra incana Cav. ；线托叶锦葵●☆

360514 Senra nubica Webb = Senra incana Cav. ●☆

360515 Senra zoes Volkens et Schweinf. ；非洲线托叶锦葵●☆

360516 Senraea Willd. = Senra Cav. ●☆

360517 Sensitiva Raf. = Mimosa L. ●■

360518 Sentis Comm. ex Brongn. = Scutia (Comm. ex DC.) Brongn. (保留属名)●

360519 Sentis F. Muell. = Pholidia R. Br. ●☆

360520 Senyumia Kiew, A. Weber et B. L. Burtt；塞尼苣苔属■☆

360521 Senyumia minutiflora (Ridl.) Kiew, A. Weber et B. L. Burtt；塞尼苣苔■☆

360522 Seorsus Rye et Trudgen = Astartea DC. ●☆

360523 Seorsus Rye et Trudgen(2008)；离生桃金娘属●☆

360524 Sepalosaccus Schltr. (1923)；囊萼兰属■☆

360525 Sepalosaccus Schltr. = Maxillaria Ruiz et Pav. ■☆

360526 Sepalosaccus humilisSchltr. ；囊萼兰■☆

360527 Sepalosiphon Schltr. = Glossorhyncha Ridl. ■☆

360528 Separotheca Waterf. = Tradescantia L. ■

360529 Sepias L. = Crassula L. ●■☆

360530 Sepikea Schltr. (1923)；新几内亚苣苔属■☆

360531 Sepikea cylindrocarpa Schltr. ；新几内亚苣苔■☆

360532 Seplimia P. V. Heath = Crassula L. ●■☆

360533 Septacanthus Wight = Leptacanthus Nees ●■

360534 Septacanthus Wight = Strobilanthes Blume ●■

360535 Septas L. = Crassula L. ●■☆

360536 Septas Lour. = Bacopa Aubl. (保留属名)■

360537 Septas Lour. = Brami Adans. (废弃属名)■

360538 Septas capensis L. = Crassula capensis (L.) Baill. ●■☆

360539 Septas globifera Sims = Crassula capensis (L.) Baill. ●■☆

360540 Septas umbella (Jacq.) Haw. = Crassula umbella Jacq. ●☆

360541 Septilia Raf. = Bacopa Aubl. (保留属名)■

360542 Septilia Raf. = Brami Adans. (废弃属名)■

360543 Septilia Raf. = Septas Lour. ■

360544 Septimetula Tiegh. = Phragmanthera Tiegh. ●☆

360545 Septimetula Tiegh. = Tapinanthus (Blume) Rchb. (保留属名) ●☆

360546 Septimetula macrosolen (Steud. ex A. Rich.) Tiegh. = Phragmanthera macrosolen (Steud. ex A. Rich.) M. G. Gilbert ●☆

360547 Septimetula rufescens (DC.) Tiegh. = Phragmanthera rufescens (DC.) Balle ●☆

360548 Septimia P. V. Heath = Crassula L. ●■☆

360549 Septimia P. V. Heath(1993)；塞普景天属■●☆

360550 Septina Nor. = ? Litsea Lam. (保留属名)●

360551 Septogarcinia Kosterm. = Garcinia L. ●

360552 Septotheca Ulbr. (1924)；秘鲁木棉属●☆

360553 Septotheca tessmannii Ulbr. ；秘鲁木棉●☆

360554 Septulina Tiegh. (1895)；南非桑寄生属●☆

360555 Septulina Tiegh. = Taxillus Tiegh. ●

360556 Septulina glauca (Thunb.) Tiegh. ；南非桑寄生●☆

360557 Septulina glauca (Thunb.) Tiegh. var. ovalis (E. Mey. ex Harv.) Balle = Septulina ovalis (E. Mey. ex Harv.) Tiegh. ●☆

360558 Septulina ovalis (E. Mey. ex Harv.) Tiegh. ；卵叶南非桑寄生 ●☆

360559 Sequencia Givnish = Brocchinia Schult. f. ■☆

360560 Sequencia Givnish(2007)；哥伦比亚凤梨属■☆

360561 Sequoia Endl. (1847) (保留属名)；北美红杉属(长叶世界爷属，红杉属)；Big Tree of California, Big-Tree, Mammoth Tree, Redwood, Sequoia, Sierra Redwood, Washingtonia, Wellingtonia ●

360562 Sequoia gigantea (Lindl.) Decne. = Sequoiadendron giganteum (Lindl.) Buchholz ●

360563 Sequoia gigantea Endl. = Sequoia sempervirens (D. Don ex Lamb.) Endl. ●

360564 Sequoia glyptostroboides (Hu et W. C. Cheng) Weide = Metasequoia glyptostroboides Hu et W. C. Cheng ●◇

360565 Sequoia sempervirens (D. Don ex Lamb.) Endl. ；北美红杉(长叶世界爷，红杉)；California Redwood, Californian Redwood, Coast Redwood, Coastal Redwood, Giant Sequoia, Mammoth Tree, Red Wood, Redwood, Sequoia ●

360566 Sequoia sempervirens (D. Don ex Lamb.) Endl. ' Adpressa '；矮生北美红杉；Coastal Redwood ●☆

360567 Sequoia sempervirens (D. Don ex Lamb.) Endl. var. adpressa Carrière = Sequoia sempervirens (D. Don ex Lamb.) Endl.

'Adpressa'●☆

360568　Sequoia sempervirens（D. Don）Endl. = Sequoia sempervirens（D. Don ex Lamb.）Endl. ●

360569　Sequoia wellingtonia Seem. = Sequoiadendron giganteum（Lindl.）Buchholz ●

360570　Sequoia wellingtonia Seem. var. pendula（Carrière）M. L. Green = Sequoiadendron giganteum（Lindl.）Buchholz 'Pendulum'●☆

360571　Sequoiaceae Arnoldi = Cupressaceae Gray（保留科名）●

360572　Sequoiaceae Arnoldi = Taxodiaceae Saporta（保留科名）●

360573　Sequoiaceae Luerss. ;北美红杉科●

360574　Sequoiaceae Luerss. = Taxodiaceae Saporta（保留科名）●

360575　Sequoiadendron J. Buchholz（1939）;巨杉属（世界爷树属,世界爷属）;Big Tree, Giant Sequoia, Giant Tree, Wellingtonia ●

360576　Sequoiadendron giganteum（Lindl.）Buchholz;巨杉（北美巨杉,世界爷）;Big Tree, Big Wood, Bigtree, California Big Tree, Californian Big Tree, Giant Redwood, Giant Sequoia, Giant Tree, Giant Tree of California, Mammoth Tree, Mammoth Tree of California, Sierra Redwood, Sierra-redwood, Wellingtonia ●

360577　Sequoiadendron giganteum（Lindl.）Buchholz 'Pendulum';垂枝巨杉;Weeping Giant Sequoia ●☆

360578　Sequoiadendron giganteum（Lindl.）Buchholz 'Powder Blue';蓝色巨杉;Blue Giant Sequoia ●☆

360579　Serangium Wood ex Salisb. = Monstera Adans.（保留属名）●■

360580　Seraphyta Fisch. et C. A. Mey. = Epidendrum L.（保留属名）■☆

360581　Serapia capensis L. = Acrolophia barbata（Thunb.）H. P. Linder ■☆

360582　Serapias L.（1753）（保留属名）;长药兰属（牛舌兰属）;Serapias, Tongue Orchid, Tongue-flowered Orchid ■☆

360583　Serapias Pers. = Epipactis Zinn（保留属名）■

360584　Serapias aculeata（L. f.）Thunb. = Eulophia aculeata（L. f.）Spreng. ☆

360585　Serapias africana（Rendle）Eaton = Epipactis africana Rendle ■☆

360586　Serapias atrorubens Hoffm. ex Bernh. = Epipactis atrorubens（Hoffm. ex Bernh.）Besser ■☆

360587　Serapias camtschatea（L.）Steud. = Neottia camtschatea（L.）Rchb. f. ■

360588　Serapias collina（Banks et Sol.）Vela et Lallain = Anacamptis collina（Banks et Sol.）R. M. Bateman, Pridgeon et Chase ■☆

360589　Serapias consimilis（D. Don）A. A. Eaton = Epipactis consimilis D. Don ■

360590　Serapias consimilis（D. Don）A. A. Eaton = Epipactis helleborine（L.）Crantz ■

360591　Serapias cordigera（Pers.）L. ;长药兰（心花舌兰）;Common Serapias, Heart-flowered Orchis, Heart-flowered Serapias ■☆

360592　Serapias cordigera L. = Serapias cordigera（Pers.）L. ■☆

360593　Serapias cordigera L. var. mauritanica（E. G. Camus）E. C. Nelson = Serapias lorenziana H. Baumann et Künkele ■☆

360594　Serapias coriophora（L.）Vela et Lallain = Anacamptis coriophora（L.）R. M. Bateman, Pridgeon et Chase ■☆

360595　Serapias damasonium Mill. = Cephalanthera damasonium（Mill.）Druce ■

360596　Serapias ensifolia Murray = Cephalanthera longifolia（L.）Fritsch ■

360597　Serapias epipogium（L.）Steud. = Epipogium aphyllum（F. W. Schmidt）Sw. ■

360598　Serapias erecta Thunb. = Cephalanthera erecta（Thunb. ex A. Murray）Blume ■

360599　Serapias erecta Thunb. ex A. Murray = Cephalanthera erecta（Thunb. ex A. Murray）Blume ■

360600　Serapias falcata Thunb. = Cephalanthera falcata（Thunb. ex A. Murray）Blume ■

360601　Serapias falcata Thunb. ex A. Murray = Cephalanthera falcata（Thunb. ex A. Murray）Blume ■

360602　Serapias grandiflora Oeder var. ensifolia L. f. = Cephalanthera longifolia（L.）Fritsch ■

360603　Serapias helleborine L. = Epipactis helleborine（L.）Crantz ■

360604　Serapias helleborine L. subsp. longifolia L. = Cephalanthera longifolia（L.）Fritsch ■

360605　Serapias helleborine L. var. latifolia L. = Epipactis helleborine（L.）Crantz ■

360606　Serapias helleborine L. var. longifolia L. = Cephalanthera longifolia（L.）Fritsch ■

360607　Serapias helleborine L. var. palustris L. = Epipactis palustris（L.）Crantz ■

360608　Serapias latifolia（L.）Huds. = Epipactis helleborine（L.）Crantz ■

360609　Serapias laxiflora（Lam.）Vela et Lallain = Anacamptis laxiflora（Lam.）R. M. Bateman, Pridgeon et Chase ■

360610　Serapias lingua L. ;舌花长药兰;Tongue Orchid, Tongue-flowered Orchis ■☆

360611　Serapias lingua L. subsp. duriaei（Rchb.）Maire = Serapias strictiflora Da Veiga ■☆

360612　Serapias lingua L. subsp. oxyglottis（Willd.）Maire et Weiller = Serapias lingua L. ■☆

360613　Serapias lingua L. subsp. stenopetala（Maire et T. Stephenson）Maire et Weiller = Serapias stenopetala Maire et T. Stephenson ■☆

360614　Serapias lingua L. var. duriaei Rchb. = Serapias strictiflora Da Veiga ■☆

360615　Serapias longibracteata（Blume）A. A. Eaton = Cephalanthera longibracteata Blume ■

360616　Serapias longifolia（L.）Huds. = Cephalanthera longifolia（L.）Fritsch ■

360617　Serapias longifolia Huds. = Cephalanthera longifolia（L.）Fritsch ■

360618　Serapias longifolia L. = Epipactis palustris（L.）Crantz ■

360619　Serapias longipetala（Ten.）Pollini;长瓣长药兰;Longpetal Serapias ■☆

360620　Serapias longipetala Pollini = Serapias vomeracea（Burm.）Briq. ■☆

360621　Serapias lorenziana H. Baumann et Künkele;滨海长药兰■☆

360622　Serapias melaleuca Thunb. = Disa bivalvata（L. f.）T. Durand et Schinz ■☆

360623　Serapias neglecta De Not. ;稀少长药兰;Scarce Serapias ■☆

360624　Serapias occultata J. Gay = Serapias parviflora Parl. ■☆

360625　Serapias olbia Verg. ;奥尔长药兰;Hybrid Serapias ■☆

360626　Serapias palustris（Jacq.）Vela et Lallain = Anacamptis palustris（Jacq.）R. M. Bateman, Pridgeon et Chase ■☆

360627　Serapias palustris L. = Epipactis palustris（L.）Crantz ■

360628　Serapias papilionacea（L.）Vela et Lallain = Anacamptis papilionacea（L.）R. M. Bateman, Pridgeon et Chase ■☆

360629　Serapias parviflora Parl. ;小花长药兰;Small Tongue Orchid ■☆

360630　Serapias parviflora Parl. subsp. occultata（Gay）Maire et Weiller ＝ Serapias parviflora Parl. ■☆

360631　Serapias patens（L. f.）Thunb. ＝ Disa tenuifolia Sw. ■☆

360632　Serapias pedicellata（L. f.）Thunb. ＝ Eulophia aculeata（L. f.）Spreng. ■☆

360633　Serapias polystachya Sw. ＝ Tropidia polystachya（Sw.）Ames ■☆

360634　Serapias repens（L.）Vill. ＝ Goodyera repens（L.）R. Br. ■

360635　Serapias salassia Steud. ＝ Liparis salassia（Pers.）Summerh. ■☆

360636　Serapias stenopetala Maire et T. Stephenson;窄瓣长药兰■☆

360637　Serapias strictiflora Da Veiga;刚叶长药兰☆

360638　Serapias tabularis（L. f.）Thunb. ＝ Eulophia tabularis（L. f.）Bolus ■☆

360639　Serapias vomeracea（Burm.）Briq.;非洲长药兰■☆

360640　Serapias vomeracea（Burm.）Briq. var. mauritanica Camus ＝ Serapias vomeracea（Burm.）Briq. ■☆

360641　Serapiastrum Kuntze ＝ Serapias L.（保留属名）■☆

360642　Serapios vomeracea（Burm.）Briq.;犁铧长药兰;Long-lipped Serapias,Ploughshare Orchid ■☆

360643　Serena Raf. ＝ Haemanthus L.

360644　Serenaea Hook. f. ＝ Serenoa Hook. f. ●☆

360645　Serenoa Hook. f.（1883）;锯齿棕属（锯柄棕属,锯箬棕属,锯叶棕属,锯棕属,蓝棕属,塞伦诺桐属,塞润桐属）;Saw Palmetto,Saw Palm ●☆

360646　Serenoa repens（Bartram）Small;锯齿棕（锯柄桐,锯箬棕,锯叶棕,匍匐蓝棕,塞润桐）;Saw Palmetto,Saw Palmetto Palm,Scrub Palmetto ●☆

360647　Serenoa serrulata（Michx.）G. Nicholson ＝ Serenoa repens（Bartram）Small ●☆

360648　Serenoa serrulata Hook. f. ＝ Serenoa repens（Bartram）Small ●☆

360649　Sererea Raf.（废弃属名）＝ Distictis Mart. ex Meisn. ●☆

360650　Sererea Raf.（废弃属名）＝ Phaedranthus Miers（保留属名）●☆

360651　Seretoberlinia P. A. Duvign. ＝ Julbernardia Pellegr. ●☆

360652　Seretoberlinia seretii（De Wild.）P. A. Duvign. ＝ Julbernardia seretii（De Wild.）Troupin ●☆

360653　Sergia Fed.（1957）;赛氏桔梗属☆

360654　Sergia regelii（Trautv.）Fed.;雷格尔赛氏桔梗■☆

360655　Sergia sewerzowii（Regel）Fed.;赛氏桔梗■☆

360656　Sergilus Gaertn. ＝ Baccharis L.（保留属名）●■☆

360657　Serialbizzia Kosterm.（1954）;丝合欢属☆

360658　Serialbizzia Kosterm. ＝ Albizia Durazz. ●

360659　Serialbizzia acle（Blanco）Kosterm.;丝合欢●☆

360660　Serialbizzia attopeuense（Pierre）Kosterm. ＝ Albizia attopeuensis（Pierre）I. C. Nielsen var. lauii（Merr.）I. C. Nielsen ●

360661　Serialbizzia splendens（Miq.）Kosterm.;纤细丝合欢●☆

360662　Seriana Willd. ＝ Seriania Schum. ●☆

360663　Seriania Plum. ex Schum. ＝ Serjania Mill. ●☆

360664　Seriania Schum. ＝ Serjania Mill. ●☆

360665　Serianthes Benth.（1844）;丝花树属●☆

360666　Serianthes grandiflora Benth.;大花丝花树●☆

360667　Serianthes tenuiflora Benth.;细花丝花树●☆

360668　Sericandra Raf. ＝ Albizia Durazz. ●

360669　Sericanthe Robbr.（1978）;丝花茜属●☆

360670　Sericanthe adamii（N. Hallé）Robbr.;阿达姆丝花茜●☆

360671　Sericanthe andongensis（Hiern）Robbr.;安东丝花茜●☆

360672　Sericanthe andongensis（Hiern）Robbr. subsp. engleri（K. Krause）Bridson;恩格勒丝花茜●☆

360673　Sericanthe andongensis（Hiern）Robbr. var. mollis Robbr.;柔软丝花茜●☆

360674　Sericanthe auriculata（Keay）Robbr.;耳形丝花茜●☆

360675　Sericanthe burundensis Robbr.;布隆迪丝花茜●☆

360676　Sericanthe chevalieri（K. Krause）Robbr.;舍瓦利耶丝花茜●☆

360677　Sericanthe chevalieri（K. Krause）Robbr. var. coffeoides（A. Chev.）Robbr.;咖啡丝花茜●☆

360678　Sericanthe chevalieri（K. Krause）Robbr. var. velutina Robbr.;绒毛丝花茜●☆

360679　Sericanthe halleana Robbr.;哈勒丝花茜●☆

360680　Sericanthe leonardii（N. Hallé）Robbr.;莱奥丝花茜●☆

360681　Sericanthe leonardii（N. Hallé）Robbr. subsp. venosa Robbr.;多脉丝花茜●☆

360682　Sericanthe odoratissima（K. Schum.）Robbr.;极香丝花茜●☆

360683　Sericanthe odoratissima（K. Schum.）Robbr. var. ulugurensis Robbr.;乌卢古尔丝花茜●☆

360684　Sericanthe pellegrinii（N. Hallé）Robbr.;佩尔格兰丝花茜●☆

360685　Sericanthe petitii（N. Hallé）Robbr.;佩蒂蒂丝花茜●☆

360686　Sericanthe raynaliorum（N. Hallé）Robbr.;雷纳尔丝花茜●☆

360687　Sericanthe roseoides（De Wild. et T. Durand）Robbr.;蔷薇丝花茜●☆

360688　Sericanthe suffruticosa（Hutch.）Robbr.;亚灌木丝花茜●☆

360689　Sericanthe testui（N. Hallé）Robbr.;泰斯图丝花茜●☆

360690　Sericanthe trilocularis（Scott-Elliot）Robbr.;三腔丝花茜●☆

360691　Sericanthe trilocularis（Scott-Elliot）Robbr. subsp. paroissei（Aubrév. et Pellegr.）Robbr.;帕罗丝花茜●☆

360692　Sericeocassia Britton ＝ Cassia L.（保留属名）●■

360693　Sericeocassia Britton ＝ Senna Mill. ●■

360694　Serichonus K. R. Thiele ＝ Stenanthemum Reissek ●☆

360695　Serichonus K. R. Thiele（2007）;澳洲狭花木属●☆

360696　Sericocactus Y. Ito ＝ Notocactus（K. Schum.）A. Berger et Backeb. ■

360697　Sericocactus Y. Ito ＝ Parodia Speg.（保留属名）●

360698　Sericocalyx Bremek.（1944）;黄球花属（丝萼爵床属）;Sericocalyx ●■

360699　Sericocalyx Bremek. ＝ Strobilanthes Blume ●■

360700　Sericocalyx chinensis（Nees）Bremek.;黄球花（白泡草,半柱花,狗泡草,火漂藤,毛虫包,毛虫药）;China Halfstyleflower,China Sericocalyx,Chinese Sericocalyx,Hemigraphis ●■

360701　Sericocalyx fluviatilis（C. B. Clarke ex W. W. Sm.）Bremek.;溪畔黄球花（岸生半柱花,溪生半柱花）;Fluvial Hemigraphis,Halfstyleflower ●■

360702　Sericocarpus Nees（1832）;丝果菊属（白顶菊属）;White-topped Aster ■☆

360703　Sericocarpus acutisquamus（Nash）Small ＝ Sericocarpus tortifolius（Michx.）Nees ■☆

360704　Sericocarpus asteroides（L.）Nees;齿叶丝果菊;Toothed White-topped Aster ■☆

360705　Sericocarpus bifoliatus（Walter）Porter ＝ Sericocarpus tortifolius（Michx.）Nees ■☆

360706　Sericocarpus bifoliatus Porter;二叶丝果菊■☆

360707　Sericocarpus californicus Durand ＝ Sericocarpus oregonensis Nutt. subsp. californicus（Durand）Ferris ■☆

360708 Sericocarpus collinsii Nutt. = Sericocarpus tortifolius (Michx.) Nees ■☆

360709 Sericocarpus linifolius Britton, Sterns et Poggenb.；亚麻叶丝果菊；Narrow-leaf White-topped Aster, Narrow-leaved White-topped Aster ■☆

360710 Sericocarpus oregonensis Nutt.；俄勒冈丝果菊；Oregon White-topped Aster ■☆

360711 Sericocarpus oregonensis Nutt. subsp. californicus (Durand) Ferris；加州丝果菊■☆

360712 Sericocarpus oregonensis Nutt. var. californicus (Durand) G. L. Nesom = Sericocarpus oregonensis Nutt. subsp. californicus (Durand) Ferris ■☆

360713 Sericocarpus rigidus Lindl.；硬丝果菊；Columbian White-topped Aster ■☆

360714 Sericocarpus solidagineus (Michx.) Nees = Sericocarpus linifolius Britton, Sterns et Poggenb. ■☆

360715 Sericocarpus tomentellus Greene = Eucephalus tomentellus (Greene) Greene ■☆

360716 Sericocarpus tortifolius (Michx.) Nees；美国南部丝果菊；Dixie white-topped aster ■☆

360717 Sericocoma Fenzl = Sericocoma Fenzl ex Endl. ■●☆

360718 Sericocoma Fenzl ex Endl. (1842)；绢毛苋属 ■●☆

360719 Sericocoma alternifolia (Schinz) C. B. Clarke = Neocentema alternifolia (Schinz) Schinz ■☆

360720 Sericocoma angustifolia (Moq.) Hook. f. = Kyphocarpa angustifolia (Moq.) Lopr. ■☆

360721 Sericocoma avolans Fenzl；澳非绢毛苋 ■☆

360722 Sericocoma bainesii Hook. f. = Leucosphaera bainesii (Hook. f.) Gilg ●☆

360723 Sericocoma capensis Moq. = Sericocoma avolans Fenzl ■☆

360724 Sericocoma capitata Moq. = Calicorema capitata (Moq.) Hook. f. ■☆

360725 Sericocoma chrysurus Meisn. = Kyphocarpa trichinoides (Fenzl) Lopr. ■☆

360726 Sericocoma denudata Hook. f. = Marcelliopsis denudata (Hook. f.) Schinz ■☆

360727 Sericocoma hereroensis Suess. et Beyerle = Kyphocarpa angustifolia (Moq.) Lopr. ■☆

360728 Sericocoma heterochiton Lopr.；异被绢毛苋 ■☆

360729 Sericocoma leucoclada Lopr. = Sericocoma pungens Fenzl ■☆

360730 Sericocoma namaensis Suess. = Sericocoma pungens Fenzl ■☆

360731 Sericocoma nelsii Schinz = Nelsia quadrangula (Engl.) Schinz ■☆

360732 Sericocoma pallida S. Moore = Sericocomopsis pallida (S. Moore) Schinz ■☆

360733 Sericocoma pungens Fenzl；刺绢毛苋 ■☆

360734 Sericocoma quadrangula Engl. = Nelsia quadrangula (Engl.) Schinz ■☆

360735 Sericocoma remotiflora (Hook.) Benth. et Hook. f. = Sericorema remotiflora (Hook.) Lopr. ●☆

360736 Sericocoma sericea Schinz = Sericorema sericea (Schinz) Lopr. ●☆

360737 Sericocoma shepperioides Schinz = Calicorema capitata (Moq.) Hook. f. ■☆

360738 Sericocoma somalensis S. Moore = Chionothrix somalensis (S. Moore) Hook. f. ●☆

360739 Sericocoma squarrosa Schinz = Calicorema squarrosa (Schinz) Schinz ■☆

360740 Sericocoma trichinioides Fenzl = Kyphocarpa trichinoides (Fenzl) Lopr. ■☆

360741 Sericocoma welwitschii Baker = Nelsia quadrangula (Engl.) Schinz ■☆

360742 Sericocoma welwitschii Hook. f. = Marcelliopsis welwitschii (Hook. f.) Schinz ■☆

360743 Sericocoma zeyheri Engl. = Sericocoma avolans Fenzl ■☆

360744 Sericocomopsis Schinz (1895)；类绢毛属 ●☆

360745 Sericocomopsis bainesii (Hook. f.) Schinz = Leucosphaera bainesii (Hook. f.) Gilg ●☆

360746 Sericocomopsis grisea Suess. = Sericocomopsis hildebrandtii Schinz ■☆

360747 Sericocomopsis hildebrandtii Schinz；类绢毛苋 ■☆

360748 Sericocomopsis lanceolata (Schinz) Peter = Cyathula lanceolata Schinz ■☆

360749 Sericocomopsis lanceolata (Schinz) Peter var. merkeri = Cyathula lanceolata Schinz ■☆

360750 Sericocomopsis lindaviana Peter = Sericocomopsis hildebrandtii Schinz ■☆

360751 Sericocomopsis meruensis Suess. = Sericocomopsis hildebrandtii Schinz ■☆

360752 Sericocomopsis orthacantha (Hochst. ex Asch.) Peter = Cyathula orthacantha (Hochst. ex Asch.) Schinz ■☆

360753 Sericocomopsis pallida (S. Moore) Schinz；苍白类绢毛苋 ■☆

360754 Sericocomopsis pallida (S. Moore) Schinz var. grandis Suess. = Sericocomopsis pallida (S. Moore) Schinz ■☆

360755 Sericocomopsis pallida (S. Moore) Schinz var. parvifolia Suess. = Sericocomopsis pallida (S. Moore) Schinz ■☆

360756 Sericocomopsis quadrangula (Engl.) Lopr. = Nelsia quadrangula (Engl.) Schinz ■☆

360757 Sericocomopsis welwitschii (Baker) Lopr. = Nelsia quadrangula (Engl.) Schinz ■☆

360758 Sericodes A. Gray (1852)；绢毛蒺藜属 ●☆

360759 Sericodes greggii A. Gray；绢毛蒺藜 ●☆

360760 Sericographis Nees = Justicia L. ●■

360761 Sericola Raf. = Miconia Ruiz et Pav. (保留属名) ●☆

360762 Sericolea Schltr. (1916)；丝鞘杜英属 ●☆

360763 Sericolea calophylla (Ridl.) Schltr.；美叶丝鞘杜英 ●☆

360764 Sericolea elegans Schltr.；雅致丝鞘杜英 ●☆

360765 Sericolea floribunda A. C. Sm.；多花丝鞘杜英 ●☆

360766 Sericolea glabra Schltr.；光丝鞘杜英 ●☆

360767 Sericolea leptophylla Kaneh. et Hatus.；细叶丝鞘杜英 ●☆

360768 Sericolea microphylla Balgooy；小叶丝鞘杜英 ●☆

360769 Sericolea pachyphylla Coode；厚叶丝鞘杜英 ●☆

360770 Sericoma Hochst. = Sericocoma Fenzl ex Endl. ■●☆

360771 Sericorema (Hook. f.) Lopr. (1899)；绢柱苋属 ■☆

360772 Sericorema Lopr. = Sericorema (Hook. f.) Lopr. ■☆

360773 Sericorema humbertiana Cavaco = Pupalia micrantha Hauman ■☆

360774 Sericorema remotiflora (Hook.) Lopr.；绢柱苋 ●☆

360775 Sericorema sericea (Schinz) Lopr.；非洲绢柱苋 ●☆

360776 Sericospora Nees (1847)；丝籽爵床属 ☆

360777 Sericospora crinita Nees；丝籽爵床 ☆

360778 Sericostachys Gilg et Lopr. (1899)；绢穗苋属 ■☆

360779 Sericostachys Gilg et Lopr. ex Lopr. = Sericostachys Gilg et Lopr. ■☆

360780　Sericostachys scandens Gilg et Lopr. ;绢穗苋■☆

360781　Sericostachys scandens Gilg et Lopr. var. tomentosa（Gilg et Lopr.）Cavaco = Sericostachys scandens Gilg et Lopr. ■☆

360782　Sericostachys tomentosa Gilg et Lopr. = Sericostachys scandens Gilg et Lopr. ■☆

360783　Sericostoma Stocks.（1848）;丝口五加属●☆

360784　Sericostoma albidum Franch. = Echiochilon persicum（Burm. f.）I. M. Johnst. ■☆

360785　Sericostoma arenarium I. M. Johnst. = Echiochilon arenarium（I. M. Johnst.）I. M. Johnst. ■☆

360786　Sericostoma calcareum（Vatke）I. M. Johnst. = Echiochilon persicum（Burm. f.）I. M. Johnst. ■☆

360787　Sericostoma pauciflorum Stocks ex Wight;丝口五加●☆

360788　Sericostoma persicum（Burm. f.）B. L. Burtt = Echiochilon persicum（Burm. f.）I. M. Johnst. ■☆

360789　Sericostoma verrucosum Beck = Echiochilon persicum（Burm. f.）I. M. Johnst. ■☆

360790　Sericotheca Raf.（废弃属名）= Holodiscus（C. Koch）Maxim.（保留属名）●☆

360791　Sericrostis Raf. = Muhlenbergia Schreb. ■

360792　Sericura Hassk. = Pennisetum Rich. ■

360793　Seridia Juss. = Centaurea L.（保留属名）●■

360794　Serigrostis Steud. = Muhlenbergia Schreb. ■

360795　Serigrostis Steud. = Sericrostis Raf. ■

360796　Seringea F. Muell. = Seringia J. Gay（保留属名）●☆

360797　Seringia J. Gay（1821）（保留属名）;塞林梧桐属●☆

360798　Seringia Spreng.（废弃属名）= Ptelidium Thouars ●☆

360799　Seringia Spreng.（废弃属名）= Seringia J. Gay（保留属名）●☆

360800　Seringia arboresceas（Aiton）Druce;塞林梧桐●☆

360801　Serinia Raf. = Krigia Schreb.（保留属名）■☆

360802　Serinia cespitosa Raf. = Krigia cespitosa（Raf.）K. L. Chambers ■☆

360803　Serinia oppositifolia（Raf.）Kuntze = Krigia cespitosa（Raf.）K. L. Chambers ■☆

360804　Seriola L. = Hypochaeris L. ■

360805　Seriola aethnensis L. = Hypochaeris achyrophora L. ■☆

360806　Seriola aethnensis L. var. hispida Bég. et Vacc. = Hypochaeris achyrophora L. ■☆

360807　Seriphidium（Besser ex Less.）Fourr.（1869）;绢蒿属（蛔蒿属,绢菊属,异蒿属）;Spunsilksage ●■

360808　Seriphidium（Besser）Poljakov = Seriphidium（Besser ex Less.）Fourr. ●■

360809　Seriphidium（Hook.）Fourr. = Seriphidium（Besser ex Less.）Fourr. ●■

360810　Seriphidium Fourr. = Artemisia L. ●■

360811　Seriphidium algeriense（Filatova）Y. R. Ling = Artemisia algeriensis Filatova ■☆

360812　Seriphidium amoenum（Poljakov）Poljakov;小针裂叶绢蒿;Lovable Spunsilksage ■

360813　Seriphidium arbusculum（Nutt.）W. A. Weber = Artemisia arbuscula Nutt. ■☆

360814　Seriphidium arbusculum（Nutt.）W. A. Weber subsp. longilobum（Osterh.）W. A. Weber = Artemisia arbuscula Nutt. subsp. longiloba（Osterh.）L. M. Shultz ■☆

360815　Seriphidium arbusculum（Nutt.）W. A. Weber var. thermopolum（Beetle）Y. R. Ling = Artemisia arbuscula Nutt. subsp. thermopola Beetle ■☆

360816　Seriphidium aucheri（Boiss.）Y. Ling et Y. R. Ling;光叶绢蒿;Smoothleaf Spunsilksage ■

360817　Seriphidium barrelieri（Besser）Soják = Artemisia barrelieri Besser ■☆

360818　Seriphidium bigelovii（A. Gray）K. Bremer et Humphries = Artemisia bigelovii A. Gray ■☆

360819　Seriphidium bolanderi（A. Gray）Y. R. Ling = Artemisia cana Pursh subsp. bolanderi（A. Gray）G. H. Ward ■☆

360820　Seriphidium borotalense（Poljakov）Y. Ling et Y. R. Ling;博洛塔绢蒿（博洛绢蒿）;Borotal Spunsilksage ■

360821　Seriphidium botschantzevii（Filatova）Y. R. Ling = Artemisia botschantzevii Filatova ■☆

360822　Seriphidium brevifolium（Wall. ex DC.）Y. Ling et Y. R. Ling;短叶绢蒿（短叶蒿）;Shortleaf Spunsilksage ■

360823　Seriphidium canum（Pursh）W. A. Weber = Artemisia cana Pursh ■☆

360824　Seriphidium canum（Pursh）W. A. Weber subsp. bolanderi（A. Gray）W. A. Weber = Artemisia cana Pursh subsp. bolanderi（A. Gray）G. H. Ward ■☆

360825　Seriphidium cinum（Berger ex Poljakov）Poljakov;蛔蒿（山道尼格,山道年草,山道年蒿,希那,沿海艾）;Levant Wormseed,Levant-wormseed,Roundworm Spunsilksage,Sea Wormwood ■

360826　Seriphidium coerulescens（L.）Y. R. Ling = Artemisia oranensis（Debeaux）Filatova ●☆

360827　Seriphidium compactum（Fisch. ex Besser）Poljakov;聚头绢蒿（聚头蒿）;Meeting Spunsilksage ■

360828　Seriphidium compactum（Fisch. ex DC.）Poljakov = Seriphidium compactum（Fisch. ex Besser）Poljakov ■

360829　Seriphidium densifolium（Filatova）Y. R. Ling = Artemisia densifolia Filatova ■☆

360830　Seriphidium fedtschenkoanum（Krasch.）Poljakov;苍绿绢蒿;Palegreen Spunsilksage ■

360831　Seriphidium ferganense（Krasch. ex Poljakov）Poljakov;费尔干绢蒿;Fergan Spunsilksage ■

360832　Seriphidium finitum（Kitag.）Y. Ling et Y. R. Ling;东北蛔蒿;Dongbei Spunsilksage ■

360833　Seriphidium finitum Kitag. = Seriphidium finitum（Kitag.）Y. Ling et Y. R. Ling ■

360834　Seriphidium glanduligerum（Krasch. ex Poljakov）Poljakov;腺体绢蒿☆

360835　Seriphidium gracilescens（Krasch. et Iljin）Poljakov;纤细绢蒿（戈壁蒿,少花蒿,纤蒿,小花蒿）;Pauciflorous Sagebrush, Thin Spunsilksage ■

360836　Seriphidium grenardii（Franch.）Y. R. Ling et Humphries;高原绢蒿;Plateau Spunsilksage ■

360837　Seriphidium heptapotamicum（Poljakov）Y. Ling et Y. R. Ling;半荒漠绢蒿;Semiaesert Spunsilksage ■

360838　Seriphidium heptapotamicum（Poljakov）Y. Ling et Y. R. Ling = Seriphidium terrae-albae（Krasch.）Poljakov ■

360839　Seriphidium herba-alba（Asso）Soják = Artemisia herba-alba Asso ■☆

360840　Seriphidium ifranense（Didier）Dobignard = Artemisia ifranensis Didier ■☆

360841　Seriphidium incanum Lam. = Seriphium cinereum L. ●☆

360842　Seriphidium incultum（Delile）Y. R. Ling = Artemisia inculta Delile ■☆

360843　Seriphidium issykkulense（Poljakov）Poljakov;伊塞克绢蒿;

Issykkol Spunsilksage ■

360844 Seriphidium issykkulense（Poljakov）Poljakov = Seriphidium fedtschenkoanum（Krasch.）Poljakov ■

360845 Seriphidium junceum（Kar. et Kir.）Poljakov；三裂叶绢蒿；Threeleft Spunsilksage ■

360846 Seriphidium junceum（Kar. et Kir.）Poljakov var. macrosciadium（Poljakov）Y. Ling et Y. R. Ling；大头三裂叶绢蒿；Bighead Threeleft Spunsilksage ■

360847 Seriphidium karatavicum（Krasch. et Abolin ex Poljakov）Y. Ling et Y. R. Ling；卡拉套绢蒿；Kalatao Spunsilksage ■

360848 Seriphidium kaschgaricum（Krasch.）Poljakov；新疆绢蒿；Kaschgar Spunsilksage ■

360849 Seriphidium kaschgaricum（Krasch.）Poljakov var. dshungaricum（Filatova）Y. R. Ling；准噶尔绢蒿；Dzungar Spunsilksage ■

360850 Seriphidium korovinii（Poljakov）Poljakov；昆仑绢蒿；Kunlun Spunsilksage ■

360851 Seriphidium lehmannianum（Bunge）Poljakov；球序绢蒿；Lehmann Spunsilksage ■

360852 Seriphidium leucotrichum K. Bremer et Humphries ex Y. R. Ling；白毛绢蒿■☆

360853 Seriphidium maritimum（L.）Poljakov = Artemisia maritima L. ■

360854 Seriphidium minchunense Y. R. Ling；民勤绢蒿（香蒿）；Minqin Spunsilksage ■

360855 Seriphidium mongolorum（Krasch.）Y. Ling et Y. R. Ling；蒙青绢蒿；Mongol Spunsilksage ■

360856 Seriphidium mongolorum Krasch. = Seriphidium mongolorum（Krasch.）Y. Ling et Y. R. Ling

360857 Seriphidium nitrosum（Weber ex Stechm.）Poljakov；西北绢蒿（新疆绢蒿）；NW. China Spunsilksage ■

360858 Seriphidium nitrosum（Weber ex Stechm.）Poljakov var. gobicum（Krasch.）Y. R. Ling；戈壁绢蒿；Gobi NW. China Spunsilksage ■

360859 Seriphidium nitrosum（Weber ex Stechm.）Poljakov var. kasakorum（Krasch.）Y. R. Ling = Seriphidium nitrosum（Weber ex Stechm.）Poljakov var. gobicum（Krasch.）Y. R. Ling ■●

360860 Seriphidium nitrosum（Weber ex Stechm.）Poljakov var. subglabrum（Krasch.）Y. R. Ling = Seriphidium nitrosum（Weber ex Stechm.）Poljakov var. gobicum（Krasch.）Y. R. Ling ■●

360861 Seriphidium novum（A. Nelson）W. A. Weber = Artemisia nova A. Nelson ■☆

360862 Seriphidium oliverianum（Gay ex Besser）K. Bremer et Humphries ex Y. R. Ling；奥里弗绢蒿■☆

360863 Seriphidium oranense（Filatova）Y. R. Ling = Artemisia oranensis（Debeaux）Filatova ●☆

360864 Seriphidium poljakovii（Filatova）Y. R. Ling = Artemisia poljakovii Filatova ■☆

360865 Seriphidium ramosum（C. Sm.）Dobignard = Artemisia ramosa C. Sm. ■☆

360866 Seriphidium rhodanthum（Rupr.）Poljakov；高山绢蒿；Alp Spunsilksage ■

360867 Seriphidium rigidum（Nutt.）W. A. Weber = Artemisia rigida（Nutt.）A. Gray ☆

360868 Seriphidium rothrockii（A. Gray）W. A. Weber = Artemisia rothrockii A. Gray ■☆

360869 Seriphidium saharae（Pomel）Y. R. Ling = Artemisia saharae Pomel ■☆

360870 Seriphidium santolinum（Schrenk）Poljakov；沙漠绢蒿；Desert Spunsilksage ■

360871 Seriphidium sawanense Y. R. Ling et Humphries；沙湾绢蒿；Shawan Spunsilksage ■

360872 Seriphidium schrenkianum（Ledeb.）Poljakov；草原绢蒿（雪岭蒿）；Grassland Spunsilksage ■

360873 Seriphidium scopiforme（Ledeb.）Poljakov；帚状绢蒿；Broomlike Spunsilksage ■

360874 Seriphidium semiaridum（Krasch. et Lavrova）Y. Ling et Y. R. Ling；半润萎绢蒿；Dryish Spunsilksage ■

360875 Seriphidium sieberi（Besser）K. Bremer et Humphries ex Y. R. Ling；西伯尔绢蒿■☆

360876 Seriphidium spiciforme（Osterh.）Y. R. Ling = Artemisia spiciformis Osterh. ■☆

360877 Seriphidium stenocephalum（Kraschen. ex Poljakov）Poljakov；窄头绢蒿■

360878 Seriphidium sublessingianum（B. Keller）Poljakov；针裂叶绢蒿（亚列兴蒿）；Acicularsplitleaf Spunsilksage ■

360879 Seriphidium terrae-albae（Krasch.）Poljakov；白茎绢蒿（白蒿）；Whitestem Spunsilksage ■

360880 Seriphidium terrae-albae（Krasch.）Poljakov var. massagetovii（Krasch.）Y. R. Ling = Seriphidium terrae-albae（Krasch.）Poljakov

360881 Seriphidium terrae-albae（Krasch.）Poljakov var. suaveolens（Poljakov）Y. R. Ling = Seriphidium terrae-albae（Krasch.）Poljakov

360882 Seriphidium thomsonianum（C. B. Clarke）Y. Ling et Y. R. Ling；西藏绢蒿（伊犁蛔蒿）；Xizang Spunsilksage ■

360883 Seriphidium transiliense（Poljakov）Poljakov；伊犁绢蒿（伊犁蛔蒿）；Yili Spunsilksage ■

360884 Seriphidium tridentatum（Nutt.）W. A. Weber；大异蒿；Big Sagebrush ●☆

360885 Seriphidium tridentatum（Nutt.）W. A. Weber = Artemisia tridentata Nutt. ■

360886 Seriphidium tridentatum（Nutt.）W. A. Weber subsp. parishii（A. Gray）W. A. Weber = Artemisia tridentata Nutt. subsp. parishii（A. Gray）H. M. Hall et Clem. ■☆

360887 Seriphidium tridentatum（Nutt.）W. A. Weber subsp. wyomingense（Beetle et A. M. Young）W. A. Weber = Artemisia tridentata Nutt. subsp. wyomingensis Beetle et A. M. Young ■☆

360888 Seriphidium tripartitum（Rydb.）W. A. Weber = Artemisia tripartita（Nutt.）Rydb. ■☆

360889 Seriphidium vaseyanum（Rydb.）W. A. Weber = Artemisia tridentata Nutt. subsp. vaseyana（Rydb.）Beetle ■☆

360890 Seriphium L.（1753）；塞里菊属●☆

360891 Seriphium L. = Stoebe L. ●■☆

360892 Seriphium adpressum DC. = Dicerothamnus rhinocerotis（L. f.）Koekemoer ■☆

360893 Seriphium alopecuroides Lam. = Stoebe alopecuroides（Lam.）Less. ●☆

360894 Seriphium asperum Pers. = Helichrysum asperum（Thunb.）Hilliard et B. L. Burtt ●☆

360895 Seriphium capitatum（P. J. Bergius）Less. = Stoebe capitata P. J. Bergius ●☆

360896 Seriphium cinereum L.；灰色塞里菊●☆

360897 Seriphium corymbiferum L. = Stoebe gomphrenoides P. J. Bergius ●☆

360898　Seriphium distichum Lam. = Trichogyne seriphioides Less. ■☆

360899　Seriphium fasciculatum Thunb. = Trichogyne seriphioides Less. ■☆

360900　Seriphium filagineum DC. = Seriphium incanum (Thunb.) Pers. ●☆

360901　Seriphium flavescens DC. = Seriphium spirale (Less.) Koekemoer ●☆

360902　Seriphium fuscum L. = Stoebe fusca (L.) Thunb. ●☆

360903　Seriphium gomphrenoides Lam. ;千日红塞里菊●☆

360904　Seriphium incanum (Thunb.) Pers. ;灰毛塞里菊●☆

360905　Seriphium juniperifolium Lam. = Stoebe aethiopica L. ●☆

360906　Seriphium kilimandscharicum (O. Hoffm.) Koekemoer = Stoebe kilimandscharica O. Hoffm. ●☆

360907　Seriphium laricifolium Lam. = Trichogyne laricifolia (Lam.) Less. ■☆

360908　Seriphium niveum Pers. = Helichrysum niveum (L.) Less. ●☆

360909　Seriphium perotrichoides Less. = Stoebe capitata P. J. Bergius ●☆

360910　Seriphium phlaeoides DC. = Stoebe phyllostachya (DC.) Sch. Bip. ●☆

360911　Seriphium phyllostachyum DC. = Stoebe phyllostachya (DC.) Sch. Bip. ●☆

360912　Seriphium plumosum L. ;羽状塞里菊●☆

360913　Seriphium plumosum L. var. canescens DC. = Seriphium plumosum L. ●☆

360914　Seriphium plumosum L. var. glabriusculum DC. = Seriphium plumosum L. ●☆

360915　Seriphium prostratum (L.) Lam. = Stoebe prostrata L. ●☆

360916　Seriphium reflexum Pers. = Trichogyne reflexa (L. f.) Less. ■☆

360917　Seriphium saxatilis (Levyns) Koekemoer;岩栖塞里菊●☆

360918　Seriphium spirale (Less.) Koekemoer;螺旋塞里菊●☆

360919　Seris Less. = Gochnatia Kunth ●

360920　Seris Less. = Richterago Kuntze ●☆

360921　Seris Willd. = Onoseris Willd. ●■☆

360922　Serissa Comm. = Serissa Comm. ex Juss. ●

360923　Serissa Comm. ex Juss. (1789);六月雪属(白马骨属,满天星属);Junesnow,Serisse,Snow Rose ●

360924　Serissa capensis Thunb. = Canthium inerme (L. f.) Kuntze ●☆

360925　Serissa democritea Baill. ex Franch. = Serissa serissoides (DC.) Druce ●

360926　Serissa foetida (L. f.) Comm. ;臭六月雪(白丁花,六月雪);Japanese Serissa,Serissa,Snowrose,Thousand Stars,Yellowrim ●

360927　Serissa foetida (L. f.) Comm. 'Flore Pleno';重瓣臭六月雪 ●☆

360928　Serissa foetida (L. f.) Comm. 'Mount Fuji';富士山臭六月雪 ●☆

360929　Serissa foetida (L. f.) Comm. 'Variegata Pink';斑叶粉臭六月雪●☆

360930　Serissa foetida (L. f.) Comm. 'Variegata';花叶六月雪●☆

360931　Serissa foetida (L. f.) Comm. = Serissa japonica (Thunb.) Thunb. ●

360932　Serissa foetida (L. f.) Lam. = Serissa japonica (Thunb.) Thunb. ●

360933　Serissa japonica (Thunb.) Thunb. ;六月雪(白丁花,白马骨,路边金,满天星,喷雪,野丁香);Japan Junesnow,Japanese Serisse,Junesnow ●

360934　Serissa japonica (Thunb.) Thunb. 'Variegata';斑叶六月雪 ●☆

360935　Serissa serissoides (DC.) Druce;白马骨(白点秤,白马里梢,白雪丹,冻米柴,光骨刺,过路黄荆,鸡骨柴,鸡骨头草,鸡脚骨,凉粉草,六月冷,六月雪,路边鸡,路边姜,路边金,路边荆,满天星,米筛花,喷雪花,千年勿大,曲节草,天星木,细牙家,野黄杨树,硬骨柴,永勿大,鱼骨刺,月月有,朱米雪);Junesnow,Serissa,Snow of June,White-flowered Serissa ●

360936　Serjania Mill. (1754);塞战藤属;Supplejack ●☆

360937　Serjania Plum. ex Schum. = Serjania Mill. ●☆

360938　Serjania Schum. = Serjania Mill. ●☆

360939　Serjania curassavica Radlk. ;加勒比塞战藤●☆

360940　Serjania mexicana (L.) Willd. ;墨西哥塞战藤●☆

360941　Serjania racemosa Schum. ;总花塞战藤●☆

360942　Serjania reticulata Cambess. ;网状塞战藤●☆

360943　Serjania schiedeana Schltdl. ;希特塞战藤●☆

360944　Serniphajus Gagnep. = Eulophia R. Br. (保留属名)■

360945　Serophyton Benth. = Argythamnia P. Browne ●☆

360946　Serpaea Gardner = Dimerostemma Cass. + Oyedaea DC. ■☆

360947　Serpenticaulis M. A. Clem. et D. L. Jones = Bulbophyllum Thouars(保留属名)■

360948　Serpenticaulis M. A. Clem. et D. L. Jones(2002);蛇茎兰属■☆

360949　Serpicula L. = Laurembergia P. J. Bergius ■☆

360950　Serpicula L. f. = Hydrilla Rich. ■

360951　Serpicula Pursh = Elodea Michx. ■☆

360952　Serpicula numidica Batt. = Laurembergia tetrandra (Schott) Kanitz ■☆

360953　Serpicula repens L. var. brachypoda Welw. ex Hiern = Laurembergia tetrandra (Schott) Kanitz var. brachypoda (Welw. ex Hiern) A. Raynal ■☆

360954　Serpicula verticillata L. f. = Hydrilla verticillata (L. f.) Royle ■

360955　Serpillaria Fabr. = Illecebrum L. ■☆

360956　Serpillaria Heist. ex Fabr. = Illecebrum L. ■☆

360957　Serpyllum Mill. = Thymus L. ●

360958　Serra Cav. = Senra Cav. ●☆

360959　Serraea Spreng. = Senra Cav. ●☆

360960　Serrafalcus Parl. (1840);假雀麦属■☆

360961　Serrafalcus Parl. = Bromus L. (保留属名)■

360962　Serrafalcus arvensis (L.) Godr. = Bromus arvensis L. ■

360963　Serrafalcus commutatus (Schrad.) Bab. = Bromus racemosus L. ■

360964　Serrafalcus grossus (Desf. ex DC.) Rouy. = Bromus grossus Desf. ex DC. et A. Camus ■

360965　Serrafalcus hordeaceus (L.) Gren. et Godr. = Bromus hordeaceus L. ■

360966　Serrafalcus hughii Tod. = Bromus intermedius Guss. ■

360967　Serrafalcus japonicus (Thunb.) Wilmott = Bromus japonicus Thunb. ■

360968　Serrafalcus lanceolatus (Roth) Parl. = Bromus lanceolatus Roth ■

360969　Serrafalcus macrostachys Parl. = Bromus lanceolatus Roth ■

360970　Serrafalcus mollis (L.) Parl. = Bromus hordeaceus L. ■

360971　Serrafalcus patulus (Mert. et Koch) Parl. = Bromus japonicus Thunb. ■

360972　Serrafalcus racemosus (L.) Parl. = Bromus racemosus L. ■

360973　Serrafalcus racemosus (L.) Parl. subsp. commutatus (Schrad.) Rouy = Bromus racemosus L. ■

360974　Serrafalcus scoparius (L.) Parl. = Bromus scoparius L. ■

360975　Serrafalcus secalinus（L.）Bab. = Bromus secalinus L. ■

360976　Serrafalcus squarrosus（L.）Bab. = Bromus squarrosus L. ■

360977　Serrafalcus squarrosus Bab. = Bromus squarrosus L. ■

360978　Serrafalcus unioloides（Kunth）Samp. = Bromus catharticus Vahl ■

360979　Serraria Adans. = Leucadendron R. Br.（保留属名）●

360980　Serraria Adans. = Serruria Burm. ex Salisb. ●☆

360981　Serrastylis Rolfe = Macradenia R. Br. ■☆

360982　Serratula L.（1753）;麻花头属（升麻属）;Sawwort,Saw-wort ■

360983　Serratula alata S. G. Gmel. = Cirsium alatum（S. G. Gmel.）Bobrov ■

360984　Serratula alatavica C. A. Mey. = Serratula alatavica C. A. Mey. ex Rupr. ■

360985　Serratula alatavica C. A. Mey. ex Rupr.;阿拉套麻花头;Alatao Sawwort ■

360986　Serratula alcalae Coss. = Klasea baetica（DC.）Holub subsp. alcalae（Coss.）Cantó et Rivas Mart. ■☆

360987　Serratula algida Iljin;全叶麻花头;Algid Sawwort,Entireleaf Sawwort ■

360988　Serratula alpina L. = Saussurea alpina（L.）DC. ■

360989　Serratula alpina L. var. angustifolia L. = Saussurea angustifolia（L.）DC. ■☆

360990　Serratula amara L. = Saussurea amara（L.）DC. ■

360991　Serratula ambiqua DC. = Jurinea multiflora（L.）B. Fedtsch. ■

360992　Serratula angulata Kar. et Kir.;窄麻花头■☆

360993　Serratula angustifolia（L.）Willd. = Saussurea angustifolia（L.）DC. ■☆

360994　Serratula aphyllopoda Iljin;无叶柄麻花头■☆

360995　Serratula arvensis L. = Cirsium arvense（L.）Scop. ■

360996　Serratula atriplicifolia（Trevis.）Benth. = Synurus deltoides（Aiton）Nakai ■

360997　Serratula atriplicifolia（Trevis.）Benth. var. excelsa Makino = Synurus deltoides（Aiton）Nakai ■

360998　Serratula atriplicifolia（Trevis.）Benth. var. incisolobata（DC.）Miyabe et Miyake = Synurus deltoides（Aiton）Nakai ■

360999　Serratula baetica DC. = Klasea baetica（DC.）Holub subsp. alcalae（Coss.）Cantó et Rivas Mart. ■☆

361000　Serratula baetica DC. subsp. alcalae（Coss.）Rouy = Klasea baetica（DC.）Holub subsp. alcalae（Coss.）Cantó et Rivas Mart. ■☆

361001　Serratula baetica DC. var. pinnatifolia Willd. = Klasea baetica（DC.）Holub subsp. alcalae（Coss.）Cantó et Rivas Mart. ■☆

361002　Serratula bracteifolia（Iljin）Stankov;具苞麻花头■☆

361003　Serratula canariensis（DC.）Sch. Bip. = Rhaponticum cynaroides Less. ■☆

361004　Serratula carduncula（Pall.）Schischk.;分枝麻花头（飞廉花头,闪光麻花头）;Branched Sawwort,Branchy Sawwort ■

361005　Serratula carthamoides（Buch.-Ham.）Kuntze = Hemistepta lyrata（Bunge）Bunge ■

361006　Serratula carthamoides Poir. = Hemistepta lyrata（Bunge）Bunge ■

361007　Serratula carthamoides Poir. = Rhaponticum carthamoides（Willd.）Iljin ■

361008　Serratula carthamoides Poir. = Stemmacantha carthamoides（Willd.）Dittrich ■

361009　Serratula caspia Pall. = Karelinia caspia（Pall.）Less. ■

361010　Serratula caucasica Boiss.;高加索麻花头■☆

361011　Serratula centauroides L.;麻花头（广东升麻,和尚头,花儿柴,华麻花头,苦郎头,麻花果）;Common Sawwort ■

361012　Serratula centauroides L. var. macrocephala Ledeb. = Serratula centauroides L. ■

361013　Serratula centauroides L. var. microcephala Ledeb. = Serratula polycephala Iljin ■

361014　Serratula chaetocarpa Ledeb. = Jurinea chaetocarpa Ledeb. ex DC. ■

361015　Serratula chanetii H. Lév.;碗苞麻花头（北京麻花头）;Bowlbract Sawwort,Chanet's Sawwort ■

361016　Serratula charbiensis A. I. Baranov et Skvortsov = Serratula cupuliformis Nakai et Kitag. ■

361017　Serratula chartacea C. Winkl.;纸质麻花头■☆

361018　Serratula chinensis S. Moore;华麻花头（广东升麻,蓝肉升麻,麻花头,升麻,鸭麻菜）;China Sawwort,Chinese Sawwort ■

361019　Serratula cichoracea（L.）DC. subsp. mucronata（Desf.）Jahand. et Maire = Klasea flavescens（L.）Holub subsp. mucronata（Desf.）Cantó et Rivas Mart. ■☆

361020　Serratula cichoracea（L.）DC. var. propinqua（Pomel）Maire = Klasea flavescens（L.）Holub subsp. mucronata（Desf.）Cantó et Rivas Mart. ■☆

361021　Serratula coriacea Fisch. et C. A. Mey.;革质麻花头■☆

361022　Serratula coronaria Pall. = Serratula coronata L. ■

361023　Serratula coronata L.;伪泥胡菜（冠麻花头,黄升麻,假升麻,升麻）;Corenate Sawwort,Fake Sawwort ■

361024　Serratula coronata L. subsp. insularis（Iljin）Kitam.;海岛伪泥胡菜■☆

361025　Serratula coronata L. subsp. insularis（Iljin）Kitam. f. albiflora（Kitam.）Kitam.;白花海岛伪泥胡菜■☆

361026　Serratula coronata L. subsp. insularis（Iljin）Kitam. var. koreana（Iljin）Kitam. = Serratula coronata L. subsp. insularis（Iljin）Kitam. ■☆

361027　Serratula coronata L. var. manshurica（Kitag.）Kitag. = Serratula coronata L. ■

361028　Serratula crupina（L.）Vill. = Crupina vulgaris Cass. ■

361029　Serratula cupuliformis Nakai et Kitag.;钟苞麻花头;Bellbract Sawwort ■

361030　Serratula cynarifolia Poir. = Rhaponticum carthamoides（Willd.）Iljin ■

361031　Serratula darrisii H. Lév. = Vernonia spirei Gand. ■

361032　Serratula davurica Adams ex Ledeb. = Saussurea davurica Adams ■

361033　Serratula deltoides（Aiton）Makino = Synurus deltoides（Aiton）Nakai ■

361034　Serratula deltoides（Aiton）Makino f. collinus Kitag. = Synurus deltoides（Aiton）Nakai ■

361035　Serratula deltoides（Aiton）Makino var. inciso-lobatus（DC.）Kitam. = Synurus deltoides（Aiton）Nakai ■

361036　Serratula deltoides（Aiton）Makino var. palmatipinnatifida Makino = Synurus deltoides（Aiton）Nakai ■

361037　Serratula demosti Boriss. = Serratula lyratifolia Schrenk ■

361038　Serratula diabolica Kitam. = Olgaea lomonosowii（Trautv.）Iljin ■

361039　Serratula dissecta Ledeb.;羽裂麻花头;Deepsplit Sawwort,Dissectedleaf Sawwort ■

361040　Serratula dissecta Ledeb. var. aperula Regel et Herder = Serratula alatavica C. A. Mey. ex Rupr. ■

361041　Serratula dissecta Steph. ex Herder = Saussurea runcinata DC. ■

361042　Serratula dissecta Steph. ex Herder var. asperula Regel et Herder = Serratula alatavica C. A. Mey. ex Rupr. ■

361043　Serratula dschungarica Iljin;密毛麻花头（准噶尔麻花头）■

361044　Serratula dschungarica Iljin = Serratula suffruticosa Schrenk ■

361045　Serratula erucifolia（L.）Boriss. ;芥叶麻花头■☆

361046　Serratula excelsus（Makino）Kitam. = Synurus deltoides（Aiton）Nakai ■

361047　Serratula flavescens（L.）Poir. = Klasea flavescens（L.）Holub ■☆

361048　Serratula flavescens（L.）Poir. subsp. mucronata（Desf.）Cantó = Klasea flavescens（L.）Holub subsp. mucronata（Desf.）Cantó et Rivas Mart. ■☆

361049　Serratula flavescens（L.）Poir. var. mucronata（Desf.）Pau et Font Quer = Klasea flavescens（L.）Holub subsp. mucronata（Desf.）Cantó et Rivas Mart. ■☆

361050　Serratula flavescens（L.）Poir. var. propinqua（Pomel）Batt. = Klasea flavescens（L.）Holub subsp. mucronata（Desf.）Cantó et Rivas Mart. ■☆

361051　Serratula flexicaulis Rupr. = Serratula procumbens Regel ■

361052　Serratula forrestii Iljin;滇麻花头；Forrest's Sawwort, Yunnan Sawwort ■

361053　Serratula geblerifolia A. I. Baranov et Skvortsov = Serratula marginata Tausch ■

361054　Serratula glauca L. = Vernonia glauca（L.）Willd. ■☆

361055　Serratula glauca Ledeb. = Serratula marginata Tausch ■

361056　Serratula gmelinii Ledeb. ex DC. = Serratula marginata Tausch ■

361057　Serratula gmelinii Tausch;格氏麻花头■☆

361058　Serratula hastifolia Kult. et Korovin;戟叶麻花头■☆

361059　Serratula heterophylla（L.）Desf. ;异叶麻花头；Diverseleaf Sawwort ■☆

361060　Serratula hondae Geibot. ? = Synurus deltoides（Aiton）Nakai ■

361061　Serratula hsingenensis Kitag. = Serratula centauroides L. ■

361062　Serratula humilis Desf. = Jurinea humilis DC. ☆

361063　Serratula incana S. G. Gmel. = Cirsium incanum（S. G. Gmel.）Fisch. ex M. Bieb. ■

361064　Serratula indica Klein ex Willd. = Goniocaulon indicum（Klein ex Willd.）C. B. Clarke ■☆

361065　Serratula inermis Gilib. ;染麻花头■☆

361066　Serratula insularis Iljin = Serratula coronata L. subsp. insularis（Iljin）Kitam. ■☆

361067　Serratula isophylla Claus;同叶麻花头■☆

361068　Serratula japonica Thunb. = Saussurea japonica（Thunb.）DC. ■

361069　Serratula kirghisorum Iljin;吉尔吉斯麻花头■☆

361070　Serratula komarovii Iljin;科马罗夫麻花头■☆

361071　Serratula komarovii Iljin = Serratula centauroides L. ■

361072　Serratula komarovii Iljin = Serratula chanetii H. Lév. ■

361073　Serratula koreana Iljin = Serratula coronata L. subsp. insularis（Iljin）Kitam. ■☆

361074　Serratula lancifolia Zakirov;剑叶麻花头■☆

361075　Serratula laxmanni Fisch. ex DC. = Serratula marginata Tausch ■

361076　Serratula lyratifolia Schrenk;无茎麻花头（琴叶麻花头, 天山麻花头）;Stemless Sawwort ■

361077　Serratula manshurica Kitag. = Serratula coronata L. ■

361078　Serratula manshuriensis W. Wang = Serratula polycephala Iljin ■

361079　Serratula marginata Tausch;薄叶麻花头（地丁叶麻花头, 球苞麻花头）;Thinleaf Sawwort ■

361080　Serratula martinii Vaniot = Serratula coronata L. ■

361081　Serratula modestii Boriss. = Serratula lyratifolia Schrenk ■

361082　Serratula mongolica Kitag. = Serratula centauroides L. ■

361083　Serratula mucronata Desf. = Klasea flavescens（L.）Holub subsp. mucronata（Desf.）Cantó et Rivas Mart. ■☆

361084　Serratula muiltiflora L. = Jurinea multiflora（L.）B. Fedtsch. ■

361085　Serratula multicaulis Wall. = Hemistepta lyrata（Bunge）Bunge ■

361086　Serratula multiflora L. = Jurinea multiflora（L.）B. Fedtsch. ■

361087　Serratula multiflora L. = Saussurea salicifolia（L.）DC. ■

361088　Serratula nitida Fisch. = Serratula carduncula（Pall.）Schischk. ■

361089　Serratula nitida Fisch. ex Spreng. = Serratula carduncula（Pall.）Schischk. ■

361090　Serratula nitida Fisch. ex Spreng. var. glauca（Ledeb.）Trautv. = Serratula marginata Tausch ■

361091　Serratula noveboracensis L. = Vernonia noveboracensis（L.）Michx. ■☆

361092　Serratula nudicaulis（L.）DC. = Klasea nudicaulis（L.）Cass. ■☆

361093　Serratula nudicaulis（L.）DC. var. albarracinensis（Pau）Maire = Klasea nudicaulis（L.）Cass. ■☆

361094　Serratula nudicaulis（L.）DC. var. ibrahimi（Iljin）Maire = Klasea nudicaulis（L.）Cass. ■☆

361095　Serratula nudicaulis（L.）DC. var. subinermis Coss. = Klasea nudicaulis（L.）Cass. ■☆

361096　Serratula nudicaulis（L.）DC. var. transiens Maire = Klasea nudicaulis（L.）Cass. ■☆

361097　Serratula ortholespis Kitag. = Serratula polycephala Iljin ■

361098　Serratula palmatopinnatifidus（Makino）Kitam. = Synurus deltoides（Aiton）Nakai ■

361099　Serratula palmatopinnatifidus（Makino）Kitam. var. indivisa Kitam. = Synurus deltoides（Aiton）Nakai ■

361100　Serratula parviflora Poir. = Saussurea parviflora（Poir.）DC. ■

361101　Serratula picris（Pall. ex Willd.）M. Bieb. = Acroptilon repens（L.）DC. ■

361102　Serratula pilosa Aiton = Liatris pilosa（Aiton）Willd. ■☆

361103　Serratula pinnatifida（Cav.）Poir. = Klasea pinnatifida（Cav.）Cass. ■☆

361104　Serratula polycephala Iljin;多花麻花头（多头麻花头）;Manyhead Sawwort ■

361105　Serratula polycephala Iljin f. leucantha Kitag. = Serratula polycephala Iljin ■

361106　Serratula polycephala Iljin var. ortholepis（Kitag.）Y. Ling ex H. C. Fu = Serratula polycephala Iljin ■

361107　Serratula polygyna A. Rich. = Laggera crispata（Vahl）Hepper et J. R. I. Wood ■

361108　Serratula potaninii Iljin = Serratula centauroides L. ■

361109　Serratula potaninii Iljin = Serratula chanetii H. Lév. ■

361110　Serratula procumbens Regel;歪斜麻花头；Crookedhead Sawwort, Slanting Sawwort ■

361111　Serratula propinqua Pomel = Klasea flavescens（L.）Holub subsp. mucronata（Desf.）Cantó et Rivas Mart. ■☆

361112　Serratula pulchella（Fisch.）Sims. = Saussurea pulchella（Fisch. ex Hornem.）Fisch. ■

361113　Serratula pulchella Sims = Saussurea pulchella（Fisch.）Fisch. ■

361114　Serratula pungens（Franch. et Sav.）Kitam. = Synurus deltoides（Aiton）Nakai ■

361115　Serratula pungens Franch. et Sav. = Synurus deltoides（Aiton）

Nakai ■

361116　Serratula pungens Franch. et Sav. var. gigantens Kitam. = Synurus deltoides（Aiton）Nakai ■

361117　Serratula quinquifolia M. Bieb. ;五叶麻花头■☆

361118　Serratula radiata（Waldst. et Kit.）M. Bieb. = Serratula centauroides L. ■

361119　Serratula rugosa Iljin;新疆麻花头;Winkled Sawwort, Xinjiang Sawwort ■

361120　Serratula salicifolia L. = Saussurea salicifolia（L.）DC. ■

361121　Serratula salicifolia Lepech. = Jurinea multiflora（L.）B. Fedtsch. ■

361122　Serratula salicina Pall. = Jurinea multiflora（L.）B. Fedtsch. ■

361123　Serratula salsa Pall. = Saussurea salsa（Pall.）Spreng. ■

361124　Serratula scariosa L. = Liatris scariosa（L.）Willd. ■☆

361125　Serratula scoanei Willk. ;矮麻花头■☆

361126　Serratula serratuloides（Fisch. et C. A. Mey.）Takht. ;齿状麻花头■☆

361127　Serratula setosa Willd. = Cirsium setosum（Willd.）M. Bieb. ■

361128　Serratula spicata L. = Liatris spicata（L.）Willd. ■☆

361129　Serratula spinosa Gilib. = Cirsium arvense（L.）Scop. ■

361130　Serratula squarrosa L. = Liatris squarrosa（L.）Michx. ■☆

361131　Serratula strangulata Iljin;缢苞麻花头（麻花头, 蕴苞麻花头）;Contracted Sawwort ■

361132　Serratula suffruticosa Schrenk = Serratula suffruticulosa Schrenk ■

361133　Serratula suffruticosa Trautv. = Serratula suffruticulosa Schrenk ■

361134　Serratula suffruticulosa Schrenk;木根麻花头;Woodroot Sawwort ■

361135　Serratula tanaitica P. A. Smirn. ;塔奈特麻花头■☆

361136　Serratula tenuifolia Bong. = Syreitschikovia tenuifolia（Bong.）Pavlov ■

361137　Serratula tianschanica Saposhn. et Nikitin = Serratula lyratifolia Schrenk ■

361138　Serratula tincta Cass. = Jurinea multiflora（L.）B. Fedtsch. ■

361139　Serratula tinctoria L. ;染色麻花头;Dyer's Plumeless Saw-wort, Dyer's Savory, Dyer's Sawwort, Dyer's Saw-wort, Dyers Serratula, Sawwort ■☆

361140　Serratula tinctoria L. = Hemistepta lyrata（Bunge）Bunge ■

361141　Serratula transcaucasica（Bornm.）Sosn. ex Grossh. ;外高加索麻花头■☆

361142　Serratula trautvetteriana Regel et Schmalh. = Serratula alatavica C. A. Mey. ex Rupr. ■

361143　Serratula uniflora Spreng. = Stemmacantha uniflora（L.）Dittrich ■

361144　Serratula xeranthemoides M. Bieb. ;干花麻花头■☆

361145　Serratula yamatsutana Kitag. = Serratula centauroides L. ■

361146　Serratula yamatsutana Kitag. var. mongolica（Kitag.）Kitag. = Serratula centauroides L. ■

361147　Serratulaceae Martinov = Asteraceae Bercht. et J. Presl（保留科名）●■

361148　Serratulaceae Martinov = Compositae Giseke（保留科名）●■

361149　Serratulaceae Martinov;麻花头科■

361150　Serronia Gaudich. = Piper L. ●■

361151　Serrulata DC. = Serratula L. ■

361152　Serrulataceae Martinov = Asteraceae Bercht. et J. Presl（保留科名）●■

361153　Serruria Adans. corr. Salisb. = Serruria Burm. ex Salisb. ●☆

361154　Serruria Burm. ex Salisb.（1807）;色罗山龙眼属（色罗里阿属）;Serruria ●☆

361155　Serruria Salisb. = Serruria Burm. ex Salisb. ●☆

361156　Serruria adscendens（Lam.）R. Br. ;上举色罗山龙眼●☆

361157　Serruria adscendens（Lam.）R. Br. var. decipiens（R. Br.）Hutch. = Serruria decipiens R. Br. ●☆

361158　Serruria aemula R. Br. ;匹敌色罗山龙眼（匹敌色罗里阿）;Rivalling Serruria ●☆

361159　Serruria aemula Salisb. ex Knight = Serruria aemula R. Br. ●☆

361160　Serruria aemula Salisb. ex Knight var. heterophylla（Meisn.）Hutch. = Serruria heterophylla Meisn. ●☆

361161　Serruria aitonii R. Br. ;澳非色罗山龙眼●☆

361162　Serruria altiscapa Rourke;高花茎色罗山龙眼●☆

361163　Serruria anethifolia Knight = Serruria triternata（Thunb.）R. Br. ●☆

361164　Serruria argentifolia E. Phillips et Hutch. = Serruria aitonii R. Br. ●☆

361165　Serruria artemisiifolia Knight = Serruria pedunculata（Lam.）R. Br. ●☆

361166　Serruria barbigera Knight = Serruria phylicoides（P. J. Bergius）R. Br. ●☆

361167　Serruria biglandulosa Schltr. = Serruria fasciflora Salisb. ex Knight ●☆

361168　Serruria bolusii E. Phillips et Hutch. ;博卢斯色罗山龙眼●☆

361169　Serruria brevifolia E. Phillips et Hutch. = Paranomus capitatus（R. Br.）Kuntze ●☆

361170　Serruria brownii Meisn. ;布朗色罗山龙眼●☆

361171　Serruria burmannii R. Br. = Serruria fasciflora Salisb. ex Knight ●☆

361172　Serruria candicans R. Br. ;纯白色罗山龙眼●☆

361173　Serruria ciliata R. Br. ;缘毛色罗山龙眼●☆

361174　Serruria ciliata R. Br. = Serruria aemula Salisb. ex Knight ●☆

361175　Serruria ciliata R. Br. var. congesta（R. Br.）Hutch. = Serruria aemula Salisb. ex Knight ●☆

361176　Serruria collina Salisb. ex Knight;山丘色罗山龙眼●☆

361177　Serruria confragosa Rourke;粗糙色罗山龙眼●☆

361178　Serruria congesta R. Br. = Serruria aemula Salisb. ex Knight ●☆

361179　Serruria cyanoides（L.）R. Br. ;蓝色色罗山龙眼●☆

361180　Serruria decipiens R. Br. ;迷惑色罗山龙眼●☆

361181　Serruria decumbens（Thunb.）R. Br. ;外倾色罗山龙眼●☆

361182　Serruria dodii E. Phillips et Hutch. ;多德色罗山龙眼●☆

361183　Serruria effusa Rourke;开展色罗山龙眼●☆

361184　Serruria elongata（P. J. Bergius）R. Br. ;伸长色罗山龙眼●☆

361185　Serruria fasciflora Salisb. ex Knight;簇花色罗山龙眼●☆

361186　Serruria flagellaris R. Br. = Serruria collina Salisb. ex Knight ●☆

361187　Serruria flagellifolia Salisb. ex Knight;鞭叶色罗山龙眼●☆

361188　Serruria flava Meisn. ;黄色罗山龙眼●☆

361189　Serruria florida（Thunb.）Salisb. ex Knight;佛罗里达色罗山龙眼（佛罗里达色罗里阿, 色罗里阿）;Blushing Bride, Blushing-bride ●☆

361190　Serruria florida Knight = Serruria florida（Thunb.）Salisb. ex Knight ●☆

361191　Serruria foeniculacea R. Br. = Serruria aemula Salisb. ex Knight ●☆

361192　Serruria furcellata R. Br. ;叉色罗山龙眼●☆

361193　Serruria glomerata（L.）R. Br. ;团集色罗山龙眼●☆

361194　Serruria gracilis Knight;纤细色罗山龙眼●☆

361195　Serruria heterophylla Meisn. ;互叶色罗山龙眼●☆

361196　Serruria hirsuta R. Br. ;粗毛色罗山龙眼●☆

361197　Serruria hyemalis Knight = Serruria decumbens (Thunb.) R. Br. ●☆

361198　Serruria inconspicua L. Guthrie et T. M. Salter;显著色罗山龙眼●☆

361199　Serruria incrassata Meisn. ;粗色罗山龙眼●☆

361200　Serruria knightii Hutch. = Serruria fasciflora Salisb. ex Knight ●☆

361201　Serruria kraussii Meisn. ;克劳斯罗山龙眼●☆

361202　Serruria lacunosa Rourke;具腔色罗山龙眼●☆

361203　Serruria leipoldtii E. Phillips et Hutch. ;莱氏色罗山龙眼●☆

361204　Serruria linearis Salisb. ex Knight;线状色罗山龙眼●☆

361205　Serruria longipes E. Phillips et Hutch. = Serruria pedunculata (Lam.) R. Br. ●☆

361206　Serruria millefolia Salisb. ex Knight;粟草叶色罗山龙眼●☆

361207　Serruria nervosa Meisn. ;多脉色罗山龙眼●☆

361208　Serruria nivenii Salisb. ex Knight;尼文色罗山龙眼●☆

361209　Serruria pauciflora E. Phillips et Hutch. = Serruria fasciflora Salisb. ex Knight ●☆

361210　Serruria pedunculata (Lam.) R. Br. ;梗花色罗山龙眼●☆

361211　Serruria phylicoides (P. J. Bergius) R. Br. ;菲利木色罗山龙眼●☆

361212　Serruria pinnata R. Br. ;羽状色罗山龙眼●☆

361213　Serruria plumosa Meisn. = Serruria nivenii Salisb. ex Knight ●☆

361214　Serruria reflexa Rourke;反折色罗山龙眼●☆

361215　Serruria rosea E. Phillips;粉红色罗山龙眼●☆

361216　Serruria rostellaris Salisb. ex Knight;喙状色罗山龙眼●☆

361217　Serruria roxburghii R. Br. ;罗氏色罗山龙眼●☆

361218　Serruria rubricaulis R. Br. ;红茎色罗山龙眼●☆

361219　Serruria scariosa R. Br. = Serruria nivenii Salisb. ex Knight ●☆

361220　Serruria scoparia R. Br. ;帚状色罗山龙眼●☆

361221　Serruria simplicifolia R. Br. = Serruria linearis Salisb. ex Knight ●☆

361222　Serruria stellata Rourke;星状色罗山龙眼●☆

361223　Serruria subsericea Hutch. = Serruria fasciflora Salisb. ex Knight ●☆

361224　Serruria trilopha Salisb. ex Knight;三冠色罗山龙眼●☆

361225　Serruria triternata (Thunb.) R. Br. ;三出色罗山龙眼●☆

361226　Serruria vallaris Knight = Serruria villosa (Lam.) R. Br. ●☆

361227　Serruria ventricosa E. Phillips et Hutch. = Serruria nervosa Meisn. ●☆

361228　Serruria villosa (Lam.) R. Br. ;长柔毛色罗山龙眼●☆

361229　Serruria viridifolia Rourke;绿叶色罗山龙眼●☆

361230　Serruria williamsii Rourke;威廉斯色罗山龙眼●☆

361231　Serruria zeyheri Meisn. ;泽赫斯色罗山龙眼●☆

361232　Sersalisia R. Br. = Lucuma Molina + Planchonella Pierre(保留属名)●

361233　Sersalisia R. Br. = Pouteria Aubl. ●

361234　Sersalisia afzelii Engl. = Synsepalum afzelii (Engl.) T. D. Penn. ●☆

361235　Sersalisia australis (R. Br.) Domin = Planchonella australis (R. Br.) Pierre ●☆

361236　Sersalisia brevipes (Baker) Baill. = Synsepalum brevipes (Baker) T. D. Penn. ●☆

361237　Sersalisia buluensis Greves = Pachystela buluensis (Greves) Aubrév. et Pellegr. ●☆

361238　Sersalisia cerasifera (Welw.) Engl. = Synsepalum cerasiferum (Welw.) T. D. Penn. ●☆

361239　Sersalisia chevalieri Engl. = Synsepalum cerasiferum (Welw.) T. D. Penn. ●☆

361240　Sersalisia disaco (Hiern) Engl. = Synsepalum cerasiferum (Welw.) T. D. Penn. ●☆

361241　Sersalisia djalonensis Aubrév. et Pellegr. = Synsepalum cerasiferum (Welw.) T. D. Penn. ●☆

361242　Sersalisia edulis S. Moore = Synsepalum cerasiferum (Welw.) T. D. Penn. ●☆

361243　Sersalisia kaessneri Engl. = Synsepalum kaessneri (Engl.) T. D. Penn. ●☆

361244　Sersalisia laurentii De Wild. = Pachystela laurentii (De Wild.) C. M. Evrard ●☆

361245　Sersalisia ledermannii Engl. et K. Krause = Synsepalum cerasiferum (Welw.) T. D. Penn. ●☆

361246　Sersalisia malchairi De Wild. = Englerophytum oblanceolatum (S. Moore) T. D. Penn. ●☆

361247　Sersalisia micrantha (A. Chev.) Aubrév. et Pellegr. = Synsepalum afzelii (Engl.) T. D. Penn. ●☆

361248　Sersalisia microphylla A. Chev. = Synsepalum afzelii (Engl.) T. D. Penn. ●☆

361249　Sersalisia obovata R. Br. = Planchonella obovata (R. Br.) Pierre ●

361250　Sersalisia usambarensis Engl. = Synsepalum cerasiferum (Welw.) T. D. Penn. ●☆

361251　Sertifera Lindl. = Sertifera Lindl. ex Rchb. f. ■☆

361252　Sertifera Lindl. et Rchb. f. = Sertifera Lindl. ex Rchb. f. ■☆

361253　Sertifera Lindl. ex Rchb. f. (1876);环花兰属■☆

361254　Sertifera colombiana Schltr. ;哥伦比亚环花兰■☆

361255　Sertifera grandifolia L. O. Williams;大叶环花兰■☆

361256　Sertifera major Schltr. ;大环花兰■☆

361257　Sertifera parviflora Schltr. ;小花环花兰■☆

361258　Sertifera purpurea Rchb. f. ;紫环花兰■☆

361259　Sertuernera Mart. = Pfaffia Mart. ■☆

361260　Sertula Kuntze = Trifolium L. ■

361261　Sertula L. = Melilotus (L.) Mill. ■

361262　Serturnera Mart. = Pfaffia Mart. ■☆

361263　Seruneum Kuntze = Wedelia Jacq. (保留属名)■●

361264　Seruneum Rumph. = Wedelia Jacq. (保留属名)■●

361265　Seruneum Rumph. ex Kuntze = Wedelia Jacq. (保留属名)■●

361266　Serveria Neck. = Doliocarpus Rol. ●☆

361267　Sesamaceae Horan. = Pedaliaceae R. Br. (保留科名)●■

361268　Sesamaceae Horan. = Sesuviaceae Horan. ■

361269　Sesamaceae R. Br. ex Bercht. = Pedaliaceae R. Br. (保留科名)●■

361270　Sesamella Rchb. = Sesamoides Ortega ■☆

361271　Sesamodes Kuntze = Astrocarpa Neck. ex Dumort. ■☆

361272　Sesamodes Kuntze = Sesamoides Ortega ■☆

361273　Sesamoides All. = Sesamoides Ortega ■☆

361274　Sesamoides Ortega(1773);拟胡麻属;Astrocarpus ■☆

361275　Sesamoides purpurascens (L.) G. López;紫色拟胡麻■☆

361276　Sesamoides purpurascens (L.) G. López subsp. spathulata (Moris) Lambinon et Kerguélen = Sesamoides pygmaea (Scheele) Kuntze ■☆

361277　Sesamoides pygmaea (Scheele) Kuntze;小拟胡麻■☆

361278　Sesamoides spathulifolia (Boreau) Rothm. ;匙叶拟胡麻■☆

361279　Sesamoides suffruticosa (Lange) Kuntze;亚灌木胡麻●☆

361280　Sesamopteris（Endl.）Meisn. = Sesamum L. ■●

361281　Sesamopteris DC. ex Meisn. = Sesamum L. ■●

361282　Sesamopteris alata（Thonn.）DC. = Sesamum alatum Thonn. ■☆

361283　Sesamopteris radiata（Schumach. et Thonn.）DC. = Sesamum radiatum Schumach. et Thonn. ■☆

361284　Sesamothamnus Welw.（1869）;壶茎麻属●☆

361285　Sesamothamnus benguellensis Welw.;本格拉壶茎麻●☆

361286　Sesamothamnus busseanus Engl.;布瑟壶茎麻●☆

361287　Sesamothamnus erlangeri Engl. = Sesamothamnus rivae Engl. ●☆

361288　Sesamothamnus guerichii（Engl.）E. A. Bruce;盖里克壶茎麻●☆

361289　Sesamothamnus lugardii N. E. Br. ex Stapf;卢格德壶茎麻●☆

361290　Sesamothamnus rivae Engl.;沟壶茎麻●☆

361291　Sesamothamnus seineri Engl. = Sesamothamnus lugardii N. E. Br. ex Stapf ●☆

361292　Sesamothamnus smithii Baker ex Stapf = Sesamothamnus rivae Engl. ●☆

361293　Sesamum Adans. = Martynia L. ■

361294　Sesamum L.（1753）;胡麻属（芝麻属,脂麻属）;Sesame ■●

361295　Sesamum abbreviatum Merxm.;缩短胡麻■☆

361296　Sesamum alatum Thonn.;翅胡麻■☆

361297　Sesamum angolense Welw.;安哥拉胡麻■☆

361298　Sesamum angustifolium（Oliv.）Engl.;窄叶胡麻■☆

361299　Sesamum antirrhinoides Welw. ex Asch. = Sesamum schinzianum Asch. ■☆

361300　Sesamum baumii Stapf = Sesamum calycinum Welw. subsp. baumii（Stapf）Seidenst. ex Ihlenf. ■☆

361301　Sesamum biapiculatum De Wild. = Sesamum radiatum Schumach. et Thonn. ■☆

361302　Sesamum caillei A. Chev. = Sesamum radiatum Schumach. et Thonn. ■☆

361303　Sesamum calycinum Welw.;萼状胡麻■☆

361304　Sesamum calycinum Welw. subsp. baumii（Stapf）Seidenst. ex Ihlenf.;鲍姆胡麻■☆

361305　Sesamum calycinum Welw. subsp. pseudoangolense Seidenst. ex Ihlenf.;假安哥拉胡麻■☆

361306　Sesamum calycinum Welw. subsp. repens（Engl. et Gilg）Seidenst. = Sesamum calycinum Welw. ■☆

361307　Sesamum calycinum Welw. var. angustifolium（Oliv.）Ihlenf. et Seidenst. = Sesamum angustifolium（Oliv.）Engl. ■☆

361308　Sesamum calycinum Welw. var. calycinum Ihlenf. et Seidenst. = Sesamum calycinum Welw. subsp. baumii（Stapf）Seidenst. ex Ihlenf. ■☆

361309　Sesamum capense Burm. f.;好望角胡麻■☆

361310　Sesamum digitaloides Welw. ex Schinz = Sesamum rigidum Peyr. ■☆

361311　Sesamum dinteri Schinz = Sesamum marlothii Engl. ■☆

361312　Sesamum gibbosum Bremek. et Oberm. = Sesamum triphyllum Welw. ex Asch. ■☆

361313　Sesamum gracile Endl. = Sesamum alatum Thonn. ■☆

361314　Sesamum grandiflorum Schinz = Sesamum triphyllum Welw. ex Asch. var. grandiflorum（Schinz）Merxm. ■☆

361315　Sesamum heudelotii Stapf = Ceratotheca sesamoides Endl. ■☆

361316　Sesamum hopkinsii Suess. = Sesamum indicum L. ■

361317　Sesamum indicum L.;芝麻（白胡麻,白油麻,白脂麻,狗虱,

黑油麻,黑芝麻,鸿藏,胡麻,机麻,交麻,巨胜,巨胜苗,麻,蔓,梦神,青蘘,乌麻,小胡麻,油麻,油子苗,芝麻花,脂麻）;Bene, Benne, Benniseed, Donegal, Gingelly, Halvah, Oilseed, Oily-grain Plant, Oriental Sesame, Sesame, Sesamum, Simsim, Tahini, Thunderbolt Plant ■

361318　Sesamum indicum L. = Sesamum orientale L. ■

361319　Sesamum indicum L. var. angustifolium Oliv. = Sesamum angustifolium（Oliv.）Engl. ■☆

361320　Sesamum indicum L. var. integerrimum Engl. = Sesamum indicum L. ■

361321　Sesamum lamiifolium Engl. = Ceratotheca triloba（Bernh.）Hook. f. ■☆

361322　Sesamum latifolium J. B. Gillett;阔叶胡麻■☆

361323　Sesamum lepidotum Schinz = Sesamum triphyllum Welw. ex Asch. ■☆

361324　Sesamum macranthum Oliv. = Sesamum angolense Welw. ■☆

361325　Sesamum marlothii Engl.;马洛斯胡麻■☆

361326　Sesamum microcarpum Engl. = Sesamum pedalioides Welw. ex Hiern ■☆

361327　Sesamum mombazense De Wild. et T. Durand = Sesamum radiatum Schumach. et Thonn. ■☆

361328　Sesamum orientale L. = Sesamum indicum L. ■

361329　Sesamum parviflorum Seidenst.;小花胡麻■☆

361330　Sesamum pedalioides Welw. ex Hiern;普通胡麻■☆

361331　Sesamum pterospermum R. Br. = Sesamum alatum Thonn. ■☆

361332　Sesamum radiatum Schumach. et Thonn.;黑胡麻■☆

361333　Sesamum repens Engl. et Gilg = Sesamum calycinum Welw. ■☆

361334　Sesamum rigidum Peyr.;硬胡麻■☆

361335　Sesamum rigidum Peyr. var. digitaloides（Welw. ex Schinz）Stapf = Sesamum rigidum Peyr. ■☆

361336　Sesamum rostratum Hochst. = Sesamum alatum Thonn. ■☆

361337　Sesamum sabulosum A. Chev. = Sesamum alatum Thonn. ■☆

361338　Sesamum schenckii Asch. = Sesamum triphyllum Welw. ex Asch. var. grandiflorum（Schinz）Merxm. ■☆

361339　Sesamum schinzianum Asch.;欣兹胡麻■☆

361340　Sesamum somalense Chiov. = Sesamum indicum L. ■

361341　Sesamum talbotii Wernham = Sesamum radiatum Schumach. et Thonn. ■☆

361342　Sesamum thonneri De Wild. et T. Durand = Sesamum radiatum Schumach. et Thonn. ■☆

361343　Sesamum triphyllum Welw. ex Asch.;三叶胡麻■☆

361344　Sesamum triphyllum Welw. ex Asch. var. grandiflorum（Schinz）Merxm.;大花三叶胡麻■☆

361345　Sesban Adans.（废弃属名）= Sesbania Scop.（保留属名）●■

361346　Sesban aculeata（Willd.）Poir. = Sesbania bispinosa（Jacq.）W. Wight ■

361347　Sesban aegyptiaca Poir. = Sesbania sesban（L.）Merr. ●

361348　Sesban grandiflora（L.）Poir. = Sesbania grandiflora（L.）Pers. ●

361349　Sesbana R. Br. = Sesbania Scop.（保留属名）●■

361350　Sesbania Scop.（1777）（保留属名）;田菁属（木田菁属,田青属）;Pea-Tree,Sesbania,Sesbanie ●■

361351　Sesbania aculeata（Willd.）Pers. = Sesbania bispinosa（Jacq.）W. Wight ■

361352　Sesbania aculeata（Willd.）Pers. var. cannabina（Retz.）Baker = Sesbania cannabina（Retz.）Poir. ■

361353　Sesbania aculeata（Willd.）Pers. var. paludosa（Roxb.）Baker

= Sesbania javanica Miq. ■

361354　Sesbania aculeata (Willd.) Poir. = Sesbania bispinosa (Jacq.) W. Wight ■

361355　Sesbania aculeata Pers. = Sesbania bispinosa (Jacq.) W. Wight ■

361356　Sesbania aculeata Pers. var. micrantha Chiov. = Sesbania bispinosa (Jacq.) W. Wight ■

361357　Sesbania aegyptiaca Pers. ;埃及田菁;Peabush ●

361358　Sesbania aegyptiaca Poir. = Sesbania sesban (L.) Merr. ●

361359　Sesbania aegyptiaca Poir. var. bicolor Wight et Arn = Sesbania sesban (L.) Merr. var. bicolor (Wight et Arn.) F. W. Andréws ●

361360　Sesbania aegyptiaca Poir. var. bicolor Wight et Arn. = Sesbania sesban (L.) Merr. ●

361361　Sesbania aegyptiaca Poir. var. concolor Wight et Arn. = Sesbania sesban (L.) Merr. ●

361362　Sesbania aegyptiaca Poir. var. picta Prain = Sesbania sesban (L.) Merr. ●

361363　Sesbania aegyptica Poir. = Sesbania sesban (L.) Merr. ●

361364　Sesbania affinis De Wild. = Sesbania wildemanii E. Phillips et Hutch. ■☆

361365　Sesbania arabica Hochst. et Steud. ex E. Phillips et Hutch. = Sesbania leptocarpa DC. ■☆

361366　Sesbania atropurpurea Taub. = Sesbania sesban (L.) Merr. var. bicolor (Wight et Arn.) F. W. Andréws ●

361367　Sesbania bispinosa (Jacq.) W. Wight;刺田菁(多刺田菁);Dhaincha,Doublespine Sesbania,Dunchi Fiber,Spiny Sesbania ■

361368　Sesbania bispinosa (Jacq.) W. Wight var. micrantha (Chiov.) J. B. Gillett = Sesbania bispinosa (Jacq.) W. Wight ■

361369　Sesbania brevipedunculata J. B. Gillett;短梗田菁●☆

361370　Sesbania cannabina (Retz.) Poir. ;田菁(铁精草,向天蜈蚣,叶顶珠);Common Sesbania,Sesbania ■

361371　Sesbania cinerascens Welw. ex Baker;浅灰田菁●☆

361372　Sesbania cochinchinensis DC. = Sesbania javanica Miq. ■

361373　Sesbania cochinchinensis Kurz = Sesbania javanica Miq. ■

361374　Sesbania coerulescens Harms;浅蓝田菁●☆

361375　Sesbania concolor Gillett;同色田菁●☆

361376　Sesbania confaloniana (Chiov.) Chiov. = Sesbania sesban (L.) Merr. var. nubica Chiov. ■☆

361377　Sesbania dalzielii E. Phillips et Hutch. ;达尔齐尔田菁■☆

361378　Sesbania drummondii (Rydb.) Cory;德拉蒙德田菁;Coffeebean,Rattlebox,Rattle-box,Rattlebrush ●☆

361379　Sesbania dummeri E. Phillips et Hutch. ;达默田菁■☆

361380　Sesbania exaltata (Raf.) Rydb. ex A. W. Hill;科罗拉多田菁;Bequilla,Colorado River Hemp,Daisha,Danglepod,Sessban ■☆

361381　Sesbania filiformis Guillaumin et Perr. = Sesbania leptocarpa DC. ■☆

361382　Sesbania goetzei Harms;格兹田菁■☆

361383　Sesbania goetzei Harms subsp. multiflora J. B. Gillett;多花格兹田菁■☆

361384　Sesbania grandiflora (L.) Pers. ;木田菁(白蝴蝶,大花田菁,大花田青,红蝴蝶,红藤,蝴蝶草,黄花马豆,疆蛇通,三叶红藤,铁马豆,小红藤);Agati,Agati Sesbania,Bakphul,Scarlet Wisteria Tree,Vegetable Hummingbird ●

361385　Sesbania grandiflora (L.) Pers. ' Alba';白木田菁(白蝴蝶);Agati Sesbania,Vegetable-hummingbird ●

361386　Sesbania grandiflora Miq. = Sesbania javanica Miq. ■

361387　Sesbania greenwayi J. B. Gillett;格林韦田菁■☆

361388　Sesbania hamata E. Phillips et Hutch. = Sesbania tetraptera

Hochst. ex Baker ■☆

361389　Sesbania hepperi J. B. Gillett;赫佩田菁■☆

361390　Sesbania herbacea (Mill.) McVaugh;草田菁;Coffeeweed ■●

361391　Sesbania hirticalyx Cronquist = Sesbania rostrata Bremek. et Oberm. ●■☆

361392　Sesbania hirtistyla J. B. Gillett;毛柱田菁■☆

361393　Sesbania hookii De Wild. = Sesbania coerulescens Harms ●☆

361394　Sesbania javanica Miq. ;爪哇田菁(越南田菁,沼生田菁);Marsh Sesbania ■

361395　Sesbania kapangensis Cronquist = Sesbania cinerascens Welw. ex Baker ●☆

361396　Sesbania keniensis J. B. Gillett;肯尼亚田菁■☆

361397　Sesbania kirkii E. Phillips et Hutch. = Sesbania tetraptera Hochst. ex Baker ■☆

361398　Sesbania leptocarpa DC. ;细果田菁■☆

361399　Sesbania leptocarpa DC. var. confaloniana Chiov. = Sesbania sesban (L.) Merr. var. nubica Chiov. ■☆

361400　Sesbania leptocarpa DC. var. minimiflora (J. B. Gillett) G. P. Lewis;小花细果田菁■☆

361401　Sesbania macowaniana Schinz;麦克欧文田菁■☆

361402　Sesbania macrantha Welw. ex E. Phillips et Hutch. ;大花田菁■☆

361403　Sesbania macrantha Welw. ex E. Phillips et Hutch. var. levis J. B. Gillett;平滑大花田菁■☆

361404　Sesbania macrocarpa Muhl. = Sesbania exaltata (Raf.) Rydb. ex A. W. Hill ■☆

361405　Sesbania madagascariensis Du Puy et Labat;马岛田菁●☆

361406　Sesbania marginata Benth. = Sesbania virgata (Cav.) Pers. ■☆

361407　Sesbania melanocaulis Bidgood et Friis;黑茎田菁●☆

361408　Sesbania microphylla Harms;小叶田菁●☆

361409　Sesbania mossambicensis Klotzsch = Sesbania leptocarpa DC. ■☆

361410　Sesbania mossambicensis Klotzsch subsp. minimiflora J. B. Gillett = Sesbania leptocarpa DC. var. minimiflora (J. B. Gillett) G. P. Lewis ■☆

361411　Sesbania multijuga Schweinf. ex Baker = Sesbania bispinosa (Jacq.) W. Wight ■

361412　Sesbania pachycarpa DC. ;厚果田菁■☆

361413　Sesbania pachycarpa DC. subsp. dinterana J. B. Gillett;丁特田菁■☆

361414　Sesbania paludosa (Roxb.) King = Sesbania javanica Miq. ■

361415　Sesbania paludosa (Roxb.) Prain = Sesbania javanica Miq. ■

361416　Sesbania paucisemina J. B. Gillett;寡籽田菁■☆

361417　Sesbania pterocarpa Welw. ex Romariz = Sesbania sphaerosperma Welw. ■☆

361418　Sesbania pubescens DC. ;短毛田菁■☆

361419　Sesbania pubescens DC. = Sesbania sericea (Willd.) Link ■☆

361420　Sesbania punctata DC. ;斑点田菁■☆

361421　Sesbania punctata DC. = Sesbania sesban (L.) Merr. subsp. punctata (DC.) J. B. Gillett ●☆

361422　Sesbania punicea (Cav.) Benth. ;橙田菁(红紫田菁);False Poinciana, Orange Wisteria Shrub, Purple Rattlebox, Purple Rattle-box, Purple Sesban, Purple Sesbane, Rattle Box, Rattlebox, Rattlebush,Scarlet Wistaria Tree,Spanish Gold ●☆

361423　Sesbania punicea Benth. = Sesbania punicea (Cav.) Benth. ●☆

361424　Sesbania quadrata J. B. Gillett;四田菁■☆

361425　Sesbania rogersii E. Phillips et Hutch. ;洛基田菁■☆

361426　Sesbania rostrata Bremek. et Oberm. ;长喙田菁●■☆

361427　Sesbania roxburghii Merr. ;罗氏田菁(田菁);Roxburgh Sesbania ■

361428　Sesbania roxburghii Merr. = Sesbania javanica Miq. ■

361429　Sesbania sericea (Willd.) Link;绢毛田菁■☆

361430　Sesbania sesban (L.) Merr. ;印度田菁(埃及田菁,臭青仔); Egyptian Riverhemp, Egyptian Sesban, India Sesbania, Indian Sesbania , Sesban ●

361431　Sesbania sesban (L.) Merr. subsp. punctata (DC.) J. B. Gillett;斑点印度田菁●☆

361432　Sesbania sesban (L.) Merr. var. bicolor (Wight et Arn.) F. W. Andréws;元江田菁(阴草树);Twocoloured Sesbania, Yuanjiang Sesbania ●

361433　Sesbania sesban (L.) Merr. var. bicolor (Wight et Arn.) F. W. Andréws = Sesbania sesban (L.) Merr. ●

361434　Sesbania sesban (L.) Merr. var. concolor (Wight et Arn.) Baquar = Sesbania sesban (L.) Merr. ●

361435　Sesbania sesban (L.) Merr. var. muricata Baquar;粗糙印度田菁■☆

361436　Sesbania sesban (L.) Merr. var. nubica Chiov. ;云雾田菁■☆

361437　Sesbania sesban (L.) Merr. var. zambesiaca J. B. Gillett = Sesbania sesban (L.) Merr. var. nubica Chiov. ■☆

361438　Sesbania sinuo-carinata Ali = Sesbania pachycarpa DC. ■☆

361439　Sesbania somalensis J. B. Gillett;索马里田菁■☆

361440　Sesbania speciosa Taub. ;美丽田菁■☆

361441　Sesbania sphaerocarpa Hiern = Sesbania sphaerosperma Welw. ■☆

361442　Sesbania sphaerosperma Welw. ;球籽田菁■☆

361443　Sesbania subalata J. B. Gillett;翅田菁■☆

361444　Sesbania sudanica J. B. Gillett;苏丹田菁■☆

361445　Sesbania tchadica A. Chev. = Sesbania sesban (L.) Merr. var. nubica Chiov. ■☆

361446　Sesbania tetraptera Hochst. ex Baker;四翅田菁■☆

361447　Sesbania tetraptera Hochst. ex Baker subsp. rogersii (E. Phillips et Hutch.) G. P. Lewis;罗杰斯田菁■☆

361448　Sesbania transvaalensis J. B. Gillett;德兰士瓦田菁■☆

361449　Sesbania tripetii (Poit.) F. T. Hubb. ;特氏田菁; Scarlet Wisteria Tree, Sesbania ●☆

361450　Sesbania tripetii (Poit.) F. T. Hubb. = Sesbania punicea (Cav.) Benth. ●☆

361451　Sesbania tripetii F. T. Hubb. = Sesbania punicea (Cav.) Benth. ●☆

361452　Sesbania tripetii F. T. Hubb. = Sesbania tripetii (Poit.) F. T. Hubb. ●☆

361453　Sesbania vesicaria (Jacq.) Elliott;囊状田菁;Bagpod, Bladder Pod , Bladderpod , Coffeebean ■☆

361454　Sesbania virgata (Cav.) Pers. ;杖田菁;Wand Riverhemp ■☆

361455　Sesbania wildemanii E. Phillips et Hutch. ;怀尔德曼田菁■☆

361456　Seseli L. (1753);西风芹属(邪蒿属);Meadow Saxifrage, Moon-carrot , Seseli ■

361457　Seseli abolinii (Korovin) Schischk. ;阿保林西风芹■☆

361458　Seseli abolinii (Korovin) Schischk. = Libanotis abolinii (Korovin) Korovin ■

361459　Seseli acaule (R. H. Shan et M. L. Sheh) V. M. Vinogr. = Libanotis acaulis R. H. Shan et M. L. Sheh ■

361460　Seseli aemulans Popov;大果西风芹;Bigfruit Seseli ■☆

361461　Seseli albescens (Franch.) Pimenov et Kljuykov = Eriocycla albescens (Franch.) H. Wolff ■

361462　Seseli alexeenkoi Lipsky;阿莱西风芹■☆

361463　Seseli altissimum Popov = Libanotis iliensis (Lipsky) Korovin ■

361464　Seseli andronakii Woronow;安德西风芹■☆

361465　Seseli annuum L. ;一年生邪蒿■☆

361466　Seseli arenarium M. Bieb. ;野邪蒿■☆

361467　Seseli asperulum (Trautv.) Schischk. ;微毛西风芹(碎叶西风芹);Roughish Seseli ■

361468　Seseli asperum (Thunb.) Sond. = Sonderina hispida (Thunb.) H. Wolff ■☆

361469　Seseli benghalense Roxb. = Oenanthe benghalensis Benth. et Hook. f. ■

361470　Seseli buchtormense (Fisch.) W. D. J. Koch = Libanotis buchtormensis (Fisch.) DC. ■

361471　Seseli caffrum Meisn. = Peucedanum caffrum (Meisn.) E. Phillips ■☆

361472　Seseli campestre Besser;平原西风芹■☆

361473　Seseli condensatum (L.) Rchb. f. = Libanotis condensata Crantz ■

361474　Seseli coreanum H. Wolff = Libanotis seseloides (Fisch. et C. A. Mey.) Turcz. ■

361475　Seseli coronatum Ledeb. ;柱冠西风芹;Stylecorolla Seseli ■

361476　Seseli coronatum Ledeb. var. asperulum Trautv. = Seseli asperulum (Trautv.) Schischk. ■

361477　Seseli cuneifolium M. Bieb. ;楔叶西风芹■☆

361478　Seseli cyclolobum (Gilli) Pimenov et Sdobnina = Libanotis buchtormensis (Fisch.) DC. ■

361479　Seseli daucifolium C. B. Clarke = Cnidium monnieri (L.) Cusson ■

361480　Seseli delavayi Franch. ;多毛西风芹(多毛邪蒿,鸡脚防风,毛果竹叶防风,云防风,竹叶防风);Hairy Seseli ■

361481　Seseli depressum (R. H. Shan et M. L. Sheh) V. M. Vinogr. = Libanotis depressa R. H. Shan et M. L. Sheh ■

361482　Seseli dichotomum Pall. ;二叉邪蒿■☆

361483　Seseli diffusum (Roxb. ex Sm.) Santapau et Wagh;印度西风芹(印度邪蒿);India Seseli ■☆

361484　Seseli dolichostylum (Schischk.) M. Hiroe = Ligusticum mucronatum (Schrenk) Leute ■

361485　Seseli elegans Schischk. ;雅致西风芹■☆

361486　Seseli eriocarpum (Schrenk) B. Fedtsch. = Libanotis eriocarpa Schrenk ■

361487　Seseli eriocephalum (Pall. ex Spreng.) Schischk. ;毛序西风芹(绵毛头邪蒿);Hairumbel Seseli ■

361488　Seseli fasciculatum Korovin;簇生西风芹■☆

361489　Seseli fedtschenkoanum Regel et Schmalh. ex Regel var. iliense Regel et Schmalh. = Libanotis iliensis (Lipsky) Korovin ■

361490　Seseli filifolium Thunb. = Itasina filifolia (Thunb.) Raf. ■☆

361491　Seseli foeniculum (L.) Koso-Pol. = Foeniculum vulgare (L.) Mill. ■

361492　Seseli foliosum (Sommier et H. Lév.) Manden. ;繁叶西风芹■☆

361493　Seseli giganteum Lipsky;巨西风芹■☆

361494　Seseli giraldii Diels = Libanotis buchtormensis (Fisch.) DC. ■

361495　Seseli glabratum Willd. ex Schult. ;膜盘西风芹(细叶邪蒿);Glabrous Seseli ■

361496　Seseli grandivittatum (Sommier et H. Lév.) Schischk. ;单纹西风芹■☆

361497　Seseli graveolens (L.) Scop. = Apium graveolens L. ■

361498　Seseli graveolens Ledeb. = Libanotis incana (Stephan) O. Fedtsch. et B. Fedtsch. ■

361499　Seseli grubovii V. M. Vinogr. et Sanchir = Libanotis grubovii (V. M. Vinogr. et Sanchir) M. L. Sheh et M. F. Watson ■

361500　Seseli gummiferum Pall. ex Sm. ;胶邪蒿■☆

361501　Seseli hippomarathrum Jacq. ;马邪蒿(苦茴香,苦邪蒿)■☆

361502　Seseli iliense (Regel et Schmalh.) Lipsky = Libanotis iliensis (Lipsky) Korovin ■

361503　Seseli iliense Lipsky = Libanotis iliensis (Lipsky) Korovin ■

361504　Seseli incanum (Stephan ex Willd.) B. Fedtsch. = Libanotis incana (Stephan) O. Fedtsch. et B. Fedtsch. ■

361505　Seseli incanum (Stephan) B. Fedtsch. = Libanotis incana (Stephan) O. Fedtsch. et B. Fedtsch. ■

361506　Seseli incisodentatum K. T. Fu;锐齿西风芹(黄花邪蒿); Sharptooth Seseli ■

361507　Seseli indicum Wight et Arn. = Seseli diffusum (Roxb. ex Sm.) Santapau et Wagh ■☆

361508　Seseli intramongolicum Ma;内蒙古西风芹(内蒙古邪蒿); Inner Mongol Seseli ■

361509　Seseli jinanense (L. C. Xu et M. D. Xu) Pimenov = Libanotis jinanensis L. C. Xu et M. D. Xu ■

361510　Seseli junatovii V. M. Vinogr. ;硬枝西风芹■

361511　Seseli karatavicum Schischk. ;卡拉塔夫西风芹■☆

361512　Seseli korovinil Schischk. ;科罗西风芹■☆

361513　Seseli lancifolium (K. T. Fu) Pimenov = Libanotis lancifolia K. T. Fu ■

361514　Seseli langshanense Y. Z. Zhao et Ma;狼山西风芹; Langshan Seseli ■

361515　Seseli langshanense Y. Z. Zhao et Ma = Libanotis abolinii (Korovin) Korovin ■

361516　Seseli lanzhouense (K. T. Fu ex R. H. Shan et M. L. Sheh) V. M. Vinogr. = Libanotis lanzhouensis K. T. Fu ex R. H. Shan et M. L. Sheh ■

361517　Seseli laserpitiifolium Palib. = Libanotis condensata Crantz ■

361518　Seseli laticalycinum (R. H. Shan et M. L. Sheh) Pimenov = Libanotis laticalycina R. H. Shan et M. L. Sheh ■

361519　Seseli ledebourii G. Don;李德氏邪蒿; Ledebour Seseli ■☆

361520　Seseli lehmannianum (Bunge) Boiss. ;李曼尼邪蒿■☆

361521　Seseli lehmannii Degen;李曼氏邪蒿; Lehmann Seseli ■☆

361522　Seseli leptocladum Woronow;细枝邪蒿■☆

361523　Seseli lessingianum Turcz. = Seseli eriocephalum (Pall. ex Spreng.) Schischk. ■

361524　Seseli lessingianum Turcz. ex Kar. et Kir. ;李辛氏邪蒿; Lessing Seseli ■

361525　Seseli lessingianum Turcz. ex Kar. et Kir. = Seseli eriocephalum (Pall. ex Spreng.) Schischk. ■

361526　Seseli libanotis (L.) W. D. J. Koch;黎巴嫩西风芹(山地邪蒿,西风芹,邪蒿);Hartwort, Moon Carrot, Mooncarrot, Moon-carrot, Mountain Meadow Seseli, Mountain Meadow-saxifrage, Mountain Meadow-seseli,Spiguel ■☆

361527　Seseli libanotis (L.) W. D. J. Koch subsp. atlanticum Maire;亚特兰大西风芹■☆

361528　Seseli libanotis (L.) W. D. J. Koch subsp. japonicum (H. Boissieu) H. Hara f. alpicola (Kitag.) Okuyama = Libanotis coreana (H. Wolff) Kitag. var. alpicola Kitag. ■☆

361529　Seseli libanotis (L.) W. D. J. Koch subsp. japonicum (H. Boissieu) H. Hara f. ugoensis (Koidz.) H. Hara = Libanotis coreana (H. Wolff) Kitag. f. ugoensis (Koidz.) Kitag. ■

361530　Seseli libanotis (L.) W. D. J. Koch subsp. japonicum (H. Boissieu) H. Hara var. alpicola (Kitag.) H. Ohba = Libanotis coreana (H. Wolff) Kitag. var. alpicola Kitag. ■☆

361531　Seseli libanotis (L.) W. D. J. Koch subsp. sibiricum (L.) Thell. = Libanotis sibirica (L.) C. A. Mey. ■

361532　Seseli libanotis (L.) W. D. J. Koch var. alpicola (Kitag.) H. Ohba;高山西风芹■☆

361533　Seseli libanotis (L.) W. D. J. Koch var. daucifolium (DC.) Franch. et Sav. = Libanotis seseloides (Fisch. et C. A. Mey.) Turcz. ■

361534　Seseli libanotis (L.) W. D. J. Koch var. japonicum H. Boissieu = Libanotis coreana (H. Wolff) Kitag. ■☆

361535　Seseli libanotis (L.) W. D. J. Koch var. sibiricum (L.) DC. = Libanotis sibirica (L.) C. A. Mey. ■

361536　Seseli libanotis (L.) W. D. J. Koch var. sibiricum DC. = Libanotis sibirica (L.) C. A. Mey. ■

361537　Seseli macrophyllum Regel et Schmalh. ;大叶西风芹■☆

361538　Seseli mairei H. Wolff;竹叶西风芹(鸡脚暗消,鸡爪防风,鸡足防风,三叶防风,西防风,西风,云防风,竹叶防风,竹叶邪蒿); Bambooleaf Seseli ■

361539　Seseli mairei H. Wolff var. simplicifolia C. Y. Wu ex R. H. Shan et M. L. Sheh;单叶西风芹■

361540　Seseli maroccanum Dobignard;摩洛哥西风芹■☆

361541　Seseli meyeri (Ledeb.) D. Dietr. = Soranthus meyeri Ledeb. ■

361542　Seseli mucronatum (Schrenk) Pimenov et Sdobnina;短尖邪蒿(短尖西风芹)■☆

361543　Seseli mucronatum (Schrenk) Pimenov et Sdobnina = Ligusticum mucronatum (Schrenk) Leute ■

361544　Seseli natalense Sond. = Peucedanum natalense (Sond.) Engl. ■☆

361545　Seseli nortonii Fedde ex H. Wolff;西藏西风芹; Tibet Seseli, Xizang Seseli ■

361546　Seseli nudum (Lindl.) Pimenov et Kljuykov. = Eriocycla nuda Lindl. ■

361547　Seseli pallasii Besser;帕拉氏邪蒿;Pallas Seseli ■☆

361548　Seseli pauciradiatum Schischk. ;稀射线西风芹■☆

361549　Seseli pelliotii (H. Boissieu) Pimenov et Kljuykov. = Eriocycla pelliotii (H. Boissieu) H. Wolff ■

361550　Seseli petraeum M. Bieb. ;石地邪蒿■☆

361551　Seseli peucedanifolium (Spreng.) Besser;前胡叶西风芹■☆

361552　Seseli peucedanoides (M. Bieb.) Koso-Pol. ;前胡西风芹■☆

361553　Seseli platyphyllum (Schrenk) O. Fedtsch. et B. Fedtsch. ;宽叶西风芹■☆

361554　Seseli ponticum Lipsky;蓬特西风芹■☆

361555　Seseli pratense Crantz = Silaum silaus (L.) Schinz et Thell. ■

361556　Seseli pratense Crantz. = Silaus pratensis (Crantz) Besser ■

361557　Seseli provostii H. Boissieu. = Eriocycla albescens (Franch.) H. Wolff ■

361558　Seseli purpureo-vaginatum R. H. Shan et M. L. Sheh;紫鞘西风芹;Purplesheath Seseli ■

361559　Seseli rigidum Waldst. et Kit. ;硬邪蒿■☆

361560　Seseli rivinianum (Ledeb.) M. Hiroe = Libanotis seseloides (Fisch. et C. A. Mey.) Turcz. ■

361561　Seseli rupicola Woronow;岩生西风芹■☆

361562　Seseli sandbergiae Fedde ex H. Wolff;山西西风芹; Shanxi Seseli ■

361563 Seseli schansiensis Fedde ex H. Wolff = Seseli sandbergiae Fedde ex H. Wolff ■

361564 Seseli schrenkianum (C. A. Mey. ex Schischk.） Pimenov et Sdobnina = Libanotis schrenkiana C. A. Mey. et Schischk. ■

361565 Seseli scopulorum C. C. Towns. ；岩栖西风芹■☆

361566 Seseli seseloides (Fisch. et C. A. Mey. ex Turcz.） M. Hiroe = Libanotis seseloides (Fisch. et C. A. Mey.） Turcz. ■

361567 Seseli seseloides (Fisch. et C. A. Mey.） M. Hiroe = Libanotis seseloides (Fisch. et C. A. Mey.） Turcz. ■

361568 Seseli seseloides (Turcz.） M. Hiroe = Libanotis seseloides (Fisch. et C. A. Mey.） Turcz. ■

361569 Seseli sessiliflorum Schrenk；无柄西风芹（无柄邪蒿）；Sessilflower Seseli ■

361570 Seseli siamicum Craib = Seseli yunnanense Franch. ■

361571 Seseli sibiricum (L.） Benth. ex C. B. Clarke = Libanotis sibirica (L.） C. A. Mey. ■

361572 Seseli sibiricum (L.） Garcke = Libanotis sibirica (L.） C. A. Mey. ■

361573 Seseli simplicifolium (C. Y. Wu ex R. H. Shan et M. L. Sheh） Pimenov et Kljukov = Seseli mairei H. Wolff var. simplicifolia C. Y. Wu ex R. H. Shan et M. L. Sheh ■

361574 Seseli sonderi M. Hiroe = Peucedanum tenuifolium Thunb. ■☆

361575 Seseli songoricum Schischk. ；准噶尔西风芹■☆

361576 Seseli songoricum Schischk. = Libanotis abolinii (Korovin） Korovin ■

361577 Seseli spodotrichoma (K. T. Fu） Pimenov = Libanotis spodotrichoma K. T. Fu ■

361578 Seseli squarrosum Schischk. ；粗鳞西风芹■☆

361579 Seseli squarrosum Schischk. = Seseli sessiliflorum Schrenk ■

361580 Seseli squarrulosum R. H. Shan et M. L. Sheh；粗糙西风芹（川防风，防风，西风）；Rugged Seseli ■

361581 Seseli striatum Thunb. = Peucedanum striatum (Thunb.） Sond. ■☆

361582 Seseli strictum Ledeb. ；劲直西风芹（直邪蒿）■

361583 Seseli tachiroei Franch. et Sav. = Ligusticum tachiroei (Franch. et Sav.） M. Hiroe et Constance ■

361584 Seseli tenuifolium Ledeb. = Seseli glabratum Willd. ex Schult. ■

361585 Seseli tenuisectum Regel et Schmalh. ；细裂西风芹■☆

361586 Seseli togasii (M. Hiroe） Pimenov et Kljukov；绒果西风芹■

361587 Seseli tortuosum L. ；扭曲邪蒿（旋扭西风芹，旋扭邪蒿）■☆

361588 Seseli trichocarpum (Schrenk） B. Fedtsch. = Stenocoelium trichocarpum Schrenk ■

361589 Seseli tschuiliense Pavlov ex Korovin；楚伊犁西风芹；Tschuili Seseli ■

361590 Seseli turbinatum Korovin；陀螺邪蒿■☆

361591 Seseli ugoense Koidz. = Libanotis coreana (H. Wolff） Kitag. ■☆

361592 Seseli ugoense Koidz. = Libanotis seseloides (Fisch. et C. A. Mey.） Turcz. ■

361593 Seseli vaginatum Ledeb. = Phlojodicarpus sibiricus (Stephan ex Spreng.） Koso-Pol. ■

361594 Seseli vaillantii H. Boissieu = Libanotis iliensis (Lipsky） Korovin ■

361595 Seseli valentinae Popov；叉枝西风芹；Forkshoot Seseli ■

361596 Seseli varium Trevir. var. atlanticum (Boiss.） Maire；北非西风芹■☆

361597 Seseli varium Trevis. ；易变邪蒿■☆

361598 Seseli varium Trevis. = Seseli pallasii Besser ■☆

361599 Seseli verticillatum Desf. = Ammoides pusilla (Brot.） Breistr. ■☆

361600 Seseli wannienchun (K. T. Fu） Pimenov = Libanotis wannienchun K. T. Fu ■

361601 Seseli wawrae H. Wolff = Peucedanum wawrae (H. Wolff） H. Y. Su ex M. L. Sheh ■

361602 Seseli webbii Coss. ；韦布邪蒿■☆

361603 Seseli yunnanense Franch. ；松叶西风芹（松叶柴胡，松叶防风，松叶西风芹，松叶邪蒿，云防风）；Pineleaf Seseli ■

361604 Seselinia G. Beck = Seseli L. ■

361605 Seselopsis Schischk. (1950）；西归芹属（假邪蒿属，天山邪蒿属）；Seselopsis ■

361606 Seselopsis tianschanica Schischk. ；西归芹（天山邪蒿，土当归）；Seselopsis ■

361607 Seshagiria Ansari et Hemadri(1971)；塞沙萝藦属☆

361608 Seshagiria sahyadrica Ansari et Hemadri；塞沙萝藦☆

361609 Seslera St. -Lag. = Sesleria Scop. ■☆

361610 Sesleria Nutt. = Buchloe Engelm. (保留属名)■

361611 Sesleria Scop. (1760)；天蓝草属（蓝禾属）；Bluc Moor-grass，Moor Grass，Moor-grass，Sesleria ■☆

361612 Sesleria albicans Schult. = Sesleria caerulea (L.） Ard. ■☆

361613 Sesleria argentea (Savi） Savi；阿根廷天蓝草■☆

361614 Sesleria autumnalis (Scop.） F. W. Schultz；秋天蓝草；Autumn Moor Grass ■☆

361615 Sesleria caerulea (L.） Ard. ；天蓝草；Blue Moor Grass，Blue Moorgrass，Blue Moor-grass，Skyblue Sesleria ■☆

361616 Sesleria cirtensis Trab. = Dactylis glomerata L. ■

361617 Sesleria dactyloides Nutt. = Buchloe dactyloides (Nutt.） Engelm. ■

361618 Sesleria gigantea Dörfl. et Hayek；大天蓝草■☆

361619 Sesleria glauca ?；灰天蓝草（灰蓝禾）；Blue-grass ■☆

361620 Sesleria gracilis Schur；纤细天蓝草■☆

361621 Sesleria heufleriana Schur；巴尔干天蓝草（巴尔干蓝禾）；Balkan Blue-grass，Blue-green Moor Grass ■☆

361622 Sesleria phleoides Stev. ex Roem. et Schult. ；梯牧草状天蓝草■☆

361623 Sesleria spicata (Willd.） Spreng. = Elytrophorus spicatus (Willd.） A. Camus ■

361624 Sesleriaceae Döll = Gramineae Juss. (保留科名)■●

361625 Sesleriaceae Döll = Poaceae Barnhart(保留科名)■●

361626 Sesleriella Deyl = Sesleria Scop. ■☆

361627 Sesquicella Alef. = Callirhoe Nutt. ■●☆

361628 Sessea Ruiz et Pav. (1794)；塞斯茄属●☆

361629 Sessea acuminata Francey；渐尖塞斯茄●☆

361630 Sessea atrovirens Benth. et Hook. f. ；墨绿塞斯茄●☆

361631 Sessea crassivenosa Bitter；粗脉塞斯茄●☆

361632 Sessea elliptica Francey；椭圆塞斯茄●☆

361633 Sessea graciliflora Bitter；细花塞斯茄●☆

361634 Sessea macrophylla Francey；大叶塞斯茄●☆

361635 Sessea multinervia Francey；多脉塞斯茄●☆

361636 Sesseopsis Hassl. = Sessea Ruiz et Pav. ●☆

361637 Sessilanthera Molseed et Cruden(1969)；矮药鸢尾属（中美鸢尾属)■☆

361638 Sessilanthera citrina Cruden；矮药鸢尾■☆

361639 Sessilanthera latifolia (Weath.） Molseed et Cruden；宽叶矮药鸢尾■☆

361640　Sessilibulbum Brieger = Scaphyglottis Poepp. et Endl. (保留属名) ■☆

361641　Sessilistigma Goldblatt = Homeria Vent. ■☆

361642　Sessilistigma Goldblatt = Moraea Mill. (保留属名) ■

361643　Sessilistigma radians Goldblatt = Moraea radians (Goldblatt) Goldblatt ☆

361644　Sessleria Spreng. = Sesleria Scop. ■☆

361645　Sestinia Boiss. = Hymenocrater Fisch. et C. A. Mey. ●■☆

361646　Sestinia Boiss. et Hohen. = Wendlandia Bartl. ex DC. (保留属名) ●

361647　Sestinia Raf. = Agrimonia L. ■

361648　Sestochilos Breda = Bulbophyllum Thouars (保留属名) ■

361649　Sestochilus Post et Kuntze = Bulbophyllum Thouars (保留属名) ■

361650　Sestochilus Post et Kuntze = Sestochilos Breda ■

361651　Sesuveriaceae Horan. ; 海马齿科 ■

361652　Sesuveriaceae Horan. = Aizoaceae Martinov (保留科名) ●■

361653　Sesuviaceae Horan. = Aizoaceae Martinov (保留科名) ●■

361654　Sesuvium L. (1759) ; 海马齿属 (滨苋属) ; Seapurslane, Sea-purslane ■

361655　Sesuvium congense Welw. ex Oliv. ; 刚果海马齿 ☆

361656　Sesuvium crithmoides Welw. ; 热带海马齿; Tropical Seapurslane ■☆

361657　Sesuvium digynum Welw. ex Oliv. var. angustifolium Schinz = Sesuvium sesuvioides (Fenzl) Verdc. var. angustifolium (Schinz) Gonc. ☆

361658　Sesuvium erectum Correll = Sesuvium verrucosum Raf. ■☆

361659　Sesuvium maritimum (Walter) Britton = Sesuvium maritimum (Walter) Britton, Stearns et Poggenb. ■☆

361660　Sesuvium maritimum (Walter) Britton, Stearns et Poggenb. ; 一年海马齿; Annual Sea-purslane, Slender Sea-purslane ■☆

361661　Sesuvium mesembryanthemoides Wawra et Peyr. ; 日中花海马齿 ■☆

361662　Sesuvium mesembryanthoides Welw. = Sesuvium crithmoides Welw. ■☆

361663　Sesuvium nyasicum (Baker) Gonc. ; 尼亚斯海马齿 ■☆

361664　Sesuvium pedunculatum Pers. = Sesuvium portulacastrum (L.) L. ■

361665　Sesuvium pentandrum Elliott = Sesuvium maritimum (Walter) Britton, Stearns et Poggenb. ■☆

361666　Sesuvium portulaca Crantz = Sesuvium portulacastrum (L.) L. ■

361667　Sesuvium portulacastrum (L.) L. ; 海马齿; Cencilla, Mboga, Seapurslane, Sea-purslane, Shoreline Sea-purslane ■

361668　Sesuvium portulacastrum (L.) L. f. majus K. Kayama; 大海马齿 ■☆

361669　Sesuvium portulacastrum (L.) L. var. griseum O. Deg. et Fosberg; 灰海马齿 ■☆

361670　Sesuvium portulacastrum (L.) L. var. griseum O. Deg. et Fosberg = Sesuvium portulacastrum (L.) L. ■

361671　Sesuvium portulacastrum (L.) L. var. tawadanum K. Nakaj. = Sesuvium portulacastrum (L.) L. var. griseum O. Deg. et Fosberg ■☆

361672　Sesuvium repens Willd. = Sesuvium portulacastrum (L.) L. ■

361673　Sesuvium revolutum Pers. = Sesuvium portulacastrum (L.) L. ■

361674　Sesuvium sesuvioides (Fenzl) Verdc. ; 普通海马齿 ■☆

361675　Sesuvium sesuvioides (Fenzl) Verdc. var. angustifolium (Schinz) Gonc. ; 窄叶普通海马齿 ■☆

361676　Sesuvium trianthemoides Correll; 得州海马齿; Texas Sea-

purslane ■☆

361677　Sesuvium verrucosum Raf. ; 西部海马齿; Western-purslane ■☆

361678　Setachna Dulac = Centaurea L. (保留属名) ●■

361679　Setaria Ach. ex Michx. (废弃属名) = Setaria P. Beauv. (保留属名) ■

361680　Setaria P. Beauv. (1812) (保留属名) ; 狗尾草属 (粟属) ; Bristle Grass, Bristlegrass, Bristle-grass, Foxtailgrass, Millet, Palmgrass, Pigeon Grass, Pigeon-grass ■

361681　Setaria × decipiens Schimp. ; 迷惑狗尾草 ■☆

361682　Setaria × pycnocoma (Steud.) Henrard ex Nakai; 密狗尾草 ■

361683　Setaria × reclinata Vill. ; 拱垂狗尾草 ■☆

361684　Setaria abyssinica Hack. = Setaria incrassata (Hochst.) Hack. ■☆

361685　Setaria acuta Stapf et C. E. Hubb. = Setaria megaphylla (Steud.) T. Durand et Schinz ☆

361686　Setaria adhaerens (Forssk.) Chiov. ; 伯尔狗尾草; Adherent Bristle-grass, Burr Bristlegrass ■☆

361687　Setaria adhaerens (Forssk.) Chiov. subsp. verticillata (L.) Belo-Corr. = Setaria verticillata (L.) P. Beauv. ■

361688　Setaria adhaerens (Forssk.) Chiov. var. antrorsa (A. Br.) Scholz = Setaria adhaerens (Forssk.) Chiov. ■☆

361689　Setaria adhaerens (Forssk.) Chiov. var. fontqueri Calduch = Setaria adhaerens (Forssk.) Chiov. ■☆

361690　Setaria aequalis Stapf = Setaria homonyma (Steud.) Chiov. ■☆

361691　Setaria albida Stapf = Setaria incrassata (Hochst.) Hack. ■☆

361692　Setaria almaspicata de Wit = Setaria sphacelata (Schumach.) Stapf et C. E. Hubb. ex M. B. Moss var. sericea (Stapf) Clayton ■☆

361693　Setaria alpestris Peter = Setaria sphacelata (Schumach.) Stapf et C. E. Hubb. ex M. B. Moss ■☆

361694　Setaria anceps Stapf = Setaria sphacelata (Schumach.) Stapf et C. E. Hubb. ex M. B. Moss var. sericea (Stapf) Clayton ■☆

361695　Setaria anceps Stapf ex R. L. Massey var. sericea Stapf = Setaria sphacelata (Schumach.) Stapf et C. E. Hubb. ex M. B. Moss var. sericea (Stapf) Clayton ■☆

361696　Setaria angustifolia Stapf = Setaria sphacelata (Schumach.) Stapf et C. E. Hubb. ex M. B. Moss ■☆

361697　Setaria angustissima Stapf = Setaria lindenbergiana (Nees) Stapf ■☆

361698　Setaria aparine (Steud.) Chiov. = Setaria verticillata (L.) P. Beauv. ■

361699　Setaria appendiculata (Hack.) Stapf; 附属物狗尾草 ■☆

361700　Setaria arenaria Kitag. ; 断穗狗尾草; Sandy Bristlegrass ■

361701　Setaria aspera Link; 粗糙狗尾草 ■☆

361702　Setaria aurea Hochst. ex A. Braun = Setaria spacelata (Schumach.) Stapf et C. E. Hubb. ex M. B. Moss var. aurea (Hochst. ex A. Braun) Clayton ■☆

361703　Setaria aurea Hochst. ex A. Braun subsp. kinsunduensis Vanderyst = Setaria sphacelata (Schumach.) Stapf et C. E. Hubb. ex M. B. Moss var. splendida (Stapf) Clayton ■☆

361704　Setaria aurea Hochst. ex A. Braun subsp. palustris Vanderyst = Setaria sphacelata (Schumach.) Stapf et C. E. Hubb. ex M. B. Moss var. sericea (Stapf) Clayton ■☆

361705　Setaria aurea Hochst. ex A. Braun var. fumigata Peter = Setaria spacelata (Schumach.) Stapf et C. E. Hubb. ex M. B. Moss var. aurea (Hochst. ex A. Braun) Clayton ■☆

361706　Setaria aurea Hochst. ex A. Braun var. latifolia Peter = Setaria spacelata (Schumach.) Stapf et C. E. Hubb. ex M. B. Moss var.

aurea（Hochst. ex A. Braun）Clayton ■☆

361707　Setaria autumnalis Ohwi = Setaria faberi R. A. W. Herrm. ■

361708　Setaria avettae Pirotta = Setaria incrassata（Hochst.）Hack. ■☆

361709　Setaria barbata（Lam.）Kunth；髯毛狗尾草；Bearded Setaria, Bristly Foxtail, East Indian Bristlegrass ■☆

361710　Setaria barbigera（Bertol.）Stapf = Setaria sagittifolia（A. Rich.）Walp. ■☆

361711　Setaria basifissa Peter = Setaria appendiculata（Hack.）Stapf ■☆

361712　Setaria bathiei A. Camus；巴西狗尾草■☆

361713　Setaria bequaertii Robyns = Setaria incrassata（Hochst.）Hack. ■☆

361714　Setaria biflora Hillebr. = Dissochondrus bifidus（Hillebr.）Kuntze ■☆

361715　Setaria bongaensis（Pilg.）Mez = Setaria homonyma（Steud.）Chiov. ■☆

361716　Setaria brachiariaeformis（Steud.）T. Durand et Schinz = Echinochloa colona（L.）Link ■

361717　Setaria braunii Peter；布劳恩狗尾草■☆

361718　Setaria breviseta Peter = Setaria incrassata（Hochst.）Hack. ■☆

361719　Setaria brevispica（Scribn. et Merr.）K. Schum. = Setaria verticillata（L.）P. Beauv. ■

361720　Setaria brevispica Schum. = Setaria verticillata（L.）P. Beauv. ■

361721　Setaria bussei Herrm. = Setaria sphacelata（Schumach.）Stapf et C. E. Hubb. ex M. B. Moss ■☆

361722　Setaria cana de Wit = Setaria sphacelata（Schumach.）Stapf et C. E. Hubb. ex M. B. Moss var. sericea（Stapf）Clayton ■☆

361723　Setaria carnei Hitchc. = Setaria verticillata（L.）P. Beauv. ■

361724　Setaria caudula Stapf = Setaria poiretiana（Schult.）Kunth ■☆

361725　Setaria chevalieri Stapf；舍瓦利耶狗尾草；Buffel Grass ■☆

361726　Setaria chevalieri Stapf = Setaria megaphylla（Steud.）T. Durand et Schinz ■☆

361727　Setaria chevalieri Stapf subsp. racemosa de Wit = Setaria megaphylla（Steud.）T. Durand et Schinz ■☆

361728　Setaria chondrachne（Steud.）Honda；莩草（松村稷）；Woody Bristlegrass ■

361729　Setaria ciliolata Stapf et C. E. Hubb. = Setaria incrassata（Hochst.）Hack. ■☆

361730　Setaria cinerea T. Koyama；灰狗尾草■☆

361731　Setaria decipiens de Wit = Setaria sphacelata（Schumach.）Stapf et C. E. Hubb. ex M. B. Moss ■☆

361732　Setaria decipiens Schimp. ex Nyman = Setaria verticillata（L.）P. Beauv. ■

361733　Setaria dioica Hochst. = Pennisetum petiolare（Hochst.）Chiov. ■☆

361734　Setaria dubia P. C. Keng et Y. K. Ma；云南莩草（云南䅟草）；Yunnan Bristlegrass ■

361735　Setaria dubia P. C. Keng et Y. K. Ma = Setaria forbesiana（Nees）Hook. f. ■

361736　Setaria erythraeae Mattei = Setaria pumila（Poir.）Roem. et Schult. ■

361737　Setaria excurrens（Trin.）Miq.；延伸狗尾草（皱叶狗尾草）；Excurrent Bristlegrass ■

361738　Setaria excurrens（Trin.）Miq. = Setaria plicata（Lam.）T. Cooke ■

361739　Setaria excurrens（Trin.）Miq. var. leviflora Keng = Setaria palmifolia（J. König）Stapf var. leviflora（Keng）S. L. Chen et G. Y. Sheng ■

361740　Setaria excurrens（Trin.）Miq. var. leviflora Keng ex S. L. Chen = Setaria palmifolia（J. König）Stapf var. leviflora（Keng）S. L. Chen et G. Y. Sheng ■

361741　Setaria excurrens（Trin.）Miq. var. leviflora Keng ex S. L. Chen = Setaria plicata（Lam.）T. Cooke var. leviflora（Keng ex S. L. Chen）S. L. Chen et S. M. Phillips ■

361742　Setaria eylesii Stapf et C. E. Hubb. = Setaria incrassata（Hochst.）Hack. ■☆

361743　Setaria faberi R. A. W. Herrm.；法氏狗尾草（大狗尾草, 狗尾巴, 狗尾草, 谷莠子）；Big Bristlegrass, Chinese Foxtail, Faber Bristlegrass, Giant Foxtail, Japanese Bristle Grass, Japanese Bristlegrass, Nodding Bristle-grass, Nodding Foxtail ■

361744　Setaria faberi R. A. W. Herrm. = Setaria glauca（L.）P. Beauv. ■

361745　Setaria flabellata Stapf = Setaria sphacelata（Schumach.）Stapf et C. E. Hubb. ex M. B. Moss ■☆

361746　Setaria flabellata Stapf subsp. natalensis de Wit = Setaria sphacelata（Schumach.）Stapf et C. E. Hubb. ex M. B. Moss var. torta（Stapf）Clayton ■☆

361747　Setaria flabelliformis de Wit = Setaria sphacelata（Schumach.）Stapf et C. E. Hubb. ex M. B. Moss var. sericea（Stapf）Clayton ■☆

361748　Setaria flaccifolia Stapf = Setaria barbata（Lam.）Kunth ■☆

361749　Setaria flavida（Retz.）Veldkamp = Paspalidium flavidum（Retz.）A. Camus ■

361750　Setaria forbesiana（Nees）Hook. f.；西南莩草（大狗尾草）；Forbes Bristlegrass ■

361751　Setaria forbesiana（Nees）Hook. f. var. breviseta S. L. Chen et G. Y. Sheng；短刺西南莩草■

361752　Setaria geminata（Forssk.）Veldkamp = Paspalidium geminatum（Forssk.）Stapf ■

361753　Setaria geniculata（Lam.）P. Beauv.；莠狗尾草（狗尾草, 光明草）；Knotroot Bristlegrass, Knot-root Bristle-grass ■

361754　Setaria geniculata（Lam.）P. Beauv. = Setaria gracilis Kunth ■

361755　Setaria geniculata（Lam.）P. Beauv. = Setaria parviflora（Poir.）Kerguélen ■

361756　Setaria geniculata P. Beauv. = Setaria geniculata（Lam.）P. Beauv. ■

361757　Setaria germanica（Mill.）P. Beauv. = Setaria italica（L.）P. Beauv. ■

361758　Setaria germanica（Mill.）P. Beauv. = Setaria italica P. Beauv. 'Major' ■

361759　Setaria germanica（Mill.）P. Beauv. = Setaria italica P. Beauv. var. germanica（Mill.）Schrad. ■

361760　Setaria gerrardii Stapf = Setaria incrassata（Hochst.）Hack. ■☆

361761　Setaria gigantea（Franch. et Sav.）Makino = Setaria viridis（L.）P. Beauv. subsp. pycnocoma（Steud.）Tzvelev ■

361762　Setaria glauca（L.）P. Beauv.；灰绿狗尾草；Yellow Foxtail ■

361763　Setaria glauca（L.）P. Beauv. = Pennisetum glaucum（L.）R. Br. ■

361764　Setaria glauca（L.）P. Beauv. = Setaria lutescens（Stuntz）F. T. Hubb. ■

361765　Setaria glauca（L.）P. Beauv. subsp. pallide-fusca（Schumach.）B. K. Simon = Setaria pumila（Poir.）Roem. et Schult. ■

361766　Setaria glauca（L.）P. Beauv. var. dura（I. C. Chung）I. C. Chung；硬稃狗尾草■

361767　Setaria glauca（L.）P. Beauv. var. dura（I. C. Chung）I. C. Chung = Setaria pumila（Poir.）Roem. et Schult. ■

361768 Setaria glauca（L.）P. Beauv. var. longispica（Honda）Makino et Nemoto = Setaria glauca（L.）P. Beauv. ■

361769 Setaria glauca（L.）P. Beauv. var. pallidefusca（Schumach.）T. Koyama = Setaria pumila（Poir.）Roem. et Schult. ■

361770 Setaria glauca（L.）P. Beauv. var. pallidefusca（Schumach.）T. Koyama = Setaria pallidifusca（Schumach.）Stapf et C. E. Hubb. ■

361771 Setaria glauca（L.）P. Beauv. var. pallidefusca（Schumach.）T. Koyama = Setaria parviflora（Poir.）Kerguélen ■

361772 Setaria gracilipes C. E. Hubb.；细梗狗尾草■☆

361773 Setaria gracilis Kunth；纤细狗尾草■

361774 Setaria gracilis Kunth = Setaria parviflora（Poir.）Kerguélen ■

361775 Setaria grandis Stapf；大狗尾草■☆

361776 Setaria grantii Stapf = Setaria kagerensis Mez ■☆

361777 Setaria guizhouensis S. L. Chen et G. Y. Sheng；贵州狗尾草；Guizhou Bristlegrass ■

361778 Setaria guizhouensis S. L. Chen et G. Y. Sheng var. paleata S. L. Chen et G. Y. Sheng；具稃贵州狗尾草■

361779 Setaria haareri Stapf et C. E. Hubb. = Setaria appendiculata（Hack.）Stapf ■☆

361780 Setaria hereroensis Herrm. = Setaria appendiculata（Hack.）Stapf ■☆

361781 Setaria hochstetteri Kunth = Pennisetum nubicum（Hochst.）K. Schum. ex Engl. ■☆

361782 Setaria holstii R. A. W. Herrm. = Setaria incrassata（Hochst.）Hack. ■☆

361783 Setaria homblei De Wild. = Setaria sphacelata（Schumach.）Stapf et C. E. Hubb. ex M. B. Moss var. torta（Stapf）Clayton ■☆

361784 Setaria homonyma（Steud.）Chiov.；同优狗尾草■☆

361785 Setaria humbertiana A. Camus；厚叶狗尾草■☆

361786 Setaria imberbis（Poir.）Roem. et Schult. = Setaria geniculata（Lam.）P. Beauv. ■

361787 Setaria incrassata（Hochst.）Hack.；粗狗尾草■☆

361788 Setaria insignis de Wit = Setaria megaphylla（Steud.）T. Durand et Schinz ■☆

361789 Setaria intermedia Roem. et Schult.；间序狗尾草；Intermixed Bristlegrass ■

361790 Setaria interpilosa Stapf et C. E. Hubb. = Setaria petiolata Stapf et C. E. Hubb. ■☆

361791 Setaria interrupta Peter = Setaria incrassata（Hochst.）Hack. ■☆

361792 Setaria ipamuensis Vanderyst = Setaria restioidea（Franch.）Stapf ■☆

361793 Setaria isalensis A. Camus；伊萨卢狗尾草■☆

361794 Setaria italica（L.）P. Beauv.；粟（八月黄，芭，白粱米，白粱粟，白苗，百日粮，贝粱，赤粱粟，赶麦黄，狗尾草，狗尾粟，古有子，古子，谷子，寒露粟，寒粟，黄粟，解粱，稞子，老军头，辽东赤粱，虋，秫，黍仔米，粟谷，籼米，籼粟，芠萁，小米，辛米，雁头青，意大利黍，硬粟，粢，粢米）；Common Millet，Fox Tail Millet，Foxtail Brisde-grass，Foxtail Bristle Grass，Foxtail Bristlegrass，Foxtail Millet，German Millet，Hungarian Millet，Italian Millet，Siberian Millet ■

361795 Setaria italica（L.）P. Beauv. var. germanica（Mill.）Schrad.；德国狗尾草■☆

361796 Setaria italica（L.）P. Beauv. var. major Ohwi = Setaria italica P. Beauv. 'Major' ■

361797 Setaria italica（L.）P. Beauv. var. stramineofructa（F. T. Hubb.）L. H. Bailey = Setaria italica（L.）P. Beauv. ■

361798 Setaria italica P. Beauv. 'Major'；粟（日本粟）■

361799 Setaria italica P. Beauv. = Setaria italica（L.）P. Beauv. ■

361800 Setaria italica P. Beauv. subsp. viridis（L.）Thell. = Setaria viridis（L.）P. Beauv. ■

361801 Setaria italica P. Beauv. var. germanica（Mill.）Schrad. = Setaria italica P. Beauv. 'Major' ■

361802 Setaria kagerensis Mez；卡盖拉狗尾草■☆

361803 Setaria kersteniana Peter = Setaria incrassata（Hochst.）Hack. ■☆

361804 Setaria kialaensis Vanderyst = Setaria homonyma（Steud.）Chiov. ■☆

361805 Setaria kinsudiensis Vanderyst；金苏迪狗尾草■☆

361806 Setaria kwamouthensis Vanderyst；夸穆特狗尾草■☆

361807 Setaria lacunosa Peter = Setaria incrassata（Hochst.）Hack. ■☆

361808 Setaria laeta de Wit；愉悦狗尾草■☆

361809 Setaria lancea Stapf ex R. L. Massey = Setaria homonyma（Steud.）Chiov. ■☆

361810 Setaria lasiothyrsa Stapf ex R. L. Massey = Setaria longiseta P. Beauv. ■☆

361811 Setaria laxispica Stapf = Setaria sphacelata（Schumach.）Stapf et C. E. Hubb. ex M. B. Moss ■☆

361812 Setaria liebmannii E. Fourn.；李布曼粟■☆

361813 Setaria lindenbergiana（Nees）Stapf；林登贝格狗尾草■☆

361814 Setaria lindiensis Pilg. = Setaria incrassata（Hochst.）Hack. ■☆

361815 Setaria longiseta P. Beauv.；长刚毛狗尾草■☆

361816 Setaria longissima Chiov. = Setaria incrassata（Hochst.）Hack. ■☆

361817 Setaria lutescens（Stuntz）F. T. Hubb. = Setaria glauca（L.）P. Beauv. ■

361818 Setaria lutescens（Stuntz）F. T. Hubb. var. dura I. C. Chung = Setaria glauca（L.）P. Beauv. var. dura（I. C. Chung）I. C. Chung ■

361819 Setaria lutescens（Weigel ex Stuntz）F. T. Hubb. = Pennisetum glaucum（L.）R. Br. ■

361820 Setaria lutescens（Weigel ex Stuntz）F. T. Hubb. var. dura I. C. Chung = Setaria pumila（Poir.）Roem. et Schult. ■

361821 Setaria lutescens（Weigel）F. T. Hubb. = Setaria glauca（L.）P. Beauv. ■

361822 Setaria lutescens（Weigel）F. T. Hubb. = Setaria pumila（Poir.）Roem. et Schult. ■

361823 Setaria lynesii Stapf et C. E. Hubb. = Setaria incrassata（Hochst.）Hack. ■☆

361824 Setaria macrophylla Andersson = Setaria megaphylla（Steud.）T. Durand et Schinz ■☆

361825 Setaria macrostachya Kunth；大穗狗尾草；Large-spike Bristlegrass ■☆

361826 Setaria magna Griseb.；巨狗尾草；Giant Fox Tail Grass，Giant Foxtail ■☆

361827 Setaria matsumurae Hack. ex Matsum. = Setaria chondrachne（Steud.）Honda ■

361828 Setaria megaphylla（Steud.）T. Durand et Schinz；大叶狗尾草；Bigleaf Bristlegrass ■☆

361829 Setaria megaphylla（Steud.）T. Durand et Schinz var. chevalieri（Stapf）Berhaut = Setaria megaphylla（Steud.）T. Durand et Schinz ■☆

361830 Setaria megaphylla T. Durand et Schinz = Setaria megaphylla（Steud.）T. Durand et Schinz ■☆

361831 Setaria merkeri Herrm. = Setaria incrassata（Hochst.）Hack. ■☆

361832　Setaria microprolepis Stapf = Setaria homonyma （Steud.） Chiov. ■☆

361833　Setaria mildbraedii C. E. Hubb. ；米尔德狗尾草■☆

361834　Setaria modesta Stapf = Setaria incrassata（Hochst.）Hack. ■☆

361835　Setaria mombassana Herrm. = Setaria incrassata（Hochst.）Hack. ■☆

361836　Setaria myosuroides Peter = Setaria spacelata（Schumach.）Stapf et C. E. Hubb. ex M. B. Moss var. aurea（Hochst. ex A. Braun）Clayton ■☆

361837　Setaria nakaiana（Honda）Ohwi = Setaria barbata（Lam.）Kunth ■☆

361838　Setaria natalensis de Wit = Setaria megaphylla（Steud.）T. Durand et Schinz ■☆

361839　Setaria neglecta de Wit = Setaria sphacelata（Schumach.）Stapf et C. E. Hubb. ex M. B. Moss ■☆

361840　Setaria nigriostris（Nees）Durand et Schinz；黑狗尾草；Black Bristlegrass ☆

361841　Setaria nigrirostris（Nees）T. Durand et Schinz var. pallida de Wit = Setaria incrassata（Hochst.）Hack. ■☆

361842　Setaria obscura de Wit；隐匿狗尾草■☆

361843　Setaria oligochaete K. Schum. = Setaria plicatilis（Hochst.）Hack. ex Engl. ■☆

361844　Setaria orthosticha Herrm. ；直列狗尾草■☆

361845　Setaria pabularis Stapf = Setaria incrassata（Hochst.）Hack. ■☆

361846　Setaria pachystachys（Franch. et Sav.）Matsum. = Setaria viridis（L.）P. Beauv. subsp. pachystachys（Franch. et Sav.）Masam. et Yanagita ■

361847　Setaria pallidifusca（Schumach.）Stapf et C. E. Hubb. ；褐毛狗尾草（狗尾草，莠狗尾草）；Bristly Foxtall Grass, Brownhair Bristlegrass, Cattail Grass ■

361848　Setaria pallidifusca（Schumach.）Stapf et C. E. Hubb. = Setaria parviflora（Poir.）Kerguélen ■

361849　Setaria pallidifusca（Schumach.）Stapf et C. E. Hubb. = Setaria pumila（Poir.）Roem. et Schult. ■

361850　Setaria palmifolia（J. König）Stapf；棕叶狗尾草（雏茅，苓草，箬叶芋，涩船草，台风草，掌叶狗尾草，竹头草，棕叶草）；Malaysian Palm Grass, New Guinea Asparagus, Palm Grass, Palmgrass, Palm-grass, Palmleaf Bristlegrass, Pitpit ■

361851　Setaria palmifolia（J. König）Stapf var. blepharoneuron（A. Braun）Veldkamp = Setaria plicata（Lam.）T. Cooke ■

361852　Setaria palmifolia（J. König）Stapf var. leviflora（Keng）S. L. Chen et G. Y. Sheng；光花狗尾草；Smoothflower Bristlegrass ■

361853　Setaria palustris Stapf = Setaria incrassata（Hochst.）Hack. ■☆

361854　Setaria paniciformis Rendle = Setaria longiseta P. Beauv. ■☆

361855　Setaria paniculifera（Steud.）E. Fourn. ex Hemsl. = Setaria palmifolia（J. König）Stapf ■

361856　Setaria parviflora（Poir.）Kerguélen；小花狗尾草（幽狗尾草）；Knotroot Bristle-grass, Knotroot Foxtail, Marsh Bristlegrass, Yellow Bristlegrass, Yellow Foxtail ■

361857　Setaria parviflora Stapf = Setaria parviflora（Poir.）Kerguélen ■

361858　Setaria perberbis Stapf ex de Wit = Setaria incrassata（Hochst.）Hack. ■☆

361859　Setaria perennis Hack. = Setaria sphacelata（Schumach.）Stapf et C. E. Hubb. ex M. B. Moss ■☆

361860　Setaria perrieri A. Camus；佩里耶狗尾草■☆

361861　Setaria petiolata Stapf et C. E. Hubb. ；柄叶狗尾草■☆

361862　Setaria phanerococca Stapf = Setaria incrassata（Hochst.）Hack. ■☆

361863　Setaria phillipsii de Wit = Setaria lindenbergiana（Nees）Stapf ■☆

361864　Setaria phleoides Stapf = Setaria incrassata（Hochst.）Hack. ■☆

361865　Setaria phragmitoides Stapf = Setaria incrassata（Hochst.）Hack. ■☆

361866　Setaria pilosa Kunth；马草■☆

361867　Setaria planifolia Stapf = Setaria sphacelata（Schumach.）Stapf et C. E. Hubb. ex M. B. Moss var. sericea（Stapf）Clayton ■☆

361868　Setaria plicata（Lam.）T. Cooke；皱叶狗尾草（胞叶草，烂衣草，马草）；Wrinkledleaf Bristlegrass ☆

361869　Setaria plicata（Lam.）T. Cooke var. leviflora（Keng ex S. L. Chen）S. L. Chen et S. M. Phillips = Setaria palmifolia（J. König）Stapf var. leviflora（Keng）S. L. Chen et G. Y. Sheng ■

361870　Setaria plicata（Lam.）T. Cooke var. leviflora（Keng）S. L. Chen et G. Y. Sheng = Setaria palmifolia（J. König）Stapf var. leviflora（Keng）S. L. Chen et G. Y. Sheng ■

361871　Setaria plicatilis（Hochst.）Hack. ex Engl. ；折叠狗尾草■☆

361872　Setaria plurinervis Stapf = Setaria incrassata（Hochst.）Hack. ■☆

361873　Setaria poiretiana（Schult.）Kunth；普瓦雷狗尾草■☆

361874　Setaria polyphylla Stapf = Setaria incrassata（Hochst.）Hack. ■☆

361875　Setaria porphyrantha Stapf = Setaria incrassata（Hochst.）Hack. ■☆

361876　Setaria pseudaristata（Peter）Pilg. ；假芒狗尾草■☆

361877　Setaria pumila（Poir.）Roem. et Schult. ；金色狗尾草（阿罗汉草，狗尾草，狗尾子，黄狗尾，金狗尾，犬尾草，洗草，小狗尾草，莠，莠草子，御谷）；Bdsde-grass, Cat's Tail Millet, Golden Bristlegrass, Little Bristlegrass, Pigeon Grass, Yellow Bristlegrass, Yellow Bristle-grass, Yellow Foxtail, Yellow Fox-tail, Yellow Giant Fox Tail Grass ■

361878　Setaria pumila（Poir.）Roem. et Schult. = Setaria glauca（L.）P. Beauv. ■

361879　Setaria pumila（Poir.）Roem. et Schult. subsp. pallidefusca（Schumach.）B. K. Simon = Setaria pumila（Poir.）Roem. et Schult. ■

361880　Setaria pumila（Poir.）Roem. et Schult. subsp. pallidefusca（Schumach.）B. K. Simon = Setaria pallidifusca（Schumach.）Stapf et C. E. Hubb. ■

361881　Setaria pumila（Poir.）Roem. et Schult. var. longispica（Honda）Masam. = Setaria pumila（Poir.）Roem. et Schult. ■

361882　Setaria punctata（Burm. f.）Veldkamp = Paspalidium punctatum（Burm. f.）A. Camus ■

361883　Setaria purpurea P. Beauv. ；紫狗尾草■☆

361884　Setaria pycnocoma（Steud.）Henrard ex Nakai = Setaria viridis（L.）P. Beauv. subsp. pycnocoma（Steud.）Tzvelev ■

361885　Setaria pycnocoma（Steud.）Henrard ex Nakai = Setaria viridis（L.）P. Beauv. var. gigantea（Franch. et Sav.）Franch. et Sav. ex Matsum. ■

361886　Setaria ramentacea Stapf = Setaria kagerensis Mez ■☆

361887　Setaria ramulosa Peter = Setaria incrassata（Hochst.）Hack. ■☆

361888　Setaria restioidea（Franch.）Stapf；绳草状狗尾草■☆

361889　Setaria rhachitricha（Hochst.）Rendle = Setaria barbata（Lam.）Kunth ■☆

361890 Setaria rigida Stapf;硬狗尾草■☆

361891 Setaria rubiginosa (Steud.) Miq. = Setaria pallidifusca (Schumach.) Stapf et C. E. Hubb. ■

361892 Setaria rubiginosa (Steud.) Miq. = Setaria pumila (Poir.) Roem. et Schult. ■

361893 Setaria rudifolia Stapf = Setaria incrassata (Hochst.) Hack. ■☆

361894 Setaria rudimentosa (Steud.) T. Durand et Schinz = Setaria sphacelata (Schumach.) Stapf et C. E. Hubb. ex M. B. Moss ■☆

361895 Setaria sagittifolia (A. Rich.) Walp. ;箭叶狗尾草■☆

361896 Setaria scalaris Peter = Setaria sphacelata (Schumach.) Stapf et C. E. Hubb. ex M. B. Moss ■☆

361897 Setaria schweinfurthii Herrm. = Setaria restioidea (Franch.) Stapf ■☆

361898 Setaria scottii (Hack.) A. Camus;司科特狗尾草■☆

361899 Setaria seriata Stapf = Setaria sphacelata (Schumach.) Stapf et C. E. Hubb. ex M. B. Moss var. sericea (Stapf) Clayton ■☆

361900 Setaria setulosa Stapf = Setaria incrassata (Hochst.) Hack. ■☆

361901 Setaria spacelata (Schumach.) Stapf et C. E. Hubb. ex M. B. Moss var. aurea (Hochst. ex A. Braun) Clayton;黄非洲狗尾草■☆

361902 Setaria sphacelata (Schumach.) Stapf et C. E. Hubb. ex M. B. Moss;非洲狗尾草; African Bristlegrass, Golden Timothy Grass, Rhodesian Timothy ■☆

361903 Setaria sphacelata (Schumach.) Stapf et C. E. Hubb. ex M. B. Moss subsp. aquamontana de Wit = Setaria sphacelata (Schumach.) Stapf et C. E. Hubb. ex M. B. Moss ■☆

361904 Setaria sphacelata (Schumach.) Stapf et C. E. Hubb. ex M. B. Moss subsp. nodosa de Wit = Setaria sphacelata (Schumach.) Stapf et C. E. Hubb. ex M. B. Moss var. sericea (Stapf) Clayton ■☆

361905 Setaria sphacelata (Schumach.) Stapf et C. E. Hubb. ex M. B. Moss subsp. pyropea de Wit = Setaria sphacelata (Schumach.) Stapf et C. E. Hubb. ex M. B. Moss var. sericea (Stapf) Clayton ■☆

361906 Setaria sphacelata (Schumach.) Stapf et C. E. Hubb. ex M. B. Moss var. splendida (Stapf) Clayton = Setaria sphacelata (Schumach.) Stapf et C. E. Hubb. ex M. B. Moss ■☆

361907 Setaria sphacelata (Schumach.) Stapf et C. E. Hubb. ex M. B. Moss var. stolonifera de Wit = Setaria sphacelata (Schumach.) Stapf et C. E. Hubb. ex M. B. Moss ■☆

361908 Setaria sphacelata (Schumach.) Stapf et C. E. Hubb. ex M. B. Moss var. sericea (Stapf) Clayton;绢毛非洲狗尾草■☆

361909 Setaria sphacelata (Schumach.) Stapf et C. E. Hubb. ex M. B. Moss var. splendida (Stapf) Clayton;纤细非洲狗尾草■☆

361910 Setaria sphacelata (Schumach.) Stapf et C. E. Hubb. ex M. B. Moss var. torta (Stapf) Clayton;回旋非洲狗尾草■☆

361911 Setaria sphacelata (Steud.) Stapf et C. E. Hubb. ex Chipp = Setaria sphacelata (Schumach.) Stapf et C. E. Hubb. ex M. B. Moss ■☆

361912 Setaria splendida Stapf = Setaria sphacelata (Schumach.) Stapf et C. E. Hubb. ex M. B. Moss ■☆

361913 Setaria splendida Stapf = Setaria sphacelata (Schumach.) Stapf et C. E. Hubb. ex M. B. Moss var. splendida (Stapf) Clayton ■☆

361914 Setaria stenantha Stapf = Setaria sphacelata (Schumach.) Stapf et C. E. Hubb. ex M. B. Moss ■☆

361915 Setaria stolzii Stapf = Setaria sphacelata (Schumach.) Stapf et C. E. Hubb. ex M. B. Moss ■☆

361916 Setaria stricta (Roth ex Roem. et Schulz) Kunth = Digitaria stricta Roth ex Roem. et Schult. ■

361917 Setaria stricta Kunth = Digitaria stricta Roth ex Roem. et Schult. ■

361918 Setaria subsetosa Stapf = Setaria lindenbergiana (Nees) Stapf ■☆

361919 Setaria sulcata Raddi;纵沟狗尾草■☆

361920 Setaria tenuiseta de Wit = Setaria pseudaristata (Peter) Pilg. ■☆

361921 Setaria tenuispica Stapf et C. E. Hubb. = Setaria sphacelata (Schumach.) Stapf et C. E. Hubb. ex M. B. Moss var. sericea (Stapf) Clayton ■☆

361922 Setaria thermitaria Chiov. = Setaria lindenbergiana (Nees) Stapf ■☆

361923 Setaria thollonii (Franch.) Stapf = Setaria homonyma (Steud.) Chiov. ■☆

361924 Setaria tomentosa (Roxb.) Kunth = Setaria intermedia Roem. et Schult. ■

361925 Setaria torta Stapf = Setaria sphacelata (Schumach.) Stapf et C. E. Hubb. ex M. B. Moss var. torta (Stapf) Clayton ■☆

361926 Setaria transiens K. Schum. = Holcolemma transiens (K. Schum.) Stapf et C. E. Hubb. ■☆

361927 Setaria trinervia Stapf = Setaria spacelata (Schumach.) Stapf et C. E. Hubb. ex M. B. Moss var. aurea (Hochst. ex A. Braun) Clayton ■☆

361928 Setaria trinervia Stapf = Setaria sphacelata (Schumach.) Stapf et C. E. Hubb. ex M. B. Moss ■☆

361929 Setaria ustilata de Wit = Setaria pumila (Poir.) Roem. et Schult. ■

361930 Setaria verticillata (L.) P. Beauv. ;倒刺狗尾草(轮生狗尾草,轮莠,轮状狗尾草); Barbed Bristle Grass, Barbed Bristlegrass, Bristly Foxtail, Bur Bristlegrass, Hooked Bristle Grass, Hooked Bristlegrass, Hooked Bristle-grass, Rough Bristle-grass, Verticillate Bristlegrass, Verticillate Foxtail, Whorled Bristlegrass ■

361931 Setaria verticillata (L.) P. Beauv. subsp. aparine (Steud.) Roem. et Schult. = Setaria adhaerens (Forssk.) Chiov. ■☆

361932 Setaria verticillata (L.) P. Beauv. var. aparine (Steud.) Asch. et Graebn. = Setaria adhaerens (Forssk.) Chiov. ■☆

361933 Setaria verticillata (L.) P. Beauv. var. pubescens Maire et Weiller = Setaria verticillata (L.) P. Beauv. ■

361934 Setaria verticilliformis Dum. Cours. = Setaria verticillata (L.) P. Beauv. ■

361935 Setaria viridis (L.) P. Beauv. ;狗尾草(阿罗汉草,狗毛尾,狗尾,狗尾半支,狗尾毛,谷莠子,光明草,光明子,金狗尾草,金毛狗尾草,毛嘟嘟,毛姑姑,毛毛草,毛娃娃,毛莠莠,犬尾草,犬尾曲,洗草,莠,莠草); Bristly Foxtail, Green Bristle Grass, Green Bristlegrass, Green Bristle-grass, Green Foxtail, Green Giant Fox Tail Grass, Green Millet ■

361936 Setaria viridis (L.) P. Beauv. f. japonica (Koidz.) Ohwi;日本狗尾草■☆

361937 Setaria viridis (L.) P. Beauv. f. misera Honda;贫弱狗尾草■☆

361938 Setaria viridis (L.) P. Beauv. f. pachystachys (Franch. et Sav.) Makino = Setaria viridis (L.) P. Beauv. var. pachystachys (Franch. et Sav.) Makino et Nemoto ■

361939 Setaria viridis (L.) P. Beauv. subsp. italica (L.) Briq. = Setaria italica (L.) P. Beauv. ■

361940 Setaria viridis (L.) P. Beauv. subsp. minor (Thunb.) T. Koyama = Setaria viridis (L.) P. Beauv. ■

361941 Setaria viridis (L.) P. Beauv. subsp. pachystachys (Franch. et Sav.) Masam. et Yanagita = Setaria viridis (L.) P. Beauv. ■

361942 Setaria viridis (L.) P. Beauv. subsp. pachystachys (Franch. et

Sav.) Masam. et Yanagita;厚穗狗尾草(海滨狗尾草)■

361943　Setaria viridis (L.) P. Beauv. subsp. pycnocoma (Steud.) Tzvelev;长穗狗尾草(巨大狗尾草)■

361944　Setaria viridis (L.) P. Beauv. subsp. pycnocoma (Steud.) Tzvelev = Setaria viridis (L.) P. Beauv. var. gigantea (Franch. et Sav.) Franch. et Sav. ex Matsum. ■

361945　Setaria viridis (L.) P. Beauv. subsp. pycnocoma (Steud.) Tzvelev = Setaria × pycnocoma (Steud.) Henrard ex Nakai ■

361946　Setaria viridis (L.) P. Beauv. var. angustifolia Y. N. Lee;狭叶狗尾草■☆

361947　Setaria viridis (L.) P. Beauv. var. breviseta (Döll) Hitchc. ;短毛狗尾草■

361948　Setaria viridis (L.) P. Beauv. var. depressa (Honda) Kitag. ;偃狗尾草■

361949　Setaria viridis (L.) P. Beauv. var. genuina Honda = Setaria viridis (L.) P. Beauv. ■

361950　Setaria viridis (L.) P. Beauv. var. giganta (Franch. et Sav.) Franch. et Sav. ex Matsum. = Setaria viridis (L.) P. Beauv. subsp. pycnocoma (Steud.) Tzvelev ■

361951　Setaria viridis (L.) P. Beauv. var. gigantea (Franch. et Sav.) Franch. et Sav. ex Matsum. = Setaria viridis (L.) P. Beauv. subsp. pycnocoma (Steud.) Tzvelev ■

361952　Setaria viridis (L.) P. Beauv. var. gigantea (Franch. et Sav.) Matsum. = Setaria pycnocoma (Steud.) Henrard ex Nakai ■

361953　Setaria viridis (L.) P. Beauv. var. japonica (Koidz.) Honda = Setaria viridis (L.) P. Beauv. f. japonica (Koidz.) Ohwi ■☆

361954　Setaria viridis (L.) P. Beauv. var. major (Gaudin) Peterm. ;巨大狗尾草; Giant Green Foxtail, Green Bristle Grass, Green Bristlegrass, Green Foxtail ■☆

361955　Setaria viridis (L.) P. Beauv. var. major (Gaudin) Peterm. = Setaria viridis (L.) P. Beauv. subsp. pycnocoma (Steud.) Tzvelev ■

361956　Setaria viridis (L.) P. Beauv. var. minor (Thunb.) Ohwi = Setaria viridis (L.) P. Beauv. ■

361957　Setaria viridis (L.) P. Beauv. var. pachystachys (Franch. et Sav.) Makino et Nemoto = Setaria viridis (L.) P. Beauv. subsp. pachystachys (Franch. et Sav.) Masam. et Yanagita ■

361958　Setaria viridis (L.) P. Beauv. var. pachystachys (Franch. et Sav.) Makino et Nemoto f. rufescens (Honda);浅红狗尾草■☆

361959　Setaria viridis (L.) P. Beauv. var. purpurascens Maxim. ;紫穗狗尾草;Purple Spike Bristlegrass ■

361960　Setaria viridis (L.) P. Beauv. var. purpurascens Maxim. = Setaria viridis (L.) P. Beauv. ■

361961　Setaria viridis (L.) P. Beauv. var. robustaalba Schreib. = Setaria viridis (L.) P. Beauv. ■

361962　Setaria viridis (L.) P. Beauv. var. robustapurpurea Schreib. = Setaria viridis (L.) P. Beauv. var. major (Gaudin) Peterm. ■☆

361963　Setaria viridis (L.) P. Beauv. var. sinica Ohwi = Setaria arenaria Kitag. ■

361964　Setaria viridis (L.) P. Beauv. var. weinmanni (Roem. et Schult.) Borbás = Setaria viridis (L.) P. Beauv. ■

361965　Setaria viridis (L.) P. Beauv. var. weinmannii (Roem. et Schult.) Heynh. = Setaria viridis (L.) P. Beauv. ■

361966　Setaria weinmannii Roem. et Schult. = Setaria viridis (L.) P. Beauv. ■

361967　Setaria welwitschii Rendle;威尔狗尾草■☆

361968　Setaria woodii Hack. = Setaria incrassata (Hochst.) Hack. ■☆

361969　Setaria woodii Hack. subsp. bechuanica de Wit = Setaria incrassata (Hochst.) Hack. ■☆

361970　Setaria woodii Hack. var. fonssalutis de Wit = Setaria incrassata (Hochst.) Hack. ■☆

361971　Setaria yunnanensis B. J. Keng et K. D. Yu ex P. C. Keng et Y. K. Ma;云南狗尾草;Yunnan Bristlegrass ■

361972　Setariopsis Scribn. ex Millsp. (1896);拟狗尾草属■☆

361973　Setariopsis latiglumis (Vasey) Scribn. ex Millsp. ;拟狗尾草■☆

361974　Setchellanthaceae Iltis(1999);夷白花菜科●☆

361975　Setchellanthus Brandegee(1909);夷白花菜属●☆

361976　Setchellanthus caeruleus Brandegee;夷白花菜■☆

361977　Setcreasea K. Schum. = Setcreasea K. Schum. et Syd. ■☆

361978　Setcreasea K. Schum. et Syd. (1901);紫竹梅属(紫露草属)■☆

361979　Setcreasea K. Schum. et Syd. = Tradescantia L. ■

361980　Setcreasea K. Schum. et Syd. = Treleasea Rose ■☆

361981　Setcreasea brevifolia (Torr.) K. Schum. et Syd. = Tradescantia brevifolia (Torr.) Rose ■☆

361982　Setcreasea buckleyi I. M. Johnst. = Tradescantia buckleyi (I. M. Johnst.) D. R. Hunt ■☆

361983　Setcreasea leiandra (Torr.) Pilg. = Tradescantia leiandra Torr. ■☆

361984　Setcreasea pallida (Rose) K. Schum. et Syd. ;苍白紫竹梅(苍白紫露草)■☆

361985　Setcreasea pallida (Rose) K. Schum. et Syd. ' Purple Heart ';紫心苍白紫竹梅(紫心苍白紫露草);Purple Heart, Wandering Jew ■☆

361986　Setcreasea pallida Rose = Tradescantia pallida (Rose) D. R. Hunt ■☆

361987　Setcreasea purpurea (Schauer) B. M. Boom;紫竹梅(紫锦草,紫露草);Purple Heart ■☆

361988　Setcreasea purpurea B. M. Boom = Setcreasea purpurea (Schauer) B. M. Boom ■☆

361989　Setcreasea purpurea B. M. Boom = Tradescantia pallida (Rose) D. R. Hunt ■☆

361990　Sethia Kunth = Erythroxylum P. Browne ●

361991　Sethia kunthiana Wall. = Erythroxylum sinense C. Y. Wu ●

361992　Setiacis S. L. Chen et Y. X. Jin = Panicum L. ■

361993　Setiacis S. L. Chen et Y. X. Jin(1988);刺毛头黍属(刺毛黍属);Setigrass ■★

361994　Setiacis diffusa (L. C. Chia) S. L. Chen et Y. X. Jin;刺毛头黍(散序凤头黍);Setigrass ■

361995　Seticereus Backeb. = Borzicactus Riccob. ●☆

361996　Seticereus Backeb. = Cleistocactus Lem. ●☆

361997　Seticleistocactus Backeb. = Cleistocactus Lem. ●☆

361998　Setiechinopsis (Backeb.) de Haas = Arthrocereus A. Berger(保留属名)●☆

361999　Setiechinopsis (Backeb.) de Haas = Echinopsis Zucc. ●

362000　Setiechinopsis (Backeb.) de Haas = Setiechinopsis Backeb. ex de Haas ☆

362001　Setiechinopsis Backeb. = Setiechinopsis Backeb. ex de Haas ■☆

362002　Setiechinopsis Backeb. ex de Haas(1940);奇想球属(刚毛刺仙人柱属,奇想丸属)■☆

362003　Setiechinopsis mirabilis (Speg.) Backeb. ex de Haas;奇想球(奇想丸)■☆

362004　Setiechinopsis mirabilis (Speg.) Backeb. ex de Haas = Echinopsis mirabilis Speg. ■☆

362005　Setilobus Baill. (1888);刚毛紫葳属●☆

362006　Setilobus bracteatus Baill. ;刚毛紫葳●☆

362007　Setirebutia Fričet Kreuz. = Rebutia K. Schum. ●

362008　Setiscapella Barnhart = Utricularia L. ■

362009　Setiscapella subulata (L.) Barnhart = Utricularia subulata L. ■☆

362010　Setosa Ewart = Chamaeraphis R. Br. ■☆

362011　Setouratea Tiegh. = Ouratea Aubl. (保留属名)●

362012　Setulocarya R. R. Mill et D. G. Long = Microcaryum I. M. Johnst. ■

362013　Setulocarya R. R. Mill et D. G. Long(1996);刺毛微果草属■☆

362014　Seubertia H. C. Watson = Bellis L. ■

362015　Seubertia Kunth = Brodiaea Sm. (保留属名)■☆

362016　Seubertia Kunth = Triteleia Douglas ex Lindl. ■☆

362017　Seubertia crocea A. W. Wood = Triteleia crocea (A. W. Wood) Greene ■☆

362018　Seubertia laxa (Benth.) Kunth = Triteleia laxa Benth. ■☆

362019　Seubertia obscura Borzí = Triteleia laxa Benth. ■☆

362020　Seutera Rchb. (1829);隔山消属■

362021　Seutera Rchb. = Cynanchum L. ●■

362022　Seutera Rchb. = Lyonia Elliott ●■

362023　Seutera Rchb. = Macbridea Raf. ●■

362024　Seutera Rchb. = Vincetoxicum Wolf ●■

362025　Seutera wilfordii (Franch. et Sav.) Pobed. = Cynanchum wilfordii (Maxim.) Hemsl. ■

362026　Seutera wilfordii (Maxim.) Pobed. = Cynanchum wilfordii (Maxim.) Hook. f. ■

362027　Seutera wilfordii Pobed. = Cynanchum wilfordii (Maxim.) Hook. f. ■

362028　Sevada Moq. (1849);坛花蓬属●☆

362029　Sevada schirnperi Moq. ;坛花蓬●☆

362030　Severinia Ten. = Atalantia Corrêa(保留属名)●

362031　Severinia Ten. = Severinia Ten. ex Endl. ●

362032　Severinia Ten. ex Endl. (1840);乌柑属(蚝壳刺属);Box-orange,Severinias ●

362033　Severinia Ten. ex Endl. = Atalantia Corrêa(保留属名)●

362034　Severinia buxifolia (Poir.) Ten. = Atalantia buxifolia (Poir.) Oliv. ●

362035　Severinia monophylla Tanaka = Atalantia buxifolia (Poir.) Oliv. ●

362036　Sewerzowia Regel et Schmalh. = Astragalus L. ●■

362037　Sexglumaceae Dulac = Juncaceae Juss. (保留科名)●■

362038　Sexilia Raf. = Polygala L. ●■

362039　Sextonia van der Werff = Ocotea Aubl. ●☆

362040　Sextonia van der Werff(1998);南美绿心樟属●☆

362041　Seychellaria Hemsl. (1907);败蕊霉草属■☆

362042　Seychellaria africana Vollesen;败蕊霉草(马岛霉草)■☆

362043　Seychellaria japonica (Makino) T. Ito = Sciaphila nana Blume ■☆

362044　Seychellaria tosaensis (Makino) T. Ito = Sciaphila secundiflora Thwaites ex Benth. ■

362045　Seymeria Pursh(1814)(保留属名);西摩尔列当属■☆

362046　Seymeria cassioides (J. Gmel.) S. F. Blake;西摩列当■☆

362047　Seymeria macrophylla Nutt. = Dasistoma macrophylla (Nutt.) Raf. ●☆

362048　Seymeriopsis Tzvelev = Seymeria Pursh(保留属名)■☆

362049　Seymeriopsis Tzvelev(1987);拟西摩列当属■☆

362050　Seymeriopsis bissei Tzvelev;拟西摩列当■☆

362051　Seymouria Sweet = Pelargonium L'Hér. ex Aiton ●■

362052　Seymouria heritieri Sweet = Pelargonium dipetalum L'Hér. ■☆

362053　Seyrigia Keraudren(1961);塞里瓜属■☆

362054　Seyrigia Rabenant. = Seyrigia Keraudren ■☆

362055　Seyrigia bosseri Rabenant. ;博瑟塞里瓜■☆

362056　Seyrigia gracilis Rabenant. ;纤细塞里瓜■☆

362057　Seyrigia humbertii Rabenant. ;亨伯特塞里瓜■☆

362058　Seyrigia multiflora Rabenant. ;多花塞里瓜■☆

362059　Shaeaceae Bertol. f. = Combretaceae R. Br. (保留科名)●

362060　Shafera Greenm. (1912);层绒菊属●☆

362061　Shafera platyphylla Greenm. ;层绒菊●☆

362062　Shaferocharis Urb. (1912);谢弗茜属☆

362063　Shaferocharis cubensis Urb. ;谢弗茜☆

362064　Shaferocharis multiflora Borhidi et O. Muniz;多花谢弗茜☆

362065　Shaferodendron Gilly = Manilkara Adans. (保留属名)●

362066　Shakua Bojer = Spondias L. ●

362067　Shallonium Raf. = Gaultheria L. ●

362068　Shangrilaia Al-Shehbaz,J. P. Yue et H. Sun(2004);拉萨荠属■★

362069　Shangrilaia nana Al-Shehbaz,J. P. Yue et H. Sun;尚礼荠■☆

362070　Shaniodendron M. B. Deng, H. T. Wei et X. K. Wang = Parrotia C. A. Mey. ●

362071　Shaniodendron subaequalis (Hung T. Chang) M. B. Deng,H. T. Wei et X. Q. Wang = Parrotia subaequalis (Hung T. Chang) R. M. Hao et H. T. Wei ●◇

362072　Shantzia Lewton = Azanza Alef. ●

362073　Shantzia Lewton = Thespesia Sol. ex Corrêa(保留属名)●

362074　Shantzia garckeana (F. Hoffm.) Lewton = Thespesia garckeana F. Hoffm. ●☆

362075　Shawia J. R. Forst. et G. Forst. (废弃属名) = Olearia Moench (保留属名)●☆

362076　Sheadendraceae G. Bertol. = Combretaceae R. Br. (保留科名)●

362077　Sheadendron G. Bertol = Combretum Loefl. (保留属名)●

362078　Sheadendron molle Klotzsch = Combretum pisoniiflorum (Klotzsch) Engl. ●☆

362079　Sheadendron pisoniiflorum Klotzsch = Combretum pisoniiflorum (Klotzsch) Engl. ●☆

362080　Sheadendron pisoniiflorum Klotzsch var. brachystachyum ? = Combretum pisoniiflorum (Klotzsch) Engl. ●☆

362081　Sheadendron pisoniiflorum Klotzsch var. macrostachyum ? = Combretum pisoniiflorum (Klotzsch) Engl. ●☆

362082　Sheareria S. Moore (1875); 虾 须 草 属; Sheareria, Shrimpfeelergrass ■★

362083　Sheareria nana S. Moore;虾须草(草麻黄,绿绿草,沙小菊); Dwarf Sheareria,Shrimpfeelergrass ■

362084　Sheareria polii Franch. = Sheareria nana S. Moore ■

362085　Sheffieldia J. R. Forst. et G. Forst. = Samolus L. ■

362086　Sheilanthera I. Williams(1981);希拉芸香属●☆

362087　Sheilanthera pubens I. Williams;希拉芸香●☆

362088　Sheperdia Raf. = Shepherdia Nutt. (保留属名)●☆

362089　Shepherdia Nutt. (1818) (保留属名);水牛果属;Buffalo Berry,Buffaloberry ●☆

362090　Shepherdia argentea (Pursh) Nutt. ;银色水牛果;Beef Suet Tree, Beef-suet Tree, Buffalo Berry, Buffaloberry, Buffalo-berry, Bullberry, Rabbitberry, Silver Buffalo Berry, Silver Buffaloberry, Silver Buffalo-berry,Silverberry ●☆

362091　Shepherdia canadensis (L.) Nutt. ; 加 拿 大 水 牛 果;

Buffaloberry，Rabbit-berry，Russet Buffaloberry，Russet Buffalo-berry，Soapberry ●☆

362092　Shepherdia rotundifolia Parry；圆叶水牛果●☆

362093　Sherarda St. -Lag. = Sherardia L. ■☆

362094　Sherardia Boehm. = Glinus L. ■

362095　Sherardia L. (1753)；野茜属(雪兰地属)；Field Madder ■☆

362096　Sherardia Mill. = Stachytarpheta Vahl(保留属名)■●

362097　Sherardia arvensis L.；野茜(阿文雪兰地，谢拉德草，雪迪亚草)；Blue Field Madder，Blue Fieldmadder，Field Madder，Purple Goose Grass，Spurwort ■☆

362098　Sherardia muralis L. = Galium murale (L.) All. ■☆

362099　Sherbournia G. Don(1855)；谢尔茜属●☆

362100　Sherbournia amaralioides (K. Schum.) Hua = Sherbournia streptocaulon (K. Schum.) Hepper ●☆

362101　Sherbournia batesii (Wernham) Hepper；贝茨谢尔茜●☆

362102　Sherbournia batesii (Wernham) Hepper subsp. kivuensis L. Pauwels et Sonké；基伍谢尔茜●☆

362103　Sherbournia bignoniiflora (Welw.) Hua；紫葳花谢尔茜●☆

362104　Sherbournia bignoniiflora (Welw.) Hua var. brazzaei (Hua) N. Hallé = Sherbournia bignoniiflora (Welw.) Hua ●☆

362105　Sherbournia brazzaei Hua = Sherbournia bignoniiflora (Welw.) Hua ●☆

362106　Sherbournia calycina (G. Don) Hua；萼状谢尔茜●☆

362107　Sherbournia curvipes (Wernham) N. Hallé；弯梗谢尔茜●☆

362108　Sherbournia foliosa G. Don = Sherbournia calycina (G. Don) Hua ●☆

362109　Sherbournia hapalophylla (Wernham) Hepper；芋叶谢尔茜●☆

362110　Sherbournia hapalophylla (Wernham) Hepper subsp. wernhamiana (N. Hallé) Sonké et L. Pauwels；沃纳姆谢尔茜●☆

362111　Sherbournia hapalophylla (Wernham) Hepper var. henrihuana N. Hallé = Sherbournia hapalophylla (Wernham) Hepper ●☆

362112　Sherbournia hapalophylla (Wernham) Hepper var. wernhamiana N. Hallé = Sherbournia hapalophylla (Wernham) Hepper subsp. wernhamiana (N. Hallé) Sonké et L. Pauwels ●☆

362113　Sherbournia millenii (Wernham) Hepper；米伦谢尔茜●☆

362114　Sherbournia streptocaulon (K. Schum.) Hepper；旋茎谢尔茜●☆

362115　Sherbournia streptocaulon (K. Schum.) Hepper var. situlunga N. Hallé = Sherbournia streptocaulon (K. Schum.) Hepper ●☆

362116　Sherbournia zenkeri Hua；岑克尔谢尔茜●☆

362117　Sherwoodia House = Shortia Torr. et A. Gray(保留属名)■

362118　Sherwoodia sinensis (Hemsl.) House = Shortia sinensis Hemsl. ■

362119　Sherwoodia uniflora House；单花谢尔茜●☆

362120　Shibataea Makino = Shibataea Makino ex Nakai ●

362121　Shibataea Makino ex Nakai(1933)；鹅毛竹属(岗姬竹属，倭竹属)；Bamboo，Shibata Bamboo，Shibataea，Wobamboo ●

362122　Shibataea chiangshanensis T. H. Wen；江山鹅毛竹(江山倭竹)；Jiangshan Shibataea，Jiangshan Wobamboo ●

362123　Shibataea chinensis Nakai；鹅毛竹(倭竹)；Chinese Bamboo，Chinese Shibataea，Goosefeather Wobamboo ●

362124　Shibataea chinensis Nakai 'Aureo-striata'；黄条纹鹅毛竹(黄条倭竹)；Large-leaved Variegated Bamboo ●

362125　Shibataea chinensis Nakai f. arueo-striata (Regel) C. H. Hu et al. = Shibataea chinensis Nakai 'Aureo-striata' ●

362126　Shibataea chinensis Nakai var. gracilis C. H. Hu；细鹅毛竹(细倭竹)；Thin Chinese Shibataea，Thin Goosefeather Wobamboo ●

362127　Shibataea fujianica Z. D. Zhu et H. Y. Zhao = Shibataea

nanpingensis Q. F. Zheng et K. F. Huang var. fujianica (C. D. Chu et H. Y. Zhou) C. H. Hu ●

362128　Shibataea fujianica Z. D. Zhu et H. Y. Zhou ex C. H. Hu, Q. F. Zheng et K. F. Huang = Shibataea nanpingensis Q. F. Zheng et K. F. Huang var. fujianica (C. D. Chu et H. Y. Zhou) C. H. Hu ●

362129　Shibataea hispida McClure；芦花竹(毛倭竹)；Hispid Bamboo，Hispid Shibataea，Reedflower Wobamboo ●

362130　Shibataea kumasasa (Zoll. ex Steud.) Makino = Shibataea kumasasa (Zoll. ex Steud.) Makino ex Nakai ●

362131　Shibataea kumasasa (Zoll. ex Steud.) Makino ex Nakai；倭竹(阿龟倭竹，丰后笹，岗姬竹，日本矮竹，日本倭竹，神乐笹，五枚笹，五叶笹，五叶竹)；Bamboo，Japanese Bamboo，Japanese Shibataea，Kuma Bamboo，Kuma Shibataea，Okame Zasa，Wobamboo ●

362132　Shibataea kumasasa (Zoll. ex Steud.) Makino ex Nakai 'Albostriata'；黄条纹倭竹；Large-leaved Variegated Bamboo ●

362133　Shibataea kumasasa (Zoll. ex Steud.) Makino ex Nakai f. arueo-striata (Regel) Sad. Suzuki = Shibataea chinensis Nakai 'Aureo-striata' ●

362134　Shibataea kumasasa (Zoll. ex Steud.) Makino ex Nakai f. arueo-striata Ohwi = Shibataea chinensis Nakai 'Aureo-striata' ●

362135　Shibataea kumasasa (Zoll. ex Steud.) Makino ex Nakai var. arueo-striata (Regel) Makino = Shibataea chinensis Nakai 'Aureo-striata' ●

362136　Shibataea lanceifolia C. H. Hu 'Lanceifolia'；狭叶倭竹(狭叶鹅毛竹)；Lance-leaved Shibataea，Narrowleaf Shibataea，Narrowleaf Wobamboo ●

362137　Shibataea lanceifolia C. H. Hu 'Smaragdina'；翡翠倭竹●

362138　Shibataea lanceifolia C. H. Hu = Shibataea lanceifolia C. H. Hu 'Lanceifolia' ●

362139　Shibataea lanceifolia C. H. Hu f. smaragdina C. H. Hu = Shibataea lanceifolia C. H. Hu 'Smaragdina' ●

362140　Shibataea nanpingensis Q. F. Zheng et K. F. Huang；南平鹅毛竹(福建倭竹，南平倭竹)；Nanping Shibataea，Nanping Wobamboo ●

362141　Shibataea nanpingensis Q. F. Zheng et K. F. Huang var. fujianica (C. D. Chu et H. Y. Zhou) C. H. Hu；福建鹅毛竹(福建倭竹)；Fujian Shibataea，Fujian Wobamboo ●

362142　Shibataea pygmaea F. Maek.；矮小倭竹●☆

362143　Shibataea ruscifolia (Siebold ex Munro) Makino = Shibataea kumasasa (Zoll. ex Steud.) Makino ex Nakai ●

362144　Shibataea ruscifolia (Siebold) Makino = Shibataea kumasasa (Zoll. ex Steud.) Makino ex Nakai ●

362145　Shibataea stellata (Maxim.) Nakai = Eranthis stellata Maxim. ■

362146　Shibataea strigosa T. H. Wen；矮雷竹(矮倭竹，雷倭竹，雷竹)；Bristly Shibataea，Dwarf Shibataea，Dwarf Wobamboo ●

362147　Shibataea tumidinoda T. H. Wen；大节倭竹；Big-nodded Shibataea ●

362148　Shibateranthis Nakai = Eranthis Salisb. (保留属名)■

362149　Shibateranthis stellata (Maxim.) Nakai = Eranthis stellata Maxim. ■

362150　Shicola M. Roem. = Eriobotrya Lindl. ●

362151　Shiia Makino = Castanopsis (D. Don) Spach(保留属名)●

362152　Shiia brachyacantha (Hayata) Kudo et Masam. = Castanopsis eyrei (Champ. ex Benth.) Tutcher ●

362153　Shiia carlesii (Hemsl.) Kudo = Castanopsis carlesii (Hemsl.) Hayata ●

362154　Shiia fissa (Champ. ex Benth.) Kudo = Castanopsis fissa (Champ. ex Benth.) Rehder et E. H. Wilson ●

362155 Shiia longicaudata（Hayata）Kudo et Masam. ex Msasam. = Castanopsis carlesii（Hemsl.）Hayata ●

362156 Shiia randaiensis（Hayata）Koidz. = Castanopsis uraiana（Hayata）Kaneh. et Hatus. ●

362157 Shiia uraiana（Hayata）Kaneh. et Hatus. = Castanopsis uraiana（Hayata）Kaneh. et Hatus. ●

362158 Shiia uraiana（Hayata）Kaneh. et Hatus. ex Kaneh. = Castanopsis uraiana（Hayata）Kaneh. et Hatus. ●

362159 Shinnersia R. M. King et H. Rob.（1970）;溪泽兰属■☆

362160 Shinnersia rivularis（A. Gray）R. M. King et H. Rob.;溪泽兰;Rio Grande Bugheal ■☆

362161 Shinnersoseris Tomb（1974）;喙骨苣属;Beaked Skeleton-weed ■☆

362162 Shinnersoseris rostrata（A. Gray）Tomb;喙骨苣;Annual Skeleton-weed,Beaked Skeleton-weed ■☆

362163 Shirakia Hurus. = Neoshirakia Esser ●

362164 Shirakia Hurus. = Sapium Jacq.（保留属名）●

362165 Shirakia aubrevillei（Léandri）Kruijt = Shirakiopsis aubrevillei（Léandri）Esser ●☆

362166 Shirakia elliptica（Hochst.）Kruijt = Shirakiopsis elliptica（Hochst.）Esser ●☆

362167 Shirakia indica（Willd.）Hurus. = Shirakiopsis indica（Willd.）Esser ●

362168 Shirakia japonica（Siebold et Zucc.）Hurus. = Sapium japonicum（Siebold et Zucc.）Pax et K. Hoffm. ●

362169 Shirakia japonica（Siebold et Zucc.）Hurus. f. macrophylla ? = Sapium japonicum（Siebold et Zucc.）Pax et K. Hoffm. ●

362170 Shirakia japonica（Siebold et Zucc.）Hurus. var. ryukyuensis ? = Sapium japonicum（Siebold et Zucc.）Pax et K. Hoffm. ●

362171 Shirakia trilocularis（Pax et K. Hoffm.）Kruijt = Shirakiopsis trilocularis（Pax et K. Hoffm.）Esser ●☆

362172 Shirakiopsis Esser（1999）;齿叶乌桕属●

362173 Shirakiopsis aubrevillei（Léandri）Esser;奥波齿叶乌桕●☆

362174 Shirakiopsis bingyricum Roxb. ex Baill. = Shirakiopsis indica（Willd.）Esser ●

362175 Shirakiopsis elliptica（Hochst.）Esser;椭圆齿叶乌桕(椭圆白木)●☆

362176 Shirakiopsis indica（Willd.）Esser;齿叶乌桕(印度白木)●

362177 Shirakiopsis trilocularis（Pax et K. Hoffm.）Esser;非洲齿叶乌桕●☆

362178 Shirleyopanax Domin = Kissodendron Seem. ●

362179 Shishindenia Makino ex Koidz. = Chamaecyparis Spach ●

362180 Shishindenia Makino ex Koidz. = Retinispora Siebold et Zucc. ●

362181 Shiuyinghua Paclt（1962）;秀英花属●

362182 Shiuyinghua silvestrii（Pamp. et Bonati）Paclt;秀英花●☆

362183 Shiuyinghua silvestrii（Pamp. et Bonati）Paclt = Paulownia silvestrii Pamp. et Bonati ●☆

362184 Shonia R. J. F. Hend. et Halford（2005）;肖恩大戟属 ☆

362185 Shorea Roxb. = Shorea Roxb. ex C. F. Gaertn.

362186 Shorea Roxb. ex C. F. Gaertn.（1805）;娑罗双属（龙脑香属,娑罗双树属）;Balau, Black Ponfianak, Chan, Doon, Illipe, Illipe Butter, Illipe Nut, Lauan, Mangasinoro, Meranti, Red Lauan, Selangan, Shorea, Yellow Seraya ●

362187 Shorea acuminata Dyer;渐尖娑罗双（红娑罗双,浅红娑罗双）;Light Red Meranti, Red Meranti ●☆

362188 Shorea albida Symington ex A. V. Thomas;沙拉瓦柳安(白柳安);Alan ●☆

362189 Shorea almon Foxw.;奥蒙娑罗双;Almon ●☆

362190 Shorea assamica Dyer;云南娑罗双;Assam Meranti, Yunnan Meranti ●◇

362191 Shorea assamica Dyer f. koordersii Brandis ex Koord. = Shorea assamica Dyer subsp. koordersii（Brandis ex Koord.）Y. K. Yang et J. K. Wu ●☆

362192 Shorea assamica Dyer subsp. globifera（Ridl.）Y. K. Yang et J. K. Wu;短隔娑罗双●☆

362193 Shorea assamica Dyer subsp. koordersii（Brandis ex Koord.）Y. K. Yang et J. K. Wu;库氏娑罗双●☆

362194 Shorea assamica Dyer subsp. yingjiangensis Y. K. Yang et J. K. Wu;盈江娑罗双;Yingjiang Meranti ●

362195 Shorea belangeran Burck;加里曼丹娑罗双;Belangeran, Kalimantan Meranti ●☆

362196 Shorea chinensis（H. Wang）H. Zhu = Parashorea chinensis H. Wang ●

362197 Shorea chinensis（Wang Hsie）H. Zhu;望天树(分界树,擎天树);Chinese Meranti, Chinese Parashorea, Guangxi Parashorea, Kwangs Parashorea i ●◇

362198 Shorea chinensis Merr. = Hopea chinensis（Merr.）Hand.-Mazz. ●◇

362199 Shorea chinensis Merr. = Hopea guangxiensis Y. K. Yang et J. K. Wu ●

362200 Shorea cochinchinensis Pierre;印支娑罗双;Cochinchina Meranti, White Meranti ●☆

362201 Shorea contorta S. Vidal;菲律宾白娑罗双;White Lauan ●☆

362202 Shorea curtisii Dyer ex King;库特斯娑罗双;Curtis Meranti, Dark Red Meranti ●☆

362203 Shorea eximia Scheff.;大柳安;Almon Lauan ●☆

362204 Shorea foxworthyi Symington;北加里曼丹娑罗双●☆

362205 Shorea glauca King;灰蓝娑罗双;Balan ●☆

362206 Shorea globifera Ridl. = Shorea assamica Dyer subsp. globifera（Ridl.）Y. K. Yang et J. K. Wu ●☆

362207 Shorea guiso（Blanco）Blume;红娑罗双;Red Balan, Selangan ●☆

362208 Shorea kunstleri King;昆氏娑罗双;Red Balan ●☆

362209 Shorea laevifolia Endert;平滑叶娑罗双;Bangkirai, Glabrousleaf Meranti ●☆

362210 Shorea leprosula Miq.;浅红娑罗双（皮屑娑罗双）;Lightred Meranti ●☆

362211 Shorea mangachapoi（Blanco）Blume = Vatica mangachapoi Blanco ●◇

362212 Shorea negrosensis Foxw.;菲律宾柳安（红柳安）;Borneo Cedar, Philippine Mahogany, Red Lauan ●☆

362213 Shorea obtusa Wall.;钝叶娑罗双（钝叶龙脑香）;Obtuse Meranti ●☆

362214 Shorea ovalis Blume;椭圆叶娑罗双;Ellipticleaf Meranti, Ovalleaf Meranti ●☆

362215 Shorea parvifolia Dyer;小叶娑罗双;Littleleaf Meranti ●☆

362216 Shorea pauciflora King;疏花娑罗双(稀花娑罗双);Dark Red Meranti, Dark Red Seraya ●☆

362217 Shorea polysperma（Blanco）Merr.;多子红柳安（多子娑罗双）;Bataan, Bataan Mahogany, Dark Red Lauan, Philippine Mahogany, Red Lauan, Tangile, Tanguile ●☆

362218 Shorea polysperma Merr. = Shorea polysperma（Blanco）Merr. ●☆

362219 Shorea robusta C. F. Gaertn. = Shorea robusta Roxb. ex C. F.

Gaertn. ●☆

362220　Shorea robusta Roxb. ex C. F. Gaertn.；娑罗双（粗壮娑罗双，沙罗树，娑罗树，娑罗双树，桫椤）；Sakhu，Sal，Sal Shorea，Sal Tree，Saul Tree，Saul-tree ●☆

362221　Shorea roxburghii G. Don；罗氏娑罗双●☆

362222　Shorea sericeiflora C. E. C. Fisch. et Hutch.；丝花娑罗双；Sericeousflower Meranti ●☆

362223　Shorea siamensis Miq. var. borealis Y. K. Yang et J. K. Wu；北缘白柳安●☆

362224　Shorea siamensis Miq. var. laevis（Pierre）Y. K. Yang et J. K. Wu；平滑白柳安●☆

362225　Shorea siamensis Miq. var. mekongensis（Pierre ex Laness.）Y. K. Yang et J. K. Wu；绒毛叶白柳安●☆

362226　Shorea squamata Benth. et Hook. f.；具鳞娑罗双；Red Lauan，White Lauan，White Meranti ●☆

362227　Shorea talura Roxb.；印南娑罗双；White Meranti Phayom ●☆

362228　Shorea wangtianshuea Y. K. Yang et J. K. Wu = Parashorea chinensis H. Wang ●

362229　Shorea wangtianshuea Y. K. Yang et J. K. Wu = Shorea chinensis（Wang Hsie）H. Zhu ●◇

362230　Shorea wangtianshuea Y. K. Yang et J. K. Wu subsp. kwangsiensis（Lin Chi）Y. K. Yang et J. K. Wu = Parashorea chinensis H. Wang var. kwangsiensis Lin Chi ●

362231　Shorea wangtianshuea Y. K. Yang et J. K. Wu subsp. kwangsiensis（Lin Chi）Y. K. Yang et J. K. Wu = Shorea chinensis（Wang Hsie）H. Zhu ●◇

362232　Shorea wangtianshuea Y. K. Yang et J. K. Wu subsp. kwangsiensis（Lin Chi）Y. K. Yang et J. K. Wu = Parashorea chinensis H. Wang ●

362233　Shorea wangtianshuea Y. K. Yang et J. K. Wu subsp. vietnamensis Y. K. Yang et J. K. Wu. = Shorea chinensis（Wang Hsie）H. Zhu ●◇

362234　Shorea wangtianshuea Y. K. Yang et J. K. Wu subsp. vietnamensis Y. K. Yang et J. K. Wu. = Parashorea chinensis H. Wang ●

362235　Shorea wangtianshuea Y. K. Yang et J. K. Wu subsp. vietnamensis Y. K. Yang et J. K. Wu；越高树●☆

362236　Shorea wangtianshuea Y. K. Yang et J. K. Wu var. chuanbanshuea Y. K. Yang et J. K. Wu = Shorea chinensis（Wang Hsie）H. Zhu ●◇

362237　Shorea wangtianshuea Y. K. Yang et J. K. Wu var. chuanbanshuea Y. K. Yang et J. K. Wu = Parashorea chinensis H. Wang ●

362238　Shorea wiesneri Schiffn.；印度娑罗双（印度娑罗双树）；Batavian Shorea，Wiesner's Meranti ●☆

362239　Shoreaceae Barldey = Dipterocarpaceae Blume（保留科名）●

362240　Shortia Raf.（废弃属名）= Arabis L. ●■

362241　Shortia Raf.（废弃属名）= Shortia Torr. et A. Gray（保留属名）■

362242　Shortia Torr. et A. Gray（1842）（保留属名）；岩扇属（裂缘花属）；Shortia ■

362243　Shortia davidii Franch. = Berneuxia thibetica Decne. ■

362244　Shortia exappendiculata Hayata；台湾岩扇（裂缘花）■

362245　Shortia exappendiculata Hayata = Shortia rotundifolia（Maxim.）Makino ■

362246　Shortia galacifolia Torr. et A. Gray；美国岩扇；Oconee Bells，Shortia ■☆

362247　Shortia ilicifolius（Maxim.）L. H. Li = Schizocodon ilicifolius

Maxim. ■☆

362248　Shortia magna（Makino）Makino = Schizocodon soldanelloides Siebold et Zucc. var. magnus（Makino）H. Hara ■☆

362249　Shortia ritoensis Hayata = Shortia exappendiculata Hayata ■

362250　Shortia ritoensis Hayata = Shortia rotundifolia（Maxim.）Makino var. ritoensis（Hayata）T. C. Huang et A. Hsiao ■

362251　Shortia rotundifolia（Maxim.）Makino；倒卵叶裂缘花（裂缘花，台湾岩扇）；Taiwan Shortia ■

362252　Shortia rotundifolia（Maxim.）Makino f. amamiana（Ohwi）T. Yamaz.；奄美倒卵叶裂缘花■☆

362253　Shortia rotundifolia（Maxim.）Makino var. amamiana Ohwi = Shortia rotundifolia（Maxim.）Makino f. amamiana（Ohwi）T. Yamaz. ■☆

362254　Shortia rotundifolia（Maxim.）Makino var. ritoensis（Hayata）T. C. Huang et A. Hsiao；李栋山裂缘花■

362255　Shortia rotundifolia（Maxim.）Makino var. subcordata（Hayata）T. C. Huang et A. Hsiao；圆叶裂缘花（心叶岩扇）；Cordate Shortia，Cordateleaf Shortia ■

362256　Shortia rotundifolia（Maxim.）Makino var. transalpina（Hayata）T. Yamaz.；高山裂缘花■

362257　Shortia sinensis Hemsl.；华岩扇；China Shortia，Chinese Shortia ■

362258　Shortia sinensis Hemsl. var. pubinervis C. Y. Wu；毛脉华岩扇；Hairyvein China Shortia ■

362259　Shortia soldanelloides（Siebold et Zucc.）Makino = Schizocodon soldanelloides Siebold et Zucc. ■☆

362260　Shortia soldanelloides（Siebold et Zucc.）Makino f. alpina（Maxim.）Makino = Schizocodon soldanelloides Siebold et Zucc. f. alpinus Maxim. ■☆

362261　Shortia soldanelloides（Siebold et Zucc.）Makino var. ilicifolia（Maxim.）Makino = Schizocodon ilicifolius Maxim. ■☆

362262　Shortia soldanelloides（Siebold et Zucc.）Makino var. intercedens Ohwi = Schizocodon ilicifolius Maxim. var. intercedens（Ohwi）T. Yamaz. ■☆

362263　Shortia soldanelloides（Siebold et Zucc.）Makino var. magna Makino = Schizocodon soldanelloides Siebold et Zucc. var. magnus（Makino）H. Hara ☆

362264　Shortia soldanelloides（Siebold et Zucc.）Makino var. minima（Makino）Masam. = Schizocodon ilicifolius Maxim. var. minimus（Makino）T. Yamaz. ■☆

362265　Shortia subcordata Hayata = Shortia exappendiculata Hayata ■

362266　Shortia subcordata Hayata = Shortia rotundifolia（Maxim.）Makino var. subcordata（Hayata）T. C. Huang et A. Hsiao ■

362267　Shortia thibetica（Decne.）Franch. = Berneuxia thibetica Decne. ■

362268　Shortia transalpina Hayata = Shortia exappendiculata Hayata ■

362269　Shortia transalpina Hayata = Shortia rotundifolia（Maxim.）Makino var. transalpina（Hayata）T. Yamaz. ■

362270　Shortia uniflora（Maxim.）Maxim.；单花岩扇（独花岩扇）；Nippon-bells ■☆

362271　Shortia uniflora（Maxim.）Maxim. f. albens Honda；白单花岩扇■☆

362272　Shortia uniflora（Maxim.）Maxim. f. albiflora Makino = Shortia uniflora（Maxim.）Maxim. f. albens Honda ■☆

362273　Shortia uniflora（Maxim.）Maxim. var. kantoensis T. Yamaz.；河原单花岩扇■☆

362274　Shortia uniflora（Maxim.）Maxim. var. kantoensis T. Yamaz. f. plena T. Yamaz.；重瓣河原单花岩扇■☆

362275　Shortia uniflora（Maxim.）Maxim. var. orbicularis Honda；圆单花岩扇■☆

362276　Shortia uniflora Maxim.‘Grandiflora’；大独花岩扇■☆

362277　Shortia uniflora Maxim. var. macrophylla？= Shortia uniflora（Maxim.）Maxim. ■☆

362278　Shortiopsis Hayata = Shortia Torr. et A. Gray（保留属名）■

362279　Shortiopsis exappendiculata Hayata = Shortia exappendiculata Hayata ■

362280　Shortiopsis exappendiculata Hayata = Shortia rotundifolia（Maxim.）Makino ■

362281　Shoshonea Evert et Constance(1982)；怀俄明属■☆

362282　Shoshonea pulvinata Evert et Constance；怀俄明草■☆

362283　Shultzia Raf.（废弃属名）= Obolaria L. ■☆

362284　Shultzia Raf.（废弃属名）= Schulzia Spreng.（保留属名）■☆

362285　Shuria Herincq = Achimenes Pers.（保留属名）■☆

362286　Shuria Hort. ex Herincq = Achimenes P. Br. ■☆

362287　Shutereia Choisy（废弃属名）= Hewittia Wight et Arn. ■

362288　Shutereia Choisy（废弃属名）= Shuteria Wight et Arn.（保留属名）■

362289　Shutereia bicolor（Vahl）Choisy = Hewittia malabarica（L.）Suresh ■

362290　Shutereia sublobata（L. f.）House = Hewittia malabarica（L.）Suresh ■

362291　Shuteria Choisy = Hewittia Wight et Arn. ■

362292　Shuteria Wight et Arn.（1834）（保留属名）；宿苞豆属；Bractbean, Shuteria ■

362293　Shuteria africana Hook. f. = Amphicarpaea africana（Hook. f.）Harms ■☆

362294　Shuteria anabaptis（Kurz）C. Y. Wu = Shuteria hirsuta Baker ■

362295　Shuteria ferruginea Baker；锈毛宿苞豆■☆

362296　Shuteria glabrata Wight et Arn. = Shuteria involucrata（Wall.）Wight et Arn. var. glabrata（Wight et Arn.）Ohashi ■

362297　Shuteria hirsuta Baker；硬毛宿苞豆（野山葛）；Hardhair Bractbean, Hisute Shuteria ■

362298　Shuteria involucrata（Wall.）Wight et Arn.；宿苞豆（铜钱麻黄,铜钱藤,野豌豆,中国宿苞豆）；Bractbean, Common Shuteria ■

362299　Shuteria involucrata（Wall.）Wight et Arn. var. glabrata（Wight et Arn.）Ohashi；光宿苞豆；Glabrous Shuteria ■

362300　Shuteria involucrata（Wall.）Wight et Arn. var. villosa（Pamp.）Ohashi；毛苞豆（草红藤,红藤,蝴蝶草,黄花马豆,铁马豆）；Pampanin Bractbean, Pampanin Shuteria ■

362301　Shuteria longipes Franch. = Hylodesmum longipes（Franch.）H. Ohashi et R. R. Mill ■

362302　Shuteria pampaniniana Hand.-Mazz. = Shuteria involucrata（Wall.）Wight et Arn. var. villosa（Pamp.）Ohashi ■

362303　Shuteria sinensis Hemsl. = Shuteria involucrata（Wall.）Wight et Arn. ■

362304　Shuteria trisperma Miq. = Amphicarpaea edgeworthii Benth. ■

362305　Shuteria vestita（Graham）Wight et Arn. = Shuteria involucrata（Wall.）Wight et Arn. var. glabrata（Wight et Arn.）Ohashi ■

362306　Shuteria vestita Wight et Arn.；西南宿苞豆（茶叶藤）；Covered Bractbean ■

362307　Shuteria vestita Wight et Arn. = Shuteria involucrata（Wall.）Wight et Arn. var. glabrata（Wight et Arn.）Ohashi ■

362308　Shuteria vestita Wight et Arn. var. glabrata（Wight et Arn.）Baker = Shuteria involucrata（Wall.）Wight et Arn. var. glabrata（Wight et Arn.）Ohashi ■

362309　Shuteria vestita Wight et Arn. var. involucrata（Wall.）Baker = Shuteria involucrata（Wall.）Wight et Arn. ■

362310　Shuteria vestita Wight et Arn. var. vilosa Pamp. = Shuteria involucrata（Wall.）Wight et Arn. var. villosa（Pamp.）Ohashi ■

362311　Shuttleworthia Steud. = Shuttlewornthia Meisn. ■●

362312　Shuttlewornthia Meisn. = Uwarowia Bunge ■●

362313　Shuttlewornthia Meisn. = Verbena L. ■●

362314　Siagonanthus Poepp. et Endl. = Maxillaria Ruiz et Pav. ■☆

362315　Siagonanthus Poepp. et Endl. = Ornithidium R. Br. ■☆

362316　Siagonanthus Pohl ex Engler = Emmotum Desv. ex Ham. ●☆

362317　Siagonarrhen Mart. ex J. A. Schmidt = Hyptis Jacq.（保留属名）●■

362318　Sialita Raf. = Dillenia L. ●

362319　Sialita Raf. = Syalita Adans. ●

362320　Sialodes Eckl. et Zeyh. = Galenia L. ●☆

362321　Siamanthus K. Larsen et Mood(1998)；泰花属■☆

362322　Siamanthus siliquosus K. Larsen et Mood；泰花■☆

362323　Siamosia K. Larsen et Pedersen(1987)；棱苞聚花苋属■☆

362324　Siamosia thailandica Larsen et Pedersen；棱苞聚花苋■☆

362325　Siapaea Pruski(1996)；匐匐尖泽兰属■●☆

362326　Siapaea liesneri Pruski；匐匐尖泽兰■☆

362327　Sibaldia L. = Potentilla L. ■●

362328　Sibaldia L. = Sibbaldia L. ■●

362329　Sibangea Oliv.（1883）；西邦大戟属●☆

362330　Sibangea Oliv. = Drypetes Vahl ●

362331　Sibangea arborescens Oliv.；西邦大戟●☆

362332　Sibangea pleioneura Radcl.-Sm.；多脉西邦大戟●☆

362333　Sibangea similis（Hutch.）Radcl.-Sm.；非洲西邦大戟●☆

362334　Sibara Greene = Cardamine L. ■

362335　Sibara Greene(1896)；假南芥属■☆

362336　Sibaropsis S. Boyd et T. S. Ross(1997)；异南芥属■☆

362337　Sibaropsis hammittii S. Boyd et T. S. Ross；异南芥■☆

362338　Sibbalda St.-Lag. = Sibbaldia L. ■

362339　Sibbaldia L.（1753）；山莓草属（山金梅属,五蕊梅属,西巴德属）；Sibbaldia, Wildberry ■

362340　Sibbaldia L. = Potentilla L. ■●

362341　Sibbaldia adpressa Bunge；伏毛山莓草（伏毛山草莓,十蕊山莓草,西山草）；Addressedhairy Wildberry, Pressedhairy Sibbaldia ■

362342　Sibbaldia altaica Laxm. = Chamaerhodos altaica（Laxm.）Bunge ●

362343　Sibbaldia aphanopetala Hand.-Mazz. = Sibbaldia procumbens L. var. aphanopetala（Hand.-Mazz.）Te T. Yu et C. L. Li ■

362344　Sibbaldia cuneata Hornem. ex Kuntze；楔叶山莓草；Cuneate Sibbaldia, Cuneate Wildberry ■

362345　Sibbaldia erecta L. = Chamaerhodos erecta（L.）Bunge ●

362346　Sibbaldia glabriuscula Te T. Yu et C. L. Li；光叶山莓草；Glabrous Sibbaldia, Glabrous Wildberry ■

362347　Sibbaldia glabriuscula Te T. Yu et C. L. Li = Potentilla glabriuscula（Te T. Yu et C. L. Li）Soják ■

362348　Sibbaldia macropetala Murav. = Sibbaldia purpurea Royle var. macropetala（Murav.）Te T. Yu et C. L. Li ■

362349　Sibbaldia macrophylla Turcz. ex Juz. = Sibbaldia procumbens L. ■

362350　Sibbaldia melinocricha Hand.-Mazz.；黄毛山莓草；Yellowhairy Sibbaldia, Yellowhairy Wildberry ■

362351　Sibbaldia melinotricha Hand.-Mazz. = Sibbaldia sikkimensis（Prain）Chatterjee ■

362352　Sibbaldia micropetala（D. Don）Hand.-Mazz.；白叶山莓草；

Smallpetal Sibbaldia，Smallpetal Wildberry ■

362353　Sibbaldia minutissima Kitam. = Sibbaldia adpressa Bunge ■

362354　Sibbaldia olgae Juz. et Ovcz. ;奥氏山莓草■☆

362355　Sibbaldia omeiensis Te T. Yu et C. L. Li;峨眉山莓草;Emei Wildberry，Omei Sibbaldia ■

362356　Sibbaldia parviflora Willd. ;小花山莓草;Smallflower Sibbaldia ■☆

362357　Sibbaldia parviflora Willd. = Sibbaldia procumbens L. var. aphanopetala（Hand. -Mazz.）Te T. Yu et C. L. Li ■

362358　Sibbaldia pentaphylla J. Krause;五 叶 山 莓 草;Fiveleaf Wildberry，Fiveleaves Sibbaldia ■

362359　Sibbaldia perpusilloides（W. W. Sm.）Hand. -Mazz.;短蕊山莓草;Shortstamen Sibbaldia，Shortstamen Wildberry ■

362360　Sibbaldia phancrophylebia Te T. Yu et C. L. Li;显脉山莓草;Manifestnerved Sibbaldia，Showvein Wildberry ■

362361　Sibbaldia procumbens L. ;山莓草（大叶山莓草,木茎山金梅, 五蕊梅）;Largeleaf Sibbaldia，Least Cinquefoil，Procumbent Sibbaldia，Procumbent Wildberry，Sibhald's Potentilla ■

362362　Sibbaldia procumbens L. = Sibbaldia cuneata Hornem. ex Kuntze ■

362363　Sibbaldia procumbens L. var. aphanopetala（Hand. -Mazz.）Te T. Yu et C. L. Li;隐瓣山莓草（木茎山金莓,木茎山金梅,匍茎五蕊莓,山莓草,五蕊莓,隐瓣山草莓,隐瓣山金莓,隐瓣山金梅）;Petalless Sibbaldia ■

362364　Sibbaldia procumbens L. var. macrophylla（Turcz. ex Juz.）Gubanov = Sibbaldia procumbens L. ■

362365　Sibbaldia procumbens L. var. valdehira Ohwi = Sibbaldia cuneata Hornem. ex Kuntze ■

362366　Sibbaldia procumbens L. var. valdehira Ohwi = Sibbaldia procumbens L. ■

362367　Sibbaldia pulvinata Te T. Yu et C. L. Li;垫状山莓草;Cushion Wildberry，Cushion-shaped Sibbaldia ■

362368　Sibbaldia pulvinata Te T. Yu et C. L. Li = Potentilla coriandrifolia D. Don var. dumosa Franch. ■

362369　Sibbaldia purpurea Royle;紫花山莓草;Purple Wildberry，Purpleflower Sibbaldia ■

362370　Sibbaldia purpurea Royle var. macropetala（Murav.）Te T. Yu et C. L. Li;大瓣紫花山莓草;Largepetal Purpleflower Sibbaldia ■

362371　Sibbaldia purpurea Royle var. pentaphylla（J. Krause）Dikshit = Sibbaldia pentaphylla J. Krause ■

362372　Sibbaldia semiglabra C. A. Mey. ;半光山莓草■☆

362373　Sibbaldia sericea（Grubov）Soják;绢 毛 山 莓 草;Silky Sibbaldia，Silky Wildberry ■

362374　Sibbaldia sericea Grubov = Sibbaldia sericea（Grubov）Soják ■

362375　Sibbaldia sikkimensis（Prain）Chatterjee;黄花山莓草■

362376　Sibbaldia sikkimensis Chatterjee = Sibbaldia melinocricha Hand. -Mazz. ■

362377　Sibbaldia sikkimensis Chatterjee = Sibbaldia sikkimensis（Prain）Chatterjee ■

362378　Sibbaldia stromatodes Melch. ;滇西山莓草■

362379　Sibbaldia taiwanensis H. L. Li = Sibbaldia cuneata Hornem. ex Kuntze ■

362380　Sibbaldia tenuis Hand. -Mazz. ;纤 细 山 莓 草;Slender Wildberry，Tenuous Sibbaldia ■

362381　Sibbaldia tetrandra Bunge = Dryadanthe tetrandra（Bunge）Juz. ■

362382　Sibbaldianthe Juz. = Sibbaldia L. ■

362383　Sibbaldianthe adpressa（Bunge）Juz. = Sibbaldia adpressa Bunge ■

Bunge ■

362384　Sibbaldianthe bifurca（L.）Kurtto et T. Erikss. = Potentilla bifurca L. ■

362385　Sibbaldianthe minutissima Kitam. = Sibbaldia adpressa Bunge ■

362386　Sibbaldianthe sericea Grubov = Sibbaldia sericea（Grubov）Soják ■

362387　Sibbaldiopsis Rydb.（1901）;拟山莓草属■☆

362388　Sibbaldiopsis Rydb. = Potentilla L. ■●

362389　Sibbaldiopsis tridentata（Aiton）Rydb. ;拟山莓草;Shrubby Five-fingers，Three-toothed Cinquefoil ■☆

362390　Sibertia Steud. = Bromus L.（保留属名）●

362391　Sibertia Steud. = Libertia Spreng.（保留属名）■☆

362392　Sibiraea Maxim.（1879）;鲜 卑 花 属;Sibiraea，Sibirea，Xianbeiflower ●

362393　Sibiraea altaiensis（Laxm.）C. K. Schneid. = Sibiraea laevigata（L.）Maxim. ●

362394　Sibiraea angustata（Rehder）Hand. -Mazz. ;窄 叶 鲜 卑 花;Narrowleaf Sibiraea，Narrowleaf Xianbeiflower，Narrow-leaved Sibiraea ●

362395　Sibiraea glaberrima K. S. Hao = Sibiraea laevigata（L.）Maxim. ●

362396　Sibiraea laevigata（L.）Maxim. ;鲜 卑 花;Smooth Sibiraea，Xianbeiflower ●

362397　Sibiraea laevigata（L.）Maxim. var. angustata Rehder = Sibiraea angustata（Rehder）Hand. -Mazz. ●

362398　Sibiraea tianschanica（Krasn.）Pojark. ;天山鲜卑花■☆

362399　Sibiraea tomentosa Diels;毛叶鲜卑花;Hairleaf Xianbeiflower，Hairyleaf Sibiraea，Tomentose Sibiraea ●

362400　Sibithorpia pinnata（Wall. ex Benth.）Benth. = Ellisiophyllum pinnatum（Wall.）Makino ■

362401　Sibithorpia pinnata Benth. = Ellisiophyllum pinnatum（Wall.）Makino ■

362402　Sibthorpia L.（1753）;鸡玄参属;Cornish Moneywort ■☆

362403　Sibthorpia africana L. ;非洲鸡玄参■☆

362404　Sibthorpia americana Sessé et Moc. ;美洲鸡玄参■☆

362405　Sibthorpia australis Hutch. = Sibthorpia europaea L. ■☆

362406　Sibthorpia europaea L. ;鸡玄参;Cornish Money Wort，Cornish Moneywort，Pennywort，Small Penny Pie ■☆

362407　Sibthorpia europaea L. var. africana Hook. = Sibthorpia europaea L. ■☆

362408　Sibthorpia peregrina L. ;外来鸡玄参■☆

362409　Sibthorpia pinnata（Wall. ex Benth.）Benth. = Ellisiophyllum pinnatum（Wall. ex Benth.）Makino ■

362410　Sibthorpia prostrata Salisb. = Sibthorpia europaea L. ■☆

362411　Sibthorpia rotundifolia（Ruiz et Pav.）Edwin;圆叶鸡玄参■☆

362412　Sibthorpiaceae D. Don = Plantaginaceae Juss.（保留科名）■

362413　Sibthorpiaceae D. Don = Scrophulariaceae Juss.（保留科名）●■

362414　Sibtorpia Scop. = Sibthorpia L. ■☆

362415　Siburatia Thouars = Maesa Forssk. ●

362416　Sicana Naudin（1862）;香蕉瓜属■

362417　Sicana atropurpurea André;暗紫香蕉瓜■☆

362418　Sicana fragrans Alain，Mejía et R. Garcia;芳香香蕉瓜■☆

362419　Sicana odorifera Naudin;香蕉瓜;Casa Banana，Curuba ■☆

362420　Siccobaccatus P. J. Braun et Esteves = Micranthocereus Backeb. ●☆

362421　Sicelium P. Browne（废弃属名）= Coccocypselum P. Browne（保留属名）●☆

362422　Sichuania M. G. Gilbert et P. T. Li（1995）;四川藤属;Sichuania ■★

362423　Sichuania alterniloba M. G. Gilbert et P. T. Li;四川藤;Sichuan Sichuania■

362424　Sickingia Willd. (1801);西金茜属(斯康吉亚属)■☆

362425　Sickingia Willd. = Simira Aubl. ■☆

362426　Sickingia cordifolia Benth. et Hook. f. ;心叶西金茜☆

362427　Sickingia fragrans (Rusby) Standl. ;芳香西金茜■☆

362428　Sickingia lancifolia Lundell;剑叶西金茜■☆

362429　Sickingia longifolia Willd. ;长叶西金茜■☆

362430　Sickingia mexicana (Bullock) Steyerm. ;墨西哥西金茜■☆

362431　Sickingia mollis Lundell;柔软西金茜■☆

362432　Sickingia oliveri K. Schum. ;奥氏西金茜☆

362433　Sickingia rubra K. Schum. ;红红西金茜■☆

362434　Sickingia viridiflora (Allemão et Saldanha) K. Schum. ;绿花西金茜■☆

362435　Sickingia williamsii Standl. ;威廉西金茜■☆

362436　Sicklera M. Roem. = Murraya J. König ex L. (保留属名)●

362437　Sicklera M. Roem. = Poechia Opiz ●

362438　Sicklera M. Roem. = Psilotrichum Blume ●■

362439　Sicklera Sendtn. = Brachistus Miers ●☆

362440　Sickmannia Nees = Ficinia Schrad. (保留属名)■☆

362441　Sickmannia radiata (L. f.) Nees = Ficinia radiata (L. f.) Kunth ■☆

362442　Sicrea (Baill.) Hallier f. (1921);落萼椴属●☆

362443　Sicrea (Pierre) Hallier f. = Sicrea (Baill.) Hallier f. ●☆

362444　Sicrea Hallier f. = Sicrea (Baill.) Hallier f. ●☆

362445　Sicrea godefroyana Hallier f. ;落萼椴●☆

362446　Sicydium A. Gray = Ibervillea Greene ■☆

362447　Sicydium A. Gray = Maximowiczia Cogn. ■☆

362448　Sicydium Schltdl. (1832);野胡瓜属■☆

362449　Sicydium lindheimeri A. Gray;美国野胡瓜■☆

362450　Sicydium lindheimeri A. Gray var. tenuisectum A. Gray = Ibervillea tenuisecta (A. Gray) Small ■☆

362451　Sicydium tamnifolium Cogn. ;野胡瓜■☆

362452　Sicyocarpus Bojer = Dregea E. Mey. (保留属名)●

362453　Sicyocarya (A. Gray) H. St. John = Sicyos L. ■

362454　Sicyocaulis Wiggins = Sicyos L. ■

362455　Sicyocodon Feer = Campanula L. ■●

362456　Sicyodea Ludw. = Sicyoides Mill. ■

362457　Sicyodea Ludw. = Sicyos L. ■

362458　Sicyoides Mill. = Sicyos L. ■

362459　Sicyomorpha Miers = Peritassa Miers ●☆

362460　Sicyos L. (1753);刺瓜藤属(西克斯属,小扁瓜属);Bur Cucumber ■

362461　Sicyos angulatus L. ;刺瓜藤(棘瓜,小扁瓜,野丝瓜);Bur Cucumber, Bur-cucumber, One-seeded Bur-cucumber, Star Cucumber,Star-cucumber,Wall Bur Cucumber,Wall Bur-cucumber ■

362462　Sicyos angulatus L. f. ohtanus Asai;棱角刺瓜藤■☆

362463　Sicyos angulatus P. J. Bergius = Kedrostis nana (Lam.) Cogn. ■☆

362464　Sicyos edulis Jacq. = Sechium edule (Jacq.) Sw. ■

362465　Sicyos fauriei H. Lév. = Momordica charantia L. ■

362466　Sicyos glandulosus Poir. = Bowlesia glandulosa (Poir.) Kuntze ■☆

362467　Sicyos laciniatus Descourt. = Sechium edule (Jacq.) Sw. ■

362468　Sicyos lobatus Michx. = Echinocystis echinata (Muhl.) Vassilcz. ■☆

362469　Sicyos polyacanthus Cogn. ;多花刺瓜藤■☆

362470　Sicyosperma A. Gray(1853);葫芦籽属■☆

362471　Sicyosperma gracile A. Gray;葫芦籽■☆

362472　Sicyus Clem. = Sicyos L. ■

362473　Sida L. (1753);黄花稔属(金午时花属);Queensland Hemp, Sida ■●

362474　Sida abutilifolia Mill. ;铺散黄花稔;Spreading Fanpetals ■☆

362475　Sida abutilon L. = Abutilon theophrastii Medik. ●■

362476　Sida abyssinica Hochst. ex D. Dietr. = Sida ovata Forssk. ●☆

362477　Sida acuminata R. Br. = Abutilon longicuspe Hochst. ex A. Rich. ■☆

362478　Sida acuta Burm. f. ;黄花稔(拔毒散,柑仔密,金午时花,披针黄花稔,扫把麻,山麻,蛇总管,细叶金午时花);Acute Sida, Narrow-leaved Sida ●■

362479　Sida acuta Burm. f. subsp. carpinifolia (L. f.) Borss. Waalk. = Sida acuta Burm. f. ●■

362480　Sida acuta Burm. f. var. carpinifolia (L. f.) K. Schum. = Sida acuta Burm. f. ●■

362481　Sida acuta Burm. f. var. garckeana (Pol.) Baker f. = Sida garckeana Pol. ●☆

362482　Sida acuta Burm. f. var. intermedia S. Y. Hu = Sida acuta Burm. f. ●■

362483　Sida acutifolia Steud. = Abutilon longicuspe Hochst. ex A. Rich. ■☆

362484　Sida affinis J. A. Schmidt = Sida alba L. ●

362485　Sida africana P. Beauv. = Sida cordifolia L. subsp. maculata (Cav.) Marais ●☆

362486　Sida alba Cav. = Sida rhombifolia L. ●

362487　Sida alba L. = Sida rhombifolia L. ●

362488　Sida albidum Willd. = Abutilon albidum (Willd.) Sweet ■☆

362489　Sida alcaeoides Michx. = Callirhoe alcaeoides (Michx.) A. Gray ●☆

362490　Sida alnifolia L. ;桤叶黄花稔(拔脓草,地膏草,地马庄,黄花草,黄花母,黄花雾,牛筋麻,牛肋筋,脓见愁,糯米药,砂宁根,小柴胡,小叶黄花稔);Alder-leaf Sida, Alder-leaved Sida ●

362491　Sida alnifolia L. = Sida rhombifolia L. subsp. retusa (L.) Borss. Waalk. ●

362492　Sida alnifolia L. = Sida rhombifolia L. ●

362493　Sida alnifolia L. var. micraphylla (Cav.) S. Y. Hu = Sida rhombifolia L. var. microphylla (Cav.) S. Y. Hu ●

362494　Sida alnifolia L. var. microphylla (Cav.) S. Y. Hu;小叶黄花稔(拔脓草,地膏草,黄花母,牛肋筋,糯米药);Littleleaf Sida ●

362495　Sida alnifolia L. var. obovata (Wall. ex Mast.) S. Y. Hu;倒卵叶黄花稔(圆齿小柴胡);Obovateleaf Sida ●

362496　Sida alnifolia L. var. obovata (Wall.) S. Y. Hu = Sida rhombifolia L. var. obvata (Wall.) S. Y. Hu ●

362497　Sida alnifolia L. var. obovata sensu Hu = Sida alba L. ●

362498　Sida alnifolia L. var. orbiculata S. Y. Hu;圆叶黄花稔;Orbicularleaf Sida,Roundleaf Sida ●

362499　Sida angustifolia Lam. = Sida rhombifolia L. ●

362500　Sida antillensis Urb. ;安的列斯黄花稔;Antilles Fanpetals ●☆

362501　Sida aurescens Ulbr. = Sida hoepfneri Gürke ●☆

362502　Sida aurita Wall. ex Link = Abutilon auritum (Wall. ex Link) Sweet ●☆

362503　Sida bakeriana Rusby = Sida garckeana Pol. ●☆

362504　Sida blepharoprion Ulbr. = Sida rhombifolia L. ●

362505　Sida bodinieri Gand. = Sida acuta Burm. f. ●■

362506　Sida calliantha Thulin;美花黄花稔●☆

362507　Sida canariensis Cav. = Sida rhombifolia L. ●

362508　Sida capitata L. = Malachra capitata (L.) L. ■☆

362509　Sida carpinifolia Bourg. ex Griseb. = Sida acuta Burm. f. ●■

362510　Sida carpinifolia L. f. = Malvastrum coromandelianum (L.) Garcke ●■

362511　Sida carpinifolia L. f. = Sida acuta Burm. f. ●■

362512　Sida carpinifolia L. f. var. acuta (Burm. f.) Kurz = Sida acuta Burm. f. ●■

362513　Sida chanetii Gand. = Sida acuta Burm. f. ●■

362514　Sida chinensis Retz. ;中华黄花稔;China Sida,Chinese Sida ●

362515　Sida chinensis Retz. = Sida rhombifolia L. ●

362516　Sida chionantha Ulbr. = Sida hoepfneri Gürke ●☆

362517　Sida chrysantha Ulbr. ;金花黄花稔●☆

362518　Sida collina Schltr. ;山丘黄花稔●☆

362519　Sida contracta Link = Wissadula contracta (Link) R. E. Fr. ■☆

362520　Sida cordata (Burm. f.) Borss. = Sida cordata (Burm. f.) Borss. Waalk. ●■

362521　Sida cordata (Burm. f.) Borss. Waalk. ;长梗黄花稔(藤本黏头婆);Longstalk Sida,Long-stalked Sida ●■

362522　Sida cordata (Burm. f.) Borss. Waalk. = Sida javensis Cav. ●■

362523　Sida cordifolia L. ;心叶黄花稔(圆叶金午时花);Cordateleaf Sida,Cordate-leaved Sida ●

362524　Sida cordifolia L. subsp. maculata (Cav.) Marais;斑心叶黄花稔●☆

362525　Sida cordifolioides K. M. Feng;湖南黄花稔;Hunan Sida ■

362526　Sida corylifolia Wall. = Sida subcordata Span. ●

362527　Sida corylifolia Wall. ex Mast. = Sida subcordata Span. ●

362528　Sida corymbosa R. E. Fr. = Sida collina Schltr. ●☆

362529　Sida corynocarpa Wall. = Sida rhombifolia L. var. corynocarpa (Wall.) S. Y. Hu ●

362530　Sida crispa L. = Abutilon crispum (L.) Medik. ■

362531　Sida crispa L. = Herissantia crispa (L.) Brizicky ■

362532　Sida cristata L. = Anoda cristata (L.) Schltdl. ■☆

362533　Sida decagyna Schumach. et Thonn. ex Schumach. = Sida cordifolia L. subsp. maculata (Cav.) Marais ●☆

362534　Sida densiflora A. Rich. = Sida urens L. ●☆

362535　Sida denticulata Fresen. = Abutilon fruticosum Guillaumin et Perr. ●☆

362536　Sida devredii Steyaert = Sida rigida (G. Don) D. Dietr. ●☆

362537　Sida dinteriana Hochr. = Sida hoepfneri Gürke ●☆

362538　Sida dregei Burtt Davy;德雷黄花稔●☆

362539　Sida elliottii Torr. et A. Gray;埃利奥特黄花稔●☆

362540　Sida fallax Walp. = Sida alnifolia L. var. orbiculata S. Y. Hu ●

362541　Sida fallax Walp. = Sida rhombifolia L. var. microphylla (Cav.) S. Y. Hu ●

362542　Sida flexuosa Burtt Davy = Sida hoepfneri Gürke ●☆

362543　Sida floccosa Thulin et Vollesen;丛卷毛黄花稔●☆

362544　Sida frutescens DC. = Sida acuta Burm. f. ●■

362545　Sida garckeana Pol. ;加尔凯黄花稔●☆

362546　Sida glauca Cav. = Abutilon pannosum (G. Forst.) Schltdl. ■☆

362547　Sida glutinosa Cav. ;胶黏黄花稔■☆

362548　Sida glutinosa Roxb. = Sida mysorensis Wight et Arn. ●■

362549　Sida gossweileri Exell = Sida rigida (G. Don) D. Dietr. ●☆

362550　Sida gracilis R. Br. = Abutilon fruticosum Guillaumin et Perr. ●☆

362551　Sida grandifolia Willd. = Abutilon grandifolium (Willd.) Sweet ●☆

362552　Sida graveolens Roxb. = Abutilon hirtum (Lam.) Sweet ■

362553　Sida graveolens Roxb. ex Hornem. = Abutilon graveolens (Roxb. ex Hornem.) Wight et Arn. ex Wight ■

362554　Sida grewioides Guillaumin et Perr. = Sida ovata Forssk. ●☆

362555　Sida grewioides Guillaumin et Perr. var. microphylla ? = Sida pakistanica Abedin ●☆

362556　Sida guineensis Schumach. = Abutilon indicum (L.) Sweet subsp. guineense (Schumach.) Borss. Waalk. ■

362557　Sida guineensis Schumach. = Abutilon indicum (L.) Sweet var. guineense (Schumach.) K. M. Feng ■

362558　Sida herbacea Cav. = Sida cordifolia L. ●

362559　Sida hirta Lam. = Abutilon hirtum (Lam.) Sweet ■

362560　Sida hirta Lam. = Sida mysorensis Wight et Arn. ●■

362561　Sida hoepfneri Gürke;赫普夫纳黄花稔●☆

362562　Sida holosericea Willd. ex Spreng. = Sida cordifolia L. ●

362563　Sida hongkongensis Gand. = Sida cordifolia L. ●

362564　Sida humilis Cav. = Sida cordata (Burm. f.) Borss. Waalk. ●■

362565　Sida humilis Cav. var. veronicifolia (Lam.) Mast. = Sida cordata (Burm. f.) Borss. Waalk. ●■

362566　Sida indica L. = Abutilon indicum (L.) Sweet ●■

362567　Sida insularis Hatus. = Sida rhombifolia L. subsp. insularis (Hatus.) Hatus. ●

362568　Sida insularis Hatus. = Sida rhombifolia L. ●

362569　Sida javensis Cav. ;爪哇黄花稔(单花黄花稔,圆叶金午时花,爪哇长梗黄花稔,爪哇金午时花);Java Sida ●■

362570　Sida lanceolata Retz. = Sida acuta Burm. f. ●■

362571　Sida lancifolia Burtt Davy = Sida dregei Burtt Davy ●☆

362572　Sida libenii Hauman;利本黄花稔●☆

362573　Sida linearifolia Thonn. = Sida linifolia Juss. ex Cav. ●☆

362574　Sida linifolia Cav. = Sida linifolia Juss. ex Cav. ●☆

362575　Sida linifolia Juss. ex Cav. ;亚麻叶黄花稔;Flaxleaf Fanpetals ●☆

362576　Sida longipedicellata Thulin;长花梗黄花稔●☆

362577　Sida longipes E. Mey. ex Harv. = Sida dregei Burtt Davy ●☆

362578　Sida longipes E. Mey. ex Harv. var. canescens Szyszyl. = Sida hoepfneri Gürke ●☆

362579　Sida maculata Cav. = Sida cordifolia L. subsp. maculata (Cav.) Marais ●☆

362580　Sida maderensis Lowe = Sida rhombifolia L. var. maderensis (Lowe) Lowe ●☆

362581　Sida massaica Vollesen;马萨黄花稔●☆

362582　Sida mauritiana Jacq. = Abutilon mauritianum (Jacq.) Medik. ■☆

362583　Sida microphylla Cav. = Sida alnifolia L. var. microphylla (Cav.) S. Y. Hu ●

362584　Sida microphylla Cav. = Sida rhombifolia L. var. microphylla (Cav.) S. Y. Hu ●

362585　Sida mollis Ortega = Abutilon grandifolium (Willd.) Sweet ●☆

362586　Sida mollis Ortega = Abutilon theophrastii Medik. ●■

362587　Sida morifolia Cav. = Sida cordata (Burm. f.) Borss. Waalk. ●■

362588　Sida multicaulis Cav. = Sida cordata (Burm. f.) Borss. Waalk. ●■

362589　Sida mutica Delile ex DC. = Abutilon pannosum (G. Forst.) Schltdl. ■☆

362590　Sida mysorensis Wight et Arn. ;黏毛黄花稔(薄叶金午时花,黄花仔,生毛虱母头,嗽血草,索仔草,吸血草);Mysor Sida,Mysore Sida,Slimyhair Sida,Slimy-haired Sida ●■

362591　Sida obovata Wall. = Sida rhombifolia L. var. obvata（Wall.）S. Y. Hu ●

362592　Sida obovata Wall. = Sida yunnanensis S. Y. Hu ●

362593　Sida ogadensis Thulin et Vollesen；欧加登黄花稔●☆

362594　Sida orientalis Cav.；东方黄花稔；Oriental Sida

362595　Sida orientalis Cav. = Sida quinquevalvacea J. L. Liu ●

362596　Sida ostryifolia Webb = Sida rhombifolia L. var. maderensis（Lowe）Lowe ●☆

362597　Sida ovata Forssk.；卵形黄花稔●☆

362598　Sida pakistanica Abedin；巴基斯坦黄花稔●☆

362599　Sida paniculata L.；圆锥花黄花稔●☆

362600　Sida pannosa G. Forst. = Abutilon pannosum（G. Forst.）Schltdl. ■☆

362601　Sida patens Andréws = Abutilon mauritianum（Jacq.）Medik. ■☆

362602　Sida periplocifolia L. = Wissadula periplocifolia（L.）C. Presl ex Thwaites ●

362603　Sida permutata Hochst. ex A. Rich. = Sida ternata L. f. ●☆

362604　Sida philippica DC. = Sida rhombifolia L. ●

362605　Sida picta Gillies ex Hook. = Abutilon striatum G. F. Dicks. ●

362606　Sida pilosa Retz. = Sida javensis Cav. ●■

362607　Sida pilosella Arw. = Sida dregei Burtt Davy ●☆

362608　Sida populifolia Lam. = Abutilon indicum（L.）Sweet subsp. albescens（Miq.）Borss. Waalk. ●■☆

362609　Sida populifolia Lam. = Abutilon indicum（L.）Sweet ●■

362610　Sida populifolia Lam. = Abutilon mauritianum（Jacq.）Medik. ■☆

362611　Sida quinquevalvacea J. L. Liu；五片黄花稔；Fivebract Sida ●

362612　Sida radiata L. = Malachra radiata（L.）L. ■☆

362613　Sida radicans Cav. = Sida cordata（Burm. f.）Borss. Waalk. ●■

362614　Sida ramosa Cav. = Abutilon ramosum（Cav.）Guillaumin et Perr. ■☆

362615　Sida retusa L. = Sida alnifolia L. ●

362616　Sida retusa L. = Sida rhombifolia L. subsp. retusa（L.）Borss. Waalk. ●

362617　Sida rhombifolia L.；白背黄花稔(拔脓消，赐米草，大地丁草，单枝落地，地膏消，地膏药，鬼柳根，黄花草，黄花地桃花，黄花猛，黄花母，黄花稔，黄花雾，金午时花，金盏花，菱叶拔毒散，麻笔，枚叶草，梅肉草，脓见愁，千斤坠，山鸡草，山木槿，生扯拢，素花草，土黄芪，细迷马桩棵，小柴胡，亚母头，枝叶草)；Broomjute Sida, Broom-jute Sida, Paddy's Lucerne, Queensland Hemp, Tea Plant, Whiteback Sida ●

362618　Sida rhombifolia L. = Sida alnifolia L. ●

362619　Sida rhombifolia L. = Sida szechuensis Matsuda ●

362620　Sida rhombifolia L. subsp. alnifolia（L.）Ugbor. = Sida rhombifolia L. ●

362621　Sida rhombifolia L. subsp. canariensis（Cav.）Pit. et Proust = Sida rhombifolia L. ●

362622　Sida rhombifolia L. subsp. insularis（Hatus.）Hatus.；恒春白背黄花稔(赐米草，恒春金午时花)●

362623　Sida rhombifolia L. subsp. insularis（Hatus.）Hatus. = Sida rhombifolia L. ●

362624　Sida rhombifolia L. subsp. maderensis（Lowe）Pit. et Proust = Sida rhombifolia L. var. maderensis（Lowe）Lowe ●☆

362625　Sida rhombifolia L. subsp. retusa（L.）Borss. Waalk. = Sida alnifolia L. ●

362626　Sida rhombifolia L. var. afroscabrida Verdc.；非洲微糙黄花稔●☆

362627　Sida rhombifolia L. var. canariensis K. Schum.；加那利白背黄花稔●☆

362628　Sida rhombifolia L. var. corynocarpa（Wall.）S. Y. Hu；棒果黄花稔；Stickfruit Sida ●

362629　Sida rhombifolia L. var. maderensis（Lowe）Lowe；梅德黄花稔●☆

362630　Sida rhombifolia L. var. microphylla（Cav.）Mast. = Sida rhombifolia L. var. microphylla（Cav.）S. Y. Hu ●

362631　Sida rhombifolia L. var. microphylla（Cav.）S. Y. Hu = Sida alnifolia L. var. microphylla（Cav.）S. Y. Hu ●

362632　Sida rhombifolia L. var. obovata Mast. = Sida szechuensis Matsuda ●

362633　Sida rhombifolia L. var. obovata Wall. ex Mast. = Sida alnifolia L. var. obovata（Wall. ex Mast.）S. Y. Hu ●

362634　Sida rhombifolia L. var. obovata Wall. ex Mast. = Sida yunnanensis S. Y. Hu ●

362635　Sida rhombifolia L. var. obvata（Wall.）Mast. = Sida rhombifolia L. var. obvata（Wall.）S. Y. Hu ●

362636　Sida rhombifolia L. var. obvata（Wall.）S. Y. Hu = Sida alnifolia L. var. obovata（Wall. ex Mast.）S. Y. Hu ●

362637　Sida rhombifolia L. var. retusa（L.）Mast. = Sida alnifolia L. ●

362638　Sida rhombifolia L. var. retusa Mast. = Sida chinensis Retz. ●

362639　Sida rhombifolia L. var. rhomboidea（Roxb. ex Fleming）Mast. = Sida rhombifolia L. ●

362640　Sida rhombifolia L. var. rhomboides（Roxb.）Mast. = Sida orientalis Cav. ●

362641　Sida rhombifolia L. var. riparia Burtt Davy；河岸白背黄花稔●☆

362642　Sida rhombifolia L. var. serratifolia（R. Wilczek et Steyaert）Verdc.；齿叶白背黄花稔●☆

362643　Sida rhombifolia Roxb. ex Fleming = Sida orientalis Cav. ●

362644　Sida rhomboidea Roxb. ex Fleming；拟金白背黄花稔●

362645　Sida rhomboidea Roxb. ex Fleming = Sida rhombifolia L. ●

362646　Sida rhomifolia L. var. petherickii Verdc.；彼瑟白背黄花稔●☆

362647　Sida rigida（G. Don）D. Dietr.；坚硬白背黄花稔●☆

362648　Sida riparia Hochst. = Sida rhombifolia L. ●

362649　Sida rostrata Schumach. = Wissadula rostrata（Schumach.）Hook. f. ●☆

362650　Sida rotundifolia Lam. = Sida cordifolia L. subsp. maculata（Cav.）Marais ●☆

362651　Sida rotundifolia Lam. ex Cav. = Sida cordifolia L. ●

362652　Sida saltii Steud. = Abutilon fruticosum Guillaumin et Perr. ●☆

362653　Sida santaremensis Monteiro；圣塔伦黄花稔；Moth Fanpetals ●☆

362654　Sida scabrida Wight et Arn. = Sida rhombifolia L. ●

362655　Sida schimperiana Hochst. ex A. Rich. = Dictyocarpus truncatus Wight ●☆

362656　Sida schweinfurthii Baker f. = Sida paniculata L. ●☆

362657　Sida scoparia Lour. = Sida acuta Burm. f. ●■

362658　Sida semicrenata Link = Sida rhombifolia L. ●

362659　Sida serratifolia R. Wilczek et Steyaert = Sida rhombifolia L. var. serratifolia（R. Wilczek et Steyaert）Verdc. ●☆

362660　Sida shinyangensis Vollesen；希尼安加黄花稔●☆

362661　Sida sonneratiana Cav. = Abutilon sonneratianum（Cav.）Sweet ●☆

362662　Sida spinosa L.；刺黄花稔；Prickly Mallow, Prickly Sida, Spiny Sida ●■

362663　Sida spinosa L. var. angustifolia Griseb. ;狭叶刺黄花稔●■

362664　Sida spinosa L. var. angustifolia Griseb. = Sida spinosa L. ●■

362665　Sida spinosa L. var. kazmii Abedin;巴基斯坦刺黄花稔●■☆

362666　Sida stauntoniana DC. = Sida acuta Burm. f. ●■

362667　Sida stipulata Cav. = Sida acuta Burm. f. ●■

362668　Sida subcordata Span. ;榛叶黄花稔(亚心叶黄花稔);Filbestleaved Sida, Subcordate Sida ●

362669　Sida subrotunda Hochst. = Sida ovata Forssk. ●☆

362670　Sida subspicata F. Muell. ex Benth. ;亚穗黄花稔●■☆

362671　Sida supina L'Hér. = Sida cordata (Burm. f.) Borss. Waalk. ●■

362672　Sida szechuensis Matsuda;四川黄花稔(巴掌叶,拔毒散,川黄花稔,肯麻尖,马庄棵,迷马桩棵,尼马庄柯,王不留行,小拔毒,小黄药,小克麻,小迷马桩,小年药,小黏药);Badusan Sida, Sichuan Sida, Szechwan Sida ●

362673　Sida tanaensis Vollesen;塔纳黄花稔●☆

362674　Sida tenuicarpa Vollesen;细果黄花稔●☆

362675　Sida ternata L. f. ;三出黄花稔●☆

362676　Sida tiliifolia Fisch. = Abutilon theophrastii Medik. ●■

362677　Sida triloba Cav. = Sida ternata L. f. ●☆

362678　Sida unilocularis L'Hér. = Sida cordata (Burm. f.) Borss. Waalk. ●■

362679　Sida urens L. ;蜇毛黄花稔●☆

362680　Sida urens L. var. bicolor Roberty;二色蜇毛黄花稔●☆

362681　Sida urens L. var. prostrata A. Chev. = Sida urens L. ●☆

362682　Sida urticifolia Wight et Arn. = Sida mysorensis Wight et Arn. ●■

362683　Sida velutina Willd. ex Spreng. = Sida cordifolia L. subsp. maculata (Cav.) Marais ●☆

362684　Sida veronicifolia Lam. ;澎湖金午时花●☆

362685　Sida veronicifolia Lam. = Sida cordata (Burm. f.) Borss. Waalk. ●■

362686　Sida veronicifolia Lam. = Sida pilosa Retz. ●☆

362687　Sida veronicifolia Lam. var. humilis (Cav.) K. Schum. = Sida cordata (Burm. f.) Borss. Waalk. ●■

362688　Sida veronicifolia Lam. var. javensis (Cav.) Baker f. = Sida javensis Cav. ●■

362689　Sida veronicifolia Lam. var. multicaulis (Cav.) Baker f. = Sida cordata (Burm. f.) Borss. Waalk. ●■

362690　Sida viscosa L. = Sida mysorensis Wight et Arn. ●■

362691　Sida vogelii Hook. f. = Sida acuta Burm. f. ●■

362692　Sida wightiana D. Dietr. = Sida mysorensis Wight et Arn. ●■

362693　Sida yunnanensis S. Y. Hu;云南黄花稔 ●

362694　Sida yunnanensis S. Y. Hu var. longistyla J. L. Liu;长柱黄花稔;Longstyle Yunnan Sida ●

362695　Sida yunnanensis S. Y. Hu var. longistyla J. L. Liu = Sida yunnanensis S. Y. Hu ●

362696　Sida yunnanensis S. Y. Hu var. viridicaulis J. L. Liu;绿茎黄花稔;Greenstem Yunnan Sida ●

362697　Sida yunnanensis S. Y. Hu var. viridicaulis J. L. Liu = Sida yunnanensis S. Y. Hu ●

362698　Sida yunnanensis S. Y. Hu var. xichangensis J. L. Liu;西昌黄花稔;Xichang Sida ●

362699　Sida yunnanensis S. Y. Hu var. xichangensis J. L. Liu = Sida yunnanensis S. Y. Hu ●

362700　Sidaceae Bercht. et Presl = Malvaceae Juss. (保留科名)●■

362701　Sidalcea A. Gray = Sidalcea A. Gray ex Benth. ■☆

362702　Sidalcea A. Gray ex Benth. (1849);双葵属(稔葵属);Checker

Mallow, Checker-Mallow, False Mallow, Greek Mallow, Prairie Mallow, Prairiemallow ■☆

362703　Sidalcea candida A. Gray ex Benth. ;白花双葵;White Prairie Mallow, White Prairiemallow ■☆

362704　Sidalcea malviflora A. Gray;双葵(双锦葵);Checkerbloom, Ckecker Bloom, Greek Mallow, Prairie Mallow ■☆

362705　Sidalcea neomexicana A. Gray;新墨西哥双葵;Checker Mallow ■☆

362706　Sidalcea oregona (Nutt. ex Torr. et A. Gray) A. Gray;俄勒冈双葵■☆

362707　Sidanoda (A. Gray) Wooton et Standl. = Anoda Cav. ■●☆

362708　Sidanoda Wooton et Standl. = Anoda Cav. ■●☆

362709　Sidasodes Fryxell et Fuertes(1992);肖锦葵属●☆

362710　Sidasodes colombiana Fryxell et J. Fuertes;脊锦葵●☆

362711　Sidasodes jamesonii (Baker f.) Fryxell et J. Fuertes;哥伦比亚脊锦葵●☆

362712　Sidastrum Baker f. (1892);小黄花稔属●☆

362713　Sidastrum Baker f. = Sida L. ■●

362714　Sidastrum paniculatum (L.) Fryxell = Sida paniculata L. ●☆

362715　Side St. -Lag. = Sida L. ●■

362716　Sideranthus Nees = Haplopappus Cass. (保留属名)■●☆

362717　Sideranthus Nutt. ex Nees = Haplopappus Cass. (保留属名) ■●☆

362718　Sideranthus aberrans (A. Nelson) Rydb. = Trinitieurybia aberrans (A. Nelson) Brouillet, Urbatsch et R. P. Roberts ■☆

362719　Sideranthus annuus Rydb. = Rayjacksonia annua (Rydb.) R. L. Hartm. et M. A. Lane ■☆

362720　Sideranthus glaberrimus Rydb. = Xanthisma spinulosum (Pursh) D. R. Morgan et R. L. Hartm. var. glaberrimum (Rydb.) D. R. Morgan et R. L. Hartm. ■☆

362721　Sideranthus gooddingii A. Nelson = Xanthisma spinulosum (Pursh) D. R. Morgan et R. L. Hartm. var. gooddingii (A. Nelson) D. R. Morgan et R. L. Hartm. ●☆

362722　Sideranthus megacephalus (Nash) Small = Rayjacksonia phyllocephala (DC.) R. L. Hartm. et M. A. Lane ■☆

362723　Sideranthus turbinellus Rydb. = Xanthisma spinulosum (Pursh) D. R. Morgan et R. L. Hartm. ●■☆

362724　Sideranthus viscidus Wooton et Standl. = Xanthisma viscidum (Wooton et Standl.) D. R. Morgan et R. L. Hartm. ■☆

362725　Siderasis Raf. (1837);锈毛草属■☆

362726　Siderasis fuscata (Lodd.) H. E. Moore;锈毛草(绒毡草);Bear Ears, Brown Spiderwort ■☆

362727　Sideria Ewart et A. H. K. Petrie = Melhania Forssk. ●■

362728　Sideritis L. (1753);毒马草属(铁尖草属);Iron Woundwort, Ironwort, Sideritis ■●

362729　Sideritis antiatlantica (Maire) Rejdali;安蒂毒马草■☆

362730　Sideritis arborescens Benth. subsp. antiatlantica (Maire) Romo = Sideritis antiatlantica (Maire) Rejdali ■☆

362731　Sideritis arborescens Benth. subsp. maireana (Font Quer) Socorro et Arreb. = Sideritis maireana Font Quer et Pau ■☆

362732　Sideritis arborescens Benth. subsp. ortonedae (Font Quer et Pau) Maire = Sideritis ortonedae (Font Quer et Pau) Obón et D. Rivera ■☆

362733　Sideritis arborescens Benth. var. angustifolia Font Quer = Sideritis ortonedae (Font Quer et Pau) Obón et D. Rivera ■☆

362734　Sideritis arborescens Benth. var. antiatlantica (Maire) Dobignard = Sideritis antiatlantica (Maire) Rejdali ■☆

362735 Sideritis arborescens Benth. var. faurei Maire = Sideritis faurei (Maire) Obón et D. Rivera ■☆

362736 Sideritis arborescens Benth. var. fontii Maire = Sideritis ortonedae (Font Quer et Pau) Obón et D. Rivera ■☆

362737 Sideritis arborescens Benth. var. kebdanensis Font Quer et Sennen = Sideritis ortonedae (Font Quer et Pau) Obón et D. Rivera ■☆

362738 Sideritis arborescens Benth. var. kerkerana Font Quer et Sennen = Sideritis ortonedae (Font Quer et Pau) Obón et D. Rivera ■☆

362739 Sideritis arborescens Benth. var. lazari Font Quer et Sennen = Sideritis ortonedae (Font Quer et Pau) Obón et D. Rivera ■☆

362740 Sideritis arborescens Benth. var. ortonedae Font Quer et Pau = Sideritis ortonedae (Font Quer et Pau) Obón et D. Rivera ■☆

362741 Sideritis arborescens Benth. var. pozasi Font Quer = Sideritis ortonedae (Font Quer et Pau) Obón et D. Rivera ■☆

362742 Sideritis arenaria Vahl var. divaricatidens (H. Lindb.) Jahand. et Maire = Stachys arenaria Vahl subsp. divaricatidens H. Lindb. ■☆

362743 Sideritis argosphacela (Webb et Berthel.) Clos;银头毒马草 ■☆

362744 Sideritis argosphacela (Webb et Berthel.) Clos var. spicata (Pit.) Bornm. = Sideritis argosphacela (Webb et Berthel.) Clos ■☆

362745 Sideritis atlantica Pomel;亚特兰大毒马草 ■☆

362746 Sideritis balansae Boiss. ;紫花毒马草;Purple-flower Sideritis ■

362747 Sideritis balansae Coss. = Sideritis cossoniana Ball ☆

362748 Sideritis bolleana Bornm. ;博勒毒马草 ■☆

362749 Sideritis brevicaulis Mend. -Heuer;短茎毒马草 ■☆

362750 Sideritis briquetiana Font Quer;布里凯毒马草 ■☆

362751 Sideritis briquetiana Font Quer var. aragonesii Font Quer et Pau = Sideritis briquetiana Font Quer ■☆

362752 Sideritis briquetiana Font Quer var. arbuscula Font Quer et Sennen = Sideritis briquetiana Font Quer ■☆

362753 Sideritis briquetiana Font Quer var. dissitiflora Sennen = Sideritis briquetiana Font Quer ■☆

362754 Sideritis briquetiana Font Quer var. foucauldiana Font Quer et Sennen = Sideritis briquetiana Font Quer ■☆

362755 Sideritis briquetiana Font Quer var. hilarii Sennen = Sideritis briquetiana Font Quer ■☆

362756 Sideritis briquetiana Font Quer var. longispica Sennen = Sideritis briquetiana Font Quer ■☆

362757 Sideritis briquetiana Font Quer var. parvifolia Font Quer et Sennen = Sideritis briquetiana Font Quer ■☆

362758 Sideritis briquetiana Font Quer var. paucidentata Font Quer et Sennen = Sideritis briquetiana Font Quer ■☆

362759 Sideritis canariensis L. ;加那利毒马草 ■☆

362760 Sideritis canariensis L. var. pannosa (Christ) Bornm. = Sideritis canariensis L. ■☆

362761 Sideritis candicans Aiton;纯白毒马草 ■☆

362762 Sideritis candicans Aiton var. crassifolia Lowe = Sideritis candicans Aiton ■☆

362763 Sideritis candicans Aiton var. multiflora (Bornm.) Mend. -Heuer = Sideritis candicans Aiton ■☆

362764 Sideritis carolipauana Peris et Stübing = Stachys fontqueri Pau ■☆

362765 Sideritis chlorostegia Juz. ;绿被毒马草;Greencovering Sideritis ■☆

362766 Sideritis ciliata Thunb. = Elsholtzia ciliata (Thunb. ex Murray) Hyl. ■

362767 Sideritis comosa (Rochel ex Benth.) Stankov;簇毛毒马草 ■☆

362768 Sideritis conferta Juz. ;密集毒马草;Crowded Sideritis ■☆

362769 Sideritis cossoniana Ball;科森毒马草 ■☆

362770 Sideritis cretica L. ;克里特毒马草 ■☆

362771 Sideritis cretica L. var. anagae (Christ) Mend. -Heuer = Sideritis cretica L. ■☆

362772 Sideritis cretica L. var. eriocephala (Clos) Mend. -Heuer = Sideritis cretica L. ■☆

362773 Sideritis cretica L. var. stricta (Webb) Mend. -Heuer = Sideritis cretica L. ■☆

362774 Sideritis curvidens Stapf;弯齿毒马草 ■☆

362775 Sideritis cystosiphon Svent. ;囊管毒马草 ■☆

362776 Sideritis dasygnaphala (Webb et Berthel.) Clos;粗毛毒马草 ■☆

362777 Sideritis debeauxii Font Quer = Sideritis pusilla (Lange) Pau ■☆

362778 Sideritis dendro-chahorra Bolle;卡岛毒马草 ■☆

362779 Sideritis dendro-chahorra Bolle var. albida (Pit.) Svent. = Sideritis dendro-chahorra Bolle ■☆

362780 Sideritis dendro-chahorra Bolle var. soluta (Clos) Svent. = Sideritis dendro-chahorra Bolle ■☆

362781 Sideritis deserti Noë = Marrubium deserti (Noë) Coss. ■☆

362782 Sideritis discolor (Noë) Bolle;异色毒马草 ■☆

362783 Sideritis edetana (Font Quer) Peris, Figuerola et Stübing;埃德毒马草 ■☆

362784 Sideritis eriocephala Marrero Rodr. ex Negrín et P. Pérez;毛头毒马草 ■☆

362785 Sideritis eriocephala Negrín et P. Pérez = Sideritis eriocephala Marrero Rodr. ex Negrín et P. Pérez ■☆

362786 Sideritis euxina Juz. ;黑海毒马草;Black Sea Sideritis ■☆

362787 Sideritis faurei (Maire) Obón et D. Rivera;福雷毒马草 ■☆

362788 Sideritis fonqueri Sennen et Mauricio = Sideritis incana L. ■☆

362789 Sideritis fontiqueriana Peris et Romo et Stübing = Sideritis subatlantica Doum. ■☆

362790 Sideritis foucauldiana Sennen et Mauricio = Sideritis briquetiana Font Quer ■☆

362791 Sideritis gaditana Rouy;加迪特毒马草 ■☆

362792 Sideritis gineslopezii Obón et D. Rivera = Sideritis subatlantica Doum. ■☆

362793 Sideritis gomeraea Bolle;戈梅拉毒马草 ■☆

362794 Sideritis gossypina Font Quer = Sideritis villosa Coss. et Balansa subsp. gossypina (Font Quer) Dobignard ■☆

362795 Sideritis gossypina Font Quer var. brevidentata Maire = Sideritis villosa Coss. et Balansa subsp. gossypina (Font Quer) Dobignard ■☆

362796 Sideritis gossypina Font Quer var. longidentata Maire = Sideritis villosa Coss. et Balansa subsp. gossypina (Font Quer) Dobignard ■☆

362797 Sideritis granatensis (Pau) Alcaraz, Peinado, Mart. Parras, J. S. Carrion et Sánchez-Gómez;格拉毒马草 ■☆

362798 Sideritis granatensis (Pau) Font Quer subsp. briquetiana (Font Quer et Pau) Socorro et Arreb. = Sideritis briquetiana Font Quer ■☆

362799 Sideritis grandiflora Benth. ;大花毒马草 ■☆

362800 Sideritis guyoniana Boiss. et Reut. ;居永毒马草 ■☆

362801 Sideritis hirsuta L. ;德国毒马草 ■☆

362802 Sideritis hirsuta L. var. latidens Font Quer = Sideritis hirsuta L. ■☆

362803 Sideritis hirsuta L. var. maroccana Coss. = Sideritis hirsuta L. ■☆

362804　Sideritis hirsuta L. var. nivalis Font Quer = Sideritis hirsuta L. ■☆

362805　Sideritis hyssopifolia Chabert subsp. atlantica (Pomel) Batt. = Sideritis atlantica Pomel ■☆

362806　Sideritis ibrahmii Obón et D. Rivera;易卜拉姆毒马草■☆

362807　Sideritis imbrex Juz.;空瓦毒马草;Tile Sideritis ■☆

362808　Sideritis imbricata H. Lindb.;覆瓦毒马草■☆

362809　Sideritis incana L.;灰毛毒马草■☆

362810　Sideritis incana L. subsp. virgata (Desf.) Malag. = Sideritis incana L. ■☆

362811　Sideritis incana L. var. albiflora Maire = Sideritis edetana (Font Quer) Peris,Figuerola et Stübing ■☆

362812　Sideritis incana L. var. altiatlantica Font Quer = Sideritis atlantica Pomel ■☆

362813　Sideritis incana L. var. aurasiaca Batt. = Sideritis atlantica Pomel ■☆

362814　Sideritis incana L. var. flavovirens Maire = Sideritis incana L. ■☆

362815　Sideritis incana L. var. guyoniana (Boiss. et Reut.) Pau = Sideritis guyoniana Boiss. et Reut. ■☆

362816　Sideritis incana L. var. henryi Maire = Sideritis edetana (Font Quer) Peris,Figuerola et Stübing ■☆

362817　Sideritis incana L. var. matris-filiae (Emb. et Maire) Maire = Sideritis incana L. ■☆

362818　Sideritis incana L. var. occidentalis Font Quer = Sideritis incana L. ■☆

362819　Sideritis incana L. var. regimonta Maire = Sideritis edetana (Font Quer) Peris,Figuerola et Stübing ■☆

362820　Sideritis incana L. var. robusta Font Quer = Sideritis incana L. ■☆

362821　Sideritis incana L. var. tomentosa Batt. et Pit. = Sideritis incana L. ■☆

362822　Sideritis jahandiezii Font Quer;贾汉毒马草■☆

362823　Sideritis jahandiezii Font Quer var. querana Emb. = Sideritis jahandiezii Font Quer ■☆

362824　Sideritis kuegleriana Bornm.;屈格勒毒马草■☆

362825　Sideritis lanata L.;铁尖草(毛萼刺草,田野水苏);Hairy Ironwort ■

362826　Sideritis leucantha Cav. subsp. mohamedii (Rejdali) Socorro et Arreb. = Sideritis jahandiezii Font Quer ■☆

362827　Sideritis lotsyi (Pit.) Bornm.;洛茨毒马草■☆

362828　Sideritis lotsyi (Pit.) Bornm. var. grandiflora Mend.-Heuer = Sideritis lotsyi (Pit.) Bornm. ■☆

362829　Sideritis lotsyi (Pit.) Bornm. var. mascaensis Svent. = Sideritis lotsyi (Pit.) Bornm. ■☆

362830　Sideritis macrostachya Poir.;大穗毒马草■☆

362831　Sideritis maireana Font Quer et Pau;迈雷毒马草■☆

362832　Sideritis maireana Font Quer et Pau var. carbonelii Sennen et Mauricio = Sideritis maireana Font Quer et Pau ■☆

362833　Sideritis marmorea Bolle;大理石毒马草■☆

362834　Sideritis maroccana (Font Quer) Obón et D. Rivera;摩洛哥毒马草■☆

362835　Sideritis marschalliana Juz.;马氏毒马草■☆

362836　Sideritis maura Noë;模糊毒马草■☆

362837　Sideritis mauritii Font Quer et Sennen = Sideritis incana L. ■☆

362838　Sideritis montana L.;毒马草;Montane Sideritis, Mountain Ironwort ■

362839　Sideritis montana L. subsp. ebracteata (Asso) Murb.;无苞毒马草■☆

362840　Sideritis montana L. var. ebracteata (Asso) Briq. = Sideritis montana L. subsp. ebracteata (Asso) Murb. ■☆

362841　Sideritis moorei Peris et al. = Sideritis hirsuta L. ■☆

362842　Sideritis mugronensis Borja;欧洲毒马草■☆

362843　Sideritis nervosa (H. Christ) Linding.;多脉毒马草■☆

362844　Sideritis nutans Svent.;俯垂毒马草■☆

362845　Sideritis occidentalis (Font Quer) Peris et al. = Sideritis incana L. ■☆

362846　Sideritis ochroleuca Willk.;淡黄白毒马草■☆

362847　Sideritis ochroleuca Willk. subsp. antiatlantica (Maire) Socorro et Arreb. = Sideritis antiatlantica (Maire) Rejdali ■☆

362848　Sideritis ochroleuca Willk. subsp. tafraoutiana (Obón et D. Rivera) Fennane = Sideritis antiatlantica (Maire) Rejdali ■☆

362849　Sideritis ochroleuca Willk. var. antiatlantica Maire = Sideritis antiatlantica (Maire) Rejdali ■☆

362850　Sideritis ochroleuca Willk. var. brevibracteata Font Quer = Sideritis ochroleuca Willk. ■☆

362851　Sideritis ochroleuca Willk. var. denticulata Font Quer = Sideritis ibrahmii Obón et D. Rivera ■☆

362852　Sideritis ochroleuca Willk. var. eremophila Maire = Sideritis ochroleuca Willk. ■☆

362853　Sideritis ochroleuca Willk. var. mairei Font Quer = Sideritis ibrahmii Obón et D. Rivera ■☆

362854　Sideritis ochroleuca Willk. var. maroccana Font Quer = Sideritis maroccana (Font Quer) Obón et D. Rivera ■☆

362855　Sideritis oromaroccana Peris et Stübing et Figuerola = Sideritis incana L. ■☆

362856　Sideritis oroteneriffae Negrín et P. Pérez var. arayae Negrin et P. Pérez = Sideritis oroteneriffae Negrín et P. Pérez ■☆

362857　Sideritis ortonedae (Font Quer et Pau) Obón et D. Rivera;奥尔毒马草■☆

362858　Sideritis penzigii (Pit.) Bornm.;彭西格毒马草■☆

362859　Sideritis pumila (H. Christ) Mend.-Heuer;矮毒马草■☆

362860　Sideritis pusilla (Lange) Pau;微小毒马草■☆

362861　Sideritis pusilla (Lange) Pau subsp. briquetiana (Font Quer) D. Rivera et al. = Sideritis briquetiana Font Quer ■☆

362862　Sideritis pycnostachys Pomel = Sideritis guyoniana Boiss. et Reut. ■☆

362863　Sideritis regimontana (Maire) Peris et Stübing et Figuerola = Sideritis edetana (Font Quer) Peris,Figuerola et Stübing ■☆

362864　Sideritis romana L.;罗马毒马草;Simplebeak Ironwort ■☆

362865　Sideritis romana L. subsp. numidica Batt.;努米底亚毒马草■☆

362866　Sideritis romoi Peris et al.;罗莫毒马草■☆

362867　Sideritis rossii Peris et al. = Sideritis hirsuta L. ■☆

362868　Sideritis sauvageana Obón et D. Rivera = Sideritis zaiana Sauvage ■☆

362869　Sideritis scardica Griseb.;希腊毒马草■☆

362870　Sideritis scordioides L. var. angustifolia Benth. = Sideritis incana L. ■☆

362871　Sideritis sierrarafolsiana Roselló et al. = Sideritis subatlantica Doum. ■☆

362872　Sideritis soluta Clos;离生毒马草■☆

362873　Sideritis spicata (Pit.) Marrero Rodr.;穗状毒马草■☆

362874　Sideritis subatlantica Doum.;拟亚特兰大毒马草■☆

362875　Sideritis subatlantica Doum. var. heterostachya Sennen = Sideritis subatlantica Doum. ■☆

362876 Sideritis subatlantica Doum. var. rhiphaea Font Quer = Sideritis subatlantica Doum. ■☆

362877 Sideritis sventenii (G. Kunkel) Mend. -Heuer;斯文顿毒马草■☆

362878 Sideritis syrinca L.;叙利亚毒马草■☆

362879 Sideritis tafraoutiana Obón et D. Rivera = Sideritis antiatlantica (Maire) Rejdali ☆

362880 Sideritis taurica Willd.;克里木毒马草☆

362881 Sideritis theezans Boiss. et Huldr.;产香毒马草■☆

362882 Sideritis tunetana (Murb.) Obón et D. Rivera;图内特毒马草■☆

362883 Sideritis villosa Coss. et Balansa;长柔毛毒马草■☆

362884 Sideritis villosa Coss. et Balansa subsp. gossypina (Font Quer) Dobignard;棉花毒马草■☆

362885 Sideritis vincentii Sennen et Mauricio = Sideritis ortonedae (Font Quer et Pau) Obón et D. Rivera ■☆

362886 Sideritis virgata Desf. = Sideritis incana L. ■☆

362887 Sideritis virgata Desf. var. lavandifolia Font Quer et Sennen = Sideritis incana L. ■☆

362888 Sideritis zaiana Sauvage;宰哈奈毒马草■☆

362889 Siderobombyx Bremek. (1947);铁蚕茜属 ☆

362890 Siderobombyx kinabaluensis Bremek.;铁蚕茜 ☆

362891 Siderocarpos Small = Acacia Mill. (保留属名)●■

362892 Siderocarpos Small = Ebenopsis Britton et Rose ●☆

362893 Siderocarpus Pierre = Pouteria Aubl. ●

362894 Siderocarpus Pierre ex L. Planch. = Planchonella Pierre(保留属名)●

362895 Siderocarpus Willis = Acacia Mill. (保留属名)●■

362896 Siderocarpus Willis = Siderocarpos Small ●■

362897 Siderodendron Roem. et Schult. = Siderodendrum Schreb. ●

362898 Siderodendrum Schreb. = Ixora L. ●

362899 Siderodendrum Schreb. = Sideroxyloides Jacq. ●

362900 Sideropogon Pichon = Arrabidaea DC. ●☆

362901 Sideroxyloides Jacq. = Ixora L. ●

362902 Sideroxylon L. (1753);铁榄属(山榄属);Jungle Plum ●☆

362903 Sideroxylon adolfi-friedericii Engl. = Pouteria adolfi-friedericii (Engl.) A. Meeuse ●☆

362904 Sideroxylon adolfi-friedericii Engl. subsp. keniensis R. E. Fr. = Pouteria adolfi-friedericii (Engl.) A. Meeuse subsp. keniensis (R. E. Fr.) L. Gaut. ●☆

362905 Sideroxylon altissimum (A. Chev.) Hutch. et Dalziel = Pouteria altissima (A. Chev.) Baehni ●☆

362906 Sideroxylon annamense (Pierre) Lecomte = Pouteria annamensis (Pierre ex Dubard) Baehni ●

362907 Sideroxylon arboreum Buch. -Ham. ex Clarke = Sarcosperma arboreum Hook. f. ●

362908 Sideroxylon atrovirens Lam. = Sideroxylon inerme L. ●☆

362909 Sideroxylon aubertii A. Chev. = Monotheca buxifolia (Falc.) A. DC. ●☆

362910 Sideroxylon aubertii A. Chev. = Sideroxylon mascatense (A. DC.) T. D. Penn. ●☆

362911 Sideroxylon aubrevillei Pellegr. = Synsepalum aubrevillei (Pellegr.) Aubrév. et Pellegr. ●☆

362912 Sideroxylon australe (R. Br.) F. Muell. = Planchonella australis (R. Br.) Pierre ●☆

362913 Sideroxylon aylmeri M. B. Scott = Neolemonniera clitandrifolia (A. Chev.) Heine ●☆

362914 Sideroxylon bakeri Scott-Elliot = Capurodendron bakeri (Scott-Elliot) Aubrév. ●☆

362915 Sideroxylon beguei Capuron ex Aubrév.;贝格铁榄●☆

362916 Sideroxylon beguei Capuron ex Aubrév. var. saboureaui Aubrév. = Sideroxylon beguei Capuron ex Aubrév. ●☆

362917 Sideroxylon bequaertii De Wild.;贝卡尔铁榄●☆

362918 Sideroxylon bodinieri H. Lév. = Handeliodendron bodinieri (H. Lév.) Rehder ●◇

362919 Sideroxylon boninense Nakai = Planchonella boninensis (Nakai) Masam. et Yanagih. ●☆

362920 Sideroxylon brevipes Baker = Synsepalum brevipes (Baker) T. D. Penn. ●☆

362921 Sideroxylon buxifolium Hutch. = Monotheca buxifolia (Falc.) A. DC. ●☆

362922 Sideroxylon buxifolium Hutch. = Sideroxylon mascatense (A. DC.) T. D. Penn. ●☆

362923 Sideroxylon capuronii Aubrév.;凯普伦铁榄●☆

362924 Sideroxylon cinereum (Pierre) Eyles = Synsepalum brevipes (Baker) T. D. Penn. ●☆

362925 Sideroxylon clemensii Lecomte = Planchonella clemensii (Lecomte) P. Royen ●

362926 Sideroxylon corradii Chiov. = Monotheca buxifolia (Falc.) A. DC. ●☆

362927 Sideroxylon corradii Chiov. = Sideroxylon mascatense (A. DC.) T. D. Penn. ●☆

362928 Sideroxylon cymosum L. f. = Olinia ventosa (L.) Cufod. ●☆

362929 Sideroxylon densiflorum Baker = Synsepalum revolutum (Baker) T. D. Penn. ●☆

362930 Sideroxylon dentatum Burm. f. = Curtisia dentata (Burm. f.) C. A. Sm. ●☆

362931 Sideroxylon diospyroides Baker = Sideroxylon inerme L. subsp. diospyroides (Baker) J. H. Hemsl. ●☆

362932 Sideroxylon discolor Radcl. -Sm.;异色铁榄●☆

362933 Sideroxylon dubium Koidz. ex H. Hara = Planchonella obovata (R. Br.) Pierre var. dubia (Koidz. ex H. Hara) Hatus. ex T. Yamaz. ●☆

362934 Sideroxylon duclitan Blanco = Planchonella duclitan (Blanco) Bakh. f. ●

362935 Sideroxylon dulcificum (Schumach. et Thonn.) A. DC. = Synsepalum dulcificum (Schumach. et Thonn.) Daniell ●☆

362936 Sideroxylon embelifolium Merr. = Xantolis longispinosa (Merr.) H. S. Lo ●

362937 Sideroxylon embeliifolium Merr. = Xantolis longispinosa (Merr.) H. S. Lo ●

362938 Sideroxylon ferrugineum Hook. et Arn. = Planchonella obovata (R. Br.) Pierre ●

362939 Sideroxylon fischeri Engl. = Manilkara fischeri (Engl.) H. J. Lam ●☆

362940 Sideroxylon foetidissimum Jacq.;臭铁榄;Barbados Mastic ●☆

362941 Sideroxylon gabonense (A. Chev.) Lecomte ex Pellegr. = Pouteria altissima (A. Chev.) Baehni ●☆

362942 Sideroxylon gamblei C. B. Clarke = Platea latifolia Blume ●

362943 Sideroxylon gerrardianum (Hook. f.) Aubrév.;杰勒德铁榄●☆

362944 Sideroxylon gillettii Hutch. et E. A. Bruce = Monotheca buxifolia (Falc.) A. DC. ●☆

362945 Sideroxylon gillettii Hutch. et E. A. Bruce = Sideroxylon mascatense（A. DC.）T. D. Penn. ●☆

362946 Sideroxylon gossweileri Greves = Tridesmostemon omphalocarpoides Engl. ●☆

362947 Sideroxylon grandifolium Wall. = Pouteria grandifolia（Wall.）Baehni ●

362948 Sideroxylon greveanum Baill. ex Aubrév. = Capurodendron greveanum Aubrév. ●☆

362949 Sideroxylon hainanense Merr. = Pouteria annamensis（Pierre ex Dubard）Baehni ●

362950 Sideroxylon inerme L. ; 无刺铁榄 ●☆

362951 Sideroxylon inerme L. subsp. diospyroides（Baker）J. H. Hemsl. ; 柿状铁榄 ●☆

362952 Sideroxylon inerme L. var. schlechteri Engl. = Sideroxylon inerme L. ●☆

362953 Sideroxylon lanuginosa Michx. ; 芽铁榄 ; Chittim Wood, False Buckthorn, Gum-elastic, Woolly Buckthorn ●☆

362954 Sideroxylon longispinosa Merr. = Xantolis longispinosa（Merr.）H. S. Lo ●

362955 Sideroxylon longispinosum Merr. = Xantolis longispinosa（Merr.）H. S. Lo ●

362956 Sideroxylon longistylum Baker = Synsepalum brevipes（Baker）T. D. Penn. ●☆

362957 Sideroxylon lycioides L. ; 卡罗来纳铁榄 ; Carolina Buckthorn, Ironwood, Southern Buckthorn ●☆

362958 Sideroxylon madagascariense Lecomte = Capurodendron madagascariense（Lecomte）Aubrév. ●☆

362959 Sideroxylon marginatum（Decne.）Cout. ; 具边铁榄 ●☆

362960 Sideroxylon marmulanum Banks ex Lowe var. edulis A. Chev. = Sideroxylon marginatum（Decne.）Cout. ●☆

362961 Sideroxylon marmulanum Banks ex Lowe var. marginatum（Decne.）A. Chev. = Sideroxylon marginatum（Decne.）Cout. ●☆

362962 Sideroxylon mascatense（A. DC.）T. D. Penn. ; 阿曼铁榄 ●☆

362963 Sideroxylon mayumbense Greves = Zeyherella mayumbensis（Greves）Aubrév. et Pellegr. ●☆

362964 Sideroxylon melanophloeos L. = Rapanea melanophleos（L.）Mez ●☆

362965 Sideroxylon microlobum Baker = Capurodendron pervillei（Engl.）Aubrév. ●☆

362966 Sideroxylon microphyllum Scott-Elliot = Capurodendron microphyllum（Scott-Elliot）Aubrév. ●☆

362967 Sideroxylon mite L. = Ilex mitis（L.）Radlk. ●☆

362968 Sideroxylon oblanceolatum S. Moore = Englerophytum oblanceolatum（S. Moore）T. D. Penn. ●☆

362969 Sideroxylon obovatum Lam. ; 倒卵形铁榄 ●☆

362970 Sideroxylon obovatum Lam. = Planchonella obovata（R. Br.）Pierre ●

362971 Sideroxylon oxyacanthum Baill. ; 尖刺铁榄 ●☆

362972 Sideroxylon perrieri Lecomte = Capurodendron perrieri（Lecomte）Aubrév. ●☆

362973 Sideroxylon perrieri Lecomte var. oblongifolium Lecomte = Capurodendron perrieri（Lecomte）Aubrév. ●☆

362974 Sideroxylon pervillei Engl. = Capurodendron pervillei（Engl.）Aubrév. ●☆

362975 Sideroxylon randii S. Moore = Englerophytum magalismontanum（Sond.）T. D. Penn. ●☆

362976 Sideroxylon revolutum Baker = Synsepalum revolutum（Baker）T. D. Penn. ●☆

362977 Sideroxylon rostrata Merr. = Xantolis boniana（Dubard）Royen var. rostrata（Merr.）Royen ●

362978 Sideroxylon rubrocostatum Jum. et H. Perrier = Capurodendron rubrocostatum（Jum. et H. Perrier）Aubrév. ●☆

362979 Sideroxylon saboureaui Capuron ex Aubrév. = Sideroxylon beguei Capuron ex Aubrév. ●☆

362980 Sideroxylon sacleuxii Baill. = Synsepalum brevipes（Baker）T. D. Penn. ●☆

362981 Sideroxylon saganeitense Schweinf. ; 瑟格呐伊提铁榄 ●☆

362982 Sideroxylon saxorum Lecomte ; 岩栖铁榄 ●☆

362983 Sideroxylon shweliense W. W. Sm. = Xantolis shweliensis（W. W. Sm.）P. Royen ●◇

362984 Sideroxylon spinosum L. = Argania spinosa（L.）Skeels ●☆

362985 Sideroxylon tampinense Lecomte = Capurodendron tampinense（Lecomte）Aubrév. ●☆

362986 Sideroxylon tenax L. ; 银色铁榄 ; Silver Buckthorn, Tough Bully, Tough Bumelia ●☆

362987 Sideroxylon tomentosum Roxb. ; 绒毛铁榄 ●☆

362988 Sideroxylon wightianum Hook. et Arn. = Sinosideroxylon wightianum（Hook. et Arn.）Aubrév. ●

362989 Sideroxylon wightianum Hook. et Arn. var. balansae Lecomte = Sinosideroxylon wightianum（Hook. et Arn.）Aubrév. ●

362990 Sideroxylon wightianum Hook. et Arn. var. tonkinense H. L. Li = Sinosideroxylon wightianum（Hook. et Arn.）Aubrév. ●

362991 Sideroxylum Salisb. = Sideroxylon L. ●☆

362992 Sidopsis Rydb. = Malvastrum A. Gray（保留属名）●■

362993 Sidopsis Rydb. = Sida L. ●■

362994 Sidopsis hispida（Pursh）Rydb. = Malvastrum hispidum（Pursh）Hochr. ■☆

362995 Sidotheca Reveal（2004）; 星苞蓼属 ; Starry Puncturebract ■☆

362996 Sidotheca caryophylloides（Parry）Reveal ; 繁缕星苞蓼 ; Chickweed Starry Puncturebract ■☆

362997 Sidotheca emarginata（H. M. Hall）Reveal ; 白边星苞蓼 ; White-margin Starry Puncturebract ■☆

362998 Sidotheca trilobata（A. Gray）Reveal ; 三裂星苞蓼 ; Three-lobed Starry Puncturebract ■☆

362999 Siebera C. Presl = Anredera Juss. ●■

363000 Siebera Hoppe = Minuartia L. ■

363001 Siebera Hoppe = Siebera J. Gay（保留属名）■☆

363002 Siebera J. Gay（1827）（保留属名）; 微刺菊属 ■☆

363003 Siebera Post et Kuntze = Sieberia Spreng.（废弃属名）■☆

363004 Siebera Rchb. = Fischera Spreng. ■☆

363005 Siebera Rchb. = Platysace Bunge ■☆

363006 Siebera punsens（Lam.）DC. ; 微刺菊 ■☆

363007 Sieberia Spreng.（废弃属名）= Coeloglossum Hartm. ■

363008 Sieberia Spreng.（废弃属名）= Leucorchis E. Mey. ■☆

363009 Sieberia Spreng.（废弃属名）= Nigritella Rich. ■☆

363010 Sieberia Spreng.（废弃属名）= Platanthera Rich.（保留属名）■

363011 Sieberia Spreng.（废弃属名）= Siebera J. Gay（保留属名）■☆

363012 Sieboldia Heynh. = Simethis Kunth（保留属名）■☆

363013 Sieboldia Hoffmanns. = Clematis L. ●■

363014 Siederella Szlach., Mytnik, Górniak et Romowicz = Oncidium Sw.（保留属名）■☆

363015 Siederella Szlach., Mytnik, Górniak et Romowicz（2006）; 墨西哥瘤瓣兰属 ■☆

363016 Siegesbeckia L. = Sigesbeckia L. ■

363017 Siegesbeckia Steud. = Sigesbeckia L. ■

363018　Siegesbeckia brachiata Roxb. = Sigesbeckia orientalis L. ■

363019　Siegesbeckia formosana Kitam. = Sigesbeckia glabrescens (Makino) Makino ■

363020　Siegesbeckia glabrescens (Makino) Makino = Sigesbeckia glabrescens (Makino) Makino ■

363021　Siegesbeckia glabrescens (Makino) Makino var. leucoclada Nakai = Sigesbeckia glabrescens Makino ■

363022　Siegesbeckia glabrescens Makino = Sigesbeckia orientalis L. ■

363023　Siegesbeckia gracilis DC. = Sigesbeckia orientalis L. ■

363024　Siegesbeckia humilis Koidz. = Sigesbeckia orientalis L. ■

363025　Siegesbeckia iberica Willd. = Sigesbeckia orientalis L. ■

363026　Siegesbeckia micorcephala DC. = Sigesbeckia orientalis L. ■

363027　Siegesbeckia occidentalis L. = Verbesina occidentalis (L.) Walter ■☆

363028　Siegesbeckia orientalis L. = Sigesbeckia orientalis L. ■

363029　Siegesbeckia orientalis L. f. angustifolia Makino = Sigesbeckia orientalis L. ■

363030　Siegesbeckia orientalis L. f. glabrescens Makino = Sigesbeckia glabrescens Makino ■

363031　Siegesbeckia orientalis L. f. pubescens Makino = Sigesbeckia pubescens (Makino) Makino ■

363032　Siegesbeckia orientalis L. subsp. glabrescens (Makino) Kitam. = Sigesbeckia glabrescens Makino ■

363033　Siegesbeckia orientalis L. subsp. pubescens (Makino) Kitam. ex H. Koyama = Sigesbeckia pubescens (Makino) Makino ■

363034　Siegesbeckia orientalis L. subsp. pubescens Kitam. = Sigesbeckia pubescens (Makino) Makino ■

363035　Siegesbeckia orientalis L. var. angustifolia Makino = Sigesbeckia orientalis L. ■

363036　Siegesbeckia orientalis L. var. glabrescens Makino = Sigesbeckia glabrescens Makino ■

363037　Siegesbeckia orientalis L. var. pubescens Makino = Sigesbeckia pubescens (Makino) Makino ■

363038　Siegesbeckia pubescens (Makino) Makino = Sigesbeckia pubescens (Makino) Makino ■

363039　Siegesbeckia pubescens Makino f. eglandulosa Y. Ling et S. M. Hwang = Sigesbeckia pubescens Makino f. eglandulosa Y. Ling et S. M. Hwang ■

363040　Siegfriedia C. A. Gardner(1933);西澳鼠李属●☆

363041　Siegfriedia darwinioides C. A. Gardner;西澳鼠李●☆

363042　Sieglingia Bernh.(废弃属名) = Danthonia DC.(保留属名)■

363043　Sieglingia decumbens (L.) Bernh. = Danthonia decumbens (L.) DC. ■☆

363044　Sieglingia decumbens (L.) Bernh. subsp. mauritanica Maire = Danthonia decumbens (L.) DC. ■☆

363045　Siella Pimenov = Berula W. D. J. Koch ■

363046　Siella erecta (Huds.) Pimenov = Berula erecta (Huds.) Coville ■

363047　Siemensia Urb.(1923);古巴茜草属●☆

363048　Siemensia pendula Urb.;古巴茜草●☆

363049　Siemssenia Steetz = Podolepis Labill.(保留属名)■☆

363050　Sieruela Raf. = Cleome L. ●■

363051　Sievekingia Rchb. f.(1871);垂序兰属■☆

363052　Sievekingia colombiana Garay;哥伦比亚垂序兰■☆

363053　Sievekingia fimbriata Rchb. f.;垂序兰■

363054　Sieveniia Willd. = Geum L. ■

363055　Sieversandreas Eb. Fisch.(1996);随氏寄生属●☆

363056　Sieversandreas madagascarianus Eb. Fisch.;随氏寄生●☆

363057　Sieversia Willd.(1811);随氏路边青属(五瓣莲属);Sieversia ●☆

363058　Sieversia elata Royle = Acomastylis elata (Royle) F. Bolle ■

363059　Sieversia elata Royle = Geum elatum Wall. ex Hook. f. ■

363060　Sieversia elata Royle var. humile Royle = Acomastylis elata (Royle) F. Bolle var. humilis (Royle) F. Bolle ■

363061　Sieversia macrantha Kearney;大花随氏路边青●☆

363062　Sieversia montana R. Br.;山地随氏路边青;Alpine Avens ●☆

363063　Sieversia pentapetala (L.) Greene;随氏路边青●☆

363064　Sieversia pentapetala (L.) Greene f. plena Miyabe et Tatew.;重瓣随氏路边青●☆

363065　Sieversia pentapetala (L.) Greene var. dilatata Takeda et Honda = Sieversia pentapetala (L.) Greene ●☆

363066　Sieversia pentapetala Greene = Sieversia pentapetala (L.) Greene ●☆

363067　Sieversia pusilla (Gaertn.) Hultén;微小随氏路边青●☆

363068　Sieversia reptans R. Br. = Geum reptans L. ■☆

363069　Sieversia triflora (Pursh) R. Br.;三花随氏路边青●☆

363070　Sieversia triflora (Pursh) R. Br. = Geum triflorum Pursh ■☆

363071　Sieversia triflora R. Br. = Sieversia triflora (Pursh) R. Br. ●☆

363072　Siflora Raf. = Sison L. ■☆

363073　Sigesbeckia L.(1753);稀莶属(西热菊属);St. Paulswort, St. Paul's-wort ■

363074　Sigesbeckia abyssinica (Sch. Bip.) Oliv. et Hiern = Micractis bojeri DC. ■☆

363075　Sigesbeckia brachiata Roxb. = Sigesbeckia orientalis L. ■

363076　Sigesbeckia discoidea (Vatke) S. F. Blake = Micractis discoidea (Vatke) D. L. Schulz ■☆

363077　Sigesbeckia emirnensis Baker = Micractis bojeri DC. ■☆

363078　Sigesbeckia filarszkyi Pit. = Eupatorium cannabinum L. ■

363079　Sigesbeckia formosana Kitam. = Sigesbeckia glabrescens Makino ■

363080　Sigesbeckia glabrescens (Makino) Makino;毛梗稀莶(肥猪草,光稀莶,棉苍狼,母猪油,疏毛稀莶,小稀莶);Hairstalk St. Paulswort ■

363081　Sigesbeckia glabrescens Makino = Sigesbeckia glabrescens (Makino) Makino ■

363082　Sigesbeckia glabrescens Makino var. leucoclada Nakai = Sigesbeckia glabrescens Makino ■

363083　Sigesbeckia gracilis DC. = Sigesbeckia orientalis L. ■

363084　Sigesbeckia gummifera ?;胶稀莶(胶西热菊,毛梗稀莶)■

363085　Sigesbeckia humilis Koidz. = Sigesbeckia orientalis L. ■

363086　Sigesbeckia iberica Willd. = Sigesbeckia orientalis L. ■

363087　Sigesbeckia jorullensis Kunth = Sigesbeckia orientalis L. ■

363088　Sigesbeckia microcephala DC. = Sigesbeckia orientalis L. ■

363089　Sigesbeckia occidentalis L. = Verbesina occidentalis (L.) Walter ■☆

363090　Sigesbeckia orientalis L.;稀莶(大接骨,大叶草,灯笼草,肥猪菜,肥猪苗,风湿草,感冒草,虎膏,虎莶,黄花草,黄花子,黄母猪,火枕草,火莶,老陈婆,老奶补补丁,老前婆,绿莶草,毛擦拉子,冇骨消,棉苍狼,棉黍棵,母猪油,黏糊菜,牛人参,四棱麻,四莶,天名精,铜锤草,土伏虱,希仙,稀莶草,虾柑草,虾疳草,亚婆针,亚婆针,野向日葵,野芝麻,油草子,粘不扎,黏苍子,黏强子,镇静草,珠草,猪膏草,猪膏莓,猪冠麻叶,猪母菜,猪屎菜);Common St. Paulswort,Eastern St. Paul's Wort,Sigesbeckia ■

363091　Sigesbeckia orientalis L. f. glabrescens Makino = Sigesbeckia glabrescens Makino ■

363092　Sigesbeckia orientalis L. subsp. glabrescens （Makino） H. Koyama = Sigesbeckia glabrescens Makino ■

363093　Sigesbeckia pubescens （Makino） Makino；腺梗稀莶（毛稀莶，棉苍狼，稀莶，野洋姜草，黏金强子，珠草）；Glandstalk St. Paulswort ■

363094　Sigesbeckia pubescens （Makino） Makino f. eglandulosa Y. Ling et S. M. Hwang；无腺稀莶（无腺腺梗稀莶）；Glandless St. Paulswort ■

363095　Sigesbeckia pubescens Makino = Sigesbeckia pubescens （Makino） Makino ■

363096　Sigesbeckia serrata ?；具齿稀莶；Western St. Paul's Wort ■☆

363097　Sigesbeckia somalensis S. Moore = Guizotia schimperi Sch. Bip. ex Walp. ■☆

363098　Sigilaria Raf. = Sigillaria Raf. ■

363099　Sigilaria Raf. = Smilacina Desf. （保留属名）■

363100　Sigillabenis Thouars = Habenaria Willd. ■

363101　Sigillaria Raf. = Maianthemum F. H. Wigg. （保留属名）■

363102　Sigillaria Raf. = Smilacina Desf. （保留属名）■

363103　Sigillum Friche-Joset et Montandon = Polygonatum Mill. ■

363104　Sigillum Montandon = Polygonatum Mill. ■

363105　Sigillum Tragus ex Montandon = Polygonatum Mill. ■

363106　Sigmatanthus Huber ex Ducke = Raputia Aubl. ●☆

363107　Sigmatanthus Huber ex Emmerich = Raputia Aubl. ●☆

363108　Sigmatochilus Rolfe = Panisea （Lindl.） Lindl. （保留属名）■

363109　Sigmatogyne Pfitzer = Panisea （Lindl.） Lindl. （保留属名）■

363110　Sigmatogyne bia Kerr = Panisea tricallosa Rolfe ■

363111　Sigmatogyne pantlingii Pfitzer = Panisea tricallosa Rolfe ■

363112　Sigmatogyne tricallosa （Rolfe） Pfitzer = Panisea tricallosa Rolfe ■

363113　Sigmatophyllum D. Dietr. = Stigmaphyllon A. Juss. ●☆

363114　Sigmatosiphon Engl. = Sesamothamnus Welw. ●☆

363115　Sigmatosiphon guerichii Engl. = Sesamothamnus guerichii （Engl.） E. A. Bruce ●☆

363116　Sigmatostalix Rchb. f. （1852）；弓柱兰属■☆

363117　Sigmatostalix mexicana L. O. Williams；墨西哥弓柱兰■☆

363118　Sigmodostyles Meisn. = Rhynchosia Lour. （保留属名）●■

363119　Sigmodostyles villosa Meisn. = Rhynchosia villosa （Meisn.） Druce ■☆

363120　Sigrnatogyne Pfitzer = Panisea （Lindl.） Lindl. （保留属名）■

363121　Sikira Raf. = Chaerophyllum L. ■

363122　Silamnus Raf. = Cephalanthus L. ●

363123　Silaum Mill. （1754）；亮叶芹属；Meadow Saxifrage, Pepper Saxifrage, Pepper-saxifrage, Pepperwort, Silaus, Sulphurwort ■

363124　Silaum besseri （DC.） Grossh.；拜斯亮叶芹■☆

363125　Silaum foliosum （Sommier et H. Lév.） Grossh.；多叶亮叶芹■☆

363126　Silaum popovii （Korovin） M. Hiroe；波波夫亮叶芹■☆

363127　Silaum rubtzovii （Schischk.） M. Hiroe；卢勃亮叶芹■☆

363128　Silaum silaus （L.） Schinz et Thell.；亮叶芹（草地亮叶芹）；Lawn Silaus, Meadow Silaus, Pepper-saxifrage ■

363129　Silaum silaus （L.） Schinz et Thell. = Silaus pratensis （Crantz） Besser ■

363130　Silaus Bernh. = Silaum Mill. ■

363131　Silaus besseri DC. = Silaum besseri （DC.） Grossh. ■☆

363132　Silaus flavescens Bernh. = Silaum silaus （L.） Schinz et Thell. ■

363133　Silaus foliosus Sommier et H. Lév. = Silaum foliosum （Sommier et H. Lév.） Grossh. ■☆

363134　Silaus popovii Korovin = Silaum popovii （Korovin） M. Hiroe ■☆

363135　Silaus pratensis （Crantz） Besser. = Silaum silaus （L.） Schinz et Thell. ■

363136　Silaus rubtzovii Schischk. = Silaum rubtzovii （Schischk.） M. Hiroe ■☆

363137　Silenaceae （DC.） Bartl. = Caryophyllaceae Juss. （保留科名） ■●

363138　Silenaceae Bartl. = Caryophyllaceae Juss. （保留科名）■●

363139　Silenanthe （Fenzl） Griseb. et Schenk = Silene L. （保留属名）■

363140　Silenanthe Griseb. et Schenk = Silene L. （保留属名）■

363141　Silene L. （1753）（保留属名）；蝇子草属（麦瓶草属，雪轮属）；Campion, Catchfly, Lamp-flower, Lychnis, Rose Campion, Silene ■

363142　Silene abietum Font Quer et Maire；冷杉蝇子草■☆

363143　Silene abyssinica （Hochst.） Neumayer = Uebelinia abyssinica Hochst. ■☆

363144　Silene acaulis L.；无茎麦瓶草（无茎蝇子草）；Cushion Pink, Cution Pink, Moss Campion, Moss Silene, Mountain Campion ■☆

363145　Silene acaulis L. = Silene davidii （Franch.） Oxelman et Lidén ■

363146　Silene acaulis L. subsp. arctica Á. Löve et D. Löve = Silene acaulis L. ■☆

363147　Silene acaulis L. subsp. exscapa （All.） DC. = Silene acaulis L. ■☆

363148　Silene acaulis L. subsp. subacaulescens （F. N. Williams） Hultén = Silene acaulis L. ■☆

363149　Silene acaulis L. var. dioica ?；异株蝇子草；Red Campion ■☆

363150　Silene adenantha Franch. = Silene asclepiadea Franch. ■

363151　Silene adenocalyx F. N. Williams；腺萼蝇子草；Glandcalyx Catchfly ■

363152　Silene adenocalyx F. N. Williams = Silene napuligera Franch. ■

363153　Silene adenopetala Raikova；腺花蝇子草■

363154　Silene aegyptiaca （L.） L. f.；埃及蝇子草■☆

363155　Silene aegyptiaca L. = Silene aegyptiaca （L.） L. f. ■☆

363156　Silene aellenii Sennen；埃伦蝇子草■☆

363157　Silene aethiopica Burm. f. = Silene burchellii Otth ■☆

363158　Silene akaisialpina （T. Yamaz.） H. Ohashi, Tateishi et H. Nakai；明石蝇子草■☆

363159　Silene akaisialpina （T. Yamaz.） H. Ohashi, Tateishi et H. Nakai f. leucantha （Takeda） H. Ohashi, Tateishi et H. Nakai；白花明石蝇子草■☆

363160　Silene alaschanica （Maxim.） Bocquet；贺兰山蝇子草（贺兰山女娄菜）；Helanshan Catchfly ■

363161　Silene alba （Mill.） E. H. L. Krause = Silene latifolia （Mill.） Rendle et Britten subsp. alba （Mill.） Greuter et Burdet ■

363162　Silene alba （Mill.） E. H. L. Krause = Silene latifolia （Mill.） Rendle et Britten ■

363163　Silene alba （Mill.） E. H. L. Krause subsp. divaricata （Rchb.） Walters = Silene latifolia （Mill.） Rendle et Britten ■

363164　Silene alba Muhl. ex Rohrb. = Silene nivea （Nutt.） Muhl. ex Otth ■☆

363165　Silene alexandrae B. Keller；斋桑蝇子草；Alexandra Catchfly ■

363166　Silene allamanii Otth = Silene laciniata Cav. ■☆

363167　Silene alpestris Willd. ex Nyman；高山生蝇子草（山生蝇子草）；Alpine Catchfly, Alpine Silene ■☆

363168　Silene alpicola Schischk；高山麦瓶草；Alpine Catchfly ■☆

363169　Silene altaica Pers.；阿尔泰蝇子草（灌丛蝇子草）；Altai Catchfly ■

363170　Silene altaica Pers. f. grandiflora C. A. Mey. = Silene alexandrae B. Keller ■

363171 Silene altaica Pers. var. grandiflora C. A. Mey. = Silene alexandrae B. Keller ■

363172 Silene altaica Pers. var. hystrix Trautv. = Silene alexandrae B. Keller ■

363173 Silene altaica Pers. var. typica Trautv. = Silene altaica Pers. ■

363174 Silene amurensis Pomel = Silene patula Desf. subsp. amurensis (Pomel) Jeanm. ■☆

363175 Silene anaglaea Maire = Silene nocturna L. ■☆

363176 Silene andersonii Clokey = Silene verecunda S. Watson ■☆

363177 Silene andryalifolia Pomel；毛托菊叶蝇子草■☆

363178 Silene anglica L. = Silene gallica L. ■

363179 Silene angustifolia Poir. = Petrorhagia illyrica (Ard.) P. W. Ball et Heywood subsp. angustifolia (Poir.) P. W. Ball et Heywood ■☆

363180 Silene anisoloba Schischk. ；同裂蝇子草■☆

363181 Silene antirrhina L. ；睡眠蝇子草；Sleepy Catchfly, Sleepy Silene ■☆

363182 Silene antirrhina L. f. apetala Farw. = Silene antirrhina L. ■☆

363183 Silene antirrhina L. f. bicolor Farw. = Silene antirrhina L. ■☆

363184 Silene antirrhina L. f. deaneana Fernald = Silene antirrhina L. ■☆

363185 Silene antirrhina L. var. confinis Fernald = Silene antirrhina L. ■☆

363186 Silene antirrhina L. var. depauperata Rydb. = Silene antirrhina L. ■☆

363187 Silene antirrhina L. var. divaricata B. L. Rob. = Silene antirrhina L. ■☆

363188 Silene antirrhina L. var. laevigata Engelm. et A. Gray = Silene antirrhina L. ■☆

363189 Silene antirrhina L. var. subglaber Engelm. et A. Gray = Silene antirrhina L. ■☆

363190 Silene antirrhina L. var. vaccarifolia Rydb. = Silene antirrhina L. ■☆

363191 Silene aomorensis M. Mizush. ；青森蝇子草■☆

363192 Silene aperta Greene；裸蝇子草；Naked Catchfly ■☆

363193 Silene apetala Willd. ；无瓣蝇子草■☆

363194 Silene apetala Willd. var. alexandrina Asch. = Silene apetala Willd. ■☆

363195 Silene apetala Willd. var. berenicea Pamp. = Silene apetala Willd. ■☆

363196 Silene apetala Willd. var. grandiflora Boiss. ；大花无瓣蝇子草■☆

363197 Silene aprica Turcz. ex Fisch. et C. A. Mey. = Melandrium apricum (Turcz. ex Fisch. et C. A. Mey.) Rohrb. ■

363198 Silene aprica Turcz. ex Fisch. et C. A. Mey. = Silene kialensis Rohrb. ■

363199 Silene aprica Turcz. ex Fisch. et C. A. Mey. = Silene nepalensis Majumdar var. kialensis (F. N. Williams) C. L. Tang ■

363200 Silene aprica Turcz. ex Fisch. et C. A. Mey. var. firma (Siebold et Zucc.) F. N. Williams = Silene firma Siebold et Zucc. ■

363201 Silene aprica Turcz. ex Fisch. et C. A. Mey. var. oldhamiana (Miq.) C. Y. Wu = Silene aprica Turcz. ex Fisch. et C. A. Mey. ■

363202 Silene aprica Turcz. ex Fisch. et C. A. Mey. var. oldhamiana (Miq.) C. Y. Wu；长冠女娄菜■

363203 Silene aprica Turcz. ex Fisch. et C. A. Mey. var. ryukyuensis T. Yamaz. ；琉球女娄菜■☆

363204 Silene aprica Turcz. var. firma (Siebold et Zucc.) F. N. Williams = Silene firma Siebold et Zucc. ■

363205 Silene arabica Boiss. ；阿拉伯蝇子草■☆

363206 Silene araratica Schischk. ；亚拉腊蝇子草■☆

363207 Silene arenaria Desf. = Silene nicaeensis All. ■☆

363208 Silene arenarioides Desf. ；拟沙地蝇子草■☆

363209 Silene arenosa K. Koch；砂蝇子草■☆

363210 Silene argillosa Munby；白土蝇子草■☆

363211 Silene argyi H. Lév. = Silene fortunei Vis. ■

363212 Silene aristidis Pomel；三芒草状蝇子草■☆

363213 Silene armena Boiss. ；亚美尼亚蝇子草■☆

363214 Silene armeria L. ；高雪轮(捕虫瞿麦,钟石竹)；Fleabane, Limewort, Lobel's Catchfly, None-so-pretty, Sweet William Catchfly, Sweet William Silene, Sweet-william Campion, Sweet-william Catchfly, Sweet-william Silene, Thriftlike Catchfly, William Sweet Catchfly ■

363215 Silene armeria L. 'Electra'；神女高雪轮■☆

363216 Silene articulata Viv. ；关节蝇子草■☆

363217 Silene artwinensis Schischk. ；阿尔特温蝇子草■☆

363218 Silene asclepiadea Franch. ；掌脉蝇子草(黑牵牛,马利筋女娄菜,腺花女娄菜)；Palmvein Catchfly ■

363219 Silene asclepiadea Franch. var. dumicola (W. W. Sm.) C. L. Tang；丛林蝇子草■

363220 Silene asclepiadea Franch. var. dumicola (W. W. Sm.) C. L. Tang = Silene viscidula Franch. ■

363221 Silene asclepiadea Franch. var. glutinosa Franch. = Silene asclepiadea Franch. ■

363222 Silene asterias Griseb. ；巴纳特蝇子草；Banat campion, Banat Silene ■☆

363223 Silene atlantica Coss. et Durieu；大西洋蝇子草■☆

363224 Silene atrocastanea Diels；栗色蝇子草(黑栗色女娄菜)；Brown Catchfly ■

363225 Silene atsaensis (C. Marquand) Bocquet；阿扎蝇子草(加查女娄菜)；Aza Catchfly ■

363226 Silene attenuata (Farr) Bocquet = Silene uralensis (Rupr.) Bocquet ■☆

363227 Silene aucheriana Boiss. ；奥氏蝇子草■☆

363228 Silene auriculifolia Pomel；耳叶蝇子草■☆

363229 Silene ayachica Humbert；阿亚希蝇子草■☆

363230 Silene baccifera (L.) Roth = Cucubalus baccifer L. ■

363231 Silene baccifera (L.) Roth var. japonica (Miq.) H. Ohashi et H. Nakai = Cucubalus baccifer L. var. japonicus Miq. ■

363232 Silene balchaschensis Schischk. ；巴尔哈什蝇子草■☆

363233 Silene baldshuanica B. Fedtsch. ；巴尔得蝇子草■☆

363234 Silene baldwinii Nutt. = Silene polypetala (Walter) Fernald et B. G. Schub. ■☆

363235 Silene banksia (Meerb.) Mabberly；剪春罗(剪红罗,剪金花,剪夏罗,金钱花,阔叶鲤鱼胆,婆婆针线包,山茶田,山田茶,山药田,碎剪罗,雄黄花,一支蒿)；Crown Campion, Gamp, Largeflowered Lychnis ■

363236 Silene barbara Humbert et Maire；外来蝇子草■☆

363237 Silene batangensis H. Limpr. ；巴塘蝇子草；Batang Catchfly ■

363238 Silene battandieriana Hochr. = Silene colorata Poir. ■☆

363239 Silene behrii (Rohrb.) F. N. Williams = Silene verecunda S. Watson ■☆

363240 Silene bellidifolia Jacq. ；雅叶蝇子草■☆

363241 Silene bellidioides Sond. ；禾鼠麹蝇子草■☆

363242 Silene benoistii Maire = Silene colorata Poir. subsp. benoistii

（Maire）Sauvage ■☆

363243   Silene bergiana Lindm. = Silene diversifolia Otth ■☆

363244   Silene bernardina S. Watson；帕默蝇子草；Palmer's Catchfly ■☆

363245   Silene bernardina S. Watson subsp. maguirei Bocquet = Silene bernardina S. Watson ■☆

363246   Silene bernardina S. Watson var. rigidula（B. L. Rob.）Tiehm = Silene bernardina S. Watson ■☆

363247   Silene bernardina S. Watson var. sierrae（C. L. Hitchc. et Maguire）Bocquet = Silene bernardina S. Watson ■☆

363248   Silene bhutanica（W. W. Sm.）Majumdar = Silene indica Roxb. ex Otth var. bhutanica（W. W. Sm.）Bocquet ■

363249   Silene biappendiculata Rohrb.；双附属物蝇子草■☆

363250   Silene bilingua W. W. Sm.；双舌蝇子草；Twintongue Catchfly ■

363251   Silene bipartita Desf. = Silene colorata Poir. ■☆

363252   Silene bobrovii Schischk.；鲍勃蝇子草■☆

363253   Silene bodinieri H. Lév. = Silene viscidula Franch. ■

363254   Silene bolanderi A. Gray = Silene hookeri Nutt. subsp. bolanderi（A. Gray）Abrams ■☆

363255   Silene bornmulleri Freyn；包尔恩蝇子草■☆

363256   Silene boryi Boiss.；鲍利蝇子草■☆

363257   Silene boryi Boiss. var. atlantica Maire = Silene boryi Boiss. ■☆

363258   Silene boryi Boiss. var. ouensae（Coss.）Maire = Silene boryi Boiss. ■☆

363259   Silene borysthenica（Gruner）Walters；小花蝇子草（小花麦瓶草）；Littleflower Silene，Smallflower Catchfly，Smallflower Silene ■

363260   Silene bourgeaei Webb ex Christ；布尔热蝇子草■☆

363261   Silene brachypetala Robill. et Castagne ex DC.；短瓣蝇子草■☆

363262   Silene bridgesii Rohrb.；布里奇斯蝇子草；Bridges' Catchfly ■☆

363263   Silene brotherana Sommier et H. Lév.；布罗特蝇子草■☆

363264   Silene bucharica Popov；布哈拉蝇子草■☆

363265   Silene bungeana（D. Don）H. Ohashi et H. Nakai；剪红纱花（邦奇蝇子草，地黄连，哈吉氏剪秋罗，哈氏剪秋罗，汉宫秋，红梅草，剪秋罗，剪秋纱，见肿消，鞠翠花，阔叶鲤鱼胆，散血沙）；Chinarose Campion，Haage Campion，Senno Campion ■

363266   Silene bungeana（D. Don）H. Ohashi et H. Nakai = Lychnis senno Siebold et Zucc. ■

363267   Silene bungei Bocquet；暗色蝇子草；Darkcolor Catchfly ■

363268   Silene bupleuroides L.；柴胡蝇子草■☆

363269   Silene burchellii Otth；伯奇尔蝇子草■☆

363270   Silene burchellii Otth var. angustifolia Sond.；窄叶伯奇尔蝇子草■☆

363271   Silene burchellii Otth var. gillettii Turrill = Silene gillettii（Turrill）M. G. Gilbert ■☆

363272   Silene burchellii Otth var. latifolia Sond.；宽叶伯奇尔蝇子草■☆

363273   Silene burchellii Otth var. macropetala Turrill；大瓣伯奇尔蝇子草■☆

363274   Silene burchellii Otth var. macrorrhiza R. E. Fr. = Silene burchellii Otth var. angustifolia Sond. ■☆

363275   Silene burchellii Otth var. pilosellifolia Sond. = Silene burchellii Otth var. angustifolia Sond. ■☆

363276   Silene burchellii Otth var. schweinfurthii（Rohrb.）Täckh. et Boulos = Silene schweinfurthii Rohrb. ■☆

363277   Silene burchellii Otth var. syngei（Turrill）Turrill = Silene syngei（Turrill）T. M. Harris et Goyder ■☆

363278   Silene caespitella F. N. Williams；丛生蝇子草；Clumpy Catchfly ■

363279   Silene caespitella F. N. Williams = Silene moorcroftiana Wall. ex Benth. ■

363280   Silene caespitosa Bureau et Franch. = Silene davidii（Franch.）Oxelman et Lidén ■

363281   Silene caespitosa Stev.；簇生蝇子草■☆

363282   Silene californica Durand；加州雪轮；California Indian Pink，Indian Pink ■☆

363283   Silene californica Durand = Silene laciniata Cav. subsp. californica（Durand）J. K. Morton ■☆

363284   Silene campanulata S. Watson；红色山地蝇子草；Red Mountain Catchfly ■☆

363285   Silene campanulata S. Watson subsp. glandulosa C. L. Hitchc. et Maguire；腺点蝇子草；Bell Catchfly ■☆

363286   Silene campanulata S. Watson subsp. greenei（S. Watson）C. L. Hitchc. et Maguire = Silene campanulata S. Watson subsp. glandulosa C. L. Hitchc. et Maguire ■☆

363287   Silene campanulata S. Watson var. angustifolia F. N. Williams = Silene campanulata S. Watson ■☆

363288   Silene campanulata S. Watson var. greenei S. Watson = Silene campanulata S. Watson subsp. glandulosa C. L. Hitchc. et Maguire ■☆

363289   Silene campanulata S. Watson var. latifolia F. N. Williams = Silene campanulata S. Watson subsp. glandulosa C. L. Hitchc. et Maguire ■☆

363290   Silene campanulata S. Watson var. orbiculata B. L. Rob. = Silene campanulata S. Watson subsp. glandulosa C. L. Hitchc. et Maguire ■☆

363291   Silene campanulata S. Watson var. petrophila Jeps. = Silene campanulata S. Watson subsp. glandulosa C. L. Hitchc. et Maguire ■☆

363292   Silene canariensis Willd.；加那利蝇子草■☆

363293   Silene canescens Ten. = Silene colorata Poir. ■☆

363294   Silene capensis Otth = Silene undulata Aiton ■

363295   Silene capitata Kom.；头序蝇子草（馒头草，馒头麦瓶草，头序麦瓶草）；Capitate Catchfly ■

363296   Silene capitellata Boiss.；小头序蝇子草■☆

363297   Silene cardiopetala Franch.；心瓣蝇子草；Cordipetal Catchfly ■

363298   Silene cardiopetala Franch. var. deqenensis C. Y. Wu = Silene monbeigii W. W. Sm. ■

363299   Silene caroliniana var. wherryi（Small）Fernald = Silene caroliniana Walter subsp. wherryi（Small）R. T. Clausen ■☆

363300   Silene caroliniana Walter；卡罗林蝇子草（加罗林雪轮）；Campion，Carolina Wild Pink，Catchfly，Hot Pink Silene，Peatpink，Wild Pink ■☆

363301   Silene caroliniana Walter subsp. pensylvanica（Michx.）R. T. Clausen；宾州蝇子草；Pennsylvania Wild Pink ■☆

363302   Silene caroliniana Walter subsp. wherryi（Small）R. T. Clausen；惠里蝇子草；Wherry's Pink ■☆

363303   Silene caroliniana Walter var. pensylvanica（Michx.）Fernald = Silene caroliniana Walter subsp. pensylvanica（Michx.）R. T. Clausen ■☆

363304   Silene caroliniana Walter var. wherryi（Small）Fernald = Silene caroliniana Walter subsp. wherryi（Small）R. T. Clausen ■☆

363305   Silene cashmeriana（Royle ex Benth.）Majumdar；克什米尔蝇子草；Kashmir Catchfly ■

363306   Silene catesbaei Walter = Silene virginica L. ■☆

363307　Silene caucasica (Bunge) Boiss. ;高加索蝇子草■☆

363308　Silene cephalantha Boiss. ;头花蝇子草■☆

363309　Silene cerastoides L. var. anomala Ball = Silene gallica L. ■

363310　Silene cerastoides L. var. dunensis Alleiz. = Silene sclerocarpa L. Dufour ■☆

363311　Silene chaetodonta Boiss. ;毛齿蝇子草■☆

363312　Silene chalcedonica (L.) E. H. L. Krause;皱叶剪秋罗(美国剪秋罗,皱叶剪夏罗);Brennende Liebe,Bridget in Her Bravery,Campion of Constantinople, Chalcedonian Lychnis, Cross of Jerusalem, Flower of Bristol, Flower of Bristow, Flower of Constantinople, Garder's Delight, Great Candlesticks, Jerusalem Cross, Knight Cross, Lampflower, London Pride, Maltese Cross, Maltese Cross Campion, Maltesecross, Maltese-cross, Nonsuch, Rose Campion, Scarlet Cross, Scarlet Lightning, Scarlet Lychnis, Yellow-green Silene ■

363313　Silene chamarensis Turcz. ;哈马尔蝇子草(皱叶剪秋箩);Maltese Cross ■☆

363314　Silene chirensis A. Rich. = Silene syngei (Turrill) T. M. Harris et Goyder ■☆

363315　Silene chlorantha Ehrh. ;绿花麦瓶草;Greenflower Catchfly, Yellowgreen Catchfly, Yellow-green Silene ■☆

363316　Silene chlorifolia Sm. ;绿叶蝇子草■☆

363317　Silene chloropetala Rupr. ;绿瓣蝇子草■☆

363318　Silene chodatii Bocquet ;球萼蝇子草;Ballcalyx Catchfly ■

363319　Silene chodatii Bocquet var. pygmaea Bocquet ;矮球萼蝇子草■

363320　Silene choulettii Coss. ;舒莱蝇子草■☆

363321　Silene chungtienensis W. W. Sm. ;中甸蝇子草;Zhongdian Catchfly ■

363322　Silene cinerea Desf. ;灰色蝇子草■☆

363323　Silene cirtensis Pomel;锡尔塔蝇子草■☆

363324　Silene clandestina Jacq. ;隐匿蝇子草■☆

363325　Silene claryi Batt. ;克拉里蝇子草■☆

363326　Silene claviformis Litv. ;棒状蝇子草■☆

363327　Silene clokeyi C. L. Hitchc. et Maguire = Silene petersonii Maguire ■☆

363328　Silene coarctata Lag. = Silene sclerocarpa L. Dufour ■☆

363329　Silene cobalticola P. A. Duvign. et Plancke;科博尔特蝇子草■☆

363330　Silene coccinea Moench = Silene virginica L. ■☆

363331　Silene coeli-rosa (L.) A. Braun;细叶麦瓶草(鞠翠花,欧洲剪秋罗,小麦剪秋罗,樱雪轮);Rose of Heaven, Rose Silene, Rose-of-heaven, Viscaria ■

363332　Silene coeli-rosa (L.) A. Braun 'Rose Angel';玫瑰天使樱雪轮☆

363333　Silene coeli-rosa (L.) Godr. = Silene coeli-rosa (L.) A. Braun ■

363334　Silene coeli-rosa (L.) Godr. var. aspera Poir. = Silene coeli-rosa (L.) A. Braun ■

363335　Silene coeli-rosa (L.) Godr. var. laevis (Poir.) Voss = Silene coeli-rosa (L.) A. Braun ■

363336　Silene coeli-rosa (L.) Godr. var. subaspera Maire = Silene coeli-rosa (L.) A. Braun ■

363337　Silene cognata (Maxim.) H. Ohashi et H. Nakai;浅裂剪秋罗(剪春罗,剪秋罗,毛缘剪秋罗,小麦剪秋罗);Cognate Campion, Lobate Campion ■

363338　Silene colorata Poir. ;着色蝇子草■☆

363339　Silene colorata Poir. subsp. benoistii (Maire) Sauvage = Silene colorata Poir. ■☆

363340　Silene colorata Poir. subsp. oliveriana (Otth) Rohrb. ;奥里弗蝇子草■☆

363341　Silene colorata Poir. subsp. pubicalycina (Fenzl) Maire = Silene colorata Poir. ■☆

363342　Silene colorata Poir. subsp. trichocalycina Fenzl;毛萼着色蝇子草■☆

363343　Silene colorata Poir. var. angustifolia Willk. = Silene colorata Poir. ■☆

363344　Silene colorata Poir. var. benoistii (Maire) Maire = Silene colorata Poir. ■☆

363345　Silene colorata Poir. var. canescens (Ten.) Soy.-Will. et Godr. = Silene colorata Poir. ■☆

363346　Silene colorata Poir. var. crassifolia Moris = Silene colorata Poir. ■☆

363347　Silene colorata Poir. var. cyrenaica E. A. Durand et Barratte = Silene colorata Poir. ■☆

363348　Silene colorata Poir. var. decumbens (Biv.) Rohrb. = Silene colorata Poir. ■☆

363349　Silene colorata Poir. var. distachya (Brot.) Rohrb. = Silene colorata Poir. ■☆

363350　Silene colorata Poir. var. lasiocalyx (Soy.-Will. et Godr.) Ball = Silene colorata Poir. ■☆

363351　Silene colorata Poir. var. monticola Murb. = Silene colorata Poir. ■☆

363352　Silene colorata Poir. var. oliveriana (Otth) Durand et Barratte = Silene colorata Poir. ■☆

363353　Silene colorata Poir. var. oliveriana (Otth) Muschl. = Silene colorata Poir. subsp. oliveriana (Otth) Rohrb. ■☆

363354　Silene colorata Poir. var. pallida Maire = Silene colorata Poir. ■☆

363355　Silene colorata Poir. var. pseudoliveriana Maire = Silene colorata Poir. ■☆

363356　Silene colorata Poir. var. pteropleura (Coss.) Batt. = Silene colorata Poir. ■☆

363357　Silene colorata Poir. var. spathulifolia (Soy.-Will. et Godr.) Batt. = Silene colorata Poir. ■☆

363358　Silene colorata Poir. var. villosissima Maire et Weiller = Silene colorata Poir. ■☆

363359　Silene colorata Poir. var. vulgaris Willk. = Silene colorata Poir. ■☆

363360　Silene commutata Guss. ;多变麦瓶草■☆

363361　Silene compacta Fisch. ex Hornem. ;紧密麦瓶草(密花麦瓶草,密花雪轮,密麦瓶草);Oriental Silene ■☆

363362　Silene concolor Greene = Silene scouleri Hook. subsp. hallii (S. Watson) C. L. Hitchc. et Maguire ■☆

363363　Silene conica L. ;圆锥麦瓶草;Conical Silene, Sand Catchfly, Striatod Catchfly, Striped Corn Catchfly ■☆

363364　Silene conica L. var. australis Maire = Silene conica L. ■☆

363365　Silene coniflora Nees ex DC. = Silene coniflora Nees ex Otth. ■☆

363366　Silene coniflora Nees ex Otth. ;锥花麦瓶草;Multinerved Catchfly ■☆

363367　Silene coniflora Otth = Silene coniflora Nees ex Otth. ■☆

363368　Silene conoidea L. ;麦瓶草(灯笼草,净瓶,梅花瓶,米瓦罐,面条菜,香炉草);Catchfly, Cone-like Silene, Conical Catchfly, Large Sand Catchfly, Weed Silene ■

363369　Silene constantia Eckl. et Zeyh. = Silene clandestina Jacq. ■☆

363370　Silene coronaria（L.）Clairv.；毛剪秋罗（毛缕，毛叶剪秋罗，醉仙翁）；Bloody William, Crown Pink, Dusty Miller, Dusty-miller, Gardener's Delight, Gardener's Eye, Gurdner's Eyes, Hairy Campion, Mullein Pink, Rose Campion, Rosecampi ■

363371　Silene coronaria（L.）Clairv. = Lychnis coronaria（L.）Desr. ■

363372　Silene corrugata Ball；皱褶蝇子草 ■☆

363373　Silene corrugata Ball subsp. adusta ? = Silene pomelii Batt. subsp. adusta（Ball）Maire ■☆

363374　Silene corrugata Ball var. adusta（Ball）Ball = Silene pomelii Batt. subsp. adusta（Ball）Maire ■☆

363375　Silene corrugata Ball var. macrosperma（Coss.）Maire = Silene corrugata Ball ■☆

363376　Silene corrugata Ball var. mogadorensis（Coss. et Balansa）Maire = Silene corrugata Ball ■☆

363377　Silene corrugata Ball var. obtusifolia（Coss.）Maire = Silene corrugata Ball ■☆

363378　Silene cossoniana Maire = Silene heterodonta F. N. Williams ■☆

363379　Silene cossoniana Maire var. rosella ? = Silene heterodonta F. N. Williams subsp. rosella（Maire）Maire ■☆

363380　Silene crassifolia L.；厚叶蝇子草 ■☆

363381　Silene cretacea Fisch.；白粉麦瓶草 ■☆

363382　Silene cretica L.；克里特蝇子草 ■☆

363383　Silene crispans Litv.；褶叶蝇子草 ■☆

363384　Silene cryptantha Diels = Psammosilene tunicoides W. C. Wu et C. Y. Wu ■

363385　Silene cserei Baumg.；切雷蝇子草；Balkan Catchfly, Biennial Campion, Glaucous Campion, Smooth Catchfly ■☆

363386　Silene cuatrecasasii Pau et Font Quer；夸特蝇子草 ■☆

363387　Silene cuatrecasasii Pau et Font Quer var. brachycarpa Font Quer = Silene cuatrecasasii Pau et Font Quer ■☆

363388　Silene cucubalus Wibel = Silene vulgaris（Moench）Garcke ■

363389　Silene cucubalus Wibel subsp. angustifolia（Guss.）Hayek = Silene vulgaris（Moench）Garcke ■

363390　Silene cucubalus Wibel var. lancifolia Rouy = Silene vulgaris（Moench）Garcke ■

363391　Silene cucubalus Wibel var. platyphylla Maire = Silene vulgaris（Moench）Garcke ■

363392　Silene cupiformis C. L. Tang；杯萼蝇子草 ■

363393　Silene cupiformis C. L. Tang = Silene atrocastanea Diels ■

363394　Silene cyrenaica Maire et Weiller；昔兰尼蝇子草 ■☆

363395　Silene cyri Schischk.；长果蝇子草（库氏麦瓶草）■

363396　Silene czerei Baumg. et Schischk.；怯氏麦瓶草 ■☆

363397　Silene dagestanica Rupr.；达吉斯坦蝇子草 ■☆

363398　Silene dasyphylla Turcz. = Silene jenisseensis Willd. ■

363399　Silene davidii（Franch.）Oxelman et Lidén；垫状蝇子草（簇生女娄菜，无茎麦瓶草）；Cution Pink, Moss Campion, Padshape Catchfly ■

363400　Silene dawoensis H. Limpr.；道孚蝇子草；Daofu Catchfly ■

363401　Silene decipiens Barceló；迷惑蝇子草 ■☆

363402　Silene deflexa Eastw. = Silene grayi S. Watson ■☆

363403　Silene delavayi Franch.；西南蝇子草（淳三七，洱源土桔梗，桔梗，凉三七，土桔梗，西南女娄菜）；Delavay Catchfly ■

363404　Silene densiflora d'Urv.；密花蝇子草（密花麦瓶草）；Denseflower Catchfly ■

363405　Silene dentipetala H. Chuang；齿瓣蝇子草；Dentipetal Catchfly ■

363406　Silene depressa M. Bieb.；凹陷蝇子草 ■☆

363407　Silene dewinteri Bocquet；德温特蝇子草 ■☆

363408　Silene dianthoides Pers.；石竹蝇子草 ■☆

363409　Silene dichotoma Ehrh.；双叉麦瓶草（二歧蝇子草）；Dichotoma Silene, Forked Catchfly, Forked Silene, Hairy Catchfly ■☆

363410　Silene dinteri Engl. = Silene burchellii Otth ■☆

363411　Silene dioica（L.）Clairv.；红剪秋罗（异株蝇子草）；Campion, Morning Campion, Red Campion, Red Catchfly, Red Cockle, Robinet Rose ■☆

363412　Silene dioica（L.）Clairv. = Lychnis dioica L. ■☆

363413　Silene dissecta Litard. et Maire；深裂蝇子草 ■☆

363414　Silene disticha Willd.；二列蝇子草 ■☆

363415　Silene divaricata Lag. = Silene aellenii Sennen ■☆

363416　Silene divaricata Lag. var. brachycalyx Pau et Font Quer = Silene aellenii Sennen ■☆

363417　Silene diversifolia Otth；异叶蝇子草 ■☆

363418　Silene dorrii Kellogg = Silene menziesii Hook. ■☆

363419　Silene douglasii Hook.；道格拉斯蝇子草；Douglas' Catchfly ■☆

363420　Silene douglasii Hook. var. brachycalyx B. L. Rob. = Silene douglasii Hook. ■☆

363421　Silene douglasii Hook. var. macounii（S. Watson）B. L. Rob. = Silene parryi（S. Watson）C. L. Hitchc. et Maguire ■☆

363422　Silene douglasii Hook. var. macrocalyx B. L. Rob. = Silene douglasii Hook. ■☆

363423　Silene douglasii Hook. var. monantha（S. Watson）B. L. Rob. = Silene douglasii Hook. ■☆

363424　Silene douglasii Hook. var. multicaulis（Nutt. ex Torr. et A. Gray）B. L. Rob. = Silene douglasii Hook. ■☆

363425　Silene douglasii Hook. var. oraria（M. Peck）C. L. Hitchc. et Maguire；奥拉蝇子草 ■☆

363426　Silene douglasii Hook. var. villosa C. L. Hitchc. et Maguire = Silene douglasii Hook. ■☆

363427　Silene drummondii Hook.；德拉蒙德蝇子草；Drummond's Catchfly, Forked Catchfly ■☆

363428　Silene drummondii Hook. subsp. striata（Rydb.）J. K. Morton；条纹蝇子草 ■☆

363429　Silene drummondii Hook. var. kruckebergii Bocquet = Silene drummondii Hook. ■☆

363430　Silene drummondii Hook. var. striata（Rydb.）Bocquet = Silene drummondii Hook. subsp. striata（Rydb.）J. K. Morton ■☆

363431　Silene dumetosa C. L. Tang；灌丛蝇子草；Shrun Catchfly ■☆

363432　Silene dumicola W. W. Sm. = Silene viscidula Franch. ■

363433　Silene duthiei Majumdar = Silene songarica（Fisch., C. A. Mey. et Avé-Lall.）Bocquet ■

363434　Silene dyris Maire；荒地蝇子草 ■☆

363435　Silene dyris Maire var. lutea Quézel = Silene dyris Maire ■☆

363436　Silene echinata Otth；具刺蝇子草 ■☆

363437　Silene eckloniana Sond.；埃氏蝇子草 ■☆

363438　Silene elisabethae Jan ex Rchb.；意大利雪轮；Large-flowered Catchfly ■☆

363439　Silene engelmannii Rohrb. = Silene bridgesii Rohrb. ■☆

363440　Silene engleri Pax；恩格勒蝇子草 ■☆

363441　Silene epilosa W. W. Sm. = Silene firma Siebold et Zucc. ■

363442　Silene eremitica Boiss.；埃雷米特蝇子草 ■☆

363443　Silene esquamata W. W. Sm.；无鳞蝇子草（无鳞麦瓶草）；Scaleless Catchfly ■

363444　Silene esquirolii H. Lév. = Swertia bimaculata（Siebold et Zucc.）Hook. f. et Thomson ex C. B. Clarke ■

363445 Silene euxina Rupr. ;黑蝇子草■☆

363446 Silene exscapa All. = Silene acaulis L. ■☆

363447 Silene filipetala Litard. et Maire;丝瓣蝇子草■☆

363448 Silene filipetala Litard. et Maire subsp. parviflora Quézel;小花丝瓣蝇子草■☆

363449 Silene filisecta M. Peck = Silene oregana S. Watson ■☆

363450 Silene fimbriata Baldwin ex Elliott = Silene polypetala (Walter) Fernald et B. G. Schub. ■☆

363451 Silene firma Siebold et Zucc. ;坚硬女娄菜(白花女娄菜,粗壮女娄菜,大叶金石榴,光萼女娄菜,女娄菜,疏毛女娄菜,无毛女娄菜,硬叶女娄菜);Hard Catchfly,Hard Melandrium ■

363452 Silene firma Siebold et Zucc. f. pubescens (Makino) M. Mizush. = Silene firma Siebold et Zucc. var. pubescens (Makino) M. Mizush. ■

363453 Silene firma Siebold et Zucc. f. pubescens (Makino) Ohwi et Ohashi = Silene firma Siebold et Zucc. ■

363454 Silene firma Siebold et Zucc. var. pubescens (Makino) M. Mizush. ;疏毛女娄菜■

363455 Silene firma Siebold et Zucc. var. pubescens (Makino) M. Mizush. = Silene firma Siebold et Zucc. ■

363456 Silene fissipetala Turcz. = Silene fortunei Vis. ■

363457 Silene flammulifolia Steud. ex A. Rich. ;焰叶蝇子草■☆

363458 Silene flavovirens C. Y. Wu;黄绿蝇子草(蓝绿蝇子草);Yellowgreen Catchfly ■

363459 Silene flavovirens C. Y. Wu = Melandrium chungtienense (W. W. Sm.) Pax et K. Hoffm. ■

363460 Silene flavovirens C. Y. Wu = Silene chungtienensis W. W. Sm. ■

363461 Silene flos-cuculi (L.) Greuter et Burdet;布谷鸟剪秋罗;Bachelor's Buttons, Billy Buttons, Bobbin Joan, Cock Robin, Cock's Comb, Crow Flower, Cuckoo, Cuckoo Flower, Cuckoo Gilliflower, Cuckooflower, Cuckoo-flower, Darmell Goddard, Drunkards, Fair Maid of France, Fair Maids of France, Gdgson, Gypsy Flower, Indian Pink, Indy, Jan the Crowder, Marsh Gilliflower, Meadow Campion, Meadow Lychnis, Meadow Pink, Meadow Spink, Meadow-lychnis, Pleasant-in-sight, Polly Baker, Poor Robin, Rag-a-tag, Ragged Jack, Ragged Robin, Ragged Urchin, Ragged Willie, Ragged-robin, Red Robin, Robin Hood, Rough Robin, Shaggy Jack, Snake's Flower, Thunder Flower, Wild Beggarman, Wild Sweet William ■☆

363462 Silene flos-jovis (L.) Greuter et Burdet;伞形剪秋罗;Flower of Jove, Flower-of-Jove, Flower-of-Jupiter, Jove's Flower ■☆

363463 Silene foliosa Maxim. ;石缝蝇子草(叶麦瓶草);Leafy Catchfly ■

363464 Silene foliosa Maxim. var. macrostyla (Maxim.) Rohrb. = Silene macrostyla Maxim. ■

363465 Silene foliosa Maxim. var. mongolica Maxim. ;小花石缝蝇子草■

363466 Silene foliosa Maxim. var. mongolica Maxim. = Silene foliosa Maxim. ■

363467 Silene fortunei Vis. ;鹤草(八月白,白柴胡,白葫芦,白花壶瓶,白花瞿麦,白接骨丹,苍蝇花,古绵草,瞿麦沙参,麦瓶草,蛇王草,水白参,土桔梗,脱力草,蚊子草,小仙桃草,小叶鲤鱼胆,野蚊子草,银柴胡,蝇子草,羽毛石竹,黏蝇草,黏蝇花);Catchfly, Crane Catchfly ■

363468 Silene fortunei Vis. var. kiruninsularis (Masam.) S. S. Ying;基隆蝇子草■

363469 Silene fruticosa L. ;西方灌丛蝇子草■☆

363470 Silene fruticosa L. subsp. cyrenaica Bég. et Vacc. ;昔兰尼灌丛蝇子草■☆

363471 Silene fruticulosa (Pall.) Schischk. ex Krylov = Silene altaica Pers. ■

363472 Silene fulgens (Fisch. ex Spreng.) E. H. L. Krause;剪秋罗(大花剪秋罗,浅裂剪秋罗);Brilliant Campion, Brilliant Lychnis ■

363473 Silene fulgens (Spreng.) E. H. L. Krause = Lychnis cognata Maxim. ■

363474 Silene fulgens (Spreng.) E. H. L. Krause = Lychnis fulgens Fisch. ■

363475 Silene furcata Raf. = Silene involucrata (Cham. et Schltdl.) Bocquet ■☆

363476 Silene fuscata Brot. ;暗棕蝇子草■☆

363477 Silene gallica L. ;蝇子草(白花蝇子草,西欧蝇子草,英国麦瓶草);Campion, Catchfly, Common Catchfly, English Catch, English Catchfly, Fly, French Catchfly, French Silene, Rose Campion, Small-flowered Catchfly ■

363478 Silene gallica L. var. agrestina Jord. et Fourr. = Silene gallica L. ■

363479 Silene gallica L. var. giraldii (Guss.) Walters;意大利蝇子草■☆

363480 Silene gallica L. var. lusitanica (L.) Willk. et Lange = Silene gallica L. ■

363481 Silene gallica L. var. minor Ball = Silene gallica L. ■

363482 Silene gallica L. var. modesta (Jord. et Fourr.) Rouy et Foucaud = Silene gallica L. ■

363483 Silene gallica L. var. quinquevulnera (L.) W. D. J. Koch;五痕蝇子草■☆

363484 Silene gallica L. var. quinquevulnera Mert. et Koch = Silene gallica L. ■

363485 Silene gallica L. var. sylvestris (Lam.) Asch. = Silene gallica L. ■

363486 Silene gasimailikensis B. Fedtsch. ;嘎西蝇子草■☆

363487 Silene gawrilowii Krasn. ;嘎氏蝇子草■☆

363488 Silene gazulensis Galán-Mera et al. ;西班牙蝇子草■☆

363489 Silene gebleriana Schrenk;线叶蝇子草■

363490 Silene getula Pomel = Silene vivianii Steud. subsp. getula (Pomel) Greuter et Burdet ■☆

363491 Silene ghiarensis Batt. ;摩洛哥蝇子草■☆

363492 Silene gibraltarica Boiss. = Silene andryalifolia Pomel ■☆

363493 Silene gibraltarica Boiss. var. papillosa Emb. = Silene andryalifolia Pomel ■☆

363494 Silene gillettii (Turrill) M. G. Gilbert;吉莱特蝇子草■☆

363495 Silene giraldii Guss. = Silene gallica L. var. giraldii (Guss.) Walters ■☆

363496 Silene githago (L.) Clairv. = Agrostemma githago L. ■

363497 Silene glabella (Ohwi) S. S. Ying et Alp. ;南湖大山蝇子草(秃玉山蝇子草)■

363498 Silene glaberrima Faure et Maire;无毛蝇子草■☆

363499 Silene glabrescens Coss. ;渐光蝇子草■☆

363500 Silene glauca Pourr. = Silene secundiflora Otth ■☆

363501 Silene glauca Pourr. subsp. macrotheca (Braun-Blanq. et Maire) Maire = Silene secundiflora Otth ■☆

363502 Silene glaucescens Schischk. ;灰绿蝇子草■☆

363503 Silene gonosperma (Rupr.) Bocquet;隐瓣蝇子草■

363504 Silene gonosperma (Rupr.) Bocquet = Melandrium apetalum (L.) Fenzl ex Ledeb. ■

363505 Silene gonosperma (Rupr.) Bocquet subsp. himalayensis (Rohrb.) Bocquet = Silene himalayensis (Edgew.) Majumdar ■

363506 Silene gonosperma (Rupr.) Bocquet var. himalayensis Bocquet = Silene himalayensis (Edgew.) Majumdar ■

363507 Silene gormanii Howell = Silene oregana S. Watson ■☆

363508　Silene gracilenta H. Chuang;纤小蝇子草;Thin Catchfly ■

363509　Silene gracilicaulis C. L. Tang;细蝇子草(滇瞿麦,九头草,细麦瓶草);Small Catchfly ■

363510　Silene gracilicaulis C. L. Tang var. longipedicellata C. L. Tang;长梗细蝇子草(长梗细麦瓶草)■

363511　Silene gracilicaulis C. L. Tang var. longipedicellata C. L. Tang = Silene gracilicaulis C. L. Tang ■

363512　Silene gracilicaulis C. L. Tang var. rubescens (Franch.) C. L. Tang;大花细蝇子草■

363513　Silene gracilicaulis C. L. Tang var. rubescens (Franch.) C. L. Tang = Silene gracilicaulis C. L. Tang ■

363514　Silene gracilis ꞈC.;纤细蝇子草■☆

363515　Silene gracillima Rohrb.;细长蝇子草■☆

363516　Silene graminifolia Otth;禾叶蝇子草(稻叶麦瓶草,滇瞿麦,禾苗蝇子草,禾叶麦瓶草,金柴胡,九头草,瞿麦,癞头参,马柴胡,毛柱蝇子草,细麦瓶草,细叶蝇子草,细蝇子草,纤细鹤草,纤细蝇子草,小九股牛,兴安旱麦瓶草,竹节防风);Grassleaf Catchfly, Hairystyle Catchfly ■

363517　Silene graminifolia Otth var. parviflora Ledeb. = Silene jenisseensis Willd. ■

363518　Silene graminoidea C. Y. Wu et C. L. Tang = Silene chodatii Bocquet var. pygmaea Bocquet ■

363519　Silene grandiflora Franch.;大花蝇子草(大花女娄菜,金蝴蝶);Bigflower Catchfly, Largeflower Melandrium ■

363520　Silene grandis Eastw. = Silene scouleri Hook. ■☆

363521　Silene grayi S. Watson;格雷蝇子草;Gray's Catchfly ■☆

363522　Silene greggii A. Gray = Silene laciniata Cav. subsp. greggii (A. Gray) C. L. Hitchc. et Maguire ■☆

363523　Silene griffithii Boiss. = Silene suaveolens Turcz. ex Kar. et Kir. ■

363524　Silene grosiana Pau et Font Quer = Silene ibosii Emb. et Maire subsp. grosiana (Pau et Font Quer) Fern. Casas ■☆

363525　Silene grossheimii Schischk.;格罗氏蝇子草■☆

363526　Silene guedirensis Pau = Silene micropetala Lag. ■☆

363527　Silene guichardii Chevassut et Quézel;吉夏尔蝇子草■☆

363528　Silene guntensis B. Fedtsch.;古恩特蝇子草■☆

363529　Silene gyirongensis L. H. Zhou;吉隆蝇子草;Jilong Catchfly ■

363530　Silene gyirongensis L. H. Zhou = Silene moorcroftiana Wall. ex Benth. ■

363531　Silene habaensis H. Chuang;哈巴蝇子草;Haba Catchfly ■

363532　Silene hallii S. Watson = Silene scouleri Hook. subsp. hallii (S. Watson) C. L. Hitchc. et Maguire ■☆

363533　Silene hellmannii Claus;亥曼氏麦瓶草■☆

363534　Silene heptapotamica Schischk.;黏蝇子草■☆

363535　Silene herbilegorum (Bocquet) Lidén et Oxelman;多裂腺毛蝇子草;Manylobed Tentacle Catchfly ■

363536　Silene heterodonta F. N. Williams;异齿蝇子草■☆

363537　Silene heterodonta F. N. Williams subsp. parvula (Coss.) Maire et Weiller = Silene heterodonta F. N. Williams ■☆

363538　Silene heterodonta F. N. Williams subsp. platycalyx (Emb. et Maire) Maire;阔萼异齿蝇子草■☆

363539　Silene heterodonta F. N. Williams subsp. rosella (Maire) Maire;粉红异齿蝇子草■☆

363540　Silene heterodonta F. N. Williams var. cossoniana Maire = Silene heterodonta F. N. Williams ■☆

363541　Silene heterodonta F. N. Williams var. platycalyx Emb. et Maire = Silene heterodonta F. N. Williams subsp. platycalyx (Emb. et Maire) Maire ■☆

363542　Silene heterodonta F. N. Williams var. thomsonii Maire = Silene heterodonta F. N. Williams ■☆

363543　Silene hidaka-alpina (Miyabe et Tatew.) Ohwi et H. Ohashi;日高山蝇子草■☆

363544　Silene himalayensis (Edgew.) Majumdar;喜马拉雅蝇子草(须弥蝇子草);Himalayas Catchfly ■

363545　Silene himalayensis (Rohrb.) Majumdar = Silene himalayensis (Edgew.) Majumdar ■

363546　Silene hirsuta Poir. = Silene bellidifolia Jacq. ■☆

363547　Silene hispida Desf. = Silene bellidifolia Jacq. ■☆

363548　Silene hitchguirei Bocquet;希氏蝇子草;Mountain Campion ■☆

363549　Silene hochstetteri Rohrb.;霍赫蝇子草■☆

363550　Silene hoggariensis Quézel = Silene lynesii Norman ■☆

363551　Silene holopetala Bunge;全缘蝇子草;Entire Catchfly ■

363552　Silene hookeri Nutt.;虎克氏雪轮;Hooker's Indian Pink ■☆

363553　Silene hookeri Nutt. subsp. bolanderi (A. Gray) Abrams;鲍氏蝇子草;Bolander's Indian Pink ■☆

363554　Silene hookeri Nutt. subsp. pulverulenta (M. Peck) C. L. Hitchc. et Maguire = Silene hookeri Nutt. ■☆

363555　Silene huguettiae Bocquet;狭果蝇子草;Narrowfruit Catchfly ■

363556　Silene huguettiae Bocquet var. pilosa C. Y. Wu et H. Chuang;无腺狭果蝇子草(毛狭瓣蝇子草)■

363557　Silene humilis C. A. Mey.;湿地蝇子草■☆

363558　Silene huochenensis X. M. Pi et X. L. Pan;霍城蝇子草;Huocheng Catchfly ■

363559　Silene hupehensis C. L. Tang;湖北蝇子草;Hubei Catchfly ■

363560　Silene hupehensis C. L. Tang var. pubescens C. L. Tang;毛湖北蝇子草;Pubescent Hubei Catchfly ■

363561　Silene iberica M. Bieb.;伊比利亚蝇子草■☆

363562　Silene ibosii Emb. et Maire subsp. grosiana (Pau et Font Quer) Fern. Casas;格罗斯蝇子草■☆

363563　Silene imbricata Desf.;覆瓦蝇子草■☆

363564　Silene incisa C. L. Tang;锐裂蝇子草(齿瓣蝇子草);Toothpetal Catchfly ■

363565　Silene incompta A. Gray = Silene bridgesii Rohrb. ■☆

363566　Silene incurvifolia Kar. et Kir.;镰叶蝇子草(内弯蝇子草);Sickleleaf Catchfly ■

363567　Silene indica Roxb. = Silene indica Roxb. ex Otth ■

363568　Silene indica Roxb. ex Otth;印度蝇子草;India Catchfly ■

363569　Silene indica Roxb. ex Otth var. bhutanica (W. W. Sm.) Bocquet;不丹蝇子草;Bhutan Catchfly ■

363570　Silene inflata (Salisb.) Sm. = Silene vulgaris (Moench) Garcke ■

363571　Silene inflata Sm.;膀胱蝇子草■☆

363572　Silene inflata Sm. = Silene vulgaris (Moench) Garcke ■

363573　Silene inflata Sm. var. angustifolia (Guss.) DC. = Silene vulgaris (Moench) Garcke ■

363574　Silene inflata Sm. var. rubriflora Ball = Silene vulgaris (Moench) Garcke ■

363575　Silene inflata Sm. var. vulgaris Turcz. = Silene vulgaris (Moench) Garcke ■

363576　Silene ingramii Tidestr. et Dayton = Silene hookeri Nutt. ■☆

363577　Silene insectivora L. F. Hend. = Silene nuda (S. Watson) C. L. Hitchc. et Maguire ■☆

363578　Silene invisa C. L. Hitchc. et Maguire;短梗蝇子草;Short-petalled Campion ■☆

363579　Silene involucrata (Cham. et Schltdl.) Bocquet;总苞蝇子草;

Arctic Campion, Greater Arctic Campion ■☆

363580　Silene involucrata（Cham. et Schltdl.）Bocquet subsp. elatior（Regel）Bocquet = Silene involucrata（Cham. et Schltdl.）Bocquet ■☆

363581　Silene involucrata（Cham. et Schltdl.）Bocquet subsp. tenella（Tolm.）Bocquet；细总苞蝇子草；Taylor's Arctic Campion ■☆

363582　Silene ispirensis Boiss.；伊斯匹尔蝇子草 ■☆

363583　Silene italica（L.）Pers.；意大利麦瓶草；Italian Catchfly, Italian Silene, Italy Catchfly, Sticky White Catchfly, Viscous Campion, Viscous Catchfly, White Sticky Catchfly ■☆

363584　Silene italica（L.）Pers. subsp. fontanesiana Maire = Silene patula Desf. ■☆

363585　Silene italica（L.）Pers. var. adenocalyx Emb. et Maire = Silene patula Desf. ■☆

363586　Silene italica（L.）Pers. var. amurensis（Pomel）Batt. = Silene patula Desf. subsp. amurensis（Pomel）Jeanm. ■☆

363587　Silene italica（L.）Pers. var. brevipes Maire = Silene patula Desf. ■☆

363588　Silene italica（L.）Pers. var. cyrenaica Maire et Weiller = Silene patula Desf. ■☆

363589　Silene italica（L.）Pers. var. denticulata Maire = Silene patula Desf. ■☆

363590　Silene italica（L.）Pers. var. hesperia Maire = Silene patula Desf. subsp. amurensis（Pomel）Jeanm. ■☆

363591　Silene italica（L.）Pers. var. leonum Maire = Silene patula Desf. ■☆

363592　Silene italica（L.）Pers. var. maura Maire = Silene patula Desf. ■☆

363593　Silene italica（L.）Pers. var. mellifera（Boiss. et Reut.）Bonnet et Barratte = Silene italica（L.）Pers. ■☆

363594　Silene italica（L.）Pers. var. patula（Desf.）Maire = Silene patula Desf. ■☆

363595　Silene italica（L.）Pers. var. pogonocalyx Svent. = Silene patula Desf. ■☆

363596　Silene italica（L.）Pers. var. rhodantha Maire = Silene patula Desf. ■☆

363597　Silene italica（L.）Pers. var. rosea Maire = Silene patula Desf. ■☆

363598　Silene italica（L.）Pers. var. wallii Maire = Silene patula Desf. ■☆

363599　Silene italica Pers. = Silene italica（L.）Pers. ■☆

363600　Silene jenissea Poir. = Silene jenisseensis Willd. ■

363601　Silene jenissea Steph. ex Bunge = Silene jenisseensis Willd. ■

363602　Silene jenissea Steph. ex Bunge var. parviflora Turcz. = Silene jenisseensis Willd. ■

363603　Silene jenisseensis Steph. ex Bunge f. parviflora（Turcz.）Schischk. = Silene jenisseensis Willd. ■

363604　Silene jenisseensis Steph. ex Bunge var. oliganthella（Nakai ex Kitag.）Y. C. Chu = Silene jenisseensis Willd. ■

363605　Silene jenisseensis Steph. ex Bunge var. vegetior Popov = Silene jenisseensis Willd. ■

363606　Silene jenisseensis Willd.；山蚂蚱草（旱麦瓶草，麦瓶草，山蚂蚱，叶尼塞蝇子草，银柴胡）；Dry Catchfly ■

363607　Silene jenisseensis Willd. f. dasyphylla（Turcz.）Schischk.；薄毛旱麦瓶草 ■

363608　Silene jenisseensis Willd. f. latifolia（Turcz.）Schischk. = Silene jenisseensis Willd. var. latifolia（Turcz.）Y. Z. Zhao ■

363609　Silene jenisseensis Willd. f. parviflora（Turcz.）Schischk.；小花旱麦瓶草（小花山蚂蚱草）；Littleflower Dry Catchfly, Smallflower Dry Catchfly ■

363610　Silene jenisseensis Willd. f. parviflora（Turcz.）Schischk. = Silene jenisseensis Willd. ■

363611　Silene jenisseensis Willd. f. setifolia（Turcz.）Schischk.；丝叶山蚂蚱草（丝叶旱麦瓶草）■

363612　Silene jenisseensis Willd. f. setifolia（Turcz.）Schischk. = Silene jenisseensis Willd. ■

363613　Silene jenisseensis Willd. var. latifolia（Turcz.）Y. Z. Zhao；宽叶旱麦瓶草；Broadleaf Dry Catchfly ■

363614　Silene jenisseensis Willd. var. oliganthella（Nakai ex Kitag.）Y. C. Chu = Silene jenisseensis Willd. ■

363615　Silene jenisseensis Willd. var. oliganthella（Nakai ex Kitag.）Y. C. Chu；长白旱麦瓶草（长白山蚂蚱草）；Changbaishan Dry Catchfly ■

363616　Silene jenisseensis Willd. var. vegetior Popov = Silene jenisseensis Willd. ■

363617　Silene jenisseensis Willd. var. viscifera Y. C. Chu；兴安旱麦瓶草（兴安麦瓶草）■

363618　Silene jenisseensis Willd. var. viscifera Y. C. Chu = Silene graminifolia Otth ■

363619　Silene kantzeensis C. L. Tang = Silene davidii（Franch.）Oxelman et Lidén ■

363620　Silene karaczukuri B. Fedtsch.；喀拉蝇子草（紫花蝇子草）；Karaczukur Catchfly ■

363621　Silene karekirii Bocquet；污色蝇子草；Karekir Catchfly ■

363622　Silene keiskei Miq.；伊藤氏蝇子草 ■☆

363623　Silene keiskei Miq. var. akaisialpina（T. Yamaz.）Ohwi et H. Ohashi f. leucantha（Takeda）Ohwi et H. Ohashi = Silene akaisialpina（T. Yamaz.）H. Ohashi, Tateishi et H. Nakai f. leucantha（Takeda）H. Ohashi, Tateishi et H. Nakai ■☆

363624　Silene keiskei Miq. var. akaisialpina（T. Yamaz.）Ohwi et H. Ohashi = Silene akaisialpina（T. Yamaz.）H. Ohashi, Tateishi et H. Nakai ■☆

363625　Silene keiskei Miq. var. minor（Takeda）Ohwi et H. Ohashi；小伊藤氏蝇子草 ■☆

363626　Silene keiskei Miq. var. minor（Takeda）Ohwi et H. Ohashi f. albescens（Takeda）H. Ohashi, Tateishi et H. Nakai；白小伊藤氏蝇子草 ■☆

363627　Silene keiskei Miq. var. minor（Takeda）Ohwi et H. Ohashi；平铺小伊藤氏蝇子草 ■☆

363628　Silene keiskei Miq. var. procumbens Takeda = Silene keiskei Miq. var. minor（Takeda）Ohwi et H. Ohashi f. procumbens（Takeda）Ohwi et H. Ohashi ■☆

363629　Silene kermesina W. W. Sm.；卡里蝇子草（红花女娄菜）；Kali Catchfly ■

363630　Silene kermesina W. W. Sm. = Silene asclepiadea Franch. ■

363631　Silene khasyana Rohrb.；卡西亚蝇子草；Khas Catchfly ■

363632　Silene kialensis（F. N. Williams）Lidén et Oxelman；甲拉蝇子草 ■

363633　Silene kialensis Rohrb. = Silene nepalensis Majumdar var. kialensis（F. N. Williams）C. L. Tang ■

363634　Silene kiiruninsularis Masam. = Silene fortunei Vis. ■

363635　Silene kilianii Maire = Silene lynesii Norman ■☆

363636　Silene kingii（S. Watson）Bocquet；金氏蝇子草；King's

Catchfly ■☆

363637 Silene kiruninsularis Masam. = Silene fortunei Vis. var. kiruninsularis（Masam.）S. S. Ying ■

363638 Silene kiusiana（Makino）H. Ohashi et H. Nakai；九州蝇子草（九州剪秋罗）■☆

363639 Silene kiusiana（Makino）H. Ohashi et H. Nakai = Lychnis kiusiana Makino ■☆

363640 Silene komarovii Schischk.；轮伞蝇子草；Komarov Catchfly ■

363641 Silene koreana Kom.；朝鲜蝇子草（朝鲜麦瓶草）；Korea Catchfly ■

363642 Silene korshinskyi Schischk.；考尔蝇子草■☆

363643 Silene kubanensis Sommier et H. Lév.；库班蝇子草■☆

363644 Silene kudrjaschevii Schischk.；库德氏蝇子草■☆

363645 Silene kumaensis F. N. Williams = Silene adenocalyx F. N. Williams ■

363646 Silene kungessana B. Fedtsch.；巩乃斯蝇子草；Gongnaisi Catchfly ■

363647 Silene kuschakewiczii Regel；库沙蝇子草■☆

363648 Silene lacera（Stev.）Sims.；撕裂蝇子草■☆

363649 Silene laciniata Cav.；裂瓣雪轮；Mexican Campion, Mexican Catchfly ■☆

363650 Silene laciniata Cav. subsp. brandegeei C. L. Hitchc. et Maguire = Silene laciniata Cav. ■☆

363651 Silene laciniata Cav. subsp. californica（Durand）J. K. Morton；加州裂瓣雪轮；California Pink ■☆

363652 Silene laciniata Cav. subsp. greggii（A. Gray）C. L. Hitchc. et Maguire；格雷格裂瓣雪轮；Gregg's Campion, Gregg's Mexican Pink ■☆

363653 Silene laciniata Cav. subsp. major C. L. Hitchc. et Maguire = Silene laciniata Cav. ■☆

363654 Silene laciniata Cav. var. angustifolia C. L. Hitchc. et Maguire = Silene laciniata Cav. ■☆

363655 Silene laciniata Cav. var. angustifolia Hitchc. et Maguire；狭叶裂瓣雪轮■☆

363656 Silene laciniata Cav. var. californica（Durand）A. Gray = Silene laciniata Cav. subsp. californica（Durand）J. K. Morton ■☆

363657 Silene laciniata Cav. var. greggii（A. Gray）S. Watson = Silene laciniata Cav. subsp. greggii（A. Gray）C. L. Hitchc. et Maguire ■☆

363658 Silene laciniata Cav. var. latifolia C. L. Hitchc. et Maguire = Silene laciniata Cav. ■☆

363659 Silene lacustris Eastw. = Silene sargentii S. Watson ■☆

363660 Silene laeta（Aiton）Godr.；愉悦蝇子草■☆

363661 Silene laeta（Aiton）Godr. var. loiseleurii Rouy et Foucaud = Silene laeta（Aiton）Godr. ■☆

363662 Silene lagrangei（Coss.）Greuter et Burdet；拉格朗热蝇子草■☆

363663 Silene lagunensis Link；拉古纳蝇子草■☆

363664 Silene lamarum C. Y. Wu；喇嘛蝇子草；Lama Catchfly ■

363665 Silene lankongensis Franch.；洱源蝇子草（滇白前，洱源瓦草，瓦草）；Lankong Catchfly ■

363666 Silene lankongensis Franch. = Silene viscidula Franch. ■

363667 Silene lasiantha K. Koch；毛花蝇子草■☆

363668 Silene lasiostyla Boiss. = Silene psammitis Spreng. subsp. lasiostyla（Boiss.）Rivas Goday ■☆

363669 Silene latifolia（Mill.）Britten et Rendle = Silene vulgaris（Moench）Garcke ■

363670 Silene latifolia（Mill.）Britten et Rendle var. pubescens（DC.）

Farw. = Silene vulgaris（Moench）Garcke ■

363671 Silene latifolia（Mill.）Rendle et Britten；叉枝蝇子草；Ben, Blandder Campion, Blandder Catchfly, Campion, Cowbell, Evening Campion, White Ben, White Campion, White Cockle ■

363672 Silene latifolia（Mill.）Rendle et Britten = Silene pratensis（Raf.）Gren. et Godr. ■

363673 Silene latifolia（Mill.）Rendle et Britten subsp. alba（Mill.）Greuter et Burdet；白叉枝蝇子草（白花剪秋罗，白花蝇子草，白剪秋罗，白女娄菜，白色蝇子草，异株女娄菜）；Bachelor's Button, Bachelor's Buttons, Billy Buttons, Bladder Campion, Bull Rattle, Butcher, Cockle, Cow Mack, Cow Rattle, Cowmack, Darmell Goddard, Evening Campion, Evening Close, Evening Lychnis, Gooseberry Pie, Granny's Nightcap, Milk Flower, Milkmaids, Plum Pudding, Poor Jane, Shades of Evening, Shades-of-evening, Shirt Buttons, Snake's Flower, Snapjack, Summer Saucers, Thunder Flower, Thunderbolts, White Bachelor's Buttons, White Campion, White Catchfly, White Cockle, White Robin, White Robin Hood ■

363674 Silene latifolia Poir. = Silene latifolia（Mill.）Rendle et Britten ■

363675 Silene latifolia Poir. subsp. alba（Mill.）Greuter et Burdet = Silene latifolia（Mill.）Rendle et Britten subsp. alba（Mill.）Greuter et Burdet ■

363676 Silene latifolia Poir. subsp. alba（Mill.）Greuter et Burdet = Silene latifolia（Mill.）Rendle et Britten ■

363677 Silene laxiflora Brot. = Silene micropetala Lag. ■☆

363678 Silene laxiflora Brot. var. fairchildiana Maire = Silene micropetala Lag. ■☆

363679 Silene laxiflora Brot. var. micropetala（Lag.）Maire = Silene micropetala Lag. ■☆

363680 Silene lazica Boiss.；拉扎蝇子草■☆

363681 Silene lemmonii S. Watson；莱蒙蝇子草；Lemmon's Catchfly ■☆

363682 Silene leptocaulis Schischk.；细茎蝇子草■☆

363683 Silene lhassana（F. N. Williams）Majumdar；拉萨蝇子草；Lasa Catchfly ■

363684 Silene lichiangensis W. W. Sm.；丽江蝇子草；Lijiang Catchfly ■

363685 Silene ligulata Viv. = Silene vivianii Steud. ■☆

363686 Silene linearifolia Otth.；丝叶蝇子草■☆

363687 Silene linearifolia Pamp. = Silene hupehensis C. L. Tang ■

363688 Silene lineariloba C. Y. Wu；线瓣蝇子草；Linearpetap Catchfly ■

363689 Silene linearis Decne.；线状蝇子草■☆

363690 Silene linicola C. C. Gmel.；亚麻地蝇子草；Flaxfield Catchfly ■☆

363691 Silene linnaeana Volosch.；林奈蝇子草（西伯利亚剪秋罗，西伯利亚麦瓶草，西伯利亚蝇子草，狭叶剪秋罗，云南剪秋罗）；Narrowleaf Campion, Siberia Catchfly, Siberian Campion ■

363692 Silene lithophila Kar. et Kir.；喜岩蝇子草■

363693 Silene littorea Brot.；滨海蝇子草■☆

363694 Silene litwinowii Schischk.；利特氏蝇子草■☆

363695 Silene lomalasinensis（Engl.）T. M. Harris et Goyder；洛马蝇子草■☆

363696 Silene lomalasinensis Engl. ex Jaeger = Silene lomalasinense（Engl.）T. M. Harris et Goyder ■☆

363697 Silene longicaulis Lag. = Silene gracilis DC. ■☆

363698 Silene longicaulis Lag. var. brachypoda Maire = Silene gracilis DC. ■☆

363699 Silene longicornuta C. Y. Wu et C. L. Tang；长角蝇子草；Longhorn Catchfly ■

363700　Silene longidens Schischk. ;长齿蝇子草■☆

363701　Silene longiflora Ehrh. ;长花麦瓶草■☆

363702　Silene longipes（Hand.-Mazz.）C. Y. Wu = Silene melanantha Franch. ■

363703　Silene longipetala Vent. ;长瓣蝇子草■☆

363704　Silene longistylis Engelm. ex S. Watson = Silene bridgesii Rohrb. ■☆

363705　Silene longitubulosa Engl. = Silene macrosolen Steud. ex A. Rich. ■☆

363706　Silene longiuscula C. Y. Wu et C. L. Tang;长花蝇子草;Longflower Catchfly ■

363707　Silene longiuscula C. Y. Wu et C. L. Tang = Silene dawoensis H. Limpr. ■

363708　Silene luisana S. Watson = Silene verecunda S. Watson ■☆

363709　Silene lusitanica L. = Silene gallica L. ■

363710　Silene lutea Franch. = Silene asclepiadea Franch. ■

363711　Silene lyallii S. Watson = Silene douglasii Hook. ■☆

363712　Silene lychnidea C. A. Mey. ;剪秋罗蝇子草■☆

363713　Silene lynesii Norman;莱恩斯蝇子草■☆

363714　Silene macounii S. Watson = Silene parryi（S. Watson）C. L. Hitchc. et Maguire ■☆

363715　Silene macrocalyx（B. L. Rob.）Howell = Silene douglasii Hook. ■☆

363716　Silene macrosolen Steud. ex A. Rich. ;大管蝇子草■☆

363717　Silene macrosperma（A. E. Porsild）Hultén = Silene uralensis（Rupr.）Bocquet subsp. porsildii Bocquet ■☆

363718　Silene macrostyla Maxim. ;长柱蝇子草（长柱麦瓶草,万年蒿）;Longstyle Catchfly ■

363719　Silene macrotheca Braun-Blanq. et Maire = Silene secundiflora Otth subsp. macrotheca（Braun-Blanq. et Maire）Greuter et Burdet ■☆

363720　Silene madens Majumdar = Silene himalayensis（Edgew.）Majumdar ■

363721　Silene maheshwarii Bocquet = Silene caespitella F. N. Williams ■

363722　Silene mairei H. Lév. = Silene viscidula Franch. ■

363723　Silene marcowiczii Schischk. ;马尔考蝇子草■☆

363724　Silene maritima Host;海滨蝇子草■☆

363725　Silene maritima With. = Silene uniflora Roth ■☆

363726　Silene markamensis L. H. Zhou;马尔康蝇子草（芒康蝇子草）;Markang Catchfly ■

363727　Silene marmarica Bég. et Vacc. ;利比亚蝇子草■☆

363728　Silene marmorensis Kruckeb. ;马莫拉蝇子草;Marble Mountain Bar Campion,Somes Bar Campion ■☆

363729　Silene maroccana Coss. = Silene vivianii Steud. subsp. getula（Pomel）Greuter et Burdet ■☆

363730　Silene marschallii C. A. Mey. ;马尔蝇子草■☆

363731　Silene mauritanica Pomel = Silene obtusifolia Willd. ■☆

363732　Silene mauritii Sennen;毛里特蝇子草■☆

363733　Silene maurorum Batt. et Pit. ;晚熟蝇子草■☆

363734　Silene maximowicziana Kozhevn. = Silene foliosa Maxim. ■

363735　Silene media（Litv.）Kleopow;中型蝇子草（中型麦瓶草）■☆

363736　Silene mekinensis Coss. ;梅金噶蝇子草■☆

363737　Silene melanantha Franch. ;黑花蝇子草（黑花女娄菜）;Blackflower Catchfly ■

363738　Silene melandriformis Maxim. = Silene aprica Turcz. ex Fisch. et C. A. Mey. ■

363739　Silene mentagensis Coss. ;门塔噶蝇子草■☆

363740　Silene mentagensis Coss. var. robusta Maire = Silene mentagensis Coss. ■☆

363741　Silene menziesii Hook. ;孟席斯蝇子草;Menzies' Catchfly ■☆

363742　Silene menziesii Hook. subsp. dorrii（Kellogg）C. L. Hitchc. et Maguire = Silene menziesii Hook. ■☆

363743　Silene menziesii Hook. subsp. williamsii（Britton）Hultén = Silene williamsii Britton ■☆

363744　Silene menziesii Hook. var. viscosa（Greene）C. L. Hitchc. et Maguire = Silene menziesii Hook. ■☆

363745　Silene menziesii Hook. var. williamsii（Britton）B. Boivin = Silene williamsii Britton ■☆

363746　Silene meruensis Engl. = Silene burchellii Otth ■☆

363747　Silene mesatlantica Maire;梅萨蝇子草■☆

363748　Silene mesatlantica Maire var. embergeri ? = Silene mesatlantica Maire ■☆

363749　Silene mesatlantica Maire var. ibrahimiana（Emb.）Maire = Silene mesatlantica Maire ■☆

363750　Silene mesatlantica Maire var. macrotricha Emb. et Maire = Silene mesatlantica Maire ■☆

363751　Silene mexicana Otth = Silene laciniata Cav. ■☆

363752　Silene meyeri Fenzl;迈氏蝇子草■☆

363753　Silene michelsonii Preobr. ;米氏蝇子草■☆

363754　Silene micropetala Lag. ;小瓣蝇子草■☆

363755　Silene miqueliana（Rohrb.）H. Ohashi et H. Nakai;女娄菜剪秋罗（全缘剪秋罗）■☆

363756　Silene miqueliana（Rohrb.）H. Ohashi et H. Nakai = Lychnis miqueliana Rohrb. ■☆

363757　Silene miqueliana（Rohrb.）H. Ohashi et H. Nakai f. argyrata（M. Mizush.）H. Ohashi et H. Nakai;白色女娄菜剪秋罗■☆

363758　Silene miqueliana（Rohrb.）H. Ohashi et H. Nakai f. plena（Makino）H. Ohashi et H. Nakai;重瓣女娄菜剪秋罗■☆

363759　Silene mirei Chevassut et Quézel;米雷蝇子草■☆

363760　Silene mogadorensis Pit. = Silene obtusifolia Willd. ■☆

363761　Silene mollissima（L.）Pers. ;柔软蝇子草■☆

363762　Silene mollissima（L.）Pers. subsp. auriculifolia（Pomel）Maire = Silene auriculifolia Pomel ■☆

363763　Silene mollissima（L.）Pers. subsp. gibraltarica（Boiss.）Maire = Silene andryalifolia Pomel ■☆

363764　Silene mollissima（L.）Pers. subsp. velutina（Pourr.）Maire = Silene andryalifolia Pomel ■☆

363765　Silene mollissima（L.）Pers. var. gibraltarica（Boiss.）Ball = Silene andryalifolia Pomel ■☆

363766　Silene mollissima（L.）Pers. var. maroccana Maire = Silene andryalifolia Pomel ■☆

363767　Silene mollissima（L.）Pers. var. oranensis Maire = Polycarpon polycarpoides（Biv.）Fiori ■☆

363768　Silene monantha S. Watson = Silene douglasii Hook. ■☆

363769　Silene monbeigii W. W. Sm. ;沧江蝇子草（滇西蝇子草）;Monbeig Catchfly ■

363770　Silene montana S. Watson;山地蝇子草;Mountain Catchfly ■☆

363771　Silene moorcroftiana Wall. ex Benth. ;冈底斯山蝇子草（西藏蝇子草）;Gangdise Catchfly ■

363772　Silene morii Hayata;台湾蝇子草■

363773　Silene morii Hayata = Silene aprica Turcz. ex Fisch. et C. A. Mey. ■

363774　Silene morrison-montana（Hayata）Ohwi et H. Ohashi;玉山蝇子草（新高山女娄菜）;Yushan Catchfly ■

363775 Silene morrison-montana（Hayata）Ohwi et H. Ohashi var. glabella（Ohwi）Ohwi et H. Ohashi = Silene glabella（Ohwi）S. S. Ying et Alp. ■

363776 Silene morrison-montana（Hayata）Ohwi et H. Ohashi var. glabella（Ohwi）Ohwi et H. Ohashi；禿玉山蝇子草■

363777 Silene morrison-montana（Hayata）Ohwi et H. Ohashi var. transalpine（Hayata）S. S. Ying = Silene transalpine（Hayata）S. S. Ying ■

363778 Silene muliensis C. Y. Wu；木里蝇子草；Muli Catchfly ■

363779 Silene multicaulis Nutt. ex Torr. et A. Gray = Silene douglasii Hook. ■☆

363780 Silene multifida（Adams）Rohrb.；多裂蝇子草■☆

363781 Silene multiflora Pers.；多花麦瓶草；Manyflower ■☆

363782 Silene multifurcata C. L. Tang；花脉蝇子草；Manyfork Catchfly ■

363783 Silene multinervia S. Watson = Silene coniflora Nees ex Otth. ■☆

363784 Silene mundiana Eckl. et Zeyh.；蒙德蝇子草■☆

363785 Silene muscipula L.；鼠夹蝇子草■☆

363786 Silene muscipula L. subsp. deserticola Murb.；荒漠蝇子草■☆

363787 Silene muscipula L. var. angustifolia Costa = Silene muscipula L. ■☆

363788 Silene muscipula L. var. arvensis（Loscos）Rouy et Foucaud = Silene muscipula L. ■☆

363789 Silene muscipula L. var. bracteosa（Bertol.）Rouy et Foucaud = Silene muscipula L. ■☆

363790 Silene muscipula L. var. malenconiana Maire = Silene muscipula L. ■☆

363791 Silene muscipula L. var. murbeckiana Maire et Weiller = Silene muscipula L. ■☆

363792 Silene muscipula L. var. oranensis（Hochr.）Maire et Weiller = Silene muscipula L. ■☆

363793 Silene mushaensis Hayata = Silene aprica Turcz. ex Fisch. et C. A. Mey. ■

363794 Silene nachlingerae Tiehm；纳氏蝇子草；Jan's Catchfly, Nachlinger's Catchfly ■☆

363795 Silene namlaensis（C. Marquand）Bocquet；墨脱蝇子草；Motuo Catchfly ■

363796 Silene nana Kar. et Kir.；矮蝇子草；Dwarf Catchfly ■

363797 Silene nangqenensis C. L. Tang；囊谦蝇子草；Nangqian Catchfly ■

363798 Silene napuligera Franch.；纺锤根蝇子草（纺锤蝇子草）；Spindleroot Catchfly ■

363799 Silene neglecta Ten.；疏忽蝇子草■☆

363800 Silene nepalensis Majumdar；尼泊尔蝇子草；Nepal Catchfly ■

363801 Silene nepalensis Majumdar var. kialensis（F. N. Williams）C. L. Tang ex C. Y. Wu = Silene kialensis（F. N. Williams）Lidén et Oxelman ■

363802 Silene nepalensis Majumdar var. kialensis（F. N. Williams）C. L. Tang ex C. Y. Wu = Silene kialensis Rohrb. ■

363803 Silene nepalensis Majumdar var. kialensis（F. N. Williams）C. L. Tang = Silene nepalensis Majumdar var. kialensis（F. N. Williams）C. L. Tang ex C. Y. Wu ■

363804 Silene nevskii Schischk.；奈氏蝇子草■☆

363805 Silene nicaeensis All.；尼卡蝇子草■☆

363806 Silene nicaeensis All. var. arenarioides（Desf.）Batt. = Silene arenarioides Desf. ■☆

363807 Silene nicaeensis All. var. arenicola（C. Presl）Bertol. = Silene nicaeensis All. ■☆

363808 Silene nicaeensis All. var. gracilis Maire = Silene nicaeensis All. ■☆

363809 Silene nicaeensis All. var. perennis Maire = Silene nicaeensis All. ■☆

363810 Silene nigrescens（Edgew.）Majumdar；变黑蝇子草（变黑女娄菜）；Blacken Catchfly ■

363811 Silene nigrescens（Edgew.）Majumdar = Silene namlaensis（C. Marquand）Bocquet ■

363812 Silene nigrescens（Edgew.）Majumdar subsp. latifolia Bocquet；宽叶变黑蝇子草；Broadleaf Blacken Catchfly ■

363813 Silene ningxiaensis C. L. Tang；宁夏蝇子草（宁夏麦瓶草）；Ningxia Catchfly ■

363814 Silene nivea（Nutt.）Muhl. ex Otth；雪白蝇子草；Evening Campion, Snowy Campion, White Campion ■

363815 Silene noctiflora L.；腋花蝇子草（夜花女娄菜，夜花蝇子草）；Nightblooming Catchfly, Night-flowering Campion, Night-flowering Catchfly, Nightflowering Silene, Night-flowering Silene, Sticky Cockle ■

363816 Silene noctiflora L. = Melandrium noctiflorum（L.）Fr. ■

363817 Silene nocturna L.；夜花蝇子草■☆

363818 Silene nocturna L. subsp. decipiens Ball = Silene nocturna L. ■☆

363819 Silene nocturna L. subsp. neglecta（Ten.）Arcang. = Silene neglecta Ten. ■☆

363820 Silene nocturna L. var. brachypetala（Robill. et Castagne ex DC.）Benth. = Silene nocturna L. ■☆

363821 Silene nocturna L. var. calva Maire = Silene nocturna L. ■☆

363822 Silene nocturna L. var. pauciflora Otth = Silene nocturna L. ■☆

363823 Silene nocturna L. var. permixta（Jord.）Rohrb. = Silene nocturna L. ■☆

363824 Silene nuda（S. Watson）C. L. Hitchc. et Maguire；裸露蝇子草（黏蝇子草）；Sticky Catchfly ■☆

363825 Silene nuda（S. Watson）C. L. Hitchc. et Maguire subsp. insectivora（L. F. Hend.）C. L. Hitchc. et Maguire = Silene nuda（S. Watson）C. L. Hitchc. et Maguire ■☆

363826 Silene nuda L. subsp. insectivora（L. F. Hend.）C. L. Hitchc. et Maguire = Silene nuda（S. Watson）C. L. Hitchc. et Maguire ■☆

363827 Silene nutans L.；俯垂麦瓶草；Eurasian Catchfly, Nodding Catchfly, Nodding Silene, Nottingham Catchfly ■☆

363828 Silene nyingchiensis L. H. Zhou = Silene lhassana（F. N. Williams）Majumdar ■

363829 Silene oblanceolata W. W. Sm.；倒披针叶蝇子草；Oblanceolate Catchfly ■

363830 Silene oblanceolata W. W. Sm. = Silene longicornuta C. Y. Wu et C. L. Tang ■

363831 Silene obovata A. E. Porsild；倒卵叶蝇子草■☆

363832 Silene obovata A. E. Porsild = Silene menziesii Hook. ■☆

363833 Silene obovata Schischk.；卵花蝇子草■☆

363834 Silene obtusidentata B. Fedtsch. et Popov；钝齿蝇子草■☆

363835 Silene obtusifolia Willd.；钝叶蝇子草■☆

363836 Silene occidentalis S. Watson；西部蝇子草；Western Catchfly ■☆

363837 Silene occidentalis S. Watson subsp. longistipitata C. L. Hitchc. et Maguire = Silene occidentalis S. Watson ■☆

363838 Silene occidentalis S. Watson var. nancta Jeps. = Silene bernardina S. Watson ■☆

363839 Silene occidentalis S. Watson var. nancta Jeps. = Silene verecunda S. Watson ■☆

363840 Silene odoratissima Bunge;香蝇子草;Aromatic Catchfly ■

363841 Silene oldhamiana Miq. = Silene aprica Turcz. ex Fisch. et C. A. Mey. ■

363842 Silene oldhamiana Miq. = Silene aprica Turcz. ex Fisch. et C. A. Mey. var. oldhamiana (Miq.) C. Y. Wu ■

363843 Silene olgiana B. Fedtsch.;奥尔嘎蝇子草■☆

363844 Silene oliganthella Nakai ex Kitag. = Silene jenisseensis Willd. ■

363845 Silene oranensis Hochr. = Silene muscipula L. subsp. deserticola Murb. ■☆

363846 Silene oraria M. Peck = Silene douglasii Hook. var. oraria (M. Peck) C. L. Hitchc. et Maguire ■☆

363847 Silene oregana S. Watson;俄勒冈蝇子草;Oregon Catchfly ■☆

363848 Silene oregana S. Watson var. filisecta (M. Peck) M. Peck = Silene oregana S. Watson ■☆

363849 Silene orgiana B. Fedtsch.;沙生蝇子草■

363850 Silene oriena Schischk.;山生蝇子草■☆

363851 Silene orientalimongolica Kozhevn.;内蒙古女娄菜;Inner Mongol Catchfly ■

363852 Silene orientalis Hook.;东麦瓶草;Oriental Catchfly ■☆

363853 Silene ornata Aiton;装饰蝇子草■☆

363854 Silene ostenfeldii (A. E. Porsild) J. K. Morton;奥斯滕蝇子草■☆

363855 Silene otites (L.) Wibel;黄雪轮;Breckland Catchfly,Earshape Catchfly,Spanish Campion,Spanish Catch Fly,Spanish Catchfly ■

363856 Silene otites (L.) Wibel var. borysthenica Gruner = Silene borysthenica (Gruner) Walters ■

363857 Silene otites (L.) Wibel var. wolgensis (Hornem.) Rohrb. = Silene wolgensis (Willd.) Besser ex Spreng. ■

363858 Silene otites Sm. = Silene otites (L.) Wibel ■

363859 Silene otodonta Franch.;耳齿蝇子草;Eartooth Catchfly, Nutate Catchfly ■

363860 Silene ovata Pursh;卵叶蝇子草;Ovate-leaved Campion,Ovate-leaved Catchfly ■☆

363861 Silene pachyneura Schischk.;粗脉蝇子草■☆

363862 Silene pachyrrhiza Franch. = Silene repens Patrin ■

363863 Silene pacifica Eastw. = Silene scouleri Hook. ■☆

363864 Silene palmeri S. Watson = Silene lemmonii S. Watson ■☆

363865 Silene pamirensis (H. Winkl.) Preobr.;帕米尔蝇子草■☆

363866 Silene pamirensis Preobr. = Silene karaczukuri B. Fedtsch. ■

363867 Silene parishii S. Watson;帕里什蝇子草;Parish's Catchfly ■☆

363868 Silene parishii S. Watson var. latifolia C. L. Hitchc. et Maguire = Silene parishii S. Watson ■☆

363869 Silene parishii S. Watson var. viscida C. L. Hitchc. et Maguire = Silene parishii S. Watson ■☆

363870 Silene parlangei Emb. = Silene muscipula L. ■☆

363871 Silene parryi (S. Watson) C. L. Hitchc. et Maguire;帕里蝇子草;Parry's Catchfly ■☆

363872 Silene parviflora (Ehrh.) Pers. = Silene borysthenica (Gruner) Walters ■

363873 Silene patula Desf.;开展蝇子草■☆

363874 Silene patula Desf. subsp. amurensis (Pomel) Jeanm.;阿穆尔开展蝇子草■☆

363875 Silene patula Desf. var. amurensis (Pomel) Jeanm. = Silene patula Desf. ■☆

363876 Silene patula Desf. var. hesperia (Maire) Jeanm. = Silene patula Desf. ■☆

363877 Silene patula Desf. var. tananorum Jeanm. = Silene patula Desf. ■☆

363878 Silene paucifolia (F. N. Williams) Nakai = Silene jenisseensis Willd. ■

363879 Silene paucifolia Ledeb.;稀叶麦瓶草■☆

363880 Silene pectinata S. Watson = Silene nuda (S. Watson) C. L. Hitchc. et Maguire ■☆

363881 Silene pectinata S. Watson var. subnuda B. L. Rob. = Silene nuda (S. Watson) C. L. Hitchc. et Maguire ■☆

363882 Silene pendula L.;大蔓樱草(矮雪轮,小町草);Drooping Catchfly, Drooping Silene, Nodding Catchfly, Pendulous-fruited Catchfly ■

363883 Silene pensylvanica Michx. = Silene caroliniana Walter subsp. pensylvanica (Michx.) R. T. Clausen ■☆

363884 Silene pentandra L. var. morisonii (Boreau) Barratte = Spergula morisonii Boreau ■☆

363885 Silene persica Boiss. subsp. moorcroftiana (Wall. ex Benth.) Chowdhuri = Silene moorcroftiana Wall. ex Benth. ■

363886 Silene petersonii Maguire;彼得森蝇子草;Peterson's Campion, Peterson's Catchfly ■☆

363887 Silene petersonii Maguire var. minor C. L. Hitchc. et Maguire = Silene petersonii Maguire ■☆

363888 Silene phoenicodonta Franch.;红齿蝇子草;Redtooth Catchfly ■☆

363889 Silene physocalyx Ledeb.;囊萼蝇子草■☆

363890 Silene pilosellifolia Cham. et Schltdl.;毛叶蝇子草■☆

363891 Silene plankii C. L. Hitchc. et Maguire;格兰德蝇子草;Grande Fire Pink,Rio Grande Fire Pink ■☆

363892 Silene platyota S. Watson = Silene verecunda S. Watson ■☆

363893 Silene platypetala Bureau et Franch. = Silene principis Oxelman et Lidén ■

363894 Silene platyphylla Franch.;宽叶蝇子草(阔叶女娄菜);Broadleaf Catchfly ■

363895 Silene platyphylla Franch. = Silene asclepiadea Franch. ■

363896 Silene platyphylla Franch. f. congesta Franch. = Silene platyphylla Franch. ■

363897 Silene platyphylla Franch. f. involucrata Franch. = Silene platyphylla Franch. ■

363898 Silene platyphylla Franch. f. paniculifera Franch. = Silene platyphylla Franch. ■

363899 Silene platyphylla Franch. var. praticola (W. W. Sm.) C. Y. Wu;草场蝇子草■

363900 Silene platyphylla Franch. var. praticola (W. W. Sm.) C. Y. Wu = Silene platyphylla Franch. ■

363901 Silene plicata S. Watson = Silene thurberi S. Watson ■☆

363902 Silene plurifolia Schischk.;繁叶蝇子草■☆

363903 Silene polaris Kleopow;极地麦瓶草■☆

363904 Silene polypetala (Walter) Fernald et B. G. Schub.;流苏蝇子草;Fringed Campion ■☆

363905 Silene polyphylla L. = Silene portensis L. ■☆

363906 Silene pomelii Batt.;波梅尔蝇子草■☆

363907 Silene pomelii Batt. subsp. adusta (Ball) Maire;煤黑蝇子草■☆

363908 Silene pomelii Batt. var. battandieri Maire = Silene pomelii Batt. ■☆

363909 Silene pomelii Batt. var. brevipedunculata Emb. et Maire = Silene pomelii Batt. ■☆

363910 Silene pomelii Batt. var. dolichocarpa Maire = Silene pomelii Batt. ■☆

363911 Silene pomelii Batt. var. rosella Maire = Silene pomelii Batt. ■☆

363912　Silene popovii Schischk. ;波波夫蝇子草■☆

363913　Silene portensis L. ;多叶蝇子草■☆

363914　Silene portensis L. subsp. maura Emb. et Maire;模糊蝇子草■☆

363915　Silene potaninii Maxim. = Silene tatarinowii Regel ■

363916　Silene pratensis（Raf.）Gren. et Godr. = Silene latifolia（Mill.）Rendle et Britten ■

363917　Silene pratensis（Raf.）Gren. et Godr. subsp. alba（Mill.）Greuter et Burdet ■

363918　Silene pratensis（Raf.）Gren. et Godr. subsp. divaricata（Rchb.）McNeill et H. C. Prent. = Melandrium boissieri Schischk. ■☆

363919　Silene pratensis（Raf.）Gren. et Godr. subsp. divaricata（Rchb.）McNeill et H. C. Prent. = Silene latifolia（Mill.）Rendle et Britten ■

363920　Silene praticola W. W. Sm. = Silene platyphylla Franch. ■

363921　Silene prilipkoana Schischk. ;普里蝇子草■☆

363922　Silene primulaeflora Eckl. et Zeyh. ;报春花蝇子草■☆

363923　Silene primulaeflora Eckl. et Zeyh. var. ciliata Sond. ;缘毛蝇子草■☆

363924　Silene principis Oxelman et Lidén;宽瓣蝇子草;Broadpetal Catchfly ■

363925　Silene pringlei S. Watson = Silene scouleri Hook. subsp. pringlei（S. Watson）C. L. Hitchc. et Maguire ■☆

363926　Silene procumbens Murray;平铺麦瓶草■☆

363927　Silene propinqua Schischk. ;邻近蝇子草■☆

363928　Silene pruinosa Boiss. ;粉蝇子草■☆

363929　Silene psammitis Spreng. subsp. lasiostyla（Boiss.）Rivas Goday;毛柱蝇子草■☆

363930　Silene pseudoatocion Desf. ;北非麦瓶草;North African Catchfly ■☆

363931　Silene pseudoatocion Desf. var. oranensis Batt. = Silene pseudoatocion Desf. ■☆

363932　Silene pseudofortunei Y. W. Tsui et C. L. Tang;团伞蝇子草;False Crane Catchfly ■

363933　Silene pseudotenuis Schischk. ;昭苏蝇子草;Zhaosu Catchfly ■

363934　Silene pseudotites Besser = Silene otites（L.）Wibel ■

363935　Silene pseudotites Besser = Silene pseudotites Besser ex Rchb. ■

363936　Silene pseudotites Besser ex Rchb. ;假耳状麦瓶草■

363937　Silene pseudotites Besser ex Rchb. = Silene otites（L.）Wibel ■

363938　Silene pseudovelutina Rothm. = Silene andryalifolia Pomel ■☆

363939　Silene pseudovestita Batt. ;假被蝇子草■☆

363940　Silene pteropleura Boiss. et Reut. = Silene stricta L. ■☆

363941　Silene pterosperma Maxim. ;长梗蝇子草（长梗细蝇子草）;Longstalk Catchfly ■

363942　Silene pubicalycina C. Y. Wu;毛萼蝇子草;Haircalyx Catchfly ■

363943　Silene pubistyla L. H. Zhou = Silene graminifolia Otth ■

363944　Silene pulchra（Willd. ex Schltdl. et Cham.）Torr. et A. Gray = Silene laciniata Cav. ■☆

363945　Silene pulverulenta M. Peck = Silene hookeri Nutt. ■☆

363946　Silene puranensis（L. H. Zhou）C. Y. Wu et H. Chuang;普兰蝇子草（普兰女娄菜）;Pulan Catchfly ■

363947　Silene purpurata Greene = Silene repens Patrin ex Pers. ■

363948　Silene pygmaea Adams;矮小蝇子草■☆

363949　Silene qiyunshanensis X. H. Guo et X. L. Liu;齐云山蝇子草;Qiyunshan Catchfly ■

363950　Silene quadriloba Turcz. ex Kar. et Kir. ;四裂蝇子草（四裂女娄菜）;Fourlobe Catchfly ■

363951　Silene quinquevulnera L. = Silene gallica L. ■

363952　Silene raddeana Trautv. ;拉德蝇子草■☆

363953　Silene radians Kar. et Kir. = Silene odoratissima Bunge ■

363954　Silene ramosissima Desf. ;多枝蝇子草■☆

363955　Silene ramosissima Desf. var. brevipes Maire et Sennen = Silene ramosissima Desf. ■☆

363956　Silene reflexa Aiton = Silene neglecta Ten. ■☆

363957　Silene regia Sims;高贵蝇子草;Royal Catchfly, Wild Pink ■☆

363958　Silene repens Patrin;蔓茎蝇子草（旱麦瓶草,葫芦草,蔓麦瓶草,毛萼麦瓶草,匍匐麦瓶草,匍茎鹤草,匍生蝇子草）;Creeping Catchfly, Pink Campion, Vine Catchfly ■

363959　Silene repens Patrin ex Pers. = Silene repens Patrin ■

363960　Silene repens Patrin f. apoiensis（H. Hara）Kitam. = Silene repens Patrin var. apoiensis H. Hara ■☆

363961　Silene repens Patrin subsp. australis C. L. Hitchc. et Maguire = Silene repens Patrin ex Pers. ■

363962　Silene repens Patrin subsp. purpurata（Greene）C. L. Hitchc. et Maguire = Silene repens Patrin ex Pers. ■

363963　Silene repens Patrin var. angustifolia Turcz. ;细叶蔓茎麦瓶草（细叶毛萼麦瓶草）;Thinleaf Creeping Catchfly ■

363964　Silene repens Patrin var. angustifolia Turcz. = Silene repens Patrin ■

363965　Silene repens Patrin var. angustifolia Turcz. ex Regel f. sinensis F. N. Williams = Silene repens Patrin ■

363966　Silene repens Patrin var. angustifolia Turcz. f. sinensis F. N. Williams = Silene repens Patrin ■

363967　Silene repens Patrin var. apoiensis H. Hara;阿伯伊蝇子草■☆

363968　Silene repens Patrin var. australis（C. L. Hitchc. et Maguire）C. L. Hitchc. = Silene repens Patrin ex Pers. ■

363969　Silene repens Patrin var. costata（Williams）B. Boivin = Silene repens Patrin ex Pers. ■

363970　Silene repens Patrin var. glandulosa Y. W. Cui et L. H. Zhou;腺蔓蝇子草;Glandulose Creeping Catchfly ■

363971　Silene repens Patrin var. latifolia Turcz. ;宽叶蔓茎蝇子草（宽叶毛萼麦瓶草,宽叶蝇子草）;Broadleaf Vine Catchfly ■

363972　Silene repens Patrin var. latifolia Turcz. = Silene repens Patrin ■

363973　Silene repens Patrin var. sinensis（F. N. Williams）C. L. Tang;线叶蔓茎蝇子草（线叶蝇子草）;Linearifolious Vine Catchfly ■

363974　Silene repens Patrin var. sinensis（F. N. Williams）C. L. Tang = Silene repens Patrin ■

363975　Silene repens Patrin var. vulgaris Turcz. = Silene repens Patrin var. sinensis（F. N. Williams）C. L. Tang ■

363976　Silene repens Patrin var. vulgaris Turcz. = Silene repens Patrin ■

363977　Silene repens Patrin var. xilingensis Y. Z. Zhao;锡林蝇子草（锡林麦瓶草）;Xilin Vine Catchfly ■

363978　Silene repens Patrin var. xilingensis Y. Z. Zhao = Silene repens Patrin ■

363979　Silene repens Patrin var. xilingensis Y. Z. Zhao = Silene repens Patrin var. sinensis（F. N. Williams）C. L. Tang ■

363980　Silene reticulata Desf. ;网状蝇子草■☆

363981　Silene reverchonii Batt. ;勒韦雄蝇子草■☆

363982　Silene rosiflora Kingdon-Ward ex W. W. Sm. ;粉花蝇子草;Rose Catchfly ■

363983　Silene rosulata Soy. -Will. et Godr. ;莲座蝇子草■☆

363984　Silene rosulata Soy. -Will. et Godr. var. adenocalyx Maire = Silene rosulata Soy. -Will. et Godr. ■☆

363985　Silene rosulata Soy. -Will. et Godr. var. ciliata Maire = Silene rosulata Soy. -Will. et Godr. ■☆

363986 Silene rosulata Soy. -Will. et Godr. var. pubescens Maire = Silene rosulata Soy. -Will. et Godr. ■☆

363987 Silene rosulata Soy. -Will. et Godr. var. reeseana (Maire) Jeanm. = Silene rosulata Soy. -Will. et Godr. ■☆

363988 Silene rosulata Soy. -Will. et Godr. var. tingitana Jeanm. = Silene rosulata Soy. -Will. et Godr. ■☆

363989 Silene rotundifolia (Oliv.) Neumayer = Uebelinia rotundifolia Oliv. ■☆

363990 Silene rotundifolia Nutt. ; 圆叶蝇子草; Round-leaved Catchfly ■☆

363991 Silene rouyana Batt. ; 卢伊蝇子草■☆

363992 Silene rubella L. subsp. segetalis (Dufour) Nyman = Silene diversifolia Otth ■☆

363993 Silene rubella L. subsp. turbinata (Guss.) Chater et Walters = Silene turbinata Guss. ■

363994 Silene rubella L. var. bifida Maire = Silene diversifolia Otth ■☆

363995 Silene rubella L. var. turbinata (Guss.) Batt. = Silene turbinata Guss. ■

363996 Silene rubella L. var. typica Fiori = Silene diversifolia Otth ■☆

363997 Silene rubicunda A. Dietr. = Silene caroliniana Walter ■☆

363998 Silene rubicunda Franch. ; 红茎蝇子草(红茎女娄菜, 红女娄菜, 九子参); Redstem Catchfly ■

363999 Silene rubicunda Franch. = Silene napuligera Franch. ■

364000 Silene rubicunda Franch. var. revoluta Franch. = Silene napuligera Franch. ■

364001 Silene rubricalyx (C. Marquand) Bocquet; 红萼蝇子草(红萼女娄菜); Redcalyx Catchfly ■

364002 Silene rupestris L. ; 岩石麦瓶草; Rock Campion, Rock Catchfly, Rupreeht's Silene ■☆

364003 Silene ruprechtii Schischk. ; 鲁氏蝇子草; Ruprecht Silene ■☆

364004 Silene sabinosae Pit. ; 萨维诺萨蝇子草■☆

364005 Silene salicifolia C. L. Tang; 柳叶蝇子草; Willowleaf Catchfly ■

364006 Silene salweenensis W. W. Sm. = Silene rosiflora Kingdon-Ward ex W. W. Sm. ■

364007 Silene samarkandensis Preobr. ; 撒马尔罕蝇子草■☆

364008 Silene saponaria Fr. = Saponaria officinalis L. ■

364009 Silene saponaria Fr. ex Willk. et Lange = Saponaria officinalis L. ■

364010 Silene sarawschanica Regel et Schmalh. ; 瑟拉蝇子草■☆

364011 Silene sargentii S. Watson; 萨金特蝇子草; Sargent's Catchfly ■☆

364012 Silene saxatilis Sims; 岩蝇子草(岩生蝇子草)■☆

364013 Silene saxifraga L. ; 针叶雪轮; Saxifrage Catchfly, Saxifrage Silene, Tufted Catchfly ■☆

364014 Silene scabrella (Nieuwl.) G. N. Jones = Silene stellata (L.) W. T. Aiton var. scabrella (Nieuwl.) E. J. Palmer et Steyerm. ■☆

364015 Silene scabrella (Nieuwl.) G. N. Jones = Silene stellata (L.) W. T. Aiton ■☆

364016 Silene scabrida Soy. -Will. et Godr. ; 微糙蝇子草■☆

364017 Silene scabriflora Brot. ; 糙花蝇子草■☆

364018 Silene scabriflora Brot. subsp. tuberculata (Ball) Talavera = Silene tuberculata (Ball) Maire et Weiller ■

364019 Silene scabriflora Brot. var. guedirensis (Pau) Maire = Silene micropetala Lag. ■☆

364020 Silene scabrifolia Kom. ; 粗叶蝇子草■☆

364021 Silene scaposa B. L. Rob. ; 罗宾逊蝇子草; Robinson's Catchfly ■☆

364022 Silene scaposa B. L. Rob. var. lobata C. L. Hitchc. et Maguire = Silene scaposa B. L. Rob. ■☆

364023 Silene schafta J. G. Gmel. ex Hohen. ; 夏弗塔雪轮; Moss Campion, Schafta Campion ■☆

364024 Silene schugnanica B. Fedtsch. ; 舒格南蝇子草■☆

364025 Silene schweinfurthii Rohrb. ; 施韦蝇子草■☆

364026 Silene sclerocarpa L. Dufour; 硬果蝇子草■☆

364027 Silene scopulorum Franch. ; 岩生蝇子草(岩生女娄菜); Saxicolous Catchfly ■

364028 Silene scouleri Hook. ; 斯考莱尔蝇子草(虎克蝇子草); Scouler's Catchfly ■☆

364029 Silene scouleri Hook. subsp. grandis (Eastw.) C. L. Hitchc. et Maguire = Silene scouleri Hook. ■☆

364030 Silene scouleri Hook. subsp. hallii (S. Watson) C. L. Hitchc. et Maguire; 霍尔蝇子草; Hall's Catchfly ■☆

364031 Silene scouleri Hook. subsp. pringlei (S. Watson) C. L. Hitchc. et Maguire; 普林格尔蝇子草; Pringle's Catchfly ■☆

364032 Silene scouleri Hook. var. concolor (Greene) C. L. Hitchc. et Maguire = Silene scouleri Hook. subsp. hallii (S. Watson) C. L. Hitchc. et Maguire ■☆

364033 Silene scouleri Hook. var. eglandulosa C. L. Hitchc. et Maguire = Silene scouleri Hook. subsp. pringlei (S. Watson) C. L. Hitchc. et Maguire ■☆

364034 Silene scouleri Hook. var. grisea C. L. Hitchc. et Maguire = Silene scouleri Hook. subsp. pringlei (S. Watson) C. L. Hitchc. et Maguire ■☆

364035 Silene scouleri Hook. var. leptophylla C. L. Hitchc. et Maguire = Silene scouleri Hook. subsp. pringlei (S. Watson) C. L. Hitchc. et Maguire ■☆

364036 Silene scouleri Hook. var. macounii (S. Watson) B. Boivin = Silene parryi (S. Watson) C. L. Hitchc. et Maguire ■☆

364037 Silene scouleri Hook. var. pacifica (Eastw.) C. L. Hitchc. = Silene scouleri Hook. ■☆

364038 Silene secundiflora Otth; 侧花蝇子草■☆

364039 Silene secundiflora Otth subsp. macrotheca (Braun-Blanq. et Maire) Greuter et Burdet; 大腔侧花蝇子草■☆

364040 Silene secundiflora Otth var. decora Maire et Wilczek = Silene secundiflora Otth ■☆

364041 Silene secundiflora Otth var. glauca (Spreng.) Maire = Silene secundiflora Otth ■☆

364042 Silene secundiflora Otth var. macrotheca Braun-Blanq. et Maire = Silene secundiflora Otth subsp. macrotheca (Braun-Blanq. et Maire) Greuter et Burdet ■☆

364043 Silene secundiflora Otth var. submacrotheca Maire = Silene secundiflora Otth ■☆

364044 Silene sedoides Poir. ; 景天蝇子草■☆

364045 Silene seelyi C. V. Morton et J. W. Thomps. ; 西利蝇子草; Seely's Catchfly, Seely's Silene ■☆

364046 Silene segetalis Neck. = Vaccaria hispanica (Mill.) Raeusch. ■☆

364047 Silene semenovii Regel; 塞麦蝇子草■☆

364048 Silene seoulensis Nakai; 汉城蝇子草; Seoul Catchfly ■

364049 Silene seoulensis Nakai var. angustata C. L. Tang; 狭叶汉城蝇子草; Narrowleaf Seoul Catchfly ■

364050 Silene sericata C. L. Tang; 绢毛蝇子草; Sericeous Catchfly ■☆

364051 Silene sericata C. L. Tang = Silene gracilicaulis C. L. Tang ■

364052 Silene serpentinicola T. W. Nelson et J. P. Nelson; 蜿蜒蝇子草; Serpentine Indian Pink ■☆

364053　Silene sersuensis Pomel = Silene vulgaris（Moench）Garcke ■

364054　Silene setacea Viv. = Silene vivianii Steud. ■☆

364055　Silene setacea Viv. subsp. getula（Pomel）Maire = Silene vivianii Steud. subsp. viscida（Boiss.）Boulos ■☆

364056　Silene setacea Viv. var. echinosperma Emb. et Maire = Silene vivianii Steud. ■☆

364057　Silene setacea Viv. var. glabrescens Pamp. = Silene vivianii Steud. ■☆

364058　Silene setacea Viv. var. macrotricha Maire = Silene vivianii Steud. ■☆

364059　Silene setacea Viv. var. maroccana（Coss.）Maire = Silene vivianii Steud. ■☆

364060　Silene setacea Viv. var. microsperma Pau = Silene vivianii Steud. ■☆

364061　Silene shockleyi S. Watson = Silene bernardina S. Watson ■☆

364062　Silene sibirica（L.）Pers.；西伯利亚蝇子草；Siberian Catchfly ■☆

364063　Silene sibirica Pers. = Silene linnaeana Volosch. ■

364064　Silene sieboldii（Van Houtte）H. Ohashi et H. Nakai；西氏蝇子草（剪春罗，希氏剪秋罗，席氏蝇子草）；Siebold Campion ■☆

364065　Silene simulans Greene = Silene laciniata Cav. ■☆

364066　Silene sinensis（Lour.）H. Ohashi et H. Nakai = Silene banksia（Meerb.）Mabberly ■

364067　Silene sinensis（Lour.）H. Ohashi et H. Nakai f. verticillata（Makino）H. Ohashi et H. Nakai；轮叶剪春罗 ■☆

364068　Silene sinowatsonii W. W. Sm. = Silene rosiflora Kingdon-Ward ex W. W. Sm. ■

364069　Silene soczaviana（Schischk.）Bocque；北方蝇子草；Boreal Catchfly ■☆

364070　Silene soczaviana（Schischk.）Bocque = Silene uralensis（Rupr.）Bocquet subsp. porsildii Bocquet ■☆

364071　Silene solenantha Trautv.；欧洲管花蝇子草 ■☆

364072　Silene songarica（Fisch.，C. A. Mey. et Avé-Lall.）Bocquet；准噶尔蝇子草（短瓣女娄菜，兴安女娄菜，准噶尔麦瓶草）；Dzungar Catchfly ■

364073　Silene sorensenis（B. Boivin）Bocquet；三花蝇子草；Three-flowered Campion ■☆

364074　Silene spaldingii S. Watson；斯波尔丁蝇子草；Spalding's Campion，Spalding's Catchfly ■☆

364075　Silene stellarioides Nutt. ex Torr. et A. Gray = Silene menziesii Hook. ■☆

364076　Silene stellata（L.）W. T. Aiton；星状蝇子草（星状麦瓶草）；Starry Campion，Widow's Frill，Widow's-frill ■☆

364077　Silene stellata（L.）W. T. Aiton var. scabrella（Nieuwl.）E. J. Palmer et Steyerm. = Silene stellata（L.）W. T. Aiton ■☆

364078　Silene stellata W. T. Aiton var. scabrella（Nieuwl.）E. J. Palmer et Steyerm. = Silene stellata（L.）W. T. Aiton ■☆

364079　Silene stenophylla Ledeb.；狭叶蝇子草 ■☆

364080　Silene stewartiana Diels；大子蝇子草；Bigseed Catchfly ■

364081　Silene stricta L.；刚直蝇子草 ■☆

364082　Silene stylosa Bunge = Silene graminifolia Otth ■

364083　Silene suaveolens Turcz. ex Kar. et Kir.；细裂蝇子草；Thinsplite Catchfly ■

364084　Silene subciliata B. L. Rob.；草原火蝇子草；Prairie-fire Pink ■☆

364085　Silene subcretacea F. N. Williams；藏蝇子草；Xizang Catchfly ■

364086　Silene succulenta Forssk.；多汁蝇子草 ■☆

364087　Silene succulenta Forssk. var. cryptantha E. A. Durand et Barratte = Silene succulenta Forssk. ■☆

364088　Silene suecica（Lodd.）Greuter et Burdet；西方山地蝇子草；Alpine Campion，Alpine Catchfly，Alpine Pink，Arctic Campion ■☆

364089　Silene suffrutescens M. Bieb.；灌木状麦瓶草 ■☆

364090　Silene suksdorfii B. L. Rob.；苏克蝇子草；Suksdorf's Catchfly ■☆

364091　Silene supina M. Bieb.；仰卧麦瓶草 ■☆

364092　Silene sveae Lidén et Oxelman；德钦蝇子草 ■

364093　Silene syngei（Turrill）T. M. Harris et Goyder；辛格麦瓶草 ■☆

364094　Silene szechuanensis F. N. Williams；四川黏萼女娄菜（白前，大牛膝，滇白前，金柴胡，九大牛，青骨藤，瓦草，瓦草参）；Sichuan Catchfly ■

364095　Silene szechuanensis F. N. Williams = Silene asclepiadea Franch. ■

364096　Silene tachtensis Franch.；冠瘤蝇子草 ■

364097　Silene tagadirtensis Murb. = Silene pomelii Batt. subsp. adusta（Ball）Maire ■☆

364098　Silene talyschensis Schischk.；塔里什蝇子草 ■☆

364099　Silene taquetii H. Lév. = Silene aprica Turcz. ex Fisch. et C. A. Mey. ■

364100　Silene tatarica Pers.；鞑靼麦瓶草；Tatar Catchfly，Tatarian Silene，Tatartan Catchfly ■☆

364101　Silene tatarica Pers. var. foliosa（Maxim.）Regel = Silene foliosa Maxim. ■

364102　Silene tatarica Pers. var. macrostyla（Maxim.）Regel = Silene macrostyla Maxim. ■

364103　Silene tatarinowii Regel；石生蝇子草（白花紫萼女娄菜，鹅耳七，连参，麦瓶草，米洋参，山女娄菜，石生麦瓶草，太子参，土洋参，瓦草，西洋参，蝇子草）；Tatarinow Catchfly，Tatarinow Melandrium ■

364104　Silene tatarinowii Regel f. albiflora（Franch.）Kitag. = Silene tatarinowii Regel var. albiflora Franch. ■

364105　Silene tatarinowii Regel f. albiflora（Franch.）Kitag. = Silene tatarinowii Regel ■

364106　Silene tatarinowii Regel var. albiflora Franch.；白花石生蝇子草 ■

364107　Silene tatarinowii Regel var. albiflora Franch. = Silene tatarinowii Regel ■

364108　Silene tatianae Schischk.；塔特蝇子草 ■☆

364109　Silene tayloriae（B. L. Rob.）Hultén = Silene involucrata（Cham. et Schltdl.）Bocquet subsp. tenella（Tolm.）Bocquet ■☆

364110　Silene tenella C. A. Mey.；柔弱麦瓶草 ■☆

364111　Silene tenuis Willd. = Silene gracilicaulis C. L. Tang ■

364112　Silene tenuis Willd. = Silene graminifolia Otth ■

364113　Silene tenuis Willd. = Silene jenisseensis Willd. ■

364114　Silene tenuis Willd. = Silene pterosperma Maxim. ■

364115　Silene tenuis Willd. f. rubescens Franch. = Silene gracilicaulis C. L. Tang ■

364116　Silene tenuis Willd. var. dentata Y. W. Cui et L. H. Zhou；具齿细蝇子草 ■

364117　Silene tenuis Willd. var. dentata Y. W. Cui et L. H. Zhou = Silene gracilicaulis C. L. Tang ■

364118　Silene tenuis Willd. var. denudata Y. W. Tsui et L. H. Zhou = Silene gracilicaulis C. L. Tang ■

364119　Silene tenuis Willd. var. jenissea Rohrb. = Silene jenisseensis Willd. ■

364120　Silene tenuis Willd. var. pauciflora F. N. Williams = Silene jenisseensis Willd. ■

364121　Silene tenuis Willd. var. rubescens（Franch.）Diels = Silene gracilicaulis C. L. Tang ■

364122　Silene tenuis Willd. var. rubescens Franch. ;癞头参■

364123　Silene tetonensis E. E. Nelson = Silene parryi（S. Watson）C. L. Hitchc. et Maguire ■☆

364124　Silene tetragyna Suksd. = Silene parryi（S. Watson）C. L. Hitchc. et Maguire ■☆

364125　Silene thirkeaua K. Koch;梯尔开蝇子草■☆

364126　Silene thunbergiana Bartl. ;通贝里蝇子草■☆

364127　Silene thurberi S. Watson;瑟伯蝇子草;Thurber's Catchfly ■☆

364128　Silene tianschanica Schischk. ;天山蝇子草;Tianshan Catchfly ■

364129　Silene tibetica Lidén et Oxelman;西藏蝇子草■

364130　Silene tokachiensis Kadota;十胜蝇子草■☆

364131　Silene tomentella Schischk. ;微毛蝇子草■☆

364132　Silene toussidana Quézel;图西德蝇子草■☆

364133　Silene trachyphylla Franch. ;糙叶蝇子草;Roughleaf Catchfly ■

364134　Silene transalpine（Hayata）S. S. Ying;高山蝇子草■

364135　Silene tridentata Desf. ;三齿蝇子草■

364136　Silene tridentata Desf. var. arenicola Sennen et Mauricio = Silene sclerocarpa L. Dufour ■☆

364137　Silene tuberculata（Ball）Maire et Weiller;多疣蝇子草

364138　Silene tubiformis C. L. Tang;剑门蝇子草;Jianmen Catchfly ■

364139　Silene tubulosa Oxelman et Lidén;管花蝇子草■

364140　Silene tunetana Batt. = Silene ghiarensis Batt. ■☆

364141　Silene tunetana Murb. ;图内特蝇子草■☆

364142　Silene turbinata Guss. ;陀螺形蝇子草■

364143　Silene turcomanica Schischk. ;土库曼蝇子草■☆

364144　Silene turgida M. Bieb. ex Bunge;膨胀蝇子草■☆

364145　Silene undulata Aiton = Silene bungei Bocquet ■

364146　Silene uniflora Roth;单花麦瓶草（滨麦瓶草,海滨麦瓶草）;Buggie Flower,Campion,Dead Man's Bells,Dead Man's Grief,Devil's Hatties,Double Bladder Campion,Grandmother's Nightcap,Oneflower Catchfly,Sea Campion,Sea Catchfly,Sea-shore Silene,Thimbles,White Snapjack,Witches Thimble,Witches' Thimbles ■☆

364147　Silene uniflora Roth ' Flore Pleno ' = Silene uniflora Roth ' Robin Whitebreast ' ■☆

364148　Silene uniflora Roth ' Robin Whitebreast ';重瓣滨麦瓶草;Double Sea Campion ■☆

364149　Silene uralensis（Rupr.）Bocquet;乌拉尔麦瓶草;Nodding Campion ■☆

364150　Silene uralensis（Rupr.）Bocquet subsp. arctica（Fr.）Bocquet = Silene uralensis（Rupr.）Bocquet ■☆

364151　Silene uralensis（Rupr.）Bocquet subsp. attenuata（Farr）McNeill = Silene uralensis（Rupr.）Bocquet ■☆

364152　Silene uralensis（Rupr.）Bocquet subsp. montana（S. Watson）McNeill = Silene hitchguirei Bocquet ■☆

364153　Silene uralensis（Rupr.）Bocquet subsp. ogilviensis（A. E. Porsild）D. F. Brunt. ;奥吉尔维蝇子草■☆

364154　Silene uralensis（Rupr.）Bocquet subsp. porsildii Bocquet;大籽乌拉尔麦瓶草;Large-seeded Nodding Campion ■☆

364155　Silene uralensis（Rupr.）Bocquet var. mollis（Cham. et Schltdl.）Bocquet = Silene uralensis（Rupr.）Bocquet ■☆

364156　Silene vallesia L. ;瑞士麦瓶草;Swiss Catchfly, Valais Catchfly ■☆

364157　Silene velutinoides Pomel;绒毛蝇子草■☆

364158　Silene venosa（Gilib.）Asch. ;膨萼蝇子草;Bladder Silene ■

364159　Silene venosa（Gilib.）Asch. = Silene vulgaris（Moench）Garcke ■

364160　Silene venosa（Gilib.）Asch. var. angustifolia（Guss.）Wirtg. = Silene vulgaris（Moench）Garcke ■

364161　Silene venosa（Gilib.）Asch. var. rhiphaea Pau et Font Quer = Silene vulgaris（Moench）Garcke subsp. glareosa（Jord.）Marsden-Jones et Turrill ■☆

364162　Silene verecunda S. Watson;旧金山蝇子草;San Francisco Campion ■☆

364163　Silene verecunda S. Watson subsp. andersonii（Clokey）C. L. Hitchc. et Maguire = Silene verecunda S. Watson ■☆

364164　Silene verecunda S. Watson subsp. platyota（S. Watson）C. L. Hitchc. et Maguire = Silene verecunda S. Watson ■☆

364165　Silene verecunda S. Watson var. eglandulosa C. L. Hitchc. et Maguire = Silene verecunda S. Watson ■☆

364166　Silene verecunda S. Watson var. platyota（S. Watson）Jeps. = Silene verecunda S. Watson ■☆

364167　Silene vespertina Retz. = Silene bellidifolia Jacq. ■☆

364168　Silene vestita Soy. -Will. et Godr. = Silene micropetala Lag. ■☆

364169　Silene villosa Forssk. ;多毛蝇子草■☆

364170　Silene villosa Forssk. var. erecta Täckh. et Boulos = Silene villosa Forssk. ■☆

364171　Silene villosa Forssk. var. graveolens Sickenb. = Silene villosa Forssk. ■☆

364172　Silene villosa Forssk. var. ismailitica Schweinf. = Silene villosa Forssk. ■☆

364173　Silene villosa Forssk. var. micropetala Batt. = Silene villosa Forssk. ■☆

364174　Silene virescens Coss. ;浅绿蝇子草■☆

364175　Silene virginica L. ;火红雪轮（弗州蝇子草）;Fire Catchfly, Fire Pink,Fire-pink,Indian Pink,Indianpink,Scarlet Catchfly ■☆

364176　Silene virginica L. var. hallensis Pickens et M. C. Pickens = Silene virginica L. ■☆

364177　Silene virginica L. var. robusta Strausbaugh et Core = Silene virginica L. ■☆

364178　Silene viridiflora L. ;欧洲绿花麦瓶草;Greenflower Catchfly ■☆

364179　Silene viscaria（L.）Jess. ;黏梗蝇子草;Clammy Campion, German Catchfly,Sticky Catchfly ■☆

364180　Silene viscaria（L.）Jess. = Lychnis viscaria L. ■☆

364181　Silene viscidula Franch. ;黏萼蝇子草（白前,大山七,大牛膝,滇白前,洱源女娄菜,金柴胡,九大牛,黏毛瓦草,青骨藤,四川精黏女娄菜,太白七,瓦草）;Adhesivecalyx Catchfly ■

364182　Silene viscidula Kom. = Silene komarovii Schischk. ■

364183　Silene viscosa（L.）Pers. f. multifida Krylov = Silene suaveolens Turcz. ex Kar. et Kir. ■

364184　Silene viscosa（L.）Pers. var. quadriloba Trautv. = Silene quadriloba Turcz. ex Kar. et Kir. ■

364185　Silene viscosa Schleich. = Silene italica（L.）Pers. ■☆

364186　Silene viscosa Schleich. f. multifida Krylov = Silene suaveolens Turcz. ex Kar. et Kir. ■

364187　Silene viscosa Schleich. var. quadriloba Trautv. = Silene quadriloba Turcz. ex Kar. et Kir. ■

364188　Silene vivianii Steud. ;维维安麦瓶草■☆

364189　Silene vivianii Steud. subsp. getula（Pomel）Greuter et Burdet;摩洛哥麦瓶草■☆

364190　Silene vivianii Steud. subsp. viscida（Boiss.）Boulos;黏维维安麦瓶草■☆

364191 Silene vlokii Masson;弗劳克麦瓶草■☆

364192 Silene volubilitana Braun-Blanq. et Maire;旋扭蝇子草■☆

364193 Silene vulgaris (Moench) Garcke;普通麦瓶草(白玉草,膀胱麦瓶草,狗筋麦瓶草,狗筋蝇子草,广布蝇子草);Adder-and-snake Plant, Ben, Billy Buster, Bird's Egg, Bird's Eggs, Bladder Bottle, Bladder Campion, Bladder Catchfly, Bladder of Lard, Bladder-campion, Bladderweed, Bletherweed, Bull Rattle, Clapweed, Cockle, Corn Pop, Cow Bells, Cow Mack, Cow Paps, Cow Rattle, Cowmack, Fat Bellies, Frothy Poppy, Hay Plant, Hay Rattle, John's Plant, Kiss-me-quick, Knap Bottle, Knap-bottle, Maidens Tears, Maiden's Tears, Maidenstears, Maiden's-tears, Malden's Tears, Pop-gun, Popper, Poppy, Ragged Robin, Rattle-bags, Rattleweed, Round Campion, Shackle-backles, Snaggs, Snappers, Spafling Poppy, Spatling Poppy, Thunderbolts, Veiny Catchfly, White Behen, White Bottle, White Cock Robin, White Cockle, White Hood, White Mintdrop, White Riding Hood, White Robin Hood ■

364194 Silene vulgaris (Moench) Garcke subsp. angustifolia Hayek = Silene vulgaris (Moench) Garcke ■

364195 Silene vulgaris (Moench) Garcke subsp. commutata (Guss.) Hayek;变异麦瓶草■☆

364196 Silene vulgaris (Moench) Garcke subsp. cratericola Franco;火山麦瓶草■☆

364197 Silene vulgaris (Moench) Garcke subsp. glareosa (Jord.) Marsden-Jones et Turrill;石砾麦瓶草■☆

364198 Silene vulgaris (Moench) Garcke subsp. macrocarpa Turrill;大果普通麦瓶草■☆

364199 Silene vulgaris (Moench) Garcke subsp. maritima (With.) Á. Löve et D. Löve = Silene uniflora Roth ■☆

364200 Silene vulgaris (Moench) Garcke var. commutata (Guss.) Good et Cullen;白玉草■

364201 Silene vulgaris (Moench) Garcke var. tenoreana (Colla) Jahand. et Maire = Silene vulgaris (Moench) Garcke ■

364202 Silene wahlbergella Chowdhuri;北亚蝇子草;Northern Catchfly ■☆

364203 Silene wahlbergella Chowdhuri = Silene uralensis (Rupr.) Bocquet ■☆

364204 Silene wahlbergella Chowdhuri subsp. arctica (Fr.) Hultén = Silene uralensis (Rupr.) Bocquet ■☆

364205 Silene wahlbergella Chowdhuri subsp. attenuata (Farr) Hultén = Silene uralensis (Rupr.) Bocquet ■☆

364206 Silene wahlbergella Chowdhuri subsp. montana (S. Watson) Hultén = Silene hitchguirei Bocquet ■☆

364207 Silene wallichiana G. Klotz = Silene vulgaris (Moench) Garcke ■

364208 Silene waltonii F. N. Williams = Silene subcretacea F. N. Williams ■

364209 Silene wardii (C. Marquand) Bocquet;林芝蝇子草(林芝女娄菜);Linzhi Catchfly ■

364210 Silene watsonii B. L. Rob. = Silene sargentii S. Watson ■☆

364211 Silene wherryi Small = Silene caroliniana Walter subsp. wherryi (Small) R. T. Clausen ■☆

364212 Silene wilczekii Sennen et Mauricio = Silene vivianii Steud. subsp. getula (Pomel) Greuter et Burdet ■☆

364213 Silene wilfordii (Regel) H. Ohashi et H. Nakai;丝瓣剪秋罗(燕尾仙翁);Wilford Campion ■

364214 Silene wilfordii (Regel) H. Ohashi et H. Nakai = Lychnis wilfordii (Regel) Maxim. ■

364215 Silene williamsii Britton;威廉斯蝇子草;Williams' Catchfly ■☆

364216 Silene wolgensis (Willd.) Besser ex Spreng.;伏尔加蝇子草;Volga Catchfly ■

364217 Silene wrightii A. Gray;赖特蝇子草;Wright's Catchfly ■☆

364218 Silene yanoei Makino;矢野蝇子草■☆

364219 Silene yetii Bocquet;腺毛蝇子草(具腺女娄菜,腺毛剪秋罗);Glandhair Campion, Tentacle Catchfly ■

364220 Silene yetii Bocquet = Silene muliensis C. Y. Wu ■

364221 Silene yetii Bocquet var. herbilegorum Bocquet = Silene herbilegorum (Bocquet) Lidén et Oxelman ■

364222 Silene yunnanensis Franch.;云南蝇子草;Yunnan Catchfly ■

364223 Silene yunnanensis Franch. = Silene trachyphylla Franch. ■

364224 Silene zangdongensis L. H. Zhou;藏东蝇子草;E. Xizang Silene ■

364225 Silene zangdongensis L. H. Zhou = Silene monbeigii W. W. Sm. ■

364226 Silene zayuensis L. H. Zhou;察隅蝇子草;Chayu Catchfly ■

364227 Silene zhongbaensis (L. H. Zhou) C. Y. Wu et C. L. Tang;仲巴蝇子草(仲巴女娄菜);Zhongba Catchfly ■

364228 Silene zhoui C. Y. Wu;耐国蝇子草(全缘瓣女娄菜);Zhou Catchfly ■

364229 Silenopsis Willk. (1847);类蝇子草属■☆

364230 Silenopsis Willk. = Petrocoptis A. Braun ex Endl. ■☆

364231 Silenopsis lagascae Willk.;类蝇子草■☆

364232 Silentvalleya V. J. Nair, Sreek., Vajr. et Bhargavan (1983);静谷草属■☆

364233 Silentvalleya nairii V. J. Nair, Sreek., Vajr. et Bhargavan;静谷草■☆

364234 Siler Mill. = Laserpitium L. ●☆

364235 Siler divaricata (Turcz.) Benth. et Hook. f. = Saposhnikovia divaricata (Turcz.) Schischk. ■

364236 Sileraceae Bercht. et J. Presl = Apiaceae Lindl. (保留科名)●■

364237 Sileraceae Bercht. et J. Presl = Umbelliferae Juss. (保留科名) ■●

364238 Sileriana Urb. et Loes. = Jacquinia L. (保留属名)●☆

364239 Silerium Raf. = Trochiscanthes W. D. J. Koch ■☆

364240 Silicularia Compton = Heliophila Burm. f. ex L. ●■☆

364241 Silicularia Compton(1953);南非角果芥属■☆

364242 Silicularia polygaloides (Schltr.) Marais = Heliophila polygaloides Schltr. ■☆

364243 Silicularia sigillata Compton = Heliophila polygaloides Schltr. ■☆

364244 Siliqua Duhamel = Ceratonia L. ●

364245 Siliquamomum Baill. (1895);长果姜属;Siliquamomum, Siliquamon ■

364246 Siliquamomum asteriscus ? var. scabrum Nutt. = Silphium asteriscus L. ■☆

364247 Siliquamomum tonkinense Baill.;长果姜;Tonkin Siliquamomum, Tonkin Siliquamon ■

364248 Siliquaria Forssk. = Cleome L. ●■

364249 Siliquaria glandulosa Forssk. = Cleome amblyocarpa Barratte et Murb. var. glandulosa (Forssk.) Botsch. ■☆

364250 Siliquastrum Duhamel = Cercis L. ●

364251 Siliybum Hassk. = Silybum Vaill. (保留属名)■

364252 Siloxerus Labill. (废弃属名) = Angianthus J. C. Wendl. (保留属名)■●☆

364253 Silphion St.-Lag. = Silphium L. ■

364254 Silphiosperma Steetz = Brachycome Cass. ●■☆

364255 Silphium L. (1753);松香草属;Compass Plant, Rosin Plant, Rosinweed ■

364256　Silphium albiflorum A. Gray；白花松香草；White Rosinweed ■☆

364257　Silphium asperrimum Hook. = Silphium asteriscus L. ■☆

364258　Silphium asperrimum Hook. = Silphium radula Nutt. ■☆

364259　Silphium asteriscus L.；星星松香草；Southern Rosinweed，Starry Rosinweed ■☆

364260　Silphium asteriscus L. var. angustatum A. Gray = Silphium asteriscus L. var. dentatum（Elliott）Chapm. ■☆

364261　Silphium asteriscus L. var. dentatum（Elliott）Chapm.；齿叶星星松香草■☆

364262　Silphium asteriscus L. var. latifolium（A. Gray）Clevinger；宽叶星星松香草■☆

364263　Silphium asteriscus L. var. simpsonii（Greene）Clevinger；辛氏星星松香草■☆

364264　Silphium asteriscus L. var. trifoliatum（L.）Clevinger；三小叶星星松香草；Whorled Rosinweed ■☆

364265　Silphium atropurpureum Retz. ex Willd. = Silphium asteriscus L. var. trifoliatum（L.）Clevinger■☆

364266　Silphium betonicifolium Hook. = Berlandiera betonicifolia（Hook.）Small ■☆

364267　Silphium brachiatum Gatt.；坎伯兰松香草；Cumberland Rosinweed ■☆

364268　Silphium compositum Michx.；肾叶松香草；Kidney-leaf Rosinweed ■☆

364269　Silphium compositum Michx. subsp. ovatifolium（Torr. et A. Gray）C. R. Sweeney et T. R. Fisher = Silphium compositum Michx. ■☆

364270　Silphium compositum Michx. subsp. reniforme（Raf. ex Nutt.）C. R. Sweeney et T. R. Fisher = Silphium compositum Michx. ■☆

364271　Silphium compositum Michx. subsp. venosum（Small）C. R. Sweeney et T. R. Fisher = Silphium compositum Michx. ■☆

364272　Silphium compositum Michx. var. reniforme（Raf. ex Nutt.）Torr. et A. Gray = Silphium compositum Michx. ■☆

364273　Silphium compositum Michx. var. venosum（Small）Kartesz et Gandhi = Silphium compositum Michx. ■☆

364274　Silphium confertifolium Small = Silphium asteriscus L. var. latifolium（A. Gray）Clevinger ■☆

364275　Silphium connatum L. = Silphium perfoliatum L. var. connatum（L.）Cronquist ■☆

364276　Silphium dentatum Elliott = Silphium asteriscus L. var. dentatum（Elliott）Chapm. ■☆

364277　Silphium dentatum Elliott var. gatesii（C. Mohr）H. E. Ahles = Silphium asteriscus L. ■☆

364278　Silphium elliottii Small = Silphium asteriscus L. var. dentatum（Elliott）Chapm. ■☆

364279　Silphium gatesii C. Mohr = Silphium asteriscus L. ■☆

364280　Silphium glabrum Eggert = Silphium asteriscus L. var. latifolium（A. Gray）Clevinger ■☆

364281　Silphium glutinosum J. R. Allison；胶松香草；Sticky Rosinweed ■☆

364282　Silphium gracile A. Gray = Silphium radula Nutt. var. gracile（A. Gray）Clevinger ■☆

364283　Silphium incisum Greene = Silphium asteriscus L. var. dentatum（Elliott）Chapm. ■☆

364284　Silphium integrifolium F. Michx. deamii（L. M. Perry）Steyerm. = Silphium integrifolium Michx. ■☆

364285　Silphium integrifolium Michx.；全缘叶松香草；Prairie Rosinweed，Rosinweed，Wholeleaf Rosinweed，Whole-leaf Rosinweed ■☆

364286　Silphium integrifolium Michx. var. deamii L. M. Perry = Silphium integrifolium Michx. ■☆

364287　Silphium integrifolium Michx. var. gattingeri L. M. Perry = Silphium asteriscus L. ■☆

364288　Silphium integrifolium Michx. var. laeve Torr. et A. Gray；平滑全缘叶松香草■☆

364289　Silphium integrifolium Michx. var. neglectum Settle et T. R. Fisher = Silphium integrifolium Michx. ■☆

364290　Silphium laciniatum L.；细裂松香草（松香草）；Compass Plant，Compass-plant，Pilot-weed，Rosinweed ■☆

364291　Silphium laciniatum L. var. robinsonii L. M. Perry = Silphium laciniatum L. ■☆

364292　Silphium laevigatum Pursh = Silphium asteriscus L. var. trifoliatum（L.）Clevinger ■☆

364293　Silphium laevigatum Pursh = Silphium asteriscus L. ■☆

364294　Silphium laevigatum Pursh = Silphium integrifolium Michx. ■☆

364295　Silphium lapsuum Small = Silphium compositum Michx. ■☆

364296　Silphium mohrii Small；蓬松松香草；Shaggy Rosinweed ■☆

364297　Silphium nodum Small = Silphium asteriscus L. var. dentatum（Elliott）Chapm. ■☆

364298　Silphium nuttallianum Torr. = Berlandiera subacaulis（Nutt.）Nutt. ■☆

364299　Silphium orae Small = Silphium compositum Michx. ■☆

364300　Silphium ovatifolium（Torr. et A. Gray）Small = Silphium compositum Michx. ■☆

364301　Silphium perfoliatum L.；串叶松香草（穿叶松香草，贯叶松香草）；Compass-plant，Cup Plant，Cup Rosinweed，Cup-plant，Indian Cup ■

364302　Silphium perfoliatum L. subsp. connatum（L.）Cruden = Silphium perfoliatum L. var. connatum（L.）Cronquist ■☆

364303　Silphium perfoliatum L. var. connatum（L.）Cronquist；合生松香草■☆

364304　Silphium peristenium Raf. = Engelmannia peristenia（Raf.）Goodman et C. A. Lawson ■☆

364305　Silphium pinnatifidum Elliott = Silphium terebinthinaceum Jacq. var. pinnatifidum（Elliott）A. Gray ■☆

364306　Silphium pumilum Michx. = Berlandiera pumila（Michx.）Nutt. ■☆

364307　Silphium radula Nutt.；糙叶松香草；Roughleaf Rosinweed ■☆

364308　Silphium radula Nutt. var. gracile（A. Gray）Clevinger；纤细糙叶松香草；Slender Rosinweed ■☆

364309　Silphium reniforme Raf. ex Nutt. = Silphium compositum Michx. ■☆

364310　Silphium reverchonii Bush = Silphium radula Nutt. ■☆

364311　Silphium rumicifolium Small = Silphium terebinthinaceum Jacq. ■☆

364312　Silphium scaberrimum Elliott = Silphium asteriscus L. ■☆

364313　Silphium simpsonii Greene = Silphium asteriscus L. var. simpsonii（Greene）Clevinger ■☆

364314　Silphium simpsonii Greene var. wrightii L. M. Perry = Silphium radula Nutt. ■☆

364315　Silphium speciosum Nutt. = Silphium integrifolium Michx. var. laeve Torr. et A. Gray ■☆

364316　Silphium subacaule Nutt. = Berlandiera subacaulis（Nutt.）Nutt. ■☆

364317　Silphium terebinthinaceum Jacq.；笃乳香状松香草；Basal-leaved Rosinweed Prairie-dock，Prairie Dock，Prairie Rosinweed ■☆

364318　Silphium terebinthinaceum Jacq. var. lucy-brauniae Steyerm. = Silphium terebinthinaceum Jacq. ■☆

364319　Silphium terebinthinaceum Jacq. var. pinnatifidum（Elliott）A. Gray；羽裂笃乳香状松香草■☆

364320　Silphium tomentosum Pursh = Berlandiera pumila（Michx.）Nutt. ■☆

364321　Silphium trifoliatum L.；三小叶松香草；Whorled Rosinweed ■☆

364322　Silphium trifoliatum L. = Silphium asteriscus L. var. trifoliatum（L.）Clevinger ■☆

364323　Silphium trifoliatum L. var. latifolium A. Gray = Silphium asteriscus L. var. latifolium（A. Gray）Clevinger ■☆

364324　Silphium trilobatum L. = Sphagneticola trilobata（L.）Pruski ■☆

364325　Silphium trilobatum L. = Wedelia trilobata（L.）Hitchc. ■☆

364326　Silphium venosum Small = Silphium compositum Michx. ■☆

364327　Silphium wasiotense Medley；阿巴拉契亚松香草；Appalachian Rosinweed ■☆

364328　Silvaea Hook. et Arn. = Trigonostemon Blume(保留属名)●

364329　Silvaea Meisn. = Mezilaurus Kuntze ex Taub. ●☆

364330　Silvaea Meisn. = Silvia Allemão ●☆

364331　Silvaea Phil. = Philippiamra Kuntze ●☆

364332　Silvalismis Thouars = Calanthe R. Br.（保留属名）■

364333　Silvia Allemão = Mezia Kuntze ●☆

364334　Silvia Allemão = Mezilaurus Kuntze ex Taub. ●☆

364335　Silvia Allemão = Neosilvia Pax ●☆

364336　Silvia Benth. = Silviella Pennell ■☆

364337　Silvia Vell. = Escobedia Ruiz et Pav. ■☆

364338　Silvianthus Hook. f.（1868）；蜘蛛花属（西威花属）；Silvianthus ●

364339　Silvianthus bracteatus Hook. f.；蜘蛛花；Common Silvianthus，Silvianthus ●

364340　Silvianthus bracteatus Hook. f. subsp. clerodendroides（Airy Shaw）H. W. Li = Silvianthus tonkinensis（Gagnep.）Ridsdale ●■

364341　Silvianthus bracteatus Hook. f. subsp. tonkinensis（Gagnep.）H. W. Li = Silvianthus tonkinensis（Gagnep.）Ridsdale ●■

364342　Silvianthus clerodendroides Airy Shaw = Silvianthus tonkinensis（Gagnep.）Ridsdale ●■

364343　Silvianthus radiciflorus C. B. Clarke = Mycetia radiciflora（C. B. Clarke）Airy Shaw ●☆

364344　Silvianthus tonkinensis（Gagnep.）Ridsdale；线萼蜘蛛花；Linearcalyx Silvianthus，Tonkin Silvianthus ●■

364345　Silviella Pennell(1928)；林列当属（林玄参属）■☆

364346　Silviella prostrata（Kunth）Pennell；林列当■☆

364347　Silvinula Pennell = Bacopa Aubl.（保留属名）■

364348　Silvorchis J. J. Sm.（1907）；森林兰属■☆

364349　Silvorchis aurea（Aver. et Averyanova）Szlach.；黄森林兰■☆

364350　Silvorchis colorata J. J. Sm.；森林兰■☆

364351　Silybon St. -Lag. = Silybum Vaill.（保留属名）■

364352　Silybum Adans. = Silybum Vaill.（保留属名）■

364353　Silybum Vaill.（1754）（保留属名）；水飞蓟属（水飞雉属）；Milk Thistle ■

364354　Silybum Vaill. ex Adans. = Silybum Vaill.（保留属名）■

364355　Silybum atriplicifolium（Trevis.）Fisch. = Synurus deltoides（Aiton）Nakai ■

364356　Silybum cernuum Germ. = Alfredia cernua（L.）Cass. ■

364357　Silybum eburneum Coss. et Durieu；象牙蓟■☆

364358　Silybum maculatum（Scop.）Moench. = Silybum marianum（L.）Gaertn. ■

364359　Silybum mariae（Crantz）Gray = Silybum marianum（L.）Gaertn. ■

364360　Silybum marianum（L.）Gaertn.；水飞蓟（斑点红花，斑水飞雉，飞雉，老鼠筋，奶蓟，水禾，小飞雉）；Blessed Mary's Thistle，Blessed Milk Thistle，Blessed Milkthistle，Blessed Thistle，Bull Thistle，Holy Thistle，Kenguel Seed，Lady's Milk，Lady's Thistle，Marian Thistle，Mary's Thistle，Mary's Titistle，Milk Thistle，Milky Thrissel，Our Lady's Milk Thistle，St. Mary Thistle，St. Mary's Thistle，Striped Milk Thistle，Striped Milky Thistle，Variegated Thistle，Virgin Mary's Thistle，Virgin's Milk，Virgin's Thistle ■

364361　Silybum marianum（L.）Gaertn. var. albiflorum Eig = Silybum marianum（L.）Gaertn. ■

364362　Silybum marianum（L.）Gaertn. var. longispina Lamotte = Silybum marianum（L.）Gaertn. ■

364363　Silybum mauclatum（Scop.）Loench = Silybum marianum（L.）Gaertn. ■

364364　Silymbrium Neck. = Sisymbrium L. ■

364365　Simaba Aubl.（1775）；苦香木属（希麻巴属，希马巴属）●☆

364366　Simaba Aubl. = Quassia L. ●☆

364367　Simaba africana Baill. = Quassia africana（Baill.）Baill. ●☆

364368　Simaba cedron Planch.；苦香木；Cedron ●☆

364369　Simaba cuspidata Spruce ex Engl.；骤尖苦香木●☆

364370　Simaba ferruginea A. St. -Hil.；锈色苦香木；Calunga-bark ●☆

364371　Simaba gabonensis（Pierre）Feuillet = Odyendyea gabonensis（Pierre）Engl. ●☆

364372　Simaba grandifolia（Engl.）Feuillet = Pierreodendron africanum（Hook. f.）Little ●☆

364373　Simaba guianensis Aubl.；圭亚那苦香木●☆

364374　Simaba multiflora A. Juss.；多花苦香木（西马巴苦木）●☆

364375　Simaba quassioides D. Don = Picrasma quassioides（D. Don）A. W. Benn. ●

364376　Simaba schweinfurthii（Oliv.）Feuillet = Quassia schweinfurthii（Oliv.）Noot. ●☆

364377　Simaba undulata Guillaumin et Perr. = Quassia undulata（Guillaumin et Perr.）F. Dietr. ●☆

364378　Simabaceae Horan.；苦香木科（水飞蓟科）●■

364379　Simabaceae Horan. = Simaroubaceae DC.（保留科名）●

364380　Simarouba Aubl.（1775）（保留属名）；苦樗属（苦木属）；Simarouba，Simaruba ●☆

364381　Simarouba Aubl.（保留属名）= Quassia L. ●☆

364382　Simarouba amara Aubl.；苦樗（苦木，类苦木，苏里南苦木）；Bitter Damson，Bitter Quassia，Bitterwood，Marupa，Mountain Damson，Orinoco Simaruba，Quassia Wood，Simaruba，South American Bitterwood，Stave-wood，Surinam Ouassia，Surinam Quassia Wood ●☆

364383　Simarouba cuspidata Spruce；尖苦樗（尖苦香木）●☆

364384　Simarouba glauca DC.；乐园树；Paradise Tree ●☆

364385　Simarouba officinalis DC.；药用苦樗（苦樗）●☆

364386　Simarouba versicolor A. St. -Hil.；变色苦樗●☆

364387　Simaroubaceae DC.（1811）（保留科名）；苦木科（樗树科）；Ailanthus Family，Quassia Family，Simaruba Family，Tree-of-heaven Family ●

364388　Simaruba Aubl. = Simarouba Aubl.（保留属名）●☆

364389　Simaruba Boehm. (废弃属名) = Bursera Jacq. ex L. (保留属名) ●☆

364390　Simaruba Boehm. (废弃属名) = Simarouba Aubl. (保留属名) ●☆

364391　Simaruba cuspidata Spruce = Simaba cuspidata Spruce ex Engl. ●☆

364392　Simarubaceae DC. = Simaroubaceae DC. (保留科名) ●

364393　Simarubopsis Engl. = Pierreodendron Engl. ●☆

364394　Simarubopsis Engl. = Quassia L. ●☆

364395　Simarubopsis kerstingii Engl. = Pierreodendron kerstingii (Engl.) Little ●☆

364396　Simblocline DC. = Diplostephium Kunth ●☆

364397　Simbuleta Forssk. (废弃属名) = Anarrhinum Desf. (保留属名) ■●☆

364398　Simbuleta arabica Poir. = Anarrhinum forskaohlii (J. F. Gmel.) Cufod. ■☆

364399　Simbuleta forskaohlii J. F. Gmel. = Anarrhinum forskaohlii (J. F. Gmel.) Cufod. ■☆

364400　Simbuleta fruticosa (Desf.) Hochr. = Anarrhinum fruticosum Desf. ■☆

364401　Simbuleta pechuelii (Kuntze) Kuntze = Diclis petiolaris Benth. ■☆

364402　Simbuleta pedata (Desf.) Hochr. = Anarrhinum pedatum Desf. ■☆

364403　Simbuleta veronicoides (A. Rich.) Kuntze = Diclis ovata Benth. ■☆

364404　Simenia Szabo = Dipsacus L. ■

364405　Simenia acaulis (Steud. ex A. Rich.) Szabó = Dipsacus pinnatifidus Steud. ex A. Rich. ■☆

364406　Simethidaceae Juss. = Simmondsiaceae Tiegh. ●☆

364407　Simethis Kunth (1843) (保留属名); 西米兹花属 (西米兹属); Kerry Lily ■☆

364408　Simethis bicolor Kunth = Simethis mattiazzii (Vand.) G. López et Jarvis ■☆

364409　Simethis mattiazzii (Vand.) G. López et Jarvis; 西米兹花; Kerry Lily ■☆

364410　Simethis planifolia (L.) Gren. et Godr. = Simethis mattiazzii (Vand.) G. López et Jarvis ■☆

364411　Simicratea N. Hallé (1983); 凹脉卫矛属 ●☆

364412　Simicratea welwitschii (Oliv.) N. Hallé; 凹脉卫矛 ●☆

364413　Simidetia humilis Raf. = Coleanthus subtilis (Tratt.) Seidl ■

364414　Similisinocarum Cauwet et Farille = Pimpinella L. ■

364415　Similisinocarum pimpinellisimulacrum Farille et S. B. Malla = Pimpinella pimpinellisimulacrum (Farille et S. B. Malla) Farille ■

364416　Simira Aubl. (1775); 西米尔茜属 ■☆

364417　Simira Raf. = Scilla L. ■

364418　Simira cordifolia (Hook. f.) Steyerm.; 心叶西米尔茜 ■☆

364419　Simira fragrans (Rusby) Steyerm.; 香西米尔茜 ■☆

364420　Simira longifolia (Willd.) Bremck.; 长叶西米尔茜 ■☆

364421　Simira nitida Poir.; 光亮西米尔茜 ■☆

364422　Simira rubra (K. Schum.) Steyerm.; 红西米尔茜 ■☆

364423　Simirestis N. Hallé = Hippocratea L. ●☆

364424　Simirestis N. Hallé (1958); 扁丝卫矛属 ●☆

364425　Simirestis andongensis (Welw. ex Oliv.) N. Hallé ex R. Wilczek = Pristimera andongensis (Welw. ex Oliv.) N. Hallé ●☆

364426　Simirestis delagoensis (Loes.) N. Hallé = Prionostemma delagoensis (Loes.) N. Hallé ●☆

364427　Simirestis dewildemaniana N. Hallé; 德怀尔德曼扁丝卫矛 ●☆

364428　Simirestis fimbriata (Exell) N. Hallé ex R. Wilczek = Prionostemma fimbriata (Exell) N. Hallé ●☆

364429　Simirestis goetzei (Loes.) N. Hallé ex R. Wilczek; 格兹扁丝卫矛 ●☆

364430　Simirestis graciliflora (Welw. ex Oliv.) N. Hallé ex R. Wilczek = Pristimera graciliflora (Welw. ex Oliv.) N. Hallé ●☆

364431　Simirestis isangiensis (De Wild.) R. Wilczek = Cuervea isangiensis (De Wild.) N. Hallé ●☆

364432　Simirestis klaineana N. Hallé; 克莱恩扁丝卫矛 ●☆

364433　Simirestis luteoviridis (Exell) N. Hallé = Pristimera luteoviridis (Exell) N. Hallé ●☆

364434　Simirestis mouilensis N. Hallé = Pristimera mouilensis (N. Hallé) N. Hallé ●☆

364435　Simirestis paniculata (Vahl) N. Hallé = Pristimera paniculata (Vahl) N. Hallé ●☆

364436　Simirestis plumbea (Blakelock et R. Wilczek) N. Hallé = Pristimera plumbea (Blakelock et R. Wilczek) N. Hallé ●☆

364437　Simirestis poggei (Loes.) R. Wilczek; 波格扁丝卫矛 ●☆

364438　Simirestis polyantha (Loes.) N. Hallé = Pristimera polyantha (Loes.) N. Hallé ●☆

364439　Simirestis preussii (Loes.) N. Hallé = Pristimera preussii (Loes.) N. Hallé ●☆

364440　Simirestis ritschardii (R. Wilczek) N. Hallé ex R. Wilczek = Prionostemma delagoensis (Loes.) N. Hallé var. ritschardii (R. Wilczek) N. Hallé ●☆

364441　Simirestis scheffleri (Loes.) N. Hallé; 谢夫勒扁丝卫矛 ●☆

364442　Simirestis staudtii (Loes.) N. Hallé; 施陶扁丝卫矛 ●☆

364443　Simirestis tisserantii N. Hallé; 蒂斯朗特扁丝卫矛 ●☆

364444　Simirestis unguiculata (Loes.) N. Hallé = Prionostemma unguiculata (Loes.) N. Hallé ●☆

364445　Simirestis welwitschii (Oliv.) N. Hallé = Simicratea welwitschii (Oliv.) N. Hallé ●☆

364446　Simlera Bubani = Leontopodium (Pers.) R. Br. ex Cass. ●■

364447　Simmondsia Nutt. (1844); 旱黄杨属 (荷荷巴属, 西蒙德木属, 希蒙德木属, 希蒙木属, 油蜡树属); Jojoba ●☆

364448　Simmondsia californica Nutt. = Simmondsia chinensis (Link) C. K. Schneid. ●☆

364449　Simmondsia chinensis (Link C. K. Schneid. = Buxus chinensis Link ●☆

364450　Simmondsia chinensis (Link) C. K. Schneid.; 旱黄杨 (荷荷巴, 藿藿巴, 加州西蒙德木, 加州希蒙, 西蒙德木, 希蒙德木, 中国西蒙德木); Deer Nut, Goat Nut, Goatnut, Goat's Nut, Jojoba, Pig Nut, Quinine Plant, Wild Hazel ●☆

364451　Simmondsiaceae (Müll. Arg.) Tiegh. ex Reveal et Hoogland (1990); 旱黄杨科 (荷荷巴科, 西蒙德木科, 希蒙德木科, 希蒙木科, 油蜡树科); Jojoba Family ●☆

364452　Simmondsiaceae (Pax) Tiegh. = Simmondsiaceae (Müll. Arg.) Tiegh. ex Reveal et Hoogland ●☆

364453　Simmondsiaceae Reveal et Hoogland = Buxaceae Dumort. (保留科名) ●■

364454　Simmondsiaceae Reveal et Hoogland = Simmondsiaceae Tiegh. ex Reveal et Hoogland ●☆

364455　Simmondsiaceae Tiegh. = Simmondsiaceae Tiegh. ex Reveal et Hoogland ●☆

364456　Simmondsiaceae Tiegh. ex Reveal et Hoogland = Buxaceae Dumort. (保留科名) ●■

364457　Simmondsiaceae Tiegh. ex Reveal et Hoogland = Simmondsiaceae（Müll. Arg.）Tiegh. ex Reveal et Hoogland ●☆

364458　Simmondslaceae（Muell. Arg.）Reveal et Hoogland = Simmondsiaceae（Müll. Arg.）Tiegh. ex Reveal et Hoogland ●☆

364459　Simocheilus Klotzsch = Erica L. ●☆

364460　Simocheilus Klotzsch（1838）；厚萼杜鹃属●☆

364461　Simocheilus acutangulus N. E. Br. = Erica glabella Thunb. subsp. laevis E. G. H. Oliv. ●☆

364462　Simocheilus albirameus N. E. Br. = Erica inaequalis（N. E. Br.）E. G. H. Oliv. ●☆

364463　Simocheilus barbiger Klotzsch = Erica uberiflora E. G. H. Oliv. ●☆

364464　Simocheilus bicolor（Klotzsch）Klotzsch = Erica inaequalis（N. E. Br.）E. G. H. Oliv. ●☆

364465　Simocheilus carneus Klotzsch = Erica uberiflora E. G. H. Oliv. ●☆

364466　Simocheilus depressus（Licht. ex Roem. et Schult.）Benth；凹陷厚萼杜鹃●☆

364467　Simocheilus depressus（Licht. ex Roem. et Schult.）Benth = Erica glabella Thunb. ●☆

364468　Simocheilus depressus（Licht. ex Roem. et Schult.）Benth var. patens（Benth.）N. E. Br. = Erica glabella Thunb. ●☆

364469　Simocheilus dispar N. E. Br. = Erica dispar（N. E. Br.）E. G. H. Oliv. ●☆

364470　Simocheilus ecklonianus Benth. = Erica inaequalis（N. E. Br.）E. G. H. Oliv. ●☆

364471　Simocheilus fourcadei（L. Guthrie）E. G. H. Oliv. = Erica angulosa E. G. H. Oliv. ●☆

364472　Simocheilus glabellus（Thunb.）Benth. = Erica glabella Thunb. ●☆

364473　Simocheilus glaber（Thunb.）Benth. = Erica inaequalis（N. E. Br.）E. G. H. Oliv. ●☆

364474　Simocheilus globiferus N. E. Br. = Erica glabella Thunb. subsp. laevis E. G. H. Oliv. ●☆

364475　Simocheilus hirsutus Benth. = Erica glabella Thunb. subsp. laevis E. G. H. Oliv. ●☆

364476　Simocheilus hirtus（Klotzsch）E. G. H. Oliv. = Erica glabella Thunb. subsp. laevis E. G. H. Oliv. ●☆

364477　Simocheilus hispidus（Klotzsch）Benth. = Erica inaequalis（N. E. Br.）E. G. H. Oliv. ●☆

364478　Simocheilus klotzschianus Benth. = Erica inaequalis（N. E. Br.）E. G. H. Oliv. ●☆

364479　Simocheilus klotzschianus Benth. var. glabrifolius N. E. Br. = Erica inaequalis（N. E. Br.）E. G. H. Oliv. ●☆

364480　Simocheilus multiflorus Klotzsch = Erica uberiflora E. G. H. Oliv. ●☆

364481　Simocheilus multiflorus Klotzsch var. atherstonei N. E. Br. = Erica uberiflora E. G. H. Oliv. ●☆

364482　Simocheilus oblongus Benth. = Erica dregei E. G. H. Oliv. ●☆

364483　Simocheilus obovatus Benth. = Erica thamnoides E. G. H. Oliv. ●☆

364484　Simocheilus patens Benth. = Erica glabella Thunb. ●☆

364485　Simocheilus patulus N. E. Br. = Erica glabella Thunb. subsp. laevis E. G. H. Oliv. ●☆

364486　Simocheilus piquetbergensis N. E. Br. = Erica piquetbergensis（N. E. Br.）E. G. H. Oliv. ●☆

364487　Simocheilus puberulus（Klotzsch）E. G. H. Oliv. = Erica

364488　Simocheilus pubescens Klotzsch = Erica uberiflora E. G. H. Oliv. ●☆

364489　Simocheilus purpureus（P. J. Bergius）Druce = Erica glabella Thunb. ●☆

364490　Simocheilus quadrifidus Benth. = Erica quadrifida（Benth.）E. G. H. Oliv. ●☆

364491　Simocheilus quadrisulcus N. E. Br. = Erica phaeocarpa E. G. H. Oliv. ●☆

364492　Simocheilus submuticus Benth. = Erica glabella Thunb. subsp. laevis E. G. H. Oliv. ●☆

364493　Simocheilus subrigidus N. E. Br. = Erica glabella Thunb. subsp. laevis E. G. H. Oliv. ●☆

364494　Simocheilus viscosus Bolus = Erica viscosissima E. G. H. Oliv. ●☆

364495　Simonenium Willis = Sinomenium Diels ●

364496　Simonisia Nees = Beloperone Nees ■☆

364497　Simonisia Nees = Justicia L. ●■

364498　Simonsia Kuntze = Beloperone Nees ■☆

364499　Simonsia Kuntze = Justicia L. ●■

364500　Simonsia Kuntze = Simonisia Nees ■☆

364501　Simphitum Neck. = Symphytum L. ■

364502　Simplicia Kirk（1897）；简禾属■☆

364503　Simplicia laxa Kirk；简禾☆

364504　Simplocarpus F. Schmidt = Symplocarpus Salisb. ex W. P. C. Barton（保留属名）■

364505　Simplocos Lex. = Symplocos Jacq. ●

364506　Simpsonia O. F. Cook = Thrinax L. f. ex Sw. ●☆

364507　Simsia Pers.（1807）；木向日葵属（西氏菊属）；Bush Sunflower,Bushsunflower ●■☆

364508　Simsia R. Br. = Stirlingia Endl. ●☆

364509　Simsia calva（A. Gray et Engelm.）A. Gray；无芒；Awnless Bush Sunflower ●■■☆

364510　Simsia exaristata A. Gray = Simsia lagasceiformis DC. ■☆

364511　Simsia frutescens A. Gray = Encelia frutescens（A. Gray）A. Gray ●☆

364512　Simsia lagasceiformis DC.；一年生西氏菊；Annual Bush Sunflower ■☆

364513　Simsia scaposa A. Gray = Encelia scaposa（A. Gray）A. Gray ■☆

364514　Simsimum Bernh. = Sesamum L. ■●

364515　Simsimum rostratum Bernh. = Sesamum alatum Thonn. ■☆

364516　Sinabraca G. H. Loos = Sinapis L. ■

364517　Sinacalia H. Rob. et Brettell（1973）；华蟹甲属（华蟹甲草属，中国千里光属）；Chinese Rag'wort,Sinacalia ■★

364518　Sinacalia caroli（C. Winkl.）C. Jeffrey et Y. L. Chen；革叶华蟹甲（华蟹甲草）；Coriaceous Sinacalia,Leatherleaf Sinacalia ■

364519　Sinacalia davidii（Franch.）H. Koyama；双花华蟹甲（双花华蟹甲草，双舌华蟹甲草）；David Sinacalia,Two Ray-florets Sinacalia ■

364520　Sinacalia henryi（Hemsl.）H. Rob. et Brettel = Sinacalia tangutica（Maxim.）B. Nord. ■

364521　Sinacalia macrocephala（H. Rob. et Brettell）C. Jeffrey et Y. L. Chen；大头华蟹甲（大花华蟹甲草）；Bighead Sinacalia,Largehead Sinacalia ■

364522　Sinacalia tangutica（Maxim.）B. Nord.；华蟹甲（登云鞋，鸡多囊，水葫芦七，水萝卜，唐古特蟹甲，唐古特蟹甲草，羊角天麻，羽裂华蟹甲草，羽裂蟹甲草，猪肚子）；Chinese Kagwort, Pinnate

Sinacalia, Pinnately Divided Sinacalia ■

364523　Sinadoxa C. Y. Wu, Z. L. Wu et R. F. Huang(1981);华福花属;Chinese Muskroot,Sinadoxa ■★

364524　Sinadoxa corydalifolia C. Y. Wu, Z. L. Wu et R. F. Huang;华福花;Chinese Muskroot,Sinadoxa ■

364525　Sinapi Dulac = Rhynchosinapis Hayek ■☆

364526　Sinapi Mill. = Sinapis L. ■

364527　Sinapidendron Lowe(1831);芥树属■●☆

364528　Sinapidendron angustifolium (DC.) Lowe;窄叶芥树■☆

364529　Sinapidendron bourgaei Webb ex Christ = Brassica bourgeaui (H. Christ) Kuntze ■☆

364530　Sinapidendron decumbens A. Chev. = Diplotaxis hirta (A. Chev.) Rustan et L. Borgen ■☆

364531　Sinapidendron frutescens (Aiton) Lowe;灌木芥树●☆

364532　Sinapidendron frutescens (Aiton) Lowe subsp. succulentus (Lowe) Rustan;多汁芥树■☆

364533　Sinapidendron frutescens (Aiton) Lowe var. succulentus Lowe = Sinapidendron frutescens (Aiton) Lowe subsp. succulentus (Lowe) Rustan ■☆

364534　Sinapidendron glaucum J. A. Schmidt = Diplotaxis glauca (J. A. Schmidt) O. E. Schulz ■☆

364535　Sinapidendron gracile Webb = Diplotaxis gracilis (Webb) O. E. Schulz ■☆

364536　Sinapidendron gymnocalyx (Lowe) Rustan;裸萼芥树●☆

364537　Sinapidendron hirtum A. Chev. = Diplotaxis hirta (A. Chev.) Rustan et L. Borgen ■☆

364538　Sinapidendron hirtum A. Chev. var. paucipilosum？= Diplotaxis hirta (A. Chev.) Rustan et L. Borgen ■☆

364539　Sinapidendron palmense (Kuntze) O. E. Schulz = Sinapis pubescens L. ■☆

364540　Sinapidendron rupestre Lowe;岩生芥树■☆

364541　Sinapidendron rupestre Lowe var. gymnocalyx Lowe = Sinapidendron gymnocalyx (Lowe) Rustan ●☆

364542　Sinapidendron sempervivifolium Menezes;常绿芥树■☆

364543　Sinapidendron vogelii Webb = Diplotaxis vogelii (Webb) Cout. ■☆

364544　Sinapis L. (1753);白芥属(芥属,欧白芥属,欧芥属);Mustard ■

364545　Sinapis alba L.;白芥(白辣菜子,胡芥,辣菜子,欧白芥,蜀芥);Mustard,Runch,Senvie,Senvy,White Mustard,Yellow Mustard ■

364546　Sinapis alba L. subsp. dissecta (Lag.) Bonnier;深裂白芥■☆

364547　Sinapis alba L. subsp. mairei (H. Lindb.) Maire;迈雷白芥■☆

364548　Sinapis alba L. var. Dittrichocarpa Maire = Sinapis alba L. ■

364549　Sinapis alba L. var. lagascana Alef. = Sinapis alba L. ■

364550　Sinapis alba L. var. latirostris O. E. Schulz = Sinapis alba L. ■

364551　Sinapis alba L. var. melanosperma Alef. = Sinapis alba L. ■

364552　Sinapis alba L. var. trichocarpa Maire = Sinapis alba L. ■

364553　Sinapis allionii Jacq. = Sinapis arvensis L. subsp. allionii (Jacq.) Baillarg. ■☆

364554　Sinapis angustifolium DC. = Sinapidendron angustifolium (DC.) Lowe ■☆

364555　Sinapis aphanoneura Maire et Weiller = Trachystoma aphanoneurum (Maire et Weiller) Maire ■☆

364556　Sinapis aristidis Pomel = Sinapis pubescens L. subsp. aristidis (Pomel) Maire et Weiller ■☆

364557　Sinapis arvensis L.;新疆白芥(田野白芥,野芥,野欧白芥);Bastard Rocket, Bazzock, Birdseed, Brashlach, Brashlagh, Brassick, Brassock, Brazzock, Bread-and-marmalade, Cabbage Flower, Cabbage Seed, Cadlock, Calf's Foot, California Rape, Callock, Careluck, Carlock, Charlick, Charlock, Charlock Mustard, Charnock, Churlick, Common Mustard, Corn Kale, Corn Mustard, Cranops, Crunchweed, Curlick, Durham Mustard, Field Kale, Field Mustard, Garlock, Goldlock, Gools, Harlock, Keblock, Kecklock, Kedlack, Kedlet, Kedlock, Kelk, Kellock, Kenry, Kerlack, Kerleck, Kerlick, Kerlock, Ketlack, Ketlock, Kettle-dock, Kilk, Kinkle, Kyerlic, Mustard Weed, Mustard Weld, Prushia, Rape, Runch, Runch-balls, Runchie, Runchik, Rungy, Scaldrick, Scalies, Senvie, Senvy, Shirt, Sinvey, Skedlock, Skellies, Skellock, Skillock, Skillog, Turnip, Warlock, Wild Charlock, Wild Cole, Wild Kale, Wild Mustard, Wild Rape, Wild Turnip, Will Kale, Xingjiang Mustard, Yellow, Yellow Top, Yellow-flower, Yellowtop, Yellow-weed, Zenry, Zenvy ■

364558　Sinapis arvensis L. subsp. allionii (Jacq.) Baillarg.;阿廖尼野芥■☆

364559　Sinapis arvensis L. var. aphanoneura (Maire et Weiller) Jahand. et Maire = Trachystoma aphanoneurum (Maire et Weiller) Maire ■☆

364560　Sinapis arvensis L. var. brachycarpa (Busch) O. E. Schulz;长喙白芥■

364561　Sinapis arvensis L. var. divaricata O. E. Schulz = Sinapis arvensis L. ■

364562　Sinapis arvensis L. var. longestylosa Sennen = Sinapis arvensis L. ■

364563　Sinapis arvensis L. var. orientalis (L.) W. D. J. Koch et Ziz;东方野芥■☆

364564　Sinapis arvensis L. var. orientalis (L.) W. D. J. Koch et Ziz = Sinapis arvensis L. ■

364565　Sinapis arvensis L. var. schkuhriana (Rchb.) Hagenb. = Sinapis arvensis L. ■

364566　Sinapis arvensis L. var. villosa Mérat = Sinapis arvensis L. ■

364567　Sinapis assurgens Delile = Diplotaxis assurgens (Delile) Thell. ■☆

364568　Sinapis bipinnata Desf. = Didesmus bipinnatus (Desf.) DC. ■☆

364569　Sinapis cernua Thunb. = Brassica juncea (L.) Czern. ■

364570　Sinapis chinensis L. var. integrifolia Stokes = Brassica juncea (L.) Czern. ■

364571　Sinapis choulettiana Coss. et Durieu = Brassica procumbens (Poir.) O. E. Schulz ■☆

364572　Sinapis circinnata Desf. = Sinapis pubescens L. ■☆

364573　Sinapis crassifolia Raf. = Diplotaxis harra (Forssk.) Boiss. subsp. crassifolia (Raf.) Maire ■☆

364574　Sinapis cuneifolia Roxb. = Brassica juncea (L.) Czern. ■

364575　Sinapis dissecta Lag. = Sinapis alba L. subsp. dissecta (Lag.) Bonnier ■☆

364576　Sinapis erucoides L. = Diplotaxis erucoides (L.) DC. ■☆

364577　Sinapis flexuosa Poir.;曲折白芥■☆

364578　Sinapis frutescens Aiton = Sinapidendron frutescens (Aiton) Lowe ●☆

364579　Sinapis geniculata Desf. = Hirschfeldia incana (L.) Lagr.-Foss. ■☆

364580　Sinapis harra Forssk. = Diplotaxis harra (Forssk.) Boiss. ■☆

364581　Sinapis hispanica L. = Erucaria hispanica (L.) Druce ■☆

364582　Sinapis hispida Schousb. = Sinapis flexuosa Poir. ■☆

364583　Sinapis incana L. = Hirschfeldia incana (L.) Lagr.-Foss. ■☆

364584　Sinapis indurata Coss. = Sinapis pubescens L. subsp. indurata (Coss.) Batt. ■☆

364585　Sinapis integrifolia West = Brassica carinata A. Braun ■☆

364586　Sinapis japonica Thunb. = Brassica juncea (L.) Czern. ■

364587　Sinapis juncea L. = Brassica juncea (L.) Czern. ■

364588　Sinapis juncea L. var. napiformis Pailleux et Bois = Brassica juncea (L.) Czern. var. napiformis (Pailleux et Bois) Kitam. ■

364589　Sinapis juncea L. var. napiformis Pailleux et Bois = Brassica napiformis (Pailleux et Bois) L. H. Bailey ■

364590　Sinapis kaber DC. = Sinapis arvensis L. ■

364591　Sinapis lanceolata DC. = Brassica juncea (L.) Czern. ■

364592　Sinapis leptopetala DC. = Erucastrum strigosum (Thunb.) O. E. Schulz ■☆

364593　Sinapis mairei H. Lindb. = Sinapis alba L. subsp. mairei (H. Lindb.) Maire ■☆

364594　Sinapis millefolia Jacq. = Descurainia millefolia (Jacq.) Webb et Berthel. ■☆

364595　Sinapis muralis (L.) R. Br. = Diplotaxis muralis (L.) DC. ■

364596　Sinapis nigra L. = Brassica nigra (L.) W. D. J. Koch ■

364597　Sinapis nudicaulis Lag. = Guenthera repanda (Willd.) Gómez-Campo subsp. africana (Maire) Gómez-Campo ■☆

364598　Sinapis patens Roxb. = Brassica juncea (L.) Czern. ■

364599　Sinapis pekinensis Lour. = Brassica campestris L. var. pekinensis (Lour.) Viehoever ■

364600　Sinapis pekinensis Lour. = Brassica rapa L. var. glabra Regel ■

364601　Sinapis pendula E. Mey. ;垂白芥■☆

364602　Sinapis philaeana Delile = Morettia philaeana (Delile) DC. ■☆

364603　Sinapis procumbens Poir. = Brassica procumbens (Poir.) O. E. Schulz ■☆

364604　Sinapis pubescens L. ;平铺白芥■☆

364605　Sinapis pubescens L. subsp. aristidis (Pomel) Maire et Weiller; 三芒草白芥■☆

364606　Sinapis pubescens L. subsp. indurata (Coss.) Batt. ;坚硬白芥 ■☆

364607　Sinapis pubescens L. var. brachyloba Coss. = Sinapis pubescens L. ■☆

364608　Sinapis pubescens L. var. brevirostrata O. E. Schulz = Sinapis pubescens L. ■☆

364609　Sinapis pubescens L. var. circinata Coss. = Sinapis pubescens L. ■☆

364610　Sinapis pubescens L. var. cyrenaica Coss. et Daveau = Sinapis pubescens L. ■☆

364611　Sinapis pubescens L. var. serrata (Huter et al.) Arcang. = Sinapis pubescens L. ■☆

364612　Sinapis pubescens L. var. tenuirostris Guss. = Sinapis pubescens L. ■☆

364613　Sinapis radicata Desf. = Brassica fruticulosa Cirillo subsp. radicata (Desf.) Batt. ■☆

364614　Sinapis recurvata All. = Coincya monensis (L.) Greuter et Burdet subsp. cheiranthos (Vill.) Aedo, Leadlay et Munoz Garm. ■☆

364615　Sinapis recurvata Desf. = Guenthera gravinae (Ten.) Gómez-Campo ■☆

364616　Sinapis retrosa Burch. ex DC. = Erucastrum strigosum (Thunb.) O. E. Schulz ■☆

364617　Sinapis rugosa Roxb. = Brassica juncea (L.) Czern. ■

364618　Sinapis schkuhriana Rchb. = Sinapis arvensis L. ■

364619　Sinapis serrata Huter et al. = Sinapis pubescens L. ■☆

364620　Sinapis virgata Cav. = Diplotaxis virgata (Cav.) DC. ■☆

364621　Sinapis weilleri Maire = Trachystoma ballii O. E. Schulz ■☆

364622　Sinapistrum Chevall. = Sinapis L. ■

364623　Sinapistrum Medik. = Gynandropsis DC. (保留属名)■

364624　Sinapistrum Mill. = Cleome L. ●■

364625　Sinapistrum Spach = Agrosinapis Fourr. ■●

364626　Sinapistrum Spach = Brassica L. ■●

364627　Sinapodendron Ball = Sinapidendron Lowe ■●☆

364628　Sinarundinaria Nakai = Fargesia Franch. emend. T. P. Yi ●

364629　Sinarundinaria Nakai = Fargesia Franch. ●

364630　Sinarundinaria Nakai = Sinoarundinaria Ohwi ●☆

364631　Sinarundinaria Nakai(1935);华桔竹属(箭竹属,四时竹属, 玉山竹属);Bamboo,China Cane,Chinacane,Fountain Bamboo ●

364632　Sinarundinaria Ohwi = Phyllostachys Siebold et Zucc. (保留属名)●

364633　Sinarundinaria acutissima (Keng) P. C. Keng = Fargesia melanostachys (Hack. ex Hand. -Mazz.) T. P. Yi ●

364634　Sinarundinaria alpina (K. Schum.) C. S. Chao et Renvoize = Arundinaria alpina K. Schum. ●☆

364635　Sinarundinaria anaurita T. H. Wen;井冈唐竹;Jinggangshan Chinacane ●

364636　Sinarundinaria anceps (Mitford) C. S. Chao et Renvoize = Arundinaria anceps Mitford ●☆

364637　Sinarundinaria andropogonoides (Hand. -Mazz.) Keng ex P. C. Kcng = Yushania andropogonoides (Hand. -Mazz.) T. P. Yi ●

364638　Sinarundinaria basihirsuta (McClure) C. D. Chu et C. S. Chao = Yushania basihirsuta (McClure) Z. P. Wang et G. H. Ye ●

364639　Sinarundinaria brevipaniculata (Hand. -Mazz.) Keng ex P. C. Keng = Yushania brevipaniculata (Hand. -Mazz.) T. P. Yi ●

364640　Sinarundinaria brevipes (McClure) Keng ex P. C. Keng = Fargesia brevipes (McClure) T. P. Yi ●

364641　Sinarundinaria chungii (Keng) P. C. Keng;大箭竹(墨竹,钟氏冷竹);Chung China Cane,Chung Chinacane ●

364642　Sinarundinaria chungii (Keng) P. C. Keng = Yushania brevipaniculata (Hand. -Mazz.) T. P. Yi ●

364643　Sinarundinaria confusa (McClure) P. C. Keng = Yushania confusa (McClure) Z. P. Wang et G. H. Ye ●

364644　Sinarundinaria cuspidata (Keng) P. C. Keng = Fargesia cuspidata (Keng) Z. P. Wang et G. H. Ye ●

364645　Sinarundinaria cuspidata (Keng) P. C. Keng = Thamnocalamus cuspidatus (Keng) P. C. Keng ●

364646　Sinarundinaria edulis T. H. Wen = Acidosasa edulis (T. H. Wen) T. H. Wen ●

364647　Sinarundinaria faberi (Rendle) P. C. Keng = Arundinaria faberi Rendle ●

364648　Sinarundinaria fangiana (A. Camus) Keng et P. C. Keng = Arundinaria faberi Rendle ●

364649　Sinarundinaria fangiana (A. Camus) Keng ex P. C. Keng = Bashania fangiana (A. Camus) P. C. Keng et T. H. Wen ●

364650　Sinarundinaria ferax (Keng) P. C. Keng = Fargesia ferax (Keng) T. P. Yi ●

364651　Sinarundinaria forrestii (Keng) P. C. Keng = Fargesia melanostachys (Hack. ex Hand. -Mazz.) T. P. Yi ●

364652　Sinarundinaria gigantea T. H. Wen = Indosasa gigantea (T. H. Wen) T. H. Wen ●

364653　Sinarundinaria glabrescens T. H. Wen;屏南唐竹;Pingnan

Chinacane ●

364654　Sinarundinaria glabrifolia T. P. Yi；光叶华橘竹；Smoothleaf Chinacane ●

364655　Sinarundinaria griffithiana（Munro）C. S. Chao et Renvoize = Chimonocalamus griffithianus（Munro）J. R. Xue et T. P. Yi ●

364656　Sinarundinaria incana T. H. Wen = Sinobambusa incana T. H. Wen ●

364657　Sinarundinaria kunishii（Hayata）Kaneh. et Hatus. = Gelidocalamus kunishii（Hayata）P. C. Keng et T. H. Wen ●

364658　Sinarundinaria latiflorus（Munro）McClure = Dendrocalamus latiflorus Munro ●

364659　Sinarundinaria longissima T. P. Yi = Yushania complanata T. P. Yi ●

364660　Sinarundinaria longissima T. P. Yi = Yushania yadongensis T. P. Yi ●

364661　Sinarundinaria longiuscula J. R. Xue et Y. Y. Dai = Fargesia longiuscula（J. R. Xue et Y. Y. Dai）J. R. Xue ●

364662　Sinarundinaria macclureana（Bor）C. S. Chao et G. Y. Yang = Fargesia macclureana（Bor）Stapleton ●

364663　Sinarundinaria mairei（Hack. ex Hand. -Mazz.）Keng ex P. C. Keng = Fargesia mairei（Hack. ex Hand. -Mazz.）T. P. Yi ●

364664　Sinarundinaria mairei（Hack.）Keng ex P. C. Keng = Fargesia mairei（Hack. ex Hand. -Mazz.）T. P. Yi ●

364665　Sinarundinaria melanostachys（Hand. -Mazz.）Keng ex P. C. Keng = Fargesia melanostachys（Hack. ex Hand. -Mazz.）T. P. Yi ●

364666　Sinarundinaria murielae（Gamble）Nakai = Fargesia murielae（Gamble）T. P. Yi ●

364667　Sinarundinaria nephroaurita C. D. Chu et C. S. Chao = Sinobambusa nephroaurita C. D. Chu et C. S. Chao ●

364668　Sinarundinaria niitakayamensis（Hayata）P. C. Keng = Yushania niitakayamensis（Hayata）P. C. Keng ●

364669　Sinarundinaria nitida（Mitford）Nakai = Fargesia nitida（Mitford ex Stapf）P. C. Keng ex T. P. Yi ●

364670　Sinarundinaria papillosa W. T. Lin = Yushania basihirsuta（McClure）Z. P. Wang et G. H. Ye ●

364671　Sinarundinaria pauciflora（Keng）P. C. Keng = Fargesia pauciflora（Keng）T. P. Yi ●

364672　Sinarundinaria puberula T. H. Wen；多毛唐竹；Manyhair Chinacane ●

364673　Sinarundinaria pulchella T. H. Wen；美丽唐竹；Beautiful Chinacane ●

364674　Sinarundinaria scabrida T. H. Wen；冬笋竹●

364675　Sinarundinaria seminuda T. H. Wen；胶南竹；Halfnaked Sinobambusa，Jiaonan Sinobambusa，Jiaonan Tangbamboo ●

364676　Sinarundinaria sichuanensis T. P. Yi = Monstruocalamus sichuanensis（T. P. Yi）T. P. Yi ●

364677　Sinarundinaria sparsiflora（Rendle）P. C. Keng = Fargesia murielae（Gamble）T. P. Yi ●

364678　Sinarundinaria striata T. H. Wen；花纹唐竹●

364679　Sinarundinaria taotsii var. dentata T. H. Wen；福鼎唐竹●

364680　Sinarundinaria vicina（Keng）P. C. Keng = Fargesia vicina（Keng）T. P. Yi ●

364681　Sinarundinaria violascens（Keng）P. C. Keng = Yushania violascens（Keng）T. P. Yi ●

364682　Sinarundinaria wilsonii（Rendle）Keng ex P. C. Keng = Indocalamus wilsonii（Rendle）C. S. Chao et C. D. Chu ●

364683　Sinarundinaria wilsonii（Rendle）P. C. Keng = Indocalamus

wilsonii（Rendle）C. S. Chao et C. D. Chu ●

364684　Sinarundinaria yunnanensis（J. R. Xue et T. P. Yi）J. R. Xue et D. Z. Li = Fargesia yunnanensis J. R. Xue et T. P. Yi ●

364685　Sincarpia Ten. = Syncarpia Ten. ●☆

364686　Sinclairea Sch. Bip. = Sinclairia Hook. et Arn. ●☆

364687　Sinclairia Hook. et Arn.（1841）；落叶黄安菊属●☆

364688　Sinclairia Hook. et Arn. = Liabum Adans. ■●☆

364689　Sinclairia adenotricha Rydb.；腺毛落叶黄安菊●☆

364690　Sinclairia angustissima（A. Gray）B. L. Turner；窄落叶黄安菊●☆

364691　Sinclairia brachypus Rydb.；短梗落叶黄安菊●☆

364692　Sinclairia discolor Hook. et Arn.；异色落叶黄安菊●☆

364693　Sinclairia glabra Rydb.；光落叶黄安菊●☆

364694　Sinclairia hypoleuca（Greenm.）Rydb.；里白落叶黄安菊●☆

364695　Sinclairiopsis Rydb. = Sinclairia Hook. et Arn. ●☆

364696　Sincoraea Ule = Orthophytum Beer ■☆

364697　Sindechites Oliv.（1888）；毛药藤属；Sindechites ●

364698　Sindechites chinensis（Merr.）Markgr. et Tsiang；坭藤；China Sindechites，Chinese Sindechites ●

364699　Sindechites esquirolii（H. Lév.）Woodson = Sindechites henryi Oliv. ●

364700　Sindechites esquirolii Woods. = Sindechites henryi Oliv. ●

364701　Sindechites henryi Oliv.；毛药藤（黄经树，蔷薇根，土牛党七）；Henry Sindechites ●

364702　Sindechites henryi Oliv. var. parvifolia Tsiang = Sindechites henryi Oliv. ●

364703　Sindora Miq.（1861）；油楠属（蚌壳树属）；Sindora ●

364704　Sindora cochinchinensis Baill.；蚌壳树（白鹤树，娥木）●☆

364705　Sindora cochinchinensis Baill. = Sindora glabra Merr. ex de Wit ●

364706　Sindora coriacea Prain；革质油楠●☆

364707　Sindora glabra Merr. ex de Wit；油楠（蚌壳树，柴油树，科楠，曲脚楠）；Oil Sindora ●

364708　Sindora javanica Backer ex K. Heyne；爪哇油楠●☆

364709　Sindora klaineana Pierre ex Pellegr.；克莱恩油楠●☆

364710　Sindora maritima Pierre；海滨油楠●

364711　Sindora parvifolia Backer ex K. Heyne；小叶油楠●☆

364712　Sindora sumatrana Miq.；苏门答腊油楠●☆

364713　Sindora supa Merr.；斯帕油楠（斯帕蚌壳树，斯帕树）●☆

364714　Sindora tonkinensis A. Chev. = Sindora tonkinensis A. Chev. ex K. Larsen et S. S. Larsen ●

364715　Sindora tonkinensis A. Chev. ex K. Larsen et S. S. Larsen；东京油楠（印支蚌壳树）；Tonkin Sindora ●

364716　Sindora tonkinensis K. Larsen et S. S. Larsen = Sindora tonkinensis A. Chev. ex K. Larsen et S. S. Larsen ●

364717　Sindora wallichii Benth.；沃利克油楠●☆

364718　Sindoropsis J. Léonard（1957）；类油楠属（赛油楠属）●☆

364719　Sindoropsis lagascae Willk.；类油楠●☆

364720　Sindoropsis letestui（Pellegr.）J. Léonard = Copaifera letestui（Pellegr.）Pellegr. ●☆

364721　Sindroa Jum. = Orania Zipp. ●☆

364722　Sindroa longisquama Jum. = Orania longisquama（Jum.）J. Dransf. et N. W. Uhl ●☆

364723　Sineoperculum Van Jaarsv. = Dorotheanthus Schwantes ■☆

364724　Sineoperculum rourkei（L. Bolus）Van Jaarsv. = Dorotheanthus rourkei L. Bolus ■☆

364725　Singana Aubl.（1775）；辛甘豆属☆

364726　Singchia Z. J. Liu et L. J. Chen（2009）；麻栗坡兰属■★

364727 Singlingia Benth. = Danthonia DC. (保留属名)■

364728 Singularybas Molloy, D. L. Jones et M. A. Clem. (2002);新西兰铠兰属■☆

364729 Singularybas Molloy, D. L. Jones et M. A. Clem. = Corybas Salisb. ■

364730 Singularybas Molloy, D. L. Jones et M. A. Clem. = Nematoceras Hook. f. ■

364731 Sinia Diels = Sauvagesia L. ●

364732 Sinia Diels(1930);辛木属(合柱金莲木属); Sinia ●★

364733 Sinia rhodoleuca Diels;辛木(合柱金莲木); Common Sinia, Sinia ●◇

364734 Sinia rhodoleuca Diels = Sauvagesia rhodoleuca (Diels) M. C. E. Amaral ●

364735 Sinistrophorum Schrank ex Endl. = Camelina Crantz ■

364736 Sinningia Nees = Rechsteineria Regel(保留属名)■☆

364737 Sinningia Nees(1825);大岩桐属(块茎苣苔属); Gloxinia, Sinningia ●■☆

364738 Sinningia × hybrida Voss;杂种大岩桐■☆

364739 Sinningia atropurpurea Hort. ex Bellair et St. -Leger;暗紫块茎苣苔■☆

364740 Sinningia barbata G. Nicholson;喉毛大岩桐(髯毛大岩桐)●☆

364741 Sinningia cardinalis (Lehm.) H. E. Moore;红花大岩桐(艳桐草); Cardinal Flower, Helmet Flower ■☆

364742 Sinningia concinna G. Nicholson;小岩桐■☆

364743 Sinningia discolor (Kunze) Sprague;异色块茎苣苔■☆

364744 Sinningia eumorpha H. E. Moore;美花大岩桐■☆

364745 Sinningia floribunda A. Dietr.;繁花大岩桐■☆

364746 Sinningia leucotricha (Hoehne) H. E. Moore;白毛大岩桐; Brazilian Edelweiss ■☆

364747 Sinningia macrorhiza (Dumort.) Wiehler;大根大岩桐(大根块茎苣苔)■☆

364748 Sinningia macrostachya (Lindl.) Chautems;大穗大岩桐(大穗块茎苣苔)■☆

364749 Sinningia polyantha (DC.) Wiehler;多花大岩桐(多花块茎苣苔)■☆

364750 Sinningia punctata Ysabeau;斑点大岩桐(斑点块茎苣苔)■☆

364751 Sinningia purpurea Hort. ex Gentil;紫大岩桐(紫块茎苣苔)■☆

364752 Sinningia regina Sprague;女王大岩桐; Cinderella-slippers, Violet Slipper Gloxinia ■☆

364753 Sinningia sanguinea Regel;血红大岩桐(血红块茎苣苔)■☆

364754 Sinningia speciosa (Lodd.) Benth. et Hook. ex Hiern;紫蓝大岩桐(大岩桐,重瓣大岩桐); Brazilian Gloxinia, Common Gloxinia, Florist's Gloxinia, Gloxinia, Violet Slipper Gloxinia ■☆

364755 Sinningia speciosa Hiern 'Berliner Rot';柏林红大岩桐■☆

364756 Sinningia speciosa Hiern 'Blut';血红紫蓝大岩桐(血红大岩桐)■☆

364757 Sinningia speciosa Hiern 'Donan';多瑙河大岩桐■☆

364758 Sinningia speciosa Hiern 'Friedrich';弗力希王大岩桐■☆

364759 Sinningia speciosa Hiern 'Gierths';吉兹蓝大岩桐■☆

364760 Sinningia speciosa Hiern 'Gratulation';庆祝大岩桐■☆

364761 Sinningia speciosa Hiern 'Lewchtfeuer';火红大岩桐■☆

364762 Sinningia speciosa Hiern 'Rhein';莱茵大岩桐■☆

364763 Sinningia tuberosa (Mart.) H. E. Moore;管状大岩桐■☆

364764 Sinningia tubiflora Fritsch;管花大岩桐(管花块茎苣苔)■☆

364765 Sinningia villosa Lindl.;柔毛大岩桐■☆

364766 Sinningia warmingii (Hieron.) Chautems;瓦氏大岩桐; Warming Sinningia ■☆

364767 Sinoadina Ridsdale (1979);鸡仔木属(水冬瓜属); Chickenwood, Sinoadina ●

364768 Sinoadina racemosa (Siebold et Zucc.) Ridsdale;鸡仔木(鸡仔木水团花,梨仔,水冬哥,水冬瓜,水团花); Chickenwood, Racemose Adina, Racemose Sinoadina, Taiwan Adina ●

364769 Sinoarundinaria Ohwi = Phyllostachys Siebold et Zucc. (保留属名)●

364770 Sinoarundinaria Ohwi(1931);华篱竹属●☆

364771 Sinoarundinaria pubescens f. nabeshimana Muroi = Phyllostachys heterocycla (Carrière) Matsum. 'Tao Kiang'●

364772 Sinobacopa D. Y. Hong = Bacopa Aubl. (保留属名)■

364773 Sinobacopa D. Y. Hong(1987);田玄参属■

364774 Sinobacopa aquatica D. Y. Hong;田玄参(假西洋菜); Sinobacopa ■

364775 Sinobacopa aquatica D. Y. Hong = Bacopa repens (Sw.) Wettst. ■

364776 Sinobaijiania C. Jeffrey et W. J. de Wilde(2006);中国白兼果属■

364777 Sinobaijiania decipiens C. Jeffrey et W. J. de Wilde;白兼果■

364778 Sinobaijiania taiwaniana (Hayata) C. Jeffrey et W. J. de Wilde;台湾白兼果(台湾罗汉果); Taiwan Balanophora ■

364779 Sinobaijiania taiwaniana (Hayata) C. Jeffrey et W. J. de Wilde = Baijiania taiwaniana (Hayata) A. M. Lu et J. Q. Li ■

364780 Sinobaijiania yunnanensis (A. M. Lu et Zhi Y. Zhang) C. Jeffrey et W. J. de Wilde = Baijiania yunnanensis (A. M. Lu et Zhi Y. Zhang) A. M. Lu et J. Q. Li ■

364781 Sinobaijiania yunnanensis (A. M. Lu et Zhi Y. Zhang) C. Jeffrey et W. J. de Wilde;云南白兼果(云南罗汉果); Yunnan Balanophora ■

364782 Sinobambusa Makino = Sinobambusa Makino ex Nakai ●

364783 Sinobambusa Makino ex Nakai(1925);唐竹属; Sinobambusa, Tangbamboo ●

364784 Sinobambusa acutiligulata W. T. Lin;尖舌唐竹; Acutiligular Sinobambusa ●

364785 Sinobambusa acutiligulata W. T. Lin = Oligostachyum hupehense (J. L. Lu) Z. P. Wang et G. H. Ye ●

364786 Sinobambusa anaurita T. H. Wen = Oligostachyum spongiosum (C. D. Chu et C. S. Chao) G. H. Ye et Z. P. Wang ●

364787 Sinobambusa dushanensis (C. D. Chu et J. Q. Zhang) T. H. Wen;独山唐竹; Dushan Sinobambusa, Dushan Tangbamboo ●

364788 Sinobambusa exaurita W. T. Lin;无耳唐竹; Earless Sinobambusa ●

364789 Sinobambusa exaurita W. T. Lin = Oligostachyum scabriflorum (McClure) Z. P. Wang et G. H. Ye ●

364790 Sinobambusa farinosa (McClure) T. H. Wen;白皮唐竹; Mealy Sinobambusa, Palegreen Tangbamboo ●

364791 Sinobambusa fimbriata T. H. Wen = Phyllostachys aurita J. L. Lu ●

364792 Sinobambusa fimbriata T. H. Wen = Phyllostachys rubromarginata McClure ●

364793 Sinobambusa gibbosa McClure = Indosasa crassiflora McClure ●

364794 Sinobambusa gigantea T. H. Wen = Indosasa gigantea (T. H. Wen) T. H. Wen ●

364795 Sinobambusa glabrescens T. H. Wen = Oligostachyum glabrescens (T. H. Wen) P. C. Keng et Z. P. Wang ●

364796 Sinobambusa glabrescens T. H. Wen = Oligostachyum glabrescens (T. H. Wen) Q. F. Zheng et Y. M. Lin ●

364797 Sinobambusa henryi (McClure) C. D. Chu et C. S. Chao;杠竹;

Henry Sinobambusa, Henry Tangbamboo ●

364798　Sinobambusa humilis McClure；竹仔●

364799　Sinobambusa incana T. H. Wen；毛环唐竹；Ashy-grey Sinobambusa, Hairring Chinacane, Hoary Sinobambusa, Palehair Tangbamboo ●

364800　Sinobambusa intermedia McClure；晾衫竹（凉衫竹）；Intermediate Sinobambusa, Sunningclothes Tangbamboo ●

364801　Sinobambusa kunishii (Hayata) Nakai = Gelidocalamus kunishii (Hayata) P. C. Keng et T. H. Wen ●

364802　Sinobambusa laeta McClure = Sinobambusa tootsik (Siebold) Makino ex Nakai var. laeta (McClure) T. H. Wen ●

364803　Sinobambusa maculata McClure = Pleioblastus maculatus (McClure) C. D. Chu et C. S. Chao ●

364804　Sinobambusa nandanensis T. H. Wen；南丹唐竹；Nandan Sinobambusa, Nandan Tangbamboo ●

364805　Sinobambusa nandanensis T. H. Wen = Sinobambusa henryi (McClure) C. D. Chu et C. S. Chao ●

364806　Sinobambusa nephroaurita C. D. Chu et C. S. Chao；肾耳唐竹；Kidneyauricle Sinobambusa, Kidneyauricle Tangbamboo, Nephroid-auriculate Sinobambusa ●

364807　Sinobambusa parvifolia T. H. Wen et S. Y. Chen = Oligostachyum sulcatum Z. P. Wang et G. H. Ye ●

364808　Sinobambusa puberula T. H. Wen = Oligostachyum puberulum (T. H. Wen) G. H. Ye et Z. P. Wang ●

364809　Sinobambusa pulchella T. H. Wen = Oligostachyum pulchellum (T. H. Wen) G. H. Ye et Z. P. Wang ●

364810　Sinobambusa pulchella T. H. Wen = Pseudosasa cantorii (Munro) P. C. Keng ex S. L. Chen et al. ●

364811　Sinobambusa rubroligula McClure；红舌唐竹；Red-ligulate Sinobambusa, Red-ligule Sinobambusa, Redtongue Tangbamboo ●

364812　Sinobambusa scabrida T. H. Wen；糙耳唐竹（冬笋竹）；Roughear Tangbamboo, Scabria Sinobambusa, Scabrous Sinobambusa ●

364813　Sinobambusa seminuda T. H. Wen = Pleioblastus hsienchuensis T. H. Wen var. subglabratus (S. Y. Chen) C. S. Chao et G. Y. Yang ●

364814　Sinobambusa sichuanensis T. P. Yi = Chimonobambusa sichuanensis (T. P. Yi) T. H. Wen ●

364815　Sinobambusa striata T. H. Wen；花箨唐竹；Striate Sinobambusa, Striate-sheath Tangbamboo ●

364816　Sinobambusa striata T. H. Wen = Indosasa longispicata W. Y. Hsiung et C. S. Chao ●

364817　Sinobambusa sulcata W. T. Lin et Z. M. Wu；沟槽唐竹；Sulcate Sinobambusa ●

364818　Sinobambusa sulcata W. T. Lin et Z. M. Wu = Oligostachyum scabriflorum (McClure) Z. P. Wang et G. H. Ye ●

364819　Sinobambusa tootsik (Makino) Makino ex Nakai var. tenuifolia (Koidz.) Sad. Suzuki；细叶唐竹●☆

364820　Sinobambusa tootsik (Siebold) Makino ex Nakai；唐竹（苦竹，疏节竹）；Chinese Cane, Chinese Sinobambusa, Long-flowered Bamboo, Tangbamboo ●

364821　Sinobambusa tootsik (Siebold) Makino ex Nakai var. dentata T. H. Wen；火管竹；Toothed Chinese Sinobambusa, Toothed Tangbamboo ●

364822　Sinobambusa tootsik (Siebold) Makino ex Nakai var. laeta (McClure) T. H. Wen；满山爆竹；Cheerful Chinese Sinobambusa, Cheerful Tangbamboo ●

364823　Sinobambusa tootsik (Siebold) Makino ex Nakai var. maeshimana Muroi ex Sugim.；光叶唐竹；Thin-leaf Chinese

Sinobambusa, Thin-leaf Chinese Tangbamboo ●

364824　Sinobambusa tootsik (Siebold) Makino ex Nakai var. maeshimana Muroi ex Sugim. = Sinobambusa tootsik (Siebold) Makino ex Nakai var. tenuifolia (Koidz.) S. Susaki ●

364825　Sinobambusa tootsik (Siebold) Makino ex Nakai var. tenuifolia (Koidz.) S. Suzuki = Sinobambusa tootsik (Siebold) Makino ex Nakai var. maeshimana Muroi ex Sugim. ●

364826　Sinobambusa urens T. H. Wen；尖头唐竹；Burning Sinobambusa, Cusp Tangbamboo, Stinging Sinobambusa ●

364827　Sinobambusa yixingensis C. S. Chao et K. S. Xiao；宜兴唐竹；Yixing Sinobambusa ●

364828　Sinoboea Chun = Ornithoboea Parish ex C. B. Clarke ■

364829　Sinoboea microcarpa Chun = Ornithoboea feddei (H. Lév.) B. L. Burtt ■

364830　Sinocalamus McClure = Dendrocalamus Nees ●

364831　Sinocalamus McClure = Neosinocalamus P. C. Keng ●★

364832　Sinocalamus affinis (Rendle) McClure = Bambusa emeiensis L. C. Chia et H. L. Fung ●

364833　Sinocalamus affinis (Rendle) McClure = Neosinocalamus affinis (Rendle) P. C. Keng ●

364834　Sinocalamus affinis (Rendle) McClure f. chrysotrichus J. R. Xue et T. P. Yi = Neosinocalamus affinis (Rendle) P. C. Keng 'Chrysotrichus' ●

364835　Sinocalamus affinis (Rendle) McClure f. flavidorivens J. R. Xue et T. P. Yi = Neosinocalamus affinis (Rendle) P. C. Keng 'Flavidorivens' ●

364836　Sinocalamus affinis (Rendle) McClure f. viridiflavus J. R. Xue et T. P. Yi = Neosinocalamus affinis (Rendle) P. C. Keng 'Viridiflavus' ●

364837　Sinocalamus beecheyana (Munro) McClure = Dendrocalamopsis beecheyana (Munro) P. C. Keng ●

364838　Sinocalamus beecheyana (Munro) McClure var. pubescens P. F. Li = Dendrocalamopsis beecheyana (Munro) P. C. Keng var. pubescens (P. F. Li) P. C. Keng ●

364839　Sinocalamus beecheyanus (Munro) McClure = Bambusa beecheyana Munro ●

364840　Sinocalamus beecheyanus (Munro) McClure var. pubescens P. F. Li = Bambusa beecheyana Munro var. pubescens (P. F. Li) W. C. Lin ●

364841　Sinocalamus bicicatricatus W. T. Lin = Bambusa bicicatricata (W. T. Lin) L. C. Chia et H. L. Fung ●

364842　Sinocalamus bicicatricatus W. T. Lin = Dendrocalamopsis bicicatricata (W. T. Lin) P. C. Keng ●

364843　Sinocalamus brandisii (Munro) P. C. Keng = Dendrocalamus brandisii (Munro) Kurz ●

364844　Sinocalamus calostachyus (Kurz) P. C. Keng = Dendrocalamus calostachyus (Kurz) Kurz ●

364845　Sinocalamus concavus W. T. Lin et Z. M. Wu = Bambusa basihirsutoides N. H. Xia ●

364846　Sinocalamus distegius Keng et P. C. Keng = Bambusa distegia (Keng et P. C. Keng) L. C. Chia et H. L. Fung ●

364847　Sinocalamus edulis (Odash.) P. C. Keng = Bambusa odashimae Hatus. ex Ohrnb. ●

364848　Sinocalamus edulis (Odash.) P. C. Keng = Dendrocalamopsis edulis (Odash.) P. C. Keng ●

364849　Sinocalamus farinosus Keng et P. C. Keng = Dendrocalamus farinosus (Keng et P. C. Keng) L. C. Chia et H. L. Fung ●

364850 Sinocalamus flagelli-fer （Munro） T. Q. Nguyen ＝ Dendrocalamus asper （Schult. et Schult. f.） Backer ex K. Heyne ●

364851 Sinocalamus giganteus （Munro） A. Camus ＝ Dendrocalamus giganteus （Wall.） Munro ●

364852 Sinocalamus hamiltonii （Nees et Arn. ex Munro） T. Q. Nguyen ＝ Dendrocalamus hamiltonii Nees et Arn. ex Munro ●

364853 Sinocalamus latiflorus （Munro） McClure ＝ Dendrocalamus latiflorus Munro ●

364854 Sinocalamus latiflorus （Munro） McClure var. magnus T. H. Wen ＝ Dendrocalamus latiflorus Munro ●

364855 Sinocalamus microphyllum J. R. Xue et T. P. Yi ＝ Drepanostachyum microphyllum （J. R. Xue et T. P. Yi） P. C. Keng ex T. P. Yi ●

364856 Sinocalamus microphyllus J. R. Xue et T. P. Yi ＝ Ampelocalamus microphyllus （J. R. Xue et T. P. Yi） J. R. Xue et T. P. Yi ●

364857 Sinocalamus minor McClure ＝ Dendrocalamus minor （McClure） L. C. Chia et H. L. Fung ●

364858 Sinocalamus minor McClure var. amoenus Q. H. Dai et C. F. Huang ＝ Dendrocalamus minor （McClure） L. C. Chia et H. L. Fung var. amoenus （Q. H. Dai et C. F. Huang） J. R. Xue et D. Z. Li ●

364859 Sinocalamus oldhamii （Munro） McClure ＝ Dendrocalamopsis oldhamii （Munro） P. C. Keng ●

364860 Sinocalamus parishii （Munro） W. T. Lin ＝ Dendrocalamus parishii Munro ●

364861 Sinocalamus pubescens （P. F. Li） P. C. Keng ＝ Bambusa beecheyana Munro var. pubescens （P. F. Li） W. C. Lin ●

364862 Sinocalamus pubescens （P. F. Li） P. C. Keng ＝ Dendrocalamopsis beecheyana （Munro） P. C. Keng var. pubescens （P. F. Li） P. C. Keng ●

364863 Sinocalamus saxatilis J. R. Xue et T. P. Yi ＝ Ampelocalamus saxatilis （J. R. Xue et T. P. Yi） J. R. Xue et T. P. Yi ●

364864 Sinocalamus saxatilis J. R. Xue et T. P. Yi ＝ Drepanostachyum saxatile （J. R. Xue et T. P. Yi） P. C. Keng ex T. P. Yi ●

364865 Sinocalamus stenoaurita W. T. Lin ＝ Dendrocalamopsis stenoaurita （W. T. Lin） P. C. Keng ex W. T. Lin ●

364866 Sinocalamus stenoauritus W. T. Lin ＝ Bambusa stenoaurita （W. T. Lin） T. H. Wen ●

364867 Sinocalamus suberosum W. T. Lin et Z. M. Wu；白沙竹●

364868 Sinocalamus variostriatus W. T. Lin ＝ Bambusa variostriata （W. T. Lin） L. C. Chia et H. L. Fung ●

364869 Sinocalycanthus （W. C. Cheng et S. Y. Chang） W. C. Cheng et S. Y. Chang ＝ Calycanthus L. （保留属名）●

364870 Sinocalycanthus chinensis W. C. Cheng et S. Y. Chang ＝ Calycanthus chinensis （W. C. Cheng et S. Y. Chang） W. C. Cheng et S. Y. Chang ex P. T. Li ◇

364871 Sinocarum H. Wolff ＝ Sinocarum H. Wolff ex R. H. Shan et F. T. Pu ■★

364872 Sinocarum H. Wolff ex R. H. Shan et F. T. Pu（1980）；小芹属；Sinocarum ■★

364873 Sinocarum bijiangense S. L. Liou ＝ Sinocarum schizopetalum （Franch.） H. Wolff ex R. H. Shan et F. T. Pu var. bijiangense （S. L. Liou） X. T. Liu ■

364874 Sinocarum caespitosum H. Wolff；贡山小芹；Gongshan Sinocarum ■

364875 Sinocarum caespitosum H. Wolff ＝ Sinocarum cruciatum （Franch.） H. Wolff ex R. H. Shan et F. T. Pu var. linearilobum （Franch.） R. H. Shan et F. T. Pu ■

364876 Sinocarum coloratum （Diels） H. Wolff ＝ Sinocarum coloratum （Diels） H. Wolff ex R. H. Shan et F. T. Pu ■

364877 Sinocarum coloratum （Diels） H. Wolff ex F. T. Pu ＝ Sinocarum coloratum （Diels） H. Wolff ex R. H. Shan et F. T. Pu ■

364878 Sinocarum coloratum （Diels） H. Wolff ex R. H. Shan et F. T. Pu；紫茎小芹；Purplestem Sinocarum ■

364879 Sinocarum cruciatum （Franch.） H. Wolff ＝ Sinocarum cruciatum （Franch.） H. Wolff ex R. H. Shan et F. T. Pu ■

364880 Sinocarum cruciatum （Franch.） H. Wolff ex F. T. Pu ＝ Sinocarum cruciatum （Franch.） H. Wolff ex R. H. Shan et F. T. Pu ■

364881 Sinocarum cruciatum （Franch.） H. Wolff ex F. T. Pu var. linearilobum （Franch.） R. H. Shan et F. T. Pu ＝ Sinocarum cruciatum （Franch.） H. Wolff ex R. H. Shan et F. T. Pu var. linearilobum （Franch.） R. H. Shan et F. T. Pu ■

364882 Sinocarum cruciatum （Franch.） H. Wolff ex R. H. Shan et F. T. Pu；钝瓣小芹；Bluntpetal Sinocarum ■

364883 Sinocarum cruciatum （Franch.） H. Wolff ex R. H. Shan et F. T. Pu var. linearilobum （Franch.） R. H. Shan et F. T. Pu；尖瓣小芹；Sharppetal Sinocarum ■

364884 Sinocarum dolichopodum （Diels） H. Wolff ＝ Sinocarum dolichopodum （Diels） H. Wolff ex R. H. Shan et F. T. Pu ■

364885 Sinocarum dolichopodum （Diels） H. Wolff ex F. T. Pu ＝ Sinocarum dolichopodum （Diels） H. Wolff ex R. H. Shan et F. T. Pu ■

364886 Sinocarum dolichopodum （Diels） H. Wolff ex R. H. Shan et F. T. Pu；长柄小芹；Longstalk Sinocarum ■

364887 Sinocarum filicinum H. Wolff；蕨叶小芹；Fernlike Sinocarum ■

364888 Sinocarum minus M. F. Watson ＝ Acronema minus （M. F. Watson） M. F. Watson et Z. H. Pan ■

364889 Sinocarum pauciradiatum R. H. Shan et F. T. Pu；少辐小芹；Fewspoke Sinocarum ■

364890 Sinocarum pityophilum （Diels） H. Wolff；松林小芹；Pineforest Sinocarum ■

364891 Sinocarum schizopetalum （Franch.） H. Wolff ex F. T. Pu ＝ Sinocarum schizopetalum （Franch.） H. Wolff ex R. H. Shan et F. T. Pu ■

364892 Sinocarum schizopetalum （Franch.） H. Wolff ex F. T. Pu var. bijiangense （S. L. Liou） X. T. Liu ＝ Sinocarum schizopetalum （Franch.） H. Wolff ex R. H. Shan et F. T. Pu var. bijiangense （S. L. Liou） X. T. Liu ■

364893 Sinocarum schizopetalum （Franch.） H. Wolff ex R. H. Shan et F. T. Pu；裂瓣小芹；Splitpetal Sinocarum ■

364894 Sinocarum schizopetalum （Franch.） H. Wolff ex R. H. Shan et F. T. Pu var. bijiangense （S. L. Liou） X. T. Liu；碧江小芹(矮小丝瓣芹)；Bijiang Sinocarum, Wolff Acronema ■

364895 Sinocarum vaginatum H. Wolff；阔鞘小芹(鸡山小芹，阔瓣小芹)；Broadsheath Sinocarum ■

364896 Sinocarum wolffianum （Fedde ex H. Wolff） R. H. Shan et F. T. Pu ＝ Sinocarum schizopetalum （Franch.） H. Wolff ex R. H. Shan et F. T. Pu var. bijiangense （S. L. Liou） X. T. Liu ■

364897 Sinochasea Keng ＝ Pseudodanthonia Bor et C. E. Hubb. ■☆

364898 Sinochasea Keng(1958)；三蕊草属(青海草属)；Sinochasea ■★

364899 Sinochasea trigyna Keng；三蕊草(青海草)；Commom Sinochasea ■

364900 Sinocitrus chachiensis C. J. Tseng ＝ Citrus reticulata Blanco ‘Chachiensis’●

364901 Sinocitrus chachiensis C. J. Tseng ＝ Citrus reticulata Blanco ●

364902　Sinocitrus erythrosa（Tanaka）C. J. Tseng = Citrus reticulata Blanco 'Erythrosa' ●

364903　Sinocitrus junos（Siebold ex Tanaka）C. J. Tseng = Citrus junos Siebold ex Tanaka ●

364904　Sinocitrus kinokuni（Tanaka）C. J. Tseng = Citrus reticulata Blanco 'Kinokuni' ●

364905　Sinocitrus nobilis（Lour.）C. J. Tseng = Citrus reticulata Blanco 'Nobilis' ●

364906　Sinocitrus poonensis（Tanaka）C. J. Tseng = Citrus reticulata Blanco 'Ponkan' ●

364907　Sinocitrus suavissima（Tanaka）C. J. Tseng = Citrus reticulata Blanco 'Suavissima' ●

364908　Sinocitrus suhuiensis（Tanaka）C. J. Tseng = Citrus reticulata Blanco 'Hanggan' ●

364909　Sinocitrus tanakan（Hayata）C. J. Tseng = Citrus reticulata Blanco 'Tankan' ●

364910　Sinocitrus tangerina（Tanaka）C. J. Tseng = Citrus reticulata Blanco 'Tangerina' ●

364911　Sinocitrus ushiu（F. P. Metcalf）C. J. Tseng = Citrus reticulata Blanco 'Unshiu' ●

364912　Sinocitrus verucosa C. J. Tseng = Citrus reticulata Blanco 'Manau Gan' ●

364913　Sinocrassula A. Berger（1930）;石莲属（华景天属,石莲花属）;Sinocrassula,Stonelotus ■

364914　Sinocrassula aliciae（Raym. -Hamet）A. Berger = Kungia aliciae（Raym. -Hamet）K. T. Fu ■

364915　Sinocrassula aliciae（Raym. -Hamet）A. Berger = Orostachys aliciae（Raym. -Hamet）H. Ohba ■

364916　Sinocrassula ambigua（Praeger）A. Berger;长萼石莲（可疑石莲）;Doubtful Sinocrassula,Longcalyx Stonelotus ■

364917　Sinocrassula densirosulata（Praeger）A. Berger;密叶石莲（立田凤）;Denseleaf Sinocrassula,Denseleaf Stonelotus ■

364918　Sinocrassula diversifolia H. Chuang;异形叶石莲■

364919　Sinocrassula indica（Decne.）A. Berger;石莲（狗牙还阳,红花岩松,景天还阳,莲花还阳,梅花狗牙瓣,山瓦松,蛇舌莲,石莲花,石莲岩松,石山莲,碎骨还阳,土三七,推山花,瓦指甲,岩松）;Indian Sinocrassula,Stonelotus ■

364920　Sinocrassula indica（Decne.）A. Berger var. forrestii（Raym. -Hamet）A. Berger;圆叶石莲;Forrest Sinocrassula ■

364921　Sinocrassula indica（Decne.）A. Berger var. luteorubra（Praeger）S. H. Fu;黄花石莲;Yellowflower Sinocrassula ■

364922　Sinocrassula indica（Decne.）A. Berger var. obtusifolia（Fröd.）S. H. Fu;钝叶石莲;Obtuseleaf Sinocrassula ■

364923　Sinocrassula indica（Decne.）A. Berger var. serrata（Raym. -Hamet）S. H. Fu;锯叶石莲;Serrateleaf Sinocrassula ■

364924　Sinocrassula indica（Decne.）A. Berger var. viridiflora K. T. Fu;绿花石莲（绿花石莲花,石灯台）;Greenflower Sinocrassula,Greenflower Stonelotus ■

364925　Sinocrassula longistyla（Praeger）S. H. Fu;长柱石莲;Longstyle Sinocrassula,Longstyle Stonelotus ■

364926　Sinocrassula paoshingensis（S. H. Fu）H. Ohba et al. = Sinocrassula indica（Decne.）A. Berger var. luteorubra（Praeger）S. H. Fu ■

364927　Sinocrassula paoshingensis H. Ohba;宝兴石莲■

364928　Sinocrassula schoenlandii（Raym. -Hamet）S. H. Fu = Kungia schoenlandii（Raym. -Hamet）K. T. Fu ■

364929　Sinocrassula stenostachya（Fröd.）S. H. Fu = Kungia schoenlandii（Raym. -Hamet）K. T. Fu var. stenostachya（Fröd.）K. T. Fu ■

364930　Sinocrassula stenostachya（Fröd.）S. H. Fu = Kungia schoenlandii（Raym. -Hamet）K. T. Fu ■

364931　Sinocrassula stenostachya（Fröd.）S. H. Fu var. integrifolia S. H. Fu = Kungia schoenlandii（Raym. -Hamet）K. T. Fu var. stenostachya（Fröd.）K. T. Fu ■

364932　Sinocrassula stenostachya（Fröd.）S. H. Fu var. lepidotricha S. H. Fu = Kungia schoenlandii（Raym. -Hamet）K. T. Fu ■

364933　Sinocrassula stenostachya（Fröd.）S. H. Fu var. lepidotricha S. H. Fu = Kungia schoenlandii（Raym. -Hamet）K. T. Fu var. stenostachya（Fröd.）K. T. Fu ■

364934　Sinocrassula techinensis（S. H. Fu）S. H. Fu;德钦石莲（德钦景天）;Deqin Sinocrassula,Deqin Stonelotus,Techin Sinocrassula ■

364935　Sinocrassula yunnanensis（Franch.）A. Berger;云南石莲（把岩香,滇石莲,四马路）;Yunnan Sinocrassula,Yunnan Stonelotus ■

364936　Sinodielsia H. Wolff = Meeboldia H. Wolff ■

364937　Sinodielsia bipinnata（R. H. Shan et F. T. Pu）Pimenov et Kljuykov. = Vicatia bipinnata R. H. Shan et F. T. Pu ■

364938　Sinodielsia cuneata（H. Wolff）Pimenov et Kljuykov. = Physospermopsis cuneata H. Wolff ■

364939　Sinodielsia delavayi（Franch.）Pimenov et Kljuykov = Peucedanum delavayi Franch. ■

364940　Sinodielsia microloba Kljuykov. = Meeboldia yunnanensis（H. Wolff）Constance et F. T. Pu ■

364941　Sinodielsia thibetica（H. Boissieu）Kljuykov et P. K. Mukh. = Vicatia thibetica H. Boissieu ■

364942　Sinodielsia yunnanensis H. Wolff = Meeboldia yunnanensis（H. Wolff）Constance et F. T. Pu ■

364943　Sinodolichos Verdc.（1970）;华扁豆属;Sinohaircot ■

364944　Sinodolichos lagopus（Dunn）Verdc.;华扁豆;Sinohaircot ■

364945　Sinofranchetia（Diels）Hemsl.（1907）;串果藤属;Sinofranchetia ●★

364946　Sinofranchetia Hemsl. = Sinofranchetia（Diels）Hemsl. ●★

364947　Sinofranchetia chinensis（Franch.）Hemsl.;串果藤（红藤）;China Sinofranchetia,Chinese Sinofranchetia ●

364948　Sinofranchetia chinensis Hemsl. = Sinofranchetia chinensis（Franch.）Hemsl. ●

364949　Sinofranchetiaceae Doweld = Lardizabalaceae R. Br.（保留科名）●

364950　Sinoga S. T. Blake = Asteromyrtus Schauer ●☆

364951　Sinojackia Hu（1928）;秤锤树属;Jacktree, Sinojackia, Weigttree ●★

364952　Sinojackia dolichocarpa C. J. Qi = Changiostyrax dolichocarpus（C. J. Qi）Tao Chen ●

364953　Sinojackia henryi（Dümmer）Merr.;棱果秤锤树;Henry Sinojackia, Henry Weigttree, Ribfruit Sinojackia ●

364954　Sinojackia microcarpa C. T. Chen et G. Y. Li;小果秤锤树●

364955　Sinojackia oblongicarpa C. T. Chen et T. R. Cao;怀化秤锤树●

364956　Sinojackia oblongicarpa C. T. Chen et T. R. Cao = Sinojackia sarcocarpa L. Q. Luo ●

364957　Sinojackia rehderiana Hu;狭果秤锤树（江西秤锤树,芮氏捷克木）;Jacktree, Narrowfruit Weigttree, Rehder Sinojackia ●

364958　Sinojackia sarcocarpa L. Q. Luo;肉果秤锤树●

364959　Sinojackia xylocarpa Hu;秤锤树（捷克木）;Jacktree, Weigttree, Xylocarpous Sinojackia ●◇

364960　Sinojackia xylocarpa Hu var. leshanensis L. Q. Luo;乐山秤锤树●

364961　Sinojohnstonia Hu（1936）；车前紫草属（琼丝东草属）；Sinojohnstonia ■★

364962　Sinojohnstonia chekiangensis（Migo）W. T. Wang；浙赣车前紫草（浙江车前紫草）；Chekiang Sinojohnstonia, Zhejiang Sinojohnstonia ■

364963　Sinojohnstonia chekiangensis（Migo）W. T. Wang ex Z. Y. Zhang = Sinojohnstonia chekiangensis（Migo）W. T. Wang ■

364964　Sinojohnstonia moupinensis（Franch.）W. T. Wang；短蕊车前紫草（宝兴车前紫草）；Muping Sinojohnstonia ■

364965　Sinojohnstonia moupinensis（Franch.）W. T. Wang ex Z. Y. Zhang = Sinojohnstonia moupinensis（Franch.）W. T. Wang ■

364966　Sinojohnstonia plantaginea Hu；车前紫草；Common Sinojohnstonia, Sinojohnstonia ■

364967　Sinoleontopodium Y. L. Chen（1985）；君范菊属；Junfandaisy ■★

364968　Sinoleontopodium lingianum Y. L. Chen；君范菊；Junfandaisy ■

364969　Sinolimprichtia H. Wolff（1922）；舟瓣芹属（华林芹属）；Sinolimptichtia ■★

364970　Sinolimprichtia alpina H. Wolff；舟瓣芹（华林芹）；Common Sinolimptichtia ■

364971　Sinolimprichtia alpina H. Wolff var. dissecta R. H. Shan et S. L. Liou；裂苞舟瓣芹；Dissected Sinolimptichtia ■

364972　Sinomalus Kdidz. = Malus Mill. ●

364973　Sinomalus honanensis（Rehder）Koidz. = Malus honanensis Rehder ●

364974　Sinomalus honanensis Koidz. = Malus honanensis Rehder ●

364975　Sinomalus toringoides（Rehder）Koidz. = Malus toringoides（Rehder）Hughes ●

364976　Sinomalus toringoides Koidz. = Malus toringoides（Rehder）Hughes ●

364977　Sinomalus transitoria（Batalin）Koidz. = Malus transitoria（Batalin）C. K. Schneid. ●

364978　Sinomalus transitoria Koidz. = Malus transitoria（Batalin）C. K. Schneid. ●

364979　Sinomanglietia Z. X. Yu = Sinomanglietia Z. X. Yu et Q. Y. Zheng ●★

364980　Sinomanglietia Z. X. Yu et Q. Y. Zheng(1994)；落叶木莲属●★

364981　Sinomanglietia Z. X. Yu et Q. Y. Zheng = Manglietia Blume ●

364982　Sinomanglietia glauca Z. X. Yu et Q. Y. Zheng = Manglietia decidua Q. Y. Zheng ●◇

364983　Sinomarsdenia P. T. Li et J. J. Chen（1997）；裂冠藤属；Sinomarsdenia ●

364984　Sinomarsdenia incisa（P. T. Li et Y. H. Li）P. T. Li et J. J. Chen；裂冠藤；Sinomarsdenia ●

364985　Sinomenium Diels(1910)；防己属（风龙属，汉防己属，青藤属）；Orientvine ●

364986　Sinomenium acutum（Thunb.）Rehder et E. H. Wilson；防己（吹风散，大风藤，防己青藤，风龙，风龙藤，海枫藤，汉防己，黑防己，毛青藤，青防己，青风藤，青藤，寻风藤）；Orientvine ●

364987　Sinomenium acutum（Thunb.）Rehder et E. H. Wilson f. nudiflorum Hiyama；裸花防己●☆

364988　Sinomenium acutum（Thunb.）Rehder et E. H. Wilson var. cinereum（Diels）Rehder et E. H. Wilson = Sinomenium acutum（Thunb.）Rehder et E. H. Wilson ●

364989　Sinomenium acutum（Thunb.）Rehder et E. H. Wilson var. cinereum（Diels）Rehder et E. H. Wilson；毛防己（吹风散，大风藤，滇防己，防己，风龙藤，汉防己，黑防己，淮通，灰毛青藤，毛汉防己，毛青藤，青防己，青风藤，青藤，寻风藤，追风散）；Hairy Orientvine ●

364990　Sinomenium acutum（Thunb.）Rehder et E. H. Wilson var. tomentosum Honda = Sinomenium acutum（Thunb.）Rehder et E. H. Wilson var. cinereum（Diels）Rehder et E. H. Wilson ●

364991　Sinomenium acutum（Thunb.）Rehder et E. H. Wilson var. tomentosum Honda；毡毛青藤；Tomentose Orientvine ●☆

364992　Sinomenium diversifolium（Miq.）Diels = Sinomenium acutum（Thunb.）Rehder et E. H. Wilson ●

364993　Sinomenium diversifolium Diels = Sinomenium acutum（Thunb.）Rehder et E. H. Wilson ●

364994　Sinomerrillia Hu = Neuropeltis Wall. ●■

364995　Sinomerrillia bracteata Hu = Neuropeltis racemosa Wall. ●■

364996　Sinopanax H. L. Li（1949）；华参属（里白八角金盘属）；Sinopanax ●★

364997　Sinopanax formosanus（Hayata）H. L. Li；华参（里白八角金盘，台湾山楸）；Formosana Sinopanax, Taiwan Sinopanax ●

364998　Sinopimelodendron Tsiang = Cleidiocarpon Airy Shaw ●

364999　Sinopimelodendron kwangsiense Tsiang = Cleidiocarpon cavalerei（H. Lév.）Airy Shaw ●◇

365000　Sinoplagiospermum Rauschert = Prinsepia Royle ●

365001　Sinoplagiospermum Rauschert(1982)；蕤核属●

365002　Sinoplagiospermum sinense（Oliv.）Rauschert；东北蕤核（扁担胡子，东北扁核木，华北扁核木，辽东扁核木，辽宁扁核木，中华扁核木）；Cherry Prinsepia ●

365003　Sinoplagiospermum uniflorum（Batalin）Rauschert；蕤核（白樱，扁核木，打枪果，打油果，单花扁核木，鸡蛋糕，李子蕤，马茹，梅花刺，蒙自扁核木，牛奶锤，炮筒果，枪子果，青刺尖，茹茹，蕤李子，蕤子，山桃，椹，孙奶子，小马茹，棫）；Hedge Prinsepia, Prinsepia ●

365004　Sinopodophyllum T. S. Ying = Podophyllum L. ■☆

365005　Sinopodophyllum T. S. Ying（1979）；桃儿七属；Chinese May-apple, Peach-seven ■

365006　Sinopodophyllum emodi（Falc. ex Royle）T. S. Ying = Sinopodophyllum hexandrum（Royle）T. S. Ying ■

365007　Sinopodophyllum emodii（Wall. ex Hook. f. et Thomson）T. S. Ying = Sinopodophyllum hexandrum（Royle）T. S. Ying ■

365008　Sinopodophyllum emodii（Wall.）T. S. Ying = Sinopodophyllum hexandrum（Royle）T. S. Ying ■

365009　Sinopodophyllum hexandrum（Royle）T. S. Ying；桃儿七（八月瓜，藏鬼臼，鬼臼，蒿果，华鬼臼，鸡素苔，墨地，桃耳七，铜筷子，西藏鬼臼，西蒙鬼臼，锡金鬼臼，小叶莲）；Chinese May Apple, Chinese May-apple, Common Peach-seven, Himalayan May Apple, Himalayan May-apple, Indian May Apple, Indian May-apple ■

365010　Sinopogonanthera H. W. Li = Paraphlomis（Prain）Prain ●■

365011　Sinopogonanthera H. W. Li(1993)；髯药草属■

365012　Sinopogonanthera cauropteris H. W. Li；翅茎髯药草■

365013　Sinopogonanthera intermedia（C. Y. Wu et H. W. Li）H. W. Li；中间髯药草■

365014　Sinopora J. Li, N. H. Xia et H. W. Li(1956)；孔药楠属●

365015　Sinopora hongkongensis（N. H. Xia et al.）J. Li et al.；孔药楠●

365016　Sinopyrenaria Hu = Pyrenaria Blume ●

365017　Sinopyrenaria cheliensis（Hu）Hu = Pyrenaria diospyricarpa Kurz ●

365018　Sinopyrenaria cheliensis Hu = Pyrenaria cheliensis Hu ●

365019　Sinopyrenaria garrettiana（Craib）Hu = Pyrenaria diospyricarpa Kurz ●

365020　Sinopyrenaria garrettiana（Craib）Hu = Pyrenaria garrettiana Craib ●

365021　Sinopyrenaria yunnanensis（Hu）Hu = Pyrenaria diospyricarpa Kurz ●

365022　Sinopyrenaria yunnanensis（Hu）Hu = Pyrenaria yunnanensis Hu ●

365023　Sinopyrenaria yunnanensis Hu = Pyrenaria yunnanensis Hu ●

365024　Sinoradlkofera F. G. Mey. = Boniodendron Gagnep. ●

365025　Sinoradlkofera minor（Hemsl.）F. G. Mey. = Boniodendron minus（Hemsl.）T. C. Chen ●

365026　Sinorchis S. C. Chen = Aphyllorchis Blume ■

365027　Sinorchis S. C. Chen = Cephalanthera Rich. ■

365028　Sinorchis S. C. Chen（1978）；梅兰属；Sinorchis ■★

365029　Sinorchis simplex（Ts. Tang et F. T. Wang）S. C. Chen；梅兰（单唇无叶兰）；Simple Sinorchis ●

365030　Sinorchis simplex（Ts. Tang et F. T. Wang）S. C. Chen = Aphyllorchis simplex Ts. Tang et F. T. Wang ■

365031　Sinosassafras（Allen）H. W. Li（1985）；华檫木属（黄脉檫木属）；Sinosassafras ●★

365032　Sinosassafras H. W. Li = Parasassafras D. G. Long ●

365033　Sinosassafras H. W. Li = Sinosassafras（Allen）H. W. Li ●★

365034　Sinosassafras flavinervia（C. K. Allen）H. W. Li；华檫木（黄脉檫木，黄脉钓樟，黄脉山胡椒，香果树）；Sinosassafras, Yellowvein Spicebush, Yellow-veined Spice-bush ●

365035　Sinosenecio B. Nord.（1978）；蒲儿根属（华千里光属，武夷千里光属）；Chinese Groundsel, Sinosenecio ■

365036　Sinosenecio bodinieri（Vaniot）B. Nord.；滇黔蒲儿根（丝带千里光，蜈蚣七，西南华千里光）；South-west Chinese Groundsel, SW. China Sinosenecio ■

365037　Sinosenecio brevior B. Nord. = Sinosenecio bodinieri（Vaniot）B. Nord. ■

365038　Sinosenecio changii（B. Nord.）B. Nord. et Pelser；莲座狗舌草（南川狗舌草）；Chang Dogtongueweed, Chang's Tephroseris ■

365039　Sinosenecio chienii（Hand.-Mazz.）B. Nord.；雨农蒲儿根（雨农华千里光）；Chien Sinosenecio, Chien's Chinese Groundsel ■

365040　Sinosenecio cortusifolius（Hand.-Mazz.）B. Nord.；齿耳蒲儿根（齿耳华千里光）；Teethear Chinese Groundsel, Toothear Sinosenecio ■

365041　Sinosenecio cyclamniifolius（Franch.）B. Nord.；仙客来蒲儿根（仙客来叶华千里光）；Cyclam Leaf Chinese Groundsel, Cyclamenleaf Sinosenecio ■

365042　Sinosenecio denticulatus J. Quan Liu；齿裂蒲儿根；Denticulate Chinese Groundsel ■

365043　Sinosenecio doryotus（Hand.-Mazz.）B. Nord. = Sinosenecio euosmus（Hand.-Mazz.）B. Nord. ■

365044　Sinosenecio dryas（Dunn）C. Jeffrey et Y. L. Chen；川鄂蒲儿根（川鄂华千里光，锦葵叶华千里光，锦葵叶千里光，岩葵，圆叶千里光）；Dryas Chinese Groundsel, Dryasalike Sinosenecio ■

365045　Sinosenecio elatior（Vaniot）B. Nord. = Sinosenecio bodinieri（Vaniot）B. Nord. ■

365046　Sinosenecio eriopodus（Cummins）C. Jeffrey et Y. L. Chen；毛柄蒲儿根（狗耳朵，毛柄华千里光，一面锣，直梗华千里光，直梗千里光）；Eriopodium Chinese Groundsel, Hairstipe Sinosenecio ■

365047　Sinosenecio euosmus（Hand.-Mazz.）B. Nord.；耳柄蒲儿根（齿裂华千里光，齿裂千里光，耳柄华千里光，槭叶千里光）；Earpetiolate Chinese Groundsel, Earstipe Sinosenecio, Winkler's Groundsel ■

365048　Sinosenecio fangianus Y. L. Chen；植夫蒲儿根（植夫华千里光）；Fang Sinosenecio, Fang's Chinese Groundsel ■

365049　Sinosenecio fanjingshanicus C. Jeffrey et Y. L. Chen；梵净蒲儿根（梵净华千里光）；Fanjingshan Chinese Groundsel, Fanjingshan Sinosenecio ■

365050　Sinosenecio globigerus（C. C. Chang）B. Nord.；匍枝蒲儿根（款冬花，莲花七，匍枝华千里光，水八角草，秃果华千里光，秃果千里光）；Barefruit Groundsel, Globose Sinosenecio, Stoloniferous Chinese Groundsel ■

365051　Sinosenecio globigerus（C. C. Chang）B. Nord. var. adenophyllus C. Jeffrey et Y. L. Chen；腺苞蒲儿根；Glandbract Chinese Groundsel ■

365052　Sinosenecio guangxiensis C. Jeffrey et Y. L. Chen；广西蒲儿根（白背青，广西华千里光，桂华千里光，走马须）；Guangxi Chinese Groundsel, Guangxi Sinosenecio ■

365053　Sinosenecio guizhouensis C. Jeffrey et Y. L. Chen；黔蒲儿根（黔华千里光）；Guizhou Chinese Groundsel, Guizhou Sinosenecio ■

365054　Sinosenecio hainanensis（C. C. Chang et Y. C. Tseng）C. Jeffrey et Y. L. Chen；海南蒲儿根（海南华千里光）；Hainan Chinese Groundsel, Hainan Sinosenecio ■

365055　Sinosenecio hederifolius（Dümmer）B. Nord.；单头蒲儿根（大寒草，单头华千里光，单头千里光，猪耳朵）；Single Head Chinese Groundsel, Singlehead Groundsel, Singlehead Sinosenecio ■

365056　Sinosenecio hederifolius（Dunn）B. Nord. var. angulatifolius Y. Ling；齿叶蒲儿根 ■

365057　Sinosenecio homogyniphyllus（Cummins）B. Nord.；肾叶蒲儿根（肾叶华千里光）；Kidneyleaf Chinese Groundsel, Kidney-leaf Sinosenecio ■

365058　Sinosenecio hunanensis（Y. Ling）B. Nord.；湖南蒲儿根（湖南华千里光）；Hunan Chinese Groundsel, Hunan Sinosenecio ■

365059　Sinosenecio jiuhuashanicus C. Jeffrey et Y. L. Chen；九华蒲儿根（九华华千里光）；Jiuhua Chinese Groundsel, Jiuhuashan Sinosenecio ■

365060　Sinosenecio koreanus（Kom.）B. Nord.；朝鲜蒲儿根（朝鲜华千里光）；Korea Sinosenecio, Korean Chinese Groundsel ■

365061　Sinosenecio koreanus（Kom.）B. Nord. = Tephroseris koreana（Kom.）B. Nord. et Pelser ■

365062　Sinosenecio latouchei（Jeffrey）B. Nord.；白背蒲儿根（白背千里光，赣闽华千里光，赣闽千里光）；Latouche's Chinese Groundsel, Latouche's Groundsel, SE. China Sinosenecio, Whiteback Groundsel ■

365063　Sinosenecio leiboensis C. Jeffrey et Y. L. Chen；雷波蒲儿根（雷波华千里光）；Leibo Chinese Groundsel, Leibo Sinosenecio ■

365064　Sinosenecio ligularioides（Hand.-Mazz.）B. Nord.；橐吾状蒲儿根（橐吾华千里光）；Goldenraylike Sinosenecio, Ligularia-like Chinese Groundsel ■

365065　Sinosenecio newcombei（Greene）Janovec et T. M. Barkley；牛氏蒲儿根 ■☆

365066　Sinosenecio oldhamianus（Maxim.）B. Nord.；蒲儿根（肥猪苗，黄菊莲，猫耳朵，野葡萄，野蒲桃）；Oldham Groundsel, Oldham Sinosenecio, Oldham's Chinese Groundsel ■

365067　Sinosenecio palmatilobus（Kitam.）C. Jeffrey et Y. L. Chen；掌裂蒲儿根（掌裂华千里光）；Palmately Leaf Chinese Groundsel, Palmsplit Sinosenecio ■

365068　Sinosenecio palmatisectus C. Jeffrey et Y. L. Chen；鄂西蒲儿根；Chinese Groundsel, Palmatisect Sinosenecio ■

365069　Sinosenecio phalacrocarpioides（C. C. Chang）B. Nord.；假果蒲儿根（矮茎华千里光，假光果千里光）；Shortstem Chinese

Groundsel, Shortstem Sinosenecio ■

365070 Sinosenecio phalacrocarpus (Hance) B. Nord.; 秃果蒲儿根（秃果华千里光）; Hairless Chinese Groundsel, Nakefruit Sinosenecio ■

365071 Sinosenecio rotundifolius Y. L. Chen; 圆叶蒲儿根（圆叶华千里光，圆叶千里光）; Round-leaf Chinese Groundsel, Roundleaf Sinosenecio ■

365072 Sinosenecio savatieri (Franch.) B. Nord. = Sinosenecio oldhamianus (Maxim.) B. Nord. ■

365073 Sinosenecio saxatilis Y. L. Chen; 岩生蒲儿根（岩生华千里光）; Chinese Groundsel ■

365074 Sinosenecio sep'ilobus (C. C. Chang) B. Nord.; 七裂蒲儿根（七裂华千里光）; Sevenlobed Sinosenecio, Sevenlobes Chinese Groundsel ■

365075 Sinosenecio subcoriaceus C. Jeffrey et Y. L. Chen; 革叶蒲儿根（近革叶华千里光，紫毛华千里光）; Nearlyleathern Sinosenecio, Subcoriaceous Chinese Groundsel ■

365076 Sinosenecio subrosulatus (Hand.-Mazz.) B. Nord.; 莲座蒲儿根（莲座华千里光）; Rosulate Chinese Groundsel, Rosulate Sinosenecio ■

365077 Sinosenecio sungpanensis (Hand.-Mazz.) B. Nord.; 松潘蒲儿根（松潘华千里光）; Songpan Chinese Groundsel, Songpan Sinosenecio ■

365078 Sinosenecio trinervius (C. C. Chang) B. Nord.; 三脉蒲儿根（三脉华千里光）; Three Veined Chinese Groundsel, Threeveins Sinosenecio ■

365079 Sinosenecio villiferus (Franch.) B. Nord.; 紫毛蒲儿根（软毛华千里光，紫毛千里光）; Purplehair Groundsel, Softhair Sinosenecio, Sost Hair Chinese Groundsel ■

365080 Sinosenecio winklerianus (Hand.-Mazz.) B. Nord. = Sinosenecio euosmus (Hand.-Mazz.) B. Nord. ■

365081 Sinosenecio wuyiensis Y. L. Chen; 武夷蒲儿根（武夷华千里光）; Wuyi Chinese Groundsel, Wuyishan Sinosenecio ■

365082 Sinosideroxylon (Engl.) Aubrév. (1963); 中国铁榄属（铁榄属）; Ironolive, Sinosideroxylon ●

365083 Sinosideroxylon (Engl.) Aubrév. = Sideroxylon L. ●☆

365084 Sinosideroxylon pedunculatum (Hemsl.) H. Chuang; 铁榄（假水石梓，山胶木）; Peduncled Sinosideroxylon ●

365085 Sinosideroxylon pedunculatum (Hemsl.) H. Chuang var. pubifolium H. Chuang; 毛叶铁榄; Hairyleaf Peduncled Sinosideroxylon ●

365086 Sinosideroxylon wightianum (Hook. et Arn.) Aubrév.; 革叶铁榄（华南羔涂木）; Wight Ironolive, Wight Sinosideroxylon ●

365087 Sinosideroxylon yunnanense (C. Y. Wu) H. Chuang; 滇铁榄（滇假水石梓）; Yunnan Sinosideroxylon ●

365088 Sinosophiopsis Al-Shehbaz(2000); 华羽芥属 ■

365089 Sinosophiopsis bartholomewii Al-Shehbaz; 华羽芥 ■

365090 Sinosophiopsis heishuiensis (W. T. Wang) Al-Shehbaz; 黑水华羽芥（黑水碎米荠）; Heishui Bittercress ■

365091 Sinosophiopsis heishuiensis (W. T. Wang) Al-Shehbaz = Cardamine heishuiensis W. T. Wang ■

365092 Sinowilsonia Hemsl. (1906); 山白树属; Wilsontree, Wilson-tree ●★

365093 Sinowilsonia henryi Hemsl.; 山白树; Henry Wilsontree, Henry Wilson-tree ●

365094 Sinowilsonia henryi Hemsl. var. glabrescens Hung T. Chang; 光叶山白树（秃山白树）; Glabrous Henry Wilsontree, Glabrous Wilsontree ●

365095 Sinthroblastes Bremek. = Strobilanthes Blume ●■

365096 Sioja Buch.-Ham. ex Lindl. = Peripterygium Hassk. ●■

365097 Siolmatra Baill. (1885); 巴西瓜属 ■☆

365098 Siolmatra amazonica Cogn.; 亚马逊巴西瓜 ■☆

365099 Siolmatra brasiliensis Baill.; 巴西瓜 ■☆

365100 Sion Adans. = Sium L. ■

365101 Siona Salisb. = Dichopogon Kunth ■☆

365102 Sipanea Aubl. (1775); 锡潘茜属 ●■☆

365103 Sipanea angustifolia A. Rich.; 窄叶西巴茜 ●☆

365104 Sipanea angustifolia A. Rich. ex DC. = Pentas angustifolia (A. Rich. ex DC.) Verdc. ●☆

365105 Sipanea biflora Cham. et Schltdl.; 双花西巴茜 ●☆

365106 Sipanea brasiliensis Wernham; 西巴茜 ●☆

365107 Sipanea colombiana Wernham; 哥伦比亚西巴茜 ●☆

365108 Sipanea elatior A. Rich. ex DC. = Otomeria elatior (A. Rich. ex DC.) Verdc. ■☆

365109 Sipanea glaberrima (Bremek.) Steyerm.; 光滑西巴茜 ●☆

365110 Sipanea ovalifolia Bremek.; 卵叶西巴茜 ●☆

365111 Sipanea saxicola J. H. Kirkbr.; 岩生西巴茜 ●☆

365112 Sipaneopsis Steyerm. (1967); 拟西巴茜属 ●☆

365113 Sipaneopsis foldatsii Steyerm.; 拟西巴茜 ●☆

365114 Sipania Seem. = Limnosipanea Hook. f. ☆

365115 Sipapoa Maguire = Diacidia Griseb. ●☆

365116 Sipapoantha Maguire et B. M. Boom(1989); 西巴龙胆属 ■☆

365117 Sipapoantha ostrina Maguire et B. M. Boom; 西巴龙胆 ■☆

365118 Siparuna Aubl. (1775); 坛罐花属（西帕木属）●☆

365119 Siparuna guianensis Aubl.; 圭亚那坛罐花（圭亚那西帕木）●☆

365120 Siparuna lindeni DC.; 林登坛罐花 ●☆

365121 Siparunaceae (A. DC.) Schodde = Monimiaceae Juss. (保留科名) ●■☆

365122 Siparunaceae Schodde = Monimiaceae Juss. (保留科名) ●■☆

365123 Siparunaceae Schodde(1970); 坛罐花科（西帕木科）●☆

365124 Siphanthemum Tiegh. = Psittacanthus Mart. ●

365125 Siphanthera Pohl ex DC. = Siphanthera Pohl ■☆

365126 Siphanthera Pohl(1828); 管药野牡丹属 ■☆

365127 Siphanthera cordifolia Gleason; 心叶管药野牡丹 ■☆

365128 Siphanthera discolor Cogn.; 异色管药野牡丹 ■☆

365129 Siphanthera foliosa (Naudin) Wurdack; 多叶管药野牡丹 ■☆

365130 Siphanthera microphylla Cogn.; 小叶管药野牡丹 ■☆

365131 Siphanthera robusta Cogn.; 粗壮管药野牡丹 ■☆

365132 Siphanthera villosa Cogn.; 毛管药野牡丹 ■☆

365133 Siphantheropsis Brade = Macairea DC. ●☆

365134 Sipharissa Post et Kuntze = Sypharissa Salisb. ■☆

365135 Sipharissa Post et Kuntze = Tenicroa Raf. ■☆

365136 Sipharissa Post et Kuntze = Urginea Steinh. ■☆

365137 Siphaulax Raf. = Nicotiana L. ●■

365138 Siphidia Raf. = Siphisia Raf. ●■

365139 Siphisia Raf. = Aristolochia L. ■●

365140 Siphisia Raf. = Isotrema Raf. ●☆

365141 Siphisia platanifolia Klotzsch = Aristolochia platanifolia Duch. ●

365142 Siphisia saccata Klotzsch = Aristolochia saccata Wall. ●

365143 Siphoboea Baill. = Clerodendrum L. ●■

365144 Siphocampylus Pohl(1830-1831); 曲管桔梗属 ●☆

365145 Siphocampylus acuminatus E. Wimm.; 渐尖曲管桔梗 ■☆

365146 Siphocampylus affinis (Mirb.) McVaugh; 近缘曲管桔梗 ■☆

365147 Siphocampylus aggregatus Rusby;聚集曲管桔梗■☆

365148 Siphocampylus albus E. Wimm.;白曲管桔梗■☆

365149 Siphocampylus angustiflorus Schlecht. et Zahlbr.;窄叶曲管桔梗■☆

365150 Siphocampylus asper Benth.;粗糙曲管桔梗■☆

365151 Siphocampylus aureus Rusby;黄曲管桔梗■☆

365152 Siphocampylus betulifolius G. Don;桦叶曲管桔梗■☆

365153 Siphocampylus bicolor G. Don;二色曲管桔梗■☆

365154 Siphocampylus boliviensis Zahlbr.;玻利维亚曲管桔梗■☆

365155 Siphocampylus brevicalyx E. Wimm.;短萼曲管桔梗■☆

365156 Siphocampylus brevidens E. Wimm.;短齿曲管桔梗■☆

365157 Siphocampylus canescens A. DC.;灰曲管桔梗■☆

365158 Siphocampylus cordifolius Otto et Dietr.;心叶曲管桔梗■☆

365159 Siphocampylus cylindricus Gleason;圆柱曲管桔梗■☆

365160 Siphocampylus densidentatus E. Wimm.;密齿曲管桔梗■☆

365161 Siphocampylus densiflorus Planch.;密花曲管桔梗■☆

365162 Siphocampylus discolor Donn. Sm.;异色曲管桔梗■☆

365163 Siphocampylus elegans Planch.;雅致曲管桔梗■☆

365164 Siphocampylus ellipticus Vatke;椭圆曲管桔梗■☆

365165 Siphocampylus fuscus G. Don;褐曲管桔梗☆

365166 Siphocampylus gracilis Britton;细曲管桔梗■☆

365167 Siphocampylus laevigatus Planch.;平滑曲管桔梗■☆

365168 Siphocampylus leptophyllus Urb.;细叶曲管桔梗■☆

365169 Siphocampylus linearifolius Léonard;线叶曲管桔梗■☆

365170 Siphocampylus longipes Vatke;长梗曲管桔梗☆

365171 Siphocampylus macranthus Pohl;大花曲管桔梗■☆

365172 Siphocampylus macrophyllus G. Don;大叶曲管桔梗■☆

365173 Siphocampylus macropodus G. Don;大梗曲管桔梗■☆

365174 Siphocampylus macrostemon A. DC.;大冠曲管桔梗■☆

365175 Siphocampylus megastoma E. Wimm.;大口曲管桔梗■☆

365176 Siphocampylus membranaceus Britton;膜质曲管桔梗■☆

365177 Siphocampylus obovatus (G. Don) E. Wimm.;倒卵曲管桔梗■☆

365178 Siphocampylus ovatus (G. Don) E. Wimm.;卵形曲管桔梗■☆

365179 Siphocampylus pallidus E. Wimm. in J. F. Macbr.;苍白曲管桔梗■☆

365180 Siphocampylus phyllobotrys E. Wimm.;叶穗曲管桔梗■☆

365181 Siphocampylus pilosus Gleason;多毛曲管桔梗■☆

365182 Siphocampylus ruber Alain;红曲管桔梗■☆

365183 Siphocodon Turcz. (1852);管花桔梗属●☆

365184 Siphocodon debilis Schltr.;非洲管花桔梗●☆

365185 Siphocodon spartioides Turcz.;管花桔梗●☆

365186 Siphocolea Baill. = Stereospermum Cham. ●

365187 Siphocolea boivinii Baill. = Stereospermum boivinii (Baill.) H. Perrier ●☆

365188 Siphocolea hildebrandtii Baill. = Stereospermum hildebrandtii (Baill.) H. Perrier ●☆

365189 Siphocolea rhoifolia Baill. = Stereospermum rhoifolium (Baill.) H. Perrier ●☆

365190 Siphocranion Kudo = Hancea Hemsl. ■★

365191 Siphocranion Kudo = Hanceola Kudo ■★

365192 Siphocranion Kudo (1929);筒冠花属(管萼草属);Siphocranion ■

365193 Siphocranion macranthum (Hook. f.) C. Y. Wu;筒冠花(草藤乌,大花筒冠花,小叶筒冠花);Bigflower Siphocranion, Smallleaf Siphocranion ■

365194 Siphocranion macranthum (Hook. f.) C. Y. Wu var.

microphyllum C. Y. Wu = Siphocranion macranthum (Hook. f.) C. Y. Wu ■

365195 Siphocranion macranthum (Hook. f.) C. Y. Wu var. prainianum (H. Lév.) C. Y. Wu et H. W. Li = Siphocranion macranthum (Hook. f.) C. Y. Wu ■

365196 Siphocranion macranthum (Hook. f.) C. Y. Wu var. prainianum (H. Lév.) C. Y. Wu et H. W. Li;长唇筒冠花;Longlip Siphocranion ■

365197 Siphocranion nudipes (Hemsl.) Kudo;光柄筒冠花;Nakedstalk Siphocranion ■

365198 Siphokentia Burret(1927);摩鹿加椰属(管鞘椰子属,马鲁古桐属,吸管堪蒂椰属)●☆

365199 Siphomeris Bojer = Lecontea A. Rich. ●■

365200 Siphomeris Bojer = Paederia L. (保留属名)●■

365201 Siphomeris Bojer ex Hook. = Lecontea A. Rich. ●■

365202 Siphomeris Bojer ex Hook. = Paederia L. (保留属名)●■

365203 Siphomeris campanulata K. Schum. = Paederia pospischilii K. Schum. ●☆

365204 Siphomeris foetens Hiern = Paederia bojeriana (A. Rich.) Drake subsp. foetens (Hiern) Verdc. ●☆

365205 Siphomeris lingun (Sweet) Bojer = Paederia bojeriana (A. Rich.) Drake ●☆

365206 Siphomeris petrophila (K. Schum.) K. Schum. = Paederia pospischilii K. Schum. ●☆

365207 Siphomeris pospischilii (K. Schum.) Engl. = Paederia pospischilii K. Schum. ●☆

365208 Siphonacanthus Nees = Ruellia L. ■●

365209 Siphonandra Klotzsch(1851);管蕊莓属(管蕊杜鹃属)●☆

365210 Siphonandra Turcz. = Chiococca P. Browne ex L. ●☆

365211 Siphonandra elliptica Klotzsch;管蕊莓●☆

365212 Siphonandraceae Klotzsch = Ericaceae Juss. (保留科名)●

365213 Siphonandrium K. Schum(1905);管蕊茜属■☆

365214 Siphonandrium intricatum K. Schum;管蕊茜■☆

365215 Siphonanthaceae Raf. = Labiatae Juss. (保留科名)●■

365216 Siphonanthaceae Raf. = Lamiaceae Martinov(保留科名)●■

365217 Siphonanthus L. (1753);管花赪桐属;Tuber Flower, Tuberflower ●☆

365218 Siphonanthus L. = Clerodendrum L. ●■

365219 Siphonanthus Schreb. ex Baill. = Hevea Aubl. ●

365220 Siphonanthus botryodes Hiern = Clerodendrum silvanum Henriq. f. botryodes (Hiern) R. Fern. ●☆

365221 Siphonanthus capitata (Willd.) S. Moore = Clerodendrum capitatum (Willd.) Schumach. ●☆

365222 Siphonanthus conglobata Hiern = Clerodendrum capitatum (Willd.) Schumach. ●☆

365223 Siphonanthus costulata Hiern = Clerodendrum silvanum Henriq. var. buchholzii (Gürke) Verdc. ●☆

365224 Siphonanthus cuneifolia Hiern = Clerodendrum buchneri Gürke ●☆

365225 Siphonanthus dumalis Hiern = Rotheca myricoides (Hochst.) Steane et Mabb. var. dumalis (Hiern) R. Fern. ●☆

365226 Siphonanthus formicarum (Gürke) Hiern = Clerodendrum formicarum Gürke ●☆

365227 Siphonanthus glabra (E. Mey.) Hiern = Clerodendrum glabrum E. Mey. ●☆

365228 Siphonanthus glabra (E. Mey.) Hiern var. vaga Hiern = Clerodendrum eriophyllum Gürke ●☆

365229 Siphonanthus indica L. = Clerodendrum indicum (L.) Kuntze ●

365230　Siphonanthus myricoides（Hochst.）Hiern = Rotheca myricoides（Hochst.）Steane et Mabb. ●☆

365231　Siphonanthus myricoides（Hochst.）Hiern var. herbacea Hiern = Rotheca luembensis（De Wild.）R. Fern. f. herbacea（Hiern）R. Fern. ■☆

365232　Siphonanthus nuxioides S. Moore = Clerodendrum silvanum Henriq. var. nuxioides（S. Moore）Verdc. ●☆

365233　Siphonanthus rotundifolius（Oliv.）S. Moore = Clerodendrum rotundifolium Oliv. ●☆

365234　Siphonanthus sanguinea Hiern = Clerodendrum poggei Gürke ●☆

365235　Siphonanthus stricta Hiern = Clerodendrum buchneri Gürke ●☆

365236　Siphonanthus trichotomum（Thunb.）Nakai = Clerodendrum trichotomum Thunb. ex A. Murray ●

365237　Siphonanthus trichotomum（Thunb.）Nakai var. fargesii（Dode）Nakai = Clerodendrum trichotomum Thunb. ex A. Murray ●

365238　Siphonanthus trichotomum Nakai = Clerodendrum trichotomum Thunb. ex A. Murray ●

365239　Siphonanthus trichotomum Nakai var. fargesii Nakai = Clerodendrum trichotomum Thunb. ex A. Murray ●

365240　Siphonanthus triphylla（Harv.）A. DC. = Rotheca hirsuta（Hochst.）R. Fern. ●☆

365241　Siphonella（A. Gray）A. Heller = Leptodactylon Hook. et Arn. ■☆

365242　Siphonella（A. Gray）A. Heller = Linanthastrum Ewan ■☆

365243　Siphonella（A. Gray）A. Heller = Linanthus Benth. ■☆

365244　Siphonella A. Heller = Siphonella（A. Gray）A. Heller ■☆

365245　Siphonella Small = Fedia Gaertn.（保留属名）■

365246　Siphonema Raf. = Nierembergia Ruiz et Pav. ■☆

365247　Siphoneranthemum（Oerst.）Kuntze = Pseuderanthemum Radlk. ●■

365248　Siphoneranthemum Kuntze = Pseuderanthemum Radlk. ●■

365249　Siphoneugena O. Berg（1856）；管蒲桃属●☆

365250　Siphonia Benth. = Lindenia Benth. ■☆

365251　Siphonia Rich. = Hevea Aubl. ●

365252　Siphonia Rich. ex Schreb. = Hevea Aubl. ●

365253　Siphonia brasiliensis Willd. ex A. Juss. = Hevea brasiliensis（Willd. ex A. Juss.）Müll. Arg. ●

365254　Siphonidium J. B. Armstr. = Euphrasia L. ■

365255　Siphoniopsis H. Karst. = Cola Schott et Endl.（保留属名）●☆

365256　Siphonochilus J. M. Wood et Franks（1911）；管唇姜属■☆

365257　Siphonochilus aethiopicus（Schweinf.）B. L. Burtt；埃塞俄比亚管唇姜■☆

365258　Siphonochilus brachystemon（K. Schum.）B. L. Burtt；短冠管唇姜■☆

365259　Siphonochilus carsonii（Baker）Lock；卡森管唇姜■☆

365260　Siphonochilus decorus（Druten）Lock；装饰管唇姜■☆

365261　Siphonochilus evae（Briq.）B. L. Burtt；埃娃管唇姜■☆

365262　Siphonochilus kilimanensis（Gagnep.）B. L. Burtt；基利马尼管唇姜■☆

365263　Siphonochilus kirkii（Hook. f.）B. L. Burtt；柯克管唇姜■☆

365264　Siphonochilus natalensis（Schltr. et K. Schum.）J. M. Wood et Franks = Siphonochilus aethiopicus（Schweinf.）B. L. Burtt ■☆

365265　Siphonochilus nigericus（Hepper）B. L. Burtt；尼日利亚管唇姜■☆

365266　Siphonochilus parvus Lock；小管唇姜■☆

365267　Siphonochilus rhodesicus（T. C. E. Fr.）Lock；罗得西亚管唇姜■☆

365268　Siphonodiscus F. Muell. = Dysoxylon Bartl. ●

365269　Siphonodiscus F. Muell. = Dysoxylum Blume ●

365270　Siphonodon Griff.（1843）；异卫矛属；Ivorywood ●☆

365271　Siphonodon australe Benth.；异卫矛；Australian Ivorywood, Ivory Wood ●☆

365272　Siphonodontaceae（Croizat）Gagnepain et Tardieu = Siphonodontaceae Gagnep. et Tardieu（保留科名）●

365273　Siphonodontaceae Gagnep. et Tardieu ex Tardieu = Celastraceae R. Br.（保留科名）●

365274　Siphonodontaceae Gagnep. et Tardieu ex Tardieu = Siphonodontaceae Gagnep. et Tardieu（保留科名）●

365275　Siphonodontaceae Gagnep. et Tardieu（1951）（保留科名）；异卫矛科●

365276　Siphonodontaceae Gagnep. et Tardieu（保留科名）= Celastraceae R. Br.（保留科名）●

365277　Siphonodontaceae Gagnep. et Tardieu（保留科名）= Sladeniaceae Airy Shaw ●

365278　Siphonoglossa Oerst.（1854）；管舌爵床属●☆

365279　Siphonoglossa leptantha（Nees）Immelman；细花管舌爵床●☆

365280　Siphonoglossa leptantha（Nees）Immelman subsp. late-ovata（C. B. Clarke）Immelman；宽卵细花管舌爵床●☆

365281　Siphonoglossa linifolia（Lindau）C. B. Clarke；亚麻叶管舌爵床●☆

365282　Siphonoglossa longiflora（Torr.）A. Gray；长花管舌爵床；Siphonoglossa ●☆

365283　Siphonoglossa longiflora A. Gray = Siphonoglossa longiflora（Torr.）A. Gray ●☆

365284　Siphonoglossa macleodiae S. Moore = Justicia ladanoides Lam. ■☆

365285　Siphonoglossa migeodii S. Moore = Justicia migeodii（S. Moore）V. A. W. Graham ●☆

365286　Siphonoglossa nkandlaensis Immelman；恩坎德拉管舌爵床●☆

365287　Siphonoglossa nummularia S. Moore = Siphonoglossa leptantha（Nees）Immelman ●☆

365288　Siphonoglossa rubra S. Moore；红管舌爵床●☆

365289　Siphonoglossa tubulosa（Nees）Baill. = Siphonoglossa leptantha（Nees）Immelman ●☆

365290　Siphonogyne Cass. = Eriocephalus L. ●☆

365291　Siphonosmanthus Stapf = Osmanthus Lour. ●

365292　Siphonosmanthus delavayi（Franch.）Stapf = Osmanthus delavayi Franch. ●

365293　Siphonosmanthus suavis（King ex C. B. Clarke）Stapf = Osmanthus suavis King ex C. B. Clarke ●

365294　Siphonosmanthus venosus（Pamp.）Knobl. = Osmanthus venosus Pamp. ●◇

365295　Siphonostegia Benth.（1835）；阴行草属；Siphonostegia ■

365296　Siphonostegia chinensis Benth. ex Hook. et Arn.；阴行草（北刘寄奴，草茵陈，除毒草，吹风草，大婆针，吊钟草，风吹草，罐儿茶，罐子草，鬼麻油，黑茵陈，壶瓶草，黄花茵陈，角茵陈，节节瓶，金壶瓶，金花屏，金钟茵陈，灵茵陈，铃茵陈，刘寄奴，漏卢，蛮老婆针，山茵陈，山油麻，山芝麻，天芝麻，铁杆茵陈，铁雨伞草，土茵陈，五毒草，徐毒草，野油麻，油罐草，油蒿菜）；China Siphonostegia，Chinese Siphonostegia ■

365297　Siphonostegia japonica（Matsum.）Matsum. ex Furumi = Siphonostegia laeta S. Moore ■

365298　Siphonostegia laeta S. Moore；腺毛阴行草（光亮阴行草）；

Glandularhair Siphonostegia ■

365299　Siphonostelma Schltr. = Brachystelma R. Br. (保留属名)■

365300　Siphonostelma stenophyllum Schltr. = Brachystelma stenophyllum (Schltr.) R. A. Dyer ■☆

365301　Siphonostema Griseb. = Ceratostema Juss. ●☆

365302　Siphonostoma Benth. et Hook. f. = Siphonostema Griseb. ●☆

365303　Siphonostylis Wern. Schulze = Iris L. ■

365304　Siphonychia Torr. et A. Gray(1838)(保留属名);管甲草属■☆

365305　Siphonychia Torr. et A. Gray(保留属名) = Paronychia Mill. ■

365306　Siphonychia americana (Nutt.) Torr. et A. Gray = Paronychia americana (Nutt.) Fenzl ex Walp. ■☆

365307　Siphonychia diffusa Chapm. = Paronychia patula Shinners ■☆

365308　Siphonychia erecta Chapm. = Paronychia erecta (Chapm.) Shinners ■☆

365309　Siphonychia interior (Small) Core = Paronychia rugelii (Chapm.) Shuttlew. ex Chapm. ■☆

365310　Siphonychia pauciflora Small = Paronychia americana (Nutt.) Fenzl ex Walp. ■☆

365311　Siphonychia rugelii Chapm. = Paronychia rugelii (Chapm.) Shuttlew. ex Chapm. ■☆

365312　Siphostigma B. D. Jacks. = Siphostima Raf. ■

365313　Siphostigma Raf. (1837);管柱鸭跖草属■☆

365314　Siphostima Raf. = Cyanotis D. Don(保留属名)■

365315　Siphostima Raf. = Tradescantia L. ■

365316　Siphotoma Raf. = Hymenocallis Salisb. ■

365317　Siphotoxis Bojer ex Benth. = Achyrospermum Blume ■●

365318　Siphotria Raf. = Renealmia L. f. (保留属名)■☆

365319　Siphyalis Raf. = Polygonatum Mill. ■

365320　Sipolisia Glaz. = Proteopsis Mart. et Zucc. ex DC. ■☆

365321　Sipolisia Glaz. ex Oliv. (1894);叉毛菊属●■☆

365322　Sipolisia lanuginosa Glaz. ex Oliv. ;叉毛菊■☆

365323　Siponima A. DC. = Ciponima Aubl. ●

365324　Siponima A. DC. = Symplocos Jacq. ●

365325　Siraitia Merr. (1934);罗汉果属;Luohanfruit, Siraitia ■

365326　Siraitia borneensis (Merr.) C. Jeffrey ex A. M. Lu et Zhi Y. Zhang;加岛罗汉果(白兼果);Scaleless Luohanfruit, Scaleless Siraitia ■

365327　Siraitia borneensis (Merr.) C. Jeffrey ex A. M. Lu et Zhi Y. Zhang var. yunnanensis A. M. Lu et Zhi Y. Zhang = Sinobaijiania yunnanensis (A. M. Lu et Zhi Y. Zhang) C. Jeffrey et W. J. de Wilde ■

365328　Siraitia borneensis (Merr.) C. Jeffrey ex A. M. Lu et Zhi Y. Zhang var. lobophylla A. M. Lu et Zhi Y. Zhang = Sinobaijiania yunnanensis (A. M. Lu et Zhi Y. Zhang) C. Jeffrey et W. J. de Wilde ■

365329　Siraitia borneensis (Merr.) C. Jeffrey ex A. M. Lu et Zhi Y. Zhang var. lobophylla A. M. Lu et Zhi Y. Zhang = Baijiania yunnanensis (A. M. Lu et Zhi Y. Zhang) A. M. Lu et J. Q. Li ■

365330　Siraitia borneensis (Merr.) C. Jeffrey ex A. M. Lu et Zhi Y. Zhang var. lobophylla A. M. Lu et Zhi Y. Zhang;裂叶罗汉果;Lobedleaf Siraitia, Splitleaf Luohanfruit ■

365331　Siraitia borneensis (Merr.) C. Jeffrey ex A. M. Lu et Zhi Y. Zhang var. yunnanensis A. M. Lu et Zhi Y. Zhang;云南罗汉果;Yunnan Luohanfruit, Yunnan Siraitia ■

365332　Siraitia grosvenorii (Swingle) C. Jeffrey ex A. M. Lu et Zhi Y. Zhang;罗汉果(戈司维若果,光果木鳖,假苦瓜,苦人参,拉汉果,罗汉表);Grosvenor Momordica, Luohanfruit, Luohanguo Momordica, Luohanguo Siraitia ■

365333　Siraitia siamensis (Craib) C. Jeffrey ex S. Q. Zhong et D. Fang;翅子罗汉果(凡力,红汞藤,山鹅);Siam Luohanfruit, Siam Siraitia ■

365334　Siraitia sikkimensis (Chakrav.) C. Jeffrey ex A. M. Lu et J. Q. Li;锡金罗汉果;Sikkim Siraitia ■

365335　Siraitia silomaradjae Merr. ;苏门答腊罗汉果;Sumatra Siraitia ■☆

365336　Siraitia taiwaniana (Hayata) C. Jeffrey ex A. M. Lu et Zhi Y. Zhang;台湾罗汉果(台湾青牛胆);Taiwan Luohanfruit, Taiwan Siraitia ■

365337　Siraitia taiwaniana (Hayata) C. Jeffrey ex A. M. Lu et Zhi Y. Zhang = Sinobaijiania taiwaniana (Hayata) C. Jeffrey et W. J. de Wilde ■

365338　Siraitos Raf. (废弃属名) = Chionographis Maxim. (保留属名)■

365339　Siraitos chinensis (K. Krause) F. T. Wang et Ts. Tang = Chionographis chinensis K. Krause ■

365340　Sirhookera Kuntze(1891);西卢兰属■☆

365341　Sirhookera lanceolata Kuntze;西卢兰■☆

365342　Sirhookera latifolia Kuntze;宽叶西卢兰■☆

365343　Sirindhornia H. A. Pedersen et Suksathan = Habenaria Willd. ■

365344　Sirindhornia H. A. Pedersen et Suksathan(2003);缅甸玉凤花属■☆

365345　Sirindhornia monophylla (Collett et Hemsl.) H. A. Pedersen et Suksathan = Ponerorchis monophylla (Collett et Hemsl.) Soó ■

365346　Sirium L. = Santalum L. ●

365347　Sirium Schreb. = Santalum L. ●

365348　Sirmuellera Kuntze = Banksia L. f. (保留属名)●☆

365349　Sirochloa S. Dransf. (2002);壕草属■☆

365350　Sirochloa parvifolia (Munro) S. Dransf. ;小花壕草■☆

365351　Siryrinchium Raf. = Sisyrinchium L. ■

365352　Sisarum Bubani = Sium L. ■

365353　Sisarum Mill. = Sium L. ■

365354　Sisarum sisaroides (DC.) Schischk. ex Krylov. = Sium sisaroides DC. ■

365355　Sisimbryum Clairv. = Sisymbrium L. ■

365356　Sismondaea Delponte = Dioscorea L. (保留属名)■

365357　Sison L. (1753);水柴胡属;Honewort, Stone Parsley ■☆

365358　Sison Wahlenb. = Apium L. ■

365359　Sison ammi Jacq. = Apium leptophyllum (Pers.) F. Muell. ex Benth. ■

365360　Sison ammi L. = Apium leptophyllum (Pers.) F. Muell. ex Benth. ■

365361　Sison ammi L. = Trachyspermum ammi (L.) Sprague ■

365362　Sison amomum L. ;田水柴胡;Honewort, Spikenard, Stone Parsley, Wild Parsley ■☆

365363　Sison anisum (L.) Spreng. = Pimpinella anisum L. ■

365364　Sison coniifolia Wall. = Vicatia coniifolia (Wall.) DC. ■

365365　Sison crinitum Pall. = Schultzia crinita (Pall.) Spreng. ■

365366　Sison inundatum L. = Apium inundatum (L.) Rchb. f. ■☆

365367　Sison ruta Burm. f. = Apium graveolens L. ■

365368　Sison tenerum Wall. = Acronema tenerum (Wall.) Edgew. ■

365369　Sisymbrella Spach = Rorippa Scop. ■

365370　Sisymbrella Spach = Sisymbrium L. ■

365371　Sisymbrella Spach(1838);姬大蒜芥属■☆

365372　Sisymbrella aspera (L.) Spach;粗糙姬大蒜芥■☆

365373　Sisymbrella aspera (L.) Spach subsp. boissieri (Coss.) Heywood = Sisymbrella aspera (L.) Spach ■☆

365374　Sisymbrella aspera (L.) Spach subsp. munbyana (Boiss. et Reut.) Greuter et Burdet;芒比粗糙姬大蒜芥■☆

365375 Sisymbriaceae Martinov = Brassicaceae Burnett(保留科名)■●

365376 Sisymbriaceae Martinov = Cruciferae Juss. (保留科名)■●

365377 Sisymbrianthus Chevall. = Rorippa Scop. ■

365378 Sisymbrion St. -Lag. = Sisymbrium L. ■

365379 Sisymbriopsis Botsch. et Tzvelev(1961); 假蒜芥属■

365380 Sisymbriopsis mollipila (Maxim.) Botsch. ; 绒毛假蒜芥■

365381 Sisymbriopsis pamirica (Y. C. Lan et C. H. An) Al-Shehbaz; 帕米尔假蒜芥(帕米尔南芥)■

365382 Sisymbriopsis shuanghuica (K. C. Kuan et C. H. An) Al-Shehbaz et al.; 双湖假蒜芥(双湖念珠芥); Shuanghu Beadcress, Shuanghu Torularia ■

365383 Sisymbriopsis shuanghuica (K. C. Kuan et C. H. An) Al-Shehbaz et al. = Torularia shuanghuica K. C. Kuan et C. H. An ■

365384 Sisymbriopsis yechengica (C. H. An) Al-Shehbaz et al.; 叶城假蒜芥(叶城小蒜芥); Yecheng Microsisymbrium, Yecheng Smallgarliccress ■

365385 Sisymbriopsis yechengica (C. H. An) Al-Shehbaz et al. = Microsisymbrium yechengicum C. H. An ■

365386 Sisymbrium L. (1753); 大蒜芥属(播娘蒿属); Garliccress, Rocket, Sisymbrium ■

365387 Sisymbrium abyssinicum E. Fourn. = Erucastrum arabicum Fisch. et C. A. Mey. ■☆

365388 Sisymbrium aculeolatum Boiss. = Torularia aculeolata (Boiss.) O. E. Schulz ■☆

365389 Sisymbrium adpressum Trautv. = Torularia dentata (Freyn et Sint.) Kitam. ■☆

365390 Sisymbrium afghanicum Gilli = Torularia afghanica (Gilli) Hedge ■☆

365391 Sisymbrium album Pall. = Smelowskia alba (Pall.) Regel ■

365392 Sisymbrium alliaria (L.) Scop. = Alliaria petiolata (M. Bieb.) Cavara et Grande ■

365393 Sisymbrium allionii Pall. = Alliaria petiolata (M. Bieb.) Cavara et Grande ■

365394 Sisymbrium alpinum (Sternb. et Hoppe) Fourn. var. aeneum (Bunge) Trautv. = Braya rosea (Turcz.) Bunge ■

365395 Sisymbrium alpinum E. Fourn. = Braya alpina Sternb. et Hoppe ■☆

365396 Sisymbrium alpinum E. Fourn. var. aeneum (Bunge) Trautv. = Braya rosea (Turcz.) Bunge ■

365397 Sisymbrium alpinum E. Fourn. var. roseum (Turcz.) Trautv. = Braya rosea (Turcz.) Bunge ■

365398 Sisymbrium altissimum L.; 大蒜芥(大叶播娘蒿, 田蒜芥); Hedge Mustard, Highest Sisymbrium, Jim Hill Mustard, Tall Rocket, Tall Sisymbrium, Tall Tumble Mustard, Tall Tumblemustard, Tall Tumble-mustard, Tumble Garliccress, Tumble Mustard, Tumbling Mustard ■

365399 Sisymbrium amphibium L. = Rorippa amphibia (L.) Besser ■☆

365400 Sisymbrium amphibium L. var. palustre L. = Rorippa islandica (Oeder) Borbás ■

365401 Sisymbrium amphibium L. var. palustre L. = Rorippa palustris (L.) Besser ■

365402 Sisymbrium amplexicaule Desf. = Guenthera amplexicaulis (Desf.) Gómez-Campo ■☆

365403 Sisymbrium asperum L. = Nasturtium asperum (L.) Coss. ■☆

365404 Sisymbrium asperum Pall. = Dontostemon pinnatifidus (Willd.) Al-Shehbaz et H. Ohba ■

365405 Sisymbrium atrovirens Hornem. = Rorippa indica (L.) Hiern ■

365406 Sisymbrium austriacum Jacq.; 奥地利大蒜芥; Austrian Rocket, Jeweled Rocket ■☆

365407 Sisymbrium austriacum Jacq. subsp. hispanicum (Jacq.) P. W. Ball et Heywood; 西班牙大蒜芥■☆

365408 Sisymbrium axillare Hook. f. et Thomson = Crucihimalaya axillaris (Hook. f. et Thomson) Al-Shehbaz, O'Kane et R. A. Price ■

365409 Sisymbrium barbaraea L. = Barbarea plantaginea DC. ■☆

365410 Sisymbrium barrelieri L. = Brassica barrelieri (L.) Janka ■☆

365411 Sisymbrium bhutanicum N. P. Balak. = Crucihimalaya lasiocarpa (Hook. f. et Thomson) Al-Shehbaz, O'Kane et R. A. Price ■

365412 Sisymbrium bilobum (K. Koch) Grossh.; 二裂大蒜芥■☆

365413 Sisymbrium bourgeanum E. Fourn. = Descurainia bourgeauana (E. Fourn.) O. E. Schulz ■☆

365414 Sisymbrium brachycarpum (N. Busch) Vassilcz.; 短果大蒜芥; Northern Tansy Mustard, Short-fruited Sisymbrium ■☆

365415 Sisymbrium brachycarpum Richardson = Descurainia pinnata (Walter) Britton subsp. brachycarpa (Rich.) Detling ■☆

365416 Sisymbrium brassiciforme C. A. Mey.; 无毛大蒜芥; Hairless Garliccress, Mustardform Sisymbrium ■

365417 Sisymbrium brevipes Kar. et Kir. = Neotorularia brevipes (Kar. et Kir.) Hedge et J. Léonard ■

365418 Sisymbrium brevipes Kar. et Kir. = Torularia brevipes (Kar. et Kir.) O. E. Schulz ■

365419 Sisymbrium briquetii Pit. = Descurainia preauxiana (Webb) O. E. Schulz ■☆

365420 Sisymbrium burchellii DC.; 伯切尔大蒜芥■☆

365421 Sisymbrium burchellii DC. var. dinteri (O. E. Schulz) Marais; 丁特大蒜芥■☆

365422 Sisymbrium burchellii DC. var. turczaninowii (Sond.) O. E. Schulz = Sisymbrium turczaninowii Sond. ■☆

365423 Sisymbrium cabulicum Hook. f. et Thomson = Olimarabidopsis cabulica (Hook. f. et Thomson) Al-Shehbaz, O'Kane et R. A. Price ■

365424 Sisymbrium canescens Benth. ; 灰大蒜芥; Dog Lime, Hedge Lime, Hedge Mustard, Pepper-grass ■☆

365425 Sisymbrium canescens Nutt. var. brachycarpa (Richardson) S. Watson = Descurainia pinnata (Walter) Britton subsp. brachycarpa (Rich.) Detling ■☆

365426 Sisymbrium capense Thunb.; 好望角大蒜芥■☆

365427 Sisymbrium catholicum L. = Diplotaxis catholica (L.) DC. ■☆

365428 Sisymbrium ceratophyllum Desf. = Nasturtiopsis coronopifolia (Desf.) Boiss. ■☆

365429 Sisymbrium cinereum Desf. = Ammosperma cinereum (Desf.) Baill. ■☆

365430 Sisymbrium columnae Jacq. = Sisymbrium orientale L. ■

365431 Sisymbrium columnae Jacq. var. orientale (L.) DC. = Sisymbrium orientale (L.) Scop. ■

365432 Sisymbrium columnae Jacq. var. stenocarpum Rouy et Foucaud = Sisymbrium orientale L. ■

365433 Sisymbrium confertum Steven ex Turcz. ; 密大蒜芥; Dense Sisymbrium ■☆

365434 Sisymbrium confusum E. Fourn. = Sisymbrium burchellii DC. ■☆

365435 Sisymbrium coronopifolium Desf. = Nasturtiopsis coronopifolia (Desf.) Boiss. ■☆

365436 Sisymbrium crassifolium Cav. ; 厚叶大蒜芥■☆

365437 Sisymbrium crassifolium Cav. var. atlanticum Maire =

Sisymbrium crassifolium Cav. ■☆

365438 Sisymbrium crassifolium Cav. var. giganteum Hochr. = Sisymbrium crassifolium Cav. ■☆

365439 Sisymbrium crassifolium Cav. var. scaposum Hochr. = Sisymbrium crassifolium Cav. ■☆

365440 Sisymbrium crassifolium Cav. var. tenuisiliqua Pomel = Sisymbrium crassifolium Cav. ■☆

365441 Sisymbrium crassifolium Cav. var. trichogynum E. Fourn. = Sisymbrium crassifolium Cav. ■☆

365442 Sisymbrium daghestanicum Vassilcz. ; 达赫斯坦大蒜芥 ■☆

365443 Sisymbrium daghestanicum Vassilcz. = Sisymbrium orientale (L.) Scop. ■

365444 Sisymbrium daghestanicum Vassilcz. = Sisymbrium orientale L. ■

365445 Sisymbrium dahuricum Turcz. = Sisymbrium heteromallum C. A. Mey. ■

365446 Sisymbrium dahuricum Turcz. ex E. Fourn. = Sisymbrium heteromallum C. A. Mey. ■

365447 Sisymbrium decipiens Bunge = Sisymbrium loeselii L. ■

365448 Sisymbrium deltoideum Hook. f. et Thomson = Eutrema deltoideum (Hook. f. et Thomson) O. E. Schulz ■

365449 Sisymbrium dentatum Torr. = Arabis shortii (Fernald) Gleason ■☆

365450 Sisymbrium dinteri O. E. Schulz = Sisymbrium burchellii DC. var. dinteri (O. E. Schulz) Marais ■☆

365451 Sisymbrium dissitiflorum O. E. Schulz; 疏花大蒜芥 ■☆

365452 Sisymbrium doumetianum Coss. = Maresia doumetiana (Coss.) Batt. ■☆

365453 Sisymbrium dubium Pers. = Rorippa dubia (Pers.) H. Hara ■

365454 Sisymbrium dubium Pers. = Rorippa heterophylla (Blume) R. O. Williams ■

365455 Sisymbrium eglandulosum DC. = Dontostemon integrifolius (L.) Ledeb. ■

365456 Sisymbrium eglandulosum DC. = Dontostemon integrifolius (L.) Ledeb. var. eglandulosus (DC.) Turcz. ■

365457 Sisymbrium elatum K. Koch; 高大蒜芥 ■☆

365458 Sisymbrium erucoides (L.) Desf. = Diplotaxis erucoides (L.) DC. ■☆

365459 Sisymbrium erysimoides Desf. ; 法国大蒜芥; French Rocket, Mediterranean Rocket ■☆

365460 Sisymbrium erysimoides Desf. var. arenarium Pit. = Sisymbrium erysimoides Desf. ■☆

365461 Sisymbrium erysimoides Desf. var. ovalifolium Webb = Sisymbrium erysimoides Desf. ■☆

365462 Sisymbrium erysimoides Desf. var. xerophilum E. Fourn. = Sisymbrium erysimoides Desf. ■☆

365463 Sisymbrium exasperatum Sond. = Sisymbrium burchellii DC. ■☆

365464 Sisymbrium falcatum (Hochst. ex A. Rich.) E. Fourn. = Oreophyton falcatum (Hochst. ex A. Rich.) O. E. Schulz ■☆

365465 Sisymbrium ferganense Korsh. = Sisymbrium brassiciforme C. A. Mey. ■

365466 Sisymbrium filifolium Willd. = Leptaleum filifolium (Willd.) DC. ■

365467 Sisymbrium flavissimum Kar. et Kir. = Sophiopsis flavissima (Kar. et Kir.) O. E. Schulz ■☆

365468 Sisymbrium foliosum Hook. f. et Thomson = Olimarabidopsis pumila (Stephan) Al-Shehbaz, O'Kane et R. A. Price ■

365469 Sisymbrium fujianense L. K. ·Ling; 福建大蒜芥; Fujian Sisymbrium, Garliccress ■

365470 Sisymbrium fujianensis L. K. Ling = Sisymbrium orientale (L.) Scop. ■

365471 Sisymbrium gallicum Willd. = Erucastrum gallicum (Willd.) O. E. Schulz ■☆

365472 Sisymbrium gariepinum Burch. ex DC. = Sisymbrium burchellii DC. ■☆

365473 Sisymbrium glabratum O. E. Schulz = Sisymbrium loeselii L. ■

365474 Sisymbrium glandulosum (Kar. et Kir.) Maxim. = Dontostemon glandulosus (Kar. et Kir.) O. E. Schulz ■

365475 Sisymbrium glandulosum (Kar. et Kir.) Maxim. var. linearifolium Maxim. = Dontostemon pinnatifidus (Willd.) Al-Shehbaz et H. Ohba subsp. linearifolius (Maxim.) Al-Shehbaz et H. Ohba ■

365476 Sisymbrium griffithianum Boiss. = Olimarabidopsis pumila (Stephan) Al-Shehbaz, O'Kane et R. A. Price ■

365477 Sisymbrium halophila C. A. Mey. = Thellungiella halophila (C. A. Mey.) O. E. Schulz ■

365478 Sisymbrium halophilum C. A. Mey. = Thellungiella halophila (C. A. Mey.) O. E. Schulz ■

365479 Sisymbrium hararense Engl. = Erucastrum arabicum Fisch. et C. A. Mey. ■☆

365480 Sisymbrium hartwegianum E. Fourn. ; 哈氏大蒜芥; Hartweg's Sisymbrium ■☆

365481 Sisymbrium hastifolium Stapf = Sisymbrium loeselii L. ■

365482 Sisymbrium heteromallum C. A. Mey. ; 垂果大蒜芥(垂果蒜芥, 弯果蒜芥); Droopingfruit Sisymbrium, Garliccress ■

365483 Sisymbrium heteromallum C. A. Mey. f. dahuricum (Turcz.) Glehn = Sisymbrium heteromallum C. A. Mey. ■

365484 Sisymbrium heteromallum C. A. Mey. f. glabrum Korsh. = Sisymbrium heteromallum C. A. Mey. ■

365485 Sisymbrium heteromallum C. A. Mey. var. dahuricum (Turcz. ex Fourn.) Glehn ex Maxim. = Sisymbrium heteromallum C. A. Mey. ■

365486 Sisymbrium heteromallum C. A. Mey. var. dahuricum (Turcz.) Glehn. = Sisymbrium heteromallum C. A. Mey. ■

365487 Sisymbrium heteromallum C. A. Mey. var. sinense O. E. Schulz; 短瓣大蒜芥; Shortpetal Sisymbrium ■

365488 Sisymbrium heteromallum C. A. Mey. var. sinense O. E. Schulz = Sisymbrium heteromallum C. A. Mey. ■

365489 Sisymbrium himalaicum Hook. f. et Thomson = Crucihimalaya himalaica (Edgew.) Al-Shehbaz, O'Kane et R. A. Price ■

365490 Sisymbrium hirsutum Lag. ex DC. ; 粗毛大蒜芥 ■☆

365491 Sisymbrium hirsutum Lag. ex DC. = Sisymbrium runcinatum DC. ■☆

365492 Sisymbrium hirtulum Regel et Schmalh. = Olimarabidopsis pumila (Stephan) Al-Shehbaz, O'Kane et R. A. Price ■

365493 Sisymbrium hispanicum Jacq. = Sisymbrium austriacum Jacq. subsp. hispanicum (Jacq.) P. W. Ball et Heywood ■☆

365494 Sisymbrium hookeri E. Fourn. = Eutrema himalaicum Hook. f. et Thomson ■

365495 Sisymbrium humile C. A. Mey. = Neotorularia humilis (C. A. Mey.) Hedge et J. Léonard ■

365496 Sisymbrium humile C. A. Mey. = Torularia humilis (C. A. Mey.) O. E. Schulz ■

365497 Sisymbrium humile C. A. Mey. var. hygrophilum E. Fourn. = Neotorularia humilis (C. A. Mey.) Hedge et J. Léonard ■

365498 Sisymbrium humile C. A. Mey. var. piasezkii (Maxim.) Maxim. =

Neotorularia humilis (C. A. Mey.) Hedge et J. Léonard ■

365499 Sisymbrium humile C. A. Mey. var. piasezkii Maxim. = Neotorularia humilis (C. A. Mey.) Hedge et J. Léonard ■

365500 Sisymbrium incisum Engelm. ex A. Gray;锐裂大蒜芥;Hedge Mustard,Tansy Mustard,Western Tansy Mustard ■☆

365501 Sisymbrium indicum L. = Rorippa indica (L.) Hiern ■

365502 Sisymbrium integrifolium L. = Dontostemon integrifolius (L.) Ledeb. ■

365503 Sisymbrium irio L. ;水蒜芥(播娘蒿,水芥菜,水蒜菜,台湾播娘蒿); Bread-leaved Hedge Mustard, London Rocket, London-rocket,Water Garliccress,Water Sisymbrium ■

365504 Sisymbrium irio L. var. dasycarpum O. E. Schulz = Sisymbrium irio L. ■

365505 Sisymbrium irio L. var. irioides (Boiss.) O. E. Schulz = Sisymbrium irio L. ■

365506 Sisymbrium irio L. var. kralikii (E. Fourn.) Batt. = Sisymbrium reboudianum Verl. ■☆

365507 Sisymbrium irio L. var. leiocarpum Maire = Sisymbrium irio L. ■

365508 Sisymbrium irio L. var. pubescens Coss. = Sisymbrium reboudianum Verl. ■☆

365509 Sisymbrium irio L. var. xerophilum E. Fourn. = Sisymbrium irio L. ■

365510 Sisymbrium irioides Boiss. = Sisymbrium irio L. ■

365511 Sisymbrium irioides Coss. = Sisymbrium reboudianum Verl. ■☆

365512 Sisymbrium iscandericum Kom. = Sisymbrium brassiciforme C. A. Mey. ■

365513 Sisymbrium isfarense Vassilcz. ;伊斯法拉大蒜芥■☆

365514 Sisymbrium islandicum Oeder = Rorippa islandica (Oeder) Borbás ■

365515 Sisymbrium japonicum Boiss. ;日本大蒜芥;Japan Garliccress, Japanese Sisymbrium ■☆

365516 Sisymbrium junceum M. Bieb. = Sisymbrium polymorphum (Murray) Roth ■

365517 Sisymbrium junceum M. Bieb. var. latifolium Korsh. = Sisymbrium polymorphum (Murray) Roth ■

365518 Sisymbrium junceum M. Bieb. var. latifolium Korsh. = Sisymbrium polymorphum (Murray) Roth var. latifolium (Korsh.) O. E. Schulz ■

365519 Sisymbrium junceum M. Bieb. var. soongaricum Regel et Herder = Sisymbrium polymorphum (Murray) Roth ■

365520 Sisymbrium junceum M. Bieb. var. soongaricum Regel et Herder = Sisymbrium polymorphum (Murray) Roth var. soongaricum (Regel et Herder) O. E. Schulz ■

365521 Sisymbrium kokanicum Regel et Schmalh. = Olimarabidopsis pumila (Stephan) Al-Shehbaz,O'Kane et R. A. Price ■

365522 Sisymbrium korolkovii Regel et Schmalh. = Neotorularia korolkowii (Regel et Schmalh.) Hedge et J. Léonard ■

365523 Sisymbrium korolkovii Regel et Schmalh. = Torularia korolkovii (Regel et Schmalh.) O. E. Schulz ■

365524 Sisymbrium korolkovii Regel et Schmalh. = Neotorularia korolkowii (Regel et Schmalh.) Hedge et J. Léonard ■

365525 Sisymbrium kralikii E. Fourn. = Sisymbrium reboudianum Verl. ■☆

365526 Sisymbrium lasiocarpum Hook. f. et Thomson = Crucihimalaya lasiocarpa (Hook. f. et Thomson) Al-Shehbaz,O'Kane et R. A. Price ■

365527 Sisymbrium limosella (Bunge) E. Fourn. = Braya rosea (Turcz.) Bunge ■

365528 Sisymbrium lipskyi N. Busch;利普斯基大蒜芥■☆

365529 Sisymbrium loeselii L. ;新疆大蒜芥(廖氏大蒜芥);False London Rocket, False London-rocket, Loesel Garliccress, Sisymbrium,Small Tumbleweed Mustard,Tall Hedge Mustard,Tall Hedge-mustard ■

365530 Sisymbrium loeselii L. var. brevicarpum C. H. An;短果新疆大蒜芥(短果大蒜芥);Shortfruit Loesel Sisymbrium ■

365531 Sisymbrium loeselii L. var. brevicarpum C. H. An = Sisymbrium loeselii L. ■

365532 Sisymbrium luteum (Maxim.) O. E. Schulz;全叶大蒜芥(黄花大蒜芥);Entire Garliccress,Yellowflower Sisymbrium ■

365533 Sisymbrium luteum (Maxim.) O. E. Schulz var. yunnanense (W. W. Sm.) O. E. Schulz;云南大蒜芥(康定南芥);Kangding Rockcress,Yunnan Yellowflower Sisymbrium ■

365534 Sisymbrium luteum (Maxim.) O. E. Schulz var. yunnanense (W. W. Sm.) O. E. Schulz = Sisymbrium yunnanense W. W. Sm. ■

365535 Sisymbrium malcolmioides Coss. et Durieu = Maresia malcolmioides (Coss. et Durieu) Pomel ■☆

365536 Sisymbrium marlothii O. E. Schulz = Sisymbrium capense Thunb. ■☆

365537 Sisymbrium maurum Maire;模糊大蒜芥■☆

365538 Sisymbrium maximowiczii J. Palib. = Berteroella maximowiczii (J. Palib.) O. E. Schulz ex Loes. ■

365539 Sisymbrium micranthum Roth = Rorippa micrantha (Roth) Jonsell ■☆

365540 Sisymbrium millefolium (Jacq.) Aiton = Descurainia millefolia (Jacq.) Webb et Berthel. ■☆

365541 Sisymbrium millefolium (Jacq.) Aiton var. brachycarpa Bornm. = Sisymbrium maurum Maire ■☆

365542 Sisymbrium millefolium (Jacq.) Aiton var. macrocarpa Pit. = Sisymbrium maurum Maire ■☆

365543 Sisymbrium minutiflorum Hook. f. et Thomson = Ianhedgea minutiflora (Hook. f. et Thomson) Al-Shehbaz et O'Kane ■

365544 Sisymbrium minutiflorum Hook. f. et Thomson = Microsisymbrium minutiflorum (Hook. f. et Thomson) O. E. Schulz ■

365545 Sisymbrium mollipilum (Maxim.) Botsch. = Sisymbriopsis mollipila (Maxim.) Botsch. ■

365546 Sisymbrium mollipilum Maxim. = Sisymbriopsis mollipila (Maxim.) Botsch. ■

365547 Sisymbrium mollissimum C. A. Mey. = Crucihimalaya mollissima (C. A. Mey.) Al-Shehbaz,O'Kane et R. A. Price ■

365548 Sisymbrium mollissimum C. A. Mey. f. pamiricum Korsh. = Crucihimalaya mollissima (C. A. Mey.) Al-Shehbaz,O'Kane et R. A. Price ■

365549 Sisymbrium monachorum W. W. Sm. = Crucihimalaya lasiocarpa (Hook. f. et Thomson) Al-Shehbaz,O'Kane et R. A. Price ■

365550 Sisymbrium monense L. = Coincya monensis (L.) Greuter et Burdet ■☆

365551 Sisymbrium mongolicum (Maxim.) Maxim. = Neotorularia korolkowii (Regel et Schmalh.) Hedge et J. Léonard ■

365552 Sisymbrium mongolicum Maxim. = Neotorularia korolkowii (Regel et Schmalh.) Hedge et J. Léonard ■

365553 Sisymbrium mongolicum Maxim. = Torularia korolkovii (Regel et Schmalh.) O. E. Schulz ■

365554 Sisymbrium murale (L.) Desf. = Diplotaxis muralis (L.) DC. ■

365555 Sisymbrium murale L. = Diplotaxis muralis (L.) DC. ■

365556 Sisymbrium nanum Bunge = Neotorularia humilis (C. A. Mey.)

Hedge et J. Léonard ■
365557 Sisymbrium nanum DC. = Maresia nana (DC.) Batt. ■☆
365558 Sisymbrium nasturtium-aquaticum L. = Nasturtium officinale R. Br. ■
365559 Sisymbrium nasturtium-aquaticum L. = Nasturtium officinale W. T. Aiton ■
365560 Sisymbrium nigrum (L.) Prantl = Brassica nigra (L.) W. D. J. Koch ■
365561 Sisymbrium nudum (Bél. ex Boiss.) Boiss. = Drabopsis verna K. Koch ■
365562 Sisymbrium nudum (Bél.) Boiss. = Drabopsis nuda (Bél.) Stapf ■
365563 Sisymbrium nudum (Bél.) Boiss. = Drabopsis verna K. Koch ■
365564 Sisymbrium officinale (L.) Scop.;药用大蒜芥(药用糖芥,钻果大蒜芥,钻果蒜芥);Bank Cress, Hedge Garliccress, Hedge Mustard, Hedge Sisymbrium, Hedgemustard, Hedge-mustard, Hedgeweed, Lucifer Matches, Poor-man's Mustard, Scrambling Rocket, Tumble Mustard, Turkey-pod, Wiry Jack, Wormseed ■
365565 Sisymbrium officinale (L.) Scop. var. leiocarpum DC.;光果药用大蒜芥■☆
365566 Sisymbrium officinale (L.) Scop. var. leiocarpum DC. = Sisymbrium officinale (L.) Scop. ■
365567 Sisymbrium officinale L. var. leiocarpum DC. = Sisymbrium officinale (L.) Scop. ■
365568 Sisymbrium orientale (L.) Scop.;东方大蒜芥(戟叶掭娘蒿,钻果大蒜芥);Eastern Rocket, Hedge Mustard, Indian Hedgemustard, Indian Hedge-mustard, Oriental Garliccress, Oriental Sisymbrium ■
365569 Sisymbrium orientale L. = Sisymbrium orientale (L.) Scop. ■
365570 Sisymbrium orientale L. subsp. macroloma (Pomel) Dvorák;大边大蒜芥■☆
365571 Sisymbrium orientale L. var. macroloma (Pomel) Halácsy = Sisymbrium orientale L. subsp. macroloma (Pomel) Dvorák ■☆
365572 Sisymbrium orientale L. var. subhastatum (Willd.) Thell. = Sisymbrium orientale L. ■
365573 Sisymbrium pachypodum Chiov. = Erucastrum pachypodum (Chiov.) Jonsell ■☆
365574 Sisymbrium pakistanicum Jafri = Torularia afghanica (Gilli) Hedge ■☆
365575 Sisymbrium palustre Leyss. = Rorippa islandica (Oeder) Borbás ■
365576 Sisymbrium panonicum Jacq. = Sisymbrium altissimum L. ■
365577 Sisymbrium parvulum (Schrenk) Lipsky = Thellungiella parvula (Schrenk) Al-Shehbaz et O'Kane ■
365578 Sisymbrium pectinatum DC. = Dontostemon pinnatifidus (Willd.) Al-Shehbaz et H. Ohba ■
365579 Sisymbrium pendulum Desf. = Diplotaxis harra (Forssk.) Boiss. ■☆
365580 Sisymbrium piasezkii Maxim. = Neotorularia humilis (C. A. Mey.) Hedge et J. Léonard ■
365581 Sisymbrium pinnatifidum (Lam.) DC. subsp. boryi (Boiss.) Font Quer = Murbeckiella boryi (Boiss.) Rothm. ■☆
365582 Sisymbrium pinnatifidum (Lam.) DC. var. longisiliqua (Font Quer) Font Quer = Murbeckiella boryi (Boiss.) Rothm. ■☆
365583 Sisymbrium pinnatifidum Forssk. = Sisymbrium irio L. ■
365584 Sisymbrium polyceratium L.;西方短果大蒜芥;Shortfruit Hedgemustard ■☆
365585 Sisymbrium polymorphum (Murray) Roth;多型大蒜芥(大蒜芥,多型蒜芥,寿蒜芥);Polymorphic Garliccress, Variousforms Sisymbrium ■
365586 Sisymbrium polymorphum (Murray) Roth var. latifolium (Korsh.) O. E. Schulz;大叶大蒜芥;Broadleaf Variousforms Sisymbrium ■
365587 Sisymbrium polymorphum (Murray) Roth var. latifolium (Korsh.) O. E. Schulz = Sisymbrium polymorphum (Murray) Roth ■
365588 Sisymbrium polymorphum (Murray) Roth var. soongaricum (Regel et Herder) O. E. Schulz = Sisymbrium polymorphum (Murray) Roth ■
365589 Sisymbrium polymorphum (Murray) Roth var. soongaricum (Regel et Herder) O. E. Schulz;准噶尔大蒜芥;Dzungar Garliccress, Songaria Variousforms Sisymbrium ■
365590 Sisymbrium preauxianum Webb = Descurainia preauxiana (Webb) O. E. Schulz ■☆
365591 Sisymbrium primulifolium Thomson = Arcyosperma primulifolium (Thomson) O. E. Schulz ■☆
365592 Sisymbrium pumilio Oliv. = Arabidopsis thaliana (L.) Heynh. ■
365593 Sisymbrium pumilum Stephan = Olimarabidopsis pumila (Stephan) Al-Shehbaz, O'Kane et R. A. Price ■
365594 Sisymbrium pumilum Stephan var. alpinum Korsh. = Olimarabidopsis cabulica (Hook. f. et Thomson) Al-Shehbaz, O'Kane et R. A. Price ■
365595 Sisymbrium reboudianum Verl.;雷博德大蒜芥■☆
365596 Sisymbrium reboudianum Verl. var. dasycarpum Maire = Sisymbrium reboudianum Verl. ■☆
365597 Sisymbrium reboudianum Verl. var. leiocarpum Maire = Sisymbrium reboudianum Verl. ■☆
365598 Sisymbrium repandum Willd. = Guenthera repanda (Willd.) Gómez-Campo ■☆
365599 Sisymbrium rigidum M. Bieb. = Neotorularia torulosa (Desf.) Hedge et J. Léonard ■
365600 Sisymbrium rigidum M. Bieb. = Torularia torulosa (Desf.) O. E. Schulz ■
365601 Sisymbrium runcinatum DC. = Sisymbrium runcinatum Lag. ex DC. ■☆
365602 Sisymbrium runcinatum DC. var. glabrum Coss. = Sisymbrium runcinatum DC. ■☆
365603 Sisymbrium runcinatum DC. var. hirsutum (DC.) Ball = Sisymbrium runcinatum DC. ■☆
365604 Sisymbrium runcinatum DC. var. intermedium Rouy et Foucaud = Sisymbrium runcinatum DC. ■☆
365605 Sisymbrium runcinatum DC. var. villosum Boiss. = Sisymbrium runcinatum DC. ■☆
365606 Sisymbrium runcinatum Lag. ex DC.;倒齿大蒜芥■☆
365607 Sisymbrium rupestre (Edgew.) Hook. f. et Thomson = Crucihimalaya himalaica (Edgew.) Al-Shehbaz, O'Kane et R. A. Price ■
365608 Sisymbrium salsugineum Pall. = Thellungiella salsuginea (Pall.) O. E. Schulz ■
365609 Sisymbrium saxatile Lam. = Guenthera repanda (Willd.) Gómez-Campo ■☆
365610 Sisymbrium schimperi Boiss. = Robeschia schimperi (Boiss.) O. E. Schulz ■☆
365611 Sisymbrium scorpiuroides Boiss. = Neotorularia torulosa (Desf.) Hedge et J. Léonard ■
365612 Sisymbrium scorpiurus Pomel = Neotorularia torulosa (Desf.) Hedge et J. Léonard ■
365613 Sisymbrium sewerzowii Regel = Arabis auriculata Lam. ■

365614　Sisymbrium simplex Viv. = Diplotaxis simplex (Viv.) Spreng. ■☆

365615　Sisymbrium sinapis Burm. f. = Rorippa indica (L.) Hiern ■

365616　Sisymbrium sinapis Burm. f. = Rorippa sinapis (Burm. f.) Keay ■☆

365617　Sisymbrium sophia L. = Descurainia sophia (L.) Webb ex Prantl ■

365618　Sisymbrium sophia L. var. schimperi (Boiss.) Hook. f. et Thomson = Robeschia schimperi (Boiss.) O. E. Schulz ■☆

365619　Sisymbrium sophioides Fisch. = Descurainia sophioides (Fisch. ex Hook.) O. E. Schulz ■

365620　Sisymbrium spectabile Hook. f. et Thomson ex E. Fourn. = Eutrema himalaicum Hook. f. et Thomson ■

365621　Sisymbrium strictissimum L. ;刚直大蒜芥;Perennial Rocket ■☆

365622　Sisymbrium stricture Hook. f. et Thomson = Crucihimalaya stricta (Cambess.) Al-Shehbaz, O'Kane et R. A. Price ■

365623　Sisymbrium strigosum Thunb. = Erucastrum strigosum (Thunb.) O. E. Schulz ■☆

365624　Sisymbrium subspinescens (Fisch. et C. A. Mey.) Bunge;细刺大蒜芥■☆

365625　Sisymbrium subtilissimum Popov;纤细大蒜芥■☆

365626　Sisymbrium subulatum E. Fourn. = Sisymbrium irio L. ■

365627　Sisymbrium sulphureum Korsh. = Neotorularia korolkowii (Regel et Schmalh.) Hedge et J. Léonard ■

365628　Sisymbrium sylvestre L. = Brassica rapa L. ■

365629　Sisymbrium sylvestre L. = Rorippa sylvestris (L.) Besser ■

365630　Sisymbrium tanacetifolium L. ;艾菊叶大蒜芥; Tansy-leaved Rocket ■☆

365631　Sisymbrium tenuifolium L. = Diplotaxis tenuifolia (L.) DC. ■☆

365632　Sisymbrium thalianum (L.) J. Gay et Monnard = Arabidopsis thaliana (L.) Heynh. ■

365633　Sisymbrium thalianum J. Gay ex Monnard = Arabidopsis thaliana (L.) Heynh. ■

365634　Sisymbrium thellungii O. E. Schulz = Erucastrum austroafricanum Al-Shehbaz et Warwick ■☆

365635　Sisymbrium thomsonii Hook. f. = Crucihimalaya mollissima (C. A. Mey.) Al-Shehbaz, O'Kane et R. A. Price ■

365636　Sisymbrium tibeticum (Hook. f. et Thomson) E. Fourn. = Braya tibetica Hook. f. et Thomson ■

365637　Sisymbrium torulosum Desf. = Neotorularia torulosa (Desf.) Hedge et J. Léonard ■

365638　Sisymbrium torulosum Desf. = Torularia torulosa (Desf.) O. E. Schulz ■

365639　Sisymbrium toxophyllum (M. Bieb.) C. A. Mey. = Pseudoarabidopsis toxophylla (M. Bieb.) Al-Shehbaz, O'Kane et R. A. Price ■

365640　Sisymbrium toxophyllum C. A. Mey. = Arabidopsis toxophylla (M. Bieb.) N. Busch ■

365641　Sisymbrium trautvetteri Lipsky = Torularia dentata (Freyn et Sint.) Kitam. ■☆

365642　Sisymbrium tripinnatum DC. = Descurainia sophia (L.) Webb ex Prantl ■

365643　Sisymbrium turcomanicum Litv. ;土库曼大蒜芥■☆

365644　Sisymbrium turczaninowii Sond. ;俄罗斯大蒜芥; Russian Rocket ■☆

365645　Sisymbrium uniflorum (Hook. f. et Thomson) E. Fourn. = Braya uniflora Hook. f. et Thomson ■

365646　Sisymbrium uniflorum (Hook. f. et Thomson) E. Fourn. = Pycnoplinthus uniflorus (Hook. f. et Thomson) O. E. Schulz ■

365647　Sisymbrium valentinum L. = Biscutella valentina (L.) Heywood ■☆

365648　Sisymbrium vimineum L. = Diplotaxis viminea (L.) DC. ■☆

365649　Sisymbrium volgense M. Bieb. ex E. Fourn. ;伏尔加大蒜芥; Russian Mustard, Volga Garliccress, Volga Sisymbrium ■☆

365650　Sisymbrium wallichii Hook. f. et Thomson = Arabidopsis wallichii (Hook. f. et Thomson) N. Busch ■

365651　Sisymbrium wallichii Hook. f. et Thomson = Crucihimalaya wallichii (Hook. f. et Thomson) Al-Shehbaz, O'Kane et R. A. Price ■

365652　Sisymbrium yunnanense W. W. Sm. = Sisymbrium luteum (Maxim.) O. E. Schulz var. yunnanense (W. W. Sm.) O. E. Schulz ■

365653　Sisymbrium zeae Spreng. = Sisymbrium erysimoides Desf. ■☆

365654　Sisyndite E. Mey. = Sisyndite E. Mey. ex Sond. ●☆

365655　Sisyndite E. Mey. ex Sond. (1860);南非蒺藜属●☆

365656　Sisyndite spartea E. Mey. ex Sond. ;南非蒺藜●☆

365657　Sisyranthus E. Mey. (1838);革花萝藦属■☆

365658　Sisyranthus anceps Schltr. ;二棱革花萝藦■☆

365659　Sisyranthus barbatus (Turcz.) N. E. Br. ;髯毛革花萝藦■☆

365660　Sisyranthus compactus N. E. Br. ;紧密革花萝藦■☆

365661　Sisyranthus expansum Schltr. = Sisyranthus trichostomus K. Schum. ■☆

365662　Sisyranthus fanniniae N. E. Br. ;范尼革花萝藦■☆

365663　Sisyranthus franksiae N. E. Br. ;弗兰克斯革花萝藦■☆

365664　Sisyranthus huttoniae (S. Moore) S. Moore;赫顿革花萝藦■☆

365665　Sisyranthus imberbis E. Mey. var. barbatus (Turcz.) Schltr. = Sisyranthus barbatus (Turcz.) N. E. Br. ■☆

365666　Sisyranthus imberbis Harv. ;无须革花萝藦■☆

365667　Sisyranthus macer (E. Mey.) Schltr. ;瘦弱革花萝藦■☆

365668　Sisyranthus randii S. Moore;兰德革花萝藦■☆

365669　Sisyranthus randii S. Moore var. abbreviatus S. Moore = Sisyranthus randii S. Moore ■☆

365670　Sisyranthus rhodesicus Weim. ;罗得西亚革花萝藦■☆

365671　Sisyranthus rotatus Schltr. = Sisyranthus trichostomus K. Schum. ■☆

365672　Sisyranthus saundersiae N. E. Br. ;桑德斯革花萝藦■☆

365673　Sisyranthus schizoglossoides N. E. Br. = Brachystelma schizoglossoides (Schltr.) N. E. Br. ■☆

365674　Sisyranthus trichostomus K. Schum. ;毛片革花萝藦■☆

365675　Sisyranthus virgatus E. Mey. ;条纹革花萝藦■☆

365676　Sisyranthus virgatus E. Mey. var. trichostomus (K. Schum.) Harv. = Sisyranthus trichostomus K. Schum. ■☆

365677　Sisyrinchium Eckl. = Aristea Sol. ex Aiton ■☆

365678　Sisyrinchium L. (1753);庭菖蒲属(豚鼻花属);Blue-eyed grass, Blue-eyed-Grass, Pig Root, Rush Lily, Satin Flower, Satinflower, Sisyrinchium ■

365679　Sisyrinchium Mill. = Gynandriris Parl. ■☆

365680　Sisyrinchium albidum Raf. ;白庭菖蒲;Common Blue-eyed-grass, Pale Blue-eyed-grass, White Blue-eyed-grass White Blue-eyed Grass ■☆

365681　Sisyrinchium alpestre E. P. Bicknell = Sisyrinchium montanum Greene ■☆

365682　Sisyrinchium amethystinum E. P. Bicknell = Sisyrinchium demissum Greene ■☆

365683　Sisyrinchium amoenum E. P. Bicknell = Sisyrinchium ensigerum E. P. Bicknell ■☆

365684　Sisyrinchium angustifolium Michx. var. mucronatum (Michx.) Baker = Sisyrinchium mucronatum Michx. ■☆

365685　Sisyrinchium angustifolium Mill. ;狭叶庭菖蒲;Blue-eyed Grass, Narrow-leaved Blue-eyed-grass, Pointed Blue-eyed-grass, Stout Blue-eyed-grass ■☆

365686　Sisyrinchium angustifolium Mill. 'Album';白花窄叶庭菖蒲(白花庭菖蒲)■

365687　Sisyrinchium angustifolium Mill. = Sisyrinchium graminoides E. P. Bicknell ■

365688　Sisyrinchium angustifolium Mill. var. bellum (S. Watson) Baker = Sisyrinchium bellum S. Watson ■☆

365689　Sisyrinchium angustifolium Mill. var. mucronatum (Michx.) Baker = Sisyrinchium mucronatum Michx. ■☆

365690　Sisyrinchium apiculatum E. P. Bicknell = Sisyrinchium atlanticum E. P. Bicknell ■

365691　Sisyrinchium apiculatum E. P. Bicknell var. mesochorum Nieuwl. = Sisyrinchium atlanticum E. P. Bicknell ■

365692　Sisyrinchium arenicola E. P. Bicknell = Sisyrinchium fuscatum E. P. Bicknell ■☆

365693　Sisyrinchium arizonicum Rothr. ;亚利桑那庭菖蒲■☆

365694　Sisyrinchium asheianum E. P. Bicknell = Sisyrinchium albidum Raf. ■☆

365695　Sisyrinchium atlanticum E. P. Bicknell;庭菖蒲;Blue-eyed Grass, Common Blue-eyed-grass, Eastern Blue-eyed-grass ■

365696　Sisyrinchium bellum S. Watson;美丽蓝眼草;Blue-eyed Grass, Pretty Satinflower ■☆

365697　Sisyrinchium bermudiana L. = Sisyrinchium angustifolium Mill. ■☆

365698　Sisyrinchium bermudiana L. var. crebrum (Fernald) B. Boivin = Sisyrinchium montanum Greene var. crebrum Fernald ■☆

365699　Sisyrinchium bermudiana L. var. mucronatum (Michx.) Baker = Sisyrinchium mucronatum Michx. ■☆

365700　Sisyrinchium bermudianum L. ;百慕大庭菖蒲;Blue-eyed Grass, Branching Blue-eyed Grass ■☆

365701　Sisyrinchium bermudianum L. var. albidum (Raf.) A. Gray = Sisyrinchium albidum Raf. ■☆

365702　Sisyrinchium bermudianum L. var. minus (Engelm. et A. Gray) Klatt = Sisyrinchium minus Engelm. et A. Gray ■☆

365703　Sisyrinchium bicknellianum Fernald = Sisyrinchium nashii E. P. Bicknell ■☆

365704　Sisyrinchium biforme E. P. Bicknell;双型庭菖蒲■☆

365705　Sisyrinchium biramewm Piper = Sisyrinchium idahoense E. P. Bicknell ■☆

365706　Sisyrinchium boreale (E. P. Bicknell) J. K. Henry = Sisyrinchium californicum Dryand. ■☆

365707　Sisyrinchium boreale J. K. Henry = Sisyrinchium californicum Dryand. ■☆

365708　Sisyrinchium brachypus (E. P. Bicknell) J. K. Henry = Sisyrinchium californicum Dryand. ■☆

365709　Sisyrinchium brachypus J. K. Henry = Sisyrinchium californicum Dryand. ■☆

365710　Sisyrinchium brayi E. P. Bicknell = Sisyrinchium pruinosum E. P. Bicknell ■☆

365711　Sisyrinchium brownei Small = Sisyrinchium rosulatum E. P. Bicknell ■☆

365712　Sisyrinchium burmudianum L. = Sisyrinchium graminoides E. P. Bicknell ■

365713　Sisyrinchium bushii E. P. Bicknell = Sisyrinchium pruinosum E. P. Bicknell ■☆

365714　Sisyrinchium californicum Dryand. ;加州蓝眼草(加州庭菖蒲); California Blue-eyed Grass, Gloden-eyed Grass, Yellow-eyed Grass ■☆

365715　Sisyrinchium campestre E. P. Bicknell;草地庭菖蒲;Prairie Blue-eyed Grass, Prairie Blue-eyed-grass, White-eyed Grass ■☆

365716　Sisyrinchium campestre E. P. Bicknell f. flaviflorum (E. P. Bicknell) Steyerm. = Sisyrinchium campestre E. P. Bicknell ■☆

365717　Sisyrinchium campestre E. P. Bicknell var. kansanum E. P. Bicknell = Sisyrinchium campestre E. P. Bicknell ■☆

365718　Sisyrinchium canbyi E. P. Bicknell = Sisyrinchium langloisii Greene ■☆

365719　Sisyrinchium capillare E. P. Bicknell;丝庭菖蒲■☆

365720　Sisyrinchium cernuum (E. P. Bicknell) Kearney;肉色庭菖蒲 ■☆

365721　Sisyrinchium corymbosum E. P. Bicknell = Sisyrinchium atlanticum E. P. Bicknell ■

365722　Sisyrinchium demissum Greene;低垂庭菖蒲■☆

365723　Sisyrinchium demissum Greene var. amethystinum (E. P. Bicknell) Kearney et Peebles = Sisyrinchium demissum Greene ■☆

365724　Sisyrinchium dichotomum E. P. Bicknell;二歧蓝眼草■☆

365725　Sisyrinchium dimorphum R. L. Oliv. = Sisyrinchium biforme E. P. Bicknell ■☆

365726　Sisyrinchium douglasii A. Dietr. ;道格拉斯蓝眼草;Douglas Blue-eyed-grass, Grass Widow, Spring Bells ■☆

365727　Sisyrinchium douglasii A. Dietr. = Olsynium douglasii (A. Dietr.) E. P. Bicknell ■☆

365728　Sisyrinchium eastwoodiae E. P. Bicknell = Sisyrinchium bellum S. Watson ■☆

365729　Sisyrinchium elmeri Greene;埃默庭菖蒲■☆

365730　Sisyrinchium ensigerum E. P. Bicknell;剑状庭菖蒲■☆

365731　Sisyrinchium exile E. P. Bicknell = Sisyrinchium rosulatum E. P. Bicknell ■☆

365732　Sisyrinchium farwellii E. P. Bicknell = Sisyrinchium fuscatum E. P. Bicknell ■☆

365733　Sisyrinchium fibrosum E. P. Bicknell = Sisyrinchium nashii E. P. Bicknell ■☆

365734　Sisyrinchium filiforme Raf. = Sisyrinchium minus Engelm. et A. Gray ■☆

365735　Sisyrinchium flaccidum E. P. Bicknell = Sisyrinchium langloisii Greene ■☆

365736　Sisyrinchium flagellum E. P. Bicknell = Sisyrinchium miamiense E. P. Bicknell ■☆

365737　Sisyrinchium flavidum Kellogg = Sisyrinchium californicum Dryand. ■☆

365738　Sisyrinchium flaviflorum E. P. Bicknell = Sisyrinchium campestre E. P. Bicknell ■☆

365739　Sisyrinchium flexile E. P. Bicknell = Sisyrinchium atlanticum E. P. Bicknell ■

365740　Sisyrinchium flexuosum Raf. = Sisyrinchium minus Engelm. et A. Gray ■☆

365741　Sisyrinchium floridanum E. P. Bicknell = Sisyrinchium nashii E. P. Bicknell ■☆

365742　Sisyrinchium floridanum Raf. = Sisyrinchium albidum Raf. ■☆

365743　Sisyrinchium fuscatum E. P. Bicknell;褐庭菖蒲(庭菖蒲); Rosette Satinflower, Rosette Sisyrinchium ■☆

365744　Sisyrinchium gramineum Curtis = Sisyrinchium angustifolium Mill. ■☆

365745　Sisyrinchium gramineum Lam. ;禾状庭菖蒲;Stout Blue-eyed Grass ■

365746　Sisyrinchium gramineum Lam. = Sisyrinchium angustifolium Mill. ■☆

365747　Sisyrinchium graminoides E. P. Bicknell;窄叶庭菖蒲(百慕大

蓝眼草,百慕大庭菖蒲,狭叶庭菖蒲,狭叶野鸢尾);Bermuda Sisyrinchium,Blue-eyed Grass,Common Blue-eyed Grass,Narrowleaf Satinflower, Narrowleaf Sisyrinchium, Narrow-leaved Blue-eyed Grass,Pointed Blue-eyed Grass,Spanish Nut,Stout Blue-eyed-grass ■

365748　Sisyrinchium graminoides E. P. Bicknell = Sisyrinchium angustifolium Mill. ■☆

365749　Sisyrinchium graminoides E. P. Bicknell = Sisyrinchium bermudianum L. ■☆

365750　Sisyrinchium grandiflorum Cav. = Tigridia grandiflora Diels ■☆

365751　Sisyrinchium grandiflorum Douglas ex Lindl. = Olsynium douglasii (A. Dietr.) E. P. Bicknell ■☆

365752　Sisyrinchium greenei E. P. Bicknell = Sisyrinchium bellum S. Watson ■☆

365753　Sisyrinchium halophilum Greene;喜盐庭菖蒲■☆

365754　Sisyrinchium hastile E. P. Bicknell = Sisyrinchium albidum Raf. ■☆

365755　Sisyrinchium helleri E. P. Bicknell = Sisyrinchium pruinosum E. P. Bicknell ■☆

365756　Sisyrinchium hesperium E. P. Bicknell = Sisyrinchium bellum S. Watson ■☆

365757　Sisyrinchium heterocarpum E. P. Bicknell = Sisyrinchium montanum Greene ■☆

365758　Sisyrinchium hibernicum Á. Löve et D. Löve = Sisyrinchium bermudianum L. ■☆

365759　Sisyrinchium hitchcockii Douglass;希契科克庭菖蒲■☆

365760　Sisyrinchium idahoense E. P. Bicknell;爱达荷庭菖蒲■☆

365761　Sisyrinchium idahoense E. P. Bicknell var. birameum (Piper) J. K. Henry = Sisyrinchium idahoense E. P. Bicknell ■☆

365762　Sisyrinchium idahoense E. P. Bicknell var. macounii (E. P. Bicknell) D. M. Hend. ;马昆蓝眼草■☆

365763　Sisyrinchium idahoense E. P. Bicknell var. occidentale (E. P. Bicknell) D. M. Hend. ;西方庭菖蒲■☆

365764　Sisyrinchium idahoense E. P. Bicknell var. segetum (E. P. Bicknell) D. M. Hend. ;田间庭菖蒲■☆

365765　Sisyrinchium incrustatum E. P. Bicknell = Sisyrinchium fuscatum E. P. Bicknell ■☆

365766　Sisyrinchium intermedium E. P. Bicknell = Sisyrinchium mucronatum Michx. ■☆

365767　Sisyrinchium iridifolium Kunth;鸢尾叶蓝眼草(黄花庭菖蒲); Irisleaf Satinflower,Spreading Blue-eyed Grass ■

365768　Sisyrinchium juncellum Greene = Sisyrinchium idahoense E. P. Bicknell var. occidentale (E. P. Bicknell) D. M. Hend. ■☆

365769　Sisyrinchium kansanum (E. P. Bicknell) Alexander = Sisyrinchium campestre E. P. Bicknell ■☆

365770　Sisyrinchium langloisii Greene;朗格庭菖蒲■☆

365771　Sisyrinchium laxum Otto ex Sims;松散庭菖蒲;Veined Yellow-eyed Grass ■☆

365772　Sisyrinchium leptocaulon E. P. Bicknell = Sisyrinchium halophilum Greene ■☆

365773　Sisyrinchium lineatum Torr. = Sisyrinchium californicum Dryand. ■☆

365774　Sisyrinchium littorale Greene;海岸庭菖蒲■☆

365775　Sisyrinchium longipedunculatum E. P. Bicknell = Sisyrinchium demissum Greene ■☆

365776　Sisyrinchium longipes (E. P. Bicknell) Kearney et Peebles;长梗庭菖蒲■☆

365777　Sisyrinchium macounii E. P. Bicknell = Sisyrinchium idahoense

E. P. Bicknell var. macounii (E. P. Bicknell) D. M. Hend. ■☆

365778　Sisyrinchium maritimum A. Heller = Sisyrinchium bellum S. Watson ■☆

365779　Sisyrinchium miamiense E. P. Bicknell;迈阿密庭菖蒲■☆

365780　Sisyrinchium micranthum Cav. ;小花庭菖蒲■☆

365781　Sisyrinchium minus Engelm. et A. Gray;小庭菖蒲■☆

365782　Sisyrinchium montanum Greene;山蓝眼草;American Blue-eyed Grass, Common Blue-eyed Grass, Mountain Blue-eyed Grass, Mountain Blue-eyed-grass, Mountain Satinflower, Strict Blue-eyed Grass ■☆

365783　Sisyrinchium montanum Greene subsp. crebrum (Fernald) Böcher = Sisyrinchium montanum Greene var. crebrum Fernald ■☆

365784　Sisyrinchium montanum Greene var. crebrum Fernald;劲直山蓝眼草;Mountain Blue-eyed Grass,Strict Blue-eyed Grass ■☆

365785　Sisyrinchium mucronatum Michx. ; 米氏庭菖蒲; Michaux's Blue-eyed Grass, Needle-tip Blue-eyed-grass, Slender Blue-eyed Grass ■☆

365786　Sisyrinchium mucronatum Michx. var. atlanticum (E. P. Bicknell) H. E. Ahles = Sisyrinchium atlanticum E. P. Bicknell ■

365787　Sisyrinchium nashii E. P. Bicknell;纳什庭菖蒲■☆

365788　Sisyrinchium niveum Raf. = Sisyrinchium albidum Raf. ■☆

365789　Sisyrinchium occidentale E. P. Bicknell = Sisyrinchium idahoense E. P. Bicknell var. occidentale (E. P. Bicknell) D. M. Hend. ■☆

365790　Sisyrinchium odoratissimum Lindl. = Olsynium biflorum (Thunb.) Goldblatt ■☆

365791　Sisyrinchium oreophilum E. P. Bicknell = Sisyrinchium idahoense E. P. Bicknell ■☆

365792　Sisyrinchium pallidum Cholewa et Douglass;苍白庭菖蒲■☆

365793　Sisyrinchium pruinosum E. P. Bicknell;簇毛庭菖蒲■☆

365794　Sisyrinchium rosulatum E. P. Bicknell;北美庭菖蒲■☆

365795　Sisyrinchium rufipes E. P. Bicknell = Sisyrinchium fuscatum E. P. Bicknell ■☆

365796　Sisyrinchium sagittiferum E. P. Bicknell;箭头庭菖蒲■☆

365797　Sisyrinchium sarmentosum Suksd. ex Greene;长匍茎庭菖蒲■☆

365798　Sisyrinchium scabrellum E. P. Bicknell = Sisyrinchium albidum Raf. ■☆

365799　Sisyrinchium scoparium E. P. Bicknell = Sisyrinchium atlanticum E. P. Bicknell ■

365800　Sisyrinchium segetum E. P. Bicknell = Sisyrinchium idahoense E. P. Bicknell var. segetum (E. P. Bicknell) D. M. Hend. ■☆

365801　Sisyrinchium septentrionale E. P. Bicknell;北方庭菖蒲■☆

365802　Sisyrinchium solstitiale E. P. Bicknell = Sisyrinchium xerophyllum Greene ■☆

365803　Sisyrinchium strictum E. P. Bicknell;智利豚鼻花(阿根廷豚鼻花,条纹庭菖蒲);Chile Satinflower,Cream Sisyrinchium,Mountain Blue-eyed-grass,Pale Yellow-eyed Grass ■☆

365804　Sisyrinchium strictum E. P. Bicknell 'Aunt May';花叶条纹庭菖蒲■☆

365805　Sisyrinchium tenellum E. P. Bicknell = Sisyrinchium fuscatum E. P. Bicknell ■☆

365806　Sisyrinchium thurowii J. M. Coult. et Fisher = Sisyrinchium minus Engelm. et A. Gray ■☆

365807　Sisyrinchium tortum E. P. Bicknell = Sisyrinchium nashii E. P. Bicknell ■☆

365808　Sisyrinchium tracyi E. P. Bicknell = Sisyrinchium atlanticum E. P. Bicknell ■

365809　Sisyrinchium varians E. P. Bicknell = Sisyrinchium pruinosum E. P. Bicknell ■☆

365810　Sisyrinchium versicolor E. P. Bicknell = Sisyrinchium mucronatum Michx. ■☆

365811　Sisyrinchium violaceum E. P. Bicknell = Sisyrinchium atlanticum E. P. Bicknell ■

365812　Sisyrinchium xerophyllum Greene;耐旱庭菖蒲■☆

365813　Sisyrocarpum Klotzsch = Sisyrocarpus Klotzsch ●■☆

365814　Sisyrocarpus Klotzsch = Capanea Decne. ex Planch. ●■☆

365815　Sisyrocarpus Post et Kuntze = Capanea Decne. ex Planch. ●■☆

365816　Sisyrolepis Radlk. (1905);革鳞无患子属●☆

365817　Sisyrolepis Radlk. = Delpya Pierre ●

365818　Sisyrolepis siamensis Radlk. ;革鳞无患子●☆

365819　Sitanion Raf. (1819);细坦麦属(单花草属)■☆

365820　Sitanion albescens Elmer;渐白细坦麦■☆

365821　Sitanion elymoides Raf. ;细坦麦☆

365822　Sitanion hystrix (Nutt.) J. G. Sm. var. brevifolium (J. G. Sm.) C. L. Hitchc. = Elymus longifolius (J. G. Sm.) Gould ■☆

365823　Sitanion longifolium J. G. Sm. = Elymus longifolius (J. G. Sm.) Gould ■☆

365824　Sitella L. H. Bailey = Waltheria L. ●■

365825　Sitilias Raf. = Pyrrhopappus DC. (保留属名)■☆

365826　Sitocodium Salisb. = Camassia Lindl. (保留属名)■☆

365827　Sitodium Banks ex Gaertn. = Radermachia Thunb. ●

365828　Sitodium Parkinson(废弃属名) = Artocarpus J. R. Forst. et G. Forst. (保留属名)●☆

365829　Sitodium altile (Banks et Sol.) Parkinson = Artocarpus communis J. R. Forst. et G. Forst. ●

365830　Sitodium altile (Banks et Sol.) Parkinson = Artocarpus incisus (Thunb.) L. f. ●

365831　Sitodium altile Parkinson = Artocarpus communis J. R. Forst. et G. Forst. ●

365832　Sitopsis (Jaub. et Spach) Á. Löve = Aegilops L. (保留属名)■

365833　Sitopsis speltoides (Tausch) Á. Löve = Aegilops bicornis (Forssk.) Jaub. et Spach ☆

365834　Sitospelos Adans. = Elymus L. ■

365835　Sium L. (1753);泽芹属(毒人参属,零余子属);Greater Water-parsnip,Skirret,Water Parsnip,Waterparsnip,Water-parsnip ■

365836　Sium angustifolium L. = Berula erecta (Huds.) Coville ■

365837　Sium asperum Thunb. = Sonderina hispida (Thunb.) H. Wolff ■☆

365838　Sium cicutifolium Schrank = Sium suave Walter ■

365839　Sium erectum Huds. = Berula erecta (Huds.) Coville ■

365840　Sium falcaria L. = Falcaria vulgaris Bernh. ■☆

365841　Sium filifolium Thunb. = Itasina filifolia (Thunb.) Raf. ■☆

365842　Sium floridanum Small = Sium suave Walter ■

365843　Sium formosanum Hayata = Sium suave Walter ■

365844　Sium frigidum Hand. -Mazz. ;滇西泽芹(冷泽芹);W. Yunnan Waterparsnip ■

365845　Sium grandiflorum Thunb. = Annesorhiza grandiflora (Thunb.) M. Hiroe ■☆

365846　Sium graveolens (L.) Vest = Apium graveolens L. ■

365847　Sium hispidum Thunb. = Sonderina hispida (Thunb.) H. Wolff ■☆

365848　Sium incisum Torr. = Berula erecta (Huds.) Coville ■

365849　Sium javanicum Blume = Oenanthe javanica (Blume) DC. ■

365850　Sium lancifolium M. Bieb. ;披针叶毒人参■☆

365851　Sium latifolium L. ;欧泽芹(宽叶毒人参,欧洲泽芹,泽芹);Broadleaf Waterparsnip, Broad-leaved Water Parsnip, Greater Water Parsnip, Greater Water-parsnip, Water Hemlock, Water-parsnip ■

365852　Sium latijugum C. B. Clarke;印巴泽芹■☆

365853　Sium matsumurae H. Boissieu = Angelica cartilaginomarginata (Makino ex Y. Yabe) Nakai ■

365854　Sium medium Fisch. et C. A. Mey. ;中亚泽芹(中泽芹);Central Asia Waterparsnip ■

365855　Sium neurophyllum (Maxim.) H. Hara = Pterygopleurum neurophyllum (Maxim.) Kitag. ■

365856　Sium neurophyllum Hara = Pterygopleurum neurophyllum (Maxim.) Kitag. ■

365857　Sium ninsi L. = Sium sisarum L. ■☆

365858　Sium nipponicum Maxim. ;日本泽芹;Japanese Waterparsnip ■

365859　Sium nipponicum Maxim. = Sium suave Walter var. nipponicum (Maxim.) H. Hara ■

365860　Sium nipponicum Maxim. = Sium suave Walter ■

365861　Sium nodiflorum L. ;裸花泽芹;Creeping Water Parsnip, Fool's Cress, Procumbent Water Parsnip, Water Parsnip, Water Skirret ■☆

365862　Sium paniculatum Thunb. = Anginon paniculatum (Thunb.) B. L. Burtt ■☆

365863　Sium patulum Thunb. = Dasispermum suffruticosum (P. J. Bergius) B. L. Burtt ■☆

365864　Sium pusillum Nutt. ex Torr. et A. Gray = Berula erecta (Huds.) Coville ■

365865　Sium repandum Welw. ex Hiern;浅波状泽芹■☆

365866　Sium rigidius L. = Oxypolis rigidior (L.) J. M. Coult. et Rose ■☆

365867　Sium serrum (Franch. et Sav.) Kitag. = Pimpinella serra Franch. et Sav. ■

365868　Sium siculum L. = Kundmannia sicula (L.) DC. ■☆

365869　Sium simense J. Gay ex A. Rich. = Oreoschimperella verrucosa (J. Gay ex A. Rich.) Rauschert ■☆

365870　Sium sisaroides DC. ;拟泽芹(阿勒泰泽芹)●■

365871　Sium sisarum L. ;小泽芹(乌苏里泽芹,西沙泽芹,泽芹);Crummock, Skerret, Skirret, Skirret Water Parsnip, Skirret Water-parsnip, Skirwits, Skirwort ■☆

365872　Sium suave Walter;泽芹(毒人参,山藁本,细叶零余子,细叶零子人参,细叶泽芹);Hemlock Waterparsnip, Hemlock Water-parsnip, Water Parsley, Water Parsnip, Water-parsnip ■

365873　Sium suave Walter subsp. nipponicum (Maxim.) Sugim. = Sium nipponicum Maxim. ■

365874　Sium suave Walter var. floridanum (Small) C. F. Reed = Sium suave Walter ■

365875　Sium suave Walter var. nipponicum (Maxim.) H. Hara = Sium nipponicum Maxim. ■

365876　Sium suave Walter var. nipponicum (Maxim.) H. Hara f. ovatum (Yatabe) Kitag. = Sium suave Walter var. ovatum (Yatabe) H. Hara ■☆

365877　Sium suave Walter var. ovatum (Yatabe) H. Hara;卵形泽芹 ■☆

365878　Sium tenue (Kom.) Kom. = Sium sisarum L. ■☆

365879　Sium thunbergii DC. = Berula erecta (Huds.) Coville subsp. thunbergii (DC.) B. L. Burtt ■☆

365880　Sium verrucosum J. Gay ex A. Rich. = Oreoschimperella verrucosa (J. Gay ex A. Rich.) Rauschert ■☆

365881　Sium villosum Thunb. = Annesorhiza grandiflora (Thunb.) M.

Hiroe ■☆

365882　Sium visnaga Stokes = Ammi visnaga（L.）Lam. ■

365883　Siumis Raf. = Sium L. ■

365884　Sivadasania N. Mohanan et Pimenov = Peucedanum L. ■

365885　Sivadasania N. Mohanan et Pimenov(2007);印度前胡属(石防风属)■☆

365886　Sixalix Raf.（1838）;肖蓝盆花属■☆

365887　Sixalix Raf. = Scabiosa L. ●■

365888　Sixalix arenaria（Forssk.）Greuter et Burdet;沙地肖蓝盆花■☆

365889　Sixalix atropurpurea（L.）Greuter et Burdet = Scabiosa atropurpurea L. ■

365890　Sixalix atropurpurea（L.）Greuter et Burdet subsp. grandiflora（Scop.）Soldano et F. Conti;大花肖蓝盆花■☆

365891　Sixalix atropurpurea（L.）Greuter et Burdet subsp. maritima（L.）Greuter et Burdet;滨海肖蓝盆花■☆

365892　Sixalix cartenniana（Pons et Quézel）Greuter et Burdet;卡尔顿肖蓝盆花■☆

365893　Sixalix daucoides（Desf.）Raf.;胡萝卜肖蓝盆花■☆

365894　Sixalix eremophila（Boiss.）Greuter et Burdet;沙漠肖蓝盆花■☆

365895　Sixalix farinosa（Coss.）Greuter et Burdet;被粉肖蓝盆花■☆

365896　Sixalix libyca（Alavi）Greuter et Burdet;利比亚肖蓝盆花■☆

365897　Sixalix parielii（Maire）Greuter et Burdet;帕里埃尔肖蓝盆花■☆

365898　Sixalix semipapposa（DC.）Greuter et Burdet;半冠毛肖蓝盆花■☆

365899　Sixalix thysdrusiana（Le Houér.）Greuter et Burdet;蒂斯肖蓝盆花■☆

365900　Sizygium Duch. = Syzygium R. Br. ex Gaertn.（保留属名）●

365901　Skapanthus C. Y. Wu = Plectranthus L'Hér.（保留属名）●■

365902　Skapanthus C. Y. Wu et H. W. Li = Isodon（Schrad. ex Benth.）Spach ●■

365903　Skapanthus C. Y. Wu et H. W. Li = Plectranthus L'Hér.（保留属名）●■

365904　Skapanthus C. Y. Wu et H. W. Li（1975）;子宫草属（龙老根属,葶花草属）;Skapanthus, Wombgrass ■★

365905　Skapanthus oreophilus（Diels）C. Y. Wu et H. W. Li;子宫草（龙老根,葶花）;Common Skapanthus, Common Wombgrass ■

365906　Skapanthus oreophilus（Diels）C. Y. Wu et H. W. Li f. albus C. Y. Wu;白子宫草（白花子宫草）;White Skapanthus, White Wombgrass ■

365907　Skapanthus oreophilus（Diels）C. Y. Wu et H. W. Li var. elonggatus（Hand.-Mazz.）C. Y. Wu et H. W. Li;茎叶子宫草（茎叶葶花）;Elonghate Wombgrass ■

365908　Skaphium Miq. = Xanthophyllum Roxb.（保留属名）●

365909　Skeptrostachys Garay = Stenorrhynchos Rich. ex Spreng. ■☆

365910　Skiatophytum L. Bolus(1927);亭花属■☆

365911　Skiatophytum tripolium（L.）L. Bolus;亭花■☆

365912　Skidanthera Raf. = Dicera J. R. Forst. et G. Forst. ●

365913　Skidanthera Raf. = Elaeocarpus L. ●

365914　Skilla Raf. = Scilla L. ■

365915　Skimmi Adans. = Illicium L. ●

365916　Skimmia Thunb.（1783）（保留属名）;茵芋属;Skimmia ●■

365917　Skimmia anquetilia N. P. Taylor et Airy Shaw;黄花茵芋●☆

365918　Skimmia arborescens T. Anderson ex Gamble;乔木茵芋;Tree Skimmia ●

365919　Skimmia arisanensis Hayata;阿里山茵芋;Alishan Skimmia ●

365920　Skimmia arisanensis Hayata = Skimmia reevesiana Fortune ●

365921　Skimmia distincte-venulosa（Hayata）C. E. Chang = Skimmia japonica Thunb. var. distincte-venulosa（Hayata）C. E. Chang ●

365922　Skimmia distincte-venulosa Hayata = Skimmia reevesiana Fortune ●

365923　Skimmia euphlebia Merr. = Skimmia arborescens T. Anderson ex Gamble ●

365924　Skimmia formosana C. E. Chang;台湾茵芋;Taiwan Skimmia ●

365925　Skimmia fortunei Mast. = Skimmia reevesiana Fortune ●

365926　Skimmia fragrans Carrière = Skimmia japonica Thunb. ●

365927　Skimmia fragrantissima T. Moore = Skimmia japonica Thunb. ●

365928　Skimmia hainanensis C. C. Huang = Skimmia reevesiana Fortune ●

365929　Skimmia japonica Thunb.;香茵芋（日本茵芋）;Japanese Skimmia ●

365930　Skimmia japonica Thunb. 'Cecilia Brown';亮叶日本茵芋●☆

365931　Skimmia japonica Thunb. 'Fructo-albo';白果茵芋（白果香茵芋）;White-fruited Skimmia ●☆

365932　Skimmia japonica Thunb. 'Nymans';尼曼斯日本茵芋●☆

365933　Skimmia japonica Thunb. 'Rubell';麻疹日本茵芋●☆

365934　Skimmia japonica Thunb. 'Snow Dwarf';小雪日本茵芋●☆

365935　Skimmia japonica Thunb. = Skimmia reevesiana Fortune ●

365936　Skimmia japonica Thunb. f. intermedia Komatsu;中型茵芋●☆

365937　Skimmia japonica Thunb. f. leucocarpa Ohwi;白果香茵芋●☆

365938　Skimmia japonica Thunb. f. longifolia H. Hara;长叶茵芋●☆

365939　Skimmia japonica Thunb. f. macrophylla？ = Skimmia japonica Thunb. ●

365940　Skimmia japonica Thunb. f. ovata（Carrière）H. Hara;卵叶茵芋●☆

365941　Skimmia japonica Thunb. f. ovata（Carrière）H. Hara = Skimmia japonica Thunb. ●

365942　Skimmia japonica Thunb. f. repens Hara;匍匐香茵芋（蔓茵芋）●☆

365943　Skimmia japonica Thunb. f. rosea Hayashi;芬香茵芋●☆

365944　Skimmia japonica Thunb. f. rosea Hayashi = Skimmia japonica Thunb. ●

365945　Skimmia japonica Thunb. f. rugosa（Yatabe）Ohwi;凸脉茵芋●☆

365946　Skimmia japonica Thunb. f. rugosa（Yatabe）Ohwi = Skimmia japonica Thunb. f. yatabei H. Ohba ■☆

365947　Skimmia japonica Thunb. f. veitchii（Carrière）H. Hara;维奇茵芋●☆

365948　Skimmia japonica Thunb. f. veitchii（Carrière）H. Hara = Skimmia japonica Thunb. ●

365949　Skimmia japonica Thunb. f. yatabei H. Ohba;谷田茵芋■☆

365950　Skimmia japonica Thunb. subsp. lutchuensis（Nakai）Kitam. = Skimmia japonica Thunb. var. lutchuensis（Nakai）Hatus. ex T. Yamaz. ●☆

365951　Skimmia japonica Thunb. var. distincte-venulosa（Hayata）C. E. Chang;显脉茵芋●

365952　Skimmia japonica Thunb. var. distincte-venulosa C. E. Chang = Skimmia reevesiana Fortune ●

365953　Skimmia japonica Thunb. var. fragrans？ = Skimmia japonica Thunb. ●

365954　Skimmia japonica Thunb. var. intermedia Komatsu;全叶香茵芋●☆

365955　Skimmia japonica Thunb. var. intermedia Komatsu f. intermedia

（Komatsu）T. Yamaz. = Skimmia japonica Thunb. var. intermedia Komatsu ●☆

365956　Skimmia japonica Thunb. var. intermedia Komatsu f. leucocarpa（Nakai）Ohwi;白果全叶香茵芋●☆

365957　Skimmia japonica Thunb. var. intermedia Komatsu f. longifolia（Makino）H. Hara;长全叶香茵芋●☆

365958　Skimmia japonica Thunb. var. intermedia Komatsu f. obovoidea（Hayashi）Hayashi;倒卵全叶香茵芋●☆

365959　Skimmia japonica Thunb. var. intermedia Komatsu f. repens（Nakai）Ohwi;匍匐全叶香茵芋●☆

365960　Skimmia japonica Thunb. var. intermedia Komatsu f. rugosa Makino = Skimmia japonica Thunb. var. intermedia Komatsu ●☆

365961　Skimmia japonica Thunb. var. intermedia Komatsu f. serrata Hayashi;具齿香茵芋●☆

365962　Skimmia japonica Thunb. var. leucocarpa Makino;日本白果茵芋●☆

365963　Skimmia japonica Thunb. var. lutchuensis（Nakai）Hatus. ex T. Yamaz.;琉球茵芋●☆

365964　Skimmia japonica Thunb. var. obovata ? = Skimmia japonica Thunb. ●

365965　Skimmia japonica Thunb. var. orthoclada Masam.;节枝茵芋;Liuqiu Skimmia ●☆

365966　Skimmia japonica Thunb. var. repens（Nakai）Ohwi = Skimmia japonica Thunb. var. intermedia Komatsu f. repens（Nakai）Ohwi ●☆

365967　Skimmia japonica Thunb. var. veitchii ? = Skimmia japonica Thunb. ●

365968　Skimmia kwangsiensis C. C. Huang = Skimmia arborescens T. Anderson ex Gamble ●

365969　Skimmia laureola（DC.）Siebold et Zucc. ex Walp.;月桂茵芋（黑果茵芋,美丽茵芋）;Laurel Skimmia ●

365970　Skimmia melanocarpa Rehder et E. H. Wilson;黑果茵芋;Blackfruit Skimmia,Black-fruited Skimmia ●

365971　Skimmia multinervia C. C. Huang;多脉茵芋;Manyvein Skimmia,Nervose Skimmia ●

365972　Skimmia oblata T. Moore = Skimmia japonica Thunb. ●

365973　Skimmia orthoclada Hayata = Skimmia reevesiana Fortune ●

365974　Skimmia reevesiana Fortune;茵芋（阿里山茵芋,卑共,卑山共,卑山竹,海南茵芋,黄山桂,山桂花,深红茵芋,莞草,卫与,香茵芋,因预,茵蒺）;Japanese Skimmia,Reeves Skimmia,Skimmia ●

365975　Skimmia repens Nakai;匍匐茵芋●☆

365976　Skimmia wallichii Hook. f. et Thoms. ex Gamble = Skimmia arborescens T. Anderson ex Gamble ●

365977　Skinnera Forssk. = Skinnera J. R. Forst. et G. Forst. ●■

365978　Skinnera J. R. Forst. et G. Forst. = Fuchsia L. ●■

365979　Skinneria Choisy = Merremia Dennst. ex Endl.（保留属名）●■

365980　Skinneria caespitosa（Roxb.）Choisy = Merremia hirta（L.）Merr. ■

365981　Skiophila Hanst. = Episcia Mart. ■☆

365982　Skiophila Hanst. = Nautilocalyx Linden ex Hanst.（保留属名）■☆

365983　Skirhophorus DC. ex Lindl. = Angianthus J. C. Wendl.（保留属名）■●☆

365984　Skirrhophorus DC. = Angianthus J. C. Wendl.（保留属名）■●☆

365985　Skirrhophorus DC. ex Lindl. = Angianthus J. C. Wendl.（保留属名）■●☆

365986　Skirrophorus C. Muell. = Angianthus J. C. Wendl.（保留属名）■●☆

365987　Skirrophorus C. Muell. = Skirhophorus DC. ex Lindl. ■●☆

365988　Skizima Raf. = Astelia Banks et Sol. ex R. Br.（保留属名）■☆

365989　Skizima Raf. = Funckia Willd.（废弃属名）■☆

365990　Skofitzia Hassk. et Kanitz = Tradescantia L. ■

365991　Skoinolon Raf. = Schoenocaulon A. Gray ■☆

365992　Skolemora Arruda = Andira Lam.（保留属名）●☆

365993　Skoliopteris Cuatrec. = Clonodia Griseb. ●☆

365994　Skoliopterys Cuatrec. = Clonodia Griseb. ●☆

365995　Skoliostigma Lauterb. = Spondias L. ●

365996　Skottsbergianthus Boelcke = Xerodraba Skottsb. ■●☆

365997　Skottsbergianthus Boelcke(1984);探险芥属■☆

365998　Skottsbergianthus colobanthoides（Skottsb.）Boelcke;探险者芥■☆

365999　Skottsbergiella Boelcke = Skottsbergianthus Boelcke ■☆

366000　Skottsbergiella Boelcke = Xerodraba Skottsb. ■●☆

366001　Skottsbergiella Epling = Cuminia Colla ●☆

366002　Skottsbergiliana H. St. John = Sicyos L. ■

366003　Skutchia Pax et K. Hoffm. = Trophis P. Browne(保留属名)●☆

366004　Skutchia Pax et K. Hoffm. ex C. V. Morton = Trophis P. Browne（保留属名）●☆

366005　Skytanthus Meyen(1834);智利夹竹桃属●☆

366006　Skytanthus acutus Meyen;智利夹竹桃●☆

366007　Skytatalanthus Endl. = Skytanthus Meyen ●☆

366008　Slackia Griff.（1848）= Decaisnea Hook. f. et Thomson（保留属名）●

366009　Slackia Griff.（1854）= Beccarinda Kuntze ■

366010　Slackia Griff.（1854）= Iguanura Blume ●☆

366011　Slackia fargesii Franch. = Decaisnea insignis（Griff.）Hook. f. et Thomson ●

366012　Slackia insignis Griff. = Decaisnea insignis（Griff.）Hook. f. et Thomson ●

366013　Slackia sinensis Chun = Beccarinda tonkinensis（Pellegr.）B. L. Burtt ■

366014　Slackia tonkinensis Pellegr. = Beccarinda tonkinensis（Pellegr.）B. L. Burtt ■

366015　Sladenia Kurz（1873）;肋果茶属（毒药树属）;Poisontree,Sladenia ●

366016　Sladenia celastrifolia Kurz;肋果茶（毒药树）;Bitteraweetleaf Poisontree,Bitteraweetleaf Sladenia,Bitteraweet-leaved Sladenia ●

366017　Sladenia integrifolia Y. M. Shui;全缘肋果茶●

366018　Sladeniaceae（Gilg et Werderm.）Airy Shaw = Theaceae Mirb.（保留科名）●

366019　Sladeniaceae（Gilg et Werderm.）Airy Shaw(1964);肋果茶科（毒药树科,独药树科）;Sladenia Family ●

366020　Sladeniaceae Airy Shaw = Sladeniaceae（Gilg et Werderm.）Airy Shaw ●

366021　Sladeniaceae Airy Shaw = Theaceae Mirb.（保留科名）●

366022　Slateria Desv. = Ophiopogon Ker Gawl.（保留属名）■

366023　Slateria japonica（L. f.）Desv. = Ophiopogon japonicus（L. f.）Ker Gawl. ■

366024　Slateria japonica Desv. = Ophiopogon japonicus（L. f.）Ker Gawl. ■

366025　Sleumeria Utteridge,Nagam. et Teo(1968);马来西亚茶茱黄属●☆

366026　Sleumerodendron Virot.（1868）;卡利登山龙眼属●☆

366027 Sleumerodendron austro-caledonicum（Brongn. et Gris）Virot；卡利登山龙眼●☆

366028 Slevogtia Rchb.（1829）；斯莱草属（斯来草属）■☆

366029 Slevogtia Rchb. = Enicostema Blume.（保留属名）■☆

366030 Slevogtia occidentalis Griseb.；西方斯来草■☆

366031 Slevogtia orientalis Griseb.；东方斯来草■☆

366032 Slevogtia verticillata D. Don；轮状斯来草■☆

366033 Sloanea Adens. = Sloanea L. ●

366034 Sloanea L.（1753）；猴欢喜属；Sloanea ●

366035 Sloanea Loefl. = Apeiba Aubl. ●☆

366036 Sloanea assamica（Benth.）Rehder et E. H. Wilson = Sloanea sterculiacea（Benth.）Rehder et E. H. Wilson var. assamica（Benth.）Coode ●

366037 Sloanea assamica Rehder et E. H. Wilson；长叶猴欢喜；Assam Sloanea, Longleaf Sloanea ●

366038 Sloanea assamica Rehder et E. H. Wilson = Sloanea sterculiacea（Benth.）Rehder et E. H. Wilson var. assamica（Benth.）Coode ●

366039 Sloanea australis（Benth.）F. Muell.；澳洲猴欢喜；Maiden's Blush ●☆

366040 Sloanea austrosinensis Hu ex Ts. Tang = Sloanea leptocarpa Diels ●

366041 Sloanea changii Coode；樟叶猴欢喜；Chang Sloanea, Cinnamonleaf Sloanea, Laurelleaf Sloanea ●

366042 Sloanea chengfengensis Hu = Sloanea hemsleyana Rehder et E. H. Wilson ●

366043 Sloanea chinensis Hu = Sloanea sinensis（Hance）Hemsl. ●

366044 Sloanea chingiana Hu；百色猴欢喜（广西猴欢喜，秦氏猴欢喜）；Baise Sloanea, Ching Sloanea ●

366045 Sloanea chingiana Hu var. integrifolia（Chun et F. C. How）Hung T. Chang = Sloanea integrifolia Chun et F. C. How ●

366046 Sloanea cordifolia K. M. Feng ex Hung T. Chang；心叶猴欢喜；Cordateleaf Sloanea, Cordate-leaved Sloanea, Heartleaf Sloanea ●

366047 Sloanea dasycarpa Hemsl.；毛果猴欢喜（薄叶猴欢喜，猴欢喜，膜叶猴欢喜）；Thick-fruitea Sloanea, Thinleaf Sloanea, Thin-leaved Sloanea ●

366048 Sloanea elegans Chun = Sloanea leptocarpa Diels ●

366049 Sloanea emeiensis W. P. Fang et P. C. Tuan；峨眉猴欢喜；Emei Sloanea ●

366050 Sloanea emeiensis W. P. Fang et P. C. Tuan = Sloanea leptocarpa Diels ●

366051 Sloanea formosana H. L. Li = Sloanea dasycarpa Hemsl. ●

366052 Sloanea forrestii W. W. Sm. = Sloanea sterculiacea（Benth.）Rehder et E. H. Wilson ●

366053 Sloanea hainanensis Merr. et Chun；海南猴欢喜；Hainan Sloanea ●

366054 Sloanea hanceana Hemsl. = Sloanea hemsleyana Rehder et E. H. Wilson ●

366055 Sloanea hemsleyana Rehder et E. H. Wilson；仿栗（药王树，油板栗）；Fake Chestnut, Hemsley Monkeyjoy, Hemsley Sloanea ●

366056 Sloanea hemsleyana Rehder et E. H. Wilson var. yunnanica Coode = Sloanea hemsleyana Rehder et E. H. Wilson ●

366057 Sloanea hongkongensis Hemsl. = Sloanea sinensis（Hance）Hemsl. ●

366058 Sloanea integrifolia Chun et F. C. How；全缘叶猴欢喜（全叶猴欢喜）；Entire-leaved Sloanea ●

366059 Sloanea integrifolia Chun et F. C. How = Sloanea chingiana Hu var. integrifolia（Chun et F. C. How）Hung T. Chang ●

366060 Sloanea kweichowensis Hu = Sloanea sinensis（Hance）Hemsl. ●

366061 Sloanea laurifolia Hung T. Chang = Sloanea changii Coode ●

366062 Sloanea leptocarpa Diels；薄果猴欢喜（北碚猴欢喜，红壳木）；Thinfruit Sloanea, Thin-fruited Sloanea ●

366063 Sloanea lordifolia K. M. Feng ex Hung T. Chang = Sloanea cordifolia K. M. Feng ex Hung T. Chang ●

366064 Sloanea mollis Gagnep.；滇越猴欢喜；Soft Sloanea ●

366065 Sloanea mollis Gagnep. = Sloanea chingiana Hu var. integrifolia（Chun et F. C. How）Hung T. Chang ●

366066 Sloanea mollis Gagnep. var. chinghsiensis Chun et F. C. How；靖西猴欢喜；Jingxi Soft Sloanea ●

366067 Sloanea mollis Gagnep. var. chinghsiensis Chun et F. C. How = Sloanea chingiana Hu var. integrifolia（Chun et F. C. How）Hung T. Chang ●

366068 Sloanea mollis Gagnep. var. chinghsiensis Chun et F. C. How = Sloanea mollis Gagnep. ●

366069 Sloanea oligophlebia Chun et K. C. Ting = Sloanea sinensis（Hance）Hemsl. ●

366070 Sloanea oligophlebia Merr. et Chun ex Gagnep. = Sloanea sinensis（Hance）Hemsl. ●

366071 Sloanea oligophlebia Merr. et Chun ex Gagnep. = Sloanea sinensis（Hance）Hemsl. ●

366072 Sloanea parvifolia Chun et F. C. How = Sloanea sinensis（Hance）Hemsl. ●

366073 Sloanea rotundifolia Hung T. Chang；圆叶猴欢喜；Roundleaf Sloanea, Round-leaved Sloanea ●

366074 Sloanea rotundifolia Hung T. Chang = Sloanea sterculiacea（Benth.）Rehder et E. H. Wilson ●

366075 Sloanea sigun（Blume）K. Schum.；斜脉猴欢喜●

366076 Sloanea sigun K. Schum. et Chun = Sloanea sinensis（Hance）Hemsl. ●

366077 Sloanea sinensis（Hance）Hemsl.；猴欢喜（狗欢喜，破布，树猬）；China Sloanea, Chinese Sloanea ●

366078 Sloanea sterculiacea（Benth.）Rehder et E. H. Wilson；苹婆叶猴欢喜（贡山猴欢喜，蒙自猴欢喜，苹婆猴欢喜）；Sterculia Sloanea, Sterculia-like Sloanea ●

366079 Sloanea sterculiacea（Benth.）Rehder et E. H. Wilson var. assamica（Benth.）Coode = Sloanea assamica Rehder et E. H. Wilson ●

366080 Sloanea tomentosa Rehder et E. H. Wilson；绒毛猴欢喜（毛猴欢喜）；Haired Sloanea, Hairy Sloanea, Tomentose Sloanea ●

366081 Sloanea tsiangiana Hu = Sloanea leptocarpa Diels ●

366082 Sloanea tsinyunensis S. S. Chien = Sloanea leptocarpa Diels ●

366083 Sloanea xichouensis K. M. Feng ex Y. Tang et Y. C. Hsu；西畴猴欢喜；Xichou Sloanea ●

366084 Sloania St. -Lag. = Sloanea L. ●

366085 Sloetia Teijsm. et Binn. = Streblus Lour. ●

366086 Sloetia Teijsm. et Binn. ex Kurz = Streblus Lour. ●

366087 Sloetiopsis Engl.（1907）；肖鹊肾树属●☆

366088 Sloetiopsis Engl. = Streblus Lour. ●

366089 Sloetiopsis usambarensis Engl. = Streblus usambarensis（Engl.）C. C. Berg ●☆

366090 Smallanthus Mack. = Smallanthus Mack. ex Small ■●

366091 Smallanthus Mack. ex Small（1933）；包果菊属（离苞果属，天山雪莲属，小花菊属）■●

366092 Smallanthus maculatus（Cav.）H. Rob.；斑点包果菊（天山雪莲，小花菊）●☆

366093 Smallanthus sonchifolius（Poeppig et Endl.）H. Rob.；菊薯■

366094　Smallanthus uvedalius（L.）Mack. = Osteospermum uvedalium L. ■☆

366095　Smallanthus uvedalius（L.）Mack. ex Small;包果菊■

366096　Smallanthus uvedalius（L.）Mack. ex Small = Osteospermum uvedalium L. ■☆

366097　Smallia Nieuwl. = Pteroglossaspis Rchb. f. ■☆

366098　Smallia Nieuwl. = Triorchos Small et Nash ■☆

366099　Smeathmannia Sol. ex R. Br.（1821）;繁柱西番莲属●☆

366100　Smeathmannia decandra Baill. = Paropsiopsis decandra（Baill.）Sleumer ●☆

366101　Smeathmannia laevigata Sol. ex R. Br. ;平滑繁柱西番莲●☆

366102　Smeathmannia laevigata Sol. ex R. Br. var. nigerica A. Chev. ex Hutch. et Dalziel;黑平滑繁柱西番莲●☆

366103　Smeathmannia pubescens Sol. ex R. Br. ;毛繁柱西番莲●☆

366104　Smeathmanniaceae Mart. ex Perleb = Passifloraceae Juss. ex Roussel(保留科名)●■

366105　Smegmadermos Ruiz et Pav. = Quillaja Molina ●☆

366106　Smegmaria Willd. = Quillaja Molina ●☆

366107　Smegmaria Willd. = Smegmadermos Ruiz et Pav. ●☆

366108　Smegmathamnium（Endl.）Rchb. = Saponaria L. ■

366109　Smegmathamnium Fenzl ex Rchb. = Saponaria L. ■

366110　Smelophyllum Radlk.（1878）;南非木属●☆

366111　Smelophyllum Radlk. = Stadmannia Lam. ●☆

366112　Smelophyllum capense（Sond.）Radlk. ;南非木●☆

366113　Smelophyllum capense Radlk. = Smelophyllum capense（Sond.）Radlk. ●☆

366114　Smelowskia C. A. Mey. = Smelowskia C. A. Mey. ex Ledebour（保留属名）■

366115　Smelowskia C. A. Mey. ex Ledebour(1830)(保留属名);芹叶荠属（芥叶荠属,裂叶芥属,裂叶荠属）;Celerycress,Smelowskia ■

366116　Smelowskia alba（Pall.）Regel;灰白芹叶荠（裂叶芥）;White Celerycress,White Smelowskia ■

366117　Smelowskia alba B. Fedtsch. = Sophiopsis sisymbrioides（Regel et Herder）O. E. Schulz ■

366118　Smelowskia annua Rupr. = Sophiopsis annua（Rupr.）O. E. Schulz ■

366119　Smelowskia aspleniifolia Turcz. ;裂叶荠;Spleenwortleaf Smelowskia ■

366120　Smelowskia aspleniifolia Turcz. = Smelowskia bifurcata（Ledeb.）Botsch. ■

366121　Smelowskia bifurcata（Ledeb.）Botsch. ;高山芹叶荠;Alpine Smelowskia,Twoforked Celerycress ■

366122　Smelowskia calycina（Stephan ex Willd.）C. A. Mey. ;芹叶荠;Calyx Smelowskia,Celerycress ■

366123　Smelowskia calycina（Stephan ex Willd.）C. A. Mey. var. densiflora O. E. Schulz = Smelowskia bifurcata（Ledeb.）Botsch. ■

366124　Smelowskia calycina（Stephan ex Willd.）C. A. Mey. var. pectinata（Bunge）B. Fedtsch. = Smelowskia calycina（Stephan ex Willd.）C. A. Mey. ■

366125　Smelowskia cineraea C. A. Mey. = Smelowskia alba（Pall.）Regel ■

366126　Smelowskia flavissima（Kar. et Kir.）Kar. et Kir. = Sophiopsis flavissima（Kar. et Kir.）O. E. Schulz ■☆

366127　Smelowskia inopinata Kom. ;意外芹叶荠■☆

366128　Smelowskia integrifolia（DC.）C. A. Mey. = Eutrema integrifolium（DC.）Bunge ■

366129　Smelowskia integrifolia C. A. Mey. = Eutrema integrifolium（DC.）Bunge ■

366130　Smelowskia koelzii（Rech. f.）Rech. f. = Smelowskia calycina（Stephan ex Willd.）C. A. Mey. ■

366131　Smelowskia pectinata（Bunge）Velichkin = Smelowskia calycina（Stephan ex Willd.）C. A. Mey. ■

366132　Smelowskia sisymbrioides（Regel et Herder）Lipsky ex Paulsen = Sophiopsis sisymbrioides（Regel et Herder）O. E. Schulz ■

366133　Smelowskia tianschania Velichkin = Smelowskia calycina（Stephan ex Willd.）C. A. Mey. ■

366134　Smelowskia tibetica（Thomson）Lipsky;西藏芹叶荠■

366135　Smelowskia tibetica（Thornson）Lipsky = Hedinia tibetica（Thomson）Ostenf. ■

366136　Smicrostigma N. E. Br.（1930）;樱龙属●☆

366137　Smicrostigma viride（Haw.）N. E. Br. ;樱龙●☆

366138　Smicrostigma viride N. E. Br. = Smicrostigma viride（Haw.）N. E. Br. ●☆

366139　Smidetia Raf. = Coleanthus Seidel(保留属名)■

366140　Smidetia Raf. = Schmidtia Tratt. ■

366141　Smilacaceae Vent.（1799）（保留科名）;菝葜科;Catbrier Family,Greenbrier Family,Smilax Family ●■

366142　Smilacina Desf.（1807）（保留属名）;鹿药属;Deerdrug,Solomonseal, Solomonplume, Starflower, False Solomon's seal, Solomon's-Plume,False Solomon's-seal ■

366143　Smilacina Desf. = Maianthemum F. H. Wigg.（保留属名）■

366144　Smilacina albiflora Wall. = Maianthemum purpureum（Wall.）LaFrankie ■

366145　Smilacina alpina Royle = Clintonia udensis Trautv. et C. A. Mey. ■

366146　Smilacina amplexicaulis Nutt. = Maianthemum racemosum（L.）Link subsp. amplexicaule（Nutt.）LaFrankie ■☆

366147　Smilacina atropurpurea（Franch.）F. T. Wang et Ts. Tang = Maianthemum atropurpureum（Franch.）LaFrankie ■

366148　Smilacina bifolia（L.）Desf. = Maianthemum bifolium（L.）F. W. Schmidt ■

366149　Smilacina bifolia Desf. = Maianthemum bifolium（L.）F. W. Schmidt ■

366150　Smilacina bootanensis Griff. = Maianthemum fuscum（Wall.）LaFrankie ■

366151　Smilacina borealis Ker Gawl. var. uniflora Menzies ex Schult. = Clintonia uniflora（Menzies ex Schult.）Kunth ■☆

366152　Smilacina canadensis（Desf.）Pursh = Maianthemum canadense Desf. ■☆

366153　Smilacina ciliata Desf. = Maianthemum racemosum（L.）Link ■☆

366154　Smilacina crassifolia Kawano = Maianthemum oleraceum（Baker）LaFrankie ■

366155　Smilacina dahurica Turcz. ex Fisch. et C. A. Mey. = Maianthemum dahuricum（Turcz. ex Fisch. et C. A. Mey.）LaFrankie ■

366156　Smilacina fargesii（Franch.）Diels = Maianthemum tubiferum（Batalin）LaFrankie ■

366157　Smilacina finitima（W. W. Sm.）F. T. Wang et Ts. Tang = Maianthemum fuscum（Wall.）LaFrankie ■

366158　Smilacina flexicaulis Wender. = Maianthemum racemosum（L.）Link ■☆

366159　Smilacina formosana Hayata = Maianthemum formosanum（Hayata）LaFrankie ■

366160　Smilacina formosana Hayata = Smilacina japonica A. Gray ■

366161　Smilacina forrestii（W. W. Sm.）Hand. -Mazz. = Maianthemum forrestii（W. W. Sm.）LaFrankie ■

366162　Smilacina forskaliana Schult. f. = Sansevieria forskaoliana（Schult. f.）Hepper et J. R. I. Wood ■☆

366163　Smilacina fusca Wall. = Maianthemum fuscum（Wall.）LaFrankie ■

366164　Smilacina fusca Wall. var. pilosa H. Hara = Maianthemum fuscum（Wall.）LaFrankie ■

366165　Smilacina fusciduliflora Kawano = Maianthemum fusciduliflorum（Kawano）S. C. Chen et Kawano ■

366166　Smilacina ginfoshanica F. T. Wang et Ts. Tang = Heteropolygonatum ginfushanicum（F. T. Wang et Ts. Tang）M. N. Tamura, S. Yun Liang et N. J. Turland ■

366167　Smilacina gongshanensis S. Yun Liang = Maianthemum gongshanense（S. Yun Liang）H. Li ■

366168　Smilacina henryi（Baker）F. T. Wang et Ts. Tang = Maianthemum henryi（Baker）LaFrankie ■

366169　Smilacina henryi（Baker）F. T. Wang et Ts. Tang var. szechuanica（F. T. Wang et Ts. Tang）F. T. Wang et Ts. Tang = Maianthemum szechuanicum（F. T. Wang et Ts. Tang）H. Li ■

366170　Smilacina henryi（Baker）H. Hara = Maianthemum henryi（Baker）LaFrankie ■

366171　Smilacina henryi（Baker）H. Hara var. szechuanica（F. T. Wang et Ts. Tang）F. T. Wang et Ts. Tang = Maianthemum szechuanicum（F. T. Wang et Ts. Tang）H. Li ■

366172　Smilacina hirta Maxim. = Maianthemum japonicum（A. Gray）LaFrankie ■

366173　Smilacina japonica A. Gray = Maianthemum japonicum（A. Gray）LaFrankie ■

366174　Smilacina japonica A. Gray var. mandshurica Maxim. = Maianthemum japonicum（A. Gray）LaFrankie ■

366175　Smilacina lichiangensis（W. W. Sm.）W. W. Sm. = Maianthemum lichiangense（W. W. Sm.）LaFrankie ■

366176　Smilacina liliacea（Greene）F. L. Wynd = Maianthemum stellatum（L.）Link ■☆

366177　Smilacina mientienensis F. T. Wang et Ts. Tang = Maianthemum oleraceum（Baker）LaFrankie ■

366178　Smilacina nanchuanensis（H. Li et J. L. Huang）S. Yun Liang = Maianthemum nanchuanense H. Li et J. L. Huang ■

366179　Smilacina nokomonticola Yamam. = Maianthemum formosanum（Hayata）LaFrankie ■

366180　Smilacina oleracea（Baker）Hook. f. et Thomson = Maianthemum oleraceum（Baker）LaFrankie ■

366181　Smilacina oleracea（Baker）Hook. f. et Thomson f. acuminata（F. T. Wang et Ts. Tang）H. Hara = Maianthemum oleraceum（Baker）LaFrankie ■

366182　Smilacina oleracea Hook. f. et Thomson ex Baker = Maianthemum oleraceum（Baker）LaFrankie ■

366183　Smilacina oleracea Hook. f. et Thomson ex Baker var. acuminatum F. T. Wang et Ts. Tang = Maianthemum oleraceum（Baker）LaFrankie ■

366184　Smilacina oleracea var. acuminata F. T. Wang et Ts. Tang = Maianthemum oleraceum（Baker）LaFrankie ■

366185　Smilacina oligophylla（Baker）Hook. f. = Maianthemum purpureum（Wall.）LaFrankie ■

366186　Smilacina pallida Royle = Maianthemum purpureum（Wall.）LaFrankie ■

366187　Smilacina paniculata（Baker）F. T. Wang et Ts. Tang = Maianthemum tatsienense（Franch.）LaFrankie ■

366188　Smilacina paniculata（Baker）F. T. Wang et Ts. Tang var. stenoloba（Franch.）F. T. Wang et Ts. Tang = Maianthemum stenolobum（Franch.）S. C. Chen et Kawano ■

366189　Smilacina prattii（Franch.）H. R. Wehrh. = Maianthemum atropurpureum（Franch.）LaFrankie ■

366190　Smilacina purpurea Wall. = Maianthemum purpureum（Wall.）LaFrankie ■

366191　Smilacina purpurea Wall. f. albiflora（Wall.）H. Hara = Maianthemum purpureum（Wall.）LaFrankie ■

366192　Smilacina purpurea Wall. f. oligophylla（Baker）H. Hara = Maianthemum purpureum（Wall.）LaFrankie ■

366193　Smilacina purpurea Wall. var. albida Wall. = Maianthemum purpureum（Wall.）LaFrankie ■

366194　Smilacina racemosa（L.）Desf. = Maianthemum racemosum（L.）Link ■☆

366195　Smilacina racemosa（L.）Desf. var. cylindrata Fernald = Maianthemum racemosum（L.）Link ■☆

366196　Smilacina racemosa（L.）Desf. var. lanceolata B. Boivin = Maianthemum racemosum（L.）Link ■☆

366197　Smilacina racemosa（L.）Desf. var. typica Fernald = Maianthemum racemosum（L.）Link ■☆

366198　Smilacina racemosa Desf.；锥花鹿药（总状鹿药）；False Solomon's Seal, False Spikenard, Panicled False Solomonseal, Scurvy-berries, Star-flowered Lily-of-the-valley, Treacle-berries ■☆

366199　Smilacina racemosa Desf. = Maianthemum racemosum（L.）Link ■☆

366200　Smilacina racemosa Desf. var. cylindrata Fernald；圆柱鹿药；Cylindrical False Solomonseal ■☆

366201　Smilacina robusta（Franch.）F. T. Wang et Ts. Tang = Maianthemum atropurpureum（Franch.）LaFrankie ■

366202　Smilacina rossii（Baker）Maxim. = Maianthemum japonicum（A. Gray）LaFrankie ■

366203　Smilacina sessilifolia Nutt. ex Baker = Maianthemum stellatum（L.）Link ■☆

366204　Smilacina smithii K. Krause = Maianthemum atropurpureum（Franch.）LaFrankie ■

366205　Smilacina souliei（Franch.）F. T. Wang et Ts. Tang = Maianthemum tubiferum（Batalin）LaFrankie ■

366206　Smilacina stellata（L.）Desf. = Maianthemum stellatum（L.）Link ■☆

366207　Smilacina stellata（L.）Desf. var. crassa Vict. = Maianthemum stellatum（L.）Link ■☆

366208　Smilacina stellata（L.）Desf. var. mollis Farw. = Maianthemum stellatum（L.）Link ■☆

366209　Smilacina stellata（L.）Desf. var. sessilifolia（Nutt. ex Baker）G. Hend. = Maianthemum stellatum（L.）Link ■☆

366210　Smilacina stellata（L.）Desf. var. sylvatica Vict. et J. Rousseau = Maianthemum stellatum（L.）Link ■☆

366211　Smilacina stellata Desf.；星果鹿药（星花鹿药）；Starflower Solomonseal, Star-flowered False Solomon's-seal, Star-flowered Lily of the Valley, Star-flowered Lily-of-the-valley, Starfruit False Solomonseal ■☆

366212　Smilacina stenoloba（Franch.）Diels = Maianthemum stenolobum（Franch.）S. C. Chen et Kawano ■

366213　Smilacina streptopoides Ledeb. = Streptopus streptopoides

（Ledeb. ）Frye et Rigg ■☆

366214 Smilacina szechuanica（F. T. Wang et Ts. Tang）H. Hara = Maianthemum szechuanicum（F. T. Wang et Ts. Tang）H. Li ■

366215 Smilacina tatsienensis（Franch. ）F. T. Wang et Ts. Tang = Maianthemum tatsienense（Franch. ）LaFrankie ■

366216 Smilacina tatsienensis（Franch. ）F. T. Wang et Ts. Tang f. stenoloba（Franch. ）H. Hara = Maianthemum stenolobum（Franch. ）S. C. Chen et Kawano ■

366217 Smilacina tatsienensis（Franch. ）F. T. Wang et Ts. Tang var. paniculata（Baker）F. T. Wang et Ts. Tang = Maianthemum tatsienense（Franch. ）LaFrankie ■

366218 Smilacina tatsienensis（Franch. ）F. T. Wang et Ts. Tang var. stenoloba（Franch. ）D. M. Liu = Maianthemum stenolobum（Franch. ）S. C. Chen et Kawano ■

366219 Smilacina tatsienensis（Franch. ）H. R. Wehrh. = Maianthemum tatsienense（Franch. ）LaFrankie ■

366220 Smilacina tatsienensis（Franch. ）H. R. Wehrh. f. stenoloba（Franch. ）H. Hara = Maianthemum stenolobum（Franch. ）S. C. Chen et Kawano ■

366221 Smilacina tatsienensis（Franch. ）H. R. Wehrh. var. paniculata（Baker）F. T. Wang et Ts. Tang = Maianthemum tatsienense（Franch. ）LaFrankie ■

366222 Smilacina tatsienensis（Franch. ）H. R. Wehrh. var. stenoloba（Franch. ）D. M. Liu = Maianthemum stenolobum（Franch. ）S. C. Chen et Kawano ■

366223 Smilacina trifolia（L. ）Desf. = Maianthemum trifolium（L. ）Slobada ■

366224 Smilacina tubifera Batalin = Maianthemum tubiferum（Batalin）LaFrankie ■

366225 Smilacina uniflora（Menzies ex Schult. ）Hook. = Clintonia uniflora（Menzies ex Schult. ）Kunth ■☆

366226 Smilacina wardii（W. W. Sm. ）F. T. Wang et Ts. Tang = Maianthemum atropurpureum（Franch. ）LaFrankie ■

366227 Smilacina yunnanensis（Franch. ）Hand. -Mazz. = Maianthemum paniculatum（Martens et Galeotti）LaFrankie ■

366228 Smilacina yunnanensis（Franch. ）Hand. -Mazz. = Maianthemum tatsienense（Franch. ）LaFrankie ■

366229 Smilacina zhongdianensis H. Li et Y. Chen = Maianthemum purpureum（Wall. ）LaFrankie ■

366230 Smilax L.（1753）;菝葜属;Catbrier, Greenbrier, Sarsaparilla, Smilax ●

366231 Smilax aberrans Gagnep. ;弯梗菝葜（毛叶菝葜）;Bent-stalk Greenbrier, Bowedstalk Greenbrier, Bowedstalk Smilax, Hairyleaf Greenbrier ●

366232 Smilax aberrans Gagnep. subsp. retroflexa（F. T. Wang et Ts. Tang）T. Koyama = Smilax retroflexa（F. T. Wang et Ts. Tang）S. C. Chen ●

366233 Smilax aberrans Gagnep. var. retroflexa F. T. Wang et Ts. Tang = Smilax retroflexa（F. T. Wang et Ts. Tang）S. C. Chen ●

366234 Smilax alba Pursh = Smilax laurifolia L. ●☆

366235 Smilax amaurophlebia Merr. = Smilax corbularia Kunth var. woodii（Merr. ）T. Koyama ●

366236 Smilax amaurophlebia Merr. = Smilax corbularia Kunth ●

366237 Smilax anceps Willd. ;二棱菝葜;Wild Sarsaparilla ●☆

366238 Smilax arisanensis（Hayata）F. T. Wang et Ts. Tang;尖叶菝葜（阿里山菝葜）;Alishan Greenbrier, Altai Mountain Greenbrier, Altai Smilax ●

366239 Smilax aristolochiifolia Mill. ;墨西哥菝葜（灰菝葜,洋菝葜）●☆

366240 Smilax asper L. ;穗菝葜（欧亚菝葜）;Eurasia Greenbrier, Eurasian Greenbrier, Eurasian Smilax, Greenbriar, Prickwind, Sharpbind ●

366241 Smilax aspera L. var. altissima Moris et De Not. = Smilax aspera L. ●

366242 Smilax aspera L. var. mauritanica（Poir. ）Gren. et Godr. ;毛里塔尼亚菝葜●☆

366243 Smilax aspericaulis Wall. ex A. DC. ;疣枝菝葜（白萆薢,糙茎菝葜）;Roughstem Greenbrier, Rough-stem Greenbrier, Roughstem Smilax, Verrucose Greenbrier ●

366244 Smilax astrosper（P. Ma）F. T. Wang et Ts. Tang;灰叶菝葜;Greyleaf Greenbrier, Grey-leaf Greenbrier, Greyleaf Smilax ●

366245 Smilax auriculata Chapm. ;耳状菝葜;Wild Bamboo, Wild-bamboo ●☆

366246 Smilax australis A. Cunn. ex A. DC. ;澳洲菝葜;Austral Sarsaparilla ●☆

366247 Smilax austrosinensis F. T. Wang et Ts. Tang = Smilax lanceifolia Roxb. var. elangata（Warb. ）F. T. Wang et Ts. Tang ●

366248 Smilax austrozhejiangensis Q. Lin;浙南菝葜;South Zhejiang Greenbrier, South Zhejiang Smilax ●

366249 Smilax balansaena H. Bonnet ex Gagnep. = Smilax corbularia Kunth var. woodii（Merr. ）T. Koyama ●

366250 Smilax banglaoensis R. H. Miao;班老菝葜;Banlao Greenbrier ●

366251 Smilax banglaoensis R. H. Miao = Smilax corbularia Kunth ●

366252 Smilax bapouensis H. Li;巴坡菝葜;Bapo Greenbrier ●

366253 Smilax basilata F. T. Wang et Ts. Tang;少花菝葜;Fewflower Greenbrier, Fewflower Smilax, Pauciflorous Greenbrier ●

366254 Smilax bauhinioides Kunth;圆叶菝葜;Roundleaf Greenbrier, Round-leaf Greenbrier ●

366255 Smilax bauhinioides T. Koyama = Smilax lunglingensis F. T. Wang et Ts. Tang ●

366256 Smilax beyrichii Kunth = Smilax auriculata Chapm. ●☆

366257 Smilax biflora Siebold ex Miq. ;小菝葜●☆

366258 Smilax biflora Siebold ex Miq. var. trinervula（Miq. ）Hatus. ex T. Koyama = Smilax trinervula Miq. ●

366259 Smilax biltmoreana（Small）J. B. Norton ex Pennell;比尔菝葜●☆

366260 Smilax blinii H. Lév. = Smilax glabra Roxb. ex C. H. Wright ●

366261 Smilax bockii Warb. = Heterosmilax japonica Kunth ●

366262 Smilax bockii Warb. ex Diels;西南菝葜（金刚藤,菝葜）;Bock Greenbrier, Bock Smilax ●

366263 Smilax bodinieri H. Lév. et Vaniot = Smilax glaucochina Warb. ex Diels ●

366264 Smilax bona-nox L. ;攀缘刺菝葜;Bullbrier, Catbrier, China Briar, Chinese Briar, Greenbrier, Saw Greenbrier, Zarzaparrilla ●☆

366265 Smilax bona-nox L. var. exauriculata Fernald = Smilax bona-nox L. ●☆

366266 Smilax bona-nox L. var. hastata（Willd. ）A. DC. = Smilax bona-nox L. ●☆

366267 Smilax bona-nox L. var. hederifolia（Beyr. ex Kunth）Fernald = Smilax bona-nox L. ●☆

366268 Smilax bona-nox L. var. hederifolia（Beyr. ）Fernald = Smilax bona-nox L. ●☆

366269 Smilax bona-nox L. var. littoralis Coker = Smilax bona-nox L. ●☆

366270　Smilax boninensis Nakai ex Tuyama = Smilax china L. var. yanagitae Honda ●☆

366271　Smilax bracteata C. Presl;圆锥菝葜（假菝葜，狭瓣菝葜）; Conical Greenbrier, Conical Smilax, False China-root Greenbrier ●

366272　Smilax bracteata C. Presl subsp. verruculosa（Merr.）T. Koyama = Smilax bracteata C. Presl var. verruculosa（Merr.）T. Koyama ●

366273　Smilax bracteata C. Presl subsp. verruculosa（Merr.）T. Koyama = Smilax aspericaulis Wall. ex A. DC. ●

366274　Smilax bracteata C. Presl var. verruculosa（Merr.）T. Koyama = Smilax aspericaulis Wall. ex A. DC. ●

366275　Smilax brevipes Warb. = Smilax scobinicaulis C. H. Wright ●■

366276　Smilax caduca L. = Smilax rotundifolia L. ●

366277　Smilax californica（A. DC.）A. Gray;加州菝葜●☆

366278　Smilax calophylla Wall. ex A. DC.;美叶菝葜●☆

366279　Smilax calophylla Wall. ex A. DC. var. concolor C. H. Wright = Smilax glabra Roxb. ex C. H. Wright ●

366280　Smilax canariensis Willd.;加那利菝葜●☆

366281　Smilax castaneiflora H. Lév. = Smilax microphylla C. H. Wright ●

366282　Smilax cavaleriei H. Lév. et Vaniot = Smilax scobinicaulis C. H. Wright ●■

366283　Smilax chapaensis Gagnep.;密疣菝葜; Chapa Greenbrier, Chapa Smilax, Densewart Greenbrier, Densewart Smilax, Densiwart Greenbrier ●

366284　Smilax china L.;菝葜（芭葜，拔谷，霸王引，鳖儿搌，豺狗刺，大溪菝葜，钉巴筲，饭巴铎，沟谷刺，红灯果，鲎壳刺，鲎壳藤，鸡肝根，假草薢，金巴筲，金巴斗，金冈拙，金刚鞭，金刚刺，金刚兜，金刚根，金刚骨，金刚树，金刚藤，金刚头，筋骨柱子，荆岗拙，净菝葜，老君须，冷饭巴，冷饭头，里白菝葜，龙爪菜，路边刷，马鞍宫，马加刺兜，马加勒，马甲，马甲刺，蓬灯果，普贴，山根儿，山菱角，算盘七，藤灯果，铁刺苓，铁菱角，铁刷子，土茯苓，王瓜草，硬饭头）; China Brier, China Greenbrier, China Root, Chinaroot Green Brier, Chinaroot Greenbrier, Chinaroot Smilax, Chinese Root ●

366285　Smilax china L. f. obltustz H. Lév. = Smilax china L. ●

366286　Smilax china L. f. xanthocarpa Sugim.;黄果菝葜●☆

366287　Smilax china L. f. yanagitae（Honda）T. Koyama = Smilax china L. var. yanagitae Honda ●☆

366288　Smilax china L. var. biflora（Siebold ex Miq.）Makino = Smilax biflora Siebold ex Miq. ●☆

366289　Smilax china L. var. brachypoda Rehder = Smilax davidiana A. DC. ●

366290　Smilax china L. var. igaensis Masam. et T. Kurok.;伊贺菝葜 ●☆

366291　Smilax china L. var. taiheiensis（Hayata）Koyama;大溪菝葜; Round-leaved Greenbrier ●

366292　Smilax china L. var. taiheiensis（Hayata）Koyama = Smilax china L. ●

366293　Smilax china L. var. trinervula（Miq.）Makino = Smilax trinervula Miq. ●

366294　Smilax china L. var. yanagitae Honda;柳田菝葜●☆

366295　Smilax chingii F. T. Wang et Ts. Tang;柔毛菝葜; Ching Greenbrier, Ching Smilax, Softhair Greenbrier ●

366296　Smilax chingii F. T. Wang et Ts. Tang var. papillosifolia J. M. Xu = Smilax chingii F. T. Wang et Ts. Tang ●

366297　Smilax cinerea Warb. = Smilax megalantha C. H. Wright ●

366298　Smilax cinnamomiifolia Small = Smilax smallii Morong ●☆

366299　Smilax cocculoides Warb. ex Diels;银叶菝葜; Silverleaf Greenbrier, Silverleaf Smilax ●

366300　Smilax cocculoides Warb. ex Diels var. lanceolata Norton = Smilax lanceifolia Roxb. var. lanceolata（Norton）T. Koyama ●

366301　Smilax corbularia Kunth;筐条菝葜（金刚，里白菝葜）; Longerpeduncle Greenbrier, Longer-peduncle Greenbrier, Longerpeduncle Smilax ●

366302　Smilax corbularia Kunth var. hypoglauca（Benth.）T. Koyama = Smilax hypoglauca Benth. ●

366303　Smilax corbularia Kunth var. hypoglauca T. Koyama = Smilax corbularia Kunth ●

366304　Smilax corbularia Kunth var. woodii（Merr.）T. Koyama;光叶菝葜; Glabrousleaf Greenbrier ●

366305　Smilax coriacea Spreng. = Smilax havanensis Jacq. ●☆

366306　Smilax coriacea Spreng. var. ilicifolia O. E. Schulz = Smilax havanensis Jacq. ●☆

366307　Smilax cyclophylla Warb.;合蕊菝葜（白茯苓藤，竹节药刺藤）; Costamen Greenbrier, Round-foliated Greenbrier, Roundfolious Smilax ●

366308　Smilax darrisii H. Lév.;平滑菝葜; Darris Smilax, Smooth Greenbrier ●

366309　Smilax davidiana A. DC.;小果菝葜; David Greenbrier, David Smilax, Smallfruit Greenbrier ●

366310　Smilax densiberbata F. T. Wang et Ts. Tang;密刺菝葜; Dense-thorned Greenbrier, Densethorny Greenbrier ●

366311　Smilax discotis Warb.;托柄菝葜（短柄菝葜，金刚豆藤，土茯苓，宜兰菝葜）; Broadsheath Greenbrier, Broad-sheath Greenbrier, Broadsheath Smilax ●

366312　Smilax discotis Warb. subsp. concolor（J. B. Norton）T. Koyama = Smilax outanscianensis Pamp. ●

366313　Smilax discotis Warb. var. concolor Norton = Smilax outanscianensis Pamp. ●

366314　Smilax diversifolia Small = Smilax lasioneura Hook. ●☆

366315　Smilax dunniana H. Lév. = Smilax glabra Roxb. ex C. H. Wright ●

366316　Smilax ecirrhata（Engelm. ex Kunth）S. Watson var. biltmoreana（Small）H. E. Ahles = Smilax biltmoreana（Small）J. B. Norton ex Pennell ●☆

366317　Smilax ecirrhata S. Watson;直立菝葜; Carrion Flower, Upright Carrion-flower ●☆

366318　Smilax ecirrhata S. Watson var. hugeri（Small）H. E. Ahles = Smilax hugeri（Small）J. B. Norton ex Pennell ●☆

366319　Smilax elegans Wall. ex Kunth;雅致菝葜（西藏菝葜）; Elagant Greenbrier ●

366320　Smilax elegans Wall. ex Kunth subsp. microphylla（C. H. Wright）Noltie = Smilax microphylla C. H. Wright ●

366321　Smilax elegans Wall. ex Kunth subsp. subrecta Noltie = Smilax longebracteolata Hook. f. ●

366322　Smilax elegantissima Gagnep.;四棱菝葜●

366323　Smilax elongatoreticulata Hayata = Smilax elongatoumbellata Hayata ●

366324　Smilax elongatoumbellata Hayata;台湾菝葜（和社菝葜，细叶菝葜）; Narrow-leaved Greenbrier, Taiwan Greenbrier ●

366325　Smilax elongatoumbellata Hayata f. elongatoreticulata（Hayata）T. Koyama = Smilax elongatoumbellata Hayata ●

366326　Smilax emeiensis J. M. Xu;峨眉菝葜; Emei Greenbrier, Emei Smilax ●

366327　Smilax erythrantha Baill. ex Gagnep. = Heterosmilax gaudichaudiana（Kunth）Maxim. ●

366328　Smilax esquirolii H. Lév. = Smilax trinervula Miq. ●

366329 Smilax excelsa L. ;高牛尾菜●☆

366330 Smilax fauri H. Lév. = Smilax china L. ●

366331 Smilax febrigula Kunth;厄瓜多尔菝葜●☆

366332 Smilax ferox Wall. ex Kunth;长托菝葜(刺草薢,大菝葜,红草薢,龙须叶,美人扇);Spiny Greenbrier, Spiny Smilax ●

366333 Smilax flaccida C. H. Wright = Smilax riparia A. DC. ●■

366334 Smilax fooningensis F. T. Wang et Ts. Tang;富宁菝葜;Fooning Greenbrier, Funing Greenbrier ●

366335 Smilax formosana (Hayata) Hayata = Smilax sieboldii Miq. ●

366336 Smilax gagnepainii T. Koyama;四翅菝葜;Fourwing Greenbrier, Four-wing Greenbrier,Fourwing Smilax ●

366337 Smilax gaudichaudiana Kunth = Heterosmilax gaudichaudiana (Kunth) Maxim. ●

366338 Smilax glabra Roxb. ex C. H. Wright;土茯苓(白葜,白土苓,白余粮,草禹余粮,川萆薢,刺猪苓,地茯苓,地胡苓,饭团根,公茯苓藤,狗朗头,狗老薯,光菝葜,光滑菝葜,光叶菝葜,过冈龙,过山龙,红萆薢,红土苓,花萆薢,花花藤,尖光头,久老薯,冷饭,冷饭块,冷饭藤,冷饭头,冷饭团,连饭,毛尾薯,木猪苓,奇良,奇粮,歧良,荣草,山地栗,山归来,山牛,山奇良,山奇粮,山遗粮,山硬硬,山猪粪,土萆薢,土太片,仙遗粮,小草薢,小红草薢,硬饭,硬饭头,硬饭头薯,禹余粮);Glabrous Greenbrier, Glabrous Smilax,Smooth Greenbrier,Tufuling ●

366339 Smilax glabra Roxb. ex C. H. Wright var. maculata Bodinier ex H. Lév. = Smilax glabra Roxb. ex C. H. Wright ●

366340 Smilax glabra Roxb. var. concolor (C. H. Wright) F. T. Wang;同色菝葜(蓝果土茯苓);Concolor Greenbrier ●

366341 Smilax glauca Walter;灰蓝菝葜;Cat Greenbrier, Catbrier, Greenbriar, Prickly Ivy, Sarsaparilla, Sawbrier, Smilax, Wild Sarsaparilla ●☆

366342 Smilax glauca Walter var. leurophylla S. F. Blake = Smilax glauca Walter ●☆

366343 Smilax glaucochina Warb. ex Diels;黑果菝葜(饭巴坨,粉菝葜,和社菝葜,红草薢,金刚菝葜,金刚刺,金刚藤,冷饭巴,鲢鱼须,龙须菜,鲇鱼须,铁菱角);Blackfruit Greenbrier, Blackfruit Smilax,Black-fruited Greenbrier ●

366344 Smilax glaucophylla Klotzsch;西藏菝葜;Tibet Greenbrier, Xizang Greenbrier ●

366345 Smilax glaucophylla Klotzsch = Smilax elegans Wall. ex Kunth ●

366346 Smilax glaucophylla Klotzsch var. randaiensis (Hayata) T. Koyama = Smilax menispermoidea A. DC. ●

366347 Smilax glaucophylla Klotzsch var. randalensis (Hayata) T. Koyama = Smilax pygmaea Merr. ●

366348 Smilax glycyphylla Hassk. ;甜叶菝葜(大豆叶菝葜);Sweet Sarsaparilla ●☆

366349 Smilax goetzeana Engl. ;格兹菝葜●☆

366350 Smilax gracillima H. Lév. et Vaniot = Smilax microphylla C. H. Wright ●

366351 Smilax gracillima Hayata = Smilax hayatae T. Koyama ●

366352 Smilax grandifolia Buckley = Smilax tamnoides L. ●☆

366353 Smilax griffithii A. DC. ;墨脱菝葜;Motuo Greenbrier ●

366354 Smilax griffithii A. DC. var. pallescens (A. DC.) T. Koyama = Smilax griffithii A. DC. ●

366355 Smilax guianensis Vitman var. subarmata O. E. Schulz = Smilax havanensis Jacq. ●☆

366356 Smilax guiyangensis C. X. Fu et C. D. Shen;花叶菝葜;Guiyang Greenbrier ●

366357 Smilax hastata Willd. = Smilax bona-nox L. ●☆

366358 Smilax hastata Willd. var. lanceolata (L.) Pursh;剑叶菝葜;Lanceleaf Greenbrier, Smilax ●☆

366359 Smilax havanensis Jacq. ;革质菝葜●☆

366360 Smilax havanensis Jacq. var. portoricensis A. DC. = Smilax havanensis Jacq. ●☆

366361 Smilax hayatae T. Koyama;菱叶菝葜(早田氏菝葜);Hayata Greenbrier, Rhombicleaf Greenbrier ●

366362 Smilax hederifolia Beyr. ex Kunth = Smilax bona-nox L. ●☆

366363 Smilax hemsleyana Craib;束丝菝葜(刺丝菝葜);Hemsley Greenbrier, Hemsley Smilax ●

366364 Smilax herbacea L. ;刺丝菝葜(白须公,草菝葜,牛尾菜,土茯苓);Carrion Flower, Carrion-flower, Carrion-flower Greenbrier, Common Carrion-flower, Jacob's Ladder, Jacob's-ladder, Smilax ●☆

366365 Smilax herbacea L. subsp. crispifolia Pennell = Smilax herbacea L. ●☆

366366 Smilax herbacea L. subsp. lasioneura (Hook.) Á. Löve et D. Löve = Smilax lasioneura Hook. ●☆

366367 Smilax herbacea L. var. acuminata C. H. Wright = Smilax riparia A. DC. var. acuminata (C. H. Wright) F. T. Wang et Ts. Tang ●■

366368 Smilax herbacea L. var. angusta C. H. Wright = Smilax riparia A. DC. ●■

366369 Smilax herbacea L. var. daibuensis Hayata = Smilax riparia A. DC. ●■

366370 Smilax herbacea L. var. ecirrata (Engelm. ex Kunth) A. DC. = Smilax ecirrhata S. Watson ●☆

366371 Smilax herbacea L. var. foetida H. Lév. = Smilax riparia A. DC. ●■

366372 Smilax herbacea L. var. heterophylla H. Lév. = Smilax riparia A. DC. ●■

366373 Smilax herbacea L. var. inodora M. E. Jones = Smilax lasioneura Hook. ●☆

366374 Smilax herbacea L. var. intermedia C. H. Wright = Smilax nipponica Miq. ●■

366375 Smilax herbacea L. var. lancilimba Merr. = Smilax riparia A. DC. ●■

366376 Smilax herbacea L. var. lasioneura (Hook.) A. DC. = Smilax lasioneura Hook. ●☆

366377 Smilax herbacea L. var. nipponica (Miq.) Maxim. = Smilax nipponica Miq. ●■

366378 Smilax herbacea L. var. oblonga C. H. Wright = Smilax nipponica Miq. ●■

366379 Smilax herbacea L. var. oldhamii (Miq.) Maxim. = Smilax sieboldii Miq. ●

366380 Smilax herbacea L. var. oldhamii Miq. = Smilax sieboldii Miq. ●

366381 Smilax herbacea L. var. peduncularis (Muhl. ex Willd.) A. DC. = Smilax herbacea L. ●☆

366382 Smilax herbacea L. var. pubescens C. H. Wright = Smilax riparia A. DC. var. pubescens (C. H. Wright) F. T. Wang et Ts. Tang ●■

366383 Smilax herbacea L. var. pulverulenta (Michx.) A. Gray = Smilax pulverulenta Michx. ●☆

366384 Smilax herbacea L. var. simsii A. DC. = Smilax herbacea L. ●☆

366385 Smilax higoensis L. var. maximowiczii (Koidz.) Kitag. = Smilax riparia A. DC. ●■

366386 Smilax higoensis L. var. ussuriensis (Regel) Kitag. = Smilax riparia A. DC. ●■

366387 Smilax hispida Muhl. ex Torr. = Smilax tamnoides L. ●☆

366388 Smilax hispida Muhl. ex Torr. var. australis Small = Smilax tamnoides L. ●☆

366389 Smilax hispida Muhl. ex Torr. var. montana Coker = Smilax tamnoides L. ●☆

366390 Smilax hongkongensis Seem. = Heterosmilax gaudichaudiana (Kunth) Maxim. ●

366391 Smilax hookeri Kunth = Smilax glabra Roxb. ex C. H. Wright ●

366392 Smilax horridiramula Hayata;刺枝菝葜（密刺菝葜）;Horrid Greenbrier,Spinybranch Greenbrier ●

366393 Smilax hugeri (Small) J. B. Norton ex Pennell;赫格菝葜●☆

366394 Smilax humilis Mill. = Smilax pumila Walter ●☆

366395 Smilax hypoglauca Benth.;粉背菝葜（大通筋,牛尾结）;Hypoglaucous Greenbrier,Hypoglaucous Smilax ●

366396 Smilax ilicifolia Desv. ex Ham. = Smilax havanensis Jacq. ●☆

366397 Smilax illinoensis Mangaly;伊利诺菝葜;Carrion Flower,Illinois Carrion-flower,Illinois Greenbrier ●☆

366398 Smilax impressinervia F. T. Wang et Ts. Tang = Smilax lanceifolia Roxb. var. impressinervia (F. T. Wang et Ts. Tang) T. Koyama ●

366399 Smilax indica Vitman = Smilax hemsleyana Craib ●

366400 Smilax inermis Walter = Smilax pseudochina Lour. ●☆

366401 Smilax insignis Kunth;显著菝葜●▪☆

366402 Smilax jamesii G. A. Wallace;詹姆斯菝葜●☆

366403 Smilax japonica (Kunth) A. Gray = Smilax china L. ●

366404 Smilax japonica A. Gray = Smilax china L. ●

366405 Smilax jiankunii H. Li;建昆菝葜;Jiankun Greenbrier ●

366406 Smilax kraussiana Meisn. = Smilax anceps Willd. ●☆

366407 Smilax kwangsiensis F. T. Wang et Ts. Tang;缘毛菝葜;Guangxi Greenbrier,Guangxi Smilax,Kwangsi Greenbrier ●

366408 Smilax kwangsiensis F. T. Wang et Ts. Tang var. setulosa F. T. Wang et Ts. Tang;小刚毛菝葜;Greenbrier,Setaceous Greenbrier ●

366409 Smilax labordei H. Lév. et Vaniot = Smilax microphylla C. H. Wright ●

366410 Smilax laevis Wall. ex A. DC. = Smilax lanceifolia Roxb. var. opaca A. DC. ●

366411 Smilax laevis Wall. ex A. DC. var. ophirensis A. DC. = Smilax lanceifolia Roxb. var. opaca A. DC. ●

366412 Smilax laevis Wall. ex A. DC. var. parkii A. DC. = Smilax lanceifolia Roxb. var. opaca A. DC. ●

366413 Smilax laevis Wall. ex A. DC. var. vanchingshanensis F. T. Wang et Ts. Tang = Smilax vanchingshanensis (F. T. Wang et Ts. Tang) F. T. Wang et Ts. Tang ●

366414 Smilax lanceifolia Roxb.;马甲菝葜（披针叶菝葜,台湾菝葜,台湾土茯苓,土茯苓）;Lanceolateleaf Smilax, Lanceolate-leaved Greenbrier,Taiwan Greenbrier,Vest Greenbrier ●

366415 Smilax lanceifolia Roxb. subsp. opaca (A. DC.) T. Koyama = Smilax lanceifolia Roxb. var. opaca A. DC. ●

366416 Smilax lanceifolia Roxb. var. elangata (Warb.) F. T. Wang et Ts. Tang;折枝菝葜;Bentshoot Greenbrier ●

366417 Smilax lanceifolia Roxb. var. impressinervia (F. T. Wang et Ts. Tang) T. Koyama;凹脉菝葜;Sunkenvein Greenbrier ●

366418 Smilax lanceifolia Roxb. var. lanceolata (Norton) T. Koyama;长叶菝葜;Longleaf Greenbrier ●

366419 Smilax lanceifolia Roxb. var. opaca A. DC.;暗色菝葜;Dull Lanceolateleaf Smilax,Dullcolor Greenbrier ●

366420 Smilax lanceifolia Roxb. var. reflexa (Norton) T. Koyama = Smilax chapaensis Gagnep. ●

366421 Smilax lanceolata Burm. f. = Smilax glabra Roxb. ex C. H. Wright ●

366422 Smilax lanceolata Engelm. ex A. DC. = Smilax rotundifolia L. ●

366423 Smilax lanceolata L. = Smilax hastata Willd. var. lanceolata (L.) Pursh ●☆

366424 Smilax lanceolata L. = Smilax laurifolia L. ●☆

366425 Smilax lanceolata Ruiz ex A. DC. = Smilax insignis Kunth ●☆

366426 Smilax lanceolata Walter = Smilax hastata Willd. ●☆

366427 Smilax lasioneura Hook.;毛脉菝葜;Carrion Flower, Common Carrion Flower, Hairy Carrion-flower ●☆

366428 Smilax laurifolia L.;月桂叶菝葜;Bamboo Vine, Blaspheme Vine,Laurel Greenbrier ●☆

366429 Smilax lebrunii H. Lév.;粗糙菝葜;Lebrun Greenbrier, Lebrun Smilax ●

366430 Smilax leptanthera Pennell = Smilax pseudochina Lour. ●☆

366431 Smilax leucocarpa H. Lév. et Vaniot = Smilax trinervula Miq. ●

366432 Smilax liukiuensis Hayata = Smilax nervo-marginata Hayata var. liukiuensis (Hayata) F. T. Wang et Ts. Tang ●

366433 Smilax longebracteolata Hook. f.;长苞菝葜（长苞片菝葜）;Longbract Greenbrier ●

366434 Smilax longipedunculata Merr. = Smilax nipponica Miq. ●▪

366435 Smilax longipes Warb.;长柄菝葜（金荞苓藤）;Longstalk Greenbrier ●

366436 Smilax loupouensis H. Lév. = Smilax megalantha C. H. Wright ●

366437 Smilax luei T. Koyama;吕氏菝葜▪

366438 Smilax lunglingensis F. T. Wang et Ts. Tang;马钱叶菝葜（白菝葜,白草薢,草薢,刺草薢,花草薢,金刚藤,铁叶菝葜）;Longling Smilax, Lungling Greenbrier, Poisonnutleaf Greenbrier ●

366439 Smilax lushuiensis S. C. Chen;泸水菝葜;Lushui Greenbrier, Lushui Smilax ●

366440 Smilax luteocaulis H. Lév. = Smilax menispermoidea A. DC. ●

366441 Smilax lyi H. Lév. = Smilax bracteata C. Presl ●

366442 Smilax maclurei T. Koyama = Heterosmilax gaudichaudiana (Kunth) Maxim. ●

366443 Smilax macrocarpa Blume = Smilax megacarpa A. DC. ●

366444 Smilax macrophylla Roxb.;大叶菝葜●☆

366445 Smilax macrophylla Roxb. = Smilax ovalifolia Roxb. ●

366446 Smilax maculata Roxb. = Smilax asper L. ●

366447 Smilax mairei H. Lév.;无刺菝葜（白草薢,草薢,川草薢,打不死,滇红草薢,红草薢,花草薢,花花藤,小草薢,小红草薢）;Maire Greenbrier, Maire Smilax ●

366448 Smilax malipoensis S. C. Chen;麻栗坡菝葜;Malipo Greenbrier, Malipo Smilax ●

366449 Smilax maritinii H. Lév. et Vaniot = Smilax scobinicaulis C. H. Wright ●▪

366450 Smilax mauritanica Poir. = Smilax aspera L. ●

366451 Smilax maximowiczii Koidz. = Smilax riparia A. DC. ●▪

366452 Smilax megacarpa A. DC.;大果菝葜;Bigfruit Greenbrier, Bigfruit Greenbrier, Bigfruit Smilax ●

366453 Smilax megalantha C. H. Wright;大花菝葜;Bigflower Greenbrier, Bigflower Smilax ●

366454 Smilax megalantha C. H. Wright var. alata F. T. Wang = Smilax megalantha C. H. Wright ●

366455 Smilax megalantha C. H. Wright var. asperata F. T. Wang = Smilax lebrunii H. Lév. ●

366456 Smilax megalantha C. H. Wright var. ferruginea F. T. Wang = Smilax chingii F. T. Wang et Ts. Tang ●

366457 Smilax megalantha C. H. Wright var. maclurei Merr. = Smilax chingii F. T. Wang et Ts. Tang ●

366458　Smilax mengmaensis R. H. Miao；勐马菝葜；Mengma Greenbrier, Mengma Smilax ●

366459　Smilax mengmaensis R. H. Miao = Smilax glabra Roxb. ex C. H. Wright ●

366460　Smilax menispermoidea A. DC.；防己叶菝葜（峦大菝葜）；Bluebead Greenbrier, Blue-bead Greenbrier, Bluebead Smilax ●

366461　Smilax menispermoidea A. DC. subsp. randaiensis（Hayata）T. Koyama = Smilax menispermoidea A. DC.

366462　Smilax menispermoidea A. DC. var. randaiensis（Hayata）T. Koyama = Smilax pygmaea Merr. ●

366463　Smilax menispermoidea A. DC. var. randaiensis（Hayata）T. Koyama = Smilax menispermoidea A. DC. ●

366464　Smilax micorpoda A. DC. = Smilax lanceifolia Roxb. ●

366465　Smilax micorpoda A. DC. var. reflexa Norton = Smilax chapaensis Gagnep. ●

366466　Smilax microphylla C. H. Wright；小叶菝葜（地茯苓藤，乌鱼刺，小菝葜）；Littleleaf Greenbrier, Little-leaf Greenbrier, Littleleaf Smilax ●

366467　Smilax microphylla C. H. Wright var. angustifolia Warb. = Smilax microphylla C. H. Wright ●

366468　Smilax microphylla C. H. Wright var. elongata T. Koyama = Smilax mairei H. Lév. ●

366469　Smilax microphylla C. H. Wright var. elongata Warb. = Smilax lanceifolia Roxb. var. elangata（Warb.）F. T. Wang et Ts. Tang ●

366470　Smilax microphylla C. H. Wright var. nigrescens Warb. = Smilax scobinicaulis C. H. Wright ●■

366471　Smilax micropoda A. DC. = Smilax lanceifolia Roxb. ●

366472　Smilax micropoda A. DC. var. reflexa J. B. Norton = Smilax chapaensis Gagnep. ●

366473　Smilax morsaniana Kunth = Smilax anceps Willd. ●☆

366474　Smilax mossambicensis Garcke = Smilax anceps Willd. ●☆

366475　Smilax munita S. C. Chen；劲直菝葜；Rigid Greenbrier, Rigid Smilax ●

366476　Smilax myrtillus A. DC.；乌饭叶菝葜；Myrtle Greenbrier ●

366477　Smilax myrtillus A. DC. var. dulongensis H. Li；独龙菝葜；Dulong Greenbrier ●

366478　Smilax myrtillus A. DC. var. dulongensis H. Li = Smilax myrtillus A. DC. ●

366479　Smilax myrtillus A. DC. var. rigida Noltie = Smilax munita S. C. Chen ●

366480　Smilax nana F. T. Wang；矮菝葜（刺瓜米草，刺梭罗）；Dwarf Greenbrier ●

366481　Smilax nantoensis T. Koyama；南投菝葜；Nantou Greenbrier ●

366482　Smilax nebelii Gilg = Smilax sieboldii Miq. ●

366483　Smilax nervo-marginata Hayata；缘脉菝葜；Marginvein Greenbrier, Marginvein Smilax, Margin-veined Greenbrier ●

366484　Smilax nervo-marginata Hayata var. liukiuensis（Hayata）F. T. Wang et Ts. Tang；无疣菝葜（琉球菝葜）；Liukiu Greenbrier, Luchu Greenbrier, Luqiu Greenbrier ●

366485　Smilax nigrescens F. T. Wang et Ts. Tang；黑叶菝葜（铁丝灵仙）；Blackleaf Greenbrier, Black-leaf Greenbrier, Blackleaf Smilax ●

366486　Smilax nipponica Miq.；白背牛尾菜（白须公，百部伸筋，长叶牛尾菜，大伸筋，大顺筋藤，大叶伸筋，老龙须，龙须草，马尾伸筋，牛尾菜，牛尾节，牛尾卷，牛尾蕨，牛尾伸筋，七星牛尾菜，日本菝葜，日本山马薯，伸筋草，水球花，水摇竹）；Japan Greenbrier, Japanese Greenbrier, Nippon Greenbrier ●■

366487　Smilax nipponica Miq. f. grandifolia H. Hara；大花白背牛尾菜 ●☆

366488　Smilax nipponica Miq. f. tenuifolia Hisauti；细叶白背牛尾菜 ●☆

366489　Smilax nipponica Miq. subsp. manshurica Kitag.；无须牛尾菜 ●

366490　Smilax oblonga（C. H. Wright）Norton ex Bailey = Smilax nipponica Miq. ●■

366491　Smilax ocreata A. DC.；抱茎菝葜（穿鞘菝葜，耳叶菝葜，红土茯苓）；Ocreate Greenbrier, Ocreate Smilax ●

366492　Smilax ocreata H. Lév. et Vaniot = Smilax scobinicaulis C. H. Wright ●■

366493　Smilax odoratissima Blume；糙茎菝葜 ●

366494　Smilax officinalis Poepp. ex A. DC.；洋菝葜；Sarsaparilla ●☆

366495　Smilax oldhami Miq. = Smilax sieboldii Miq. ●

366496　Smilax oldhami Miq. var. daibuensis（Hayata）T. Koyama = Smilax riparia A. DC. ●■

366497　Smilax oldhami Miq. var. ussuriensis（Regel）A. DC. = Smilax riparia A. DC. ●■

366498　Smilax oldhamii Miq. = Smilax sieboldii Miq. ●

366499　Smilax opaca（A. DC.）J. B. Norton = Smilax lanceifolia Roxb. var. opaca A. DC. ●

366500　Smilax opaca（A. DC.）J. B. Norton = Smilax lanceifolia Roxb. ●

366501　Smilax ornata Hook. = Smilax regelii Killip et Morton ●☆

366502　Smilax outanscianensis Pamp.；武当菝葜；Wudangshan Greenbrier, Wudangshan Smilax, Wutang Mountain Greenbrier ●

366503　Smilax ovalifolia Roxb.；卵叶菝葜 ●

366504　Smilax ovalifolia Roxb. = Smilax macrophylla Roxb. ●☆

366505　Smilax ovata Pursh = Smilax smallii Morong ●☆

366506　Smilax ovatorotunda Hayata = Smilax riparia A. DC. ●■

366507　Smilax ovatorotunda Hayata var. ussuriensis（Regel）H. Hara = Smilax riparia A. DC. ●■

366508　Smilax ovatorotunda Hayata var. ussuriensis（Regel）H. Hara f. stenophylla H. Hara = Smilax riparia A. DC. var. ussuriensis（Regel）H. Hara et T. Koyama f. stenophylla（H. Hara）T. Koyama ●☆

366509　Smilax oxyphylla T. Koyama = Smilax arisanensis（Hayata）F. T. Wang et Ts. Tang ●

366510　Smilax oxyphylla Wall. ex Kunth = Smilax arisanensis（Hayata）F. T. Wang et Ts. Tang ●

366511　Smilax pachysandroides T. Koyama；川鄂菝葜（湖北菝葜）；Hubei Greenbrier ●

366512　Smilax pallescens A. DC. = Smilax griffithii A. DC. ●

366513　Smilax papyracea Duhamel；纸质菝葜 ●☆

366514　Smilax parviflora Wall. ex Hook. f. = Smilax glaucophylla Klotzsch ●

366515　Smilax parvifolia Wall. ex Hook. f. = Smilax elegans Wall. ex Kunth ●

366516　Smilax peduncularis Muhl. ex Willd. = Smilax herbacea L. ●☆

366517　Smilax pekingensis A. DC. = Smilax stans Maxim. ●

366518　Smilax perfoliata Lour.；穿鞘菝葜（白草薢，穿耳菝葜，穿叶菝葜，大托叶菝葜，多育菝葜，耳叶菝葜，金刚藤）；Auricled Greenbrier, Perfoliate Greenbrier ●

366519　Smilax perolifera Wall. ex Roxb. = Smilax perfoliata Lour. ●

366520　Smilax perulata H. Lév. et Vaniot = Smilax ocreata A. DC. ●

366521　Smilax pinfaensis H. Lév. et Vaniot；平伐菝葜；Pingfa Greenbrier ●

366522　Smilax planipedunculata Hayata = Heterosmilax japonica Kunth ●

366523　Smilax planipes F. T. Wang et Ts. Tang；扁柄菝葜；Flatstalk

Greenbrier, Flat-stalk Greenbrier, Flatstalk Smilax ●

366524 Smilax polycephala F. T. Wang et Ts. Tang；多头菝葜；Fourangular Greenbrier, Four-angular Greenbrier, Fourridge Greenbrier ●☆

366525 Smilax polycephala F. T. Wang et Ts. Tang = Smilax elegantissima Gagnep. ●

366526 Smilax polycolea Warb. ex Diels；红果菝葜（金姜豆藤）；Redfruit Greenbrier, Redfruit Smilax, Red-fruited Greenbrier ●

366527 Smilax polycolea Warb. ex Diels var. acuminata Warb. = Smilax cocculoides Warb. ex Diels ●

366528 Smilax pottingeri Prain；纤柄菝葜（纤柄肖菝葜）；Pottinger Greenbrier, Pottinger Heterosmilax ●

366529 Smilax prolifera Roxb. = Smilax perfoliata Lour. ●

366530 Smilax prolifera Wall. ex Roxb. = Smilax perfoliata Lour. ●

366531 Smilax pseudochina Lour.；竹叶菝葜●☆

366532 Smilax pteropus Miq. = Smilax china L. ●

366533 Smilax pubera Michx. = Smilax pumila Walter ●☆

366534 Smilax puberula Kunth = Smilax pumila Walter ●☆

366535 Smilax pulverulenta Michx.；被粉菝葜；Carrion Flower ●☆

366536 Smilax pumila Walter；矮小菝葜；Dwarf Greenbrier, Sarsaparilla Vine ●☆

366537 Smilax pygmaea Merr.；峦大菝葜；Luanda Greenbrier ●

366538 Smilax quadrangularis Muhl. ex Willd. = Smilax rotundifolia L. ●

366539 Smilax quadrata A. DC.；方枝菝葜；Squarestem Greenbrier, Square-stem Greenbrier ●

366540 Smilax randaiensis Hayata = Smilax menispermoidea A. DC. ●

366541 Smilax randaiensis Hayata = Smilax pygmaea Merr. ●

366542 Smilax regelii Killip et Morton；洪都拉斯菝葜；Beautiful-flower Greenbrier, Jamaica Sarsaparilla, Sarsaparilla ●☆

366543 Smilax renifolia Small = Smilax bona-nox L. ●☆

366544 Smilax retroflexa (F. T. Wang et Ts. Tang) S. C. Chen；弯柄菝葜（苍白菝葜，弯梗菝葜）；Bowedstalk Greenbrier, Bowedstalk Smilax ●

366545 Smilax rigida Wall. ex Kunth = Smilax munita S. C. Chen ●

366546 Smilax rigida Wall. ex Kunth subsp. myrtillus (A. DC.) T. Koyama = Smilax myrtillus A. DC. ●

366547 Smilax rigida Wall. ex Kunth var. myrtillus (A. DC.) T. Koyama = Smilax myrtillus A. DC. ●

366548 Smilax riparia A. DC.；牛尾菜（白须公，草菝葜，春根藤，大伸筋，大伸筋草，大武牛尾菜，过江蕨，尖叶牛尾菜，金刚豆藤，老龙须，鲤鱼须，龙须牛尾菜，马氏菝葜，马尾伸根，马尾伸筋，牛尾结，牛尾蕨，七层楼，千层塔，软叶菝葜，山豇豆，山竹花，土春根，乌苏里山马薯）；Oxtail Greenbrier, Riparian Greenbrier, Ussuri Greenbrier ●■

366549 Smilax riparia A. DC. f. acuta (Hiyama) M. Kobay.；河岸菝葜 ●☆

366550 Smilax riparia A. DC. f. ovatorotunda (Hayata) T. Koyama = Smilax riparia A. DC. ●■

366551 Smilax riparia A. DC. f. sadoensis (Honda) T. Koyama；佐渡菝葜●☆

366552 Smilax riparia A. DC. subsp. ussuriensis (Regel) Kitag. = Smilax riparia A. DC. ●■

366553 Smilax riparia A. DC. var. acuminata (C. H. Wright) F. T. Wang et Ts. Tang；尖叶牛尾菜；Sharp Riparian Greenbrier, Sharpleaf Oxtail Greenbrier ●■

366554 Smilax riparia A. DC. var. pubescens (C. H. Wright) F. T. Wang et Ts. Tang；毛牛尾菜（金刚藤，圆叶菝葜）；Common Greenbrier,

Hairy Oxtail Greenbrier, Horse-brier Riparian Greenbrier ●■

366555 Smilax riparia A. DC. var. pubescens Miq. = Smilax riparia A. DC. var. pubescens (C. H. Wright) F. T. Wang et Ts. Tang ●■

366556 Smilax riparia A. DC. var. ussuriensis (Regel) H. Hara et T. Koyama f. maximowiczii (Koidz.) T. Koyama = Smilax riparia A. DC. ●■

366557 Smilax riparia A. DC. var. ussuriensis (Regel) H. Hara et T. Koyama f. stenophylla (H. Hara) T. Koyama；狭叶牛尾菜●☆

366558 Smilax riparia A. DC. var. ussuriensis (Regel) H. Hara et T. Koyama = Smilax riparia A. DC. ●■

366559 Smilax rotundifolia L.；金刚菝葜（金刚藤）；Broadleaf Greenbrier, Bull Brier, Bullbrier, Cat Brier Catbrier, Catbrier, Common Catbrier, Common Greenbrier, Green Briar, Greenbrier, Horse Brier, Horsebrier, Round-leaved Greenbrier ●

366560 Smilax rotundifolia L. var. californica A. DC. = Smilax californica (A. DC.) A. Gray ●☆

366561 Smilax rotundifolia L. var. crenulata Small et A. Heller = Smilax rotundifolia L. ●

366562 Smilax rotundifolia L. var. quadrangularis (Muhl. ex Willd.) A. W. Wood = Smilax rotundifolia L. ●

366563 Smilax rubiflora Rehder = Smilax menispermoidea A. DC. ●

366564 Smilax saluberrima ?；健身菝葜●☆

366565 Smilax sarumame Ohwi = Smilax biflora Siebold ex Miq. var. trinervula (Miq.) Hatus. ex T. Koyama ●

366566 Smilax scobinicaulis C. H. Wright；短梗菝葜（黑刺菝葜，金刚刺，金刚藤，威灵仙）；Scabrous-stem Greenbrier, Scabrousstem Smilax, Shortstalk Greenbrier ●■

366567 Smilax scobinicaulis C. H. Wright var. brevipes (Warb.) Hand.-Mazz. = Smilax scobinicaulis C. H. Wright ●■

366568 Smilax sebeana T. Koyama = Smilax china L. ●

366569 Smilax sebeana T. Koyama var. glaucochina (Warb.) T. Koyama = Smilax glaucochina Warb. ex Diels ●

366570 Smilax sempervirens F. T. Wang = Smilax nervo-marginata Hayata ●

366571 Smilax setiramula F. T. Wang et Ts. Tang；密刚毛菝葜；Denseseta Greenbrier, Densesetaceous Greenbrier, Dense-setaceous Greenbrier ●

366572 Smilax siderophylla Hand.-Mazz. = Smilax lunglingensis F. T. Wang et Ts. Tang ●

366573 Smilax sieboldii Miq.；华东菝葜（奥氏菝葜，倒钩刺，金刚藤，鲢鱼须，龙须菜，鲇鱼须，鲇鱼须草，黏鱼须，黏鱼须菝葜，山何首乌，台湾山马薯）；Siebold Greenbrier, Siebold Smilax ●

366574 Smilax sieboldii Miq. f. inermis (Nakai) H. Hara；无刺华东菝葜●☆

366575 Smilax sieboldii Miq. f. inermis (Nakai) H. Hara = Smilax sieboldii Miq. ●

366576 Smilax sieboldii Miq. var. formosana Hayata = Smilax sieboldii Miq. ●

366577 Smilax sieboldii Miq. var. inermis Nakai = Smilax sieboldii Miq. ●

366578 Smilax sieboldii Miq. var. scobinicaulis (C. H. Wright) T. Koyama = Smilax scobinicaulis C. H. Wright ●■

366579 Smilax simadai Masam. = Smilax nipponica Miq. ●■

366580 Smilax smallii Morong；斯莫尔菝葜；Cantaque, Jacksonbrier, Smilax ●☆

366581 Smilax spinosa Mill.；多刺菝葜●☆

366582 Smilax spinulosa Sm. = Smilax glauca Walter ●☆

366583 Smilax stans Maxim.；鞘柄菝葜（北京菝葜，薄叶菝葜，具鞘菝

蒉,鞘菝葜,玉山菝葜);SHeathed Greenbrier, Sheathstipe Greenbrier,Sheathstipe Smilax,Sheath-stiped Greenbrier ●

366584　Smilax stans Maxim. var. verruculosifolia J. M. Xu = Smilax trachypoda J. B. Norton ●

366585　Smilax stemonifolia H. Lév. et Vaniot = Heterosmilax japonica Kunth ●

366586　Smilax stenopetala A. Gray = Smilax bracteata C. Presl ●

366587　Smilax subarmata (O. E. Schulz) O. E. Schulz = Smilax havanensis Jacq. ●☆

366588　Smilax synandra Gagnep. ;筒被菝葜(直立肖菝葜); Erect Heterosmilax ●

366589　Smilax taiheiensis Hayata = Smilax china L. var. taiheiensis (Hayata) Koyama ●

366590　Smilax takaoensis Hayata = Smilax riparia A. DC. ●■

366591　Smilax tamnifolia Michx. = Smilax pseudochina Lour. ●☆

366592　Smilax tamnoides L. ;硬毛菝葜; Bristly Greenbrier, Catbrier, China Root,Hellfetter, Hispid Greenbrier ●☆

366593　Smilax tamnoides L. var. hispida (Muhl. ex Torr.) Fernald = Smilax tamnoides L. ●☆

366594　Smilax tamnoides L. var. hispida (Muhl.) Fernald = Smilax hispida Muhl. ex Torr. ●☆

366595　Smilax tenuis Small = Smilax lasioneura Hook. ●☆

366596　Smilax tenuissima Hayata = Smilax stans Maxim. ●

366597　Smilax tenuissima Hayata = Smilax vaginata Decne. ●

366598　Smilax tequetii H. Lév. = Smilax china L. ●

366599　Smilax tetraptera Gagnep. = Smilax gagnepainii T. Koyama ●

366600　Smilax tortipetiolata H. Lév. et Vaniot = Smilax lanceifolia Roxb. var. elangata (Warb.) F. T. Wang et Ts. Tang ●

366601　Smilax tortuosus Diels = Smilax ferox Wall. ex Kunth ●

366602　Smilax tortuosus Diels = Smilax megalantha C. H. Wright ●

366603　Smilax trachyclada Hayata = Smilax aspericaulis Wall. ex A. DC. ●

366604　Smilax trachyclada Hayata = Smilax bracteata C. Presl var. verruculosa (Merr.) T. Koyama ●

366605　Smilax trachypoda J. B. Norton;糙柄菝葜(粗柄菝葜); Roughstalk Greenbrier, Rough-stalk Greenbrier ●

366606　Smilax trigona Warb. ex Diels = Smilax glabra Roxb. ex C. H. Wright ●

366607　Smilax trinervula Miq. ;三脉菝葜(山梨儿); Threenerve Greenbrier,Three-nerve Greenbrier,Threenerve Smilax ●

366608　Smilax trinervula Miq. = Smilax biflora Siebold ex Miq. var. trinervula (Miq.) Hatus. ex T. Koyama ●

366609　Smilax tsaii F. T. Wang = Smilax aberrans Gagnep. ●

366610　Smilax tsaii F. T. Wang et Ts. Tang = Heterosmilax japonica Kunth ●

366611　Smilax tsinchengshanensis F. T. Wang;青城菝葜; Qingcheng Mountain Greenbrier,Qingcheng Smilax,Qingchengshan Greenbrier ●

366612　Smilax umbrosa J. M. Xu;荫生菝葜;Shady Greenbrier ●

366613　Smilax umbrosa J. M. Xu = Smilax pachysandroides T. Koyama ●

366614　Smilax vaginata Decne. = Smilax stans Maxim. ●

366615　Smilax vaginata Decne. var. pekingensis (A. DC.) Koyama = Smilax stans Maxim. ●

366616　Smilax vaginata Decne. var. stans (Maxim.) T. Koyama = Smilax stans Maxim. ●

366617　Smilax vanchingshanensis (F. T. Wang et Ts. Tang) F. T. Wang et Ts. Tang;梵净山菝葜; Fanching Mountain Greenbrier, Fanjing Mountain Greenbrier,Fanjingshan Greenbrier,Fanjingshan Smilax ●

366618　Smilax variegata Walter = Smilax bona-nox L. ●☆

366619　Smilax verruculosa Merr. = Smilax aspericaulis Wall. ex A. DC. ●

366620　Smilax verruculosa Merr. = Smilax bracteata C. Presl var. verruculosa (Merr.) T. Koyama ●

366621　Smilax walteri Pursh;沃尔特菝葜; Red-berried Bamboo, Red-berried Greenbrier ●☆

366622　Smilax woodii Merr. = Smilax corbularia Kunth var. woodii (Merr.) T. Koyama ●

366623　Smilax yunnanensis S. C. Chen;云南菝葜;Yunnan Greenbrier ●

366624　Smilax zeylanica L. ;锡兰菝葜;Ceylon Greenbrier ●☆

366625　Smilax zeylanica L. subsp. hemsleyana (Craib) TKoyama = Smilax hemsleyana Craib ●

366626　Smirnovia Bunge(1876);没药豆属●☆

366627　Smirnovia turkestana Bunge;没药豆(土耳其豆)●☆

366628　Smirnowia Bunge = Smirnovia Bunge ●☆

366629　Smithanthe Szlach. et Marg. = Habenaria Willd. ■

366630　Smithanthe rhodocheila (Hance) Szlach. et Marg. = Habenaria rhodochelia Hance ■

366631　Smithatris W. J. Kress et K. Larsen(2001);泰国姜属■☆

366632　Smithia Aiton(1789)(保留属名);坡油甘属(合叶豆属,施密草属,施氏豆属,史密豆属);Smithia ●■

366633　Smithia J. F. Gmel. = Humbertia Comm. ex Lam. ●☆

366634　Smithia Scop. (废弃属名) = Clusia L. ●☆

366635　Smithia Scop. (废弃属名) = Quapoya Aubl. ●☆

366636　Smithia Scop. (废弃属名) = Smithia Aiton(保留属名)●■

366637　Smithia abyssinica (A. Rich.) Verdc. ;阿比西尼亚坡油甘●☆

366638　Smithia aeschynomenoides Welw. ex Baker = Kotschya aeschynomenoides (Welw. ex Baker) Dewit et P. A. Duvign. ●☆

366639　Smithia bequaertii De Wild. = Kotschya africana Endl. var. bequaertii (De Wild.) Verdc. ■☆

366640　Smithia bernieri Baill. = Aeschynomene uniflora E. Mey. ●☆

366641　Smithia bingilensis Micheli ex Pellegr. = Kotschya ochreata (Taub.) Dewit et P. A. Duvign. ■☆

366642　Smithia blanda Wall. = Smithia blanda Wall. ex Wight et Arn. ●

366643　Smithia blanda Wall. ex Wight et Arn. ;黄花合叶豆(艳丽施氏豆);Yellow Smithia, Yellow-flowered Smithia ●

366644　Smithia blanda Wall. ex Wight et Arn. var. paniculata (Wight et Arn.) Baker = Smithia blanda Wall. ex Wight et Arn. ●

366645　Smithia blanda Wall. ex Wight et Arn. var. racemosa (Wight et Arn.) Baker = Smithia blanda Wall. ex Wight et Arn. ●

366646　Smithia bodinieri H. Lév. = Smithia blanda Wall. ex Wight et Arn. ●

366647　Smithia burttii Baker f. = Kotschya capitulifera (Welw. ex Baker) Dewit et P. A. Duvign. ■☆

366648　Smithia capitulifera Welw. ex Baker = Kotschya capitulifera (Welw. ex Baker) Dewit et P. A. Duvign. ■☆

366649　Smithia carsonii Baker = Kotschya carsonii (Baker) Dewit et P. A. Duvign. ■☆

366650　Smithia cavaleriei H. Lév. = Smithia ciliata Royle ■

366651　Smithia chamaecrista Benth. = Kotschya africana Endl. ■☆

366652　Smithia chamaecrista Benth. var. genuina R. Vig. = Kotschya africana Endl. ☆

366653　Smithia chamaecrista Benth. var. stipulata R. Vig. = Kotschya africana Endl. ■☆

366654　Smithia ciliata Royle;缘毛合叶豆(薄萼坡油甘); Ciliate Smithia ■

366655　Smithia conferta Sm. ;密节坡油甘(密节膜苞豆,密节施氏豆,蛇头草);Densenode Smithia ■

366656　Smithia conferta var. geminiflora（Roth）Cooke = Smithia conferta Sm. ■

366657　Smithia congesta Baker = Kotschya recurvifolia（Taub.）White ■☆

366658　Smithia dichotoma Dalzell ex Baker = Smithia salsuginea Hance ■

366659　Smithia drepanophylla Baker = Kotschya recurvifolia（Taub.）White ■☆

366660　Smithia elaphroxylon Baill. = Aeschynomene elaphroxylon（Guillaumin et Perr.）Taub. ■☆

366661　Smithia elliotii Baker f.；埃利坡油甘■☆

366662　Smithia elliotii Baker f. var. sparse-strigosa Verdc.；稀糙伏毛坡油甘●☆

366663　Smithia erubescens（E. Mey.）Baker f.；变红坡油甘●☆

366664　Smithia eurycalyx Harms = Kotschya eurycalyx（Harms）Dewit et P. A. Duvign. ■☆

366665　Smithia geminiflora Roth = Smithia conferta Sm. ■

366666　Smithia geminiflora Roth var. coferta Baker = Smithia conferta Sm. ■

366667　Smithia geminiflora Roth var. conferta（Smith）Baker = Smithia conferta Sm. ■

366668　Smithia goetzei Harms = Kotschya goetzei（Harms）Verdc. ■☆

366669　Smithia grandidieri Baill. = Aeschynomene elaphroxylon（Guillaumin et Perr.）Taub. ■☆

366670　Smithia japonica Maxim.；日本坡油甘；Japan Smithia，Japanese Smithia ■☆

366671　Smithia japonica Maxim. = Smithia ciliata Royle ■

366672　Smithia javanica Benth. = Smithia sensitiva Aiton ■

366673　Smithia kotschyi Benth. = Kotschya africana Endl. ■☆

366674　Smithia lutea Portères = Kotschya lutea（Portères）Hepper ■☆

366675　Smithia megalophylla Harms = Humularia welwitschii（Taub.）P. A. Duvign. var. lundaensis（P. A. Duvign.）Verdc. ■☆

366676　Smithia micrantha Harms = Kotschya micrantha（Harms）Hepper ■☆

366677　Smithia mildbraedii Harms = Kotschya aeschynomenoides（Welw. ex Baker）Dewit et P. A. Duvign. ●☆

366678　Smithia nagasawai Hayata = Smithia ciliata Royle ■

366679　Smithia nodulosa Baker = Aeschynomene nodulosa（Baker）Baker f. ■☆

366680　Smithia ochreata Taub. = Kotschya ochreata（Taub.）Dewit et P. A. Duvign. ■☆

366681　Smithia oubanguiensis Tisser. = Kotschya oubanguiensis（Tisser.）Verdc. ■☆

366682　Smithia paniculata Arn. = Smithia blanda Wall. ex Wight et Arn. ●

366683　Smithia parvifolia Burtt Davy = Kotschya parvifolia（Burtt Davy）Verdc. ■☆

366684　Smithia perrieri R. Vig. = Kotschya perrieri（R. Vig.）Verdc. ■☆

366685　Smithia platyphylla Brenan = Kotschya platyphylla（Brenan）Verdc. ■☆

366686　Smithia prittwitzii Harms = Kotschya prittwitzii（Harms）Verdc. ■☆

366687　Smithia racemosa Wight et Arn. = Smithia blanda Wall. ex Wight et Arn. ●

366688　Smithia recurvifolia Taub. = Kotschya recurvifolia（Taub.）White ■☆

366689　Smithia reflexa Portères = Kotschya carsonii（Baker）Dewit et

P. A. Duvign. subsp. reflexa（Portères）Verdc. ■☆

366690　Smithia ringoetii De Wild. = Kotschya africana Endl. var. ringoetii（De Wild.）Dewit et P. A. Duvign. ■☆

366691　Smithia riparia R. E. Fr. = Kotschya africana Endl. var. bequaertii（De Wild.）Verdc. ■☆

366692　Smithia rosea R. Vig. = Smithia elliotii Baker f. ■☆

366693　Smithia rubrofarinacea Taub. = Aeschynomene rubrofarinacea（Taub.）F. White ●☆

366694　Smithia ruwenzoriensis Baker f. = Kotschya aeschynomenoides（Welw. ex Baker）Dewit et P. A. Duvign. ●☆

366695　Smithia salsuginea Hance；盐碱土坡油甘（盐碱坡油甘，盐碱土膜苞豆，盐碱土施氏豆）；Saline Smithia ■

366696　Smithia scaberrima Taub. = Kotschya scaberrima（Taub.）Wild ■☆

366697　Smithia schweinfurthii Taub. = Kotschya schweinfurthii（Taub.）Dewit et P. A. Duvign. ■☆

366698　Smithia sensitiva Aiton；坡油甘（玻油甘，黄花儿，敏感施氏豆，施氏豆，水百足，水老虎，田唇乌蝇翼，田基豆，田基黄）；Sensitive Smithia ■

366699　Smithia sensitiva Aiton var. abyssinica A. Rich. = Smithia abyssinica（A. Rich.）Verdc. ●☆

366700　Smithia setosissima Harms = Kotschya carsonii（Baker）Dewit et P. A. Duvign. ■☆

366701　Smithia speciosa Hutch. = Kotschya speciosa（Hutch.）Hepper ■☆

366702　Smithia sphaerocephala Baker = Kotschya aeschynomenoides（Welw. ex Baker）Dewit et P. A. Duvign. ●☆

366703　Smithia stolonifera Brenan = Kotschya stolonifera（Brenan）Dewit et P. A. Duvign. ■☆

366704　Smithia strigosa Benth. = Kotschya strigosa（Benth.）Dewit et P. A. Duvign. ■☆

366705　Smithia strobilantha Welw. ex Baker = Kotschya strobilantha（Welw. ex Baker）Dewit et P. A. Duvign. ■☆

366706　Smithia thouinia J. F. Gmel. = Humbertia madagascariensis Lam. ●☆

366707　Smithia trochainii Berhaut = Aeschynomene crassicaulis Harms ■☆

366708　Smithia uguenensis Taub. = Kotschya uguenensis（Taub.）White ■☆

366709　Smithia uniflora A. Chev. = Kotschya uniflora（A. Chev.）Hepper ■☆

366710　Smithia volkensii Taub. = Kotschya aeschynomenoides（Welw. ex Baker）Dewit et P. A. Duvign. ●☆

366711　Smithia welwitschii Taub. = Humularia welwitschii（Taub.）P. A. Duvign. ■☆

366712　Smithia yunnanensis Franch. = Smithia blanda Wall. ex Wight et Arn. ●

366713　Smithia yunnanensis Franch. = Smithia blanda Wall. ●

366714　Smithiantha Kuntze（1891）；庙铃苔属（绒桐草属）；Temple Bells，Temple-bells ■☆

366715　Smithiantha cinnabarina（Linden）Kuntze；红庙铃苔（绒桐草，天鹅绒桐花）；Red Temple-bells，Temple Bells ■☆

366716　Smithiantha laui Wiehler；楼氏庙铃苔；Lau Temple-bells ■☆

366717　Smithiantha multiflora Fritsch；多花庙铃苔；Naegelia ■☆

366718　Smithiantha punctata Kuntze；斑点庙铃苔；Punctate Temple-bells ■☆

366719　Smithiantha zebrina Kuntze；条斑庙铃苔（斑叶绒桐草）；

Zebra-striped Temple-bells ■☆

366720　Smithiella Dunn = Aboriella Bennet ■

366721　Smithiella myriantha Dunn = Pilea myriantha (Dunn) C. J. Chen ■

366722　Smithiodendron Hu = Broussonetia L'Hér. ex Vent.（保留属名）●

366723　Smithiodendron artocarpioideum Hu = Broussonetia papyrifera (L.) L'Hér. ex Vent. ●

366724　Smithorchis Ts. Tang et F. T. Wang（1936）；反唇兰属；Smithorchis ■★

366725　Smithorchis calceoliformis (W. W. Sm.) Ts. Tang et F. T. Wang；反唇兰；Common Smithorchis, Smithorchis ■

366726　Smithsonia C. J. Saldanha(1974)；史密森兰属■☆

366727　Smithsonia maculata (Dalzell) C. J. Saldanha；斑点史密森兰■☆

366728　Smithsonia straminea C. J. Saldanha；史密森兰■☆

366729　Smithsonia viridiflora (Dalzell) C. J. Saldanha；绿花史密森兰■☆

366730　Smitinandia Holttum(1969)；盖喉兰属；Smitinandia ■

366731　Smitinandia micrantha (Lindl.) Holttum；盖喉兰；Smitinandia ■

366732　Smodingium E. Mey. = Smodingium E. Mey. ex Sond. ●☆

366733　Smodingium E. Mey. ex Sond.（1860）；肿漆属●☆

366734　Smodingium argutum E. Mey. ex Sond.；肿漆●☆

366735　Smyrniaceae Burnett = Apiaceae Lindl.（保留科名）●■

366736　Smyrniaceae Burnett = Umbelliferae Juss.（保留科名）■●

366737　Smyrniopsis Boiss.（1844）；肖没药属■☆

366738　Smyrniopsis armena Schischk.；肖没药■☆

366739　Smyrnium L.（1753）；类没药属（马芹属，美味芹属，亚历山大草属，异叶芹属）；Alexanders ■☆

366740　Smyrnium aureum L. = Zizia aurea (L.) W. D. J. Koch ■☆

366741　Smyrnium cordifolium Boiss.；心叶类没药■☆

366742　Smyrnium laterale Thunb. = Apium graveolens L. ■

366743　Smyrnium olusatrum L.；类没药（小美味芹）；Alexanders, Alexandrian Parsley, Alick, Alisanders, Alizanders, Alshinders, Black Lovage, Black Potherb, Hellroot, Helrut, Horse Parsley, Horse-nut Tree, Lovage, Macedonian Parsley, Megweed, Skeet, Skeets, Skit, Stanmarch, Wild Celery, Wild Parsley ■☆

366744　Smyrnium perfoliatum (L.) Mill.；穿叶芹；Perfoliate Alexanders ■☆

366745　Smyrnium perfoliatum (L.) Mill. subsp. rotundifolium (L.) Mill. = Smyrnium rotundifolium Mill. ■☆

366746　Smyrnium perfoliatum L. = Smyrnium perfoliatum (L.) Mill. ■☆

366747　Smyrnium rotundifolium Mill.；圆叶类没药■☆

366748　Smythea Seem.（1862）；扁果藤属；Smythea ●☆

366749　Smythea nitida Merr.；扁果藤；Shining Smythea ●☆

366750　Smythea nitida Merr. = Ventilago leiocarpa Benth. ●

366751　Snowdenia C. E. Hubb.（1929）；斯诺登草属●☆

366752　Snowdenia gracilis (Hochst.) Pilg. = Snowdenia mutica (Hochst.) Pilg. ■☆

366753　Snowdenia microcarpha C. E. Hubb.；小果斯诺登草■☆

366754　Snowdenia mutica (Hochst.) Pilg.；斯诺登草■☆

366755　Snowdenia petitiana (A. Rich.) C. E. Hubb.；非洲斯诺登草■☆

366756　Snowdenia polystachya (Fresen.) Pilg.；多穗斯诺登草■☆

366757　Snowdenia scabra (Pilg.) Pilg. = Snowdenia petitiana (A. Rich.) C. E. Hubb. ■☆

366758　Soala Blanco = Cyathocalyx Champ. ex Hook. f. et Thomson ●

366759　Soaresia Allemão（废弃属名）= Clarisia Ruiz et Pav.（保留属名）●☆

366760　Soaresia Allemão（废弃属名）= Soaresia Sch. Bip.（保留属名）●☆

366761　Soaresia Sch. Bip.（1863）（保留属名）；纵脉菊属●☆

366762　Soaresia velutina Sch. Bip.；纵脉菊■☆

366763　Sobennikoffia Schltr.（1925）；苏本兰属■☆

366764　Sobennikoffia fournieriana (André) Schltr.；苏本兰■☆

366765　Sobennikoffia humbertiana H. Perrier；亨伯特苏本兰■☆

366766　Sobennikoffia poissoniana H. Perrier；普瓦松苏本兰■☆

366767　Sobennikoffia robusta (Schltr.) Schltr.；粗壮苏本兰■☆

366768　Soberbaea D. Dietr. = Sowerbaea Sm. ■☆

366769　Sobisco Merr. = Sobiso Raf. ●■

366770　Sobiso Raf. = Salvia L. ●■

366771　Sobolewskia M. Bieb.（1832）；索包草属■☆

366772　Sobolewskia caucasica Busch；高加索索包草■☆

366773　Sobolewskia clavata Fenzl；棒状索包草■☆

366774　Sobolewskia lithophila M. Bieb.；索包草■☆

366775　Sobolewskia sibirica (Willd.) P. W. Ball；西伯利亚索包草■☆

366776　Sobralia Ruiz et Pav.（1794）；折叶兰属；Sobralia ■☆

366777　Sobralia × hybrida Hort.；杂交竹叶兰■☆

366778　Sobralia biflora Ruiz et Pav.；双花折叶兰■☆

366779　Sobralia cattleya Rchb. f.；卡特兰状折叶兰■☆

366780　Sobralia chrysantha Lindl.；黄花折叶兰■☆

366781　Sobralia decora Bateman；美丽折叶兰；Beautiful Sobralia ■☆

366782　Sobralia dichotoma Ruiz et Pav.；二岐折叶兰；Dichotomous Sobralia ■☆

366783　Sobralia fragrans Lindl.；芳香折叶兰；Fragrant Sobralia ■☆

366784　Sobralia leucoxantha Rchb. f.；乳白折叶兰；Whiteyellow Sobralia ■☆

366785　Sobralia lowii Rolfe；楼氏折叶兰■☆

366786　Sobralia macrantha Lindl.；大花折叶兰（苏伯兰）；Largeflower Sobralia ■☆

366787　Sobralia macrantha Lindl. var. alba Lindl.；白色大花折叶兰；White Largeflower Sobralia ■☆

366788　Sobralia macrantha Lindl. var. purpurea Lindl.；紫色大花折叶兰；Purple Largeflower Sobralia ■☆

366789　Sobralia macrophylla Rchb. f.；大叶折叶兰；Largeleaf Sobralia ■☆

366790　Sobralia parviflora L. O. Williams；小花折叶兰；Smallflower Sobralia ■☆

366791　Sobralia rosea Poepp. et Endl.；红花折叶兰；Redflower Sobralia ■☆

366792　Sobralia sessilis Lindl.；无柄折叶兰；Sessile Sobralia ■☆

366793　Sobralia violacea Linden ex Lindl.；堇色折叶兰；Violet Sobralia ■☆

366794　Sobralia virginalis Peeters et Cogn.；处女折叶兰■☆

366795　Sobralia xantholeuca Williams；黄白折叶兰；Yellow-white Sobralia ■☆

366796　Sobreyra Ruiz et Pav. = Enydra Lour. ■

366797　Sobrya Pers. = Sobreyra Ruiz et Pav. ■

366798　Socotora Balf. f. = Periploca L. ●

366799　Socotora aphylla Balf. f. = Periploca visciformis (Vatke) K. Schum. ●☆

366800　Socotranthus Kuntze = Cochlanthus Balf. f. ●☆

366801　Socotranthus Kuntze(1903)；螺花藤属●☆

366802　Socotranthus socotranus (Balf. f.) Bullock = Cochlanthus socotranus Balf. f. ●☆

366803　Socotrella Bruyns et A. G. Mill. (2002);索科特拉萝藦属■☆

366804　Socotrella dolichocnema Bruyns;索科特拉萝藦■☆

366805　Socotria G. M. Levin = Punica L. ●

366806　Socratea H. Karst.(1857);高跷椰属(高根柱椰属,高跷桐属,苏格拉底棕属,苏格椰子属,苏快特桐属)●☆

366807　Socratea exorrhiza （C. Mart.） H. Wendl.;高跷椰;Paxiuba Palm ●☆

366808　Socratesia Klotzsch = Cavendishia Lindl.(保留属名)●☆

366809　Socratina Balle(1964);索克寄生属●☆

366810　Socratina bemarivensis (Lecomte) Balle;索克寄生●☆

366811　Socratina keraudreniana Balle;马岛索克寄生●☆

366812　Soda (Dumort.) Fourr. = Salsola L. ●■

366813　Soda Fourr. = Salsola L. ●■

366814　Sodada Forssk. = Capparis L. ●

366815　Sodada decidua Forssk. = Capparis decidua (Forssk.) Edgew. ●

366816　Soderstromia C. V. Morton(1966);矮草原花属■☆

366817　Soderstromia mexicana (Scribn.) C. V. Morton;矮草原花■☆

366818　Sodiroa André = Guzmania Ruiz et Pav. ■☆

366819　Sodiroella Schltr. = Stellilabium Schltr. ■☆

366820　Soehrensia (Backeb.) Backeb.(1938);炮弹仙人球属■☆

366821　Soehrensia (Backeb.) Backeb. = Echinopsis Zucc. ●

366822　Soehrensia Backeb. = Echinopsis Zucc. ●

366823　Soehrensia Backeb. = Soehrensia (Backeb.) Backeb. ■☆

366824　Soehrensia bruchii (Britton et Rose) Backeb.;炮弹仙人球(湘阳丸)■☆

366825　Soehrensia formosa (Pfeiff.) Backeb.;丽刺玉■☆

366826　Soehrensia grandis (Britton et Rose) Backeb.;巨黄龙■☆

366827　Soehrensia korethroides (Werderm.) Backeb.;狂魔玉■☆

366828　Soejatmia K. M. Wong(1993);苏亚竹属■☆

366829　Soejatmia ridleyi (Gamble) K. M. Wong;苏亚竹■☆

366830　Soelanthus Raf. = Cissus L. ●

366831　Soelanthus Raf. = Saelanthus Forssk. ●

366832　Soemmeringia Mart.(1828);永花豆属(常花豆属)●☆

366833　Soemmeringia psittacorhyncha Webb = Bryaspis lupulina (Planch.) P. A. Duvign. ☆

366834　Soemmeringia semperflorens Mart.;永花豆(常花豆)●☆

366835　Sofianthe Tzvelev = Lychnis L.(废弃属名)■

366836　Sofianthe Tzvelev = Silene L.(保留属名)■

366837　Sofianthe Tzvelev(2001);索菲石竹属■☆

366838　Sogalgina Cass. = Tridax L. ■●

366839　Sogaligna Steud. = Sogalgina Cass. ■●

366840　Sogaligna Steud. = Tridax L. ■●

366841　Sogerianthe Danser(1933);索花属●☆

366842　Sogerianthe ferruginea Danser;索花●☆

366843　Sohnreyia K. Krause = Spathelia L.(保留属名)●☆

366844　Sohnsia Airy Shaw(1965);白霜叶属■☆

366845　Sohnsia filifolia (E. Fourn.) Airy Shaw;白霜叶■☆

366846　Sohrea Steud. = Shorea Roxb. ex C. F. Gaertn. ●

366847　Soja Moench(废弃属名) = Glycine Willd.(保留属名)■

366848　Soja hispida Moench = Glycine max (L.) Merr. ■

366849　Soja max (L.) Piper = Glycine max (L.) Merr. ■

366850　Sokolofia Raf. = Salix L.(保留属名)●

366851　Solaenacanthus Oerst. = Ruellia L. ■●

366852　Solanaceae Adans. = Solanaceae Juss.(保留科名)●■

366853　Solanaceae Juss.(1789)(保留科名);茄科;Nightshade Family ●■

366854　Solanandra Pers. = Galax Sims(保留属名)●☆

366855　Solanandra Pers. = Solenandria P. Beauv. ex Vent. ■☆

366856　Solanastrum Fabr. = Solanum L. ●■

366857　Solanastrum Heist. ex Fabr. = Solanum L. ●■

366858　Solandera Kuntze = Centella L. ■

366859　Solandera Kuntze = Solandra Sw.(保留属名)●☆

366860　Solandra L.(废弃属名) = Centella L. ■

366861　Solandra L.(废弃属名) = Solandra Sw.(保留属名)●☆

366862　Solandra Murray = Hibiscus L.(保留属名)●■

366863　Solandra Sw.(1787)(保留属名);金盏藤属(苏兰茄属,苏南花属);Chalice Vine,Chalice-vine,Trumpet Flower ●☆

366864　Solandra capensis L. = Centella capensis (L.) Domin ■☆

366865　Solandra decipens L.;迷金盏藤(迷苏兰茄)●☆

366866　Solandra grandiflora Sw.;大花金盏藤;Silver Cup Vine ●☆

366867　Solandra guttata D. Don;金喇叭●☆

366868　Solandra hartwegii N. E. Br.;哈氏金盏藤●☆

366869　Solandra lobata Murray = Hibiscus lobatus (Murray) Kuntze ■

366870　Solandra longiflora Tussac;长花金盏藤;Gabriel's Trumpet ●☆

366871　Solandra maxima (Sessé et Moc.) P. S. Green;金盏藤(大苏南,金杯花);Capa de Oro,Chalice Vine,Cup of Gold,Cup of Gold Vine,Cup-of-gold,Gold Cup,Golden-chalice Vine ●☆

366872　Solandra nitida Zucc.;金杯藤●☆

366873　Solandra ternata Cav. = Hibiscus sidiformis Baill. ●☆

366874　Solandra viridiflora Sims;绿花金盏藤●☆

366875　Solanecio (Sch. Bip.) Walp.(1846);盘花千里光属■●☆

366876　Solanecio (Sch. Bip.) Walp. = Senecio L. ■●

366877　Solanecio angulatus (Vahl) C. Jeffrey;棱角盘花千里光■☆

366878　Solanecio buchwaldii (O. Hoffm.) C. Jeffrey;布赫盘花千里光●☆

366879　Solanecio cydoniifolius (O. Hoffm. ex Engl.) C. Jeffrey;榅桲叶千里光■☆

366880　Solanecio cydoniifolius (O. Hoffm.) C. Jeffrey = Solanecio cydoniifolius (O. Hoffm. ex Engl.) C. Jeffrey ■☆

366881　Solanecio epidendricus (Mattf.) C. Jeffrey;柱瓣兰盘花千里光■☆

366882　Solanecio gigas (Vatke) C. Jeffrey;巨大盘花千里光■☆

366883　Solanecio goetzei (O. Hoffm.) C. Jeffrey;格兹盘花千里光■☆

366884　Solanecio gymnocarpus C. Jeffrey;裸果盘花千里光■☆

366885　Solanecio gynuroides C. Jeffrey;菊三七盘花千里光■☆

366886　Solanecio harennensis Mesfin;哈伦千里光■☆

366887　Solanecio kanzibiensis (Humbert et Staner) C. Jeffrey;刚果千里光■☆

366888　Solanecio mannii (Hook. f.) C. Jeffrey;曼氏盘花千里光■☆

366889　Solanecio mirabilis (Muschl.) C. Jeffrey;奇异盘花千里光■☆

366890　Solanecio nandensis (S. Moore) C. Jeffrey;南德千里光■☆

366891　Solanecio tuberosus (Sch. Bip. ex A. Rich.) C. Jeffrey;块状盘花千里光■☆

366892　Solanecio tuberosus (Sch. Bip. ex A. Rich.) C. Jeffrey var. pubescens Mesfin;短毛块状盘花千里光■☆

366893　Solanoa Greene = Asclepias L. ■

366894　Solanoana Kuntze = Solanoa Greene ■

366895　Solanocharis Bitter = Solanum L. ●■

366896　Solanoides Mill. = Rivina L. ●

366897　Solanoides Tourn. ex Moench = Rivina L. ●

366898　Solanoides Tourn. ex Moench(1794);茄商陆属●☆

366899　Solanopsis Bitter = Solanum L. ●■

366900　Solanopsis Börner = Battata Hill ●■

366901　Solanopsis Börner = Solanum L. + Lycopersicon Mill. ●■

366902　Solanum L.（1753）;茄属;Dragon Mallow,Eggplant,Egg-plant, Nightshade,Solanum ●■

366903　Solanum acanthocalyx Klotzsch = Solanum richardii Dunal ●☆

366904　Solanum acanthoideum Drège ex Dunal = Solanum macrocarpum L. ■☆

366905　Solanum acaule Bitter;无梗茄●☆

366906　Solanum aculeastrum Dunal;鬼茄; Apple of Sodom, Bitter Apple, Devil's Apple,Soda Apple,Sodaapple ●☆

366907　Solanum aculeastrum Dunal var. albifolium（C. H. Wright） Bitter;白花无梗茄●☆

366908　Solanum aculeastrum Dunal var. conraui（Dammer）Bitter = Solanum aculeastrum Dunal var. albifolium（C. H. Wright）Bitter ●☆

366909　Solanum aculeastrum Dunal var. exarmatum Bitter = Solanum aculeastrum Dunal var. albifolium（C. H. Wright）Bitter ●☆

366910　Solanum aculeatissimum Jacq.;喀西茄（刺茄,刺茄子,刺天 茄,大苦葛,狗茄子,谷雀蛋,金银茄,苦颠茄,苦茄子,苦天茄,天 茄果,添钱果,哑口子,印度茄）;Khasi Nightshade,Sodom apple ●■

366911　Solanum aculeatissimum Jacq. = Solanum capsicoides All. ■

366912　Solanum aculeatissimum Jacq. = Solanum ciliatum Lam. ●■

366913　Solanum aculeatissimum Jacq. = Solanum surattense Burm. f. ●■

366914　Solanum aculeatissimum Jacq. = Solanum virginianum L. ●■

366915　Solanum aculeatissimum Jacq. var. purpureum A. Chev. = Solanum capsicoides All. ■

366916　Solanum acutilobatum Dammer = Solanum richardii Dunal var. acutilobatum（Dammer）A. E. Gonc. ●☆

366917　Solanum adelense Delile;阿代尔茄■☆

366918　Solanum adoense Hochst. ex A. Br.;阿多茄■☆

366919　Solanum adoense Hochst. ex A. Br. var. schweinfurthii Engl. = Solanum hastifolium Hochst. ex Dunal ■☆

366920　Solanum aethiopicum L.;红茄; African Eggplant, Chinese Scarlet Egg-plant, Ethiopia Eggplant, Ethiopian Egg-plant, Tomato Egg-plant ●■

366921　Solanum africanum Dunal;非洲茄■☆

366922　Solanum afzelii Dunal = Solanum dasyphyllum Schumach. et Thonn. ■☆

366923　Solanum aggerum Dunal = Solanum africanum Dunal ■☆

366924　Solanum aggregatum Jacq.;聚生茄■☆

366925　Solanum aggregatum Jacq. = Solanum guineense L. ■☆

366926　Solanum alatum Moench = Solanum americanum Mill. ■

366927　Solanum alatum Moench = Solanum nigrum L. var. villosum L. ■

366928　Solanum alatum Moench = Solanum villosum Mill. subsp. miniatum（Bernh. ex Willd.）Edmonds ■

366929　Solanum alatum Moench = Solanum villosum Mill. ■

366930　Solanum albicaule Kotschy ex Dunal = Solanum forskalii Dunal ●☆

366931　Solanum albidum De Wild.;微白茄■☆

366932　Solanum albiflorum De Wild.;白花茄■☆

366933　Solanum albifolium C. H. Wright = Solanum aculeastrum Dunal var. albifolium（C. H. Wright）Bitter ●☆

366934　Solanum americanum Mill.;光果龙葵（白花菜,打卜子,耳坠 子,古纽子,古钮菜,扣子草,美洲茄,鸟疔草,七粒扣,少花龙葵,乌 疔草,五地茄,衣扣草,痣草）;American Black Nightshade,American Nightshade, Apple of Sodom, Night Blooming Nightshade, Popolokikania,Shiningfruit Dragon Mallow,Shiningfruit Nightshade ■

366935　Solanum americanum Mill. = Solanum ptycanthum Dunal ex DC. ■☆

366936　Solanum americanum Mill. subsp. nodiflorum（Jacq.）R. J. F. Hend. = Solanum americanum Mill. var. patulum（L.）Edmonds ■☆

366937　Solanum americanum Mill. var. nodiflorum（Jacq.）A. Gray = Solanum americanum Mill. ■

366938　Solanum americanum Mill. var. nodiflorum（Jacq.）Edmonds = Solanum americanum Mill. var. patulum（L.）Edmonds ■☆

366939　Solanum americanum Mill. var. patulum（L.）Edmonds;张开光 果龙葵■☆

366940　Solanum anguivi Lam.;安古茄●☆

366941　Solanum angulatum Ruiz et Pav.;棱角茄;Cow Pops ☆

366942　Solanum angustifolium Lanza = Solanum lanzae J. -P. Lebrun et Stork ■☆

366943　Solanum angustifolium Mill.;狭叶茄■

366944　Solanum angustispinosum De Wild.;狭刺茄☆

366945　Solanum anisantherum Dammer = Solanum somalense Franch. ■☆

366946　Solanum anodontum H. Lév. et Vaniot = Tubocapsicum anomalum（Franch. et Sav.）Makino ■

366947　Solanum anomalum Thonn.;非洲异茄; Children's Tomato, Children's Tomatoes ■☆

366948　Solanum arabicum Dunal = Solanum surattense Burm. f. ●■

366949　Solanum aranoideum Dammer = Solanum supinum Dunal ■☆

366950　Solanum armatum Forssk. = Solanum surattense Burm. f. ●■

366951　Solanum arrebenta Vell. = Solanum capsicoides All. ■

366952　Solanum arundo Mattei;阿伦多茄■☆

366953　Solanum asiae-mediae Pojark.;中亚茄■☆

366954　Solanum asperum Sieber ex Dunal;粗糙茄■☆

366955　Solanum atriplicifolium Desp. = Solanum nigrum L. ■

366956　Solanum aurantiacobaccatum De Wild.;橙黄茄■☆

366957　Solanum auriculatum Aiton;耳状茄●☆

366958　Solanum auriculatum Aiton = Solanum mauritianum Scop. ●☆

366959　Solanum auriculatum Mart. ex Dunal = Solanum auriculatum Aiton ●☆

366960　Solanum aviculare G. Forst. = Solanum laciniatum Aiton ●

366961　Solanum bagamojense Bitter et Dammer;巴加莫约茄☆

366962　Solanum baidoense Chiov. = Solanum tettense Klotzsch ■☆

366963　Solanum balbisii Dunal = Solanum sisymbriifolium Lam. ■

366964　Solanum bansoense Dammer = Solanum terminale Forssk. ●☆

366965　Solanum bansoense Dammer subsp. sanaganum Bitter = Solanum terminale Forssk. ●☆

366966　Solanum bansoense Dammer var. episporadotrichum Bitter = Solanum terminale Forssk. ●☆

366967　Solanum barbisetum Nees;刺苞茄;Spinebract Nightshade ●■

366968　Solanum barbisetum Nees var. griffithii Prain = Solanum griffithii （Prain）C. Y. Wu et S. C. Huang ●■

366969　Solanum batangense Dammer;巴坦加茄☆

366970　Solanum batoides D'Arcy et Rakot.;肉穗果茄●☆

366971　Solanum benadirense Chiov.;贝纳迪尔茄●☆

366972　Solanum benguelense Peyr. = Solanum delagoense Dunal ☆

366973　Solanum beniense De Wild.;贝尼茄☆

366974　Solanum bequaertii De Wild.;贝卡尔茄●☆

366975　Solanum betaceum Cav. = Cyphomandra betacea（Cav.） Sendtn. ●

366976　Solanum biflorum Lour.;双花龙葵（耳钩草,红头耳钩草）●

366977　Solanum biflorum Lour. = Lycianthes biflora（Lour.）Bitter ●

366978　Solanum biflorum Lour. var. glabrum Hatus. = Lycianthes biflora （Lour.）Bitter ●

366979　Solanum biflorum Lour. var. glabrum Koidz. ex Hatus. = Lycianthes boninensis Bitter ●

366980　Solanum biflorum Lour. var. kotoense Y. C. Liu et C. H. Ou = Lycianthes laevis（Dunal）Bitter var. kotoensis（L. C. Liu et C. H. Ou）T. Yamaz. ●☆

366981　Solanum biflorum Lour. var. kotoensis Y. C. Liu et C. H. Ou = Lycianthes biflora（Lour.）Bitter ●

366982　Solanum bifurcatum Hochst. ex A. Rich. = Solanum terminale Forssk. ●☆

366983　Solanum bifurcum Hochst. ex Dunal = Solanum terminale Forssk. ●☆

366984　Solanum bigeminatum Nees；二对茄■

366985　Solanum bilabiatum Dammer；双唇茄☆

366986　Solanum bipinnatipartitum Dunal = Solanum sisymbriifolium Lam. ■

366987　Solanum bodinieri H. Lév. = Solanum capsicoides All. ■

366988　Solanum boerhaviifolium Sendtn. ；黄细心叶茄☆

366989　Solanum bojeri Dunal = Solanum incanum L. ●■

366990　Solanum bojeri Dunal var. deckenii（Dammer）Bitter = Solanum incanum L. ●■

366991　Solanum bonariense L. ；博纳里茄☆

366992　Solanum boninense Nakai ex Tuyama = Lycianthes boninensis Bitter ●

366993　Solanum boreali-sinense C. Y. Wu et S. C. Huang = Solanum kitagawae Schönb. -Tem. ●■

366994　Solanum brieyi De Wild. ；布里茄☆

366995　Solanum buchwaldii Dammer；布赫茄☆

366996　Solanum bumeliifolium Dunal；榄叶茄☆

366997　Solanum burbankii Bitter = Solanum retroflexum Dunal ■☆

366998　Solanum burchellii Dunal；伯切尔茄■☆

366999　Solanum burtt-davyi Dunkley = Solanum richardii Dunal var. burtt-davyi（Dunkley）A. E. Gonc. ☆

367000　Solanum bussei Dammer；布瑟茄☆

367001　Solanum butaguense De Wild. = Solanum terminale Forssk. ●☆

367002　Solanum calleryanum Dunal = Lycianthes biflora（Lour.）Bitter ●

367003　Solanum callium C. T. White ex R. J. F. Hend. ；美茄☆

367004　Solanum campylacanthum Dunal；弯刺茄☆

367005　Solanum capense L. ；好望角茄●☆

367006　Solanum capsica Link；辣茄■☆

367007　Solanum capsicastrum Link ex Schauer = Solanum diphyllum L. ●

367008　Solanum capsicastrum Link ex Schauer = Solanum pseudocapsicum L. var. diflorum（Vell.）Bitter ●

367009　Solanum capsicastrum Link. ex Schauer = Solanum diflorum Vell. ●

367010　Solanum capsiciforme（Domin）G. T. S. Baylis；辣椒茄■☆

367011　Solanum capsicoides All. ；牛茄子（刺茄）■

367012　Solanum cardiophyllum Dunal；心叶茄；Heartleaf Horsenettle ☆

367013　Solanum carense Dunal；阿拉伯茄☆

367014　Solanum carolinense L. ；北美水茄（卡罗来纳茄，卡罗林茄，美洲野茄）；Bull Nettle，Carolina Horse-nettle，Carolina Nettle，Carolina Nightshade，Devil's Potato，Devil's Tomato，Horse Nettle，Horse-nettle，Nettles Bull，Sand Brier ■

367015　Solanum carolinense L. f. albiflorum（Kuntze）Benke；白花卡罗来纳水茄（白花北美水茄）■☆

367016　Solanum carolinense L. f. albiflorum（Kuntze）Benke = Solanum carolinense L. ■

367017　Solanum carterianum Rock = Solanum mauritianum Scop. ●☆

367018　Solanum cathayanum C. Y. Wu et S. C. Huang = Solanum lyratum Thunb. ■

367019　Solanum catombelense Peyr. ；卡顿茄☆

367020　Solanum caulorhizum Dunal = Lycianthes lysimachioides（Wall.）Bitter var. caulorrhiza（Dunal）Bitter ■

367021　Solanum cavaleriei H. Lév. et Vaniot = Solanum aculeatissimum Jacq. ●■

367022　Solanum cerasiferum Dunal；樱桃茄☆

367023　Solanum cerasiferum Dunal subsp. crepinii（Van Heurck）Bitter；克雷潘茄☆

367024　Solanum cerasiferum Dunal subsp. duchartrei（Heckel）Bitter = Solanum cerasiferum Dunal subsp. crepinii（Van Heurck）Bitter ☆

367025　Solanum cerasiferum Dunal var. garuense Bitter = Solanum cerasiferum Dunal ☆

367026　Solanum chacoense Bitter；卡茄■☆

367027　Solanum chariense A. Chev. ；沙里茄☆

367028　Solanum chenopodioides Lam. ；浅裂茄；Tall Nightshade ●☆

367029　Solanum chinense Dunal = Solanum violaceum Ortega ●

367030　Solanum chiovendae Lanza；基奥文达茄☆

367031　Solanum chloranthum DC. = Solanum viarum Dunal ■

367032　Solanum chlorocarpum（Spenn.）Schur = Solanum humile Bernh. ex Willd. ■

367033　Solanum chondropetalum Dammer = Solanum tettense Klotzsch var. renschii（Vatke）A. E. Gonc. ■☆

367034　Solanum chousboe var. merrillianum（Liou）C. Y. Wu. et S. C. Huang = Solanum merrillianum Liou ●

367035　Solanum chrysotrichum Schltdl. ；多裂水茄；Manylobed Water Nightshade ●

367036　Solanum cicatricosum Chiov. = Solanum benadirense Chiov. ●☆

367037　Solanum ciliare Willd. = Solanum capsicoides All. ■

367038　Solanum ciliatum Lam. = Solanum aculeatissimum Jacq. ●■

367039　Solanum ciliatum Lam. = Solanum capsicoides All. ■

367040　Solanum cinereum R. Br. ；灰色茄；Narrawa Burr ☆

367041　Solanum cirsioides A. Chev. ；蓟茄☆

367042　Solanum citrullifolium A. Br. ；西瓜叶茄■☆

367043　Solanum clerodendroides Hutch. et Dalziel = Solanum madagascariense Dammer ●☆

367044　Solanum coagulans Forssk. ；非洲野茄；Wild Nightshade ●■☆

367045　Solanum coagulans Forssk. = Solanum incanum L. ●■

367046　Solanum coagulans Forssk. = Solanum thruppii C. H. Wright ■☆

367047　Solanum commersonii Dunal ex Poir. ；康氏茄（康密索茄，克默森茄）■☆

367048　Solanum comorense Dammer = Solanum terminale Forssk. ●☆

367049　Solanum congense Link；刚果茄●☆

367050　Solanum congestiflorum Dunal；密花茄■☆

367051　Solanum conraui Dammer = Solanum aculeastrum Dunal var. albifolium（C. H. Wright）Bitter ●☆

367052　Solanum cordatum Forssk. ；心形茄●☆

367053　Solanum cornigerum André = Solanum mammosum L. ■

367054　Solanum cornutum Lam. = Solanum angustifolium Mill. ■

367055　Solanum cornutum Lam. = Solanum rostratum Dunal ■

367056　Solanum crassifolium Lam. = Solanum africanum Dunal ■☆

367057　Solanum crassifolium Ortega = Cyphomandra betacea（Cav.）Sendtn. ●

367058　Solanum crassifolium Salisb. = Solanum macrocarpum L. ■☆

367059　Solanum crassipetalum Wall. = Solanum crassipetalum Wight ex C. B. Clarke ■☆

367060　Solanum crassipetalum Wight ex C. B. Clarke;厚瓣茄■☆

367061　Solanum crepinii Van Heurck = Solanum cerasiferum Dunal subsp. crepinii (Van Heurck) Bitter ☆

367062　Solanum crispum Ruiz et Pav.;智利藤茄(皱波茄);Chilean Potato Tree ●☆

367063　Solanum crispum Ruiz et Pav. 'Glasnevin';秋花智利藤茄●☆

367064　Solanum cufodontii Lanza;卡佛茄 ☆

367065　Solanum cumingii Dunal. = Solanum undatum Lam. ■

367066　Solanum cyaneopurpureum De Wild.;蓝紫茄☆

367067　Solanum cymbalariifolium Chiov.;舟叶茄☆

367068　Solanum cynanchoides Chiov. = Solanum hastifolium Hochst. ex Dunal ■☆

367069　Solanum damarense Bitter;达马尔茄☆

367070　Solanum darassumense Dammer = Solanum cordatum Forssk. ●☆

367071　Solanum darbandense A. Chev.;达尔班德茄☆

367072　Solanum dasyphyllum Schumach. et Thonn.;毛叶茄(粗毛叶茄)■☆

367073　Solanum dasyphyllum Schumach. et Thonn. var. decaisneanum Bitter = Solanum dasyphyllum Schumach. et Thonn. ■☆

367074　Solanum dasyphyllum Schumach. et Thonn. var. inerme Bitter = Solanum dasyphyllum Schumach. et Thonn. ■☆

367075　Solanum dasyphyllum Schumach. et Thonn. var. semiglabrum (C. H. Wright) Bitter = Solanum dasyphyllum Schumach. et Thonn. ■☆

367076　Solanum dasypus Drège ex Dunal = Solanum guineense L. ■☆

367077　Solanum dasytrichum Bitter;密毛茄☆

367078　Solanum daturifolium Dunal = Solanum torvum Sw. ●

367079　Solanum debilissimum Merr. = Lycianthes lysimachioides (Wall.) Bitter var. caulorrhiza (Dunal) Bitter ■

367080　Solanum decaisneanum (Bitter) Schimp. = Solanum dasyphyllum Schumach. et Thonn. ■☆

367081　Solanum decemdentatum Roxb. = Lycianthes biflora (Lour.) Bitter ●

367082　Solanum decemfidum Nees = Lycianthes biflora (Lour.) Bitter ●

367083　Solanum decipiens Opiz;易混茄■☆

367084　Solanum deckenii Dammer = Solanum incanum L. ●■

367085　Solanum decurrens Balb. = Solanum sisymbriifolium Lam. ■

367086　Solanum decurrens Wall. ex Dunal;下延茄■☆

367087　Solanum deflexicarpum C. Y. Wu et S. C. Huang;苦刺茄(苦刺,苦果);Bitterspine Nightshade,Deflective-fruited Nightshade ●

367088　Solanum delagoense Dunal;迪拉果茄☆

367089　Solanum delagoense Dunal subsp. omahekense (Dammer) Bitter = Solanum delagoense Dunal ☆

367090　Solanum delagoense Dunal var. benguelense (Peyr.) Bitter = Solanum delagoense Dunal ☆

367091　Solanum delpierrei De Wild.;戴尔皮埃尔茄☆

367092　Solanum demissum Lindl.;垂茄■☆

367093　Solanum dennekense Dammer;代呐科茄■☆

367094　Solanum denticulatum Blume;细齿茄■

367095　Solanum depilatum Bitter = Solanum americanum Mill. ■

367096　Solanum depilatum Kitag. = Solanum kitagawae Schönb.-Tem. ●■

367097　Solanum depressum Bitter = Solanum thruppii C. H. Wright ■☆

367098　Solanum dewildemanianum Robyns;德怀尔德曼茄■☆

367099　Solanum dichroanthum Dammer;二色花茄■☆

367100　Solanum diffusum Roxb. = Solanum surattense Burm. f. ●■

367101　Solanum diflorum Vell. = Solanum pseudocapsicum L. var. diflorum (Vell.) Bitter ●

367102　Solanum dimidiatum Raf.;西部马茄;Western Horse Nettle ■☆

367103　Solanum dimorphum Matsum. = Solanum macrocarpum L. ■

367104　Solanum dinklagei Dammer;丁克茄☆

367105　Solanum dinteri Bitter;丁特茄☆

367106　Solanum diphyllum Forssk. = Solanum pseudocapsicum L. ●

367107　Solanum diphyllum L.;玛瑙珠(黄果龙葵);Twoleaf Nightshade,Two-leaf Nightshade ●

367108　Solanum diplacanthum Dammer = Solanum arundo Mattei ■☆

367109　Solanum diplocincinnum Dammer = Solanum tettense Klotzsch var. renschii (Vatke) A. E. Gonc. ■☆

367110　Solanum diporum ?;假银叶茄;False Jerusalem Cherry ■☆

367111　Solanum distichum Schumach. et Thonn.;二列茄■☆

367112　Solanum distichum Schumach. et Thonn. var. halophilum (Pax) Cufod. = Solanum distichum Schumach. et Thonn. ■☆

367113　Solanum distichum Schumach. et Thonn. var. modicearmatum Bitter = Solanum distichum Schumach. et Thonn. ■☆

367114　Solanum domesticum A. Chev.;土著茄☆

367115　Solanum donianum Walp.;毛蕊花茄;Mullein Nightshade ☆

367116　Solanum douglasii Dunal;道氏茄■☆

367117　Solanum dregei Dunal = Solanum capense L. ●☆

367118　Solanum dubium Fresen. = Solanum thruppii C. H. Wright ■☆

367119　Solanum duchartrei Heckel = Solanum cerasiferum Dunal subsp. crepinii (Van Heurck) Bitter ☆

367120　Solanum dulcamara L.;欧白英(白草,茯,谷菜,鬼目,苦茄,来甘,六甲草,毛和尚草,排风子,千年不烂心,蜀羊泉,天抛子,天泡草);Bitter Nightshade, Bitter Night-shade, Bitter Sweet, Bittersweet, Bittersweet Nightshade, Blue Bindweed, Climbing Nightshade,Cron-reish, Deadly Nightshade, Devil's Cherries, Devil's Cherry,Dogwood, Europe Nightshade, European Bitter Night Shade, European Bittersweet, Fellon-wood, Fellon-wort, Felonewood, Fool's Cap, Granny's Nightcap, Halfwood, Lady's Umbrella, Mad Dog's Berries, Mortal, Nightshade, Poison Berry, Poison Flower, Poisoning Berries, Poisonous Tea Plant, Purple Nightshade, Robin-run-the-hedge, Shady Night, Snake's Food, Snake's Meat, Snake's Poison-food,Terrydevil, Terryddidle, Tether Devil, Tether-devil, Wild Potato Flower, Witchelower, Wood Nightshade, Woodbine, Woody Nightshade ●■

367121　Solanum dulcamara L. f. albiflorum House = Solanum dulcamara L. ●■

367122　Solanum dulcamara L. var. canescens Farw. = Solanum dulcamara L. ●■

367123　Solanum dulcamara L. var. chinense Dunal = Solanum lyratum Thunb. ■

367124　Solanum dulcamara L. var. heterophyllum Makino = Solanum japonense Nakai ■

367125　Solanum dulcamara L. var. indivisum Boiss. = Solanum dulcamara L. ●■

367126　Solanum dulcamara L. var. lyratum (Thunb.) Bonati = Solanum lyratum Thunb. ■

367127　Solanum dulcamara L. var. marinum ?;海滨欧白英;Sea Bittersweet ■☆

367128　Solanum dulcamara L. var. pubescens Blume = Solanum lyratum Thunb. ■

367129　Solanum dulcamara L. var. pubescens Roem. et Schult. = Solanum dulcamara L. ●■

367130　Solanum dulcamara L. var. villosissimum Desv.;攀缘欧白英;

Climbing Nightshade ●■☆

367131　Solanum dulcamara L. var. villosissimum Desv. = Solanum dulcamara L. ●■

367132　Solanum dulcamarum St. -Lag. = Solanum dulcamara L. ●■

367133　Solanum dulmacara L. var. heterophyllum Makino;异叶欧白英☆

367134　Solanum dulmacara L. var. heterophyllum Makino = Solanum japonense Nakai ■

367135　Solanum dunnianum H. Lév. = Solanum pseudocapsicum L. var. diflorum (Vell.) Bitter ●

367136　Solanum duplosinuatum Klotzsch = Solanum dasyphyllum Schumach. et Thonn. ■☆

367137　Solanum duplosinuatum Klotzsch var. semiglabrum C. H. Wright = Solanum dasyphyllum Schumach. et Thonn. ■☆

367138　Solanum eickii Dammer;艾克茄☆

367139　Solanum elaeagnifolium Cav. ;银叶茄;Bull Nettle, Jerusalem Cherry, Nettles Bull, Silver Horse Nettle, Silverleaf Nightshade, Tropillo, White Horse Nettle, White Horsenettle ■☆

367140　Solanum elaeagnifolium Cav. f. albiflorum Cockerell;白花银叶茄■☆

367141　Solanum ellenbeckii Dammer = Solanum thruppii C. H. Wright ■☆

367142　Solanum ellenbeckii Dammer var. oligopilum Bitter = Solanum thruppii C. H. Wright ■☆

367143　Solanum ellipticum R. Br. ;沙地茄;Desert Raisin ☆

367144　Solanum emarginatum L. f. ex Engl. = Solanum marginatum L. f. ●☆

367145　Solanum endlichii Dammer;恩德茄☆

367146　Solanum erianthum D. Don;假烟叶树(臭鹏木,臭枇杷,臭屎花,臭烟,大发散,大黄叶,大毛叶,大王叶,黄带来仔,黄毛茄,假烟叶,酱权树,老公须,毛叶树,绵毛苗,暖叶根,茄树,三权树,三姐妹,三姊妹,山烟,山烟草,山烟头,生毛将军,石烟,树茄,天蓬草,土臭烟,土烟,土烟叶,洗碗叶,袖扭果,野茄树,野烟叶);Mountain Tobacco, Mullein Nightshade, Potato Tree, Tobacco Tree ●

367147　Solanum erianthum D. Don var. adulterinum (Ham. ex G. Don) Baker et Simmonds = Solanum erianthum D. Don ●

367148　Solanum erythracanthum Bojer ex Dunal;红花茄●☆

367149　Solanum esculentum Dunal = Solanum melongena L. ●■

367150　Solanum esculentum Dunal var. inerme ? = Solanum melongena L. ●■

367151　Solanum exasperatum Drège ex Dunal = Solanum africanum Dunal ■☆

367152　Solanum farinosum Wall. ex Roxb. = Solanum giganteum Jacq. ■

367153　Solanum fastigiatum Willd. ;帚状茄☆

367154　Solanum ferox L. = Solanum lasiocarpum Dunal ■●

367155　Solanum ferrugineum Jacq. = Solanum torvum Sw. ●

367156　Solanum ficifolium Ortega = Solanum torvum Sw. ●

367157　Solanum ficifolium Pav. ex Dunal;榕叶茄;Small Red Trubba, Trowie Girse, Trowie Gliv ☆

367158　Solanum filicaule Dammer;线茎茄☆

367159　Solanum flagelliferum Baker = Solanum erythracanthum Bojer ex Dunal ●☆

367160　Solanum flamignii De Wild. ;弗拉米尼茄☆

367161　Solanum flavum Kit. = Solanum humile Bernh. ex Willd. ■

367162　Solanum floccosistellatum Bitter = Solanum incanum L. ●■

367163　Solanum florulentum Bitter;多花茄☆

367164　Solanum fontanesianum Schrank;丰塔茄■☆

367165　Solanum forskalii Dunal;福氏茄●☆

367166　Solanum forsythii Dammer = Solanum erythracanthum Bojer ex Dunal ●☆

367167　Solanum francoisii Dammer ex Dinter = Solanum delagoense Dunal ☆

367168　Solanum furcatum Dunal;叉茄;Forked Nightshade ☆

367169　Solanum ganchouenense H. Lév. = Solanum americanum Mill. ■

367170　Solanum gayanum F. Phil. ;加亚茄;Chilean Nightshade ■☆

367171　Solanum geminifolium Thonn. = Solanum aethiopicum L. ●■

367172　Solanum geniculatum Drège ex Dunal = Solanum africanum Dunal ■☆

367173　Solanum giganteum Jacq. ;大茄■

367174　Solanum gillettii Hutch. et E. A. Bruce = Solanum dennekense Dammer ☆

367175　Solanum gilo Raddi = Solanum aethiopicum L. ●■

367176　Solanum giorgii De Wild. ;乔治茄☆

367177　Solanum glabratum Dunal = Solanum sepicula Dunal ■☆

367178　Solanum glaucophyllum Desf. ;粉绿叶茄;Waxyleaf Nightshade ●☆

367179　Solanum glaucum Bertol. = Solanum glaucophyllum Desf. ●☆

367180　Solanum globiferum Dunal = Solanum mammosum L. ■

367181　Solanum goetzei Dammer;格兹茄☆

367182　Solanum gracile Dunal = Solanum chenopodioides Lam. ■☆

367183　Solanum gracile Otto ex Baxter;纤细茄;Graceful Nightshade ■☆

367184　Solanum gracilescens Nakai = Solanum japonense Nakai ■

367185　Solanum gracilescens Nakai ex Makino = Solanum japonense Nakai ■

367186　Solanum gracilipes Decne = Solanum cordatum Forssk. ●☆

367187　Solanum gracilipes Decne. ;细柄茄■

367188　Solanum grandiflorum Ruiz et Pav. = Solanum wrightii Benth. ●■

367189　Solanum grewioides Lanza;扁担杆茄☆

367190　Solanum griffithii (Prain) C. Y. Wu et S. C. Huang;膜萼茄;Griffith Nightshade ●■

367191　Solanum grossedentatum A. Rich. = Solanum memphiticum J. F. Gmel. var. abyssinicum (Dunal) Cufod. ●☆

367192　Solanum grotei Dammer;格罗特茄☆

367193　Solanum guineense (L.) Lam. = Solanum guineense Lam. ■☆

367194　Solanum guineense (L.) Lam. = Solanum nigrum L. ■

367195　Solanum guineense L. ;几内亚茄■☆

367196　Solanum guineense Lam. = Solanum guineense L. ■☆

367197　Solanum guineense Lam. = Solanum intrusum J. Soria ■☆

367198　Solanum hadaq Deflers = Solanum cordatum Forssk. ●☆

367199　Solanum hainanense Hance = Solanum procumbens Lour. ●

367200　Solanum halophilum Pax = Solanum distichum Schumach. et Thonn. ■☆

367201　Solanum hastifolium Hochst. ex Dunal;戟叶茄■☆

367202　Solanum herculeum Bohs = Triguera ambrosiaca Cav. ■☆

367203　Solanum hermannioides Schinz = Solanum pseudocapsicum L. ●

367204　Solanum heterandrum Pursh = Solanum rostratum Dunal ■

367205　Solanum heterodoxum Dun. ;瓜叶茄;Melon-leaved Nightshade ■☆

367206　Solanum heudelotii Dunal = Solanum forskalii Dunal ●☆

367207　Solanum heudesii H. Lév. = Solanum angustifolium Mill. ■

367208　Solanum hidetaroi Masam. ;台白英(玉山茄);Taiwan Nightshade ■

367209　Solanum hidetaroi Masam. = Solanum pittosporifolium Hemsl. ●

367210　Solanum hierochunticum Dun. = Solanum incanum L. ●■

367211　Solanum hildebrandtii A. Br. et Bouché;希尔德茄☆

367212　Solanum hindsianum Benth. ;海因兹茄;Blue Solanum Shrub ●☆

367213　Solanum hirsuticaule Werderm. ;粗毛茎茄☆

367214　Solanum hirsutum Dunal var. abyssinicum Dunal = Solanum memphiticum J. F. Gmel. var. abyssinicum (Dunal) Cufod. ●☆

367215　Solanum hirtulum Steud. ex A. Rich. ;多毛茄☆

367216　Solanum holstii Dammer ex Engl. = Solanum incanum L. ●■

367217　Solanum holtzii Dammer = Solanum incanum L. ●■

367218　Solanum homblei De Wild. ;洪布勒茄☆

367219　Solanum humblotii Dammer;安布洛茄●☆

367220　Solanum humile Bernh. ex Willd. = Solanum nigrum L. var. humile (Bernh.) C. Y. Wu et S. C. Huang ■

367221　Solanum humile Bernh. ex Willd. = Solanum villosum Mill. ■

367222　Solanum humistratum R. Shaw ?;平卧茄■☆

367223　Solanum hybridum Jacq. ;杂种茄☆

367224　Solanum hypoleucum (Standl.) Morton;白背茄■☆

367225　Solanum hypomalacophyllum Bitter ex Pittier;软叶背茄■☆

367226　Solanum imamense Dunal;伊玛目茄●☆

367227　Solanum imamense Dunal var. grandiflora Dunal = Solanum imamense Dunal ●☆

367228　Solanum imerinense Bitter = Solanum americanum Mill. ■

367229　Solanum immane Hance = Solanum lasiocarpum Dunal ■●

367230　Solanum immane Hance ex Walp. = Solanum lasiocarpum Dunal ■●

367231　Solanum inaequilaterale Merr. = Solanum macaonense Dunal ●■

367232　Solanum inaequiradians Bitter = Solanum lamprocarpum Bitter ■☆

367233　Solanum incanum L. ;黄水茄(灰茄);Apple of Sodom, Bitter Tomato, Gray Nightshade, Hoary Nightshade, Jericho Potato, Palestine Nightshade ●■

367234　Solanum incanum L. = Solanum coagulans Forssk. ●■☆

367235　Solanum incanum L. = Solanum undatum Lam. ■

367236　Solanum incanum L. subsp. horridescens Bitter;多刺黄水茄●☆

367237　Solanum incanum L. subsp. schoanum Bitter;绍氏黄水茄●☆

367238　Solanum incanum L. var. unguiculatum (A. Rich.) Bitter = Solanum incanum L. ●■

367239　Solanum incertum Dunal;可疑茄■☆

367240　Solanum inconstans C. H. Wright = Solanum terminale Forssk. ●☆

367241　Solanum indicum L. ;印度茄(巴山虎,刺茄,刺天茄,黄面仔,黄水茄,鸡刺子,假茄子,金扣,金扣钮,金钮刺,金钮头,苦果,勒矮瓜,满天星,天茄子,天星子,五角颠茄,五宅茄,细颠茄,细黄茄,细纽扣,小颠茄,小闹杨,紫花茄);India Nightshade, Indian Nightshade ●

367242　Solanum indicum L. = Solanum violaceum Ortega ●

367243　Solanum indicum L. f. album C. Y. Wu et S. C. Huang;白花刺天茄;Whiteflower Indian Nightshade ●

367244　Solanum indicum L. subsp. distichum (Schumach. et Thonn.) Bitter = Solanum distichum Schumach. et Thonn. ■☆

367245　Solanum indicum L. subsp. rohrii (C. H. Wright) Bitter = Solanum rohrii C. H. Wright ■☆

367246　Solanum indicum L. var. grandemunitum Bitter = Solanum distichum Schumach. et Thonn. ■☆

367247　Solanum indicum L. var. halophilum (Pax) Bitter = Solanum distichum Schumach. et Thonn. ■☆

367248　Solanum indicum L. var. modicearmatum Bitter = Solanum distichum Schumach. et Thonn. ■☆

367249　Solanum indicum L. var. recurvatum C. Y. Wu et S. C. Huang;弯柄刺天茄;Curved Indian Nightshade ●

367250　Solanum indicum L. var. recurvatum C. Y. Wu et S. C. Huang = Solanum violaceum Ortega ●

367251　Solanum indicum L. var. uollense Chiov. = Solanum uollense (Chiov.) Pic. Serm. ●☆

367252　Solanum insanum L. = Solanum melongena L. ●■

367253　Solanum insidiosum Mart. ;坐生茄■☆

367254　Solanum integrifolium Poir. = Solanum aethiopicum L. ●■

367255　Solanum integrifolium Poir. = Solanum melongena L. var. incanum (L.) Kuntze ■☆

367256　Solanum intrusum J. Soria;内嵌茄;Garden Huckleberry, Wonderberry ■☆

367257　Solanum iodes Dammer = Solanum incanum L. ●■

367258　Solanum ivohibe D'Arcy et Rakot. ;伊武希贝茄●☆

367259　Solanum jacquini Miq. ;杰昆茄■☆

367260　Solanum jacquini Willd. = Solanum surattense Burm. f. ●■

367261　Solanum jaegeri Dammer;耶格茄☆

367262　Solanum japonense Nakai;野海茄(毛九里光,山茄);Jamiaca Nightshade, Japan Nightshade, Japanese Nightshade ■

367263　Solanum japonense Nakai f. xanthocarpum H. Hara;黄果野海茄■☆

367264　Solanum japonense Nakai var. takaoyamense (Makino) H. Hara = Solanum japonense Nakai ■

367265　Solanum jasminoides Paxton;素馨叶白英(茉莉状茄,素馨茄,野海茄);Jasmine Nightshade, Potato Vine ●

367266　Solanum jasminoides Paxton 'Album';白花素馨茄●☆

367267　Solanum jemense Bitter = Solanum carense Dunal ☆

367268　Solanum jubae Bitter;朱巴茄☆

367269　Solanum juciri Mart. ex Sendtn. ;朱氏茄(朱塞尔茄)■☆

367270　Solanum judaicum Besser;犹太茄■☆

367271　Solanum kagehense Dammer;科格茄☆

367272　Solanum kandtii Dammer;坎德茄☆

367273　Solanum kayamae T. Yamaz. ;加山茄■☆

367274　Solanum keniense Standl. = Solanum dewildemanianum Robyns ■☆

367275　Solanum keniense Turrill;肯尼亚茄☆

367276　Solanum kerrii Bonati = Solanum seaforthianum Andréws ●■

367277　Solanum khasianum C. B. Clarke = Solanum aculeatissimum Jacq. ●■

367278　Solanum khasianum C. B. Clarke = Solanum myriacanthum Dunal ■☆

367279　Solanum khasianum C. B. Clarke var. chatterjeeanum Sengupta = Solanum viarum Dunal ■

367280　Solanum kibweziense Dammer = Solanum tettense Klotzsch var. renschii (Vatke) A. E. Gonc. ■☆

367281　Solanum kieseritzkii C. A. Mey. ;基氏茄■☆

367282　Solanum kitagawae Schönb. -Tem. ;光白英;N. China Nightshade, North-China Bittersweet, North-China Nightshade ●■

367283　Solanum koniortodes Dammer = Solanum tettense Klotzsch var. renschii (Vatke) A. E. Gonc. ■☆

367284　Solanum kurzii Brace ex Prain;库鲁茄■☆

367285　Solanum kwebense N. E. Br. ex C. H. Wright = Solanum tettense Klotzsch var. renschii (Vatke) A. E. Gonc. ■☆

367286　Solanum kwebense N. E. Br. ex C. H. Wright var.

chondropetalum（Dammer）Bitter = Solanum tettense Klotzsch var. renschii（Vatke）A. E. Gonc. ■☆

367287   Solanum kwebense N. E. Br. ex C. H. Wright var. luderitzii（Schinz）Bitter = Solanum tettense Klotzsch var. renschii（Vatke）A. E. Gonc. ■☆

367288   Solanum lachneion Dammer = Solanum incanum L. ●■

367289   Solanum laciniatum Aiton；澳洲茄（大澳洲茄，锯边茄，裂叶茄，条裂叶茄）；Australia Nightshade, Australian Nightshade, Kangaroo Apple, Kangaroo-apple, Large Kangaroo Apple, Large Poroporo, New Zealand Nightshade, Porcupine Plant, Poro Poro, Poroporo ●

367290   Solanum laeve Dunal = Lycianthes laevis（Dunal）Bitter ■

367291   Solanum lamprocarpum Bitter；亮果茄■☆

367292   Solanum lanceolatum Cav. ；橘果茄；Orangeberry Nightshade ●☆

367293   Solanum lanzae J. -P. Lebrun et Stork；兰扎茄■☆

367294   Solanum largifolium C. T. White = Solanum torvum Sw. ●

367295   Solanum lasiocarpum Dunal；毛茄（大样颠茄，大叶毛刺茄，毛果茄，毛茄树，羊不食）；Hairy Nightshade, Hairyfruited Nightshade ■●

367296   Solanum lasiocarpum Dunal var. velutinum Dunal = Solanum lasiocarpum Dunal ■●

367297   Solanum lasiophyllum Humb. et Bonpl. ex Dunal；绵毛叶茄■☆

367298   Solanum lasiostylum（L. C. Liu et C. H. Ou）Tawada = Solanum macaonense Dunal ●■

367299   Solanum laurentii Dammer；洛朗茄☆

367300   Solanum lavae Dunal；光茄☆

367301   Solanum laxum Spreng. ；疏松茄☆

367302   Solanum lichtensteinii Willd. ；利希滕茄☆

367303   Solanum lidii Sunding；利德茄☆

367304   Solanum lignosum Werderm. = Solanum ulugurense Holub ■☆

367305   Solanum linnaeanum Hepper et P. M. Jaeger；林奈茄；Sodom Apple ■☆

367306   Solanum litorale Raab；滨海茄☆

367307   Solanum longipedicellatum De Wild. = Solanum dewildemanianum Robyns ■☆

367308   Solanum longistamineum Dammer = Solanum hastifolium Hochst. ex Dunal ■☆

367309   Solanum luederitzii Schinz = Solanum tettense Klotzsch var. renschii（Vatke）A. E. Gonc. ■☆

367310   Solanum luteovirescens J. F. Gmel. = Solanum humile Bernh. ex Willd. ■

367311   Solanum luteum Mill. ；黄茄；Yellow Nightshade, Yellow-berried Nightshade ■

367312   Solanum luteum Mill. = Solanum nigrum L. var. villosum L. ■

367313   Solanum luteum Mill. = Solanum villosum Mill. ■

367314   Solanum luteum Mill. subsp. alatum（Moench）Dostal = Solanum nigrum L. var. villosum L. ■

367315   Solanum luteum Mill. subsp. alatum（Moench）Dostál = Solanum villosum Mill. subsp. miniatum（Bernh. ex Willd.）Edmonds ■

367316   Solanum luteum Mill. subsp. alatum Mill. ；翅黄茄■☆

367317   Solanum luzoniense Merr. ；吕宋茄；Luzon Nightshade ●

367318   Solanum lycopersicoides Dunal；番茄状茄☆

367319   Solanum lycopersicum L. = Lycopersicon esculentum Mill. ■

367320   Solanum lycopersicum L. var. cerasiforme（Dunal）D. M. Spooner, G. J. Anderson et R. K. Jansen = Lycopersicon esculentum

Mill. var. cerasiforme（Dunal）A. Gray ■

367321   Solanum lyratifolium Dammer = Solanum supinum Dunal ■☆

367322   Solanum lyratum Thunb. ；白英（白草，白毛藤，白幕，北风藤，耳坠菜，苻，谷菜，鬼目，鬼目菜，鬼目草，和尚头草，红道士，红麦禾，胡毛藤，葫芦草，假辣椒，金线绿毛龟，金线绿毛龟草，耒甘，龙毛龟，曼茄，蔓茄，毛风藤，毛和尚，毛和尚草，毛老人，毛母猪藤，毛千里光，毛藤，毛藤果，毛秀才，毛燕仔，钮仔黄，排风，排风藤，排风子，千年不烂心，山甜菜，生毛鸡屎藤，生毛梢，蜀羊泉，酸尖菜，天灯笼，天抛子，舔菜，土防风，望冬红，小儿拳，羊仔耳，野猫耳朵）；Bittersweet, China Nightshade, Chinese Nightshade, Garden Tomato, Nightshade ■

367323   Solanum lyratum Thunb. f. leucanthum（Nakai）Sugim. ；白花白英■

367324   Solanum lyratum Thunb. f. leucanthum（Nakai）Sugim. = Solanum lyratum Thunb. var. leucanthemum Nakai ■

367325   Solanum lyratum Thunb. f. purpuratum Konta et Katsuyama；紫花白英■☆

367326   Solanum lyratum Thunb. f. xanthocarpum（Makino）H. Hara；黄果白英■

367327   Solanum lyratum Thunb. f. xanthocarpum（Makino）H. Hara = Solanum lyratum Thunb. var. xanthocarpium Makino ■

367328   Solanum lyratum Thunb. var. filamentosum Hayashi；丝状白英■☆

367329   Solanum lyratum Thunb. var. leucanthemum Nakai = Solanum lyratum Thunb. f. leucanthum（Nakai）Sugim. ■

367330   Solanum lyratum Thunb. var. maruyamanum Honda；丸山白英■☆

367331   Solanum lyratum Thunb. var. xanthocarpium Makino = Solanum lyratum Thunb. f. xanthocarpum（Makino）H. Hara ■

367332   Solanum lysimachioides Wall. ；蔓茄■

367333   Solanum lysimachioides Wall. = Lycianthes lysimachioides（Wall.）Bitter ■

367334   Solanum macaonense Dunal；山茄（大丁茄子，毛万桃花，毛柱万桃花）；Hairy-styled Water Nightshade, Macao Nightshade ●■

367335   Solanum macilentum A. Rich. = Solanum forskalii Dunal ●☆

367336   Solanum macinae A. Chev. = Solanum aculeatissimum Jacq. ●■

367337   Solanum macowanii Fourc. = Solanum capsicoides All. ■

367338   Solanum macracanthum A. Rich. ；大刺茄☆

367339   Solanum macranthum Carrière = Solanum wrightii Benth. ●■

367340   Solanum macranthum Dunal = Solanum megacarpum Koidz. ■☆

367341   Solanum macranthum Dunal = Solanum wrightii Benth. ●■

367342   Solanum macrocarpon L. = Solanum macrocarpum L. ■☆

367343   Solanum macrocarpon L. var. calvum Bitter = Solanum macrocarpon L. ■☆

367344   Solanum macrocarpum L. ；大果茄；African Eggplant ■☆

367345   Solanum macrodon Wall. = Lycianthes macrodon（Wall.）Bitter ●

367346   Solanum macrodon Wall. ex Nees = Lycianthes macrodon（Wall.）Bitter ●

367347   Solanum macrodon Wall. ex Nees var. lysimachioides（Wall.）C. B. Clarke = Lycianthes lysimachioides（Wall.）Bitter ■

367348   Solanum macrodon Wall. var. lysimachioides（Wall.）C. B. Clarke = Lycianthes lysimachioides（Wall.）Bitter ■

367349   Solanum macrodon Wall. var. lysimachioides（Wall.）C. B. Clarke = Lycianthes biflora（Lour.）Bitter subsp. lysimachioides（Wall.）Debeaux ■

367350   Solanum macrosepalum Dammer = Solanum incanum L. ●■

367351   Solanum madagascariense Dammer；马岛茄●☆

367352　Solanum magdalenae Dammer = Solanum incanum L. ●■

367353　Solanum magnusianum Dammer = Solanum richardii Dunal ●☆

367354　Solanum mahoriensis D'Arcy et Rakot. ;马赫里茄●☆

367355　Solanum mairei H. Lév. = Solanum virginianum L. ●■

367356　Solanum malacochlamys Bitter = Solanum incanum L. ●■

367357　Solanum mammosum L. ;乳茄(北美乳茄,五角茄,五指丁茄,五指茄,五子登科);Amoise, Bachelor's Pear, Jumby-bubby, Macaw Bush, Macaw-bush, Papillate Nightshade ■

367358　Solanum mannii C. H. Wright = Solanum torvum Sw. ●

367359　Solanum mannii C. H. Wright var. compactum (C. H. Wright) C. H. Wright = Solanum torvum Sw. ●

367360　Solanum maranguense Bitter = Solanum incanum L. ●■

367361　Solanum margaritense J. R. Johnst. ;边茄■☆

367362　Solanum marginatum L. f. ;绿边茄;Purple African Nightshade ●☆

367363　Solanum marojejy D'Arcy et Rakot. ;马罗茄●☆

367364　Solanum mauritianum Scop. ;毛里求斯野烟树(阿根廷野烟树);Earleaf Nightshade, Tree Tobaco, Wild Tobaco, Woolly Nightshade ●☆

367365　Solanum maximowiczii Koidz. ;席氏山茄;Maximowicz Nightshade ■☆

367366　Solanum mayanum Lundell = Solanum torvum Sw. ●

367367　Solanum megacarpum Koidz. ;马铃薯茄(大果茄)■☆

367368　Solanum melanocerasum All. ;黑樱桃茄;Huckleberry ■☆

367369　Solanum melastomoides C. H. Wright;野牡丹茄●☆

367370　Solanum melongena L. ;茄(矮瓜,白茄,草鳖甲,长弯茄,大圆茄,癫茄,吊菜子,东风草,红茄,鸡蛋茄,昆仑瓜,酪酥,六苏,落苏,牛心茄,茄子,茄子花,圆茄,猪胆茄,紫茄);Aubergine, Badinjan, Begoon, Bringal, Brinjal, Brinjaul, Brown-Jolly, Common Eggplant, Egg Fruit, Egg Plant, Eggplant, Garden Egg, Garden Eggplant, Jew's Apple, Mad Apple, Mad-apple ●■

367371　Solanum melongena L. ovigerum ?;绵毛茄■☆

367372　Solanum melongena L. var. depressum L. = Solanum melongena L. ●■

367373　Solanum melongena L. var. depressum L. H. Bailey = Solanum melongena L. ●■

367374　Solanum melongena L. var. esculentum (Dunal) Nees = Solanum melongena L. ●■

367375　Solanum melongena L. var. incanum (L.) Kuntze;灰茄■☆

367376　Solanum melongena L. var. inerme (Dunal) Hiern = Solanum melongena L. ●■

367377　Solanum melongena L. var. serpentinum L. = Solanum melongena L. ●■

367378　Solanum melongena L. var. serpentinum L. H. Bailey = Solanum melongena L. ●■

367379　Solanum memphiticum J. F. Gmel. ;孟斐茄■☆

367380　Solanum memphiticum J. F. Gmel. var. abyssinicum (Dunal) Cufod. ;阿比西尼亚茄●☆

367381　Solanum merkeri Dammer = Solanum incanum L. ●■

367382　Solanum merrillianum Liou;光枝木龙葵;Glabrous Subshrub Nightshade ●

367383　Solanum mesadenium Bitter = Solanum benadirense Chiov. ●☆

367384　Solanum mesomorphum Bitter = Solanum incanum L. ●■

367385　Solanum meyeri-johannis Dammer;迈尔约翰茄☆

367386　Solanum micranthum Willd. ex Roem. et Schult. ;小花茄■☆

367387　Solanum microcarpum Vahl = Solanum pseudocapsicum L. ●

367388　Solanum mildbraedii Dammer;米尔德茄☆

367389　Solanum miniatum Bernh. = Solanum villosum Mill. ■

367390　Solanum miniatum Bernh. ex Willd. = Solanum nigrum L. var. villosum L. ■

367391　Solanum miniatum Bernh. ex Willd. = Solanum villosum Mill. ■

367392　Solanum miyakojimense T. Yamaz. et Takushi;宫古茄■☆

367393　Solanum molliusculum Bitter = Solanum pseudospinosum C. H. Wright ■☆

367394　Solanum monactinanthum Dammer;单花茄☆

367395　Solanum monticola Dunal = Solanum guineense L. ■☆

367396　Solanum morrisonense Hayata = Solanum hidetaroi Masam. ■

367397　Solanum mors-elephantum Dammer = Solanum macrocarpum L. ■☆

367398　Solanum muansense Dammer = Solanum kagehense Dammer ☆

367399　Solanum multiglandulosum Bitter;多腺茄☆

367400　Solanum muricatum Aiton;茄瓜(粗茎茄);Melon Pear, Pepino, Pepino Dulce Melon ■☆

367401　Solanum myoxotrichum Baker;多刺茄■☆

367402　Solanum myriacanthum Dunal = Solanum myoxotrichum Baker ■☆

367403　Solanum myrsinoides D'Arcy et Rakot. ;铁仔茄●☆

367404　Solanum nakurense C. H. Wright;呐古尔茄●☆

367405　Solanum namaquense Dammer;纳马夸茄●☆

367406　Solanum naumannii Engl. = Solanum aethiopicum L. ●■

367407　Solanum ndellense A. Chev. ;恩代尔茄●☆

367408　Solanum neesianum Wall. ex Nees = Lycianthes neesiana (Wall. ex Nees) D'Arcy et Zhi Y. Zhang ●

367409　Solanum neumannii Dammer = Solanum incanum L. ●■

367410　Solanum neumannii Dammer var. schoense Bitter = Solanum incanum L. ●■

367411　Solanum newtonii Dammer;纽敦茄☆

367412　Solanum nguelense Dammer;恩盖尔茄☆

367413　Solanum nienkui Merr. et Chun;疏刺茄;Laxspine Bittersweet, Laxspine Nightshade, Nienku Nightshade ●

367414　Solanum nigrescens Mart. et Aellen;变黑龙葵●☆

367415　Solanum nigriviolaceum Bitter = Solanum sessilistellatum Bitter☆

367416　Solanum nigrum L. ;龙葵(白花菜,滨藜叶龙葵,灯笼草,地胡草,地葫草,地泡子,耳坠菜,耳坠子,飞天龙,古钮菜,黑蛋蛋棵,黑姑娘,黑辣椒,黑茄,黑茄子,黑天地棵,黑天棵,黑天天,黑天天棵,黑星星,后红子,假灯笼菜,救儿草,扣子草,苦菜,苦葵,乌鸦酸浆草,老鸦眼睛草,龙眼草,钮仔菜,七粒扣,惹子草,山海椒,山辣椒,少花龙葵,石海椒,水茄,天地豆,天抛子,天泡草,天泡果,天泡子,天砲,天茄菜,天茄苗儿,天茄子,天天茄,甜豆茄棵,甜甜,乌疔草,乌归菜,五地茄,小果果,小苦菜,野海椒,野海角,野辣虎,野辣椒,野辣椒树,野辣子,野葡萄,野茄秧,野茄子,野伞子);Black Nightshade, Common Nightshade, Deadly Nightshade, Dragon Mallow, Dwale, Fellon-wort, Garden Nightshade, Houndsberry, Morel, Petty Morel, Saltbushleaf Nightshade, Stubbleberry, Wild Tomato, Woody Nightshade ■☆

367417　Solanum nigrum L. = Solanum americanum Mill. ■

367418　Solanum nigrum L. subsp. humile (Bernh. ex Willd.) Hartm. = Solanum humile Bernh. ex Willd. ■

367419　Solanum nigrum L. subsp. miniatum (Bernh. ex Willd.) Hartm. = Solanum villosum Mill. subsp. miniatum (Bernh. ex Willd.) Edmonds ■

367420　Solanum nigrum L. subsp. puniceum Kirschl. = Solanum villosum Mill. subsp. miniatum (Bernh. ex Willd.) Edmonds ■

367421　Solanum nigrum L. subsp. villosum (L.) Ball = Solanum villosum Mill. ■

367422　Solanum nigrum L. var. atriplicifolium（Desp.）G. Mey. = Solanum nigrum L. ■

367423　Solanum nigrum L. var. atriplicifolium G. Mey. = Solanum nigrum L. ■

367424　Solanum nigrum L. var. aurantium Maire = Solanum nigrum L. ■

367425　Solanum nigrum L. var. chlorocarpum（Spenn.）Boiss. = Solanum nigrum L. ■

367426　Solanum nigrum L. var. chlorocarpum（Spenn.）Koch = Solanum humile Bernh. ex Willd. ■

367427　Solanum nigrum L. var. elbaensis Täckh. et Boulos = Solanum nigrum L. ■

367428　Solanum nigrum L. var. flavovirens S. Z. Liou et W. Q. Wang；黄果龙葵■

367429　Solanum nigrum L. var. hirsutum Vahl = Solanum memphiticum J. F. Gmel. ■☆

367430　Solanum nigrum L. var. humile（Bernh. ex Willd.）Boiss. = Solanum humile Bernh. ex Willd. ■

367431　Solanum nigrum L. var. humile（Bernh. ex Willd.）C. Y. Wu et S. C. Huang = Solanum villosum Mill. ■

367432　Solanum nigrum L. var. humile（Bernh.）C. Y. Wu et S. C. Huang；矮株龙葵（矮龙葵）；Dwarf Nightshade ■

367433　Solanum nigrum L. var. incisum Täckh. et Boulos = Solanum nigrum L. ■

367434　Solanum nigrum L. var. induratum Boiss. = Solanum nigrum L. ■

367435　Solanum nigrum L. var. miniatum（Bernh. ex Willd.）Fr. = Solanum villosum Mill. ■

367436　Solanum nigrum L. var. miniatum（Willd.）Mert. et Koch = Solanum villosum Mill. subsp. miniatum（Bernh. ex Willd.）Edmonds ■

367437　Solanum nigrum L. var. miniatum Hook.；红色龙葵■☆

367438　Solanum nigrum L. var. nodiflorum（Jacq.）A. Gray = Solanum americanum Mill. var. patulum（L.）Edmonds ■☆

367439　Solanum nigrum L. var. pauciflorum Liou；少花龙葵■☆

367440　Solanum nigrum L. var. pauciflorum Liou = Solanum americanum Mill. ■

367441　Solanum nigrum L. var. pterocaulos（Rchb.）Batt. = Solanum nigrum L. ■

367442　Solanum nigrum L. var. sarrachoides Sendtn.；毛龙葵■

367443　Solanum nigrum L. var. suffruticosum（Schousb. ex Willd.）Moris = Solanum nigrum L. ■

367444　Solanum nigrum L. var. villosissimum（Guss.）Fiori = Solanum nigrum L. ■

367445　Solanum nigrum L. var. villosum L. = Solanum luteum Mill. ■

367446　Solanum nigrum L. var. villosum L. = Solanum villosum Mill. ■

367447　Solanum nigrum L. var. violaceum F. H. Chen = Solanum photeinocarpum Nakam. et Odash. var. violaceum（F. H. Chen ex Wessely）C. Y. Wu et S. C. Huang ■

367448　Solanum nigrum L. var. violaceum F. H. Chen ex Wessely = Solanum photeinocarpum Nakam. et Odash. var. violaceum（F. H. Chen ex Wessely）C. Y. Wu et S. C. Huang ■

367449　Solanum nigrum L. var. viride Neilr. = Solanum humile Bernh. ex Willd. ■

367450　Solanum nigrum L. var. vulgare？ = Solanum nigrum L. ■

367451　Solanum nipponense Makino = Solanum japonense Nakai ■

367452　Solanum nitidibaccatum Bitter = Solanum physalifolium Rusby ■☆

367453　Solanum nivalomontanum C. Y. Wu et S. C. Huang = Solanum violaceum Ortega ●

367454　Solanum niveum Thunb. = Solanum giganteum Jacq. ■

367455　Solanum nodiflorum Jacq.；节花茄；Black Nightshade ■☆

367456　Solanum nodiflorum Jacq. = Solanum americanum Mill. ■

367457　Solanum nossibeense Vatke = Solanum erythracanthum Bojer ex Dunal ●☆

367458　Solanum nossibeense Vatke var. elongatius Bitter = Solanum erythracanthum Bojer ex Dunal ●☆

367459　Solanum nossibeense Vatke var. robustius Bitter = Solanum erythracanthum Bojer ex Dunal ●☆

367460　Solanum numile Bernh. ex Willd. = Solanum villosum Mill. ■

367461　Solanum nummulifolium Chiov. = Solanum cordatum Forssk. ●☆

367462　Solanum obbiadense Chiov. = Solanum cordatum Forssk. ●☆

367463　Solanum ochroleucum Bastard；淡黄茄■☆

367464　Solanum ochroleucum Bastard = Solanum humile Bernh. ex Willd. ■

367465　Solanum ochroleucum Bastard var. flavum（Kit.）Dunal = Solanum humile Bernh. ex Willd. ■

367466　Solanum ogadense Bitter = Solanum dennekense Dammer ■☆

367467　Solanum oleraceum Dunal ex Poir.；橄榄茄■☆

367468　Solanum olgae Pojark.；乌恰茄■☆

367469　Solanum olivaceum Dammer；橄榄绿茄☆

367470　Solanum omahekense Dammer = Solanum delagoense Dunal ☆

367471　Solanum omitiomirense Dammer = Solanum delagoense Dunal ☆

367472　Solanum opacum A. Br. et Bouché；深暗茄■☆

367473　Solanum orthocarpum Pic. Serm.；直果茄☆

367474　Solanum osbeckii Dunal = Lycianthes biflora（Lour.）Bitter ●

367475　Solanum osbeckii Dunal var. stauntonii Dunal = Lycianthes biflora（Lour.）Bitter ●

367476　Solanum ovatifolium De Wild.；卵叶茄☆

367477　Solanum ovigerum Dunal = Solanum melongena L. ●■

367478　Solanum paaschenianum H. J. P. Winkl. = Solanum aethiopicum L. ●■

367479　Solanum palmetorum Dunal = Solanum sepicula Dunal ■☆

367480　Solanum pampaninii Chiov.；潘帕尼尼茄☆

367481　Solanum panduriforme E. Mey. ex Dunal；琴茄；Apple of Sodom，Bitter Apple，Poison Apple ☆

367482　Solanum paniculatum L.；锥花茄■☆

367483　Solanum parcebarbatum Bitter；稀髯毛茄■☆

367484　Solanum patens Lowe；铺展茄■☆

367485　Solanum pauperum Wright = Solanum anomalum Thonn. ■☆

367486　Solanum pectinatum Dunal；篦齿茄■☆

367487　Solanum pedunculatum Roem. et Schult. = Solanum muricatum Aiton ■☆

367488　Solanum peikuoensis S. S. Ying；白狗大山茄；Baigoudashan Nightshade ●

367489　Solanum pembae Bitter = Solanum incanum L. ●■

367490　Solanum penduliflorum Dammer；垂花茄■☆

367491　Solanum pentagonocalyx Bitter；五角萼花茄☆

367492　Solanum pentapetaloides Hornem. = Solanum capsicoides All. ■

367493　Solanum persicum Willd. ex Roem. et Schult.；波斯茄■☆

367494　Solanum peruvianum L.；秘鲁茄；Peruvian Nightshade ☆

367495　Solanum pharmacum Klotzsch；药茄■☆

367496　Solanum photeinocarpum Nakam. et Odash.；少花龙葵■☆

367497　Solanum photeinocarpum Nakam. et Odash. = Solanum americanum Mill. ■

367498　Solanum photeinocarpum Nakam. et Odash. var. violaceum（F. H. Chen ex Wessely）C. Y. Wu et S. C. Huang；紫少花龙葵；Purple Shiningfruit Dragon Mallow，Violet Nightshade ■

367499　Solanum physalifolium Rusby；光果茄；Green Nightshade，Hoe Nightshade ■☆

367500　Solanum physalifolium Rusby var. nitidibaccatum（Bitter）Edmonds = Solanum physalifolium Rusby ■☆

367501　Solanum phytolaccoides C. H. Wright = Solanum terminale Forssk. ●☆

367502　Solanum pimpinellifolium L. = Lycopersicon pimpinellifolium（L.）Mill. ■☆

367503　Solanum pittosporifolium Hemsl.；海桐叶白英（疏毛海桐叶白英，玉山茄）；Currant Tomato，Pilose Pittosporumleaf Bittersweet，Pilose Pittosporumleaf Nightshade，Pilose Seatung Nightshade，Pittosporumleaf Bittersweet，Pittosporum-leaved Bittersweet，Seatung Nightshade ●

367504　Solanum pittosporifolium Hemsl. var. pilosum C. Y. Wu et S. C. Huang = Solanum pittosporifolium Hemsl. ●

367505　Solanum pittosporifolium Nakam. et Odash. var. pilosum C. Y. Wu et S. C. Huang = Solanum pittosporifolium Hemsl. ●

367506　Solanum platanifolium Hook.；悬铃木叶茄■

367507　Solanum platanifolium Hook. = Solanum mammosum L. ■

367508　Solanum plebeium A. Rich.；普通茄●☆

367509　Solanum plebeium A. Rich. var. brachysepalum Bitter = Solanum plebeium A. Rich. ●☆

367510　Solanum plebeium A. Rich. var. grossedentatum（A. Rich.）Chiov. = Solanum memphiticum J. F. Gmel. var. abyssinicum（Dunal）Cufod. ●☆

367511　Solanum plebeium A. Rich. var. subtile Bitter = Solanum plebeium A. Rich. ●☆

367512　Solanum plousianthemum Dammer = Solanum terminale Forssk. ●☆

367513　Solanum pluviale Standl. = Solanum chrysotrichum Schltdl. ●

367514　Solanum poggei Dammer；波格茄☆

367515　Solanum polyanthemum Hochst. ex A. Rich.；繁花茄☆

367516　Solanum praematurum Dammer；早熟茄☆

367517　Solanum preussii Dammer；普罗伊斯茄☆

367518　Solanum procumbens Lour.；海南茄（卜古雀，耳环草，鸡公箣子，金纽头，衫纽藤，细颠茄，小丁茄）；Hainan Eggplant，Hainan Nightshade ●

367519　Solanum protodasypogon Bitter = Solanum aculeastrum Dunal ●☆

367520　Solanum pruinosum Dunal；白粉茄●☆

367521　Solanum pruinosum Dunal var. pilosulum ? = Solanum pruinosum Dunal ●☆

367522　Solanum pseudocapsicum L.；珊瑚樱（冬珊瑚，红珊瑚，吉祥子，吉杏，珊瑚豆，珊瑚子，万寿果，洋辣子，野海椒，野辣茄，玉珊瑚，玉珊瑚茄）；Christmas Cherry，Jerusalem Cherry，Jerusalem Nightshade，Jerusalemcherry，Jerusalem-cherry，Winter Cherry ●

367523　Solanum pseudocapsicum L. 'Ballon'；气球珊瑚樱●☆

367524　Solanum pseudocapsicum L. 'Fancy'；幻想珊瑚樱●☆

367525　Solanum pseudocapsicum L. 'Red Giant'；红巨珊瑚樱●☆

367526　Solanum pseudocapsicum L. 'Snowfire'；雪火珊瑚樱●☆

367527　Solanum pseudocapsicum L. subsp. diflorum（Vell.）Hassl. = Solanum pseudocapsicum L. var. diflorum（Vell.）Bitter ●

367528　Solanum pseudocapsicum L. var. diflorum（Vell.）Bitter；珊瑚豆（陈龙茄，刺石榴，冬珊瑚，观音莲，海茄子，玛瑙珠，毛叶冬珊

瑚，青杞，蜀羊泉，我立丁子花，岩海椒，洋海椒，野枸杞，野海椒，玉珊瑚）；Capsicum，Christmas Cherry，False Jerusalem Cherry，Tree Nightshade，Twoflower Jerusalemcherry，Winter Cherry ●

367529　Solanum pseudoflavum Pojark.；假黄茄■☆

367530　Solanum pseudogeminifolium Dammer；假对叶茄☆

367531　Solanum pseudopersicum Pojark.；假波斯茄☆

367532　Solanum pseudoquinum A. St.-Hil.；假五叶茄■☆

367533　Solanum pseudospinosum C. H. Wright；假刺茄■☆

367534　Solanum psilostylum Dammer = Solanum incanum L. ●■

367535　Solanum ptycanthum Dunal ex DC.；美洲茄；Black Nightshade，Nightshade ■☆

367536　Solanum ptychanthum Dunal = Solanum ptycanthum Dunal ex DC. ■☆

367537　Solanum pubescens Willd.；柔毛茄■

367538　Solanum pulverulentum L.；粉末茄☆

367539　Solanum pynaertii De Wild. = Solanum chrysotrichum Schltdl. ●

367540　Solanum pyracanthos Lam. = Solanum pyracanthum Jacq. ●☆

367541　Solanum pyracanthum Jacq.；焰刺茄；Spiny Solanum ●☆

367542　Solanum quadrangulare Thunb. ex L. f. = Solanum africanum Dunal ■☆

367543　Solanum quitoense Lam.；奎东茄（基多茄）；Lulo，Naranjilla，Quito Orange ●☆

367544　Solanum racemosum Mill.；总花茄；Bitter-berry ☆

367545　Solanum radicans L. f.；生根茄■☆

367546　Solanum rangei Dammer = Solanum burchellii Dunal ☆

367547　Solanum rantonnei Carrière = Lycianthes rantonnei（Carrière）Bitter ●■☆

367548　Solanum rantonnei Carrière ex Lesc. = Lycianthes rantonnei（Carrière）Bitter ●■☆

367549　Solanum rantonnetii Carrière ex Lesc.；巴拉圭茄（南茄）；Blue Potato Bush，Blue Potato-bush，Paraguay Nightshade，Potato Bush ●☆

367550　Solanum rantonnetii Carrière ex Lesc. 'Royal Robe'；皇袍南茄 ●☆

367551　Solanum rautanenii Schinz = Solanum catombelense Peyr. ☆

367552　Solanum rederi Dammer；雷德茄☆

367553　Solanum reflexum Schrank = Solanum aculeatissimum Jacq. ●■

367554　Solanum reichenbachii Vatke = Withania reichenbachii（Vatke）Bitter ■☆

367555　Solanum renschii Vatke = Solanum tettense Klotzsch var. renschii（Vatke）A. E. Gonc. ■☆

367556　Solanum repandifrons Bitter = Solanum incanum L. ●■

367557　Solanum reticulatum Willd. ex Roem. et Schult.；网脉茄☆

367558　Solanum retroflexum Dunal；下弯茄；Sun Berry，Wonderberry ■☆

367559　Solanum rhodesianum Dammer；罗得西亚茄☆

367560　Solanum richardii Dunal；理查德茄●☆

367561　Solanum richardii Dunal var. acutilobatum（Dammer）A. E. Gonc.；尖裂片理查德茄●☆

367562　Solanum richardii Dunal var. burtt-davyi（Dunkley）A. E. Gonc.；伯特-戴维茄☆

367563　Solanum robecchii Bitter et Dammer；罗贝克茄☆

367564　Solanum robustum H. Wendl.；灌木茄；Shrubby Nightshade ●☆

367565　Solanum rogersii S. Moore = Solanum sisymbriifolium Lam. ■

367566　Solanum rohrii C. H. Wright；勒尔茄■☆

367567　Solanum rostratum Dunal；壶萼刺茄（黄花刺茄）；Buffalo Bur，Buffalo-bur，Buffalobur Nightshade，Buffalo-bur Nightshade，Kansas Thistle，Prickly Nightshade，Rostratum Nightshade，Sand Bur，Texas

Thistle ■

367568　Solanum roxburghii Dun. = Solanum alatum Moench ■

367569　Solanum ruandae Bitter;卢旺达茄☆

367570　Solanum rugosum Dunal;皱褶茄☆

367571　Solanum rugulosum De Wild. ;稍皱茄☆

367572　Solanum runsoriense C. H. Wright;伦索里茄☆

367573　Solanum ruwenzoriense De Wild. = Solanum distichum Schumach. et Thonn. ■☆

367574　Solanum sakarense Dammer;萨卡尔茄☆

367575　Solanum sambiranense D'Arcy et Rakot. ;桑比朗茄●☆

367576　Solanum sanctum L. = Solanum coagulans Forssk. ●■☆

367577　Solanum sanctum L. = Solanum incanum L. ●■

367578　Solanum sanitwongsei Craib. ;萨氏茄■☆

367579　Solanum sapiaceum Dammer = Solanum aculeastrum Dunal ●☆

367580　Solanum sapinii De Wild. = Solanum macrocarpum L. ■☆

367581　Solanum saponaceum Welw. = Solanum aculeastrum Dunal var. albifolium (C. H. Wright) Bitter ●☆

367582　Solanum sarmentosum Nees;匍匐茄■

367583　Solanum sarrachoides Sendtn. ;黏毛茄;Green Nightshade,Hairy Nightshade,Leafy-fruited Nightshade,Viscid Nightshade ■☆

367584　Solanum scabridum Dunal;糙茄■☆

367585　Solanum scabrum Mill. ;木龙葵;Garden-huckleberry ■

367586　Solanum scabrum Mill. subsp. nigericum Gbile？;黑木龙葵■☆

367587　Solanum scaforthianum Andréws;星茄(南青杞)●

367588　Solanum scalare C. H. Wright = Solanum distichum Schumach. et Thonn. ■☆

367589　Solanum schaeferi Dammer = Solanum burchellii Dunal ■☆

367590　Solanum scheffleri Dammer;谢夫勒茄☆

367591　Solanum schimperianum Hochst. = Solanum schimperianum Hochst. ex A. Rich. ■☆

367592　Solanum schimperianum Hochst. ex A. Rich. ;荷花茄■☆

367593　Solanum schimperianum Hochst. ex A. Rich. var. polyanthemum (Hochst. ex A. Rich.) Bitter = Solanum schimperianum Hochst. ex A. Rich. ■☆

367594　Solanum schliebenii Werderm. ;施利本茄■☆

367595　Solanum schumannianum Dammer;舒曼茄■☆

367596　Solanum seaforthianum Andréws;南青杞(海茄,玲珑茄,南青茄,葡萄茄,藤茄,星茄);Brazilian Nightshade,Kerr Nightshade,Potato Creeper ●■

367597　Solanum secedens Dammer;塞采德茄●☆

367598　Solanum senegambicum Dunal = Solanum anguivi Lam. ●☆

367599　Solanum sennii Chiov. = Solanum campylacanthum Dunal ☆

367600　Solanum sepiaceum Dammer;篱笆茄■☆

367601　Solanum sepicula Dunal;庭院茄■☆

367602　Solanum sepicula Dunal var. calvifrons Bitter = Solanum sepicula Dunal ■☆

367603　Solanum sepicula Dunal var. microlepis Bitter = Solanum sepicula Dunal ■☆

367604　Solanum septemlobum Bunge;青杞(狗杞子,红葵,裂叶龙葵,蜀羊泉,药鸡豆,药人豆,野枸杞,野辣子,野茄,野茄子,野苏子);Sevanlobed Nightshade ■

367605　Solanum septemlobum Bunge var. indutum Hand.-Mazz. ;茄子蒿;Eggplant Wormwood ■

367606　Solanum septemlobum Bunge var. ovoideocarpum C. Y. Wu et S. C. Huang;卵果青杞;Ovatefruit Sevanlobed Nightshade ■

367607　Solanum septemlobum Bunge var. ovoideocarpum C. Y. Wu et S. C. Huang = Solanum septemlobum Bunge ■

367608　Solanum septemlobum Bunge var. subintegrifolium C. Y. Wu et S. C. Huang = Solanum septemlobum Bunge ■

367609　Solanum septemlobum Bunge var. subintegrifolium C. Y. Wu et S. C. Huang;单叶青杞;Singleleaf Sevanlobed Nightshade ■

367610　Solanum seretii De Wild. ;赛雷茄☆

367611　Solanum sessiliflorum Dunal;无柄花茄(奎东茄);Naranjilla ☆

367612　Solanum sessilistellatum Bitter;无梗星状茄☆

367613　Solanum setaceum Dammer;刚毛茄☆

367614　Solanum sinaicum Boiss. ;西奈茄☆

367615　Solanum sinaicum Boiss. = Solanum forskalii Dunal ●☆

367616　Solanum sinuatifolium Vell. = Solanum capsicoides All. ■

367617　Solanum sisymbrifolium Lam. ;蒜芥茄(二列星毛刺茄,二裂星毛刺茄,拟刺茄,蒜介茄);Galiccressleaf Nightshade, Morelle De Balbis, Red Buffalo-bur, Sisymbriumleaf Nightshade, Sticky Nightshade ■

367618　Solanum sodomaeum L. var. hermannii Dunal = Solanum linnaeanum Hepper et P. M. Jaeger ■☆

367619　Solanum sodomaeum L. var. mediterraneum Dunal = Solanum linnaeanum Hepper et P. M. Jaeger ■☆

367620　Solanum sodomeum L. ;索多米茄;Apple of Sodom, Dead Sea Apple,Popolo ■☆

367621　Solanum sodomeum L. = Solanum anguivi Lam. ●☆

367622　Solanum sodomeum L. = Solanum incanum L. ●■

367623　Solanum somalense Franch. ;索马里茄■☆

367624　Solanum somalense Franch. var. anisantherum (Dammer) Bitter = Solanum somalense Franch. ■☆

367625　Solanum somalense Franch. var. parvifrons Bitter = Solanum somalense Franch. ■☆

367626　Solanum somalense Franch. var. withaniifolium (Dammer) Bitter = Solanum somalense Franch. ■☆

367627　Solanum sordidescens Bitter = Solanum kagehense Dammer ☆

367628　Solanum sparsespinosum De Wild. ;稀刺茄☆

367629　Solanum sparsiflorum Dammer = Solanum catombelense Peyr. ☆

367630　Solanum spathotrichum Dammer;毛苞茄☆

367631　Solanum sphaerocarpon Moric. = Solanum capsicoides All. ■

367632　Solanum spirale Roxb. ;旋花茄(白条花,百两金,大苦溜溜,倒提壶,滴打稀,敷药,苦凉菜,理肺散,螺旋茄,帕笠,山烟木,四萼旋花茄);Coiledflower Nightshade, Coil-flowered Nightshade, Tetrasepaled Nightshade ●

367633　Solanum spirale Roxb. var. tetrasepalum H. Chu = Solanum spirale Roxb. ●

367634　Solanum stellativillosum Bitter = Solanum incanum L. ●■

367635　Solanum stolzii Dammer;斯托尔兹茄☆

367636　Solanum stramoniifolium Jacq. ;曼佗罗叶茄■☆

367637　Solanum stramoniifolium Jacq. = Solanum torvum Sw. ●

367638　Solanum subcoriaceum T. Durand et H. Durand;革质茄☆

367639　Solanum suberosum Dammer = Solanum terminale Forssk. ●☆

367640　Solanum subhastatum De Wild. = Solanum aculeastrum Dunal ●☆

367641　Solanum sublobatum Roem. et Schult. ;微裂茄☆

367642　Solanum sublobatum Willd. ex Roem. et Schult. = Solanum chenopodioides Lam. ■☆

367643　Solanum subsessile De Wild. ;近无柄茄■☆

367644　Solanum subtruncatum Wall. = Lycianthes neesiana (Wall. ex Nees) D'Arcy et Zhi Y. Zhang ●

367645　Solanum subtruncatum Wall. ex Dunal = Lycianthes neesiana (Wall. ex Nees) D'Arcy et Zhi Y. Zhang ●

367646   Solanum subulatum Wright;钻形茄☆

367647   Solanum subuniflorum Bitter;亚单花茄☆

367648   Solanum subviscidum Schrank = Solanum sisymbriifolium Lam. ■

367649   Solanum sudanense Hammerstein;苏丹茄☆

367650   Solanum suffruticosum Schousb. = Solanum merrillianum Liou ●

367651   Solanum suffruticosum Schousb. ex Willd.;白花木龙葵（白花仔草，黄果茄，木龙葵）;Subshrub Dragon Mallow, Subshrub Nightshade ●

367652   Solanum suffruticosum Schousb. ex Willd. = Solanum merrillianum Liou ●

367653   Solanum suffruticosum Schousb. ex Willd. = Solanum nigrum L. ■

367654   Solanum suffruticosum Schousb. ex Willd. var. merrillianum (Liou) C. Y. Wu = Solanum merrillianum Liou ●

367655   Solanum supinum Dunal;仰卧茄■☆

367656   Solanum surattense Burm. f. = Solanum virginianum L. ●■

367657   Solanum tabaccifolium Vell. = Solanum mauritianum Scop. ●☆

367658   Solanum tabacicolor Dammer = Solanum incanum L. ●■

367659   Solanum taitense Vatke;泰塔茄■☆

367660   Solanum takaoyamense Makino = Solanum japonense Nakai ■

367661   Solanum tampicense Dunal;湿地茄;Aquatic Soda Apple, Scambling Nightshade, Wetland Night Shade, Wetland Nightshade ☆

367662   Solanum tanganikense Bitter = Solanum cyaneopurpureum De Wild. ☆

367663   Solanum tarolinense ?;塔罗林茄;Horse Nettle ☆

367664   Solanum teitense Vatke = Solanum taitense Vatke ■☆

367665   Solanum tenuiramosum Dammer = Solanum tettense Klotzsch var. renschii (Vatke) A. E. Gonc. ■☆

367666   Solanum terminale Forssk.;顶生茄●☆

367667   Solanum terminale Forssk. subsp. inconstans (C. H. Wright) Heine = Solanum terminale Forssk. ●☆

367668   Solanum terminale Forssk. subsp. sanaganum (Bitter) Heine = Solanum terminale Forssk. ●☆

367669   Solanum terminale Forssk. subsp. welwitschii (C. H. Wright) Heine = Solanum welwitschii C. H. Wright ■☆

367670   Solanum tetrachondrum Bitter;四粒茄☆

367671   Solanum tettense Klotzsch;泰特茄☆

367672   Solanum tettense Klotzsch var. renschii (Vatke) A. E. Gonc.;伦施茄■☆

367673   Solanum texanum Ten.;金银茄;Tomatofruited Eggplant ■

367674   Solanum texense Engelm. et A. Gray = Solanum aethiopicum L. ●■

367675   Solanum thruppii C. H. Wright;斯拉普茄■☆

367676   Solanum togoense Dammer = Solanum terminale Forssk. ●☆

367677   Solanum tomentosum L.;绒毛茄☆

367678   Solanum tomentosum L. var. burchellii (Dunal) Wright = Solanum burchellii Dunal ☆

367679   Solanum tomentosum L. var. coccineum (Jacq.) Willd.;绯红茄☆

367680   Solanum torreanum A. E. Gonc.;托尔茄☆

367681   Solanum torreyi A. Gray = Solanum dimidiatum Raf. ■☆

367682   Solanum torvum Sw.;水茄（刺番茄，刺茄，大金扣，金纽扣，金钮头，金衫扣，拦路虎，扭茄木，青茄，山颠茄，山烟草，天茄子，万桃花，乌凉，小登茄，鸭卡，洋毛辣，野茄，野茄子，一面针）;Susumber, Turkey Berry, Water Nightshade ●

367683   Solanum torvum Sw. var. compactum C. H. Wright = Solanum torvum Sw. ●

367684   Solanum torvum Sw. var. lasiostylum C. Y. Liu et C. H. Ou =

367685   Solanum torvum Sw. var. lasiostylum Y. C. Liu et C. H. Ou;毛柱万桃花●■

367686   Solanum torvum Sw. var. pleiotomum C. Y. Wu et S. C. Huang = Solanum chrysotrichum Schltdl. ●

367687   Solanum transcaucasicum Pojark.;外高加索茄☆

367688   Solanum trichopetiolatum D'Arcy et Rakot.;毛梗茄●☆

367689   Solanum triflorum Nutt.;三花茄;Cutleaf Nightshade, Cut-leaf Nightshade, Cut-leaved Nightshade, Small Nightshade, Three-flowered Nightshade ■☆

367690   Solanum trifolium Dunal;三叶茄;Wild Tomato ☆

367691   Solanum trilobatum L.;三裂水茄（三裂茄）■

367692   Solanum tripartitum Dunal;三深裂茄■☆

367693   Solanum trisectum Dunal;三全裂茄☆

367694   Solanum truncatum Standl. et C. V. Morton = Solanum chrysotrichum Schltdl. ●

367695   Solanum truncicola Bitter = Solanum humblotii Dammer ●☆

367696   Solanum tuberosum L.;马铃薯（地蛋，甘同，荷兰薯，山洋芋，山药蛋，山药豆，薯仔，土豆，土豆儿，喜旧花，阳芋，洋番薯，洋山芋，洋芋）;Chats, Cheddies, Chiddies, Crokers, Edible Nightshade, Eggs-and-bacon, English Arrowroot, European Potato, Fata, Frata, Indian Potato, Irish Potato, Murfeys, Poltate, Potato, Potato Crisps, Prase, Pratie, Pridda, Spud, Taters, Tateys, Tetty, Tiddy, Tuberous Nightshade, Uala-kahiki, Vodka, White Potato ■

367697   Solanum tuntula De Wild. = Solanum delagoense Dunal ☆

367698   Solanum ubanghense A. Chev.;乌班吉茄☆

367699   Solanum ueleense De Wild.;韦莱茄☆

367700   Solanum ukerewense Bitter;乌凯雷韦茄☆

367701   Solanum ulugurense Dammer = Solanum schumannianum Dammer ■☆

367702   Solanum ulugurense Holub;乌卢古尔茄■☆

367703   Solanum umbellatum Mill.;伞茄■☆

367704   Solanum uncinatum R. Br.;钩茄☆

367705   Solanum undatum Jacq. = Solanum aethiopicum L. ●■

367706   Solanum undatum Lam.;野茄（刺颠茄，大颠茄，颠茄树，丁茄，菲岛茄，黄刺茄，黄果珊瑚，黄水茄，黄天茄，苦天茄，牛茄子，衫纽果，衫钮果，野颠茄，野海茄，中茄子）;Philippine Nightshade ■

367707   Solanum unguiculatum A. Rich. = Solanum incanum L. ●■

367708   Solanum uollense (Chiov.) Pic. Serm.;沃尔茄●☆

367709   Solanum upingtoniae Schinz = Solanum tettense Klotzsch var. renschii (Vatke) A. E. Gonc. ■☆

367710   Solanum urbanianum Dammer = Solanum incanum L. ●■

367711   Solanum urosepalum Dammer = Solanum anguivi Lam. ●☆

367712   Solanum usambarense Bitter et Dammer;乌桑巴拉茄☆

367713   Solanum vagum Hayne;无定形茄■

367714   Solanum variegatum Ruiz et Pav. = Solanum muricatum Aiton ■☆

367715   Solanum verapazense Standl. et Steyerm. = Solanum torvum Sw. ●

367716   Solanum verbascifolium L. = Solanum erianthum D. Don ●

367717   Solanum verbascifolium L. var. adulterinum Ham. ex G. Don = Solanum erianthum D. Don ●

367718   Solanum verbascifolium L. var. auriculatum (Aiton) Kuntze = Solanum mauritianum Scop. ●☆

367719   Solanum verbascifolium L. var. exstipulatum Kuntze = Solanum erianthum D. Don ●

367720   Solanum verbascifrons Bitter = Solanum incanum L. ●■

367721   Solanum vespertilio Aiton;夕茄☆

Solanum macaonense Dunal ●■

367722　Solanum viarum Dunal；毛果茄；Tropical Soda Apple ■

367723　Solanum villosissimum Zucc. = Solanum mammosum L. ■

367724　Solanum villosum（L.）Moench = Solanum nigrum L. var. villosum L. ■

367725　Solanum villosum Forssk. = Solanum forskalii Dunal ●☆

367726　Solanum villosum Mill.；红果龙葵（光果龙葵，红葵）；Alate Nightshade, Hairy Nightshade, Redfruit Dragon Mallow, Red-fruited Nightshade, Yellow Nightshade, Yellow-berried Nightshade ■

367727　Solanum villosum Mill. = Solanum luteum Mill. ■

367728　Solanum villosum Mill. = Solanum nigrum L. var. villosum L. ■

367729　Solanum villosum Mill. subsp. miniatum（Bernh. ex Willd.）Edmonds = Solanum villosum Mill. ■

367730　Solanum villosum Mill. subsp. puniceum（Kirschl.）Edmonds = Solanum villosum Mill. subsp. miniatum（Bernh. ex Willd.）Edmonds ■

367731　Solanum villosum Mill. subsp. puniceum（Kirschl.）Edmonds = Solanum villosum Mill. ■

367732　Solanum villosum Moench = Solanum nigrum L. var. villosum L. ■

367733　Solanum violaceum Ortega；刺天茄（巴山虎，刺茄，颠茄，丁茄子，钉茄，黄面仔，黄木荞，黄茄花，黄水荞，黄水茄，鸡刺子，假茄子，金吊柳，金吊钮，金扣，金扣拦路虎，金扣钮，金纽扣，金钮刺，金钮头，苦果，苦天茄，勒矮瓜，满天星，南天茄，扭茄木，生刺矮瓜，生宅茄，天茄子，天星子，五角颠茄，五提茄，五宅茄，细颠茄，细黄茄，小颠茄，小闹telephone，袖扣果，雪山茄，野海椒，印度茄，紫花茄）；Indian Nightshade, Snowmountain Bittersweet, Snowmountain Eggplant, Snowmountain Nightshade, Snow-mountain Nightshade ●

367734　Solanum virginianum Jacq. = Solanum surattense Burm. f. ●■

367735　Solanum virginianum L.；黄果茄（刺丁茄，刺茄，刺天果，大颠茄，大丁茄，大苦果，大苦菜，颠茄，颠茄子，丁茄，番鬼茄，鬼茄，红颠茄，红果丁茄，红水茄，黄打破碗，黄贡茄，黄果珊瑚，黄茄果，黄水茄，黄天茄，假茄子，金银茄，笋丁茄，马刺，毛果茄，牛子，钮茄根，山马铃，天茄子，小颠茄，哑口子，野颠茄，野番茄，野茄果，油辣果）；Gold Silver Nightshade, Soda-apple Nightshade, Surattense Nightshade, Yellowfruit Nightshade ●■

367736　Solanum viridiflorum Schltdl. = Solanum viarum Dunal ■

367737　Solanum viridimaculatum Gilli；绿斑茄●☆

367738　Solanum volkensii Dammer = Solanum incanum L. ●■

367739　Solanum vulgatum Willd. ex Steud. var. chlorocarpum Spenn. = Solanum humile Bernh. ex Willd. ■

367740　Solanum warneckianum Dammer = Solanum anomalum Thonn. ■☆

367741　Solanum warzcewiczii Huber = Solanum chrysotrichum Schltdl. ●

367742　Solanum welwitschii C. H. Wright；韦尔茄■☆

367743　Solanum welwitschii C. H. Wright var. oblongum？= Solanum welwitschii C. H. Wright ■☆

367744　Solanum welwitschii C. H. Wright var. strictum？= Solanum welwitschii C. H. Wright ■☆

367745　Solanum wendlandii Hook. f.；温南茄（薯仔藤）；Costa Rican Nightshade, Giant Potato Creeper, Giant Potatocreeper, Paradise Flower ■☆

367746　Solanum withaniifolium Dammer = Solanum somalense Franch. ■☆

367747　Solanum wittei Robyns = Solanum tettense Klotzsch ■☆

367748　Solanum worouowii Pojark.；沃氏茄■☆

367749　Solanum wrightii Benth.；大花茄（水柿）；Brazilian Potato Tree, Potato Tree, Wright Nightshade ●■

367750　Solanum xanthocarpum Schrad. et H. Wendl. = Solanum ciliatum Lam. ●■

367751　Solanum xanthocarpum Schrad. et H. Wendl. = Solanum surattense Burm. f. ●■

367752　Solanum xanthocarpum Schrad. et H. Wendl. = Solanum virginianum L. ●■

367753　Solanum xanti A. Gray；紫茄；Purple Nightshade, Purple Robe ●☆

367754　Solanum yangambiense De Wild.；扬甘比茄☆

367755　Solanum yolense Hutch. et Dalziel = Solanum cerasiferum Dunal ☆

367756　Solanum zanzibarense Vatke；桑给巴尔茄●☆

367757　Solanum zelenetzkii Pojark.；泽氏茄■☆

367758　Solaria Phil.（1858）；喜阳葱属■☆

367759　Solaria atropurpurea（Phil.）Ravenna；暗紫喜阳葱■☆

367760　Solaria major Reiche；大喜阳葱■☆

367761　Solaria miersioides Phil.；喜阳葱■☆

367762　Soldanella L.（1753）；圆币草属（高山钟花属，雪铃花属）；Alpenclock, Blue Moonwort, Gravel Bind, Gravel-bind, Moon-wort, Snowbell ■☆

367763　Soldanella alpina L.；圆币草（高山雪铃花，高山钟花，圆叶高山樱草）；Alpine Snowbell, Blue Moonwort, Glacier Alpenclock ■☆

367764　Soldanella austriaca Vierh.；澳大利亚圆币草；Austrian Snowbell ■☆

367765　Soldanella minima Hoppe ex Sturm；小圆币草；Least Bell, Least Snowbell ■☆

367766　Soldanella montana Willd.；山圆币草；Alpine Tassel Flower, Greater Alpenclock, Mountain Snowbell, Mountain Tassel, Mountain Tassel Flower ■☆

367767　Soldanella pusilla Baumg.；微小圆币草；Dwarf Snowbell ■☆

367768　Soldanella villosa Darracq；密毛圆币草■☆

367769　Soldevilla Lag. = Hispidella Barnadez ex Lam. ■☆

367770　Solea Spreng. = Hybanthus Jacq.（保留属名）●■

367771　Soleirolia Gaudich.（1830）；金钱麻属；Baby's Tears, Baby's-tears, Mind-your-own-business, Mother of Thousands ■☆

367772　Soleirolia soleirolii（Req.）Dandy；金钱麻（绿珠草）；Angel's Tears, Baby's Tears, Helxine, Lady's Tears, Mind-your-own-business, Mother of Thousands, Mother-of-thousands, Oliver Cromwell's Creeping Companion ■☆

367773　Solena Lour.（1790）；茅瓜属；Solena ■

367774　Solena Willd. = Posoqueria Aubl. ●☆

367775　Solena amplexicaulis（Lam.）Gandhi；茅瓜（抱瓜，抱茎马㼎儿，变叶马㼎儿，波瓜公，地苦胆，滇藏茅瓜，杜瓜，狗黄瓜，狗屎瓜，耗子瓜，金丝瓜，老鼠瓜，老鼠黄瓜，老鼠拉冬瓜，老鼠香瓜，牛奶子，山鸡仔，山天瓜，山熊胆，天瓜，土白蔹，小鸡黄瓜，野黄瓜，异叶马㼎儿，银丝莲）；Claspingstem Solena, Delavay Solena Telfairia, Solena ■

367776　Solena bowieana（A. Cunn. ex Hook.）D. Dietr. = Euclinia longiflora Salisb. ●☆

367777　Solena delavayi（Cogn.）C. Y. Wu = Solena heterophylla Lour. ■

367778　Solena heterophylla（Lour.）Cogn. = Solena amplexicaulis（Lam.）Gandhi ■

367779　Solena heterophylla Lour. = Solena amplexicaulis（Lam.）Gandhi ■

367780　Solena heterophylla Lour. subsp. napaulensis（Ser.）W. J. de Wilde et Duyfjes；西藏茅瓜■

367781　Solena latifolia Rudge = Posoqueria latifolia（Rudge）Roem. et Schult. ●☆

367782　Solena longistyla（DC.）D. Dietr. = Macrosphyra longistyla（DC.）Hiern ●☆

367783　Solena macrantha（Schult.）D. Dietr. = Euclinia longiflora Salisb. ●☆

367784　Solena maculata（DC.）D. Dietr. = Rothmannia longiflora Salisb. ●☆

367785　Solena madagascariensis（Lam.）D. Dietr. = Hyperacanthus madagascariensis（Lam.）Rakotonas. et A. P. Davis ●☆

367786　Solenacanthus C. Muell. = Ruellia L. ■●

367787　Solenacanthus C. Muell. = Solaenacanthus Oerst. ■●

367788　Solenachne Steud. = Spartina Schreb. ex J. F. Gmel. ■

367789　Solenandra（Reissek）Kuntze = Stenanthemum Reissek ●☆

367790　Solenandra Benth. et Hook. f. = Galax Sims（保留属名）■☆

367791　Solenandra Benth. et Hook. f. = Solenandria P. Beauv. ex Vent. ■☆

367792　Solenandra Hook. f. = Exostema（Pers.）Humb. et Bonpl. ●☆

367793　Solenandra Kuntze = Stenanthemum Reissek ●☆

367794　Solenandria P. Beauv. ex Vent. = Galax Sims（保留属名）■☆

367795　Solenangis Schltr.（1918）；沟管兰属■☆

367796　Solenangis angustifolia Summerh. = Solenangis conica（Schltr.）L. Jonss. ■☆

367797　Solenangis aphylla（Thouars）Summerh. = Microcoelia aphylla（Thouars）Summerh. ■☆

367798　Solenangis clavata（Rolfe）Schltr. ；棒状沟管兰■☆

367799　Solenangis conica（Schltr.）L. Jonss. ；沟管兰■☆

367800　Solenangis scandens（Schltr.）Schltr. ；非洲沟管兰■☆

367801　Solenangis wakefieldii（Rolfe）P. J. Cribb et J. Stewart；瓦氏沟管兰■☆

367802　Solenantha G. Don = Hymenanthera R. Br. ●☆

367803　Solenantha G. Don = Melicytus J. R. Forst. et G. Forst. ●☆

367804　Solenanthus Ledeb.（1829）；长蕊琉璃草属（长筒琉璃草属，管花属）；Glazegrass, Solenanthus ■☆

367805　Solenanthus Steud. ex Klatt = Acidanthera Hochst. ■

367806　Solenanthus Steud. ex Klatt = Gladiolus L. ■

367807　Solenanthus amplifolius Boiss. = Solenanthus circinnatus Ledeb. ■☆

367808　Solenanthus atlanticus Pit. = Cynoglossum pitardianum Greuter et Burdet ●☆

367809　Solenanthus biebersteinii DC. ；毕氏长蕊琉璃草■☆

367810　Solenanthus brachystemon Fisch. et C. A. Mey. ；短长蕊琉璃草■☆

367811　Solenanthus circinnatus Ledeb. ；长蕊琉璃草；Circinate Glazegrass, Circinate Solenanthus ■

367812　Solenanthus coronatus Regel = Solenanthus circinnatus Ledeb. ■

367813　Solenanthus hirsutus Regel；毛长蕊琉璃草■☆

367814　Solenanthus hupehensis R. R. Mill；湖北长蕊琉璃草■

367815　Solenanthus karateginus Lipsky；卡拉特金长蕊琉璃草■☆

367816　Solenanthus kokanicus Regel；浩罕长蕊琉璃草■☆

367817　Solenanthus lanatus（L.）DC. = Cynoglossum mathezii Greuter et Burdet ■☆

367818　Solenanthus lanatus（L.）DC. var. glabrescens Batt. = Cynoglossum tubiflorum（Murb.）Greuter et Burdet ■☆

367819　Solenanthus nigricans Schrenk ex Fisch. et C. A. Mey. = Lindelofia stylosa（Kar. et Kir.）Brand ■

367820　Solenanthus nigricans Schrenk ex Fisch. et C. A. Mey. var. pterocarpus Rupr. = Lindelofia stylosa（Kar. et Kir.）Brand subsp. pterocarpa（Rupr.）Kamelin ■

367821　Solenanthus petiolaris DC. = Solenanthus circinnatus Ledeb. ■

367822　Solenanthus plantaginifolius Lipsky；车前叶长蕊琉璃草■☆

367823　Solenanthus pteiolaris DC. = Solenanthus circinnatus Ledeb. ■

367824　Solenanthus rumicifolius Boiss. = Solenanthus circinnatus Ledeb. ■

367825　Solenanthus stamineus（Desf.）Wettst. ；雄长蕊琉璃草■☆

367826　Solenanthus tubiflorus Murb. = Cynoglossum tubiflorum（Murb.）Greuter et Burdet ■☆

367827　Solenanthus tubiflorus Murb. var. glabrescens（Batt.）Maire = Cynoglossum tubiflorum（Murb.）Greuter et Burdet ■☆

367828　Solenanthus turkestanicus（Regel et Smirn.）Kusn. ；土耳其斯坦长蕊琉璃草■☆

367829　Solenanthus watieri Batt. et Maire = Cynoglossum watieri（Batt. et Maire）Braun-Blanq. et Maire ■☆

367830　Solenarium Dulac = Gagea Salisb. ■

367831　Solenidiopsis Senghas（1986）；拟小管兰属■☆

367832　Solenidiopsis flavobrunnea Senghas；拟小管兰■☆

367833　Solenidium Lindl.（1846）；小管兰属■☆

367834　Solenidium racemosum Lindl. ；小管兰■☆

367835　Solenipedium Beer = Selenipedium Rchb. f. ■☆

367836　Soleniscia DC. = Styphelia（Sol. ex G. Forst.）Sm. ●☆

367837　Solenisia Steud. = Soleniscia DC. ●☆

367838　Solenixora Baill. = Coffea L. ●

367839　Solenocalyx Tiegh. = Psittacanthus Mart. ●☆

367840　Solenocarpus Wight et Arn. = Spondias L. ●

367841　Solenocentrum Schltr.（1911）；管距兰属■☆

367842　Solenocentrum costaricense Schltr. ；管距兰■☆

367843　Solenochasma Fenzl = Dicliptera Juss.（保留属名）■

367844　Solenochasma Fenzl = Justicia L. ●■

367845　Solenogyne Cass.（1828）；短喙菊属■●☆

367846　Solenogyne Cass.（1828）= Lagenifera Cass. ■●

367847　Solenogyne Cass.（1897）= Eriocephalus L. ●☆

367848　Solenogyne mikadoi Koidz. ；短喙菊■☆

367849　Solenolantana（Nakai）Nakai = Viburnum L. ●

367850　Solenomeles T. Durand et Jacks. = Solenomelus Miers ●☆

367851　Solenomelus Miers（1842）；管鸢尾属■☆

367852　Solenomelus biflorus Baker；双花管鸢尾■☆

367853　Solenomelus chilensis Miers；管鸢尾■☆

367854　Solenophora Benth.（1840）；管梗苣苔属●☆

367855　Solenophora australis C. V. Morton；巴拿马管梗苣苔●☆

367856　Solenophora coccinea Benth. ；管梗苣苔●☆

367857　Solenophora maculata D. N. Gibson；斑点管梗苣苔●☆

367858　Solenophyllum Baill. = Monanthochloe Engelm. ■☆

367859　Solenophyllum Nutt. ex Baill. = Monanthochloe Engelm. ■☆

367860　Solenopsis C. Presl = Laurentia Neck. ■☆

367861　Solenopsis C. Presl（1836）；茅瓜桔梗属；Shrub-harebell ■☆

367862　Solenopsis bicolor（Batt.）Greuter et Burdet；二色茅瓜桔梗■☆

367863　Solenopsis laurentia（L.）C. Presl；茅瓜桔梗■☆

367864　Solenoruellia Baill. = Tetramerium Nees（保留属名）●☆

367865　Solenospermum Zoll. = Lophopetalum Wight ex Arn. ●☆

367866　Solenostemma Hayne（1825）；筒冠萝藦属（狭冠花属）●☆

367867　Solenostemma argel（Delile）Hayne；筒冠萝藦（狭冠花）●☆

367868　Solenostemma argel Hayne = Solenostemma argel（Delile）Hayne ●☆

367869　Solenostemma oleifolium（Nectoux）Bullock et E. A. Bruce ex Maire = Solenostemma argel（Delile）Hayne ●☆

367870　Solenostemon Thonn.（1827）；鞘蕊属（管蕊花属，五彩苏属）；

Flame Nettle , Coleus ■

367871　Solenostemon Thonn. = Plectranthus L'Hér. (保留属名) ●■

367872　Solenostemon africanus Briq. ; 非洲鞘蕊 ■☆

367873　Solenostemon africanus Briq. = Plectranthus monostachyus (P. Beauv.) B. J. Pollard ■☆

367874　Solenostemon autranii (Briq.) J. K. Morton ; 林斑点鞘蕊 ■☆

367875　Solenostemon bernieri (Briq.) Guillaumet et Cornet = Plectranthus bojeri (Benth.) Hedge ■☆

367876　Solenostemon bojeri (Benth.) Guillaumet et Cornet = Plectranthus bojeri (Benth.) Hedge ■☆

367877　Solenostemon bullatus Briq. = Coleus bullulatus Briq. ■☆

367878　Solenostemon calaminthoides Baker ; 新风轮鞘蕊 ■☆

367879　Solenostemon chevalieri Briq. = Plectranthus chevalieri (Briq.) B. J. Pollard et A. J. Paton ■☆

367880　Solenostemon collinus (Lebrun et L. Touss.) Troupin ; 山丘鞘蕊 ■☆

367881　Solenostemon cymosus (Baker) Guillaumet et Cornet = Plectranthus bojeri (Benth.) Hedge ■☆

367882　Solenostemon cymosus Baker = Plectranthus occidentalis B. J. Pollard ■☆

367883　Solenostemon decumbens (Hook. f.) Baker = Plectranthus decumbens Hook. f. ■☆

367884　Solenostemon giorgii (De Wild.) Champl. ; 乔治鞘蕊 ■☆

367885　Solenostemon godefroyae N. E. Br. ; 戈德曼鞘蕊 ■☆

367886　Solenostemon gouanensis A. Chev. = Plectranthus monostachyus (P. Beauv.) B. J. Pollard subsp. marrubiifolius (Brenan) B. J. Pollard ■☆

367887　Solenostemon goudotii (Briq.) Guillaumet et Cornet = Plectranthus persoonii (Benth.) Hedge ■☆

367888　Solenostemon gracilifolius (Briq.) Guillaumet et Cornet = Plectranthus bojeri (Benth.) Hedge ■☆

367889　Solenostemon graniticola A. Chev. = Plectranthus saxicola B. J. Pollard et A. J. Paton ■☆

367890　Solenostemon graniticola J. K. Morton = Plectranthus monostachyus (P. Beauv.) B. J. Pollard subsp. marrubiifolius (Brenan) B. J. Pollard ■☆

367891　Solenostemon koualensis (A. Chev. ex Hutch. et Dalziel) J. K. Morton = Plectranthus koualensis (A. Chev. ex Hutch. et Dalziel) B. J. Pollard et A. J. Paton ■☆

367892　Solenostemon latericola A. Chev. = Plectranthus monostachyus (P. Beauv.) B. J. Pollard subsp. latericola (A. Chev.) B. J. Pollard ■☆

367893　Solenostemon latifolius (Hochst. ex Benth.) J. K. Morton ; 宽叶鞘蕊 ■☆

367894　Solenostemon latifolius (Hochst. ex Benth.) J. K. Morton = Plectranthus bojeri (Benth.) Hedge ■☆

367895　Solenostemon linearifolius J. K. Morton = Plectranthus linearifolius (J. K. Morton) B. J. Pollard et A. J. Paton ☆

367896　Solenostemon mannii (Hook. f.) Baker = Plectranthus occidentalis B. J. Pollard ■☆

367897　Solenostemon minor J. K. Morton = Plectranthus decumbens Hook. f. ■☆

367898　Solenostemon monostachyus (P. Beauv.) Briq. ; 单穗鞘蕊 ■☆

367899　Solenostemon monostachyus (P. Beauv.) Briq. subsp. latericola (A. Chev.) J. K. Morton = Plectranthus monostachyus (P. Beauv.) B. J. Pollard subsp. latericola (A. Chev.) B. J. Pollard ■☆

367900　Solenostemon monostachyus (P. Beauv.) Briq. subsp.

367901　Solenostemon monostachyus (P. Beauv.) Briq. var. granaticola (A. Chev.) Brenan = Plectranthus saxicola B. J. Pollard et A. J. Paton ■☆

367902　Solenostemon monostachyus (P. Beauv.) Briq. var. marrubiifolius Brenan = Plectranthus monostachyus (P. Beauv.) B. J. Pollard subsp. marrubiifolius (Brenan) B. J. Pollard ■☆

367903　Solenostemon niveus Hiern ; 雪白鞘蕊 ■☆

367904　Solenostemon ocymoides Schumach. et Thonn. ; 罗勒鞘蕊 (罗勒管蕊花) ■☆

367905　Solenostemon ocymoides Schumach. et Thonn. = Plectranthus monostachyus (P. Beauv.) B. J. Pollard ■☆

367906　Solenostemon ocymoides Schumach. et Thonn. var. monostachyus (P. Beauv.) Baker = Plectranthus monostachyus (P. Beauv.) B. J. Pollard ■☆

367907　Solenostemon paniculatus (Pers.) Guillaumet et Cornet = Plectranthus persoonii (Benth.) Hedge ■☆

367908　Solenostemon platostomoides (Robyns et Lebrun) Troupin ; 平口花鞘蕊 ■☆

367909　Solenostemon porpeodon (Baker) J. K. Morton ; 小叶鞘蕊 ■☆

367910　Solenostemon pumilus ? = Coleus pumilus Blanco ■

367911　Solenostemon repens (Gürke) J. K. Morton = Plectranthus epilithicus B. J. Pollard ■☆

367912　Solenostemon robustus Hiern = Holostylon robustum (Hiern) G. Taylor ☆

367913　Solenostemon rotundifolius (Poir.) J. K. Morton ; 圆叶鞘蕊 ■☆

367914　Solenostemon rutenbergianus (Vatke) Guillaumet et Cornet = Plectranthus rutenbergianus Vatke ■☆

367915　Solenostemon scutellarioides L. = Coleus scutellarioides (L.) Benth. ■

367916　Solenostemon shirensis (Gürke) Codd = Solenostemon autranii (Briq.) J. K. Morton ■☆

367917　Solenostemon sylvaticus (Gürke) Agnew = Solenostemon autranii (Briq.) J. K. Morton ■☆

367918　Solenostemon thyrsiflorus (Lebrun et L. Touss.) Troupin ; 聚伞鞘蕊 ■☆

367919　Solenostemon zambesiacus Baker = Solenostemon autranii (Briq.) J. K. Morton ■☆

367920　Solenosterigma Klotasch ex K. Krause = Philodendron Schott (保留属名) ■●

367921　Solenostigma Endl. = Celtis L. ●

367922　Solenostigma Klotzsch ex Walp. = Retzia Thunb. ●☆

367923　Solenostigma consimile Blume = Celtis philippensis Blanco var. consimilis (Blume) Leroy ●

367924　Solenostigma consimile Blume = Celtis philippensis Blanco var. wightii (Planch.) Soepadmo ●

367925　Solenostyles Hort. ex Pasq. (1867) ; 鞘柱爵床属 ■☆

367926　Solenostyles Pasq. = Solenostyles Hort. ex Pasq. ■☆

367927　Solenostyles aurantiacus Hort. ex Pasq. ; 鞘柱爵床 ■☆

367928　Solenotheca Nutt. = Tagetes L. ■●

367929　Solenotinus (DC.) Spach = Viburnum L. ●

367930　Solenotinus Oerst. = Viburnum L. ●

367931　Solenotinus Spach = Viburnum L. ●

367932　Solenotinus erubescens (Wall.) Oerst. = Viburnum erubescens Wall. ●

367933 Solenotinus nervosus (D. Don) Oerst. = Viburnum nervosum D. Don ●

367934 Solenotinus nervosus Oerst. = Viburnum grandiflorum Wall. ex DC. ●

367935 Solenotus (Steven) Steven = Astragalus L. ●■

367936 Solenotus Steven = Astragalus L. ●■

367937 Solfia Rech. = Drymophloeus Zipp. ●☆

367938 Solia Noronha = Premna L. (保留属名) ●■

367939 Solidago L. (1753) ; 一枝黄花属; Golden Rod, Goldenrod, Solidago ■

367940 Solidago Mill. = Senecio L. ■●

367941 Solidago ' Goldkind ' ; 金娃娃一枝黄花; Golden Baby Goldenrod ■☆

367942 Solidago × raymondii J. Rousseau = Solidago simplex Kunth var. racemosa (Greene) G. S. Ringius ■☆

367943 Solidago aestivalis E. P. Bicknell = Solidago rugosa Mill. var. sphagnophila C. Graves ■☆

367944 Solidago alba Mill. = Solidago bicolor L. ■☆

367945 Solidago albopilosa E. L. Braun; 白茅一枝黄花; White-haired Goldenrod ■☆

367946 Solidago algida Piper = Solidago multiradiata Aiton ■☆

367947 Solidago alleghaniensis House = Solidago roanensis Porter ■☆

367948 Solidago alpestris Waldst. et Kit. ; 亚高山一枝黄花 ■☆

367949 Solidago altiplanities C. E. S. Taylor et R. J. Taylor; 高原一枝黄花; High-plains Goldenrod ■☆

367950 Solidago altissima L. ; 高大一枝黄花; Late Goldenrod, Tall Goldenrod ■☆

367951 Solidago altissima L. = Solidago canadensis L. var. scabra Torr. et A. Gray ■☆

367952 Solidago altissima L. = Solidago canadensis L. ■

367953 Solidago altissima L. subsp. gilvocanescens (Rydb.) Semple; 灰一枝黄花; Great Plains late Goldenrod ■☆

367954 Solidago altissima L. var. gilvocanescens (Rydb.) Semple = Solidago canadensis L. var. gilbocanescens Rydb. ■☆

367955 Solidago altissima L. var. gilvocanscens (Rydb.) Semple = Solidago altissima L. subsp. gilvocanescens (Rydb.) Semple ■☆

367956 Solidago altissima L. var. pluricephala M. C. Johnst. = Solidago canadensis L. var. scabra Torr. et A. Gray ■☆

367957 Solidago altissima L. var. procera (Aiton) Fernald = Solidago canadensis L. var. scabra Torr. et A. Gray ■☆

367958 Solidago ambigua Aiton var. curtisii (Torr. et A. Gray) A. W. Wood = Solidago curtisii Torr. et A. Gray ■☆

367959 Solidago ambigua Aiton var. lancifolia Torr. et A. Gray = Solidago lancifolia (Torr. et A. Gray) Chapm. ■☆

367960 Solidago amplexicaulis M. Martens = Solidago riddellii Frank ex Riddell ■☆

367961 Solidago amplexicaulis Torr. et A. Gray ex Chapm. = Solidago auriculata Shuttlew. ex S. F. Blake ■☆

367962 Solidago angusta Torr. et A. Gray = Solidago petiolaris Aiton ■☆

367963 Solidago anticostensis Fernald = Solidago simplex Kunth var. racemosa (Greene) G. S. Ringius ■☆

367964 Solidago arenicola B. R. Keener et Král; 南部山地一枝黄花; Southern Racemose Goldenrod ■☆

367965 Solidago arguta Aiton; 尖一枝黄花(狭叶一枝黄花) ; Atlantic Goldenrod, Cut-leaf Goldenrod, Goldenrod, Sharp-leaved Goldenrod ■☆

367966 Solidago arguta Aiton subsp. boottii (Hook.) G. H. Morton =

Solidago arguta Aiton var. boottii (Hook.) E. J. Palmer et Steyerm. ■☆

367967 Solidago arguta Aiton subsp. pseudoyadkinensis G. H. Morton = Solidago arguta Aiton var. caroliniana A. Gray ■☆

367968 Solidago arguta Aiton var. boottii (Hook.) E. J. Palmer et Steyerm. ; 布特一枝黄花; Boott's Goldenrod ■☆

367969 Solidago arguta Aiton var. caroliniana A. Gray; 卡罗来纳一枝黄花 ■☆

367970 Solidago arguta Aiton var. harrisii (E. S. Steele) Cronquist; 哈里斯一枝黄花; Harris' Goldenrod, Shale-barren Goldenrod ■☆

367971 Solidago arguta Aiton var. juncea (Aiton) Torr. et A. Gray = Solidago juncea Aiton ■☆

367972 Solidago arguta Aiton var. scabrella Torr. et A. Gray = Solidago juncea Aiton ■☆

367973 Solidago arguta Aiton var. strigosa (Small) Steyerm. = Solidago ludoviciana (A. Gray) Small ■☆

367974 Solidago arizonica (A. Gray) Wooton et Standl. = Solidago velutina DC. subsp. sparsiflora (A. Gray) Semple ■☆

367975 Solidago aspera Aiton = Solidago rugosa Aiton var. aspera (Aiton) Fernald ■☆

367976 Solidago aspericaulis A. H. Moore = Solidago fistulosa Mill. ■☆

367977 Solidago asterifolia Small = Solidago curtisii Torr. et A. Gray var. flaccidifolia (Small) R. E. Cook et Semple ■☆

367978 Solidago asteroides Semple = Solidago ptarmicoides (Torr. et A. Gray) B. Boivin ■☆

367979 Solidago aureola Greene = Solidago simplex Kunth ■☆

367980 Solidago auriculata Shuttlew. ex S. F. Blake; 耳状一枝黄花; Clasping Goldenrod, Claspingleaf Goldenrod, Eared Goldenrod ■☆

367981 Solidago austrina Small = Solidago stricta Aiton subsp. gracillima (Torr. et A. Gray) Semple ■☆

367982 Solidago axillaris Pursh = Solidago caesia L. ■☆

367983 Solidago bellidiflora Greene = Solidago simplex Kunth var. nana (A. Gray) G. S. Ringius ■☆

367984 Solidago bernardii B. Boivin; 伯纳德一枝黄花; Bernard's Goldenrod, Yellow Stiff Aster ■☆

367985 Solidago bernardii B. Boivin = Solidago ptarmicoides (Torr. et A. Gray) B. Boivin ■☆

367986 Solidago bicolor L. ; 一枝银花; Late Golden Rod, Pale Goldenrod, Silver Rod, Silverrod, Silver-rod, White Golden Rod, White Goldenrod ■☆

367987 Solidago bicolor L. var. concolor Torr. et A. Gray = Solidago hispida Muhl. ex Willd. ■☆

367988 Solidago bicolor L. var. hispida (Muhl. ex Willd.) Britton, Sterns et Poggenb. = Solidago hispida Muhl. ex Willd. ■☆

367989 Solidago bicolor L. var. lanata (Hook.) A. Gray = Solidago hispida Muhl. ex Willd. ■☆

367990 Solidago bicolor L. var. luteola Farw. = Solidago hispida Muhl. ex Willd. ■☆

367991 Solidago bicolor L. var. ovalis Farw. = Solidago hispida Muhl. ex Willd. ■☆

367992 Solidago bicolor L. var. spathulata Farw. = Solidago hispida Muhl. ex Willd. ■☆

367993 Solidago bicolor L. var. tonsa (Fernald) B. Boivin = Solidago hispida Muhl. ex Willd. ■☆

367994 Solidago bigelovii A. Gray = Solidago wrightii A. Gray ■☆

367995 Solidago bigelovii A. Gray var. wrightii (A. Gray) A. Gray = Solidago wrightii A. Gray ■☆

367996 Solidago bombycinum (Lunell) Friesner = Solidago rigida L. subsp. humilis (Porter) S. B. Heard et Semple ■☆

367997 Solidago boottii Hook. = Solidago arguta Aiton var. boottii (Hook.) E. J. Palmer et Steyerm. ■☆

367998 Solidago boottii Hook. var. brachyphylla (Chapm. ex Torr. et A. Gray) A. Gray = Solidago brachyphylla Chapm. ex Torr. et A. Gray ■☆

367999 Solidago boottii Hook. var. caroliniana (A. Gray) Cronquist = Solidago arguta Aiton var. caroliniana A. Gray ■☆

368000 Solidago boottii Hook. var. ludoviciana A. Gray = Solidago ludoviciana (A. Gray) Small ■☆

368001 Solidago brachyphylla Chapm. ex Torr. et A. Gray；短叶一枝黄花；Dixie Goldenrod ■☆

368002 Solidago buckleyi Torr. et A. Gray；巴克利一枝黄花；Buckley's Goldenrod , Goldenrod ■☆

368003 Solidago caesia L. ；蓝茎一枝黄花；Axillary Goldenrod , Blue-stemmed Goldenrod , Woodland Goldenrod , Wreath Golden Rod , Wreath Goldenrod ■☆

368004 Solidago caesia L. var. axillaris (Pursh) A. Gray = Solidago caesia L. ■☆

368005 Solidago caesia L. var. curtisii (Torr. et A. Gray) C. E. S. Taylor et R. J. Taylor = Solidago curtisii Torr. et A. Gray ■☆

368006 Solidago caesia L. var. hispida A. Wood = Solidago curtisii Torr. et A. Gray ■☆

368007 Solidago caesia L. var. paniculata A. Gray = Solidago caesia L. ■☆

368008 Solidago californica Nutt. = Solidago velutina DC. subsp. californica (Nutt.) Semple ■☆

368009 Solidago californica Nutt. var. aperta L. F. Hend. = Solidago velutina DC. subsp. californica (Nutt.) Semple ■☆

368010 Solidago californica Nutt. var. nevadensis A. Gray = Solidago velutina DC. subsp. sparsiflora (A. Gray) Semple ■☆

368011 Solidago camporum (Greene) A. Nelson = Euthamia gymnospermoides Greene ■☆

368012 Solidago camporum Greene var. tricostata Lunell = Euthamia graminifolia (L.) Nutt. ■☆

368013 Solidago canadensis L. ；加拿大一枝黄花（北美一枝黄花, 高茎一枝黄花, 金棒草, 美洲一枝黄花）；Canada Goldenrod , Canadian Golden Rod , Canadian Goldenrod , Common Goldenrod , Goldenrod , Meadow Goldenrod , Rock Golden Rod , Tall Golden Rod , Tall Goldenrod ■

368014 Solidago canadensis L. subsp. altissima (L.) O. Bolòs et Vigo = Solidago altissima L. ■☆

368015 Solidago canadensis L. subsp. elongata (Nutt.) D. D. Keck = Solidago canadensis L. var. salebrosa (Piper) M. E. Jones ■☆

368016 Solidago canadensis L. subsp. elongata (Nutt.) D. D. Keck = Solidago elongata Nutt. ■☆

368017 Solidago canadensis L. subsp. gilbocanescens (Rydb.) Á. Löve et D. Löve = Solidago canadensis L. var. gilbocanescens Rydb. ■☆

368018 Solidago canadensis L. subsp. gilvocanescens (Rydb.) Á. Löve et D. Löve = Solidago altissima L. subsp. gilvocanescens (Rydb.) Semple ■☆

368019 Solidago canadensis L. subsp. salebrosa (Piper) D. D. Keck = Solidago canadensis L. var. salebrosa (Piper) M. E. Jones ■☆

368020 Solidago canadensis L. subsp. salebrosa (Piper) D. D. Keck = Solidago lepida DC. var. salebrosa (Piper) Semple ■☆

368021 Solidago canadensis L. var. arizonica A. Gray = Solidago velutina DC. subsp. sparsiflora (A. Gray) Semple ■☆

368022 Solidago canadensis L. var. canescens A. Gray = Solidago juliae G. L. Nesom ■☆

368023 Solidago canadensis L. var. canescens A. Gray = Solidago nemoralis Aiton subsp. decemflora (DC.) Brammall ex Semple ■☆

368024 Solidago canadensis L. var. elongata (Nutt.) M. Peck = Solidago canadensis L. var. salebrosa (Piper) M. E. Jones ■☆

368025 Solidago canadensis L. var. fallax (Fernald) Beaudry = Solidago lepida DC. subsp. fallax (Fernald) Semple ■☆

368026 Solidago canadensis L. var. gilbocanescens Rydb. ；灰黄一枝黄花；Great Plains Canadian Goldenrod , Short-hair Goldenrod ■☆

368027 Solidago canadensis L. var. gilvocanescens Rydb. = Solidago altissima L. ■☆

368028 Solidago canadensis L. var. gilvocanescens Rydb. = Solidago altissima L. subsp. gilvocanescens (Rydb.) Semple ■☆

368029 Solidago canadensis L. var. hargeri Fernald；哈格一枝黄花；Canadian Goldenrod , Harger's Goldenrod ■☆

368030 Solidago canadensis L. var. lepida (DC.) Cronquist = Solidago lepida DC. ■☆

368031 Solidago canadensis L. var. rupestris (Raf.) Porter = Solidago rupestris Raf. ■☆

368032 Solidago canadensis L. var. salebrosa (Piper) M. E. Jones；加拿大岩地一枝黄花；Canadian Goldenrod , Rocky Mountains Canada Goldenrod , Salebrosa Goldenrod ■☆

368033 Solidago canadensis L. var. salebrosa (Piper) M. E. Jones = Solidago lepida DC. var. salebrosa (Piper) Semple ■☆

368034 Solidago canadensis L. var. scabra (Muhl. ex Willd.) Torr. et A. Gray = Solidago altissima L. ■☆

368035 Solidago canadensis L. var. scabra Torr. et A. Gray；高加拿大一枝黄花；Canadian Goldenrod , Common Goldenrod , Tall Goldenrod ■☆

368036 Solidago canescens (Rydb.) Friesner = Solidago rigida L. subsp. humilis (Porter) S. B. Heard et Semple ■☆

368037 Solidago cantonensis Lour. = Solidago decurrens Lour. ■

368038 Solidago caroliniana Britton , Sterns et Poggenb. = Euthamia tenuifolia (Pursh) Nutt. ■☆

368039 Solidago castrensis E. S. Steele = Solidago speciosa Nutt. ■☆

368040 Solidago caurina Piper = Solidago elongata Nutt. ■☆

368041 Solidago chandonnettii E. S. Steele = Solidago speciosa Nutt. var. rigidiuscula Torr. et A. Gray ■☆

368042 Solidago chapmanii A. Gray = Solidago odora Aiton subsp. chapmanii (A. Gray) Semple ■☆

368043 Solidago chinensis Osbeck = Wedelia chinensis (Osbeck) Merr. ■

368044 Solidago chlorolepis Fernald = Solidago simplex Kunth var. chlorolepis (Fernald) G. S. Ringius ■☆

368045 Solidago chrysolepis Fernald = Solidago uliginosa Nutt. ■☆

368046 Solidago chrysopsis Small = Solidago stricta Aiton ■☆

368047 Solidago chrysothamnoides (Greene) Bush = Euthamia gymnospermoides Greene ■☆

368048 Solidago cleliae DC. = Solidago gigantea Aiton ■☆

368049 Solidago compacta Turcz. ；密集一枝黄花■☆

368050 Solidago concinna A. Nelson = Solidago missouriensis Nutt. ■☆

368051 Solidago conferta Mill. = Solidago speciosa Nutt. ■☆

368052 Solidago confinis A. Gray；毗邻一枝黄花；Southern Goldenrod ■☆

368053 Solidago confinis A. Gray var. luxurians (H. M. Hall) Jeps. = Solidago confinis A. Gray ■☆

368054　Solidago cordata Short et R. Peter = Solidago sphacelata Raf. ■☆

368055　Solidago corymbosa Elliott = Solidago rigida L. subsp. glabrata (E. L. Braun) S. B. Heard et Semple ■☆

368056　Solidago cuprea Juz. ;铜色一枝黄花■☆

368057　Solidago curtisii Torr. et A. Gray;库特斯一枝黄花;Curtis' Goldenrod ■☆

368058　Solidago curtisii Torr. et A. Gray var. flaccidifolia (Small) R. E. Cook et Semple;软叶一枝黄花;Appalachian Goldenrod, Mountain Goldenrod ■☆

368059　Solidago curtisii Torr. et A. Gray var. monticola Torr. et A. Gray = Solidago roanensis Porter ■☆

368060　Solidago curtisii Torr. et A. Gray var. pubens (M. A. Curtis ex Torr. et A. Gray) A. Gray = Solidago bicolor L. ■☆

368061　Solidago cusickii Piper = Solidago multiradiata Aiton ■☆

368062　Solidago cuspidata Wall. = Duhaldea cuspidata (DC.) Anderb. ●

368063　Solidago cuspidata Wall. = Inula cuspidata C. B. Clarke ●

368064　Solidago cutleri Fernald;高山一枝黄花;Alpine Goldenrod ■☆

368065　Solidago cutleri Fernald = Solidago leiocarpa DC. ☆

368066　Solidago dahurica Kitag. = Solidago virgaurea L. var. dahurica Kitag. ■

368067　Solidago deamii Fernald = Solidago simplex Kunth var. gillmanii (A. Gray) G. S. Ringius ■☆

368068　Solidago decemflora A. Gray = Solidago radula Nutt. ■☆

368069　Solidago decemflora DC. = Solidago nemoralis Aiton subsp. decemflora (DC.) Brammall ex Semple ■☆

368070　Solidago decumbens Greene = Solidago simplex Kunth ■☆

368071　Solidago decumbens Greene var. chlorolepis (Fernald) Beaudry = Solidago simplex Kunth var. chlorolepis (Fernald) G. S. Ringius ■☆

368072　Solidago decumbens Greene var. oreophila (Rydb.) Fernald = Solidago simplex Kunth ■☆

368073　Solidago decurrens Lour. ;一枝黄花(百根草,百条根,朝天一炷香,大败毒,大叶七星剑,红柴胡,红胶苦菜,黄柴胡,黄花草,黄花儿,黄花马兰,黄花细辛,黄花一枝香,黄花仔,见血飞,金柴胡,金锁匙,老虎尿,满山黄,黏糊菜,破布叶,千根癀,洒金花,洒金兰,山边半枝香,山厚合,蛇头黄,蛇头王,铁金拐,土泽兰,苍子草,小白龙须,野黄菊,一枝箭,一枝枪,一枝香);Common Goldenrod ■☆

368074　Solidago decurrens Lour. = Solidago virgaurea L. subsp. leiocarpa (Benth.) Hultén ■

368075　Solidago decurrens Lour. f. padudosa (Honda) Kitam. = Solidago decurrens Lour. ■

368076　Solidago decurrens Lour. var. gigantea (Nakai) Ohwi = Solidago virgaurea L. subsp. gigantea (Nakai) Kitam. ■☆

368077　Solidago delicatula Small;光榆叶一枝黄花;Smooth Elm-leaf Goldenrod ■☆

368078　Solidago dilatata A. Nelson = Solidago multiradiata Aiton ■☆

368079　Solidago discoidea (Elliott) Torr. et A. Gray = Brintonia discoidea (Elliott) Greene ■☆

368080　Solidago drummondii Torr. et A. Gray;德拉蒙德一枝黄花;Drummond's Goldenrod, Goldenrod ■☆

368081　Solidago dumetorum Lunell = Solidago gigantea Aiton ■☆

368082　Solidago duriuscula Greene = Solidago missouriensis Nutt. ■☆

368083　Solidago earlei Small = Solidago hispida Muhl. ex Willd. ■☆

368084　Solidago edisoniana Mack. = Solidago latissimifolia Mill. ■☆

368085　Solidago elliottii Torr. et A. Gray = Solidago latissimifolia Mill. ■☆

368086　Solidago elliottii Torr. et A. Gray var. ascendens Fernald = Solidago latissimifolia Mill. ■☆

368087　Solidago elliottii Torr. et A. Gray var. divaricata Fernald = Solidago latissimifolia Mill. ■☆

368088　Solidago elliottii Torr. et A. Gray var. edisoniana (Mack.) Fernald = Solidago latissimifolia Mill. ■☆

368089　Solidago elliottii Torr. et A. Gray var. pedicellata Fernald = Solidago latissimifolia Mill. ■☆

368090　Solidago elliptica Aiton = Solidago latissimifolia Mill. ■☆

368091　Solidago elongata Nutt. ; 长一枝黄花; Cascade Canada Goldenrod, West Coast Goldenrod ■☆

368092　Solidago elongata Nutt. = Solidago canadensis L. var. salebrosa (Piper) M. E. Jones ■☆

368093　Solidago elongata Nutt. var. fallax (Fernald) G. N. Jones = Solidago lepida DC. subsp. fallax (Fernald) Semple ■☆

368094　Solidago elongata Nutt. var. microcephala Kellogg = Solidago elongata Nutt. ■☆

368095　Solidago erecta Banks ex Pursh;直立一枝黄花;Erect Goldenrod, Slender Goldenrod ■☆

368096　Solidago faucibus Wieboldt;山谷一枝黄花;Gorge Goldenrod ■☆

368097　Solidago fistulosa Mill. ;松林一枝黄花;Pine-barren Goldenrod ■☆

368098　Solidago flaccidifolia Small = Solidago curtisii Torr. et A. Gray var. flaccidifolia (Small) R. E. Cook et Semple ■☆

368099　Solidago flavovirens Chapm. = Solidago stricta Aiton ■☆

368100　Solidago flexicaulis L. ;宽叶一枝黄花;Broadleaf Goldenrod, Broad-leaved Golden Rod, Broadleaved Goldenrod, Zigzag Golden Rod, Zigzag Goldenrod, Zig-zag Goldenrod ■☆

368101　Solidago flexicaulis L. var. ciliata DC. = Solidago flexicaulis L. ■☆

368102　Solidago flexicaulis L. var. latifolia (L.) Pursh = Solidago flexicaulis L. ■☆

368103　Solidago garrettii Rydb. = Solidago velutina DC. subsp. sparsiflora (A. Gray) Semple ■☆

368104　Solidago gattingeri Chapm. ex A. Gray;加氏一枝黄花;Gattinger's Goldenrod, Goldenrod ■☆

368105　Solidago gebleri Juz. ;格布勒一枝黄花■☆

368106　Solidago gigantea Aiton;巨大一枝黄花(晚花一枝黄花);Blue Mountain Tea, Early Golden Rod, Early Goldenrod, Giant Goldenrod, Goldenrod, Late Goldenrod, November Goldenrod, Smooth Goldenrod, Smooth Three-ribbed Golden Rod, Tall Goldenrod ■☆

368107　Solidago gigantea Aiton subsp. serotina (Kuntze) McNeill = Solidago gigantea Aiton ■☆

368108　Solidago gigantea Aiton var. leiophylla Fernald = Solidago gigantea Aiton subsp. serotina (Kuntze) McNeill ■☆

368109　Solidago gigantea Aiton var. leiophylla Fernald = Solidago gigantea Aiton ■☆

368110　Solidago gigantea Aiton var. pitcheri (Nutt.) Shinners = Solidago gigantea Aiton ■☆

368111　Solidago gigantea Aiton var. salebrosa (Piper) Friesner = Solidago lepida DC. var. salebrosa (Piper) Semple ■☆

368112　Solidago gigantea Aiton var. serotina (Kuntze) Cronquist = Solidago gigantea Aiton ☆

368113　Solidago gigantea Aiton var. shinnersii Beaudry = Solidago gigantea Aiton ■☆

368114　Solidago gillmanii (A. Gray) E. S. Steele = Solidago simplex

Kunth var. gillmanii（A. Gray）G. S. Ringius ■☆

368115　Solidago gilvocanescens（Rydb.）Smyth = Solidago altissima L. subsp. gilvocanescens（Rydb.）Semple ■☆

368116　Solidago gilvocanescens（Rydb.）Smyth = Solidago canadensis L. var. gilbocanescens Rydb. ■☆

368117　Solidago glaberrima M. Martens = Solidago missouriensis Nutt. var. fasciculata Holz. ■☆

368118　Solidago glaberrima M. Martens = Solidago missouriensis Nutt. ■☆

368119　Solidago glaberrima M. Martens var. montana（A. Gray）Lunell = Solidago missouriensis Nutt. ■☆

368120　Solidago glaberrima M. Martens var. moritura（E. S. Steele）E. J. Palmer et Steyerm. = Solidago missouriensis Nutt. ■☆

368121　Solidago glaberrima M. Martens var. moritura（E. S. Steele）E. J. Palmer et Steyerm. = Solidago missouriensis Nutt. var. fasciculata Holz. ■☆

368122　Solidago glaucophylla Rydb. = Solidago missouriensis Nutt. ■☆

368123　Solidago glomerata Michx.；丛生一枝黄花；Clustered Goldenrod，Skunk Goldenrod ■☆

368124　Solidago glutinosa Nutt. = Solidago simplex Kunth ■☆

368125　Solidago glutinosa Nutt. subsp. randii（Porter）Cronquist = Solidago simplex Kunth subsp. randii（Porter）G. S. Ringius ■☆

368126　Solidago glutinosa Nutt. var. chlorolepis（Fernald）G. S. Ringius = Solidago simplex Kunth var. chlorolepis（Fernald）G. S. Ringius ■☆

368127　Solidago glutinosa Nutt. var. gillmanii（A. Gray）Cronquist = Solidago simplex Kunth var. gillmanii（A. Gray）G. S. Ringius ■☆

368128　Solidago glutinosa Nutt. var. nana（A. Gray）Cronquist = Solidago simplex Kunth var. nana（A. Gray）G. S. Ringius ■☆

368129　Solidago glutinosa Nutt. var. ontarioensis G. S. Ringius = Solidago simplex Kunth var. ontarioensis（G. S. Ringius）G. S. Ringius ■☆

368130　Solidago glutinosa Nutt. var. racemosa（Greene）Cronquist = Solidago simplex Kunth var. racemosa（Greene）G. S. Ringius ■☆

368131　Solidago glutinosa Nutt. var. randii（Porter）Cronquist = Solidago simplex Kunth subsp. randii（Porter）G. S. Ringius ■☆

368132　Solidago gracilis Poir. = Solidago caesia L. ■☆

368133　Solidago gracillima Torr. et A. Gray = Solidago stricta Aiton subsp. gracillima（Torr. et A. Gray）Semple ■☆

368134　Solidago graminifolia（L.）Salisb.；禾叶一枝黄花；Fragrant Golden Rod，Grass-leaved Goldenrod，Lance-leaved Goldenrod，Narrow-leaved Goldenrod ■☆

368135　Solidago graminifolia（L.）Salisb. = Euthamia graminifolia（L.）Nutt. ■☆

368136　Solidago graminifolia（L.）Salisb. var. gymnospermoides（Greene）Croat = Euthamia gymnospermoides Greene ■☆

368137　Solidago graminifolia（L.）Salisb. var. major（Michx.）Fernald = Euthamia graminifolia（L.）Nutt. ■☆

368138　Solidago graminifolia（L.）Salisb. var. media（Greene）S. K. Harris = Euthamia gymnospermoides Greene ■☆

368139　Solidago graminifolia（L.）Salisb. var. nuttallii（Greene）Fernald = Euthamia graminifolia（L.）Nutt. ■☆

368140　Solidago graminifolia（L.）Salisb. var. polycephala（Fernald）Fernald = Euthamia graminifolia（L.）Nutt. ■☆

368141　Solidago graminifolia（L.）Salisb. var. septentrionalis Fernald = Euthamia graminifolia（L.）Nutt. ■☆

368142　Solidago graminifolia（L.）Salisb. var. typica Rosend. et Cronquist = Euthamia graminifolia（L.）Nutt. ■☆

368143　Solidago grandiflora Raf. = Solidago rigida L. ■☆

368144　Solidago guiradonis A. Gray；吉拉多一枝黄花；Guirado's Goldenrod ■☆

368145　Solidago guiradonis A. Gray var. spectabilis D. C. Eaton = Solidago spectabilis（D. C. Eaton）A. Gray ■☆

368146　Solidago gymnospermoides（Greene）Fernald = Euthamia gymnospermoides Greene ■☆

368147　Solidago gymnospermoides（Greene）Fernald var. callosa S. K. Harris = Euthamia gymnospermoides Greene ■☆

368148　Solidago hachijoensis Nakai = Solidago virgaurea L. subsp. leiocarpa（Benth.）Hultén var. praeflorens Nakai ■☆

368149　Solidago hapemaniana Rydb. = Solidago missouriensis Nutt. ■☆

368150　Solidago harperi Mack. = Solidago speciosa Nutt. ■☆

368151　Solidago harperi Mack. ex Small = Solidago petiolaris Aiton ■☆

368152　Solidago harperi Mack. ex Small = Solidago speciosa Nutt. ■☆

368153　Solidago harrisii E. S. Steele = Solidago arguta Aiton var. harrisii（E. S. Steele）Cronquist ■☆

368154　Solidago helleri Small = Solidago ulmifolia Muhl. ex Willd. ■☆

368155　Solidago hesperia Howell = Solidago simplex Kunth var. nana（A. Gray）G. S. Ringius ■☆

368156　Solidago heterophylla Nutt. = Solidago multiradiata Aiton ■☆

368157　Solidago heterotricha Wall. = Inula eupatorioides DC. ●

368158　Solidago hirsuta Nutt. = Solidago hispida Muhl. ex Willd. ■☆

368159　Solidago hirsutissima Mill. = Solidago canadensis L. var. scabra Torr. et A. Gray ■☆

368160　Solidago hirtella（Greene）Bush = Euthamia graminifolia（L.）Nutt. ■☆

368161　Solidago hirtipes Fernald = Euthamia graminifolia（L.）Nutt. ■☆

368162　Solidago hispida Muhl. ex Willd.；硬毛一枝黄花；Goldenrod，Hairy Goldenrod ■☆

368163　Solidago hispida Muhl. ex Willd. var. arnoglossa Fernald；软毛一枝黄花；Hairy Goldenrod ■☆

368164　Solidago hispida Muhl. ex Willd. var. tonsa Fernald = Solidago hispida Muhl. ex Willd. ■☆

368165　Solidago hispida Nutt. var. arnoglossa Fernald = Solidago hispida Muhl. ex Willd. ■☆

368166　Solidago hispida Nutt. var. disjuncta Fernald = Solidago hispida Muhl. ex Willd. ■☆

368167　Solidago hispida Nutt. var. huronensis Semple = Solidago hispida Muhl. ex Willd. ■☆

368168　Solidago hispida Nutt. var. lanata（Hook.）Fernald = Solidago hispida Muhl. ex Willd. ■☆

368169　Solidago hispida Nutt. var. tonsa Fernald = Solidago hispida Muhl. ex Willd. ■☆

368170　Solidago houghtonii Torr. et A. Gray；霍顿一枝黄花；Houghton's Goldenrod ■☆

368171　Solidago howellii Wooton et Standl. = Solidago velutina DC. subsp. sparsiflora（A. Gray）Semple ■☆

368172　Solidago humilis Pursh = Solidago uliginosa Nutt. ■☆

368173　Solidago humilis Pursh var. abbei B. Boivin = Solidago uliginosa Nutt. ■☆

368174　Solidago humilis Pursh var. gillmanii A. Gray = Solidago simplex Kunth var. gillmanii（A. Gray）G. S. Ringius ■☆

368175　Solidago humilis Pursh var. microcephala Porter = Solidago uliginosa Nutt. ■☆

368176　Solidago humilis Pursh var. nana A. Gray = Solidago simplex

Kunth var. nana（A. Gray）G. S. Ringius ■☆

368177　Solidago humilis Pursh var. peracuta Fernald = Solidago uliginosa Nutt. ■☆

368178　Solidago humilis Pursh var. reducta Farw. = Solidago uliginosa Nutt. ■☆

368179　Solidago hybrida Hort. ;杂种一枝黄花;Goldenrod ■☆

368180　Solidago incana Torr. et A. Gray = Solidago mollis Bartl. ■☆

368181　Solidago jacksonii（Kuntze）Fernald = Solidago rigida L. subsp. glabrata（E. L. Braun）S. B. Heard et Semple ■☆

368182　Solidago jacksonii（Kuntze）Fernald var. humilis（Porter）Beaudry = Solidago rigida L. subsp. humilis（Porter）S. B. Heard et Semple ■☆

368183　Solidago jailarum Juz. ;亚林一枝黄花■☆

368184　Solidago japonica Kitam. = Solidago virgaurea L. subsp. asiatica（Nakai ex H. Hara）Kitam. ex H. Hara ■☆

368185　Solidago japonica Kitam. var. paludosa Honda = Solidago decurrens Lour. ■

368186　Solidago jejunifolia E. S. Steele = Solidago speciosa Nutt. ■☆

368187　Solidago juliae G. L. Nesom;朱莉亚一枝黄花;Julia's Goldenrod ■☆

368188　Solidago juncea Aiton;早一枝黄花;Early Goldenrod ■☆

368189　Solidago juncea Aiton f. scabrella（Torr. et A. Gray）Fernald = Solidago juncea Aiton ■☆

368190　Solidago juncea Aiton var. neobohemica Fernald = Solidago juncea Aiton ■☆

368191　Solidago juncea Aiton var. ramosa Porter et Britton = Solidago juncea Aiton ■☆

368192　Solidago juncea Aiton var. scabrella（Torr. et A. Gray）A. Gray = Solidago juncea Aiton ■☆

368193　Solidago klughii Fernald = Solidago uliginosa Nutt. ■☆

368194　Solidago kralii Semple;克拉尔一枝黄花;Kral's Goldenrod ■☆

368195　Solidago krotkovii B. Boivin;克劳特考夫一枝黄花;Krotkov's Goldenrod ■☆

368196　Solidago kuhistanica Popov;库西斯坦一枝黄花■☆

368197　Solidago laeta Greene = Solidago radula Nutt. ■☆

368198　Solidago lanata Hook. = Solidago hispida Muhl. ex Willd. ■☆

368199　Solidago lanceolata L. = Euthamia graminifolia（L.）Nutt. ■☆

368200　Solidago lanceolata L. var. minor Michx. = Euthamia caroliniana（L.）Greene ex Porter et Britton ■☆

368201　Solidago lancifolia（Torr. et A. Gray）Chapm. ;剑叶一枝黄花;Lance-leaf Goldenrod ■☆

368202　Solidago lateriflora L. ;侧花一枝黄花■☆

368203　Solidago lateriflora L. = Symphyotrichum lateriflorum（L.）Á. Löve et D. Löve ■☆

368204　Solidago lateriflora Raf. ex DC. = Solidago caesia L. ■☆

368205　Solidago latifolia L. = Solidago flexicaulis L. ■☆

368206　Solidago latifolia Mill. ;阔叶一枝黄花;Broad-leaved Goldenrod ■☆

368207　Solidago latissimifolia Mill. ;埃利奥特一枝黄花;Elliott's Goldenrod ■☆

368208　Solidago leavenworthii Torr. et A. Gray;北美一枝黄花;Leavenworth's Goldenrod ■☆

368209　Solidago leiocarpa DC. ;卡特勒一枝黄花;Cutler's Alpine Goldenrod ■☆

368210　Solidago leiophallax Friesner = Solidago gigantea Aiton ■☆

368211　Solidago lepida DC. ;雅鳞一枝黄花;Elegant Goldenrod, Western Canada Goldenrod ■☆

368212　Solidago lepida DC. subsp. fallax（Fernald）Semple;假雅鳞一枝黄花■☆

368213　Solidago lepida DC. var. caurina（Piper）M. Peck = Solidago elongata Nutt. ■☆

368214　Solidago lepida DC. var. elongata（Nutt.）Fernald = Solidago canadensis L. var. salebrosa（Piper）M. E. Jones ■☆

368215　Solidago lepida DC. var. elongata（Nutt.）Fernald = Solidago elongata Nutt. ■☆

368216　Solidago lepida DC. var. fallax Fernald = Solidago canadensis L. var. salebrosa（Piper）M. E. Jones ■☆

368217　Solidago lepida DC. var. fallax Fernald = Solidago lepida DC. subsp. fallax（Fernald）Semple ■☆

368218　Solidago lepida DC. var. molina Fernald = Solidago lepida DC. subsp. fallax（Fernald）Semple ■☆

368219　Solidago lepida DC. var. salebrosa（Piper）Semple;美国一枝黄花■☆

368220　Solidago leptocephala Torr. et A. Gray = Euthamia leptocephala（Torr. et A. Gray）Greene ■☆

368221　Solidago lindheimeriana Scheele = Solidago petiolaris Aiton ■☆

368222　Solidago linoides Torr. et A. Gray = Solidago uliginosa Nutt. ■☆

368223　Solidago longipetiolata Mack. et Bush = Solidago nemoralis Aiton subsp. decemflora（DC.）Brammall ex Semple ■☆

368224　Solidago ludoviciana（A. Gray）Small;路易斯安娜一枝黄花;Louisiana Goldenrod ■☆

368225　Solidago lunellii Rydb. = Solidago canadensis L. var. scabra Torr. et A. Gray ■☆

368226　Solidago macrophylla Banks ex Pursh;大叶一枝黄花;Large-leaved Goldenrod ■☆

368227　Solidago macrophylla Banks ex Pursh var. thyrsoidea（E. Mey.）Fernald = Solidago macrophylla Banks ex Pursh ■☆

368228　Solidago marshallii Rothr. = Solidago missouriensis Nutt. ■☆

368229　Solidago maxonii Pollard = Solidago roanensis Porter ■☆

368230　Solidago media（Greene）Bush = Euthamia gymnospermoides Greene ■☆

368231　Solidago mensalis Fernald = Solidago macrophylla Banks ex Pursh ■☆

368232　Solidago mexicana L. = Solidago sempervirens L. subsp. mexicana（L.）Semple ■☆

368233　Solidago microphylla（Greene）Bush;小叶一枝黄花;Large-leaved Goldenrod ■☆

368234　Solidago microphylla（Greene）Bush = Euthamia tenuifolia（Pursh）Nutt. ■☆

368235　Solidago microphylla Engelm. ex Small = Solidago delicatula Small ■☆

368236　Solidago milleriana Mack. ex Small = Solidago petiolaris Aiton ■☆

368237　Solidago minutissima（Makino）Kitam. ;微小一枝黄花■☆

368238　Solidago mirabilis Kitam. = Solidago virgaurea L. subsp. gigantea（Nakai）Kitam. ■☆

368239　Solidago mirabilis Small = Solidago latissimifolia Mill. ■☆

368240　Solidago missouriensis Nutt. ;密苏里一枝黄花;Goldenrod, Missouri Goldenrod ■☆

368241　Solidago missouriensis Nutt. var. extraria A. Gray = Solidago missouriensis Nutt. ■☆

368242　Solidago missouriensis Nutt. var. fasciculata Holz. ;簇生密苏里一枝黄花;Missouri Goldenrod ■☆

368243　Solidago missouriensis Nutt. var. fasciculata Holz. = Solidago missouriensis Nutt. ■☆

368244 Solidago missouriensis Nutt. var. glaberrima （M. Martens） Rosend. et Cronquist = Solidago missouriensis Nutt. var. fasciculata Holz. ■☆

368245 Solidago missouriensis Nutt. var. glaberrima （M. Martens） Rosend. et Cronquist = Solidago missouriensis Nutt. ■☆

368246 Solidago missouriensis Nutt. var. montana A. Gray = Solidago missouriensis Nutt. ■☆

368247 Solidago missouriensis Nutt. var. pumila Chapm. = Solidago gattingeri Chapm. ex A. Gray ■☆

368248 Solidago missouriensis Nutt. var. tenuissima （Wooton et Standl.） C. E. S. Taylor et R. J. Taylor = Solidago missouriensis Nutt. ■☆

368249 Solidago missouriensis Nutt. var. tolmieana （A. Gray） Cronquist = Solidago missouriensis Nutt. ■☆

368250 Solidago mollis Bartl.；柔软一枝黄花；Ashly Goldenrod, Soft Goldenrod, Velvet Goldenrod, Velvety Goldenrod ■☆

368251 Solidago mollis Bartl. var. angustata Shinners = Solidago mollis Bartl. ■☆

368252 Solidago mollis Rothr. = Solidago velutina DC. subsp. sparsiflora （A. Gray） Semple ■☆

368253 Solidago monticola （Torr. et A. Gray） Chapm. = Solidago roanensis Porter ■☆

368254 Solidago moritura E. S. Steele = Solidago missouriensis Nutt. ■☆

368255 Solidago moritura Steele = Solidago missouriensis Nutt. var. fasciculata Holz. ■☆

368256 Solidago moseleyi Fernald = Euthamia caroliniana （L.） Greene ex Porter et Britton ■☆

368257 Solidago moseleyi Fernald = Euthamia gymnospermoides Greene ■☆

368258 Solidago multiradiata Aiton；库西克一枝黄花；Northern Goldenrod, Rocky Mountain Goldenrod ■☆

368259 Solidago multiradiata Aiton var. arctica （DC.） Fernald = Solidago multiradiata Aiton ■☆

368260 Solidago multiradiata Aiton var. neomexicana A. Gray = Solidago simplex Kunth ■☆

368261 Solidago multiradiata Aiton var. scopulorum A. Gray = Solidago multiradiata Aiton ■☆

368262 Solidago nana Nutt.；矮小一枝黄花；Baby Goldenrod, Dwarf Goldenrod ■☆

368263 Solidago neglecta Torr. et A. Gray；疏忽一枝黄花■☆

368264 Solidago neglecta Torr. et A. Gray = Solidago uliginosa Nutt. ■☆

368265 Solidago neglecta Torr. et A. Gray var. linoides （Torr. et A. Gray） A. Gray = Solidago uliginosa Nutt. ■☆

368266 Solidago neglecta Torr. et A. Gray var. simulata Farw. = Solidago uliginosa Nutt. ■☆

368267 Solidago neglecta Torr. et A. Gray var. uniligulata （DC.） Britton, Stern et Poggenb. = Solidago uliginosa Nutt. ■☆

368268 Solidago nemoralis Aiton；灰茎一枝黄花（灰一枝黄花）；Common Goldenrod, Dyersweed, Dyer's-weed Goldenrod, Field Goldenrod, Gray Goldenrod, Gray-stemmed Goldenrod, Old-field Goldenrod ■☆

368269 Solidago nemoralis Aiton subsp. decemflora （DC.） Brammall = Solidago nemoralis Aiton subsp. decemflora （DC.） Brammall ex Semple ■☆

368270 Solidago nemoralis Aiton subsp. decemflora （DC.） Brammall ex Semple；十花一枝黄花；Field Goldenrod, Old-field Goldenrod ■☆

368271 Solidago nemoralis Aiton subsp. haleana （Fernald） G. W. Douglas = Solidago nemoralis Aiton ■☆

368272 Solidago nemoralis Aiton subsp. longipetiolata （Mack. et Bush） G. W. Douglas = Solidago nemoralis Aiton subsp. decemflora （DC.） Brammall ■☆

368273 Solidago nemoralis Aiton var. arenicola Burgess = Solidago nemoralis Aiton ■☆

368274 Solidago nemoralis Aiton var. decemflora （DC.） Fernald = Solidago nemoralis Aiton subsp. decemflora （DC.） Brammall ex Semple ■☆

368275 Solidago nemoralis Aiton var. elongata Peck = Solidago nemoralis Aiton ■☆

368276 Solidago nemoralis Aiton var. haleana Fernald = Solidago nemoralis Aiton ■☆

368277 Solidago nemoralis Aiton var. incana （Torr. et A. Gray） A. Gray = Solidago mollis Bartl. ■☆

368278 Solidago nemoralis Aiton var. longipetiolata （Mack. et Bush） E. J. Palmer et Steyerm. = Solidago nemoralis Aiton subsp. decemflora （DC.） Brammall ex Semple ■☆

368279 Solidago nemoralis Aiton var. mollis （Bartl.） A. Gray = Solidago mollis Bartl. ■☆

368280 Solidago nemoralis Aiton var. typica Rosend. et Cronquist = Solidago nemoralis Aiton ■☆

368281 Solidago nemoralis Dryand. var. decemflora （DC.） Fernald = Solidago nemoralis Aiton ■☆

368282 Solidago nemorosa ?；林生一枝黄花；Wood Goldenrod ■☆

368283 Solidago neomexicana （A. Gray） Wooton et Standl. = Solidago simplex Kunth ■☆

368284 Solidago nitida Torr. et A. Gray；光亮一枝黄花；Shiny Goldenrod ■☆

368285 Solidago nivea Rydb. = Solidago nana Nutt. ■☆

368286 Solidago notabilis Mack. ex Small = Solidago auriculata Shuttlew. ex S. F. Blake ■☆

368287 Solidago nuttallii （Greene） Bush = Euthamia graminifolia （L.） Nutt. ■☆

368288 Solidago occidentalis （Nutt.） Torr. et A. Gray = Euthamia occidentalis Nutt. ■☆

368289 Solidago occidentalis Torr. et Gray；西方一枝黄花；Western Goldenrod ■☆

368290 Solidago odora Aiton；芳香一枝黄花（甜一枝黄花，香一枝黄花）；Anise-scented Goldenrod, Fragrant Goldenrod, Sweet Goldenrod ■☆

368291 Solidago odora Aiton subsp. chapmanii （A. Gray） Semple；查普曼一枝黄花；Chapman's Goldenrod ■☆

368292 Solidago odora Aiton var. chapmanii （A. Gray） Cronquist = Solidago odora Aiton subsp. chapmanii （A. Gray） Semple ■☆

368293 Solidago odora Aiton var. inodora A. Gray = Solidago odora Aiton ■☆

368294 Solidago ohioensis Riddell；俄亥俄一枝黄花；Ohio Goldenrod ■☆

368295 Solidago oreophila Rydb. = Solidago simplex Kunth ■☆

368296 Solidago ouachitensis C. E. S. Taylor et R. J. Taylor；欧山一枝黄花；Ouachita Mountains Goldenrod ■☆

368297 Solidago pacifica Juz.；钝苞一枝黄花（朝鲜一枝黄花）；Bluntbract Goldenrod, Pacific Goldenrod ■☆

368298 Solidago pallescens C. Mohr = Solidago brachyphylla Chapm. ex Torr. et A. Gray ■☆

368299　Solidago pallida（Porter）Rydb. = Solidago speciosa Nutt. subsp. pallida（Porter）Semple ■☆

368300　Solidago palmata Pall. = Senecio cannabifolius Less. ■

368301　Solidago paramuschirensis Barkalov = Solidago virgaurea L. subsp. leiocarpa（Benth.）Hultén ■

368302　Solidago parryi（A. Gray）Greene = Oreochrysum parryi（A. Gray）Rydb. ■☆

368303　Solidago parvirigida Beaudry = Solidago rigida L. subsp. humilis（Porter）S. B. Heard et Semple ■☆

368304　Solidago patula Muhl. ex Willd. ；糙叶一枝黄花；Rough-leaved Goldenrod, Round-leaved Goldenrod, Swamp Goldenrod ■☆

368305　Solidago patula Muhl. ex Willd. subsp. strictula（Torr. et A. Gray）Semple；紧缩一枝黄花■☆

368306　Solidago patula Muhl. ex Willd. var. macra Farw. = Solidago patula Muhl. ex Willd. ■☆

368307　Solidago patula Muhl. ex Willd. var. strictula Torr. et A. Gray = Solidago patula Muhl. ex Willd. subsp. strictula（Torr. et A. Gray）Semple ■☆

368308　Solidago pauciflosculosa Michx. = Chrysoma pauciflosculosa（Michx.）Greene ●☆

368309　Solidago pendula Small = Solidago radula Nutt. ■☆

368310　Solidago perglabra Friesner = Euthamia gymnospermoides Greene ■☆

368311　Solidago perlonga Fernald = Solidago stricta Aiton subsp. gracillima（Torr. et A. Gray）Semple ■☆

368312　Solidago petiolaris Aiton；哈珀一枝黄花；Downy Ragged Goldenrod, Goldenrod ■☆

368313　Solidago petiolaris Aiton var. angusta（Torr. et A. Gray）A. Gray = Solidago petiolaris Aiton ■☆

368314　Solidago petiolaris Aiton var. squarrulosa Torr. et A. Gray = Solidago petiolaris Aiton ■☆

368315　Solidago petiolaris Aiton var. wardii（Britton）Fernald = Solidago petiolaris Aiton ■☆

368316　Solidago petradoria S. F. Blake；石茅一枝黄花■☆

368317　Solidago pinetorum Small；斯莫尔一枝黄花；Small's Goldenrod ■☆

368318　Solidago pitcheri Nutt. = Solidago gigantea Aiton ■☆

368319　Solidago plumosa Small；羽状一枝黄花；Plumed Goldenrod, Plumose Goldenrod ■☆

368320　Solidago polycephala Fernald = Euthamia graminifolia（L.）Nutt. ■☆

368321　Solidago porteri Small = Solidago erecta Banks ex Pursh ■☆

368322　Solidago procera Aiton = Solidago canadensis L. var. scabra Torr. et A. Gray ■☆

368323　Solidago pruinosa Greene = Solidago canadensis L. var. gilbocanescens Rydb. ■☆

368324　Solidago ptarmicoides（Nees）B. Boivin；丘陵一枝黄花（白平头一枝黄花）；Prairie Goldenrod, Sneezewort Aster, Upland White Aster, Upland White Goldenrod, White Flat-top Goldenrod, White Upland Aster ■☆

368325　Solidago ptarmicoides（Torr. et A. Gray）B. Boivin = Solidago ptarmicoides（Nees）B. Boivin ■☆

368326　Solidago pterocaulon（Franch. et Sav.）Maxim. var. calvescens Pamp. = Anaphalis aureopunctata Lingelsh. et Borza ■

368327　Solidago pterocaulon（Franch. et Sav.）Maxim. var. intermedia Pamp. = Anaphalis aureopunctata Lingelsh. et Borza ■

368328　Solidago pubens M. A. Curtis ex Torr. et A. Gray = Solidago bicolor L. ■☆

368329　Solidago puberula Nutt. ；柔毛一枝黄花；Downy Golden Rod, Downy Goldenrod ■☆

368330　Solidago puberula Nutt. subsp. pulverulenta（Nutt.）Semple；粉状柔毛一枝黄花■☆

368331　Solidago puberula Nutt. var. borealis Vict. = Solidago puberula Nutt. ■☆

368332　Solidago puberula Nutt. var. monticola Porter = Solidago simplex Kunth var. monticola（Porter）G. S. Ringius ■☆

368333　Solidago puberula Nutt. var. pulverulenta（Nutt.）Chapm. = Solidago puberula Nutt. subsp. pulverulenta（Nutt.）Semple ■☆

368334　Solidago pubescens Wall. = Solidago decurrens Lour. ■

368335　Solidago pulcherrima A. Nelson = Solidago nemoralis Aiton subsp. decemflora（DC.）Brammall ex Semple ■☆

368336　Solidago pulchra Small；美丽一枝黄花；Carolina Goldenrod ■☆

368337　Solidago pulverulenta Nutt. = Solidago puberula Nutt. subsp. pulverulenta（Nutt.）Semple ■☆

368338　Solidago purshii Porter；珀什一枝黄花；Bog Goldenrod ■☆

368339　Solidago purshii Porter = Solidago uliginosa Nutt. ■☆

368340　Solidago purshii Porter var. gillmanii（A. Gray）Farw. = Solidago simplex Kunth var. gillmanii（A. Gray）G. S. Ringius ■☆

368341　Solidago purshii Porter var. nana（A. Gray）Farw. = Solidago simplex Kunth var. nana（A. Gray）G. S. Ringius ■☆

368342　Solidago purshii Porter var. racemosa（Greene）Farw. = Solidago simplex Kunth var. racemosa（Greene）G. S. Ringius ■☆

368343　Solidago racemosa Greene = Solidago simplex Kunth var. racemosa（Greene）G. S. Ringius ■☆

368344　Solidago racemosa Greene var. gillmanii（A. Gray）Fernald = Solidago simplex Kunth var. gillmanii（A. Gray）G. S. Ringius ■☆

368345　Solidago radula Nutt. ；西部糙叶一枝黄花；Rough Goldenrod, Western Rough Goldenrod ■☆

368346　Solidago radula Nutt. var. laeta（Greene）Fernald = Solidago radula Nutt. ■☆

368347　Solidago radula Nutt. var. rotundifolia（DC.）A. Gray = Solidago radula Nutt. ■☆

368348　Solidago radula Nutt. var. stenolepis Fernald = Solidago radula Nutt. ■☆

368349　Solidago randii（Porter）Britton = Solidago simplex Kunth subsp. randii（Porter）G. S. Ringius ■☆

368350　Solidago randii（Porter）Britton var. monticola（Porter）Fernald = Solidago simplex Kunth var. monticola（Porter）G. S. Ringius ■☆

368351　Solidago remota（Greene）Friesner = Euthamia tenuifolia（Pursh）Nutt. ■☆

368352　Solidago riddellii Frank = Solidago riddellii Frank ex Riddell ■☆

368353　Solidago riddellii Frank ex Riddell；里德尔一枝黄花；Goldenrod, Riddell's Goldenrod ■☆

368354　Solidago rigida L. ；坚硬一枝黄花；Hardleaf Golden Rod, Rigid Goldenrod, Stiff Goldenrod, Stiff-leaved Goldenrod ■☆

368355　Solidago rigida L. subsp. glabrata（E. L. Braun）S. B. Heard et Semple；无毛坚硬一枝黄花■☆

368356　Solidago rigida L. subsp. humilis（Porter）S. B. Heard et Semple；荫地坚硬一枝黄花；Rigid Goldenrod, Stiff Goldenrod ■☆

368357　Solidago rigida L. var. canescens（Rydb.）Breitung = Solidago rigida L. subsp. humilis（Porter）S. B. Heard et Semple ■☆

368358　Solidago rigida L. var. glabrata E. L. Braun = Solidago rigida L. subsp. glabrata（E. L. Braun）S. B. Heard et Semple ■☆

368359　Solidago rigida L. var. humilis Porter = Solidago rigida L. subsp.

humilis (Porter) S. B. Heard et Semple ■☆

368360　Solidago rigida L. var. laevicaulis Shinners = Solidago rigida L. subsp. glabrata (E. L. Braun) S. B. Heard et Semple ■☆

368361　Solidago rigida L. var. magna Clute = Solidago rigida L. ■☆

368362　Solidago rigida L. var. microcephala DC. = Solidago rigida L. ■☆

368363　Solidago rigidiuscula (Torr. et A. Gray) Porter = Solidago speciosa Nutt. var. rigidiuscula Torr. et A. Gray ■☆

368364　Solidago rigidiuscula (Torr. et A. Gray) Porter = Solidago spectabilis (D. C. Eaton) A. Gray ■☆

368365　Solidago rigidula Bosc ex DC. ;稍硬一枝黄花■☆

368366　Solidago roanensis Porter ;罗昂一枝黄花; Roan Mountain Goldenrod ■☆

368367　Solidago roanensis Porter var. monticola (Torr. et A. Gray) Fernald = Solidago roanensis Porter ■☆

368368　Solidago rotundifolia DC. = Solidago radula Nutt. ■☆

368369　Solidago rubra Rydb. = Solidago multiradiata Aiton ■☆

368370　Solidago rubricaulis Wall. = Inula rubricaulis (DC.) Benth. et Hook. f. ●

368371　Solidago rugosa Aiton ' Fireworks ';烟火糙茎一枝黄花; Rough-stemmed Goldenrod ■☆

368372　Solidago rugosa Aiton var. aspera (Aiton) Fernald ;粗糙一枝黄花■☆

368373　Solidago rugosa Mill. ;糙茎一枝黄花(糙叶一枝黄花,长毛一枝黄花); Rough Goldenrod, Rough Leaved Goldenrod, Rough-leaved Goldenrod, Rough-stemmed Goldenrod, Wrinkled Goldenrod, Wrinkle-leaf Goldenrod, Wrinkle-leaved Goldenrod ■☆

368374　Solidago rugosa Mill. var. glabrata Farw. = Solidago rugosa Mill. ■☆

368375　Solidago rugosa Mill. var. sphagnophila C. Graves ;夏一枝黄花 ■☆

368376　Solidago rugosa Mill. var. villosa (Pursh) Fernald = Solidago rugosa Mill. ■☆

368377　Solidago rupestris Raf. ;岩地一枝黄花; Rock Goldenrod ■☆

368378　Solidago salebrosa (Piper) Rydb. = Solidago lepida DC. var. salebrosa (Piper) Semple ■☆

368379　Solidago salicifolia Wall. = Aster albescens (DC.) Wall. ex Hand. -Mazz. ●

368380　Solidago sarothrae Pursh = Gutierrezia sarothrae (Pursh) Britton et Rusby ☆

368381　Solidago scaberrima Torr. et A. Gray = Solidago radula Nutt. ■☆

368382　Solidago scabra Muhl. ex Willd. = Solidago altissima L. ■☆

368383　Solidago sciaphila E. S. Steele ;喜阴一枝黄花; Cliff Goldenrod, Driftless Area Goldenrod, Shadowy Goldenrod ■☆

368384　Solidago scopulorum (A. Gray) A. Nelson = Solidago multiradiata Aiton ■☆

368385　Solidago scrophulariifolia Mill. = Solidago flexicaulis L. ■☆

368386　Solidago sempervirens L. ;海滨一枝黄花; Seaside Golden Rod, Seaside Goldenrod ■☆

368387　Solidago sempervirens L. subsp. mexicana (L.) Semple ;墨西哥海滨一枝黄花■☆

368388　Solidago sempervirens L. var. mexicana (L.) Fernald = Solidago sempervirens L. subsp. mexicana (L.) Semple ■☆

368389　Solidago serotina Aiton = Solidago gigantea Aiton ■☆

368390　Solidago serotina Aiton var. gigantea (Aiton) A. Gray = Solidago gigantea Aiton ■☆

368391　Solidago serotina Aiton var. minor Hook. = Solidago gigantea

Aiton ■☆

368392　Solidago serotina Aiton var. salebrosa Piper = Solidago canadensis L. var. salebrosa (Piper) M. E. Jones ■☆

368393　Solidago serotina Aiton var. salebrosa Piper = Solidago lepida DC. var. salebrosa (Piper) Semple ■☆

368394　Solidago serotinoides Á. Löve et D. Löve = Solidago gigantea Aiton ■☆

368395　Solidago serra Rydb. = Solidago lepida DC. var. salebrosa (Piper) Semple ■☆

368396　Solidago shinnersii (Beaudry) Beaudry = Solidago gigantea Aiton ■☆

368397　Solidago shortii Torr. et A. Gray ;肖特一枝黄花; Short's Goldenrod ■☆

368398　Solidago simplex Kunth ;黏一枝黄花; Decumbent Goldenrod, Mountain Goldenrod, Mt. Albert Goldenrod, Sticky Goldenrod ■☆

368399　Solidago simplex Kunth subsp. randii (Porter) G. S. Ringius ;兰德一枝黄花; Rand's Goldenrod ■☆

368400　Solidago simplex Kunth subsp. randii (Porter) G. S. Ringius var. gillmanii (A. Gray) G. S. Ringius = Solidago simplex Kunth var. gillmanii (A. Gray) G. S. Ringius ■☆

368401　Solidago simplex Kunth var. chlorolepis (Fernald) G. S. Ringius ;绿鳞一枝黄花; Mt. Albert Goldenrod ■☆

368402　Solidago simplex Kunth var. gillmanii (A. Gray) G. S. Ringius ;吉尔曼一枝黄花; Dune Goldenrod, Gillman's Goldenrod, Sticky Goldenrod ■☆

368403　Solidago simplex Kunth var. monticola (Porter) G. S. Ringius ;山地黏一枝黄花; Rand's Goldenrod ■☆

368404　Solidago simplex Kunth var. nana (A. Gray) G. S. Ringius ;矮黏一枝黄花; Dwarf Goldenrod ■☆

368405　Solidago simplex Kunth var. ontarioensis (G. S. Ringius) G. S. Ringius ;安大略黏一枝黄花; Ontario Goldenrod ■☆

368406　Solidago simplex Kunth var. racemosa (Greene) G. S. Ringius ;总状黏一枝黄花; Racemose Goldenrod ■☆

368407　Solidago simplex Kunth var. spathulata (DC.) Cronquist = Solidago spathulata DC. ■☆

368408　Solidago simulans Fernald = Solidago uliginosa Nutt. ■☆

368409　Solidago sinica Hance subsp. intermedia (Pamp.) Kitam. = Anaphalis aureopunctata Lingelsh. et Borza ■

368410　Solidago sinica Hance var. calvescens (Pamp.) S. Y. Hu = Anaphalis aureopunctata Lingelsh. et Borza ■

368411　Solidago somesii Rydb. = Solidago gigantea Aiton ■☆

368412　Solidago sparsiflora A. Gray = Solidago velutina DC. subsp. sparsiflora (A. Gray) Semple ■☆

368413　Solidago sparsiflora A. Gray var. subcinerea A. Gray = Solidago velutina DC. subsp. sparsiflora (A. Gray) Semple ■☆

368414　Solidago spathulata DC. ;海岸一枝黄花; Coast Goldenrod ■☆

368415　Solidago spathulata DC. subsp. glutinosa (Nutt.) D. D. Keck = Solidago simplex Kunth ■☆

368416　Solidago spathulata DC. subsp. randii (Porter) Cronquist ex Gleason = Solidago simplex Kunth subsp. randii (Porter) G. S. Ringius ■☆

368417　Solidago spathulata DC. var. gillmanii (A. Gray) Cronquist = Solidago simplex Kunth var. gillmanii (A. Gray) G. S. Ringius ■☆

368418　Solidago spathulata DC. var. gillmanii (A. Gray) Cronquist ex Gleason = Solidago simplex Kunth var. gillmanii (A. Gray) G. S. Ringius ■☆

368419　Solidago spathulata DC. var. nana (A. Gray) Cronquist =

Solidago simplex Kunth var. nana (A. Gray) G. S. Ringius ■☆

368420 Solidago spathulata DC. var. neomexicana (A. Gray) Cronquist = Solidago simplex Kunth ■☆

368421 Solidago spathulata DC. var. racemosa (Greene) Cronquist ex Gleason = Solidago simplex Kunth var. racemosa (Greene) G. S. Ringius ■☆

368422 Solidago speciosa Nutt. ;优雅一枝黄花;Noble Goldenrod, Prairie Goldenrod,Showy Goldenrod ■☆

368423 Solidago speciosa Nutt. subsp. pallida (Porter) Semple;苍白优雅一枝黄花■☆

368424 Solidago speciosa Nutt. var. angustata Torr. et A. Gray = Solidago speciosa Nutt. ■☆

368425 Solidago speciosa Nutt. var. angustata Torr. et A. Gray = Solidago spectabilis (D. C. Eaton) A. Gray ■☆

368426 Solidago speciosa Nutt. var. erecta (Banks ex Pursh) MacMill. = Solidago erecta Banks ex Pursh ■☆

368427 Solidago speciosa Nutt. var. jejunifolia (E. S. Steele) Cronquist = Solidago speciosa Nutt. ■☆

368428 Solidago speciosa Nutt. var. pallida Porter = Solidago speciosa Nutt. subsp. pallida (Porter) Semple ■☆

368429 Solidago speciosa Nutt. var. rigidiuscula Torr. et A. Gray;硬壳一枝黄花■☆

368430 Solidago spectabilis (D. C. Eaton) A. Gray;内华达一枝黄花;Basin Goldenrod,Nevada Goldenrod,Showy Goldenrod ■☆

368431 Solidago spectabilis A. Gray var. confinis (A. Gray) Cronquist = Solidago confinis A. Gray ■☆

368432 Solidago sphacelata Raf. ;秋一枝黄花;Autumn Goldenrod, False Goldenrod,Goldenrod ■☆

368433 Solidago spiciformis Torr. et A. Gray = Solidago spathulata DC. ■☆

368434 Solidago spithamaea M. A. Curtis ex A. Gray;蓝脊一枝黄花;Blue Ridge Goldenrod,Skunk Goldenrod ■☆

368435 Solidago squarrosa Muhl. ;糙一枝黄花;Rugged Goldenrod, Stout Goldenrod ■☆

368436 Solidago squarrosa Muhl. var. ramosa Peck = Solidago squarrosa Muhl. ■☆

368437 Solidago squarrulosa (Torr. et A. Gray) A. W. Wood = Solidago petiolaris Aiton ■☆

368438 Solidago stricta Aiton;棒状一枝黄花;Wand Goldenrod, Wandlike Goldenrod, Wand-like Goldenrod, Willow-leaf Goldenrod ■☆

368439 Solidago stricta Aiton subsp. gracillima (Torr. et A. Gray) Semple;细棒状一枝黄花■☆

368440 Solidago stricta Aiton var. angustifolia (Elliott) A. Gray = Solidago stricta Aiton ■☆

368441 Solidago strigosa Small = Solidago ludoviciana (A. Gray) Small ■☆

368442 Solidago taurica Juz. ;克里木一枝黄花■☆

368443 Solidago tenuifolia Pursh = Euthamia caroliniana (L.) Greene ex Porter et Britton ■☆

368444 Solidago tenuifolia Pursh = Euthamia tenuifolia (Pursh) Nutt. ■☆

368445 Solidago tenuifolia Pursh var. pycnocephala Fernald = Euthamia caroliniana (L.) Greene ex Porter et Britton ■☆

368446 Solidago tenuissima Wooton et Standl. = Solidago missouriensis Nutt. ■☆

368447 Solidago terrae-novae Torr. et A. Gray = Solidago uliginosa Nutt. ■☆

368448 Solidago texensis Friesner = Euthamia gymnospermoides Greene ■☆

368449 Solidago thyrsoidea E. Mey. = Solidago macrophylla Banks ex Pursh ■☆

368450 Solidago tolmieana A. Gray = Solidago missouriensis Nutt. ■☆

368451 Solidago tortifolia Elliott;扭叶一枝黄花;Twist-leaf Goldenrod ■☆

368452 Solidago trinervata Greene = Solidago velutina DC. subsp. sparsiflora (A. Gray) Semple ■☆

368453 Solidago turfosa Woronow ex Grossh. ;泥炭一枝黄花■☆

368454 Solidago uliginosa Nutt. ;沼地一枝黄花;Bog Goldenrod, Marsh Goldenrod,Northern Bog Goldenrod,Swamp Goldenrod ■☆

368455 Solidago uliginosa Nutt. var. jejunifolia (E. S. Steele) B. Boivin = Solidago speciosa Nutt. ■☆

368456 Solidago uliginosa Nutt. var. levipes (Fernald) Fernald = Solidago uliginosa Nutt. ■☆

368457 Solidago uliginosa Nutt. var. linoides (Torr. et A. Gray) Fernald = Solidago uliginosa Nutt. ■☆

368458 Solidago uliginosa Nutt. var. neglecta (Torr. et A. Gray) Fernald = Solidago uliginosa Nutt. ■☆

368459 Solidago uliginosa Nutt. var. peracuta (Fernald) Friesner = Solidago uliginosa Nutt. ■☆

368460 Solidago uliginosa Nutt. var. terrae-novae (Torr. et A. Gray) Fernald = Solidago uliginosa Nutt. ■☆

368461 Solidago ulmifolia Muhl. ex Willd. ;榆叶一枝黄花;Elm-leaved Goldenrod ■☆

368462 Solidago ulmifolia Muhl. ex Willd. var. microphylla A. Gray = Solidago delicatula Small ■☆

368463 Solidago ulmifolia Muhl. ex Willd. var. palmeri Cronquist;帕默一枝黄花;Palmer's elm-leaf Goldenrod ■☆

368464 Solidago uniligulata (DC.) Porter = Solidago uliginosa Nutt. ■☆

368465 Solidago uniligulata (DC.) Porter var. levipes Fernald = Solidago uliginosa Nutt. ■☆

368466 Solidago uniligulata (DC.) Porter var. neglecta (Torr. et A. Gray) Fernald = Solidago uliginosa Nutt. ■☆

368467 Solidago uniligulata (DC.) Porter var. terrae-novae (Torr. et A. Gray) Fernald = Solidago uliginosa Nutt. ■☆

368468 Solidago urticifolia Mill. = Calea urticifolia (Mill.) DC. ■☆

368469 Solidago vaseyi A. Heller = Solidago arguta Aiton var. caroliniana A. Gray ■☆

368470 Solidago velutina DC. ;三脉一枝黄花;Three-nerved Goldenrod,Velvety Goldenrod ■☆

368471 Solidago velutina DC. subsp. californica (Nutt.) Semple;加州一枝黄花;California Goldenrod ■☆

368472 Solidago velutina DC. subsp. sparsiflora (A. Gray) Semple;散花一枝黄花■☆

368473 Solidago velutina DC. var. nevadensis (A. Gray) C. E. S. Taylor et R. J. Taylor = Solidago velutina DC. subsp. sparsiflora (A. Gray) Semple ■☆

368474 Solidago venulosa Greene = Solidago speciosa Nutt. var. rigidiuscula Torr. et A. Gray ■☆

368475 Solidago verna M. A. Curtis ex Torr. et A. Gray;春一枝黄花;Spring-flowering Goldenrod ☆

368476 Solidago vespertina Piper = Solidago simplex Kunth ■☆

368477 Solidago victorinii Fernald = Solidago simplex Kunth var. racemosa (Greene) G. S. Ringius ■☆

368478   Solidago villosa Pursh = Solidago rugosa Mill. ■☆

368479   Solidago villosicarpa LeBlond；腺枝一枝黄花；Glandular Wand Goldenrod ■☆

368480   Solidago virgaurea L.；毛果一枝黄花（金柴胡，新疆一枝黄花，一枝黄花）；Aaron's Rod, Cast-the-spear, Common Goldenrod, Europe Goldenrod, European Goldenrod, Farewell-summer, Golden Dust, Golden Glow, Golden Rod, Golden Wings, Goldenrod, St. Joseph's Staff, Woundweed, Woundwort ■

368481   Solidago virgaurea L. = Solidago decurrens Lour. ■

368482   Solidago virgaurea L. subsp. alpestris（Waldst. et Kit.）Gremli；高山毛果一枝黄花■☆

368483   Solidago virgaurea L. subsp. asiatica（Nakai ex H. Hara）Kitam. ex H. Hara；亚洲一枝黄花■☆

368484   Solidago virgaurea L. subsp. asiatica Kitam. ex H. Hara var. insularis（Kitam.）H. Hara；海岛毛果一枝黄花■☆

368485   Solidago virgaurea L. subsp. gigantea（Nakai）Kitam.；大毛果一枝黄花■☆

368486   Solidago virgaurea L. subsp. leiocarpa（Benth.）Hultén；光果一枝黄花■

368487   Solidago virgaurea L. subsp. leiocarpa（Benth.）Hultén = Solidago decurrens Lour. ■

368488   Solidago virgaurea L. subsp. leiocarpa（Benth.）Hultén f. japonalpestris Kitam. = Solidago virgaurea L. subsp. leiocarpa（Benth.）Hultén ■

368489   Solidago virgaurea L. subsp. leiocarpa（Benth.）Hultén f. paludosa（Honda）Kitam. ex Ohwi = Solidago decurrens Lour. ■

368490   Solidago virgaurea L. subsp. leiocarpa（Benth.）Hultén f. paludosa（Honda）Kitam. ex Ohwi；沼泽光果一枝黄花■

368491   Solidago virgaurea L. subsp. leiocarpa（Benth.）Hultén var. ovata（Honda）Ohba；卵叶光果一枝黄花■☆

368492   Solidago virgaurea L. subsp. leiocarpa（Benth.）Hultén var. paludosa Honda = Solidago virgaurea L. subsp. leiocarpa（Benth.）Hultén f. paludosa（Honda）Kitam. ex Ohwi ■

368493   Solidago virgaurea L. subsp. leiocarpa（Benth.）Hultén var. paludosa Honda = Solidago decurrens Lour. ■

368494   Solidago virgaurea L. subsp. leiocarpa（Benth.）Hultén var. praeflorens Nakai；早花光果一枝黄花■☆

368495   Solidago virgaurea L. subsp. leiocarpa Hultén = Solidago decurrens Lour. ■

368496   Solidago virgaurea L. subsp. minuta（L.）Arcang. = Solidago virgaurea L. subsp. alpestris（Waldst. et Kit.）Gremli ■☆

368497   Solidago virgaurea L. var. alpina Bigelow = Solidago leiocarpa DC. ■☆

368498   Solidago virgaurea L. var. arctica DC. = Solidago multiradiata Aiton ■☆

368499   Solidago virgaurea L. var. asiatica Nakai = Solidago virgaurea L. subsp. asiatica（Nakai ex H. Hara）Kitam. ex H. Hara ■☆

368500   Solidago virgaurea L. var. confertiflora（DC.）Kurtz = Solidago simplex Kunth ■☆

368501   Solidago virgaurea L. var. coreana Nakai = Solidago pacifica Juz. ■☆

368502   Solidago virgaurea L. var. dahurica Kitag.；寡毛一枝黄花（寡毛毛果一枝黄花，宽毛毛果一枝黄花，兴安一枝黄花）；Dahur Goldenrod, Dahurian Goldenrod ■

368503   Solidago virgaurea L. var. deanei Porter = Solidago simplex Kunth var. monticola（Porter）G. S. Ringius ■☆

368504   Solidago virgaurea L. var. gigantea Nakai = Solidago virgaurea L. subsp. gigantea（Nakai）Kitam. ■☆

368505   Solidago virgaurea L. var. gillmanii（A. Gray）Porter = Solidago simplex Kunth var. gillmanii（A. Gray）G. S. Ringius ■☆

368506   Solidago virgaurea L. var. glabriuscula C. B. Clarke = Solidago decurrens Lour. ■

368507   Solidago virgaurea L. var. humilis A. Gray = Solidago uliginosa Nutt. ■☆

368508   Solidago virgaurea L. var. leiocarpa（Benth.）A. Gray = Solidago decurrens Lour. ■

368509   Solidago virgaurea L. var. minutussima Makino = Solidago minutissima（Makino）Kitam. ■☆

368510   Solidago virgaurea L. var. multiradiata（Aiton）Torr. et A. Gray = Solidago multiradiata Aiton ■☆

368511   Solidago virgaurea L. var. paludosa Honda = Solidago decurrens Lour. ■

368512   Solidago virgaurea L. var. pubescens（Wall.）C. B. Clarke = Solidago decurrens Lour. ■

368513   Solidago virgaurea L. var. randii Porter = Solidago simplex Kunth subsp. randii（Porter）G. S. Ringius ■☆

368514   Solidago virgaurea L. var. redfieldii Porter = Solidago simplex Kunth var. monticola（Porter）G. S. Ringius ■☆

368515   Solidago virgaurea L. var. yakusimensis Nakai = Solidago minutissima（Makino）Kitam. ■☆

368516   Solidago wardii Britton = Solidago petiolaris Aiton ■☆

368517   Solidago wrightii A. Gray；赖特一枝黄花；Wright's Goldenrod ■☆

368518   Solidago wrightii A. Gray var. adenophora S. F. Blake = Solidago wrightii A. Gray ■☆

368519   Solidago yakusimensis（Nakai）Masam. = Solidago minutissima（Makino）Kitam. ■☆

368520   Solidago yokusaiana Makino；八草一枝黄花■☆

368521   Soliera Clos = Kurzamra Kuntze ■☆

368522   Soliera Gay = Kurzamra Kuntze ■☆

368523   Solisia Britton et Rose = Mammillaria Haw.（保留属名）●

368524   Solisia Britton et Rose（1923）；白斜子属；Solisia ■☆

368525   Solisia mexicana DC.；墨西哥白斜子■☆

368526   Solisia pectinata Britton et Rose；白斜子；Common Solisia ■☆

368527   Soliva Ruiz et Pav.（1794）；裸柱菊属（根头菊属，假吐金菊属，梭氏菊属）；Burrweed, Soliva ■

368528   Soliva anthemifolia（Juss.）R. Br.；裸柱菊（鹅草，根头菊，假吐金菊，九龙吐珠，七星菊，七星坠地，座地菊）；Button Burrweed, Camomileleaf Soliva, False Corianda ■

368529   Soliva anthemifolia（Juss.）Sweet = Soliva anthemifolia（Juss.）R. Br. ■

368530   Soliva daucifolia Nutt. = Soliva sessilis Ruiz et Pav. ■☆

368531   Soliva mutisii Kunth；穆氏裸柱菊；Mutis' Burrweed ■☆

368532   Soliva mutisii Kunth = Soliva anthemifolia（Juss.）R. Br. ■

368533   Soliva nasturtifolia（Juss.）DC. = Soliva stolonifera（Brot.）Sweet ■☆

368534   Soliva pterosperma（Juss.）Less.；翅果假吐金菊■

368535   Soliva pterosperma（Juss.）Less. = Soliva sessilis Ruiz et Pav. ■☆

368536   Soliva sessilis Ruiz et Pav.；无柄裸柱菊；Field Burrweed, Jo-Jo Weed, Lawn Bburweed ■☆

368537   Soliva stolonifera（Brot.）Sweet；匍匐裸柱菊；Carpet Burrweed ■☆

368538   Solivaea Cass. = Soliva Ruiz et Pav. ■

368539   Sollya Lindl.（1832）；蓝钟藤属（梭利藤属，索里亚属）；Bluebell, Bluebell Creeper ●☆

368540　Sollya angustifolia Lindl. ;狭叶蓝钟藤;Australian Bluebell Creeper, Narrowleaf Bluebell Creeper ●☆

368541　Sollya heterophylla Lindl. ;蓝钟藤（索里亚）;Australian Bluebell, Australian Bluebell Creeper, Blue Beli, Blue Beli Creeper, Bluebell Creeper ●☆

368542　Solmsia Baill. (1871);佐尔木属●☆

368543　Solmsia calophylla Baill. ;佐尔木●☆

368544　Solmsia chrysophylla Baill. ;金叶佐尔木●☆

368545　Solmsiella Borbas = Capsella Medik. (保留属名)■

368546　Solms-Laubachia Muschl. (1912);丛菔属;Shrubcress, Solms-Laubachia ■★

368547　Solms-Laubachia carnosifolia C. H. An = Braya scharnhorstii Regel et Schmalh. ■

368548　Solms-Laubachia ciliaris (Bureau et Franch.) Botsch. ;睫毛丛菔■

368549　Solms-Laubachia ciliaris (Bureau et Franch.) Botsch. = Solms-Laubachia pulcherrima Muschl. ■

368550　Solms-Laubachia dolichocarpa Y. C. Lan et T. Y. Cheo;长果丛菔■

368551　Solms-Laubachia dolichocarpa Y. C. Lan et T. Y. Cheo = Solms-Laubachia eurycarpa (Maxim.) Botsch. ■

368552　Solms-Laubachia eurycarpa (Maxim.) Botsch. ;宽果丛菔(巴蓼草,长果丛菔,短柄丛菔,宽叶丛菔,梭罗加博);Broadfruit Shrubcress, Broadfruit Solms-Laubachia, Broadleaf Shrubcress, Broadleaf Solms-Laubachia, Longfruit Shrubcress, Longfruit Solms-Laubachia, Shortstalk Broadfruit Shrubcress, Shortstalk Broadfruit Solms-Laubachia ■

368553　Solms-Laubachia eurycarpa (Maxim.) Botsch. var. brevistipes Y. C. Lan et T. Y. Cheo = Solms-Laubachia eurycarpa (Maxim.) Botsch. ■

368554　Solms-Laubachia eurycarpa (Maxim.) Botsch. var. brevistipes Y. C. Lan et T. Y. Cheo;短柄丛菔■

368555　Solms-Laubachia eurycarpa (Maxim.) Botsch. var. lasiophylla R. F. Huang = Solms-Laubachia eurycarpa (Maxim.) Botsch. ■

368556　Solms-Laubachia floribunda Y. C. Lan et T. Y. Cheo;多花丛菔; Manyflorous Shrubcress, Manyflowers Solms-Laubachia ■

368557　Solms-Laubachia gamosepala Al-Shehbaz;合萼丛菔■

368558　Solms-Laubachia glabra Y. C. Lan et T. Y. Cheo = Pycnoplinthus uniflorus (Hook. f. et Thomson) O. E. Schulz ■

368559　Solms-Laubachia haranensis (Al-Shehbaz) J. P. Yue, Al-Shehbaz et H. Sun;尼泊尔丛菔■☆

368560　Solms-Laubachia lanata Botsch. ;绵毛丛菔;Woolly Shrubcress, Woolly Solms-Laubachia ■

368561　Solms-Laubachia latifolia (O. E. Schulz) Y. C. Lan et T. Y. Cheo;宽叶丛菔■

368562　Solms-Laubachia latifolia (O. E. Schulz) Y. C. Lan et T. Y. Cheo = Solms-Laubachia eurycarpa (Maxim.) Botsch. ■

368563　Solms-Laubachia linearifolia (W. W. Sm.) O. E. Schulz;线叶丛菔(光果丛菔,鸡掌,鸡掌七,线形叶丛菔);Linearleaf Shrubcress, Linearleaf Solms-Laubachia, Smoothfruit Linearleaf Solms-Laubachia ■

368564　Solms-Laubachia linearifolia (W. W. Sm.) O. E. Schulz var. leiocarpa O. E. Schulz;光果丛菔■

368565　Solms-Laubachia linearifolia (W. W. Sm.) O. E. Schulz var. leiocarpa O. E. Schulz = Solms-Laubachia linearifolia (W. W. Sm.) O. E. Schulz ■

368566　Solms-Laubachia minor Hand. -Mazz. ;细叶丛菔(短叶丛菔); Fineleaf Shrubcress, Fineleaf Solms-Laubachia ■

368567　Solms-Laubachia orbiculata Y. C. Lan et T. Y. Cheo;圆叶丛菔■

368568　Solms-Laubachia orbiculata Y. C. Lan et T. Y. Cheo = Solms-Laubachia platycarpa (Hook. f. et Thomson) Botsch. ■

368569　Solms-Laubachia pamirica C. H. An;帕米尔丛菔;Pamir Solms-Laubachia ■

368570　Solms-Laubachia platycarpa (Hook. f. et Thomson) Botsch. ;总状丛菔(圆叶丛菔);Broadfruit Solms-Laubachia, Orbiculate Shrubcress, Orbiculate Solms-Laubachia, Racemoce Shrubcress ■

368571　Solms-Laubachia pulcherrima Muschl. ;丛菔(睫毛丛菔,狭叶丛菔);Beautiful Shrubcress, Beautiful Solms-Laubachia, Ciliate Shrubcress, Ciliate Solms-Laubachia, Narrowleaf Beautiful Shrubcress, Narrowleaf Beautiful Solms-Laubachia ■

368572　Solms-Laubachia pulcherrima Muschl. f. angustifolia O. E. Schulz;狭叶丛菔■

368573　Solms-Laubachia pulcherrima Muschl. f. angustifolia O. E. Schulz = Solms-Laubachia pulcherrima Muschl. ■

368574　Solms-Laubachia pulcherrima Muschl. f. atrichophylla Hand. -Mazz. = Solms-Laubachia pulcherrima Muschl. ■

368575　Solms-Laubachia pulcherrima Muschl. var. latifolia O. E. Schulz = Solms-Laubachia eurycarpa (Maxim.) Botsch. ■

368576　Solms-Laubachia pulcherrima Muschl. var. latifolia O. E. Schulz = Solms-Laubachia latifolia (O. E. Schulz) Y. C. Lan et T. Y. Cheo ■

368577　Solms-Laubachia pumila (Kurz) Dvorák = Desideria pumila (Kurz) Al-Shehbaz ■

368578　Solms-Laubachia retropilosa Botsch. ;倒毛丛菔;Backwardpilose Solms-laubachia, Retropilose Shrubcress ■

368579　Solms-Laubachia xerophyta (W. W. Sm.) Comber;旱生丛菔; Dry Shrubcress, Dry Solms-Laubachia ■

368580　Solms-Laubachia zhongdianensis J. P. Yue, Al-Shehbaz et H. Sun;中甸丛菔■

368581　Solonia Urb. (1922);索伦紫金牛属●☆

368582　Solonia reflexa Urb. ;索伦紫金牛●☆

368583　Solori Adans. (废弃属名) = Derris Lour. (保留属名)●

368584　Solstitiaria Hill = Centaurea L. (保留属名)●■

368585　Soltmannia Klotzsch ex Naudin = Miconia Ruiz et Pav. (保留属名)●☆

368586　Solulus Kuntze = Diphaca Lour. (废弃属名)●

368587　Solulus Kuntze = Ormocarpum P. Beauv. (保留属名)●

368588　Somalia Oliv. (1886);索马里玄参属■☆

368589　Somalia diffusa Oliv. = Barleria diffusa (Oliv.) Lindau ●☆

368590　Somalluma Plowes = Caralluma R. Br. ■

368591　Somalluma Plowes(1995);索马里水牛角属■☆

368592　Somalluma baradii (Lavranos) Plowes = Caralluma baradii Lavranos ■☆

368593　Somera Salisb. = Hyacinthoides Medik. ■☆

368594　Somera Salisb. = Scilla L. ■

368595　Someraura Hoppe = Minuartia L. ■

368596　Sommea Bory = Acicarpha Juss. ■☆

368597　Sommera Schltdl. (1835);萨默茜属●☆

368598　Sommera acuminata Oerst. ;尖萨默茜●☆

368599　Sommera arborescens Schltdl. ;萨默茜●☆

368600　Sommera fusca Oerst. ;褐萨默茜●☆

368601　Sommera grandis Standl. ;大萨默茜●☆

368602　Sommera lanceolata K. Krause;披针叶萨默茜●☆

368603　Sommerauera Endl. = Minuartia L. ■

368604　Sommerauera Endl. = Somerauera Hoppe ■

368605　Sommerfeldtia Schumach. = Drepanocarpus G. Mey. ●☆

368606　Sommerfeltia Flörke ex Sommerf.（废弃属名）= Sommerfeltia Less.（保留属名）■●☆

368607　Sommerfeltia Less.（1832）（保留属名）;柄腺层菀属■●☆

368608　Sommerfeltia spinulosa Less. ;柄腺层菀■●☆

368609　Sommeringio Lindl. = Soemmeringia Mart. ●☆

368610　Sommiera Benth. et Hook. f. = Sommieria Becc. ●☆

368611　Sommieria Becc.（1877）;瘤果椰属（白叶椰属，苏米阿椰属）●☆

368612　Sommieria affinis Becc. ;近缘瘤果椰●☆

368613　Sommieria elegans Becc. ;雅致瘤果椰●☆

368614　Sommieria leucophylla Becc. ;瘤果椰●☆

368615　Somphocarya Torr. ex Steud. = Scirpus L.（保留属名）■

368616　Somphoxylon Eichler = Odontocarya Miers ●☆

368617　Sonchella Sennikov = Crepis L. ■

368618　Sonchella Sennikov（2008）;小苦苣菜属■

368619　Sonchidium Pomel = Sonchoseris Fourr. ■

368620　Sonchidium Pomel. = Sonchus L. ■

368621　Sonchidium maritimum（L.）Pomel = Sonchus maritimus L. ■☆

368622　Sonchidium palustre（L.）Pomel = Sonchus palustris L. ■

368623　Sonchos St. -Lag. = Sonchus L. ■

368624　Sonchoseris Fourr. = Sonchus L. ■

368625　Sonchus L.（1753）;苦苣菜属;Milk Thistle, Sow Thistle, Sow-thistle, Sowthistle ■

368626　Sonchus abbreviatus Link = Sonchus congestus Willd. ■☆

368627　Sonchus acaulis Dum. Cours. ;无茎苦苣菜■☆

368628　Sonchus acidus Schousb. = Sonchus pinnatifidus Cav. ■☆

368629　Sonchus afromontanus R. E. Fr. ;非洲山生苦苣菜■☆

368630　Sonchus angustifolius Desf. = Launaea angustifolia（Desf.）Kuntze ■☆

368631　Sonchus angustissimus Hook. f. ;极窄苦苣菜■☆

368632　Sonchus aquatilis Pourr. = Sonchus maritimus L. ■☆

368633　Sonchus arboreus DC. ;树状苦苣菜（细头苦苣菜）■☆

368634　Sonchus arenicola Vorosch. = Sonchus brachyotus DC. ■

368635　Sonchus arvensis L. ;苣荬菜（滇苦荬菜，短耳苣荬菜，怀特苦苣菜，苣菜，苦菜，苦葛麻，苦苣菜，苦荬菜，裂叶苦苣菜，荬菜，牛舌头，匍茎苦菜，取麻菜，山苦荬，台湾苣荬菜，田野苦荬菜，甜苣，野苦菜，野苦荬）;Corn Sow Thistle, Corn Sow-thistle, Creeping Sow Thistle, Field Milk Thistle, Field Milky Thistle, Field Sow Thistle, Field Sowthistle, Field Sow-thistle, Perennial Sow Thistle, Perennial Sow-thistle, Rosecampi, Sow Thistle, Swine Thistle, Tree Sow Thistle ■

368636　Sonchus arvensis L. = Sonchus brachyotus DC. ■

368637　Sonchus arvensis L. f. brachyotus（DC.）Kirp. = Sonchus brachyotus DC. ■

368638　Sonchus arvensis L. f. glabrescens（Günther, Grab. et Wimm.）Kirp. = Sonchus uliginosus M. Bieb. ■

368639　Sonchus arvensis L. subsp. brachyotus〔DC.）Kitam. = Sonchus brachyotus DC. ■

368640　Sonchus arvensis L. subsp. uliginosus（M. Bieb.）Nyman = Sonchus uliginosus M. Bieb. ■

368641　Sonchus arvensis L. var. glabrescens Günther, Grab. et Wimm. = Sonchus uliginosus M. Bieb. ■

368642　Sonchus arvensis L. var. glabrescens Günther, Grab. et Wimm. = Sonchus arvensis L. ■

368643　Sonchus arvensis L. var. laevipes Koch = Sonchus brachyotus DC. ■

368644　Sonchus arvensis L. var. mauritanicus（Boiss. et Reut.）Batt. = Sonchus mauritanicus Boiss. et Reut. ■☆

368645　Sonchus arvensis L. var. shumovichii B. Boivin = Sonchus arvensis L. ■

368646　Sonchus arvensis L. var. typicus Beck = Sonchus arvensis L. ■

368647　Sonchus arvensis L. var. uliginosus（M. Bieb.）Trautv. = Sonchus uliginosus M. Bieb. ■

368648　Sonchus arvensis L. var. uliginosus Trautv. = Sonchus uliginosus M. Bieb. ■

368649　Sonchus asper（L.）Hill;花叶滇苦菜（白石头，败酱草，大叶苣荬菜，滇古菜，滇苦荬菜，鬼苦苣菜，苦马菜，续断，续断菊，圆耳苦苣菜）;Prickly Sow Thistle, Prickly Sowthistle, Prickly Sow-thistle, Spiny Leaf Sow-thistle, Spiny Milk Thistle, Spiny Sow Thistle, Spiny Sowthistle, Spiny Sow-thistle, Spiny-leaf Sow-thistle, Spiny-leaved Sow Thistle, Spiny-leaved Sow-thistle ■

368650　Sonchus asper（L.）Hill f. glandulosus Beckh. ;腺花叶滇苦菜■☆

368651　Sonchus asper（L.）Hill f. glandulosus Beckh. = Sonchus asper（L.）Hill ■

368652　Sonchus asper（L.）Hill f. inermis（Bisch.）Beck = Sonchus asper（L.）Hill ■

368653　Sonchus asper（L.）Hill subsp. glaucescens（Jord.）Ball;灰绿滇苦菜■☆

368654　Sonchus asper（L.）Hill subsp. nymanii（Tineo et Guss.）Hegi = Sonchus asper（L.）Hill subsp. glaucescens（Jord.）Ball ■☆

368655　Sonchus asper（L.）Hill var. glaucescens（Jord.）Batt. = Sonchus asper（L.）Hill subsp. glaucescens（Jord.）Ball ■☆

368656　Sonchus asper（L.）Hill var. inermis Bisch. = Sonchus asper（L.）Hill ■

368657　Sonchus asper（L.）Hill var. inermis Bisch. f. gracilis A. F. Schwarz = Sonchus asper（L.）Hill ■

368658　Sonchus asper（L.）Hill var. pungens Bisch. = Sonchus asper（L.）Hill ■

368659　Sonchus asper Vill. = Sonchus asper（L.）Hill ■

368660　Sonchus azureus Ledeb. = Cicerbita azurea（Ledeb.）Beauverd ■

368661　Sonchus bequaertii De Wild. = Sonchus bipontini Asch. var. glanduligerus（R. E. Fr.）Robyns ■☆

368662　Sonchus biennis Moench = Lactuca biennis（Moench）Fernald ■☆

368663　Sonchus bipontini Asch. ;非洲滇苦菜■☆

368664　Sonchus bipontini Asch. f. glanduligerus R. E. Fr. = Sonchus bipontini Asch. var. glanduligerus（R. E. Fr.）Robyns ■☆

368665　Sonchus bipontini Asch. f. luxurians R. E. Fr. = Sonchus luxurians（R. E. Fr.）C. Jeffrey ■☆

368666　Sonchus bipontini Asch. var. exauriculatus Oliv. et Hiern = Launaea cornuta（Hochst. ex Oliv. et Hiern）C. Jeffrey ■☆

368667　Sonchus bipontini Asch. var. glanduligerus（R. E. Fr.）Robyns;腺体苦苣菜■☆

368668　Sonchus bipontini Asch. var. louisii Robyns = Sonchus bipontini Asch. var. glanduligerus（R. E. Fr.）Robyns ■☆

368669　Sonchus bipontini Asch. var. pinnatifidus Oliv. et Hiern = Launaea cornuta（Hochst. ex Oliv. et Hiern）C. Jeffrey ■☆

368670　Sonchus bornmuelleri Pit. ;博恩苦苣菜■☆

368671　Sonchus bourgeaui Sch. Bip. ;布尔苦苣菜■☆

368672　Sonchus bourgeaui Sch. Bip. var. imbricatus（Svent.）Boulos = Sonchus bourgeaui Sch. Bip. ■☆

368673　Sonchus brachylobus Webb et Berthel. ;短裂苦苣菜■☆

368674　Sonchus brachylobus Webb et Berthel. var. canariae（Pit.）Boulos = Sonchus brachylobus Webb et Berthel. ■☆

368675　Sonchus brachyotus DC. ;长裂苦苣菜（败酱,败酱草,苣菜,苣荬菜,苦葛麻,苦苦菜,苦荬菜,荬菜,曲麻菜,取麻菜,小蓟,野苦菜,野苦荬）;Longlobed Sowthistle

368676　Sonchus briquetianus Gand. ;布里凯苦苣菜■☆

368677　Sonchus brunneri（Webb）Oliv. et Hiern = Launaea brunneri（Webb）Amin ex Boulos ■☆

368678　Sonchus bulbosus（L.）N. Kilian et Greuter = Aetheorhiza bulbosa（L.）Cass. ■☆

368679　Sonchus bupleuroides（Font Quer）N. Kilian et Greuter;柴胡苦苣菜■☆

368680　Sonchus camporum（R. E. Fr.）Boulos ex C. Jeffrey;弯苦苣菜■☆

368681　Sonchus canariae Pit. = Sonchus brachylobus Webb et Berthel. ■☆

368682　Sonchus canariensis（Sch. Bip.）Boulos;加那利苦苣菜■☆

368683　Sonchus capillaris Svent. ;发状苦苣菜■☆

368684　Sonchus capitatus Spreng. = Launaea capitata（Spreng.）Dandy ■☆

368685　Sonchus cassianus Jaub. et Spach = Launaea mucronata（Forssk.）Muschl. subsp. cassiana（Jaub. et Spach）N. Kilian ■☆

368686　Sonchus caucasicus Biehler = Crepis sibirica L. ■

368687　Sonchus cavaleriei H. Lév. = Sonchus brachyotus DC. ■

368688　Sonchus charmelii Sennen et Mauricio = Sonchus tenerrimus L. ■☆

368689　Sonchus chevalieri（O. Hoffm. et Muschl.）Dandy = Launaea brunneri（Webb）Amin ex Boulos ■☆

368690　Sonchus chinensis Fisch. = Sonchus brachyotus DC. ■

368691　Sonchus chondrilloides Desf. = Launaea fragilis（Asso）Pau ■☆

368692　Sonchus ciliatus Lam. = Sonchus oleraceus L. ■

368693　Sonchus congestus Willd. ;密集苦苣菜■☆

368694　Sonchus congestus Willd. var. gibbosus（Svent.）G. Kunkel = Sonchus congestus Willd. ■☆

368695　Sonchus cornutus Hochst. ex Oliv. et Hiern = Launaea cornuta（Hochst. ex Oliv. et Hiern）C. Jeffrey ■☆

368696　Sonchus cubanguensis S. Moore = Lactuca cubanguensis（S. Moore）C. Jeffrey ■☆

368697　Sonchus cyaneus D. Don = Chaetoseris cyanea（D. Don）C. Shih ■

368698　Sonchus daltonii Webb;多尔顿苦苣菜■☆

368699　Sonchus delagoensis Thell. = Sonchus integrifolius Harv. ■☆

368700　Sonchus dianthoseris Chiov. var. rueppellii（Sch. Bip. ex Oliv. et Hiern）Chiov. = Launaea rueppellii（Sch. Bip. ex Oliv. et Hiern）Amin ex Boulos ■☆

368701　Sonchus dianthoseris Chiov. var. schimperi Sch. Bip. ex A. Rich. = Dianthoseris schimperi Sch. Bip. ex A. Rich. ☆

368702　Sonchus divaricatus Desf. ;叉开苦苣菜■☆

368703　Sonchus dregeanus DC. ;德雷苦苣菜■☆

368704　Sonchus ecklonianus DC. = Sonchus dregeanus DC. ■☆

368705　Sonchus elliotianus Hiern = Launaea nana（Baker）Chiov. ■☆

368706　Sonchus exauriculatus（Oliv. et Hiern）O. Hoffm. = Launaea cornuta（Hochst. ex Oliv. et Hiern）C. Jeffrey ■☆

368707　Sonchus fauriei H. Lév. = Sonchus brachyotus DC. ■

368708　Sonchus fauriei H. Lév. et Vaniot = Sonchus brachyotus DC. ■

368709　Sonchus filifolius N. Kilian et Greuter;线叶苦苣菜■☆

368710　Sonchus fischeri O. Hoffm. = Launaea rarifolia（Oliv. et Hiern）Boulos ■☆

368711　Sonchus flexuosus Ledeb. = Crepis sibirica L. ■

368712　Sonchus floridanus L. = Lactuca floridana（L.）Gaertn. ■☆

368713　Sonchus fragilis Ball;脆苦苣菜■☆

368714　Sonchus freynianus Huter = Launaea arborescens（Batt.）Murb. ■☆

368715　Sonchus friesii Boulos;福瑞苦苣菜■☆

368716　Sonchus friesii Boulos var. integer G. V. Pope;全缘福瑞苦苣菜■☆

368717　Sonchus fruticosus L. f. ;灌丛苦苣菜■☆

368718　Sonchus gigas Boulos ex Humbert = Sonchus asper（L.）Hill ■

368719　Sonchus gigas Boulos ex Humbert subsp. medius ? = Sonchus asper（L.）Hill ■

368720　Sonchus glaucescens Jord. = Sonchus asper（L.）Hill subsp. glaucescens（Jord.）Ball ■☆

368721　Sonchus gomerensis Boulos;戈梅拉苦苣菜■☆

368722　Sonchus goraeensis Lam. = Launaea intybacea（Jacq.）Beauverd ■☆

368723　Sonchus gorgadensis Bolle = Launaea gorgadensis（Bolle）Kilian ■☆

368724　Sonchus gracilis Sennen;细苦苣菜■☆

368725　Sonchus gummifer Link;产胶苦苣菜■☆

368726　Sonchus hastatus Less. = Prenanthes alata（Hook.）D. Dietr. ■☆

368727　Sonchus heterophyllus（Boulos）U. Reifenb. et A. Reifenb. ;互叶苦苣菜■☆

368728　Sonchus hierrensis（Pit.）Boulos;耶罗苦苣菜■☆

368729　Sonchus hierrensis（Pit.）Boulos var. benehoavensis Svent. = Sonchus hierrensis（Pit.）Boulos ■☆

368730　Sonchus hispidus Gilib. = Sonchus arvensis L. ■

368731　Sonchus integrifolius Harv. ;全缘苦苣菜■☆

368732　Sonchus integrifolius Harv. f. lobatus R. E. Fr. = Sonchus integrifolius Harv. ■☆

368733　Sonchus integrifolius Harv. var. schlechteri R. E. Fr. = Sonchus integrifolius Harv. ■☆

368734　Sonchus jacottetianus Thell. ;贾科泰苦苣菜■☆

368735　Sonchus jacquinii DC. = Sonchus congestus Willd. ■☆

368736　Sonchus jacquinii DC. var. hierrensis Pit. = Sonchus hierrensis（Pit.）Boulos ■☆

368737　Sonchus kabarensis De Wild. = Launaea cornuta（Hochst. ex Oliv. et Hiern）C. Jeffrey ■☆

368738　Sonchus lactucoides Sch. Bip. ex A. Rich. = Sonchus bipontini Asch. ■☆

368739　Sonchus lakouensis S. Y. Hu = Paramicrorhynchus procumbens（Roxb.）Kirp. ■

368740　Sonchus lanifer Dinter = Launaea rarifolia（Oliv. et Hiern）Boulos ■☆

368741　Sonchus lasiorhizus O. Hoffm. = Lactuca lasiorhiza（O. Hoffm.）C. Jeffrey ■☆

368742　Sonchus ledermannii R. E. Fr. = Lactuca lasiorhiza（O. Hoffm.）C. Jeffrey ■☆

368743　Sonchus leptocephalus Cass. ;细头苦苣菜■☆

368744　Sonchus lidii Boulos;利德苦苣菜■☆

368745　Sonchus lingianus C. Shih;南苦苣菜（南苦荬菜）;Ling's Sowthistle ■

368746　Sonchus ludovicianus Nutt. = Lactuca ludoviciana（Nutt.）Riddell ■☆

368747　Sonchus luxurians（R. E. Fr.）C. Jeffrey；茂盛苦苣菜■☆

368748　Sonchus macer S. Moore = Launaea rarifolia（Oliv. et Hiern）Boulos ■☆

368749　Sonchus macrocarpus Boulos et C. Jeffrey；大果苦苣菜■☆

368750　Sonchus maculigerus H. Lindb. ；斑点苦苣菜■☆

368751　Sonchus mairei H. Lév. = Launaea sarmentosa（Willd.）Kuntze ■

368752　Sonchus mairei H. Lév. = Paramicrorhynchus procumbens（Roxb.）Kirp. ■

368753　Sonchus mairei H. Lév. = Sonchus oleraceus L. ■

368754　Sonchus maritimus L. ；滨海苦苣菜■☆

368755　Sonchus maritimus L. subsp. aquatilis（Pourr.）Nyman = Sonchus maritimus L. ■☆

368756　Sonchus maritimus L. subsp. cartilagineus Maire et Sennen = Sonchus maritimus L. ■☆

368757　Sonchus maritimus L. var. canescens Maire et Sennen = Sonchus maritimus L. ■☆

368758　Sonchus masguindalii Pau et Font Quer；马斯苦苣菜■☆

368759　Sonchus massauensis（Fresen.）Sch. Bip. = Launaea massavensis（Fresen.）Sch. Bip. ex Kuntze ☆

368760　Sonchus mauritanicus Boiss. et Reut. ；毛里塔尼亚苦苣菜■☆

368761　Sonchus melanolepis Fresen. ；黑鳞苦苣菜■☆

368762　Sonchus melanolepis Fresen. f. stramineus R. E. Fr. = Sonchus melanolepis Fresen. ■☆

368763　Sonchus melanolepis Fresen. var. linearis R. E. Fr. = Sonchus melanolepis Fresen. ■☆

368764　Sonchus microcarpus（Boulos）U. Reifenb. et A. Reifenb. ；小果苦苣菜■☆

368765　Sonchus nanellus R. E. Fr. = Launaea cabrae（De Wild.）N. Kilian subsp. nanella（R. E. Fr.）N. Kilian ■☆

368766　Sonchus nanus O. Hoffm. = Launaea nana（Baker）Chiov. ■☆

368767　Sonchus neglectus Pit. ；疏忽苦苣菜■☆

368768　Sonchus nudicaulis（L.）Sch. Bip. = Launaea nudicaulis（L.）Hook. f. ■☆

368769　Sonchus nymannii Guss. = Sonchus asper（L.）Hill ■

368770　Sonchus obtusilobus R. E. Fr. ；钝裂苦苣菜■☆

368771　Sonchus oleraceus L. ；苦苣菜（苦苣苦菜，滇苦菜，滇苦苣菜，滇苦荬菜，鹅菜，鹅仔菜，堇菜，空心苦马菜，苦菜，苦菜花，苦地胆，苦滇菜，苦苣，苦马菜，苦买菜，苦荬，苦荬菜，苦荬麻，老鸦荬，麻苦苣，奶浆草，苣，青菜，天香菜，荼，荼草，小鹅菜，野苦马，野苦荬，游冬，紫苦菜）；Annual Sow-thistle, Common Sow Thistle, Common Sowthistle, Common Sow-thistle, Dindle, Dog's Thistle, Hare's Colewort, Hare's Lettuce, Hare's Palace, Hare's Thistle, Hate's Colewort, Hate's Thistle, Mary's Seed, Milk Thistle, Milkweed, Milkwort, Milky Dashel, Milky Dassel, Milky Dicel, Milky Dickle, Milky Disle, Milky Tassel, Rabbit's Meat, Rabbit's Victuals, Smooth Sow Thistle, Smooth Sow-thistle, Sow Bread, Sow Dingle, Sow Flower, Sow Thistle, Sow Thristle, Sowbread, Sowthristle, Swine Thistle, Swinies, Thistle, Virgin's Milk, Wild Thistle ■

368772　Sonchus oleraceus L. subsp. angustissimus H. Lindb. = Sonchus oleraceus L. ■

368773　Sonchus oleraceus L. var. asper L. = Sonchus asper（L.）Hill ■

368774　Sonchus oleraceus L. var. integrifolius Wallr. = Sonchus oleraceus L. ■

368775　Sonchus oleraceus L. var. lacerus Wallr. = Sonchus oleraceus L. ■

368776　Sonchus oleraceus L. var. triangularis Wallr. = Sonchus oleraceus L. ■

368777　Sonchus oliveri-hiernii Boulos = Launaea cornuta（Hochst. ex Oliv. et Hiern）C. Jeffrey ■☆

368778　Sonchus oliveri-hiernii Boulos var. luxurians（R. E. Fr.）Boulos = Sonchus luxurians（R. E. Fr.）C. Jeffrey ■☆

368779　Sonchus otaviensis Dinter = Sonchus maritimus L. ■☆

368780　Sonchus palmensis（Sch. Bip.）Boulos；帕尔马苦苣菜■☆

368781　Sonchus palustris L. ；沼生苦苣菜；Fen Sow-thistle, Hogweed, Marsh Sow Thistle, Marsh Sow-thistle, Marshy Sowthistle, Tree Sow Thistle ■

368782　Sonchus pauciflorus Baker = Launaea rarifolia（Oliv. et Hiern）Boulos ■☆

368783　Sonchus pectinatus DC. = Sonchus tenerrimus L. ■☆

368784　Sonchus pendulus（Sch. Bip.）Sennikov；下垂苦苣菜■☆

368785　Sonchus pendulus（Sch. Bip.）Sennikov subsp. flaccidus（Svent.）N. Kilian et Greuter；柔软下垂苦苣菜■☆

368786　Sonchus picris H. Lév. et Vaniot = Sonchus arvensis L. ■

368787　Sonchus pinnatifidus Cav. ；羽裂苦苣菜■☆

368788　Sonchus pinnatus Aiton；双羽裂苦苣菜■☆

368789　Sonchus pinnatus Aiton var. canariensis Sch. Bip. = Sonchus canariensis（Sch. Bip.）Boulos ■☆

368790　Sonchus pinnatus Aiton var. palmensis Sch. Bip. = Sonchus palmensis（Sch. Bip.）Boulos ■☆

368791　Sonchus pitardii Boulos；皮塔德苦苣菜■☆

368792　Sonchus platylepis Webb；宽鳞苦苣菜■☆

368793　Sonchus prenanthoides Oliv. et Hiern = Launaea brunneri（Webb）Amin ex Boulos ■☆

368794　Sonchus pulchellus Pursh = Lactuca pulchella（Pursh）DC. ■☆

368795　Sonchus pulchellus Pursh = Mulgedium pulchellum（Pursh）G. Don ■☆

368796　Sonchus pustulatus Willk. ；泡状苦苣菜■☆

368797　Sonchus pycnocephalus R. E. Fr. = Launaea rogersii（Humb.）Humb. et Boulos ■☆

368798　Sonchus quercifolius Desf. = Launaea quercifolia（Desf.）Pamp. ■☆

368799　Sonchus quercifolius Philipson = Lactuca lasiorhiza（O. Hoffm.）C. Jeffrey ■☆

368800　Sonchus radicatus Aiton；具根苦苣菜■☆

368801　Sonchus radicatus Aiton var. glaucus DC. = Sonchus radicatus Aiton ■☆

368802　Sonchus rarifolius Oliv. et Hiern = Launaea rarifolia（Oliv. et Hiern）Boulos ■☆

368803　Sonchus regis-jubae Pit. ；加那利硬苦苣菜■☆

368804　Sonchus rueppellii（Sch. Bip. ex Oliv. et Hiern）R. E. Fr. = Launaea rueppellii（Sch. Bip. ex Oliv. et Hiern）Amin ex Boulos ■☆

368805　Sonchus scapiforme Thell. = Lactuca tysonii（E. Phillips）C. Jeffrey ■☆

368806　Sonchus schweinfurthii Muschl. = Sonchus afromontanus R. E. Fr. ■☆

368807　Sonchus schweinfurthii Oliv. et Hiern；施韦苦苣菜■☆

368808　Sonchus schweinfurthii Oliv. et Hiern var. camporum R. E. Fr. = Sonchus camporum（R. E. Fr.）Boulos ex C. Jeffrey ■☆

368809　Sonchus schweinfurthii Oliv. et Hiern var. violaceus Hiern = Sonchus schweinfurthii Oliv. et Hiern ■☆

368810　Sonchus septenensis Gand. = Sonchus tenerrimus L. ■☆

368811　Sonchus shzucinianus Turcz. ex Herder = Sonchus brachyotus DC. ■

368812　Sonchus sibiricus L. = Lagedium sibiricum（L.）Soják ■

368813　Sonchus sosnowskyi Schchian；索氏苦苣菜■☆

368814 Sonchus spinosus Lam. = Sonchus asper（L.）Hill ■

368815 Sonchus spinulifoius Sennen = Sonchus oleraceus L. ■

368816 Sonchus stenophyllus R. E. Fr.；窄叶苦苣菜■☆

368817 Sonchus taquetii H. Lév. = Sonchus brachyotus DC. ■

368818 Sonchus taraxacifolius Willd. = Launaea taraxacifolia（Willd.）Amin ex C. Jeffrey ■☆

368819 Sonchus tataricus L. = Mulgedium tataricum（L.）DC. ■

368820 Sonchus tectifolius Svent.；拱叶苦苣菜■☆

368821 Sonchus tenerrimus L.；纤细苦苣菜；Slender Sowthistle，Slender Sow-thistle ■☆

368822 Sonchus tenerrimus L. subsp. arborescens（Salzm.）Batt. = Sonchus tenerrimus L. ■☆

368823 Sonchus tenerrimus L. subsp. pustulatus（Willk.）Batt. = Sonchus pustulatus Willk. ■☆

368824 Sonchus tenerrimus L. subsp. tuberculatus（Ball）Batt. = Sonchus tenerrimus L. ■☆

368825 Sonchus tenerrimus L. var. adenobasis Maire = Sonchus tenerrimus L. ■☆

368826 Sonchus tenerrimus L. var. amicus Maire et Wilczek = Sonchus tenerrimus L. ■☆

368827 Sonchus tenerrimus L. var. angustissimus（H. Lindb.）Jahand. et Maire = Sonchus tenerrimus L. ■☆

368828 Sonchus tenerrimus L. var. arborescens（Salzm.）Ball = Sonchus tenerrimus L. ■☆

368829 Sonchus tenerrimus L. var. briquetianus Gand. = Sonchus briquetianus Gand. ■☆

368830 Sonchus tenerrimus L. var. laevigatus Lange = Sonchus tenerrimus L. ■☆

368831 Sonchus tenerrimus L. var. maritimus Ball = Sonchus tenerrimus L. ■☆

368832 Sonchus tenerrimus L. var. pallidulus Maire = Sonchus tenerrimus L. ■☆

368833 Sonchus tenerrimus L. var. pectinatus（DC.）Coss. = Sonchus tenerrimus L. ■☆

368834 Sonchus tenerrimus L. var. perennis DC. = Sonchus tenerrimus L. ■☆

368835 Sonchus tenerrimus L. var. septenensis（Gand.）Maire = Sonchus tenerrimus L. ■☆

368836 Sonchus tenerrimus L. var. tuberculatus Ball = Sonchus tenerrimus L. ■☆

368837 Sonchus tibesticus Quézel = Sonchus asper（L.）Hill ■

368838 Sonchus transcaspicus Nevski；全叶苦苣菜■

368839 Sonchus tuberifer Svent.；块茎苦苣菜■☆

368840 Sonchus tuberifer Svent. var. latisectus？ = Sonchus tuberifer Svent. ■☆

368841 Sonchus tysonii E. Phillips = Lactuca tysonii（E. Phillips）C. Jeffrey ■☆

368842 Sonchus uliginosus M. Bieb.；湿生苦苣菜（短裂苦苣菜）；Shortlobed Sowthistle ■

368843 Sonchus uliginosus M. Bieb. = Sonchus arvensis L. subsp. uliginosus（M. Bieb.）Nyman ■

368844 Sonchus ustulatus Lowe；马岛泡状苦苣菜■☆

368845 Sonchus ustulatus Lowe subsp. maderensis Anderw.；梅德苦苣菜■☆

368846 Sonchus verdickii（De Wild.）R. E. Fr. = Launaea verdickii（De Wild.）Boulos ■☆

368847 Sonchus violaceus O. Hoffm. = Launaea violacea（O. Hoffm.）Boulos ■☆

368848 Sonchus wallichianus DC. = Sonchus uliginosus M. Bieb. ■

368849 Sonchus webbii Sch. Bip.；韦布苦苣菜■☆

368850 Sonchus welwitschii（Scott-Elliot）Chiov. = Launaea rarifolia（Oliv. et Hiern）Boulos ■☆

368851 Sonchus welwitschii（Scott-Elliot）Chiov. ex S. Moore = Launaea rarifolia（Oliv. et Hiern）Boulos ■☆

368852 Sonchus wightianus DC. = Sonchus arvensis L. ■

368853 Sonchus wightianus DC. subsp. wallichianus（DC.）Boulos = Sonchus uliginosus M. Bieb. ■

368854 Sonchus wildpretii U. Reifenb. et A. Reifenb.；维尔德苦苣菜■☆

368855 Sonchus wilmsii R. E. Fr.；维尔姆斯苦苣菜■☆

368856 Sondera Lehm. = Drosera L. ■

368857 Sonderina H. Wolff（1927）；桑德尔草属■☆

368858 Sonderina didyma（Sond.）Adamson = Stoibrax capense（Lam.）B. L. Burtt ■☆

368859 Sonderina hispida（Thunb.）H. Wolff；硬毛桑德尔草■☆

368860 Sonderina humilis（Meisn.）H. Wolff；低矮桑德尔草■☆

368861 Sonderina streyi Merxm. = Anginon streyi（Merxm.）I. Allison et B. -E. van Wyk ■☆

368862 Sonderina tenuis（Sond.）H. Wolff；细桑德尔草■☆

368863 Sonderothamnus R. Dahlgren（1968）；桑德尔木属●☆

368864 Sonderothamnus petraeus（W. F. Barker）R. Dahlgren；桑德尔木●☆

368865 Sonderothamnus speciosus（Sond.）R. Dahlgren；非洲桑德尔木●☆

368866 Sondottia P. S. Short（1989）；光鼠麹属■☆

368867 Sondottia glabrata P. S. Short；光鼠麹■☆

368868 Soneratiaceae Engl. et Gilg = Sonneratiaceae Engl.（保留科名）■

368869 Sonerila Roxb.（1820）（保留属名）；蜂斗草属（地胆属）；Sonerila ●■

368870 Sonerila alata Chun et F. C. How ex C. Chen；翅茎蜂斗草；Alate Sonerila，Winged-stem Sonerila，Winged-stemmed Sonerila ■

368871 Sonerila alata Chun et F. C. How ex C. Chen = Sonerila plagiocardia Diels ■

368872 Sonerila alata Chun et F. C. How ex C. Chen var. triangula C. Chen；短萼蜂斗草（短茎蜂斗草，山风穿筋藤，田螺掩）；Short-calyx Winged-stem Sonerila，Threeangle Sonerila ■

368873 Sonerila alata Chun et F. C. How ex C. Chen var. triangula C. Chen = Sonerila plagiocardia Diels ■

368874 Sonerila cantonensis Stapf；蜂斗草（地胆，喉痧药，尖尾痧，桑簕草，四大天王）；Canton Sonerila，Guangdong Sonerila，Stapf Sonerila ■

368875 Sonerila cantonensis Stapf var. strigosa C. Chen；毛蜂斗草（桑叶草）；Hairy Canton Sonerila ■

368876 Sonerila cantonensis Stapf var. strigosa C. Chen = Sonerila cantonensis Stapf ■

368877 Sonerila cavaleriei H. Lév. = Oxyspora paniculata（D. Don）DC. ●

368878 Sonerila cheliensis H. L. Li；景洪蜂斗草（景洪地胆）；Cheli Sonerila，Jinghong Sonerila ■

368879 Sonerila cheliensis H. L. Li = Sonerila erecta Jack ■

368880 Sonerila epilobioides Stapf et King；柳叶菜蜂斗草（柳叶菜地胆）；Willowweed Sonerila，Willow-weed Sonerila ■

368881 Sonerila epilobioides Stapf et King = Sonerila erecta Jack ■

368882 Sonerila erecta Jack；直立蜂斗草■

368883　Sonerila esquirolii H. Lév. = Plagiopetalum esquirolii（H. Lév.）Rehder ●

368884　Sonerila fordii Oliv. = Fordiophyton fordii（Oliv.）Krasser ●■

368885　Sonerila hainanensis Merr.；海南桑叶草；Hainan Sonerila ■

368886　Sonerila henryi Kraenzl. = Plagiopetalum esquirolii（H. Lév.）Rehder ●

368887　Sonerila laeta Stapf；小蜂斗草（彩斑桑勒草，彩斑桑筋草，地胆，花花草，花叶叶，小花草）；Bright-coloured Sonerila, Small Sonerila ■

368888　Sonerila laeta Stapf = Sonerila maculata Roxb. ■

368889　Sonerila maculata Roxb.；溪边桑勒草■

368890　Sonerila margaritacea Lindl.；珍珠地胆■☆

368891　Sonerila margaritacea Lindl.‘Argentea’；银叶珍珠地胆■☆

368892　Sonerila margaritacea Lindl.‘Hendersonii’；银斑珍珠地胆■☆

368893　Sonerila peperomiifolia Oliv. = Fordiophyton peperomiifolium（Oliv.）Hansen ■

368894　Sonerila peperomiifolia Oliv. = Stapfiophyton peperomiifolium（Oliv.）H. L. Li ■

368895　Sonerila picta Korth.；地胆（彩斑桑筋草，花花草）；Canton Sonerila ■☆

368896　Sonerila plagiocardia Diels；海棠叶蜂斗草（海棠叶地胆）；Begonialeaf Sonerila, Grabappleaf Sonerila ■

368897　Sonerila primuloides C. Y. Wu ex C. Chen；报春蜂斗草（报春地胆）；Primroseleaf Sonerila, Primroselike Sonerila ■

368898　Sonerila rivularis Cogn.；溪边蜂斗草（溪边地胆，溪边桑勒草，溪边桑筋草）；Brooklet Sonerila, Riparian Sonerila, River-sida Sonerila ■

368899　Sonerila rivularis Cogn. = Sonerila maculata Roxb. ■

368900　Sonerila shanlinensis C. Chen；上林蜂斗草；Shanglin Sonerila ■

368901　Sonerila shanlinensis C. Chen = Sonerila erecta Jack ■

368902　Sonerila tenera Royle；三蕊蜂斗草（短药地胆，三蕊草）；Shortanther Sonerila, Threestamen Sonerila, Triander Sonerila ■

368903　Sonerila tenera Royle = Sonerila erecta Jack ■

368904　Sonerila yunnanensis Jeffrey = Sonerila cantonensis Stapf ■

368905　Sonerila yunnanensis Jeffrey ex W. W. Sm.；毛叶蜂斗草（毛叶地胆）；Hairleaf Sonerila, Yunnan Sonerila ■

368906　Soninnia Kostel. = Diplolepis R. Br. ●

368907　Soninnia Kostel. = Sonninia Rchb. ●

368908　Sonnea Greene = Plagiobothrys Fisch. et C. A. Mey. ■☆

368909　Sonneratia Comm. ex Endl. = Celastrus L.（保留属名）●

368910　Sonneratia L. f.（1782）（保留属名）；海桑属；Seamulberry, Sonneratia ●

368911　Sonneratia × gulngai N. C. Duke；拟海桑；False Sonneratia ●

368912　Sonneratia acida L. f. = Sonneratia caseolaris（L.）Engl. ●

368913　Sonneratia alba Sm.；杯萼海桑（海桑，环萼海桑，枷果，剪包树，剪刀树）；Cupcalyx Seamulberry, Cupcalyx Sonneratia, Cupcalyxed Sonneratia ●

368914　Sonneratia apetala Buch.-Ham.；无瓣海桑●

368915　Sonneratia caseolaris（L.）Engl.；海桑；Common Seamulberry, Common Sonneratia ●

368916　Sonneratia evenia Blume = Sonneratia caseolaris（L.）Engl. ●

368917　Sonneratia hainanensis Merr. et Chun；海南海桑；Hainan Seamulberry, Hainan Sonneratia ●◇

368918　Sonneratia iriomotensis Masam. = Sonneratia alba Sm. ●

368919　Sonneratia mossambicensis Klotzsch = Sonneratia alba Sm. ●

368920　Sonneratia mossambicensis Klotzsch ex Peters = Sonneratia alba Sm. ●

368921　Sonneratia neglecta Blume = Sonneratia caseolaris（L.）Engl. ●

368922　Sonneratia obovata Blume = Sonneratia caseolaris（L.）Engl. ●

368923　Sonneratia ovalis Korth. = Sonneratia caseolaris（L.）Engl. ●

368924　Sonneratia ovata Backer；桑海桑●

368925　Sonneratia paracaseolaris W. C. Ko et al. = Sonneratia × gulngai N. C. Duke et Jackes ●

368926　Sonneratiaceae Engl.（1897）（保留科名）；海桑科；Seamulberry Family, Soneratia Family ●

368927　Sonneratiaceae Engl.（保留科名）= Lythraceae J. St.-Hil.（保留科名）■●

368928　Sonneratiaceae Engl. et Gilg = Sonneratiaceae Engl.（保留科名）●

368929　Sonninia Rchb. = Diplolepis R. Br. ●

368930　Sonraya Engl. = Sonzaya Marchand ●

368931　Sonzaya Marchand = Canarium L. ●

368932　Sonzeya Engl. = Sonzaya Marchand ●

368933　Sooja Pócs = Epiclastopelma Lindau ■☆

368934　Sooja Siebold = Cassia L.（保留属名）●■

368935　Sooja Siebold = Chamaecrista Moench ■●

368936　Sooja nomame Siebold = Cassia nomame（Siebold）Kitag. ●

368937　Sooja nomame Siebold = Senna nomame（Siebold）T. Chen ●

368938　Sophandra Meisn. = Erica L. ●☆

368939　Sophandra Meisn. = Lophandra D. Don ●☆

368940　Sophia Adans.（废弃属名）= Descurainia Webb et Berthel.（保留属名）■

368941　Sophia L. = Pachira Aubl. ●

368942　Sophia brachycarpa（Richardson）Rydb. = Descurainia pinnata（Walter）Britton subsp. brachycarpa（Rich.）Detling ■☆

368943　Sophia sophia（L.）Britton = Descurainia sophia（L.）Webb ex Prantl ■

368944　Sophiopsis O. E. Schulz（1924）；羽裂荠属（假播娘蒿属，羽裂芥属，羽裂叶荠属）；Sophiopsis ■

368945　Sophiopsis annua（Rupr.）O. E. Schulz；中亚羽裂荠；Central Asia Sophiopsis ■

368946　Sophiopsis annua（Rupr.）O. E. Schulz var. fontinalis O. E. Schulz = Sophiopsis annua（Rupr.）O. E. Schulz ■

368947　Sophiopsis flavissima（Kar. et Kir.）O. E. Schulz；黄羽裂荠 ■☆

368948　Sophiopsis flavissima O. E. Schulz = Sophiopsis flavissima（Kar. et Kir.）O. E. Schulz ■☆

368949　Sophiopsis micrantha Botsch. et Vved.；小花羽裂荠■☆

368950　Sophiopsis mongolica（Kom.）N. Busch；蒙古羽裂荠■☆

368951　Sophiopsis sisymbrioides（Regel et Herder）O. E. Schulz；羽裂荠；Galiccress-like Sophiopsis ■

368952　Sophisteques Comm. ex Endl. = Ochna L. ●

368953　Sophoclesia Klotzsch = Sphyrospermum Poepp. et Endl. ●☆

368954　Sophonodon Miq. = Siphonodon Griff. ●☆

368955　Sophora L.（1753）；槐属（苦参属）；Pagoda Tree, Pagodatree, Sophora ●■

368956　Sophora acuminata Benth. ex Baker = Sophora benthamii Steenis ●

368957　Sophora acuminata Desv. = Sophora benthamii Steenis ●

368958　Sophora affinis Torr. et Gray；得州槐（紫果槐）；Coralbean, Eves-necklace, Purple-fruit Sophora, Texas Sophora ●☆

368959　Sophora albescens（Rehder）C. Y. Ma；白花槐（白花灰毛槐）；Whiteflower Pagodatree, Whiteflower Sophora, White-flower Velvety Sophora, White-flowered Sophora ●

368960　Sophora albescens Jaume = Sophora albescens（Rehder）C. Y.

Ma ●

368961 Sophora albo-petiolulata Léonard;多米尼加槐;White-petiole Sophora ●☆

368962 Sophora alopecuroides L.;苦豆子(白头蒿子,草槐,哥培尔槐,狐尾槐,看麦娘苦参,苦豆根,苦甘草);Bitterbean Pagodatree, Foxtail-like Sophora ■●

368963 Sophora alopecuroides L. subsp. tomentosa (Boiss.) Yakovlev = Sophora alopecuroides L. var. tomentosa (Boiss.) Ponert ●

368964 Sophora alopecuroides L. var. tomentosa (Boiss.) Bornm. = Sophora alopecuroides L. var. tomentosa (Boiss.) Ponert ●

368965 Sophora alopecuroides L. var. tomentosa (Boiss.) Ponert;毛苦豆子(绒毛苦豆子);Tomentose Sophora ●

368966 Sophora alpina Pall. = Thermopsis alpina (Pall.) Ledeb. ■

368967 Sophora ambigua P. C. Tsoong;贝哈利槐;Doubtful Sophora ●☆

368968 Sophora angustifolia Siebold et Zucc. = Sophora flavescens Aiton ●■

368969 Sophora argentea Pall. = Ammodendron argenteum (Pall.) Kuntze ●◇

368970 Sophora argentea Pall. = Ammodendron bifolium (Pall.) Yakovlev ●◇

368971 Sophora arizonica S. Watson;亚利桑那槐;Arizona Sophora ●☆

368972 Sophora arizonica S. Watson var. formosa (Kearney et Peebles) P. C. Tsoong;美丽槐;Arizona Necklacepod, Beautiful Sophora, Gila Sophora ●☆

368973 Sophora aurea Aiton = Calpurnia aurea (Aiton) Benth. ■☆

368974 Sophora bakeri C. B. Clarke ex Baker;印度槐;Baker Sophora ●☆

368975 Sophora benthamii Steenis;尾叶槐(短绒槐);Acuminate Sophora, Bentham Sophora ●

368976 Sophora bhudaannica Ohashi;不丹槐;Bhutan Sophora ●☆

368977 Sophora bifolia Pall. = Ammodendron argenteum (Pall.) Kuntze ●◇

368978 Sophora bifolia Pall. = Ammodendron bifolium (Pall.) Yakovlev ●◇

368979 Sophora brachygyna C. Y. Ma;短蕊槐;Short-stamen Pagodatree, Short-stamen Sophora, Short-stamened Sophora ●

368980 Sophora buxifolia Retz. = Podalyria buxifolia (Retz.) Willd. ●☆

368981 Sophora calyptrata Retz. = Podalyria calyptrata (Retz.) Willd. ●☆

368982 Sophora capensis L. = Virgilia oroboides (P. J. Bergius) T. M. Salter ●☆

368983 Sophora cavaleriei H. Lév. = Sophora velutina Lindl. var. cavaleriei (H. Lév.) Brummitt et Gillett ●

368984 Sophora ceylonica Trimen;斯里兰卡槐;Ceylon Sophora ●☆

368985 Sophora chathamica Cockayne;奥克兰槐;Chatham Sophora ●☆

368986 Sophora chinensis G. Don;中华槐;China Pagodatree, Chinese Sophora ●

368987 Sophora chinensis Zabel = Sophora japonica L. var. pubescens (Tausch.) Bosse ●

368988 Sophora chrysophylla (Salisb.) Seem.;黄叶槐(金叶槐);Mamane Sophora ●☆

368989 Sophora chrysophylla (Salisb.) Seem. var. glabrata Rock;无毛黄叶槐;Glabrous Sophora ●☆

368990 Sophora conzattii Standl.;墨西哥槐●☆

368991 Sophora davidii (Franch.) Pavol. = Sophora davidii (Franch.) Skeels ●

368992 Sophora davidii (Franch.) Pavol. var. chuansiensis C. Y. Ma = Sophora davidii (Franch.) Skeels var. chuansiensis C. Y. Ma ●

368993 Sophora davidii (Franch.) Pavol. var. liangshanensis C. Y. Ma = Sophora davidii (Franch.) Skeels var. liangshanensis C. Y. Ma ●

368994 Sophora davidii (Franch.) Skeels;白刺槐(白刺花,白花刺,白刺针,蚕豆叶槐,戴维槐,苦刺,苦刺花,苦豆刺,狼牙刺,狼牙槐,马鞭采,马皮采,马蹄针,铁马胡烧);David Sophora, David's Mountain Laurel, Vetchleaf Pagodatree, Vetchleaf Sophora, Whitepine Pagodatree ●

368995 Sophora davidii (Franch.) Skeels var. chuansiensis C. Y. Ma;川西白刺花;W. Sichuan Pagodatree, West Sichuan Sophora ●

368996 Sophora davidii (Franch.) Skeels var. liangshanensis C. Y. Ma;凉山白刺花;Liangshan Pagodatree, Liangshan Sophora ●

368997 Sophora denudata Bory;裸槐;Dentate Sophora ●☆

368998 Sophora denudata Prain. = Sophora dunnii Prain ●

368999 Sophora dispar Craib = Sophora dunnii Prain ●

369000 Sophora duclouxii Gagnep. = Sophora prazeri Prain var. mairei (Pamp.) P. C. Tsoong ●

369001 Sophora duclouxii Gagnep. = Sophora prazeri Prain ●

369002 Sophora dunnii Prain;柳叶槐(凹脉槐);Dunn Sophora, Willowleaf Pagodatree ●

369003 Sophora exigua Craib;稀见槐(弱小槐);Little Sophora ●☆

369004 Sophora exigua Craib var. elatior P. C. Tsoong;缅甸稀见槐(缅甸槐);Tall Little Sophora ●☆

369005 Sophora fabacea Pall. = Thermopsis lupinoides (L.) Link ●■

369006 Sophora fernandeziana Skottsb.;弗尔南德斯槐;Fernandez Sophora ●☆

369007 Sophora fernandeziana Skottsb. f. glacilior Skottsb.;硬毛槐;Hard-hair Sophora ●☆

369008 Sophora fernandeziana Skottsb. var. glacilior Skottsb. = Sophora fernandeziana Skottsb. f. glacilior Skottsb. ●☆

369009 Sophora fernandeziana Skottsb. var. reedeana Skottsb.;圆瓣槐(圆叶槐)●☆

369010 Sophora flavescens Aiton;苦参(拔麻,白萼,白茎,百本,川参,戴椹,地骨,地槐,独椹,凤凰爪,好汉枝,虎卷扁府,虎麻,黄耆,芰草,骄槐,苦参麻,苦参树,苦参炭,苦豆,苦骨,苦槐,苦辛,苦藏,陵郎,鹿白,禄白,绿白,牛参,牛苦参,芩茎,软肉虾,山豆根,山槐,山槐树,蜀脂,水槐,菟槐,王孙,野槐,沼水槐,蘽苦骨);Bittergiseng, Lightyellow Sophora, Light-yellow Sophora ●■

369011 Sophora flavescens Aiton f. angustifolia (Siebold et Zucc.) Yakovlev = Sophora flavescens Aiton ●■

369012 Sophora flavescens Aiton f. purpurascens (Makino) Sugim.;紫苦参●■☆

369013 Sophora flavescens Aiton var. angustifolia (Siebold et Zucc.) Kitag. = Sophora flavescens Aiton ●■

369014 Sophora flavescens Aiton var. galegoides (Pall.) DC.;红花苦参;Redflower Sophora ■

369015 Sophora flavescens Aiton var. kronei (Hance) C. Y. Ma;毛苦参;Downy Lightyellow Sophora ■

369016 Sophora flavescens Aiton var. stenophylla Hayata = Sophora flavescens Aiton ●■

369017 Sophora formosa Kearney et Peebles = Sophora arizonica S. Watson var. formosa (Kearney et Peebles) P. C. Tsoong ●☆

369018 Sophora franchetiana Dunn;闽槐;Franchet Sophora, Fujian Pagodatree ●

369019 Sophora fraseri Benth.;澳大利亚槐;Fraser Sophora ●☆

369020 Sophora galegioides Debeaux = Sophora vestita Nakai ●

369021　Sophora galegoides Pall. = Sophora flavescens Aiton var. galegoides（Pall.）DC. ■

369022　Sophora galioides P. J. Bergius = Cyclopia galioides（P. J. Bergius）DC. ●☆

369023　Sophora genistoides L. = Cyclopia genistoides（L.）R. Br. ●☆

369024　Sophora gibbosa Kuntze；驼曲苦豆子（驼风苦豆子）；Gibbous Sophora ●☆

369025　Sophora glauca Lesch. = Sophora velutina Lindl. ●

369026　Sophora glauca Lesch. ex DC. = Sophora velutina Lindl. ●

369027　Sophora glauca Lesch. ex DC. var. albescens Rehder = Sophora albescens（Rehder）C. Y. Ma ●

369028　Sophora glauca Lesch. var. albescens Rehder = Sophora albescens（Rehder）C. Y. Ma ●

369029　Sophora glauca Lesch. var. albescens Rehder et E. H. Wilson = Sophora velutina Lindl. var. albescens（Rehder et E. H. Wilson）P. C. Tsoong ●

369030　Sophora griffithii Stocks = Sophora mollis（Royle）Baker var. griffithii（Stocker）P. C. Tsoong ●

369031　Sophora gypsophila B. L. Turner et A. M. Powell；奇瓦瓦槐●☆

369032　Sophora gypsophila B. L. Turner et A. M. Powell var. guabalupensis B. L. Turner et A. M. Powell；大叶奇瓦瓦槐●☆

369033　Sophora hirsuta Aiton = Podalyria hirsuta（Aiton）Willd. ●☆

369034　Sophora howinsula（Oliv.）P. C. Tsoong；腺鳞果槐；Gland-scaleleaffruit Sophora ●☆

369035　Sophora inhambanensis Klotzsch；因地槐；South Africac Sophora ●☆

369036　Sophora interrupta Bedd. ；间断槐；Interrupted Sophora ●☆

369037　Sophora japonica L. ；槐树（白槐，豆槐，国槐，黑槐，护房树，怀槐，槐，槐花，槐花木，槐花树，槐角，槐米，槐木，槐实，槐树芽，槐子，家槐，金药树，守宫槐，细叶槐）；Chinese Scholar Tree，Chinese Scholatree，Japan Pagodatree，Japanese Pagoda，Japanese Pagoda Tree，Japanese Pagodatree，Pagoda Tree，Pagodatree，Scholar Tree，Scholars' Tree，Scholar-tree，Waifa ●

369038　Sophora japonica L. 'Pendula'；龙爪槐（倒栽槐，盘槐，蟠槐）；Pendent Japanese Pagodatree，Pendulous Pagodatree，Weeping Pagoda Tree ●

369039　Sophora japonica L. 'Princeton Upright'；直立槐●

369040　Sophora japonica L. 'Regent'；摄政王槐●☆

369041　Sophora japonica L. 'Violacea'；堇花槐（玫瑰紫花槐，紫花槐，紫堇槐）；Violetflower Japanese Pagodatree ●

369042　Sophora japonica L. = Styphnolobium japonicum（L.）Schott ●

369043　Sophora japonica L. f. columnalis Schwer. ；圆柱槐●

369044　Sophora japonica L. f. hybrida Carrière；杂蟠槐；Hybrid Pagodatree ●

369045　Sophora japonica L. f. oligophylla Franch. ；五叶槐；Fiveleaves Japanese Pagodatree ●

369046　Sophora japonica L. f. pendula Loudon = Sophora japonica L. 'Pendula' ●

369047　Sophora japonica L. var. praecox Schwer. ；早生槐●

369048　Sophora japonica L. var. praecox Schwer. f. columnalis Schwer. ；柱花槐●

369049　Sophora japonica L. var. pubescens（Tausch.）Bosse；毛叶槐（柔毛槐，紫花槐）；Pubescent Japanese Pagodatree ●

369050　Sophora japonica L. var. tomentosa Hort. = Sophora japonica L. var. pubescens（Tausch.）Bosse ●

369051　Sophora japonica L. var. vestita Rehder；宜昌槐；Yichang Japanese Pagodatree ●

369052　Sophora japonica L. var. violacea Carrière = Sophora japonica L. 'Violacea' ●

369053　Sophora jaubertii Spach ex Jaub. et Spach；土耳其槐；Jaubert Sophora ●☆

369054　Sophora koreensis Nakai；朝鲜狼牙刺；Korean Sophora ●☆

369055　Sophora kronei Hance = Sophora flavescens Aiton var. kronei（Hance）C. Y. Ma ■

369056　Sophora lehmannii（Bunge）Yakovlev；曲果苦豆子；Lehmann Sophora ●☆

369057　Sophora linearifolia Griseb. ；线叶槐；Linearleaf Sophora ●☆

369058　Sophora longipes Merr. ；长柄槐；Longstalk Sophora ●☆

369059　Sophora lupinoides L. = Thermopsis lupinoides（L.）Link ●■

369060　Sophora macrocarpa Sm. ；大果槐；Bigfruit Sophora ●☆

369061　Sophora mairei H. Lév. = Sophora japonica L. ●

369062　Sophora mairei Pamp = Sophora prazeri Prain ●

369063　Sophora mairei Pamp. = Sophora prazeri Prain var. mairei（Pamp.）P. C. Tsoong ●

369064　Sophora mairei Pamp. = Sophora wilsonii Craib ●

369065　Sophora masafuerana Skottsb. ；智利槐；Chilean Sophora ●☆

369066　Sophora microcarpa C. Y. Ma；细果槐（小果槐）；Little Fruit Sophora，Little-fruited Sophora，Smallfruit Pagodatree ●

369067　Sophora microphylla Sol. ex Aiton；小叶槐；Kowhai，Little-leaf Sophora ●☆

369068　Sophora microphylla Sol. ex Aiton var. longicarinata（Simpson）Allen；长瓣小叶槐●☆

369069　Sophora microphyllafulvida Allen；多小叶槐●☆

369070　Sophora mollis（Royle）Baker；翅果槐（翅果苦参，云南槐树）；Himalayan Laburnum，Soft Sophora，Winged-fruit Sophora，Wingfruit Pagodatree，Wing-fruited Sophora ●

369071　Sophora mollis（Royle）Baker subsp. duthiei（Prain）Ali = Sophora mollis（Royle）Baker var. duthiei Prain ●

369072　Sophora mollis（Royle）Baker subsp. griffithii（Stocks）Ali = Sophora mollis（Royle）Baker var. griffithii（Stocker）P. C. Tsoong ●

369073　Sophora mollis（Royle）Baker var. duthiei Prain；无翅果槐；Wingless Fruit Sophora ●

369074　Sophora mollis（Royle）Baker var. duthiei Prain = Sophora mollis（Royle）Baker subsp. duthiei（Prain）Ali ●

369075　Sophora mollis（Royle）Baker var. griffithii（Stocker）P. C. Tsoong；硬脊槐；Griffith Sophora ●

369076　Sophora mollis（Royle）Baker var. hydaspidis Baker；毛翅果槐；Hairywinged Fruit Sophora ●

369077　Sophora mollis（Royle）Baker var. hydaspidis Baker = Sophora mollis（Royle）Baker var. griffithii（Stocker）P. C. Tsoong ●

369078　Sophora moocroftiana（Graham）Benth. ex Baker；沙生槐树（蓟瓦，狼牙刺，沙生槐，砂生槐）；Moorcroft Sophora，Sandliving Sophora，Sandy Pagodatree ●

369079　Sophora moocroftiana（Graham）Benth. ex Baker subsp. viciifolia（Hance）Yakovlev = Sophora davidii（Franch.）Pavol. ●

369080　Sophora moocroftiana（Graham）Benth. ex Baker var. davidii Franch. = Sophora davidii（Franch.）Pavol. ●

369081　Sophora moocroftiana（Benth.）Baker var. davidii Franch. = Sophora davidii（Franch.）Skeels ●

369082　Sophora moocroftiana Kanitz = Sophora davidii（Franch.）Pavol. ●

369083　Sophora moocroftiana Kanitz subsp. viciifolia（Hance）Yakovlev = Sophora davidii（Franch.）Pavol. ●

369084　Sophora moocroftiana Kanitz var. davidii Franch. = Sophora davidii（Franch.）Pavol.●

369085 Sophora myrtillifolia Retz. = Podalyria myrtillifolia（Retz.）Willd. ●☆

369086 Sophora nitens Harv. = Sophora inhambanensis Klotzsch ●☆

369087 Sophora nitens Schumach. = Sophora tomentosa L. subsp. occidentalis（L.）Brummitt ●☆

369088 Sophora nuttalliana B. L. Turner；丝毛槐；Silkyhair Sophora ●☆

369089 Sophora oblongata P. C. Tsoong；赫布里底槐；Oblong Sophora ●☆

369090 Sophora occidentalis L.；西方槐（紫花绒毛槐）；Occidental Sophora ●☆

369091 Sophora occidentalis L. = Sophora tomentosa L. subsp. occidentalis（L.）Brummitt ●☆

369092 Sophora oligophylla Baker = Angylocalyx oligophyllus（Baker）Baker f. ■☆

369093 Sophora orientalis Pall. = Sophora alopecuroides L. ■●

369094 Sophora oroboides P. J. Bergius = Virgilia oroboides（P. J. Bergius）T. M. Salter ●☆

369095 Sophora pachycarpa Schrenk ex C. A. Mey.；甘肃槐树（厚果哥培尔槐，厚果槐，胖果苦参）；Thick-fruit Sophora, Thickpod Pagodatree ●

369096 Sophora pallida Salisb. = Sophora alopecuroides L. ■●

369097 Sophora pendula Spach. = Sophora japonica L. 'Pendula' ●

369098 Sophora philippinensis Merr.；菲律宾槐；Philippine Platycarpa ●☆

369099 Sophora platycarpa Maxim. = Cladrastis platycarpa（Maxim.）Makino ●

369100 Sophora polyphylla Urb.；古巴槐；Manyleaves Sophora ●☆

369101 Sophora praetorulosa Chun et T. C. Chen；疏节槐；Loose-node Sophora, Loose-noded Sophora, Scatternode Pagodatree ●

369102 Sophora praetorulosa Chun et T. C. Chen var. mairei（Pamp.）P. C. Tsoong = Sophora prazeri Prain ●

369103 Sophora praetorulosa Chun et T. C. Chen var. mairei（Pamp.）P. C. Tsoong = Sophora prazeri Prain var. mairei（Pamp.）P. C. Tsoong ●

369104 Sophora prazeri Prain；锈毛槐；Maire Sophora, Prazer Sophora, Rusthair Pagodatree, Rustyhair Sophora ●

369105 Sophora prazeri Prain subsp. mairei（Pamp.）Yakovlev = Sophora prazeri Prain var. mairei（Pamp.）P. C. Tsoong ●

369106 Sophora prazeri Prain var. burkei P. C. Tsoong；小叶锈毛槐；Small-leaf Sophora ●☆

369107 Sophora prazeri Prain var. mairei（Pamp.）P. C. Tsoong；西南槐（短槐，墨里疏节槐，山豆根，乌豆根，西南槐树）；Maire Loose-node Sophora, Maire Pagodatree, Maire Sophora ●

369108 Sophora prazeri Prain var. mairei（Pamp.）P. C. Tsoong = Sophora wilsonii Craib ●

369109 Sophora prazeri Prain var. micrantha P. C. Tsoong；小花锈毛槐；Littleflower Sophora ●☆

369110 Sophora prodanii E. S. Anderson；普罗槐●☆

369111 Sophora prostrata Buchholz；新西兰槐；Dwarf Kowhai, Prostrate Sophora ●☆

369112 Sophora pubescnes Tausch. = Sophora japonica L. var. pubescens（Tausch.）Bosse ●

369113 Sophora purpusii Brandegee；紫花槐；Purpleflower Sophora ●☆

369114 Sophora rubriflora P. C. Tsoong；红花槐；Redflower Sophora ●☆

369115 Sophora secundiflora（Ortega）DC. = Sophora secundiflora（Ortega）Lag. ex DC. ●☆

369116 Sophora secundiflora（Ortega）Lag. ex DC.；得克萨斯槐（侧花槐，侧花槐树，红豆槐，黄子侧花槐，偏花槐）；Coral Bean, Evergreen Coral Bean, Frijolillo, Frijolko, Medicinal Mescal Bean, Mescal Bean, Mescalbean, Mountain Laurel, Red Bean, Secud-flower Sophora, Texas Mountain Laurel, Texas Mountain-laurel ●☆

369117 Sophora secundiflora（Ortega）Lag. ex DC. f. xanthosperma Rehder；黄籽偏花槐（黄花紫果槐）；Yellowseed Sophora ●☆

369118 Sophora senegalensis Deless. ex DC. = Requienia obcordata（Lam. ex Poir.）DC. ■☆

369119 Sophora sericea Andréws = Podalyria sericea（Andréws）R. Br. ex W. T. Aiton ●☆

369120 Sophora somalensis Chiov.；索马里兰槐；Somalen Sophora ●☆

369121 Sophora somalensis Chiov. = Millettia usaramensis Taub. ●☆

369122 Sophora songarica Schrenk；准噶尔苦豆子（弯果苦豆子）；Soongarian Sophora ●

369123 Sophora sororia Hance = Sophora flavescens Aiton ●■

369124 Sophora speciosa Torr. = Sophora secundiflora（Ortega）Lag. ex DC. ●☆

369125 Sophora stenophylla A. Gray；狭叶槐；Narrowleaf Sophora ●☆

369126 Sophora subprostrata Chun et T. C. Chen；柔枝槐●

369127 Sophora subprostrata Chun et T. C. Chen = Sophora tonkinensis Gagnep. ●

369128 Sophora sylvatica Burch. = Calpurnia aurea（Aiton）Benth. ■☆

369129 Sophora tetragonocarpa Hayata = Sophora flavescens Aiton ●■

369130 Sophora tetraptera J. S. Muell.；四翅槐（羽实槐）；Four-wing Sophora, Fourwings Pagodatree, Fourwings Sophora, Kowhai, New Zealand Kowhai, New Zealand Laburnum, Yellow Kowhai ●

369131 Sophora tetraptera J. S. Muell. microphylla ? = Sophora microphylla Sol. ex Aiton ●☆

369132 Sophora tomentosa Dippel = Sophora japonica L. var. pubescens（Tausch.）Bosse ●

369133 Sophora tomentosa Drake = Sophora tonkinensis Gagnep. ●

369134 Sophora tomentosa L.；绒毛槐（巴哈马绒毛槐，海南槐树，黄芪，岭南槐，岭南槐树，毛苦参）；Doulwnny Sophora, Hainan Sophora, Silverbush, Tomentose Pagodatree, Tomentose Sophora ●

369135 Sophora tomentosa L. f. glabra Steenis；光叶绒毛槐；Glabrous Leaf Sophora, Glabrous Tomentose Sophora ●☆

369136 Sophora tomentosa L. subsp. occidentalis（L.）Brummitt = Sophora occidentalis L. ●☆

369137 Sophora tomentosa L. var. bahamensis P. C. Tsoong；巴哈马绒毛槐；Bahama Sophora ●☆

369138 Sophora tomentosa L. var. glabra Steenis = Sophora tomentosa L. f. glabra Steenis ●☆

369139 Sophora tomentosa L. var. occidentalis（L.）Brummitt = Sophora occidentalis L. ●☆

369140 Sophora tomentosa L. var. occidentalis（L.）Isely = Sophora occidentalis L. ●☆

369141 Sophora tonkinensis Gagnep.；越南槐树（东京槐，广豆根，黄结，金锁匙，苦豆根，南豆根，柔枝槐，柔枝槐树，山大豆根，土豆根，小黄连，岩黄连，越南槐，云豆根）；Subprostrate Pagodatree, Subprostrate Sophora, Tonkin Sophora, Vietnam Pagodatree ●

369142 Sophora tonkinensis Gagnep. = Cephalostigmaton tonkinensis（Gagnep.）Yakovlev ●

369143 Sophora tonkinensis Gagnep. var. polyphylla S. C. Huang et Z. C. Zhou；多叶越南槐；Many-leaves Pagodatree, Many-leaves Sophora ●

369144 Sophora tonkinensis Gagnep. var. purpurescens C. Y. Ma；紫花

越南槐；Purple-flower Pagodatree, Purple-flower Sophora ●

369145 Sophora toromiro Skottsb.；倾卧槐（南美槐）；Lateral Prostrate Sophora ●☆

369146 Sophora unifoliata（Rock）O. Deg. et Sherff；单叶槐；Simle Leaf Sophora ●☆

369147 Sophora unifoliata（Rock）O. Deg. et Sherff var. elliptica（Chock）O. Deg. et Sherff；椭圆叶槐●☆

369148 Sophora unifoliata（Rock）O. Deg. et Sherff var. kanaioensis（Chock）O. Deg. et Sherff；尖瓣槐●☆

369149 Sophora velutina Lindl.；短绒槐（黑豆根，灰毛槐角，灰毛槐树，小苦参，贼骨头，紫花苦参）；Downy Pagodatree, Greyblue Sophora, Greyhair Pagodatree, Velvety Sophora ●

369150 Sophora velutina Lindl. subsp. cavaleriei（H. Lév.）Yakovlev = Sophora velutina Lindl. var. cavaleriei（H. Lév.）Brummitt et Gillett ●

369151 Sophora velutina Lindl. subsp. zimbabweensis J. B. Gillett et Brummitt；津巴布韦槐●☆

369152 Sophora velutina Lindl. var. albescens（Rehder et E. H. Wilson）P. C. Tsoong；白花灰毛槐（白花灰毛槐树，千层皮，山豆根）；Whiteflower Greyblue Sophora, Whiteflower Greyhair Pagodatree ●

369153 Sophora velutina Lindl. var. albescens（Rehder）P. C. Tsoong et C. Y. Ma = Sophora albescens（Rehder）C. Y. Ma ●

369154 Sophora velutina Lindl. var. cavaleriei（H. Lév.）Brummitt et Gillett；光叶短绒槐（毛叶短绒槐）；Cavalerie Velvety Sophora ●

369155 Sophora velutina Lindl. var. dolichopoda C. Y. Ma；长颈槐；Long-stalked Velvety Sophora ●

369156 Sophora velutina Lindl. var. multifoliolata C. Y. Ma；多叶槐（多叶短绒槐）；Climbing Sophora, Many-leaflets Sophora ●

369157 Sophora velutina Lindl. var. scandens C. Y. Ma；攀缘槐；Climbing Pagodatree, Climbing Sophora ●

369158 Sophora vestita Nakai；曲阜槐●

369159 Sophora viciifolia Hance；白刺花●

369160 Sophora viciifolia Hance = Sophora davidii（Franch.）Pavol. ●

369161 Sophora wightii Baker；长梗槐；Wight Sophora ●☆

369162 Sophora wilsonii Craib；瓦山槐；E. H. Wilson Pagodatree, E. H. Wilson Sophora, Wilson Sophora ●

369163 Sophora xanthoantha C. Y. Ma；黄花槐；Yellowflower Pagodatree, Yellowflower Sophora, Yellow-flowered Sophora ●

369164 Sophora yunnanensis C. Y. Ma；云南槐；Yunnan Pagodatree, Yunnan Sophora ●

369165 Sophora zambesiaca Baker = Xanthocercis zambesiaca（Baker）Dumaz-le-Grand ■☆

369166 Sophoraceae Bercht. et J. Presl = Fabaceae Lindl.（保留科名）●■

369167 Sophoraceae Bercht. et J. Presl = Leguminosae Juss.（保留科名）●■

369168 Sophorocapnos Turcz. = Corydalis DC.（保留属名）■

369169 Sophorocapnos pallida（Thunb.）Turcz. = Corydalis pallida（Thunb.）Pers. ■

369170 Sophronanthe Benth. = Gratiola L. ■

369171 Sophronia Licht. ex Roem. et Schult. = Lapeirousia Pourr. ■☆

369172 Sophronia Lindl. = Sophronitis Lindl. ■☆

369173 Sophronia Roem. et Schult. = Lapeirousia Pourr. ■☆

369174 Sophronia caespitosa Licht. = Lapeirousia plicata（Jacq.）Diels ■☆

369175 Sophronitella Schltr.（1925）；小丑角兰属■☆

369176 Sophronitella Schltr. = Isabelia Barb. Rodr. ■☆

369177 Sophronitella violacea（Lindl.）Schltr.；小丑角兰■☆

369178 Sophronitis Lindl.（1828）；丑角兰属（贞兰属）；Sophronitis ■☆

369179 Sophronitis cernua Lindl.；垂头丑角兰；Nodding Sophronitis ■☆

369180 Sophronitis coccinea（Lindl.）Rchb. f.；绯红丑角兰；Scarlet Sophronitis ■☆

369181 Sophronitis lowii Lindl.；楼氏丑角兰■☆

369182 Sophronitis pterocarpa Lindl.；翼果丑角兰■☆

369183 Sopropis Britton et Rose = Prosopis L. ●

369184 Sopubia Buch.-Ham. = Sopubia Buch.-Ham. ex D. Don ■

369185 Sopubia Buch.-Ham. ex D. Don（1825）；短冠草属（短冠花属）；Sopubia ■

369186 Sopubia aemula S. Moore；匹敌短冠草■☆

369187 Sopubia angolensis Engl.；安哥拉短冠草■☆

369188 Sopubia argentea Hiern；银色短冠草■☆

369189 Sopubia buchneri Engl.；布赫纳短冠草■

369190 Sopubia cana Harv. var. glabrescens Diels；渐光短冠草■

369191 Sopubia candeoi A. Terracc. = Merremia candeoi（A. Terracc.）Sebsebe ■☆

369192 Sopubia carsonii V. Naray. = Sopubia lanata Engl. ■

369193 Sopubia comosa（Bonati）T. Yamaz. = Petitmenginia comosa Bonati ■

369194 Sopubia conferta S. Moore；密集短冠草■

369195 Sopubia conferta S. Moore var. congensis（S. Moore）Mielcarek；刚果短冠草■☆

369196 Sopubia congensis S. Moore = Sopubia conferta S. Moore var. congensis（S. Moore）Mielcarek ■☆

369197 Sopubia densiflora V. Naray. = Sopubia lanata Engl. var. densiflora（V. Naray.）O. J. Hansen ■☆

369198 Sopubia dregeana（Benth. ex Hochst.）Benth. = Sopubia simplex（Hochst.）Hochst. ■☆

369199 Sopubia dregeana（Benth. ex Hochst.）Benth. var. tenuifolia Engl. et Gilg = Sopubia mannii V. Naray. var. tenuifolia（Engl. et Gilg）Hepper ■☆

369200 Sopubia duvigneaudiana H. P. Hofm. et Eb. Fisch.；迪维尼奥短冠草■

369201 Sopubia eenii S. Moore = Gomphostigma virgatum（L. f.）Baill. ●☆

369202 Sopubia elatior Pilg.；高短冠草■

369203 Sopubia eminii Engl. = Sopubia parviflora Engl. subsp. eminii（Engl.）Mielcarek ■☆

369204 Sopubia fastigiata Hiern = Sopubia karaguensis Oliv. ■☆

369205 Sopubia filiformis（Schumach. et Thonn.）G. Don = Micrargeria filiformis（Schumach. et Thonn.）Hutch. et Dalziel ■☆

369206 Sopubia filiformis Hiern = Micrargeria filiformis（Schumach. et Thonn.）Hutch. et Dalziel ■☆

369207 Sopubia formosana Hayata = Melasma arvense（Benth.）Hand.-Mazz. ■

369208 Sopubia graminicola Exell；草莺短冠草■☆

369209 Sopubia hildebrandtii Vatke = Pseudosopubia hildebrandtii（Vatke）Engl. ■☆

369210 Sopubia karaguensis Oliv.；卡拉古短冠草■☆

369211 Sopubia karaguensis Oliv. var. macrocalyx O. J. Hansen；大萼短冠草■☆

369212 Sopubia karaguensis Oliv. var. welwitschii（Engl.）O. J. Hansen；韦尔短冠草■☆

369213 Sopubia kassneri Pilg. = Sopubia lanata Engl. var. densiflora

（V. Naray.）O. J. Hansen ■☆

369214　Sopubia kituiensis Vatke = Pseudosopubia kituiensis（Vatke）Engl. ■☆

369215　Sopubia lanata Engl. ；绵毛短冠草■

369216　Sopubia lanata Engl. var. densiflora（V. Naray.）O. J. Hansen；密花绵毛短冠草■☆

369217　Sopubia lasiocarpa P. C. Tsoong；毛果短冠草（钟山草）；Hairfruit Sopubia ■

369218　Sopubia latifolia Engl. ；宽叶短冠草■☆

369219　Sopubia laxior S. Moore = Sopubia ramosa（Hochst.）Hochst. ■☆

369220　Sopubia lejolyana Mielcarek；勒若利短冠草■☆

369221　Sopubia leprosa S. Moore = Gomphostigma virgatum（L. f.）Baill. ●☆

369222　Sopubia mannii V. Naray. ；曼氏短冠草■☆

369223　Sopubia mannii V. Naray. var. linearifolia O. J. Hansen；线叶曼氏短冠草■☆

369224　Sopubia mannii V. Naray. var. metallorum（P. A. Duvign.）Mielcarek；光泽曼氏短冠草■☆

369225　Sopubia mannii V. Naray. var. tenuifolia（Engl. et Gilg）Hepper；细叶曼氏短冠草■☆

369226　Sopubia matsumurae（T. Yamaz.）C. Y. Wu；中南短冠草■

369227　Sopubia menglianensis Y. Y. Qian；孟连短冠草■

369228　Sopubia metallorum P. A. Duvign. = Sopubia mannii V. Naray. var. metallorum（P. A. Duvign.）Mielcarek ■☆

369229　Sopubia monteiroi V. Naray. = Sopubia lanata Engl. ■

369230　Sopubia neptunii P. A. Duvign. et Van Bockstal；内普丘恩短冠草■☆

369231　Sopubia obtusifolia（Benth.）G. Don = Harveya obtusifolia（Benth.）Vatke ■☆

369232　Sopubia parviflora Engl. ；小花短冠草■☆

369233　Sopubia parviflora Engl. subsp. eminii（Engl.）Mielcarek；埃明小花短冠草■☆

369234　Sopubia ramosa（Hochst.）Hochst. ；分枝短冠草■☆

369235　Sopubia scabra（L. f.）G. Don = Graderia scabra（L. f.）Benth. ■☆

369236　Sopubia scaettae Staner = Sopubia lanata Engl. ■

369237　Sopubia scopiformis（Klotzsch）Vatke = Micrargeria filiformis（Schumach. et Thonn.）Hutch. et Dalziel ■☆

369238　Sopubia similis V. Naray. = Sopubia ramosa（Hochst.）Hochst. ■☆

369239　Sopubia simplex（Hochst.）Hochst. ；简单短冠草■☆

369240　Sopubia stricta G. Don；坚挺短冠草；Strict Sopubia ■

369241　Sopubia trifida Buch. -Ham. ；短冠草（虫药，小伸筋草，英雄草）；Trifid Sopubia ■

369242　Sopubia trifida Buch. -Ham. ex D. Don var. ramosa（Hochst.）Engl. = Sopubia ramosa（Hochst.）Hochst. ■☆

369243　Sopubia trifida Buch. -Ham. ex D. Don f. humilis Engl. et Gilg = Sopubia mannii V. Naray. var. tenuifolia（Engl. et Gilg）Hepper ■☆

369244　Sopubia ugandensis S. Moore；乌干达短冠草■☆

369245　Sopubia welwitschii Engl. = Sopubia karaguensis Oliv. var. welwitschii（Engl.）O. J. Hansen ■☆

369246　Sopubia welwitschii Engl. var. micrantha ？= Sopubia karaguensis Oliv. ■☆

369247　Soramia Aubl. = Doliocarpus Rol. ●☆

369248　Soramiaceae Martinov = Dilleniaceae Salisb.（保留科名）●■

369249　Soranthe Salisb. = Sorocephalus R. Br.（保留属名）●☆

369250　Soranthe Salisb. ex Knight（废弃属名）= Sorocephalus R. Br.（保留属名）●☆

369251　Soranthe clavigera Salisb. ex Knight = Sorocephalus clavigerus（Salisb. ex Knight）Hutch. ●☆

369252　Soranthe pinifolia Salisb. ex Knight = Sorocephalus pinifolius（Salisb. ex Knight）Rourke ●☆

369253　Soranthe rupestris Salisb. ex Knight = Sorocephalus clavigerus（Salisb. ex Knight）Hutch. ●☆

369254　Soranthus Ledeb.（1829）；簇花芹属（束花属）；Soranthus ■

369255　Soranthus Ledeb. = Ferula L. ●

369256　Soranthus meyeri Ledeb. ；簇花芹；Soranthus ■

369257　Sorbaceae Brenner = Rosaceae Juss.（保留科名）●■

369258　Sorbaria（DC.）A. Braun = Sorbaria（Ser. ex DC.）A. Braun（保留属名）●

369259　Sorbaria（Ser. ex DC.）A. Braun（1860）（保留属名）；珍珠梅属（珍珠花属）；False Spiraea, Falsespiraea, False-spiraea ●

369260　Sorbaria（Ser.）A. Braun = Sorbaria（Ser. ex DC.）A. Braun（保留属名）●

369261　Sorbaria（Ser.）A. Braun ex Asch. = Sorbaria（Ser. ex DC.）A. Braun（保留属名）●

369262　Sorbaria A. Braun = Sorbaria（Ser. ex DC.）A. Braun（保留属名）●

369263　Sorbaria aitchisonii Hemsl. = Sorbaria tomentosa（Lindl.）Rehder ●

369264　Sorbaria amurensis Koehne；阿穆尔珍珠梅；Amur Falsespiraea ●☆

369265　Sorbaria arborea Schneid. ；高丛珍珠梅（八木条，花儿杆，山高粱，野生珍珠梅，珍珠杆，珍珠梅）；Tree Falsespiraea, Tree False-spiraea ●

369266　Sorbaria arborea Schneid. = Sorbaria kirilowii（Regel）Maxim. ●

369267　Sorbaria arborea Schneid. var. glabrata Rehder；光叶高丛珍珠梅（光叶珍珠梅）；Glabrous Tree Falsespiraea ●

369268　Sorbaria arborea Schneid. var. subtomentosa Rehder；毛叶高丛珍珠梅（毛叶珍珠梅）；Hairy Tree Falsespiraea ●

369269　Sorbaria assurgens Vilm. et Bois = Sorbaria kirilowii（Regel）Maxim. ●

369270　Sorbaria diabolica Koidz. ；红绒毛珍珠梅●☆

369271　Sorbaria grandiflora Maxim. ；毛珍珠梅●☆

369272　Sorbaria kirilowii（Regel）Maxim. ；华北珍珠梅（花楸珍珠梅，吉氏珍珠梅，珍珠梅）；Chinese Sorbaria, False Spirea, Giant False Spiraea, Kirilow Falsespiraea, Kirilow False-spiraea ●

369273　Sorbaria lindleyana Maxim. = Sorbaria tomentosa（Lindl.）Rehder ●

369274　Sorbaria olgae Zinserl. ；奥氏珍珠梅●☆

369275　Sorbaria pallasii（G. Don）Pojark. ；帕氏珍珠梅●☆

369276　Sorbaria rhoifolia Kom. ；盐肤木叶珍珠梅●☆

369277　Sorbaria sibirica Hedl. ；西伯利亚珍珠梅；Siberian Falsespiraea ●☆

369278　Sorbaria sorbifolia（L.）A. Brau var. kirilowii（Regel et Tiling）Ito = Sorbaria kirilowii（Regel）Maxim. ●

369279　Sorbaria sorbifolia（L.）A. Braun；珍珠梅（八本条，东北珍珠梅，高楷子，花儿杆，花楸珍珠梅，华楸珍珠梅，毛漆，山高粱，山高粱条子，珍珠杆，珍珠花，走马蓁）；False Spiraea, Mountain-ash False Spiraea, Mountainash Falsespiraea, Mountainash False-spiraea, Sorbaria, Ural False Spiraea, Ural Falsespiraea, Ural False-spiraea ●

369280　Sorbaria sorbifolia（L.）A. Braun f. incerta（C. K. Schneid.）

Kitag. ;虾夷珍珠梅●☆

369281　Sorbaria sorbifolia（L.）A. Braun var. kirilowii Ito = Sorbaria kirilowii（Regel）Maxim. ●

369282　Sorbaria sorbifolia（L.）A. Braun var. stellipila Maxim. ;星毛珍珠梅（穗形七度灶,星毛花楸珍珠梅,星毛华楸珍珠梅,走马蓁）;Starshaped Hair Ural Falsespiraea ●

369283　Sorbaria sorbifolia（L.）A. Braun var. typica Schneid. = Sorbaria sorbifolia（L.）A. Braun ●

369284　Sorbaria sorbifolia A. Braun = Sorbaria kirilowii（Regel）Maxim. ●

369285　Sorbaria stellipila（Maxim.）C. K. Schneid. = Sorbaria sorbifolia（L.）A. Braun var. stellipila Maxim. ●

369286　Sorbaria stellipila（Maxim.）C. K. Schneid. var. incerta C. K. Schneid. = Sorbaria sorbifolia（L.）A. Braun f. incerta（C. K. Schneid.）Kitag. ●☆

369287　Sorbaria stellipila C. K. Schneid. = Sorbaria sorbifolia（L.）A. Braun var. stellipila Maxim. ●

369288　Sorbaria tomentosa（Lindl.）Rehder;西藏珍珠梅（大花珍珠梅）;Himalayan Sorbaria, Lindley False Spirea, Tibet Falsespiraea ●

369289　Sorbaria tomentosa（Lindl.）Rehder var. angustifolia（Wenz.）Rehder;狭叶西藏珍珠梅（狭叶大花珍珠梅,狭叶珍珠梅）●☆

369290　Sorbus L.（1753）;花楸属（花楸树属）;Dogberry, Mountain Ash, Mountainash, Mountain-ash, Rowan, Rowan Tree, Scorb, Whitebeam ●

369291　Sorbus × arnoldiana Rehder;阿诺德花楸●☆

369292　Sorbus × arnoldiana Rehder 'Carpet of Gold';金毯阿诺德花楸●☆

369293　Sorbus × arnoldiana Rehder 'Chamois Glow';黄果阿诺德花楸●☆

369294　Sorbus × arnoldiana Rehder 'Golden Wonder';金色奇迹阿诺德花楸●☆

369295　Sorbus × arnoldiana Rehder 'Kirsten Pink';柯尔斯顿粉阿诺德花楸●☆

369296　Sorbus × kawashiroi Koji Ito ex Murata;川诚花楸●☆

369297　Sorbus × thuringiaca（Hedl.）Fritsch;栎叶花楸（锐齿花楸）;Bastard Service-tree, Oak-leafed Mountain Ash ●☆

369298　Sorbus × thuringiaca（Hedl.）Fritsch 'Fastigiana';帚枝栎叶花楸（帚枝锐齿花楸）●☆

369299　Sorbus × uzenensis Koidz. ;羽前花楸●☆

369300　Sorbus aestivalis Koehne = Sorbus prattii Koehne var. aestivalis（Koehne）Te T. Yu ●

369301　Sorbus albopilosa Te T. Yu et L. T. Lu;白毛花楸;Whitehair Mountainash ●

369302　Sorbus albovii Zinserl. ;阿氏花楸●☆

369303　Sorbus alnifolia（Siebold et Zucc.）K. Koch;水榆花楸●

369304　Sorbus alnifolia（Siebold et Zucc.）K. Koch f. lobulata Koidz. = Aria alnifolia（Siebold et Zucc.）Decne. var. lobulata Koidz. ●

369305　Sorbus alnifolia（Siebold et Zucc.）K. Koch var. angulata S. B. Liang;棱果花楸;Angulate Densehead Mountainash ●

369306　Sorbus alnifolia（Siebold et Zucc.）K. Koch var. lobulata（Koidz.）Rehder;裂叶水榆花楸;Lobed Densehead Mountainash, Lobed Mountain Ash ●

369307　Sorbus alnifolia（Siebold et Zucc.）K. Koch var. submollis Rehder;稍软水榆花楸●☆

369308　Sorbus alnifolia（Siebold et Zucc.）K. Koch var. tiliifolia Hisauti;椴叶花楸●

369309　Sorbus amabilis W. C. Cheng ex Te T. Yu;黄山花楸;

Huangshan Mountainash, Huangshan Mountain-ash ●◇

369310　Sorbus amabilis W. C. Cheng ex Te T. Yu var. wuyishanensis Z. X. Yu;武夷山花楸;Wuyishan Mountainash, Wuyishan Mountain-ash ●

369311　Sorbus amabilis W. C. Cheng ex Te T. Yu var. wuyishanensis Z. X. Yu = Sorbus amabilis W. C. Cheng ex Te T. Yu ●◇

369312　Sorbus ambrozyana C. K. Schneid. = Sorbus coronata Hedl. var. ambrozzyana（C. K. Schneid.）L. T. Lu ●

369313　Sorbus americana Marshall;美洲花楸（美国花楸）;America Mountainash, American Mountain Ash, American Mountain-ash, American Rowan-tree, Mountain Ash, Roundwood ●☆

369314　Sorbus americana Marshall subsp. japonica（Maxim.）Kitam. = Sorbus commixta Hedl. ●

369315　Sorbus americana Marshall subsp. japonica（Maxim.）Kitam. var. rufoferruginea（C. K. Schneid.）Kitam. = Sorbus commixta Hedl. var. rufoferruginea C. K. Schneid. ●☆

369316　Sorbus amurensis Koehne = Sorbus pohuashanensis（Hance）Hedl. ●

369317　Sorbus anadyrensis Kom. ;阿纳代尔花楸●☆

369318　Sorbus anglica Hedl. ;英国花楸●☆

369319　Sorbus aperta Koehne = Sorbus hupehensis C. K. Schneid. ●

369320　Sorbus arbutifolia（L.）Heynh. var. atropurpurea（Britton）C. K. Schneid. = Aronia prunifolia（Marshall）Rehder ●☆

369321　Sorbus arguta Te T. Yu;锐齿花楸;Sharptooth Mountainash, Sharp-toothed Mountain-ash ●

369322　Sorbus aria（L.）Crantz;白花楸（白背花楸,白面子树）; Beam Tree, Chaw-leaf, Chess Apple, Chess-apple, Common Whitebeam, Cumberland Hawthorn, Hen Apple, Hoar Withy, Iron Pear, Iron-pear, Lot-tree, Mulberry, Quickbeam, Service-berry, Whipbeam, Whipcrop, Whitbin Pear, Whitcbeam Mountain Ash, White Beam, White Beamtree, White Rice, Whitebeam, Whitebeam Mountain Ash, Whiteleaf Tree, Whithin Pear, Whitten, Widbin Pear, Widbin Pear Tree, Wild Pear, Winterbeam ●☆

369323　Sorbus aria（L.）Crantz 'Aurea';金叶白花楸（金叶白面子树）●☆

369324　Sorbus aria（L.）Crantz 'Chrysophylla';黄叶白花楸（黄叶白面子树,金叶白背花楸）●☆

369325　Sorbus aria（L.）Crantz 'Decne. ana' = Sorbus aria（L.）Crantz 'Majestica' ●☆

369326　Sorbus aria（L.）Crantz 'Lutescens';土黄白花楸（淡黄白花楸,土黄白面子树）●☆

369327　Sorbus aria（L.）Crantz 'Majestica';壮丽白花楸（大叶白背花楸,壮丽白面子树）●☆

369328　Sorbus aria（L.）Crantz 'Quercoides';栎叶白花楸（栎叶白面子树）●☆

369329　Sorbus aria（L.）Crantz subsp. meridionalis（Guss.）= Sorbus umbellata（Desf.）Fritsch ●☆

369330　Sorbus aria（L.）Crantz var. cyclophylla C. K. Schneid. = Sorbus aria（L.）Crantz ☆

369331　Sorbus aria（L.）Crantz var. incisa Rchb. = Sorbus aria（L.）Crantz ●☆

369332　Sorbus aria（L.）Crantz var. mairei H. Lév. = Sorbus keissleri（C. K. Schneid.）Rehder ●

369333　Sorbus aria Crantz = Sorbus aria（L.）Crantz ●☆

369334　Sorbus armeniaca Hedl. ;亚美尼亚花楸●☆

369335　Sorbus aronioides Rehder;毛背花楸;Hairyback Mountainash, Hairy-backed Mountain-ash ●

369336　Sorbus astateria（Cardot）Hand. -Mazz. ;多变花楸;Variable

Mountainash, Variable Mountain-ash ●

369337　Sorbus atrosanguinea Te T. Yu et H. T. Tsai = Sorbus thibetica (Cardot) Hand. -Mazz. ●

369338　Sorbus atrovirens Hort. ex K. Koch;墨绿花楸●☆

369339　Sorbus aucuparia L.;欧洲花楸(北欧花楸,鹿梨,鸟花楸,鸟梨,鸟梨花楸,欧亚花楸,野槐);Caorthann, Care-tree, Carr Care, Chit-chat, Cock-drunks, Cuirn, Devil's Hate, Dog Berries, Eowler's Service, Eurasian Mountain-ash, Europe Mountainash, European Mountain Ash, European Mountainash, European Mountain-ash, Field Ash, Fowler's Service, Hen Drunks, Hicken, Keer, Keirn, Kern, Kieran, Kitty Keys, Mountain Ash, Poison Berry, Quickbane, Quickbeam, Quicken-berry, Quicken-tree, Quickenwood, Quick-tree, Rantle-tree, Ran-tree, Rantry, Ranty-berry, Rauntree, Raven-tree, Rawntree, Roan-thee, Rodden, Roddin, Roddon, Roden, Roden-quicken-rowan, Rodin, Rodon, Rone, Roundberry, Roun-tree, Rowan, Rowan Tree, Rowntree, Sap Tree, Serb, Shepherd's Friend, Sip-sap, Twickband, Twig Bean, Twig-bean, Twlckbine, Whicky, Whistlewood, White Ash, Whitten-tree, Whitty, Whltty-tree, Wicken Tree, Wickenwood, Wiggan, Wiggen, Wiggin, Wiggy, Wild Ash, Wild Service, Wild Sorb, Witch Elm, Witch Hazel, Witch Wicken, Witch Wiggin, Witchbeam, Witchen, Witchen-tree, Witchin Tree, Witchwood, Withen, Withwine, Witty, Wltchbane, Wychen ●

369340　Sorbus aucuparia L. 'Aspleniifolia';铁角蕨叶欧洲花楸●☆

369341　Sorbus aucuparia L. 'Cardinal Royal';主教欧洲花楸●☆

369342　Sorbus aucuparia L. 'Fastigiata';帚状欧洲花楸(帚状鸟花楸);Upright European Mountain-ash ●☆

369343　Sorbus aucuparia L. 'Fructu Luteo';乳黄果欧洲花楸(黄果鸟花楸)●☆

369344　Sorbus aucuparia L. 'Pendula';垂枝欧洲花楸●☆

369345　Sorbus aucuparia L. 'Rossiva Major';大果鸟花楸●☆

369346　Sorbus aucuparia L. 'SheerwaterbSeedling';希尔沃特实生鸟花楸●☆

369347　Sorbus aucuparia L. 'Variegata';斑叶欧洲花楸●☆

369348　Sorbus aucuparia L. 'Xanthocarpa';黄果欧洲花楸●☆

369349　Sorbus baldaccii Degen et Fritsch;巴尔花楸●☆

369350　Sorbus boissieri C. K. Schneid.;布瓦西耶花楸●☆

369351　Sorbus bushiana Zinserl.;布氏花楸●☆

369352　Sorbus caloneura (Stapf) Rehder;美脉花楸(川花楸,山黄果);Browndot Mountainash, Browndot Mountain-ash ●

369353　Sorbus caloneura (Stapf) Rehder var. kwangtungensis Te T. Yu;广东美脉花楸;Guangdong Browndot Mountainash, Kwangtung Browndot Mountainash ●

369354　Sorbus carpinifolia Te T. Yu et L. T. Lu;鹅耳枥叶花楸;Hornbeam Mountain-ash ●

369355　Sorbus carpinifolia Te T. Yu et L. T. Lu = Sorbus yunnanensis L. T. Lu ●

369356　Sorbus cashmiriana Hedl.;克什米尔花楸;Kashmir Rowan ●☆

369357　Sorbus caucasica Zinserl.;高加索花楸●☆

369358　Sorbus chamaemespilus Crantz;矮生白面子树;Dwarf Whitebeam, False Medlar ●☆

369359　Sorbus chengii C. J. Qi;大叶石灰树;Cheng's Mountainash ●

369360　Sorbus chengii C. J. Qi = Sorbus folgneri (C. K. Schneid.) Rehder var. duplicatodentata Te T. Yu et A. M. Lu ●

369361　Sorbus colchica Zinserl.;黑海花楸●☆

369362　Sorbus commixta Hedl.;朝鲜花楸(日本花楸,杂色花楸);Japanese Rowan, Korean Sorbus ●☆

369363　Sorbus commixta Hedl. 'Embley';恩波利朝鲜花楸(恩波利

日本花楸,红叶杂色花楸);Japanese Rowan ●☆

369364　Sorbus commixta Hedl. 'Ethel's Gold';埃塞尔金朝鲜花楸(埃塞尔金日本花楸);Japanese Rowan ●☆

369365　Sorbus commixta Hedl. 'Jermyns';杰明斯朝鲜花楸(杰明斯日本花楸);Japanese Rowan ●☆

369366　Sorbus commixta Hedl. var. rufoferruginea C. K. Schneid.;褐毛朝鲜花楸;Brown-haired Korean Sorbus ●☆

369367　Sorbus commixta Hedl. var. sachalinensis Koidz.;库页花楸;Sachalin Mountainash ●☆

369368　Sorbus commixta Hedl. var. sachalinensis Koidz. = Sorbus commixta Hedl. ●☆

369369　Sorbus commixta Hedl. var. wilfordii (Koehne) Sugim.;威尔朝鲜花楸●☆

369370　Sorbus conradinae Koehne = Sorbus esserteauiana Koehne ●

369371　Sorbus coronata (Cardot) Te T. Yu et H. T. Tsai;冠萼花楸;Coronate Mountainash ●

369372　Sorbus coronata Hedl. var. ambrozzyana (C. K. Schneid.) L. T. Lu;少脉冠萼花楸;Few-veins Coronate Mountainash ●

369373　Sorbus coronata Hedl. var. glabrescens Te T. Yu et L. T. Lu;脱毛冠萼花楸;Glabrescent Coronate Mountainash ●

369374　Sorbus corymbifera (Miq.) T. H. Nguyên et Yakovlev;疣果花楸;Granular Mountainash, Granular Mountain-ash ●

369375　Sorbus cuspidata (Spach) Hedl.;白叶花楸(喜马拉雅花楸);Cuspidate Mountainash, Himalayan Whitebeam, Sharp-tipped Mountain-ash, Whiteleaf Mountainash ●

369376　Sorbus cuspidata (Spach) Hedl. = Sorbus vestita (Wall. ex G. Don) S. Schauer ●

369377　Sorbus decora (Sarg.) C. K. Schneid.;美丽花楸(复花楸叶);Northern Mountain-ash, Showy Mountain Ash, Showy Mountainash, Showy Mountain-ash, Shrub Mountain-ash ●

369378　Sorbus detergibilis Merr. = Sorbus epidendron Hand. -Mazz. ●

369379　Sorbus discolor (Maxim.) Maxim.;北京花楸(白果臭山槐,白果花楸,北平花楸树,红叶花楸,黄果臭山槐);Beijing Mountainash, Mountain-ash, Snowberry Mountain Ash, Snowberry Mountainash, Snowberry Mountain-ash ●

369380　Sorbus discolor (Maxim.) Maxim. var. paucijuga D. K. Zang et P. C. Huang = Sorbus hupehensis C. K. Schneid. var. paucijuga (D. K. Zang et P. C. Huang) L. T. Lu ●

369381　Sorbus domestica L.;欧亚花楸(红果花楸,花楸果,花楸树叶,家种花楸);Cultivate Mountainash, Right Service, Service Tree, Service-tree, Service-tree Mountain Ash, Sorb Apple, Sorb-tree, True Service, True Service-tree, Whitty Pear, Whitty-pear, Withy Pear, Witten Pear, Witten Pear Tree ●☆

369382　Sorbus dunnii Rehder;棕脉花楸;Dunn Mountainash, Dunn Mountain-ash ●

369383　Sorbus epidendron Hand. -Mazz.;附生花楸;Epiphytic Mountainash, Epiphytic Mountain-ash ●

369384　Sorbus esserteauiana Koehne;麻叶花楸;Hempleaf Mountainash, Hemp-leaved Mountain-ash ●

369385　Sorbus expansa Koehne = Sorbus wilsoniana C. K. Schneid. ●

369386　Sorbus febrifuga A. Juss. = Soymida febrifuga (Roxb.) A. Juss. ●☆

369387　Sorbus ferruginea (Wenz.) Rehder;锈色花楸;Rusty Mountainash, Rusty Mountain-ash ●

369388　Sorbus filipes Hand. -Mazz.;纤细花楸;Fine Mountainash, Fine Mountain-ash ●

369389　Sorbus folgneri (C. K. Schneid.) Rehder;石灰花楸(白绵子

树,翻白树,反白树,粉背叶,傅氏花楸,华盖木,毛桲子,石灰树,石灰条子);Folgner Mountainash,Folgner Mountain-ash,Lime Mountainash ●

369390 Sorbus folgneri (C. K. Schneid.) Rehder var. duplicatodentata Te T. Yu et A. M. Lu;密齿石灰花楸(齿叶石灰花楸);Dense-toothed Folgner Mountainash ●

369391 Sorbus foliolosa (Wall.) Spach;尼泊尔花楸(美叶花楸);Nepal Mountainash,Wallich Mountainash,Wallich Mountain-ash ●

369392 Sorbus foliolosa (Wall.) Spach var. pluripinnata C. K. Schneid. = Sorbus scalaris Koehne ●

369393 Sorbus foliolosa (Wall.) Spach var. ursina Wenz. = Sorbus ursina (Wenz.) Hedl. ●

369394 Sorbus formosanus (Koidz.) S. S. Ying;台湾海棠●

369395 Sorbus forrestii McAll. et Gillham;福莱斯花楸●☆

369396 Sorbus giraldiana C. K. Schneid. = Sorbus tapashana C. K. Schneid. ●

369397 Sorbus glabrata Hedl. ;无毛花楸;Glabrous Mountainash ●☆

369398 Sorbus glabrescens (Cardot) Hand. -Mazz. = Sorbus oligodonta (Cardot) Hand. -Mazz. ●

369399 Sorbus globosa Te T. Yu et H. T. Tsai;圆果花楸;Roundfruit Mountainash,Round-fruited Mountain-ash ●

369400 Sorbus glomerulata Koehne;球穗花楸;Ballraceme Mountainash, Glomerulate Mountainash,Glomerulate Mountain-ash ●

369401 Sorbus gracilis (Siebold et Zucc.) K. Koch;细花楸(纤细花楸);Slender Mountainash ●☆

369402 Sorbus gracilis K. Koch = Sorbus gracilis (Siebold et Zucc.) K. Koch ●☆

369403 Sorbus graeca (Spach) Hedl. ;希腊花楸;Greece Mountainash ●☆

369404 Sorbus graeca Lodd. ex Steud. = Sorbus graeca (Spach) Hedl. ●☆

369405 Sorbus granulosa (Bertol.) Rehder = Sorbus corymbifera (Miq.) T. H. Nguyên et Yakovlev ●

369406 Sorbus guanxianensis T. C. Ku;灌县花楸;Guanxian Mountainash,Guanxian Mountain-ash ●

369407 Sorbus harrowiana (Balf. f. et W. W. Sm.) Rehder;巨叶花楸;Harrow Mountainash,Harrow Mountain-ash ●

369408 Sorbus harrowiana (Balf. f. et W. W. Sm.) Rehder = Sorbus insignis (Hook. f.) Hedl. ●

369409 Sorbus hedlundii C. K. Schneid. ;海德伦花楸●☆

369410 Sorbus helenae Koehne;钝齿花楸;Obtusetooth Mountainash, Obtuse-toothed Mountain-ash ●

369411 Sorbus helenae Koehne f. rufidula Koehne = Sorbus helenae Koehne ●

369412 Sorbus helenae Koehne f. subglabra Koehne = Sorbus helenae Koehne ●

369413 Sorbus helenae Koehne var. argutiserrata Te T. Yu;川西钝齿花楸(尖齿花楸,锐齿花楸);Sharptooth Mountainash ●

369414 Sorbus hemsleyi (C. K. Schneid.) Rehder;江南花楸;Hemsley Mountainash,Hemsley Mountain-ash ●

369415 Sorbus henryi Rehder = Sorbus hemsleyi (C. K. Schneid.) Rehder ●

369416 Sorbus hoii Fang = Sorbus prattii Koehne ●

369417 Sorbus hostii Heynh. ;锐尖花楸●☆

369418 Sorbus hunanica C. J. Qi;湖南花楸;Hunan Mountainash, Hunan Mountain-ash ●

369419 Sorbus hunanica C. J. Qi = Sorbus zahlbruckneri C. K. Schneid. ●

369420 Sorbus hupehensis C. K. Schneid. ;湖北花楸(雪压花);Hubei Mountain-ash,Hupeh Mountain-ash,Hupeh Rowan ●

369421 Sorbus hupehensis C. K. Schneid. ' Coral Rire';珊瑚火湖北花楸●☆

369422 Sorbus hupehensis C. K. Schneid. ' Rosea';绯果湖北花楸●☆

369423 Sorbus hupehensis C. K. Schneid. var. aperta (Koehne) C. K. Schneid. = Sorbus hupehensis C. K. Schneid. ●

369424 Sorbus hupehensis C. K. Schneid. var. laxiflora (Koehne) C. K. Schneid. = Sorbus hupehensis C. K. Schneid. ●

369425 Sorbus hupehensis C. K. Schneid. var. obtusa C. K. Schneid. = Sorbus hupehensis C. K. Schneid. ●

369426 Sorbus hupehensis C. K. Schneid. var. obtusa C. K. Schneid. = Sorbus oligodonta (Cardot) Hand. -Mazz. ●

369427 Sorbus hupehensis C. K. Schneid. var. paucijuga (D. K. Zang et P. C. Huang) L. T. Lu;少叶花楸;Few-leaves Mountainash ●

369428 Sorbus hupehensis C. K. Schneid. var. syncarpa Koehne = Sorbus hupehensis C. K. Schneid. ●

369429 Sorbus hybrida L. ;杂种花楸;Bastard Service Tree, Bastard Service-tree, Cornish Whitebeam, Finnish Whitebeam, Oakleaf Mountain Ash, Oak-leaf Mountain Ash, Service Tree of Fontainebleau,Swedish Service-tree ●☆

369430 Sorbus hypoglauca (Cardot) Hand. -Mazz. = Sorbus rehderiana Koehne ●

369431 Sorbus insignis (Hook. f.) Hedl. ;卷边花楸;Revelute Mountainash,Revelute Mountain-ash ●

369432 Sorbus intermedia Pers. ;瑞典花楸;Cut-leaved Whitebeam, Scots Whitebeam, Swedish Mountain Ash, Swedish White Beam, Swedish Whitebeam ●☆

369433 Sorbus japonica (Decne.) Hedl. = Aria japonica Decne. ●☆

369434 Sorbus japonica (Decne.) Hedl. f. denudata (Nakai) Hiyama = Aria japonica Decne. f. denudata (Nakai) Yonek. ●☆

369435 Sorbus japonica (Decne.) Hedl. var. calocarpa Rehder = Aria japonica Decne. f. calocarpa (Rehder) Yonek. ●☆

369436 Sorbus kamtschatcensis Kom. ;勘察加花楸●☆

369437 Sorbus keissleri (C. K. Schneid.) Rehder;毛序花楸(凯旋花);Keissler Mountainash,Keissler Mountain-ash ●

369438 Sorbus kewensis K. J. W. Hensen;邱园花楸●☆

369439 Sorbus kiukiangensis Te T. Yu;俅江花楸;Kiukiang Mountainash,Kiukiang Mountain-ash,Qiujiang Mountainash ●

369440 Sorbus kiukiangensis Te T. Yu = Malus kansuensis (Batalin) C. K. Schneid. ●

369441 Sorbus kiukiangensis Te T. Yu var. crossotocalyx (Cardot) C. Y. Wu = Sorbus kiukiangensis Te T. Yu ●

369442 Sorbus kiukiangensis Te T. Yu var. glabrescens Te T. Yu;无毛俅江花楸;Glabrous Qiujiang Mountainash ●

369443 Sorbus koehneana C. K. Schneid. ;陕甘花楸(昆氏花楸);Koehne Mountainash,Koehne Mountain-ash,Shangan Mountainash ●

369444 Sorbus kusnetzovii Zinserl. ;库兹花楸●☆

369445 Sorbus lanpingensis L. T. Lu;兰坪花楸;Lanping Mountainash ●

369446 Sorbus latifolia (Lam.) Pers. ;宽叶花楸;Broad-leaved Whitebeam, French Hales, Service-tree-of-fontainebleau ●☆

369447 Sorbus latifolia Pers. ' French';法国花楸;French Hales ●☆

369448 Sorbus latifolia Pers. = Sorbus latifolia (Lam.) Pers. ●☆

369449 Sorbus laxiflora Koehne = Sorbus hupehensis C. K. Schneid. ●

369450 Sorbus macrantha Merr. ;大花花楸(疏序花楸);Bigflower Mountainash,Tree of Fountainbleu ●

369451 Sorbus maderensis (Lowe) Dode;梅德花楸●☆

369452 Sorbus mairei Rehder et H. Lév. = Sorbus keissleri（C. K. Schneid.）Rehder ●

369453 Sorbus manshuriensis Kitag. = Sorbus pohuashanensis（Hance）Hedl. ●

369454 Sorbus matsumurana（Makino）Koehne；松村氏花楸；Japanese Mountain Ash ●☆

369455 Sorbus matsumurana（Makino）Koehne f. pseudogracilis（Koidz.）Ohwi；大托叶松村氏花楸●☆

369456 Sorbus medogensis L. T. Lu et T. C. Ku；墨脱花楸●

369457 Sorbus megalocarpa Rehder；大果花楸（砂糖果）；Bigfruit Mountainash，Big-fruited Mountain-ash，Large-fruited Whitebeam ●

369458 Sorbus megalocarpa Rehder var. cuneata Rehder；楔叶大果花楸（圆大果花楸，圆果大果花楸）；Cuneate Bigfruit Mountainash ●

369459 Sorbus melanocarpa（Michx.）Heynh.；黑果花楸；Blackfruit Mountainash ●☆

369460 Sorbus melanocarpa（Michx.）Heynh. = Aronia melanocarpa（Michx.）Elliott ●☆

369461 Sorbus meliosmifolia Rehder；泡吹叶花楸；Meliosmaleaf Mountainash，Meliosma-leaved Mountain-ash ●

369462 Sorbus micrantha Koidz.；小花花楸；Smallflower Mountainash ●☆

369463 Sorbus microcarpa Pursh = Sorbus americana Marshall ●☆

369464 Sorbus microphylla Wenz.；小叶花楸；Littleleaf Mountainash ●

369465 Sorbus monbeigii（Cardot）Te T. Yu；维西花楸；Monbeig Mountainash，Monbeig Mountain-ash ●

369466 Sorbus mougeotti Godr. et Soy.-Will.；莫高花楸；Edible Mountain Ash ●☆

369467 Sorbus multijuga Koehne；多对花楸；Multijugous Mountainash，Multijugous Mountain-ash ●

369468 Sorbus multijuga Koehne var. microdenta Koehne = Sorbus multijuga Koehne ●

369469 Sorbus multijuga Koehne var. microdonta Koehne = Sorbus koehneana C. K. Schneid. ●

369470 Sorbus munda Koehne = Sorbus prattii Koehne ●

369471 Sorbus munda Koehne f. subarachnoidea Koehne = Sorbus prattii Koehne ●

369472 Sorbus munda Koehne f. tatsienensis Koehne = Sorbus prattii Koehne ●

369473 Sorbus nubium Hand.-Mazz. = Sorbus folgneri（C. K. Schneid.）Rehder ●

369474 Sorbus obsoletidentata（Cardot）Te T. Yu；宾川花楸；Binchuan Mountainash，Obsoletedentate Mountainash，Obsoletedentate Mountain-ash ●

369475 Sorbus ochracea（Hand.-Mazz.）J. E. Vidal；褐毛花楸；Brownhair Mountainash，Ochre-yellow Mountain-ash ●

369476 Sorbus ochrocarpa Rehder = Sorbus pallescens Rehder ●

369477 Sorbus oligodonta（Cardot）Hand.-Mazz.；少齿花楸；Fewteeth Mountainash，Few-teethed Mountain-ash ●

369478 Sorbus pallescens Rehder；灰叶花楸；Greyleaf Mountainash，Grey-leaved Mountain-ash ●

369479 Sorbus paniculata Te T. Yu et H. T. Tsai = Sorbus rhamnoides（Decne.）Rehder ●

369480 Sorbus pekinensis Koehne = Sorbus discolor（Maxim.）Maxim. ●

369481 Sorbus persica Hedl.；波斯花楸●☆

369482 Sorbus pluripinnata（C. K. Schneid.）Koehne = Sorbus scalaris Koehne ●

369483 Sorbus pogonopetala Koehne = Sorbus prattii Koehne ●

369484 Sorbus pohuashanensis（Hance）Hedl.；花楸树（百花山花楸，百华花楸，百华山花楸，黑龙江花楸，红果臭山槐，花楸，马加木，马家木，绒花树，绒毛树，山槐子）；Amur Mountain Ash，Amur Mountainash，Mountain-ash，Pohuashan Mountain Mountain-ash，Pohuashan Mountainash，Pohuashan Mountain-ash ●

369485 Sorbus pohuashanensis（Hance）Hedl. var. amurensis（Koehne）Y. L. Chou et S. L. Tung = Sorbus pohuashanensis（Hance）Hedl. ●

369486 Sorbus pohuashanensis（Hance）Hedl. var. manshurensis（Kitag.）Y. C. Chu；东北花楸●☆

369487 Sorbus pohuashanensis（Hance）Hedl. var. manshuriensis（Kitag.）Y. C. Chu = Sorbus pohuashanensis（Hance）Hedl. ●

369488 Sorbus poteriifolia Hand.-Mazz.；侏儒花楸；Dwarf Mountainash，Dwarf Mountain-ash ●

369489 Sorbus poteriifolia Hand.-Mazz. = Sorbus filipes Hand.-Mazz. ●

369490 Sorbus prattii Koehne；西康花楸（川滇花楸，独椒，蒲氏花楸，爪瓣花楸）；Pratt Mountainash，Pratt Mountain-ash，Xikang Mountainash ●

369491 Sorbus prattii Koehne var. aestivalis（Koehne）Te T. Yu；多对西康花楸●

369492 Sorbus prattii Koehne var. tatsienensis C. K. Schneid. = Sorbus prattii Koehne ●

369493 Sorbus pseudofennica E. F. Warb.；阿兰花楸；Arran Service Tree，Arran Service-tree ●☆

369494 Sorbus pseudogracilis（C. K. Schneid.）Koehne = Sorbus sambucifolia（Cham. et Schltdl.）M. Roem. var. pseudogracilis C. K. Schneid. ●☆

369495 Sorbus pteridophylla Hand.-Mazz.；蕨叶花楸；Fernleaf Mountainash，Fern-leaved Mountain-ash ●

369496 Sorbus pteridophylla Hand.-Mazz. var. tephroclada Hand.-Mazz.；灰毛蕨叶花楸（毛叶蕨叶花楸）；Grey-hair Fernleaf Mountainash ●

369497 Sorbus pymaea Hutch. = Sorbus poteriifolia Hand.-Mazz. ●

369498 Sorbus randaiensis（Hayata）Koidz.；台湾花楸（峦大花楸，峦大山花楸）；Taiwan Mountainash，Taiwan Mountain-ash ●

369499 Sorbus reducta Diels；铺地花楸（矮丛花楸）；Decumbent Mountainash，Decumbent Mountain-ash，Dwarf Chinese Mountain Ash ●

369500 Sorbus reducta Diels var. pubescens L. T. Lu；毛萼铺地花楸；Pubescent Decumbent Mountainash ●

369501 Sorbus rehderiana Koehne；西南花楸（芮德花楸）；Rehder Mountainash，Rehder Mountain-ash ●

369502 Sorbus rehderiana Koehne var. cupreonitens Hand.-Mazz.；锈毛西南花楸；Rusty-hair Rehder Mountainash ●

369503 Sorbus rehderiana Koehne var. grosseserrata Koehne；巨齿西南花楸；Bigtooth Rehder Mountainash ●

369504 Sorbus rhamnoides（Decne.）Rehder；鼠李叶花楸；Buckthornleaf Mountainash，Buckthorn-leaved Mountain-ash ●

369505 Sorbus rhombifolia C. J. Qi et K. W. Liu；菱叶花楸；Rhombifolious Mountainash ●

369506 Sorbus rubiginosa Te T. Yu = Sorbus ochracea（Hand.-Mazz.）J. E. Vidal ●

369507 Sorbus rufoferrugineus（Shiras.）Koidz. var. trilocularis（Hayata）Koidz. = Sorbus randaiensis（Hayata）Koidz. ●

369508 Sorbus rufoferrugineus C. K. Schneid. var. trilocularis Koidz. = Sorbus randaiensis（Hayata）Koidz. ●

369509 Sorbus rufopilosa C. K. Schneid.；红毛花楸；Redhair Mountainash，Redpilose Mountainash，Red-pilosed Mountainash ●

369510 Sorbus rufopilosa C. K. Schneid. var. stenophylla Koehne;狭叶花楸 ●

369511 Sorbus salwinensis Te T. Yu et L. T. Lu;怒江花楸;Nujiang Mountainash,Nujiang Mountain-ash ●

369512 Sorbus sambucifolia（Cham. et Schltdl.）M. Roem.;高岑花楸;Siberian Mountain Ash ●☆

369513 Sorbus sambucifolia（Cham. et Schltdl.）M. Roem. var. pseudogracilis C. K. Schneid.;小高岑花楸●☆

369514 Sorbus sambucifolia Roem.;接骨木叶花楸●☆

369515 Sorbus sargentiana Koehne;晚绿花楸（山麻柳,佘坚花楸,晚绣花球,晚绣球）;Sargent Mountainash,Sargent Mountain-ash,Sargent's Rowan ●

369516 Sorbus scalaris Koehne;梯叶花楸;Scalar Mountainash,Scalar Mountain-ash ●

369517 Sorbus scandica Fr. = Sorbus intermedia Pers. ●☆

369518 Sorbus schemachensis Zinserl.;谢马花楸●☆

369519 Sorbus scopulina Greene;格林花楸;Greene Mountainash,Western Mountain Ash,Western Mountainash,Western Mountain-ash ●☆

369520 Sorbus setschwanensis（C. K. Schneid.）Koehne;四川花楸;Sichuan Mountainash,Szechwan Mountain-ash ●

369521 Sorbus sibirica Hedl.;西伯利亚花楸;Siberia Mountainash ●

369522 Sorbus sikkimensis Wenz. = Sorbus granulosa（Bertol.）Rehder ●

369523 Sorbus sikkimensis Wenz. var. ferruginea Wenz. = Sorbus ferruginea（Wenz.）Rehder ●

369524 Sorbus sikkimensis Wenz. var. oblongifolia ? = Sorbus rhamnoides（Decne.）Rehder ●

369525 Sorbus sitchensis Roem.;太平洋花楸;Ocean Mountainash,Pacific Mountainash,Sitka Mountain Ash,Sitka Mountainash ●

369526 Sorbus subfusca（Ledeb.）Boiss.;浅棕花楸●☆

369527 Sorbus subochracea Te T. Yu et L. T. Lu;尾叶花楸;Caudateleaf Mountainash,Subochre-yellow Mountain-ash,Tailleaf Mountainash ●

369528 Sorbus subtomentusa（Albov）Zinserl.;亚毛花楸●☆

369529 Sorbus taishanensis F. Z. Li et X. D. Chen;泰山花楸（山东花楸）;Taishan Mountainash,Taishan Mountain-ash ●

369530 Sorbus taishanensis F. Z. Li et X. D. Chen = Sorbus pohuashanensis（Hance）Hedl. ●

369531 Sorbus tapashana C. K. Schneid.;太白花楸;Taibai Mountainash,Taibaishan Mountainash,Taibaishan Mountain-ash ●

369532 Sorbus taurica Zinserl.;克里木花楸●☆

369533 Sorbus thibetica（Cardot）Hand.-Mazz.;康藏花楸;Kangzang Mountainash,Tibet Mountainash,Tibet Mountain-ash ●

369534 Sorbus thibetica（Cardot）Hand.-Mazz.'John Mitchell';约翰·米切尔康藏花楸●☆

369535 Sorbus thomsonii（King）Rehder;滇缅花楸;Sino-Burma Mountainash,Thomson Mountainash,Thomson Mountain-ash ●

369536 Sorbus tianschanica Rupr.;天山花楸（花楸）;Tianshan Mountain Mountain-ash,Tianshan Mountainash,Tianshan Mountain-ash ●

369537 Sorbus tianschanica Rupr. var. integrifoliata Te T. Yu;全缘叶天山花楸;Entire-leaf Tianshan Mountainash ●

369538 Sorbus tianschanica Rupr. var. tomentosa Chang Y. Yang et Y. L. Han;天山毛花楸;Tomentose Tianshan Mountainash ●

369539 Sorbus tianschanica Rupr. var. tomentosa Chang Y. Yang et Y. L. Han = Sorbus tapashana C. K. Schneid. ●

369540 Sorbus toringo K. Koch = Malus sieboldii（Regel）Rehder ●

369541 Sorbus torminalis（L.）Crantz;红叶花楸（野花楸,治疝花

楸）;Checker Tree,Checker-tree Mountain Ash,Checker-tree Mountain-ash,Chequer Tree,Chequers,Chequer-tree,Chequer-wood,Choke Pear,Corme,Hagberry,Lezzory,Lizzory,Maple Service,Maple Tree,Redleaf Mountainash,Serb,Serb Apple,Service Tree,Sherves,Shir,Surry,Swallow Pear,Swallow-pear,Whitty Pear,Whitty-bush,Whitty-pear,Wild Service,Wild Service Tree,Wild Service-tree ●

369542 Sorbus torminalis（L.）Crantz var. mollis（Beck）Asch. et Graebn. = Sorbus torminalis（L.）Crantz ●

369543 Sorbus torminalis（L.）Crantz var. perincisa（Borbás et Fekete）C. K. Schneid. = Sorbus torminalis（L.）Crantz ●

369544 Sorbus torminalis（L.）Crantz var. pinnatifida Boiss. = Sorbus torminalis（L.）Crantz ●

369545 Sorbus torminalis Crantz = Sorbus torminalis（L.）Crantz ●

369546 Sorbus trilocularis（Hayata）Masam. = Sorbus randaiensis（Hayata）Koidz. ●

369547 Sorbus tsinlingensis C. L. Tang;秦岭花楸;Qinling Mountainash,Qinling Mountain-ash ●

369548 Sorbus turcica Zinserl.;土耳其花楸;Turkey Mountainash ●☆

369549 Sorbus turkestanica（Franch.）Hedl.;新疆花楸（土耳其斯坦花楸）●

369550 Sorbus umbellata（Desf.）Fritsch;小伞花楸●☆

369551 Sorbus unguiculata Koehne = Sorbus prattii Koehne ●

369552 Sorbus ursina（Wenz.）Hedl.;美叶花楸●

369553 Sorbus ursina（Wenz.）Hedl. var. wenzigiana C. K. Schneid.;西藏美叶花楸●

369554 Sorbus valbrayi H. Lév. = Sorbus koehneana C. K. Schneid. ●

369555 Sorbus variabilis Zabel = Photinia villosa（Thunb.）DC. ●

369556 Sorbus velutina（Albov）Schneid.;绒毛花楸●☆

369557 Sorbus vestita（Wall. ex G. Don）S. Schauer = Sorbus cuspidata（Spach）Hedl. ●

369558 Sorbus vilmorinii C. K. Schneid.;川滇花楸;Vilmorin Mountainash,Vilmorin Mountain-ash ●

369559 Sorbus vilmorinii C. K. Schneid. var. setschwanensis C. K. Schneid. = Sorbus setschwanensis（C. K. Schneid.）Koehne ●

369560 Sorbus vilmorinii C. K. Schneid. var. typica Schneid. = Sorbus vilmorinii C. K. Schneid. ●

369561 Sorbus viminalis Koidz.;柳条状花楸●☆

369562 Sorbus wallichii（Hook. f.）Te T. Yu = Sorbus foliolosa（Wall.）Spach ●

369563 Sorbus wardii Merr. = Sorbus thibetica（Cardot）Hand.-Mazz. ●

369564 Sorbus wenzigiana（C. K. Schneid.）Koehne = Sorbus ursina（Wenz.）Hedl. var. wenzigiana C. K. Schneid. ●

369565 Sorbus wilsoniana C. K. Schneid.;华西花楸（威氏花楸）;E. H. Wilson Mountainash,Wilson Mountain-ash ●

369566 Sorbus woronowii Zinserl.;沃氏花楸●☆

369567 Sorbus xanthoneura Rehder;黄脉花楸;Yellowvein Mountainash,Yellow-veined Mountain-ash ●

369568 Sorbus xanthoneura Rehder = Sorbus hemsleyi（C. K. Schneid.）Rehder ●

369569 Sorbus yuana Spongberg;鄂西花楸（神农架花楸）;Yu Mountainash ●

369570 Sorbus yunnanensis L. T. Lu;栎叶花楸;Hornbeamleaf Mountainash ●

369571 Sorbus zahlbruckneri C. K. Schneid.;长果花楸;Longfruit Mountainash,Zahlbruckner Mountainash,Zahlbruckner Mountain-ash ●

369572 Sorbus zayuensis Te T. Yu et L. T. Lu;察隅花楸;Chayu Mountainash,Chayu Mountain-ash ●

369573　Sorema Lindl. = Nolana L. ex L. f. ■☆

369574　Sorema Lindl. = Periloba Raf. ■☆

369575　Sorghastrum Nash（1901）；假高粱属；Indian Grass，Indiangrass ■☆

369576　Sorghastrum avenaceum（Michx.）Nash = Sorghastrum nutans（L.）Nash ■☆

369577　Sorghastrum bipennatum（Hack.）Pilg.；双羽假高粱■☆

369578　Sorghastrum friesii（Pilg.）Pilg. = Sorghastrum nudipes Nash ■☆

369579　Sorghastrum fuscescens（Pilg.）Clayton；浅棕假高粱■☆

369580　Sorghastrum micratherum（Stapf）Pilg.；小药假高粱■☆

369581　Sorghastrum nudipes Nash；光梗假高粱■☆

369582　Sorghastrum nutans（L.）Nash；黄假高粱；Indian Grass，Nodding Indian-grass，Yellow Indian Grass，Yellow Indiangrass ■☆

369583　Sorghastrum pogonostachyum（Stapf）Clayton；须穗假高粱■☆

369584　Sorghastrum rigidifolium（Stapf）Chippind. ex Pole-Evans = Sorghastrum stipoides（Kunth）Nash ■☆

369585　Sorghastrum stipoides（Kunth）Nash；针叶假高粱；Needle Indiangrass ■☆

369586　Sorghastrum stipoides Nash = Sorghastrum stipoides（Kunth）Nash ■☆

369587　Sorghastrum tisserantii Clayton；蒂斯朗特假高粱■☆

369588　Sorghastrum trichopus（Stapf）Pilg. = Sorghastrum stipoides（Kunth）Nash ■☆

369589　Sorghastrum trollii Pilg. = Sorghastrum stipoides（Kunth）Nash ■☆

369590　Sorghum Moench（1794）（保留属名）；高粱属（蜀黍属）；Gaoliang，Guinea Corn，Indian Millet，Millet，Sorghum ■

369591　Sorghum × drummondii（Nees ex Steud.）Millsp. et Chase = Sorghum sudanense（Piper）Stapf ■

369592　Sorghum aethiopicum（Hack.）Stapf；埃塞俄比亚高粱■☆

369593　Sorghum almum Parodi；杂高粱；Almum Sorghum，Columbus Grass，Mixed Gaoliang，Mixed Sorghum ■

369594　Sorghum almum Parodi = Sorghum halepense（L.）Pers. ■

369595　Sorghum ankolib（Hack.）Stapf = Sorghum bicolor（L.）Moench ■

369596　Sorghum annuum（Trab.）Maire；一年高粱■☆

369597　Sorghum arundinaceum（Desv.）Stapf；普通野高粱；Common Wild Sorghum ■☆

369598　Sorghum aterrimum Stapf；黑高粱■☆

369599　Sorghum basutorum Snowden = Sorghum bicolor（L.）Moench ■

369600　Sorghum bicolor（L.）Moench；高粱（荻粱，二色高粱，高粱七，红高粱，两色蜀黍，芦穄，扫帚高粱，蜀秫，蜀黍，爪龙）；Broom Corn，Broom-corn，Chicken Corn，Durra，Feterita，Florence Whisk，Grain Sorghum，Great Millet，Guinea Corn，Imphee，Jowar，Kaffir Corn，Kafir Corn，Millet，Shattercane Gaoliang，Sorgho，Sorghum，Sudan Grass，Two-coloured Sorghum ■

369601　Sorghum bicolor（L.）Moench 'Cernuum'；弯头高粱（垂穗高粱草，高粱，黄绒毛草，蜀黍）；Nodding Sorghum，Nutant Gaoliang，White Durra ■

369602　Sorghum bicolor（L.）Moench 'Dochna' = Sorghum dochna（Forssk.）Snowden ■

369603　Sorghum bicolor（L.）Moench 'Nervosum' = Sorghum nervosum Besser ex Schult. ■

369604　Sorghum bicolor（L.）Moench nothosubsp. drummondii（Nees ex Steud.）de Wet ex Davidse = Sorghum × drummondii（Nees ex Steud.）Millsp. et Chase ■

369605　Sorghum bicolor（L.）Moench subsp. arundinaceum（Desv.）de Wet et Harlan；菁状高粱■

369606　Sorghum bicolor（L.）Moench subsp. drummondii（Nees ex Steud.）de Wet；德拉蒙德高粱■☆

369607　Sorghum bicolor（L.）Moench var. aethiopicum（Hack.）de Wet et Huckaby = Sorghum aethiopicum（Hack.）Stapf ■☆

369608　Sorghum bicolor（L.）Moench var. arundinaceum（Desv.）de Wet et Huckaby = Sorghum arundinaceum（Desv.）Stapf ■☆

369609　Sorghum bicolor（L.）Moench var. caffrorum（Retz.）Mohlenbr. = Sorghum bicolor（L.）Moench ■

369610　Sorghum bicolor（L.）Moench var. drummondii（Steud.）Mohlenbr. = Sorghum bicolor（L.）Moench ■

369611　Sorghum bicolor（L.）Moench var. picigutta Snowden；元江高粱■

369612　Sorghum bicolor（L.）Moench var. subglobosum（Hack.）Snowden；球果高粱■

369613　Sorghum bicolor（L.）Moench var. verticilliforum（Steud.）de Wett et Huckaby = Sorghum arundinaceum（Desv.）Stapf ■☆

369614　Sorghum bipennatum（Hack.）Kuntze = Sorghastrum bipennatum（Hack.）Pilg. ■☆

369615　Sorghum bracteata（Humb. et Bonpl. ex Willd.）Kuntze = Hyparrhenia bracteata（Humb. et Bonpl. ex Willd.）Stapf ■

369616　Sorghum brevicarinatum Snowden = Sorghum arundinaceum（Desv.）Stapf ■☆

369617　Sorghum caffrorum（Retz.）P. Beauv. = Sorghum bicolor（L.）Moench ■

369618　Sorghum caffrorum（Retz.）P. Beauv. = Sorghum caffrorum（Thunb.）P. Beauv. ■

369619　Sorghum caffrorum（Thunb.）P. Beauv.；卡佛尔高粱；Caffron Gaoliang，Caffron Sorghum ■

369620　Sorghum castaneum C. E. Hubb. et Snowden = Sorghum arundinaceum（Desv.）Stapf ■☆

369621　Sorghum caucasicum（Trin.）Griseb. = Bothriochloa bladhii（Retz.）S. T. Blake ■

369622　Sorghum caudatum（Hack.）Stapf = Sorghum bicolor（L.）Moench ■

369623　Sorghum caudatum Stapf = Sorghum bicolor（L.）Moench ■

369624　Sorghum centroplicatum Chiov.；折扇高粱■☆

369625　Sorghum cernuum（Ard.）Host = Sorghum bicolor（L.）Moench 'Cernuum' ■

369626　Sorghum cernuum（Ard.）Host = Sorghum bicolor（L.）Moench ■

369627　Sorghum comosum Kuntze = Hyparrhenia coleotricha（Steud.）Andersson ex Clayton ■☆

369628　Sorghum conspicuum Snowden = Sorghum bicolor（L.）Moench ■

369629　Sorghum coriaceum Snowden = Sorghum bicolor（L.）Moench ■

369630　Sorghum cymbarium（L.）Kuntze = Hyparrhenia cymbaria（L.）Stapf ■☆

369631　Sorghum dimidiatum Stapf = Sorghum purpureo-sericeum（Hochst. ex A. Rich.）Asch. et Schweinf. ■☆

369632　Sorghum diplandrum（Hack.）Kuntze = Hyparrhenia diplandra（Hack.）Stapf ■

369633　Sorghum dochna（Forssk.）Snowden；甜高粱；Sweet Gaoliang，Sweet Sorghum ■

369634　Sorghum dochna（Forssk.）Snowden = Sorghum bicolor（L.）Moench ■

369635　Sorghum dochna（Forssk.）Snowden var. technicum（Körn.）

Snowden = Sorghum bicolor（L.）Moench ■

369636　Sorghum dochna（Forssk.）Snowden var. techicum（Körn.）Snowden；工艺高粱；Broomcorn ■

369637　Sorghum drummondii（Nees ex Steud.）Millsp. et Chase = Sorghum bicolor（L.）Moench ■

369638　Sorghum drummondii（Steud.）Millsp. et Chase = Sorghum bicolor（L.）Moench ■

369639　Sorghum dulcicaule Snowden = Sorghum bicolor（L.）Moench ■

369640　Sorghum durra（Forssk.）Stapf；硬秆高粱（硬高粱草）；Brown Durra，Durra，Hard Sorghum，Hardculm Gaoliang ■

369641　Sorghum durra（Forssk.）Stapf = Sorghum bicolor（L.）Moench ■

369642　Sorghum elliotii Stapf = Sorghum drummondii（Steud.）Millsp. et Chase ■

369643　Sorghum exsertum Snowden = Sorghum bicolor（L.）Moench ■

369644　Sorghum fasciculare（Roxb.）Haines = Pseudosorghum fasciculare（Roxb. A. Camus ■

369645　Sorghum filipendulum（Hochst.）Kuntze = Hyparrhenia filipendula（Hochst.）Stapf ■

369646　Sorghum fulvum（R. Br.）P. Beauv. = Sorghum nitidum（Vahl）Pers. ■

369647　Sorghum fulvum P. Beauv. = Sorghum nitidum（Vahl）Pers. ■

369648　Sorghum gambicum Snowden = Sorghum bicolor（L.）Moench ■

369649　Sorghum giganteum Edgew. = Sorghum halepense（L.）Pers. ■

369650　Sorghum guineense Stapf；几内亚高粱■

369651　Sorghum guineense Stapf = Sorghum bicolor（L.）Moench ■

369652　Sorghum halepense（L.）Pers.；石茅（阿拉伯高粱，琼生草，石茅高粱，亚勒伯高粱，约翰逊草，詹森草）；Aleppo Grass，Aleppo-grass，Guinea Grass，Guinea-grass，Helep Gaoliang，Johnson Grass，Johnsongrass，Johnson-grass ■

369653　Sorghum halepense（L.）Pers. f. muticum（Hack.）C. E. Hubb.；秃石茅■☆

369654　Sorghum halepense（L.）Pers. subsp. annuum Trab. = Sorghum annuum（Trab.）Maire ■☆

369655　Sorghum halepense（L.）Pers. subsp. sativum（Hack.）Trab. = Sorghum bicolor（L.）Moench ■

369656　Sorghum halepense（L.）Pers. var. genuinus Honda = Sorghum halepense（L.）Pers. ■

369657　Sorghum halepense（L.）Pers. var. muticum（Hack.）Honda = Sorghum halepense（L.）Pers. f. muticum（Hack.）C. E. Hubb. ■☆

369658　Sorghum halepense（L.）Pers. var. propinquum（Kunth）Ohwi = Sorghum propinquum（Kunth）Hitchc. ■

369659　Sorghum heteroclitum（Roxb.）Kuntze = Pseudanthistria heteroclita（Roxb.）Hook. f. ■

369660　Sorghum japonicum（Hack.）Roshev. = Sorghum nervosum Besser ex Schult. ■

369661　Sorghum japonicum Roshev. = Sorghum nervosum Besser ex Schult. ■

369662　Sorghum lanceolatum Stapf = Sorghum arundinaceum（Desv.）Stapf ■☆

369663　Sorghum margaritiferum Stapf；珍珠高粱■☆

369664　Sorghum melaleucum Stapf；黑白高粱■☆

369665　Sorghum melanocarpum Huber = Sorghum bicolor（L.）Moench ■

369666　Sorghum mellitum Snowden = Sorghum bicolor（L.）Moench ■

369667　Sorghum membranaceum Chiov. = Sorghum bicolor（L.）Moench ■

369668　Sorghum micratherum Stapf = Sorghastrum micratherum（Stapf）Pilg. ■☆

369669　Sorghum miliaceum（Roxb.）Snowden = Sorghum halepense（L.）Pers. ■

369670　Sorghum miliaceum（Roxb.）Snowden var. parvispiculum Snowden = Sorghum halepense（L.）Pers. ■

369671　Sorghum nervosum Besser ex Schult.；多脉高粱（凸脉高粱）；Kaliang，Nerved Sorghum，Nervose Gaoliang ■

369672　Sorghum nervosum Besser ex Schult. = Sorghum bicolor（L.）Moench ■

369673　Sorghum nervosum Besser ex Schult. var. flexibile Snowden；散穗高粱■

369674　Sorghum nervosum Chiov. = Sorghum nervosum Besser ex Schult. ■

369675　Sorghum newtonii（Hack.）Kuntze = Hyparrhenia newtonii（Hack.）Stapf ■

369676　Sorghum nigricans（Ruiz et Pav.）Snowden = Sorghum bicolor（L.）Moench ■

369677　Sorghum nitidum（Vahl）Pers.；光高粱（草蜀黍，芦稼茅）；Shine Gaoliang，Shining Sorghum ■

369678　Sorghum nitidum（Vahl）Pers. f. aristatum C. E. Hubb.；具芒光高粱（小光高粱）■☆

369679　Sorghum nitidum（Vahl）Pers. f. aristatum C. E. Hubb. = Sorghum nitidum（Vahl）Pers. ■

369680　Sorghum nitidum（Vahl）Pers. subsp. dichroanthum（Steud.）T. Koyama = Sorghum nitidum（Vahl）Pers. var. dichroanthum（Steud.）Ohwi ■☆

369681　Sorghum nitidum（Vahl）Pers. var. dichroanthum（Steud.）Ohwi；两色花光高粱■☆

369682　Sorghum nitidum（Vahl）Pers. var. fulvus（R. Br.）Hand. - Mazz. = Sorghum nitidum（Vahl）Pers. ■

369683　Sorghum notabile Snowden = Sorghum bicolor（L.）Moench ■

369684　Sorghum pallidum（Chiov.）Chiov.；苍白高粱■☆

369685　Sorghum panicoides Stapf = Sorghum bicolor（L.）Moench subsp. arundinaceum（Desv.）de Wet et Harlan ■

369686　Sorghum parviflorum（R. Br.）P. Beauv.；小花高粱■☆

369687　Sorghum parviflorum P. Beauv. = Sorghum parviflorum（R. Br.）P. Beauv. ■☆

369688　Sorghum pogonostachyum Stapf = Sorghastrum pogonostachyum（Stapf）Clayton ■☆

369689　Sorghum propinquum（Kunth）Hitchc.；拟高粱（大茅根，高粱七，山高粱）；Nearly Gaoliang，Resembling Sorghum ■

369690　Sorghum propinquum（Kunth）Hitchc. = Sorghum halepense（L.）Pers. ■

369691　Sorghum pugionifolium Snowden = Sorghum arundinaceum（Desv.）Stapf ■☆

369692　Sorghum purpureo-sericeum（Hochst. ex A. Rich.）Asch. et Schweinf.；紫绢毛高粱■☆

369693　Sorghum quartinianum（A. Rich.）Hack. = Capillipedium parviflorum（R. Br.）Stapf ■

369694　Sorghum rigidifolium Stapf = Sorghastrum stipoides（Kunth）Nash ■☆

369695　Sorghum rigidifolium Stapf var. microstachyum？ = Sorghastrum stipoides（Kunth）Nash ■☆

369696　Sorghum rigidum Snowden = Sorghum bicolor（L.）Moench ■

369697　Sorghum rollii Chiov.；罗尔高粱■☆

369698　Sorghum roxburghii Stapf = Sorghum bicolor（L.）Moench ■

369699　Sorghum saccharatum（L.）Moench = Sorghum bicolor（L.）Moench ■

369700　Sorghum saccharatum Poir.；甜糖高粱（获蔗,芦粟,甜秆高粱,甜高粱）；Broom Corn, Broomcorn, Chinese Sugar Maple, Sorgho, Sugar Sorghum, Sweet Sorghum ■

369701　Sorghum schimperi（Hochst. ex A. Rich.）Kuntze = Hyparrhenia schimperi（Hochst. ex A. Rich.）Andersson ex Stapf ■☆

369702　Sorghum serratum（Thunb.）Kuntze = Sorghum nitidum（Vahl）Pers. ■

369703　Sorghum simulans Snowden = Sorghum bicolor（L.）Moench ■

369704　Sorghum somaliense Snowden = Sorghum arundinaceum（Desv.）Stapf ■☆

369705　Sorghum stapfii（Hook. f.）Fisch. = Sorghum arundinaceum（Desv.）Stapf ■☆

369706　Sorghum subglabrescens Schweinf. et Asch. = Sorghum bicolor（L.）Moench ■

369707　Sorghum sudanense（Piper）Stapf；苏丹草（杜氏高粱）；Dragon's-teeth, Shattercane, Sudan Gaoliang, Sudan Grass, Sudangrass, Sudan-grass ■

369708　Sorghum sudanense（Piper）Stapf = Sorghum × drummondii（Nees ex Steud.）Millsp. et Chase ■

369709　Sorghum sudanense（Piper）Stapf = Sorghum bicolor（L.）Moench ■

369710　Sorghum sudanense（Piper）Stapf = Sorghum drummondii（Steud.）Millsp. et Chase ■

369711　Sorghum technicum（Körn.）Roshev. = Sorghum bicolor（L.）Moench ■

369712　Sorghum technicum（Körn.）Roshev. = Sorghum dochna（Forssk.）Snowden var. techicum（Körn.）Snowden ■

369713　Sorghum trichopus（Stapf）Stapf = Sorghastrum stipoides（Kunth）Nash ■☆

369714　Sorghum triticeum（R. Br.）Kuntze = Heteropogon triticeus（R. Br.）Stapf ex Craib ■

369715　Sorghum usambarense Snowden = Sorghum arundinaceum（Desv.）Stapf ■☆

369716　Sorghum versicolor Andersson；变色黍■☆

369717　Sorghum verticilliflorum（Steud.）Stapf = Sorghum arundinaceum（Desv.）Stapf ■☆

369718　Sorghum verticilliflorum（Steud.）Stapf = Sorghum bicolor（L.）Moench subsp. arundinaceum（Desv.）de Wet et Harlan ■

369719　Sorghum virgatum（Hack.）Stapf；条纹黍■☆

369720　Sorghum vulgare（L.）Pers.；蜀黍（获粱,番麦,番黍,高粱,高粱七,瓜龙,薥粱,红高粱,芦穄,芦黍,芦粟,木稷,蜀秫）；Broom Corn, Broomcorn, Chicken Corn, Durra, Egyptian Millet, Gaoliang, Guinea Corn, Kaffir Corn, Millet, Sorghum ■

369721　Sorghum vulgare（L.）Pers. = Sorghum bicolor（L.）Moench ■

369722　Sorghum vulgare（L.）Pers. subsp. bicolor（L.）Maire et Weiller = Sorghum bicolor（L.）Moench ■

369723　Sorghum vulgare（L.）Pers. subsp. cernuum（Ard.）Maire et Weiller = Sorghum cernuum（Ard.）Host ■

369724　Sorghum vulgare（L.）Pers. subsp. durra（Forssk.）Maire et Weiller = Sorghum durra（Forssk.）Stapf ■

369725　Sorghum vulgare（L.）Pers. subsp. saccharatumm（L.）Maire et Weiller = Sorghum saccharatum（L.）Moench ■

369726　Sorghum vulgare（L.）Pers. var. caffrorum（Retz.）F. T. Hubb. et Rehder = Sorghum bicolor（L.）Moench ■

369727　Sorghum vulgare（L.）Pers. var. durra（Forssk.）F. T. Hubb. et Rehder = Sorghum bicolor（L.）Moench ■

369728　Sorghum vulgare（L.）Pers. var. durra（Forssk.）F. T. Hubb. et Rehder = Sorghum durra（Forssk.）Stapf ■

369729　Sorghum vulgare（L.）Pers. var. nervosum？ = Sorghum nervosum Besser ex Schult. ■

369730　Sorghum vulgare（L.）Pers. var. roxburghii（Stapf）Haines = Sorghum bicolor（L.）Moench ■

369731　Sorghum vulgare（L.）Pers. var. saccharatum（L.）Boerl. = Sorghum bicolor（L.）Moench ■

369732　Sorghum vulgare（L.）Pers. var. saccharatum（L.）Boerl. = Sorghum dochna（Forssk.）Snowden ■

369733　Sorghum vulgare（L.）Pers. var. saccharatum（L.）Boerl. = Sorghum saccharatum Poir. ■

369734　Sorghum vulgare（L.）Pers. var. sudanense（Piper）Hitchc. = Sorghum sudanense（Piper）Stapf ■

369735　Sorghum vulgare（L.）Pers. var. sudanense Hitchc. = Sorghum sudanense（Piper）Stapf ■

369736　Sorghum vulgare（L.）Pers. var. technicum？ = Sorghum technicum（Körn.）Roshev. ■

369737　Sorghum zizanioides（L.）Kuntze = Vetiveria zizanioides（L.）Nash ■

369738　Sorghum zollingeri（Steud.）Kuntze = Pseudosorghum fasciculare（Roxb. A. Camus ■

369739　Sorgum Adans.（废弃属名）= Holcus L.（保留属名）■

369740　Sorgum Adans.（废弃属名）= Sorghum Moench（保留属名）■

369741　Sorgum Kuntze = Andropogon L.（保留属名）■

369742　Soria Adans.（废弃属名）= Euclidium W. T. Aiton（保留属名）■

369743　Soridium Miers ex Henfrey = Peltophyllum Gardner ■☆

369744　Soridium Miers（1850）；丘霉草属■☆

369745　Soridium spruceanum Miers；丘霉草■☆

369746　Sorindeia Thouars（1806）；索林漆属●☆

369747　Sorindeia acutifolia Engl. = Sorindeia grandifolia Engl. ●☆

369748　Sorindeia adolfi-fredericii Engl. et Brehmer；弗里德里西索林漆●☆

369749　Sorindeia africana（Engl.）Van der Veken；非洲索林漆●☆

369750　Sorindeia africana（Engl.）Van der Veken var. lastoursvillensis（Pellegr.）Van der Veken；拉斯图维尔索林漆●☆

369751　Sorindeia afzelii Engl. = Sorindeia juglandifolia（A. Rich.）Planch. ex Oliv. ●☆

369752　Sorindeia albiflora Engl. et K. Krause；白花索林漆●☆

369753　Sorindeia batekeensis Lecomte；巴泰凯索林漆●☆

369754　Sorindeia befalensis Van der Veken；贝法莱索林漆●☆

369755　Sorindeia calantha Mildbr.；美花索林漆●☆

369756　Sorindeia claessensii De Wild.；克莱森斯索林漆●☆

369757　Sorindeia claessensii De Wild. var. monticola Van der Veken；山地索林漆●☆

369758　Sorindeia collina Keay；丘陵索林漆●☆

369759　Sorindeia crassifolia Engl. et K. Krause；厚叶索林漆●☆

369760　Sorindeia deliciosa A. Chev. ex Hutch. et Dalziel = Santiria trimera（Oliv.）Aubrév. ●☆

369761　Sorindeia doeringii Engl. et K. Krause = Ekebergia capensis Sparrm. ●☆

369762　Sorindeia ferruginea Engl.；锈色索林漆●☆

369763　Sorindeia gabonensis Bourobou et Breteler；加蓬索林漆●☆

369764　Sorindeia gilletii De Wild.；吉勒特索林漆●☆

369765 Sorindeia gossweileri Exell;戈斯索林漆●☆

369766 Sorindeia goudotii Briq. = Sophora tomentosa L. ●

369767 Sorindeia heterophylla Hook. f. = Sorindeia juglandifolia (A. Rich.) Planch. ex Oliv. ●☆

369768 Sorindeia immersinervia Engl. et Brehmer;水中索林漆●☆

369769 Sorindeia juglandifolia (A. Rich.) Planch. ex Oliv.;胡桃叶索林漆●☆

369770 Sorindeia katangensis Van der Veken;加丹加索林漆●☆

369771 Sorindeia kimuenzae De Wild. = Sorindeia gilletii De Wild. ●☆

369772 Sorindeia lagdoensis Engl. et K. Krause = Lannea lagdoensis (Engl. et K. Krause) Mildbr. ●☆

369773 Sorindeia lamprophylla Engl. et K. Krause;亮叶索林漆●☆

369774 Sorindeia lastoursvillensis Pellegr. = Sorindeia africana (Engl.) Van der Veken var. lastoursvillensis (Pellegr.) Van der Veken ●☆

369775 Sorindeia ledermannii Engl. et K. Krause;莱德索林漆●☆

369776 Sorindeia letestui Pellegr.;莱泰斯图索林漆●☆

369777 Sorindeia longifolia (Hook. f.) Oliv. = Trichoscypha longifolia (Hook. f.) Engl. ●☆

369778 Sorindeia longipetiolulata Engl. et Brehmer;长梗索林漆●☆

369779 Sorindeia lundensis Exell et Mendonça;隆德索林漆●☆

369780 Sorindeia macrophylla Planch. ex Oliv.;大叶索林漆●☆

369781 Sorindeia madagascariensis DC.;马岛索林漆●☆

369782 Sorindeia maxima Vermoesen = Sorindeia gilletii De Wild. ●☆

369783 Sorindeia mayumbensis Van der Veken;马永巴索林漆●☆

369784 Sorindeia mildbraedii Engl. et Brehmer;米尔德索林漆●☆

369785 Sorindeia multifoliolata Van der Veken;多小叶索林漆●☆

369786 Sorindeia ngounyensis Pellegr.;恩戈尼亚索林漆●☆

369787 Sorindeia nitida Engl.;光亮索林漆●☆

369788 Sorindeia nitidula Engl.;稍亮索林漆●☆

369789 Sorindeia obliquifoliolata Engl.;斜叶索林漆●☆

369790 Sorindeia obtusifoliolata Engl. = Sorindeia madagascariensis DC. ●☆

369791 Sorindeia obtusifoliolata Engl. var. parvifoliolata ? = Pseudospondias microcarpa (A. Rich.) Engl. ●☆

369792 Sorindeia ochracea Engl.;淡黄褐索林漆●☆

369793 Sorindeia oxyandra Bourobou et Breteler;尖蕊索林漆●☆

369794 Sorindeia patens Oliv. = Trichoscypha patens (Oliv.) Engl. ●☆

369795 Sorindeia poggei Engl.;波格索林漆●☆

369796 Sorindeia protioides Engl. et K. Krause;马蹄果索林漆●☆

369797 Sorindeia reticulata Engl. et Brehmer;网脉索林漆●☆

369798 Sorindeia revoluta Engl. et Brehmer;外卷索林漆●☆

369799 Sorindeia rhodesica R. Fern. et A. Fern.;罗得西亚索林漆●☆

369800 Sorindeia ripicola Champl.;岩地索林漆●☆

369801 Sorindeia rubriflora Engl.;红花索林漆●☆

369802 Sorindeia schroederi Engl. et K. Krause = Sorindeia warneckei Engl. ●☆

369803 Sorindeia schweinfurthii Engl.;施韦索林漆●☆

369804 Sorindeia somalensis (Chiov.) Chiov.;索马里索林漆●☆

369805 Sorindeia submontana Van der Veken;亚山生索林漆●☆

369806 Sorindeia tchibangensis Pellegr.;奇班加索林漆●☆

369807 Sorindeia tessmannii Engl.;泰斯曼索林漆●☆

369808 Sorindeia tholloni Lecomte;托伦索林漆●☆

369809 Sorindeia trimera Oliv. = Santiria trimera (Oliv.) Aubrév. ●☆

369810 Sorindeia undulata R. Fern. et A. Fern.;波状索林漆●☆

369811 Sorindeia usambarensis Engl. = Sorindeia madagascariensis DC. ●☆

369812 Sorindeia warneckei Engl.;沃内克索林漆●☆

369813 Sorindeia winkleri Engl.;温克勒索林漆●☆

369814 Sorindeia zenkeri Engl.;岑克尔索林漆●☆

369815 Sorindeiopsis Engl. = Sorindeia Thouars ●☆

369816 Sorocea A. St. -Hil. (1821);堆桑属●☆

369817 Sorocea affinis Hemsl.;近缘堆桑●☆

369818 Sorocea amazonica Miq.;亚马孙堆桑●☆

369819 Sorocea grandis Warb.;大堆桑●☆

369820 Sorocea macrophylla Gaudich.;大叶堆桑●☆

369821 Sorocea micranthera Warb.;小花堆桑●☆

369822 Sorocea nitida Warb.;光亮堆桑●☆

369823 Sorocea stenophylla Standl.;窄叶堆桑●☆

369824 Sorocea sylvicola Chodat;林地堆桑●☆

369825 Sorocephalus R. Br. (1810)(保留属名);丘头山龙眼属●☆

369826 Sorocephalus alopecurus Rourke;看麦娘丘头山龙眼●☆

369827 Sorocephalus capitatus Rourke;头状丘头山龙眼●☆

369828 Sorocephalus clavigerus (Salisb. ex Knight) Hutch.;珊瑚丘头山龙眼●☆

369829 Sorocephalus crassifolius Hutch.;厚叶丘头山龙眼●☆

369830 Sorocephalus diversifolius (Roem. et Schult.) R. Br. = Paranomus longicaulis Salisb. ex Knight ●☆

369831 Sorocephalus imberbis R. Br. = Sorocephalus pinifolius (Salisb. ex Knight) Rourke ●☆

369832 Sorocephalus imberbis R. Br. var. longifolius Meisn. = Sorocephalus pinifolius (Salisb. ex Knight) Rourke ●☆

369833 Sorocephalus imbricatus (Thunb.) R. Br.;覆瓦丘头山龙眼 ●☆

369834 Sorocephalus lanatus (Thunb.) R. Br.;绵毛丘头山龙眼●☆

369835 Sorocephalus lanatus (Thunb.) R. Br. var. teretifolius Meisn. = Sorocephalus teretifolius (Meisn.) E. Phillips ●☆

369836 Sorocephalus longifolius (Meisn.) E. Phillips = Sorocephalus pinifolius (Salisb. ex Knight) Rourke ●☆

369837 Sorocephalus palustris Rourke;沼泽丘头山龙眼●☆

369838 Sorocephalus phylicoides Meisn. = Sorocephalus lanatus (Thunb.) R. Br. ●☆

369839 Sorocephalus pinifolius (Salisb. ex Knight) Rourke;松叶丘头山龙眼●☆

369840 Sorocephalus rupestris (Salisb. ex Knight) E. Phillips = Sorocephalus clavigerus (Salisb. ex Knight) Hutch. ●☆

369841 Sorocephalus salsoloides R. Br. = Spatalla salsoloides (R. Br.) Rourke ●☆

369842 Sorocephalus scabridus Meisn.;微糙丘头山龙眼●☆

369843 Sorocephalus schlechteri E. Phillips = Sorocephalus lanatus (Thunb.) R. Br. ●☆

369844 Sorocephalus setaceus R. Br. = Spatalla setacea (R. Br.) Rourke ●☆

369845 Sorocephalus spatalloides R. Br. = Spatalla thyrsiflora Salisb. ex Knight ●☆

369846 Sorocephalus tenuifolius R. Br.;细叶丘头山龙眼●☆

369847 Sorocephalus teretifolius (Meisn.) E. Phillips;柱叶丘头山龙眼●☆

369848 Sorocephalus tulbaghensis E. Phillips = Spatalla tulbaghensis (E. Phillips) Rourke ●☆

369849 Soroseris Stebbins(1940);绢毛苣属(绢毛菊属,兔苣属);Soroseris ■

369850 Soroseris bellidifolia (Hand. -Mazz.) Stebbins = Soroseris glomerata (Decne.) Stebbins ■

369851 Soroseris chrysocephala C. Shih = Syncalathium chrysocephalum

（C. Shih）S. W. Liu■

369852　Soroseris deasyi（S. Moore）Stebbins = Soroseris glomerata（Decne.）Stebbins■

369853　Soroseris depressa（Hook. f. et Thomson）Stebbins = Soroseris glomerata（Decne.）Stebbins■

369854　Soroseris erysimoides（Hand.-Mazz.）C. Shih；空桶参（绢毛苣,空洞参,空空参,啦吧花,糖芥绢毛菊）■

369855　Soroseris gillii（S. Moore）Stebbins；金沙绢毛苣（金沙绢毛菊,绢毛菊,绢毛苣,空洞参,空空参,空桶参,啦吧花,喇叭花,搜空瓦）；Gill Soroseris,Hooker Hawksbeard■

369856　Soroseris gillii（S. Moore）Stebbins subsp. handelii Stebbins = Soroseris hirsuta（J. Anthony）C. Shih■

369857　Soroseris gillii（S. Moore）Stebbins subsp. hirsuta（Anthony）Stebbins = Soroseris hirsuta（J. Anthony）C. Shih■

369858　Soroseris gillii（S. Moore）Stebbins subsp. occidentalis Stebbins = Soroseris hirsuta（J. Anthony）C. Shih■

369859　Soroseris gillii（S. Moore）Stebbins subsp. typica Stebbins = Soroseris gillii（S. Moore）Stebbins■

369860　Soroseris glomerata（Decne.）Stebbins；绢毛苣（绢毛菊,莲状绢毛菊,条参,团花绢毛苣）；Glomerate Soroseris■

369861　Soroseris hirsuta（J. Anthony）C. Shih；羽裂绢毛苣（硬毛金沙绢毛菊）；Hirsute Soroseris■

369862　Soroseris hookeriana（C. B. Clarke）Stebbins；皱叶绢毛苣（虎克绢毛菊,绢毛菊,空洞参,空空参,空桶参,喇叭花）；Hooker Soroseris■

369863　Soroseris hookeriana（C. B. Clarke）Stebbins = Soroseris gillii（S. Moore）Stebbins■

369864　Soroseris hookeriana（C. B. Clarke）Stebbins subsp. erysimoides（Hand.-Mazz.）Stebbins；糖芥绢毛菊（绢毛菊,绢毛苣）■

369865　Soroseris hookeriana（C. B. Clarke）Stebbins subsp. erysimoides（Hand.-Mazz.）Stebbins = Soroseris erysimoides（Hand.-Mazz.）C. Shih■

369866　Soroseris hookeriana（C. B. Clarke）Stebbins subsp. typica Stebbins = Soroseris hookeriana（C. B. Clarke）Stebbins■

369867　Soroseris pumila Stebbins = Soroseris glomerata（Decne.）Stebbins■

369868　Soroseris qinghaiensis C. Shih = Syncalathium qinghaiense（C. Shih）C. Shih■

369869　Soroseris rosularis（Diels）Stebbins = Soroseris glomerata（Decne.）Stebbins■

369870　Soroseris teres C. Shih；柱序绢毛苣（柱序绢毛菊）■

369871　Soroseris umbrella（Franch.）Stebbins = Stebbinsia umbrella（Franch.）Lipsch.■

369872　Soroseris umbrellata（Franch.）Stebbins；伞花绢毛菊■

369873　Sorostachys Steud. = Cyperus L.■

369874　Sorostachys kyllingioides Steud. = Cyperus pulchellus R. Br.■☆

369875　Sorostachys pulchellus（R. Br.）Lye = Cyperus pulchellus R. Br.■☆

369876　Sosnovskya Takht. = Centaurea L.（保留属名）●■

369877　Soterosanthus Lehm. ex Jenny（1986）；丘花兰属■☆

369878　Soterosanthus shepheardii（Rolfe）Jenny；丘花兰■☆

369879　Sotor Fenzl = Kigelia DC.●

369880　Sotor aethiopum Fenzl = Kigelia africana（Lam.）Benth.●☆

369881　Sotrophola Buch.-Ham. = Chukrasia A. Juss.●

369882　Sotularia Raf. = Catu-Adamboe Adans.●

369883　Sotularia Raf. = Lagerstroemia L.●

369884　Souari Endl. = Caryocar F. Allam. ex L.●☆

369885　Souari Endl. = Saouari Aubl.●☆

369886　Soubeyrania Neck. = Barleria L.●■

369887　Soulamea Lam.（1785）；苦苦木属●☆

369888　Soulamea amara Lam.；苦苦木●☆

369889　Soulameaceae Endl. = Simaroubaceae DC.（保留科名）●

369890　Soulangia Brongn. = Phylica L.●☆

369891　Soulangia lutescens Eckl. et Zeyh. = Phylica axillaris Lam. var. lutescens（Eckl. et Zeyh.）Pillans●☆

369892　Soulangia microphylla Eckl. et Zeyh. = Phylica axillaris Lam. var. microphylla（Eckl. et Zeyh.）Pillans●☆

369893　Souleyetia Gaudich. = Pandanus Parkinson ex Du Roi●■

369894　Souliea Franch.（1898）；黄三七属；Souliea■

369895　Souliea vaginata（Maxim.）Franch.；黄三七（长果升麻,太白黄连,土黄连）；Common Souliea■

369896　Souroubea Aubl.（1775）；距苞藤属●☆

369897　Souroubea amazonica Delp.；亚马孙距苞藤●☆

369898　Souroubea auriculata Delp.；小耳距苞藤●☆

369899　Souroubea bicolor（Benth.）de Roon；二色距苞藤●☆

369900　Souroubea crassipes Wittm.；粗梗距苞藤●☆

369901　Souroubea dasystachya Gilg ex Ule；毛穗距苞藤●☆

369902　Souroubea fragilis de Roon；脆距苞藤●☆

369903　Souroubea guianensis Aubl.；圭亚那距苞藤●☆

369904　Souroubea intermedia de Roon；间型距苞藤●☆

369905　Souroubea micrantha Standl. et Steyerm.；小花距苞藤●☆

369906　Souroubea pachyphylla Gilg；厚叶距苞藤●☆

369907　Souroubea triandra Lundell；三蕊距苞藤●☆

369908　Souroubea venosa Schery；黏距苞藤●☆

369909　Southwellia Salisb. = Sterculia L.●

369910　Souza Vell. = Sisyrinchium L.■

369911　Sovara Raf. = Polygonum L.（保留属名）■●

369912　Sowerbaea Sm.（1798）；三雄兰属■☆

369913　Sowerbaea juncea Sm.；三雄兰■☆

369914　Sowerbaea laxiflora Lindl.；疏花三雄兰■☆

369915　Sowerbaea multicaulis E. Pritz.；多茎三雄兰■☆

369916　Sowerbea Dum. Cours. = Sowerbaea Sm.■☆

369917　Sowerbia Andrews = Sowerbaea Sm.■☆

369918　Soya Benth. = Glycine Willd.（保留属名）■

369919　Soyauxia Oliv.（1882）；索亚花属●☆

369920　Soyauxia bipindensis Gilg ex Hutch. et Dalziel = Soyauxia gabonensis Oliv.●☆

369921　Soyauxia floribunda Hutch.；繁花索亚花●☆

369922　Soyauxia gabonensis Oliv.；加蓬索亚花●☆

369923　Soyauxia glabrescens Engl.；光索亚花●☆

369924　Soyauxia grandifolia Gilg et Stapf；大叶索亚花●☆

369925　Soyauxia ledermannii Sleumer；莱德索亚花●☆

369926　Soyauxia talbotii Baker f.；塔尔博特索亚花●☆

369927　Soyauxia velutina Hutch. et Dalziel；绒毛索亚花●☆

369928　Soyauxiaceae Barkley = Flacourtiaceae Rich. ex DC.（保留科名）●

369929　Soyera St.-Lag. = Soyeria Monnier■

369930　Soyeria Monnier = Crepis L.■

369931　Soyeria chrysantha（Ledeb.）D. Dietr. = Crepis chrysantha（Ledeb.）Turcz.■

369932　Soyeria sibirica（L.）Monnier = Crepis sibirica L.■

369933　Soymida A. Juss.（1830）；印度红木属●☆

369934　Soymida febrifuga（Roxb.）A. Juss.；印度红木（解热桃花心

木）；Bastard Cedar, Indian Mahogany, Indian Redwood, Rohan Soymida ●☆

369935　Soymida febrifuga（Roxb.）A. Juss. = Swietenia febrifuga Roxb. ●☆

369936　Soymida roupalifolia Schweinf. = Pseudocedrela kotschyi （Schweinf.）Harms ●☆

369937　Spachea A. Juss.（1838）；斯帕木属●☆

369938　Spachea correa Cuatrec. et Croat；斯帕木●☆

369939　Spachea elegans A. Juss. ；雅致斯帕木●☆

369940　Spachea tenuifolia Griseb. ；细叶斯帕木●☆

369941　Spachea tricarpa A. Juss. ；三果斯帕木●☆

369942　Spachelodes Y. Kimura = Hypericum L. ■●

369943　Spachia Lilja = Fuchsia L. ●■

369944　Spadactis Cass. = Atractylis L. ■☆

369945　Spadicaceae Dulac = Araceae Juss.（保留科名）■●

369946　Spadonia Less. = Moquinia DC.（保留属名）●☆

369947　Spadostyles Benth. = Pultenaea Sm. ●☆

369948　Spaendoncea Desf. = Cadia Forssk. ●■☆

369949　Spaendoncea Desf. ex Usteri = Cadia Forssk. ●■☆

369950　Spalanthus Walp. = Quisqualis L. ●

369951　Spalanthus Walp. = Sphalanthus Jack ●

369952　Spallanzania DC. = Mussaenda L. ●■

369953　Spallanzania Neck. = Gustavia L.（保留属名）●☆

369954　Spallanzania Pollini = Aremonia Neck. ex Nestl.（保留属名）■☆

369955　Spananthe Jacq.（1791）；寡花草属■☆

369956　Spananthe paniculata Jacq. ；圆锥寡花草■☆

369957　Spaniopappus B. L. Rob.（1926）；疏毛泽兰属（疏泽兰属）■☆

369958　Spaniopappus ekmanii B. L. Rob. ；疏毛泽兰■☆

369959　Spanioptilon Less. = Cirsium Mill. ■

369960　Spanioptilon lineare（Thunb.）Less. = Cirsium lineare （Thunb.）Sch. Bip. ■

369961　Spanioptilon lineare Less. = Cirsium lineare（Thunb.）Sch. Bip. ■

369962　Spanizium Griseb. = Saponaria L. ■

369963　Spanoghea Blume = Alectryon Gaertn. ●☆

369964　Spanotrichum E. Mey. ex DC. = Osmites L.（废弃属名）●■

369965　Sparattanthelium Mart.（1841）；疏花桐属●☆

369966　Sparattanthelium amazonum Mart. ；疏花桐●☆

369967　Sparattosperma Mart. ex DC. = Bignonia L.（保留属名）●

369968　Sparattosperma Mart. ex DC. = Sparattosperma Mart. ex Meisn. ●☆

369969　Sparattosperma Mart. ex Meisn.（1840）；裂紫葳属●☆

369970　Sparattosperma Mart. ex Meisn. = Bignonia L.（保留属名）●

369971　Sparattosperma lithontripticum Mart. ；裂紫葳●☆

369972　Sparattosyce Bureau（1869）；假榕属●☆

369973　Sparattosyce dioica Bureau；假榕●☆

369974　Sparattothamnella Steenls = Spartothamnella Briq. ●☆

369975　Sparaxis Ker Gawl.（1802）；魔杖花属（裂缘莲属，芒苞菖属）；Harlequin Flower, Wand Flower, Wandflower, Wand-flower ■☆

369976　Sparaxis albiflora Eckl. = Sparaxis bulbifera（L.）Ker Gawl. ■☆

369977　Sparaxis atropurpurea Klatt = Sparaxis grandiflora（D. Delaroche）Ker Gawl. ■☆

369978　Sparaxis auriculata Goldblatt et J. C. Manning；耳状魔杖花■☆

369979　Sparaxis bicolor（Thunb.）Ker Gawl. = Sparaxis villosa （Burm. f.）Goldblatt ■☆

369980　Sparaxis bulbifera（L.）Ker Gawl. ；鳞茎魔杖花；Bulbiferous Wand-flower ■☆

369981　Sparaxis bulbifera（L.）Ker Gawl. var. violacea（Eckl.）Baker sensu Baker = Sparaxis grandiflora（D. Delaroche）Ker Gawl. subsp. violacea（Eckl.）Goldblatt ■☆

369982　Sparaxis cana Eckl. = Sparaxis grandiflora（D. Delaroche）Ker Gawl. subsp. violacea（Eckl.）Goldblatt ■☆

369983　Sparaxis caryophyllacea Goldblatt；石竹状魔杖花■☆

369984　Sparaxis cuprea（Sweet）Klatt = Sparaxis elegans（Sweet）Goldblatt ■☆

369985　Sparaxis elegans（Sweet）Goldblatt；雅致魔杖花■☆

369986　Sparaxis fimbriata（Lam.）Ker Gawl. = Sparaxis grandiflora （D. Delaroche）Ker Gawl. subsp. fimbriata（Lam.）Goldblatt ■☆

369987　Sparaxis fragrans（Jacq.）Ker Gawl. ；芳香魔杖花；Fragrant Wandflower ■☆

369988　Sparaxis fragrans Ker Gawl. = Sparaxis fragrans（Jacq.）Ker Gawl. ■☆

369989　Sparaxis galeata Ker Gawl. ；盔形魔杖花■☆

369990　Sparaxis grandiflora（D. Delaroche）Ker Gawl. ；大花魔杖花；Fragrant Wandflower, Plain Harlequin Flower ■☆

369991　Sparaxis grandiflora（D. Delaroche）Ker Gawl. subsp. acutiloba Goldblatt；尖裂大花魔杖花■☆

369992　Sparaxis grandiflora（D. Delaroche）Ker Gawl. subsp. fimbriata （Lam.）Goldblatt；流苏大花魔杖花■☆

369993　Sparaxis grandiflora（D. Delaroche）Ker Gawl. subsp. violacea （Eckl.）Goldblatt；堇色大花魔杖花■☆

369994　Sparaxis grandiflora（D. Delaroche）Ker Gawl. var. liliago （DC.）Ker Gawl. = Sparaxis grandiflora（D. Delaroche）Ker Gawl. subsp. fimbriata（Lam.）Goldblatt ■☆

369995　Sparaxis grandiflora（D. Delaroche）Ker Gawl. var. striata Sweet = Sparaxis grandiflora（D. Delaroche）Ker Gawl. subsp. fimbriata （Lam.）Goldblatt ■☆

369996　Sparaxis grandiflora Ker Gawl. ；大魔杖花；Big Wand-flower, Large Wand-flower ■☆

369997　Sparaxis grandiflora Ker Gawl. = Sparaxis grandiflora（D. Delaroche）Ker Gawl. ■☆

369998　Sparaxis liliago（DC.）Sweet = Sparaxis grandiflora（D. Delaroche）Ker Gawl. subsp. fimbriata（Lam.）Goldblatt ■☆

369999　Sparaxis lutea Eckl. = Sparaxis grandiflora（D. Delaroche）Ker Gawl. subsp. acutiloba Goldblatt ■☆

370000　Sparaxis maculosa Goldblatt；斑点魔杖花■☆

370001　Sparaxis metelerkampiae（L. Bolus）Goldblatt et J. C. Manning；梅泰魔杖花■☆

370002　Sparaxis miniata Klatt = Sparaxis grandiflora（D. Delaroche）Ker Gawl. ■☆

370003　Sparaxis monanthos（D. Delaroche）N. E. Br. = Sparaxis grandiflora（D. Delaroche）Ker Gawl. subsp. acutiloba Goldblatt ■☆

370004　Sparaxis orchidiflora Lodd. = Sparaxis variegata（Sweet）Goldblatt ■☆

370005　Sparaxis parviflora（G. J. Lewis）Goldblatt；小花美丽魔杖花■☆

370006　Sparaxis pillansii L. Bolus；皮朗斯魔杖花■☆

370007　Sparaxis pulcherrima Hook. f. = Dierama pulcherrimum（Hook. f.）Baker ■☆

370008　Sparaxis roxburghii（Baker）Goldblatt；罗氏美丽魔杖花■☆

370009　Sparaxis tricolor（Schneev.）Ker Gawl. ；三色魔杖花（彩眼花）；Harlequin Flower, Threecolor Wand-flower, Velvet Flower,

Wandflower ■☆

370010　Sparaxis tricolor Ker Gawl. = Sparaxis tricolor（Schneev.）Ker Gawl. ■☆

370011　Sparaxis variegata（Sweet）Goldblatt;杂色魔杖花■☆

370012　Sparaxis variegata（Sweet）Goldblatt subsp. metelerkampiae（L. Bolus）Goldblatt = Sparaxis metelerkampiae（L. Bolus）Goldblatt et J. C. Manning ■☆

370013　Sparaxis villosa（Burm. f.）Goldblatt;新诺鸢尾■☆

370014　Sparaxis violacea Eckl. = Sparaxis grandiflora（D. Delaroche）Ker Gawl. subsp. violacea（Eckl.）Goldblatt ■☆

370015　Sparaxis walthamii Hort. = Sparaxis variegata（Sweet）Goldblatt ■☆

370016　Sparaxis wattii Harv. = Sparaxis variegata（Sweet）Goldblatt ■☆

370017　Sparganiaceae F. Rudolphi = Sparganiaceae Hanin(保留科名)■

370018　Sparganiaceae Hanin(1811)(保留科名);黑三棱科;Burreed Family, Bur-reed Family ■

370019　Sparganiaceae Schultz-Sch. = Sparganiaceae Hanin(保留科名)■

370020　Sparganion Adans. = Sparganium L. ■

370021　Sparganium L.（1753）;黑三棱属;Bur Reed, Bur Weed, Burreed, Bur-reed, Burr-reed ■

370022　Sparganium acaule（Beeby ex Macoun）Rydb. = Sparganium emersum Rehmann ■

370023　Sparganium affine Schnizl.;近亲黑三棱■☆

370024　Sparganium affine Schnizl. = Sparganium angustifolium Michx. ■

370025　Sparganium americanum Nutt.;美国黑三棱;American Burreed, American Bur-reed, Bur-reed, Lesser Bur-reed ■☆

370026　Sparganium americanum Nutt. var. androcladum（Engelm.）Fernald et Eames = Sparganium androcladum（Engelm.）Morong ■☆

370027　Sparganium amplexicaulium D. Yu;抱茎黑三棱;Amplexicaul Burreed ■

370028　Sparganium androcladum（Engelm.）Morong;雄枝黑三棱;Branched Bur-reed, Melabranch Burreed ■☆

370029　Sparganium androcladum（Engelm.）Morong var. fluctuans Engelm. ex Morong = Sparganium fluctuans（Engelm. ex Morong）B. L. Rob. ■☆

370030　Sparganium angustifolium Michx.;线叶黑三棱(狭叶黑三棱);Bur-reed, Floating Bur-reed, Narrowleaf Burreed, Narrowleaf Bur-reed, Narrow-leaved Bur-reed, Threedleaf Burreed ■

370031　Sparganium angustifolium Michx. subsp. emersum（Rehmann）Brayshaw var. multipedunculatum（Morong）Reveal = Sparganium angustifolium Michx. ■

370032　Sparganium angustifolium Michx. subsp. emersum（Rehmann）Brayshaw = Sparganium emersum Rehmann ■

370033　Sparganium angustifolium Michx. var. multipedunculatum（Morong）Brayshaw = Sparganium angustifolium Michx. ■

370034　Sparganium arcuscaulis D. Yu et G. T. Yang;弓杆黑三棱;Arcuscaul Burreed ■

370035　Sparganium arcuscaulis D. Yu et G. T. Yang = Sparganium stoloniferum（Graebn.）Buch. -Ham. ex Juz. ■

370036　Sparganium californicum Greene = Sparganium eurycarpum Engelm. ■☆

370037　Sparganium chlorocarpum Rydb.;绿果黑三棱;Green Burreed, Green-fruit Burreed ■☆

370038　Sparganium chlorocarpum Rydb. = Sparganium emersum Rehmann ■

370039　Sparganium chlorocarpum Rydb. f. acaule（Beeby ex Macoun）E. G. Voss = Sparganium emersum Rehmann ■

370040　Sparganium chlorocarpum Rydb. var. acaule（Beeby ex Macoun）Fernald = Sparganium emersum Rehmann ■

370041　Sparganium choui D. Yu = Sparganium stoloniferum（Graebn.）Buch. -Ham. ex Juz. subsp. choui（D. Yu）K. Sun

370042　Sparganium confertum Y. D. Chen;穗状黑三棱（密集黑三棱）;Spike Burreed ■

370043　Sparganium coreanum H. Lév. = Sparganium erectum L. ■

370044　Sparganium emersum Rehmann;欧洲黑三棱;European Burreed, Narrow-leaved Bur-reed, Unbranched Bur-reed ■

370045　Sparganium emersum Rehmann = Sparganium simplex Huds. ■

370046　Sparganium emersum Rehmann subsp. acaule（Beeby ex Macoun）C. D. K. Cook et Nicholls = Sparganium emersum Rehmann ■

370047　Sparganium emersum Rehmann subsp. acaule（Beeby ex Macoun）C. D. K. Cook et M. S. Nicholls = Sparganium emersum Rehmann ■

370048　Sparganium emersum Rehmann var. angustifolium（Michx.）R. L. Taylor et MacBryde = Sparganium angustifolium Michx. ■

370049　Sparganium emersum Rehmann var. multipedunculatum（Morong）Reveal = Sparganium angustifolium Michx. ■

370050　Sparganium erectum L.;直立黑三棱(多枝黑三棱,黑三棱);Bead Sedge, Branched Bur-reed, Bur Flag, Bur-reed ■

370051　Sparganium erectum L. subsp. neglectum（Beeby）K. Richt.;隐黑三棱■☆

370052　Sparganium erectum L. subsp. polyedrum（Asch. et Graebn.）Schinz et Thell. = Sparganium erectum L. ■

370053　Sparganium erectum L. subsp. stoloniferum（Buch. -Ham. ex Graebn.）C. D. K. Cook et M. S. Nicholls = Sparganium eurycarpum Engelm. ■☆

370054　Sparganium erectum L. subsp. stoloniferum（Graebn.）H. Hara = Sparganium erectum L. ■

370055　Sparganium erectum L. var. coreanum（H. Lév.）H. Hara = Sparganium erectum L. ■

370056　Sparganium erectum L. var. glomeratum Beurl. ex Laest. = Sparganium glomeratum（Beurl. ex Laest.）Neuman ■

370057　Sparganium erectum L. var. macrocarpum（Makino）H. Hara;大果直立黑三棱■☆

370058　Sparganium eurycarpum Engelm.;巨大黑三棱;Broad-fruit Burreed, Broad-fruited Bur-reed, Common Bur-reed, Giant Burreed, Giant Bur-reed ■☆

370059　Sparganium eurycarpum Engelm. subsp. coreanum（H. Lév.）C. D. K. Cook et Nicholls = Sparganium erectum L. ■

370060　Sparganium eurycarpum Engelm. var. greenei（Morong）Graebn. = Sparganium eurycarpum Engelm. ■☆

370061　Sparganium fallax Graebn.;曲轴黑三棱（东亚黑三棱）;Bentaxis Burreed ■

370062　Sparganium fluctuans（Engelm. ex Morong）B. L. Rob.;水黑三棱（变叶黑三棱）;Floating Bur-reed, Floating-leaved Bur-reed, Water Burreed ■☆

370063　Sparganium friesii Beurl. = Sparganium gramineum Georgi ■☆

370064　Sparganium glehnii Meinsh.;旋序黑三棱■

370065　Sparganium glomeratum（Beurl. ex Laest.）Neuman = Sparganium glomeratum Laest. ex Beurl. ■

370066　Sparganium glomeratum Laest. ex Beurl.;短序黑三棱（密黑三棱,密序黑三棱,球状黑三棱）;Clustered Bur-reed, Glomerate Burreed, Northern Bur-reed, Shortspike Burreed ■

370067 Sparganium glomeratum Laest. ex Beurl. var. angustifolium Graebn. ;狭叶短序黑三棱■☆

370068 Sparganium gramineum Georgi;弗瑞氏黑三棱■☆

370069 Sparganium greenei Morong = Sparganium eurycarpum Engelm. ■☆

370070 Sparganium hyperboreum Beurl. ex Laest. ;北方黑三棱■☆

370071 Sparganium hyperboreum Laest. ex Beurl. ;无柱黑三棱(北方黑三棱);Northern Bur-reed,Styleless Burreed ■

370072 Sparganium japonicum Rothert;日本黑三棱;Japan Burreed ■☆

370073 Sparganium kawakamii H. Hara = Sparganium angustifolium Michx. ■

370074 Sparganium limosum Y. D. Chen;沼生黑三棱;Paludal Burreed ■☆

370075 Sparganium longifolium Turcz. ;长叶黑三棱(黑三棱);Longleaf Burreed ■☆

370076 Sparganium lucidum Fernald et Eames = Sparganium androcladum (Engelm.) Morong ■☆

370077 Sparganium macrocarpum Makino = Sparganium erectum L. var. macrocarpum (Makino) H. Hara ■☆

370078 Sparganium manshuricum D. Yu;东北黑三棱; NE. China Burreed ■

370079 Sparganium manshuricum D. Yu = Sparganium glomeratum Laest. ex Beurl. ■

370080 Sparganium microcarpum Celak. ;小果黑三棱(京三棱);Smallfruit Burreed ■

370081 Sparganium minimum (Hartm.) Fr. = Sparganium natans Pursh ■

370082 Sparganium minimum Hill = Sparganium natans Pursh ■

370083 Sparganium minimum Wallr. = Sparganium natans Pursh ■

370084 Sparganium multipedunculatum (Morong) Rydb. ;多梗黑三棱;Many-pedunded Bur-reed ■☆

370085 Sparganium multipedunculatum (Morong) Rydb. = Sparganium angustifolium Michx. ■

370086 Sparganium multiporcatum D. Yu;多脊黑三棱;Manyrib Burreed ■

370087 Sparganium multiporcatum D. Yu = Sparganium stoloniferum (Graebn.) Buch. -Ham. ex Juz. ■

370088 Sparganium natans Pursh;矮黑三棱(短黑三棱,小黑三棱);Dwarf Burreed,Least Bur Reed,Least Burreed,Least Bur-reed,Small Bur-reed ■

370089 Sparganium neglectum Beeby = Sparganium erectum L. subsp. neglectum (Beeby) K. Richt. ■☆

370090 Sparganium polyedrum Asch. et Graebn. ;塔果黑三棱(多面黑三棱)■

370091 Sparganium ramosum Huds. = Sparganium erectum L. ■

370092 Sparganium ramosum Huds. = Sparganium stoloniferum (Graebn.) Buch. -Ham. ex Juz. ■

370093 Sparganium ramosum Huds. subsp. neglectum (Beeby) Schinz et Thell. = Sparganium erectum L. subsp. neglectum (Beeby) K. Richt. ■☆

370094 Sparganium ramosum Huds. subsp. polyedrum Asch. et Graebn. = Sparganium erectum L. ■

370095 Sparganium ramosum Huds. subsp. stoloniferum Graebn. = Sparganium stoloniferum (Graebn.) Buch. -Ham. ex Juz. ■

370096 Sparganium simplex Huds. ;小黑三棱(单枝黑三棱,三棱);Little Burreed,Small Burreed,Small Bur-reed ■

370097 Sparganium simplex Huds. = Sparganium emersum Rehmann ■

370098 Sparganium simplex Huds. var. androcladum Engelm. = Sparganium androcladum (Engelm.) Morong ■☆

370099 Sparganium simplex Huds. var. multipedunculatum Morong = Sparganium angustifolium Michx. ■

370100 Sparganium stenophyllum Maxim. ex Meinsh. ;狭叶黑三棱(细叶黑三棱);Narrowleaf Burreed ■

370101 Sparganium stenophyllum Maxim. ex Meinsh. = Sparganium subglobosum Morong ■☆

370102 Sparganium stoloniferum (Graebn.) Buch. -Ham. ex Juz. ;黑三棱(草三棱,醋三棱,光三棱,红蒲根,鸡爪三棱,京三棱,荆三棱,理三棱,泡三棱,三棱,三棱草,山棱,山林,细叶黑三棱,小黑三棱);Bur Reed,Burreed,Common Burreed ■

370103 Sparganium stoloniferum (Graebn.) Buch. -Ham. ex Juz. = Sparganium erectum L. ■

370104 Sparganium stoloniferum (Graebn.) Buch. -Ham. ex Juz. subsp. choui (D. Yu) K. Sun;周氏黑三棱;Chou's Burreed ■

370105 Sparganium stoloniferum Buch. -Ham. = Sparganium stoloniferum (Graebn.) Buch. -Ham. ex Juz. ■

370106 Sparganium subglobosum Morong;亚球形黑三棱■☆

370107 Sparganium tenuicaule D. Yu et L. H. Liu;细茎黑三棱;Thinstem Burreed ■

370108 Sparganium yamatense Makino ex H. Hara = Sparganium fallax Graebn. ■

370109 Sparganium yunnanense Y. D. Chen;云南黑三棱;Yunnan Burreed ■

370110 Sparganophoros Adans. = Sparganophorus Boehm. ■☆

370111 Sparganophoros Vaill. = Struchium P. Browne ■☆

370112 Sparganophoros sparganophora (L.) Kuntze = Sparganophorus sparganophorus (L.) C. Jeffrey ■☆

370113 Sparganophorus Boehm. (1760) ('Spharganophorus');骨冠斑鸠菊属(带菊属)■☆

370114 Sparganophorus Boehm. = Struchium P. Browne ■☆

370115 Sparganophorus Vaill. ex Boehm. = Struchium P. Browne ■☆

370116 Sparganophorus Vaill. ex Crantz = Struchium P. Browne ■☆

370117 Sparganophorus africanus Steud. = Sparganophorus sparganophorus (L.) C. Jeffrey ■☆

370118 Sparganophorus sparganophorus (L.) C. Jeffrey;骨冠斑鸠菊(带菊)■☆

370119 Sparganophorus sparganophorus (L.) C. Jeffrey = Struchium sparganophorum (L.) Kuntze ■☆

370120 Sparganophorus vaillantii Crantz;维氏骨冠斑鸠菊(维氏带菊)■☆

370121 Sparganophorus vaillantii Crantz = Sparganophorus sparganophorus (L.) C. Jeffrey ■☆

370122 Sparmannia Buc' hoz(废弃属名) = Rehmannia Libosch. ex Fisch. et C. A. Mey. (保留属名)■★

370123 Sparmannia Buc' hoz(废弃属名) = Sparrmannia L. f. (保留属名)●☆

370124 Sparmannia L. f. = Sparrmannia L. f. (保留属名)●☆

370125 Sparmanniaceae J. Agardh = Malvaceae Juss. (保留科名)●■

370126 Sparmanniaceae J. Agardh = Tiliaceae Juss. (保留科名)●■

370127 Sparrea Hunz. et Dottori = Celtis L. ●

370128 Sparrea schippii (Standl.) Hunz. et Dottori = Celtis schippii Standl. ●☆

370129 Sparrmania L. ex B. D. Jacks. = Melanthium L. ■☆

370130 Sparrmannia L. f. (1782) ('Sparmannia') (保留属名);庭院椴属(垂蕾树属,斯珀曼木属);African Hemp, House Lime, Sparmannia ●☆

370131 Sparrmannia abyssinica A. Rich. var. fischeri Engl. = Sparrmannia ricinocarpa（Eckl. et Zeyh.）Kuntze ●☆

370132　Sparrmannia abyssinica A. Rich. var. hirsuta Oliv. = Sparrmannia ricinocarpa（Eckl. et Zeyh.）Kuntze ●☆

370133　Sparrmannia abyssinica A. Rich. var. micrantha Burret = Sparrmannia ricinocarpa（Eckl. et Zeyh.）Kuntze ●☆

370134　Sparrmannia abyssinica Hochst. ex A. Rich. = Sparrmannia ricinocarpa（Eckl. et Zeyh.）Kuntze ●☆

370135　Sparrmannia abyssinica Hochst. ex A. Rich. var. concolor Chiov. = Sparrmannia ricinocarpa（Eckl. et Zeyh.）Kuntze ●☆

370136　Sparrmannia africana L. f.；非洲庭院椴（垂蕾树，庭院椴）；African Hemp, African Sparmannla, African Windflower, House Lime, Stock Rose ●☆

370137　Sparrmannia discolor Baker；马岛庭院椴●☆

370138　Sparrmannia macrocarpa Ulbr. = Sparrmannia ricinocarpa（Eckl. et Zeyh.）Kuntze var. macrocarpa（Ulbr.）Weim. ●☆

370139　Sparrmannia palmata E. Mey. ex C. Presl；掌叶庭院椴●☆

370140　Sparrmannia palmata E. Mey. ex Harv. = Sparrmannia ricinocarpa（Eckl. et Zeyh.）Kuntze ●☆

370141　Sparrmannia ricinocarpa（Eckl. et Zeyh.）Kuntze；蓖麻果庭院椴●☆

370142　Sparrmannia ricinocarpa（Eckl. et Zeyh.）Kuntze subsp. hirsuta Weim. = Sparrmannia ricinocarpa（Eckl. et Zeyh.）Kuntze ●☆

370143　Sparrmannia ricinocarpa（Eckl. et Zeyh.）Kuntze subsp. micrantha（Burret）Weim. = Sparrmannia ricinocarpa（Eckl. et Zeyh.）Kuntze ●☆

370144　Sparrmannia ricinocarpa（Eckl. et Zeyh.）Kuntze var. abyssinica（Hochst. ex A. Rich.）Weim. = Sparrmannia ricinocarpa（Eckl. et Zeyh.）Kuntze ●☆

370145　Sparrmannia ricinocarpa（Eckl. et Zeyh.）Kuntze var. cinerea Weim.；灰蓖麻果庭院椴●☆

370146　Sparrmannia ricinocarpa（Eckl. et Zeyh.）Kuntze var. fischeri（Engl.）Weim. = Sparrmannia ricinocarpa（Eckl. et Zeyh.）Kuntze ●☆

370147　Sparrmannia ricinocarpa（Eckl. et Zeyh.）Kuntze var. macrocarpa（Ulbr.）Weim.；大果蓖麻果庭院椴●☆

370148　Sparrmannia subpalmata Baker；掌状庭院椴●☆

370149　Sparteum P. Beauv. = Stipa L. ■

370150　Spartianthus Link = Spartium L. ●

370151　Spartidium Pomel（1874）；撒哈拉染料木属●☆

370152　Spartidium Porael = Genista L. ●

370153　Spartidium saharae（Coss. et Durieu）Pomel；撒哈拉染料木 ●☆

370154　Spartina Schreb. = Spartina Schreb. ex J. F. Gmel. ■

370155　Spartina Schreb. ex J. F. Gmel.（1789）；米草属（大米草属，绳草属，网茅属）；Cord Grass, Cordgrass, Cord-grass, Marsh Grass, Rice Grass, Spartina ■

370156　Spartina alterniflora Loisel.；互花米草（平滑网茅）；Alternate-flowered Spartina, Atlantic Cordgrass, Smooth Cord Grass, Smooth Cordgrass, Smooth Cord-grass ■

370157　Spartina angelica C. E. Hubb.；大米草；Common Cordgrass, Common Cord-grass ■

370158　Spartina bakeri Merr.；巴氏米草；Baker's Cord Grass ■☆

370159　Spartina capensis Nees = Spartina maritima（Curtis）Fernald ■☆

370160　Spartina cynosuroides Roth；盐地禾；Big Cord-grass ■☆

370161　Spartina densiflora Brongn.；密花米草；Denseflower Cordgrass ■☆

370162　Spartina durieui Parl. = Spartina versicolor E. Fabre ■☆

370163　Spartina glabra Muhl. = Spartina alterniflora Loisel. ■

370164　Spartina glabra Muhl. ex Elliott var. alterniflora（Loisel.）Merr. = Spartina alterniflora Loisel. ■

370165　Spartina juncea（Michx.）Willd. = Spartina versicolor E. Fabre ■☆

370166　Spartina maritima（Curtis）Fernald；海岸米草；Cord-grass, Lesser Cord-grass, Small Cordgrass, Small Cord-grass ■☆

370167　Spartina maritima（Curtis）Fernald subsp. glabra（Muhl.）St. - Yves；光滑米草■☆

370168　Spartina maritima（Curtis）Fernald subsp. stricta（Roth）St. - Yves；直米草■☆

370169　Spartina maritima（Curtis）Fernald var. alterni-flora（Loisel.）St. -Yves = Spartina townsendii H. Groves et J. Groves ■☆

370170　Spartina michauxiana Hitchc. = Spartina pectinata Bosc ex Link ■☆

370171　Spartina patens（Aiton）Muhl.；狐米草（伸展网茅）；Marshhay Cordgrass, Salt Marsh Cord Grass, Spreading Spartina ■☆

370172　Spartina patens（Aiton）Muhl. var. juncea（Michx.）Hitchc. = Spartina versicolor E. Fabre ■☆

370173　Spartina pectinata Bosc ex Link；草原米草（草原网茅）；Pectinate Spartina, Prairie Cord Grass, Prairie Cord-grass, Slough Grass, Slough-grass ■☆

370174　Spartina pectinata Bosc ex Link 'Aureoarginata'；黄边草原米草（黄边草原网茅）；Prairie Cord Grass, Slough Grass ☆

370175　Spartina pectinata Bosc ex Link var. suttei（Farw.）Fernald = Spartina pectinata Bosc ex Link ■☆

370176　Spartina pectinata Link = Spartina pectinata Bosc ex Link ■☆

370177　Spartina pectinata Link var. suttiei（Farw.）Fernald = Spartina pectinata Bosc ex Link ■☆

370178　Spartina schreberi J. F. Gmel.；美洲米草■☆

370179　Spartina spartinae（Trin.）Merr. ex Hitchc.；普通米草；Gulf Cord Grass ■☆

370180　Spartina stricta（Aiton）Roth = Spartina maritima（Curtis）Fernald ■☆

370181　Spartina stricta Roth var. alterniflora（Loisel.）A. Gray = Spartina alterniflora Loisel. ■

370182　Spartina townsendii H. Groves et J. Groves；稻米草；Cord-grass, Rice Grass, Townsend's Cord-grass ■☆

370183　Spartina townsendii H. Groves et J. Groves var. anglica（C. E. Hubb.）Lambinon et Maquet = Spartina anglica C. E. Hubb. ■

370184　Spartina versicolor E. Fabre；变色米草■☆

370185　Spartinaceae Burnett = Gramineae Juss.（保留科名）■●

370186　Spartinaceae Burnett = Poaceae Barnhart（保留科名）■●

370187　Spartinaceae Burnett；米草科■

370188　Spartinaceae Link = Gramineae Juss.（保留科名）■●

370189　Spartinaceae Link = Poaceae Barnhart（保留科名）■●

370190　Spartium Duhamel = Genista L. ●

370191　Spartium L.（1753）；鹰爪豆属（无叶豆属）；Broom, Eagleclawbean, Spanish Broom, Weaver's Broom, Weaversbroom, Weaver's-broom ●

370192　Spartium albicans Cav. = Teline canariensis（L.）Webb et Berthel. ●☆

370193　Spartium album Desf. = Cytisus multiflorus（L'Hér.）Sweet ●☆

370194　Spartium arboreum L. = Cytisus arboreus（Desf.）DC. ●☆

370195　Spartium aspalathoides（Lam.）Desf. = Genista aspalathoides Lam. ●☆

370196　Spartium biflorum Desf. = Cytisus fontanesii Spach ●☆

370197　Spartium capense L. = Rafnia capensis（L.）Schinz ■☆

370198　Spartium capitatum Cav. = Genista clavata Poir. ●☆

370199　Spartium cuspidatum Cav. = Genista hirsuta Vahl ●☆

370200　Spartium ferox Poir. = Genista ferox（Poir.）Dum. Cours. ●☆

370201　Spartium interruptum Cav. = Genista triacanthos Brot. ●☆

370202　Spartium junceum L.；鹰爪豆（无叶豆，莺织柳，鹰爪，鹰爪花）；Eagleclawbean，Spanish Broom，Weaver's Broom，Weaver's-broom，Weaversbroom ●

370203　Spartium junceum L. f. flore-plenum Collins；重瓣鹰爪豆；Double Eagleclawbean，Double Weaversbroom ●

370204　Spartium junceum L. f. ochroleucum（Spreng.）Rehder；黄花鹰爪豆（白花鹰爪豆）；Whiteflower Eagleclawbean，White-flower Weaversbroom ●

370205　Spartium junceum L. f. odoraratissimum Sweet；小花鹰爪豆●

370206　Spartium lanigerum Desf. = Calycotome villosa（Poir.）Link ●☆

370207　Spartium linifolium L. = Teline linifolia（L.）Webb et Berthel. ●☆

370208　Spartium microphyllum Cav. = Adenocarpus foliolosus（Aiton）DC. ●☆

370209　Spartium molle Cav. = Chamaecytisus mollis（Cav.）Greuter et Burdet ●☆

370210　Spartium monospermum L. = Retama monosperma（L.）Boiss. ●☆

370211　Spartium ovatum P. J. Bergius = Rafnia capensis（L.）Schinz subsp. ovata（P. J. Bergius）G. J. Campb. et B. -E. van Wyk ●☆

370212　Spartium persicum（Burm. f.）Willd. = Crotalaria persica（Burm. f.）Merr. ●☆

370213　Spartium ramosissimum Desf. = Genista cinerea（Vill.）DC. subsp. speciosa Rivas Mart. et al. ●☆

370214　Spartium rigidum Viv. = Calycotome rigida（Viv.）Maire et Weiller ●☆

370215　Spartium scoparium L. = Cytisus scoparius（L.）Link ●

370216　Spartium scorpius L. = Genista scorpius（L.）DC. ●☆

370217　Spartium sophoroides P. J. Bergius = Hypocalyptus sophoroides（P. J. Bergius）Baill. ●☆

370218　Spartium sphaerocarpum L. = Retama sphaerocarpa（L.）Boiss. ●☆

370219　Spartium spinosum L. = Calycotome spinosa（L.）Link ●☆

370220　Spartium tricuspidatum Cav. = Genista tricuspidata Desf. ●☆

370221　Spartium tridens Cav. = Genista tridens（Cav.）DC. ●☆

370222　Spartium umbellatum L'Hér. = Genista umbellata（L'Hér.）Poir. ●☆

370223　Spartium villosum Poir. = Calycotome villosa（Poir.）Link ●☆

370224　Spartochloa C. E. Hubb.（1952）；金雀枝草属■☆

370225　Spartochloa scirpoidea（Steud.）C. E. Hubb.；金雀枝草■☆

370226　Spartocysus Willk. et Lange = Spartocytisus Webb et Berthel. ●

370227　Spartocytisus Webb et Berthel. = Cytisus Desf.（保留属名）●

370228　Spartocytisus Webb et Berthel. ex Presl = Spartocytisus Webb et Berthel. ●

370229　Spartothamnella Briq.（1895）；小索灌属●☆

370230　Spartothamnella juncea（Walp.）Briq.；小索灌●☆

370231　Spartothamnus A. Cunn. = Spartothamnella Briq. ●☆

370232　Spartothamnus A. Cunn. ex Walp. = Spartothamnella Briq. ●☆

370233　Spartothamnus Walp. = Cytisus Desf.（保留属名）●

370234　Spartothamnus Walp. = Spartocytisus Webb et Berthel. ●

370235　Spartothamnus Walp. = Spartothamnus Webb et Berthel. ●

370236　Spartothamnus Webb et Berthel. = Spartothamnus Webb et Berthel. ex C. Presl ●

370237　Spartothamnus Webb et Berthel. ex C. Presl = Spartothamnella Briq. ●☆

370238　Spartothamnus Webb et Berthel. ex C. Presl = Spartothamnus A. Cunn. ex Walp. ●☆

370239　Spartum P. Beauv. = Lygeum L. ■☆

370240　Spatalanthus Sweet = Romulea Maratti（保留属名）■☆

370241　Spatalla Salisb.（1807）；南非少花山龙眼属●☆

370242　Spatalla argentea Rourke；银白南非少花山龙眼●☆

370243　Spatalla barbigera Salisb. ex Knight；髯毛南非少花山龙眼●☆

370244　Spatalla brachyloba E. Phillips = Spatalla mollis R. Br. ●☆

370245　Spatalla burchellii E. Phillips = Spatalla barbigera Salisb. ex Knight ●☆

370246　Spatalla caudata（Thunb.）R. Br.；尾状南非少花山龙眼●☆

370247　Spatalla caudiflora Salisb. ex Knight = Spatalla caudata（Thunb.）R. Br. ●☆

370248　Spatalla colorata Meisn.；着色南非少花山龙眼●☆

370249　Spatalla confusa（E. Phillips）Rourke；混乱南非少花山龙眼●☆

370250　Spatalla curvifolia Salisb. ex Knight；折叶南非少花山龙眼●☆

370251　Spatalla cylindrica E. Phillips = Spatalla longifolia Salisb. ex Knight ●☆

370252　Spatalla ericifolia Salisb. ex Knight = Spatalla caudata（Thunb.）R. Br. ●☆

370253　Spatalla ericoides E. Phillips；石南状南非少花山龙眼●☆

370254　Spatalla galpinii E. Phillips = Spatalla curvifolia Salisb. ex Knight ●☆

370255　Spatalla gracilis Salisb. ex Knight = Spatalla racemosa（L.）Druce ●☆

370256　Spatalla incurva（Thunb.）R. Br.；内折南非少花山龙眼●☆

370257　Spatalla longifolia Salisb. ex Knight；长叶南非少花山龙眼●☆

370258　Spatalla mollis R. Br.；柔软南非少花山龙眼●☆

370259　Spatalla mucronifolia E. Phillips = Spatalla incurva（Thunb.）R. Br. ●☆

370260　Spatalla nubicola Rourke；云雾南非少花山龙眼●☆

370261　Spatalla parilis Salisb. ex Knight；相似南非少花山龙眼●☆

370262　Spatalla procera Salisb. ex Knight = Spatalla incurva（Thunb.）R. Br. ●☆

370263　Spatalla prolifera（Thunb.）Salisb. ex Knight；多育南非少花山龙眼●☆

370264　Spatalla propinqua R. Br.；邻近南非少花山龙眼●☆

370265　Spatalla racemosa（L.）Druce；总状南非少花山龙眼●☆

370266　Spatalla salsoloides（R. Br.）Rourke；猪毛菜状南非少花山龙眼●☆

370267　Spatalla sericea R. Br. = Spatalla barbigera Salisb. ex Knight ●☆

370268　Spatalla setacea（R. Br.）Rourke；刚毛南非少花山龙眼●☆

370269　Spatalla squamata Meisn.；鳞南非少花山龙眼●☆

370270　Spatalla thyrsiflora Salisb. ex Knight；聚伞南非少花山龙眼●☆

370271　Spatalla tulbaghensis（E. Phillips）Rourke；塔尔巴赫南非少花山龙眼●☆

370272　Spatalla wallichii E. Phillips = Spatalla incurva（Thunb.）R. Br. ●☆

370273　Spatallopsis E. Phillips = Spatalla Salisb. ●☆

370274　Spatallopsis begleyi E. Phillips = Spatalla setacea（R. Br.）

Rourke ●☆

370275　Spatallopsis caudata（Thunb.）E. Phillips = Spatalla caudata（Thunb.）R. Br. ●☆

370276　Spatallopsis caudiflora（Salisb. ex Knight）E. Phillips = Spatalla caudata（Thunb.）R. Br. ●☆

370277　Spatallopsis confusa E. Phillips = Spatalla confusa（E. Phillips）Rourke ●☆

370278　Spatallopsis ericifolia（Salisb. ex Knight）E. Phillips = Spatalla caudata（Thunb.）R. Br. ●☆

370279　Spatallopsis propinqua（R. Br.）E. Phillips = Spatalla propinqua R. Br. ●☆

370280　Spatanthus Juss. = Spathanthus Desv. ■☆

370281　Spatela Adans. = Spathelia L.（保留属名）●☆

370282　Spatellaria Rchb. = Amphirrhox Spreng.（保留属名）■☆

370283　Spatellaria Rchb. = Spathularia A. St. -Hil. ■☆

370284　Spatha Post et Kuntze = Spathe P. Browne ●☆

370285　Spatha Post et Kuntze = Spathelia L.（保留属名）●☆

370286　Spathacanthus Baill.（1891）;扁刺爵床属●☆

370287　Spathacanthus hahnianus Baill. ;扁刺爵床●☆

370288　Spathacanthus parviflorus Léonard;小花扁刺爵床●☆

370289　Spathaceae Dulac = Iridaceae Juss.（保留科名）■●

370290　Spathalea L. = Spathe P. Browne ●☆

370291　Spathalea L. = Spathelia L.（保留属名）●☆

370292　Spathandra Guill. et Perr.（1833）;鞘蕊野牡丹属●☆

370293　Spathandra Guill. et Perr. = Memecylon L. ●

370294　Spathandra barteri（Hook. f.）Jacq. -Fél. = Lijndenia barteri（Hook. f.）K. Bremer ●☆

370295　Spathandra blakeoides（G. Don）Jacq. -Fél. ;鞘蕊野牡丹●☆

370296　Spathandra blakeoides（G. Don）Jacq. -Fél. var. fleuryi（Jacq. -Fél.）Jacq. -Fél. ;弗勒里鞘蕊野牡丹●☆

370297　Spathandra fascicularis Planch. ex Benth. = Warneckea fascicularis（Planch. ex Benth.）Jacq. -Fél. ●☆

370298　Spathandra memecyloides Benth. = Warneckea memecyloides（Benth.）Jacq. -Fél. ●☆

370299　Spathandus Steud. = Spathanthus Desv. ■☆

370300　Spathantheum Schott（1859）;鞘花南星属■☆

370301　Spathantheum orbignyanum Schott;鞘花南星■☆

370302　Spathanthus Desv.（1828）;长穗草属■☆

370303　Spathanthus unilateralis Desv. ;长穗草■☆

370304　Spathe P. Browne = Spathelia L.（保留属名）●☆

370305　Spathe P. Browne et Boehm. = Spathelia L.（保留属名）●☆

370306　Spathelia L.（1762）（保留属名）;苞芸香属●☆

370307　Spathelia simplex L. ;简单苞芸香●☆

370308　Spathelia sorbifolia（L.）Fawc. et Rendle;苞芸香;Mountain Pride ●☆

370309　Spatheliaceae J. Agardh = Rutaceae Juss.（保留科名）●■

370310　Spathestigma Hook. et Arn. = Adenosma R. Br. ■

370311　Spathia Ewart（1917）;佛焰苞草属■☆

370312　Spathia neurosa Ewart et M. E. L. Archer;佛焰苞草■☆

370313　Spathicalyx J. C. Gomes（1956）;匙萼紫葳属●☆

370314　Spathicalyx kuhlmannii J. C. Gomes;匙萼紫葳●☆

370315　Spathicarpa Hook.（1831）;匙果南星属■☆

370316　Spathicarpa sagittifolia Schott;匙果南星;Caterpillar Plant, Fruit Sheath Plant ■☆

370317　Spathichlamys R. Parker（1931）;缅甸茜属■☆

370318　Spathichlamys oblonga R. Parker;缅甸茜■☆

370319　Spathidolepis Schltr.（1905）;薄鳞萝藦属■☆

370320　Spathidolepis torricelliensis Schltr. ;薄鳞萝藦■☆

370321　Spathiger Small = Epidendrum L.（保留属名）■☆

370322　Spathiger rigidus（Jacq.）Small = Epidendrum rigidum Jacq. ■☆

370323　Spathiger strobiliferus（Rchb. f.）Small = Epidendrum strobiliferum Rchb. f. ■☆

370324　Spathionema Taub.（1895）;匙蕊豆属（窄线豆属）■☆

370325　Spathionema kilimandscharicum Taub. ;匙蕊豆（窄线豆）■☆

370326　Spathiostemon Blume（1826）;匙蕊大戟属●☆

370327　Spathiostemon javensis Blume;匙蕊大戟●☆

370328　Spathipappus Tzvelev = Tanacetum L. ■●

370329　Spathiphyllopsis Teijsm. et Binn.（1863）;类苞叶芋属■☆

370330　Spathiphyllopsis Teijsm. et Binn. = Spathiphyllum Schott ■☆

370331　Spathiphyllopsis minahassae Teijsm. et Binn. ;类苞叶芋■☆

370332　Spathiphyllum Schott（1832）;苞叶芋属（白鹤芋属,匙芋叶属）;Madonna Lily, Peace Lily, Peace-Lily, Spathe Flower, Spathiphyllum ■☆

370333　Spathiphyllum 'Perfume';香水白掌■☆

370334　Spathiphyllum 'Supreme';绿巨人■☆

370335　Spathiphyllum brevirostre Schott;短喙苞叶芋■☆

370336　Spathiphyllum cannifolium（Dryand.）Schott;苞叶芋;Canna-leaf Spathiphyllum ■☆

370337　Spathiphyllum cochlearispathum Engl. ;匙状苞叶芋■☆

370338　Spathiphyllum floribundum（Lindl. et André）N. E. Br. ;翼柄苞叶芋（多花苞叶芋,翼柄白鹤芋）;Peace Lily, Snow Flower, Snowflower, Spathe Flower ■☆

370339　Spathiphyllum hybridum N. E. Br. ;杂种苞叶芋;Hybrid Spathiphyllum ■☆

370340　Spathiphyllum kochii Engl. et Krause;白鹤芋;Spathe Flower ■☆

370341　Spathiphyllum laeve Engl. ;平滑苞叶芋■☆

370342　Spathiphyllum longirostre Schott;长喙苞叶芋■☆

370343　Spathiphyllum minus G. S. Bunting;小苞叶芋■☆

370344　Spathiphyllum patinii（Hogg）N. E. Br. ;披针叶苞叶芋;Lanceolate-leaf Spathiphyllum, Peace Lily, White Sails ■☆

370345　Spathiphyllum wallisii Regel;矮小苞叶芋（白鹤芋,瓦氏白鹤芋）;Dwarf Madonna Lily, Peace Lily, Spathiphyllum, White Sails ■☆

370346　Spathirachis Klotzsch ex Klatt = Sisyrinchium L. ■

370347　Spathium Edgew. = Aponogeton L. f.（保留属名）■

370348　Spathium Lour. = Saururus L. ■

370349　Spathium chinense Lour. = Saururus chinensis（Lour.）Baill. ■

370350　Spathocarpus Post et Kuntze = Spathicarpa Hook. ■☆

370351　Spathodea P. Beauv.（1805）;火焰树属（苞萼木属,火焰木属）;African Tulip Tree, Flambeau Tree, Flambeautree, Flambeau-tree, Flamtree, Spathodea ●

370352　Spathodea acuminata Klotzsch = Markhamia zanzibarica（Bojer ex DC.）K. Schum. ●☆

370353　Spathodea adenantha G. Don = Newbouldia laevis（P. Beauv.）Seem. ex Bureau ●☆

370354　Spathodea adenophylla DC. = Fernandoa adenophylla（Wall. ex G. Don）Steenis ●☆

370355　Spathodea alba Sim;白火焰树●☆

370356　Spathodea alba Sim = Dolichandrone alba（Sim）Sprague ●☆

370357　Spathodea alternifolia R. Br. ;互叶火焰树●☆

370358　Spathodea campanulata P. Beauv. ;火焰树（火焰木,喷泉树,钟形火焰树）;African Tulip Tree, African Tuliptree, African Tuliptree，Bell Flambeau Tree，Bell Flambeautree，Bell Flambeau-tree，Flambeau Tree，Flambeau-tree，Flame of the Forest，Flame Tree，Flame-of-the-forest，Fountain Tree，Nandi Flame，Tulip Tree ●

370359　Spathodea campanulata P. Beauv. subsp. congolana Bidgood；刚果火焰树●☆

370360　Spathodea campanulata P. Beauv. subsp. nilotica（Seem.）Bidgood；尼罗河火焰树●☆

370361　Spathodea cauda-felina Hance = Dolichandrone caudafelina（Hance）Benth. et Hook. f. ●

370362　Spathodea cauda-felina Hance = Dolichandrone stipulata（Wall.）Benth. et Hook. f. var. kerrii（Sprague）C. Y. Wu et W. C. Yin ●

370363　Spathodea cauda-felina Hance = Markhamia caudafelina（Hance）Craib ●

370364　Spathodea cauda-felina Hance = Markhamia stipulata（Wall.）Seem. ex K. Schum. var. kerrii Sprague ●

370365　Spathodea caude-felina Hance = Dolichandrone caudafelina（Hance）Benth. et Hook. f. ●

370366　Spathodea danckelmaniana Büttner = Spathodea campanulata P. Beauv. ●

370367　Spathodea glandulosa Blume = Radermachera glandulosa（Blume）Miq. ●

370368　Spathodea igneum Kurz = Mayodendron igneum（Kurz）Kurz ●

370369　Spathodea jenischii Sond. = Newbouldia laevis（P. Beauv.）Seem. ex Bureau ●☆

370370　Spathodea laevis P. Beauv.；平滑火焰树●☆

370371　Spathodea laevis P. Beauv. = Newbouldia laevis（P. Beauv.）Seem. ex Bureau ●☆

370372　Spathodea longiflora P. Beauv.；长花火焰树●☆

370373　Spathodea lutea Benth. = Markhamia lutea（Benth.）K. Schum. ●☆

370374　Spathodea nilotica P. Beauv.；尼罗火焰树（火烧花，火焰树，喷泉树）；Flambeautree，Flamtree ●

370375　Spathodea nilotica P. Beauv. = Spathodea campanulata P. Beauv. ●

370376　Spathodea nilotica Seem. = Spathodea campanulata P. Beauv. subsp. nilotica（Seem.）Bidgood ●☆

370377　Spathodea pentandra Hook.；五雄蕊火焰树●☆

370378　Spathodea pentandra Hook. = Newbouldia laevis（P. Beauv.）Seem. ex Bureau ●☆

370379　Spathodea puberula Klotzsch = Markhamia zanzibarica（Bojer ex DC.）K. Schum. ●☆

370380　Spathodea speciosa Brongn. = Newbouldia laevis（P. Beauv.）Seem. ex Bureau ●☆

370381　Spathodea stenocarpa Welw. = Markhamia zanzibarica（Bojer ex DC.）K. Schum. ●☆

370382　Spathodea stipulata Wall. = Dolichandrone stipulata（Wall.）Benth. et Hook. f. ●

370383　Spathodea stipulata Wall. = Markhamia stipulata（Wall.）Seem. ex K. Schum. ●

370384　Spathodea tomentosa Benth. = Markhamia tomentosa（Benth.）K. Schum. ex Engl. ●☆

370385　Spathodea tulipifera（Thonn.）G. Don = Spathodea campanulata P. Beauv. ●

370386　Spathodea velutina Kurz = Dolichandrone stipulata（Wall.）Benth. et Hook. f. ●

370387　Spathodea velutina Kurz = Markhamia stipulata（Wall.）Seem.

ex K. Schum. ●

370388　Spathodea zanzibarica Bojer ex DC. = Markhamia zanzibarica（Bojer ex DC.）K. Schum. ●☆

370389　Spathodeopsis Dop = Fernandoa Welw. ex Seem. ●

370390　Spathodeopsis Dop（1930）；拟火焰树属●☆

370391　Spathodeopsis rossignolii Dop；拟火焰树●☆

370392　Spathodithyros Hassk. = Commelina L. ■

370393　Spathoglottis Blume（1825）；苞舌兰属（黄花独蒜属，药兰属，紫兰属）；Spathoglottis ■

370394　Spathoglottis aurea Lindl.；黄苞舌兰；Yellow Spathoglottis ■☆

370395　Spathoglottis chrysantha Ames；菲律宾苞舌兰（菲律宾黄苞舌兰）；Philippine Yellow Spathoglottis ■☆

370396　Spathoglottis fortunei Lindl. = Spathoglottis pubescens Lindl. ■

370397　Spathoglottis grandifolia Schltr.；大叶苞舌兰；Largeleaf Spathoglottis ■☆

370398　Spathoglottis ixioides（D. Don）Lindl.；少花苞舌兰；Foorflower Spathoglottis ■

370399　Spathoglottis kimballiana Sand.；肯氏苞舌兰；Kimball Spathoglottis ■☆

370400　Spathoglottis petri Rchb. f.；彼得苞舌兰■☆

370401　Spathoglottis plicata Blume；紫花苞舌兰（剑叶苞舌兰，紫苞舌兰）；Malayan Ground Orchid，Philippine Ground Orchid，Plicate Spathoglottis，Purplr Spathoglottis ■

370402　Spathoglottis pubescens Lindl.；苞舌兰（冰梨子，弗氏苞舌兰，黄花独蒜，老鸦蒜，牛油杯，土白芨）；Fortune Spathoglottis，Pubescent Spathoglottis ■

370403　Spathoglottis tomentosa Lindl.；毛苞舌兰；Tomentose Spathoglottis ■☆

370404　Spathoglottis vieillardii Rchb. f.；维氏苞舌兰■☆

370405　Spatholirion Ridl.（1896）；竹叶吉祥草属；Luckyweed ■

370406　Spatholirion elegans（Cherfils）C. Y. Wu；矩叶吉祥草■

370407　Spatholirion longifolium（Gagnep.）Dunn；竹叶吉祥草（白龙须，缠百合，马耳草，马耳朵草，秦归，珊瑚草，竹叶菜，竹叶凤，竹叶红参，竹叶藤参）；Bambooleaf Luckyweed，Longleaf Luckyweed ■

370408　Spatholirion longifolium Gagnep. = Spatholirion longifolium（Gagnep.）Dunn ■

370409　Spatholirion ornatum Ridl.；泰国吉祥草■☆

370410　Spatholirion scandens Dunn = Spatholirion longifolium（Gagnep.）Dunn ■

370411　Spatholobus Hassk.（1842）；密花豆属（翅豆藤属）；Spatholobus ●

370412　Spatholobus biauritus C. F. Wei；双耳密花豆；Double-ear Spatholobus，Two-auricled Spatholobus ●

370413　Spatholobus discolor C. F. Wei；变色密花豆；Discolor Spatholobus，Variable-coloured Spatholobus，Variant-coloured Spatholobus ●

370414　Spatholobus gengmaensis C. F. Wei；耿马密花豆；Gengma Spatholobus ●

370415　Spatholobus harmandii Gagnep.；光叶密花豆（大样荔枝藤）；Harmand Spatholobus ●

370416　Spatholobus parviflorus Kuntze = Spatholobus suberectus Dunn ●

370417　Spatholobus pulcher Dunn；美丽密花豆；Beautiful Spatholobus，Spiffy Spatholobus ●

370418　Spatholobus roseus Prain；老贯藤（大红藤，瑰花密花豆，老涩藤玫）；Rose Spatholobus ●

370419　Spatholobus roxburghii Benth.；罗氏密花豆（红花密花豆）；Red Spatholobus，Roxburgh Spatholobus ●

370420　Spatholobus roxburghii Benth. var. denudatus Baker;显脉密花豆;Conspicuous-veined Spatholobus ●

370421　Spatholobus sinensis Chun et T. C. Chen;红血藤(华密花豆);Bloodvine Spatholobus,Chinese Spatholobus ●

370422　Spatholobus suberectus Dunn;密花豆(大活血,大血藤,丰城鸡血藤,贯肠血藤,光叶崖豆藤,过岗龙,过山龙,鸡血藤,九层风,苦藤,亮叶岩豆藤,马鹿花,密花豆藤,三叶鸡血藤,山鸡血藤,香花岩豆藤,血风,血风藤,血筋藤,血龙藤,血藤,野奶豆,猪婆藤,紫梗藤);Spatholobus,Suberect Spatholobus ●

370423　Spatholobus uniauritus C. F. Wei;单耳密花豆;Monoear Spatholobus,One-auricled Spatholobus,Single-ear Spatholobus ●

370424　Spatholobus varians Dunn;云南密花豆(变异密花豆,栗色密花豆);Variable Spatholobus,Yunnan Spatholobus ●

370425　Spathophyllopsis Post et Kuntze = Spathiphyllopsis Teijsm. et Binn. ■☆

370426　Spathophyllopsis Post et Kuntze = Spathiphyllum Schott ■☆

370427　Spathophyllum Post et Kuntze = Spathiphyllum Schott ■☆

370428　Spathorachis Post et Kuntze = Sisyrinchium L. ■

370429　Spathorachis Post et Kuntze = Spathirachis Klotzsch ex Klatt ■

370430　Spathoscaphe Oerst. = Chamaedorea Willd. (保留属名)●☆

370431　Spathostigma Post et Kuntze = Adenosma R. Br. ■

370432　Spathostigma Post et Kuntze = Spathestigma Hook. et Arn. ■

370433　Spathotecoma Bureau = Newbouldia Seem. ex Bureau ●☆

370434　Spathula (Tausch) Fourr. = Iris L. ■

370435　Spathula Fourr. = Iris L. ■

370436　Spathularia A. St.-Hil. = Amphirrhox Spreng. (保留属名)■☆

370437　Spathularia DC. = Saxifraga L. ■

370438　Spathularia DC. = Spatularia Haw. ■

370439　Spathulata (Boriss.) Á. Löve et D. Löve = Sedum L. ●■

370440　Spathulopetalum Chiov. = Caralluma R. Br. ■

370441　Spathulopetalum arachnoideum (P. R. O. Bally) Plowes = Caralluma arachnoidea (P. R. O. Bally) M. G. Gilbert ■☆

370442　Spathulopetalum congestiflora (P. R. O. Bally) Plowes = Caralluma congestiflora P. R. O. Bally ■☆

370443　Spathulopetalum edwardsiae (M. G. Gilbert) Plowes = Caralluma edwardsiae (M. G. Gilbert) M. G. Gilbert ■☆

370444　Spathulopetalum gracilipes (K. Schum.) Plowes = Caralluma gracilipes K. Schum. ■☆

370445　Spathulopetalum longiflorum (M. G. Gilbert) Plowes = Caralluma longiflora M. G. Gilbert ■☆

370446　Spathulopetalum moniliforme (P. R. O. Bally) Plowes = Caralluma moniliformis P. R. O. Bally ■☆

370447　Spathulopetalum peckii (P. R. O. Bally) Plowes = Caralluma peckii P. R. O. Bally ■☆

370448　Spathulopetalum priogonium (K. Schum.) Plowes = Caralluma priogonium K. Schum. ■☆

370449　Spathulopetalum turneri (E. A. Bruce) Plowes = Caralluma turneri E. A. Bruce ■☆

370450　Spathulopetalum vaduliae (Lavranos) Plowes = Caralluma vaduliae Lavranos ■☆

370451　Spathyema Raf. = Symplocarpus Salisb. ex W. P. C. Barton(保留属名)■

370452　Spathyema foetida (L.) Raf. = Symplocarpus foetidus (L.) Salisb. ex W. P. C. Barton ■

370453　Spatularia Haw. = Hydatica Neck. ex Gray ■

370454　Spatularia Haw. = Saxifraga L. ■

370455　Spatulima Raf. = Lathyrus L. ■

370456　Specklinia Lindl. = Pleurothallis R. Br. ■☆

370457　Spectaculum Luer = Masdevallia Ruiz et Pav. ■☆

370458　Speculantha D. L. Jones et M. A. Clem. (2002);镜花兰属■☆

370459　Speculantha D. L. Jones et M. A. Clem. = Pterostylis R. Br. (保留属名)■☆

370460　Specularia A. DC. = Legousia T. Durand ●■☆

370461　Specularia Heist. = Legousia T. Durand ●■☆

370462　Specularia Heist. ex A. DC. = Legousia T. Durand ●■☆

370463　Specularia Heist. ex Fabr. = Legousia T. Durand ●■☆

370464　Specularia biflora (Ruiz et Pav.) Fisch. et C. A. Mey. = Triodanis biflora (Ruiz et Pav.) Greene ■

370465　Specularia castellana Lange = Legousia falcata (Ten.) Janch. subsp. castellana (Lange) Jauzein ●☆

370466　Specularia castellana Lange var. grandiflora Willk. = -Legousia falcata (Ten.) Janch. subsp. castellana (Lange) Jauzein ●☆

370467　Specularia castellana Lange var. maroccana (Pau et Font Quer) Maire = Legousia falcata (Ten.) Janch. subsp. castellana (Lange) Jauzein ●☆

370468　Specularia falcata (Ten.) A. DC. = Legousia falcata (Ten.) Janch. ●☆

370469　Specularia holzingeri (McVaugh) Fernald = Triodanis holzingeri McVaugh ■☆

370470　Specularia hybrida (L.) A. DC. = Legousia hybrida (L.) Delarbre ■☆

370471　Specularia juliani Batt. = Legousia juliani (Batt.) Briq. ●☆

370472　Specularia lamprosperma (McVaugh) Fernald = Triodanis lamprosperma McVaugh ■☆

370473　Specularia leptocarpa (Nutt.) A. Gray = Triodanis leptocarpa (Nutt.) Nieuwl. ■☆

370474　Specularia perfoliata (L.) A. DC. = Triodanis perfoliata (L.) Nieuwl. ■

370475　Specularia perfoliata (L.) A. DC. f. alba (Voigt) Steyerm. = Triodanis perfoliata (L.) Nieuwl. f. alba Voigt ■☆

370476　Specularia perfoliata (L.) A. DC. var. alba (Voigt) Steyerm. = Triodanis perfoliata (L.) Nieuwl. ■

370477　Specularia speculum L. = Legousia speculum-veneris (L.) Chaix ●☆

370478　Specularia speculum-veneris (L.) Caruel = Legousia speculum-veneris (L.) Chaix ●☆

370479　Speea Loes. (1927);斯皮葱属■☆

370480　Speea humilis Loes. ;斯皮葱■☆

370481　Spegazzinia Backeb. = Rebutia K. Schum. ●

370482　Spegazzinia Backeb. = Weingartia Werderm. ■☆

370483　Spegazziniophytum Esser = Colliguaja Molina ●☆

370484　Spegazziniophytum Esser(2001);巴塔哥尼亚大戟属●☆

370485　Spegazziniophytum patagonicum (Speg.) Esser;巴塔哥尼亚大戟●☆

370486　Spegpzinia Backeb. = Gymnocalycium Sweet ex Mittler ●

370487　Spegpzinia Backeb. = Weingartia Werderm. ■☆

370488　Speirantha Baker (1875);白穗花属;Speirantha,Whitespike ■★

370489　Speirantha convallarioides Baker = Speirantha gardenii (Hook.) Baill. ■

370490　Speirantha gardenii (Hook.) Baill. ;白穗花;Common Speirantha,Whitespike ■

370491　Speiranthes Hassk. = Spiranthes Rich. (保留属名)■

370492　Speirema Hook. f. et Thomson = Pratia Gaudich. ■

370493　Speirema montanum Hook. f. et Thomson = Pratia montana（Reinw. ex Blume）Hassk. ■

370494　Speirodela S. Watson = Spirodela Schleid. ■

370495　Speirostyla Baker = Christiana DC. ●☆

370496　Speirostyla tiliifolia Baker = Christiana africana DC. ●☆

370497　Spelaeanthus Kiew, A. Weber et B. L. Burtt（1998）;小岩苣苔属■☆

370498　Spelaeanthus chinii Kiew, A. Weber et B. L. Burtt;小岩苣苔■☆

370499　Spelta Wolf = Triticum L. ■

370500　Spencera Stapf = Spenceria Trimen ■★

370501　Spenceria Trimen（1879）;马蹄黄属;Spenceria ■★

370502　Spenceria parviflora Stapf = Spenceria ramalana Trimen ■

370503　Spenceria ramalana Trimen;马蹄黄（白地榆,黄地榆,黄总花草,小地榆）;Spenceria, Common Spenceria ■

370504　Spenceria ramalana Trimen var. parviflora（Stapf）Kitam. ;小花马蹄黄■

370505　Spenceria ramalana Trimen var. parviflora（Stapf）Kitam. = Spenceria ramalana Trimen ■

370506　Spennera Mart. ex DC. = Aciotis D. Don ☆

370507　Spenocarpus B. D. Jacks. = Magnolia L. ●

370508　Spenocarpus B. D. Jacks. = Sphenocarpus Korovin ■☆

370509　Spenotoma G. Don = Dracophyllum Labill. ●☆

370510　Spenotoma G. Don = Sphenotoma R. Br. ex Sweet ●☆

370511　Speranskia Baill.（1858）;地构叶属;Speranskia ■★

370512　Speranskia cantonensis（Hance）Pax et K. Hoffm. ;广东地构叶（白花蛋不老,旦不老,蛋不老,地胡椒,挂裂搽,广州地构叶,华南地构叶,黄鸡胆,六月雪,南地构叶,仁砂草,透骨草）;Guangzhou Speranskia ■

370513　Speranskia henryi Oliv. = Speranskia cantonensis（Hance）Pax et K. Hoffm. ■

370514　Speranskia pekinensis Pax et K. Hoffm. = Speranskia tuberculata（Bunge）Baill. ■

370515　Speranskia tuberculata（Bunge）Baill. ;地构叶（地构菜,海地透骨草,瘤果地构叶,透骨草,疣果地构菜,疣果地构叶,珍珠透骨草）;Speranskia ■

370516　Speranskia yunnanensis S. M. Hwang;云南地构叶（云南地构木）;Yunnan Speranskia ■

370517　Spergella Rchb. = Sagina L. ■

370518　Spergella caespitosa（J. Vahl）Á. Löve et D. Löve = Sagina caespitosa（J. Vahl）Lange ■☆

370519　Spergella decumbens Elliott = Sagina decumbens（Elliott）Torr. et A. Gray ■☆

370520　Spergella intermedia（Fenzl ex Ledeb. ）Á. Löve et D. Löve = Sagina nivalis（Lindblom）Fr. ■☆

370521　Spergula L.（1753）;大爪草属;Spurrey, Spurry ■

370522　Spergula arvensis L. ;大爪草（欧大爪草,普通大爪草）;Beggarweed, Bottle Brush, Bottle-brush, Carran, Cat's Hair, Corn Spurry, Cowquake, Devil's Flower, Dodder, Dother, Farmer's Ruin, Franck Spurry, Francke Spurry, Francking Spurwort, Franke, Granyagh, Guano-weed, Lousy Grass, Mountain Flax, Pickpocket, Pick-purse, Poverty, Sandweed, Spurrey, Spurry, Starwort, Stickwort, Tailor's Needle, Toadflax, Toad's Brass, Yarr ■

370523　Spergula arvensis L. subsp. chieusseana（Pomel）Briq. ;希厄斯大爪草■☆

370524　Spergula arvensis L. var. glutinosa Lange = Spergula arvensis L. ■

370525　Spergula arvensis L. var. laevis Chabert = Spergula arvensis L. ■

370526　Spergula arvensis L. var. maxima（Weihe）Mert. et W. D. J. Koch;高大爪草■

370527　Spergula arvensis L. var. maxima（Weihe）Mert. et W. D. J. Koch = Spergula arvensis L. ■

370528　Spergula arvensis L. var. pomeliana Maire et Weiller = Spergula arvensis L. ■

370529　Spergula arvensis L. var. sativa（Boenn. ）Mert. et W. D. J. Koch = Spergula arvensis L. ■

370530　Spergula chieusseana Pomel = Spergula arvensis L. subsp. chieusseana（Pomel）Briq. ■☆

370531　Spergula diandra（Guss. ）Murb. = Spergularia diandra（Guss. ）Heldr. et Sartori ■

370532　Spergula diandra（Guss. ）Murb. var. leiosperma（Bunge）Asch. et Schweinf. = Spergularia diandra（Guss. ）Heldr. et Sartori ■

370533　Spergula diandra（Guss. ）Murb. var. maura Pau et Sennen = Spergularia diandra（Guss. ）Heldr. et Sartori ■

370534　Spergula fallax（Lowe）E. H. L. Krause;迷惑大爪草■☆

370535　Spergula fimbriata（Boiss. et Reut. ）Murb. = Spergularia fimbriata Boiss. et Reut. ■☆

370536　Spergula fimbriata（Boiss. et Reut. ）Murb. var. condensata Ball = Spergularia fimbriata Boiss. et Reut. ■☆

370537　Spergula fimbriata（Boiss. et Reut. ）Murb. var. hebephylla Maire et Weiller = Spergularia fimbriata Boiss. et Reut. ■☆

370538　Spergula fimbriata（Boiss. et Reut. ）Murb. var. pulvinata Sauvage = Spergularia fimbriata Boiss. et Reut. ■☆

370539　Spergula fimbriata（Boiss. et Reut. ）Murb. var. tenuis Ball = Spergularia fimbriata Boiss. et Reut. ■☆

370540　Spergula flaccida Asch. = Spergula fallax（Lowe）E. H. L. Krause ■☆

370541　Spergula fontenellei Maire = Spergularia microsperma（Kindb. ）Vved. subsp. fontenellei（Maire）Greuter et Burdet ■☆

370542　Spergula fontinalis（Short et R. Peter）Dietrich = Stellaria fontinalis（Short et R. Peter）B. L. Rob. ■☆

370543　Spergula gamostyla（Pomel）Maire = Spergularia media（L. ）C. Presl ex Griseb. subsp. sauvagei（P. Monnier）Lambinon et Dobignard ■☆

370544　Spergula japonica Sw. = Sagina japonica（Sw. ex Steud. ）Ohwi ■

370545　Spergula laricina L. = Minuartia laricina（L. ）Mattf. ■

370546　Spergula linicola Boreau;麻生大爪草■☆

370547　Spergula linicola Boreau ex Nyman = Spergula arvensis L. ■

370548　Spergula linnaei C. Presl = Sagina linnaei C. Presl ■

370549　Spergula linnaei C. Presl = Sagina saginoides（L. ）H. Karst. ■

370550　Spergula longipes（Lange）Murb. = Spergularia purpurea（Pers. ）G. Don ■☆

370551　Spergula marginata（DC. ）Murb. = Spergularia media（L. ）C. Presl ex Griseb. ■

370552　Spergula marginata（DC. ）Murb. var. battandieri（Foucaud）Batt. = Spergularia media（L. ）C. Presl ex Griseb. subsp. sauvagei（P. Monnier）Lambinon et Dobignard ■☆

370553　Spergula marginata（DC. ）Murb. var. glandulosissima Faure et Maire = Spergularia media（L. ）C. Presl ex Griseb. ■

370554　Spergula marginata（DC. ）Murb. var. intermedia Maire = Spergularia maritima（All. ）Chiov. subsp. intermedia（Maire）Greuter et Burdet ■☆

370555　Spergula marginata（DC. ）Murb. var. kralikii Foucaud = Spergularia media（L. ）C. Presl ex Griseb. ■

370556　Spergula marginata（DC. ）Murb. var. latifolia Pau = Spergularia media（L.）C. Presl ex Griseb. ■

370557　Spergula marginata（DC.）Murb. var. munbyana（Pomel）Maire = Spergularia munbyana Pomel ■☆

370558　Spergula marginata（DC.）Murb. var. reverchonii Batt. = Spergularia media（L.）C. Presl ex Griseb. subsp. sauvagei（P. Monnier）Lambinon et Dobignard ■☆

370559　Spergula marginata（DC.）Murb. var. roberti（Foucaud）Batt. = Spergularia media（L.）C. Presl ex Griseb. subsp. sauvagei（P. Monnier）Lambinon et Dobignard ■☆

370560　Spergula marginata（DC.）Murb. var. vulgaris Clavaud = Spergularia media（L.）C. Presl ex Griseb. ■

370561　Spergula maxima Weihe = Spergula arvensis L. ■

370562　Spergula maxima Weihe ex Boenn. = Spergula arvensis L. ■

370563　Spergula micrantha（Bunge）Fernald = Sagina saginoides（L.）H. Karst. ■

370564　Spergula micrantha Bunge = Sagina saginoides（L.）H. Karst. ■

370565　Spergula morisonii Boreau；莫氏大爪草；Morison's Spurry, Pearlwort Spurrey ■☆

370566　Spergula morisonii Boreau = Spergula vernalis Willd. ■☆

370567　Spergula nodosa L. = Sagina nodosa（L.）Fenzl ■☆

370568　Spergula pentandra L.；翼茎大爪草；Wing-seeded Spurry, Wingstem Spurry ■☆

370569　Spergula pitardiana（Pit.）Maire = Spergularia pitardiana Pit. ■☆

370570　Spergula pitardiana（Pit.）Maire var. villosa Emb. et Maire = Spergularia pitardiana Pit. ■☆

370571　Spergula pycnorrhiza（Batt.）Maire = Spergularia media（L.）C. Presl ex Griseb. subsp. sauvagei（P. Monnier）Lambinon et Dobignard ■☆

370572　Spergula rubra（L.）D. Dietr. = Spergularia purpurea（Pers.）G. Don ■☆

370573　Spergula rubra（L.）D. Dietr. subsp. atheniensis（Asch.）Maire = Spergularia bocconei（Scheele）Asch. et Graebn. ■☆

370574　Spergula rubra（L.）D. Dietr. subsp. campestris（L.）Maire = Spergularia rubra（L.）J. Presl et C. Presl ■

370575　Spergula rubra（L.）D. Dietr. subsp. longipes（Lange）Maire = Spergularia purpurea（Pers.）G. Don ■☆

370576　Spergula rubra（L.）D. Dietr. subsp. nicaeensis（Burnat）Maire = Spergularia nicaeensis Burnat ■☆

370577　Spergula rubra（L.）D. Dietr. subsp. oreophila Litard. et Maire = Spergularia microsperma（Kindb.）Vved. subsp. oreophila（Litard. et Maire）P. Monnier ■☆

370578　Spergula rubra（L.）D. Dietr. subsp. radiata Maire = Spergularia echinosperma（Celak.）Asch. et Graebn. ■☆

370579　Spergula rubra（L.）D. Dietr. subsp. tenuifolia（Pomel）Maire = Spergularia tenuifolia Pomel ■☆

370580　Spergula rubra（L.）D. Dietr. var. decipiens（Burnat）Maire et Weiller = Spergularia purpurea（Pers.）G. Don ■☆

370581　Spergula rubra（L.）D. Dietr. var. jallui Maire = Spergularia purpurea（Pers.）G. Don ■☆

370582　Spergula rubra（L.）D. Dietr. var. marceli（Sennen）Maire = Spergularia purpurea（Pers.）G. Don ■☆

370583　Spergula rubra（L.）D. Dietr. var. rouyana Cout. = Spergularia purpurea（Pers.）G. Don ■☆

370584　Spergula saginoides L. = Sagina saginoides（L.）H. Karst. ■

370585　Spergula saginoides L. var. hesperia Fernald = Sagina saginoides（L.）H. Karst. ■

370586　Spergula saginoides L. var. nivalis Lindblom = Sagina nivalis（Lindblom）Fr. ■☆

370587　Spergula salina（J. Presl et C. Presl）D. Dietr. = Spergularia salina J. Presl et C. Presl ■☆

370588　Spergula salina（J. Presl et C. Presl）D. Dietr. var. australis（Lebel）Maire = Spergularia marina（L.）Besser ■

370589　Spergula salina（J. Presl et C. Presl）D. Dietr. var. faurei Maire = Spergularia marina（L.）Besser ■

370590　Spergula salina（J. Presl et C. Presl）D. Dietr. var. halophila（Bunge）Maire = Spergularia marina（L.）Besser ■

370591　Spergula salina（J. Presl et C. Presl）D. Dietr. var. leiosperma（Kindb.）Maire et Weiller = Spergularia marina（L.）Besser ■

370592　Spergula salina（J. Presl et C. Presl）D. Dietr. var. sperguloides（Lehm.）Gürke = Spergularia marina（L.）Besser ■

370593　Spergula salina（J. Presl et C. Presl）D. Dietr. var. stenopetala（A. Chev.）Maire et Weiller = Spergularia marina（L.）Besser ■

370594　Spergula salina（J. Presl et C. Presl）D. Dietr. var. urbica（Leffler）Gürke = Spergularia marina（L.）Besser ■

370595　Spergula sativa Boenn.；栽培大爪草 ■☆

370596　Spergula sativa Boenn. = Spergula arvensis L. ■

370597　Spergula seminulifera（Hy）Maire = Spergularia salina J. Presl et C. Presl ■☆

370598　Spergula stricta Sw. = Minuartia stricta（Sw.）Hiern ■☆

370599　Spergula subulata Sw. = Sagina subulata（Sw.）C. Presl ■☆

370600　Spergula tunetana Maire = Spergularia media（L.）C. Presl ex Griseb. subsp. tunetana（Maire）Lambinon et Dobignard ■☆

370601　Spergula tunetana Maire var. chevallieri（Hy）Maire et Weiller = Spergularia media（L.）C. Presl ex Griseb. subsp. tunetana（Maire）Lambinon et Dobignard ■☆

370602　Spergula tunetana Maire var. mixta？ = Spergularia media（L.）C. Presl ex Griseb. subsp. tunetana（Maire）Lambinon et Dobignard ■☆

370603　Spergula tunetana Maire var. pseudopycnorrhiza？ = Spergularia media（L.）C. Presl ex Griseb. subsp. tunetana（Maire）Lambinon et Dobignard ■☆

370604　Spergula vernalis Willd.；春大爪草；Spring Spurry ■☆

370605　Spergula villosa Pers. = Spergularia villosa（Pers.）Cambess. ■☆

370606　Spergula vulgaris Boenn. = Spergula arvensis L. ■

370607　Spergulaceae Bartl. = Caryophyllaceae Juss.（保留科名）■●

370608　Spergulaceae Tzvelev = Caryophyllaceae Juss.（保留科名）■●

370609　Spergularia（Pers.）J. Presl et C. Presl（1819）（保留属名）；拟漆姑草属（假漆姑草属，拟漆姑草属，牛漆姑草属）；Sand Spurrey, Sand Spurry, Sandspurry, Sand-spurry, Sea-spurrey, Spergularia, Spurrey ■

370610　Spergularia J. Presl et C. Presl = Spergularia（Pers.）J. Presl et C. Presl（保留属名）■

370611　Spergularia alata Wiegand = Spergularia marina（L.）Griseb. ■

370612　Spergularia amurensis Pomel = Spergularia diandra（Guss.）Heldr. et Sartori ■

370613　Spergularia atheniensis Asch. et Schweinf. = Spergularia bocconei（Scheele）Asch. et Graebn. ■☆

370614　Spergularia atrosperma R. Rossbach；暗籽拟漆姑草；Black-seed Sand-spurrey ■☆

370615　Spergularia bocconei（Scheele）Asch. et Graebn.；博氏拟漆姑草；Boccone's Sand-spurrey, Boccone's Sandspurry, Greek Sand Spurrey, Greek Sea-spurrey, Red Sand Spurrey ■☆

370616　Spergularia bocconii（Scheele）Foucaud ex Merino = Spergularia bocconei（Scheele）Asch. et Graebn. ■☆

370617　Spergularia bourgeaui Lebel;布尔拟漆姑■☆

370618　Spergularia campestris（L.）Asch. = Spergularia rubra（L.）J. Presl et C. Presl ■

370619　Spergularia canadensis（Pers.）G. Don;加拿大拟漆姑;Canada Sand-spurry,Canadian Sand-spurrey ■☆

370620　Spergularia canadensis（Pers.）G. Don var. occidentalis R. Rossbach;西方拟漆姑■☆

370621　Spergularia chieusseana Pomel = Spergula arvensis L. subsp. chieusseana（Pomel）Briq. ■☆

370622　Spergularia diandra（Guss.）Boiss. var. tenuifolia（Pomel）Batt. = Spergularia tenuifolia Pomel ■☆

370623　Spergularia diandra（Guss.）Heldr. = Spergularia diandra（Guss.）Heldr. et Sartori ■

370624　Spergularia diandra（Guss.）Heldr. et Sartori;二蕊拟漆姑（二雄蕊拟漆姑）;Alkali Sand-spurrey,Diandra Sandspurry,Twoanther Sandspurry ■

370625　Spergularia dillenii Lebel = Spergularia marina（L.）Griseb. ■

370626　Spergularia doumerguei P. Monnier;杜梅格拟漆姑■☆

370627　Spergularia echinosperma（Celak.）Asch. et Graebn.;刺子拟漆姑;Bristle-seed Sand-spurrey,Bristleseed Sandspurry ■☆

370628　Spergularia echinosperma Celak. = Spergularia echinosperma（Celak.）Asch. et Graebn. ■☆

370629　Spergularia embergeri P. Monnier;恩贝格尔拟漆姑■☆

370630　Spergularia fallax Lowe = Spergula fallax（Lowe）E. H. L. Krause ■☆

370631　Spergularia fimbriata Boiss. et Reut. ;流苏拟漆姑■☆

370632　Spergularia fimbriata Boiss. et Reut. subsp. condensata（Ball）H. Lindb. = Spergularia fimbriata Boiss. et Reut. ■☆

370633　Spergularia fimbriata Boiss. et Reut. var. condensata Ball = Spergularia fimbriata Boiss. et Reut. subsp. condensata（Ball）H. Lindb. ■☆

370634　Spergularia fimbriata Boiss. et Reut. var. interclusa Svent. = Spergularia fimbriata Boiss. et Reut. ■☆

370635　Spergularia fimbriata Boiss. et Reut. var. tenue Ball = Spergularia fimbriata Boiss. et Reut. ■☆

370636　Spergularia fontenellei（Maire）Ozenda = Spergularia microsperma（Kindb.）Vved. subsp. fontenellei（Maire）Greuter et Burdet ■☆

370637　Spergularia hanoverensis Simon;汉诺威拟漆姑■☆

370638　Spergularia heldreichii Foucaud;赫德拟漆姑■☆

370639　Spergularia heterosperma（Guss.）Lebel = Spergularia marina（L.）Besser ■

370640　Spergularia leiosperma（Kindb.）Schmidt = Spergularia marina（L.）Griseb. ■

370641　Spergularia longicaulis Pomel = Spergularia salina J. Presl et C. Presl ■☆

370642　Spergularia longipes（Lange）Rouy = Spergularia purpurea（Pers.）G. Don ■☆

370643　Spergularia macrotheca（Hornem. ex Cham. et Schltdl.）Heynh. ;黏拟漆姑;Sticky Sand-spurrey ■☆

370644　Spergularia macrotheca（Hornem. ex Cham. et Schltdl.）Heynh. var. leucantha（Greene）B. L. Rob.;白花黏拟漆姑■☆

370645　Spergularia macrotheca（Hornem. ex Cham. et Schltdl.）Heynh. var. longistyla R. Rossbach;长柱黏拟漆姑■☆

370646　Spergularia marceli Sennen = Spergularia bocconei（Scheele）Asch. et Graebn. ■☆

370647　Spergularia marginata（DC.）Kitt. = Spergularia media（L.）

370648　Spergularia marginata（DC.）Kitt. subsp. chevallieri Pott. - Alap. = Spergularia media（L.）C. Presl ex Griseb. subsp. sauvagei（P. Monnier）Lambinon et Dobignard ■☆

370649　Spergularia marginata（DC.）Kitt. subsp. sauvagei P. Monnier = Spergularia media（L.）C. Presl ex Griseb. subsp. sauvagei（P. Monnier）Lambinon et Dobignard ■☆

370650　Spergularia marginata（DC.）Kitt. subsp. tunetana（Maire）P. Monnier = Spergularia media（L.）C. Presl ex Griseb. subsp. tunetana（Maire）Lambinon et Dobignard ■☆

370651　Spergularia marginata（DC.）Kitt. subsp. vulgaris（Clavaud）P. Monnier = Spergularia media（L.）C. Presl ex Griseb. ■

370652　Spergularia marginata（DC.）Kitt. var. battandieri（Foucaud）Batt. = Spergularia media（L.）C. Presl ex Griseb. subsp. sauvagei（P. Monnier）Lambinon et Dobignard ■☆

370653　Spergularia marginata（DC.）Kitt. var. grandiflora Pit. = Spergularia media（L.）C. Presl ex Griseb. ■

370654　Spergularia marginata（DC.）Kitt. var. munbyana（Pomel）Sennen = Spergularia munbyana Pomel ■☆

370655　Spergularia marginata（DC.）Kitt. var. pedicellaris Pit. = Spergularia media（L.）C. Presl ex Griseb. ■

370656　Spergularia marginata（DC.）Kitt. var. reverchonii（Foucaud）Batt. = Spergularia media（L.）C. Presl ex Griseb. subsp. sauvagei（P. Monnier）Lambinon et Dobignard ■☆

370657　Spergularia marginata（DC.）Kitt. var. roberti（Foucaud）Batt. = Spergularia media（L.）C. Presl ex Griseb. subsp. sauvagei（P. Monnier）Lambinon et Dobignard ■☆

370658　Spergularia marginata（DC.）Murb. var. crassa Pit. = Spergularia media（L.）C. Presl ex Griseb. ■

370659　Spergularia marginata Kitt. = Spergularia media（L.）C. Presl ex Griseb. ■

370660　Spergularia marginata Kitt. subsp. intermedia（Maire）Monnier = Spergularia media（L.）C. Presl ex Griseb. subsp. intermedia（Maire）Lambinon et Dobignard ■☆

370661　Spergularia marina（L.）Griseb. ;拟漆姑（牛漆姑草,盐地拟漆姑）; Lesser Sea Spurrey, Lesser Sea-spurrey, Salt Sandspurry, Saltmarsh Sand Spurrey, Salt-Marsh Sand Spurry, Saltmarsh Sand-spurry, Sea Plantain, Sea Shore Spergularia, Sea Spurrey, Shore Sand Spurrey ■

370662　Spergularia marina（L.）Griseb. = Spergularia salina J. Presl et C. Presl ■☆

370663　Spergularia marina（L.）Griseb. var. asiatica（H. Hara）H. Hara = Spergularia marina（L.）Griseb. ■

370664　Spergularia marina（L.）Griseb. var. simonii O. Deg. et I. Deg. = Spergularia marina（L.）Griseb. ■

370665　Spergularia marina（L.）Griseb. var. tenuis（Greene）R. Rossbach = Spergularia salina J. Presl et C. Presl ■☆

370666　Spergularia maritima（All.）Chiov. = Spergularia media（L.）C. Presl ex Griseb. ■

370667　Spergularia maritima（All.）Chiov. subsp. angustata（Clavaud）Greuter et Burdet = Spergularia media（L.）C. Presl ex Griseb. subsp. angustata（Clavaud）Kerguélen et Lambinon ■☆

370668　Spergularia maritima（All.）Chiov. subsp. intermedia（Maire）Greuter et Burdet = Spergularia media（L.）C. Presl ex Griseb. subsp. intermedia（Maire）Lambinon et Dobignard ■☆

370669　Spergularia maritima（All.）Chiov. subsp. occidentalis P. Monnier = Spergularia media（L.）C. Presl ex Griseb. subsp.

occidentalis (P. Monnier) Lambinon et Dobignard ■☆

370670 Spergularia maritima (All.) Chiov. subsp. sauvagei P. Monnier = Spergularia media (L.) C. Presl ex Griseb. subsp. sauvagei (P. Monnier) Lambinon et Dobignard ■☆

370671 Spergularia maritima (All.) Chiov. subsp. tunetana (Maire) Greuter et Burdet = Spergularia media (L.) C. Presl ex Griseb. subsp. tunetana (Maire) Lambinon et Dobignard ■☆

370672 Spergularia maritima (L.) Chiov. = Spergularia media (L.) C. Presl ex Griseb. ■

370673 Spergularia maritima Chiov. = Spergularia media (L.) C. Presl ex Griseb. ■

370674 Spergularia media (L.) C. Presl = Spergularia media (L.) C. Presl ex Griseb. ■

370675 Spergularia media (L.) C. Presl ex Griseb. ;缘翅拟漆姑(海滨拟漆姑,具缘拟漆姑);Greater Sand-spurrey,Greater Sea Spurrey, Greater Sea-spurrey,Marginate Sand Spurrey,Marginate Sand-spurrcy,Media Sandspurry,Median Sandspurry ■

370676 Spergularia media (L.) C. Presl ex Griseb. subsp. angustata (Clavaud) Kerguélen et Lambinon;窄叶缘翅拟漆姑■☆

370677 Spergularia media (L.) C. Presl ex Griseb. subsp. intermedia (Maire) Lambinon et Dobignard;间型缘翅拟漆姑■☆

370678 Spergularia media (L.) C. Presl ex Griseb. subsp. occidentalis (P. Monnier) Lambinon et Dobignard;西方缘翅拟漆姑■☆

370679 Spergularia media (L.) C. Presl ex Griseb. subsp. sauvagei (P. Monnier) Lambinon et Dobignard;索瓦热缘翅拟漆姑■☆

370680 Spergularia media (L.) C. Presl ex Griseb. subsp. tunetana (Maire) Lambinon et Dobignard;图奈缘翅拟漆姑■☆

370681 Spergularia microsperma (Kindb.) Vved. ;小籽拟漆姑■☆

370682 Spergularia microsperma (Kindb.) Vved. subsp. fontenellei (Maire) Greuter et Burdet;丰特小籽拟漆姑■☆

370683 Spergularia microsperma (Kindb.) Vved. subsp. oreophila (Litard. et Maire) P. Monnier;喜沙小籽拟漆姑■☆

370684 Spergularia munbyana Pomel;芒比拟漆姑■☆

370685 Spergularia nicaeensis Burnat;尼西亚拟漆姑■☆

370686 Spergularia pentandra L. var. intermedia Boiss. = Spergula fallax (Lowe) E. H. L. Krause ■☆

370687 Spergularia pitardiana Pit. ;皮塔德拟漆姑■☆

370688 Spergularia pitardiana Pit. var. villosa Emb. et Maire = Spergularia pitardiana Pit. ■☆

370689 Spergularia platensis (Cambess.) Fenzl;平原拟漆姑■☆

370690 Spergularia purpurea (Pers.) G. Don;紫拟漆姑;Purple Sandspurry ■☆

370691 Spergularia purpurea G. Don = Spergularia purpurea (Pers.) G. Don ■☆

370692 Spergularia pycnantha R. Rossbach;密花拟漆姑■☆

370693 Spergularia pycnorrhiza Batt. ;密根拟漆姑■☆

370694 Spergularia radicans C. Presl = Spergularia rubra (L.) J. Presl et C. Presl ■

370695 Spergularia ramosa Cambess. ;多枝拟漆姑■☆

370696 Spergularia rubra (L.) J. Presl et C. Presl;无翅拟漆姑(田野拟漆姑,野拟漆姑);Field Sandspurry, Purple Sand-spurry, Red Sand Spurry,Red Sand-spurry,Red Sandspurry,Red Sandwort,Red Spurrey,Roadside Sand Spurry,Sand Spurrey,Sand-spurrey ■

370697 Spergularia rubra (L.) J. Presl et C. Presl subsp. longipes (Lange) Briq. = Spergularia purpurea (Pers.) G. Don ■☆

370698 Spergularia rubra (L.) J. Presl et C. Presl subsp. marceli Maire et Sennen = Spergularia bocconei (Scheele) Asch. et Graebn. ■☆

370699 Spergularia rubra (L.) J. Presl et C. Presl subsp. oreophila Litard. et Maire = Spergularia microsperma (Kindb.) Vved. subsp. oreophila (Litard. et Maire) P. Monnier ■☆

370700 Spergularia rubra (L.) J. Presl et C. Presl var. amurensis (Pomel) Batt. = Spergularia rubra (L.) J. Presl et C. Presl ■

370701 Spergularia rubra (L.) J. Presl et C. Presl var. campestris (Willk.) Batt. = Spergularia rubra (L.) J. Presl et C. Presl ■

370702 Spergularia rubra (L.) J. Presl et C. Presl var. perennans (Kindb.) B. L. Rob. = Spergularia rubra (L.) J. Presl et C. Presl ■

370703 Spergularia rubra (L.) J. Presl et C. Presl var. pinguis (Fenzl) Ball = Spergularia marina (L.) Besser ■

370704 Spergularia rubra (L.) J. Presl et C. Presl var. sperguloides (Lehm.) Ball = Spergularia marina (L.) Besser ■

370705 Spergularia rupicola Lebel;岩生拟漆姑;Cliff Spurrey, Rock Sand Spurrey,Rock Sea-spurrey ■☆

370706 Spergularia salina J. Presl et C. Presl;盐沼拟漆姑;Lesser Sea-spurrey,Salt-marsh Sand-spurrey ■☆

370707 Spergularia salina J. Presl et C. Presl = Spergularia marina (L.) Griseb. ■

370708 Spergularia salina J. Presl et C. Presl subsp. microcarpa (Batt.) P. Monnier = Spergularia tangerina P. Monnier ■☆

370709 Spergularia salina J. Presl et C. Presl var. halophila (Bunge) Maire = Spergularia marina (L.) Besser ■

370710 Spergularia salina J. Presl et C. Presl var. heterosperma (Guss.) Halácsy = Spergularia marina (L.) Besser ■

370711 Spergularia salina J. Presl et C. Presl var. leiosperma Kit. = Spergularia marina (L.) Besser ■

370712 Spergularia salina J. Presl et C. Presl var. tenuis (Greene) Jeps. = Spergularia salina J. Presl et C. Presl ■☆

370713 Spergularia salsuginea (Bunge) Fenzl = Spergularia diandra (Guss.) Heldr. et Sartori ■

370714 Spergularia salsuginea Fenzl = Spergularia diandra (Guss.) Heldr. ■

370715 Spergularia sativa Roem. ;牛漆姑草;Spurry ■☆

370716 Spergularia segetalis (L.) G. Don;谷地漆姑草■☆

370717 Spergularia seminulifera Hy = Spergularia salina J. Presl et C. Presl ■☆

370718 Spergularia sparsiflora (Greene) A. Nelson = Spergularia marina (L.) Griseb. ■

370719 Spergularia tangerina P. Monnier;丹吉尔漆姑草■☆

370720 Spergularia tenuifolia Pomel;细叶漆姑草■☆

370721 Spergularia tenuis Greene = Spergularia salina J. Presl et C. Presl ■☆

370722 Spergularia tibestica P. Monnier et Quézel = Spergularia microsperma (Kindb.) Vved. ■☆

370723 Spergularia uliginosa Pomel = Spergularia marina (L.) Besser ■

370724 Spergularia villosa (Pers.) Cambess. ;毛漆姑草;Hairy Sand-spurrey,Hairy Sandspurry ■☆

370725 Spergularia villosa Cambess. = Spergularia villosa (Pers.) Cambess. ■☆

370726 Spergulastrum Michx. = Arenaria L. ■

370727 Spergulastrum Rich. = Stellaria L. ■

370728 Spergulastrum lanceolatum Michx. = Stellaria borealis Bigelow ■☆

370729 Spergulus Brot. ex Steud. = Drosophyllum Link ●☆

370730 Sperlingia Vahl = Hoya R. Br. ●

370731 Spermabolus Teijsm. et Binn. = Anaxagorea A. St. -Hil. ●

370732　Spermachiton Llanos = Sporobolus R. Br. ■

370733　Spermacocaceae Bercht. et J. Presl = Rubiaceae Juss. （保留科名）●■

370734　Spermacoce Dill. ex L. = Spermacoce L. ●■

370735　Spermacoce Gaertn. = Spermacoce L. ●■

370736　Spermacoce L. （1753）；拟鸭舌癀舅属（丰花草属，仔熟茜属）；Button Weed ●■

370737　Spermacoce abyssinica Kuntze；阿比西尼亚拟鸭舌癀舅●☆

370738　Spermacoce adscendens Pav. ex DC. ；上举鸭舌癀舅●☆

370739　Spermacoce ampliata （A. Rich.）Oliv. = Spermacoce sphaerostigma （A. Rich.）Vatke ■☆

370740　Spermacoce andongensis （Hiern）R. D. Good；安东鸭舌癀舅■☆

370741　Spermacoce annua Verdc. ；一年鸭舌癀舅■☆

370742　Spermacoce apiculata Willd. ex Roem. et Schult. = Diodia apiculata （Willd. ex Roem. et Schult.）K. Schum. ■☆

370743　Spermacoce aprica （Hiern）Govaerts；向阳鸭舌癀舅■☆

370744　Spermacoce articularis （L. f.）G. Mey. ；鸭舌癀舅（糙叶丰花草）■

370745　Spermacoce articularis （L. f.）G. Mey. = Borreria articularis （L. f.）F. N. Williams ■

370746　Spermacoce articularis L. f. = Spermacoce articularis （L. f.）G. Mey. ■

370747　Spermacoce articularis L. f. = Spermacoce hispida L. ■☆

370748　Spermacoce arvensis （Hiern）R. D. Good；田野鸭舌癀舅■☆

370749　Spermacoce assurgens Ruiz et Pav. ；光叶鸭舌癀舅■

370750　Spermacoce azurea Verdc. ；天蓝鸭舌癀舅■☆

370751　Spermacoce bambusicola （Berhaut）J. -P. Lebrun et Stork；邦布鸭舌癀舅■☆

370752　Spermacoce bangweolensis （R. E. Fr.）Verdc. ；班韦鸭舌癀舅■☆

370753　Spermacoce bequaertii （De Wild.）Verdc. ；贝卡尔鸭舌癀舅■☆

370754　Spermacoce bisepala Verdc. ；双萼鸭舌癀舅■☆

370755　Spermacoce brachyantha Verdc. ；短花鸭舌癀舅■☆

370756　Spermacoce buchneri （K. Schum.）Govaerts；布赫纳鸭舌癀舅■☆

370757　Spermacoce calycoptera Decne. = Pterogaillonia calycoptera （Decne.）Lincz. ■☆

370758　Spermacoce capitata Ruiz et Pav. ；头状鸭舌癀舅；Baldhead False Buttonweed ■☆

370759　Spermacoce cassuangensis R. D. Good；卡苏鸭舌癀舅■☆

370760　Spermacoce chaetocephala DC. ；毛头鸭舌癀舅■☆

370761　Spermacoce chaetocephala DC. var. minor （Hepper）Puff；小毛头鸭舌癀舅■☆

370762　Spermacoce compacta Hochst. ex Hiern = Spermacoce chaetocephala DC. ■☆

370763　Spermacoce compressa Hiern = Spermacoce hepperana Verdc. ■☆

370764　Spermacoce congensis （Bremek.）Verdc. ；刚果鸭舌癀舅■☆

370765　Spermacoce costata Roxb. = Hedyotis costata （Roxb.）Kurz ■

370766　Spermacoce decandollei Deb et R. M. Dutta = Borreria ocymoides （Burm. f.）DC. ●☆

370767　Spermacoce decandollei Deb et R. M. Dutta = Spermacoce mauritiana Gideon ■

370768　Spermacoce deserti N. E. Br. ；荒漠鸭舌癀舅■☆

370769　Spermacoce dibrachiata Oliv. ；异枝鸭舌癀舅■☆

370770　Spermacoce ericifolia Licht. ex Roem. et Schult. = Anthospermum ericifolium （Licht. ex Roem. et Schult.）Kuntze ●☆

370771　Spermacoce exilis （L. O. Williams）C. D. Adams = Spermacoce exilis （L. O. Williams）C. D. Adams ex W. C. Burger et C. M. Taylor ■☆

370772　Spermacoce exilis （L. O. Williams）C. D. Adams ex W. C. Burger et C. M. Taylor = Borreria exilis L. O. Williams ■☆

370773　Spermacoce exilis （L. O. Williams）C. D. Adams ex W. C. Burger et C. M. Taylor；太平洋鸭舌癀舅；Pacific False Buttonweed ■☆

370774　Spermacoce filifolia （Schumach. et Thonn.）J. -P. Lebrun et Stork；线叶鸭舌癀舅●☆

370775　Spermacoce filifolia Perr. et Lepr. ex DC. = Spermacoce filifolia （Schumach. et Thonn.）J. -P. Lebrun et Stork ●☆

370776　Spermacoce filiformis Hiern；线形鸭舌癀舅■☆

370777　Spermacoce filituba （K. Schum.）Verdc. ；线管鸭舌癀舅■☆

370778　Spermacoce flexuosa Lour. = Borreria articularis （L. f.）F. N. Williams ■

370779　Spermacoce flexuosa Lour. = Spermacoce articularis （L. f.）G. Mey. ■

370780　Spermacoce galeopsidis DC. = Borreria galeopsidis （DC.）Berhaut ■☆

370781　Spermacoce garuensis （K. Krause）Govaerts；加鲁鸭舌癀舅■☆

370782　Spermacoce glabra Michx. ；光鸭舌癀舅；Smooth Buttonweed ■☆

370783　Spermacoce globosa Schumach. et Thonn. = Spermacoce verticillata L. ●☆

370784　Spermacoce hebecarpa （A. Rich.）Oliv. var. major Hiern = Spermacoce chaetocephala DC. ■☆

370785　Spermacoce hebecarpa （Hochst. ex A. Rich.）Oliv. = Spermacoce chaetocephala DC. var. minor （Hepper）Puff ■☆

370786　Spermacoce hedyotidea DC. = Hedyotis hedyotidea （DC.）Merr. ●■

370787　Spermacoce hepperana Verdc. ；赫佩鸭舌癀舅■☆

370788　Spermacoce heteromorpha Dessein；异形鸭舌癀舅■☆

370789　Spermacoce hirta L. = Mitracarpus hirtus （L.）DC. ■

370790　Spermacoce hirta L. = Mitracarpus villosus （Sw.）DC. ■

370791　Spermacoce hispida L. ；硬毛鸭舌癀舅■☆

370792　Spermacoce hockii （De Wild.）Dessein；霍克鸭舌癀舅●☆

370793　Spermacoce homblei （De Wild.）Govaerts；洪布勒鸭舌癀舅■☆

370794　Spermacoce huillensis （Hiern）R. D. Good；威拉鸭舌癀舅●☆

370795　Spermacoce intricans （Hepper）H. M. Burkill；缠结鸭舌癀舅●☆

370796　Spermacoce ivorensis Govaerts；伊沃里鸭舌癀舅●☆

370797　Spermacoce kirkii （Hiern）Verdc. ；柯克鸭舌癀舅■☆

370798　Spermacoce kotschyana Oliv. = Spermacoce chaetocephala DC. ■☆

370799　Spermacoce laevis Lam. ；平滑鸭舌癀舅■☆

370800　Spermacoce laevis Lam. = Spermacoce tenuior L. ■☆

370801　Spermacoce lancea （Hiern）Govaerts；披针状鸭舌癀舅■☆

370802　Spermacoce latifolia Aubl. ；阔叶鸭舌癀舅（宽叶鸭舌癀舅，阔叶丰花草）；Oval-leaf False Buttonweed ■

370803　Spermacoce latifolia Aubl. = Borreria latifolia （Aubl.）K. Schum. ■

370804　Spermacoce latituba （K. Schum.）Verdc. ；宽管鸭舌癀舅■☆

370805 Spermacoce ledermannii（K. Krause）Govaerts；莱德鸭舌癀舅 ●☆

370806 Spermacoce leucadea Hochst. ex Hiern = Spermacoce stachydea DC. ●☆

370807 Spermacoce macrantha（Hepper）H. M. Burkill = Spermacoce ivorensis Govaerts ●☆

370808 Spermacoce malacophylla（K. Schum.）Govaerts；软叶鸭舌癀舅 ●☆

370809 Spermacoce manikensis Dessein；马尼科鸭舌癀舅 ●☆

370810 Spermacoce mauritiana Gideon；蔓鸭舌癀舅（二萼丰花草）■

370811 Spermacoce milnei Verdc.；米尔恩鸭舌癀舅 ●☆

370812 Spermacoce minutiflora（K. Schum.）Verdc.；微花鸭舌癀舅 ●☆

370813 Spermacoce molleri（Gand.）Govaerts = Spermacoce verticillata L. ●☆

370814 Spermacoce natalensis Hochst.；纳塔尔鸭舌癀舅 ●☆

370815 Spermacoce octodon（Hepper）J. -P. Lebrun et Stork；八齿鸭舌癀舅 ●☆

370816 Spermacoce ocymoides Burm. f. = Borreria ocymoides（Burm. f.）DC. ●☆

370817 Spermacoce ovalifolia Hemsl.；卵叶鸭舌癀舅；Broadleaf False Buttonweed ■☆

370818 Spermacoce palmetorum DC. = Spermacoce ruelliae DC. ■☆

370819 Spermacoce paolii（Chiov.）Verdc.；保尔鸭舌癀舅 ●☆

370820 Spermacoce perennis Verdc.；多年生鸭舌癀舅 ■☆

370821 Spermacoce philippensis Willd. ex Spreng. = Hedyotis philippensis（Willd. ex Spreng.）Merr. ex C. B. Rob. ■

370822 Spermacoce phyllocephala DC. = Spermacoce stachydea DC. var. phyllocephala（DC.）J. -P. Lebrun et Stork ● ☆

370823 Spermacoce phyteuma Schweinf. ex Hiern；牧根草鸭舌癀舅 ■☆

370824 Spermacoce phyteumoides Verdc.；拟牧根草鸭舌癀舅 ■☆

370825 Spermacoce phyteumoides Verdc. var. caerulea Verdc.；天蓝拟牧根草鸭舌癀舅 ■☆

370826 Spermacoce pilosa（Schumach. et Thonn.）DC. = Diodia sarmentosa Sw. ■☆

370827 Spermacoce princeae（K. Schum.）Verdc. var. mwinilungae Verdc.；穆维尼鸭舌癀舅 ●☆

370828 Spermacoce princeae（K. Schum.）Verdc. var. pubescens（Hepper）Verdc.；短柔毛鸭舌癀舅 ●☆

370829 Spermacoce prostrata Aubl.；平卧鸭舌癀舅 ■☆

370830 Spermacoce prostrata R. D. Good = Lelya prostrata（R. D. Good）W. H. Lewis ■☆

370831 Spermacoce pusilla Wall.；小鸭舌癀舅 ■

370832 Spermacoce quadrisulcata（Bremek.）Verdc.；五纵沟鸭舌癀舅 ■☆

370833 Spermacoce radiata（DC.）Hiern；辐射沟鸭舌癀舅 ■☆

370834 Spermacoce repens（DC.）Fosberg et J. M. Powell = Spermacoce exilis（L. O. Williams）C. D. Adams ex W. C. Burger et C. M. Taylor ■☆

370835 Spermacoce rigida Willd. ex Roem. et Schult. = Diodia apiculata（Willd. ex Roem. et Schult.）K. Schum. ■☆

370836 Spermacoce ruelliae DC.；鲁氏鸭舌癀舅 ■☆

370837 Spermacoce samfya Verdc.；萨姆菲亚鸭舌癀舅 ■☆

370838 Spermacoce schlechteri K. Schum. ex Verdc.；施莱鸭舌癀舅 ■☆

370839 Spermacoce senensis（Klotzsch）Hiern；塞纳鸭舌癀舅 ■☆

370840 Spermacoce serrulata P. Beauv. = Diodia serrulata（P. Beauv.）G. Taylor ■☆

370841 Spermacoce shangdongensis（F. Z. Li et X. D. Chen）X. D. Chen et al.；山东丰花草；Shandong Borreria ●

370842 Spermacoce somalica（K. Schum.）Govaerts = Spermacoce sphaerostigma（A. Rich.）Vatke ■☆

370843 Spermacoce spermacocina（K. Schum.）Bridson et Puff；尖籽鸭舌癀舅 ■☆

370844 Spermacoce sphaerostigma（A. Rich.）Vatke；球柱头鸭舌癀舅 ■☆

370845 Spermacoce stachydea DC.；热非鸭舌癀舅 ●☆

370846 Spermacoce stachydea DC. var. phyllocephala（DC.）J. -P. Lebrun et Stork；叶头鸭舌癀舅 ●☆

370847 Spermacoce stricta L. f.；丰花草（波利亚草，长叶鸭舌癀，假蛇舌草，破帽草，破帽花，四方枝节花，乌骨四方枝节节花，叶里藏珠）；Small Borreria，Strict Borreria ■

370848 Spermacoce stricta L. f. = Borreria stricta（L. f.）G. Mey. ●

370849 Spermacoce subvulgata（K. Schum.）J. G. Garcia；广布鸭舌癀舅 ●☆

370850 Spermacoce subvulgata（K. Schum.）J. G. Garcia var. quadrisepala Verdc.；四萼鸭舌癀舅 ■☆

370851 Spermacoce taylorii Verdc.；泰勒鸭舌癀舅 ■☆

370852 Spermacoce tenuior L.；细鸭舌癀舅 ■☆

370853 Spermacoce tenuior L. var. commersonii Verdc.；科梅逊鸭舌癀舅 ■☆

370854 Spermacoce tenuissima Hiern；极细鸭舌癀舅 ●☆

370855 Spermacoce terminaliflora R. D. Good；顶花鸭舌癀舅 ■☆

370856 Spermacoce thymoidea（Hiern）Verdc.；百里香鸭舌癀舅 ■☆

370857 Spermacoce verticillata L.；灌木鸭舌癀舅（轮生仔熟茜）；Shrubby False Buttonweed ●☆

370858 Spermacoce villosa Sw. = Mitracarpus hirtus（L.）DC. ■

370859 Spermacoce villosa Sw. = Mitracarpus villosus（Sw.）DC. ■

370860 Spermacoceaceae Bercht. et J. Presl = Rubiaceae Juss.（保留科名）●■

370861 Spermacoceodes Kuntze = Spermacoce L. ●■

370862 Spermacon Raf. = Spermacoce L. ●■

370863 Spermadictyon Roxb.（1815）；网纹茜属（香茜草属，香叶木属）；Hamiltonia ●

370864 Spermadictyon Roxb. = Hamiltonia Roxb. ●☆

370865 Spermadictyon azureum Wall. = Spermadictyon suaveolens Roxb. ●

370866 Spermadictyon mysorense Steud. = Spermadictyon suaveolens Roxb. ●

370867 Spermadictyon pilosum Spreng.；毛网纹茜（毛香叶木）●☆

370868 Spermadictyon scabrum Spreng. = Spermadictyon suaveolens Roxb. ●

370869 Spermadictyon suaveolens Roxb.；网纹茜（香花木，香叶木）；Aromatic Hamiltonia ●

370870 Spermadon Post et Kuntze = Rhynchospora Vahl（保留属名）●☆

370871 Spermadon Post et Kuntze = Spermodon P. Beauv. ex T. Lestib. ●

370872 Spermaphyllum Post et Kuntze = Spermophylla Neck. ●■☆

370873 Spermaphyllum Post et Kuntze = Ursinia Gaertn.（保留属名）●■☆

370874 Spermatochiton Pilg. = Spermachiton Llanos ■

370875 Spermatochiton Pilg. = Sporobolus R. Br. ■

370876 Spermatococe Clem. = Spermacoce L. ●■

370877 Spermatolepis Clem. = Arillastrum Pancher ex Baill. ●☆

370878 Spermatolepis Clem. = Myrtomera B. C. Stone ●☆

370879　Spermatolepis Clem. = Spermolepis Brongn. et Gris ●☆

370880　Spermatolepis Clem. = Stereocaryum Burret ●☆

370881　Spermatura Rchb. = Osmorhiza Raf.（保留属名）■

370882　Spermatura Rchb. = Uraspermum Nutt.（废弃属名）■

370883　Spermaulaxen Raf. = Polygonum L.（保留属名）■●

370884　Spermaxyron Steud. = Olax L. ●

370885　Spermaxyron Steud. = Spermaxyrum Labill. ●

370886　Spermaxyrum Labill. = Olax L. ●

370887　Spermodon P. Beauv. ex T. Lestib. = Rhynchospora Vahl（保留属名）■

370888　Spermodon eximius Nees = Rhynchospora eximia （Nees）Boeck. ■☆

370889　Spermolepis Brongn. et Gris = Arillastrum Pancher ex Baill. ●☆

370890　Spermolepis Brongn. et Gris = Stereocaryum Burret ●☆

370891　Spermolepis Raf.（1825）；鳞籽草属■☆

370892　Spermolepis divaricata （Walter）Britton；叉枝鳞籽草（叉枝鳞果草）；Forked Scale-seed ■☆

370893　Spermolepis echinata （Nutt.）A. Heller；刺鳞籽草（刺鳞果草）；Scale-seed ■☆

370894　Spermolepis hawaiiensis H. Wolff；夏威夷鳞籽草（夏威夷鳞果草）■☆

370895　Spermolepis inermis （Nutt.）Mathias et Constance；无刺鳞籽草（无刺鳞果草）；Scale-seed ■☆

370896　Spermophylla Neck. = Ursinia Gaertn.（保留属名）●■☆

370897　Sperulastrum lanuginosum Michx. = Arenaria lanuginosa （Michx.）Rohrb. ■☆

370898　Sperulastrum lanuginosum Michx. subsp. saxosum （A. Gray）W. A. Weber = Arenaria lanuginosa （Michx.）Rohrb. var. saxosa （A. Gray）Zarucchi ■☆

370899　Spetaea Wetschnig et Pfosser（2003）；斯氏风信子属☆

370900　Sphacanthus Benoist（1939）；楔刺爵床属☆

370901　Sphacanthus brillantaisia Benoist；楔刺爵床☆

370902　Sphacanthus humbertii Benoist；马岛楔刺爵床☆

370903　Sphacele Benth.（1829）（保留属名）；热美鳞翅草属●■☆

370904　Sphacele Benth.（保留属名）= Lepechinia Willd. ●■☆

370905　Sphacophyllum Benth. = Anisopappus Hook. et Arn. ■

370906　Sphacophyllum Benth. = Epallage DC. ■

370907　Sphacophyllum africanum （Oliv.）O. Hoffm. = Anisopappus chinensis Hook. et Arn. subsp. oliveranus （Wild）S. Ortiz, Paiva et Rodr. Oubina ■☆

370908　Sphacophyllum bojeri Benth. = Anisopappus salviifolius （DC.）Wild ■☆

370909　Sphacophyllum buchwaldii O. Hoffm. = Anisopappus chinensis Hook. et Arn. var. buchwaldii （O. Hoffm.）S. Ortiz, Paiva et Rodr. Oubina ■☆

370910　Sphacophyllum candelabrum O. Hoffm. = Anisopappus chinensis Hook. et Arn. ■

370911　Sphacophyllum flexuosum Hutch. = Anisopappus kirkii （Oliv.）Brenan ■☆

370912　Sphacophyllum gossweileri S. Moore = Anisopappus grangeoides （Vatke et Höpfner ex Klatt）Merxm. ■☆

370913　Sphacophyllum helenae Buscal. et Muschl. = Anisopappus chinensis Hook. et Arn. var. buchwaldii （O. Hoffm.）S. Ortiz, Paiva et Rodr. Oubina ■☆

370914　Sphacophyllum holstii O. Hoffm. = Anisopappus holstii （O. Hoffm.）Wild ■☆

370915　Sphacophyllum kirkii Oliv. = Anisopappus kirkii （Oliv.）Brenan ■☆

370916　Sphacophyllum lastii O. Hoffm. = Anisopappus chinensis Hook. et Arn. var. dentatus （DC.）S. Ortiz, Paiva et Rodr. Oubina ■☆

370917　Sphacophyllum madagascariense Benth. = Anisopappus salviifolius （DC.）Wild ■☆

370918　Sphacophyllum pinnatifidum O. Hoffm. = Anisopappus grangeoides （Vatke et Höpfner ex Klatt）Merxm. ■☆

370919　Sphacophyllum pumilum Hiern = Anisopappus pumilus （Hiern）Wild ■☆

370920　Sphacophyllum spilanthoides S. Moore = Anisopappus anemonifolius （DC.）G. Taylor ■☆

370921　Sphacophyllum stuhlmannii O. Hoffm. = Anisopappus chinensis Hook. et Arn. var. buchwaldii （O. Hoffm.）S. Ortiz, Paiva et Rodr. Oubina ■☆

370922　Sphacophyllum tenerum S. Moore = Anisopappus chinensis Hook. et Arn. var. dentatus （DC.）S. Ortiz, Paiva et Rodr. Oubina ■☆

370923　Sphacophyllum welwitschii O. Hoffm. = Anisopappus chinensis Hook. et Arn. var. buchwaldii （O. Hoffm.）S. Ortiz, Paiva et Rodr. Oubina ■☆

370924　Sphacopsis Briq. = Salvia L. ●■

370925　Sphaenodesma Schauer = Sphenodesme Jack ●

370926　Sphaenolobium Pimenov = Selinum L.（保留属名）■

370927　Sphaenolobium Pimenov（1975）；楔片草属■☆

370928　Sphaenolobium tenuisectum （Korovin）Pimenov；楔片草■☆

370929　Sphaeradenia Harling（1954）；球腺草属■☆

370930　Sphaeradenia alba R. Erikss.；白球腺草■☆

370931　Sphaeradenia alleniana Harling；阿伦球腺草■☆

370932　Sphaeradenia amazonica Harling；亚马孙球腺草■☆

370933　Sphaeradenia angustifolia （Ruiz et Pav.）Harling；窄叶球腺草■☆

370934　Sphaeralcea A. St.-Hil.（1827）；球葵属；Globe Mallow, Globemallow, Globe-mallow, Negrita ●●☆

370935　Sphaeralcea ambigua A. Gray；球葵；Desert Globemallow, Desert Hollyhock, Desert Mallow, Globe Mallow ■☆

370936　Sphaeralcea americana （L.）Metz = Malvastrum americanum （L.）Torr. ■

370937　Sphaeralcea angusta （A. Gray）Fernald = Malvastrum hispidum （Pursh）Hochr. ■☆

370938　Sphaeralcea angustifolia （Cav.）G. Don；狭叶球葵；Narrowleaf Globemallow, Scarlet Globe Mallow ■☆

370939　Sphaeralcea angustifolia （Cav.）G. Don subsp. cuspidata （A. Gray）A. E. Murray；骤尖球葵；Scarlet Globe Mallow ■☆

370940　Sphaeralcea bonariensis （Cav.）Griseb.；博纳里球葵；Latin Globemallow ■☆

370941　Sphaeralcea bonariensis Griseb. = Sphaeralcea bonariensis （Cav.）Griseb. ■☆

370942　Sphaeralcea coccinea （Nutt.）Rydb.；猩红球葵；Globe Mallow, Scarlet Globe Mallow, Scarlet Globemallow ■☆

370943　Sphaeralcea coulteri A. Gray；库尔特球葵；Coulter's Globemallow ■☆

370944　Sphaeralcea digitata Rydb.；指裂球葵；Juniper Globemallow ■☆

370945　Sphaeralcea dregeana （C. Presl）Harv. = Anisodontea anomala （Link et Otto）Bates ●☆

370946　Sphaeralcea elegans （Cav.）G. Don = Anisodontea elegans （Cav.）Bates ●☆

370947　Sphaeralcea fendleri A. Gray；芬德勒球葵；Fendler's Globemallow，Globe Mallow ■☆

370948　Sphaeralcea incana Torr. ex A. Gray；灰球葵；Gray Globe Mallow ■☆

370949　Sphaeralcea julii（Burch. ex DC.）Baker f. = Anisodontea julii（Burch. ex DC.）Bates ●☆

370950　Sphaeralcea malvastroides Baker f. = Anisodontea malvastroides（Baker f.）Bates ■☆

370951　Sphaeralcea munroana Spach；小果球葵■☆

370952　Sphaeralcea pannosa Bolus = Anisodontea julii（Burch. ex DC.）Bates subsp. pannosa（Bolus）Bates ●☆

370953　Sphaeralcea prostrata（Harv.）Baker f. var. mollis = Anisodontea julii（Burch. ex DC.）Bates subsp. prostrata（E. Mey. ex Turcz.）Bates ●☆

370954　Sphaeralcea prostrata（Turcz.）Baker f. = Anisodontea julii（Burch. ex DC.）Bates subsp. prostrata（E. Mey. ex Turcz.）Bates ●☆

370955　Sphaeranthoides A. Cunn. ex DC. = Pterocaulon Elliott ■

370956　Sphaeranthus L.（1753）；戴星草属；Sphaeranthus ■

370957　Sphaeranthus abyssinicus Steetz = Sphaeranthus suaveolens（Forssk.）DC. ■☆

370958　Sphaeranthus africanus L.；非洲戴星草（戴星草，三点花，田艾草）；Africa Sphaeranthus，African Sphaeranthus ■

370959　Sphaeranthus africanus L. var. suberiflorus（Hayata）Yamam. = Sphaeranthus africanus L. ■

370960　Sphaeranthus angolensis O. Hoffm.；安哥拉戴星草■☆

370961　Sphaeranthus angustifolius DC.；狭叶戴星草■☆

370962　Sphaeranthus brachystachys O. Hoffm. ex Cufod. = Sphaeranthus ukambensis Vatke et O. Hoffm. ■☆

370963　Sphaeranthus brounae Robyns = Sphaeranthus flexuosus O. Hoffm. ■☆

370964　Sphaeranthus bullatus Mattf.；泡状戴星草■☆

370965　Sphaeranthus calcareus Robyns = Sphaeranthus flexuosus O. Hoffm. ■☆

370966　Sphaeranthus cochinchinensis Lour. = Sphaeranthus africanus L. ■

370967　Sphaeranthus confertifolius Robyns；密叶戴星草■☆

370968　Sphaeranthus cotuloides DC.；马岛戴星草■☆

370969　Sphaeranthus cristatus O. Hoffm.；冠状戴星草■☆

370970　Sphaeranthus cufodontii Lanza = Sphaeranthus suaveolens（Forssk.）DC. ■☆

370971　Sphaeranthus cylindraceus Cufod. = Sphaeranthus ukambensis Vatke et O. Hoffm. ■☆

370972　Sphaeranthus cylindricus O. Hoffm. ex Engl. = Sphaeranthus ukambensis Vatke et O. Hoffm. ■☆

370973　Sphaeranthus dauensis Cufod. = Sphaeranthus ukambensis Vatke et O. Hoffm. var. dauensis（Cufod.）Beentje ■☆

370974　Sphaeranthus dinteri Muschl. = Sphaeranthus flexuosus O. Hoffm. ■☆

370975　Sphaeranthus fischeri O. Hoffm.；菲舍尔戴星草■☆

370976　Sphaeranthus flexuosus O. Hoffm.；曲折戴星草■☆

370977　Sphaeranthus foliosus Ross-Craig；多叶戴星草■☆

370978　Sphaeranthus gallensis Sacleux = Sphaeranthus bullatus Mattf. ■☆

370979　Sphaeranthus gazaensis Bremek. = Sphaeranthus senegalensis DC. ■

370980　Sphaeranthus glaber DC. = Sphaeranthus africanus L. ■

370981　Sphaeranthus gomphrenoides O. Hoffm. = Sphaeranthus steetzii Oliv. et Hiern ■☆

370982　Sphaeranthus gracilis Oliv. = Athroisma gracile（Oliv.）Mattf. ●☆

370983　Sphaeranthus greenwayi Ross-Craig；格林韦戴星草■☆

370984　Sphaeranthus hildebrandtii Baker = Sphaeranthus africanus L. ■

370985　Sphaeranthus hirtus Willd. = Sphaeranthus indicus L. ■

370986　Sphaeranthus hirtus Willd. = Sphaeranthus senegalensis DC. ■

370987　Sphaeranthus humilis O. Hoffm. = Sphaeranthus flexuosus O. Hoffm. ■☆

370988　Sphaeranthus incisus Robyns = Sphaeranthus peduncularis DC. ■☆

370989　Sphaeranthus indicus L.；绒毛戴星草（戴星草，麻腊干，印度戴星草）；India Sphaeranthus，Indian Sphaeranthus ■

370990　Sphaeranthus johnstonii Robyns = Sphaeranthus kirkii Oliv. et Hiern ■☆

370991　Sphaeranthus kalahariensis Bremek. et Oberm. = Sphaeranthus flexuosus O. Hoffm. ■☆

370992　Sphaeranthus keniensis Robyns = Sphaeranthus steetzii Oliv. et Hiern ■☆

370993　Sphaeranthus kirkii Oliv. et Hiern；柯克戴星草■☆

370994　Sphaeranthus kirkii Oliv. et Hiern var. cyathuloides（O. Hoffm.）Beentje；杯苋戴星草■☆

370995　Sphaeranthus kotschyi Sch. Bip. ex Schweinf. = Sphaeranthus suaveolens（Forssk.）DC. ■☆

370996　Sphaeranthus lecornteanus O. Hoffm. et Muschl. = Sphaeranthus senegalensis DC. ■☆

370997　Sphaeranthus lelyi Robyns = Sphaeranthus angustifolius DC. ■☆

370998　Sphaeranthus madagascariensis Robyns = Sphaeranthus angustifolius DC. ■☆

370999　Sphaeranthus madagascariensis Robyns = Sphaeranthus peduncularis DC. subsp. rogersii（N. E. Br.）Wild ■☆

371000　Sphaeranthus microcephalus Vatke = Sphaeranthus africanus L. ■

371001　Sphaeranthus mimetes Ross-Craig；相似戴星草■☆

371002　Sphaeranthus mollis Roxb. = Sphaeranthus indicus L. ■

371003　Sphaeranthus mozambiquensis Steetz；莫桑比克戴星草■☆

371004　Sphaeranthus napierae Ross-Craig = Sphaeranthus suaveolens（Forssk.）DC. ■☆

371005　Sphaeranthus neglectus R. E. Fr.；疏忽戴星草■☆

371006　Sphaeranthus neglectus R. E. Fr. var. lanatus？ = Sphaeranthus peduncularis DC. subsp. rogersii（N. E. Br.）Wild ■☆

371007　Sphaeranthus nubicus Sch. Bip. ex Oliv. et Hiern = Sphaeranthus angustifolius DC. ■☆

371008　Sphaeranthus oppositifolius Ross-Craig；对叶戴星草■☆

371009　Sphaeranthus ovalis Steetz = Sphaeranthus africanus L. ■

371010　Sphaeranthus peduncularis DC.；梗花戴星草■☆

371011　Sphaeranthus peduncularis DC. subsp. rogersii（N. E. Br.）Wild；罗杰斯戴星草■☆

371012　Sphaeranthus polycephalus Oliv. et Hiern = Sphaeranthus senegalensis DC. ■☆

371013　Sphaeranthus pusillus Sacleux = Sphaeranthus spathulatus Peter ■☆

371014　Sphaeranthus ramosus（Klatt）Mesfin；分枝戴星草■☆

371015　Sphaeranthus randii S. Moore；兰德戴星草■☆

371016　Sphaeranthus randii S. Moore var. bibracteata Ross-Craig；双苞戴星草■☆

371017　Sphaeranthus rogersii N. E. Br. = Sphaeranthus peduncularis DC. subsp. rogersii (N. E. Br.) Wild ■☆

371018　Sphaeranthus salinarum Symoens;柳戴星草■☆

371019　Sphaeranthus samburuensis Beentje;桑布鲁戴星草■☆

371020　Sphaeranthus senegalensis DC. ;塞内加尔戴星草（非洲戴星草）;Senegal Sphaeranthus ■

371021　Sphaeranthus setulosus R. E. Fr. = Sphaeranthus flexuosus O. Hoffm. ■☆

371022　Sphaeranthus similis Kers;近似戴星草■☆

371023　Sphaeranthus spathulatus Peter;匙形戴星草■☆

371024　Sphaeranthus sphenocleoides Oliv. et Hiern = Sphaeranthus africanus L. ■

371025　Sphaeranthus steetzii Oliv. et Hiern;斯蒂兹戴星草■☆

371026　Sphaeranthus stenostachys Chiov. = Sphaeranthus ukambensis Vatke et O. Hoffm. ■☆

371027　Sphaeranthus strobilaceus Peter = Sphaeranthus kirkii Oliv. et Hiern ■☆

371028　Sphaeranthus stuhlmannii O. Hoffm. ;斯图尔曼戴星草■☆

371029　Sphaeranthus suaveolens (Forssk.) DC. ;芳香戴星草■☆

371030　Sphaeranthus suaveolens (Forssk.) DC. var. abyssinicus (Steetz) Ross-Craig = Sphaeranthus suaveolens (Forssk.) DC. ■☆

371031　Sphaeranthus suaveolens (Forssk.) DC. var. angustifolius Oliv. = Sphaeranthus suaveolens (Forssk.) DC. ■☆

371032　Sphaeranthus suaveolens (Forssk.) DC. var. tetraphyllus (S. Moore) Ross-Craig = Sphaeranthus suaveolens (Forssk.) DC. ■☆

371033　Sphaeranthus suberiflorus Hayata = Sphaeranthus africanus L. ■

371034　Sphaeranthus talbotii S. Moore = Sphaeranthus steetzii Oliv. et Hiern ■☆

371035　Sphaeranthus tenuis R. E. Fr. = Sphaeranthus flexuosus O. Hoffm. ■☆

371036　Sphaeranthus tetraphyllus S. Moore = Sphaeranthus suaveolens (Forssk.) DC. ■☆

371037　Sphaeranthus tomentellus Mattf. = Sphaeranthus confertifolius Robyns ■☆

371038　Sphaeranthus ukambensis Vatke et O. Hoffm. ;短穗戴星草■☆

371039　Sphaeranthus ukambensis Vatke et O. Hoffm. var. dauensis (Cufod.) Beentje;达瓦戴星草■☆

371040　Sphaeranthus variabilis Robyns = Sphaeranthus angolensis O. Hoffm. ■☆

371041　Sphaeranthus wattii Giess ex Merxm. ;瓦特戴星草■☆

371042　Sphaeranthus zavattarii Cufod. ;扎瓦戴星草■☆

371043　Sphaerantia Peter G. Wilson et B. Hyland(1988);澳洲球金娘属●☆

371044　Sphaerantia chartacea Peter G. Wilson et B. Hyland;澳洲球金娘●☆

371045　Sphaerantia discolor Peter G. Wilson et B. Hyland;杂色澳洲球金娘●☆

371046　Sphaerella Bubani = Airopsis Desv. ■☆

371047　Sphaereupatorium (O. Hoffm.) Kuntze ex B. L. Rob. (1920);球泽兰属■☆

371048　Sphaereupatorium Kuntze = Sphaereupatorium (O. Hoffm.) Kuntze ex B. L. Rob. ■☆

371049　Sphaereupatorium scandens (Gardner) R. M. King et H. Rob. ;球泽兰■☆

371050　Sphaereupaturium (O. Hoffm.) B. L. Rob. = Sphaereupatorium (O. Hoffm.) Kuntze ex B. L. Rob. ■☆

371051　Sphaeridiophora Benth. et Hook. f. = Indigofera L. ●■

371052　Sphaeridiophora Benth. et Hook. f. = Sphaeridtophorum Desv. ●■

371053　Sphaeridiophorum Desv. = Indigofera L. ●■

371054　Sphaeridiophorum abyssinicum Jaub. et Spach = Indigofera linifolia (L. f.) Retz. ●

371055　Sphaeridiophorum linifolium (L. f.) Desv. = Indigofera linifolia (L. f.) Retz. ●

371056　Sphaerine Herb. = Bomarea Mirb. ■☆

371057　Sphaeritis Eckl. et Zeyh. = Crassula L. ●■☆

371058　Sphaeritis biconvexa Eckl. et Zeyh. = Crassula pubescens Thunb. ●☆

371059　Sphaeritis incana Eckl. et Zeyh. = Crassula subaphylla (Eckl. et Zeyh.) Harv. ■☆

371060　Sphaeritis margaritifera Eckl. et Zeyh. = Crassula mollis Thunb. ■☆

371061　Sphaeritis paucifolia Eckl. et Zeyh. = Crassula mesembryanthoides (Haw.) D. Dietr. subsp. hispida (Haw.) Toelken ■☆

371062　Sphaeritis puberula Eckl. et Zeyh. = Crassula subaphylla (Eckl. et Zeyh.) Harv. ■☆

371063　Sphaeritis setigera Eckl. et Zeyh. = Crassula tomentosa Thunb. ■☆

371064　Sphaeritis subaphylla Eckl. et Zeyh. = Crassula subaphylla (Eckl. et Zeyh.) Harv. ■☆

371065　Sphaeritis tomentosa (Thunb.) Eckl. et Zeyh. = Crassula tomentosa Thunb. ■☆

371066　Sphaeritis trachysantha Eckl. et Zeyh. = Crassula mesembryanthoides (Haw.) D. Dietr. ●☆

371067　Sphaeritis typica Eckl. et Zeyh. = Crassula subulata L. ■☆

371068　Sphaerium Kuntze = Coix L. ●■

371069　Sphaerobambos S. Dransf. (1989);球籽竹属（球子竹属）●☆

371070　Sphaerobambos hirsuta S. Dransf. ;球籽竹●☆

371071　Sphaerobambos philippinensis (Gamble) S. Dransf. ;菲律宾球籽竹●☆

371072　Sphaerocardamum Nees et Schauer(1847);球形碎米荠属■☆

371073　Sphaerocardamum S. Schauer = Sphaerocardamum Nees et Schauer ■☆

371074　Sphaerocardamum nesliiforme S. Schauer;球形碎米荠■☆

371075　Sphaerocarpos J. F. Gmel. = Globba L. ■

371076　Sphaerocarpos J. F. Gmel. = Manitia Giseke ■

371077　Sphaerocarpum Nees ex Steud. = Sphaerocaryum Nees ex Hook. f. ■

371078　Sphaerocarpum Steud. = Sphaerocaryum Nees ex Hook. f. ■

371079　Sphaerocarpus Fabr. = Neslia Desv. (保留属名)■

371080　Sphaerocarpus Rich. = Laguncularia C. F. Gaertn. ●☆

371081　Sphaerocarpus Steud. = Laguncularia C. F. Gaertn. ●☆

371082　Sphaerocarpus Steud. = Sphenocarpus Korovin ■☆

371083　Sphaerocarya Dalzell ex DC. = Strombosia Blume ●☆

371084　Sphaerocarya Wall. = Pyrularia Michx. ●

371085　Sphaerocarya edulis Wall. = Pyrularia edulis (Wall.) A. DC. ●

371086　Sphaerocarya wallichiana Wight et Arn. = Scleropyrum wallichianum (Wight et Arn.) Arn. ●

371087　Sphaerocaryum Nees ex Hook. f. (1896);稃荩属（秆荩属,圆柱草属）;Sphaerocaryum ■

371088　Sphaerocaryum Nees ex Steud. = Sphaerocaryum Nees ex Hook. f. ■

371089　Sphaerocaryum elegans (Wight et Arn.) Nees ex Steud. =

Sphaerocaryum malaccense (Trin.) Pilg. ■

371090　Sphaerocaryum malaccense (Trin.) Pilg. ;稗荩(稃荩,稃荩);
Malacca Sphaerocaryum ■

371091　Sphaerocaryum pulchella (Roth) A. Camus = Sphaerocaryum
malaccense (Trin.) Pilg. ■

371092　Sphaerocaryum pulchellum (Roth) Merr. = Isachne pulchella
Roth ■

371093　Sphaerocephala Hill = Centaurea L. (保留属名)●■

371094　Sphaerocephalus Kuntze = Echinops L. ■

371095　Sphaerocephalus Lag. ex DC. = Nassauvia Comm. ex Juss. ●☆

371096　Sphaerocephalus dahuricus (Fisch.) Kuntze ex Kom. =
Echinops latifolius Tausch ■

371097　Sphaerochloa P. Beauv. ex Desv. = Eriocaulon L. ●

371098　Sphaerochloa compressa P. Beauv. = Eriocaulon compressum
Lam. ■☆

371099　Sphaeroclinium (DC.) Sch. Bip. = Cotula L. ■

371100　Sphaeroclinium (DC.) Sch. Bip. = Matricaria L. ■

371101　Sphaeroclinium Sch. Bip. = Matricaria L. ■

371102　Sphaeroclinium Sch. Bip. = Sphaeroclinium (DC.) Sch. Bip. ■

371103　Sphaeroclinium nigellifolium (DC.) Sch. Bip. = Cotula
nigellifolia (DC.) K. Bremer et Humphries ■☆

371104　Sphaerocodon Benth. (1876);球冠萝藦属●☆

371105　Sphaerocodon acutifolius Schum. = Sphaerocodon
longipedunculatus K. Schum. ●☆

371106　Sphaerocodon angolensis S. Moore;安哥拉球冠萝藦●☆

371107　Sphaerocodon caffer (Meisn.) Schltr. ;开菲尔球冠萝藦●☆

371108　Sphaerocodon longipedunculatus K. Schum. = Tylophora
heterophylla A. Rich. ●☆

371109　Sphaerocodon melananthus N. E. Br. ;黑花球冠萝藦●☆

371110　Sphaerocodon natalensis Benth. = Sphaerocodon caffer
(Meisn.) Schltr. ●☆

371111　Sphaerocodon obtusifolius Benth. = Sphaerocodon caffer
(Meisn.) Schltr. ●☆

371112　Sphaerocodon platypodus K. Schum. ex De Wild. ;阔足球冠萝
藦●☆

371113　Sphaerocoma T. Anderson(1861);聚果指甲木属●☆

371114　Sphaerocoma aucheri Boiss. ;奥氏聚果指甲木●☆

371115　Sphaerocoma hookeri T. Anderson;聚果指甲木●☆

371116　Sphaerocoma hookeri T. Anderson subsp. intermedia J. B.
Gillett;中间聚果指甲木●☆

371117　Sphaerocoryne (Boerl.) Ridl. = Melodorum Lour. ●☆

371118　Sphaerocoryne Scheff. = Melodorum Lour. ●☆

371119　Sphaerocoryne Scheff. ex Ridl. = Melodorum Lour. ●☆

371120　Sphaerocoryne gracilis (Engl. et Diels) Verdc. ;纤细金帽花
●☆

371121　Sphaerocoryne gracilis (Engl. et Diels) Verdc. subsp. engleriana
(Exell et Mendonça) Verdc. ;恩氏纤细金帽花●☆

371122　Sphaerocyperus Lye(1972);球莎草属■☆

371123　Sphaerocyperus erinaceus (Ridl.) Lye;球莎草■☆

371124　Sphaerodendron Seem. = Cussonia Thunb. ●☆

371125　Sphaerodendron angolense Seem. = Cussonia angolensis
(Seem.) Hiern ●☆

371126　Sphaerodiscus Nakai = Euonymus L. (保留属名)●

371127　Sphaerogyne Naudin = Tococa Aubl. ●☆

371128　Sphaerolobium Sm. (1805);澳洲球豆属■☆

371129　Sphaerolobium gracile Benth. ;细澳洲球豆■☆

371130　Sphaerolobium grandiflorum Benth. ;大花澳洲球豆■☆

371131　Sphaerolobium stenopterum Meisn. ;窄翅澳洲球豆■☆

371132　Sphaeroma (DC.) Schltdl. = Phymosia Desv. ex Ham. ●☆

371133　Sphaeroma Schltdl. = Phymosia Desv. ex Ham. ●☆

371134　Sphaeroma divaricatum Kuntze = Anisodontea anomala (Link et
Otto) Bates ●☆

371135　Sphaeroma julii (Burch. ex DC.) Harv. = Anisodontea julii
(Burch. ex DC.) Bates ●☆

371136　Sphaeroma pannosum (Bolus) Kuntze = Anisodontea julii
(Burch. ex DC.) Bates subsp. pannosa (Bolus) Bates ●☆

371137　Sphaeroma prostratum (E. Mey. ex Turcz.) Harv. =
Anisodontea julii (Burch. ex DC.) Bates subsp. prostrata (E. Mey.
ex Turcz.) Bates ●☆

371138　Sphaeroma prostratum (Turcz.) Harv. var. molle Harv. =
Anisodontea julii (Burch. ex DC.) Bates subsp. prostrata (E. Mey.
ex Turcz.) Bates ●☆

371139　Sphaeromariscus E. G. Camus = Cyperus L. ■

371140　Sphaeromariscus microcephalus E. G. Camus = Mariscus
compactus (Retz.) Druce ■

371141　Sphaeromeria Nutt. (1841);球序蒿属;False Sagebrush,
Chickensage ■☆

371142　Sphaeromeria Nutt. = Tanacetum L. ■●

371143　Sphaeromeria argentea Nutt. ;银色球序蒿;Nuttall's False
Sagebrush, Silver Chickensage ■☆

371144　Sphaeromeria cana (D. C. Eaton) A. Heller;灰球序蒿;Gray
Chickensage ■☆

371145　Sphaeromeria capitata Nutt. ;球序蒿■☆

371146　Sphaeromeria compacta (H. M. Hall) A. H. Holmgren, L. M.
Shultz et Lowrey;紧凑球序蒿;Charleston Tansy, Compact
Chickensage ■☆

371147　Sphaeromeria diversifolia (D. C. Eaton) Rydb. ;异叶球序蒿;
False Sagebrush, Separateleaf Chickensage ■☆

371148　Sphaeromeria potentilloides (A. Gray) A. Heller;五指球序蒿;
Cinquefoil False Sagebrush, Fivefinger Chickensage ■☆

371149　Sphaeromeria potentilloides (A. Gray) A. Heller var. nitrophila
(Cronquist) A. H. Holmgren, L. M. Shultz et Lowrey;喜硝球序蒿
■☆

371150　Sphaeromeria ruthiae A. H. Holmgren;锡安山球序蒿;Zion
Chickensage, Zion Tansy ■☆

371151　Sphaeromeria simplex (A. Nelson) A. Heller;单茎球序蒿;
Laramie Chickensage ■☆

371152　Sphaeromorphaea DC. = Epaltes Cass. ■

371153　Sphaeromorphaea australis (Less.) Kitam. = Epaltes australis
Less. ■

371154　Sphaeromorphaea centipeda DC. = Centipeda minima (L.) A.
Braun et Asch. ■

371155　Sphaeromorphaea centipeda DC. = Centipeda orbicularis Lour. ■

371156　Sphaeromorphaea russeliana DC. = Epaltes australis Less. ■

371157　Sphaerophora Blume = Morinda L. ●■

371158　Sphaerophora Sch. Bip. = Eremanthus Less. ●☆

371159　Sphaerophora Sch. Bip. = Paralychnophora MacLeish ●☆

371160　Sphaerophysa DC. (1825);苦马豆属;Bitterhorsebean,
Globepea, Sphaerophysa, Swainsona ■●

371161　Sphaerophysa salsula (Pall.) DC. ;苦马豆(爆竹花,红花苦豆
子,红花土豆子,红苦豆,红苦豆子,苦黑子,铃当草,马皮泡,尿
泡草,泡泡豆,盐生苦马豆,羊吹泡,羊卵蛋,羊卵泡,羊萝泡,羊
奶奶,羊尿泡);Alkali Swainsonpea, Bitterhorsebean, Saline
Sphaerophysa, Saline Swainsona, Salt Globepea, Swainsonpea ●■

371162 Sphaeropus Boock. = Scleria P. J. Bergius ■

371163 Sphaerorhizon Hook. f. = Scybalium Schott et Endl. ■☆

371164 Sphaerorrhiza Roalson et Boggan（2005）；球根莒苔■☆

371165 Sphaerorrhiza burchellii（S. M. Phillips）Roalson et Boggan；球根莒苔■☆

371166 Sphaerorrhiza sarmentiana（Gardner ex Hook.）Roalson et Boggan；长匍茎球根莒苔■☆

371167 Sphaerosacme Wall. = Amoora Roxb. ●

371168 Sphaerosacme Wall. ex Roem.（1846）；喜马拉雅楝属●☆

371169 Sphaerosacme Wall. ex Roem. = Lansium Jacq. ●

371170 Sphaerosacme Wall. ex Roxb. = Sphaerosacme Wall. ex Royle ●☆

371171 Sphaerosacme decandra（Wall.）T. D. Penn.；喜马拉雅楝●☆

371172 Sphaerosacme decandra Wall. = Sphaerosacme decandra（Wall.）T. D. Penn. ●☆

371173 Sphaeroschoenus Arn. = Haplostylis Nees ■

371174 Sphaeroschoenus Arn. = Rhynchospora Vahl（保留属名）■

371175 Sphaeroschoenus Nees = Rhynchospora Vahl（保留属名）■

371176 Sphaerosciadium Pimenov et Kljuykov（1981）；球伞芹属■☆

371177 Sphaerosciadium denaense（Schischk.）Pimenov et Kljuykov；球伞芹■☆

371178 Sphaerosepalaceae（Warb.）Tiegh. ex Bullock.（1959）；球萼树科（刺果萼树科，球形萼科，圆萼树科）●☆

371179 Sphaerosepalaceae（Warb.）Tiegh. ex Bullock. = Ochnaceae DC.（保留科名）●☆

371180 Sphaerosepalaceae Tiegh. = Ochnaceae DC.（保留科名）●■

371181 Sphaerosepalaceae Tiegh. = Sphaerosepalaceae Tiegh. ex Bullock. ●☆

371182 Sphaerosepalaceae Tiegh. ex Bullock. = Ochnaceae DC.（保留科名）●■

371183 Sphaerosepalaceae Tiegh. ex Bullock. = Sphaerosepalaceae（Warb.）Tiegh. ex Bullock. ●☆

371184 Sphaerosepalum Baker = Rhopalocarpus Bojer ●☆

371185 Sphaerosepalum Baker（1884）；球萼树属（球形萼属）●☆

371186 Sphaerosepalum alternifolium Baker = Rhopalocarpus alternifolius（Baker）Capuron ●☆

371187 Sphaerosepalum coriaceum Scott-Elliot = Rhopalocarpus alternifolius（Baker）Capuron ●☆

371188 Sphaerosepalum louvelii Danguy = Rhopalocarpus louvelii（Danguy）Capuron ●☆

371189 Sphaerosepalum madagascariense Danguy = Rhopalocarpus similis Hemsl. ●☆

371190 Sphaerosicyos Hook. f. = Lagenaria Ser. ■

371191 Sphaerosicyos meyeri Hook. f. = Lagenaria sphaerica（Sond.）Naudin ■☆

371192 Sphaerosicyos sphaericus（Sond.）Cogn. = Lagenaria sphaerica（Sond.）Naudin ■☆

371193 Sphaerosicyos sphaericus（Sond.）Hook. f. = Lagenaria sphaerica（Sond.）Naudin ■☆

371194 Sphaerosicyus Post et Kuntze = Lagenaria Ser. ■

371195 Sphaerospora Klatt = Acidanthera Hochst. ■

371196 Sphaerospora Klatt = Gladiolus L. ■

371197 Sphaerospora Sweet = Gladiolus L. ■

371198 Sphaerospora flexuosa（L. f.）Klatt = Tritoniopsis flexuosa（L. f.）G. J. Lewis ■☆

371199 Sphaerospora gigantea Klatt = Gladiolus murielae Kelway ■☆

371200 Sphaerostachys Miq. = Piper L. ●■

371201 Sphaerostema Blume = Schisandra Michx.（保留属名）●

371202 Sphaerostema elongatum Blume = Schisandra elongata Hook. f. et Thomson ●

371203 Sphaerostema propinquum Blume = Schisandra propinqua（Wall.）Baill. ●

371204 Sphaerostemma Rchb. = Schisandra Michx.（保留属名）●

371205 Sphaerostigma（Ser.）Fisch. et C. A. Mey. = Camissonia Link ■☆

371206 Sphaerostigma Fisch. et C. A. Mey. = Camissonia Link ■☆

371207 Sphaerostylis Baill.（1858）；球柱大戟属●☆

371208 Sphaerostylis anomala（Prain）Croizat = Tragiella anomala（Prain）Pax et K. Hoffm. ●☆

371209 Sphaerostylis natalensis（Sond.）Croizat = Tragiella natalensis（Sond.）Pax et K. Hoffm. ●☆

371210 Sphaerostylis tulasneana Baill.；球柱大戟●☆

371211 Sphaerotele C. Presl = Phycella Lindl. ■☆

371212 Sphaerotele C. Presl = Stenomesson Herb. ■☆

371213 Sphaerotele Link = Urceolina Rchb.（保留属名）■☆

371214 Sphaerothalamus Hook. f. = Polyalthia Blume ●

371215 Sphaerotheca Cham. et Schltdl. = Conobea Aubl. ■☆

371216 Sphaerothele Benth. et Hook. f. = Sphaerotele C. Presl ■☆

371217 Sphaerothele Benth. et Hook. f. = Stenomesson Herb. ■☆

371218 Sphaerothylax Bisch. = Sphaerothylax Bisch. ex Krauss ■☆

371219 Sphaerothylax Bisch. ex Krauss（1844）；球囊苔草属■☆

371220 Sphaerothylax abyssinica（Wedd.）Warm.；阿比西尼亚球囊苔草■☆

371221 Sphaerothylax heteromorpha Baill. = Macropodiella heteromorpha（Baill.）C. Cusset ■☆

371222 Sphaerothylax pusilla Warm. = Ledermanniella pusilla（Warm.）C. Cusset ■☆

371223 Sphaerothylax pygmaea Pellegr. = Ledermanniella pygmaea（Pellegr.）C. Cusset ■☆

371224 Sphaerothylax sanguinea Chiov. = Sphaerothylax abyssinica（Wedd.）Warm. ■☆

371225 Sphaerothylax sphaerocarpa（Engl.）G. Taylor；球囊苔草■☆

371226 Sphaerothylax warmingiana Gilg = Ledermanniella warmingiana（Gilg）C. Cusset ■☆

371227 Sphaerotorrhiza（O. E. Schulz）Khokhr. = Cardamine L. ■

371228 Sphaerotorrhiza trifida（Lam. ex Poir.）Khokhr. = Cardamine trifida（Lam. ex Poir.）B. M. G. Jones ■

371229 Sphaerotylos C. J. Chen = Sarcochlamys Gaudich. ●

371230 Sphaerotylos medogensis C. J. Chen = Sarcochlamys pulcherrima Gaudich. ●

371231 Sphaerula W. Anderson ex Hook. f. = Acaena L. ■●☆

371232 Sphagneticola O. Hoffm.（1900）；薜菊属■☆

371233 Sphagneticola trilobata（L.）Pruski；三裂薜菊（乳甲菊）；Bay Biscayne，Creeping-oxeye，Singapore Daisy，Wedelia ■☆

371234 Sphalanthus Jack = Quisqualis L. ●

371235 Sphalanthus confertus Jack = Quisqualis conferta（Jack）Exell ●

371236 Sphallerocarpus Bess. = Sphallerocarpus Besser ex DC. ■

371237 Sphallerocarpus Besser ex DC.（1830）；迷果芹属；Losefruit，Sphallerocarpus ■

371238 Sphallerocarpus coniifolius（Wall. ex DC.）Koso-Pol. = Vicatia coniifolia Wall. ex DC. ■

371239 Sphallerocarpus cyminum Besser ex DC. = Sphallerocarpus gracilis（Trevir.）Koso-Pol. ■

371240 Sphallerocarpus gracilis（Trevir.）Koso-Pol.；迷果芹（达扭，东北迷

果芹,小叶山红萝卜);Thin Losefruit,Thin Sphallerocarpus ■

371241   Sphallerocarpus longilobus Kar. et Kir. ;长裂片迷果芹■

371242   Sphallerocarpus millefolius (Klotzsch) Koso-Pol. = Vicatia coniifolia Wall. ex DC. ■

371243   Sphalmanthus N. E. Br. = Phyllobolus N. E. Br. ●☆

371244   Sphalmanthus abbreviatus (L. Bolus) L. Bolus = Phyllobolus abbreviatus (L. Bolus) Gerbaulet ●☆

371245   Sphalmanthus acocksii L. Bolus = Phyllobolus saturatus (L. Bolus) Gerbaulet ●☆

371246   Sphalmanthus acuminatus (Haw.) L. Bolus = Phyllobolus splendens (L.) Gerbaulet ●☆

371247   Sphalmanthus albertensis (L. Bolus) L. Bolus = Phyllobolus pumilus (L. Bolus) Gerbaulet ●☆

371248   Sphalmanthus albicaulis (Haw.) L. Bolus = Phyllobolus splendens (L.) Gerbaulet ●☆

371249   Sphalmanthus anguineus (L. Bolus) L. Bolus = Phyllobolus oculatus (N. E. Br.) Gerbaulet ●☆

371250   Sphalmanthus arenicola (L. Bolus) L. Bolus = Phyllobolus oculatus (N. E. Br.) Gerbaulet ●☆

371251   Sphalmanthus auratus (Sond.) L. Bolus = Phyllobolus nitidus (Haw.) Gerbaulet ●☆

371252   Sphalmanthus baylissii L. Bolus = Phyllobolus saturatus (L. Bolus) Gerbaulet ●☆

371253   Sphalmanthus bijliae (N. E. Br.) L. Bolus = Phyllobolus splendens (L.) Gerbaulet ●☆

371254   Sphalmanthus blandus (L. Bolus) L. Bolus = Phyllobolus splendens (L.) Gerbaulet ●☆

371255   Sphalmanthus brevisepalus (L. Bolus) L. Bolus = Phyllobolus spinuliferus (Haw.) Gerbaulet ●☆

371256   Sphalmanthus brevisepalus (L. Bolus) L. Bolus var. ferus ? = Phyllobolus spinuliferus (Haw.) Gerbaulet ●☆

371257   Sphalmanthus calycinus L. Bolus = Phyllobolus canaliculatus (Haw.) Bittrich ●☆

371258   Sphalmanthus canaliculatus (Haw.) N. E. Br. = Phyllobolus canaliculatus (Haw.) Bittrich ●☆

371259   Sphalmanthus carneus (Haw.) N. E. Br. = Phyllobolus spinuliferus (Haw.) Gerbaulet ●☆

371260   Sphalmanthus caudatus (L. Bolus) N. E. Br. = Phyllobolus caudatus (L. Bolus) Gerbaulet ●☆

371261   Sphalmanthus celans (L. Bolus) L. Bolus = Phyllobolus splendens (L.) Gerbaulet ●☆

371262   Sphalmanthus commutatus (A. Berger) N. E. Br. = Phyllobolus grossus (Aiton) Gerbaulet ■☆

371263   Sphalmanthus congestus (L. Bolus) L. Bolus = Phyllobolus congestus (L. Bolus) Gerbaulet ●☆

371264   Sphalmanthus constrictus (L. Bolus) L. Bolus = Phyllobolus splendens (L.) Gerbaulet ●☆

371265   Sphalmanthus crassus L. Bolus = Phyllobolus prasinus (L. Bolus) Gerbaulet ●☆

371266   Sphalmanthus deciduus (L. Bolus) L. Bolus = Phyllobolus deciduus (L. Bolus). Gerbaulet ●☆

371267   Sphalmanthus decurvatus (L. Bolus) L. Bolus = Phyllobolus decurvatus (L. Bolus) Gerbaulet ●☆

371268   Sphalmanthus decussatus (Thunb.) L. Bolus = Aptenia geniculiflora (L.) Bittrich ex Gerbaulet ■☆

371269   Sphalmanthus defoliatus (Haw.) L. Bolus = Aridaria noctiflora (L.) Schwantes subsp. defoliata (Haw.) Gerbaulet ●☆

371270   Sphalmanthus delus (L. Bolus) L. Bolus = Phyllobolus delus (L. Bolus) Gerbaulet ■☆

371271   Sphalmanthus dinteri (L. Bolus) L. Bolus = Phyllobolus melanospermus (Dinter et Schwantes) Gerbaulet ●☆

371272   Sphalmanthus dyeri (L. Bolus) L. Bolus = Phyllobolus splendens (L.) Gerbaulet ●☆

371273   Sphalmanthus englishiae (L. Bolus) L. Bolus = Prenia englishiae (L. Bolus) Gerbaulet ■☆

371274   Sphalmanthus flexuosus (Haw.) L. Bolus = Phyllobolus splendens (L.) Gerbaulet ●☆

371275   Sphalmanthus fourcadei (L. Bolus) L. Bolus = Phyllobolus splendens (L.) Gerbaulet ●☆

371276   Sphalmanthus fragilis N. E. Br. = Phyllobolus oculatus (N. E. Br.) Gerbaulet ●☆

371277   Sphalmanthus framesii (L. Bolus) L. Bolus = Phyllobolus spinuliferus (Haw.) Gerbaulet ●☆

371278   Sphalmanthus geniculiflorus (L.) L. Bolus = Aptenia geniculiflora (L.) Bittrich ex Gerbaulet ■☆

371279   Sphalmanthus glanduliferus (L. Bolus) L. Bolus = Phyllobolus sinuosus (L. Bolus) Gerbaulet ●☆

371280   Sphalmanthus godmaniae (L. Bolus) L. Bolus = Phyllobolus sinuosus (L. Bolus) Gerbaulet ●☆

371281   Sphalmanthus gratiae (L. Bolus) L. Bolus = Phyllobolus grossus (Aiton) Gerbaulet ■☆

371282   Sphalmanthus grossus (Aiton) N. E. Br. = Phyllobolus grossus (Aiton) Gerbaulet ■☆

371283   Sphalmanthus gydouwensis L. Bolus = Phyllobolus grossus (Aiton) Gerbaulet ■☆

371284   Sphalmanthus hallii L. Bolus = Phyllobolus delus (L. Bolus) Gerbaulet ■☆

371285   Sphalmanthus herbertii N. E. Br. = Phyllobolus herbertii (N. E. Br.) Gerbaulet ●☆

371286   Sphalmanthus herrei L. Bolus = Phyllobolus prasinus (L. Bolus) Gerbaulet ●☆

371287   Sphalmanthus humilis L. Bolus = Phyllobolus herbertii (N. E. Br.) Gerbaulet ●☆

371288   Sphalmanthus latipetalus (L. Bolus) L. Bolus = Phyllobolus latipetalus (L. Bolus) Gerbaulet ●☆

371289   Sphalmanthus laxipetalus (L. Bolus) L. Bolus = Phyllobolus grossus (Aiton) Gerbaulet ■☆

371290   Sphalmanthus laxus (L. Bolus) N. E. Br. = Phyllobolus decurvatus (L. Bolus) Gerbaulet ●☆

371291   Sphalmanthus leipoldtii L. Bolus = Phyllobolus grossus (Aiton) Gerbaulet ■☆

371292   Sphalmanthus leptopetalus (L. Bolus) L. Bolus = Phyllobolus splendens (L.) Gerbaulet ●☆

371293   Sphalmanthus lignescens L. Bolus = Phyllobolus lignescens (L. Bolus) Gerbaulet ●☆

371294   Sphalmanthus ligneus (L. Bolus) L. Bolus = Phyllobolus melanospermus (Dinter et Schwantes) Gerbaulet ●☆

371295   Sphalmanthus littlewoodii L. Bolus = Phyllobolus nitidus (Haw.) Gerbaulet ●☆

371296   Sphalmanthus longipapillatus L. Bolus = Phyllobolus oculatus (N. E. Br.) Gerbaulet ●☆

371297   Sphalmanthus longispinulus (Haw.) N. E. Br. = Phyllobolus grossus (Aiton) Gerbaulet ■☆

371298   Sphalmanthus longitubus (L. Bolus) L. Bolus = Phyllobolus tenuiflorus (Jacq.) Gerbaulet ●☆

371299 Sphalmanthus macrosiphon (L. Bolus) L. Bolus = Phyllobolus tenuiflorus (Jacq.) Gerbaulet ●☆

371300 Sphalmanthus melanospermus Dinter et Schwantes = Phyllobolus melanospermus (Dinter et Schwantes) Gerbaulet ●☆

371301 Sphalmanthus micans L. Bolus = Phyllobolus resurgens (Kensit) Schwantes ●☆

371302 Sphalmanthus nanus L. Bolus = Phyllobolus tenuiflorus (Jacq.) Gerbaulet ●☆

371303 Sphalmanthus nitidus (Haw.) L. Bolus = Phyllobolus nitidus (Haw.) Gerbaulet ●☆

371304 Sphalmanthus nothus (N. E. Br.) Schwantes = Phyllobolus splendens (L.) Gerbaulet ●☆

371305 Sphalmanthus obtusus (L. Bolus) L. Bolus = Phyllobolus decurvatus (L. Bolus) Gerbaulet ●☆

371306 Sphalmanthus oculatus (N. E. Br.) N. E. Br. = Phyllobolus oculatus (N. E. Br.) Gerbaulet ●☆

371307 Sphalmanthus olivaceus (Schltr.) L. Bolus = Phyllobolus tenuiflorus (Jacq.) Gerbaulet ●☆

371308 Sphalmanthus oubergensis (L. Bolus) L. Bolus = Phyllobolus pumilus (L. Bolus) Gerbaulet ●☆

371309 Sphalmanthus pentagonus (L. Bolus) L. Bolus = Phyllobolus splendens (L.) Gerbaulet subsp. pentagonus (L. Bolus) Gerbaulet ●☆

371310 Sphalmanthus pentagonus (L. Bolus) L. Bolus var. occidentalis ? = Phyllobolus splendens (L.) Gerbaulet subsp. pentagonus (L. Bolus) Gerbaulet ●☆

371311 Sphalmanthus platysepalus (L. Bolus) L. Bolus = Phyllobolus grossus (Aiton) Gerbaulet ■☆

371312 Sphalmanthus plenifolius (N. E. Br.) L. Bolus = Phyllobolus splendens (L.) Gerbaulet ●☆

371313 Sphalmanthus pomonae (L. Bolus) L. Bolus = Phyllobolus oculatus (N. E. Br.) Gerbaulet ●☆

371314 Sphalmanthus praecox L. Bolus = Phyllobolus sinuosus (L. Bolus) Gerbaulet ●☆

371315 Sphalmanthus prasinus (L. Bolus) L. Bolus = Phyllobolus prasinus (L. Bolus) Gerbaulet ●☆

371316 Sphalmanthus primulinus (L. Bolus) L. Bolus = Phyllobolus splendens (L.) Gerbaulet ●☆

371317 Sphalmanthus pumulis (L. Bolus) L. Bolus = Phyllobolus pumilus (L. Bolus) Gerbaulet ●☆

371318 Sphalmanthus quarternus (L. Bolus) L. Bolus = Phyllobolus spinuliferus (Haw.) Gerbaulet ●☆

371319 Sphalmanthus quarziticus (L. Bolus) L. Bolus = Phyllobolus quartziticus (L. Bolus) Gerbaulet ●☆

371320 Sphalmanthus rabiei (L. Bolus) N. E. Br. = Phyllobolus rabiei (L. Bolus) Gerbaulet ●☆

371321 Sphalmanthus rabiesbergensis (L. Bolus) L. Bolus = Phyllobolus splendens (L.) Gerbaulet ●☆

371322 Sphalmanthus radicans (L. Bolus) L. Bolus = Prenia radicans (L. Bolus) Gerbaulet ●☆

371323 Sphalmanthus recurvus (L. Bolus) L. Bolus = Phyllobolus sinuosus (L. Bolus) Gerbaulet ●☆

371324 Sphalmanthus reflexus (Haw.) L. Bolus = Phyllobolus splendens (L.) Gerbaulet ●☆

371325 Sphalmanthus rejuvenalis L. Bolus = Phyllobolus spinuliferus (Haw.) Gerbaulet ●☆

371326 Sphalmanthus resurgens (Kensit) L. Bolus = Phyllobolus resurgens (Kensit) Schwantes ●☆

371327 Sphalmanthus rhodandrus (L. Bolus) L. Bolus = Phyllobolus nitidus (Haw.) Gerbaulet ●☆

371328 Sphalmanthus roseus (L. Bolus) L. Bolus = Phyllobolus splendens (L.) Gerbaulet ●☆

371329 Sphalmanthus salmoneus (Haw.) N. E. Br. = Phyllobolus canaliculatus (Haw.) Bittrich ●☆

371330 Sphalmanthus saturatus (L. Bolus) L. Bolus = Phyllobolus saturatus (L. Bolus) Gerbaulet ●☆

371331 Sphalmanthus scintillans (Dinter) Dinter et Schwantes = Phyllobolus oculatus (N. E. Br.) Gerbaulet ●☆

371332 Sphalmanthus sinuosus (L. Bolus) L. Bolus = Phyllobolus sinuosus (L. Bolus) Gerbaulet ●☆

371333 Sphalmanthus spinuliferus (Haw.) L. Bolus = Phyllobolus spinuliferus (Haw.) Gerbaulet ●☆

371334 Sphalmanthus splendens (L.) L. Bolus = Phyllobolus splendens (L.) Gerbaulet ●☆

371335 Sphalmanthus stayneri L. Bolus = Phyllobolus delus (L. Bolus) Gerbaulet ■☆

371336 Sphalmanthus straminicolor (L. Bolus) L. Bolus = Phyllobolus sinuosus (L. Bolus) Gerbaulet ●☆

371337 Sphalmanthus striatus (L. Bolus) L. Bolus = Phyllobolus splendens (L.) Gerbaulet ●☆

371338 Sphalmanthus strictus (L. Bolus) L. Bolus = Phyllobolus spinuliferus (Haw.) Gerbaulet ●☆

371339 Sphalmanthus suaveolens (L. Bolus) H. Jacobsen = Phyllobolus lignescens (L. Bolus) Gerbaulet ●☆

371340 Sphalmanthus subaequans (L. Bolus) L. Bolus = Phyllobolus splendens (L.) Gerbaulet subsp. pentagonus (L. Bolus) Gerbaulet ●☆

371341 Sphalmanthus subpatens (L. Bolus) L. Bolus = Phyllobolus splendens (L.) Gerbaulet subsp. pentagonus (L. Bolus) Gerbaulet ●☆

371342 Sphalmanthus subpetiolatus (L. Bolus) L. Bolus = Phyllobolus grossus (Aiton) Gerbaulet ■☆

371343 Sphalmanthus suffusus (L. Bolus) L. Bolus = Prenia tetragona (Thunb.) Gerbaulet ■☆

371344 Sphalmanthus sulcatus (Haw.) L. Bolus = Phyllobolus splendens (L.) Gerbaulet ●☆

371345 Sphalmanthus tenuiflorus (Jacq.) N. E. Br. = Phyllobolus tenuiflorus (Jacq.) Gerbaulet ●☆

371346 Sphalmanthus tetragonus (Thunb.) L. Bolus = Prenia tetragona (Thunb.) Gerbaulet ■☆

371347 Sphalmanthus tetramerus (L. Bolus) L. Bolus var. parviflorus ? = Phyllobolus trichotomus (Thunb.) Gerbaulet ●☆

371348 Sphalmanthus tetramerus (L. Bolus) L. Bolus var. tetramerus ? = Phyllobolus trichotomus (Thunb.) Gerbaulet ●☆

371349 Sphalmanthus trichotomus (Thunb.) L. Bolus = Phyllobolus trichotomus (Thunb.) Gerbaulet ●☆

371350 Sphalmanthus umbelliflorus (Jacq.) L. Bolus = Phyllobolus splendens (L.) Gerbaulet ●☆

371351 Sphalmanthus vanheerdei L. Bolus = Phyllobolus roseus (L. Bolus) Gerbaulet ●☆

371352 Sphalmanthus varians (L. Bolus) L. Bolus = Phyllobolus oculatus (N. E. Br.) Gerbaulet ●☆

371353 Sphalmanthus vernalis (L. Bolus) L. Bolus = Phyllobolus splendens（L.）Gerbaulet ●☆

371354 Sphalmanthus vigilans （ L. Bolus ） L. Bolus = Mesembryanthemum longistylum DC. ■☆

371355 Sphalmanthus viridiflorus（Aiton）N. E. Br. = Phyllobolus viridiflorus（Aiton）Gerbaulet ●☆

371356 Sphalmanthus watermeyeri（L. Bolus）L. Bolus = Phyllobolus spinuliferus（Haw.）Gerbaulet ●☆

371357 Sphalmanthus willowmorensis（L. Bolus）L. Bolus = Phyllobolus grossus（Aiton）Gerbaulet ■☆

371358 Sphalmium B. G. Briggs，B. Hyland et L. A. S. Johnson（1975）；澳龙眼属●☆

371359 Sphalmium racemosum（C. T. White）B. G. Briggs，B. Hyland et L. A. S. Johnson；澳龙眼●☆

371360 Sphanellolepis Cogn. = Sphanellopsis Steud. ex Naudin ●☆

371361 Sphanellopsis Steud. ex Naudin = Adelobotrys DC. ●☆

371362 Spharganophorus Boehm. = Sparganophorus Boehm. ■☆

371363 Spharganophorus Boehm. = Struchium P. Browne ■☆

371364 Sphedamnocarpus Planch. ex Benth.（1862）；楔果金虎尾属 ●☆

371365 Sphedamnocarpus Planch. ex Benth. et Hook. f. = Sphedamnocarpus Planch. ex Benth. ●☆

371366 Sphedamnocarpus angolensis（A. Juss.）Planch. ex Oliv.；安哥拉楔果金虎尾●☆

371367 Sphedamnocarpus angolensis（A. Juss.）Planch. ex Oliv. var. pulcherrimus（Engl. et Gilg）Nied. = Sphedamnocarpus pruriens（A. Juss.）Szyszyl. ●☆

371368 Sphedamnocarpus barbosae Launert；非洲楔果金虎尾●☆

371369 Sphedamnocarpus galphimiifolius（A. Juss.）Szyszyl.；金英叶楔果金虎尾●☆

371370 Sphedamnocarpus galphimiifolius（A. Juss.）Szyszyl. subsp. rehmannii（Szyszyl.）Launert = Sphedamnocarpus pruriens（A. Juss.）Szyszyl. subsp. galphimiifolius（A. Juss.）P. D. de Villiers et D. J. Botha ●☆

371371 Sphedamnocarpus latifolius（Engl.）Nied. = Sphedamnocarpus pruriens（A. Juss.）Szyszyl. ●☆

371372 Sphedamnocarpus pruriens（A. Juss.）Szyszyl.；刺痒楔果金虎尾●☆

371373 Sphedamnocarpus pruriens （ A. Juss. ） Szyszyl. f. brevipedunculatus Nied. = Sphedamnocarpus pruriens（A. Juss.）Szyszyl. ●☆

371374 Sphedamnocarpus pruriens （ A. Juss. ） Szyszyl. f. wilmsii（Engl.）Nied. = Sphedamnocarpus pruriens（A. Juss.）Szyszyl. ●☆

371375 Sphedamnocarpus pruriens （ A. Juss. ） Szyszyl. subsp. galphimiifolius （ A. Juss. ） P. D. de Villiers et D. J. Botha = Sphedamnocarpus galphimiifolius（A. Juss.）Szyszyl. ●☆

371376 Sphedamnocarpus pruriens（A. Juss.）Szyszyl. var. lanceolatus Launert；剑叶刺痒楔果金虎尾●☆

371377 Sphedamnocarpus pruriens（A. Juss.）Szyszyl. var. latifolius Engl. = Sphedamnocarpus latifolius（Engl.）Nied. ●☆

371378 Sphedamnocarpus pruriens（A. Juss.）Szyszyl. var. platypterus Arènes = Sphedamnocarpus pruriens（A. Juss.）Szyszyl. ●☆

371379 Sphedamnocarpus pulcherrimus Engl. et Gilg = Sphedamnocarpus pruriens（A. Juss.）Szyszyl. ●☆

371380 Sphedamnocarpus rehmannii Szyszyl. = Sphedamnocarpus pruriens（A. Juss.）Szyszyl. subsp. galphimiifolius（A. Juss.）P. D. de Villiers et D. J. Botha ●☆

371381 Sphedamnocarpus rogersii Burtt Davy = Sphedamnocarpus pruriens（A. Juss.）Szyszyl. subsp. galphimiifolius（A. Juss.）P. D. de Villiers et D. J. Botha ●☆

371382 Sphedamnocarpus transvalicus （ Kuntze ） Burtt Davy = Sphedamnocarpus pruriens（A. Juss.）Szyszyl. subsp. galphimiifolius （A. Juss.）P. D. de Villiers et D. J. Botha ●☆

371383 Sphedamnocarpus wilmsii Engl. = Sphedamnocarpus pruriens（A. Juss.）Szyszyl. ●☆

371384 Sphedamnocarpus woodianus Arènes = Sphedamnocarpus pruriens（A. Juss.）Szyszyl. subsp. galphimiifolius（A. Juss.）P. D. de Villiers et D. J. Botha ●☆

371385 Sphenandra Benth. = Sutera Roth ■●☆

371386 Sphenandra cinerea Engl. = Sutera halimifolia（Benth.）Kuntze ■☆

371387 Sphenandra coerulea（L. f.）Kuntze = Sutera caerulea（L. f.）Hiern ■☆

371388 Sphenandra viscosa（Aiton）Benth. = Sutera caerulea（L. f.）Hiern ■☆

371389 Sphenantha Schrad. = Cucurbita L. ■

371390 Sphenanthera Hassk.（1856）；楔药秋海棠属■☆

371391 Sphenanthera Hassk. = Begonia L. ●■

371392 Sphenanthera multangula Klotzsch = Begonia multangula Blume ■☆

371393 Sphenanthera robusta Hassk. = Begonia robusta Blume ■☆

371394 Sphendamnocarpus Baker = Sphedamnocarpus Planch. ex Benth. ●☆

371395 Spheneria Kuhlm.（1922）；假颖草属■☆

371396 Spheneria setifolia Kuhlm.；假颖草■☆

371397 Sphenista Raf. = Cosmibuena Ruiz et Pav.（保留属名）●☆

371398 Sphenista Raf. = Hirtella L. ●☆

371399 Sphenocarpus Korovin = Seseli L. ■

371400 Sphenocarpus Korovin（1947）；楔果芹属■☆

371401 Sphenocarpus Rich. = Laguncularia C. F. Gaertn. ●☆

371402 Sphenocarpus Wall. = Magnolia L. ●

371403 Sphenocarpus eryngioides Korovin；楔果芹■☆

371404 Sphenocentrum Pierre（1898）；楔心藤属●☆

371405 Sphenocentrum jollyanum Pierre；楔心藤●☆

371406 Sphenoclea Gaertn.（1788）（保留属名）；楔瓣花属（尖瓣花属，密穗桔梗属）；Sphenoclea ■

371407 Sphenoclea dalzielii N. E. Br.；戴尔楔瓣花■☆

371408 Sphenoclea pongatia DC. = Sphenoclea zeylanica Gaertn. ■

371409 Sphenoclea zeylanica Gaertn.；楔瓣花（尖瓣花，蜜穗桔梗）；Celon Sphenoclea，Chicken Spike，Chickenspike，Gooseweed ■

371410 Sphenocleaceae（Lindl.）Mart. ex DC. = Sphenocleaceae T. Baskerv.（保留科名）■

371411 Sphenocleaceae Lindl = Sphenocleaceae T. Baskerv.（保留科名）■

371412 Sphenocleaceae Mart. ex DC. = Sphenocleaceae T. Baskerv.（保留科名）■

371413 Sphenocleaceae T. Baskerv.（1839）（保留科名）；楔瓣花科（尖瓣花科，蜜穗桔梗科）■

371414 Sphenocleaceae T. Baskerv.（保留科名）= Campanulaceae Juss.（保留科名）■●

371415 Sphenodesma Griff. = Sphenodesme Jack ●

371416 Sphenodesme Jack（1820）；楔翅藤属；Sphenodesma ●

371417 Sphenodesme annamitica Dop = Sphenodesme mollis Craib ●

371418 Sphenodesme floribunda Chun et F. C. How；多花楔翅藤；

Flowery Sphenodesma, Manyflower Sphenodesma ●

371419 Sphenodesme involucrata (C. Presl) B. L. Rob. ;爪楔翅藤(楔翅藤);Involucrate Sphenodesma ●

371420 Sphenodesme mollis Craib;毛楔翅藤;Hairy Sphenodesma ●

371421 Sphenodesme pentandra Jack;楔翅藤 (五蕊楔翅藤); Fivestamen Sphenodesma,Sphenodesma ●☆

371422 Sphenodesme pentandra Jack var. wallichiana (Schauer) Munir; 山白藤(楔起翅);Pentandrous Sphenodesma, Wallich Fivestamen Sphenodesma ●

371423 Sphenodesme unguiculatum Kurz. = Sphenodesme involucrata (C. Presl) B. L. Rob. ●

371424 Sphenodesme wallichiana Schauer = Sphenodesme pentandra Jack var. wallichiana (Schauer) Munir ●

371425 Sphenogyne R. Br. = Ursinia Gaertn. (保留属名)●■☆

371426 Sphenogyne abrotanifolia R. Br. = Ursinia abrotanifolia (R. Br.) Spreng. ■☆

371427 Sphenogyne adonidifolia DC. = Ursinia anthemoides (L.) Poir. ■☆

371428 Sphenogyne anethifolia Less. = Ursinia paleacea (L.) Moench ●☆

371429 Sphenogyne anethoides DC. = Ursinia anethoides (DC.) N. E. Br. ■☆

371430 Sphenogyne anethoides DC. var. brachyglossa ? = Ursinia punctata (Thunb.) N. E. Br. ●☆

371431 Sphenogyne anethoides DC. var. ramossima ? = Ursinia punctata (Thunb.) N. E. Br. ●☆

371432 Sphenogyne anthemoides (L.) R. Br. = Ursinia anthemoides (L.) Poir. ■☆

371433 Sphenogyne anthemoides (L.) R. Br. var. versicolor (DC.) Harv. = Ursinia anthemoides (L.) Poir. subsp. versicolor (DC.) Prassler ■☆

371434 Sphenogyne brachyloba Kunze = Ursinia brachyloba (Kunze) N. E. Br. ■☆

371435 Sphenogyne brachypoda Harv. = Ursinia hispida (DC.) N. E. Br. ●☆

371436 Sphenogyne brevifolia DC. = Ursinia discolor (Less.) N. E. Br. ■☆

371437 Sphenogyne chamomillifolia DC. = Ursinia nudicaulis (Thunb.) N. E. Br. ●☆

371438 Sphenogyne chamomillifolia DC. var. elongata Harv. = Ursinia nudicaulis (Thunb.) N. E. Br. ●☆

371439 Sphenogyne chrysanthemoides Less. = Ursinia chrysanthemoides (Less.) Harv. ■☆

371440 Sphenogyne ciliaris DC. = Ursinia tenuifolia (L.) Poir. subsp. ciliaris (DC.) Prassler ●☆

371441 Sphenogyne concolor Harv. = Ursinia punctata (Thunb.) N. E. Br. ●☆

371442 Sphenogyne coronopifolia Less. = Ursinia coronopifolia (Less.) N. E. Br. ■☆

371443 Sphenogyne crithmifolia R. Br. = Ursinia paleacea (L.) Moench ●☆

371444 Sphenogyne crithmifolia R. Br. var. grandiflora Harv. = Ursinia paleacea (L.) Moench ●☆

371445 Sphenogyne crithmifolia R. Br. var. trifurcata DC. = Ursinia paleacea (L.) Moench ●☆

371446 Sphenogyne dentata (L.) R. Br. = Ursinia dentata (L.) Poir. ■☆

371447 Sphenogyne dentata (L.) R. Br. var. setigera (DC.) Harv. = Ursinia dentata (L.) Poir. ■☆

371448 Sphenogyne discolor Less. = Ursinia discolor (Less.) N. E. Br. ■☆

371449 Sphenogyne dregeana DC. = Ursinia dregeana (DC.) N. E. Br. ●☆

371450 Sphenogyne eckloniana Sond. = Ursinia eckloniana (Sond.) N. E. Br. ■☆

371451 Sphenogyne filipes E. Mey. ex DC. = Ursinia filipes (E. Mey. ex DC.) N. E. Br. ●☆

371452 Sphenogyne foeniculacea (Jacq.) Less. = Ursinia anthemoides (L.) Poir. ■☆

371453 Sphenogyne gracilis DC. = Ursinia punctata (Thunb.) N. E. Br. ●☆

371454 Sphenogyne grandiflora DC. = Ursinia paleacea (L.) Moench ●☆

371455 Sphenogyne heterodonta DC. = Ursinia heterodonta (DC.) N. E. Br. ●☆

371456 Sphenogyne hispida DC. = Ursinia hispida (DC.) N. E. Br. ●☆

371457 Sphenogyne incisa DC. = Ursinia serrata (L. f.) Poir. ●☆

371458 Sphenogyne leptoglossa DC. = Ursinia trifida (Thunb.) N. E. Br. ■☆

371459 Sphenogyne macropoda DC. = Ursinia macropoda (DC.) N. E. Br. ●☆

371460 Sphenogyne microcephala DC. = Ursinia anthemoides (L.) Poir. ■☆

371461 Sphenogyne natalensis Sch. Bip. = Ursinia tenuiloba DC. ■☆

371462 Sphenogyne nudicaulis (Thunb.) Less. = Ursinia nudicaulis (Thunb.) N. E. Br. ●☆

371463 Sphenogyne nudicaulis (Thunb.) Less. var. alpina Harv. = Ursinia nudicaulis (Thunb.) N. E. Br. ●☆

371464 Sphenogyne nudicaulis (Thunb.) Less. var. gracilior Harv. = Ursinia nudicaulis (Thunb.) N. E. Br. ●☆

371465 Sphenogyne paleacea (Thunb.) Less. = Ursinia subflosculosa (DC.) Prassler ●☆

371466 Sphenogyne pallida DC. = Ursinia nana DC. ■☆

371467 Sphenogyne pallida DC. var. immarginata ? = Ursinia nana DC. ■☆

371468 Sphenogyne pauciloba DC. = Ursinia punctata (Thunb.) N. E. Br. ●☆

371469 Sphenogyne pilifera (P. J. Bergius) DC. = Ursinia pilifera (P. J. Bergius) Poir. ■☆

371470 Sphenogyne pusilla DC. = Ursinia anthemoides (L.) Poir. ■☆

371471 Sphenogyne quinquepartita DC. = Ursinia quinquepartita (DC.) N. E. Br. ●☆

371472 Sphenogyne rigidula DC. = Ursinia rigidula (DC.) N. E. Br. ●☆

371473 Sphenogyne scapiformis DC. = Ursinia nudicaulis (Thunb.) N. E. Br. ●☆

371474 Sphenogyne scariosa (Aiton) R. Br. = Ursinia scariosa (Aiton) Poir. ●☆

371475 Sphenogyne sericea (Thunb.) Less. = Ursinia sericea (Thunb.) N. E. Br. ●☆

371476 Sphenogyne serrata (L. f.) DC. = Ursinia serrata (L. f.) Poir. ●☆

371477 Sphenogyne setigera DC. = Ursinia dentata (L.) Poir. ■☆

371478 Sphenogyne speciosa Knowles et Westc. = Ursinia anthemoides

(L.) Poir. ■☆

371479　Sphenogyne subflosculosa DC. = Ursinia subflosculosa (DC.) Prassler ●☆

371480　Sphenogyne subhirsuta DC. = Ursinia scariosa (Aiton) Poir. subsp. subhirsuta (DC.) Prassler ●☆

371481　Sphenogyne tenuifolia (L.) DC. = Ursinia tenuifolia (L.) Poir. ●☆

371482　Sphenogyne tenuifolia (L.) DC. var. heterochroma Harv. = Ursinia tenuifolia (L.) Poir. ●☆

371483　Sphenogyne trifida (Thunb.) Less. = Ursinia trifida (Thunb.) N. E. Br. ■☆

371484　Sphenogyne trifurca Harv. = Ursinia tenuifolia (L.) Poir. subsp. ciliaris (DC.) Prassler ●☆

371485　Sphenogyne triloba DC. = Ursinia trifida (Thunb.) N. E. Br. ■☆

371486　Sphenogyne tripartita DC. = Ursinia quinquepartita (DC.) N. E. Br. ●☆

371487　Sphenogyne versicolor DC. = Ursinia anthemoides (L.) Poir. subsp. versicolor (DC.) Prassler ■☆

371488　Sphenopholis Scribn. (1906);革颖草属(楔鳞茅属);Wedge Grass ■☆

371489　Sphenopholis intermedia (Rydb.) Rydb.;纤细革颖草;Slender Wedge Grass ■☆

371490　Sphenopholis intermedia (Rydb.) Rydb. var. pilosa Dore = Sphenopholis intermedia (Rydb.) Rydb. ■☆

371491　Sphenopholis intermedia Rydb. = Sphenopholis obtusata (Michx.) Scribn. ■☆

371492　Sphenopholis nitida (Biehler) Scribn.;光亮革颖草;Shining Wedge Grass ■☆

371493　Sphenopholis obtusata (Michx.) Scribn.;草原革颖草;Obtuse Sphenopholis, Prairie Wedge Grass, Prairie Wedge Scale, Prairie Wedgescale, Wedge Grass ■☆

371494　Sphenopholis obtusata (Michx.) Scribn. var. lobata (Trin.) Scribn. = Sphenopholis obtusata (Michx.) Scribn. ■☆

371495　Sphenopholis obtusata (Michx.) Scribn. var. major (Torr.) Erdman = Sphenopholis intermedia (Rydb.) Rydb. ■☆

371496　Sphenopholis obtusata (Michx.) Scribn. var. pubescens (Scribn. et Merr.) Scribn. = Sphenopholis obtusata (Michx.) Scribn. ■☆

371497　Sphenopholis pallens Scribn.;苍白革颖草;Pale Sphenopholis ■☆

371498　Sphenopholis pensylvanica (L.) Hitchc.;沼泽革颖草;Swamp Oat, Swamp Oats ■☆

371499　Sphenopus Trin. (1820);楔梗禾属■☆

371500　Sphenopus divaricatus (Gouan) Rchb. subsp. syrticus Murb. = Sphenopus ehrenbergii Hausskn. ■☆

371501　Sphenopus divaricatus (Gouan) Rchb. var. ehrenbergii (Hausskn.) Bég. et Vacc. = Sphenopus ehrenbergii Hausskn. ■☆

371502　Sphenopus divaricatus (Gouan.) Rchb.;叉枝楔梗禾■☆

371503　Sphenopus ehrenbergii Hausskn.;楔梗禾■☆

371504　Sphenopus gouani Trin. = Sphenopus divaricatus (Gouan) Rchb. ■☆

371505　Sphenopus syrticus (Murb.) Batt. et Trab. = Sphenopus ehrenbergii Hausskn. ■☆

371506　Sphenosciadium A. Gray(1865);楔伞芹属■☆

371507　Sphenosciadium capitellatum A. Gray;楔伞芹■☆

371508　Sphenostemon Baill. (1875);楔药花属●☆

371509　Sphenostemon balansae Baill.;楔药花●☆

371510　Sphenostemonaceae P. Royen et Airy Shaw = Aquifoliaceae Bercht. et J. Presl(保留科名)●

371511　Sphenostemonaceae P. Royen et Airy Shaw(1972);楔药花科 ●☆

371512　Sphenostigma Baker = Gelasine Herb. ■☆

371513　Sphenostigma Baker(1877);楔点鸢尾属■☆

371514　Sphenostigma boliviense Baker;玻利维亚楔点鸢尾■☆

371515　Sphenostigma caeruleum Klatt;蓝楔点鸢尾■☆

371516　Sphenostigma coelestina (W. Bartram) R. C. Foster = Calydorea coelestina (W. Bartram) Goldblatt et Henrich ■☆

371517　Sphenostigma gracile Benth. et Hook. f.;细楔点鸢尾■☆

371518　Sphenostigma mexicanum R. C. Foster;墨西哥楔点鸢尾■☆

371519　Sphenostylis E. Mey. (1836);楔柱豆属■☆

371520　Sphenostylis angustifolia Sond.;窄叶楔柱豆■☆

371521　Sphenostylis briartii (De Wild.) Baker f.;布里亚特楔柱豆 ■☆

371522　Sphenostylis calantha Harms = Nesphostylis holosericea (Baker) Verdc. ■☆

371523　Sphenostylis congensis A. Chev. = Sphenostylis stenocarpa (Hochst. ex A. Rich.) Harms ■☆

371524　Sphenostylis erecta (Baker f.) Hutch. ex Baker f.;直立楔柱豆 ■☆

371525　Sphenostylis erecta (Baker f.) Hutch. ex Baker f. subsp. obtusifolia (Harms) Potter et Doyle;钝叶直立楔柱豆■☆

371526　Sphenostylis gossweileri Baker f. = Sphenostylis erecta (Baker f.) Hutch. ex Baker f. subsp. obtusifolia (Harms) Potter et Doyle ■☆

371527　Sphenostylis holosericea (Baker) Harms = Nesphostylis holosericea (Baker) Verdc. ■☆

371528　Sphenostylis homblei De Wild. = Sphenostylis erecta (Baker f.) Hutch. ex Baker f. ■☆

371529　Sphenostylis katangensis (De Wild.) Harms = Sphenostylis stenocarpa (Hochst. ex A. Rich.) Harms ■☆

371530　Sphenostylis kerstingii Harms = Nesphostylis holosericea (Baker) Verdc. ■☆

371531　Sphenostylis marginata E. Mey.;具边楔柱豆■☆

371532　Sphenostylis marginata E. Mey. subsp. erecta (Baker f.) Verdc. = Sphenostylis erecta (Baker f.) Hutch. ex Baker f. ■☆

371533　Sphenostylis marginata E. Mey. subsp. obtusifolia (Harms) Verdc. = Sphenostylis erecta (Baker f.) Hutch. ex Baker f. subsp. obtusifolia (Harms) Potter et Doyle ■☆

371534　Sphenostylis obtusifolia Harms;钝叶楔柱豆■☆

371535　Sphenostylis obtusifolia Harms = Sphenostylis erecta (Baker f.) Hutch. ex Baker f. subsp. obtusifolia (Harms) Potter et Doyle ■☆

371536　Sphenostylis ornata A. Chev. = Sphenostylis stenocarpa (Hochst. ex A. Rich.) Harms ■☆

371537　Sphenostylis ringoetii De Wild. = Sphenostylis erecta (Baker f.) Hutch. ex Baker f. ■☆

371538　Sphenostylis schweinfurthii Harms;施韦楔柱豆■☆

371539　Sphenostylis schweinfurthii Harms subsp. benguellensis Torre;本格拉楔柱豆■☆

371540　Sphenostylis stenocarpa (Hochst. ex A. Rich.) Harms;狭果楔柱豆;Girigiri, Yam Bean ■☆

371541　Sphenostylis stenocarpa Harms = Sphenostylis stenocarpa (Hochst. ex A. Rich.) Harms ■☆

371542　Sphenostylis wildemaniana Baker f. = Sphenostylis briartii (De

Wild.）Baker f.■☆

371543　Sphenostylis zimbabweensis Mithen;津巴布韦楔柱豆■☆

371544　Sphenotoma（R. Br.）Sweet = Sphenotoma R. Br. ex Sweet ●☆

371545　Sphenotoma R. Br. = Sphenotoma R. Br. ex Sweet ●☆

371546　Sphenotoma R. Br. ex Sweet（1828）;报春石南属●☆

371547　Sphenotoma Sweet = Dracophyllum Labill. ●☆

371548　Sphenotoma Sweet = Sphenotoma R. Br. ex Sweet ●☆

371549　Sphenotoma capitata Lindl.;头状报春石南●☆

371550　Sphenotoma gracilis Sweet;细报春石南●☆

371551　Sphenotoma parviflora（Benth.）F. Muell.;小花报春石南●☆

371552　Spheranthus Hill = Sphaeranthus L. ■

371553　Sphinctacanthus Benth.（1876）;韧喉花属（断穗爵床属,小苞爵床属）;Sphinctacanthus ■☆

371554　Sphinctacanthus griffithii Benth.;韧喉花（小苞爵床）;Sphinctacanthus ■☆

371555　Sphinctacanthus siamensis C. B. Clarke ex Hosseus;缅甸韧喉花■☆

371556　Sphinctanthus Benth.（1841）;束花茜属●☆

371557　Sphinctanthus rupestris Benth.;束花茜●☆

371558　Sphincterostigma Schott = Philodendron Schott（保留属名）■●

371559　Sphincterostigma Schott ex B. D. Jacks. = Philodendron Schott（保留属名）■●

371560　Sphincterostoma Stschegl. = Andersonia Buch. -Ham. ex Wall. ●☆

371561　Sphinctolobium Vogel = Lonchocarpus Kunth（保留属名）●■☆

371562　Sphinctospermum Rose = Tephrosia Pers.（保留属名）●■

371563　Sphinctospermum Rose（1906）;沙漏灰毛豆属■☆

371564　Sphinctospermum constrictum（S. Watson）Rose;沙漏灰毛豆■☆

371565　Sphinga Barneby et J. W. Grimes = Acacia Mill.（保留属名）●■

371566　Sphinga Barneby et J. W. Grimes（1996）;束豆属●☆

371567　Sphingiphila A. H. Gentry（1990）;束紫葳属●☆

371568　Sphingiphila tetramera A. H. Gentry;束紫葳●☆

371569　Sphingium E. Mey. = Melolobium Eckl. et Zeyh. ■☆

371570　Sphingium canaliculatum E. Mey. = Melolobium stipulatum（Thunb.）Harv. ■☆

371571　Sphingium canescens E. Mey. = Melolobium canescens（E. Mey.）Benth. ■☆

371572　Sphingium decumbens E. Mey. = Melolobium microphyllum（L. f.）Eckl. et Zeyh. ■☆

371573　Sphingium spicatum E. Mey. = Melolobium aethiopicum（L.）Druce ■☆

371574　Sphingium spicatum E. Mey. var. hirsutiusculum ? = Melolobium humile Eckl. et Zeyh. ■☆

371575　Sphingium spicatum E. Mey. var. orthotrichum ? = Melolobium humile Eckl. et Zeyh. ■☆

371576　Sphingium velutinum E. Mey. = Melolobium candicans（E. Mey.）Eckl. et Zeyh. ■☆

371577　Sphingium viscidulum E. Mey. = Melolobium adenodes Eckl. et Zeyh. ■☆

371578　Sphondylantha Endl. = Spondylantha C. Presl ●

371579　Sphondylantha Endl. = Vitis L. ●

371580　Sphondylastrum Rchb. = Myriophyllum L. ■

371581　Sphondylium Adans. = Sphondylium Mill. ■

371582　Sphondylium Mill. = Heracleum L. ■

371583　Sphondylococca Willd. = Bergia L. ●■

371584　Sphondylococca Willd. ex Schult. = Bergia L. ●■

371585　Sphondylococcum Schauer = Callicarpa L. ●

371586　Sphragidia Thwaites = Drypetes Vahl ●

371587　Sphyranthera Hook. f.（1887）;槌药大戟属●☆

371588　Sphyranthera capitellata Hook. f.;槌药大戟●☆

371589　Sphyrarhynchus Mansf.（1935）;锤喙兰属■☆

371590　Sphyrarhynchus schliebenii Mansf.;锤喙兰■☆

371591　Sphyrastylis Schltr.（1920）;槌柱兰属■☆

371592　Sphyrastylis hoppii Schltr.;槌柱兰■☆

371593　Sphyrospermum Poepp. et Endl.（1835）;槌籽莓属（提灯莓属）●☆

371594　Sphyrospermum buxifolium Poepp. et Endl.;槌籽莓●☆

371595　Spicillaria A. Rich. = Hypobathrum Blume ●☆

371596　Spicillaria A. Rich. = Petunga DC. ●☆

371597　Spiciviscum Engelm. = Phoradendron Nutt. ●☆

371598　Spiculaea Lindl.（1840）;矛兰属■☆

371599　Spiculaea ciliata Lindl.;矛兰■☆

371600　Spielmannia Cuss. ex Juss. = Trinia Hoffm.（保留属名）■☆

371601　Spielmannia Medik.（1775）;异玄参属■☆

371602　Spielmannia Medik. = Oftia Adans. ●■☆

371603　Spielmannia africana（L.）Willd. = Oftia africana（L.）Bocq. ●☆

371604　Spielmannia decurrens Moench = Oftia africana（L.）Bocq. ●☆

371605　Spielmannia desertorum Eckl. et Zeyh. ex Schauer = Oftia revoluta（E. Mey.）Bocq. ●☆

371606　Spielmannia jasminum Medik. = Oftia africana（L.）Bocq. ●☆

371607　Spielmannia revoluta E. Mey. = Oftia revoluta（E. Mey.）Bocq. ●☆

371608　Spielmanniaceae J. Agardh = Myoporaceae R. Br.（保留科名）●

371609　Spielmanniaceae J. Agardh = Oftiaceae Takht. et Reveal ●☆

371610　Spielmanniaceae J. Agardh;异玄参科●☆

371611　Spiesia Neck. = Oxytropis DC.（保留属名）●■

371612　Spigelia L.（1753）;驱虫草属;Pink-root, Spigelia ■☆

371613　Spigelia P. Browne = Andira Lam.（保留属名）●☆

371614　Spigelia anthelmia L.;驱虫草;Demerara Pinkroot, Pink-root, West Indian Spigelia ■☆

371615　Spigelia anthelmia L. var. nervosa（Steud.）Progel = Spigelia anthelmia L. ■☆

371616　Spigelia marilandica L.;赤根驱虫草;Carolina Pink, Indian Pink, Maryland Pinkroot, Pinkroot, Woodland Pinkroot, Wormgrass ■☆

371617　Spigelia multispica Steud. = Spigelia anthelmia L. ■☆

371618　Spigelia multispica Steud. var. discolor Progel = Spigelia anthelmia L. ■☆

371619　Spigelia nervosa Steud. = Spigelia anthelmia L. ■☆

371620　Spigeliaceae Bercht. et J. Presl = Loganiaceae R. Br. ex Mart.（保留科名）●■

371621　Spigeliaceae Bercht. et J. Presl = Spigeliaceae C. Mart. ■●

371622　Spigeliaceae Mart.（1827）;驱虫草科（度量草科）■●

371623　Spigeliaceae Mart. = Loganiaceae R. Br. ex Mart.（保留科名）●■

371624　Spigeliaceae Mart. = Strychnaceae Link ●■

371625　Spilacron Cass. = Centaurea L.（保留属名）●■

371626　Spiladocorys Ridl. = Pentasacme Wall. ex Wight et Arn. ■

371627　Spilanthes Jacq.（1760）;金钮扣属（千日菊属,小铜钟属）;Goldenbutton, Spot Flower, Spotflower ■

371628　Spilanthes Jacq. = Acmella Rich. ex Pers. ■

371629　Spilanthes abyssinica Sch. Bip. ex A. Rich. = Acmella caulirhiza

Delile ■☆

371630　Spilanthes acmella（L.）Dalzell et A. Gibson = Spilanthes paniculata Wall. ex DC. ■

371631　Spilanthes acmella（L.）Dalzell et A. Gibson var. oleracea（Jacq.）Baker = Spilanthes oleracea L. ■

371632　Spilanthes acmella（L.）Murray；铁拳头（金钮扣）■

371633　Spilanthes acmella（L.）Murray = Acmella caulirhiza Delile ■☆

371634　Spilanthes acmella（L.）Murray var. calva（DC.）C. B. Clarke ex Hook. f. = Acmella calva（DC.）R. K. Jansen ■

371635　Spilanthes acmella（L.）Murray var. paniculata（Wall. ex DC.）Hook. f. = Acmella paniculata（Wall. ex DC.）R. K. Jansen ■

371636　Spilanthes acmella L. = Spilanthes paniculata Wall. ex DC. ■

371637　Spilanthes acmella L. var. boninensis Nakai = Acmella uliginosa（Sw.）Cass. ■☆

371638　Spilanthes acmella L. var. oleracea（L.）C. B. Clarke = Acmella oleracea（L.）R. K. Jansen ■

371639　Spilanthes acmella L. var. oleracea（L.）C. B. Clarke = Spilanthes oleracea L. ■

371640　Spilanthes acmella L. var. oleracea（L.）C. B. Clarke ex Hook. f. = Acmella oleracea（L.）R. K. Jansen ■

371641　Spilanthes africana DC. = Acmella caulirhiza Delile ■☆

371642　Spilanthes alpestris Griseb. ；高山金钮扣■☆

371643　Spilanthes americana（L. f.）Hieron. var. repens（Walter）A. H. Moore = Acmella oppositifolia（Lam.）R. K. Jansen ■☆

371644　Spilanthes americana（Mutis ex L. f.）Hieron. ex Sodiro = Acmella oppositifolia（Lam.）R. K. Jansen var. repens（Walter）R. K. Jansen ■☆

371645　Spilanthes callimorpha A. H. Moore；美形金钮扣（遍地红，过海龙，黄花草，铜锤草，乌龙过江，细麻药，小麻药，小铜锤）；Beautifulform Goldenbutton ■

371646　Spilanthes callimorpha A. H. Moore = Acmella calva（DC.）R. K. Jansen ■

371647　Spilanthes calva DC. = Acmella calva（DC.）R. K. Jansen ■

371648　Spilanthes caulirhiza（Delile）DC. = Acmella caulirhiza Delile ■☆

371649　Spilanthes caulirhiza（Delile）DC. var. madagascariensis DC. = Acmella caulirhiza Delile ■☆

371650　Spilanthes costata Benth. ；单脉金钮扣■☆

371651　Spilanthes decumbens（Sm.）A. H. Moore；外倾金钮扣■☆

371652　Spilanthes exasperatus Jacq. = Acmella radicans（Jacq.）R. K. Jansen ■☆

371653　Spilanthes filicaulis（Schumach. et Thonn.）C. D. Adams；丝茎金钮扣■☆

371654　Spilanthes filicaulis（Schumach. et Thonn.）C. D. Adams = Acmella caulirhiza Delile ■☆

371655　Spilanthes grandifolia Miq. ；大花金钮扣■☆

371656　Spilanthes iabadicensis A. H. Moore = Acmella uliginosa（Sw.）Cass. ■☆

371657　Spilanthes mauritiana（A. Rich. ex Pers.）DC. ；毛里求斯金钮扣■☆

371658　Spilanthes mauritiana（A. Rich. ex Pers.）DC. = Acmella caulirhiza Delile ■☆

371659　Spilanthes mauritiana（A. Rich. ex Pers.）DC. f. madagascariensis（DC.）A. H. Moore = Acmella caulirhiza Delile ■☆

371660　Spilanthes mauritiana（A. Rich.）DC. = Acmella caulirhiza Delile ■☆

371661　Spilanthes mauritiana DC. = Spilanthes mauritiana（A. Rich. ex Pers.）DC. ■☆

371662　Spilanthes mauritiana DC. f. madagascariensis（DC.）A. H. Moore = Acmella caulirhiza Delile ■☆

371663　Spilanthes nervosa Chodat；多脉金钮扣■☆

371664　Spilanthes ocymifolia（Lam.）A. H. Moore；罗勒叶金钮扣■☆

371665　Spilanthes oleracea（L.）Murray = Spilanthes oleracea L. ■

371666　Spilanthes oleracea Clarke；印度金钮扣■

371667　Spilanthes oleracea Jacq. = Spilanthes oleracea L. ■

371668　Spilanthes oleracea L. = Acmella oleracea（L.）R. K. Jansen ■

371669　Spilanthes oppositifolia（Lam.）D' Arcy；对叶金钮扣■☆

371670　Spilanthes paniculata Wall. ex DC. = Acmella paniculata（Wall. ex DC.）R. K. Jansen ■

371671　Spilanthes pusilla Hook. et Arn. = Acmella pusilla（Hook. et Arn.）R. K. Jansen ■☆

371672　Spilanthes radicans Jacq. = Acmella radicans（Jacq.）R. K. Jansen ■☆

371673　Spilanthes radicans Schrad. ex DC. = Acmella oleracea（L.）R. K. Jansen ■

371674　Spilanthes tinctoria Lour. = Adenostemma lavenia（L.）Kuntze ■

371675　Spilanthes uliginosa Sw. = Acmella uliginosa（Sw.）Cass. ■☆

371676　Spilanthus L. = Spilanthes Jacq. ■

371677　Spilocarpus Lem. = Tournefortia L. ●■

371678　Spilorchis D. L. Jones et M. A. Clem.（2005）；斑兰属■☆

371679　Spilorchis weinthalii（R. S. Rogers）D. L. Jones et M. A. Clem. ；斑兰■☆

371680　Spilotantha Luer = Masdevallia Ruiz et Pav. ■☆

371681　Spilotantha Luer（2006）；斑花细瓣兰属■☆

371682　Spiloxene Salisb.（1866）；南非仙茅属■☆

371683　Spiloxene acida（Nel）Garside；针形南非仙茅■☆

371684　Spiloxene aemulans（Nel）Garside；匹敌南非仙茅■☆

371685　Spiloxene alba（Thunb.）Fourc. ；白南非仙茅■☆

371686　Spiloxene aquatica（L. f.）Fourc. ；水生南非仙茅■☆

371687　Spiloxene canaliculata Garside；具沟南非仙茅■☆

371688　Spiloxene capensis（L.）Garside；好望角南非仙茅■☆

371689　Spiloxene curculigoides（Bolus）Garside；仙茅状南非仙茅■☆

371690　Spiloxene cuspidata（Nel）Garside = Spiloxene ovata（L. f.）Garside ■☆

371691　Spiloxene declinata（Nel）Garside = Spiloxene curculigoides（Bolus）Garside ■☆

371692　Spiloxene dielsiana（Nel）Garside；迪尔斯南非仙茅■☆

371693　Spiloxene flaccida（Nel）Garside；柔弱南非仙茅■☆

371694　Spiloxene gracilipes（Schltr.）Garside = Spiloxene ovata（L. f.）Garside ■☆

371695　Spiloxene linearis（Andréws）Garside = Spiloxene serrata（Thunb.）Garside ■☆

371696　Spiloxene maximiliani（Schltr.）Garside = Spiloxene umbraticola（Schltr.）Garside ■☆

371697　Spiloxene minuta（L.）Fourc. ；小南非仙茅■☆

371698　Spiloxene monophylla（Schltr. ex Baker）Garside；单叶南非仙茅■☆

371699　Spiloxene nana Snijman；矮小南非仙茅■☆

371700　Spiloxene ovata（L. f.）Garside；卵形南非仙茅■☆

371701　Spiloxene pusilla Snijman；微小南非仙茅■☆

371702　Spiloxene schlechteri（Bolus）Garside；施莱南非仙茅■☆

371703　Spiloxene scullyi（Baker）Garside；斯卡里南非仙茅■☆

371704　Spiloxene serrata（Thunb.）Garside；具齿南非仙茅■☆

371705　Spiloxene serrata（Thunb.）Garside var. albiflora（Nel）

Garside;白花具齿南非仙茅■☆

371706 Spiloxene stellata（L. f.）Salisb. = Spiloxene capensis（L.）Garside ■☆

371707 Spiloxene umbraticola（Schltr.）Garside;荫蔽南非仙茅■☆

371708 Spinacea Schur = Spinacia L. ■

371709 Spinachia Hill = Spinacia L. ■

371710 Spinacia L.（1753）;菠菜属（菠薐菜属,菠蓤属）;Spinach, Spinage ■

371711 Spinacia divaricata Turcz. ex Moq. = Atriplex fera（L.）Bunge ●

371712 Spinacia fera L. = Atriplex fera（L.）Bunge ●

371713 Spinacia oleracea L.;菠菜（波棱菜,波斯菜,菠棱,菠棱菜,菠蓤,菠薐菜,菠菱菜,赤根菜,角菜,颇菱,鼠根菜,苋菜,苋草,鹦鹉菜）;Common Spinach, Prickly-seeded Spinach, Spinach, Spinage, Spinnage, Winter Spinach ■

371714 Spinacia oleracea L. var. ineracea Peterm.;无刺菠菜■☆

371715 Spinacia spinosa Moench. = Spinacia oleracea L. ■

371716 Spinacia tetrandra Steven ex M. Bieb.;四蕊菠菜■☆

371717 Spinacia turkestanica Iljin;土耳其斯坦菠菜■☆

371718 Spinaciaceae Menge = Amaranthaceae Juss.（保留科名）●■

371719 Spinaciaceae Menge = Chenopodiaceae Vent.（保留科名）●■

371720 Spingula Noronha = Hygrophila R. Br. ●■

371721 Spingula Noronha = Springula Noronha ●■

371722 Spinicalycium Frič = Acanthocalycium Backeb. ●■☆

371723 Spinifex L.（1771）;鼠芳属（滨草属,鬣刺属）;Spinifex ■

371724 Spinifex littoreus（Burm. f.）Merr.;老鼠芳（滨刺草,腊刺,鬣刺）;Littoral Spinifex ■

371725 Spinifex longifolius R. Br.;长叶鼠芳草（长叶鬣刺）■☆

371726 Spinifex squarrosus L. = Spinifex littoreus（Burm. f.）Merr. ■

371727 Spiniluma（Baill.）Aubrév. = Sideroxylon L. ●☆

371728 Spiniluma Baill. = Sideroxylon L. ●☆

371729 Spiniluma Baill. ex Aubrév. = Sideroxylon L. ●☆

371730 Spiniluma buxifolia（Hutch.）Aubrév. = Monotheca buxifolia（Falc.）A. DC. ●☆

371731 Spiniluma buxifolia（Hutch.）Aubrév. = Sideroxylon mascatense（A. DC.）T. D. Penn. ●☆

371732 Spiniluma discolor（Radcl.-Sm.）Friis = Sideroxylon discolor Radcl.-Sm. ●☆

371733 Spiniluma oxyacantha（Baill.）Aubrév. = Sideroxylon oxyacanthum Baill. ●☆

371734 Spinovitis Rom. Caill. = Vitis L. ●

371735 Spinovitis davidii Rom. Caill. = Vitis davidii（Rom. Caill.）Foëx ●

371736 Spipa nakaii Honda = Achnatherum nakaii（Honda）Tateoka ■

371737 Spipa pubicalyx Ohwi = Achnatherum pubicalyx（Ohwi）Keng ex P. C. Kuo ■

371738 Spirabutilon Krapov.（2009）;螺苘麻属■☆

371739 Spirabutilon citrinum Krapov.;螺苘麻■☆

371740 Spiracantha Kunth（1818）;螺刺菊属（旋花菊属）■☆

371741 Spiracantha cornifolia Kunth;螺刺菊（旋花菊）■☆

371742 Spiradiclis Blume（1827）;螺序草属;Spiradiclis ■●

371743 Spiradiclis arunachanensis Deb et R. C. Rout = Spiradiclis caespitosa Blume f. subimmersa H. S. Lo ■

371744 Spiradiclis baishaiensis X. X. Chen et W. L. Sha;百色螺序草;Baise Spiradiclis ■

371745 Spiradiclis bifida Wall. ex Kurz;大叶螺序草;Bigleaf Spiradiclis ■

371746 Spiradiclis caespitosa Blume;螺序草;Spiradiclis ■

371747 Spiradiclis caespitosa Blume f. cylindrica（Wall. ex Hook. f.）

H. S. Lo;柱花螺序草（尖叶螺序草）;Cylindric Spiradiclis, Sharpleaf Spiradiclis ■

371748 Spiradiclis caespitosa Blume f. subimmersa H. S. Lo;柳叶螺序草;Willowleaf Spiradiclis ■

371749 Spiradiclis chuniana R. J. Wang;陈氏螺序草■

371750 Spiradiclis coccinea H. S. Lo;红花螺序草;Red Spiradiclis ■

371751 Spiradiclis cordata H. S. Lo et W. L. Sha;心叶螺序草;Heartleaf Spiradiclis ■

371752 Spiradiclis corymbosa W. L. Sha et X. X. Chen ex H. S. Lo;密花螺序草;Denseflor Spiradiclis ■

371753 Spiradiclis cylindrica Wall. ex Hook. f. = Spiradiclis caespitosa Blume f. cylindrica（Wall. ex Hook. f.）H. S. Lo ■

371754 Spiradiclis cylindrica Wall. ex Hook. f. = Spiradiclis caespitosa Blume ■

371755 Spiradiclis emeiensis H. S. Lo;峨眉螺序草;Emei Spiradiclis ■

371756 Spiradiclis emeiensis H. S. Lo var. yunnanensis H. S. Lo;河口螺序草;Hekou Spiradiclis ■

371757 Spiradiclis ferruginea D. Fang et D. H. Qin;锈茎螺序草;Ruststem Spiradiclis ■

371758 Spiradiclis fusca H. S. Lo;两广螺序草;Brown Spiradiclis ■

371759 Spiradiclis guangdongensis H. S. Lo;广东螺序草（春根藤）;Guangdong Spiradiclis, Kwangtung Spiradiclis ■

371760 Spiradiclis hainanensis H. S. Lo;海南螺序草;Hainan Spiradiclis ■

371761 Spiradiclis howii H. S. Lo;宽昭螺序草;Kuanzhao Spiradiclis ■

371762 Spiradiclis laxiflora W. L. Sha et X. X. Chen;疏花螺序草;Scatterflor Spiradiclis ■

371763 Spiradiclis leptobotrya Pit. var. longiflora Merr. = Spiradiclis corymbosa W. L. Sha et X. X. Chen ex H. S. Lo ■

371764 Spiradiclis loana R. J. Wang;献瑞螺序草■

371765 Spiradiclis longibracteata S. Y. Liu et S. J. Wei;长苞螺序草;Longbract Spiradiclis ■

371766 Spiradiclis longipedunculata W. L. Sha et X. X. Chen;长梗螺序草;Longpedicel Spiradiclis ■

371767 Spiradiclis longzhouensis H. S. Lo;龙州螺序草;Longzhou Spiradiclis ■

371768 Spiradiclis luochengensis H. S. Lo et W. L. Sha;桂北螺序草;N. Guangxi Spiradiclis ■

371769 Spiradiclis malipoensis H. S. Lo;滇南螺序草;Malipo Spiradiclis ■

371770 Spiradiclis micrantha（Drake）H. S. Lo = Lerchea micrantha（Drake）H. S. Lo ■

371771 Spiradiclis microcarpa H. S. Lo;小果螺序草;Smallfruit Spiradiclis ■

371772 Spiradiclis micropllylla H. S. Lo;小叶螺序草;Littleleaf Spiradiclis ■

371773 Spiradiclis napoensis D. Fang et Z. M. Xie;那坡螺序草;Napo Spiradiclis ■

371774 Spiradiclis oblanceolata W. L. Sha et X. X. Chen;长叶螺序草;Longleaf Spiradiclis ■

371775 Spiradiclis petrophila H. S. Lo;石生螺序草;Petrophila Spiradiclis ■

371776 Spiradiclis pseuducaespitosa Chun et F. C. How;假螺序草;Falsetufuted Spiradiclis ■

371777 Spiradiclis purpureocoerulea H. S. Lo;紫花螺序草;Purple Spiradiclis ■

371778 Spiradiclis rubescens H. S. Lo;红叶螺序草;Redleaf Spiradiclis ■

371779 Spiradiclis scabrida D. Fang et D. H. Qin;糙边螺序草(粗边螺序草);Coarse-edge Spiradiclis ■

371780 Spiradiclis spathulata X. X. Chen et C. C. Huang;匙叶螺序草;Spoonleaf Spiradiclis ■

371781 Spiradiclis tomentosa D. Fang et D. H. Qin;黏毛螺序草;Adhesivehair Spiradiclis ■

371782 Spiradiclis umbelliformis H. S. Lo;伞花螺序草;Umbel Spiradiclis ■

371783 Spiradiclis villosa X. X. Chen et W. L. Sha;毛螺序草;Hair Spiradiclis ■

371784 Spiradiclis xizangensis H. S. Lo;西藏螺序草;Xizang Spiradiclis ■

371785 Spiraea L. (1753);绣线菊属(珍珠梅属);Bridal Wreath, Bridalwreath, Bridal-wreath, Bridewort, Meadow Sweet, Meadowsweet,Spiraea,Spirea ●

371786 Spiraea × cinerea Zabel;格雷绣线菊;Grefsheim Spirea ●☆

371787 Spiraea × cinerea Zabel 'Compacta';紧凑格雷绣线菊;Dwarf Garland Spirea ●☆

371788 Spiraea × cinerea Zabel 'Grefsheim';格雷福塞姆绣线菊;Grefsheim Spirea ●☆

371789 Spiraea × hayatae Koidz.;早田氏绣线菊●☆

371790 Spiraea × pseudosalicifolia Silverside;假柳叶绣线菊;Billard's Bridewort,Billiard Spirea,Confused Bridewort ●☆

371791 Spiraea × rubella Dippel;微红绣线菊;Bridewort ●☆

371792 Spiraea aemiliana C. K. Schneid. = Spiraea betulifolia Pall. var. aemiliana (C. K. Schneid.) Koidz. ●☆

371793 Spiraea aemulans Rehder;酷似绣线菊;Likest Spiraea ●

371794 Spiraea aemulans Rehder = Spiraea sargentiana Rehder ●

371795 Spiraea alba Du Roi;白花柳叶绣线菊(宽叶绣线菊);Broad-leaved Meadowsweet,Large-leaved Meadowsweet,Meadowsweet,Pale Bridewort,Quaker Lady,White Meadowsweet ●☆

371796 Spiraea alba Du Roi var. latifolia (Aiton) H. E. Ahles;阔叶白花绣线菊●☆

371797 Spiraea albiflora C. K. Schneid.;日本白绣线菊;Japanese White Spirea ●☆

371798 Spiraea alpina Pall.;高山绣线菊;Alpine Spiraea,Alpine Spirea ●

371799 Spiraea alpina Pall. var. dahurica Rupr. = Spiraea dahurica Maxim. ●

371800 Spiraea altaica Pall. = Sibiraea laevigata (L.) Maxim. ●

371801 Spiraea altaiensis Laxm. = Sibiraea laevigata (L.) Maxim. ●

371802 Spiraea amurensis Maxim. = Physocarpus amurensis (Maxim.) Maxim. ●

371803 Spiraea angulata Fritsch ex C. K. Schneid. = Spiraea fritschiana C. K. Schneid. var. angulata Rehder ●

371804 Spiraea angulata Fritschex Schneid. = Spiraea fritschiana C. K. Schneid. var. angulata Rehder ●

371805 Spiraea angustiloba Turcz. = Filipendula angustiloba (Turcz.) Maxim. ■

371806 Spiraea anomala Batalin;异常绣线菊●

371807 Spiraea aquilegiifolia Pall.;楼斗菜叶绣线菊(楼斗叶绣线菊);Columbineleaf Spiraea,Columbine-leaved Spiraea ●

371808 Spiraea aquilegiifolia Pall. var. vanhouttei Briot = Spiraea vanhouttei (Briot) Zabel ●

371809 Spiraea arborea (C. K. Schneid.) Bean = Sorbaria arborea Schneid. var. glabrata Rehder ●

371810 Spiraea arborea (C. K. Schneid.) Bean = Sorbaria arborea Schneid. ●

371811 Spiraea arborea (C. K. Schneid.) Bean var. glabrata Bean = Sorbaria arborea Schneid. var. glabrata Rehder ●

371812 Spiraea arborea Bean = Sorbaria arborea Schneid. var. glabrata Rehder ●

371813 Spiraea arborea Bean var. glabrata (Rehder) Bean = Sorbaria arborea Schneid. var. glabrata Rehder ●

371814 Spiraea arcuata Hook. f.;拱枝绣线菊;Arcuate Spiraea ●

371815 Spiraea arguta Zabel;锐绣线菊(尖绣线菊);Bridal Wreath, Bridal-spray,Foam of May,Foam-of-may,Gariand Spirea,Thunberg Spiraea ●

371816 Spiraea aruncus L. = Aruncus sylvester Kostel. ex Maxim. ■

371817 Spiraea atemnophylla H. Lév. = Spiraea veitchii Hemsl. ●

371818 Spiraea baldschuanica B. Fedtsch.;巴尔德绣线菊●☆

371819 Spiraea barbara Wall. = Astilbe rivularis Buch. -Ham. ex D. Don ■

371820 Spiraea beauverdiana C. K. Schneid. = Spiraea betulifolia Pall. var. aemiliana (C. K. Schneid.) Koidz. ●☆

371821 Spiraea bella Sims;藏南绣线菊(美丽绣线菊);Beautiful Spiraea,S. Xizang Spiraea ●

371822 Spiraea bella Sims var. pubicarpa Te T. Yu et A. M. Lu;毛果藏南绣线菊;Hairy-fruit Beautiful Spiraea ●

371823 Spiraea betulifolia Pall.;桦叶绣线菊;Birchleaf Spiraea,Birch-leaved Spirea ●☆

371824 Spiraea betulifolia Pall. f. grandifolia (Nakai) Kitam. ex Kawano;大叶桦叶绣线菊●☆

371825 Spiraea betulifolia Pall. f. grandifolia Kitam. ex Kawano = Spiraea betulifolia Pall. f. grandifolia (Nakai) Kitam. ex Kawano ●☆

371826 Spiraea betulifolia Pall. f. oblanceolata (Tatew.) Kitam.;倒披针桦叶绣线菊●☆

371827 Spiraea betulifolia Pall. subsp. aemiliana (C. K. Schneid.) H. Hara = Spiraea betulifolia Pall. var. aemiliana (C. K. Schneid.) Koidz. ●☆

371828 Spiraea betulifolia Pall. var. aemiliana (C. K. Schneid.) Koidz.;阔桦叶绣线菊●☆

371829 Spiraea betulifolia Pall. var. aemiliana (C. K. Schneid.) Koidz. f. glabra (H. Hara) Ohwi;光阔桦叶绣线菊●☆

371830 Spiraea betulifolia Pall. var. glabra (H. Hara) H. Hara;光桦叶绣线菊;Glabrous Birchleaf Spiraea ●☆

371831 Spiraea betulifolia Pall. var. glabra (H. Hara) H. Hara = Spiraea betulifolia Pall. var. aemiliana (C. K. Schneid.) Koidz. f. glabra (H. Hara) Ohwi ●☆

371832 Spiraea betulifolia Pall. var. glabra H. Hara = Spiraea betulifolia Pall. var. aemiliana (C. K. Schneid.) Koidz. f. glabra (H. Hara) Ohwi ●☆

371833 Spiraea betulifolia Pall. var. oblanceolata Tatew. = Spiraea betulifolia Pall. f. oblanceolata (Tatew.) Kitam. ●☆

371834 Spiraea billardii Meehan;毕氏绣线菊(比拉尔绣线菊,杂种绣线菊);Billard Spirea,Billard's Spiraea ●☆

371835 Spiraea billardii Meehan 'Triumphans';胜利比拉尔绣线菊(狂欢杂种绣线菊)●☆

371836 Spiraea blumei G. Don;绣球绣线菊(补氏绣线菊,翠蓝茶,麻叶绣球,山茴香,碎米丫,碎米桠,绣球,珍珠梅,珍珠绣球,珍珠绣线菊);Blume Spiraea ●

371837 Spiraea blumei G. Don f. amabilis (Koidz.) Sugim.;秀丽绣球绣线菊●☆

371838 Spiraea blumei G. Don f. obtusa (Nakai) Kitam. = Spiraea blumei G. Don var. obtusa (Nakai) Sugim. ●☆

371839 Spiraea blumei G. Don var. hayatae (Koidz.) Ohwi = Spiraea ×

hayatae Koidz. ●☆

371840　Spiraea blumei G. Don var. hirsuta Hemsl. = Spiraea hirsuta (Hemsl.) C. K. Schneid. ●

371841　Spiraea blumei G. Don var. latipetala Hemsl.；宽瓣绣球绣线菊；Broadpetal Blume Spiraea ●

371842　Spiraea blumei G. Don var. maximowicziana (C. K. Schneid.) Dunn = Spiraea hirsuta (Hemsl.) C. K. Schneid. var. rotundifolia (Hemsl.) Rehder ●

371843　Spiraea blumei G. Don var. microphylla Rehder；小叶绣球绣线菊；Little-leaf Blume Spiraea ●

371844　Spiraea blumei G. Don var. obtusa (Nakai) Sugim.；钝叶绣球绣线菊●☆

371845　Spiraea blumei G. Don var. pubescens (Koidz.) Ohwi = Spiraea × hayatae Koidz. ●☆

371846　Spiraea blumei G. Don var. pupicarpa W. C. Cheng；毛果绣球绣线菊；Hairyfruit Blume Spiraea ●

371847　Spiraea blumei G. Don var. rotundifolia Hemsl. = Spiraea hirsuta (Hemsl.) C. K. Schneid. var. rotundifolia (Hemsl.) Rehder ●

371848　Spiraea bodinieri H. Lév. = Spiraea japonica L. f. var. acuminata Franch. ●

371849　Spiraea bodinieri H. Lév. var. concolor H. Lév. = Spiraea japonica L. f. var. acuminata Franch. ●

371850　Spiraea brachybotrys Lange；短穗绣线菊；Lange's Spiraea ●☆

371851　Spiraea bracteata Raf. = Sibiraea laevigata (L.) Maxim. ●

371852　Spiraea bumalda Burv.；日本绣线菊；Japanese Spiraea ●☆

371853　Spiraea caespitosa Nutt.；丛生绣线菊；Turf Spiraea ●☆

371854　Spiraea calcicola W. W. Sm.；石灰岩绣线菊；Calcarious Spiraea，Limestone Spiraea ●

371855　Spiraea callosa Thunb. = Spiraea japonica L. f. ●

371856　Spiraea callosa Thunb. = Spiraea robusta (Hook. f. et Thomson) Hand. -Mazz. ●

371857　Spiraea callosa Thunb. var. glabra Regel = Spiraea japonica L. f. var. glabra (Regel) Koidz. ●

371858　Spiraea callosa Thunb. var. robusta Hook. f. et Thomson = Spiraea robusta (Hook. f. et Thomson) Hand. -Mazz. ●

371859　Spiraea camtschatica Pall. var. himalensis Lindl. = Filipendula vestita (Wall.) Maxim. ■

371860　Spiraea canescens D. Don；楔叶绣线菊（刺杨，铁刷子）；Cunealleaf Spiraea，Himalayan Spiraea，Hoary Spiraea ●

371861　Spiraea canescens D. Don var. glabra Hook. f. et Thomson = Spiraea arcuata Hook. f. ●

371862　Spiraea canescens D. Don var. glaucophylla Franch.；粉背楔叶绣线菊（粉楔叶绣线菊）；Glaucousleaf Hoary Spiraea ●

371863　Spiraea canescens D. Don var. glaucophylla Franch. = Spiraea myrtilloides Rehder ●

371864　Spiraea canescens D. Don var. myrtifolia Zabel = Spiraea canescens D. Don var. glaucophylla Franch. ●

371865　Spiraea canescens D. Don var. oblanceolata Rehder = Spiraea canescens D. Don ●

371866　Spiraea canescens D. Don var. oblonceolata Rehder；窄楔叶绣线菊（窄叶楔叶绣线菊）；Narrowleaf Hoary Spiraea ●

371867　Spiraea canescens D. Don var. sulphurea Batalin = Spiraea canescens D. Don var. glaucophylla Franch. ●

371868　Spiraea cantoniensis Lour.；麻叶绣线菊（广州绣线菊，麻球，麻叶绣球，麻叶绣球绣线菊，毛萼麻叶绣线菊，石棒子，粤绣线菊，珍珠花）；Double Reeves Spirea，Double White Spirea，Hempleaf Spiraea，Reeves' Meadowsweet，Reeves Spiraea，Reeves'

Spiraea ●

371869　Spiraea cantoniensis Lour. 'Flore Pleno' = Spiraea cantoniensis Lour. var. lanceata Zabel ●

371870　Spiraea cantoniensis Lour. 'Lanceata' = Spiraea cantoniensis Lour. var. lanceata Zabel ●

371871　Spiraea cantoniensis Lour. f. plena (Koidz.) Okuyama；日本重瓣麻叶绣线菊（重瓣麻叶绣线菊）●☆

371872　Spiraea cantoniensis Lour. var. jiangxiensis (Z. X. Yu) L. T. Lu；江西绣线菊；Jiangxi Spiraea ●

371873　Spiraea cantoniensis Lour. var. jiangxiensis (Z. X. Yu) L. T. Lu = Spiraea jiangxiensis Z. X. Yu ●

371874　Spiraea cantoniensis Lour. var. lanceata Zabel；重瓣麻叶绣线菊（重瓣麻叶，重瓣麻叶绣球）；Doubleflower Reeves Spiraea ●

371875　Spiraea cantoniensis Lour. var. pilosa Te T. Yu；毛萼麻叶绣线菊（毛萼绣线菊）；Pilose Reeves Spiraea ●

371876　Spiraea cavaleriei H. Lév.；独山绣线菊；Dushan Spiraea ●

371877　Spiraea chamaedryfolia L.；石蚕叶绣线菊（大叶绣线菊，乌苏里绣线菊，兴安绣线菊，榆叶绣线菊）；Elm-leaved Spiraea，Germander Meadowsweet，Germander Spiraea ●

371878　Spiraea chamaedryfolia L. = Spiraea elegans Pojark. ●

371879　Spiraea chamaedryfolia L. var. flexuosa (Fisch. ex Cambess.) Maxim. = Spiraea flexuosa Fisch. ex Cambess. ●

371880　Spiraea chamaedryfolia L. var. flexuosa Maxim. = Spiraea flexuosa Fisch. ex Cambess. ●

371881　Spiraea chamaedryfolia L. var. pilosa (Nakai) H. Hara；毛石蚕叶绣线菊●☆

371882　Spiraea chamaedryfolia L. var. pubescens H. Hara = Spiraea chamaedryfolia L. var. pilosa (Nakai) H. Hara ●☆

371883　Spiraea chamaedryfolia L. var. ulmifolia (Scop.) J. Duvign.；榆叶绣线菊；Germander Meadowsweet ●☆

371884　Spiraea chinensis Maxim.；中华绣线菊（华绣线菊，铁黑汉条）；China Spiraea，Chinese Spiraea，Chinese Spirea ●

371885　Spiraea chinensis Maxim. var. erecticarpa Y. Q. Zhu et X. W. Li；直果绣线菊；Erectfruit China Spiraea ■

371886　Spiraea chinensis Maxim. var. grandiflora Te T. Yu；大花中华绣线菊（岩刷子）；Largeflower Chinese Spiraea ●

371887　Spiraea compsophylla Hand. -Mazz.；粉叶绣线菊；Glaucous Spiraea，Grey-blue-leaved Spiraea，Powderleaf Spiraea ●

371888　Spiraea confusa Regel et Körn. var. sericea (Turcz.) Regel = Spiraea sericea Turcz. ●

371889　Spiraea confusa Regel et Körn. var. serieea Regel = Spiraea sericea Turcz. ●

371890　Spiraea crenata L.；圆齿绣线菊；Snow Spirea ●☆

371891　Spiraea crenata L. = Spiraea thunbergii Siebold ex Blume ●

371892　Spiraea crenifolia C. A. Mey. var. mongolica Maxim. = Spiraea mongolica Maxim. ●

371893　Spiraea cuneifolia Wall. = Spiraea canescens D. Don ●

371894　Spiraea cuntoniensis Lour.；麻叶绣球（奥绣线菊，麻叶绣线菊，石棒子）；Reeves Spiraea ●

371895　Spiraea dahurica Maxim.；窄叶绣线菊；Dahurian Spiraea，Narrowleaf Spiraea ●

371896　Spiraea daochengensis L. T. Lu；稻城绣线菊；Daocheng Spiraea ●

371897　Spiraea dasyantha Bunge；毛花绣线菊（凹脉绣线菊，翠蓝茶，筷棒，筷子木，绒毛绣线菊，石崩子）；Hairyflower Spiraea，Hairy-flowered Spiraea ●

371898　Spiraea densiflora Nutt. ex Rydb.；密花绣线菊●☆

371899　Spiraea densiflora Nutt. ex Rydb. var. splendens (Baumann ex

K. Koch) C. L. Hitchc. ;美丽密花绣线菊●☆

371900 Spiraea digitata Willd. = Filipendula palmata (Pall.) Maxim. ■

371901 Spiraea digitata Willd. var. intermedia Glehn = Filipendula intermedia (Glehn) Juz. ■

371902 Spiraea douglasii Hook. ;道格拉斯绣线菊(灰背绣线菊); Douglas Spirea, Hardhack, Spiraea, Steeplebush, Western Spiraea ●☆

371903 Spiraea elegans Pojark. ;美丽绣线菊(丽绣线菊); Elegant Spiraea ●

371904 Spiraea elegans Pojark. = Spiraea flexuosa Fisch. ex Cambess. ●

371905 Spiraea esquirolii H. Lév. = Spiraea japonica L. f. var. acuminata Franch. ●

371906 Spiraea faurieana C. K. Schneid. ;法氏绣线菊;Faurioe Spiraea ●☆

371907 Spiraea ferganensis Pojark. ;费尔干绣线菊●☆

371908 Spiraea filipendula L. = Filipendula filipendula (L.) Voss ■☆

371909 Spiraea filipendula L. = Filipendula vulgaris Moench ■☆

371910 Spiraea flexuosa Fisch. ex Cambess. ;曲萼绣线菊;Bowedsepal Spiraea, Bowed-sepaled Spiraea ●

371911 Spiraea flexuosa Fisch. ex Cambess. var. pubescens Liou;柔毛曲萼绣线菊;Pubescent Bowedsepal Spiraea ●

371912 Spiraea formosana Hayata;台湾绣线菊(光叶绣线菊);Taiwan Spiraea ●

371913 Spiraea formosana Hayata var. brevistyla Hayata = Spiraea hayatana H. L. Li ●

371914 Spiraea fortunei Planch. = Spiraea japonica L. f. var. fortunei (Planch.) Rehder ●

371915 Spiraea fritschiana C. K. Schneid. ;华北绣线菊(弗氏绣线菊);Fritschiana Spirea, Korean Spiraea ●

371916 Spiraea fritschiana C. K. Schneid. var. angulata Rehder;大叶华北绣线菊(叫驴腿);Largeleaf Fritsch Spiraea ●

371917 Spiraea fritschiana C. K. Schneid. var. latifolia Liou = Spiraea fritschiana C. K. Schneid. var. angulata Rehder ●

371918 Spiraea fritschiana C. K. Schneid. var. parvifolia Liou;小叶华北绣线菊;Little-leaf FritschSpiraea ●

371919 Spiraea fritschiana C. K. Schneid. var. pilosula Rehder;毛叶华北绣线菊(毛叶长蕊绣线菊)●

371920 Spiraea fritschiana C. K. Schneid. var. villosa Y. Q. Zhu et D. K. Zang;长毛华北绣线菊;Villose FritschSpiraea ●

371921 Spiraea fritschiana C. K. Schneid. var. villosa Y. Q. Zhu et D. K. Zang = Spiraea fritschiana C. K. Schneid. ●

371922 Spiraea fulvescens Rehder = Spiraea martinii H. Lév. ●

371923 Spiraea gemmata Zabel = Spiraea mongolica Maxim. ●

371924 Spiraea gracilis Maxim. ;细弱绣线菊;Slenser Spiraea ●☆

371925 Spiraea grandiflora Sweet = Exochorda racemosa (Lindl.) Rehder ●

371926 Spiraea hailarensis Liou;海拉尔绣线菊;Hailar Spiraea ●

371927 Spiraea hayatana H. L. Li;假绣线菊;Hayata Spiraea ●

371928 Spiraea henryi Hemsl. ex Forbes et Hemsl. ;翠蓝绣线菊(翠蓝茶,亨利绣线菊);Henry Spiraea ●

371929 Spiraea henryi Hemsl. ex Forbes et Hemsl. var. glabrata Te T. Yu et L. T. Lu;光叶翠蓝绣线菊;Glabrous Henry Spiraea ●

371930 Spiraea henryi Hemsl. ex Forbes et Hemsl. var. omeiensis Te T. Yu;峨眉翠蓝绣线菊;Emei Henry Spiraea, Omei Henry Spiraea ●

371931 Spiraea henryi var. glabrata Te T. Yu et L. T. Lu = Spiraea henryi Hemsl. ex Forbes et Hemsl. var. omeiensis Te T. Yu ●

371932 Spiraea hingshanensis Te T. Yu et L. T. Lu;兴山绣线菊; Xingshan Spiraea ●

371933 Spiraea hirsuta (Hemsl.) C. K. Schneid. ;疏毛绣线菊;Hirsute Spiraea ●

371934 Spiraea hirsuta (Hemsl.) C. K. Schneid. var. rotundifolia (Hemsl.) Rehder;圆叶疏毛绣线菊(圆疏毛绣线菊);Roundleaf Hirsute Spiraea ●

371935 Spiraea hispanica (Willd.) Hoffmanns. et Link = Galium concatenatum Coss. ■☆

371936 Spiraea holorhodantha H. Lév. = Rodgersia sambucifolia Hemsl. ■

371937 Spiraea humilis Pojark. ;矮小绣线菊●☆

371938 Spiraea hypericifolia L. ;金丝桃叶绣线菊;Hypericifoliate Spiraea, Iberian Spirea, St. John's Wortleaf Spiraea, St. John's Wort-leaved Spiraea ●

371939 Spiraea hypericifolia L. subsp. obovata (Willd.) Huber;倒卵叶绣线菊●☆

371940 Spiraea hypericifolia L. var. hupehensis Rehder = Spiraea prunifolia Siebold et Zucc. var. hupehensis (Rehder) Rehder ●

371941 Spiraea hypericifolia L. var. thalictroides (Pall.) Ledeb. = Spiraea aquilegiifolia Pall. ●

371942 Spiraea hypericifolia L. var. thalictroides Ledeb. = Spiraea aquilegiifolia Pall. ●

371943 Spiraea incisa Thunb. = Stephanandra incisa (Thunb. ex A. Murray) Zabel ●

371944 Spiraea japonica L. f. ;粉花绣线菊(火烧尖,蚂蟥梢,日本绣线菊,土黄连,绣线菊,野鞘火树,杂干树);Feather Fern, Japanese Meadowsweet, Japanese Spiraea, Pink Spiraea, Powderflower Spirea ●

371945 Spiraea japonica L. f. 'Alpina';高山粉花绣线菊;Alpine Spirea, Dwarf Japanese Spirea ●☆

371946 Spiraea japonica L. f. 'Anthony Watwrer';安东尼·沃特尔粉花绣线菊(安东尼水粉花绣线菊)●☆

371947 Spiraea japonica L. f. 'Bullata';布拉塔粉花绣线菊●☆

371948 Spiraea japonica L. f. 'Bumalda';布玛尔达粉花绣线菊●☆

371949 Spiraea japonica L. f. 'Dart's Red';红镖粉花绣线菊●☆

371950 Spiraea japonica L. f. 'Fire Light';火光粉花绣线菊●☆

371951 Spiraea japonica L. f. 'Golden Princess';金色公主粉花绣线菊;Japanese Spirea ●☆

371952 Spiraea japonica L. f. 'Goldflame';金焰粉花绣线菊●☆

371953 Spiraea japonica L. f. 'Little Princess';小公主粉花绣线菊;Japanese Spirea ●☆

371954 Spiraea japonica L. f. 'Monhup';毛胡布粉花绣线菊●☆

371955 Spiraea japonica L. f. 'Nana';矮生粉花绣线菊●☆

371956 Spiraea japonica L. f. 'Neon Flash';氖光粉花绣线菊;Japanese Spirea ●☆

371957 Spiraea japonica L. f. f. albiflora (Miq.) Kitam. ;白粉花绣线菊(白花日本绣线菊);Whiteflower Japanese Spiraea ●☆

371958 Spiraea japonica L. f. f. alpina (Maxim.) Makino = Spiraea japonica L. f. 'Alpina' ●☆

371959 Spiraea japonica L. f. f. bullata (Maxim.) Kitam. = Spiraea japonica L. f. 'Bullata' ●☆

371960 Spiraea japonica L. f. f. hypoglauca (Koidz.) Kitam. ;粉绿背绣线菊●☆

371961 Spiraea japonica L. f. f. ibukiensis Makino;伊吹山绣线菊●☆

371962 Spiraea japonica L. f. f. pubescens Kitam. ;茸毛粉花绣线菊(茸毛日本绣线菊);Pubescent Japanese Spiraea ●☆

371963 Spiraea japonica L. f. subsp. glabra var. fortunei Koidz. = Spiraea japonica L. f. var. fortunei (Planch.) Rehder ●

371964 Spiraea japonica L. f. var. acuminata Franch. ;狭叶绣线菊(吹火筒,火筒花,尖叶绣线菊,渐尖粉花绣线菊,渐尖叶粉花绣线

菊,狭叶粉花绣线菊）；Taperleaf Japanese Spiraea, Taperleaf Spiraea ●

371965　Spiraea japonica L. f. var. acuta Te T. Yu；急尖叶粉花绣线菊（急尖粉花绣线菊）；Acuteleaf Japanese Spiraea ●

371966　Spiraea japonica L. f. var. albiflora（Miq.）Z. Wei et Y. B. Chang = Spiraea japonica L. f. f. albiflora（Miq.）Kitam. ●☆

371967　Spiraea japonica L. f. var. bullata（Maxim.）Makino = Spiraea japonica L. f. 'Bullata' ●☆

371968　Spiraea japonica L. f. var. formosana（Hayata）Masam. = Spiraea formosana Hayata ●

371969　Spiraea japonica L. f. var. formosana（Hayata）Masam. subvar. brevistyla（Hayata）Masam. = Spiraea hayatana H. L. Li ●

371970　Spiraea japonica L. f. var. formosana Masam. = Spiraea formosana Hayata ●

371971　Spiraea japonica L. f. var. fortunei（Planch.）Rehder；光叶粉花绣线菊（大粉红绣线菊,大绣线菊,光叶绣线菊,红花绣线菊,绣线菊）；Fortune Japanese Spiraea, Fortune Meadowsweet ●

371972　Spiraea japonica L. f. var. glabra（Regel）Koidz.；无毛粉花绣线菊（红绣线菊,无毛绣线菊）；Glabrous Japanese Spiraea ●

371973　Spiraea japonica L. f. var. hypoglauca（Koidz.）Kitam. = Spiraea japonica L. f. f. hypoglauca（Koidz.）Kitam. ●☆

371974　Spiraea japonica L. f. var. incisa Te T. Yu；裂叶粉花绣线菊；Incisedleaf Spiraea ●

371975　Spiraea japonica L. f. var. leucantha Koidz.；白花粉花绣线菊（白花日本绣线菊）●☆

371976　Spiraea japonica L. f. var. morrisonicola（Hayata）Kitam. = Spiraea morrisonicola Hayata ●

371977　Spiraea japonica L. f. var. ovalifolia Franch.；卵叶绣线菊（椭圆叶粉花绣线菊）；Ovalleaf Japanese Spiraea ●

371978　Spiraea japonica L. f. var. pinnatifida Te T. Yu et L. T. Lu；羽叶粉花绣线菊；Pinnate-leaf Japanese Spiraea ●

371979　Spiraea japonica L. f. var. pubescens Kitam.；毛叶粉花绣线菊；Hairleaf Japanese Spiraea ●☆

371980　Spiraea japonica L. f. var. ripensis Kitam.；河岸粉花绣线菊■☆

371981　Spiraea japonica L. f. var. stellaris Rehder；白升麻（千颗米,筛子花,星花绣线菊）；Stellate Japanese Spiraea ●

371982　Spiraea japonica L. f. var. typica Gilg = Spiraea japonica L. f. ●

371983　Spiraea japonica L. f. var. typica Gilg f. glabra Schneid. = Spiraea japonica L. f. var. glabra（Regel）Koidz. ●

371984　Spiraea jiangxiensis Z. X. Yu = Spiraea cantoniensis Lour. var. jiangxiensis（Z. X. Yu）L. T. Lu ●

371985　Spiraea kamtschatica Wall. var. himalensis Lindl. = Filipendula vestita（Wall.）Maxim. ■

371986　Spiraea kirilowii Regel = Sorbaria kirilowii（Regel）Maxim. ●

371987　Spiraea kirilowii Regel et Tiling = Sorbaria kirilowii（Regel）Maxim. ●

371988　Spiraea kwangsiensis Te T. Yu；广西绣线菊（珍珠梅）；Guaangxi Spiraea, Kwangsi Spiraea ●

371989　Spiraea kweichouensis Te T. Yu et L. T. Lu；贵州绣线菊；Guizhou Spiraea ●

371990　Spiraea laeta Rehder；华西绣线菊；W. China Spiraea, West China Spiraea ●

371991　Spiraea laeta Rehder var. subpubescens Rehder；毛叶华西绣线菊；Hairyleaf West China Spiraea ●

371992　Spiraea laeta Rehder var. tenuis Rehder；细叶华西绣线菊；Fineleaf West China Spiraea ●

371993　Spiraea laevigata L. = Sibiraea laevigata（L.）Maxim. ●

371994　Spiraea lasiocarpa Kar. et Kir.；欧洲毛果绣线菊●☆

371995　Spiraea latifolia Borkh. = Spiraea alba Du Roi ●☆

371996　Spiraea laucheana Koehne = Spiraea pubescens Turcz. ●

371997　Spiraea lichiangensis W. W. Sm.；丽江绣线菊；Lijiang Spiraea ●

371998　Spiraea lobata Jacq. = Filipendula rubra（Hill）B. L. Rob. ■☆

371999　Spiraea lobulata Te T. Yu et L. T. Lu；裂叶绣线菊；Lobed Spiraea, Lobulate Spiraea ●

372000　Spiraea longigemmis Maxim.；长芽绣线菊；Longbud Spiraea, Longibuded Spiraea, Spiraea ●

372001　Spiraea mairei（H. Lév.）L. T. Lu；长毛绣线菊（麦地绣线菊）；Long-hair Spiraea ●

372002　Spiraea martinii H. Lév.；毛枝绣线菊；Martin Spiraea ●

372003　Spiraea martinii H. Lév. var. pubescens Te T. Yu；长梗毛枝绣线菊；Longstalk Martin Spiraea ●

372004　Spiraea martinii H. Lév. var. tomentosa Te T. Yu；毛叶毛枝绣线菊（绒叶毛枝绣线菊,绒毛枝绣线菊）；Hairyleaf Martin Spiraea ●

372005　Spiraea maximowicziana C. K. Schneid. = Spiraea hirsuta（Hemsl.）C. K. Schneid. var. rotundifolia（Hemsl.）Rehder ●

372006　Spiraea media F. Schmidt；欧亚绣线菊（石棒绣线菊,石棒子）；Eurasiatic Spiraea, Oriental Spiraea, Oriental Spirea ●

372007　Spiraea media F. Schmidt var. monbetsuensis（Franch.）Cardot ex Nakai = Spiraea media F. Schmidt var. sericea（Turcz.）Regel ex Maxim. ●☆

372008　Spiraea media F. Schmidt var. sericea（Turcz.）Maxim. = Spiraea sericea Turcz. ●

372009　Spiraea media F. Schmidt var. sericea（Turcz.）Regel ex Maxim.；虾夷小绣线菊●☆

372010　Spiraea media F. Schmidt var. sericea（Turcz.）Regel ex Maxim. = Spiraea sericea Turcz. ●

372011　Spiraea microphylla H. Lév. = Spiraea myrtilloides Rehder ●

372012　Spiraea miyabei Koidz.；长蕊绣线菊（虾夷白花绣线菊）；Miyabe Spiraea ●

372013　Spiraea miyabei Koidz. var. glabrata Rehder；无毛长蕊绣线菊；Glabrous Miyabe Spiraea ●

372014　Spiraea miyabei Koidz. var. pilosula Rehder；毛叶长蕊绣线菊（疏毛绣线菊）；Pilose Miyabe Spiraea ●

372015　Spiraea miyabei Koidz. var. tenuifolia Rehder；细叶长蕊绣线菊（细叶绣线菊）；Fineleaf Miyabe Spiraea ●

372016　Spiraea mollifolia Rehder；毛叶绣线菊（丝毛叶绣线菊）；Hairyleaf Spiraea, Hairy-leaved Spiraea ●

372017　Spiraea mollifolia Rehder var. glabrata Te T. Yu et L. T. Lu；光秃绣线菊；Glabrous Hairyleaf Spiraea ●

372018　Spiraea mongolica Maxim.；蒙古绣线菊；Mongol Spiraea, Mongolian Spiraea ●

372019　Spiraea mongolica Maxim. var. pubescens Y. Z. Zhao et T. J. Wang；毛枝蒙古绣线菊；Pubescent Mongolian Spiraea ●

372020　Spiraea mongolica Maxim. var. tomentulosa Te T. Yu = Spiraea tomentulosa（Te T. Yu）Y. Z. Zhao ●

372021　Spiraea morrisonicola Hayata；玉山绣线菊（新高山绣线菊）；Morrison Mountain Spiraea, Morrison Spiraea ●

372022　Spiraea muliensis Te T. Yu et L. T. Lu；木里绣线菊；Muli Spiraea ●

372023　Spiraea myrtilloides Rehder；细枝绣线菊；Fineshoot Spiraea, Fine-shooted Spiraea ●

372024　Spiraea myrtilloides Rehder var. pubicarpa Te T. Yu et L. T. Lu；毛果细枝绣线菊；Hairy-fruit Fineshoot Spiraea ●

372025　Spiraea nervosa Franch. et Sav. = Spiraea dasyantha Bunge ●

372026　Spiraea nervosa Franch. et Sav. var. angustifolia（Yatabe）Ohwi = Spiraea dasyantha Bunge ●

372027　Spiraea ningshiaensis Te T. Yu et L. T. Lu；宁夏绣线菊；Ningxia Spiraea ●

372028　Spiraea ningshiaensis Te T. Yu et L. T. Lu = Spiraea tomentulosa（Te T. Yu）Y. Z. Zhao ●

372029　Spiraea nipponica Maxim.；东瀛绣线菊（日本绣线菊）；Nippon Spirea ●☆

372030　Spiraea nipponica Maxim. 'Halward's Silver'；哈尔沃德银东瀛绣线菊（哈尔沃德银日本绣线菊）；Nippon Spirea ●☆

372031　Spiraea nipponica Maxim. 'Rotundifolia'；圆叶东瀛绣线菊（圆叶日本绣线菊）●☆

372032　Spiraea nipponica Maxim. 'Snowmound'；雪堆东瀛绣线菊（雪堆日本绣线菊）；Snowmound Spirea ●☆

372033　Spiraea nipponica Maxim. f. oblanceolata（Nakai）Ohwi = Spiraea nipponica Maxim. ●☆

372034　Spiraea nipponica Maxim. f. rotundifolia（G. Nicholson）Makino = Spiraea nipponica Maxim. 'Rotundifolia' ●☆

372035　Spiraea nipponica Maxim. f. rotundifolia（G. Nicholson）Makino = Spiraea nipponica Maxim. ●☆

372036　Spiraea nipponica Maxim. var. ogawae（Nakai）T. Yamanaka；小川绣线菊；Ogawa Spiraea ●☆

372037　Spiraea nipponica Maxim. var. tosaensis（Yatabe）Makino = Spiraea tosaensis Yatabe ●☆

372038　Spiraea nishimurae Kitag.；金州绣线菊；Chinchow Spiraea，Jinzhou Spiraea ●

372039　Spiraea obtusa Nakai = Spiraea blumei G. Don ●

372040　Spiraea ogawai Nakai = Spiraea nipponica Maxim. var. ogawae（Nakai）T. Yamanaka ●☆

372041　Spiraea opulifolia L. = Physocarpus opulifolius（L.）Maxim. ●☆

372042　Spiraea ouensanensis H. Lév. = Spiraea pubescens Turcz. ●

372043　Spiraea ovalis Rehder；广椭绣线菊；Oval Spiraea ●

372044　Spiraea palmata Pall. = Filipendula palmata（Pall.）Maxim. ■

372045　Spiraea papillosa Rehder；乳突绣线菊；Papillose Spiraea ●

372046　Spiraea papillosa Rehder var. yunnanensis Te T. Yu；云南乳突绣线菊；Yunnan Papillose Spiraea ●

372047　Spiraea pilosa Franch.；柔毛绣线菊●☆

372048　Spiraea prattii C. K. Schneid. = Spiraea rosthornii E. Pritz. ex Diels ●

372049　Spiraea prostrata Maxim.；平卧绣线菊；Prostrate Spiraea ●

372050　Spiraea prunifolia Siebold et Zucc.；李叶绣线菊（单瓣绣线菊，李叶笑靥花，蚬花，小米花，小叶米筛草，笑靥花，玉屑，珍珠绣线菊，中华绣线菊）；Bridal Wreath，Bridal Wreath Spiraea，Bridalwreath，Bridal-wreath Spiraea，Bridalwreath Spirea，Bridal-wreath Spirea，Plumleaf Spiraea，Shoe Button Spiraea ●

372051　Spiraea prunifolia Siebold et Zucc. 'Plena'；重瓣李叶绣线菊●

372052　Spiraea prunifolia Siebold et Zucc. f. pseudoprunifolia（Hayata ex Nakai）H. L. Li = Spiraea prunifolia Siebold et Zucc. f. pseudoprunifolia（Hayata ex Nakai）Kitam. ●

372053　Spiraea prunifolia Siebold et Zucc. f. pseudoprunifolia（Hayata ex Nakai）Kitam.；多毛李叶绣线菊（假笑靥花，笑靥花）；Hairy Bridalwreath Spiraea ●

372054　Spiraea prunifolia Siebold et Zucc. f. simpliciflora Nakai = Spiraea prunifolia Siebold et Zucc. var. simpliciflora Nakai ●

372055　Spiraea prunifolia Siebold et Zucc. var. hupehensis（Rehder）Rehder；无毛李叶绣线菊●

372056　Spiraea prunifolia Siebold et Zucc. var. plena C. K. Schneid. = Spiraea prunifolia Siebold et Zucc. ●

372057　Spiraea prunifolia Siebold et Zucc. var. plena C. K. Schneid. = Spiraea prunifolia Siebold et Zucc. 'Plena' ●

372058　Spiraea prunifolia Siebold et Zucc. var. prunifolia f. pseudoprunifolia（Hayata ex Nakai）Kitam. = Spiraea prunifolia Siebold et Zucc. var. pseudoprunifolia（Hayata ex Nakai）H. L. Li ●

372059　Spiraea prunifolia Siebold et Zucc. var. pseudoprunifolia（Hayata ex Nakai）H. L. Li = Spiraea prunifolia Siebold et Zucc. f. pseudoprunifolia（Hayata ex Nakai）Kitam. ●

372060　Spiraea prunifolia Siebold et Zucc. var. pseudoprunifolia（Hayata）H. L. Li = Spiraea prunifolia Siebold et Zucc. f. pseudoprunifolia（Hayata ex Nakai）Kitam. ●

372061　Spiraea prunifolia Siebold et Zucc. var. pseudoprunifolia（Hayata）Kitam. = Spiraea prunifolia Siebold et Zucc. f. pseudoprunifolia（Hayata ex Nakai）H. L. Li ●

372062　Spiraea prunifolia Siebold et Zucc. var. simpliciflora Nakai；单瓣李叶绣线菊（单瓣笑靥花）；Simpleflower Bridalwreath Spiraea ●

372063　Spiraea prunifolia Siebold et Zucc. var. simplicifolia Nakai = Spiraea prunifolia Siebold et Zucc. ●

372064　Spiraea prunifolia Siebold et Zucc. var. typica C. K. Schneid. = Spiraea prunifolia Siebold et Zucc. var. pseudoprunifolia（Hayata）H. L. Li ●

372065　Spiraea pseudoprunifolia Hayata ex Nakai = Spiraea prunifolia Siebold et Zucc. f. pseudoprunifolia（Hayata ex Nakai）Kitam. ●

372066　Spiraea pseudoprunifolia Hayata ex Nakai = Spiraea prunifolia Siebold et Zucc. var. pseudoprunifolia（Hayata）H. L. Li ●

372067　Spiraea pubescens Turcz.；土庄绣线菊（蚂蚱腿，柔毛绣线菊，石蒡子，土庄花，小叶石棒子）；Pubescent Spiraea ●

372068　Spiraea pubescens Turcz. var. lasiocarpa Nakai；毛果土庄绣线菊；Woollyfruit Pubescent Spiraea ●

372069　Spiraea purpurea Hand.-Mazz.；紫花绣线菊；Purpleflower Spiraea，Purple-flowered Spiraea ●☆

372070　Spiraea reevesiana Lindl. = Spiraea cantoniensis Lour. ●

372071　Spiraea robusta（Hook. f. et Thomson）Hand.-Mazz.；粗壮绣线菊；Robust Spiraea ●

372072　Spiraea rosthornii E. Pritz. ex Diels；南川绣线菊（罗氏绣线菊）；Nanchuan Spiraea，Rosthorn Spiraea，Rosthorn Spirea ●

372073　Spiraea rotundifolia Lindl. = Spiraea canescens D. Don ●

372074　Spiraea rubiacea Wall. = Neillia rubiflora D. Don ●

372075　Spiraea salicifolia L.；柳叶绣线菊（空心柳，马尿溲，绣线菊，珍珠梅）；Aaron's Beard，Aaron's-beard，Bridewort，Bridewort Spiraea，Meadowsweet，Pink Bridewort Spirea，Willow Spiraea，Willowleaf Meadowsweet，Willowleaf Spiraea，Willowleaved Spiraea，Willow-leaved Spiraea ●☆

372076　Spiraea salicifolia L. var. cligodonta Te T. Yu；贫齿柳叶绣线菊（贫齿绣线菊，疏齿绣线菊）；Fewtooth Willowleaf Spiraea ●

372077　Spiraea salicifolia L. var. grosserserrata Liou et Y. X. Liou；巨齿柳叶绣线菊（巨齿绣线菊）；Grosstooth Willowleaf Spiraea ●

372078　Spiraea sargentiana Rehder；茂纹绣线菊（佘坚绣线菊）；Sargent Spiraea ●

372079　Spiraea schneideriana Rehder；川滇绣线菊；Schneider Spiraea ●

372080　Spiraea schneideriana Rehder var. amphidoxa Rehder；无毛川滇绣线菊；Glabrous Schneider Spiraea ●

372081　Spiraea schochiana Rehder；滇中绣线菊；Schoch Spiraea ●

372082　Spiraea schrenkiana（Fisch. et C. A. Mey.）Maxim.；施雷绣线菊●

372083　Spiraea sericea Turcz. ;绢毛绣线菊;Sericeous Spiraea ●

372084　Spiraea sericea Turcz. = Spiraea media F. Schmidt var. sericea（Turcz.）Regel ex Maxim. ●☆

372085　Spiraea sericea Turcz. = Spiraea media F. Schmidt ●

372086　Spiraea siccanea（W. W. Sm.）Rehder;干地绣线菊;Dryground Spiraea,Dry-ground Spiraea ●

372087　Spiraea silvestris Nakai = Spiraea miyabei Koidz. ●

372088　Spiraea simpliciflora（Nakai）Nakai = Spiraea prunifolia Siebold et Zucc. var. simpliciflora Nakai ●

372089　Spiraea sinobrahuica W. W. Sm. = Spiraea yunnanensis Franch. ●

372090　Spiraea sinobrahuica W. W. Sm. var. aridicola W. W. Sm. = Spiraea yunnanensis Franch. ●

372091　Spiraea sorbifolia L. = Sorbaria kirilowii（Regel）Maxim. ●

372092　Spiraea sorbifolia L. = Sorbaria sorbifolia（L.）A. Braun ●

372093　Spiraea sublobata Hand.-Mazz. ;浅裂绣线菊;Sublobed Spiraea ●

372094　Spiraea tarokoensis Hayata;太鲁阁绣线菊（大罗口绣线菊）;Daluokou Spiraea,Taroko Spiraea ●

372095　Spiraea teniana Rehder;伏毛绣线菊;Hairy Spiraea,Ten Spiraea ●

372096　Spiraea teniana Rehder var. mairei H. Lév. = Spiraea mairei（H. Lév.）L. T. Lu ●

372097　Spiraea teretiuscula C. K. Schneid. ;圆枝绣线菊●

372098　Spiraea thalictroides Pall. = Spiraea aquilegiifolia Pall. ●

372099　Spiraea thunbergii Siebold et Zucc. = Spiraea arguta Zabel ●

372100　Spiraea thunbergii Siebold et Zucc. = Spiraea thunbergii Siebold ex Blume ●

372101　Spiraea thunbergii Siebold ex Blume;珍珠绣线菊（喷雪花,绣线菊,雪柳,珍珠花）;Japanese Spirea,Thunberg Spirea ●

372102　Spiraea thunbergii Siebold ex Blume 'Mountain Fuji';富士山绣线菊;White Japanese Spirea ●☆

372103　Spiraea thyrsiflora K. Koch = Neillia thyrsiflora D. Don ●◇

372104　Spiraea tianschanica Pojark. ;天山绣线菊;Tianshan Spiraea ●

372105　Spiraea tibetica Te T. Yu et L. T. Lu = Spiraea xizangensis L. T. Lu ●

372106　Spiraea tomentosa L. ;美国绒毛绣线菊（绒毛绣线菊）;Hard Hack,Hardhack,Hard-hack,Hardhack Spiraea,Silver-leaf,Steeple Bush,Steeplebush,Tomentose Meadowsweet,White Cap,Whiteleaf ●☆

372107　Spiraea tomentosa L. var. rosea（Raf.）Fernald;粉美国绒毛绣线菊;Hard-hack,Steeplebush ●☆

372108　Spiraea tomentosa L. var. rosea（Raf.）Fernald = Spiraea tomentosa L. ●☆

372109　Spiraea tomentulosa（Te T. Yu）Y. Z. Zhao;回折绣线菊（毛枝蒙古绣线菊）;Tomentose Mongolian Spiraea,Tomentose Spiraea ●

372110　Spiraea tomentulosa（Te T. Yu）Y. Z. Zhao = Spiraea mongolica Maxim. var. pubescens Y. Z. Zhao et T. J. Wang ●

372111　Spiraea tortuosa Rehder = Spiraea yunnanensis Franch. f. tortuosa（Rehder）Rehder ●

372112　Spiraea tosaensis Yatabe;土佐绣线菊●☆

372113　Spiraea tosaensis Yatabe = Spiraea nipponica Maxim. var. tosaensis（Yatabe）Makino ●☆

372114　Spiraea trichocarpa Nakai;毛果绣线菊（石蚌树）;Korea Spiraea,Korean Spiraea ●

372115　Spiraea trichocarpa Nakai 'Snow White';雪白花绣线菊●☆

372116　Spiraea triloba L. = Spiraea trilobata L. ●

372117　Spiraea trilobata L. ;三裂绣线菊（三裂叶绣线菊,三桠绣球,石棒子,石朋子,团叶绣球）;Asian Meadowsweet,Threelobe Spiraea,

Three-lobe Spirea,Threelobed Spiraea,Three-lobed Spirea ●

372118　Spiraea trilobata L. 'Fairy Queen';美丽女王三裂绣线菊;Three-lobe Spirea ●☆

372119　Spiraea trilobata L. 'Swan Lake';天鹅湖三裂绣线菊;Three-lobe Spirea ●☆

372120　Spiraea trilobata L. var. pubescens Te T. Yu;毛叶三裂绣线菊;Hairy Threelobed Spiraea ●

372121　Spiraea ulmaria L. = Filipendula ulmaria（L.）Maxim. ■

372122　Spiraea ulmifolia Scop. = Spiraea chamaedryfolia L. var. ulmifolia（Scop.）J. Duvign. ●

372123　Spiraea uratensis Franch. ;乌拉特绣线菊;Urat Spiraea ●

372124　Spiraea ussuriensis Pojark. = Spiraea chamaedryfolia L. var. pilosa（Nakai）H. Hara ●☆

372125　Spiraea ussuriensis Pojark. = Spiraea chamaedryfolia L. ●

372126　Spiraea vanhouttei（Briot）Zabel;菱叶绣线菊（范氏绣线菊）;Bridal Wreath,Bridal-wreath,Bridalwreath Spiraea,Trapaleaf Spiraea,Van Houtte Spiraea,Van Houtte Spirea,Van Houtte's Spiraea,Vanhoutte Spiraea ●

372127　Spiraea veitchii Hemsl. ;鄂西绣线菊（魏忌绣线菊）;Veitch Spiraea,W. Hubei Spiraea ●

372128　Spiraea velutina Franch. ;绒毛绣线菊;Velutinous Spiraea,Velvety Spiraea ●

372129　Spiraea velutina Franch. = Spiraea robusta（Hook. f. et Thomson）Hand.-Mazz. ●

372130　Spiraea velutina Franch. var. glabrescens Te T. Yu et L. T. Lu;脱毛绣线菊;Glabrescent Velvety Spiraea ●

372131　Spiraea venusta Franch. = Filipendula rubra（Hill）B. L. Rob. ■☆

372132　Spiraea vestita Wall. ex Hook. f. = Filipendula vestita（Wall.）Maxim. ■

372133　Spiraea virgata Franch. = Spiraea myrtilloides Rehder ●

372134　Spiraea virginiana Britton;弗吉尼亚绣线菊;Virginia Spirea ●☆

372135　Spiraea wilsonii Duthie ex J. H. Veitch;陕西绣线菊（威氏绣线菊）;E. H. Wilson Spiraea,Shaanxi Spiraea,Wilson Spiraea,Wilson Spirea ●

372136　Spiraea xizangensis L. T. Lu;西藏绣线菊;Tibet Spiraea,Xizang Spiraea ●

372137　Spiraea yatabei Nakai = Spiraea nipponica Maxim. var. tosaensis（Yatabe）Makino ●☆

372138　Spiraea yunnanensis Franch. ;云南绣线菊;Yunnan Spiraea ●

372139　Spiraea yunnanensis Franch. f. tortuosa（Rehder）Rehder;曲枝云南绣线菊;Twisted-branch Yunnan Spiraea ●

372140　Spiraea yunnanensis Franch. var. siccanea W. W. Sm. = Spiraea siccanea（W. W. Sm.）Rehder ●

372141　Spiraea zabeliana Schneid. ;北美绣线菊;Zabel Spirea ●☆

372142　Spiraeaceae Bartl. ;绣线菊科●

372143　Spiraeaceae Bartl. = Rosaceae Juss. (保留科名）●■

372144　Spiraeaceae Bertuch = Spiraeaceae Bartl. ●

372145　Spiraeanthemaceae Doweld = Cunoniaceae R. Br. (保留科名）●☆

372146　Spiraeanthemum A. Gray（1854）;螺花树属●☆

372147　Spiraeanthemum densiflorum Brongn. et Gris;密花螺花树●☆

372148　Spiraeanthemum ellipticum Vieill. ex Pamp. ;椭圆螺花树●☆

372149　Spiraeanthemum lanceolatum L. M. Perry;披针叶螺花树●☆

372150　Spiraeanthemum pubescens Pamp. ;毛螺花树●☆

372151　Spiraeanthus（Fisch. et C. A. Mey.）Maxim. (1879）;螺花蔷薇属（绣线花属）■

372152　Spiraeanthus Maxim. = Spiraeanthus (Fisch. et C. A. Mey.) Maxim. ■

372153　Spiraeanthus schrenckianus Maxim. ;螺花蔷薇(绣线花)■

372154　Spiraeopsis Miq. (1856);拟绣线菊属●☆

372155　Spiraeopsis Miq. = Caldcluvia D. Don ●☆

372156　Spiraeopsis Miq. = Dirhynchosia Blume ●☆

372157　Spiraeopsis celebica Miq. ;拟绣线菊●☆

372158　Spiragyne Neck. = Gentiana L. ■

372159　Spiraia Raf. = Spiraea L. ●

372160　Spiralepis D. Don = Helichrysum Mill. (保留属名)●■

372161　Spiralepis D. Don = Leontonyx Cass. ●■

372162　Spiralepis declinata (L. f.) D. Don = Gnaphalium declinatum L. f. ■☆

372163　Spiralepis squarrosa D. Don = Helichrysum spiralepis Hilliard et B. L. Burtt ●☆

372164　Spiralluma Plowes(1995);螺牛角属●☆

372165　Spiralluma mouretii (A. Chev.) Plowes = Caudanthera edulis (Edgew.) Meve et Liede ■☆

372166　Spirantha Post et Kuntze = Speirantha Baker ■★

372167　Spiranthera A. St. -Hil. (1823);螺药芸香属■☆

372168　Spiranthera Bojer = Merremia Dennst. ex Endl. (保留属名)●■

372169　Spiranthera Hook. = Pronaya Hügel ex Endl. ●☆

372170　Spiranthera Raf. = Eustrephus R. Br. ●☆

372171　Spiranthera aegyptia (L.) Roberty = Merremia aegyptia (L.) Urb. ■☆

372172　Spiranthera odoratissima A. St. -Hil. ;螺药芸香■☆

372173　Spiranthera parviflora Sandwith;小花螺药芸香■☆

372174　Spiranthera turpethum (L.) Bojer. = Convolvulus turpethum L. ■

372175　Spiranthera turpethum (L.) Bojer. = Operculina turpetha (L.) Silva Manso ■

372176　Spiranthes Rich. (1817)(保留属名);绶草属(盘龙参属); Ladies'-tresses, Lady's Tresses, Lady's Tresses Orchid, Lady's-tresses,Ladytress,Spiranthes ■

372177　Spiranthes adnata (Sw.) Benth. ex Fawc. = Pelexia adnata (Sw.) Spreng. ■☆

372178　Spiranthes aestivalis (Poir.) A. Rich. ;夏花绶草; Summer Lady's Tresses ■☆

372179　Spiranthes aestivalis A. Rich. = Spiranthes aestivalis (Poir.) A. Rich. ■☆

372180　Spiranthes africana Lindl. = Benthamia perularioides Schltr. ●☆

372181　Spiranthes amoena (M. Bieb.) Spreng. = Spiranthes sinensis (Pers.) Ames ■

372182　Spiranthes australis (R. Br.) Lindl. = Spiranthes sinensis (Pers.) Ames ■

372183　Spiranthes australis Lindl. = Spiranthes sinensis (Pers.) Ames ■

372184　Spiranthes australis Lindl. var. suishaensis Hayata = Spiranthes sinensis (Pers.) Ames ■

372185　Spiranthes autumnalis (Balb.) Rich. = Spiranthes spiralis (L.) Chevall. ■☆

372186　Spiranthes autumnalis Rich. = Spiranthes spiralis (L.) Chevall. ■☆

372187　Spiranthes beckii (Lindl.) House = Spiranthes tuberosa Raf. ■☆

372188　Spiranthes beckii Lindl. = Spiranthes lacera (Raf.) Raf. ■☆

372189　Spiranthes brevilabris Lindl. var. floridana (Wherry) Luer = Spiranthes floridana (Wherry) Cory ■☆

372190　Spiranthes calcarata (Sw.) Jimenez = Eltroplectris calcarata (Sw.) Garay et H. R. Sweet ■☆

372191　Spiranthes casei Catling et Cruise;凯斯绶草; Case's Lady's-tresses ■☆

372192　Spiranthes cernua (L.) Rich. ;蜡色绶草(俯花绶草);Autumn Tresess, Common Ladies' Tresses, Nodding Ladies' Tresess,Nodding Lady's Tresses, Nodding Lady's-tresses, Nodding Tresess, Screw Auger,Screw Augers ■☆

372193　Spiranthes cernua (L.) Rich. var. incurva Jenn. = Spiranthes cernua (L.) Rich. ■☆

372194　Spiranthes cernua (L.) Rich. var. ochroleuca (Rydb.) Ames = Spiranthes ochroleuca (Rydb.) Rydb. ■☆

372195　Spiranthes cernua (L.) Rich. var. odorata (Nutt.) Correll = Spiranthes odorata (Nutt.) Lindl. ■☆

372196　Spiranthes cernua (L.) Rich. var. parviflora Chapm. = Spiranthes ovalis Lindl. ■☆

372197　Spiranthes cinnabarina (La Llave et Lex.) Hemsl. = Dichromanthus cinnabarinus (La Llave et Lex.) Garay ■☆

372198　Spiranthes cinnabarina Hemsl. ;朱红绶草■☆

372199　Spiranthes confusa (Garay) Kartesz et Gandhi = Deiregyne confusa Garay ■☆

372200　Spiranthes costaricensis Rchb. f. = Beloglottis costaricensis (Rchb. f.) Schltr. ■☆

372201　Spiranthes cranichoides (Griseb.) Cogn. = Cyclopogon cranichoides (Griseb.) Schltr. ■☆

372202　Spiranthes diuretica Lindl. ;利尿绶草■☆

372203　Spiranthes eatonii Ames ex P. M. Br. ;伊顿绶草■☆

372204　Spiranthes elata (Sw.) Rich. = Cyclopogon elatus (Sw.) Schltr. ■☆

372205　Spiranthes exigua Rolfe = Chamaegastrodia vaginata (Hook. f.) Seidenf. ■

372206　Spiranthes floridana (Wherry) Cory;佛罗里达绶草■☆

372207　Spiranthes gracilis (Bigelow) L. C. Beck = Spiranthes lacera (Raf.) Raf. ■☆

372208　Spiranthes gracilis (Bigelow) L. C. Beck var. floridana (Wherry) Correll = Spiranthes floridana (Wherry) Cory ■☆

372209　Spiranthes gracilis Bigelow = Spiranthes lacera (Raf.) Raf. ■☆

372210　Spiranthes grandis (Blume) Hassk. = Goodyera rubicunda (Rchb. f.) J. J. Sm. ■

372211　Spiranthes grayi Ames = Spiranthes tuberosa Raf. ■☆

372212　Spiranthes hongkongensis S. Y. Hu et Barretto;香港绶草■

372213　Spiranthes jaliscana S. Watson = Sacoila lanceolata (Aubl.) Garay ■☆

372214　Spiranthes lacera (Raf.) Raf. ;纤细绶草; Northern Slender Lady's-tresses, Slender Ladies' Tresses, Slender Ladies'-tresses, Slender Lady's Tresses ■☆

372215　Spiranthes lacera (Raf.) Raf. = Spiranthes simpsonii Catling et Sheviak ■☆

372216　Spiranthes lacera (Raf.) Raf. var. gracilis (Bigelow) Luer = Spiranthes lacera (Raf.) Raf. ■☆

372217　Spiranthes lacera (Raf.) Raf. var. gracilis (Bigelow) Luer = Spiranthes gracilis Bigelow ■☆

372218　Spiranthes laciniata (Small) Ames;鹰爪绶草■☆

372219　Spiranthes lancea (Thunb. ex Sw.) Baker, Bakh. et Steenis = Spiranthes sinensis (Pers.) Ames ■☆

372220　Spiranthes lancea (Thunb. ex Sw.) Bakh. et Steenis = Herminium lanceum (Thunb. ex Sw.) Vuijk ■

372221　Spiranthes lancea (Thunb.) Backer = Spiranthes sinensis

（Pers.）Ames ■

372222　Spiranthes lancea（Thunb.）Bakh. et Steenis = Herminium lanceum（Thunb. ex Sw.）Vuijk ■

372223　Spiranthes lancea Backer = Spiranthes sinensis（Pers.）Ames ■

372224　Spiranthes lanceolata（Aubl.）León = Sacoila lanceolata（Aubl.）Garay ■☆

372225　Spiranthes lanceolata（Aubl.）León var. paludicola Luer = Sacoila lanceolata（Aubl.）Garay var. paludicola（Luer）Sauleda, Wunderlin et B. F. Hansen ■☆

372226　Spiranthes lucayana（Britton）Cogn. = Mesadenus lucayanus（Britton）Schltr. ■☆

372227　Spiranthes lucida（H. H. Eaton）Ames；光亮绶草；Broad-leaved Lady's Tresses, Shampoo Orchid, Shining Ladies'-tresses, Shining Lady's Tresses, Wide-leaved Ladies'-tresses, Wide-leaved Lady's Tresses, Yellow-lipped Ladies' Tresses ■☆

372228　Spiranthes macrophylla（D. Don）Spreng. = Herminium macrophyllum（D. Don）Dandy ■

372229　Spiranthes magnicamporum Sheviak；草原绶草；Great Plains Ladies' Tresses, Great Plains lady's-tresses ■☆

372230　Spiranthes obliqua J. J. Sm. = Pelexia obliqua（J. J. Sm.）Garay ■

372231　Spiranthes ochroleuca（Rydb.）Rydb.；浅黄褐绶草■☆

372232　Spiranthes odorata（Nutt.）Lindl.；香绶草■☆

372233　Spiranthes orchioides（Sw.）A. Rich.；红门兰绶草■☆

372234　Spiranthes orchioides（Sw.）A. Rich. = Sacoila lanceolata（Aubl.）Garay ■☆

372235　Spiranthes ovalis Lindl.；十月绶草；October Lady's-tresses, Oval Ladies' Tresses, Oval Lady's-tresses ■☆

372236　Spiranthes ovalis Lindl. var. erostellata Catling；不齐十月绶草；October Lady's-tresses, Oval Lady's-tresses ■☆

372237　Spiranthes parksii Correll；帕克斯绶草■☆

372238　Spiranthes parviflora（Chapm.）Ames = Spiranthes ovalis Lindl. ■☆

372239　Spiranthes plantaginea Raf. = Spiranthes lucida（H. H. Eaton）Ames ■☆

372240　Spiranthes porrifolia Lindl.；展叶绶草■☆

372241　Spiranthes praecox（Walter）S. Watson；早熟绶草■☆

372242　Spiranthes romanzoffiana Cham.；美洲绶草；American Lady's Tresses, Cork Lady's Tresses, Hooded Ladies'-tresses, Hooded Lady's-tresses, Irish Lady's Tresses, Irish Lady's-tresses, Romanzofi's Ladies'-tresses ■☆

372243　Spiranthes romanzoffiana Cham. = Spiranthes simpsonii Catling et Sheviak ■☆

372244　Spiranthes romanzoffiana Cham. var. porrifolia（Lindl.）Ames et Correll = Spiranthes porrifolia Lindl. ■☆

372245　Spiranthes simplex A. Gray = Spiranthes tuberosa Raf. ■☆

372246　Spiranthes simpsonii Catling et Sheviak；辛普森绶草；Lady's-tresses ■☆

372247　Spiranthes sinensis（Pers.）Ames；绶草（爱绶草，大叶青，反皮索，过水龙，海珠草，红龙盘柱，金龙盘树，九龙蛇，懒蛇上树，鲤鱼草，镰刀草，龙抱柱，龙缠柱，马牙七，扭兰，扭扭兰，盘龙参，盘龙草，盘龙棍，盘龙花，盘龙箭，蛇崽草，胜杖草，双瑚草，小猪獠参，笑天龙，一线香，一叶一枝花，一枝枪，䖳，猪鞭草，猪獠子，猪辽参，猪牙参）；China Ladytress, Chinese Ladiesstresses, Chinese Spiranthes, Northern Slender Lady's-tresses, Pink Lady's-tresses, Southern Ladiesstresses ■

372248　Spiranthes sinensis（Pers.）Ames = Spiranthes sinensis

（Pers.）Ames ■

372249　Spiranthes sinensis（Pers.）Ames subsp. australis（R. Br.）Kitam. = Spiranthes sinensis（Pers.）Ames var. amoena（M. Bieb.）H. Hara ■

372250　Spiranthes sinensis（Pers.）Ames var. amoena（M. Bieb.）H. Hara；南盘龙参■

372251　Spiranthes sinensis（Pers.）Ames var. amoena（M. Bieb.）H. Hara = Spiranthes sinensis（Pers.）Ames ■

372252　Spiranthes sinensis（Pers.）Ames var. amoena（M. Bieb.）H. Hara f. albescens（Honda）Honda；白花南盘龙参■☆

372253　Spiranthes sinensis（Pers.）Ames var. amoena（M. Bieb.）H. Hara f. autumnus Tsukaya；秋南盘龙参■☆

372254　Spiranthes sinensis（Pers.）Ames var. amoena（M. Bieb.）H. Hara f. gracilis F. Maek.；纤细南盘龙参■☆

372255　Spiranthes sinensis（Pers.）Ames var. amoena（M. Bieb.）H. Hara f. viridiflora（Makino）Ohwi；绿花南盘龙参■☆

372256　Spiranthes sinensis（Pers.）Ames var. australis（R. Br.）H. Hara et Kitam. = Spiranthes sinensis（Pers.）Ames var. amoena（M. Bieb.）H. Hara ■

372257　Spiranthes sinensis（Pers.）Ames var. australis（R. Br.）H. Hara et S. Kitam. = Spiranthes sinensis（Pers.）Ames ■

372258　Spiranthes speciosa（Jacq.）A. Rich.；美花绶草■☆

372259　Spiranthes spiralis（L.）Chevall.；秋绶草；Autumn Lady's Tresses, Autumn Lady's-tresses, Autumn Ladytress, Common Lady's-tresses, Lady Traces, Lady's Tresses, Lady's-tresses, Stander-grass, Sweet Ballocks, Sweet Cods, Sweet Cullins ■☆

372260　Spiranthes squamulosa（Kunth）León = Sacoila squamulosa（Kunth）Garay ■☆

372261　Spiranthes steigeri Correll = Spiranthes ochroleuca（Rydb.）Rydb. ■☆

372262　Spiranthes storeri Chapm. = Cyclopogon cranichoides（Griseb.）Schltr. ■☆

372263　Spiranthes strateumatica（L.）Lindl. = Pecteilis susannae（L.）Raf. ■

372264　Spiranthes strateumatica（L.）Lindl. = Zeuxine strateumatica（L.）Schltr. ■

372265　Spiranthes stricta（House）A. Nelson = Spiranthes romanzoffiana Cham. ■☆

372266　Spiranthes stricta（Rydb.）A. Nelson = Spiranthes romanzoffiana Cham. ■☆

372267　Spiranthes stylites Lindl. = Spiranthes sinensis（Pers.）Ames ■

372268　Spiranthes suishaensis（Hayata）Hayata = Spiranthes sinensis（Pers.）Ames ■

372269　Spiranthes suishaensis（Hayata）Schltr. = Spiranthes sinensis（Pers.）Ames ■

372270　Spiranthes suishaensis Schltr. = Spiranthes sinensis（Pers.）Ames ■

372271　Spiranthes sunii Boufford et Wen H. Zhang；宋氏绶草■

372272　Spiranthes torta（Thunb.）Garay et H. R. Sweet；旋扭绶草■☆

372273　Spiranthes tortilis（Sw.）A. Rich. = Spiranthes torta（Thunb.）Garay et H. R. Sweet ■☆

372274　Spiranthes tuberosa Raf.；小绶草；Little Ladies'-tresses ■☆

372275　Spiranthes tuberosa Raf. var. gracilis ？；纤小绶草；Little Ladies' Tresses ■☆

372276　Spiranthes tuberosa Raf. var. grayi（Ames）Fernald = Spiranthes tuberosa Raf. ■☆

372277　Spiranthes vernalis Engelm. et Gray；春花绶草；Grass-leaved

Ladies'-tresses, Spring Ladies'-tresses, Spring Lady's-tresses, Twisted Ladies' Tresses ■☆

372278 Spirastigma L'Hér. ex Schult. f. = Pitcairnia L'Hér.（保留属名）■☆

372279 Spirea Pall. = Spiraea L. ●

372280 Spirea Piarre = Aspilia Thouars ■☆

372281 Spirella Costantin(1912);小螺旋萝藦属☆

372282 Spirella robinsonii Costantin;小螺旋萝藦☆

372283 Spirema Benth. = Pratia Gaudich. ■

372284 Spirema Benth. = Speirema Hook. f. et Thomson ■

372285 Spirenia Raf. = Spiraea L. ●

372286 Spiridanthus Fenzl = Monolopia DC. ■☆

372287 Spiridanthus Fenzl ex Endl. = Monolopia DC. ■☆

372288 Spirillus J. Gay = Potamogeton L. ■

372289 Spirocarpus（Ser.）Opiz = Medicago L.（保留属名）●■

372290 Spirocarpus Opiz = Medicago L.（保留属名）●■

372291 Spiroceratium H. Wolff = Pimpinella L. ■

372292 Spirochaeta Turcz. = Elephantopus L. ■

372293 Spirochloe Lunell = Schedonnardus Steud. ■☆

372294 Spiroconus Steven = Trichodesma R. Br.（保留属名）●■

372295 Spirodela Schleid.（1839）;紫萍属（浮萍属）;Duck's-meat, Ducksmet, Greater Duckweed ■

372296 Spirodela maxima McCann = Spirodela polyrrhiza（L.）Schleid. ■

372297 Spirodela oligorhiza（Kurz）Hegelm. = Spirodela punctata（G. Mey.）C. H. Thomps. ■

372298 Spirodela oligorrhiza（Kurz）Hegelm. = Landoltia punctata（G. Mey.）Les et D. J. Crawford ■

372299 Spirodela oligorrhiza（Kurz）Hegelm. = Spirodela punctata（G. Mey.）C. H. Thomps. ■

372300 Spirodela polyrrhiza（L.）Schleid.;紫萍（氽头蕰草,浮瓜叶,浮漂草,浮萍,浮萍草,九子萍,藻,苹,萍,萍子草,水白,水花,水帘,水萍,水萍草,水苏,水薸,田萍,小浮萍,小萍子,紫背浮萍）;Big Duckweed, Common Ducksmeat, Duckmeat, Giant Duckweed, Great Duckweed, Greater Duckweed, Larger Duckweed ■

372301 Spirodela polyrrhiza（L.）Schleid. var. masonii Daubs = Spirodela polyrrhiza（L.）Schleid. ■

372302 Spirodela punctata（G. Mey.）C. H. Thomps. = Landoltia punctata（G. Mey.）Les et D. J. Crawford ■

372303 Spirodela punctata（G. Mey.）Schleid. = Spirodela punctata（G. Mey.）C. H. Thomps. ■

372304 Spirodela sichuanensis M. G. Liu et K. M. Xie;四川紫萍;Sichuan Ducksmeat ■

372305 Spirodiclis Post et Kuntze = Spiradiclis Blume ■●

372306 Spirogardnera Stauffer(1968);螺檀香属☆

372307 Spirogardnera rubescens Stauffer;螺檀香☆

372308 Spirogyna Post et Kuntze = Gentiana L. ■

372309 Spirogyna Post et Kuntze = Spiragyne Neck. ■

372310 Spirolepia Post et Kuntze = Leontonyx Cass. ●■

372311 Spirolepia Post et Kuntze = Spiralepis D. Don ●■

372312 Spiroloba Raf. = Pithecellobium Mart.（保留属名）●

372313 Spirolobae Link = Sapindaceae Juss.（保留科名）●■

372314 Spirolobium Baill.（1889）（保留属名）;旋片木属●☆

372315 Spirolobium Orb. = Prosopis L. ●

372316 Spirolobium cambodianum Baill.;旋片木●☆

372317 Spironema Hochst. = Clerodendrum L. ■

372318 Spironema Lindl. = Callisia Loefl. ■☆

372319 Spironema Lindl. = Rectanthera O. Deg. ■☆

372320 Spironema Raf. = Cassytha L. ■●

372321 Spironema Raf. = Volutella Forssk. ■●

372322 Spironema fragrans Lindl. = Callisia fragrans（Lindl.）Woodson ■☆

372323 Spironema myricoides Hochst. = Rotheca myricoides（Hochst.）Steane et Mabb. ●☆

372324 Spiropetalum Gilg = Rourea Aubl.（保留属名）●

372325 Spiropetalum erythrocarpum Gilg = Rourea solanderi Baker ●☆

372326 Spiropetalum heterophyllum（Baker）Gilg = Rourea solanderi Baker ●☆

372327 Spiropetalum liberosepalum Baker f. = Rourea myriantha Baill. ●☆

372328 Spiropetalum odoratum Gilg = Rourea solanderi Baker ●☆

372329 Spiropetalum polyanthum Gilg = Rourea solanderi Baker ●☆

372330 Spiropetalum reynoldsii（Stapf）G. Schellenb. = Rourea solanderi Baker ●☆

372331 Spiropetalum solanderi（Baker）Gilg = Rourea solanderi Baker ●☆

372332 Spiropodium F. Muell. = Pluchea Cass. ●■

372333 Spirorhynchus Kar. et Kir.（1842）;螺果荠属（螺喙芥属,螺喙荠属）;Spirorrhynchus ■

372334 Spirorhynchus bulleri（Burkill）O. E. Schulz = Spirorhynchus sabulosus Kar. et Kir. ■

372335 Spirorhynchus sabulosus Kar. et Kir.;螺喙荠;Common Spirorrhynchus ■

372336 Spirorrhynchus bulleri（Burkill）O. E. Schulz = Spirorhynchus sabulosus Kar. et Kir. ■

372337 Spirorrhynchus sabulosus Kar. et Kir. = Spirorhynchus sabulosus Kar. et Kir. ■

372338 Spirosatis Thouars = Habenaria Willd. ■

372339 Spirosatis Thouars = Satyrium Sw.（保留属名）■

372340 Spiroseris Rech. f.（1977）;叶苞苣属■☆

372341 Spiroseris phyllocephala Rech. f.;叶苞苣■☆

372342 Spirospatha Raf. = Homalomena Schott ■

372343 Spirospatha occulta（Lour.）Raf. = Homalomena occulta（Lour.）Schott ■

372344 Spirospermum Thouars(1806);旋子藤属●☆

372345 Spirospermum penduliflorum DC. = Spirospermum penduliflorum Thouars ●☆

372346 Spirospermum penduliflorum Thouars;旋子藤●☆

372347 Spirostachys S. Watson = Allenrolfea Kuntze ●☆

372348 Spirostachys Sond.（1850）;螺穗戟属■☆

372349 Spirostachys Ung. -Sternb. = Heterostachys Ung. -Sternb. ●☆

372350 Spirostachys africana Sond.;非洲螺穗戟;African Sandalwood, Tambootie ■☆

372351 Spirostachys madagascariensis Baill. = Excoecaria madagascariensis（Baill.）Müll. Arg. ●☆

372352 Spirostalis Raf. = Spirostylis Raf. ■☆

372353 Spirostalis Raf. = Thalia L. ■☆

372354 Spirostegia Ivanina(1955);螺盖参属■☆

372355 Spirostegia bucharica（B. Fedtsch.）Ivanina;螺盖参■☆

372356 Spirostemon Griff.（1854）（'Spirastemon'）;螺蕊夹竹桃属●☆

372357 Spirostemon Griff. = Parsonsia R. Br.（保留属名）●

372358 Spirostemon spiralis Griff.;螺蕊夹竹桃●☆

372359 Spirostigma Nees(1847);螺柱头爵床属■☆

372360 Spirostigma Post et Kuntze = Pitcairnia L'Hér.（保留属名）■☆

372361　Spirostigma Post et Kuntze = Spirastigma L'Hér. ex Schult. f. ■☆

372362　Spirostigma hirsutissima Nees；螺柱头爵床■☆

372363　Spirostylis C. Presl ex Schult. et Schult. f. = Struthanthus Mart.（保留属名）●☆

372364　Spirostylis C. Presl（废弃属名）= Loranthus Jacq.（保留属名）●

372365　Spirostylis C. Presl（废弃属名）= Struthanthus Mart.（保留属名）●☆

372366　Spirostylis Nees ex Mart. = Willdenowia Thunb. ■☆

372367　Spirostylis Nees ex Mast. = Willdenowia Thunb. ■☆

372368　Spirostylis Post et Kuntze = Christiana DC. ●☆

372369　Spirostylis Post et Kuntze = Speirostyla Baker ●☆

372370　Spirostylis Raf. = Thalia L. ■☆

372371　Spirotecoma（Baill.）Dalla Torre et Harms（1904）；螺凌霄属 ●☆

372372　Spirotecoma Baill. = Spirotecoma（Baill.）Dalla Torre et Harms ●☆

372373　Spirotecoma Baill. ex Dalla Torre et Harms = Spirotecoma（Baill.）Dalla Torre et Harms ●☆

372374　Spirotecoma rubriflora（Léonard）Alain；红花螺凌霄●☆

372375　Spirotheca Ulbr. = Ceiba Mill. ●

372376　Spirotheros Raf. = Heteropogon Pers. ■

372377　Spirotropis Tul.（1844）；螺骨豆属（圭亚那豆属）■☆

372378　Spirotropis candollei Tul.；螺骨豆■☆

372379　Spitgelia Sch. Bip. = Picris L. ■

372380　Spitzelia aspera Pomel = Leontodon hispanicus Poir. ■☆

372381　Spitzelia aviorum Pomel = Picris asplenioides L. subsp. saharae（Coss. et Kralik）Dobignard ■☆

372382　Spitzelia coronopifolia（Desf.）Sch. Bip. = Picris asplenioides L. ■☆

372383　Spitzelia coronopifolia（Desf.）Sch. Bip. var. pilosa（Delile）Asch. et Schweinf. = Picris asplenioides L. ■☆

372384　Spitzelia cupuligera Durieu = Picris cupuligera（Durieu）Walp. ■☆

372385　Spitzelia getula Pomel = Picris asplenioides L. subsp. saharae（Coss. et Kralik）Dobignard ■☆

372386　Spitzelia saharae Coss. et Kralik = Picris asplenioides L. subsp. saharae（Coss. et Kralik）Dobignard ■☆

372387　Spitzelia saharae Coss. et Kralik var. cyrenaica Pamp. = Picris asplenioides L. subsp. saharae（Coss. et Kralik）Dobignard ■☆

372388　Spitzelia saharae Coss. et Kralik var. getula（Pomel）Batt. = Picris asplenioides L. subsp. saharae（Coss. et Kralik）Dobignard ■☆

372389　Spitzelia willkommii Sch. Bip. = Picris willkommii（Sch. Bip.）Nyman ■☆

372390　Spixia Leandro = Pera Mutis ●☆

372391　Spixia Schrank = Centratherum Cass. ■☆

372392　Splitgerbera Miq. = Boehmeria Jacq. ●

372393　Splitgerbera macrostachya Wight. = Boehmeria macrophylla Hornem. ●■

372394　Spodiadaceae Hassk. = Anacardiaceae R. Br.（保留科名）●

372395　Spodiadaceae Hassk. = Spondiadaceae Kunth ●

372396　Spodias Hassk. = Spondias L. ●

372397　Spodiopogon Fourn. = Erianthus Michx. ■

372398　Spodiopogon Trin.（1820）；大油芒属（大荻属，油芒属）；Greyawngrass，Spodiopogon ■

372399　Spodiopogon angustifolius Trin. = Eulaliopsis binata（Retz.）C. E. Hubb. ■

372400　Spodiopogon arcuatus Nees = Ischaemum fasciculatum Brongn. ■☆

372401　Spodiopogon aureum Hook. et Arn. = Ischaemum aureum（Hook. et Arn.）Hack. ■

372402　Spodiopogon baiyuensis L. Liou；白玉大油芒；Baiyu Greyawngrass，Baiyu Spodiopogon ■

372403　Spodiopogon bambusoides（P. C. Keng）S. M. Phillips et S. L. Chen；竹油芒；Bamboo Eccoilopus，Bamboo Oilawn ■

372404　Spodiopogon bambusoides Keng = Eccoilopus bambusoides P. C. Keng ex L. Liou ■

372405　Spodiopogon binatus（Retz.）Roberty = Eulaliopsis binata（Retz.）C. E. Hubb. ■

372406　Spodiopogon cotulifer（Thunb.）Hack.；油芒（秋茅，山高粱）；Articulation-bearing，Oilawn ■

372407　Spodiopogon cotulifer（Thunb.）Hack. = Eccoilopus cotulifer（Thunb.）A. Camus ■

372408　Spodiopogon depauperatus Hack.；萎缩大油芒■☆

372409　Spodiopogon depauperatus Hack. var. purpurascens Honda = Spodiopogon sibiricus Trin. ■

372410　Spodiopogon dubius Hack.；绒毛大油芒；Floss Greyawngrass，Floss Spodiopogon ■

372411　Spodiopogon duclouxii A. Camus；滇大油芒（杜氏油芒）；Ducloux Spodiopogon，Yunnan Greyawngrass ■

372412　Spodiopogon formosanus Rendle；台湾油芒；Taiwan Eccoilopus，Taiwan Oilawn，Taiwan Spodiopogon ■

372413　Spodiopogon formosanus Rendle = Eccoilopus formosanus（Rendle）A. Camus ■

372414　Spodiopogon gracilis Honda = Spodiopogon tainanensis Hayata ■

372415　Spodiopogon grandiflorus L. Liou；长花大油芒；Longflower Greyawngrass，Longflower Spodiopogon ■

372416　Spodiopogon hayatae Honda = Spodiopogon tainanensis Hayata ■

372417　Spodiopogon hogo-ensis Hayata = Spodiopogon tainanensis Hayata ■

372418　Spodiopogon kawakamii Hayata = Eccoilopus cotulifer（Thunb.）A. Camus ■

372419　Spodiopogon kawakamii Hayata = Spodiopogon formosanus Rendle ■

372420　Spodiopogon kawakamii Hayata var. sativus Honda = Spodiopogon formosanus Rendle ■

372421　Spodiopogon ludingensis L. Liou；泸定大油芒；Luding Greyawngrass，Luding Spodiopogon ■

372422　Spodiopogon obliquivalvis Nees var. villosus Benth. = Ischaemum ciliare Retz. ■

372423　Spodiopogon paucistachyus L. Liou；寡穗大油芒；Fewspike Greyawngrass，Fewspike Spodiopogon ■

372424　Spodiopogon petiolaris Trin. = Microstegium petiolare（Trin.）Bor ■

372425　Spodiopogon ramosus Keng；分枝大油芒；Arrowleaf Greyawngrass，Branched Spodiopogon，Ramose Greyawngrass ■

372426　Spodiopogon ramosus Keng = Spodiopogon tainanensis Hayata ■

372427　Spodiopogon sagittifolius Rendle；箭叶大油芒（茨菇草，催生草，大油芒，灵芝草）；Sagittateleaf Spodiopogon ■

372428　Spodiopogon sibiricus Trin.；大油芒（大荻，山黄菅，油芒）；Frost Grass，Greyawngrass，Siberian Graybeard，Siberian Spodiopogon，Spodiopogon ■

372429　Spodiopogon sibiricus Trin. var. grandiflorus L. Liou ex S. M. Phillips et S. L. Chen；大花大油芒■

372430　Spodiopogon sibiricus Trin. var. purpurascens（Honda）Honda ＝ Spodiopogon sibiricus Trin. ■

372431　Spodiopogon sibiricus Trin. var. tenuis（Kitag.）Kitag.；细大油芒■

372432　Spodiopogon sibiricus Trin. var. tenuis（Kitag.）Kitag. ＝ Spodiopogon sibiricus Trin. ■

372433　Spodiopogon sibiricus Trin. var. tomentosus Koidz. ＝ Spodiopogon sibiricus Trin. ■

372434　Spodiopogon tainanensis Hayata；台南大油芒；Tainan Greyawngrass，Tainan Spodiopogon ■

372435　Spodiopogon tainanensis Hayata f. hayatai（Honda）C. C. Hsu；双梗大油芒；Hayata Tainan Greyawngrass ■

372436　Spodiopogon tainanensis Hayata f. hogoensis（Hayata）C. C. Hsu；短叶大油芒；Shortleaf Tainan Greyawngrass ■

372437　Spodiopogon tainanensis Hayata f. takeoi（Hayata）C. C. Hsu；无脉大油芒；Takeo Tainan Greyawngrass ■

372438　Spodiopogon tainanensis Hayata var. hogoensis（Hayata）Ohwi ＝ Spodiopogon tainanensis Hayata ■

372439　Spodiopogon tainanensis Hayata var. takeoi（Hayata）Honda ＝ Spodiopogon tainanensis Hayata ■

372440　Spodiopogon takeoi Hayata ＝ Spodiopogon tainanensis Hayata ■

372441　Spodiopogon tenuis Kitag. ＝ Spodiopogon sibiricus Trin. ■

372442　Spodiopogon tohoensis Hayata ＝ Eccoilopus formosanus（Rendle）A. Camus ■

372443　Spodiopogon tohoensis Hayata ＝ Spodiopogon formosanus Rendle ■

372444　Spodiopogon villosus L. Liou ＝ Spodiopogon dubius Hack. ■

372445　Spodiopogon villosus Nees ＝ Ischaemum ciliare Retz. ■

372446　Spodiopogon villosus Nees ＝ Ischaemum indicum（Houtt.）Merr. ■

372447　Spodiopogon yuexiensis S. L. Zhong；越西大油芒（白玉大油芒）；Yuexi Greyawngrass，Yuexi Spodiopogon ■

372448　Spogopsis Raf. ＝ Gilia Ruiz et Pav. ■●☆

372449　Spogopsis Raf. ＝ Ipomopsis Michx. ■☆

372450　Sponcopsis Raf. ＝ Spogopsis Raf. ■●☆

372451　Spondiaceae Martinov ＝ Anacardiaceae R. Br.（保留科名）●

372452　Spondiadaceae Kunth ＝ Anacardiaceae R. Br.（保留科名）●

372453　Spondiadaceae Martinov ＝ Anacardiaceae R. Br.（保留科名）●

372454　Spondianthus Engl.（1905）；梅花大戟属●☆

372455　Spondianthus glaber Engl. ＝ Spondianthus preussii Engl. subsp. glaber（Engl.）J. Léonard et Nkounkou●☆

372456　Spondianthus preussii Engl.；梅花大戟●☆

372457　Spondianthus preussii Engl. subsp. glaber（Engl.）J. Léonard et Nkounkou；光梅花大戟●☆

372458　Spondianthus preussii Engl. var. glaber（Engl.）Engl. ＝ Spondianthus preussii Engl. subsp. glaber（Engl.）J. Léonard et Nkounkou●☆

372459　Spondias L.（1753）；槟榔青属；Hog Plum，Imbu，Mombin，Otaheite Apple ●

372460　Spondias acuminata Roxb. ＝ Spondias pinnata（L. f.）Kurz ●

372461　Spondias amara Lam.；苦槟榔青●☆

372462　Spondias axillaris Roxb. ＝ Choerospondias axillaris（Roxb.）B. L. Burtt et A. W. Hill ●

372463　Spondias axillaris Roxb. var. pubinervis Rehder et E. H. Wilson ＝ Choerospondias axillaris（Roxb.）B. L. Burtt et A. W. Hill var. pubinervis（Rehder et E. H. Wilson）Burtt et Hill ●

372464　Spondias birrea A. Rich. ＝ Sclerocarya birrea（A. Rich.）Hochst. ●☆

372465　Spondias bivenomarginalis K. M. Feng et P. I. Mao；皮蛋果●

372466　Spondias chinensis（Merr.）F. P. Metcalf ＝ Spondias lakonensis Pierre ●

372467　Spondias cytherea Sonn.；甜槟榔青（加椰芒，食用槟榔青，太平洋楹梓）；Ambarella，Amboina-berry，Golden Apple，Golden-apple Tree，Hog Plum，Jew Plum，Jew's Plum，Mombin，Otaheite Apple，Otaheite-apple，Tahitian Quince，Vi-apple ●☆

372468　Spondias dulcis Sol. ex G. Forst. ＝ Spondias cytherea Sonn. ●☆

372469　Spondias haplophylla Airy Shaw et Forman；单叶槟榔青；Singleleaf Mombin，Single-leaved Mombin ●

372470　Spondias klaineana（Pierre）Engl. ＝ Antrocaryon klaineanum Pierre ●☆

372471　Spondias lakonensis Pierre；岭南槟榔青（假酸枣，岭南酸枣）；Canton Mombin，Lingnan Mombin ●

372472　Spondias lakonensis Pierre var. hirsuta C. Y. Wu et T. L. Ming；毛叶岭南槟榔青（毛叶岭南酸枣）；Hairy-leaf Canton Mombin ●

372473　Spondias lutea Engl. ＝ Choerospondias axillaris（Roxb.）B. L. Burtt et A. W. Hill ●

372474　Spondias lutea L. ＝ Spondias mombin L. ●

372475　Spondias mangifera Willd. ＝ Spondias pinnata（L. f.）Kurz ●

372476　Spondias microcarpa A. Rich. ＝ Pseudospondias microcarpa（A. Rich.）Engl. ●☆

372477　Spondias mombin L.；黄槟榔青（黄酸枣，蒙滨槟榔青，深黄槟榔青）；Caja Fruit，Hog Plum，Jamaica Plum，Jobo，Leather-coatplum，Monbin，Red Mombin，Spanish Plum，Yellow Mombin，Yellow Mombin Tree，Yellow Spanish Plum，Yucatan Plum ●

372478　Spondias pinnata（L. f.）Kurz；槟榔青（爪哇楹梓）；Amra，Andaman Mombin，Hog-plum ●

372479　Spondias purpurea L.；紫色槟榔青；Hog Plum，Purple Mombin，Red Mombin，Spanish Plum ●☆

372480　Spondias radlkoferi J. D. Sm.；拉得槟榔青●☆

372481　Spondias soyauxii Engl. ＝ Antrocaryon klaineanum Pierre ●☆

372482　Spondias tuberosa Arruda；块状槟榔青；Imbu，Umbu ●☆

372483　Spondias venulosa Mart. ex Engl.；细脉槟榔青●☆

372484　Spondiodes Kuntze ＝ Cnestis Juss. ●

372485　Spondiodes Kuntze ＝ Spondioides Smeathman ex Lam. ●☆

372486　Spondioides Smeathman ex Lam.（1789）；拟槟榔青属●☆

372487　Spondioides Smeathman ex Lam. ＝ Cnestis Juss. ●

372488　Spondioides pruriens Smeathman ex Lam.；拟槟榔青●☆

372489　Spondiopsis Engl.（1895）；类槟榔青属●☆

372490　Spondiopsis Engl. ＝ Commiphora Jacq.（保留属名）●

372491　Spondiopsis trifoliata Engl.；类槟榔青●☆

372492　Spondiopsis trifoliolata Engl. ＝ Commiphora eminii Engl. subsp. trifoliolata（Engl.）J. B. Gillett ●☆

372493　Spondogona Raf.（废弃属名）＝ Dipholis A. DC.（保留属名）●☆

372494　Spondogona Raf.（废弃属名）＝ Sideroxylon L. ●☆

372495　Spondylantha C. Presl ＝ Vitis L. ●

372496　Spondylococcos Mitch. ＝ Callicarpa L. ●

372497　Spondylococcus Rchb. ＝ Bergia L. ●■

372498　Spondylococcus Rchb. ＝ Sphondylococca Willd. ex Schult. ●■

372499　Spongiocarpella Yakovlev et N. Ulziykh. ＝ Chesneya Lindl. ex Endl. ●

372500　Spongiocarpella Yakovlev et N. Ulziykh. ＝ Spongiocarpella Yakovlev et N. Ulziykh. ex Yakovlev et Sviaz. ●

372501　Spongiocarpella Yakovlev et N. Ulziykh. ex Yakovlev et Sviaz.（1987）；海绵豆属●

372502 Spongiocarpella Yakovlev et N. Ulziykh. ex Yakovlev et Sviaz. = Chesneya Lindl. ex Endl. ●

372503 Spongiocarpella grubovii (N. Ulziykh.) Yakovlev;红花海绵豆●

372504 Spongiocarpella grubovii (N. Ulziykh.) Yakovlev = Chesneya macrantha H. S. Cheng ex H. C. Fu ●■

372505 Spongiocarpella intermedia Yakovlev = Chesneya nubigena (D. Don) Ali ●☆

372506 Spongiocarpella intermedia Yakovlev et N. Ulziykh. ex Yakovlev et Sviaz. = Chesneya intermedia (Yakovlev et N. Ulziykh. ex Yakovlev et Sviaz.) Z. G. Qian ●

372507 Spongiocarpella nubigena (D. Don) Ali = Chesneya nubigena (D. Don) Ali ●☆

372508 Spongiocarpella nubigena (D. Don) Ali subsp. kumaoensis Yakovlev = Chesneya nubigena (D. Don) Ali ●☆

372509 Spongiocarpella paucifoliolata Yakovlev = Chesneya nubigena (D. Don) Ali ●☆

372510 Spongiocarpella paucifoliolata Yakovlev et N. Ulziykh. ex Yakovlev et Sviaz. = Chesneya paucifoliolata (Yakovlev et N. Ulziykh. ex Yakovlev et Sviaz.) Z. G. Qian ■

372511 Spongiocarpella paucifoliolata Yakovlev et N. Ulziykh. ex Yakovlev et Sviaz. = Chesneya nubigena (D. Don) Ali ●☆

372512 Spongiocarpella polystichoides (Hand. -Mazz.) Yakovlev = Chesneya polystichoides (Hand. -Mazz.) Ali ●

372513 Spongiocarpella polystichoides (Hand. -Mazz.) Yakovlev et N. Ulziykh. ex Yakovlev et Sviaz. = Chesneya polystichoides (Hand. -Mazz.) Ali ●

372514 Spongiocarpella potaninii Yakovlev = Chesneya macrantha S. H. Cheng ex H. C. Fu ●

372515 Spongiocarpella purpurea (P. C. Li) Yakovlev = Chesneya nubigena (D. Don) Ali subsp. purpurea (P. C. Li) X. Y. Zhu ●

372516 Spongiocarpella purpurea (P. C. Li) Yakovlev et N. Ulziykh. ex Yakovlev et Sviaz. = Chesneya nubigena (D. Don) Ali subsp. purpurea (P. C. Li) X. Y. Zhu ●

372517 Spongiocarpella spinosa (P. C. Li) Yakovlev = Chesneya spinosa P. C. Li ●

372518 Spongiocarpella yunnanensis Yakovlev = Chesneya nubigena (D. Don) Ali ●☆

372519 Spongiocarpella yunnanensis Yakovlev et N. Ulziykh. ex Yakovlev et Sviaz. = Chesneya yunnanensis (Yakovlev et N. Ulziykh. ex Yakovlev et Sviaz.) Z. G. Qian ●☆

372520 Spongiocarpella yunnanensis Yakovlev et N. Ulziykh. ex Yakovlev et Sviaz. = Chesneya nubigena (D. Don) Ali ●☆

372521 Spongiola J. J. Wood et A. L. Lamb(2009);海绵兰属■☆

372522 Spongiola lohokii J. J. Wood et A. L. Lamb;海绵兰●☆

372523 Spongiosperma Zarucchi(1988);绵籽夹竹桃属●☆

372524 Spongiosperma cataractarum Zarucchi;绵籽夹竹桃●☆

372525 Spongiosperma grandiflorum (Huber) Zarucchi;大花绵籽夹竹桃●☆

372526 Spongiosperma macrophyllum (Müll. Arg.) Zarucchi;大叶绵籽夹竹桃●☆

372527 Spongiosyndesmus Gilli = Ladyginia Lipsky ■☆

372528 Spongopyrena Tiegh. = Ouratea Aubl. (保留属名)●

372529 Spongopyrena elongata (Oliv.) Tiegh. = Campylospermum elongatum (Oliv.) Tiegh. ●☆

372530 Spongostemma (Rchb.) Rchb. = Scabiosa L. ●■

372531 Spongostemma Rchb. = Scabiosa L. ●■

372532 Spongostemma Tiegh. = Scabiosa L. ●■

372533 Spongotrichum Nees = Olearia Moench(保留属名)●☆

372534 Sponia Comm. ex Decne. = Trema Lour. ●

372535 Sponia Comm. ex Lam. = Trema Lour. ●

372536 Sponia africana Planch. = Trema orientalis (L.) Blume ●

372537 Sponia amboinensis (Willd.) Decne. = Trema cannabina Lour. ●

372538 Sponia amboinensis (Willd.) Decne. = Trema tomentosa (Roxb.) H. Hara ●

372539 Sponia angustifolia Planch. = Trema angustifolia (Planch.) Blume ●

372540 Sponia argentea Planch. = Trema orientalis (L.) Blume ●

372541 Sponia bracteolata Hochst. = Trema orientalis (L.) Blume ●

372542 Sponia glomerata Hochst. = Trema orientalis (L.) Blume ●

372543 Sponia guineensis (Schumach. et Thonn.) Planch. = Trema orientalis (L.) Blume ●

372544 Sponia hochstetteri Planch. = Trema orientalis (L.) Blume ●

372545 Sponia micrantha (L.) Decne. = Trema micrantha (L.) Blume ●☆

372546 Sponia nitens Planch. = Trema orientalis (L.) Blume ●

372547 Sponia orientalis (L.) Decne. = Trema orientalis (L.) Blume ●

372548 Sponia orientalis (L.) Planch. = Trema orientalis (L.) Blume ●

372549 Sponia orientalis (L.) Planch. var. asperata Solms = Trema orientalis (L.) Blume ●

372550 Sponia politoria Planch. = Trema politoria (Planch.) Blume ●☆

372551 Sponia sampsonii Hance = Trema angustifolia (Planch.) Blume ●

372552 Sponia strigosa Planch. = Trema orientalis (L.) Blume ●

372553 Sponia timorensis Kurz = Trema cannabina Lour. ●

372554 Sponia tomentosa (Roxb.) Planch. = Trema tomentosa (Roxb.) H. Hara ●

372555 Sponia velutina Planch. = Trema tomentosa (Roxb.) H. Hara ●

372556 Sponia virgata Planch. = Trema cannabina Lour. ●

372557 Sponia wightii Planch. = Trema orientalis (L.) Blume ●

372558 Sporabolus Hassk. = Sporobolus R. Br. ■

372559 Sporadanthus F. Muell. = Sporadanthus F. Muell. ex J. Buch. ■☆

372560 Sporadanthus F. Muell. ex J. Buch. (1874);散花帚灯草属■☆

372561 Sporadanthus caudatus (L. A. S. Johnson et O. D. Evans) B. G. Briggs et L. A. S. Johnson;尾状散花帚灯草■☆

372562 Sporadanthus ferrugineus de Lange, Heenan et B. D. Clarkson;锈色散花帚灯草■☆

372563 Sporadanthus gracilis (R. Br.) B. G. Briggs et L. A. S. Johnson;细散花帚灯草■☆

372564 Sporadanthus strictus (R. Br.) B. G. Briggs et L. A. S. Johnson;坚挺散花帚灯草■☆

372565 Sporichloe Pilg. = Schedonnardus Steud. ■☆

372566 Sporichloe Pilg. = Spirochloe Lunell ■☆

372567 Sporledera Bernh. = Ceratotheca Endl. ■●☆

372568 Sporledera kraussiana Bernh. = Ceratotheca triloba (Bernh.) Hook. f. ■☆

372569 Sporledera triloba Bernh. = Ceratotheca triloba (Bernh.) Hook. f. ■☆

372570 Sporobolaceae (Stapf) Herter = Sporobolaceae Herter ■

372571 Sporobolaceae Herter = Gramineae Juss. (保留科名)■●

372572 Sporobolaceae Herter = Poaceae Barnhart(保留科名)■●

372573 Sporobolaceae Herter;鼠尾粟科■

372574 Sporobolus R. Br. (1810);鼠尾粟属;Dropseed, Dropseedgrass, Rush Grass ■

372575　Sporobolus acinifolius Stapf;葡萄叶鼠尾粟■☆

372576　Sporobolus aeneus（Trin.）Kunth;黄铜色鼠尾粟;Brass-coloured Dropseedgrass ■☆

372577　Sporobolus aequiglumis Stapf ex A. Chev. = Sporobolus microprotus Stapf ■☆

372578　Sporobolus affinis A. Rich. = Sporobolus confinis（Steud.）Chiov. ■☆

372579　Sporobolus africanus（Poir.）Robyns et Tournay;非洲鼠尾粟;African Dropseed, Parramatta Grass, Rat-tail Grass ■☆

372580　Sporobolus agrostoides Chiov. ;剪股颖鼠尾粟■☆

372581　Sporobolus airoides（Torr.）Torr. ;碱鼠尾粟;Alkali Dropseed, Alkali Sacaton, Finetop Salt Grass ■☆

372582　Sporobolus albicans Nees;微白鼠尾粟■☆

372583　Sporobolus albomarginatus Stent et J. M. Rattray = Sporobolus cordofanus（Hochst. ex Steud.）Henriq. ex Coss. ■☆

372584　Sporobolus alpicola Hochst. ex A. Rich. = Agrostis sclerophylla C. E. Hubb. ■☆

372585　Sporobolus andongensis Rendle = Sporobolus welwitschii Rendle ■☆

372586　Sporobolus angustifolius A. Rich. ;窄叶鼠尾粟■☆

372587　Sporobolus angustifolius De Wild. = Sporobolus myrianthus Benth. ■☆

372588　Sporobolus angustus Mez ex Peter = Sporobolus subulatus Hack. ■☆

372589　Sporobolus arabicus Boiss. = Sporobolus ioclados（Nees ex Trin.）Nees ■☆

372590　Sporobolus arabicus Boiss. var. littoralis Peter = Sporobolus ioclados（Trin.）Nees ■☆

372591　Sporobolus arenarius（Gouan）Duval-Jouve;沙生鼠尾粟;Sandy Dropseedgrass ■☆

372592　Sporobolus arenarius（Gouan）Duval-Jouve = Sporobolus pungens（Schreb.）Kunth ■☆

372593　Sporobolus argutus（Nees）Kunth = Sporobolus coromandelianus（Retz.）Kunth ■

372594　Sporobolus artus Stent = Sporobolus subulatus Hack. ■☆

372595　Sporobolus asper（Michx.）Kunth = Sporobolus compositus（Michx.）Kunth ■☆

372596　Sporobolus asper（Michx.）Kunth var. canovirens（Nash）Shinners = Sporobolus clandestinus（Biehler）Hitchc. ■☆

372597　Sporobolus asper（Michx.）Kunth var. clandestinus（Biehler）Shinners = Sporobolus clandestinus（Biehler）Hitchc. ■☆

372598　Sporobolus asper（Michx.）Kunth var. hookeri（Trin.）Vasey = Sporobolus compositus（Michx.）Kunth ■☆

372599　Sporobolus asper（P. Beauv.）Kunth = Sporobolus compositus（Michx.）Kunth ■☆

372600　Sporobolus asper（P. Beauv.）Kunth var. clandestinus（Biehler）Shinners = Sporobolus clandestinus（Biehler）Hitchc. ■☆

372601　Sporobolus asper（P. Beauv.）Kunth var. drummondii（Trin.）Vasey = Sporobolus compositus（Michx.）Kunth ■☆

372602　Sporobolus asper（P. Beauv.）Kunth var. hookeri（Trin.）Vasey = Sporobolus compositus（Michx.）Kunth ■☆

372603　Sporobolus asper（P. Beauv.）Kunth var. macer（Trin.）Shinners = Sporobolus compositus（Michx.）Kunth ■☆

372604　Sporobolus asper（P. Beauv.）Kunth var. pilosus（Vasey）Hitchc. = Sporobolus compositus（Michx.）Kunth ■☆

372605　Sporobolus asperifolius Nees et Meyen ex Trin. = Muhlenbergia asperifolia（Nees et Meyen ex Trin.）Parodi ■☆

372606　Sporobolus assakae Caball. = Sporobolus robustus Kunth ■☆

372607　Sporobolus barbigerus Franch. = Sporobolus subtilis Kunth ■☆

372608　Sporobolus batesii A. Chev. = Sporobolus africanus（Poir.）Robyns et Tournay ■☆

372609　Sporobolus baumianus Pilg. = Sporobolus welwitschii Rendle ■☆

372610　Sporobolus bechuanicus Gooss. ;贝专鼠尾粟■☆

372611　Sporobolus bequaertii De Wild. = Sporobolus sanguineus Rendle ■☆

372612　Sporobolus bianoensis De Wild. = Sporobolus welwitschii Rendle ■☆

372613　Sporobolus blephariphyllus A. Rich. = Sporobolus discosporus Nees ■☆

372614　Sporobolus braunii Mez = Sporobolus sanguineus Rendle ■☆

372615　Sporobolus breviglumis Hack. ex De Wild. = Sporobolus festivus Hochst. ex A. Rich. ■☆

372616　Sporobolus canovirens Nash = Sporobolus clandestinus（Biehler）Hitchc. ■☆

372617　Sporobolus capensis（P. Beauv.）Kunth = Sporobolus africanus（Poir.）Robyns et Tournay ■☆

372618　Sporobolus capensis（P. Beauv.）Kunth var. laxus Nees = Sporobolus natalensis（Steud.）T. Durand et Schinz ■☆

372619　Sporobolus capillaris A. Chev. = Sporobolus pectinellus Mez ■☆

372620　Sporobolus centrifugus（Trin.）Nees;离心鼠尾粟■☆

372621　Sporobolus centrifugus（Trin.）Nees var. filifolius（Stent）Gooss. = Sporobolus centrifugus（Trin.）Nees ■☆

372622　Sporobolus centrifugus（Trin.）Nees var. laxivaginatus（Stent）Gooss. = Sporobolus centrifugus（Trin.）Nees ■☆

372623　Sporobolus ciliatus De Wild. = Sporobolus uniglumis Stent et J. M. Rattray ■☆

372624　Sporobolus ciliatus De Wild. var. japonicus（Steud.）Hack. = Sporobolus pilifer（Trin.）Kunth ■

372625　Sporobolus clandestinus（Biehler）Hitchc. ;粗糙鼠尾粟;Hidden Drop-seed, Rough Drop-seed, Rough Rush Grass ■☆

372626　Sporobolus clandestinus（Biehler）Hitchc. var. canovirens（Nash）Steyerm. et C. Kucera = Sporobolus clandestinus（Biehler）Hitchc. ■☆

372627　Sporobolus commutatus（Trin.）Kunth = Sporobolus coromandelianus（Retz.）Kunth ■

372628　Sporobolus compactus Clayton;紧密鼠尾粟■☆

372629　Sporobolus compositus（Michx.）Kunth;高大鼠尾粟;Flag Grass, Meadow Drop-seed, Rough Dropseed, Rough Drop-seed, Tall Drop-seed ■☆

372630　Sporobolus compositus（Poir.）Merr. ;牧场鼠尾粟;Flag Grass, Meadow Drop-seed, Rough Dropseed, Rough Drop-seed, Tall Drop-seed ■☆

372631　Sporobolus compositus（Poir.）Merr. = Sporobolus compositus（Michx.）Kunth ■☆

372632　Sporobolus compositus（Poir.）Merr. var. clandestinus（Biehler）Wipff et S. D. Jones = Sporobolus clandestinus（Biehler）Hitchc. ■☆

372633　Sporobolus confertiflorus A. Rich. = Eragrostis japonica（Thunb.）Trin. ■

372634　Sporobolus confertus J. A. Schmidt = Sporobolus minutus Link subsp. confertus（J. A. Schmidt）Lobin et N. Kilian et Leyens ■☆

372635　Sporobolus confinis（Steud.）Chiov. ;邻近鼠尾粟■☆

372636　Sporobolus congoensis Franch. ;刚果鼠尾粟■☆

372637　Sporobolus conrathii Chiov. ;康拉特鼠尾粟■☆

372638　Sporobolus consimilis Fresen. ;相似鼠尾粟■☆

372639　Sporobolus cordofanus（Hochst. ex Steud.）Henriq. ex Coss. ;庙宇鼠尾粟■☆

372640　Sporobolus coromandelianus（Retz.）Kunth;卡鲁满德鼠尾粟■☆

372641　Sporobolus creber De Nardi;紧密细鼠尾粟;Slender Dropseed■☆

372642　Sporobolus cryptandrus（Torr.）A. Gray;隐雄鼠尾粟;Cryptandrous Dropseed,Sand Dropseed,Sand Drop-seed■☆

372643　Sporobolus cryptandrus（Torr.）A. Gray subsp. fuscicola（Hook.）E. K. Jones et Fassett = Sporobolus cryptandrus（Torr.）A. Gray var. fuscicola（Hook.）Pohl■☆

372644　Sporobolus cryptandrus（Torr.）A. Gray var. fuscicola（Hook.）Pohl;沙地隐雄鼠尾粟;Sand Drop-seed■☆

372645　Sporobolus cryptandrus（Torr.）A. Gray var. fusicolor（Hook.）R. W. Pohl = Sporobolus cryptandrus（Torr.）A. Gray■☆

372646　Sporobolus cryptandrus（Torr.）A. Gray var. occidentalis E. K. Jones et Fassett = Sporobolus cryptandrus（Torr.）A. Gray■☆

372647　Sporobolus cryptandrus var. fusicolor（Hook.）R. W. Pohl = Sporobolus cryptandrus（Torr.）A. Gray■☆

372648　Sporobolus densissimus Pilg. ;极密鼠尾粟■☆

372649　Sporobolus deschampsioides P. A. Duvign. = Sporobolus subulatus Hack. ■☆

372650　Sporobolus diander（Retz.）P. Beauv. ;双蕊鼠尾粟;Doubleanther Dropseed, Indian Dropseed, Tussock Dropseed, Twoanther Dropseedgrass, West Indies Smutgrass■

372651　Sporobolus diander（Retz.）P. Beauv. = Sporobolus indicus（L.）R. Br. var. flaccidus（Roem. et Schult.）R. Br. ■

372652　Sporobolus diander P. Beauv. = Sporobolus diander（Retz.）P. Beauv. ■

372653　Sporobolus diffusus Clayton;松散鼠尾粟■☆

372654　Sporobolus dinklagei Mez;丁克鼠尾粟■☆

372655　Sporobolus discosporus Nees;盘籽鼠尾粟■☆

372656　Sporobolus effusus Franch. ;开展鼠尾粟■☆

372657　Sporobolus eichingeri Mez = Sporobolus nervosus Hochst. ■☆

372658　Sporobolus elatior Bosser;高鼠尾粟■☆

372659　Sporobolus elongatus R. Br. ;沟稃草（鼠尾粟）;Australian Dropseedgrass, Elongate Dropseed■☆

372660　Sporobolus elongatus R. Br. = Sporobolus fertilis（Steud.）Clayton■

372661　Sporobolus elongatus R. Br. var. purpureosuffusus（Ohwi）Koyama = Sporobolus fertilis（Steud.）Clayton■

372662　Sporobolus elongatus R. Br. var. purpureosuffusus Ohwi = Sporobolus fertilis（Steud.）Clayton■

372663　Sporobolus engleri Pilg. ;恩格勒鼠尾粟■☆

372664　Sporobolus eylesii Stent et J. M. Rattray = Sporobolus congoensis Franch. ■☆

372665　Sporobolus faucicola Peter = Sporobolus spicatus（Vahl）Kunth■☆

372666　Sporobolus fertilis（Steud.）Clayton;鼠尾粟（长鼠尾粟,老鼠尾,牛夭草,鼠尾粟牛顿草,双蕊鼠尾粟,西印度鼠尾粟,线香草）;Australian Smut-grass, Dropseed, Dropseedgrass, Purple Indian Dropseedgrass,Smut Grass,Smutgrass, West Indian Dropseed■

372667　Sporobolus fertilis（Steud.）Clayton = Sporobolus indicus（L.）R. Br. var. major（Büse）Baaijens■

372668　Sporobolus fertilis（Steud.）Clayton var. pallidior（T. Koyama）Hatus. = Sporobolus fertilis（Steud.）Clayton var. purpureosuffusus

（Ohwi）Ohwi■

372669　Sporobolus fertilis（Steud.）Clayton var. pallidior（T. Koyama）Hatus. = Sporobolus fertilis（Steud.）Clayton var. purpureosuffusus（Ohwi）P. C. Keng et X. S. Shen■

372670　Sporobolus fertilis（Steud.）Clayton var. purpureosuffusus（Ohwi）Ohwi = Sporobolus fertilis（Steud.）Clayton■

372671　Sporobolus fertilis（Steud.）Clayton var. purpureosuffusus（Ohwi）P. C. Keng et X. S. Shen = Sporobolus fertilis（Steud.）Clayton■

372672　Sporobolus festivus Hochst. ex A. Rich. ;华美鼠尾粟■☆

372673　Sporobolus festivus Hochst. ex A. Rich. var. dilloniana Schweinf. = Sporobolus festivus Hochst. ex A. Rich. ■☆

372674　Sporobolus festivus Hochst. ex A. Rich. var. fibrosus Stapf = Sporobolus festivus Hochst. ex A. Rich. ■☆

372675　Sporobolus festivus Hochst. ex A. Rich. var. stuppeus Stapf = Sporobolus stapfianus Gand. ■☆

372676　Sporobolus fibrosus Cope;纤维质鼠尾粟■☆

372677　Sporobolus filifolius Stent = Sporobolus centrifugus（Trin.）Nees■☆

372678　Sporobolus filipes Stapf ex Napper = Sporobolus agrostoides Chiov. ■☆

372679　Sporobolus fimbriatus（Trin.）Nees;流苏鼠尾粟;Fringed Dropseed■☆

372680　Sporobolus fimbriatus（Trin.）Nees var. latifolius Stent = Sporobolus fimbriatus（Trin.）Nees■☆

372681　Sporobolus fimbriatus Nees = Sporobolus fimbriatus（Trin.）Nees■☆

372682　Sporobolus flagelliferus Peter = Sporobolus helvolus（Trin.）T. Durand et Schinz■☆

372683　Sporobolus fourcadei Stent;富尔卡德鼠尾粟■☆

372684　Sporobolus fruticulosus Stapf = Sporobolus ruspolianus Chiov. ■☆

372685　Sporobolus geminatus Clayton;双生鼠尾粟■☆

372686　Sporobolus genalensis Chiov. = Sporobolus ioclados（Trin.）Nees■☆

372687　Sporobolus geniculatus（Nees ex Steud.）Aitch. = Sporobolus tremulus（Willd.）Kunth■☆

372688　Sporobolus ghikae Schweinf. et Volkens;吉卡鼠尾粟■☆

372689　Sporobolus gillii Stent = Sporobolus ioclados（Trin.）Nees■☆

372690　Sporobolus glaucifolius（Hochst. ex Steud.）T. Durand et Schinz = Sporobolus helvolus（Trin.）T. Durand et Schinz■☆

372691　Sporobolus glaucifolius（Steud.）T. Durand et Schinz = Sporobolus helvolus（Trin.）T. Durand et Schinz■☆

372692　Sporobolus glaucus Mez;灰绿鼠尾粟■☆

372693　Sporobolus granularis Mez = Sporobolus stolzii Mez■☆

372694　Sporobolus greenwayi Napper = Sporobolus macranthelus Chiov. ■☆

372695　Sporobolus halophilus Bosser;喜盐鼠尾粟■☆

372696　Sporobolus hancei Rendle;广州鼠尾粟（韩氏鼠尾粟）;Guangzhou Dropseed,Hance Dropseedgrass■

372697　Sporobolus helvolus（Trin.）T. Durand et Schinz;苍白鼠尾粟■☆

372698　Sporobolus heterolepis（A. Gray）A. Gray;异鳞鼠尾粟;Northern Dropseed, Northern Drop-seed, Prairie Droopseed, Prairie Drop-seed,Short-scale Dropseed■☆

372699　Sporobolus hockii De Wild. = Sporobolus sanguineus Rendle■☆

372700 Sporobolus homblei De Wild. = Sporobolus sanguineus Rendle ■☆

372701 Sporobolus hubbardii A. Chev. = Sporobolus subulatus Hack. ■☆

372702 Sporobolus humifusus (Kunth) Kunth var. cordofanus (Hochst. ex Steud.) Massey = Sporobolus cordofanus (Hochst. ex Steud.) Henriq. ex Coss. ■☆

372703 Sporobolus hypseloteros Chiov. = Sporobolus pyramidalis P. Beauv. ■☆

372704 Sporobolus inconspicuus Hack. = Sporobolus rangei Pilg. ■☆

372705 Sporobolus indicus (L.) R. Br. = Sporobolus fertilis (Steud.) Clayton ■

372706 Sporobolus indicus (L.) R. Br. f. pyramidalis (P. Beauv.) Peter = Sporobolus pyramidalis P. Beauv. ■☆

372707 Sporobolus indicus (L.) R. Br. f. spiciformis Kouama = Sporobolus fertilis (Steud.) Clayton ■

372708 Sporobolus indicus (L.) R. Br. subsp. pallidior (T. Koyama) T. Koyama = Sporobolus fertilis (Steud.) Clayton var. purpureosuffusus (Ohwi) Ohwi ■

372709 Sporobolus indicus (L.) R. Br. subsp. pallidior (T. Koyama) T. Koyama = Sporobolus fertilis (Steud.) Clayton ■

372710 Sporobolus indicus (L.) R. Br. subsp. pallidior (T. Koyama) T. Koyama = Sporobolus fertilis (Steud.) Clayton var. purpureosuffusus (Ohwi) P. C. Keng et X. S. Shen ■

372711 Sporobolus indicus (L.) R. Br. subsp. purpureosuffusus (Ohwi) T. Koyama = Sporobolus fertilis (Steud.) Clayton var. purpureosuffusus (Ohwi) Ohwi ■

372712 Sporobolus indicus (L.) R. Br. subsp. purpureosuffusus (Ohwi) T. Koyama = Sporobolus fertilis (Steud.) Clayton var. purpureosuffusus (Ohwi) P. C. Keng et X. S. Shen ■

372713 Sporobolus indicus (L.) R. Br. subsp. purpureosuffusus (Ohwi) T. Koyama = Sporobolus fertilis (Steud.) Clayton ■

372714 Sporobolus indicus (L.) R. Br. var. africanus (Poir.) Jovet et Guédès = Sporobolus africanus (Poir.) Robyns et Tournay ■☆

372715 Sporobolus indicus (L.) R. Br. var. capensis Engl. = Sporobolus africanus (Poir.) Robyns et Tournay ■☆

372716 Sporobolus indicus (L.) R. Br. var. diandrus (Retz.) Jovet et Guédès = Sporobolus indicus (L.) R. Br. var. flaccidus (Roem. et Schult.) Veldkamp ■

372717 Sporobolus indicus (L.) R. Br. var. flaccidus (Roem. et Schult.) R. Br. = Sporobolus indicus (L.) R. Br. var. flaccidus (Roem. et Schult.) Veldkamp ■

372718 Sporobolus indicus (L.) R. Br. var. flaccidus (Roem. et Schult.) Veldkamp = Sporobolus diander (Retz.) P. Beauv. ■

372719 Sporobolus indicus (L.) R. Br. var. flaccidus (Roem. et Schult.) Veldkamp = Sporobolus fertilis (Steud.) Clayton ■

372720 Sporobolus indicus (L.) R. Br. var. flaccidus (Roth ex Roem. et Schult.) Veldkamp = Sporobolus diander (Retz.) P. Beauv. ■

372721 Sporobolus indicus (L.) R. Br. var. laxus (Nees) Stapf = Sporobolus natalensis (Steud.) T. Durand et Schinz ■☆

372722 Sporobolus indicus (L.) R. Br. var. major (Büse) Baaijens = Sporobolus fertilis (Steud.) Clayton ■

372723 Sporobolus indicus (L.) R. Br. var. pallidior T. Koyama = Sporobolus fertilis (Steud.) Clayton var. purpureosuffusus (Ohwi) Ohwi ■

372724 Sporobolus indicus (L.) R. Br. var. pallidior T. Koyama = Sporobolus fertilis (Steud.) Clayton ■

372725 Sporobolus indicus (L.) R. Br. var. pellucidus (Hochst.) Chiov. = Sporobolus pellucidus Hochst. ■☆

372726 Sporobolus indicus (L.) R. Br. var. purpureo-suffusus (Ohwi) T. Koyama = Sporobolus fertilis (Steud.) Clayton ■

372727 Sporobolus indicus (L.) R. Br. var. pyramidalis (P. Beauv.) Veldkamp = Sporobolus pyramidalis P. Beauv. ■☆

372728 Sporobolus indicus (L.) R. Br. var. saxicola Sosef et Ngok;岩生鼠尾粟■☆

372729 Sporobolus infirmus Mez;柔弱鼠尾粟■☆

372730 Sporobolus insulanus Parl. = Sporobolus robustus Kunth ■☆

372731 Sporobolus iocladoides Chiov. = Sporobolus nervosus Hochst. ■☆

372732 Sporobolus ioclados (Nees ex Trin.) Nees;阿拉伯鼠尾粟■☆

372733 Sporobolus ioclados (Trin.) Nees = Sporobolus ioclados (Nees ex Trin.) Nees ■☆

372734 Sporobolus ioclados (Trin.) Nees var. usitatus (Stent) Chippind. = Sporobolus ioclados (Trin.) Nees ■☆

372735 Sporobolus ioclados Hook. f. = Sporobolus nervosus Hochst. ■☆

372736 Sporobolus jacquemontii Kunth = Sporobolus pyramidalis P. Beauv. ■☆

372737 Sporobolus japonicus (Steud.) Maxim. ex Rendle;日本鼠尾粟■☆

372738 Sporobolus japonicus (Steud.) Maxim. ex Rendle = Sporobolus pilifer (Trin.) Kunth ■

372739 Sporobolus kentrophyllus (K. Schum.) Clayton;刺叶鼠尾粟■☆

372740 Sporobolus kentrophyllus (K. Schum.) Clayton = Sporobolus ioclados (Trin.) Nees ■☆

372741 Sporobolus kenyensis Chiov. = Sporobolus angustifolius A. Rich. ■☆

372742 Sporobolus kimuenzaensis Vanderyst = Sporobolus myrianthus Benth. ■☆

372743 Sporobolus kwiluensis Vanderyst = Sporobolus natalensis (Steud.) T. Durand et Schinz ■☆

372744 Sporobolus laetevirens Coss. = Sporobolus ioclados (Trin.) Nees ■☆

372745 Sporobolus lampranthus Pilg. = Sporobolus nervosus Hochst. ■☆

372746 Sporobolus lanuginosus De Wild. = Sporobolus subulatus Hack. ■☆

372747 Sporobolus laxivaginatus Stent = Sporobolus centrifugus (Trin.) Nees ■☆

372748 Sporobolus ledermannii Mez = Sporobolus subulatus Hack. ■☆

372749 Sporobolus leptostachys Ficalho et Hiern = Mosdenia leptostachys (Ficalho et Hiern) Clayton ■☆

372750 Sporobolus littoralis (Lam.) Kunth = Sporobolus virginicus (L.) Kunth ■

372751 Sporobolus longibrachiatus Stapf = Sporobolus nervosus Hochst. ■☆

372752 Sporobolus longifolius (Torr.) A. W. Wood = Sporobolus clandestinus (Biehler) Hitchc. ■☆

372753 Sporobolus ludwigii Hochst.;路德维格鼠尾粟■☆

372754 Sporobolus macranthelus Chiov.;多小花鼠尾粟■☆

372755 Sporobolus macrothrix Pilg. = Sporobolus welwitschii Rendle ■☆

372756 Sporobolus marginatus Hochst. ex A. Rich. = Sporobolus ioclados (Nees ex Trin.) Nees ■☆

372757 Sporobolus marginatus Hochst. ex A. Rich. var. anceps Chiov. =

Sporobolus ioclados (Trin.) Nees ■☆

372758 Sporobolus marginatus Hochst. ex A. Rich. var. scabrifolius Chiov. = Sporobolus ioclados (Trin.) Nees ■☆

372759 Sporobolus marginatus Hochst. ex A. Rich. var. usitatus (Stent) Chippind. = Sporobolus ioclados (Trin.) Nees ■☆

372760 Sporobolus marlothii Hack. = Sporobolus fimbriatus (Trin.) Nees ■☆

372761 Sporobolus matrella Nees = Sporobolus virginicus (L.) Kunth ■

372762 Sporobolus mauritianus (Steud.) T. Durand et Schinz;毛里求斯鼠尾粟■☆

372763 Sporobolus mayumbensis Franch. = Sporobolus molleri Hack. ■☆

372764 Sporobolus menyharthii Hack. = Sporobolus festivus Hochst. ex A. Rich. ■☆

372765 Sporobolus micranthus (Steud.) T. Durand et Schinz;小花鼠尾粟■☆

372766 Sporobolus micranthus Conrath et Hack. = Sporobolus welwitschii Rendle ■☆

372767 Sporobolus microprotus Stapf = Sporobolus scabriflorus Stapf ex R. L. Massey ■☆

372768 Sporobolus mildbraedii Pilg. ;米尔德鼠尾粟■☆

372769 Sporobolus minimus Cope;极小鼠尾粟■☆

372770 Sporobolus minutiflorus (Trin.) Link = Sporobolus tenuissimus (Mart. ex Schrank) Kuntze ■

372771 Sporobolus minutus Link;微小鼠尾粟■☆

372772 Sporobolus minutus Link subsp. confertus (J. A. Schmidt) Lobin et N. Kilian et Leyens;密集微小鼠尾粟■☆

372773 Sporobolus mobigiensis Vanderyst = Sporobolus congoensis Franch. ■☆

372774 Sporobolus modestus Peter = Sporobolus helvolus (Trin.) T. Durand et Schinz ■☆

372775 Sporobolus molleri Hack. ;摩尔鼠尾粟■☆

372776 Sporobolus montanus Engl. ;山地鼠尾粟■☆

372777 Sporobolus myrianthus Benth. ;多花鼠尾粟■☆

372778 Sporobolus myriostachyus Peter = Sporobolus festivus Hochst. ex A. Rich. ■☆

372779 Sporobolus myxosperma Stapf ex Hutch. et Dalziel = Sporobolus paniculatus (Trin.) T. Durand et Schinz ■☆

372780 Sporobolus natalensis (Steud.) T. Durand et Schinz;纳塔尔鼠尾粟■☆

372781 Sporobolus nebulosus Hack. ;星云鼠尾粟■☆

372782 Sporobolus neglectus Nash;小鼠尾粟;Neglected Dropseed, Poverty Grass, Puff-sheath Drop-seed, Small Drop-seed ■☆

372783 Sporobolus neglectus Nash var. ozarkanus (Fernald) Steyerm. et C. Kucera = Sporobolus ozarkanus Fernald ■☆

372784 Sporobolus nervosus Hochst. ;多脉鼠尾粟■☆

372785 Sporobolus niamensis Mez = Sporobolus sanguineus Rendle ■☆

372786 Sporobolus nitens Stent;光亮鼠尾粟■☆

372787 Sporobolus nogalensis Chiov. = Sporobolus consimilis Fresen. ■☆

372788 Sporobolus olivaceus Napper;橄榄绿鼠尾粟■☆

372789 Sporobolus oxylepsis Mez = Sporobolus sanguineus Rendle ■☆

372790 Sporobolus oxyphyllus Fish;尖叶鼠尾粟■☆

372791 Sporobolus ozarkanus Fernald;奥扎克鼠尾粟;Ozark Dropseed, Ozark Poverty Grass ■☆

372792 Sporobolus pallidus (Nees ex Trin.) Boiss. = Sporobolus ioclados (Nees ex Trin.) Nees ■☆

372793 Sporobolus panicoides A. Rich. ;黍鼠尾粟■☆

372794 Sporobolus paniculatus (Trin.) T. Durand et Schinz;锥形鼠尾粟■☆

372795 Sporobolus paniculatus T. Durand et Schinz = Sporobolus paniculatus (Trin.) T. Durand et Schinz ■☆

372796 Sporobolus parvulus Stent = Sporobolus coromandelianus (Retz.) Kunth ■

372797 Sporobolus patulus Hack. = Sporobolus paniculatus (Trin.) T. Durand et Schinz ■☆

372798 Sporobolus pauciflorus A. Chev. ;少花鼠尾粟■☆

372799 Sporobolus pectinatus Hack. ;篦状鼠尾粟■☆

372800 Sporobolus pectinellus Mez;篦齿鼠尾粟■☆

372801 Sporobolus pellucidus Hochst. ;透明鼠尾粟■☆

372802 Sporobolus phyllotrichus Hochst. = Sporobolus confinis (Steud.) Chiov. ■☆

372803 Sporobolus pilifer (Trin.) Kunth;毛鼠尾粟;Barundi Dropseed, Hairy Dropseed, Hairy Dropseedgrass ■

372804 Sporobolus podotrichus Chiov. = Sporobolus helvolus (Trin.) T. Durand et Schinz ■☆

372805 Sporobolus poiretii (Roem. et Schult.) Hitchc. = Sporobolus fertilis (Steud.) Clayton ■☆

372806 Sporobolus poiretii (Roem. et Schult.) Hitchc. = Sporobolus indicus (L.) R. Br. ■

372807 Sporobolus polycyclus Berhaut = Sporobolus paniculatus (Trin.) T. Durand et Schinz ■☆

372808 Sporobolus praecox A. Chev. = Sporobolus pilifer (Trin.) Kunth ■

372809 Sporobolus psammophilus Stent et J. M. Rattray = Sporobolus micranthus (Steud.) T. Durand et Schinz ■☆

372810 Sporobolus pulvinatus Swallen;具枕鼠尾粟;Cashionshape Dropseedgrass, Cashiony Dropseed ■

372811 Sporobolus pungens (Schreb.) Kunth;刺鼠尾粟■☆

372812 Sporobolus pyramidalis (Lam.) Hitchc. = Sporobolus pyramidalis P. Beauv. ■☆

372813 Sporobolus pyramidalis P. Beauv. ;塔形鼠尾粟(螺纹鼠尾粟); Whorled Dropseed ■☆

372814 Sporobolus pyramidalis P. Beauv. var. jacquemontii Jovet et Guédès = Sporobolus pyramidalis P. Beauv. ■☆

372815 Sporobolus quadratus Clayton;四出鼠尾粟■☆

372816 Sporobolus rangei Pilg. ;朗格鼠尾粟■☆

372817 Sporobolus regularis Mez = Sporobolus micranthus (Steud.) T. Durand et Schinz ■☆

372818 Sporobolus rehmannii Hack. = Sporobolus fimbriatus (Trin.) Nees ■☆

372819 Sporobolus rehmannii Hack. var. hirsutus Peter = Sporobolus fimbriatus (Trin.) Nees ■☆

372820 Sporobolus rhodesiensis Stent et J. M. Rattray = Sporobolus sanguineus Rendle ■☆

372821 Sporobolus robustus Kunth;粗壮鼠尾粟■☆

372822 Sporobolus rueppellianus Fresen. = Sporobolus pyramidalis P. Beauv. ■☆

372823 Sporobolus ruspolianus Chiov. ;鲁斯波利鼠尾粟■☆

372824 Sporobolus sanguineus Rendle;血红鼠尾粟■☆

372825 Sporobolus scabriflorus Stapf ex R. L. Massey;糙花鼠尾粟■☆

372826 Sporobolus schlechteri Schweick. = Sporobolus centrifugus (Trin.) Nees ■☆

372827 Sporobolus schliebenii Pilg. = Sporobolus sanguineus Rendle ■☆

372828 Sporobolus schnellii A. Chev. = Sporobolus infirmus Mez ■☆

372829 Sporobolus schweinfurthii Stapf = Sporobolus sanguineus Rendle ■☆

372830 Sporobolus scitulus Clayton;绮丽鼠尾粟■☆

372831 Sporobolus secundispiculus Chiov. = Sporobolus molleri Hack. ■☆

372832 Sporobolus seineri Mez = Sporobolus ioclados (Trin.) Nees ■☆

372833 Sporobolus semisterilis Chiov. = Sporobolus sanguineus Rendle ■☆

372834 Sporobolus senegalensis Chiov. = Sporobolus robustus Kunth ■☆

372835 Sporobolus senegalensis Chiov. var. glaucifolius (Steud.) Chiov. = Sporobolus helvolus (Trin.) T. Durand et Schinz ■☆

372836 Sporobolus senegalensis Chiov. var. microstachyus ? = Sporobolus helvolus (Trin.) T. Durand et Schinz ■☆

372837 Sporobolus senegalensis Chiov. var. podotrichus (Chiov.) Chiov. = Sporobolus helvolus (Trin.) T. Durand et Schinz ■☆

372838 Sporobolus setarioides Peter = Sacciolepis indica (L.) Chase ■

372839 Sporobolus setifolius Peter = Sporobolus subulatus Hack. ■☆

372840 Sporobolus setulosus (Trin.) N. Terracc. = Urochondra setulosa (Trin.) C. E. Hubb. ■☆

372841 Sporobolus sindicus Stapf ex T. Cooke = Sporobolus tourneuxii Coss. ■☆

372842 Sporobolus sladenianus F. Bolus = Sporobolus nervosus Hochst. ■☆

372843 Sporobolus smutsii Stent = Sporobolus ioclados (Trin.) Nees ■☆

372844 Sporobolus somalensis Chiov. ;索马里鼠尾粟■☆

372845 Sporobolus spicatus (Vahl) Kunth;长穗鼠尾粟■☆

372846 Sporobolus stachyanthus A. Rich. = Sporobolus pilifer (Trin.) Kunth ■

372847 Sporobolus stapfianus Gand. ;施塔普夫鼠尾粟■☆

372848 Sporobolus stelliger P. A. Duvign. et Kiwak = Sporobolus congoensis Franch. ■☆

372849 Sporobolus stenostachyus Peter = Sporobolus consimilis Fresen. ■☆

372850 Sporobolus stocksii Bor = Sporobolus nervosus Hochst. ■☆

372851 Sporobolus stolzii Mez;斯托尔兹鼠尾粟■☆

372852 Sporobolus strictus Franch. = Sporobolus paniculatus (Trin.) T. Durand et Schinz ■☆

372853 Sporobolus stuppeus (Stapf) Stent = Sporobolus stapfianus Gand. ■☆

372854 Sporobolus subglobosus A. Chev. ;亚球形鼠尾粟■☆

372855 Sporobolus subtilis Kunth;纤细鼠尾粟■☆

372856 Sporobolus subulatus Hack. ;钻形鼠尾粟■☆

372857 Sporobolus tenellus (A. Spreng.) Kunth;柔软鼠尾粟■☆

372858 Sporobolus tenuis Stapf ex A. Chev. ;细鼠尾粟■☆

372859 Sporobolus tenuissimus (Mart. ex Schrank) Kuntze;热带鼠尾粟■

372860 Sporobolus tourneuxii Coss. ;图尔鼠尾粟■☆

372861 Sporobolus transvaalensis Gooss. = Sporobolus albicans Nees ■☆

372862 Sporobolus tremulus (Willd.) Kunth;颤鼠尾粟■☆

372863 Sporobolus tremulus Kunth = Sporobolus virginicus (L.) Kunth ■

372864 Sporobolus trichophorus Gand. = Sporobolus helvolus (Trin.) T. Durand et Schinz ■☆

372865 Sporobolus tysonii Stent = Sporobolus centrifugus (Trin.) Nees ■☆

372866 Sporobolus uniflorus (Muhl.) Scribn. et Merr. ;单花鼠尾粟;One-flowered Dropseed ■☆

372867 Sporobolus uniflorus (Muhl.) Scribn. et Merr. = Muhlenbergia uniflora (Muhl.) Fernald ■☆

372868 Sporobolus uniglumis Stent et J. M. Rattray;单颖鼠尾粟■☆

372869 Sporobolus usitatus Stent = Sporobolus ioclados (Trin.) Nees ■☆

372870 Sporobolus vaginiflorus (Torr. ex A. Gray) A. W. Wood;鞘花鼠尾粟;Poverty Drop-seed, Poverty Grass, Poverty-grass, Sheathed Dropseed, Sheathed Drop-seed ■☆

372871 Sporobolus vaginiflorus (Torr. ex A. Gray) A. W. Wood var. inaequalis Fernald = Sporobolus vaginiflorus (Torr. ex A. Gray) A. W. Wood ■☆

372872 Sporobolus vaginiflorus (Torr. ex A. Gray) A. W. Wood var. neglectus (Nash) Scribn. = Sporobolus neglectus Nash ■☆

372873 Sporobolus vaginiflorus (Torr. ex A. Gray) A. W. Wood var. ozarkanus (Fernald) Shinners = Sporobolus ozarkanus Fernald ■☆

372874 Sporobolus vaginifolius (Torr. ex A. Gray) A. W. Wood;鞘叶鼠尾粟■☆

372875 Sporobolus variegatus Stapf = Sporobolus somalensis Chiov. ■☆

372876 Sporobolus verdcourtii Napper = Sporobolus ioclados (Trin.) Nees ■☆

372877 Sporobolus verdcourtii Napper = Sporobolus kentrophyllus (K. Schum.) Clayton ■☆

372878 Sporobolus verticillatus Peter;轮生鼠尾粟■☆

372879 Sporobolus virgatus Mez ex Peter = Sporobolus myrianthus Benth. ■☆

372880 Sporobolus virginicus (L.) Kunth;盐地鼠尾粟;Seashore Dropseed, Seashore Dropseedgrass ■

372881 Sporobolus virginicus (L.) Kunth var. arenarius (Gouan) Maire = Sporobolus pungens (Schreb.) Kunth ■☆

372882 Sporobolus virginicus (L.) Kunth var. gaditanus (Boiss. et Reut.) Kerguélen = Sporobolus pungens (Schreb.) Kunth ■☆

372883 Sporobolus virginicus (L.) Kunth var. major Franch. = Sporobolus virginicus (L.) Kunth ■

372884 Sporobolus vryburgensis Stent = Sporobolus ioclados (Trin.) Nees ■☆

372885 Sporobolus wallichii Munro ex Trimen;瓦丽鼠尾粟■

372886 Sporobolus welwitschii Rendle;韦尔鼠尾粟■☆

372887 Sporobolus wildemannii Chiov. = Sporobolus myrianthus Benth. ■☆

372888 Sporobolus wrightii Munro ex Vasey;赖氏鼠尾粟(瓦丽鼠尾粟);Giant Sacaton ■

372889 Sporoxeia W. W. Sm. (1917);八蕊花属;Sporoxeia ●★

372890 Sporoxeia clavicalcarata C. Chen;棒距八蕊花(娘阿拔翠);Clavate Sporoxeia, Clavatespur Sporoxeia, Clubspur Sporoxeia ●

372891 Sporoxeia fengii S. Y. Hu = Sporoxeia latifolia (H. L. Li) C. Y. Wu et Y. C. Huang ex C. Chen var. fengii (S. Y. Hu) C. Chen ●

372892 Sporoxeia hirsuta (H. L. Li) C. Y. Wu;毛萼八蕊花;Haircalyx Sporoxeia, Hirsute Sporoxeia ●

372893 Sporoxeia hirsuta (H. L. Li) C. Y. Wu = Sporoxeia sciadophila W. W. Sm. ●

372894 Sporoxeia latifolia (H. L. Li) C. Y. Wu et Y. C. Huang = Sporoxeia sciadophila W. W. Sm. ●

372895 Sporoxeia latifolia (H. L. Li) C. Y. Wu et Y. C. Huang ex C. Chen;尖叶八蕊花(宽叶八蕊花);Broadleaf Sporoxeia, Broad-leaved Sporoxeia, Sharpleaf Sporoxeia ●

372896 Sporoxeia latifolia（H. L. Li）C. Y. Wu et Y. C. Huang ex C. Chen var. fengii（S. Y. Hu）C. Chen；光萼八蕊花；Feng Broadleaf Sporoxeia ●

372897 Sporoxeia latifolia（H. L. Li）C. Y. Wu et Y. C. Huang var. fengii（S. Y. Hu）C. Chen = Sporoxeia sciadophila W. W. Sm. ●

372898 Sporoxeia sciadophila W. W. Sm.；八蕊花；Shadiloving Sporoxeia，Shad-loving Sporoxeia，Sporoxeia ●

372899 Sportella Hance = Pyracantha M. Roem. ●

372900 Sportella atalantioides Hance = Pyracantha atalantioides（Hance）Stapf ●

372901 Spraguea Torr.（1851）；斯普马齿苋属■☆

372902 Spraguea Torr. = Cistanthe Spach ■☆

372903 Spraguea monosperma（Greene）Rydb. = Cistanthe monosperma（Greene）Hershk. ■☆

372904 Spraguea pulchella Eastw. = Cistanthe pulchella（Eastw.）Hershk. ■☆

372905 Spraguea pulcherrima A. Heller = Cistanthe monosperma（Greene）Hershk. ■☆

372906 Spraguea umbellata Torr. = Cistanthe umbellata（Torr.）Hershk. ■☆

372907 Spraguea umbellata Torr. var. caudicifera A. Gray = Cistanthe umbellata（Torr.）Hershk. ■☆

372908 Spragueanella Balle（1954）；斯普寄生属●☆

372909 Spragueanella curta Wiens et Polhill；斯普寄生●☆

372910 Spragueanella rhamnifolia（Engl.）Balle；鼠李叶斯普寄生●☆

372911 Sprekelia Heist. = Sprekelia Herb. ■

372912 Sprekelia Herb.（1755）；龙头花属（燕水仙属）；Aztec Lily，Azteclily，Dragonheadflower，Jacobean Lily，Jacobeanlily，St. James Lily ■

372913 Sprekelia formosissima（L.）Herb.；龙头花（火燕兰，燕水仙）；Aztec Lily，Azteclily，Dragonheadflower，Jacobean Amaryllis，Jacobean Lily，Jacobeanlily，Malta Lily，St. James Lily，St. James' Lily ■

372914 Sprengalia Steud. = Sprengelia Sm. ●☆

372915 Sprengelia Schult. = Melhania Forssk. ●■

372916 Sprengelia Sm.（1794）；昙石南属（湿生石南属）●☆

372917 Sprengelia incarnata Sm.；粉红昙石南（粉红湿生石南）●☆

372918 Sprengeria Greene = Lepidium L. ■

372919 Springalia DC. = Sprengelia Sm. ●☆

372920 Springia Heurck et Müll. Arg. = Ichnocarpus R. Br.（保留属名）●■

372921 Springia Van Heurck et Müll. Arg. = Ichnocarpus R. Br.（保留属名）●■

372922 Springula Noronha = Hygrophila R. Br. ●■

372923 Sprucea Benth. = Simira Aubl. ■☆

372924 Spruceanthus Sleumer = Hasseltia Kunth ●☆

372925 Spruceanthus Sleumer = Neosprucea Sleumer ●☆

372926 Sprucella Pierre = Micropholis（Griseb.）Pierre ●☆

372927 Sprucina Nied. = Diplopterys A. Juss. ●☆

372928 Sprucina Nied. = Jubelina A. Juss. ●☆

372929 Sprunera Sch. Bip. ex Hochst. = Sphaeranthus L. ■

372930 Sprunnera Sch. Bip. = Codonocephalum Fenzl ■☆

372931 Sprunnera Sch. Bip. = Inula L. ●■

372932 Spryginia Popov = Orychophragmus Bunge ■

372933 Spryginia Popov（1923）；斯皮里芥属■☆

372934 Spryginia winkleri Popov；斯皮里芥■☆

372935 Spuricianthus Szlach. et Marg.（2001）；假钻花兰属■☆

372936 Spuricianthus Szlach. et Marg. = Acianthus R. Br. ■☆

372937 Spuriodaucus C. Norman（1930）；异萝卜属■☆

372938 Spuriodaucus asper C. Norman；异萝卜■☆

372939 Spuriodaucus atropurpureus C. Norman = Physotrichia atropurpurea（C. Norman）Cannon ■☆

372940 Spurionucaceae Dulac = Ambrosiaceae Martinov ●■

372941 Spuriopimpinella（H. Boissieu）Kitag.（1941）；大叶芹属■

372942 Spuriopimpinella（H. Boissieu）Kitag. = Pimpinella L. ■

372943 Spuriopimpinella Kitag. = Spuriopimpinella（H. Boissieu）Kitag. ■

372944 Spuriopimpinella brachycarpa（Kom.）Kitag. = Pimpinella brachycarpa（Kom.）Nakai ■

372945 Spuriopimpinella calycina（Maxim.）Kitag. = Pimpinella calycina Maxim. ■

372946 Spuriopimpinella komarovii Kitag. = Pimpinella komarovii（Kitag.）R. H. Shan et F. T. Pu ■

372947 Spuriopimpinella koreana（Nakai）Kitag. = Pimpinella koreana（Y. Yabe）Nakai ■

372948 Spuriopimpinella koreana（Y. Yabe）Kitag. = Pimpinella koreana（Y. Yabe）Nakai ■

372949 Spuriopimpinella nikoensis（Y. Yabe ex Makino et Nemoto）Kitag. = Spuriopimpinella koreana（Y. Yabe）Kitag. ■

372950 Spyridauthus Wirtst. = Monolopia DC. ■☆

372951 Spyridauthus Wirtst. = Spiridanthus Fenzl ex Endl. ■☆

372952 Spyridium Fenzl（1837）；篮鼠李属●☆

372953 Spyridium glaucum Rye；篮鼠李●☆

372954 Squamaria Ludw.（1757）；鳞玄参属■☆

372955 Squamaria Ludw. = Anblatum Hill ■

372956 Squamaria Ludw. = Lathraea L. ■

372957 Squamaria orobanche Scop.；鳞玄参■☆

372958 Squamataxus J. Nelson = Saxegothaea Lindl.（保留属名）●☆

372959 Squamellaria Becc.（1886）；小鳞茜属☆

372960 Squamellaria imberbis Becc.；小鳞茜☆

372961 Squamopappus R. K. Jansen, N. A. Harriman et Urbatsch（1982）；冠鳞菊属●☆

372962 Squamopappus skutchii（S. F. Blake）R. K. Jansen；冠鳞菊■☆

372963 Squibbia Raf. = Sesuvium L. ●☆

372964 Squilla Steinh. = Charybdis Speta ■☆

372965 Squilla Steinh. = Urginea Steinh. ■☆

372966 Sredinskya（Stein ex Kusn.）Fed.（1950）；大报春属■☆

372967 Sredinskya（Stein ex Kusn.）Fed. = Primula L. ■☆

372968 Sredinskya（Stein）Fed. = Sredinskya（Stein ex Kusn.）Fed. ■☆

372969 Sredinskya grandis（Trautv.）Fed.；大报春■☆

372970 Sredinskya grandis（Trautv.）Fed. = Primula grandis Trautv. ■☆

372971 Sreemadhavana Rauschert = Aphelandra R. Br. ●■☆

372972 Srutanthus Pritz. = Sruthanthus DC. ●☆

372973 Srutanthus Pritz. = Struthanthus Mart.（保留属名）●☆

372974 Sruthanthus DC. = Struthanthus Mart.（保留属名）●☆

372975 Staavia Dahl（1787）；斯塔树属●☆

372976 Staavia adenandrifolia Eckl. et Zeyh. = Staavia capitella（Thunb.）Sond. ●☆

372977 Staavia brownii Dümmer；布朗斯塔树●☆

372978 Staavia capitella（Thunb.）Sond.；小头斯塔树●☆

372979 Staavia comosa Colozza = Staavia capitella（Thunb.）Sond. ●☆

372980 Staavia dodii Bolus；多德斯塔树●☆

372981　Staavia dregeana C. Presl;德雷斯塔树●☆

372982　Staavia glaucescens E. Mey. = Staavia glutinosa（P. J. Bergius）Dahl ●☆

372983　Staavia globosa Sond. = Staavia capitella（Thunb.）Sond. ●☆

372984　Staavia glutinosa（P. J. Bergius）Dahl;黏性斯塔树●☆

372985　Staavia lateriflora Colozza = Staavia capitella（Thunb.）Sond. ●☆

372986　Staavia nuda Brongn. = Staavia verticillata（L. f.）Pillans ●☆

372987　Staavia phylicoides Pillans;菲利木斯塔树●☆

372988　Staavia radiata（L.）Dahl;辐射斯塔树●☆

372989　Staavia radiata（L.）Dahl var. glabra Sond. = Staavia dregeana C. Presl ●☆

372990　Staavia rupestris Eckl. et Zeyh. = Staavia capitella（Thunb.）Sond. ●☆

372991　Staavia trichotoma（Thunb.）Pillans = Staavia capitella（Thunb.）Sond. ●☆

372992　Staavia verticillata（L. f.）Pillans;轮生斯塔树●☆

372993　Staavia zeyheri Sond. ;泽赫斯塔树●☆

372994　Staberoha Kunth（1841）;纸苞帚灯草属■☆

372995　Staberoha aemula（Kunth）Pillans;匹敌纸苞帚灯草■☆

372996　Staberoha banksii Pillans;班克斯纸苞帚灯草■☆

372997　Staberoha caricina（Mast.）T. Durand et Schinz = Thamnochortus erectus（Thunb.）Mast. ●☆

372998　Staberoha cernua（L. f.）T. Durand et Schinz;俯垂纸苞帚灯草■☆

372999　Staberoha distachyos（Rottb.）Kunth;双穗纸苞帚灯草■☆

373000　Staberoha disticha（Rottb.）T. Durand et Schinz = Restio distichus Rottb. ■☆

373001　Staberoha gracilis（Mast.）T. Durand et Schinz = Thamnochortus gracilis Mast. ●☆

373002　Staberoha imbricata（Thunb.）Kunth = Staberoha distachyos（Rottb.）Kunth ■☆

373003　Staberoha imbricata（Thunb.）Kunth var. stenoptera（Kunth）T. Durand et Schinz = Staberoha aemula（Kunth）Pillans ■☆

373004　Staberoha multispicula Pillans;多细刺纸苞帚灯草■☆

373005　Staberoha ornata Esterh. ;装饰纸苞帚灯草■☆

373006　Staberoha remota Pillans;稀疏纸苞帚灯草■☆

373007　Staberoha stenoptera Kunth = Staberoha aemula（Kunth）Pillans ■☆

373008　Staberoha stokoei Pillans;斯托克纸苞帚灯草■☆

373009　Staberoha vaginata（Thunb.）Pillans;具鞘纸苞帚灯草■☆

373010　Stachiopsis Ikonn. -Gal. = Stachyopsis Popov et Vved. ■

373011　Stachiopsis Popow et Vved. ex Ikonn. -Gal. = Stachyopsis Popov et Vved. ■

373012　Stachis Neck. = Stachys L. ●■

373013　Stachyacanthus Nees（1847）;刺穗爵床属☆

373014　Stachyandra Leroy ex Radcl. -Sm.（1990）;穗蕊大戟属●☆

373015　Stachyandra imberbis（Airy Shaw）Radcl. -Sm. ;穗蕊大戟●☆

373016　Stachyandra viticifolia（Airy Shaw）Radcl. -Sm. ;葡萄叶穗蕊大戟●☆

373017　Stachyanthemum Klotzsch = Cyrilla Garden ex L. ●☆

373018　Stachyanthus Blume = Bulbophyllum Thouars（保留属名）■

373019　Stachyanthus Blume = Phyllorkis Thouars（废弃属名）■

373020　Stachyanthus DC.（废弃属名）= Argyrothamnia Müll. Arg. ●☆

373021　Stachyanthus DC.（废弃属名）= Argyrovernonia MacLeish（废弃属名）■●☆

373022　Stachyanthus DC.（废弃属名）= Argythamnia P. Browne ●☆

373023　Stachyanthus DC.（废弃属名）= Eremanthus Less. ●☆

373024　Stachyanthus DC.（废弃属名）= Stachyanthus Engl.（保留属名）●☆

373025　Stachyanthus Engl.（1897）（保留属名）;穗花茱萸属●☆

373026　Stachyanthus Engl.（保留属名）= Bulbophyllum Thouars（保留属名）■

373027　Stachyanthus cuneatus Sleumer;楔形穗花茱萸●☆

373028　Stachyanthus devredii Boutique;德夫雷穗花茱萸●☆

373029　Stachyanthus donisii（Boutique）Boutique;多尼斯穗花茱萸■☆

373030　Stachyanthus nigeriensis S. Moore;尼日利亚穗花茱萸●☆

373031　Stachyanthus obovatus S. Moore;倒卵穗花茱萸●☆

373032　Stachyanthus occidentalis（Keay et J. Miège）Boutique;西方穗花茱萸●☆

373033　Stachyanthus zenkeri Engl. ;岑克尔穗花茱萸●☆

373034　Stachyarpagophora M. Gómez = Achyranthes L.（保留属名）■

373035　Stachyarpagophora Vaill. ex M. Gómez = Achyranthes L.（保留属名）■

373036　Stachyarrhena Hook. f.（1870）;雄穗茜属☆

373037　Stachyarrhena acuminata Standl. ;渐尖雄穗茜☆

373038　Stachyarrhena acutiloba Steyerm. ;尖裂雄穗茜☆

373039　Stachyarrhena longifolia Hook. f. ;长叶雄穗茜☆

373040　Stachyarrhena reticulata Steyerm. ;网脉雄穗茜☆

373041　Stachyarrhena spicata Hook. f. ;雄穗茜☆

373042　Stachycarpus（Endl.）Tiegh. = Podocarpus Pers.（保留属名）●

373043　Stachycarpus Ticgh. = Podocarpus Pers.（保留属名）●

373044　Stachycephalum Sch. Bip. ex Benth.（1872）;穗头菊属■●☆

373045　Stachycephalum mexicanum Sch. Bip. ex Benth. ;穗头菊■●☆

373046　Stachychrysum Bojer = Piptadenia Benth. ●☆

373047　Stachycnida Post et Kuntze = Pouzolzia Gaudich. ●■

373048　Stachycnida Post et Kuntze = Stachyocnide Blume ●■

373049　Stachycrater Turcz. = Osmelia Thwaites ●☆

373050　Stachydaceae Döll = Labiatae Juss.（保留科名）●■

373051　Stachydaceae Döll = Lamiaceae Martinov（保留科名）●■

373052　Stachydaceae Salisb. = Labiatae Juss.（保留科名）●■

373053　Stachydaceae Salisb. = Lamiaceae Martinov（保留科名）●■

373054　Stachydeoma Small（1903）;北美穗灌属●☆

373055　Stachydeoma angulata Tharp;窄北美穗灌●☆

373056　Stachydeoma ciliata Small;北美穗灌●☆

373057　Stachydesma Willis = Stachydeoma Small ●☆

373058　Stachyobium Rchb. f. = Dendrobium Sw.（保留属名）■

373059　Stachyocnide Blume = Pouzolzia Gaudich. ●■

373060　Stachyococcus Standl.（1936）;穗果茜属●☆

373061　Stachyococcus adinanthus（Standl.）Standl. ;穗果茜●☆

373062　Stachyophorbe（Liebm. ex Mart.）Liebm. = Chamaedorea Willd.（保留属名）●☆

373063　Stachyophorbe Liebm. = Chamaedorea Willd.（保留属名）●☆

373064　Stachyopogon Klotz. = Aletris L. ■

373065　Stachyopogon Klotzsch = Aletris L. ■

373066　Stachyopogon pauciflorus G. Klotz. = Aletris pauciflora（G. Klotz）Franch. ■

373067　Stachyopogon spicata G. Klotz. = Aletris pauciflora（G. Klotz）Franch. var. khasiana（Hook. f.）F. T. Wang et Ts. Tang ■

373068　Stachyopsis Popov et Vved.（1923）;假水苏属;Falsebetony ■

373069　Stachyopsis lamiiflora（Rupr.）Popov et Vved. ;心叶假水苏;Heartleaf Falsebetony ■

373070　Stachyopsis marrubioides（Regel）Ikonn. -Gal. ;多毛假水苏;

Hairy Falsebetony ■

373071 Stachyopsis oblongata (Schrenk ex Fisch. et C. A. Mey.) Popov et Vved.;假水苏;Falsebetony,Oblong Falsebetony ■

373072 Stachyopsis oblongata (Schrenk) Popov et Vved. = Stachyopsis oblongata (Schrenk ex Fisch. et C. A. Mey.) Popov et Vved. ■

373073 Stachyopsis oblongata (Schrenk) Popov et Vved. var. canescens (Regel) Popov et Vved. = Stachyopsis marrubioides (Regel) Ikonn. -Gal. ■

373074 Stachyothyrsus Harms(1897);大穗苏木属■☆

373075 Stachyothyrsus germainii (R. Wilczek) J. Léonard = Stachyothyrsus staudtii Harms ■☆

373076 Stachyothyrsus stapfiana (A. Chev.) J. Léonard et Voorh.;大穗苏木■☆

373077 Stachyothyrsus staudtii Harms;施陶大穗苏木■☆

373078 Stachyothyrsus tessmannii Harms = Pachyelasma tessmannii (Harms) Harms ●☆

373079 Stachyphrynium K. Schum. (1902);竹花柊叶属(穗花柊叶属)■

373080 Stachyphrynium sinense H. Li;竹花柊叶(穗花柊叶)■

373081 Stachyphyllum Tiegh. = Antidaphne Poepp. et Endl. ●☆

373082 Stachypitys A. V. Bobrov et Melikyan = Podocarpus Pers. (保留属名)●

373083 Stachypogon Post et Kuntze = Aletris L. ■

373084 Stachypogon Post et Kuntze = Stachyopogon Klotz. ■

373085 Stachys L. (1753);水苏属;Betony,Hedge Nettle,Hedgenettle,Stachys,Woundwort ●■

373086 Stachys × ambigua Sm.;可疑水苏;Hybrid Woundwort ■☆

373087 Stachys aberdarica T. C. E. Fr. = Stachys aculeolata Hook. f. ■☆

373088 Stachys aculeolata Hook.f.;小皮刺水苏■☆

373089 Stachys aculeolata Hook.f. var. afromontana T. C. E. Fr. = Stachys aculeolata Hook.f. ■☆

373090 Stachys aculeolata Hook.f. var. camerunensis T. C. E. Fr. = Stachys aculeolata Hook.f. ■☆

373091 Stachys aculeolata Hook.f. var. inermis Y. B. Harv.;无刺水苏■☆

373092 Stachys adulterina Hemsl.;少毛甘露子(蚕子);Fewhair Betony ■

373093 Stachys adulterina Hemsl. var. malacotricha Hand.-Mazz.;柔软少毛甘露子■

373094 Stachys aegyptiaca Pers.;埃及水苏■☆

373095 Stachys aethiopica L.;埃塞俄比亚水苏■☆

373096 Stachys aethiopica L. var. glandulifera V. Naray. = Stachys aethiopica L. ■☆

373097 Stachys aethiopica L. var. hispidissima Benth. = Stachys aethiopica L. ■☆

373098 Stachys aethiopica L. var. parviflora V. Naray. = Stachys aethiopica L. ■☆

373099 Stachys aethiopica L. var. tenella Kuntze = Stachys cymbalaria Briq. ■☆

373100 Stachys affinis Bunge;近缘水苏;China Artichoke, China Crosnes, Chinese Artichoke, Crosnes, Japan Artichoke, Japanese Artichoke ■

373101 Stachys affinis Bunge = Stachys sieboldii Miq. ■

373102 Stachys albiflora N. E. Br.;白花水苏■☆

373103 Stachys albotomentosa Ramamoorthy;白毛水苏;Hidalgo ■☆

373104 Stachys alopecuros Benth.;狐尾水苏;Yellow Betony ■☆

373105 Stachys alpigena T. C. E. Fr.;山生水苏■☆

373106 Stachys alpigena T. C. E. Fr. subsp. longipetala Sebsebe;长瓣山生水苏■☆

373107 Stachys alpina L.;高山水苏;Alpine Betony, Alpine Woundwort,Limestone Woundwort ■☆

373108 Stachys andongensis Hiern;安东水苏■☆

373109 Stachys angustifolia M. Bieb.;窄叶水苏;Narrowleaf Betony ■☆

373110 Stachys annua (L.) L.;一年生水苏;Annual Hedgenettle, Annual Yellow Woundwort, Annual Yellow-woundwort, Hedge-nettle Betuny ■☆

373111 Stachys arachnoidea Codd;蛛网水苏■☆

373112 Stachys arenaria Vahl;沙地水苏■☆

373113 Stachys arenaria Vahl subsp. divaricatidens H. Lindb.;叉开沙地水苏■☆

373114 Stachys arenaria Vahl subsp. mollis (Benth.) Gómiz = Stachys arenaria Vahl ■☆

373115 Stachys arenaria Vahl var. angustifolia (Sennen) Guarr. = Stachys arenaria Vahl ■☆

373116 Stachys arenaria Vahl var. latifolia Maire = Stachys arenaria Vahl ■☆

373117 Stachys arenaria Vahl var. maroccana Batt. = Stachys arenaria Vahl ■☆

373118 Stachys arenaria Vahl var. peduncularis (Pau) Maire = Stachys arenaria Vahl ■☆

373119 Stachys arenaria Vahl var. villosissima Andr. = Stachys arenaria Vahl ■☆

373120 Stachys argilacea Pomel = Stachys duriaei Noë ■☆

373121 Stachys argillicola Sebsebe;白土水苏■☆

373122 Stachys arrecta L. H. Bailey;蜗儿菜(宝塔菜,地蚕);Erect Betony ■

373123 Stachys arrecta L. H. Bailey = Stachys geobombycis C. Y. Wu ■

373124 Stachys artemisia Lour. = Leonurus japonicus Houtt. ■

373125 Stachys arvensis L.;田野水苏(草石蚕,甘露子,野水苏);Corn Woundwort, Field Betony, Field Nettle, Field Woundwort, Fieldnettle Betony,Field-nettle Betony,Staggerweed,Stagger-weed ■

373126 Stachys aspera Michx.;粗糙水苏(水苏);Hedge Nettle, Hyssop Hedge Nettle,Rough Hedge-nettle ■

373127 Stachys aspera Michx. = Stachys geobombycis C. Y. Wu ■

373128 Stachys aspera Michx. var. baicalensis (Fisch. ex Benth.) Maxim. = Stachys baicalensis Fisch. ex Benth. ■

373129 Stachys aspera Michx. var. baicalensis Maxim. = Stachys baicalensis Fisch. ex Benth. ■

373130 Stachys aspera Michx. var. chinensis (Bunge ex Benth.) Maxim. = Stachys chinensis Bunge ex Benth. ■

373131 Stachys aspera Michx. var. chinensis (Bunge ex Benth.) Maxim. f. glabrata Nakai = Stachys japonica Miq. ■

373132 Stachys aspera Michx. var. hispidula (Regel) Vorosch. = Stachys baicalensis Fisch. ex Benth. var. hispidula (Regel) Nakai ■

373133 Stachys aspera Michx. var. japonica (Miq.) Maxim. = Stachys japonica Miq. ■

373134 Stachys asperata Hedge;微糙水苏■☆

373135 Stachys aspericaulis Pax = Stachys aculeolata Hook. f. ■☆

373136 Stachys attenuata V. Naray. = Stachys aethiopica L. ■☆

373137 Stachys aurea Benth.;黄水苏■☆

373138 Stachys bachmannii Gürke = Stachys sessilifolia E. Mey. ex Benth. ■☆

373139 Stachys baicalensis Fisch. ex Benth.;毛水苏(陈痧草,鸡苏,芥

苴,芥蒩,劳蒩,龙脑薄荷,朋头草,山升麻,水鸡苏,水苏,水苏草,天芝麻,望江青,乌雷公,香苏,野香苏,野紫苏); Baikal Betony ■

373140   Stachys baicalensis Fisch. ex Benth. = Stachys japonica Miq. ■

373141   Stachys baicalensis Fisch. ex Benth. f. intermedia Kudo = Stachys baicalensis Fisch. ex Benth. var. hispidula (Regel) Nakai ■

373142   Stachys baicalensis Fisch. ex Benth. var. angustifolia Honda;狭叶毛水苏;Narrowleaf Baikal Betony ■

373143   Stachys baicalensis Fisch. ex Benth. var. angustifolia Honda = Stachys riederi Cham. var. hispidula (Regel) H. Hara f. angustifolia (Honda) H. Hara ■

373144   Stachys baicalensis Fisch. ex Benth. var. chinensis (Bunge ex Benth.) Kom. = Stachys chinensis Bunge ex Benth. ■

373145   Stachys baicalensis Fisch. ex Benth. var. hispida (L.) Nakai = Stachys baicalensis Fisch. ex Benth. ■

373146   Stachys baicalensis Fisch. ex Benth. var. hispida (Ledeb.) Nakai = Stachys baicalensis Fisch. ex Benth. ■

373147   Stachys baicalensis Fisch. ex Benth. var. hispidula (Regel) Nakai;小刚毛水苏;Hispid Baikal Betony ■

373148   Stachys baicalensis Fisch. ex Benth. var. hispidula (Regel) Nakai = Stachys aspera Michx. var. hispidula (Regel) Vorosch. ■

373149   Stachys baicalensis Fisch. ex Benth. var. japonica (Miq.) Kom. = Stachys aspera Michx. var. japonica (Miq.) Maxim. ■

373150   Stachys baicalensis Fisch. ex Benth. var. japonica Kom. = Stachys japonica Miq. ■

373151   Stachys balansae Boiss. et Kotschy ex Boiss.;巴兰水苏■☆

373152   Stachys balensis Sebsebe;巴莱水苏■☆

373153   Stachys bambuseti T. C. E. Fr. = Stachys aculeolata Hook. f. ■☆

373154   Stachys betonica Benth. = Betonica officinalis L. ■

373155   Stachys betoniciflora Rupr.;拟水苏■☆

373156   Stachys bizensis Schweinf. ex Baker;比扎水苏■☆

373157   Stachys boissieri Kapeller;布瓦西耶水苏■☆

373158   Stachys bolusii V. Naray.;博卢斯水苏■☆

373159   Stachys brachiata Bojer ex Benth.;短水苏■☆

373160   Stachys brachyclada Noë;短穗水苏■☆

373161   Stachys brachyclada Noë var. immaculata Maire et Wilczek = Stachys brachyclada Noë ■☆

373162   Stachys brachyclada Noë var. punctata Maire = Stachys brachyclada Noë ■☆

373163   Stachys bullata Benth.;泡状水苏(泡水苏);Hedge Nettle ■☆

373164   Stachys burchelliana Launert;布尔水苏■☆

373165   Stachys burchellii Benth. = Stachys burchelliana Launert ■☆

373166   Stachys byzantina K. Koch;绵毛水苏(奥林匹克水苏,厚毛水苏,毛草石蚕,绵水苏,土耳其水苏); Betony, Blanket-leaf, Bunnies' Ears, Bunny's Ear, Bunny's Ears, Donkey's Ear, Donkey's Ears, Lamb's Ear, Lamb's Ears, Lamb's Tongue, Lamb's-ear, Lamb's-tongue, Lanate Betony, Mouse Ear, Mouse-car Betony, Rabbit's Ear, Rabbit's Ears, Saviour's Blanket, Sheep's Ear, Sheep's Ears, Silverleaf, Stachys, Woolly Betony, Woolly Hedeg-nettle, Woolly Hedgenettle, Woolly Woundwort ■

373167   Stachys byzantina K. Koch 'Primrose Heron';黄叶海伦绵毛水苏■☆

373168   Stachys byzantina K. Koch 'Silver Carpet';银毯绵毛水苏■☆

373169   Stachys caffra E. Mey. ex Benth.;开菲尔水苏■☆

373170   Stachys capensis C. Presl = Stachys aethiopica L. ■☆

373171   Stachys cardiophylla Prain ex Dunn = Stachys kouyangensis (Vaniot) Dunn ■

373172   Stachys chanetii H. Lév. = Stachys chinensis Bunge ex Benth. ■

373173   Stachys chinensis Bunge ex Benth.;华水苏(水苏);China Betony, Chinese Betony ■

373174   Stachys chinensis Bunge var. albiflora C. Y. Li;白花华水苏;Whiteflower Chinese Betony ■

373175   Stachys chrysotrichos Gürke = Stachys simplex Schltr. ■☆

373176   Stachys circinata L'Hér.;卷须水苏■☆

373177   Stachys circinata L'Hér. subsp. numidica (Pomel) Batt.;努米底亚水苏■☆

373178   Stachys circinata L'Hér. subsp. zaiana Emb. et Maire;宰哈奈水苏■☆

373179   Stachys coccinea J. Jacq.;红花水苏;Great Hedge-nettle, Green Hedge-nettle, Scarlet Hedge Nettle, Scarlet Hedgenettle, Scarlet Sage, Seven-up Plant, Texas Betony ■☆

373180   Stachys coerulea Burch. ex Benth. = Stachys hyssopoides Burch. ex Benth. ■☆

373181   Stachys comosa Codd;簇毛水苏■☆

373182   Stachys cooperi V. Naray. = Stachys graciliflora C. Presl ■☆

373183   Stachys cordifolia K. Koch = Stachys kouyangensis (Vaniot) Dunn ■

373184   Stachys crenulata Briq. = Stachys rugosa Aiton ■☆

373185   Stachys cretica Sibth. et Sm.;克里特水苏;Cret Betony ■☆

373186   Stachys cuneata Banks ex Benth.;楔形水苏■☆

373187   Stachys cymbalaria Briq.;船状水苏■☆

373188   Stachys cymbalaria Briq. var. alba V. Naray. = Stachys cymbalaria Briq. ■☆

373189   Stachys czernjaevii Des. -Shost.;蔡氏水苏■☆

373190   Stachys denticulata Burch. ex Benth. = Stachys cuneata Banks ex Benth. ■☆

373191   Stachys didymantha Brenan;双花水苏■☆

373192   Stachys dinteri Launert;丁特水苏■☆

373193   Stachys dolichodeira Briq. = Stachys tubulosa MacOwan ■☆

373194   Stachys dregeana Benth.;德雷水苏■☆

373195   Stachys dregeana Benth. var. lasiocalyx (Schltr.) V. Naray. = Stachys dregeana Benth. ■☆

373196   Stachys dregeana Benth. var. tenuior V. Naray. = Stachys dregeana Benth. ■☆

373197   Stachys durandiana Coss.;杜朗水苏■☆

373198   Stachys durandiana Coss. var. grantii (Batt.) Garrigues = Stachys durandiana Coss. ■☆

373199   Stachys duriaei Noë;杜里奥水苏■☆

373200   Stachys duriaei Noë var. argilacea (Pomel) Batt. = Stachys duriaei Noë ■☆

373201   Stachys duriaei Noë var. purpurea Batt. = Stachys duriaei Noë ■☆

373202   Stachys erectiuscula Gürke;稍直立水苏■☆

373203   Stachys erectiuscula Gürke var. natalensis V. Naray. = Stachys erectiuscula Gürke ■☆

373204   Stachys erioleuca Pomel = Stachys circinata L'Hér. ■☆

373205   Stachys filifolia Hedge;线叶水苏■☆

373206   Stachys flavescens Benth.;浅黄水苏■☆

373207   Stachys flexuosa V. Naray.;曲折水苏■☆

373208   Stachys floccosa Benth.;丛卷毛水苏■☆

373209   Stachys floridana Shuttlew. ex Benth.;银苗;Florida Betony ■☆

373210   Stachys foeniculum Pursh = Agastache foeniculum (Pursh) Kuntze ■☆

373211   Stachys foliosa Benth. = Stachys dregeana Benth. ■☆

373212 Stachys fominii Sosn. ;福明水苏■☆

373213 Stachys fontqueri Pau ;丰特水苏■☆

373214 Stachys forsythii Hedge ;福赛斯水苏■☆

373215 Stachys foucauldiana Sennen = Stachys rifana Font Quer et Maire ■☆

373216 Stachys franchetiana H. Lév. = Stachys kouyangensis （Vaniot） Dunn var. franchetiana （H. Lév.） C. Y. Wu ■

373217 Stachys franchetiana H. Lév. = Stachys kouyangensis （Vaniot） Dunn ■

373218 Stachys fruticetorum Briq. = Stachys aethiopica L. ■☆

373219 Stachys fruticulosa M. Bieb. ;小灌木状水苏■☆

373220 Stachys galpinii Briq. = Stachys natalensis Hochst. var. galpinii （Briq.） Codd ■☆

373221 Stachys gariepina Benth. = Stachys flavescens Benth. ■☆

373222 Stachys geobombycis C. Y. Wu ;地蚕（白虫草，白冬虫草，冬虫草，冬虫夏草，肺痨草，黄花地纽菜，土虫草，土冬虫草，五眼草，野麻子）;Earth-silkworm Betony ■

373223 Stachys geobombycis C. Y. Wu var. alba C. Y. Wu et H. W. Li ;白花地蚕（白花地纽菜）;Whiteflower Earth-silkworm Betony ■

373224 Stachys germanica L. ;德国水苏;Downy Woundwort, German Betony, German Hedgenettle, Mouse-ear Betony ■☆

373225 Stachys germanica L. var. canariensis Font Quer et Svent. = Stachys germanica L. ■☆

373226 Stachys glabra Riddell = Stachys tenuifolia Willd. ■☆

373227 Stachys glandulibracteata Y. B. Harv. ;腺苞水苏■☆

373228 Stachys glandulosa Hutch. et E. A. Bruce ;具腺水苏■☆

373229 Stachys glutinosa L. ;胶质水苏■☆

373230 Stachys gossweileri G. Taylor ;戈斯水苏■☆

373231 Stachys graciliflora C. Presl ;细花水苏■☆

373232 Stachys grandiflora Benth. ;广布大花水苏;Big Betony ■☆

373233 Stachys grantii Batt. = Stachys durandiana Coss. ■☆

373234 Stachys grossheimii Kapeller ;格罗氏水苏■☆

373235 Stachys guyoniana Batt. ;居永水苏■☆

373236 Stachys hantamensis Vatke = Stachys aurea Benth. ■☆

373237 Stachys harveyi V. Naray. = Stachys aethiopica L. ■☆

373238 Stachys heterodonta Zefir. ;异齿水苏■☆

373239 Stachys hildebrandtii Vatke ;希尔德水苏■☆

373240 Stachys hirta L. = Stachys ocymastrum （L.） Briq. ■☆

373241 Stachys hirta L. var. hirtula （Pomel） Batt. = Stachys ocymastrum （L.） Briq. ■☆

373242 Stachys hirtula Pomel = Stachys ocymastrum （L.） Briq. ■☆

373243 Stachys hispida （Thunb.） Briq. = Stachys thunbergii Benth. ■☆

373244 Stachys hispida Pursh ;细叶毛水苏;Hairy Hedge Nettle ■☆

373245 Stachys hispida Pursh = Stachys tenuifolia Willd. ■☆

373246 Stachys hispidula Hochst. = Stachys aethiopica L. ■☆

373247 Stachys hissarica Regel ;希萨尔水苏■☆

373248 Stachys homotricha （Fernald） Rydb. = Stachys palustris L. subsp. arenicola （Britton） Gill ■☆

373249 Stachys huillensis Hiern ;威拉水苏■☆

373250 Stachys humifusa Burch. ex Benth. ;平伏水苏■☆

373251 Stachys hydrophila Boiss. ;喜水水苏■☆

373252 Stachys hypoleuca Hochst. ex A. Rich. ;白背水苏■☆

373253 Stachys hyssopifolia Michx. ;北美水苏;Hedge Nettle, Hyssop Hedge Nettle, Hyssop Hedge-nettle ■☆

373254 Stachys hyssopifolia Michx. var. ambigua A. Gray = Stachys aspera Michx. ■

373255 Stachys hyssopoides Burch. ex Benth. ;神香草水苏■☆

373256 Stachys iberica M. Bieb. ;伊比利亚水苏■☆

373257 Stachys imaii Nakai = Stachys oblongifolia Benth. ■

373258 Stachys inflata Benth. ;膀胱水苏■☆

373259 Stachys integrifolia Vahl ex Benth. = Stachys aurea Benth. ■☆

373260 Stachys intermedia Aiton ;间型水苏■☆

373261 Stachys iranica Rchb. f. ;伊朗水苏■☆

373262 Stachys japonica Miq. ;水苏（白根草，白马蓝，还精草，鸡苏，宽叶水苏，泥灯心，日本粗水苏，水鸡苏，天芝麻，望江青，血见愁，野地蚕，银脚鹭鸶，玉蒁草，元宝草，芝麻草）;Japan Betony, Japanese Betony, Water Betony ■

373263 Stachys japonica Miq. = Stachys aspera Michx. var. japonica （Miq.） Maxim. ■

373264 Stachys japonica Miq. f. angustifolia Miq. = Stachys baicalensis Fisch. ex Benth. var. angustifolia Honda ■

373265 Stachys japonica Miq. f. glabrata Matsum. et Kudo = Stachys japonica Miq. ■

373266 Stachys japonica Miq. f. glabrata Matsum. et Kudo ex Kudo = Stachys japonica Miq. ■

373267 Stachys japonica Miq. f. villosa Kudo = Stachys baicalensis Fisch. ex Benth. ■

373268 Stachys japonica Miq. var. intermedia （Kudo） Ohwi = Stachys aspera Michx. var. hispidula （Regel） Vorosch. ■

373269 Stachys japonica Miq. var. intermedia （Kudo） Ohwi f. angustifolia （Honda） Kitag. = Stachys riederi Cham. var. hispidula （Regel） H. Hara f. angustifolia （Honda） H. Hara ■

373270 Stachys japonica Miq. var. villosa （Kudo） Ohwi = Stachys aspera Michx. ■

373271 Stachys jijigaensis Sebsebe ;吉吉加水苏■☆

373272 Stachys karasmontana Dinter = Stachys spathulata Burch. ex Benth. ■☆

373273 Stachys komarovii Knorring ;科马罗夫水苏■☆

373274 Stachys kouyangensis （Vaniot） Dunn ;西南水苏（白根药，冬虫草，麻布草，猫猫菜，破布草，山波罗子，水苏，铁骡子，土石蚕，野甘露）;Guiyang Betony, Kweiyang Betony ■

373275 Stachys kouyangensis （Vaniot） Dunn = Stachys geobombycis C. Y. Wu ■

373276 Stachys kouyangensis （Vaniot） Dunn var. franchetiana （H. Lév.） C. Y. Wu ;粗齿西南水苏（黄狼鼠花）;Thicktooth Guiyang Betony ■

373277 Stachys kouyangensis （Vaniot） Dunn var. leptodon （Dunn） C. Y. Wu ;细齿西南水苏;Thintooth Guiyang Betony ■

373278 Stachys kouyangensis （Vaniot） Dunn var. tuberculata （Hand. - Mazz.） C. Y. Wu ;具瘤西南水苏;Tuberculate Guiyang Betony ■

373279 Stachys kouyangensis （Vaniot） Dunn var. vilosissima C. Y. Wu ;柔毛西南水苏;Softhair Guiyang Betony ■

373280 Stachys kryptantha Range ;隐花水苏■☆

373281 Stachys kulalensis Sebsebe ;库拉尔水苏■☆

373282 Stachys kuntzei Gürke ;孔策水苏■☆

373283 Stachys lagascae Caball. = Stachys ocymastrum （L.） Briq. ■☆

373284 Stachys lamarckii Benth. ;拉马克水苏■☆

373285 Stachys lamiiflora Rupr. = Stachyopsis lamiiflora （Rupr.） Popov et Vved. ■

373286 Stachys lanata Jacq. = Stachys byzantina K. Koch ■

373287 Stachys lasiocalyx Schltr. = Stachys dregeana Benth. ■☆

373288 Stachys lavandulifolia Vahl ;薰衣草叶水苏■☆

373289 Stachys leptoclada Briq. = Stachys natalensis Hochst. ■☆

373290　Stachys leptodon Dunn = Stachys kouyangensis（Vaniot）Dunn var. leptodon（Dunn）C. Y. Wu ■

373291　Stachys leptopoda Hayata = Stachys oblongifolia Benth. var. leptopoda（Hayata）C. Y. Wu ■

373292　Stachys lindblomiana T. C. E. Fr. = Stachys aculeolata Hook. f. ■☆

373293　Stachys linearis Burch. ex Benth. ;线状水苏■☆

373294　Stachys longespicata Boiss. et Kotschy ex Boiss. ; 长穗水苏; Longspike Hedgenettle ■☆

373295　Stachys lupulina Briq. = Stachys natalensis Hochst. var. galpinii（Briq.）Codd ■☆

373296　Stachys lyallii Benth. ;莱尔水苏■☆

373297　Stachys macilenta E. Mey. ex Benth. = Stachys hyssopoides Burch. ex Benth. ■☆

373298　Stachys macrantha（K. Koch）Stearn;大花水苏;Big-sage ■☆

373299　Stachys macrantha（K. Koch）Stearn 'Superba';华丽大花水苏■☆

373300　Stachys macrophylla Albov;大叶水苏●☆

373301　Stachys mairei H. Lév. = Isodon sculponeatus（Vaniot）Kudo ■

373302　Stachys malacophylla V. Naray. ;软叶水苏■☆

373303　Stachys maritima Gouan;海滨水苏;Seashore Betony ■☆

373304　Stachys marrubiifolia Viv. ;夏至草叶水苏■☆

373305　Stachys martini Vaniot = Stachys oblongifolia Benth. ■

373306　Stachys maweana Ball = Stachys saxicola Coss. subsp. maweana（Ball）Maire ■☆

373307　Stachys melissifolia Benth. ;多枝水苏;Melissaleaf Betony ■

373308　Stachys micrantha K. Koch;小花水苏☆

373309　Stachys minima Gürke = Stachys spathulata Burch. ex Benth. ■☆

373310　Stachys modica Hance = Stachys oblongifolia Benth. ■

373311　Stachys mollis Benth. = Stachys arenaria Vahl ■☆

373312　Stachys mouretii Batt. et Pit. ;穆雷水苏■☆

373313　Stachys multiflora Benth. = Stachys rugosa Aiton ■☆

373314　Stachys natalensis Hochst. ;纳塔尔水苏■☆

373315　Stachys natalensis Hochst. var. galpinii（Briq.）Codd;盖尔水苏■☆

373316　Stachys neglecta Klokov;湮没水苏■☆

373317　Stachys nemorivaga Briq. ;尼莫水苏■☆

373318　Stachys nigricans Benth. ;变黑水苏■☆

373319　Stachys numidica Pomel = Stachys circinata L'Hér. subsp. numidica（Pomel）Batt. ■☆

373320　Stachys nutans Benth. = Stachys lamarckii Benth. ■☆

373321　Stachys oblongifolia Benth. ;针筒菜（长叶草石蚕,长叶水苏,长圆叶水苏,地参,方形叶水苏,牛舌条,千密灌,水茴香,野蚕蛹子,野油麻,针筒果）;Oblongleaf Betony ■

373322　Stachys oblongifolia Benth. = Stachys melissifolia Benth. ■

373323　Stachys oblongifolia Benth. f. leptopoda（Hayata）Kudo = Stachys oblongifolia Benth. var. leptopoda（Hayata）C. Y. Wu ■

373324　Stachys oblongifolia Benth. var. leptopoda（Hayata）C. Y. Wu; 细柄针筒菜（臭草）;Thinstalk Oblongleaf Betony ■

373325　Stachys obtusifolia MacOwan;钝叶水苏■☆

373326　Stachys obtusifolia MacOwan var. angustifolia V. Naray. = Stachys tysonii V. Naray. ☆

373327　Stachys obtusifolia MacOwan var. flanaganii V. Naray. = Stachys obtusifolia MacOwan ■☆

373328　Stachys ochroleuca Pomel = Stachys duriaei Noë ■☆

373329　Stachys ocymastrum（L.）Briq. ; 意大利水苏; Italian

Hedgenettle ■☆

373330　Stachys ocymastrum（L.）Briq. var. bicolor Faure et Maire = Stachys ocymastrum（L.）Briq. ■☆

373331　Stachys ocymastrum（L.）Briq. var. lagascae（Caball.）Sennen et Mauricio = Stachys ocymastrum（L.）Briq. ■☆

373332　Stachys ocymastrum（L.）Briq. var. violascens Faure et Maire = Stachys ocymastrum（L.）Briq. ■☆

373333　Stachys ocymastrum Briq. = Stachys ocymastrum（L.）Briq. ■☆

373334　Stachys odontophylla Freyn;齿叶水苏■☆

373335　Stachys officinalis（L.）Franch. = Betonica officinalis L. ■

373336　Stachys officinalis（L.）Trevis. = Betonica officinalis L. ■

373337　Stachys officinalis（L.）Trevis. ex Briq. = Betonica officinalis L. ■

373338　Stachys officinalis（L.）Trevis. subsp. algeriensis（Noë）Franco;阿尔及利亚水苏■☆

373339　Stachys officinalis（L.）Trevis. var. tangerina Pau = Stachys officinalis（L.）Trevis. ■

373340　Stachys oligantha Baker;寡花水苏■☆

373341　Stachys olympica Poir. = Stachys byzantina K. Koch ■

373342　Stachys oreophila Hedge;喜山水苏■☆

373343　Stachys pachycalamna Briq. = Stachys spathulata Burch. ex Benth. ■☆

373344　Stachys palaestina L. var. hypoleuca（Hochst. ex A. Rich.）Benth. = Stachys hypoleuca Hochst. ex A. Rich. ■☆

373345　Stachys palaestina L. var. schimperi（Vatke）Baker = Stachys schimperi Vatke ■☆

373346　Stachys palustris L. ;沼生水苏（白根草,白马兰,光叶水苏,泥灯心,天芝麻,望江青,野地蚕,银脚鹭鸶）;Allheal, Clown's Allheal, Clown's Woundwort, Dea Nettle, Hedge Nettle, Hedge-nettle, Hound's Tongue, Husbandman's Weatherglass, Husbandman's Weather-warner, Husbandman's Woundwort, Marsh Betony, Marsh Hedge-nettle, Marsh Woundwort, Marshy Betony, Maskert, Roughweed, Sheep's Brisken, Swine Arnut, Swine's Beads, Swine's Maskert, Swine's Mosscorts, Swine's Murrill, Swine's Murrills, Woundwort ■

373347　Stachys palustris L. = Stachys pilosa Nutt. ■

373348　Stachys palustris L. subsp. arenicola（Britton）Gill;旱沙水苏; Hedge-nettle, Marsh Hedge-nettle ■☆

373349　Stachys palustris L. subsp. pilosa（Nutt.）Epling = Stachys palustris L. var. pilosa（Nutt.）Fernald ■

373350　Stachys palustris L. subsp. pilosa（Nutt.）Epling = Stachys pilosa Nutt. ■

373351　Stachys palustris L. var. baicalensis Turcz. = Stachys baicalensis Fisch. ex Benth. var. angustifolia Honda ■

373352　Stachys palustris L. var. hispida（Pursh）B. Boivin = Stachys tenuifolia Willd. ■☆

373353　Stachys palustris L. var. hispida Ledeb. = Stachys baicalensis Fisch. ex Benth. var. angustifolia Honda ■

373354　Stachys palustris L. var. hispida Ledeb. = Stachys baicalensis Fisch. ex Benth. ■

373355　Stachys palustris L. var. hispidula Regel = Stachys baicalensis Fisch. ex Benth. var. hispidula（Regel）Nakai ■

373356　Stachys palustris L. var. homotricha Fernald = Stachys palustris L. subsp. arenicola（Britton）Gill ■☆

373357　Stachys palustris L. var. homotricha Fernald = Stachys pilosa Nutt. ■

373358　Stachys palustris L. var. imaii（Nakai）Nakai = Stachys oblongifolia Benth. ■

373359　Stachys palustris L. var. imaii Nakai = Stachys oblongifolia Benth. ■

373360　Stachys palustris L. var. macrocalyx Jenn. = Stachys palustris L. ■

373361　Stachys palustris L. var. nipigonensis Jenn. = Stachys palustris L. ■

373362　Stachys palustris L. var. nipigonensis Jenn. = Stachys pilosa Nutt. ■

373363　Stachys palustris L. var. phaneropoda Weath. = Stachys palustris L. subsp. arenicola（Britton）Gill ■☆

373364　Stachys palustris L. var. phaneropoda Weath. = Stachys pilosa Nutt. ■

373365　Stachys palustris L. var. pilosa（Nutt.）Fernald；毛沼生水苏；Hedge-nettle，Marsh Betony，Marsh Hedge-nettle，Woundwort ■

373366　Stachys palustris L. var. pilosa（Nutt.）Fernald = Stachys palustris L. subsp. pilosa（Nutt.）Epling ■

373367　Stachys palustris L. var. pilosa（Nutt.）Fernald = Stachys pilosa Nutt. ■

373368　Stachys parilis N. E. Br. = Stachys natalensis Hochst. var. galpinii（Briq.）Codd ■☆

373369　Stachys parviflora Benth. = Phlomidoschema parviflorum（Benth.）Vved. ■☆

373370　Stachys pascuicola Briq. = Stachys simplex Schltr. ■☆

373371　Stachys paulii Grossh. ；帕氏水苏■☆

373372　Stachys peduncularis Pau = Stachys arenaria Vahl ■☆

373373　Stachys persica S. G. Gmel. ex C. A. Mey. ；波斯水苏■☆

373374　Stachys petrogenes Briq. = Stachys kuntzei Gürke ■☆

373375　Stachys pilosa Nutt. = Stachys palustris L. subsp. pilosa（Nutt.）Epling ■

373376　Stachys pilosa Nutt. = Stachys palustris L. var. pilosa（Nutt.）Fernald ■

373377　Stachys pilosa Nutt. = Stachys palustris L. ■

373378　Stachys pilosa Nutt. var. arenicola（Britton）G. A. Mulligan et D. B. Munro = Stachys palustris L. subsp. arenicola（Britton）Gill ■☆

373379　Stachys priorii V. Naray. = Stachys scabrida V. Naray. ■☆

373380　Stachys pseudoarenaria Sennen = Stachys arenaria Vahl ■☆

373381　Stachys pseudofloccosa Knorring；假丛卷毛水苏■☆

373382　Stachys pseudohumifusa Sebsebe；拟平伏水苏■☆

373383　Stachys pseudohumifusa Sebsebe subsp. minutiflora Y. B. Harv. ；微花平伏水苏■☆

373384　Stachys pseudonigricans Gürke；假变黑水苏■☆

373385　Stachys pseuophlomis C. Y. Wo；狭齿水苏；Narrowteeth Betony ■

373386　Stachys pubescens Ten. ；短柔毛水苏；Pubescent Betony ■☆

373387　Stachys pulchra Pomel = Stachys duriaei Noë ■☆

373388　Stachys pyramidalis J. K. Morton；塔状水苏■☆

373389　Stachys recta L. ；直水苏；Perennial Yellow-woundwort，Stiff Hedgenettle，Upright Hedge Nettle，Upright Hedge-nettle ■☆

373390　Stachys recurva Gürke = Stachys linearis Burch. ex Benth. ■☆

373391　Stachys rehmannii V. Naray. ；拉赫曼水苏■☆

373392　Stachys reptans Hedge；匍匐水苏■☆

373393　Stachys reticulata Codd；网状水苏■☆

373394　Stachys riederi Cham. = Stachys baicalensis Fisch. ex Benth. ■

373395　Stachys riederi Cham. ex Benth. = Stachys baicalensis Fisch. ex Benth. ■

373396　Stachys riederi Cham. ex Benth. var. hispida（Ledeb.）H. Hara

= Stachys baicalensis Fisch. ex Benth. ■

373397　Stachys riederi Cham. ex Benth. var. hispida（Ledeb.）H. Hara = Stachys riederi Cham. ex Benth. ■

373398　Stachys riederi Cham. ex Benth. var. hispidula（Regel）H. Hara = Stachys baicalensis Fisch. ex Benth. var. hispidula（Regel）Nakai ■

373399　Stachys riederi Cham. ex Benth. var. hispidula（Regel）H. Hara f. angustifolia（Honda）Hara = Stachys baicalensis Fisch. ex Benth. var. angustifolia Honda ■

373400　Stachys riederi Cham. ex Benth. var. hispidula（Regel）H. Hara f. leucantha Hiyama；白花硬毛水苏■☆

373401　Stachys riederi Cham. var. hispida（Ledeb.）H. Hara = Stachys aspera Michx. ■

373402　Stachys riederi Cham. var. hispida（Ledeb.）H. Hara = Stachys baicalensis Fisch. ex Benth. ■

373403　Stachys riederi Cham. var. hispidula（Regel）H. Hara = Stachys aspera Michx. var. hispidula（Regel）Vorosch. ■

373404　Stachys riederi Cham. var. hispidula（Regel）H. Hara = Stachys baicalensis Fisch. ex Benth. var. hispidula（Regel）Nakai ■

373405　Stachys riederi Cham. var. hispidula（Regel）H. Hara f. angustifolia（Honda）H. Hara = Stachys baicalensis Fisch. ex Benth. var. angustifolia Honda ■

373406　Stachys riederi Cham. var. intermedia（Kudo）Kitam. = Stachys aspera Michx. var. hispidula（Regel）Vorosch. ■

373407　Stachys riederi Cham. var. japonica（Miq.）H. Hara = Stachys aspera Michx. var. japonica（Miq.）Maxim. ■

373408　Stachys riederi Cham. var. japonica（Miq.）H. Hara = Stachys japonica Miq. ■

373409　Stachys riederi Cham. var. villosa（Kudo）Kitam. = Stachys aspera Michx. ■

373410　Stachys rifana Font Quer et Maire；里夫水苏■☆

373411　Stachys rifana Font Quer et Maire var. foucauldiana（Sennen）Font Quer et Sennen = Stachys rifana Font Quer et Maire ■☆

373412　Stachys ringens Oett. en = Stachys baicalensis Fisch. ex Benth. ■

373413　Stachys rivularis J. M. Wood and M. S. Evans；溪边水苏■☆

373414　Stachys rosea（Desf.）Boiss. ；粉红水苏■☆

373415　Stachys rubella Hedge；微红水苏■☆

373416　Stachys rudatisii V. Naray. ；鲁达蒂斯水苏■☆

373417　Stachys rugosa Aiton；褶皱水苏■☆

373418　Stachys rugosa Aiton var. foliosa（Benth.）V. Naray. = Stachys dregeana Benth. ■☆

373419　Stachys rugosa Aiton var. linearis（Burch. ex Benth.）V. Naray. = Stachys linearis Burch. ex Benth. ■☆

373420　Stachys saxicola Coss. ；岩栖水苏■☆

373421　Stachys saxicola Coss. subsp. chelifensis Quézel et Simonn. ；谢利夫水苏■☆

373422　Stachys saxicola Coss. subsp. laxa Faure et Maire；疏松岩栖水苏■☆

373423　Stachys saxicola Coss. subsp. maweana（Ball）Maire；马韦岩栖水苏■☆

373424　Stachys saxicola Coss. subsp. platyodon Maire；阔齿岩栖水苏■☆

373425　Stachys saxicola Coss. subsp. villosissima（Ball）Maire；长毛岩栖水苏■☆

373426　Stachys saxicola Coss. var. villosissima Ball = Stachys saxicola Coss. subsp. villosissima（Ball）Maire ■☆

373427　Stachys scaberula Vatke；糙水苏；Scabrous Betony ■☆

373428　Stachys Stachys schiedeana Schltdl. ；谢得水苏■☆

373429 Stachys schimperi Vatke;欣珀水苏■☆

373430 Stachys schlechteri Gürke = Stachys rivularis J. M. Wood et M. S. Evans ■☆

373431 Stachys sericea Wall. ;绢毛水苏;Seriseous Betony ■☆

373432 Stachys serrulata Burch. ex Benth. = Stachys aethiopica L. ■☆

373433 Stachys sessilifolia E. Mey. ex Benth. ;无柄叶水苏■☆

373434 Stachys sessilis Gürke;无柄水苏■☆

373435 Stachys setifera C. A. Mey. ;刚毛水苏■☆

373436 Stachys sidamoensis Gürke = Stachys aculeolata Hook. f. ■☆

373437 Stachys sidamoensis Gürke. f. neumannii T. C. E. Fr. = Stachys aculeolata Hook. f. ■☆

373438 Stachys sieboldi Miq. var. tuberculata Hand. -Mazz. = Stachys kouyangensis (Vaniot) Dunn var. tuberculata (Hand. -Mazz.) C. Y. Wu ■

373439 Stachys sieboldii Miq. ;甘露子(宝塔菜,草石蚕,长圆叶水苏,滴露,地参,地蚕,地牯牛,地牯牛草,地瓜儿,地果,地母,地钮,地蕊,甘露儿,旱螺狮,块茎水苏,露水果,罗汉菜,螺狮菜,螺丝菜,米累累,千蜜罐,水苗香,土虫草,土人参,土蛹,蜗儿菜,小地梨,野油麻,益母膏);Artichoke Betony, Artichoke Betony Chorogi,China Artichoke,Chinese Artichoke,Chorogi,Choroki,Japan Artichoke,Japanese Artichoke ■

373440 Stachys sieboldii Miq. var. glabrescens Hemsl. ;近无毛甘露子;Glabros Betony ■

373441 Stachys sieboldii Miq. var. malacotricha Hand. -Mazz. ;软毛甘露子;Softhair Betony ■

373442 Stachys simplex Schltr. ;简单水苏■☆

373443 Stachys sosnowskyi Kapeller;索斯水苏■☆

373444 Stachys spathulata Burch. ex Benth. ;匙形水苏■☆

373445 Stachys spectabiliformis Kapeller;拟壮观水苏■☆

373446 Stachys spectabilis Choisy ex DC. ;壮观水苏■☆

373447 Stachys splendens Wall. = Stachys melissifolia Benth. ■

373448 Stachys steingroeveri Briq. = Acrotome pallescens Benth. ■☆

373449 Stachys strictiflora C. Y. Wu;直花水苏;Standflower Betony, Strictflower Betony ■

373450 Stachys strictiflora C. Y. Wu var. latidens C. Y. Wu et H. W. Li;宽齿直花水苏;Broadtooth Strictflower Betony ■

373451 Stachys subargentea Hayata = Stachys oblongifolia Benth. ■

373452 Stachys sublobata V. Naray. ;微裂水苏■☆

373453 Stachys subrenifolia T. C. E. Fr. = Stachys aculeolata Hook. f. ■☆

373454 Stachys subsessilis Burch. ex Benth. = Stachys humifusa Burch. ex Benth. ■☆

373455 Stachys sylvatica L. ;林地水苏;Archangel, Blind Nettle, Clown's Allheal, Cow's Weatherwind, Cow's Weather-wind, Cow's Withwind, Cow's Withywind, Deye Nettle, Forest Betony, Grass Nettle, Hedge Dead Nettle, Hedge Nettle, Hedge Woundwort, Husbandman's Weatherglass, Husbandman's Weather-warner, Husbandman's Woundwort, Nettleeoot, Red Archangel, Whitespot, Whitespot Betony, Wild Grass Nettle, Wood Woundwort, Woundwort ■

373456 Stachys taliensis C. Y. Wu;大理水苏;Dali Betony, Tali Betony ■

373457 Stachys talyschensis Kapeller;塔里什水苏■☆

373458 Stachys tenella V. Naray. = Stachys humifusa Burch. ex Benth. ■☆

373459 Stachys tenuifolia Willd. ;细叶水苏;Common Hedge-nettle, Hedge Nettle, Narrow-leaved Hedge-nettle, Slenderleaf Betony, Smooth Hedge-nettle,Thinleaf Betony ■☆

373460 Stachys tenuifolia Willd. var. hispida (Pursh) Fernald = Stachys hispida Pursh ■☆

373461 Stachys tenuifolia Willd. var. hispida (Pursh) Fernald = Stachys tenuifolia Willd. ■☆

373462 Stachys tenuifolia Willd. var. platyphylla Fernald = Stachys tenuifolia Willd. ■☆

373463 Stachys terekensis Knorring;捷列克水苏■☆

373464 Stachys teres V. Naray. = Stachys aurea Benth. ■☆

373465 Stachys thunbergii Benth. ;通贝里水苏■☆

373466 Stachys tibetica Vatke;西藏水苏(藏水苏);Tibet Betony, Xizang Betony ●

373467 Stachys tournefortii Poir. ;图内福尔水苏■☆

373468 Stachys transvaalensis Gürke = Stachys natalensis Hochst. ■☆

373469 Stachys trapezuntea Boiss. ;四边形水苏■☆

373470 Stachys trichophylla Baker;毛叶水苏■☆

373471 Stachys trinervis Aitch. et Hemsl. ;三脉水苏■☆

373472 Stachys tschatkalensis Knorring;查特卡尔水苏■☆

373473 Stachys tuberifera Naudin = Stachys sieboldii Miq. ■

373474 Stachys tubulosa MacOwan;管状水苏■☆

373475 Stachys turcomanica Trautv. ;土库曼水苏■☆

373476 Stachys turkestanica (Regel) Popov;土耳其斯坦水苏■☆

373477 Stachys tysonii V. Naray. ;泰森水苏■☆

373478 Stachys villosissima H. M. L. Forbes = Stachys natalensis Hochst. var. galpinii (Briq.) Codd ■☆

373479 Stachys vincentii Sennen;文森特水苏■☆

373480 Stachys volgensis Willems;伏尔加水苏;Volga Betony ■☆

373481 Stachys woronowii Maleev;沃氏水苏■☆

373482 Stachys xanthantha C. Y. Wu;黄花地钮菜(地钮菜,黄花水苏);Yellowflower Betony ■

373483 Stachys xanthantha C. Y. Wu var. gracilis C. Y. Wu et H. W. Li;柔弱黄花地钮菜;Weak Yellowflower Betony ■

373484 Stachys xanthantha C. Y. Wu var. gracilis C. Y. Wu et H. W. Li = Stachys xanthantha C. Y. Wu ■

373485 Stachys zeyheri V. Naray. ;泽赫水苏■☆

373486 Stachystemon Planch. (1845);穗雄大戟属■☆

373487 Stachystemon brevifolius Grüning;短叶穗雄大戟■☆

373488 Stachystemon vermicularis Planch. ;澳洲穗雄大戟■☆

373489 Stachytarpha Link = Stachytarpheta Vahl (保留属名)■●

373490 Stachytarpheta Vahl(1804)(保留属名);假马鞭属(假败酱属,假马鞭草属,木马鞭属,玉龙鞭属);Falsevalerian ■●

373491 Stachytarpheta angolensis Moldenke;安哥拉假马鞭●☆

373492 Stachytarpheta angustifolia (Mill.) Vahl;狭叶假马鞭;Narrowleaf Falsevalerian ●☆

373493 Stachytarpheta angustifolia (Mill.) Vahl = Stachytarpheta indica (L.) Vahl ●

373494 Stachytarpheta australis Moldenke;南方猿尾木;Branched Porterweed ●☆

373495 Stachytarpheta cayeanensis (Rich.) Vahl;卡也假马鞭(卡也假败酱,蓝蝶猿尾木)●■☆

373496 Stachytarpheta cayennensis (Rich.) Vahl = Stachytarpheta urticifolia (Salisb.) Sims ●☆

373497 Stachytarpheta dichotoma (Ruiz et Pav.) Vahl;叉枝猿尾木;Oi,Owi ●☆

373498 Stachytarpheta dichotoma (Ruiz et Pav.) Vahl = Stachytarpheta cayennensis (Rich.) Vahl ●☆

373499 Stachytarpheta dichotoma A. Chev. = Stachytarpheta fallax A. E. Gonc. ●☆

373500 Stachytarpheta elegans Welw. ;雅致假马鞭●☆

373501　Stachytarpheta fallax A. E. Gonc. ;迷惑假马鞭●☆

373502　Stachytarpheta hildebrandtii Vatke = Chascanum hildebrandtii (Vatke) J. B. Gillett ●☆

373503　Stachytarpheta indica (L.) Vahl = Stachytarpheta jamaicensis (L.) Vahl ●

373504　Stachytarpheta indica C. B. Clarke = Stachytarpheta jamaicensis (L.) Vahl ●

373505　Stachytarpheta indica Vahl = Stachytarpheta jamaicensis (L.) Vahl ●

373506　Stachytarpheta jabassensis H. Winkl. = Stachytarpheta indica (L.) Vahl ●

373507　Stachytarpheta jamaicensis (L.) Vahl;假马鞭（长穗木,大蓝草,大种马鞭草,倒扣藤,倒困蛇,倒团蛇,假败酱,马鞭草,木马鞭,牛鞭草,蛇尾草,铁马鞭,万灵草,牙买加长穗木,玉郎鞭,玉龙鞭）; Blue Porterweed, Blue Snakeweed, Jamaica Falsevalerian, Jamaica Vervain,Oi,Owi,Porterweed ●

373508　Stachytarpheta jamaicensis Vahl = Stachytarpheta urticifolia (Salisb.) Sims ●☆

373509　Stachytarpheta laevis Moldenke;平滑假马鞭●☆

373510　Stachytarpheta mutabilis (Jacq.) J. Vahl;粉红假马鞭; Changeable Velvetberry,Pink Snakeweed ●☆

373511　Stachytarpheta urticifolia (Salisb.) Sims;长穗木;Nettle-leaf Porterweed, Nettleleaf Velvetberry, Nettle-leaved Vervain, Oi, Owi ●☆

373512　Stachytarpheta urticifolia (Salisb.) Sims = Stachytarpheta cayeanensis (Rich.) Vahl ●☆

373513　Stachytarpheta urticifolia Sims = Stachytarpheta urticifolia (Salisb.) Sims ●☆

373514　Stachythyrsus Post et Kuntze = Stachyothyrsus Harms ■☆

373515　Stachyuraceae J. Agardh (1858) (保留科名);旌节花科;Stachyurus Family ●

373516　Stachyurus Siebold et Zucc. (1836);旌节花属;Stachyurus ●

373517　Stachyurus brachystachyus (C. Y. Wu et S. K. Chen) Y. C. Tang et Y. L. Cao = Stachyurus himalaicus Hook. f. et Thomson ex Benth. ●

373518　Stachyurus brachystachyus (C. Y. Wu et S. K. Chen) Y. C. Tang et Y. L. Cao = Stachyurus chinensis Franch. var. brachystachyus C. Y. Wu et S. K. Chen ●

373519　Stachyurus calcareus Hung T. Chang ex Z. R. Xu;石山旌节花;Calcicolous Stachyurus,Lime-loving Stachyurus ●

373520　Stachyurus callosus C. Y. Wu ex S. K. Chen;椭圆叶旌节花;Callous Stachyurus,Elliptic-leaf Stachyurus ●

373521　Stachyurus callosus C. Y. Wu ex S. K. Chen = Stachyurus yunnanensis Franch. ●

373522　Stachyurus caudatilimbus C. Y. Wu et S. K. Chen = Stachyurus chinensis Franch. var. latus H. L. Li ●

373523　Stachyurus caudatilimbus C. Y. Wu et S. K. Chen = Stachyurus chinensis Franch. ●

373524　Stachyurus chinensis Franch.;中国旌节花（旌节花,萝卜药,实心通草,水凉子,通草,通草棍,通花,小通花,鱼泡通,中华旌节花）;China Stachyurus,Chinese Stachyurus ●

373525　Stachyurus chinensis Franch. subsp. brachystachyus (C. Y. Wu et S. K. Chen) Y. C. Tang et Y. L. Cao = Stachyurus himalaicus Hook. f. et Thomson ex Benth. ●

373526　Stachyurus chinensis Franch. subsp. cuspidatus (H. L. Li) Y. C. Tang et Y. L. Cao = Stachyurus chinensis Franch. ●

373527　Stachyurus chinensis Franch. subsp. cuspidatus (H. L. Li) Y. C. Tang et Y. L. Cao = Stachyurus chinensis Franch. var. cuspidatus H. L. Li ●

373528　Stachyurus chinensis Franch. subsp. latus (H. L. Li) Y. C. Tang et Y. L. Cao = Stachyurus chinensis Franch. ●

373529　Stachyurus chinensis Franch. subsp. latus (H. L. Li) Y. C. Tang et Y. L. Cao = Stachyurus chinensis Franch. var. latus H. L. Li ●

373530　Stachyurus chinensis Franch. var. brachystachyus C. Y. Wu et S. K. Chen = Stachyurus himalaicus Hook. f. et Thomson ex Benth. ●

373531　Stachyurus chinensis Franch. var. brachystachyus C. Y. Wu et S. K. Chen;短穗旌节花（短轴旌节花）; Short-stachys Chinese Stachyurus ●

373532　Stachyurus chinensis Franch. var. cuspidatus H. L. Li;尖尾叶旌节花(尖叶旌节花,骤尖叶旌节花);Cuspidate Chinese Stachyurus ●

373533　Stachyurus chinensis Franch. var. cuspidatus H. L. Li = Stachyurus chinensis Franch. ●

373534　Stachyurus chinensis Franch. var. lancifolius C. Y. Wu;披针旌节花;Lanceolate Chinese Stachyurus ●

373535　Stachyurus chinensis Franch. var. latus H. L. Li;宽叶旌节花(通草);Broadleaf China Stachyurus,Broadleaf Stachyurus ●

373536　Stachyurus chinensis Franch. var. latus H. L. Li = Stachyurus chinensis Franch. ●

373537　Stachyurus cordatulus Merr. ;滇缅旌节花（心叶旌节花）;Cordate Stachyurus,Heart-leaved Stachyurus ●

373538　Stachyurus duclouxii Pit. ex Chung = Stachyurus chinensis Franch. ●

373539　Stachyurus esquirolii H. Lév. = Stachyurus yunnanensis Franch. ●

373540　Stachyurus himalaicus Hook. f. et Thomson ex Benth. ;西域旌节花(翅柄旌节花,空藤杆,毛轴旌节花,实心通草,通草,通草棍,通草树,通棍,通条柳,通条木,通条树,喜马拉雅旌节花,喜马山旌节花,小通草,小通花,小叶旌节花,鱼泡通);Hairyrachis Himalayan Stachyurus, Himalayan Stachyurus, Himalayas Stachyurus, Small-leaf Himalayan Stachyurus ●

373541　Stachyurus himalaicus Hook. f. et Thomson ex Benth. subsp. purpureus Y. P. Zhu et Z. Y. Zhang = Stachyurus himalaicus Hook. f. et Thomson ex Benth. ●

373542　Stachyurus himalaicus Hook. f. et Thomson ex Benth. subsp. purpureus Y. P. Zhu et Z. Y. Zhang;紫红旌节花●

373543　Stachyurus himalaicus Hook. f. et Thomson ex Benth. var. alatipes C. Y. Wu ex S. K. Chen = Stachyurus himalaicus Hook. f. et Thomson ex Benth. ●

373544　Stachyurus himalaicus Hook. f. et Thomson ex Benth. var. alatipes C. Y. Wu = Stachyurus himalaicus Hook. f. et Thomson ex Benth. ●

373545　Stachyurus himalaicus Hook. f. et Thomson ex Benth. var. dasyrachis C. Y. Wu = Stachyurus himalaicus Hook. f. et Thomson ex Benth. ●

373546　Stachyurus himalaicus Hook. f. et Thomson ex Benth. var. microphyllus C. Y. Wu = Stachyurus himalaicus Hook. f. et Thomson ex Benth. ●

373547　Stachyurus littoralis Nakai;四国旌节花;Sikoku Stachyurus ●☆

373548　Stachyurus macrocarpus Koidz. = Stachyurus praecox Siebold et Zucc. var. macrocarpus (Koidz.) Tuyama ex H. Ohba ●☆

373549　Stachyurus macrocarpus Koidz. var. prunifolius Tuyama = Stachyurus praecox Siebold et Zucc. var. macrocarpus (Koidz.) Tuyama ex H. Ohba ●☆

373550　Stachyurus oblongifolius F. T. Wang et Ts. Tang;矩圆叶旌节花（长圆旌节花,长圆叶旌节花）;Oblong-leaf Stachyurus, Oblong-leaved Stachyurus ●

373551　Stachyurus oblongifolius F. T. Wang et Ts. Tang = Stachyurus yunnanensis Franch. ●

373552　Stachyurus obovatus (Rehder) H. L. Li = Stachyurus oblongifolius F. T. Wang et Ts. Tang ●

373553　Stachyurus obovatus (Rehder) H. L. Li = Stachyurus obovatus (Rehder) Hand. -Mazz. ●

373554　Stachyurus obovatus (Rehder) Hand. -Mazz. ;倒卵叶旌节花(卵叶旌节花);Obovateleaf Stachyurus, Obovate-leaved Stachyurus ●

373555　Stachyurus obovatus (Rehder) W. C. Cheng = Stachyurus obovatus (Rehder) Hand. -Mazz. ●

373556　Stachyurus praecox Siebold et Zucc. ;旌节花(通草,通条花,通条叶,小通草,早春旌节花,早花旌节花,早开旌节花,早生旌节花);Precocious Stachyurus, Stachyurus ●

373557　Stachyurus praecox Siebold et Zucc. 'Magpie';鹊旌节花●☆

373558　Stachyurus praecox Siebold et Zucc. = Stachyurus chinensis Franch. var. latus H. L. Li ●

373559　Stachyurus praecox Siebold et Zucc. f. bicolor Sakata;二色旌节花●☆

373560　Stachyurus praecox Siebold et Zucc. f. marginatus Hiyama;花边旌节花●☆

373561　Stachyurus praecox Siebold et Zucc. f. microphyllus (Nakai) H. Hara;小叶旌节花●☆

373562　Stachyurus praecox Siebold et Zucc. f. rotundifolius (Tuyama) Tuyama ex H. Hara;圆叶旌节花●☆

373563　Stachyurus praecox Siebold et Zucc. var. lancifolius Koidz. ;狭叶旌节花;Narrowleaf Precocious Stachyurus ●☆

373564　Stachyurus praecox Siebold et Zucc. var. leucotrichus Hayashi;白毛旌节花●☆

373565　Stachyurus praecox Siebold et Zucc. var. macrocarpus (Koidz.) Tuyama ex H. Ohba;大果旌节花●☆

373566　Stachyurus praecox Siebold et Zucc. var. microphyllus Nakai = Stachyurus praecox Siebold et Zucc. f. microphyllus (Nakai) H. Hara ●☆

373567　Stachyurus praecox Siebold et Zucc. var. parviflorus Makino ex H. Hara;小花旌节花●☆

373568　Stachyurus retusus Yen C. Yang;凹叶旌节花(小通草);Concave-leaved Stachyurus, Retuse Stachyurus ●◇

373569　Stachyurus salicifolius Franch. ;柳叶旌节花(铁泡桐,通花,小通花);Willowleaf Stachyurus, Willow-leaved Stachyurus ●

373570　Stachyurus salicifolius Franch. f. lancifolius (C. Y. Wu ex S. K. Chen) Y. C. Tang et al. = Stachyurus salicifolius Franch. var. lancifolius C. Y. Wu ex S. K. Chen ●

373571　Stachyurus salicifolius Franch. subsp. lancifolius (C. Y. Wu ex S. K. Chen) Y. C. Tang et Y. L. Cao = Stachyurus salicifolius Franch. ●

373572　Stachyurus salicifolius Franch. var. lancifolius C. Y. Wu ex S. K. Chen = Stachyurus salicifolius Franch. ●

373573　Stachyurus salicifolius Franch. var. lancifolius C. Y. Wu ex S. K. Chen;披针叶旌节花;Lanceleaf Stachyurus ●

373574　Stachyurus sigeyosii Masam. = Stachyurus chinensis Franch. ●

373575　Stachyurus sigeyosii Masam. = Stachyurus himalaicus Hook. f. et Thomson ex Benth. ●

373576　Stachyurus szechuanensis W. P. Fang;四川旌节花;Sichuan Stachyurus ●◇

373577　Stachyurus szechuanensis W. P. Fang = Stachyurus retusus Yen C. Yang ●◇

373578　Stachyurus yunnanensis Franch. ;云南旌节花(滇旌节花,通草);Yunnan Stachyurus ●

373579　Stachyurus yunnanensis Franch. var. obovata Rehder = Stachyurus oblongifolius F. T. Wang et Ts. Tang ●

373580　Stachyurus yunnanensis Franch. var. obovata Rehder = Stachyurus obovatus (Rehder) Hand. -Mazz. ●

373581　Stachyurus yunnanensis Franch. var. pedicellatus Rehder;长柄旌节花(长梗旌节花,具柄旌节花,具梗旌节花);Longpedicel Stachyurus ●

373582　Stachyurus yunnanensis Franch. var. pedicellatus Rehder = Stachyurus yunnanensis Franch. ●

373583　Stachyus St. -Lag. = Stachys L. ●■

373584　Stackhousia Sm. (1798);异雄蕊属(木根草属)■☆

373585　Stackhousia monogyna Labill. ;单蕊异雄蕊■☆

373586　Stackhousia tryonii Bailey;特利异雄蕊■☆

373587　Stackhousiaceae R. Br. (1814)(保留科名);异雄蕊科(木根草科)■☆

373588　Stackhousiaceae R. Br. (保留科名) = Celastraceae R. Br. (保留科名)●

373589　Stacyella Szlach. = Oncidium Sw. (保留属名)■☆

373590　Stadiochilus R. M. Sm. (1980);缅甸姜属■☆

373591　Stadmania Lam. = Stadmannia Lam. ●☆

373592　Stadmania acuminata Capuron;渐尖缅甸姜■☆

373593　Stadmania excelsa Capuron;高大缅甸姜■☆

373594　Stadmania glauca Capuron;灰绿缅甸姜■☆

373595　Stadmania leandrii Capuron;利安缅甸姜■☆

373596　Stadmania serrulata Capuron;细齿缅甸姜■☆

373597　Stadmannia Lam. (1794);斯达无患子属●☆

373598　Stadmannia oppositifolia Lam. ;对叶斯达无患子●☆

373599　Stadmannia oppositifolia Lam. subsp. rhodesica Exell;罗得西亚斯达无患子●☆

373600　Stadmannia sideroxylon DC. = Stadmannia oppositifolia Lam. ●☆

373601　Stadtmannia Walp. = Stadmannia Lam. ●☆

373602　Staebe Hill(1762);马岛矢车菊属■☆

373603　Staebe Juss. = Stoebe L. ●■☆

373604　Staebe aethiopica Hill;马岛矢车菊■☆

373605　Staeblorhiza Dur. = Streblorrhiza Endl. ■☆

373606　Staehelina L. (1753);卷翅菊属(长冠菊属,斯泰赫菊属)●■☆

373607　Staehelina Raf. = Helipterum DC. ex Lindl. ■☆

373608　Staehelina alpina (L.) Crantz;高山卷翅菊;Alpine Bartsia, Poly-mountain, Velvet Bells ■☆

373609　Staehelina alpina (L.) Crantz = Bartsia alpina L. ■☆

373610　Staehelina centauroides L. = Athanasia crenata (L.) L. ●☆

373611　Staehelina chamaepeuce Viv. = Ptilostemon gnaphaloides (Cirillo) Soják ■☆

373612　Staehelina corymbosa L. f. = Vernonia tigna Klatt ●☆

373613　Staehelina dubia L. ;卷翅菊(斯泰赫菊)■☆

373614　Staehelina dubia L. var. macrocephala Faure et Maire = Staehelina dubia L. ■☆

373615　Staehelina elegans Walter = Liatris elegans (Walter) Michx. ■☆

373616　Staehelina fasciculata Thunb. = Lachnospermum fasciculatum (Thunb.) Baill. ●☆

373617　Staehelina gnaphaloides L. = Syncarpha gnaphaloides (L.) DC. ■☆

373618　Staehelina hastata Vahl = Vernonia spatulata (Forssk.) Sch. Bip. ■☆

373619 Staehelina imbricata P. J. Bergius = Lachnospermum imbricatum (P. J. Bergius) Hilliard ●☆

373620 Staehelinia Crantz = Bartsia L. (保留属名)■●☆

373621 Staehelinia Haller = Bartsia L. (保留属名)■●☆

373622 Staehelinoides Loefl. = Ludwigia L. ●■

373623 Staelia Cham. et Schltdl. (1828);施泰茜属☆

373624 Staelia thymoides Cham. et Schltdl.;施泰茜☆

373625 Staflinus Raf. = Daucus L. ■

373626 Stagmaria Jack = Gluta L. ●

373627 Stagmaria Jack(1823);肖胶漆树属●☆

373628 Stagmaria verniciflua Jack;肖胶漆树;Rengas ●☆

373629 Stahelia Jonker = Tapeinostemon Benth. ●☆

373630 Stahlia Bello(1881);单籽苏木属●☆

373631 Stahlia monosperma (Tul.) Urb.;单籽苏木■☆

373632 Stahlia monosperma Urb. = Stahlia monosperma (Tul.) Urb. ■☆

373633 Stahlianthus Kuntze (1891);土田七属(姜三七属);Local Tianqi,Stahlianthus ■

373634 Stahlianthus involucratus (King ex Baker) Craib = Stahlianthus involucratus (King ex Baker) Craib ex Loes. ■

373635 Stahlianthus involucratus (King ex Baker) Craib ex Loes.;土田七(打不死,峨参,姜七,姜三七,姜田七,姜叶三七,内消子,三七姜,土三七,竹叶三七);Involucrate Local Tianqi, Involucrate Stahlianthus ■

373636 Stahlianthus rubromarginatus S. Q. Tong;红缘土田七;Redmarginate Stahlianthus ■

373637 Stahlianthus rubromarginatus S. Q. Tong = Kaempferia parviflora Wall. ex Baker ■

373638 Stahlianthus thorelii Gagnep.;索氏土田七(土田七)■☆

373639 Stahycarpus (Endl.) Tiegh. = Prumnopitys Phil. ●☆

373640 Staintoniella H. Hara = Aphragmus Andrz. ex DC. ■

373641 Staintoniella H. Hara = Taphrospermum C. A. Mey. ■

373642 Staintoniella H. Hara (1974);无隔荠属(无隔芥属);Nowallcress,Staintoniella ■

373643 Staintoniella verticillata (Jeffrey et W. W. Sm.) H. Hara;轮叶无隔荠(轮叶柏蕾荠);Veticillate Nowallcress, Veticillate Staintoniella ■

373644 Staintoniella verticillata (Jeffrey et W. W. Sm.) H. Hara = Taphrospermum verticillatum (Jeffrey et W. W. Sm.) Al-Shehbaz ■

373645 Stalagmites Miq. = Cratoxylum Blume ●

373646 Stalagmites Murray = Garcinia L. ●

373647 Stalagmites Spreng. = Stalagmites Murray ●

373648 Stalagmites erosipetala Miq. = Cratoxylum cochinchinense (Lour.) Blume ●

373649 Stalagmitis Murray = Garcinia L. ●

373650 Stalkya Garay(1982);委内瑞拉兰属■☆

373651 Stalkya muscicola (Garay et Dunst.) Garay;委内瑞拉兰■☆

373652 Stammarium Willd. ex DC. = Pectis L. ■☆

373653 Stamnorchis D. L. Jones et M. A. Clem. (2002);瓶兰属■☆

373654 Stamnorchis D. L. Jones et M. A. Clem. = Pterostylis R. Br. (保留属名)■☆

373655 Standleya Brade(1932);斯坦茜属(巴西草属)■☆

373656 Standleya prostrata (Schumach.) Brade;斯坦茜■☆

373657 Standleyacanthus Léonard(1952);斯坦爵床属☆

373658 Standleyacanthus costaricanus Léonard;斯坦爵床☆

373659 Standleyanthus R. M. King et H. Rob. (1971);三叶泽兰属●☆

373660 Standleyanthus triptychus (B. L. Rob.) R. M. King et H. Rob.;三叶泽兰■☆

373661 Stanfieldia Small = Haplopappus Cass. (保留属名)■●☆

373662 Stanfieldiella Brenan(1960);光花草属■☆

373663 Stanfieldiella axillaris J. K. Morton;腋花光花草■☆

373664 Stanfieldiella brachycarpa (Gilg et Ledermann ex Mildbr.) Brenan;短果光花草■☆

373665 Stanfieldiella brachycarpa (Gilg et Ledermann ex Mildbr.) Brenan var. hirsuta (Brenan) Brenan;毛短果光花草■☆

373666 Stanfieldiella imperforata (C. B. Clarke) Brenan;无孔光花草■☆

373667 Stanfieldiella imperforata (C. B. Clarke) Brenan var. glabrisepala (De Wild.) Brenan;光萼光花草■☆

373668 Stanfieldiella oligantha (Mildbr.) Brenan;寡花光花草■☆

373669 Stanfordia S. Watson = Caulanthus S. Watson ■

373670 Stangea Graebn. (1906);施坦格草属■☆

373671 Stangea Graebn. = Valeriana L. ●■

373672 Stangea emiliae Graebn.;施坦格草■☆

373673 Stangeria T. Moore(1853);托叶苏铁属(蕨苏铁属,托叶铁属);Fern Cycad ●☆

373674 Stangeria eriopus (Kunze) Nash;托叶苏铁(蕨苏铁,蕨铁,绵柄蕨苏铁,托叶铁)●☆

373675 Stangeria paradoxa T. Moore;奇异托叶苏铁●☆

373676 Stangeriaceae (Pilg.) L. A. S. Johnson = Zamiaceae Rchb. ●☆

373677 Stangeriaceae L. A. S. Johnson = Zamiaceae Rchb. ●☆

373678 Stangeriaceae L. A. S. Johnson (1959);托叶苏铁科(托叶铁科)●☆

373679 Stangeriaceae Schimp. et Schenk = Stangeriaceae L. A. S. Johnson ●☆

373680 Stanggeria Stevens = Stangeria T. Moore ●☆

373681 Stanhopea J. Frost ex Hook. (1829);老虎兰属(奇唇兰属);Stanhopea ■☆

373682 Stanhopea J. Frost. = Stanhopea J. Frost ex Hook. ■☆

373683 Stanhopea atropurpurea Lodd. ex Planch.;暗紫老虎兰■☆

373684 Stanhopea aurea Lodd. ex Lindl.;金黄老虎兰■☆

373685 Stanhopea bicolor K. Koch;二色老虎兰■☆

373686 Stanhopea bucephalus Lindl.;牛头老虎兰■☆

373687 Stanhopea devoniensis Lindl.;德文郡老虎兰;Devon Stanhopea ■☆

373688 Stanhopea eburnea Lindl.;象牙白老虎兰(象牙色老虎兰);Ivorywhite Stanhopea ■☆

373689 Stanhopea ecornuta Lem.;白花老虎兰;Whiteflower Stanhopea ■☆

373690 Stanhopea gibbosa Rchb. f.;圆形老虎兰■☆

373691 Stanhopea grandiflora (Kunth) Rchb. f.;大花老虎兰;Largeflower Stanhopea ■☆

373692 Stanhopea graveolens Lindl.;烈味老虎兰;Scented Stanhopea ■☆

373693 Stanhopea hernandezii Schltr.;亨氏老虎兰;Hernandez Stanhopea, Lynx Flower ■☆

373694 Stanhopea insignis Frost;美花老虎兰;Notable Stanhopea ■☆

373695 Stanhopea lowii Rolfe;娄氏老虎兰;Low Stanhopea ■☆

373696 Stanhopea martiana Lindl.;马氏老虎兰■☆

373697 Stanhopea nigripes Rolfe;黑梗老虎兰■☆

373698 Stanhopea oculata Lindl.;眼状老虎兰;Oculate Stanhopea ■☆

373699 Stanhopea platyceras Rchb. f.;平角老虎兰■☆

373700 Stanhopea pulla Rchb. f.;黑老虎兰;Black Stanhopea ■☆

373701 Stanhopea quadricornis Lindl.;四角老虎兰;Fourcorners

Stanhopea ■☆

373702　Stanhopea saccata Bateman;香奇唇兰■☆

373703　Stanhopea tigrina Bateman;老虎兰(虎斑奇唇兰,金鱼兰,奇船兰);Tiger Stanhopea■☆

373704　Stanhopea wardii Lodd. ex Lindl.;瓦氏老虎兰(金鱼兰,奇唇兰);Ward Stanhopea■☆

373705　Stanhopeastrum Rchb. f. = Stanhopea J. Frost ex Hook.■☆

373706　Stanley L. Welsh = Atriplex L.■●

373707　Stanleya Nutt.(1818);长药芥属■☆

373708　Stanleya albescens M. E. Jones;微白长药芥;White Desert Plume■☆

373709　Stanleya pinnata (Pursh) Britton;羽状长药芥;Desert Plume, Golden Prince's Plume,Paiute Cabbage,Prince's Plume■☆

373710　Stanleya tomentosa Parry;毛长药芥■☆

373711　Stanleya viridiflora Nutt.;长药芥■☆

373712　Stanleyaceae Nutt. = Brassicaceae Burnett(保留科名)■●

373713　Stanleyaceae Nutt. = Cruciferae Juss.(保留科名)■●

373714　Stanleyella Rydb. = Thelypodium Endl.■☆

373715　Stanmarkia Almeda(1993);斯坦野牡丹属●☆

373716　Stanmarkia medialis (Standl. et Steyerm.) Almeda;斯坦野牡丹●☆

373717　Stanmarkia spectabilis Almeda;壮观斯坦野牡丹●☆

373718　Stannia H. Karst. = Posoqueria Aubl.●☆

373719　Stapelia L.(1753)(保留属名);豹皮花属(狗皮花属,国章属,魔星花属,五星国徽属,犀角属);Carrion Flower, Carrionflower,Leopardflower,Stapelia■

373720　Stapelia acuminata Masson;渐尖豹皮花■☆

373721　Stapelia acuminata Masson var. brevicuspis N. E. Br. = Stapelia acuminata Masson■☆

373722　Stapelia adscendens Roxb. = Caralluma adscendens (Roxb.) Haw.■☆

373723　Stapelia affinis N. E. Br. = Stapelia hirsuta L.■☆

373724　Stapelia albipilosa Giess = Tridentea marientalensis (Nel) L. C. Leach subsp. albipilosa (Giess) L. C. Leach■☆

373725　Stapelia albocastanea Marloth = Orbea albocastanea (Marloth) Bruyns■☆

373726　Stapelia ambigua Masson = Stapelia grandiflora Masson■

373727　Stapelia ambigua Masson var. fulva Sweet = Stapelia grandiflora Masson■

373728　Stapelia anguinea Jacq. = Orbea variegata (L.) Haw.■☆

373729　Stapelia aperta Masson = Tromotriche aperta (Masson) Bruyns■☆

373730　Stapelia arenosa C. A. Lückh.;沙豹皮花■☆

373731　Stapelia arida Masson = Quaqua arida (Masson) Bruyns■☆

373732　Stapelia arnotii N. E. Br.;阿诺特豹皮花■☆

373733　Stapelia articulata Aiton = Pectinaria articulata (Aiton) Haw.■☆

373734　Stapelia asterias Masson = Stapelia hirsuta L.■☆

373735　Stapelia asterias Masson var. gibba N. E. Br. = Stapelia hirsuta L.■☆

373736　Stapelia asterias Masson var. lucida (DC.) N. E. Br. = Stapelia hirsuta L.■☆

373737　Stapelia atrata Tod.;黑豹皮花■☆

373738　Stapelia atropurpurea Salm-Dyck = Orbea variegata (L.) Haw.■☆

373739　Stapelia atrosanguinea N. E. Br. = Piaranthus atrosanguineus (N. E. Br.) Bruyns■☆

373740　Stapelia auobensis Nel = Tridentea marientalensis (Nel) L. C. Leach■☆

373741　Stapelia aurea Dinter = Tridentea virescens (N. E. Br.) L. C. Leach■☆

373742　Stapelia ausana Dinter et A. Berger ex Dinter = Tridentea jucunda (N. E. Br.) L. C. Leach■☆

373743　Stapelia barbata Masson = Huernia barbata (Masson) Haw.■☆

373744　Stapelia baylissii L. C. Leach = Stapelia hirsuta L. var. baylissii (L. C. Leach) Bruyns■☆

373745　Stapelia bergeriana Dinter = Stapelia schinzii A. Berger et Schltr. var. bergeriana (Dinter) L. C. Leach■☆

373746　Stapelia beukmanii C. A. Lückh. = Stapelia arenosa C. A. Lückh.■☆

373747　Stapelia bisulca Schult. = Orbea variegata (L.) Haw.■☆

373748　Stapelia bufonia (Haw.) Sims = Orbea variegata (L.) Haw.■☆

373749　Stapelia bufonia Jacq. = Orbea variegata (L.) Haw.■☆

373750　Stapelia cactiformis Hook. = Larryleachia cactiformis (Hook.) Plowes●☆

373751　Stapelia caespitosa Masson;丛生豹皮花■☆

373752　Stapelia caespitosa Masson = Duvalia caespitosa (Masson) Haw.■☆

373753　Stapelia caespitosa Masson var. hirtella (Jacq.) Loudon = Duvalia caespitosa (Masson) Haw.■☆

373754　Stapelia campanulata Masson = Huernia barbata (Masson) Haw.■☆

373755　Stapelia caroli-schmidtii Dinter et A. Berger = Orbea albocastanea (Marloth) Bruyns■☆

373756　Stapelia caudata Thunb. = Brachystelma tuberosum (Meerb.) R. Br. ex Sims■☆

373757　Stapelia cedrimontana Frandsen;山地豹皮花■☆

373758　Stapelia chinensis Lour. = Hoya carnosa (L. f.) R. Br.●

373759　Stapelia choanantha Lavranos et H. Hall = Tromotriche choanantha (Lavranos et H. Hall) Bruyns■☆

373760　Stapelia ciliata Thunb. = Orbea ciliata (Thunb.) L. C. Leach■☆

373761　Stapelia ciliolata Tod. = Orbea variegata (L.) Haw.■☆

373762　Stapelia ciliolulata Tod. ex Rust = Orbea variegata (L.) Haw.■☆

373763　Stapelia cincta Marloth = Tridentea jucunda (N. E. Br.) L. C. Leach■☆

373764　Stapelia clavata Willd. = Trichocaulon clavatum (Willd.) H. Huber■☆

373765　Stapelia clavigera Jacq. = Huernia barbata (Masson) Haw.■☆

373766　Stapelia clypeata Jacq. = Orbea variegata (L.) Haw.■☆

373767　Stapelia comata Jacq. = Stapelia hirsuta L.■☆

373768　Stapelia compacta (Haw.) Schult. = Duvalia caespitosa (Masson) Haw.■☆

373769　Stapelia concinna Masson;整洁豹皮花■☆

373770　Stapelia concinna Masson var. paniculata (Willd.) N. E. Br. = Stapelia paniculata Willd.■☆

373771　Stapelia concolor Salm-Dyck = Duvalia concolor (Salm-Dyck) Schltr.■☆

373772　Stapelia conformis N. E. Br. = Stapelia grandiflora Masson var. conformis (N. E. Br.) Bruyns■☆

373773　Stapelia conformis N. E. Br. var. abrosa ? = Stapelia grandiflora Masson var. conformis (N. E. Br.) Bruyns■☆

373774　Stapelia congestiflora Delile;密花豹皮花■☆

373775　Stapelia conspurcata Willd. = Orbea variegata（L.）Haw.■☆

373776　Stapelia cooperi N. E. Br. = Orbea cooperi（N. E. Br.）L. C. Leach■☆

373777　Stapelia corderoyi Hook. f. = Duvalia corderoyi（Hook. f.）N. E. Br.■☆

373778　Stapelia crassa Donn ex Haw. = Huernia guttata（Masson）Haw. subsp. reticulata（Masson）Bruyns■☆

373779　Stapelia cylista C. A. Lückh. = Stapelia gigantea N. E. Br.■

373780　Stapelia decaisneana（Lem.）A. Chev. = Orbea decaisneana（Lehm.）Bruyns■☆

373781　Stapelia decora Masson = Piaranthus geminatus（Masson）N. E. Br. subsp. decorus（Masson）Bruyns■☆

373782　Stapelia deflexa Jacq.；外折豹皮花■☆

373783　Stapelia depressa Jacq. = Stapelia hirsuta L.■☆

373784　Stapelia desmetiana N. E. Br. = Stapelia grandiflora Masson■

373785　Stapelia desmetiana N. E. Br. var. apicalis？= Stapelia grandiflora Masson■

373786　Stapelia desmetiana N. E. Br. var. fergusoniae R. A. Dyer = Stapelia grandiflora Masson■

373787　Stapelia desmetiana N. E. Br. var. pallida？= Stapelia grandiflora Masson■

373788　Stapelia dinteri A. Berger = Tridentea jucunda（N. E. Br.）L. C. Leach■☆

373789　Stapelia dinteri A. Berger var. capensis C. A. Lückh. = Tridentea jucunda（N. E. Br.）L. C. Leach■☆

373790　Stapelia dinteri A. Berger var. pseudocapensis C. A. Lückh. = Tridentea jucunda（N. E. Br.）L. C. Leach■☆

373791　Stapelia discoidea Oberm.；盘状豹皮花■☆

373792　Stapelia discoidea Oberm. = Orbea semota（N. E. Br.）L. C. Leach■☆

373793　Stapelia divaricata Masson；叉开豹皮花■☆

373794　Stapelia divergens N. E. Br.；黄豹皮花■☆

373795　Stapelia dummeri N. E. Br.；杜默豹皮花■☆

373796　Stapelia dummeri N. E. Br. = Orbea dummeri（N. E. Br.）Bruyns■☆

373797　Stapelia duodecimfida Jacq. = Huernia barbata（Masson）Haw.■☆

373798　Stapelia dwequensis C. A. Lückh. = Tridentea dwequensis（C. A. Lückh.）L. C. Leach■☆

373799　Stapelia elegans Masson = Duvalia elegans（Masson）Haw.■☆

373800　Stapelia engleriana Schltr.；星天角■☆

373801　Stapelia erectiflora N. E. Br.；直立花豹皮花■☆

373802　Stapelia europea Guss. = Apteranthes europaea（Guss.）Plowes■☆

373803　Stapelia fasciculata Thunb.；簇生豹皮花■☆

373804　Stapelia fissirostris Jacq. = Stapelia rufa Masson■☆

373805　Stapelia flavicomata Haw.；黄束毛豹皮花■☆

373806　Stapelia flavirostris N. E. Br.；妖星角（黑犀角）■☆

373807　Stapelia flavirostris N. E. Br. = Stapelia grandiflora Masson■

373808　Stapelia flavopurpurea Marloth；紫黄豹皮花；Carrion Plant■☆

373809　Stapelia flavopurpurea Marloth var. fleckii（A. Berger et Schltr.）White = Stapelia flavopurpurea Marloth■☆

373810　Stapelia fleckii A. Berger et Schltr. = Stapelia flavopurpurea Marloth■☆

373811　Stapelia forcipis E. Phillips et Letty = Stapelia hirsuta L. var. tsomoensis（N. E. Br.）Bruyns■☆

373812　Stapelia fucosa N. E. Br. = Orbea verrucosa（Masson）L. C. Leach■☆

373813　Stapelia furcata N. E. Br. = Orbea melanantha（Schltr.）Bruyns■☆

373814　Stapelia fuscata Jacq. = Tromotriche revoluta（Masson）Haw.■☆

373815　Stapelia gariepensis Pillans = Stapelia hirsuta L. var. gariepensis（Pillans）Bruyns■☆

373816　Stapelia geminata Masson = Piaranthus geminatus（Masson）N. E. Br.■☆

373817　Stapelia gemmiflora Masson = Tridentea gemmiflora（Masson）Haw.■☆

373818　Stapelia gemmiflora Masson var. densa（N. E. Br.）N. E. Br. = Tridentea gemmiflora（Masson）Haw.■☆

373819　Stapelia gemmiflora Masson var. hircosa（Jacq.）N. E. Br. = Tridentea gemmiflora（Masson）Haw.■☆

373820　Stapelia gettliffei R. Pott；格特豹皮花■☆

373821　Stapelia gigantea N. E. Br.；大豹皮花（大犀角，帝王魔星花，王犀角）；Carrion Flower, Carrion Plant, Gigantic Carrionflower, Gigantic Leopardflower, Zulu Giant■

373822　Stapelia gigantea N. E. Br. var. pallida E. Phillips = Stapelia gigantea N. E. Br.■

373823　Stapelia glabricaulis N. E. Br. = Stapelia hirsuta L. var. tsomoensis（N. E. Br.）Bruyns■☆

373824　Stapelia glanduliflora Masson；具腺豹皮花■☆

373825　Stapelia glanduliflora Masson var. emarginata N. E. Br. = Stapelia glanduliflora Masson■☆

373826　Stapelia gordonii Masson = Hoodia gordonii（Masson）Sweet ex Decne.■☆

373827　Stapelia grandiflora Masson；大花魔星花（大花犀角）；Carrion Plant, Largeflower Carrionflower■

373828　Stapelia grandiflora Masson var. conformis（N. E. Br.）Bruyns；同形豹皮花■☆

373829　Stapelia grandiflora Masson var. lineata N. E. Br. = Stapelia grandiflora Masson■

373830　Stapelia guttata Masson = Huernia guttata（Masson）Haw.■☆

373831　Stapelia hamata Jacq. = Stapelia hirsuta L.■☆

373832　Stapelia hanburyana Berger et Rüst；拟牛角■☆

373833　Stapelia herrei Nel = Tromotriche herrei（Nel）Bruyns■☆

373834　Stapelia hircosa Jacq. = Tridentea gemmiflora（Masson）Haw.■☆

373835　Stapelia hircosa Jacq. var. densa N. E. Br. = Tridentea gemmiflora（Masson）Haw.■☆

373836　Stapelia hirsuta L.；魔星花（犀角，英犀角，硬毛豹皮花）；Carrion Plant■☆

373837　Stapelia hirsuta L. var. affinis（N. E. Br.）N. E. Br. = Stapelia hirsuta L.■☆

373838　Stapelia hirsuta L. var. baylissii（L. C. Leach）Bruyns；贝利斯豹皮花■☆

373839　Stapelia hirsuta L. var. comata（Jacq.）N. E. Br. = Stapelia hirsuta L.■☆

373840　Stapelia hirsuta L. var. depressa（Jacq.）N. E. Br. = Stapelia hirsuta L.■☆

373841　Stapelia hirsuta L. var. gariepensis（Pillans）Bruyns；加里豹皮花■☆

373842　Stapelia hirsuta L. var. grata N. E. Br. = Stapelia hirsuta L.■☆

373843　Stapelia hirsuta L. var. longirostris（N. E. Br.）N. E. Br. =

Stapelia hirsuta L. ■☆

373844 Stapelia hirsuta L. var. lutea N. E. Br. = Stapelia hirsuta L. ■☆

373845 Stapelia hirsuta L. var. patula (Willd.) N. E. Br. = Stapelia hirsuta L. ■☆

373846 Stapelia hirsuta L. var. tsomoensis (N. E. Br.) Bruyns;措莫豹皮花■☆

373847 Stapelia hirsuta L. var. unguipetala (N. E. Br.) N. E. Br. = Stapelia hirsuta L. ■☆

373848 Stapelia hirsuta L. var. vetula (Masson) Bruyns;老翁豹皮花■☆

373849 Stapelia hirtella Jacq. = Duvalia caespitosa (Masson) Haw. ☆

373850 Stapelia hispida Horn ex Rust = Orbea variegata (L.) Haw. ☆

373851 Stapelia hispidula Hornem. = Stapelia glanduliflora Masson ■☆

373852 Stapelia horizontalis N. E. Br. = Orbea variegata (L.) Haw. ☆

373853 Stapelia humilis Masson = Huernia humilis (Masson) Haw. ■☆

373854 Stapelia hystrix Hook. f. = Huernia hystrix (Hook. f.) N. E. Br. ■☆

373855 Stapelia immelmaniae Pillans = Stapelia paniculata Willd. ■☆

373856 Stapelia incarnata L. f. = Quaqua incarnata (L. f.) Bruyns ☆

373857 Stapelia indocta Nel = Stapelia acuminata Masson ■☆

373858 Stapelia inodora (Haw.) Decne. = Orbea variegata (L.) Haw. ■☆

373859 Stapelia intermedia N. E. Br. = Caralluma intermedia (N. E. Br.) Schltr. ■☆

373860 Stapelia irrorata Masson = Orbea verrucosa (Masson) L. C. Leach ■☆

373861 Stapelia jacquiniana Schult. = Duvalia elegans (Masson) Haw. ■☆

373862 Stapelia johni-lavrani Halda = Stapelia hirsuta L. var. gariepensis (Pillans) Bruyns ■☆

373863 Stapelia jucunda N. E. Br. = Tridentea jucunda (N. E. Br.) L. C. Leach ■☆

373864 Stapelia jucunda N. E. Br. var. deficiens ? = Tridentea jucunda (N. E. Br.) L. C. Leach ■☆

373865 Stapelia juttae Dinter = Stapelia similis N. E. Br. ■☆

373866 Stapelia juvencula Jacq. = Stapelia hirsuta L. var. vetula (Masson) Bruyns ■☆

373867 Stapelia kagerensis J. -P. Lebrun et Taton = Orbea semota (N. E. Br.) L. C. Leach ☆

373868 Stapelia knobelii E. Phillips = Orbea knobelii (E. Phillips) Bruyns ■☆

373869 Stapelia kougabergensis L. C. Leach = Stapelia paniculata Willd. subsp. kougabergensis (L. C. Leach) Bruyns ■☆

373870 Stapelia kwebensis N. E. Br. ;奎波豹皮花■☆

373871 Stapelia kwebensis N. E. Br. var. longipedicellata A. Berger = Stapelia kwebensis N. E. Br. ■☆

373872 Stapelia laevis Decne. = Tromotriche pedunculata (Masson) Bruyns ■☆

373873 Stapelia lanifera Haw. = Stapelia hirsuta L. ■☆

373874 Stapelia leendertziae N. E. Br. ;钟楼角■☆

373875 Stapelia lentiginosa Sims = Huernia guttata (Masson) Haw. ■☆

373876 Stapelia lepida Jacq. = Orbea variegata (L.) Haw. ■☆

373877 Stapelia longidens N. E. Br. ;长齿豹皮花■☆

373878 Stapelia longidens N. E. Br. = Orbea longidens (N. E. Br.) L. C. Leach ■☆

373879 Stapelia longii C. A. Lückh. = Orbea longii (C. A. Lückh.) Bruyns ■☆

373880 Stapelia longipedicellata (A. Berger) N. E. Br. = Stapelia kwebensis N. E. Br. ■☆

373881 Stapelia longipes C. A. Lückh. ;长柄豹皮花■☆

373882 Stapelia longipes C. A. Lückh. = Tromotriche pedunculata (Masson) Bruyns subsp. longipes (C. A. Lückh.) Bruyns ■☆

373883 Stapelia longipes C. A. Lückh. var. namaquensis ? = Tromotriche pedunculata (Masson) Bruyns subsp. longipes (C. A. Lückh.) Bruyns ■☆

373884 Stapelia lucida DC. = Stapelia hirsuta L. ■☆

373885 Stapelia macloughlinii I. Verd. = Orbea macloughlinii (I. Verd.) L. C. Leach ■☆

373886 Stapelia macowanii N. E. Br. = Stapelia grandiflora Masson var. conformis (N. E. Br.) Bruyns ■☆

373887 Stapelia macowanii N. E. Br. var. conformis (N. E. Br.) L. C. Leach = Stapelia grandiflora Masson var. conformis (N. E. Br.) Bruyns ■☆

373888 Stapelia macrocarpa A. Rich. = Huernia macrocarpa (A. Rich.) Spreng. ■☆

373889 Stapelia mammillaris L. = Quaqua mammillaris (L.) Bruyns ■☆

373890 Stapelia margarita B. Sloane = Stapelia hirsuta L. ■☆

373891 Stapelia marginata Willd. = Orbea variegata (L.) Haw. ☆

373892 Stapelia marientalensis Nel = Tridentea marientalensis (Nel) L. C. Leach ■☆

373893 Stapelia marlothii N. E. Br. = Stapelia gigantea N. E. Br. ■

373894 Stapelia marmorata Jacq. = Orbea variegata (L.) Haw. ■☆

373895 Stapelia mastodes Jacq. = Duvalia caespitosa (Masson) Haw. ■☆

373896 Stapelia melanantha Schltr. = Orbea melanantha (Schltr.) Bruyns ■☆

373897 Stapelia miscella N. E. Br. = Orbea miscella (N. E. Br.) Meve ■☆

373898 Stapelia mixta Masson = Orbea variegata (L.) Haw. ■☆

373899 Stapelia molonyae A. C. White et B. Sloane = Orbea semota (N. E. Br.) L. C. Leach ■☆

373900 Stapelia montana L. C. Leach = Stapelia cedrimontana Frandsen ■☆

373901 Stapelia montana L. C. Leach var. grossa L. C. Leach = Stapelia cedrimontana Frandsen ■☆

373902 Stapelia namaquensis N. E. Br. = Orbea namaquensis (N. E. Br.) L. C. Leach ■☆

373903 Stapelia namaquensis N. E. Br. var. tridentata ? = Orbea namaquensis (N. E. Br.) L. C. Leach ■☆

373904 Stapelia natalensis Rust = Orbea variegata (L.) Haw. ■☆

373905 Stapelia neliana A. C. White et B. Sloane = Tromotriche herrei (Nel) Bruyns ■☆

373906 Stapelia noachabibensis C. A. Lückh. = Stapelia similis N. E. Br. ■☆

373907 Stapelia nobilis N. E. Br. = Stapelia gigantea N. E. Br. ■

373908 Stapelia nobilis N. E. Br. ex Hook. f. ;帝王犀角(富丽豹皮花) ■☆

373909 Stapelia normalis Jacq. = Orbea variegata (L.) Haw. ■☆

373910 Stapelia nouhuysii E. Phillips = Stapelia paniculata Willd. ■☆

373911 Stapelia nudiflora Pillans;光花豹皮花■☆

373912 Stapelia nudiflora Pillans = Stapelia hirsuta L. var. vetula (Masson) Bruyns ■☆

373913 Stapelia obducta L. C. Leach;包被豹皮花■☆

373914 Stapelia obliqua Willd. = Orbea variegata（L.）Haw. ■☆

373915 Stapelia obscura N. E. Br. = Stapelia grandiflora Masson ■

373916 Stapelia ocellata Jacq. = Huernia guttata（Masson）Haw. ■☆

373917 Stapelia olivacea N. E. Br. ；紫水角■☆

373918 Stapelia ophiuncula Haw. = Orbea variegata（L.）Haw. ■☆

373919 Stapelia orbicularis Andréws = Orbea variegata（L.）Haw. ■☆

373920 Stapelia pachyrrhiza Dinter = Tridentea pachyrrhiza（Dinter）L. C. Leach ■☆

373921 Stapelia pallida H. Wendl. = Stapelia divaricata Masson ■☆

373922 Stapelia paniculata Willd. ；圆锥豹皮花■☆

373923 Stapelia paniculata Willd. subsp. kougabergensis（L. C. Leach）Bruyns；库卡圆锥豹皮花■☆

373924 Stapelia paniculata Willd. subsp. scitula（L. C. Leach）Bruyns；纤弱圆锥豹皮花■☆

373925 Stapelia parviflora Masson = Quaqua parviflora（Masson）Bruyns ■☆

373926 Stapelia parvipuncta N. E. Br. ；小斑豹皮花■☆

373927 Stapelia parvipuncta N. E. Br. var. truncata C. A. Lückh. = Tridentea parvipuncta（N. E. Br.）L. C. Leach subsp. truncata（C. A. Lückh.）Bruyns ■☆

373928 Stapelia parvula Kers；较小豹皮花■☆

373929 Stapelia patentirostris N. E. Br. = Stapelia hirsuta L. ■☆

373930 Stapelia patula Willd. = Stapelia hirsuta L. ■☆

373931 Stapelia patula Willd. var. depressa（Jacq.）N. E. Br. = Stapelia hirsuta L. ■☆

373932 Stapelia patula Willd. var. longirostris N. E. Br. = Stapelia hirsuta L. ■☆

373933 Stapelia pearsonii N. E. Br. ；皮尔逊豹皮花■☆

373934 Stapelia peculiaris C. A. Lückh. = Tridentea peculiaris（C. A. Lückh.）L. C. Leach ■☆

373935 Stapelia pedunculata Masson = Tromotriche pedunculata（Masson）Bruyns ■☆

373936 Stapelia peglerae N. E. Br. = Stapelia hirsuta L. var. tsomoensis（N. E. Br.）Bruyns ■☆

373937 Stapelia picta Donn ex Sims = Orbea variegata（L.）Haw. ■☆

373938 Stapelia pilifera L. f. = Hoodia pilifera（L. f.）Plowes ■☆

373939 Stapelia pillansii N. E. Br. ；非洲豹皮花■☆

373940 Stapelia pillansii N. E. Br. var. attenuata ？ = Stapelia pillansii N. E. Br. ■☆

373941 Stapelia pillansii N. E. Br. var. fontinalis Nel = Stapelia pillansii N. E. Br. ■☆

373942 Stapelia planiflora Jacq. = Orbea variegata（L.）Haw. ■☆

373943 Stapelia portae-taurinae Dinter et A. Berger = Stapelia similis N. E. Br. ■☆

373944 Stapelia praetermissa L. C. Leach = Stapelia hirsuta L. var. baylissii（L. C. Leach）Bruyns ■☆

373945 Stapelia praetermissa L. C. Leach var. luteola L. C. Leach = Stapelia hirsuta L. var. baylissii（L. C. Leach）Bruyns ■☆

373946 Stapelia pruinosa Masson = Quaqua pruinosa（Masson）Bruyns ■☆

373947 Stapelia pulchella Masson；豹皮花；Beautiful Carrionflower, Beautiful Leopardflower ■

373948 Stapelia pulchella Masson = Orbea pulchella（Masson）L. C. Leach ■☆

373949 Stapelia pulchra（Haw.）Schult. = Orbea verrucosa（Masson）L. C. Leach ■☆

373950 Stapelia pulla Aiton = Quaqua mammillaris（L.）Bruyns ■☆

373951 Stapelia pulvinata Masson = Stapelia hirsuta L. ■☆

373952 Stapelia punctata Masson = Piaranthus punctatus（Masson）R. Br. ■☆

373953 Stapelia putida A. Berger = Orbea variegata（L.）Haw. ■☆

373954 Stapelia radiata Jacq. = Duvalia elegans（Masson）Haw. ■☆

373955 Stapelia radiata Sims = Duvalia caespitosa（Masson）Haw. ■☆

373956 Stapelia ramosa Masson = Quaqua ramosa（Masson）Bruyns ■☆

373957 Stapelia reclinata Masson = Duvalia caespitosa（Masson）Haw. ■☆

373958 Stapelia reflexa Haw. ；反折豹皮花■☆

373959 Stapelia remota R. A. Dyer；稀疏豹皮花■☆

373960 Stapelia replicata Jacq. = Duvalia caespitosa（Masson）Haw. ■☆

373961 Stapelia reticulata Masson = Huernia guttata（Masson）Haw. subsp. reticulata（Masson）Bruyns ■☆

373962 Stapelia reticulata Schult. var. deformis Jacq. = Huernia guttata（Masson）Haw. subsp. reticulata（Masson）Bruyns ■☆

373963 Stapelia retusa Schult. = Orbea variegata（L.）Haw. ■☆

373964 Stapelia revoluta Masson；反卷豹皮花■☆

373965 Stapelia revoluta Masson = Tromotriche revoluta（Masson）Haw. ■☆

373966 Stapelia rogersii L. Bolus = Orbea rogersii（L. Bolus）Bruyns ■☆

373967 Stapelia roriflua Jacq. = Orbea verrucosa（Masson）L. C. Leach ■☆

373968 Stapelia rubiginosa Nel；锈红豹皮花■☆

373969 Stapelia rufa Masson；赤褐豹皮花■☆

373970 Stapelia rufa Masson var. attenuata N. E. Br. = Stapelia rufa Masson ■☆

373971 Stapelia rufa Masson var. fissirostris（Jacq.）A. C. White et B. Sloane = Stapelia rufa Masson ■☆

373972 Stapelia rufescens Salm-Dyck = Stapelia rufa Masson ■☆

373973 Stapelia rugosa Jacq. = Orbea variegata（L.）Haw. ■☆

373974 Stapelia ruschiana Dinter = Tromotriche ruschiana（Dinter）Bruyns ■☆

373975 Stapelia schinzii A. Berger et Schltr. ；虎犀角■☆

373976 Stapelia schinzii A. Berger et Schltr. var. angolensis Kers；安哥拉豹皮花■☆

373977 Stapelia schinzii A. Berger et Schltr. var. bergeriana（Dinter）L. C. Leach；贝格尔豹皮花■☆

373978 Stapelia scitula L. C. Leach = Stapelia paniculata Willd. subsp. scitula（L. C. Leach）Bruyns ■☆

373979 Stapelia scutellata Tod. = Orbea variegata（L.）Haw. ■☆

373980 Stapelia scylla Sprenger = Orbea variegata（L.）Haw. ■☆

373981 Stapelia semota N. E. Br. ；分离豹皮花■☆

373982 Stapelia senilis N. E. Br. ；翁犀角■☆

373983 Stapelia senilis N. E. Br. = Stapelia grandiflora Masson ■

373984 Stapelia serrulata Jacq. = Piaranthus geminatus（Masson）N. E. Br. subsp. decorus（Masson）Bruyns ■☆

373985 Stapelia similis N. E. Br. ；相似豹皮花■☆

373986 Stapelia simsii（Haw.）Schult. = Stapelia hirsuta L. var. vetula（Masson）Bruyns ■☆

373987 Stapelia sororia Masson = Stapelia hirsuta L. ■☆

373988 Stapelia spectabilis Haw. = Stapelia grandiflora Masson ■

373989 Stapelia stellaris Haw. = Stapelia hirsuta L. ■☆

373990 Stapelia stricta Sims = Stapelia divaricata Masson ■☆

373991 Stapelia stultitioides C. A. Lückh. = Stapelia arenosa C. A.

Lückh. ■☆

373992 Stapelia stygia (Haw.) Schult. = Tridentea gemmiflora (Masson) Haw. ■☆

373993 Stapelia subulata Forssk. = Caralluma adscendens (Roxb.) Haw. ■☆

373994 Stapelia surrecta N. E. Br. ;直立豹皮花■☆

373995 Stapelia surrecta N. E. Br. var. primosii C. A. Lückh. = Stapelia surrecta N. E. Br. ■☆

373996 Stapelia tapscottii I. Verd. = Orbea tapscottii (I. Verd.) L. C. Leach ■☆

373997 Stapelia thudichumii Pillans = Tromotriche thudichumii (Pillans) L. C. Leach ■☆

373998 Stapelia thuretii (F. Cels) Croucher = Huernia thuretii F. Cels ■☆

373999 Stapelia tigridia Decne. = Tromotriche revoluta (Masson) Haw. ■☆

374000 Stapelia tigrina Nel = Tromotriche herrei (Nel) Bruyns ■☆

374001 Stapelia tridentata (N. E. Br.) Rüst = Orbea namaquensis (N. E. Br.) L. C. Leach ■☆

374002 Stapelia trisulca Donn ex Jacq. = Orbea variegata (L.) Haw. ■☆

374003 Stapelia tsomoensis N. E. Br. = Stapelia hirsuta L. var. tsomoensis (N. E. Br.) Bruyns ■☆

374004 Stapelia tuberosa Meerb. = Brachystelma tuberosum (Meerb.) R. Br. ex Sims ■☆

374005 Stapelia umbonata Pillans = Tridentea pachyrrhiza (Dinter) L. C. Leach ■☆

374006 Stapelia uncinata J. Jacq. ;具钩豹皮花□☆

374007 Stapelia unguipetala N. E. Br. = Stapelia hirsuta L. ■☆

374008 Stapelia unicornis C. A. Lückh. ;单角豹皮花■☆

374009 Stapelia vaga N. E. Br. = Orbea lutea (N. E. Br.) Bruyns subsp. vaga (N. E. Br.) Bruyns ■☆

374010 Stapelia variegata (L.) Haw. var. pallida N. E. Br. = Orbea variegata (L.) Haw. ■☆

374011 Stapelia variegata (L.) Haw. var. prometheus Dammann ex Rust = Orbea variegata (L.) Haw. ■☆

374012 Stapelia variegata Forssk. = Ceropegia variegata (Forssk.) Decne. ■☆

374013 Stapelia variegata L. ;杂色豹皮花(牛角);Carrion Flower, Starfish Flower, Star-flower, Toad Flower ■☆

374014 Stapelia variegata L. = Orbea variegata (L.) Haw. ■☆

374015 Stapelia variegata L. var. brevicornis N. E. Br. = Orbea variegata (L.) Haw. ■☆

374016 Stapelia venusta Masson = Huernia guttata (Masson) Haw. ■☆

374017 Stapelia venusta Masson var. minor Jacq. = Huernia guttata (Masson) Haw. ■☆

374018 Stapelia verrucosa Haw. ;姬牛角(姬犀角,野牛角)■☆

374019 Stapelia verrucosa Masson = Orbea verrucosa (Masson) L. C. Leach ■☆

374020 Stapelia verrucosa Masson var. pulchra (Haw.) N. E. Br. = Orbea verrucosa (Masson) L. C. Leach ■☆

374021 Stapelia verrucosa Masson var. robusta ?;大姬牛角;Warty Carrion Flower, Warty Carrion-flower ■☆

374022 Stapelia vetula Masson = Stapelia hirsuta L. var. vetula (Masson) Bruyns ☆

374023 Stapelia vetula Masson var. juvencula (Jacq.) A. Berger = Stapelia hirsuta L. var. vetula (Masson) Bruyns ☆

374024 Stapelia vetula Masson var. simsii (Haw.) N. E. Br. = Stapelia hirsuta L. var. vetula (Masson) Bruyns ■☆

374025 Stapelia villosa N. E. Br. = Stapelia hirsuta L. ■☆

374026 Stapelia virescens N. E. Br. = Tridentea virescens (N. E. Br.) L. C. Leach ■☆

374027 Stapelia wendlandiana Schult. = Orbea verrucosa (Masson) L. C. Leach ■☆

374028 Stapelia wilmaniae C. A. Lückh. = Stapelia leendertziae N. E. Br. ■☆

374029 Stapelia woodii N. E. Br. = Orbea woodii (N. E. Br.) L. C. Leach ■☆

374030 Stapelia woodii N. E. Br. var. westii R. A. Dyer = Orbea woodii (N. E. Br.) L. C. Leach ■☆

374031 Stapelia youngii N. E. Br. ;蛮犀角■☆

374032 Stapelia youngii N. E. Br. = Stapelia gigantea N. E. Br. ■

374033 Stapeliaceae Horan. ;豹皮花科■

374034 Stapeliaceae Horan. = Apocynaceae Juss. (保留科名)●■

374035 Stapeliaceae Horan. = Asclepiadaceae Borkh. (保留科名)●■

374036 Stapelianthus Choux = Stapelianthus Choux ex A. C. White et B. Sloane ■☆

374037 Stapelianthus Choux ex A. C. White et B. Sloane(1933);海葵萝藦属■☆

374038 Stapelianthus baylissii L. C. Leach = Tromotriche baylissii (L. C. Leach) Bruyns ■☆

374039 Stapelianthus choananthus (Lavranos and A. V. Hall) R. A. Dyer = Tromotriche choanantha (Lavranos et H. Hall) Bruyns ■☆

374040 Stapelianthus decaryi Choux;海葵萝藦(雷角)■☆

374041 Stapelianthus madagascariensis (Choux) Choux;臭钟花(雷角)■☆

374042 Stapeliopsis Choux = Stapelianthus Choux ex A. C. White et B. Sloane ■☆

374043 Stapeliopsis E. Phillips = Stultitia E. Phillips ■☆

374044 Stapeliopsis E. Pillans(1928);拟豹皮花属■☆

374045 Stapeliopsis ballyi Marn. -Lap. = Echidnopsis ballyi (Marn. -Lap.) P. R. O. Bally ■☆

374046 Stapeliopsis breviloba (R. A. Dyer) Bruyns;短裂拟豹皮花■☆

374047 Stapeliopsis cooperi (N. E. Br.) E. Phillips = Orbea cooperi (N. E. Br.) L. C. Leach ■☆

374048 Stapeliopsis exasperata (Bruyns) Bruyns;粗糙拟豹皮花■☆

374049 Stapeliopsis khamiesbergensis Bruyns;卡米拟豹皮花■☆

374050 Stapeliopsis neronis Pillans;拟豹皮花■☆

374051 Stapeliopsis pillansii (N. E. Br.) Bruyns;皮朗斯拟豹皮花■☆

374052 Stapeliopsis saxatilis (N. E. Br.) Bruyns;岩地拟豹皮花■☆

374053 Stapeliopsis saxatilis (N. E. Br.) Bruyns subsp. stayneri (M. B. Bayer) Bruyns = Stapeliopsis stayneri (M. B. Bayer) Bruyns ■☆

374054 Stapeliopsis stayneri (M. B. Bayer) Bruyns;斯泰纳拟豹皮花■☆

374055 Stapeliopsis urniflora Lavranos;单花拟豹皮花■☆

374056 Stapfia Burtt Davy(1898) = Davyella Hack. ■☆

374057 Stapfia Burtt Davy(1898) = Neostapfia Burtt Davy ■☆

374058 Stapfiella Gilg(1913);热非时钟花属●☆

374059 Stapfiella claoxyloides Gilg;白桐树热非时钟花●☆

374060 Stapfiella lucida Robyns;亮热非时钟花●☆

374061 Stapfiella lucida Robyns var. pubescens Verdc. ;短柔毛热非时钟花●☆

374062 Stapfiella muricata Staner;粗糙热非时钟花●☆

374063 Stapfiella ulugurica Mildbr. ;乌卢古尔热非时钟花●☆

374064　Stapfiella usambarica J. Lewis;乌桑巴拉热非时钟花●☆

374065　Stapfiella zambesiensis R. Fern. ;赞比西热非时钟花●☆

374066　Stapfiella zambesiensis R. Fern. f. grandifolia R. Fern. ;大叶热非时钟花●☆

374067　Stapfiola Kuntze = Desmostachya (Stapf) Stapf ■

374068　Stapfiola bipinnata (L.) Kuntze = Desmostachya bipinnata (L.) Stapf ■

374069　Stapfiophyton H. L. Li = Fordiophyton Stapf ●■★

374070　Stapfiophyton H. L. Li(1944);无距花属(熊掌草属,异药花属);Nospurflower, Stapfiophyton ■★

374071　Stapfiophyton breviscapum C. Chen = Fordiophyton breviscapum (C. Chen) Y. F. Deng et T. L. Wu ■

374072　Stapfiophyton degeneratum C. Chen = Fordiophyton degeneratum (C. Chen) Y. F. Deng et T. L. Wu ■

374073　Stapfiophyton elattandra (Diels) H. L. Li = Phyllagathis elattandra Diels ■

374074　Stapfiophyton elattandrum (Diels) H. L. Li = Phyllagathis elattandra Diels ■

374075　Stapfiophyton erectum S. Y. Hu = Phyllagathis erecta (S. Y. Hu) C. Y. Wu ex C. Chen ●■

374076　Stapfiophyton peperomiifolium (Oliv.) H. L. Li;无距花(岩娇草);Common Nospurflower, Common Stapfiophyton ■

374077　Stapfiophyton peperomiifolium (Oliv.) H. L. Li = Fordiophyton peperomiifolium (Oliv.) Hansen ■

374078　Stapfiophyton tetrandrum (Diels) H. L. Li = Phyllagathis tetrandra Diels ■

374079　Stapfochloa H. Scholz = Chloris Sw. ●■

374080　Stapfochloa H. Scholz(2004);苏丹草属■☆

374081　Staphidiastrum Naudin = Clidemia D. Don ●☆

374082　Staphidiastrum Naudin = Sagraea DC. ●☆

374083　Staphidium Naudin = Clidemia D. Don ●☆

374084　Staphilea Medik. = Staphylea L. ●

374085　Staphisagria Hill = Delphinium L. ■

374086　Staphisagria Hill = Staphysagria (DC.) Spach ■

374087　Staphylea L. (1913);省沽油属;Bladder Nut, Bladdernut, Bladder-nut ●

374088　Staphylea 'Elegans';雅致省沽油;Elegans Bladdernut ●☆

374089　Staphylea bumalda (Thunb.) DC. ;省沽油(三叶空木,双蝴蝶,水条,珍珠花);Bladdernut, Bumalda Bladder-nut, Japanese Bladdernut ●

374090　Staphylea bumalda (Thunb.) DC. f. kobotokensis Hayashi;讲武省沽油●☆

374091　Staphylea bumalda (Thunb.) DC. f. rosea Okuyama;粉省沽油●☆

374092　Staphylea bumalda (Thunb.) DC. f. stenophylla (Honda) Okuyama = Staphylea bumalda (Thunb.) DC. ●

374093　Staphylea bumalda (Thunb.) DC. f. viridis (Nakai) H. Hara;白花省沽油●☆

374094　Staphylea bumalda (Thunb.) DC. var. glabra Nakai = Staphylea bumalda (Thunb.) DC. ●

374095　Staphylea bumalda (Thunb.) DC. var. pubescens N. Li et Y. H. He;毛省沽油;Pubescent Bumalda Bladder-nut ●

374096　Staphylea bumalda (Thunb.) DC. var. robustior ? = Staphylea bumalda (Thunb.) DC. ●

374097　Staphylea bumalda (Thunb.) DC. var. viridis ? = Staphylea bumalda (Thunb.) DC. ●

374098　Staphylea colchica Steven;科尔切斯省沽油(高加索省沽油);Caucasian Bladdernut, Colchis Bladdernut ●☆

374099　Staphylea colchica Steven var. coulombieri (André) Zabel;法国省沽油;French Bladdernut ●☆

374100　Staphylea emodii Brandis;印巴省沽油;Himalayan Bladdernut ●☆

374101　Staphylea emodii Wall. = Staphylea emodii Brandis ●☆

374102　Staphylea forrestii Balf. f. ;嵩明省沽油(枫树);Forrest Bladdernut, Forrest Bladder-nut ●◇

374103　Staphylea holocarpa Hemsl. ;大果省沽油(白凉子,膀胱果,凉子树,泡泡树);Bladdernut, Chinese Bladdernut, Chinese Bladder-nut ●

374104　Staphylea holocarpa Hemsl. 'Rosea' = Staphylea holocarpa Hemsl. var. rosea Rehder et E. H. Wilson ●

374105　Staphylea holocarpa Hemsl. var. rosea Rehder et E. H. Wilson;玫红省沽油(粉花膀胱果);Rose Bladdernut ●

374106　Staphylea indica Burm. f. = Leea indica (Burm. f.) Merr. ●

374107　Staphylea pinnata L. ;羽状省沽油(欧洲省沽油,羽裂省沽油);Bladder Nut, Bladdernut, Europe Bladdernut, European Bladdernut, St. Anthony's Nut, Wild Pistacia ●☆

374108　Staphylea shweliensis W. W. Sm. ;腺齿省沽油;Ruili Bladdernut, Ruili Bladder-nut, Shwelien Bladdernut, Shwelien Bladder-nut ●

374109　Staphylea simplicifolea Gardner et Champ. = Turpinia arguta (Lindl.) Seem. ●

374110　Staphylea trifolia L. ;美洲省沽油(美国省沽油,三叶省沽油);American Bladder Nut, American Bladdernut, Bladdernut, Eastern Bladdernut ●☆

374111　Staphylea yuanjiangensis K. M. Feng et T. Z. Hsu;元江省沽油;Yuanjiang Bladdernut ●

374112　Staphyleaceae (DC.) Lindl. = Staphyleaceae Martinov(保留科名)●

374113　Staphyleaceae Lindl. = Staphyleaceae Martinov(保留科名)●

374114　Staphyleaceae Martinov (1820) (保留科名);省沽油科;Bladdernut Family, Bladder-nut Family ●

374115　Staphylis St. -Lag. = Staphylea L. ●

374116　Staphyllaea Scop. = Staphylea L. ●

374117　Staphyllodendron Scop. = Staphylea L. ●

374118　Staphyllodendron Scop. = Staphylodendron Mill. ●

374119　Staphyllum Dumort. (nomen) = ? Daucus L. ■

374120　Staphylodendron Mill. = Staphylea L. ●

374121　Staphylodendrum Moench = Staphylodendron Mill. ●

374122　Staphylorhodos Turcz. = ? Azara Ruiz et Pav. ●☆

374123　Staphylosyce Hook. f. = Coccinia Wight et Arn. ■

374124　Staphylosyce barteri Hook. f. = Coccinia barteri (Hook. f.) Keay ■☆

374125　Staphysagria (DC.) Spach = Delphinium L. ■

374126　Staphysagria Spach = Delphinium L. ■

374127　Staphysora Pierre = Maesobotrya Benth. ●☆

374128　Staphysora Pierre ex Pax = Maesobotrya Benth. ●☆

374129　Staphysora albida Pierre ex Pax = Maesobotrya longipes (Pax) Hutch. ●☆

374130　Staphysora dusenii Pax = Maesobotrya klaineana (Pierre) J. Léonard ●☆

374131　Staphysora klaineana Pierre = Maesobotrya klaineana (Pierre) J. Léonard ●☆

374132　Staphysora sapinii De Wild. = Maesobotrya bertramiana Büttner ●☆

374133　Starbia Thouars = Alectra Thunb. ■

374134　Starkea Willd. = Liabum Adans. ■●☆

374135　Starkia Juss. ex Steud. = Starkea Willd. ■●☆

374136　Starkia Steud. = Starkea Willd. ■■●☆

374137　Stathmostelma K. Schum. (1893);尺冠萝藦属■■☆

374138　Stathmostelma angustatum K. Schum. 狭尺冠萝藦■☆

374139　Stathmostelma angustatum K. Schum. subsp. vomeriforme (S. Moore) Goyder;犁头尺冠萝藦■☆

374140　Stathmostelma bicolor K. Schum. = Stathmostelma gigantiflorum K. Schum. ■☆

374141　Stathmostelma chironiodes K. Schum. ex De Wild. et T. Durand = Stathmostelma welwitschii Britten et Rendle ■☆

374142　Stathmostelma crassinerve (N. E. Br.) Bullock = Asclepias crassinervis N. E. Br. ■☆

374143　Stathmostelma diversifolium Goyder;异叶尺冠萝藦■☆

374144　Stathmostelma fornicatum (N. E. Br.) Bullock;拱形尺冠萝藦 ■☆

374145　Stathmostelma fornicatum (N. E. Br.) Bullock subsp. tridentatum Goyder;三齿尺冠萝藦■☆

374146　Stathmostelma frommii Schltr. = Stathmostelma spectabile (N. E. Br.) Schltr. subsp. frommii (Schltr.) Goyder ■☆

374147　Stathmostelma gigantiflorum K. Schum. ;巨花尺冠萝藦■☆

374148　Stathmostelma globuliflorum K. Schum. = Stathmostelma pedunculatum (Decne.) K. Schum. ●☆

374149　Stathmostelma incarnatum K. Schum. ;肉色尺冠萝藦■☆

374150　Stathmostelma katangense (De Wild.) Goyder;加丹加尺冠萝藦■☆

374151　Stathmostelma laurentianum Dewèvre = Stathmostelma welwitschii Britten et Rendle ■☆

374152　Stathmostelma macranthum (Hochst. ex Oliv.) Schltr. = Stathmostelma pedunculatum (Decne.) K. Schum. ●☆

374153　Stathmostelma macropetalum Schltr. et K. Schum. = Stathmostelma spectabile (N. E. Br.) Schltr. ■☆

374154　Stathmostelma nuttii (N. E. Br.) Bullock;纳特尺冠萝藦■☆

374155　Stathmostelma odoratum K. Schum. = Stathmostelma spectabile (N. E. Br.) Schltr. ■☆

374156　Stathmostelma pachycladum K. Schum. = Stathmostelma spectabile (N. E. Br.) Schltr. ■☆

374157　Stathmostelma pauciflorum (Klotzsch) K. Schum. ;寡花尺冠萝藦●☆

374158　Stathmostelma pedunculatum (Decne.) K. Schum. ;梗花尺冠萝藦●☆

374159　Stathmostelma praetermissum Bullock = Stathmostelma gigantiflorum K. Schum. ■☆

374160　Stathmostelma propinquum (N. E. Br.) Schltr. ;邻近尺冠萝藦■☆

374161　Stathmostelma reflexum Britten et Rendle = Stathmostelma pauciflorum (Klotzsch) K. Schum. ●☆

374162　Stathmostelma spectabile (N. E. Br.) Schltr. ;壮观尺冠萝藦■☆

374163　Stathmostelma spectabile (N. E. Br.) Schltr. subsp. frommii (Schltr.) Goyder;弗罗姆尺冠萝藦■☆

374164　Stathmostelma thomasii Bullock = Stathmostelma angustatum K. Schum. subsp. vomeriforme (S. Moore) Goyder ■☆

374165　Stathmostelma verdickii De Wild. ;韦氏尺冠萝藦■☆

374166　Stathmostelma welwitschii Britten et Rendle;韦尔尺冠萝藦■☆

374167　Stathmostelma welwitschii Britten et Rendle var. bagshawei (S. Moore) Goyder;巴格肖尺冠萝藦■☆

374168　Stathmostelma wildemanianum Durand;怀尔德曼尺冠萝藦■☆

374169　Staticaceae Cassel = Plumbaginaceae Juss. (保留科名)●■

374170　Staticaceae Hoffmanns. et Link ex Gray = Plumbaginaceae Juss. (保留科名)●■

374171　Statice L. (废弃属名) = Armeria Willd. (保留属名)■☆

374172　Statice L. (废弃属名) = Armeria Willd. + Limonium Mill. ■

374173　Statice alliacea Cav. var. apollinaris Sennen = Armeria alliacea (Cav.) Hoffmanns. et Link ■☆

374174　Statice allioides (Boiss.) Kuntze = Armeria alliacea (Cav.) Hoffmanns. et Link ■☆

374175　Statice allioides (Boiss.) Kuntze var. yebalica (Pau) Maire = Armeria alliacea (Cav.) Hoffmanns. et Link ■☆

374176　Statice alpinifolia (Pau et Font Quer) Maire = Armeria alpinifolia Pau et Font Quer ■☆

374177　Statice amoena C. H. Wright = Afrolimon amoenum (C. H. Wright) Lincz. ●☆

374178　Statice amplifoliata (Pau) Maire = Armeria simplex Pomel ■☆

374179　Statice anthericoides Schltr. = Limonium anthericoides (Schltr.) R. A. Dyer ■☆

374180　Statice aphylla Poir. = Limonium coralloides (Tausch) Lincz. ■

374181　Statice arbuscula Maxim. = Limonium wrightii (Hance) Kuntze ●■

374182　Statice argentea Pall. ex Siev. = Goniolimon callicomum (C. A. Mey.) Boiss. ■

374183　Statice armeria L. = Armeria maritima (Mill.) Willd. ■☆

374184　Statice asparagoides Batt. = Limonium asparagoides (Batt.) Maire ■☆

374185　Statice aurea L. = Limonium aureum (L.) Hill ex Kuntze ■

374186　Statice aurea L. = Limonium potaninii Ikonn. -Gal. ■

374187　Statice avenacea C. H. Wright = Limonium scabrum (Thunb.) Kuntze var. avenaceum (C. H. Wright) R. A. Dyer ■☆

374188　Statice axillaris Forssk. = Limonium axillare (Forssk.) Kuntze ■☆

374189　Statice beaumieriana Maire = Limonium sinuatum (L.) Mill. subsp. beaumierianum (Maire) Sauvage et Vindt ■☆

374190　Statice beaumieriana Maire var. akkensis ? = Limonium sinuatum (L.) Mill. subsp. beaumierianum (Maire) Sauvage et Vindt ■☆

374191　Statice beaumieriana Maire var. tripeaui ? = Limonium sinuatum (L.) Mill. subsp. beaumierianum (Maire) Sauvage et Vindt ■☆

374192　Statice bicolor Bunge = Limonium bicolor (Bunge) Kuntze ■

374193　Statice bicolor Bunge = Limonium sinense (Girard) Kuntze ■

374194　Statice bicolor Bunge var. laxiflora Bunge = Limonium bicolor (Bunge) Kuntze ■

374195　Statice bonduellei T. Lestib. = Limonium sinuatum (L.) Mill. var. bonduellei (T. Lestib.) Sauvage et Vindt ■☆

374196　Statice bonduellei T. Lestib. var. leucocalyx Maire = Limonium sinuatum (L.) Mill. subsp. bonduellei (F. Lestib.) Sauvage et Vindt ■☆

374197　Statice bonduellei T. Lestib. var. xanthocalyx Maire = Limonium sinuatum (L.) Mill. var. bonduellei (T. Lestib.) Sauvage et Vindt ■☆

374198　Statice braunii Bolle = Limonium braunii (Bolle) A. Chev. ■☆

374199　Statice brunneri Webb = Limonium brunneri (Webb) Kuntze ■☆

374200 Statice bungeana Boiss. = Limonium bicolor (Bunge) Kuntze ■

374201 Statice cabulica Boiss. = Limonium cabulicum (Boiss.) Kuntze ■☆

374202 Statice californica Boiss. = Limonium californicum (Boiss.) A. Heller ■☆

374203 Statice callicoma C. A. Mey. = Goniolimon callicomum (C. A. Mey.) Boiss. ■

374204 Statice capensis L. Bolus = Afrolimon capense (L. Bolus) Lincz. ■☆

374205 Statice carinensis Chiov. = Limonium xipholepis (Baker) Hutch. et E. A. Bruce ■☆

374206 Statice caroliniana Walter = Limonium carolinianum (Walter) Britton ■☆

374207 Statice chazaliei Boissieu = Limonium chazaliei (Boissieu) Maire ■☆

374208 Statice chrysocephala Regel = Limonium chrysocomum (Kar. et Kir.) Kuntze ●■

374209 Statice chrysocoma Kar. et Kir. = Limonium chrysocomum (Kar. et Kir.) Kuntze ●■

374210 Statice congesta Ledeb. = Limonium congestum (Ledeb.) Kuntze ■

374211 Statice coralloides Tausch = Limonium coralloides (Tausch) Lincz. ■

374212 Statice corymbulosa Boiss. = Limonium scabrum (Thunb.) Kuntze var. corymbulosum (Boiss.) R. A. Dyer ■☆

374213 Statice cylindrifolium Forssk. = Limonium cylindrifolium (Forssk.) Verdc. ■☆

374214 Statice cymulifera Boiss. = Limonium cymuliferum (Boiss.) Sauvage et Vindt ■☆

374215 Statice cymulifera Boiss. subsp. mauritii Sennen = Limonium cymuliferum (Boiss.) Sauvage et Vindt ■☆

374216 Statice cyrenaica Rouy = Limonium cyrenaicum (Rouy) Brullo ■☆

374217 Statice cyrtostachia Girard = Limonium cyrtostachyum (Girard) Brullo ■☆

374218 Statice decipiens Ledeb. = Limonium coralloides (Tausch) Lincz. ■

374219 Statice decumbens Boiss. = Limonium decumbens (Boiss.) Kuntze ■☆

374220 Statice delicatula Girard = Limonium delicatulum (Girard) Kuntze ■☆

374221 Statice delicatula Girard var. minor (Boiss.) Bég. et Vacc. = Limonium delicatulum (Girard) Kuntze ■☆

374222 Statice delicatula Girard var. subrotundifolia Bég. et Vacc. = Limonium delicatulum (Girard) Kuntze ■☆

374223 Statice desipiens Ledeb. = Limonium coralloides (Tausch) Lincz. ■

374224 Statice dichroantha Rupr. = Limonium dichoroanthum (Rupr.) Ikonn. -Gal. ex Lincz. ■

374225 Statice dielsiana Wangerin = Limonium dielsianum (Wangerin) Kamelin ■

374226 Statice dregeana C. Presl = Limonium dregeanum (C. Presl) Kuntze ■☆

374227 Statice dschungarica Regel = Goniolimon dschungaricum (Regel) O. Fedtsch. et B. Fedtsch. ■

374228 Statice duriaei Girard = Limonium duriaei (Girard) Kuntze ■☆

374229 Statice ebracteata (Pomel) Maire = Armeria ebracteata Pomel ■☆

374230 Statice ebracteata (Pomel) Maire var. laevis Maire = Armeria ebracteata Pomel ■☆

374231 Statice echioides L. = Limonium echioides (L.) Mill. ■☆

374232 Statice echioides L. var. exaristata Murb. = Limonium echioides (L.) Mill. ■☆

374233 Statice equisetina Boiss. = Limonium equisetinum (Boiss.) R. A. Dyer ■☆

374234 Statice equisetina Boiss. var. depauperata Boiss. = Limonium depauperatum (Boiss.) R. A. Dyer ■☆

374235 Statice eximia Schrenk = Goniolimon eximium (Schrenk) Boiss. ■

374236 Statice eximia Schrenk var. turkestanica Regel = Goniolimon eximium (Schrenk) Boiss. ■

374237 Statice faustii Sennen et Mauricio = Limonium cossonianum Kuntze ■☆

374238 Statice ferulacea L. = Myriolepis ferulacea (L.) Lledò et Erben et Crespo ■☆

374239 Statice filicaulis (Boiss.) Maire = Armeria filicaulis (Boiss.) Boiss. ■☆

374240 Statice filicaulis (Boiss.) Maire var. maroccana Pau et Font Quer = Armeria filicaulis (Boiss.) Boiss. ■☆

374241 Statice flexuosa L. = Limonium flexuosum (L.) Kuntze ■

374242 Statice florida Kitag. = Limonium bicolor (Bunge) Kuntze ■

374243 Statice fortunei Lindl. = Limonium sinense (Girard) Kuntze ■

374244 Statice fradiniana Pomel = Limonium densiflorum (Guss.) Kuntze ■☆

374245 Statice franchetii Debeaux = Limonium franchetii (Debeaux) Kuntze ■

374246 Statice gavilae Sennen et Mauricio = Limonium cossonianum Kuntze ■☆

374247 Statice glauca Willd. ex Schult. = Limonium gmelinii (Willd.) Kuntze ■

374248 Statice glauca Willd. ex Schult. = Limonium suffruticosum (L.) Kuntze ●

374249 Statice globulariifolia Desf. = Limonium ramosissimum (Poir.) Maire ■☆

374250 Statice gmelinii M. Bieb. = Limonium tomentellum (Boiss.) Kuntze ■☆

374251 Statice gmelinii Willd. = Limonium flexuosum (L.) Kuntze ■

374252 Statice gmelinii Willd. = Limonium gmelinii (Willd.) Kuntze ■

374253 Statice gmelinii Willd. var. scoparia (Pall. ex Willd.) Schmalh. = Limonium gmelinii (Willd.) Kuntze ■

374254 Statice gomezi-jordanae Sennen = Limonium cossonianum Kuntze ■☆

374255 Statice gougetiana Girard = Limonium gougetianum (Girard) Kuntze ■☆

374256 Statice griffithii Aitch. et Hemsl. = Limonium cabulicum (Boiss.) Kuntze ■☆

374257 Statice gummifera Boiss. et Reut. = Limonium battandieri Greuter et Burdet ■☆

374258 Statice gummifera Durieu = Limonium cymuliferum (Boiss.) Sauvage et Vindt ■☆

374259 Statice gummifera Durieu var. corymbulosa Coss. = Limonium cossonianum Kuntze ■☆

374260 Statice holtzeri Regel = Limonium kaschgaricum (Rupr.) Ikonn. -Gal. ■

374261 Statice imbricata Girard = Limonium imbricatum (Girard) C. E. Hubb. ■☆

374262　Statice japonica Siebold et Zucc. = Limonium sinense (Girard) Kuntze ■

374263　Statice japonica Siebold et Zucc. = Limonium tetragonum (Thunb.) Bullock ■

374264　Statice kaschgarica Rupr. = Limonium kaschgaricum (Rupr.) Ikonn. -Gal. ■

374265　Statice kaufmanniana Regel = Ikonnikovia kaufmanniana (Regel) Lincz. ●◇

374266　Statice kossmatii R. Wagner et Vierh. = Limonium kossmatii (R. Wagner et Vierh.) Verdc. et Hemming ex Cufod. ■☆

374267　Statice kraussiana Buchinger ex Boiss. = Limonium kraussianum (Buchinger ex Boiss.) Kuntze ■☆

374268　Statice lachnolepis (Pomel) Maire = Armeria choulettiana Pomel ■☆

374269　Statice lacostei Danguy = Limonium aureum (L.) Hill ex Kuntze ■

374270　Statice lacostei Danguy = Limonium lacostei (Danguy) Kamelin ●■

374271　Statice laeta Ball = Limonium ornatum (Ball) Kuntze ■☆

374272　Statice latifolia Sm. = Limonium latifolium (Sm.) Kuntze ■☆

374273　Statice latissima Kar. et Kir. = Limonium myrianthum (Schrenk) Kuntze ■

374274　Statice lepidorachis Pomel = Limonium cymuliferum (Boiss.) Sauvage et Vindt ■☆

374275　Statice leptoloba Regel = Limonium leptolobum (Regel) Kuntze ■

374276　Statice leptoloba Regel var. subaphylla Regel = Limonium leptolobum (Regel) Kuntze ■

374277　Statice leptostachys Pomel = Limonium delicatulum (Girard) Kuntze ■☆

374278　Statice letourneuxii Batt. = Limonium letourneuxii (Batt.) Greuter et Burdet ■☆

374279　Statice limbata (Small) K. Schum. = Limonium limbatum Small ■☆

374280　Statice limonium L. = Limonium narbonense Mill. ■☆

374281　Statice lingua Pomel = Limonium cymuliferum (Boiss.) Sauvage et Vindt ■☆

374282　Statice linifolia L. f. = Limonium linifolium (L. f.) Kuntze ■☆

374283　Statice linifolia L. f. var. maritima Eckl. et Zeyh. ex Boiss. = Limonium linifolium (L. f.) Kuntze var. maritimum (Eckl. et Zeyh. ex Boiss.) R. A. Dyer ■☆

374284　Statice longiaristata Boiss. var. cuspidata Faure et Maire = Armeria atlantica Pomel ■☆

374285　Statice longifolia Thunb. = Afrolimon longifolium (Thunb.) Lincz. ■☆

374286　Statice lusitanica Poir. = Armeria mauritanica Wallr. ■☆

374287　Statice lychnidifolia Girard = Limonium auriculiursifolium (Pourr.) Druce ■☆

374288　Statice lycopodioides Girard = Acantholimon lycopodioides Boiss. ●

374289　Statice macrorrhabdos Boiss. = Limonium macrorhabdon (Boiss.) Kuntze ■☆

374290　Statice maritima Mill. = Armeria maritima (Mill.) Willd. ■☆

374291　Statice molesii Sennen = Limonium cossonianum Kuntze ■☆

374292　Statice monopetala L. = Limoniastrum monopetalum (L.) Boiss. ●☆

374293　Statice mouretii Pit. = Limonium mouretii (Pit.) Maire ■☆

374294　Statice mucronata L. f. = Limonium mucronatum (L. f.) Chaz. ■☆

374295　Statice multiceps Pomel = Limonium gougetianum (Girard) Kuntze ■☆

374296　Statice myriantha Schrenk = Limonium myrianthum (Schrenk) Kuntze ■

374297　Statice nogalensis Chiov. = Limonium xipholepis (Baker) Hutch. et E. A. Bruce ■☆

374298　Statice occidentalis Lloyd = Limonium binervosum (G. E. Sm.) C. E. Salmon ■☆

374299　Statice ochrantha Kar. et Kir. = Goniolimon speciosum (L.) Boiss. ■

374300　Statice oleifolia Scop. var. fradiniana (Pomel) Batt. = Limonium densiflorum (Guss.) Kuntze ■☆

374301　Statice ornata Ball = Limonium ornatum (Ball) Kuntze ■☆

374302　Statice otolepis Schrenk = Limonium otolepis (Schrenk) Kuntze ■

374303　Statice ovalifolia Poir. = Limonium ovalifolium (Poir.) Kuntze ■☆

374304　Statice perezii Stapf = Limonium perezii (Stapf) F. T. Hubb. ●☆

374305　Statice perigrina P. J. Bergius = Afrolimon peregrinum (P. J. Bergius) Lincz. ■☆

374306　Statice plantaginea All. subsp. choulettiana (Pomel) Maire = Armeria choulettiana Pomel ■☆

374307　Statice plantaginea All. subsp. leucantha (Boiss.) Maire = Armeria atlantica Pomel ■☆

374308　Statice plantaginea All. subsp. medians Maire = Armeria atlantica Pomel ■☆

374309　Statice plantaginea All. var. atlantica (Pomel) Maire = Armeria atlantica Pomel ■☆

374310　Statice plantaginea All. var. barbata Maire = Armeria choulettiana Pomel ■☆

374311　Statice plantaginea All. var. brachylepis (Batt.) Maire = Armeria choulettiana Pomel ■☆

374312　Statice plantaginea All. var. cuspidata (Faure et Maire) Maire = Armeria atlantica Pomel ■☆

374313　Statice plantaginea All. var. djurdjurae Maire = Armeria choulettiana Pomel ■☆

374314　Statice plantaginea All. var. masguindalii (Pau) Maire = Armeria atlantica Pomel ■☆

374315　Statice plantaginea All. var. microcephala Maire = Armeria choulettiana Pomel ■☆

374316　Statice plantaginea All. var. subcuspidata Maire = Armeria atlantica Pomel ■☆

374317　Statice plantaginea All. var. zaianica (Emb. et Maire) Maire = Armeria choulettiana Pomel ■☆

374318　Statice pruinosa L. = Limonium pruinosum (L.) Chaz. ■☆

374319　Statice pruinosa L. var. hirtiflora Cavara = Limonium pruinosum (L.) Chaz. ■☆

374320　Statice psiloclada Boiss. = Limonium pescadense Greuter et Burdet ■☆

374321　Statice psiloclada Boiss. = Limonium ramosissimum (Poir.) Maire ■☆

374322　Statice purpurata L. = Afrolimon purpuratum (L.) Lincz. ■☆

374323　Statice pycnantha K. Koch = Limonium gmelinii (Willd.) Kuntze ■

374324　Statice pyrrholepis Pomel = Limonium delicatulum (Girard) Kuntze ■☆

374325　Statice ramosissima Poir. = Limonium ramosissimum (Poir.)

Maire ■☆

374326 Statice rosea Sm. = Afrolimon peregrinum (P. J. Bergius) Lincz. ■☆

374327 Statice sanjurjoi Sennen et Mauricio = Limonium cossonianum Kuntze ■☆

374328 Statice scabra Thunb. = Limonium scabrum (Thunb.) Kuntze ■☆

374329 Statice schrenkiana Fisch. et C. A. Mey. = Limonium aureum (L.) Hill ex Kuntze ■

374330 Statice schrenkiana Fisch. et C. A. Mey. = Limonium chrysocomum (Kar. et Kir.) Kuntze ●■

374331 Statice scoparia Pall. ex Willd. = Limonium gmelinii (Willd.) Kuntze ■

374332 Statice sebkarum Pomel = Limonium cymuliferum (Boiss.) Sauvage et Vindt ■☆

374333 Statice sebkarum Pomel var. mauritii Sennen = Limonium cymuliferum (Boiss.) Sauvage et Vindt ■☆

374334 Statice sedoides Regel = Limonium chrysocomum (Kar. et Kir.) Kuntze ●■

374335 Statice semenowii Herder = Limonium chrysocomum (Kar. et Kir.) Kuntze var. semenowii (Herder) T. H. Peng ■

374336 Statice sinensis Girard = Limonium sinense (Girard) Kuntze ■

374337 Statice sinensium Gand. = Limonium bicolor (Bunge) Kuntze ■

374338 Statice sinuata L. = Limonium sinuatum (L.) Mill. ■☆

374339 Statice sinuata L. var. emarginata Boiss. = Limonium sinuatum (L.) Mill. ■☆

374340 Statice sinuata L. var. integrifolia Boiss. = Limonium sinuatum (L.) Mill. ■☆

374341 Statice somalorum Vierh. = Limonium axillare (Forssk.) Kuntze ■☆

374342 Statice spathulata Desf. = Limonium spathulatum (Desf.) Kuntze ■☆

374343 Statice speciosa L. = Goniolimon speciosum (L.) Boiss. ■

374344 Statice speciosa L. var. crispa Regel = Goniolimon eximium (Schrenk) Boiss. ■

374345 Statice speciosa L. var. lanceolata Regel = Goniolimon speciosum (L.) Boiss. ■

374346 Statice speciosa L. var. stricta Regel = Goniolimon speciosum (L.) Boiss. var. strictum (Regel) T. H. Peng ■

374347 Statice spicata Willd. = Psylliostachys spicatus (Willd.) Nevski ■☆

374348 Statice spinulosa (Boiss.) Maire = Armeria spinulosa Boiss. ■☆

374349 Statice stocksii Boiss. = Limonium stocksii (Boiss.) Kuntze ■☆

374350 Statice subaphylla Regel = Limonium leptolobum (Regel) Kuntze ■

374351 Statice subfruticosum L. = Limonium suffruticosum (L.) Kuntze ●

374352 Statice suffruticosa L. = Limonium suffruticosum (L.) Kuntze ●

374353 Statice suworowii Regel = Psylliostachys suworowii (Regel) Roshkova ■☆

374354 Statice tchefouensis Gand. = Limonium franchetii (Debeaux) Kuntze ■

374355 Statice tenella Turcz. = Limonium tenellum (Turcz.) Kuntze ■

374356 Statice teretifolia Baker ex Oliv. = Limonium cylindrifolium (Forssk.) Verdc. ■☆

374357 Statice tetragona Thunb. = Limonium tetragonum (Thunb.)

Bullock ■

374358 Statice thouinii Viv. = Limonium lobatum (L. f.) Chaz. ■☆

374359 Statice tuberculata Boiss. = Limonium tuberculatum (Boiss.) Kuntze ■☆

374360 Statice tubiflora Delile = Limonium tubiflorum (Delile) Kuntze ■☆

374361 Statice tubiflora Delile var. zanonii Pamp. = Limonium tubiflorum (Delile) Kuntze ■☆

374362 Statice tunetana Bonnet et Barratte = Limonium tunetanum (Bonnet et Barratte) Maire ■☆

374363 Statice varia Hance = Limonium bicolor (Bunge) Kuntze ■

374364 Statice villosa (Girard) Maire = Armeria choulettiana Pomel ■☆

374365 Statice virgata Willd. = Limonium virgatum (Willd.) Fourr. ■☆

374366 Statice wrightii Hance = Limonium wrightii (Hance) Kuntze ●■

374367 Statice xipholepis Baker = Limonium xipholepis (Baker) Hutch. et E. A. Bruce ■☆

374368 Statiotes acoroides L. f. = Enhalus acoroides (L. f.) Rich. ex Steud. ■

374369 Statiotes alismoides L. = Ottelia alismoides (L.) Pers. ■

374370 Staudtia Warb. (1897) ; 非洲蔻木属 (斯托木属) ●☆

374371 Staudtia congoensis Vermoesen = Staudtia kamerunensis Warb. var. gabonensis (Warb.) Fouilloy ●☆

374372 Staudtia gabonensis Warb. = Staudtia kamerunensis Warb. var. gabonensis (Warb.) Fouilloy ●☆

374373 Staudtia gabonensis Warb. var. macrocarpa G. C. C. Gilbert et Troupin = Staudtia kamerunensis Warb. ●☆

374374 Staudtia kamerunensis Warb. ; 喀麦隆非洲蔻木 (喀麦隆斯托木) ●☆

374375 Staudtia kamerunensis Warb. var. gabonensis (Warb.) Fouilloy ; 加蓬非洲蔻木 (加蓬斯托木) ●☆

374376 Staudtia pterocarpa (Warb.) Warb. ; 翅果非洲蔻木 (翅果斯托木) ●☆

374377 Staudtia pterocarpa Warb. = Staudtia pterocarpa (Warb.) Warb. ●☆

374378 Staudtia stipitata Warb. ; 具柄非洲蔻木 (具柄斯托木 , 有柄斯托木) ; Niove ●☆

374379 Staudtia stipitata Warb. = Staudtia kamerunensis Warb. var. gabonensis (Warb.) Fouilloy ●☆

374380 Staufferia Z. S. Rogers , Nickrent et Malécot (2008) ; 斯氏檀香属 ●☆

374381 Stauntonia DC. (1817) ; 野木瓜属 ; Sausage Vine ●

374382 Stauntonia alata Merr. = Stauntonia decora (Dunn) C. Y. Wu ●

374383 Stauntonia angustifolia Wall. = Holboellia angustifolia Wall. ●

374384 Stauntonia brachyanthera Hand. -Mazz. ; 黄蜡果 (黄药野木瓜 , 黄腊果 , 假木瓜 , 萝卜藤 , 蒙姑芦) ; Short-anthered Staunton-vine , Short-flower Stauntonvine ●

374385 Stauntonia brachyanthera Hand. -Mazz. var. minor Diels ex C. Y. Wu ; 小黄蜡果 (小果黄腊果) ; Small Short-flower Stauntonvine ●

374386 Stauntonia brachyanthera Hand. -Mazz. var. minor Diels ex C. Y. Wu = Stauntonia obovatifoliola Hayata subsp. urophylla (Hand. -Mazz.) H. N. Qin ●

374387 Stauntonia brachybotrya T. C. Chen ; 短序野木瓜 ; Shortraceme Stauntonvine , Shortraceme Wild Quince , Short-racemed Staunton-vine ●

374388 Stauntonia brachybotrya T. C. Chen = Stauntonia obovatifoliola Hayata subsp. urophylla (Hand. -Mazz.) H. N. Qin ●

374389 Stauntonia brevipes Hemsl. = Holboellia coriacea Diels ●

374390 Stauntonia brunoniana Wall.；三叶野木瓜（印度野木瓜）；Brunon Stauntonvine，Brunon Staunton-vine，Trifoliate Wild Quince ●

374391 Stauntonia cavalerieana Gagnep.；西南野木瓜（八叶瓜，九叶木通，六月瓜）；Cavalerie Stauntonvine，Cavalerie Staunton-vine ●

374392 Stauntonia chinensis DC.；野木瓜（假荔枝，假荔枝牛藤，木通七叶莲，拿绳，那藤，牛藤，牛芽标，七叶莲，绕绕藤，沙藤，沙引藤，山芭蕉，似荔枝，土牛藤，五叶木通，五爪金龙，五爪野金龙，鸭脚莲，鸭脚藤）；Chinese Stauntonia，Chinese Stauntonvine，Chinese Staunton-vine，False Lychee，Wild Quince ●

374393 Stauntonia conspicua R. H. Chang；腺脉野木瓜（三叶绳，显脉野木瓜）；Distinctvein Stauntonvine，Distinctvein Wild Quince ●

374394 Stauntonia crassipes T. C. Chen；粗柄野木瓜；Thick-stiped Staunton-vine，Thick-stslk Stauntonvine，Thick-stslk Wild Quince ●

374395 Stauntonia crassipes T. C. Chen = Stauntonia trinervia Merr. ●

374396 Stauntonia decora（Dunn）C. Y. Wu；翅野木瓜（大酸藤，猪腰子果）；Decored Stauntonvine，Ornate Staunton-vine ●

374397 Stauntonia dielsiana C. Y. Wu = Stauntonia brachyanthera Hand.-Mazz. ●

374398 Stauntonia duclouxii Gagnep.；羊瓜藤（野木瓜，云南野木瓜）；Ducloux Stauntonvine，Ducloux Staunton-vine，Ducloux Wild Quince ●

374399 Stauntonia elliptica Hemsl.；牛藤果；Elliptic Stauntonvine，Elliptical Staunton-vine ●

374400 Stauntonia formosana Hayata；台湾野木瓜；Taiwan Staunton-vine ●

374401 Stauntonia formosana Hayata = Stauntonia obovata Hemsl. ●

374402 Stauntonia glauca Merr. et F. P. Metcalf；粉叶野木瓜（白背野木瓜）；Glaucous Stauntonvine，Grey-blue Staunton-vine ●

374403 Stauntonia hainanensis T. C. Chen；海南野木瓜；Hainan Stauntonvine，Hainan Staunton-vine，Hainan Wild Quince ●

374404 Stauntonia hainanensis T. C. Chen = Stauntonia chinensis DC. ●

374405 Stauntonia hebandra Hayata = Stauntonia obovata Hemsl. ●

374406 Stauntonia hebandra Hayata var. angustata C. Y. Wu = Stauntonia obovata Hemsl. ●

374407 Stauntonia hedandra Hayata var. angustata C. Y. Wu = Stauntonia obovata Hemsl. var. angustata（C. Y. Wu）H. L. Li ●

374408 Stauntonia hexaphylla（Thunb. ex Murray）Decne. = Stauntonia hexaphylla（Thunb.）Decne. ●

374409 Stauntonia hexaphylla（Thunb.）Decne.；那藤（假荔枝，六叶野木瓜，牛藤，七姐妹藤，日本野木瓜，石月，鸭脚莲，野木瓜，野人瓜）；Hexaphyllous Staunton-vine，Japanese Staunton Vine，Japanese Staunton-vine，Sechsblattrige Staunton-vine，Sixleaves Stauntonvine ●

374410 Stauntonia hexaphylla（Thunb.）Decne. f. cordata H. L. Li = Stauntonia obovatifoliola Hayata ●

374411 Stauntonia hexaphylla（Thunb.）Decne. f. intermedia C. Y. Wu = Stauntonia obovatifoliola Hayata subsp. intermedia（C. Y. Wu）T. C. Chen ●

374412 Stauntonia hexaphylla（Thunb.）Decne. f. intermedia C. Y. Wu = Stauntonia obovatifoliola Hayata subsp. urophylla（Hand. -Mazz.）H. N. Qin ●

374413 Stauntonia hexaphylla（Thunb.）Decne. f. obovata Y. C. Wu；倒卵叶六叶野木瓜（变尾叶石月，倒卵叶那藤）；Obovate Japanese Staunton-vine ●

374414 Stauntonia hexaphylla（Thunb.）Decne. f. rotundata C. Y. Wu；椭圆叶石月（石月，野人瓜，郁子，圆叶那藤）；Rotundate Japanese Staunton-vine ●

374415 Stauntonia hexaphylla（Thunb.）Decne. f. urophylla（Hand. -Mazz.）C. Y. Wu = Stauntonia obovatifoliola Hayata subsp. urophylla（Hand. -Mazz.）H. N. Qin ●

374416 Stauntonia hexaphylla（Thunb.）Decne. var. urophylla Hand. -Mazz. = Stauntonia obovatifoliola Hayata subsp. urophylla（Hand. -Mazz.）H. N. Qin ●

374417 Stauntonia keitaoensis Hayata；阿里山野木瓜（溪头野木瓜）；Keitao Stauntonvine，Keitao Staunton-vine ●

374418 Stauntonia keitaoensis Hayata = Stauntonia obovata Hemsl. ●

374419 Stauntonia latifolia Wall. = Holboellia latifolia Wall. ●

374420 Stauntonia leucantha Diels ex C. Y. Wu；短药野木瓜（艾口藤，八月瓜，钝药野木瓜，九月黄，绕绕藤，五叶木通，芽曲藤）；White-flower Stauntonvine，White-flower Wild Quince，White-flowered Staunton-vine ●

374421 Stauntonia libera H. N. Qin；离丝野木瓜；Liberated Stauntonvine，Liberated Wild Quince ●

374422 Stauntonia longipes Hemsl. = Holboellia angustifolia Wall. ●

374423 Stauntonia maculata Merr.；斑叶野木瓜；Maculate Stauntonvine，Spotted Staunton-vine ●

374424 Stauntonia obcordatilimba C. Y. Wu et S. H. Huang；倒心叶野木瓜；Obcordate Stauntonvine，Obcordate Staunton-vine ●

374425 Stauntonia obovata Hemsl.；倒卵叶野木瓜（倒卵野木瓜，钝药野木瓜，四叶野木瓜，台湾野木瓜，圆叶野木瓜）；Formosan Staunton-vine，Obovate Staunton-vine，Obovateleaf Stauntonvine，Taiwan Stauntonvine，Taiwan Wild Quince ●

374426 Stauntonia obovata Hemsl. var. angustata（C. Y. Wu）H. L. Li；狭萼倒卵叶野木瓜（狭萼钝药野木瓜，狭叶倒卵叶野木瓜，小圆叶野木瓜）；Narrowly Obovateleaf Stauntonvine ●

374427 Stauntonia obovata Hemsl. var. angustata（C. Y. Wu）H. L. Li = Stauntonia obovata Hemsl. ●

374428 Stauntonia obovatifoliola Hayata；石月（椭圆叶石月，心基六叶野木瓜，心叶那藤，心叶石月）；Cordate Japanese Staunton-vine，Obovate-leaf Stauntonvine ●

374429 Stauntonia obovatifoliola Hayata subsp. intermedia（C. Y. Wu）T. C. Chen；五指那藤（木通七叶莲，那藤，牛藤，七叶莲，七叶木通，山木通）；Intermediate Sixleaves Stauntonvine ●

374430 Stauntonia obovatifoliola Hayata subsp. urophylla（Hand. -Mazz.）H. N. Qin；尾叶那藤（尾叶野木瓜）；Caudate-leaf Sixleaves Stauntonvine ●

374431 Stauntonia obovatifoliola Hayata var. pinninervis Hayata = Stauntonia obovatifoliola Hayata ●

374432 Stauntonia obscurinervia T. C. Chen；隐脉野木瓜；Obscure-nerved Staunton-vine ●

374433 Stauntonia oligophylla Merr. et Chun；少叶野木瓜；Few-leaves Stauntonvine，Oligophyllous Staunton-vine ●

374434 Stauntonia parviflora Hemsl. = Holboellia parviflora（Hemsl.）Gagnep. ●

374435 Stauntonia pseudomaculata C. Y. Wu et S. H. Huang；假斑点野木瓜（假斑叶野木瓜）；False Spotted Staunton-vine，False-spotted Stauntonvine ●

374436 Stauntonia purpurea Y. C. Liu et F. Y. Lu；紫花野木瓜；Purple Stauntonvine ●

374437 Stauntonia sinii Diels = Stauntonia decora（Dunn）C. Y. Wu ●

374438 Stauntonia trifoliata Griff. = Stauntonia brunoniana Wall. ●

374439 Stauntonia trinervia Merr.；三脉野木瓜（炮仗花藤）；Trinerves Stauntonvine，Trinervious Staunton-vine ●

374440 Stauntonia yaoshanensis F. N. Wei et S. L. Mo；瑶山野木瓜（瑶山七姐妹）；Yaoshan Stauntonvine，Yaoshan Staunton-vine ●

374441 Stauracanthus Link(1807);十字豆属(十字角荆豆属)●☆

374442 Stauracanthus boivinii (Webb) Samp.；布氏十字豆●☆

374443 Stauracanthus genistoides (Brot.) Samp. subsp. spectabilis (Webb) Rothm. = Stauracanthus spectabilis Webb ●☆

374444 Stauracanthus spectabilis Webb；非洲十字豆●☆

374445 Stauranthera Benth. (1835);十字苣苔属；Stauranthera ■

374446 Stauranthera chiritiflora Oliv. = Whytockia chiritiflora (Oliv.) W. W. Sm. ■

374447 Stauranthera grandiflora Benth.；大花十字苣苔；Bigflower Stauranthera ■

374448 Stauranthera tsiangiana Hand. -Mazz. = Whytockia tsiangiana (Hand. -Mazz.) A. Weber ■

374449 Stauranthera umbrosa (Griff.) C. B. Clarke；十字苣苔(山地蓝);Shady Stauranthera ■

374450 Stauranthus Liebm. (1854);十字花芸香属●☆

374451 Stauranthus perforatus Liebm.；十字花芸香●☆

374452 Stauregton Fourr. = Lemna L. ■

374453 Stauregton Fourr. = Staurogeton Rchb. ■

374454 Staurites Rchb. f. = Phalaenopsis Blume ■

374455 Stauritis Rchb. f. = Phalaenopsis Blume ■

374456 Staurochilus Ridl. = Staurochilus Ridl. ex Pfitzer ■

374457 Staurochilus Ridl. ex Pfitzer(1900);掌唇兰属(豹纹兰属,十字唇兰属,十字唇兰属);Staurochilus ■

374458 Staurochilus dawsonianus (Rchb. f.) Schltr.；掌唇兰；Staurochilus ■

374459 Staurochilus loratus (Rolfe ex Downie) Seidenf.；小掌唇兰；Small Staurochilus ■

374460 Staurochilus luchuensis (Rolfe) Fukuy.；豹纹掌唇兰(豹纹兰);Leopard Staurochilus ■

374461 Staurochilus lutchuensis (Rolfe) Fukuy. = Trichoglottis lutchuensis (Rolfe) Garay et H. R. Sweet ■☆

374462 Staurochilus lutchuensis (Rolfe) Fukuy. var. grossemaculata (Koidz.) Masam. = Trichoglottis lutchuensis (Rolfe) Garay et H. R. Sweet ■☆

374463 Staurochlamys Baker(1889);裂舌菊属■☆

374464 Staurochlamys burchellii Baker;裂舌菊■☆

374465 Staurogeton Rchb. = Lemna L. ■

374466 Stauroglottis Schauer = Phalaenopsis Blume ■

374467 Stauroglottis equestris Schauer = Phalaenopsis equestris (Schauer) Rchb. f. ■

374468 Staurogyne Wall. (1831);叉柱花属(哈哼花属);Forkstyleflower ■

374469 Staurogyne alboviolacea Benoist = Anisosepalum alboviolaceum (Benoist) E. Hossain ■☆

374470 Staurogyne alboviolacea Benoist subsp. grandiflora Napper = Anisosepalum alboviolaceum (Benoist) E. Hossain ■☆

374471 Staurogyne alboviolaceum Benoist var. gracilior Heine = Anisosepalum alboviolaceum (Benoist) E. Hossain subsp. gracilis (Heine) Champl. ■☆

374472 Staurogyne bicolor (Mildbr.) Champl.；双色叉柱花■☆

374473 Staurogyne brachystachya Benoist；短穗叉柱花；Shortstachys Forkstyleflower ■

374474 Staurogyne capitata E. A. Bruce；头状叉柱花■☆

374475 Staurogyne chapaensis Benoist;弯花叉柱花■

374476 Staurogyne concinnula (Hance) Kuntze;叉柱花(糙叶叉柱花,哈哼花);Forkstyleflower ■

374477 Staurogyne congoensis S. Moore = Staurogyne letestuana Benoist ■☆

374478 Staurogyne debilis (T. Anderson) C. B. Clarke;菲律宾叉柱花(菲律宾哈哼花)■

374479 Staurogyne dolichocalyx E. Hossain = Staurogyne sesamoides (Hand. -Mazz.) B. L. Burtt ■

374480 Staurogyne hainanensis C. Y. Wu et H. S. Lo;海南叉柱花；Hainan Forkstyleflower ■

374481 Staurogyne humbertii Mildbr. = Anisosepalum humbertii (Mildbr.) E. Hossain ■☆

374482 Staurogyne hypoleuca Benoist;灰背叉柱花;Greyback Forkstyleflower ■☆

374483 Staurogyne kamerunensis (Engl.) Benoist;喀麦隆叉柱花■☆

374484 Staurogyne kamerunensis (Engl.) Benoist subsp. calabarensis Champl.；卡拉巴尔叉柱花■☆

374485 Staurogyne lebrunii (Staner) B. L. Burtt = Saintpauliopsis lebrunii Staner ■☆

374486 Staurogyne letestuana Benoist;莱泰叉柱花■☆

374487 Staurogyne longicuneata H. S. Lo;楔叶叉柱花;Cuneateleaf Forkstyleflower ■

374488 Staurogyne merguensis Kuntze;马来叉柱花■☆

374489 Staurogyne paludosa (Mangenot et Aké Assi) Heine = Staurogyne capitata E. A. Bruce ■☆

374490 Staurogyne paotingensis C. Y. Wu et H. S. Lo;保亭叉柱花；Baoting Forkstyleflower ■

374491 Staurogyne pseudocapitata Champl.；假头状叉柱花■☆

374492 Staurogyne rivularis Merr.；瘦叉柱花;Thin Forkstyleflower ■

374493 Staurogyne sesamoides (Hand. -Mazz.) B. L. Burtt;大花叉柱花;Big-flowered Forkstyleflower ■

374494 Staurogyne sichuanica H. S. Lo;四川叉柱花(金长莲);Sichuan Forkstyleflower ■

374495 Staurogyne sinica C. Y. Wu et H. S. Lo;中华叉柱花；China Forkstyleflower,Chinese Forkstyleflower ■

374496 Staurogyne stenophylla Merr. et Chun;狭叶叉柱花;Narrowleaf Forkstyleflower ■

374497 Staurogyne strigosa C. Y. Wu et H. S. Lo;琼海叉柱花;Strigose Forkstyleflower ■

374498 Staurogyne yunnanensis H. S. Lo;云南叉柱花；Yunnan Forkstyleflower ■

374499 Staurogynopsis Mangenot et Aké Assi = Staurogyne Wall. ■

374500 Staurogynopsis Mangenot et Aké Assi(1959);类叉柱花属■☆

374501 Staurogynopsis capitata (E. A. Bruce) Mangenot et Aké Assi = Staurogyne capitata E. A. Bruce ■☆

374502 Staurogynopsis maiana Mangenot et Aké Assi = Staurogyne capitata E. A. Bruce ■☆

374503 Staurogynopsis paludosa Mangenot et Aké Assi;类叉柱花■☆

374504 Staurogynopsis paludosa Mangenot et Aké Assi = Staurogyne capitata E. A. Bruce ■☆

374505 Stauromatum Endl. = Sauromatum Schott ■

374506 Staurophragma Fisch. et C. A. Mey. = Verbascum L. ■●

374507 Stauropsis Rchb. f. (1860);船唇兰属■

374508 Stauropsis Rchb. f. = Trichoglottis Blume ■

374509 Stauropsis alpina (Lindl.) Ts. Tang et F. T. Wang = Vanda alpina Lindl. ■

374510 Stauropsis championii (Lindl. ex Benth.) Ts. Tang et F. T. Wang = Diploprora championii (Lindl. ex Benth.) Hook. f. ■

374511 Stauropsis championii (Lindl.) Ts. Tang et F. T. Wang = Diploprora championii (Lindl.) Hook. f. ■

374512 Stauropsis chinensis Rolfe = Vandopsis gigantea (Lindl.) Pfitzer ■

374513　Stauropsis gigantea（Lindl.）Benth. ex Pfitzer = Vandopsis gigantea（Lindl.）Pfitzer ■

374514　Stauropsis kusukusensis（Hayata）Ts. Tang et F. T. Wang = Diploprora championii（Lindl. ex Benth.）Hook. f. ■

374515　Stauropsis kusukusensis（Hayata）Ts. Tang et F. T. Wang = Diploprora championii（Lindl.）Hook. f. ■

374516　Stauropsis luchuensis Rolfe = Staurochilus luchuensis（Rolfe）Fukuy. ■

374517　Stauropsis parishii（Rchb. f.）Rolfe = Hygrochilus parishii（Veitch et Rchb. f.）Pfitzer ■

374518　Stauropsis polyantha W. W. Sm. = Vandopsis undulata（Lindl.）J. J. Sm. ■

374519　Stauropsis undulata（Lindl.）Benth. ex Hook. f. ;船唇兰■

374520　Stauropsis undulata（Lindl.）Benth. ex Hook. f. = Vandopsis undulata（Lindl.）J. J. Sm. ■

374521　Staurospermum Thonn. = Mitracarpus Zucc. ■

374522　Staurospermum verticillatum Schumach. et Thonn. = Mitracarpus hirtus（L.）DC. ■

374523　Staurospermum verticillatum Schumach. et Thonn. = Mitracarpus villosus（Sw.）DC. ■

374524　Staurostigma Scheidw. = Asterostigma Fisch. et C. A. Mey. ■☆

374525　Staurothylax Griff. = Cieca Adans. ●■☆

374526　Staurothylax Griff. = Phyllanthus L. ●■

374527　Staurothyrax Griff. = Staurothylax Griff. ●■

374528　Stavta Thunb. = Staavia Dahl ●☆

374529　Stawellia F. Muell.（1870）;凸花草属■☆

374530　Stawellia dimorphantha F. Muell. ;凸花草■☆

374531　Stawellia gymnocephala Diels;裸头凸花草■☆

374532　Stayneria L. Bolus（1960）;镰玉树属●☆

374533　Stayneria littlewoodii L. Bolus = Stayneria neilii（L. Bolus）L. Bolus ●☆

374534　Stayneria neilii（L. Bolus）L. Bolus;镰玉树●☆

374535　Stearodendron Engl. = Allanblackia Oliv. ex Benth. ●☆

374536　Stearodendron stuhlmannii Engl. = Allanblackia stuhlmannii（Engl.）Engl. ●☆

374537　Stebbinsia Lipsch.（1956）;肉菊属;Stebbinsia ■★

374538　Stebbinsia Lipsch. = Soroseris Stebbins ■

374539　Stebbinsia umbrella（Franch.）Lipsch. ;肉菊（红茶参,红条参,莲状绢毛菊,伞花绢毛菊,条参,雪条参）;Stebbinsia, Umbrellate Soroseris ■

374540　Stebbinsoseris K. L. Chambers = Microseris D. Don ■☆

374541　Stebbinsoseris K. L. Chambers（1991）;斯特宾斯菊属■☆

374542　Stebbinsoseris decipiens（K. L. Chambers）K. L. Chambers;斯特宾斯菊■☆

374543　Stebbinsoseris heterocarpa（Nutt.）K. L. Chambers;异果斯特宾斯菊;Derived Silverpuffs, Grassland Silverpuffs ■☆

374544　Stechmannia DC. = Jurinea Cass. ●■

374545　Stechys Boiss. = Stachys L. ●■

374546　Steegia Steud. = Stegia DC. ■●

374547　Steenhamera Kostel. = Steenhammera Rchb. ■

374548　Steenhammera Rchb. = Mertensia Roth（保留属名）■

374549　Steenhammera sibirica（L.）Turcz. = Mertensia sibirica（L.）G. Don ■

374550　Steenisia Bakh. f.（1952）;斯地茜属■☆

374551　Steenisia borneensis（Valeton）Bakh. f. ;斯地茜■☆

374552　Steentsia Kuprian. = Nothofagus Blume（保留属名）●☆

374553　Steerbeckia J. F. Gmel. = Singana Aubl. ☆

374554　Steerbeckia J. F. Gmel. = Sterbeckia Schreb. ☆

374555　Steetzia Sond. = Olearia Moench（保留属名）●☆

374556　Stefaninia Chiov. = Reseda L. ■

374557　Stefaninia telephiifolia Chiov. = Reseda telephiifolia（Chiov.）Abdallah et de Wit ■☆

374558　Stefanoffia H. Wolff（1925）;斯特草属■☆

374559　Stefanoffia aurea（Boiss.）Pimenov et Kljuykov;黄斯特草■☆

374560　Stefanoffia daucoides H. Wolff;斯特草■☆

374561　Steffensia Kunth = Piper L. ●■

374562　Steganotaenia Hochst.（1844）;五加前胡属■☆

374563　Steganotaenia araliacea Hochst. ;五加前胡■☆

374564　Steganotaenia commiphoroides Thulin;没药五加前胡■☆

374565　Steganotaenia hockii（C. Norman）C. Norman;霍克五加前胡■☆

374566　Steganotropis Lehm.（废弃属名）= Centrosema（DC.）Benth.（保留属名）●■☆

374567　Steganotus Cass. = Arctotis L. ●■☆

374568　Steganthera Perkins（1898）;闭药桂属●☆

374569　Steganthera alpina Perkins;高山闭药桂●☆

374570　Steganthera australiana C. T. White;澳洲闭药桂●☆

374571　Steganthera brassii（A. C. Sm.）Kaneh. et Hatus. ;巴西闭药桂●☆

374572　Steganthera laxiflora（Benth.）Whiffin et Foreman;疏花闭药桂●☆

374573　Steganthera ledermannii（Perkins）Kaneh. et Hatus. ;莱德曼闭药桂●☆

374574　Steganthera oblongiflora Perkins;矩圆闭药桂●☆

374575　Steganthera oligantha（Perkins）Kaneh. et Hatus. ;寡花闭药桂●☆

374576　Steganthera villosa Kaneh. et Hatus. = Faika villosa（Kaneh. et Hatus.）Philipson ●☆

374577　Steganthus Knobl. = Olea L. ●

374578　Steganthus welwitschii（Knobl.）Knobl. = Olea capensis L. subsp. welwitschii（Knobl.）Friis et P. S. Green ●☆

374579　Stegastrum Tiegh. = Lepeostegeres Blume ●☆

374580　Stegia DC. = Lavatera L. ■●

374581　Stegia trimestris（L.）Aké Assi et Devesa = Lavatera trimestris L. ■

374582　Stegitrio Post et Kuntze = Halimium（Dunal）Spach ●☆

374583　Stegitrio Post et Kuntze = Stegitris Raf. ●☆

374584　Stegitris Raf. = Halimium（Dunal）Spach ●☆

374585　Stegnocarpus（DC.）Torr. = Coldenia L. ■

374586　Stegnocarpus Torr. = Coldenia L. ■

374587　Stegnocarpus Torr. et A. Gray = Tiquilia Pers. ■☆

374588　Stegnosperma Benth.（1844）;白籽树属（白子树属,闭籽花属）●☆

374589　Stegnosperma alimifolia Benth. ;白籽树●☆

374590　Stegnospermataceae（A. Rich.）Nakai（1942）;白籽树科（闭籽花科）●☆

374591　Stegnospermataceae（H. Walter）Nakai = Stegnospermataceae（A. Rich.）Nakai ●☆

374592　Stegnospermataceae Nakai = Stegnospermataceae（A. Rich.）Nakai ●☆

374593　Stegocedrus Doweld = Libocedrus Endl. ●☆

374594　Stegocedrus Doweld（2001）;盖柏属●☆

374595　Stegolepis Klotzsch ex Körn.（1872）;鳞盖草属■☆

374596　Stegolepis albiflora Steyerm. ;白花鳞盖草■☆

374597　Stegolepis allenii Steyerm. ;阿伦鳞盖草■☆

374598　Stegolepis angustata Gleason;窄鳞盖草■☆

374599　Stegolepis ferruginea Baker;锈色鳞盖草■☆

374600　Stegolepis grandis Maguire;大鳞盖草■☆

374601　Stegolepis linearis Gleason;线形鳞盖草■☆

374602　Stegolepis membranacea Maguire;膜质鳞盖草■☆

374603　Stegolepis microcephala Maguire;小头鳞盖草■☆

374604　Stegolepis pauciflora Gleason;少花鳞盖草■☆

374605　Stegolepis pulchella Maguire;美丽鳞盖草■☆

374606　Stegonotus Cass. = Steganotus Cass. ●■☆

374607　Stegonotus Post et Kuntze = Arctotis L. ●■☆

374608　Stegonotus Post et Kuntze = Steganotus Cass. ●■☆

374609　Stegosia Lour. = Rottboellia L. f. (保留属名)■

374610　Stegosia cochinchinensis Lour. = Rottboellia cochinchinensis (Lour.) Clayton ■

374611　Stegostyla D. L. Jones et M. A. Clem. (2001);盖柱兰属■☆

374612　Stegostyla D. L. Jones et M. A. Clem. = Caladenia R. Br. ■☆

374613　Steigeria Müll. Arg. = Baloghia Endl. ●■☆

374614　Steinbachiella Harms = Diphysa Jacq. ■☆

374615　Steinchisma Raf. (1830);无柄黍属■☆

374616　Steinchisma Raf. = Panicum L. ■

374617　Steinchisma hians (Elliott) Nash = Panicum hians Elliott ■☆

374618　Steinchisma hians (Elliott) Nash et Small = Panicum hians Elliott ☆

374619　Steinchisma hians Raf. ;无柄黍■☆

374620　Steinchisma laxa (Sw.) Zuloaga;松散无柄黍■☆

374621　Steinchisma stenophylla Zuloaga et Morrone;窄无柄黍■☆

374622　Steineria Klotzsch = Begonia L. ●■

374623　Steinhauera Post et Kuntze = Sequoia Endl. + Sequoiadendron Buchholz ●

374624　Steinhauera gigantea (Lindl.) Kuntze ex Voss = Sequoiadendron giganteum (Lindl.) Buchholz ●

374625　Steinheilia Decne. = Odontanthera Wight ex Lindl. ■☆

374626　Steinheilia radians (Forssk.) Decne. = Odontanthera radians (Forssk.) D. V. Field ■☆

374627　Steinitzia Gand. = Anthemis L. ■

374628　Steinmannia F. Phil. = Garaventia Looser ■☆

374629　Steinmannia F. Phil. = Tristagma Poepp. ■☆

374630　Steinmannia Opiz = Rumex L. ■●

374631　Steinreitera Opiz = Thesium L. ■

374632　Steinschisma Steud. = Panicum L. ■

374633　Steinschisma Steud. = Steinchisma Raf. ■☆

374634　Steirachne Ekman(1911);南美毛枝草属■☆

374635　Steirachne diandra Ekman;南美毛枝草■☆

374636　Steiractinia S. F. Blake(1915);斑实菊属●☆

374637　Steiractinia glandulosa S. F. Blake;斑实菊●☆

374638　Steiractis DC. = Olearia Moench(保留属名)●☆

374639　Steiractis Raf. = Layia Hook. et Arn. ex DC. (保留属名)■☆

374640　Steiractis Raf. = Oxyura DC. ■☆

374641　Steiranisia Raf. = Heteresia Raf. ■

374642　Steiranisia Raf. = Micranthes Haw. ☆

374643　Steiranisia Raf. = Saxifraga L. ■

374644　Steirema Benth. et Hook. f. = Steiremis Raf. ■

374645　Steiremis Raf. = Telanthera R. Br. ■

374646　Steirexa Raf. = Trichopus Gaertn. ■☆

374647　Steirexa B. D. Jacks. = Steirexa Raf. ■☆

374648　Steireya Raf. (1838);施泰薯蓣属■☆

374649　Steireya Raf. = Trichopus Gaertn. ■☆

374650　Steirocoma (DC.) Rchb. = Dicoma Cass. ●☆

374651　Steirocoma Rchb. = Dicoma Cass. ●☆

374652　Steiroctis Raf. = Lachnaea L. + Cryptadenia Meisn. ●☆

374653　Steirodiscus Less. (1832);黄窄叶菊属■☆

374654　Steirodiscus Less. = Psilothonna E. Mey. ex DC. ■☆

374655　Steirodiscus capillaceus (L. f.) Less. ;细毛黄窄叶菊☆

374656　Steirodiscus linearilobus DC. ;线裂片黄窄叶菊■☆

374657　Steirodiscus schlechteri Bolus ex Schltr. ;施莱黄窄叶菊■☆

374658　Steirodiscus speciosus (Pillans) B. Nord. ;美丽黄窄叶菊■☆

374659　Steirodiscus tagetes (L.) Schltr. ;万寿黄窄叶菊■☆

374660　Steiroglossa DC. = Brachycome Cass. ●■☆

374661　Steironema Raf. (1821);肖珍珠菜属■☆

374662　Steironema Raf. = Lysimachia L. ●■

374663　Steironema ciliatum (L.) Baudo = Lysimachia ciliata L. ■☆

374664　Steironema heterophyllum (Michx.) Raf. = Lysimachia lanceolata Walter ■☆

374665　Steironema hybridum (Michx.) Raf. ex B. D. Jacks. ;杂种肖珍珠菜☆

374666　Steironema laevigatum Howell;光滑肖珍珠菜■☆

374667　Steironema lanceolatum (Walter) A. Gray;披针形肖珍珠菜; Lance-leaved Steironema ■☆

374668　Steironema lanceolatum (Walter) A. Gray = Lysimachia lanceolata Walter ■☆

374669　Steironema pumilum Greene = Lysimachia ciliata L. ■☆

374670　Steironema quadriflorum (Sims) Hitchc. = Lysimachia quadriflora Sims ☆

374671　Steirosanchezia Lindau = Sanchezia Ruiz et Pav. ●■

374672　Steirostemon (Griseb.) Phil. = Samolus L. ■

374673　Steirostemon Phil. = Samolus L. ■

374674　Steirotis Raf. = Struthanthus Mart. (保留属名)●☆

374675　Stekhovia de Vriese = Goodenia Sm. ●■☆

374676　Stelanthes Stokes = Alangium Lam. (保留属名)●

374677　Stelanthes Stokes = Marlea Roxb. ●

374678　Stelanthes Stokes = Stylidium Lour. (废弃属名)●

374679　Stelbophyllum D. L. Jones et M. A. Clem. = Stilbophyllum D. L. Jones et M. A. Clem. ■☆

374680　Stelechanteria Thouars ex Baill. = Drypetes Vahl ●

374681　Stelechanteria thouarsiana Baill. = Drypetes thouarsiana (Baill.) Capuron ●☆

374682　Stelechantha Bremek. (1940);根茎花茜属■☆

374683　Stelechantha arcuata S. E. Dawson;拱根茎花茜■☆

374684　Stelechantha cauliflora (R. D. Good) Bremek. ;茎花根茎花茜 ■☆

374685　Stelechantha ziamaeana (Jacq. -Fél.) N. Hallé;济阿马根茎花茜■☆

374686　Stelechocarpus (Blume) Hook. f. et Thomson(1855);茎花玉盘属●☆

374687　Stelechocarpus Hook. f. et Thomson = Stelechocarpus (Blume) Hook. f. et Thomson ●☆

374688　Stelechocarpus burahol (Blume) Hook. f. et Thomson;茎花玉盘;Keppel ●☆

374689　Stelechospermum Blume = Mischocarpus Blume(保留属名)●

374690　Steleocodon Gilli = Phalacraea DC. ■☆

374691　Steleostemma Schltr. (1906);把子花属■☆

374692　Steleostemma pulchellum Schltr. = Amblystigma pulchellum (Schltr.) T. Mey. ■☆

374693　Stelephuros Adans. = Phleum L. ■

374694　Stelestylis Drude(1881);把柱草属■☆

374695　Stelestylis anomala Harling;把柱草■☆

374696　Stelin Bubani = Viscum L. ●

374697　Steliopsis Brieger = Stelis Sw. (保留属名)■☆

374698　Stelis Loefl. = Oryctanthus (Griseb.) Eichler + Struthanthus Mart. (保留属名)●☆

374699　Stelis Sw. (1800)(保留属名);微花兰属;Stelis ■☆

374700　Stelis barbata Rolfe;毛萼微花兰■☆

374701　Stelis bidentata Schltr. ;二齿微花兰;Twoteeth Stelis ■☆

374702　Stelis bruchmuelleri Rchb. f. ex Hook. f. ;布氏微花兰■☆

374703　Stelis canaliculata Rchb. f. ;沟叶微花兰■☆

374704　Stelis ciliaris Lindl. ;缘毛微花兰;Ciliate Stelis ■☆

374705　Stelis endresii Rchb. f. ;尹氏微花兰;Endres Stelis ■☆

374706　Stelis gracilis Ames;纤细微花兰;Slender Stelis ■☆

374707　Stelis grandiflora Lindl. ;大花微花兰■☆

374708　Stelis hirta Sm. = Bulbophyllum hirtum (Sm.) Lindl. ■

374709　Stelis hymenantha Schltr. ;膜花微花兰;Membrane-flower Stelis ■☆

374710　Stelis micrantha (Sw.) Sw. ;小花微花兰;Littleflower Stelis ■☆

374711　Stelis mucronata D. Don = Oberonia mucronata (D. Don) Ormerod et Seidenf. ■

374712　Stelis odoratissima Sm. = Bulbophyllum odoratissimum (Sm.) Lindl. ■

374713　Stelis puberula Barb. Rodr. ;微毛微花兰;Giant Chickweed ■☆

374714　Stelis purpurascens A. Rich. ;紫红微花兰;Purple-red Stelis ■☆

374715　Stelis racemosa Sm. = Sunipia scariosa Lindl. ■

374716　Stelis rubens Schltr. ;红微花兰;Red Stelis ■☆

374717　Stelistylis Drude;巴西环花属■☆

374718　Stelitaceae Dulac = Primulaceae Batsch ex Borkh. (保留科名)●■

374719　Stella Medik. = Astragalus L. ●■

374720　Stellandria Brickell(废弃属名) = Schisandra Michx. (保留属名)●

374721　Stellandria glabra Brickell = Schisandra glabra (Brickell) Rehder ●☆

374722　Stellara Fisch. ex Reut. = Boschniakia C. A. Mey. ex Bong. ■

374723　Stellaria Hill = Corispermum L. ■

374724　Stellaria L. (1753);繁缕属;Chickweed,Starwort,Stitch Grass, Stitchwort ■

374725　Stellaria Ség. = Callitriche L. ■

374726　Stellaria Zinn = Callitriche L. ■

374727　Stellaria alaschanica Y. Z. Zhao;贺兰山繁缕;Helanshan Chickweed ■

374728　Stellaria alaskana Hultén;阿拉斯加繁缕;Alaska Starwort ■☆

374729　Stellaria alatavica Popov;阿拉套繁缕;Alatai Chickweed ■

374730　Stellaria alexeenkoana Schischk. ;阿氏繁缕■☆

374731　Stellaria alsine Grimm;雀舌草(滨繁缕,瓜子草,寒草,金线吊葫芦,泥潭繁缕,雀舌繁缕,天蓬草,葶苈子,吴檀,雪里花,雪里开花, 指甲草); Alsine Starwort, Birdtongue Chickweed, Bog Chickweed,Bog Starwort,Bog Stitchwort,Slender Starwort ■

374732　Stellaria alsine Grimm = Stellaria uliginosa Murray ■

374733　Stellaria alsine Grimm var. alpina (Schur) Hand. -Mazz. = Stellaria uliginosa Murray var. alpina (Schur) Hand. -Mazz. ■

374734　Stellaria alsine Grimm var. apetala (Rchb.) Hausm. = Stellaria

374735　Stellaria alsine Grimm var. atlantica Jahand. et Maire = Stellaria alsine Grimm ■

374736　Stellaria alsine Grimm var. phaenopetala Hand. -Mazz. = Stellaria alsine Grimm ■

374737　Stellaria alsine Grimm var. undulata (Thunb.) Ohwi = Stellaria alsine Grimm ■

374738　Stellaria alsine Grimm var. undulata (Thunb.) Ohwi = Stellaria uliginosa Murray var. undulata (Thunb.) Fenzl ■

374739　Stellaria alsinoides Boiss. et Buhse;假雀舌草■☆

374740　Stellaria amblyosepala Schrenk; 钝萼繁缕; Obtusecalyx Chickweed ■

374741　Stellaria americana (Porter ex B. L. Rob.) Standl. ;美国繁缕 (美洲繁缕);American Starwort ■☆

374742　Stellaria anagalloides C. A. Mey. ex Rupr. ;拟琉璃繁缕■☆

374743　Stellaria anhweiensis Migo;安徽繁缕;Anhui Chickweed ■

374744　Stellaria anhweiensis Migo = Moehringia trinervia (L.) Clairv. ■

374745　Stellaria anhweiensis Migo = Stellaria pallida (Dumort.) Pire ■

374746　Stellaria apetala Ucria = Stellaria pallida (Dumort.) Piré ■

374747　Stellaria apetala Ucria ex Roem. = Stellaria media (L.) Vill. ■

374748　Stellaria apetala Ucria ex Roem. = Stellaria pallida (Dumort.) Pire ■

374749　Stellaria aquatica (L.) Scop. ;水繁缕;Giant Chickweed,Water Chickweed ■☆

374750　Stellaria aquatica (L.) Scop. = Myosoton aquaticum (L.) Moench ■

374751　Stellaria arctica Schischk. ;北极繁缕■☆

374752　Stellaria arctica Schischk. = Stellaria longipes Goldie ■☆

374753　Stellaria arenaria Maxim. = Stellaria arenarioides Shi L. Chen et al. ■

374754　Stellaria arenaria Salzm. = Arenaria hispanica Spreng. ■☆

374755　Stellaria arenarioides Shi L. Chen et al. ;沙生繁缕;Sandy Chickweed ■

374756　Stellaria arenicola Raup = Stellaria longipes Goldie subsp. arenicola (Raup) C. C. Chinnappa et J. K. Morton ■☆

374757　Stellaria arisanensis (Hayata) Hayata;阿里山繁缕;Alishan Chickweed ■

374758　Stellaria arisanensis (Hayata) Hayata var. leptophylla Hayata; 大叶阿里山繁缕;Bigleaf Alishan Chickweed ■

374759　Stellaria arisanensis (Hayata) Hayata var. leptophylla Hayata = Stellaria arisanensis (Hayata) Hayata ■

374760　Stellaria atrata (J. W. Moore) B. Boivin = Stellaria longifolia Muhl. ex Willd. ■

374761　Stellaria atrata (J. W. Moore) B. Boivin = Stellaria longifolia Muhl. ex Willd. var. atrata J. W. Moore ■☆

374762　Stellaria atrata (J. W. Moore) B. Boivin var. eciliata B. Boivin = Stellaria longifolia Muhl. ex Willd. ■

374763　Stellaria biflora L. = Minuartia biflora (L.) Schinz et Thell. ■

374764　Stellaria bistyla Y. Z. Zhao;二柱繁缕;Bistyle Chickweed ■

374765　Stellaria bistylata W. Z. Di et Y. Ren = Stellaria bistyla Y. Z. Zhao ■

374766　Stellaria boraeana Jord. = Stellaria pallida (Dumort.) Pire ■

374767　Stellaria borealis Bigelow;北方繁缕;Boreal Starwort,Northern Starwort,Northern Stitchwort ■☆

374768　Stellaria borealis Bigelow subsp. bongardiana (Fernald) Piper et Beattie = Stellaria borealis Bigelow subsp. sitchana (Steud.) Piper et Beattie ■☆

374769　Stellaria borealis Bigelow subsp. sitchana（Steud.）Piper et Beattie；锡特卡繁缕；Sitka Starwort ■☆

374770　Stellaria borealis Bigelow var. bongardiana Fernald = Stellaria borealis Bigelow subsp. sitchana（Steud.）Piper et Beattie ■☆

374771　Stellaria borealis Bigelow var. crispa（Cham. et Schltdl.）Fenzl ex Torr. et A. Gray = Stellaria crispa Cham. et Schltdl. ■☆

374772　Stellaria borealis Bigelow var. floribunda Fernald = Stellaria borealis Bigelow ■☆

374773　Stellaria borealis Bigelow var. isophylla Fernald = Stellaria borealis Bigelow ■☆

374774　Stellaria borealis Bigelow var. simcoei（Howell）Fernald = Stellaria calycantha（Ledeb.）Bong. ■☆

374775　Stellaria borealis Bigelow var. sitchana（Steud.）Fernald = Stellaria borealis Bigelow subsp. sitchana（Steud.）Piper et Beattie ■☆

374776　Stellaria brachypetala Bong. = Stellaria borealis Bigelow subsp. sitchana（Steud.）Piper et Beattie ■☆

374777　Stellaria brachypetala Bunge；短瓣繁缕；Shortpetal Chickweed ■☆

374778　Stellaria brauniana Engl. ex Robyns = Stellaria sennii Chiov. ■☆

374779　Stellaria bubbosa Wulfen = Pseudostellaria himalaica（Franch.）Pax ■

374780　Stellaria bungeana Fenzl；长瓣繁缕（林繁缕）；Longpetal Chickweed ■

374781　Stellaria bungeana Fenzl var. stubendorfii（Regel）Y. C. Chu；林繁缕；Woodland Chickweed ■

374782　Stellaria bungeana Fenzl var. stubendorfii（Regel）Y. C. Chu = Stellaria monosperma Buch. -Ham. ex D. Don var. japonica Maxim. ■

374783　Stellaria calycantha（Ledeb.）Bong.；杯花繁缕；Northern Starwort ■☆

374784　Stellaria calycantha（Ledeb.）Bong. subsp. interior Hultén = Stellaria borealis Bigelow ■☆

374785　Stellaria calycantha（Ledeb.）Bong. var. bongardiana（Fernald）Fernald = Stellaria borealis Bigelow subsp. sitchana（Steud.）Piper et Beattie ■☆

374786　Stellaria calycantha（Ledeb.）Bong. var. floribunda（Fernald）Fernald = Stellaria borealis Bigelow ■☆

374787　Stellaria calycantha（Ledeb.）Bong. var. isophylla（Fernald）Fernald = Stellaria borealis Bigelow ■☆

374788　Stellaria calycantha（Ledeb.）Bong. var. latifolia B. Boivin = Stellaria borealis Bigelow ■☆

374789　Stellaria calycantha（Ledeb.）Bong. var. laurentiana Fernald = Stellaria borealis Bigelow ■☆

374790　Stellaria calycantha（Ledeb.）Bong. var. simcoei（Howell）Fernald = Stellaria calycantha（Ledeb.）Bong. ■☆

374791　Stellaria calycantha（Ledeb.）Bong. var. sitchana（Steud.）Fernald = Stellaria borealis Bigelow subsp. sitchana（Steud.）Piper et Beattie ■☆

374792　Stellaria capillipes（Franch.）C. Y. Wu = Stellaria petiolaris Hand. -Mazz. ■

374793　Stellaria cerastoides L. = Cerastium cerastoides（L.）Britton ■

374794　Stellaria cherleriae（Fisch. ex Ser.）F. N. Williams；兴安繁缕（东北繁缕，绿花繁缕）；Xing'an Chickweed ■

374795　Stellaria cherleriae（Fisch. ex Ser.）F. N. Williams = Stellaria petraea Bunge ■

374796　Stellaria cherleriae（Fisch. ex Ser.）F. N. Williams var. alpina（Bunge）Schischk. = Stellaria petraea Bunge ■

374797　Stellaria cherleriae（Fisch. ex Ser.）F. N. Williams var. alpina

374798　Schischk. = Stellaria petraea Bunge ■

374798　Stellaria cherleriae（Fisch. ex Ser.）F. N. Williams var. minor（Edgew. et Hook. f.）Majumdar = Stellaria decumbens Edgew. ■

374799　Stellaria cherleriae（Fisch. ex Ser.）F. N. Williams var. polyantha（Edgew. et Hook. f.）R. R. Stewart = Stellaria decumbens Edgew. var. polyantha Edgew. et Hook. f. ■

374800　Stellaria cherleriae（Fisch. ex Ser.）F. N. Williams var. typica Williams = Stellaria petraea Bunge ■

374801　Stellaria cherleriae（Fisch. ex Ser.）F. N. Williams var. uniflora Willams = Stellaria decumbens Edgew. ■

374802　Stellaria chinensis Regel；中国繁缕（华繁缕，蛇舌草，鸭雀子窝）；China Chickweed，Chinese Star Wort ■

374803　Stellaria ciliatosepala Trautv.；毛瓣繁缕；■☆

374804　Stellaria ciliatosepala Trautv. = Stellaria longipes Goldie ■☆

374805　Stellaria congestiflora H. Hara；密花繁缕；Denseflower Chickweed ■

374806　Stellaria corei Shinners；田纳西繁缕；Tennessee Chickweed ■☆

374807　Stellaria crassifolia Ehrh.；叶苞繁缕（厚叶繁缕）；Fleshy Stitchwort，Thickleaf Chickweed，Thick-leaved Starwort ■

374808　Stellaria crassifolia Ehrh. var. eriocalycina Schischk. = Stellaria crassifolia Ehrh. ■

374809　Stellaria crassifolia Ehrh. var. linearis Fenzl；线形叶苞繁缕（叶苞繁缕）；Linear Thickleaf Chickweed ■

374810　Stellaria crassifolia Ehrh. var. linearis Fenzl = Stellaria crassifolia Ehrh. ■

374811　Stellaria crassipes Hultén = Stellaria longipes Goldie ■☆

374812　Stellaria crispa Cham. et Schltdl.；脆繁缕；Crisp Starwort ■☆

374813　Stellaria crispata Wall. = Stellaria monosperma Buch. -Ham. ex D. Don var. japonica Maxim. ■

374814　Stellaria crispata Wall. ex D. Don = Brachystemma calycinum D. Don ■

374815　Stellaria crispata Wall. ex D. Don = Stellaria monosperma Buch. -Ham. ex D. Don ■

374816　Stellaria cuonaensis L. = Stellaria decumbens Edgew. var. arenarioides L. H. Zhou ■

374817　Stellaria cuonaensis L. H. Zhou = Stellaria decumbens Edgew. var. arenarioides L. H. Zhou ■

374818　Stellaria cupaniana（Jord. et Fourr.）Bég.；库潘繁缕■☆

374819　Stellaria cuspidata Willd. ex Schltdl.；墨西哥繁缕；Mexican Chickweed ■☆

374820　Stellaria cuspidata Willd. ex Schltdl. subsp. prostrata（Baldwin）J. K. Morton；平卧绥草■☆

374821　Stellaria dahurica Willd. ex Schltdl.；达乌尔繁缕■☆

374822　Stellaria davidii（Franch.）Hemsl. = Pseudostellaria davidii（Franch.）Pax ■

374823　Stellaria davidii（Franch.）Hemsl. var. himalaica Franch. = Pseudostellaria himalaica（Franch.）Pax ■

374824　Stellaria davidii（Franch.）Hemsl. var. sessilifolia Franch. = Pseudostellaria heterantha（Maxim.）Pax ■

374825　Stellaria davidii Hemsl. var. himalaica Franch. = Pseudostellaria himalaica（Franch.）Pax ■

374826　Stellaria decumbens Edgew.；偃卧繁缕；Supine Chickweed ■

374827　Stellaria decumbens Edgew. var. arenarioides L. H. Zhou；错那繁缕（雪灵芝状繁缕）；Cuona Chickweed ■

374828　Stellaria decumbens Edgew. var. edgeworthii Edgew. et Hook. f. = Stellaria decumbens Edgew. ■

374829　Stellaria decumbens Edgew. var. minor Edgew. et Hook. f. =

Stellaria decumbens Edgew. ■

374830 Stellaria decumbens Edgew. var. polyantha Edgew. et Hook. f. ；多花偃卧繁缕；Manyflower Supine Chickweed ■

374831 Stellaria decumbens Edgew. var. pulvinata Edgew.；垫状偃卧繁缕（垫状繁缕）；Pulvinate Supine Chickweed ■

374832 Stellaria delavayi Franch.；大叶繁缕（川滇繁缕，大花繁缕，青葙子，西南繁缕）；Bigleaf Chickweed ■

374833 Stellaria depressa Em. Schmid；凹陷繁缕；Depressed Chickweed ■

374834 Stellaria dianthifolia F. N. Williams；石竹叶繁缕；Pinkleaf Chickweed ■

374835 Stellaria dichasioides F. N. Williams = Stellaria infracta Maxim. ■

374836 Stellaria dichotoma L.；叉歧繁缕（叉繁缕，叉枝繁缕，歧枝繁缕，双歧繁缕）；Dichotomal Chickweed, Dichotomous Star Wort ■

374837 Stellaria dichotoma L. f. lanceolata（Bunge）Kitag. = Stellaria dichotoma L. var. lanceolata Bunge ■

374838 Stellaria dichotoma L. var. americana Porter ex B. L. Rob. = Stellaria americana（Porter ex B. L. Rob.）Standl. ■☆

374839 Stellaria dichotoma L. var. cordifolia Bunge = Stellaria dichotoma L. ■

374840 Stellaria dichotoma L. var. heterophylla Fenzl. = Stellaria dichotoma L. var. lanceolata Bunge ■

374841 Stellaria dichotoma L. var. lanceolata Bunge；银柴胡（白根子，鳖血银柴胡，长枝繁缕，马蹄踏菜根，牛肚根，披针叶叉繁缕，披针叶繁缕，歧繁缕，沙参儿，沙地繁缕，沙根子，山菜，山马踏菜，土参，西银柴胡，狭叶歧繁缕，霞草状繁缕，银胡，云菁）；Lanceolate Dichotomous Star Wort ■

374842 Stellaria dichotoma L. var. linearis Fenzl；线叶繁缕（条叶叉歧繁缕，线叶叉繁缕）；Linearleaf Dichotomous Star Wort ■

374843 Stellaria dichotoma L. var. linearis Fenzl = Stellaria amblyosepala Schrenk ■

374844 Stellaria dichotoma L. var. rigida Bunge = Stellaria amblyosepala Schrenk ■

374845 Stellaria dichotoma L. var. stephaniana（Willd. ex Schltdl.）Regel = Stellaria dichotoma L. var. lanceolata Bunge ■

374846 Stellaria dicranoides（Cham. et Schltdl.）Fenzl；绒毛繁缕；Chamisso's Starwort ■☆

374847 Stellaria dicranoides Fenzl = Stellaria dicranoides（Cham. et Schltdl.）Fenzl ■☆

374848 Stellaria diffusa Willd. ex Schltdl.；铺散繁缕 ■

374849 Stellaria diffusa Willd. ex Schltdl. = Stellaria longifolia Muhl. ex Willd. ■

374850 Stellaria diffusa Willd. ex Schltdl. f. ciliolata Kitag. = Stellaria longifolia Muhl. ex Willd. ■

374851 Stellaria diffusa Willd. ex Schltdl. f. ciliolata Kitag. = Stellaria longifolia Muhl. ex Willd. f. ciliolata（Kitag.）Y. C. Chu ■

374852 Stellaria diffusa Willd. ex Schltdl. var. ciliolata（Kitag.）Kitag. = Stellaria longifolia Muhl. ex Willd. ■

374853 Stellaria diffusa Willd. ex Schltdl. Willd. ex Schltdl. = Stellaria longifolia Muhl. ex Willd. ■

374854 Stellaria diffusa Willd. ex Schltr. = Stellaria longifolia Muhl. ex Willd. ■

374855 Stellaria diffusa Willd. ex Schltr. f. ciliolata Kitag. = Stellaria longifolia Muhl. ex Willd. ■

374856 Stellaria diffusa Willd. ex Schltr. var. ciliolata（Kitag.）Kitag. = Stellaria longifolia Muhl. ex Willd. ■

374857 Stellaria dilleniana Moench = Stellaria palustris Ehrh. ex Hoffm. ■

374858 Stellaria discolor Turcz.；翻白繁缕（异色繁缕）；Bicolor Chickweed, Different Colours Star Wort ■

374859 Stellaria diversiflora Maxim.；异型花繁缕（泽繁缕）；Diversiflorous Star Wort ■

374860 Stellaria diversiflora Maxim. f. angustifolia M. Mizush.；狭叶异型花繁缕 ■☆

374861 Stellaria diversiflora Maxim. f. robusta M. Mizush.；钝叶异型花繁缕 ■☆

374862 Stellaria diversiflora Maxim. f. yakumontana（Masam.）M. Mizush. = Stellaria diversiflora Maxim. var. yakumontana（Masam.）Masam. ■☆

374863 Stellaria diversiflora Maxim. var. gymnandra Franch. = Stellaria neglecta Weihe ex Bluff et Fingerh. ■

374864 Stellaria diversiflora Maxim. var. gymnandra Franch. = Stellaria nemorum L. ■

374865 Stellaria diversiflora Maxim. var. leptophylla（Hayata）Mizush. = Stellaria arisanensis（Hayata）Hayata var. leptophylla Hayata ■

374866 Stellaria diversiflora Maxim. var. leptophylla（Hayata）Mizush. = Stellaria arisanensis（Hayata）Hayata ■

374867 Stellaria diversiflora Maxim. var. robusta（M. Mizush.）Okuyama = Stellaria diversiflora Maxim. f. robusta M. Mizush. ■☆

374868 Stellaria diversiflora Maxim. var. yakumontana（Masam.）Masam.；屋久岛繁缕 ■☆

374869 Stellaria diversifolia Maxim. = Stellaria nemorum L. ■

374870 Stellaria dubia Bastard = Cerastium dubium（Bastard）Guépin ■☆

374871 Stellaria dulcis Gervais = Stellaria longipes Goldie ■☆

374872 Stellaria ebracteata Kom.；无苞繁缕；Bractless Chickweed ■

374873 Stellaria edwardsii R. Br.；爱德华兹繁缕 ■☆

374874 Stellaria edwardsii R. Br. = Stellaria longipes Goldle ■☆

374875 Stellaria edwardsii R. Br. var. arctica（Schischk.）Hultén = Stellaria longipes Goldie ■☆

374876 Stellaria edwardsii R. Br. var. crassipes（Hultén）B. Boivin = Stellaria longipes Goldie ■☆

374877 Stellaria erlangeriana Engl. ex Pax et K. Hoffm.；厄兰格繁缕 ■☆

374878 Stellaria eschschohziana Fenzl；埃绍氏繁缕 ■☆

374879 Stellaria fenzliana Klotzsch = Stellaria patens D. Don ■

374880 Stellaria fenzlii Regel；范氏繁缕 ■☆

374881 Stellaria fenzlii Regel f. glaberrima（H. Hara）M. Mizush.；无毛范氏繁缕 ■☆

374882 Stellaria filicaulis Makino；线茎繁缕（细叶繁缕，鸭绿繁缕）；Thinleaf Chickweed ■

374883 Stellaria filicaulis Makino f. jaluana（Nakai）Kitag. = Stellaria filicaulis Makino ■

374884 Stellaria fischeriana Ser.；菲舍尔繁缕 ■☆

374885 Stellaria florida Fisch. = Stellaria nipponica Ohwi ■

374886 Stellaria florida Fisch. ex DC. var. angustifolia Maxim. = Stellaria nipponica Ohwi ■

374887 Stellaria florida Fisch. var. angustifolia Maxim. = Stellaria nipponica Ohwi ■

374888 Stellaria fontana Popov；泉繁缕 ■☆

374889 Stellaria fontinalis（Short et R. Peter）B. L. Rob.；肯塔基繁缕；Kentucky Starwort ■☆

374890 Stellaria friesiana Ser. = Stellaria longifolia Muhl. ex Willd. ■

374891 Stellaria glandulifera Klotzsch = Stellaria monosperma Buch.-Ham. ex D. Don ■

374892 Stellaria glauca With. = Stellaria graminea L. ■

374893　Stellaria glauca With. = Stellaria gyangtseensis F. N. Williams ■

374894　Stellaria glauca With. = Stellaria palustris Ehrh. ex Hoffm. ■

374895　Stellaria gonomischa B. Boivin = Stellaria umbellata Turcz. ■

374896　Stellaria gracilis Richardson = Stellaria crassifolia Ehrh. ■

374897　Stellaria graminea L.；禾叶繁缕（草状繁缕）；Babes-in-the-wood，Break Jack，Break-Jack，Common Starwort，Common Stitchwort，Granny's Nightcap，Grassleaf Chickweed，Grass-leaved Starwort，Grass-leaved Stitchwort，Grasslike Starwort，Grass-like Starwort，Heath Stitchwort，Lady's Milking Stools，Lady's Milking-stools，Lesser Stitchwort，Little Starwort，Snake's Flower，Snappers，Starwort，White Sunday ■

374898　Stellaria graminea L. var. brachypetala（Bunge）Regel = Stellaria brachypetala Bunge ■

374899　Stellaria graminea L. var. chinensis Maxim.；中华禾叶繁缕；China Grassleaf Chickweed，Chinese Chickweed ■

374900　Stellaria graminea L. var. lanceolata Fenzl = Stellaria graminea L. var. chinensis Maxim. ■

374901　Stellaria graminea L. var. pilosula Maxim.；毛禾叶繁缕；Pilose Chinese Chickweed ■

374902　Stellaria graminea L. var. pilosula Maxim. = Stellaria patens D. Don ■

374903　Stellaria graminea L. var. viridescens Maxim.；常绿禾叶繁缕；Evergreen Chinese Chickweed ■

374904　Stellaria gramineoides Hazit = Stellaria graminea L. ■

374905　Stellaria groenlandica Retz. = Minuartia groenlandica（Retz.）Ostenf. ■☆

374906　Stellaria gyangtseensis F. N. Williams；江孜繁缕；Jiangzi Chickweed ■

374907　Stellaria gyirongensis L. H. Zhou；吉隆繁缕；Jilong Chickweed ■

374908　Stellaria gypsophiloides Fenzl = Stellaria dichotoma L. var. lanceolata Bunge ■

374909　Stellaria gypsophiloides Fenzl var. lanceolata（Bunge）Kozhevn. = Stellaria dichotoma L. var. lanceolata Bunge ■

374910　Stellaria hamiltoniana Majumdar = Stellaria vestita Kurz ■

374911　Stellaria hamiltoniana Majumdar var. vestita（Kurz）Majumdar = Stellaria vestita Kurz ■

374912　Stellaria hassiana Loes. = Stellaria chinensis Regel ■

374913　Stellaria hebecalyx Fenzl；毛萼繁缕（钝萼繁缕）；Obtusesepal Chickweed ■☆

374914　Stellaria henryi F. N. Williams；湖北繁缕（续筋草）；Hubei Chickweed ■

374915　Stellaria heterophylla（Miq.）Hemsl. = Pseudostellaria heterophylla（Miq.）Pax ■

374916　Stellaria holostea L.；披针叶繁缕（复活节钟草，复活节钟花）；Adder's Meat，Adder's Spit，Addersmeat，Adder's-meat，Agworm Flower，Agworm-flower，All-bone，Baalem's Smite，Bachelor's Buttons，Biddy's Eyes，Billy Bachelor's Buttons，Billy Buttons，Billy White's Buttons，Bird's Eye，Bird's Tongue，Brandy Snap，Brandy Snaps，Break Jack，Breakbones，Break-Jack，Crow's Foot，Cuckoo's Meat，Cuckoo's Victuals，Darnell Goddard，Dead Man's Bones，Devil's Corn，Devil's Ear，Devil's Eye，Devil's Eyes，Devil's Flower，Devil's Nightcap，Devil's Shirt Buttons，Easter Bell，Easter Bells，Easter Flower，Easter-bell，Easter-bell Starwort，Eyebright，Granny's Nightcap，Greater Stitchwort，Hagworm Flower，Headache，Hedge Stitchwort，Lady's Lint，Jack Snaps，Jack Sprat，Jack-in-the-hedge，Jack-in-the-lantern，Jack-snaps，Lady's Buttons，Lady's Chemise，Lady's Embroidery，Lady's Lint，Lady's Needlework，Lady's Smock，Lady's Thimble，Lady's Thimbles，Lady's White Petticoats，Little John，Looking-glass，Mary-at-the-cottage-gate，May-flower，May-grass，Milk Maidens，Milkcans，Milkpans，Miller's Star，Moon Flower，Moonwort，Morning Star，Mother Shimble's Snick-needles，Mother's Thimble，Mother's Thimbles，Nancy，Nancy Pretty，Nancy-pretty，Nightingale，Nits，Old Lad's Corn，Old Man's Shirt，One-o'clock，Paigle，Pickpocket，Pins-and-needles，Pisgie-flower，Piskies，Pixies，Pixy Lily，Pop Jack，Pop-gun，Pop-Jack，Popper，Poppy，Pops，Pretty Nancy，Sailor Buttons，Sailor's Buttons，Satin Flower，Scurvy-grass，Sgreat Tarwort，Shepherd's Weatherglass，Shimmies，Shimmies-and-shirts，Skirt Buttons，Smock，Smock Frock，Snake-grass，Snake's Flower，Snakeweed，Snap Jack，Snapcracker，Snapjack，Snapper-flower，Snaps，Snapstalks，Snapwort，Snow，Snowflake，Snow-on-the-mountain，Soldier Buttons，Soldier's Buttons，St. Leger-gordon Pixylily，Star of Bethlehem，Star of the Wood，Star Wort，Star-flower，Star-grass，Starwort，Stepmother，Stitchwort，Sugar Basin，Sugar Basins，Sunday Whites，Sweet hearts，Sweet Nance，Sweet Nancy，Thunder Flower，Thunderbolts，Twinkle Star，Twinkle-star，Unglepigle，Watches，Wedding-flower，White Bells，White Bird's Eye，White Bird's Eyes，White Bobby's Eye，White Bobby's Eyes，White Lady's Petticoats，White Robin's Eye，White Robin's Eyes，White Sunday，Whiteflower，White-flowered Grass ■

374917　Stellaria hsinganensis Kitag. = Stellaria palustris Ehrh. ex Hoffm. ■

374918　Stellaria hultenii B. Boivin = Stellaria longipes Goldie ■☆

374919　Stellaria humifusa Rottb.；扩散繁缕；Low Starwort，Salt-marsh Starwort ■☆

374920　Stellaria humifusa Rottb. var. marginata Fenzl = Stellaria humifusa Rottb. ■☆

374921　Stellaria humifusa Rottb. var. oblongifolia Fenzl = Stellaria humifusa Rottb. ■☆

374922　Stellaria humifusa Rottb. var. suberecta B. Boivin = Stellaria humifusa Rottb. ■☆

374923　Stellaria imbricata Bunge；覆瓦繁缕；Imbricate Chickweed ■

374924　Stellaria infracta Maxim.；内弯繁缕（内曲繁缕）；Incurved Star Wort，Infracted Chickweed ■

374925　Stellaria infracta Maxim. var. ovatolanceolata Mattf. = Stellaria infracta Maxim. ■

374926　Stellaria irrigua Bunge；冻原繁缕（贮水繁缕）；Altai Chickweed，Altai Starwort，Watery Chickweed ■

374927　Stellaria jaluana Nakai = Stellaria filicaulis Makino ■

374928　Stellaria jamesiana Torr. = Pseudostellaria jamesiana（Torr.）W. A. Weber et R. L. Hartm. ■☆

374929　Stellaria karatavica（Lipsch.）Schischk.；卡拉塔夫繁缕■☆

374930　Stellaria kingii S. Watson = Arenaria kingii（S. Watson）M. E. Jones ■☆

374931　Stellaria kingii S. Watson = Eremogone kingii（S. Watson）Ikonn. ■☆

374932　Stellaria kotschyana Fenzl；考奇繁缕■☆

374933　Stellaria kotschyana Fenzl ex Boiss.；光萼繁缕；Glabrous-calyx Chickweed ■

374934　Stellaria laeta Richardson = Stellaria longipes Goldie ■☆

374935　Stellaria laeta Richardson var. altocaulis（Hultén）B. Boivin = Stellaria longipes Goldie ■☆

374936　Stellaria lanata Hook. f. ex Edgew. et Hook. f.；绵毛繁缕；Cotton Chickweed ■

374937　Stellaria lanipes C. Y. Wu et H. Chuang；绵柄繁缕（绵毛繁

缕）；Flossstalk Chickweed ■

374938　Stellaria laxa Merr. = Stellaria vestita Kurz ■

374939　Stellaria laxmannii Fisch. = Stellaria longipes Goldie ■☆

374940　Stellaria laxmannii Fisch. ex Ser.；拉克斯曼繁缕■☆

374941　Stellaria littoralis Torr.；海滩繁缕；Beach Chickweed, Beach Starwort ■☆

374942　Stellaria longifolia Muhl. = Stellaria filicaulis Makino ■

374943　Stellaria longifolia Muhl. = Stellaria longifolia Muhl. ex Willd. ■

374944　Stellaria longifolia Muhl. ex Willd.；长叶繁缕（多叉繁缕，睫毛长叶繁缕，睫伞繁缕，铺散繁缕，伞繁缕）；Ciliate Longleaf Chickweed, Diffuse Star Wort, Eyebright, Longleaf Chickweed, Long-leaved Starwort, Long-leaved Stitchwort ■

374945　Stellaria longifolia Muhl. ex Willd. f. ciliolata（Kitag.）Y. C. Chu = Stellaria longifolia Muhl. ex Willd. ■

374946　Stellaria longifolia Muhl. ex Willd. var. atrata J. W. Moore；黑长叶繁缕；Long-leaved Stitchwort ■☆

374947　Stellaria longifolia Muhl. ex Willd. var. atrata J. W. Moore = Stellaria longifolia Muhl. ex Willd. ■

374948　Stellaria longifolia Muhl. ex Willd. var. eciliata（B. Boivin）B. Boivin = Stellaria longifolia Muhl. ex Willd. ■

374949　Stellaria longifolia Muhl. ex Willd. var. laeta（Richardson）S. Watson = Stellaria longipes Goldie ■☆

374950　Stellaria longifolia Muhl. ex Willd. var. legitima Regel = Stellaria longifolia Muhl. ex Willd. ■

374951　Stellaria longipes Goldie；具梗繁缕（爱氏繁缕）；Edwards Chickweed, Goldie's Starwort, Long-leaved Starwort, Long-stalked Sitchwort, Long-stalked Starwort ■☆

374952　Stellaria longipes Goldie subsp. arenicola（Raup）C. C. Chinnappa et J. K. Morton；沙丘绥草；Lake Athabasca starwort ■☆

374953　Stellaria longipes Goldie subsp. monantha（Hultén）W. A. Weber = Stellaria longipes Goldie ■☆

374954　Stellaria longipes Goldie subsp. stricta（Richardson）W. A. Weber = Stellaria longipes Goldie ■☆

374955　Stellaria longipes Goldie var. altocaulis（Hultén）C. L. Hitchc. = Stellaria longipes Goldie ■☆

374956　Stellaria longipes Goldie var. arenicola（Raup）B. Boivin = Stellaria longipes Goldie subsp. arenicola（Raup）C. C. Chinnappa et J. K. Morton ■☆

374957　Stellaria longipes Goldie var. edwardsii（R. Br.）A. Gray = Stellaria longipes Goldie ■☆

374958　Stellaria longipes Goldie var. laeta（Richardson）S. Watson = Stellaria longipes Goldie ■☆

374959　Stellaria longipes Goldie var. minor Hook. = Stellaria longipes Goldie ■☆

374960　Stellaria longipes Goldie var. monantha（Hultén）S. L. Welsh = Stellaria longipes Goldie ■☆

374961　Stellaria longipes Goldie var. subvestita（Greene）Polunin = Stellaria longipes Goldie ■☆

374962　Stellaria longissima Wall. ex Edgew. et Hook. f. = Stellaria patens D. Don ■

374963　Stellaria mainlingensis L. H. Zhou；米林繁缕；Milin Chickweed ■

374964　Stellaria mannii Hook. f.；曼氏繁缕■☆

374965　Stellaria martjanovii Krylov；长裂繁缕；Lonsplit Chickweed ■

374966　Stellaria maximowiczii Kozhevn. = Stellaria decumbens Edgew. var. pulvinata Edgew. ■

374967　Stellaria media（L.）Cirillo = Stellaria media（L.）Vill. ■

374968　Stellaria media（L.）Vill.；繁缕（蔜，鹅肠菜，鹅儿肠菜，鹅儿伸筋，鹅馄饨，蘩蒌，狗蚤菜，合筋草，鸡肠菜，鸡肠草，鸡儿肠，五爪龙，小被单草，小鸡草，园酸菜，滋草）；Arvi Arva, Biddy's Eyes, Bird's Eye, Chick Wittles, Chicken's Evergreen, Chicken's Meat, Chickenweed, Chickenwort, Chicknyweed, Chickweed, Chuckenwort, Cickenwort, Cluckenweed, Cluckenwort, Cluckweed, Common Chickweed, Craches, Cuckenwort, Cyrillo, Flig, Foxtail, Hen's Evergreen, Mischievous Jack, Murren, Murun, Pokeweed, Ram's Claws, Satin Flower, Skirt Buttons, Star Chickweed, Starweed, Starwort, Tongue-grass, White Bird's Eye, White Bird's Eyes, Winterweed ■

374969　Stellaria media（L.）Vill. f. apetala Rouy et Foucaud = Stellaria pallida（Dumort.）Pire ■

374970　Stellaria media（L.）Vill. f. glaberrima（Beck）?；秃繁缕■☆

374971　Stellaria media（L.）Vill. f. pallida（Dumort.）Asch. et Graebn. = Stellaria pallida（Dumort.）Pire ■

374972　Stellaria media（L.）Vill. subsp. apetala（Ucria）Gaudin = Stellaria pallida（Dumort.）Pire ■

374973　Stellaria media（L.）Vill. subsp. cupaniana（Jord. et Fourr.）Nyman = Stellaria cupaniana（Jord. et Fourr.）Bég. ■☆

374974　Stellaria media（L.）Vill. subsp. major（Koch）Arcang. = Stellaria neglecta Weihe ex Bluff et Fingerh. ■☆

374975　Stellaria media（L.）Vill. subsp. neglecta（Weihe）Gremli；疏忽繁缕；Common Chickweed ■☆

374976　Stellaria media（L.）Vill. subsp. neglecta（Weihe）Gremli = Stellaria neglecta Weihe ex Bluff et Fingerh. ■

374977　Stellaria media（L.）Vill. subsp. pallida Dumort. = Stellaria pallida（Dumort.）Pire ■

374978　Stellaria media（L.）Vill. var. apetala（Ucria）Gaudin = Stellaria pallida（Dumort.）Pire ■

374979　Stellaria media（L.）Vill. var. boraeana（Jord.）Petit = Stellaria media（L.）Vill. ■

374980　Stellaria media（L.）Vill. var. candollei Briq. = Stellaria media（L.）Vill. ■

374981　Stellaria media（L.）Vill. var. decandra Fenzl = Stellaria neglecta Weihe ex Bluff et Fingerh. ■

374982　Stellaria media（L.）Vill. var. glabella（Jord. et Fourr.）Briq. = Stellaria media（L.）Vill. ■

374983　Stellaria media（L.）Vill. var. gymnocalyx Trautv. = Stellaria media（L.）Vill. ■

374984　Stellaria media（L.）Vill. var. latifolia Pau = Stellaria media（L.）Vill. ■

374985　Stellaria media（L.）Vill. var. major（Koch）Ball = Stellaria neglecta Weihe ex Bluff et Fingerh. ■☆

374986　Stellaria media（L.）Vill. var. micrantha（Hayata）T. Sh. Liu et S. S. Ying；小花繁缕；Littleflower Chickweed, Smallflower Chickweed ■

374987　Stellaria media（L.）Vill. var. micropetala Batt. = Stellaria media（L.）Vill. ■

374988　Stellaria media（L.）Vill. var. neglecta Mert. et Koch = Stellaria media（L.）Vill. ■

374989　Stellaria media（L.）Vill. var. oligandra Fenzl = Stellaria media（L.）Vill. ■

374990　Stellaria media（L.）Vill. var. procera Klatt et Richt. = Stellaria media（L.）Vill. ■

374991　Stellaria media（L.）Vill. var. procera Klatt et Richt. = Stellaria neglecta Weihe ex Bluff et Fingerh. ■

374992　Stellaria micrantha Hayata = Stellaria media（L.）Vill. var. micrantha（Hayata）T. Sh. Liu et S. S. Ying ■

374993　Stellaria mollis Klotzsch = Stellaria patens D. Don ■

374994　Stellaria monantha Hultén = Stellaria longipes Goldie ■☆

374995　Stellaria monantha Hultén subsp. atlantica Hultén = Stellaria longipes Goldie ■☆

374996　Stellaria monantha Hultén var. altocaulis Hultén = Stellaria longipes Goldie ■☆

374997　Stellaria monantha Hultén var. atlantica（Hultén）B. Boivin = Stellaria longipes Goldie ■☆

374998　Stellaria monogyna D. Don = Stellaria media（L.）Vill. ■

374999　Stellaria monosperma（F. N. Williams）Kozhevn. = Arenaria monosperma F. N. Williams ■

375000　Stellaria monosperma Buch. -Ham. ex D. Don;独籽繁缕（寸金草,大鹅肠,大种鹅儿肠,独子繁缕,黑牵牛,藤牛膝,通经草）;Singleseed Chickweed ■

375001　Stellaria monosperma Buch. -Ham. ex D. Don f. paniculata Mizush. = Stellaria monosperma Buch. -Ham. ex D. Don var. paniculata Majumdar ■

375002　Stellaria monosperma Buch. -Ham. ex D. Don var. japonica Maxim. ;皱叶繁缕（大繁缕）;Japanese Singleseed Chickweed ■

375003　Stellaria monosperma Buch. -Ham. ex D. Don var. paniculata Majumdar;锥花繁缕;Paniculate Singleseed Chickweed ■

375004　Stellaria montana Rose = Cerastium texanum Britton ■☆

375005　Stellaria muscorum Fassett = Minuartia muscorum（Fassett）Rabeler ■☆

375006　Stellaria neglecta Weihe = Stellaria neglecta Weihe ex Bluff et Fingerh. ■

375007　Stellaria neglecta Weihe ex Bluff et Fingerh. ;鸡肠繁缕（鹅肠繁缕,鸡肚肠草,绿繁缕,赛繁缕,小鸡草,鱼肚肠草）;Chickintestine Chickweed,Greater Chickweed ■

375008　Stellaria nemorum L. ;腺毛繁缕（森林繁缕）;Glandhair Chickweed,Wood Chickweed,Wood Starwort,Wood Stitchwort ■

375009　Stellaria nemorum L. = Stellaria bungeana Fenzl var. stubendorfii（Regel）Y. C. Chu ■

375010　Stellaria nemorum L. = Stellaria bungeana Fenzl ■

375011　Stellaria nemorum L. var. bracteata Fenzl = Stellaria nemorum L. ■

375012　Stellaria nemorum L. var. stubendorfii Regel = Stellaria bungeana Fenzl var. stubendorfii（Regel）Y. C. Chu ■

375013　Stellaria nemorum L. var. subebracteata Fenzl. = Stellaria nemorum L. ■

375014　Stellaria neopalustris Kitag. = Stellaria filicaulis Makino ■

375015　Stellaria neotomentosa Mizush. ex H. Ohba = Stellaria nepalensis Majumdar et Vartak ■

375016　Stellaria neotomentosa Mizush. ex Ohba = Stellaria nepalensis Majumdar et Vartak ■

375017　Stellaria nepalensis Majumdar et Vartak;尼泊尔繁缕;Nepal Chickweed ■

375018　Stellaria nipponica Ohwi;多花繁缕（白花蛇舌草）;Flowery Chickweed ■

375019　Stellaria nipponica Ohwi f. yezoensis（H. Hara）Toyok. = Stellaria nipponica Ohwi var. yezoensis H. Hara ■☆

375020　Stellaria nipponica Ohwi var. yezoensis H. Hara;北海道繁缕 ■☆

375021　Stellaria nitens Nutt. ;光亮繁缕;Shining Starwort ■☆

375022　Stellaria nutans F. N. Williams = Stellaria infracta Maxim. ■

375023　Stellaria nuttallii Torr. et A. Gray = Minuartia drummondii（Shinners）McNeill ■☆

375024　Stellaria nyalamensis L. H. Zhou;聂拉木繁缕;Nielamu Chickweed ■

375025　Stellaria nyalamensis L. H. Zhou = Stellaria monosperma Buch. -Ham. ex D. Don ■

375026　Stellaria obtusa Engelm. ;落基山繁缕;Blunt-sepaled Starwort,Rocky Mountain Starwort ■☆

375027　Stellaria octandra Pobed. = Stellaria neglecta Weihe ex Bluff et Fingerh. ■

375028　Stellaria omeiensis C. Y. Wu et Y. W. Tsui ex P. Ke;峨眉繁缕（大鹅儿肠,双蝴蝶）;Emei Chickweed,Omei Chickweed ■

375029　Stellaria ovatifolia（Mizush.）Mizush. ;卵叶繁缕;Ooleaf Chickweed ■

375030　Stellaria oxycoccoides Kom. ;莓苔状繁缕;Cranberry Chickweed ■

375031　Stellaria oxyphylla B. L. Rob. = Pseudostellaria oxyphylla（B. L. Rob.）R. L. Hartm. et Rabeler ■☆

375032　Stellaria pallida（Dumort.）Crépin = Stellaria pallida（Dumort.）Pire ■

375033　Stellaria pallida（Dumort.）Pire;苍白繁缕（无瓣繁缕）;Common Chickweed, Lesser Chickweed, Pale Chickweed, Petalless Chickweed ■

375034　Stellaria palmeri（Rydb.）Tidestr. = Stellaria longipes Goldie ■☆

375035　Stellaria paludicola Fernald et B. G. Schub. = Minuartia godfreyi（Shinners）McNeill ■☆

375036　Stellaria palustris Ehrh. ex Hoffm. ;沼泽繁缕（灰蓝繁缕,沼繁缕,沼生繁缕）;Bog Chickweed, European Chickweed, Marsh Stitchwort,Marshy Chickweed,Meadow Starwort ■

375037　Stellaria palustris Ehrh. ex Hoffm. var. imbricata Krylov = Stellaria imbricata Bunge ■

375038　Stellaria palustris Ehrh. var. imbricata（Bunge）Krylov = Stellaria imbricata Bunge ■

375039　Stellaria palustris Retz. = Stellaria palustris Ehrh. ex Hoffm. ■

375040　Stellaria paniculata Edgew. = Stellaria monosperma Buch. -Ham. ex D. Don var. paniculata Majumdar ■

375041　Stellaria paniculigera Makino;大繁缕（寸金草,大鹅儿肠,大种鹅儿肠,黑牵牛,老鹳精,通经草）■

375042　Stellaria paniculigera Makino = Stellaria monosperma Buch. -Ham. ex D. Don var. japonica Maxim. ■

375043　Stellaria parva Pedersen;微小繁缕;Pygmy Starwort, Small Starwort ■☆

375044　Stellaria parviumbellata Y. Z. Zhao;小伞花繁缕;Smallumbel Chickweed ■

375045　Stellaria patens D. Don;白毛繁缕;Whitehair Chickweed ■

375046　Stellaria patens D. Don = Stellaria lanata Hook. f. ex Edgew. et Hook. f. ■

375047　Stellaria patentifolia Kitag. = Stellaria graminea L. ■

375048　Stellaria peduncularis Bunge = Stellaria longipes Goldie ■☆

375049　Stellaria persica Boiss. ;波斯繁缕 ■☆

375050　Stellaria petiolaris Hand. -Mazz. ;细柄繁缕;Thinstalk Chickweed ■

375051　Stellaria petraea Bunge;岩生繁缕（绿花繁缕）;Saxicolous Chickweed ■

375052　Stellaria petraea Bunge = Stellaria decumbens Edgew. ■

375053　Stellaria petraea Bunge var. alpina（Bunge）Turcz. = Stellaria petraea Bunge ■

375054　Stellaria petraea Bunge var. alpina Turcz. = Stellaria petraea Bunge ■

375055　Stellaria petraea Bunge var. fasciculata Bunge ex Turcz. =

Stellaria cherleriae (Fisch. ex Ser.) F. N. Williams ■

375056　Stellaria petraea Bunge var. imbricata Fenzl = Stellaria petraea Bunge ■

375057　Stellaria petraea Bunge var. vegeta Fenzl = Stellaria cherleriae (Fisch. ex Ser.) F. N. Williams ■

375058　Stellaria pilosa Franch. = Stellaria pilosoides Shi L. Chen et al. ■

375059　Stellaria pilosa Franch. var. capillipes (Franch.) Hand. -Mazz. = Stellaria petiolaris Hand. -Mazz. ■

375060　Stellaria pilosoides Shi L. Chen et al. ;长柔毛繁缕(长毛繁缕,长毛箐姑草);Longhair Chickweed ■

375061　Stellaria porsildii C. C. Chinnappa;波尔繁缕;Porsild's Starwort ■☆

375062　Stellaria potaninii Krylov = Stellaria amblyosepala Schrenk ■

375063　Stellaria praecox A. Nelson = Stellaria nitens Nutt. ■☆

375064　Stellaria prostrata Baldwin = Stellaria cuspidata Willd. ex Schltdl. subsp. prostrata (Baldwin) J. K. Morton ■☆

375065　Stellaria pseudosaxatilis Hand. -Mazz. = Stellaria vestita Kurz ■

375066　Stellaria pterosperma Ohwi;翅籽繁缕■☆

375067　Stellaria pubera Michx. ;北美繁缕; Giant Chickweed, Star Chickweed ■☆

375068　Stellaria pubera Michx. subsp. silvatica Bég. = Stellaria corei Shinners ■☆

375069　Stellaria pubera Michx. var. silvatica (Bég.) Weath. = Stellaria corei Shinners ■☆

375070　Stellaria pusilla Em. Schmid;小繁缕;Small Chickweed ■

375071　Stellaria radians L. ;缞瓣繁缕(瓣繁缕,垂梗繁缕);Radiation Chickweed ■

375072　Stellaria radians L. f. fimbriata (Ledeb.) Kitag. = Stellaria radians L. ■

375073　Stellaria radians L. var. ovatooblonga Koidz. = Stellaria radians L. ■

375074　Stellaria reticulivena Hayata;网脉繁缕;Netvein Chickweed ■

375075　Stellaria rhaphanorrhiza Hemsl. = Pseudostellaria heterophylla (Miq.) Pax ■

375076　Stellaria rugegensis Engl. = Stellaria mannii Hook. f. ■☆

375077　Stellaria rugegensis Engl. f. parvipetala Balle = Stellaria mannii Hook. f. ■☆

375078　Stellaria ruscifolia Pall. ex Schltdl. ;假叶树叶繁缕;Prickly-leaved Starwort ■☆

375079　Stellaria ruscifolia Pall. ex Schltdl. subsp. aleutica Hultén = Stellaria ruscifolia Pall. ex Schltdl. ■☆

375080　Stellaria ruscifolia Willd. ex Schltdl. = Stellaria ruscifolia Pall. ex Schltdl. ■☆

375081　Stellaria ruscifolia Willd. ex Schltdl. subsp. aleutica Hultén = Stellaria ruscifolia Willd. ex Schltdl. ■☆

375082　Stellaria sachalinensis (Regel) Takeda = Stellaria fenzlii Regel ■☆

375083　Stellaria salicifolia Y. W. Tsui ex P. Ke;柳叶繁缕;Willowleaf Chickweed ■

375084　Stellaria saxatilis Buch. -Ham. ex D. Don = Stellaria vestita Kurz ■

375085　Stellaria saxatilis Buch. -Ham. ex D. Don f. petiolata Mizush. = Stellaria vestita Kurz ■

375086　Stellaria saxatilis Buch. -Ham. ex D. Don var. amplexicaulis Hand. -Mazz. = Stellaria vestita Kurz var. amplexicaulis (Hand. -Mazz.) C. Y. Wu ■

375087　Stellaria saxatilis Buch. -Ham. ex D. Don var. capillipes Franch. = Stellaria petiolaris Hand. -Mazz. ■

375088　Stellaria schimperi Engl. = Cerastium indicum Wight et Arn. ■☆

375089　Stellaria schugnanica Schischk. ;舒格南繁缕■☆

375090　Stellaria semivestita Edgew. var. brevipetala L. H. Zhou;短瓣绵毛繁缕■

375091　Stellaria semivestita Edgew. var. brevipetala L. H. Zhou = Stellaria congestiflora H. Hara ■

375092　Stellaria sennii Chiov. ;森恩繁缕■☆

375093　Stellaria sessiliflora Y. Yabe;无柄繁缕■☆

375094　Stellaria siberia (Regel et Tiling) Schischk. ;西伯利亚繁缕;Siberian Chickweed ■☆

375095　Stellaria silvatica (Bég.) Maguire = Stellaria corei Shinners ■☆

375096　Stellaria simcoei (Howell) C. L. Hitchc. = Stellaria calycantha (Ledeb.) Bong. ■☆

375097　Stellaria sitchana (Ledeb.) Bong. var. bongardiana (Fernald) Hultén = Stellaria borealis Bigelow subsp. sitchana (Steud.) Piper et Beattie ■☆

375098　Stellaria sitchana Steud. = Stellaria borealis Bigelow subsp. sitchana (Steud.) Piper et Beattie ■☆

375099　Stellaria sitchana Steud. var. bongardiana (Fernald) Hultén = Stellaria borealis Bigelow subsp. sitchana (Steud.) Piper et Beattie ■☆

375100　Stellaria soongorica Roshev. ;准噶尔繁缕;Dzungar Chickweed ■

375101　Stellaria souliei F. N. Williams;康定繁缕(苏氏繁缕);Kangding Chickweed ■

375102　Stellaria stellarioides L. = Cerastium cerastoides (L.) Britton ■

375103　Stellaria stellatopilosa Hayata = Stellaria vestita Kurz ■

375104　Stellaria stephaniana Willd. ex Schltdl. = Stellaria dichotoma L. var. lanceolata Bunge ■

375105　Stellaria stricta Richardson = Stellaria longipes Goldie ■☆

375106　Stellaria strongylosepala Hand. -Mazz. ;圆萼繁缕;Roundcalyx Chickweed ■

375107　Stellaria subumbellata Edgew. ;亚伞花繁缕;Subumbel Chickweed ■

375108　Stellaria subumbellata Edgew. = Stellaria umbellata Turcz. ■

375109　Stellaria subumbellata Edgew. f. villosa H. Huang;毛拟伞花繁缕■

375110　Stellaria subvestita Greene = Stellaria longipes Goldie ■☆

375111　Stellaria sylvatica (Maxim.) Maxim. ex Regel = Pseudostellaria sylvatica (Maxim.) Pax ■

375112　Stellaria sylvatica (Maxim.) Regel = Pseudostellaria sylvatica (Maxim.) Pax ■

375113　Stellaria tennesseensis (C. Mohr) Strausbaugh et Core = Stellaria corei Shinners ■☆

375114　Stellaria tibetica Kurz;西藏繁缕;Tibet Chickweed, Xizang Chickweed ■

375115　Stellaria tomentella Ohwi = Stellaria uchiyamana Makino var. apetala (Kitam.) Ohwi ■☆

375116　Stellaria turkestanica Schischk. ;土耳其斯坦繁缕■☆

375117　Stellaria uchiyamana Makino;日本山繁缕■☆

375118　Stellaria uchiyamana Makino f. apetala Kitam. = Stellaria uchiyamana Makino var. apetala (Kitam.) Ohwi ■☆

375119　Stellaria uchiyamana Makino var. apetala (Kitam.) Ohwi;无瓣日本山繁缕■☆

375120　Stellaria uda F. N. Williams;湿地繁缕;Wet Chickweed ■

375121　Stellaria uda F. N. Williams var. pubescens Y. W. Chi et L. H. Zhou = Stellaria alaschanica Y. Z. Zhao ■

375122　Stellaria uliginosa Murray = Stellaria alsine Grimm ■

375123　Stellaria uliginosa Murray var. alpina（Schur）Gürke = Stellaria alsine Grimm var. alpina（Schur）Hand. -Mazz. ■

375124　Stellaria uliginosa Murray var. alpina（Schur）Hand. -Mazz. ；高山雀舌草；Alpine Birdtongue Chickweed ■

375125　Stellaria uliginosa Murray var. atlantica Jahand. et Maire = Stellaria alsine Grimm ■

375126　Stellaria uliginosa Murray var. undulata（Thunb.）Fenzl；天蓬草(雀舌草)■

375127　Stellaria uliginosa Murray var. undulata（Thunb.）Fenzl = Stellaria alsine Grimm ■

375128　Stellaria umbellata Turcz. ；伞花繁缕；Umbel Chickweed，Umbellate Star Wort，Umbellate Starwort ■

375129　Stellaria umbellata Turcz. = Stellaria parviumbellata Y. Z. Zhao ■

375130　Stellaria umbellata Turcz. = Stellaria subumbellata Edgew. ■

375131　Stellaria undulata Thunb. = Stellaria alsine Grimm var. undulata（Thunb.）Ohwi ■

375132　Stellaria undulata Thunb. = Stellaria alsine Grimm ■

375133　Stellaria undulata Thunb. = Stellaria chinensis Regel ■

375134　Stellaria uniflora Walter = Minuartia uniflora（Walter）Mattf. ■☆

375135　Stellaria valida（Goodd.）Coult. et A. Nelson = Stellaria longipes Goldie ■☆

375136　Stellaria vestita Kurz；箐姑草(白筋骨草，白老鸦草，白老鸦肠，被单草，抽筋草，大鹅肠草，单背叶，地精草，滇繁缕，假石生繁缕，接筋草，金缠菜，筋骨菜，筋骨草，菁姑草，青姑草，青骨草，石灰草，石生繁缕，疏花繁缕，星毛繁缕)；Cliff Chickweed，Jinggu Chickweed，Rocky Star Wort ■

375137　Stellaria vestita Kurz var. amplexicaulis（Hand. -Mazz.）C. Y. Wu；抱茎箐姑草(贯叶繁缕)；Amplexicaul Jinggu Chickweed ■

375138　Stellaria viridescens（Maxim.）Kozhevn. = Stellaria graminea L. var. viridescens Maxim. ■

375139　Stellaria viridiflora Pax et K. Hoffm. = Stellaria cherleriae（Fisch. ex Ser.）F. N. Williams ■

375140　Stellaria viridula（Piper）St. John = Stellaria obtusa Engelm. ■☆

375141　Stellaria washingtoniana B. L. Rob. = Stellaria obtusa Engelm. ■☆

375142　Stellaria weberi B. Boivin = Stellaria umbellata Turcz. ■

375143　Stellaria williamsiana Kozhevn. = Arenaria monosperma F. N. Williams ■

375144　Stellaria winkleri（Briq.）Schischk. ；帕米尔繁缕；Pamir Chickweed ■

375145　Stellaria wushaneensis F. N. Williams；巫山繁缕（武冈繁缕）；Wushan Chickweed ■

375146　Stellaria wushaneensis F. N. Williams var. trientaloides Hand. -Mazz. = Stellaria wushaneensis F. N. Williams ■

375147　Stellaria wushaneensis F. N. Williams var. trientaloides Hand. -Mazz. ；武冈繁缕；Wugang Chickweed ■

375148　Stellaria wushanensis F. N. Williams var. trientaloides Hand. -Mazz. = Stellaria wushaneensis F. N. Williams ■

375149　Stellaria wutaica Hand. -Mazz. = Stellaria umbellata Turcz. ■

375150　Stellaria yesoalpina Nakai = Stellaria calycantha（Ledeb.）Bong. ■☆

375151　Stellaria yunnanensis Franch. ；云南繁缕(大鹅肠菜，筋骨草，麦参，麦冬，千针万线草)；Yunnan Chickweed ■

375152　Stellaria yunnanensis Franch. f. villosa C. Y. Wu ex P. Ke =

375153　Stellaria yunnanensis Franch. var. villosa C. Y. Wu ex P. Ke；密柔毛繁缕；Villose Yunnan Chickweed ■

375154　Stellaria yunnanensis Franch. var. villosa C. Y. Wu ex P. Ke = Stellaria dianthifolia F. N. Williams ■

375155　Stellaria yunnanensts Franch. = Stellaria dianthifolia F. N. Williams ■

375156　Stellaria zangnanensis L. H. Zhou；藏南繁缕；S. Xizang Chickweed ■

375157　Stellariaceae Bercht. et J. Presl = Caryophyllaceae Juss.（保留科名）■●

375158　Stellariaceae Dumort. = Caryophyllaceae Juss.（保留科名）■●

375159　Stellariaceae MacMill. = Callitrichaceae Link(保留科名)■

375160　Stellariaceae MacMill. = Caryophyllaceae Juss.（保留科名）■●

375161　Stellarioides Medik.（1790）；类繁缕属■☆

375162　Stellarioides Medik. = Anthericum L. ■☆

375163　Stellarioides canaliculata Medik. ；类繁缕■☆

375164　Stellarioides sessiliflora（Desf.）Speta；无花梗类繁缕■☆

375165　Stellariopsis（Baill.）Rydb.（1898）；拟繁缕属■☆

375166　Stellariopsis（Baill.）Rydb. = Ivesia Torr. et A. Gray ■☆

375167　Stellariopsis（Baill.）Rydb. = Potentilla L. ■●

375168　Stellariopsis Rydb. = Potentilla L. ■●

375169　Stellariopsis santolinoides（A. Gray）Rydb. ；拟繁缕■☆

375170　Stellaris Dill. ex Moench = Scilla L. ■

375171　Stellaris Fabr. = Scilla L. ■

375172　Stellaris Moench = Scilla L. ＋ Ornithogalum L. ＋ Gagea Salisb. ■

375173　Stellaris Moench = Scilla L. ■

375174　Stellaster Fabr. = Scilla L. ■

375175　Stellaster Heist. = Stellaster Heist. ex Fabr. ■

375176　Stellaster Heist. ex Fabr. = Scilla L. ■

375177　Stellaster Heist. ex Fabr. = Stellaris Fabr. ■

375178　Stellatae Batsch = Rubiaceae Juss.（保留科名）●■

375179　Stellera L.（1753）；似狼毒属；Stellera ■●

375180　Stellera L. = Wikstroemia Endl. ＋ Dendrostellera（C. A. Mey.）Tiegh. ■

375181　Stellera Turcz. = Rellesta Turcz. ■

375182　Stellera Turcz. = Swertia L. ■

375183　Stellera alberti Regel；阿尔波特狼毒●☆

375184　Stellera altaica Thieb. -Bern. = Stelleropsis altaica（Thieb. -Bern.）Pobed. ■

375185　Stellera annua Salisb. = Thymelaea passerina（L.）Coss. et Germ. ■

375186　Stellera bodinieri H. Lév. = Stellera chamaejasme L. ■

375187　Stellera chamaejasme L. ；似狼毒（矮茉莉荛花，拔萝卜，白狼毒，川狼毒，大将军，大猫眼草，断肠草，甘遂，红火柴头花，红狼毒，黄皮狼毒，鸡肠狼毒，狼毒疙瘩，馒头花，猫眼根，棉大戟，千里马，瑞香狼毒，山萝卜，顺水龙，西北狼毒，狭叶甘遂，小狼毒，续毒，燕子花，一把香）；Chinese Stellera，Dwarf Stringbush，Narrowleaf Stellera，Stellera ■

375188　Stellera chamaejasme L. f. angustifolia Diels = Stellera chamaejasme L. ■

375189　Stellera chamaejasme L. f. chrysantha S. C. Huang；黄花甘遂；Yellowflower Chinese Stellera ■

375190　Stellera chamaejasme L. f. chrysantha S. C. Huang = Stellera chamaejasme L. ■

375191　Stellera chamaejasme L. var. angustifolia Diels；狭叶瑞香狼毒

（瑞香狼毒）■

375192　Stellera chamaejasme L. var. angustifolia Diels = Stellera chamaejasme L. ■

375193　Stellera chinensis Lecomte = Daphne rosmarinifolia Rehder ●

375194　Stellera circinata Lecomte = Wikstroemia dolichantha Diels ●

375195　Stellera circinata Lecomte var. divaricata Lecomte = Wikstroemia dolichantha Diels ●

375196　Stellera diffusa Lecomte = Daphne rosmarinifolia Rehder ●

375197　Stellera fargesii Lecomte = Wikstroemia fargesii（Lecomte）Domke ●

375198　Stellera formosana Hayata ex H. L. Li；台湾狼毒（矮瑞香）；Taiwan Stellera ●

375199　Stellera japonica（Siebold et Zucc.）Meisn. = Wikstroemia trichotoma（Thunb.）Makino ●

375200　Stellera japonica Siebold = Diplomorpha trichotoma（Thunb.）Nakai ●

375201　Stellera lessertii（Wikstr.）C. A. Mey.；莱塞特狼毒●☆

375202　Stellera mairei Lecomte = Daphne esquirolii H. Lév. ●

375203　Stellera passerina L. = Thymelaea passerina（L.）Coss. et Germ. ■

375204　Stellera rosea Nakai = Stellera chamaejasme L. ■

375205　Stellera tenuiflora（Bureau et Franch.）Lecomte = Daphne tenuiflora Bureau et Franch. ●

375206　Stellera tenuiflora（Bureau et Franch.）Lecomte var. legendrei Lecomte = Daphne tenuiflora Bureau et Franch. var. legendrei（Lecomte）Hamaya ●

375207　Stelleropsis Pobed.（1950）；假狼毒属；Fakestellera ■

375208　Stelleropsis Pobed. = Diarthron Turcz. ●■

375209　Stelleropsis altaica（Thieb. -Bern.）Pobed.；阿尔泰假狼毒（假狼毒）；Altai Fakestellera ■

375210　Stelleropsis antoninae Pobed.；安氏假狼毒■☆

375211　Stelleropsis caucasica Pobed.；高加索假狼毒■☆

375212　Stelleropsis iranica Pobed.；伊朗假狼毒■☆

375213　Stelleropsis issykkulensis Pobed.；伊塞克假狼毒■☆

375214　Stelleropsis magakjanii（Sosn.）Pobed.；马氏假狼毒■☆

375215　Stelleropsis tianschanica Pobed.；天山假狼毒；Tianshan Fakestellera ■

375216　Stelleropsis turcomanica（Czerniak.）Pobed.；土库曼假狼毒■☆

375217　Stellia Noronha = Tarenna Gaertn. ●

375218　Stelligera A. J. Scott = Sclerolaena R. Br. ●☆

375219　Stellilabium Schltr.（1914）；星唇兰属■☆

375220　Stellilabium alticola Dodson et R. Escobar；高原星唇兰■☆

375221　Stellilabium atropurpureum P. Ortiz；暗紫星唇兰■☆

375222　Stellilabium microglossum（Schltr.）Dodson；小舌星唇兰■☆

375223　Stellilabium minutiflorum（Kraenzl.）Garay；多花星唇兰■☆

375224　Stellimia Raf. = Pectis L. ■☆

375225　Stellina Bubani = Callitriche L. ■

375226　Stellix Noronha = Psychotria L.（保留属名）●

375227　Stellorchis Thouars = Nervilia Comm. ex Gaudich.（保留属名）■

375228　Stellorkis Thouars（废弃属名）= Nervilia Comm. ex Gaudich.（保留属名）■

375229　Stellorkis aplostellis Thouars = Nervilia simplex（Thouars）Schltr. ■☆

375230　Stellularia Benth. = Buchnera L. ■

375231　Stellularia Hill = Stellaria L. ■

375232　Stellularia inflata De Wild. = Buchnera inflata（De Wild.）V. Naray. ■☆

375233　Stellularia nigricans Benth. = Buchnera nigricans（Benth.）V. Naray. ■☆

375234　Stelmacrypton Baill.（1889）；须药藤属（生藤属，须叶藤属，隐冠萝藦属）；Stelmatocrypton ●

375235　Stelmacrypton Baill. = Pentanura Blume ●

375236　Stelmacrypton khasianum（Benth.）Baill. = Stelmacrypton khasianum（Kurz）Baill. ●

375237　Stelmacrypton khasianum（Kurz）Baill.；须药藤（大花藤，冷水发汗，生藤，水逼药，香根藤，小杜仲，羊角藤）；Common Stelmatocrypton ●☆

375238　Stelmacrypton khasianum（Kurz）Baill. var. major C. Y. Wu；大须药藤（大羊奶藤）●

375239　Stelmacrypton khasianum Baill. = Stelmacrypton khasianum（Kurz）Baill. ●☆

375240　Stelmagonum Baill.（1890）；膝冠萝藦属☆

375241　Stelmagonum hahnianum Baill.；膝冠萝藦☆

375242　Stelmanis Raf.（1836）= Heterotheca Cass. ■☆

375243　Stelmanis Raf.（1840）= Anistelma Raf. ●■

375244　Stelmanis Raf.（1840）= Hedyotis L.（保留属名）●■

375245　Stelmanis Raf. = Oldenlandia L. ●■

375246　Stelmanis scabra Raf. = Heterotheca subaxillaris（Lam.）Britton et Rusby ■☆

375247　Stelmation E. Fourn. = Metastelma R. Br. ●☆

375248　Stelmatocodon Schltr.（1906）；钟冠萝藦属■☆

375249　Stelmatocodon fiebrigii Schltr.；钟冠萝藦■☆

375250　Stelmatocrypton Baill. = Pentanura Blume ●

375251　Stelmatocrypton Baill. = Stelmacrypton Baill. ●

375252　Stelmatocrypton khasianum（Kurz）Baill. = Stelmacrypton khasianum（Kurz）Baill. ●

375253　Stelmatogonum K. Schum. = Stelmagonum Baill. ☆

375254　Stelmesus Raf. = Allium L. ■

375255　Stelmotis Raf. = Anistelma Raf. ●■

375256　Stelmotis Raf. = Hedyotis L.（保留属名）●■

375257　Stelmotis Raf. = Oldenlandia L. ●■

375258　Stelmotis Raf. = Stelmanis Raf. ●■

375259　Stelophurus Post et Kuntze = Phleum L. ■

375260　Stelophurus Post et Kuntze = Stelephuros Adans. ■

375261　Stelostylis Post et Kuntze = Stelestylis Drude ■☆

375262　Stemelena Raf. = Krameria L. ex Loefl. ●■☆

375263　Stemmacantha Cass.（1817）；祁州漏芦属（刺冠菊属，漏芦属）；Swiss Centaury，Swisscentaury ■

375264　Stemmacantha Cass. = Cirsium Mill. ■

375265　Stemmacantha Cass. = Leuzea DC. ■☆

375266　Stemmacantha Cass. = Rhaponticum Adans. ■

375267　Stemmacantha acaulis（L.）Dittrich = Rhaponticum acaule（L.）DC. ■☆

375268　Stemmacantha carthamoides（Willd.）Dittrich；鹿草（鹿根）；Safflowerlike Swisscentaury ■

375269　Stemmacantha carthamoides（Willd.）Dittrich = Rhaponticum carthamoides（Willd.）Iljin ■

375270　Stemmacantha cynaroides（Less.）Dittrich = Rhaponticum cynaroides Less. ■☆

375271　Stemmacantha exaltata（Cutanda）Dittrich = Rhaponticum exaltatum（Willk.）Greuter ■☆

375272　Stemmacantha longifolia（Hoffmanns. et Link）Dittrich = Rhaponticum longifolium（Hoffmanns. et Link）Dittrich ■☆

375273　Stemmacantha longifolia（Hoffmanns. et Link）Dittrich var. ericeticola（Font Quer）Dittrich = Rhaponticum longifolium（Hoffmanns. et Link）Dittrich subsp. ericeticola（Font Quer）Greuter ■☆

375274　Stemmacantha uniflora（L.）Dittrich;漏芦（打锣锤,大口袋花,大脑袋花,大头翁,单花矢车菊,独花山牛蒡,鬼油麻,和尚头,郎头花,狼头花,老虎爪,馒头草,牛蒡头,牛馒头花,祁州漏芦,土烟叶,野兰）;Common Swisscentaury,Swisscentaury ■

375275　Stemmacantha uniflora（L.）Dittrich = Rhaponticum uniflorum（L.）DC. ■

375276　Stemmadenia Benth.（1845）;腺冠夹竹桃属●☆

375277　Stemmadenia glabra Benth.;腺冠夹竹桃●☆

375278　Stemmadenia macrophylla Greenm.;大叶腺冠夹竹桃●☆

375279　Stemmadenia minima A. H. Gentry;小腺冠夹竹桃●☆

375280　Stemmadenia obovata Schum.;倒卵腺冠夹竹桃●☆

375281　Stemmadenia pauciflora Woodson;少花腺冠夹竹桃●☆

375282　Stemmadenia pubescens Benth. ;毛腺冠夹竹桃●☆

375283　Stemmatella Wedd. ex Benth. = Galinsoga Ruiz et Pav. ■●

375284　Stemmatella Wedd. ex Benth. et Hook. f. = Galinsoga Ruiz et Pav. ■●

375285　Stemmatella Wedd. ex Sch. Bip. = Galinsoga Ruiz et Pav. ■●

375286　Stemmatium Phil. = Leucocoryne Lindl. ■☆

375287　Stemmatium Phil. = Tristagma Poepp. ■☆

375288　Stemmatodaphne Gamble = Alseodaphne Nees ●

375289　Stemmatophyllum Tiegh. = Amyema Tiegh. ●☆

375290　Stemmatophysum Steud. = Stemmatosiphum Pohl ●

375291　Stemmatosiphon Meisn. = Stemmatosiphum Pohl ●

375292　Stemmatosiphum Pohl = Symplocos Jacq. ●

375293　Stemmatospermum P. Beauv. = Nastus Juss. ●☆

375294　Stemmodontia Cass. = Wedelia Jacq.（保留属名）■●

375295　Stemodia L.（1759）（保留属名）;离药草属■☆

375296　Stemodia L.（保留属名）= Unanuea Ruiz et Pav. ex Pennell ■☆

375297　Stemodia ceratophylloides（Hiern）K. Schum. = Limnophila ceratophylloides（Hiern）V. Naray. ■☆

375298　Stemodia chilensis Benth.;智利离药草■☆

375299　Stemodia floribunda（R. Br.）Roberty;多花离药草■☆

375300　Stemodia floribunda（R. Br.）Roberty = Bacopa floribunda（R. Br.）Wettst. ■

375301　Stemodia glabra Spreng. ;光离药草■☆

375302　Stemodia grandiflora Buch. -Ham. = Lindenbergia grandiflora Benth. ■

375303　Stemodia grandiflora Buch. -Ham. ex D. Don = Lindenbergia grandiflora（Buch. -Ham. ex D. Don）Benth. ■

375304　Stemodia grossa Benth. ;大齿离药草■☆

375305　Stemodia hirsuta Heyne ex Benth. = Limnophila chinensis（Osbeck）Merr. ■

375306　Stemodia hypericifolia Benth. = Limnophila connata（Buch. -Ham. ex D. Don）Hand. -Mazz. ■

375307　Stemodia lanceolata Benth. ;剑叶离药草■☆

375308　Stemodia lythrifolia Benth. ;千屈菜离药草■☆

375309　Stemodia multifida（Michx.）Spreng. ;多裂离药草■☆

375310　Stemodia muraria Roxb. = Lindenbergia muraria（Roxb. ex D. Don）Brühl ■

375311　Stemodia muraria Roxb. ex D. Don = Lindenbergia muraria（Roxb. ex D. Don）Brühl ■

375312　Stemodia parviflora Aiton = Stemodia verticillata（Mill.）Bold. ■☆

375313　Stemodia philippensis Cham. et Schltdl. = Lindenbergia philippensis（Cham. et Schltdl.）Benth. ■

375314　Stemodia repens Benth. = Limnophila repens（Benth.）Benth. ■

375315　Stemodia ruderalis Vahl = Lindenbergia muraria（Roxb. ex D. Don）Brühl ■

375316　Stemodia senegalensis Desf. ;塞内加尔离药草■☆

375317　Stemodia serrata Benth. ;具齿离药草■☆

375318　Stemodia tenera（Hiern）K. Schum. = Dopatrium tenerum（Hiern）Eb. Fisch. ■☆

375319　Stemodia tenuifolia Minod;细叶离药草■☆

375320　Stemodia verticillata（Mill.）Bold. ;轮生离药草■☆

375321　Stemodia viscosa Roxb. ;黏质离药草■☆

375322　Stemodiacra Kuntze = Limnophila R. Br.（保留属名）■

375323　Stemodiacra P. Browne（废弃属名）= Stemodia L.（保留属名）■☆

375324　Stemodiacra ceratophylloides Hiern = Limnophila ceratophylloides（Hiern）V. Naray. ■☆

375325　Stemodiacra tenera Hiern = Dopatrium tenerum（Hiern）Eb. Fisch. ■☆

375326　Stemodiopsis Engl.（1898）;拟离药草属■☆

375327　Stemodiopsis buchananii V. Naray. ;布坎南拟离药草■☆

375328　Stemodiopsis buchananii V. Naray. var. pubescens Philcox;短柔毛拟离药草■☆

375329　Stemodiopsis eylesii S. Moore;艾尔斯拟离药草■☆

375330　Stemodiopsis glandulosa Philcox;具腺拟离药草■☆

375331　Stemodiopsis humilis V. Naray. = Stemodiopsis rivae Engl. ■☆

375332　Stemodiopsis linearis S. Moore;线状拟离药草■☆

375333　Stemodiopsis rivae Engl. ;沟拟离药草■☆

375334　Stemodiopsis ruandensis Eb. Fisch. ;卢旺达拟离药草■☆

375335　Stemodoxis Raf. = Allium L. ■

375336　Stemona Lour.（1790）;百部属;Roxburghia,Stemona ■

375337　Stemona acuta C. H. Wright = Stemona tuberosa Lour. ■

375338　Stemona argyi（H. Lév. et Vaniot）H. Lév. = Stemona japonica（Blume）Miq. ■

375339　Stemona collinsae Craib;小丘百部■☆

375340　Stemona erecta C. H. Wright = Stemona sessilifolia（Miq.）Miq. ■

375341　Stemona filifolia Schltr. = Stemona mairei（H. Lév.）Krause ■

375342　Stemona gloriosoides Voigt = Stemona tuberosa Lour. ■

375343　Stemona japonica（Blume）Miq. ;百部（百奶,百条根,大叶百部,九虫根,九丛根,九十九条根,九重根,蔓生百部,闹虱药,牛虱鬼,婆妇草,山百部,虱婆草,嗽药,药虱药,野天门冬,一窝虎）;Japan Stemona,Japanese Stemona ■

375344　Stemona jinshanjiangensis X. D. Cong et G. J. Xu;金沙江百部;Jinshajiang Stemona ■

375345　Stemona kerrii Craib;克氏百部■

375346　Stemona mairei（H. Lév.）Krause;云南百部（丽江百部,狭叶百部,线叶百部）;Maire Stemona,Yunnan Stemona ■

375347　Stemona ovata Nakai = Stemona japonica（Blume）Miq. ■

375348　Stemona ovata Nakai ex Kishida et Matsuno = Stemona japonica（Blume）Miq. ■

375349　Stemona parviflora C. H. Wright;细花百部（大百部,披针叶百部,细叶百部,小花百部）;Littleflower Stemona ■

375350　Stemona saxorum Gagnep. = Stemona kerrii Craib ■

375351　Stemona sessilifolia（Miq.）Miq. ;直立百部（百部,百部袋,百奶,百条根,九虫根,九丛根,九十九条根,九重根,闹虱药,牛虱鬼,婆妇草,山百部,嗽药,药虱药,野天门冬,一窝虎）;Sessile

Stemona, Stand Stemona ■
375352 Stemona shandongensis D. K. Zang; 山东百部; Shandong Stemona ■
375353 Stemona stenophylla Diels = Stemona mairei (H. Lév.) Krause ■
375354 Stemona stenophylla Diels ex Schltr. = Stemona mairei (H. Lév.) Krause ■
375355 Stemona tuberosa Lour.;大百部(百部,百部根,百蔀,百奶,百条根,大春根药,对叶百部,九虫根,九丛根,九十九条根,九重根,闹虱药,牛虱鬼,山百部,嗽药,药虱药,野天门冬,野天门冬根,一窝虎);Large Stemona ■
375356 Stemona vagula W. W. Sm. = Stemona mairei (H. Lév.) Krause ■
375357 Stemona wardii W. W. Sm. = Stemona mairei (H. Lév.) Krause ■
375358 Stemonacanthus Nees = Ruellia L. ■●
375359 Stemonaceae Caruel (1878) (保留科名); 百部科; Stemona Family ■
375360 Stemonaceae Caruel(保留科名) = Croomiaceae Nakai ■
375361 Stemonaceae Engl. = Stemonaceae Caruel(保留科名) ■
375362 Stemone Franch. et Sav. = Stemona Lour. ■
375363 Stemonix Raf. = Eurycles Salisb. ■☆
375364 Stemonocoleus Harms(1905);鞘蕊苏木属(小花苏木属)●☆
375365 Stemonocoleus micranthus Harms;小花苏木●☆
375366 Stemonoporus Thwaites(1854);孔雄蕊香属●☆
375367 Stemonoporus acuminatus Bedd.;尖孔雄蕊香●☆
375368 Stemonoporus nervosus Trim.;多脉孔雄蕊香●☆
375369 Stemonoporus rigidus Pierre;硬孔雄蕊香●☆
375370 Stemonoporus roseus Trim.;粉红孔雄蕊香●☆
375371 Stemonuraceae Kårehed(2001);尾药木科(金檀木科)●
375372 Stemonurus Blume(1826);尾药木属(粗丝木属,毛蕊木属);Stemonurus ●
375373 Stemonurus chingianus Hand.-Mazz. = Gomphandra tetrandra (Wall.) Sleumer ●
375374 Stemonurus foetidus Wight = Nothapodytes foetida (Wight) Sleumer ●
375375 Stemonurus foetidus Wight = Nothapodytes nimmoniana (J. Graham) Mabb. ●
375376 Stemonurus hainanensis (Merr.) Hu = Gomphandra tetrandra (Wall.) Sleumer ●
375377 Stemonurus luzoniensis Merr. = Gomphandra luzoniensis (Merr.) Merr. ●
375378 Stemonurus mollis (Merr.) Howard ex Dahl = Gomphandra mollis Merr. ●
375379 Stemonurus secundiflorus Blume;尾药木●☆
375380 Stemonurus yunnanensis Hu = Pittosporopsis kerrii Craib ●
375381 Stemoptera Miers = Apteria Nutt. ■☆
375382 Stemotis Raf. = Rhododendron L. ●
375383 Stemotria Wettst. et Harms = Stemotria Wettst. et Harms ex Engl. ●☆
375384 Stemotria Wettst. et Harms ex Engl. (1899);秘鲁玄参属●☆
375385 Stemotria triandra (Cav.) Govaerts;秘鲁玄参●☆
375386 Stenachaenium Benth. (1873);长尾菊属■☆
375387 Stenachaenium adenanthum Krasch.;腺花长尾菊■☆
375388 Stenachaenium macrocephalum Benth. ex Benth. et Hook. f.;大头长尾菊■☆
375389 Stenactis Cass. = Erigeron L. ■●
375390 Stenactis annuus (L.) Cass. = Erigeron annuus (L.) Pers. ■
375391 Stenactis annuus Cass. = Erigeron annuus (L.) Pers. ■
375392 Stenactis annuus Nees = Erigeron annuus (L.) Pers. ■
375393 Stenactis beyrichii Fisch. et C. A. Mey. = Erigeron strigosus Muhl. ex Willd. ■☆
375394 Stenactis multiradiatus Lindl. ex DC. = Erigeron multiradiatus (Lindl. ex DC.) Benth. ■
375395 Stenactis septentrionalis (Fernald et Wiegand) Holub = Erigeron strigosus Muhl. ex Willd. var. septentrionalis (Fernald et Wiegand) Fernald ■☆
375396 Stenactis speciosus Lindl. = Erigeron speciosus (Lindl.) DC. ■☆
375397 Stenactis strigosus (Muhl. ex Willd.) DC. = Erigeron strigosus Muhl. ex Willd. ■☆
375398 Stenadenium Pax = Monadenium Pax ■☆
375399 Stenadenium spinescens Pax = Euphorbia neospinescens Bruyns ●☆
375400 Stenandriopsis S. Moore = Crossandra Salisb. ●
375401 Stenandriopsis S. Moore = Stenandrium Nees(保留属名) ■☆
375402 Stenandriopsis S. Moore(1906);类狭蕊爵床属■☆
375403 Stenandriopsis afromontana (Mildbr.) Benoist = Stenandrium afromontanum (Mildbr.) Vollesen ■☆
375404 Stenandriopsis buntingii (S. Moore) Heine = Stenandrium buntingii (S. Moore) Vollesen ■☆
375405 Stenandriopsis gabonica (Benoist) Heine = Stenandrium gabonicum (Benoist) Vollesen ■☆
375406 Stenandriopsis guineensis (Nees) Benoist = Stenandrium guineense (Nees) Vollesen ■☆
375407 Stenandriopsis talbotii (S. Moore) Heine = Stenandrium talbotii (S. Moore) Vollesen ■☆
375408 Stenandriopsis thomensis (Milne-Redh.) Heine = Stenandrium thomense (Milne-Redh.) Vollesen ■☆
375409 Stenandriopsis thompsoni S. Moore;类狭蕊爵床■☆
375410 Stenandriopsis warneckei (S. Moore) Napper = Stenandrium warneckei (S. Moore) Vollesen ■☆
375411 Stenandrium Nees (1836) (保留属名); 狭蕊爵床属; False Foxglove ■☆
375412 Stenandrium afromontanum (Mildbr.) Vollesen;非洲山生狭蕊爵床■☆
375413 Stenandrium buntingii (S. Moore) Vollesen;邦廷狭蕊爵床■☆
375414 Stenandrium gabonicum (Benoist) Vollesen;加蓬狭蕊爵床■☆
375415 Stenandrium grandiflorum Vollesen;大花狭蕊爵床■☆
375416 Stenandrium guineense (Nees) Vollesen;几内亚狭蕊爵床■☆
375417 Stenandrium leptostachyum (Benoist) Vollesen;细穗狭蕊爵床■☆
375418 Stenandrium lindenii N. E. Br.;林登狭蕊爵床■☆
375419 Stenandrium pauciflorum Vollesen;少花狭蕊爵床■☆
375420 Stenandrium talbotii (S. Moore) Vollesen;塔尔博特狭蕊爵床■☆
375421 Stenandrium thomense (Milne-Redh.) Vollesen; 毛狭蕊爵床■☆
375422 Stenandrium warneckei (S. Moore) Vollesen; 瓦尔狭蕊爵床■☆
375423 Stenanona Standl. (1929);狭瓣花属●☆
375424 Stenanona cauliflora (J. W. Walker) G. E. Schatz;茎花狭瓣花●☆
375425 Stenanona panamensis Standl.;巴拿马狭瓣花●☆
375426 Stenanona stenopetala (Donn. Sm.) G. E. Schatz;狭瓣花●☆
375427 Stenanthella Rydb. (1900);小狭被莲属■☆
375428 Stenanthella Rydb. = Stenanthium (A. Gray) Kunth(保留属

名)■☆

375429 Stenanthella occidentalis（A. Gray）Rydb.；小狭被莲■☆

375430 Stenanthella occidentalis（A. Gray）Rydb. = Stenanthium occidentale A. Gray ■☆

375431 Stenanthemum Reissek = Cryptandra Sm. ●☆

375432 Stenanthemum Reissek（1858）；狭花木属●☆

375433 Stenanthemum gracilipes Diels；细梗狭花木●☆

375434 Stenanthemum pomaderroides Reissek；狭花木●☆

375435 Stenanthemum tridentatum Reissek；三齿狭花木●☆

375436 Stenanthera（Oliv.）Engl. et Diels（1900）；窄药花属●☆

375437 Stenanthera Engl. et Diels = Neostenanthera Exell ●☆

375438 Stenanthera Engl. et Diels = Stenanthera（Oliv.）Engl. et Diels ●☆

375439 Stenanthera R. Br. = Astroloma R. Br. ●☆

375440 Stenanthera bakuana A. Chev. ex Hutch. et Dalziel = Neostenanthera gabonensis（Engl. et Diels）Exell ●☆

375441 Stenanthera gabonensis（Engl. et Diels）Engl. et Diels = Neostenanthera gabonensis（Engl. et Diels）Exell ●☆

375442 Stenanthera gabonensis Engl. et Diels；窄药花●☆

375443 Stenanthera hamata（Benth.）Engl. et Diels = Neostenanthera hamata（Benth.）Exell ●☆

375444 Stenanthera macrantha Mildbr. et Diels；大花窄药花●☆

375445 Stenanthera macrantha Mildbr. et Diels = Boutiquea platypetala（Engl. et Diels）Le Thomas ●☆

375446 Stenanthera myristicifolia（Oliv.）Engl. et Diels = Neostenanthera myristicifolia（Oliv.）Exell ●☆

375447 Stenanthera neurosericea Diels = Neostenanthera gabonensis（Engl. et Diels）Exell ●☆

375448 Stenanthera platypetala Engl. et Diels；宽瓣窄药花●☆

375449 Stenanthera platypetala Engl. et Diels = Boutiquea platypetala（Engl. et Diels）Le Thomas ●☆

375450 Stenanthera pluriflora De Wild. = Neostenanthera myristicifolia（Oliv.）Exell ●☆

375451 Stenanthium（A. Gray）Kunth（1843）（保留属名）；狭被莲属（瘦花属）；Stenanthium ■☆

375452 Stenanthium Kunth = Stenanthium（A. Gray）Kunth（保留属名）■☆

375453 Stenanthium gramineum Morong；禾叶狭被莲；Feather Bells，Featherbells，Feather-bells，Featherfleece，Feather-fleece ■☆

375454 Stenanthium gramineum Morong var. micranthum Fernald = Stenanthium gramineum Morong ■☆

375455 Stenanthium gramineum Morong var. robustum（S. Watson）Fernald = Stenanthium gramineum Morong ■☆

375456 Stenanthium occidentale A. Gray；西方狭被莲；Bronze-bells，Mountain-bells ■☆

375457 Stenanthium rhombipetalum Suksd. = Stenanthium occidentale A. Gray ■☆

375458 Stenanthium robustum S. Watson = Stenanthium gramineum Morong ■☆

375459 Stenanthium sachalinense F. Schmidt；库页狭被莲■☆

375460 Stenanthus Oerst. ex Hanst.（1854）；狭花苣苔属■☆

375461 Stenanthus Oerst. ex Hanst. = Columnea L. ●■☆

375462 Stenanthus heterophyllus Oerst. = Stenanthus heterophyllus Oerst. ex Hanst. ■☆

375463 Stenanthus heterophyllus Oerst. ex Hanst.；狭花苣苔■☆

375464 Stenaphia A. Rich. = Stephania Lour. ●■

375465 Stenaria（Raf.）Terrell（2001）；窄石竹属■☆

375466 Stenaria Raf. = Houstonia L. ■☆

375467 Stenaria Raf. ex Steud. = Houstonia L. ■☆

375468 Stenaria nigricans（Lam.）Terrell；窄石竹■☆

375469 Stenarrhena D. Don = Salvia L. ●■

375470 Stengelia Neck. = Mourera Aubl. ■☆

375471 Stengelia Sch. Bip. = Stengelia Sch. Bip. ex Steetz ■☆

375472 Stengelia Sch. Bip. = Vernonia Schreb.（保留属名）●■

375473 Stengelia Sch. Bip. ex Steetz = Baccharoides Moench ●■

375474 Stengelia Sch. Bip. ex Steetz（1841）；斯滕菊属■☆

375475 Stengelia adoensis Sch. Bip.；斯滕菊●☆

375476 Stengelia calvoana Hook. f. = Vernonia calvoana（Hook. f.）Hook. f. ●☆

375477 Stengelia insignis Hook. f. = Vernonia calvoana（Hook. f.）Hook. f. ●☆

375478 Stenhammaria Nyman = Mertensia Roth（保留属名）■

375479 Stenhammaria Nyman = Steenhammera Rchb. ■

375480 Stenia Lindl.（1837）；狭团兰属■☆

375481 Stenia angustilabia D. E. Benn. et Christenson；窄叶狭团兰■☆

375482 Stenia caudata（Ackerman）Dodson et D. E. Benn.；尾状狭团兰■☆

375483 Stenia pallida Lindl.；苍白狭团兰■☆

375484 Stenocactus（K. Schum.）A. Berger（1929）；薄棱玉属（多棱球属）●☆

375485 Stenocactus（K. Schum.）A. W. Hill = Stenocactus（K. Schum.）A. Berger ●☆

375486 Stenocactus A. Berger = Echinofossulocactus Lawr. ■

375487 Stenocactus A. Berger = Stenocactus（K. Schum.）A. Berger ●☆

375488 Stenocactus albatus（A. Dietr.）F. M. Knuth = Echinocactus albatus A. Dietr. ●☆

375489 Stenocactus coptonogonus A. Berger = Echinofossulocactus coptonogonus（Lem.）Lawr. ■☆

375490 Stenocactus crispatus（DC.）A. Berger = Echinocactus crispatus DC. ●☆

375491 Stenocactus crispatus（DC.）A. Berger = Ferocactus crispatus（DC.）N. P. Taylor ■☆

375492 Stenocactus hastatus A. Berger = Echinocactus hastatus Hopffer ●☆

375493 Stenocactus lamellosus A. Berger；龙舌玉●☆

375494 Stenocactus multicostatus A. Berger；多脉薄棱玉；Brain Cactus ■☆

375495 Stenocactus multicostatus A. Berger 'Zacatecasensis'；洒卡特卡斯薄棱玉；Brain Cactus，Wave Cactus ●☆

375496 Stenocactus obvallatus A. Berger；有栅玉（瑞晃龙，太刀岚）■☆

375497 Stenocactus pentacanthus（Lem.）A. Berger = Echinocactus pentacanthus Lem. ●☆

375498 Stenocactus zacatecasensis A. Berger = Echinofossulocactus zacatecasensis Britton et Rose ■

375499 Stenocaelium Benth. et Hook. f. = Stenocoelium Ledeb. ■

375500 Stenocaelium Ledeb. et Hook. f. = Stenocoelium Ledeb. ■

375501 Stenocalyx O. Berg = Eugenia L. ●

375502 Stenocalyx Turcz. = Diplopterys A. Juss. ●☆

375503 Stenocalyx Turcz. = Mezia Schwacke ex Nied. ●☆

375504 Stenocalyx involuta Turcz. = Diplopterys involuta Nied. ●☆

375505 Stenocalyx michelii O. Berg = Eugenia uniflora L. ●

375506 Stenocalyx uniflorus（L.）Kausel = Eugenia uniflora L. ●

375507 Stenocarpha S. F. Blake（1915）；狭苞菊属■☆

375508 Stenocarpha filiformis S. F. Blake；狭苞菊■☆

375509　Stenocarpha filipes S. F. Blake；<u>丝梗狭苞菊</u>■☆

375510　Stenocarpus R. Br.（1810）（保留属名）；火轮树属（狭果树属）；Fircwhcel Tree，Fire-wheel Tree ●☆

375511　Stenocarpus salignus R. Br.；柳状火轮树（柳叶火焰树）；Beef Wood，Beefsteak，Red Silkyoak，Red Sliky Oak，Scrub Beefwood，Willow Fircwheel Tree，Willow Fire-wheel Tree ●☆

375512　Stenocarpus sinuatus（Loudon）Endl.；深波火轮树（波叶狭果树，火轮木）；Australian Fireweheel Tree，Fire Tree，Fire Wheel Tree，Firewheel Tree，Fire-wheel Tree，Queensland Fire-wheel Tree，Queenssland Firewheel Tree ●☆

375513　Stenocephalum Sch. Bip.（1863）；窄头斑鸠菊属■☆

375514　Stenocephalum Sch. Bip. = Vernonia Schreb.（保留属名）●■

375515　Stenocephalum monticola（Mart. ex DC.）Sch. Bip.；山地窄头斑鸠菊■☆

375516　Stenocereus（A. Berger）Riccob.（1909）（保留属名）；狭花柱属（新绿柱属）；Pitaya ●☆

375517　Stenocereus（A. Berger）Riccob. = Lemaireocereus Britton et Rose ●☆

375518　Stenocereus Riccob. = Stenocereus（A. Berger）Riccob.（保留属名）●☆

375519　Stenocereus alamosensis（J. M. Coult.）A. C. Gibson et K. E. Horak；阿拉莫斯狭花柱；Octopus Cactus ●☆

375520　Stenocereus beneckei（Ehrenb.）A. Berger et Buxb.；雷斧阁 ●☆

375521　Stenocereus dumortieri（Scheidw.）Buxb.；碧塔（杜氏新绿柱）●☆

375522　Stenocereus eruca（Brandegee）A. C. Gibson et K. E. Horak；虫狭花柱；Chirinole，Creeping Devil，Creeping Devil Cactus ●☆

375523　Stenocereus griseus（Haw.）Buxb.；灰狭花柱；Pitayo De Mayo ●☆

375524　Stenocereus gummosus（Engelm.）A. C. Gibson et K. E. Horak；产胶狭花柱；Pitaya Agria ●☆

375525　Stenocereus kerberi（K. Schum.）A. C. Gibson et K. E. Horak；克伯尔狭花柱●☆

375526　Stenocereus longispinus（Britton et Rose）Buxb.；长角狭花柱（白云角）●☆

375527　Stenocereus marginatus（DC.）A. Berger et Buxb. = Marginatocereus marginatus（DC.）Backeb. ●☆

375528　Stenocereus martinezii（J. G. Ortega）Bravo；马丁内斯狭花柱 ●☆

375529　Stenocereus montanus（Britton et Rose）Buxb.；山地狭花柱；Mountain Organ Pipe，Pitaya Colorada，Saguira ●☆

375530　Stenocereus pruinosus（Otto）Buxb.；朝雾阁（白粉狭花柱）；Gray Ghost Organ Pipe，Pitayo ●☆

375531　Stenocereus queretaroensis（F. A. C. Weber）Buxb.；克雷塔罗狭花柱；Pitahaya De Queretaro ●☆

375532　Stenocereus standleyi（Ortega）Buxb.；斯坦德狭花柱●☆

375533　Stenocereus stellatus Riccob.；新绿柱（星状狭花柱）；Jonocostle，Pitaya，Xoconochtli ●☆

375534　Stenocereus thurberi（Engelm.）Buxb.；大王阁（茶柱）；Organ Pipe Cactus，Pitahaya Dulce ●☆

375535　Stenochasma Griff. = Hornstedtia Retz. ■

375536　Stenochasma Miq. = Broussonetia L'Hér. ex Vent.（保留属名）●

375537　Stenochilum Willd. ex Gham. et Schltdl. = Lamourouxia Kunth ■☆

375538　Stenochilus Post et Kuntze = Lamourouxia Kunth ■☆

375539　Stenochilus Post et Kuntze = Stenochilum Willd. ex Chum. et Schltdl. ■☆

375540　Stenochilus R. Br. = Eremophila R. Br. ●☆

375541　Stenochloa Nutt. = Dissanthelium Trin. ■☆

375542　Stenocline DC.（1838）；多头鼠麴木属●☆

375543　Stenocline bracteifera DC. = Helichrysum bracteiferum（DC.）Humbert ●☆

375544　Stenocline ericoides DC.；多头鼠麴木●☆

375545　Stenocline ferruginea Baker = Helichrysum myriocephalum Humbert ●☆

375546　Stenocline filaginoides DC. = Helichrysum filaginoides（DC.）Humbert ●☆

375547　Stenocline fruticosa Baker = Helichrysum baronii Humbert ●☆

375548　Stenocline gymnocephala DC. = Helichrysum gymnocephalum（DC.）Humbert ●☆

375549　Stenocline incana Baker = Helichrysum gymnocephalum（DC.）Humbert ●☆

375550　Stenocline tomentosula Klatt = Helichrysum tomentosulum（Klatt）Merxm. ●☆

375551　Stenocoelium Ledeb.（1829）；狭腔芹属（细腔属）；Stenocoelium ■

375552　Stenocoelium athamantoides（M. Bieb.）Ledeb.；狭腔芹；Common Stenocoelium ■

375553　Stenocoelium divaricatum Turcz. = Saposhnikovia divaricata（Turcz.）Schischk. ■

375554　Stenocoelium popovii V. M. Vinogr. et Fedor.；波氏狭腔芹■

375555　Stenocoelium trichocarpum Schrenk；毛果狭腔芹；Hairfruit Stenocoelium ■

375556　Stenocoelium villosum（Turcz. ex Fisch. et C. A. Mey.）Koso-Pol. = Phlojodicarpus villosus（Turcz. ex Fisch. et C. A. Mey.）Turcz. ex Ledeb. ■

375557　Stenocoelium villosum（Turcz.）Koso-Pol. = Phlojodicarpus villosus（Turcz. ex Fisch. et C. A. Mey.）Turcz. ex Ledeb. ■

375558　Stenocoryne Lindl.（1843）；狭棒兰属■☆

375559　Stenocoryne Lindl. = Bifrenaria Lindl. ■☆

375560　Stenocoryne longicornis Lindl.；狭棒兰■☆

375561　Stenodiptera Koso-Pol. = Caropodium Stapf et Wettst. ■☆

375562　Stenodiscus Reissek = Spyridium Fenzl ●☆

375563　Stenodon Naudin（1844）；细齿野牡丹属☆

375564　Stenodon suberosus Naudin；细齿野牡丹☆

375565　Stenodraba O. E. Schulz = Weberbauera Gilg et Muschl. ■☆

375566　Stenodrepanum Harms（1921）；窄镰苏木属（阿根廷苏木属）■☆

375567　Stenodrepanum bergii Harms；窄镰苏木■☆

375568　Stenofestuca（Honda）Nakai = Bromus L.（保留属名）■

375569　Stenofestuca pauciflora（Thunb.）Nakai = Bromus remotiflorus（Steud.）Ohwi ■

375570　Stenogastra Hanst. = Almana Raf. ●■☆

375571　Stenogastra Hanst. = Sinningia Nees ●■☆

375572　Stenoglossum Kunth = Epidendrum L.（保留属名）■☆

375573　Stenoglottis Lindl.（1837）；狭舌兰属■☆

375574　Stenoglottis calcarata Rchb. f. = Cynorkis anacamptoides Kraenzl. ■☆

375575　Stenoglottis fimbriata Lindl.；狭舌兰■☆

375576　Stenoglottis fimbriata Lindl. var. saxicola Schltr. ex Kraenzl. = Stenoglottis fimbriata Lindl. ■☆

375577　Stenoglottis longifolia Hook. f.；长叶狭舌兰■☆

375578　Stenoglottis macloughlinii（L. Bolus）G. McDonald = Stenoglottis woodii Schltr. ■☆

375579 Stenoglottis woodii Schltr. ;伍得狭舌兰■☆

375580 Stenoglottis zambesiaca Rolfe;赞比西狭舌兰■☆

375581 Stenogonum Nutt. (1848);双轮蓼属;Two-whorl Buckwheat ■☆

375582 Stenogonum Nutt. = Eriogonum Michx. ●■☆

375583 Stenogonum flexum (M. E. Jones) Reveal et J. T. Howell;弯双轮蓼;Bent Two-whorl Buckwheat ■☆

375584 Stenogonum salsuginosum Nutt. ;光双轮蓼;Smooth Two-whorl Buckwheat ■☆

375585 Stenogtossum Kunth = Epidendrum L. (保留属名)■☆

375586 Stenogyne Benth. (1830) (保留属名);狭蕊藤属■☆

375587 Stenogyne Cass. = Eriocephalus L. ●☆

375588 Stenogyne alba H. St. John;白狭蕊藤■☆

375589 Stenogyne biflora (Sherff) H. St. John;双花狭蕊藤■☆

375590 Stenolirion Baker = Ammocharis Herb. ■☆

375591 Stenolobium Benth. = Calopogonium Desv. ●

375592 Stenolobium D. Don = Cybistax Mart. ex Meisn. ●☆

375593 Stenolobium D. Don = Tecoma Juss. ●

375594 Stenolobium brachycarpum Benth. = Calopogonium mucunoides Desv. ●

375595 Stenolobium stans (L.) D. Don = Tecoma stans (L.) Juss. ex Kunth ●☆

375596 Stenoloma Cass. = Centaurea L. (保留属名)●■

375597 Stenomeria Turcz. (1852);块茎藤属■☆

375598 Stenomeria decalepis Turcz. ;块茎藤■☆

375599 Stenomeridaceae J. Agardh(1858) (保留科名);块茎藤科(丝瓣藤科)■☆

375600 Stenomeridaceae J. Agardh(保留科名) = Dioscoreaceae R. Br. (保留科名)●■

375601 Stenomeridaceae J. Agardh(保留科名) = Sterculiaceae Vent. (保留科名)●■

375602 Stenomeris Planch. (1852);多子薯蓣属■☆

375603 Stenomeris dioscoreifolia Planch. ;多子薯蓣■☆

375604 Stenomesson Herb. (1821);狭管石蒜属(狭管蒜属)■☆

375605 Stenomesson elwesii (Baker) J. F. Macbr. ;埃尔威斯狭管石蒜■☆

375606 Stenomesson flavum Herb. ;黄色狭管石蒜(黄色狭管蒜)■☆

375607 Stenomesson incarnatum Baker;肉色狭管石蒜(肉色狭管蒜);Stenomesson ■☆

375608 Stenomesson incarum Kraenzl. = Stenomesson variegatum (Ruiz et Pav.) J. F. Macbr. ■☆

375609 Stenomesson miniatum (Herb.) Ravenna;出蕊狭管石蒜(出蕊狭管蒜)■☆

375610 Stenomesson variegatum (Ruiz et Pav.) J. F. Macbr. ;五彩狭管石蒜(五彩狭管蒜)■☆

375611 Stenonema Hook. = Draba L. ■

375612 Stenonema Hook. ex Benth. et Hook. f. = Dolichostylis Turcz. ■

375613 Stenonema Hook. ex Benth. et Hook. f. = Draba L. ■

375614 Stenonia Baill. = Cleistanthus Hook. f. ex Planch. ●

375615 Stenonia Baill. = Stenoniella Kuntze ●

375616 Stenonia Didr. = Argythamnia P. Browne ●☆

375617 Stenonia Didr. = Ditaxis Vahl ex A. Juss. ●☆

375618 Stenoniella Kuntze = Cleistanthus Hook. f. ex Planch. ●

375619 Stenoniella Post et Kuntze = Cleistanthus Hook. f. ex Planch. ●

375620 Stenopadus S. F. Blake(1931);绛菊木属●☆

375621 Stenopadus affinis Maguire, Steyerm. et Wurdack;近缘绛菊木●☆

375622 Stenopadus talaumifolius S. F. Blake;绛菊木●☆

375623 Stenopetalum R. Br. ex DC. (1821);狭瓣芥属■☆

375624 Stenopetalum album E. Pritz. ;白狭瓣芥■☆

375625 Stenopetalum australis C. Muell. ;澳洲狭瓣芥■☆

375626 Stenopetalum filifolium Benth. ;线叶狭瓣芥■☆

375627 Stenopetalum gracile Bunge;细狭瓣芥■☆

375628 Stenopetalum robustum Endl. ;粗壮狭瓣芥■☆

375629 Stenopetalum velutinum F. Muell. ;黏狭瓣芥■☆

375630 Stenophalium Anderb. (1991);光果彩鼠麹属■☆

375631 Stenophalium chionaeum (DC.) Anderb. ;光果彩鼠麹■☆

375632 Stenophragma Celak. = Arabidopsis Heynh. (保留属名)■

375633 Stenophragma Celak. = Arabis L. ●■

375634 Stenophragma glandulosum (Kar. et Kir.) B. Fedtsch. = Dontostemon glandulosus (Kar. et Kir.) O. E. Schulz ■

375635 Stenophragma griffithianum (Boiss.) B. Fedtsch. = Olimarabidopsis pumila (Stephan) Al-Shehbaz, O'Kane et R. A. Price ■

375636 Stenophragma halophilum (C. A. Mey.) B. Fedtsch. = Thellungiella halophila (C. A. Mey.) O. E. Schulz ■

375637 Stenophragma mollipilum (Maxim.) B. Fedtsch. = Sisymbriopsis mollipila (Maxim.) Botsch. ■

375638 Stenophragma mollissimum (C. A. Mey.) B. Fedtsch. = Crucihimalaya mollissima (C. A. Mey.) Al-Shehbaz, O'Kane et R. A. Price ■

375639 Stenophragma nudum (Bél.) B. Fedtsch. = Drabopsis nuda (Bél.) Stapf ■

375640 Stenophragma nudum (Bél.) B. Fedtsch. = Drabopsis verna K. Koch ■

375641 Stenophragma parvulum (Schrenk) B. Fedtsch. = Thellungiella parvula (Schrenk) Al-Shehbaz et O'Kane ■

375642 Stenophragma pumilum (Stephan) B. Fedtsch. = Olimarabidopsis pumila (Stephan) Al-Shehbaz, O'Kane et R. A. Price ■

375643 Stenophragma salusgineum (Pall.) Prantl = Thellungiella salsuginea (Pall.) O. E. Schulz ■

375644 Stenophragma thalianum (L.) Celak. = Arabidopsis thaliana (L.) Heynh. ■

375645 Stenophragma toxophyllum (M. Bieb.) B. Fedtsch. = Pseudoarabidopsis toxophylla (M. Bieb.) Al-Shehbaz, O'Kane et R. A. Price ■

375646 Stenophyllum Sch. Bip. ex Benth. et Hook. f. = Calea L. ●■☆

375647 Stenophyllus Raf. (废弃属名) = Bulbostylis Kunth(保留属名)■☆

375648 Stenophyllus capillaris (L.) Britton = Bulbostylis capillaris (L.) Kunth ex C. B. Clarke ■☆

375649 Stenophyllus capillaris (L.) Britton = Scirpus capillaris L. ■☆

375650 Stenophyllus carteri Britton = Bulbostylis ciliatifolia (Elliott) Torr. var. coarctata (Elliott) Král ■☆

375651 Stenophyllus cespitosus (Muhl.) Raf. = Bulbostylis stenophylla (Elliott) C. B. Clarke ■☆

375652 Stenophyllus ciliatifolius (Elliott) C. Mohr = Bulbostylis ciliatifolia (Elliott) Fernald ■☆

375653 Stenophyllus coarctatus (Elliott) Britton = Bulbostylis ciliatifolia (Elliott) Torr. var. coarctata (Elliott) Král ■☆

375654 Stenophyllus floridanus Britton ex Nash = Bulbostylis barbata (Rottb.) C. B. Clarke ■

375655 Stenophyllus warei (Torr.) Britton = Bulbostylis warei (Torr.) C. B. Clarke ■☆

375656　Stenopolen Raf. = Stenia Lindl. ■☆

375657　Stenops B. Nord. (1978);窄叶菊属■☆

375658　Stenops helodes B. Nord. ;窄叶菊■☆

375659　Stenops zairensis (Lisowski) B. Nord. ;扎伊尔窄叶菊■☆

375660　Stenoptera C. Presl(1827);狭翅兰属☆

375661　Stenoptera brachystachys (Schltr.) L. O. Williams;短穗狭翅兰■☆

375662　Stenoptera ciliaria C. Schweinf. ;睫毛狭翅兰■☆

375663　Stenoptera elata Schltr. ;高狭翅兰■☆

375664　Stenoptera elegans Kraenzl. ;雅致狭翅兰■☆

375665　Stenoptera gracilis (Schltr.) L. O. Williams;细狭翅兰■☆

375666　Stenoptera lancipetala (Schltr.) Garay;剑瓣狭翅兰■☆

375667　Stenoptera laxiflora C. Schweinf. ;疏花狭翅兰■☆

375668　Stenoptera longifolia Rolfe;长叶狭翅兰■☆

375669　Stenoptera macrostachya Rchb. f. ;大穗狭翅兰■☆

375670　Stenoptera montana C. Schweinf. ;山地狭翅兰■☆

375671　Stenoptera parviflora (C. Schweinf.) C. Schweinf. ;小花狭翅兰■☆

375672　Stenoptera viscosa Rchb. f. ;毛狭翅兰■☆

375673　Stenorhynchus Lindl. = Stenorhyncus Lindl. ■☆

375674　Stenorhynchus Lindl. = Stenorrhynchos Rich. ex Spreng. ■☆

375675　Stenorhyncus Lindl. = Stenorrhynchos Rich. ex Spreng. ■☆

375676　Stenorrhynchium Rchb. = Stenorrhynchos Rich. ex Spreng. ■☆

375677　Stenorrhynchos Rchb. = Stenorrhynchos Rich. ex Spreng. ■☆

375678　Stenorrhynchos Rich. ex Spreng. (1826);狭喙兰属■☆

375679　Stenorrhynchos Spreng. = Stenorrhynchos Rich. ex Spreng. ■☆

375680　Stenorrhynchos apetalum Kraenzl. ;无瓣狭喙兰■☆

375681　Stenorrhynchos aphyllum Lindl. ;无叶狭喙兰■☆

375682　Stenorrhynchos australe Lindl. ;澳洲狭喙兰■☆

375683　Stenorrhynchos calcaratus (Sw.) Rich. = Eltroplectris calcarata (Sw.) Garay et H. R. Sweet ■☆

375684　Stenorrhynchos cinnabarinum (La Llave et Lex.) Lindl. var. paludicola (Luer) W. J. Schrenk;沼泽剑叶狭喙兰;Swamp beaked orchid ■☆

375685　Stenorrhynchos cinnabarinus (La Llave et Lex.) Lindl. = Dichromanthus cinnabarinus (La Llave et Lex.) Garay ■☆

375686　Stenorrhynchos densiflorum (C. Schweinf.) Szlach. ;密花狭喙兰■☆

375687　Stenorrhynchos densum Hauman;密集狭喙兰■☆

375688　Stenorrhynchos flavum (Sw.) Spreng. ;黄狭喙兰■☆

375689　Stenorrhynchos foliosum Schltr. ;多叶狭喙兰■☆

375690　Stenorrhynchos lanceolatum (Willd.) Rich. ;披针叶狭喙兰(剑叶囊唇兰);Leafless Beaked Orchid ■☆

375691　Stenorrhynchos lanceolatum Rich. = Stenorrhynchos lanceolatum (Willd.) Rich. ■☆

375692　Stenorrhynchos laxum Poepp. et Endl. ;松散狭喙兰■☆

375693　Stenorrhynchos longifolium Cogn. ;长叶狭喙兰■☆

375694　Stenorrhynchos macranthum (Rchb. f.) Cogn. ;大花狭喙兰■☆

375695　Stenorrhynchos montanum Lindl. ;山地狭喙兰■☆

375696　Stenorrhynchos pauciflorum Rchb. f. ;少花狭喙兰■☆

375697　Stenorrhynchos pilosum Cogn. ;柔毛狭喙兰■☆

375698　Stenorrhynchos polystachyon (Sw.) Spreng. = Tropidia polystachya (Sw.) Ames ■☆

375699　Stenorrhynchos speciosum (Jacq.) Spreng. ;狭喙兰■☆

375700　Stenorrhynchos squamulosum (Kunth) Spreng. ;灰白狭喙兰;Hoary Beaked Orchid ■☆

375701　Stenorrhynchos squamulosum (Kunth) Spreng. = Sacoila squamulosa (Kunth) Garay ■☆

375702　Stenorynchus Rich. = Stenorrhynchos Rich. ex Spreng. ■☆

375703　Stenoschista Bremek. = Ruellia L. ■●

375704　Stenoschista togoensis (Lindau) Bremek. = Ruellia togoensis (Lindau) Heine ■☆

375705　Stenoselenium Popov = Stenosolenium Turcz. ■

375706　Stenosemis E. Mey. ex Harv. et Sond. = Annesorhiza Cham. et Schltdl. ■☆

375707　Stenosemis caffra (Eckl. et Zeyh.) Sond. = Annesorhiza caffra (Eckl. et Zeyh.) Schönland ■☆

375708　Stenosemis caffra (Eckl. et Zeyh.) Sond. = Krubera caffra Eckl. et Zeyh. ■☆

375709　Stenosepala C. Perss. (2000);狭萼茜属☆

375710　Stenoseris C. Shih(1991);细莴苣属;Stenoseris ■

375711　Stenoseris auriculiformis C. Shih;抱茎细莴苣(耳叶细莴苣);Ear-leaved Stenoseris ■

375712　Stenoseris graciliflora (Wall. ex DC.) C. Shih;细莴苣(细花莴苣);Thin Stenoseris ■

375713　Stenoseris leptantha C. Shih;景东细莴苣;Jingdong Stenoseris ■

375714　Stenoseris taliensis (Franch.) C. Shih;大理细莴苣;Dali Stenoseris ■

375715　Stenoseris tenuis C. Shih;全叶细莴苣(三花盘果菊);Tenuis Stenoseris ■

375716　Stenoseris triflora C. C. Chang et C. Shih;栉齿细莴苣;Threeflower Stenoseris ■

375717　Stenosiphanthus A. Samp. = Arrabidaea DC. ●☆

375718　Stenosiphon Spach(1835);窄管柳叶菜属■☆

375719　Stenosiphon linifolius (Nutt.) Heynh. ;窄管柳叶菜■☆

375720　Stenosiphonium Nees(1832);窄管爵床属■☆

375721　Stenosiphonium confertum Nees;窄管爵床■☆

375722　Stenosiphonium diandrum Wight;二蕊窄管爵床■☆

375723　Stenosiphonium parviflorum T. Anderson;小花窄管爵床■☆

375724　Stenosolen (Müll. Arg.) Markgr. = Tabernaemontana L. ●

375725　Stenosolenium Turcz. (1840);紫筒草属(狭管紫草属);Stenosolenium ■

375726　Stenosolenium saxatile (Pall.) Turcz. ;紫筒草(白毛草,伏地蜈蚣草,紫根根);Cliff Stenosolenium ■

375727　Stenospermation Schott(1858);窄籽南星属■☆

375728　Stenospermation angustifolium Hemsl. ;窄叶窄籽南星■☆

375729　Stenospermation arborescens Madison;树状窄籽南星■☆

375730　Stenospermation brachypodum Sodiro;短梗窄籽南星■☆

375731　Stenospermation dictyoneurum Croat et Acebey;指脉窄籽南星■☆

375732　Stenospermation flavescens Engl. ;浅黄窄籽南星■☆

375733　Stenospermation flavum Croat et D. C. Bay;黄窄籽南星■☆

375734　Stenospermation gracile Sodiro;细窄籽南星■☆

375735　Stenospermation latifolium Engl. ;宽叶窄籽南星■☆

375736　Stenospermation longifolium Engl. ;长叶窄籽南星■☆

375737　Stenospermation maximum Engl. ;大窄籽南星■☆

375738　Stenospermatium Schott = Stenospermation Schott ☆

375739　Stenospermum Sweet = Kunzea Rchb. (保留属名)●☆

375740　Stenospermum Sweet = Metrosideros Banks ex Gaertn. (保留属名)●☆

375741　Stenosperrnum Sweet ex Heynh. = Kunzea Rchb. (保留属名)●☆

375742　Stenostachys Turcz. (1862);狭穗草属■☆

375743　Stenostachys Turcz. = Hystrix Moench ■

375744　Stenostachys gracilis（Hook. f.）Connor;细狭穗草■☆

375745　Stenostachys laevis（Petrie）Connor;平滑狭穗草■☆

375746　Stenostachys narduroides Turcz.;狭穗草■☆

375747　Stenostegia A. R. Bean（1998）;狭盖桃金娘属●☆

375748　Stenostegia congesta A. R. Bean;狭盖桃金娘●☆

375749　Stenostelma Schltr.（1894）;细冠萝藦属■☆

375750　Stenostelma capense Schltr.;好望角细冠萝藦■☆

375751　Stenostelma corniculatum（E. Mey.）Bullock;圆锥细冠萝藦■☆

375752　Stenostelma eminens（Harv.）Bullock = Asclepias eminens（Harv.）Schltr.■☆

375753　Stenostelma umbelluliferum（Schltr.）S. P. Bester et Nicholas;伞花细冠萝藦■☆

375754　Stenostephanus Nees（1847）;窄冠爵床属■☆

375755　Stenostephanus atropurpureus（Lindau）J. R. I. Wood;暗紫窄冠爵床■☆

375756　Stenostephanus bolivianus Rusby;玻利维亚窄冠爵床■☆

375757　Stenostephanus lasiostachyus Nees;毛穗窄冠爵床■☆

375758　Stenostephanus laxus（Wassh.）Wassh.;松散窄冠爵床■☆

375759　Stenostomum C. F. Gaertn. = Antirhea Comm. ex Juss. ●

375760　Stenotaenia Boiss.（1844）;窄带芹属■☆

375761　Stenotaenia daralaghezica（Takht.）Schischk.;窄带芹■☆

375762　Stenotalis B. G. Briggs et L. A. S. Johnson（1998）;窄花帚灯草属（寡小花帚灯草属）■☆

375763　Stenotalis ramosissima（Gilg）B. G. Briggs et L. A. S. Johnson;窄花帚灯草■☆

375764　Stenotaphrum Trin.（1820）;钝叶草属（窄沟草属）;Bluntleaf grass,Stenotaphrum ■

375765　Stenotaphrum americanum Schrank;美洲钝叶草■☆

375766　Stenotaphrum americanum Schrank = Stenotaphrum secundatum（Walter）Kuntze ■

375767　Stenotaphrum dimidiatum（L.）Brongn.;光钝叶草;Glabrous Stenotaphrum ■

375768　Stenotaphrum diplotaphrum Pilg. = Stenotaphrum micranthum（Desv.）C. E. Hubb. ■

375769　Stenotaphrum glabrum Trin. = Stenotaphrum dimidiatum（L.）Brongn. ■

375770　Stenotaphrum helferi Munro ex Hook. f.;钝叶草（鸭口草,苡米草,薏米草）;Helfer Bluntleaf grass,Helfer Stenotaphrum ■

375771　Stenotaphrum lepturoides Hensl. = Stenotaphrum micranthum（Desv.）C. E. Hubb. ■

375772　Stenotaphrum madagascariense Kunth = Stenotaphrum dimidiatum（L.）Brongn. ■

375773　Stenotaphrum micranthum（Desv.）C. E. Hubb.;锥穗钝叶草;Awlshaped Stenotaphrum,Sharpspike Bluntleaf Grass ■

375774　Stenotaphrum secundatum（Walter）Kuntze;侧钝叶草（钝叶草）;Buffalo Grass,Saint Augustine Grass,St. Augustine Grass ■

375775　Stenotaphrum secundatum（Walter）Kuntze 'Variegatum';条纹钝叶草（白纹钝叶草）;Variegated St. Augustine Grass ■☆

375776　Stenotaphrum subulatum Trin. = Stenotaphrum micranthum（Desv.）C. E. Hubb. ■

375777　Stenotheca Monn. = Hieracium L. ■

375778　Stenotheca tristis（Willd. ex Spreng.）Schljakov = Hieracium triste Willd. ex Spreng. ■☆

375779　Stenothyrsus C. B. Clarke（1908）;细茎爵床属■☆

375780　Stenothyrsus ridleyi C. B. Clarke;细茎爵床■☆

375781　Stenotis Terrell = Hedyotis L.（保留属名）●■

375782　Stenotium Presl ex Steud. = Lobelia L. ●■

375783　Stenotopsis Rydb. = Ericameria Nutt. ●☆

375784　Stenotopsis Rydb. = Haplopappus Cass.（保留属名）■●☆

375785　Stenotropis Hassk. = Erythrina L. ●■

375786　Stenotus Nutt.（1840）;窄黄花属;Goldenweed Goldenweed,Mock Goldenweed ■☆

375787　Stenotus Nutt. = Haplopappus Cass.（保留属名）■●☆

375788　Stenotus acaulis（Nutt.）Nutt.;无茎窄黄花;Stemless Goldenweed ■☆

375789　Stenotus acaulis（Nutt.）Nutt. var. kennedyi Jeps. = Stenotus acaulis（Nutt.）Nutt. ■☆

375790　Stenotus andersonii Rydb. = Stenotus lanuginosus（A. Gray）Greene var. andersonii（Rydb.）C. A. Morse ■☆

375791　Stenotus armerioides Nutt.;海石竹窄黄花;Thrifty Goldenweed ■☆

375792　Stenotus armerioides Nutt. var. gramineus（S. L. Welsh et F. J. Sm.）Kartesz et Gandhi;禾状海石竹窄黄花■☆

375793　Stenotus falcatus Rydb. = Stenotus acaulis（Nutt.）Nutt. ■☆

375794　Stenotus lanuginosus（A. Gray）Greene;毛窄黄花;Woolly Goldenweed ■☆

375795　Stenotus lanuginosus（A. Gray）Greene var. andersonii（Rydb.）C. A. Morse;安氏窄黄花■☆

375796　Stenotus macleanii（Brandegee）A. Heller = Nestotus macleanii（Brandegee）Urbatsch,R. P. Roberts et Neubig ●☆

375797　Stenotus multicaulis Nutt. = Oonopsis multicaulis（Nutt.）Greene ■☆

375798　Stenotus pygmaeus Torr. et A. Gray = Tonestus pygmaeus（Torr. et A. Gray）A. Nelson ☆

375799　Stenotus stenophyllus（A. Gray）Greene = Nestotus stenophyllus（A. Gray）Urbatsch,R. P. Roberts et Neubig ●☆

375800　Stenotyla Dressler = Chondrorhyncha Lindl. ■☆

375801　Stenotyla Dressler（2005）;狭节兰属■☆

375802　Stenouratea Tiegh. = Ouratea Aubl.（保留属名）●

375803　Stenurus Salisb. = Biarum Schott（保留属名）■☆

375804　Stephalea Raf. = Campanula L. ■●

375805　Stephanachne Keng（1934）;冠毛草属;Pappcsedge,Stephanachne ■

375806　Stephanachne monandra（P. C. Kuo et S. L. Lu）P. C. Kao et S. L. Lu;单蕊冠毛草;Onestamen Stephanachne,Singleanther Papposedge ■

375807　Stephanachne nigrescens Keng;黑穗茅;Blackspike Papposedge,Blackspike Stephanachne ■

375808　Stephanachne nigrescens Keng var. monandra P. C. Kuo et S. L. Lu = Stephanachne monandra（P. C. Kuo et S. L. Lu）P. C. Kao et S. L. Lu ■

375809　Stephanachne pappophorea（Hack.）Keng;冠毛草（索草）;Common Papposedge,Common Stephanachne ■

375810　Stephanachne pappophorea（Hack.）Keng var. monandra P. C. Kuo et S. L. Lu = Stephanachne monandra（P. C. Kuo et S. L. Lu）P. C. Kao et S. L. Lu ■

375811　Stephanandra Siebold et Zucc.（1843）;小米空木属（冠蕊木属,野珠兰属）;Lace Shrub,Stephanandra ●

375812　Stephanandra chinensis Hance;华空木（滴滴金,凤尾米筛花,野珠兰,中国小米空木）;China Stephanandra,Chinese Stephanandra ●

375813　Stephanandra flexuosa Siebold et Zucc. = Stephanandra incisa

(Thunb. ex A. Murray) Zabel ●

375814 Stephanandra flexuosa Siebold et Zucc. var. chinensis (Hance) Pamp. = Stephanandra chinensis Hance ●

375815 Stephanandra flexuosa Siebold et Zucc. var. chinensis Pamp. = Stephanandra chinensis Hance ●

375816 Stephanandra gracilis Franch. et Sav. ;细野珠兰●☆

375817 Stephanandra incisa (Thunb. ex A. Murray) Zabel;小米空木（冠蕊木,檬子树青阳,稀米菜,小野珠兰）;Catleaf Stephanandra, Cat-leaved Stephanandra, Cut-leaf Stephanandra, Cut-leaf Stephanandre, Cutwafed Stephanandra, Lace Shrub, Laceshrub ●

375818 Stephanandra incisa (Thunb.) Zabel 'Crispa';波缘叶小米空木;Cutleaf Stephanandra, Dwarf Cutleaf Stephanandra, Lace Shrub ●☆

375819 Stephanandra incisa (Thunb.) Zabel = Stephanandra incisa (Thunb. ex A. Murray) Zabel ●

375820 Stephanandra incisa (Thunb.) Zabel var. macrophylla Hid. Takah. ;大叶小野珠兰●☆

375821 Stephanandra tanakae Franch. et Sav. ;日本小米空木（日本野珠兰,田中氏野珠兰）;Tanaka Stephanandra ●☆

375822 Stephanangaceae Dulac = Valerianaceae Batsch(保留科名)●■

375823 Stephananthus Lehm. = Baccharis L. (保留属名)●■☆

375824 Stephanella (Engl.) Tiegh. = Dichapetalum Thouars ●

375825 Stephanella Tiegh. = Dichapetalum Thouars ●

375826 Stephania Kuntze = Astephania Oliv. ■☆

375827 Stephania Lour. (1790);千金藤属;Stephania ●■

375828 Stephania Willd. = Steriphoma Spreng. (保留属名)●☆

375829 Stephania abyssinica (Quart. -Dill. et A. Rich.) Walp. ;阿比西尼亚千金藤●☆

375830 Stephania abyssinica (Quart.-Dill. et A. Rich.) Walp. var. tomentella (Oliv.) Diels;毛阿比西尼亚千金藤●☆

375831 Stephania brachyandra Diels;白线薯（地不容,短蕊千斤藤,短蕊千金藤,山乌龟）;Short-stemmed Stephania, Whitethread Yam ●■

375832 Stephania brevipedunculata C. Y. Wu et D. D. Tao;短梗地不容（短梗千金藤）;Short-peduncled Stephania, Shortstalk Stephania ●■

375833 Stephania cephalantha Hayata ex Yamam. ;金线吊乌龟（白虾蟆,白药,白药脂,白药子,大还魂,地苦胆,独脚乌桕,荕萎,金丝吊蛤蟆,金线吊鳖,金线吊蛤蟆,金线吊葫芦,盘花地不容,青藤,山乌龟,台湾千金藤,铁秤砣,头花千金藤,细三角藤,玉笑葛藤）;Oriental Stephania ●■

375834 Stephania chingtungensis H. S. Lo;景东千金藤（山乌龟）;Jingdong Stephania ■

375835 Stephania cyanantha Welw. ex Hiern;蓝花千金藤■☆

375836 Stephania delavayi Diels;一文钱（白地胆,抱母鸡,地不荣,地不容,地胆,地芙蓉,地乌龟,荷叶暗消,解毒子,金不换,金钱暗消,金钱寒药,金丝荷叶,金线吊乌龟,青藤,山乌龟,藤子暗消,藤子内消,铜钱暗消,铜钱根,乌龟抱蛋,乌龟梢,小寒药,小黑藤）;Delavay Stephania, Farthing Stephania ●■

375837 Stephania delavayi Diels = Stephania epigaea H. S. Lo ●■

375838 Stephania dentifolia H. S. Lo et M. Yang;齿叶地不容;Toothleaf Stephania ■

375839 Stephania dicentrinifera H. S. Lo et M. Yang;荷包地不容（寒地山乌龟）;Pouch Stephania, Stephania ■

375840 Stephania dielsiana C. Y. Wu;血散薯（独角乌桕,金不换,黔桂千金藤,山乌龟,一滴血,一点血）;Diels Stephania ●■

375841 Stephania dinklagei (Engl.) Diels;丁克拉千金藤;Dinklage Stephania ●☆

375842 Stephania dinklagei (Engl.) Diels var. axillaris Troupin =

375843 Stephania dinklagei (Engl.) Diels ●☆

375844 Stephania disciflora Hand. -Mazz. = Stephania cephalantha Hayata ex Yamam. ●■

375845 Stephania dolichopoda Diels;大叶地不容;Bigleaf Stephania ■

375846 Stephania ebracteata S. Y. Zhao et H. S. Lo;川南地不容;Bractless Stephania ■

375847 Stephania elegans Hook. f. et Thomson;雅丽千金藤（千金藤,山豆根,小山豆根,雅致千金藤）;Elegant Stephania ■

375848 Stephania epigaea H. S. Lo;地不容（白地胆,地乌龟,荷叶暗消,金不换,金丝荷叶,山乌龟,乌龟抢蛋）;Epigeal Stephania ●■

375849 Stephania erecta Craib;直立千金藤●☆

375850 Stephania excentrica H. S. Lo;江南地不容（夜牵牛）;Jiangnan Stephania, Slant Stephania ●■

375851 Stephania fastosa Miers = Stephania abyssinica (Quart. -Dill. et A. Rich.) Walp. ●☆

375852 Stephania forsteri (DC.) A. Gray;光千金藤;Forster Stephania ●■

375853 Stephania forsteri (DC.) A. Gray = Stephania japonica (Thunb.) Miers var. timoriensis (DC.) Forman ●■

375854 Stephania glabra (Roxb.) Miers;西藏地不容（光叶地不容,无花千金藤,西藏千金藤）;Glabrous Stephania, Xizang Stephania ■

375855 Stephania glandulifera Miers;具腺千金藤●☆

375856 Stephania gracilenta Miers;纤细千金藤;Fine Stephania ■

375857 Stephania graciliflora Yamam. ;纤花千金藤（荷叶暗消,金钱暗消,藤子暗消,铜钱暗消,小黑藤）;Thin-flower Stephania ●■

375858 Stephania graciliflora Yamam. = Stephania delavayi Diels ●■

375859 Stephania hainanensis H. S. Lo et Y. Tsoong;海南地不容（海南金不换,金不换）;Hainan Stephania ●■

375860 Stephania herbacea Gagnep. ;草质千金藤（铜锣七,乌龟七,乌龟梢,乌龟条）;Herb Stephania, Herbaceous Stephania ●■

375861 Stephania hernandifolia (Willd.) Walp. ;桐叶千金藤（地不容,吊金龟,华千金藤,解毒子,金龟莲草,金线吊乌龟,莲叶桐叶千金藤,毛背千金藤,毛千金藤,牛金藤,千金藤,汝兰,山乌龟,乌龟梢）;Hernandia-leaed Stephania, Hernandialeaf Stephania, Tungleaf Stephania ●■

375862 Stephania hernandifolia (Willd.) Walp. var. tomentella Oliv. = Stephania abyssinica (Quart. -Dill. et A. Rich.) Walp. var. tomentella (Oliv.) Diels ●☆

375863 Stephania hernandiifolia (Willd.) Walp. = Stephania japonica (Thunb.) Miers var. discolor (Blume) Forman ●■

375864 Stephania hernandiifolia (Willd.) Walp. var. discolor (Blume) Miq. = Stephania japonica (Thunb.) Miers var. discolor (Blume) Forman ●■

375865 Stephania hispidula (Yamam.) Yamam. ;毛千金藤●■

375866 Stephania hispidula (Yamam.) Yamam. = Stephania japonica (Thunb.) Miers var. hispidula Yamam. ●■

375867 Stephania hispidula (Yamam.) Yamam. = Stephania longa Lour. ■

375868 Stephania intermedia H. S. Lo;河谷地不容;Intermediate Stephania, Rivervalley Stephania ●■

375869 Stephania japonica (Thunb. ex A. Murray) Miers = Stephania japonica (Thunb.) Miers ●■

375870 Stephania japonica (Thunb.) Miers;千金藤（白药子,爆竹消,朝天膏药,粉防己,刚毛千金藤,公老鼠藤,古藤,合钹草,金盆寒药,金丝荷叶,金线吊青蛙,金线吊乌龟,金线钓乌龟,毛千金藤,日本地不容,山乌龟,天药膏,铁板膏药,土番薯,土广木香,乌虎藤,小青藤,野薯藤,野桃草);Hispidulous Stephania, Japan Stephania,Japanese Stephania ●■

375871 Stephania japonica(Thunb.)Miers var. australis Hatus. = Stephania japonica(Thunb.)Miers ●■

375872 Stephania japonica(Thunb.)Miers var. discolor(Blume)Forman;桐叶千斤藤●■

375873 Stephania japonica(Thunb.)Miers var. discolor(Blume)Forman = Stephania hernandifolia(Willd.)Walp. ●■

375874 Stephania japonica(Thunb.)Miers var. hispidula Yamam. = Stephania hispidula(Yamam.)Yamam. ●■

375875 Stephania japonica(Thunb.)Miers var. hispidula Yamam. = Stephania japonica(Thunb.)Miers ●■

375876 Stephania japonica(Thunb.)Miers var. hispidula Yamam. = Stephania longa Lour. ■

375877 Stephania japonica(Thunb.)Miers var. timoriensis(DC.)Forman;光叶千金藤●■

375878 Stephania japonica(Thunb.)Miers var. timoriensis(DC.)Forman = Stephania forsteri(DC.)A. Gray ●■

375879 Stephania kuinanensis H. S. Lo et M. Yang;桂南地不容(山乌龟);Guinan Stephania ●■

375880 Stephania kwangsiensis H. S. Lo;广西地不容(地不容,金不换,山乌龟);Guangxi Stephania,Kwangsi Stephania ●■

375881 Stephania laetificata(Miers)Benth. = Perichasma laetificata Miers ●☆

375882 Stephania laevigata Miers = Stephania abyssinica(Quart.-Dill. et A. Rich.)Walp. ●☆

375883 Stephania lincangensis H. S. Lo et M. Yang;临沧地不容;Lincang Stephania ■

375884 Stephania longa Lour.;粪箕笃(畚箕草,飞天雷公,粪箕藤,膏药草,蛤蟆草,雷钵嘴,犁壁藤,黎壁叶,七厘,七厘藤,千金藤,田鸡草,铁板膏药草,铁线藤);Long Stephania ■

375885 Stephania longa Lour. = Stephania japonica(Thunb. ex A. Murray)Miers ●■

375886 Stephania longipes H. S. Lo;长柄地不容(长柄千金藤);Longstalk Stephania,Long-stalked Stephania ●■

375887 Stephania macrantha H. S. Lo et M. Yang;大花地不容;Bigflower Stephania ■

375888 Stephania mashanica H. S. Lo et B. N. Chang;马山地不容(山乌龟);Mashan Stephania ●■◇

375889 Stephania merrillii Diels;兰屿千金藤●

375890 Stephania micrantha H. S. Lo et M. Yang;小花地不容(山乌龟);Microflower Stephania,Smallflower Stephania ●■

375891 Stephania mildbraedii Diels;米尔德地不容■☆

375892 Stephania miyiensis S. Y. Zhao et H. S. Lo;米易地不容;Miyi Stephania ■

375893 Stephania officinarum H. S. Lo et M. Yang;药用地不容;Medicine Stephania ■

375894 Stephania pierrei Diels;皮氏千金藤●☆

375895 Stephania praelata Miers = Stephania abyssinica(Quart.-Dill. et A. Rich.)Walp. var. tomentella(Oliv.)Diels ●☆

375896 Stephania rotunda Lour.;圆叶千金藤(独脚乌柏,防己,乌柏茹,乌柏薯);Round-leaf Stephania,Saboo Leard ●

375897 Stephania sasakii Hayata ex Yamam.;台湾千金藤(兰屿千金藤,佐佐木千金藤);Taiwan Stephania ●■

375898 Stephania sasakii Hayata ex Yamam. = Stephania merrillii Diels ●

375899 Stephania sinica Diels;汝兰(地不容,地乌龟,吊金龟,独脚乌柏,华千金藤,金不换,金线吊乌龟,山乌龟,石蟾薯,石琴薯);China Stephania,Chinese Stephania ●■

375900 Stephania suberosa Forman;软木千金藤●☆

375901 Stephania subpeltata H. S. Lo;西南千金藤(高原千金藤);Plateau Stephania,Subpeltate Stephania ●■

375902 Stephania succifera H. S. Lo et Y. Tsoong;小叶地不容(金不换);Littleleaf Stephania, Little-leaved Stephania ●■

375903 Stephania sutchuenensis H. S. Lo;四川千金藤(山豆根,小山豆根);Sichuan Stephania ●■

375904 Stephania tetrandra S. Moore;粉防己(白木香,蟾蜍薯,长根金不换,倒地拱,独脚蟾蜍,防己,防杞,房苑,粉寸己,汉防己,解离,金丝吊鳖,金线吊蛤蟆,千年薯,山乌龟,石蟾蜍,石解,土防己,乌龟梢,载君行,猪大肠,猪屎碌);Fourstamen Stephania, Powder Stephania,Tetrandous Stephania ●■

375905 Stephania tetrandra S. Moore var. glabra Maxim. = Stephania cephalantha Hayata ex Yamam. ●■

375906 Stephania venosa Spreng.;具脉千金藤●☆

375907 Stephania viridiflavens H. S. Lo et M. Yang;黄叶地不容(金不换,山乌龟);Greenyellow Stephania, Green-yellow Stephania, Yellowleaf Stephania ●■

375908 Stephania yunnanensis H. S. Lo;云南地不容(地不容,红藤,山乌龟,一滴血);Yunnan Stephania ●■

375909 Stephania yunnanensis H. S. Lo var. trichocalyx H. S. Lo et M. Yang;毛萼地不容;Haircalyx Yunnan Stephania ●■

375910 Stephaniscus Tiegh. = Englerina Tiegh. ●☆

375911 Stephaniscus Tiegh. = Tapinanthus(Blume)Rchb.(保留属名)●☆

375912 Stephaniscus gabonensis(Engl.)Tiegh. = Englerina gabonensis(Engl.)Balle ●☆

375913 Stephaniscus lecomtei Tiegh. = Englerina gabonensis(Engl.)Balle ●☆

375914 Stephanium Schreb. = Palicourea Aubl. ●☆

375915 Stephanocarpus Spach = Cistus L. ●

375916 Stephanocaryum Popov(1951);冠果紫草属■☆

375917 Stephanocaryum olgae(B. Fedtsch.)Popov;冠果紫草■☆

375918 Stephanocereus A. Berger(1926);毛环柱属(毛环翁属,毛环翁柱属)●☆

375919 Stephanocereus leucostele(Gürke)A. Berger;毛环柱(毛环翁,毛环翁柱)●☆

375920 Stephanochilus Coss. et Durieu ex Benth. = Centaurea L.(保留属名)●■

375921 Stephanochilus Coss. et Durieu ex Benth. et Hook. f. = Volutarella Cass. ■☆

375922 Stephanochilus Coss. et Durieu ex Maire(1935);冠唇菊属■☆

375923 Stephanochilus omphalodes(Benth. et Hook. f.)Maire;冠唇菊■☆

375924 Stephanochilus omphalodes(Coss.)Maire var. flavescens Le Houér. = Stephanochilus omphalodes(Benth. et Hook. f.)Maire ■☆

375925 Stephanococcus Bremek.(1952);冠果茜属■☆

375926 Stephanococcus crepinianus(K. Schum.)Bremek.;冠果茜■☆

375927 Stephanocoma Less. = Berkheya Ehrh.(保留属名)●■☆

375928 Stephanocoma atriplicifolia urcz. ex Ledeb. = Synurus deltoides(Aiton)Nakai ■

375929 Stephanocoma atriplicifolium(Trev.)Turcz. ex Ledeb. = Synurus deltoides(Aiton)Nakai ■

375930 Stephanocoma carduoides Less. = Berkheya carduoides(Less.)

Hutch. ■☆

375931　Stephanodaphne Baill.（1875）；冠瑞香属●☆

375932　Stephanodaphne capitata（Léandri）Léandri = Stephanodaphne geminata H. Perrier ex Léandri ●☆

375933　Stephanodaphne cremostachya Baill. subsp. capitata Léandri = Stephanodaphne geminata H. Perrier ex Léandri ●☆

375934　Stephanodaphne cremostachya Baill. subsp. cuspidata Léandri = Stephanodaphne cuspidata（Léandri）Léandri ●☆

375935　Stephanodaphne cuspidata（Léandri）Léandri；骤尖冠瑞香●☆

375936　Stephanodaphne geminata H. Perrier ex Léandri；双冠瑞香●☆

375937　Stephanodaphne humbertii Léandri；亨伯特冠瑞香●☆

375938　Stephanodaphne oblongifolia Léandri = Stephanodaphne geminata H. Perrier ex Léandri ●☆

375939　Stephanodaphne pedicellata Z. S. Rogers；梗花冠瑞香●☆

375940　Stephanodaphne perrieri Léandri；佩里耶冠瑞香●☆

375941　Stephanodaphne pilosa Z. S. Rogers；疏毛冠瑞香●☆

375942　Stephanodaphne pulchra Léandri = Stephanodaphne geminata H. Perrier ex Léandri ●☆

375943　Stephanodaphne schatzii Z. S. Rogers；沙茨冠瑞香●☆

375944　Stephanodoria Greene（1895）；短舌黄头菊属■☆

375945　Stephanodoria tomentella（B. L. Rob.）Greene；短舌黄头菊 ■☆

375946　Stephanogastra H. Karat. et Triana = Centronia D. Don ●☆

375947　Stephanogastra Triana = Centronia D. Don ●☆

375948　Stephanogyna Post et Kuntze = Mitragyna Korth.（保留属名）●

375949　Stephanogyna Post et Kuntze = Stephegyne Korth. ●

375950　Stephanolepis S. Moore = Erlangea Sch. Bip. ■☆

375951　Stephanolepis centauroides S. Moore = Erlangea centauroides（S. Moore）S. Moore ■☆

375952　Stephanolirion Baker = Tristagma Poepp. ■☆

375953　Stephanoluma Baill. = Sideroxylon L. ●☆

375954　Stephanolurna Baill. = Micropholis（Griseb.）Pierre ●☆

375955　Stephanomeria Nutt.（1841）（保留属名）；线莴苣属；Skeletonweed,Stickweed,Wirelettuce ●■☆

375956　Stephanomeria blairii Munz et I. M. Johnst. = Munzothamnus blairii（Munz et I. M. Johnst.）P. H. Raven ●☆

375957　Stephanomeria carotifera Hoover = Stephanomeria exigua Nutt. subsp. carotifera（Hoover）Gottlieb ■☆

375958　Stephanomeria cichoriacea A. Gray；菊苣叶线莴苣；Chicoryleaf Wirelettuce ■☆

375959　Stephanomeria cinerea（S. F. Blake）S. F. Blake = Stephanomeria pauciflora（Torr.）A. Nelson ■☆

375960　Stephanomeria coronaria Greene = Stephanomeria exigua Nutt. subsp. coronaria（Greene）Gottlieb ■☆

375961　Stephanomeria diegensis Gottlieb；圣地亚哥线莴苣；San Diego Wirelettuce ■☆

375962　Stephanomeria elata Nutt.；纳托尔线莴苣；Nuttall's Wirelettuce ■☆

375963　Stephanomeria exigua Greene var. coronaria（Greene）Jeps. = Stephanomeria exigua Nutt. subsp. coronaria（Greene）Gottlieb ■☆

375964　Stephanomeria exigua Nutt.；弱小线莴苣；Skeletonplant,Small Wirelettuce ■☆

375965　Stephanomeria exigua Nutt. subsp. carotifera（Hoover）Gottlieb；胡佛线莴苣；Hoover's Wirelettuce ■☆

375966　Stephanomeria exigua Nutt. subsp. coronaria（Greene）Gottlieb；小冠线莴苣；Small Crown Wirelettuce ■☆

375967　Stephanomeria exigua Nutt. subsp. deanei（J. F. Macbr.）Gottlieb；迪恩线莴苣；Deane's Wirelettuce ■☆

375968　Stephanomeria exigua Nutt. subsp. macrocarpa Gottlieb.；大果线莴苣；Large Seed Wirelettuce ■☆

375969　Stephanomeria exigua Nutt. var. deanei J. F. Macbr. = Stephanomeria exigua Nutt. subsp. deanei（J. F. Macbr.）Gottlieb ■☆

375970　Stephanomeria exigua Nutt. var. pentachaeta（D. C. Eaton）H. M. Hall = Stephanomeria exigua Nutt. ■☆

375971　Stephanomeria fluminea Gottlieb.；河岸线莴苣；Creekside Wirelettuce ■☆

375972　Stephanomeria lygodesmoides M. E. Jones ex L. F. Hend. = Stephanomeria pauciflora（Torr.）A. Nelson ■☆

375973　Stephanomeria malheurensis Gottlieb.；马尔线莴苣；Malheur Wirelettuce ■☆

375974　Stephanomeria myrioclada D. C. Eaton = Stephanomeria tenuifolia（Raf.）H. M. Hall ■☆

375975　Stephanomeria neomexicana（Greene）Cory = Stephanomeria tenuifolia（Raf.）H. M. Hall ■☆

375976　Stephanomeria paniculata Nutt.；硬枝线莴苣；Stiff-branched Wirelettuce ■☆

375977　Stephanomeria parryi A. Gray；帕里线莴苣；Parry's Wirelettuce ■☆

375978　Stephanomeria pauciflora（Torr.）A. Nelson；少花线莴苣；Brownplume Wirelettuce, Few-flowered Wirelettuce, Prairie Skeletonplant ■☆

375979　Stephanomeria pauciflora（Torr.）A. Nelson var. parishii（Jeps.）Munz = Stephanomeria pauciflora（Torr.）A. Nelson ■☆

375980　Stephanomeria pentachaeta D. C. Eaton = Stephanomeria exigua Nutt. ■☆

375981　Stephanomeria runcinata Nutt.；沙漠线莴苣；Desert Wirelettuce,Sawtooth Wirelettuce ■☆

375982　Stephanomeria schottii（A. Gray）A. Gray = Stephanomeria exigua Nutt. ■☆

375983　Stephanomeria spinosa（Nutt.）Tomb = Pleiacanthus spinosus（Nutt.）Rydb. ■☆

375984　Stephanomeria tenuifolia（Raf.）H. M. Hall；细叶线莴苣；Narrow-leaved Skeletonplant,Slender Wirelettuce ■☆

375985　Stephanomeria tenuifolia（Raf.）H. M. Hall var. myrioclada（D. C. Eaton）Cronquist = Stephanomeria tenuifolia（Raf.）H. M. Hall ■☆

375986　Stephanomeria tenuifolia（Raf.）H. M. Hall var. uintaensis Goodrich et S. L. Welsh = Stephanomeria tenuifolia（Raf.）H. M. Hall ■☆

375987　Stephanomeria thurberi A. Gray；瑟伯线莴苣；Skeletonplant,Thurber's Wirelettuce ■☆

375988　Stephanomeria tomentosa Greene = Stephanomeria virgata Benth. ■☆

375989　Stephanomeria virgata Benth.；棒状线莴苣；Virgate Wirelettuce ■☆

375990　Stephanomeria virgata Benth. subsp. pleurocarpa（Greene）Gottlieb；歪果线莴苣；Wand Wirelettuce ■☆

375991　Stephanomeria virgata Benth. var. tomentosa（Greene）Munz = Stephanomeria virgata Benth. ■☆

375992　Stephanomeria wrightii A. Gray = Stephanomeria tenuifolia（Raf.）H. M. Hall ■☆

375993　Stephanopappus Less. = Nestlera Spreng. ■☆

375994　Stephanopholis S. F. Blake = Chromolepis Benth. ■☆

375995　Stephanophorum Dulac = Narcissus L. ■

375996　Stephanophyllum Guill. = Paepalanthus Kunth（保留属名）■☆

375997　Stephanophysum Pohl = Ruellia L. ■●

375998　Stephanopodium Poepp. = Stephanopodium Poepp. et Endl. ●☆

375999　Stephanopodium Poepp. et Endl. (1843)；冠足毒鼠子属●☆

376000　Stephanopodium peruvianum Poepp. et Endl. ；冠足毒鼠子●☆

376001　Stephanorossia Chiov. = Oenanthe L. ■

376002　Stephanorossia elliotii Clark = Oenanthe procumbens（H. Wolff）C. Norman ■☆

376003　Stephanorossia palustris Chiov. = Oenanthe palustris（Chiov.）C. Norman ■☆

376004　Stephanosiphon Boiv. ex C. DC. = Turraea L. ●

376005　Stephanostachys（Klotzsch）Klotzsch ex O. E. Schulz = Chamaedorea Willd.（保留属名）●☆

376006　Stephanostachys Klotzsch ex Oerst. = Chamaedorea Willd.（保留属名）●☆

376007　Stephanostegia Baill. (1888)；顶冠夹竹桃属●

376008　Stephanostegia brevis Markgr. = Stephanostegia capuronii Markgr. ●☆

376009　Stephanostegia capuronii Markgr. ；凯普伦顶冠夹竹桃●☆

376010　Stephanostegia hildebrandtii Baill. ；马岛顶冠夹竹桃●☆

376011　Stephanostegia holophaea Pichon = Stephanostegia hildebrandtii Baill. ●☆

376012　Stephanostegia holophaea Pichon var. parvifolia（Pichon）Markgr. = Stephanostegia hildebrandtii Baill. ●☆

376013　Stephanostegia megalocarpa Markgr. = Stephanostegia hildebrandtii Baill. ●☆

376014　Stephanostegia parvifolia Pichon = Stephanostegia hildebrandtii Baill. ●☆

376015　Stephanostema K. Schum. (1904)；冠蕊夹竹桃属●☆

376016　Stephanostema stenocarpum K. Schum. ；冠蕊夹竹桃●☆

376017　Stephanotella E. Fourn. (1885)；小黑鳗藤属●☆

376018　Stephanotella E. Fourn. = Marsdenia R. Br. (保留属名)●

376019　Stephanotella glaziovii E. Fourn. ；小黑鳗藤●☆

376020　Stephanothelys Garay(1977)；女人兰属■☆

376021　Stephanothelys colombiana Garay；女人兰■☆

376022　Stephanotis Thouars(1806)(废弃属名)；黑鳗藤属(冠豆藤属,千金子藤属,舌瓣花属)；Madagascar Jasmine, Stephanotis ●

376023　Stephanotis Thouars(废弃属名) = Jasminanthes Blume ●

376024　Stephanotis Thouars(废弃属名) = Marsdenia R. Br. (保留属名)●

376025　Stephanotis chinensis Champ. ex Benth. = Jasminanthes mucronata（Blanco）W. D. Stevens et P. T. Li ●

376026　Stephanotis chinensis Champ. ex Benth. = Stephanotis mucronata（Blanco）Merr. ●

376027　Stephanotis chunii Tsiang = Jasminanthes chunii（Tsiang）W. D. Stevens et P. T. Li ●

376028　Stephanotis floribunda（R. Br.）Brongn. = Marsdenia floribunda（Brongn.）Schltr. ●☆

376029　Stephanotis grandiflora Decne. = Dramsenia grandiflora（Norman）Bullock ex Aké Assi ■☆

376030　Stephanotis japonica Makino ex Nakai = Jasminanthes mucronata（Blanco）W. D. Stevens et P. T. Li ●

376031　Stephanotis japonica Makino ex Nakai = Stephanotis mucronata（Blanco）Merr. ●

376032　Stephanotis lutchuensis Koidz. = Jasminanthes mucronata（Blanco）W. D. Stevens et P. T. Li ●

376033　Stephanotis lutchuensis Koidz. = Stephanotis mucronata（Blanco）Merr. ●

376034　Stephanotis lutchuensis Koidz. var. japonica（Makino ex Nakai）Hatus. = Jasminanthes mucronata（Blanco）W. D. Stevens et P. T. Li ●

376035　Stephanotis lutchuensis Koidz. var. japonica（Makino ex Nakai）Hatus. = Stephanotis mucronata（Blanco）Merr. ●

376036　Stephanotis mucronata（Blanco）Merr. = Jasminanthes mucronata（Blanco）W. D. Stevens et P. T. Li ●

376037　Stephanotis mucronata（Blanco）Merr. = Jasminanthes pilosa（Kerr）W. D. Stevens et P. T. Li ●

376038　Stephanotis nana P. T. Li = Marsdenia stenantha Hand. -Mazz. ●

376039　Stephanotis pilosa Kerr = Jasminanthes pilosa（Kerr）W. D. Stevens et P. T. Li ●

376040　Stephanotis saxatilis Tsiang et P. T. Li = Jasminanthes saxatilis（Tsiang et P. T. Li）W. D. Stevens et P. T. Li ●

376041　Stephanotis yunnanensis H. Lév. = Marsdenia stenantha Hand. -Mazz. ●

376042　Stephanotrichum Naudin = Clidemia D. Don ●☆

376043　Stephegyne Korth. = Mitragyna Korth. (保留属名)●

376044　Stephegyne africana（Willd.）Walp. = Mitragyna inermis（Willd.）K. Schum. ●☆

376045　Stephegyne parvifolia（Roxb.）Korth. = Mitragyna parvifolia（Roxb.）Korth. ●☆

376046　Stephensonia Hort. = Phoenicophorium H. Wendl. ●☆

376047　Stephensonia Hort. = Stevensonia Duncan ex Balf. f. ●☆

376048　Stephensonia Hort. ex Van Houtte = Phoenicophorium H. Wendl. ●☆

376049　Stephensonia Hort. ex Van Houtte = Stevensonia Duncan ex Balf. f. ●☆

376050　Steptium Boiss. = Priva Adans. ■☆

376051　Steptium Boiss. = Streptium Roxb. ■☆

376052　Steptorhamphus Bunge(1852)；线嘴苣属■☆

376053　Steptorhamphus crambifolius Bunge；线嘴苣■☆

376054　Steptorhamphus crassicaulis（Trautv.）Kirp. ；粗茎线嘴苣■☆

376055　Steptorhamphus czerepanovii Kirp. ；契氏线嘴苣■☆

376056　Steptorhamphus linczevskii Kirp. ；林氏线嘴苣■☆

376057　Steptorhamphus persicus O. Fedtsch. et B. Fedtsch. ；波斯线嘴苣■☆

376058　Steptorhamphus petraeus（Fisch. et C. A. Mey.）Grossh. ；岩生线嘴苣■☆

376059　Steptorhamphus tuberosus（Jacq.）Grossh. ；瘤线嘴苣■☆

376060　Steptorhamphus tuberosus（L.）Grossh. = Steptorhamphus tuberosus（Jacq.）Grossh. ■☆

376061　Stera Ewart = Cratystylis S. Moore ■☆

376062　Sterbeckia Schreb. = Singana Aubl. ☆

376063　Sterculia L. (1753)；苹婆属；Bottle Tree, Bottle-Tree, Sterculia ●

376064　Sterculia acerifolia A. Cunn. = Brachychiton acerifolium Macarthur et C. Moore ●

376065　Sterculia acuminata P. Beauv. = Cola acuminata（Brenan）Schott et Endl. ●☆

376066　Sterculia affinis Mast. ；近缘苹婆；Affined Sterculia ●☆

376067　Sterculia africana（Lour.）Fiori；非洲苹婆；African Star Chestnut, Illawarra Flame Tree, Mopopaja Tree ●☆

376068　Sterculia africana（Lour.）Fiori var. rivae（K. Schum.）Cufod. = Sterculia rhynchocarpa K. Schum. ●☆

376069　Sterculia africana（Lour.）Fiori var. socotrana（K. Schum.）Fiori = Sterculia africana（Lour.）Fiori ●☆

376070　Sterculia alata Roxb. ; 有翅苹婆（翅果苹婆, 翅子苹婆, 海南苹婆）; Winged Seed Sterculia, Winged Sterculia ●

376071　Sterculia alata Roxb. = Pterygota alata（Roxb.）R. Br. ●

376072　Sterculia alexandri Harv. ; 亚历山大苹婆●☆

376073　Sterculia ambacensis Welw. ex Hiern = Sterculia subviolacea K. Schum. ●☆

376074　Sterculia ankaranensis Arènes = Hildegardia ankaranensis（Arènes）Kosterm. ●☆

376075　Sterculia apetala（Jacq.）H. Karst. ; 拉美苹婆（巴拿马苹婆）; Apetalous Sterculia, Bellota, Panama Tree ●☆

376076　Sterculia appendiculata K. Schum. ; 附属物苹婆●☆

376077　Sterculia armata Mast. ; 有刺苹婆; Spiny Sterculia ●☆

376078　Sterculia armata Mast. = Sterculia villosa Roxb. et G. Don ●

376079　Sterculia balansae A. DC. = Sterculia lanceolata Cav. ●

376080　Sterculia barteri Mast. = Hildegardia barteri（Mast.）Kosterm. ●☆

376081　Sterculia bequaertii De Wild. ; 贝卡尔苹婆●☆

376082　Sterculia bicolor Mast. ; 二色苹婆; Bicolor Sterculia ●☆

376083　Sterculia bidwillii（Hook.）Benth. = Sterculia bidwillii Hook. ex Benth. ●☆

376084　Sterculia bidwillii Hook. ex Benth. ; 披威利苹婆; Bidwilli Sterculia ●☆

376085　Sterculia bodinieri H. Lév. = Phyllanthus bodinieri（H. Lév.）Rehder ●

376086　Sterculia brevipetiolata H. T. Tsai et P. I Mao = Sterculia brevissima H. H. Hsue ex Y. Tang ●

376087　Sterculia brevissima H. H. Hsue = Sterculia brevissima H. H. Hsue ex Y. Tang ●

376088　Sterculia brevissima H. H. Hsue ex Y. Tang; 短柄苹婆（骂良王）; Short-pediceled Sterculia, Shortstalk Sterculia, Shortstipe Sterculia ●

376089　Sterculia campanulata Wall. ex Mast. ; 钟状苹婆; Bellshaped Sterculia ●☆

376090　Sterculia carthaginensis Cav. = Sterculia apetala（Jacq.）H. Karst. ●☆

376091　Sterculia cauliflora（Mast.）Roberty = Cola cauliflora Mast. ●☆

376092　Sterculia ceramica R. Br. ; 台湾苹婆（兰屿苹婆）; Lanyu Sterculia, Taiwan Sterculia ●

376093　Sterculia cinerea A. Rich. ; 灰色苹婆; Tartar Gum ●☆

376094　Sterculia cinnamomifolia H. T. Tsai et P. I. Mao; 樟叶苹婆; Cinnamon-leaf Sterculia, Cinnamon-leaved Sterculia ●

376095　Sterculia coccinea Roxb. et G. Don; 绯红苹婆; Red Sterculia ●☆

376096　Sterculia colorata Roxb. = Erythropsis colorata（Roxb.）Burkill ●

376097　Sterculia colorata Roxb. = Firmiana colorata（Roxb.）R. Br. ●

376098　Sterculia cordifolia Cav. = Cola cordifolia（Cav.）R. Br. ●☆

376099　Sterculia cubensis Urb. ; 古巴苹婆●☆

376100　Sterculia dawei Sprague; 道氏苹婆●☆

376101　Sterculia digitata（Mast.）Roberty = Cola digitata Mast. ●☆

376102　Sterculia diversifolia G. Don; 异叶苹婆（白杨叶瓶木, 瓶子木, 杨叶桐, 异叶瓶木, 掌叶酒瓶树）; Bottle Tree, Brachychiton, Kurrajong, Kurrajong Bottle Tree, Laceback Kurrajong, Whiteflower Kurrajong ●☆

376103　Sterculia diversifolia G. Don = Brachychiton populneum（Schott et Endl.）R. Br. ●☆

376104　Sterculia diversifolia G. Don var. occidentalis Benth. ; 西方异叶苹婆●☆

376105　Sterculia elegantiflora Hutch. et Dalziel = Eribroma oblongum（Mast.）Pierre ex A. Chev. ■☆

376106　Sterculia ensifolia Mast. ; 剑叶苹婆; Swordleaf Sterculia ●☆

376107　Sterculia erythrosiphon Baill. = Hildegardia erythrosiphon（Baill.）Kosterm. ●☆

376108　Sterculia euosma W. W. Sm. ; 粉苹婆; Powder Sterculia, Wellflavoured Sterculia, Well-flavoured Sterculia ●

376109　Sterculia firmiana J. F. Gmel. = Firmiana simplex（L.）W. Wight ●

376110　Sterculia foetida L. ; 香苹婆（臭苹婆, 裂叶苹婆, 粟苹婆, 香草婆, 掌叶苹婆）; Fragrant Sterculia, Hazel Bottle Tree, Hazel Sterculia, Java Olive ●

376111　Sterculia foetida L. = Sterculia pexa Pierre ●

376112　Sterculia fulgens Wall. ; 光亮苹婆; Shining Sterculia ●☆

376113　Sterculia gengmaensis H. H. Hsue = Sterculia gengmaensis H. H. Hsue ex Y. Tang ●

376114　Sterculia gengmaensis H. H. Hsue ex Y. Tang; 绿花苹婆; Green Sterculia, Greenflower Sterculia, Green-flowered Sterculia ●

376115　Sterculia geniculata Miq. = Botryophora geniculata（Miq.）Beumee ex Airy Shaw ●☆

376116　Sterculia guangxiensis S. J. Xu et P. T. Li; 广西苹婆（桂苹婆）; Guangxi Sterculia ●

376117　Sterculia guerichii K. Schum. = Sterculia africana（Lour.）Fiori ●☆

376118　Sterculia guttata Roxb. et G. Don; 斑点苹婆; Spotted Sterculia ●☆

376119　Sterculia hainanensis Merr. et Chun; 海南苹婆（小苹婆）; Hainan Sterculia ●

376120　Sterculia hartmanniana Schweinf. = Sterculia cinerea A. Rich. ●☆

376121　Sterculia henryi Hemsl. ; 蒙自苹婆; Henry Sterculia, Mengzi Sterculia ●

376122　Sterculia henryi Hemsl. var. cuneata Chun et H. H. Hsue; 大围山苹婆; Cuneate Sterculia ●

376123　Sterculia heterophylla P. Beauv. = Cola heterophylla（P. Beauv.）Schott et Endl. ●☆

376124　Sterculia hymenocalyx K. Schum. ; 膜萼苹婆; Membranecalyx Sterculia, Membranous-calyxed Sterculia ●

376125　Sterculia impressinervis H. H. Hsue; 凹脉苹婆; Concavevein Sterculia, Sunkenvein Sterculia, Sunken-veined Sterculia ●

376126　Sterculia kingtungensis H. H. Hsue = Sterculia kingtungensis H. H. Hsue ex Y. Tang ●

376127　Sterculia kingtungensis H. H. Hsue ex Y. Tang; 大叶苹婆; Bigleaf Sterculia, Jingdong Sterculia ●

376128　Sterculia laevis Wall. ; 平滑苹婆; Smooth Sterculia ●☆

376129　Sterculia lanceifolia Roxb. et G. Don; 西蜀苹婆; Lanceleaf Sterculia, Lance-leaved Sterculia, W. Sichuan Sterculia ●

376130　Sterculia lanceolata Cav. ; 假苹婆（狗麻, 红郎伞, 鸡冠木, 鸡冠皮, 苹婆, 赛苹婆, 山木棉）; Fake Sterculia, Scarlet Sterculia ●

376131　Sterculia lanceolata Cav. var. principis（Gagnep.）Phengklai = Sterculia principis Gagnep. ●

376132　Sterculia lantsangensis Hu = Sterculia villosa Roxb. et G. Don ●

376133　Sterculia lastoursvillensis M. Bodard et Pellegr. = Chlamydocola lastoursvillensis（M. Bodard et Pellegr.）N. Hallé ●☆

376134　Sterculia lindensis Engl. = Sterculia appendiculata K. Schum. ●☆

376135　Sterculia linguifolia Mast. ; 舌状叶苹婆; Tobguelikeleaf

Sterculia ●☆

376136　Sterculia luzonica Warb. = Sterculia ceramica R. Br. ●

376137　Sterculia lychnophora Hance；胖大海（安南子，大洞果，大发，大海，大海子，胡大海，裂叶胖大海，通大海）●☆

376138　Sterculia macrophylla Vent. ；巨叶苹婆（大叶苹婆）；Bigleaf Sterculia ●☆

376139　Sterculia maingayi Mast. ；马音加苹婆；Maingay Sterculia ●☆

376140　Sterculia malvacea H. Lév. = Eriolaena spectabilis （DC.）Planch. ex Hook. f. ●

376141　Sterculia megaphylla H. T. Tsai et P. I Mao = Sterculia kingtungensis H. H. Hsue ex Y. Tang ●

376142　Sterculia micrantha Chun et H. H. Hsue；小花苹婆；Littleflower Sterculia，Smallflower Sterculia，Small-flowered Sterculia ●

376143　Sterculia mirabilis （A. Chev.）Roberty = Chlamydocola chlamydantha （K. Schum.）M. Bodard ●☆

376144　Sterculia monosperma Vent. ；苹婆（凤眼果，富贵子，红皮果，九层皮，罗晃子，罗望子，潘安果，频婆果，苹婆果，萍婆，七姐果）；China Chestnut，China Chest-nut，Chinese Chest-nut，Common Sterculia，Noble Bottle-tree，Pimpon，Ping-pong，Sterculia ●

376145　Sterculia monosperma Vent. var. subspontanea （H. H. Hsue et S. J. Xu）Y. Tang；野生苹婆 ●

376146　Sterculia nitida Vent. = Cola nitida （Vent.）Schott et Endl. ●☆

376147　Sterculia nobilis R. Br. = Sterculia nobilis Sm. ●

376148　Sterculia nobilis Sm. = Sterculia monosperma Vent. ●

376149　Sterculia nobilis Sm. var. subspontanea J. R. Xue et S. J. Xu = Sterculia monosperma Vent. var. subspontanea （H. H. Hsue et S. J. Xu）Y. Tang ●

376150　Sterculia oblonga Mast. ；西非黄苹婆；Eyong，W. Africa Sterculia，Yellow Sterculia ●☆

376151　Sterculia oblonga Mast. = Eribroma oblongum （Mast.）Pierre ex A. Chev. ■☆

376152　Sterculia ornata Wall. ex Kurz = Sterculia villosa Roxb. et G. Don ●

376153　Sterculia ovalifolia Wall. = Sterculia lanceifolia Roxb. et G. Don ●

376154　Sterculia pallens Wall. ex King = Firmiana pallens （Wall. ex King）Stearn ●☆

376155　Sterculia parviflora Roxb. et G. Don；小叶苹婆（小花苹婆）；Smallflower Sterculia，Small-leaf Sterculia ●☆

376156　Sterculia pexa Pierre；家麻树（哥波，家麻桐，九层皮，绵毛苹婆，千层皮）；Woolly Sterculia ●

376157　Sterculia pexa var. yunnanensis （Hu）H. H. Hsue = Sterculia pexa Pierre ●

376158　Sterculia pinbienensis H. T. Tsai et P. I. Mao；屏边苹婆；Pingbian Sterculia ●

376159　Sterculia plantanifolia L. f. = Firmiana simplex （L.）W. Wight ●

376160　Sterculia plantanifolia L. f. var. major W. W. Sm. = Firmiana major （W. W. Sm.）Hand. -Mazz. ●◇

376161　Sterculia platanifolia L. = Firmiana simplex （L.）W. Wight ●

376162　Sterculia platanifolia L. f. = Firmiana platanifolia （L. f.）Marsili ●

376163　Sterculia platanifolia L. f. = Firmiana simplex （L.）W. Wight ●

376164　Sterculia platanifolia L. f. var. major W. W. Sm. = Firmiana major （W. W. Sm.）Hand. -Mazz. ●◇

376165　Sterculia populifolia Roxb. ex Wall. ；杨叶苹婆；Poplarleaf Sterculia ●☆

376166　Sterculia principis Gagnep. ；基苹婆；True Sterculia ●

376167　Sterculia pubescens Mast. ；柔毛苹婆（毛苹婆）；Pubescent

376168　Sterculia purpurea Exell. ；紫苹婆 ■☆

376169　Sterculia pyriformis Bunge = Firmiana platanifolia （L. f.）Marsili ●

376170　Sterculia pyriformis Bunge = Firmiana simplex （L.）W. Wight ●

376171　Sterculia quadrifida R. Br. ；四裂苹婆；Peanut Tree ●☆

376172　Sterculia quinqueloba （Garcke）K. Schum. ；五裂苹婆 ●☆

376173　Sterculia reticulata （A. Chev.）Roberty = Cola reticulata A. Chev. ●☆

376174　Sterculia rhinopetala K. Schum. ；象鼻黄苹婆；Aye, Brown Sterculia ●☆

376175　Sterculia rhynchocarpa K. Schum. ；喙果苹婆 ●☆

376176　Sterculia richardiana Baill. = Sterculia ceramica R. Br. ●

376177　Sterculia rogersii N. E. Br. ；罗杰斯苹婆 ●☆

376178　Sterculia roxburghii Wall. = Sterculia lanceifolia Roxb. et G. Don ●

376179　Sterculia rubiginosa Vent. ；褐赤苹婆；Brownred Sterculia ●☆

376180　Sterculia rupestris （Lindl.）Benth；岩生苹婆；Barrel Bottle Tree ●☆

376181　Sterculia rupestris （Lindl.）Benth. = Brachychiton rupestris （Lindl.）K. Schum. ●☆

376182　Sterculia rupestris Benth = Sterculia rupestris （Lindl.）Benth. ●☆

376183　Sterculia scandens Hemsl. ；河口苹婆；Hekou Sterculia ●

376184　Sterculia scaphigera Wall. ；圆粒苹婆（安南子，大洞果，大发，大海，大海榄，大海子，胡大海，胖大海，蓬大海，膨大海，通大海，星大海，圆粒胖大海，舟状苹婆）；Boat Sterculia ●

376185　Sterculia schliebenii Mildbr. ；施利本苹婆 ●☆

376186　Sterculia setigera Delile；刚毛苹婆 ●☆

376187　Sterculia simaoensis Y. Y. Qian；思茅苹婆；Simao Sterculia ●

376188　Sterculia simplex （L.）Druce = Firmiana simplex （L.）W. Wight ●

376189　Sterculia spectabilis （Welw.）Roberty = Octolobus spectabilis Welw. ●☆

376190　Sterculia stenocarpa H. Winkl. ；窄果苹婆 ■☆

376191　Sterculia striatiflora Mast. ；条纹花苹婆（条纹苹婆）；Striateflower Sterculia ●☆

376192　Sterculia subnobilis H. H. Hsue；罗浮苹婆；Lofu Sterculia, Luofu Sterculia ●

376193　Sterculia subracemosa Chun et H. H. Hsue；信宜苹婆；Hsinyi Sterculia，Xinyi Sterculia ●

376194　Sterculia subviolacea K. Schum. ；堇色苹婆 ●☆

376195　Sterculia thwaitesii Mast. ；特维特氏苹婆；Thwaites Sterculia ●☆

376196　Sterculia tiliacea H. Lév. = Grewia abutilifolia Vent. ex Juss. ●

376197　Sterculia tomentosa Thunb. = Firmiana simplex （L.）W. Wight ●

376198　Sterculia tonkinensis A. DC. ；北越苹婆；Tonkin Sterculia ●

376199　Sterculia tragacantha Lindl. ；丰花苹婆（胶梧桐）；African Tragacanth ●☆

376200　Sterculia tragacanthoides Engl. ；拟丰花苹婆 ●☆

376201　Sterculia trichosiphon Benth. = Brachychiton australis （Schott et Endl.）A. Terracc. ●☆

376202　Sterculia triphaca R. Br. = Sterculia africana （Lour.）Fiori ●☆

376203　Sterculia triphaca R. Br. var. rivae K. Schum. = Sterculia rhynchocarpa K. Schum. ●☆

376204　Sterculia tubulata Mast. ；有管苹婆；Tubulate Sterculia ●☆

376205　Sterculia urens Roxb. ；炴毛梧桐；Bottle Tree, Karaya, Kataya

Gum, Kutira Gum ●☆

376206 Sterculia urens sensu Qureshi et Saeed = Firmiana simplex (L.) W. Wight ●

376207 Sterculia versicolor Wall. ;变色苹婆;Versicolor Sterculia ●☆

376208 Sterculia verticillata Thonn. = Cola verticillata (Thonn.) Stapf ex A. Chev. ●☆

376209 Sterculia villosa Roxb. = Sterculia villosa Roxb. et G. Don ●

376210 Sterculia villosa Roxb. et G. Don;绒毛苹婆(白椰皮,长毛苹婆,考西抱,椰皮树,色白告);Villous Sterculia ●

376211 Sterculia viridiflora H. T. Tsai et P. I Mao = Sterculia gengmaensis H. H. Hsue ex Y. Tang ●

376212 Sterculia yuanjiangensis H. H. Hsue et S. J. Xu;元江苹婆;Yuanjiang Sterculia ●

376213 Sterculia yunnanensis Hu = Sterculia pexa Pierre ●

376214 Sterculia zastrowiana Engl. = Sterculia quinqueloba (Garcke) K. Schum. ●☆

376215 Sterculiaceae (DC.) Bartl. = Sterculiaceae Vent. (保留科名) ●■

376216 Sterculiaceae Bartl. = Sterculiaceae Vent. (保留科名) ●■

376217 Sterculiaceae DC. = Sterculiaceae Vent. (保留科名) ●■

376218 Sterculiaceae Vent. (1807)(保留科名);梧桐科;Chocolate Family, Sterculia Family ●■

376219 Sterculiaceae Vent. (保留科名) = Malvaceae Juss. (保留科名) ●■

376220 Sterculiaceae Vent. (保留科名) = Stilaginaceae C. Agardh ●

376221 Stereimis Raf. = Alternanthera Forssk. ■

376222 Stereimis Raf. = Steiremis Raf. ■

376223 Stereimis Raf. = Telanthera R. Br. ■

376224 Stereocarpus (Pierre) Hallier f. = Camellia L. ●

376225 Stereocarpus Hallier f. = Camellia L. ●

376226 Stereocaryum Burret(1941);坚果桃金娘属 ●☆

376227 Stereocaryum neocaledonicum (Brongn. et Gris) Burret;坚果桃金娘 ●☆

376228 Stereochilus Lindl. (1858);坚唇兰属 ■

376229 Stereochilus Lindl. = Sarcanthus Lindl. (废弃属名) ■

376230 Stereochilus brevirachis Christenson;短轴坚唇兰 ■

376231 Stereochilus dalatensis (Guillaumin) Garay;坚唇兰 ■

376232 Stereochlaena Hack. (1908);小翼轴草属 ■☆

376233 Stereochlaena annua Clayton;一年小翼轴草 ■☆

376234 Stereochlaena caespitosa Clayton;小翼轴草 ■☆

376235 Stereochlaena cameronii (Stapf) Pilg. ;非洲小翼轴草 ■☆

376236 Stereochlaena foliacea Clayton = Baptorhachis foliacea (Clayton) Clayton ■☆

376237 Stereochlaena jeffreysii Hack. = Stereochlaena cameronii (Stapf) Pilg. ■☆

376238 Stereochlaena tridentata Clayton;三齿小翼轴草 ■☆

376239 Stereoderma Blume = Olea L. ●

376240 Stereoderma Blume = Pachyderma Blume ●

376241 Stereoderma Blume ex Endl. = Olea L. ●

376242 Stereoderma Blume ex Endl. = Pachyderma Blume ●

376243 Stereosandra Blume(1856);肉药兰属;Stereosandra ■

376244 Stereosandra javanica Blume;肉药兰;Java Stereosandra ■

376245 Stereosandra javanica Blume var. papuana J. J. Sm. = Stereosandra javanica Blume ■

376246 Stereosandra koidsumiana Ohwi = Stereosandra javanica Blume ■

376247 Stereosandra liukiuensis Tuyama = Stereosandra javanica Blume ■

376248 Stereosandra pendula Kraenzl. = Stereosandra javanica Blume ■

376249 Stereosandra schinziana (Kraenzl.) Garay = Epipogium roseum (D. Don) Lindl. ■

376250 Stereosanthus Franch. = Nannoglottis Maxim. ■●★

376251 Stereosanthus delavayi Franch. = Nannoglottis delavayi (Franch.) Y. Ling et Y. L. Chen ■

376252 Stereosanthus gynura (C. Winkl.) Hand. -Mazz. = Nannoglottis gynura (C. Winkl.) Y. Ling et Y. L. Chen ■

376253 Stereosanthus hieraciifolius Diels = Nannoglottis gynura (C. Winkl.) Y. Ling et Y. L. Chen ■

376254 Stereosanthus souliei Franch. = Nannoglottis gynura (C. Winkl.) Y. Ling et Y. L. Chen ■

376255 Stereosanthus souliei Franch. = Nannoglottis macrocarpa Y. Ling et Y. L. Chen ■

376256 Stereosanthus yunnanensis Franch. = Nannoglottis latisquama Y. Ling et Y. L. Chen ■

376257 Stereospermum Cham. (1833);羽叶楸属;Padri Tree, Padritree ●

376258 Stereospermum acuminatissimum K. Schum. ;渐尖羽叶楸 ●☆

376259 Stereospermum arcuatum H. Perrier;拱羽叶楸 ●☆

376260 Stereospermum arguezana A. Rich. = Stereospermum kunthianum Cham. var. dentatum (A. Rich.) Fiori ●☆

376261 Stereospermum arnoldianum De Wild. = Stereospermum kunthianum Cham. var. dentatum (A. Rich.) Fiori ●☆

376262 Stereospermum boivinii (Baill.) H. Perrier;博伊文羽叶楸 ●☆

376263 Stereospermum bracteosum K. Schum. = Stereospermum zenkeri K. Schum. ex De Wild. ●☆

376264 Stereospermum chelonoides DC. ;龟头花羽叶楸 ●☆

376265 Stereospermum cinereo-viride K. Schum. = Stereospermum kunthianum Cham. var. dentatum (A. Rich.) Fiori ●☆

376266 Stereospermum colais (Buch. -Ham. ex Dillwyn) Mabb. ;羽叶楸(钝刀木,四角夹子树,四角羽叶楸,咸沙木);Common Padritree, Fourangular Padritree, Four-angular Padritree, Padri ●

376267 Stereospermum colais (Buch. -Ham. ex Dillwyn) Mabb. var. puberula (Dop) D. D. Tao = Stereospermum colais (Buch. -Ham. ex Dillwyn) Mabb. ●

376268 Stereospermum colais (Buch. -Ham. ex Dillwyn) Mabb. var. puberula (Dop) D. D. Tao;广西羽叶楸;Guangxi Padritree, Kwangsi Padritree ●

376269 Stereospermum dentatum A. Rich. = Stereospermum kunthianum Cham. var. dentatum (A. Rich.) Fiori ●☆

376270 Stereospermum discolor K. Schum. = Stereospermum kunthianum Cham. var. dentatum (A. Rich.) Fiori ●☆

376271 Stereospermum euphorioides (Bojer) A. DC. ;龙眼羽叶楸 ●☆

376272 Stereospermum ghorta (Buch. -Ham. ex G. Don) C. B. Clarke = Pauldopia ghorta (Buch. -Ham. ex G. Don) Steenis ●

376273 Stereospermum glandulosum Miq. = Radermachera glandulosa (Blume) Miq. ●

376274 Stereospermum harmsianum K. Schum. ;哈姆斯羽叶楸 ●☆

376275 Stereospermum hildebrandtii (Baill.) H. Perrier;希尔德羽叶楸 ●☆

376276 Stereospermum integrifolium A. Rich. = Stereospermum kunthianum Cham. var. dentatum (A. Rich.) Fiori ●☆

376277 Stereospermum katangense De Wild. = Stereospermum harmsianum K. Schum. ●☆

376278 Stereospermum kunthianum Cham. ;孔特羽叶楸 ●☆

376279 Stereospermum kunthianum Cham. var. dentatum (A. Rich.) Fiori;齿孔特羽叶楸 ●☆

376280 Stereospermum longiflorum Capuron;长花羽叶楸 ●☆

376281 Stereospermum molle K. Schum. = Stereospermum kunthianum Cham. var. dentatum (A. Rich.) Fiori ●☆

376282 Stereospermum nematocarpum A. DC.；线果羽叶楸●☆

376283 Stereospermum neuranthum Kurz；毛叶羽叶楸；Hairyleaf Padritree, Hairy-leaved Padritree ●

376284 Stereospermum personatum (Hassk.) Chatterjee；假面羽叶楸（羽叶楸）●

376285 Stereospermum personatum (Hassk.) Chatterjee = Stereospermum colais (Buch. -Ham. ex Dillwyn) Mabb. ●

376286 Stereospermum personatum (Hassk.) Chatterjee var. puberula Dop = Stereospermum colais (Buch. -Ham. ex Dillwyn) Mabb. ●

376287 Stereospermum personatum (Hassk.) Chatterjee var. puberula Dop = Stereospermum colais (Buch. -Ham. ex Dillwyn) Mabb. var. puberula (Dop) D. D. Tao ●

376288 Stereospermum rhoifolium (Baill.) H. Perrier；盐肤木叶羽叶楸●☆

376289 Stereospermum senegalense Miq. = Stereospermum kunthianum Cham. ●☆

376290 Stereospermum sinicum Hance = Radermachera sinica (Hance) Hemsl. ●

376291 Stereospermum strigillosum C. Y. Wu et W. C. Yin；伏毛萼羽叶楸（伏毛羽叶楸，毛萼羽叶楸）；Strigillose Padritree, Strigosecalyx Padritree ●

376292 Stereospermum suaveolens DC.；香羽叶楸；Fragrant Padritree ●

376293 Stereospermum suaveolens DC. = Stereospermum colais (Buch. -Ham. ex Dillwyn) Mabb. ●

376294 Stereospermum tetragonum (Wall.) DC. = Stereospermum colais (Buch. -Ham. ex Dillwyn) Mabb. ●

376295 Stereospermum tetragonum DC. = Stereospermum colais (Buch. -Ham. ex Dillwyn) Mabb. ●

376296 Stereospermum tomentosum H. Perrier；绒毛羽叶楸●☆

376297 Stereospermum variabile H. Perrier；易变羽叶楸●☆

376298 Stereospermum verdickii De Wild. = Stereospermum harmsianum K. Schum. ●☆

376299 Stereospermum xylocarpum Benth. et Hook. f.；木果羽叶楸；Woody-fruit Stereospermum ●☆

376300 Stereospermum zenkeri K. Schum. ex De Wild.；岑克尔羽叶楸●☆

376301 Stereoxylon Ruiz et Pav. = Escallonia Mutis ex L. f. ●☆

376302 Sterigma DC. = Sterigmostemum M. Bieb. ■

376303 Sterigma purpurescens Boiss. = Sterigmostemum purpurescens (Boiss.) Parsa ■☆

376304 Sterigmanthe Klotzsch et Garcke = Euphorbia L. ●■

376305 Sterigmanthe Klotzsch et Garcke = Lacanthis Raf. ●■

376306 Sterigmanthe bojeri Klotzsch et Garcke = Euphorbia milii Des Moul. ●

376307 Sterigmapetalum Kuhlm. (1925)；南美红树属●☆

376308 Sterigmapetalum chrysophyllum Aymard et N. Cuello；金叶南美红树●☆

376309 Sterigmapetalum guianense Steyerm.；圭亚那南美红树●☆

376310 Sterigmapetalum heterodoxum Steyerm. et Liesner；异齿南美红树●☆

376311 Sterigmapetalum obovatum Kuhlm.；南美红树●☆

376312 Sterigmostemon Kuntze = Sterigmostemum M. Bieb. ■

376313 Sterigmostemon Poir. = Sterigmostemum M. Bieb. ■

376314 Sterigmostemum M. Bieb. (1819)；棒果芥属（棒果芥属，小梗属）；Clubfruitcress, Sterigmostemum ■

376315 Sterigmostemum acanthocarpum Fisch. et C. A. Mey.；尖果棒果芥■☆

376316 Sterigmostemum caspicum (Lam.) Rupr.；棒果芥■

376317 Sterigmostemum eglandulosum (Botsch.) H. L. Yang = Oreoloma eglandulosum Botsch. ■

376318 Sterigmostemum fuhaiense H. L. Yang；福海棒果芥；Fuhai Sterigmostemum ■

376319 Sterigmostemum fuhaiense H. L. Yang = Oreoloma violaceum Botsch. ■

376320 Sterigmostemum grandiflorum K. C. Kuan；大花棒果芥；Bigflower Clubfruitcress, Bigflower Sterigmostemum ■

376321 Sterigmostemum grandiflorum K. C. Kuan = Oreoloma eglandulosum Botsch. ■

376322 Sterigmostemum incanum M. Bieb.；灰毛棒果芥；Greyhair Clubfruitcress, Greyhair Sterigmostemum ■

376323 Sterigmostemum matthioloides (Franch.) Botsch.；紫花棒果芥■

376324 Sterigmostemum matthioloides (Franch.) Botsch. = Oreoloma matthioloides (Franch.) Botsch. ■

376325 Sterigmostemum purpurescens (Boiss.) Parsa；紫棒果芥■☆

376326 Sterigmostemum rhodanthum Rech. f. = Sterigmostemum purpurescens (Boiss.) Parsa ■☆

376327 Sterigmostemum sulfureum (Banks et Sol.) Bornm.；黄花棒果芥；Yellow-flowered Sterigmostemum ■

376328 Sterigmostemum tomentosum (Willd.) M. Bieb. = Oreoloma violaceum Botsch. ■

376329 Sterigmostemum tomentosum (Willd.) M. Bieb. = Sterigmostemum caspicum (Lam.) Rupr. ■

376330 Sterigmostemum torulosum (M. Bieb.) Stapf = Sterigmostemum incanum M. Bieb. ■

376331 Sterigmostemum violaceum (Botsch.) H. L. Yang = Oreoloma violaceum Botsch. ■

376332 Sterigrnanthe Klotzsch et Garcke = Euphorbia L. ●■

376333 Steripha Banks ex Gaertn. = Dichondra J. R. Forst. et G. Forst. ■

376334 Steripha reniformia Gaertn. = Dichondra repens J. R. Forst. et G. Forst. ■

376335 Steriphe Phil. = Haplopappus Cass. (保留属名) ■●☆

376336 Steriphoma Spreng. (1827) (保留属名)；硬点山柑属●☆

376337 Steriphoma cleomoides Spreng.；硬点山柑●☆

376338 Steriphoma elliptica Spreng.；椭圆硬点山柑●☆

376339 Steriphoma macrantha Standl.；大花硬点山柑●☆

376340 Steriphoma paradoxa Endl.；奇异硬点山柑●☆

376341 Steris Adans. (废弃属名) = Silene L. (保留属名) ■

376342 Steris L. = Hydrolea L. (保留属名) ■

376343 Steris alpina (L.) Sourkova = Silene suecica (Lodd.) Greuter et Burdet ■☆

376344 Steris aquatica Burm. f. = Hydrolea zeylanica (L.) J. Vahl ■

376345 Steris javanica L. = Hydrolea zeylanica (L.) J. Vahl ■

376346 Steris viscaria (L.) Raf. = Silene viscaria (L.) Jess. ■☆

376347 Sterisia Raf. = Steris L. ■

376348 Sternbeckia Pers. = Singana Aubl. ☆

376349 Sternbeckia Pers. = Sterbeckia Schreb. ☆

376350 Sternbergia Waldst. et Kit. (1804)；黄韭兰属（黄花石蒜属，斯坦堡属，斯坦恩伯格属）；Sternbergia, Winter Daffodil ■☆

376351 Sternbergia candida B. Mathew et T. Baytop；白韭兰■☆

376352 Sternbergia clusiana Ker Gawl. ex Schult.；大花黄韭兰■☆

376353 Sternbergia colchiciflora Waldst. et Kit.；秋水仙黄韭兰■☆

376354 Sternbergia fischeriana (Herb.) Roem.；菲舍尔黄韭兰■☆

376355　Sternbergia lutea（L.）Ker Gawl. = Sternbergia lutea（L.）Ker Gawl. ex Roem. et Schult. ■☆

376356　Sternbergia lutea（L.）Ker Gawl. ex Roem. et Schult. ;黄花韭兰;Fall Daffodil, Fall-daffodil, Winter Daffodil, Yellow Star Flower ■☆

376357　Sternbergia sicula Tineo ex Guss. ;西西里黄韭兰（斯坦堡）■☆

376358　Sterrhymenia Griseb. = Sclerophylax Miers ■☆

376359　Sterropetalum N. E. Br. = Nelia Schwantes ●■☆

376360　Sterropetalum pillansii N. E. Br. = Nelia pillansii（N. E. Br.）Schwantes ■☆

376361　Stethoma Raf. = Justicia L. ●■

376362　Stethoma pectoralis（Jacq.）Raf. = Justicia pectoralis Jacq. ■☆

376363　Stetsonia Britton et Rose（1920）;近卫柱属（仙影掌属）●☆

376364　Stetsonia coryne（Salm-Dyck）Britton et Rose;近卫柱（近卫，仙影掌）;Toothpick Cactus ●☆

376365　Stetsonia coryne Britton et Rose = Stetsonia coryne（Salm-Dyck）Britton et Rose ●☆

376366　Steuarta Catesb. ex Mill. = Stewartia L. ●

376367　Steuartia Catesb. ex Mill. = Stewartia L. ●

376368　Steudelago Kuntze = Exostema（Pers.）Humb. et Bonpl. ●☆

376369　Steudelago Kuntze = Solenandra Hook. f. ●☆

376370　Steudelella Honda = Sphaerocaryum Nees ex Hook. f. ■

376371　Steudelella pulchella（Roth）Honda = Isachne pulchella Roth ■

376372　Steudelia C. Presl = Adenogramma Rchb. ■☆

376373　Steudelia Mart. = Leonia Ruiz et Pav. ●☆

376374　Steudelia Sprong. = Erythroxylum P. Browne ●

376375　Steudelia capillaris Eckl. et Zeyh. = Adenogramma capillaris（Eckl. et Zeyh.）Druce ■☆

376376　Steudelia diffusa Eckl. et Zeyh. = Adenogramma lichtensteiniana（Schult.）Druce ■☆

376377　Steudelia sylvatica Eckl. et Zeyh. = Adenogramma sylvatica（Eckl. et Zeyh.）Fenzl ■☆

376378　Steudelia viridis Gand. = Adenogramma sylvatica（Eckl. et Zeyh.）Fenzl ■☆

376379　Steudnera C. Koch = Steudnera K. Koch ■

376380　Steudnera K. Koch（1862）;泉七属（香芋属）;Steudnera ■

376381　Steudnera colocasiifolia Engl. = Steudnera griffithii Schott ■

376382　Steudnera colocasiifolia K. Koch;泉七（团芋，湾洪，香芋，小毒芋）;Henry Steudnera, Steudnera, Taro-leaf Steudnera ■

376383　Steudnera discolor Bull. ;异色泉七（异色岩芋）;Discolor Steudnera ■

376384　Steudnera griffithii Schott;全缘泉七;Entire Steudnera ■

376385　Steudnera henryana Engl. = Steudnera colocasiifolia K. Koch ■

376386　Stevena Andrz. ex DC. = Alyssum L. ■●

376387　Stevenia Adams et Fisch.（1817）;曙南芥属（念珠南芥属）;Stevenia ■

376388　Stevenia alyssoides Adams et Fisch. ;庭荠曙南芥■☆

376389　Stevenia cheiranthoides DC. ;曙南芥（山葶苈，施第芥）;Lipfern Stevenia, Montane Draba, Montane Whitlowgrass ■

376390　Steveniella Schltr.（1918）;史蒂文兰属■☆

376391　Steveniella satyrioides（Stev.）Schltr. ;史蒂文兰;Hooded Orchid ☆

376392　Stevenorchis Wankow et Kraenzl. = Steveniella Schltr. ■☆

376393　Stevensia Poit.（1802）;史蒂茜属●☆

376394　Stevensia Poit. = Rondeletia L. ●

376395　Stevensia buxifolia Poit. ;史蒂茜●☆

376396　Stevensia grandiflora Alain;大花史蒂茜●☆

376397　Stevensia minutifolia Alain;小叶史蒂茜●☆

376398　Stevensia ovatifolia Urb. et Ekman;卵叶史蒂茜●☆

376399　Stevensonia Duncan = Stevensonia Duncan ex Balf. f. ●☆

376400　Stevensonia Duncan ex Balf. f.（1877）;凤凰椰属;Stevenson Palm ●☆

376401　Stevensonia Duncan ex Balf. f. = Phoenicophorium H. Wendl. ●☆

376402　Stevensonia borsigiana L. H. Bailey = Phoenicophorium borsigianum Stuntz ●☆

376403　Stevensonia grandiflora Duncan;大花凤凰椰●☆

376404　Stevensonia viridifolia Duncan;绿叶凤凰椰●☆

376405　Stevia Cav.（1797）;甜叶菊属（甜菊属）;Candyleaf, Stevia ■●☆

376406　Stevia callosa Nutt. = Palafoxia callosa（Nutt.）Torr. et A. Gray ■☆

376407　Stevia ivifolia Willd. ;伊瓦菊叶甜叶菊■☆

376408　Stevia lemmonii A. Gray;莱蒙甜叶菊;Lemmon's Candyleaf ■☆

376409　Stevia micrantha Lag. ;小花甜叶菊;Annual Candyleaf ■☆

376410　Stevia ovata Lag. ;卵叶甜叶菊（卵叶甜菊）;Round-leaf Candyleaf ■☆

376411　Stevia ovata Willd. var. texana Grashoff;得州甜叶菊;Texas Candyleaf ■☆

376412　Stevia plummerae A. Gray;普氏甜叶菊;Plummer's Candyleaf ■☆

376413　Stevia plummerae A. Gray var. alba A. Gray = Stevia plummerae A. Gray ■☆

376414　Stevia purpurea Pers. ;紫色甜叶菊■☆

376415　Stevia rebaudiana（Bertoni）Bertoni;甜叶菊（瑞宝泽兰，蛇菊，甜菊）;Caa-ehe, Stevia, Sugar Plant, Sweet Honey Leaf ■☆

376416　Stevia salicifolia Cav. ;柳叶甜叶菊;Willow-leaf Candyleaf ■☆

376417　Stevia serrata Cav. ;齿叶甜叶菊;Saw-tooth Candyleaf ■☆

376418　Stevia serrata Cav. var. haplopappa B. L. Rob. = Stevia serrata Cav. ■☆

376419　Stevia serrata Cav. var. ivifolia（Willd.）B. L. Rob. = Stevia serrata Cav. ■☆

376420　Stevia sphacelata Nutt. ex Torr. = Palafoxia sphacelata（Nutt. ex Torr.）Cory ■☆

376421　Stevia viscida Kunth;黏甜叶菊;Viscid Candyleaf ■☆

376422　Steviopsis R. M. King et H. Rob.（1971）;轮叶修泽兰属■☆

376423　Steviopsis fendleri（A. Gray）B. L. Turner = Brickelliastrum fendleri（A. Gray）R. M. King et H. Rob. ■☆

376424　Steviopsis rapunculoides（DC.）R. M. King et H. Rob. ;轮叶修泽兰■☆

376425　Stevogtia Neck. = Convolvulus L. ■●

376426　Stevogtia Neck. ex Raf. = Phacelia Juss. ■☆

376427　Stevogtia Raf. = Phacelia Juss. ■☆

376428　Stewartia L.（1753）;紫茎属（旃檀属）;Earl of Bute, Purplestem, Stewarta, Stewartia ●

376429　Stewartia L. = Stuartia L'Hér. ●

376430　Stewartia acutisepala P. L. Chiu et G. R. Zhong = Stewartia gemmata S. S. Chien et W. C. Cheng ●

376431　Stewartia acutisepala P. L. Chiu et G. R. Zhong = Stewartia sinensis Rehder et E. H. Wilson var. acutisepala（P. L. Chiu et G. R. Zhong）T. L. Ming et J. Li ●

376432　Stewartia brevicalyx S. Z. Yan = Stewartia sinensis Rehder et E. H. Wilson var. brevicalyx（S. Z. Yan）T. L. Ming et J. Li ●

376433　Stewartia calcicola T. L. Ming et J. Li;云南紫茎（云南折柄茶，

云南舟柄茶）；Yunnan Hartia，Yunnan Purplestem ●

376434　Stewartia calcicola T. L. Ming et J. Li＝Hartia yunnanensis Hu ●

376435　Stewartia cordifolia（H. L. Li）J. Li et T. L. Ming；心叶紫茎（贵州折柄茶，心叶折柄茶）；Guizhou Hartia，Heartleaf Hartia，Heart-leaved Hartia ●

376436　Stewartia cordifolia（H. L. Li）J. Li et T. L. Ming＝Hartia cordifolia H. L. Li ●

376437　Stewartia crassifolia（S. Z. Yan）J. Li et T. L. Ming；厚叶紫茎（厚叶折柄茶，圆萼折柄茶）；Round-calyx Hartia，Roundsepal Hartia ●

376438　Stewartia crassifolia（S. Z. Yan）J. Li et T. L. Ming＝Hartia crassifolia S. Z. Yan ●

376439　Stewartia damingshanica J. Li et T. L. Ming＝Stewartia rubiginosa Hung T. Chang var. damingshanica（J. Li et T. L. Ming）T. L. Ming ●

376440　Stewartia densivillosa（Hu ex Hung T. Chang et C. X. Ye）J. Li et T. L. Ming＝Hartia densivillosa Hu ex Hung T. Chang et C. X. Ye ●

376441　Stewartia densivillosa（Hu ex Hung T. Chang et C. X. Ye）J. Li et T. L. Ming；狭萼紫茎（狭萼折柄茶）；Narrow-calyx Hartia，Narrowsepal Hartia，Narrow-sepaled Hartia ●

376442　Stewartia gemmata S. S. Chien et W. C. Cheng；天目紫茎（尖萼紫茎）；Chinese Stewartia，Tianmu Mountain Stewartia，Tianmu Stewartia，Tianmushan Purplestem ●

376443　Stewartia gemmata S. S. Chien et W. C. Cheng＝Stewartia sinensis Rehder et E. H. Wilson ●

376444　Stewartia glabra S. Z. Yan；秃房紫茎（光紫茎）；Glabrous Stewartia，Nakeovary Purplestem ●

376445　Stewartia glabra S. Z. Yan＝Stewartia rostrata Spongberg ●

376446　Stewartia koreana Nakai ex Rehder；朝鲜紫茎；Korean Stewartia ●☆

376447　Stewartia laotica（Gagnep.）J. Li et T. L. Ming；老挝紫茎（细柄折柄茶，小萼折柄茶，总状折柄茶）；Little-caalyx Hartia，Raceme Hartia，Slender Hartia，Slender Yunnan Hartia ●

376448　Stewartia laotica（Gagnep.）J. Li et T. L. Ming＝Hartia laotica Gagnep. ●

376449　Stewartia longibracteata Hung T. Chang；长苞紫茎；Longbract Purplestem，Longbract Stewartia，Long-bracted Stewartia ●

376450　Stewartia longibracteata Hung T. Chang＝Pyrenaria diospyricarpa Kurz ●

376451　Stewartia longibracteata Hung T. Chang＝Pyrenaria yunnanensis Hu ●

376452　Stewartia malacodendron L.；圆果紫茎（弗吉尼亚紫茎，绢毛紫茎）；Round-fruit Stewartia，Sewartia，Silky Camellia，Silky-camellia，Virginia Stewartia ●☆

376453　Stewartia mangshanica C. X. Ye＝Stewartia rubiginosa Hung T. Chang ●

376454　Stewartia medogensis J. Li et T. L. Ming；墨脱紫茎（短萼折柄茶）；Shortcalyx Hartia ●

376455　Stewartia micrantha（Chun）Sealy；小花紫茎（小花折柄茶）；Littleflower Hartia，Smallflower Hartia，Small-flowered Hartia ●

376456　Stewartia micrantha（Chun）Sealy＝Hartia micrantha Chun ●

376457　Stewartia monadelpha Siebold et Zucc.；大紫茎（单体蕊紫茎，马榴光，日本紫茎）；Japanese Stewartia，Tall Stewartia ●☆

376458　Stewartia monadelpha Siebold et Zucc. f. sericea（Nakai）H. Hara；绢毛大紫茎●☆

376459　Stewartia nanlingensis S. Z. Yan；南岭紫茎；Nanling Purplestem，Nanling Stewartia ●

376460　Stewartia nanlingensis S. Z. Yan＝Stewartia sinensis Rehder et E. H. Wilson ●

376461　Stewartia oblongifolia Hu ex S. Z. Yan；长叶紫茎；Longleaf Purplestem，Longleaf Stewartia ●

376462　Stewartia oblongifolia Hu ex S. Z. Yan＝Stewartia rubiginosa Hung T. Chang ●

376463　Stewartia obovata（Chun ex Hung T. Chang）J. Li et T. L. Ming；钝叶紫茎（钝叶赫德木，钝叶折柄茶，密脉折柄茶）；Densenerve Hartia，Obvate Hartia，Obvate-leaf Hartia ●

376464　Stewartia obovata（Chun ex Hung T. Chang）J. Li et T. L. Ming＝Hartia obovata Chun ex Hung T. Chang ●

376465　Stewartia ovata（Cav.）Weath.；山紫茎；Angle-fruit Stewartia，Mountain Camellia，Mountain Purplestem，Mountain Stewartia，Mountain-camellia ●

376466　Stewartia pseudocamellia Maxim.；日本紫茎（大花紫茎，假山茶，沙罗树，娑罗花，夏椿）；Japanese Stewarta，Japanese Stewartia ●☆

376467　Stewartia pteropetiolata W. C. Cheng；翅柄紫茎（折柄茶，舟柄茶）；China Hartia，Chinese Hartia，Wingedpetiole Purplestem，Wingedpetiole Stewartia ●

376468　Stewartia pteropetiolata W. C. Cheng＝Hartia sinensis Dunn ●

376469　Stewartia rostrata Spongber；长柱紫茎（长喙紫茎，尖嘴紫茎）；Longstyle Purplestem，Long-styled Stewartia，Rostrate Stewartia，Upright Stewarta ●

376470　Stewartia rostrata Spongber＝Stewartia sinensis Spongber var. rostrata（Spongber）Hung T. Chang ●

376471　Stewartia rubiginosa Hung T. Chang；红皮紫茎；Redbark Stewartia，Red-barked Stewartia，Rust Purplestem ●

376472　Stewartia rubiginosa Hung T. Chang var. damingshanica（J. Li et T. L. Ming）T. L. Ming；大明紫茎●

376473　Stewartia serrata Maxim.；髯毛紫茎；Chinese Stewartia ●☆

376474　Stewartia serrata Maxim. var. epitricha（Nakai）Ohwi＝Stewartia serrata Maxim. ●☆

376475　Stewartia shensiensis Hung T. Chang＝Stewartia sinensis Rehder et E. H. Wilson var. shensiensis（Hung T. Chang）T. L. Ming et J. Li ●

376476　Stewartia sichuanensis（S. Z. Yan）J. Li et T. L. Ming；四川紫茎（四川折柄茶）；Sichuan Hartia ●

376477　Stewartia sichuanensis（S. Z. Yan）J. Li et T. L. Ming＝Hartia sichuanensis S. Z. Yan ●

376478　Stewartia sinensis Rehder et E. H. Wilson；紫茎（马骝光，帽兰，天目紫茎，野茶子）；China Purplestem，Chinese Stewartia ●

376479　Stewartia sinensis Rehder et E. H. Wilson var. acutisepala（P. L. Chiu et G. R. Zhong）T. L. Ming et J. Li；尖萼紫茎 ●

376480　Stewartia sinensis Rehder et E. H. Wilson var. brevicalyx（S. Z. Yan）T. L. Ming et J. Li；短萼紫茎；Shortcalyx Purplestem，Shortcalyx Stewartia ●

376481　Stewartia sinensis Rehder et E. H. Wilson var. rostrata（Spongber）Hung T. Chang＝Stewartia rostrata Spongber ●

376482　Stewartia sinensis Rehder et E. H. Wilson var. shensiensis（Hung T. Chang）T. L. Ming et J. Li；陕西紫茎；Shaanxi Purplestem，Shaanxi Stewartia ●

376483　Stewartia sinensis Spongber＝Stewartia gemmata S. S. Chien et W. C. Cheng ●

376484　Stewartia sinensis Spongber var. rostrata（Spongber）Hung T. Chang；长喙紫茎；Rostrate Chinese Purplestem，Rostrate Chinese Stewartia ●

376485 Stewartia sinii（Y. C. Wu）Sealy；黄毛紫茎（黄毛折柄茶）；Sin Hartia，Yellow-hair Hartia ●

376486 Stewartia sinii（Y. C. Wu）Sealy = Hartia sinii Y. C. Wu ●

376487 Stewartia villosa Merr.；柔毛紫茎（毛折柄茶，南昆折柄茶）；Hair Hartia，Nankun Hartia，Villous Hartia ●

376488 Stewartia villosa Merr. = Hartia villosa（Merr.）Merr. ●

376489 Stewartia villosa Merr. var. grandifolia（Chun）J. Li et T. L. Ming = Stewartia villosa Merr. var. kwangtungensis（Chun）J. Li et T. L. Ming ●

376490 Stewartia villosa Merr. var. kwangtungensis（Chun）J. Li et T. L. Ming；广东柔毛紫茎（大叶毛折柄茶，贴毛折柄茶）；Bigleaf Hartia，Guangdong Hartia ●

376491 Stewartia villosa Merr. var. kwangtungensis（Chun）J. Li et T. L. Ming = Hartia villosa（Merr.）Merr. var. kwangtungensis（Chun）Hung T. Chang ●

376492 Stewartia villosa Merr. var. serrata（Hu）T. L. Ming；齿叶柔毛紫茎 ●

376493 Stewartia yunnanensis Hung T. Chang = Pyrenaria diospyricarpa Kurz ●

376494 Stewartiella Nasir（1972）；斯图阿魏属■☆

376495 Stewartiella baluchistanica Nasir；斯图阿魏■☆

376496 Stewartiella crucifolia（Gilli）Hedge et Lamond；阿富汗斯图阿魏■☆

376497 Steyerbromelia L. B. Sm.（1987）；施泰凤梨属■☆

376498 Steyerbromelia deflexa L. B. Sm. et H. Rob.；施泰凤梨■☆

376499 Steyermarkia Standl.（1940）；斯泰茜属☆

376500 Steyermarkia guatemalensis Standl.；斯泰茜☆

376501 Steyermarkina R. M. King et H. Rob.（1971）；长瓣亮泽兰属 ●☆

376502 Steyermarkina pyrifolia（DC.）R. M. King et H. Rob.；梨叶长瓣亮泽兰 ●☆

376503 Steyermarkina triflora R. M. King et H. Rob.；三花长瓣亮泽兰 ●☆

376504 Steyermarkochloa Davidse et R. P. Ellis（1985）；单叶草属■☆

376505 Steyermarkochloa angustifolia（Sprengel）Judz.；窄单叶草■☆

376506 Steyermarkochloa unifolia Davidse et R. P. Ellis；单叶草■☆

376507 Sthaelina Lag. = Staehelina L. ●☆

376508 Stibadotheca Klotzsch = Begonia L. ●■

376509 Stibas Comm. ex DC. = Levenhookia R. Br. ■☆

376510 Stiburus Stapf = Eragrostis Wolf ■

376511 Stiburus Stapf（1900）；肖画眉草属■☆

376512 Stiburus alopecuroides（Hack.）Stapf；肖画眉草■☆

376513 Stiburus conrathii Hack.；非洲肖画眉草■☆

376514 Stichianthus Valeton et Bremek. = Stichianthus Valeton ☆

376515 Stichianthus Valeton（1920）；单列花属☆

376516 Stichianthus minutiflorus Valeton；单列花☆

376517 Stichoneuron Hook. f.（1883）；单列脉属■☆

376518 Stichoneuron membranaceum Hook. f. et Thomson；单列脉■☆

376519 Stichophyllum Phil. = Pycnophyllum J. Rémy ■☆

376520 Stichorchis Thouars = Liparis Rich.（保留属名）■

376521 Stichorkis Thouars = Liparis Rich.（保留属名）■

376522 Stichorkis latifolia（Lindl.）Pfitzer = Liparis latifolia（Blume）Lindl. ■

376523 Stickmannia Neck. = Dichorisandra J. C. Mikan（保留属名）■☆

376524 Stictocardia Hallier f.（1893）；腺叶藤属（大萼旋花属）；Stictocardia ●■

376525 Stictocardia beraviensis（Vatke）Hallier f.；贝拉维腺叶藤●☆

376526 Stictocardia beraviensis（Vatke）Hallier f. subsp. laxiflora（Baker）Verdc. = Stictocardia laxiflora（Baker）Hallier f. ●☆

376527 Stictocardia campanulata（L.）House = Stictocardia tiliifolia（Desr.）Hallier f. ●■

376528 Stictocardia incomta（Hallier f.）Hallier f.；贫弱腺叶藤■☆

376529 Stictocardia laxiflora（Baker）Hallier f.；疏花腺叶藤●☆

376530 Stictocardia laxiflora（Baker）Hallier f. var. woodii（N. E. Br.）Verdc. = Stictocardia laxiflora（Baker）Hallier f. ●☆

376531 Stictocardia lutambensis（Schulze-Menz）Verdc.；卢塔腺叶藤■☆

376532 Stictocardia macalusoi（Mattei）Verdc.；马卡卢索腺叶藤●☆

376533 Stictocardia multiflora Hallier f.；多花腺叶藤●☆

376534 Stictocardia tiliifolia（Choisy）Hallier f. = Stictocardia tiliifolia（Desr.）Hallier f. ●■

376535 Stictocardia tiliifolia（Choisy）Hallier f. subsp. macalusoi（Mattei）Verdc. = Stictocardia macalusoi（Mattei）Verdc. ●☆

376536 Stictocardia tiliifolia（Desr.）Hallier f.；腺叶藤（大萼旋花，椴叶白鹤藤，椴叶牵牛）；Common Stictocardia，Linden-leaf Asiaglory，Stictocardia ●■

376537 Stictocardia woodii（N. E. Br.）Hallier f. = Stictocardia laxiflora（Baker）Hallier f. ●☆

376538 Stictophyllorchis Carnevali et Dodson = Stictophyllorchis Dodson et Carnevali ■☆

376539 Stictophyllorchis Dodson et Carnevali（1993）；类斑叶兰属■☆

376540 Stictophyllorchis pygmaea（Cogn.）Dodson et Carnevali；类斑叶兰■☆

376541 Stictophyllum Dodson et M. W. Chase = Stictophyllorchis Dodson et Carnevali ■☆

376542 Stictophyllum Edgew. = Tricholepis DC. ■

376543 Stiefia Medik. = Salvia L. ●■

376544 Stifftia J. C. Mikan（1820）（保留属名）；斯迪菊属（亮毛菊属）●☆

376545 Stifftia axillaris Vinha；腋生斯迪菊（腋生亮毛菊）●☆

376546 Stifftia chrysantha Mikan；斯迪菊（金花亮毛菊）●☆

376547 Stifftia chrysantha Mikan var. oligantha Baker；寡花斯迪菊●☆

376548 Stifftia fruticosa（Vell.）D. J. N. Hind et J. Semir；灌木斯迪菊（灌木亮毛菊）●☆

376549 Stifftia uniflora Ducke；单花斯迪菊（单花亮毛菊）●☆

376550 Stiftia Cass. = Stifftia J. C. Mikan（保留属名）●☆

376551 Stiftia Pohl ex Nees = Ebermaiera Nees ■

376552 Stigmamblys Kuntze = Amblystigma Benth. ☆

376553 Stigmanthus Lour. = Morinda L. ●■

376554 Stigmaphyllon A. Juss.（1833）；蕊叶藤属（刺叶藤属）；Amazonvine，Brazilian Golden Vine，Orchid Vine ●☆

376555 Stigmaphyllon albiflorum Cuatrec.；白花蕊叶藤●☆

376556 Stigmaphyllon angustifolium Griseb.；狭叶蕊叶藤●☆

376557 Stigmaphyllon ciliatum A. Juss.；蕊叶藤（巴西土地库买伦）；Fringed Amazonvine，Golden Vine，Orchid Vine ●☆

376558 Stigmaphyllon fulgens A. Juss.；光亮蕊叶藤（光亮刺叶）●☆

376559 Stigmaphyllon littorale A. Juss.；海岸蕊叶藤；Orchid Vine ●☆

376560 Stigmaphyllon microphyllum Griseb.；小叶蕊叶藤●☆

376561 Stigmaphyllon orientale Cuatrec.；东方蕊叶藤●☆

376562 Stigmaphyllon ovatum（Cav.）Nied.；卵状蕊叶藤●☆

376563 Stigmaphyllon pseudopuberum Nied.；假毛蕊叶藤（假毛刺叶）●☆

376564 Stigmaphyllon sinuatum A. Juss.；深波刺叶●☆

376565 Stigmaphyllon strigosum（Poepp.）A. Juss.；糙毛蕊叶藤（糙毛

刺叶) ●☆

376566　Stigmarosa Hook. f. et Thomson = Stigmarota Lour. ●

376567　Stigmarota Lour. = Flacourtia Comm. ex L'Hér. ●

376568　Stigmarota africana Lour. = Flacourtia ramontchii L'Hér. ●

376569　Stigmarota jangomas Lour. = Flacourtia jangomas (Lour.) Raeusch. ●

376570　Stigmatanthus Roem. et Schult. = Morinda L. ●■

376571　Stigmatanthus Roem. et Schult. = Stigmanthus Lour. ●■

376572　Stigmatella Eig = Eigia Soják ■☆

376573　Stigmatocarpum L. Bolus = Dorotheanthus Schwantes ■☆

376574　Stigmatococca Willd. = Ardisia Sw. (保留属名) ●■

376575　Stigmatococca Willd. ex Schult. = Ardisia Sw. (保留属名) ●■

376576　Stigmatodactylus Maxim. ex Makino(1891);指柱兰属(腐指柱兰属);Digitstyleorchis ●

376577　Stigmatodactylus palawensis Tuyama = Disperis neilgherrensis Wight ■

376578　Stigmatodactylus sikokianus Maxim. ex Makino;指柱兰;Digitstyleorchis ■

376579　Stigmatophyllon Meisn. = Stigmaphyllon A. Juss. ●☆

376580　Stigmatophyllum Spach = Stigmatophyllon Meisn. ●☆

376581　Stigmatorhynchus Schltr. (1913);喙柱萝藦属●☆

376582　Stigmatorhynchus hereroensis Schltr. ;赫雷罗喙柱萝藦●☆

376583　Stigmatorhynchus stelostigma (K. Schum.) Schltr. = Dregea stelostigma (K. Schum.) Bullock ●☆

376584　Stigmatorhynchus umbelliferus (K. Schum.) Schltr. ;伞花喙柱萝藦●☆

376585　Stigmatorthos M. W. Chase et D. E. Benn. (1993);直柱兰属■☆

376586　Stigmatorthos peruviana M. W. Chase et D. E. Benn. ;直柱兰■☆

376587　Stigmatosema Garay(1982);显柱兰属■☆

376588　Stigmatosema polyaden (Vell.) Garay;显柱兰■☆

376589　Stigmatotheca Sch. Bip. = Argyranthemum Webb ex Sch. Bip. ●

376590　Stigmatotheca Sch. Bip. = Leucopoa Griseb. ■

376591　Stigraatocarpum L. Bolus = Dorotheanthus Schwantes ■☆

376592　Stilaginaceae C. Agardh = Euphorbiaceae Juss. (保留科名) ●■

376593　Stilaginaceae C. Agardh = Phyllanthaceae J. Agardh ●■

376594　Stilaginaceae C. Agardh;五月茶科●

376595　Stilaginella Tul. = Hieronima Allemão ●☆

376596　Stilago L. = Antidesma L. ●

376597　Stilago bunius L. = Antidesma bunius (L.) Spreng. ●

376598　Stilago diandra Roxb. = Antidesma acidum Retz. ●

376599　Stilago lanceolaria Roxb. = Antidesma acidum Retz. ●

376600　Stilbaceae Kunth(1831)(保留科名);密穗草科●☆

376601　Stilbanthus Hook. f. (1879);巨苋藤属(巨藤苋属,巨苋属);Stilbanthus ●

376602　Stilbanthus scandens Hook. f. ;巨苋藤●

376603　Stilbe P. J. Bergius(1767);密穗草属●☆

376604　Stilbe albiflora E. Mey. ;白花密穗草●☆

376605　Stilbe cernua L. f. = Campylostachys cernua (L. f.) Kunth ●☆

376606　Stilbe chorisepala Suess. = Kogelbergia verticillata (Eckl. et Zeyh.) Rourke ●☆

376607　Stilbe ericoides (L.) L. ;石南状密穗草●☆

376608　Stilbe mucronata N. E. Br. = Kogelbergia verticillata (Eckl. et Zeyh.) Rourke ●☆

376609　Stilbe mucronata N. E. Br. var. cuspidata H. Pearson = Kogelbergia verticillata (Eckl. et Zeyh.) Rourke ●☆

376610　Stilbe overbergensis Rourke;奥沃贝格密穗草●☆

376611　Stilbe phylicoides A. DC. = Kogelbergia phylicoides (A. DC.) Rourke ●☆

376612　Stilbe rupestris Compton;岩生密穗草●☆

376613　Stilbe serrulata (Hochst.) Rourke;细齿密穗草●☆

376614　Stilbe verticillata (Eckl. et Zeyh.) Moldenke = Kogelbergia verticillata (Eckl. et Zeyh.) Rourke ●☆

376615　Stilbe verticillata (Eckl. et Zeyh.) Moldenke var. cuspidata (H. Pearson) Moldenke = Kogelbergia verticillata (Eckl. et Zeyh.) Rourke ●☆

376616　Stilbe vestita P. J. Bergius;包被密穗草●☆

376617　Stilbe zeyheri Gand. = Stilbe albiflora E. Mey. ●☆

376618　Stilbeaceae Bullock = Stilbaceae Kunth(保留科名) ●☆

376619　Stilbocarpa (Hook. f.) A. Gray = Stilbocarpa (Hook. f.) Decne. et Planch. ●☆

376620　Stilbocarpa (Hook. f.) Decne. et Planch. (1854);槭果五加属●☆

376621　Stilbocarpa A. Gray = Stilbocarpa (Hook. f.) Decne. et Planch. ●☆

376622　Stilbocarpa polaris (Hook. f.) A. Gray;槭果五加●☆

376623　Stilbophyllum D. L. Jones et M. A. Clem. (2002);槭叶兰属■☆

376624　Stilifolium Königer et D. Pongratz(1997);柱叶兰属■☆

376625　Stilingia Raf. = Stillingia Garden ex L. ●■☆

376626　Stillengia Torr. = Stillingia Garden ex L. ●■☆

376627　Stillingfleetia Bojer = Sapium Jacq. (保留属名) ●

376628　Stillingfleetia sebifera Bojer = Sapium sebiferum (L.) Roxb. ●

376629　Stillingia Garden ex L. (1767);假乌桕属(皇后根属,柿苓属) ●■☆

376630　Stillingia L. = Stillingia Garden ex L. ●■☆

376631　Stillingia africana (Sond.) Müll. Arg. = Spirostachys africana Sond. ■☆

376632　Stillingia agallocha Müll. Arg. = Excoecaria agallocha L. ●

376633　Stillingia baccata (Roxb.) Baill. = Sapium baccatum Roxb. ●

376634　Stillingia discolor Champ. ex Benth. = Sapium discolor (Champ. ex Benth.) Müll. Arg. ●

376635　Stillingia diversifolia Miq. = Shirakiopsis indica (Willd.) Esser ●

376636　Stillingia elliptica (Hochst.) Baill. = Shirakiopsis elliptica (Hochst.) Esser ●☆

376637　Stillingia himalayensis Klotzsch = Excoecaria acerifolia Didr. ●

376638　Stillingia integerrima (Hochst.) Baill. = Sclerocroton integerrimus Hochst. ●☆

376639　Stillingia japonica Siebold et Zucc. = Sapium japonicum (Siebold et Zucc.) Pax et K. Hoffm. ●

376640　Stillingia laurifolia A. Rich. ;月桂叶假乌桕●☆

376641　Stillingia paucidentata S. Watson;齿叶假乌桕;Tooth-leaf ●■☆

376642　Stillingia sebifera Michx. = Sapium sebiferum (L.) Roxb. ●

376643　Stillingia sinensis Müll. Arg. = Sapium sebiferum (L.) Roxb. ●

376644　Stillingia sylvatica L. = Sapium sylvaticum Torr. ●☆

376645　Stillingia treculiana I. M. Johnst. ;银叶假乌桕(皇后根);Queen's Delight,Queen's Root,Silver-leaf,Yawroot ●■☆

376646　Stilopus Hook. = Geum L. ■

376647　Stilopus Hook. = Stylypus Raf. ■☆

376648　Stilpnogyne DC. (1838);耳雏菊属■☆

376649　Stilpnogyne bellidioides DC. ;耳雏菊■☆

376650　Stilpnolepis Krasch. (1946);百花蒿属;Stilpnolepis ■

376651　Stilpnolepis centiflora (Maxim.) Krasch. ;百花蒿;Manyflower Stilpnolepis,Stilpnolepis ■

376652　Stilpnolepis centiflora（Maxim.）Krasch. var. pilifera（Y. Ling）H. C. Fu = Stilpnolepis centiflora（Maxim.）Krasch. ■

376653　Stilpnolepis intricata（Franch.）C. Shih = Elachanthemum intricatum（Franch.）Y. Ling et Y. R. Ling ■

376654　Stilpnopappus Mart. = Stilpnopappus Mart. ex DC. ●■☆

376655　Stilpnopappus Mart. ex DC.（1836）；芒冠斑鸠菊属●■☆

376656　Stilpnopappus bicolor Mart. ex Baker；二色芒冠斑鸠菊●■☆

376657　Stilpnopappus tomentosus Mart. ex DC. ；芒冠斑鸠菊●■☆

376658　Stilpnophleum Nevski = Calamagrostis Adans. ■

376659　Stilpnophleum Nevski = Deyeuxia Clarion ■

376660　Stilpnophleum Nevski（1937）；亮梯牧草属■☆

376661　Stilpnophleum anthoxanthoides（Munro）Nevski；黄花亮梯牧草☆

376662　Stilpnophleum laguroides（Regel）Nevski；亮梯牧草■☆

376663　Stilpnophyllum（Endl.）Drury = Ficus L. ■

376664　Stilpnophyllum Hook. f.（1873）；亮叶茜属●☆

376665　Stilpnophyllum grandifolium L. Andersson；大叶亮叶茜●☆

376666　Stilpnophyllum lineatum Hook. f. ；亮叶茜●☆

376667　Stilpnophyton Less. = Athanasia L. ●☆

376668　Stilpnophyton axillare（Thunb.）Less. = Athanasia minuta（L. f.）Källersjö ●☆

376669　Stilpnophyton inopinatum Hutch. = Athanasia inopinata（Hutch.）Källersjö ●☆

376670　Stilpnophyton linifolium（L. f.）Less. = Athanasia linifolia Burm. ●☆

376671　Stilpnophyton linifolium（L. f.）Less. var. longifolium（Thunb.）Harv. = Athanasia linifolia Burm. ●☆

376672　Stilpnophyton longifolium（Thunb.）Less. = Athanasia linifolia Burm. ●☆

376673　Stilpnophyton oocephalum DC. = Athanasia oocephala（DC.）Källersjö ●☆

376674　Stilpnophytum Less. = Athanasia L. ●☆

376675　Stimegas Raf.（废弃属名）= Cypripedium L. ■

376676　Stimegas Raf.（废弃属名）= Paphiopedilum Pfitzer（保留属名）■

376677　Stimegas venustum（Wall. ex Sims）Raf. = Paphiopedilum venustum（Wall. ex Sims）Pfitzer ■

376678　Stimenes Raf.（废弃属名）= Nierembergia Ruiz et Pav. ■☆

376679　Stimomphis Raf. = Calibrachoa Cerv. ■☆

376680　Stimomphis Raf. = Salpiglossis Ruiz et Pav. ■☆

376681　Stimoryne Raf. = Petunia Juss.（保留属名）■

376682　Stimpsonia C. Wright ex A. Gray（1858）；假婆婆纳属（施丁草属）；Stimpsonia ■

376683　Stimpsonia chamaedryoides Wright ex A. Gray；假婆婆纳（施丁草）；Common Stimpsonia ■

376684　Stimpsonia chamaedryoides Wright ex A. Gray f. rubriflora J. Z. Shao；红花假婆婆纳；Redflower Stimpsonia ■

376685　Stimpsonia crispidens Hance = Lysimachia crispidens（Hance）Hemsl. ■

376686　Stingana B. D. Jacks. = Singana Aubl. ☆

376687　Stipa L.（1753）；针茅属（羽茅属）；Corkscrew Grass, Feather Grass, Feathergrass, Feather-grass, Grass, Needdle Grass, Needlegrass, Needle-grass, Spear Grass ■

376688　Stipa africana Burm. f. ；非洲针茅■☆

376689　Stipa aliena Keng；异针茅；Foreign Feathergrass, Foreign Needlegrass ■

376690　Stipa alpina（F. Schmidt）Petr. ；高山针茅■☆

376691　Stipa anomala P. A. Smirn. ；异常针茅■☆

376692　Stipa antiatlantica Barrena, D. Rivera, Alcaraz et Obón = Macrochloa antatlantica（Barrena, D. Rivera, Alcaraz et Obón）H. Scholz et Valdés ■☆

376693　Stipa apertifolia Martinovsky；阔叶针茅■☆

376694　Stipa apertifolia Martinovsky subsp. longiglumis（Scholz）R. Vásquez et Devesa；长颖针茅■☆

376695　Stipa arabica Trin. et Rupr. ；图尔盖针茅；Turgai Feathergrass, Turgai Needlegrass ■

376696　Stipa arabica Trin. et Rupr. subsp. caspia（K. Koch）Tzvelev = Stipa arabica Trin. et Rupr. ■

376697　Stipa arabica Trin. et Rupr. subsp. caspia（K. Koch）Tzvelev = Stipa szowitsiana Trin. ex Hohen. ■

376698　Stipa arabica Trin. et Rupr. subsp. caspia（K. Koch）Tzvelev = Stipa turgaica Roshev. ■

376699　Stipa arabica Trin. et Rupr. var. szovitsiana Trin. = Stipa arabica Trin. et Rupr. ■

376700　Stipa arabica Trin. et Rupr. var. turgaica（Roshev.）Tzvelev = Stipa arabica Trin. et Rupr. ■

376701　Stipa araxensis Grossh. ；阿拉克西针茅■☆

376702　Stipa arenaria Brot. = Macrochloa arenaria（Brot.）Kunth ■☆

376703　Stipa aristella L. = Achnatherum bromoides（L.）P. Beauv. ■☆

376704　Stipa armeniaca P. A. Smirn. ；亚美尼亚针茅■☆

376705　Stipa arundinacea Benth. ；芦针茅；Pheasant Grass, Pheasant's Tail Grass, Pheasant-grass ■☆

376706　Stipa atlantica Smirn. ；大西洋针茅■☆

376707　Stipa avenacea Hook. et Arn. ；燕麦针茅；Black Oat Grass ■☆

376708　Stipa avenoides Honda = Achnatherum sibiricum（L.）Keng ■

376709　Stipa badachschanica Roshev. ；巴达针茅■☆

376710　Stipa baicalensis Roshev. ；狼针草（贝加尔针茅, 狼针茅）；Baical Feathergrass, Baical Needlegrass ■

376711　Stipa balansae Scholz；巴兰萨针茅■☆

376712　Stipa barbata Desf. ；髯毛针茅■☆

376713　Stipa barbata Desf. subsp. brevipila（Coss. et Durieu）R. Vásquez et Devesa；短髯毛针茅■☆

376714　Stipa barbata Desf. var. brevipila Coss. et Durieu = Stipa barbata Desf. subsp. brevipila（Coss. et Durieu）R. Vásquez et Devesa ■☆

376715　Stipa basiplumosa Munro ex Hook. f. = Stipa subsessiliflora（Rupr.）Roshev. var. basiplumosa（Munro ex Hook. f.）P. C. Kuo et Y. H. Sun ■

376716　Stipa basiplumosa Munro ex Hook. f. = Stipa subsessiliflora（Rupr.）Roshev. ■

376717　Stipa basiplumosa Munro ex Hook. f. var. longearistata Munro ex Hook. f. = Stipa roborowskyi Roshev. ■

376718　Stipa bella Drobow；雅丽针茅■☆

376719　Stipa brandisii Mez = Stipa sibirica（L.）Lam. ■

376720　Stipa breviflora Griseb. ；短花针茅；Shortflower Feathergrass, Shortflower Needlegrass ■

376721　Stipa bromoides（L.）Dörfl. ；雀麦状针茅；Bromus-like Feathergrass ■☆

376722　Stipa bromoides（L.）Dörfl. = Achnatherum bromoides（L.）P. Beauv. ■☆

376723　Stipa bungeana Trin. ex Bunge；本氏针茅（长芒草）；Bunge Feathergrass, Bunge Needlegrass ■

376724　Stipa calamagrostis（L.）Wahlenb. = Achnatherum calamagrostis（L.）P. Beauv. ■

376725　Stipa calamagrostis Wahlenb. ；拂子茅状针茅■☆

376726 Stipa canadensis Poir. = Oryzopsis canadensis (Poir.) Torr. ■☆

376727 Stipa canescens P. A. Smirn.; 灰针茅■☆

376728 Stipa capensis Thunb.; 地中海针茅; Mediterranean Needle-grass, Pheasant-grass ■☆

376729 Stipa capensis Thunb. var. pubescens (Ball) Breistr. = Stipa capensis Thunb. ■☆

376730 Stipa capillacea Keng; 丝颖针茅; Capillary Feathergrass, Silkglume Needlegrass ■

376731 Stipa capillacea Keng var. parviflora N. X. Zhao et M. F. Li; 小花丝颖针茅; Smallflower Capillary Feathergrass ■

376732 Stipa capillata L.; 针茅; Bridal-veil-grass, Needlegrass ■

376733 Stipa capillata L. var. coronata Roshev. = Stipa sareptana Becker var. krylovii (Roshev.) P. C. Kuo et Y. H. Sun ■

376734 Stipa capillata L. var. sareptana (A. K. Becker) Schmalh. = Stipa sareptana Becker ■

376735 Stipa caragana Trin. et Rupr. = Achnatherum caraganum (Trin. et Rupr.) Nevski ■

376736 Stipa caspia K. Koch = Stipa arabica Trin. et Rupr. ■

376737 Stipa caucasica Schmalh.; 镰芒针茅; Caucasus Feathergrass, Silkleawn Needlegrass ■

376738 Stipa caucasica Schmalh. f. desertorum Roshev. = Stipa caucasica Schmalh. subsp. glareosa (P. A. Sm.) Tzvelev ■

376739 Stipa caucasica Schmalh. subsp. desertorum (Roshev.) Tzvelev; 荒漠镰芒针茅■

376740 Stipa caucasica Schmalh. subsp. desertorum (Roshev.) Tzvelev = Stipa caucasica Schmalh. subsp. glareosa (P. A. Sm.) Tzvelev ■

376741 Stipa caucasica Schmalh. subsp. glareosa (P. A. Sm.) Tzvelev = Stipa glareosa C. C. Davis ■

376742 Stipa caucasica Schmalh. var. desertorum (Roshev.) Tzvelev = Stipa caucasica Schmalh. subsp. glareosa (P. A. Sm.) Tzvelev ■

376743 Stipa celakovskyi Martinovsky = Stipa juncea L. ■☆

376744 Stipa charruana Arechav.; 卡瑞针茅; Charruana Feathergrass ■☆

376745 Stipa chingii Hitchc. = Achnatherum chingii (Hitchc.) Keng ex P. C. Kuo ■

376746 Stipa chitralensis Bor; 吉德拉尔针茅■☆

376747 Stipa clandestina Hack. = Achnatherum clandestinum (Hack.) Barkworth ■☆

376748 Stipa clausa (Trab.) R. Vàsquez Pardo et Devesa; 云雾针茅 ■☆

376749 Stipa comata Trin. et Rupr.; 种缨针茅; Needle and Thread Grass, Needle-and-thread ■☆

376750 Stipa comata Trin. et Rupr. subsp. intonsa Piper = Stipa comata Trin. et Rupr. ■☆

376751 Stipa comata Trin. et Rupr. var. suksdorfii H. St. John = Stipa comata Trin. et Rupr. ■☆

376752 Stipa concinna Hook. f. = Ptilagrostis concinna (Hook. f.) Roshev. ■

376753 Stipa consanguinea Trin. et Rupr.; 宜红针茅 (近亲针茅) ■

376754 Stipa coreana Honda; 朝鲜针茅■☆

376755 Stipa coreana Honda var. japonica (Hack.) Y. N. Lee; 日本针茅■☆

376756 Stipa coreana Honda var. kengii (Ohwi) Ohwi = Stipa coreana Honda ■☆

376757 Stipa crassiculmis P. A. Smirn.; 粗秆针茅■☆

376758 Stipa daghestanica Grossh.; 达赫斯坦针茅■☆

376759 Stipa dasyphylla Czern.; 毛叶针茅■☆

376760 Stipa dasyvaginata Martinovsky = Stipa apertifolia Martinovsky ■☆

376761 Stipa dasyvaginata Martinovsky subsp. longiglumis Scholz = Stipa apertifolia Martinovsky subsp. longiglumis (Scholz) R. Vásquez et Devesa ■☆

376762 Stipa densiflora P. A. Smirn.; 密花针茅■☆

376763 Stipa diminuta Mez = Aristida diminuta (Mez) C. E. Hubb. ■☆

376764 Stipa dregeana Steud.; 德雷针茅■☆

376765 Stipa dregeana Steud. var. elongata (Nees) Stapf; 伸长针茅■☆

376766 Stipa duthiei Hook. f. = Achnatherum duthiei (Hook. f.) P. C. Kuo et S. L. Lu ■

376767 Stipa effusa (Maxim.) Nakai ex Honda = Achnatherum extremiorientale (H. Hara) Keng ex P. C. Kuo ■

376768 Stipa elegantissima Labill.; 优雅针茅■☆

376769 Stipa elongata (Nees) Steud. = Stipa keniensis (Pilg.) Freitag ■☆

376770 Stipa extremiorientalis H. Hara = Achnatherum extremiorientale (H. Hara) Keng ex P. C. Kuo ■

376771 Stipa extremiorientalis H. Hara = Stipa pekinensis Hance ■

376772 Stipa filiculmis Delile; 丝秆针茅; Filiculm Feathergrass ■☆

376773 Stipa gabesensis Moraldo, Raffaelli et Ricceri = Macrochloa tenacissima (L.) Kunth ■☆

376774 Stipa gigantea Lag. = Stipa lagascae Roem. et Schult. ■☆

376775 Stipa gigantea Lag. subsp. lagascae Trab. = Stipa lagascae Roem. et Schult. ■☆

376776 Stipa gigantea Lag. subsp. letourneuxii (Trab.) Trab. = Stipa letourneuxii Trab. ■☆

376777 Stipa gigantea Lag. var. clausa (Trab.) Pau et Font Quer = Stipa clausa (Trab.) R. Vàsquez Pardo et Devesa ■☆

376778 Stipa gigantea Lag. var. pubescens Trab. = Stipa clausa (Trab.) R. Vàsquez Pardo et Devesa ■☆

376779 Stipa gigantea Link; 巨大针茅; Giant Feather Grass, Giant Feathergrass, Golden Oats ■☆

376780 Stipa gigantea Link = Macrochloa arenaria (Brot.) Kunth ■☆

376781 Stipa gigantea Link subsp. donyanae R. Vàsquez Pardo et Devesa = Macrochloa arenaria (Brot.) Kunth ■☆

376782 Stipa gigantea Link subsp. maroccana (Pau et Font Quer) R. Vásquez et Devesa = Macrochloa arenaria (Brot.) Kunth ■☆

376783 Stipa gigantea Link var. maroccana Pau et Font Quer = Macrochloa arenaria (Brot.) Kunth ■☆

376784 Stipa gigantea Link var. mesatlantica Andr. = Macrochloa arenaria (Brot.) Kunth ■☆

376785 Stipa glareosa C. C. Davis; 沙生针茅; Sandy Feathergrass, Sandy Needlegrass ■

376786 Stipa glareosa C. C. Davis var. langshanica Y. Z. Zhao; 狼山针茅; Langshan Feathergrass ■

376787 Stipa glareosa P. A. Smirn. = Stipa caucasica Schmalh. subsp. glareosa (P. A. Sm.) Tzvelev ■

376788 Stipa glareosa P. A. Smirn. var. lang-shanica Y. Z. Zhao = Stipa caucasica Schmalh. subsp. glareosa (P. A. Sm.) Tzvelev ■

376789 Stipa gobica Roshev. = Stipa tianschanica Roshev. subsp. gobica (Roshev.) D. F. Cui ■

376790 Stipa gobica Roshev. = Stipa tianschanica Roshev. var. gobica (Roshev.) P. C. Kuo et Y. H. Sun ■

376791 Stipa gobica Roshev. var. wulateica Y. Z. Zhao; 乌拉特针茅; Wulate Feathergrass ■

376792 Stipa gobica Roshev. var. wulateica Y. Z. Zhao = Stipa tianschanica Roshev. subsp. gobica (Roshev.) D. F. Cui ■

376793　Stipa gracilis Roshev. ;纤细针茅■☆

376794　Stipa grandifolium Keng = Orthoraphium grandifolium (Keng) Keng ex P. C. Kuo ■

376795　Stipa grandis C. C. Davis; 大针茅（高针茅）; Large Feathergrass, Large Needlegrass ■

376796　Stipa himalaica Roshev. ;喜马拉雅针茅■

376797　Stipa hoggarensis Chrtek et Martinovsky;霍加尔针茅■☆

376798　Stipa hohenackeriana Trin. et Rupr. ;浩瀚针茅■

376799　Stipa holosericea Trin. ;全毛针茅■☆

376800　Stipa hookeri Stapf = Trikeraia hookeri (Stapf) Bor ■

376801　Stipa hyalina Nees;透明针茅;Hyaline Feathergrass ■☆

376802　Stipa hymenoides Roem. et Schult. = Achnatherum hymenoides (Roem. et Schult.) Barkworth ■☆

376803　Stipa hymenoides Roem. et Schult. = Oryzopsis hymenoides (Roem. et Schult.) Ricker et Piper ■

376804　Stipa iljinii Roshev. ;伊里金针茅■☆

376805　Stipa inebrians Hance = Achnatherum inebrians (Hance) Keng ■

376806　Stipa jacquemontii Jaub. et Spach = Achnatherum jacquemontii (Jaub. et Spach) P. C. Kuo et S. L. Lu ■

376807　Stipa jagnobica Ovcz. et Czukav. ;雅格诺勃针茅■☆

376808　Stipa japonica (Hack.) Hack. ex Nakai = Stipa coreana Honda var. japonica (Hack.) Y. N. Lee ☆

376809　Stipa juncea L. ;灯芯草针茅■☆

376810　Stipa karataviensis Rusher. ;卡拉塔夫针茅■☆

376811　Stipa kelibiae Moraldo et Raffaelli et Riccceri = Macrochloa tenacissima (L.) Kunth ■☆

376812　Stipa keniensis (Pilg.) Freitag;肯尼亚针茅■☆

376813　Stipa keniensis (Pilg.) Freitag subsp. somalensis Freitag;索马里针茅■☆

376814　Stipa kirghisorum C. C. Davis;大羽针茅（长羽针茅）; Long Feathergrass, Longfeather Needlegrass ■

376815　Stipa klemenzii Roshev. = Stipa tianschanica Roshev. var. klemenzii (Roshev.) Norl. ■

376816　Stipa koelzii R. R. Stewart = Stipa capillata L. ■

376817　Stipa koelzii R. R. Stewart = Stipa consanguinea Trin. et Rupr. ■

376818　Stipa kokonorica K. S. Hao = Achnatherum splendens (Trin.) Nevski ■

376819　Stipa korshinskyi Roshev. ;考尔针茅■☆

376820　Stipa kralifii Moraldo et Raffaelli et Riccceri = Macrochloa tenacissima (L.) Kunth ■☆

376821　Stipa kraschenirnnikowii Roshev. ;克拉赛针茅■■☆

376822　Stipa krylovii Roshev. = Stipa sareptana Becker var. krylovii (Roshev.) P. C. Kuo et Y. H. Sun ■

376823　Stipa lagascae Roem. et Schult. ;拉加针茅■☆

376824　Stipa lagascae Roem. et Schult. subsp. letourneuxii (Trab.) Maire = Stipa letourneuxii Trab. ■☆

376825　Stipa lagascae Roem. et Schult. subsp. normalis Maire = Stipa lagascae Roem. et Schult. ■☆

376826　Stipa lagascae Roem. et Schult. var. clausa Trab. = Stipa clausa (Trab.) R. Vàsquez Pardo et Devesa ■☆

376827　Stipa lagascae Roem. et Schult. var. embergeri Maire;恩贝格尔针茅■☆

376828　Stipa lagascae Roem. et Schult. var. hackelii Fiori = Stipa lagascae Roem. et Schult. ■☆

376829　Stipa lagascae Roem. et Schult. var. letourneuxii Trab. = Stipa letourneuxii Trab. ■☆

376830　Stipa lagascae Roem. et Schult. var. malvana Brichan et Sauvage = Stipa lagascae Roem. et Schult. ■☆

376831　Stipa lagascae Roem. et Schult. var. oropediorum Maire = Stipa lagascae Roem. et Schult. ■☆

376832　Stipa lagascae Roem. et Schult. var. pubescens Maire et Weiller = Stipa letourneuxii Trab. subsp. pellita (Trin. et Rupr.) H. Scholz ■☆

376833　Stipa lagascae Roem. et Schult. var. trabutii Maire = Stipa clausa (Trab.) R. Vàsquez Pardo et Devesa ■☆

376834　Stipa langshanica (Y. Z. Zhao) Y. Z. Zhao = Stipa caucasica Schmalh. subsp. glareosa (P. A. Sm.) Tzvelev ■

376835　Stipa laxiflora Keng = Stipa penicillata Hand. -Mazz. ■

376836　Stipa lessingiana Trin. et Rupr. ;细叶针茅;Slenderleaf Needlegrass, Tenous Feathergrass ■

376837　Stipa letourneuxii Trab. ;勒图尔针茅■☆

376838　Stipa letourneuxii Trab. subsp. pellita (Trin. et Rupr.) H. Scholz;遮皮针茅■☆

376839　Stipa letourneuxii Trab. subsp. tunetana (H. Scholz) H. Scholz;图呐特针茅■☆

376840　Stipa lingua Junge;舌状针茅■☆

376841　Stipa lipskyi Roshev. ;利普斯基针茅■☆

376842　Stipa lithophila P. A. Smirn. ;喜石针茅■☆

376843　Stipa littorea Burm. f. = Spinifex littoreus (Burm. f.) Merr. ■

376844　Stipa longiplumosa Roshev. ;长羽状针茅■☆

376845　Stipa macroglossa C. C. Davis; 长舌针茅; Long-glosse Feathergrass, Longtongue Needlegrass ■

376846　Stipa magnifica Junge;华丽针茅■☆

376847　Stipa maroccana Scholz;摩洛哥针茅■☆

376848　Stipa megapotamia Spreng. ex Trin. ; 河生针茅; River Feathergrass ■☆

376849　Stipa melanosperma J. Presl;黑籽针茅;Blackseed Feathergrass ■☆

376850　Stipa meridionalis R. Vàsquez Pardo et Devesa;南方针茅■☆

376851　Stipa mongholica Turcz. ex Trin. = Ptilagrostis mongholica (Turcz. ex Trin.) Griseb. ■

376852　Stipa mongolorum Tzvelev; 蒙古针茅; Mongol Needlegrass, Mongolian Feathergrass ■

376853　Stipa nakaii Honda = Achnatherum nakaii (Honda) Tateoka ■

376854　Stipa namaquensis Pilg. = Stipagrostis anomala De Winter ■☆

376855　Stipa neesiana Trin. et Rupr. ;尼氏针茅; American Needle-grass, Nees Feathergrass ■☆

376856　Stipa neesiana Trin. et Rupr. = Nassella neesiana (Trin. et Rupr.) Barkworth ■☆

376857　Stipa nitens Ball;光亮针茅■☆

376858　Stipa offneri Breistr. ;奥夫纳针茅■☆

376859　Stipa orientalis Trin. ex Ledeb. ; 东 方 针 茅; Eastern Feathergrass, Oriental Needlegrass ■

376860　Stipa orientalis Trin. var. grandiflora Rupr. = Stipa caucasica Schmalh. ■

376861　Stipa orientalis Trin. var. persica Trin. = Stipa arabica Trin. et Rupr. ■

376862　Stipa ovcxinnikovii Roshev. ;奥氏针茅■☆

376863　Stipa paleacea Poir. = Themeda triandra Forssk. ■

376864　Stipa pamirica Roshev. ;帕米尔针茅■☆

376865　Stipa pappiformis Keng = Trikeraia pappiformis (Keng) P. C. Kuo et S. L. Lu ■

376866　Stipa papposa Nees;毛针茅;Hairy Feathergrass ■☆

376867　Stipa paradoxa（Junge）P. A. Smirn. ;奇异针茅■☆

376868　Stipa parviflora Desf. ;小花针茅■☆

376869　Stipa parviflora Desf. var. mareotica Chrtek et Martinovsky = Stipa parviflora Desf. ■☆

376870　Stipa parviflora Desf. var. pilosa Chrtek et Martinovsky = Stipa parviflora Desf. ■☆

376871　Stipa parvula Nees = Aristida parvula（Nees）De Winter ■☆

376872　Stipa pekinensis Hance = Achnatherum pekinense（Hance）Ohwi ■

376873　Stipa pelliotii Danguy = Ptilagrostis pelliotii（Danguy）Grubov ■

376874　Stipa pellita（Tineo et Rupr.）Tzvelev = Stipa letourneuxii Trab. subsp. pellita（Trin. et Rupr.）H. Scholz ■☆

376875　Stipa penicillata Hand.-Mazz.；疏花针茅；Looseflower Needlegrass,Loose-flowered Feathergrass ■

376876　Stipa penicillata Hand.-Mazz. var. hirsuta P. C. Kuo et Y. H. Sun ex C. P. Wang et X. L. Yang;毛疏花针茅;Hirsute Feathergrass ■

376877　Stipa pennata L. ;羽状针茅（欧洲针茅）;European Feather Grass, European Feathergrass, European Feather-grass, European Needle Grass,European Needle-grass,Feather Grass ■

376878　Stipa pennata L. = Stipa himalaica Roshev. ■

376879　Stipa pennata L. subsp. kirghisorum（P. A. Smirn.）Freitag = Stipa kirghisorum C. C. Davis ■

376880　Stipa pennata L. subsp. pulcherrima（K. Koch）Á. Löve et D. Löve = Stipa pulcherrima K. Koch ■☆

376881　Stipa pennata L. subsp. pulcherrima（K. Koch）Freitag = Stipa pulcherrima K. Koch ■☆

376882　Stipa pennata L. var. breviglumis Maire = Stipa atlantica Smirn. ■☆

376883　Stipa pilgeriana K. S. Hao = Stipa purpurea Griseb. ■

376884　Stipa pontica P. A. Smirn. ;蓬特针茅■☆

376885　Stipa potaninii Roshev. = Stipa tianschanica Roshev. ■

376886　Stipa prolifera Steud. = Stipagrostis uniplumis（Licht. ex Roem. et Schult.）De Winter ■☆

376887　Stipa przewalskyi Roshev. ;甘青针茅（勃氏针茅）;Ganqing Needlegrass,Przewalsky Feathergrass,Przewalsky Needlegrass ■

376888　Stipa pseudocapillata Roshev. ;假细毛针茅■☆

376889　Stipa pubicalyx Ohwi = Achnatherum pubicalyx（Ohwi）Keng ex P. C. Kuo ■

376890　Stipa pulcherrima K. Koch;美丽针茅;Beautiful Feathergrass ■☆

376891　Stipa pulchra Hitchc. ;北美丽针茅;Purple Needle-grass ■☆

376892　Stipa purpurascens Hitchc. = Stipa regeliana Hack. ■

376893　Stipa purpurea Griseb. ;紫花针茅;Purpleflower Feathergrass, Purpleflower Needlegrass ■

376894　Stipa purpurea Griseb. = Ptilagrostis purpurea（Griseb.）Roshev. ■

376895　Stipa purpurea Griseb. subsp. arenosa（Tzvelev）D. F. Cui = Stipa purpurea Griseb. var. arenosa Tzvelev ■

376896　Stipa purpurea Griseb. subsp. arenosa（Tzvelev）D. F. Cui = Stipa purpurea Griseb. ■

376897　Stipa purpurea Griseb. var. arenosa Tzvelev;大紫花针茅（大颖紫花针茅）;Sandy Purpleflower Feathergrass ■

376898　Stipa purpurea Griseb. var. arenosa Tzvelev = Stipa purpurea Griseb. ■

376899　Stipa regeliana Hack. ;狭穗针茅（紫花芨芨草）;Narrowspike Needlegrass,Regel Feathergrass ■

376900　Stipa retorta Cav. = Stipa capensis Thunb. ■☆

376901　Stipa retorta Cav. var. pubescens（Ball）Jahand. et Maire = Stipa capensis Thunb. ■☆

376902　Stipa richteriana Kar. et Kir. ;瑞氏针茅■

376903　Stipa roborowskyi Roshev. ;昆仑针茅;Kunlun Needlegrass, Roborowsky Feathergrass ■

376904　Stipa robusta Nutt. ex Trin. et Rupr. ;醉马羽茅（睡眠草）■☆

376905　Stipa robusta Nutt. ex Trin. et Rupr. 'Sleepy';睡眠草;Sleepy Grass ■☆

376906　Stipa roylei（Nees）Duthie = Orthoraphium roylei Nees ■

376907　Stipa rubens C. C. Davis = Stipa zalesskii Wilensky ex Grossh. ■

376908　Stipa sareptana Becker;新疆针茅;Xinjiang Feathergrass, Xinjiang Needlegrass ■

376909　Stipa sareptana Becker subsp. krylovii（Roshev.）D. F. Cui = Stipa krylovii Roshev. ■

376910　Stipa sareptana Becker var. krylovii（Roshev.）P. C. Kuo et Y. H. Sun;西北针茅（阿尔泰针茅,克氏针茅）;Krylov Feathergrass ■

376911　Stipa sareptana Becker var. krylovii（Roshev.）P. C. Kuo et Y. H. Sun = Stipa krylovii Roshev. ■

376912　Stipa sareptana subsp. krylovii（Roshev.）D. F. Cui = Stipa sareptana Becker var. krylovii（Roshev.）P. C. Kuo et Y. H. Sun ■

376913　Stipa scabra Lindl. ;粗糙针茅;Rough Spear-grass ■☆

376914　Stipa sibirica（L.）Lam. = Achnatherum sibiricum（L.）Keng ■

376915　Stipa sibirica（L.）Lam. var. effusa Maxim. = Achnatherum extremiorientale（H. Hara）Keng ex P. C. Kuo ■

376916　Stipa sinomongholica Ohwi = Stipa tianschanica Roshev. subsp. gobica（Roshev.）D. F. Cui ■

376917　Stipa spartea Trin. ;豪猪针茅;Needle Grass, Needle-grass, Porcupine Grass ■☆

376918　Stipa spicata L. f. = Trachypogon spicatus（L. f.）Kuntze ■☆

376919　Stipa spinifex L. = Spinifex littoreus（Burm. f.）Merr. ■

376920　Stipa spiridonovii Roshev. ;斯皮里针茅■☆

376921　Stipa splendens Trin. = Achnatherum splendens（Trin.）Nevski ■

376922　Stipa splendens Trin. var. gracilis Bor = Achnatherum splendens（Trin.）Nevski ■

376923　Stipa stenophylla（Lindl.）Trautv. ;窄叶针茅■☆

376924　Stipa subsessiliflora（Rupr.）Roshev. ;座花针茅;Subsessile-flower Feathergrass,Subsessileflower Needlegrass ■

376925　Stipa subsessiliflora（Rupr.）Roshev. subsp. basiplumosa（Munro ex Hook. f.）D. F. Cui = Stipa subsessiliflora（Rupr.）Roshev. ■

376926　Stipa subsessiliflora（Rupr.）Roshev. var. basiplumosa（Munro ex Hook. f.）P. C. Kuo et Y. H. Sun = Stipa subsessiliflora（Rupr.）Roshev. ■

376927　Stipa subsessiliflora（Rupr.）Roshev. var. basiplumosa（Munro ex Hook. f.）P. C. Kuo et Y. H. Sun;羽柱针茅;Plumose Feathergrass ■

376928　Stipa szovitsiana（Trin.）Griseb. = Stipa arabica Trin. et Rupr. ■

376929　Stipa szowitsiana（Trin.）Griseb. = Stipa arabica Trin. et Rupr. ■

376930　Stipa szowitsiana Trin. ex Hohen. ;拟长舌针茅（图尔盖针茅,伊犁针茅）■

376931　Stipa tenacissima L. ;西班牙纸草;Algerian Grass, Esparto Grass, Esparto Needle Grass, Esparto Needle-grass, Esparto-grass, Halfa,Mexican Feather Grass,Needle Grass ■☆

376932　Stipa tenacissima L. subsp. gabesensis Moraldo et Raffaelli et Riccceri = Macrochloa tenacissima（L.）Kunth subsp. gabesensis（Moraldo et al.）H. Scholz et Valdés ■☆

376933　Stipa tenacissima L. var. villosiuscula H. Lindb. = Macrochloa tenacissima (L.) Kunth ■☆

376934　Stipa tenuissima Trin.；细茎针茅；Finestem Feathergrass, Mexican Feather Grass ■☆

376935　Stipa tenuissima Trin. = Nassella tenuissima (Trin.) Barkworth ■☆

376936　Stipa tianschanica Roshev.；天山针茅；Tianshan Feathergrass, Tianshan Needlegrass ■

376937　Stipa tianschanica Roshev. subsp. gobica (Roshev.) D. F. Cui；戈壁针茅；Gobi Tianshan Feathergrass ■

376938　Stipa tianschanica Roshev. subsp. gobica (Roshev.) D. F. Cui = Stipa klemenzii Roshev. ■

376939　Stipa tianschanica Roshev. var. gobica (Roshev.) P. C. Kuo et Y. H. Sun = Stipa gobica Roshev. ■

376940　Stipa tianschanica Roshev. var. klemenzii (Roshev.) Norl.；内蒙针茅(克里门茨针茅,石生针茅,小叶针茅)；Klemenz Feathergrass ■

376941　Stipa tianschanica Roshev. var. klemenzii (Roshev.) Norl. = Stipa klemenzii Roshev. ■

376942　Stipa tibestica Maire；提贝斯提针茅■☆

376943　Stipa tibetica Mez = Ptilagrostis mongholica (Turcz. ex Trin.) Griseb. ■

376944　Stipa tigrensis Chiov.；蒂格雷针茅■☆

376945　Stipa tortilis Desf. = Stipa capensis Thunb. ■☆

376946　Stipa tortilis Desf. subsp. nitens Ball = Stipa nitens Ball ■☆

376947　Stipa tortilis Desf. var. pilosa Trab. = Stipa capensis Thunb. ■☆

376948　Stipa tortilis Desf. var. pubescens Ball = Stipa capensis Thunb. ■☆

376949　Stipa trichoides P. A. Smirn.；毛状针茅■☆

376950　Stipa trichotoma Nees；分枝针茅；Trichotomous Feathergrass ■☆

376951　Stipa trichotoma Nees = Nassella trichotoma (Nees) Hack. ex Arechav. ■☆

376952　Stipa tunetana H. Scholz = Stipa letourneuxii Trab. subsp. tunetana (H. Scholz) H. Scholz ■☆

376953　Stipa turcomanica P. A. Smirn.；土库曼针茅■☆

376954　Stipa turgaica Roshev. = Stipa arabica Trin. et Rupr. ■

376955　Stipa turgaica Roshev. = Stipa szowitsiana Trin. ex Hohen. ■

376956　Stipa turkestanica Hack.；土耳其斯坦针茅■☆

376957　Stipa turkestanica Hack. subsp. trichoides (P. A. Smirn.) Tzvelev = Stipa trichoides P. A. Smirn. ■☆

376958　Stipa ucrainica P. A. Smirn.；乌克兰针茅■☆

376959　Stipa variabilis Hughes；易变针茅■☆

376960　Stipa viridula Trin.；绿针茅(美洲醉马草,睡眠草)；Green Feathergrass,Green Needle Grass ■☆

376961　Stipa wulateica (Y. Z. Zhao) Y. Z. Zhao = Stipa tianschanica Roshev. subsp. gobica (Roshev.) D. F. Cui ■

376962　Stipa zalesskii Wilensky ex Grossh.；红针茅■

376963　Stipaceae Bercht. et J. Presl = Gramineae Juss. (保留科名)■●

376964　Stipaceae Bercht. et J. Presl = Poaceae Barnhart(保留科名)■●

376965　Stipaceae Burnett = Gramineae Juss. (保留科名)■●

376966　Stipaceae Burnett = Poaceae Barnhart(保留科名)■●

376967　Stipaceae Burnett；针茅科■

376968　Stipagrostis Ness = Aristida L. ■

376969　Stipagrostis Ness(1832)；针茅草属(针禾属)■

376970　Stipagrostis acutiflora (Trin. et Rupr.) De Winter；尖花针茅草(尖花针禾)■☆

376971　Stipagrostis acutiflora (Trin. et Rupr.) De Winter subsp. algeriensis (Henrard) H. Scholz；阿尔及利亚针茅草■☆

376972　Stipagrostis affinis H. Scholz；近缘针茅草(近缘针禾)■☆

376973　Stipagrostis amabilis (Schweick.) De Winter；秀丽针茅草■☆

376974　Stipagrostis anomala De Winter；异常针茅草■☆

376975　Stipagrostis brachyathera (Coss. et Balansa) De Winter；短药针茅草■☆

376976　Stipagrostis brachypoda (Tausch) De Winter = Stipagrostis plumosa (L.) Munro ex T. Anderson ■☆

376977　Stipagrostis brevifolia (Nees) De Winter；短叶针茅草(短叶针禾)■☆

376978　Stipagrostis ciliata (Desf.) De Winter；睫毛针茅草(睫毛针禾)■☆

376979　Stipagrostis ciliata (Desf.) De Winter var. capensis (Trin. et Rupr.) De Winter；好望角针茅草■☆

376980　Stipagrostis damarensis (Mez) De Winter；达马尔针茅草(达马尔针禾)■☆

376981　Stipagrostis dinteri (Hack.) De Winter；丁特针茅草■☆

376982　Stipagrostis dregeana Nees；德雷针茅草■☆

376983　Stipagrostis fastigiata (Hack.) De Winter；帚状针茅草■☆

376984　Stipagrostis foexiana (Maire et Wilczek) De Winter；福埃针茅草■☆

376985　Stipagrostis garubensis (Pilg.) De Winter；加鲁布针茅草■☆

376986　Stipagrostis geminifolia Nees；对叶针茅草■☆

376987　Stipagrostis giessii Kers；吉斯针茅草(吉斯针禾)■☆

376988　Stipagrostis gonatostachys (Pilg.) De Winter；膝穗针茅草■☆

376989　Stipagrostis grandiglumis (Roshev.) Tzvelev；大颖针禾(大颖三芒草)；Bigglume Threeawngrass, Bigglume Triawn ■

376990　Stipagrostis grandiglumis (Roshev.) Tzvelev = Aristida grandiglumis Roshev. ■

376991　Stipagrostis hermannii (Mez) De Winter；赫尔曼针茅草■☆

376992　Stipagrostis hirtigluma (Steud. ex Trin. et Rupr.) De Winter；毛颖针茅草■☆

376993　Stipagrostis hirtigluma (Steud. ex Trin. et Rupr.) De Winter subsp. patula (Hack.) De Winter；张开针茅草■☆

376994　Stipagrostis hirtigluma (Steud. ex Trin. et Rupr.) De Winter subsp. pearsonii (Henrard) De Winter；皮尔逊针茅草■☆

376995　Stipagrostis hirtigluma (Steud. ex Trin. et Rupr.) De Winter var. patula (Hack.) De Winter = Stipagrostis hirtigluma (Steud. ex Trin. et Rupr.) De Winter subsp. patula (Hack.) De Winter ■☆

376996　Stipagrostis hochstetteriana (Beck ex Hack.) De Winter；霍赫针茅草(霍赫针禾)■☆

376997　Stipagrostis lanata (Forssk.) De Winter；绵毛针茅草(绵毛针禾)■☆

376998　Stipagrostis lanipes (Mez) De Winter；毛梗针茅草(毛梗针禾)■☆

376999　Stipagrostis libyca (H. Scholz) H. Scholz；利比亚针茅草(利比亚针禾)■☆

377000　Stipagrostis libyca (H. Scholz) H. Scholz subsp. darfurensis H. Scholz；达尔富尔针茅草(达尔富尔针禾)■☆

377001　Stipagrostis lutescens (Nees) De Winter；淡黄针茅草(淡黄针禾)■☆

377002　Stipagrostis lutescens (Nees) De Winter var. marlothii (Hack.) De Winter；马洛斯针茅草(马洛斯针禾)■☆

377003　Stipagrostis multinerva H. Scholz；多脉针茅草(多脉针禾)■☆

377004　Stipagrostis namaquensis (Nees) De Winter；纳马夸针茅草(纳马夸针禾)■☆

377005　Stipagrostis namibensis De Winter；纳米布针茅草(纳米布针禾)■☆

377006　Stipagrostis obtusa (Delile) Nees;钝针茅草(钝针禾)■☆

377007　Stipagrostis obtusa Delile var. foexiana Le Houér. = Stipagrostis foexiana (Maire et Wilczek) De Winter ■☆

377008　Stipagrostis oranensis (Henrard) De Winter;奥兰针茅草(奥兰针禾)■☆

377009　Stipagrostis papposa (Trin. et Rupr.) De Winter = Stipagrostis uniplumis (Licht. ex Roem. et Schult.) De Winter ■☆

377010　Stipagrostis paradisea (Edgew.) De Winter;公园针茅草(公园针禾)■☆

377011　Stipagrostis pennata (Trin.) De Winter;羽毛针茅草(羽裂三芒草,羽毛三芒草,羽毛针禾);Pennate Threeawngrass, Pennate Triawn ■

377012　Stipagrostis pennata (Trin.) De Winter = Aristida pennata Trin. ■

377013　Stipagrostis plumosa (L.) Munro ex T. Anderson;羽状针茅草(羽状针禾)■☆

377014　Stipagrostis plumosa (L.) Munro ex T. Anderson subsp. seminuda (Trin. et Rupr.) H. Scholz;半裸羽状针禾■☆

377015　Stipagrostis plumosa (L.) Munro ex T. Anderson subsp. syrtica (Maire et Weiller) H. Scholz;瑟尔特针禾■☆

377016　Stipagrostis plumosa (L.) Munro ex T. Anderson var. aethiopica (Trin. et Rupr.) Täckh. = Stipagrostis plumosa (L.) Munro ex T. Anderson subsp. seminuda (Trin. et Rupr.) H. Scholz ■☆

377017　Stipagrostis plumosa (L.) Munro ex T. Anderson var. alexandrina (Trin. et Rupr.) Täckh. = Stipagrostis plumosa (L.) Munro ex T. Anderson subsp. seminuda (Trin. et Rupr.) H. Scholz ■☆

377018　Stipagrostis plumosa (L.) Munro ex T. Anderson var. brachypoda (Tausch) Bor = Stipagrostis plumosa (L.) Munro ex T. Anderson ■☆

377019　Stipagrostis pogonoptila (Jaub. et Spach) De Winter;髯毛针茅草(髯毛针禾)■☆

377020　Stipagrostis pogonoptila (Jaub. et Spach) De Winter subsp. tibestica Maire = Stipagrostis pogonoptila (Jaub. et Spach) De Winter subsp. tibestica (Maire) J. -P. Lebrun et Stork ■☆

377021　Stipagrostis pogonoptila (Jaub. et Spach) De Winter subsp. tibestica (Maire) J. -P. Lebrun et Stork;提贝斯提针禾■☆

377022　Stipagrostis proxima (Steud.) De Winter;近基针茅草(近基针禾)■☆

377023　Stipagrostis pungens (Desf.) De Winter;锐尖针茅草(锐尖针禾)■☆

377024　Stipagrostis pungens (Desf.) De Winter subsp. pubescens (Henrard) H. Scholz;短柔毛针茅草(短柔毛针禾)■☆

377025　Stipagrostis pungens (Desf.) De Winter subsp. transiens (Maire) H. Scholz;中间针茅草(中间针禾)■☆

377026　Stipagrostis raddiana (Savi) De Winter;分枝针茅草(分枝针禾)■☆

377027　Stipagrostis ramulosa De Winter = Stipagrostis raddiana (Savi) De Winter ■☆

377028　Stipagrostis rigidifolia H. Scholz;硬叶针茅草(硬叶针禾)■☆

377029　Stipagrostis sabulicola (Pilg.) De Winter;砂地针茅草(砂地针禾)■☆

377030　Stipagrostis sahelica (Trab.) De Winter;萨赫勒针茅草(萨赫勒针禾)■☆

377031　Stipagrostis schaeferi (Mez) De Winter;谢夫针茅草(谢夫针禾)■☆

377032　Stipagrostis scoparia (Trin. et Rupr.) De Winter;刷状针茅草(帚状针禾)■☆

377033　Stipagrostis shawii (H. Scholz) H. Scholz;范肖针茅草(范肖针禾)■☆

377034　Stipagrostis subacaulis (Nees) De Winter;近无茎针茅草(近无茎针禾)■☆

377035　Stipagrostis tenuirostris (Henrard) De Winter;细喙针茅草(细喙针禾)■☆

377036　Stipagrostis uniplumis (Licht. ex Roem. et Schult.) De Winter;单羽针茅草(单羽针禾)■☆

377037　Stipagrostis uniplumis (Licht. ex Roem. et Schult.) De Winter subsp. papposa (Trin. et Rupr.) Bourreil = Stipagrostis uniplumis (Licht. ex Roem. et Schult.) De Winter ■☆

377038　Stipagrostis uniplumis (Licht. ex Roem. et Schult.) De Winter var. intermedia (Schweick.) De Winter;间型针禾■☆

377039　Stipagrostis uniplumis (Licht. ex Roem. et Schult.) De Winter var. neesii (Trin. et Rupr.) De Winter;尼斯针禾■☆

377040　Stipagrostis uniplumis (Licht.) De Winter = Stipagrostis uniplumis (Licht. ex Roem. et Schult.) De Winter ■☆

377041　Stipagrostis xylosa Cope;木质针禾■☆

377042　Stipagrostis zeyheri (Nees) De Winter;泽赫针茅草(泽赫针禾)■☆

377043　Stipagrostis zeyheri (Nees) De Winter subsp. barbata (Stapf) De Winter;髯毛泽赫针茅草(髯毛泽赫针禾)■☆

377044　Stipagrostis zeyheri (Nees) De Winter subsp. macropus ?;大足泽赫针禾■☆

377045　Stipagrostis zeyheri (Nees) De Winter subsp. sericans (Hack.) De Winter;绢毛泽赫针茅草(绢毛泽赫针禾)■☆

377046　Stipagrostis zittelii (Asch.) De Winter;齐特尔针茅草(齐特尔针禾)■☆

377047　Stipavena Vierh. = Helictotrichon Besser ex Schult. et Schult. f. ■

377048　Stipecoma Müll. Arg. (1860);毛梗夹竹桃属●☆

377049　Stipecoma peltigera Müll. Arg.;毛梗夹竹桃●☆

377050　Stipellaria Benth. = Alchornea Sw. ●

377051　Stipellaria mollis Benth. = Alchornea mollis (Benth.) Müll. Arg. ●

377052　Stipellaria tiliifolia Benth. = Alchornea tiliifolia (Benth.) Müll. Arg. ●

377053　Stipellaria trewioides Benth. = Alchornea trewioides (Benth.) Müll. Arg. ●

377054　Stiphonia Hemsl. = Rhus L. ●

377055　Stiphonia Hemsl. = Styphonia Nutt. ●

377056　Stipocoma Post et Kuntze = Stipecoma Müll. Arg. ●☆

377057　Stiptanthus (Benth.) Briq. (1897);多穗香属■☆

377058　Stiptanthus (Benth.) Briq. = Anisochilus Wall. ex Benth. ●■

377059　Stiptanthus Briq. = Anisochilus Wall. ex Benth. ●■

377060　Stiptanthus polystachyus Briq.;多穗香■☆

377061　Stipularia Dalpino = Piuttia Mattei ■

377062　Stipularia Dalpino = Thalictrum L. ■

377063　Stipularia Haw. = Spergularia (Pers.) J. Presl et C. Presl(保留属名)■

377064　Stipularia P. Beauv. (1810);托叶茜属■☆

377065　Stipularia P. Beauv. = Sabicea Aubl. ●☆

377066　Stipularia africana P. Beauv.;非洲托叶茜■☆

377067　Stipularia efulenensis Hutch. = Sabicea efulenensis (Hutch.) Hepper ●☆

377068　Stipularia elliptica Schweinf. ex Hiern;托叶茜■☆

377069　Stipularia gabonica Hiern = Sabicea gabonica (Hiern) Hepper ●☆

377070 Stipularia mollis Wernham = Sabicea lanata Hepper ●☆

377071 Stipulicida Michx. (1803);齿托草属■☆

377072 Stipulicida Rich. = Stipulicida Michx. ■☆

377073 Stipulicida filiformis Nash = Stipulicida setacea Michx. ■☆

377074 Stipulicida setacea Michx. ;齿托草;Pineland Scaly-pink ■☆

377075 Stipulicida setacea Michx. var. filiformis (Nash) D. B. Ward = Stipulicida setacea Michx. ■☆

377076 Stipulicida setacea Michx. var. lacerata C. W. James;古巴齿托草■☆

377077 Stiractis Post et Kuntze = Layia Hook. et Arn. ex DC. (保留属名)■☆

377078 Stiractis Post et Kuntze = Steiractis Raf. ■☆

377079 Stiradotheca Klotzseh = Begonia L. ●■

377080 Stiradotheca Klotzseh = Stibadotheca Klotzsch ●■

377081 Stiranisia Post et Kuntze = Saxifraga L. ■

377082 Stiranisia Post et Kuntze = Steiranisia Raf. ■☆

377083 Stireja Post et Kuntze = Steirexa Raf. ■☆

377084 Stireja Post et Kuntze = Steireya Raf. ■☆

377085 Stireja Post et Kuntze = Trichopus Gaertn. ■☆

377086 Stiremis Post et Kuntze = Steiremis Raf. ■

377087 Stiremis Post et Kuntze = Telanthera R. Br. ■

377088 Stirlingia Endl. (1837);斯迪林木属●☆

377089 Stirlingia latifolia (R. Br.) Steud. ;阔叶斯迪林木●☆

377090 Stiroctis Post et Kuntze = Lachnaea L. + Cryptadenia Meisn. ●☆

377091 Stiroctis Post et Kuntze = Steiroctis Raf. ●☆

377092 Stirodiscus Post et Kuntze = Psilothonna E. Mey. ex DC. ■☆

377093 Stirodiscus Post et Kuntze = Steirodiscus Less. ■☆

377094 Stiroglossa Post et Kuntze = Brachycome Cass. ●■☆

377095 Stiroglossa Post et Kuntze = Steiroglossa DC. ■☆

377096 Stironema Post et Kuntze = Lysimachia L. ●■

377097 Stironema Post et Kuntze = Steironema Raf. ■☆

377098 Stironeuron Radlk. = Synsepalum (A. DC.) Daniell ●☆

377099 Stironeurum Radlk. = Synsepalum (A. DC.) Daniell ●☆

377100 Stironeurum Radlk. ex De Wild. et T. Durand = Synsepalum (A. DC.) Daniell ●☆

377101 Stironeurum stipulatum Radlk. = Synsepalum stipulatum (Radlk.) Engl. ●☆

377102 Stirostemon Post et Kuntze = Samolus L. ■

377103 Stirostemon Post et Kuntze = Steirostemon Phil. ■

377104 Stirotis Post et Kuntze = Steirotis Raf. ●☆

377105 Stirtonanthus B. -E. van Wyk et A. L. Schutte(1995);肖香豆木属●☆

377106 Stirtonanthus chrysanthus (Adamson) B. -E. van Wyk et A. L. Schutte;金花肖香豆木●☆

377107 Stirtonanthus insignis (Compton) B. -E. van Wyk et A. L. Schutte;显著肖香豆木●☆

377108 Stirtonanthus taylorianus (L. Bolus) B. -E. van Wyk et A. L. Schutte;泰勒肖香豆木●☆

377109 Stirtonia B. -E. van Wyk et A. L. Schutte = Stirtonanthus B. -E. van Wyk et A. L. Schutte ●☆

377110 Stirtonia chrysantha (Adamson) B. -E. van Wyk et A. L. Schutte = Stirtonanthus chrysanthus (Adamson) B. -E. van Wyk et A. L. Schutte ●☆

377111 Stirtonia insignis (Compton) B. -E. van Wyk et A. L. Schutte = Stirtonanthus insignis (Compton) B. -E. van Wyk et A. L. Schutte ●☆

377112 Stirtonia tayloriana (L. Bolus) B. -E. van Wyk et A. L. Schutte = Stirtonanthus taylorianus (L. Bolus) B. -E. van Wyk et A. L. Schutte ●☆

377113 Stissera Giseke = Curcuma L. (保留属名)■

377114 Stissera Heist. ex Fabr. = Stapelia L. (保留属名)■

377115 Stissera Kuntze = Stapelia L. (保留属名)■

377116 Stisseria Fabr. = Stapelia L. (保留属名)■

377117 Stisseria Heist. ex Fabr. = Stapelia L. (保留属名)■

377118 Stisseria Scop. = Manilkara Adans. (保留属名)●

377119 Stisseria Scop. = Mimusops L. ●☆

377120 Stixaceae Doweld(2008);斑果藤科(六萼藤科,罗志藤科)●

377121 Stixis Lour. (1790);斑果藤属(斑果藤属,六萼藤属,罗志藤属);Stixis ●

377122 Stixis fasciculata (King) Gagnep. = Stixis ovata (Korth.) Hallier f. subsp. fasciculata (King) Jacobs ●

377123 Stixis ovata (Korth.) Hallier f. subsp. fasciculata (King) Jacobs;锥序斑果藤●

377124 Stixis scandens Lour. ;闭脉斑果藤●

377125 Stixis suaveolens (Roxb.) Pierre;斑果藤(六萼藤,罗志藤);Fragrant Stixis,Stixis ●

377126 Stiza E. Mey. = Lebeckia Thunb. ■☆

377127 Stizolobium P. Browne(废弃属名) = Mucuna Adans. (保留属名)●■

377128 Stizolobium capitatum (Sweet) Kuntze = Mucuna pruriens (L.) DC. var. utilis (Wall. ex Wight) Baker ex Burck ■

377129 Stizolobium cochinchinensis (Lour.) Ts. Tang et F. T. Wang = Mucuna pruriens (L.) DC. var. utilis (Wall. ex Wight) Baker ex Burck ■

377130 Stizolobium hassjoo Piper et Tracy = Mucuna pruriens (L.) DC. var. utilis (Wall. ex Wight) Baker ex Burck ■

377131 Stizolobium imbricatum Kuntze = Mucuna nigricans (Lour.) Steud. ●■

377132 Stizolobium niveum (Roxb.) Kuntze = Mucuna pruriens (L.) DC. var. utilis (Wall. ex Wight) Baker ex Burck ■

377133 Stizolobium pruriens (L.) Medik. = Mucuna pruriens (L.) DC. ●■

377134 Stizolobium pruriens (L.) Medik. var. hassjoo (Piper et Tracy) Makino = Mucuna pruriens (L.) DC. var. utilis (Wall. ex Wight) Baker ex Burck ■

377135 Stizolobium utile (Wall.) Ditmer = Mucuna pruriens (L.) DC. var. utilis (Wall. ex Wight) Baker ex Burck ■

377136 Stizolobium venulosum Piper = Mucuna bracteata DC. ex Kurz ■

377137 Stizolophus Cass. (1826);纤刺菊属■☆

377138 Stizolophus Cass. = Centaurea L. (保留属名)●■

377139 Stizolophus balsamita (Lam.) Cass. ex Takht. = Centaurea balsamita Lam. ■☆

377140 Stizolophus coronopifolius (Lam.) Cass. ;鸟足叶纤刺菊■☆

377141 Stizophyllum Miers(1863);刺叶紫葳属●☆

377142 Stizophyllum adspersum Miers;刺叶紫葳●☆

377143 Stizophyllum affinis Miers;近缘刺叶紫葳●☆

377144 Stobaea Thunb. = Berkheya Ehrh. (保留属名)●■☆

377145 Stobaea acanthopoda DC. = Berkheya acanthopoda (DC.) Rössler ●☆

377146 Stobaea acarnoides DC. = Berkheya discolor (DC.) O. Hoffm. et Muschl. ■☆

377147 Stobaea aristosa DC. = Berkheya rhapontica (DC.) Hutch. et Burtt Davy var. aristosa (DC.) Rössler ■☆

377148　Stobaea atractyloides（L.）Thunb. var. carlinoides（Thunb.）DC. = Berkheya onobromoides（DC.）O. Hoffm. et Muschl. var. carlinoides（Thunb.）Rössler ■☆

377149　Stobaea biloba DC. = Berkheya heterophylla（Thunb.）O. Hoffm. ■☆

377150　Stobaea bipinnatifida Harv. = Berkheya bipinnatifida（Harv.）Rössler ■☆

377151　Stobaea cardopatifolia DC. = Berkheya cardopatifolia（DC.）Rössler ■☆

377152　Stobaea carlinifolia DC. = Berkheya carlinifolia（DC.）Rössler ■☆

377153　Stobaea carlinoides Thunb. = Berkheya onobromoides（DC.）O. Hoffm. et Muschl. var. carlinoides（Thunb.）Rössler ■☆

377154　Stobaea cirsiifolia DC. = Berkheya cirsiifolia（DC.）Rössler ■☆

377155　Stobaea cruciata（Houtt.）Harv. = Berkheya cruciata（Houtt.）Willd. ■☆

377156　Stobaea discolor DC. = Berkheya discolor（DC.）O. Hoffm. et Muschl. ■☆

377157　Stobaea echinacea Harv. = Berkheya echinacea（Harv.）O. Hoffm. ex Burtt Davy ■☆

377158　Stobaea echinopoda DC. = Berkheya decurrens（Thunb.）Willd. ■☆

377159　Stobaea epitrachys DC. = Berkheya pinnatifida（Thunb.）Thell. ■☆

377160　Stobaea eriobasis DC. = Berkheya eriobasis（DC.）Rössler ■☆

377161　Stobaea erysithales DC. = Berkheya erysithales（DC.）Rössler ■☆

377162　Stobaea gerrardii Harv. = Berkheya echinacea（Harv.）O. Hoffm. ex Burtt Davy ■☆

377163　Stobaea glabrata Thunb. = Berkheya glabrata（Thunb.）Fourc. ■☆

377164　Stobaea glabriuscula DC. = Berkheya decurrens（Thunb.）Willd. ■☆

377165　Stobaea glomerata（Thunb.）Spreng. = Platycarpha glomerata（Thunb.）Less. ■☆

377166　Stobaea grandifolia DC. = Berkheya multijuga（DC.）Rössler ■☆

377167　Stobaea helianthiflora DC. = Berkheya decurrens（Thunb.）Willd. ■☆

377168　Stobaea heterophylla Thunb. = Berkheya heterophylla（Thunb.）O. Hoffm. ■☆

377169　Stobaea heterophylla Thunb. var. radiata DC. = Berkheya heterophylla（Thunb.）O. Hoffm. var. radiata（DC.）Rössler ■☆

377170　Stobaea insignis Harv. = Berkheya insignis（Harv.）Thell. ■☆

377171　Stobaea membranifolia DC. = Berkheya decurrens（Thunb.）Willd. ■☆

377172　Stobaea microcephala DC. = Berkheya discolor（DC.）O. Hoffm. et Muschl. ■☆

377173　Stobaea multijuga DC. = Berkheya multijuga（DC.）Rössler ■☆

377174　Stobaea onobromoides DC. = Berkheya onobromoides（DC.）O. Hoffm. et Muschl. ■☆

377175　Stobaea onopordifolia DC. = Berkheya onopordifolia（DC.）O. Hoffm. ex Burtt Davy ■☆

377176　Stobaea oppositifolia DC. = Berkheya spinosissima（Thunb.）Willd. var. namaensis Rössler ■☆

377177　Stobaea petiolata DC. = Berkheya decurrens（Thunb.）Willd. ■☆

377178　Stobaea pinnata Thunb. = Heterorhachis aculeata（Burm. f.）Rössler ●☆

377179　Stobaea pinnatifida Thunb. = Berkheya pinnatifida（Thunb.）Thell. ■☆

377180　Stobaea platyptera Harv. = Berkheya rhapontica（DC.）Hutch. et Burtt Davy subsp. platyptera（Harv.）Rössler ■☆

377181　Stobaea polyacantha DC. = Berkheya decurrens（Thunb.）Willd. ■☆

377182　Stobaea purpurea DC. = Berkheya purpurea（DC.）Mast. ■☆

377183　Stobaea radula Harv. = Berkheya radula（Harv.）De Wild. ■☆

377184　Stobaea rigida Thunb. = Berkheya rigida（Thunb.）Bolus et Wolley-Dod ex Adamson et T. M. Salter ■☆

377185　Stobaea scolymoides DC. = Berkheya decurrens（Thunb.）Willd. ■☆

377186　Stobaea seminivea DC. = Berkheya bipinnatifida（Harv.）Rössler ■☆

377187　Stobaea sonchifolia Harv. = Berkheya erysithales（DC.）Rössler ■☆

377188　Stobaea speciosa DC. = Berkheya speciosa（DC.）O. Hoffm. ■☆

377189　Stobaea sphaerocephala DC. = Berkheya sphaerocephala（DC.）Rössler ■☆

377190　Stobaea viscosa DC. = Berkheya viscosa（DC.）Hutch. ■☆

377191　Stobaea zeyheri Sond. et Harv. = Berkheya zeyheri Oliv. et Hiern ■☆

377192　Stocksia Benth.（1853）;斯托无患子属●☆

377193　Stocksia brahuica Benth. ;斯托无患子●☆

377194　Stocksia brahuica Benth. = Stracheya tibetica Benth. ●■◇

377195　Stockwellia D. J. Carr, S. G. M. Carr et B. Hyland（2002）;斯托克木属●☆

377196　Stockwellia quadrifida D. J. Carr, S. G. M. Carr et B. Hyland;斯托克木●☆

377197　Stoebe L.（1753）;帚鼠麴属●■☆

377198　Stoebe aethiopica L. ;埃塞俄比亚帚鼠麴●☆

377199　Stoebe aethiopica Sieber ex DC. = Seriphium incanum（Thunb.）Pers. ●☆

377200　Stoebe affinis S. Moore = Stoebe capitata P. J. Bergius ●☆

377201　Stoebe alopecuroides（Lam.）Less. ;看麦娘帚鼠麴●☆

377202　Stoebe aspera Thunb. = Helichrysum asperum（Thunb.）Hilliard et B. L. Burtt ●☆

377203　Stoebe biotoides Baker = Stoebe cryptophylla Baker ●☆

377204　Stoebe bruniades（Rchb.）Levyns = Stoebe capitata P. J. Bergius ●☆

377205　Stoebe burchellii Levyns = Seriphium plumosum L. ●☆

377206　Stoebe capitata P. J. Bergius;头状帚鼠麴●☆

377207　Stoebe cernua Thunb. = Dicerothamnus rhinocerotis（L. f.）Koekemoer ■☆

377208　Stoebe cinerea（L.）Thunb. = Seriphium cinereum L. ●☆

377209　Stoebe cinerea Thunb. var. plumosa（Less.）Harv. = Seriphium plumosum L. ●☆

377210　Stoebe copholepis Sch. Bip. = Stoebe phyllostachya（DC.）Sch. Bip. ●☆

377211　Stoebe cryptophylla Baker;隐叶帚鼠麴●☆

377212　Stoebe cupressina Rchb. = Dicerothamnus rhinocerotis（L. f.）Koekemoer ■☆

377213 Stoebe cyathuloides Schltr. ;杯苋帚鼠麴●☆

377214 Stoebe disticha L. f. = Trichogyne ambigua (L.) Druce ■☆

377215 Stoebe elgonensis Mattf. = Seriphium kilimandscharicum (O. Hoffm.) Koekemoer ●☆

377216 Stoebe ensorii Compton = Stoebe phyllostachya (DC.) Sch. Bip. ●☆

377217 Stoebe ericoides P. J. Bergius = Disparago ericoides (P. J. Bergius) Gaertn. ●☆

377218 Stoebe fasciculata Cass. = Seriphium plumosum L. ●☆

377219 Stoebe fasciculata Thunb. = Trichogyne ambigua (L.) Druce ■☆

377220 Stoebe filaginea (DC.) Sch. Bip. ex Harv. = Seriphium spirale (Less.) Koekemoer ●☆

377221 Stoebe fusca (L.) Thunb. ;棕色帚鼠麴●☆

377222 Stoebe gnaphalodes Thunb. = Metalasia pulcherrima Less. ●☆

377223 Stoebe gomphrenoides P. J. Bergius ;千日红帚鼠麴●☆

377224 Stoebe incana Thunb. = Seriphium incanum (Thunb.) Pers. ●☆

377225 Stoebe kilimandscharica O. Hoffm. = Seriphium kilimandscharicum (O. Hoffm.) Koekemoer ●☆

377226 Stoebe kilimandscharica O. Hoffm. var. densiflora ? = Seriphium kilimandscharicum (O. Hoffm.) Koekemoer ●☆

377227 Stoebe leiocarpa Sch. Bip. = Seriphium incanum (Thunb.) Pers. ●☆

377228 Stoebe leucocephala DC. ;白头帚鼠麴●☆

377229 Stoebe microphylla DC. ;小叶帚鼠麴●☆

377230 Stoebe montana Schltr. ex Levyns;山地帚鼠麴●☆

377231 Stoebe mossii S. Moore = Stoebe capitata P. J. Bergius ●☆

377232 Stoebe muirii Levyns ;缪里帚鼠麴●☆

377233 Stoebe nivea Thunb. = Dolichothrix ericoides (Lam.) Hilliard et B. L. Burtt ■☆

377234 Stoebe pachyclada Humbert ;粗枝帚鼠麴●☆

377235 Stoebe pentheri O. Hoffm. = Stoebe rosea Wolley-Dod ●☆

377236 Stoebe phlaeoides (DC.) Sch. Bip. = Stoebe phyllostachya (DC.) Sch. Bip. ●☆

377237 Stoebe phylicoides Thunb. = Stoebe aethiopica L. ●☆

377238 Stoebe phyllostachya (DC.) Sch. Bip. ;叶穗帚鼠麴●☆

377239 Stoebe plumosa (L.) Thunb. = Seriphium plumosum L. ●☆

377240 Stoebe prostrata L. ;平卧帚鼠麴●☆

377241 Stoebe reflexum L. f. ;反折帚鼠麴●☆

377242 Stoebe rhinocerotis L. f. = Dicerothamnus rhinocerotis (L. f.) Koekemoer ■☆

377243 Stoebe rosea Wolley-Dod;蔷薇帚鼠麴●☆

377244 Stoebe rugulosa Harv. ;稍皱帚鼠麴●☆

377245 Stoebe salteri Levyns = Stoebe schultzii Levyns ●☆

377246 Stoebe saxatilis Levyns = Seriphium saxatilis (Levyns) Koekemoer ●☆

377247 Stoebe schultzii Levyns;舒尔茨帚鼠麴●☆

377248 Stoebe spiralis Less. = Seriphium spirale (Less.) Koekemoer ●☆

377249 Stoebe spiralis Less. var. flavescens (DC.) Harv. = Seriphium spirale (Less.) Koekemoer ●☆

377250 Stoebe squarrosa Harv. = Stoebe leucocephala DC. ●☆

377251 Stoebe tortilis DC. = Gongyloglossa tortilis (DC.) Koekemoer ■☆

377252 Stoebe virgata Thunb. = Seriphium plumosum L. ●☆

377253 Stoebe vulgaris Levyns = Seriphium plumosum L. ●☆

377254 Stoeberia Dinter et Schwantes(1927);松菊树属●☆

377255 Stoeberia apetala L. Bolus = Stoeberia beetzii (Dinter) Dinter et Schwantes ●☆

377256 Stoeberia arborea Van Jaarsv. ;树状松菊树●☆

377257 Stoeberia beetzii (Dinter) Dinter et Schwantes;贝茨松菊树●☆

377258 Stoeberia beetzii (Dinter) Dinter et Schwantes var. arborescens Friedrich = Stoeberia beetzii (Dinter) Dinter et Schwantes ●☆

377259 Stoeberia carpii Friedrich;澳非松菊树●☆

377260 Stoeberia frutescens (L. Bolus) Van Jaarsv. ;灌木松菊树●☆

377261 Stoeberia gigas (Dinter) Dinter et Schwantes;巨大松菊树●☆

377262 Stoeberia gigas (Dinter) Dinter et Schwantes = Ruschianthemum gigas (Dinter) Friedrich ●☆

377263 Stoeberia hallii L. Bolus = Phiambolia hallii (L. Bolus) Klak ■☆

377264 Stoeberia littlewoodii L. Bolus = Lampranthus mutatus (G. D. Rowley) H. E. K. Hartmann ■☆

377265 Stoeberia porphyrea H. E. K. Hartmann = Stoeberia arborea Van Jaarsv. ●☆

377266 Stoeberia rupis-arcuatae (Dinter) Dinter et Schwantes = Amphibolia rupis-arcuatae (Dinter) H. E. K. Hartmann ●☆

377267 Stoeberia utilis (L. Bolus) Van Jaarsv. ;有用松菊树●☆

377268 Stoechadomentha Kunze = Adenosma R. Br. ■

377269 Stoechas Gueldenst. = Helichrysum Mill. (保留属名)●■

377270 Stoechas Gueldenst. ex Ledeb. = Helichrysum Mill. (保留属名)●■

377271 Stoechas Mill. = Lavandula L. ●■

377272 Stoechas Rumph. = Adenosma R. Br. ■

377273 Stoechas Tourn. ex L. = Lavandula L. ●■

377274 Stoechas citrina Gueldenst. = Helichrysum arenarium (L.) Moench ●■

377275 Stoehelina Benth. = Bartsia L. (保留属名)■●☆

377276 Stoehelina Benth. = Staehelina L. ●☆

377277 Stoerkea Baker = Stoerkia Crantz ●■

377278 Stoerkia Crantz = Dracaena Vand. ex L. ●■

377279 Stoibrax Raf. (1840);斯托草属■☆

377280 Stoibrax Raf. = Carum L. ■

377281 Stoibrax capense (Lam.) B. L. Burtt;好望角斯托草■☆

377282 Stoibrax dichotomum (L.) Raf. ;二歧斯托草■☆

377283 Stoibrax hanotei (Maire) B. L. Burtt;哈诺特斯托草■☆

377284 Stoibrax involucratum (Braun-Blanq. et Maire) B. L. Burtt;筒鞘斯托草■☆

377285 Stoibrax pomelianum (Maire) B. L. Burtt;波梅尔斯托草■☆

377286 Stokesia L'Hér. (1789);琉璃菊属; Stokes Aster, Stokesia, Stokes' Aster ■☆

377287 Stokesia cyanea L'Hér. ; 蓝 琉 璃 菊; Cornflower Aster, Cornflower-aster,Stoke's Aster ■☆

377288 Stokesia laevis (Hill) Greene;琉璃菊（美国蓝菊）;Stokes Aster,Stokesia,Stokes' Aster,White Stokes' Aster ■☆

377289 Stokesia laevis (Hill) Greene 'Alba';白琉璃菊;White Stokes Aster ■☆

377290 Stokesia laevis (Hill) Greene 'Blue Star';蓝星琉璃菊■☆

377291 Stokesia laevis (Hill) Greene ' Mary Gregory ';黄琉璃菊; Yellow Stokes Aster ■☆

377292 Stokesia laevis (Hill) Greene 'Omega Skyrocket';高琉璃菊; Tall Stokes Aster ■☆

377293 Stokesia laevis (Hill) Greene ' Purple Parasols ';紫琉璃菊; Purple Stokes Aster ■☆

377294 Stokesia laevis (Hill) Greene ' Silver Moon ';银月琉璃菊; White Stokes Aster ■☆

377295 Stokoeanthus E. G. H. Oliv. (1976);斯托花属●☆

377296 Stokoeanthus E. G. H. Oliv. = Erica L. ●☆

377297 Stokoeanthus chionophilus E. G. H. Oliv. ;斯托花●☆

377298 Stokoeanthus chionophilus E. G. H. Oliv. = Erica stokoeanthus E. G. H. Oliv. ●☆

377299 Stolidia Baill. = Badula Juss. ●☆

377300 Stollaea Schltr. = Caldcluvia D. Don ●☆

377301 Stolzia Schltr. (1915);司徒兰属■☆

377302 Stolzia angustifolia Mansf. ;窄叶司徒兰■☆

377303 Stolzia atrorubra Mansf. ;深红司徒兰■☆

377304 Stolzia compacta P. J. Cribb;紧密司徒兰■☆

377305 Stolzia compacta P. J. Cribb subsp. iringana P. J. Cribb;伊林加司徒兰■☆

377306 Stolzia compacta P. J. Cribb subsp. purpurata P. J. Cribb;紫司徒兰■☆

377307 Stolzia cupuligera (Kraenzl.) Summerh. ;杯状司徒兰■☆

377308 Stolzia denticulata P. J. Cribb et Stévart;细齿司徒兰■☆

377309 Stolzia diffusa Summerh. = Stolzia cupuligera (Kraenzl.) Summerh. ■☆

377310 Stolzia grandiflora P. J. Cribb;大花司徒兰■☆

377311 Stolzia leedalii P. J. Cribb;利达尔司徒兰■☆

377312 Stolzia moniliformis P. J. Cribb;串珠状司徒兰■☆

377313 Stolzia nyassana Schltr. ;尼亚萨司徒兰■☆

377314 Stolzia oligantha Mansf. ;寡花司徒兰■☆

377315 Stolzia peperomioides (Kraenzl.) Summerh. ;草胡椒司徒兰■☆

377316 Stolzia repens (Rolfe) Summerh. ;匍匐司徒兰■☆

377317 Stolzia repens (Rolfe) Summerh. var. obtusata G. Will. ;钝司徒兰■☆

377318 Stolzia thomensis Stévart et P. J. Cribb;爱岛司徒兰■☆

377319 Stolzia viridis P. J. Cribb;绿司徒兰■☆

377320 Stolzia williamsonii P. J. Cribb;威廉森司徒兰■☆

377321 Stomadena Raf. = Ipomoea L. (保留属名)●■

377322 Stomandra Standl. (1947);口蕊茜属■☆

377323 Stomandra costaricensis Standl. ;口蕊茜■☆

377324 Stomarrhena DC. = Astroloma R. Br. ●☆

377325 Stomatanthes R. M. King et H. Rob. (1970);口泽兰属■●☆

377326 Stomatanthes africanus (Oliv. et Hiern) R. M. King et H. Rob. ;非洲口泽兰■☆

377327 Stomatanthes meyeri R. M. King et H. Rob. ;口泽兰■☆

377328 Stomatanthes zambiensis R. M. King et H. Rob. ;赞比亚口泽兰■☆

377329 Stomatechium B. D. Jacks. = Anchusa L. ■

377330 Stomatechium B. D. Jacks. = Stomotechium Lehm. ■

377331 Stomatium Schwantes(1926);齿舌叶属(史都草属)■☆

377332 Stomatium acutifolium L. Bolus;尖齿齿舌叶■☆

377333 Stomatium agninum (Haw.) Schwantes;全缘齿舌叶■☆

377334 Stomatium agninum (Haw.) Schwantes var. integrifolium (Salm-Dyck) Volk = Stomatium agninum (Haw.) Schwantes ■☆

377335 Stomatium agninum Schwantes = Stomatium agninum (Haw.) Schwantes ■☆

377336 Stomatium alboroseum L. Bolus;粉白齿舌叶■☆

377337 Stomatium angustifolium L. Bolus;窄叶齿舌叶■☆

377338 Stomatium beaufortense L. Bolus;博福特齿舌叶■☆

377339 Stomatium bolusiae Schwantes;博卢斯齿舌叶■☆

377340 Stomatium braunsii L. Bolus;布劳恩齿舌叶■☆

377341 Stomatium difforme L. Bolus;异形齿舌叶■☆

377342 Stomatium duthieae L. Bolus;笹舟玉●☆

377343 Stomatium fulleri L. Bolus;浮城●☆

377344 Stomatium gerstneri L. Bolus;格斯齿舌叶■☆

377345 Stomatium grandidens L. Bolus;大齿齿舌叶■☆

377346 Stomatium integrum L. Bolus;全缘叶齿舌叶■☆

377347 Stomatium jamesii L. Bolus;詹姆斯齿舌叶■☆

377348 Stomatium latifolium L. Bolus;宽叶齿舌叶■☆

377349 Stomatium lesliei (Schwantes) Volk;莱斯利齿舌叶■☆

377350 Stomatium loganii L. Bolus;洛根齿舌叶■☆

377351 Stomatium meyeri L. Bolus;迈尔齿舌叶■☆

377352 Stomatium middelburgense L. Bolus;米德尔堡齿舌叶■☆

377353 Stomatium murinum (Haw.) Schwantes;鼠色齿舌叶■☆

377354 Stomatium musculinum Schwantes = Chasmatophyllum musculinum (Haw.) Dinter et Schwantes ●☆

377355 Stomatium mustelinum (Haw.) Schwantes = Chasmatophyllum musculinum (Haw.) Dinter et Schwantes ●☆

377356 Stomatium niveum L. Bolus = Stomatium alboroseum L. Bolus ■☆

377357 Stomatium patulum H. Jacobsen = Stomatium patulum L. Bolus ex H. Jacobsen ■☆

377358 Stomatium patulum L. Bolus ex H. Jacobsen;齿舌叶■☆

377359 Stomatium paucidens L. Bolus;少齿齿舌叶■☆

377360 Stomatium peersii L. Bolus;浮舟玉●☆

377361 Stomatium pluridens L. Bolus;多齿齿舌叶■☆

377362 Stomatium pyrodorum (Diels) L. Bolus = Stomatium mustelinum (Haw.) Schwantes ●☆

377363 Stomatium resedolens L. Bolus;木犀草齿舌叶■☆

377364 Stomatium ronaldii L. Bolus;罗纳德齿舌叶■☆

377365 Stomatium rouxii L. Bolus;楠舟齿舌叶●☆

377366 Stomatium ryderae L. Bolus;吕德齿舌叶■☆

377367 Stomatium suaveolens Schwantes;芳香齿舌叶■☆

377368 Stomatium trifarium L. Bolus;三列齿舌叶■☆

377369 Stomatium villetii L. Bolus;维莱齿舌叶■☆

377370 Stomatium viride L. Bolus;绿齿舌叶■☆

377371 Stomatocalyx Müll. Arg. = Pimelodendron Hassk. ●

377372 Stomatochaeta (S. F. Blake) Maguire et Wurdack(1957);毛菊木属●☆

377373 Stomatochaeta acuminata Pruski;渐尖毛菊木●☆

377374 Stomatochaeta colveei Steyerm. ;科氏毛菊木●☆

377375 Stomatochaeta cylindrica Maguire et Wurdack = Stomatochaeta cymbifolia (S. F. Blake) Maguire et Wurdack ●☆

377376 Stomatochaeta cymbifolia (S. F. Blake) Maguire et Wurdack;毛菊木●☆

377377 Stomatostemma N. E. Br. (1902);口冠萝藦属■☆

377378 Stomatostemma monteiroae (Oliv.) N. E. Br. ;口冠萝藦■☆

377379 Stomatostemma monteiroae N. E. Br. = Stomatostemma monteiroae (Oliv.) N. E. Br. ■☆

377380 Stomatostemma pendulina Venter et D. V. Field;非洲口冠萝藦■☆

377381 Stomatotechium Spach = Anchusa L. ■

377382 Stomatotechium Spach = Stomotechium Lehm. ■

377383 Stomoisia Raf. = Utricularia L. ■

377384 Stomoisia cornuta (Michx.) Raf. = Utricularia cornuta Michx. ■☆

377385 Stomotechium Lehm. = Anchusa L. ■

377386 Stonekenya Raf. = Honkenya Ehrh. ■☆

377387 Stonesia G. Taylor(1953);斯通草属■☆

377388 Stonesia fascicularis G. Taylor;斯通草■☆

377389 Stonesia gracilis G. Taylor;纤细斯通草■☆

377390 Stonesia heterospathella G. Taylor;异苞斯通草■☆

377391 Stonesia taylorii C. Cusset;泰勒斯通草■☆

377392 Stonesiella Crisp et P. H. Weston = Pultenaea Sm. ●☆

377393 Stonesiella Crisp et P. H. Weston(1999);澳洲灌木豆属●☆

377394 Stongylocaryum Burret = Strongylocaryum Burret ●☆

377395 Stooria Neck. = Lobelia L. ●■

377396 Stooria Neck. ex T. Post et Kuntze = Lobelia L. ●■

377397 Stopinaca Raf. = Polygonella Michx. ■☆

377398 Storckiella Seem. (1861);名材豆属●☆

377399 Storckiella australiensis J. H. Ross et B. Hyland;澳洲名材豆 ●☆

377400 Storckiella vitiensis Seem. ;名材豆●☆

377401 Stormia S. Moore = Cardiopetalum Schltdl. ●☆

377402 Storthocalyx Radlk. (1879);尖萼无患子属●☆

377403 Storthocalyx chryseus Radlk. ;金色尖萼无患子●☆

377404 Storthocalyx leioneurus Radlk. ;光脉尖萼无患子●☆

377405 Storthocalyx pancheri Radlk. ;尖萼无患子●☆

377406 Storthocalyx sordidus Radlk. ;污浊尖萼无患子●☆

377407 Strabonia DC. = Pulicaria Gaertn. ■●

377408 Strabonia gnaphaloides DC. = Pulicaria gnaphaloides (Vent.) Boiss. ■

377409 Stracheya Benth. (1853);藏豆属;Stracheya,Zangbean ●■

377410 Stracheya Benth. = Hedysarum L. (保留属名)●■

377411 Stracheya tibetica Benth. ;藏豆;Tibet Stracheya,Zangbean ●■◇

377412 Stracheya tibetica Benth. = Hedysarum tibeticum (Benth.) B. H. Choi et H. Ohashi ●■◇

377413 Strailia T. Durand = Lecythis Loefl. ●☆

377414 Strakaea C. Presl = Apama Lam. ●

377415 Strakaea C. Presl = Thottea Rottb. ●

377416 Stramentopappus H. Rob. et V. A. Funk(1987);黄冠单毛菊属 ●☆

377417 Stramentopappus pooleae (B. L. Turner) H. Rob. et V. A. Funk;黄冠单毛菊■☆

377418 Stramonium Mill. = Datura L. ●■

377419 Strangalis Dulac = Hirschfeldia Moench ■☆

377420 Strangea Meisn. (1855);斯特山龙眼属●☆

377421 Strangea linearis Meisn. ;斯特山龙眼●☆

377422 Strangeveia Baker = Hyacinthus L. ■☆

377423 Strangeveia Baker = Strangweja Bertol. ■☆

377424 Strangula Noronha = Ardisia Sw. (保留属名)●■

377425 Strangwaysia Post et Kuntze = Hyacinthus L. ■☆

377426 Strangwaysia Post et Kuntze = Strangweja Bertol. ■☆

377427 Strangwaysia Post et Kuntze = Stranvaesia Lindl. ●

377428 Strangweia Baker = Strangweja Bertol. ■☆

377429 Strangweja Bertol. = Bellevalia Lapeyr. (保留属名)■☆

377430 Strangweja Bertol. = Hyacinthus L. ■☆

377431 Strangweya Benth. et Hook. f. = Strangweja Bertol. ■☆

377432 Strania Noronha = Canarium L. ●

377433 Stranvaesia Lindl. (1837);红果树属(假花楸属,斯脱兰木 属,斯脱木属,夏皮楠属);Stranvaesia ●

377434 Stranvaesia Lindl. = Aronia Medik. (保留属名)●☆

377435 Stranvaesia Lindl. = Photinia Lindl. ●

377436 Stranvaesia amphidoxa C. K. Schneid. ;毛萼红果树;Hairy Stranvaesia,Hairycalyx Stranvaesia,Hairy-calyxed Stranvaesia ●

377437 Stranvaesia amphidoxa C. K. Schneid. = Photinia amphidoxa (C. K. Schneid.) Rehder et E. H. Wilson ●

377438 Stranvaesia amphidoxa C. K. Schneid. var. amphileia (Hand. - Mazz.) Te T. Yu;湖南红果树(光萼红果树,无毛毛萼红果树); Smooth Hairy Stranvaesia ●

377439 Stranvaesia amphidoxa C. K. Schneid. var. kwangsiensis Metcalf; 广西毛萼红果树●

377440 Stranvaesia argyi H. Lév. = Photinia serrulata Lindl. ●

377441 Stranvaesia benthamiana (Hance) Merr. = Photinia benthamiana Hance ●

377442 Stranvaesia calleryana (Decne.) Decne. = Photinia calleryana (Decne.) Cardot ●

377443 Stranvaesia calleryana Decne. = Photinia benthamiana Hance ●

377444 Stranvaesia davidiana Decne. ;红果树(红枫子,椤木石楠,斯 脱兰威木); Chinese Photinia, Chinese Stranvaesia, David Stranvaesia,Stranvaesia ●

377445 Stranvaesia davidiana Decne. = Photinia davidiana (Decne.) Cardot ●

377446 Stranvaesia davidiana Decne. var. salicifolia (Hutch.) Rehder; 柳叶红果树(夏皮楠,玉山假沙梨);Taiwan Stranvaesia, Willowleaf Stranvaesia ●

377447 Stranvaesia davidiana Decne. var. salicifolia (Hutch.) Rehder = Photinia davidiana (Decne.) Cardot var. formosana (Cardot) H. Ohashi et Iketani ●

377448 Stranvaesia davidiana Decne. var. salicifolia (Hutch.) Rehder = Photinia niitakayamensis Hayata ●

377449 Stranvaesia davidiana Decne. var. salicifolia (Hutch.) Rehder = Stranvaesia davidiana Decne. ●

377450 Stranvaesia davidiana Decne. var. suoxiyuensis C. J. Qi et C. L. Peng;索溪峪红果树;Suoxiyu David Stranvaesia ●

377451 Stranvaesia davidiana Decne. var. suoxiyuensis C. J. Qi et C. L. Peng = Stranvaesia davidiana Decne. var. undulata (Decne.) Rehder et E. H. Wilson ●

377452 Stranvaesia davidiana Decne. var. undulata (Decne.) Rehder et E. H. Wilson;波叶红果树;Undulate Stranvaesia ●

377453 Stranvaesia glaucescens Lindl. = Stranvaesia nussia (Buch. - Ham. ex D. Don) Decne. ●

377454 Stranvaesia glaucescens Lindl. var. yunnanensis Franch. = Photinia lasiogyna (Franch.) C. K. Schneid. ●

377455 Stranvaesia henryi Diels = Stranvaesia davidiana Decne. ●

377456 Stranvaesia impressivena (Hayata) Masam. = Photinia impressivena Hayata ●

377457 Stranvaesia niitakayamensis (Hayata) Hayata;新高山斯脱兰木 (玉山假沙梨);Morrison Stranvaesia ●

377458 Stranvaesia niitakayamensis (Hayata) Hayata = Photinia davidiana (Decne.) Cardot var. formosana (Cardot) H. Ohashi et Iketani ●

377459 Stranvaesia niitakayamensis (Hayata) Hayata = Photinia niitakayamensis Hayata ●

377460 Stranvaesia niitakayamensis (Hayata) Hayata = Stranvaesia davidiana Decne. ●

377461 Stranvaesia niitakayamensis (Hayata) Hayata = Stranvaesia davidiana Decne. var. salicifolia (Hutch.) Rehder ●

377462 Stranvaesia nussia (Buch. -Ham. ex D. Don) Decne. ;印缅红果 树;India-Burma Stranvaesia ●

377463 Stranvaesia nussia (Buch. -Ham.) Decne. = Stranvaesia nussia (Buch. -Ham. ex D. Don) Decne. ●

377464 Stranvaesia nussia (Buch. -Ham.) Decne. var. oblanceolata Rehder et E. H. Wilson = Stranvaesia oblanceolata (Rehder et E. H.

Wilson）Stapf ●

377465　Stranvaesia nussia Buch.-Ham. ex D. Don var. oblanceolata Rehder et E. H. Wilson = Stranvaesia oblanceolata（Rehder et E. H. Wilson）Stapf ●

377466　Stranvaesia oblanceolata（Rehder et E. H. Wilson）Stapf；滇南红果树；Oblanceolate Stranvaesia，S. Yunnan Stranvaesia ●

377467　Stranvaesia salicifolia Hutch. = Stranvaesia davidiana Decne. var. salicifolia（Hutch.）Rehder ●

377468　Stranvaesia salicifolia Hutch. = Stranvaesia davidiana Decne. ●

377469　Stranvaesia scandens（Stapf）Hand.-Mazz. = Photinia integrifolia Lindl. ●

377470　Stranvaesia tomentosa Te T. Yu et T. C. Ku；绒毛红果树；Tomentose Stranvaesia ●

377471　Stranvaesia undulata Decne. = Stranvaesia davidiana Decne. var. undulata（Decne.）Rehder et E. H. Wilson ●

377472　Strasburgeria Baill.（1876）；栓皮果属●☆

377473　Strasburgeria calliantha Baill. ；栓皮果●☆

377474　Strasburgeriaceae Engl. et Gilg = Strasburgeriaceae Tiegh.（保留科名）●☆

377475　Strasburgeriaceae Tiegh.（1908）（保留科名）；栓皮果科●☆

377476　Strateuma Raf. = Zeuxine Lindl.（保留属名）■

377477　Strateuma Salisb. = Orchis L. ■

377478　Stratioites Gilib. = Stratiotes L. ■☆

377479　Stratiotaceae Link = Hydrocharitaceae Juss.（保留科名）■

377480　Stratiotaceae Link；水剑叶科■

377481　Stratiotaceae Schultz Sch. = Stratiotaceae Link ■

377482　Stratiotes L.（1753）；水剑叶属（斯特藻属）；Water Soldier ■☆

377483　Stratiotes acoroides L. f. = Enhalus acoroides（L. f.）Royle ■

377484　Stratiotes alismoides L. = Ottelia alismoides（L.）Pers. ■

377485　Stratiotes aloides L. ；水剑叶；Crab's Claws，Freshwater Soldiers，Knigbt's Pondweed，Knight's Pondweed，Sea Green，Soldier's Arrow，Soldier's Yarrow，Water Aloe，Water Houseleek，Water Parsnip，Water Pine，Water Sengren，Water Soldier，Water Soldiers ■☆

377486　Stratiotes nymphoides Willd. = Hydrocleys nymphoides（Willd.）Buchenau ■☆

377487　Straussia（DC.）A. Gray = Psychotria L.（保留属名）●

377488　Straussia A. Gray = Psychotria L.（保留属名）●

377489　Straussiella Hausskn.（1897）；伊朗芥属■☆

377490　Straussiella iranica Hausskn. ；伊朗芥■☆

377491　Stravadia Pers. = Stravndium Juss. ●

377492　Stravndium Juss. = Barringtonia J. R. Forst. et G. Forst.（保留属名）●

377493　Strebanthus Raf. = Eryngium L. ■

377494　Strebanthus Raf. = Streblanthus Raf. ■

377495　Streblacanthus Kuntze（1891）；大刺爵床属●☆

377496　Streblacanthus boliviensis Lindau；玻利维亚大刺爵床●☆

377497　Streblacanthus cordifolius T. F. Daniel；心叶大刺爵床●☆

377498　Streblacanthus longiflorus Cufod. ；长花大刺爵床●☆

377499　Streblacanthus macrophyllus Lindau；大叶大刺爵床●☆

377500　Streblacanthus monospermus Kuntze；单籽大刺爵床●☆

377501　Streblacanthus parviflorus Léonard；小花大刺爵床●☆

377502　Streblacanthus roseus（Radlk.）B. L. Burtt；粉红大刺爵床●☆

377503　Streblanthera Steud. = Trichodesma R. Br.（保留属名）●■

377504　Streblanthera Steud. ex A. Rich. = Trichodesma R. Br.（保留属名）●■

377505　Streblanthera oleifolia A. Rich. = Trichodesma trichodesmoides（Bunge）Gürke ■☆

377506　Streblanthus Raf. = Eryngium L. ■

377507　Streblidia Link = Schoenus L. ■

377508　Streblina Raf. = Nyssa L. ●

377509　Streblocarpus Arn. = Maerua Forssk. ●☆

377510　Streblocarpus angustifolia（A. Rich.）Endl. ex Walp. = Maerua oblongifolia（Forssk.）A. Rich. ●☆

377511　Streblocarpus fenzlii Parl. = Cadaba farinosa Forssk. ●☆

377512　Streblocarpus oblongifolius（Forssk.）Endl. ex Walp. = Maerua oblongifolia（Forssk.）A. Rich. ●☆

377513　Streblocarpus pubescens Klotzsch = Maerua triphylla A. Rich. var. pubescens（Klotzsch）DeWolf ●☆

377514　Streblocarpus scandens Klotzsch = Maerua scandens（Klotzsch）Gilg ●☆

377515　Streblochaeta Benth. et Hook. f. = Streblochaete Hochst. ex A. Rich. ■☆

377516　Streblochaete Hochst. ex A. Rich. = Streblochaete Hochst. ex Pilg. ■☆

377517　Streblochaete Hochst. ex Pilg.（1906）；长芒草属■☆

377518　Streblochaete Pilg. = Streblochaete Hochst. ex Pilg. ■☆

377519　Streblochaete longiarista（A. Rich.）Pilg. ；长芒草■☆

377520　Streblorrhiza Endl.（1833）；绞根耀花豆属■☆

377521　Streblorrhiza speciosa Endl. ；绞根耀花豆■☆

377522　Streblosa Korth.（1851）；马来茜属●☆

377523　Streblosa axilliflora Merr. ；腋花马来茜●☆

377524　Streblosa bracteata Ridl. ；大苞马来茜●☆

377525　Streblosa bracteolata Merr. ；小苞马来茜●☆

377526　Streblosa glabra Valeton；光马来茜●☆

377527　Streblosa hirta Ridl. ；粗毛马来茜●☆

377528　Streblosa leiophylla Bremek. ；光叶马来茜●☆

377529　Streblosa maxima Bremek. ；大马来茜●☆

377530　Streblosa microcarpa Ridl. ；小果马来茜●☆

377531　Streblosa multiglandulosa Merr. ；多腺马来茜●☆

377532　Streblosa pubescens Ridl. ；毛马来茜●☆

377533　Streblosa undulata Korth. ；波缘马来茜●☆

377534　Streblosa wallichii Merr. ；瓦氏马来茜●☆

377535　Streblosiopsis Valeton（1910）；假马来茜属●☆

377536　Streblosiopsis cupulata Valeton；假马来茜●☆

377537　Streblus Lour.（1790）；鹊肾树属；Streblus ●

377538　Streblus asper Lour. ；鹊肾树（鸡仔，万里果，莺哥果）；Paperbark，Rough Streblus ●

377539　Streblus brunonianus F. Muell. ；澳洲鹊肾树；Whalebone Tree ●☆

377540　Streblus ilicifolius（S. Vidal）Corner；刺桑（赤回）；Hollyleaf Streblus，Jungle Holly ●

377541　Streblus indicus（Bureau）Corner；假鹊肾树（滑叶跌打，青树跌打，清水跌打，止血树）；Indian Pseudostreblus，Indian Streblus ●

377542　Streblus macrophyllus Blume；大叶假刺桑（刺桑，双果桑）；Bigleaf Streblus，Big-leaved Spine-mulberry ●

377543　Streblus taxoides（K. Heyne）Kurz；叶被木（酒饼树）；Yew-like Phyllochlamys，Yew-like Streblus ●

377544　Streblus tonkinensis（Eberh. et Dubard）Corner；米扬噎（米农液，米杨噎，霜降胶木，条龙胶树，条隆胶）；Tongking Rubber，Tonkin Streblus，Tonkin Teonongia ●

377545　Streblus usambarensis（Engl.）C. C. Berg；乌桑巴拉鹊肾树●☆

377546　Streblus zeylanicus（Thwaites）Kurz；尾叶刺桑；Ceylon Streblus ●

377547   Streckera Sch. Bip. = Leontodon L. (保留属名)■☆

377548   Streleskia Hook. f. = Wahlenbergia Schrad. ex Roth(保留属名)■●

377549   Strelitsia Thunb. = Strelitzia Aiton ●

377550   Strelitzia Aiton(1789);鸟蕉属(鹤望兰属,扇芭蕉属);Bird of Paradise Flower, Bird's Tongue Flower, Bird-of-paradise, Bird-of-paradise Flower,Strelitzia ●

377551   Strelitzia Banks = Strelitzia Aiton ●

377552   Strelitzia Banks ex Dryand. = Strelitzia Aiton ●

377553   Strelitzia Dryand = Strelitzia Aiton ●

377554   Strelitzia alba (L. f.) Skeels;扇芭蕉;White Bird-of-paradise, White Bird-of-paradise Flower ●

377555   Strelitzia angustifolia W. T. Aiton = Strelitzia reginae Banks ex Aiton ●

377556   Strelitzia augusta Thunb. = Strelitzia alba (L. f.) Skeels ●

377557   Strelitzia caudata R. A. Dyer;尾状鸟蕉●☆

377558   Strelitzia farinosa W. T. Aiton = Strelitzia reginae Banks ex Aiton ●

377559   Strelitzia gigantea J. Kern = Strelitzia reginae Banks ex Aiton ●

377560   Strelitzia glauca Rich. = Strelitzia reginae Banks ex Aiton ●

377561   Strelitzia humilis Link = Strelitzia reginae Banks ex Aiton ●

377562   Strelitzia juncea Link;灯心草鹤望兰●☆

377563   Strelitzia nicolai Regel et Körn.;大鹤望兰(白花鹤望兰,尼古拉鹤望兰,扇蕉);Big Bird-of-paradise, Big Bird-of-paradise Flower, Giant Bird of Paradise,Natal Wild Banana ●

377564   Strelitzia ovata W. T. Aiton = Strelitzia reginae Banks ex Aiton ●

377565   Strelitzia parvifolia Dryand.;小叶鹤望兰●☆

377566   Strelitzia parvifolia W. T. Aiton var. juncea Ker Gawl. = Strelitzia juncea Link ●☆

377567   Strelitzia quensonii Lem. = Strelitzia nicolai Regel et Körn. ●

377568   Strelitzia regalis Salisb. = Strelitzia reginae Banks ex Aiton ●

377569   Strelitzia reginae Aiton = Strelitzia reginae Banks ex Aiton ●

377570   Strelitzia reginae Aiton var. farinosa (W. T. Aiton) Baker = Strelitzia reginae Banks ex Aiton ●

377571   Strelitzia reginae Aiton var. glauca (Rich.) Baker = Strelitzia reginae Banks ex Aiton ●

377572   Strelitzia reginae Aiton var. humilis (Link) Baker = Strelitzia reginae Banks ex Aiton ●

377573   Strelitzia reginae Aiton var. juncea (Ker Gawl.) H. E. Moore = Strelitzia juncea Link ●☆

377574   Strelitzia reginae Aiton var. ovata (W. T. Aiton) Baker = Strelitzia reginae Banks ex Aiton ●

377575   Strelitzia reginae Aiton var. rutilans (C. Morren) K. Schum. = Strelitzia reginae Banks ex Aiton ●

377576   Strelitzia reginae Banks ex Aiton;鹤望兰(极乐鸟,鸟蕉,天堂鸟,天堂鸟蕉);Bird of Paradise, Bird-of-paradise, Bird-of-paradise Flower, Bird's Tongue, Crane Flower, Crane Lily, Queen Bird-of-paradise Flower,Strelitzia ●

377577   Strelitzia rutilans C. Morren = Strelitzia reginae Banks ex Aiton ●

377578   Strelitziaceae (K. Schum.) Hutch. = Musaceae Juss. (保留科名)■

377579   Strelitziaceae (K. Schum.) Hutch. = Strelitziaceae Hutch. ●■

377580   Strelitziaceae Hutch. (1934)(保留科名);鹤望兰科(旅人蕉科);Strelitzia Family ●■

377581   Strelitziaceae Hutch. (保留科名) = Musaceae Juss. (保留科名)■

377582   Strempelia A. Rich. = Psychotria L. (保留属名)●

377583   Strempelia A. Rich. = Strempelia A. Rich. ex DC. ●☆

377584   Strempelia A. Rich. ex DC. (1830);施特茜属●☆

377585   Strempelia A. Rich. ex DC. = Psychotria L. (保留属名)●

377586   Strempelia ceratopetala (Donn. Sm.) Bremek.;角瓣施特茜●☆

377587   Strempelia guianensis A. Rich.;圭亚那施特茜●☆

377588   Strempeliopsis Benth. (1876);施特夹竹桃属●☆

377589   Strempeliopsis arborea Urb.;乔木施特夹竹桃●☆

377590   Strempeliopsis cubensis M. Gómez;古巴施特夹竹桃●☆

377591   Strempeliopsis strempelioides Benth.;施特夹竹桃●☆

377592   Strepalon Raf. = Hypericum L. ■●

377593   Strepalon Raf. = Streptalon Raf. ■●

377594   Strephium Nees = Raddia Bertol. ■☆

377595   Strephium Schrad. ex Nees = Olyra L. ■☆

377596   Strephium Schrad. ex Nees = Raddia Bertol. ■☆

377597   Strephonema Hook. f. (1867);扭丝使君子属●☆

377598   Strephonema apolloniense Clark = Strephonema pseudocolum A. Chev. ●☆

377599   Strephonema gilletii De Wild. = Strephonema sericeum Hook. f. ●☆

377600   Strephonema klaineanum Pierre = Strephonema sericeum Hook. f. ●☆

377601   Strephonema mannii Hook. f.;曼氏扭丝使君子●☆

377602   Strephonema polybotryum Mildbr. = Strephonema sericeum Hook. f. ●☆

377603   Strephonema pseudocolum A. Chev.;假圆柱扭丝使君子●☆

377604   Strephonema sericeum Hook. f.;绢毛使君子●☆

377605   Strephonema tessmannii Mildbr. = Strephonema mannii Hook. f. ●☆

377606   Strephonema tessmannii Mildbr. var. micranthum ? = Strephonema mannii Hook. f. ●☆

377607   Strephonemataceae Venkat. et Prak. Rao = Combretaceae R. Br. (保留科名)●

377608   Strepsanthera Raf. = Anthurium Schott ■

377609   Strepsia Steud. = Tillandsia L. ■☆

377610   Strepsiloba Raf. = Strepsilobus Raf. ●

377611   Strepsilobus Raf. = Entada Adans. (保留属名)●

377612   Strepsimela Raf. = Helixanthera Lour. ●

377613   Strepsiphigla Krause = Drimia Jacq. ex Willd. ■☆

377614   Strepsiphigla Krause = Strepsiphyla Raf. ■☆

377615   Strepsiphus Raf. = Peristrophe Nees ■

377616   Strepsiphyla Raf. = Drimia Jacq. ex Willd. ■☆

377617   Streptachne Kunth = Aristida L. ■

377618   Streptachne R. Br. = Aristida L. ■●

377619   Streptachne domingensis Spreng. ex Schult. = Schizachyrium sanguineum (Retz.) Alston ■

377620   Streptalon Raf. = Hypericum L. ■●

377621   Streptanthella Rydb. (1917);长喙提琴芥属■☆

377622   Streptanthella longirostris (S. Watson) Rydb.;长喙提琴芥■☆

377623   Streptanthella longirostris Rydb. = Streptanthella longirostris (S. Watson) Rydb. ■☆

377624   Streptanthera Sweet = Sparaxis Ker Gawl. ■☆

377625   Streptanthera Sweet(1827);扭药花属;Streptanthera, Twisted Anther Flower ■☆

377626   Streptanthera cuprea Sweet;扭药花(旋药花);Common Streptanthera ■☆

377627   Streptanthera cuprea Sweet = Sparaxis elegans (Sweet) Goldblatt ■☆

377628　Streptanthera cuprea Sweet var. non-picta L.；红扭药花；Red Streptanthera ■☆

377629　Streptanthera cuprea Sweet var. non-picta L. Bolus = Sparaxis elegans（Sweet）Goldblatt ■☆

377630　Streptanthera elegans Sweet；雅致扭药花■☆

377631　Streptanthera elegans Sweet = Sparaxis elegans（Sweet）Goldblatt ■☆

377632　Streptanthera tricolor（Schneev.）Klatt = Sparaxis tricolor（Schneev.）Ker Gawl. ■☆

377633　Streptanthus Nutt.（1825）；扭花芥属■☆

377634　Streptanthus arizonicus S. Watson；亚利桑那扭花芥；Arizona Jewelflower ■☆

377635　Streptanthus campestris S. Watson；加利福尼亚扭花芥■☆

377636　Streptanthus campestris var. jacobaeus（Greene）Jeps.；雅各扭花芥■☆

377637　Streptanthus carinatus Wright ex A. Gray；龙骨状扭花芥；Lyreleaf Jewelflower, Pecos Twistflower ■☆

377638　Streptanthus cordatus Nutt. ex Torr. et A. Gray；心形扭花芥；Heartleaf Jewelflower ■☆

377639　Streptanthus tortuosus Kellogg；山地扭花芥；Mountain Jewelflower ■☆

377640　Streptia Döll = Streptogyna P. Beauv. ■☆

377641　Streptia Rich. ex Hook. f. = Streptogyna P. Beauv. ■☆

377642　Streptilon Raf. = Geum L. ■

377643　Streptima Raf. = Frankenia L. ●■

377644　Streptium Roxb. = Priva Adans. ■☆

377645　Streptocalyx Beer = Aechmea Ruiz et Pav.（保留属名）■☆

377646　Streptocalyx Beer（1854）；扭萼凤梨属（塔花凤梨属，塔花属，旋萼属）；Streptocalyx ■☆

377647　Streptocalyx angustifolius Mez；窄叶扭萼凤梨■☆

377648　Streptocalyx biflorus L. B. Sm.；双花扭萼凤梨■☆

377649　Streptocalyx brachystachys Harms；短穗扭萼凤梨■☆

377650　Streptocalyx floribunda Mez；多花扭萼凤梨■☆

377651　Streptocalyx laxiflora Baker；疏花扭萼凤梨■☆

377652　Streptocarpus Lindl.（1828）；扭果花属（好望角苣苔属，旋果花属）；Cape Primrose ■☆

377653　Streptocarpus 'Constant Nymph' = Streptocarpus rexii Lindl. ■☆

377654　Streptocarpus × kewensis Hort.；邱园扭果花■☆

377655　Streptocarpus albiflorus Engl. = Streptocarpus nobilis C. B. Clarke ■☆

377656　Streptocarpus albus（E. A. Bruce）I. Darbysh.；白扭果花■☆

377657　Streptocarpus albus（E. A. Bruce）I. Darbysh. subsp. edwardsii（Weigend）I. Darbysh.；爱德华兹扭果花■☆

377658　Streptocarpus andohahelensis Humbert；安杜扭果花■☆

377659　Streptocarpus arcuatus Hilliard et B. L. Burtt；拱扭果花■☆

377660　Streptocarpus armitagei Baker f. et S. Moore = Streptocarpus dunnii Hook. f. ■☆

377661　Streptocarpus atroviolaceus Engl. = Streptocarpus nobilis C. B. Clarke ■☆

377662　Streptocarpus balsaminoides Engl. = Streptocarpus nobilis C. B. Clarke ■☆

377663　Streptocarpus balsaminoides Engl. var. tenuifolius ? = Streptocarpus nobilis C. B. Clarke ■☆

377664　Streptocarpus bambuseti B. L. Burtt；邦布塞特扭果花■☆

377665　Streptocarpus baudertii L. L. Britten；鲍德扭果花■☆

377666　Streptocarpus beampingaratrensis Humbert；贝安扭果花■☆

377667　Streptocarpus benguellensis Welw. ex C. B. Clarke = Streptocarpus monophyllus Welw. ■☆

377668　Streptocarpus bequaertii De Wild. = Streptocarpus glandulosissimus Engl. ■☆

377669　Streptocarpus boinensis Humbert；博伊纳扭果花■☆

377670　Streptocarpus bolusii C. B. Clarke；博卢斯扭果花■☆

377671　Streptocarpus brachynema Hilliard et B. L. Burtt；短丝扭果花■☆

377672　Streptocarpus breviflos（C. B. Clarke）C. B. Clarke；短花扭果花■☆

377673　Streptocarpus brevistamineus Humbert；短线扭果花■☆

377674　Streptocarpus buchananii C. B. Clarke；布坎南扭果花■☆

377675　Streptocarpus bullatus Mansf.；泡状扭果花■☆

377676　Streptocarpus burttianus Pócs；伯特扭果花■☆

377677　Streptocarpus burundianus Hilliard et B. L. Burtt；布隆迪扭果花■☆

377678　Streptocarpus caeruleus Hilliard et B. L. Burtt；天蓝扭果花■☆

377679　Streptocarpus caeruleus Hilliard et B. L. Burtt subsp. longiflorus Hilliard et B. L. Burtt = Streptocarpus longiflorus（Hilliard et B. L. Burtt）T. J. Edwards ■☆

377680　Streptocarpus campanulatus B. L. Burtt；风铃扭果花■☆

377681　Streptocarpus candidus Hilliard；纯白扭果花■☆

377682　Streptocarpus capuronii Humbert；凯普伦扭果花■☆

377683　Streptocarpus caulescens Vatke；具茎旋果花（蓝蝶旋果花）■☆

377684　Streptocarpus caulescens Vatke var. ovatus C. B. Clarke = Streptocarpus holstii Engl. ■☆

377685　Streptocarpus caulescens Vatke var. pallescens Engl. = Streptocarpus pallidiflorus C. B. Clarke ■☆

377686　Streptocarpus chariensis A. Chev. = Streptocarpus nobilis C. B. Clarke ■☆

377687　Streptocarpus chinensis Franch. = Rhabdothamnopsis sinensis Hemsl. ■

377688　Streptocarpus clarkeanus（Hemsl.）Hill. et Burtt = Boea clarkeana Hemsl. ■

377689　Streptocarpus compressus B. L. Burtt；扁扭果花■☆

377690　Streptocarpus comptonii Mansf. = Streptocarpus polyanthus Hook. subsp. comptonii（Mansf.）Hilliard ■☆

377691　Streptocarpus confusus Hilliard；混乱扭果花■☆

377692　Streptocarpus confusus Hilliard subsp. lebomboensis Hilliard et B. L. Burtt；莱邦博扭果花■☆

377693　Streptocarpus cooksonii B. L. Burtt；库克逊扭果花■☆

377694　Streptocarpus cooperi C. B. Clarke；库珀扭果花■☆

377695　Streptocarpus cordifolius Humbert；心叶扭果花■☆

377696　Streptocarpus coursii Humbert；库尔斯扭果花■☆

377697　Streptocarpus cyanandrus B. L. Burtt；蓝蕊扭果花■☆

377698　Streptocarpus cyaneus S. Moore；蓝扭果花■☆

377699　Streptocarpus cyaneus S. Moore subsp. nigridens Weigend et T. J. Edwards；黑齿蓝蕊扭果花■☆

377700　Streptocarpus cyaneus S. Moore subsp. polackii（B. L. Burtt）Weigend et T. J. Edwards；波拉克扭果花■☆

377701　Streptocarpus daviesii N. E. Br. ex C. B. Clarke；戴维斯扭果花■☆

377702　Streptocarpus davyi S. Moore；戴维扭果花■☆

377703　Streptocarpus decipiens Hilliard et B. L. Burtt；迷惑扭果花■☆

377704　Streptocarpus denticulatus Engl. = Streptocarpus insularis Hutch. et Dalziel ■☆

377705　Streptocarpus denticulatus Turrill;细齿扭果花■☆

377706　Streptocarpus dunnii Hook. f. ;邓恩扭果花■☆

377707　Streptocarpus elongatus Engl. ;伸长扭果花■☆

377708　Streptocarpus elongatus Engl. var. glabrescens = Streptocarpus elongatus Engl. ■☆

377709　Streptocarpus erubescens Hilliard et B. L. Burtt;变红扭果花■☆

377710　Streptocarpus exsertus Hilliard et B. L. Burtt;伸出扭果花■☆

377711　Streptocarpus eylesii S. Moore;艾尔斯扭果花■☆

377712　Streptocarpus eylesii S. Moore subsp. brevistylus Hilliard et B. L. Burtt;短柱扭果花■☆

377713　Streptocarpus eylesii S. Moore subsp. chalensis I. Darbysh. ;查拉扭果花■☆

377714　Streptocarpus eylesii S. Moore subsp. silvicola Hilliard et B. L. Burtt;森林扭果花■☆

377715　Streptocarpus fanniniae Harv. ex C. B. Clarke;范尼扭果花■☆

377716　Streptocarpus fanniniae Harv. ex C. B. Clarke var. minor C. B. Clarke = Streptocarpus fanniniae Harv. ex C. B. Clarke ■☆

377717　Streptocarpus fasciatus T. J. Edwards et Kunhardt;带状扭果花■☆

377718　Streptocarpus fenestradei Weigend et T. J. Edwards;窗孔扭果花■☆

377719　Streptocarpus floribundus Weigend et T. J. Edwards;多花孔扭果花■☆

377720　Streptocarpus formosus (Hilliard et B. L. Burtt) T. J. Edwards;美丽花孔扭果花■☆

377721　Streptocarpus galpinii Hook. f. ;盖尔扭果花■☆

377722　Streptocarpus glandulosissimus Engl. ;多腺扭果花■☆

377723　Streptocarpus glandulosissimus Engl. var. longiflorus Mansf. = Streptocarpus bambuseti B. L. Burtt ■☆

377724　Streptocarpus goetzei Engl. ;格兹扭果花■☆

377725　Streptocarpus gonjaensis Engl. ;贡贾扭果花■☆

377726　Streptocarpus gracilis B. L. Burtt = Streptocarpus prolixus C. B. Clarke ■☆

377727　Streptocarpus grandis N. E. Br. ;大扭果花(大苣苔)■☆

377728　Streptocarpus grandis N. E. Br. subsp. septentrionalis Hilliard et B. L. Burtt;北方扭果花■☆

377729　Streptocarpus haygarthii N. E. Br. ex C. B. Clarke;海加斯扭果花■☆

377730　Streptocarpus heckmannianus (Engl.) I. Darbysh. ;赫克曼扭果花■☆

377731　Streptocarpus heckmannianus (Engl.) I. Darbysh. subsp. gracilis (E. A. Bruce) I. Darbysch. ;纤细赫克曼扭果花■☆

377732　Streptocarpus hildebrandtii Vatke;希尔德扭果花■☆

377733　Streptocarpus hilsenbergii R. Br. ;希尔森扭果花■☆

377734　Streptocarpus hirsutissimus Bruce;粗毛扭果花■☆

377735　Streptocarpus holstii Engl. ;霍尔扭果花■☆

377736　Streptocarpus huamboensis Hilliard et B. L. Burtt;万博扭果花■☆

377737　Streptocarpus ibityensis Humbert;伊比提扭果花■☆

377738　Streptocarpus inflatus B. L. Burtt;膨胀扭果花■☆

377739　Streptocarpus insignis B. L. Burtt = Streptocarpus primulifolius Gand. ■☆

377740　Streptocarpus insularis Hutch. et Dalziel;海岛扭果花■☆

377741　Streptocarpus itremensis B. L. Burtt;伊特雷穆扭果花■☆

377742　Streptocarpus johannis L. L. Britten;约翰扭果花■☆

377743　Streptocarpus junodii P. Beauv. = Streptocarpus cyaneus S. Moore ■☆

377744　Streptocarpus katangensis De Wild. et T. Durand;加丹加扭果花■☆

377745　Streptocarpus kentaniensis L. L. Britten et Story;肯塔尼扭果花■☆

377746　Streptocarpus kerstingii Engl. = Streptocarpus nobilis C. B. Clarke ■☆

377747　Streptocarpus kirkii Hook. f. ;吉氏扭果花■☆

377748　Streptocarpus kungwensis Hilliard et B. L. Burtt;昆圭扭果花■☆

377749　Streptocarpus kunhardtii T. J. Edwards;孔哈特扭果花■☆

377750　Streptocarpus lagosensis C. B. Clarke = Streptocarpus nobilis C. B. Clarke ■☆

377751　Streptocarpus lanatus MacMaster;绵毛扭果花■☆

377752　Streptocarpus latens Hilliard et B. L. Burtt;隐匿扭果花■☆

377753　Streptocarpus ledermannii Engl. = Streptocarpus nobilis C. B. Clarke ■☆

377754　Streptocarpus leptopus Hilliard et B. L. Burtt;细梗扭果花■☆

377755　Streptocarpus levis B. L. Burtt;平滑扭果花■☆

377756　Streptocarpus lilacinus Engl. = Streptocarpus buchananii C. B. Clarke ■☆

377757　Streptocarpus longiflorus (Hilliard et B. L. Burtt) T. J. Edwards;长花扭果花■☆

377758　Streptocarpus lujai De Wild. = Streptocarpus goetzei Engl. ■☆

377759　Streptocarpus luteus C. B. Clarke = Streptocarpus candidus Hilliard ■☆

377760　Streptocarpus macropodus B. L. Burtt;大足扭果花■☆

377761　Streptocarpus mahonii Hook. = Streptocarpus goetzei Engl. ■☆

377762　Streptocarpus makabengensis Hilliard;马卡邦扭果花■☆

377763　Streptocarpus masisiensis De Wild. ;马西西扭果花■☆

377764　Streptocarpus mbeyensis I. Darbysh. ;姆贝扭果花■☆

377765　Streptocarpus meyeri B. L. Burtt;迈尔扭果花■☆

377766　Streptocarpus michelmorei B. L. Burtt;米氏扭果花■☆

377767　Streptocarpus micranthus C. B. Clarke;小花扭果花■☆

377768　Streptocarpus milanjianus Hilliard et B. L. Burtt;米兰吉扭果花■☆

377769　Streptocarpus milbraedii Engl. = Streptocarpus glandulosissimus Engl. ■☆

377770　Streptocarpus minutiflorus Mansf. = Streptocarpus bullatus Mansf. ■☆

377771　Streptocarpus modestus L. L. Britten;适度扭果花■☆

377772　Streptocarpus molweniensis Hilliard;莫尔韦尼扭果花■☆

377773　Streptocarpus monophyllus Welw. ;单叶扭果花■☆

377774　Streptocarpus montanus Oliv. ;山地扭果花■☆

377775　Streptocarpus montigena L. L. Britten;山生扭果花■☆

377776　Streptocarpus muddii C. B. Clarke = Streptocarpus wilmsii Engl. ■☆

377777　Streptocarpus muscicola Engl. ;苔地扭果花■☆

377778　Streptocarpus muscosus C. B. Clarke;苔藓扭果花■☆

377779　Streptocarpus myoporoides Hilliard et B. L. Burtt;苦槛蓝扭果花■☆

377780　Streptocarpus nobilis C. B. Clarke;名贵扭果花■☆

377781　Streptocarpus occultus Hilliard;隐蔽扭果花■☆

377782　Streptocarpus ovatus (C. B. Clarke) C. B. Clarke = Streptocarpus holstii Engl. ■☆

377783　Streptocarpus pallidiflorus C. B. Clarke;苍白扭果花■☆

377784　Streptocarpus papangae Humbert;帕潘加扭果花■☆

377785　Streptocarpus parviflorus E. Mey. ex C. B. Clarke = Streptocarpus meyeri B. L. Burtt ■☆

377786　Streptocarpus parviflorus Hook. f. subsp. soutpansbergensis Weigend et T. J. Edwards;索特潘扭果花■☆

377787　Streptocarpus paucispiralis Engl. = Streptocarpus rhodesianus S. Moore ■☆

377788　Streptocarpus pentherianus Fritsch;彭泰尔扭果花■☆

377789　Streptocarpus phaeotrichus B. L. Burtt;褐毛扭果花■☆

377790　Streptocarpus pogonites Hilliard et B. L. Burtt;髯毛扭果花■☆

377791　Streptocarpus polackii B. L. Burtt = Streptocarpus cyaneus S. Moore subsp. polackii (B. L. Burtt) Weigend et T. J. Edwards ■☆

377792　Streptocarpus pole-evansii I. Verd.;埃文斯扭果花■☆

377793　Streptocarpus polyanthus Hook.;多花扭果花■☆

377794　Streptocarpus polyanthus Hook. subsp. comptonii (Mansf.) Hilliard;康普顿扭果花■☆

377795　Streptocarpus polyanthus Hook. subsp. verecundus Hilliard;羞涩扭果花■☆

377796　Streptocarpus porphyrostachys Hilliard;紫穗扭果花■☆

377797　Streptocarpus primulifolius Gand.;报春叶扭果花■☆

377798　Streptocarpus primulifolius Gand. subsp. formosus Hilliard et B. L. Burtt = Streptocarpus formosus (Hilliard et B. L. Burtt) T. J. Edwards ■☆

377799　Streptocarpus princeps Engl. et Mildbr. = Streptocarpus nobilis C. B. Clarke ■☆

377800　Streptocarpus prolixus C. B. Clarke;伸展扭果花■☆

377801　Streptocarpus prostratus (Humbert) B. L. Burtt;平卧扭果花■☆

377802　Streptocarpus pumilus B. L. Burtt;矮小扭果花■☆

377803　Streptocarpus pusillus Harv. ex C. B. Clarke;小扭果花■☆

377804　Streptocarpus rexii (Bowie ex Hook.) Lindl.;旋果花;Cape Primrose ■☆

377805　Streptocarpus rexii Lindl. = Streptocarpus rexii (Bowie ex Hook.) Lindl. ■☆

377806　Streptocarpus reynoldsii I. Verd. = Streptocarpus haygarthii N. E. Br. ex C. B. Clarke ■☆

377807　Streptocarpus rhodesianus S. Moore;罗得西亚扭果花■☆

377808　Streptocarpus rhodesianus S. Moore subsp. grandiflorus I. Darbysh.;大花扭果花■☆

377809　Streptocarpus rhodesianus S. Moore var. perlanatus P. A. Duvign. = Streptocarpus rhodesianus S. Moore ■☆

377810　Streptocarpus rimicola Story;隙居扭果花■☆

377811　Streptocarpus rivularis Engl. = Streptocarpus caulescens Vatke ■☆

377812　Streptocarpus roseo-albus Weigend et T. J. Edwards;红白扭果花■☆

377813　Streptocarpus rungwensis Engl. = Streptocarpus goetzei Engl. ■☆

377814　Streptocarpus rungwensis Engl. var. latifolius？= Streptocarpus goetzei Engl. ■☆

377815　Streptocarpus ruwenzoriensis Baker = Streptocarpus glandulosissimus Engl. ■☆

377816　Streptocarpus saundersii Hook.;桑德斯扭果花■☆

377817　Streptocarpus saundersii Hook. var. breviflos C. B. Clarke = Streptocarpus breviflos (C. B. Clarke) C. B. Clarke ■☆

377818　Streptocarpus saxorum Engl.;圆叶旋果花;Cape Primrose, False African Violet ■☆

377819　Streptocarpus schliebenii Mansf.;施利本扭果花■☆

377820　Streptocarpus silvaticus Hilliard;林地扭果花■☆

377821　Streptocarpus smithii C. B. Clarke = Streptocarpus glandulosissimus Engl. ■☆

377822　Streptocarpus solenanthus Mansf.;管花扭果花■☆

377823　Streptocarpus subscandens (B. L. Burtt) I. Darbysh.;亚攀缘扭果花■☆

377824　Streptocarpus suffruticosus Humbert;亚灌木扭果花●☆

377825　Streptocarpus tanala Humbert;塔纳尔扭果花●☆

377826　Streptocarpus tchenzemae Gilli = Streptocarpus glandulosissimus Engl. ■☆

377827　Streptocarpus thomensis Exell = Streptocarpus elongatus Engl. ■☆

377828　Streptocarpus thysanotus Hilliard et B. L. Burtt;流苏扭果花■☆

377829　Streptocarpus trabeculatus Hilliard;横条扭果花■☆

377830　Streptocarpus tsaratananensis Humbert ex B. L. Burtt;察拉塔纳纳扭果花■☆

377831　Streptocarpus tubiflos C. B. Clarke = Streptocarpus grandis N. E. Br. ■☆

377832　Streptocarpus umtaliensis B. L. Burtt;乌穆塔利旋果花■☆

377833　Streptocarpus vandeleurii Baker f. et S. Moore;范德旋果花■☆

377834　Streptocarpus venosus B. L. Burtt;显脉旋果花■☆

377835　Streptocarpus violascens Engl. = Streptocarpus nobilis C. B. Clarke ■☆

377836　Streptocarpus violascens Engl. f. nanus？= Streptocarpus nobilis C. B. Clarke ■☆

377837　Streptocarpus volkensii Engl. = Streptocarpus glandulosissimus Engl. ■☆

377838　Streptocarpus wendlandii Dammann;牛舌旋果花■☆

377839　Streptocarpus wendlandii Spreng.;文德兰旋果花■☆

377840　Streptocarpus wilmsii Engl.;维尔姆斯旋果花■☆

377841　Streptocarpus wittei De Wild.;维特旋果花■☆

377842　Streptocarpus woodii C. B. Clarke = Streptocarpus fanniniae Harv. ex C. B. Clarke ■☆

377843　Streptocarpus zimmermannii Engl.;齐默尔曼旋果花■☆

377844　Streptocaulon Wight et Arn. (1834);马莲鞍属(古羊藤属);Streptocaulon ■

377845　Streptocaulon calophyllum Wight = Periploca calophylla (Wight) Falc. ●

377846　Streptocaulon chinense (Spreng.) G. Don = Cryptolepis sinensis (Lour.) Merr. ●

377847　Streptocaulon chinensis G. Don = Cryptolepis sinensis (Lour.) Merr. ●

377848　Streptocaulon cochinchinense (Lour.) G. Don = Calotropis gigantea (L.) Dryand. ex W. T. Aiton ●

377849　Streptocaulon divaricata G. Don = Strophanthus divaricatus (Lour.) Hook. et Arn. ●

377850　Streptocaulon extensum Wight = Myriopteron extensum (Wight et Arn.) K. Schum. ●

377851　Streptocaulon extensum Wight et Arn. = Myriopteron extensum (Wight et Arn.) K. Schum. ●

377852　Streptocaulon extensum Wight et Arn. var. paniculatum (Griff.) Kurz = Myriopteron extensum (Wight et Arn.) K. Schum. ●

377853　Streptocaulon extensum Wight et Arn. var. paniculatum Kurz = Myriopteron extensum (Wight et Arn.) K. Schum. ●

377854　Streptocaulon griffithii Hook. f. = Streptocaulon juventas (Lour.) Merr. ■

377855　Streptocaulon horsfieldii Miq. = Myriopteron extensum (Wight et Arn.) K. Schum. ●

377856 Streptocaulon juventas (Lour.) Merr. ; 马莲鞍(暗消藤,白花鸡矢藤,地苦参,古羊藤,红马莲鞍,红藤,虎阴藤,苦羊藤,老鸦嘴,马达,马连鞍,毛青才,奶藤,南苦参,山暗消,藤苦参,小暗消,有毛老鸦嘴);Griffith Streptocaulon,Infant Streptocaulon ■

377857 Streptocaulon tomentosum Wight et Arn. = Streptocaulon juventas (Lour.) Merr. ■

377858 Streptochaeta Schrad. = Streptochaeta Schrad. ex Nees ■☆

377859 Streptochaeta Schrad. ex Nees(1829);槟芒笄属■☆

377860 Streptochaetaceae Nakai = Gramineae Juss. (保留科名)■●

377861 Streptochaetaceae Nakai = Poaceae Barnhart(保留科名)■●

377862 Streptochaetaceae Nakai;槟芒笄科

377863 Streptodesmia A. Gray = Adesmia DC. (保留属名)■☆

377864 Streptoglossa Steetz = Streptoglossa Steetz ex F. Muell. ■●☆

377865 Streptoglossa Steetz ex F. Muell. (1863);紫蓬菊属■●☆

377866 Streptoglossa Steetz ex F. Muell. = Allopterigeron Dunlop ■☆

377867 Streptoglossa Steetz ex F. Muell. = Oliganthemum F. Muell. ■☆

377868 Streptoglossa odora (F. Muell.) Dunlop;香紫蓬菊■●☆

377869 Streptoglossa steetzii F. Muell. ;紫蓬菊■☆

377870 Streptoglossa tenuiflora Dunlop;细叶紫蓬菊■●☆

377871 Streptogloxinia hort. = Sinningia Nees ●■☆

377872 Streptogyna P. Beauv. (1812);槟果禾属■☆

377873 Streptogyna crinita P. Beauv. ;槟果禾■☆

377874 Streptogyna gerontogaea Hook. f. = Streptogyna crinita P. Beauv. ■☆

377875 Streptogyne (Rchb.) Rchb. = Satyrium L. (废弃属名)■

377876 Streptogyne Poir. = Streptogyna P. Beauv. ■☆

377877 Streptolirion Edgew. (1845);竹叶子属;Streptolirion ■

377878 Streptolirion cordifolium (Griff.) Kuntze = Streptolirion volubile Edgew. ■

377879 Streptolirion duclouxii H. Lév. et Vaniot = Streptolirion volubile Edgew. ■

377880 Streptolirion elegans Cherfils = Spatholirion elegans (Cherfils) C. Y. Wu ■

377881 Streptolirion lineare Fukuoka et N. Kurosaki = Streptolirion volubile Edgew. ■

377882 Streptolirion longifolium Gagnep. = Spatholirion longifolium (Gagnep.) Dunn ■

377883 Streptolirion mairei H. Lév. = Streptolirion volubile Edgew. ■

377884 Streptolirion volubile Edgew. ;竹叶子(露水草,水百步还魂,猪草,猪耳草,猪耳朵,猪伢草,竹皮空藤,竹叶子草);Twining Streptolirion ■

377885 Streptolirion volubile Edgew. subsp. khasianum (C. B. Clarke) D. Y. Hong;红毛竹叶子(猪肚子草);Khas Streptolirion ■

377886 Streptolirion volubile Edgew. subsp. subalpinum C. Y. Wu = Streptolirion volubile Edgew. ■

377887 Streptolirion volubile Edgew. var. khasianum C. B. Clarke = Streptolirion volubile Edgew. subsp. khasianum (C. B. Clarke) D. Y. Hong ■

377888 Streptoloma Bunge(1847);曲缘芥属(拧缘芥属)■☆

377889 Streptoloma desertorum Bunge;曲缘芥■☆

377890 Streptolophus Hughes(1923);攀缘箭叶草属■☆

377891 Streptolophus sagittifolius Hughes;攀缘箭叶草■☆

377892 Streptomanes K. Schum. = Streptomanes K. Schum. ex Schltr. ■☆

377893 Streptomanes K. Schum. ex Schltr. (1905);扭杯萝藦属(新几内亚萝藦属)■☆

377894 Streptomanes nymanii Schum. ;扭杯萝藦(新几内亚萝藦)■☆

377895 Streptopetalum Hochst. (1841);扭瓣时钟花属■☆

377896 Streptopetalum arenarium Thulin;沙地扭瓣时钟花■☆

377897 Streptopetalum graminifolium Urb. ;禾叶扭瓣时钟花■☆

377898 Streptopetalum hildebrandtii Urb. ;希尔扭瓣时钟花■☆

377899 Streptopetalum luteoglandulosum R. Fern. ;非洲扭瓣时钟花■☆

377900 Streptopetalum serratum Hochst. ;齿扭瓣时钟花■☆

377901 Streptopetalum wittei Staner;威特扭瓣时钟花■☆

377902 Streptopus Michx. (1803);扭柄花属(算盘七属);Mandarin,Scootberry,Twistedstalk,Twisted-stalk ■

377903 Streptopus Rich. = Streptopus Michx. ■

377904 Streptopus ajanensis Tiling var. koreanus Kom. = Streptopus koreanus Ohwi ■

377905 Streptopus amplexifolius (L.) DC. ;抱茎扭柄花;Clasp-leaf Twisted-stalk,Twisted-stalk,White Mandarin ■☆

377906 Streptopus amplexifolius (L.) DC. = Streptopus obtusatus Fassett ■

377907 Streptopus amplexifolius (L.) DC. subsp. americanus (Schult.) Á. Löve et D. Löve = Streptopus amplexifolius (L.) DC. ■☆

377908 Streptopus amplexifolius (L.) DC. var. americanus Schult. = Streptopus amplexifolius (L.) DC. ■☆

377909 Streptopus amplexifolius (L.) DC. var. chalazatus Fassett = Streptopus amplexifolius (L.) DC. ■☆

377910 Streptopus amplexifolius (L.) DC. var. denticulatus Fassett = Streptopus amplexifolius (L.) DC. ■☆

377911 Streptopus amplexifolius (L.) DC. var. grandiflorus Fassett = Streptopus amplexifolius (L.) DC. ■☆

377912 Streptopus amplexifolius (L.) DC. var. papillatus Ohwi;乳头抱茎扭柄花;Claspleaf Twistedstalk ■☆

377913 Streptopus amplexifolius C. H. Wright = Streptopus obtusatus Fassett ■

377914 Streptopus brevipes Baker = Streptopus streptopoides (Ledeb.) Frye et Rigg ■☆

377915 Streptopus chinensis (Ker Gawl.) Sm. = Disporum cantoniense (Lour.) Merr. ■

377916 Streptopus curvipes Vail = Streptopus lanceolatus (Aiton) Reveal ■☆

377917 Streptopus fassettii Á. Löve et D. Löve = Streptopus amplexifolius (L.) DC. ■☆

377918 Streptopus geniculatus F. T. Wang et Ts. Tang = Streptopus obtusatus Fassett ■

377919 Streptopus koreanus Ohwi;丝梗扭柄花;Korea Twistedstalk,Korean Twistedstalk ■

377920 Streptopus lanceolatus (Aiton) Reveal;披针状扭柄花;Twisted-stalk ■☆

377921 Streptopus lanceolatus (Aiton) Reveal var. curvipes (Vail) Reveal = Streptopus lanceolatus (Aiton) Reveal ■☆

377922 Streptopus lanceolatus (Aiton) Reveal var. longipes (Fernald) Reveal = Streptopus lanceolatus (Aiton) Reveal ■☆

377923 Streptopus lanceolatus (Aiton) Reveal var. longipes (Fernald) Reveal;长柄披针状扭柄花;Twisted-stalk ■☆

377924 Streptopus lanceolatus (Aiton) Reveal var. roseus (Michx.) Reveal = Streptopus lanceolatus (Aiton) Reveal ■☆

377925 Streptopus lanuginosus Michx. = Prosartes lanuginosa (Michx.) D. Don ■☆

377926　Streptopus longipes Fernald ＝ Streptopus lanceolatus （Aiton）Reveal var. longipes （Fernald） Reveal ■☆

377927　Streptopus longipes Fernald ＝ Streptopus lanceolatus （Aiton）Reveal ■☆

377928　Streptopus maculatus Buckley ＝ Prosartes maculata （Buckley）A. Gray ■☆

377929　Streptopus mairei H. Lév. ＝ Streptopus parviflorus Franch. ■

377930　Streptopus obtusatus Fassett；扭柄花（抱茎叶扭柄花，钝叶算盘七，曲柄算盘七，算盘七）；Clasping Twistedstalk，Clasping-leaved Streptopus，Claspleaf Twistedstalk，Clasp-leaf Twisted-stalk，Obtuse Twistedstalk，Twistedstalk，Twisted-stalk，White Mandarin ■

377931　Streptopus ovalis （Ohwi） F. T. Wang et Y. C. Tang；卵叶扭柄花（金钢草）；Ovalleaf Twistedstalk，Ovateleaf Twistedstalk ■

377932　Streptopus paniculatus Baker ＝ Maianthemum tatsienense （Franch.） LaFrankie ■

377933　Streptopus parviflorus Franch.；小花扭柄花（高山竹林梢，小花算盘七）；Littleflower Twistedstalk ■

377934　Streptopus roseus Michx.；无柄叶扭柄花；Pink Streptopus，Rose Mandarin，Rosybetls，Row Twisted-stalk，Sessile Twistedstalk ■☆

377935　Streptopus roseus Michx. ＝ Streptopus lanceolatus （Aiton）Reveal ■☆

377936　Streptopus roseus Michx. subsp. curvipes （Vail） Hultén ＝ Streptopus lanceolatus （Aiton） Reveal ■☆

377937　Streptopus roseus Michx. var. curvipes （Vail） Fassett ＝ Streptopus lanceolatus （Aiton） Reveal ■☆

377938　Streptopus roseus Michx. var. longipes （Fernald） Fassett ＝ Streptopus lanceolatus （Aiton） Reveal ■☆

377939　Streptopus roseus Michx. var. longipes （Fernald） Fassett ＝ Streptopus lanceolatus （Aiton） Reveal var. longipes （Fernald） Reveal ■☆

377940　Streptopus roseus Michx. var. perspectus Fassett ＝ Streptopus lanceolatus （Aiton） Reveal ■☆

377941　Streptopus simplex D. Don；腋花扭柄花（草茎算盘七，单茎算盘草，单茎算盘七，鸡爪参，女楂，算盘花，野参须，竹林梢，竹林消）；Axillaryflower Twistedstalk ■

377942　Streptopus streptopoides （Ledeb.） Frye et Rigg；北美扭柄花 ■☆

377943　Streptopus streptopoides （Ledeb.） Frye et Rigg subsp. brevipes （Baker） Calder et R. L. Taylor ＝ Streptopus streptopoides （Ledeb.） Frye et Rigg ■☆

377944　Streptopus streptopoides （Ledeb.） Frye et Rigg var. brevipes （Baker） Fassett ＝ Streptopus streptopoides （Ledeb.） Frye et Rigg ■☆

377945　Streptopus streptopoides （Ledeb.） Nelson et J. F. Macbr. ＝ Streptopus streptopoides （Ledeb.） Frye et Rigg ■☆

377946　Streptorhamphus Regel ＝ Steptorhamphus Bunge ■☆

377947　Streptosema C. Presl ＝ Aspalathus L. ●☆

377948　Streptosiphon Mildbr. （1935）；扭管爵床属 ☆

377949　Streptosiphon hirsutus Mildbr.；扭管爵床 ☆

377950　Streptosolen Miers（1850）；橙茄属（扭管花属）●☆

377951　Streptosolen jamesonii （Benth.） Miers；橙茄（果酱木，扭管花）；Marmalade Bush ●☆

377952　Streptosolen jamesonii Miers ＝ Streptosolen jamesonii （Benth.） Miers ●☆

377953　Streptostachys Desv. （1812）；弯穗黍属 ■☆

377954　Streptostachys acuminata Renvoize；渐尖弯穗黍 ■☆

377955　Streptostachys asperifolia Desv.；弯穗黍 ■☆

377956　Streptostachys rigidifolia Filg.，Morrone et Zuloaga；硬叶弯穗黍 ■☆

377957　Streptostachys robusta Renvoize；粗壮弯穗黍 ■☆

377958　Streptostigma Regel ＝ Cacabus Bernh. ■☆

377959　Streptostigma Regel ＝ Exodeconus Raf. ■☆

377960　Streptostigma Thwaites ＝ Harpullia Roxb. ●

377961　Streptothamnus F. Muell. （1862）；扭风灌属 ●☆

377962　Streptothamnus beckleri F. Muell.；扭风灌 ●☆

377963　Streptotrachelus Greenm. ＝ Laubertia A. DC. ●☆

377964　Streptoura Luer ＝ Masdevallia Ruiz et Pav. ■☆

377965　Streptoura Luer（2006）；扭尾细瓣兰属 ■☆

377966　Streptylis Raf.（废弃属名）＝ Murdannia Royle（保留属名）■

377967　Striangis Thouars ＝ Angraecum Bory ■

377968　Stricklandia Baker ＝ Phaedranassa Herb. ■☆

377969　Striga Lour. （1790）；独脚金属；Striga ■

377970　Striga aequinoctialis A. Chev. ex Hutch. et Dalziel；昼夜独脚金 ■

377971　Striga angolensis Mohamed et Musselman；安哥拉独脚金 ■☆

377972　Striga angustifolia （D. Don） C. J. Saldanha；狭叶独脚金 ■

377973　Striga asiatica （L.） Kuntze；独脚金（矮脚子，地丁草，地连枝，地莲芝，独脚柑，干草，疳积草，黄独脚金，黄花草，黄花甘，金锁匙，宽叶独脚金，鹿草，马佬含菊，细独角马骝，消米虫）；Asia Striga，Asiatic Striga，Asiatic Witchweed，Broadleaf Striga，Witch Weed，Yellow Striga ■

377974　Striga asiatica （L.） Kuntze var. humilis （Benth.） D. Y. Hong ＝ Striga asiatica （L.） Kuntze ■

377975　Striga aspera （Willd.） Benth.；粗糙独脚金 ■☆

377976　Striga aspera （Willd.） Benth. var. filiformis Benth. ＝ Striga aspera （Willd.） Benth. ■☆

377977　Striga barteri Engl. ＝ Striga bilabiata （Thunb.） Kuntze subsp. barteri （Engl.） Hepper ■☆

377978　Striga baumannii Engl.；鲍曼独脚金 ■☆

377979　Striga bilabiata （Thunb.） Kuntze；双唇独脚金 ■☆

377980　Striga bilabiata （Thunb.） Kuntze subsp. barteri （Engl.） Hepper；巴特独脚金 ■☆

377981　Striga bilabiata （Thunb.） Kuntze subsp. jaegeri Hepper；耶格独脚金 ■☆

377982　Striga bilabiata （Thunb.） Kuntze subsp. ledermannii （Pilg.） Hepper；莱德独脚金 ■☆

377983　Striga bilabiata （Thunb.） Kuntze subsp. linearifolia （Schumach. et Thonn.） Mohamed；线叶双唇独脚金 ■☆

377984　Striga bilabiata （Thunb.） Kuntze subsp. rowlandii （Engl.） Hepper；罗兰独脚金 ■☆

377985　Striga brachycalyx Engl. ex V. Naray.；短萼独脚金 ■☆

377986　Striga brouilletii Mielcarek ＝ Striga bilabiata （Thunb.） Kuntze subsp. jaegeri Hepper ■☆

377987　Striga buettneri Engl. ＝ Striga macrantha （Benth.） Benth. ■☆

377988　Striga canescens Engl. ＝ Striga bilabiata （Thunb.） Kuntze subsp. linearifolia （Schumach. et Thonn.） Mohamed ■☆

377989　Striga chloroleuca Dinter ＝ Striga gesnerioides （Willd.） Vatke ■☆

377990　Striga chrysantha A. Raynal；金花独脚金 ■☆

377991　Striga coccinea （Hook.） Benth. ＝ Striga asiatica （L.） Kuntze ■

377992　Striga curvata G. Don；弯花独脚金 ■☆

377993　Striga dalzielii Hutch.；达尔齐尔独脚金 ■☆

377994　Striga densiflora Benth.；密花独脚金；Denseflower Striga ■

377995　Striga diversifolia Pires de Lima；异叶独脚金 ■☆

377996　Striga elegans Benth. ;雅致独脚金■☆

377997　Striga ellenbergeri A. Raynal;埃伦独脚金■☆

377998　Striga esquirolii H. Lév. = Cancsora diffusa（Vahl）R. Br. ex Roem. et Schuldt ■

377999　Striga euphrasioides Benth. ;大米草状独脚金;Eyebrightlike Striga ■☆

378000　Striga forbesii Benth. :福布斯独脚金■☆

378001　Striga fulgens（Engl.）Hepper;光亮独脚金■☆

378002　Striga gastonii A. Raynal;加斯顿独脚金■☆

378003　Striga gesnerioides（Willd.）Vatke;豇豆独脚金;Cowpea Witchweed ■☆

378004　Striga gesnerioides Vatke ex Engl. = Striga gesnerioides （Willd.）Vatke ■☆

378005　Striga glandulifera Engl. = Striga bilabiata（Thunb.）Kuntze subsp. barteri（Engl.）Hepper ■☆

378006　Striga glumacea A. Raynal;壳独脚金■☆

378007　Striga gracillima Melch. ;细长独脚金■☆

378008　Striga hallei A. Raynal;哈勒独脚金■☆

378009　Striga hermonthica（Delile）Benth. ;赫尔独脚金■☆

378010　Striga hermonthica（Delile）Benth. subsp. senegalensis （Benth.）Maire = Striga hermonthica（Delile）Benth. ■☆

378011　Striga hirsuta（L.）Kuntze var. humilis Benth. = Striga asiatica （L.）Kuntze var. humilis（Benth.）D. Y. Hong ■

378012　Striga hirsuta Benth. = Striga asiatica（L.）Kuntze ■

378013　Striga hirsuta Benth. var. humilis Benth. = Striga asiatica（L.）Kuntze ■

378014　Striga humifusa（Forssk.）Benth. = Cycniopsis humifusa （Forssk.）Engl. ■☆

378015　Striga humilis Hochst. ex A. Rich. = Cycniopsis humifusa （Forssk.）Engl. ■☆

378016　Striga junodii Schinz;朱诺德独脚金■☆

378017　Striga klingii（Engl.）V. Naray. ;金氏独脚金■☆

378018　Striga latericea Vatke;侧生独脚金■☆

378019　Striga ledermannii Pilg. = Striga bilabiata（Thunb.）Kuntze subsp. ledermannii（Pilg.）Hepper ■☆

378020　Striga linearifolia（Schumach. et Thonn.）Hepper = Striga bilabiata（Thunb.）Kuntze subsp. linearifolia（Schumach. et Thonn.）Mohamed ■☆

378021　Striga lutea Lour. = Striga asiatica（L.）Kuntze ■

378022　Striga lutea Lour. var. coccinea（Hook.）Kuntze = Striga asiatica（L.）Kuntze ■

378023　Striga macrantha（Benth.）Benth. = Striga micrantha A. Rich. ■☆

378024　Striga masuria（Buch. -Ham. ex Benth.）Benth. ;大独脚金(干草,高雄独脚金,小白花草,小白花苏);Big Striga ■

378025　Striga micrantha A. Rich. ;大花独脚金■☆

378026　Striga orobanchoides（R. Br.）Benth. = Striga gesnerioides （Willd.）Vatke ■☆

378027　Striga orobanchoides Benth. ;列当状独脚金;Broomrap-like Striga ■☆

378028　Striga orobanchoides Benth. = Striga gesnerioides Vatke ex Engl. ■☆

378029　Striga passargei Engl. ;帕萨独脚金■☆

378030　Striga pinnatifida Getachew;羽裂独脚金■☆

378031　Striga primuloides A. Chev. ;报春独脚金■☆

378032　Striga pubiflora Klotzsch;短毛花独脚金■☆

378033　Striga pusilla Hochst. ex Benth. = Striga asiatica（L.）Kuntze ■

378034　Striga rowlandii Engl. = Striga bilabiata（Thunb.）Kuntze subsp. rowlandii（Engl.）Hepper ■☆

378035　Striga schimperiana Hochst. ex A. Rich. = Buchnera hispida Buch. -Ham. ex D. Don ■

378036　Striga senegalensis Benth. ;塞内加尔独脚金■☆

378037　Striga senegalensis Benth. = Striga hermonthica（Delile）Benth. ■☆

378038　Striga senegalensis Thomson = Buchnera usuiensis Oliv. ■☆

378039　Striga somaliensis V. Naray. = Striga latericea Vatke ■☆

378040　Striga strictissima V. Naray. = Striga bilabiata（Thunb.）Kuntze subsp. linearifolia（Schumach. et Thonn.）Mohamed ■☆

378041　Striga strigosa Good;索马里独脚金■☆

378042　Striga sulphurea Dalzell et A. Gibson;硫黄独脚金;Sulphur Striga ■☆

378043　Striga thunbergii Benth. = Striga bilabiata（Thunb.）Kuntze ■☆

378044　Striga warneckei Engl. ex V. Naray. = Striga brachycalyx Engl. ex V. Naray. ■☆

378045　Striga welwitschii Engl. = Striga bilabiata（Thunb.）Kuntze ■☆

378046　Striga yemenica Musselman et Hepper;也门独脚金■☆

378047　Striga zanzibarensis Vatke = Striga pubiflora Klotzsch ■☆

378048　Strigilia Cav. = Styrax L. ●

378049　Strigilia ferruginea（Nees et C. Mart.）Miers = Styrax ferrugineus Nees et Mart. ●☆

378050　Strigilia florida（Pohl）Miers = Styrax ferrugineus Nees et Mart. ●☆

378051　Strigilia guyanensis（A. DC.）Miers = Styrax guyanensis A. DC. ●☆

378052　Strigilia nervosa（A. DC.）Miers = Styrax ferrugineus Nees et Mart. ●☆

378053　Strigilia oblonga（Ruiz et Pav.）DC. = Styrax oblongus（Ruiz et Pav.）A. DC. ●☆

378054　Strigilia parvifolia（Pohl）Miers = Styrax ferrugineus Nees et Mart. ●☆

378055　Strigilia pohlii（A. DC.）Miers = Styrax pohlii A. DC. ●☆

378056　Strigilia punctata（A. DC.）Miers = Styrax pohlii A. DC. ●☆

378057　Strigilia reticulata（C. Mart.）Miers = Styrax ferrugineus Nees et Mart. ●☆

378058　Strigina Engl. = Lindernia All. ■

378059　Strigosella Boiss. = Malcolmia W. T. Aiton(保留属名)■

378060　Strigosella africana（L.）Botsch. = Malcolmia africana（L.）R. Br. ■

378061　Strigosella africana（L.）Botsch. var. laxa（Lam.）Botsch. = Malcolmia africana（L.）R. Br. ■

378062　Strigosella brevipes（Bunge）Botsch. = Malcolmia karelinii Lipsky ■

378063　Strigosella brevipes（Kar. et Kir.）Botsch. = Neotorularia brevipes（Kar. et Kir.）Hedge et J. Léonard ■

378064　Strigosella hispida（Litv.）Botsch. = Malcolmia hispida Litv. ■

378065　Strigosella scorpioides（Bunge）Botsch. = Malcolmia scorpioides （Bunge）Boiss. ■

378066　Strigosella stenopetala（Bernh. ex Fisch. et C. A. Mey.）Botsch. = Malcolmia africana（L.）R. Br. ■

378067　Strigosella stenopetala（Bernh.）Botsch. = Malcolmia africana （L.）R. Br. var. stenopetala Claus ■

378068　Strigosella trichocarpa（Boiss. et Buhse）Botsch. = Malcolmia africana（L.）R. Br. var. trichocarpa（Boiss. et Buhse）Boiss. ■

378069 Strigosella trichocarpa（Boiss. et Buhse）Botsch. = Malcolmia africana（L.）R. Br. ■

378070 Striolaria Ducke(1945)；条纹茜属（亚马孙茜草属）■☆

378071 Striolaria amazonica Ducke；条纹茜■☆

378072 Strobidia Miq. = Alpinia Roxb.（保留属名）■

378073 Strobila G. Don = Arnebia Forssk. ●■

378074 Strobila Noronha = Nicolaia Horan.（保留属名）■☆

378075 Strobilacanthus Griseb.（1858）；球刺爵床属☆

378076 Strobilacanthus lepidospermus Griseb. ；球刺爵床☆

378077 Strobilaceae Dulac = Cannabaceae Martinov(保留科名)■

378078 Strobilanthes Blume et Bremek. = Strobilanthes Blume ●■

378079 Strobilanthes Blume(1826)；马蓝属（紫云菜属，紫云英属）；Conehead ●■

378080 Strobilanthes acrocephala T. Anderson = Tarphochlamys affinis（Griff.）Bremek. ■

378081 Strobilanthes aenobarba W. W. Sm. = Pteracanthus aenobarbus（W. W. Sm.）C. Y. Wu et C. C. Hu ■

378082 Strobilanthes affinis（Griff.）Y. C. Tang = Tarphochlamys affinis（Griff.）Bremek. ■

378083 Strobilanthes alatiramosa H. S. Lo et D. Fang = Pteracanthus alatiramosus（H. S. Lo et D. Fang）C. Y. Wu et C. C. Hu ■

378084 Strobilanthes alatus Nees = Strobilanthes urticifolia Wall. ex Kuntze ●

378085 Strobilanthes anisandra Benoist = Paragutzlaffia lyi（H. Lév.）H. P. Tsui ■

378086 Strobilanthes anisophylla T. Anderson；喜马拉雅马蓝；Goldfussia ■☆

378087 Strobilanthes aprica（Hance）T. Anderson = Gutzlaffia aprica Hance ■

378088 Strobilanthes aprica（Hance）T. Anderson var. glabra Imlay = Gutzlaffia aprica Hance var. glabra（Imlay）H. S. Lo ■

378089 Strobilanthes atropurpurea Nees；深紫马蓝■☆

378090 Strobilanthes attenuata Nees = Strobilanthes urticifolia Wall. ex Kuntze ●

378091 Strobilanthes auriculata（Wall.）Nees；耳叶马蓝（耳形马蓝）■

378092 Strobilanthes auriculata（Wall.）Nees = Perilepta auriculata（Wall.）Bremek. ■

378093 Strobilanthes auriculata（Wall.）Nees var. edgewworthianus（Nees）C. B. Clarke = Strobilanthes auriculata（Wall.）Nees ■

378094 Strobilanthes auriculata（Wall.）Nees var. edgewworthianus（Nees）C. B. Clarke = Perilepta edgewworthiana（Nees）Bremek. ■

378095 Strobilanthes auriculata Nees = Perilepta auriculata（Wall.）Bremek. ■

378096 Strobilanthes auriculata Nees var. edgeworthianus（Nees）C. B. Clarke = Perilepta edgeworthiana（Nees）Bremek. ■

378097 Strobilanthes auriculata Nees var. siamensis C. B. Clarke = Perilepta siamensis（C. B. Clarke）Bremek. ■

378098 Strobilanthes austinii C. B. Clarke ex W. W. Sm. = Goldfussia austinii（C. B. Clarke ex W. W. Sm.）Bremek. ■

378099 Strobilanthes balansae Lindau = Baphicacanthus cusia（Ness）Bremek. ●

378100 Strobilanthes blinii H. Lév. = Aechmanthera gossypina（Nees）Nees ●

378101 Strobilanthes bonatiana H. Lév. = Championella japonica（Thunb.）Bremek. ●■

378102 Strobilanthes botryantha D. Fang et H. S. Lo = Pteracanthus botryanthus（D. Fang et H. S. Lo）C. Y. Wu et C. C. Hu ●

378103 Strobilanthes breviceps Benoist；短头马蓝■☆

378104 Strobilanthes capitata T. Anderson = Goldfussia capitata Nees ●■

378105 Strobilanthes cavaleriei H. Lév. = Aechmanthera gossypina（Nees）Nees ●

378106 Strobilanthes chaffanjonii H. Lév. = Goldfussia pentastemonoides（Wall.）Nees ■

378107 Strobilanthes championi T. Anderson ex Benth. = Baphicacanthus cusia（Ness）Bremek. ●■

378108 Strobilanthes claviculatus C. B. Clarke ex W. W. Sm. = Pteracanthus clavicalatus（C. B. Clarke ex W. W. Sm.）C. Y. Wu ■

378109 Strobilanthes cognata Benoist = Pteracanthus cognatus（Benoist）C. Y. Wu et C. C. Hu ■

378110 Strobilanthes colorata（Nees）T. Anderson = Diflugossa colorata（Nees）Bremek. ■

378111 Strobilanthes compacta D. Fang et H. S. Lo；密苞紫云菜●

378112 Strobilanthes congesta Terao = Pteracanthus congestus（Terao）C. Y. Wu et C. C. Hu ●

378113 Strobilanthes cuneifolia Benoist；楔叶马蓝■☆

378114 Strobilanthes curviflora C. B. Clarke = Pteracanthus cyphanthus（Diels）C. Y. Wu et C. C. Hu ●

378115 Strobilanthes cusia（Ness）Kuntze = Baphicacanthus cusia（Ness）Bremek. ●

378116 Strobilanthes cyclus C. B. Clarke ex W. W. Sm. ；环状马蓝（观音座莲，环毛紫云菜）；Cycleshape Conehead ■

378117 Strobilanthes cyphanthus Diels = Pteracanthus cyphanthus（Diels）C. Y. Wu et C. C. Hu ●

378118 Strobilanthes dalzielii（W. W. Sm.）Benoist = Pterolobium punctatum Hemsl. ex Forbes et Hemsl. ●

378119 Strobilanthes dalziellii（W. W. Sm.）Benoist var. inaequalis Benoist = Pteroptychia dalziellii（Sm.）H. S. Lo ●■

378120 Strobilanthes darrisi H. Lév. = Tarphochlamys darrisii（H. Lév.）E. Hossain ■

378121 Strobilanthes debilis Hemsl. = Championella tetrasperma（Champ. ex Benth.）Bremek. ●

378122 Strobilanthes densa Benoist；密花紫云菜；Denseflower Conehead ■

378123 Strobilanthes dielsiana W. W. Sm. = Gutzlaffia aprica Hance ■

378124 Strobilanthes dimorphotricha Hance = Goldfussia pentastemonoides（Wall.）Nees ■

378125 Strobilanthes divaricata（Nees）T. Anderson = Diflugossa divaricata（Nees）Bremek. ■

378126 Strobilanthes divaricata T. Anderson；疏花马蓝■

378127 Strobilanthes dryadum Benoist = Pteracanthus dryadum（C. B. Clarke ex Benoist）C. Y. Wu et C. C. Hu ■

378128 Strobilanthes duclouxii C. B. Clarke ex Benoist = Pteracanthus duclouxii（C. B. Clarke ex Benoist）C. Y. Wu et C. C. Hu ■

378129 Strobilanthes dyeriana Mast. = Perilepta dyeriana（Mast.）Bremek. ●■

378130 Strobilanthes edgeworthiana Nees = Perilepta edgewworthiana（Nees）Bremek. ■

378131 Strobilanthes equitans H. Lév. = Goldfussia equitans（H. Lév.）E. Hossain ■

378132 Strobilanthes esquirolii H. Lév. = Tetragoga esquirolii（H. Lév.）E. Hossain ●■

378133 Strobilanthes extensa Nees = Pteracanthus extensus（Nees）Bremek. ●■

378134 Strobilanthes fauriei Benoist = Championella fauriei（Benoist）

C. Y. Wu et C. C. Hu ●

378135 Strobilanthes feddei H. Lév. = Goldfussia feddei (H. Lév.) E. Hossain ■

378136 Strobilanthes ferruginea D. Fang et H. S. Lo = Perilepta ferruginea (D. Fang et H. S. Lo) C. Y. Wu et C. C. Hu ■

378137 Strobilanthes flaccidifolia Nees = Baphicacanthus cusia (Ness) Bremek. ●

378138 Strobilanthes flexicaulis Hayata = Parachampionella flexicaulis (Hayata) C. F. Hsieh et T. C. Huang ●■

378139 Strobilanthes flexicaulis Hayata var. tashiroi (Hayata) T. Yamaz. = Strobilanthes tashiroi Hayata ■☆

378140 Strobilanthes flexus Benoist = Pteracanthus flexus (Benoist) C. Y. Wu et C. C. Hu ■

378141 Strobilanthes formosana S. Moore = Goldfussia formosana (S. Moore) C. F. Hsieh et T. C. Huang ■

378142 Strobilanthes forrestii Diels = Pteracanthus forrestii (Diels) C. Y. Wu ●■

378143 Strobilanthes fulvihispida D. Fang et H. S. Lo = Championella fulvihispida (D. Fang et H. S. Lo) C. Y. Wu et C. C. Hu ●

378144 Strobilanthes gentiliana H. Lév. = Sesamum indicum L. ■

378145 Strobilanthes gigantodes Lindau = Tetraglochidium gigantodes (Lindau) C. Y. Wu et C. C. Hu ●

378146 Strobilanthes glandibracteata D. Fang et H. S. Lo = Goldfussia glandibracteata (D. Fang et H. S. Lo) C. Y. Wu ■

378147 Strobilanthes glandulifera Hatus. = Strobilanthes flexicaulis Hayata ●■

378148 Strobilanthes glomerata (Wall.) T. Anderson = Goldfussia glomerata (Wall.) Nees ●■

378149 Strobilanthes glutinosa Nees;黏性马蓝■☆

378150 Strobilanthes glutinosa Nees = Pseudaechmanthera glutinosa (Nees) Bremek. ■

378151 Strobilanthes gracilicaulis Benoist = Dyschoriste gracilicaulis (Benoist) Benoist ■

378152 Strobilanthes grossa C. B. Clarke;大紫云菜(味牛膝);Big Conehead ■

378153 Strobilanthes guangxiensis S. Z. Huang = Pteracanthus guangxiensis (S. Z. Huang) C. Y. Wu et C. C. Hu ■

378154 Strobilanthes hancockii C. B. Clarke ex W. W. Sm. = Goldfussia austinii (C. B. Clarke ex W. W. Sm.) Bremek. ■

378155 Strobilanthes helictus T. Anderson = Pteracanthus calycinus (Nees) Bremek. ●

378156 Strobilanthes henryi Hemsl. = Paragutzlaffia henryi (Hemsl.) H. P. Tsui ●■

378157 Strobilanthes heterochroa Hand.-Mazz. = Pyrrothrix heterochrous (Hand.-Mazz.) C. Y. Wu et C. C. Hu ■

378158 Strobilanthes heteroclita D. Fang et H. S. Lo;异序紫云菜(异序马蓝)●■

378159 Strobilanthes hispidula Baker = Dyschoriste hispidula (Baker) Benoist ■☆

378160 Strobilanthes hossei C. B. Clarke = Pyrrothrix hossei (C. B. Clarke) C. Y. Wu et C. C. Hu ■

378161 Strobilanthes humblotii Benoist;洪布紫云菜●☆

378162 Strobilanthes hupehensis W. W. Sm. = Goldfussia pentastemonoides (Wall.) Nees ■

378163 Strobilanthes hygrophiloides C. B. Clarke ex W. W. Sm. = Pteracanthus hygrophiloides (C. B. Clarke ex W. W. Sm.) H. W. Li ●

378164 Strobilanthes inflatus T. Anderson = Pteracanthus inflatus (T.

Anderson) Bremek. ■

378165 Strobilanthes isoglossoides Lindau = Mimulopsis madagascariensis (Baker) Benoist ■☆

378166 Strobilanthes isophylla T. Anderson;异叶马蓝■☆

378167 Strobilanthes japonica (Thunb.) Miq. = Championella japonica (Thunb.) Bremek. ■

378168 Strobilanthes jugorum Benoist = Tetraglochidium jugorum (Benoist) Bremek. ■

378169 Strobilanthes labordei H. Lév. = Championella labordei (H. Lév.) E. Hossain ●■

378170 Strobilanthes lactucifolia H. Lév.;莴苣叶紫云菜■

378171 Strobilanthes lamius C. B. Clarke ex W. W. Sm. = Pteracanthus lamius (C. B. Clarke ex W. W. Sm.) C. Y. Wu et C. C. Hu ■

378172 Strobilanthes larium Hand.-Mazz.;闭花紫云菜;Conehead ■

378173 Strobilanthes latisepala Hemsl. = Pteracanthus alatus (Nees) Bremek. ●■

378174 Strobilanthes laxiocalyx Hayata = Goldfussia pentastemonoides (Wall.) Nees ■

378175 Strobilanthes leucocephala Craib = Goldfussia leucocephala (Craib) C. Y. Wu ■

378176 Strobilanthes leucotricha Benoist = Pteracanthus leucotrichus (Benoist) C. Y. Wu et C. C. Hu ■

378177 Strobilanthes limprichtii Diels;雅安紫云菜;Ya'an Conehead ●■

378178 Strobilanthes lofouensis H. Lév. = Echinacanthus lofuensis (H. Lév.) J. R. I. Wood ●

378179 Strobilanthes longespicatus Hayata = Semnostachya longispicata (Hayata) C. F. Hsieh et C. C. Huang ●■

378180 Strobilanthes longgangensis D. Fang et H. S. Lo = Perilepta longgangensis (D. Fang et H. S. Lo) C. Y. Wu et C. C. Hu ■

378181 Strobilanthes longiflora Benoist = Championella longiflora (Benoist) C. Y. Wu et C. C. Hu ●

378182 Strobilanthes longispicata Hayata = Semnostachya longispicata (Hayata) C. F. Hsieh et C. C. Huang ●■

378183 Strobilanthes longzhouensis H. S. Lo et D. Fang = Perilepta longzhouensis (H. S. Lo et D. Fang) C. Y. Wu et C. C. Hu ■

378184 Strobilanthes maclurei Merr. = Championella maclurei (Merr.) C. Y. Wu et H. S. Lo ●

378185 Strobilanthes madagascariensis Baker;马岛紫云菜●☆

378186 Strobilanthes mahongensis H. Lév. = Goldfussia austinii (C. B. Clarke ex W. W. Sm.) Bremek. ■

378187 Strobilanthes mairei H. Lév. = Gutzlaffia aprica Hance ■

378188 Strobilanthes marchandii H. Lév. = Goldfussia pentastemonoides (Wall.) Nees ■

378189 Strobilanthes martinii H. Lév.;镇宁紫云菜■

378190 Strobilanthes mediocris Benoist;中位紫云菜■☆

378191 Strobilanthes mekongensis W. W. Sm. = Pteracanthus mekongensis (W. W. Sm.) C. Y. Wu et C. C. Hu ●

378192 Strobilanthes monadelpha Nees = Sympagis monadelpha (Nees) Bremek. ●■

378193 Strobilanthes mucronatoproducta Lindau;尾苞紫云菜●

378194 Strobilanthes myriostachya D. Fang et H. S. Lo = Pteracanthus botryanthus (D. Fang et H. S. Lo) C. Y. Wu et C. C. Hu ●

378195 Strobilanthes myura Benoist;鼠尾紫云菜●

378196 Strobilanthes nemorosa Benoist = Pteracanthus nemorosus (Benoist) C. Y. Wu et C. C. Hu ●

378197 Strobilanthes ningmingensis D. Fang et H. S. Lo = Goldfussia ningmingensis（D. Fang et H. S. Lo）C. Y. Wu ●

378198 Strobilanthes oligantha Miq. = Championella oligantha（Miq.）Bremek. ■

378199 Strobilanthes oresbius W. W. Sm. = Pteracanthus oresbius（W. W. Sm.）C. Y. Wu et C. C. Hu ■

378200 Strobilanthes ovatibracteata H. S. Lo et D. Fang = Goldfussia ovatibracteata（H. S. Lo et D. Fang）C. Y. Wu ●■

378201 Strobilanthes pandurata Hand. -Mazz. = Pteracanthus panduratus（Hand. -Mazz.）C. Y. Wu et C. C. Hu ■

378202 Strobilanthes panpienkaiensis H. Lév. = Pteracanthus forrestii（Diels）C. Y. Wu ●■

378203 Strobilanthes penstemonoides（Nees）T. Anderson = Goldfussia pentastemonoides（Wall.）Nees ■

378204 Strobilanthes perrieri Benoist；佩里耶紫云菜■☆

378205 Strobilanthes petelotii Benoist；沙坝紫云菜■

378206 Strobilanthes petiolaris C. B. Clarke = Sympagis petiolaris（Nees）Bremek. ●■

378207 Strobilanthes petiolaris Nees = Sympagis petiolaris（Nees）Bremek. ●■

378208 Strobilanthes pinetorum W. W. Sm. = Diflugossa pinetorum（W. W. Sm.）C. Y. Wu et C. C. Hu ■

378209 Strobilanthes pinnatifida C. Z. Zheng = Pteracanthus pinnatifidus（C. Z. Zheng）C. Y. Wu et C. C. Hu ■

378210 Strobilanthes polyneuros C. B. Clarke ex W. W. Sm. ；多脉紫云菜；Manyvein Conehead ■

378211 Strobilanthes prionophylla Hayata = Parachampionella flexicaulis（Hayata）C. F. Hsieh et T. C. Huang ●■

378212 Strobilanthes psilostachys C. B. Clarke ex W. W. Sm. = Goldfussia psilostachys（C. B. Clarke ex W. W. Sm.）Bremek. ■

378213 Strobilanthes pteroclada Benoist = Hymenochlaena pteroclada（Benoist）C. Y. Wu et C. C. Hu ■

378214 Strobilanthes radicans T. Anderson ex Benth. = Championella tetrasperma（Champ. ex Benth.）Bremek. ●

378215 Strobilanthes ramiflora Benoist；枝花紫云菜■☆

378216 Strobilanthes rankanensis Hayata = Parachampionella rankanensis（Hayata）Bremek. ■

378217 Strobilanthes refracta D. Fang, Y. G. Wei et J. Murata = Perilepta refracta（D. Fang, Y. G. Wei et J. Murata）C. Y. Wu et C. C. Hu ●

378218 Strobilanthes retusa D. Fang = Perilepta retusa（D. Fang）C. Y. Wu et C. C. Hu ■

378219 Strobilanthes ridleyi sensu Yamam. = Semnostachya longispicata（Hayata）C. F. Hsieh et C. C. Huang ●■

378220 Strobilanthes rotundifolia Benoist = Pteracanthus rotundifolius（D. Don）Bremek. ■

378221 Strobilanthes rufo-hirta C. B. Clarke ex W. W. Sm. = Pyrrothrix rufohirtus（C. B. Clarke）C. Y. Wu et C. C. Hu ■

378222 Strobilanthes sarcorrhiza（C. Ling）C. Z. Zheng = Championella sarcorrhiza C. Ling ●

378223 Strobilanthes sarcorrhiza（C. Ling）H. S. Lo = Championella sarcorrhiza C. Ling ●

378224 Strobilanthes scoriarum W. W. Sm. = Diflugossa scoriarum（W. W. Sm.）E. Hossain ■

378225 Strobilanthes scrobiculata Dalz. ex C. B. Clarke = Supushpa scrobiculata（Dalz. ex C. B. Clarke）Suryan. ■☆

378226 Strobilanthes seguinii H. Lév. = Goldfussia seguinii（H. Lév.）C. Y. Wu et C. C. Hu ●

378227 Strobilanthes shweliensis W. W. Sm. = Diflugossa scoriarum（W. W. Sm.）E. Hossain ■

378228 Strobilanthes siamensis C. B. Clarke = Perilepta siamensis（C. B. Clarke）Bremek. ■

378229 Strobilanthes stolonifer Benoist；匍枝紫云菜；Stolonifer Conehead ●

378230 Strobilanthes straminea W. W. Sm. = Goldfussia straminea（W. W. Sm.）C. Y. Wu et C. C. Hu ■

378231 Strobilanthes tashiroi Hayata；琉球兰嵌马蓝（田代氏马蓝）；Tashiro Parachampionella ■☆

378232 Strobilanthes tashiroi Hayata = Parachampionella tashiroi（Hayata）Bremek. ■☆

378233 Strobilanthes tashiroi Hayata = Strobilanthes flexicaulis Hayata var. tashiroi（Hayata）T. Yamaz. ■☆

378234 Strobilanthes tetraspermus（Champ. ex Benth.）Druce = Championella tetrasperma（Champ. ex Benth.）Bremek. ●

378235 Strobilanthes thirionnii H. Lév. = Tarphochlamys affinis（Griff.）Bremek. ■

378236 Strobilanthes thomsonii T. Anderson = Strobilanthes triflora Y. C. Tang ●■

378237 Strobilanthes tibetica J. R. I. Wood = Pteracanthus tibeticus（J. R. I. Wood）C. Y. Wu et C. C. Hu ●

378238 Strobilanthes torrentium Benoist；急流紫云菜■

378239 Strobilanthes triflora Y. C. Tang = Pteracanthus alatus（Wall.）Bremek. ●■

378240 Strobilanthes truncata D. Fang et H. S. Lo；截头紫云菜；Truncate Conehead ■

378241 Strobilanthes urophylla Nees = Pteracanthus urophyllus（Nees）Bremek. ●

378242 Strobilanthes urticifolia Kuntze = Pteracanthus urticifolius（Kuntze）Bremek. ●

378243 Strobilanthes urticifolia Wall. ex Kuntze = Pteracanthus urticifolius（Kuntze）Bremek. ●

378244 Strobilanthes versicolor Diels = Pteracanthus versicolor（Diels）H. W. Li ●■

378245 Strobilanthes wallichii Nees = Pteracanthus alatus（Wall.）Bremek. ●■

378246 Strobilanthes wallichii Nees var. microphylla Nees = Pteracanthus alatus（Wall.）Bremek. ●■

378247 Strobilanthes wallichii Nees var. microphylla Nees = Strobilanthes wallichii Nees ●■

378248 Strobilanthes xanthantha Diels = Championella xanthantha（Diels）Bremek. ●

378249 Strobilanthes yangzekiangensis H. Lév. = Cystacanthus yangtsekiangensis（H. Lév.）Rehder ●

378250 Strobilanthes yunnanensis Diels = Pteracanthus yunnanensis（Diels）C. Y. Wu et C. C. Hu ●

378251 Strobilanthopsis H. Lév. = Strobilanthes Blume ●■

378252 Strobilanthopsis S. Moore（1900）；类马蓝属●☆

378253 Strobilanthopsis glutinifolia（Lindau）S. Moore = Strobilanthopsis linifolia（T. Anderson ex C. B. Clarke）Milne-Redh. ●☆

378254 Strobilanthopsis hircina S. Moore = Strobilanthopsis linifolia（T. Anderson ex C. B. Clarke）Milne-Redh. ●☆

378255 Strobilanthopsis linifolia（T. Anderson ex C. B. Clarke）Milne-Redh. ；亚麻叶类马蓝●☆

378256 Strobilanthopsis prostrata Milne-Redh. ；平卧类马蓝●☆

378257 Strobilanthopsis rogersii S. Moore = Strobilanthopsis linifolia（T.

Anderson ex C. B. Clarke) Milne-Redh. ●☆

378258　Strobilanthos St. -Lag. = Strobilanthus Rchb. ●■

378259　Strobilanthus Rchb. = Strobilanthes Blume ●■

378260　Strobilanthus lofuensis H. Lév. = Echinacanthus lofuensis（H. Lév.）J. R. I. Wood ●

378261　Strobilocarpos Benth. et Hook. f. = Strobilocarpus Klotzsch ●☆

378262　Strobilocarpus Klotzsch = Grubbia P. J. Bergius ●☆

378263　Strobilopanax R. Vig. = Meryta J. R. Forst. et G. Forst. ●☆

378264　Strobilopsis Hilliard et B. L. Burtt(1977);球果玄参属■☆

378265　Strobilopsis wrightii Hilliard et B. L. Burtt;球果玄参■☆

378266　Strobilorhachis Klotzsch = Aphelandra R. Br. ●■☆

378267　Strobllanthes bodinieri H. Lév. = Aechmanthera gossypina（Nees）Nees ●

378268　Strobocalyx（Blume ex DC.）Spach = Vernonia Schreb.（保留属名）●■

378269　Strobocalyx Sch. Bip. = Vernonia Schreb.（保留属名）●■

378270　Strobon Raf. = Cistus L. + Halimium（Dunal）Spach ●☆

378271　Strobopetalum N. E. Br.（1894）;扭瓣萝藦属■☆

378272　Strobopetalum bentii N. E. Br.;扭瓣萝藦■☆

378273　Strobopetalum bentii N. E. Br. = Pentatropis bentii（N. E. Br.）Liede ■☆

378274　Strobus（Endl.）Opiz = Pinus L. ●

378275　Strobus（Sweet ex Spach）Opiz = Pinus L. ●

378276　Strobus（Sweet）Opiz = Pinus L. ●

378277　Strobus Opiz = Pinus L. ●

378278　Strobus koraiensis（Siebold et Zucc.）Moldenke = Pinus koraiensis Siebold et Zucc. ●◇

378279　Strobus koraiensis Moldenke = Pinus koraiensis Siebold et Zucc. ●◇

378280　Strobus monticola（Douglas ex D. Don）Rydb. = Pinus monticola Douglas ex D. Don ●☆

378281　Strobus strobus（L.）Small = Pinus strobus L. ●

378282　Stroemeria Roxb. = Stroemia Vahl ●☆

378283　Stroemia Vahl = Cadaba Forssk. ●☆

378284　Stroemia farinosa（Forssk.）Vahl = Cadaba farinosa Forssk. ●☆

378285　Stroemia trifoliata Schumach. et Thonn. = Euadenia trifoliata（Schumach. et Thonn.）Oliv. ●☆

378286　Stroganovia Kar. et Kir. = Stroganowia Kar. et Kir. ●☆

378287　Stroganowia Kar. et Kir.（1841）;革叶荠属（革叶芥属）;Leathercress, Stroganowia ■

378288　Stroganowia Kar. et Kir. = Lepidium L. ■

378289　Stroganowia affghana（Boiss.）Pavlov;阿富汗革叶荠■☆

378290　Stroganowia brachyota Kar. et Kir.；革叶荠（革叶芥）;Leathercress, Shortauricle Stroganowia ■

378291　Stroganowia brachyota Kar. et Kir. = Lepidium brachyotum（Kar. et Kir.）Al-Shehbaz ■☆

378292　Stroganowia cardiophylla Pavlov;心叶革叶荠■☆

378293　Stroganowia desertorum（Schrenk）Botsch. = Stroganowia brachyota Kar. et Kir. ■

378294　Stroganowia gracilis Pavlov;纤细革叶荠■☆

378295　Stroganowia intermedia Kar. et Kir.;间型革叶荠■☆

378296　Stroganowia litvinoyii Lipsky;里特革叶荠■☆

378297　Stroganowia paniculata Regel et Schmalh.;圆锥革叶荠■☆

378298　Stroganowia persica N. Busch;波斯革叶荠■☆

378299　Stroganowia robusta Pavlov;粗壮革叶荠■☆

378300　Stroganowia sagittata Kar. et Kir.;箭头革叶荠■☆

378301　Stroganowia subalpina（Kom.）Thell.;亚高山革叶荠■☆

378302　Stroganowia trautvetteri Botsch.;特劳特革叶荠■☆

378303　Strogylodon T. Durand et Jacks. = Strongylodon Vogel ●☆

378304　Stromadendrum Pav. ex Bur. = Broussonetia L'Hér. ex Vent.（保留属名）●

378305　Stromanthe Sond.（1849）;紫背竹芋属■☆

378306　Stromanthe porteana Gris;波特氏紫背竹芋■☆

378307　Stromanthe sanguinea Sond.;紫背竹芋（红里蕉）■☆

378308　Stromatocactus Karw. ex Foerst. = Ariocarpus Scheidw. ●

378309　Stromatocactus Karw. ex Rümpler = Ariocarpus Scheidw. ●

378310　Stromatocarpus Rümpler = Ariocarpus Scheidw. ●

378311　Stromatocarpus Rümpler = Stromatocactus Karw. ex Rümpler ●

378312　Strombocactus Britton et Rose(1922);鳞茎玉属（独乐球属，菊水属）●☆

378313　Strombocactus disciformis Britton et Rose;鳞茎玉（独乐玉，菊水，鳞茎仙人球）■☆

378314　Strombocarpa（Benth.）A. Gray = Prosopis L. ●

378315　Strombocarpa A. Gray = Prosopis L. ●

378316　Strombocarpus Benth. et Hook. f. = Strombocarpa A. Gray ●

378317　Strombodurtus Steud. = Pentarrhaphis Kunth ■☆

378318　Strombodurus Willd. ex Steud. = Pentarrhaphis Kunth ■☆

378319　Strombosia Blume(1827);陀螺树属;Strombosia ●☆

378320　Strombosia cyanescens Mildbr. = Strombosia nigropunctata Louis et J. Léonard ●☆

378321　Strombosia fleuryana Breteler;弗勒里陀螺树●☆

378322　Strombosia glaucescens Engl. = Strombosia pustulata Oliv. ●☆

378323　Strombosia glaucescens Engl. var. lucida J. Léonard = Strombosia pustulata Oliv. var. lucida（J. Léonard）Villiers ●☆

378324　Strombosia gossweileri S. Moore;戈斯陀螺树●☆

378325　Strombosia grandifolia Hook. f.;大叶陀螺树●☆

378326　Strombosia klaineana Pierre = Strombosia grandifolia Hook. f. ●☆

378327　Strombosia majuscula S. Moore = Diogoa zenkeri（Engl.）Exell et Mendonça ●☆

378328　Strombosia mannii Engl. = Strombosia zenkeri Engl. ●☆

378329　Strombosia minor Engl. = Strombosia scheffleri Engl. ●☆

378330　Strombosia nigropunctata Louis et J. Léonard;黑斑陀螺树●☆

378331　Strombosia pustulata Oliv.;泡状陀螺树●☆

378332　Strombosia pustulata Oliv. var. lucida（J. Léonard）Villiers;光亮泡状陀螺树●☆

378333　Strombosia retevenia S. Moore = Diogoa retivenia（S. Moore）Breteler ●☆

378334　Strombosia scheffleri Engl.;谢夫勒陀螺树●☆

378335　Strombosia toroensis S. Moore = Strombosia scheffleri Engl. ●☆

378336　Strombosia zenkeri Engl.;岑克尔陀螺树●☆

378337　Strombosiaceae Tiegh. = Erythropalaceae Planch. ex Miq.（保留科名）●

378338　Strombosiaceae Tiegh. = Olacaceae R. Br.（保留科名）●

378339　Strombosiopsis Engl.（1897）;拟陀螺树属●☆

378340　Strombosiopsis buxifolia S. Moore = Diospyros hoyleana F. White ●☆

378341　Strombosiopsis congolensis De Wild. et T. Durand = Strombosiopsis tetrandra Engl. ●☆

378342　Strombosiopsis nana Breteler;矮小拟陀螺树●☆

378343　Strombosiopsis tetrandra Engl.;四蕊拟陀螺树●☆

378344　Strombosiopsis zenkeri Engl. = Diogoa zenkeri（Engl.）Exell et Mendonça ●☆

378345　Strongylocalyx Blume = Syzygium R. Br. ex Gaertn.（保留属名）●

378346　Strongylocaryum Burret = Ptychosperma Labill. ●☆

378347　Strongylodon Vogel（1836）；圆萼藤属（玉花豆属）●☆

378348　Strongylodon campenonii Drake = Strongylodon craveniae Baron et Baker ●☆

378349　Strongylodon catatii Drake = Strongylodon madagascariensis Baker ●☆

378350　Strongylodon craveniae Baron et Baker；马达加斯加圆萼藤●☆

378351　Strongylodon lantzianus（Baill.）Drake = Strongylodon madagascariensis Baker ●☆

378352　Strongylodon lastellianus Baill. = Strongylodon madagascariensis Baker ●☆

378353　Strongylodon macrobotrys A. Gray；圆萼藤（绿玉藤）；Emerald Creeper，Jade Vine，Philippin Jade Vine ●☆

378354　Strongylodon madagascariensis Baker；马岛圆萼藤●☆

378355　Strongylodon perrieri R. Vig. = Sylvichadsia perrieri（R. Vig.）Du Puy et Labat ●☆

378356　Strongyloma DC. = Nassauvia Comm. ex Juss. ●☆

378357　Strongylomopsis Speg. = Nassauvia Comm. ex Juss. ☆

378358　Strongylomopsis Speg. = Strongyloma DC. ●☆

378359　Strongylosperma Less. = Cotula L. ■

378360　Stropha Noronha = Chloranthus Sw. ■●

378361　Strophacanthus Lindau = Isoglossa Oerst.（保留属名）■★

378362　Strophades Boiss. = Erysimum L. ■●

378363　Strophanthus DC.（1802）；羊角拗属（毒毛旋花属，旋花羊角拗属）；Strophanthus ●☆

378364　Strophanthus amboensis（Schinz）Engl. et Pax；安博羊角拗●☆

378365　Strophanthus arnoldianus De Wild. et T. Durand；阿诺德羊角拗 ●☆

378366　Strophanthus asper Oliv. ex Planch. = Strophanthus nicholsonii Holmes ●☆

378367　Strophanthus bariba Boyé et Béréni = Strophanthus hispidus DC. ●

378368　Strophanthus barteri Franch.；巴特羊角拗●☆

378369　Strophanthus bequaertii Staner et Michotte；贝卡尔羊角拗●☆

378370　Strophanthus bracteatus Franch.；苞片羊角拗●☆

378371　Strophanthus bracteatus Franch. = Strophanthus preussii Engl. et Pax ●☆

378372　Strophanthus bullenianus Mast.；布伦羊角拗●☆

378373　Strophanthus capensis A. DC. = Strophanthus speciosus（Ward et Harv.）Reber ●☆

378374　Strophanthus caudatus（Burm. f.）Kurz = Strophanthus caudatus（L.）Kurz ●

378375　Strophanthus caudatus（L.）Kurz；卵萼羊角拗（金龙花，羊肝狼头草）；Malay Strophanthus，Ovatesepal Strophanthus，Ovate-sepaled Strophanthus ●

378376　Strophanthus chinensis（Hunter ex Roxb.）G. Don = Strophanthus divaricatus（Lour.）Hook. et Arn. ●

378377　Strophanthus chinensis G. Don = Strophanthus divaricatus（Lour.）Hook. et Arn. ●

378378　Strophanthus congoensis Franch.；刚果羊角拗●☆

378379　Strophanthus courmontii Sacleux ex Franch.；考蒙羊角拗●☆

378380　Strophanthus courmontii Sacleux ex Franch. var. fallax Holmes = Strophanthus courmontii Sacleux ex Franch. ●☆

378381　Strophanthus cumingii A. DC.；卡明羊角拗●☆

378382　Strophanthus demeusei Dewèvre = Strophanthus amboensis（Schinz）Engl. et Pax ●☆

378383　Strophanthus dewevrei De Wild. = Strophanthus parviflorus Franch. ●☆

378384　Strophanthus dichotomus DC. = Strophanthus caudatus（Burm. f.）Kurz ●

378385　Strophanthus dichotomus DC. var. chinensis Ker Gawl. = Strophanthus divaricatus（Lour.）Hook. et Arn. ●

378386　Strophanthus divaricatus（Lour.）Hook. et Arn.；羊角拗（布渣叶，打破碗花，打破碗碗，大羊角扭蕹，倒钓笔，断肠草，黄葛扭，金龙花，金龙角，鲤鱼橄榄，沥口花，菱角扭，牛角橹，牛角藤，山羊角，武靴藤，羊角，羊角崩，羊角果，羊角黎，羊角捩，羊角柳，羊角墓，羊角扭，羊角纽，羊角藕，羊角树，羊角藤，阳角右藤，猪屎壳）；Divaricate Strophanthus ●

378387　Strophanthus divergens G. G. Graham = Strophanthus divaricatus（Lour.）Hook. et Arn. ●

378388　Strophanthus ecaudatus Rolfe = Strophanthus welwitschii（Baill.）K. Schum. ●☆

378389　Strophanthus emini Asch. ex Pax；埃明羊角拗（伊朗羊角拗）●☆

378390　Strophanthus emini Asch. ex Pax var. wittei（Staner）Staner = Strophanthus emini Asch. ex Pax ●☆

378391　Strophanthus erythroleucus Gilg = Strophanthus bullenianus Mast. ●☆

378392　Strophanthus fischeri Asch. et K. Schum. ex Holmes = Strophanthus eminii Asch. et Pax ●☆

378393　Strophanthus gerrardii Stapf；杰勒德羊角拗●☆

378394　Strophanthus gilletii De Wild. = Strophanthus welwitschii（Baill.）K. Schum. ●☆

378395　Strophanthus glabriflorus（Monach.）Monach. = Strophanthus sarmentosus DC. var. glabriflorus Monach. ●☆

378396　Strophanthus gossweileri H. E. Hess = Strophanthus amboensis（Schinz）Engl. et Pax ●☆

378397　Strophanthus gracilis K. Schum. et Pax；纤细羊角拗●☆

378398　Strophanthus grandiflorus（N. E. Br.）Gilg = Strophanthus petersianus Klotzsch ●☆

378399　Strophanthus grandiflorus Gilg；大花羊角拗；Corkscrew Flower ●☆

378400　Strophanthus gratus（Hook.）Franch. = Strophanthus gratus（Wall. et Hook. ex Benth.）Baill. ●

378401　Strophanthus gratus（Wall. et Hook. ex Benth.）Baill.；旋花羊角拗（非洲羊角拗，苦毒毛旋花）；Gream Fruit, Pleasant Strophanthus, Spinyflower Strophanthus, Spiny-flowered Strophanthus ●

378402　Strophanthus gratus（Wall. et Hook.）Baill. = Strophanthus gratus（Wall. et Hook. ex Benth.）Baill. ●

378403　Strophanthus gratus（Wall. et Hook.）Franch. = Strophanthus gratus（Wall. et Hook. ex Benth.）Baill. ●

378404　Strophanthus hirsutus H. E. Hess = Strophanthus amboensis（Schinz）Engl. et Pax ●☆

378405　Strophanthus hispidus DC.；箭毒羊角拗（棕毒毛旋花，棕羊角拗）；Transvasl Strophanthus ●

378406　Strophanthus hispidus DC. var. bosere De Wild. = Strophanthus hispidus DC. ●

378407　Strophanthus hispidus DC. var. latistigmatica Schnell = Strophanthus hispidus DC. ●

378408　Strophanthus hispidus DC. var. lobatistigmatica Schnell = Strophanthus hispidus DC. ●

378409　Strophanthus hispidus DC. var. parvistigmatica Schnell = Strophanthus hispidus DC. ●

378410　Strophanthus hispidus DC. var. seidenii ? = Strophanthus hispidus DC. ●

378411 Strophanthus holosericeus K. Schum. et Gilg;绢毛羊角拗●☆

378412 Strophanthus hypoleucos Stapf;白背羊角拗●☆

378413 Strophanthus intermedius Pax = Strophanthus amboensis (Schinz) Engl. et Pax ●☆

378414 Strophanthus intermedius Pax var. bieleri De Wild. = Strophanthus congoensis Franch. ●☆

378415 Strophanthus katangensis Staner = Strophanthus welwitschii (Baill.) K. Schum. ●☆

378416 Strophanthus klainei De Wild. = Strophanthus gracilis K. Schum. et Pax ●☆

378417 Strophanthus kombe Oliv. ;绿羊角拗(毒毛旋花子,绿毒毛旋花);Kombe ●

378418 Strophanthus laurifolius DC. = Strophanthus sarmentosus DC. ●

378419 Strophanthus ledienii Stein;莱丁羊角拗●☆

378420 Strophanthus letei Merr. ex Wiells et Garcia;李特羊角拗●☆

378421 Strophanthus longicalyx H. E. Hess = Strophanthus amboensis (Schinz) Engl. et Pax ●☆

378422 Strophanthus luteolus Codd;淡黄羊角拗●☆

378423 Strophanthus minor Blondel;较小羊角拗●☆

378424 Strophanthus mirabilis Gilg;奇异羊角拗●☆

378425 Strophanthus mortehanii De Wild. ;莫特汉羊角拗●☆

378426 Strophanthus nicholsonii Holmes;尼氏羊角拗●☆

378427 Strophanthus ogovensis Franch. = Strophanthus sarmentosus DC. ●

378428 Strophanthus ouabaio Holmes = Strophanthus gratus (Wall. et Hook.) Baill. ●

378429 Strophanthus paroissei Franch. = Strophanthus sarmentosus DC. ●

378430 Strophanthus parviflorus Franch. ;小花羊角拗●☆

378431 Strophanthus parvifolius K. Schum. = Strophanthus welwitschii (Baill.) K. Schum. ●☆

378432 Strophanthus paxii H. E. Hess = Strophanthus amboensis (Schinz) Engl. et Pax ●☆

378433 Strophanthus pendulus Hook. = Strophanthus sarmentosus DC. ●

378434 Strophanthus perrotii A. Chev. = Strophanthus gratus (Wall. et Hook.) Baill. ●

378435 Strophanthus petersianus Klotzsch;彼得斯羊角拗●☆

378436 Strophanthus petersianus Klotzsch var. amboensis Schinz = Strophanthus amboensis (Schinz) Engl. et Pax ●☆

378437 Strophanthus petersianus Klotzsch var. grandiflorus N. E. Br. = Strophanthus petersianus Klotzsch ●☆

378438 Strophanthus pierreanus De Wild. = Strophanthus thollonii Franch. ●☆

378439 Strophanthus preussii Engl. et Pax = Strophanthus preussii Engl. et Pax ex Pax ●☆

378440 Strophanthus preussii Engl. et Pax ex Pax;普罗羊角拗●☆

378441 Strophanthus preussii Engl. et Pax var. brevifolius De Wild. = Strophanthus preussii Engl. et Pax ●☆

378442 Strophanthus preussii Engl. et Pax var. scabridulus Monach. = Strophanthus preussii Engl. et Pax ●☆

378443 Strophanthus puncti ferus A. Chev. = Strophanthus sarmentosus DC. ●

378444 Strophanthus radcliffei S. Moore = Cryptolepis sanguinolenta (Lindl.) Schltr. ●☆

378445 Strophanthus sarmentosus DC. ;西非羊角拗(蔓茎毒毛旋花,蔓茎羊角拗);Arrowpoison Strophanthus, Arrow-poison Strophanthus ●

378446 Strophanthus sarmentosus DC. f. paroissei (Franch.) Chiov. = Strophanthus sarmentosus DC. ●

378447 Strophanthus sarmentosus DC. f. senegambiae (A. DC.) A. Chev. = Strophanthus sarmentosus DC. ●

378448 Strophanthus sarmentosus DC. var. glabriflorus Monach. ;光花羊角拗●☆

378449 Strophanthus sarmentosus DC. var. major Dewèvre = Strophanthus sarmentosus DC. ●

378450 Strophanthus sarmentosus DC. var. pendulus (Hook.) Pax = Strophanthus sarmentosus DC. ●

378451 Strophanthus sarmentosus DC. var. pubescens Staner et Michotte = Strophanthus sarmentosus DC. ●

378452 Strophanthus sarmentosus DC. var. verrucosus Pax = Strophanthus petersianus Klotzsch ●☆

378453 Strophanthus scaber Pax = Strophanthus gracilis K. Schum. et Pax ●☆

378454 Strophanthus schlechteri K. Schum. et Gilg = Strophanthus bullenianus Mast. ●☆

378455 Strophanthus schuchardtii Pax = Strophanthus amboensis (Schinz) Engl. et Pax ●☆

378456 Strophanthus schultzei Mildbr. ;舒尔茨羊角拗●☆

378457 Strophanthus senegambiae A. DC. = Strophanthus sarmentosus DC. ●

378458 Strophanthus speciosus (Ward et Harv.) Reber;美丽羊角拗●☆

378459 Strophanthus standleyanus Wall. et Hook. = Strophanthus gratus (Wall. et Hook.) Baill. ●

378460 Strophanthus tchabe Boyé et Béréni = Strophanthus hispidus DC. ●

378461 Strophanthus thierreanus K. Schum. et Gilg = Strophanthus hispidus DC. ●

378462 Strophanthus thollonii Franch. ;索郎羊角拗●☆

378463 Strophanthus vanderijstii Staner;范德羊角拗●☆

378464 Strophanthus verdickii De Wild. = Strophanthus welwitschii (Baill.) K. Schum. ●☆

378465 Strophanthus verdickii De Wild. var. latisepalus ? = Strophanthus welwitschii (Baill.) K. Schum. ●☆

378466 Strophanthus verrucosus Stapf;疣状羊角拗●☆

378467 Strophanthus verrucosus Stapf = Strophanthus petersianus Klotzsch ●☆

378468 Strophanthus wallichii A. DC. ;云南羊角拗(羊角拗);Wallich Strophanthus ●

378469 Strophanthus welwitschii (Baill.) K. Schum. ;韦尔羊角拗●☆

378470 Strophanthus wildemanianus Gilg = Strophanthus bullenianus Mast. ●☆

378471 Strophanthus wittei Staner = Strophanthus emini Asch. ex Pax ●☆

378472 Strophanthus zimmermannianus Monach. ;齐默尔曼羊角拗●☆

378473 Strophioblachia Boerl. (1900);宿萼木属(腺萼木属);Strophioblachia ●

378474 Strophioblachia fimbricalyx Boerl. ;宿萼木(施巴蜡,腺萼木);Common Strophioblachia ●

378475 Strophioblachia fimbricalyx Boerl. var. efimbriata Airy Shaw;广西宿萼木;Guangxi Strophioblachia, Kwangsi Strophioblachia ●

378476 Strophioblachia glandulosa Pax;越南宿萼木;Viatnam ●

378477 Strophioblachia glandulosa Pax var. cordifolia Airy Shaw;心叶宿萼木;Cordateleaf Strophioblachia ●

378478 Strophioblachia glandulosa Pax var. tonkinensis Gagnep. = Strophioblachia fimbricalyx Boerl. ●

378479 Strophiodiscus Choux = Plagioscyphus Radlk. ●☆

378480 Strophiodiscus jumellei Choux = Plagioscyphus jumellei (Choux) Capuron ●☆

378481 Strophiostoma Turcz. = Myosotis L. ■

378482 Strophiostoma sparsiflorum (J. C. Mikan) Turcz. = Myosotis sparsiflora J. C. Mikan ■

378483 Strophiostoma sparsiflorum Turcz. = Myosotis sparsiflora J. C. Mikan ■

378484 Strophis Salisb. = Dioscorea L. (保留属名)■

378485 Strophium Dulac = Moehringia L. ■

378486 Strophocactus Britton et Rose = Selenicereus (A. Berger) Britton et Rose ●

378487 Strophocactus Britton et Rose(1913);百足柱属●☆

378488 Strophocactus wittii Britton et Rose;百足柱●☆

378489 Strophocaulos (G. Don) Small = Convolvulus L. ■●

378490 Strophocaulos Small = Convolvulus L. ■●

378491 Strophocaulos arvensis (L.) Small = Convolvulus arvensis L. ■

378492 Strophocereus Fričet Kreuz. = Selenicereus (A. Berger) Britton et Rose ●

378493 Stropholirion Torr. = Dichelostemma Kunth ■☆

378494 Stropholirion californicum Torr. = Dichelostemma volubile (Kellogg) A. Heller ■☆

378495 Strophopappus DC. = Stilpnopappus Mart. ex DC. ●■☆

378496 Strophostyles E. Mey. = Vigna Savi(保留属名)■

378497 Strophostyles Elliott(1823)(保留属名);扭柱豆属(曲瓣菜豆属);Wild Bean ■☆

378498 Strophostyles Elliott(保留属名) = Glycine Willd. (保留属名)■

378499 Strophostyles Elliott(保留属名) = Phaseolus L. ■

378500 Strophostyles angulosa (Willd.) Elliott var. missouriensis S. Watson = Strophostyles helvula (L.) Elliott ■☆

378501 Strophostyles capensis E. Mey. var. ovatus E. Mey. = Vigna vexillata (L.) A. Rich. var. ovata (E. Mey.) B. J. Pienaar ■☆

378502 Strophostyles helvula (L.) Elliott;一年生扭柱豆;Annual Woolly Bean, Trailing Fuzzy Bean, Trailing Wild Bean, Wild Bean ■☆

378503 Strophostyles helvula (L.) Elliott var. missouriensis (S. Watson) Britton = Strophostyles helvula (L.) Elliott ■☆

378504 Strophostyles leiosperma (Torr. et A. Gray) Piper;光籽扭柱豆;Slickseed Bean, Slick-seed Fuzzy Bean, Small-flowered Wild Bean ■☆

378505 Strophostyles pauciflora (Benth.) S. Watson = Strophostyles leiosperma (Torr. et A. Gray) Piper ■☆

378506 Strophostyles umbellata (Muhl. ex Willd.) Britton;粉色扭柱豆;Pink Wild Bean, Wild Bean ■☆

378507 Strotheria B. L. Turner(1972);岩丘菊属●☆

378508 Strotheria gypsophila B. L. Turner;岩丘菊■☆

378509 Struchium P. Browne = Sparganophoros Vaill. ■☆

378510 Struchium africanum (Steud.) P. Beauv. = Sparganophorus sparganophorus (L.) C. Jeffrey ■☆

378511 Struchium africanum P. Beauv. = Struchium sparganophorum (L.) Kuntze ■☆

378512 Struchium sparganophorum (L.) Kuntze = Sparganophorus sparganophorus (L.) C. Jeffrey ■☆

378513 Struckeria Steud. = Strukeria Vell. ●☆

378514 Strukeria Vell. = Vochysia Aubl. (保留属名)●☆

378515 Strumaria Jacq. (1797);疣石蒜属■☆

378516 Strumaria Jacq. ex Willd. = Strumaria Jacq. ■☆

378517 Strumaria aestivalis Snijman;夏疣石蒜■☆

378518 Strumaria angustifolia Jacq. = Strumaria truncata Jacq. ■☆

378519 Strumaria barbarae Oberm. ;外来疣石蒜■☆

378520 Strumaria bidentata Schinz;双齿疣石蒜■☆

378521 Strumaria chaplinii (W. F. Barker) Snijman;哈普林疣石蒜■☆

378522 Strumaria discifera Marloth ex Snijman subsp. bulbifera Snijman;鳞茎疣石蒜■☆

378523 Strumaria gemmata Ker Gawl. ;具芽疣石蒜■☆

378524 Strumaria hardyana D. Müll. -Doblies et U. Müll. -Doblies;哈迪疣石蒜■☆

378525 Strumaria karooica (W. F. Barker) Snijman;卡卢疣石蒜■☆

378526 Strumaria karoopoortensis (D. Müll. -Doblies et U. Müll. -Doblies) Snijman;卡卢普尔疣石蒜■☆

378527 Strumaria leipoldtii (L. Bolus) Snijman;莱波尔德疣石蒜■☆

378528 Strumaria linguifolia Jacq. = Strumaria truncata Jacq. ■☆

378529 Strumaria luteoloba Snijman;黄裂片疣石蒜■☆

378530 Strumaria massoniella (D. Müll. -Doblies et U. Müll. -Doblies) Snijman;马森疣石蒜■☆

378531 Strumaria merxmuelleriana (D. Müll. -Doblies et U. Müll. -Doblies) Snijman;梅尔疣石蒜■☆

378532 Strumaria perryae Snijman;佩里疣石蒜■☆

378533 Strumaria picta W. F. Barker;着色疣石蒜■☆

378534 Strumaria prolifera Snijman;多育疣石蒜■☆

378535 Strumaria pubescens W. F. Barker;短柔毛疣石蒜■☆

378536 Strumaria pygmaea Snijman;矮小疣石蒜■☆

378537 Strumaria rubella Jacq. = Strumaria truncata Jacq. ■☆

378538 Strumaria salteri W. F. Barker;索尔特疣石蒜■☆

378539 Strumaria speciosa Snijman;美丽疣石蒜■☆

378540 Strumaria spiralis L'Hér. ;螺旋疣石蒜■☆

378541 Strumaria tenella (L. f.) Snijman;柔软疣石蒜■☆

378542 Strumaria tenella (L. f.) Snijman subsp. orientalis Snijman;东方柔软疣石蒜■☆

378543 Strumaria truncata Jacq. ;平截疣石蒜■☆

378544 Strumaria undulata Jacq. ;波状疣石蒜■☆

378545 Strumaria unguiculata (W. F. Barker) Snijman;爪状疣石蒜■☆

378546 Strumaria villosa Snijman;长柔毛疣石蒜■☆

378547 Strumaria watermeyeri L. Bolus;沃特迈耶疣石蒜■☆

378548 Strumaria watermeyeri L. Bolus subsp. botterkloofensis (D. Müll. -Doblies et U. Müll. -Doblies) Snijman;博泰尔疣石蒜■☆

378549 Strumariaceae Salisb. = Amaryllidaceae J. St. -Hil. (保留科名)■●

378550 Strumariaceae Salisb. = Poaceae Barnhart(保留科名)■●

378551 Strumarium Raf. = Xanthium L. ■

378552 Strumpfia Jacq. (1760);斯特茜属☆

378553 Strumpfia maritima Jacq. ;斯特茜☆

378554 Strusiola Raf. = Struthiola L. (保留属名)●☆

378555 Struthanthus Mart. (1830)(保留属名);鸵鸟花属(驼花属)●☆

378556 Struthanthus cassythoides Millsp. ex Standl. ;无根藤驼花●☆

378557 Struthanthus flexicaulis Mart. ;弯茎驼花●☆

378558 Struthia Boehm. = Gnidia L. ●☆

378559 Struthia L. = Gnidia L. ●☆

378560 Struthiola L. (1767)(保留属名);鸵鸟木属●☆

378561 Struthiola albersii H. Pearson = Struthiola thomsonii Oliv. ●☆

378562 Struthiola amabilis Gilg = Struthiola thomsonii Oliv. ●☆

378563 Struthiola angustifolia Lam. = Struthiola ciliata (L.) Lam. ●☆

378564　Struthiola angustiloba B. Peterson et Hilliard;窄裂鸵鸟木●☆

378565　Struthiola anomala Hilliard;异常鸵鸟木●☆

378566　Struthiola argentea Lehm.;银白鸵鸟木●☆

378567　Struthiola bachmanniana Gilg;巴克曼鸵鸟木●☆

378568　Struthiola cicatricosa C. H. Wright;疤痕鸵鸟木●☆

378569　Struthiola ciliata (L.) Lam.;睫毛鸵鸟木●☆

378570　Struthiola concava S. Moore;凹鸵鸟木●☆

378571　Struthiola confusa C. H. Wright;混乱鸵鸟木●☆

378572　Struthiola congesta C. H. Wright = Struthiola pondoensis Gilg ex C. H. Wright ●☆

378573　Struthiola dodecandra (L.) Druce;十二雄蕊鸵鸟木(全裂鸵鸟木)●☆

378574　Struthiola eckloniana Meisn.;埃氏鸵鸟木●☆

378575　Struthiola epacridioides C. H. Wright = Struthiola hirsuta Wikstr. ●☆

378576　Struthiola erecta Lam. = Struthiola dodecandra (L.) Druce ●☆

378577　Struthiola ericina Gilg = Struthiola thomsonii Oliv. ●☆

378578　Struthiola ericoides C. H. Wright;石南状鸵鸟木●☆

378579　Struthiola fasciata C. H. Wright;带状鸵鸟木●☆

378580　Struthiola flavescens Gilg ex C. H. Wright = Struthiola ciliata (L.) Lam. ●☆

378581　Struthiola floribunda C. H. Wright;繁花鸵鸟木●☆

378582　Struthiola fourcadei Compton = Struthiola martiana Meisn. ●☆

378583　Struthiola galpinii C. H. Wright;盖尔鸵鸟木●☆

378584　Struthiola garciana C. H. Wright;加西亚鸵鸟木●☆

378585　Struthiola gilgiana H. Pearson = Struthiola thomsonii Oliv. ●☆

378586　Struthiola hirsuta Wikstr.;粗毛鸵鸟木●☆

378587　Struthiola kilimandscharica Gilg = Struthiola thomsonii Oliv. ●☆

378588　Struthiola leiosiphon Gilg ex C. H. Wright = Struthiola martiana Meisn. ●☆

378589　Struthiola leptantha Bolus;细花鸵鸟木●☆

378590　Struthiola linealriloba Meisn.;线裂片鸵鸟木●☆

378591　Struthiola longiflora Lam. = Struthiola ciliata (L.) Lam. ●☆

378592　Struthiola longifolia C. H. Wright;长叶鸵鸟木●☆

378593　Struthiola lucens Lam. = Struthiola ciliata (L.) Lam. ●☆

378594　Struthiola macowanii C. H. Wright;麦克欧文鸵鸟木●☆

378595　Struthiola martiana Meisn.;马氏鸵鸟木●☆

378596　Struthiola montana B. Peterson;山地鸵鸟木●☆

378597　Struthiola mundtii Eckl. ex Meisn.;蒙特鸵鸟木●☆

378598　Struthiola myrsinites Lam.;铁仔鸵鸟木●☆

378599　Struthiola nana L. f. = Gnidia nana (L. f.) Wikstr. ●☆

378600　Struthiola ovata Thunb. = Struthiola myrsinites Lam. ●☆

378601　Struthiola parviflora Bartl. ex Meisn.;小花分枝鸵鸟木●☆

378602　Struthiola pentheri S. Moore = Struthiola hirsuta Wikstr. ●☆

378603　Struthiola pillansii Hutch. = Struthiola ciliata (L.) Lam. ●☆

378604　Struthiola pondoensis Gilg ex C. H. Wright;庞多鸵鸟木●☆

378605　Struthiola ramosa C. H. Wright;分枝鸵鸟木●☆

378606　Struthiola recta C. H. Wright;直立鸵鸟木●☆

378607　Struthiola rhodesiana B. Peterson;罗得西亚鸵鸟木●☆

378608　Struthiola rigida Meisn.;坚硬鸵鸟木●☆

378609　Struthiola rustiana Gilg = Struthiola ciliata (L.) Lam. ●☆

378610　Struthiola salteri Levyns;索尔特鸵鸟木●☆

378611　Struthiola schlechteri Gilg ex C. H. Wright = Struthiola ciliata (L.) Lam. ●☆

378612　Struthiola striata Lam.;条纹鸵鸟木●☆

378613　Struthiola stuhlmannii Gilg = Struthiola thomsonii Oliv. ●☆

378614　Struthiola tetralepis Schltr.;四鳞鸵鸟木●☆

378615　Struthiola tetralepis Schltr. var. glabricaulis ?;光茎鸵鸟木●☆

378616　Struthiola thomsonii Oliv.;托马森鸵鸟木●☆

378617　Struthiola tomentosa Andréws;绒毛鸵鸟木●☆

378618　Struthiola usambarensis Engl. = Struthiola thomsonii Oliv. ●☆

378619　Struthiola virgata L. = Struthiola ciliata (L.) Lam. ●☆

378620　Struthiola volkensii H. Winkl. = Struthiola thomsonii Oliv. ●☆

378621　Struthiolopsis E. Phillips = Gnidia L. ●☆

378622　Struthiolopsis bolusii E. Phillips = Gnidia linearifolia (Wikstr.) B. Peterson ●☆

378623　Struthiolopsis pulvinata (Bolus) E. Phillips = Gnidia nana (L. f.) Wikstr. ●☆

378624　Struvea Rchb. = Torreya Arn. ●

378625　Strychnaceae DC. ex Perleb = Loganiaceae R. Br. ex Mart. (1827)(保留科名)●

378626　Strychnaceae Link = Loganiaceae R. Br. ex Mart.(保留科名)●■

378627　Strychnaceae Perleb = Strychnaceae DC ex Perleb ●

378628　Strychnodaphne Nees = Ocotea Aubl. ●☆

378629　Strychnodaphne Nees et Mart. = Ocotea Aubl. ●☆

378630　Strychnodaphne Nees et Mart. ex Nees = Ocotea Aubl. ●☆

378631　Strychnopsis Baill. (1885);马钱藤属●☆

378632　Strychnopsis thouarsii Baill.;马钱藤●☆

378633　Strychnos L. (1753);马钱属(马钱子属); Monkey Apple, Poison Nut, Poisonnut, Poison-nut, Strychnos ●

378634　Strychnos abyssinica Hochst. = Acokanthera schimperi (A. DC.) Schweinf. ☆

378635　Strychnos aculeata Soler.;皮刺马钱●☆

378636　Strychnos acutissima Gilg = Strychnos splendens Gilg ●☆

378637　Strychnos adolfi-friderici Gilg = Strychnos mitis S. Moore ●☆

378638　Strychnos adolphi-fridericii Gilg = Strychnos mitis S. Moore ●☆

378639　Strychnos afzelii Gilg;阿氏马钱(阿芙泽尔马钱); Afzel Strychnos ●☆

378640　Strychnos albersii Gilg et Busse = Strychnos henningsii Gilg ●☆

378641　Strychnos alnifolia Baker = Strychnos innocua Delile ●☆

378642　Strychnos amazonica Krukoff;亚马孙马钱●☆

378643　Strychnos angolensis Gilg;安哥拉马钱●☆

378644　Strychnos angolensis Gilg var. lacourtiana (De Wild.) P. A. Duvign. = Strychnos angolensis Gilg ●☆

378645　Strychnos angolensis Gilg var. latifolia P. A. Duvign. = Strychnos angolensis Gilg ●☆

378646　Strychnos angolensis Gilg var. tanganykae P. A. Duvign. = Strychnos angolensis Gilg ●☆

378647　Strychnos angolensis Gilg var. tisserantii P. A. Duvign. = Strychnos angolensis Gilg ●☆

378648　Strychnos angustiflora Benth.;牛眼马钱(车前树,钩梗树,牛目周,牛眼睛,牛眼球,牛眼珠,狭花马钱,狭叶马钱); Narrowflower Poison-nut, Narrowflower Strychnos, Narrow-flowered Poisonnut ●

378649　Strychnos asterantha Leeuwenb.;星花马钱●☆

378650　Strychnos atherstonei Harv. = Strychnos decussata (Pappe) Gilg ●☆

378651　Strychnos axillaris Colebr.;腋花马钱; Axillary Poisonnut, Axillary Strychnos, Axillaryflower Poisonnut ●

378652　Strychnos bakanko Bourquelet et Hérissey = Strychnos madagascariensis Poir. ●☆

378653　Strychnos balansae Hill. = Strychnos ignatii Berger ●

378654　Strychnos barbata Chiov. = Strychnos henningsii Gilg ●☆

378655　Strychnos baronii Baker = Strychnos madagascariensis Poir. ●☆

378656 Strychnos barteri Soler. ;巴特马钱●☆

378657 Strychnos behrensiana Gilg et Busse = Strychnos madagascariensis Poir. ●☆

378658 Strychnos bequaertii De Wild. = Strychnos angolensis Gilg ●☆

378659 Strychnos bicirrifera Dunkley = Strychnos panganensis Gilg ●☆

378660 Strychnos bifurcata Leeuwenb. ;双叉马钱●☆

378661 Strychnos boinensis Jum. et H. Perrier = Strychnos decussata (Pappe) Gilg ●☆

378662 Strychnos bourdillonii Brandis = Strychnos wallichiana Steud. ex A. DC. ●◇

378663 Strychnos brazzavillensis A. Chev. = Strychnos pungens Soler. ●☆

378664 Strychnos brevicymosa De Wild. = Strychnos longicaudata Gilg ●☆

378665 Strychnos buettneri Gilg = Strychnos spinosa Lam. ●☆

378666 Strychnos burtonii Baker = Strychnos madagascariensis Poir. ●☆

378667 Strychnos caespitosa Good = Strychnos gossweileri Exell ●☆

378668 Strychnos campicola Gilg ex Leeuwenb. ;平原马钱●☆

378669 Strychnos camptoneura Gilg et Busse ;弯脉马钱●☆

378670 Strychnos canthioides Leeuwenb. ;鱼骨木马钱●☆

378671 Strychnos cardiophylla Gilg et Busse = Strychnos spinosa Lam. ●☆

378672 Strychnos carvalhoi Gilg = Strychnos spinosa Lam. ●☆

378673 Strychnos caryophyllus A. Chev. = Strychnos afzelii Gilg ●☆

378674 Strychnos castelnaeana Baill. ;卡斯马钱●☆

378675 Strychnos cathayensis Merr. ;华马钱(百节藤,登欧梅罗,亨利马钱,牛目椒,三脉马钱,台湾马钱);Cathay Poisonnut,Chinese Poisonnut,Chinese Strychnos,Henry Snakewood ●

378676 Strychnos cathayensis Merr. var. spinata P. T. Li ;小刺马钱;Spiny Poisonnut,Spiny Strychnos ●

378677 Strychnos cheliensis Hu = Strychnos nitida G. Don ●

378678 Strychnos chlorocarpa Gilg = Strychnos densiflora Baill. ●☆

378679 Strychnos chromatoxylon Leeuwenb. ;色材马钱●☆

378680 Strychnos chrysocarpa Baker = Strychnos splendens Gilg ●☆

378681 Strychnos chrysophylla Gilg;金叶马钱●☆

378682 Strychnos ciliicalyx Gilg et Busse = Strychnos nigritana Baker ●☆

378683 Strychnos cinnabarina Gilg ex Hutch. et Dalziel = Strychnos angolensis Gilg ●☆

378684 Strychnos cinnabarina Gilg ex Hutch. et Dalziel var. ctenotricha P. A. Duvign. = Strychnos angolensis Gilg ●☆

378685 Strychnos cinnamomifolia Thwaites = Strychnos wallichiana Steud. ex A. DC. ●◇

378686 Strychnos cinnamomifolia Thwaites var. wightii Hill. = Strychnos wallichiana Steud. ex A. DC. ●◇

378687 Strychnos cirrhosa Stokes = Strychnos wallichiana Steud. ex A. DC. ●◇

378688 Strychnos cocculoides Baker;小凯菲尔马钱;Small Kaffir Orange,Wild Orange ●☆

378689 Strychnos colubrina L. = Strychnos wallichiana Steud. ex A. DC. ●◇

378690 Strychnos confertiflora Merr. et Chun = Strychnos ovata A. W. Hill ●

378691 Strychnos congolana Gilg;刚果马钱●☆

378692 Strychnos cooperi Hutch. et M. B. Moss = Strychnos usambarensis Gilg ex Engl. ●☆

378693 Strychnos corymbifera Gilg ex P. A. Duvign. = Strychnos panganensis Gilg ●☆

378694 Strychnos courteti A. Chev. = Strychnos spinosa Lam. ●☆

378695 Strychnos cuneata Schinz ex Klein et Herndlhofer = Diospyros quiloensis (Hiern) F. White ●☆

378696 Strychnos cuneifolia Gilg et Busse = Strychnos spinosa Lam. ●☆

378697 Strychnos dale De Wild. ;戴尔豆马钱●☆

378698 Strychnos decorsei A. Chev. = Strychnos moandaensis De Wild. ●☆

378699 Strychnos decussata (Pappe) Gilg;海角马钱(对生马钱,海角柚木,南非马钱);Cape Teak,Chaka's Wood,Unhlamalala ●☆

378700 Strychnos decussata Gilg = Strychnos decussata (Pappe) Gilg ●☆

378701 Strychnos dekindtiana Gilg = Strychnos cocculoides Baker ●☆

378702 Strychnos densiflora Baill. ;密花马钱●☆

378703 Strychnos dewevrei Gilg = Strychnos icaja Baill. ●☆

378704 Strychnos diaboli Sandwith;戴氏马钱;Diabol Strychnos ●☆

378705 Strychnos dinklagei Gilg;丁克马钱●☆

378706 Strychnos diplotricha Leeuwenb. ;双毛马钱●☆

378707 Strychnos djalonis A. Chev. = Strychnos spinosa Lam. ●☆

378708 Strychnos dolichothyrsa Gilg ex Onochie et Hepper;长花序马钱 ●☆

378709 Strychnos dschurica (Gilg) Gilg = Strychnos innocua Delile ●☆

378710 Strychnos dubia De Wild. = Strychnos scheffleri Gilg ●☆

378711 Strychnos dulcis A. Chev. = Strychnos spinosa Lam. ●☆

378712 Strychnos dundusanensis De Wild. = Strychnos icaja Baill. ●☆

378713 Strychnos dysophylla Benth. ;斜叶马钱●☆

378714 Strychnos dysophylla Benth. = Strychnos madagascariensis Poir. ●☆

378715 Strychnos dysophylla Benth. subsp. engleri (Gilg) E. A. Bruce et Lewis = Strychnos madagascariensis Poir. ●☆

378716 Strychnos edulis Schweinf. = Strychnos innocua Delile ●☆

378717 Strychnos eketensis S. Moore = Strychnos chrysophylla Gilg ●☆

378718 Strychnos elliottii Gilg et Busse = Strychnos henningsii Gilg ●☆

378719 Strychnos emarginata Baker = Strychnos spinosa Lam. ●☆

378720 Strychnos engleri Gilg = Strychnos madagascariensis Poir. ●☆

378721 Strychnos erythrocarpa Gilg = Strychnos afzelii Gilg ●☆

378722 Strychnos esquirolii H. Lév. = Ziziphus incurva Roxb. ●

378723 Strychnos esquirolii H. Lév. = Ziziphus pubinervis Rehder ●

378724 Strychnos euryphylla Gilg et Busse = Strychnos spinosa Lam. ●☆

378725 Strychnos excellens Gilg = Strychnos scheffleri Gilg ●☆

378726 Strychnos fallax Leeuwenb. ;迷惑马钱●☆

378727 Strychnos fernandiae P. A. Duvign. = Strychnos usambarensis Gilg ex Engl. ●☆

378728 Strychnos fischeri Gilg = Strychnos innocua Delile ●☆

378729 Strychnos flacurtii Desv. ex Dubuisson et Thouars = Strychnos spinosa Lam. ●☆

378730 Strychnos fleuryana A. Chev. = Strychnos nigritana Baker ●☆

378731 Strychnos floribunda Gilg;繁叶马钱●☆

378732 Strychnos gauthierana Pierre ex Dop = Strychnos wallichiana Steud. ex A. DC. ●◇

378733 Strychnos gerrardii N. E. Br. = Strychnos madagascariensis Poir. ●☆

378734 Strychnos gilletii De Wild. = Strychnos spinosa Lam. ●☆

378735 Strychnos goetzei Gilg = Strychnos cocculoides Baker ●☆

378736 Strychnos gonioides P. A. Duvign. = Strychnos johnsonii Hutch. et M. B. Moss ●☆

378737 Strychnos gossweileri Exell;戈斯马钱●☆

378738 Strychnos gracillima Gilg = Strychnos spinosa Lam. ●☆

378739 Strychnos gracillima Gilg var. paucispinosa De Wild. = Strychnos spinosa Lam. ●☆

378740 Strychnos greveana Baill. ex Pernet = Strychnos decussata (Pappe) Gilg ●☆

378741 Strychnos guerkeana Gilg = Strychnos panganensis Gilg ●☆

378742 Strychnos guineensis Schumach. et Thonn. ex Didr. = Ancylobotrys scandens (Schumach. et Thonn.) Pichon ●☆

378743 Strychnos hainanensis Merr. et Chun = Strychnos ignatii Berger ●

378744 Strychnos hankei H. J. P. Winkl. ex Guinea = Strychnos malacoclados C. H. Wright ●☆

378745 Strychnos harmsii Gilg et Busse = Strychnos spinosa Lam. ●☆

378746 Strychnos henningsii Gilg;亨氏马钱(硬梨木);Umnonono ●☆

378747 Strychnos henriquesiana Baker = Strychnos pungens Soler. ●☆

378748 Strychnos henriquesiana Gilg = Strychnos floribunda Gilg ●☆

378749 Strychnos henryi Merr. et Yamam. ex Yamam. = Strychnos cathayensis Merr. ●

378750 Strychnos heterodoxa Gilg = Strychnos potatorum L. f. ●☆

378751 Strychnos hippocrateoides Gilg = Strychnos angolensis Gilg ●☆

378752 Strychnos hirsutostylosa De Wild. = Strychnos densiflora Baill. ●☆

378753 Strychnos holstii Gilg = Strychnos henningsii Gilg ●☆

378754 Strychnos holstii Gilg f. condensata P. A. Duvign. = Strychnos henningsii Gilg ●☆

378755 Strychnos holstii Gilg f. laxiuscula P. A. Duvign. = Strychnos henningsii Gilg ●☆

378756 Strychnos holstii Gilg var. procera (Gilg et Busse) P. A. Duvign. = Strychnos henningsii Gilg ●☆

378757 Strychnos holstii Gilg var. reticulata (Burt-Davy et Honore) J. Duvign. = Strychnos henningsii Gilg ●☆

378758 Strychnos huillensis Gilg et Busse = Strychnos innocua Delile ●☆

378759 Strychnos icaja Baill. = Strychnos icaya Baill. ●☆

378760 Strychnos icaya Baill. ;伊卡亚马钱●☆

378761 Strychnos ignatii Berger;吕宋果(宝豆、海南马钱、解热豆、金马长子,苦果,苦果子,吕宋豆,椭圆叶马钱,云海马钱);Hainan Poisonnut,Hainan Strychnos,Ignat Poisonnut,Ignatius Bean,Luzon Fruit,Luzon Strychnos,Saint Ignatius' Bean,St. Ignatius Poison Nut, St. Ignatius Poison-nut,Upas Climber ●

378762 Strychnos imbricata A. W. Hill ex P. A. Duvign. = Strychnos nigritana Baker ●☆

378763 Strychnos innocua Delile;无害马钱●☆

378764 Strychnos innocua Delile subsp. burtonii (Baker) E. A. Bruce et Lewis = Strychnos madagascariensis Poir. ●☆

378765 Strychnos innocua Delile subsp. dysophylla (Benth.) I. Verd. = Strychnos madagascariensis Poir. ●☆

378766 Strychnos innocua Delile subsp. gerrardii (N. E. Br.) I. Verd. = Strychnos gerrardii N. E. Br. ●☆

378767 Strychnos innocua Delile var. glabra E. A. Bruce et Lewis = Strychnos madagascariensis Poir. ●☆

378768 Strychnos innocua Delile var. pubescens Soler. = Strychnos innocua Delile ●☆

378769 Strychnos isabellina Gilg = Strychnos tricalysioides Hutch. et M. B. Moss ●☆

378770 Strychnos johnsonii Hutch. et M. B. Moss;约翰斯顿马钱●☆

378771 Strychnos jollyana Pierre ex A. Chev. = Strychnos soubrensis Hutch. et Dalziel ●☆

378772 Strychnos kasengaensis De Wild. ;卡森加马钱●☆

378773 Strychnos kerrii A. W. Hill = Strychnos nitida G. Don ●

378774 Strychnos kipapa Gilg = Strychnos icaja Baill. ●☆

378775 Strychnos kongofera Thoms = Strychnos innocua Delile ●☆

378776 Strychnos lacourtiana De Wild. = Strychnos angolensis Gilg ●☆

378777 Strychnos laxa Soler. = Strychnos spinosa Lam. ●☆

378778 Strychnos lecomtei A. Chev. ex Hutch. et Dalziel = Strychnos congolana Gilg ●☆

378779 Strychnos leiocarpa Gilg et Busse = Strychnos madagascariensis Poir. ●☆

378780 Strychnos leiosepala Gilg et Busse = Strychnos spinosa Lam. ●☆

378781 Strychnos lethalis Barb. Rodr. ;毒马钱●☆

378782 Strychnos ligustroides Gossw. et Mendonça = Strychnos henningsii Gilg ●☆

378783 Strychnos likimiensis De Wild. = Strychnos angolensis Gilg ●☆

378784 Strychnos limbogeton H. J. P. Winkl. = Strychnos malacoclados C. H. Wright ●☆

378785 Strychnos littoralis A. Chev. ex Hutch. et Dalziel = Strychnos floribunda Gilg ●☆

378786 Strychnos loandensis Baker = Strychnos lucens Baker ●☆

378787 Strychnos lokua A. Rich. = Strychnos spinosa Lam. ●☆

378788 Strychnos longicaudata Gilg;长尾马钱●☆

378789 Strychnos longicaudata Gilg var. niamniamensis ? = Strychnos longicaudata Gilg ●☆

378790 Strychnos lucens Baker;光亮马钱●☆

378791 Strychnos lucida R. Br. ;腺叶马钱●

378792 Strychnos macrorhiza Pierre ex P. A. Duvign. = Strychnos tricalysioides Hutch. et M. B. Moss ●☆

378793 Strychnos madagascariensis Poir. ;马岛马钱●☆

378794 Strychnos malaccensis Benth. ;马六甲马钱●☆

378795 Strychnos malaccensis Benth. = Strychnos wallichiana Steud. ex A. DC. ●◇

378796 Strychnos malacoclados C. H. Wright;软枝马钱●☆

378797 Strychnos malchairii De Wild. ;马尔谢里马钱●☆

378798 Strychnos malifolia Baker = Strychnos floribunda Gilg ●☆

378799 Strychnos marquesii Baker = Strychnos floribunda Gilg ●☆

378800 Strychnos martreti A. Chev. = Strychnos nigritana Baker ●☆

378801 Strychnos matopensis S. Moore;马托马钱●☆

378802 Strychnos medeola Sagot ex Progel;美洲马钱●☆

378803 Strychnos megalocarpa Gilg et Busse = Strychnos spinosa Lam. ●☆

378804 Strychnos melastomatoides Gilg;野牡丹马钱●☆

378805 Strychnos melinoniana Baill. ;梅林马钱●☆

378806 Strychnos mellodora S. Moore;蜜味马钱●☆

378807 Strychnos melonicarpa Gilg et Busse = Strychnos madagascariensis Poir. ●☆

378808 Strychnos memecyloides S. Moore;谷木马钱●☆

378809 Strychnos memecyloides S. Moore var. effusior ? = Strychnos memecyloides S. Moore ●☆

378810 Strychnos micans S. Moore = Strychnos usambarensis Gilg ex Engl. ●☆

378811 Strychnos microcarpa Baker = Strychnos floribunda Gilg ●☆

378812 Strychnos mildbraedii Gilg = Strychnos icaja Baill. ●☆

378813 Strychnos milneredheadii P. A. Duvign. et Staquet = Strychnos lucens Baker ●☆

378814 Strychnos mimfiensis Gilg ex Leeuwenb. ;明菲马钱●☆

378815 Strychnos miniungansamba Gilg = Strychnos spinosa Lam. ●☆

378816　Strychnos minor Dennst. ;小马钱;Snake Wood ●☆

378817　Strychnos mitis S. Moore;柔软马钱●☆

378818　Strychnos moandaensis De Wild. ;莫安达马钱●☆

378819　Strychnos mocquerysii Aug. DC. = Strychnos madagascariensis Poir. ●☆

378820　Strychnos moloneyi Baker = Strychnos floribunda Gilg ●☆

378821　Strychnos mongonda De Wild. = Strychnos angolensis Gilg ●☆

378822　Strychnos mortehanii De Wild. = Strychnos aculeata Soler. ●☆

378823　Strychnos mostueoides Leeuwenb. ;拟摩斯马钱●☆

378824　Strychnos mueghe Chiov. = Strychnos spinosa Lam. ●☆

378825　Strychnos myrcioides S. Moore = Strychnos henningsii Gilg ●☆

378826　Strychnos myrtoides Gilg et Busse;香桃木马钱●☆

378827　Strychnos nauphylla P. A. Duvign. = Strychnos angolensis Gilg ●☆

378828　Strychnos ndengensis Pellegr. ;恩登马钱●☆

378829　Strychnos ngouniensis Pellegr. ;恩古涅马钱●☆

378830　Strychnos niamniamensis（Gilg）Gilg = Strychnos longicaudata Gilg ●☆

378831　Strychnos nigritana Baker;尼格里塔马钱●☆

378832　Strychnos nigrovillosa De Wild. = Strychnos longicaudata Gilg ●☆

378833　Strychnos nitida G. Don;毛柱马钱（车里马钱,滇南马钱,马钱子,云南马钱）;Hairstyle Poisonnut, S. Yunnan Poisonnut, Shining Poisonnut,Shiny Strychnos,South Yunnan Strychnos ●

378834　Strychnos nux-blanda A. W. Hill;山马钱（柔毛番木鳖,柔毛马钱）;Mountain Poisonnut, Mountain Strychnos, Nux-vomica Poison Nut,Wild Poisonnut ●

378835　Strychnos nux-blanda A. W. Hill var. hirsuta Hill. = Strychnos nux-blanda A. W. Hill ●

378836　Strychnos nux-vomica L. ;马钱子（大方八,番木鳖,火失刻把都,苦实,苦实巴豆,苦实把豆儿,马前,马钱,马钱木,马钱树,牛眼,印度马钱）;Koochla Tree, Nux Vomica, Nux-vomica, Nuxvomica Poisonnut, Nux-vomica Poison-nut, Nux-vomica Tree, Poison Nut, Poisonnut, Quaker Buttons, Snake Wood, Snakewood, Strychnine, Strychnine Tree,Strychnos ●

378837　Strychnos nux-vomica L. var. grandifolia Dop = Strychnos nux-blanda A. W. Hill ●

378838　Strychnos nux-vomica L. var. oligosperma Dop = Strychnos nux-vomica L. ●

378839　Strychnos oblongifolia Hochst. = Acokanthera oblongifolia （Hochst. ）Codd ●☆

378840　Strychnos occidentalis Soler. = Strychnos pungens Soler. ●☆

378841　Strychnos odorata A. Chev. ;芳香马钱●☆

378842　Strychnos omphalocarpa Gilg et Busse = Strychnos spinosa Lam. ●☆

378843　Strychnos ovalifolia Wall. = Strychnos ignatii Berger ●

378844　Strychnos ovalifolia Wall. ex G. Don = Strychnos ignatii Berger ●

378845　Strychnos ovata A. W. Hill;卵形马钱（密花马钱）;Denseflower Poisonnut,Denseflower Strychnos,Ovate Poisonnut,Ovate Strychnos ●

378846　Strychnos pachyphylla Gilg et Busse = Strychnos madagascariensis Poir. ●☆

378847　Strychnos panganensis Gilg;潘甘马钱●☆

378848　Strychnos paniculata Champ. ex Benth. = Strychnos umbellata （Lour. ）Merr. ●

378849　Strychnos pansa S. Moore = Strychnos malacoclados C. H. Wright ●☆

378850　Strychnos paralleloneura Gilg et Busse = Strychnos cocculoides Baker ●☆

378851　Strychnos pauciflora Gilg = Strychnos henningsii Gilg ●☆

378852　Strychnos penduliflora Baker = Strychnos innocua Delile ●☆

378853　Strychnos penninervis A. Chev. ;羽脉马钱●☆

378854　Strychnos pentantha Leeuwenb. ;五花马钱●☆

378855　Strychnos phaeopoda Gilg ex De Wild. = Strychnos ngouniensis Pellegr. ●☆

378856　Strychnos phaeotricha Gilg;褐毛马钱●☆

378857　Strychnos pierriana A. W. Hill = Strychnos wallichiana Steud. ex A. DC. ●◇

378858　Strychnos pluvialis A. Chev. = Strychnos spinosa Lam. ●☆

378859　Strychnos polyphylla Gilg et Busse = Strychnos madagascariensis Poir. ●☆

378860　Strychnos potatorum L. f. ;马铃马钱（马铃番木鳖,饮粉马钱子）;Clearing Nut,Kataka,Water-filter Nut ●☆

378861　Strychnos procera Gilg et Busse = Strychnos henningsii Gilg ●☆

378862　Strychnos pseudojollyana A. Chev. = Strychnos aculeata Soler. ●☆

378863　Strychnos pseudoquina A. St. -Hil. ;假奎马钱（假金鸡纳马钱,假金马钱）●☆

378864　Strychnos pungens Gagnep. = Carissa spinarum L. ●

378865　Strychnos pungens Soler. ;刺马钱●☆

378866　Strychnos pusilliflora S. Moore = Strychnos icaja Baill. ●☆

378867　Strychnos quadrangularis Mildbr. = Strychnos johnsonii Hutch. et M. B. Moss ●☆

378868　Strychnos quaqua Gilg = Strychnos madagascariensis Poir. ●☆

378869　Strychnos radiosperma Gilg et Busse = Strychnos spinosa Lam. ●☆

378870　Strychnos randiaeformis Baill. = Strychnos madagascariensis Poir. ●☆

378871　Strychnos reticulata Burtt Davy et Honoré = Strychnos henningsii Gilg ●☆

378872　Strychnos retinervis Leeuwenb. ;网脉马钱●☆

378873　Strychnos reygartii De Wild. = Strychnos talbotiae S. Moore ●☆

378874　Strychnos rheedei C. B. Clarke = Strychnos wallichiana Steud. ex A. DC. ◇

378875　Strychnos rhombifolia Gilg et Busse = Strychnos spinosa Lam. ●☆

378876　Strychnos samba P. A. Duvign. ;桑巴马钱●☆

378877　Strychnos sansibariensis Gilg = Strychnos spinosa Lam. ●☆

378878　Strychnos sapini De Wild. = Strychnos pungens Soler. ●☆

378879　Strychnos scaberrima Gilg ex Pellegr. = Strychnos phaeotricha Gilg ●☆

378880　Strychnos scandens Schumach. et Thonn. = Ancylobotrys scandens （Schumach. et Thonn. ）Pichon ●☆

378881　Strychnos scheffleri Gilg;谢夫勒马钱●☆

378882　Strychnos scheffleri Gilg var. expansa E. A. Bruce = Strychnos scheffleri Gilg ●☆

378883　Strychnos schumanniana Gilg = Strychnos cocculoides Baker ●☆

378884　Strychnos schweinfurthii Gilg = Strychnos spinosa Lam. ●☆

378885　Strychnos sennensis Baker = Strychnos henningsii Gilg ●☆

378886　Strychnos simiarum （Hochst. ）Gilg ex A. Chev. = Strychnos innocua Delile ●☆

378887　Strychnos soubrensis Hutch. et Dalziel;苏布雷马钱●☆

378888　Strychnos spinosa Lam. ;多刺马钱;Elephant Orange, Kaffir Orange,Monkey Apple,Monkey Balls,Natal Orange ●☆

378889　Strychnos spinosa Lam. subsp. volkensii （Gilg）E. A. Bruce = Strychnos spinosa Lam. ●☆

378890　Strychnos spinosa Lam. var. lokua （A. Rich. ）E. A. Bruce =

Strychnos spinosa Lam. ●☆

378891　Strychnos spinosa Lam. var. pubescens Baker = Strychnos spinosa Lam. ●☆

378892　Strychnos spireana Dop = Strychnos nux-vomica L. ●

378893　Strychnos splendens Gilg;亮马钱●☆

378894　Strychnos staudtii Gilg;施陶马钱●☆

378895　Strychnos stenoneura Gilg et Busse = Strychnos madagascariensis Poir. ●☆

378896　Strychnos stuhlmannii Gilg = Strychnos potatorum L. f. ●☆

378897　Strychnos suaveolens Gilg = Strychnos densiflora Baill. ●☆

378898　Strychnos subaquatica De Wild. = Strychnos scheffleri Gilg ●☆

378899　Strychnos suberifera Gilg et Busse = Strychnos cocculoides Baker ●☆

378900　Strychnos suberosa De Wild. = Strychnos cocculoides Baker ●☆

378901　Strychnos suberosa Sim = Strychnos cocculoides Baker ●☆

378902　Strychnos subscandens Baker = Strychnos floribunda Gilg ●☆

378903　Strychnos sumbensis Good = Strychnos scheffleri Gilg ●☆

378904　Strychnos syringiflora A. Chev. = Strychnos melastomatoides Gilg ●☆

378905　Strychnos talbotiae S. Moore;塔尔博特马钱●☆

378906　Strychnos tchibangensis Pellegr. ;奇班加马钱●☆

378907　Strychnos ternata Gilg ex Leeuwenb. ;三出马钱●☆

378908　Strychnos tetragyna Gilg et Busse var. glabrrima Hung T. Chang;河口马钱;Hekou Poisonnut,Hekou Strychnos ●

378909　Strychnos thomsiana Gilg et Busse = Strychnos cocculoides Baker ●☆

378910　Strychnos thomsiana Gilg et Busse var. elegans = Strychnos cocculoides Baker ●☆

378911　Strychnos thyrsiflora Gilg = Strychnos phaeotricha Gilg ●☆

378912　Strychnos togoensis Gilg et Busse = Strychnos nigritana Baker ●☆

378913　Strychnos tonga Gilg = Strychnos spinosa Lam. ●☆

378914　Strychnos toxifera R. H. Schomb. ex Benth. ;箭毒马钱(毒马钱,南美箭毒树);Curare ●

378915　Strychnos transiens Gilg = Strychnos congolana Gilg ●☆

378916　Strychnos tricalysioides Hutch. et M. B. Moss;三萼马钱●☆

378917　Strychnos trichoneura Leeuwenb. ;毛脉马钱●☆

378918　Strychnos triclisioides Baker = Strychnos innocua Delile ●☆

378919　Strychnos trillesiana Pierre ex A. Chev. = Strychnos dale De Wild. ●☆

378920　Strychnos tubiflora A. W. Hill = Strychnos wallichiana Steud. ex A. DC. ●◇

378921　Strychnos tuvungasala P. A. Duvign. = Strychnos angolensis Gilg ●☆

378922　Strychnos umbellata (Lour.) Merr. ;伞花马钱(牛目椒,牛目周,三脉马钱);Umbellaflower Strychnos,Umbellaflowered Poisonnut,Umbella-flowered Poisonnut ●

378923　Strychnos unguacha A. Rich. = Strychnos innocua Delile ●☆

378924　Strychnos unguacha A. Rich. var. dschurica Gilg = Strychnos innocua Delile ●☆

378925　Strychnos unguacha A. Rich. var. dysophylla (Benth.) Gilg = Strychnos madagascariensis Poir. ●☆

378926　Strychnos unguacha A. Rich. var. grandifolia Gilg = Strychnos innocua Delile ●☆

378927　Strychnos unguacha A. Rich. var. micrantha Gilg = Strychnos madagascariensis Poir. ●☆

378928　Strychnos unguacha A. Rich. var. microcarpa Gilg = Strychnos innocua Delile ●☆

378929　Strychnos unguacha A. Rich. var. obovata De Wild. = Strychnos innocua Delile ●☆

378930　Strychnos unguacha A. Rich. var. polyantha Gilg = Strychnos innocua Delile ●☆

378931　Strychnos unguacha A. Rich. var. pubescens (Soler.) Gilg = Strychnos innocua Delile ●☆

378932　Strychnos unguacha A. Rich. var. retusa Chiov. = Strychnos spinosa Lam. ●☆

378933　Strychnos unguacha A. Rich. var. steudneri Gilg = Strychnos innocua Delile ●☆

378934　Strychnos unguacha A. Rich. var. typica Gilg = Strychnos innocua Delile ●☆

378935　Strychnos urceolata Leeuwenb. ;坛状马钱●☆

378936　Strychnos usambarensis Gilg = Strychnos usambarensis Gilg ex Engl. ●☆

378937　Strychnos usambarensis Gilg ex Engl. ;乌桑巴拉山马钱●☆

378938　Strychnos usitata Pierre ex Dop var. cirrosa Dop = Strychnos angustiflora Benth. ●

378939　Strychnos vacacoua Baill. = Strychnos madagascariensis Poir. ●☆

378940　Strychnos vanderystii De Wild. = Strychnos kasengaensis De Wild. ●☆

378941　Strychnos variabilis De Wild. ;易变马钱●☆

378942　Strychnos venulosa Hutch. et M. B. Moss = Strychnos icaja Baill. ●☆

378943　Strychnos viridescens Gilg ex Mildbr. = Strychnos afzelii Gilg ●☆

378944　Strychnos viridiflora De Wild. = Strychnos congolana Gilg ●☆

378945　Strychnos vogelii Baker = Strychnos nigritana Baker ●☆

378946　Strychnos volkensii Gilg = Strychnos spinosa Lam. ●☆

378947　Strychnos wakefieldii Baker = Strychnos madagascariensis Poir. ●☆

378948　Strychnos wallichiana Steud. ex A. DC. ;长籽马钱(大方八,冬绿马钱,毒胡桃,番木鳖,方八,苦实,苦实把豆儿,马前,马钱,马钱藤,闹狗药,牛银,皮氏马钱,生马钱,尾叶马钱,云南马钱,制马钱);Snakewood,Wallich Poisonnut,Wallich Strychnos ●◇

378949　Strychnos wallichiana Steud. ex A. DC. var. intermiedia A. W. Hill = Strychnos nitida G. Don ●

378950　Strychnos wallichiana Steud. ex A. DC. var. ovata A. W. Hill = Strychnos nitida G. Don ●

378951　Strychnos welwitschii Gilg = Strychnos floribunda Gilg ●☆

378952　Strychnos xantha Leeuwenb. ;黄马钱●☆

378953　Strychnos xerophila Baker = Strychnos innocua Delile ●☆

378954　Strychnos xylophylla Gilg;木叶马钱●☆

378955　Strychnos yunnanensis S. Y. Pao = Strychnos nitida G. Don ●

378956　Strychnos zenkeri Gilg ex Baker;岑克尔马钱●☆

378957　Strychnos zizyphoides Baker = Strychnos afzelii Gilg ●☆

378958　Strychnus Post et Kuntze = Strychnos L. ●

378959　Stryphnodendron Mart. (1837);涩树属;Alum Bark Tree, Alum-bark Tree ●☆

378960　Stryphnodendron adstringens (C. Mart.) Coville;巴西涩树;Barbatimao Alum Bark Tree,Barbatimao Alum-bark Tree ●☆

378961　Stryphnodendron barbatiman C. Mart. = Stryphnodendron adstringens (C. Mart.) Coville ●☆

378962　Stryphnodendron coriaceum Benth. ;草质涩树●☆

378963　Stryphnodendron discolor Benth. ;异色涩树●☆

378964　Stryphnodendron floribundum Benth. ;多花涩树●☆

378965 Stryphnodendron guianense Benth. ;圭亚那涩树●☆

378966 Stryphnodendron microstachyum Poepp. et Endl. ;小穗花涩树 ●☆

378967 Stryphnodendron obovatum Benth. ;倒卵涩树●☆

378968 Stryphnodendron polystachyum Kleinh. ;多穗涩树●☆

378969 Stryphnodendron purpureum Ducke;紫涩树●☆

378970 Strzeleckya F. Muell. = Flindersia R. Br. ●

378971 Stuartia L'Hér. = Stewartia L. ●

378972 Stuartia monadelpha Siebold et Zucc. = Stewartia monadelpha Siebold et Zucc. ●☆

378973 Stuartina Sond. (1853);无冠紫绒草属■☆

378974 Stuartina hamata Philipson;澳洲无冠紫绒草■☆

378975 Stuartina muelleri Sond. ;无冠紫绒草■☆

378976 Stubendorffia Schrenk ex Fisch. , C. A. Mey. et Avé-Lall. (1844);施图芥属(斯图芥属)■☆

378977 Stubendorffia aptera Lipsky;无翅施图芥■☆

378978 Stubendorffia lipskyi Busch;离氏施图芥■☆

378979 Stubendorffia orientalis Schrenk ex Fisch. ;东方施图芥■☆

378980 Stubendorffia subdidyma N. Busch;施图芥■☆

378981 Stuckenia Börner = Potamogeton L. ■

378982 Stuckenia Börner(1912);施图肯草属;Potamot■☆

378983 Stuckenia amblyphylla (C. A. Mey.) Holub;钝叶菹草; Obtuseleaf Curly Pondweed ■

378984 Stuckenia filiformis (Pers. ex Nolte) Börner = Potamogeton filiformis Pers. ■

378985 Stuckenia filiformis (Pers. ex Nolte) Börner = Stuckenia filiformis (Pers.) Börner ■

378986 Stuckenia filiformis (Pers.) Börner;丝叶眼子菜(纤眼子菜); Filiform Pondweed, Fine-leaved Pondweed, Silkleaf Pondweed, Slender-leaved Pondweed, Thread-leaf Pondweed, Thread-leaved Pondweed ■

378987 Stuckenia filiformis (Pers.) Börner = Potamogeton filiformis Pers. ■

378988 Stuckenia filiformis (Pers.) Börner subsp. alpina (Blytt) R. R. Haynes,Les et Král;高山丝叶眼子菜(高山施图肯草); Fine-leaved Pondweed, Thread-leaf Pondweed, Thread-leaved Pondweed ■☆

378989 Stuckenia filiformis (Pers.) Börner subsp. occidentalis (J. W. Robbins) R. R. Haynes,Les et Král;西部丝叶眼子菜; Fine-leaved Pondweed,Thread-leaf Pondweed,Thread-leaved Pondweed ■☆

378990 Stuckenia pamirica (Baagoe) Z. Kaplan;长鞘菹草■☆

378991 Stuckenia pectinata (L.) Börner;篦齿眼子菜(红线草,红线儿菹,龙须草,龙须眼子菜,马尾巴草,松毛草,酸水草,线形眼子菜,眼子菜); Comb Pondweed, Fennel Pondweed, Fennelleaf Pondweed, Fennel-leaved Pondweed, Labyrinth Pondweed, Pectinate Pondweed,Sago Pondweed,Sago-pondweed ■

378992 Stuckenia pectinata (L.) Börner = Potamogeton pectinatus L. ■

378993 Stuckenia striata (Ruiz et Pav.) Holub;条纹施图肯草; Nevada-pondweed ■☆

378994 Stuckenia vaginata (Turcz.) Holub;大鞘施图肯草;Big-sheath Pondweed,Sheathed Pondweed ■☆

378995 Stuckertia Kuntze(1903);施图萝藦属●☆

378996 Stuckertia stuckertiana Kuntze;施图萝藦☆

378997 Stuckertiella Beauverd(1913);联冠紫绒草属■☆

378998 Stuckertiella capitata Beauverd;头状联冠紫绒草■☆

378999 Stuckertiella peregrina Beauverd;联冠紫绒草■☆

379000 Stuebelia Pax = Belencita H. Karst. ●☆

379001 Stuessya B. L. Turner = Stuessya B. L. Turner et F. G. Davies ■●☆

379002 Stuessya B. L. Turner et F. G. Davies(1980);芒苞菊属■●☆

379003 Stuessya michoacana B. L. Turner et F. Davies;芒苞菊■●☆

379004 Stuhlmannia Taub. (1895);斯图云实属(东非云实属)●■☆

379005 Stuhlmannia moavi Taub. ;东非云实●☆

379006 Stultitia E. Phillips = Orbea Haw. ■☆

379007 Stultitia E. Phillips(1933);神鹿殿属■☆

379008 Stultitia conjuncta A. C. White et B. Sloane = Orbea conjuncta (A. C. White et B. Sloane) Bruyns ■☆

379009 Stultitia cooperi (N. E. Br.) E. Phillips = Orbea cooperi (N. E. Br.) L. C. Leach ■☆

379010 Stultitia hardyi R. A. Dyer = Orbea hardyi (R. A. Dyer) Bruyns ■☆

379011 Stultitia miscella (N. E. Br.) C. A. Lückh. = Orbea miscella (N. E. Br.) Meve ■☆

379012 Stultitia paradoxa I. Verd. = Orbea paradoxa (I. Verd.) L. C. Leach ■☆

379013 Stultitia tapscottii (I. Verd.) E. Phillips;神鹿殿(神鹿草)■☆

379014 Stultitia tapscottii (I. Verd.) E. Phillips = Orbea tapscottii (I. Verd.) L. C. Leach ■☆

379015 Stultitia umbracula M. D. Hend. = Orbea umbracula (M. D. Hend.) L. C. Leach ■☆

379016 Stupa Asch. = Stipa L. ■

379017 Sturmia C. F. Gaertn. = Antirhea Comm. ex Juss. ●

379018 Sturmia Hoppe = Mibora Adans. ■☆

379019 Sturmia Rchb. = Liparis Rich. (保留属名)■

379020 Sturmia Rchb. f. = Liparis Rich. (保留属名)■

379021 Sturmia abbyssinica (A. Rich.) Rchb. f. = Liparis abyssinica A. Rich. ■☆

379022 Sturmia bituberculata Rchb. f. = Liparis nervosa (Thunb. ex A. Murray) Lindl. ■

379023 Sturmia capensis (Lindl.) Sond. = Liparis capensis Lindl. ■☆

379024 Sturmia longipes (Lindl.) Rchb. f. = Liparis viridiflora (Blume) Lindl. ■

379025 Sturmia nervosa (Thunb. ex A. Murray) Rchb. f. = Liparis nervosa (Thunb. ex A. Murray) Lindl. ■

379026 Sturmia nervosa (Thunb.) Rchb. f. = Liparis nervosa (Thunb.) Lindl. ■

379027 Sturmia paludosa (L.) Rchb. = Malaxis paludosa (L.) Sw. ■☆

379028 Sturtia R. Br. = Gossypium L. ●■

379029 Stussenia C. Hansen(1985);南亚野牡丹属●■

379030 Stussenia membranifolia (H. L. Li) C. Hansen;南亚野牡丹●■

379031 Stussenia membranifolia (H. L. Li) C. Hansen = Blastus membranifolius H. L. Li ●

379032 Stussenia membranifolia (H. L. Li) C. Hansen = Neodriessenia membranifolia (H. L. Li) C. Hansen ●

379033 Stutzeria F. Muell. = Pullea Schltr. ●☆

379034 Styasasia S. Moore = Asystasia Blume ●■

379035 Styasasia africana (S. Moore) S. Moore = Asystasia africana (S. Moore) C. B. Clarke ■☆

379036 Stychophyllum Phil. = Pycnophyllum J. Rémy ■☆

379037 Stygiaria Ehrh. = Juncus L. ■

379038 Stygiopsis Gand. = Juncus L. ■

379039 Stygnanthe Hanst. = Columnea L. ●■☆

379040 Stygnanthe Hanst. = Pentadenia (Planch.) Hanst. ●☆

379041　Stylago Salisb. = Strumaria Jacq. ■☆

379042　Stylagrostis Mez = Calamagrostis Adans. ■

379043　Stylagrostis Mez = Deyeuxia Clarion ■

379044　Stylandra Nutt. = Podostigma Elliott ■

379045　Stylanthus Rchb. et Zoll. = Mallotus Lour. ●

379046　Stylanthus Rchb. f. et Zoll. = Mallotus Lour. ●

379047　Stylaptera Benth. et Hook. f. = Stylapterus A. Juss. ●☆

379048　Stylapterus A. Juss. (1846); 翼柱管萼木属 ●☆

379049　Stylapterus A. Juss. = Penaea L. ●☆

379050　Stylapterus barbatus A. Juss. ; 髯毛翼柱管萼木 ●☆

379051　Stylapterus candolleanus (Stephens) R. Dahlgren; 康氏翼柱管萼木 ●☆

379052　Stylapterus dubius (Stephens) R. Dahlgren; 可疑翼柱管萼木 ●☆

379053　Stylapterus ericifolius (A. Juss.) R. Dahlgren; 毛叶翼柱管萼木 ●☆

379054　Stylapterus ericoides A. Juss. ; 石南状翼柱管萼木 ●☆

379055　Stylapterus ericoides A. Juss. subsp. pallidus R. Dahlgren; 苍白石南状翼柱管萼木 ●☆

379056　Stylapterus fruticulosus (L. f.) A. Juss. ; 灌木翼柱管萼木 ●☆

379057　Stylapterus micranthus R. Dahlgren; 小花翼柱管萼木 ●☆

379058　Stylapterus sulcatus R. Dahlgren; 纵沟翼柱管萼木 ●☆

379059　Stylarthropus Baill. = Whitfieldia Hook. ■☆

379060　Stylarthropus brazzei Baill. = Whitfieldia brazzae (Baill.) C. B. Clarke ■☆

379061　Stylarthropus laurentii Lindau = Whitfieldia laurentii (Lindau) C. B. Clarke ■☆

379062　Stylarthropus preussii Lindau = Whitfieldia preussii (Lindau) C. B. Clarke ■☆

379063　Stylarthropus stuhlmannii Lindau = Whitfieldia stuhlmannii (Lindau) C. B. Clarke ■☆

379064　Stylarthropus tenuiflora Baill. = Whitfieldia brazzae (Baill.) C. B. Clarke ■☆

379065　Stylarthropus thollonii Baill. = Whitfieldia thollonii (Baill.) Benoist ■☆

379066　Stylbocarpa Decne. et Planch. = Stilbocarpa (Hook. f.) Decne. et Planch. ●☆

379067　Styledium Andrews = Stylidium Sw. ex Willd. (保留属名) ■

379068　Stylesia Nutt. = Bahia Lag. ■☆

379069　Styleurodon Raf. = Stylodon Raf. ■☆

379070　Stylexia Raf. = Caylusea A. St. -Hil. (保留属名) ■☆

379071　Stylidiaceae R. Br. (1810)(保留科名); 花柱草科(丝滴草科); Stylegrass Family, Stylidium Family ●■

379072　Stylidium Lour. (废弃属名) = Alangium Lam. (保留属名) ●

379073　Stylidium Lour. (废弃属名) = Pautsauvia Juss. ●

379074　Stylidium Lour. (废弃属名) = Stylidium Sw. ex Willd. (保留属名) ■

379075　Stylidium Sw. = Stylidium Sw. ex Willd. (保留属名) ■

379076　Stylidium Sw. ex Willd. (1790)(保留属名); 花柱草属(丝滴草属); Stylegrass, Stylewort, Stylidium, Trigger Plant, Triggerplant ■

379077　Stylidium bulbiferum Benth. ; 珠芽花柱草; Circus Triggerplant ■☆

379078　Stylidium chinense Lour. = Alangium chinense (Lour.) Harms ●

379079　Stylidium graminifolium Sw. ex Willd. ; 禾叶花柱草; Grass Triggerplant, Trigger Plant, Trigger Stylewort ■☆

379080　Stylidium sinicum Hance = Stylidium uliginosum Sw. ■

379081　Stylidium tenellum Sw. ; 狭叶花柱草(柔丝滴草); Narrowleaf Stylegrass, Narrowleaf Stylidium ■

379082　Stylidium uliginosum Sw. ; 花柱草(红口锁); Marshy Stylegrass, Marshy Stylidium ■

379083　Stylimnus Raf. = Pluchea Cass. ●■

379084　Stylipus Raf. = Geum L. ■

379085　Stylipus Raf. = Stylypus Raf. ■

379086　Stylis Poir. = Alangium Lam. (保留属名) ●

379087　Stylis Poir. = Stylidium Lour. (废弃属名) ■

379088　Stylis chinensis (Lour.) Poir. = Alangium chinense (Lour.) Harms ●

379089　Stylis chinensis Poir. = Alangium chinense (Lour.) Harms ●

379090　Stylisma Raf. (1818); 尖柱旋花属; Stylisma ■☆

379091　Stylisma pickeringii (Torr.) A. Gray; 尖柱旋花; Stylisma ■☆

379092　Stylismus Spach = Stylisma Raf. ■☆

379093　Stylista Raf. = Cleome L. ●■

379094　Stylobasiaceae J. Agardh = Surianaceae Arn. (保留科名) ●

379095　Stylobasiaceae J. Agardh; 过柱花科 ●☆

379096　Stylobasium Desf. (1819); 过柱花属 ●☆

379097　Stylobasium lineare Nees; 线叶过柱花 ●☆

379098　Stylobasium spathulatum Desf. ; 过柱花 ●☆

379099　Stylocarpum Noulet = Rapistrum Crantz (保留属名) ■☆

379100　Styloceras A. Juss. = Styloceras Kunth ex A. Juss. ●☆

379101　Styloceras Kunth ex A. Juss. (1824); 尖角黄杨属(柱角木属) ●☆

379102　Styloceras kunthianum A. Juss. ; 尖角黄杨 ●☆

379103　Stylocerataceae (Pax) Baill. ex Reveal et Hoogland = Stylocerataceae Baill. ex Reveal et Hoogland ●☆

379104　Stylocerataceae (Pax) Reveal et Hoogland = Buxaceae Dumort. (保留科名) ●■

379105　Stylocerataceae Baill. = Buxaceae Dumort. (保留科名) ●■

379106　Stylocerataceae Baill. = Euphorbiaceae Juss. (保留科名) ●■

379107　Stylocerataceae Baill. = Stylocerataceae Baill. ex Reveal et Hoogland ●☆

379108　Stylocerataceae Baill. = Styracaceae DC. et Spreng. (保留科名) ●

379109　Stylocerataceae Baill. ex Reveal et Hoogland = Buxaceae Dumort. (保留科名) ●■

379110　Stylocerataceae Baill. ex Reveal et Hoogland = Euphorbiaceae Juss. (保留科名) ●■

379111　Stylocerataceae Baill. ex Reveal et Hoogland (1990); 尖角黄杨科 ●☆

379112　Stylocerataceae Rev. et Hoogland = Buxaceae Dumort. (保留科名) ●■

379113　Stylocerataceae Rev. et Hoogland = Euphorbiaceae Juss. (保留科名) ●■

379114　Stylocerataceae Rev. et Hoogland = Stylocerataceae Baill. ex Reveal et Hoogland ●☆

379115　Stylocerataceae Takht. ex Reveal et Hoogland = Stylocerataceae Baill. ex Reveal et Hoogland ●☆

379116　Stylochaeton Lepr. (1834); 毛柱南星属 ■☆

379117　Stylochaeton angolensis Engl. ; 安哥拉毛柱南星 ■☆

379118　Stylochaeton angustifolius Peter = Stylochaeton borumensis N. E. Br. ■☆

379119　Stylochaeton baguirmiense A. Chev. = Stylochaeton hypogaeus Lepr. ■☆

379120　Stylochaeton barteri N. E. Br. = Stylochaeton hypogaeus Lepr. ■☆

379121 Stylochaeton bogneri Mayo;博格纳毛柱南星■☆

379122 Stylochaeton borumensis N. E. Br.;博鲁姆南星■☆

379123 Stylochaeton chevalieri Engl. = Stylochaeton lancifolius Kotschy et Peyr. ■☆

379124 Stylochaeton crassispathus Bogner;厚苞毛柱南星■☆

379125 Stylochaeton dalzielii N. E. Br. = Stylochaeton lancifolius Kotschy et Peyr. ■☆

379126 Stylochaeton euryphyllus Mildbr.;宽叶毛柱南星■☆

379127 Stylochaeton fissus Peter = Anchomanes boehmii Engl. ■☆

379128 Stylochaeton gabonicus N. E. Br. = Stylochaeton zenkeri Engl. ■☆

379129 Stylochaeton gazensis Rendle = Stylochaeton natalensis Schott ■☆

379130 Stylochaeton grandis N. E. Br.;大毛柱南星■☆

379131 Stylochaeton hennigii Engl. = Stylochaeton natalensis Schott ■☆

379132 Stylochaeton heterophyllus Peter = Stylochaeton borumensis N. E. Br. ■☆

379133 Stylochaeton hostiifolius Engl. = Stylochaeton lancifolius Kotschy et Peyr. ■☆

379134 Stylochaeton hypogaeus Lepr.;地下毛柱南星■☆

379135 Stylochaeton kerensis N. E. Br.;盖拉毛柱南星■☆

379136 Stylochaeton kornasii Malaisse et Bamps;科纳斯毛柱南星■☆

379137 Stylochaeton lancifolius Kotschy et Peyr.;剑叶毛柱南星■☆

379138 Stylochaeton lobatus N. E. Br. = Stylochaeton borumensis N. E. Br. ■☆

379139 Stylochaeton maximus Engl. =Stylochaeton natalensis Schott ■☆

379140 Stylochaeton milneanus Mayo;米尔恩毛柱南星■☆

379141 Stylochaeton natalensis Schott;纳塔尔毛柱南星■☆

379142 Stylochaeton obliquinervis Peter = Stylochaeton borumensis N. E. Br. ■☆

379143 Stylochaeton oligocarpus Riedl;寡果毛柱南星■☆

379144 Stylochaeton puberulus N. E. Br.;微毛毛柱南星■☆

379145 Stylochaeton rogersii N. E. Br. = Stylochaeton borumensis N. E. Br. ■☆

379146 Stylochaeton shabaensis Malaisse et Bamps;沙巴毛柱南星■☆

379147 Stylochaeton similis N. E. Br. =Stylochaeton hypogaeus Lepr. ■☆

379148 Stylochaeton tenuinervis Peter =Stylochaeton natalensis Schott ■☆

379149 Stylochaeton tubulosus Peter = Stylochaeton borumensis N. E. Br. ■☆

379150 Stylochaeton warneckei Engl. = Stylochaeton lancifolius Kotschy et Peyr. ■☆

379151 Stylochaeton zenkeri Engl.;岑克尔毛柱南星■☆

379152 Stylochiton Lepr. = Stylochaeton Lepr. ■☆

379153 Stylochiton Schott = Stylochaeton Lepr. ■☆

379154 Stylocline Nutt. (1840);筑巢草属;Neststraw ■☆

379155 Stylocline Nutt. = Cymbolaena Smoljan. ☆

379156 Stylocline acaulis Kellogg = Hesperevax acaulis (Kellogg) Greene ■☆

379157 Stylocline amphibola (A. Gray) J. T. Howell = Micropus amphibolus A. Gray ■☆

379158 Stylocline citroleum Morefield.;油筑巢草;Oil Neststraw ■☆

379159 Stylocline filaginea (A. Gray) A. Gray = Ancistrocarphus filagineus A. Gray ■☆

379160 Stylocline gnaphaloides Nutt.;山地筑巢草;Everlasting Neststraw,Mountain Neststraw ■☆

379161 Stylocline griffithii A. Gray = Cymbolaena griffithii (A. Gray) Wagenitz ■☆

379162 Stylocline intertexta Morefield.;莫哈韦筑巢草;Mojave Neststraw,Morefield Neststraw ■☆

379163 Stylocline masonii Morefield.;梅森筑巢草;Mason Neststraw ■☆

379164 Stylocline micropoides A. Gray;毛头筑巢草;Desert Fanbract,Woollyhead Fanbract ■☆

379165 Stylocline psilocarphoides M. Peck;裸筑巢草;Baretwig Neststraw,Peck Neststraw ■☆

379166 Stylocline sonorensis Wiggins;索诺拉筑巢草;Mesquite Neststraw,Sonoran Neststraw ■☆

379167 Styloconus Baill. = Blancoa Lindl. ■☆

379168 Stylocoryna Cav. = Aidia Lour. ●

379169 Stylocoryna Cav. = Tarenna Gaertn. ●

379170 Stylocoryna attenuata Voigt =Tarenna attenuata (Voigt) Hutch. ●

379171 Stylocoryna barbertonensis Bremek. = Coptosperma supra-axillare (Hemsl.) Degreef ●☆

379172 Stylocoryna conferta Benth. =Tarenna conferta (Benth.) Hiern ●☆

379173 Stylocoryna grandiflora Benth. = Tarenna grandiflora (Benth.) Hiern ●☆

379174 Stylocoryna mollissima Walp. = Tarenna mollissima (Hook. et Arn.) Rob. ●

379175 Stylocoryna neurophylla (S. Moore) Bremek. = Coptosperma neurophyllum (S. Moore) Degreef ●☆

379176 Stylocoryna nitidula Benth. =Tarenna nitidula (Benth.) Hiern ●☆

379177 Stylocoryne Wight et Arn. = Aidia Lour. ●

379178 Stylodiscus Benn. = Bischofia Blume ●

379179 Stylodiscus trifoliatus Benn. = Bischofia javanica Blume ●

379180 Stylodon Raf. (1825);柱齿马鞭草属■☆

379181 Stylodon Raf. = Verbena L. ■●

379182 Stylodon carneus (Medik.) Moldenke;柱齿马鞭草■☆

379183 Styloglossura Breda = Calanthe R. Br. (保留属名)■

379184 Stylogyne A. DC. (1841);柱蕊紫金牛属●☆

379185 Stylogyne amazonica Mez;亚马逊柱蕊紫金牛●☆

379186 Stylogyne atra Mez;暗柱蕊紫金牛●☆

379187 Stylogyne brasiliensis (A. DC.) Mez;巴西柱蕊紫金牛●☆

379188 Stylogyne cauliflora (Mart. et Miq.) Mez;茎花柱蕊紫金牛●☆

379189 Stylogyne laevis (Oerst.) Mez;平滑柱蕊紫金牛●☆

379190 Stylogyne lateriflora (Sw.) Mez;侧花柱蕊紫金牛●☆

379191 Stylogyne latipes Imkhan.;粗梗柱蕊紫金牛●☆

379192 Stylogyne laxiflora Mez;疏花柱蕊紫金牛●☆

379193 Stylogyne leptantha (Miq.) Mez;细花柱蕊紫金牛●☆

379194 Stylogyne micrantha (Kunth) Mez;小花柱蕊紫金牛●☆

379195 Stylogyne micrantha Mez = Stylogyne leptantha (Miq.) Mez ●☆

379196 Stylogyne nigricans Mez;黑柱蕊紫金牛●☆

379197 Stylogyne pauciflora Mez;少花柱蕊紫金牛●☆

379198 Stylogyne perpunctata Lundell;紫斑柱蕊紫金牛●☆

379199 Stylogyne sordida Mez;污浊柱蕊紫金牛●☆

379200 Stylogyne tenuifolia Britton;细叶柱蕊紫金牛●☆

379201 Stylogyne viridis (Lundell) Ricketson et Pipoly;绿柱蕊紫金牛●☆

379202 Stylolepis Lehm. = Podolepis Labill. (保留属名)■☆

379203 Styloma O. F. Cook = Eupritchardia Kuntze ●☆

379204 Styloma O. F. Cook = Pritchardia Seem. et H. Wendl. (保留属名)●☆

379205 Stylomecon Benth. = Hylomecon Maxim. ■

379206 Stylomecon Benth. = Stylophorum Nutt. ■

379207 Stylomecon G. Taylor(1930);火焰罂粟属(细柱罂粟属)■☆

379208 Stylomecon heterophylla G. Taylor;火焰罂粟■☆

379209　Styloncerus Labill. = Angianthus J. C. Wendl. (保留属名)■●☆

379210　Styloncerus Spreng. = Angianthus J. C. Wendl. (保留属名)■●☆

379211　Styloncerus Spreng. = Siloxerus Labill. (废弃属名)■●☆

379212　Stylonema (DC.) Kuntze = Syrenia Andrz. ex DC. ■

379213　Stylonema Kuntze = Syrenia Andrz. ex DC. ■

379214　Stylopappus Nutt. = Troximon Gaertn. ■☆

379215　Stylopappus elatus Nutt. = Agoseris elata (Nutt.) Greene ■☆

379216　Stylopappus grandiflorus Nutt. = Agoseris grandiflora (Nutt.) Greene ■☆

379217　Stylopappus laciniatus Nutt. = Agoseris elata (Nutt.) Greene ■☆

379218　Stylopappus laciniatus Nutt. var. longifolius Nutt. = Agoseris elata (Nutt.) Greene ■☆

379219　Stylophorum Nutt. (1818);金罂粟属(刺罂粟属,人血草属);Goldenpoppy, Stylophorum, Celandine Poppy ■

379220　Stylophorum diphyllum (Michx.) Nutt.;二叶金罂粟(二叶苞罂粟,金罂粟,美国金罂粟);Celandine Poppy, Celandine-poppy, Mock Poppy, Wood Poppy, Wood-poppy, Yellow Wood Poppy, Yellow-poppy ■☆

379221　Stylophorum diphyllum Nutt. = Stylophorum diphyllum (Michx.) Nutt. ■☆

379222　Stylophorum japonicum (Murray) Miq. = Hylomecon japonica (Thunb.) Prantl et Kündig ■

379223　Stylophorum japonicum (Thunb.) Prantl et Kündig var. dissectum Franch. et Sav. = Hylomecon japonica (Thunb.) Prantl et Kündig var. dissecta (Franch. et Sav.) Fedde ■

379224　Stylophorum japonicum Miq. = Hylomecon japonica (Thunb.) Prantl et Kündig ■

379225　Stylophorum japonicum Miq. var. dissectum Franch. et Sav. = Hylomecon japonica (Thunb.) Prantl et Kündig var. dissecta (Franch. et Sav.) Fedde ■

379226　Stylophorum lactucoides (Hook. f. et Thomson) Baill. = Dicranostigma lactucoides Hook. f. et Thomson ■

379227　Stylophorum lactucoides (Hook. f. et Thomson) Prain = Dicranostigma lactucoides Hook. f. et Thomson ■

379228　Stylophorum lasiocarpum (Oliv.) Fedde;金罂粟(大金盘,大金盆,大人血七,豆叶七,人血草,人血七,野人血草);Goldenpoppy, Woollyfruit Stylophorum ■

379229　Stylophorum nepalense (DC.) Spreng. = Meconopsis paniculata (D. Don) Prain ■

379230　Stylophorum nepalense (DC.) Spring = Meconopsis paniculata (D. Don) Prain ■

379231　Stylophorum ohiense Spreng. = Stylophorum diphyllum (Michx.) Nutt. ■☆

379232　Stylophorum pratense Froebel = Stylophorum diphyllum (Michx.) Nutt. ■☆

379233　Stylophorum sutchuense (Franch.) Fedde;四川金罂粟(天青地白);Sichuan Goldenpoppy, Sichuan Stylophorum ■

379234　Stylophyllum Britton et Rose = Dudleya Britton et Rose ■☆

379235　Stylopus Hook. = Geum L. ■

379236　Stylopus Hook. = Stylypus Raf. ■

379237　Stylosanthes Sw. (1788);笔花豆属(笔豆属);Penflower, Stylo, Stylosanthes ●■

379238　Stylosanthes angustifolia Vogel;狭叶笔花豆☆

379239　Stylosanthes biflora (L.) Britton, Sterns et Poggenb.;双花笔花豆(二花笔花豆);Pencil Flower, Pencilflower, Pencil-flower, Sidebark Pencil-flower ■☆

379240　Stylosanthes bojeri Vogel = Stylosanthes fruticosa (Retz.) Alston ●☆

379241　Stylosanthes erecta P. Beauv.;直立笔花豆;Nigerian Stylo ■☆

379242　Stylosanthes flavicans Baker = Stylosanthes fruticosa (Retz.) Alston ●☆

379243　Stylosanthes fruticosa (Retz.) Alston;灌木笔花豆;Shrubby Pencilflower ●☆

379244　Stylosanthes gracilis Kunth = Stylosanthes guianensis (Aubl.) Sw. ■●

379245　Stylosanthes guianensis (Aubl.) Sw.;圭亚那笔花豆(笔花豆);Brazilian Lucerne, Penflower Stylosanthes, Slender Stylosanthes, Stylo ■●

379246　Stylosanthes guineensis Schumach. et Thonn. = Stylosanthes erecta P. Beauv. ■☆

379247　Stylosanthes hamata (L.) Taub.;牙买加笔花豆■☆

379248　Stylosanthes humilis Kunth;小笔花豆;Townsville Stylo ■☆

379249　Stylosanthes longiseta Micheli;长毛笔花豆■☆

379250　Stylosanthes macrocarpa S. F. Blake;大果笔花豆■☆

379251　Stylosanthes mexicana Taub.;墨西哥笔花豆■☆

379252　Stylosanthes montevidensis Vogel;蒙得维的亚笔花豆■☆

379253　Stylosanthes mucronata Willd. = Stylosanthes fruticosa (Retz.) Alston ●☆

379254　Stylosanthes suborbiculata Chiov.;亚圆笔花豆■☆

379255　Stylosanthes sundaica Taub. = Stylosanthes humilis Kunth ■☆

379256　Stylosanthes viscosa Sw.;黏笔花豆■☆

379257　Stylosiphonia Brandegee(1914);管柱茜属☆

379258　Stylosiphonia glabra Brandegee;管柱茜☆

379259　Stylosiphonia salvadorensis Standl.;萨尔瓦多管柱茜☆

379260　Stylotrichium Mattf. (1923);毛柱柄泽兰属●☆

379261　Stylotrichium corymbosum Mattf.;毛柱柄泽兰●☆

379262　Stylotrichium rotundifolium Mattf.;圆叶毛柱柄泽兰●☆

379263　Stylurus Raf. = Ranunculus L. ■

379264　Stylurus Salisb. = Grevillea R. Br. ex Knight(保留属名)●

379265　Stylurus Salisb. ex Knight(废弃属名) = Grevillea R. Br. ex Knight(保留属名)●

379266　Stylvianthes Raf. = Stylosanthes Sw. ●■

379267　Stylypus Raf. = Geum L. ■

379268　Stylypus vernus Raf. = Geum vernum (Raf.) Torr. et A. Gray ■☆

379269　Stypa Garcke = Stipa L. ■

379270　Stypandra R. Br. (1810);粗雄花属(干花属)■☆

379271　Stypandra imbricata R. Br.;覆瓦粗雄花(覆瓦干花);Blind Grass ■☆

379272　Styphania C. Muell. = Stephania Lour. ●■

379273　Styphelia (Sol. ex G. Forst.) Sm. (1795);垂钉石南属(斯迪菲木属);Five-corners ●☆

379274　Styphelia Sm. = Styphelia (Sol. ex G. Forst.) Sm. ●☆

379275　Styphelia adscendens R. Br.;匍匐垂钉石南(匍匐斯迪菲木)●☆

379276　Styphelia behrii (Schltdl.) Sleumer;垂钉石南;Flame Heath ●☆

379277　Styphelia tubiflora Sm.;狭叶垂钉石南(狭叶斯迪菲木);Red Five-corners ●☆

379278　Styphelia viridis Andréws;绿垂钉石南(绿斯迪菲木);Green Five Corners, Green Five-corners ●☆

379279　Stypheliaceae Horan. = Epacridaceae R. Br. (保留科名)●☆

379280　Stypheliaceae Horan. = Ericaceae Juss. (保留科名)●

379281　Styphnolobium (Schott) P. C. Tsoong = Sophora L. ●■

379282　Styphnolobium Schott = Sophora L. ●■

379283　Styphnolobium Schott ex Endl. = Sophora L. ●■

379284　Styphnolobium japonicum (L.) Schott = Sophora japonica L. ●

379285　Styphnolobium japonicum（L.）Schott var. pubescens Hort. = Sophora japonica L. var. pubescens（Tausch.）Bosse ●

379286　Styphnolobium japonicum Schott var. pubescens Kirsch = Sophora japonica L. var. pubescens（Tausch.）Bosse ●

379287　Styphonia Medik. = Lavandula L. ●■

379288　Styphonia Nutt. = Rhus L. ●

379289　Styphonia Nutt. ex Torr. et A. Gray = Rhus L. ●

379290　Styphorrhiza Ehrh. = Polygonum L.（保留属名）■●

379291　Styponema Salisb. = Stypandra R. Br. ■☆

379292　Stypostylis Raf. = Geum L. ■

379293　Styppeiochloa De Winter（1966）；纤维鞘草属■☆

379294　Styppeiochloa gynoglossa（Gooss.）De Winter；纤维鞘草■☆

379295　Styppeiochloa hitchcockii（A. Camus）Cope；希契科克纤维鞘草■☆

379296　Styracaceae DC. et Spreng.（1821）（保留科名）；安息香科（齐墩果科，野茉莉科）；Storax Family ●

379297　Styracaceae Dumort. = Styracaceae DC. et Spreng.（保留科名）●

379298　Styracaceae Spreng. = Styracaceae L. C. Rich. + Ebenaceae Gürke（保留科名）●

379299　Styrandra Raf. = Maianthemum F. H. Wigg.（保留属名）■

379300　Styrax L.（1753）；安息香属（野茉莉属）；Friar's Balsam, Snowbell, Snowdrop Bush, Storax, Styrax ●

379301　Styrax agrestis（Lour.）G. Don；喙果安息香（南粤安息香，南粤野茉莉）；Field Storax ●

379302　Styrax ambiguus Seub. = Styrax pohlii A. DC. ●☆

379303　Styrax ambiguus Seub. var. apiculatus Chodat et Hassl. = Styrax pohlii A. DC. ●☆

379304　Styrax americanus Lam.；美洲安息香；American Snowbell, American Storax, Mock Orange, Snowbell, Storax ●☆

379305　Styrax americanus Lam. var. pulverulentus（Michx.）Perkins = Styrax americanus Lam. var. pulverulentus（Michx.）Rehder ●☆

379306　Styrax americanus Lam. var. pulverulentus（Michx.）Rehder；绒毛美洲安息香（星毛美国野茉莉）；Downy American Snowbell, Downy Snowbell ●☆

379307　Styrax argenteus Presl；银白安息香●☆

379308　Styrax argentifolius H. L. Li；银叶安息香；Siverleaf Snowbell, Siverleaf Storax, Siver-leaved Storax ●

379309　Styrax argyi H. Lév. = Styrax dasyanthus Perkins ●

379310　Styrax bashanensis S. Z. Qu et K. Y. Wang；巴山安息香；Bashan Snowbell ●

379311　Styrax benzoides Craib；滇南安息香（泰国安息香，暹罗安息香）；Benjamin Tree, Benzoin, Benzoin Laurel, Gum Benjamin, S. Yunnan Snowbell, South Yunnan Storax, Sumatra Snowbell, Yunnan Storax ●

379312　Styrax benzoin Dryand.；安息香（安息香树）；Benjamin Tree, Sumatra Snowbell ●

379313　Styrax biaristatus W. W. Sm. = Huodendron biaristatum（W. W. Sm.）Rehder ●

379314　Styrax bodinieri H. Lév. = Styrax japonicus Siebold et Zucc. ●

379315　Styrax bracteolatus Guillaumin = Styrax roseus Dunn ●

379316　Styrax buchtienii Sleumer = Styrax pentlandianus J. Rémy ☆

379317　Styrax burchellii Perkins = Styrax sieberi Perkins ●☆

379318　Styrax burchellii Perkins var. longifolius Perkins = Styrax ferrugineus Nees et Mart. ●☆

379319　Styrax californicus Torr.；加州安息香；California Snowbell ●☆

379320　Styrax caloneurus Perkins = Styrax suberifolius Hook. et Arn. ●

379321　Styrax calvescens Perkins；灰叶安息香（变秃安息香，灰叶野

茉莉，毛垂珠花，突变安息香）；Glabrescent Storax ●

379322　Styrax camporum Pohl；田园安息香●☆

379323　Styrax cavaleriei H. Lév. = Pterostyrax psilophyllus Diels ex Perkins ●

379324　Styrax cavaleriei H. Lév. = Styrax grandiflorus Griff. ●

379325　Styrax chinensis Hu et S. Ye Liang；中华安息香（大果安息香，大籽安息香，米哥蚊，山柿）；China Snowbell, Chinese Storax ●

379326　Styrax chrysocarpus H. L. Li；黄果安息香；Yellowfruit Snowbell, Yellowfruit Storax, Yellow-fruited Storax ●

379327　Styrax confusus Hemsl.；赛山梅（白花龙，白扣子，白山龙，高山梅，猛骨子，乌蚊子，油榨果）；Confused Storax, Muddy Snowbell, Philadelphia Snowbell ●

379328　Styrax confusus Hemsl. var. microphyllus Perkins；小叶赛山梅；Littleleaf Confused Storax, Littleleaf Muddy Snowbell ●

379329　Styrax confusus Hemsl. var. superbus（Chun）S. M. Hwang；华丽赛山梅；Elegant Confused Storax ●

379330　Styrax crotonoides C. B. Clarke；巴豆叶安息香●☆

379331　Styrax dasyanthus Perkins；垂珠花（白花树，白克马，小叶硬田螺）；Henry-flowered Storax, Roughhairflower Storax ●

379332　Styrax dasyanthus Perkins var. cinerascens Rehder = Styrax calvescens Perkins ●

379333　Styrax discolor M. F. Silva = Styrax pohlii A. DC. ●☆

379334　Styrax duclouxii Perkins = Styrax grandiflorus Griff. ●

379335　Styrax esquirolii H. Lév. = Deutzia esquirolii（H. Lév.）Rehder ●

379336　Styrax faberi Perkins；白花龙（白花朵，白花笼，白条龙，扣子柴，梦童子，棉子树，扫酒树，响铃子）；Faber Snowbell, Faber Storax, White-dragon Storax ●

379337　Styrax faberi Perkins var. acutiserratus Perkins = Styrax faberi Perkins ●

379338　Styrax faberi Perkins var. amplexifolius Chun et F. C. How ex S. M. Hwang；抱茎叶白花龙（抱茎安息香）；Amplexicaul-leaf White-dragon Storax ●

379339　Styrax faberi Perkins var. formosanus（Matsum.）S. M. Hwang = Styrax formosanus Matsum. ●

379340　Styrax faberi Perkins var. matsumurae（Perkins）S. M. Hwang = Styrax faberi Perkins var. formosanus（Matsum.）S. M. Hwang ●

379341　Styrax faberi Perkins var. matsumurae（Perkins）S. M. Hwang = Styrax matsumurae Perkins ●

379342　Styrax faberi Perkins var. matsumuraei（Perkins）S. M. Hwang = Styrax formosanus Matsum. ●

379343　Styrax ferax J. F. Macbr. = Styrax pentlandianus J. Rémy ●☆

379344　Styrax ferrugineus Nees et Mart.；锈色安息香●☆

379345　Styrax ferrugineus Pohl = Styrax ferrugineus Nees et Mart. ●☆

379346　Styrax ferrugineus Pohl var. grandifolius Perkins = Styrax ferrugineus Nees et Mart. ●☆

379347　Styrax formosanus Matsum.；台湾安息香（白树，奋起湖野茉莉，苗栗白花龙，乌鸡母，乌皮九芎，叶下白）；Formosan Snowbell, Taiwan Snowbell, Taiwan Storax ●

379348　Styrax formosanus Matsum. var. hayataianus（Perkins）H. L. Li；恒春野茉莉（台北安息香，早田安息香，早田氏红皮，早田野茉莉）；Hayata Corkleaf Storax, Hayata Storax, Hayata's Snowbell ●

379349　Styrax formosanus Matsum. var. hayataianus（Perkins）H. L. Li = Styrax suberifolius Hook. et Arn. var. hayataianus（Perkins）Mori ●

379350　Styrax formosanus Matsum. var. hirtus S. M. Hwang；长柔毛安息香；Hairy Taiwan Storax ●

379351　Styrax formosanus Matsum. var. matsumuraei（Perkins）Y. C. Liu = Styrax faberi Perkins var. formosanus（Matsum.）S. M. Hwang

379352 Styrax formosanus Matsum. var. matsumuraei（Perkins）Y. C. Liu ＝ Styrax formosanus Matsum. ●

379353 Styrax formosanus Matsum. var. matsumurai（Perkins）Y. C. Liu ＝ Styrax matsumurae Perkins ●

379354 Styrax fukiensis Mori ＝ Styrax formosanus Matsum. ●

379355 Styrax fukiensis W. W. Sm. et Jeffrey ＝ Styrax confusus Hemsl. ●

379356 Styrax funkikensis Mori ＝ Styrax formosanus Matsum. ●

379357 Styrax grandiflorus Griff.；大花安息香（大花野茉莉，兰屿安息香，兰屿野茉莉）；Bigflower Snowbell, Bigflower Storax, Big-flowered Storax, Bigleaf Snowbell, Lanyu Snowbell, Snowbell, Storax ●

379358 Styrax grandifolius Aiton；大叶安息香（大叶野茉莉，兰屿安息香，兰屿野茉莉）；Bigleaf Snowbell, Big-leafed Snowbell ● ☆

379359 Styrax guyanensis A. DC.；圭亚那安息香● ☆

379360 Styrax guyanensis A. DC. var. japurensis Seub. ＝ Styrax guyanensis A. DC. ● ☆

379361 Styrax hainanensis F. C. How；海南安息香（海南野茉莉，厚叶安息香）；Hainan Snowbell, Hainan Storax, Thick-leaf Storax ● ◇

379362 Styrax hayataianus Perkins ＝ Styrax formosanus Matsum. var. hayataianus（Perkins）H. L. Li ●

379363 Styrax hayataianus Perkins ＝ Styrax suberifolius Hook. et Arn. var. hayataianus（Perkins）Mori ●

379364 Styrax hemsleyanus Diels；老鸹铃（赫斯利野茉莉）；Hemsle Storax ●

379365 Styrax hemsleyanus Diels var. griseus Rehder ＝ Styrax hemsleyanus Diels ●

379366 Styrax henryi Perkins ＝ Styrax formosanus Matsum. ●

379367 Styrax henryi Perkins var. microcalyx Perkins ＝ Styrax formosanus Matsum. ●

379368 Styrax hookeri C. B. Clarke var. yunnanensis Perkins ＝ Styrax grandiflorus Griff. ●

379369 Styrax hookeri Perkins var. yunnanensis Perkins ＝ Styrax grandiflorus Griff. ●

379370 Styrax huanus Rehder；墨泡（麦泡）；Hu Storax ● ◇

379371 Styrax hypoglaucus Perkins ＝ Styrax tonkinensis（Pierre）Craib ex Hartwich ●

379372 Styrax iopilina Diels ＝ Styrax faberi Perkins ●

379373 Styrax iopolinus Diels ＝ Styrax dasyanthus Perkins ●

379374 Styrax japonicus Siebold et Zucc.；日本安息香（安息香，耳完桃，黑茶花，候风藤，君迁子，买子木，茉莉包，茉莉苞，木橘子，齐墩果，野花楀，野茉莉）；Japan Snowbell, Japanese Snowbell, Japanese Snowdrop Tree, Japanese Styrax, Snowbell Tree ●

379375 Styrax japonicus Siebold et Zucc. 'Carillon'；俯垂日本安息香；Weeping Snowbell ● ☆

379376 Styrax japonicus Siebold et Zucc. 'Fargesii'；法尔格斯安息香● ☆

379377 Styrax japonicus Siebold et Zucc. 'Pink Chimes'；粉钟安息香；Pink Japanese Snowbell ● ☆

379378 Styrax japonicus Siebold et Zucc. 'Snowcone'；雪球日本安息香；Snowbell ● ☆

379379 Styrax japonicus Siebold et Zucc. ＝ Styrax grandiflorus Griff. ●

379380 Styrax japonicus Siebold et Zucc. f. angustifolia（Koidz.）Sugim.；狭叶日本安息香● ☆

379381 Styrax japonicus Siebold et Zucc. f. jippei-kawakamii（Yanagita）T. Yamaz.；川上日本安息香● ☆

379382 Styrax japonicus Siebold et Zucc. f. parviflora Y. Kimura；小花日本安息香● ☆

379383 Styrax japonicus Siebold et Zucc. f. pendula T. Yamaz.；垂枝日本安息香● ☆

379384 Styrax japonicus Siebold et Zucc. f. rubicalyx Satomi；红萼日本安息香● ☆

379385 Styrax japonicus Siebold et Zucc. var. calycothrix Gilg；毛萼野茉莉；Hairycalyx Storax ●

379386 Styrax japonicus Siebold et Zucc. var. jippei-kawakamii（Yanagita）H. Hara ＝ Styrax japonicus Siebold et Zucc. f. jippei-kawakamii（Yanagita）T. Yamaz. ● ☆

379387 Styrax japonicus Siebold et Zucc. var. kotoensis（Hayata）Masam. et T. Suzuki ＝ Styrax grandiflorus Griff. ●

379388 Styrax japonicus Siebold et Zucc. var. kotoensis（Hayata）Masam. et T. Suzuki f. tomentosa（Hatus.）T. Yamaz.；毛兰屿野茉莉●

379389 Styrax japonicus Siebold et Zucc. var. kotoensis（Hayata）Masam. et T. Suzuki；兰屿野茉莉（兰屿安息香）●

379390 Styrax japonicus Siebold et Zucc. var. longipedunculata Z. Y. Zhang；长梗野茉莉；Longipedunculate Hairycalyx Storax ●

379391 Styrax japonicus Siebold et Zucc. var. nervilosa Z. Y. Zhang；毛脉野茉莉；Hairy-nerve Storax ●

379392 Styrax japonicus Siebold et Zucc. var. tomentosa Hatus. ＝ Styrax japonicus Siebold et Zucc. var. kotoensis（Hayata）Masam. et T. Suzuki f. tomentosa（Hatus.）T. Yamaz. ●

379393 Styrax juncuda Diels ＝ Styrax confusus Hemsl. ●

379394 Styrax kotoensis Hayata ＝ Styrax grandiflorus Griff. ●

379395 Styrax kotoensis Hayata ＝ Styrax japonicus Siebold et Zucc. var. kotoensis（Hayata）Masam. et Suzuki ●

379396 Styrax lacei W. W. Sm. ＝ Parastyrax lacei（W. W. Sm.）W. W. Sm. ● ◇

379397 Styrax langkongensis W. W. Sm. ＝ Styrax limprichtii Lingelsh. et Borza ●

379398 Styrax leptactinosus Cuatrec. ＝ Styrax pentlandianus J. Rémy ● ☆

379399 Styrax leveillei Fedde ex H. Lév. ＝ Pterostyrax psilophyllus Diels ex Perkins ●

379400 Styrax limprichtii Lingelsh. et Borza；楚雄安息香（楚雄野茉莉，兰贡安息香）；Langong Snowbell, Limpricht Storax ●

379401 Styrax longifolius Standl. ＝ Styrax guyanensis A. DC. ● ☆

379402 Styrax macranthus Perkins；逢春安息香（禄春安息香）；Big-flowered Styrax, Largeflowered Styrax ●

379403 Styrax macrocarpus W. C. Cheng；大果安息香；Bigfruit Snowbell, Bigfruit Storax, Big-fruited Storax ● ◇

379404 Styrax macrothyrsus Perkins；青山安息香（禄春安息香，青山安息香树）；Bigflower Snowbell, Big-thrse Styrax ●

379405 Styrax macrothyrsus Perkins ＝ Styrax tonkinensis（Pierre）Craib ex Hartwich ●

379406 Styrax mallotifolia C. Y. Wu；桐叶野茉莉；Mallotus-leaf Storax ●

379407 Styrax martii Seub.；马蒂安息香● ☆

379408 Styrax matsumurae Perkins；台湾野茉莉（苗栗白花龙，松村氏野茉莉）；Matsumura White-dragon Storax, Matsumura's Snowbell ●

379409 Styrax matsumurae Perkins ＝ Styrax faberi Perkins var. formosanus（Matsum.）S. M. Hwang ●

379410 Styrax matsumurae Perkins ＝ Styrax formosanus Matsum. ●

379411 Styrax megalocarpus Hu et S. Ye Liang；广西安息香（大果安息香，大籽安息香，山柿）；Guangxi Storax ●

379412 Styrax mollis Dunn ＝ Styrax confusus Hemsl. ●

379413 Styrax nervosus A. DC. var. elongatus Seub. ＝ Styrax ferrugineus Nees et Mart. ● ☆

379414 Styrax obassia Siebold et Zucc.；玉铃花安息香（白云木，分松子，老丹皮，老开皮，山榛子，玉铃花）；Big-leaf Storax, Big-leafed

Storax, Fragrant Snowbell, Fragrant Storax ●

379415　Styrax oblongus（Ruiz et Pav.）A. DC.；矩圆安息香●☆

379416　Styrax odoratissimus Champ. = Styrax odoratissimus Champ. ex Benth. ●

379417　Styrax odoratissimus Champ. ex Benth.；芬芳安息香（白木，芳香安息香，乳白野茉莉，野茉莉，郁香安息香，郁香野茉莉）；Sweetscented Snowbell, Sweetscented Storax, Sweet-scented Storax ●

379418　Styrax odoratissirnus Champ. = Styrax confusus Hemsl. ●

379419　Styrax officinalis Walter；药用安息香（南欧安息香，药用苏合香）；Drug Snowbell, Snowbells, Snowdrop-bush, Storax, Styrax ●☆

379420　Styrax oligophlebis Merr. ex H. L. Li = Styrax tonkinensis（Pierre）Craib ex Hartwich ●

379421　Styrax pachyphyllus Merr. et Chun = Styrax hainanensis F. C. How ●◇

379422　Styrax pachyphyllus Pilg. = Styrax pohlii A. DC. ●☆

379423　Styrax paralleloneurus Perkins；苏门答腊安息香；Sumatra Storax ●☆

379424　Styrax parviflorum Merr. = Huodendron biaristatum（W. W. Sm.）Rehder var. parviflorum（Merr.）Rehder ●

379425　Styrax pearcei Perkins var. bolivianus Perkins = Styrax sieberi Perkins ●☆

379426　Styrax pentlandianus J. Rémy；彭特兰安息香●☆

379427　Styrax perkinsiae Rehder；瓦山安息香（瑞丽安息香）；Washan Snowbell, Washan Storax ●

379428　Styrax philadelphoides Perkins = Styrax confusus Hemsl. ●

379429　Styrax philadelphoides Perkins var. superbus Chun = Styrax confusus Hemsl. var. superbus（Chun）S. M. Hwang ●

379430　Styrax philippinensis Merr. et Quisumb. = Styrax grandiflorus Griff. ●

379431　Styrax platanifolia Engelm.；梧桐叶安息香；Sycamoreleaf Snowbell ●☆

379432　Styrax platanifolia Engelm. var. stellata（Engelm.）Cory；毛梧桐叶安息香；Hairy Sycamoreleaf Snowbell ●☆

379433　Styrax pohlii A. DC.；波尔安息香●☆

379434　Styrax pohlii A. DC. f. calvescens Perkins = Styrax pohlii A. DC. ●☆

379435　Styrax polyspermus C. B. Clarke = Bruinsmia polysperma（C. B. Clarke）Steenis ●

379436　Styrax prunifolius Perkins = Styrax odoratissimus Champ. ex Benth. ●

379437　Styrax punctatus A. DC. = Styrax pohlii A. DC. ●☆

379438　Styrax reticulatus Mart.；网状安息香●☆

379439　Styrax roseus Dunn；粉花安息香（粉花野茉莉）；Roseflower Storax, Rose-flowered Storax ●

379440　Styrax rostratus Hosok. = Styrax agrestis（Lour.）G. Don ●

379441　Styrax rubifolius Guillaumin = Styrax dasyanthus Perkins ●

379442　Styrax rugosus Kurz；皱叶安息香（皱叶野茉莉）；Wrinkledleaf Storax, Wrinkle-leaved Storax ●

379443　Styrax rugosus Kurz var. formosanus Matsum. = Styrax faberi Perkins var. formosanus（Matsum.）S. M. Hwang ●

379444　Styrax rugosus Kurz var. formosanus Matsum. = Styrax formosanus Matsum. ●

379445　Styrax serrulatus Roxb. = Styrax grandiflorus Griff. ●

379446　Styrax serrulatus Roxb. = Styrax japonicus Siebold et Zucc. ●

379447　Styrax serrulatus Roxb. var. agrestis？= Styrax serrulatus Roxb. ●

379448　Styrax serrulatus Roxb. var. vesdrum Hemsl. = Styrax confusus Hemsl. ●

379449　Styrax serrulatus Wall.；齿叶安息香（白花木，滑蚁木，茉莉包，野茉莉）；Serrate-leaf Storax, Serrate-leaved Storax, Toothleaf Snowbell ●

379450　Styrax shiraianus Makino；白井安息香；Strigila Snowbell ●☆

379451　Styrax shiraianus Makino f. discolor（Nakai）Sugim.；异色白井安息香●☆

379452　Styrax shiraianus Makino var. discolor Nakai = Styrax shiraianus Makino f. discolor（Nakai）Sugim. ●☆

379453　Styrax shweliensis W. W. Sm. = Styrax perkinsiae Rehder ●

379454　Styrax sieberi Perkins；西伯尔安息香●☆

379455　Styrax socialis J. F. Macbr. = Styrax pentlandianus J. Rémy ●☆

379456　Styrax suberenatus Hand.-Mazz. = Styrax agrestis（Lour.）G. Don ●

379457　Styrax suberifolius Hook. et Arn.；栓皮安息香（赤皮，赤血仔，赤仔尾，椆树，滇红皮，红皮，红皮树，狐狸公，栓叶安息香，铁甲子，叶下白，粘高树）；Corkleaf Snowbell, Corkleaf Storax, Corky-leaved Storax, Prettynerved Storax ●

379458　Styrax suberifolius Hook. et Arn. var. caloneurus Perkins = Styrax suberifolius Hook. et Arn. ●

379459　Styrax suberifolius Hook. et Arn. var. fargesii Perkins = Styrax suberifolius Hook. et Arn. ●

379460　Styrax suberifolius Hook. et Arn. var. hayataianus（Perkins）Mori = Styrax formosanus Matsum. var. hayataianus（Perkins）H. L. Li ●

379461　Styrax subheterotrichus Herzog = Styrax pentlandianus J. Rémy ●☆

379462　Styrax subniveus Merr. et Chun = Styrax tonkinensis（Pierre）Craib ex Hartwich ●

379463　Styrax sumatranus J. J. Sm. = Styrax paralleloneurus Perkins ●☆

379464　Styrax supaii Chun et F. Chun；裂叶安息香；Lobate-leaf Storax, Splitleaf Snowbell, Supa Storax ●◇

379465　Styrax suzukii Mori = Styrax formosanus Matsum. ●

379466　Styrax sweliensis Sm.；瑞丽安息香；Ruili Storax, Sweli Storax ●☆

379467　Styrax tafelbergensis Maguire = Styrax pohlii A. DC. ●☆

379468　Styrax tarapotensis Perkins = Styrax oblongus（Ruiz et Pav.）A. DC. ●☆

379469　Styrax texanus Cory；得州安息香；Texas Snowbell ●☆

379470　Styrax tibeticus J. Anthony = Huodendron tibeticum（J. Anthony）Rehder ●

379471　Styrax tonkinensis（Pierre）Craib ex Hartwich；越南安息香（安息香，安悉香，八翻龙，白背安息香，白花椆，白花椆树，白花木，白花树，白脉安息香，白叶安息香，白叶野茉莉，大青山安息香，滇桂安息香，滇桂野茉莉，粉背安息香树，牛奶树，牛油树，青山安息香，泰国安息香，姊永）；Hypoglaucous Storax, Tonkin Snowbell, Tonkin Storax ●

379472　Styrax touchanensis H. Lév. = Styrax grandiflorus Griff. ●

379473　Styrax veitchiorum Hemsl. et E. H. Wilson = Styrax odoratissimus Champ. ex Benth. ●

379474　Styrax wilsonii Rehder；小叶安息香（矮茉莉，小叶野茉莉）；China Snowbell, Chinese Snowbell, Chinese Storax ●

379475　Styrax wuchanensis H. Lév. = Styrax grandiflorus Griff. ●

379476　Styrax wuyuanensis S. M. Hwang；婺源安息香；Wuyuan Snowbell, Wuyuan Storax ●

379477　Styrax youngae Cory；杨氏安息香；Yang Snowbell ●

379478　Styrax zhejiangensis S. M. Hwang et L. L. Yu；浙江安息香；Zhejiang Snowbell, Zhejiang Storax ●

379479　Styrophyton S. Y. Hu = Allomorphia Blume ●

379480　Styrophyton S. Y. Hu（1952）；长穗花属（长尾花属）；

Styrophyton ●★

379481　Styrophyton caudatum（Diels）S. Y. Hu；长穗花（假欧八竹）；Caudate Styrophyton ●◇

379482　Styrosinia Raf.（废弃属名）=Rechsteineria Regel（保留属名）■☆

379483　Styrosinia Raf.（废弃属名）=Sinningia Nees ●■☆

379484　Styssopus Raf.（废弃属名）=Hyssopus L. ●■

379485　Suaeda Forssk. = Suaeda Forssk. ex J. F. Gmel.（保留属名）●■

379486　Suaeda Forssk. ex J. F. Gmel.（1776）（保留属名）；碱蓬属（翼花蓬属）；Barilla，Sea-blite，Seepweed ●■

379487　Suaeda Forssk. ex Scop. = Suaeda Forssk. ex J. F. Gmel.（保留属名）●■

379488　Suaeda Scop. = Suaeda Forssk. ex J. F. Gmel.（保留属名）●■

379489　Suaeda acuminata（C. A. Mey.）Moq.；刺毛碱蓬；Spinyhair Seepweed ■

379490　Suaeda aegyptiaca（Hasselq.）Zohary；埃及碱蓬■☆

379491　Suaeda albescens Lázaro Ibiza；白碱蓬■☆

379492　Suaeda altissima（L.）Pall.；高碱蓬；Lorty Seepweed ■

379493　Suaeda americana（Pers.）Fernald = Suaeda calceoliformis（Hook.）Moq. ■☆

379494　Suaeda ampullacea Bunge = Borsczowia aralocaspica Bunge ■

379495　Suaeda arcuata Bunge；五蕊碱蓬；Arcuate Seepweed ■

379496　Suaeda arguinensis Maire；阿尔金碱蓬●☆

379497　Suaeda articulata Aellen；关节碱蓬●☆

379498　Suaeda asparagoides（Miq.）Makino = Suaeda glauca（Bunge）Bunge ■

379499　Suaeda australis（R. Br.）Moq.；南方碱蓬；Seablite，Southern Seepweed ●

379500　Suaeda baccata Forssk. ex J. F. Gmel. = Suaeda aegyptiaca（Hasselq.）Zohary ■☆

379501　Suaeda baccifera Pall.；浆果碱蓬■☆

379502　Suaeda caespitosa Wolley-Dod；丛生碱蓬■☆

379503　Suaeda calceoliformis（Hook.）Moq.；草地碱蓬；Horned Sea-blite，Plains Sea-blite，Pursh's Seepweed，Pursh's Seep-weed，Sea Blite ■☆

379504　Suaeda californica S. Watson；加州碱蓬；California Sea-blite ■☆

379505　Suaeda californica S. Watson var. pubescens Jeps. = Suaeda taxifolia（Standl.）Standl. ■☆

379506　Suaeda californica S. Watson var. taxifolia（Standl.）Munz = Suaeda taxifolia（Standl.）Standl. ■☆

379507　Suaeda conferta（Small）I. M. Johnst.；簇生碱蓬；Tufted Sea-blite ■☆

379508　Suaeda confusa Iljin；乱碱蓬■☆

379509　Suaeda corniculata（C. A. Mey.）Bunge；角果碱蓬；Hornyfruit Seepweed ■

379510　Suaeda corniculata（C. A. Mey.）Bunge var. microcarpa Fuh et W. Wang；小果角碱蓬；Smallfruit Hornyfruit Seepweed ■

379511　Suaeda corniculata（C. A. Mey.）Bunge var. olufsenii（Paulsen）G. L. Chu；西藏角果碱蓬（藏角果碱蓬）；Tibet Seepweed，Xizang Seepweed ■

379512　Suaeda corniculata Bunge var. microcarpa P. Y. Fu et W. Wang = Suaeda corniculata（C. A. Mey.）Bunge ■

379513　Suaeda crassifolia Pall.；镰叶碱蓬；Sickleleaf Seepweed ■

379514　Suaeda dendroides（C. A. Mey.）Moq.；木碱蓬（灌木型碱蓬）；Arborescent Seepweed ●

379515　Suaeda depressa（Pursh）S. Watson = Suaeda calceoliformis（Hook.）Moq. ■☆

379516　Suaeda depressa（Pursh）S. Watson var. erecta S. Watson = Suaeda calceoliformis（Hook.）Moq. ■

379517　Suaeda drepanophylla Litv. = Suaeda crassifolia Pall. ■

379518　Suaeda duripes I. M. Johnst. = Suaeda nigra（Raf.）J. F. Macbr. ■☆

379519　Suaeda eltonica Iljin；埃尔塘碱蓬■☆

379520　Suaeda erecta（S. Watson）A. Nelson = Suaeda calceoliformis（Hook.）Moq. ■☆

379521　Suaeda esteroa Ferren et S. Whitmore；河口碱蓬；Estuary Sea-blite ■☆

379522　Suaeda fernaldii（Standl.）Standl. = Suaeda maritima（L.）Dumort. ■☆

379523　Suaeda fruticosa（L.）Forssk. = Suaeda fruticosa Forssk. ex J. F. Gmel. ●☆

379524　Suaeda fruticosa（L.）Forssk. subsp. vera（Forssk.）Maire et Weiller = Suaeda vera Forssk. ex J. F. Gmel. ●☆

379525　Suaeda fruticosa（L.）Forssk. var. ambigua Maire = Suaeda vera Forssk. ex J. F. Gmel. ●☆

379526　Suaeda fruticosa（L.）Forssk. var. brevifolia Moq. = Suaeda vera Forssk. ex J. F. Gmel. ●☆

379527　Suaeda fruticosa（L.）Forssk. var. longifolia（Koch）Fenzl = Suaeda vera Forssk. ex J. F. Gmel. ●☆

379528　Suaeda fruticosa Forssk. ex J. F. Gmel.；灌丛碱蓬●☆

379529　Suaeda fruticosa Forssk. ex J. F. Gmel. = Suaeda vera Forssk. ex J. F. Gmel. ●☆

379530　Suaeda fruticosa Forssk. ex J. F. Gmel. = Suaeda vermiculata Forssk. ex J. F. Gmel. ■☆

379531　Suaeda glauca（Bunge）Bunge；碱蓬（断肠草，灰绿碱蓬，碱蒿子，霜草，相思草，小螺草，盐蒿子，盐蓬，猪尾巴草）；Common Seepweed ■

379532　Suaeda glauca（Bunge）Bunge var. coneriflora H. C. Fu et Z. Y. Chu；密花碱蓬■

379533　Suaeda heterophylla（Kar. et Kir.）Bunge；盘果碱蓬（异型叶碱蓬）；Dishfruit Seepweed ■

379534　Suaeda heteroptera Kitag. = Suaeda salsa（L.）Pall. ■

379535　Suaeda heteroptera Kitag. var. tenuiramea P. Y. Fu et W. Wang = Suaeda salsa（L.）Pall. ■

379536　Suaeda hortensis Forssk. ex J. F. Gmel. = Suaeda aegyptiaca（Hasselq.）Zohary ■☆

379537　Suaeda hyssopifolia Pall. = Bassia hyssopifolia（Pall.）Kuntze ■

379538　Suaeda ifniensis Caball. ex Maire；伊夫尼碱蓬●☆

379539　Suaeda inflata Aellen；膨胀碱蓬●☆

379540　Suaeda insularis（Britton）Urb. et Ekman = Suaeda conferta（Small）I. M. Johnst. ■☆

379541　Suaeda intermedia S. Watson = Suaeda nigra（Raf.）J. F. Macbr. ■☆

379542　Suaeda japonica Makino；日本碱蓬；Japan Seepweed，Japanese Seepweed ■☆

379543　Suaeda kossinskyi Iljin；肥叶碱蓬（柯辛氏碱蓬）；Fatleaf Seepweed，Kossinsky Seepweed ■

379544　Suaeda laevissima Kitag.；光碱蓬；Laevigate Seepweed ■

379545　Suaeda liaotungensis Kitag.；辽宁碱蓬；Liaoning Seepweed ■

379546　Suaeda linearis（Elliott）Moq.；南方高碱蓬；Southern Sea-blite，Tall Sea-blite ■☆

379547　Suaeda linifolia Pall.；亚麻叶碱蓬；Flaxleaf Seepweed，Pin-leaf Seepweed ■

379548　Suaeda lipskyi Litv.；利普斯基碱蓬■☆

379549　Suaeda lipskyi Litv. = Suaeda arcuata Bunge ■

379550　Suaeda longifolia K. Koch ＝ Suaeda vera Forssk. ex J. F. Gmel. ●☆

379551　Suaeda malacosperma H. Hara；阔籽碱蓬■☆

379552　Suaeda maris-mortui Post ＝ Suaeda aegyptiaca（Hasselq.）Zohary ■☆

379553　Suaeda maritima（L.）Dumort.；裸花碱蓬（海滨碱蓬）；Annual Sea-blite，Herbaceous Seepweed，Sand Spurrey，Sea Blite，Sea Seepweed，Seablite，Seashore Seepweed，Seepweed，White Sea-blite ■

379554　Suaeda maritima（L.）Dumort. ＝ Suaeda prostrata Pall. ■

379555　Suaeda maritima（L.）Dumort. subsp. richii（Fernald）Bassett et Crompton ＝ Suaeda maritima（L.）Dumort. ■

379556　Suaeda maritima（L.）Dumort. subsp. salsa（L.）Soó ＝ Suaeda salsa（L.）Pall. ■

379557　Suaeda maritima（L.）Dumort. var. americana（Pers.）B. Boivin ＝ Suaeda calceoliformis（Hook.）Moq. ■☆

379558　Suaeda maritima（L.）Dumort. var. australis（R. Br.）Domin ＝ Suaeda australis（R. Br.）Moq. ●

379559　Suaeda maritima（L.）Dumort. var. malacosperma（H. Hara）Kitam. ＝ Suaeda malacosperma H. Hara ■☆

379560　Suaeda maritima（L.）Dumort. var. perennans Maire ＝ Suaeda salsa（L.）Pall. ■

379561　Suaeda maritima（L.）Dumort. var. vulgaris Moq. ＝ Suaeda prostrata Pall. ■

379562　Suaeda maritima Dumort. ＝ Suaeda prostrata Pall. ■

379563　Suaeda merxmuelleri Aellen；梅尔碱蓬■☆

379564　Suaeda mesopotamica Eig ＝ Suaeda vermiculata Forssk. ex J. F. Gmel. ■☆

379565　Suaeda micromeris Brenan；小碱蓬■☆

379566　Suaeda microphylla（C. A. Mey.）Pall.；小叶碱蓬；Smallleaf Seepweed ●

379567　Suaeda microsperma（（C. A. Mey.）Fenzl；小籽碱蓬■☆

379568　Suaeda minutiflora S. Watson ＝ Suaeda calceoliformis（Hook.）Moq. ■☆

379569　Suaeda mollis（Desf.）Delile ＝ Suaeda vermiculata Forssk. ex J. F. Gmel. ■☆

379570　Suaeda monodiana Maire ＝ Suaeda vermiculata Forssk. ex J. F. Gmel. ■☆

379571　Suaeda monoica Forssk. ex J. F. Gmel.；同株碱蓬■☆

379572　Suaeda moquinii（Torr.）Greene ＝ Suaeda nigra（Raf.）J. F. Macbr. ■☆

379573　Suaeda nigra（Raf.）J. F. Macbr.；黑碱蓬；Bush Seepweed，Seepweed ■☆

379574　Suaeda nigrescens I. M. Johnst. ＝ Suaeda nigra（Raf.）J. F. Macbr. ■☆

379575　Suaeda nudiflora（Willd.）Moq. ＝ Suaeda maritima（L.）Dumort. ■

379576　Suaeda occidentalis（S. Watson）S. Watson；西方碱蓬；Western Seepweed ■☆

379577　Suaeda olufsenii Paulsen；奥卢碱蓬■☆

379578　Suaeda olufsenii Paulsen ＝ Suaeda corniculata（C. A. Mey.）Bunge var. olufsenii（Paulsen）G. L. Chu ■

379579　Suaeda palaestina Eig et Zohary；燥地碱蓬■☆

379580　Suaeda paradoxa Bunge；奇异碱蓬；Wonderful Seepweed ■

379581　Suaeda paulayana Vierh. ＝ Suaeda vermiculata Forssk. ex J. F. Gmel. ■☆

379582　Suaeda physophora Pall.；囊果碱蓬（气囊果碱蓬）；Alashan Seepweed，Saccatefruit Seepweed ●

379583　Suaeda plumosa Aellen；羽状碱蓬■☆

379584　Suaeda prostrata Pall.；平卧碱蓬；Prostrate Seepweed ■

379585　Suaeda pruinosa Lange；白粉碱蓬■☆

379586　Suaeda pruinosa Lange var. kochii（Tod.）Maire et Weiller ＝ Salsola kochii Guss. ex Tod. ■☆

379587　Suaeda pruinosa Lange var. kochii（Tod.）Maire et Weiller ＝ Suaeda pruinosa Lange ■☆

379588　Suaeda pruinosa Lange var. solmsiana Maire et Weiller ＝ Suaeda pruinosa Lange ■☆

379589　Suaeda przewalskii Bunge；阿拉善碱蓬（茄叶碱蓬，水杏，水珠子）；Przewalski Seepweed ■☆

379590　Suaeda pterantha（Kar. et Kir.）Bunge；纵翅碱蓬；Longitudinalwing Seepweed ■

379591　Suaeda pygmaea（Kar. et Kir.）Iljin ＝ Suaeda pterantha（Kar. et Kir.）Bunge ■

379592　Suaeda ramosissima（Standl.）I. M. Johnst. ＝ Suaeda nigra（Raf.）J. F. Macbr. ■☆

379593　Suaeda richii Fernald ＝ Suaeda maritima（L.）Dumort. ■☆

379594　Suaeda rigida H. W. Kung et G. L. Chu；硬枝碱蓬（土耳其斯坦碱蓬）；Rigid Seepweed ●

379595　Suaeda roborowskii Iljin ＝ Suaeda pterantha（Kar. et Kir.）Bunge ■

379596　Suaeda rolandii Bassett et Crompton；罗兰碱蓬；Roland's Sea-blite ■☆

379597　Suaeda salina B. Nord.；盐生碱蓬■☆

379598　Suaeda salsa（L.）Pall.；盐地碱蓬（翅碱蓬，黄须菜，碱葱，汊棚，乌苏里碱蓬）；Saline Seepweed ■

379599　Suaeda schimperi（Moq.）Martelli ＝ Sevada schimperi Moq. ■☆

379600　Suaeda sieversiana Pall. ＝ Kochia scoparia（L.）Schrad. var. sieversiana（Pall.）Ulbr. ex Asch. et Graebn. ■

379601　Suaeda spicata（Willd.）Moq.；长穗亮绿碱蓬■☆

379602　Suaeda splendens（Pourr.）Gren. et Godr.；光亮碱蓬■☆

379603　Suaeda stauntonii Moq. ＝ Suaeda glauca（Bunge）Bunge ■

379604　Suaeda stellatiflora G. L. Chu；星花碱蓬（星果碱蓬）；Stellateflower Seepweed ■

379605　Suaeda suffrutescens S. Watson ＝ Suaeda nigra（Raf.）J. F. Macbr. ■☆

379606　Suaeda suffrutescens S. Watson var. detonsa I. M. Johnst. ＝ Suaeda nigra（Raf.）J. F. Macbr. ■☆

379607　Suaeda tampicensis（Standl.）Standl.；坦皮科碱蓬；Tampico Sea-blite ■☆

379608　Suaeda taxifolia（Standl.）Standl.；绵毛碱蓬；Woolly Sea-blite ■☆

379609　Suaeda tomentosa Lowe ＝ Bassia tomentosa（Lowe）Maire et Weiller ■☆

379610　Suaeda torreyana S. Watson 'Ramosissima'；多枝托雷碱蓬；Inkweed，Iodine Weed ■☆

379611　Suaeda torreyana S. Watson ＝ Suaeda nigra（Raf.）J. F. Macbr. ■☆

379612　Suaeda torreyana S. Watson var. ramosissima（Standl.）Munz ＝ Suaeda nigra（Raf.）J. F. Macbr. ■☆

379613　Suaeda transoxana（Bunge）Boiss.；外阿穆达尔碱蓬■☆

379614　Suaeda turkestanica Litv. ＝ Suaeda rigida H. W. Kung et G. L. Chu ●

379615　Suaeda ussuriensis Iljin ＝ Suaeda salsa（L.）Pall. ■

379616　Suaeda vera Forssk. ex J. F. Gmel.；灌木碱蓬；Inkbush，Shrubby Saltwort，Shrubby Seablite，Shrubby Sea-blite ●☆

379617　Suaeda vera J. F. Gmel. subsp. longifolia（Koch）O. Bolòs et

Vigo ＝ Suaeda vera Forssk. ex J. F. Gmel. ●☆

379618　Suaeda vera J. F. Gmel. subsp. pruinosa（Lange）O. Bolòs et Vigo ＝ Suaeda pruinosa Lange ■☆

379619　Suaeda vermiculata Forssk. ex J. F. Gmel.；虫状碱蓬■☆

379620　Suaeda vermiculata Forssk. ex J. F. Gmel. var. puberula C. B. Clarke ＝ Sevada schimperi Moq. ■☆

379621　Suaeda vesceritensis L. Chevall. ＝ Suaeda vera Forssk. ex J. F. Gmel. ●☆

379622　Suaeda volkensii C. B. Clarke ＝ Suaeda vermiculata Forssk. ex J. F. Gmel. ■☆

379623　Suarda Nocca ex Steud. ＝ Eugenia L. ●

379624　Suardia Schrank ＝ Melinis P. Beauv. ■

379625　Suardia picta Schrank ＝ Melinis minutiflora P. Beauv. ■

379626　Suarezia Dodson（1989）；苏阿兰属■☆

379627　Suarezia ecuadorana Dodson；苏阿兰■☆

379628　Suber Mill. ＝ Quercus L. ●

379629　Suberanthus Borhidi et M. Fernandez（1982）；栓花茜属●☆

379630　Suberanthus yamuriensis（Britton）Borhidi et M. Fernandez；栓花茜●☆

379631　Subertia Wood ＝ Brodiaea Sm.（保留属名）■☆

379632　Subertia Wood ＝ Seubertia Kunth ■☆

379633　Sublimia Comm. ex Mart. ＝ Hyophorbe Gaertn. ●

379634　Submatucana Backeb.（1959）；黄仙玉属■☆

379635　Submatucana Backeb. ＝ Borzicactus Riccob. ■☆

379636　Submatucana Backeb. ＝ Matucana Britton et Rose ●☆

379637　Submatucana Backeb. ＝ Oreocereus（A. Berger）Riccob. ●

379638　Submatucana aurantiaca（Vaupel）Backeb.；黄仙玉；Aurantiaceous Matucana ■☆

379639　Submatucana aurantiaca（Vaupel）Backeb. ＝ Borzicactus acanthurus（Vaupel）Britton et Rose ■☆

379640　Submatucana currundayensis（F. Ritter）Backeb. ＝ Matucana currundayensis F. Ritter ■☆

379641　Submatucana formosa（F. Ritter）Backeb. ＝ Borzicactus formosus（F. Ritter）Donald ■☆

379642　Submatucana ritteri（Buining）Backeb. ＝ Matucana ritteri Buining ■☆

379643　Subpilocereus Backeb. ＝ Cephalocereus Pfeiff. ●

379644　Subpilocereus Backeb. ＝ Cereus Mill. ●

379645　Subrisia Raf. ＝ Ehretia P. Browne ●

379646　Subscariosaceae Dulac ＝ Amaranthaceae Juss.（保留科名）●■

379647　Subularia Boehm. ＝ Littorella P. J. Bergius ■☆

379648　Subularia Forssk. ＝ Schouwia DC.（保留属名）■☆

379649　Subularia L.（1753）；锥叶芥属（钻状荠属）；Awlwort ■☆

379650　Subularia aquatica L.；水锥叶芥；Awlwort, Awl-wort, Subularia, Water Awlwort ■☆

379651　Subularia monticola A. Br. ex Schweinf. ；山生锥叶芥■☆

379652　Subularia purpurea Forssk. ＝ Schouwia purpurea（Forssk.）Schweinf. ■☆

379653　Subulatopuntia Frič et Schelle ＝ Opuntia Mill. ●

379654　Subulatopuntia Frič et Schelle ex Kreuz. ＝ Opuntia Mill. ●

379655　Succisa Haller（1768）；断草属；Devil's-bit Scabious ■☆

379656　Succisa australis Borbás ＝ Scabiosa australis Wulfen ■☆

379657　Succisa australis Rchb.；南方断草（南方蓝盆花）；Southern Succisa ■☆

379658　Succisa australis Rchb. ＝ Scabiosa australis Wulfen ■☆

379659　Succisa kamerunensis Engl. ex Mildbr. ＝ Succisa trichotocephala Baksay ■☆

379660　Succisa pratensis Moench；草原断草；Bachelor's Buttons, Blue Bonnets, Blue Buttons, Blue Head, Blue Kiss, Blue Top, Bund, Bundweed, Curldoddy, Devil's Bit, Devil's Bit Scabious, Devil's Bite, Devil's Button, Devil's Guts, Devilsbit, Devil's-bit, Devil's-bit Scabious, For-bete, Forbitten More, Forebit, Gentleman's Buttons, Gentlemen's Buttons, Gypsy Rose, Hardhead, Hog-a-back, Lamb's Ear, Meadow Scabious, Meadow Succisa, Ofbit, Ofbiten, Pincushion, Premorse Scabious, Remcope, Sailor Buttons, Sailor's Buttons, Stinking Nancy, Woolly Hardhead ■☆

379661　Succisa trichotocephala Baksay；热非断草■☆

379662　Succisella Beck（1893）；小断草属；Succisella ■☆

379663　Succisella inflexa（Kluk）Beck；南方小断草；Southern Succisella ■☆

379664　Succisocrepis Fourr. ＝ Crepis L. ■

379665　Succisocrepis Fourr. ＝ Wibelia P. Gaertn. ,B. Mey. et Scherb. ■

379666　Succosaria Raf. ＝ Aloe L. ●■

379667　Succovia Desv. ＝ Succowia Medik. ■☆

379668　Succowia Dennst. ＝ Hiptage Gaertn.（保留属名）●

379669　Succowia Medik.（1792）；苏氏芥属■☆

379670　Succowia balearica（L.）Medik.；苏氏芥■☆

379671　Succuta Des Moul. ＝ Cuscuta L. ■

379672　Suchtelenia Kar. ＝ Suchtelenia Kar. ex Meisn. ■☆

379673　Suchtelenia Kar. ex Meisn.（1840）；苏合草属■☆

379674　Suchtelenia calycina（C. A. Mey.）DC.；苏合草■☆

379675　Suckleya A. Gray（1876）；异被滨藜属■☆

379676　Suckleya suckleyana（Torr.）Rydb.；异被滨藜；Poison Suckleya ■☆

379677　Sucrea Soderstr.（1981）；糖禾属（苏克蕾禾属）■☆

379678　Sucrea monophylla Soderstr.；巴西苏克蕾禾■☆

379679　Sucrea sampaiana（Hitchc.）Soderstr. ；苏克蕾禾■☆

379680　Sudamerlycaste Archila ＝ Lycaste Lindl. ■☆

379681　Suddia Renvoize（1984）；苏德禾属（箭叶苏丹草属，苏丹禾属）■☆

379682　Suddia sagittifolia Renvoize；苏丹禾■☆

379683　Sueda Edgew. ＝ Suaeda Forssk. ex J. F. Gmel.（保留属名）●■

379684　Suensonia Gaudich. ＝ Piper L. ●■

379685　Suensonia Gaudich. ex Miq. ＝ Piper L. ●■

379686　Suessenguthia Merxm.（1953）；苏氏爵床属■☆

379687　Suessenguthia multisetosa（Rusby）Wassh. et J. R. I. Wood；多刚毛苏氏爵床■☆

379688　Suessenguthia trochilophila Merxm. ；苏氏爵床■☆

379689　Suessenguthiella Friedrich（1955）；针叶粟草属■☆

379690　Suessenguthiella caespitosa Friedrich；簇生针叶粟草■☆

379691　Suessenguthiella scleranthoides（Sond.）Friedrich；针叶粟草■☆

379692　Suffrenia Bellardi ＝ Rotala L. ■

379693　Suffrenia capensis Harv. ＝ Rotala capensis（Harv.）A. Fern. et Diniz ■☆

379694　Suffrenia dichotoma Miq. ＝ Ammannia multiflora Roxb. ■

379695　Suffrenia filiformis Bellardi ＝ Rotala filiformis（Bellardi）Hiern ■☆

379696　Sugerokia Miq. ＝ Heloniopsis A. Gray（保留属名）■

379697　Sugerokia acutifolia（Hayata）Koidz. ＝ Heloniopsis umbellata（Baker）N. Tanaka ■

379698　Sugerokia arisanensis（Hayata ex Honda）Koidz. ＝ Heloniopsis umbellata（Baker）N. Tanaka ■

379699　Sugerokia umbellata（Baker）Koidz. ＝ Heloniopsis umbellata（Baker）N. Tanaka ■

379700　Sugillaria Salisb. ＝ Scilla L. ■

379701 Suida Opiz ＝ Swida Opiz ●

379702 Suitenia Stokes ＝ Swietenia Jacq. ●

379703 Suitramia Rchb. ＝ Svitramia Cham. ●☆

379704 Sukaminea Raf. ＝ Chlorophora Gaudich. ●☆

379705 Sukana Adans. ＝ Celosia L. ■

379706 Suksdorfia A. Gray（1880）（保留属名）；苏克草属■☆

379707 Suksdorfia violacea A. Gray；苏克草■☆

379708 Sukunia A. C. Sm.（1936）；苏昆茜属●☆

379709 Sukunia longipes A. C. Sm.；长梗苏昆茜●☆

379710 Sukunia pentagonioides（Seem.）A. C. Sm.；苏昆茜●☆

379711 Sulaimania Hedge et Rech. f.（1982）；苏赖曼草属●☆

379712 Sulaimania otostegioides（Prain）Hedge et Rech. f.；苏赖曼草●☆

379713 Sulamea K. Schum. et Lauterb. ＝ Soulamea Lam. ●☆

379714 Sulcanux Raf. ＝ Geophila D. Don（保留属名）■

379715 Sulcorebutia Backeb.（1951）；沟宝山属（有沟宝山属）●☆

379716 Sulcorebutia Backeb. ＝ Rebutia K. Schum. ●

379717 Sulcorebutia steinbachii（Werderm.）Backeb.；沟宝山（斯氏沟宝山）●☆

379718 Sulipa Blanco ＝ Gardenia Ellis（保留属名）●

379719 Sulitia Merr.（1926）；苏利特茜属●☆

379720 Sulitia longiflora Merr.；苏利特茜●☆

379721 Sulitra Medik.（废弃属名）＝ Lessertia DC.（保留属名）●■☆

379722 Sulla Medik. ＝ Hedysarum L.（保留属名）●■

379723 Sulla capitata（Desf.）B. H. Choi et H. Ohashi ＝ Hedysarum glomeratum F. Dietr. ■☆

379724 Sulla carnosa（Desf.）B. H. Choi et H. Ohashi ＝ Hedysarum carnosum Desf. ■☆

379725 Sulla coronaria（L.）Medik. ＝ Hedysarum coronarium L. ■☆

379726 Sulla flexuosa（L.）Medik. ＝ Hedysarum flexuosum L. ■☆

379727 Sulla glomerata（F. Dietr.）B. H. Choi et H. Ohashi ＝ Hedysarum glomeratum F. Dietr. ■☆

379728 Sulla pallida（Desf.）B. H. Choi et H. Ohashi ＝ Hedysarum pallidum Desf. ■☆

379729 Sulla spinosissima（L.）B. H. Choi et H. Ohashi ＝ Hedysarum spinosissimum L. ■☆

379730 Sullivantia Torr. et A. Gray（1842）；沙氏虎耳草属■☆

379731 Sullivantia ohionis Torr. et A. Gray ex A. Gray ＝ Sullivantia sullivantii（Torr. et A. Gray）Britton ■☆

379732 Sullivantia renifolia Rosend. ＝ Sullivantia sullivantii（Torr. et A. Gray）Britton ■☆

379733 Sullivantia sullivantii（Torr. et A. Gray）Britton；沙氏虎耳草；Sullivantia，Sullivant's Cool-wort ■☆

379734 Sulpitia Raf. ＝ Encyclia Hook. ■☆

379735 Sulpitia Raf. ＝ Exophya Raf. ■☆

379736 Sulzeria Roem. et Schult. ＝ Faramea Aubl. ●☆

379737 Sumachiaceae DC. ex Perleb ＝ Anacardiaceae R. Br.（保留科名）●

379738 Sumachium Raf. ＝ Rhus L. ●

379739 Sumacrus R. Hedw. ＝ Rhus L. ●

379740 Sumacrus R. Hedw. ＝ Sumachium Raf. ●

379741 Sumacus Raf. ＝ Rhus L. ●

379742 Sumacus Raf. ＝ Sumachium Raf. ●

379743 Sumatroscirpus Oteng-Yeb.（1974）；苏门答腊莎草属■☆

379744 Sumatroscirpus junghuhnii（Miq.）Oteng-Yeb.；苏门答腊莎草■☆

379745 Sumbavia Baill. ＝ Doryxylon Zoll. ●☆

379746 Sumbavia macrophylla Müll. Arg. ＝ Sumbaviopsis albicans（Blume）J. J. Sm. ●

379747 Sumbaviopsis J. J. Sm.（1910）；狭瓣木属（缅桐属）；Sumbaviopsis ●

379748 Sumbaviopsis albicans（Blume）J. J. Sm.；狭瓣木（缅桐）；Common Sumbaviopsis ●

379749 Sumbulus H. Reinsch ＝ Ferula L. ■

379750 Sumbulus moschatus H. Reinsch ＝ Ferula moschata（H. Reinsch）Koso-Pol. ■

379751 Sumbulus moschatus H. Reinsch ＝ Ferula sumbul（Kauffm.）Hook. f. ■

379752 Summerhayesia P. J. Cribb（1977）；萨默兰属（赛姆兰属）■☆

379753 Summerhayesia laurentii（De Wild.）P. J. Cribb；洛朗萨默兰■☆

379754 Summerhayesia rwandensis Geerinck；卢旺达萨默兰■☆

379755 Summerhayesia zambesiaca P. J. Cribb；赞比西萨默兰■☆

379756 Sumnera Nieuwl. ＝ Physocarpum Bercht. et J. Presl ■

379757 Sumnera Nieuwl. ＝ Thalictrum L. ■

379758 Sunania Raf. ＝ Antenoron Raf. ■

379759 Sunania filiformis（Thunb.）Raf. ＝ Antenoron filiforme（Thunb.）Rob. et Vautier ■

379760 Sunania neofiliformis（Nakai）H. Hara ＝ Antenoron filiforme（Thunb.）Rob. et Vautier var. neofiliforme（Nakai）A. J. Li ■

379761 Sunaptea Griff. ＝ Vatica L. ●

379762 Sunapteopsis Heim ＝ Stemonoporus Thwaites ●☆

379763 Sunapteopsis Heim ＝ Vateria L. ●☆

379764 Sundacarpus（J. Buchholz et N. E. Gray）C. N. Page（1989）；苦味罗汉松属●☆

379765 Sundacarpus C. N. Page ＝ Sundacarpus（J. Buchholz et N. E. Gray）C. N. Page ●☆

379766 Sundacarpus amara（Blume）C. N. Page；苦味罗汉松；Black Pine ●☆

379767 Sundacarpus amara（Blume）C. N. Page ＝ Podocarpus amara Blume ●☆

379768 Sunipia Buch. -Ham. ex Lindl.（1826）；大苞兰属（宝石兰属）；Sunipia ■

379769 Sunipia Buch. -Ham. ex Sm. ＝ Sunipia Buch. -Ham. ex Lindl. ■

379770 Sunipia Lindl. ＝ Sunipia Buch. -Ham. ex Lindl. ■

379771 Sunipia andersonii（King et Pantl.）P. F. Hunt；黄花大苞兰（黄花菫兰，绿花宝石兰，台湾菫兰）；Yellow Sunipia ■

379772 Sunipia annamensis（Ridl.）P. F. Hunt；绿花大苞兰■

379773 Sunipia bicolor Lindl.；二色大苞兰（二色卷瓣兰）；Bicolor Bulbophyllum，Bicolor Curlylip-orchis，Twocolor Sunipia ■

379774 Sunipia bicolor Lindl. ＝ Bulbophyllum bicolor（Lindl.）Hook. f. ■

379775 Sunipia bifurcatoflorens（Fukuy.）P. F. Hunt ＝ Sunipia andersonii（King et Pantl.）P. F. Hunt ■

379776 Sunipia candida（Lindl.）P. F. Hunt；白花大苞兰（白花菫兰）；White Sunipia ■

379777 Sunipia cirrhata（Lindl.）P. F. Hunt；云南大苞兰■

379778 Sunipia hainanensis Z. H. Tsi；海南大苞兰；Hainan Sunipia ■

379779 Sunipia intermedia（King et Pantl.）P. F. Hunt；少花大苞兰；Foorflower Sunipia ■

379780 Sunipia racemosa（Sm.）Ts. Tang et F. T. Wang ＝ Sunipia scariosa Lindl. ■

379781 Sunipia rimannii（Rchb. f.）Seidenf.；圆瓣大苞兰；Rimann Sunipia ■

379782 Sunipia salweenensis（Phil. et W. W. Sm.）P. F. Hunt ＝ Sunipia rimannii（Rchb. f.）Seidenf. ■

379783 Sunipia sasakii（Hayata）P. F. Hunt ＝ Sunipia andersonii（King et Pantl.）P. F. Hunt ■

379784 Sunipia scariosa Lindl.;大苞兰(独叶果,果上叶,一叶七,总序大苞兰);Sunipia ■

379785 Sunipia soidaoensis (Seidenf.) P. F. Hunt;苏瓣大苞兰;Tassel Sunipia ■

379786 Sunipia thailandica (Seidenf. et Smitinand) P. F. Hunt;光花大苞兰;Tailand Sunipia ■

379787 Superbangis Thouars = Angraecum Bory ■

379788 Suprago Gaertn. = Vemonia Edgew. ●■

379789 Suprago Gaertn. = Vernonia Schreb. (保留属名)●■

379790 Supushpa Suryan. (1970);西印度爵床属■☆

379791 Supushpa scrobiculata (Dalz. ex C. B. Clarke) Suryan.;西印度爵床■☆

379792 Suregada Roxb. ex Rottler (1803);白树属(饼树属);Suregada,Whitetree ●

379793 Suregada Willd. = Suregada Roxb. ex Rottler ●

379794 Suregada aequorea (Hance) Seem.;台湾白树(白树仔);Taiwan Suregada,Taiwan Whitetree,White Suregada ●

379795 Suregada aequorea (Hance) Seem. = Gelonium aequoreum Hance ●

379796 Suregada africana (Sond.) Kuntze;非洲白树●☆

379797 Suregada angolensis (Prain) Croizat = Tetrorchidium didymostemon (Baill.) Pax et K. Hoffm. ●☆

379798 Suregada ceratophora Baill. = Suregada africana (Sond.) Kuntze ●☆

379799 Suregada congoensis (S. Moore) Croizat = Suregada gossweileri (S. Moore) Croizat ●☆

379800 Suregada glomerulata (Blume) Baill.;白树(饼树);Glamerule Whitetree,Glomerulate Suregada ●

379801 Suregada gossweileri (S. Moore) Croizat;戈斯白树●☆

379802 Suregada ivorensis (Aubrév. et Pellegr.) J. Léonard;伊沃里白树●☆

379803 Suregada lithoxyla (Pax et K. Hoffm.) Croizat;是木白树●☆

379804 Suregada multiflorum Baill.;多花白树●☆

379805 Suregada occidentalis (Hoyle) Croizat;西部白树●☆

379806 Suregada procera (Prain) Croizat;高大白树●☆

379807 Suregada zanzibariensis Baill.;桑给巴尔白树●☆

379808 Surenus Kuntze = Cedrela P. Browne ●

379809 Surenus Kuntze = Toona (Endl.) M. Roem. ●

379810 Surfacea Moldenke = Premna L. (保留属名)●■

379811 Suriana Dombey et Cav. ex D. Don = Ercilla A. Juss. ●☆

379812 Suriana L. (1753);海人树属;Suriana,Seamantree ●

379813 Suriana maritima L.;海人树(滨樗);Seamantree, Seashore Suriana,Sea-shore Suriana ●

379814 Surianaceae Arn. (1834)(保留科名);海人树科●

379815 Suringaria Pierre = Symplocos Jacq. ●

379816 Surubea J. St. -Hil. = Ruyschia Jacq. ●☆

379817 Surubea J. St. -Hil. = Souroubea Aubl. ●☆

379818 Surwala M. Roem. = Walsura Roxb. ●

379819 Susanna E. Phillips = Amellus L. (保留属名)●■☆

379820 Susanna dinteri E. Phillips = Felicia namaquana (Harv.) Merxm. ■☆

379821 Susanna epaleacea (O. Hoffm.) E. Phillips = Amellus epaleaceus O. Hoffm. ■☆

379822 Susanna microglossa (DC.) E. Phillips = Amellus microglossus DC. ■☆

379823 Susarium Phil. = Symphyostemon Miers ex Klatt ■☆

379824 Susilkumara Bennet = Alajja Ikonn. ■

379825 Sussea Gaudich. = Pandanus Parkinson ex Du Roi ●■

379826 Sussodia Buch. -Ham. ex D. Don = Colebrookia Donn ex T. Lestib. ■

379827 Susum Blume = Hanguana Blume ■☆

379828 Susum Blume ex Schult. et Schult. f. = Hanguana Blume ■☆

379829 Sutera Hort. ex Steud. = Lessertia DC. (保留属名)●■☆

379830 Sutera Roth(1807);裂口花属■●☆

379831 Sutera Roth(1821) = Jamesbrittenia Kuntze ■●☆

379832 Sutera accrescens Hiern = Jamesbrittenia accrescens (Hiern) Hilliard ■☆

379833 Sutera acutiloba (Pilg.) Overkott ex Rössler = Jamesbrittenia acutiloba (Pilg.) Hilliard ■☆

379834 Sutera adpressa Dinter = Jamesbrittenia adpressa (Dinter) Hilliard ■☆

379835 Sutera aethiopica (L.) Kuntze;埃塞俄比亚裂口花■☆

379836 Sutera affinis (Bernh.) Kuntze;近缘裂口花■☆

379837 Sutera albiflora I. Verd. = Jamesbrittenia albiflora (I. Verd.) Hilliard ■☆

379838 Sutera altoplana Hiern = Jamesbrittenia tysonii (Hiern) Hilliard ■☆

379839 Sutera amplexicaulis (Benth.) Hiern = Jamesbrittenia amplexicaulis (Benth.) Hilliard ■☆

379840 Sutera annua (Schltr. ex Hiern) Hiern var. laxa Hiern = Manulea paucibarbata Hilliard ■☆

379841 Sutera annua Hiern = Manulea altissima L. f. subsp. longifolia (Benth.) Hilliard ■☆

379842 Sutera antirrhinoides (L. f.) Hiern = Lyperia antirrhinoides (L. f.) Hilliard ■☆

379843 Sutera archeri Compton;阿谢尔裂口花■☆

379844 Sutera arcuata Hiern = Sutera floribunda (Benth.) Kuntze ■☆

379845 Sutera argentea (L. f.) Hiern = Jamesbrittenia argentea (L. f.) Hilliard ■☆

379846 Sutera asbestina Hiern = Jamesbrittenia integerrima (Benth.) Hilliard ■☆

379847 Sutera aspalathoides (Benth.) Hiern = Jamesbrittenia aspalathoides (Benth.) Hilliard ■☆

379848 Sutera atrocaerulea Fourc. = Jamesbrittenia tenuifolia (Bernh.) Hilliard ■☆

379849 Sutera atrocaerulea Fourc. var. latifolia ? = Jamesbrittenia tenuifolia (Bernh.) Hilliard ■☆

379850 Sutera atropurpurea (Benth.) Hiern = Jamesbrittenia atropurpurea (Benth.) Hilliard ■☆

379851 Sutera aurantiaca (Burch.) Hiern = Jamesbrittenia aurantiaca (Burch.) Hilliard ■☆

379852 Sutera ausana Dinter ex Range = Jamesbrittenia integerrima (Benth.) Hilliard ■☆

379853 Sutera batlapina Hiern = Jamesbrittenia integerrima (Benth.) Hilliard ■☆

379854 Sutera bicolor Dinter = Jamesbrittenia bicolor (Dinter) Hilliard ■☆

379855 Sutera blantyrensis V. Naray. = Jamesbrittenia micrantha (Klotzsch) Hilliard ■☆

379856 Sutera bolusii Hiern = Jamesbrittenia micrantha (Klotzsch) Hilliard ■☆

379857 Sutera brachiata Roth = Sutera hispida (Thunb.) Druce ■☆

379858 Sutera bracteolata Hiern = Sutera cooperi Hiern ■☆

379859 Sutera breviflora (Schltr.) Hiern = Jamesbrittenia breviflora (Schltr.) Hilliard ■☆

379860 Sutera brunnea Hiern = Jamesbrittenia huillana (Diels) Hilliard ■☆

379861　Sutera brunnea Hiern var. macrophylla ? = Jamesbrittenia accrescens（Hiern）Hilliard ■☆

379862　Sutera burchellii Hiern = Sutera griquensis Hiern ■☆

379863　Sutera burkeana（Benth.）Hiern = Jamesbrittenia burkeana（Benth.）Hilliard ■☆

379864　Sutera caerulea（L. f.）Hiern；天蓝裂口花■☆

379865　Sutera caerulea（L. f.）Hiern = Manulea viscosa（Aiton）Willd. ■☆

379866　Sutera caerulea（L. f.）Kuntze = Sutera caerulea（L. f.）Hiern ■☆

379867　Sutera calciphila Hilliard；喜岩裂口花■☆

379868　Sutera calycina（Benth.）Kuntze；尊状裂口花■☆

379869　Sutera calycina（Benth.）Kuntze var. laxiflora（Benth.）Hiern = Sutera calycina（Benth.）Kuntze ■☆

379870　Sutera campanulata（Benth.）Kuntze；风铃草状裂口花■☆

379871　Sutera canariensis（Webb et Berthel.）Sunding et G. Kunkel；加那利裂口花■☆

379872　Sutera canescens（Benth.）Hiern var. laevior Dinter = Jamesbrittenia canescens（Benth.）Hilliard var. laevior（Dinter）Hilliard ■☆

379873　Sutera carvalhoi（Engl.）V. Naray. = Jamesbrittenia carvalhoi（Engl.）Hilliard ■☆

379874　Sutera cephalotes（Thunb.）Kuntze = Manulea cephalotes Thunb. ■☆

379875　Sutera cephalotes Hiern = Sutera aethiopica（L.）Kuntze ■☆

379876　Sutera cephalotes Hiern var. glabrata ? = Sutera aethiopica（L.）Kuntze ■☆

379877　Sutera cinerea Hilliard；灰色裂口花■☆

379878　Sutera compta Hiern = Sutera floribunda（Benth.）Kuntze ■☆

379879　Sutera comptonii Hilliard；康普顿裂口花■☆

379880　Sutera concinna Hiern = Jamesbrittenia concinna（Hiern）Hilliard ■☆

379881　Sutera cooperi Hiern；库珀裂口花■☆

379882　Sutera cordata（Thunb.）Kuntze；心叶裂口花；Bacopa ■☆

379883　Sutera cordata（Thunb.）Kuntze 'Snowflake'；雪花裂口花；Bacopa 'Snowflake' ■☆

379884　Sutera cordata（Thunb.）Kuntze var. hirsutior Hiern = Sutera cooperi Hiern ■☆

379885　Sutera corymbosa（Marloth et Engl.）Hiern = Camptoloma rotundifolia Benth. ■☆

379886　Sutera corymbosa（Marloth et Engl.）Hiern var. huillana（Diels）Hiern = Camptoloma rotundifolia Benth. ■☆

379887　Sutera crassicaulis（Benth.）Hiern = Jamesbrittenia crassicaulis（Benth.）Hilliard ■☆

379888　Sutera crassicaulis（Benth.）Hiern var. purpurea Hiern = Jamesbrittenia pristisepala（Hiern）Hilliard ■☆

379889　Sutera cuneata（Benth.）Kuntze = Sutera hispida（Thunb.）Druce ■☆

379890　Sutera cymbalarifolia Chiov. = Stemodiopsis buchananii V. Naray. ■☆

379891　Sutera cymulosa Hiern = Sutera cooperi Hiern ■☆

379892　Sutera debilis Hutch. ；弱小裂口花■☆

379893　Sutera decipiens Hilliard；迷惑裂口花■☆

379894　Sutera densifolia Hiern = Jamesbrittenia microphylla（L. f.）Hilliard ■☆

379895　Sutera dentatisepala Overkott = Jamesbrittenia dentatisepala（Overkott）Hilliard ■☆

379896　Sutera denudata（Benth.）Kuntze；裸露裂口花■☆

379897　Sutera dielsiana Hiern = Jamesbrittenia racemosa（Benth.）Hilliard ■☆

379898　Sutera dioritica Dinter = Jamesbrittenia canescens（Benth.）Hilliard ■☆

379899　Sutera dissecta（Delile）Walp. = Jamesbrittenia dissecta（Delile）Kuntze ■☆

379900　Sutera divaricata（Diels）Hiern = Manulea annua（Hiern）Hilliard ■☆

379901　Sutera dubia V. Naray. = Manulea dubia（V. Naray.）Overkott ex Rössler ■☆

379902　Sutera elegantissima（Schinz）V. Naray. = Jamesbrittenia elegantissima（Schinz）Hilliard ■☆

379903　Sutera elliotensis Hiern = Manulea paniculata Benth. ■☆

379904　Sutera esculenta Bond = Jamesbrittenia incisa（Thunb.）Hilliard ●☆

379905　Sutera fastigiata（Benth.）Druce = Sutera aethiopica（L.）Kuntze ■☆

379906　Sutera filicaulis（Benth.）Hiern = Jamesbrittenia filicaulis（Benth.）Hilliard ■☆

379907　Sutera fissifolia S. Moore = Jamesbrittenia micrantha（Klotzsch）Hilliard ■☆

379908　Sutera flexuosa Hiern = Jamesbrittenia tenella（Hiern）Hilliard ■☆

379909　Sutera floribunda（Benth.）Kuntze；繁花裂口花■☆

379910　Sutera foetida Roth；臭裂口花■☆

379911　Sutera foliolosa（Benth.）Hiern = Jamesbrittenia foliolosa（Benth.）Hilliard ■☆

379912　Sutera fragilis Pilg. = Jamesbrittenia fragilis（Pilg.）Hilliard ■☆

379913　Sutera fraterna Hiern = Jamesbrittenia thunbergii（G. Don）Hilliard ■☆

379914　Sutera fruticosa（Benth.）Hiern = Jamesbrittenia fruticosa（Benth.）Hilliard ■☆

379915　Sutera gariusana Dinter = Jamesbrittenia sessilifolia（Diels）Hilliard ■☆

379916　Sutera glabrata（Benth.）Kuntze；光滑裂口花■☆

379917　Sutera glandulifera Hilliard；腺点裂口花■☆

379918　Sutera glandulosa Roth = Jamesbrittenia dissecta（Delile）Kuntze ■☆

379919　Sutera gossweileri V. Naray. = Jamesbrittenia heucherifolia（Diels）Hilliard ■☆

379920　Sutera gracilis（Diels）Hiern = Jamesbrittenia thunbergii（G. Don）Hilliard ■☆

379921　Sutera grandiflora（Galpin）Hiern；大花裂口花(紫裂口花)●■☆

379922　Sutera grandiflora（Galpin）Hiern = Jamesbrittenia grandiflora（Galpin）Hilliard ■☆

379923　Sutera grandiflora Hiern. = Sutera grandiflora（Galpin）Hiern ●■☆

379924　Sutera griquensis Hiern；格里夸裂口花■☆

379925　Sutera halimifolia（Benth.）Kuntze；滨藜叶裂口花■☆

379926　Sutera henrici Hiern = Jamesbrittenia filicaulis（Benth.）Hilliard ■☆

379927　Sutera hereroensis（Engl.）V. Naray. = Jamesbrittenia hereroensis（Engl.）Hilliard ■☆

379928　Sutera heucherifolia（Diels）Hiern = Jamesbrittenia heucherifolia（Diels）Hilliard ■☆

379929　Sutera hispida（Thunb.）Druce；硬毛裂口花■☆

379930　Sutera huillana（Diels）Hiern = Jamesbrittenia huillana（Diels）Hilliard ■☆

379931　Sutera humifusa Hiern = Sutera platysepala Hiern ■☆

379932　Sutera impedita Hilliard;赘裂口花■☆

379933　Sutera incisa （Thunb.） Hiern ＝ Jamesbrittenia incisa （Thunb.） Hilliard ●☆

379934　Sutera infundibuliformis Schinz ＝ Sutera polyantha （Benth.） Kuntze ■☆

379935　Sutera integerrima （Benth.） Hiern ＝Jamesbrittenia integerrima （Benth.） Hilliard ■☆

379936　Sutera integrifolia （L. f.） Kuntze;全叶裂口花■☆

379937　Sutera integrifolia （L. f.） Kuntze var. parvifolia Hiern ＝ Sutera hispida （Thunb.） Druce ■☆

379938　Sutera intertexta Hiern ＝Sutera campanulata （Benth.） Kuntze ■☆

379939　Sutera kraussiana （Bernh.） Hiern ＝ Jamesbrittenia kraussiana （Bernh.） Hilliard ■☆

379940　Sutera kraussiana （Bernh.） Hiern var. latifolia ？ ＝ Jamesbrittenia argentea （L. f.） Hilliard ■☆

379941　Sutera langebergensis Hilliard;朗厄山裂口花■☆

379942　Sutera latifolia Hiern ＝ Sutera cooperi Hiern ■☆

379943　Sutera laxiflora （Benth.） Kuntze ＝ Sutera halimifolia （Benth.） Kuntze ■☆

379944　Sutera levis Hiern;平滑裂口花■☆

379945　Sutera lilacina Dinter ＝ Jamesbrittenia fruticosa （Benth.） Hilliard ■☆

379946　Sutera linifolia （Thunb.） Kuntze ＝ Sutera uncinata （Desr.） Hilliard ■☆

379947　Sutera linifolia （Thunb.） Kuntze var. heterophylla （Kuntze） Hiern ＝Jamesbrittenia albiflora （I. Verd.） Hilliard ■☆

379948　Sutera litoralis （Schinz） Hiern ＝ Jamesbrittenia fruticosa （Benth.） Hilliard ■☆

379949　Sutera longipedicellata Hilliard;长梗裂口花■☆

379950　Sutera longituba （Dinter） Range ＝ Jamesbrittenia huillana （Diels） Hilliard ■☆

379951　Sutera luteiflora Hiern ＝ Jamesbrittenia micrantha （Klotzsch） Hilliard ■☆

379952　Sutera lychnidea （L.） Hiern ＝Lyperia lychnidea （L.） Druce ■☆

379953　Sutera lyperiiflora （Vatke） V. Naray. ＝ Camptoloma lyperiiflorum （Vatke） Hilliard ■●☆

379954　Sutera lyperioides （Engl.） Engl. ex Range ＝ Jamesbrittenia lyperioides （Engl.） Hilliard ■☆

379955　Sutera macleana Hiern ＝Sutera floribunda （Benth.） Kuntze ■☆

379956　Sutera macrantha Codd ＝ Jamesbrittenia macrantha （Codd） Hilliard ■☆

379957　Sutera macrosiphon （Schltr.） Hiern;大管裂口花■☆

379958　Sutera major （Pilg.） Range ＝ Jamesbrittenia major （Pilg.） Hilliard ■☆

379959　Sutera marifolia （Benth.） Kuntze;芋叶裂口花■☆

379960　Sutera maritima Hiern ＝ Jamesbrittenia maritima （Hiern） Hilliard ■☆

379961　Sutera maxii Hiern ＝Jamesbrittenia maxii （Hiern） Hilliard ■☆

379962　Sutera merxmuelleri Rössler ＝ Jamesbrittenia merxmuelleri （Rössler） Hilliard ■☆

379963　Sutera microphylla （L. f.） Hiern ＝Jamesbrittenia microphylla （L. f.） Hilliard ■☆

379964　Sutera mollis （Benth.） Hiern ＝Jamesbrittenia pinnatifida （L. f.） Hilliard ■☆

379965　Sutera montana （Diels） S. Moore ＝ Jamesbrittenia montana （Diels） Hilliard ■☆

379966　Sutera multiramosa Hilliard;多枝裂口花■☆

379967　Sutera natalensis （Bernh.） Kuntze ＝ Sutera floribunda （Benth.） Kuntze ■☆

379968　Sutera neglecta （J. M. Wood et M. S. Evans） Hiern;疏忽裂口花■☆

379969　Sutera noodsbergensis Hiern ＝ Sutera floribunda （Benth.） Kuntze ■☆

379970　Sutera ochracea Hiern ＝Lyperia antirrhinoides （L. f.） Hilliard ■☆

379971　Sutera oppositiflora （Vent.） Kuntze ＝ Sutera hispida （Thunb.） Druce ■☆

379972　Sutera pallescens Hiern ＝Sutera platysepala Hiern ■☆

379973　Sutera pallida （Pilg.） Overkott ex Rössler ＝ Jamesbrittenia pallida （Pilg.） Hilliard ■☆

379974　Sutera palustris Hiern ＝ Sutera levis Hiern ■☆

379975　Sutera paniculata Hilliard;圆锥裂口花■☆

379976　Sutera pauciflora （Benth.） Kuntze;少花裂口花■☆

379977　Sutera pedunculata （Andréws） Hiern ＝ Jamesbrittenia argentea （L. f.） Hilliard ■☆

379978　Sutera pedunculosa （Benth.） Kuntze ＝Jamesbrittenia pedunculosa （Benth.） Hilliard ■☆

379979　Sutera phlogiflora （Benth.） Hiern ＝ Jamesbrittenia argentea （L. f.） Hilliard ■☆

379980　Sutera pilgeriana （Dinter） Range ＝ Jamesbrittenia pilgeriana （Dinter） Hilliard ■☆

379981　Sutera pinnatifida Hiern ＝ Jamesbrittenia filicaulis （Benth.） Hilliard ■☆

379982　Sutera placida Hilliard;温柔裂口花■☆

379983　Sutera platysepala Hiern;宽瓣裂口花■☆

379984　Sutera polelensis Hiern;波莱尔裂口花■☆

379985　Sutera polelensis Hiern subsp. fraterna Hilliard;兄弟裂口花■☆

379986　Sutera polyantha （Benth.） Kuntze;多花裂口花■☆

379987　Sutera polysepala Hiern ＝Sutera calycina （Benth.） Kuntze ■☆

379988　Sutera primuliflora （Thell.） Range ＝ Jamesbrittenia primuliflora （Thell.） Hilliard ■☆

379989　Sutera pristisepala Hiern ＝ Jamesbrittenia pristisepala （Hiern） Hilliard ■☆

379990　Sutera procumbens （Benth.） Kuntze ＝ Sutera polyantha （Benth.） Kuntze ■☆

379991　Sutera pulchra Norl. ＝ Sutera floribunda （Benth.） Kuntze ■☆

379992　Sutera pumila （Benth.） Kuntze ＝ Sutera halimifolia （Benth.） Kuntze ■☆

379993　Sutera racemosa （Benth.） Kuntze;总花裂口花■☆

379994　Sutera ramosissima Hiern ＝ Jamesbrittenia ramosissima （Hiern） Hilliard ■☆

379995　Sutera remotiflora Dinter ex Range ＝ Manulea robusta Pilg. ■☆

379996　Sutera revoluta （Thunb.） Kuntze;外卷裂口花■☆

379997　Sutera revoluta （Thunb.） Kuntze var. pubescens Hiern ＝ Sutera paniculata Hilliard ■☆

379998　Sutera rhombifolia Schinz ＝ Jamesbrittenia argentea （L. f.） Hilliard ■☆

379999　Sutera rigida L. Bolus ＝ Antherothamnus pearsonii N. E. Br. ●☆

380000　Sutera roseoflava Hiern;粉黄裂口花■☆

380001　Sutera rotundifolia （Benth.） Kuntze;圆叶裂口花■☆

380002　Sutera septentrionalis Hilliard;北方裂口花■☆

380003　Sutera sessilifolia （Diels） Hiern ＝ Jamesbrittenia sessilifolia （Diels） Hilliard ■☆

380004　Sutera silenoides Hilliard ＝ Jamesbrittenia silenoides （Hilliard） Hilliard ■☆

380005　Sutera squarrosa（Pilg.）Hiern ex Range ＝ Jamesbrittenia integerrima（Benth.）Hilliard ■☆

380006　Sutera stenopetala（Diels）Hiern ＝ Jamesbrittenia incisa（Thunb.）Hilliard ●☆

380007　Sutera stenophylla Hiern ＝ Sutera subnuda（N. E. Br.）Hiern ■☆

380008　Sutera subnuda（N. E. Br.）Hiern；亚裸裂口花■☆

380009　Sutera subsessilis Hilliard；近无柄裂口花■☆

380010　Sutera subspicata（Benth.）Kuntze；亚长穗裂口花■☆

380011　Sutera tenella Hiern ＝ Jamesbrittenia tenella（Hiern）Hilliard ■☆

380012　Sutera tenuicaulis Hilliard；细茎裂口花■☆

380013　Sutera tenuiflora（Benth.）Hiern ＝ Lyperia tenuiflora Benth. ■☆

380014　Sutera tenuifolia（Bernh.）Fourc. ＝ Jamesbrittenia tenuifolia（Bernh.）Hilliard ■☆

380015　Sutera tenuis Pilg. ＝ Jamesbrittenia concinna（Hiern）Hilliard ■☆

380016　Sutera tomentosa Hiern ＝ Jamesbrittenia glutinosa（Benth.）Hilliard ■☆

380017　Sutera tortuosa（Benth.）Hiern ＝ Jamesbrittenia tortuosa（Benth.）Hilliard ■☆

380018　Sutera tristis（L. f.）Hiern ＝ Lyperia tristis（L. f.）Benth. ■☆

380019　Sutera tristis（L. f.）Hiern var. montana（Diels）Hiern ＝ Lyperia tristis（L. f.）Benth. ■☆

380020　Sutera tysonii Hiern ＝ Jamesbrittenia tysonii（Hiern）Hilliard ■☆

380021　Sutera uncinata（Desr.）Hilliard；具钩裂口花■☆

380022　Sutera violacea（Schltr.）Hiern；堇色裂口花■☆

380023　Sutera virgulosa Hiern ＝ Jamesbrittenia filicaulis（Benth.）Hilliard ■☆

380024　Sutera welwitschii V. Naray. ＝ Camptoloma rotundifolia Benth. ■☆

380025　Sutera zambesica（R. E. Fr.）R. E. Fr. ＝ Jamesbrittenia zambesica（R. E. Fr.）Hilliard ■☆

380026　Suteria DC. ＝ Psychotria L.（保留属名）●

380027　Sutheriandia R. Br. ex Aiton ＝ Sutherlandia R. Br.（保留属名）●☆

380028　Sutherlandia J. F. Gmel.（废弃属名）＝ Heritiera Aiton ●

380029　Sutherlandia J. F. Gmel.（废弃属名）＝ Sutherlandia R. Br.（保留属名）●☆

380030　Sutherlandia R. Br.（1812）（保留属名）；气球豆属（纸荚豆属）●☆

380031　Sutherlandia R. Br. ex W. T. Aiton ＝ Sutherlandia R. Br.（保留属名）●☆

380032　Sutherlandia frutescens（L.）R. Br.；气球豆（纸荚豆）；Ballon Pea, Cancer Bush, Cape Bladder Pea, Duck Plant ●☆

380033　Sutherlandia frutescens R. Br. ＝ Sutherlandia frutescens（L.）R. Br. ●☆

380034　Sutherlandia humilis E. Phillips et R. A. Dyer；小气球豆●☆

380035　Sutherlandia microphylla Burch. ex DC.；小叶气球豆●☆

380036　Sutherlandia montana E. Phillips et R. A. Dyer；山地气球豆●☆

380037　Sutherlandia speciosa E. Phillips et R. A. Dyer；美丽气球豆●☆

380038　Sutherlandia tomentosa Eckl. et Zeyh.；毛气球豆●☆

380039　Sutrina Lindl.（1842）；苏特兰属■☆

380040　Sutrina bicolor Lindl.；苏特兰■☆

380041　Suttonia A. Rich. ＝ Rapanea Aubl. ●

380042　Suttonia Mez ＝ Suttonia A. Rich. ●

380043　Suxifraga turfosa Engl. et Irmsch. ＝ Saxifraga glaucophylla Franch. ■

380044　Suzukia Kudo（1930）；台钱草属（铃木草属，台连钱属）；Suzukia ■★

380045　Suzukia luchuensis Kudo；齿唇台钱草（琉球铃木草）；Luchuen Suzukia, Luchun Suzukia, Toothlip Suzukia ■

380046　Suzukia shikikunensis Kudo；台钱草（假马蹄草，铃木草）；Taiwan Suzukia ■

380047　Suzygium P. Browne（废弃属名）＝ Calyptranthes Sw.（保留属名）●☆

380048　Suzygium P. Browne（废弃属名）＝ Syzygium R. Br. ex Gaertn.（保留属名）●

380049　Svenhedinia Urb. ＝ Talauma Juss. ●

380050　Svenkoeltzia Burns-Bal. ＝ Funkiella Schltr. ■☆

380051　Svenkoeltzia Burns-Bal. ＝ Stenorrhynchos Rich. ex Spreng. ■☆

380052　Svensonia Moldenke ＝ Chascanum E. Mey.（保留属名）●☆

380053　Svensonia laeta（Walp.）Moldenke ＝ Chascanum laetum Walp. ●☆

380054　Svensonia moldenkei J. B. Gillett ＝ Chascanum moldenkei（J. B. Gillett）Sebsebe et Verdc. ●☆

380055　Sventenia Font Quer ＝ Sonchus L. ■

380056　Sventenia Font Quer（1949）；斯文菊属■☆

380057　Sventenia bupleuroides Font Quer ＝ Sonchus bupleuroides（Font Quer）N. Kilian et Greuter ■☆

380058　Svida Opiz ＝ Cornus L. ●

380059　Svida Small ＝ Cornus L. ●

380060　Svida Small ＝ Swida Opiz ●

380061　Svida sanguinea（L.）Opiz. ＝ Swida sanguinea（L.）Opiz ●

380062　Svitramia Cham.（1835）；斯维野牡丹属●☆

380063　Svitramia pulchra Cham.；斯维野牡丹●☆

380064　Svjda Opiz ＝ Swida Opiz ●

380065　Swainsona Salisb.（1806）；澳洲苦马豆属（苦马豆属，年豆属，斯氏豆属，斯万森木属，枝弯豆属）；Swainson Pea, Swainson-pea, Glory Pea ●■☆

380066　Swainsona formosa（G. Don）Joy Thomps.；美丽澳洲苦马豆（美丽斯氏豆）；Desett Pea, Glory Pea, Sturt Desert Pea, Sturt's Desert Pea ●☆

380067　Swainsona galegifolia（Andréws）R. Br.；山羊豆叶澳洲苦马豆（光叶苦马豆，山羊豆叶斯氏豆，山羊豆叶斯万森木）；Darling Pea, Smooth Darling Pea ●☆

380068　Swainsona greyana Lindl.；格雷澳洲苦马豆（格雷斯氏豆）；Darling pea ●☆

380069　Swainsona salsula（Pall.）Taub. ＝ Sphaerophysa salsula（Pall.）DC. ●■

380070　Swainsona sejuncta Joy Thomps.；脱落澳洲苦马豆（美丽澳洲苦马豆，美丽斯氏豆，美丽斯万森木）●☆

380071　Swainsonia Salisb. ＝ Swainsona Salisb. ●■☆

380072　Swainsonia Spreng. ＝ Swainsona Salisb. ●■☆

380073　Swainsonia salsula（Pall.）Taub. ＝ Sphaerophysa salsula（Pall.）DC. ●■

380074　Swallenia Soderstr. et H. F. Decker（1963）；斯沃伦草属（斯窝伦草属）■☆

380075　Swallenia alexandrae（Swallen）Soderstr. et H. F. Decker；斯沃伦草■☆

380076　Swallenochloa McClure ＝ Chusquea Kunth ●☆

380077　Swammerdamia DC. ＝ Helichrysum Mill.（保留属名）●■

380078　Swanalloia Horq ex Walp. ＝ Juanulloa Ruiz et Pav. ●☆

380079　Swantia Alef. ＝ Vicia L. ■

380080　Swartsia J. F. Gmel. ＝ Solandra Sw.（保留属名）●☆

380081　Swartzia Schreb.（1791）（保留属名）；铁木豆属●☆

380082　Swartzia fistuloides Harms ＝ Bobgunnia fistuloides（Harms）J. H. Kirkbr. et Wiersema ●☆

380083　Swartzia madagascariensis Desv.；马岛铁木豆；Ironheart Tree,

Snake Bean Tree ●☆

380084　Swartzia madagascariensis Desv. = Bobgunnia madagascariensis（Desv.）J. H. Kirkbr. et Wiersema ●☆

380085　Swartzia madagascariensis Desv. f. glabrescens G. C. C. Gilbert et Boutique = Bobgunnia madagascariensis（Desv.）J. H. Kirkbr. et Wiersema ●☆

380086　Swartzia madagascariensis Desv. f. grandifoliolata G. C. C. Gilbert et Boutique = Bobgunnia madagascariensis（Desv.）J. H. Kirkbr. et Wiersema ●☆

380087　Swartzia marginata Benth. = Bobgunnia madagascariensis（Desv.）J. H. Kirkbr. et Wiersema ●☆

380088　Swartzia sapinii De Wild. = Bobgunnia madagascariensis（Desv.）J. H. Kirkbr. et Wiersema ●☆

380089　Swartziaceae（DC.）Bartl. = Fabaceae Lindl.（保留科名）●■

380090　Swartziaceae（DC.）Bartl. = Leguminosae Juss.（保留科名）●■

380091　Swartziaceae Bani. = Fabaceae Lindl.（保留科名）●■

380092　Swartziaceae Bani. = Leguminosae Juss.（保留科名）●■

380093　Swartziaceae Bartl. = Fabaceae Lindl.（保留科名）●■

380094　Swartziaceae Bartl. = Leguminosae Juss.（保留科名）●■

380095　Swarzia Retz. = Costus L. ■

380096　Swarzia Retz. = Hellenia Retz. ■

380097　Sweertia Post et Kuntze = Swertia Boehm. ●■☆

380098　Sweertia Post et Kuntze = Tolpis Adans. ●■☆

380099　Sweertia W. D. J. Koch = Swertia L. ■

380100　Sweetia DC. = Galactia P. Browne ■

380101　Sweetia Spreng.（1825）（保留属名）;斯威特豆属;Sucupira ●☆

380102　Sweetia fruticosa Spreng. ;斯威特豆 ●☆

380103　Sweetiopsis Chodat et Hassk. = Riedeliella Harms ■☆

380104　Swertia Boehm. = Tolpis Adans. ●■☆

380105　Swertia L.（1753）;獐牙菜属（当药属）;Ferworth,Swertia ■

380106　Swertia aberdarica T. C. E. Fr. = Swertia kilimandscharica Engl. ■☆

380107　Swertia abyssinica Hochst. ;阿比西尼亚獐牙菜 ■☆

380108　Swertia adolfi-friderici Mildbr. et Gilg;弗里德里西獐牙菜 ■☆

380109　Swertia affinis C. B. Clarke = Swertia angustifolia Buch. -Ham. ex D. Don var. pulchella（D. Don）Burkill ■

380110　Swertia alata Hayata = Swertia arisanensis Hayata ■

380111　Swertia alba T. N. Ho et S. W. Liu;白花獐牙菜;White Swertia, Narrowleaved Swertia ■

380112　Swertia alboviolacea H. Lév. = Swertia cincta Burkill ■

380113　Swertia angustifolia Buch. -Ham. ex D. Don;狭叶獐牙菜（滇獐牙菜,肝炎草,华南当药,美丽獐牙菜,青叶丹,青叶胆,思茅獐牙菜,土疸药,小青鱼胆,粤北獐牙菜,云南当药）●■

380114　Swertia angustifolia Buch. -Ham. ex D. Don var. hamiltoniana Burkill = Swertia angustifolia Buch. -Ham. ex D. Don ■

380115　Swertia angustifolia Buch. -Ham. ex D. Don var. pulchella（D. Don）Burkill;美丽獐牙菜（肝炎草,青叶胆,青鱼草,水黄连,土疸药,小青鱼胆,粤北獐牙菜）;Beautiful Swertia ■

380116　Swertia anomala Nakai = Swertia tetrapetala Pall. ■

380117　Swertia arisanensis Hayata;阿里山獐牙菜（阿里山当药）;Alishan Swertia ■

380118　Swertia asarifolia Franch. ;细辛叶獐牙菜;Wildgingerleaf Swertia ■

380119　Swertia asterocalyx T. N. Ho et S. W. Liu;星萼獐牙菜;Starcalyx Swertia ■

380120　Swertia asterocalyx T. N. Ho et S. W. Liu = Swertia cuneata Wall. ex D. Don ■

380121　Swertia atroviolacea Harry Sm. ;黑紫獐牙菜;Darkpurple Swertia ■

380122　Swertia atroviolacea Harry Sm. = Swertia asarifolia Franch. ■

380123　Swertia aucheri Boiss. ;奥氏獐牙菜 ■☆

380124　Swertia bella Hemsl. = Lomatogonium bellum（Hemsl.）Harry Sm. ■

380125　Swertia biauriculata H. Lév. = Swertia bimaculata（Siebold et Zucc.）Hook. f. et Thomson ex C. B. Clarke ■

380126　Swertia bifolia Batalin;二叶獐牙菜（乌金草,异花獐牙菜）;Twoleaved Swertia ■

380127　Swertia bifolia Batalin var. wardii（C. Marquand）T. N. Ho et S. W. Liu;少花二叶獐牙菜;Fewflower Twoleaved Swertia ■

380128　Swertia bimaculata（Siebold et Zucc.）Hook. f. et Thomson ex C. B. Clarke f. impunctata（Makino）Satake;无斑獐牙菜 ■☆

380129　Swertia bimaculata（Siebold et Zucc.）Hook. f. et Thomson ex C. B. Clarke var. arisanensis（Hayata）S. S. Ying = Swertia arisanensis Hayata ■

380130　Swertia bimaculata（Siebold et Zucc.）Hook. f. et Thomson ex C. B. Clarke var. macrocarpa Nakai = Swertia bimaculata（Siebold et Zucc.）Hook. f. et Thomson ex C. B. Clarke ■

380131　Swertia bimaculata（Siebold et Zucc.）Hook. f. et Thomson ex C. B. Clarke;獐牙菜（大苦草,黑节苦草,黑药黄,双斑獐牙菜,双点獐牙菜,蓑衣草,紫花青叶胆,走胆草）;Swertia,Twospot Swertia ■

380132　Swertia binchuanensis T. N. Ho et S. W. Liu;滨川獐牙菜;Binchuan Swertia ■

380133　Swertia bonatiana Burkill = Lomatogonium forrestii（Balf. f.）Fernald var. bonatianum（Burkill）T. N. Ho ■

380134　Swertia brachyanthera（C. B. Clarke）Knobl. = Lomatogonium brachyantherum（C. B. Clarke）Fernald ■

380135　Swertia brevipedicellata Gilg ex N. E. Br. = Swertia abyssinica Hochst. ■☆

380136　Swertia brownii Shah;热非獐牙菜 ■☆

380137　Swertia burkilliana W. W. Sm. = Veratrilla burkilliana（W. W. Sm.）Harry Sm. ■

380138　Swertia caerulea A. Chev. = Djaloniella ypsilostyla P. Taylor ■☆

380139　Swertia caerulea Royle = Lomatogonium caeruleum（Royle）Harry Sm. ex B. L. Burtt ■

380140　Swertia calycina Franch. ;叶萼獐牙菜（抱萼獐牙菜,草黄连）;Leaflikecalyx Swertia ■

380141　Swertia calycina Franch. var. major Diels = Swertia calycina Franch. ■

380142　Swertia calycina N. E. Br. = Swertia brownii Shah ■☆

380143　Swertia carinthiaca Wulfen = Lomatogonium carinthiacum（Wulfen）Rchb. ■

380144　Swertia carinthiaca Wulfen var. afghanica Burkill = Lomatogonium brachyantherum（C. B. Clarke）Fernald ■

380145　Swertia caroliniensis（Walter）Kuntze = Frasera caroliniensis Walter ■☆

380146　Swertia caroliniensis（Walter）Kunze;美洲獐牙菜;Amer, American Columbo,Columbo,Green Gentian,Monument Plant ■☆

380147　Swertia cavaleriei H. Lév. = Swertia nervosa（G. Don）Wall. ex C. B. Clarke ■

380148　Swertia cavbaleriei H. Lév. = Swertia nervosa（G. Don）Wall. ex C. B. Clarke ■

380149　Swertia changii S. Z. Yang,C. F. Chen et C. H. Chen;张氏獐牙菜 ■

380150　Swertia chinensis Franch. = Lomatogonium forrestii（Balf. f.）Fernald var. bonatianum（Burkill）T. N. Ho ■

380151 Swertia chinensis Franch. ex Hemsl. = Swertia diluta（Turcz.）Benth. et Hook. f. ■

380152 Swertia chinensis Franch. f. grandiflora Franch. = Swertia pseudochinensis H. Hara ■

380153 Swertia chinensis Franch. f. stenopetala Franch. = Swertia diluta（Turcz.）Benth. et Hook. f. ■

380154 Swertia chinensis Franch. f. violacea Makino = Swertia pseudochinensis H. Hara ■

380155 Swertia chinensis Franch. var. tosaensis Makino = Swertia diluta（Turcz.）Benth. et Hook. f. var. tosaensis（Makino）H. Hara ■

380156 Swertia chinensis Franch. var. tosaensis Makino = Swertia tosaensis Makino ■

380157 Swertia chirayita Buch.-Ham. ex Wall.；印度獐牙菜（印度龙胆,斋瑞塔獐牙菜）；Chiretta, Indian Balmony, Indian Gentian ■☆

380158 Swertia chumbica Burkill = Lomatogonium chumbicum（Burkill）Harry Sm. ■

380159 Swertia ciliata（D. Don ex G. Don）B. L. Burtt；普兰獐牙菜（带紫獐牙菜,蒂达）；Pulan Swertia ■

380160 Swertia cincta Burkill；西南獐牙菜（大青叶胆）；Surrounded Swertia ■

380161 Swertia clarenceana Hook. f. = Swertia abyssinica Hochst. ■☆

380162 Swertia clarkei Knobl. = Lomatogonium brachyantherum（C. B. Clarke）Fernald ■

380163 Swertia coerulea Royle = Lomatogonium coeruleum（Burkill）Harry Sm. ■☆

380164 Swertia conaensis T. N. Ho et S. W. Liu；错那獐牙菜；Cuona Swertia ■

380165 Swertia connata Schrenk；短筒獐牙菜（合生獐牙菜）；Shorttube Swertia ■

380166 Swertia cordata（G. Don）Wall. ex C. B. Clarke；心叶獐牙菜；Heartleaf Swertia ■

380167 Swertia cordata Wall. = Swertia cordata（G. Don）Wall. ex C. B. Clarke ■

380168 Swertia corniculata L. = Halenia corniculata（L.）Cornaz ■

380169 Swertia crassiuscula Gilg；稍粗獐牙菜■☆

380170 Swertia crassiuscula Gilg subsp. robusta Sileshi；粗壮稍粗獐牙菜■☆

380171 Swertia crassiuscula Gilg var. leucantha（T. C. E. Fr.）Sileshi；白花稍粗獐牙菜■☆

380172 Swertia cuneata Wall. ex D. Don；楔叶獐牙菜；Cuneateleaf Swertia ■

380173 Swertia curtioides Gilg = Swertia usambarensis Engl. var. curtioides（Gilg）Sileshi ■☆

380174 Swertia cuspidata（Maxim.）Kitag. = Swertia perennis L. subsp. cuspidata（Maxim.）H. Hara ■☆

380175 Swertia davidii Franch.；川东獐牙菜（金盆,青鱼胆草,水黄连,水灵芝,鱼胆草）；E. Sichuan Swertia ■

380176 Swertia decora Franch.；观赏獐牙菜；Decorative Swertia ■

380177 Swertia decussata Nimmo；对生獐牙菜■☆

380178 Swertia deflexa Sm. = Halenia deflexa（Sm.）Griseb. ■☆

380179 Swertia delavayi Franch.；丽江獐牙菜（青叶胆）；Delevay Swertia, Lijiang Swertia ■

380180 Swertia deltoides Burkill = Lomatogonium macranthum（Diels et Gilg）Fernald ■

380181 Swertia dichotoma L.；歧伞獐牙菜（类海绿,歧伞当药,腺鳞草）；Forked Swertia ■

380182 Swertia dichotoma L. var. punctata T. N. He et J. X. Yang；紫斑歧伞獐牙菜（斑点歧伞獐牙菜）■

380183 Swertia dilatata C. B. Clarke = Swertia paniculata Wall. ■

380184 Swertia diluta（Turcz.）Benth. et Hook. f.；北方獐牙菜（淡花当药,淡味当药,淡味獐牙菜,当药,苦草,水黄莲,水灵芝,乌金散,小方杆,兴安獐牙菜,獐牙菜,中国当药）；Diluted Swertia ■

380185 Swertia diluta（Turcz.）Benth. et Hook. f. var. tosaensis（Makino）H. Hara = Swertia tosaensis Makino ■

380186 Swertia dimorpha Batalin = Swertia tetraptera Pall. ■

380187 Swertia dissimilis N. E. Br. = Swertia abyssinica Hochst. ■☆

380188 Swertia divaricata Harry Sm.；叉序獐牙菜；Divaricate Swertia ■

380189 Swertia duclouxii Burkill = Swertia punicea Hemsl. ■

380190 Swertia duemmeriana T. C. E. Fr. = Swertia crassiuscula Gilg ■☆

380191 Swertia elata Harry Sm.；高獐牙菜；Tall Swertia ■

380192 Swertia ellenbeckiana Gilg ex Engl. = Swertia tischeri Chiov. ■☆

380193 Swertia elongata T. N. Ho et S. W. Liu = Swertia elongata T. N. Ho et S. W. Liu ex J. X. Yang ■

380194 Swertia elongata T. N. Ho et S. W. Liu = Swertia kouitchensis Franch. ■

380195 Swertia elongata T. N. Ho et S. W. Liu ex J. X. Yang；伸梗獐牙菜；Elongate Swertia ■

380196 Swertia elongata T. N. Ho et S. W. Liu ex J. X. Yang = Swertia kouitchensis Franch. ■

380197 Swertia emeiensis Ma ex T. N. Ho et S. W. Liu；峨眉獐牙菜（一匹瓦）；Emei Swertia ■

380198 Swertia eminii Engl.；埃明獐牙菜■☆

380199 Swertia endotricha Harry Sm.；直毛獐牙菜；Erecthair Swertia ■

380200 Swertia engleri Gilg；恩格勒獐牙菜■☆

380201 Swertia engleri Gilg var. woodii（J. Shan）Sileshi；伍得獐牙菜■☆

380202 Swertia erosula N. E. Br.；啮蚀状獐牙菜■☆

380203 Swertia erosula N. E. Br. = Swertia kilimandscharica Engl. ■☆

380204 Swertia erythraea Chiov. = Swertia abyssinica Hochst. ■☆

380205 Swertia erythrosticta Maxim.；红直獐牙菜（红点獐牙菜,红直当药）；Redspot Swertia ■

380206 Swertia erythrosticta Maxim. var. epunctata T. N. Ho et S. W. Liu；素色獐牙菜；Epunctate Swertia ■

380207 Swertia esquirolii H. Lév. = Swertia angustifolia Buch.-Ham. ex D. Don var. pulchella（D. Don）Burkill ■

380208 Swertia fasciculata T. N. Ho et S. W. Liu；簇花獐牙菜；Clusterflowered Swertia ■

380209 Swertia filicaulis Gilg；线茎獐牙菜■☆

380210 Swertia fimbriata（Hochst.）Cufod.；流苏獐牙菜■☆

380211 Swertia forrestii Harry Sm.；紫萼獐牙菜；Purplecalyx Swertia ■

380212 Swertia franchetiana Harry Sm.；抱茎獐牙菜；Amlexicaul Swertia, Clasping Swertia ■

380213 Swertia fwambensis N. E. Br. = Swertia quartiniana A. Rich. ■☆

380214 Swertia gamosepala Burkill = Lomatogonium gamosepalum（Burkill）Harry Sm. ■

380215 Swertia gentianifolia Chiov. = Swertia kilimandscharica Engl. ■☆

380216 Swertia gentianoides Franch. = Gentianella gentianoides（Franch.）Harry Sm. ■

380217 Swertia graciliflora Gontsch.；细花獐牙菜；Smallflowered Swertia ■

380218 Swertia gracilis Franch. = Swertia tenuis T. N. Ho et S. W. Liu ■

380219 Swertia granvikii T. C. E. Fr. = Swertia crassiuscula Gilg ■☆

380220 Swertia guibeiensis C. Z. Gao；桂北獐牙菜；Guibei Swertia ■

380221 Swertia gyacaensis T. N. Ho et S. W. Liu；加查獐牙菜；Jiacha Swertia ■

380222 Swertia handeliana Harry Sm. ;矮獐牙菜;Dwarf Swertia ■

380223 Swertia heterantha Y. Ling = Swertia bifolia Batalin ■

380224 Swertia heterosepala Gilg = Swertia pleurogynoides Baker ■☆

380225 Swertia hickinii Burkill;浙江獐牙菜;Zhejiang Swertia ■

380226 Swertia hispidicalyx Burkill;毛萼獐牙菜;Hairycalyx Swertia ■

380227 Swertia hispidicalyx Burkill f. subglabra Marquand = Swertia hispidicalyx Burkill var. minima Burkill ■

380228 Swertia hispidicalyx Burkill var. major Burkill = Swertia hispidicalyx Burkill var. minima Burkill ■

380229 Swertia hispidicalyx Burkill var. minima Burkill;小毛萼獐牙菜;Small Hairycalyx Swertia ■

380230 Swertia hookeri C. B. Clarke;粗壮獐牙菜;Strong Swertia ■

380231 Swertia hypericoides Diels = Swertia fasciculata T. N. Ho et S. W. Liu ■

380232 Swertia intermixta A. Rich. ;混杂獐牙菜■☆

380233 Swertia janssensii De Wild. = Swertia kilimandscharica Engl. ■☆

380234 Swertia japonica (Schult.) Makino;日本当药(日本獐牙菜);Japan Swertia,Japanese Swertia ■☆

380235 Swertia japonica (Schult.) Makino f. chionantha F. Maek. ;雪花獐牙菜■☆

380236 Swertia japonica (Schult.) Makino f. littoralis Hid. Takah. ;滨海獐牙菜■☆

380237 Swertia japonica (Schult.) Makino f. plena (Makino) Satake;重瓣日本当药■☆

380238 Swertia japonica (Schult.) Makino var. latifolia Konta;宽叶日本当药■☆

380239 Swertia jiandeensis Y. Y. Fang;建德獐牙菜;Jiande Swertia ■

380240 Swertia johnsonii N. E. Br. = Swertia fimbriata (Hochst.) Cufod. ■☆

380241 Swertia kanasimi Satake = Swertia tashiroi (Maxim.) Makino ■☆

380242 Swertia keniensis T. C. E. Fr. = Swertia volkensii Gilg ■☆

380243 Swertia kilimandscharica Engl. ;基利獐牙菜■☆

380244 Swertia kingii Hook. f. ;黄花獐牙菜;Yellow Swertia ■

380245 Swertia kouitchensis Franch. ;贵州獐牙菜;Guizhou Swertia ■

380246 Swertia kuroiwae Makino = Swertia tashiroi (Maxim.) Makino ■☆

380247 Swertia kuroiwai Makino var. condensata ? = Swertia tashiroi (Maxim.) Makino ■☆

380248 Swertia kuroiwai Makino var. laxa ? = Swertia tashiroi (Maxim.) Makino ■☆

380249 Swertia kuroiwai Makino var. shintenensis (Hayata) Satake = Swertia shintenensis Hayata ■

380250 Swertia lactea Bunge;乳白獐牙菜■☆

380251 Swertia lastii Engl. = Swertia abyssinica Hochst. ■☆

380252 Swertia leducii Franch. ;蒙自獐牙菜(肝炎草,苦胆草,青叶胆,青鱼胆,土疸药,小青鱼胆);Mengzi Swertia, Mile Swertia, Millen Swertia ■

380253 Swertia leucantha T. C. E. Fr. = Swertia crassiuscula Gilg var. leucantha (T. C. E. Fr.) Sileshi ■☆

380254 Swertia lloydioides Burkill = Lomatogonium brachyantherum (C. B. Clarke) Fernald ■

380255 Swertia lloydioides Burkill = Lomatogonium lloydioides (Burkill) Harry Sm. ex Chater ■

380256 Swertia longipes Franch. ;长梗獐牙菜;Longstalk Swertia ■

380257 Swertia longipes Franch. = Swertia davidii Franch. ■

380258 Swertia lovenii T. C. E. Fr. = Swertia kilimandscharica Engl. ■☆

380259 Swertia lubahniana (Vatke) Engl. = Swertia rosulata (Baker) Klack. ■☆

380260 Swertia lugardae Bullock;卢格德獐牙菜■☆

380261 Swertia luquanensis S. W. Liu;禄劝獐牙菜■

380262 Swertia macrosepala Gilg;大萼獐牙菜■☆

380263 Swertia macrosepala Gilg subsp. microsperma Sileshi;小籽獐牙菜■☆

380264 Swertia macrosperma (C. B. Clarke) C. B. Clarke;大籽獐牙菜(大籽当药,峦大当药);Bigseed Swertia ■

380265 Swertia maculata ?;斑点獐牙菜■☆

380266 Swertia mairei H. Lév. = Swertia bimaculata (Siebold et Zucc.) Hook. f. et Thomson ex C. B. Clarke ■

380267 Swertia makinoana F. Maek. ;牧野氏獐牙菜■☆

380268 Swertia mannii Hook. f. ;曼氏獐牙菜■☆

380269 Swertia manshurica (Kom.) Kitag. ;东北獐牙菜■

380270 Swertia manshurica (Kom.) Kitag. = Swertia perennis L. ■

380271 Swertia marginata Schrenk;膜边獐牙菜(边膜獐牙菜);Membraneedge Swertia ■

380272 Swertia marginata Schrenk. = Swertia wolfgangiana Grüning ■

380273 Swertia matsudae Hayata ex Satake;细叶獐牙菜(松田獐牙菜,细叶当药)■

380274 Swertia matsudae Hayata ex Satake = Swertia tozanensis Hayata ■

380275 Swertia mattirolii Chiov. ;马蒂奥里獐牙菜■☆

380276 Swertia mearnsii De Wild. = Swertia crassiuscula Gilg ■☆

380277 Swertia mekongensis Balf. f. et Forrest = Veratrilla baillonii Franch. ■

380278 Swertia membranifolia Franch. ;膜叶獐牙菜;Membraneleaf Swertia ■

380279 Swertia mildbraedii Gilg = Swertia adolfi-friderici Mildbr. et Gilg ■☆

380280 Swertia mileensis T. N. Ho et W. L. Shih = Swertia leducii Franch. ■

380281 Swertia minima Gilg;极小獐牙菜■☆

380282 Swertia multicaulis D. Don;多茎獐牙菜;Pluristem Swertia ■

380283 Swertia multicaulis D. Don var. umbellifera T. N. Ho et S. W. Liu;伞花獐牙菜■

380284 Swertia multicaulis Engl. ex Gilg = Swertia engleri Gilg ■☆

380285 Swertia mussotii Franch. ;川西獐牙菜;Mussot Swertia, W. Sichuan Swertia ■

380286 Swertia mussotii Franch. var. flavescens T. N. Ho et S. W. Liu;黄花川西獐牙菜(地鉴,黄花药);Yellowflower Mussot Swertia ■

380287 Swertia nervosa (G. Don) Wall. ex C. B. Clarke;显脉獐牙菜(四棱草,翼梗獐牙菜);Nervate Swertia ■

380288 Swertia nervosa Wall. = Swertia nervosa (G. Don) Wall. ex C. B. Clarke ■

380289 Swertia nummularifolia Baker = Exacum spathulatum Baker ■☆

380290 Swertia obmsa Ledeb. var. qunheensis T. N. Ho et S. W. Liu = Swertia connata Schrenk ■

380291 Swertia obtusa Ledeb. ;互叶獐牙菜(钝獐牙菜);Obtuse Swertia ■

380292 Swertia obtusa Ledeb. var. quingheensis T. N. Ho et S. W. Liu;清河獐牙菜■

380293 Swertia obtusa Ledeb. var. quingheensis T. N. Ho et S. W. Liu = Swertia connata Schrenk ■

380294 Swertia obtusipetala Grüning = Swertia wolfgangiana Grüning ■

380295 Swertia oculata Hemsl. ;鄂西獐牙菜;W. Hubei Swertia ■

380296 Swertia paniculata Wall. ;宽丝獐牙菜;Widefilament Swertia ■

380297 Swertia parnassiiflora T. C. E. Fr. = Swertia crassiuscula Gilg var. leucantha (T. C. E. Fr.) Sileshi ■☆

380298 Swertia patens Burkill;斜茎獐牙菜(广展獐牙菜,金沙青叶胆,金

沙獐牙菜,小儿腹痛草,斜茎青叶胆);Slantingstem Swertia ■

380299 Swertia patula Harry Sm. ;开展獐牙菜;Spread Swertia ■

380300 Swertia pauciflora Harry Sm. = Swertia souliaei Burkill ■

380301 Swertia pauciflora Harry Sm. = Swertia tibetica Batalin ■

380302 Swertia perennis L. ;宿根獐牙菜(北温带獐牙菜,东北獐牙菜,多年生獐牙菜,多年獐牙菜);Alpine-bog Swertla, Elkweed, Felwort,Green Gentian, Marsh Felwort, NE. China Swertia, Perennial Swertia ■

380303 Swertia perennis L. subsp. cuspidata (Maxim.) H. Hara;骤尖宿根獐牙菜■☆

380304 Swertia perennis L. subsp. obtusa (Ledeb.) Hara = Swertia obtusa Ledeb. ■

380305 Swertia perennis L. var. cuspidata Maxim. = Swertia perennis L. subsp. cuspidata (Maxim.) H. Hara ■☆

380306 Swertia perennis L. var. manshurica Kom. = Swertia perennis L. ■

380307 Swertia petiolata Royle ex D. Don;长柄獐牙菜(叶柄獐牙菜);Petiolate Swertia ■

380308 Swertia petiolata Royle ex D. Don = Swertia souliaei Burkill ■

380309 Swertia petitiana A. Rich. ;佩蒂那獐牙菜■☆

380310 Swertia phragmitiphylla T. N. Ho et S. W. Liu = Swertia wardii C. Marquand ■

380311 Swertia phragmitiphylla T. N. Ho et S. W. Liu var. rigida T. N. Ho et S. W. Liu = Swertia wardii C. Marquand var. rigida (T. N. Ho et S. W. Liu) T. N. Ho ■

380312 Swertia phragmitiphylla T. N. Ho et S. W. Liu var. rigida T. N. Ho et S. W. Liu = Swertia wardii C. Marquand ■

380313 Swertia pianmaensis T. N. Ho et S. W. Liu;片马獐牙菜;Pianma Swertia ■

380314 Swertia platyphylla Merr. = Swertia bimaculata (Siebold et Zucc.) Hook. f. et Thomson ex C. B. Clarke ■

380315 Swertia pleurogynoides Baker;异萼獐牙菜■☆

380316 Swertia polyantha Gilg = Swertia usambarensis Engl. ■☆

380317 Swertia porphyrantha Baker = Swertia abyssinica Hochst. ■☆

380318 Swertia przewalskii Pissjauk. ;祁连獐牙菜;Qilianshan Swertia ■

380319 Swertia pseudochinensis H. Hara;瘤毛獐牙菜(大苦草,当药,黑节骨草,黑菊黄,蓑衣莲,獐牙菜,紫花当药,紫花青叶胆,走胆草);False Chinese Swertia,Tumaorhair Swertia ■

380320 Swertia pseudochinensis H. Hara f. grandiflora (Franch.) H. Hara = Swertia pseudochinensis H. Hara ■

380321 Swertia pubescens Franch. ;毛獐牙菜;Pubescent Swertia ■

380322 Swertia pulchella Buch. -Ham. ex Wall. = Swertia angustifolia Buch. -Ham. ex D. Don var. pulchella (D. Don) Burkill ■

380323 Swertia pumila Hochst. ;低矮獐牙菜■☆

380324 Swertia punicea Hemsl. ;紫红獐牙菜(草龙胆,苦胆草,青叶胆,山飘儿草,水黄连,土黄连);Scarlet Swertia ■

380325 Swertia punicea Hemsl. var. lutescens Franch. ex Harry Sm. = Swertia punicea Hemsl. var. lutescens Franch. ex T. N. Ho ■

380326 Swertia punicea Hemsl. var. lutescens Franch. ex T. N. Ho;淡黄獐牙菜;Yellowish Scarlet Swertia ■

380327 Swertia purpurascens (Wall. ex D. Don) C. B. Clarke var. violaceocincta Franch. = Swertia cincta Burkill ■

380328 Swertia purpurascens Wall. = Swertia ciliata (D. Don ex G. Don) B. L. Burtt ■

380329 Swertia purpurascens Wall. ex C. B. Clarke = Swertia ciliata (D. Don ex G. Don) B. L. Burtt ■

380330 Swertia purpurascens Wall. ex C. B. Clarke var. violaceocincta Franch. = Swertia cincta Burkill ■

380331 Swertia pusilla Diels = Swertia tetraptera Pall. ■

380332 Swertia quartiniana A. Rich. ;夸尔廷獐牙菜■☆

380333 Swertia racemosa (Griseb.) Wall. ex C. B. Clarke;藏獐牙菜;Raceme Swertia ■

380334 Swertia racemosa (Wall. ex Griseb.) C. B. Clarke = Swertia racemosa (Griseb.) Wall. ex C. B. Clarke ■

380335 Swertia racemosa Wall. = Swertia racemosa (Griseb.) Wall. ex C. B. Clarke ■

380336 Swertia radiata (Kellogg) Kuntze;鹿耳獐牙菜;Deer's Ear, Deer's Ears ■☆

380337 Swertia randaiensis Hayata = Swertia macrosperma (C. B. Clarke) C. B. Clarke ■

380338 Swertia richardii Engl. ;理查德獐牙菜■☆

380339 Swertia rosea Burkill = Swertia decora Franch. ■

380340 Swertia rosularis T. N. Ho et S. W. Liu;莲座獐牙菜;Rosulate Swertia ■

380341 Swertia rosulata (Baker) Klack. ;马岛獐牙菜■☆

380342 Swertia rotata L. = Lomatogonium rotatum (L.) Fr. ex Fernald ■

380343 Swertia rotata Thunb. = Swertia pseudochinensis H. Hara ■

380344 Swertia rotundiglandula T. N. Ho et S. W. Liu;圆腺獐牙菜;Roundglandula Swertia ■

380345 Swertia sattimae T. C. E. Fr. = Swertia crassiuscula Gilg ■☆

380346 Swertia scandens H. Lév. = Swertia macrosperma (C. B. Clarke) C. B. Clarke ■

380347 Swertia scandens T. C. E. Fr. ;攀缘獐牙菜■☆

380348 Swertia scapiformis T. N. Ho et S. W. Liu;花葶獐牙菜;Scape Swertia ■

380349 Swertia schimperi (Hochst.) Griseb. = Swertia fimbriata (Hochst.) Cufod. ■☆

380350 Swertia schliebenii Mildbr. ;施利本獐牙菜■☆

380351 Swertia scottii J. Shah;司科特獐牙菜■☆

380352 Swertia sharpei N. E. Br. = Swertia welwitschii Engl. ■☆

380353 Swertia shigucao Z. Y. Zhu;石骨草(小疝气草);Shigucao Swertia ■

380354 Swertia shintenensis Hayata;新店獐牙菜(新店当药);Xindian Swertia ■

380355 Swertia sikkimensis Burkill = Lomatogonium sikkimense (Burkill) Harry Sm. ■

380356 Swertia silenifolia T. C. E. Fr. = Swertia kilimandscharica Engl. ■☆

380357 Swertia souliaei Burkill;康定獐牙菜;Kangding Swertia ■

380358 Swertia splendens Harry Sm. ;光亮獐牙菜;Bright Swertia ■

380359 Swertia squamigera Sileshi;鳞獐牙菜■☆

380360 Swertia stapfii Burkill = Lomatogonium stapfii (Burkill) Harry Sm. ■

380361 Swertia stellarioides Ficalho ex Hiern = Swertia welwitschii Engl. ■☆

380362 Swertia stricta Franch. = Swertia franchetiana Harry Sm. ■

380363 Swertia subalpina N. E. Br. = Swertia abyssinica Hochst. ■☆

380364 Swertia subnivalis T. C. E. Fr. ;雪獐牙菜■☆

380365 Swertia subspeciosa Burkill = Swertia bifolia Batalin ■

380366 Swertia subspeciosa Burkill = Swertia souliaei Burkill ■

380367 Swertia swertopsis Makino;拟獐牙菜(当药)■☆

380368 Swertia tashiroi (Maxim.) Makino;田代氏獐牙菜■☆

380369 Swertia tashiroi (Maxim.) Makino f. immaculata Sakata;无斑田代氏獐牙菜■☆

380370 Swertia tashiroi (Maxim.) Makino var. cruciata F. Maek. = Swertia tashiroi (Maxim.) Makino ■☆

380371 Swertia tashiroi Makino = Swertia makinoana F. Maek. ■☆

380372 Swertia tenuis T. N. Ho et S. W. Liu;细瘦獐牙菜;Thin Swertia ■

380373 Swertia tetragona C. B. Clarke = Swertia kouitchensis Franch. ■

380374 Swertia tetragona R. H. Miao;棱茎獐牙菜(四数獐牙菜)■

380375 Swertia tetrandra Hochst.;四蕊獐牙菜■■

380376 Swertia tetrapetala Pall.;千岛獐牙菜(卵叶獐牙菜)■

380377 Swertia tetrapetala Pall. f. albiflora Tatew.;白花千岛獐牙菜■☆

380378 Swertia tetrapetala Pall. f. variegata Tatew.;斑点千岛獐牙菜■☆

380379 Swertia tetrapetala Pall. subsp. micrantha (Takeda) Kitam.;小花千岛獐牙菜■☆

380380 Swertia tetrapetala Pall. subsp. micrantha (Takeda) Kitam. f. leucantha (Hid. Takah.) T. Shimizu;白小花千岛獐牙菜■☆

380381 Swertia tetrapetala Pall. subsp. micrantha (Takeda) Kitam. var. happoensis Hid. Takah. ex T. Shimizu;八风獐牙菜■☆

380382 Swertia tetrapetala Pall. var. chrysantha (Honda et Tatew.) Sugim.;黄花千岛獐牙菜■☆

380383 Swertia tetrapetala Pall. var. yezoalpina (H. Hara) H. Hara;北海道山地獐牙菜■☆

380384 Swertia tetraptera Maxim. = Swertia tetraptera Pall. ■

380385 Swertia tetraptera Pall.;四数獐牙菜(藏茵陈,二型腺鳞草);Fourtimes Swertia ■

380386 Swertia thomsonii C. B. Clarke;汤氏獐牙菜;Thomson Swertia ■☆

380387 Swertia tibetica Batalin;大药獐牙菜(西藏獐牙菜);Tibet Swertia,Xizang Swertia ■

380388 Swertia tischeri Chiov.;蒂舍尔獐牙菜■☆

380389 Swertia tosaensis Makino;日本獐牙菜■

380390 Swertia tosaensis Makino = Swertia diluta (Turcz.) Benth. et Hook. f. var. tosaensis (Makino) H. Hara

380391 Swertia tozanensis Hayata;搭山獐牙菜(高山当药,松田獐牙菜,细叶当药,细叶獐牙菜);Dashan Swertia,Thinleaf Swertia ■

380392 Swertia tscherskyi Kom.;切尔斯基獐牙菜■☆

380393 Swertia tshitirungensis De Wild. = Swertia eminii Engl. ■☆

380394 Swertia umbellata Gilg = Swertia swertopsis Makino ■☆

380395 Swertia uniflora Mildbr.;单花獐牙菜■☆

380396 Swertia usambarensis Engl.;乌桑巴拉獐牙菜■☆

380397 Swertia usambarensis Engl. var. curtioides (Gilg) Sileshi;短獐牙菜■☆

380398 Swertia vacillans (Hance) Maxim. = Swertia angustifolia Buch. -Ham. ex D. Don ■

380399 Swertia vacillans Maxim. = Swertia angustifolia Buch. -Ham. ex D. Don var. pulchella (D. Don) Burkill ■

380400 Swertia veratroides Maxim. ex Kom.;藜芦獐牙菜;Falsehelleborelike Swertia ■

380401 Swertia verticillifolia T. N. Ho et S. W. Liu;轮叶獐牙菜;Verticillateleaf Swertia ■

380402 Swertia virescens Harry Sm.;绿花獐牙菜;Greenflower Swertia ■

380403 Swertia volkensii Gilg;福尔獐牙菜■☆

380404 Swertia volkensii Gilg var. baleensis Sileshi;巴莱獐牙菜■☆

380405 Swertia wardii C. Marquand;苇叶獐牙菜(坚梗獐牙菜)■

380406 Swertia wardii C. Marquand var. rigida (T. N. Ho et S. W. Liu) T. N. Ho;硬秆獐牙菜■

380407 Swertia wardii C. Marquand var. rigida T. N. Ho et S. W. Liu = Swertia wardii C. Marquand var. rigida (T. N. Ho et S. W. Liu) T. N. Ho ■

380408 Swertia wellbyi N. E. Br. = Swertia abyssinica Hochst. ■☆

380409 Swertia welwitschii Engl.;韦尔獐牙菜■☆

380410 Swertia whytei N. E. Br. = Swertia usambarensis Engl. ■☆

380411 Swertia wilfordii (A. Kern.) Kom.;卵叶獐牙菜;Ovateleaf Swertia ■

380412 Swertia wilfordii A. Kern. = Swertia tetrapetala Pall. ■

380413 Swertia wilfordii A. Kern. = Swertia wilfordii (A. Kern.) Kom. ■

380414 Swertia wojerensis N. E. Br.;沃耶拉特獐牙菜■☆

380415 Swertia wolfgangiana Grüning;华北獐牙菜(乌氏当药);N. China Swertia ■

380416 Swertia woodii J. Shan = Swertia engleri Gilg var. woodii (J. Shan) Sileshi ■☆

380417 Swertia younghusbandii Burkill;少花獐牙菜;Fewflowered Swertia ■

380418 Swertia yunnanensis Burkill;云南獐牙菜(滇獐牙菜,肝炎草,苦草,荞杆草,青叶丹,青叶胆,青鱼胆,小苦胆草,小龙胆草,紫花苦胆草,走胆药);Yunnan Swertia ■

380419 Swertia zayuensis T. N. Ho et S. W. Liu;察隅獐牙菜;Chayu Swertia ■

380420 Swertopsis Makino = Swertia L. ■

380421 Swertopsis umbellata Makino = Swertia swertopsis Makino ■☆

380422 Swertya Steud. = Swertia Boehm. ●■☆

380423 Swertya Steud. = Tolpis Adans. ●■☆

380424 Swida Opiz = Cornus L. ●

380425 Swida Opiz(1838);梾木属;Cornel,Dogwood ●

380426 Swida alba (L.) Opiz = Cornus alba L. ●

380427 Swida alpina (W. P. Fang et W. K. Hu) W. P. Fang et W. K. Hu;高山梾木;Alpine Dogwood ●

380428 Swida alpina (W. P. Fang et W. K. Hu) W. P. Fang et W. K. Hu = Cornus macrophylla Wall. ●

380429 Swida alsophila (W. W. Sm.) Holub;凉山梾木(云南四照花);Alsophila Dogwood,Cold Dogwood,Shady Dogwood ●

380430 Swida alsophila (W. W. Sm.) Holub = Cornus hemsleyi C. K. Schneid. et Wangerin ●

380431 Swida alternifolia (L. f.) Small = Cornus alternifolia L. f. ●☆

380432 Swida austrosinensis (W. P. Fang et W. K. Hu) W. P. Fang et W. K. Hu = Cornus austrosinensis W. P. Fang et W. K. Hu ●

380433 Swida baileyi (J. M. Coult. et W. H. Evans) Rydb. = Cornus stolonifera Michx. ●☆

380434 Swida bretschneideri (L. Henry) Soják;沙梾(毛山茱萸);Bretschneider Dogwood,Sand Dogwood ●

380435 Swida bretschneideri (L. Henry) Soják = Cornus bretschneideri L. Henry ●

380436 Swida bretschneideri (L. Henry) Soják var. crispa (W. P. Fang et W. K. Hu) W. P. Fang et W. K. Hu = Cornus bretschneideri L. Henry var. crispa W. P. Fang et W. K. Hu ●

380437 Swida bretschneideri (L. Henry) Soják var. gracilis (Wangerin) W. K. Hu. = Cornus bretschneideri L. Henry ●

380438 Swida controversa (Hemsl. ex Prain) Soják = Bothrocaryum controversum (Hemsl. ex Prain) Pojark. ●

380439 Swida controversa (Hemsl. ex Prain) Soják var. alpina (Wangerin) H. Hara ex Noshiro;高山灯台树●☆

380440 Swida controversa (Hemsl. ex Prain) Soják var. shikokumontana (Hiyama) H. Hara ex Noshiro;四国梾木●☆

380441 Swida controversa (Hemsl.) Holub = Bothrocaryum controversum (Hemsl. ex Prain) Pojark. ●

380442 Swida controversa (Hemsl.) S. S. Ying = Bothrocaryum controversum (Hemsl. ex Prain) Pojark. ●

380443 Swida controversa (Hemsl.) Soják. = Cornus controversa Hemsl. ex Prain ●

380444 Swida coreana (Wangerin) Soják;朝鲜梾木;Dogwood,Korean

Dogwood ●

380445　Swida coreana（Wangerin）Soják. = Cornus coreana Wangerin ●

380446　Swida daijinensis（W. P. Fang et W. K. Hu）W. P. Fang et W. K. Hu；大金梾木；Dajin Dogwood ●◇

380447　Swida daijinensis（W. P. Fang et W. K. Hu）W. P. Fang et W. K. Hu = Cornus schindleri Wangerin ●

380448　Swida darvasica（Pojark.）Soják；中亚梾木●☆

380449　Swida fulvescens（W. P. Fang et W. K. Hu）W. P. Fang et W. K. Hu；黄褐毛梾木；Brown-haired Dogwood, Fulvescent Dogwood, Fulvous Dogwood ●

380450　Swida fulvescens（W. P. Fang et W. K. Hu）W. P. Fang et W. K. Hu = Cornus schindleri Wangerin ●

380451　Swida hemsleyi（C. K. Schneid. et Wangerin）Soják；红椋子（娘子木，青构）；Hemsley Dogwood ●

380452　Swida hemsleyi（C. K. Schneid. et Wangerin）Soják = Cornus hemsleyi C. K. Schneid. et Wangerin ●

380453　Swida hemsleyi（C. K. Schneid. et Wangerin）Soják var. gracilipes（W. P. Fang et W. K. Hu）W. P. Fang et W. K. Hu = Cornus hemsleyi C. K. Schneid. et Wangerin ●

380454　Swida hemsleyi（C. K. Schneid. et Wangerin）Soják var. gracilipes（W. P. Fang et W. K. Hu）W. P. Fang et W. K. Hu；细梗红椋子；Slender Hemsle Dogwood ●

380455　Swida hemsleyi（C. K. Schneid. et Wangerin）Soják var. longistyla（W. P. Fang et W. K. Hu）W. P. Fang et W. K. Hu = Cornus hemsleyi C. K. Schneid. et Wangerin ●

380456　Swida hemsleyi（C. K. Schneid. et Wangerin）Soják var. longistyla（W. P. Fang et W. K. Hu）W. P. Fang et W. K. Hu；长花柱红椋子（长柱红椋子）；Long-style Hemsle Dogwood ●

380457　Swida instolonea（A. Nelson）Rydb. = Cornus stolonifera Michx. ●☆

380458　Swida interior Rydb. = Cornus stolonifera Michx. ●☆

380459　Swida koehneana（Wangerin）Soják；川陕梾木；Koehne Dogwood ●

380460　Swida koehneana（Wangerin）Soják = Cornus koehneana Wangerin ●

380461　Swida macrophylla（Wall.）Soják；梾木（冬青果，光皮树，棶，凉木，凉子，凉子木，椋子，椋子木，毛梾棶木，松杨，松杨木，台灯树）；Common Dogwood, Largeleaf Dogwood, Largeleaved Dogwood, Large-leaved Dogwood, Macrophyllous Dogwood ●

380462　Swida macrophylla（Wall.）Soják = Cornus macrophylla Wall. ●

380463　Swida macrophylla（Wall.）Soják var. longipedunculata（W. P. Fang et W. K. Hu）W. P. Fang et W. K. Hu = Cornus macrophylla Wall. ●

380464　Swida macrophylla（Wall.）Soják var. longipedunculata（W. P. Fang et W. K. Wu）W. P. Fang et W. K. Wu = Swida macrophylla（Wall.）Soják ●

380465　Swida monbeigii（Hemsl.）Soják；曲瓣梾木（滇藏梾木）；Monbeig Dogwood ●

380466　Swida monbeigii（Hemsl.）Soják = Cornus schindleri Wangerin ●

380467　Swida monbeigii（Hemsl.）Soják var. crassa（W. P. Fang et W. K. Hu）W. P. Fang et W. K. Hu = Cornus schindleri Wangerin ●

380468　Swida monbeigii（Hemsl.）Soják var. crassa（W. P. Fang et W. K. Hu）W. P. Fang et W. K. Hu；粗壮曲瓣梾木；Strong Monbeig Dogwood ●

380469　Swida monbeigii（Hemsl.）Soják var. popolufolia（W. P. Fang et W. K. Hu）W. P. Fang et W. K. Hu；杨叶曲瓣梾木；Poplarleaf Monbeig Dogwood ●

380470　Swida monbeigii（Hemsl.）Soják var. populifolia（W. P. Fang et W. K. Hu）W. P. Fang et W. K. Hu = Cornus schindleri Wangerin ●

380471　Swida monbeigii（Hemsl.）Soják var. xanthotricha（W. P. Fang et W. K. Hu）W. P. Fang et W. K. Hu = Cornus schindleri Wangerin ●

380472　Swida monbeigii（Hemsl.）Soják var. xanthotricha（W. P. Fang et W. K. Hu）W. P. Fang et W. K. Hu；黄毛曲瓣梾木；Yellowhair Monbeig Dogwood ●

380473　Swida muchuanensis（Z. Y. Zhu）Holub = Cornus oblonga Wall. ex Roxb. ●☆

380474　Swida oblonga（Wall.）Soják；长圆叶梾木（臭条子，粉帕树，黑皮楠，矩圆叶梾木）；Oblong-leaf Dogwood, Oblong-leaved Dogwood ●

380475　Swida oblonga（Wall.）Soják = Cornus oblonga Wall. ex Roxb. ●☆

380476　Swida oblonga（Wall.）Soják var. glabrescena（W. P. Fang et W. K. Hu）W. P. Fang et W. K. Hu；无毛长圆叶梾木；Smooth Oblongleaf Dogwood ●

380477　Swida oblonga（Wall.）Soják var. glabrescens（W. P. Fang et W. K. Hu）W. P. Fang et W. K. Hu = Cornus oblonga Wall. var. glabrescena W. P. Fang et W. K. Hu ●

380478　Swida oblonga（Wall.）Soják var. griffithii（C. B. Clarke）W. K. Hu；毛叶梾木；Griffith Dogwood ●

380479　Swida oblonga（Wall.）Soják var. griffithii（C. B. Clarke）W. K. Hu = Cornus oblonga Wall. var. griffithii C. B. Clarke ●

380480　Swida oligophlebia（Merr.）W. K. Hu；樟叶梾木；Cinnamonleaf Dogwood, Laurel-leaved Dogwood ●

380481　Swida oligophlebia（Merr.）W. K. Hu = Cornus oligophlebia Merr. ●

380482　Swida papillosa（W. P. Fang et W. K. Hu）W. P. Fang et W. K. Hu；乳突梾木；Papillose Dogwood ●

380483　Swida papillosa（W. P. Fang et W. K. Hu）W. P. Fang et W. K. Hu = Cornus papillosa W. P. Fang et W. K. Hu ●

380484　Swida parviflora（S. S. Chien）Holub；小花梾木（贵州梾木，贵州四照花）；Smallflower Dogwood, Small-flowered Dogwood ●

380485　Swida parviflora（S. S. Chien）Holub. = Cornus parviflora S. S. Chien ●

380486　Swida paucinervis（Hance）Soják；小梾木（臭黄荆，穿鱼藤，大穿鱼草，南天种，疏脉山茱萸，水杨柳，小叶梾木）；Littleleaf Dogwood, Little-leaved Dogwood, Small Dogwood ●

380487　Swida paucinervis（Hance）Soják = Cornus quinquenervis Franch. ●

380488　Swida poliophylla（C. K. Schneid. et Wangerin）Soják；灰叶梾木（黑椋子）；Greyleaf Dogwood, Grey-leaved Dogwood ●

380489　Swida poliophylla（C. K. Schneid. et Wangerin）Soják = Cornus schindleri Wangerin subsp. poliophylla（C. K. Schneid. et Wangerin）Q. Y. Xiang ●

380490　Swida poliophylla（C. K. Schneid. et Wangerin）Soják var. malifolia（W. P. Fang et W. K. Hu）W. P. Fang et W. K. Hu = Cornus schindleri Wangerin ●

380491　Swida poliophylla（C. K. Schneid. et Wangerin）Soják var. malifolia（W. P. Fang et W. K. Wu）W. P. Fang et W. K. Wu；海棠叶梾木●

380492　Swida poliophylla（C. K. Schneid. et Wangerin）Soják var. praelonga（W. P. Fang et W. K. Hu）W. P. Fang et W. K. Hu = Cornus schindleri Wangerin ●

380493　Swida poliophylla（C. K. Schneid. et Wangerin）Soják var. praelonga（W. P. Fang et W. K. Hu）W. P. Fang et W. K. Hu；高大灰叶梾木；Tall Greyleaf Dogwood ●

380494 Swida polyantha（W. P. Fang et W. K. Hu）W. P. Fang et W.
K. Hu；多花楝木；Manyflower Dogwood，Multiflrous Dogwood ●

380495 Swida polyantha（W. P. Fang et W. K. Hu）W. P. Fang et W.
K. Hu. = Cornus hemsleyi C. K. Schneid. et Wangerin ●

380496 Swida priceae（Small）Small = Cornus drummondii C. A. Mey. ●☆

380497 Swida racemosa（Lam.）Moldenke = Cornus racemosa Lam. ●☆

380498 Swida rugosa（Lam.）Rydb. = Cornus rugosa Lam. ●☆

380499 Swida sanguinea（L.）Opiz；欧洲红瑞木（黑果红瑞木，红楝
木，欧亚红花山茱萸，欧洲山茱萸）；Bloodtwig Dogwood，Blood-twig
Dogwood，Bloodwig Shrub Dogwood，Bloodwing Dogwood，Blood-wing
Dogwood，Bloody Rod，Bloody Twig，Catteridge Tree，Cat-tree，
Catwood，Common Dogwood，Cornalee，Cornel，Cornelian，Cornwood，
Dog Cherry，Dog Tree，Dogberry，Dog's Timber，Dogwood，European
Dogwood，Female Cornel，Gadrise，Gaiter，Gaiter-tree，Gatten-tree，
Gatter-bush，Gatteridge-tree，Gatter-tree，Houndberry Tree，Hound's
Tree，Pegwood，Prick-timber，Prick-tree，Prick-wood，Red Dogwood，
Skewerwood，Skiver，Skiver-tree，Skiver-wood，Skiwer，Snake's
Cherries，Snake's Cherry，Swamp Dog Wood，Widbin，Wild Cornel ●

380500 Swida sanguinea Opiz = Swida sanguinea（L.）Opiz ●

380501 Swida scabrida（Franch.）Holub；宝兴楝木；Baoxing
Dogwood，Rugged Dogwood ●

380502 Swida scabrida（Franch.）Holub = Cornus schindleri Wangerin ●

380503 Swida schindleri（Wangerin）Soják；康定楝木；Kangding
Dogwood ●

380504 Swida schindleri（Wangerin）Soják = Cornus schindleri
Wangerin ●

380505 Swida schindleri（Wangerin）Soják var. lixianensis（W. P. Fang et
W. K. Hu）W. P. Fang et W. K. Hu = Cornus schindleri Wangerin ●

380506 Swida schindleri（Wangerin）Soják var. lixianensis（W. P. Fang
et W. K. Hu）W. P. Fang et W. K. Hu；理县楝木；Lixian Dogwood ●

380507 Swida sericea（L.）Holub = Cornus stolonifera Michx. ●☆

380508 Swida stolonifera（Michx.）Rydb. = Cornus stolonifera Michx. ●☆

380509 Swida stracheyi（C. B. Clarke）Soják = Cornus macrophylla
Wall. var. stracheyi C. B. Clarke ●

380510 Swida ulotricha（C. K. Schneid. et Wangerin）Soják；卷毛楝木
（川萼楝木，西蜀楝木，西蜀四照花）；Rollinghair Dogwood，
Sichuan Dogwood，Szechwan Dogwood ●

380511 Swida ulotricha（C. K. Schneid. et Wangerin）Soják = Cornus
ulotricha C. K. Schneid. et Wangerin ●

380512 Swida ulotricha（C. K. Schneid. et Wangerin）Soják var. leptophylla
W. K. Hu = Cornus ulotricha C. K. Schneid. et Wangerin ●

380513 Swida ulotricha（C. K. Schneid. et Wangerin）Soják var.
leptophylla W. K. Hu；细叶卷毛楝木（薄叶卷毛楝木）；Thin-leaf
Sichuan Dogwood ●

380514 Swida walteri（Wangerin）Soják；毛楝（八树，车梁木，红梗山
茱萸，红零子，癞树，椋子木，小六谷，油树）；Hair Dogwood，Walter
Dogwood ●

380515 Swida walteri（Wangerin）Soják = Cornus walteri Wangerin ●

380516 Swida walteri（Wangerin）Soják var. confertiflora（W. P. Fang
et W. K. Hu）W. P. Fang et W. K. Hu = Cornus walteri Wangerin ●

380517 Swida walteri（Wangerin）Soják var. confertiflora（W. P. Fang
et W. K. Hu）W. P. Fang et W. K. Hu = Swida walteri（Wangerin）
Soják ●

380518 Swida walteri（Wangerin）Soják var. insignis（W. P. Fang et
W. K. Hu）W. P. Fang et W. K. Hu = Cornus walteri Wangerin ●

380519 Swida walteri（Wangerin）Soják var. insignis（W. P. Fang et
W. K. Hu）W. P. Fang et W. K. Hu = Swida walteri（Wangerin）
Soják ●

380520 Swida wilsoniana（Wangerin）Soják；光皮楝木（狗骨木，光皮
树，花皮树，马林光）；E. H. Wilson Dogwood，Wilson Dogwood ●

380521 Swida wilsoniana（Wangerin）Soják = Cornus wilsoniana
Wangerin ●

380522 Swietenia Jacq.（1760）；桃花心木属；Mahogany ●

380523 Swietenia angolensis Welw. = Entandrophragma angolense
（Welw.）C. DC. ●☆

380524 Swietenia candollea Pittier；美洲桃花心木；American
Mahogany，Venezuelan Mahogany ●☆

380525 Swietenia febrifuga Roxb. = Soymida febrifuga（Roxb.）A.
Juss. ●☆

380526 Swietenia humilis Zucc.；小桃花心木（矮桃花心木，墨西哥桃
花心木）；Mexican Mahogany，Pacific Mahogany ●☆

380527 Swietenia macrophylla King；大叶桃花心木（洪都拉斯桃花心
木，美洲桃花心木，桃花心木，真桃花心木，正宗桃花心木）；
Acajou，American Mahogany，Baywood，Big Leaf Mahogany，Bigleaf
Mahogany，Central American Mahogany，Colombian Mahogany，
Genuine Mahogany，Honduras Mahogany，Tabasco Mahogany，True
Mahogany ●

380528 Swietenia mahagoni（L.）Jacq.；桃花心木（马哈贡尼桃花心
木，小叶桃花心木）；Baywood，Cuba Mahogany，Cuban Mahogany，
Honduras Mahogany，Mahogany，Spanish Mahogany，St. Domingo
Mahogany，W. India Mahogany，West Indian Mahogany ●

380529 Swietenia mahagoni Jacq. = Swietenia mahagoni（L.）Jacq. ●

380530 Swietenia mahagoni Lam. = Swietenia mahagoni（L.）Jacq. ●

380531 Swietenia senegalensis Desr. = Khaya senegalensis（Desr.）A.
Juss. ●

380532 Swietenia tessmannii Harms；巴西桃花心木；Brazilian
Mahogany ●☆

380533 Swieteniaceae Bercht. et J. Presl = Meliaceae Juss.（保留科名）●

380534 Swieteniaceae Kirchn. = Meliaceae Juss.（保留科名）●

380535 Swinburnia Ewart = Neotysonia Dalla Torre et Harms ■☆

380536 Swingera Dunal = Nolana L. ex L. f. ■☆

380537 Swingera Dunal = Zwingera Hofer ■☆

380538 Swinglea Merr.（1927）；菲律宾木橘属（菲律宾木桔属）；
Swinglea ●☆

380539 Swinglea glutinosa（Blanco）Merr.；菲律宾木橘；Stycky
Swinglea ●☆

380540 Swintonia Griff.（1846）；斯温顿漆属 ●☆

380541 Swintonia acuminata Merr.；渐尖斯温顿漆 ●☆

380542 Swintonia acuta Engl.；尖斯温顿漆 ●☆

380543 Swintonia floribunda Griff.；多花斯温顿漆 ●☆

380544 Swintonia glauca Engl.；灰蓝斯温顿漆 ●☆

380545 Swintonia griffithii Kurz；格氏斯温顿漆 ●☆

380546 Swintonia lurida King；亮斯温顿漆 ●☆

380547 Swintonia minuta Evrard；小斯温顿漆 ●☆

380548 Swintonia obtusifolia Engl.；钝叶斯温顿漆 ●☆

380549 Swjda Opiz = Swida Opiz ●

380550 Swynnertonia S. Moore（1908）；斯温萝藦属 ■☆

380551 Swynnertonia cardinea S. Moore；斯温萝藦 ■☆

380552 Swynnertonia cardinea S. Moore = Neoschumannia cardinea（S.
Moore）Meve ■☆

380553 Syagrus Mart.（1824）；金山葵属（凤尾棕属，皇后葵属，皇后
椰属，女王椰子属，射古椰子属，西雅椰子属，下个棕属）；
Jinshanpalm，Queen Palm，Syagrus ●

380554 Syagrus amara Mart.；沟金山葵（马提尼桐）●☆

380555　Syagrus comosa Mart.；褐斑金山葵（褐斑皇后椰）；Candy Palm ●☆

380556　Syagrus coronata（Mart.）Becc.；冠金山葵（西雅棕）；Nicuri，Nicuri Palm Nut，Ouricuri，Ouricuri Palm，Urucury Wax ●☆

380557　Syagrus romanzoffiana（Cham.）Glassman ＝ Arecastrum romanzoffianum（Cham.）Becc. ●

380558　Syagrus romanzoffiana（Cham.）Glassman var. australe Becc. ＝ Arecastrum romanzoffianum（Cham.）Becc. var. australe（Mart.）Becc. ●

380559　Syagrus schizophylla（Mart.）Glassman；裂叶金山葵 ●☆

380560　Syagrus tessmannii Burret；德氏金山葵（德森西雅棕）●☆

380561　Syagrus weddellianus（H. Wendl.）Becc.；韦氏金山葵；Weddell Palm ●☆

380562　Syalita Adans. ＝ Dillenia L. ●

380563　Syama Jones ＝ Pupalia Juss.（保留属名）■☆

380564　Sycamorus Oliv. ＝ Ficus L. ●

380565　Sycamorus Oliv. ＝ Sycomorus Gasp. ●

380566　Sychinium Desv. ＝ Dorstenia L. ■●☆

380567　Sychnosepalum Eichl. ＝ Sciadotaenia Benth. ●☆

380568　Sychnosepalum Eichl. ＝ Sciadotenia Miers ●☆

380569　Sycios Medik. ＝ Sicyos L. ■

380570　Sycocarpus Britton ＝ Guarea F. Allam.（保留属名）●☆

380571　Sycodendron Rojas ＝ Ficus L. ●

380572　Sycodendron Rojas Acosta ＝ Ficus L. ●

380573　Sycodium Pomel ＝ Anvillea DC. ●☆

380574　Sycodium radiatum（Coss. et Durieu）Pomel ＝ Anvillea garcinii（Burm. f.）DC. subsp. radiata（Coss. et Durieu）Anderb. ●☆

380575　Sycomorphe Miq. ＝ Ficus L. ●

380576　Sycomorus Gasp. ＝ Ficus L. ●

380577　Sycomorus gnaphalocarpa Miq. ＝ Ficus sycomorus L. subsp. gnaphalocarpa（Miq.）C. C. Berg ●☆

380578　Sycomorus guineensis Miq. ＝ Ficus sur Forssk. ●☆

380579　Sycomorus riparia Miq. ＝ Ficus sur Forssk. ●☆

380580　Sycomorus trachyphylla Miq. ＝ Ficus sycomorus L. subsp. gnaphalocarpa（Miq.）C. C. Berg ●☆

380581　Sycomorus vogeliana Miq. ＝ Ficus vogeliana（Miq.）Miq. ●☆

380582　Sycophila Welw. ex Tiegh. ＝ Helixanthera Lour. ●

380583　Sycophila combretoides Welw. ex Tiegh. ＝ Helixanthera mannii（Oliv.）Danser ●☆

380584　Sycophila mannii（Oliv.）Tiegh. ＝ Helixanthera mannii（Oliv.）Danser ●☆

380585　Sycophila ternata Tiegh. ＝ Helixanthera mannii（Oliv.）Danser ●☆

380586　Sycopsis Oliv.（1860）；水丝梨属；Fighazel ●

380587　Sycopsis chungii F. P. Metcalf ＝ Distylium chungii（F. P. Metcalf）W. C. Cheng ●

380588　Sycopsis dunnii Hemsl. ＝ Distyliopsis dunnii（Hemsl.）P. K. Endress ●

380589　Sycopsis formosana（Kaneh.）Kaneh. et Hatus. ＝ Sycopsis sinensis Oliv. ●

380590　Sycopsis formosana（Kaneh.）Kaneh. et Hatus. ex Hatus.；台湾水丝梨；Formosan Fig Hazel ●

380591　Sycopsis formosana（Kaneh.）Kaneh. et Hatus. ex Hatus. ＝ Sycopsis sinensis Oliv. ●

380592　Sycopsis formosana Kaneh. et Hatus. ＝ Sycopsis sinensis Oliv. ●

380593　Sycopsis griffithiana Oliv.；喀西亚水丝梨；Griffith Fighazel，Khasia Fighazel ●

380594　Sycopsis griffithiana Oliv. ＝ Eustigma lenticellatum C. Y. Wu ●

380595　Sycopsis griffithiana Rehder et E. H. Wilson ＝ Eustigma lenticellatum C. Y. Wu ●

380596　Sycopsis laurifolia Hemsl. ＝ Distyliopsis laurifolia（Hemsl.）P. K. Endress ●

380597　Sycopsis loii（S. S. Ying）S. S. Ying ＝ Sycopsis sinensis Oliv. ●

380598　Sycopsis oblanceolata Hung T. Chang ＝ Distyliopsis tutcheri（Hemsl.）P. K. Endress ●

380599　Sycopsis philippinensis Hemsl.；菲律宾水丝梨；Philippine Fighazel ●☆

380600　Sycopsis pingpienensis Hu ＝ Distylium pingpienense（Hu）Walker ●

380601　Sycopsis salicifolia H. L. Li ＝ Distyliopsis salicifolia（H. L. Li）P. K. Endress ●

380602　Sycopsis sinensis Oliv.；水丝梨（南湖水丝梨，水私梨）；Chinese Fig Hazel，Chinese Fighazel，Fighazel ●

380603　Sycopsis sinensis Oliv. var. integrifolia Diels ＝ Sycopsis sinensis Oliv. ●

380604　Sycopsis triplinervia Hung T. Chang；三脉水丝梨；Threenerve Fighazel，Trinerved Fighazel ●

380605　Sycopsis tutcheri Hemsl. ＝ Distyliopsis tutcheri（Hemsl.）P. K. Endress ●

380606　Sycopsis yunnanensis Hung T. Chang ＝ Distyliopsis yunnanensis（Hung T. Chang）C. Y. Wu ●

380607　Syderitis All. ＝ Sideritis L. ■●

380608　Syena Schreb. ＝ Mayaca Aubl. ■☆

380609　Sykesia Arn. ＝ Gaertnera Lam. ●

380610　Sykesia hongkongensis（Seem.）Kuntze ＝ Tsiangia hongkongensis（Seem.）But，H. H. Hsue et P. T. Li ●

380611　Sykesia vaginans（DC.）Kuntze ＝ Gaertnera vaginans（DC.）Merr. ●☆

380612　Sykoraea Opiz ＝ Campanula L. ■●

380613　Sylitra E. Mey. ＝ Ptycholobium Harms ■☆

380614　Sylitra angolensis Baker ＝ Ptycholobium biflorum（E. Mey.）Brummitt subsp. angolensis（Baker）Brummitt ■☆

380615　Sylitra biflora E. Mey. ＝ Ptycholobium biflorum（E. Mey.）Brummitt ■☆

380616　Sylitra contorta（N. E. Br.）Baker f. ＝ Ptycholobium contortum（N. E. Br.）Brummitt ■☆

380617　Syllepis E. Fourn. ＝ Imperata Cyrillo ■

380618　Syllepis E. Fourn. ex Benth. et Hook. f. ＝ Imperata Cyrillo ■

380619　Syllisium Endl. ＝ Syzygium R. Br. ex Gaertn.（保留属名）●

380620　Syllisium buxifolium（Hook. et Arn.）Meyen et Schauer ＝ Syzygium buxifolium Hook. et Arn. ●

380621　Syllisium buxifolium Meyen et Schauer ＝ Syzygium buxifolium Hook. et Arn. ●

380622　Syllysium Meyen et Schauer ＝ Syzygium R. Br. ex Gaertn.（保留属名）●

380623　Sylvalismis Dalla Torre et Harms ＝ Calanthe R. Br.（保留属名）■

380624　Sylvalismis Dalla Torre et Harms ＝ Sylvalismis Thouars

380625　Sylvalismis Dalla Torre et Harms ＝ Sylvalismus Post et Kuntze ■

380626　Sylvalismis Thouars ＝ Calanthe R. Br.（保留属名）■

380627　Sylvalismus Post et Kuntze ＝ Calanthe R. Br.（保留属名）■

380628　Sylvalismus Post et Kuntze ＝ Sylvalismis Thouars ■

380629　Sylvia Lindl. ＝ Silvia Benth. ●☆

380630　Sylvia Lindl. ＝ Silviella Pennell ■☆

380631　Sylvichadsia Du Puy et Labat（1998）；林灌豆属 ●☆

380632　Sylvichadsia grandidieri（Baill.）Du Puy et Labat；格氏西尔豆 ●☆

380633 Sylvichadsia grandifolia (R. Vig.) Du Puy et Labat;大叶西尔豆●☆

380634 Sylvichadsia macrophylla (R. Vig.) Du Puy et Labat;西尔豆●☆

380635 Sylvichadsia perrieri (R. Vig.) Du Puy et Labat;佩里耶西尔豆●☆

380636 Sylvipoa Soreng, L. J. Gillespie et S. W. L. Jacobs(2009);昆士兰禾属■☆

380637 Sylvipoa queenslandica (C. E. Hubb.) Soreng, L. J. Gillespie et S. W. L. Jacobs;昆士兰禾■☆

380638 Sylvorchis Schltr. = Silvorchis J. J. Sm. ■☆

380639 Symbasiandra Steud. = Hilaria Kunth ■☆

380640 Symbasiandra Willd. ex Steud. = Hilaria Kunth ■☆

380641 Symbegonia Warb. (1894);类秋海棠属■☆

380642 Symbegonia fulvo-villosa Warb.;黄毛类秋海棠■☆

380643 Symbegonia hirta Ridl.;粗毛类秋海棠■☆

380644 Symbegonia sanguinea Warb.;血红类秋海棠■☆

380645 Symbegonia strigosa Warb.;直类秋海棠■☆

380646 Symblomeria Nutt. = Albertinia Spreng. ●☆

380647 Symbolanthus G. Don(1837);热美龙胆属●☆

380648 Symbolanthus aureus Struwe et V. A. Albert;黄热美龙胆●☆

380649 Symbolanthus australis Struwe;澳洲热美龙胆●☆

380650 Symbolanthus latifolius Gilg;宽叶热美龙胆●☆

380651 Symbolanthus macranthus (Benth.) Moldenke;大花热美龙胆●☆

380652 Symbolanthus microphyllus Gilg;小叶热美龙胆●☆

380653 Symbolanthus pauciflorus Gilg;少花热美龙胆●☆

380654 Symbolanthus tricolor Gilg;三色热美龙胆●☆

380655 Symbryon Griseb. = Lunania Hook. (保留属名)●☆

380656 Symea Baker = Solaria Phil. ■☆

380657 Symethus Raf. = Convolvulus L. ■●

380658 Symingtonia Steenis = Exbucklandia R. W. Br. ●

380659 Symingtonia Steenis(1952);异马蹄荷属●

380660 Symingtonia populnea (R. Br. ex Griff.) Steenis = Exbucklandia populnea (R. Br. ex Griff.) R. W. Br. ●

380661 Symingtonia populnea (R. Br.) Steenis = Exbucklandia populnea (R. Br. ex Griff.) R. W. Br. ●

380662 Symingtonia tonkinensis (Lecomte) Steenis = Exbucklandia tonkinensis (Lecomte) Hung T. Chang ●

380663 Symingtonia tonkinensis (Lecomte) Steenis ex Vink = Exbucklandia tonkinensis (Lecomte) Hung T. Chang ●

380664 Symmeria Benth. (1845);多蕊蓼树属●☆

380665 Symmeria Hook. f. = Habenaria Willd. ■

380666 Symmeria Hook. f. = Synmeria Nimmo ■

380667 Symmeria paniculata Benth.;圆锥多蕊蓼树●☆

380668 Symmetria Blume = Carallia Roxb. (保留属名)●

380669 Symonanthus Haegi = Isandra F. Muell. ■☆

380670 Symonanthus Haegi(1981);西蒙茄属■☆

380671 Symonanthus bancroftii (F. Muell.) Haegi;西蒙茄■☆

380672 Sympa Ravenna(1981);巴西鸢尾属■☆

380673 Sympa riograndensis Ravenna;巴西鸢尾■☆

380674 Sympachue Steud. = Eriocaulon L. ■

380675 Sympachue Steud. = Symphachne P. Beauv. ■

380676 Sympagis (Nees) Bremek. (1944);合页草属●■

380677 Sympagis (Nees) Bremek. = Strobilanthes Blume ●■

380678 Sympagis Bremek. = Sympagis (Nees) Bremek. ●■

380679 Sympagis monadelpha (Nees) Bremek.;合页草●■

380680 Sympagis petiolaris (Nees) Bremek.;具柄合页草●■

380681 Sympegma Bunge(1879);合头草属●■

380682 Sympegma regelii Bunge;合头草(合头藜,黑柴,列氏合头草);Regel Sympegma,Sympegma●■

380683 Sympetalandra Stapf(1891);东南亚苏木属●☆

380684 Sympetalandra borneensis Stapf;东南亚苏木●☆

380685 Sympetalandra densiflora (Elm.) Steenis;密花东南亚苏木●☆

380686 Sympetaleia A. Gray = Eucnide Zucc. ■☆

380687 Symphachne P. Beauv. = Eriocaulon L. ■

380688 Symphachne P. Beauv. ex Desv. = Eriocaulon L. ■

380689 Symphachne xyrioides (L.) P. Beauv. = Eriocaulon decangulare L. ■☆

380690 Symphiandra Steud. = Symphyandra A. DC. ■☆

380691 Symphionema R. Br. (1810);合丝山龙眼属●☆

380692 Symphionema montanum R. Br.;合丝山龙眼●☆

380693 Symphipappus Klatt = Cadiscus E. Mey. ex DC. ■☆

380694 Symphitum Neck. = Symphytum L. ■

380695 Symphocoronis Dur. = Scyphocoronis A. Gray ■☆

380696 Symphonia L. f. (1782);合声木属;Symphonia ●☆

380697 Symphonia gabonensis (Vesque) Pierre;加布合声木;Gabon Symphonia ●☆

380698 Symphonia gabonensis (Vesque) Pierre = Symphonia globulifera L. f. ●☆

380699 Symphonia globulifera L. f.;小球合声木●☆

380700 Symphonia macrocarpa Jum.;大果合声木●☆

380701 Symphonia microphylla R. E. Schult.;小叶合声木●☆

380702 Symphonia rhodosepala Jum.;粉瓣合声木●☆

380703 Symphoniaceae (C. Presl) Barnhart = Clusiaceae Lindl. (保留科名)●■

380704 Symphoniaceae (C. Presl) Barnhart = Guttiferae Juss. (保留科名)●■

380705 Symphoniaceae Barnhart = Clusiaceae Lindl. (保留科名)●■

380706 Symphoniaceae Barnhart = Guttiferae Juss. (保留科名)●■

380707 Symphoniaceae C. Presl = Clusiaceae Lindl. (保留科名)●■

380708 Symphoniaceae C. Presl = Guttiferae Juss. (保留科名)●■

380709 Symphoranthera T. Durand et Jacks. = Dialypetalum Benth. ●■☆

380710 Symphoranthera T. Durand et Jacks. = Synphoranthera Bojer ●■☆

380711 Symphoranthus Mitch. = Polypremum L. ■☆

380712 Symphorema Roxb. (1805);六苞藤属;Symphorema ●

380713 Symphorema involucratum Roxb.;六苞藤;Involucrete Symphorema ●

380714 Symphorema jackianum Kurz = Sphenodesme pentandra Jack ●☆

380715 Symphorema unguiculatum Kurz = Sphenodesme involucrata (C. Presl) B. L. Rob. ●

380716 Symphoremataceae (Meisn.) Reveal et Hoogland = Symphoremataceae Moldenke ex Reveal et Hoogland ●

380717 Symphoremataceae (Meisn.) Reveal et Hoogland = Verbenaceae J. St. -Hil. (保留科名)●■

380718 Symphoremataceae Moldenke ex Reveal et Hoogland = Symphoremataceae Reveal et Hoogland ●

380719 Symphoremataceae Moldenke ex Reveal et Hoogland = Verbenaceae J. St. -Hil. (保留科名)●■

380720 Symphoremataceae Moldenke ex Reveal et Hoogland;六苞藤科(伞序材科)●

380721 Symphoremataceae Reveal et Hoogland = Symphoremataceae Moldenke ex Reveal et Hoogland ●

380722 Symphoremataceae Reveal et Hoogland = Verbenaceae J. St. -Hil. (保留科名)●■

380723 Symphoremataceae Tiegh. = Labiatae Juss. (保留科名)●■

380724 Symphoremataceae Tiegh. = Lamiaceae Martinov(保留科名)●■

380725 Symphoremataceae Tiegh. = Symphoremataceae Reveal et Hoogland ●

380726 Symphoremataceae Tiegh. = Symplocaceae Desf. (保留科名)●

380727 Symphoremataceae Tiegh. = Verbenaceae J. St. -Hil. (保留科名)●■

380728 Symphoremataceae Wight = Labiatae Juss. (保留科名)●■

380729 Symphoremataceae Wight = Lamiaceae Martinov(保留科名)●■

380730 Symphorerna unguiculata Kurz = Sphenodesme involucrata (C. Presl) B. L. Rob. ●

380731 Symphoria Pers. = Symphoricarpos Duhamel ●

380732 Symphoricarpa Neck. = Symphoricarpos Duhamel ●

380733 Symphoricarpos Dill. ex Juss. = Symphoricarpos Duhamel ●

380734 Symphoricarpos Duhamel(1755);毛核木属(雪果属,雪莓属);Coralberry,Snowberry,St. Peter's Wort ●

380735 Symphoricarpos Juss. = Symphoricarpos Duhamel ●

380736 Symphoricarpos albus (L.) S. F. Blake;白毛核木(毛核木,雪果,雪果毛核木,雪晃木);Common Snowberry,Snowberry,Thin-leaved Snowberry,Waxberry,White Coralberry ●☆

380737 Symphoricarpos albus (L.) S. F. Blake var. laevigatus (Fernald) S. F. Blake = Symphoricarpos albus (L.) S. F. Blake ●☆

380738 Symphoricarpos albus (L.) S. F. Blake var. laevigatus (Fernald) S. F. Blake;光滑毛核木(花叶雪果,平滑毛核木);Snowberry,Western Snowberry ●☆

380739 Symphoricarpos albus (L.) S. F. Blake var. laevigatus S. F. Blake = Symphoricarpos albus (L.) S. F. Blake var. laevigatus (Fernald) S. F. Blake ●☆

380740 Symphoricarpos albus (L.) S. F. Blake var. pauciflorus (W. J. Robbins ex A. Gray) S. F. Blake = Symphoricarpos albus (L.) S. F. Blake ●☆

380741 Symphoricarpos chenaultii Rehder;查纳尔特毛核木(匍枝毛核木);Chenault Coralberry,Pink Snowberry ●☆

380742 Symphoricarpos chenaultii Rehder 'Hancock';汉考克匍枝毛核木;Chenault Coralberry ●☆

380743 Symphoricarpos longiflorus A. Gray;长花毛核木;Long-flowered Snowberry ●☆

380744 Symphoricarpos mexicanus Hort. ex K. Koch;墨西哥毛核木 ●☆

380745 Symphoricarpos mexicanus K. Koch = Symphoricarpos mexicanus Hort. ex K. Koch ●☆

380746 Symphoricarpos mollis Nutt. ex Torr. et Gray;柔毛核木(柔毛雪果);Creeping Snowberry,Hairy Snowberry ●☆

380747 Symphoricarpos montanus S. Watson;山生毛核木 ●☆

380748 Symphoricarpos occidentalis Hook.;西方毛核木;Buck Brush,Western Snowberry,Western Wolfberry,Wolfberry ●☆

380749 Symphoricarpos orbiculatus Moench;小花毛核木(圆果毛核木,圆叶毛核木,圆叶雪果);Buck Brush,Buckbrush,Buck-brush,Common Snowberry,Coral Beauty,Coral Berry,Coralberry,Coral-berry,India-currant,India-currant Snowberry,Indian Currant,Indian Currant Coralberry,Indian-currant,Indian-currant Coralberry,Indian-currant Snowberry,Red Woleberry,Stagberry,Waxberry ●☆

380750 Symphoricarpos orbiculatus Moench 'Foliis Variegatis';花叶圆果毛核木 ●☆

380751 Symphoricarpos oreophilus A. Gray;山地毛核木;Mountain Snowberry ●☆

380752 Symphoricarpos pauciflorus W. J. Robbins ex A. Gray = Symphoricarpos albus (L.) S. F. Blake ●☆

380753 Symphoricarpos racemosus Michx. ;加拿大毛核木 ●☆

380754 Symphoricarpos racemosus Michx. = Symphoricarpos albus (L.) S. F. Blake ●☆

380755 Symphoricarpos racemosus Michx. var. pauciflorus W. J. Robbins ex A. Gray = Symphoricarpos albus (L.) S. F. Blake ●☆

380756 Symphoricarpos rivularis Suksd. ;溪边毛核木;Egg Plant,Snotterberries,Snowball,Snowberry,Squirters,St. Peter's Wort,Tea Tree ●☆

380757 Symphoricarpos rivularis Suksd. = Symphoricarpos albus (L.) S. F. Blake ●☆

380758 Symphoricarpos sinensis Rehder;毛核木(雪果,雪莓);China Snowberry,Chinese Snowberry ●☆

380759 Symphoricarpos symphoricarpos (L.) MacMill. = Symphoricarpos orbiculatus Moench ●☆

380760 Symphoricarpos vulgaris Michx. = Symphoricarpos orbiculatus Moench ●☆

380761 Symphoricarpus Kunth = Symphoricarpos Duhamel ●

380762 Symphostemon Hiern = Plectranthus L'Her. (保留属名)●■

380763 Symphostemon Hiern(1900);合蕊草属■☆

380764 Symphostemon articulatus I. M. Johnst. ;合蕊草■☆

380765 Symphostemon insolitus (C. H. Wright) Hiern = Plectranthus insolitus C. H. Wright ■☆

380766 Symphostemon strictus Klotzsch = Cleome stricta (Klotzsch) R. A. Graham ■☆

380767 Symphachna Post et Kuntze = Eriocaulon L. ■

380768 Symphachna Post et Kuntze = Symphachne P. Beauv. ex Desv. ■

380769 Symphyandra A. DC. (1830);联药花属(共药花属)■☆

380770 Symphyandra A. DC. = Campanula L. ■●

380771 Symphyandra armena (Stev.) A. DC. ;艳丽联药花■☆

380772 Symphyandra asiatica Nakai;亚洲联药花■☆

380773 Symphyandra hofmannii Pant. ;霍氏联药花■☆

380774 Symphyandra lazica Boiss. et Balansa ex Boiss. ;拉扎联药花■☆

380775 Symphyandra pendula A. DC. ;联药花(共药花)■☆

380776 Symphyandra stylosa Royle = Asyneuma thomsonii (Hook. f.) Bornm. ■☆

380777 Symphyandra transcaucasica (Sommier et H. Lév.) Grossh. ;外高加索联药花■☆

380778 Symphyandra wanneri Heuff. ;阿尔卑斯联药花■☆

380779 Symphyandra zangezura Lipsky;赞格祖尔联药花■☆

380780 Symphydolon Salisb. = Gladiolus L. ■

380781 Symphyglossum Schltr. (1919)(保留属名);密舌兰属■☆

380782 Symphyglossum sanguineum Schltr. 血红密舌兰■☆

380783 Symphyglossum strictum Schltr. ;密舌兰■☆

380784 Symphyglossum umbrosum (Rchb. f.) Garay et Dunst. ;伞密舌兰■☆

380785 Symphyllanthus Vahl = Dichapetalum Thouars ●

380786 Symphyllarion Gagnep. = Hedyotis L. (保留属名)●■

380787 Symphyllia Baill. = Epiprinus Griff. ●

380788 Symphyllia siletiana Baill. = Epiprinus siletianus (Bailey) Croizat ●

380789 Symphyllium Post et Kuntze = Curanga Juss. ■☆

380790 Symphyllium Post et Kuntze = Synphyllium Griff. ■☆

380791 Symphyllocarpus Maxim. (1859);合苞菊属(含苞草属);Symphyllocarpus ■

380792 Symphyllocarpus exilis Maxim. ;合苞菊(含苞草,合苞草);Common Symphyllocarpus ■

380793 Symphyllochlamys Willis = Symphyochlamys Gürke ●☆

380794 Symphyllophyton Gilg(1897);合叶龙胆属■☆

380795　Symphyllophyton caprifolioides Gilg;合叶龙胆■☆

380796　Symphyloma Steud. = Symphyoloma C. A. Mey. ☆

　　380797　Symphyobasis K. Krause = Goodenia Sm. ●■☆

380798　Symphyochaeta (DC.) Skottsb. = Robinsonia DC.(保留属名)●☆

380799　Symphyochlamys Gürke(1903);合被锦葵属●☆

380800　Symphyochlamys erlangeri Gürke;合被锦葵●☆

380801　Symphyochlamys erlangeri Gürke = Hibiscus erlangeri (Gürke) Thulin ●☆

380802　Symphyodolon Baker = Gladiolus L. ■

380803　Symphyodolon Baker = Symphydolon Salisb. ■

380804　Symphyoglossum Turcz.(废弃属名) = Symphyglossum Schltr. (保留属名)■☆

380805　Symphyoglossum hastatum (Bunge) Turcz. = Cynanchum bungei Decne. ●■

380806　Symphyoglossum hastatum Turcz. = Cynanchum bungei Decne. ●■

380807　Symphyogyne Burret = Liberbaileya Furtado ●☆

380808　Symphyogyne Burret = Maxburretia Furtado ●☆

380809　Symphyoloma C. A. Mey.(1831);合缘芹属☆

380810　Symphyoloma graveolens C. A. Mey. ;合缘芹☆

380811　Symphyomera Hook. f. = Cotula L. ■

380812　Symphyomyrtus Schauer = Eucalyptus L'Her. ●

380813　Symphyomyrtus lehmanii Schauer = Eucalyptus lehmannii (Schauer) Benth. ●☆

380814　Symphyonema R. Br. = Symphionema R. Br. ●☆

380815　Symphyonema Spreng. = Symphionema R. Br. ●☆

380816　Symphyopappus Post et Kuntze = Cadiscus E. Mey. ex DC. ■

380817　Symphyopappus Post et Kuntze = Symphipappus Klatt ■☆

380818　Symphyopappus Turcz.(1848);合冠菊属●☆

380819　Symphyopappus angustifolius Cabrera;窄叶合冠菊●☆

380820　Symphyopappus brasiliensis (Gardner) R. M. King et H. Rob. ; 巴西合冠菊●☆

380821　Symphyopappus decussatus Turcz. ;合冠菊●☆

380822　Symphyopetaion J. Drumm. ex Harv. = Nematolepis Turcz. ●☆

380823　Symphyosepalum Hand. -Mazz. = Neottianthe (Rchb.) Schltr. ■

380824　Symphyosepalum gymnadenioides Hand. -Mazz. = Neottianthe cucullata (W. W. Sm.) Schltr. var. calcicola (W. W. Sm.) Soó ■

380825　Symphyosepalum gymnadenioides Hand. -Mazz. = Neottianthe gymnadenioides (Hand. -Mazz.) K. Y. Lang et S. C. Chen ■

380826　Symphyostemon Klotzsch = Cleome L. ●■

380827　Symphyostemon Miers = Phaiophleps Raf. ■☆

380828　Symphyostemon Miers = Symphyostemon Miers ex Klatt ■☆

380829　Symphyostemon Miers et Klatt = Symphyostemon Miers ■☆

380830　Symphyostemon Miers ex Klatt = Olsynium Raf. ■☆

380831　Symphyostemon Miers ex Klatt = Phaiophleps Raf. ■☆

380832　Symphyotrichum Nees = Aster L. ●■

380833　Symphyotrichum Nees(1832);卷舌菊属;Aster ■☆

380834　Symphyotrichum adnatum (Nutt.) G. L. Nesom;贴生卷舌菊; Scaleleaf Aster ■☆

380835　Symphyotrichum amethystinum (Nutt.) G. L. Nesom;紫晶卷舌菊;Amethyst Aster ■☆

380836　Symphyotrichum amethystinum (Nutt.) G. L. Nesom = Aster amethystinus Nutt. ■☆

380837　Symphyotrichum anomalum (Engelm. ex Torr. et A. Gray) G. L. Nesom;多射线卷舌菊;Manyray Aster ■☆

380838　Symphyotrichum anticostense (Fernald) G. L. Nesom;安蒂科斯蒂卷舌菊;Anticosti Aster ■☆

380839　Symphyotrichum ascendens (Lindl.) G. L. Nesom;上升卷舌菊;Intermountain Aster,Long-leaved Aster,Western Aster ■☆

380840　Symphyotrichum attenuatum (Lindl.) Semple = Symphyotrichum laeve (L.) G. L. Nesom var. purpuratum (Nees) G. L. Nesom ■☆

380841　Symphyotrichum bahamense (Britton) G. L. Nesom = Symphyotrichum subulatum (Michx.) G. L. Nesom var. elongatum (Bosser. ex A. G. Jones et Lowry) S. D. Sundb. ■☆

380842　Symphyotrichum boreale (Torr. et A. Gray) Á. Löve et D. Löve;北方卷舌菊;Northern Bog Aster,Rush Aster,Slender White Aster ■☆

380843　Symphyotrichum bracei (Britton) G. L. Nesom = Symphyotrichum tenuifolium (L.) G. L. Nesom var. aphyllum (R. W. Long) S. D. Sundb. ■☆

380844　Symphyotrichum bracteolatum (Nutt.) G. L. Nesom = Symphyotrichum eatonii (A. Gray) G. L. Nesom ■☆

380845　Symphyotrichum campestre (Nutt.) G. L. Nesom;平原卷舌菊; Western Meadow Aster ■☆

380846　Symphyotrichum campestre (Nutt.) G. L. Nesom var. bloomeri (A. Gray) G. L. Nesom = Symphyotrichum campestre (Nutt.) G. L. Nesom ■☆

380847　Symphyotrichum carolinianum (Walter) Wunderlin et B. F. Hansen = Ampelaster carolinianus (Walter) G. L. Nesom ■☆

380848　Symphyotrichum chapmanii (Torr. et A. Gray) Semple et Brouillet;查普曼卷舌菊;Savanna Aster ■☆

380849　Symphyotrichum chilense (Nees) G. L. Nesom;智利卷舌菊■☆

380850　Symphyotrichum ciliatum (Ledeb.) G. L. Nesom;缘毛卷舌菊; Rayless Alkali Aster,Rayless Annual Aster ■☆

380851　Symphyotrichum ciliatum (Ledeb.) G. L. Nesom = Brachyactis ciliata (Ledeb.) Ledeb. ■

380852　Symphyotrichum ciliolatum (Lindl.) Á. Löve et D. Löve;林氏卷舌菊;Fringed Blue Aster,Lindley's Aster,Northern Heart-leaved Aster ■☆

380853　Symphyotrichum ciliolatum (Lindl.) Á. Löve et D. Löve = Aster ciliolatus Lindl. ■☆

380854　Symphyotrichum concolor (L.) G. L. Nesom;东部银卷舌菊(银紫菀);Eastern Silvery Aster ■☆

380855　Symphyotrichum concolor (L.) G. L. Nesom var. plumosum (Small) Wunderlin et B. F. Hansen = Symphyotrichum plumosum (Small) Semple ■☆

380856　Symphyotrichum cordifolium (L.) G. L. Nesom;心叶卷舌菊; Common Blue Wood Aster,Heartleaf Aster ■☆

380857　Symphyotrichum cordifolium (L.) G. L. Nesom = Aster cordifolius L. ■☆

380858　Symphyotrichum cordifolium (L.) G. L. Nesom = Aster sagittifolius Wedem. ex Willd. ■☆

380859　Symphyotrichum cusickii (A. Gray) G. L. Nesom;库西克卷舌菊;Cusick's Aster ■☆

380860　Symphyotrichum defoliatum (Parish) G. L. Nesom;圣贝尔纳多卷舌菊;San Bernardino Aster ■☆

380861　Symphyotrichum depauperatum (Fernald) G. L. Nesom;蜒蜓卷舌菊;Serpentine Aster,Starved Aster ■☆

380862　Symphyotrichum divaricatum (Nutt.) G. L. Nesom = Symphyotrichum subulatum (Michx.) G. L. Nesom var. ligulatum S. D. Sundb. ■☆

380863　Symphyotrichum drummondii (Lindl.) G. L. Nesom;德拉蒙德卷舌菊(德氏紫菀);Drummond Aster,Drummond's Aster ■☆

380864　Symphyotrichum drummondii (Lindl.) G. L. Nesom var. parviceps (Shinners) G. L. Nesom = Symphyotrichum drummondii

（Lindl.）G. L. Nesom var. texanum（E. S. Burgess）G. L. Nesom ■☆

380865　Symphyotrichum drummondii（Lindl.）G. L. Nesom var. texanum（E. S. Burgess）G. L. Nesom;得州卷舌菊;Texas Aster ■☆

380866　Symphyotrichum dumosum（L.）G. L. Nesom;灌丛卷舌菊;Aster, Bushy Aster, Long-stalked Aster, Rice-button Aster, Wood's Aster ■☆

380867　Symphyotrichum dumosum（L.）G. L. Nesom = Aster dumosus L. ■☆

380868　Symphyotrichum dumosum（L.）G. L. Nesom var. strictior（Torr. et A. Gray）G. L. Nesom = Aster dumosus L. var. strictior Torr. et A. Gray ■☆

380869　Symphyotrichum dumosum（L.）G. L. Nesom var. strictior（Torr. et A. Gray）G. L. Nesom;直立灌丛卷舌菊;Bushy Aster, Long-stalked Aster, Rice-button Aster ■☆

380870　Symphyotrichum eatonii（A. Gray）G. L. Nesom;伊顿卷舌菊;Eaton's Aster ■☆

380871　Symphyotrichum elliottii（Torr. et A. Gray）G. L. Nesom;埃利奥特卷舌菊;Elliott's Aster ■☆

380872　Symphyotrichum ericoides（L.）G. L. Nesom;白心卷舌菊（毛紫菀）;Downy Aster, Heath Aster, Many-flowered Aster, Snow Flurry Aster, Squarrose White Aster, White Aster, White Heath, White Heath Aster, White Old-field Aster, White Prairie Aster, Wreath Aster ■☆

380873　Symphyotrichum ericoides（L.）G. L. Nesom = Aster ericoides L. ■☆

380874　Symphyotrichum ericoides（L.）G. L. Nesom subsp. pansum（S. F. Blake）Semple = Symphyotrichum ericoides（L.）G. L. Nesom var. pansum（S. F. Blake）G. L. Nesom ■☆

380875　Symphyotrichum ericoides（L.）G. L. Nesom var. ericoides = Symphyotrichum ericoides（L.）G. L. Nesom ■☆

380876　Symphyotrichum ericoides（L.）G. L. Nesom var. pansum（S. F. Blake）G. L. Nesom;铺散卷舌菊■☆

380877　Symphyotrichum ericoides（L.）G. L. Nesom var. prostratum（Kuntze）G. L. Nesom = Aster ericoides L. var. prostratus（Kuntze）S. F. Blake ■☆

380878　Symphyotrichum ericoides（L.）G. L. Nesom var. prostratum（Kuntze）G. L. Nesom = Symphyotrichum ericoides（L.）G. L. Nesom ■☆

380879　Symphyotrichum ericoides（L.）G. L. Nesom var. prostratum（Kuntze）G. L. Nesom;平卧白心卷舌菊;Prostrate Heath Aster ■☆

380880　Symphyotrichum ericoides（L.）G. L. Nesom var. stricticaule（Torr. et A. Gray）G. L. Nesom = Symphyotrichum ericoides（L.）G. L. Nesom var. pansum（S. F. Blake）G. L. Nesom ■☆

380881　Symphyotrichum eulae（Shinners）G. L. Nesom;欧拉卷舌菊;Eula's Aster ■☆

380882　Symphyotrichum expansum（Poepp. ex Spreng.）G. L. Nesom = Symphyotrichum subulatum（Michx.）G. L. Nesom var. parviflorum（Nees）S. D. Sundb. ■☆

380883　Symphyotrichum falcatum（Lindl.）G. L. Nesom;西部心叶卷舌菊;Cluster Aster, Western Heath Aster, White Prairie Aster ■☆

380884　Symphyotrichum falcatum（Lindl.）G. L. Nesom = Aster falcatus Lindl. ■☆

380885　Symphyotrichum falcatum（Lindl.）G. L. Nesom subsp. commutatum（Torr. et A. Gray）Semple = Symphyotrichum falcatum（Lindl.）G. L. Nesom var. commutatum（Torr. et A. Gray）G. L. Nesom ■☆

380886　Symphyotrichum falcatum（Lindl.）G. L. Nesom var. commutatum（Torr. et A. Gray）= Aster falcatus Lindl. var. commutatus（Torr. et A. Gray）A. G. Jones ■☆

380887　Symphyotrichum falcatum（Lindl.）G. L. Nesom var. commutatum（Torr. et A. Gray）G. L. Nesom;变化卷舌菊;Cluster Aster, White Prairie Aster ■☆

380888　Symphyotrichum falcatum（Lindl.）G. L. Nesom var. crassulum（Torr. et A. Gray）G. L. Nesom = Symphyotrichum falcatum（Lindl.）G. L. Nesom var. commutatum（Torr. et A. Gray）G. L. Nesom ■☆

380889　Symphyotrichum fendleri（A. Gray）G. L. Nesom;芬德勒卷舌菊;Fendler's Aster ■☆

380890　Symphyotrichum firmum（Nees）G. L. Nesom;亮叶卷舌菊;Glossy-leaved Aster, Shining Aster, Shiny-leaved Aster ■☆

380891　Symphyotrichum firmum（Nees）G. L. Nesom = Aster firmus Nees ■☆

380892　Symphyotrichum foliaceum（Lindl. ex DC.）G. L. Nesom;叶苞卷舌菊（叶苞紫菀）;Alpine Leafybract Aster, Leafy Aster, Leafy-bract Aster, Leafy-bracted Aster ■☆

380893　Symphyotrichum foliaceum（Lindl. ex DC.）G. L. Nesom = Aster foliaceus Lindl. ex DC. ■☆

380894　Symphyotrichum foliaceum（Lindl. ex DC.）G. L. Nesom var. apricum（A. Gray）G. L. Nesom;喜光卷舌菊■☆

380895　Symphyotrichum foliaceum（Lindl. ex DC.）G. L. Nesom var. canbyi（A. Gray）G. L. Nesom;康比卷舌菊■☆

380896　Symphyotrichum foliaceum（Lindl. ex DC.）G. L. Nesom var. parryi（D. C. Eaton）G. L. Nesom;帕里卷舌菊■☆

380897　Symphyotrichum fontinale（Alexander）G. L. Nesom;佛罗里达卷舌菊;Florida water Aster ■☆

380898　Symphyotrichum frondosum（Nutt.）G. L. Nesom;碱地卷舌菊;Short-rayed Alkali Aster ■☆

380899　Symphyotrichum georgianum（Alexander）G. L. Nesom;乔治亚卷舌菊;Georgia Aster ■☆

380900　Symphyotrichum grandiflorum（L.）G. L. Nesom;大花卷舌菊（大花紫菀,大紫菀）;Christmas Daisy, Great Aster, Large-flowered Aster ■☆

380901　Symphyotrichum grandiflorum（L.）G. L. Nesom = Aster grandiflorus L. ■☆

380902　Symphyotrichum greatae（Parish）G. L. Nesom;格氏卷舌菊;Greata's Aster ■☆

380903　Symphyotrichum hallii（A. Gray）G. L. Nesom;霍尔卷舌菊;Hall's Aster ■☆

380904　Symphyotrichum hendersonii（Fernald）G. L. Nesom;亨德森卷舌菊;Henderson's Aster ■☆

380905　Symphyotrichum hesperium（A. Gray）Á. Löve and D. Löve = Symphyotrichum lanceolatum（Willd.）G. L. Nesom var. hesperium（A. Gray）G. L. Nesom ■☆

380906　Symphyotrichum jessicae（Piper）G. L. Nesom;杰西卡卷舌菊;Jessica's Aster ■☆

380907　Symphyotrichum laeve（L.）Á. Löve et D. Löve;光滑卷舌菊;Glaucous Michaelmas-daisy, Smooth Aster, Smooth Blue Aster ■☆

380908　Symphyotrichum laeve（L.）Á. Löve et D. Löve = Aster laevis L. ■☆

380909　Symphyotrichum laeve（L.）G. L. Nesom var. concinnum（Willd.）G. L. Nesom;雅致卷舌菊■☆

380910　Symphyotrichum laeve（L.）G. L. Nesom var. geyeri（A. Gray）G. L. Nesom;盖耶氏卷舌菊;Geyer's Aster ■☆

380911　Symphyotrichum laeve（L.）G. L. Nesom var. purpuratum（Nees）G. L. Nesom;紫色光滑卷舌菊■☆

380912 Symphyotrichum lanceolatum（Willd.）G. L. Nesom；剑叶卷舌菊（剑叶紫菀）；Eastern Lined Aster, Lance-leaved Aster, Marsh Aster, Narrow-leaved Michaelmas-daisy Panicled Aster, Panicled Aster,Tall White Aster,White Panicle Aster,White Panicled Aster ■☆

380913 Symphyotrichum lanceolatum（Willd.）G. L. Nesom subsp. hesperium（A. Gray）G. L. Nesom ＝Symphyotrichum lanceolatum（Willd.）G. L. Nesom var. hesperium（A. Gray）G. L. Nesom ■☆

380914 Symphyotrichum lanceolatum（Willd.）G. L. Nesom subsp. lanceolatum var. interior（Wiegand）G. L. Nesom ＝Symphyotrichum lanceolatum（Willd.）G. L. Nesom var. interior（Wiegand）G. L. Nesom ■☆

380915 Symphyotrichum lanceolatum（Willd.）G. L. Nesom var. hesperium（A. Gray）G. L. Nesom ＝Aster hesperius A. Gray ■☆

380916 Symphyotrichum lanceolatum（Willd.）G. L. Nesom var. hesperium（A. Gray）G. L. Nesom；金星剑叶卷舌菊；Western Lined Aster ■☆

380917 Symphyotrichum lanceolatum（Willd.）G. L. Nesom var. hirsuticaule（Semple et Chmiel.）G. L. Nesom ＝Aster lanceolatus Willd. var. hirsuticaulis Semple et Chmiel. ■☆

380918 Symphyotrichum lanceolatum（Willd.）G. L. Nesom var. hirsuticaule（Semple et Chmiel.）G. L. Nesom；毛茎剑叶卷舌菊；White Panicle Aster ■☆

380919 Symphyotrichum lanceolatum（Willd.）G. L. Nesom var. interior（Wiegand）G. L. Nesom；内地剑叶卷舌菊；Inland Panicled Aster, Panicled Aster,White Panicle Aster ■☆

380920 Symphyotrichum lanceolatum（Willd.）G. L. Nesom var. latiflorum（Semple et Chmiel.）G. L. Nesom ＝Aster lanceolatus Willd. var. latifolius Semple et Chmiel. ■☆

380921 Symphyotrichum lanceolatum（Willd.）G. L. Nesom var. latifolium（Semple et Chmiel.）G. L. Nesom；宽剑叶卷舌菊；White Panicle Aster ■☆

380922 Symphyotrichum lateriflorum（L.）Á. Löve et D. Löve；宽花卷舌菊；Calico Aster, CalicoAster, One-sided Aster, Starved Aster, White Woodland Aster ■☆

380923 Symphyotrichum lateriflorum（L.）Á. Löve et D. Löve ＝Aster lateriflorus（L.）Britton ■☆

380924 Symphyotrichum laurentianum（Fernald）G. L. Nesom；圣劳伦斯卷舌菊；Gulf of St. Lawrence Aster ■☆

380925 Symphyotrichum lentum（Greene）G. L. Nesom；湿地卷舌菊；Suisun Marsh Aster ■☆

380926 Symphyotrichum longifolium（Lam.）G. L. Nesom；长叶卷舌菊；Long-leaved Aster,Long-leaved Blue Aster ■☆

380927 Symphyotrichum longifolium（Lam.）G. L. Nesom ＝Aster longifolius Lam. ■☆

380928 Symphyotrichum molle（Rydb.）G. L. Nesom；软卷舌菊；Soft Aster ■☆

380929 Symphyotrichum nahanniense（Cody）Semple；那汗卷舌菊；Nahanni Aster ■☆

380930 Symphyotrichum novae-angliae（L.）G. L. Nesom；新英格兰卷舌菊（红花紫菀，美国紫菀）；Autumn's Welcome, Hairy Michaelmas-daisy, Michaelmas Daisy, New England Aster, New-England-staudenaster,Starwort ■☆

380931 Symphyotrichum novae-angliae（L.）G. L. Nesom ＝Aster novi-belgii L. ■

380932 Symphyotrichum novi-belgii（L.）G. L. Nesom ＝Aster novi-belgii L. ■

380933 Symphyotrichum novi-belgii（L.）G. L. Nesom ＝

Symphyotrichum longifolium（Lam.）G. L. Nesom ■☆

380934 Symphyotrichum novi-belgii（L.）G. L. Nesom var. crenifolium（Fernald）Labrecque et Brouillet ＝Aster foliaceus Lindl. var. crenifolius Fernald ■☆

380935 Symphyotrichum novi-belgii（L.）G. L. Nesom var. novi-belgii ? ＝Symphyotrichum longifolium（Lam.）G. L. Nesom ■☆

380936 Symphyotrichum novi-belgii（L.）G. L. Nesom var. villicaule（A. Gray）Labrecque et Brouillet ＝Aster longifolius Lam. var. villicaulis A. Gray ■☆

380937 Symphyotrichum oblongifolium（Nutt.）G. L. Nesom；矩圆叶卷舌菊；Aromatic Aster, Oblong-leaved Aster ■☆

380938 Symphyotrichum oblongifolium（Nutt.）G. L. Nesom ＝Aster oblongifolius Nutt. ■☆

380939 Symphyotrichum ontarionis（Wiegand）G. L. Nesom；安大略卷舌菊（安大略紫菀）；Bottomland Aster,Ontario Aster ■☆

380940 Symphyotrichum ontarionis（Wiegand）G. L. Nesom ＝Aster ontarionis Wiegand ■☆

380941 Symphyotrichum ontarionis（Wiegand）G. L. Nesom var. glabratum（Semple）Brouillet et Bouchard；光安大略卷舌菊■☆

380942 Symphyotrichum oolentagiense（Riddell）G. L. Nesom ＝Symphyotrichum oolentangiense（Riddell）G. L. Nesom ■☆

380943 Symphyotrichum oolentangiense（Riddell）G. L. Nesom；天蓝卷舌菊；Azure Aster, Blue Devil, Prairie Heart-leaved Aster, Skyblue Aster,Sky-blue Aster ■☆

380944 Symphyotrichum oolentangiense（Riddell）G. L. Nesom ＝Aster oolentangiensis Riddell ■☆

380945 Symphyotrichum parviceps（E. S. Burgess）G. L. Nesom；小柄卷舌菊；Small White Aster,Smallhead Aster ■☆

380946 Symphyotrichum parviceps（E. S. Burgess）G. L. Nesom ＝Aster parviceps（E. S. Burgess）Mack. et Bush ■☆

380947 Symphyotrichum patens（Aiton）G. L. Nesom；晚熟卷舌菊（迟紫菀）；Late Aster,Late Purple Aster,Purple Aster,Spreading Aster ■☆

380948 Symphyotrichum patens（Aiton）G. L. Nesom ＝Aster patens Aiton ■☆

380949 Symphyotrichum patens（Aiton）G. L. Nesom var. gracile（Hook.）G. L. Nesom；纤细晚熟卷舌菊■☆

380950 Symphyotrichum patens（Aiton）G. L. Nesom var. patentissimum（Lindl. ex DC.）G. L. Nesom；极晚熟卷舌菊■☆

380951 Symphyotrichum patulum（Lam.）Karlsson；北美洲卷舌菊■☆

380952 Symphyotrichum phlogifolium（Muhl. ex Willd.）G. L. Nesom；细叶晚熟卷舌菊；Thin-leaf Late Purple Aster ■☆

380953 Symphyotrichum pilosum（Willd.）G. L. Nesom；林地卷舌菊；Awl Aster, Frost Aster, Frost Weed Aster, Hairy Aster, Hairy White Oldfield Aster, Oldfield Aster, White Heath Aster, White Oldfield Aster,White Old-field Aster ■☆

380954 Symphyotrichum pilosum（Willd.）G. L. Nesom ＝Aster pilosus Willd. ■☆

380955 Symphyotrichum pilosum（Willd.）G. L. Nesom var. pringlei（A. Gray）G. L. Nesom；普林格尔卷舌菊；Awl Aster, Frost Aster, Hairy Aster,Pringle's Aster,White Old-field Aster ■☆

380956 Symphyotrichum plumosum（Small）Semple；羽状卷舌菊■☆

380957 Symphyotrichum porteri（A. Gray）G. L. Nesom；波特卷舌菊；Porter's Aster,Smooth White Aster ■☆

380958 Symphyotrichum potosinum（A. Gray）G. L. Nesom；圣丽塔卷舌菊；Santa Rita Mountain Aster ■☆

380959 Symphyotrichum praealtum（Poir.）G. L. Nesom；柳叶卷舌菊；Veiny Lined Aster, Willow Aster, Willowleaf Aster, Willow-leaved

Aster ■☆

380960　Symphyotrichum praealtum（Poir.）G. L. Nesom ＝ Aster praealtus Poir. ■☆

380961　Symphyotrichum praealtum（Poir.）G. L. Nesom var. angustior（Wiegand）G. L. Nesom ＝ Aster praealtus Poir. var. angustior Wiegand ■☆

380962　Symphyotrichum praealtum（Poir.）G. L. Nesom；窄柳叶卷舌菊；Veiny Lined Aster，Willow Aster，Willow-leaved Aster ■☆

380963　Symphyotrichum praealtum（Poir.）G. L. Nesom var. texicola（Wiegand）G. L. Nesom ＝ Aster praealtus Poir. var. coerulescens（DC.）A. G. Jones ■☆

380964　Symphyotrichum praealtum（Poir.）G. L. Nesom var. texicola（Wiegand）G. L. Nesom；天蓝柳叶卷舌菊；Veiny Lined Aster，Willow Aster，Willow-leaved Aster ■☆

380965　Symphyotrichum pratense（Raf.）G. L. Nesom；荒地卷舌菊；Barrens Silky Aster ■☆

380966　Symphyotrichum prenanthoides（Muhl. ex Willd.）G. L. Nesom；弯茎卷舌菊；Crooked Aster，Crookedstem Aster，Crooked-stem Aster，Crooked-stemmed Aster，Zigzag Aster ■☆

380967　Symphyotrichum prenanthoides（Muhl. ex Willd.）G. L. Nesom ＝ Aster prenanthoides Muhl. ex Willd. ■☆

380968　Symphyotrichum prenanthoides（Muhl. ex Willd.）G. L. Nesom ＝ Symphyotrichum pratense（Raf.）G. L. Nesom ■☆

380969　Symphyotrichum priceae（Britton）G. L. Nesom；普来氏卷舌菊；Lavender Oldfield Aster，Price's Aster ■☆

380970　Symphyotrichum puniceum（L.）Á. Löve et D. Löve；红茎卷舌菊（红茎紫菀，深红紫菀）；Bristly Aster，Glossy-leaved Aster，Purplestem Aster，Purple-stem Aster，Red-stalked Aster，Red-stemmed Aster，Swamp Aster ■☆

380971　Symphyotrichum puniceum（L.）Á. Löve et D. Löve ＝ Aster puniceus L. ■☆

380972　Symphyotrichum puniceum（L.）Á. Löve et D. Löve var. calderi（B. Boivin）G. L. Nesom ＝ Symphyotrichum puniceum（L.）Á. Löve et D. Löve ■☆

380973　Symphyotrichum puniceum（L.）Á. Löve et D. Löve var. scabricaule（Shinners）G. L. Nesom；糙茎卷舌菊；Roughstem Aster ■☆

380974　Symphyotrichum pygmaeum（Lindl.）Brouillet et S. Selliah；微小卷舌菊；Pygmy Aster ■☆

380975　Symphyotrichum racemosum（Elliott）G. L. Nesom；小白卷舌菊；Small White Aster，Smooth White Oldfield Aster ■☆

380976　Symphyotrichum racemosum（Elliott）G. L. Nesom var. subdumosum（Wiegand）G. L. Nesom ＝ Aster fragilis Willd. var. subdumosus（Wiegand）A. G. Jones ■☆

380977　Symphyotrichum retroflexum（Lindl. ex DC.）G. L. Nesom；硬白头卷舌菊；Curtis Aster，Rigid Whitetop Aster ■☆

380978　Symphyotrichum rhiannon Weakley et Govus；卷舌菊；Rhiannon's Aster ■☆

380979　Symphyotrichum robynsianum（J. Rousseau）Brouillet et Labrecque；罗宾卷舌菊；Robyns' Aster ■☆

380980　Symphyotrichum sericeum（Vent.）G. L. Nesom；西部银色卷舌菊；Western Silvery Aster ■☆

380981　Symphyotrichum sericeum（Vent.）G. L. Nesom ＝ Aster sericeus Vent. ■☆

380982　Symphyotrichum sericeum（Vent.）G. L. Nesom var. microphyllum（DC.）Wunderlin et B. F. Hansen ＝ Symphyotrichum pratense（Raf.）G. L. Nesom ■☆

380983　Symphyotrichum shortii（Lindl.）G. L. Nesom；肖特卷舌菊；Midwestern Blue Heart-leaved Aster，Short's Aster ■☆

380984　Symphyotrichum shortii（Lindl.）G. L. Nesom ＝ Aster shortii Lindl. ■☆

380985　Symphyotrichum simmondsii（Small）G. L. Nesom；西蒙卷舌菊；Simmonds' Aster ■☆

380986　Symphyotrichum simplex（Willd.）Á. Löve et D. Löve；单枝卷舌菊；Branched Panicled Aster，Panicled Aster，White Panicle Aster ■☆

380987　Symphyotrichum simplex（Willd.）Á. Löve et D. Löve ＝ Aster lanceolatus Willd. var. simplex（Willd.）A. G. Jones ■☆

380988　Symphyotrichum simplex（Willd.）Á. Löve et D. Löve ＝ Symphyotrichum lanceolatum（Willd.）G. L. Nesom ■☆

380989　Symphyotrichum spathulatum（Lindl.）G. L. Nesom；西部山地卷舌菊；Western Mountain Aster ■☆

380990　Symphyotrichum spathulatum（Lindl.）G. L. Nesom var. intermedium（A. Gray）G. L. Nesom；间型卷舌菊 ■☆

380991　Symphyotrichum spathulatum（Lindl.）G. L. Nesom var. yosemitanum（A. Gray）G. L. Nesom；西部沼泽卷舌菊；Western Bog Aster ■☆

380992　Symphyotrichum squamatum（Spreng.）G. L. Nesom ＝ Symphyotrichum subulatum（Michx.）G. L. Nesom var. squamatum（Spreng.）S. D. Sundb. ■☆

380993　Symphyotrichum subspicatum（Nees）G. L. Nesom；道格拉斯卷舌菊；Douglas' Aster ■☆

380994　Symphyotrichum subulatum（Michx.）G. L. Nesom ＝ Aster subulatus Michx. ■

380995　Symphyotrichum subulatum（Michx.）G. L. Nesom var. elongatum（Bosser. ex A. G. Jones et Lowry）S. D. Sundb.；巴哈曼卷舌菊；Bahaman Aster ■☆

380996　Symphyotrichum subulatum（Michx.）G. L. Nesom var. ligulatum S. D. Sundb.；盐沼卷舌菊；Saltmarsh Aster，Southern Annual Saltmarsh Aster ■☆

380997　Symphyotrichum subulatum（Michx.）G. L. Nesom var. parviflorum（Nees）S. D. Sundb.；西南一年生卷舌菊；Southwestern Annual Saltmarsh Aster ■☆

380998　Symphyotrichum subulatum（Michx.）G. L. Nesom var. squamatum（Spreng.）S. D. Sundb.；南部一年生卷舌菊；Southeastern Annual Saltmarsh Aster ■☆

380999　Symphyotrichum tenuifolium（L.）G. L. Nesom；盐沼多年生卷舌菊；Large Sahmarsh Aster，Perennial Saltmarsh Aster，Saltmarsh Aster ■☆

381000　Symphyotrichum tenuifolium（L.）G. L. Nesom ＝ Aster tenuifolius L. ■☆

381001　Symphyotrichum tenuifolium（L.）G. L. Nesom var. aphyllum（R. W. Long）S. D. Sundb.；无叶盐沼卷舌菊；Brace's Aster ■☆

381002　Symphyotrichum texanum（E. S. Burgess）Semple ＝ Symphyotrichum drummondii（Lindl.）G. L. Nesom var. texanum（E. S. Burgess）G. L. Nesom ■☆

381003　Symphyotrichum tradescantii（L.）G. L. Nesom；海滨卷舌菊；Shore Aster，Tradescant's Aster ■☆

381004　Symphyotrichum tradescantii（L.）G. L. Nesom ＝ Aster lateriflorus（L.）Britton var. hirsuticaulis（Lindl. ex DC.）Porter ■☆

381005　Symphyotrichum turbinellum（Lindl.）G. L. Nesom；小陀螺卷舌菊（小陀螺紫菀）；Prairie Aster ■☆

381006　Symphyotrichum undulatum（L.）G. L. Nesom；波叶卷舌菊；Wavyleaf Aster ■☆

381007　Symphyotrichum urophyllum（Lindl. ex DC.）G. L. Nesom；白

箭叶卷舌菊（箭叶紫菀）；Arrow Aster，Arrowleaf Aster，Arrow-leaved Aster，Blue Wood Aster，Heart Leaved Aster，White Arrowleaf Aster ■☆

381008　Symphyotrichum urophyllum (Lindl. ex DC.) G. L. Nesom ＝ Aster sagittifolius Wedem. ex Willd. ■☆

381009　Symphyotrichum urophyllum (Lindl. ex DC.) G. L. Nesom ＝ Aster urophyllus Lindl. ex DC. ■☆

381010　Symphyotrichum walteri (Alexander) G. L. Nesom；沃尔特卷舌菊；Walter's Aster ■☆

381011　Symphyotrichum welshii (Cronquist) G. L. Nesom；威尔士卷舌菊；Welsh's Aster ■☆

381012　Symphyotrichum yukonense (Cronquist) G. L. Nesom；育空卷舌菊；Yukon Aster ■☆

381013　Symphysia C. Presl(1827)；合囊莓属（西印度杜鹃花属）●☆

381014　Symphysia martinicensis C. Presl；合囊莓●☆

381015　Symphysicarpus Hassk. ＝ Heterostemma Wight et Arn. ●

381016　Symphysodaphne A. Rich. ＝ Licaria Aubl. ●☆

381017　Symphytonema Schltr. ＝ Camptocarpus Decne.（保留属名）●■☆

381018　Symphytonema Schltr. ＝ Tanulepis Balf. f. ●☆

381019　Symphytonema acuminatum Choux ＝ Camptocarpus acuminatus (Choux) Venter ●☆

381020　Symphytonema crassifolium (Decne.) Choux ＝ Camptocarpus crassifolius Decne. ●☆

381021　Symphytonema lineare (Decne.) Choux ＝ Camptocarpus crassifolius Decne. ●☆

381022　Symphytonema madagascariense Schltr. ＝ Camptocarpus crassifolius Decne. ●☆

381023　Symphytosiphon Harms ＝ Trichilia P. Browne(保留属名)●

381024　Symphytum L. (1753)；聚合草属（合生草属，合生花属，聚生草属，块根紫芹属，西门肺草属）；Collectivegrass，Comfrey ■

381025　Symphytum asperrimum Donn ＝ Symphytum asperum Lepech. ■☆

381026　Symphytum asperum Lepech.；糙聚合草（糙叶康复力）；Prickly Comfrey，Rough Comfrey ■☆

381027　Symphytum bohemicum F. W. Schmidt；波希米亚聚合草■☆

381028　Symphytum bulbosum K. F. Schimp.；鳞茎聚合草■☆

381029　Symphytum caucasicum M. Bieb.；高加索西门肺草（高加索聚合草，辛菲草）；Blue Comfrey，Caucasia Collectivegrass，Caucasian Comfrey ■☆

381030　Symphytum cordatum Waldst. et Kit.；心叶聚合草；Heartleaf Collectivegrass ■☆

381031　Symphytum grandiflorum DC.；大花聚合草（伊比利亚聚合草）；Creeping Comfrey ■☆

381032　Symphytum ibericum Steven ex M. Bieb. ' Jubilee ' ＝ Symphytum ibericum Steven ex M. Bieb. ' Variegatum ' ■☆

381033　Symphytum ibericum Steven ex M. Bieb. ' Variegatum '；花叶聚合草■☆

381034　Symphytum ibericum Steven ex M. Bieb. ＝ Symphytum grandiflorum DC. ■☆

381035　Symphytum officinale L.；聚合草（爱国草，雏菊，康复力，块茎紫草，西门肺草，药用聚合草，友谊草）；Alum，Ass Ear，Ass-ear，Backwort，Blackroot，Blackwort，Boneset，Briswort，Bruisewort，Church Bells，Coffee Flower，Comfrey，Comfrey Consound，Common Comfrey，Common Comh'ey，Consound，Cumfurt，Galla，Galluc，Gooseberry Pie，Great Consound，Gum Plant，Heartsease，Knitback，Knitbone，Knotbone，Medicinal Collectivegrass，Medicinal Comfrey，Needle-cases，Nipbone，Pigweed，Shop-consound，Slippery Root，Snake，Suckers，Sweet Suckers，Swinesnap，Woundwort，Yalluc ■

381036　Symphytum officinale L. subsp. uliginosum (J. Kern.) Nyman ＝ Symphytum officinale L. ■

381037　Symphytum orientale L.；东方聚合草；Soft Comfrey，Turkish Comfrey，White Comfrey ■☆

381038　Symphytum peregrinum Ledeb.；俄西门肺草；Blue Comfrey，Russian Comfrey ■☆

381039　Symphytum tauricum Willd.；克里木聚合草；Crimean Comfrey ■☆

381040　Symphytum tuberosum L.；管聚合草；Tuberous Comfrey ■☆

381041　Symphytum uliginosum J. Kern. ＝ Symphytum officinale L. ■

381042　Symphytum uplandicum Nyman；俄罗斯聚合草；Russian Comfrey ■☆

381043　Sympieza Licht. ex Roem. et Schult. (1818)；好望角杜鹃花属●☆

381044　Sympieza Licht. ex Roem. et Schult. ＝ Erica L. ●☆

381045　Sympieza brachyphylla Benth. ＝ Erica labialis Salisb. ●☆

381046　Sympieza breviflora N. E. Br. ＝ Erica labialis Salisb. ●☆

381047　Sympieza capitellata Licht. ex Roem. et Schult.；小头好望角杜鹃花●☆

381048　Sympieza capitellata Licht. ex Roem. et Schult. ＝ Erica labialis Salisb. ●☆

381049　Sympieza capitellata Licht. ex Roem. et Schult. var. angustata N. E. Br. ＝ Erica labialis Salisb. ●☆

381050　Sympieza capitellata Licht. ex Roem. et Schult. var. crassistigma N. E. Br. ＝ Erica labialis Salisb. ●☆

381051　Sympieza eckloniana Klotzsch ＝ Erica ecklonii E. G. H. Oliv. ●☆

381052　Sympieza gracilis (Bartl.) E. G. H. Oliv. ＝ Erica benthamiana E. G. H. Oliv. ●☆

381053　Sympieza kunthii Klotzsch ＝ Erica benthamiana E. G. H. Oliv. ●☆

381054　Sympieza kunthii Klotzsch var. brachyphylla Benth. ＝ Erica benthamiana E. G. H. Oliv. ●☆

381055　Sympieza kunthii Klotzsch var. hispida Benth. ＝ Erica benthamiana E. G. H. Oliv. ●☆

381056　Sympieza labialis (Salisb.) Druce ＝ Erica labialis Salisb. ●☆

381057　Sympieza pallescens N. E. Br. ＝ Erica labialis Salisb. ●☆

381058　Sympieza tenuiflora Benth. ＝ Erica labialis Salisb. ●☆

381059　Sympieza vestita N. E. Br. ＝ Erica labialis Salisb. ●☆

381060　Symplectochilus Lindau ＝ Anisotes Nees(保留属名)●☆

381061　Symplectochilus formosissimus (Klotzsch) Lindau ＝ Anisotes formosissimus (Klotzsch) Milne-Redh. ●☆

381062　Symplectrodia Lazarides(1985)；根茎三齿稃属■☆

381063　Symplectrodia gracilis Lazarides；细根茎三齿稃■☆

381064　Symplectrodia lanosa Lazarides；根茎三齿稃■☆

381065　Sympleura Miers ＝ Barberina Vell. ●

381066　Sympleura Miers ＝ Symplocos Jacq. ●

381067　Symplocaceae Desf. (1820)（保留科名）；山矾科（灰木科）；Sweetleaf Family，Symplocos Family ●

381068　Symplocarpus Salisb. ＝ Symplocarpus Salisb. ex W. P. C. Barton（保留属名）■

381069　Symplocarpus Salisb. ex Nutt. ＝ Symplocarpus Salisb. ex W. P. C. Barton（保留属名）■

381070　Symplocarpus Salisb. ex W. P. C. Barton(1817)（保留属名）；臭菘属；Skunk Cabbage，Skunkcabbage，Skunk-cabbage，Symplocarpus ■

381071　Symplocarpus foetidus (L.) Salisb. ＝ Symplocarpus foetidus (L.) Salisb. ex W. P. C. Barton ■

381072　Symplocarpus foetidus (L.) Salisb. ex Nutt. ＝ Symplocarpus foetidus (L.) Salisb. ex W. P. C. Barton ■

381073　Symplocarpus foetidus (L.) Salisb. ex W. P. C. Barton；臭菘

（地涌金莲，黑瞎子白菜）；Bad-smell Symplocarpus, Meadow Cabbage, Polecat-weed, Skunk Cabbage, Skunkcabbage, Skunk-cabbage, Skunkweed ■

381074 Symplocarpus foetidus（L.）Salisb. ex W. P. C. Barton var. latissimus H. Hara = Symplocarpus renifolius Schott ex Tzvelev ■☆

381075 Symplocarpus foetidus Nutt. = Symplocarpus foetidus（L.）Salisb. ex W. P. C. Barton ■

381076 Symplocarpus foetidus Salisb. ex W. P. C. Barton = Symplocarpus foetidus（L.）Salisb. ex W. P. C. Barton ■

381077 Symplocarpus foetidus Salisb. ex W. P. C. Barton var. latissimus H. Hara = Symplocarpus renifolius Schott ex Tzvelev ■☆

381078 Symplocarpus nabekuraensis Otsuka et K. Inoue;锅仓臭菘■☆

381079 Symplocarpus nipponicus Makino;姬坐禅草■☆

381080 Symplocarpus nipponicus Makino f. variegatus T. Koyama;斑点姬坐禅草■☆

381081 Symplocarpus nipponicus Makino f. viridispathus J. Ohara;绿苞姬坐禅草■☆

381082 Symplocarpus renifolius Schott ex Tzvelev;肾叶臭菘（地涌金莲,坐禅草）;Skunk Cabbage ■☆

381083 Symplocarpus renifolius Schott ex Tzvelev = Symplocarpus foetidus（L.）Salisb. ex W. P. C. Barton var. latissimus H. Hara ■☆

381084 Symplococarpon Airy Shaw（1937）;合果山茶属●☆

381085 Symplococarpon australe Sandwith et Kobuski;哥伦比亚合果山茶●☆

381086 Symplococarpon flaviflium Lundell;黄叶合果山茶●☆

381087 Symplococarpon hintonii（Bullock）Airy Shaw;合果山茶●☆

381088 Symplococarpon lucidum Lundell;亮合果山茶●☆

381089 Symplococarpon lucidum Lundell = Symplococarpon hintonii（Bullock）Airy Shaw ●☆

381090 Symplocos Jacq.（1760）;山矾属（灰木属）;Sweetleaf, Symplocos ●

381091 Symplocos acuminata（Blume）Miq.;大里力灰木（黄奶树,泡花子,月桂山矾）●

381092 Symplocos acutangula Brand = Symplocos lucida（Thunb.）Siebold et Zucc. ●

381093 Symplocos acutangula Brand = Symplocos setchuensis Brand ex Diels ●

381094 Symplocos adenophylla Wall. et G. Don;腺叶山矾（琼中山矾）;Glandularleaf Sweetleaf, Glandular-leaved Sweetleaf, Machure Sweetleaf ●

381095 Symplocos adenopus Hance;腺柄山矾（被毛腺柄山矾,赤牙木）;Clothed Glandularstipe Sweetleaf, Glandularstipe Sweetleaf, Glandular-stiped Sweetleaf ●

381096 Symplocos adenopus Hance var. vestita C. C. Huang et Y. F. Wu;被毛腺柄山矾●

381097 Symplocos adenopus Hance var. vestita C. C. Huang et Y. F. Wu = Symplocos adenopus Hance ●

381098 Symplocos adinandrifolia Hayata = Symplocos congesta Benth. ●

381099 Symplocos adinandrifolia Hayata var. theifolia Hayata = Symplocos congesta Benth. ●

381100 Symplocos aenea Hand. -Mazz.;铜绿山矾（无量山山矾）;Verdigris Sweetleaf, Wuliangshan Sweetleaf ●

381101 Symplocos aenea Hand. -Mazz. = Symplocos stellaris Brand var. aenea（Hand. -Mazz.）Noot. ●

381102 Symplocos alata Brand = Symplocos anomala Brand ●

381103 Symplocos anastomosans Chun ex Tanaka et Odash. = Symplocos adenophylla Wall. et G. Don ●

381104 Symplocos anomala Brand;薄叶山矾(薄叶冬青,盆若虎亮,山桂花,台湾山矾,玉山灰木,竹山灰木)；Doi's Sweetleaf, Morrison Sweetleaf, Taiwan Sweetleaf, Thin-leaf Sweetleaf, Thin-leaved Sweetleaf ●

381105 Symplocos anomala Brand var. fusonii（Merr.）Hand. -Mazz. = Symplocos anomala Brand ●

381106 Symplocos anomala Brand var. fusonii（Merr.）Hand. -Mazz. et Peter-Stibbal = Symplocos anomala Brand ●

381107 Symplocos anomala Brand var. morrisonicola（Hayata）S. S. Ying = Symplocos anomala Brand ●

381108 Symplocos anomala Brand var. morrisonicola（Hayata）S. S. Ying f. kiraishiensis（Hayata）S. S. Ying = Symplocos anomala Brand ●

381109 Symplocos anomala Brand var. morrisonicola（Hayata）S. S. Ying f. matudai（Hatus.）S. S. Ying = Symplocos morrisonicola Hayata ●

381110 Symplocos anomala Brand var. morrisonicola（Hayata）S. S. Ying f. matudai（Hatus.）S. S. Ying = Symplocos anomala Brand ●

381111 Symplocos anomala Brand var. nitida H. L. Li = Symplocos anomala Brand ●

381112 Symplocos argentea Brand = Symplocos anomala Brand ●

381113 Symplocos argutidens Nakai = Symplocos coreana（H. Lév.）Ohwi ●☆

381114 Symplocos argyi H. Lév. = Symplocos setchuensis Brand ex Diels ●

381115 Symplocos arisanensis Hayata;阿里山灰木●

381116 Symplocos arisanensis Hayata = Symplocos lancifolia Siebold et Zucc. ●

381117 Symplocos ascidiiformis Y. F. Wu;瓶核山矾●

381118 Symplocos ascidiiformis Y. F. Wu = Symplocos viridissima Brand ●

381119 Symplocos atriolivacea Merr. et Chun ex H. L. Li;橄榄山矾;Olivaceous Sweetleaf, Olive Sweetleaf ●◇

381120 Symplocos aurea H. Lév. = Symplocos lancifolia Siebold et Zucc. ●

381121 Symplocos austroisinensis Hand. -Mazz.;南国山矾;Australis Sweetleaf, S. China Sweetleaf, South China Sweetleaf ●

381122 Symplocos austrosinensis Hand. -Mazz. = Symplocos yaoshanensis Chun et K. C. Ting ●

381123 Symplocos balfourii H. Lév. = Symplocos cochinchinensis（Lour.）S. Moore var. laurina（Retz.）Raizada ●

381124 Symplocos boninensis Rehder et E. H. Wilson;小笠原山矾●☆

381125 Symplocos botryantha Franch.;总状山矾●

381126 Symplocos botryantha Franch. = Symplocos sumuntia Buch. -Ham. ex D. Don ●

381127 Symplocos botryantha Franch. var. stenophylla Brand = Symplocos sumuntia Buch. -Ham. ex D. Don ●

381128 Symplocos caerulea H. Lév. = Symplocos sumuntia Buch. -Ham. ex D. Don ●

381129 Symplocos cauclata Wall. ex A. DC. var. macrocalyx Hand. -Mazz. = Symplocos sumuntia Buch. -Ham. ex D. Don ●

381130 Symplocos caudata Wall. ex A. DC. = Symplocos sumuntia Buch. -Ham. ex D. Don ●

381131 Symplocos caudata Wall. ex A. DC. var. macrantha Hand. -Mazz. = Symplocos sumuntia Buch. -Ham. ex D. Don ●

381132 Symplocos cavaleriei H. Lév.;葫芦果山矾●

381133 Symplocos cavaleriei H. Lév. = Symplocos sumuntia Buch. -Ham. ex D. Don ●

381134 Symplocos celastrinea Mart. ex Miq. et Mart.；南山矾●☆

381135 Symplocos chinense（Lour.）Merr. = Symplocos paniculata（Thunb.）Miq. ●

381136 Symplocos chinensis（Lour.）Druce = Symplocos paniculata（Thunb.）Miq. ●

381137 Symplocos chinensis（Lour.）Druce subsp. pilosa（Nakai）Kitag.；毛华山矾（白檀，白檀山矾）；Pilose Chinese Sweetleaf ●

381138 Symplocos chinensis（Lour.）Druce subsp. pilosa（Nakai）Kitag. = Symplocos sawafutagi Nagam. ●

381139 Symplocos chinensis（Lour.）Druce var. leucocarpa（Nakai）Ohwi f. pilosa（Nakai）Ohwi = Symplocos sawafutagi Nagam. ●

381140 Symplocos chinensis（Lour.）Druce var. vestita（Hemsl.）Hand.-Mazz. = Symplocos paniculata（Thunb.）Miq. ●

381141 Symplocos chunii Merr.；十棱山矾●

381142 Symplocos chunii Merr. = Symplocos poilanei Guillaumin ●

381143 Symplocos cochinchinensis（Lour.）S. Moore；越南山矾（大叶灰木，大叶山矾，火织树，火样灰树，日本山矾，铁锈叶灰木，铁锈叶山矾，锈叶灰木，越南灰木）；Cochinchina Sweetleaf, Cochin-China Sweetleaf, Downy Sweetleaf, Japenese Sweetleaf, Vietnam Sweetleaf ●

381144 Symplocos cochinchinensis（Lour.）S. Moore subsp. laurina（Retz.）Noot. ex Ramam. = Symplocos cochinchinensis（Lour.）S. Moore var. laurina（Retz.）Noot. ex Ramam. ●

381145 Symplocos cochinchinensis（Lour.）S. Moore var. angustifolia（Guillaumin）Noot；狭叶山矾；Angustifoliate Sweetleaf, Narrowleaf Sweetleaf ●

381146 Symplocos cochinchinensis（Lour.）S. Moore var. javanica（Blume）S. S. Ying = Symplocos cochinchinensis（Lour.）S. Moore ●

381147 Symplocos cochinchinensis（Lour.）S. Moore var. laurina（Retz.）Noot. ex Ramam. = Symplocos cochinchinensis（Lour.）S. Moore var. laurina（Retz.）Raizada ●

381148 Symplocos cochinchinensis（Lour.）S. Moore var. laurina（Retz.）Raizada；黄牛奶树（白门，大叶白矾，短穗花山矾，短序花山矾，花香木，火灰山矾，火灰树，苦山矾，泡花子，莐花叶山矾，散风木，山猪肝，水冬瓜，台东山矾，狭叶黄牛奶树，小西氏灰木，小西氏山矾，月桂叶山矾）；Cinderlike Sweetleaf, Cinder-like Sweetleaf, Divaricated-veined Sweetleaf, Divaricate-vein Sweetleaf, Formosan Sweetleaf, Konishi Sweetleaf, Laurel Sweetleaf, Laurelleaf Sweetleaf, Narrowleaf Laurel Sweetleaf, Short-flower Sweetleaf, Shortspike Sweetleaf ●

381149 Symplocos cochinchinensis（Lour.）S. Moore var. philippinensis（Brand）Noot.；兰屿山矾（兰屿锈叶灰木）；Lanyu Sweetleaf, Philippine Cochinchina Sweetleaf ●

381150 Symplocos cochinchinensis（Lour.）S. Moore var. puberula C. C. Huang et Y. F. Wu = Symplocos cochinchinensis（Lour.）S. Moore ●

381151 Symplocos cochinchinensis（Lour.）S. Moore var. puberula C. C. Huang et Y. F. Wu；微毛越南山矾（越南山矾）；Hairy Cochinchina Sweetleaf ●

381152 Symplocos confusa Brand = Symplocos pendula Wight var. hirtistylus（C. B. Clarke）Noot. ●

381153 Symplocos confusa Brand var. lysiostemon Hand.-Mazz. = Symplocos pendula Wight var. hirtistylus（C. B. Clarke）Noot. ●

381154 Symplocos congesta Benth.；密花山矾（杨桐叶灰木，杨桐叶山矾）；Adenandra-leaf Sweetleaf, Denseflower Sweetleaf, Dense-flowered Sweetleaf ●

381155 Symplocos congesta Benth. var. glomeratifolia（Hayata）S. S. Ying = Symplocos setchuensis Brand ex Diels ●

381156 Symplocos congesta Benth. var. theifolia（Hayata）Yuen P. Yang et S. Y. Lu；茶叶灰木●

381157 Symplocos cordatifolia H. L. Li = Symplocos fordii Hance ●

381158 Symplocos coreana（H. Lév.）Ohwi；朝鲜山矾●☆

381159 Symplocos coronigera H. Lév. = Symplocos lucida（Thunb.）Siebold et Zucc ●

381160 Symplocos courtoisii H. Lév. = Ilex chinensis Sims ●

381161 Symplocos crassifolia Benth.；厚皮灰木●

381162 Symplocos crassifolia Benth. = Symplocos lucida（Thunb.）Siebold et Zucc. ●

381163 Symplocos crassilimba Merr.；厚叶山矾（白布果）；Thickleaf Sweetleaf, Thick-leaved Sweetleaf ●

381164 Symplocos crataegoides Buch.-Ham. ex D. Don；山楂山矾●☆

381165 Symplocos crataegoides Buch.-Ham. ex D. Don = Symplocos coreana（H. Lév.）Ohwi ●☆

381166 Symplocos crataegoides Buch.-Ham. ex D. Don = Symplocos paniculata（Thunb.）Miq. ●

381167 Symplocos crataegoides Buch.-Ham. ex D. Don = Symplocos sawafutagi Nagam. ●

381168 Symplocos crataegoides Buch.-Ham. ex D. Don f. major ? = Symplocos coreana（H. Lév.）Ohwi ●☆

381169 Symplocos crataegoides Buch.-Ham. ex D. Don f. major ? = Symplocos paniculata（Thunb.）Miq. ●

381170 Symplocos crataegoides Buch.-Ham. ex D. Don var. glabra ? = Symplocos paniculata（Thunb.）Miq. ●

381171 Symplocos crataegoides Buch.-Ham. ex D. Don var. glabrifolia ? = Symplocos paniculata（Thunb.）Miq. ●

381172 Symplocos crataegoides Buch.-Ham. ex D. Don var. leucocarpa ? = Symplocos sawafutagi Nagam. ●

381173 Symplocos crataegoides Buch.-Ham. ex D. Don var. pallida ? = Symplocos paniculata（Thunb.）Miq. ●

381174 Symplocos crenatifolia（Yamam.）Makino et Nemoto = Symplocos nokoensis（Hayata）Kaneh. ●

381175 Symplocos cuspidata Brand = Symplocos congesta Benth. ●

381176 Symplocos cuspidata Brand var. doii（Hayata）S. S. Ying = Symplocos morrisonicola Hayata ●

381177 Symplocos decora Hance；小泉氏灰木（美山矾）●

381178 Symplocos decora Hance = Symplocos sumuntia Buch.-Ham. ex D. Don ●

381179 Symplocos delavayi Brand = Symplocos dryophila C. B. Clarke ●

381180 Symplocos dielsii H. Lév. = Symplocos anomala Brand ●

381181 Symplocos discolor Brand = Symplocos lucida（Thunb.）Siebold et Zucc. ●

381182 Symplocos divaricativena Hayata；短穗花山矾●

381183 Symplocos divaricativena Hayata = Symplocos cochinchinensis（Lour.）S. Moore var. laurina（Retz.）Raizada ●

381184 Symplocos divaricativena Hayata = Symplocos heishanensis Hayata ●

381185 Symplocos doii Hayata = Symplocos anomala Brand ●

381186 Symplocos dolichostylosa Y. F. Wu = Symplocos sumuntia Buch.-Ham. ex D. Don ●

381187 Symplocos dolichotricha Merr.；长毛山矾（土白术）；Longhair Sweetleaf, Long-haired Sweetleaf ●

381188 Symplocos dryophila C. B. Clarke；坚木山矾；Dryophyla Sweetleaf, Hardwood Sweetleaf, Oak-like Sweetleaf ●

381189 Symplocos dung Eberh. et Dubard；火灰山矾●

381190 Symplocos dung Eberh. et Dubard = Symplocos cochinchinensis (Lour.) S. Moore var. laurina (Retz.) Raizada ●

381191 Symplocos dunniana H. Lév. = Symplocos stellaris Brand ●

381192 Symplocos eriobotryifolia Hayata = Symplocos stellaris Brand ●

381193 Symplocos eriostroma Hayata；薄叶灰木●

381194 Symplocos eriostroma Hayata = Symplocos modesta Brand ●

381195 Symplocos ernestii Dunn = Symplocos lucida (Thunb.) Siebold et Zucc. ●

381196 Symplocos ernestii Dunn = Symplocos phyllocalyx C. B. Clarke ●

381197 Symplocos esquirolii H. Lév. = Symplocos anomala Brand ●

381198 Symplocos euryifolia Masam. et Syozi = Symplocos euryoides Hand. -Mazz. ●

381199 Symplocos euryoides Hand. -Mazz. ；柃叶山矾；Euryaleaf Sweetleaf , Eurya-leaved Sweetleaf ●

381200 Symplocos fasciculata Zoll. var. chinensis Brand = Symplocos ramosissima Wall. et G. Don ●

381201 Symplocos fasciculiflora Merr. = Symplocos poilanei Guillaumin ●

381202 Symplocos ferruginea Roxb. = Symplocos cochinchinensis (Lour.) S. Moore ●

381203 Symplocos ferruginea Roxb. var. philippinensis Brand = Symplocos cochinchinensis (Lour.) S. Moore var. philippinensis (Brand) Noot. ●

381204 Symplocos ferruginifolia Kaneh. = Symplocos cochinchinensis (Lour.) S. Moore ●

381205 Symplocos fordii Hance；三裂山矾；Ford Sweetleaf ●

381206 Symplocos formosana Brand；台湾灰木（花莲灰木，台湾山矾）；Formosan Sweetleaf ●

381207 Symplocos formosana Brand = Symplocos lancifolia Siebold et Zucc. ●

381208 Symplocos forrestii W. W. Sm. = Symplocos dryophila C. B. Clarke ●

381209 Symplocos fuboensis M. Y. Fang = Symplocos sumuntia Buch. -Ham. ex D. Don ●

381210 Symplocos fukienensis Y. Ling；福建山矾；Fujian Sweetleaf ●

381211 Symplocos fulvipes (C. B. Clarke) Brand = Symplocos lancifolia Siebold et Zucc. ●

381212 Symplocos fusonii Merr. = Symplocos anomala Brand ●

381213 Symplocos gardneriana Wight；加氏山矾；Gardner Sweetleaf ●☆

381214 Symplocos glandulifera Brand；腺缘山矾●

381215 Symplocos glandulifera Brand = Symplocos sulcata Kurz ●

381216 Symplocos glandulosopunctata Y. F. Wu = Symplocos sulcata Kurz ●

381217 Symplocos glauca (Thunb.) Koidz. ；羊舌树（大山矾，大叶灰木，大叶山矾，粉叶山矾，恒春灰木，恒春山矾，山羊耳）；Bigleaf Sweetleaf , Big-leaved Sweetleaf , Glaucous Sweetleaf , Hengchun Sweetleaf , Pale Sweetleaf ●

381218 Symplocos glauca (Thunb.) Koidz. var. epapillata Noot. ；无乳突羊舌树（倒披针叶山矾）；Oblanceleaf Sweetleaf , Oblanceolate Sweetleaf , Oblanceolate-leaved Sweetleaf ●

381219 Symplocos glauca (Thunb.) Koidz. var. koshunensis (Kaneh.) S. S. Ying = Symplocos koshunensis Kaneh. ●

381220 Symplocos glauca (Thunb.) Koidz. var. tashiroi (Matsum.) E. Walker = Symplocos stellaris Brand ●

381221 Symplocos glomerata King ex C. B. Clarke；团花山矾（团伞花山矾，文山山矾，宜章山矾）；Club-shaped Sweetleaf , Glomerateflower Sweetleaf , Glomerule Sweetleaf , Groupflower Sweetleaf , Wenshan Sweetleaf , Yizhang Sweetleaf ●

381222 Symplocos glomerata King ex C. B. Clarke subsp. congesta (Benth.) Noot. = Symplocos congesta Benth. ●

381223 Symplocos glomerata King ex C. B. Clarke subsp. congesta (Benth.) Noot. var. congesta Noot = Symplocos congesta Benth. ●

381224 Symplocos glomerata King ex C. B. Clarke subsp. congesta (Benth.) Noot. var. poilanei (Guillaumin) Noot. = Symplocos poilanei Guillaumin ●

381225 Symplocos glomerata King ex C. B. Clarke subsp. glomerata var. adenopus (Hance) Noot. = Symplocos adenopus Hance ●

381226 Symplocos glomerata King ex C. B. Clarke var. adenopus (Hance) Noot. = Symplocos adenopus Hance ●

381227 Symplocos glomerata King ex C. B. Clarke var. congesta Noot. = Symplocos congesta Benth. ●

381228 Symplocos glomerata King ex C. B. Clarke var. glomeratifolia (Hayata) S. S. Ying = Symplocos setchuensis Brand ex Diels ●

381229 Symplocos glomerata King ex C. B. Clarke var. poilanei (Guillaumin) Noot. = Symplocos poilanei Guillaumin ●

381230 Symplocos glomeratifolia Hayata = Symplocos setchuensis Brand ex Diels ●

381231 Symplocos grandis Hand. -Mazz. = Symplocos glauca (Thunb.) Koidz. ●

381232 Symplocos groffii Merr. ；毛山矾；Groff Sweetleaf , Hairy Sweetleaf ●

381233 Symplocos hainanensis Merr. et Chun ex H. L. Li；海南山矾；Hainan Sweetleaf ●

381234 Symplocos hayatae Mori = Symplocos congesta Benth. ●

381235 Symplocos heishanensis Hayata；海桐山矾（平遮那灰木，平遮那山矾）；Heishan Sweetleaf , Pittosporumleaf Sweetleaf ●

381236 Symplocos henryi Brand = Symplocos lucida (Thunb.) Siebold et Zucc ●

381237 Symplocos henschelii Benth. ex Clarke = Symplocos pendula Wight var. hirtistylus (C. B. Clarke) Noot. ●

381238 Symplocos henschelii Benth. ex Clarke var. hirtistylis C. B. Clarke = Symplocos pendula Wight var. hirtistylus (C. B. Clarke) Noot. ●

381239 Symplocos hiraishiensis Hayata = Symplocos anomala Brand ●

381240 Symplocos hookeri C. B. Clarke；滇南山矾；Hooker Sweetleaf , S. Yunnan Sweetleaf ●

381241 Symplocos hookeri C. B. Clarke var. tomentosa Y. F. Wu；绒毛滇南山矾；Hairy Hooker Sweetleaf ●

381242 Symplocos howii Merr. et Chun ex H. L. Li = Symplocos lucida (Thunb.) Siebold et Zucc ●

381243 Symplocos hualiensis S. S. Ying；花莲灰木●

381244 Symplocos hualiensis S. S. Ying = Symplocos formosana Brand ●

381245 Symplocos hunanensis Hand. -Mazz. = Symplocos paniculata (Thunb.) Miq. ●

381246 Symplocos ilicifolia Hayata = Symplocos lucida (Thunb.) Siebold et Zucc ●

381247 Symplocos ilicifolia Hayata = Symplocos setchuensis Brand ex Diels ●

381248 Symplocos indochinensis H. L. Li = Symplocos dolichotricha Merr. ●

381249 Symplocos intermedia Brand = Symplocos racemosa Roxb. ●

381250 Symplocos intermedia Brand var. trichantha Hand. -Mazz. = Symplocos racemosa Roxb. ●

381251 Symplocos iteophylla Miq. = Symplocos adenophylla Wall. et G. Don ●

381252 Symplocos japonica A. DC. ;日本山矾;Japanese Sweetleaf ●

381253 Symplocos japonica A. DC. = Symplocos cochinchinensis (Lour.) S. Moore ●

381254 Symplocos javanica Kurz = Symplocos cochinchinensis (Lour.) S. Moore ●

381255 Symplocos junshanensis Q. X. Liu;君山山矾;Junshan Sweetleaf ●

381256 Symplocos kawakamii Hayata;川上氏灰木;Kawakami Sweetleaf ●

381257 Symplocos kiratishiensis Hayata = Symplocos anomala Brand ●

381258 Symplocos kiratishiensis Hayata = Symplocos morrisonicola Hayata ●

381259 Symplocos koidzumiana Tatew. et B. Yoshim. = Symplocos decora Hance ●

381260 Symplocos konishii Hayata;小西氏灰木(台东山矾)●

381261 Symplocos konishii Hayata = Symplocos cochinchinensis (Lour.) S. Moore var. laurina (Retz.) Raizada ●

381262 Symplocos koshunensis Kaneh. ;恒春灰木●

381263 Symplocos koshunensis Kaneh. = Symplocos glauca (Thunb.) Koidz. ●

381264 Symplocos kotoensis (Hayata) Yamam. = Symplocos cochinchinensis (Lour.) S. Moore var. philippinensis (Brand) Noot. ●

381265 Symplocos kotoensis Hayata = Symplocos cochinchinensis (Lour.) S. Moore ●

381266 Symplocos kotoensis Hayata = Symplocos cochinchinensis (Lour.) S. Moore var. philippinensis (Brand) Noot. ●

381267 Symplocos kudoi Mori = Symplocos congesta Benth. ●

381268 Symplocos kwangsiensis Merr. ex H. L. Li;广西山矾●

381269 Symplocos kwangsiensis Merr. ex H. L. Li = Symplocos lancifolia Siebold et Zucc. ●

381270 Symplocos kwangsiensis Merr. ex H. L. Li = Symplocos trichoclada Hayata ●

381271 Symplocos kwangtungensis H. L. Li = Symplocos dolichotricha Merr. ●

381272 Symplocos lancifolia Siebold et Zucc. ;光叶山矾(阿里山灰木,潮州山矾,刀灰树,广西山矾,褐毛灰木,滑叶常山,滑叶山矾,剑叶灰木,卵叶山矾,密毛山矾,披针叶山矾,甜茶);Alishan Sweetleaf, Brownhaie Sweetleaf, Chaozhou Sweetleaf, Guangxi Sweetleaf, Kwangsi Sweetleaf, Ovateleaf Sweetleaf, Ovate-leaved Sweetleaf, Smoothleaf Sweetleaf, Smooth-leaved Sweetleaf, Soft-haired Sweetleaf, Softhairy Sweetleaf ●

381273 Symplocos lancifolia Siebold et Zucc. var. cryptostachya ？ = Symplocos lancifolia Siebold et Zucc. ●

381274 Symplocos lancifolia Siebold et Zucc. var. fulvipes C. B. Clarke = Symplocos lancifolia Siebold et Zucc. ●

381275 Symplocos lancifolia Siebold et Zucc. var. hualiensis (S. S. Ying) S. S. Ying = Symplocos formosana Brand ●

381276 Symplocos lancifolia Siebold et Zucc. var. leptostachys ？ = Symplocos lancifolia Siebold et Zucc. ●

381277 Symplocos lancifolia Siebold et Zucc. var. microcarpa (Champ.) Hand. -Mazz. = Symplocos lancifolia Siebold et Zucc. ●

381278 Symplocos lancifolia Siebold et Zucc. var. taiheizanensis (Mori) S. S. Ying = Symplocos formosana Brand ●

381279 Symplocos lancilimba Merr. ;披针叶山矾;Lanceleaf Sweetleaf, Lanceolate-leaved Sweetleaf ●

381280 Symplocos lancilimba Merr. = Symplocos viridissima Brand ●

381281 Symplocos latouchei W. W. Sm. ex Hand. -Mazz. = Symplocos lancifolia Siebold et Zucc. ●

381282 Symplocos laurina (Retz.) Wall. et G. Don = Symplocos acuminata (Blume) Miq. ●

381283 Symplocos laurina (Retz.) Wall. et G. Don = Symplocos cochinchinensis (Lour.) S. Moore var. laurina (Retz.) Raizada ●

381284 Symplocos laurina (Retz.) Wall. var. bodinieri (Brand) Hand. -Mazz. = Symplocos cochinchinensis (Lour.) S. Moore var. laurina (Retz.) Raizada ●

381285 Symplocos leucophylla Brand = Symplocos sumuntia Buch. -Ham. ex D. Don ●

381286 Symplocos limprichtii H. Winkl. = Symplocos stellaris Brand ●

381287 Symplocos lithocarpoides Nakai = Symplocos cochinchinensis (Lour.) S. Moore ●

381288 Symplocos liukiuensis Matsum. ;琉球山矾●

381289 Symplocos liukiuensis Matsum. var. iriomotensis Nagam. ;西表山矾●☆

381290 Symplocos longipetiolata Rehder = Symplocos dryophila C. B. Clarke ●

381291 Symplocos lucida (Thunb.) Siebold et Zucc. ;光亮山矾(茶条果,多倍山矾,过冬青,厚皮灰木,厚叶冬青,棱角山矾,留春树,蒙自山矾,日本灰木,叶萼山矾,枝穗山矾);Calyx-leaved Sweetleaf, Ernest Sweetleaf, Forkspike Sweetleaf, Fork-spiked Sweetleaf, Fourangle Sweetleaf, Henry Sweetleaf, Leafcalyx Sweetleaf, Mengzi Sweetleaf, Phyllocalyx Sweetleaf, Tetragonal Sweetleaf, Thickbark Sweetleaf, Thick-barked Sweetleaf ●

381292 Symplocos lucida (Thunb.) Siebold et Zucc. = Symplocos setchuensis Brand ●

381293 Symplocos lungtauensis Merr. = Symplocos groffii Merr. ●

381294 Symplocos maclurei Merr. ;琼中山矾●

381295 Symplocos maclurei Merr. = Symplocos adenophylla Wall. et G. Don ●

381296 Symplocos macrophylla Wall. ex DC. subsp. sulcata (Kurz) Noot. var. glandulifera (Brand) Noot. = Symplocos glandulifera Brand ●

381297 Symplocos macrostachya Brand = Symplocos racemosa Roxb. ●

381298 Symplocos macrostachya Brand var. leducii Brand = Symplocos racemosa Roxb. ●

381299 Symplocos macrostroma Hayata;大花灰木●

381300 Symplocos macrostroma Hayata = Symplocos caudata Wall. ex A. DC. ●

381301 Symplocos macrostroma Hayata = Symplocos sumuntia Buch. -Ham. ex D. Don ●

381302 Symplocos mairei H. Lév. = Symplocos adenopus Hance ●

381303 Symplocos martinii H. Lév. ;贵阳山矾●

381304 Symplocos menglianensis Y. Y. Qian;孟连山矾;Menglian Sweetleaf ●

381305 Symplocos microcalyx Hayata = Symplocos formosana Brand ●

381306 Symplocos microcalyx Hayata var. taiheizanensis Mori = Symplocos formosana Brand ●

381307 Symplocos microcarpa Champ. ex Benth. = Symplocos lancifolia Siebold et Zucc. ●

381308 Symplocos microtricha Hand. -Mazz. = Symplocos wikstroemiifolia Hayata ●

381309 Symplocos migoi Nagam. ;拟日本灰木;Migo Sweetleaf ●

381310 Symplocos modesta Brand;长梗山矾(薄叶灰木,小叶白笔);Long-stalked Sweetleaf, Taiwan Sweetleaf ●

381311 Symplocos mollifolia Dunn;潮州山矾●

381312 Symplocos mollifolia Dunn = Symplocos arisanensis Hayata ●

381313　Symplocos mollifolia Dunn ＝ Symplocos lancifolia Siebold et Zucc. ●

381314　Symplocos mollipila H. L. Li ＝ Symplocos groffii Merr. ●

381315　Symplocos morrisonicola Hayata；台湾山矾（玉山灰木）●☆

381316　Symplocos morrisonicola Hayata ＝ Symplocos anomala Brand ●

381317　Symplocos morrisonicola Hayata var. kiraishiensis（Hayata）S. S. Ying ＝ Symplocos morrisonicola Hayata ●

381318　Symplocos morrisonicola Hayata var. matudai（Hatus.）S. S. Ying ＝ Symplocos morrisonicola Hayata ●

381319　Symplocos multipes Brand；枝穗山矾●

381320　Symplocos multipes Brand ＝ Symplocos lucida（Thunb.）Siebold et Zucc. ●

381321　Symplocos myriadena Merr. ＝ Symplocos adenopus Hance ●

381322　Symplocos myriantha Rehder ＝ Symplocos ramosissima Wall. et G. Don ●

381323　Symplocos myrtacea Siebold et Zucc. ；香桃木山矾●☆

381324　Symplocos myrtacea Siebold et Zucc. f. latifolia（Hatus.）Sugim. ex Ohwi et Kitag. ＝ Symplocos myrtacea Siebold et Zucc. var. latifolia Hatus. ●☆

381325　Symplocos myrtacea Siebold et Zucc. f. pubescens（Uyeki et Tokui）Sugim. ex Ohwi et Kitag. ；短柔毛山矾●☆

381326　Symplocos myrtacea Siebold et Zucc. var. latifolia Hatus. ；宽叶山矾●☆

381327　Symplocos nakaharae（Hayata）Masam. ；中原氏山矾●☆

381328　Symplocos nakaii Hayata ＝ Symplocos congesta Benth. ●

381329　Symplocos neriifolia Siebold et Zucc. ＝ Symplocos glauca（Thunb.）Koidz. ●

381330　Symplocos nokoensis（Hayata）Kaneh. ；能高山矾（能高山灰木，能高山灰树）；Nenggaoshan Sweetleaf, Nengkao Sweetleaf, Nenkao Sweetleaf, Noko Sweetleaf ●

381331　Symplocos oblanceolata Y. F. Wu ＝ Symplocos glauca（Thunb.）Koidz. var. epapillata Noot. ●

381332　Symplocos okinawensis Matsum. ＝ Symplocos anomala Brand ●

381333　Symplocos oreades Guillaumin ＝ Symplocos heishanensis Hayata ●

381334　Symplocos orisanensis Hayata ＝ Symplocos lancifolia Siebold et Zucc. ●

381335　Symplocos ovalifolia Hand. -Mazz. ；卵叶山矾●

381336　Symplocos ovalifolia Hand. -Mazz. ＝ Symplocos lancifolia Siebold et Zucc. ●

381337　Symplocos ovatibracteata Y. F. Wu；卵苞山矾●

381338　Symplocos ovatibracteata Y. F. Wu ＝ Symplocos sumuntia Buch. -Ham. ex D. Don ●

381339　Symplocos ovatilobata Noot. ；卵裂山矾（单花山矾）；Singleflower, Single-flowered Sweetleaf ●◇

381340　Symplocos pallida Franch. et Sav. ＝ Symplocos paniculata（Thunb.）Miq. ●

381341　Symplocos paniculata（Thunb.）Miq. ；华山矾（白矾，白花丹，白檀，白檀山矾，瓜瓜叶，大米仔花，地胡椒，地黄木，钉地黄，豆豉果，蛤蟆涎，贡檀兜，狗檬树，狗屎木，华灰木，黄檀树，黄檀子木，灰木，降痰黄，降痰王，雷公针，毛柴子，毛壳子树，檬子柴，米碎花木，牛特木，糯米树，膨药，砒霜子，七针，水泡木，思茅山矾，碎米子树，檀花青，土常山，土黄柴，乌子树，羊仔屎，羊子屎，渣子树，止血树，中华白檀，猪婆柴）；Asiatic Sweetleaf, China Sweetleaf, Chinese Sweetleaf, Sapphire Berry, Sapphireberry, Sapphire-berry, Sapphireberry Sweetleaf, Simao Sweetleaf ●

381342　Symplocos paniculata（Thunb.）Miq. var. glabra Makino ＝ Symplocos paniculata（Thunb.）Miq. ●

381343　Symplocos paniculata（Thunb.）Miq. var. glabrifolia ？ ＝ Symplocos paniculata（Thunb.）Miq. ●

381344　Symplocos paniculata（Thunb.）Miq. var. parvifolia ？ ＝ Symplocos paniculata（Thunb.）Miq. ●

381345　Symplocos paniculata（Thunb.）Miq. var. pubescens（Nakai）Ohwi ＝ Symplocos tanakae Matsum. ●☆

381346　Symplocos paniculata ？ ＝ Symplocos coreana（H. Lév.）Ohwi ●☆

381347　Symplocos paniculata Wall. ＝ Symplocos paniculata（Thunb.）Miq. ●

381348　Symplocos paniculata Wall. ex D. Don ＝ Symplocos sawafutagi Nagam. ●

381349　Symplocos paniculata Wall. ex D. Don var. leucocarpa Nakai ＝ Symplocos sawafutagi Nagam. ●

381350　Symplocos papyracea ？ ＝ Symplocos coreana（H. Lév.）Ohwi ●☆

381351　Symplocos patens C. Presl ＝ Symplocos cochinchinensis（Lour.）S. Moore ●

381352　Symplocos pauciflora Wight ＝ Symplocos pendula Wight ●

381353　Symplocos paucinervia Noot. ；少脉山矾；Few-nerved Sweetleaf, Fewveined Sweetleaf, Paucinerved Sweetleaf ●

381354　Symplocos pendula Wight；吊钟山矾（垂枝灰木，南岭灰木）；Pendulous Sweetleaf ●

381355　Symplocos pendula Wight var. hirtistylus（C. B. Clarke）Noot. ；南岭山矾（南岭灰木）；Asiatic Sweetleaf, Confuse Sweetleaf ●

381356　Symplocos pergracilis（Nakai）T. Yamaz. ；纤细山矾●☆

381357　Symplocos permicrophylla Merr. et Chun ＝ Symplocos euryoides Hand. -Mazz. ●

381358　Symplocos permicrophylla Merr. et Chun ex H. L. Li ＝ Symplocos euryoides Hand. -Mazz. ●

381359　Symplocos persistens C. C. Huang et Y. F. Wu ＝ Symplocos sulcata Kurz ●

381360　Symplocos phaeophylla Hayata ＝ Symplocos congesta Benth. ●

381361　Symplocos phyllocalyx C. B. Clarke；叶萼山矾●

381362　Symplocos phyllocalyx C. B. Clarke ＝ Symplocos lucida（Thunb.）Siebold et Zucc. ●

381363　Symplocos pilosa Rehder；柔毛山矾；Pilose Sweetleaf ●

381364　Symplocos pinfaensis H. Lév. ；平伐山矾●

381365　Symplocos pittosporifolia Hand. -Mazz. ＝ Symplocos heishanensis Hayata ●

381366　Symplocos poilanei Guillaumin；丛花山矾（上身眉，十棱山矾，乌脚木）；Chun Sweetleaf, Clusteredflower Sweetleaf, Cluster-flowered Sweetleaf ●

381367　Symplocos potanini Gontsch. ＝ Symplocos lucida（Thunb.）Siebold et Zucc ●

381368　Symplocos prainii H. Lév. ＝ Symplocos adenopus Hance ●

381369　Symplocos propinqua Hance ＝ Symplocos racemosa Roxb. ●

381370　Symplocos prunifolia Siebold et Zucc. ＝ Symplocos sumuntia Buch. -Ham. ex D. Don ●

381371　Symplocos prunifolia Siebold et Zucc. var. tawadae Nagam. ；田和代氏山矾●☆

381372　Symplocos pseudobarberina Gontsch. ；铁山矾（白花木，白檀木，刀灰树）；Iron Sweetleaf ●

381373　Symplocos pseudolancifolia（Hatus.）Hand. -Mazz. ＝ Symplocos lancifolia Siebold et Zucc. ●

381374　Symplocos punctata Brand ＝ Symplocos sumuntia Buch. -Ham. ex D. Don ●

381375　Symplocos punctomarginata A. Chev. ex Guillaumin ＝ Symplocos adenophylla Wall. et G. Don ●

381376 Symplocos punctulata Masam. et Syozi = Symplocos pendula Wight ●

381377 Symplocos punctulata Masam. et Syozi = Symplocos sumuntia Buch. -Ham. ex D. Don ●

381378 Symplocos pyrifolia Wall. et G. Don；梨叶山矾；Pearleaf Sweetleaf, Pear-leaved Sweetleaf ●

381379 Symplocos racemosa Roxb.；珠仔山矾(山麻栗,乌口木,乌口树,珠仔树,总花山矾,总序山矾,总状花灰木)；Lodh Bark, Racemose Sweetleaf ●

381380 Symplocos rachitricha Y. F. Wu = Symplocos sumuntia Buch. -Ham. ex D. Don ●

381381 Symplocos ramosissima Wall. et G. Don；多花山矾；Flowery Sweetleaf, Much-flowered Sweetleaf, Multiflorous Sweetleaf, Stapf Sweetleaf ●

381382 Symplocos rarnosissima Wall. et G. Don var. salwinensis Hand. -Mazz. = Symplocos ramosissima Wall. et G. Don ●

381383 Symplocos risekiensis Hayata = Symplocos heishanensis Hayata ●

381384 Symplocos sasakii Hayata；佐佐木氏灰木 ●

381385 Symplocos sasakii Hayata = Symplocos sumuntia Buch. -Ham. ex D. Don ●

381386 Symplocos sawafutagi Nagam.；东北山矾 ●

381387 Symplocos schaefferae Merr. = Symplocos cochinchinensis (Lour.) S. Moore var. laurina (Retz.) Raizada ●

381388 Symplocos seguinii H. Lév. = Eriobotrya seguinii (H. Lév.) Cardot ex Guillaumin ●

381389 Symplocos setchuensis Brand = Symplocos lucida (Thunb.) Siebold et Zucc. ●

381390 Symplocos setchuensis Brand ex Diels；四川山矾(冬青叶山矾,日本灰木,四川灰木)；Setchuan Sweetleaf, Sichuan Sweetleaf, Sinuate Sweetleaf ●

381391 Symplocos setchuensis Brand ex Diels = Symplocos lucida (Thunb.) Siebold et Zucc. ●

381392 Symplocos shilanensis Y. C. Liu et F. Y. Lu；希兰灰木 ●

381393 Symplocos shilanensis Y. C. Liu et F. Y. Lu = Symplocos lucida (Thunb.) Siebold et Zucc. ●

381394 Symplocos simaoensis Y. Y. Qian = Symplocos paniculata (Thunb.) Miq. ●

381395 Symplocos singuliflora Guillaumin；单花山矾；Oneflower Sweetleaf ●☆

381396 Symplocos singuliflora Guillaumin = Symplocos ovatilobata Noot. ●◇

381397 Symplocos sinica Ker Gawl. = Symplocos chinensis (Lour.) Druce ●

381398 Symplocos sinica Ker Gawl. = Symplocos paniculata (Thunb.) Miq. ●

381399 Symplocos sinica Ker Gawl. var. vestita Hemsl. = Symplocos paniculata (Thunb.) Miq. ●

381400 Symplocos sinuata Brand = Symplocos lucida (Thunb.) Siebold et Zucc. ●

381401 Symplocos sinuata Brand = Symplocos setchuensis Brand ex Diels ●

381402 Symplocos somai Hayata = Symplocos caudata Wall. ex A. DC. ●

381403 Symplocos somai Hayata = Symplocos sumuntia Buch. -Ham. ex D. Don ●

381404 Symplocos sonoharae Koidz.；南岭灰木 ●

381405 Symplocos sozanensis Hayata = Symplocos caudata Wall. ex A. DC. ●

381406 Symplocos sozanensis Hayata = Symplocos sumuntia Buch. -Ham. ex D. Don ●

381407 Symplocos spathulata H. L. Li = Symplocos poilanei Guillaumin ●

381408 Symplocos spectabilis Brand；禄春山矾(绿春山矾,清秀山矾)●

381409 Symplocos spicata Roxb. = Symplocos cochinchinensis (Lour.) S. Moore var. laurina (Retz.) Raizada ●

381410 Symplocos stapfiana H. Lév. = Symplocos ramosissima Wall. et G. Don ●

381411 Symplocos stapfiana H. Lév. var. leiocalyx Hand. -Mazz. = Symplocos ramosissima Wall. et G. Don ●

381412 Symplocos stellaris Brand；枇杷叶山矾(老鼠刺,老鼠矢,枇杷叶灰木)；Loquatleaf Sweetleaf, Loquat-leaf Sweetleaf, Starshape Sweetleaf, Star-shape Sweetleaf ●

381413 Symplocos stellaris Brand var. aenea (Hand. -Mazz.) Noot. = Symplocos aenea Hand. -Mazz. ●

381414 Symplocos stenophylla Merr. et Chun ex H. L. Li = Symplocos cochinchinensis (Lour.) S. Moore var. angustifolia (Guillaumin) Noot. ●

381415 Symplocos stenostachys Hayata = Symplocos cochinchinensis (Lour.) S. Moore var. laurina (Retz.) Raizada ●

381416 Symplocos stenostachys Hayata = Symplocos theophrastifolia Siebold et Zucc. ●

381417 Symplocos stewardii Sleumer = Symplocos adenophylla Wall. et G. Don ●

381418 Symplocos stnuata Brand = Symplocos setchuensis Brand ex Diels ●

381419 Symplocos subconnata Hand. -Mazz. = Symplocos sumuntia Buch. -Ham. ex D. Don ●

381420 Symplocos suishariensis Hayata = Symplocos formosana Brand ●

381421 Symplocos suishariensis Hayata = Symplocos lancifolia Siebold et Zucc. ●

381422 Symplocos sulcata Kurz；沟槽山矾(宿苞山矾,腺斑山矾,腺点山矾)；Bract-persisted Sweetleaf, Bractpersistent Sweetleaf, Glandspot Sweetleaf, Glandular-dotted Sweetleaf, Glandular-punctated Sweetleaf, Persistent Sweetleaf ●

381423 Symplocos sumuntia Buch. -Ham. ex D. Don；山矾(长花柱山矾,珤花,春桂,大萼山矾,吊钟山矾,福宝山矾,黑厚皮柴,葫芦果山矾,黄仔叶柴,卵苞山矾,毛轴山矾,美丽山矾,美山矾,七里香,三月桂,山桂花,十里香,坛果山矾,田螺柴,土白芷,尾叶灰木,尾叶山矾,小泉氏山矾,小元柴,银色山矾,樱叶山矾,芸香,柘花,槮关花,总状山矾,佐佐木氏灰木)；Argentate Sweetleaf, Beautiful Sweetleaf, Calabash Sweetleaf, Calabash-fruited Sweetleaf, Caudate Sweetleaf, Cavaler Sweetleaf, Common Sweetleaf, Fubao Sweetleaf, Hairaxi Sweetleaf, Hairyrachis Sweetleaf, Hairy-rachised Sweetleaf, Large-calyx Sweetleaf, Longstyled Sweetleaf, Long-styled Sweetleaf, Ovatebract Sweetleaf, Ovate-bracted Sweetleaf, Punctulate Sweetleaf, Raceme Sweetleaf, Sasaki Sweetleaf, Sumntia Sweetleaf, Sweetleaf, Tailed-leaf Sweetleaf, Urceolar Sweetleaf, Urnshapedfruit Sweetleaf ●

381424 Symplocos swinhoeana Hance = Symplocos sumuntia Buch. -Ham. ex D. Don ●

381425 Symplocos taiheizanensis (Mori) Mori = Symplocos formosana Brand ●

381426 Symplocos taiheizanensis (Mori) Mori = Symplocos lancifolia Siebold et Zucc. ●

381427 Symplocos taiheizanensis Mori = Symplocos taiheizanensis (Mori) Mori ●

381495 Synadenium piscatorium Pax = Synadenium pereskiifolium (Baill.) Guillaumin ●☆

381496 Synadenium umbellatum Pax = Euphorbia pseudograntii Pax ●☆

381497 Synadenium umbellatum Pax var. puberulum N. E. Br. = Euphorbia pseudograntii Pax ●☆

381498 Synaecia Pritz. = Ficus L. ●

381499 Synaecia Pritz. = Synoecia Miq. ●

381500 Synaedris Steud. = Lithocarpus Blume ●

381501 Synaedrys Lindl. = Lithocarpus Blume ●

381502 Synaedrys amygdalifolia (V. Naray.) Koidz. = Lithocarpus amygdalifolius (V. Naray. ex Forbes et Hemsl.) Hayata ●

381503 Synaedrys amygdalifolia (V. Naray.) Koidz. f. castanopsifolia (Hayata) Kudo = Lithocarpus lepidocarpus (Hayata) Hayata ●

381504 Synaedrys attenuata (V. Naray.) Koidz. = Lithocarpus attenuatus (V. Naray.) Rehder ●

381505 Synaedrys balansae (Drake) Koidz. = Lithocarpus balansae (Drake) A. Camus ●

381506 Synaedrys baviensis (Drake) Koidz. = Lithocarpus truncatus (King) Rehder var. baviensis (Drake) A. Camus ●

381507 Synaedrys brachyacantha (Hayata) Koidz. = Castanopsis eyrei (Champ.) Tutcher ●

381508 Synaedrys brevicaudata (V. Naray.) Koidz. = Lithocarpus brevicaudatus (V. Naray.) Hayata ●

381509 Synaedrys brevicaudata (V. Naray.) Koidz. var. pinnativena Yamam. = Lithocarpus brevicaudatus (V. Naray.) Hayata ●

381510 Synaedrys calathiformis (V. Naray.) Koidz. = Castanopsis calathiformis (V. Naray.) Rehder et P. Wilson ●

381511 Synaedrys carlesii (Hemsl.) Koidz. = Castanopsis carlesii (Hemsl.) Hayata ●

381512 Synaedrys cathayana (Seem.) Koidz. = Lithocarpus truncatus (King) Rehder ●

381513 Synaedrys cavaleriei (H. Lév. et Vaniot) Koidz. = Castanopsis eyrei (Champ.) Tutcher ●

381514 Synaedrys cleistocarpa (Seem.) Koidz. = Lithocarpus cleistocarpus (Seemen) Rehder et E. H. Wilson ●

381515 Synaedrys cornea (Lour.) Koidz. = Lithocarpus corneus (Lour.) Rehder ●

381516 Synaedrys cyrtocarpa (Drake) Koidz. = Lithocarpus cyrtocarpus (Drake) A. Camus ●

381517 Synaedrys dealbata (DC.) Koidz. = Lithocarpus dealbatus (Hook. f. et Thomson ex DC.) Rehder ●

381518 Synaedrys dealbata (Hook. f. et Thomson ex Miq.) Koidz. = Lithocarpus dealbatus (Hook. f. et Thomson ex Miq.) Rehder ●

381519 Synaedrys delavayi (Franch.) Koidz. = Castanopsis delavayi Franch. ●

381520 Synaedrys elaeagnifolia (Seemen) Koidz. = Lithocarpus elaeagnifolius (Seemen) Chun ●

381521 Synaedrys elizabethae (Tutcher) Kudo = Lithocarpus elizabethae (Tutcher) Rehder ●

381522 Synaedrys fenestrata (Roxb.) Koidz. = Lithocarpus fenestratus (Roxb.) Rehder ●

381523 Synaedrys fissa (Champ. ex Benth.) Koidz. = Castanopsis fissa (Champ. ex Benth.) Rehder et E. H. Wilson ●

381524 Synaedrys fordiana (Hemsl.) Koidz. = Lithocarpus fordianus (Hemsl.) Chun ●

381525 Synaedrys formosana (Hayata) Koidz. f. dodonaeifolia (Hayata) Kudo = Lithocarpus dodonaeifolius (Hayata) Hayata ●

381526 Synaedrys formosana (V. Naray.) Koidz. = Lithocarpus formosanus (V. Naray. ex Forbes et Hemsl.) Hayata ●

381527 Synaedrys glabra (Thunb.) Koidz. = Lithocarpus glaber (Thunb.) Nakai ●

381528 Synaedrys hancei (Benth.) Koidz. = Lithocarpus hancei (Benth.) Rehder ●

381529 Synaedrys harlandii (Hance ex Walp.) Koidz. = Lithocarpus harlandii (Hance) Rehder ●

381530 Synaedrys harlandii (Hance) Koidz. = Lithocarpus harlandii (Hance) Rehder ●

381531 Synaedrys hemisphaerica (Drake) Koidz. = Lithocarpus corneus (Lour.) Rehder var. zonatus C. C. Huang et Y. T. Chang ●

381532 Synaedrys impressivena (Hayata) Masam. = Lithocarpus brevicaudatus (V. Naray.) Hayata ●

381533 Synaedrys irwinii (Hance) Koidz. = Lithocarpus irwinii (Hance) Rehder ●◇

381534 Synaedrys iteaphylla (Hance) Koidz. = Lithocarpus iteaphyllus (Hance) Rehder ●

381535 Synaedrys kawakamii (Hayata) Koidz. = Lithocarpus kawakamii (Hayata) Hayata ●

381536 Synaedrys koidaikoensis (Hayata) Kudo = Lithocarpus corneus (Lour.) Rehder ●

381537 Synaedrys konishii (Hayata) Koidz. = Lithocarpus konishii (Hayata) Hayata ●

381538 Synaedrys kuarunensis Tomiya = Lithocarpus hancei (Benth.) Rehder ●

381539 Synaedrys lepidocarpa (Hayata) Koidz. = Lithocarpus lepidocarpus (Hayata) Hayata ●

381540 Synaedrys litseifolia (Hance) Koidz. = Lithocarpus litseifolius (Hance) Chun ●

381541 Synaedrys lycoperdon (V. Naray.) Koidz. = Lithocarpus lycoperdon (V. Naray.) A. Camus ●

381542 Synaedrys mairei (Schottky) Koidz. = Lithocarpus mairei (Schottky) Rehder ●

381543 Synaedrys matsudai (Hayata) Kudo = Lithocarpus hancei (Benth.) Rehder ●

381544 Synaedrys naiadarum (Hance) Koidz. = Lithocarpus naiadarum (Hance) Chun ●

381545 Synaedrys nakaii (Hayata) Kudo = Lithocarpus taitoensis (Hayata) Hayata ●

381546 Synaedrys nantoensis (Hayata) Koidz. = Lithocarpus nantoensis (Hayata) Hayata ●

381547 Synaedrys nariakii (Hayata) Kudo = Lithocarpus silvicolarum (Hance) Chun ●

381548 Synaedrys pachyphylla (Kurz) Koidz. = Lithocarpus pachyphyllus (Kurz) Rehder ●

381549 Synaedrys rhombocarpa (Hayata) Kudo = Lithocarpus taitoensis (Hayata) Hayata ●

381550 Synaedrys rhombocarpa (Hayata) Kudo = Lithocarpus taitoensis C. C. Huang et Y. T. Chang ●

381551 Synaedrys rhombocarpa (Hayata) Kudo f. suishaensis (Kaneh. et Yamam.) Kudo = Lithocarpus taitoensis C. C. Huang et Y. T. Chang ●

381552 Synaedrys rosthornii (Schottky) Koidz. = Lithocarpus rosthornii (Schottky) Barnett ●

381553 Synaedrys sclerophylla (Lindl. et Paxton) Koidz. = Castanopsis sclerophylla (Lindl.) Schottky ●

381554　Synaedrys sclerophylla（Lindl.）Koidz. = Castanopsis sclerophylla（Lindl.）Schottky ●

381555　Synaedrys shinsuiensis（Hayata et Kaneh.）Kudo = Lithocarpus shinsuiensis Hayata et Kaneh. ●

381556　Synaedrys silvicolarum（Hance）Koidz. = Lithocarpus silvicolarum（Hance）Chun ●

381557　Synaedrys stephrocarpa（Drake）Koidz. = Lithocarpus tephrocarpus（Drake）A. Camus ●

381558　Synaedrys taitoensis（Hayata）Koidz. = Lithocarpus taitoensis（Hayata）Hayata ●

381559　Synaedrys taitoensis（Hayata）Koidz. = Lithocarpus taitoensis C. C. Huang et Y. T. Chang ●

381560　Synaedrys tephrocarpa（Drake）Koidz. = Lithocarpus tephrocarpus（Drake）A. Camus ●

381561　Synaedrys ternaticupula（Hayata）Koidz. = Lithocarpus hancei（Benth.）Rehder ●

381562　Synaedrys thomsonii（Miq.）Koidz. = Lithocarpus thomsonii（Miq.）Rehder ●

381563　Synaedrys tunkinensis（Drake）Koidz. = Castanopsis fissa（Champ. ex Benth.）Rehder et E. H. Wilson ●

381564　Synaedrys uraiana（Hayata）Koidz. = Castanopsis uraiana（Hayata）Kaneh. et Hatus. ●

381565　Synaedrys uvariifolia（Hance）Koidz. = Lithocarpus urariifolius（Hance）Rehder ●

381566　Synaedrys variolosa（Franch.）Koidz. = Lithocarpus variolosus（Franch.）Chun ●

381567　Synaedrys viridis（Schottky）Koidz. = Lithocarpus litseifolius（Hance）Chun ●

381568　Synaedrys wilsonii（Seem.）Koidz. = Lithocarpus cleistocarpus（Seemen）Rehder et E. H. Wilson ●

381569　Synaedrys xylocarpa（Kurz）Koidz. = Lithocarpus xylocarpus（Kurz）Markgr. ●

381570　Synallodia Raf. = Swertia L. ■

381571　Synandra Nutt.（1818）；聚雄草属■☆

381572　Synandra Schrad. = Aphelandra R. Br. ●■☆

381573　Synandra Schrad. = Stenandrium Nees（保留属名）■☆

381574　Synandra grandiflora Nutt.；聚雄草■☆

381575　Synandrina Standl. et L. O. Williams = Casearia Jacq. ●

381576　Synandrodaphne Gilg（1915）（保留属名）；联蕊木属●☆

381577　Synandrodaphne Meisn.（废弃属名）= Ocotea Aubl. ●☆

381578　Synandrodaphne Meisn.（废弃属名）= Rhodostemonodaphne Rohwer et Kubitzki ●☆

381579　Synandrodaphne Meisn.（废弃属名）= Synandrodaphne Gilg（保留属名）●☆

381580　Synandrodaphne paradoxa Gilg；联蕊木●☆

381581　Synandrogyne Buchet = Arophyton Jum. ■☆

381582　Synandrogyne rhizomatosum Buchet = Arophyton rhizomatosum（Buchet）Bogner ■☆

381583　Synandropus A. C. Sm.（1931）；聚药藤属●☆

381584　Synandropus membranaceus A. C. Sm.；聚药藤●☆

381585　Synandrospadix Engl.（1883）；合蕊南星属■☆

381586　Synandrospadix vermitoxica（Griseb.）Engl.；合蕊南星■☆

381587　Synantheraceae Cass. = Asteraceae Bercht. et J. Presl（保留科名）●■

381588　Synantheraceae Cass. = Compositae Giseke（保留科名）●■

381589　Synantherias Schott = Amorphophallus Blume ex Decne.（保留属名）●■

381590　Synanthes Burns-Bal.，H. Rob. et M. S. Foster（1985）；合花兰属■☆

381591　Synanthes bertonii Burns-Bal.，H. Rob. et M. S. Foster；合花兰■☆

381592　Synanthes borealis（A. H. Heller）Burns-Bal.，H. Rob. et M. S. Foster；北方合花兰■☆

381593　Synaphe Dulac = Catapodium Link ■☆

381594　Synaphe Dulac = Scleropoa Griseb. ■☆

381595　Synaphea R. Br.（1810）；合龙眼属●☆

381596　Synaphea acutiloba Meisn.；尖裂合龙眼●☆

381597　Synaphea brachyceras R. Butcher；短角合龙眼●☆

381598　Synaphea cuneata A. S. George；楔形合龙眼●☆

381599　Synaphea gracillima Lindl.；纤细合龙眼●☆

381600　Synaphea grandis A. S. George；大合龙眼●☆

381601　Synaphea macrophylla A. S. George；大叶合龙眼●☆

381602　Synaphea oligantha A. S. George；寡花合龙眼●☆

381603　Synaphea reticulata Druce；网脉合龙眼●☆

381604　Synaphea stenoloba A. S. George；细裂合龙眼●☆

381605　Synapisma Steud. = Codiaeum A. Juss.（保留属名）●

381606　Synapisma Steud. = Synaspisma Endl. ●

381607　Synapsis Griseb.（1866）；古巴参木属●☆

381608　Synapsis ilicifolia Griseb.；古巴参木●☆

381609　Synaptantha Hook. f.（1873）；合花茜属■☆

381610　Synaptantha tillaeacea（F. Muell.）Hook. f.；合花茜■☆

381611　Synaptanthe Willis = Synaptantha Hook. f. ■☆

381612　Synaptanthera K. Schum. = Synaptantha Hook. f. ■☆

381613　Synaptea Griff. = Sunaptea Griff. ●

381614　Synaptea Griff. = Vatica L. ●

381615　Synaptea Kurz = Vatica L. ●

381616　Synapteopsis Post et Kuntze = Stemonoporus Thwaites ●☆

381617　Synapteopsis Post et Kuntze = Sunapteopsis Heim ●☆

381618　Synapteopsis Post et Kuntze = Vateria L. ●☆

381619　Synaptera Willis = Synaptea Griff. ●

381620　Synaptera Willis = Vatica L. ●

381621　Synaptolepis Oliv.（1870）；合鳞瑞香属●☆

381622　Synaptolepis alternifolia Oliv.；互叶合鳞瑞香●☆

381623　Synaptolepis angolensis Domke ex Nolde；安哥拉合鳞瑞香●☆

381624　Synaptolepis kirkii Oliv.；吉尔合鳞瑞香●☆

381625　Synaptolepis oliveriana Gilg；合鳞瑞香●☆

381626　Synaptolepis perrieri Léandri；佩里耶合鳞瑞香●☆

381627　Synaptolepis retusa H. Pearson；非洲合鳞瑞香●☆

381628　Synaptophyllum N. E. Br.（1925）；合叶日中花属■☆

381629　Synaptophyllum juttae（Dinter et A. Berger）N. E. Br.；合叶日中花■☆

381630　Synaptophyllum sladenianum（L. Bolus）N. E. Br. = Prenia sladeniana（L. Bolus）L. Bolus ■☆

381631　Synardisia（Mez）Lundell = Ardisia Sw.（保留属名）●■

381632　Synardisia（Mez）Lundell（1963）；合金牛属●☆

381633　Synardisia venosa（Mast. ex Donn. Sm.）Lundell；合金牛●☆

381634　Synarmosepalum Garay，Hamer et Siegerist = Bulbophyllum Thouars（保留属名）■

381635　Synarmosepalum Garay，Hamer et Siegerist（1994）；聚萼兰属■☆

381636　Synarrhena F. Muell. = Saurauia Willd.（保留属名）●

381637　Synarrhena Fisch. et C. A. Mey. = Manilkara Adans.（保留属名）●

381638　Synarthron Benth. et Hook. f. = Senecio L. ■●

381639　Synarthron Benth. et Hook. f. = Synarthrum Cass. ●■

381640　Synarthrum Cass. = Senecio L. ■●

381641　Synaspisma Endl. = Codiaeum A. Juss. (保留属名)●

381642　Synassa Lindl. = Sauroglossum Lindl. ☆

381643　Synastemon F. Muell. = Sauropus Blume ●■

381644　Syncalathium Lipsch. (1956);合头菊属;Syncalathium ■★

381645　Syncalathium chrysocephalum (C. Shih) C. Shih = Syncalathium chrysocephalum (C. Shih) S. W. Liu ■

381646　Syncalathium chrysocephalum (C. Shih) S. W. Liu;黄花合头菊;Yellowflower Syncalathium ■

381647　Syncalathium disciforme (Mattf.) Y. Ling;盘状合头菊;Dishshaped Syncalathium ■

381648　Syncalathium kawaguchii (Kitam.) Y. Ling;合头菊;Common Syncalathium ■

381649　Syncalathium orbiculariforme C. Shih;圆叶合头菊;Roundleaf Syncalathium ■

381650　Syncalathium pilosum (Y. Ling) C. Shih;柔毛合头菊;Pilose Syncalathium ■

381651　Syncalathium porphyreum (C. Marquand et Airy Shaw) Y. Ling;紫花合头菊;Purpleflower Syncalathium ■

381652　Syncalathium porphyreum (C. Marquand et Airy Shaw) Y. Ling = Syncalathium kawaguchii (Kitam.) Y. Ling ■

381653　Syncalathium qinghaiense (C. Shih) C. Shih;青海合头菊(青海绢毛菊,青海绢毛苣);Qinghai Soroseris,Qinghai Syncalathium ■

381654　Syncalathium roseum Y. Ling;红花合头菊;Rose Syncalathium ■

381655　Syncalathium souliei (Franch.) Y. Ling;康滇合头菊;Soulie Syncalathium ■

381656　Syncalathium sukaczevii Lipsch. = Syncalathium kawaguchii (Kitam.) Y. Ling ■

381657　Syncalathium sukaczevii Lipsch. var. pilosum Y. Ling = Syncalathium pilosum (Y. Ling) C. Shih ■

381658　Syncarpha DC. (1810);小麦秆菊属■☆

381659　Syncarpha DC. = Helipterum DC. ex Lindl. ☆

381660　Syncarpha affinis (B. Nord.) B. Nord.;近缘小麦秆菊☆

381661　Syncarpha argentea (Thunb.) B. Nord.;银色小麦秆菊■☆

381662　Syncarpha argyropsis (DC.) B. Nord.;拟银色小麦秆菊■☆

381663　Syncarpha aurea B. Nord.;金黄小麦秆菊☆

381664　Syncarpha canescens (L.) B. Nord.;灰小麦秆菊■☆

381665　Syncarpha canescens (L.) B. Nord. subsp. leucolepis (DC.) B. Nord.;白鳞黄小麦秆菊■☆

381666　Syncarpha canescens (L.) B. Nord. subsp. tricolor (DC.) B. Nord.;三色小麦秆菊☆

381667　Syncarpha chlorochrysum (DC.) B. Nord.;绿穗小麦秆菊☆

381668　Syncarpha dregeana (DC.) B. Nord.;德雷小麦秆菊■☆

381669　Syncarpha dykei (Bolus) B. Nord.;戴克小麦秆菊☆

381670　Syncarpha eximia (L.) B. Nord.;优异小麦秆菊■☆

381671　Syncarpha ferruginea (Lam.) B. Nord.;锈色小麦秆菊■☆

381672　Syncarpha flava (Compton) B. Nord.;黄小麦秆菊■☆

381673　Syncarpha gnaphaloides (L.) DC.;鼠曲草小麦秆菊■☆

381674　Syncarpha lepidopodium (Bolus) B. Nord.;鳞足小麦秆菊☆

381675　Syncarpha loganiana (Compton) B. Nord.;洛根小麦秆菊☆

381676　Syncarpha marlothii (Schltr.) B. Nord.;马洛斯小麦秆菊■☆

381677　Syncarpha milleflora (L. f.) B. Nord.;蜜花小麦秆菊■☆

381678　Syncarpha montana (B. Nord.) B. Nord.;山地小麦秆菊■☆

381679　Syncarpha mucronata (P. J. Bergius) B. Nord.;短尖小麦秆菊☆

381680　Syncarpha paniculata (L.) B. Nord.;圆锥小麦秆菊☆

381681　Syncarpha recurvata (L. f.) B. Nord.;反折小麦秆菊■☆

381682　Syncarpha sordescens (DC.) B. Nord.;污浊小麦秆菊☆

381683　Syncarpha speciosissima (L.) B. Nord.;极美小麦秆菊■☆

381684　Syncarpha speciosissima (L.) B. Nord. subsp. angustifolia (DC.) B. Nord.;窄叶极美小麦秆菊■☆

381685　Syncarpha staehelina (L.) B. Nord.;澳非小麦秆菊■☆

381686　Syncarpha striata (Thunb.) B. Nord.;条纹小麦秆菊☆

381687　Syncarpha variegata (P. J. Bergius) B. Nord.;斑叶小麦秆菊☆

381688　Syncarpha vestita (L.) B. Nord.;包被小麦秆菊☆

381689　Syncarpha virgata (P. J. Bergius) B. Nord.;线纹小麦秆菊■☆

381690　Syncarpha zeyheri (Sond.) B. Nord.;泽赫小麦秆菊■☆

381691　Syncarpia Ten. (1839);合生果树属●☆

381692　Syncarpia glomulifera (Sm.) Nied.;球花合生果树(球花辛尔卡木);Turpentine,Turpentine Tree,Turpentine Wood ●☆

381693　Syncarpia laurifolia Ten. = Syncarpia glomulifera (Sm.) Nied. ●☆

381694　Syncephalantha Bartl. = Dyssodia Cav. ■☆

381695　Syncephalanthus Benth. et Hook. f. = Dyssodia Cav. ■☆

381696　Syncephalanthus Benth. et Hook. f. = Syncephalantha Bartl. ■☆

381697　Syncephalum DC. (1838);合头鼠麹木属●☆

381698　Syncephalum arbutifolium (Baker) Humbert;浆果鹃叶合头鼠麹木●☆

381699　Syncephalum bojeri DC. = Syncephalum arbutifolium (Baker) Humbert ●☆

381700　Syncephalum candidum Humbert;纯白合头鼠麹木●☆

381701　Syncephalum perrieri Humbert = Syncephalum arbutifolium (Baker) Humbert ●☆

381702　Syncephalum stenoclinoides Humbert;狭合头鼠麹木●☆

381703　Syncephalum suborbiculare Humbert;亚圆合头鼠麹木●☆

381704　Syncephalum tsinjoarivense Humbert;钦祖阿里武鼠麹木●☆

381705　Synchaeta Kirp. = Gnaphalium L. ■

381706　Synchaeta norvegica (Gunn.) Kirp. = Gnaphalium norvegicum Gunnerus ■

381707　Synchaeta sylvatica (L.) Kirp. = Gnaphalium sylvaticum L. ■

381708　Synchodendron Bojer ex DC. = Brachylaena R. Br. ●☆

381709　Synchodendron senegalense Klatt = Elephantopus senegalensis (Klatt) Oliv. et Hiern ■☆

381710　Synchoriste Baill. = Lasiocladus Bojer ex Nees ●☆

381711　Synclinostyles Farille et Lachard(2002);合柱草属■☆

381712　Synclinostyles denisjordanii Farille et Lachard;戴尼斯合柱草■☆

381713　Synclinostyles exadversum Farille et Lachard;合柱草■☆

381714　Synclisia Benth. (1862);合被藤属●☆

381715　Synclisia delagoensis N. E. Br. = Albertisia delagoensis (N. E. Br.) Forman ●☆

381716　Synclisia ferruginea (Diels) Hutch. et Dalziel = Albertisia ferruginea (Diels) Forman ●☆

381717　Synclisia junodii Schinz = Albertisia delagoensis (N. E. Br.) Forman ●☆

381718　Synclisia leonensis Scott-Elliot = Tiliacora leonensis (Scott-Elliot) Diels ●☆

381719　Synclisia scabrida Miers;非洲合被藤●☆

381720　Synclisia villosa Exell = Albertisia villosa (Exell) Forman ●☆

381721　Synclisia zambesiaca N. E. Br. = Albertisia delagoensis (N. E. Br.) Forman ●☆

381722　Syncodium Raf. = Honorius Gray ■☆

381723　Syncodium Raf. = Myogalum Link ■

381724　Syncodium Raf. = Ornithogalum L. ■

381725　Syncodon Fourr. = Campanula L. ■●

381726　Syncolostemon E. Mey. = Syncolostemon E. Mey. ex Benth. ●■☆

381727　Syncolostemon E. Mey. ex Benth. (1838);杂蕊草属●■☆

381728　Syncolostemon argenteus N. E. Br.;银白杂蕊草●☆

381729 Syncolostemon comptonii Codd;康普顿杂蕊草●☆

381730 Syncolostemon concinnus N. E. Br. ;整洁杂蕊草●☆

381731 Syncolostemon cooperi Briq. = Syncolostemon parviflorus E. Mey. ex Benth. var. lanceolatus (Gürke) Codd ●☆

381732 Syncolostemon densiflorus Benth. ;密花杂蕊草●☆

381733 Syncolostemon dissitiflorus E. Mey. ex Benth. = Syncolostemon parviflorus E. Mey. ex Benth. ●☆

381734 Syncolostemon eriocephalus I. Verd. ;红头杂蕊草●☆

381735 Syncolostemon flabellifolius (S. Moore) A. J. Paton;扇叶杂蕊草●☆

381736 Syncolostemon lanceolatus Gürke;剑叶杂蕊草●☆

381737 Syncolostemon lanceolatus Gürke = Syncolostemon parviflorus E. Mey. ex Benth. var. lanceolatus (Gürke) Codd ●☆

381738 Syncolostemon lanceolatus Gürke var. cooperi (Briq.) N. E. Br. = Syncolostemon parviflorus E. Mey. ex Benth. var. lanceolatus (Gürke) Codd ●☆

381739 Syncolostemon lanceolatus Gürke var. grandiflorus N. E. Br. = Syncolostemon parviflorus E. Mey. ex Benth. var. lanceolatus (Gürke) Codd ●☆

381740 Syncolostemon latidens (N. E. Br.) Codd;宽齿杂蕊草●☆

381741 Syncolostemon macranthus (Gürke) M. Ashby;大花杂蕊草●☆

381742 Syncolostemon macrophyllus Gürke = Hemizygia macrophylla (Gürke) Codd ●☆

381743 Syncolostemon madagascariensis (A. J. Paton et Hedge) D. F. Otieno;马岛杂蕊草●☆

381744 Syncolostemon parviflorus E. Mey. ex Benth. ;小花杂蕊草●☆

381745 Syncolostemon parviflorus E. Mey. ex Benth. var. dissitiflorus (Benth.) N. E. Br. =Syncolostemon parviflorus E. Mey. ex Benth. ●☆

381746 Syncolostemon parviflorus E. Mey. ex Benth. var. lanceolatus (Gürke) Codd;披针形杂蕊草●☆

381747 Syncolostemon ramulosus E. Mey. ex Benth. ;多枝杂蕊草●☆

381748 Syncolostemon rotundifolius E. Mey. ex Benth. ;圆叶杂蕊草●☆

381749 Syncretocarpus S. F. Blake(1916);油果菊属■●☆

381750 Syncretocarpus sericeus S. F. Blake;油果菊■●☆

381751 Syndechites T. Durand et Jacks. = Sindechites Oliv. ●

381752 Syndesmanthus Klotzsch = Erica L. ●☆

381753 Syndesmanthus Klotzsch = Scyphogyne Brongn. ●☆

381754 Syndesmanthus Klotzsch(1838);集带花属●☆

381755 Syndesmanthus articulatus (L.) Klotzsch;关节集带花●☆

381756 Syndesmanthus articulatus (L.) Klotzsch = Erica similis (N. E. Br.) E. G. H. Oliv. ●☆

381757 Syndesmanthus articulatus (L.) Klotzsch var. fasciculatus N. E. Br. = Erica similis (N. E. Br.) E. G. H. Oliv. ●☆

381758 Syndesmanthus articulatus (L.) Klotzsch var. hirtus Benth. = Erica similis (N. E. Br.) E. G. H. Oliv. ●☆

381759 Syndesmanthus breviflorus N. E. Br. = Erica brownii E. G. H. Oliv. ●☆

381760 Syndesmanthus ciliatus (Klotzsch) Benth. = Erica paucifolia (J. C. Wendl.) E. G. H. Oliv. subsp. ciliata (Klotzsch) E. G. H. Oliv. ●☆

381761 Syndesmanthus elimensis N. E. Br. ;埃利姆集带花●☆

381762 Syndesmanthus elimensis N. E. Br. = Erica similis (N. E. Br.) E. G. H. Oliv. ●☆

381763 Syndesmanthus elimensis N. E. Br. var. incertus ? = Erica similis (N. E. Br.) E. G. H. Oliv. ●☆

381764 Syndesmanthus erinus (Klotzsch ex Benth.) N. E. Br. ;绵毛集带花●☆

381765 Syndesmanthus erinus (Klotzsch ex Benth.) N. E. Br. = Erica erina (Klotzsch ex Benth.) E. G. H. Oliv. ●☆

381766 Syndesmanthus erinus (Klotzsch ex Benth.) N. E. Br. var. validus N. E. Br. = Erica erina (Klotzsch ex Benth.) E. G. H. Oliv. ●☆

381767 Syndesmanthus fasciculatus Klotzsch = Erica similis (N. E. Br.) E. G. H. Oliv. ●☆

381768 Syndesmanthus glaucus Klotzsch = Erica similis (N. E. Br.) E. G. H. Oliv. ●☆

381769 Syndesmanthus globiceps N. E. Br. = Erica globiceps (N. E. Br.) E. G. H. Oliv. ●☆

381770 Syndesmanthus gracilis (Benth.) N. E. Br. = Erica globiceps (N. E. Br.) E. G. H. Oliv. subsp. gracilis (Benth.) E. G. H. Oliv. ●☆

381771 Syndesmanthus nivenii N. E. Br. = Erica niveniana E. G. H. Oliv. ●☆

381772 Syndesmanthus paucifolius (J. C. Wendl.) Benth. = Erica paucifolia (J. C. Wendl.) E. G. H. Oliv. ●☆

381773 Syndesmanthus pulchellus N. E. Br. = Erica pulchelliflora E. G. H. Oliv. ●☆

381774 Syndesmanthus scaber Klotzsch = Erica similis (N. E. Br.) E. G. H. Oliv. ●☆

381775 Syndesmanthus scaber Klotzsch var. gracilis Benth. = Erica globiceps (N. E. Br.) E. G. H. Oliv. subsp. gracilis (Benth.) E. G. H. Oliv. ●☆

381776 Syndesmanthus similis N. E. Br. = Erica similis (N. E. Br.) E. G. H. Oliv. ●☆

381777 Syndesmanthus squarrosus Benth. = Erica paucifolia (J. C. Wendl.) E. G. H. Oliv. subsp. squarrosa (Benth.) E. G. H. Oliv. ●☆

381778 Syndesmanthus sympiezoides N. E. Br. = Erica globiceps (N. E. Br.) E. G. H. Oliv. ●☆

381779 Syndesmanthus venustus N. E. Br. = Erica venustiflora E. G. H. Oliv. ●☆

381780 Syndesmanthus viscosus (Bolus) N. E. Br. = Erica viscosissima E. G. H. Oliv. ●☆

381781 Syndesmanthus zeyheri Bolus = Erica globiceps (N. E. Br.) E. G. H. Oliv. ●☆

381782 Syndesmis Wall. = Gluta L. ●

381783 Syndesmon (Hoffmanns. ex Endl.) Britton = Anemonella Spach ■☆

381784 Syndesmon Hoffmanns. = Anemonella Spach ■☆

381785 Syndesmon thalictroides (L.) Hoffmanns. = Anemonella thalictroides (L.) Spach ■☆

381786 Syndesmon thalictroides (L.) Hoffmanns. = Thalictrum thalictroides (L.) A. J. Eames et B. Boivin ■☆

381787 Syndiaspermaceae Dulac = Orobanchaceae Vent. (保留科名)●■

381788 Syndiclis Hook. f. (1886);油果樟属 (油果楠属);Oilfruitcamphor,Syndiclis ●

381789 Syndiclis Hook. f. = Potameia Thouars ●☆

381790 Syndiclis anlungensis H. W. Li;安龙油果樟;Anlong Oilfruitcamphor,Anlong Syndiclis ●

381791 Syndiclis chinensis C. K. Allen;油果樟(白面柴,油樟);Chinese Oilfruitcamphor,Chinese Syndiclis ●

381792 Syndiclis fooningensis H. W. Li;富宁油果樟;Funing Oilfruitcamphor,Funing Syndiclis ●

381793 Syndiclis furfuracea H. W. Li;鳞秕油果樟;Furturaceous Syndiclis,Scurfy Fruit Syndiclis,Scurfy Oilfruitcamphor ●

381794 Syndiclis hongkongensis N. H. Xia et al. = Sinopora hongkongensis (N. H. Xia et al.) J. Li et al. ●

381795 Syndiclis kwangsiensis (Kosterm.) H. W. Li;广西油果樟;

Guangxi Oilfruitcamphor，Guangxi Syndiclis，Kwangsi Oilfruitcamphor ●

381796 Syndiclis lotungensis S. K. Lee；乐东油果樟（乐东油樟）；Ledong Oilfruitcamphor，Ledong Syndiclis ●

381797 Syndiclis marlipoensis H. W. Li；麻栗坡油果樟；Malipo Oilfruitcamphor，Malipo Syndiclis ●

381798 Syndiclis pingbianensis H. W. Li；屏边油果樟；Pingbian Oilfruitcamphor，Pingbian Syndiclis ●

381799 Syndiclis sichouensis H. W. Li；西畴油果樟；Xichou Oilfruitcamphor，Xichou Syndiclis ●

381800 Syndyophyllum K. Schum. et Lauterb. = Syndyophyllum Lauterb. et K. Schum. ☆ 381801 Syndyophyllum Lauterb. et K. Schum. (1900)；双叶大戟属 ☆ 381802 Syndyophyllum excelsum K. Schum. et Lauterb.；双叶大戟 ☆ 381803 Synecanthus H. Wendl. = Synechanthus H. Wendl. ●☆

381804 Synechanthaceae O. F. Cook = Arecaceae Bercht. et J. Presl(保留科名) ●

381805 Synechanthaceae O. F. Cook = Palmae Juss. (保留科名) ●

381806 Synechanthus H. Wendl. (1858)；聚花椰属(簇羽棕属,簇棕属,合生花棕属,巧椰属)；Synechanthus ●☆

381807 Synechanthus panamensis H. E. Moore；聚花椰(合生花棕) ●☆

381808 Synedrella Gaertn. (1791)(保留属名)；金腰箭属(破伞菊属)；Synedrella ■

381809 Synedrella nodiflora (L.) Gaertn.；金腰箭(苞壳菊,苦草,水慈姑,猪毛草)；Axile Arrow，Nodalflower Synedrella，Synedrella ■

381810 Synedrella nodiflora Gaertn. = Synedrella nodiflora (L.) Gaertn. ■

381811 Synedrellopsis Hieron. et Kuntze(1898)；黄腰箭属(类金腰箭属) ■☆

381812 Synedrellopsis grisebachii Hieron. et Kuntze；黄腰箭(类金腰箭) ■☆

381813 Syneilesis Maxim. (1859)；兔儿伞属；Syneilesis ■

381814 Syneilesis aconitifolia (Bunge) Maxim.；兔儿伞(观音伞,龙头七,帽头菜,南天扇,破阳伞,七里麻,伞把草,伞草,水鹅掌,贴骨伞,贴骨散,铁凉伞,兔打伞,小鬼伞,雪里伞,鸭脚莲,一把伞,雨伞菜,雨伞草)；Aconiteleaf Syneilesis ■

381815 Syneilesis aconitifolia (Bunge) Maxim. var. longilepis Kitam.；长梗兔儿伞 ■☆

381816 Syneilesis australis Y. Ling；南方兔儿伞；Southern Syneilesis ■

381817 Syneilesis hayatae Kitam. = Syneilesis intermedia (Hayata) Kitam. ■

381818 Syneilesis intermedia (Hayata) Kitam.；台湾兔儿伞(台湾破伞菊)；Taiwan Syneilesis ■

381819 Syneilesis intermedia Kitam. = Syneilesis intermedia (Hayata) Kitam. ■

381820 Syneilesis palmata (Thunb.) Maxim.；掌状兔儿伞(兔儿伞,小吴风草)；Palm Syneilesis ■☆

381821 Syneilesis palmata Maxim. = Syneilesis palmata (Thunb.) Maxim. ■☆

381822 Syneilesis subglabrata (Yamam. et Sasaki) Kitam.；高山兔儿伞(高山破伞菊) ■

381823 Syneilesis tagawae (Kitam.) Kitam.；田川氏兔儿伞 ■☆

381824 Syneilesis tagawae (Kitam.) Kitam. var. latifolia H. Koyama；宽叶田川氏兔儿伞 ■☆

381825 Synekosciadium Boiss. = Tordylium L. ■☆

381826 Synema Dulac = Mercurialis L. ■

381827 Synepilaena Baill. = Kohleria Regel ●■☆

381828 Synexemia Raf. = Andrachne L. ●☆

381829 Synexemia Raf. = Phyllanthus L. ●■

381830 Syngeneticae Horan. = Compositae Giseke(保留科名) ●■

381831 Syngonanthus Ruhland(1900)；合瓣花属；Shoe-buttons ■☆

381832 Syngonanthus angolensis H. E. Hess；安哥拉合瓣花 ■☆

381833 Syngonanthus bianoensis Kimpouni；比亚诺合瓣花 ■☆

381834 Syngonanthus chevalieri Lecomte = Syngonanthus wahlbergii (Wikstr. ex Körn.) Ruhland ■☆

381835 Syngonanthus elegans (Kunth) Ruhland；雅致合瓣花 ■☆

381836 Syngonanthus exilis S. M. Phillips；瘦小合瓣花 ■☆

381837 Syngonanthus flavidulus (Michx.) Ruhland；浅黄合瓣花 ■☆

381838 Syngonanthus hessii Moldenke = Syngonanthus angolensis H. E. Hess ■☆

381839 Syngonanthus hybridus Moldenke = Syngonanthus angolensis H. E. Hess ■☆

381840 Syngonanthus lisowskii Kimpouni；利索合瓣花 ■☆

381841 Syngonanthus longibracteatus Kimpouni；长苞合瓣花 ■☆

381842 Syngonanthus manikaensis Kimpouni；马尼科合瓣花 ■☆

381843 Syngonanthus mwinilungensis S. M. Phillips；穆维尼合瓣花 ■☆

381844 Syngonanthus ngoweensis Lecomte；恩戈韦合瓣花 ■☆

381845 Syngonanthus paleaceus S. M. Phillips；膜片合瓣花 ■☆

381846 Syngonanthus poggeanus Ruhland；波格合瓣花 ■☆

381847 Syngonanthus robinsonii Moldenke；罗氏合瓣花 ■☆

381848 Syngonanthus schlechteri Ruhland；施莱合瓣花 ■☆

381849 Syngonanthus schlechteri Ruhland subsp. appendiculata Kimpouni；附属物合瓣花 ■☆

381850 Syngonanthus upembaensis Kimpouni；乌彭巴合瓣花 ■☆

381851 Syngonanthus wahlbergii (Wikstr. ex Körn.) Ruhland；瓦尔贝里合瓣花 ■☆

381852 Syngonanthus wahlbergii (Wikstr. ex Körn.) Ruhland var. sinkabolensis S. M. Phillips；辛卡波尔合瓣花 ■☆

381853 Syngonanthus welwitschii (Rendle) Ruhland；韦尔合瓣花 ■☆

381854 Syngonium Schott ex Endl. = Syngonium Schott ■☆

381855 Syngonium Schott(1829)；合果芋属(箭头藤属)；Arrowhead，Fivefingers，Syngonium ■☆

381856 Syngonium auritum (L.) Schott；五指合果芋(长耳合果芋)；Five Fingers，Five Fingers Syngonium，Fivefingers ■☆

381857 Syngonium erythrophyllum Birdsey ex G. S. Bunting；红叶合果芋 ■☆

381858 Syngonium hoffmannii Schott；霍夫曼合果芋 ■☆

381859 Syngonium macrophyllum Engl.；大叶合果芋 ■☆

381860 Syngonium mauroanum Birdsey；三裂合果芋 ■☆

381861 Syngonium podophyllum Schott；合果芋(长柄合果芋,合果梗芋,箭头藤)；African Evergreen，African Syngonium，American Evergreen，Arrowhead Vine，Goosefoot Plant，Nephthytis ■☆

381862 Syngonium podophyllum Schott 'Emerald Gem'Variegated'；翠玉合果芋 ■☆

381863 Syngonium podophyllum Schott 'Fresh Marble'；白斑合果芋 ■☆

381864 Syngonium podophyllum Schott 'Jenny'；珍妮合果芋 ■☆

381865 Syngonium podophyllum Schott 'Pinky'；粉红合果芋 ■☆

381866 Syngonium podophyllum Schott 'Pixie'；翠绿合果芋 ■☆

381867 Syngonium podophyllum Schott 'Trileaf Wonder'；银脉合果芋 ■☆

381868 Syngonium podophyllum Schott 'Variegata'；白蝴蝶 ■☆

381869 Syngonium podophyllum Schott var. albolineatum Engl.；白纹合果芋；White Veind Arrowhead Vine ■☆

381870 Syngonium podophynum Schott 'Alborirens'；白缘合果芋 ■☆

381871 Syngonium wendlandii Schott；绒叶合果芋(银脉合果芋) ■☆

381872 Synima Radlk. (1879)；合生无患子属 ●☆

381873　Synima cordierorum（F. Muell.）Radlk. ;合生无患子●☆

381874　Synisoon Baill. = Retiniphyllum Humb. et Bonpl. ●☆

381875　Synmeria Nimmo = Habenaria Willd. ■

381876　Synnema Benth. = Hygrophila R. Br. ●■

381877　Synnema abyssinicum（Hochst. ex Nees）Bremek. = Hygrophila abyssinica（Hochst. ex Nees）T. Anderson ●☆

381878　Synnema acinos S. Moore = Hygrophila acinos（S. Moore）Heine ■☆

381879　Synnema africanum（T. Anderson）Kuntze = Hygrophila africana（T. Anderson）Heine ■☆

381880　Synnema angolense S. Moore = Hygrophila angolensis（S. Moore）Heine ■☆

381881　Synnema borellii（Lindau）Benoist = Hygrophila borellii（Lindau）Heine ■☆

381882　Synnema brevitubum Burkill = Hygrophila brevituba（Burkill）Heine ■☆

381883　Synnema diffusum J. K. Morton = Hygrophila borellii（Lindau）Heine ■☆

381884　Synnema gossweileri S. Moore = Hygrophila gossweileri（S. Moore）Heine ■☆

381885　Synnema hygrophiloides Lindau = Hygrophila hygrophiloides（Lindau）Heine ■☆

381886　Synnema limnophiloides S. Moore = Hygrophila limnophiloides（S. Moore）Heine ■☆

381887　Synnema origanoides（Lindau）Bremek. = Hygrophila origanoides（Lindau）Heine ■☆

381888　Synnema prunelloides（S. Moore）Bremek. = Hygrophila prunelloides（S. Moore）Heine ■☆

381889　Synnema schweinfurthii（S. Moore）Bremek. = Hygrophila abyssinica（Hochst. ex Nees）T. Anderson ●☆

381890　Synnema tenerum（Lindau）Bremek. = Hygrophila tenera（Lindau）Heine ■☆

381891　Synnema triflorum Kuntze = Hygrophila difformis（L. f.）Blume ■☆

381892　Synnotia Sweet = Sparaxis Ker Gawl. ■☆

381893　Synnotia Sweet（1827）;漏斗莲属■☆

381894　Synnotia bicolor（Thunb.）Sweet = Sparaxis villosa（Burm. f.）Goldblatt ■☆

381895　Synnotia bicolor（Thunb.）Sweet var. roxburghii Baker = Sparaxis roxburghii（Baker）Goldblatt ■☆

381896　Synnotia bicolor Pole-Evans = Sparaxis galeata Ker Gawl. ■☆

381897　Synnotia galeata（Ker Gawl.）Sweet = Sparaxis galeata Ker Gawl. ■☆

381898　Synnotia metelerkampiae L. Bolus = Sparaxis metelerkampiae（L. Bolus）Goldblatt et J. C. Manning ■☆

381899　Synnotia parviflora G. J. Lewis = Sparaxis parviflora（G. J. Lewis）Goldblatt ■☆

381900　Synnotia roxburghii（Baker）G. J. Lewis = Sparaxis roxburghii（Baker）Goldblatt ■☆

381901　Synnotia variegata Sweet;漏斗莲■☆

381902　Synnotia variegata Sweet var. metelerkampiae（L. Bolus）G. J. Lewis = Sparaxis metelerkampiae（L. Bolus）Goldblatt et J. C. Manning ■☆

381903　Synnotia villosa（Burm. f.）N. E. Br. = Sparaxis villosa（Burm. f.）Goldblatt ☆

381904　Synnottia Baker = Synnotia Sweet ■☆

381905　Synodon Raf. = Conostegia D. Don ■☆

381906　Synoecia Miq. = Ficus L. ●

381907　Synoliga Raf. = Xyris L. ■

381908　Synoplectris Raf. = Sarcoglottis C. Presl ■☆

381909　Synoptera Raf. = Miconia Ruiz et Pav.（保留属名）●☆

381910　Synosma Raf. = Cacalia L. ●■

381911　Synosma Raf. ex Britton = Hasteola Raf. ■☆

381912　Synosma Raf. ex Britton et A. Br. = Hasteola Raf. ■☆

381913　Synosma suaveolens（L.）Raf. ex Britton = Hasteola suaveolens（L.）Pojark. ■☆

381914　Synostemon F. Muell.（1858）;假叶下珠属（艾堇属，合蕊木属）●■

381915　Synostemon F. Muell. = Sauropus Blume ●■

381916　Synostemon bacciforme（L.）Webster;假叶下珠（艾堇，艾堇守宫木，艾茎守宫木，红果草）;Berry-shaped Sauropus ■

381917　Synostemon bacciforme（L.）Webster = Sauropus bacciformis（L.）Airy Shaw ■

381918　Synotis（C. B. Clarke）C. Jeffrey et Y. L. Chen（1984）;合耳菊属（尾药菊属，尾药千里光属）;Synotis, Tailanther ●

381919　Synotis acuminata（Wall. ex DC.）C. Jeffrey et Y. L. Chen;尾尖合耳菊（尾尖尾药菊）;Acuminate Synotis, Sharp Tailanther ■

381920　Synotis ainsliifolia C. Jeffrey et Y. L. Chen;宽翅合耳菊（宽翅尾药菊）;Broadwing Tailanther, Wide Wing Synotis ■

381921　Synotis alata（Wall. ex DC.）C. Jeffrey et Y. L. Chen;翅柄合耳菊（翅柄千里光，翅柄尾药菊，聚伞千里光）;Winged Petiole Synotis, Wingstipe Tailanther ■

381922　Synotis atractylidifolia（Y. Ling）C. Jeffrey et Y. L. Chen;术叶合耳菊（术叶菊，术叶千里光，术叶尾药菊）;Atractylodeleaf Groundsel, Atractylodes-leaf Synotis ■

381923　Synotis auriculata C. Jeffrey et Y. L. Chen;耳叶合耳菊（耳叶尾药菊）;Ear-laef Synotis, Earlaef Tailanther ■

381924　Synotis austro-yunnanensis C. Jeffrey et Y. L. Chen;滇南合耳菊（滇南尾药菊）;S. Yunnan Tailanther, South Yunnan Synotis ■

381925　Synotis birmanica C. Jeffrey et Y. L. Chen;缅甸合耳菊（缅甸尾药菊）;Burma Tailanther, Burmese Synotis ■

381926　Synotis blevipappa C. Jeffrey et Y. L. Chen;短缨合耳菊（短缨尾药菊）;Short Pappus Synotis, Shortpappo Tailanther ■

381927　Synotis calocephala C. Jeffrey et Y. L. Chen;美头合耳菊（美头尾药菊，鞋头千里光）;Beautiful Head Synotis, Beautifulhead Tailanther ■

381928　Synotis cappa（Buch.-Ham. ex D. Don）C. Jeffrey et Y. L. Chen;密花合耳菊（白叶火草，密花千里光，密花尾药菊，密花尾药千里光）;Dense Flower Synotis, Denseflower Groundsel, Densehead Tailanther ■

381929　Synotis cavaleriei（H. Lév.）C. Jeffrey et Y. L. Chen;昆明合耳菊（西南尾药菊）;Cavalerie Tailanther, Cavalerie's Synotis ■

381930　Synotis changiana Y. L. Chen;肇骞合耳菊（肇骞尾药菊）;Zhaoqian Synotis, Zhaoqian Tailanther ■

381931　Synotis chingiana C. Jeffrey et Y. L. Chen;子农合耳菊（子农尾药菊）;Ching Tailanther, Ching's Synotis ■

381932　Synotis cordiflora Y. L. Chen;心叶尾药菊;Heartleaf Synotis, Heartleaf Tailanther ■

381933　Synotis damiaoshanica C. Jeffrey et Y. L. Chen;大苗山合耳菊（大苗山尾药菊）;Damiaoshan Synotis, Damiaoshan Tailanther ■

381934　Synotis duclouxii（Dunn）C. Jeffrey et Y. L. Chen;滇东合耳菊（滇车，滇东千里光，滇尾药菊，杜氏千里光，金毛草，千里光，血当归，银毛草）;Ducloux Tailanther, Ducloux's Synotis ■

381935　Synotis erythropappa（Bureau et Franch.）C. Jeffrey et Y. L. Chen;红缨合耳菊（红冠尾药菊，红毛千里光，红缨尾药菊，双花

千里光,双花尾药千里光,一扫光);Redpappo Tailanther, Red-pappus Synotis,Twinflower Groundsel ■

381936 Synotis fulvipes (Y. Ling) C. Jeffrey et Y. L. Chen;褐柄合耳菊(褐头尾药菊,湖南千里光);Brown Head Synotis, Brownhead Tailanther ■

381937 Synotis glomerata (Jeffrey) C. Jeffrey et Y. L. Chen;聚花合耳菊(团聚尾药菊);Glamerated Synotis, Glamerated Tailanther ■

381938 Synotis guizhouensis C. Jeffrey et Y. L. Chen;黔合耳菊(黔尾药菊);Guizhou Synotis, Guizhou Tailanther ■

381939 Synotis hieraciifolia (H. Lév.) C. Jeffrey et Y. L. Chen;矛叶合耳菊(毛叶合耳菊,山柳菊叶尾药菊);Hawkweed Leaf Synotis, Hawkweedleaf Tailanther ■

381940 Synotis ionodasys (Hand. -Mazz.) C. Jeffrey et Y. L. Chen;紫毛合耳菊(锈毛尾药菊,紫毛千里光);Brown Hair Synotis, Rusthair Tailanther ■

381941 Synotis longipes C. Jeffrey et Y. L. Chen;长柄合耳菊(长梗尾药菊);Long-peduncled Synotis, Longstalk Tailanther ■

381942 Synotis lucorum (Franch.) C. Jeffrey et Y. L. Chen;丽江合耳菊(丽江尾药菊);Lijiang Synotis, Lijiang Tailanther ■

381943 Synotis muliensis Y. L. Chen;木里合耳菊(木里尾药菊);Muli Synotis, Muli Tailanther ■

381944 Synotis nagensis (C. B. Clarke) C. Jeffrey et Y. L. Chen;锯叶合耳菊(白背艾,白千里光,白叶火草,大白叶子火草,大叶艾,火门艾,锯叶千里光,锯叶尾药菊,满山香,拿嘎千里光,拿嘎尾药千里光,舒嘎千里光);Nagens Groundsel, Nagesi Groundsel, Sawleaf Groundsel, Serrated Synotis, Serrated Tailanther ■

381945 Synotis nayongensis C. Jeffrey et Y. L. Chen;纳拥合耳菊(纳拥尾药菊);Nayong Synotis, Nayong Tailanther ■

381946 Synotis otophylla Y. L. Chen;耳柄合耳菊(耳柄尾药菊);Ear-stalk Synotis ■

381947 Synotis palmatisecta Y. L. Chen et J. D. Liu;掌裂合耳菊(掌裂尾药菊);Palmateleaf Synotis ■

381948 Synotis pseudoalata (C. C. Chang) C. Jeffrey et Y. L. Chen;紫背合耳菊(假翅柄千里光,紫背尾药菊);Purple Beneath Synotis, Purpleback Tailanther ■

381949 Synotis reniformis Y. L. Chen;肾叶合耳菊;Kidneyleaf Synotis ■

381950 Synotis saluenensis (Diels) C. Jeffrey et Y. L. Chen;腺毛合耳菊(怒江千里光,腺毛尾药菊);Glandhair Tailanther, Glandular Synotis ■

381951 Synotis sciatrephes (W. W. Sm.) C. Jeffrey et Y. L. Chen;林荫合耳菊(林荫千里光,狭翅尾药菊);Narrowwing Tailanther, Narrow-winged Synotis ■

381952 Synotis setchuanensis (Franch.) C. Jeffrey et Y. L. Chen;四川合耳菊(四川尾药菊);Sichuan Synotis, Sichuan Tailanther ■

381953 Synotis sinica (Diels) C. Jeffrey et Y. L. Chen;华合耳菊(华尾药菊);China Tailanther, Chinese Synotis ■

381954 Synotis solidaginea (Hand. -Mazz.) C. Jeffrey et Y. L. Chen;川西合耳菊(川西千里光,川西尾药菊);Goldenrod-like Synotis, W. Sichuan Tailanther ■

381955 Synotis tetrantha (DC.) C. Jeffrey et Y. L. Chen;四花合耳菊(四头尾药菊);Four Flower Synotis, Fourhead Tailanther ■

381956 Synotis triligulata (Buch. -Ham. ex D. Don) C. Jeffrey et Y. L. Chen;三舌合耳菊(三舌千里光,三舌尾药菊,三舌尾药千里光,山东风);Three Tongues Synotis, Threetongue Tailanther ■

381957 Synotis vaniotii (H. Lév.) C. Jeffrey et Y. L. Chen;羽裂合耳菊(药山千里光,羽裂尾药菊);Pinnatifid Synotis, Vaniot Tailanther ■

381958 Synotis wallichii (DC.) C. Jeffrey et Y. L. Chen;合耳菊(无翅

尾药菊);Wingless Synotis, Wingless Tailanther ■

381959 Synotis xantholeuca (Hand. -Mazz.) C. Jeffrey et Y. L. Chen;黄白合耳菊(黄白千里光,黄白尾药菊);Yellow-white Synotis, Yellowwish Tailanther ■

381960 Synotis yakoensis (Jeffrey ex Diels) C. Jeffrey et Y. L. Chen;丫口合耳菊(丫口千里光,丫口尾药菊,雅各尾药菊);Yake Synotis, Yakou Tailanther ■

381961 Synotis yui C. Jeffrey et Y. L. Chen;蔓生合耳菊(季川尾药菊);Yu Tailanther, Yu's Synotis ■

381962 Synotoma (G. Don) R. Schulz = Physoplexis (Endl.) Schur ■☆

381963 Synotoma R. Schulz = Physoplexis (Endl.) Schur ■☆

381964 Synoum A. Juss. (1830);东澳楝属●☆

381965 Synoum glandulosum (Sm.) A. Juss.;东澳楝;Bastard Rosewood, Scentless Rosewood, Snake's Food ●☆

381966 Synphoranthera Bojer = Dialypetalum Benth. ●■☆

381967 Synphoranthera Bojer ex A. Zahlbr. = Dialypetalum Benth. ●■☆

381968 Synphyllium Griff. = Curanga Juss. ■☆

381969 Synptera Llanos = Trichoglottis Blume ■

381970 Synsepalum (A. DC.) Baill. = Synsepalum (A. DC.) Daniell ●☆

381971 Synsepalum (A. DC.) Daniell(1852);神秘果属●☆

381972 Synsepalum Baill. = Synsepalum (A. DC.) Daniell ●☆

381973 Synsepalum afzelii (Engl.) T. D. Penn.;阿芙泽尔神秘果●☆

381974 Synsepalum attenuatum Hutch. et Dalziel = Synsepalum seretii (De Wild.) T. D. Penn. ●☆

381975 Synsepalum aubrevillei (Pellegr.) Aubrév. et Pellegr.;奥布神秘果●☆

381976 Synsepalum batesii (A. Chev.) Aubrév. et Pellegr.;贝茨神秘果●☆

381977 Synsepalum bequaertii De Wild.;贝卡尔神秘果●☆

381978 Synsepalum brenanii (Heine) T. D. Penn.;布雷南神秘果●☆

381979 Synsepalum brevipes (Baker) T. D. Penn.;短梗神秘果●☆

381980 Synsepalum cerasiferum (Welw.) T. D. Penn.;樱桃神秘果●☆

381981 Synsepalum congolense Lecomte;刚果神秘果●☆

381982 Synsepalum dulcificum (Schumach. et Thonn.) Daniell;神秘果;Daniell, Miraculous Berry, Sweetberry ●☆

381983 Synsepalum dulcificum Daniell = Synsepalum dulcificum (Schumach. et Thonn.) Daniell ●☆

381984 Synsepalum fleuryanum A. Chev.;弗勒里神秘果●☆

381985 Synsepalum gabonense (Aubrév. et Pellegr.) T. D. Penn.;加蓬神秘果●☆

381986 Synsepalum kaessneri (Engl.) T. D. Penn.;卡斯纳神秘果●☆

381987 Synsepalum kemoense (Dubard) Aubrév.;凯莫神秘果●☆

381988 Synsepalum letestui Aubrév. et Pellegr.;莱泰斯图神秘果●☆

381989 Synsepalum letouzeyi Aubrév.;勒图神秘果●☆

381990 Synsepalum longecuneatum De Wild.;长楔形神秘果●☆

381991 Synsepalum msolo (Engl.) T. D. Penn.;热非神秘果●☆

381992 Synsepalum muelleri (Kupicha) T. D. Penn.;梅尔神秘果●☆

381993 Synsepalum nyangense (Pellegr.) McPherson et L. J. T. White;尼扬加神秘果●☆

381994 Synsepalum passargei (Engl.) T. D. Penn.;帕萨神秘果●☆

381995 Synsepalum pobeguinianum (Pierre ex Lecomte) Aké Assi et L. Gaut.;波别神秘果●☆

381996 Synsepalum revolutum (Baker) T. D. Penn.;外卷神秘果●☆

381997 Synsepalum seretii (De Wild.) T. D. Penn.;赛雷神秘果●☆

381998 Synsepalum stipulatum (Radlk.) Engl.;托叶神秘果●☆

381999 Synsepalum subcordatum De Wild.;近心形神秘果●☆

382000 Synsepalum subverticillatum (E. A. Bruce) T. D. Penn.;近轮

生神秘果●☆

382001 Synsepalum ulugurense (Engl.) Engl.;乌卢古尔神秘果●☆

382002 Synsepalum zenkeri Engl. ex Aubrév. et Pellegr.;岑克尔神秘果●☆

382003 Synsiphon Regel = Colchicum L. ■

382004 Systemon Botsch. (1959);连蕊芥属(合蕊草属);Systemon ■★

382005 Systemon Taub. = Synostemon F. Muell. ●■

382006 Systemon deserticola Y. Z. Zhao = Systemon petrovii Botsch. ■

382007 Systemon linearifolium C. H. An = Dontostemon integrifolius (L.) Ledeb. ■

382008 Systemon linearifolium C. H. An = Dontostemon integrifolius (L.) Ledeb. var. eglandulosus (DC.) Turcz. ■

382009 Systemon linearifolium C. H. An = Dontostemon integrifolius (L.) Ledeb. ■

382010 Systemon lulianlianus Al-Shehbaz et al.;陆氏连蕊芥■

382011 Systemon petrovii Botsch.;连蕊芥(荒漠连蕊芥);Petrov Systemon,Sand Systemon,Systemon ■

382012 Systemon petrovii Botsch. var. pilosus Botsch.;柔毛连蕊芥;Pilose Petrov Systemon,Pilose Systemon ■

382013 Systemon petrovii Botsch. var. pilosus Botsch. = Systemon petrovii Botsch. ■

382014 Systemon petrovii Botsch. var. xinglongicus C. H. An = Systemon petrovii Botsch. ■

382015 Systemon petrovii Botsch. var. xinglonica C. H. An;兴隆连蕊芥;Xinglong Petrov Systemon,Xinglong Systemon ■

382016 Systemonanthus Botsch. = Systemon Botsch. ■★

382017 Systemonanthus petrovii (Botsch.) Botsch. = Systemon petrovii Botsch. ■

382018 Systemonanthus petrovii (Botsch.) Botsch. var. pilosus (Botsch.) Botsch. = Systemon petrovii Botsch. ■

382019 Synstima Raf. = Ilex L. ●

382020 Synstylis C. Cusset = Hydrobryum Endl. ■

382021 Syntherisma Walter = Digitaria Haller(保留属名)■

382022 Syntherisma argillacea Hitchc. et Chase = Digitaria argillacea (Hitchc. et Chase) Fernald ■☆

382023 Syntherisma chinensis (Nees) Hitchc. = Digitaria violascens Link ■

382024 Syntherisma ciliaris (Retz.) Schrad. = Digitaria ciliaris (Retz.) Koeler ■

382025 Syntherisma formosana (Rendle) Honda = Digitaria radicosa (J. Presl et C. Presl) Miq. ■

382026 Syntherisma formosana (Rendle) Honda var. hirsuta Honda = Digitaria radicosa (J. Presl et C. Presl) Miq. ■

382027 Syntherisma fusca (C. Presl) Scribn. = Digitaria violascens Link ■

382028 Syntherisma glabra Schrad. = Digitaria ischaemum (Schreb.) Schreb. ex Muhl. ■

382029 Syntherisma hayatae Honda = Digitaria mollicoma (Kunth) Henrard ■

382030 Syntherisma hayatae Honda var. magna Honda = Digitaria mollicoma (Kunth) Henrard ■

382031 Syntherisma henryi (Rendle) Newbold = Digitaria henryi Rendle ■

382032 Syntherisma humifusa (Pers.) Rydb. = Digitaria ischaemum (Schreb.) Schreb. ex Muhl. ■

382033 Syntherisma ischaemum (Schreb.) Nash = Digitaria ischaemum (Schreb.) Schreb. ex Muhl. ■

382034 Syntherisma longiflora (Retz.) Skeels = Digitaria longiflora (Retz.) Pers. ■

382035 Syntherisma longiflora Skeels = Digitaria longiflora (Retz.) Pers. ■

382036 Syntherisma magna Honda = Digitaria mollicoma (Kunth) Henrard ■

382037 Syntherisma microbachne (J. Presl) Hitchc. = Digitaria microbachne (J. Presl) Henrard ■

382038 Syntherisma microbachne (J. Presl) Hitchc. = Digitaria setigera Roth ex Roem. et Schult. ■

382039 Syntherisma nodosa (Parl.) Newbold = Digitaria nodosa Parl. ■☆

382040 Syntherisma royleana (Nees) Newbold = Digitaria stricta Roth ex Roem. et Schult. ■

382041 Syntherisma sanguinalis (L.) Dulac = Digitaria sanguinalis (L.) Scop. ■

382042 Syntherisma sasakii Honda = Digitaria henryi Rendle ■

382043 Syntherisma sericea Honda = Digitaria ciliaris (Retz.) Koeler ■

382044 Syntherisma tenuispica Keng = Digitaria radicosa (J. Presl et C. Presl) Miq. ■

382045 Syntherisma ternata (Hochst. ex A. Rich.) Newbold = Digitaria ternata (Hochst. ex A. Rich.) Stapf ex Dyer ■

382046 Syntherisma ternata (Hochst.) Newbold = Digitaria ternata (Hochst. ex A. Rich.) Stapf ex Dyer ■

382047 Synthlipsis A. Gray(1849);合集芥属■☆

382048 Synthlipsis berlandieri A. Gray;合集芥■☆

382049 Synthlipsis densiflora Rollins;密花合集芥■☆

382050 Synthlipsis lepidota Rose;细齿合集芥■☆

382051 Synthyris Benth. (1846);美洲玄参属(猫尾草属)■☆

382052 Synthyris bullii (Eaton) A. Heller = Besseya bullii (Eaton) Rydb. ■☆

382053 Synthyris reniformis Benth.;肾叶美洲玄参(肾叶猫尾草);Snow Queen ■☆

382054 Synthyris stellata Pennell;重齿美洲玄参(重齿猫尾草)■☆

382055 Syntriandrium Engl. (1899);三蕊藤属(三叶藤属)●☆

382056 Syntriandrium cordatum Hutch. et Dalziel = Syntriandrium preussii Engl. ●☆

382057 Syntriandrium dinklagei Engl. = Syntriandrium preussii Engl. ●☆

382058 Syntriandrium edentatum Engl. ex Diels = Syntriandrium preussii Engl. ●☆

382059 Syntriandrium preussii Engl.;三蕊藤(三叶藤)●☆

382060 Syntrichopappus A. Gray(1857);集毛菊属;Fremont's-gold ■☆

382061 Syntrichopappus fremontii A. Gray;集毛菊;Yellow Syntrichopappus,Yellowray Fremont's-gold ■☆

382062 Syntrichopappus lemmonii (A. Gray) A. Gray;莱蒙集毛菊;Lemmon's Syntrichopappus,Pinkray Fremont's-gold ■☆

382063 Syntrinema H. Pfeiff. = Rhynchospora Vahl(保留属名)■

382064 Syntrinema Radlk. = Syntrinema H. Pfeiff. ■

382065 Syntrophe Ehrenb. = Caylusea A. St. -Hil. (保留属名)■☆

382066 Syntrophe Ehrenb. ex Müll. Arg. = Caylusea A. St. -Hil. (保留属名)■☆

382067 Synurus Iljin(1926);山牛蒡属;Synurus ■

382068 Synurus atriplicifolius (Trevis.) Iljin = Synurus deltoides (Aiton) Nakai ■

382069 Synurus deltoides (Aiton) Nakai;山牛蒡(老鼠愁);Deltoid Synurus ■

382070 Synurus deltoides (Aiton) Nakai var. incisolobata (DC.) Kitam. = Synurus deltoides (Aiton) Nakai ■

382071 Synurus deltoides (Aiton) Nakai var. incisolobatus (Miyabe)

Kitam. ;锐裂山牛蒡■

382072 Synurus diabolicus (Kitam.) Kitam. = Olgaea lomonosowii (Trautv.) Iljin ■

382073 Synurus excelsus (Makino) Kitam. ;高大山牛蒡■☆

382074 Synurus excelsus (Makino) Kitam. = Synurus deltoides (Aiton) Nakai ■

382075 Synurus hondae Kitag. = Synurus deltoides (Aiton) Nakai ■

382076 Synurus palmatopinnatifidus (Makino) Kitam. ;掌裂山牛蒡■☆

382077 Synurus palmatopinnatifidus (Makino) Kitam. = Synurus deltoides (Aiton) Nakai ■

382078 Synurus palmatopinnatifidus (Makino) Kitam. var. indivisus Kitam. ;不裂山牛蒡■☆

382079 Synurus pungens (Franch. et Sav.) Kitam. ;锐尖山牛蒡■☆

382080 Synurus pungens (Franch. et Sav.) Kitam. = Synurus deltoides (Aiton) Nakai ■

382081 Synurus pungens (Franch. et Sav.) Kitam. var. giganens Kitam. = Synurus deltoides (Aiton) Nakai ■

382082 Synurus pungens (Franch. et Sav.) Kitam. var. gigantens Kitam. = Synurus deltoides (Aiton) Nakai ■

382083 Synzistachium Raf. = Heliotropium L. ●■

382084 Synzyganthera Ruiz et Pav. = Lacistema Sw. ●☆

382085 Syoctonum Bernh. = Chenopodium L. ■●

382086 Syorhynchium Hoffmanns. = Sisyrinchium L. ■

382087 Sypharissa Salisb. = Tenicroa Raf. ■☆

382088 Sypharissa Salisb. = Urginea Steinh. ■☆

382089 Sypharissa exuviata (Jacq.) Salisb. ex Oberm. = Drimia exuviata (Jacq.) Jessop ■☆

382090 Sypharissa filifolia (Jacq.) Salisb. ex Oberm. = Drimia filifolia (Jacq.) J. C. Manning et Goldblatt ■☆

382091 Sypharissa fragrans (Jacq.) Salisb. ex Oberm. = Drimia fragrans (Jacq.) J. C. Manning et Goldblatt ■☆

382092 Sypharissa multifolia (G. J. Lewis) Oberm. = Drimia multifolia (G. J. Lewis) Jessop ■☆

382093 Syphocampylos Hort. Belg. ex Hook. = Siphocampylus Pohl ■●☆

382094 Syphomeris Steud. = Grewia L. ●

382095 Syphomeris Steud. = Siphomeris Bojer ●■

382096 Syreitschikovia Pavlov (1933) ; 疆 菊 属 ; Syreitschikovia, Xinjiangdaisy ●

382097 Syreitschikovia spinulosa (Franch.) Pavlov;疆菊■☆

382098 Syreitschikovia tenuifolia (Bong.) Pavlov;细叶疆菊;Thinleaf Syreitschikovia, Xinjiangdaisy ■

382099 Syreitschikovia tenuis (Bunge) Botsch. = Syreitschikovia tenuifolia (Bong.) Pavlov ■

382100 Syrenia Andrz. ex Besser = Erysimum L. ■●

382101 Syrenia Andrz. ex DC. (1821) ;棱果芥属(茜兰芥属,赛糖芥属) ;Syrenia ■

382102 Syrenia Andrz. ex DC. = Erysimum L. ■●

382103 Syrenia angustifolia Rchb. ; 窄叶棱果芥(窄叶赛糖芥) ; Narrowleaf Syrenia ■☆

382104 Syrenia macrocarpa V. N. Vassil. ; 大果棱果芥; Largefruit Syrenia ■

382105 Syrenia sessiliflora Ledeb. ;无柄花棱果芥(无柄花棱果芥,无柄茜兰芥,无柄赛糖芥) ;Sessile Syrenia ■☆

382106 Syrenia siliculosa (M. Bieb.) Andrz. = Erysimum siliculosum (M. Bieb.) DC.

382107 Syrenia siliculosa (M. Bieb.) Andrz. ex DC. ;棱果芥(短荚赛糖芥) ;Cilicle Syrenia ■

382108 Syrenia talliewii Klokov;塔氏棱果芥(塔氏赛糖芥) ;Taliew Syrenia ■☆

382109 Syreniopsis H. P. Fuchs = Acachmena H. P. Fuchs(废弃属名)●■

382110 Syreniopsis H. P. Fuchs = Erysimum L. ■●

382111 Syreniopsis H. P. Fuchs(1959) ;拟棱果芥属■☆

382112 Syreniopsis cuspidata (M. Bieb.) H. P. Fuchs;拟棱果芥■☆

382113 Syrenopsis Jaub. et Spach = Thlaspi L. ■

382114 Syringa L. (1753) ;丁香属(丁香花属) ;Lilac,Syringa ●

382115 Syringa Mill. = Philadelphus L. ●

382116 Syringa ' Betsy Ross ' ;早熟丁香;Early Lilac Hybrid ●☆

382117 Syringa ' Red Pixie ' ;红精灵丁香;Lilac ●☆

382118 Syringa adamiana Balf. f. et W. W. Sm. = Syringa tomentella Bureau et Franch. ●

382119 Syringa adamiana Balfour f. et W. W. Sm. = Syringa tomentella Bureau et Franch. ●

382120 Syringa affinis L. = Syringa oblata Lindl. ex Carrière ●

382121 Syringa affinis L. var. giraldi Schneid. = Syringa oblata Lindl. ex Carrière ●

382122 Syringa afghanica C. K. Schneid. ; 阿富汗丁香●☆

382123 Syringa alborosea N. E. Br. = Syringa tomentella Bureau et Franch. ●

382124 Syringa amurensis Rupr. = Syringa reticulata (Blume) H. Hara subsp. amurensis (Rupr.) P. S. Green et M. C. Chang ●

382125 Syringa amurensis Rupr. var. japonica ? = Syringa japonica Decne. ●☆

382126 Syringa amurensis Rupr. var. mandshurica (Maxim.) Korsh. = Syringa reticulata (Blume) H. Hara subsp. amurensis (Rupr.) P. S. Green et M. C. Chang ●

382127 Syringa amurensis Rupr. var. pekiensis (Rupr.) Maxim. = Syringa reticulata (Blume) H. Hara subsp. pekinensis (Rupr.) P. S. Green et M. C. Chang ●

382128 Syringa angustifolia Salisb. = Syringa persica L. ●☆

382129 Syringa bretschneideri Lemoine = Syringa villosa Vahl ●

382130 Syringa buxifolia Nakai = Syringa protolaciniata P. S. Green et M. C. Chang ●

382131 Syringa caerulea Jonst. = Syringa vulgaris L. ●

382132 Syringa chinensis Schmidt;什锦丁香(华丁香) ;China Lilac, Chinese Lilac,Rouen Lilac ●

382133 Syringa chinensis Schmidt ' Alba ' ;白花什锦丁香;White-flowered Chinese Lilac ●

382134 Syringa chinensis Schmidt ' Lilac Sunday ' ;星期日丁香; Chinese Lilac ●☆

382135 Syringa chinensis Schmidt f. alba (Kirchn.) Schelle = Syringa chinensis Schmidt ' Alba ' ●

382136 Syringa chinensis Schmidt f. duplex (Lemoine) Schelle;重瓣什锦丁香;Doubleflower Chinese Lilac ●

382137 Syringa chinensis Schmidt var. alba (Kirchn.) Rehder = Syringa chinensis Schmidt ' Alba ' ●

382138 Syringa chinensis Schmidt var. duplex (Lemoine) Rehder = Syringa chinensis Schmidt f. duplex (Lemoine) Schelle ●

382139 Syringa chuanxiensis S. Z. Qu et X. L. Chen;川西丁香;W. Sichuan Lilac ●☆

382140 Syringa chuanxiensis S. Z. Qu et X. L. Chen = Syringa mairei (H. Lév.) Rehder ●

382141 Syringa delavayi Franch. ;德丁香(山桂花) ●

382142 Syringa didymopetalus P. S. Green;离瓣木犀●

382143 Syringa dielsiana C. K. Schneid. = Syringa pubescens Turcz.

subsp. microphylla (Diels) M. C. Chang et X. L. Chen ●

382144　Syringa dilatata Nakai 'Lacera';裂叶朝鲜丁香●

382145　Syringa dilatata Nakai = Syringa oblata Lindl. ex Carrière subsp. dilatata (Nakai) P. S. Green et M. C. Chang ●

382146　Syringa dilatata Nakai f. alba (W. Wang et Skvortsov) S. D. Zhao = Syringa oblata Lindl. ex Carrière subsp. dilatata (Nakai) P. S. Green et M. C. Chang ●

382147　Syringa dilatata Nakai var. alba W. Wang et Skvortsov = Syringa oblata Lindl. ex Carrière subsp. dilatata (Nakai) P. S. Green et M. C. Chang ●

382148　Syringa dilatata Nakai var. longituba W. Wang et Skvortsov = Syringa oblata Lindl. ex Carrière subsp. dilatata (Nakai) P. S. Green et M. C. Chang ●

382149　Syringa dilatata Nakai var. pubescens S. D. Zhao = Syringa oblata Lindl. ex Carrière subsp. dilatata (Nakai) P. S. Green et M. C. Chang ●

382150　Syringa dilatata Nakai var. rubra W. Wang et Skvortsov = Syringa oblata Lindl. ex Carrière subsp. dilatata (Nakai) P. S. Green et M. C. Chang ●

382151　Syringa dilatata Nakai var. violacea W. Wang et Skvortsov = Syringa oblata Lindl. ex Carrière subsp. dilatata (Nakai) P. S. Green et M. C. Chang ●

382152　Syringa dubia Pers. = Syringa chinensis Schmidt ●

382153　Syringa emodi Wall. et G. Don;喜马拉雅丁香(西蜀丁香);Himalayan Lilac, Rouen Lilac ●

382154　Syringa emodi Wall. ex Royle = Syringa emodi Wall. et G. Don ●

382155　Syringa emodii Wall. et G. Don var. pilosissima C. K. Schneid. = Syringa tomentella Bureau et Franch. ●

382156　Syringa emodii Wall. et G. Don var. rosea Cornu = Syringa villosa Vahl ●

382157　Syringa emodii Wall. ex G. Don = Syringa tibetica P. Y. Bai ●

382158　Syringa emodi-rosea Cornu = Syringa villosa Vahl ●

382159　Syringa fauriei H. Lév. = Syringa pubescens Turcz. subsp. patula (Palib.) M. C. Chang et X. L. Chen ●

382160　Syringa fauriei H. Lév. var. lactea (Nakai) Nakai = Syringa pubescens Turcz. subsp. patula (Palib.) M. C. Chang et X. L. Chen ●

382161　Syringa formosissima Nakai = Syringa wolfii C. K. Schneid. ●

382162　Syringa formosissima Nakai var. hirsuta (C. K. Schneid.) Nakai = Syringa wolfii C. K. Schneid. ●

382163　Syringa geraldiana Sarg. = Syringa oblata Lindl. ex Carrière ●

382164　Syringa giraldiana C. K. Schneid.;秦岭丁香;Girald Lilac ●

382165　Syringa giraldiana C. K. Schneid. = Syringa pubescens Turcz. subsp. microphylla (Diels) M. C. Chang et X. L. Chen ●

382166　Syringa giraldii Lemoine = Syringa oblata Lindl. ex Carrière ●

382167　Syringa glabra (C. K. Schneid.) Lingelsh. = Syringa komarowii C. K. Schneid. ●

382168　Syringa henryi C. K. Schneid.;亨利丁香●☆

382169　Syringa hirsuta (C. K. Schneid.) Nakai = Syringa wolfii C. K. Schneid. ●

382170　Syringa hirsuta (C. K. Schneid.) Nakai var. formosissima (Nakai) Nakai = Syringa wolfii C. K. Schneid. ●

382171　Syringa hyacinthiflora Rehder;早花丁香;American Hybrid Lilac, Early Flowering Lilac, French Hybrid Lilac, Hyacinth Lilac, Skinner Hybrid ●☆

382172　Syringa hyacinthiflora Rehder 'Blue Hyacinth';蓝紫早花丁香(蓝风信子丁香)●☆

382173　Syringa hyacinthiflora Rehder 'Charles Nordine';查尔斯·诺尔丁早花丁香●☆

382174　Syringa hyacinthiflora Rehder 'Clarke's Giant';克拉克巨人丁香●☆

382175　Syringa hyacinthiflora Rehder 'Cora Brandt';布朗特丁香●☆

382176　Syringa hyacinthiflora Rehder 'Laurentian';劳伦斯早花丁香●☆

382177　Syringa indica Royle ex Lindl. = Syringa emodi Wall. ex Royle ●

382178　Syringa japonica Decne.;日本丁香;Japanese Tree Lilac ●☆

382179　Syringa josikaea J. Jacq. = Syringa josikaea J. Jacq. ex Rchb. ●☆

382180　Syringa josikaea J. Jacq. ex Rchb.;匈牙利丁香(反折丁香,胶希丁香);Hungarian Lilac ●☆

382181　Syringa josikaea J. Jacq. ex Rchb. 'Anna Amhoff';安娜·阿姆霍夫胶希丁香●☆

382182　Syringa josikaea J. Jacq. ex Rchb. 'Bellicent';美丽反折丁香●☆

382183　Syringa josikaea J. Jacq. ex Rchb. 'Elaine';伊兰胶希丁香●☆

382184　Syringa josikaea J. Jacq. ex Rchb. 'Lynette';利纳特胶希丁香●☆

382185　Syringa josikaea J. Jacq. ex Rchb. 'Royalty';王权胶希丁香●☆

382186　Syringa julianae C. K. Schneid. 'Hers Variety';垂枝匈牙利丁香;Weeping Hers Lilac ●☆

382187　Syringa julianae C. K. Schneid. = Syringa pubescens Turcz. subsp. julianae (C. K. Schneid.) M. C. Chang et X. L. Chen ●

382188　Syringa kamibayashi Nakai = Syringa pubescens Turcz. subsp. patula (Palib.) M. C. Chang et X. L. Chen ●

382189　Syringa koehneana C. K. Schneid. = Syringa pubescens Turcz. subsp. patula (Palib.) M. C. Chang et X. L. Chen ●

382190　Syringa komarowii C. K. Schneid.;西蜀丁香(垂丝丁香,柯氏丁香);Komarov Lilac, Nodding Lilac, Reflexed Komarov Lilac ●

382191　Syringa komarowii C. K. Schneid. subsp. reflexa (C. K. Schneid.) P. S. Green et M. C. Chang = Syringa komarowii C. K. Schneid. ●

382192　Syringa komarowii C. K. Schneid. subsp. reflexa (C. K. Schneid.) P. S. Green et M. C. Chang;垂丝丁香●

382193　Syringa komarowii C. K. Schneid. var. reflexa (C. K. Schneid.) Z. P. Jien ex M. C. Chang = Syringa komarowii C. K. Schneid. subsp. reflexa (C. K. Schneid.) P. S. Green et M. C. Chang ●

382194　Syringa komarowii C. K. Schneid. var. sargentiana C. K. Schneid. = Syringa komarowii C. K. Schneid. ●

382195　Syringa laciniata Mill.;裂叶丁香;Cutleaf Lilac, Cut-leaf Lilac ●☆

382196　Syringa latifolia Salisb. = Syringa vulgaris L. ●

382197　Syringa mairei (H. Lév.) Rehder;皱叶丁香;Maire Lilac ●

382198　Syringa marginata (Champ. ex Benth.) Hemsl.;月桂丁香●

382199　Syringa media Hort. = Syringa chinensis Schmidt ●

382200　Syringa meyeri C. K. Schneid.;蓝丁香(南子香,细管丁香);Blue Syzygium, Dwarf Korean Lilac, Dwarf Lilac, Meyer Lilac ●

382201　Syringa meyeri C. K. Schneid. 'Palibin';帕利宾蓝丁香;Meyer Lilac ●

382202　Syringa meyeri C. K. Schneid. 'Palibin' = Syringa velutina Kom. ●

382203　Syringa meyeri C. K. Schneid. var. spontanea (M. C. Chang) X. K. Qin f. alba (W. Wang, Fuh et H. C. Chao) M. C. Chang = Syringa spontanea (M. C. Chang) X. K. Qin f. alba (W. Wang, Fuh et H. C. Chao) X. K. Qin ●

382204　Syringa meyeri C. K. Schneid. var. spontanea f. alba (W. Wang, Fuh, et H. C. Chao) M. C. Chang = Syringa meyeri C. K. Schneid. var. spontanea M. C. Chang ●

382205　Syringa meyeri C. K. Schneid. var. spontanea M. C. Chang;小叶蓝丁香;Little-leaf Meyer Lilac ●

382206　Syringa meyeri C. K. Schneid. var. spontanea M. C. Chang =

Syringa meyeri C. K. Schneid. 'Palibin' ●

382207　Syringa micrantha Nakai ＝ Syringa pubescens Turcz. subsp. patula（Palib.）M. C. Chang et X. L. Chen ●

382208　Syringa microphylla Diels ＝ Syringa pubescens Turcz. subsp. microphylla（Diels）M. C. Chang et X. L. Chen ●

382209　Syringa microphylla Diels Diels ＝ Syringa pubescens Turcz. subsp. microphylla（Diels）M. C. Chang et X. L. Chen ●

382210　Syringa microphylla Diels f. alba（W. Wang, Fuh et H. C. Chao）Kitag. ＝ Syringa meyeri C. K. Schneid. var. spontanea M. C. Chang ●

382211　Syringa microphylla Diels f. alba（W. Wang, Fuh et H. C. Chao）Kitag. ＝ Syringa spontanea（M. C. Chang）X. K. Qin f. alba（W. Wang, Fuh et H. C. Chao）X. K. Qin ●

382212　Syringa microphylla Diels minor Dropmore ＝ Syringa meyeri C. K. Schneid. var. spontanea M. C. Chang ●

382213　Syringa microphylla Diels var. alba W. Wang ＝ Syringa meyeri C. K. Schneid. var. spontanea M. C. Chang ●

382214　Syringa microphylla Diels var. alba W. Wang, Fuh et H. C. Chao ＝ Syringa spontanea（M. C. Chang）X. K. Qin f. alba（W. Wang, Fuh et H. C. Chao）X. K. Qin ●

382215　Syringa microphylla Diels var. alba W. Wang, Fuh, et H. C. Chao ＝ Syringa meyeri C. K. Schneid. var. spontanea M. C. Chang ●

382216　Syringa microphylla Diels var. flavoanthera X. L. Chen ＝ Syringa pubescens Turcz. var. flavoanthera（X. L. Chen）M. C. Chang ●

382217　Syringa microphylla Diels var. giraldiana（C. K. Schneid.）S. Z. Qu et X. L. Chen ＝ Syringa pubescens Turcz. subsp. microphylla（Diels）M. C. Chang et X. L. Chen ●

382218　Syringa microphylla Diels var. glabriuscula C. K. Schneid. ＝ Syringa pubescens Turcz. subsp. microphylla（Diels）M. C. Chang et X. L. Chen ●

382219　Syringa oblata Lindl. ＝ Syringa oblata Lindl. ex Carrière ●

382220　Syringa oblata Lindl. ex Carrière;紫丁香（丁香,丁香花,华北紫丁香）;Broadleaf Lilac, Early Lilac ●

382221　Syringa oblata Lindl. ex Carrière 'Luolanzi';罗蓝紫丁香（罗蓝紫）●

382222　Syringa oblata Lindl. ex Carrière 'Xiangxue';香雪丁香（香雪）●

382223　Syringa oblata Lindl. ex Carrière 'Ziyun';紫云丁香（紫云）●

382224　Syringa oblata Lindl. ex Carrière subsp. dilatata（Nakai）P. S. Green et M. C. Chang;朝鲜丁香（朝阳丁香,大紫丁香）;Korean Early Lilac ●

382225　Syringa oblata Lindl. ex Carrière var. affinis（Henry）Lingelsh.;白花丁香（白丁香）;White Early Lilac ●

382226　Syringa oblata Lindl. ex Carrière var. affinis（Henry）Lingelsh. ＝ Syringa oblata Lindl. ex Carrière ●

382227　Syringa oblata Lindl. ex Carrière var. alba Rehder;白丁香（白花紫丁香）;White Early Lilac ●

382228　Syringa oblata Lindl. ex Carrière var. alba Rehder ＝ Syringa oblata Lindl. ex Carrière ●

382229　Syringa oblata Lindl. ex Carrière var. dilatata（Nakai）Rehder ＝ Syringa oblata Lindl. ex Carrière subsp. dilatata（Nakai）P. S. Green et M. C. Chang ●

382230　Syringa oblata Lindl. ex Carrière var. giraldii（Lemoine）Rehder;毛紫丁香（紫萼丁香,紫萼紫丁香）;Purple Early Lilac ●

382231　Syringa oblata Lindl. ex Carrière var. giraldii（Lemoine）Rehder ＝ Syringa oblata Lindl. ex Carrière ●

382232　Syringa oblata Lindl. ex Carrière var. hupehensis Pamp. ;湖北丁香;Hubei Early Lilac ●

382233　Syringa oblata Lindl. ex Carrière var. hupehensis Pamp. ＝

Syringa oblata Lindl. ex Carrière ●

382234　Syringa oblata Lindl. ex Carrière var. typica f. alba Lingelsh. ＝ Syringa pubescens Turcz. subsp. patula（Palib.）M. C. Chang et X. L. Chen ●

382235　Syringa oblata Lindl. ex Carrière var. typica f. alba Lingelsh. ＝ Syringa oblata Lindl. ex Carrière ●

382236　Syringa palibiniana Nakai ＝ Syringa pubescens Turcz. subsp. patula（Palib.）M. C. Chang et X. L. Chen ●

382237　Syringa palibiniana Nakai var. lactea（Nakai）Nakai ＝ Syringa pubescens Turcz. subsp. patula（Palib.）M. C. Chang et X. L. Chen ●

382238　Syringa patula（Palib.）Nakai ＝ Syringa pubescens Turcz. subsp. patula（Palib.）M. C. Chang et X. L. Chen ●

382239　Syringa patulum Palib. ＝ Syringa pubescens Turcz. subsp. patula（Palib.）M. C. Chang et X. L. Chen ●

382240　Syringa pekinensis Rupr. ＝ Syringa reticulata（Blume）H. Hara subsp. pekinensis（Rupr.）P. S. Green et M. C. Chang ●

382241　Syringa persica L. ;花叶丁香（波斯丁香,历细,野丁香）;Peach Lilac, Persian Lilac, Rouen Lilac ●☆

382242　Syringa persica L. 'Alba';白花花叶丁香（白花波斯丁香）;Whiteflower Persian Lilac ●

382243　Syringa persica L. f. alba（Weston）Voss ＝ Syringa persica L. 'Alba' ●

382244　Syringa persica L. var. alba（Weston）Voss ＝ Syringa persica L. 'Alba' ●

382245　Syringa persica L. var. typica f. alba Lingelsh. ＝ Syringa persica L. 'Alba' ●

382246　Syringa pinetorum W. W. Sm. ;松林丁香;Pinewood Syzygium, Piney Lilac ●

382247　Syringa pinnatifolia Hemsl. ;羽叶丁香（复叶丁香,山沉香）;Pinnate Lilac, Pinnateleaf Lilac, Pinnate-leavedf Lilac ●

382248　Syringa pinnatifolia Hemsl. var. alashanensis Ma et S. Q. Zhou;贺兰山丁香（山沉香）;Alashan Lilac ●

382249　Syringa pinnatifolia Hemsl. var. alashanensis Ma et S. Q. Zhou ＝ Syringa pinnatifolia Hemsl. ●

382250　Syringa potanini Schneid. ＝ Syringa pubescens Turcz. subsp. microphylla（Diels）M. C. Chang et X. L. Chen ●

382251　Syringa prestoniae McKelvey;普瑞斯顿丁香;Canadian Hybrid Lilac, Nodding Lilac, Preston Lilac ●☆

382252　Syringa prestoniae McKelvey 'Donald Wyman';唐纳德·维曼普雷斯顿丁香●☆

382253　Syringa prestoniae McKelvey 'Isabella';伊莎贝拉普雷斯顿丁香●☆

382254　Syringa prestoniae McKelvey 'Miss Canada';加拿大小姐丁香;Preston Lilac ●☆

382255　Syringa prestoniae McKelvey 'Nocturne';小夜曲普雷斯顿丁香●☆

382256　Syringa protolaciniata P. S. Green et M. C. Chang;华丁香（甘肃丁香）;China Lilac, Laciniate-leaved Lilac, Protolaciniate Lilac ●

382257　Syringa pubescens Turcz. ;巧玲花（毛丁香,毛叶丁香,小叶丁香）;Hairy Lilac ●

382258　Syringa pubescens Turcz. 'excellens';优秀巧玲花●☆

382259　Syringa pubescens Turcz. 'Miss Kim';金小姐巧玲花;Manchurian Lilac, Miss Kim Lilac ●☆

382260　Syringa pubescens Turcz. 'Sarah Sands';萨拉·杉兹巧玲花●☆

382261　Syringa pubescens Turcz. 'Superba';卓越巧玲花●☆

382262　Syringa pubescens Turcz. ＝ Syringa villosa Vahl ●

382263　Syringa pubescens Turcz. f. alba S. D. Zhao;白花巧玲花;White

Hairy Lilac ●

382264 Syringa pubescens Turcz. f. hirsuta（Skvortsov et W. Wang）Kitag. = Syringa pubescens Turcz. subsp. patula（Palib.）M. C. Chang et X. L. Chen ●

382265 Syringa pubescens Turcz. subsp. julianae（C. K. Schneid.）M. C. Chang et X. L. Chen；光萼巧玲花（毛序紫丁香）；Juliana Lilac ●

382266 Syringa pubescens Turcz. subsp. microphylla（Diels）M. C. Chang et X. L. Chen；小叶巧玲花（四季丁香，小叶丁香）；Littleleaf Lilac ●

382267 Syringa pubescens Turcz. subsp. patula（Palib.）M. C. Chang = Syringa pubescens Turcz. subsp. patula（Palib.）M. C. Chang et X. L. Chen ●

382268 Syringa pubescens Turcz. subsp. patula（Palib.）M. C. Chang et X. L. Chen；关东巧玲花（关东丁香，红丁香，舒展巧玲花）；Korean Lilac，Late Lilac，Manchurian Lilac，Spread Hairy Lilac，Velvety Lilac ●

382269 Syringa pubescens Turcz. var. flavoanthera（X. L. Chen）M. C. Chang；黄药小叶巧玲花；Yellow-anther Hairy Lilac ●

382270 Syringa pubescens Turcz. var. hirsuta Skvortsov et W. Wang = Syringa pubescens Turcz. subsp. patula（Palib.）M. C. Chang et X. L. Chen ●

382271 Syringa pubescens Turcz. var. tibetica Batalin = Syringa pubescens Turcz. subsp. microphylla（Diels）M. C. Chang et X. L. Chen ●

382272 Syringa pubescens Turcz. var. typica Batalin f. pilosa Schneid. = Syringa pubescens Turcz. ●

382273 Syringa reflexa C. K. Schneid. = Syringa komarowii C. K. Schneid. subsp. reflexa（C. K. Schneid.）P. S. Green et M. C. Chang ●

382274 Syringa reflexa C. K. Schneid. = Syringa komarowii C. K. Schneid. ●

382275 Syringa rehderiana C. K. Schneid. = Syringa tomentella Bureau et Franch. ●

382276 Syringa reticulata（Blume）H. Hara；丁香（暴马丁香，青杠子，日本丁香，网叶丁香）；Japan Lilac，Japanese Tree Lilac ●☆

382277 Syringa reticulata（Blume）H. Hara subsp. amurensis（Rupr.）P. S. Green et M. C. Chang；暴马丁香（白丁香，棒棒木，暴马子，荷花丁香，青杠子）；Amur Lilac，Japanese Lilac，Japanese Tree Lilac，Manchurian Lilac ●

382278 Syringa reticulata（Blume）H. Hara subsp. pekinensis（Rupr.）P. S. Green et M. C. Chang；北京丁香；Beijing Lilac，Chinese Tree Lilac，Peking Lilac，Peking Tree Lilac ●

382279 Syringa reticulata（Blume）H. Hara var. amurensis（Rupr.）J. S. Pringle = Syringa reticulata（Blume）H. Hara subsp. amurensis（Rupr.）P. S. Green et M. C. Chang ●

382280 Syringa reticulata（Blume）H. Hara var. mandshurica（Maxim.）H. Hara = Syringa reticulata（Blume）H. Hara subsp. amurensis（Rupr.）P. S. Green et M. C. Chang ●

382281 Syringa reticulata（Blume）H. Hara var. tatewakiana（Yanagita）H. Hara；馆肋丁香 ●☆

382282 Syringa robusta Nakai；粗壮丁香 ●☆

382283 Syringa robusta Nakai = Syringa wolfii C. K. Schneid. ●

382284 Syringa robusta Nakai f. glabra Nakai = Syringa wolfii C. K. Schneid. ●

382285 Syringa robusta Nakai f. subhirsuta Nakai = Syringa wolfii C. K. Schneid. ●

382286 Syringa robusta Nakai var. rupestris A. I. Baranov et Skvortsov = Syringa wolfii C. K. Schneid. ●

382287 Syringa rothomagensis Hort. = Syringa chinensis Schmidt ●

382288 Syringa rotundifolia Decne. = Syringa reticulata（Blume）H. Hara subsp. amurensis（Rupr.）P. S. Green et M. C. Chang ●

382289 Syringa rugulosa McKelvey = Syringa mairei（H. Lév.）Rehder ●

382290 Syringa sargentiana C. K. Schneid. = Syringa komarowii C. K. Schneid. ●

382291 Syringa schneideri Lingelsh. = Syringa pubescens Turcz. subsp. microphylla（Diels）M. C. Chang et X. L. Chen ●

382292 Syringa sempervirens Franch. = Ligustrum sempervirens（Franch.）Lingelsh. ●

382293 Syringa spontanea（M. C. Chang）X. K. Qin；山丁香 ●

382294 Syringa spontanea（M. C. Chang）X. K. Qin = Syringa meyeri C. K. Schneid. var. spontanea M. C. Chang ●

382295 Syringa spontanea（M. C. Chang）X. K. Qin f. alba（W. Wang，Fuh et H. C. Chao）X. K. Qin；白花山丁香（白花小叶蓝丁香）●

382296 Syringa suspensa（Thunb.）Thunb. ex Murray = Forsythia suspensa（Thunb.）Vahl ●

382297 Syringa suspensa Thunb. = Forsythia suspensa（Thunb.）Vahl ●

382298 Syringa sweginzowii Koehne et Lingelsh.；四川丁香（细枝丁香）；Sichuan Lilac，Szechwan Lilac ●

382299 Syringa tetanoloba C. K. Schneid. = Syringa sweginzowii Koehne et Lingelsh. ●

382300 Syringa tibetica P. Y. Bai；藏南丁香；S. Xizang Lilac，Tibet Lilac，Xizang Lilac ●

382301 Syringa tigerstedtii Harry Sm. = Syringa sweginzowii Koehne et Lingelsh. ●

382302 Syringa tomentella Bureau et Franch.；毛丁香（毛叶丁香，绒毛丁香）；Ferty Lilac，Hair Lilac，Tomentose Lilac ●

382303 Syringa tomentella Bureau et Franch. var. rehderiana（C. K. Schneid.）Rehder = Syringa tomentella Bureau et Franch. ●

382304 Syringa trichophylla Ts. Tang = Syringa pubescens Turcz. subsp. microphylla（Diels）M. C. Chang et X. L. Chen ●

382305 Syringa tsinlingensana C. K. Schneid. = Syringa pubescens Turcz. subsp. microphylla（Diels）M. C. Chang et X. L. Chen ●

382306 Syringa velutina Bureau et Franch. = Syringa tomentella Bureau et Franch. ●

382307 Syringa velutina Kom. = Syringa pubescens Turcz. subsp. patula（Palib.）M. C. Chang et X. L. Chen ●

382308 Syringa venosa Nakai；凸脉丁香 ●☆

382309 Syringa venosa Nakai = Syringa pubescens Turcz. subsp. patula（Palib.）M. C. Chang et X. L. Chen ●

382310 Syringa venosa Nakai var. lactea（Nakai）Nakai = Syringa pubescens Turcz. subsp. patula（Palib.）M. C. Chang et X. L. Chen ●

382311 Syringa verrucosa C. K. Schneid. = Syringa pubescens Turcz. subsp. julianae（C. K. Schneid.）M. C. Chang et X. L. Chen ●

382312 Syringa villosa Vahl；红丁香；Hairy Lilac，Late Lilac，Villous Lilac ●

382313 Syringa villosa Vahl = Syringa pubescens Turcz. subsp. microphylla（Diels）M. C. Chang et X. L. Chen ●

382314 Syringa villosa Vahl = Syringa pubescens Turcz. subsp. patula（Palib.）M. C. Chang et X. L. Chen ●

382315 Syringa villosa Vahl = Syringa pubescens Turcz. ●

382316 Syringa villosa Vahl var. emodi（Wall. ex Royle）Rehder = Syringa emodi Wall. ex Royle ●

382317 Syringa villosa Vahl var. giraldi Spreng. = Syringa oblata Lindl. ex Carrière ●

382318 Syringa villosa Vahl var. glabra C. K. Schneid. = Syringa komarowii C. K. Schneid. ●

382319　Syringa villosa Vahl var. hirsuta C. K. Schneid. = Syringa wolfii C. K. Schneid. ●

382320　Syringa villosa Vahl var. lactea Nakai = Syringa pubescens Turcz. subsp. patula（Palib.）M. C. Chang et X. L. Chen ●

382321　Syringa villosa Vahl var. limprichtii Lingelsh. = Syringa villosa Vahl ●

382322　Syringa villosa Vahl var. ovalifolia DC. = Syringa pubescens Turcz. ●

382323　Syringa villosa Vahl var. pubescens Anomymous = Syringa pubescens Turcz. ●

382324　Syringa villosa Vahl var. rosea（Cornu）Schneid. = Syringa villosa Vahl ●

382325　Syringa villosa Vahl var. rosea Cornu ex Rehder = Syringa villosa Vahl ●

382326　Syringa villosa Vahl var. typica C. K. Schneid. = Syringa villosa Vahl ●

382327　Syringa villosa Vahl var. typica C. K. Schneid. f. glabra C. K. Schneid. = Syringa komarowii C. K. Schneid. ●

382328　Syringa villosa Vahl var. typica C. K. Schneid. f. subhirsuta C. K. Schneid. = Syringa villosa Vahl ●

382329　Syringa vulgaris L.;欧丁香（丁香花,欧洲丁香,西洋丁香,洋丁香,志金花,紫丁香,紫丁香花）;Ash,Blow-pipe Tree,Blue Ash,Blue Pipe,Common Lilac,Duck's Bill,French Lilac,Laylock,Lilac,Lily Oak,Lily-oak,May Lily,May-flower,May-lily,Oyster,Pipe Privit,Pipe Tree,Pipe-privit,Pipe-tree,Prince of Wales' Feathers,Prince's Feathers,Queen Flower,Queen's Feather,Queen's Feathers,Roman Willow,Soldier's Feathers,Spanish Ash,Spanishash,Tree Blow-pipe,Whitsuntide ●

382330　Syringa vulgaris L. 'Ambassadeur';大使欧丁香●☆

382331　Syringa vulgaris L. 'Ami Schott';艾米·斯科特欧丁香●☆

382332　Syringa vulgaris L. 'Andenken an Ludwig Spaeth';斯帕斯欧丁香●☆

382333　Syringa vulgaris L. 'Belle de Nancy';贝勒·德·南希欧丁香（丽人欧丁香）●☆

382334　Syringa vulgaris L. 'Charles X';查里十世欧丁香●☆

382335　Syringa vulgaris L. 'Charles Joly';查尔斯·朱莉欧丁香（乔里欧丁香）;Red Hybrid Lilac ●☆

382336　Syringa vulgaris L. 'Chunge';春阁欧丁香●

382337　Syringa vulgaris L. 'Congo';刚果欧丁香●☆

382338　Syringa vulgaris L. 'Decne. ';德凯逊欧丁香●☆

382339　Syringa vulgaris L. 'Edith Cavell';艾迪斯·卡维尔欧丁香●☆

382340　Syringa vulgaris L. 'Ellen Willmott';埃伦·维尔姆特欧丁香●☆

382341　Syringa vulgaris L. 'Jan van Tol';范托尔欧丁香●☆

382342　Syringa vulgaris L. 'Katherine Havemeyer';哈夫迈尔欧丁香●☆

382343　Syringa vulgaris L. 'Madame Abel Chatenay';艾贝尔·查特纳夫人欧丁香●☆

382344　Syringa vulgaris L. 'Madame Antoine Buchner';安托万·毕希纳夫人欧丁香（巴克勒欧丁香）●☆

382345　Syringa vulgaris L. 'Madame Charles Souchet';查尔斯·苏查特夫人欧丁香●☆

382346　Syringa vulgaris L. 'Madame F. Morel';莫里尔欧丁香●☆

382347　Syringa vulgaris L. 'Madame Florent Stepman';史蒂普曼欧丁香●☆

382348　Syringa vulgaris L. 'Madame Lemoine';莱蒙尼夫人欧丁香（勒蒙利欧丁香）●☆

382349　Syringa vulgaris L. 'Mare Chal Foch';玛勒查尔·福奇欧丁香（弗奇欧丁香）●☆

382350　Syringa vulgaris L. 'Massena';马森那欧丁香●☆

382351　Syringa vulgaris L. 'Maud Notcutt';莫德·诺特卡特欧丁香（那特卡特欧丁香）●☆

382352　Syringa vulgaris L. 'Michel Buchner';布克勒欧丁香●☆

382353　Syringa vulgaris L. 'Monge';蒙吉欧丁香●☆

382354　Syringa vulgaris L. 'Monique Lemoine';莫尼奇·莱蒙尼欧丁香●☆

382355　Syringa vulgaris L. 'Mrs Edward Harding';爱德华·哈定夫人欧丁香（哈丁欧丁香）●☆

382356　Syringa vulgaris L. 'Olivier de Serres';奥利维尔·德·塞雷斯欧丁香●☆

382357　Syringa vulgaris L. 'Paul Thirion';保罗·斯利奥欧丁香（舍里翁）●☆

382358　Syringa vulgaris L. 'President Grevy';格雷维总统欧丁香（格里维）●☆

382359　Syringa vulgaris L. 'Primrose';报春花欧丁香（淡黄）●☆

382360　Syringa vulgaris L. 'Reaumur';利姆尔欧丁香●☆

382361　Syringa vulgaris L. 'Sensation';知觉欧丁香;Picotee Lilac ●☆

382362　Syringa vulgaris L. 'Souvenir d' Alice Harding';艾利丝·哈丁的纪念品欧丁香●☆

382363　Syringa vulgaris L. 'Souvenir de Louis Spaeth';路易斯·斯帕锡的纪念品欧丁香●☆

382364　Syringa vulgaris L. 'Vestale';修女欧丁香●☆

382365　Syringa vulgaris L. 'Victor Lemoine';维克托·莱蒙尼欧丁香●☆

382366　Syringa vulgaris L. 'Volcan';火山欧丁香●☆

382367　Syringa vulgaris L. 'William Robinson';威廉·罗宾逊欧丁香●☆

382368　Syringa vulgaris L. f. alba（Weston）Voss;白花欧丁香（白丁香,白花丁香）;White-flower Common Lilac ●

382369　Syringa vulgaris L. f. albipleniflora S. D. Zhao;白花重瓣洋丁香●

382370　Syringa vulgaris L. f. coerulea（Weston）Schelle;蓝花欧丁香;Blue-flower Common Lilac ●

382371　Syringa vulgaris L. f. plena（Oudin）Rehder;重瓣欧丁香;Double Common Lilac,Doubleflower Common Lilac ●

382372　Syringa vulgaris L. f. purpurea（Weston）Schelle;紫花欧丁香;Purple-flower Common Lilac ●

382373　Syringa vulgaris L. var. alba Weston = Syringa vulgaris L. f. alba（Weston）Voss ●

382374　Syringa vulgaris L. var. coerulea Weston = Syringa vulgaris L. f. coerulea（Weston）Schelle ●

382375　Syringa vulgaris L. var. emodi（Wall. ex Royle）Jaub. = Syringa emodi Wall. ex Royle ●

382376　Syringa vulgaris L. var. oblata（Lindl.）Franch. = Syringa oblata Lindl. ex Carrière ●

382377　Syringa vulgaris L. var. oblata Franch. = Syringa oblata Lindl. ex Carrière ●

382378　Syringa vulgaris L. var. plena Oudin = Syringa vulgaris L. f. plena（Oudin）Rehder ●

382379　Syringa vulgaris L. var. purpurea Weston = Syringa vulgaris L. f. purpurea（Weston）Schelle ●

382380　Syringa vulgaris Lam. = Syringa vulgaris L. ●

382381　Syringa wardii W. W. Sm.;圆叶丁香 ●

382382　Syringa wilsonii C. K. Schneid. = Syringa tomentella Bureau et Franch. ●

382383　Syringa wolfii C. K. Schneid. ;辽东丁香;Liaodong Lilac, Wolf Lilac ●

382384　Syringa wolfii C. K. Schneid. = Syringa pubescens Turcz. subsp. patula（Palib.）M. C. Chang et X. L. Chen ●

382385　Syringa wolfii C. K. Schneid. var. hirsuta（C. K. Schneid.）Hatus. = Syringa wolfii C. K. Schneid. ●

382386　Syringa wulingensis Skvortsov et W. Wang；雾灵丁香；Wuling Lilac ●

382387　Syringa wulingensis Skvortsov et W. Wang = Syringa pubescens Turcz. ●

382388　Syringa yunnanensis Franch.；云南丁香（野桂花）；Yunnan Lilac ●

382389　Syringa yunnanensis Franch. f. pubicalyx（Z. P. Jien ex P. Y. Bai）M. C. Chang = Syringa yunnanensis Franch. ●

382390　Syringa yunnanensis Franch. f. pubicalyx（Z. P. Jien ex P. Y. Bai）M. C. Chang；毛萼云南丁香；Pubicalyx Lilac ●

382391　Syringa yunnanensis Franch. var. pubicalyx Z. P. Jien ex P. Y. Bai = Syringa yunnanensis Franch. ●

382392　Syringaceae Horan.；丁香科●

382393　Syringaceae Horan. = Oleaceae Hoffmanns. et Link（保留科名）●

382394　Syringantha Standl.（1930）；管花茜●☆

382395　Syringantha loranthoides Standl.；管花茜●☆

382396　Syringidium Lindau = Habracanthus Nees ●☆

382397　Syringidium Lindau = Kalbreyeracanthus Wassh. ●☆

382398　Syringodea D. Don（废弃属名）= Erica L. ●☆

382399　Syringodea D. Don（废弃属名）= Syringodea Hook. f.（保留属名）■☆

382400　Syringodea Hook. f.（1873）（保留属名）；管花鸢尾属■☆

382401　Syringodea bicolor Baker = Syringodea bifucata M. P. de Vos ■☆

382402　Syringodea bicolor Baker var. concolor = Syringodea concolor（Baker）M. P. de Vos ■☆

382403　Syringodea bifucata M. P. de Vos；二色管花鸢尾■☆

382404　Syringodea concolor（Baker）M. P. de Vos；同色管花鸢尾■☆

382405　Syringodea derustensis M. P. de Vos；德卢斯特管花鸢尾■☆

382406　Syringodea filifolia Baker = Syringodea longituba（Klatt）Kuntze ■☆

382407　Syringodea flanaganii Baker；弗拉纳根管花鸢尾■☆

382408　Syringodea latifolia Klatt = Hesperantha latifolia（Klatt）M. P. de Vos ■☆

382409　Syringodea leipoldtii L. Bolus = Syringodea longituba（Klatt）Kuntze ■☆

382410　Syringodea linifolia E. Phillips = Duthiastrum linifolium（E. Phillips）M. P. de Vos ●☆

382411　Syringodea longituba（Klatt）Kuntze；长管花鸢尾■☆

382412　Syringodea longituba（Klatt）Kuntze var. violacea M. P. de Vos；堇色管花鸢尾■☆

382413　Syringodea luteonigra Baker = Romulea macowanii Baker ■☆

382414　Syringodea montana Klatt = Syringodea longituba（Klatt）Kuntze ■☆

382415　Syringodea pulchella Hook. f.；美丽管花鸢尾■☆

382416　Syringodea saxatilis M. P. de Vos；岩石管花鸢尾■☆

382417　Syringodea unifolia Goldblatt；单花管花鸢尾■☆

382418　Syringodium Kuntze（1939）；针叶藻属；Manatee-grass, Needlealga ■

382419　Syringodium filiforme Kurtz；丝状针叶藻■☆

382420　Syringodium isoetifolium（Asch.）Dandy；针叶藻；Needlealga ■

382421　Syringodium isoetifolium（Asch.）Dandy = Cymodocea isoetifolia Asch. ■

382422　Syringosma Mart. ex Lindl. = Forsteronia G. Mey. ●☆

382423　Syringosma Mart. ex Rchb. = Forsteronia G. Mey. ●☆

382424　Syrium Steud. = Santalum L. ●

382425　Syrium Steud. = Sirium Schreb. ●

382426　Syrmatium Vogel = Hosackia Douglas ex Benth. ■☆

382427　Syrrheonema Miers（1867）；西非丝藤属●☆

382428　Syrrheonema boukokoense Tisser. = Syrrheonema welwitschii（Hiern）Diels ●☆

382429　Syrrheonema fasciculatum Miers；簇生西非丝藤●☆

382430　Syrrheonema hexastamineum Keay；六雄蕊西非丝藤●☆

382431　Syrrheonema welwitschii（Hiern）Diels；韦尔西非丝藤●☆

382432　Syrrhonema Miers = Syrrheonema Miers ●☆

382433　Sysepalum Post et Kuntze = Synsepalum（A. DC.）Daniell ●☆

382434　Sysimbrium Pall. = Sisimbrium L. ■

382435　Sysiphon Post et Kuntze = Colchicum L. ■

382436　Sysiphon Post et Kuntze = Synsiphon Regel ■

382437　Sysirinchium Raf. = Sisyrinchium L. ■

382438　Syspone Griseb. = Chamaespartium Adans. ●

382439　Syspone Griseb. = Genistella Moench ●

382440　Systellantha B. C. Stone（1992）；聚四花属●☆

382441　Systellantha brookeae B. C. Stone；聚四花●☆

382442　Systellantha fruticosa（B. C. Stone）B. C. Stone；灌木聚四花●☆

382443　Systeloglossum Schltr.（1923）；合舌兰属■☆

382444　Systeloglossum costaricense Schltr.；合舌兰■☆

382445　Systemon Post et Kuntze = Synostemon F. Muell. ●■

382446　Systemon Regel = Galipea Aubl. ●☆

382447　Systemonodaphne Mez（1889）；合蕊樟属●☆

382448　Systemonodaphne geminiflora Mez；合蕊樟●☆

382449　Systemonodaphne macrantha Kosterm.；大花合蕊樟●☆

382450　Systenotheca Reveal et Hardham = Chorizanthe R. Br. ex Benth. ■●☆

382451　Systenotheca Reveal et Hardham（1989）；齿苞蓼属；Vortriede's Spinyherb ■☆

382452　Systenotheca vortriedei（Brandegee）Reveal et Hardham；齿苞蓼■☆

382453　Systigma Post et Kuntze = Ilex L. ●

382454　Systigma Post et Kuntze = Synstima Raf. ●

382455　Systrepha Burch. = Ceropegia L. ■

382456　Systrepha filiforme Burch. = Ceropegia filiformis（Burch.）Schltr. ■☆

382457　Systrephia Benth. et Hook. f. = Systrepha Burch. ■

382458　Syziganthus Steud. = Gahnia J. R. Forst. et G. Forst. ■

382459　Syzistachyum Post et Kuntze = Heliotropium L. ●■

382460　Syzistachyum Post et Kuntze = Synzistachium Raf. ●■

382461　Syziganthera Post et Kuntze = Lacistema Sw. ●☆

382462　Syziganthera Post et Kuntze = Synzyganthera Ruiz et Pav. ●☆

382463　Syzigiopsis Ducke = Pouteria Aubl. ●

382464　Syzygium Gaertn. = Syzygium R. Br. ex Gaertn.（保留属名）●

382465　Syzygium R. Br. ex Gaertn.（1788）（保留属名）；蒲桃属（赤楠属）；Eugenia, Jambu, Syzygium ●

382466　Syzygium abidjanense Aubrév. et Pellegr. = Syzygium rowlandii Sprague ●☆

382467　Syzygium acuminatissimum（Blume）DC. = Acmena acuminatissima（Blume）Merr. et L. M. Perry ●

382468　Syzygium acutisepalum（Hayata）Mori；尖萼蒲桃；Acute-sepaled Syzygium ●

382469　Syzygium acutisepalum（Hayata）Mori = Syzygium formosanum（Hayata）Mori ●

382470　Syzygium album Q. F. Zheng；白果蒲桃●

382471　Syzygium angkolanum Miq. = Cleistocalyx operculatus（Roxb.）Merr. et L. M. Perry ●

382472　Syzygium angkolanum Miq. = Syzygium nervosum DC. ●

382473　Syzygium angustinii Merr. et L. M. Perry = Syzygium toddalioides（Wight）Walp. ●

382474　Syzygium aqueum（Burm. f.）Alston；西方水蒲桃；Water Apple，Water Rose Apple ●☆

382475　Syzygium aqueum Alston = Syzygium aqueum（Burm. f.）Alston ●☆

382476　Syzygium araiocladum Merr. et L. M. Perry；纤枝蒲桃（红三门，上思蒲桃，线枝蒲桃，小叶红营）；Fineshoot Syzygium，Fine-shooted Syzygium，Thinshoot Syzygium ●

382477　Syzygium aromaticum（L.）Merr. et L. M. Perry；丁香蒲桃（百里馨，雌丁香，大花，大花丁香，大叶丁香，丁香，丁香树，丁子，丁子香，公丁，公丁香，鸡舌香，母丁香，如宇香，瘦香娇，亭炅独生，香蒲桃，小叶丁香，雄丁香，支解香，紫丁香）；Aromatic Syzygium，Clove，Clove Tree，Cloves，Clove-tree，Zanzibar Redheads ●

382478　Syzygium aromaticum（L.）Merr. et L. M. Perry = Eugenia caryophyllata Thunb. ●

382479　Syzygium augustinii Merr. et L. M. Perry = Syzygium toddalioides（Wight）Walp. ●

382480　Syzygium aurantiacum（H. Perrier）Labat et G. E. Schatz；橘色蒲桃●☆

382481　Syzygium australe（H. Wendl. ex Link）B. Hyland；澳洲蒲桃（澳洲圆锥花番樱桃，圆锥花番樱桃）；Australian Brush-cherry，Australian Brush-cherry Eugenia，Brush Cherry，Brush-chery，Brychchery Eugenia，Lillypilly，Magenta Cherry ●☆

382482　Syzygium australe（Link）B. Hyland = Syzygium australe（H. Wendl. ex Link）B. Hyland ●☆

382483　Syzygium austrosinense（Merr. et L. M. Perry）Hung T. Chang et R. H. Miao；华南蒲桃（南方蒲桃，小山稔）；S. China Syzygium，South China Syzygium ●

382484　Syzygium austroyunnanense Hung T. Chang et R. H. Miao；滇南蒲桃（八家妙）；S. Yunnan Syzygium，South Yunnan Syzygium ●

382485　Syzygium balsameum Wall.；香胶蒲桃；Balsam Syzygium ●

382486　Syzygium baronii Labat et G. E. Schatz；巴隆蒲桃●☆

382487　Syzygium baviense（Gagnep.）Merr. et L. M. Perry；短棒蒲桃；Bavi Syzygium，Shortstalk Syzygium ●

382488　Syzygium benguellense（Welw. ex Hiern）Engl. ex Engl. et Gilg；本格拉蒲桃●☆

382489　Syzygium bernieri（Drake）Labat et G. E. Schatz；伯尼尔蒲桃●☆

382490　Syzygium boissianum（Gagnep.）Merr. et L. M. Perry；无柄蒲桃（保亭蒲桃）；Bois Syzygium，Sessile Syzygium ●

382491　Syzygium brachyantherum Merr. et L. M. Perry；短药蒲桃（麻里果，砂糖树果，野冬青果）；Shortanther Syzygium，Short-anthered Syzygium ●

382492　Syzygium brachyantherum Merr. et L. M. Perry = Syzygium globiflorum（Craib）Chantar. et J. Parn. ●

382493　Syzygium brachythyrsum Merr. et L. M. Perry；短序蒲桃（野冬青果）；Shortspike Syzygium，Short-spiked Syzygium ●

382494　Syzygium bracteatum Korth. = Syzygium zeylanicum（L.）DC. ●

382495　Syzygium brazzavillense Aubrév. et Pellegr.；布拉柴维尔蒲桃●☆

382496　Syzygium bubengense C. Chen；补崩蒲桃●

382497　Syzygium bullockii（Hance）Merr. et L. M. Perry；黑嘴蒲桃（黑嘴树，碎叶子）；Bullock Syzygium ●

382498　Syzygium buxifolioideum Hung T. Chang et R. H. Miao；假赤楠；Boxifoliateoid Syzygium，Fake Boxleaf Syzygium，False Boxleaf Syzygium ●

382499　Syzygium buxifolium Hook. et Arn.；赤楠（赤兰，赤楠蒲桃，瓜子柴，瓜子木，假黄杨，金牛子，轮叶赤楠，牛金子，千年树，山石榴，山乌珠，石枔，细叶紫陵树，细子莲，小号犁头树，小叶赤楠，小叶蒲桃，鱼鳞木）；Boxleaf Eugenia，Boxleaf Syzygium，Box-leaved Eugenia，Box-leaved Syzygium ●

382500　Syzygium buxifolium Hook. et Arn. var. austrosinense Merr. et L. M. Perry = Syzygium austrosinense（Merr. et L. M. Perry）Hung T. Chang et R. H. Miao ●

382501　Syzygium buxifolium Hook. et Arn. var. cleyerifolium（Yatabe）Tateishi = Syzygium cleyerifolium（Yatabe）Makino ●☆

382502　Syzygium campicola Mildbr. = Syzygium guineense（Willd.）DC. subsp. macrocarpum（Engl.）F. White ●☆

382503　Syzygium caryophyllifolium DC.；石竹叶蒲桃●

382504　Syzygium cathayense Merr. et L. M. Perry；华夏蒲桃；China Syzygium，Chinese Syzygium ●

382505　Syzygium cerasoides（Roxb.）Raizada = Syzygium nervosum DC. ●

382506　Syzygium championii（Benth.）Merr. et L. M. Perry；子凌蒲桃（子凌树）；Champion Syzygium ●

382507　Syzygium chunianum Merr. et L. M. Perry；密脉蒲桃；Chun Syzygium ●

382508　Syzygium cinereum Wall.；钝叶蒲桃；Ash-coloured Syzygium，Bluntleaf Syzygium ●

382509　Syzygium claviflorum（Roxb.）A. M. Cowan et Cowan var. oblongifolium（Hayata）Mori = Syzygium taiwanicum Hung T. Chang et R. H. Miao ●

382510　Syzygium claviflorum（Roxb.）Wall. = Syzygium claviflorum（Roxb.）Wall. ex A. M. Cowan et Cowan ●

382511　Syzygium claviflorum（Roxb.）Wall. = Syzygium taiwanicum Hung T. Chang et R. H. Miao ●

382512　Syzygium claviflorum（Roxb.）Wall. ex A. M. Cowan et Cowan；棒花蒲桃（棒花赤楠）；Clavate-flower Syzygium，Club-flowered Syzygium，Stickleaf Syzygium ●

382513　Syzygium claviflorum（Roxb.）Wall. ex A. M. Cowan et Cowan var. oblongifolium（Hayata）Mori = Syzygium taiwanicum Hung T. Chang et R. H. Miao ●

382514　Syzygium cleyerifolium（Yatabe）Makino；杨桐蒲桃●☆

382515　Syzygium condensatum（Baker）Labat et G. E. Schatz；密集蒲桃■☆

382516　Syzygium congestiflorum Hung T. Chang et R. H. Miao；团花蒲桃；Congested-flowered Syzygium，Globose-flower Syzygium，Groupflower Syzygium ●

382517　Syzygium congolense Vermoesen；刚果蒲桃■☆

382518　Syzygium conspersipunctatum（Merr. et L. M. Perry）Craven et Biffin；散点蒲桃●

382519　Syzygium cordatum Hochst. ex C. Krauss；心叶蒲桃；Water Berry ■☆

382520　Syzygium cordatum Hochst. ex C. Krauss subsp. shimbaense Verdc.；欣巴蒲桃●

382521　Syzygium cordatum Hochst. ex C. Krauss var. gracile Amshoff；细心叶蒲桃■☆

382522　Syzygium cumini（L.）Skeels；乌墨（海南蒲桃，黑墨树，假紫荆，董宝莲，肯氏蒲桃，密脉蒲桃，乌贯木，乌口树，乌楣，乌墨蒲桃，乌木，野冬青果，印度蒲桃）；Duhat，Jaman，Jambolan，Jambolanplum，Jambolan-plum，Jambul，Java Plum，Java-plum，Jembolan Plum，Roseapple，Rose-apple ●

382523　Syzygium cumini（L.）Skeels var. caryophyllifolium（Duthie）R. R. Stewart；阔叶董宝莲●

382524　Syzygium cumini（L.）Skeels var. tsoi（Merr. et Chun）Hung T. Chang et R. H. Miao；长萼乌墨；Long-calyx Duhat ●

382525　Syzygium cuspidatoobovatum（Hayata）Mori ＝ Acmena acuminatissima（Blume）Merr. et L. M. Perry ●

382526　Syzygium cuspidatoovatum（Hayata）Mori ＝ Acmena acuminatissima（Blume）Merr. et L. M. Perry

382527　Syzygium cymiferum E. Mey. ex C. Presl ＝ Syzygium cordatum Hochst. ex C. Krauss ■☆

382528　Syzygium danguyanum（H. Perrier）Labat et G. E. Schatz；当吉蒲桃 ●☆

382529　Syzygium densinervium（Merr.）Merr.；菲律宾蒲桃 ●☆

382530　Syzygium densinervium（Merr.）Merr. var. insulare Hung T. Chang；岛生蒲桃（密脉赤楠）；Dense Nerve Eugenia ●

382531　Syzygium emirnense（Baker）Labat et G. E. Schatz；埃米蒲桃 ●☆

382532　Syzygium euonymifolium（F. P. Metcalf）Merr. et L. M. Perry；卫矛叶蒲桃；Euonymusleaf Syzygium, Euonymus-leaved Syzygium, Evonymusleaf Syzygium ●

382533　Syzygium euphlebium（Hayata）Mori；细叶蒲桃（细脉赤楠,细脉蒲桃）；Beautiful Nerve Eugenia, Beautiful-nerved Syzygium, Thinvein Syzygium ●

382534　Syzygium fluviatile（Hemsl.）Merr. et L. M. Perry；水竹蒲桃（方竹蒲桃,柳叶蒲桃,水边蒲桃）；Fluvial Syzygium, Waterbamboo Syzygium ●

382535　Syzygium formosanum（Hayata）Mori；台湾蒲桃（赤兰,大号犁头树,番樱桃,尖萼蒲桃,台湾赤楠）；Acutesepal Syzygium, Sharpcalyx Syzygium, Taiwan Eugenia, Taiwan Syzygium ●

382536　Syzygium forrestii Merr. et L. M. Perry；滇边蒲桃；Forrest Syzygium ●

382537　Syzygium francisii（F. M. Bailey）L. A. S. Johnson；弗朗西斯蒲桃；Francis' Water-gum, Giant Water Gum ●☆

382538　Syzygium fruticosum DC.；簇花蒲桃（鼻虫窝树,灌木蒲桃,黑果树,黑叶蒲桃,水榕树）；Clusterflower Syzygium, Shrubby Syzygium ●

382539　Syzygium garcinioides（Ridl.）Merr. et L. M. Perry；拟加尔桑蒲桃 ●☆

382540　Syzygium germainii Amshoff；吉曼蒲桃 ●☆

382541　Syzygium gerrardii（Harv.）Burtt Davy；水蒲桃；Waterpear ●

382542　Syzygium gerrardii Burtt Davy ＝ Syzygium gerrardii（Harv.）Burtt Davy ●

382543　Syzygium gilletii De Wild.；吉勒特蒲桃 ●☆

382544　Syzygium globiflorum（Craib）Chantar. et J. Parn.；球花蒲桃（短药蒲桃）●

382545　Syzygium gongshanense P. Y. Bai；贡山蒲桃；Gongshan Syzygium ●

382546　Syzygium gracilentum（Hance）Hu ＝ Decaspermum gracilentum（Hance）Merr. et L. M. Perry ●

382547　Syzygium gracilentum Hu ＝ Decaspermum gracilentum（Hance）Merr. et L. M. Perry ●

382548　Syzygium grijsii（Hance）Merr. et L. M. Perry；轮叶蒲桃（赤兰,枸铃子,构柃子,三叶赤楠,山乌珠,紫藤子）；Three-leaved Syzygium, Whorled-leaved Syzygium, Whorlleaf Syzygium ●

382549　Syzygium guangxiensis Hung T. Chang et R. H. Miao；广西蒲桃；Guangxi Syzygium, Kwangsi Syzygium ●

382550　Syzygium guineense（Willd.）DC. subsp. afromontanum F. White；非洲山生几内亚蒲桃 ●☆

382551　Syzygium guineense（Willd.）DC. subsp. bamendae F. White；巴门达蒲桃 ●☆

382552　Syzygium guineense（Willd.）DC. subsp. barotsense F. White；巴罗策蒲桃 ●☆

382553　Syzygium guineense（Willd.）DC. subsp. gerrardii（Harv.）F. White ＝ Syzygium gerrardii（Harv.）Burtt Davy ●

382554　Syzygium guineense（Willd.）DC. subsp. huillense（Hiern）F. White；威拉蒲桃 ●☆

382555　Syzygium guineense（Willd.）DC. subsp. littorale（Keay）Boutique ＝ Syzygium guineense（Willd.）DC. ●☆

382556　Syzygium guineense（Willd.）DC. subsp. macrocarpum（Engl.）F. White；大果几内亚蒲桃 ●☆

382557　Syzygium guineense（Willd.）DC. subsp. obovatum F. White；倒卵几内亚蒲桃 ●☆

382558　Syzygium guineense（Willd.）DC. subsp. occidentale F. White；西部几内亚蒲桃 ●☆

382559　Syzygium guineense（Willd.）DC. subsp. parvifolium（Engl.）F. White；小花几内亚蒲桃 ●☆

382560　Syzygium guineense（Willd.）DC. subsp. urophyllum（Welw. ex Hiern）Amshoff；尾叶几内亚蒲桃 ●☆

382561　Syzygium guineense（Willd.）DC. var. huillensis Hiern ＝ Syzygium guineense（Willd.）DC. subsp. huillense（Hiern）F. White ●☆

382562　Syzygium guineense（Willd.）DC. var. littorale Keay ＝ Syzygium guineense（Willd.）DC. ●☆

382563　Syzygium guineense（Willd.）DC. var. macrocarpum Engl. ＝ Syzygium guineense（Willd.）DC. subsp. macrocarpum（Engl.）F. White ●☆

382564　Syzygium guineense（Willd.）DC. var. parvifolium Engl. ＝ Syzygium guineense（Willd.）DC. subsp. parvifolium（Engl.）F. White ●☆

382565　Syzygium guineense（Willd.）DC. var. staudtii Engl. ＝ Syzygium staudtii（Engl.）Mildbr. ●☆

382566　Syzygium guineense（Willd.）DC. var. sudanicum（A. Chev.）Roberty；苏丹蒲桃 ●☆

382567　Syzygium guineense DC.；几内亚蒲桃；Guinea Syzygium, Guinean Syzygium, Water Berry ●☆

382568　Syzygium guineensis（Willd.）DC. var. palustre Aubrév. ＝ Syzygium owariense（P. Beauv.）Benth. ●☆

382569　Syzygium hainanense Hung T. Chang et R. H. Miao；海南蒲桃（山蒲桃,乌墨,乌木）；Hainan Syzygium ●

382570　Syzygium hancei Merr. et L. M. Perry；红鳞蒲桃（白车木,车辕木,红车木,红鳞树,磨推树,蘑堆树,小花蒲桃）；Hance Syzygium ●

382571　Syzygium handelii Merr. et L. M. Perry；贵州蒲桃；Handel Syzygium ●

382572　Syzygium howii Merr. et L. M. Perry；万宁蒲桃；How Syzygium, Wanning Syzygium ●

382573　Syzygium huillense（Hiern）Engl. ＝ Syzygium guineense（Willd.）DC. subsp. huillense（Hiern）F. White ●☆

382574　Syzygium imitans Merr. et L. M. Perry；桂南蒲桃；Guinan Syzygium, South Guangxi Syzygium ●

382575　Syzygium impressum N. H. Xia；香港蒲桃 ●

382576　Syzygium infra-rubiginosum Hung T. Chang et R. H. Miao；褐背蒲桃；Brown-back Syzygium, Brown-backed Syzygium ●

382577　Syzygium intermediiforme Hung T. Chang et R. H. Miao；中间型蒲桃；Intermediate Syzygium ●

382578　Syzygium intermedium Engl. et Brehmer；间型蒲桃 ●☆

382579　Syzygium jambolanum（Lam.）DC.；菫宝莲 ●

382580　Syzygium jambolanum（Lam.）DC. ＝ Syzygium cumini（L.）

Skeels ●

382581 Syzygium jambolanum DC. = Syzygium cumini（L.）Skeels ●

382582 Syzygium jambos（L.）Alston；蒲桃（风鼓，南无，撑布，攘波，水葡桃，水蒲桃，水石榴，香果，檐木）；Jambos，Jambu，Malabar Plum，Malay Apple，Pomarrosa，Rose Apple，Roseapple，Rose-apple ●

382583 Syzygium jambos（L.）Alston var. linearilimbum Hung T. Chang et R. H. Miao；线叶蒲桃（带叶蒲桃）；Linearleaf Syzygium ●

382584 Syzygium jambos（L.）Alston var. sylvaticum（Gagnep.）Merr. et L. M. Perry = Syzygium jambos（L.）Alston ●

382585 Syzygium jambos（L.）Alston var. tripinnatum（Blanco）C. Chen；大花赤楠；Tripinnate Eugenia ●

382586 Syzygium jienfunicum Hung T. Chang et R. H. Miao；尖峰蒲桃；Jianfeng Syzygium，Jienfun Syzygium ●

382587 Syzygium kashotoense（Hayata）Mori = Syzygium paucivenium（C. B. Rob.）Merr. ●

382588 Syzygium kerstingii Engl.；克斯廷蒲桃●☆

382589 Syzygium kusukusuense（Hayata）Mori；恒春蒲桃（高士佛赤楠）；Hengchun Eugenia，Hengchun Syzygium ●

382590 Syzygium kwangsiense Hung T. Chang et R. H. Miao = Syzygium guangxiensis Hung T. Chang et R. H. Miao ●

382591 Syzygium kwangtungense（Merr.）Merr. et L. M. Perry；广东蒲桃；Guangdong Syzygium，Kwangtung Syzygium ●

382592 Syzygium lanyuense E. C. Chang = Syzygium simile（Merr.）Merr. ●

382593 Syzygium laosense（Gagnep.）Merr. et L. M. Perry；老挝蒲桃（野苹果）●

382594 Syzygium laosense（Gagnep.）Merr. et L. M. Perry var. quocense（Gagnep.）Hung T. Chang et R. H. Miao；少花老挝蒲桃（老挝蒲桃，野苹果）；Laos Syzygium ●

382595 Syzygium lasianthifolium Hung T. Chang et R. H. Miao；粗叶蒲桃（粗叶木蒲桃）；Lasianthus-leaf Syzygium，Lasianthus-leaved Syzygium，Roughleaf Syzygium ●

382596 Syzygium latilimbum（Merr.）Merr. et L. M. Perry = Syzygium megacarpum（Craib）Rathakrishnan et N. C. Nair ●

382597 Syzygium leptanthum（Wight）Nied.；纤花蒲桃；Fineflower Syzygium，Fine-flowered Syzygium ●

382598 Syzygium leptanthum（Wight）Nied. = Syzygium claviflorum（Roxb.）Wall. ex A. M. Cowan et Cowan ●

382599 Syzygium levinei（Merr.）Merr. et L. M. Perry；山蒲桃（白车，红营，红营蒲桃，万宁蒲桃）；Levine Syzygium，Wild Syzygium ●

382600 Syzygium linearilimbum C. Y. Wu = Syzygium jambos（L.）Alston var. linearilimbum Hung T. Chang et R. H. Miao ●

382601 Syzygium lineatum（DC.）Merr. et L. M. Perry；长花蒲桃；Guava Berry，Lineate Syzygium，Long-flower Syzygium ●

382602 Syzygium littorale Aubrév. = Syzygium guineense（Willd.）DC. ●☆

382603 Syzygium longiflorum C. Presl = Syzygium lineatum（DC.）Merr. et L. M. Perry ●

382604 Syzygium luehmannii（F. Muell.）L. A. S. Johnson；小叶蒲桃；Brush Cherry，Riberry，Small-leafed Lillypilly ●☆

382605 Syzygium lugubre（H. Perrier）Labat et G. E. Schatz；暗淡蒲桃●☆

382606 Syzygium malaccense（L.）Merr. et L. M. Perry；马六甲蒲桃（莲雾，麻六甲蒲桃，马来番樱桃，马来蒲桃，辇雾）；Jambos，Large-fruited Rose-apple，Malay Apple，Malayapple，Malay-apple，Malaysian Apple，Malaysian Eugenia，Mountain Apple，Mountain-apple，Ohia，Ohla，Otaheite Apple，Pomarack，Pomerac，Rose Apple ●

382607 Syzygium marounzense Pellegr. = Syzygium staudtii（Engl.）

Mildbr. ●☆

382608 Syzygium masukuense（Baker）R. E. Fr.；马苏克蒲桃●☆

382609 Syzygium masukuense（Baker）R. E. Fr. subsp. pachyphyllum F. White；厚叶蒲桃●☆

382610 Syzygium megacarpum（Craib）Rathakrishnan et N. C. Nair；阔叶蒲桃（大叶蒲桃，水梅）；Broadleaf Syzygium ●

382611 Syzygium melanophyllum Hung T. Chang et R. H. Miao；黑长叶蒲桃；Black-leaved Syzygium，Black-long-leaf Syzygium ●

382612 Syzygium micklethwaitii Verdc. var. subcordatum Verdc.；近心形蒲桃●☆

382613 Syzygium microphyllum Gamble = Syzygium buxifolium Hook. et Arn. ●

382614 Syzygium micropodum（Baker）Labat et G. E. Schatz；小梗蒲桃●☆

382615 Syzygium montanum Aubrév. = Syzygium staudtii（Engl.）Mildbr. ●☆

382616 Syzygium moorei（F. Muell.）L. A. S. Johnson；莫尔蒲桃；Coolamon，Durobby，Rose Apple，Rose-apple ●☆

382617 Syzygium mumbwaense Greenway = Syzygium guineense（Willd.）DC. subsp. huillense（Hiern）F. White ●☆

382618 Syzygium muriculatum Hung T. Chang et R. H. Miao；瘤突蒲桃；Muriculate Syzygium ●

382619 Syzygium myrsinifolium（Hance）Merr. et L. M. Perry；竹叶蒲桃（黄杨叶番樱桃）；Bambooleaf Syzygium，Boxleaf Eugenia，Myrsine-leaf Syzygium，Myrsine-leaved Syzygium ●

382620 Syzygium myrsinifolium（Hance）Merr. et L. M. Perry var. grandiflorum Hung T. Chang et R. H. Miao；大花竹叶蒲桃；Bigflower Myrsine-leaf Syzygium ●

382621 Syzygium myrtifolium Miq. = Syzygium zeylanicum（L.）DC. ●

382622 Syzygium nanpingense Y. Y. Qian；南屏蒲桃；Nanping Syzygium ●

382623 Syzygium nervosum DC.；水翁蒲桃●

382624 Syzygium nervosum DC. = Cleistocalyx operculatus（Roxb.）Merr. et L. M. Perry ●

382625 Syzygium nienkui Merr. et L. M. Perry = Syzygium tetragonum Wall. ex Wight ●

382626 Syzygium nigrans（Gagnep.）Hung T. Chang et R. H. Miao；变黑蒲桃；Blacken Syzygium ●

382627 Syzygium nodosum Miq. = Cleistocalyx operculatus（Roxb.）Merr. et L. M. Perry ●

382628 Syzygium nodosum Miq. = Syzygium nervosum DC. ●

382629 Syzygium oblancilimbum Hung T. Chang et R. H. Miao；倒披针叶蒲桃；Oblanceleaf Syzygium，Oblanceolate Syzygium，Oblanceolateleaf Syzygium ●

382630 Syzygium oblatum（Roxb.）Wall. ex A. M. Cowan et Cowan = Syzygium oblatum（Roxb.）Wall. ex Steud. ●

382631 Syzygium oblatum（Roxb.）Wall. ex Steud.；高檐蒲桃；Oblate Syzygium ●

382632 Syzygium obudai Mori = Syzygium tripinnatum（Blanco）Merr. ●

382633 Syzygium odoratum（Lour.）DC.；香蒲桃（香花蒲桃）；Fragrant Syzygium ●

382634 Syzygium okudae Mori = Syzygium jambos（L.）Alston var. tripinnatum（Blanco）C. Chen ●

382635 Syzygium okudae Mori = Syzygium jambos（L.）Alston ●

382636 Syzygium oleosum（F. Muell.）B. Hyland；澳洲蓝蒲桃；Blue Lillypilly ●☆

382637 Syzygium oleosum（F. Muell.）B. Hyland = Eugenia oleosa F. Muell. ●☆

382638 Syzygium onivense（H. Perrier）Labat et G. E. Schatz；乌尼韦

蒲桃●☆

382639 Syzygium operculatum （ Roxb. ） Nied. = Cleistocalyx operculatus （ Roxb. ） Merr. et L. M. Perry ●

382640 Syzygium operculatum （ Roxb. ） Nied. = Syzygium nervosum DC. ●

382641 Syzygium operculatum Nied. = Cleistocalyx operculatus （ Roxb. ） Merr. et L. M. Perry ●

382642 Syzygium owariense （ P. Beauv. ） Benth. ；尾张蒲桃（奥瓦蒲桃）●☆

382643 Syzygium owariense Benth. = Syzygium owariense （ P. Beauv. ） Benth. ●☆

382644 Syzygium paniculatum Gaertn. ；紫红蒲桃（圆锥蒲桃）；Brush Cherry，Magenta Brush Cherry，Magenta Cherry ●☆

382645 Syzygium paniculatum Gaertn. = Syzygium australe （ H. Wendl. ex Link ） B. Hyland ●☆

382646 Syzygium parkeri （ Baker ） Labat et G. E. Schatz；帕克蒲桃●☆

382647 Syzygium parvifolium （ Engl. ） Mildbr. = Syzygium guineense （ Willd. ） DC. subsp. parvifolium （ Engl. ） F. White ●☆

382648 Syzygium parvulum Mildbr. ex Amshoff；较小赤楠●☆

382649 Syzygium paucivenium （ C. B. Rob. ） Merr. ；疏脉赤楠（绿岛赤楠，圆顶蒲桃）；Kashot Syzygium, Lutao Eugenia, Roundtop Syzygium ●

382650 Syzygium phillyreifolium （ Baker ） Labat et G. E. Schatz；欧女贞叶蒲桃●☆

382651 Syzygium poluanthum （ Wright ） Masam. ；多花蒲桃；Many-flower Syzygium，Multiflorous Syzygium ●

382652 Syzygium polypetaloideum Merr. et L. M. Perry；假多瓣蒲桃；Manypetal-like Syzygium，Polypetaloid Syzygium ●

382653 Syzygium pondoense Engl. ；庞多蒲桃●☆

382654 Syzygium rehderianum Merr. et L. M. Perry；红枝蒲桃；Rehder Syzygium ●

382655 Syzygium rockii Merr. et L. M. Perry；滇西蒲桃；Rock Syzygium ●

382656 Syzygium rowlandii Sprague；罗兰蒲桃●☆

382657 Syzygium rysopodum Merr. et L. M. Perry；皱萼蒲桃；Ruogh Syzygium，Wrinklecalyx Syzygium，Wrinkled Syzygium ●

382658 Syzygium sakalavarum （ H. Perrier ） Labat et G. E. Schatz；萨卡拉瓦蒲桃●☆

382659 Syzygium salwinense Merr. et L. M. Perry；怒江蒲桃；Salwin Syzygium ●

382660 Syzygium samarangense （ Blume ） Merr. et L. M. Perry；洋蒲桃（金山蒲桃，莲雾，南洋蒲桃，爪哇番樱桃，爪哇蒲桃）；Curacao Apple，Jambosa，Java Apple，Java Eugenia，Makopa，Mankil Apple，Mankil Rose-apple，Pini Jambu，Rose Apple，Samarang Rose-apple，Samaranga Syzygium，Sang Rose Apple，Wax Jambu，Wax-apple ●

382661 Syzygium sambiranense （ H. Perrier ） Labat et G. E. Schatz = Syzygium samarangense （ Blume ） Merr. et L. M. Perry ●

382662 Syzygium saxatile Hung T. Chang et R. H. Miao；石生蒲桃；Rocky Syzygium，Saxatile Syzygium ●

382663 Syzygium seylanicum （ L. ） DC. = Syzygium zeylanicum （ L. ） DC. ●

382664 Syzygium simile （ Merr. ） Merr. ；兰屿赤楠；Lanyu Eugenia ●

382665 Syzygium somae （ Hayata ） Mori = Syzygium buxifolium Hook. et Arn. ●

382666 Syzygium staudtii （ Engl. ） Mildbr. ；施陶蒲桃●☆

382667 Syzygium stenocladum Merr. et L. M. Perry；细枝蒲桃；Thin-branch Syzygium，Thin-branched Syzygium ●

382668 Syzygium sterrophyllum Merr. et L. M. Perry；硬叶蒲桃；

Rigidleaf Syzygium，Rigid-leaved Syzygium ●

382669 Syzygium subdecurrens Miq. = Acmena acuminatissima （ Blume ） Merr. et L. M. Perry ●

382670 Syzygium szechuanense Hung T. Chang et R. H. Miao；四川蒲桃；Sichuan Syzygium ●

382671 Syzygium szemaoense Merr. et L. M. Perry；思茅蒲桃（赤兰,赤兰营）；Simao Syzygium，Szemao Syzygium ●

382672 Syzygium taiwanicum Hung T. Chang et R. H. Miao；台湾棒花蒲桃（棒萼赤楠，兰屿赤楠）；Club-shaped Flower Eugenia，Lanyu Eugenia，Taiwan Stickflower Syzygium，Taiwan Syzygium ●

382673 Syzygium tenuirhachis Hung T. Chang et R. H. Miao；细轴蒲桃；Thinaxis Syzygium，Thin-rachis Syzygium，Thin-rachised Syzygium ●

382674 Syzygium tephrodes （ Hance ） Merr. et L. M. Perry；方枝蒲桃；Squarebranch Syzygium，Square-branched Syzygium ●

382675 Syzygium tetragonum Wall. ex Wight；四角蒲桃（大树果，棱翅蒲桃，时令树果）；Fourangular Syzygium，Four-angular Syzygium ●

382676 Syzygium thumra （ Roxb. ） Merr. et L. M. Perry；黑叶蒲桃；Black-leaf Syzygium，Black-leaved Syzygium ●

382677 Syzygium toddalioides （ Wight ） Walp. ；假乌墨；Angustin Syzygium，Angustine Syzygium，False Duhat ●

382678 Syzygium tripinnatum （ Blanco ） Merr. = Syzygium jambos （ L. ） Alston var. tripinnatum （ Blanco ） C. Chen ●

382679 Syzygium tsoongii （ Merr. ） Merr. et L. M. Perry；狭叶蒲桃；Tsoong Syzygium ●

382680 Syzygium verrucatoglandurosum K. M. Feng；腺瘤蒲桃；Verrucato-glandurose Syzygium ●

382681 Syzygium vestitum Merr. et L. M. Perry；毛脉蒲桃；Hairy-nerved Syzygium，Hairy-vein Syzygium ●

382682 Syzygium wenshanense Hung T. Chang et R. H. Miao；文山蒲桃；Wenshan Syzygium ●

382683 Syzygium wilsonii （ F. Muell. ） B. Hyland；威尔逊蒲桃；Powderpuff Lillypilly ●☆

382684 Syzygium xizangense Hung T. Chang et R. H. Miao；西藏蒲桃；Xizang Syzygium ●

382685 Syzygium yunnanense Merr. et L. M. Perry；云南蒲桃；Yunnan Syzygium ●

382686 Syzygium zeylanicum （ L. ） DC. ；锡兰蒲桃（两广蒲桃）；Ceylon Syzygium ●

382687 Szechenyia Kanitz = Gagea Salisb. ■

382688 Szechenyia lloydioides Kanitz = Gagea pauciflora Turcz. ■

382689 Szegleewia C. Muell. = Sczegleewia Turcz. ●☆

382690 Szegleewia C. Muell. = Symphorema Roxb. ●

382691 Szlachetkoella Mytnik = Polystachya Hook. （保留属名）■

382692 Szlachetkoella Mytnik（2007）；热非多穗兰属■☆

382693 Szovitsia （ Fisch. et C. A. Mey. ） Drude = Aphanopleura Boiss. ■

382694 Szovitsia Fisch. et C. A. Mey. （1835）；绍维草属■☆

382695 Szovitsia Fisch. et C. A. Mey. = Aphanopleura Boiss. ■

382696 Szovitsia callicarpa Fisch. et C. A. Mey. ；绍维草■☆

382697 Szowitsia Fisch. et C. A. Mey. = Szovitsia Fisch. et C. A. Mey. ■☆

382698 Tabacina Rchb. = Tabacum Gilib. ●■

382699 Tabacum （ Gilib. ） Opiz = Nicotiana L. ●■

382700 Tabacum Gilib. = Nicotiana L. ●■

382701 Tabacus Moench = Nicotiana L. ●■

382702 Tabacus Moench = Tabacum Gilib. ●■

382703 Tabascina Baill. = Justicia L. ●■

382704 Tabebuia Gomes ex DC. （1838）；黄钟木属（风铃木属，喇叭木属，皮蚁木属，蚁木属，钟花树属）；Lapacho, Lapachol, Trumpet

Tree ●☆

382705　Tabebuia Gomez = Tabebuia Gomes ex DC. ●☆

382706　Tabebuia argentea Britton；银叶喇叭木（钟花树）；Lapacho, Paraguayan Siver Trumpet Tree, Siver Trumpet Tree, Tree of Gold ●☆

382707　Tabebuia aurea Benth. et Hook. f. ex S. Moore；加勒比钟花树；Caribbean Trumpet Tree, Caribbean Trumpet-tree ●☆

382708　Tabebuia avellanedae Lorentz ex Griseb.；褐色钟花树（南美蚁木）；New World Trumpet Tree ●☆

382709　Tabebuia avellanedae Lorentz ex Griseb. = Tabebuia impetiginosa（Mart. ex DC.）Standl. ●☆

382710　Tabebuia cassinoides DC.；钟花树●☆

382711　Tabebuia chrysantha（Jacq.）G. Nicholson；黄钟木（黄花风铃木,黄花喇叭木）；Golden Trumpet Tree, Roble Amarillo ●☆

382712　Tabebuia chrysotricha（Mart. ex DC.）Standl.；红果喇叭木（毛黄钟花）；Golden Trumpet Tree ●☆

382713　Tabebuia donnell-smithii Rose = Cybistax donnell-smithii（Rose）Seibert ●☆

382714　Tabebuia guayacan Hemsl.；中美洲蚁木●☆

382715　Tabebuia heterophylla（DC.）Britton；粉红喇叭木；Pink Trumpet Tree ●☆

382716　Tabebuia heterophylla Britton = Tabebuia heterophylla（DC.）Britton ●☆

382717　Tabebuia ilicifolia（Pers.）A. DC. = Phyllarthron ilicifolium（Pers.）H. Perrier ●☆

382718　Tabebuia impetiginosa（Mart. ex DC.）Standl.；斑疹钟花树（斑绣球,风铃木,光皮喇叭木）；Lapacho, Lapachol, Pink Trumpet Tree ●☆

382719　Tabebuia insignis（Miq.）Sandwith；显著钟花树●☆

382720　Tabebuia pallida（Lindl.）Miers；古巴喇叭木（灰白蚁木）；Cuba Pink Tree, Cuba Pink Trumpet Tree ●☆

382721　Tabebuia pallida（Lindl.）Miers = Tabebuia heterophylla（DC.）Britton ●☆

382722　Tabebuia pentaphylla（L.）Hemsl.；五叶黄钟木（五叶钟花树）；Hawaiian Cherry, Pink Tecoma Tree ●☆

382723　Tabebuia rosea（Bertol.）DC.；掌叶黄钟木（红蚁木,玫瑰喇叭木,洋红风铃木,掌叶木）；Pink Poui, Pink Trumpet Tree, Pink Trumpet-tree ●☆

382724　Tabebuia serratifolia Nicolson；齿叶黄钟木（齿叶蚁木）；Washiba Wood, Yellow Poui ●☆

382725　Taberna（DC.）Miers = Tabernaemontana L. ●

382726　Taberna Miers = Tabernaemontana L. ●

382727　Tabernaemontana L.（1753）；红月桂属（狗牙花属,假金橘属,马蹄花属,山辣椒属,山马茶属）；Ervatamia, Rejoua, Tabernaemontana ●

382728　Tabernaemontana Plum. ex L. = Tabernaemontana L. ●

382729　Tabernaemontana accedens Müll. Arg. 相近山马茶●☆

382730　Tabernaemontana affinis Müll. Arg. 近山马茶●☆

382731　Tabernaemontana africana Hook.；非洲红月桂●☆

382732　Tabernaemontana alba Mill.；白红月桂；White Milkwood ●☆

382733　Tabernaemontana amblyblasta S. F. Blake；钝胚山马茶●☆

382734　Tabernaemontana angolensis Stapf = Tabernaemontana pachysiphon Stapf ●☆

382735　Tabernaemontana anisophylla Pichon = Pandaca anisophylla（Pichon）Markgr. ●☆

382736　Tabernaemontana arborea Rose ex Donn. Sm.；乔木山马茶●☆

382737　Tabernaemontana australis Müll. Arg.；南方红月桂；Sapieangy, Zapieandi ●☆

382738　Tabernaemontana barteri Hook. f. = Callichilia barteri（Hook. f.）Stapf ●☆

382739　Tabernaemontana bouquetii（Boiteau）Leeuwenb.；布凯红月桂●☆

382740　Tabernaemontana bovina Lour.；药用狗牙花；Medicinal Cogflower, Medicinal Ervatamia ●

382741　Tabernaemontana brachyantha Stapf；短花红月桂●☆

382742　Tabernaemontana brachypoda K. Schum. = Tabernaemontana eglandulosa Stapf ●☆

382743　Tabernaemontana bufalina Lour.；尖蕾狗牙花（艾角青,澄江狗牙花,单根木,独根木,海南狗牙花,鸡爪花,角状狗牙花,山辣椒树,震天雷）；Chengjiang Cogflower, Chengjiang Ervatamia, Jianlei Tabernaemontana ●

382744　Tabernaemontana calcarea Pichon；石灰红月桂●☆

382745　Tabernaemontana capuronii Leeuwenb.；凯普伦狗牙花●☆

382746　Tabernaemontana celastroides（Kerr）P. T. Li；南蛇藤叶狗牙花●☆

382747　Tabernaemontana ceratocarpa（Kerr）P. T. Li = Tabernaemontana bufalina Lour. ●

382748　Tabernaemontana chartacea Pichon；纸质山马茶●☆

382749　Tabernaemontana chartacea Pichon = Tabernaemontana eglandulosa Stapf ●☆

382750　Tabernaemontana chengkiangensis（Tsiang）P. T. Li = Tabernaemontana bufalina Lour. ●

382751　Tabernaemontana chinensis Merr. = Tabernaemontana corymbosa Roxb. ex Wall. ●

382752　Tabernaemontana chippii（Stapf）Pichon = Tabernaemontana africana Hook. ●☆

382753　Tabernaemontana chrysocarpa S. F. Blake；黄果山马茶●☆

382754　Tabernaemontana ciliata Pichon；睫毛狗牙花●☆

382755　Tabernaemontana citrifolia DC.；柠檬叶山马茶●☆

382756　Tabernaemontana coffeoides Bojer ex A. DC.；咖啡狗牙花●☆

382757　Tabernaemontana continentalis（Tsiang）P. T. Li = Tabernaemontana corymbosa Roxb. ex Wall. ●

382758　Tabernaemontana continentalis（Tsiang）P. T. Li var. pubiflora（Tsiang）P. T. Li = Tabernaemontana corymbosa Roxb. ex Wall. ●

382759　Tabernaemontana contorta Stapf；缠扭狗牙花●☆

382760　Tabernaemontana coronaria（Jacq.）Willd. = Tabernaemontana divaricata（L.）R. Br. ex Roem. et Schult. ●

382761　Tabernaemontana coronaria R. Br. = Tabernaemontana divaricata（L.）R. Br. ex Roem. et Schult. ●

382762　Tabernaemontana coronaria Wild. = Ervatamia coronaria（Jacq.）Stapf ●

382763　Tabernaemontana coronaria Willd.；冠状山辣椒；Adam's Apple, Broad-leaved Rosebay, Crape Jasmine, East Indian Rosebay, Nero's Crown, Wax-flower ●☆

382764　Tabernaemontana coronaria Willd. = Ervatamia coronaria（Jacq.）Stapf ●

382765　Tabernaemontana coronaria Willd. var. flore-pleno hort；重瓣冠状山辣椒；Butterfly Gardenia ●☆

382766　Tabernaemontana corymbosa Roxb. ex Wall.；伞房狗牙花（大陆狗牙花,广西狗牙花,贵州狗牙花,毛瓣狗牙花,毛大陆狗牙花,纤花狗牙花,异尊云南狗牙花,云南狗牙花,中国狗牙花）；China Cogflower, Chinese Ervatamia, Combose Tabernaemontana, Continent Ervatamia, Continental Cogflower, Guangxi Cogflower, Guangxi Ervatamia, Guizhou Cogflower, Guizhou Ervatamia, Hairypetal Cogflower, Hairypetal Ervatamia, Heterosepal Cogflower, Heterosepal Ervatamia, Tenuousflower Ervatamia, Thin Cogflower,

Yunnan Cogflower, Yunnan Ervatamia ●

382767　Tabernaemontana crassa Benth. ; 马岛厚叶狗牙花●☆

382768　Tabernaemontana crassifolia Pichon; 厚叶狗牙花●☆

382769　Tabernaemontana crispiflora K. Schum. = Tabernaemontana eglandulosa Stapf ●☆

382770　Tabernaemontana cuneata Pichon = Tabernaemontana ciliata Pichon ●☆

382771　Tabernaemontana debrayi（Markgr.）Leeuwenb. ; 德布雷狗牙花●☆

382772　Tabernaemontana dichotoma Roxb. = Rejoua dichotoma（Roxb.）Gamble ●

382773　Tabernaemontana dichotoma Roxb. ex Wall. = Rejoua dichotoma（Roxb.）Gamble ●

382774　Tabernaemontana divaricata（L.）R. Br. ex Roem. et Schult. 'Flore Pleno'; 重瓣狗牙花●☆

382775　Tabernaemontana divaricata（L.）R. Br. ex Roem. et Schult. 'Grandifolia'; 大叶狗牙花●☆

382776　Tabernaemontana divaricata（L.）R. Br. ex Roem. et Schult. = Ervatamia divaricata（L.）Burkill ●

382777　Tabernaemontana durissima Stapf = Tabernaemontana crassa Benth. ●☆

382778　Tabernaemontana eglandulosa Stapf; 无腺狗牙花●☆

382779　Tabernaemontana eglandulosa Stapf var. macrocalyx = Tabernaemontana eglandulosa Stapf ●☆

382780　Tabernaemontana elegans Stapf; 美丽红月桂; Toad Tree ●☆

382781　Tabernaemontana elliptica Thunb. = Amsonia elliptica（Thunb.）Roem. et Schult. ■

382782　Tabernaemontana elliptica Thunb. et Murray = Alstonia elliptica（Thunb.）Roem. et Schult. ●☆

382783　Tabernaemontana erythrophthalma K. Schum. = Alafia erythrophthalma（K. Schum.）Leeuwenb. ●☆

382784　Tabernaemontana eusepala Aug. DC. ; 马达加斯加山辣椒●☆

382785　Tabernaemontana eusepaloides（Markgr.）Leeuwenb. ; 拟马达加斯加山辣椒●☆

382786　Tabernaemontana flabelliformis（Tsiang）P. T. Li = Tabernaemontana divaricata（L.）R. Br. ex Roem. et Schult. ●

382787　Tabernaemontana glandulosa（Stapf）Pichon; 腺质山辣椒●☆

382788　Tabernaemontana grandiflora Hook. = Tabernaemontana africana Hook. ●☆

382789　Tabernaemontana grandiflora Jacq. ; 大花狗牙花; Carthage Crape Jasmine ●☆

382790　Tabernaemontana guangdongensis P. T. Li = Tabernaemontana pandacaqui Lam. ●

382791　Tabernaemontana hainanensis（Tsiang）P. T. Li = Tabernaemontana bufalina Lour. ●

382792　Tabernaemontana hallei（Boiteau）Leeuwenb. ; 哈勒狗牙花●☆

382793　Tabernaemontana heyneana Wall. ; 海尼山辣椒（山马茶）●☆

382794　Tabernaemontana holstii K. Schum. ; 霍氏山马茶●☆

382795　Tabernaemontana holstii K. Schum. = Tabernaemontana pachysiphon Stapf ●☆

382796　Tabernaemontana humblotii（Baill.）Pichon; 安布洛狗牙花●☆

382797　Tabernaemontana inconspicua Stapf; 显著狗牙花●☆

382798　Tabernaemontana jasminoides Tsiang = Tabernaemontana bufalina Lour. ●

382799　Tabernaemontana johnstonii（Stapf）Pichon; 约翰山马茶（肯尼亚山辣椒树）●☆

382800　Tabernaemontana johnstonii（Stapf）Pichon = Tabernaemontana

stapfiana Britten ●☆

382801　Tabernaemontana kwangsiensis（Tsiang）P. T. Li = Tabernaemontana corymbosa Roxb. ex Wall. ●

382802　Tabernaemontana kweichowensis（Tsiang）P. T. Li = Tabernaemontana corymbosa Roxb. ex Wall. ●

382803　Tabernaemontana latifolia（Stapf）Pichon = Tabernaemontana eglandulosa Stapf ●☆

382804　Tabernaemontana letestui（Pellegr.）Pichon; 赖特狗牙花●☆

382805　Tabernaemontana longiflora Benth. = Tabernaemontana africana Hook. ●☆

382806　Tabernaemontana longituba Pichon = Tabernaemontana ciliata Pichon ●☆

382807　Tabernaemontana macrocarpa Jack = Ervatamia macrocarpa（Jack）Merr. ●

382808　Tabernaemontana malaccensis Hook. f. ; 马六甲山马茶●☆

382809　Tabernaemontana membranacea A. DC. = Tabernaemontana coffeoides Bojer ex A. DC. ●☆

382810　Tabernaemontana minutiflora Pichon = Pentopetia ovalifolia（Costantin et Gallaud）Klack. ●☆

382811　Tabernaemontana mocquerysii Aug. DC. ; 莫克狗牙花●☆

382812　Tabernaemontana modesta Baker = Tabernaemontana coffeoides Bojer ex A. DC. ●☆

382813　Tabernaemontana mollis Hook. et Arn. = Tabernaemontana pandacaqui Lam. ●

382814　Tabernaemontana monopodialis K. Schum. = Callichilia monopodialis（K. Schum.）Stapf ●☆

382815　Tabernaemontana mucronata Merr. = Tabernaemontana pandacaqui Lam. ●

382816　Tabernaemontana nitida Stapf = Picralima nitida（Stapf）T. Durand et H. Durand ●☆

382817　Tabernaemontana noronhiana Bojer ex A. DC. = Tabernaemontana retusa（Lam.）Palacky ●☆

382818　Tabernaemontana ochrascens Pichon = Tabernaemontana ciliata Pichon ●☆

382819　Tabernaemontana odoratissima（Stapf）Leeuwenb. ; 极香狗牙花●☆

382820　Tabernaemontana officinalis（Tsiang）P. T. Li = Tabernaemontana bovina Lour. ●

382821　Tabernaemontana pachysiphon Stapf; 粗管狗牙花; Dodo Cloth ●☆

382822　Tabernaemontana pandacaqui Lam. ; 平脉狗牙花（尖果狗牙花, 兰屿马蹄花, 毛叶狗牙花, 南洋马蹄花, 潘达卡红月桂, 台湾狗牙花, 泰国狗牙花）; Hairyleaf Cogflower, Hairyleaf Ervatamia, Lanyu Cogflower, Mucronatefruit Ervatamia, Pandacaqu Tabernaemontana, Sharpfruit Cogflower, Taiwan Cogflower, Taiwan Ervatamia, Thailand Cogflower ●

382823　Tabernaemontana pandacaqui Poir. = Tabernaemontana pandacaqui Lam. ●

382824　Tabernaemontana parvifolia Pichon = Tabernaemontana mocquerysii Aug. DC. ●☆

382825　Tabernaemontana penduliflora K. Schum. ; 垂花狗牙花●☆

382826　Tabernaemontana peninsularis（Kerr）P. T. Li; 岛生狗牙花●☆

382827　Tabernaemontana phymata Leeuwenb. ; 肿胀狗牙花●☆

382828　Tabernaemontana polyantha（Blume）Miq. = Ichnocarpus polyanthus（Blume）P. I. Forst. ●

382829　Tabernaemontana polyantha（Blume）Miq. = Microchites polyantha（Blume）Miq. ●

382830　Tabernaemontana polyantha Blume = Ichnocarpus polyanthus（Blume）P. I. Forst. ●

382831 Tabernaemontana polyantha Blume = Microchites polyantha (Blume) Miq. ●

382832 Tabernaemontana retusa (Lam.) Palacky;马岛狗牙花●☆

382833 Tabernaemontana riedelii Müll. Arg.;瑞得山马茶●☆

382834 Tabernaemontana rigida (Miers) Leeuwenb.;坚硬山马茶●☆

382835 Tabernaemontana rotensis (Kaneh.) P. T. Li;马里安狗牙花●

382836 Tabernaemontana salicifolia Wall. = Hunteria zeylanica (Retz.) Gardner ex Thwaites ●

382837 Tabernaemontana sambiranensis Pichon;桑比朗狗牙花●☆

382838 Tabernaemontana sessilifolia Baker;无柄狗牙花●☆

382839 Tabernaemontana smithii Stapf = Tabernaemontana crassa Benth. ●☆

382840 Tabernaemontana stapfiana Britten;施塔普夫狗牙花●☆

382841 Tabernaemontana stellata Pichon;星形狗牙花●☆

382842 Tabernaemontana stenosiphon Stapf;窄管狗牙花●☆

382843 Tabernaemontana subglobosa Merr. = Tabernaemontana pandacaqui Lam. ●

382844 Tabernaemontana subsessilis Benth. = Callichilia subsessilis (Benth.) Stapf ●☆

382845 Tabernaemontana thailandensis P. T. Li = Tabernaemontana pandacaqui Lam. ●

382846 Tabernaemontana thonneri T. Durand et De Wild. ex Stapf = Tabernaemontana crassa Benth. ●☆

382847 Tabernaemontana thouarsii (Roem. et Schult.) Palacky = Voacanga thouarsii Roem. et Schult. ●☆

382848 Tabernaemontana tonkinensis Pierre ex Pit. = Tabernaemontana bovina Lour. ●

382849 Tabernaemontana tsiangiana P. T. Li = Tabernaemontana corymbosa Roxb. ex Wall. ●

382850 Tabernaemontana undulata Perr. ex A. DC.;波叶山马茶●☆

382851 Tabernaemontana usambarensis K. Schum. ex Engl. = Tabernaemontana ventricosa Hochst. ex A. DC. ●☆

382852 Tabernaemontana utilis Arn.;有用狗牙花;Cow Tree Tabernaemontana,Cow-tree Tabernaemontana ●☆

382853 Tabernaemontana vanheurckii Müll. Arg.;范氏山马茶(范浩山马茶)●☆

382854 Tabernaemontana ventricosa Hochst. ex A. DC.;偏肿山马茶●☆

382855 Tabernaemontana verrucosa Blume = Chonemorpha verrucosa (Blume) D. J. Middleton ●

382856 Tabernaemontana volkensii K. Schum. = Rauvolfia volkensii (K. Schum.) Stapf ●☆

382857 Tabernaemontana yunnanensis (Tsiang) P. T. Li = Tabernaemontana corymbosa Roxb. ex Wall. ●

382858 Tabernaemontana yunnanensis (Tsiang) P. T. Li var. heterosepala (Tsiang) P. T. Li = Tabernaemontana corymbosa Roxb. ex Wall. ●

382859 Tabernaemontanaceae Baum.-Bod.;红月桂科●

382860 Tabernaemontanaceae Baum.-Bod. = Apocynaceae Juss. (保留科名)●■

382861 Tabernanthe Baill. (1889);马山茶属(塔拜尔木属)●☆

382862 Tabernanthe albiflora Stapf = Tabernanthe iboga Baill. ●☆

382863 Tabernanthe bocca Stapf = Tabernanthe iboga Baill. ●☆

382864 Tabernanthe elliptica (Stapf) Leeuwenb.;椭圆马山茶●☆

382865 Tabernanthe iboga Baill.;伊波加木;Iboga ●☆

382866 Tabernanthe mannii Stapf = Tabernanthe iboga Baill. ●☆

382867 Tabernanthe pubescens Pichon = Tabernanthe iboga Baill. ●☆

382868 Tabernanthe subsessilis Stapf = Tabernanthe iboga Baill. ●☆

382869 Tabernanthe tenuiflora Stapf = Tabernanthe iboga Baill. ●☆

382870 Tabernaria Raf. = Tabernaemontana L. ●

382871 Tacamahaca Mill. = Populus L. ●

382872 Tacarcuna Huft(1989);塔卡大戟属☆

382873 Tacarcuna amanoifolia Huft;塔卡大戟☆

382874 Tacazzea Decne. (1844);塔卡萝藦属●☆

382875 Tacazzea africana (Schltr.) N. E. Br. = Schlechterella africana (Schltr.) K. Schum. ●☆

382876 Tacazzea amplifolia S. Moore = Mondia whitei (Hook. f.) Skeels ●☆

382877 Tacazzea apiculata Oliv.;细尖塔卡萝藦●☆

382878 Tacazzea apiculata Oliv. var. benedicta Scott-Elliot = Tacazzea apiculata Oliv. ●☆

382879 Tacazzea apiculata Oliv. var. glabra K. Schum. = Tacazzea apiculata Oliv. ●☆

382880 Tacazzea bagshawei S. Moore = Tacazzea apiculata Oliv. ●☆

382881 Tacazzea bagshawei S. Moore var. occidentalis C. Norman = Tacazzea apiculata Oliv. ●☆

382882 Tacazzea barteri Baill. = Tacazzea apiculata Oliv. ●☆

382883 Tacazzea brazzaeana Baill. = Tacazzea apiculata Oliv. ●☆

382884 Tacazzea conferta N. E. Br.;密集塔卡萝藦●☆

382885 Tacazzea floribunda K. Schum. = Tacazzea conferta N. E. Br. ●☆

382886 Tacazzea galactagoga Bullock = Tacazzea conferta N. E. Br. ●☆

382887 Tacazzea kirkii N. E. Br. = Tacazzea apiculata Oliv. ●☆

382888 Tacazzea laxiflora Engl. = Mondia whitei (Hook. f.) Skeels ●☆

382889 Tacazzea martini Baill. = Tacazzea venosa Decne. ●☆

382890 Tacazzea natalensis (Schltr.) N. E. Br. = Ischnolepis natalensis (Schltr.) Venter ■☆

382891 Tacazzea nigritana N. E. Br. = Tacazzea apiculata Oliv. ●☆

382892 Tacazzea oleander S. Moore = Tacazzea rosmarinifolia (Decne.) N. E. Br. ●☆

382893 Tacazzea pedicellata K. Schum. = Zacateza pedicellata (K. Schum.) Bullock ●☆

382894 Tacazzea pedicellata K. Schum. var. occidentalis N. E. Br. = Zacateza pedicellata (K. Schum.) Bullock ●☆

382895 Tacazzea racemosus N. E. Br. = Zacateza pedicellata (K. Schum.) Bullock ●☆

382896 Tacazzea rosmarinifolia (Decne.) N. E. Br.;迷迭香叶塔卡萝藦●☆

382897 Tacazzea salicina Schltr. = Tacazzea rosmarinifolia (Decne.) N. E. Br. ●☆

382898 Tacazzea stipularis N. E. Br. = Tacazzea apiculata Oliv. ●☆

382899 Tacazzea thollonii Baill. = Tacazzea apiculata Oliv. ●☆

382900 Tacazzea tomentosa E. A. Bruce = Buckollia tomentosa (E. A. Bruce) Venter et R. L. Verh. ●☆

382901 Tacazzea venosa Decne.;多脉塔卡萝藦●☆

382902 Tacazzea venosa Decne. subsp. rosmarinifolia (Decne.) Bullock = Tacazzea rosmarinifolia (Decne.) N. E. Br. ●☆

382903 Tacazzea venosa Decne. var. martini (Baill.) N. E. Br. = Tacazzea venosa Decne. ●☆

382904 Tacazzea verticillata K. Schum. = Tacazzea apiculata Oliv. ●☆

382905 Tacazzea viridis A. Chev. ex Hutch. et Dalziel = Mondia whitei (Hook. f.) Skeels ●☆

382906 Tacazzea volubilis (Schltr.) N. E. Br. = Buckollia volubilis (Schltr.) Venter et R. L. Verh. ●☆

382907 Tacazzea welwitschii Baill. = Tacazzea apiculata Oliv. ●☆

382908 Tacca J. R. Forst. et G. Forst. (1775)(保留属名);蒟蒻薯属

（箭根薯属）；Tacca ■

382909　Tacca ankaranensis Bard. -Vauc. ;安卡兰蒟蒻薯■☆

382910　Tacca artocarpifolia Seem. = Tacca leontopetaloides（L.）Kuntze ■

382911　Tacca birkinshawii ?;伯金肖蒟蒻薯●☆

382912　Tacca chantrieri André;箭根薯（碧石雷，蝙蝠花，长须果，大水田七，大叶屈头鸡，黑冬叶，蒟蒻薯，老虎花，老虎须，山大黄，水狗仔）；Arrowroot Tacca, Bat Flower, Batflower, Bat-flower, Cat's Whiskers, Chantrier Tacca, Devil-flower ■

382913　Tacca chantrieri André var. macrantha Limpr. ;大花箭根薯■☆

382914　Tacca cristata Jack.;冠状蒟蒻薯■☆

382915　Tacca cristata Jack. = Tacca integrifolia Ker Gawl. ■

382916　Tacca esquirolii（H. Lév.）Rehder = Tacca chantrieri André ■

382917　Tacca gaogao Blanco = Tacca leontopetaloides（L.）Kuntze ■

382918　Tacca hawaiiensis H. Limpr. ;夏威夷蒟蒻薯; Hawaiian Arrowroot ■☆

382919　Tacca hawaiiensis H. Limpr. = Tacca leontopetaloides（L.）Kuntze ■

382920　Tacca integrifolia Ker Gawl. ;丝须蒟蒻薯（丝蒟蒻薯）; Entireleaf Tacca ■

382921　Tacca involucrata Schum. et Thonn. = Tacca leontopetaloides（L.）Kuntze ■

382922　Tacca laevis Roxb. = Tacca integrifolia Ker Gawl. ■

382923　Tacca leontopetaloides（L.）Kuntze;蒟蒻薯（田代薯，羽裂蒟蒻薯，蛛丝草）; Afraca Arrowroot, Arrowroot Tacca, East Indian Arrowroot, Fiji Arrowroot, Hawaiian Arrowroot, Otaheite Salap-plant, Pia, Polynesian Arrowroot, South Sea Arrowroot, South Sea Arrowroot-plant, Tacca, Tahiti Arrowroot ■

382924　Tacca madagascariensis Bojer = Tacca leontopetaloides（L.）Kuntze ■

382925　Tacca minor Ridl. = Tacca chantrieri André ■

382926　Tacca palmata Blume. ;掌叶蒟蒻薯■☆

382927　Tacca paxiana H. Limpr. = Tacca chantrieri André ■

382928　Tacca pinnatifida J. R. Forst. et G. Forst. = Tacca leontopetaloides（L.）Kuntze ■

382929　Tacca pinnatifida J. R. Forst. et G. Forst. subsp. madagascariensis Limpr. = Tacca leontopetaloides（L.）Kuntze ■

382930　Tacca plantaginea（Hance）Drenth = Schizocapsa plantaginea Hance ■

382931　Tacca subflabellata P. P. Ling et C. C. Ting;扇苞蒟蒻薯; Fanbract Tacca ■

382932　Tacca umbrarum Jum. et H. Perrier = Tacca artocarpifolia Seem. ■

382933　Taccaceae Bercht. et J.Presl = Taccaceae Dumort.（保留科名）■

382934　Taccaceae Dumort.（1829）（保留科名）;蒟蒻薯科（箭根薯科，蛛丝草科）;Pia Family, Tacca Family ■

382935　Taccaceae Dumort.（保留科名）= Dioscoreaceae R. Br.（保留科名）●■

382936　Taccarum Brongn. = Taccarum Brongn. ex Schott ■☆

382937　Taccarum Brongn. ex Schott（1857）;箭根南星属■☆

382938　Taccarum caudatum Rusby;尾状箭根南星■☆

382939　Taccarum cylindricum Arcangeli;圆柱箭根南星■☆

382940　Taccarum weddellianum Brongn. ex Schott;箭根南星■☆

382941　Tachia Aubl.（1775）;圭亚那龙胆属（大吉阿属）●☆

382942　Tachia Pers. = Tachigali Aubl. ●☆

382943　Tachia guianensis Aubl. ;圭亚那龙胆（圭亚那大吉阿）●☆

382944　Tachiadenus Griseb.（1838）;腺龙胆属●■☆

382945　Tachiadenus boivinii Humbert ex Klack. ;博伊文腺龙胆●☆

382946　Tachiadenus carinatus（Desr.）Griseb. ;龙骨状腺龙胆●☆

382947　Tachiadenus continentalis Baker = Sebaea teuszii（Schinz）P. Taylor ■☆

382948　Tachiadenus elatus Hemsl. = Tachiadenus tubiflorus（Thouars ex Roem. et Schult.）Griseb. ●☆

382949　Tachiadenus gracilis Griseb. ;纤细腺龙胆●☆

382950　Tachiadenus longiflorus Griseb. ;长花腺龙胆●☆

382951　Tachiadenus longifolius Scott-Elliot;长叶腺龙胆●☆

382952　Tachiadenus mechowianus（Vatke ex Schinz）Hill;梅休腺龙胆●☆

382953　Tachiadenus pervillei Humbert ex Klack. ;佩尔腺龙胆●☆

382954　Tachiadenus platypterus Baker;宽翅腺龙胆●☆

382955　Tachiadenus trinervis（Desr.）Griseb. = Ornichia trinervis（Desr.）Klack. ●☆

382956　Tachiadenus tubiflorus（Thouars ex Roem. et Schult.）Griseb. ;管花腺龙胆●☆

382957　Tachiadenus umbellatus Klack. ;小伞腺龙胆●☆

382958　Tachiadenus vohimavensis Humbert;武希腺龙胆●☆

382959　Tachibota Aubl. = Hirtella L. ●☆

382960　Tachigalea Griseb. = Amasonia L. f.（保留属名）●■☆

382961　Tachigali Aubl.（1775）;塔奇苏木属●☆

382962　Tachigali Juss. = Tachigali Aubl. ●☆

382963　Tachigali alba Ducke. ;白塔奇苏木●☆

382964　Tachigali angustifolia Miq. ;窄叶塔奇苏木●☆

382965　Tachigali argyrophylla Ducke;银叶塔奇苏木●☆

382966　Tachigali aurea Tul. ;黄塔奇苏木●☆

382967　Tachigali colombiana Dwyer;哥伦比亚塔奇苏木●☆

382968　Tachigali densiflora（Benth.）L. F. Gomes da Silva et H. C. Lima;密花塔奇苏木●☆

382969　Tachigali eriocalyx Tul. ;毛萼塔奇苏木●☆

382970　Tachigali floribunda Steud. ;繁花塔奇苏木●☆

382971　Tachigali fusca van der Werff;褐塔奇苏木●☆

382972　Tachigali glauca Tul. ;灰蓝塔奇苏木●☆

382973　Tachigali grandiflora Huber;大花塔奇苏木●☆

382974　Tachigali leiocalyx（Ducke）L. F. Gomes da Silva et H. C. Lima;光萼塔奇苏木●☆

382975　Tachigali longiflora Ducke;塔奇苏木●☆

382976　Tachigali macropetala（Ducke）L. F. Gomes da Silva et H. C. Lima;大瓣塔奇苏木●☆

382977　Tachigali macrostachya Huber;大穗塔奇苏木●☆

382978　Tachigali melanocarpa（Ducke）van der Werff;黑果塔奇苏木●☆

382979　Tachigali micropetala（Ducke）Zarucchi et Pipoly;小瓣塔奇苏木●☆

382980　Tachigali polyphylla Poepp. et Endl. ;多叶塔奇苏木●☆

382981　Tachigali pulchra Dwyer;美丽塔奇苏木●☆

382982　Tachigali purpurea Rich. ;紫塔奇苏木●☆

382983　Tachigali rigida Ducke;硬塔奇苏木●☆

382984　Tachigali rugosa（Mart. ex Benth.）Zarucchi et Pipoly;皱塔奇苏木●☆

382985　Tachigali sericea Tul. ;绢毛塔奇苏木●☆

382986　Tachigali tinctoria（Benth.）Zarucchi et Herend. ;染色塔奇苏木●☆

382987　Tachigali urbaniana（Harms）L. F. Gomes da Silva et H. C. Lima;乌尔班塔奇苏木●☆

382988　Tachigali vulgaris L. F. Gomes da Silva et H. C. Lima;普通塔奇苏木●☆

382989　Tachites Sol. ex Gaertn. = Melicytus J. R. Forst. et G. Forst. ●☆

382990　Tachytes Steud. = Tachites Sol. ex Gaertn. ●☆

382991 Tacinga Britton et Rose(1919);长蕊掌属(白林属)●☆

382992 Tacinga funalis Britton et Rose;狼烟台■☆

382993 Tacitus Moran = Graptopetalum Rose ■●☆

382994 Tacitus Moran(1974);齐花山景天属■☆

382995 Tacitus bellus Moran et J. Meyrán;齐花山景天■☆

382996 Tacoanthus Baill.(1890);玻利维亚爵床属■☆

382997 Tacoanthus pearcei Baill.;玻利维亚爵床■☆

382998 Tacsonia Juss. = Passiflora L. ●■

382999 Tacsonia manicata Juss. = Passiflora manicata (Juss.) Pers. ●☆

383000 Tacsonia mollissima Kunth = Passiflora mollissima (Kunth) L. H. Bailey ●☆

383001 Tacsonia van-volxemii Hook. = Passiflora antioquiensis H. Karst. ●☆

383002 Tadeastrum Szlach.(2007);巴西老虎兰属■☆

383003 Tadeastrum Szlach. = Stanhopea J. Frost ex Hook. ■☆

383004 Tadehagi (Schindl.) H. Ohashi = Pteroloma Desv. ex Benth. ●

383005 Tadehagi (Schindl.) H. Ohashi = Tadehagi H. Ohashi ●

383006 Tadehagi H. Ohashi(1973);葫芦茶属;Tadehagi,Calabash ●

383007 Tadehagi pseudotriquetra (DC.) H. Ohashi;蔓茎葫芦茶(葫芦茶,假葫芦茶,龙舌黄,一条根);False Calabash, False Tadehagi, Prostrte Tadehagi, Sham Calabash ●

383008 Tadehagi pseudotriquetra (DC.) Yen C. Yang et P. H. Huang = Tadehagi pseudotriquetra (DC.) H. Ohashi ●

383009 Tadehagi triquetra (L.) H. Ohashi;葫芦茶(百劳舌,宝剑草,虫草,地马桩,金葫芦,金剑草,懒狗舌,鲮鲤舌,龙舌茶,龙舌癀,麻草,牛虫草,迫颈草,螳螂草,剃刀柄,田刀柄,土豆,咸虾茶,咸鱼草,钊板茶);Calabash, Triquetrous Tadehagi ●

383010 Tadehagi triquetra (L.) H. Ohashi subsp. alata (DC.) H. Ohashi;宽果葫芦茶●

383011 Tadehagi triquetra (L.) H. Ohashi subsp. pseudotriquetra (DC.) H. Ohashi = Tadehagi pseudotriquetra (DC.) H. Ohashi ●

383012 Tadtm Moran = Graptopetalum Rose ■●☆

383013 Taeckholmia Boulos = Atalanthus D. Don ■☆

383014 Taeckholmia Boulos = Sonchus L. ■

383015 Taeckholmia arborea (DC.) Boulos = Sonchus arboreus DC. ■☆

383016 Taeckholmia canariensis (Svent.) Boulos = Sonchus filifolius N. Kilian et Greuter ■☆

383017 Taeckholmia capillaris (Svent.) Boulos = Sonchus capillaris Svent. ■☆

383018 Taeckholmia filifolia G. Kunkel = Sonchus filifolius N. Kilian et Greuter ■☆

383019 Taeckholmia heterophylla Boulos = Sonchus heterophyllus (Boulos) U. Reifenb. et A. Reifenb. ■☆

383020 Taeckholmia microcarpa Boulos = Sonchus microcarpus (Boulos) U. Reifenb. et A. Reifenb. ■☆

383021 Taeckholmia pinnata (Aiton) Boulos = Sonchus pinnatus Aiton ■☆

383022 Taeckholmia regis-jubae (Pit.) Boulos = Sonchus regis-jubae Pit. ■☆

383023 Taenais Salisb. = Crinum L. ■

383024 Taenia Post et Kuntze = Tainia Blume ■

383025 Taeniandra Bremek. = Strobilanthes Blume ●■

383026 Taenianthera Burret = Geonoma Willd. ■☆

383027 Taenianthera Burret(1930);带药椰属■☆

383028 Taenianthera acaulis Burret;无茎带药椰●☆

383029 Taenianthera gracilis Burret;细带药椰●☆

383030 Taenianthera macrostachys Burret;大穗带药椰●☆

383031 Taenianthera minor Burret;小带药椰●☆

383032 Taenianthera multisecta Burret;多刚毛带药椰●☆

383033 Taenianthera oligosticha Burret;寡列带药椰●☆

383034 Taeniatherum Nevski(1934);带药禾属;Medusahead ■☆

383035 Taeniatherum asperum (Simonk.) Nevski;粗糙带药禾■☆

383036 Taeniatherum caput-medusae (L.) Nevski = Taeniatherum crinitum (Schreb.) Nevski ■☆

383037 Taeniatherum caput-medusae (L.) Nevski subsp. asperum (Simonk.) Melderis = Taeniatherum asperum (Simonk.) Nevski ■☆

383038 Taeniatherum caput-medusae (L.) Nevski subsp. crinitum (Schreb.) Melderis = Taeniatherum caput-medusae (L.) Nevski ■☆

383039 Taeniatherum caput-medusae (L.) Nevski var. crinitum (Schreb.) Humphries = Taeniatherum caput-medusae (L.) Nevski ■☆

383040 Taeniatherum crinitum (Schreb.) Nevski;带药禾;Medusahead ■☆

383041 Taenidia (Torr. et A. Gray) Drude(1898);太尼草属☆

383042 Taenidia Drude = Taenidia (Torr. et A. Gray) Drude ☆

383043 Taenidia integerrima (L.) Drude;太尼草;Yellow Pimpernel, Yellow-pimpernel☆

383044 Taenidium Targ. Tozz. = Posidonia K. D. König(保留属名)■

383045 Taeniocarpum Desv. = Pachyrhizus Rich. ex DC. (保留属名)■

383046 Taeniochlaena Hook. f. = Rourea Aubl. (保留属名)●

383047 Taeniochlaena Hook. f. = Roureopsis Planch. ●

383048 Taeniola Salisb. = Ornithogalum L. ■

383049 Taenionema Post et Kuntze = Tainionema Schltr. ●☆

383050 Taeniopetalum Vis. (1850);纹瓣花属■☆

383051 Taeniopetalum Vis. = Peucedanum L. ■

383052 Taeniopetalum neumayeri Vis. ;纹瓣花(带瓣花)■☆

383053 Taeniopetalum peucedanoides Bunge;高纹瓣花■☆

383054 Taeniophyllum Blume (1825);带叶兰属(蜘蛛兰属);Taeniophyllum ■

383055 Taeniophyllum aphyllum (Makino) Makino = Taeniophyllum glandulosum Blume ■

383056 Taeniophyllum aphyllum Makino = Taeniophyllum glandulosum Blume ■

383057 Taeniophyllum breviscapum J. J. Sm.;短花茎带叶兰;Shortscape Taeniophyllum ■☆

383058 Taeniophyllum chitouense S. S. Ying = Taeniophyllum glandulosum Blume ■

383059 Taeniophyllum compactum Ames = Microtatorchis compacta (Ames) Schltr. ■

383060 Taeniophyllum complanatum Fukuy. ;扁根带叶兰(扁带兰,扁带叶兰,扁蜘蛛兰,厚脚蜘蛛兰)■

383061 Taeniophyllum crassipes Fukuy. = Taeniophyllum complanatum Fukuy. ■

383062 Taeniophyllum filiforme J. J. Sm. ;丝状带叶兰;Thread-shaped Taeniophyllum ■☆

383063 Taeniophyllum formosanum Hayata;台湾蜘蛛兰■

383064 Taeniophyllum glandulosum Blume;带叶兰(蜘蛛兰);Leafless Taeniophyllum,Taeniophyllum ■

383065 Taeniophyllum obtusum Blume = Taeniophyllum pusillum (Willd.) Seidenf. et Ormerod ■

383066 Taeniophyllum pusillum (Willd.) Seidenf. et Ormerod;兜唇带叶兰;Obtuse Taeniophyllum ■

383067 Taenioplehrum T. Durand et Jacks. = Taeniopleurum J. M. Coult. et Rose ■☆

383068 Taeniopleurum J. M. Coult. et Rose = Perideridia Rchb. ■☆

383069 Taeniorhachis Cope(1993);带刺禾属■☆

383070 Taeniorhachis repens Cope;匍匐带刺禾■☆

383071　Taeniorrhiza Summerh. (1943)；带根兰属■☆

383072　Taeniorrhiza gabonensis Summerh. ；带根兰■☆

383073　Taeniosapium Müll. Arg. = Excoecaria L. ●

383074　Taeniosapium Müll. Arg. = Sapium Jacq. (保留属名)●

383075　Taeniostema Spach = Crocanthemum Spach ●☆

383076　Taenosapium Benth. et Hock. f. = Excoecaria L. ●

383077　Taenosapium Benth. et Hock. f. = Taeniosapium Müll. Arg. ●

383078　Taetsia Medik. (废弃属名) = Cordyline Comm. ex R. Br. (保留属名)●

383079　Taetsia ferrea (L.) Medik. = Cordyline fruticosa (L.) A. Chev. ●

383080　Taetsia fruticosa (L.) Merr. = Cordyline fruticosa (L.) A. Chev. ●

383081　Taetsia terminalis (L.) W. Wight ex Saff. = Cordyline fruticosa (L.) A. Chev. ●

383082　Tafalla D. Don = Loricaria Wedd. ●☆

383083　Tafalla Ruiz et Pav. = Hedyosmum Sw. ●■

383084　Tafallaea Kuntze = Tafalla Ruiz et Pav. ●■

383085　Tagera Raf. = Cassia L. (保留属名)●■

383086　Tagera Raf. = Chamaecrista Moench ■●

383087　Tagetes L. (1753)；万寿菊属；Marigold，Tagetes ■●

383088　Tagetes apetala Posada-Ar. ；无舌万寿菊；Rayless Marigold ■☆

383089　Tagetes biflora Cabrera；双花万寿菊■☆

383090　Tagetes erecta L. ；万寿菊(柏花，臭芙蓉，臭芙蓉花，大芙蓉，大万寿菊，蜂窝菊，芙蓉花，黑苦艾，红花，黄菊，金花菊，金鸡菊，金菊，金盏菊，军旗花，里苦艾，千寿菊，伞段花，西番菊)；African Marigold，American Marigold，Aztec Marigold，Big Marigold，Common Marigold，French Marigold，Marigold，Tall Marigold，Turkey Gilliflower，Upright Velvetflower ■

383091　Tagetes filifolia Lag. ；丝叶万寿菊■☆

383092　Tagetes gilletii De Wild. ；吉勒特万寿菊■☆

383093　Tagetes glandulifera Schrank = Tagetes minuta L. ■☆

383094　Tagetes glandulosa Link = Tagetes minuta L. ■☆

383095　Tagetes graveolens L. ；刺鼻万寿菊■☆

383096　Tagetes lemmonii A. Gray；山地万寿菊；Copper Canyon Daisy，Mount Lemmon Marigold，Mountain Marigold ■☆

383097　Tagetes lucida Cav. ；香万寿菊(光亮万寿菊，透明万寿菊，香红黄草，香叶万寿菊)；Mexican Marigold，Mexican Mint Marigold，Mexican Tarragon，Sweet Marigold，Sweetscented Marigold，Sweet-scented Marigold，Sweet-scented Mexican Marigold ■☆

383098　Tagetes maxima Kuntze；大万寿菊■☆

383099　Tagetes micrantha Cav. ；小花万寿菊(小万寿菊)；Little Marigold ■☆

383100　Tagetes minuta L. ；弱小万寿菊；Miniature Marigold，Muster John Henry，Southern Marigold，Stinking Roger ■☆

383101　Tagetes multiflora Kunth；多花万寿菊■☆

383102　Tagetes papposa Vent. = Dyssodia papposa (Vent.) Hitchc. ■☆

383103　Tagetes patula L. ；孔雀草(步步登高花，臭菊花，缎子花，缎子菊，红黄草，红黄万寿菊，黄菊花，孔雀菊，藤菊，万寿菊，西番菊，小芙蓉花，小万寿菊)；French Marigold，Marigold，Spreading Marigold ■

383104　Tagetes patula L. = Tagetes erecta L. ■

383105　Tagetes pusilla Kunth；微小万寿菊；Lesser Marigold ■☆

383106　Tagetes rotundifolia Mill. = Tithonia rotundifolia (Mill.) S. F. Blake ■☆

383107　Tagetes signata Bartl. ；小万寿菊(姬红黄草，姬孔雀草，金星菊，墨西哥万寿菊，细叶红黄草，细叶孔雀草，细叶万寿菊)；

French Spotted Marigold，Mexican Marigold，Signet Marigold，Striped Mexican Marigold ■☆

383108　Tagetes tenuifolia Cav. = Tagetes erecta L. ■

383109　Tagetes tenuifolia Cav. = Tagetes signata Bartl. ■☆

383110　Taguaria Raf. = Gaiadendron G. Don ●☆

383111　Tahina J. Dransf. et Rakotoarin. (1926)；马岛棕属●☆

383112　Tahina spectabilis J. Dransf. et Rakotoarin. ；马岛棕●☆

383113　Tahitia Burret = Berrya Roxb. (保留属名)●

383114　Tahitia Burret = Christiana DC. ●☆

383115　Taihangia Te T. Yu et C. L. Li(1980)；太行花属；Taihang Flower，Taihangia ■★

383116　Taihangia Te T. Yu et C. L. Li = Geum L. ■

383117　Taihangia rupestris Te T. Yu et C. L. Li；太行花；Cliff Taihangia，Taihang Flower，Taihangia ■

383118　Taihangia rupestris Te T. Yu et C. L. Li var. ciliata Te T. Yu et C. L. Li；缘毛太行花；Ciliate Cliff Taihangia，Ciliate Taihang flower ■

383119　Tainia Blume(1825)；带唇兰属(杜鹃属)；Tainia ■

383120　Tainia angustifolia (Lindl.) Benth. et Hook. f. ；狭叶带唇兰(狭叶安兰)；Narrowleaf Tainia ■

383121　Tainia balansae Gagnep. = Collabium chinense (Rolfe) Ts. Tang et F. T. Wang ■

383122　Tainia barbara Lindl. = Eriodes barbata (Lindl.) Rolfe ■

383123　Tainia bilamellata (Fukuy.) S. S. Ying = Oreorchis bilamellata Fukuy. ■

383124　Tainia chapaensis Gagnep. = Collabium formosanum Hayata ■

383125　Tainia chinensis (Rolfe) Gagnep. = Collabium chinense (Rolfe) Ts. Tang et F. T. Wang ■

383126　Tainia cordata Hook. f. = Tainia latifolia (Lindl.) Rchb. f. ■

383127　Tainia cordifolia Hook. f. ；心叶带唇兰(葵兰，心叶葵兰，心叶球柄兰)；Heartleaf Mischobulbum，Mischobulbum ■

383128　Tainia cordifolia Hook. f. = Mischobulbum cordifolium (Hook. f.) Schltr. ■

383129　Tainia cristata (Rolfe) Gagnep. = Nephelaphyllum tenuiflorum Blume ■

383130　Tainia dalavayi Gagnep. = Collabium formosanum Hayata ■

383131　Tainia delavayi Gagnep. = Collabium delavayi (Gagnep.) Seidenf. ■

383132　Tainia dunnii Rolfe；带唇兰(长叶杜鹃兰，小杜鹃兰)；Dunn Tainia ■

383133　Tainia elata (Schltr.) P. F. Hunt = Tainia ruybarrettoi (S. Y. Hu et Barretto) Z. H. Tsi ■

383134　Tainia elliptica Fukuy. ；竹东杜鹃兰■

383135　Tainia elliptica Fukuy. = Tainia dunnii Rolfe ■

383136　Tainia elliptica Fukuy. = Tainia latifolia (Lindl.) Rchb. f. ■

383137　Tainia emeiensis (K. Y. Lang) Z. H. Tsi；峨眉带唇兰(峨眉球柄兰)；Emei Mischobulbum，Emei Tainia ■

383138　Tainia fauriei Schltr. = Mischobulbum cordifolium (Hook. f.) Schltr. ■

383139　Tainia fauriei Schltr. = Tainia cordifolia Hook. f. ■

383140　Tainia flabellilobata C. L. Tso = Tainia dunnii Rolfe ■

383141　Tainia gokanzanensis Masam. ；拟山兰■

383142　Tainia gokanzanensis Masam. = Oreorchis indica (Lindl.) Hook. f. ■

383143　Tainia gracilis C. L. Tso = Tainia dunnii Rolfe ■

383144　Tainia hastata (Lindl.) Hook. f. = Tainia latifolia (Lindl.) Rchb. f. ■

383145　Tainia hongkongensis Rolfe；香港带唇兰(香港安兰)；Hongkong Tainia ■

383146 Tainia hookeriana（King et Pantl.）Ts. Tang et F. T. Wang ex Summerh. = Tainia hookeriana King et Pantl. ■

383147 Tainia hookeriana King et Pantl. = Ania hookeriana（King et Pantl.）Ts. Tang et F. T. Wang ex Summerh. ■

383148 Tainia hookeriana King et Pantl. = Tainia penangiana Hook. f. ■

383149 Tainia hualienia S. S. Ying；花莲带唇兰（华丽带唇兰，黄花小杜鹃兰）；Hualian Tainia ■

383150 Tainia khasiana Hook. f. = Tainia latifolia（Lindl.）Rchb. f. ■

383151 Tainia latifolia（Lindl.）Rchb. f.；阔叶带唇兰（大花邓兰，阔叶杜鹃兰，圆叶小杜鹃兰，竹东杜鹃兰）；Broadleaf Tainia ■

383152 Tainia latilingua Hook. f.；宽舌带唇兰■☆

383153 Tainia laxiflora Makino；疏花带唇兰■

383154 Tainia laxiflora Makino var. piyananensis（Fukuy.）Masam. = Tainia laxiflora Makino ■

383155 Tainia laxiflora Makino var. shimadae（Hayata）M. Hiroe = Tainia dunnii Rolfe ■

383156 Tainia longiscapa（Seidenf. ex H. Turner）J. J. Wood et A. L. Lamb；卵叶带唇兰；Ooleaf Tainia ■

383157 Tainia macrantha Hook. f.；大花带唇兰（大花球柄兰）；Largeflower Mischobulbum, Largeflower Tainia ■

383158 Tainia maculata（Thwaites）Trimen = Chrysoglossum ornatum Blume ■

383159 Tainia minor Hook. f.；滇南带唇兰；S. Yunnan Tainia ■

383160 Tainia minor Hook. f. var. laxiflora（Makino）T. Hashim. = Tainia laxiflora Makino ■

383161 Tainia ovifolia Z. H. Tsi et S. C. Chen = Tainia longiscapa（Seidenf. ex H. Turner）J. J. Wood et A. L. Lamb ■

383162 Tainia parvifolia C. L. Tso = Tainia dunnii Rolfe ■

383163 Tainia penangiana Hook. f.；绿花带唇兰（绿花安兰，马来带唇兰）■

383164 Tainia penangiana Hook. f. = Ania penangiana（Hook. f.）Summerh. ■

383165 Tainia piyananensis Fukuy. = Tainia dunnii Rolfe ■

383166 Tainia piyananensis Fukuy. = Tainia laxiflora Makino ■

383167 Tainia procera Senghas = Tainia dunnii Rolfe ■

383168 Tainia quadriloba Summerh. = Tainia dunnii Rolfe ■

383169 Tainia ruybarrettoi（S. Y. Hu et Barretto）Z. H. Tsi；南方带唇兰；Southern Tainia ■

383170 Tainia shimadae Hayata；长叶杜鹃兰■

383171 Tainia shimadae Hayata = Tainia dunnii Rolfe ■

383172 Tainia shimadae Hayata var. elliptica（Fukuy.）S. S. Ying = Tainia latifolia（Lindl.）Rchb. f. ■

383173 Tainia shimadai Hayata var. elliptica（Fukuy.）S. S. Ying = Tainia latifolia（Lindl.）Rchb. f. ■

383174 Tainia stellata（Lindl.）Pfitzer = Eria javanica（Sw.）Blume ■

383175 Tainia taiwaniana S. S. Ying = Tainia hookeriana King et Pantl. ■

383176 Tainia taiwaniana S. S. Ying = Tainia penangiana Hook. f. ■

383177 Tainia tenuiflora（Blume）Gagnep. = Nephelaphyllum tenuiflorum Blume ■

383178 Tainia unguiculata Hayata = Acanthephippium striatum Lindl. ■

383179 Tainia viridifusca（Hook.）Benth. et Hook. f.；高褶带唇兰（滇粤安兰，五脊安兰）；Greenbrown Tainia ■

383180 Tainionema Schltr.（1899）；带蕊萝藦属●☆

383181 Tainionema occidentale Schltr.；带蕊萝藦●☆

383182 Tainiopsis Hayata = Tainia Blume ■

383183 Tainiopsis Schltr. = Eriodes Rolfe ■

383184 Tainiopsis barbata（Lindl.）Schltr. = Eriodes barbata（Lindl.）Rolfe ■

383185 Tainiopsis barbata Schltr. = Eriodes barbata（Lindl.）Rolfe ■

383186 Taitonia Yamam. = Gomphostemma Wall. ex Benth. ●■

383187 Taitonia callicarpoides Yamam. = Gomphostemma callicarpoides（Yamam.）Masam. ●■

383188 Taiwania Hayata（1906）；台湾杉属（秃杉属）；Taiwania ●★

383189 Taiwania cryptomerioides Hayata；台湾杉（松萝，台杉，台湾松，秃杉，土杉，亚杉）；Cryptomeria-like Taiwania, Flous Taiwania, Taiwan Cryptomeria, Taiwania ●◇

383190 Taiwania cryptomerioides Hayata var. flousiana（Gaussen）Silba = Taiwania cryptomerioides Hayata ●◇

383191 Taiwania flousiana Gaussen；秃杉●

383192 Taiwania flousiana Gaussen = Taiwania cryptomerioides Hayata ●◇

383193 Taiwania yunnanensis Koidz. = Taiwania cryptomerioides Hayata ●◇

383194 Taiwania yunnanensis Koidz. = Taiwania flousiana Gaussen ●

383195 Taiwaniaceae Hayata = Cupressaceae Gray（保留科名）●

383196 Taiwaniaceae Hayata = Taxodiaceae Saporta（保留科名）●

383197 Taiwaniaceae Hayata；台湾杉科●

383198 Taiwanites Hayata = Taiwania Hayata ●★

383199 Takaikazuchia Kitag. = Olgaea Iljin ■

383200 Takaikazuchia Kitag. et Kitam. = Olgaea Iljin ■

383201 Takaikazuchia lomonosowii（Trautv.）Kitag. et Kitam. = Olgaea lomonosowii（Trautv.）Iljin ■

383202 Takasagoya Y. Kimura = Hypericum L. ■●

383203 Takasagoya acutisepala（Hayata）Y. Kimura = Hypericum geminiflorum Hemsl. ●

383204 Takasagoya formosana（Maxim.）Y. Kimura = Hypericum formosanum Maxim. ●

383205 Takasagoya geminiflora（Hesml.）Y. Kimura = Hypericum geminiflorum Hemsl. ●

383206 Takasagoya nakamurae Masam. = Hypericum nakamurai（Masam.）N. Robson ●

383207 Takasagoya simplicistyla（Hayata）Y. Kimura = Hypericum geminiflorum Hemsl. var. simplicistylum（Hayata）N. Robson ●

383208 Takasagoya subalata（Hayata）Y. Kimura = Hypericum subalatum Hayata ●

383209 Takasagoya trinervia（Hemsl.）Y. Kimura = Hypericum geminiflorum Hemsl. ●

383210 Takeikadzuchia Kitag. et Kitam.（1934）；鳍蓟属（猬菊属）■

383211 Takeikadzuchia Kitag. et Kitam. = Olgaea Iljin ■

383212 Takeikadzuchia bonosowii（Trautv.）Kitag. et Kitam. = Olgaea lomonosowii（Trautv.）Iljin ■

383213 Takeikadzuchia lomonossowii（Trautv.）Kitag. et Kitam.；鳍蓟■

383214 Takeikatzukia Kitag. et Kitam. = Olgaea Iljin ■

383215 Takhtajania Baranova et J. -F. Leroy（1978）；塔氏林仙属■☆

383216 Takhtajania perrieri（Capuron）Baranova et J. -F. Leroy；塔氏林仙●☆

383217 Takhtajaniaceae（J. -F. Leroy）J. -F. Leroy = Winteraceae R. Br. ex Lindl.（保留科名）●

383218 Takhtajaniaceae J. -F. Leroy = Winteraceae R. Br. ex Lindl.（保留科名）●

383219 Takhtajaniantha Nazarova = Scorzonera L. ■

383220 Takhtajanianthus A. B. De = Rhanteriopsis Rauschert ●☆

383221 Takhtajaniella V. E. Avet. = Alyssoides Mill. ■●☆

383222 Takhtajaniella V. E. Avet. = Alyssum L. ■●

383223 Takulumena Szlach.（2006）；厄瓜多尔柱瓣兰属■☆

383224 Takulumena Szlach. = Epidendrum L.（保留属名）■☆

383225 Tala Blanco = Limnophila R. Br. (保留属名)■

383226 Talamancalia H. Rob. et Cuatrec. (1994);翅柄千里光属■●☆

383227 Talamancalia boquetensis (Standl.) H. Rob. et Cuatrec.;翅柄千里光■☆

383228 Talanelis Raf. = Campanula L. ■●

383229 Talangninia Chapel. ex DC. = Randia L. ●

383230 Talasium Spreng. = Panicum L. ■

383231 Talasium Spreng. = Thalasium Spreng. ■

383232 Talassia Korovin = Ferula L. ■

383233 Talassia Korovin(1962);伊犁芹属;Yilicelery ■

383234 Talassia transiliensis (Herder) Korovin = Talassia transiliensis (Regel et Herder) Korovin ■

383235 Talassia transiliensis (Regel et Herder) Korovin;伊犁芹;Yilicelery Talassia ■

383236 Talauma Juss. (1789);盖裂木属(盖裂木兰属,脱轴木属);Talauma ●

383237 Talauma Juss. = Magnolia L. ●

383238 Talauma coco (Lour.) Merr. = Lirianthe coco (Lour.) N. H. Xia et C. Y. Wu ●

383239 Talauma coco Merr. = Magnolia coco (Lour.) DC. ●

383240 Talauma dubia Eichler;可疑盖裂木●☆

383241 Talauma elegans Miq.;南洋盖裂木(南洋玉兰)●☆

383242 Talauma fistulosa Finet et Gagnep. = Lirianthe fistulosa (Finet et Gagnep.) N. H. Xia et C. Y. Wu ●

383243 Talauma fistulosa Finet et Gagnep. = Magnolia paenetalauma Dandy ●

383244 Talauma gioi A. Chev. = Michelia gioi (A. Chev.) Sima et Hong Yu ●◇

383245 Talauma gitingensis Elmer = Magnolia paenetalauma Dandy ●

383246 Talauma hodgsonii Hook. f. et Thomson;盖裂木;Hodgson Talauma ●

383247 Talauma kerrii Craib = Lirianthe henryi (Dunn) N. H. Xia et C. Y. Wu ●◇

383248 Talauma kerrii Craib = Magnolia henryi Dunn ●◇

383249 Talauma mexicana G. Don;墨西哥盖裂木(墨西哥达老玉兰)●☆

383250 Talauma mutabilis Blume;易变盖裂木●☆

383251 Talauma ovata A. St.-Hil.;卵圆盖裂木;Ovate Talauma ●

383252 Talauma phellocarpa King = Paramichelia baillonii (Pierre) Hu ●◇

383253 Talauma pumila (Andréws) Blume = Lirianthe coco (Lour.) N. H. Xia et C. Y. Wu ●

383254 Talauma spongiocarpa King = Paramichelia baillonii (Pierre) Hu ●◇

383255 Talbotia Balf. = Vellozia Vand. ■☆

383256 Talbotia Balf. = Xerophyta Juss. ●■☆

383257 Talbotia S. Moore = Afrofittonia Lindau ■☆

383258 Talbotia elegans Balf.;塔若翠■☆

383259 Talbotia radicans S. Moore = Afrofittonia silvestris Lindau ■☆

383260 Talbotiella Baker f. (1914);塔氏豆属■☆

383261 Talbotiella batesii Baker f.;非洲塔氏豆■☆

383262 Talbotiella eketensis Baker f.;塔氏豆■☆

383263 Talbotiopsis L. B. Sm. = Talbotia Balf. ■☆

383264 Talechium Hill = Campanula L. ■●

383265 Talechium Hill = Trachelium Hill ■●

383266 Talguenea Miers = Talguenea Miers ex Endl. ●☆

383267 Talguenea Miers ex Endl. (1840);智利鼠李属●☆

383268 Talguenea costata Miers;智利鼠李●☆

383269 Tali Adans. = Connarus L. ●

383270 Taliera Mart. = Corypha L. ●

383271 Taligalea Aubl. (废弃属名) = Amasonia L. f. (保留属名)●■☆

383272 Talinaceae Doweld = Portulacaceae Juss. (保留科名)■●

383273 Talinaceae Doweld(2001);土人参科●

383274 Talinaria Brandegee = Talinum Adans. (保留属名)■●

383275 Talinaria Brandegee(1908);肖土人参属■☆

383276 Talinaria palmeri Brandegee;肖土人参■☆

383277 Talinella Baill. (1886);小土人参属●☆

383278 Talinella albidiflora Appleq.;白花小土人参●☆

383279 Talinella ankaranensis Appleq.;安卡兰小土人参●☆

383280 Talinella boiviniana Baill.;博伊文小土人参●☆

383281 Talinella bosseri Appleq.;博瑟小土人参●☆

383282 Talinella dauphinensis Scott-Elliot;多芬小土人参●☆

383283 Talinella grevei Danguy;格雷弗小土人参●☆

383284 Talinella humbertii Appleq.;亨伯特小土人参●☆

383285 Talinella latifolia Appleq.;宽叶小土人参●☆

383286 Talinella microphylla Eggli;小叶小土人参●☆

383287 Talinella pachypoda Eggli;粗梗小土人参●☆

383288 Talinella tsitondroinensis Appleq.;马岛小土人参●☆

383289 Talinella xerophila Appleq.;沙地小土人参●☆

383290 Talinium Raf. = Talinum Adans. (保留属名)■●

383291 Talinopsis A. Gray(1852);树土人参属●☆

383292 Talinopsis frutescens A. Gray;树土人参●☆

383293 Talinum Adans. (1763)(保留属名);土人参属;Fame Flower, Fameflower, Localginseng, Talinum ■●

383294 Talinum angustissimum (Engelm.) Wooton et Standl. = Phemeranthus aurantiacus (Engelm.) Kiger ■☆

383295 Talinum appalachianum W. Wolf = Phemeranthus parviflorus (Nutt.) Kiger ■☆

383296 Talinum arnotii Hook. f.;阿诺特土人参■☆

383297 Talinum aurantiacum Engelm. = Phemeranthus aurantiacus (Engelm.) Kiger ■☆

383298 Talinum aurantiacum Engelm. var. angustissimum Engelm. = Phemeranthus aurantiacus (Engelm.) Kiger ■☆

383299 Talinum brachypodum S. Watson = Phemeranthus brevifolius (Torr.) Hershk. ■☆

383300 Talinum brevicaule S. Watson = Phemeranthus brevicaulis (S. Watson) Kiger ■☆

383301 Talinum brevifolium Torr. = Phemeranthus brevifolius (Torr.) Hershk. ■☆

383302 Talinum caffrum (Thunb.) Eckl. et Zeyh.;开菲尔土人参■☆

383303 Talinum calcaricum S. Ware = Phemeranthus calcaricus (S. Ware) Kiger ■☆

383304 Talinum calycinum Engelm.;大花土人参;Fame Flower, Flame Flower, Large Flower-of-an-hour, Largeflower Fameflower, Rock Pink ■☆

383305 Talinum calycinum Engelm. = Phemeranthus calycinus (Engelm.) Kiger ■

383306 Talinum chrysanthum Rose et Standl. = Talinum paniculatum (Jacq.) Gaertn. ■

383307 Talinum ciliatum Lindl. = Phemeranthus teretifolius (Pursh) Raf. ■☆

383308 Talinum ciliatum Ruiz et Pav. = Calandrinia ciliata DC. ■☆

383309 Talinum confertiflorum Greene = Phemeranthus parviflorus (Nutt.) Kiger ■☆

383310 Talinum crassifolium Willd. = Talinum triangulare (Jacq.) Willd. ●☆

383311 Talinum cuneifolium (Vahl) Willd. = Talinum portulacifolium

（Forssk.）Asch. ex Schweinf. ■☆

383312　Talinum dinteri Poelln. = Talinum tenuissimum Dinter ■☆

383313　Talinum esculentum Dinter et G. Schellenb. = Talinum caffrum（Thunb.）Eckl. et Zeyh. ■☆

383314　Talinum eximium A. Nelson = Phemeranthus brevicaulis（S. Watson）Kiger ■☆

383315　Talinum fallax Poelln. = Phemeranthus parviflorus（Nutt.）Kiger ■☆

383316　Talinum fruticosum（L.）Juss.；灌木状土人参■☆

383317　Talinum gooddingii P. Wilson = Phemeranthus parviflorus（Nutt.）Kiger ■☆

383318　Talinum humile Greene = Phemeranthus humilis（Greene）Kiger ■☆

383319　Talinum longipes Wooton et Standl. = Phemeranthus longipes（Wooton et Standl.）Kiger ■☆

383320　Talinum marginatum Greene = Phemeranthus marginatus（Greene）Kiger ■☆

383321　Talinum mengesii W. Wolf = Phemeranthus mengesii（W. Wolf）Kiger ■☆

383322　Talinum menziesii Hook. = Calandrinia ciliata（Ruiz et Pav.）DC. ■☆

383323　Talinum okanoganense English = Phemeranthus sediformis（Poelln.）Kiger ■☆

383324　Talinum paniculatum（Jacq.）Gaertn.；土人参（波世兰,波丝兰,参草,东洋参,飞来参,福参,高丽参,红参,红芍药,假人参,力参,卢兰,栌兰,申时花,水人参,桃参,土参,土高丽参,土红参,土花旗参,土洋参,瓦参,瓦坑头,煮饭花,锥花土人参,紫人参）；Jewels of Opar, Panicled Fameflower, Panicled Localginseng, Pink Baby-breath, Pink Baby's Breath ■

383325　Talinum paniculatum（Jacq.）Gaertn. var. sarmentosum（Engelm.）Poelln. = Talinum paniculatum（Jacq.）Gaertn. ■

383326　Talinum parviflorum Nutt. = Phemeranthus parviflorus（Nutt.）Kiger ■☆

383327　Talinum parviflorum Nutt. ex Torr. et A. Gray；小花土人参；Prairie Fame Flower, Rock Pink, Small Flower-of-an-hour ■☆

383328　Talinum patens（L.）Willd. = Talinum paniculatum（Jacq.）Gaertn. ■

383329　Talinum patens Juss. = Talinum paniculatum（Jacq.）Gaertn. ■

383330　Talinum patens Willd. = Talinum paniculatum（Jacq.）Gaertn. ■

383331　Talinum portulacifolium（Forssk.）Asch. ex Schweinf.；马齿苋叶土人参■☆

383332　Talinum pulchellum Wooton et Standl. = Phemeranthus brevicaulis（S. Watson）Kiger ■☆

383333　Talinum pygmaeum A. Gray = Lewisia pygmaea（A. Gray）B. L. Rob. ■☆

383334　Talinum rugospermum Holz. = Phemeranthus rugospermus（Holz.）Kiger ■☆

383335　Talinum sarmentosum Engelm. = Talinum paniculatum（Jacq.）Gaertn. ■

383336　Talinum sediforme Poelln. = Phemeranthus sediformis（Poelln.）Kiger ■☆

383337　Talinum spathulatum Engelm. ex A. Gray = Talinum paniculatum（Jacq.）Gaertn. ■

383338　Talinum spinescens Torr. = Phemeranthus spinescens（Torr.）Hershk. ■☆

383339　Talinum taitense Pax et Vatke = Calyptrotheca taitense（Pax et Vatke）Brenan ●☆

383340　Talinum tenuissimum Dinter；极细土人参■☆

383341　Talinum teretifolium Pursh；柱叶土人参■☆

383342　Talinum teretifolium Pursh = Phemeranthus teretifolius（Pursh）Raf. ■☆

383343　Talinum thompsonii N. D. Atwood et S. L. Welsh = Phemeranthus thompsonii（N. D. Atwood et S. L. Welsh）Kiger ■☆

383344　Talinum transvaalense Poelln. = Talinum tenuissimum Dinter ■☆

383345　Talinum triangulare（Jacq.）Willd.；棱轴土人参（栌兰,土人参）；Ceylon Spinach, Fameflower, Waterleaf ■☆

383346　Talinum triangulare（Jacq.）Willd. = Talinum fruticosum（L.）Juss. ■☆

383347　Talinum validulum Greene = Phemeranthus validulus（Greene）Kiger ■☆

383348　Talinum wayae Eastw. = Phemeranthus sediformis（Poelln.）Kiger ■☆

383349　Talinum whitei I. M. Johnst. = Phemeranthus aurantiacus（Engelm.）Kiger ■☆

383350　Talinum youngiae C. H. Mull. = Phemeranthus brevicaulis（S. Watson）Kiger ■☆

383351　Talipariti Fryxell = Hibiscus L.（保留属名）●■

383352　Talipariti Fryxell（2001）；五片果木槿属●☆

383353　Talipariti hamabo（Siebold et Zucc.）Fryxell = Hibiscus hamabo Siebold et Zucc. ●

383354　Talipariti tiliaceum（L.）Fryxell = Hibiscus tiliaceus L. ●

383355　Talipulia Raf. = Aneilema R. Br. ■☆

383356　Talisia Aubl.（1775）；塔利木属（塔利西属）●☆

383357　Talisia amazonica Guarim；亚马逊塔利木●☆

383358　Talisia angustifolia Radlk.；窄叶塔利木●☆

383359　Talisia caudata Steyerm.；尾状塔利木●☆

383360　Talisia dasyclada Radlk.；毛枝塔利木●☆

383361　Talisia erecta Radlk.；直立塔利木●☆

383362　Talisia glabra DC.；光塔利木●☆

383363　Talisia grandifolia Cuatrec.；大叶塔利木●☆

383364　Talisia laxiflora Benth.；疏花塔利木●☆

383365　Talisia micrantha Radlk.；小花塔利木●☆

383366　Talisia multinervis Radlk.；多脉塔利木●☆

383367　Talisia pachycarpa Radlk.；粗果塔利木●☆

383368　Talisia rosea Vahl；粉红塔利木●☆

383369　Talisia squarrosa Radlk.；粗塔利木●☆

383370　Talisiopsis Radlk. = Zanha Hiern ●☆

383371　Talmella Dur. = Talinella Baill. ●☆

383372　Talpa Raf. = Catalpa Scop. ●

383373　Talpinaria H. Karst. = Pleurothallis R. Br. ■☆

383374　Taltalia Ehr. Bayer = Alstroemeria L. ■☆

383375　Taltalia Ehr. Bayer（1998）；智利六出花属■☆

383376　Taltalia graminea（Phil.）Ehr. Bayer；智利六出花■☆

383377　Tamaceae Bercht. et J. Presl = Dioscoreaceae R. Br.（保留科名）●■

383378　Tamaceae Gray = Dioscoreaceae R. Br.（保留科名）●■

383379　Tamaceae Martinov = Dioscoreaceae R. Br.（保留科名）●■

383380　Tamala Raf. = Persea Mill.（保留属名）●

383381　Tamala borbonia（L.）Raf. = Persea borbonia（L.）Spreng. ●☆

383382　Tamala humilis（Nash）Small = Persea humilis Nash ●☆

383383　Tamala littoralis（Small）Small = Persea borbonia（L.）Spreng. ●☆

383384　Tamala palustris Raf. = Persea palustris（Raf.）Sarg. ●☆

383385　Tamala pubescens（Pursh）Small = Persea palustris（Raf.）

Sarg. ●☆

383386　Tamamschjanella Pimenov et Kljuykov(1996);小塔马草属■☆

383387　Tamamschjanella rhizomatica(Hartvig)Pimenov et Kljuykov;小塔马草■☆

383388　Tamamschjanella rubella(E. A. Busch)Pimenov et Kljuykov;红小塔马草■☆

383389　Tamamschjania Pimenov et Kljuykov(1981);塔马草属■☆

383390　Tamamschjania lazica(Boiss. et Balansa)Pimenov et Kljuykov;塔马草■☆

383391　Tamananthus V. M. Badillo(1985);肿叶菊属■●☆

383392　Tamananthus crinitus V. M. Badillo;肿叶菊■●☆

383393　Tamania Cuatrec. = Espeletia Mutis ex Humb. et Bonpl. ●☆

383394　Tamara Roxb. ex Steud. = Nelumbo Adans. ■

383395　Tamaricaceae Bercht. et J. Presl = Tamaricaceae Link(保留科名)●■

383396　Tamaricaceae Gray = Tamaricaceae Link(保留科名)●■

383397　Tamaricaceae Link(1821)(保留科名);柽柳科;Tamarisk Family,Tamarix Family ●

383398　Tamaricaria Qaiser et Ali. = Myricaria Desv. ●

383399　Tamaricaria Qaiser et Ali. = Myrtama Ovcz. et Kinzik. ●

383400　Tamaricaria elegans(Royle)Qaiser et Ali = Myricaria elegans Royle ●

383401　Tamarindaceae Bercht. et J. Presl = Fabaceae Lindl. (保留科名)●■

383402　Tamarindaceae Bercht. et J. Presl = Leguminosae Juss. (保留科名)●■

383403　Tamarindaceae Bercht. et J. Presl;酸豆科

383404　Tamarindus L. (1753);酸豆属(罗晃子属,罗望子属);Sour Bean,Sourbean,Tamarind ●

383405　Tamarindus Tourn. ex L. = Tamarindus L. ●

383406　Tamarindus erythraea Mattei = Tamarindus indica L. ●

383407　Tamarindus indica L. ;酸豆(罗晃子,罗望子,曼姆,酸果树,酸角,酸饺,酸梅,通血图,通血香,乌梅,油楠);Indian Date,Sour Bean, Sourbean, Tamarind Date, Tamarind Tree, Tamarindo,Tamarind-tree ●

383408　Tamarindus officinalis Hook. = Tamarindus indica L. ●

383409　Tamarindus somalensis Mattei = Tamarindus indica L. ●

383410　Tamariscaceae A. St. -Hil. = Tamaricaceae Link(保留科名)●■

383411　Tamariscus Mill. = Tamarix L. ●

383412　Tamarix L. (1753);柽柳属(红柳属);Salt Cedar, Saltcedar, Tamarisk,Tamarisk Salt Cedar,Tamarix ●

383413　Tamarix affinis Bunge = Tamarix gracilis Willd. ●

383414　Tamarix africana Poir. ;非洲柽柳;African Tamarisk ●☆

383415　Tamarix africana Poir. var. bimamillata(Trab.)Maire = Tamarix africana Poir. ●☆

383416　Tamarix africana Poir. var. brevistyla Trab. = Tamarix africana Poir. ●☆

383417　Tamarix africana Poir. var. brivesii Maire et Trab. = Tamarix africana Poir. ●☆

383418　Tamarix africana Poir. var. ducellieri Maire = Tamarix africana Poir. ●☆

383419　Tamarix africana Poir. var. faurei(Sennen)Maire = Tamarix africana Poir. ●☆

383420　Tamarix africana Poir. var. fluminensis(Maire)Baum = Tamarix africana Poir. ●☆

383421　Tamarix africana Poir. var. macrostachya Gay = Tamarix africana Poir. ●☆

383422　Tamarix africana Poir. var. mairei(Trab.)Maire = Tamarix africana Poir. ●☆

383423　Tamarix africana Poir. var. microstachys Maire = Tamarix africana Poir. ●☆

383424　Tamarix africana Poir. var. mogadorensis Maire et Trab. = Tamarix africana Poir. ●☆

383425　Tamarix africana Poir. var. palasyana Maire = Tamarix africana Poir. ●☆

383426　Tamarix africana Poir. var. rifana Maire et Trab. = Tamarix africana Poir. ●☆

383427　Tamarix africana Poir. var. weilleri Maire = Tamarix africana Poir. ●☆

383428　Tamarix africana Poir. var. wilczekii Maire = Tamarix africana Poir. ●☆

383429　Tamarix allorgeana Sennen et Mauricio = Tamarix africana Poir. ●☆

383430　Tamarix amplexicaulis Ehrenb. ;抱茎柽柳●☆

383431　Tamarix androssowii Litv. ;紫杆柽柳(白花柽柳,直立紫杆柽柳,中亚柽柳,紫茎柽柳);Adrossow Tamarisk, White Tamarisk ●

383432　Tamarix androssowii Litv. var. transcaucassica(Bunge)Qaiser;外高加索柽柳●☆

383433　Tamarix anglica Webb;英国柽柳;Brummel, English Tamarisk, Tamarisk ●☆

383434　Tamarix anglica Webb = Tamarix gallica L. ●☆

383435　Tamarix angolensis Nied. = Tamarix usneoides E. Mey. ex Bunge ●☆

383436　Tamarix angustifolia Ledeb. = Tamarix gracilis Willd. ●

383437　Tamarix aphylla(L.)H. Karst. ;无叶柽柳(节枝柽柳,亚非柽柳);Athel, Athel Tamarisk, Athel Tree, Evergreen Tamarisk, Leafless Tamarisk, Tamarisk, Tamarisk Galls ●

383438　Tamarix aphylla H. karst. = Tamarix aphylla(L.)H. Karst. ●

383439　Tamarix arabica Bunge = Tamarix nilotica(Ehrenb.)Bunge ●☆

383440　Tamarix aralensis Bunge;俄罗斯柽柳;Russian Tamarisk ●☆

383441　Tamarix araratica(Bunge)Gorschk. = Tamarix kotschyi Bunge ●☆

383442　Tamarix arborea(Sieber ex Ehrenb.)Bunge;树状柽柳●☆

383443　Tamarix arborea(Sieber ex Ehrenb.)Bunge var. subvelutina Boiss. = Tamarix arborea(Sieber ex Ehrenb.)Bunge ●☆

383444　Tamarix arceuthoides Bunge;密花柽柳;Denseflower Tamarisk, Dense-flowered Tamarisk ●

383445　Tamarix articulata Vahl = Tamarix aphylla(L.)H. Karst. ●

383446　Tamarix articulata Wall. = Tamarix dioica Roxb. ●☆

383447　Tamarix askabadansis Freyn = Tamarix arceuthoides Bunge ●

383448　Tamarix austro-africana Schinz = Tamarix usneoides E. Mey. ex Bunge ●☆

383449　Tamarix austromongolica Nakai;甘蒙柽柳;S. Mongol Tamarisk, South Mongolia Tamarisk, South Mongolian Tamarisk ●

383450　Tamarix balansae Batt. = Tamarix amplexicaulis Ehrenb. ●☆

383451　Tamarix balansae J. Gay = Tamarix passernioides Desv. var. macrocarpa Ehrenb. ●☆

383452　Tamarix balansae J. Gay var. longipes Maire = Tamarix amplexicaulis Ehrenb. ●☆

383453　Tamarix balansae J. Gay var. micrantha Maire et Trab. = Tamarix canariensis Willd. ●☆

383454　Tamarix balansae J. Gay var. microstyla Maire = Tamarix amplexicaulis Ehrenb. ●☆

383455　Tamarix balansae J. Gay var. oxysepala Trab. = Tamarix passerinoides Desv. ●☆

383456　Tamarix balansae J. Gay var. squarrosa Maire = Tamarix amplexicaulis Ehrenb. ●☆

383457　Tamarix bengalensis Baum ＝ Tamarix indica Willd. ●☆

383458　Tamarix bounopaea Batt. ＝ Tamarix boveana Bunge ●☆

383459　Tamarix boveana Bunge；博韦柽柳●☆

383460　Tamarix boveana Bunge var. longipes（Trab.）Maire ＝ Tamarix boveana Bunge ●☆

383461　Tamarix brachystilis Batt. ＝ Tamarix canariensis Willd. ●☆

383462　Tamarix brachystilis J. Gay var. fluminensis Maire ＝ Tamarix africana Poir. ●☆

383463　Tamarix brachystilis J. Gay var. geyrii（Diels）Maire ＝ Tamarix canariensis Willd. ●☆

383464　Tamarix brachystilis J. Gay var. littoralis Pau et Font Quer ＝ Tamarix canariensis Willd. ●☆

383465　Tamarix brachystilis J. Gay var. stenonema Maire et Trab. ＝ Tamarix canariensis Willd. ●☆

383466　Tamarix bungei Boiss.；邦奇柽柳●☆

383467　Tamarix bungei Boiss. ＝ Tamarix arceuthoides Bunge ●

383468　Tamarix canariensis Willd.；加那利柽柳；Canary Island Tamarisk ●☆

383469　Tamarix chinensis Lour.；柽柳（长寿仙人柳，柽，赤柽，赤柽柳，赤柳，赤杨，垂丝柳，春柳，观音柳，河柳，红筋条，红柳，华北柽柳，桧柽柳，桧叶柽柳，菩萨柳，人柳，三川柳，三春柳，三眠柳，山川柳，蜀柳，丝柳，西河柳，西湖柳，殷柽，雨师，雨师柳，雨丝，御柳，中国柽柳）；Chinese Tamarisk，Chinese Tamarix，Fivestamen Tamarisk，Five-stamen Tamarisk，Juniper Tamarisk，Salt Cedar，Saltcedar，Tamarisk ●

383470　Tamarix chinensis Lour. subsp. austromongolica（Nakai）S. Q. Zhou ＝ Tamarix austromongolica Nakai ●

383471　Tamarix cretica Bunge ＝ Tamarix parviflora DC. ●

383472　Tamarix cupressiformis Ledeb. ＝ Tamarix gracilis Willd. ●

383473　Tamarix davurica Willd. ＝ Myricaria davurica（Willd.）Ehrenb. ●

383474　Tamarix deserti Boiss. ＝ Tamarix tetragyna Ehrenb. var. meyeri（Boiss.）Boiss. ●☆

383475　Tamarix dioica Roxb. ＝ Tamarix dioica Roxb. ex Roch ●☆

383476　Tamarix dioica Roxb. ex Roch；印缅柽柳●☆

383477　Tamarix edessana Steven；奥杰沙柽柳●☆

383478　Tamarix elegans Spach ＝ Tamarix chinensis Lour. ●

383479　Tamarix elongata Ledeb.；长穗柽柳（长柽柳，红柳）；Long Spike Tamarisk，Longspike Tamarisk，Long-spiked Tamarisk ●

383480　Tamarix engleri Arendt ＝ Tamarix usneoides E. Mey. ex Bunge ●☆

383481　Tamarix epacroides Sm. ＝ Tamarix indica Willd. ●☆

383482　Tamarix ericoides Rottler ＝ Tamarix ericoides Rottler et Willd. ●☆

383483　Tamarix ericoides Rottler et Willd.；欧石楠柽柳●☆

383484　Tamarix eversmannii C. Presl ex Ledeb.；埃威氏柽柳；Eversmann's Tamarisk ●☆

383485　Tamarix eversmannii C. Presl ex Ledeb. ＝ Tamarix ramosissima Ledeb. ●

383486　Tamarix florida Bunge；繁花柽柳●☆

383487　Tamarix florida Bunge var. albifore Bunge ＝ Tamarix arceuthoides Bunge ●

383488　Tamarix florida Bunge var. kotschyi Bunge ＝ Tamarix arceuthoides Bunge ●

383489　Tamarix florida Bunge var. rosea Bunge ＝ Tamarix smyrnensis Bunge ●☆

383490　Tamarix fontqueri Maire et Trab. ＝ Tamarix africana Poir. ●☆

383491　Tamarix gallica L.；法国柽柳（五倍子柽柳）；Common Tamarisk，French Tamarisk，French Tree，Heath，Ling，Manna Plant，Salt Cedar，Tamarisk，Tamarisk Galls ●☆

383492　Tamarix gallica L. ＝ Tamarix indica Willd. ●☆

383493　Tamarix gallica L. subsp. epidiscina Maire ＝ Tamarix canariensis Willd. ●☆

383494　Tamarix gallica L. subsp. leucocharis（Maire）Maire ＝ Tamarix canariensis Willd. ●☆

383495　Tamarix gallica L. subsp. speciosa Ball. ＝ Tamarix africana Poir. ●☆

383496　Tamarix gallica L. var. arborea Sieber ex Ehrenb. ＝ Tamarix arborea（Sieber ex Ehrenb.）Bunge ●☆

383497　Tamarix gallica L. var. brevibracteata Maire ＝ Tamarix gallica L. ●☆

383498　Tamarix gallica L. var. chinensis（Lour.）Ehrenb. ＝ Tamarix chinensis Lour. ●

383499　Tamarix gallica L. var. indica（Willd.）Ehrenb. ＝ Tamarix gallica L. ●☆

383500　Tamarix gallica L. var. indica（Willd.）Ehrenb. ＝ Tamarix indica Willd. ●☆

383501　Tamarix gallica L. var. lagunae（Caball.）Maire ＝ Tamarix canariensis Willd. ●☆

383502　Tamarix gallica L. var. littoralis（Pau et Font Quer）Maire ＝ Tamarix gallica L. ●☆

383503　Tamarix gallica L. var. longibracteata Maire ＝ Tamarix gallica L. ●☆

383504　Tamarix gallica L. var. micrantha Ledeb. ＝ Tamarix leptostachys Bunge ●

383505　Tamarix gallica L. var. micrantha Ledeb. ＝ Tamarix ramosissima Ledeb. ●

383506　Tamarix gallica L. var. monodiana Maire ＝ Tamarix senegalensis DC. ●☆

383507　Tamarix gallica L. var. nilotica Ehrenb. ＝ Tamarix nilotica（Ehrenb.）Bunge ●☆

383508　Tamarix gallica L. var. pallasii ? ＝ Tamarix ramosissima Ledeb. ●

383509　Tamarix gallica L. var. pleiandra Maire ＝ Tamarix africana Poir. ●☆

383510　Tamarix gallica L. var. pyconostachys Ledeb. ＝ Tamarix smyrnensis Bunge ●☆

383511　Tamarix gallica L. var. submutica Maire et Trab. ＝ Tamarix gallica L. ●☆

383512　Tamarix gallica L. var. wallii Maire ＝ Tamarix gallica L. ●☆

383513　Tamarix gallica sensu Wight et Arn. ＝ Tamarix indica Willd. ●☆

383514　Tamarix gansuensis H. Z. Zhang ex P. Y. Zhang et M. T. Liu；甘肃柽柳；Gansu Tamarisk ●

383515　Tamarix geyrii Diels ＝ Tamarix canariensis Willd. ●☆

383516　Tamarix gracilis Willd.；翠枝柽柳（异花柽柳）；Greenbranch Tamarisk，Slender Tamarisk，Thin Tamarisk ●

383517　Tamarix hispida Willd.；刚毛柽柳（粗毛柽柳，粗硬毛柽柳，毛柽柳，毛红柳）；Bristle Tamarisk，Kaschgar Tamarisk ●

383518　Tamarix hispida Willd. var. karelinii（Bunge）B. R. Baum ＝ Tamarix karelinii Bunge ●

383519　Tamarix hohenackeri Bunge；多花柽柳（贺恒氏柽柳，霍氏柽柳）；Hohenacker's Tamarisk，Many Flower Tamarisk，Manyflower Tamarisk，Multiflorous Tamarisk ●

383520　Tamarix indica Willd.；印度柽柳●☆

383521　Tamarix jintaensis P. Y. Zhang et M. T. Liu；金塔柽柳（金穗柽柳）；Jinta Tamarisk ●

383522　Tamarix juniperina Bunge ＝ Tamarix chinensis Lour. ●

383523　Tamarix karakalensis Freyn；卡拉卡利柽柳●☆

383524　Tamarix karakalensis Freyn var. myriantha Freyn ＝ Tamarix arceuthoides Bunge ●

383525　Tamarix karakalensis Freyn var. verrucifera Freyn　= Tamarix arceuthoides Bunge ●

383526　Tamarix karelinii Bunge;盐地柽柳(短毛柽柳);Karelin's Tamarisk,Saline Tamarisk ●

383527　Tamarix karelinii Bunge var. densior Trautv. = Tamarix karelinii Bunge ●

383528　Tamarix komarovii Gorschk.;科马罗夫柽柳●☆

383529　Tamarix korolkowii Regel et Schmalh. = Tamarix korolkowii Regel et Schmalh. ex Regel ●

383530　Tamarix korolkowii Regel et Schmalh. ex Regel;垂枝柽柳; Korolkow Tamarisk,Nutantbranch Tamarisk ●

383531　Tamarix kotschyi Bunge;科奇柽柳●☆

383532　Tamarix kotschyi Bunge var. rosea Litv. = Tamarix kotschyi Bunge ●☆

383533　Tamarix ladachensis B. R. Baum = Myricaria elegans Royle ●

383534　Tamarix lagunae Caball. = Tamarix canariensis Willd. ●☆

383535　Tamarix laxa Willd.;短穗柽柳(短穗红柳,毛穗柽柳,疏花柽柳); Short Spike Tamarisk,Shortspike Tamarisk,Short-spiked Tamarisk ●

383536　Tamarix laxa Willd. var. araratica Bunge = Tamarix kotschyi Bunge ●☆

383537　Tamarix laxa Willd. var. parvifora Litv. = Tamarix androssowii Litv. var. transcaucassica(Bunge)Qaiser ●☆

383538　Tamarix laxa Willd. var. polystachya(Ledeb.)Bunge;伞花短穗柽柳;Many-stachys Tamarisk ●

383539　Tamarix laxa Willd. var. subspicata Ehrenb. = Tamarix parviflora DC. ●

383540　Tamarix laxa Willd. var. transcaucassica Bunge = Tamarix androssowii Litv. var. transcaucassica(Bunge)Qaiser ●☆

383541　Tamarix leptopetala Bunge = Tamarix mascatensis Bunge ●☆

383542　Tamarix leptostachys Bunge;细穗柽柳;Thinspike Tamarisk, Thin-spiked Tamarisk ●

383543　Tamarix leptostachys Bunge = Tamarix korolkowii Regel et Schmalh. ex Regel ●

383544　Tamarix leptostachys sensu Russ = Tamarix korolkowii Regel et Schmalh. ex Regel ●

383545　Tamarix lessingii Persl. ex Bunge = Tamarix smyrnensis Bungee ●☆

383546　Tamarix leucocharis Maire = Tamarix canariensis Willd. ●☆

383547　Tamarix litwinowii Gorschk.;利特氏柽柳●☆

383548　Tamarix litwinowii Gorschk. = Tamarix androssowii Litv. var. transcaucassica(Bunge)Qaiser ●☆

383549　Tamarix longe-pedunculata Blatt. et Hallb. = Tamarix dioica Roxb. ex Roch ●☆

383550　Tamarix lucronensis Sennen et Elias = Tamarix parviflora DC. ●

383551　Tamarix macrocarpa(Ehrenb.)Bunge = Tamarix passernioides Desv. var. macrocarpa Ehrenb. ●☆

383552　Tamarix macrocarpa(Ehrenb.)Bunge var. micrantha(Corti)Baum = Tamarix macrocarpa(Ehrenb.)Bunge ●☆

383553　Tamarix mannifera Ehrenb. ex Bunge;甘露柽柳;Mannatamarisk ●☆

383554　Tamarix martana Kom. = Tamarix korolkowii Regel et Schmalh. ex Regel ●

383555　Tamarix mascatensis Bunge;细瓣柽柳●☆

383556　Tamarix mauritii Sennen = Tamarix africana Poir. ●☆

383557　Tamarix meyeri Boiss.;梅氏柽柳;Meyer Tamarisk ●☆

383558　Tamarix meyeri Boiss. = Tamarix tetragyna Ehrenb. var. meyeri(Boiss.)Boiss. ●☆

383559　Tamarix mongolica Nied.;蒙古柽柳;Mongol Tamarisk, Mongolian Tamarisk ●

383560　Tamarix mucronata Sm. = Tamarix ericoides Rottler et Willd. ●☆

383561　Tamarix muluyana Sennen = Tamarix canariensis Willd. ●☆

383562　Tamarix murbeckii Sennen = Tamarix canariensis Willd. ●☆

383563　Tamarix nilotica(Ehrenb.)Bunge;阿拉伯柽柳;Nile Tamarisk ●☆

383564　Tamarix octandra Bunge;八雄蕊柽柳●☆

383565　Tamarix odessana Stev. ex Bunge = Tamarix ramosissima Ledeb. ●

383566　Tamarix orientalis Forssk. = Tamarix aphylla(L.)H. Karst. ●

383567　Tamarix pakistanica Qaiser;巴基斯坦柽柳●☆

383568　Tamarix pallasii Desv. = Tamarix laxa Willd. ●

383569　Tamarix pallasii Desv. = Tamarix ramosissima Ledeb. ●

383570　Tamarix pallasii Desv. var. albiflora Grint. = Tamarix smyrnensis Bunge ●☆

383571　Tamarix pallasii Desv. var. brachystachys Bunge = Tamarix ramosissima Ledeb. ●

383572　Tamarix pallasii Desv. var. effusa Bunge = Tamarix smyrnensis Bunge ●☆

383573　Tamarix pallasii Desv. var. macrostemon Bunge = Tamarix smyrnensis Bunge ●☆

383574　Tamarix pallasii Desv. var. minutijlora Bunge = Tamarix ramosissima Ledeb. ●

383575　Tamarix pallasii Desv. var. moldavica Bunge = Tamarix smyrnensis Bunge ●☆

383576　Tamarix pallasii Desv. var. ramosissima(Ledeb.)Bunge = Tamarix ramosissima Ledeb. ●

383577　Tamarix pallasii Desv. var. ramosissima Kar. = Tamarix karelinii Bunge ●

383578　Tamarix pallasii Desv. var. smyrnensis(Bunge)Boiss. = Tamarix smyrnensis Bunge ●☆

383579　Tamarix pallasii Desv. var. tigrensis Bunge = Tamarix ramosissima Ledeb. ●

383580　Tamarix paniculata Stev. ex DC. = Tamarix gracilis Willd. ●

383581　Tamarix parviflora DC.;小花柽柳(浅色柽柳,四蕊柽柳); Early Tamarisk,Four-stamen Tamarisk,Little-flower Tamarisk, Smallflower Tamarisk,Small-flower Tamarisk,Spring-flowering Tamarisk ●

383582　Tamarix parviflora DC. = Tamarix ramosissima Ledeb. ●

383583　Tamarix parviflora DC. var. rubella(Batt.)Maire = Tamarix parviflora DC. ●

383584　Tamarix passerinoides Delile = Tamarix passerinoides Desv. ●☆

383585　Tamarix passerinoides Desv.;雀柽柳●☆

383586　Tamarix passerinoides Desv. var. macrocarpa Ehrenb. = Tamarix passerinoides Desv. ●☆

383587　Tamarix passernioides Desv. var. macrocarpa Ehrenb.;大果柽柳●☆

383588　Tamarix pauciovulata J. Gay = Tamarix amplexicaulis Ehrenb. ●☆

383589　Tamarix pauciovulata J. Gay ex Coss. var. micrantha Corti = Tamarix passernioides Desv. var. macrocarpa Ehrenb. ●☆

383590　Tamarix paviflora DC. var. cretica Boiss. = Tamarix parviflora DC. ●

383591　Tamarix pentandra(Desv.)Pall.;五蕊柽柳(柽柳);Fivestamen Tamarisk,Salt Cedar ●

383592　Tamarix pentandra(Desv.)Pall. = Tamarix ramosissima Ledeb. ●

383593　Tamarix pentandra Pall. = Tamarix ramosissima Ledeb. ●

383594　Tamarix polystachya Ledeb. = Tamarix laxa Willd. var. polystachya(Ledeb.)Bunge ●

383595　Tamarix ramosissima Ledeb.;多枝柽柳(俄罗斯柽柳,分枝柽柳,红柳,五蕊柽柳,西河柳);Branchy Tamarisk, Five-stamen Tamarisk, Late Tamarisk, Odessa Tamarisk, Ramosest Tamarisk, Salt Cedar, Saltcedar, Tamarisk, Tamarix ●

383596　Tamarix ramosissima Ledeb. 'Pink Cascade';粉红瀑布柽柳●☆

383597　Tamarix ramosissima Ledeb. 'Rosea';玫瑰红柽柳●☆

383598　Tamarix ramosissima Ledeb. 'Rubra';品红柽柳;Summer Glow Tamarix ●☆

383599　Tamarix ramosissima Ledeb. 'Summer Glow';夏日时光柽柳●☆

383600　Tamarix rubella Batt. = Tamarix parviflora DC. ●

383601　Tamarix sachuensis P. Y. Zhang et M. T. Liu;莎车柽柳;Sachu Tamarisk, Shanche Tamarisk ●

383602　Tamarix salina Baum = Tamarix pakistanica Qaiser ●☆

383603　Tamarix salina sensu Baum = Tamarix pakistanica Qaiser ●☆

383604　Tamarix scebelensis Chiov. = Tamarix nilotica (Ehrenb.) Bunge ●☆

383605　Tamarix senegalensis DC.;塞内加尔柽柳●☆

383606　Tamarix serotina Bunge ex Boiss. = Tamarix karelinii Bunge ●

383607　Tamarix smyrnensis Bunge;欧亚柽柳●☆

383608　Tamarix soongarica Pall. = Reaumuria soongarica (Pall.) Maxim. ●

383609　Tamarix speciosa (Ball) Ball = Tamarix africana Poir. ●☆

383610　Tamarix speciosa (Ball) Ball subsp. luribunda (Maire) Maire = Tamarix africana Poir. ●☆

383611　Tamarix speciosa (Ball) Ball var. acutibracteata Maire = Tamarix africana Poir. ●☆

383612　Tamarix speciosa (Ball) Ball var. calarantha (Pau) Maire = Tamarix africana Poir. ●☆

383613　Tamarix speciosa (Ball) Ball var. fontqueri (Maire et Trab.) Maire = Tamarix africana Poir. ●☆

383614　Tamarix speciosa (Ball) Ball var. longicollis (Batt.) Maire = Tamarix africana Poir. ●☆

383615　Tamarix speciosa (Ball) Ball var. tingitana (Pau) Maire = Tamarix africana Poir. ●☆

383616　Tamarix spiridonowii Fedtsch. = Tamarix gracilis Willd. ●

383617　Tamarix stricta Boiss.;刚直柽柳●☆

383618　Tamarix taklamakanensis M. T. Liu;沙生柽柳(塔克拉玛干柽柳);Takelamagan Tamarisk, Taklamakan Tamarisk ●◇

383619　Tamarix tarimensis P. Y. Zhang et M. T. Liu;塔里木柽柳;Talimu Tamarisk, Tarim Tamarisk ●

383620　Tamarix tenacissima Buch.-Ham. ex Wall. = Tamarix ericoides Rottler et Willd. ●☆

383621　Tamarix tenuifolia Maire et Trab. = Tamarix passerinoides Desv. ●☆

383622　Tamarix tenuissima Nakai;纤细柽柳(细柽柳);Slender Tamarisk, Tenuis Tamarisk ●

383623　Tamarix tetragyna Ehrenb.;四蕊柽柳●☆

383624　Tamarix tetragyna Ehrenb. var. deserti (Boiss.) Zohary = Tamarix tetragyna Ehrenb. var. meyeri (Boiss.) Boiss. ●☆

383625　Tamarix tetragyna Ehrenb. var. meyeri (Boiss.) Boiss.;迈尔柽柳●☆

383626　Tamarix tetrandra Pall. ex M. Bieb. = Tamarix parviflora DC. ●

383627　Tamarix tetrandra Pall. ex M. Bieb. var. parviflora Boiss. ex Bunge = Tamarix kotschyi Bunge ●☆

383628　Tamarix trabutii Maire = Tamarix amplexicaulis Ehrenb. ●☆

383629　Tamarix troupii Hole;特鲁柽柳●☆

383630　Tamarix troupii Hole = Tamarix indica Willd. ●☆

383631　Tamarix usneoides E. Mey. ex Bunge;苔藓柽柳●☆

383632　Tamarix valdesquamigera Sennen = Tamarix canariensis Willd. ●☆

383633　Tamarix weyleri Pau = Tamarix canariensis Willd. ●☆

383634　Tamatavia Hook. f. = Chapelieria A. Rich. ex DC. ●☆

383635　Tamaulipa R. M. King et H. Rob. (1971);天泽兰属●☆

383636　Tamaulipa azurea (DC.) R. M. King et H. Rob.;天泽兰;Blue Boneset, Blueweed ■☆

383637　Tamayoa V. M. Badillo = Lepidesmia Klatt ■☆

383638　Tamayorkis Szlach. (1995);墨西哥小柱兰属■☆

383639　Tamayorkis Szlach. = Microstylis (Nutt.) Eaton(保留属名)■☆

383640　Tambourissa Sonn. (1782);马岛甜桂属●☆

383641　Tambourissa alternifolia A. DC.;异叶马岛甜桂●☆

383642　Tambourissa cordifolia Lorence;心叶马岛甜桂●☆

383643　Tambourissa elliptica A. DC.;椭圆马岛甜桂●☆

383644　Tambourissa gracilis Perkins;细马岛甜桂●☆

383645　Tambourissa leptophylla A. DC.;细叶马岛甜桂●☆

383646　Tambourissa longicarpa Lorence;长果马岛甜桂●☆

383647　Tambourissa madagascariensis Cavaco;马岛甜桂●☆

383648　Tambourissa microphylla Perkins;小叶马岛甜桂●☆

383649　Tambourissa obovata A. DC.;倒卵马岛甜桂●☆

383650　Tambourissa paradoxa Perkins;奇异马岛甜桂●☆

383651　Tambourissa purpurea A. DC.;紫马岛甜桂●☆

383652　Tambourissa trichophylla Baker;毛叶马岛甜桂●☆

383653　Tamia Ravenna = Calydorea Herb. ■☆

383654　Tamia Ravenna(2001);南美矛鞘鸢尾属■☆

383655　Tamia pallens (Griseb.) Ravenna;南美矛鞘鸢尾■☆

383656　Tamijia S. Sakai et Nagam. (2000);加岛姜属■☆

383657　Tamijia flagellaris S. Sakai et Nagam.;加岛姜■☆

383658　Tamilnadia Tirveng. et Sastre(1979);塔米茜属●☆

383659　Tamilnadia uliginosa (Retz.) Tirveng. et Sastre;塔米茜●☆

383660　Tammsia H. Karst. (1861)(保留属名);塔姆斯茜属☆

383661　Tammsia anomala H. Karst.;塔姆斯茜☆

383662　Tamnaceae J. Kickx f. = Dioscoreaceae R. Br. (保留科名)●■

383663　Tamnus Mill. = Tamus L. ■☆

383664　Tamonea Aubl. (1775);塔蒙草属■●☆

383665　Tamonea Aubl. = Ghinia Schreb. ■●☆

383666　Tamonea Aubl. = Miconia Ruiz et Pav. (保留属名)●☆

383667　Tamonea Aubl. ex Krasser = Miconia Ruiz et Pav. (保留属名)●☆

383668　Tamonea Krasn. = Miconia Ruiz et Pav. (保留属名)●☆

383669　Tamonea arabica Mirb. = Priva adhaerens (Forssk.) Chiov. ■☆

383670　Tamonea boxiana (Moldenke) R. A. Howard;塔蒙草;Crow Broom ■☆

383671　Tamonea spicata Aubl.;穗花塔蒙草(穗花马鞭草)■☆

383672　Tamonopsis Griseb. = Lantana L. (保留属名)●

383673　Tampoa Aubl. = Salacia L. (保留属名)●

383674　Tamus L. (1753);浆果薯蓣属(达马薯蓣属);Black Bryony ■☆

383675　Tamus communis L.;浆果薯蓣(达马薯蓣);Adder's Meat, Adder's Poison, Bead-bind, Big-root, Black Bindweed, Black Bryony, Black Vine, Blackeye Root, Chilblain-berry, Devil's Berries, Hedge Ivy, Isle of Wight Vine, Lady's Seal, Lady's Signet, Lady's-seal, Mandrake, Murrain Berries, Murren-berries, Our Lady's Vine, Oxberry, Poison Berry, Rowberry, Snake's Food, Snake's Meat, Snakeweed, Tetter-berries, Tetterwort, Wild Hop, Wild Nep, Wild Vine, With Ywinny, Withywiny, Wound Root ☆

383676　Tamus communis L. = Dioscorea communis (L.) Caddick et Wilkin ■☆

383677　Tamus communis L. var. subtriloba Guss. = Dioscorea communis (L.) Caddick et Wilkin ■☆

383678　Tamus edulis Lowe；可食浆果薯蓣■☆

383679　Tamus edulis Lowe. = Dioscorea communis（L.）Caddick et Wilkin ■☆

383680　Tamus elephantipes L'Hér. = Dioscorea elephantipes（L'Hér.）Engl. ■☆

383681　Tana B. -E. van Wyk = Peucedanum L. ■

383682　Tana bojeriana（Baker）B. -E. van Wyk；塔纳草■☆

383683　Tana bojeriana（Baker）B. -E. van Wyk = Peucedanum bojerianum Baker ■☆

383684　Tanacetaceae Vest = Asteraceae Bercht. et J. Presl（保留科名）●■

383685　Tanacetaceae Vest = Compositae Giseke（保留科名）●■

383686　Tanacetaceae Vest；菊蒿科●■

383687　Tanacetopsis（Tzvelev）Kovalevsk.（1962）；类菊蒿属■●☆

383688　Tanacetopsis afghanica（Gilli）K. Bremer et Humphries；阿富汗类菊蒿■☆

383689　Tanacetopsis eriobasis（Rech. f.）Kovalevsk. ；类菊蒿■☆

383690　Tanacetum L.（1753）；菊蒿属（艾菊属）；Tansy ■●

383691　Tanacetum abrotanifolium Druce；美叶菊蒿■☆

383692　Tanacetum achilleifolium Sch. Bip. ；蓍草叶菊蒿■☆

383693　Tanacetum achilleoides（Turcz.）DC. = Ajania achilleoides（Turcz.）Poljakov ex Grubov ●

383694　Tanacetum achilleoides（Turcz.）Hand. -Mazz. = Ajania achilleoides（Turcz.）Poljakov ex Grubov ●

383695　Tanacetum achilloides（Turcz.）DC. = Ajania achilloides（Turcz.）Poljakov ex Grubov ■

383696　Tanacetum acutilobum DC. = Myxopappus acutilobus（DC.）Källersjö ■☆

383697　Tanacetum adenanthum Diels = Ajania adenantha（Diels）Y. Ling et C. Shih ■

383698　Tanacetum aegyptiacum Juss. ex Jacq. = Grangea maderaspatana（L.）Poir. ■

383699　Tanacetum aegytiacum Juss. ex Jacq. = Grangea maderaspatana（L.）Poir. ■

383700　Tanacetum akinfiewii（Alex.）Tzvelev；阿氏菊蒿■☆

383701　Tanacetum alashanense Y. Ling = Hippolytia alashanensis（Y. Ling）C. Shih ●◇

383702　Tanacetum alashanense Y. Ling = Hippolytia kaschgarica（Krasch.）Poljakov ■

383703　Tanacetum alatavicum Herder = Pyrethrum alatavicum（Herder）O. Fedtsch. et B. Fedtsch. ■

383704　Tanacetum albidum DC. = Foveolina dichotoma（DC.）Källersjö ■☆

383705　Tanacetum annuum L. ；一年菊蒿■☆

383706　Tanacetum argenteum Willd. ；银白菊蒿■☆

383707　Tanacetum argyreum DC. = Schistostephium flabelliforme Less. ●☆

383708　Tanacetum argyrophyllum（K. Koch）Tzvelev；银叶菊蒿■☆

383709　Tanacetum arisanense Kitam. = Dendranthema arisanense（Hayata）Y. Ling et C. Shih ■

383710　Tanacetum aureoglobosum W. W. Sm. et Farrer = Ajania fruticulosa（Ledeb.）Poljakov ●

383711　Tanacetum axillare Thunb. = Athanasia minuta（L. f.）Källersjö ●☆

383712　Tanacetum balsamita L. ；香脂菊蒿（艾菊，香脂茼蒿）；Alecost, Alecost Costmary, Atllspice, Balsam Herb, Balsamint, Camphor, Camphor Plant, Costmary, Costmary Chrysanthemum, Goose Tongue, Goose-tongue, Herb Mary, Maudeline, Mint Geranium, Mint-geranium, Our Lady's Mint, St. Mary ■☆

383713　Tanacetum balsamita L. = Balsamita major Desf. ■☆

383714　Tanacetum barclayanum DC. ；阿尔泰菊蒿；Altai Tansy ■

383715　Tanacetum bipinnatum（L.）Sch. Bip. subsp. huronense（Nutt.）Breitung = Tanacetum huronense Nutt. ■☆

383716　Tanacetum bipinnatum Sch. Bip. ；羽状菊蒿■☆

383717　Tanacetum bipinnatum Sch. Bip. subsp. huronense（Nutt.）Breitung = Tanacetum bipinnatum（L.）Sch. Bip. ■☆

383718　Tanacetum boreale Fisch. ex DC. = Tanacetum vulgare L. var. boreale（Fisch. ex DC.）Trautv. et C. A. Mey. ■☆

383719　Tanacetum boreale Fisch. ex DC. = Tanacetum vulgare L. ■

383720　Tanacetum brachanthemoides（C. Winkl.）Krasch. = Kaschgaria brachanthemoides（C. Winkl.）Poljakov ●■

383721　Tanacetum bulbosum Hand. -Mazz. = Hippolytia delavayi（Franch. ex W. W. Sm.）C. Shih ●■

383722　Tanacetum burchellii DC. = Pentzia punctata Harv. ■☆

383723　Tanacetum camphoratum Less. = Tanacetum bipinnatum（L.）Sch. Bip. ■☆

383724　Tanacetum canum D. C. Eaton = Sphaeromeria cana（D. C. Eaton）A. Heller ■☆

383725　Tanacetum capitatum（Nutt.）Torr. et A. Gray = Sphaeromeria capitata Nutt. ■☆

383726　Tanacetum capitatum Torr. et Gray；岩生菊蒿（岩生艾菊）；Rock-living Tansy ■

383727　Tanacetum chiliophyllum Sch. Bip. ；千叶菊蒿■☆

383728　Tanacetum chinense A. Gray ex Maxim. = Crossostephium chinense（L.）Makino ●

383729　Tanacetum cinerariifolium（Trevir.）Sch. Bip. = Pyrethrum cinerariifolium Trevis. ■

383730　Tanacetum cinereum（Delile）DC. = Cotula cinerea Delile ■☆

383731　Tanacetum coccineum（Willd.）Grierson 'Brenda'；紫舌红花除虫菊■☆

383732　Tanacetum coccineum（Willd.）Grierson 'Eileen May Robinson'；淡舌红花除虫菊■☆

383733　Tanacetum coccineum（Willd.）Grierson 'James Kelway'；变色红花除虫菊■☆

383734　Tanacetum coccineum（Willd.）Grierson = Pyrethrum coccineum（Willd.）Vorosch. ■

383735　Tanacetum compactum H. M. Hall = Sphaeromeria compacta（H. M. Hall）A. H. Holmgren, L. M. Shultz et Lowrey ■☆

383736　Tanacetum consanguineum DC. = Schistostephium crataegifolium（DC.）Fenzl ex Harv. ●☆

383737　Tanacetum corymbosum（L.）Sch. Bip. ；伞花除虫菊；Corymbflower Tansy ■☆

383738　Tanacetum corymbosum（L.）Sch. Bip. var. tenuifolium（Willd.）Ledeb. = Tanacetum corymbosum（L.）Sch. Bip. ■☆

383739　Tanacetum corymbosum（L.）Sch. Bip. var. webbianum（Coss.）Maire = Tanacetum corymbosum（L.）Sch. Bip. ■☆

383740　Tanacetum crassipes（Stschegl.）Tzvelev；密头菊蒿；Densehead Tansy ■

383741　Tanacetum crataegifolium DC. = Schistostephium crataegifolium（DC.）Fenzl ex Harv. ●☆

383742　Tanacetum crispum Steud. = Tanacetum vulgare L. ■

383743　Tanacetum davidii Krasch. = Ajania parviflora（Grüning）Y. Ling ■

383744　Tanacetum delavayi（Franch. ex W. W. Sm.）Hand. -Mazz. = Hippolytia delavayi（Franch. ex W. W. Sm.）C. Shih ●■

383745　Tanacetum delavayi Franch. ex W. W. Sm. = Hippolytia delavayi

(Franch. ex W. W. Sm.) C. Shih ●■

383746 Tanacetum densum Sch. Bip. ;密集菊蒿;Partridge Feather ■☆

383747 Tanacetum diversifolium D. C. Eaton = Sphaeromeria diversifolia (D. C. Eaton) Rydb. ■☆

383748 Tanacetum dolichophyllum (Kitam.) Kitam. ;长叶菊蒿■☆

383749 Tanacetum douglasii DC. = Tanacetum bipinnatum (L.) Sch. Bip. ■☆

383750 Tanacetum elegantulum W. W. Sm. = Ajania elegantyla (W. W. Sm.) C. Shih ■

383751 Tanacetum emodii K. R. Khan;喜马拉雅菊蒿■☆

383752 Tanacetum facatolobatum Krasch. = Cancrinia maximowiczii C. Winkl. ●

383753 Tanacetum ferulaceum (Webb) Sch. Bip. ;手杖菊蒿■☆

383754 Tanacetum ferulaceum (Webb) Sch. Bip. var. latipinnum (Svent.) G. Kunkel = Tanacetum ferulaceum (Webb) Sch. Bip. ■☆

383755 Tanacetum frutescens L. = Hippia frutescens (L.) L. ●☆

383756 Tanacetum fruticulosum Ledeb. = Ajania fruticulosa (Ledeb.) Poljakov ●

383757 Tanacetum glabriusculum W. W. Sm. = Dendranthema glabriusculum (W. W. Sm.) C. Shih ■

383758 Tanacetum gossypinum C. B. Clarke = Hippolytia gossypina (C. B. Clarke) C. Shih ■

383759 Tanacetum grandiflorum Thunb. = Oncosiphon grandiflorum (Thunb.) Källersjö ■☆

383760 Tanacetum griffithii (C. B. Clarke) Muradyan;格氏菊蒿■☆

383761 Tanacetum griseum Harv. = Schistostephium griseum (Harv.) Hutch. ●☆

383762 Tanacetum haradjanii (Rech. f.) Grierson;西亚菊蒿■☆

383763 Tanacetum heptalobum DC. = Schistostephium heptalobum (DC.) Oliv. et Hiern ●☆

383764 Tanacetum herderi Regel et Schmalh. = Hippolytia herderi (Regel et Schmalh.) Poljakov ■

383765 Tanacetum heterophyllum Boiss. ;互叶菊蒿■☆

383766 Tanacetum hippiifolium DC. = Schistostephium hippiifolium (DC.) Hutch. ●☆

383767 Tanacetum hispidum DC. = Cotula hispida (DC.) Harv. ■☆

383768 Tanacetum huronense Nutt. ;休伦湖菊蒿;Eastern Tansy, Lake Huron Tansy ■☆

383769 Tanacetum huronense Nutt. = Tanacetum bipinnatum (L.) Sch. Bip. ■☆

383770 Tanacetum huronense Nutt. var. bifarium Fernald = Tanacetum bipinnatum (L.) Sch. Bip. ■☆

383771 Tanacetum huronense Nutt. var. bifarium Fernald = Tanacetum huronense Nutt. ■☆

383772 Tanacetum huronense Nutt. var. floccosum Raup = Tanacetum bipinnatum (L.) Sch. Bip. ■☆

383773 Tanacetum huronense Nutt. var. floccosum Raup = Tanacetum huronense Nutt. ■☆

383774 Tanacetum huronense Nutt. var. johannense Fernald = Tanacetum bipinnatum (L.) Sch. Bip. ■☆

383775 Tanacetum huronense Nutt. var. johannense Fernald = Tanacetum huronense Nutt. ■☆

383776 Tanacetum huronense Nutt. var. terrae-novae Fernald = Tanacetum bipinnatum (L.) Sch. Bip. ■☆

383777 Tanacetum huronense Nutt. var. terrae-novae Fernald = Tanacetum huronense Nutt. ■☆

383778 Tanacetum indicum (L.) Sch. Bip. = Chrysanthemum indicum L. ■

383779 Tanacetum indicum (L.) Sch. Bip. = Dendranthema indicum (L.) Des Moul. ■

383780 Tanacetum karelinii Tzvelev;中亚菊蒿(三裂菊蒿,准噶尔菊蒿);Carelin Tansy ■

383781 Tanacetum kaschgaricum Krasch. = Hippolytia kaschgarica (Krasch.) Poljakov ■

383782 Tanacetum kennedyi Dunn = Hippolytia kennedyi (Dunn) Y. Ling ■

383783 Tanacetum khartense Dunn = Ajania khartensis (Dunn) C. Shih ■

383784 Tanacetum kittaryanum (C. A. Mey.) Tzvelev;基氏菊蒿■☆

383785 Tanacetum komarovii Krasch. et N. I. Rubtzov = Kaschgaria komalovii (Krasch. et N. I. Rubtzov) Poljakov ●

383786 Tanacetum ledebourii Sch. Bip. = Cancrinia discoides (Ledeb.) Poljakov ex Tzvelev ■

383787 Tanacetum leucanthemum (L.) Sch. Bip. = Leucanthemum vulgare Lam. ■

383788 Tanacetum leucophyllum Regel = Hippolytia herderi (Regel et Schmalh.) Poljakov ■

383789 Tanacetum lineare Kitam. = Leucanthemella linearis (Matsum.) Tzvelev ■

383790 Tanacetum linearilobum DC. = Cotula lineariloba (DC.) Hilliard ■☆

383791 Tanacetum linifolium (L. f.) Thunb. = Athanasia linifolia Burm. ●☆

383792 Tanacetum longifolium Thunb. = Athanasia linifolia Burm. ●☆

383793 Tanacetum longipedunculatum (Sosn.) Tzvelev;长柄菊蒿■☆

383794 Tanacetum macrophyllum Simonk. ;大叶菊蒿;Rayed Tansy ■☆

383795 Tanacetum mairei H. Lév. = Ajania myriantha (Bureau et Franch.) Y. Ling ex C. Shih ■

383796 Tanacetum meyerianum Sch. Bip. = Tanacetum tanacetoides (DC.) Tzvelev ■

383797 Tanacetum microphyllum DC. ;小叶菊蒿■☆

383798 Tanacetum millefolium (L.) Tzvelev = Anthemis millefolia L. ■☆

383799 Tanacetum monanthos L. = Anacyclus monanthos (L.) Thell. ■☆

383800 Tanacetum morifolium (Ramat.) Kitam. = Dendranthema morifolium (Ramat.) Tzvelev ■

383801 Tanacetum mutellinum Hand. -Mazz. = Ajania khartensis (Dunn) C. Shih ■

383802 Tanacetum myrianthum Bureau et Franch. = Ajania myriantha (Bureau et Franch.) Y. Ling ex C. Shih ■

383803 Tanacetum myrianthum Franch. = Ajania myriantha (Franch.) Y. Ling ex C. Shih ■

383804 Tanacetum myrianthum Franch. var. wardii Marquand et Shaw = Ajania myriantha (Franch.) Y. Ling ex C. Shih ■

383805 Tanacetum myrianthum Franch. var. wardii Marquand et Shaw = Ajania myriantha (Bureau et Franch.) Y. Ling ex C. Shih ■

383806 Tanacetum myriophyllum Willd. ;密叶菊蒿■☆

383807 Tanacetum nubigenum (Wall.) DC. = Ajania nubigena (Wall.) C. Shih ■

383808 Tanacetum nuttallii Torr. et A. Gray = Sphaeromeria argentea Nutt. ■☆

383809 Tanacetum obtusum Thunb. = Oncosiphon piluliferum (L. f.) Källersjö ■☆

383810 Tanacetum oligocephalum Sch. Bip. ;寡头菊蒿■☆

383811 Tanacetum oresbium W. W. Sm. = Ajania myriantha (Bureau et Franch.) Y. Ling ex C. Shih ■

383812 Tanacetum oresbium W. W. Sm. = Ajania myriantha (Franch.) Y. Ling ex C. Shih ■

383813 Tanacetum paczoskii (Zefir.) Tzvelev;帕氏菊蒿■☆

383814 Tanacetum pallasianum (Fisch. ex Besser) Trautv. et C. A. Mey. = Ajania pallasiana (Fisch. ex Besser) Poljakov ■

383815 Tanacetum pallasianum (Fisch. ex Besser) Trautv. et C. A. Mey. = Chrysanthemum pallasianum (Fisch. ex Besser) Kom. ■

383816 Tanacetum pallasianum (Fisch. ex Besser) Trautv. et C. A. Mey. = Dendranthema pacificum (Nakai) Kitam. ■

383817 Tanacetum pallasianum (Fisch. ex Besser) Trautv. et C. A. Mey. var. brevilobum Franch. ex Diels = Ajania breviloba (Franch. ex Hand. -Mazz.) Y. Ling et C. Shih ■

383818 Tanacetum pallasianum (Fisch.) Trautv. et C. A. Mey. var. brevilobum Franch. ex Diels = Ajania breviloba (Franch. ex Hand. -Mazz.) Y. Ling et C. Shih ■

383819 Tanacetum pallidum (Mill.) Maire = Leucanthemopsis pallida (Mill.) Heywood ■☆

383820 Tanacetum pallidum (Mill.) Maire subsp. longipectinatum (Font Quer) Maire = Leucanthemopsis longipectinata (Pau) Heywood ■☆

383821 Tanacetum parthenifolium (Willd.) Sch. Bip. = Pyrethrum parthenifolium Willd. ■

383822 Tanacetum parthenium (L.) Sch. Bip. ' Aureum';金叶短舌菊蒿;Golden Feather ■☆

383823 Tanacetum parthenium (L.) Sch. Bip. = Chrysanthemum parthenium (L.) Benth. ■

383824 Tanacetum parthenium (L.) Sch. Bip. = Pyrethrum parthenium (L.) Sm. ■

383825 Tanacetum parviflorum (Grüning) H. W. Kung = Ajania parviflora (Grüning) Y. Ling ■

383826 Tanacetum pilosum P. J. Bergius = Hippia pilosa (P. J. Bergius) Druce ●☆

383827 Tanacetum potaninii Krasch. = Ajania potaninii (Krasch.) Poljakov ■●

383828 Tanacetum potaninii Krasch. var. nanum Krasch. = Ajania potaninii (Krasch.) Poljakov ■●

383829 Tanacetum potaninii Krasch. var. suffruticosum Krasch. = Ajania potaninii (Krasch.) Poljakov ■●

383830 Tanacetum potentilloides (A. Gray) A. Gray = Sphaeromeria potentilloides (A. Gray) A. Heller ■☆

383831 Tanacetum potentilloides (A. Gray) A. Gray var. nitrophilum Cronquist = Sphaeromeria potentilloides (A. Gray) A. Heller var. nitrophila (Cronquist) A. H. Holmgren, L. M. Shultz et Lowrey ■☆

383832 Tanacetum pseudachillea C. Winkl. ;假蓍草菊蒿■☆

383833 Tanacetum pulchrum (Ledeb.) Sch. Bip. = Pyrethrum pulchrum Ledeb. ■

383834 Tanacetum purpureum Buch. -Ham. = Cyathocline purpurea (Buch. -Ham. ex De Don) Kuntze ■

383835 Tanacetum purpureum Buch. -Ham. ex D. Don = Cyathocline purpurea (Buch. -Ham. ex D. Don) Kuntze ■

383836 Tanacetum quercifolium W. W. Sm. = Ajania quercifolia (W. W. Sm.) Y. Ling et C. Shih ■

383837 Tanacetum rockii Mattf. ex Rehder et Kobuski = Ajania potaninii (Krasch.) Poljakov ■●

383838 Tanacetum rotundifolium DC. = Schistostephium rotundifolium (DC.) Fenzl ex Harv. ●☆

383839 Tanacetum roylei (DC.) K. R. Khan;罗伊尔菊蒿■☆

383840 Tanacetum salicifolium Mattf. = Ajania salicifolia (Mattf. ex Rehder et Kobuski) Poljakov ■

383841 Tanacetum salicifolium Mattf. ex Rehder et Kobuski = Ajania salicifolia (Mattf. ex Rehder et Kobuski) Poljakov ●

383842 Tanacetum santolina C. Winkl. ;散头菊蒿 (散花菊蒿); Scattered-flower Tansy ■

383843 Tanacetum saxicola (Krasch.) Tzvelev;岩地菊蒿■☆

383844 Tanacetum scharnhorstii Regel et Schmalh. = Ajania scharnhorstii (Regel et Schmalh.) Tzvelev ■

383845 Tanacetum sclerophyllum (Krasch.) Tzvelev;硬叶菊蒿■☆

383846 Tanacetum scopulorum (Krasch.) Tzvelev; 岩菊蒿; Rocky Tansy ■

383847 Tanacetum senecionis (Jacq. ex Besser) J. Gay = Hippolytia senecionis (Jacquem. ex Besser) Poljakov ex Tzvelev ■

383848 Tanacetum senecionis (Jacq. ex Besser) J. Gay ex DC. = Hippolytia senecionis (Jacquem. ex Besser) Poljakov ex Tzvelev ■

383849 Tanacetum serotinum (L.) Sch. Bip. = Leucanthemella serotina (L.) Tzvelev ■☆

383850 Tanacetum sibiricum L. = Artemisia sibirica (L.) Maxim. ■☆

383851 Tanacetum sibiricum L. = Filifolium sibiricum (L.) Kitam. ■

383852 Tanacetum simplex A. Nelson = Sphaeromeria simplex (A. Nelson) A. Heller ■☆

383853 Tanacetum sinense (Sabine) Des Moul. = Dendranthema morifolium (Ramat.) Tzvelev ■

383854 Tanacetum suaveolens (Pursh) Hook. = Matricaria discoidea DC. ■

383855 Tanacetum suffruticosum L. = Oncosiphon schlechteri (Bolus ex Schltr.) Källersjö ■☆

383856 Tanacetum suffruticosum L. = Oncosiphon suffruticosum (L.) Källersjö ■☆

383857 Tanacetum tanacetoides (DC.) Tzvelev;伞房菊蒿; Corymb Tansy, Umbei Tansy ■

383858 Tanacetum tenuifolium Jacq. = Ajania tenuifolia (Jacq.) Tzvelev ■

383859 Tanacetum tenuifolium Jacquem. ex DC. = Ajania tenuifolia (Jacq.) Tzvelev ■

383860 Tanacetum tibeticum Hook. f. et Thomson = Ajania tibetica (Hook. f. et Thomson ex C. B. Clarke) Tzvelev ■

383861 Tanacetum tibeticum Hook. f. et Thomson ex C. B. Clarke = Ajania tibetica (Hook. f. et Thomson ex C. B. Clarke) Tzvelev ●

383862 Tanacetum tomentosum DC. = Hippolytia tomentosa (DC.) Tzvelev ■

383863 Tanacetum tortuosum DC. = Pentzia tortuosa (DC.) Fenzl ex Harv. ■☆

383864 Tanacetum transiliense Herder = Pyrethrum transiliense (Herder) Regel et Schmalh. ■

383865 Tanacetum trifidum (Turcz.) DC. = Hippolytia trifida (Turcz.) Poljakov ●

383866 Tanacetum trifidum DC. = Ajania parviflora (Grüning) Y. Ling ■

383867 Tanacetum turcomanicum (Krasch.) Tzvelev;土库曼菊蒿■☆

383868 Tanacetum turlanicum (Pavlov) Tzvelev = Tanacetum barclayanum DC. ■

383869 Tanacetum ulutavicum Tzvelev;乌鲁塔夫菊蒿■☆

383870 Tanacetum umbellatum Gilib. = Tanacetum vulgare L. ■

383871 Tanacetum uniflorum Sch. Bip. ;单花菊蒿■☆

383872 Tanacetum uralense (Krasch.) Tzvelev;乌拉尔菊蒿■☆

383873 Tanacetum vestitum Thunb. = Athanasia vestita (Thunb.)

Druce ●☆

383874 Tanacetum vulgare L.；菊蒿（艾菊，普通菊蒿）；Bachelor's Buttons, Bitter Buttons, Buttons, Cheese, Common Tansy, English Cost, English Costmary, Ginger, Ginger-plant, Golden Buttons, Golden-buttons, Goose Grass, Hindheel, Joynson's Remedy Cheese, Parsley Fern, Scented Daisy, Scented Fern, She Bulkishawn, She-bulkishawn, Stinking Elshander, Stinking Willie, Tansy, Trnvener's Rest, Weebow, Yellow Buttons ■

383875 Tanacetum vulgare L. f. crispum (L.) Fernald = Tanacetum vulgare L. ■

383876 Tanacetum vulgare L. var. boreale (Fisch. ex DC.) Trautv. et C. A. Mey.；北方菊蒿■☆

383877 Tanacetum vulgare L. var. boreale (Fisch. ex DC.) Trautv. et C. A. Mey. = Tanacetum vulgare L. ■

383878 Tanacetum vulgare L. var. crispum L.；卷曲菊蒿；Curled Tansy ■☆

383879 Tanacetum vulgare L. var. crispum L. = Tanacetum vulgare L. ■

383880 Tanacetum yunnanense Jeffrey = Hippolytia yunnanensis (Jeffrey) C. Shih ■

383881 Tanaecium Sw. (1788)；塔纳葳属■☆

383882 Tanaecium albiflorum DC.；白花塔纳葳■☆

383883 Tanaecium nocturnum Bureau et K. Schum.；塔纳葳■☆

383884 Tanaecium ovatum Bureau et K. Schum.；卵形塔纳葳■☆

383885 Tanaecium pinnatum Willd. = Kigelia africana (Lam.) Benth. ●☆

383886 Tanaesium Raf. = Tanaecium Sw. ■☆

383887 Tanakaea Franch. et Sav. (1878)；峨屏草属（岩雪下属）；Tanakaea ■

383888 Tanakaea omeiensis Nakai = Tanakaea radicans Franch. et Sav. ■

383889 Tanakaea omeiensis Nakai var. nanchuanensis W. T. Wang = Tanakaea radicans Franch. et Sav. ■

383890 Tanakaea radicans Franch. et Sav.；峨屏草（峨眉岩下雪，日本峨屏草，岩雪下）；Emei Tanakaea, Omei Mountain Tanakaea ■

383891 Tanakea Franch. et Sav. = Tanakaea Franch. et Sav. ■

383892 Tanaosolen N. E. Br. = Tritoniopsis L. Bolus ■☆

383893 Tanarius Kuntze = Macaranga Thouars ■

383894 Tanarius Rumph. = Macaranga Thouars ●

383895 Tanarius Rumph. ex Kuntze = Macaranga Thouars ●

383896 Tanarius glaber Kuntze = Macaranga tanarius (L.) Müll. Arg. ●

383897 Tanarius gmelinifolius (King ex Hook. f.) Kuntze = Macaranga pustulata King ex Hook. f. ●

383898 Tanarius kurzii Kuntze = Macaranga kurzii (Kuntze) Pax et K. Hoffm. ●

383899 Tanarius pustulatus (King ex Hook. f.) Kuntze = Macaranga pustulata King ex Hook. f. ●

383900 Tanaxion Raf. = Pluchea Cass. ●■

383901 Tandonia Baill. = Tannodia Baill. ■☆

383902 Tandonia Moq. = Anredera Juss. ●■

383903 Tangaraca Adans. = Hamelia Jacq. ●

383904 Tanghekolli Adans. = Crinum L. ■

383905 Tanghinia Thouars = Cerbera L. ●

383906 Tanghinia Thouars(1806)；坦杙果属●☆

383907 Tanghinia venenifera Poir.；坦杙果；Ordeal Tree, Tangin Tree ●☆

383908 Tangtsinia S. C. Chen = Cephalanthera Rich. ■

383909 Tangtsinia S. C. Chen(1965)；金佛山兰属（金佛兰属，金兰属）；Tangtsinia ■★

383910 Tangtsinia japonica (Thunb.) Szyszył. = Ternstroemia japonica Thunb. ●

383911 Tangtsinia nanchuanica S. C. Chen；金佛山兰（金兰）；Nanchuan Tangtsinia, Tangtsinia ■

383912 Tanibouca Aubl. = Terminalia L. （保留属名）●

383913 Tankervillia Link = Phaius Lour. ■

383914 Tannodia Baill. (1861)；塔诺大戟属■☆

383915 Tannodia congolensis J. Léonard；刚果塔诺大戟■☆

383916 Tannodia sessiliflora (Pax) Prain = Tannodia tenuifolia (Pax) Prain var. glabrata Prain ■☆

383917 Tannodia swynnertonii (S. Moore) Prain；塔诺大戟■☆

383918 Tannodia tenuifolia (Pax) Prain；细叶塔诺大戟■☆

383919 Tannodia tenuifolia (Pax) Prain var. glabrata Prain；光细叶塔诺大戟■☆

383920 Tanquana H. E. K. Hartm. et Liede(1986)；黄指玉属■☆

383921 Tanquana archeri (L. Bolus) H. E. K. Hartmann et Liede；非洲黄指玉■☆

383922 Tanquana hilmarii (L. Bolus) H. E. K. Hartmann et Liede；希尔黄指玉■☆

383923 Tanquana prismatica (Schwantes) H. E. K. Hartmann et Liede；黄指玉■☆

383924 Tanroujou Juss. = Hymenaea L. ●

383925 Tansaniochloa Rauschert = Setaria P. Beauv. （保留属名）■

383926 Tantalus Noronha ex Thouars = Sarcolaena Thouars ●☆

383927 Tanulepis Balf. f. (1877)；大鳞萝藦属●☆

383928 Tanulepis Balf. f. ex Baker = Camptocarpus Decne. （保留属名）●■☆

383929 Tanulepis acuminata (Choux) Choux = Camptocarpus acuminatus (Choux) Venter ●☆

383930 Tanulepis crassifolia (Decne.) Choux = Camptocarpus crassifolius Decne. ●☆

383931 Tanulepis decaryi Choux = Camptocarpus decaryi (Choux) Venter ●☆

383932 Tanulepis linearis (Decne.) Choux = Camptocarpus crassifolius Decne. ●☆

383933 Tanulepis madagascariensis (Schltr.) Choux = Camptocarpus crassifolius Decne. ●☆

383934 Tanulepis sphenophylla Balf. f.；大鳞萝藦●☆

383935 Taonabo Aubl. （废弃属名） = Ternstroemia Mutis ex L. f. （保留属名）●

383936 Taonabo fragrans (Champ.) Choisy = Ternstroemia japonica (Thunb.) Thunb. ●

383937 Taonabo japonica (Thunb.) Szyszył. = Ternstroemia japonica (Thunb.) Thunb. ●

383938 Taonabo mokof Nakai = Ternstroemia japonica (Thunb.) Thunb. ●

383939 Tapagomea Kuntze = Cephaelis Sw. （保留属名）●

383940 Tapagomea Kuntze = Tapogomea Aubl. （废弃属名）●

383941 Tapanava Adans. = Pothos L. ●■

383942 Tapanava chinensis Raf. = Pothos chinensis (Raf.) Merr. ■

383943 Tapanawa Hassk. = Tapanava Adans. ●■

383944 Tapeinaegle Herb. = Braxireon Raf. ■

383945 Tapeinaegle Herb. = Tapeinanthus Herb. （废弃属名）●☆

383946 Tapeinanthus Boiss. ex Benth. = Thuspeinanta T. Durand ■☆

383947 Tapeinanthus Herb. （废弃属名） = Braxireon Raf. ■

383948 Tapeinanthus Herb. （废弃属名） = Narcissus L. ■

383949 Tapeinanthus Herb. （废弃属名） = Tapinanthus (Blume) Rchb. （保留属名）●☆

383950 Tapeinanthus humilis (Cav.) Herb. = Narcissus cavanillesii

Barra et G. López ■☆

383951　Tapeinia Comm. ex Juss. (1789);低鸢尾属■☆

383952　Tapeinia F. Dietrich = Tritonia Ker Gawl. ■

383953　Tapeinia Juss. = Tapeinia Comm. ex Juss. ■☆

383954　Tapeinia magellanica J. F. Gmel.;低鸢尾■☆

383955　Tapeinocheilos Miq. = Tapeinochilos Miq. (保留属名)■☆

383956　Tapeinochilos Miq. (1869) (保留属名);小唇姜属■☆

383957　Tapeinochilos ananassae (Hassk.) K. Schum.;小唇姜■☆

383958　Tapeinochilus Benth. et Hook. f. = Tapeinocheilos Miq. ■☆

383959　Tapeinoglossum Schltr. = Bulbophyllum Thouars(保留属名)■

383960　Tapeinophallus Baill. = Amorphophallus Blume ex Decne. (保留属名)■●

383961　Tapeinosperma Hook. f. (1876);小籽金牛属●☆

383962　Tapeinosperma glandulosum Guillaumin;多腺小籽金牛●☆

383963　Tapeinosperma gracile Mez;细小籽金牛●☆

383964　Tapeinosperma grandiflorum Guillaumin;大花小籽金牛●☆

383965　Tapeinosperma laeve Mez;平滑小籽金牛●☆

383966　Tapeinosperma lanceifolium (Merr.) Philipson;披针叶小籽金牛●☆

383967　Tapeinosperma laurifolium Mez = Tapeinosperma glandulosum Guillaumin ●☆

383968　Tapeinosperma megaphyllum Mez;大叶小籽金牛●☆

383969　Tapeinosperma minutum Mez;小小籽金牛●☆

383970　Tapeinosperma multiflorum (Gillespie) A. C. Sm.;小花小籽金牛●☆

383971　Tapeinosperma multipunctatum Guillaumin;多斑小籽金牛●☆

383972　Tapeinosperma nitidum Mez;光亮小籽金牛●☆

383973　Tapeinosperma pauciflorum Mez;少花小籽金牛●☆

383974　Tapeinosperma pulchellum Mez;美丽小籽金牛●☆

383975　Tapeinosperma sessilifolium Mez;无柄小籽金牛●☆

383976　Tapeinostelma Schltr. = Brachystelma R. Br. (保留属名)●

383977　Tapeinostelma caffrum Schltr. = Brachystelma caffrum (Schltr.) N. E. Br. ■☆

383978　Tapeinostemon Benth. (1854);小雄蕊龙胆属■☆

383979　Tapeinostemon capitatum Benth.;头状小雄蕊龙胆■☆

383980　Tapeinostemon longiflorum Maguire et Steyerm.;长花小雄蕊龙胆■☆

383981　Tapeinostemon rugosum Maguire et Steyerm.;粗糙小雄蕊龙胆■☆

383982　Tapeinotes DC. = Sinningia Nees ●■☆

383983　Tapeinotes DC. = Tapina Mart. ●■☆

383984　Tapesia C. F. Gaertn. = Hamelia Jacq. ●

383985　Tapheocarpa Conran(1994);隐果草属■☆

383986　Tapheocarpa calandrinioides (F. Muell.) Conran;隐果草■☆

383987　Taphrogiton Friche-Joset et Montandon = Scirpus L. (保留属名)■

383988　Taphrogiton Montandon = Scirpus L. (保留属名)■

383989　Taphrospermum C. A. Mey. (1831);沟子荠属;Furrowcress, Taphropermum ■

383990　Taphrospermum altaicum C. A. Mey.;沟子荠(大果沟子荠);Altai Furrowcress, Altai Taphropermum, Bigfruit Altai Taphropermum,Bigfruit Furrowcress ■

383991　Taphrospermum altaicum C. A. Mey. var. macrocarpum C. H. An = Taphrospermum altaicum C. A. Mey. ■

383992　Taphrospermum fontana (Maxim.) Al-Shehbaz et G. Yang;泉沟子荠(双脊荠)■

383993　Taphrospermum fontana (Maxim.) Al-Shehbaz et G. Yang subsp. microspermum Al-Shehbaz et G. Yang;小籽泉沟子荠(毛果

双脊荠小籽沟子荠)■

383994　Taphrospermum himalaicum (Hook. f. et Thomson) Al-Shehbaz et G. Yang;须弥沟子荠■

383995　Taphrospermum lowndesii (H. Hara) Al-Shehbaz;郎氏沟子荠■

383996　Taphrospermum platypetalum Schrenk;宽瓣沟子荠■☆

383997　Taphrospermum tibeticum (O. E. Schulz) Al-Shehbaz;西藏沟子荠(西藏蛇头荠); Tibet Dipoma, Xizang Dipoma, Xizang Snakeheadcress ■

383998　Taphrospermum tomentosum (Willd.) M. Bieb. = Sterigmostemum caspicum (Lam.) Rupr. ■

383999　Taphrospermum verticillatum (Jeffrey et W. W. Sm.) Al-Shehbaz;轮叶沟子荠■

384000　Tapia Mill. = Crateva L. ●

384001　Tapina Mart. = Sinningia Nees ●■☆

384002　Tapinaegle Post et Kuntze = Tapeinaegle Herb. ■

384003　Tapinaegle Post et Kuntze = Tapeinanthus Herb. (废弃属名)●☆

384004　Tapinanthus (Blume) Blume = Tapinanthus (Blume) Rchb. (保留属名)●☆

384005　Tapinanthus (Blume) Rchb. (1841) (保留属名);大岩桐寄生属●☆

384006　Tapinanthus Blume = Tapinanthus (Blume) Rchb. (保留属名)●☆

384007　Tapinanthus Post et Kuntze = Tapinanthus (Blume) Rchb. (保留属名)●☆

384008　Tapinanthus acacietorum (Bullock) Danser = Oncocalyx fischeri (Engl.) M. G. Gilbert ●☆

384009　Tapinanthus adolfi-friderici (Engl. et K. Krause) Danser = Englerina woodfordioides (Schweinf.) Balle ●☆

384010　Tapinanthus alatus Danser = Tapinanthus dependens (Engl.) Danser ●☆

384011　Tapinanthus albizziae (De Wild.) Danser = Phragmanthera usuiensis (Oliv.) M. G. Gilbert ●☆

384012　Tapinanthus alboannulatus (Engl. et K. Krause) Danser = Agelanthus brunneus (Engl.) Balle et N. Hallé ●☆

384013　Tapinanthus angiensis (De Wild.) Danser = Tapinanthus buvumae (Rendle) Danser ●☆

384014　Tapinanthus angolensis (Engl.) Danser = Phragmanthera polycrypta (Didr.) Balle ●☆

384015　Tapinanthus annulatus (Engl. et K. Krause) Danser = Agelanthus fuellebornii (Engl.) Polhill et Wiens ●☆

384016　Tapinanthus apodanthus (Sprague) Danser;无梗花大岩桐寄生●☆

384017　Tapinanthus aurantiacus Danser = Agelanthus scassellatii (Chiov.) Polhill et Wiens ●☆

384018　Tapinanthus bagshawei (Rendle) Danser = Englerina woodfordioides (Schweinf.) Balle ●☆

384019　Tapinanthus bangwensis (Engl. et K. Krause) Danser;班乌大岩桐寄生●☆

384020　Tapinanthus batangae (Engl.) Danser = Phragmanthera batangae (Engl.) Balle ●☆

384021　Tapinanthus batesii (S. Moore et Sprague) Danser = Tapinanthus ogowensis (Engl.) Danser ●☆

384022　Tapinanthus baumii (Engl. et Gilg) Danser = Phragmanthera baumii (Engl. et Gilg) Polhill et Wiens ●☆

384023　Tapinanthus belvisii (DC.) Danser;贝尔维斯大岩桐寄生●☆

384024　Tapinanthus bemarivensis (Lecomte) Danser = Socratina bemarivensis (Lecomte) Balle ●☆

384025　Tapinanthus bequaertii (De Wild.) Danser = Phragmanthera usuiensis（Oliv.）M. G. Gilbert ●☆

384026　Tapinanthus berliniicola（Engl.）Danser ＝ Phragmanthera usuiensis（Oliv.）M. G. Gilbert subsp. sigensis（Engl.）Polhill et Wiens ●☆

384027　Tapinanthus blantyreanus（Engl.）Danser ＝ Agelanthus pungu（De Wild.）Polhill et Wiens ●☆

384028　Tapinanthus bogoroensis（De Wild.）Danser ＝ Phragmanthera usuiensis（Oliv.）M. G. Gilbert ●☆

384029　Tapinanthus bolusii（Sprague）Danser ＝ Oncocalyx bolusii（Sprague）Wiens et Polhill ●☆

384030　Tapinanthus boonei（De Wild.）Danser ＝ Agelanthus djurensis（Engl.）Polhill et Wiens ●☆

384031　Tapinanthus brachyanthus（Peter）Danser ＝ Englerina triplinervia（Baker et Sprague）Polhill et Wiens ●☆

384032　Tapinanthus brachyphyllus（Peter）Danser ＝ Englerina woodfordioides（Schweinf.）Balle ●☆

384033　Tapinanthus brazzavillensis（De Wild. et T. Durand）Danser ＝ Tapinanthus constrictiflorus（Engl.）Danser ●☆

384034　Tapinanthus brevilobus（Engl. et K. Krause）Danser ＝ Agelanthus heteromorphus（A. Rich.）Polhill et Wiens ●☆

384035　Tapinanthus brieyi（De Wild.）Danser ＝ Phragmanthera brieyi（De Wild.）Polhill et Wiens ●☆

384036　Tapinanthus brunneus（Engl.）Danser ＝ Agelanthus brunneus（Engl.）Balle et N. Hallé ●☆

384037　Tapinanthus brunneus（Engl.）Danser subsp. buchholzii（Engl.）Balle ＝ Agelanthus brunneus（Engl.）Balle et N. Hallé ●☆

384038　Tapinanthus brunneus（Engl.）Danser subsp. djurensis（Engl.）Balle ＝ Agelanthus djurensis（Engl.）Polhill et Wiens ●☆

384039　Tapinanthus brunneus（Engl.）Danser subsp. krausei（Engl.）Balle ＝ Agelanthus krausei（Engl.）Polhill et Wiens ●☆

384040　Tapinanthus buchneri（Engl.）Danser；布赫纳大岩桐寄生●☆

384041　Tapinanthus bulawayensis（Engl.）Danser ＝ Agelanthus pungu（De Wild.）Polhill et Wiens ●☆

384042　Tapinanthus buntingii（Sprague）Danser；邦廷大岩桐寄生●☆

384043　Tapinanthus bussei（Sprague）Danser ＝ Oliverella bussei（Sprague）Polhill et Wiens ●☆

384044　Tapinanthus butaguensis（De Wild.）Danser ＝ Tapinanthus constrictiflorus（Engl.）Danser ●☆

384045　Tapinanthus buvumae（Rendle）Danser；布夫大岩桐寄生●☆

384046　Tapinanthus campestris（Engl.）Danser ＝ Oliverella hildebrandtii（Engl.）Tiegh. ●☆

384047　Tapinanthus capitatus（Spreng.）Danser ＝ Phragmanthera capitata（Spreng.）Balle ●☆

384048　Tapinanthus carsonii（Baker et Sprague）Danser ＝ Agelanthus pungu（De Wild.）Polhill et Wiens ●☆

384049　Tapinanthus ceciliae（N. E. Br.）Danser ＝ Agelanthus pungu（De Wild.）Polhill et Wiens ●☆

384050　Tapinanthus celtidifolius Danser ＝ Agelanthus scassellatii（Chiov.）Polhill et Wiens ●☆

384051　Tapinanthus chunguensis（R. E. Fr.）Danser ＝ Agelanthus subulatus（Engl.）Polhill et Wiens ●☆

384052　Tapinanthus cinereus（Engl.）Danser ＝ Phragmanthera cinerea（Engl.）Balle ●☆

384053　Tapinanthus cistoides（Welw. ex Engl.）Danser ＝ Phragmanthera glaucocarpa（Peyr.）Balle ●☆

384054　Tapinanthus constrictiflorus（Engl.）Danser；缩花大岩桐寄生●☆

384055　Tapinanthus copaiferae（Sprague）Danser ＝ Agelanthus

384056　Tapinanthus cordifolius Polhill et Wiens；心叶大岩桐寄生●☆

384057　Tapinanthus cornetii（Dewèvre）Danser ＝ Phragmanthera cornetii（Dewèvre）Polhill et Wiens ●☆

384058　Tapinanthus coronatus（Tiegh.）Danser；花冠大岩桐寄生●☆

384059　Tapinanthus crassicaulis（Engl.）Danser ＝ Phragmanthera crassicaulis（Engl.）Balle ●☆

384060　Tapinanthus crassifolius Wiens ＝ Agelanthus crassifolius（Wiens）Polhill et Wiens ●☆

384061　Tapinanthus crataevae（Sprague）Danser ＝ Agelanthus toroensis（Sprague）Polhill et Wiens ●☆

384062　Tapinanthus crispatulus（Sprague）Danser ＝ Tapinanthus buchneri（Engl.）Danser ●☆

384063　Tapinanthus dekindtianus（Engl.）Danser ＝ Agelanthus molleri（Engl.）Polhill et Wiens ●☆

384064　Tapinanthus deltae（Baker et Sprague）Danser ＝ Agelanthus deltae（Baker et Sprague）Polhill et Wiens ●☆

384065　Tapinanthus demeusei（Engl.）Danser ＝ Agelanthus djurensis（Engl.）Polhill et Wiens ●☆

384066　Tapinanthus dependens（Engl.）Danser；悬垂大岩桐寄生●☆

384067　Tapinanthus dichrous Danser ＝ Agelanthus dichrous（Danser）Polhill et Wiens ●☆

384068　Tapinanthus discolor（Schinz）Danser ＝ Agelanthus discolor（Schinz）Balle ●☆

384069　Tapinanthus djurensis（Engl.）Danser ＝ Agelanthus djurensis（Engl.）Polhill et Wiens ●☆

384070　Tapinanthus dodoneifolius（DC.）Danser ＝ Agelanthus dodoneifolius（DC.）Polhill et Wiens ●☆

384071　Tapinanthus dodoneifolius（DC.）Danser subsp. glaucoviridis（Engl.）Balle ＝ Agelanthus glaucoviridis（Engl.）Polhill et Wiens ●☆

384072　Tapinanthus dombeyae（K. Krause et Dinter）Danser ＝ Phragmanthera dombeyae（K. Krause et Dinter）Polhill et Wiens ●☆

384073　Tapinanthus dschallensis（Engl.）Danser ＝ Phragmanthera dschallensis（Engl.）M. G. Gilbert ●☆

384074　Tapinanthus ehlersii（Schweinf.）Danser ＝ Englerina woodfordioides（Schweinf.）Balle ●☆

384075　Tapinanthus elegantissimus（Schinz）Danser ＝ Oncocalyx welwitschii（Engl.）Polhill et Wiens ●☆

384076　Tapinanthus elegantulus（Engl.）Danser ＝ Agelanthus elegantulus（Engl.）Polhill et Wiens ●☆

384077　Tapinanthus eminii（Engl.）Danser ＝ Phragmanthera eminii（Engl.）Polhill et Wiens ●☆

384078　Tapinanthus engleri（Hiern）Danser ＝ Phragmanthera engleri（Hiern）Polhill et Wiens ●☆

384079　Tapinanthus englerianus（K. Krause et Dinter）Danser ＝ Oncocalyx welwitschii（Engl.）Polhill et Wiens ●☆

384080　Tapinanthus entebbensis（Sprague）Danser ＝ Agelanthus entebbensis（Sprague）Polhill et Wiens ●☆

384081　Tapinanthus erectotruncatus Balle ex Polhill et Wiens；截形大岩桐寄生●☆

384082　Tapinanthus erianthus（Sprague）Danser；毛花大岩桐寄生●☆

384083　Tapinanthus erythraeus（Sprague）Danser ＝ Phragmanthera erythraea（Sprague）M. G. Gilbert ●☆

384084　Tapinanthus eucalyptoides Danser ＝ Englerina woodfordioides（Schweinf.）Balle ●☆

384085　Tapinanthus eylesii（Sprague）Danser ＝ Agelanthus fuellebornii（Engl.）Polhill et Wiens ●☆

384086　Tapinanthus falcifolius（Sprague）Danser ＝ Agelanthus falcifolius（Sprague）Polhill et Wiens ●☆

384087　Tapinanthus findens（Sprague）Danser ＝ Tapinanthus constrictiflorus（Engl.）Danser ●☆

384088　Tapinanthus fischeri（Engl.）Danser ＝ Oncocalyx fischeri（Engl.）M. G. Gilbert ●☆

384089　Tapinanthus flamignii（De Wild.）Danser ＝ Agelanthus unyorensis（Sprague）Polhill et Wiens ●☆

384090　Tapinanthus forbesii（Sprague）Wiens；福布斯大岩桐寄生●☆

384091　Tapinanthus fragilis（Sprague）Danser ＝ Oncocalyx welwitschii（Engl.）Polhill et Wiens ●☆

384092　Tapinanthus friesiorum（K. Krause）Danser ＝ Oncocalyx sulfureus（Engl.）Wiens et Polhill ●☆

384093　Tapinanthus fuellebornii（Engl.）Danser ＝ Agelanthus fuellebornii（Engl.）Polhill et Wiens ●☆

384094　Tapinanthus gabonensis（Engl.）Danser ＝ Englerina gabonensis（Engl.）Balle ●☆

384095　Tapinanthus ghikae（Volkens et Schweinf.）Danser ＝ Oncocalyx ghikae（Volkens et Schweinf.）M. G. Gilbert ●☆

384096　Tapinanthus gilgii（Engl.）Danser ＝ Agelanthus glomeratus（Engl.）Polhill et Wiens ●☆

384097　Tapinanthus glabratus（Engl.）Danser ＝ Oncocalyx glabratus（Engl.）M. G. Gilbert ●☆

384098　Tapinanthus glaucescens（Engl. et K. Krause）Danser ＝ Agelanthus pungu（De Wild.）Polhill et Wiens ●☆

384099　Tapinanthus glaucocarpus（Peyr.）Danser ＝ Phragmanthera glaucocarpa（Peyr.）Balle ●☆

384100　Tapinanthus glaucophyllus（Engl.）Danser；灰绿大岩桐寄生●☆

384101　Tapinanthus glaucoviridis（Engl.）Danser ＝ Agelanthus glaucoviridis（Engl.）Polhill et Wiens ●☆

384102　Tapinanthus globiferus（A. Rich.）Tiegh.；球大岩桐寄生●☆

384103　Tapinanthus globiferus（A. Rich.）Tiegh. subsp. apodanthus（Sprague）Balle ＝ Tapinanthus apodanthus（Sprague）Danser ●☆

384104　Tapinanthus globiferus（A. Rich.）Tiegh. subsp. bangwensis（Engl. et K. Krause）Balle ＝ Tapinanthus bangwensis（Engl. et K. Krause）Danser ●☆

384105　Tapinanthus globiferus（A. Rich.）Tiegh. subsp. bornuensis（Sprague）Balle ＝ Tapinanthus globiferus（A. Rich.）Tiegh. ●☆

384106　Tapinanthus globiferus（A. Rich.）Tiegh. subsp. letouzeyi Balle ＝ Tapinanthus letouzeyi（Balle）Polhill et Wiens ●☆

384107　Tapinanthus glomeratus（Engl.）Danser ＝ Agelanthus glomeratus（Engl.）Polhill et Wiens ●☆

384108　Tapinanthus goetzei（Sprague）Danser ＝ Agelanthus fuellebornii（Engl.）Polhill et Wiens ●☆

384109　Tapinanthus gracilis Toelken et Wiens ＝ Agelanthus gracilis（Toelken et Wiens）Polhill et Wiens ●☆

384110　Tapinanthus guerichii（Engl.）Danser ＝ Phragmanthera guerichii（Engl.）Balle ●☆

384111　Tapinanthus guttatus（Sprague）Danser ＝ Tapinanthus dependens（Engl.）Danser ●☆

384112　Tapinanthus heckmannianus（Engl.）Danser ＝ Englerina heckmanniana（Engl.）Polhill et Wiens ●☆

384113　Tapinanthus henriquesii（Engl.）Danser ＝ Agelanthus glomeratus（Engl.）Polhill et Wiens ●☆

384114　Tapinanthus heteromorphus（A. Rich.）Danser ＝ Agelanthus heteromorphus（A. Rich.）Polhill et Wiens ●☆

384115　Tapinanthus heteromorphus（A. Rich.）Danser subsp. dichrous（Danser）Balle ＝ Agelanthus dichrous（Danser）Polhill et Wiens ●☆

384116　Tapinanthus hildebrandtii（Engl.）Danser ＝ Oliverella hildebrandtii（Engl.）Tiegh. ●☆

384117　Tapinanthus holstii（Engl.）Danser ＝ Englerina holstii（Engl.）Tiegh. ●☆

384118　Tapinanthus holtzii（Engl.）Danser ＝ Agelanthus elegantulus（Engl.）Polhill et Wiens ●☆

384119　Tapinanthus homblei（De Wild.）Danser ＝ Tapinanthus dependens（Engl.）Danser ●☆

384120　Tapinanthus igneus Danser ＝ Agelanthus igneus（Danser）Polhill et Wiens ●☆

384121　Tapinanthus inaequilaterus（Engl.）Danser ＝ Englerina inaequilatera（Engl.）Gilli ●☆

384122　Tapinanthus irangensis（Engl.）Danser ＝ Agelanthus irangensis（Engl.）Polhill et Wiens ●☆

384123　Tapinanthus irebuensis（De Wild.）Danser ＝ Phragmanthera batangae（Engl.）Balle ●☆

384124　Tapinanthus ituriensis（De Wild.）Danser ＝ Tapinanthus constrictiflorus（Engl.）Danser ●☆

384125　Tapinanthus kagehensis（Engl.）Danser ＝ Englerina kagehensis（Engl.）Polhill et Wiens ●☆

384126　Tapinanthus kamerunensis（Engl.）Danser ＝ Phragmanthera kamerunensis（Engl.）Balle ●☆

384127　Tapinanthus kayseri（Engl.）Danser ＝ Agelanthus kayseri（Engl.）Polhill et Wiens ●☆

384128　Tapinanthus keilii（Engl. et K. Krause）Danser ＝ Agelanthus keilii（Engl. et K. Krause）Polhill et Wiens ●☆

384129　Tapinanthus kelleri（Engl.）Danser ＝ Oncocalyx kelleri（Engl.）M. G. Gilbert ●☆

384130　Tapinanthus keniae（K. Krause）Danser ＝ Agelanthus pennatulus（Sprague）Polhill et Wiens ●☆

384131　Tapinanthus kerstingii（Engl.）Balle ＝ Tapinanthus globiferus（A. Rich.）Tiegh. ●☆

384132　Tapinanthus keudelii（Engl.）Danser ＝ Agelanthus elegantulus（Engl.）Polhill et Wiens ●☆

384133　Tapinanthus kihuirensis（Engl.）Danser ＝ Oncocalyx kelleri（Engl.）M. G. Gilbert ●☆

384134　Tapinanthus kilimandscharicus（Engl.）Danser ＝ Agelanthus elegantulus（Engl.）Polhill et Wiens ●☆

384135　Tapinanthus kimuenzae（De Wild.）Danser ＝ Tapinanthus constrictiflorus（Engl.）Danser ●☆

384136　Tapinanthus kisaguka（Engl. et K. Krause）Danser ＝ Phragmanthera usuiensis（Oliv.）M. G. Gilbert ●☆

384137　Tapinanthus kisantuensis（De Wild. et T. Durand）Danser ＝ Tapinanthus constrictiflorus（Engl.）Danser ●☆

384138　Tapinanthus krausei（Engl.）Danser ＝ Agelanthus krausei（Engl.）Polhill et Wiens ●☆

384139　Tapinanthus kraussianus（Meisn.）Tiegh.；克劳斯大岩桐寄生●☆

384140　Tapinanthus kraussianus（Meisn.）Tiegh. ＝ Agelanthus kraussianus（Meisn.）Polhill et Wiens ●☆

384141　Tapinanthus kraussianus（Meisn.）Tiegh. subsp. transvaalensis（Sprague）Wiens ＝ Agelanthus transvaalensis（Sprague）Polhill et Wiens ●☆

384142　Tapinanthus kwaiensis（Engl.）Danser ＝ Englerina kwaiensis（Engl.）Polhill et Wiens ●☆

384143　Tapinanthus laciniatus（Engl.）Danser ＝ Agelanthus elegantulus（Engl.）Polhill et Wiens ●☆

384144　Tapinanthus lamborayi（De Wild. ）Danser ＝ Tapinanthus globiferus（A. Rich. ）Tiegh. ●☆

384145　Tapinanthus landanaensis（De Wild. ）Danser ＝ Tapinanthus buchneri（Engl. ）Danser ●☆

384146　Tapinanthus lapathifolius（Engl. et K. Krause）Danser ＝ Phragmanthera capitata（Spreng. ）Balle ●☆

384147　Tapinanthus lateritiostriatus（Engl. et K. Krause）Danser ＝ Agelanthus nyasicus（Baker et Sprague）Polhill et Wiens ●☆

384148　Tapinanthus latibracteatus（Engl. ）Danser ＝ Agelanthus subulatus（Engl. ）Polhill et Wiens ●☆

384149　Tapinanthus lecardii（Engl. ）Danser ＝ Englerina lecardii（Engl. ）Balle ●☆

384150　Tapinanthus lecomtei（Tiegh. ）Danser ＝ Englerina gabonensis（Engl. ）Balle ●☆

384151　Tapinanthus leendertziae（Sprague）Wiens ＝ Tapinanthus quequensis（Weim. ）Polhill et Wiens ●☆

384152　Tapinanthus lenticellatus（De Wild. et T. Durand）Danser ＝ Agelanthus djurensis（Engl. ）Polhill et Wiens ●☆

384153　Tapinanthus leonensis（Sprague）Danser ＝ Phragmanthera leonensis（Sprague）Balle ●☆

384154　Tapinanthus letouzeyi（Balle）Polhill et Wiens；勒图大岩桐寄生●☆

384155　Tapinanthus loandensis（Engl. et K. Krause）Danser ＝ Agelanthus brunneus（Engl. ）Balle et N. Hallé ●☆

384156　Tapinanthus longiflorus Polhill et Wiens；长花大岩桐寄生●☆

384157　Tapinanthus longifolius（Peter）Danser ＝ Englerina woodfordioides（Schweinf. ）Balle ●☆

384158　Tapinanthus longipes（Baker et Sprague）Danser ＝ Agelanthus longipes（Baker et Sprague）Polhill et Wiens ●☆

384159　Tapinanthus lugardii（N. E. Br. ）Danser ＝ Agelanthus lugardii（N. E. Br. ）Polhill et Wiens ●☆

384160　Tapinanthus lujaei（De Wild. et T. Durand）Danser ＝ Agelanthus djurensis（Engl. ）Polhill et Wiens ●☆

384161　Tapinanthus lukwangulensis（Engl. ）Danser ＝ Englerina inaequilatera（Engl. ）Gilli ●☆

384162　Tapinanthus luluensis（Engl. ）Danser ＝ Englerina luluensis（Engl. ）Polhill et Wiens ●☆

384163　Tapinanthus luteiflorus（Engl. et K. Krause）Danser ＝ Agelanthus pungu（De Wild. ）Polhill et Wiens ●☆

384164　Tapinanthus luteoaurantiacus（De Wild. ）Danser ＝ Englerina schubotziana（Engl. et K. Krause）Polhill et Wiens ●☆

384165　Tapinanthus luteostriatus（Engl. et K. Krause）Danser ＝ Agelanthus nyasicus（Baker et Sprague）Polhill et Wiens ●☆

384166　Tapinanthus luteovittatus（Engl. et K. Krause）Danser ＝ Phragmanthera luteovittata（Engl. et K. Krause）Polhill et Wiens ●☆

384167　Tapinanthus macrosolen（Steud. ex A. Rich. ）Danser ＝ Phragmanthera macrosolen（Steud. ex A. Rich. ）M. G. Gilbert ●☆

384168　Tapinanthus malacophyllus（Engl. et K. Krause）Danser；软叶大岩桐寄生●☆

384169　Tapinanthus mangheensis（De Wild. ）Danser ＝ Tapinanthus constrictiflorus（Engl. ）Danser ●☆

384170　Tapinanthus marginatus Danser ＝ Phragmanthera crassicaulis（Engl. ）Balle ●☆

384171　Tapinanthus mbogaensis（De Wild. ）Danser ＝ Agelanthus djurensis（Engl. ）Polhill et Wiens ●☆

384172　Tapinanthus mechowii（Engl. ）Tiegh. ；梅休大岩桐寄生●☆

384173　Tapinanthus menyharthii（Engl. et Schinz ex Schinz）Danser ＝ Actinanthella menyharthii（Engl. et Schinz ex Schinz）Balle ●☆

384174　Tapinanthus micrantherus（Engl. ）Danser ＝ Englerina gabonensis（Engl. ）Balle ●☆

384175　Tapinanthus microphyllus（Engl. ）Danser ＝ Oncocalyx ugogensis（Engl. ）Wiens et Polhill ●☆

384176　Tapinanthus minor（Harv. ）Danser ＝ Agelanthus natalitius（Meisn. ）Polhill et Wiens subsp. zeyheri（Harv. ）Polhill et Wiens ●☆

384177　Tapinanthus minor（Sprague）Danser ＝ Agelanthus gracilis（Toelken et Wiens）Polhill et Wiens ●☆

384178　Tapinanthus molleri（Engl. ）Danser ＝ Agelanthus molleri（Engl. ）Polhill et Wiens ●☆

384179　Tapinanthus mollissimus（Engl. ）Danser；柔软大岩桐寄生●☆

384180　Tapinanthus moorei（Sprague）Danser ＝ Agelanthus natalitius（Meisn. ）Polhill et Wiens ●☆

384181　Tapinanthus mortehanii（De Wild. ）Danser ＝ Tapinanthus apodanthus（Sprague）Danser ●☆

384182　Tapinanthus muerensis（Engl. ）Danser ＝ Englerina muerensis（Engl. ）Polhill et Wiens ●☆

384183　Tapinanthus musozensis（Rendle）Danser ＝ Agelanthus musozensis（Rendle）Polhill et Wiens ●☆

384184　Tapinanthus myrsinifolius（Engl. et K. Krause）Danser ＝ Agelanthus myrsinifolius（Engl. et K. Krause）Polhill et Wiens ●☆

384185　Tapinanthus namaquensis（Harv. ）Tiegh. ＝ Tapinanthus oleifolius（J. C. Wendl. ）Danser ●☆

384186　Tapinanthus natalitius（Meisn. ）Danser ＝ Agelanthus natalitius（Meisn. ）Polhill et Wiens ●☆

384187　Tapinanthus natalitius（Meisn. ）Danser subsp. zeyheri（Harv. ）Wiens ＝ Agelanthus natalitius（Meisn. ）Polhill et Wiens subsp. zeyheri（Harv. ）Polhill et Wiens ●☆

384188　Tapinanthus nigritanus（Hook. f. ex Benth. ）Danser ＝ Phragmanthera nigritana（Hook. f. ex Benth. ）Balle ●☆

384189　Tapinanthus nitidulus（Sprague）Danser ＝ Phragmanthera polycrypta（Didr. ）Balle ●☆

384190　Tapinanthus nyasicus（Baker et Sprague）Danser ＝ Agelanthus nyasicus（Baker et Sprague）Polhill et Wiens ●☆

384191　Tapinanthus obovatus（Peter）Danser ＝ Agelanthus zizyphifolius（Engl. ）Polhill et Wiens subsp. vittatus（Engl. ）Polhill et Wiens ●☆

384192　Tapinanthus ochroleucus（Engl. et K. Krause）Danser ＝ Englerina ochroleuca（Engl. et K. Krause）Balle ●☆

384193　Tapinanthus oedostemon Danser ＝ Englerina oedostemon（Danser）Polhill et Wiens ●☆

384194　Tapinanthus oehleri（Engl. ）Danser ＝ Agelanthus oehleri（Engl. ）Polhill et Wiens ●☆

384195　Tapinanthus ogowensis（Engl. ）Danser；奥古大岩桐寄生●☆

384196　Tapinanthus ogowensis（Engl. ）Danser var. batesii（S. Moore et Sprague）Balle ＝ Tapinanthus ogowensis（Engl. ）Danser ●☆

384197　Tapinanthus oleifolius（J. C. Wendl. ）Danser；木犀榄叶大岩桐寄生●☆

384198　Tapinanthus ophiodes（Sprague）Danser；蛇状大岩桐寄生●☆

384199　Tapinanthus pallideviridis（Engl. et K. Krause）Danser ＝ Agelanthus zizyphifolius（Engl. ）Polhill et Wiens subsp. vittatus（Engl. ）Polhill et Wiens ●☆

384200　Tapinanthus parviflorus（Engl. ）Danser ＝ Englerina parviflora（Engl. ）Balle ●☆

384201　Tapinanthus pennatulus（Sprague）Danser ＝ Agelanthus pennatulus（Sprague）Polhill et Wiens ●☆

384202　Tapinanthus pentagonia（DC. ）Tiegh. ；五节大岩桐寄生●☆

384203 Tapinanthus pentagonia（DC.）Tiegh. var. guineensis Balle ＝ Tapinanthus pentagonia（DC.）Tiegh. ●☆

384204 Tapinanthus pentagonia（DC.）Tiegh. var. senegalensis（De Wild.）Balle ＝Tapinanthus pentagonia（DC.）Tiegh. ●☆

384205 Tapinanthus penteneurus Danser ＝ Agelanthus djurensis（Engl.）Polhill et Wiens ●☆

384206 Tapinanthus petiolatus（De Wild.）Danser ＝ Agelanthus djurensis（Engl.）Polhill et Wiens ●☆

384207 Tapinanthus platyphyllus（Hochst. ex A. Rich.）Danser ＝ Agelanthus platyphyllus（Hochst. ex A. Rich.）Balle ●☆

384208 Tapinanthus poggei（Engl.）Danser ＝ Tapinanthus constrictiflorus（Engl.）Danser ●☆

384209 Tapinanthus polycryptus（Didr.）Danser ＝ Phragmanthera polycrypta（Didr.）Balle ●☆

384210 Tapinanthus polygonifolius（Engl.）Danser ＝ Agelanthus polygonifolius（Engl.）Polhill et Wiens ●☆

384211 Tapinanthus preussii（Engl.）Tiegh.；普罗伊斯大岩桐寄生●☆

384212 Tapinanthus proteicola（Engl.）Danser ＝ Phragmanthera proteicola（Engl.）Polhill et Wiens ●☆

384213 Tapinanthus prunifolius（E. Mey. ex Harv.）Tiegh. ＝ Agelanthus prunifolius（E. Mey. ex Harv.）Polhill et Wiens ●☆

384214 Tapinanthus prunifolius（E. Mey. ex Harv.）Tiegh. subsp. combreticola（Lebrun et L. Touss.）Balle ＝ Agelanthus combreticola（Lebrun et L. Touss.）Polhill et Wiens ●☆

384215 Tapinanthus prunifolius（E. Mey. ex Harv.）Tiegh. subsp. keilii（Engl. et K. Krause）Balle ＝Agelanthus keilii（Engl. et K. Krause）Polhill et Wiens ●☆

384216 Tapinanthus prunifolius（E. Mey. ex Harv.）Tiegh. subsp. musozensis（Rendle）Balle ＝ Agelanthus musozensis（Rendle）Polhill et Wiens ●☆

384217 Tapinanthus pseudonymus Danser ＝ Phragmanthera polycrypta（Didr.）Balle ●☆

384218 Tapinanthus pubiflorus（Sprague）Danser ＝ Tapinanthus belvisii（DC.）Danser ●☆

384219 Tapinanthus pungu（De Wild.）Danser ＝ Agelanthus pungu（De Wild.）Polhill et Wiens ●☆

384220 Tapinanthus quequensis（Weim.）Polhill et Wiens；克克大岩桐寄生●☆

384221 Tapinanthus quinquangulus（Engl. et Schinz）Danser ＝ Tapinanthus oleifolius（J. C. Wendl.）Danser ●☆

384222 Tapinanthus quinquenervius（Hochst.）Danser ＝ Oncocalyx quinquenervius（Hochst.）Wiens et Polhill ●☆

384223 Tapinanthus ramulosus（Sprague）Danser ＝Englerina ramulosa（Sprague）Polhill et Wiens ●☆

384224 Tapinanthus redingii（De Wild.）Danser ＝ Phragmanthera capitata（Spreng.）Balle ●☆

384225 Tapinanthus regularis（Steud. ex Sprague）Danser ＝Phragmanthera regularis（Steud. ex Sprague）M. G. Gilbert ☆

384226 Tapinanthus remotus（Baker et Sprague）Danser ＝ Vanwykia remota（Baker et Sprague）Wiens ●☆

384227 Tapinanthus reygaertii（De Wild.）Danser ＝ Tapinanthus ogowensis（Engl.）Danser ●☆

384228 Tapinanthus rhamnifolius（Engl.）Danser ＝ Spragueanella rhamnifolia（Engl.）Balle ●☆

384229 Tapinanthus rigidilobus（K. Krause）Danser ＝ Phragmanthera usuiensis（Oliv.）M. G. Gilbert ●☆

384230 Tapinanthus rondensis（Engl.）Danser ＝ Agelanthus rondensis（Engl.）Polhill et Wiens ●☆

384231 Tapinanthus rosiflorus（Engl. et K. Krause）Danser ＝ Tapinanthus globiferus（A. Rich.）Tiegh. ●☆

384232 Tapinanthus rubiginosus（De Wild.）Danser ＝ Phragmanthera cornetii（Dewèvre）Polhill et Wiens ●☆

384233 Tapinanthus rubromarginatus（Engl.）Danser；红边大岩桐寄生●☆

384234 Tapinanthus rubroviridis（Tiegh.）Danser ＝Oliverella rubroviridis Tiegh. ●☆

384235 Tapinanthus rubrovittatus（Engl. et K. Krause）Danser ＝ Tapinanthus globiferus（A. Rich.）Tiegh. ●☆

384236 Tapinanthus rufescens（DC.）Danser ＝ Phragmanthera rufescens（DC.）Balle ●☆

384237 Tapinanthus rugegensis（Engl. et K. Krause）Danser ＝ Englerina woodfordioides（Schweinf.）Balle ●☆

384238 Tapinanthus rugulosus Danser ＝ Tapinanthus buvumae（Rendle）Danser ●☆

384239 Tapinanthus sacleuxii（Tiegh.）Danser ＝ Oliverella hildebrandtii（Engl.）Tiegh. ●☆

384240 Tapinanthus sakarensis（Engl.）Danser ＝Agelanthus tanganyikae（Engl.）Polhill et Wiens ●☆

384241 Tapinanthus sambesiacus（Engl. et Schinz）Danser ＝ Agelanthus sambesiacus（Engl. et Schinz）Polhill et Wiens ●☆

384242 Tapinanthus sankuruensis（De Wild.）Danser ＝ Agelanthus brunneus（Engl.）Balle et N. Hallé ●☆

384243 Tapinanthus sansibarensis（Engl.）Danser ＝ Agelanthus sansibarensis（Engl.）Polhill et Wiens ●☆

384244 Tapinanthus sapinii（De Wild.）Danser ＝ Agelanthus djurensis（Engl.）Polhill et Wiens ●☆

384245 Tapinanthus scassellatii（Chiov.）Danser ＝ Agelanthus scassellatii（Chiov.）Polhill et Wiens ●☆

384246 Tapinanthus schimperi（Hochst. ex A. Rich.）Danser ＝ Oncocalyx schimperi（Hochst. ex A. Rich.）M. G. Gilbert ●☆

384247 Tapinanthus schlechteri（Engl.）Danser ＝ Englerina schlechteri（Engl.）Polhill et Wiens ●☆

384248 Tapinanthus schubotzianus（Engl. et K. Krause）Danser ＝ Englerina schubotziana（Engl. et K. Krause）Polhill et Wiens ●☆

384249 Tapinanthus schweinfurthii（Engl.）Danser ＝ Agelanthus schweinfurthii（Engl.）Polhill et Wiens ●☆

384250 Tapinanthus senegalensis（De Wild.）Danser ＝ Tapinanthus pentagonia（DC.）Tiegh. ●☆

384251 Tapinanthus seretii（De Wild.）Danser ＝Phragmanthera seretii（De Wild.）Balle ●☆

384252 Tapinanthus sessilifolius（P. Beauv.）Tiegh.；无梗大岩桐寄生●☆

384253 Tapinanthus sigensis（Engl.）Danser ＝ Phragmanthera usuiensis（Oliv.）M. G. Gilbert subsp. sigensis（Engl.）Polhill et Wiens ●☆

384254 Tapinanthus sphaericocompressus（De Wild.）Danser ＝ Agelanthus brunneus（Engl.）Balle et N. Hallé ●☆

384255 Tapinanthus sterculiae（Hiern）Danser ＝ Phragmanthera sterculiae（Hiern）Polhill et Wiens ●☆

384256 Tapinanthus stolzii（Engl. et K. Krause）Danser ＝ Agelanthus nyasicus（Baker et Sprague）Polhill et Wiens ●☆

384257 Tapinanthus stuhlmannii（Engl.）Danser ＝ Oncocalyx fischeri（Engl.）M. G. Gilbert ●☆

384258 Tapinanthus subquadrangularis（De Wild.）Danser ＝ Englerina subquadrangularis（De Wild.）Polhill et Wiens ●☆

384259 Tapinanthus subulatus (Engl.) Danser = Agelanthus subulatus (Engl.) Polhill et Wiens ●☆

384260 Tapinanthus sulfureus (Engl.) Danser = Oncocalyx sulfureus (Engl.) Wiens et Polhill ●☆

384261 Tapinanthus swynnertonii (Sprague) Danser = Englerina swynnertonii (Sprague) Polhill et Wiens ●☆

384262 Tapinanthus syringifolius (Tiegh.) Danser = Tapinanthus constrictiflorus (Engl.) Danser ●☆

384263 Tapinanthus talbotiorum (Sprague) Danser = Phragmanthera talbotiorum (Sprague) Balle ●☆

384264 Tapinanthus tambermensis (Engl. et K. Krause) Danser = Agelanthus heteromorphus (A. Rich.) Polhill et Wiens ●☆

384265 Tapinanthus tanganyikae (Engl.) Danser = Agelanthus tanganyikae (Engl.) Polhill et Wiens ●☆

384266 Tapinanthus tenuifolius (Engl.) Danser = Englerina inaequilatera (Engl.) Gilli ●☆

384267 Tapinanthus terminaliae (Engl. et Gilg) Danser = Agelanthus terminaliae (Engl. et Gilg) Polhill et Wiens ●☆

384268 Tapinanthus thollonii (Tiegh.) Danser = Phragmanthera capitata (Spreng.) Balle ●☆

384269 Tapinanthus thonningii Danser = Tapinanthus bangwensis (Engl. et K. Krause) Danser ●☆

384270 Tapinanthus thorei (K. Krause) Danser = Agelanthus sansibarensis (Engl.) Polhill et Wiens subsp. montanus Polhill et Wiens ●☆

384271 Tapinanthus toroensis (Sprague) Danser = Agelanthus toroensis (Sprague) Polhill et Wiens ●☆

384272 Tapinanthus tricolor (Peter) Danser = Agelanthus sansibarensis (Engl.) Polhill et Wiens ●☆

384273 Tapinanthus trinervius (Engl.) Danser = Agelanthus djurensis (Engl.) Polhill et Wiens ●☆

384274 Tapinanthus triplinervius (Baker et Sprague) Danser = Englerina triplinervia (Baker et Sprague) Polhill et Wiens ●☆

384275 Tapinanthus truncatus (Engl.) Danser = Tapinanthus belvisii (DC.) Danser ●☆

384276 Tapinanthus tschertscherensis (Pax) Danser = Englerina woodfordioides (Schweinf.) Balle ●☆

384277 Tapinanthus tschintschochensis (Engl.) Danser = Tapinanthus buchneri (Engl.) Danser ●☆

384278 Tapinanthus ugogensis (Engl.) Danser = Oncocalyx ugogensis (Engl.) Wiens et Polhill ●☆

384279 Tapinanthus uhehensis (Engl.) Danser = Agelanthus uhehensis (Engl.) Polhill et Wiens ●☆

384280 Tapinanthus umbelliflorus (De Wild.) Danser = Englerina woodfordioides (Schweinf.) Balle ●☆

384281 Tapinanthus unyorensis (Sprague) Danser =Agelanthus unyorensis (Sprague) Polhill et Wiens ●☆

384282 Tapinanthus usambarensis (Engl.) Danser = Agelanthus subulatus (Engl.) Polhill et Wiens ●☆

384283 Tapinanthus usuiensis (Oliv.) Danser =Phragmanthera usuiensis (Oliv.) M. G. Gilbert ●☆

384284 Tapinanthus vanderystii (De Wild.) Danser = Tapinanthus ogowensis (Engl.) Danser ●☆

384285 Tapinanthus variifolius (De Wild.) Danser = Agelanthus brunneus (Engl.) Balle et N. Hallé ●☆

384286 Tapinanthus verrucosus (Engl.) Tiegh. = Tapinanthus globiferus (A. Rich.) Tiegh. ●☆

384287 Tapinanthus verschuerenii (De Wild.) Danser = Tapinanthus buchneri (Engl.) Danser ●☆

384288 Tapinanthus villosiflorus (Engl.) Danser =Agelanthus villosiflorus (Engl.) Polhill et Wiens ●☆

384289 Tapinanthus viminalis (Engl. et K. Krause) Danser = Englerina woodfordioides (Schweinf.) Balle ●☆

384290 Tapinanthus vittatus (Engl.) Danser = Agelanthus zizyphifolius (Engl.) Polhill et Wiens subsp. vittatus (Engl.) Polhill et Wiens ●☆

384291 Tapinanthus voltensis Tiegh. ex Balle = Tapinanthus globiferus (A. Rich.) Tiegh. ●☆

384292 Tapinanthus warneckei (Engl.) Danser = Tapinanthus sessilifolius (P. Beauv.) Tiegh. ●☆

384293 Tapinanthus welwitschii (Engl.) Danser = Oncocalyx welwitschii (Engl.) Polhill et Wiens ●☆

384294 Tapinanthus wentzelianus (Engl.) Danser = Phragmanthera usuiensis (Oliv.) M. G. Gilbert subsp. sigensis (Engl.) Polhill et Wiens ●☆

384295 Tapinanthus woodfordioides (Schweinf.) Danser = Englerina woodfordioides (Schweinf.) Balle ●☆

384296 Tapinanthus wyliei (Sprague) Danser = Actinanthella wyliei (Sprague) Wiens ●☆

384297 Tapinanthus xanthanthus Danser = Phragmanthera capitata (Spreng.) Balle ●☆

384298 Tapinanthus zeyheri (Harv.) Danser = Agelanthus natalitius (Meisn.) Polhill et Wiens subsp. zeyheri (Harv.) Polhill et Wiens ●☆

384299 Tapinanthus zizyphifolius (Engl.) Danser = Agelanthus zizyphifolius (Engl.) Polhill et Wiens ●☆

384300 Tapinanthus zygiarum (Hiern) Danser = Phragmanthera zygiarum (Hiern) Polhill et Wiens ●☆

384301 Tapinia Post et Kuntze =Tapeinia Comm. ex Juss. ■☆

384302 Tapinia Steud. = Tapirira Aubl. ●☆

384303 Tapinocarpus Dalzell = Theriophonum Blume ■☆

384304 Tapinochilus Post et Kuntze =Tapeinochilos Miq. (保留属名)■☆

384305 Tapinopentas Bremek. = Otomeria Benth. ■☆

384306 Tapinopentas cameronica Bremek. = Otomeria cameronica (Bremek.) Hepper ■☆

384307 Tapinopentas latifolia Verdc. = Otomeria cameronica (Bremek.) Hepper ■☆

384308 Tapinopentas ulugurica Verdc. = Pentas ulugurica (Verdc.) Hepper ■☆

384309 Tapinophallus Post et Kuntze = Amorphophallus Blume ex Decne. (保留属名)■●

384310 Tapinophallus Post et Kuntze =Tapeinophallus Baill. ■●

384311 Tapinosperma Post et Kuntze =Tapeinosperma Hook. f. ●☆

384312 Tapinostemma (Benth.) Tiegh. = Plicosepalus Tiegh. ●☆

384313 Tapinostemma Tiegh. =Plicosepalus Tiegh. ●☆

384314 Tapinostemma acaciae (Zucc.) Tiegh. = Plicosepalus acaciae (Zucc.) Wiens et Polhill ●☆

384315 Tapinostemma meridianum Danser = Plicosepalus meridianus (Danser) Wiens et Polhill ●☆

384316 Tapinostemma nummulariifolium (Franch.) Tiegh. = Plicosepalus nummulariifolius (Franch.) Wiens et Polhill ●☆

384317 Tapiphyllum Robyns(1928);热非茜属■☆

384318 Tapiphyllum burnettii Tennant;伯内特热非茜■☆

384319 Tapiphyllum cinerascens (Hiern) Robyns;浅灰热非茜■☆

384320 Tapiphyllum cinerascens (Hiern) Robyns subsp. inaequale (Robyns) Verdc. = Tapiphyllum cinerascens (Hiern) Robyns var.

inaequale（Robyns）Verdc. ■☆

384321 Tapiphyllum cinerascens（Hiern）Robyns var. inaequale（Robyns）Verdc. ;不等浅灰热非茜■☆

384322 Tapiphyllum cinerascens（Hiern）Robyns var. laetum（Robyns）Verdc. ;愉悦浅灰热非茜■☆

384323 Tapiphyllum cinerascens（Hiern）Robyns var. laevius（K. Schum.）Verdc. ;无毛浅灰热非茜■☆

384324 Tapiphyllum cinerascens（Hiern）Robyns var. richardsiae（Robyns）Verdc. ;理查浅灰热非茜■☆

384325 Tapiphyllum cinerascens（Welw. ex Hiern）Robyns subsp. cinerascens Verdc. = Tapiphyllum cinerascens（Hiern）Robyns ■☆

384326 Tapiphyllum cistifolium（Welw. ex Hiern）Robyns = Vangueria cistifolia（Welw. ex Hiern）Lantz ■☆

384327 Tapiphyllum confertiflorum Robyns = Tapiphyllum discolor（De Wild.）Robyns ■☆

384328 Tapiphyllum discolor（De Wild.）Robyns;异色热非茜■☆

384329 Tapiphyllum fadogia Bullock = Tapiphyllum discolor（De Wild.）Robyns ■☆

384330 Tapiphyllum floribundum Bullock = Tapiphyllum obtusifolium（K. Schum.）Robyns ■☆

384331 Tapiphyllum fulvum Robyns;黄褐热非茜■☆

384332 Tapiphyllum glomeratum Robyns = Tapiphyllum cinerascens（Hiern）Robyns var. richardsiae（Robyns）Verdc. ■☆

384333 Tapiphyllum gossweileri Robyns;戈斯热非茜■☆

384334 Tapiphyllum grandiflorum Bullock = Tapiphyllum discolor（De Wild.）Robyns ■☆

384335 Tapiphyllum griseum Robyns = Tapiphyllum cinerascens（Hiern）Robyns var. laevius（K. Schum.）Verdc. ■☆

384336 Tapiphyllum herbaceum Robyns = Tapiphyllum discolor（De Wild.）Robyns ■☆

384337 Tapiphyllum inaequale Robyns = Tapiphyllum cinerascens（Hiern）Robyns var. inaequale（Robyns）Verdc. ■☆

384338 Tapiphyllum kaessneri（S. Moore）Robyns = Tapiphyllum cinerascens（Hiern）Robyns var. inaequale（Robyns）Verdc. ■☆

384339 Tapiphyllum laetum Robyns = Tapiphyllum cinerascens（Hiern）Robyns var. laetum（Robyns）Verdc. ■☆

384340 Tapiphyllum molle Robyns;柔软热非茜■☆

384341 Tapiphyllum mucronulatum Robyns = Pachystigma schumannianum（Robyns）Bridson et Verdc. subsp. mucronulatum（Robyns）Bridson et Verdc. ●☆

384342 Tapiphyllum oblongifolium Robyns = Tapiphyllum discolor（De Wild.）Robyns ■☆

384343 Tapiphyllum obtusifolium（K. Schum.）Robyns;钝叶热非茜■☆

384344 Tapiphyllum pachyanthum Robyns;粗花热非茜■☆

384345 Tapiphyllum parvifolium（Sond.）Robyns = Vangueria parvifolia Sond. ■☆

384346 Tapiphyllum psammophilum（K. Schum.）Robyns;喜沙热非茜■☆

384347 Tapiphyllum quarrei Robyns;卡雷热非茜■☆

384348 Tapiphyllum rhodesiacum（Tennant）Bridson;罗得西亚热非茜■☆

384349 Tapiphyllum richardsii Robyns = Tapiphyllum cinerascens（Hiern）Robyns var. richardsiae（Robyns）Verdc. ■☆

384350 Tapiphyllum schliebenii Verdc. ;施利本热非茜■☆

384351 Tapiphyllum schumannianum Robyns = Pachystigma schumannianum（Robyns）Bridson et Verdc. ●☆

384352 Tapiphyllum schumannianum Robyns subsp. mucronulatum（Robyns）Verdc. = Pachystigma schumannianum（Robyns）Bridson et Verdc. subsp. mucronulatum（Robyns）Bridson et Verdc. ●☆

384353 Tapiphyllum tanganyikense Robyns = Tapiphyllum cinerascens（Hiern）Robyns var. laevius（K. Schum.）Verdc. ■☆

384354 Tapiphyllum velutinum（Hiern）Robyns;绒毛热非茜■☆

384355 Tapiphyllum velutinum（Hiern）Robyns var. laevius（K. Schum.）Robyns = Tapiphyllum cinerascens（Hiern）Robyns var. laevius（K. Schum.）Verdc. ■☆

384356 Tapiphyllum verticillatum Robyns;轮生热非茜■☆

384357 Tapiphyllum vestitum Robyns = Tapiphyllum velutinum（Hiern）Robyns ■☆

384358 Tapiria Hook. f. = Pegia Colebr. ●

384359 Tapiria Juss. = Tapirira Aubl. ●☆

384360 Tapiria hirsuta（Roxb.）Hook. f. = Pegia nitida Colebr. ●

384361 Tapirira Aubl.（1775）;塔皮木属●☆

384362 Tapirira edulis Brandegee;可食塔皮木●☆

384363 Tapirira extensa（Wall.）Hook. f. ex Marchand = Pegia nitida Colebr. ●

384364 Tapirira hirsula（Roxb.）Hu = Pegia nitida Colebr. ●

384365 Tapirira marchandii Engl. ;马钱塔皮木●☆

384366 Tapirira mexicana Marchand;墨西哥塔皮木●☆

384367 Tapirocarpus Sagot（1882）;厚果橄榄属●☆

384368 Tapirocarpus talisia Sagot;厚果橄榄●☆

384369 Tapiscia Oliv.（1890）;瘿椒树属（银雀树属,银鹊树属）;False Pistache,Falsepistache,False-pistache ●★

384370 Tapiscia lichunensis W. C. Cheng et C. D. Chu;利川瘿椒树（利川银鹊树）;Lichuan Falsepistache,Lichuan False-pistache ●◇

384371 Tapiscia sinensis Oliv. ;瘿椒树（丹树,接骨,泡花,皮巴风,银雀树,银鹊树,瘿漆树）;Chinese False-pistache,Falsepistache ●◇

384372 Tapiscia sinensis Oliv. var. concolor W. C. Cheng = Tapiscia sinensis Oliv. ●◇

384373 Tapiscia sinensis Oliv. var. macrocarpa T. Z. Hsu;大果瘿椒树;Bigfruit Falsepistache,Big-fruited Falsepistache ●

384374 Tapiscia sinensis Rehder et E. H. Wilson = Tapiscia yunnanensis W. C. Cheng et C. D. Chu ●

384375 Tapiscia yunnanensis W. C. Cheng et C. D. Chu;云南瘿椒树;Yunnan Falsepistache,Yunnan False-pistache ●

384376 Tapisciaceae（Pax）Takht.（1987）;瘿椒树科●

384377 Tapisciaceae（Pax）Takht. = Staphyleaceae Martinov（保留科名）●

384378 Tapisciaceae Takht. = Staphyleaceae Martinov（保留科名）●

384379 Tapisciaceae Takht. = Tapisciaceae（Pax）Takht. ●

384380 Taplinia Lander（1989）;岩鼠麹属■☆

384381 Taplinia saxatilis Lander;岩鼠麹■☆

384382 Tapogamea Raf. = Tapogomea Aubl.（废弃属名）●

384383 Tapogomea Aubl.（废弃属名）= Cephaelis Sw.（保留属名）●

384384 Tapogomea Aubl.（废弃属名）= Psychotria L.（保留属名）●

384385 Tapoides Airy Shaw（1960）;加岛大戟属●☆

384386 Tapoides villamilii（Merr.）Airy Shaw;加岛大戟●☆

384387 Tapomana Adans. = Connarus L. ●

384388 Taprobanea Christenson（1992）;匙树兰属（小匙兰属）■☆

384389 Taprobanea spathulata（L.）Christenson;匙树兰（小匙兰）■☆

384390 Tapura Aubl.（1775）;塔普木属●☆

384391 Tapura africana Oliv. ;非洲塔普木●☆

384392 Tapura arachnoidea Breteler;蛛网塔普木●☆

384393 Tapura bouquetiana N. Hallé et Heine;布凯塔普木●☆

384394 Tapura carinata Breteler;龙骨状塔普木●☆

384395 Tapura fischeri Engl. ;菲舍尔塔普木●☆

384396 Tapura fischeri Engl. var. pubescens Verdc. et Torre = Tapura fischeri Engl. ●☆

384397　Tapura guianensis Aubl. ;圭亚那塔普木●☆

384398　Tapura ivorensis Breteler;伊沃里塔普木●☆

384399　Tapura le-testui Pellegr. ;勒泰斯蒂塔普木●☆

384400　Tapura lujae De Wild. = Tapura fischeri Engl. ●☆

384401　Tapura neglecta N. Hallé et Heine;疏忽塔普木●☆

384402　Tara Molina = Caesalpinia L. ●

384403　Tara spinosa（Molina）Britton et Rose = Caesalpinia spinosa （Molina）Kuntze ●☆

384404　Taraktogenos Hassk. = Hydnocarpus Gaertn. ●

384405　Taraktogenos annamensis Gagnep. = Hydnocarpus annamensis （Gagnep.）Lescot et Sleumer ●◇

384406　Taraktogenos hainanensis Merr. = Hydnocarpus hainanensis （Merr.）Sleumer ●◇

384407　Taraktogenos kurzii King = Hydnocarpus kurzii（King）Warb. ●

384408　Taraktogenos merrilliana（H. L. Li）C. Y. Wu = Hydnocarpus annamensis（Gagnep.）Lescot et Sleumer ●◇

384409　Taraktogenos merrilliana C. Y. Wu = Hydnocarpus annamensis （Gagnep.）Lescot et Sleumer ●◇

384410　Taralea Aubl.（废弃属名）= Dipteryx Schreb.（保留属名）●☆

384411　Taramea Raf. = Faramea Aubl. ●☆

384412　Tarasa Phil.（1891）;星毛卷萼锦属■●☆

384413　Tarasa alberti Phil. ;星毛卷萼锦■●☆

384414　Tarasa capitata（Cav.）D. M. Bates;头状星毛卷萼锦■☆

384415　Tarasa tenuis Krapov. ;细星毛卷萼锦■☆

384416　Tarasa trisecta（Griseb.）Krapov. ;三星毛卷萼锦■☆

384417　Taravalia Greene = Ptelea L. ●

384418　Taraxaconastrum Guett. = Hyoseris L. ■☆

384419　Taraxaconoides Guett. = Leontodon L.（保留属名）■☆

384420　Taraxacum F. H. Wigg.（1780）（保留属名）;蒲公英属; Dandelion, Gowan ■

384421　Taraxacum Weber ex F. H. Wigg. = Taraxacum F. H. Wigg.（保留属名）■

384422　Taraxacum Weber. = Taraxacum F. H. Wigg.（保留属名）■

384423　Taraxacum Zinn（废弃属名）= Leontodon L.（保留属名）■☆

384424　Taraxacum Zinn（废弃属名）= Taraxacum F. H. Wigg.（保留属名）■

384425　Taraxacum alaicum Schischk. ;阿莱蒲公英■☆

384426　Taraxacum alaskanum Rydb. ;阿拉斯加蒲公英; Alaska Dandelion ■☆

384427　Taraxacum alatavicum Schischk. ;阿拉套蒲公英■☆

384428　Taraxacum alatopetiolum D. T. Zhai et C. H. An;冀柄蒲公英; Wingpetal Dandelion ■

384429　Taraxacum albescens Dahlst. ;微白蒲公英■☆

384430　Taraxacum albidum Dahlst. ;白蒲公英（白鼓钉）■☆

384431　Taraxacum albiflorum Koidz. = Taraxacum albidum Dahlst. ■☆

384432　Taraxacum albomarginatum Kitam. = Taraxacum platypecidium Diels ex H. Limpr. ■

384433　Taraxacum almaatense Schischk. = Taraxacum officinale F. H. Wigg. ■

384434　Taraxacum alpicola Kitam. ;高山生蒲公英（山地蒲公英）■☆

384435　Taraxacum alpicola Kitam. var. shiroumense（H. Koidz.）Kitam. ;白马岳蒲公英■☆

384436　Taraxacum alpigenum Dshanaeva = Taraxacum goloskokovii Schischk. ■

384437　Taraxacum alpinum Hegetschw. et Heer. ;高山蒲公英■☆

384438　Taraxacum altaicum Schischk. ;阿尔泰蒲公英（阿勒泰蒲公英）;Altai Dandelion ■

384439　Taraxacum altune D. T. Zhai et C. H. An;阿尔金蒲公英;Aerjin Dandelion ■

384440　Taraxacum alukense R. Doll – Taraxacum lapponicum Kihlm. ex Hand. -Mazz. ■☆

384441　Taraxacum ambigens Fernald var. fultius Fernald = Taraxacum ceratophorum（Ledeb.）DC. ■☆

384442　Taraxacum anadyrense Nakai et Koidz. ;阿纳代尔蒲公英■☆

384443　Taraxacum andersovii Schischk. ;安得蒲公英■☆

384444　Taraxacum angulatum G. E. Haglund = Taraxacum ceratophorum （Ledeb.）DC. ■☆

384445　Taraxacum angustifolium Greene;窄叶蒲公英■☆

384446　Taraxacum apargiiforme Dahlst. ;天全蒲公英; Tianquan Dandelion ■

384447　Taraxacum arakii Kitam. ;荒木蒲公英■☆

384448　Taraxacum arcticum Dahlst. ;北极蒲公英■☆

384449　Taraxacum arctogenum Dahlst. = Taraxacum ceratophorum （Ledeb.）DC. ■☆

384450　Taraxacum argute-denticulatum Nakai et Koidz. = Taraxacum mongolicum Hand. -Mazz. et Dahlst. ■

384451　Taraxacum armeniacum Schischk. ;亚美尼亚蒲公英■☆

384452　Taraxacum aschabadense Schischk. ;阿沙巴得蒲公英■☆

384453　Taraxacum asiaticum Dahlst. ;亚洲蒲公英■

384454　Taraxacum asiaticum Dahlst. = Taraxacum leucanthum （Ledeb.）Ledeb. ■

384455　Taraxacum asiaticum Dahlst. f. lonchophyllum（Kitag.）Kitag. = Taraxacum asiaticum Dahlst. ■

384456　Taraxacum asiaticum Dahlst. f. lonchophyllum（Kitag.）Kitag. = Taraxacum leucanthum（Ledeb.）Ledeb. ■

384457　Taraxacum asiaticum Dahlst. var. lonchophyllum Kitag. = Taraxacum asiaticum Dahlst. ■

384458　Taraxacum atlanticola H. Lindb. ;北非蒲公英■☆

384459　Taraxacum atlanticum Pomel;大西洋蒲公英■☆

384460　Taraxacum atrans Schischk. ;光冠蒲公英■☆

384461　Taraxacum atratum G. E. Haglund = Taraxacum pseudoatratum Orazova ■

384462　Taraxacum atratum Schischk. ;俄罗斯黑蒲公英■☆

384463　Taraxacum aurantiacum Dahlst. ;黄花蒲公英■☆

384464　Taraxacum badachschanicum Schischk. = Taraxacum luridum G. E. Haglund ex Persson ■

384465　Taraxacum baicalense Schischk. = Taraxacum dissectum （Ledeb.）Ledeb. ■

384466　Taraxacum balticum Dahlst. ;波罗的海蒲公英■☆

384467　Taraxacum bessarabicum（Hornem.）Hand. -Mazz. ;窄苞蒲公英（别比萨拉比亚蒲公英,厚叶蒲公英,西疆蒲公英）; Narrowbract Dandelion ■

384468　Taraxacum bhutanicum Soest = Taraxacum parvulum（Wall.）DC. ■

384469　Taraxacum bicolor DC. ;双色蒲公英■☆

384470　Taraxacum bicolor DC. = Taraxacum borealisinense Kitam. ■

384471　Taraxacum bicolor DC. = Taraxacum parvulum（Wall.）DC. ■

384472　Taraxacum bicorne Dahlst. ;双角蒲公英（两角蒲公英）; Koksaghyz, Kok-saghyz Rubber, Russian Dandelion, Twohorn Dandelion ■

384473　Taraxacum borealisinense Kitam. ;华蒲公英（白鼓丁,孛孛地丁,孛孛丁,孛孛丁菜,鹁鸪英,卜地蜈蚣,卜公英,残飞坠,倒披针叶蒲公英,耳瘢草,凫公英,古古丁,鬼灯笼,黄狗头,黄花草黄花郎，黄花郎草，黄花苗，黄花三七，灰花蒲公英，碱地蒲公英，耩耨草，金簪草，金簪花，茅萝卜，奶汁草，婆婆丁，仆公英，仆公婴，扑灯儿，蒲公丁，双英卜地）；China Dandelion, Chinese Dandelion ■

384474　Taraxacum botschantzevii Schischk.；鲍氏蒲公英■

384475　Taraxacum brachyceras Dahlst.；短角蒲公英■☆

384476　Taraxacum brachycranum Dahlst.；短头蒲公英■☆

384477　Taraxacum brachyglossum（Dahlst.）Dahlst.；短舌蒲公英■☆

384478　Taraxacum brachyglossum Dahlst. = Taraxacum brachyglossum（Dahlst.）Dahlst.■☆

384479　Taraxacum brassicifolium Kitag.；芥叶蒲公英；Mustardleaf Dandelion ■

384480　Taraxacum brevicorne Dahlst.；北极短角蒲公英■☆

384481　Taraxacum brevicorniculatum Korol. = Taraxacum kok-saghyz L. E. Rodin ■

384482　Taraxacum brevirostre Hand.-Mazz.；短喙蒲公英；Shortbeak Dandelion ■

384483　Taraxacum breviscapum A. J. Richards；短花茎蒲公英■☆

384484　Taraxacum calanthodium Dahlst.；大头蒲公英（丽花蒲公英）；Bighead Dandelion ■

384485　Taraxacum calareum Korol.；碱地蒲公英■

384486　Taraxacum californicum Munz et I. M. Johnst.；加州蒲公英；California Dandelion ■☆

384487　Taraxacum canitiosum Dahlst. = Taraxacum calanthodium Dahlst.■

384488　Taraxacum carneocoloratum A. Nelson；粉色蒲公英；Pink Dandelion ■☆

384489　Taraxacum carthamopsis A. E. Porsild = Taraxacum ceratophorum（Ledeb.）DC.■☆

384490　Taraxacum caucasicum（Stev.）DC.；高加索蒲公英■☆

384491　Taraxacum centrasiaticum D. T. Zhai et C. H. An；中亚蒲公英；Centrol Asia Dandelion ■

384492　Taraxacum ceratolepis Kitam.；角鳞蒲公英■☆

384493　Taraxacum ceratophorum（Ledeb.）DC.；角蒲公英（粗糙蒲公英，角状蒲公英）；Horned Dandelion ■☆

384494　Taraxacum ceratophorum（Ledeb.）DC. = Taraxacum officinale F. H. Wigg. ■

384495　Taraxacum ceratophorum（Ledeb.）DC. var. bernardinum Jeps. = Taraxacum californicum Munz et I. M. Johnst. ■☆

384496　Taraxacum chamissonis Greene = Taraxacum ceratophorum（Ledeb.）DC. ■☆

384497　Taraxacum chionophilum Dahlst.；川西蒲公英；W. Sichuan Dandelion ■

384498　Taraxacum chirieanum Kitam.；西里蒲公英■☆

384499　Taraxacum ciscaucasicum Schischk.；北高加索蒲公英■☆

384500　Taraxacum collinum DC.；丘陵蒲公英■

384501　Taraxacum compactum Schischk.；堆叶蒲公英（密叶蒲公英）；Compactleaf Dandelion ■

384502　Taraxacum complicatum Soest；折叠蒲公英■☆

384503　Taraxacum confusum Schischk.；混乱蒲公英■☆

384504　Taraxacum connectens Dahlst. = Taraxacum lugubre Dahlst. ■

384505　Taraxacum coreanum Nakai；朝鲜蒲公英（白花蒲公英）；Korea Dandelion ■

384506　Taraxacum coverum R. Doll = Taraxacum ceratophorum（Ledeb.）DC. ■☆

384507　Taraxacum crepidiforme DC.；还阳参蒲公英■☆

384508　Taraxacum croceum Dahlst.；镉黄蒲公英■☆

384509　Taraxacum cuspidatum Dahlst.；凸尖蒲公英（雾灵蒲公英）■

384510　Taraxacum cuspidatum Dahlst. = Taraxacum asiaticum Dahlst. ■

384511　Taraxacum cuspidatum Dahlst. = Taraxacum leucanthum（Ledeb.）Ledeb. ■

384512　Taraxacum cuspidatum Dahlst. var. lonchophyllum Kitag. = Taraxacum asiaticum Dahlst. ■

384513　Taraxacum czuense Schischk.；丘蒲公英■☆

384514　Taraxacum dantungense Kitag.；丹东蒲公英（阴山蒲公英）；Dandong Dandelion ■

384515　Taraxacum dasypodum Soest；丽江蒲公英；Lijiang Dandelion ■

384516　Taraxacum dealbatum Hand.-Mazz.；粉绿蒲公英；Darkgreen Dandelion ■

384517　Taraxacum decipiens Raunk. et G. E. Haglund；迷惑蒲公英■☆

384518　Taraxacum decolorans Dahlst.；褪色蒲公英■☆

384519　Taraxacum dens-leonis Dahlst.；狮齿蒲公英；Dandelion ■☆

384520　Taraxacum denudatum H. Koidz.；裸露蒲公英■☆

384521　Taraxacum desertorum Schischk.；沙漠蒲公英■☆

384522　Taraxacum dilatatum H. Lindb.；膨大蒲公英■☆

384523　Taraxacum dilutum Dahlst.；稀薄蒲公英■☆

384524　Taraxacum dissectum（Ledeb.）Ledeb.；多裂蒲公英（裂叶蒲公英）；Dissected Dandelion ■

384525　Taraxacum disseminatum G. E. Haglund = Taraxacum laevigatum（Willd.）DC.■

384526　Taraxacum dissimile Dahlst.；不似蒲公英■☆

384527　Taraxacum dumetorum Greene = Taraxacum ceratophorum（Ledeb.）DC.■☆

384528　Taraxacum duplex Jacot Guillaumin；山东蒲公英■

384529　Taraxacum ecornutum Kovalevsk.；无角蒲公英；Hornless Dandelion ■

384530　Taraxacum elatum Kitam.；高蒲公英■☆

384531　Taraxacum elatum Kitam. var. deforme Kitam. = Taraxacum elatum Kitam. ■☆

384532　Taraxacum elatum Kitam. var. ibukiense Kitam. = Taraxacum elatum Kitam. ■☆

384533　Taraxacum elongatum Kovalevsk.；伸展蒲公英■

384534　Taraxacum eriobasis Kovalevsk.；毛基蒲公英■☆

384535　Taraxacum eriophorum Rydb. = Taraxacum ceratophorum（Ledeb.）DC.■☆

384536　Taraxacum eriopodum（D. Don）DC.；毛柄蒲公英（毛葶蒲公英）；Hairstalk Dandelion ■

384537　Taraxacum erypodon Dahlst.；折苞蒲公英■

384538　Taraxacum erythropodium Kitag.；红梗蒲公英■

384539　Taraxacum erythropodium Kitag. = Taraxacum variegatum Kitag. ■

384540　Taraxacum erythrospermum Andrz. ex Besser；红果蒲公英（红籽蒲公英）；Redfruit Dandelion, Red-seeded Dandelion ■

384541　Taraxacum erythrospermum Andrz. ex Besser = Taraxacum laevigatum（Willd.）DC. ■

384542　Taraxacum erythrospermum Andrz. ex Besser subsp. brachyglossum Dahlst. = Taraxacum brachyglossum（Dahlst.）Dahlst. ■☆

384543　Taraxacum erythrospermum Andrz. ex Besser var. bessarabicum（Hornem.）DC. = Taraxacum bessarabicum（Hornem.）Hand.-Mazz. ■

384544　Taraxacum erythrospermum Andrz. var. bessarabicum（Hornem.）DC. = Taraxacum bessarabicum（Hornem.）Hand.-Mazz. ■

384545　Taraxacum erythrospermum Besser = Taraxacum erythrospermum Andrz. ex Besser ■

384546 Taraxacum eurylepium Dahlst. = Taraxacum ceratophorum (Ledeb.) DC. ■☆

384547 Taraxacum evittatum Dahlst.；无线蒲公英■☆

384548 Taraxacum falcilobum Kitag.；兴安蒲公英■

384549 Taraxacum falcilobum Kitag. = Taraxacum asiaticum Dahlst. ■

384550 Taraxacum falcilobum Kitag. = Taraxacum leucanthum (Ledeb.) Ledeb. ■

384551 Taraxacum fedtschenkoi Hand.-Mazz.；费氏蒲公英■☆

384552 Taraxacum fendleri var. wrightii (A. Gray) Trel. = Thalictrum fendleri Engelm. ex A. Gray ■

384553 Taraxacum fontanum Hand.-Mazz.；泉蒲公英■☆

384554 Taraxacum formosanum Kitam.；台湾蒲公英■

384555 Taraxacum formosanum Kitam. = Taraxacum mongolicum Hand.-Mazz. et Dahlst. ■

384556 Taraxacum forrestii Soest；网苞蒲公英；Netbract Dandelion ■

384557 Taraxacum fulvipilis Harv. = Taraxacum bessarabicum (Hornem.) Hand.-Mazz. ■

384558 Taraxacum fulvum Raunk.；黄蒲公英■☆

384559 Taraxacum glabellum Schischk.；光蒲公英■☆

384560 Taraxacum glabrum DC.；光果蒲公英（无毛蒲公英）；Nakedfruit Dandelion ■

384561 Taraxacum glaucanthum DC. = Taraxacum borealisinense Kitam. ■

384562 Taraxacum glaucivirens Schischk.；灰绿蒲公英■☆

384563 Taraxacum glaucophyllum Soest；苍叶蒲公英；Glaucousleaf Dandelion ■

384564 Taraxacum goloskokovii Schischk.；小叶蒲公英（高劳蒲公英，高氏蒲公英）；Littleleaf Dandelion ■

384565 Taraxacum grandisquamatum H. Koidz.；大壳蒲公英■☆

384566 Taraxacum groenlandicum Dahlst. = Taraxacum ceratophorum (Ledeb.) DC. ■☆

384567 Taraxacum grossheimii Schischk.；格罗蒲公英■☆

384568 Taraxacum grupodon Dahlst.；反苞蒲公英■

384569 Taraxacum gymnanthum DC.；无被蒲公英■☆

384570 Taraxacum hamatiforme Dahlst.；顶钩蒲公英■☆

384571 Taraxacum hangchouense H. Koidz. = Taraxacum mongolicum Hand.-Mazz. et Dahlst. ■

384572 Taraxacum heterolepis Nakai et Koidz. ex Kitag.；异苞蒲公英（细裂蒲公英）；Differbract Dandelion ■

384573 Taraxacum hideoi Nakai ex H. Koidz.；秀雄蒲公英■☆

384574 Taraxacum himalaicum Soest = Taraxacum parvulum (Wall.) DC. ■

384575 Taraxacum hirsutum (Hook.) Torr. et A. Gray = Agoseris hirsuta (Hook.) Greene ■☆

384576 Taraxacum hjeltii Dahlst.；耶尔特蒲公英■☆

384577 Taraxacum holmenianum Sahlin；霍尔曼蒲公英；Holmen's Dandelion ■☆

384578 Taraxacum holophyllum Schischk. = Taraxacum monochlamydeum Hand.-Mazz. et G. E. Haglund ■

384579 Taraxacum hondae Nakai et Koidz. = Taraxacum mongolicum Hand.-Mazz. et Dahlst. ■

384580 Taraxacum huhhoticum Z. Xu et H. C. Fu = Taraxacum mongolicum Hand.-Mazz. et Dahlst. ■

384581 Taraxacum hultenii Dahlst.；胡尔滕蒲公英■

384582 Taraxacum humbertii Maire；亨伯特蒲公英■☆

384583 Taraxacum hyparcticum Dahlst.；高寒蒲公英；High-arctic Dandelion ■☆

384584 Taraxacum hyperboreum Dahlst. = Taraxacum ceratophorum (Ledeb.) DC. ■☆

384585 Taraxacum ikonnikovii Schischk. = Taraxacum luridum G. E. Haglund ex Persson ■

384586 Taraxacum inaequilobum Pomel = Taraxacum megalorrhizon (Forssk.) Hand.-Mazz. ■☆

384587 Taraxacum indicum H. Lindb.；印度蒲公英；India Dandelion ■

384588 Taraxacum integratiforme R. Doll = Taraxacum ceratophorum (Ledeb.) DC. ■☆

384589 Taraxacum integratum G. E. Haglund = Taraxacum ceratophorum (Ledeb.) DC. ■☆

384590 Taraxacum intercedens Markl.；中间蒲公英■☆

384591 Taraxacum japonicum Koidz.；日本蒲公英；Japan Dandelion, Japanese Dandelion ■☆

384592 Taraxacum junpeianum Kitam.；长春蒲公英■

384593 Taraxacum junpeianum Kitam. = Taraxacum licentii Soest ■

384594 Taraxacum junpeianum Kitam. = Taraxacum ohwianum Kitam. ■

384595 Taraxacum juzepczukii Schischk.；尤氏蒲公英■☆

384596 Taraxacum kamtchaticum Dahlst.；勘察加蒲公英■☆

384597 Taraxacum kamtchaticum Dahlst. = Taraxacum alaskanum Rydb. ■☆

384598 Taraxacum kansuense Nakai ex Koidz. = Taraxacum mongolicum Hand.-Mazz. et Dahlst. ■

384599 Taraxacum karatavicum Pavlov；卡拉塔夫蒲公英■☆

384600 Taraxacum kawaguchii Kitam. = Taraxacum parvulum (Wall.) DC. ■

384601 Taraxacum ketoiense Tatew. et Kitam.；计吐夷蒲公英■☆

384602 Taraxacum kirghizicum Schischk.；基尔吉斯蒲公英■☆

384603 Taraxacum kljutschevskoanum Kom.；克柳奇蒲公英■☆

384604 Taraxacum klokovii Litv.；克劳氏蒲公英■☆

384605 Taraxacum kojimae Kitam.；光岛蒲公英■☆

384606 Taraxacum kok-saghyz L. E. Rodin；橡胶草（罗胶草）；Chewroot, Rubber Dandelion, Russian Dandelion ■

384607 Taraxacum koraginense Kom.；科拉金蒲公英■☆

384608 Taraxacum koraginicola Kom.；科拉金生蒲公英■☆

384609 Taraxacum kozlovii Tzvelev；褐果蒲公英（大刺蒲公英，科兹洛夫蒲公英）；Kozlov Dandelion ■

384610 Taraxacum kuzakaiense Kitam.；岩手蒲公英■☆

384611 Taraxacum lacerum Greene = Taraxacum ceratophorum (Ledeb.) DC. ■☆

384612 Taraxacum lacistophyllum (Dahlst.) Raunk. = Taraxacum laevigatum (Willd.) DC. ■

384613 Taraxacum laetum Dahlst.；愉悦蒲公英■☆

384614 Taraxacum laevigatum (Willd.) DC.；无毛蒲公英（普通蒲公英）；Lesser Dandelion, Red-fruited Dandelion, Red-seed Dandelion, Red-seeded Dandelion, Rock Dandelion, Small Dandelion, Smooth Dandelion ■

384615 Taraxacum laevigatum (Willd.) DC. = Taraxacum erythrospermum Andrz. ex Besser ■

384616 Taraxacum laevigatum (Willd.) DC. var. erythrospermum (Andrz. ex Besser) J. Weiss = Taraxacum erythrospermum Andrz. ex Besser ■

384617 Taraxacum laevigatum Maire var. maroccanum (H. Lindb.) Maire = Taraxacum maroccanum H. Lindb. ■☆

384618 Taraxacum lamprolepis Kitag.；光苞蒲公英；Brightbract Dandelion ■

384619 Taraxacum lanigerum Soest；多毛蒲公英；Cottony Dandelion ■

384620　Taraxacum lapponicum Kihlm. ex Hand. -Mazz. ；拉普兰蒲公英；Lapland Dandelion ■☆

384621　Taraxacum lateritium Dahlst. ；砖红蒲公英■☆

384622　Taraxacum lateritium Dahlst. = Taraxacum ceratophorum（Ledeb.）DC. ■☆

384623　Taraxacum latilobum DC. ；宽裂蒲公英；Large-lobed Dandelion ■☆

384624　Taraxacum latisquameum Dahlst. ；远东蒲公英■☆

384625　Taraxacum laurentianum Fernald；洛朗蒲公英；Gulf of St. Lawrence Dandelion ■☆

384626　Taraxacum leptocephalum Rchb. = Taraxacum bessarabicum（Hornem.）Hand. -Mazz. ■

384627　Taraxacum leptoceras Dahlst. ；细角蒲公英■☆

384628　Taraxacum leucanthum（Ledeb.）Ledeb. ；白花蒲公英（戟叶蒲公英，凸尖蒲公英，狭戟片蒲公英，兴安蒲公英，亚洲蒲公英）；Asia Dandelion，White Dandelion，Whiteflower Dandelion ■

384629　Taraxacum liaotungense Kitag. ；辽东蒲公英■

384630　Taraxacum liaotungense Kitag. = Taraxacum mongolicum Hand. -Mazz. et Dahlst. ■

384631　Taraxacum liaotungense Kitag. f. lobulatum Kitag. = Taraxacum mongolicum Hand. -Mazz. et Dahlst. ■

384632　Taraxacum licentii Soest；山西蒲公英；Shanxi Dandelion ■

384633　Taraxacum lilacinum Krasn. ex Schischk. ；紫花蒲公英；Lilac Dandelion ■

384634　Taraxacum lincevskyi Schischk. ；林克柴夫斯基蒲公英■☆

384635　Taraxacum linguatifrons Markl. ；舌叶蒲公英■☆

384636　Taraxacum lipskyi Schischk. ；小果蒲公英；Smallfruit Dandelion ■☆

384637　Taraxacum lissocarpum Dahlst. ；欧洲光果蒲公英■☆

384638　Taraxacum litvinovii Schischk. ；里特蒲公英■

384639　Taraxacum longeappendiculatum Nakai = Taraxacum platycarpum Dahlst. var. longeappendiculatum（Nakai）Morita ■☆

384640　Taraxacum longicorne Dahlst. ；欧洲长角蒲公英■☆

384641　Taraxacum longii Fernald = Taraxacum ceratophorum（Ledeb.）DC. ■☆

384642　Taraxacum longipes Kom. ；长梗蒲公英■☆

384643　Taraxacum longipyramidatum Schischk. ；长锥蒲公英（长塔蒲公英，长嘴蒲公英）；Lobgawl Dandelion ■

384644　Taraxacum longirostre Schischk. = Taraxacum longipyramidatum Schischk. ■

384645　Taraxacum ludlowii Soest；林周蒲公英；Ludlow Dandelion ■

384646　Taraxacum lugubre Dahlst. ；川甘蒲公英；Chuangan Dandelion ■☆

384647　Taraxacum luridum G. E. Haglund；红角蒲公英；Redhorn Dandelion ■

384648　Taraxacum luridum G. E. Haglund ex Persson = Taraxacum luridum G. E. Haglund ■

384649　Taraxacum lyratum DC. ；大头羽裂蒲公英■☆

384650　Taraxacum macilentum Dahlst. ；弱小蒲公英■☆

384651　Taraxacum mackenziense A. E. Porsild = Taraxacum ceratophorum（Ledeb.）DC. ■☆

384652　Taraxacum macrocarpum Dahlst. ；川藏蒲公英■

384653　Taraxacum macrolepis Schischk. ；大鳞蒲公英■☆

384654　Taraxacum magnum Korol. ；大花蒲公英■

384655　Taraxacum majus Schischk. ；大叶蒲公英■

384656　Taraxacum malaisei Dahlst. ；马氏蒲公英■☆

384657　Taraxacum malteanum Dahlst. ex G. E. Haglund = Taraxacum ceratophorum（Ledeb.）DC. ■☆

384658　Taraxacum mandshuricum Nakai ex Koidz. = Taraxacum licentii Soest ■

384659　Taraxacum mandshuricum Nakai ex Koidz. = Taraxacum ohwianum Kitam. ■

384660　Taraxacum maracandicum Kovalevsk. ；马拉坎达蒲公英■☆

384661　Taraxacum marginatum Dahlst. ；具边蒲公英■☆

384662　Taraxacum maroccanum H. Lindb. ；摩洛哥蒲公英■☆

384663　Taraxacum maruyamanum Kitam. ；丸山蒲公英■☆

384664　Taraxacum maurocarpum Dahlst. ；灰果蒲公英（川藏蒲公英，蒲公英）；Blackfruit Dandelion，Greyfruit Dandelion ■

384665　Taraxacum maurolepium G. E. Haglund = Taraxacum ceratophorum（Ledeb.）DC. ■☆

384666　Taraxacum megalorrhizon（Forssk.）Hand. -Mazz. ；大根蒲公英；Krim-saghyz Rubber ■☆

384667　Taraxacum megalorrhizon（Forssk.）Hand. -Mazz. subsp. obovatum（Willd.）Braun-Blanq. et Wilczek = Taraxacum obovatum（Willd.）DC. ■☆

384668　Taraxacum messanense H. Lindb. ；梅桑蒲公英■☆

384669　Taraxacum mexicanum DC. ；墨西哥蒲公英■☆

384670　Taraxacum microcephalum Pomel；小头蒲公英■☆

384671　Taraxacum microcephalum Pomel var. atlanticum（Pomel）Batt. = Taraxacum atlanticum Pomel ■☆

384672　Taraxacum microcerum R. Doll = Taraxacum ceratophorum（Ledeb.）DC. ■☆

384673　Taraxacum microlobum Markl. ex Eklund et Markl. ；微裂蒲公英■☆

384674　Taraxacum microspermum Schischk. ；小籽蒲公英■☆

384675　Taraxacum minimum（Guss.）N. Terracc. ；非洲小蒲公英■☆

384676　Taraxacum minutilobum Popov ex Kovalevsk. ；毛叶蒲公英；Hairleaf Dandelion ■

384677　Taraxacum mitalii Soest；亚东蒲公英；Yadong Dandelion ■

384678　Taraxacum miyakei Kitam. ；三宅蒲公英■☆

384679　Taraxacum modestum Schischk. ；适度蒲公英■☆

384680　Taraxacum mongolicum Hand. -Mazz. = Taraxacum mongolicum Hand. -Mazz. et Dahlst. ■

384681　Taraxacum mongolicum Hand. -Mazz. et Dahlst. ；蒲公英（把儿英，白鼓丁，白花蒲公英，白婆婆丁，波波丁，孛孛丁地丁，孛孛丁，孛孛丁菜，鹁鸪英，卜地蜈蚣，卜公英，不知奈，残飞坠，大丁草，淡黄花蒲公英，灯笼草，地丁，地丁草，地丁花，东北蒲公英，多裂蒲公英，耳瘢草，凫公英，公英，狗乳草，姑姑丁，姑姑英，古丁，古古丁，鼓钉，鬼灯笼，红梗蒲公英，黄草地丁，黄狗头，黄花草，黄花地丁，黄花郎，黄花郎草，黄花苗，黄花三七，碱地蒲公英，浆浆菜，耩耨草，角状蒲公英，金鼓草，金簪草，金簪花，辽东蒲公英，陆英，满地金钱，茅萝卜，蒙古蒲公英，奶奶草，奶汁草，婆补丁，婆婆丁，仆公英，仆公婴，蒲公草，蒲公丁，蒲蒲丁，蒲英，热河蒲公英，石生长，双英卜地，台湾蒲公英，太奈，雾灵蒲公英，小瘤蒲公英，亚洲蒲公英，羊奶奶草，异苞蒲公英）；Formosan Dandelion，Mongol Dandelion，Mongolian Dandelion ■

384682　Taraxacum mongolicum Hand. -Mazz. var. formosanum（Kitam.）Kitam. = Taraxacum mongolicum Hand. -Mazz. et Dahlst. ■

384683　Taraxacum monochlamydeum Hand. -Mazz. et G. E. Haglund；荒漠蒲公英（单被蒲公英）；Desert Dandelion ■

384684　Taraxacum montanum DC. ；山地蒲公英■☆

384685　Taraxacum multiscaposum Schischk. ；多葶蒲公英；Manyscape Dandelion ■

384686　Taraxacum multisectum Kitag. = Taraxacum heterolepis Nakai et Koidz. ex Kitag. ■

384687　Taraxacum muricatum Schischk. ；粗糙蒲公英■☆

384688　Taraxacum murmanicum N. I. Orlova；穆尔曼蒲公英■☆

384689　Taraxacum nairoense H. Koidz. ;奈罗蒲公英■☆

384690　Taraxacum natschikense Kom. 纳奇克蒲公英■☆

384691　Taraxacum neglectum Nakai et Koidz. ;疏忽蒲公英■☆

384692　Taraxacum neosachalinense H. Koidz. ;新库页蒲公英■☆

384693　Taraxacum nevskii Juz. ;涅氏蒲公英■☆

384694　Taraxacum nigricans Rchb. ;浅黑蒲公英■☆

384695　Taraxacum nikitinii Schischk. ;尼氏蒲公英■☆

384696　Taraxacum nivale Lange ex Kihlm. ;雪线蒲公英■☆

384697　Taraxacum norvegicum Dahlst. ;挪威蒲公英■☆

384698　Taraxacum novae-zemliae Holmboe ;新泽米尔蒲公英■☆

384699　Taraxacum nuratavicum Schischk. ;努拉套蒲公英■☆

384700　Taraxacum nutans Dahlst. ;垂头蒲公英;Nutant Dandelion ■

384701　Taraxacum oblancifolium D. Z. Ma = Taraxacum borealisinense Kitam. ■

384702　Taraxacum obliquum Dahlst. ;斜蒲公英■☆

384703　Taraxacum obovatum (Willd.) DC. ;倒卵蒲公英■☆

384704　Taraxacum obovatum (Willd.) DC. subsp. ochrocarpum Soest = Taraxacum atlanticola H. Lindb. ■☆

384705　Taraxacum obovatum (Willd.) DC. var. atlanticola (H. Lindb.) Jahand. et Maire = Taraxacum atlanticola H. Lindb. ■☆

384706　Taraxacum obovatum (Willd.) DC. var. rifeum Sennen et Mauricio = Taraxacum obovatum (Willd.) DC. ■☆

384707　Taraxacum officinale F. H. Wigg. = Taraxacum officinale Weber ex F. H. Wigg. ■

384708　Taraxacum officinale F. H. Wigg. var. affine Rouy = Taraxacum officinale F. H. Wigg. ■

384709　Taraxacum officinale F. H. Wigg. var. depressum Barratte = Taraxacum megalorrhizon (Forssk.) Hand. -Mazz. ☆

384710　Taraxacum officinale F. H. Wigg. var. erythrospermum (Andrz. ex Besser) Bab. = Taraxacum erythrospermum Andrz. ex Besser ■

384711　Taraxacum officinale F. H. Wigg. var. obovatum (Willd.) Batt. = Taraxacum obovatum (Willd.) DC. ■☆

384712　Taraxacum officinale F. H. Wigg. var. scopulorum A. Gray = Taraxacum scopulorum (A. Gray) Rydb. ■☆

384713　Taraxacum officinale Weber = Taraxacum officinale Weber ex F. H. Wigg. ■

384714　Taraxacum officinale Weber ex F. H. Wigg. ;药用蒲公英(白白菜,白鼓钉,鹁鸪英,粗糙蒲公英,黄花郎,黄花苗,节蒲公英,苦菜,苦马菜,满地金,欧蒲公英,欧洲蒲公英,婆婆丁花孛孛丁菜,仆公罂,蒲公草,蒲公英,西洋蒲公英,洋蒲公英,药蒲公英); Bashels, Bitferwort, Bitter Aks, Blowball, Blower, Bum Pipe, Bum-pipe, Burning Fire, Canker, Children's Clock, Clock, Clock Posy, Clock-flower, Clock-posy, Clocks-and-watches, Comb-and-hairpins, Combs-and-hairpins, Common Dandelion, Conquer-more, Crow Parsnip, Dandelion, Dashel, Dashel Flower, Day Lily, Dazzle, Dentelion, Dentilinn, Dentylion, Devil's Milk Plant, Devil's Milkpail, Devil's Milkplant, Dindle, Dirt-a-bed, Dog Posy, Doonhead Clock, Doon-head Clock, Eksis-girse, Fairy Clock, Farmer's Clock, Fluffy-puffy, Fortune-teller, Four-o'clock, Golden Sun, Gowan, Granfer Griggle-sticks, Grumsel, Hawkweed Gowan, Heart Clover, Heart Fever Grass, Horse Gowan, Irish Daisy, Jinny Joes, Lady of Spring, Lay-a-bed, Lion's Teeth, Little Flame of God, Male, Medicine Dandelion, Mess-a-bed, Milk Gowan, Old Man's Clock, One-o'clock, One-two-three, Peasant's Clock, Pee-a-bed, Pee-bed, Pee-in-bed, Piddly-bed, Pishamoolag, Pismire, Pissibed, Pissimire, Pittle-bed, Priest's Crown, Puff Clocks, Rabbit's Meat, Rough Dandelion, Schoolboy's Clock, Shepherd's Clock, Shit-a-bed, Snowball, St. Bride's Forerunner, Stink Davie, Swine's Snout, Tell-time, Three Pee-abed, Three Two One, Time Flower, Time Teller, Tlme Table, Waiteweed, Watches-and-clocks, Weather Clock, Wet-a-bed, Wet-the-bed, Wetweed, What o'clock, Wiggers, Wild William, Wishes, Witch Gowan, Yellow Gowan ■

384715　Taraxacum officinale Weber ex F. H. Wigg. var. palustre Blytt = Taraxacum officinale Weber ex F. H. Wigg. ■

384716　Taraxacum officinale Weber ex F. H. Wigg. var. parvulum ? = Taraxacum officinale Weber ex F. H. Wigg. ■

384717　Taraxacum officinale Weber subsp. vulgare (Lam.) Schinz et R. Keller;普通蒲公英;Common Dandelion ■☆

384718　Taraxacum ohirense S. Watan. et Morita;大平蒲公英■☆

384719　Taraxacum ohwianum Kitam. ; 东北蒲公英(长春蒲公英); Manchurian Dandelion, NE. China Dandelion ■

384720　Taraxacum oliganthum Schott et Kotschy ex Hand. -Mazz. ;少花蒲公英■☆

384721　Taraxacum oschense Schischk. ;奥什蒲公英■☆

384722　Taraxacum otagirianum Koidz. ex Kitam. = Taraxacum platypecidium Diels ex H. Limpr. ■

384723　Taraxacum ovinum Greene = Taraxacum ceratophorum (Ledeb.) DC. ■☆

384724　Taraxacum pachypodum H. Lindb. ;粗足蒲公英■☆

384725　Taraxacum paludosum Schltr. = Taraxacum borealisinense Kitam. ■

384726　Taraxacum paludosum Schltr. = Taraxacum suecicum G. E. Haglund ■☆

384727　Taraxacum palustre (Lyons) Symons;沼地蒲公英■☆

384728　Taraxacum palustre (Willd.) DC. = Taraxacum suecicum G. E. Haglund ■☆

384729　Taraxacum pamiricum Schischk. ;帕米尔蒲公英■☆

384730　Taraxacum parlovii Orazova;帕洛蒲公英■☆

384731　Taraxacum parvilobum Dahlst. ;小裂片蒲公英■☆

384732　Taraxacum parvulum (Wall.) DC. ;小花蒲公英;Smallflower Dandelion ■

384733　Taraxacum parvulum DC. = Taraxacum parvulum (Wall.) DC. ■

384734　Taraxacum pectinatum Kitam. ;篦状蒲公英■☆

384735　Taraxacum pellianum A. E. Porsild = Taraxacum ceratophorum (Ledeb.) DC. ■☆

384736　Taraxacum perlatesens Dahlst. ;极宽蒲公英■☆

384737　Taraxacum perpusillum Schischk. ;极小蒲公英■

384738　Taraxacum phymatocarpum J. Vahl;北部蒲公英;Northern Dandelion ■☆

384739　Taraxacum pineticola Klokov;松林蒲公英■☆

384740　Taraxacum pingue Schischk. ; 尖角蒲公英(肥滑蒲公英); Tinehorn Dandelion ■

384741　Taraxacum platycarpum Dahlst. f. alboflavescens H. Koidz. ;浅黄宽果蒲公英■☆

384742　Taraxacum platycarpum Dahlst. subsp. hondoense (Nakai ex Koidz.) Morita;本州宽果蒲公英■☆

384743　Taraxacum platycarpum Dahlst. subsp. maruyamanum (Kitam.) Morita = Taraxacum maruyamanum Kitam. ■☆

384744　Taraxacum platycarpum Dahlst. var. longeappendiculatum (Nakai) Morita;长附属物蒲公英■☆

384745　Taraxacum platycarpum Dahlst. var. variabile (Kitam.) H. Koidz. = Taraxacum variabile Kitam. ■☆

384746　Taraxacum platycarpum Hayata;宽果蒲公英(蒲公英)■

384747　Taraxacum platyceras Dahlst. ;宽角蒲公英■☆

384748　Taraxacum platylepium Dahlst. ;宽鳞蒲公英■☆

384749　Taraxacum platypecidium Diels ex H. Limpr. ;白缘蒲公英(河北蒲公英,热河蒲公英,山海蒲公英);Whiteedge Dandelion ■

384750　Taraxacum platypecidum Diels ＝ Taraxacum platypecidium Diels ex H. Limpr. ■

384751　Taraxacum platypecidum Diels ex H. Limpr. var. angustibracteatum Y. Ling;狭苞白缘蒲公英(狭苞蒲公英);Narrowbract Whiteedge Dandelion ■

384752　Taraxacum pojarkoviae Schischk. ;波氏蒲公英■☆

384753　Taraxacum porphyranthum Boiss. ;紫蒲公英■☆

384754　Taraxacum potaninii Tzvelev;波塔宁蒲公英(宽边蒲公英,玫瑰色蒲公英);Potanin Dandelion ■

384755　Taraxacum praecox Schischk. ;早熟蒲公英■☆

384756　Taraxacum praticola Schischk. ;草原蒲公英■☆

384757　Taraxacum printzii Dahlst. ex Printz;普林氏蒲公英■☆

384758　Taraxacum proximum Dahlst. ;近基蒲公英■☆

384759　Taraxacum przevalskii Tzvelev;普氏蒲公英(亮红蒲公英);Przevalski Dandelion ■

384760　Taraxacum pseudoalbidum Kitag. ;假白蒲公英(白花蒲公英,拟白花蒲公英)■

384761　Taraxacum pseudoalbidum Kitag. ＝Taraxacum coreanum Nakai ■

384762　Taraxacum pseudoalbidum Kitag. f. lutescens Kitag. ＝ Taraxacum coreanum Nakai ■

384763　Taraxacum pseudoalpinum Schischk. ex Orazova;假高山蒲公英(草甸蒲公英,山地蒲公英,亚高山蒲公英);Montane Dandelion ■

384764　Taraxacum pseudoatratum Orazova;窄边蒲公英;Narrowedge Dandelion ■

384765　Taraxacum pseudodissectum Nakai et Koidz. ＝ Taraxacum mongolicum Hand. -Mazz. et Dahlst. ■

384766　Taraxacum pseudofulvum H. Lindb. ex Florstr. ;假黄褐蒲公英■☆

384767　Taraxacum pseudoglabrum Dahlst. ;亚光蒲公英■☆

384768　Taraxacum pseudokamtschaticum Jurtzev ＝ Taraxacum alaskanum Rydb. ■☆

384769　Taraxacum pseudominutilobum Kovalevsk. ;葱岭蒲公英(杂叶蒲公英);Congling Dandelion ■

384770　Taraxacum pseudonigricans Hand. -Mazz. ;亚黑蒲公英■☆

384771　Taraxacum pseudonorvegicum Dahlst. ex G. E. Haglund ＝ Taraxacum ceratophorum (Ledeb.) DC. ■☆

384772　Taraxacum pseudoroseum Schischk. ;绯红蒲公英;Bright-red Dandelion ■

384773　Taraxacum pseudostenoceras Soest;长角蒲公英;Longhorn Dandelion ■

384774　Taraxacum pumilum Dahlst. ＝Taraxacum holmenianum Sahlin ■☆

384775　Taraxacum pycnodes H. Lindb. ;密蒲公英■☆

384776　Taraxacum pyropappum Boiss. et Reut. ;密冠毛蒲公英■☆

384777　Taraxacum qirae D. T. Zhai et C. H. An;策勒蒲公英;Cele Dandelion ■

384778　Taraxacum repandum Pavlov;皿果蒲公英(浅波蒲公英);Bloodfruit Dandelion ■

384779　Taraxacum roborovskyi Tzvelev;罗氏蒲公英(黑柱蒲公英,膜叶蒲公英);Roborovsky Dandelion ■

384780　Taraxacum robtzovii Schischk. ;洛氏网苞蒲公英■

384781　Taraxacum roseoflavescens Tzvelev;刺果蒲公英(二色蒲公英)■

384782　Taraxacum rubidum Schischk. ;浅红蒲公英■☆

384783　Taraxacum rubtzovii Schischk. ;鲁氏蒲公英■☆

384784　Taraxacum rufum Dahlst. ;浅黄蒲公英■☆

384785　Taraxacum sachalinense Kitam. ;库页蒲公英■☆

384786　Taraxacum sachalinense Kitam. ＝ Taraxacum ceratophorum (Ledeb.) DC. ■☆

384787　Taraxacum sachalinense Koidz. ＝ Taraxacum platypecidium Diels ex H. Limpr. ■

384788　Taraxacum sagittifolium H. Lindb. ;箭叶蒲公英■☆

384789　Taraxacum saposhnikovii Schischk. ;沙氏蒲公英■

384790　Taraxacum scanicum Dahlst. ＝ Taraxacum erythrospermum Andrz. ex Besser ■

384791　Taraxacum scanicum Dahlst. ＝ Taraxacum laevigatum (Willd.) DC. ■

384792　Taraxacum schelkovnikovii Schischk. ;谢尔蒲公英■☆

384793　Taraxacum schischkinii Korol. ;希施蒲公英(斯氏蒲公英)■

384794　Taraxacum schugnanicum Schischk. ;舒格南蒲公英■☆

384795　Taraxacum scopulorum (A. Gray) Rydb. ;美洲高山蒲公英;Alpine Dandelion ■☆

384796　Taraxacum sendaicum Kitam. ＝Taraxacum platycarpum Dahlst. ■

384797　Taraxacum seravschanicum Schischk. ;塞拉夫蒲公英■☆

384798　Taraxacum serotinum (Waldst. et Kit.) Poir. ;晚熟蒲公英■☆

384799　Taraxacum serotinum Poir. ;秋蒲公英;Autumn Dandelion ■☆

384800　Taraxacum sherriffii Soest;拉萨蒲公英;Lasa Dandelion ■

384801　Taraxacum shikokianum Kitam. ;四国蒲公英■☆

384802　Taraxacum shikotanense Kitam. ;色丹蒲公英■☆

384803　Taraxacum shimushirense Tatew. et Kitam. ;西姆希尔蒲公英■☆

384804　Taraxacum shumushuense Kitam. ;舒姆蒲公英■☆

384805　Taraxacum sibiricum Dahlst. ;西伯利亚蒲公英■☆

384806　Taraxacum sikkimense Hand. -Mazz. ;锡金蒲公英;Sikkim Dandelion ■

384807　Taraxacum sinense Dahlst. ＝ Taraxacum borealisinense Kitam. ■

384808　Taraxacum sinense Poir. ＝ Taraxacum borealisinense Kitam. ■

384809　Taraxacum sinicum Kitag. ＝ Taraxacum borealisinense Kitam. ■

384810　Taraxacum sinomongolicum Kitag. ＝ Taraxacum asiaticum Dahlst. ■

384811　Taraxacum sinomongolicum Kitag. ＝ Taraxacum leucanthum (Ledeb.) Ledeb. ■

384812　Taraxacum songoricum Schischk. ;准噶尔蒲公英■☆

384813　Taraxacum spectabile Dahlst. ;宽叶沼生蒲公英;Broadleaved Dandelion , Broad-leaved Marsh Dandelion, Showy Dandelion ■

384814　Taraxacum stanjukoviczii Schischk. ;和田蒲公英;Hetian Dandelion ■

384815　Taraxacum staticifolium Soest ＝ Taraxacum parvulum (Wall.) DC. ■

384816　Taraxacum stenoceras Dahlst. ;角苞蒲公英;Hornbract Dandelion ■

384817　Taraxacum stenolepium Hand. -Mazz. ;南亚窄苞蒲公英■☆

384818　Taraxacum stenolobum Stschegl. ;深裂蒲公英;Deepsplit Dandelion ■

384819　Taraxacum stereni DC. ;独公英■☆

384820　Taraxacum strobilocephalum Kovalevsk. ;球头蒲公英■☆

384821　Taraxacum subcoronatum Tzvelev;黑柱蒲公英(亚心蒲公英)■

384822　Taraxacum suberiopodum Soest;滇北蒲公英;N. Yunnan Dandelion ■

384823　Taraxacum subglaciale Schischk. ;寒生蒲公英;Coldish Dandelion ■

384824　Taraxacum suecicum G. E. Haglund;沼生蒲公英(沼泽蒲公英);Lesser Marsh Dandelion, Marsh Dandelion, Narroeleved Marsh Dandelion, Narrow-leaved Marsh Dandelion ■☆

384825　Taraxacum sumnevsczii Schischk. ;紫果蒲公英;Purplefruit

Dandelion ■

384826 Taraxacum sylvanicum R. Doll = Taraxacum officinale Weber ex F. H. Wigg. ■

384827 Taraxacum syriacum Boiss. ;西亚蒲公英(序利亚蒲公英)●■

384828 Taraxacum tadshicorum Ovcz. ;塔什克蒲公英■☆

384829 Taraxacum tatewakii Kitam. ;馆肋蒲公英■☆

384830 Taraxacum tauricum Kotov;克里木蒲公英■☆

384831 Taraxacum tenuisectum Sommier et H. Lév. ;细裂蒲公英■☆

384832 Taraxacum tianschanicum Pavlov; 天 山 蒲 公 英; Tianshan Dandelion ■

384833 Taraxacum tibetanum Hand. -Mazz. ;藏蒲公英(西藏蒲公英); Tibet Dandelion,Xizang Dandelion ■

384834 Taraxacum togakushiense H. Koidz. ;户隐山蒲公英■☆

384835 Taraxacum trigonolobum Dahlst. = Taraxacum ceratophorum (Ledeb.) DC. ■☆

384836 Taraxacum tujuksuense Orazova;托可逊蒲公英■

384837 Taraxacum tundricola Hand. -Mazz. ;冻原蒲公英■☆

384838 Taraxacum turfosum (Sch. Bip.) Soest = Taraxacum palustre (Lyons) Symons ■☆

384839 Taraxacum turgaicum Schischk. ;图尔嘎蒲公英■☆

384840 Taraxacum turiense N. I. Orlova;图里蒲公英■☆

384841 Taraxacum tzvelevii Schischk. ;茨氏蒲公英■☆

384842 Taraxacum umbriniforme R. Doll = Taraxacum ceratophorum (Ledeb.) DC. ■☆

384843 Taraxacum umbrinum Dahlst. ex G. E. Haglund = Taraxacum ceratophorum (Ledeb.) DC. ■☆

384844 Taraxacum urbanum Kitag. = Taraxacum antungense Kitag. ■

384845 Taraxacum urdzharense Orazova;乌扎蒲公英■

384846 Taraxacum ussuriense Kom. ;乌苏里蒲公英;Ussuri Dandelion ■☆

384847 Taraxacum variabile Kitam. ;变色蒲公英■☆

384848 Taraxacum variegatum Kitag. ;斑叶蒲公英(红梗蒲公英); Variegate Dandelion ■

384849 Taraxacum varsobicum Schischk. ;瓦尔索波蒲公英■☆

384850 Taraxacum vassilczenkoi Schischk. ;瓦西里蒲公英■☆

384851 Taraxacum venustum H. Koidz. ;雅致蒲公英■☆

384852 Taraxacum vernale Schischk. = Taraxacum monochlamydeum Hand. -Mazz. et G. E. Haglund ■

384853 Taraxacum vitalii Orazova;锥塔蒲公英■

384854 Taraxacum voronovii Schischk. ;沃氏蒲公英■☆

384855 Taraxacum vulcanorum H. Koidz. ;火蒲公英■☆

384856 Taraxacum vulgare Schkuhr = Taraxacum officinale F. H. Wigg. ■

384857 Taraxacum wattii Hook. f. = Taraxacum eriopodum (D. Don) DC. ■

384858 Taraxacum wutaishanense Kitag. = Taraxacum parvulum (Wall.) DC. ■

384859 Taraxacum xerophilum Markl. ex H. Lindb. ;旱生蒲公英■☆

384860 Taraxacum xinyuanicum D. T. Zhai et C. H. An;新源蒲公英; Xinyuan Dandelion ■

384861 Taraxacum yamamotoi Koidz. ex Kitam. ;山本蒲公英■☆

384862 Taraxacum yatsugatakense H. Koidz. ;富士蒲公英■☆

384863 Taraxacum yesoalpinum Nakai ex H. Koidz. = Taraxacum ceratophorum (Ledeb.) DC. ■☆

384864 Taraxacum yetrofuense Kitam. ;耶特洛夫蒲公英■☆

384865 Taraxacum yinshanicum Z. Xu et H. C. Fu;阴山蒲公英; Yinshan Dandelion ■

384866 Taraxacum yinshanicum Z. Xu et H. C. Fu = Taraxacum antungense Kitag. ■

384867 Taraxacum yuparense H. Koidz. ;高岭蒲公英■☆

384868 Taraxacum yuparense H. Koidz. var. grandisquamatum Toyok. ; 大鳞高岭蒲公英■☆

384869 Taraxia (Nutt. ex Torr. et A. Gray) Raim. = Camissonia Link ■☆

384870 Taraxia (Nutt.) Raim. = Camissonia Link ■☆

384871 Taraxia Nutt. ex Torr. et A. Gray = Camissonia Link ■☆

384872 Taraxis B. G. Briggs et L. A. S. Johnson(1998);多枝帚灯草属■☆

384873 Taraxis grossa B. G. Briggs et L. A. S. Johnson;多枝帚灯草■☆

384874 Tarchonanthus L. (1753);平柱菊属●☆

384875 Tarchonanthus abyssinicus Sch. Bip. ex Schweinf. et Asch. = Tarchonanthus camphoratus L. ●☆

384876 Tarchonanthus angustissimus DC. = Tarchonanthus minor Less. ●☆

384877 Tarchonanthus camphoratus L. ;平 柱 菊; Camphor Wood, Camphorwood ●☆

384878 Tarchonanthus camphoratus L. var. litakunensis (DC.) Harv. = Tarchonanthus camphoratus L. ●☆

384879 Tarchonanthus dentatus Eckl. et Zeyh. ex DC. = Brachylaena neriifolia (L.) R. Br. ●☆

384880 Tarchonanthus dentatus Thunb. = Brachylaena glabra (L. f.) Druce ●☆

384881 Tarchonanthus ellipticus Thunb. = Brachylaena elliptica (Thunb.) DC. ●☆

384882 Tarchonanthus ericoides L. f. = Eriocephalus ericoides (L. f.) Druce ●☆

384883 Tarchonanthus galpinii Hutch. et E. Phillips = Tarchonanthus trilobus DC. var. galpinii (Hutch. et E. Phillips) Paiva ■☆

384884 Tarchonanthus glaber L. f. = Brachylaena glabra (L. f.) Druce ●☆

384885 Tarchonanthus lanceolatus Thunb. = Brachylaena neriifolia (L.) R. Br. ●☆

384886 Tarchonanthus litakunensis DC. = Tarchonanthus camphoratus L. ●☆

384887 Tarchonanthus littoralis Herman;滨海平柱菊■☆

384888 Tarchonanthus minor Less. ;小平柱菊■☆

384889 Tarchonanthus obovatus DC. ;倒卵平柱菊■☆

384890 Tarchonanthus parvicapitulatus Herman;小头平柱菊■☆

384891 Tarchonanthus procerus Salisb. = Tarchonanthus camphoratus L. ●☆

384892 Tarchonanthus racemosus Thunb. = Brachylaena ilicifolia (Lam.) E. Phillips et Schweick. ●☆

384893 Tarchonanthus trilobus DC. ;三裂平柱菊■☆

384894 Tarchonanthus trilobus DC. var. galpinii (Hutch. et E. Phillips) Paiva;盖尔平柱菊■☆

384895 Tardavel Adans. (废弃属名) = Borreria G. Mey. (保留属名)●■

384896 Tardavel Adans. (废弃属名) = Spermacoce L. ●■

384897 Tardavel andongensis Hiern = Spermacoce andongensis (Hiern) R. D. Good ■☆

384898 Tardavel arvensis Hiern = Spermacoce arvensis (Hiern) R. D. Good ■☆

384899 Tardavel huillensis Hiern = Spermacoce huillensis (Hiern) R. D. Good ●☆

384900 Tardavel kaessneri S. Moore = Spermacoce subvulgata (K. Schum.) J. G. Garcia ●☆

384901 Tardavel ocymoidés (Burm. f.) Hiern = Borreria ocymoides (Burm. f.) DC. ●☆

384902 Tardavel scabra (Schumach. et Thonn.) Hiern = Spermacoce ruelliae DC. ■☆

384903 Tardavel stricta Hiern = Spermacoce quadrisulcata (Bremek.)

Verdc. ■☆

384904　Tardavel thymoidea Hiern = Spermacoce thymoidea（Hiern） Verdc. ■☆

384905　Tardiella Gagnep. = Casearia Jacq. ●

384906　Tarenaya Raf.（1838）;醉蝶花属■

384907　Tarenaya Raf. = Cleome L. ●■

384908　Tarenaya hassleriana（Chodat） Iltis;醉蝶花■

384909　Tarenna Gaertn.（1788）;乌口树属（玉心花属）;Tarenna ●

384910　Tarenna acutisepala F. C. How ex W. C. Chen;尖萼乌口树; Sharpsepal Tarenna,Sharp-sepaled Tarenna,Tinesepal Tarenna ●

384911　Tarenna adamii Schnell = Tarenna bipindensis（K. Schum.） Bremek. ●☆

384912　Tarenna adamii Schnell var. nigeriana ? = Tarenna bipindensis （K. Schum.） Bremek. ●☆

384913　Tarenna affinis（K. Schum.） S. Moore = Tarenna pavettoides （Harv.） Sim subsp. affinis（K. Schum.） Bridson ●☆

384914　Tarenna alleizettei（Dubard et Dop） De Block;阿雷乌口树●☆

384915　Tarenna angolensis Hiern = Crossopteryx febrifuga（Afzel. ex G. Don） Benth. ■☆

384916　Tarenna asteriscus（K. Schum.） Bremek. = Nichallea soyauxii （Hiern） Bridson ●☆

384917　Tarenna attenuata（Voigt） Hutch.;假桂乌口树（茶山虫,杀山虫,山乌口树,树节,土五味子,乌口树,小林乌口树）;Taper Tarenna,Tapered Tarenna,Woody Tarenna ●

384918　Tarenna attenuata（Voigt） Hutch. var. puberula Chun et F. C. How;毛萼乌口树●

384919　Tarenna attenuata（Voigt） Hutch. var. puberula Chun et F. C. How = Tarenna lancilimba W. C. Chen ●

384920　Tarenna austrosinensis Chun et F. C. How ex W. C. Chen;华南乌口树;S. China Tarenna,South China Tarenna ●

384921　Tarenna baconoides Wernham var. nephrosperma N. Hallé;肾籽乌口树●☆

384922　Tarenna barbertonensis（Bremek.） Bremek. = Coptosperma supra-axillare（Hemsl.） Degreef ●☆

384923　Tarenna bipindensis（K. Schum.） Bremek.;比平迪乌口树●☆

384924　Tarenna boranensis Cufod.;博兰乌口树●☆

384925　Tarenna boranensis Cufod. = Coptosperma graveolens（S. Moore） Degreef ●☆

384926　Tarenna boranensis Cufod. subsp. arabica ? = Coptosperma graveolens（S. Moore） Degreef subsp. arabicum（Cufod.） Degreef ●☆

384927　Tarenna brachypoda Homolle = Tarenna grevei（Drake） Homolle ●☆

384928　Tarenna brachysiphon（Hiern） Keay;短管乌口树●☆

384929　Tarenna brunnea R. D. Good = Aidia ochroleuca（K. Schum.） E. M. Petit ●☆

384930　Tarenna burttii Bridson;伯特乌口树●☆

384931　Tarenna calliblepharis N. Hallé;美毛乌口树●☆

384932　Tarenna capuroniana De Block;凯普伦乌口树●☆

384933　Tarenna catocensi R. D. Good = Tarenna gossweileri S. Moore ●☆

384934　Tarenna conferta（Benth.） Hiern;密集乌口树●☆

384935　Tarenna congensis Hiern;刚果乌口树●☆

384936　Tarenna depauperata Hutch.;白皮乌口树（白骨木,狗骨头,假枝子）;Whitebark Tarenna,White-barked Tarenna ●

384937　Tarenna drummondii Bridson;德拉蒙德乌口树●☆

384938　Tarenna edgardii Chiov. = Coptosperma graveolens（S. Moore） Degreef ●☆

384939　Tarenna eketensis Wernham;埃凯特乌口树●☆

384940　Tarenna eketensis Wernham var. auricluna N. Hallé;耳乌口树●☆

384941　Tarenna flavo-fusca（K. Schum.） S. Moore = Tarenna fusco-flava（K. Schum.） S. Moore ●☆

384942　Tarenna flexilis A. Chev. = Tarenna thomasii Hutch. et Dalziel ●☆

384943　Tarenna friesiorum（K. Krause） Bremek. = Tarenna pavettoides （Harv.） Sim subsp. friesiorum（K. Krause） Bridson ●☆

384944　Tarenna funebris（Bremek.） N. Hallé;墓地乌口树●☆

384945　Tarenna fusco-flava（K. Schum.） N. Hallé = Tarenna fusco-flava（K. Schum.） S. Moore ●☆

384946　Tarenna fusco-flava（K. Schum.） S. Moore;棕黄乌口树●☆

384947　Tarenna gilletii（De Wild. et T. Durand） N. Hallé ex Gereau;吉勒特乌口树●☆

384948　Tarenna gossweileri S. Moore;戈斯乌口树●☆

384949　Tarenna gossweileri S. Moore var. brevituba Bridson;短管戈斯乌口树●☆

384950　Tarenna gracilipes（Hayata） Ohwi;薄叶乌口树（薄叶玉心花）;Thinleaf Tarenna ●

384951　Tarenna gracilipes（Hayata） Ohwi var. kotoensis（Hayata） T. Yamaz.;湖东乌口树●☆

384952　Tarenna gracilipes（Hayata） Ohwi var. yaeyamensis（Masam.） T. Yamaz. = Tarenna gracilipes（Hayata） Ohwi var. kotoensis （Hayata） T. Yamaz. ●☆

384953　Tarenna gracilis（Stapf） Keay;纤细乌口树●☆

384954　Tarenna grandiflora（Benth.） Hiern;大花乌口树●☆

384955　Tarenna graveolens（S. Moore） Bremek. = Coptosperma graveolens（S. Moore） Degreef ●☆

384956　Tarenna graveolens（S. Moore） Bremek. subsp. arabica （Cufod.） Bridson = Coptosperma graveolens（S. Moore） Degreef subsp. arabicum（Cufod.） Degreef ●☆

384957　Tarenna graveolens（S. Moore） Bremek. var. impolita Bridson = Coptosperma graveolens（S. Moore） Degreef var. impolitum （Bridson） Degreef ●☆

384958　Tarenna grevei（Drake） Homolle = Ixora brachypoda DC. ●☆

384959　Tarenna gyokushinkwa Ohwi = Tarenna gracilipes（Hayata） Ohwi ●

384960　Tarenna hayataiana Kaneh. = Tarenna gracilipes（Hayata） Ohwi ●

384961　Tarenna hutchinsonii Bremek.;哈钦森乌口树●☆

384962　Tarenna incana Diels = Tarenna mollissima（Hook. et Arn.） Rob. ●

384963　Tarenna incerta Koord. et Valeton = Randia wallichii Hook. f. ●

384964　Tarenna incerta Koord. et Valeton = Tarennoidea wallichii （Hook. f.） Tirveng. ●

384965　Tarenna junodii（Schinz） Bremek.;朱诺德乌口树●☆

384966　Tarenna kibuwae Bridson = Coptosperma kibuwae（Bridson） Degreef ●☆

384967　Tarenna kotoensis（Hayata） Kaneh. et Sasaki = Tarenna zeylanica Gaertn. ●

384968　Tarenna kotoensis（Hayata） Masam.;兰屿玉心花●

384969　Tarenna kotoensis（Hayata） Masam. = Tarenna zeylanica Gaertn. ●

384970　Tarenna kwangxiensis Hand.-Mazz. = Pavetta hongkongensis Bremek. ●

384971　Tarenna lagosensis Hutch. et Dalziel = Nichallea soyauxii （Hiern） Bridson ●☆

384972　Tarenna lanceolata Chun et F. C. How ex W. C. Chen;广西乌口树;Guangxi Tarenna,Kwangsi Tarenna ●

384973　Tarenna lancifolia（Hayata） Kaneh. et Sasaki = Tarenna gracilipes（Hayata）Ohwi ●

384974　Tarenna lancilimba W. C. Chen；披针叶乌口树（毛萼乌口树）；Hairycalyx Tarenna，Lanceleaf Tarenna，Lanceolate Tarenna ●

384975　Tarenna lasiorhachis（K. Schum. et K. Krause）Bremek.；毛轴乌口树●☆

384976　Tarenna laticorymbosa Chun et F. C. How ex W. C. Chen；宽序乌口树；Broadcorymb Tarenna，Broad-flower Tarenna ●

384977　Tarenna laui Merr.；崖州乌口树（毛冠乌口树）；Lau Tarenna，Yazhou Tarenna ●

384978　Tarenna laurentii（De Wild.）J. G. Garcia；洛朗乌口树●☆

384979　Tarenna laurentii（De Wild.）J. G. Garcia var. longipedicellata J. G. Garcia ＝ Tarenna longipedicellata（J. G. Garcia）Bridson ●☆

384980　Tarenna leonardii N. Hallé；莱奥乌口树●☆

384981　Tarenta letestui Pellegr. ＝ Tarenna conferta（Benth.）Hiern ●☆

384982　Tarenna littoralis（Hiern）Bridson ＝ Coptosperma littorale（Hiern）Degreef ●☆

384983　Tarenna longifolia Ridl.；长叶乌口树；Long-leaved Tarenna ●

384984　Tarenna longipedicellata（J. G. Garcia）Bridson；长柄乌口树●☆

384985　Tarenna luhomeroensis Bridson；卢赫梅罗乌口树●☆

384986　Tarenna luteola（Stapf）Bremek.；黄乌口树●☆

384987　Tarenna mangenotii Schnell ＝ Tarenna bipindensis（K. Schum.）Bremek. ●☆

384988　Tarenna mollissima（Hook. et Arn.）Rob.；白花苦灯笼（白花鬼点灯，白花乌口树，白青乌心，达仑木，达伦木，黑虎，鸡公辣，毛乌木树，密毛蒿香，密毛乌口树，青柞树，乌口树，乌木，小肠枫）；Whiteflower Tarenna，White-flowered Tarenna ●

384989　Tarenna mossambicensis Hiern ＝ Crossopteryx febrifuga（Afzel. ex G. Don）Benth. ■☆

384990　Tarenna neurophylla（S. Moore）Bremek. ＝ Coptosperma neurophyllum（S. Moore）Degreef ●☆

384991　Tarenna nigrescens（Hook. f.）Hiern ＝ Coptosperma nigrescens Hook. f. ●☆

384992　Tarenna nigrescens R. D. Good ＝ Nichallea soyauxii（Hiern）Bridson ●☆

384993　Tarenna nigroviridis R. D. Good ＝ Nichallea soyauxii（Hiern）Bridson ●☆

384994　Tarenna nilotica Hiern；尼罗河乌口树●☆

384995　Tarenna nimbana Schnell ＝ Tarenna nitidula（Benth.）Hiern ●☆

384996　Tarenna nitidula（Benth.）Hiern；光亮乌口树●☆

384997　Tarenna nitiduloides G. Taylor；拟光亮乌口树●☆

384998　Tarenna odora（K. Krause）Cufod. ＝ Coptosperma neurophyllum（S. Moore）Degreef ●☆

384999　Tarenna oligantha（K. Schum. et K. Krause）Cufod. ＝ Coptosperma graveolens（S. Moore）Degreef ●☆

385000　Tarenna ombrophila Schnell ＝ Tarenna brachysiphon（Hiern）Keay ●☆

385001　Tarenna pallida（Franch. ex Brandis）Hutch. ＝ Randia wallichii Hook. f. ●

385002　Tarenna pallida（Franch. ex Brandis）Hutch. ＝ Tarennoidea wallichii（Hook. f.）Tirveng. ●

385003　Tarenna pallidula Hiern；苍白乌口树●☆

385004　Tarenna pallidula Hiern var. oligoneura（K. Schum.）N. Hallé；寡脉苍白乌口树●☆

385005　Tarenna papyracea Burtt Davy；纸质乌口树●☆

385006　Tarenna patens S. Moore；铺展乌口树●☆

385007　Tarenna patula Hutch. et Dalziel ＝ Tarenna pallidula Hiern ●☆

385008　Tarenna pavettoides（Harv.）Sim；大沙叶乌口树●☆

385009　Tarenna pavettoides（Harv.）Sim subsp. affinis（K. Schum.）Bridson；近缘乌口树●☆

385010　Tarenna pavettoides（Harv.）Sim subsp. friesiorum（K. Krause）Bridson；弗里斯乌口树●☆

385011　Tarenna pavettoides（Harv.）Sim subsp. gillmanii Bridson；吉尔曼乌口树●☆

385012　Tarenna peteri Bridson ＝ Coptosperma peteri（Bridson）Degreef ●☆

385013　Tarenna petitii N. Hallé；佩蒂蒂乌口树●☆

385014　Tarenna pobeguinii Pobeg. ＝ Rutidea parviflora DC. ●☆

385015　Tarenna polysperma Chun et F. C. How ex W. C. Chen；多籽乌口树；Manyseed Tarenna，Seedy Tarenna ●

385016　Tarenna pubinervis Hutch.；滇南乌口树（毛脉乌口树）；Hairynerve Tarenna，Hairy-nerved Tarenna ●

385017　Tarenna punctata（K. Krause）Bremek.；斑点乌口树●☆

385018　Tarenna quadrangularis Bremek.；四棱乌口树●☆

385019　Tarenna rhodesiaca Bremek. ＝ Tarenna gossweileri S. Moore ●☆

385020　Tarenna richardii Drake ex Verdc. ＝ Tarenna grevei（Drake）Homolle ●☆

385021　Tarenna richardii Verdc.；理查德乌口树●☆

385022　Tarenna richardii Verdc. ＝ Tarenna grevei（Drake）Homolle ●☆

385023　Tarenna rwandensis Bridson；卢旺达乌口树●☆

385024　Tarenna saxatilis（Scott-Elliot）Homolle ＝ Paracephaelis saxatilis（Scott-Elliot）De Block ●☆

385025　Tarenna scandens R. D. Good；攀缘乌口树●☆

385026　Tarenna sinica W. C. Chen；长梗乌口树；China Tarenna，Chinese Tarenna ●

385027　Tarenna soyauxii（Hiern）Bremek. ＝ Nichallea soyauxii（Hiern）Bridson ●☆

385028　Tarenna subsessilis（A. Gray）T. Ito；近无柄乌口树●☆

385029　Tarenna supra-axillaris（Hemsl.）Bremek. ＝ Coptosperma supra-axillare（Hemsl.）Degreef ●☆

385030　Tarenna supra-axillaris（Hemsl.）Bremek. subsp. barbertonensis（Bremek.）Bridson ＝ Coptosperma supra-axillare（Hemsl.）Degreef ●☆

385031　Tarenna supra-axillaris（Hemsl.）Bremek. var. supra-axillaris ＝ Coptosperma supra-axillare（Hemsl.）Degreef ●☆

385032　Tarenna sylvestris Hutch. ＝ Tarenna attenuata（Voigt）Hutch. ●

385033　Tarenna talbotii Wernham ＝ Tarenna conferta（Benth.）Hiern ●☆

385034　Tarenna tetramera Hiern ＝ Pavetta tetramera（Hiern）Bremek. ●☆

385035　Tarenna thomasii Hutch. et Dalziel；托马斯氏乌口树●☆

385036　Tarenna thouarsiana（Drake）Homolle；图氏乌口树●☆

385037　Tarenna tomensis Schnell ＝ Tarenna brachysiphon（Hiern）Keay ●☆

385038　Tarenna trichantha（Baker）Bremek. ＝ Paracephaelis trichantha（Baker）De Block ●☆

385039　Tarenna tsangii Merr.；海南乌口树（光冠乌口树）；Hainan Tarenna ●

385040　Tarenna tsangii Merr. f. elliptica Chun et F. C. How；椭圆叶乌口树；Elliptic Tarenna ●

385041　Tarenna uniflora（Drake）Homolle；单花乌口树●☆

385042　Tarenna uzungwaensis Bridson；乌尊季沃乌口树●☆

385043　Tarenna verdcourtiana Fosberg；韦尔德乌口树●☆

385044　Tarenna vignei Hutch. et Dalziel；维涅乌口树●☆

385045　Tarenna vignei Hutch. et Dalziel var. subglabra Keay；近光乌口树●☆

385046　Tarenna wajirensis Bridson ＝ Coptosperma wajirense（Bridson）Degreef ●☆

385047　Tarenna wangii Chun et F. C. How ex W. C. Chen；王氏乌口树

（长叶乌口树）；Longleaf Tarenna，Wang Tarenna ●

385048　Tarenna yunnanensis F. C. How ex W. C. Chen；云南乌口树；Yunnan Tarenna ●

385049　Tarenna zeylanica Gaertn.；锡兰乌口树（兰屿玉心花，锡兰玉心花）；Ceylon Tarenna ●

385050　Tarenna zimbabwensis Bridson；罗得西亚乌口树●☆

385051　Tarenna zygoon Bridson；对称乌口树●☆

385052　Tarennoidea Tirveng. et Sastre（1979）；岭罗麦属；Tarennoidea ●

385053　Tarennoidea wallichii（Hook. f.）Tirveng. = Randia wallichii Hook. f. ●

385054　Tarennoidea wallichii（Hook. f.）Tirveng. = Tarennoidea wallichii（Hook. f.）Tirveng. et Sastre ●

385055　Tarennoidea wallichii（Hook. f.）Tirveng. et Sastre；岭罗麦（大果玉心花，蒿香）；Tarennoidea，Wallich Randia ●

385056　Tarigidia Stent（1932）；等颖草属■☆

385057　Tarigidia aequiglumis（Gooss.）Stent；等颖草■☆

385058　Tariri Aubl.（废弃属名）= Picramnia Sw.（保留属名）●☆

385059　Tarlmounia H. Rob.，S. C. Keeley，Skvarla et R. Chan = Vernonia Schreb.（保留属名）■●

385060　Tarlmounia H. Rob.，S. C. Keeley，Skvarla et R. Chan（2008）；椭圆斑鸠菊属■☆

385061　Tarlmounia elliptica（DC.）H. Rob.，S. C. Keeley，Skvarla et R. Chan；椭圆斑鸠菊■☆

385062　Tarlmounia elliptica（DC.）H. Rob.，S. C. Keeley，Skvarla et R. Chan = Vernonia elliptica DC. ■☆

385063　Tarnaricaria Qaiser et Ali = Myrtama Ovcz. et Kinzik. ●

385064　Tarpheta Raf. = Stachytarpheta Vahl（保留属名）■●

385065　Tarphochlamys Bremek.（1944）；肖笼鸡属（顶头马兰属，顶头马蓝属）；Tarphochlamys ■

385066　Tarphochlamys Bremek. = Strobilanthes Blume ●■

385067　Tarphochlamys affinis（Griff.）Bremek.；肖笼鸡（顶头草，顶头马兰，顶头马蓝，汗斑草，尖头马蓝）；Affined Conehead，Tarphochlamys ■

385068　Tarphochlamys darrisii（H. Lév.）E. Hossain；贵州肖笼鸡（肖笼鸡，兴义顶头马兰）；Darris Tarphochlamys ■

385069　Tarrietia Blume = Heritiera Aiton ●

385070　Tarrietia Blume = Hildegardia Schott et Endl. ●

385071　Tarrietia erythrosiphon（Baill.）Hochr. = Hildegardia erythrosiphon（Baill.）Kosterm. ●☆

385072　Tarrietia javanica Blume = Heritiera javanica（Blume）Kosterm. ●☆

385073　Tarrietia parvifolia（Merr.）Merr. et Chun = Heritiera parvifolia Merr. ●◇

385074　Tarrietia perrieri Hochr. = Hildegardia perrieri（Hochr.）Arènes ●

385075　Tarrietia utilis（Sprague）Sprague = Heritiera utilis（Sprague）Sprague ●☆

385076　Tarsina Noronha = Lepeostegeres Blume ●☆

385077　Tartagalia（A. Robyns）T. Mey. = Eriotheca Schott et Endl. ●☆

385078　Tartagalia Capurro = Eriotheca Schott et Endl. ●☆

385079　Tartonia Raf. = Thymelaea Mill.（保留属名）●■

385080　Tashiroa Willis = Tashiroea Matsum. ●■

385081　Tashiroea Matsum. = Bredia Blume ●■

385082　Tashiroea Matsum. ex T. Ito et Matsum. = Bredia Blume ●■

385083　Tashiroea okinawaensis Matsum. = Pachycentria formosana Hayata ●

385084　Tashiroea sinensis Diels = Bredia sinensis（Diels）H. L. Li ●

385085　Tasmannia DC. = Drimys J. R. Forst. et G. Forst.（保留属名）●☆

385086　Tasmannia DC. = Tasmannia R. Br. ex DC. ●☆

385087　Tasmannia R. Br. = Drimys J. R. Forst. et G. Forst.（保留属名）●☆

385088　Tasmannia R. Br. ex DC.（1817）；澳洲林仙属（塔司马尼木属）●☆

385089　Tasmannia R. Br. ex DC. = Drimys J. R. Forst. et G. Forst.（保留属名）●☆

385090　Tasmannia insipida R. Br. ex DC.；厚叶澳洲林仙（厚叶塔司马尼木）；Brush Pepperbush，Pepper Bush ●☆

385091　Tasmannia xerophila（Parm.）M. Gray；高山澳洲林仙（高山塔司马尼木）；Alpine Pepperbush ●☆

385092　Tasoba Raf. = Polygonum L.（保留属名）■●

385093　Tassadia Decne.（1844）；热美萝藦属●☆

385094　Tassadia angustifolia Malme；窄叶热美萝藦●☆

385095　Tassadia capitata W. D. Stevens；头状热美萝藦●☆

385096　Tassadia lanceolata Decne.；剑形热美萝藦●☆

385097　Tassadia minutiflora Malme；小花热美萝藦●☆

385098　Tassadia ovalifolia（E. Fourn.）Fontella；卵叶热美萝藦●☆

385099　Tassadia pilosula K. Schum.；柔毛热美萝藦●☆

385100　Tassia Rich. = Tachigalia Aubl. ●☆

385101　Tassia Rich. ex DC. = Tachigalia Aubl. ●☆

385102　Tassilicyparis A. V. Bobrov et Melikyan（2006）；撒哈拉柏属●☆

385103　Tassilicyparis dupreziana（A. Camus）A. V. Bobrov et Melikyan；撒哈拉柏●☆

385104　Tassilicyparis dupreziana（A. Camus）A. V. Bobrov et Melikyan = Cupressus dupreziana A. Camus ●☆

385105　Tatea F. Muell. = Premna L.（保留属名）●■

385106　Tatea F. Muell. = Pygmaeopremna Merr. ●

385107　Tatea Seem. = Bikkia Reinw.（保留属名）●☆

385108　Tatea herbacea（Roxb.）Junell = Premna herbacea Roxb. ●

385109　Tatea herbacea（Roxb.）Junell = Pygmaeopremna herbacea（Roxb.）Moldenke ●

385110　Tatea humilis（Merr.）Junell = Premna herbacea Roxb. ●

385111　Tatea humilis（Merr.）Junell = Pygmaeopremna herbacea（Roxb.）Moldenke ●

385112　Tateanthus Gleason（1931）；威尼斯野牡丹属☆

385113　Tateanthus duidae Gleason；威尼斯野牡丹☆

385114　Tatianyx Zuloaga et Soderstr.（1985）；稠毛草属■☆

385115　Tatianyx arnacites（Trin.）Zuloaga et Soderstr.；稠毛草■☆

385116　Tatina Raf. = Ilex L. ●

385117　Tatina Raf. = Sideroxylon L. ●☆

385118　Tattia Scop. = Homalium Jacq. ●

385119　Tattia Scop. = Napimoga Aubl. ●

385120　Taubertia K. Schum. = Disciphania Eichler ●☆

385121　Taubertia K. Schum. ex Taub. = Disciphania Eichler ●☆

385122　Taumastos Raf. = Libertia Spreng.（保留属名）■☆

385123　Tauroceras Britton et Rose = Acacia Mill.（保留属名）●■

385124　Taurophthalmum Duchass. ex Griseb. = Dioclea Kunth ■☆

385125　Taurostalix Rchb. f. = Bulbophyllum Thouars（保留属名）■

385126　Taurostalix herminiostachys Rchb. f. = Bulbophyllum pumilum（Sw.）Lindl. ■☆

385127　Taurrettia Raeusch. = Tourrettia Foug.（保留属名）■☆

385128　Tauscheria Fisch. = Tauscheria Fisch. ex DC. ■

385129　Tauscheria Fisch. ex DC.（1821）；舟果荠属；Boatcress，Tauscheria ■

385130　Tauscheria desertorum Ledeb. = Tauscheria lasiocarpa Fisch. ex DC. ■

385131 Tauscheria gymnocarpa Fisch. ex DC. = Tauscheria lasiocarpa Fisch. ex DC. ■

385132 Tauscheria lasiocarpa Fisch. ex DC.；舟果荠；Hairyfruit Boatcress, Hairyfruit Tauscheria ■

385133 Tauscheria lasiocarpa Fisch. ex DC. var. gymnocarpa（Fisch. ex DC.）Boiss. = Tauscheria lasiocarpa Fisch. ex DC. ■

385134 Tauscheria lasiocarpa Fisch. ex DC. var. gymnocarpa（Fisch. ex DC.）Boiss.；光果舟果荠；Smoothfruit Tauscheria ■

385135 Tauscheria oblonga Vassilcz. = Tauscheria lasiocarpa Fisch. ex DC. ■

385136 Tauschia Preissler（废弃属名）= Symphysia C. Presl ●☆

385137 Tauschia Preissler（废弃属名）= Tauschia Schltdl.（保留属名）■☆

385138 Tauschia Schltdl.（1835）（保留属名）；陶施草属■☆

385139 Tauschia bicolor Constance et Bye；二色陶施草■☆

385140 Tauschia edulis J. M. Coult. et Rose；可食陶施草■☆

385141 Tauschia nudicaulis Schltdl.；裸茎陶施草■☆

385142 Tauschia pubescens J. F. Macbr.；毛陶施草■☆

385143 Tavalla Pars. = Hedyosmum Sw. ●

385144 Tavalla Pars. = Tafalla Ruiz et Pav. ●■

385145 Tavaresia Welw. = Tavaresia Welw. ex N. E. Br. ●■☆

385146 Tavaresia Welw. ex N. E. Br.（1854）；丽钟角属；Tavaresia ●■☆

385147 Tavaresia Welw. ex N. E. Br. = Decabelone Decne. ●■☆

385148 Tavaresia angolensis Welw.；安哥拉丽钟角■☆

385149 Tavaresia barklyi（Dyer）N. E. Br.；巴氏丽钟角；Barkly's Tavaresia ■☆

385150 Tavaresia barklyi N. E. Br. = Tavaresia barklyi（Dyer）N. E. Br. ■☆

385151 Tavaresia grandiflora（K. Schum.）A. Berger；丽钟角（丽钟阁）；Largeflower Tavaresia, Largeflowered Tavaresia ●■☆

385152 Tavaresia grandiflora（K. Schum.）A. Berger = Tavaresia barklyi（Dyer）N. E. Br. ■☆

385153 Tavaresia grandiflora（K. Schum.）A. Berger var. recta van Son = Tavaresia barklyi（Dyer）N. E. Br. ■☆

385154 Tavernaria Rchb. = Taverniera DC. ■☆

385155 Taverniera DC.（1825）；塔韦豆属■☆

385156 Taverniera abyssinica A. Rich.；阿比西尼亚塔韦豆■☆

385157 Taverniera aegyptiaca Boiss.；埃及塔韦豆■☆

385158 Taverniera albida Thulin；白塔韦豆■☆

385159 Taverniera cuneifolia（Roth）Arn.；楔叶塔韦豆■☆

385160 Taverniera ephedroides Jaub. et Spach = Taverniera glabra Boiss. ■☆

385161 Taverniera floribunda Schweinf.；多花塔韦豆■☆

385162 Taverniera glabra Boiss.；光塔韦豆■☆

385163 Taverniera gonoclada Jaub. et Spach = Taverniera spartea（Burm. f.）DC. ■☆

385164 Taverniera lappacea（Forssk.）DC.；钩毛塔韦豆■☆

385165 Taverniera longisetosa Thulin；长刚毛塔韦豆■☆

385166 Taverniera multinoda Thulin；多节塔韦豆■☆

385167 Taverniera oligantha（Franch.）Thulin；少花塔韦豆■☆

385168 Taverniera schimperi Jaub. et Spach；欣班塔韦豆■☆

385169 Taverniera schimperi Jaub. et Spach var. oligantha Franch. = Taverniera oligantha（Franch.）Thulin ■☆

385170 Taverniera spartea（Burm. f.）DC.；鹰爪塔韦豆■☆

385171 Taverniera stefaninii Chiov. = Taverniera lappacea（Forssk.）DC. ■☆

385172 Taverniera stocksii Boiss. = Taverniera cuneifolia（Roth）Arn. ■☆

385173 Taveunia Burret = Cyphosperma H. Wendl. ex Benth. et Hook. f. ●☆

385174 Tavomyta Vitman = Tovomita Aubl. ●☆

385175 Taxaceae Gray（1822）（保留科名）；红豆杉科（紫杉科）；Yew Family, Taxus Family ●

385176 Taxandria（Benth.）J. R. Wheeler et N. G. Marchant = Agonis（DC.）Sweet（保留属名）●☆

385177 Taxandria（Benth.）J. R. Wheeler et N. G. Marchant（2007）；齐蕊木属●☆

385178 Taxandria angustifolia（Schauer）J. R. Wheeler et N. G. Marchant；窄叶齐蕊木●☆

385179 Taxandria spathulata（Schauer）J. R. Wheeler et N. G. Marchant；匙叶齐蕊木●☆

385180 Taxanthema Neck. = Armeria Willd.（保留属名）■☆

385181 Taxanthema Neck. = Statice L.（废弃属名）■☆

385182 Taxanthema Neck. ex R. Br. = Armeria Willd.（保留属名）■☆

385183 Taxanthema Neck. ex R. Br. = Statice L.（废弃属名）■☆

385184 Taxanthema R. Br. = Armeria Willd.（保留属名）■☆

385185 Taxanthema R. Br. = Statice L.（废弃属名）■☆

385186 Taxillus Tiegh.（1895）；钝果寄生属（钝果桑寄生属，杨寄生属）；Taxillus ●

385187 Taxillus balansae（Lecomte）Danser；栗毛钝果寄生（栗毛寄生）；Chestnus Taxillus, Chestnut Taxillus ●

385188 Taxillus balfourianus（Diels）Danser = Taxillus delavayi（Tiegh.）Danser ●

385189 Taxillus caloreas（Diels）Danser；松柏钝果寄生（松寄生，松田寄生，油杉寄生）；Matsuda Scurrula, Matsuda Taxillus, Pine Taxillus ●

385190 Taxillus caloreas（Diels）Danser var. fargesii（Lecomte）H. S. Kiu；显脉钝果寄生（显脉柏寄生，显脉松寄生）；Farges Pine Taxillus ●

385191 Taxillus cavaleriei（H. Lév.）Danser = Taxillus limprichtii（Grüning）H. S. Kiu ●

385192 Taxillus chinensis（DC.）Danser；广寄生（冰粉树，冬青，冻青，蠹心宝，广钝果寄生，华桑寄生，槐花寄生，黄皮寄生，寄生草，寄生茶，寄生树，寄屑，苦楝寄生，梨寄生，茑，茑木，枇杷寄生，桑寄生，桑上寄生，桑树寄生，松树桑寄生，酸柚寄生，桃寄生，宛童，梧州寄生茶，相思寄生，相思树寄生，寓木）；China Scurrula, Chinese Scurrula, Chinese Taxillus ●

385193 Taxillus collapsus（Lecomte）Danser = Bakerella collapsa（Lecomte）Balle ●☆

385194 Taxillus delavayi（Tiegh.）Danser；柳树寄生（花椒寄生，火把树寄生，柳寄生，柳树钝果寄生，柳叶钝果寄生，桑寄生）；Delavay Taxillus, Willow Scurrula, Willowleaf Scurrula ●

385195 Taxillus delavayi（Tiegh.）Danser var. barbatum W. L. Cheng；髯毛钝果寄生；Barbate Delavay Taxillus ●

385196 Taxillus delavayi（Tiegh.）Danser var. yanjingensis W. L. Cheng；盐井钝果寄生；Yanjing Delavay Taxillus ●

385197 Taxillus diplocrater（Baker）Danser = Bakerella diplocrater（Baker）Tiegh. ●☆

385198 Taxillus duclouxii（Lecomte）Danser = Taxillus sutchuenensis（Lecomte）Danser var. duclouxii（Lecomte）H. S. Kiu ●

385199 Taxillus estipitatus（Stapf）Danser = Taxillus chinensis（DC.）Danser ●

385200 Taxillus glaucus（Thunb.）Danser = Septulina glauca（Thunb.）Tiegh. ●☆

385201 Taxillus gonocladus（Baker）Danser = Bakerella gonoclada（Baker）Balle ●☆

385202　Taxillus kaempferi（DC.）Danser；小叶钝果寄生（华东松寄生，茑萝松，松胡颓子，松寄生）；Kaempfer Taxillus●

385203　Taxillus kaempferi（DC.）Danser var. grandiflorus H. S. Kiu；黄松钝果寄生（黄杉钝果寄生，四川松寄生）；Largeflower Kaempfer Taxillus●

385204　Taxillus kaempferi（DC.）Danser var. obovata Hatus.；倒卵叶钝果寄生●☆

385205　Taxillus kwangtungensis（Merr.）Danser = Taxillus limprichtii（Grüning）H. S. Kiu var. liquidambaricola（Hayata）H. S. Kiu●

385206　Taxillus kwangtungensis（Merr.）Danser = Taxillus limprichtii（Grüning）H. S. Kiu●

385207　Taxillus levinei（Merr.）H. S. Kiu；锈毛钝果寄生（李寄生，连江寄生，锈毛寄生）；Rustyhair Taxillus，Rusty-haired Taxillus●

385208　Taxillus limprichtii（Grüning）H. S. Kiu；木兰钝果寄生（枫木寄生，李栋山桑寄生，木兰寄生，木兰桑寄生，粤桑寄生）；Limpricht Taxillus，Magnolia Taxillus，Ritozan Scurrula●

385209　Taxillus limprichtii（Grüning）H. S. Kiu var. liquidambaricola（Hayata）= Taxillus liquidambaricola（Hayata）Hosok.●

385210　Taxillus limprichtii（Grüning）H. S. Kiu var. longiflorus（Lecomte）H. S. Kiu；亮叶木兰钝果寄生（亮叶寄生，亮叶木兰寄生）；Longflower Limpricht Taxillus●

385211　Taxillus liquidambaricola（Hayata）Hosok.；阆阓果寄生（赤柯寄生，大叶枫寄生，大叶桑寄生，螃蟹脚，显脉寄生，显脉木兰钝果寄生，显脉木兰寄生）；Distinctvein Taxillus，Sweetgum-loving Scurrula●

385212　Taxillus liquidambaricola（Hayata）Hosok. = Taxillus limprichtii（Grüning）H. S. Kiu var. liquidambaricola（Hayata）H. S. Kiu●

385213　Taxillus liquidambaricola（Hayata）Hosok. var. neriifolius H. S. Kiu；狭叶钝果寄生（狭叶果寄生）●

385214　Taxillus lonicerifolius（Hayata）S. T. Chiu = Scurrula lonicerifolia（Hayata）Danser●

385215　Taxillus lonicerifolius（Hayata）S. T. Chiu = Taxillus nigrans（Hance）Danser●

385216　Taxillus lonicerifolius（Hayata）S. T. Chiu var. longifolius P. S. Chiu；大花忍冬叶桑寄生●

385217　Taxillus lonicerifolius（Hayata）S. T. Chiu var. longifolius S. T. Chiu = Taxillus nigrans（Hance）Danser●

385218　Taxillus madagascaricus（Hochr.）Danser = Bakerella gonoclada（Baker）Balle●☆

385219　Taxillus matsudae（Hayata）Danser = Taxillus caloreas（Diels）Danser●

385220　Taxillus microcuspis（Baker）Danser = Bakerella microcuspis（Baker）Tiegh.●☆

385221　Taxillus nigrans（Hance）Danser；毛叶钝果寄生（杜鹃寄生，杜鹃桑寄生，寄生泡，借母怀胎，毛叶寄生，忍冬叶桑寄生，桑寄生，柿寄生）；Hairyleaf Taxillus，Hairy-leaved Taxillus，Rhododendron-inhabiting Scurrula●

385222　Taxillus notothixoides（Hance）Danser = Scurrula notothixoides（Hance）Danser●

385223　Taxillus ovalis（E. Mey. ex Harv.）Danser = Septulina ovalis（E. Mey. ex Harv.）Tiegh.●☆

385224　Taxillus parasiticus（L.）S. T. Chiu = Scurrula parasitica L.●

385225　Taxillus pseudochinensis（Yamam.）Danser；高雄钝果寄生（恒春桑寄生）；False Chinese Taxillus，Gaoxiong Taxillus●

385226　Taxillus pseudochinensis Yamam. = Taxillus pseudochinensis（Yamam.）Danser●

385227　Taxillus renii H. S. Kiu；油杉钝果寄生●

385228　Taxillus rhododendricola（Hayata）S. T. Chiu = Taxillus nigrans（Hance）Danser●

385229　Taxillus rhododendricola Hayata = Taxillus nigrans（Hance）Danser●

385230　Taxillus ritozanensis（Hayata）S. T. Chiu；李栋山桑寄生●

385231　Taxillus ritozanensis（Hayata）S. T. Chiu = Taxillus limprichtii（Grüning）H. S. Kiu●

385232　Taxillus rutilus Danser = Taxillus levinei（Merr.）H. S. Kiu●

385233　Taxillus sericeus Danser；龙陵钝果寄生（龙陵寄生）；Longling Taxillus，Silky Taxillus●

385234　Taxillus sutchuenensis（Lecomte）Danser；四川寄生（板栗寄生，冬青，冻青，寄生，毛叶寄生，桑寄生，桑上寄生，四川桑寄生）；Sichuan Taxillus，Szechwan Taxillus●

385235　Taxillus sutchuenensis（Lecomte）Danser var. duclouxii（Lecomte）H. S. Kiu；灰毛桑寄生（湖北桑寄生，灰毛寄生）；Ducloux Sichuan Taxillus●

385236　Taxillus tandrokensis（Lecomte）Danser = Bakerella tandrokensis（Lecomte）Balle●☆

385237　Taxillus theifer（Hayata）H. S. Kiu；台湾钝果寄生（埔姜桑寄生）；Taiwan Taxillus●

385238　Taxillus thibetensis（Lecomte）Danser；滇藏钝果寄生（藏寄生，寄生草，金沙江寄生，梨寄生，牛筋刺寄生）；Jinshajiang Taxillus，Tibetan Taxillus，Xizang Taxillus●

385239　Taxillus thibetensis（Lecomte）Danser var. albus J. R. Wu；白毛寄生；Whitehair Xizang Taxillus●

385240　Taxillus thibetensis（Lecomte）Danser var. albus J. R. Wu = Taxillus thibetensis（Lecomte）Danser●

385241　Taxillus thibetensis Danser = Taxillus thibetensis（Lecomte）Danser●

385242　Taxillus tomentosus（Roth）Tiegh.；钝果寄生；Taxillus●☆

385243　Taxillus tricostatus（Lecomte）Danser = Bakerella tricostata（Lecomte）Balle●☆

385244　Taxillus tsaii S. T. Chiu；莲花池寄生；Tsai Taxillus●

385245　Taxillus umbellifer（Schult.）Danser；伞花钝果寄生；Umbel Taxillus，Umbrellateflower Taxillus●

385246　Taxillus vestitus（Wall.）Danser；短梗钝果寄生（短果钝果寄生，怒江寄生）；Clothed Taxillus，Nujiang Taxillus，Shortstalk Taxillus，Short-stalked Taxillus●

385247　Taxillus vestitus（Wall.）Danser. = Loranthus vestitus Wall.●

385248　Taxillus wiensii Polhill；温斯钝果寄生●☆

385249　Taxillus yadoriki（Siebold ex Maxim.）Danser = Scurrula yadoriki（Siebold ex Maxim.）Danser●☆

385250　Taxodiaceae Saporta（1865）（保留科名）；杉科；Bald Cypress Family，Baldcypress Family，Redwood Family，Swamp Cypress Family，Taxodium Family●

385251　Taxodiaceae Saporta（保留科名）= Cupressaceae Gray（保留科名）●

385252　Taxodiaceae Warm. = Taxodiaceae Saporta（保留科名）●

385253　Taxodium Rich.（1810）；落羽杉属（落羽松属）；Bald Cypress，Baldcypress，Bald-cypress，Cypress，Deciduous Swamp Cypress，Deciduous-swamp-cypress，Deciduous-yew-cypress，Pond Baldcypress，Swamp Cypress，Swamp-cypress●

385254　Taxodium ascendens Brongn.'Nutans'；垂枝池杉；Deciduous Cypress，Pond Bald-cypress，Pond Cypress，Swamp Cypress，Sweeping Pond Cypress●

385255　Taxodium ascendens Brongn.'Xianyechisha'；线叶池杉；Lineleaf Pond Cypress ●

385256 Taxodium ascendens Brongn. ' Yuyechishan '；羽叶池杉；
Pinnate-leaf Pond Cypress ●

385257 Taxodium ascendens Brongn. ' Zhuiyechishan '；锥叶池杉；
Panicleleaf Pond Cypress ●

385258 Taxodium ascendens Brongn. = Taxodium distichum （L.）
Rich. var. imbricatum （Nutt.）Croom ●

385259 Taxodium ascendens Brongn. f. nutans （Aiton）Rehder =
Taxodium ascendens Brongn. ' Nutans ' ●

385260 Taxodium ascendens Brongn. var. nutans （Aiton）Rehder =
Taxodium ascendens Brongn. ' Nutans ' ●

385261 Taxodium ascendens Brongn. var. nutans （Aiton）Rehder =
Taxodium distichum （L.）Rich. var. imbricatum （Nutt.）Croom ●

385262 Taxodium distichum （L.）Rich.；落羽杉(落羽松,美国水杉,
美国水松)；Bald Cypress, Baldcypress, Bald-cypress, Common
Baldcypress, Cypress, Cypress Knees, Cypress-knees, Deciduous
Cypress, Gulf Cypress, Louisiana Cypress, Montezuma Cypress, Red
Cypress, Southern Cypress, Swamp Cypress, Tidewater Red Cypress ●

385263 Taxodium distichum （L.）Rich. f. confusum E. J. Palmer et
Steyerm. = Taxodium distichum （L.）Rich. ●

385264 Taxodium distichum （L.）Rich. var. imbricatum （Nutt.）
Croom；池杉(池柏,沼落羽松)；Pond Baldcypress, Pond Cypress,
Pondcypress ●

385265 Taxodium distichum （L.）Rich. var. mexicanum （Carrière）
Gordon = Taxodium mucronatum Ten. ●

385266 Taxodium distichum （L.）Rich. var. mexicanum Gordon =
Taxodium mucronatum Ten. ●

385267 Taxodium distichum （L.）Rich. var. mucronatum （Ten.）A.
Henry = Taxodium mucronatum Ten. ●

385268 Taxodium distichum （L.）Rich. var. nutans （Aiton）Sweet =
Taxodium distichum （L.）Rich. ●

385269 Taxodium distichum （L.）Rich. var. nutans （Aiton）Sweet =
Taxodium distichum （L.）Rich. var. imbricatum （Nutt.）Croom ●

385270 Taxodium distichum （L.）Rich. var. nutans Carrière =
Taxodium distichum （L.）Rich. ●

385271 Taxodium distichum （L.）Rich. var. nutans Sweet = Taxodium
ascendens Brongn. ' Nutans ' ●

385272 Taxodium distichum （L.）Rich. var. nutans Sweet = Taxodium
distichum （L.）Rich. ●

385273 Taxodium distichum （L.）Rich. var. pendulum Carrière；垂枝
落羽杉(垂枝落叶松)●

385274 Taxodium heterophyllum Brongn. = Glyptostrobus pensilis
（Staunton ex D. Don）K. Koch ●◇

385275 Taxodium imbricatum （Nutt.）R. M. Harper = Taxodium
distichum （L.）Rich. var. imbricatum （Nutt.）Croom ●

385276 Taxodium japonicum （L. f.）Brongn. = Cryptomeria japonica
（L. f.）D. Don ●

385277 Taxodium japonicum （L. f.）Brongn. = Glyptostrobus pensilis
（Staunton ex D. Don）K. Koch ●◇

385278 Taxodium japonicum （L. f.）Brongn. var. heterophyllum
Brongn. = Glyptostrobus pensilis （Staunton ex D. Don）K. Koch ●◇

385279 Taxodium japonicum （Thunb. ex L. f.）Brongn. var.
heterophyllum Brongn. = Glyptostrobus pensilis （Staunton ex D.
Don）K. Koch ●◇

385280 Taxodium mexicanum Carrière = Taxodium mucronatum Ten. ●

385281 Taxodium mucronatum Ten.；墨西哥落羽杉(尖叶落羽杉,墨
西哥落羽松)；Bald Cypress, Mexican Bald Cypress, Mexican

Baldcypress, Mexican Bald-cypress, Mexican Cypress, Mexican
Swamp Cypress, Montezuma Bald Cypress, Montezuma Baldcypress,
Montezuma Bald-cypress, Montezuma Cypress, Sabino ●

385282 Taxodium mucronatum Ten. = Taxus wallichiana Zucc. ●◇

385283 Taxodium nutans （Aiton）Sweet = Taxodium distichum （L.）
Rich. var. imbricatum （Nutt.）Croom ●

385284 Taxodium sempervirens D. Don = Sequoia sempervirens （D. Don
ex Lamb.）Endl. ●

385285 Taxodium sempervirens D. Don ex Lamb. = Sequoia
sempervirens （D. Don ex Lamb.）Endl. ●

385286 Taxodium sinense J. Forbes = Glyptostrobus pensilis （Staunton
ex D. Don）K. Koch ●◇

385287 Taxodium sinense Nois. ex Gordon = Taxodium ascendens
Brongn. ' Nutans ' ●

385288 Taxotrophis Blume = Streblus Lour. ●

385289 Taxotrophis Blume(1856)；刺桑属●☆

385290 Taxotrophis aquifolioides W. C. Ko = Streblus ilicifolius （S.
Vidal）Corner ●

385291 Taxotrophis balansae Hutch. = Streblus macrophyllus Blume ●

385292 Taxotrophis caudata Hutch. = Streblus zeylanicus （Thwaites）
Kurz ●

385293 Taxotrophis ilicifolia （S. Vidal）Corner = Streblus zeylanicus
（Thwaites）Kurz ●

385294 Taxotrophis ilicifolia S. Vidal = Streblus ilicifolius （S. Vidal）
Corner ●

385295 Taxotrophis longispina Merr. et Chun = Streblus ilicifolius （S.
Vidal）Corner ●

385296 Taxotrophis macrophylla Boerl. = Streblus ilicifolius （S. Vidal）
Corner ●

385297 Taxotrophis obtusa Elmer = Streblus ilicifolius （S. Vidal）
Corner ●

385298 Taxotrophis triapiculata Gamble = Streblus ilicifolius （S. Vidal）
Corner ●

385299 Taxotrophis zeylanica （Thwaites）Thwaites = Streblus
zeylanicus （Thwaites）Kurz ●

385300 Taxotrophis zeylanica Thwaites = Streblus zeylanicus
（Thwaites）Kurz ●

385301 Taxus L. (1753)；红豆杉属(紫杉属)；Taxol, Yew ●

385302 Taxus baccata L.；欧洲红豆杉(观音杉,红豆杉,红头杉,浆
果紫杉,欧洲紫杉,欧紫杉,西洋红豆杉,紫杉)；Common Yew,
Cup-of-wine, English Yew, Eugh, Europe Yew, European Yew, Ewe,
Ewgh, Florence Court Yew, Hebon, Love-in-Idleness Ife, Palm, Poison
Bush, Shooter Yew, Shoter, Snatberry, Snoder Gill, Snod-goggles, Snot
Berry, Snots, Snotter Gall, Snotterberries, Snotty Gog, Snottygobbles,
Vew, Vewe, View, Wire Thorn, Yeugh, Yew, Youe, Yugh ●☆

385303 Taxus baccata L. ' Adpresa '；灌丛欧洲红豆杉(紧贴欧洲红豆
杉)；Dense-shrub Common Yew ●☆

385304 Taxus baccata L. ' Aurea '；金黄欧洲红豆杉；Golden Common
Yew, Golden Yew ●☆

385305 Taxus baccata L. ' Dovastoniana '；道氏垂枝欧洲红豆杉；
Dovaston Yew, Westfeltan Yew, Westfelton Yew ●☆

385306 Taxus baccata L. ' Dovastonii Aurea '；道氏金色欧洲红豆杉●☆

385307 Taxus baccata L. ' Dovastonii '；道氏欧洲红豆杉；Wesffelton
Yew ●☆

385308 Taxus baccata L. ' Dwarf White '；矮白欧洲红豆杉●☆

385309 Taxus baccata L. ' Elegantissima '；极美欧洲红豆杉●☆

385310 Taxus baccata L. ' Erecta '；直立欧洲红豆杉；Erect Common

Yew,Fulham Yew,Neidpath Yew ●☆

385311　Taxus baccata L. 'Fastigiata Aurea'；金帚欧洲红豆杉(帚形黄叶欧洲红豆杉)；Columnar English Yew ●☆

385312　Taxus baccata L. 'Fastigiata Aureomarginata'；金边帚形欧洲红豆杉；Golden Irish Yew ●☆

385313　Taxus baccata L. 'Fastigiata Robusta'；粗帚欧洲红豆杉；Upright Yew ●☆

385314　Taxus baccata L. 'Fastigiata'；帚状欧洲红豆杉(帚形欧洲红豆杉)；Churchyard Yew,Florence Court Yew,Irish Yew ●☆

385315　Taxus baccata L. 'Lutea'；黄果欧洲红豆杉；Yellow-berried Yew ●☆

385316　Taxus baccata L. 'Nana'；矮生欧洲红豆杉●☆

385317　Taxus baccata L. 'Nutans'；低垂欧洲红豆杉●☆

385318　Taxus baccata L. 'Repandens Aurea'；金黄垂枝欧洲红豆杉；Golden Spreading English Yew ●☆

385319　Taxus baccata L. 'Repandens'；垂枝欧洲红豆杉(展枝欧洲红豆杉)；Repandens Yew,Spreading English Yew ●☆

385320　Taxus baccata L. 'Semperaurea'；金叶斯坦迪西欧洲红豆杉(四季黄欧洲红豆杉)●☆

385321　Taxus baccata L. 'Standishii'；金欧洲红豆杉；Golden Yew ●☆

385322　Taxus baccata L. 'Stricta Aurea'；金黄爱尔兰红豆杉；Golden Irish Yew ●☆

385323　Taxus baccata L. 'Stricta Variegata'；银斑爱尔兰欧洲红豆杉；Variegated Irish Yew ●☆

385324　Taxus baccata L. 'Stricta'；爱尔兰欧洲垂枝红豆杉；Irish Yew ●☆

385325　Taxus baccata L. = Taxus wallichiana Zucc. var. chinensis (Pilg.) Florin ●◇

385326　Taxus baccata L. subsp. brevifolia (Nutt.) Pilg. = Taxus brevifolia Nutt. ●☆

385327　Taxus baccata L. subsp. canadensis (Marshall) Pilg. = Taxus canadensis Marshall ●☆

385328　Taxus baccata L. subsp. cuspidata (Siebold et Zucc.) Pilg. = Taxus cuspidata Siebold et Zucc. ●◇

385329　Taxus baccata L. subsp. cuspidata (Siebold et Zucc.) Pilg. var. chinensis Pilg. = Taxus wallichiana Zucc. var. chinensis (Pilg.) Florin ●◇

385330　Taxus baccata L. subsp. cuspidata (Siebold et Zucc.) Pilg. var. latifolia Pilg. = Taxus cuspidata Siebold et Zucc. ●◇

385331　Taxus baccata L. subsp. wallichiana (Zucc.) Pilg. = Taxus wallichiana Zucc. ●◇

385332　Taxus baccata L. var. brevifolia (Nutt.) Koehne = Taxus brevifolia Nutt. ●☆

385333　Taxus baccata L. var. canadensis Benth. = Taxus brevifolia Nutt. ●☆

385334　Taxus baccata L. var. chinensis Henry = Taxus chinensis (Pilg.) Rehder ●◇

385335　Taxus baccata L. var. cuspidata Carrière = Taxus cuspidata Siebold et Zucc. ●◇

385336　Taxus baccata L. var. floridana (Nutt. ex Chapm.) Pilg. = Taxus floridana Nutt. ex Chapm. ●☆

385337　Taxus baccata L. var. microcarpa Trautv. = Taxus cuspidata Siebold et Zucc. ●◇

385338　Taxus baccata L. var. minor Michx. = Taxus canadensis Marshall ●☆

385339　Taxus baccata L. var. sinensis A. Henry = Taxus wallichiana Zucc. var. chinensis (Pilg.) Florin ●◇

385340　Taxus baccata L. var. stricta ? = Taxus baccata L. 'Stricta'●☆

385341　Taxus bourcieri Carrière = Taxus brevifolia Nutt. ●☆

385342　Taxus brevifolia Nutt.；短叶红豆杉(短叶紫杉，红豆杉)；

American Yew,Californian Yew,Oregon,Oregon Yew,Pacific Yew,Shortleaf Yew,Western Yew,Yew ●☆

385343　Taxus caespitosa Nakai = Taxus cuspidata Siebold et Zucc. ●◇

385344　Taxus canadensis Marshall；加拿大红豆杉；American Yew,Canada Yew,Canadian Yew,Ground Hemlock,Ground-hemlock ●☆

385345　Taxus celebica (Warb.) H. L. Li = Taxus mairei (Lemée et H. Lév.) S. Y. Hu ●

385346　Taxus chienii W. C. Cheng = Pseudotaxus chienii (W. C. Cheng) W. C. Cheng ●

385347　Taxus chinensis (Pilg.) Rehder = Taxus wallichiana Zucc. var. chinensis (Pilg.) Florin ●◇

385348　Taxus chinensis (Pilg.) Rehder var. mairei (Lemée et H. Lév.) W. C. Cheng et L. K. Fu = Taxus wallichiana Zucc. var. mairei (Lemée et H. Lév.) L. K. Fu et Nan Li ●◇

385349　Taxus chinensis (Pilg.) Rehder var. mairei (Lemée et H. Lév.) W. C. Cheng et L. K. Fu = Taxus mairei (Lemée et H. Lév.) S. Y. Hu ●

385350　Taxus chinensis Rehder var. mairei (Lemée et H. Lév.) W. C. Cheng et L. K. Fu = Taxus wallichiana Zucc. var. mairei (Lemée et H. Lév.) L. K. Fu et Nan Li ●◇

385351　Taxus chinensis Roxb. = Podocarpus macrophyllus (Thunb. ex A. Murray) D. Don ●

385352　Taxus chinensis Zucc. = Taxus wallichiana Zucc. ●◇

385353　Taxus cuspidata Nakai var. latifolia (Pilg.) Nakai = Taxus cuspidata Siebold et Zucc. ●◇

385354　Taxus cuspidata Nakai var. microcarpa (Trautv.) Kolesn. = Taxus cuspidata Siebold et Zucc. ●◇

385355　Taxus cuspidata Siebold et Zucc.；东北红豆杉(赤柏木，赤柏松，宽叶紫杉，米树，小果柳杉，小果紫杉，紫柏松，紫杉)；Japan Yew,Japanese Yew,Little Fruit Japanese Yew ●◇

385356　Taxus cuspidata Siebold et Zucc. 'Aurescens'；杂黄叶东北红豆杉●☆

385357　Taxus cuspidata Siebold et Zucc. 'Capitata'；塔紫杉(头状东北红豆杉，头状紫杉)；Pyramid Japanese Yew ●

385358　Taxus cuspidata Siebold et Zucc. 'Densa'；密集紫杉(紧密东北红豆杉)●☆

385359　Taxus cuspidata Siebold et Zucc. 'Densiformis'；丛生紫杉(浓密紫杉)；Dense Japanese Yew ●

385360　Taxus cuspidata Siebold et Zucc. 'Minima'；极小紫杉●☆

385361　Taxus cuspidata Siebold et Zucc. 'Nana'；矮紫杉(矮生紫杉)；Dwarf Japanese Yew ●

385362　Taxus cuspidata Siebold et Zucc. = Taxus wallichiana Zucc. var. mairei (Lemée et H. Lév.) L. K. Fu et Nan Li ●◇

385363　Taxus cuspidata Siebold et Zucc. f. luteobaccata (Miyabe et Tatew.) Hayashi；札幌红豆杉●☆

385364　Taxus cuspidata Siebold et Zucc. var. caespitosa (Nakai) Q. L. Wang；大山伽罗木●

385365　Taxus cuspidata Siebold et Zucc. var. chinensis (Pilg.) C. K. Schneid. ex Silva = Taxus chinensis (Pilg.) Rehder ●◇

385366　Taxus cuspidata Siebold et Zucc. var. chinensis (Pilg.) C. K. Schneid. = Taxus wallichiana Zucc. var. chinensis (Pilg.) Florin ●◇

385367　Taxus cuspidata Siebold et Zucc. var. latifolia (Pilg.) Nakai = Taxus cuspidata Siebold et Zucc. ●◇

385368　Taxus cuspidata Siebold et Zucc. var. luteobaccata Miyabe et Tatew. = Taxus cuspidata Siebold et Zucc. f. luteobaccata (Miyabe et Tatew.) Hayashi ●☆

385369　Taxus cuspidata Siebold et Zucc. var. microcarpa (Trautv.) S.

Y. Hu = Taxus cuspidata Siebold et Zucc. ●◇

385370 Taxus cuspidata Siebold et Zucc. var. nana Rehder；小红豆杉（矮紫杉）●

385371 Taxus cuspidata Siebold et Zucc. var. nana Rehder 'Aurescens' = Taxus cuspidata Siebold et Zucc. 'Aurescens' ●☆

385372 Taxus cuspidata Siebold et Zucc. var. umbraculifera Makino；伞状红豆杉●☆

385373 Taxus cuspidata var. chinensis （Pilg.）Schneid. ex Silva = Taxus wallichiana Zucc. var. chinensis （Pilg.）Florin ●◇

385374 Taxus floridana Chapm. = Taxus floridana Nutt. ex Chapm. ●☆

385375 Taxus floridana Nutt. ex Chapm.；佛罗里达红豆杉；Florida Yew ●☆

385376 Taxus fuana Nan Li et R. R. Mill；密叶红豆杉（喜马拉雅密叶红豆杉）；Dense-leaved Yew ●☆

385377 Taxus globosa Schltdl.；球果红豆杉●☆

385378 Taxus harringtonia Knight ex Forbes = Cephalotaxus harringtonia （Knight ex J. Forbes）K. Koch ●☆

385379 Taxus lindleyana A. Murray bis = Taxus brevifolia Nutt. ●☆

385380 Taxus macrophylla Thunb. = Podocarpus macrophyllus （Thunb. ex A. Murray）D. Don ●

385381 Taxus macrophylla Thunb. = Podocarpus macrophyllus （Thunb.）D. Don ●

385382 Taxus mairei （Lemée et H. Lév.）S. Y. Hu = Taxus wallichiana Zucc. var. mairei （Lemée et H. Lév.）L. K. Fu et Nan Li ●◇

385383 Taxus mairei （Lemée et H. Lév.）S. Y. Hu ex T. S. Liu = Taxus wallichiana Zucc. var. mairei （Lemée et H. Lév.）L. K. Fu et Nan Li ●◇

385384 Taxus media Rehder；杂种红豆杉（间型红豆杉，杂种紫杉）；Anglojap，Anglo-jap Yew，Hybrid Yew，Yew ●☆

385385 Taxus media Rehder 'Brownii'；密生杂种紫杉（布朗间型红豆杉，布鲁尼杂种紫杉）；Dense Hybrid Yew ●☆

385386 Taxus media Rehder 'Dark Green Spreder'；暗绿杂种紫杉●☆

385387 Taxus media Rehder 'Densiformis'；紧密间型红豆杉●☆

385388 Taxus media Rehder 'Evelow'；艾维鲁杂种紫杉●☆

385389 Taxus media Rehder 'Hartfieldii'；塔形杂种紫杉（哈特菲尔德杂种紫杉）；Pyramid Hybrid Yew ●☆

385390 Taxus media Rehder 'Hicksii'；直立杂种紫杉（西科斯杂种紫杉，希克斯间型红豆杉）；Columnar Yew，Erect Hybrid Yew ●☆

385391 Taxus media Rehder 'Hillii'；希尔间型红豆杉●☆

385392 Taxus media Rehder 'Nigra'；黑人杂种紫杉●☆

385393 Taxus media Rehder 'Wardii'；沃德间型红豆杉●☆

385394 Taxus minor （Michx.）Britton = Taxus canadensis Marshall ●☆

385395 Taxus nucifera L. = Torreya nucifera （L.）Siebold et Zucc. ●

385396 Taxus procumbens Lodd. = Taxus canadensis Marshall ●☆

385397 Taxus sieboldii Hort. = Taxus cuspidata Siebold et Zucc. ●◇

385398 Taxus speciosa Florin = Taxus mairei （Lemée et H. Lév.）S. Y. Hu ●

385399 Taxus speciosa Florin = Taxus wallichiana Zucc. var. mairei （Lemée et H. Lév.）L. K. Fu et Nan Li ●◇

385400 Taxus sumatrana （Miq.）de Laub.；南洋红豆杉（苏门答腊红豆杉）●☆

385401 Taxus tomentosa Thunb. = Grubbia tomentosa （Thunb.）Harms ●☆

385402 Taxus verticillata Thunb. = Sciadopitys verticillata （Thunb.）Siebold et Zucc. ●

385403 Taxus virgata Wall. = Taxus wallichiana Zucc. ●◇

385404 Taxus wallichiana Zucc.；西藏红豆杉（西南红豆杉，喜马拉雅红豆杉，须弥红豆杉，云南红豆杉）；Himalayan Yew，Xizang Yew ●◇

385405 Taxus wallichiana Zucc. var. chinensis （Pilg.）Florin；红豆杉（扁柏，观音杉，红豆树，卷柏，血柏）；China Yew，Chinese Yew ●◇

385406 Taxus wallichiana Zucc. var. chinensis （Pilg.）Florin = Taxus chinensis （Pilg.）Rehder ●◇

385407 Taxus wallichiana Zucc. var. mairei （Lemée et H. Lév.）L. K. Fu et Nan Li；南方红豆杉（赤椎，臭榧，榧子木，海罗松，红榧，红叶水杉，华紫杉，美丽红豆杉，杉公子，台湾红豆杉，血榧）；Chinese Yew，Maire Yew，Southern Yew，Taiwan Yew ●◇

385408 Taxus wallichiana Zucc. var. yunnanensis （W. C. Cheng et L. K. Fu）C. T. Kuan = Taxus wallichiana Zucc. ●◇

385409 Taxus yunnanensis W. C. Cheng et L. K. Fu；云南红豆杉（土榧子，西南红豆杉）；Yunnan Yew ●◇

385410 Taxus yunnanensis W. C. Cheng et L. K. Fu = Taxus wallichiana Zucc. ●◇

385411 Tayloriophyton Nayar（1968）；塔氏木属●☆

385412 Tayloriophyton glabrum Nayar；塔氏木●☆

385413 Tayloriophyton longisetosum （Ridl.）Nayar；长刚毛塔氏木●☆

385414 Tayotum Blanco = Geniostoma J. R. Forst. et G. Forst. ●

385415 Tchihatchewia Boiss. = Neotchihatchewia Rauschert ■☆

385416 Teagueia （Luer）Luer（1991）；蒂格兰属■☆

385417 Teagueia teaguei （Luer）Luer；蒂格兰■☆

385418 Teclea Delile（1843）（保留属名）；柚木芸香属●☆

385419 Teclea Delile（保留属名）= Vepris Comm. ex A. Juss. ●☆

385420 Teclea afzelii Engl. = Vepris afzelii （Engl.）Mziray ●☆

385421 Teclea alexandrae （Chiov.）Senni = Vepris eugeniifolia （Engl.）I. Verd. ●☆

385422 Teclea amaniensis Engl. = Vepris amaniensis （Engl.）Mziray ●☆

385423 Teclea borenensis M. G. Gilbert = Vepris borenensis （M. G. Gilbert）Mziray ●☆

385424 Teclea campestris Engl.；田野柚木芸香●☆

385425 Teclea crenulata （Engl.）Engl.；细圆齿柚木芸香●☆

385426 Teclea diversifolia Lanza；异叶柚木芸香●☆

385427 Teclea ebolowensis Engl. = Toddaliopsis ebolowensis Engl. ●☆

385428 Teclea eggelingii Kokwaro = Vepris eggelingii （Kokwaro）Mziray ●☆

385429 Teclea engleriana De Wild.；恩格勒柚木芸香●☆

385430 Teclea evodioides Chiov.；吴茱萸柚木芸香●☆

385431 Teclea ferruginea A. Chev.；锈色柚木芸香●☆

385432 Teclea fischeri （Engl.）Engl.；菲舍尔柚木芸香●☆

385433 Teclea gerrardii I. Verd.；杰勒德柚木芸香●☆

385434 Teclea glomerata （F. Hoffm.）I. Verd. = Vepris glomerata （F. Hoffm.）Engl. ●☆

385435 Teclea gossweileri I. Verd. = Vepris gossweileri （I. Verd.）Mziray ●☆

385436 Teclea grandifolia Engl. = Vepris grandifolia （Engl.）Mziray ●☆

385437 Teclea hanangensis Kokwaro = Vepris borenensis （M. G. Gilbert）Mziray ●☆

385438 Teclea hanangensis Kokwaro = Vepris hanangensis （Kokwaro）Mziray ●☆

385439 Teclea hanangensis Kokwaro var. unifoliolata Kokwaro = Vepris hanangensis （Kokwaro）Mziray ●☆

385440 Teclea heterophylla Engl. = Vepris heterophylla （Engl.）Letouzey ●☆

385441 Teclea macedoi Exell et Mendonça = Vepris macedoi （Exell et Mendonça）Mziray ●☆

385442 Teclea myrei Exell et Mendonça = Vepris myrei （Exell et Mendonça）Mziray ●☆

385443 Teclea natalensis （Sond.）Engl.；纳塔尔柚木芸香●☆

385444 Teclea nobilis Delile = Vepris nobilis (Delile) Mziray ●☆

385445 Teclea oubanguiensis Aubrév. et Pellegr. ;乌班吉柚木芸香●☆

385446 Teclea pilosa (Engl.) I. Verd. ;疏毛柚木芸香●☆

385447 Teclea punctata I. Verd. = Vepris punctata (I. Verd.) Mziray ●☆

385448 Teclea rogersii Mendonça = Vepris rogersii (Mendonça) Mziray ●☆

385449 Teclea simplicifolia (Engl.) I. Verd. = Vepris simplicifolia (Engl.) Mziray ●☆

385450 Teclea suaveolens Engl. = Vepris suaveolens (Engl.) Mziray ●☆

385451 Teclea sudanica A. Chev. ;苏丹柚木芸香●☆

385452 Teclea sudanica A. Chev. = Vepris heterophylla (Engl.) Letouzey ●☆

385453 Teclea swynnertonii Baker f. = Oricia bachmannii (Engl.) I. Verd. ●☆

385454 Teclea trichocarpa (Engl.) Engl. = Vepris trichocarpa (Engl.) Mziray ●☆

385455 Teclea utilis Engl. ;有用柚木芸香●☆

385456 Teclea verdoorniana Exell et Mendonça = Vepris verdoorniana (Engl. et Mendonça) Mziray ●☆

385457 Tecleopsis Hoyle et Leakey = Vepris Comm. ex A. Juss. ●☆

385458 Tecleopsis glandulosa Hoyle et Leakey = Vepris glandulosa (Hoyle et Leakey) Kokwaro ●☆

385459 Tecmarsis DC. = Vernonia Schreb. (保留属名)●■

385460 Tecmarsis bojeri DC. = Vernonia speiracephala Baker ●☆

385461 Tecoma Juss. (1789);黄钟花属（硬骨凌霄属）; Tecoma, Trumpet Flower,Trumpetbush,Trumpetcreeper,Yellowtrumpet ●

385462 Tecoma × smithii Hort. ;史密斯黄钟花;Orange Bells ●☆

385463 Tecoma africana (Lam.) G. Don = Kigelia africana (Lam.) Benth. ●☆

385464 Tecoma alata DC. ;翅黄钟花;Orange Bells ●☆

385465 Tecoma australis R. Br. = Bignonia pandorana Andréws ●☆

385466 Tecoma australis R. Br. = Pandorea pandorana Steenis ●☆

385467 Tecoma bipinnata Collett et Hemsl. = Pauldopia ghorta (Buch. -Ham. ex G. Don) Steenis ●

385468 Tecoma brycei N. E. Br. = Podranea brycei (N. E. Br.) Sprague ●☆

385469 Tecoma capensis (Thunb.) Lindl. = Tecomaria capensis (Thunb.) Spach ●

385470 Tecoma capensis (Thunb.) Spach = Tecomaria capensis (Thunb.) Spach ●

385471 Tecoma capensis Lindl. = Tecomaria capensis (Thunb.) Spach ●

385472 Tecoma castanifolia (D. Don) Melch. ;栗叶黄钟花; Chestnutleaf Trumpetbush ●☆

385473 Tecoma cavaleriei H. Lév. = Staphylea holocarpa Hemsl. ●

385474 Tecoma chinensis (Lam.) K. Koch = Campsis grandiflora (Thunb.) K. Schum. ●

385475 Tecoma chinensis K. Koch = Campsis grandiflora (Thunb.) Loisel. ●

385476 Tecoma dendrophila Blume = Tecomanthe dendrophila (Blume) K. Schum. ■☆

385477 Tecoma garrocha Hieron. ;卡拉查黄钟花;Argentine Tecoma, Garrocha,Guaran Colorado ●☆

385478 Tecoma grandiflora Loisel. = Campsis grandiflora (Thunb.) K. Schum. ●

385479 Tecoma impetiginosa Mart. ex DC. = Tabebuia impetiginosa (Mart. ex DC.) Standl. ●☆

385480 Tecoma ipe Mart. ex K. Schum. ;伊佩黄钟花●☆

385481 Tecoma jasminoides Lindl. ;洋凌霄●

385482 Tecoma mairei H. Lév. = Incarvillea delavayi Bureau et Franch. ■

385483 Tecoma mairei H. Lév. = Incarvillea mairei (H. Lév.) Grierson ■

385484 Tecoma nyassae Oliv. = Tecomaria capensis (Thunb.) Spach subsp. nyassae (Oliv.) Brummitt ●☆

385485 Tecoma nyikensis Baker = Tecomaria capensis (Thunb.) Spach subsp. nyassae (Oliv.) Brummitt ●☆

385486 Tecoma pentaphylla (L.) A. DC. = Tabebuia heterophylla (DC.) Britton ●☆

385487 Tecoma radicans (L.) Juss. = Campsis radicans (L.) Seem. ●

385488 Tecoma radicans Juss. = Campsis radicans (L.) Seem. ●

385489 Tecoma radicans Juss. ex Spreng. = Campsis radicans (L.) Seem. ex Bureau ●

385490 Tecoma ricasoliana Tanfani = Podranea ricasoliana (Tanfani) Sprague ●

385491 Tecoma shirensis Baker = Tecomaria capensis (Thunb.) Spach subsp. nyassae (Oliv.) Brummitt ●☆

385492 Tecoma stans (L.) Griseb. = Tecoma stans (L.) Juss. ex Kunth ●☆

385493 Tecoma stans (L.) Juss. ex Kunth;直立黄钟花（黄钟花,直立钟花）; Florida Yellowtrumpet, Trumpet Bush, Trumpet Flower, Trumpetcreeper,Yellow Bells, Yellow Elder, Yellow Trumpet Bush, Yellow Trumpet Flower,Yellow Trumpet Tree,Yellowbells ●☆

385494 Tecoma stans (L.) Juss. ex Kunth var. velutina DC. ;绒毛黄钟花●☆

385495 Tecoma stans (L.) Kunth = Tecoma stans (L.) Juss. ex Kunth ●☆

385496 Tecoma stans Juss. = Tecoma stans (L.) Juss. ex Kunth ●☆

385497 Tecoma tenuiflora (DC.) Fabris;细花黄钟花●☆

385498 Tecoma undulata (Roxb.) G. Don = Tecomella undulata (Roxb.) Seem. ●☆

385499 Tecoma undulata G. Don = Tecomella undulata (Sm.) Seem. ●☆

385500 Tecoma whytei C. H. Wright = Tecomaria capensis (Thunb.) Spach subsp. nyassae (Oliv.) Brummitt ●☆

385501 Tecomanthe Baill. (1888);南洋凌霄属■☆

385502 Tecomanthe dendrophila (Blume) K. Schum. ;喜树南洋凌霄●☆

385503 Tecomanthe speciosa W. R. B. Oliv. et J. A. Hunter;南洋凌霄■☆

385504 Tecomaria (Endl.) Spach = Tecoma Juss. ●

385505 Tecomaria (Endl.) Spach (1840);硬骨凌霄属; Cape Honeysuckle,Honeysuckle,Tecomaria,Yellowbells ●

385506 Tecomaria Spach = Tecoma Juss. ●

385507 Tecomaria Spach = Tecomaria (Endl.) Spach ●

385508 Tecomaria capensis (Thunb.) Spach;硬骨凌霄(得克马树,南非凌霄花,铁角凌霄,硬枝爆竹花,竹林标);Cape Honey Suckle, Cape Honeysuckle ●

385509 Tecomaria capensis (Thunb.) Spach 'Aurea';黄花硬骨凌霄●☆

385510 Tecomaria capensis (Thunb.) Spach subsp. nyassae (Oliv.) Brummitt;尼萨硬骨凌霄●☆

385511 Tecomaria capensis (Thunb.) Spach var. flava Verdc. ;黄硬骨凌霄●☆

385512 Tecomaria krebsii Klotzsch = Tecoma capensis (Thunb.) Lindl. ●

385513 Tecomaria nyassae (Oliv.) K. Schum. = Tecomaria capensis (Thunb.) Spach subsp. nyassae (Oliv.) Brummitt ●☆

385514 Tecomaria petersii Klotzsch = Tecoma capensis (Thunb.) Lindl. ●

385515 Tecomaria rupium Bullock = Tecomaria capensis (Thunb.) Spach subsp. nyassae (Oliv.) Brummitt ●☆

385516 Tecomaria shirensis (Baker) K. Schum. = Tecomaria capensis (Thunb.) Spach subsp. nyassae (Oliv.) Brummitt ●☆

385517　Tecomella Seem. (1863);小黄钟花属●☆

385518　Tecomella undulata (Roxb.) Seem. = Tecomella undulata (Sm.) Seem. ●☆

385519　Tecomella undulata (Sm.) Seem.;小黄钟花●☆

385520　Tecophilaea Bertero ex Colla(1836);蒂可花属;Blue Crocus, Chilean Blue Crocus,Chilean Crocus ■☆

385521　Tecophilaea cyanocrocus Leyb.;蓝蒂可花(智利蓝番红花); Chilean Crocus ■☆

385522　Tecophilaeaceae Leyb. (1862)(保留科名);蒂可花科(百鸢科,基叶草科)■☆

385523　Tecophilea Herb. = Tecophilaea Bertero ex Colla ■☆

385524　Tectania Spreng. = Tectona L. f. (保留属名)●

385525　Tecticornia Hook. f. (1880);肉被盐角草属■☆

385526　Tecticornia cinerea Hook. f.;肉被盐角草■☆

385527　Tectiphiala H. E. Moore. (1978);碟苞椰属(披针叶刺椰属)●☆

385528　Tectiphiala ferox H. E. Moore;披针刺椰●☆

385529　Tectona L. f. (1782)(保留属名);柚木属;Teak ●

385530　Tectona australis W. Hill;澳洲柚木;Australian Beech ●☆

385531　Tectona grandis L. f.;柚木(麻栗,埋桑,乌楠,硬木树,脂树,紫柚木);Burma Teak, Common Teak, Indian Teak-tree, Moulmein Teak,Rangoon,Teak ●

385532　Tectona philippinensis Benth. et Hook.;菲岛柚木;Philippin Teak ●☆

385533　Tectona theka Lour. = Tectona grandis L. f. ●

385534　Tectonia Spreng. = Tectona L. f. (保留属名)●

385535　Tecunumania Standl. et Steyerm. (1944);特库瓜属☆

385536　Tecunumania quetzalteca Standl. et Steyerm.;特库瓜☆

385537　Tedingea D. Müll. -Doblies et U. Müll. -Doblies = Strumaria Jacq. ■☆

385538　Tedingea D. Müll. -Doblies et U. Müll. -Doblies(1985);特丁石蒜属■☆

385539　Tedingea pygmaea (Snijman) D. Müll. -Doblies et U. Müll. -Doblies = Strumaria pygmaea Snijman ■☆

385540　Tedingea spiralis (F. M. Leight.) D. Müll. -Doblies et U. Müll. -Doblies = Strumaria pygmaea Snijman ■☆

385541　Tedingea tenella (L. f.) D. Müll. -Doblies et U. Müll. -Doblies = Strumaria tenella (L. f.) Snijman ■☆

385542　Tedingea transkarooica D. Müll. -Doblies et U. Müll. -Doblies = Strumaria tenella (L. f.) Snijman subsp. orientalis Snijman ■☆

385543　Teedea Post et Kuntze = Teedia Rudolphi ●■☆

385544　Teedia Rudolphi(1800);梯玄参属●■☆

385545　Teedia lucida (Sol.) Rudolphi;光亮梯玄参●☆

385546　Teedia pentheri Gand. = Teedia pubescens Burch. ■☆

385547　Teedia pubescens Burch.;短柔毛梯玄参■☆

385548　Teesdalea Asch. = Teesdalia R. Br. ■☆

385549　Teesdalia R. Br. (1812);野屈曲花属;Shepherd's Cress ■☆

385550　Teesdalia W. T. Aiton = Guepinia Bastard ■☆

385551　Teesdalia W. T. Aiton = Teesdalia R. Br. ■☆

385552　Teesdalia coronopifolia (Steud.) Thell.;小野屈曲花;Lesser Shepherd's Cress, Lesser Shepherdscress ■☆

385553　Teesdalia iberis DC. = Teesdalia nudicaulis (L.) R. Br. ■☆

385554　Teesdalia lepidium DC. = Teesdalia nudicaulis (L.) R. Br. ■☆

385555　Teesdalia nudicaulis (L.) R. Br.;野屈曲花;Barestem Teesdalia, Naked-stalked Candytuft, Naked-stalked Rock-cress, Shepherd's Cress ■☆

385556　Teesdaliopsis (Willk.) Rothm. (1940);密生屈曲花属■☆

385557　Teesdaliopsis conferta (Lag.) Rothm.;密生屈曲花■☆

385558　Teganium Schmidel = Nolana L. ex L. f. ■☆

385559　Teganocharis Hochst. = Tenagocharis Hochst. ■☆

385560　Tegicornia Paul G. Wilson(1980);异株盐角草属■☆

385561　Tegicornia uniflora Paul G. Wilson;异株盐角草■☆

385562　Tegneria Lilja = Calandrinia Kunth(保留属名)■☆

385563　Tehuana Panero et Villasenor(1997);长托菊属■☆

385564　Tehuana calzadae Panero et Villasenor;长托菊■☆

385565　Teichmeyeria Scop. = Gustavia L. (保留属名)●☆

385566　Teichmeyeria Scop. = Japarandiba Adans. (废弃属名)●☆

385567　Teichostemma R. Br. = Vernonia Schreb. (保留属名)●■

385568　Teijsmannia Post et Kuntze = Pottsia Hook. et Arn. ●

385569　Teijsmannia Post et Kuntze = Teysmannia Miq. ●

385570　Teijsmanniodendron Koord. (1904);泰树属●☆

385571　Teijsmanniodendron bogoriense Koord.;泰树●☆

385572　Teijsmanniodendron glabrum Merr.;光泰树●☆

385573　Teijsmanniodendron longifolium Merr.;长叶泰树●☆

385574　Teijsmanniodendron monophyllum Kurata;单叶泰树●☆

385575　Teijsmanniodendron pteropodum Bakh.;翅梗泰树●☆

385576　Teijsmanniodendron unifoliolatum (Merr.) Moldenke;单小叶泰树●☆

385577　Teinosolen Hook. f. = Heterophyllaea Hook. f. ●☆

385578　Teinostachyum Munro = Schizostachyum Nees ●

385579　Teinostachyum Munro(1868);长穗竹属(疏穗竹属)●☆

385580　Teinostachyum attenuatum Munro;环长穗竹●☆

385581　Teinostachyum griffithii Munro;长穗竹●☆

385582　Teinostachyum maculatum Trim.;斑点长穗竹●☆

385583　Teinostachyum schisostachyoides Kurz;裂穗长穗竹●☆

385584　Teixeiranthus R. M. King et H. Rob. (1980);点腺菊属■☆

385585　Teixeiranthus foliosus (Gardn.) R. M. King et H. Rob.;多叶点腺菊■☆

385586　Teixeiranthus pohlii (Baker) R. M. King et H. Rob.;点腺菊■☆

385587　Tekel Adans. (废弃属名) = Libertia Spreng. (保留属名)■☆

385588　Tekelia Adans. ex Kuntze = Tekel Adans. (废弃属名)■☆

385589　Tekelia Kuntze = Tekel Adans. (废弃属名)■☆

385590　Tekelia Scop. = Argania Roem. et Schult. (保留属名)●☆

385591　Tektona L. f. = Tectona L. f. (保留属名)●

385592　Telanthera R. Br. (1818);织锦苋属■

385593　Telanthera R. Br. = Alternanthera Forssk. ■

385594　Telanthera amoena (Lem.) Regel = Alternanthera ficoidea (L.) Sm. var. amoena (Lem.) L. B. Sm. et Downs ■

385595　Telanthera amoena Regel = Alternanthera ficoidea (L.) Sm. var. amoena (Lem.) L. B. Sm. et Downs ■

385596　Telanthera bettzickiana Regel = Alternanthera bettzickiana (Regel) G. Nicholson ■

385597　Telanthera bettzickiana Regel = Alternanthera paronychioides A. St. -Hil. var. bettzickiana (Regel) Fosberg ■☆

385598　Telanthera ficoidea (L.) Moq. var. versicolor Lem. = Alternanthera bettzickiana (Regel) G. Nicholson ■

385599　Telanthera maritima (Mart.) Moq. var. sparmannii Moq. = Alternanthera littoralis P. Beauv. var. sparmannii (Moq.) Pedersen ■☆

385600　Telanthera philoxeroides (C. Mart.) Moq. = Alternanthera philoxeroides (Mart.) Griseb. ■

385601　Telanthera versicolor (Lem.) Regel = Alternanthera versicolor (Lem.) Regel ■

385602　Telanthera versicolor Regel = Alternanthera versicolor (Lem.) Regel ■

385603　Telanthophora H. Rob. et Brettell(1974);顶叶千里光属●☆

385604　Telanthophora andrieuxii (DC.) H. Rob. et Brettell;顶叶千里光●☆

385605　Telanthophora arborescens (Steetz) H. Rob. et Brettell;树顶叶千里光●☆

385606　Telanthophora grandifolia (Less.) H. Rob. et Brettell;墨西哥顶叶千里光●☆

385607　Telectadium Baill.(1889);东南亚杠柳属●☆

385608　Telectadium edule Baill.;东南亚杠柳●☆

385609　Telectadium linearicarpum Pierre;线果东南亚杠柳●☆

385610　Teleiandra Nees et Meyen = Ocotea Aubl.●☆

385611　Teleiandra Nees et Meyen ex Nees = Ocotea Aubl.●☆

385612　Teleianthera Endl. = Alternanthera Forssk.■

385613　Teleianthera Endl. = Telanthera R. Br.■

385614　Telekia Baumg.(1816);泰氏菊属(蒂立菊属,特勒菊属);Oxeye,Yellow Oxeye■☆

385615　Telekia africana Hook. f. = Anisopappus chinensis Hook. et Arn. subsp. africanus (Hook. f.) S. Ortiz et Paiva■☆

385616　Telekia speciosa (Schreb.) Baumg.;泰氏菊(蒂立菊,美丽特勒菊,心叶牛眼菊);Heartleaf Oxeye, Heart-leaved Oxeye, Oxeye, Sunwheel, Yellow Oxeye■☆

385617　Telekia speciosa (Schreb.) Baumg. = Buphthalmum speciosum Schreb.■☆

385618　Telelophus Dulac = Dethawia Endl.●☆

385619　Telelophus Dulac = Seseli L.■

385620　Telelophus Dulac = Wallrothia Spreng.■☆

385621　Telemachia Urb. = Elaeodendron J. Jacq.●☆

385622　Telephiaceae Link = Aizoaceae Martinov(保留科名)●■

385623　Telephiaceae Link = Caryophyllaceae Juss. (保留科名)■●

385624　Telephiaceae Martinov = Caryophyllaceae Juss. (保留科名)●■

385625　Telephiastrum Fabr. = Anacampseros L. (保留属名)■☆

385626　Telephioides Ortega = Andrachne L.●☆

385627　Telephium Hill = Hylotelephium H. Ohba■

385628　Telephium Hill = Sedum L.●■

385629　Telephium L.(1753);耳托指甲草属;Orpine■☆

385630　Telephium barbeyanum Bornm.;巴比耳托指甲草■☆

385631　Telephium compressa (Desf.) Fisher et C. A. Mey. = Petrorhagia illyrica (Ard.) P. W. Ball et Heywood subsp. angustifolia (Poir.) P. W. Ball et Heywood■☆

385632　Telephium compressa (Desf.) Fisher et C. A. Mey. var. australis Batt. = Petrorhagia illyrica (Ard.) P. W. Ball et Heywood subsp. angustifolia (Poir.) P. W. Ball et Heywood■☆

385633　Telephium exiguum Batt. = Telephium sphaerospermum Boiss.■☆

385634　Telephium imperati L.;非洲耳托指甲草■☆

385635　Telephium imperati L. subsp. orientale (Boiss.) Nyman = Telephium orientale Boiss.■☆

385636　Telephium imperati L. var. pseudorientale Maire = Telephium imperati L.■☆

385637　Telephium oligospermum Boiss.;寡籽耳托指甲草■☆

385638　Telephium orientale Boiss.;东方耳托指甲草■☆

385639　Telephium sphaerospermum Boiss.;圆籽耳托指甲草■☆

385640　Telephium sphaerospermum Boiss. var. barbeyanum (Bornm.) Maire et Weiller = Telephium barbeyanum Bornm.■☆

385641　Telephium sphaerospermum Boiss. var. boissieri Maire et Weiller = Telephium sphaerospermum Boiss.■☆

385642　Telesia Raf. = Zexmenia La Llave●■☆

385643　Telesilla Klotzsch(1849);特莱斯萝藦属■☆

385644　Telesilla cynanchioides Klotzsch;特莱斯萝藦■☆

385645　Telesmia Raf. = Salix L. (保留属名)●

385646　Telesonix Raf. = Boykinia Nutt. (保留属名)●■☆

385647　Telestria Raf. = Bauhinia L.●

385648　Telfairia Hook. = Telfairia Newman ex Hook.■☆

385649　Telfairia Newm. ex Hook. = Byttneria Loefl. (保留属名)●

385650　Telfairia Newman ex Hook.(1827);特非瓜属(太肥瓜属,特费瓜属);Oil Vine, Oil-vine■☆

385651　Telfairia africana (Delile) A. Chev.;非洲特非瓜■☆

385652　Telfairia africana (Delile) A. Chev. = Telfairia pedata (Sims) Hook.■☆

385653　Telfairia batesii Rabenant.;贝茨特非瓜■☆

385654　Telfairia occidentalis Hook. f.;西部特非瓜;Fluted Pumpkin, Krobonko, Oyster-nut Vine■☆

385655　Telfairia pedata (Sims) Hook.;鸟足特非瓜(鸟足特费瓜);Kweme Nut, Oyster Nut, Oysternut, Zanzibar Oil Vine■☆

385656　Telfairia pedata Hook. = Telfairia pedata (Sims) Hook.■☆

385657　Telfairia volubilis Newman ex Hook.;特非瓜(太肥瓜)■☆

385658　Telina Dttr. et Jacks. = Chamaespartium Adans.●

385659　Telina Dttr. et Jacks. = Teline Medik.●☆

385660　Telina E. Mey. = Lotononis (DC.) Eckl. et Zeyh. (保留属名)■

385661　Telina brevifolia Eckl. et Zeyh. ex Drège = Lotononis stricta (Eckl. et Zeyh.) B.-E. van Wyk■☆

385662　Telina capnitidis E. Mey. = Lotononis capnitidis (E. Mey.) Benth. ex Walp.■☆

385663　Telina cytisoides E. Mey. = Lotononis stricta (Eckl. et Zeyh.) B.-E. van Wyk■☆

385664　Telina eriocarpa E. Mey. = Lotononis eriocarpa (E. Mey.) B.-E. van Wyk■☆

385665　Telina genuflexa E. Mey. = Lotononis divaricata (Eckl. et Zeyh.) Benth.■☆

385666　Telina varia E. Mey. = Lotononis varia (E. Mey.) Steud.■☆

385667　Telina villosa E. Mey. = Lotononis villosa (E. Mey.) Steud.■☆

385668　Telinaria C. Presl = Teline Medik.●☆

385669　Teline Medik.(1786);同金雀花属●☆

385670　Teline Medik. = Chamaespartium Adans.●

385671　Teline Medik. = Cytisus Desf. (保留属名)●

385672　Teline Medik. = Cytisus L. (废弃属名)●

385673　Teline Webb = Genista L.●

385674　Teline canariensis (L.) Webb et Berthel.;加那利同金雀花●☆

385675　Teline linifolia (L.) Webb et Berthel.;亚麻叶同金雀花●☆

385676　Teline maderensis Webb et Berthel.;梅德同金雀花●☆

385677　Teline maderensis Webb et Berthel. var. paivae (Lowe) del Arco = Teline maderensis Webb et Berthel.●☆

385678　Teline microphylla (DC.) Gibbs et Dingwall;小叶同金雀花●☆

385679　Teline monspessulana (L.) K. Koch = Cytisus monspessulanus L.●☆

385680　Teline monspessulana (L.) K. Koch = Genista monspessulana (L.) L. A. S. Johnson●☆

385681　Teline nervosa (Esteve) A. Hansen et Sunding;多脉同金雀花●☆

385682　Teline osmarensis (Coss.) Gibbs et Dingwall;奥马同金雀花●☆

385683　Teline osyroides (Svent.) Gibbs et Dingwall;沙针同金雀花●☆

385684　Teline osyroides (Svent.) Gibbs et Dingwall subsp. sericea (Kuntze) del Arco et Acebes;绢毛同金雀花●☆

385685　Teline patens (DC.) Talavera et Gibbs;铺展同金雀花●☆

385686　Teline rosmarinifolia Webb et Berthel.;迷迭香叶同金雀花●☆

385687　Teline rosmarinifolia Webb et Berthel. var. eurifolia del Arco =

Teline rosmarinifolia Webb et Berthel. ●☆

385688　Teline salsoloides del Arco et Acebes；猪毛菜同金雀花●☆

385689　Teline segonnei (Maire) Raynaud；塞贡同金雀花●☆

385690　Teline splendens (Webb et Berthel.) del Arco；光亮同金雀花●☆

385691　Teline stenopetala (Webb et Berthel.) Webb et Berthel.；窄瓣同金雀花●☆

385692　Teline stenopetala (Webb et Berthel.) Webb et Berthel. var. microphylla (Pit. et Proust) Gibbs et Dingwall = Teline stenopetala (Webb et Berthel.) Webb et Berthel. ●☆

385693　Teline stenopetala (Webb et Berthel.) Webb et Berthel. var. pauciovulata del Arco = Teline stenopetala (Webb et Berthel.) Webb et Berthel. ●☆

385694　Teline stenopetala (Webb et Berthel.) Webb et Berthel. var. sericea (Pit. et Proust) del Arco = Teline stenopetala (Webb et Berthel.) Webb et Berthel. ●☆

385695　Teline stenopetala (Webb et Berthel.) Webb et Berthel. var. spachiana (Webb) del Arco = Teline stenopetala (Webb et Berthel.) Webb et Berthel. ●☆

385696　Teliostachya Nees(1847)；冬穗爵床属■☆

385697　Teliostachya alopecuroidea (Vahl) Nees = Lepidagathis alopecuroides (Vahl) R. Br. ex Griseb. ■☆

385698　Teliostachya cataractae Nees；冬穗爵床☆

385699　Teliostachya hyssopifolia Benth. = Lepidagathis alopecuroides (Vahl) R. Br. ex Griseb. ■☆

385700　Teliostachya laguroidea Nees = Lepidagathis alopecuroides (Vahl) R. Br. ex Griseb. ■☆

385701　Telipodus Raf. = Philodendron Schott(保留属名)■●

385702　Telipogon Kunth = Telipogon Mutis ex Kunth ■☆

385703　Telipogon Mutis ex Kunth(1816)；毛顶兰属■☆

385704　Telipogon alberti Rchb. f. ；阿氏毛顶兰■☆

385705　Telipogon angustifolius Kunth；窄叶毛顶兰■☆

385706　Telipogon atropurpureus D. E. Benn. et Ric. Fernández；暗紫毛顶兰■☆

385707　Telipogon aureus Lindl. ；黄毛顶兰■☆

385708　Telipogon australis Dodson et Hirtz；南方毛顶兰■☆

385709　Telipogon gracilipes Schltr. ；细梗毛顶兰■☆

385710　Telipogon gracilis Schltr. ；细毛顶兰■☆

385711　Telipogon minutiflorus Kraenzl. ；小花毛顶兰■☆

385712　Telipogon monticola L. O. Williams；山地毛顶兰■☆

385713　Telipogon obovatus Lindl. ；倒卵毛顶兰■☆

385714　Telipogon polyneuros Rchb. f. ex Kraenzl. ；多脉毛顶兰■☆

385715　Telis Kuntze = Trigonella L. ■

385716　Telis cachemiriana Kuntze = Trigonella cachemiriana Cambess. ■

385717　Telitoxicum Moldenke(1938)；矛毒藤属●☆

385718　Telitoxicum minutiflorum (Diels) Moldenke；矛毒藤●☆

385719　Tellima R. Br. (1823)；穗杯花属(饰缘花属,特罗马属,新唢呐草属)；Fringe Cups, Fringecups ☆

385720　Tellima grandiflora (Pursh) Douglas ex Lindl. ；大穗杯花(饰缘花,新唢呐草)；Big Flower Tellima, Fringe Cups, Fringecups ☆

385721　Tellima grandiflora R. Br. = Tellima grandiflora (Pursh) Douglas ex Lindl. ☆

385722　Telmatophace Schleid. = Lemna L. ■

385723　Telmatophila Ehrh. = Scheuchzeria L. ■

385724　Telmatophila Mart. ex Baker(1873)；少花瘦片菊属■☆

385725　Telmatophila scolymastrum Mart. ex Baker；少花瘦片菊■☆

385726　Telmatosphace Ball = Lemna L. ■

385727　Telmatosphace Ball = Telmatophace Schleid. ■

385728　Telminostelma E. Fourn. (1885)；鹅绒藤萝藦属■☆

385729　Telminostelma roulinioides Fourn. ；鹅绒藤萝藦■☆

385730　Telmissa Fenzl = Sedum L. ●■

385731　Telogyne Baill. = Trigonostemon Blume(保留属名)●

385732　Telopaea Parkinson = Aleurites J. R. Forst. et G. Forst. ●

385733　Telopaea Sol. ex Parkinson = Aleurites J. R. Forst. et G. Forst. ●

385734　Telopea R. Br. (1810)(保留属名)；蒂罗花属(泰洛帕属)；Waratah ●☆

385735　Telopea Sol. ex Baill. = Aleurites J. R. Forst. et G. Forst. ●

385736　Telopea mongaensis Cheel；大蒂罗花(大花泰洛帕)●☆

385737　Telopea oreades F. Muell. ；湿生蒂罗花(湿生泰洛帕)；Gippsland Waratah ●☆

385738　Telopea speciosissima (Sm.) R. Br. ；蒂罗花(极美泰洛帕,瓦格塔)；Kiwi Rose, New South Wales Waratah, Waratah, Warratau ●☆

385739　Telopea speciosissima (Sm.) R. Br. 'Corroboree'；庆典蒂罗花(庆典极美泰洛帕)●☆

385740　Telopea speciosissima (Sm.) R. Br. 'Flaming Beacon'；灯塔极美泰洛帕(闪烁的灯塔极美泰洛帕)●☆

385741　Telopea speciosissima (Sm.) R. Br. 'Olympic Flame'；奥运圣火蒂罗花(奥运圣火极美泰洛帕)●☆

385742　Telopea speciosissima (Sm.) R. Br. 'Wirrimbirra White'；维林比拉白蒂罗花(维林比拉白极美泰洛帕)●☆

385743　Telopea truncata (Labill.) R. Br. ；截叶蒂罗花(截形泰洛帕,塔斯马尼亚泰洛帕)；Tasmanian Waratah ●☆

385744　Telophyllum Tiegh. = Myzodendron Sol. ex DC. ●☆

385745　Telopogon Mutis ex Spreng. = Telipogon Mutis ex Kunth ■☆

385746　Telosiphonia (Woodson) Henrickson = Echites P. Browne ●☆

385747　Telosiphonia (Woodson) Henrickson；远管木属●☆

385748　Telosiphonia macrosiphon (Torr.) Henrickson；大管远管木●☆

385749　Telosma Coville(1905)；夜来香属(夜香花属)；Telosma ●

385750　Telosma africana (N. E. Br.) N. E. Br. ；非洲夜来香●☆

385751　Telosma cathayensis Merr. = Telosma procumbens (Blanco) Merr. ●

385752　Telosma cordata (Burm. f.) Merr. ；夜来香(卵叶尖槐藤,浅色夜香花,夜兰香,夜香花)；Chinese Violet, Cordate Telosma, Fragrant Telosma, Ovateleaf Oxystelma, Telosma, Tonkin Creeper ●

385753　Telosma minor (Andréws) Craib = Telosma cordata (Burm. f.) Merr. ●

385754　Telosma minor Craib = Telosma cordata (Burm. f.) Merr. ●

385755　Telosma odoratissima (Lour.) Coville = Telosma cordata (Burm. f.) Merr. ●

385756　Telosma odoratissima Coville = Telosma cordata (Burm. f.) Merr. ●

385757　Telosma pallida (Roxb.) Craib；台湾夜来香(浅色夜香花,夜香花)；Taiwan Telosma ●

385758　Telosma pallida (Roxb.) Craib = Telosma cordata (Burm. f.) Merr. ●

385759　Telosma procumbens (Blanco) Merr. ；卧茎夜来香(华南夜香)；Creeping Telosma, S. China Telosma, South China Telosma ●

385760　Telosma unyorensis S. Moore；乌尼奥尔夜来香●☆

385761　Telotia Pierre = Pycnarrhena Miers ex Hook. f. et Thomson ●

385762　Telotia nodiflora Pierre = Pycnarrhena lucida (Teijsm. et Binn.) Miq. ●

385763　Teloxis Rchb. = Teloxys Moq. ■

385764　Teloxys Moq. (1834)；针藜属(针尖藜属)■

385765　Teloxys Moq. = Chenopodium L. ■●

385766　Teloxys Moq. = Dysphania R. Br. ■

385767　Teloxys ambrosioides（L.）W. A. Weber ＝ Dysphania ambrosioides（L.）Mosyakin et Clemants ■

385768　Teloxys aristata（L.）Moq. ＝ Chenopodium aristatum L. ■

385769　Teloxys aristata（L.）Moq. ＝ Dysphania aristata（L.）Mosyakin et Clemants ■

385770　Teloxys aristata Moq. ＝ Chenopodium aristatum L. ■

385771　Teloxys asiatica（L.）Moq.；针藜■

385772　Teloxys botrys（L.）W. A. Weber ＝ Dysphania botrys（L.）Mosyakin et Clemants ■

385773　Teloxys foetida Kitag. ＝ Chenopodium foetidum Schrad. ■

385774　Teloxys foetida Kitag. ＝ Dysphania schraderiana（Roem. et Schult.）Mosyakin et Clemants ■

385775　Teloxys graveolens（Willd.）W. A. Weber ＝ Dysphania graveolens（Willd.）Mosyakin et Clemants ■☆

385776　Teloxys multifida（L.）W. A. Weber ＝ Dysphania multifidum（L.）Mosyakin et Clemants ■☆

385777　Teloxys pumilio（R. Br.）W. A. Weber ＝ Dysphania pumilio（R. Br.）Mosyakin et Clemants ■☆

385778　Teloxys schraderiana（Roem. et Schult.）W. A. Weber ＝ Dysphania schraderiana（Roem. et Schult.）Mosyakin et Clemants ■

385779　Telukrama Raf. ＝ Swida Opiz ●

385780　Telukrama Raf. ＝ Thelycrania（Dumort.）Fourr. ●☆

385781　Tema Adans.（废弃属名）＝ Echinochloa P. Beauv.（保留属名）■

385782　Tema Adans.（废弃属名）＝ Setaria P. Beauv.（保留属名）■

385783　Temburongia S. Dransf. et K. M. Wong（1996）；加岛禾属■☆

385784　Temburongia simplex S. Dransf. et K. M. Wong；加岛禾■☆

385785　Temenia O. F. Cook ＝ Maximiliana Mart.（保留属名）●

385786　Temminekia de Vriese ＝ Scaevola L.（保留属名）●■

385787　Temmodaphne Kosterm.（1973）；泰樟属●☆

385788　Temmodaphne Kosterm. ＝ Cinnamomum Schaeff.（保留属名）●

385789　Temmodaphne thailandica Kosterm.；泰樟●☆

385790　Temnadenia Miers et Woodson ＝ Temnadenia Miers ●☆

385791　Temnadenia Miers（1878）；割腺夹竹桃属●☆

385792　Temnadenia annularis Miers；割腺夹竹桃●☆

385793　Temnadenia tomentosa Miers；毛割腺夹竹桃●☆

385794　Temnemis Raf. ＝ Carex L. ■

385795　Temnocalyx Robyns ＝ Temnocalyx Robyns ex Ridl. ☆

385796　Temnocalyx Robyns ex Ridl.（1928）；坦桑尼亚茜草属☆

385797　Temnocalyx ancylanthus（Hiern）Robyns var. puberulus Robyns ＝ Fadogia ancylantha Schweinf. ●☆

385798　Temnocalyx ancylanthus（Schweinf.）Robyns ＝ Fadogia ancylantha Schweinf. ●☆

385799　Temnocalyx fuchsioides（Oliv.）Robyns ＝ Fadogia fuchsioides Oliv. ●☆

385800　Temnocalyx nodulosus Robyns；多节坦桑尼亚茜草●☆

385801　Temnocalyx obovatus（N. E. Br.）Robyns ＝ Fadogia ancylantha Schweinf. ●☆

385802　Temnocalyx verdickii（De Wild. et T. Durand）Robyns ＝ Fadogia verdickii De Wild. et T. Durand ●☆

385803　Temnocydia Mart. ex DC. ＝ Bignonia L.（保留属名）●

385804　Temnolepis Baker ＝ Epallage DC. ■

385805　Temnolepis Baker（1887）；割鳞菊属■☆

385806　Temnolepis scrophulariifolia Baker；割鳞菊☆

385807　Temnopteryx Hook. f.（1873）；割翅茜属☆

385808　Temnopteryx sericea Hook. f.；割翅茜☆

385809　Temochloa S. Dransf.（2000）；泰草属■☆

385810　Templetonia R. Br. ＝ Templetonia R. Br. ex W. T. Aiton ●☆

385811　Templetonia R. Br. ex W. T. Aiton（1812）；傲慢木属（珊瑚豆属）●☆

385812　Templetonia retusa R. Br.；傲慢木（珊瑚豆）；Cockies' Tongues，Coral Bush，Coral Plant ●☆

385813　Temu O. Berg ＝ Blepharocalyx O. Berg ●☆

385814　Temus Molina ＝ ? Drimys J. R. Forst. et G. Forst.（保留属名）●☆

385815　Temus Molina（1782）；泰木属●☆

385816　Tenacistachya L. Liou（1847）；坚轴草属（新禾草属）；Hardgrass ■

385817　Tenacistachya minor L. Liou；小坚轴草；Minor Hardgrass ■

385818　Tenacistachya sichuanensis L. Liou；坚轴草；Hardgrass ■

385819　Tenageia（Rchb.）Rchb. ＝ Juncus L. ■

385820　Tenageia Ehrh. ＝ Juncus L. ■

385821　Tenagocharis Hochst. ＝ Butomopsis Kunth ■☆

385822　Tenagocharis latifolia（D. Don）Buchenau ＝ Butomopsis latifolia（D. Don）Kunth ■

385823　Tenaris E. Mey.（1838）；泰纳萝藦属■☆

385824　Tenaris browniana S. Moore；布朗泰纳萝藦■☆

385825　Tenaris chlorantha Schltr. ＝ Brachystelma chloranthum（Schltr.）Peckover ■☆

385826　Tenaris filifolia N. E. Br. ＝ Brachystelma filifolium（N. E. Br.）Peckover ■☆

385827　Tenaris rostrata N. E. Br. ＝ Brachystelma rubellum（E. Mey.）Peckover ■☆

385828　Tenaris rubella E. Mey. ＝ Brachystelma rubellum（E. Mey.）Peckover ■☆

385829　Tenaris schultzei（Schltr.）E. Phillips ＝ Brachystelma schultzei（Schltr.）Bruyns ■☆

385830　Tenaris simulans N. E. Br. ＝ Brachystelma rubellum（E. Mey.）Peckover ■☆

385831　Tenaris somalensis（Schltr.）N. E. Br. ＝ Caralluma priogonium K. Schum. ■☆

385832　Tenaris subaphylla（K. Schum.）N. E. Br. ＝ Caralluma edulis（Edgew.）Benth. ■☆

385833　Tenaris volkensii K. Schum. ＝ Brachystelma rubellum（E. Mey.）Peckover ■☆

385834　Tendana Rchb. ＝ Piperella（C. Presl ex Rchb.）Spach ●

385835　Tendana Rchb. f. ＝ Micromeria Benth.（保留属名）■●

385836　Tengia Chun（1946）；世纬苣苔属（黔苣苔属）；Tengia ■★

385837　Tengia potiflora S. Z. He ＝ Tengia scopulorum Chun var. potiflora（S. Z. He）W. T. Wang ■

385838　Tengia scopulorum Chun；世纬苣苔（黔苣苔）；Guizhou Tengia，Tengia ■

385839　Tengia scopulorum Chun var. potiflora（S. Z. He）W. T. Wang；壶花世纬苣苔（壶花黔苣苔）■

385840　Tenicroa Raf.（1837）；泰尼风信子属■☆

385841　Tenicroa Raf. ＝ Drimia Jacq. ex Willd. ■☆

385842　Tenicroa Raf. ＝ Urginea Steinh. ■☆

385843　Tenicroa exuviata（Jacq.）Speta ＝ Drimia exuviata（Jacq.）Jessop ■☆

385844　Tenicroa filifolia（Jacq.）Oberm. ＝ Drimia filifolia（Jacq.）J. C. Manning et Goldblatt ■☆

385845　Tenicroa fragrans（Jacq.）Raf. ＝ Drimia fragrans（Jacq.）J. C. Manning et Goldblatt ■☆

385846　Tenicroa multifolia（G. J. Lewis）Oberm. ＝ Drimia multifolia（G. J. Lewis）Jessop ■☆

385847 Tennantia Verdc. (1981);特南茜属●☆

385848 Tennantia sennii (Chiov.) Verdc. et Bridson;特南茜●☆

385849 Tenorea C. Koch = Bupleurum L. ●■

385850 Tenorea C. Koch = Tenoria Spreng. ●■

385851 Tenorea Colla = Trixis P. Browne ■●☆

385852 Tenorea Gasp. = Ficus L. ●

385853 Tenorea K. Koch = Bupleurum L. ●■

385854 Tenorea K. Koch = Tenoria Spreng. ●■

385855 Tenorea Raf. = Zanthoxylum L. ●

385856 Tenoria Dehnh. et Giord. = Hygrophila R. Br. ●■

385857 Tenoria Spreng. = Bupleurum L. ●■

385858 Tenrhynea Hilliard et B. L. Burtt(1981);密头紫绒草属■☆

385859 Tenrhynea phylicifolia (DC.) Hilliard et B. L. Burtt;密头紫绒草■☆

385860 Teonongia Stapf = Streblus Lour. ●

385861 Teonongia tonkinensis (Dubard et Eberh.) Stapf = Streblus tonkinensis (Eberh. et Dubard) Corner ●

385862 Teonongia tonkinensis (Eberh. et Dubard) Stapf = Streblus tonkinensis (Eberh. et Dubard) Corner ●

385863 Tepesia C. F. Gaertn. = Hamelia Jacq. ●

385864 Tephea Delile = Olinia Thunb. ●☆

385865 Tephea aequipetala Delile = Olinia rochetiana Juss. ●☆

385866 Tephis Adans. = Polygonum L. (保留属名)■●

385867 Tephis Raf. = Atraphaxis L. ●

385868 Tephranthus Neck. = Phyllanthus L. ●■

385869 Tephras E. Mey. ex Harv. et Sond. = Galenia L. ●☆

385870 Tephrocactus Lem. (1868);灰球掌属(球形节仙人掌属)■☆

385871 Tephrocactus Lem. = Opuntia Mill. ●

385872 Tephrocactus alexanderi (Britton et Rose) Backeb.;蛮将殿■☆

385873 Tephrocactus articulatus Backeb.;仙人结;Paper Spine Cactus ■☆

385874 Tephrocactus articulatus Backeb. var. diadematus Backeb.;饰冠灰球掌;Paper Spine Cactus, Spruce Cone Cholla ■☆

385875 Tephrocactus articulatus Backeb. var. papyacanthus Backeb.;银刺仙人结;Paper Spine Cactus ■☆

385876 Tephrocactus glomeratus (Haw.) Speg.;姬武藏野■☆

385877 Tephrocactus glomeratus (Haw.) Speg. var. fulvispinus (Lem.) Backeb.;九万王■☆

385878 Tephrocactus kuehnrichianus (Werderm. et Backeb.) Backeb.;白骨城■☆

385879 Tephrocactus kuehnrichianus var. applanatus (Werderm. et Backeb.) Backeb.;平展白骨城■☆

385880 Tephrocactus molinensis (Speg.) Backeb.;蛸壶■☆

385881 Tephrocactus rauhii Backeb.;大酋长■☆

385882 Tephrocactus sphaericus (C. F. Först.) Backeb.;寿星仙人掌■☆

385883 Tephrocactus udonis (Weing.) Backeb.;塞翁团扇■☆

385884 Tephroseris (Rchb.) Rchb. (1841);狗舌草属;Dogtongueweed, Fleawort, Tephroseris ■

385885 Tephroseris Rchb. = Senecio L. ■●

385886 Tephroseris Rchb. = Tephroseris (Rchb.) Rchb. ■

385887 Tephroseris adenolepis C. Jeffrey et Y. L. Chen;腺苞狗舌草;Glandbract Dogtongueweed, Glandular Phyllaries Tephroseris ■

385888 Tephroseris atropurpurea (Ledeb.) Holub subsp. frigida (Richardson) Á. Löve et D. Löve = Tephroseris frigida (Richardson) Holub ■☆

385889 Tephroseris atropurpurea (Ledeb.) Holub subsp. tomentosa (Kjellm.) Á. Löve et D. Löve = Tephroseris kjellmanii (A. E. Porsild) Holub ■☆

385890 Tephroseris birubonensis (Kitam.) B. Nord. = Tephroseris phaeantha (Nakai) C. Jeffrey et Y. L. Chen ■

385891 Tephroseris changii B. Nord. = Sinosenecio changii (B. Nord.) B. Nord. et Pelser ■

385892 Tephroseris flammea (Turcz. ex DC.) Holub;红轮狗舌草(红轮千里光);Flamecolored Groundsel, Orange Dogtongueweed, Orange Ligulate Tephroseris ■

385893 Tephroseris flammea (Turcz. ex DC.) Holub subsp. glabrifolia (Cufod.) B. Nord. = Tephroseris flammea (Turcz. ex DC.) Holub ■

385894 Tephroseris flammea (Turcz. ex DC.) Holub subsp. glabrifolia (Cufod.) B. Nord.;光叶红轮狗舌草■☆

385895 Tephroseris flammea (Turcz. ex DC.) Holub var. chaerocarhe (C. Jeffrey et Y. L. Chen) Y. M. Yuan = Tephroseris rufa (Hand.-Mazz.) B. Nord. var. chaetocarpa C. Jeffrey et Y. L. Chen ■

385896 Tephroseris flammea (Turcz. ex DC.) Holub var. chaetocarpa (C. Jeffrey et Y. L. Chen) Y. M. Yuan = Tephroseris rufa (Hand.-Mazz.) B. Nord. var. chaetocarpa C. Jeffrey et Y. L. Chen ■

385897 Tephroseris frigida (Richardson) Holub;冷地狗舌草■☆

385898 Tephroseris furusei (Kitam.) B. Nord.;古施狗舌草■☆

385899 Tephroseris integrifolia (L.) Holub;全叶狗舌草;Field Fleawort, Fleawori ■☆

385900 Tephroseris integrifolia (L.) Holub subsp. kirilowii (Turcz. ex DC.) B. Nord. = Tephroseris kirilowii (Turcz. ex DC.) Holub ■

385901 Tephroseris integrifolia (L.) Holub var. spathulata (Miq.) H. Ohba = Tephroseris integrifolia (L.) Holub subsp. kirilowii (Turcz. ex DC.) B. Nord. ■

385902 Tephroseris kawakamii (Makino) Holub;川上氏狗舌草■☆

385903 Tephroseris kirilowii (Turcz. ex DC.) Holub;狗舌草(白火丹草,草地狗舌草,朝阳花,肥猪苗,狗舌头草,黄菊莲,猫耳朵,糯米青,平地茱窝,蒲儿根,丘狗舌草,全缘叶狗舌草,田野狗舌草,铜交杯,铜盘一枝香,一枝花);Dogtongueweed, Field Fleawort, Field Groundsel, Kirilow's Groundsel, Kirilow's Tephroseris, Welsh Ragwort ■

385904 Tephroseris kirilowii (Turcz. ex DC.) Holub = Tephroseris integrifolia (L.) Holub subsp. kirilowii (Turcz. ex DC.) B. Nord. ■

385905 Tephroseris kjellmanii (A. E. Porsild) Holub;谢尔曼狗舌草■☆

385906 Tephroseris koreana (Kom.) B. Nord. et Pelser = Sinosenecio koreanus (Kom.) B. Nord. ■

385907 Tephroseris lindstroemii (Ostenf.) Á. Löve et D. Löve;林德狗舌草■☆

385908 Tephroseris palustris (L.) Fourr.;湿生狗舌草(沼泽狗舌草);Marsh Fleawort, Marshy Tephroseris, Wet Dogtongueweed ■

385909 Tephroseris palustris (L.) Fourr. subsp. congesta (R. Br.) Holub = Tephroseris palustris (L.) Fourr. ■

385910 Tephroseris palustris (L.) Fourr. var. congestus (R. Br.) Kom. = Tephroseris palustris (L.) Fourr. ■

385911 Tephroseris palustris (L.) Rchb. = Tephroseris palustris (L.) Fourr. ■

385912 Tephroseris paraticola (Schischk. et Serg.) Holub = Tephroseris praticola (Schischk. et Serg.) Holub ■

385913 Tephroseris phaeantha (Nakai) C. Jeffrey et Y. L. Chen;长白狗舌草;Changbaishan Dogtongueweed, Changbaishan Tephroseris, Changpeishan Tephroseris ■

385914 Tephroseris pierotii (Miq.) Holub;江浙狗舌草(皮氏千里光);Pierot Dogtongueweed, Pierot's Tephroseris ■

385915 Tephroseris praticola (Schischk. et Serg.) Holub;草原狗舌草(亚洲千里光);Grassland Dogtongueweed, Parture Tephroseris ■

385916　Tephroseris pseudosonchus（Vaniot）C. Jeffrey et Y. L. Chen；黔狗舌草（抱茎狗舌草，朝阳花）；Guizhou Dogtongueweed，Pseudosowthisitle Tephroseris ■

385917　Tephroseris rufa（Hand. -Mazz.）B. Nord.；橙舌狗舌草（橙舌千里光，橙叶千里光，红舌狗舌草，红舌千里光，头状千里光）；Orange Groundsel，Red Dogtongueweed，Red Groundsel，Red-ligulate Tephroseris ■

385918　Tephroseris rufa（Hand. -Mazz.）B. Nord. var. chaetocarpa C. Jeffrey et Y. L. Chen；毛果橙舌狗舌草；Hairfruit Red-ligulate Tephroseris ■

385919　Tephroseris stolonifera（Cufod.）Holub；匍枝狗舌草；Stoloniferous Dogtongueweed，Stoloniferous Tephroseris ■

385920　Tephroseris subdentata（Bunge）Holub；尖齿狗舌草（近全缘狗舌草）；Nearly Entire Tephroseris，Nearlyentireleaf Dogtongueweed ■

385921　Tephroseris taitoensis（Hayata）Holub；台东狗舌草（台东黄菀，台湾狗舌草）；Taiwan Dogtongueweed，Taiwan Tephroseris ■

385922　Tephroseris tundricola（Tolm.）Holub；冻原狗舌草■☆

385923　Tephroseris tundricola（Tolm.）Holub subsp. lindstroemii（Ostenf.）Wiebe = Tephroseris lindstroemii（Ostenf.）Á. Löve et D. Löve ■☆

385924　Tephroseris turczaninowii（DC.）Holub；天山狗舌草（灰千里光）；Tianshan Dogtongueweed，Turczaninow's Tephroseris ■

385925　Tephroseris yukonensis（A. E. Porsild）Holub；育空狗舌草■☆

385926　Tephrosia Pers.（1807）（保留属名）；灰毛豆属（灰叶豆属，灰叶属）；Hoary Pea，Hoarypea，Tephrosia ●■

385927　Tephrosia acaciifolia Welw. ex Baker；合欢叶灰毛豆■☆

385928　Tephrosia aemula（E. Mey.）Harv. = Tephrosia macropoda（E. Mey.）Harv. var. diffusa（E. Mey.）Schrire ■☆

385929　Tephrosia aequilata Baker；相等灰毛豆■☆

385930　Tephrosia aequilata Baker subsp. mlanjeana Brummitt；姆兰杰灰毛豆■☆

385931　Tephrosia aequilata Baker subsp. namuliana Brummitt；纳木里灰毛豆■☆

385932　Tephrosia aequilata Baker subsp. nyasae（Baker f.）Brummitt；尼亚萨灰毛豆■☆

385933　Tephrosia aequilata Baker var. meyeri-johannis（Taub.）Brummitt = Tephrosia aequilata Baker ■☆

385934　Tephrosia alba Du Puy et Labat；白花灰毛豆●☆

385935　Tephrosia albifoliolis Nongon. et Sarr；白毛灰毛豆●☆

385936　Tephrosia albissima H. M. L. Forbes；极白灰毛豆●☆

385937　Tephrosia albissima H. M. L. Forbes subsp. zuluensis（H. M. L. Forbes）Schrire；祖卢灰毛豆●☆

385938　Tephrosia aldabrensis J. R. Drumm. et J. H. Hemsl. = Tephrosia pumila（Lam.）Pers. var. aldabrensis（J. R. Drumm. et Hemsl.）Brummitt ●☆

385939　Tephrosia alpestris Taub.；高山灰毛豆●☆

385940　Tephrosia amoena E. Mey. = Tephrosia kraussiana Meisn.●☆

385941　Tephrosia andongensis Welw. ex Baker；安东灰毛豆●☆

385942　Tephrosia angulata E. Mey. = Tephrosia capensis（Jacq.）Pers. ■☆

385943　Tephrosia angustissima Engl. = Tephrosia longipes Meisn.■☆

385944　Tephrosia anomala Thulin；异常灰毛豆●☆

385945　Tephrosia ansellii Hook. f. = Tephrosia platycarpa Guillaumin et Perr. ●☆

385946　Tephrosia anthylloides Hochst. ex Webb = Tephrosia uniflora Pers. ■☆

385947　Tephrosia apiculata H. M. L. Forbes = Tephrosia natalensis H. M. L. Forbes ■☆

385948　Tephrosia apollinea（Delile）Link = Tephrosia purpurea（L.）Pers. subsp. apollinea（Delile）Hosni et El Karemy ■☆

385949　Tephrosia apollinea Link = Tephrosia purpurea（L.）Pers. subsp. apollinea（Delile）Hosni et El Karemy ■☆

385950　Tephrosia arabica（Boiss.）Martelli；阿拉伯灰毛豆●☆

385951　Tephrosia argyrotricha Harms；银毛灰毛豆●☆

385952　Tephrosia argyrotricha Harms var. burttii（Baker f.）J. B. Gillett = Tephrosia argyrotricha Harms ●☆

385953　Tephrosia armitageana Chiov. = Tephrosia macropoda（E. Mey.）Harv.■☆

385954　Tephrosia athiensis Baker f.；阿西灰毛豆●☆

385955　Tephrosia atroviolacea Baker f. = Tephrosia interrupta Hochst. et Steud. ex Engl. subsp. mildbraedii（Harms）J. B. Gillett ●☆

385956　Tephrosia aurantiaca Harms；橘色灰毛豆●☆

385957　Tephrosia aurantiaca Harms subsp. hirsutostylosa Dewit = Tephrosia hockii De Wild. subsp. hirsutostylosa（Dewit）J. B. Gillett ●☆

385958　Tephrosia aurantiaca Harms var. brevifolia Dewit = Tephrosia hockii De Wild. ●☆

385959　Tephrosia aurantiaca Harms var. longifolia Dewit = Tephrosia subpraecox Cronquist ■☆

385960　Tephrosia aurantiaca Harms. f. albescens Dewit = Tephrosia hockii De Wild. ●☆

385961　Tephrosia aurantiaca Harms. f. cinerea Dewit = Tephrosia hockii De Wild. ●☆

385962　Tephrosia aurantiaca Harms. f. fulvescens Dewit = Tephrosia hockii De Wild. ●☆

385963　Tephrosia avasmontana Dinter = Dolichos angustissimus E. Mey.■☆

385964　Tephrosia bachmannii Harms；巴克曼灰毛豆●☆

385965　Tephrosia barbigera Welw. ex Baker = Tephrosia nana Kotschy ex Schweinf. ●☆

385966　Tephrosia barclayi（Telfair ex Hook.）Baill. = Mundulea barclayi（Telfair ex Hook.）R. Vig. ex Du Puy et Labat ●☆

385967　Tephrosia bequaertii De Wild. = Tephrosia elata Deflers ●☆

385968　Tephrosia betsileensis（R. Vig.）Du Puy et Labat；贝齐尔灰毛豆●☆

385969　Tephrosia bibracteolata（Dumaz-le-Grand）Du Puy et Labat；双苞片灰毛豆●☆

385970　Tephrosia boiviniana Baill.；博伊文灰毛豆●☆

385971　Tephrosia boiviniana Baill. f. annua R. Vig. = Tephrosia boiviniana Baill. ●☆

385972　Tephrosia boiviniana Baill. f. typica R. Vig. = Tephrosia boiviniana Baill. ●☆

385973　Tephrosia bojeri Baill. = Tephrosia lyallii Baker ■☆

385974　Tephrosia boranensis Chiov. = Tephrosia hildebrandtii Vatke ■☆

385975　Tephrosia brachyloba E. Mey. = Tephrosia dregeana E. Mey. ●☆

385976　Tephrosia bracteolata Guillaumin et Perr.；小苞片灰毛豆●☆

385977　Tephrosia bracteolata Guillaumin et Perr. var. strigulosa Brummitt；硬毛灰毛豆●☆

385978　Tephrosia brummittii Schrire；布鲁米特灰毛豆●☆

385979　Tephrosia brunnea Baker = Tephrosia dichroocarpa Steud. ex A. Rich. ●☆

385980　Tephrosia burchellii Burtt Davy；伯切尔灰毛豆●☆

385981　Tephrosia burttii Baker f. = Tephrosia argyrotricha Harms ●☆

385982　Tephrosia butayei De Wild. et T. Durand = Tephrosia dasyphylla Welw. ex Baker subsp. butayei（De Wild. et T. Durand）Brummitt ●☆

385983　Tephrosia caerulea Baker f.；天蓝灰毛豆●☆

385984　Tephrosia caerulea Baker f. subsp. otaviensis（Dinter）A.

Schreib. et Brummitt;奥塔维灰毛豆●☆

385985　Tephrosia candida（Roxb.）DC.;白灰毛豆（白花灰叶,白花铁富豆,短萼灰毛豆,短萼灰叶,山毛豆,印度豆）;Boga Medaloa, White Hoarypea,White Tephrosia ●

385986　Tephrosia candida DC. = Tephrosia candida（Roxb.）DC. ●

385987　Tephrosia canescens E. Mey. = Tephrosia purpurea（L.）Pers. subsp. canescens（E. Mey.）Brummitt ●☆

385988　Tephrosia capensis（Jacq.）Pers.;好望角灰毛豆■☆

385989　Tephrosia capensis（Jacq.）Pers. var. acutifolia E. Mey.;尖叶好望角灰毛豆●☆

385990　Tephrosia capensis（Jacq.）Pers. var. angustifolia E. Mey.;窄叶灰毛豆●☆

385991　Tephrosia capensis（Jacq.）Pers. var. hirsuta Harv.;粗毛灰毛豆●☆

385992　Tephrosia capensis（Jacq.）Pers. var. longipetiolata H. M. L. Forbes;长梗好望角灰毛豆●☆

385993　Tephrosia capillipes Welw. ex Baker = Tephrosia dregeana E. Mey. var. capillipes（Welw. ex Baker）Torre ●

385994　Tephrosia capitata Verdc.;头状灰毛豆●☆

385995　Tephrosia carvalhoi Taub. = Tephrosia reptans Baker ●☆

385996　Tephrosia cephalantha Welw. ex Baker;头花灰毛豆●☆

385997　Tephrosia cephalantha Welw. ex Baker var. decumbens ?;外倾灰毛豆●☆

385998　Tephrosia chimanimaniana Brummitt;奇马尼曼灰毛豆●☆

385999　Tephrosia cineria（L.）Pers.;灰色灰叶■☆

386000　Tephrosia coccinea Wall.;红灰毛豆;Red Hoarypea ■☆

386001　Tephrosia coccinea Wall. var. stenophylla Hosok.;狭叶红灰毛豆（狭叶红花灰叶）;Narrow Leaf Red Tephrosia, Narrowleaf Red Hoarypea ■

386002　Tephrosia commersonii Scott-Elliot = Tephrosia pumila（Lam.）Pers. ■

386003　Tephrosia concinna Baker = Tephrosia bracteolata Guillaumin et Perr. ●☆

386004　Tephrosia confertiflora Benth. = Tephrosia luzonensis Vogel ■

386005　Tephrosia congestiflora Harms = Tephrosia nyikensis Baker subsp. victoriensis Brummitt et J. B. Gillett ●☆

386006　Tephrosia contorta N. E. Br. = Ptycholobium contortum（N. E. Br.）Brummitt ■☆

386007　Tephrosia cordata Hutch. et Burtt Davy;心形灰毛豆●☆

386008　Tephrosia cordatistipulata J. B. Gillett;心托叶灰毛豆●☆

386009　Tephrosia cordofana Hochst. = Tephrosia subtriflora Hochst. ex Baker ●☆

386010　Tephrosia coronilloides Welw. ex Baker;小冠花灰毛豆●☆

386011　Tephrosia crotalarioides Klotzsch = Indigofera crotalarioides（Klotzsch）Baker ●☆

386012　Tephrosia curvata De Wild.;内折灰毛豆●☆

386013　Tephrosia damarensis Engl. = Tephrosia dregeana E. Mey. ●☆

386014　Tephrosia dasyphylla Baker = Tephrosia dasyphylla Welw. ex Baker ■☆

386015　Tephrosia dasyphylla Baker subsp. amplissima Brummitt = Tephrosia dasyphylla Welw. ex Baker subsp. amplissima Brummitt ●☆

386016　Tephrosia dasyphylla Baker subsp. butayei（De Wild. et T. Durand）Brummitt = Tephrosia dasyphylla Welw. ex Baker subsp. butayei（De Wild. et T. Durand）Brummitt ●☆

386017　Tephrosia dasyphylla Baker subsp. youngii（Torre）Brummitt = Tephrosia dasyphylla Welw. ex Baker subsp. youngii（Torre）Brummitt ●☆

386018　Tephrosia dasyphylla Welw. ex Baker;毛叶灰毛豆■☆

386019　Tephrosia dasyphylla Welw. ex Baker subsp. amplissima Brummitt;膨大灰毛豆●☆

386020　Tephrosia dasyphylla Welw. ex Baker subsp. butayei（De Wild. et T. Durand）Brummitt;布塔耶毛叶灰毛豆●☆

386021　Tephrosia dasyphylla Welw. ex Baker subsp. youngii（Torre）Brummitt;扬氏灰毛豆●☆

386022　Tephrosia dawsonii Baker f. = Tephrosia longipes Meisn. ■☆

386023　Tephrosia decaryana（Dumaz-le-Grand）Du Puy et Labat;德卡里灰毛豆●☆

386024　Tephrosia decidua A. Rich. = Tephrosia pentaphylla（Roxb.）G. Don ●☆

386025　Tephrosia decora Baker;装饰灰毛豆●☆

386026　Tephrosia decorticans Taub. = Tephrosia dura Baker ●☆

386027　Tephrosia deflexa Baker;外折灰毛豆●☆

386028　Tephrosia delagoensis H. M. L. Forbes = Tephrosia purpurea（L.）Pers. var. delagoensis（H. M. L. Forbes）Brummitt ●☆

386029　Tephrosia delicata Baker f. = Tephrosia decora Baker ●☆

386030　Tephrosia densiflora Hook. f.;密花灰毛豆●☆

386031　Tephrosia desertorum Scheele;荒漠灰毛豆●☆

386032　Tephrosia dichotoma Desv. = Tephrosia luzonensis Vogel ■

386033　Tephrosia dichotoma Desv. = Tephrosia pumila（Lam.）Pers. ■

386034　Tephrosia dichroocarpa Steud. ex A. Rich.;二色果灰毛豆●☆

386035　Tephrosia diffusa（E. Mey.）Harv. = Tephrosia macropoda（E. Mey.）Harv. var. diffusa（E. Mey.）Schrire ■☆

386036　Tephrosia diffusa Roxb. = Tephrosia purpurea（L.）Pers. ●■

386037　Tephrosia digitata DC. = Tephrosia lupinifolia DC. ●☆

386038　Tephrosia dimorphophylla Welw. ex Baker = Tephrosia paniculata Welw. ex Baker ●☆

386039　Tephrosia dinteri Schinz = Tephrosia dregeana E. Mey. ●☆

386040　Tephrosia discolor E. Mey. = Tephrosia linearis（Willd.）Pers. ●☆

386041　Tephrosia disperma Welw. ex Baker;双籽灰毛豆●☆

386042　Tephrosia dissitiflora Baker = Tephrosia elongata E. Mey. ●☆

386043　Tephrosia djalonica A. Chev. ex Hutch. et Dalziel;贾隆灰毛豆●☆

386044　Tephrosia doggettii Baker f. = Tephrosia interrupta Hochst. et Steud. ex Engl. subsp. mildbraedii（Harms）J. B. Gillett ●☆

386045　Tephrosia downsonii Baker f. = Tephrosia lurida Sond. ■☆

386046　Tephrosia dregeana E. Mey.;德雷灰毛豆●☆

386047　Tephrosia dregeana E. Mey. var. capillipes（Welw. ex Baker）Torre;细毛灰毛豆●☆

386048　Tephrosia drepanocarpa Welw. ex Baker;镰果灰毛豆●☆

386049　Tephrosia dura Baker;硬灰毛豆●☆

386050　Tephrosia ehrenbergiana Schweinf. = Tephrosia villosa（L.）Pers. var. ehrenbergiana（Schweinf.）Brummitt ■☆

386051　Tephrosia elata Deflers;高灰毛豆●☆

386052　Tephrosia elata Deflers subsp. heckmanniana（Harms）Brummitt = Tephrosia heckmanniana Harms ●☆

386053　Tephrosia elata Deflers var. tomentella Brummitt;软毛高灰毛豆●☆

386054　Tephrosia elegans Schumach.;雅致灰毛豆●☆

386055　Tephrosia elongata E. Mey.;伸长灰毛豆●☆

386056　Tephrosia elongata E. Mey. var. glabra Sond. = Tephrosia elongata E. Mey. ●☆

386057　Tephrosia elongata E. Mey. var. lasiocaulos Brummitt;绵毛茎灰毛豆●☆

386058　Tephrosia elongata E. Mey. var. pubescens Harv. = Tephrosia elongata E. Mey. ●☆

386059　Tephrosia elongata E. Mey. var. tzaneenensis（H. M. L. Forbes）

Brummitt;察嫩灰毛豆●☆

386060 Tephrosia emarginato-foliolata De Wild. = Tephrosia heckmanniana Harms ●☆

386061 Tephrosia encoptosperma Schweinf. = Tephrosia subtriflora Hochst. ex Baker ●☆

386062 Tephrosia ensifolia Harv. = Tephrosia elongata E. Mey. ●☆

386063 Tephrosia eriosemoides Oliv. = Tephrosia paniculata Welw. ex Baker ●☆

386064 Tephrosia evansii Hutch. et Burtt Davy = Tephrosia rhodesica Baker f. var. evansii (Hutch. et Burtt Davy) Brummitt ●☆

386065 Tephrosia eylesii Baker f. = Tephrosia stormsii De Wild. ●☆

386066 Tephrosia falcata (Thunb.) Pers. ;镰状灰毛豆●☆

386067 Tephrosia fasciculata Hook. f. = Tephrosia bracteolata Guillaumin et Perr. ●☆

386068 Tephrosia faulknerae Brummitt;福克纳灰毛豆●☆

386069 Tephrosia filiflora Chiov. ;细花灰毛豆■☆

386070 Tephrosia filipes Benth. ; 细梗灰毛豆; Slender-pedicel Tephrosia, Thinstalk Hoarypea ■

386071 Tephrosia flava Thulin;黄花灰毛豆■☆

386072 Tephrosia flexuosa G. Don = Tephrosia platycarpa Guillaumin et Perr. ●☆

386073 Tephrosia forbesii Baker;福布斯灰毛豆■☆

386074 Tephrosia forbesii Baker subsp. inhacensis Brummitt;伊尼亚卡灰毛豆●☆

386075 Tephrosia forbesii Baker subsp. interior Brummitt;间型灰毛豆●☆

386076 Tephrosia franchetii Hutch. et E. A. Bruce = Tephrosia heterophylla Vatke ●☆

386077 Tephrosia fulvinervis Hochst. ex A. Rich. ;黄褐脉灰毛豆●☆

386078 Tephrosia galpinii H. M. L. Forbes;盖尔灰毛豆●☆

386079 Tephrosia genistoides (Dumaz-le-Grand) Du Puy et Labat;金雀灰毛豆●☆

386080 Tephrosia glomeruliflora Meisn. ;团花灰毛豆●☆

386081 Tephrosia glomeruliflora Meisn. subsp. meisneri (Hutch. et Burtt Davy) Schrire;梅斯纳灰毛豆●☆

386082 Tephrosia gobensis Brummitt;戈贝灰毛豆●☆

386083 Tephrosia godmaniae Baker f. = Tephrosia reptans Baker ●☆

386084 Tephrosia gorgonea Cout. = Tephrosia pedicellata Baker ●☆

386085 Tephrosia gossweileri Baker f. ;戈斯灰毛豆●☆

386086 Tephrosia gracilipes Guillaumin et Perr. ;丝梗灰毛豆●☆

386087 Tephrosia graminifolia Chiov. = Tephrosia subtriflora Hochst. ex Baker ●☆

386088 Tephrosia grandibracteata Merxm. ;大苞灰毛豆●☆

386089 Tephrosia grandiflora (Aiton) Pers. = Tephrosia grandiflora (L'Her. ex Aiton) Pers. ●☆

386090 Tephrosia grandiflora (L'Her. ex Aiton) Pers. ;大花灰毛豆, Tephrosia ●☆

386091 Tephrosia granitica R. Vig. = Tephrosia reptans Baker ●☆

386092 Tephrosia hamiltonii J. R. Drumm. = Tephrosia purpurea (L.) Pers. ●■

386093 Tephrosia harmsii Hochr. = Tephrosia elegans Schumach. ●☆

386094 Tephrosia heckmanniana Harms;赫克曼灰毛豆●☆

386095 Tephrosia heterophylla Vatke;互叶灰毛豆●☆

386096 Tephrosia hildebrandtii Vatke;希尔德灰毛豆●☆

386097 Tephrosia hirsuta Schumach. et Thonn. = Tephrosia pumila (Lam.) Pers. ■

386098 Tephrosia hochstetteri Chiov. ;霍赫灰毛豆●☆

386099 Tephrosia hockii De Wild. ;霍克灰毛豆●☆

386100 Tephrosia hockii De Wild. subsp. hirsutostylosa (Dewit) J. B. Gillett;毛柱霍克灰毛豆●☆

386101 Tephrosia holosericea Nutt. = Tephrosia virginiana (L.) Pers. ■☆

386102 Tephrosia holstii Taub. ;霍尔斯灰毛豆■☆

386103 Tephrosia hookeriana var. amoena Prain = Tephrosia noctiflora Bojer ex Baker ■

386104 Tephrosia hookeriana Wight et Arn. var. amoena Prain = Tephrosia noctiflora Bojer ex Baker ■

386105 Tephrosia huillensis Welw. ex Baker;威拉灰毛豆●☆

386106 Tephrosia huillensis Welw. ex Baker var. grandiflora Baker;大花威拉灰毛豆●☆

386107 Tephrosia humbertii Dumaz-le-Grand;亨伯特灰毛豆●☆

386108 Tephrosia humilis Guillaumin et Perr. ;低矮灰毛豆●☆

386109 Tephrosia ibityensis (R. Vig.) Du Puy et Labat;伊比提灰毛豆●☆

386110 Tephrosia ilorinensis J. R. Drumm. ex Baker f. = Tephrosia pedicellata Baker ●☆

386111 Tephrosia inandensis H. M. L. Forbes;伊南德灰毛豆●☆

386112 Tephrosia incarnata Brummitt = Tephrosia glomeruliflora Meisn. subsp. meisneri (Hutch. et Burtt Davy) Schrire ●☆

386113 Tephrosia indigofera Bertol. = Tephrosia purpurea (L.) Pers. var. delagoensis (H. M. L. Forbes) Brummitt ●☆

386114 Tephrosia interrupta Hochst. et Steud. ex Engl. ;间断灰毛豆●☆

386115 Tephrosia interrupta Hochst. et Steud. ex Engl. subsp. elongatiflora J. B. Gillett;长花间断灰毛豆●☆

386116 Tephrosia interrupta Hochst. et Steud. ex Engl. subsp. mildbraedii (Harms) J. B. Gillett;米尔德灰毛豆●☆

386117 Tephrosia ionophlebia Hayata;台湾灰毛豆■

386118 Tephrosia ionophlebia Hayata = Tephrosia purpurea (L.) Pers. ●■

386119 Tephrosia iringae Baker f. ;伊林加灰毛豆●☆

386120 Tephrosia isaloensis Du Puy et Labat;伊萨卢灰毛豆●☆

386121 Tephrosia jelfiae Baker f. = Tephrosia ringoetii Baker f. ●☆

386122 Tephrosia junodii De Wild. = Tephrosia forbesii Baker ■☆

386123 Tephrosia kalamboensis Brummitt et J. B. Gillett;卡兰博灰毛豆●☆

386124 Tephrosia karkarensis Thulin;卡尔卡尔灰毛豆●☆

386125 Tephrosia kasikiensis Baker f. ;卡西基灰毛豆●☆

386126 Tephrosia kassasii Boulos;卡萨斯灰毛豆●☆

386127 Tephrosia kassneri Baker f. = Tephrosia holstii Taub. ■☆

386128 Tephrosia katangensis De Wild. ;加丹加灰毛豆●☆

386129 Tephrosia kerrii J. R. Drumm. et Craib;银灰毛豆(银毛灰叶);Kerr Tephrosia, Silver Hoarypea ■

386130 Tephrosia kirkii Baker = Tephrosia reptans Baker ●☆

386131 Tephrosia kraussiana Meisn. ;克劳斯灰毛豆●☆

386132 Tephrosia lactea Schinz = Tephrosia oxygona Welw. ex Baker subsp. lactea (Schinz) A. Schreib. ●☆

386133 Tephrosia laevigata Welw. ex Baker = Tephrosia lupinifolia DC. ●☆

386134 Tephrosia lateritia Merxm. = Tephrosia decora Baker ●☆

386135 Tephrosia lathyroides Guillaumin et Perr. ;山黧豆状灰毛豆●☆

386136 Tephrosia latidens (Small) Standl. ;宽齿灰毛豆(宽灰叶,宽叶灰叶)■☆

386137 Tephrosia latidens (Small) Standl. = Tephrosia virginiana (L.) Pers. ■☆

386138 Tephrosia laurentii De Wild. = Tephrosia purpurea (L.) Pers. ●■

386139 Tephrosia laxiflora R. E. Fr. = Tephrosia lurida Sond. ■☆

386140 Tephrosia lebrunii Cronquist;勒布伦灰毛豆●☆

386141 Tephrosia lelyi Baker f. =Tephrosia paniculata Welw. ex Baker ●☆

386142 Tephrosia lepida Baker f. ;小鳞灰毛豆●☆

386143 Tephrosia lepida Baker f. subsp. nigrescens Brummitt;变黑小鳞

灰毛豆●☆

386144　Tephrosia leptostachya DC. ;细穗花灰叶■☆

386145　Tephrosia leptostachya DC. = Tephrosia purpurea (L.) Pers. var. leptostachya (DC.) Brummitt ■☆

386146　Tephrosia lessertioides Baker f. = Tephrosia curvata De Wild. ●☆

386147　Tephrosia letestui Tisser. ;莱泰斯图灰毛豆●☆

386148　Tephrosia leucoclada Scott-Elliot = Tephrosia purpurea (L.) Pers. subsp. dunensis Brummitt ●☆

386149　Tephrosia limpopoensis J. B. Gillett;林波波灰毛豆●☆

386150　Tephrosia linearis (Willd.) Pers. ;线状灰毛豆●☆

386151　Tephrosia linearis (Willd.) Pers. subsp. discolor (E. Mey.) J. B. Gillett = Tephrosia linearis (Willd.) Pers. ●☆

386152　Tephrosia linearis (Willd.) Pers. var. discolor (E. Mey.) Brummitt = Tephrosia linearis (Willd.) Pers. ●☆

386153　Tephrosia lineata Schumach. et Thonn. = Tephrosia purpurea (L.) Pers. var. leptostachya (DC.) Brummitt ■☆

386154　Tephrosia longana Harms = Tephrosia coronilloides Welw. ex Baker ●☆

386155　Tephrosia longipes Meisn. ;长梗灰毛豆■☆

386156　Tephrosia longipes Meisn. subsp. swynnertonii (Baker f.) Brummitt;斯温纳顿灰毛豆■☆

386157　Tephrosia longipes Meisn. var. lurida (Sond.) J. B. Gillett = Tephrosia lurida Sond. ■☆

386158　Tephrosia longipes Meisn. var. ringoetii (Baker f.) J. B. Gillett = Tephrosia ringoetii Baker f. ●☆

386159　Tephrosia longipes Meisn. var. uncinata Harv. ;具钩灰毛豆●☆

386160　Tephrosia lortii Baker f. ;洛特灰毛豆●☆

386161　Tephrosia luembensis De Wild. = Tephrosia dasyphylla Welw. ex Baker ■☆

386162　Tephrosia lupinifolia DC. ;羽扇豆灰叶●☆

386163　Tephrosia lurida Sond. = Tephrosia longipes Meisn. ■☆

386164　Tephrosia lurida Sond. var. drummondii Brummitt = Tephrosia longipes Meisn. ■☆

386165　Tephrosia lurida Sond. var. lissocarpa Brummitt = Tephrosia longipes Meisn. ■☆

386166　Tephrosia lutea R. E. Fr. = Tephrosia hockii De Wild. ●☆

386167　Tephrosia luzonensis Vogel;西沙灰毛豆(西沙灰叶);Luzon Tephrosia,Xisha Hoarypea ■

386168　Tephrosia lyallii Baker;莱尔灰毛豆●☆

386169　Tephrosia macropoda (E. Mey.) Harv. ;大柄灰叶;Lazane ■☆

386170　Tephrosia macropoda (E. Mey.) Harv. var. diffusa (E. Mey.) Schrire;铺散大柄灰毛豆■☆

386171　Tephrosia malvina Brummitt;锦葵灰毛豆●☆

386172　Tephrosia manikensis De Wild. ;马尼科灰毛豆■☆

386173　Tephrosia manikensis De Wild. var. albosericea Brummitt;白绢毛灰毛豆■☆

386174　Tephrosia marginella H. M. L. Forbes;小边灰毛豆■☆

386175　Tephrosia maxima (L.) Pers. ;最大灰叶●☆

386176　Tephrosia maxima Pers. = Tephrosia maxima (L.) Pers. ●☆

386177　Tephrosia mbogaensis De Wild. = Tephrosia elata Deflers ●☆

386178　Tephrosia medleyi H. M. L. Forbes = Tephrosia shiluwanensis Schinz ●☆

386179　Tephrosia megalantha Micheli = Tephrosia vogelii Hook. f. ●■

386180　Tephrosia meisneri Hutch. et Burtt Davy = Tephrosia glomeruliflora Meisn. subsp. meisneri (Hutch. et Burtt Davy) Schrire ●☆

386181　Tephrosia melanocalyx Welw. ex Baker;黑萼灰毛豆●☆

386182　Tephrosia meyeri-johannis Taub. = Tephrosia aequilata Baker ■☆

386183　Tephrosia micrantha J. B. Gillett;小花灰毛豆●☆

386184　Tephrosia mildbraedii Harms = Tephrosia interrupta Hochst. et Steud. ex Engl. subsp. mildbraedii (Harms) J. B. Gillett ●☆

386185　Tephrosia miranda Brummitt;奇异灰毛豆●☆

386186　Tephrosia mohrii (Rydb.) Godfrey = Tephrosia virginiana (L.) Pers. ■☆

386187　Tephrosia monantha Baker = Pyranthus monantha (Baker) Du Puy et Labat ●☆

386188　Tephrosia monophylla Schinz;单叶灰毛豆●☆

386189　Tephrosia montana Brummitt;山地灰毛豆●☆

386190　Tephrosia mossambicensis Schinz = Tephrosia uniflora Pers. ■☆

386191　Tephrosia mossiensis A. Chev. ;莫西灰毛豆●☆

386192　Tephrosia multiflora Blatt. et Hallb. = Tephrosia subtriflora Hochst. ex Baker ●☆

386193　Tephrosia multijuga R. G. N. Young;多对灰毛豆●☆

386194　Tephrosia multinervis Baker f. = Tephrosia heckmanniana Harms ●☆

386195　Tephrosia nana Kotschy ex Schweinf. ;矮小灰毛豆●☆

386196　Tephrosia natalensis H. M. L. Forbes;纳塔尔灰毛豆●☆

386197　Tephrosia natalensis H. M. L. Forbes subsp. pseudocapitata (H. M. L. Forbes) Schrire;假头状纳塔尔灰毛豆■☆

386198　Tephrosia newtoniana Torre;纽敦灰毛豆■☆

386199　Tephrosia noctiflora Bojer ex Baker;长序灰毛豆(黄花铁富豆,夜花灰毛豆);Longspike Hoarypea,South African Hoarypea ■

386200　Tephrosia nseleensis De Wild. ;恩塞勒灰毛豆■☆

386201　Tephrosia nubica (Boiss.) Baker;云雾灰毛豆■☆

386202　Tephrosia nubica (Boiss.) Baker var. abissinica Schweinf. = Tephrosia nubica (Boiss.) Baker ■☆

386203　Tephrosia nubica (Boiss.) Baker var. polyphylla Chiov. = Tephrosia polyphylla (Chiov.) J. B. Gillett ■☆

386204　Tephrosia nyassae Baker f. = Tephrosia aequilata Baker subsp. nyasae (Baker f.) Brummitt ■☆

386205　Tephrosia nyikensis Baker;尼卡灰毛豆■☆

386206　Tephrosia nyikensis Baker subsp. victoriensis Brummitt et J. B. Gillett;维多利亚灰毛豆●☆

386207　Tephrosia obbiadensis Chiov. ;奥比亚德灰毛豆●☆

386208　Tephrosia obcordata (Lam. ex Poir.) Baker = Requienia obcordata (Lam. ex Poir.) DC. ■☆

386209　Tephrosia oblongifolia E. Mey. = Ophrestia oblongifolia (E. Mey.) H. M. L. Forbes ■☆

386210　Tephrosia obovata Merr. ;卵叶灰毛豆(台湾灰毛豆);Obovate Leaf Tephrosia,Obovateleaf Hoarypea ■

386211　Tephrosia oligantha Drake = Pyranthus pauciflora (Baker) Du Puy et Labat ●☆

386212　Tephrosia oraria Hance = Millettia oraria (Hance) Dunn ●

386213　Tephrosia orientalis Baker f. = Tephrosia hildebrandtii Vatke ■☆

386214　Tephrosia otaviensis Dinter = Tephrosia caerulea Baker f. subsp. otaviensis (Dinter) A. Schreib. et Brummitt ●☆

386215　Tephrosia oubanguiensis Tisser. ;乌班吉灰毛豆●☆

386216　Tephrosia oxygona Welw. ex Baker;尖角灰毛豆●☆

386217　Tephrosia oxygona Welw. ex Baker subsp. lactea (Schinz) A. Schreib. ;乳白尖角灰毛豆●☆

386218　Tephrosia oxygona Welw. ex Baker var. obcordata Torre;倒心形尖角灰毛豆●☆

386219　Tephrosia pallida H. M. L. Forbes;苍白灰毛豆●☆

386220　Tephrosia paniculata Welw. ex Baker;圆锥灰毛豆●☆

386221　Tephrosia paniculata Welw. ex Baker subsp. holstii (Taub.) Brummitt = Tephrosia holstii Taub. ■☆

386222　Tephrosia paniculata Welw. ex Baker var. schizocalyx（Taub.）J. B. Gillett ＝ Tephrosia paniculata Welw. ex Baker ●☆

386223　Tephrosia paradoxa Brummitt ＝ Tephrosia stormsii De Wild. ●☆

386224　Tephrosia parvifolia（R. Vig.）Du Puy et Labat;小叶灰毛豆●☆

386225　Tephrosia pauciflora Wall. ＝ Tephrosia subtriflora Hochst. ex Baker ●☆

386226　Tephrosia paucijuga Harms;少轭灰毛豆●☆

386227　Tephrosia pearsonii Baker f. ;皮尔逊灰毛豆●☆

386228　Tephrosia pedicellata Baker;梗花灰毛豆●☆

386229　Tephrosia pentaphylla（Roxb.）G. Don;五叶灰毛豆●☆

386230　Tephrosia periculosa Baker ＝ Tephrosia vogelii Hook. f. ●■

386231　Tephrosia perrieri R. Vig. ;佩里耶灰毛豆■☆

386232　Tephrosia petrosa Blatt. et Hallb. ＝ Tephrosia uniflora Pers. subsp. petrosa（Blatt. et Hallb.）J. B. Gillett et Ali ●☆

386233　Tephrosia phaeosperma F. Muell. ex Benth. ;褐色灰毛豆■☆

386234　Tephrosia pinifolia Du Puy et Labat;松叶灰毛豆●☆

386235　Tephrosia pinnata（Thunb.）Pers. ;羽状灰毛豆■☆

386236　Tephrosia platycarpa Guillaumin et Perr. ;阔果灰毛豆●☆

386237　Tephrosia plicata Oliv. ＝ Ptycholobium plicatum（Oliv.）Harms ■☆

386238　Tephrosia polyphylla（Chiov.）J. B. Gillett;多叶灰毛豆■☆

386239　Tephrosia polysperma Baker ＝ Tephrosia nana Kotschy ex Schweinf. ●☆

386240　Tephrosia polystachya E. Mey. ;多穗灰毛豆■☆

386241　Tephrosia polystachya E. Mey. var. hirta Harv. ;毛多穗灰毛豆■☆

386242　Tephrosia polystachya E. Mey. var. latifolia Harv. ;宽叶多穗灰毛豆■☆

386243　Tephrosia polystachya E. Mey. var. longidens H. M. L. Forbes;长齿多穗灰毛豆■☆

386244　Tephrosia polystachyoides Baker f. ＝ Tephrosia rhodesica Baker f. var. polystachyoides（Baker f.）Brummitt ●☆

386245　Tephrosia pondoensis（Codd）Schrire;庞多灰毛豆●☆

386246　Tephrosia preussii Taub. ＝ Tephrosia paniculata Welw. ex Baker ●☆

386247　Tephrosia procumbens（Buch. -Ham.）Benth. ＝ Tephrosia pumila（Lam.）Pers. ■

386248　Tephrosia pseudocapitata H. M. L. Forbes ＝ Tephrosia natalensis H. M. L. Forbes subsp. pseudocapitata（H. M. L. Forbes）Schrire ■☆

386249　Tephrosia pseudosphaerosperma Schinz ＝ Requienia pseudosphaerosperma（Schinz）Brummitt ■☆

386250　Tephrosia pulchella Hook. f. ＝ Tephrosia linearis（Willd.）Pers. ●☆

386251　Tephrosia pulchra Colebr. ＝ Millettia pulchra（Benth.）Kurz ●

386252　Tephrosia pumila（Lam.）Pers. ;矮灰毛豆;Dwarf Hoarypea, Dwarf Tephrosia ■

386253　Tephrosia pumila（Lam.）Pers. subsp. aldabrensis（J. R. Drumm. et Hemsl.）Bosman et A. J. P. de Haas ＝Tephrosia pumila（Lam.）Pers. var. aldabrensis（J. R. Drumm. et Hemsl.）Brummitt ●☆

386254　Tephrosia pumila（Lam.）Pers. var. aldabrensis（J. R. Drumm. et Hemsl.）Brummitt;阿尔达布拉灰毛豆●☆

386255　Tephrosia punctata J. B. Gillett;斑点灰毛豆●☆

386256　Tephrosia pungens（R. Vig.）Du Puy et Labat;锐尖灰毛豆●☆

386257　Tephrosia purpurea（L.）Pers. ;灰毛豆（红花灰叶,灰叶,假靛青,假蓝靛,山青,台湾灰毛豆,野蓝,野蓝靛,野青树,野青子,紫叶灰毛豆）;Bastard Indigo, Bluish-purple Tephrosia, Fishpoison, Purple Hoarypea, Purple Tephrosia, Taiwan Hoarypea, Wild Indigo ●■

386258　Tephrosia purpurea（L.）Pers. subsp. altissima Brummitt;极高灰毛豆●☆

386259　Tephrosia purpurea（L.）Pers. subsp. apollinea（Delile）Hosni

et El Karemy;阿波林灰毛豆（阿波林灰叶）■☆

386260　Tephrosia purpurea（L.）Pers. subsp. canescens（E. Mey.）Brummitt;灰白灰毛豆●☆

386261　Tephrosia purpurea（L.）Pers. subsp. dunensis Brummitt;砂丘灰毛豆●☆

386262　Tephrosia purpurea（L.）Pers. subsp. leptostachya（DC.）Brummitt ＝ Tephrosia purpurea（L.）Pers. var. leptostachya（DC.）Brummitt ■☆

386263　Tephrosia purpurea（L.）Pers. var. delagoensis（H. M. L. Forbes）Brummitt;迪拉果灰毛豆●☆

386264　Tephrosia purpurea（L.）Pers. var. glabra Hosok. ;秃净灰毛豆（秃灰毛豆）;Glabrous Tephrosia ■

386265　Tephrosia purpurea（L.）Pers. var. leptostachya（DC.）Brummitt;细穗灰毛豆■☆

386266　Tephrosia purpurea（L.）Pers. var. maxima（L.）Baker;滇川灰毛豆●■

386267　Tephrosia purpurea（L.）Pers. var. pubescens Baker ＝ Tephrosia purpurea（L.）Pers. ●■

386268　Tephrosia purpurea（L.）Pers. var. pumila（Lam.）Baker ＝ Tephrosia pumila（Lam.）Pers. ■

386269　Tephrosia purpurea（L.）Pers. var. yunnanensis Z. Wei;云南灰毛豆;Yunnan Tephrosia ■

386270　Tephrosia pusilla（Thunb.）Pers. ;微小灰毛豆■☆

386271　Tephrosia quartiniana Cufod. ＝ Tephrosia uniflora Pers. ■☆

386272　Tephrosia quartiniana Cufod. var. inflexa（Chiov.）Cufod. ＝ Tephrosia pumila（Lam.）Pers. ■

386273　Tephrosia radicans Welw. ex Baker;具根灰毛豆■☆

386274　Tephrosia remotiflora F. Muell. ex Benth. ;疏花灰毛豆■☆

386275　Tephrosia rensburgii Verdc. ＝ Tephrosia radicans Welw. ex Baker ■☆

386276　Tephrosia reptans Baker;匍匐灰毛豆●☆

386277　Tephrosia reptans Baker var. arenicola Brummitt et J. B. Gillett;沙生灰毛豆■☆

386278　Tephrosia reptans Baker var. microfoliata（Pires de Lima）Brummitt;小小叶灰毛豆■☆

386279　Tephrosia retamoides（Baker）Soler. ;勒塔木灰毛豆●☆

386280　Tephrosia retamoides（Baker）Soler. var. genuina R. Vig. ＝ Tephrosia retamoides（Baker）Soler. ●☆

386281　Tephrosia retamoides（Baker）Soler. var. glabrescens R. Vig. ＝ Tephrosia viguieri Du Puy et Labat ●☆

386282　Tephrosia retusa Burtt Davy;微凹灰毛豆●☆

386283　Tephrosia rhizomatosa De Wild. ＝ Tephrosia hockii De Wild. ●☆

386284　Tephrosia rhodesica Baker f. ;罗得西亚灰毛豆●☆

386285　Tephrosia rhodesica Baker f. var. evansii（Hutch. et Burtt Davy）Brummitt;埃文斯灰毛豆●☆

386286　Tephrosia rhodesica Baker f. var. polystachyoides（Baker f.）Brummitt;多穗罗得西亚灰毛豆●☆

386287　Tephrosia richardsiae J. B. Gillett;理查兹灰毛豆●☆

386288　Tephrosia rigida Baker ＝ Tephrosia elata Deflers ●☆

386289　Tephrosia rigidula Welw. ex Baker;稍硬灰毛豆●☆

386290　Tephrosia ringoetii Baker f. ;林戈灰毛豆●☆

386291　Tephrosia rivae Taub. ex Harms ＝ Tephrosia holstii Taub. ■☆

386292　Tephrosia robinsoniana Brummitt;鲁滨逊灰毛豆●☆

386293　Tephrosia rosea F. Muell. ex Benth. ;玫瑰灰毛豆■☆

386294　Tephrosia rupicola J. B. Gillett;岩生灰毛豆●☆

386295　Tephrosia rupicola J. B. Gillett subsp. dreweana Brummitt;德鲁灰毛豆●☆

386296 Tephrosia rutenbergiana Vatke ＝ Tephrosia linearis（Willd.）Pers. ●☆

386297 Tephrosia salicifolia Schinz ＝ Tephrosia acaciifolia Welw. ex Baker ■☆

386298 Tephrosia sambesiaca Taub. ＝ Tephrosia elongata E. Mey. ●☆

386299 Tephrosia schizocalyx Harms ＝ Tephrosia paniculata Welw. ex Baker ●☆

386300 Tephrosia schweinfurthii Deflers ＝Tephrosia heterophylla Vatke ●☆

386301 Tephrosia scopulata Thulin；岩栖灰毛豆●☆

386302 Tephrosia secunda Welw. ex Baker ＝Tephrosia longipes Meisn. ■☆

386303 Tephrosia semiglabra Sond.；半光灰毛豆●☆

386304 Tephrosia seminuda Bojer ex Baker ＝ Tephrosia linearis（Willd.）Pers. ●☆

386305 Tephrosia sengaensis Baker f.；森加灰毛豆●☆

386306 Tephrosia senna Kunth；番泻灰叶■☆

386307 Tephrosia sericea Baker ＝Dolichos kilimandscharicus Taub. ■☆

386308 Tephrosia sessiliflora Hassl.；无柄花灰叶；Sessileflower Hoarypea ■☆

386309 Tephrosia shiluwanensis Schinz；希卢灰毛豆●☆

386310 Tephrosia similis Chiov. ＝ Tephrosia pentaphylla（Roxb.）G. Don ●☆

386311 Tephrosia simplicifolia Franch. ＝Tephrosia heterophylla Vatke ●☆

386312 Tephrosia sinapou（Buchoz）A. Chev.；芥灰毛豆；Yarroconalli ●☆

386313 Tephrosia sparsiflora H. M. L. Forbes ＝ Tephrosia purpurea（L.）Pers. var. pubescens Baker ●■

386314 Tephrosia spathacea Hutch. et Burtt Davy ＝ Tephrosia shiluwanensis Schinz ●☆

386315 Tephrosia sphaerosperma（DC.）Baker ＝ Requienia sphaerosperma DC. ■☆

386316 Tephrosia spicata Torr. et A. Gray；褐毛灰叶；Brownhair Tephrosia ■☆

386317 Tephrosia stormsii De Wild.；斯托姆斯灰毛豆●☆

386318 Tephrosia stormsii De Wild. var. pilosa Brummitt；疏毛灰毛豆●☆

386319 Tephrosia stricta（L. f.）Pers. ＝ Indigofera stricta L. f. ●☆

386320 Tephrosia strigosa（Dalzell）Santapau et Maheshwari；糙伏毛灰毛豆●☆

386321 Tephrosia stuhlmannii Taub. ＝ Tephrosia stormsii De Wild. ●☆

386322 Tephrosia subalpina A. Chev. ＝ Dolichos dinklagei Harms ■☆

386323 Tephrosia subaphylla R. Vig. ex Du Puy et Labat；亚无叶灰毛豆●☆

386324 Tephrosia suberosa DC. ＝ Mundulea sericea（Willd.）A. Chev. ●☆

386325 Tephrosia subfalcato-stipulata De Wild. ＝ Tephrosia dasyphylla Welw. ex Baker ■☆

386326 Tephrosia subpraecox Cronquist；较早生灰毛豆■☆

386327 Tephrosia subtriflora Hochst. ex Baker；亚三花灰毛豆●☆

386328 Tephrosia subulata Hutch. et Burtt Davy；钻形灰毛豆●☆

386329 Tephrosia sulphurea Chiov. ＝ Tephrosia subtriflora Hochst. ex Baker ●☆

386330 Tephrosia swynnertonii Baker f. ＝ Tephrosia longipes Meisn. subsp. swynnertonii（Baker f.）Brummitt ■☆

386331 Tephrosia sylitroides Baker f.；拟异灰毛豆●☆

386332 Tephrosia sylviae Berhaut；西尔维亚灰毛豆●☆

386333 Tephrosia tanganicensis De Wild.；坦噶尼喀灰毛豆●☆

386334 Tephrosia ternatifolia R. G. N. Young ＝ Tephrosia capensis（Jacq.）Pers. ■☆

386335 Tephrosia timoriensis DC. ＝ Tephrosia pumila（Lam.）Pers. ■

386336 Tephrosia tinctoria Perrs. var. coccinea Baker ＝ Tephrosia coccinea Wall. ■☆

386337 Tephrosia totta（Thunb.）Pers.；托特灰毛豆■☆

386338 Tephrosia toxicaria（Sav.）Pers.；毒灰毛豆■☆

386339 Tephrosia toxicaria（Sw.）Pers. ＝ Tephrosia sinapou（Buchoz）A. Chev. ●☆

386340 Tephrosia transjubensis Chiov. ＝ Tephrosia uniflora Pers. ■☆

386341 Tephrosia transvaalensis Hutch. ex Burtt Davy ＝ Tephrosia purpurea（L.）Pers. var. pubescens Baker ●■

386342 Tephrosia tricolor（Hook.）Sweet；三色灰毛豆■☆

386343 Tephrosia triphylla Harms；三叶灰毛豆■☆

386344 Tephrosia tundavalensis Bamps；通达灰毛豆■☆

386345 Tephrosia tutcheri Dunn ＝ Millettia pulchra（Benth.）Kurz ●

386346 Tephrosia tzaneenensis H. M. L. Forbes ＝ Tephrosia elongata E. Mey. var. tzaneenensis（H. M. L. Forbes）Brummitt ●☆

386347 Tephrosia uniflora Pers.；单花灰叶■☆

386348 Tephrosia uniflora Pers. subsp. petrosa（Blatt. et Hallb.）J. B. Gillett et Ali；岩生单花灰叶■☆

386349 Tephrosia uniflora Pers. var. parviflora Baker f. ＝ Tephrosia uniflora Pers. ■☆

386350 Tephrosia unifolia H. M. L. Forbes ＝ Tephrosia albissima H. M. L. Forbes subsp. zuluensis（H. M. L. Forbes）Schrire ●☆

386351 Tephrosia varians（F. M. Bailey）C. T. White；变异灰叶■☆

386352 Tephrosia verdickii De Wild.；韦尔灰毛豆■☆

386353 Tephrosia vestita Vogel；黄灰毛豆（狐狸射草，黄毛灰毛豆，黄毛灰叶，假鸟豆）；Yellow Hoarypea, Yellowhair Tephrosia ●■

386354 Tephrosia vicioides A. Rich. ＝ Tephrosia uniflora Pers. ■☆

386355 Tephrosia vicioides A. Rich. var. inflexa Chiov. ＝ Tephrosia pumila（Lam.）Pers. ■

386356 Tephrosia viguieri Du Puy et Labat；维基耶灰毛豆●☆

386357 Tephrosia vilersii Drake ＝ Pyranthus monantha（Baker）Du Puy et Labat ●☆

386358 Tephrosia villosa（L.）Pers.；长毛灰叶（弗州灰毛豆）■

386359 Tephrosia villosa（L.）Pers. subsp. ehrenbergiana（Schweinf.）Brummitt ＝ Tephrosia villosa（L.）Pers. var. ehrenbergiana（Schweinf.）Brummitt ■☆

386360 Tephrosia villosa（L.）Pers. var. daviesii Brummitt；戴维斯灰毛豆■☆

386361 Tephrosia villosa（L.）Pers. var. ehrenbergiana（Schweinf.）Brummitt；爱伦堡灰毛豆■☆

386362 Tephrosia villosa（L.）Pers. var. incana（Roxb.）Baker ＝ Tephrosia villosa（L.）Pers. var. ehrenbergiana（Schweinf.）Brummitt ■☆

386363 Tephrosia villosa Pers. ＝ Tephrosia villosa（L.）Pers. ■

386364 Tephrosia virgata H. M. L. Forbes；条纹灰毛豆■☆

386365 Tephrosia virginiana（L.）Pers.；弗州灰毛豆；Cat Gut, Catgut, Devil's Shoestrings, Dolly Varden, Goat's Rue, Goat's-rue, Hoary Pea, Rabbit Pea, Rabbit-pea, Turkey Pea, Virginia Tephrosia ■☆

386366 Tephrosia virginiana（L.）Pers. var. glabra Nutt. ＝ Tephrosia virginiana（L.）Pers. ■☆

386367 Tephrosia virginiana（L.）Pers. var. holosericea（Nutt.）Torr. et A. Gray ＝ Tephrosia virginiana（L.）Pers. ■☆

386368 Tephrosia vogelii Hook. f.；西非灰毛豆（灰毛豆，灰叶豆，沃氏灰毛豆）；Fish Poison Bean, Vogel Hoarypea, Vogel Tephrosia ●■

386369 Tephrosia vohimenaensis Du Puy et Labat；武希梅纳灰毛豆●☆

386370 Tephrosia wallichii Graham ＝Tephrosia purpurea（L.）Pers. ●■

386371 Tephrosia whyteana Baker f.；怀特灰毛豆■☆

386372 Tephrosia whyteana Baker f. subsp. gemina Brummitt；双生灰毛

豆■☆

386373　Tephrosia wittei Baker f. = Tephrosia curvata De Wild. ●☆

386374　Tephrosia woodii Burtt Davy = Tephrosia multijuga R. G. N. Young ●☆

386375　Tephrosia wyliei H. M. L. Forbes = Tephrosia shiluwanensis Schinz ●☆

386376　Tephrosia youngii Torre = Tephrosia dasyphylla Welw. ex Baker subsp. youngii (Torre) Brummitt ●☆

386377　Tephrosia zambesiaca Taub. = Tephrosia elongata E. Mey. ●☆

386378　Tephrosia zambiana Brummitt;赞比亚软毛豆■☆

386379　Tephrosia zombensis Baker = Tephrosia aequilata Baker subsp. nyasae (Baker f.) Brummitt ■☆

386380　Tephrosia zoutpansbergensis Bremek.;佐特灰毛豆■☆

386381　Tephrosia zuluensis H. M. L. Forbes = Tephrosia albissima H. M. L. Forbes subsp. zuluensis (H. M. L. Forbes) Schrire ●☆

386382　Tephrothamnus Sch. Bip. (1863);灰灌菊属●☆

386383　Tephrothamnus Sch. Bip. = Critoniopsis Sch. Bip. ●■☆

386384　Tephrothamnus Sch. Bip. = Vernonia Schreb. (保留属名)●■

386385　Tephrothamnus Sweet = Argyrolobium Eckl. et Zeyh. (保留属名)●☆

386386　Tephrothamnus calophyllus Sch. Bip.;美叶灰灌菊●☆

386387　Tephrothamnus paradoxus Sch. Bip.;奇异灰灌菊●☆

386388　Tephrothamnus pycnanthus Sch. Bip.;密花灰灌菊●☆

386389　Tepion Adans. = Verbesina L. (保留属名)●■☆

386390　Tepso Raf. = Agostana Bute ex Gray ●■

386391　Tepso Raf. = Bupleurum L. ●■

386392　Tepualia Griseb. (1854);塔普桃金娘属●☆

386393　Tepualia philippiana Griseb.;塔普桃金娘●☆

386394　Tepuia Camp(1939);腺白珠属●☆

386395　Tepuia floribunda Camp;繁花腺白珠●☆

386396　Tepuia intermedia Steyerm.;间型腺白珠●☆

386397　Tepuia speciosa A. C. Sm.;美丽腺白珠●☆

386398　Tepuianthaceae Maguire et Steyerm. (1981);苦皮树科(绢毛果科)●☆

386399　Tepuianthaceae Maguire et Steyerm. = Thymelaea Mill. (保留属名)●■

386400　Tepuianthus Maguire et Steyerm. (1981);苦皮树属●☆

386401　Tepuianthus auyantepuiensis Maguire et Steyerm.;苦皮树●☆

386402　Tepuianthus colombianus Maguire et Steyerm.;哥伦比亚苦皮树●☆

386403　Teramnus P. Browne(1756);软荚豆属(钩豆属,毛豆属,野黄豆属);Softbean,Teramnus ●

386404　Teramnus andongensis (Welw. ex Baker) Baker f. = Teramnus uncinatus (L.) Sw. subsp. ringoetii (De Wild.) Verdc. ■☆

386405　Teramnus angustifolius Merr. = Teramnus labialis (L. f.) Spreng. ■

386406　Teramnus axilliflorus (Kotschy) Baker f. = Teramnus uncinatus (L.) Sw. subsp. axilliflorus (Kotschy) Verdc. ■☆

386407　Teramnus buettneri (Harms) Baker f.;比特纳软荚豆■☆

386408　Teramnus gilletii (De Wild.) Baker f. = Teramnus uncinatus (L.) Sw. subsp. axilliflorus (Kotschy) Verdc. ■☆

386409　Teramnus gracilis Chiov. = Teramnus repens (Taub.) Baker f. subsp. gracilis (Chiov.) Verdc. ■☆

386410　Teramnus labialis (L. f.) Spreng.;软荚豆(钩豆,广叶南洋豆,野黄豆);Common Teramnus,Softbean ●

386411　Teramnus labialis (L. f.) Spreng. subsp. arabicus Verdc.;阿拉伯软荚豆■☆

386412　Teramnus labialis (L. f.) Spreng. var. abyssinicus (Hochst. ex A. Rich.) Verdc.;阿比西尼亚软荚豆■☆

386413　Teramnus labialis (L. f.) Spreng. var. mollis Baker = Teramnus mollis Benth. ●☆

386414　Teramnus lanceolifoliatus (De Wild.) Baker f. = Teramnus uncinatus (L.) Sw. subsp. ringoetii (De Wild.) Verdc. ■☆

386415　Teramnus micans (Welw. ex Baker) Baker f.;弱光泽软荚豆■☆

386416　Teramnus micans (Welw. ex Baker) Baker f. var. cyaneus (De Wild.) Hauman;蓝色软荚豆■☆

386417　Teramnus mollis Benth.;柔软软荚豆●☆

386418　Teramnus obcordatus Baill. = Galactia tenuiflora (Klein ex Willd.) Wight et Arn. ■

386419　Teramnus repens (Taub.) Baker f.;匍匐软荚豆■☆

386420　Teramnus repens (Taub.) Baker f. subsp. gracilis (Chiov.) Verdc.;纤细匍匐软荚豆■☆

386421　Teramnus stolzii Baker f. = Teramnus micans (Welw. ex Baker) Baker f. var. cyaneus (De Wild.) Hauman ■☆

386422　Teramnus uncinatus (L.) Sw.;具钩软荚豆■☆

386423　Teramnus uncinatus (L.) Sw. subsp. axilliflorus (Kotschy) Verdc.;腋花软荚豆■☆

386424　Teramnus uncinatus (L.) Sw. subsp. ringoetii (De Wild.) Verdc.;林戈软荚豆■☆

386425　Terana La Llave(1884);太拉菊属☆

386426　Terana lanceolata La Llave;太拉菊☆

386427　Teranta Beriand. = Leucophyllum Bonpl. ●☆

386428　Terauchia Nakai(1913);朝鲜百合属■☆

386429　Terebintaceae Juss. = Anacardiaceae R. Br. (保留科名)●

386430　Terebinthaceae Juss. = Anacardiaceae R. Br. (保留科名) + Pistaciaceae + Connaraceae R. Br. (保留科名) + Burserae + Rutaceae Juss. (保留科名)●■

386431　Terebinthina Kuntze = Ambuli Adans. (废弃属名)■

386432　Terebinthina Kuntze = Limnophila R. Br. (保留属名)■

386433　Terebinthina Rumph. = Limnophila R. Br. (保留属名)■

386434　Terebinthina Rumph. ex Kuntze = Limnophila R. Br. (保留属名)■

386435　Terebinthus Mill. = Pistacia L. + Bursera Jacq. ex L. (保留属名)●☆

386436　Terebinthus P. Browne = Bursera Jacq. ex L. (保留属名)●☆

386437　Terebraria Kuntze = Neolaugeria Nicolson ●☆

386438　Terebraria Sessé ex DC. = Neolaugeria Nicolson ●☆

386439　Terebraria Sessé ex Kunth = Neolaugeria Nicolson ●☆

386440　Tereianthes Raf. = Reseda L. ■

386441　Tereianthus Fourr. = Tereianthes Raf. ■

386442　Tereietra Raf. = Calonyction Choisy ■

386443　Tereietra Raf. = Ipomoea L. (保留属名)●■

386444　Tereiphas Raf. = Scabiosa L. ●■

386445　Teremis Raf. = Lycium L. ●

386446　Terepis Raf. = Salvia L. ●■

386447　Terera Dombey ex Naudin = Miconia Ruiz et Pav. (保留属名)●☆

386448　Terminalia L. (1767) (保留属名);榄仁树属(诃子属,榄仁属);Ashanti Gum,Myrobalan,Myrobalans,Subakh,Terminalia ●

386449　Terminalia acuta Walp. = Terminalia chebula Retz. ●

386450　Terminalia adamanensis Engl. et Diels = Terminalia macroptera Guillaumin et Perr. ●☆

386451　Terminalia aemula Diels = Terminalia sambesiaca Engl. et Diels ●☆

386452　Terminalia alata Roth;有翅榄仁树;Asna,Indian Laurel ●☆

386453　Terminalia albida Scott-Elliot;微白榄仁树●☆

386454　Terminalia altissima A. Chev. = Terminalia superba Engl. et Diels ●☆

386455　Terminalia amazonia（J. F. Gmel.）Exell；亚马逊榄仁树●☆

386456　Terminalia angolensis O. Hoffm. = Terminalia sericea Burch. ex DC. ●☆

386457　Terminalia angolensis Welw. ex Ficalho = Terminalia sericea Burch. ex DC. ●☆

386458　Terminalia angustifolia Jack. ；狭叶榄仁树●☆

386459　Terminalia argyrophylla Engl. et Diels = Terminalia albida Scott-Elliot ●☆

386460　Terminalia argyrophylla King et Prain；银叶诃子（小诃子）；Silverleaf Terminalia，Silver-leaved Terminalia ●

386461　Terminalia argyrophylla King et Prain = Terminalia chebula Retz. var. tomentella（Kurz）C. B. Clarke ●

386462　Terminalia arjuna（Roxb. ex DC.）Wight et Arn. ；印度榄仁树（阿江榄仁，阿江榄仁树，诃子，三果木）；Arjun，Arjun Terminalia，Kahua Kahua Bark，Kumbuk ●☆

386463　Terminalia arjuna（Roxb.）Wight et Arn. = Terminalia arjuna（Roxb. ex DC.）Wight et Arn. ●☆

386464　Terminalia arjuna Wight et Arn. = Terminalia arjuna（Roxb. ex DC.）Wight et Arn. ●☆

386465　Terminalia arostrata Ewart et O. B. Davies；无喙榄仁树；Nut Wood ●☆

386466　Terminalia attenuata Edgew. = Terminalia bellerica（Gaertn.）Roxb. ●

386467　Terminalia avicennioides Guillaumin et Perr. ；海榄雌榄仁树●☆

386468　Terminalia badamia Tul. = Terminalia catappa L. ●

386469　Terminalia balladellii Chiov. = Terminalia brevipes Pamp. ●☆

386470　Terminalia basilei Chiov. ；基生榄仁树●☆

386471　Terminalia baumannii Engl. et Diels = Terminalia schimperiana Hochst. ●☆

386472　Terminalia baumii Engl. et Gilg = Terminalia brachystemma Welw. ex Hiern ●☆

386473　Terminalia bellerica（Gaertn.）Roxb. ；红果榄仁树（川楝，毛诃子，毗梨筋，毗黎筋，油榄仁）；Bahera Nut，Bastard Myrobalan，Bedda Nut，Belilaj，Bellir Terminalia，Myrobalan ●

386474　Terminalia bellerica（Gaertn.）Roxb. var. laurinoides（Teijsm. et Binn.）C. B. Clarke = Terminalia bellerica（Gaertn.）Roxb. ●

386475　Terminalia bellerica（Gaertn.）Roxb. var. laurinoides C. B. Clarke = Terminalia bellerica（Gaertn.）Roxb. ●

386476　Terminalia benguellensis Welw. ex Hiern = Terminalia prunioides M. A. Lawson ●☆

386477　Terminalia benguellensis Welw. ex Hiern var. ovalis Hiern = Terminalia prunioides M. A. Lawson ●☆

386478　Terminalia bialata Kurz；银灰榄仁树；Silvergrey Data，White Chuglam，White Chuglam Wood ●☆

386479　Terminalia bimarginata H. Perrier；双边榄仁树●☆

386480　Terminalia bispinosa Schweinf. et Volkens = Terminalia spinosa Engl. ●☆

386481　Terminalia boivinii Tul. ；博伊文榄仁树●☆

386482　Terminalia boivinii Tul. var. chlorophylla Tul. = Terminalia boivinii Tul. ●☆

386483　Terminalia boivinii Tul. var. microcarpa Tul. = Terminalia boivinii Tul. ●☆

386484　Terminalia brachystemma Welw. ex Hiern；短冠榄仁树●☆

386485　Terminalia brachystemma Welw. ex Hiern subsp. sessilifolia（R. E. Fr.）Wickens；无梗短冠榄仁树●☆

386486　Terminalia brevipes Pamp. ；短梗榄仁树●☆

386487　Terminalia brosigiana Engl. et Diels = Terminalia sericea Burch. ex DC. ●☆

386488　Terminalia brownii Fresen. ；布氏榄仁树（细叶榄仁树）；Brown's Myrobalan ●☆

386489　Terminalia brownii Fresen. var. albertensis Bagsh. et Baker f. = Terminalia brownii Fresen. ●☆

386490　Terminalia brownii Fresen. var. gallaensis Engl. ex Diels = Terminalia brownii Fresen. ●☆

386491　Terminalia brownii Fresen. var. stenocarpa Fiori = Terminalia brownii Fresen. ●☆

386492　Terminalia bubu De Wild. et Ledoux = Terminalia sericea Burch. ex DC. ●☆

386493　Terminalia calamansanai（Blanco）Rolfe；马尼拉榄仁树●☆

386494　Terminalia calcicola H. Perrier；钙生榄仁树●☆

386495　Terminalia calophylla Tul. ；美叶榄仁树●☆

386496　Terminalia canescens Engl. = Swertia kilimandscharica Engl. ■☆

386497　Terminalia canescens Engl. = Terminalia kilimandscharica Engl. ●☆

386498　Terminalia capuronii H. Perrier；凯普伦榄仁树●☆

386499　Terminalia carpentariae C. T. White；卡奔他榄仁树（榄仁树）●☆

386500　Terminalia catappa L. ；榄仁树（榄仁，枇杷树，山枇杷树，戌士）；Almond，Amendoeira，Barbados Almond，Bastard Almond，Chebule，Country Almond，India-almond，Indian Almond，Indian Swede，Indian Turnip，Katamba，Malabar Almond，Myrobalan，Oliver-bark Tree，Sea Almond，Tropical Almond，Wild Almond ●

386501　Terminalia catappa L. var. rhodocarpa Hassk. = Terminalia catappa L. ●

386502　Terminalia catappa L. var. subcordata（Humb. et Bonpl. ex Willd.）DC. = Terminalia catappa L. ●

386503　Terminalia cephalota McPherson；头状榄仁树●☆

386504　Terminalia chebula Retz. ；诃子（藏青果，诃梨勒，诃黎，诃黎勒，诃黎筋，诃仔，呵子，卡西，苛子，桄，码蜡，码漏，毗梨勒，婆罗得，婆罗勒，三果，涩翁，随风子，西藏青果，西青果）；Chebula Terminalia，Chebulic Myrobalan，Firda-tree，Medicine Terminalia，Myrobalan，Myrobalans，Myrobalanwood，Terminalia ●

386505　Terminalia chebula Retz. var. gangetica（Roxb.）C. B. Clarke；恒河诃子；Henghe Terminalia ●

386506　Terminalia chebula Retz. var. tomentella（Kurz）C. B. Clarke；微毛呵子（呵子，绒毛呵子）；Tomentose Medicine Terminalia ●

386507　Terminalia chevalieri Diels = Terminalia laxiflora Engl. ●☆

386508　Terminalia chlorocarpa H. Perrier；绿果榄仁●☆

386509　Terminalia circumalata F. Muell. ；环翅榄仁●☆

386510　Terminalia citrina Roxb. ex Fleming；柠檬榄仁树●☆

386511　Terminalia claessensii De Wild. = Terminalia macroptera Guillaumin et Perr. ●☆

386512　Terminalia comintana Merr. ；科明榄仁树●☆

386513　Terminalia confertifolia Steud. ex A. Rich. = Terminalia brownii Fresen. ●☆

386514　Terminalia crenata Tul. ；圆齿榄仁树●☆

386515　Terminalia crenulata Roth；印锡榄仁树●☆

386516　Terminalia cycloptera R. Br. = Terminalia brownii Fresen. ●☆

386517　Terminalia dawei Rolfe = Terminalia macroptera Guillaumin et Perr. ●☆

386518　Terminalia dewevrei De Wild. et T. Durand = Terminalia mollis M. A. Lawson ●☆

386519　Terminalia dictyoneura Engl. et Diels = Terminalia avicennioides

Guillaumin et Perr. ●☆

386520　Terminalia dielsii Engl. = Terminalia stenostachya Engl. et Diels ●☆

386521　Terminalia divaricata H. Perrier;叉枝榄仁树●☆

386522　Terminalia diversipilosa H. Perrier;异毛榄仁树●☆

386523　Terminalia dukouensis W. P. Fang et P. C. Kao;渡口榄仁; Dukou Terminalia ●

386524　Terminalia dukouensis W. P. Fang et P. C. Kao = Terminalia franchetii Gagnep. ●

386525　Terminalia edulis Blanco;食用榄仁树●☆

386526　Terminalia eglandulosa Roxb. ex C. B. Clarke = Terminalia bellerica (Gaertn.) Roxb. ●

386527　Terminalia elliotii Engl. et Diels = Terminalia laxiflora Engl. ●☆

386528　Terminalia erici-rosenii R. E. Fr. ;欧石南榄仁树●☆

386529　Terminalia erythrocarpa F. Muell. ;红果榄仁●☆

386530　Terminalia erythrophylla Burch. = Combretum erythrophyllum (Burch.) Sond. ●☆

386531　Terminalia excelsior A. Chev. = Terminalia schimperiana Hochst. ●☆

386532　Terminalia exsculpta Tul. ;雕刻榄仁树●☆

386533　Terminalia fischeri Engl. = Terminalia sericea Burch. ex DC. ●☆

386534　Terminalia flava Engl. = Terminalia schimperiana Hochst. ●☆

386535　Terminalia flavicans Boivin ex Tul. ;浅黄榄仁树●☆

386536　Terminalia foetens Engl. = Terminalia sambesiaca Engl. et Diels ●☆

386537　Terminalia franchetii Gagnep. ;滇榄仁(夫兰氏榄仁,黄心树, 毛叶榄仁);Franchet Terminalia ●

386538　Terminalia franchetii Gagnep. var. glabra Exell;光叶滇榄仁(光叶夫兰氏榄仁,牛板筋,牛筋树,牛荆,无毛滇榄仁);Glabrous Franchet Terminalia ●

386539　Terminalia franchetii Gagnep. var. glabra Exell = Terminalia franchetii Gagnep. ●

386540　Terminalia franchetii Gagnep. var. intricata (Hand. -Mazz.) Turland et C. Chen;错枝榄仁(云南榄仁,云南榄仁树);Intricate Terminalia ●

386541　Terminalia franchetii Gagnep. var. membranifolia A. C. Chao = Terminalia franchetii Gagnep. ●

386542　Terminalia franchetii Gagnep. var. membranifolia H. C. Chao;薄叶滇榄仁(薄叶夫兰氏榄仁,膜叶滇榄仁);Thinleaf Franchet Terminalia ●

386543　Terminalia franchetii Gagnep. var. tomentosa Nanakorn = Terminalia franchetii Gagnep. ●

386544　Terminalia gangetica Roxb. = Terminalia chebula Retz. var. gangetica (Roxb.) C. B. Clarke ●

386545　Terminalia gangetica Roxb. = Terminalia chebula Retz. ●

386546　Terminalia gazensis Baker f. ;加兹榄仁树●☆

386547　Terminalia gella Dalzell = Terminalia bellerica (Gaertn.) Roxb. ●

386548　Terminalia glabra Wight et Arn. = Terminalia arjuna (Roxb. ex DC.) Wight et Arn. ●☆

386549　Terminalia glandulipetiolata De Wild. = Terminalia chebula Retz. ●

386550　Terminalia glandulosa De Wild. = Terminalia mollis M. A. Lawson ●☆

386551　Terminalia glaucescens Planch. ex Benth. = Terminalia schimperiana Hochst. ●☆

386552　Terminalia gossweileri Exell et J. G. Garcia;戈斯榄仁树●☆

386553　Terminalia griffithsiana Liben;格氏榄仁树●☆

386554　Terminalia hainanensis Exell = Terminalia nigrovenulosa Pierre ●

386555　Terminalia hararensis Engl. ex Diels;哈拉雷榄仁树●☆

386556　Terminalia hecistocarpa Engl. et Diels;寡果榄仁树●☆

386557　Terminalia hemignosta Steud. ex A. Rich. = Terminalia brownii Fresen. ●☆

386558　Terminalia hildebrandtii Engl. = Swertia kilimandscharica Engl. ■☆

386559　Terminalia holstii Engl. = Terminalia prunioides M. A. Lawson ●☆

386560　Terminalia horrida Steud. ;刺毛榄仁树●☆

386561　Terminalia intermedia Bertero ex Spreng. = Terminalia catappa L. ●

386562　Terminalia intricata Hand. -Mazz. = Terminalia franchetii Gagnep. var. intricata (Hand. -Mazz.) Turland et C. Chen ●

386563　Terminalia ivorensis A. Chev. ;象牙海岸榄仁;Black Afara, Blackbark, Brimstone-wood, Emeri, Idigbo, Ivory Coast Almond, Satinwood, Yellow Terminalia ●☆

386564　Terminalia kelleri Engl. et Diels = Terminalia polycarpa Engl. et Diels ●☆

386565　Terminalia kerstingii Engl. = Terminalia mollis M. A. Lawson ●☆

386566　Terminalia kilimandscharica Engl. ;基尔曼榄仁树●☆

386567　Terminalia kouytchensis H. Lév. = Gouania javanica Miq. ●

386568　Terminalia latifolia Blanco = Terminalia catappa L. ●

386569　Terminalia laurinoides Teijsm. et Binn. = Terminalia bellerica (Gaertn.) Roxb. ●

386570　Terminalia laxiflora Engl. ;疏花榄仁●☆

386571　Terminalia leandriana H. Perrier;利安榄仁●☆

386572　Terminalia lecardii Engl. et Diels = Terminalia avicennioides Guillaumin et Perr. ●☆

386573　Terminalia longipes Engl. = Terminalia schimperiana Hochst. ●☆

386574　Terminalia macroptera Guillaumin et Perr. ;大翅榄仁树●☆

386575　Terminalia mairei H. Lév. = Combretum wallichii DC. ●

386576　Terminalia mauritiana Lam. = Terminalia catappa L. ●

386577　Terminalia melanocarpa F. Muell. ;黑果榄仁●☆

386578　Terminalia menezesii Mendes et Exell;梅内塞斯榄仁树●☆

386579　Terminalia micans Hand. -Mazz. = Terminalia franchetii Gagnep. ●

386580　Terminalia mildbreadii Gilg ex Mildbr. = Terminalia mollis M. A. Lawson ●☆

386581　Terminalia mollis M. A. Lawson;柔软榄仁树●☆

386582　Terminalia moluccana Lam. = Terminalia catappa L. ●

386583　Terminalia monoceros H. Perrier;单角榄仁树●☆

386584　Terminalia muelleri Benth. ;卵果榄仁(莫氏榄仁);Australian Almond ●☆

386585　Terminalia myriocarpa Van Heurck et Müll. Arg. ;千果榄仁(大马缨子花,多果榄仁,多果榄仁树,红花树,千红花树);Bayberry Waxmyrtle-fruit, Bayberry Waxmyrtlefruit Terminalia, East Indian Almond, Hollock ●◇

386586　Terminalia myriocarpa Van Heurck et Müll. Arg. var. hirsuta Craib;硬毛千果榄仁(硬千果榄仁);Hirsute Bayberry Waxmyrtlefruit Terminalia ●

386587　Terminalia myrobalana Roth = Terminalia catappa L. ●

386588　Terminalia namorokensis H. Perrier;纳莫鲁克榄仁树●☆

386589　Terminalia nigrovenulosa Pierre;海南榄仁(海南金针木,黑脉榄仁树,鸡尖,鸡占,鸡针木,鸡珍);Hainan Terminalia ●

386590　Terminalia nitens C. Presl;光亮榄仁树●☆

386591　Terminalia nyassensis Engl. = Terminalia sericea Burch. ex DC. ●☆

386592　Terminalia obcordiformis H. Perrier;倒心形榄仁树●☆

386593　Terminalia obliqua Craib = Terminalia nigrovenulosa Pierre ●

386594　Terminalia oblonga (Ruiz et Pav.) Steud. ;秘鲁榄仁树;

Peruvian Almond ●☆

386595   Terminalia oblonga Steud. = Terminalia oblonga (Ruiz et Pav.) Steud. ●☆

386596   Terminalia obovata Sim = Terminalia sambesiaca Engl. et Diels ●☆

386597   Terminalia ombrophila H. Perrier;喜雨榄仁树●☆

386598   Terminalia orbicularis Engl. et Diels;圆形榄仁树●☆

386599   Terminalia orbicularis Engl. et Diels var. macroptera ? = Terminalia orbicularis Engl. et Diels ●☆

386600   Terminalia ovatifolia Noronha = Terminalia catappa L. ●

386601   Terminalia paniculata Roth;锥花榄仁树●☆

386602   Terminalia paraensis Mart. = Terminalia catappa L. ●

386603   Terminalia parviflora Thwaites = Terminalia chebula Retz. ●

386604   Terminalia parvula Pamp. ;较小榄仁树●☆

386605   Terminalia passargei Engl. ex Engl. et Diels = Terminalia schimperiana Hochst. ●☆

386606   Terminalia pauciflora Tul. ;少花榄仁树●☆

386607   Terminalia petersii Engl. = Terminalia prunioides M. A. Lawson ●☆

386608   Terminalia phanerophlebia Engl. et Diels;显脉榄仁树●☆

386609   Terminalia poliotricha Diels = Terminalia trichopoda Diels ●☆

386610   Terminalia polycarpa Engl. et Diels;多果花榄仁树●☆

386611   Terminalia porphyrocarpa Schinz = Terminalia prunioides M. A. Lawson ●☆

386612   Terminalia praecox Engl. et Diels = Terminalia orbicularis Engl. et Diels ●☆

386613   Terminalia procera Roxb. ; 高 大 榄 仁 树; Badam, White Bombway ●☆

386614   Terminalia procera Roxb. = Terminalia catappa L. ●

386615   Terminalia prunioides M. A. Lawson;粉榄仁树●☆

386616   Terminalia pteleopsoides Exell = Pteleopsis pteleopsoides (Exell) Vollesen ●☆

386617   Terminalia pumila Thouars ex Tul. ;矮榄仁树●☆

386618   Terminalia punctata Roth = Terminalia bellerica (Gaertn.) Roxb. ●

386619   Terminalia randii Baker f. ;兰德榄仁树●☆

386620   Terminalia rautanenii Schinz = Terminalia prunioides M. A. Lawson ●☆

386621   Terminalia repanda A. Chev. = Terminalia laxiflora Engl. ●☆

386622   Terminalia reticulata Engl. = Terminalia mollis M. A. Lawson ●☆

386623   Terminalia reticulata Roth = Terminalia chebula Retz. ●

386624   Terminalia rhodesica R. E. Fr. = Terminalia stenostachya Engl. et Diels ●☆

386625   Terminalia riparia Engl. et Diels = Terminalia sambesiaca Engl. et Diels ●☆

386626   Terminalia robecchii Chiov. = Terminalia spinosa Engl. ●☆

386627   Terminalia roseo-grisea Gilg et Ledermann ex Engl. = Terminalia laxiflora Engl. ●☆

386628   Terminalia rubrigemmis Tul. = Terminalia catappa L. ●

386629   Terminalia ruspolii Engl. et Diels = Terminalia orbicularis Engl. et Diels ●☆

386630   Terminalia ruspolii Engl. et Diels var. macroptera Pamp. = Terminalia orbicularis Engl. et Diels ●☆

386631   Terminalia saja Steud. = Terminalia myriocarpa Van Heurck et Müll. Arg. ●◇

386632   Terminalia salicifolia Schweinf. = Terminalia schimperiana Hochst. ●☆

386633   Terminalia sambesiaca Engl. et Diels;热非榄仁树●☆

386634   Terminalia schimperiana Hochst. ;欣珀榄仁树●☆

386635   Terminalia schweinfurthii Engl. et Diels = Terminalia laxiflora Engl. ●☆

386636   Terminalia scutifera Planch. ex M. A. Lawson;盾状榄仁树●☆

386637   Terminalia semlikiensis De Wild. = Terminalia brownii Fresen. ●☆

386638   Terminalia sericea Burch. ex DC. ;丝质榄仁树(如丝榄仁树); Assegai Wood, Mang We, Yellow-wood ●☆

386639   Terminalia sericea Burch. ex DC. var. angolensis Hiern = Terminalia sericea Burch. ex DC. ●☆

386640   Terminalia sericea Burch. ex DC. var. huillensis Hiern = Terminalia sericea Burch. ex DC. ●☆

386641   Terminalia sessilifolia R. E. Fr. = Terminalia brachystemma Welw. ex Hiern subsp. sessilifolia (R. E. Fr.) Wickens ●☆

386642   Terminalia sokodensis Engl. = Terminalia laxiflora Engl. ●☆

386643   Terminalia spekei Rolfe = Terminalia mollis M. A. Lawson ●☆

386644   Terminalia spinosa Engl. ;多刺榄仁树●☆

386645   Terminalia splendida Engl. et Diels = Terminalia stenostachya Engl. et Diels ●☆

386646   Terminalia stenostachya Engl. et Diels;细穗榄仁树●☆

386647   Terminalia stuhlmannii Engl. ;斯图尔曼榄仁树●☆

386648   Terminalia subcordata Humb. et Bonpl. ex Willd. = Terminalia catappa L. ●

386649   Terminalia suberosa R. E. Fr. = Terminalia mollis M. A. Lawson ●☆

386650   Terminalia subserrata H. Perrier;具齿榄仁树●☆

386651   Terminalia sulcata Tul. ;纵沟榄仁树●☆

386652   Terminalia superba Engl. et Diels;非洲榄仁树(艳榄仁); Afara, Aiara Terminaila, Akom, Limba, Ofram, White Afara ●☆

386653   Terminalia thomasii Engl. et Diels = Terminalia sambesiaca Engl. et Diels ●☆

386654   Terminalia togoensis Engl. et Diels = Terminalia schimperiana Hochst. ●☆

386655   Terminalia tomentella Kurz = Terminalia chebula Retz. var. tomentella (Kurz) C. B. Clarke ●

386656   Terminalia tomentosa Bedd. ;毛榄仁树(毛榄仁);Rokfa,Sain, Woolly Terminalia ●☆

386657   Terminalia torulosa F. Hoffm. = Terminalia mollis M. A. Lawson ●☆

386658   Terminalia trichopoda Diels;毛梗榄仁树●☆

386659   Terminalia triptera Franch. = Terminalia franchetii Gagnep. ●

386660   Terminalia triptera Stapf = Terminalia nigrovenulosa Pierre ●

386661   Terminalia tripteroides Craib = Terminalia nigrovenulosa Pierre ●

386662   Terminalia ulexoides H. Perrier;荆豆榄仁树●☆

386663   Terminalia urjan Royle = Terminalia arjuna (Roxb. ex DC.) Wight et Arn. ●☆

386664   Terminalia urschii H. Perrier;乌尔施榄仁树●☆

386665   Terminalia velutina Rolfe = Terminalia schimperiana Hochst. ●☆

386666   Terminalia worensis ?;沃尔榄仁树;Shingle-wood ●☆

386667   Terminalia zeylanica Van Heurck et Müll. Arg. = Terminalia chebula Retz. ●

386668   Terminaliaceae J. St. -Hil. ;榄仁树科;Myrobalan Family ●

386669   Terminaliaceae J. St. -Hil. = Combretaceae R. Br. (保留科名)●

386670   Terminaliopsis Danguy(1923);类榄仁树属●☆

386671   Terminaliopsis tetrandrus Danguy;类榄仁树●☆

386672   Terminalis Kuntze = Cordyline Comm. ex R. Br. (保留属名)●

386673   Terminalis Medik. = Dracaena Vand. ex L. ●■

386674   Terminalis Rumph. = Cordyline Comm. ex R. Br. (保留属名)●

386675   Terminalis Rumph. ex Kuntze = Cordyline Comm. ex R. Br. (保留属名)●

386676   Terminthia Bernh. (1938);三叶漆属;Threeleaf Lacquertree,

Trileaved Lacquer-tree ●

386677 Terminthia Bernh. = Rhus L. ●

386678 Terminthia paniculata（Wall. ex G. Don）C. Y. Wu et T. L. Ming；三叶漆（扁果，野荔枝）；Paniculate Threeleaf Lacquertree，Paniculate Trileaved Lacquer-tree，Threeleaf Lacquertree ●

386679 Terminthodia Ridl. = Tetractomia Hook. f. ●☆

386680 Terminthos St. -Lag. = Pistacia L. ●

386681 Termontis Raf.（1815）= Linaria Mill. ■

386682 Termontis Raf.（1840）= Antirrhinum L. ●■

386683 Ternatea Mill. = Clitoria L. ●

386684 Terniola Tul. = Dalzellia Wight ■

386685 Terniola Tul. = Lawia Tul. ■

386686 Terniopsis H. C. Chao = Dalzellia Wight ■

386687 Terniopsis sessilis H. C. Chao = Dalzellia sessilis（H. C. Chao）C. Cusset et G. Cusset ■

386688 Ternstroemia L. f. = Ternstroemia Mutis ex L. f.（保留属名）●

386689 Ternstroemia Mutis ex L. f.（1782）（保留属名）；厚皮香属；Hoferia，Ternstroemia ●

386690 Ternstroemia africana Melch. ；非洲厚皮香●☆

386691 Ternstroemia biangulipes Hung T. Chang；角柄厚皮香；Biangular-stalk Ternstroemia ●

386692 Ternstroemia coniocarpa Hu ex L. K. Ling；锥果厚皮香；Conicalfruit Ternstroemia，Conical-fruited Ternstroemia ●

386693 Ternstroemia discolor Hung T. Chang et S. H. Shi；异色厚皮香；Discolor Ternstroemia ●

386694 Ternstroemia dubia（Champ.）Choisy = Ternstroemia japonica（Thunb.）Thunb. ●

386695 Ternstroemia fragrans（Champ.）Choisy = Ternstroemia japonica Thunb. ●

386696 Ternstroemia gymnanthera（Wight et Arn.）Bedd. ；厚皮香（白花果，称杆红，称杆血，秤杆木，杆红，红柴，红果树，气血藤，日本厚皮香，猪血柴）；Bare-anthered Ternstroemia，Cleyera，Glabrous Ternstroemia，Japanese Ternstroemia，Nakedanther Ternstroemia，Ternstroemia ●

386697 Ternstroemia gymnanthera（Wight et Arn.）Bedd. ' Variegata'；花叶厚皮香●

386698 Ternstroemia gymnanthera（Wight et Arn.）Bedd. f. subserrata（Makino）Ohwi；具齿厚皮香●☆

386699 Ternstroemia gymnanthera（Wight et Arn.）Bedd. var. wightii（Choisy）Hand. -Mazz. = Ternstroemia gymnanthera（Wight et Arn.）Bedd. ●

386700 Ternstroemia gymnanthera（Wight et Arn.）Bedd. var. wightii（Choisy）Hand. -Mazz. ；凹脉厚皮香（阔叶厚皮香，梅木）；Wight Nakedanther Ternstroemia ●

386701 Ternstroemia gymnanthera（Wight et Arn.）Sprague = Ternstroemia gymnanthera（Wight et Arn.）Bedd. ●

386702 Ternstroemia hainanensis Hung T. Chang；海南厚皮香；Hainan Ternstroemia ●

386703 Ternstroemia insignis Y. C. Wu；大果厚皮香；Largefruit Ternstroemia，Large-fruited Ternstroemia ●

386704 Ternstroemia japonica（Thunb.）Thunb. ；日本厚皮香●

386705 Ternstroemia japonica（Thunb.）Thunb. var. wightii（Choisy）Dyer = Ternstroemia gymnanthera（Wight et Arn.）Bedd. var. wightii（Choisy）Hand. -Mazz. ●

386706 Ternstroemia japonica Siebold et Zucc. = Ternstroemia gymnanthera（Wight et Arn.）Bedd. ●

386707 Ternstroemia japonica Thunb. = Cleyera japonica Thunb. ●

386708 Ternstroemia japonica Thunb. = Ternstroemia gymnanthera（Wight et Arn.）Bedd. ●

386709 Ternstroemia japonica Thunb. var. wightii Dyer = Ternstroemia gymnanthera（Wight et Arn.）Bedd. ●

386710 Ternstroemia japonica Thunb. var. wigtii Choisy = Ternstroemia gymnanthera（Wight et Arn.）Bedd. var. wightii（Choisy）Hand. -Mazz. ●

386711 Ternstroemia khasyna Choisy = Illicium griffithii Hook. f. et Thomson ex Walp. ●

386712 Ternstroemia kwangtungensis Merr. ；厚叶厚皮香（广东厚皮香，华南厚皮香，小红木头树，圆叶厚皮香）；Guangdong Ternstroemia，Kwangtung Ternstroemia，Subroundleaf Ternstroemia，Thickleaf Ternstroemia，Thick-leaved Ternstroemia ●

386713 Ternstroemia longipedicellata L. K. Ling；长梗厚皮香；Longpedicel Ternstroemia，Long-pediceled Ternstroemia ●

386714 Ternstroemia longipedicellata L. K. Ling = Ternstroemia biangulipes Hung T. Chang ●

386715 Ternstroemia longipes Hu；长柄厚皮香；Longstalk Ternstroemia ●

386716 Ternstroemia lushia Buch. -Ham. ex D. Don = Cleyera japonica Thunb. var. wallichiana（DC.）Sealy ●

386717 Ternstroemia lushia Ham. ex D. Don = Cleyera japonica Thunb. ●

386718 Ternstroemia luteoflora Hu ex L. K. Ling；尖萼厚皮香；Sharpsepal Ternstroemia，Sharp-sepaled Ternstroemia ●

386719 Ternstroemia mabianensis L. K. Ling；条苞厚皮香；Mabian Ternstroemia ●

386720 Ternstroemia microphylla Merr. ；小叶厚皮香；Littleleaf Ternstroemia，Little-leaved Ternstroemia ●

386721 Ternstroemia mokof Nakai = Ternstroemia japonica Thunb. ●

386722 Ternstroemia nitida Merr. ；亮叶厚皮香；Shinyleaf Ternstroemia，Shiny-leaved Ternstroemia ●

386723 Ternstroemia oblancilimba Hung T. Chang = Ternstroemia microphylla Merr. ●

386724 Ternstroemia pachyphylla L. K. Ling = Ternstroemia kwangtungensis Merr. ●

386725 Ternstroemia parvifolia Hu = Ternstroemia gymnanthera（Wight et Arn.）Bedd. ●

386726 Ternstroemia polypetala Melch. ；多瓣厚皮香●☆

386727 Ternstroemia pringlei Standl. ；平氏厚皮香●☆

386728 Ternstroemia pseudomicrophylla Hung T. Chang = Ternstroemia gymnanthera（Wight et Arn.）Bedd. ●

386729 Ternstroemia pseudoverticillata Merr. et Chun = Ternstroemia microphylla Merr. ●

386730 Ternstroemia sichuanensis L. K. Ling；四川厚皮香；Sichuan Ternstroemia ●

386731 Ternstroemia simaoensis L. K. Ling；思茅厚皮香；Simao Ternstroemia ●

386732 Ternstroemia subrotundifolia Hung T. Chang = Ternstroemia kwangtungensis Merr. ●

386733 Ternstroemia tepezapote Cham. et Schltdl. ；特佩厚皮香●☆

386734 Ternstroemia yunnanensis L. K. Ling；云南厚皮香；Yunnan Ternstroemia ●

386735 Ternstroemiaceae Mirb. = Pentaphylacaceae Engl.（保留科名）●

386736 Ternstroemiaceae Mirb. = Ternstroemiaceae Mirb. ex DC. ●

386737 Ternstroemiaceae Mirb. = Theaceae Mirb.（保留科名）●

386738 Ternstroemiaceae Mirb. ex DC.（1816）；厚皮香科●

386739 Ternstroemiaceae Mirb. ex DC. = Theaceae Mirb.（保留科名）●

386740 Ternstroemiopsis Urb. = Eurya Thunb. ●

386741 Ternu O. Berg ＝ Blepharocalyx O. Berg ●☆

386742 Terobera Steud. ＝ Cladium P. Browne ■

386743 Terogia Raf. ＝ Ortegia L. ■☆

386744 Terpnanthus Nees et Mart. ＝ Spiranthera A. St. -Hil. ■☆

386745 Terpnophyllum Thwaites ＝ Garcinia L. ●

386746 Terranea Colla ＝ Erigeron L. ■●

386747 Terrella Nevski ＝ Elymus L. ■

386748 Terrella Nevski ＝ Terrellia Lunell ■

386749 Terrellia Lunell ＝ Elymus L. ■

386750 Terrellia villosa （Muhl. ex Willd.）Baum ＝ Elymus villosus Muhl. ex Willd. ■☆

386751 Terrentia Vell. ＝ Ichthyothere Mart. ■●☆

386752 Tersonia Moq. （1849）；泰尔森环蕊木属■☆

386753 Tersonia brevipes Moq. ；泰尔森环蕊木■☆

386754 Tertrea DC. ＝ Machaonia Bonpl. ■☆

386755 Tertria Schrank ＝ Polygala L. ●■

386756 Tertria Schrank ＝ Polygaloides Haller ●☆

386757 Terua Standl. et F. J. Herm. ＝ Lonchocarpus Kunth（保留属名）●■☆

386758 Teruncius Lunell ＝ Thlaspi L. ■

386759 Terustroemia Jack ＝ Ternstroemia Mutis ex L. f. （保留属名）●

386760 Tesmannia Willis ＝ Tessmannia Harms ●☆

386761 Tesota C. Muell. ＝ Olneya A. Gray ●☆

386762 Tessarandra Miers ＝ Linociera Sw. ex Schreb. （保留属名）●

386763 Tessaranthium Kellogg ＝ Frasera Walter ■☆

386764 Tessaria Ruiz et Pav. （1794）；单树菊属（特萨菊属）；Arrow-Weed ●☆

386765 Tessaria dodoneifolia （Hook. et Arn.）Cabrera；单树菊（特萨菊）●☆

386766 Tessaria integrifolia Ruiz et Pav. ；全缘叶单树菊（全缘叶特萨菊）●☆

386767 Tessaria redolens Less. ＝ Pterocaulon redolens （G. Forst.）Fern. -Vill. ■

386768 Tessaria sericea （Nutt.）Shinners ＝ Pluchea sericea （Nutt.）Coville ■☆

386769 Tessenia Bubani ＝ Erigeron L. ■●

386770 Tesserantherum Curran ＝ Frasera Walter ■☆

386771 Tesserantherum Curran ＝ Tessaranthium Kellogg ■☆

386772 Tesserantherum Curran ＝ Tesseranthium Pritz. ■☆

386773 Tesseranthium Kellogg ＝ Frasera Walter ■☆

386774 Tesseranthium Pritz. ＝ Frasera Walter ■☆

386775 Tesseranthium Pritz. ＝ Tessaranthium Kellogg ■☆

386776 Tessiera DC. ＝ Spermacoce L. ●■

386777 Tessmannia Harms（1910）；特斯曼苏木属●☆

386778 Tessmannia africana Harms；非洲特斯曼苏木●☆

386779 Tessmannia anomala （Micheli）Harms；异常特斯曼苏木●☆

386780 Tessmannia baikiaeoides Hutch. et Dalziel；拟红苏木●☆

386781 Tessmannia burttii Harms；伯氏特斯曼苏木●☆

386782 Tessmannia claessensi De Wild. ＝ Tessmannia africana Harms ●☆

386783 Tessmannia dawei J. Léonard；道氏特斯曼苏木●☆

386784 Tessmannia densiflora Harms；密花特斯曼苏木●☆

386785 Tessmannia dewildemaniana Harms；德怀尔苏木●☆

386786 Tessmannia lescrauwaetii （De Wild.）Harms；莱斯特斯曼苏木●☆

386787 Tessmannia martiniana Harms；马丁特斯曼苏木●☆

386788 Tessmannia martiniana Harms var. pauloi J. Léonard；保罗特斯曼苏木●☆

386789 Tessmannia parvifolia Harms ＝ Tessmannia anomala （Micheli）Harms ●☆

386790 Tessmannia yangambiensis Louis ex J. Léonard；扬甘比苏木●☆

386791 Tessmanniacanthus Mildbr. （1926）；特斯曼爵床属☆

386792 Tessmanniacanthus chlamydocardioides Mildbr. ；特斯曼爵床☆

386793 Tessmannianthus Markgr. （1927）；特斯曼野牡丹属●☆

386794 Tessmannianthus heterostemon Markgr. ；特斯曼野牡丹●☆

386795 Tessmanniodoxa Burret ＝ Chelyocarpus Dammer ●☆

386796 Tessmanniophoenix Burret ＝ Chaetopappa DC. ■☆

386797 Tessmanniophoenix Burret ＝ Chelyocarpus Dammer ●☆

386798 Testudinaria Salisb. （1824）；肖薯蓣属■☆

386799 Testudinaria Salisb. ＝ Dioscorea L. （保留属名）■

386800 Testudinaria elephantipes Lindl. ＝ Dioscorea elephantipes （L'Hér.）Engl. ■☆

386801 Testudinaria elephantipes Salisb. ；肖薯蓣■☆

386802 Testudinaria macrostachya （Benth.）G. D. Rowley；大穗肖薯蓣●☆

386803 Testudinaria montana Burch. ；山地肖薯蓣■☆

386804 Testudinaria montana Burch. ＝ Dioscorea elephantipes （L'Hér.）Engl. ■☆

386805 Testudinaria montana Eckl. et Zeyh. ＝ Dioscorea sylvatica Eckl. ■☆

386806 Testudinaria multiflora Marloth；多花肖薯蓣■☆

386807 Testudinaria multiflora Marloth ＝ Dioscorea sylvatica Eckl. var. multiflora （Marloth）Burkill ■☆

386808 Testudinaria paniculata Dümmer；圆锥肖薯蓣■☆

386809 Testudinaria paniculata Dümmer ＝ Dioscorea sylvatica Eckl. var. paniculata （Dümmer）Burkill ■☆

386810 Testudinaria sylvatica （Eckl.）Kunth ＝ Dioscorea sylvatica Eckl. ■☆

386811 Testudipes Markgr. ＝ Tabernaemontana L. ●

386812 Testulea Pellegr. （1924）；泰斯木属●☆

386813 Testulea gabonensis Pellegr. ；加蓬泰斯木●☆

386814 Teta Roxb. ＝ Peliosanthes Andréws ●

386815 Tetanosia Rich. ex M. Roem. ＝ Opilia Roxb. ●

386816 Tetaris Chesney ＝ Arnebia Forssk. ●■

386817 Tetaris Chesney ＝ Tetaris Lindl. ●■

386818 Tetaris Lindl. ＝ Arnebia Forssk. ●■

386819 Tetilla DC. （1830）；智利虎耳草属■☆

386820 Tetilla hydrocotylifolia DC. ；智利虎耳草●☆

386821 Tetraberlinia （Harms）Hauman（1952）；四鞋木属●☆

386822 Tetraberlinia bifoliolata （Harms）Hauman；二小叶四鞋木●☆

386823 Tetraberlinia korupensis Wieringa；科鲁普四鞋木●☆

386824 Tetraberlinia longiracemosa （A. Chev.）Wieringa；长序四鞋木●☆

386825 Tetraberlinia microphylla （Troupin）Aubrév. ＝ Michelsonia microphylla （Troupin）Hauman ●☆

386826 Tetraberlinia moreliana Aubrév. ；默勒尔四鞋木●☆

386827 Tetraberlinia polyphylla （Harms）J. Léonard；多叶四鞋木●☆

386828 Tetracanthus A. Rich. ＝ Pectis L. ■☆

386829 Tetracanthus C. Wright ex Griseb. ＝ Pinillosia Ossa ex DC. ■☆

386830 Tetracarpaea Benth. ＝ Anisophyllea R. Br. ex Sabine ●☆

386831 Tetracarpaea Benth. ＝ Tetracrypta Gardner ●☆

386832 Tetracarpaea Hook. （1840）；四果木属●☆

386833 Tetracarpaea Hook. f. ＝ Tetracarpaea Hook. ●☆

386834 Tetracarpaea tasmannica Hook. ；四果木●☆

386835 Tetracarpaeaceae Nakai ＝ Grossulariaceae DC. （保留科名）●

386836 Tetracarpaeaceae Nakai（1943）；四果木科●☆

386837 Tetracarpidium Pax（1899）；四锥木属●☆

386838 Tetracarpidium conophorum （Müll. Arg.）Hutch. et Dalziel；四

锥木;Awusa Nut,Conophor Nut ●☆

386839　Tetracarpidium staudtii Pax ＝ Tetracarpidium conophorum（Müll. Arg.）Hutch. et Dalziel ●☆

386840　Tetracarpidium tenuifolium（Pax et K. Hoffm.）Pax et K. Hoffm. ;细叶四锥木●☆

386841　Tetracarpum Moench ＝Schkuhria Roth(保留属名)■☆

386842　Tetracarpus Post et Kuntze ＝ Tetracarpaea Hook. ●☆

386843　Tetracarya Dur. ＝ Microula Benth. ■

386844　Tetracarya Dur. ＝ Tretocarya Maxim. ■

386845　Tetracellion Turcz. ex Fisch. et C. A. Mey. ＝ Rorippa Scop. ■

386846　Tetracellion Turcz. ex Fisch. et C. A. Mey. ＝ Tetrapoma Turcz. ex Fisch. et C. A. Mey. ■

386847　Tetracentraceae A. C. Sm.（1945）(保留科名);水青树科;Tetracentron Family ●

386848　Tetracentraceae A. C. Sm.（保留科名）＝ Trochodendraceae Eichler(保留科名)●

386849　Tetracentraceae Tiegh. ＝ Tetracentraceae A. C. Sm.（保留科名）●

386850　Tetracentron Oliv.(1889);水青树属;Tetracentron ●

386851　Tetracentron sinense Oliv. ;水青树;Tetracentron ●

386852　Tetracentron sinense Oliv. var. himalense H. Hara et Kanai ＝ Tetracentron sinense Oliv. ●

386853　Tetracera L.（1753）;锡叶藤属（第伦桃属,涩叶藤属）;Tetracera ●

386854　Tetracera affinis Hutch. ;近缘锡叶藤●☆

386855　Tetracera alnifolia Willd. ;桤叶锡叶藤●☆

386856　Tetracera alnifolia Willd. subsp. dinklagei（Gilg）Kubitzki;丁克锡叶藤●☆

386857　Tetracera alnifolia Willd. var. demeusei De Wild. et T. Durand ＝ Tetracera alnifolia Willd. ●☆

386858　Tetracera alnifolia Willd. var. podotricha（Gilg）Staner ＝ Tetracera alnifolia Willd. ●☆

386859　Tetracera asiatica（Lour.）Hoogland ＝ Tetracera sarmentosa（L.）Vahl ●

386860　Tetracera aspera（Lour.）Hoogland var. boliviana Kuntze ＝ Tetracera parviflora（Rusby）Sleumer ●☆

386861　Tetracera boiviniana Baill. ;博伊文锡叶藤●☆

386862　Tetracera bussei Gilg;布瑟锡叶藤●☆

386863　Tetracera calothyrsa Gilg et Ledermann ex Mildbr. ;美序锡叶藤●☆

386864　Tetracera claessensii De Wild. ＝ Tetracera masuiana De Wild. et T. Durand ●☆

386865　Tetracera demeusei（De Wild. et T. Durand）De Wild. ＝ Tetracera alnifolia Willd. ●☆

386866　Tetracera dinklagei Gilg ＝ Tetracera alnifolia Willd. subsp. dinklagei（Gilg）Kubitzki ●☆

386867　Tetracera djalonica A. Chev. ex Hutch. et Dalziel ＝ Tetracera alnifolia Willd. ●☆

386868　Tetracera edentata H. Perrier;无齿锡叶藤●☆

386869　Tetracera eriantha（Oliv.）Hutch. ;毛花锡叶藤●☆

386870　Tetracera fragrans De Wild. et T. Durand ＝ Tetracera poggei Gilg ●☆

386871　Tetracera gilletii De Wild. ＝ Tetracera masuiana De Wild. et T. Durand ●☆

386872　Tetracera gilletii De Wild. ＝ Tetracera rosiflora Gilg ●☆

386873　Tetracera guineensis A. Chev. ＝ Tetracera leiocarpa Stapf ●☆

386874　Tetracera humilis A. Chev. ＝ Tetracera masuiana De Wild. et T. Durand ●☆

386875　Tetracera hydrophila Triana et Planch. ;喜水锡叶藤●☆

386876　Tetracera indica Merr. ;印度锡叶藤●☆

386877　Tetracera leiocarpa Stapf;光果锡叶藤●☆

386878　Tetracera levinei Merr. ＝ Tetracera asiatica（Lour.）Hoogland ●

386879　Tetracera levinei Merr. ＝ Tetracera sarmentosa（L.）Vahl ●

386880　Tetracera litoralis Gilg;海边锡叶藤●☆

386881　Tetracera macrophylla A. Chev. ＝ Tetracera alnifolia Willd. ●☆

386882　Tetracera madagascariensis Willd. ex Schltdl. ;马岛锡叶藤●☆

386883　Tetracera malangensis Exell ＝ Tetracera poggei Gilg ●☆

386884　Tetracera marquesii Gilg ＝ Tetracera poggei Gilg ●☆

386885　Tetracera masuiana De Wild. et T. Durand;马苏锡叶藤●☆

386886　Tetracera masuiana De Wild. et T. Durand var. sapinii De Wild. ＝ Tetracera masuiana De Wild. et T. Durand ●☆

386887　Tetracera mayumbensis Exell ＝ Tetracera rosiflora Gilg ●☆

386888　Tetracera obtusata Planch. ex Oliv. ＝ Tetracera potatoria Afzel. ex G. Don ●☆

386889　Tetracera obtusata Planch. ex Oliv. var. eriantha Oliv. ＝ Tetracera eriantha（Oliv.）Hutch. ●☆

386890　Tetracera parviflora（Rusby）Sleumer;小花锡叶藤●☆

386891　Tetracera pauciflora Baker ＝ Tetracera rutenbergii Buchenau ●☆

386892　Tetracera podotricha Gilg ＝ Tetracera alnifolia Willd. ●☆

386893　Tetracera poggei Gilg;波格锡叶藤●☆

386894　Tetracera potatoria Afzel. ex G. Don ＝ Tetracera alnifolia Willd. ●☆

386895　Tetracera rosiflora Gilg;粉花锡叶藤●☆

386896　Tetracera rutenbergii Buchenau;鲁氏锡叶藤●☆

386897　Tetracera sarmentosa（L.）Vahl;锡叶藤（大涩沙,狗舌藤,老糠藤,涩谷藤,涩沙藤,涩藤,涩叶藤,砂藤,水车藤,锡叶）;Asian Tetracera ●

386898　Tetracera sarmentosa Vahl ＝ Tetracera sarmentosa（L.）Vahl ●

386899　Tetracera sarmentosa Vahl ＝ Tetracera scandens（L.）Merr. ●

386900　Tetracera scandens（L.）Merr. ;毛果锡叶藤;Hairy-fruit Tetracera,Hairy-fruited Tetracera ●

386901　Tetracera senegalensis DC. ＝ Tetracera alnifolia Willd. ●☆

386902　Tetracera strigillosa A. Chev. ＝ Tetracera masuiana De Wild. et T. Durand ●☆

386903　Tetracera strigillosa Gilg ＝ Tetracera masuiana De Wild. et T. Durand ●☆

386904　Tetracera stuhlmanniana Gilg;斯图尔曼锡叶藤●☆

386905　Tetracera triceras Thouars ex Baill. ＝ Tetracera madagascariensis Willd. ex Schltdl. ●☆

386906　Tetracera willdenowiana Steud. ;威氏锡叶藤●☆

386907　Tetraceraceae Baum. -Bod. ;锡叶藤科●

386908　Tetraceraceae Baum. -Bod. ＝ Dilleniaceae Salisb.（保留科名）●■

386909　Tetraceras Post et Kuntze ＝ Tetracera L. ●

386910　Tetraceras Webb ＝ Tetraceratium（DC.）Kuntze ■

386911　Tetraceratium（DC.）Kuntze ＝ Tetracme Bunge ■

386912　Tetraceratium Kuntze ＝ Tetracme Bunge ■

386913　Tetrachaete Chiov.（1903）;胶鳞禾状草属■☆

386914　Tetrachaete elionuroides Chiov. ;胶鳞禾状草■☆

386915　Tetracheilos Lehm. ＝ Acacia Mill.（保留属名）●■

386916　Tetrachne Nees(1841);四秤禾属(高原牧场草属)■☆

386917　Tetrachne aristulata Hack. et Rendle ex Scott-Elliot ＝ Entoplocamia aristulata（Hack. et Rendle ex Scott-Elliot）Stapf ■☆

386918　Tetrachne dregei Nees;四秤禾(高原牧场草)■☆

386919　Tetrachondra Petrie ＝ Tetrachondra Petrie ex Oliv. ■☆

386920　Tetrachondra Petrie ex Oliv.（1892）;四粉草属(四核草属)■☆

386921　Tetrachondra hamiltonii Petrie;四粉草■☆

386922 Tetrachondraceae Skottsb. = Labiatae Juss.(保留科名)●■

386923 Tetrachondraceae Skottsb. = Lamiaceae Martinov(保留科名)●■

386924 Tetrachondraceae Skottsb. = Tetrachondraceae Skottsb. ex R. W. Sanders et P. D. Cantino ■☆

386925 Tetrachondraceae Skottsb. ex R. W. Sanders et P. D. Cantino = Labiatae Juss.(保留科名)●■

386926 Tetrachondraceae Skottsb. ex R. W. Sanders et P. D. Cantino = Lamiaceae Martinov(保留科名)●■

386927 Tetrachondraceae Skottsb. ex R. W. Sanders et P. D. Cantino (1984);四粉草科■☆

386928 Tetrachondraceae Wettst. = Labiatae Juss.(保留科名)●■

386929 Tetrachondraceae Wettst. = Lamiaceae Martinov(保留科名)●■

386930 Tetrachyron Schltdl.(1847);四芒菊属■☆

386931 Tetrachyron Schltdl. = Calea L. ●■☆

386932 Tetrachyron discolor (A. Gray) Wussow et Urbatsch;杂色四芒菊■☆

386933 Tetrachyron grayi (Klatt) J. R. Wussow et Urbatsch;格雷四芒菊■☆

386934 Tetrachyron manicatum Schltdl.;四芒菊■☆

386935 Tetraclea A. Gray(1853);四封草属●☆

386936 Tetraclea coulteri A. Gray;四封草●☆

386937 Tetraclea viscida Lundell = Trichostema brachiatum L. ●☆

386938 Tetraclinaceae Hayata = Cupressaceae Gray(保留科名)●

386939 Tetraclinaceae Hayata;山达木科●

386940 Tetraclinidaceae Hayata = Cupressaceae Gray(保留科名)●

386941 Tetraclinis Mast.(1892);山达木属(方楔柏属,香漆柏属)●☆

386942 Tetraclinis articulata (Vahl) Mast.;山达木(北非山达树,方苞澳洲柏,方苞非洲柏,节状方楔柏);Afar Tree,Alerce,Arar,Arar Tree,Arar-tree,Cypress Pine,Great Juniper,Gum-tree,Pounce,Sandarac,Sandrach,Spanish Juniper,Thuya ●☆

386943 Tetraclinis articulata Mast. = Tetraclinis articulata (Vahl) Mast. ●☆

386944 Tetraclis Hiern = Diospyros L. ●

386945 Tetracma Post et Kuntze = Tetracme Bunge ■

386946 Tetracme Bunge(1838);四齿芥属(四齿荠属);Tetracme ■

386947 Tetracme appressa Rech. f. = Tetracme pamirica Vassilcz. ■☆

386948 Tetracme bucharica (Korsh.) O. E. Schulz;布赫四齿芥■☆

386949 Tetracme contorta Boiss.;扭果四齿芥;Contorted Tetracme ■

386950 Tetracme elongata Kitam. = Tetracme quadricornis (Stephan) Bunge ■

386951 Tetracme glochidiata (Botsch. et Vved.) Pachom.;钩四齿芥■☆

386952 Tetracme pamirica Vassilcz.;帕米尔四齿芥■☆

386953 Tetracme quadricornis (Stephan) Bunge;四齿芥;Fourtooth Tetracme ■

386954 Tetracme quadricornis (Stephan) Bunge var. longicornis Regel = Tetracme quadricornis (Stephan) Bunge ■

386955 Tetracme recurvata Bunge;弯角四齿芥;Recurvate Tetracme ■

386956 Tetracme recurvata Bunge var. integrifolia Gillies = Tetracme contorta Boiss. ■

386957 Tetracme secunda Boiss.;单侧四齿芥■☆

386958 Tetracme stocksii Boiss.;斯托克斯四齿芥■☆

386959 Tetracmidion Korsh. = Tetracme Bunge ■

386960 Tetracmidion bucharicum Korsh. = Tetracme bucharica (Korsh.) O. E. Schulz ■☆

386961 Tetracmidion glochidiatum Botsch. et Vved. = Tetracme glochidiata (Botsch. et Vved.) Pachom. ■☆

386962 Tetracoccus Engelm. ex Parry(1885);四仁大戟属●☆

386963 Tetracoccus capensis (I. M. Johnst.) Croizat;好望角四仁大戟●☆

386964 Tetracoccus hallii Brandegee;霍尔四仁大戟;Purple-bush ●☆

386965 Tetracoccus ilicifolius Coville et Gilman;冬青叶四仁大戟;Holly-lenf Spurge ●☆

386966 Tetracocyne Turcz. = Patrisa Rich.(废弃属名)●☆

386967 Tetracocyne Turcz. = Ryania Vahl(保留属名)●☆

386968 Tetracoilanthus Rappa et Camarrone = Aptenia N. E. Br. ●☆

386969 Tetracoilanthus anatomicus (Haw.) Rappa et Camarrone = Sceletium emarcidum (Thunb.) L. Bolus ex H. Jacobsen ●☆

386970 Tetracoilanthus concavus (Haw.) Rappa et Camarrone = Sceletium tortuosum (L.) N. E. Br. ●☆

386971 Tetracoilanthus cordifolius (L. f.) Rappa et Camarrone = Aptenia cordifolia (L. f.) Schwantes ■

386972 Tetracronia Pierre = Glycosmis Corrêa(保留属名)●

386973 Tetracronia cymosa Pierre = Glycosmis montana Pierre ●

386974 Tetracrypta Gardner = Anisophyllea R. Br. ex Sabine ●☆

386975 Tetracrypta Gardner et Champ. = Anisophyllea R. Br. ex Sabine ●☆

386976 Tetractinostigma Hassk. = Aporusa Blume ●

386977 Tetractomia Hook. f.(1875);四片芸香属●☆

386978 Tetractomia acuminata Merr.;渐尖四片芸香●☆

386979 Tetractomia lanceolata (Lauterb.) Merr. et L. M. Perry;披针叶四片芸香●☆

386980 Tetractomia latifolia Ridl.;宽叶四片芸香●☆

386981 Tetractomia majus Hook. f.;大四片芸香●☆

386982 Tetractomia montana Ridl.;山地四片芸香●☆

386983 Tetractomia obovata Merr.;倒卵四片芸香●☆

386984 Tetractomia oppositifolia (Ridl.) Merr. et L. M. Perry;对叶四片芸香●☆

386985 Tetractomia pachyphylla Merr.;厚叶四片芸香●☆

386986 Tetractomia parviflora Ridl.;小花四片芸香●☆

386987 Tetractomia philippinensis Elmer;菲律宾四片芸香●☆

386988 Tetractomia rotundifolia (Ridl.) Merr. et L. M. Perry;圆叶四片芸香●☆

386989 Tetracustelma Baill. = Matelea Aubl. ●☆

386990 Tetradapa Osbeck = Erythrina L. ●■

386991 Tetradema Schltr. = Agalmyla Blume ●☆

386992 Tetradenia Benth.(1830);四腺木姜子属(南非木姜子属)●☆

386993 Tetradenia Nees = Neolitsea (Benth. et Hook. f.) Merr.(保留属名)●

386994 Tetradenia acuminatissima Hayata = Neolitsea acuminatissima (Hayata) Kaneh. et Sasaki ●

386995 Tetradenia acutivena (Hayata) Nemoto = Litsea acutivena Hayata ●

386996 Tetradenia acutotrinervia Hayata = Neolitsea aciculata (Blume) Koidz. ●

386997 Tetradenia acutotrinervia Hayata = Neolitsea acutotrinervia (Hayata) Kaneh. et Sasaki ●

386998 Tetradenia akoensis (Hayata) Nemoto = Litsea akoensis Hayata ●

386999 Tetradenia akoensis (Hayata) Nemoto ex Makino et Nemoto = Litsea akoensis Hayata ●

387000 Tetradenia aurata (Hayata) Hayata = Neolitsea aurata (Hayata) Koidz. ●

387001 Tetradenia aurata Hayata = Neolitsea aurata (Hayata) Koidz. ●

387002 Tetradenia barberae (N. E. Br.) Codd;巴尔四腺木姜子●☆

387003 Tetradenia brevispicata (N. E. Br.) Codd;短穗四腺木姜子(短穗南非木姜子)●☆

387004 Tetradenia clementiana Phillipson;圣克利门蒂四腺木姜子(圣

克利门蒂木姜子)●☆

387005　Tetradenia consimilis Nees ＝Neolitsea pallens（D. Don）Momiy. et H. Hara ●

387006　Tetradenia cordata Phillipson;心叶四腺木姜子(心叶南非木姜子)●☆

387007　Tetradenia dolichocarpa（Hayata）Makino et Nemoto ＝Actinodaphne acuminata（Blume）Meisn. ●

387008　Tetradenia falafa Phillipson;镰叶四腺木姜子(镰叶南非木姜子)●☆

387009　Tetradenia fruticosa Benth. ;四腺木姜子(南非木姜子)●☆

387010　Tetradenia glauca（Siebold）Matsum. ＝Neolitsea sericea（Blume）Koidz. ◇

387011　Tetradenia hayatae Nemoto ＝Neolitsea kotoensis（Hayata）Kaneh. et Sasaki ●

387012　Tetradenia hayatae Nemoto ＝Neolitsea villosa（Blume）Merr. ●

387013　Tetradenia herbacea Phillipson;草本四腺木姜子■☆

387014　Tetradenia hildebrandtii Briq. ＝Tetradenia fruticosa Benth. ●☆

387015　Tetradenia hypophaea（Hayata）Makino et Nemoto. ＝Litsea hypophaea Hayata ●

387016　Tetradenia isaloensis Phillipson;伊萨卢四腺木姜子●☆

387017　Tetradenia kaokoensis Van Jaarsv. et A. E. van Wyk;卡奥科四腺木姜子●☆

387018　Tetradenia kawakamii（Hayata）Nemoto ＝Litsea garciae Vidal ●

387019　Tetradenia kawakamii（Hayata）Nemoto ex Makino et Nemoto ＝Litsea garciae Vidal ●

387020　Tetradenia konishii（Hayata）Hayata ＝Neolitsea konishii（Hayata）Kaneh. et Sasaki ●

387021　Tetradenia kotoensis Hayata ＝Neolitsea kotoensis（Hayata）Kaneh. et Sasaki ●

387022　Tetradenia kotoensis Hayata ＝Neolitsea villosa（Blume）Merr. ●

387023　Tetradenia lanuginosa（Wall. ex Nees）Nees ＝Neolitsea cuipala（D. Don）Kosterm. ●☆

387024　Tetradenia nakaii（Hayata）Makino et Nemoto ＝Litsea acutivena Hayata ●

387025　Tetradenia nervosa Codd;多脉南非木姜子●☆

387026　Tetradenia obovata（Hayata）Nemoto ex Makino et Nemoto ＝Litsea akoensis Hayata ●

387027　Tetradenia obovata（Hayata）Nemoto ex Makino et Nemoto ＝Litsea hayatae Kaneh. ●

387028　Tetradenia obovata Nees ＝Actinodaphne obovata（Nees）Blume ●

387029　Tetradenia obovata Nemoto ＝Litsea hayatae Kaneh. ●

387030　Tetradenia pallens D. Don ＝Neolitsea pallens（D. Don）Momiy. et H. Hara ●

387031　Tetradenia parvigemma Hayata ＝Neolitsea parvigemma（Hayata）Kaneh. et Sasaki ●

387032　Tetradenia riparia（Hochst.）Codd;河岸四腺木姜子(河岸南非木姜子);Ginger Bush,Moschosma,Nutmeg Bush ●☆

387033　Tetradenia variabillima Hayata ＝Neolitsea aciculata（Blume）Koidz. var. variabilima（Hayata）J. C. Liao ●

387034　Tetradenia variabillima Hayata ＝Neolitsea variabilima Hayata ●

387035　Tetradenia variabillima Hayata ＝Neolitsea variabilis（Hayata）Kaneh. et Sasaki ●

387036　Tetradenia zeylanica（Nees et T. Nees）Nees ＝Neolitsea zeylanica（Nees et T. Nees）Merr. ●

387037　Tetradenia zeylanica Nees ＝Neolitsea zeylanica（Nees et T. Nees）Merr. ●

387038　Tetradia R. Br. ＝Pterygota Schott et Endl. ●

387039　Tetradia Thouars ex Tul. ＝Tetrataxis Hook. f. ●☆

387040　Tetradiclidaceae Takht. (1986);旱霸王科●☆

387041　Tetradiclidaceae Takht. ＝Nitrariaceae Bercht et J. Presl ●

387042　Tetradiclidaceae Takht. ＝Tetragoniaceae Lindl. (保留科名)●■☆

387043　Tetradiclidaceae Takht. ＝Zygophyllaceae R. Br. (保留科名)●■

387044　Tetradiclidaeeae（Engl.）Takht. ＝Tetradiclidaceae Takht. ●■

387045　Tetradiclis Steven ＝Tetradiclis Steven ex M. Bieb. ■☆

387046　Tetradiclis Steven ex M. Bieb. (1819);旱霸王属■☆

387047　Tetradiclis eversmannii Bunge ＝Tetradiclis tenella（Ehrenb.）Litv. ■☆

387048　Tetradiclis salsa C. A. Mey. ＝Tetradiclis tenella（Ehrenb.）Litv. ■☆

387049　Tetradiclis tenella（Ehrenb.）Litv. ;旱霸王■☆

387050　Tetradium Dulac ＝Evodia J. R. Forst. et G. Forst. ●

387051　Tetradium Dulac ＝Rhodiola L. ■

387052　Tetradium Dulac ＝Sedum L. ●■

387053　Tetradium Lour. ＝Evodia J. R. Forst. et G. Forst. ●

387054　Tetradium austrosinense（Hand. -Mazz.）T. G. Hartley ＝Evodia austrosinensis Hand. -Mazz. ●

387055　Tetradium calcicola（Chun ex C. C. Huang）T. G. Hartley ＝Evodia calcicola Chun ex C. C. Huang ●

387056　Tetradium daniellii（A. W. Benn. ex Daniell）T. G. Hartley ＝Evodia daniellii（A. W. Benn.）Hemsl. ex Forbes et Hemsl. ●

387057　Tetradium daniellii（A. W. Benn.）Hartley ＝Evodia daniellii（A. W. Benn.）Hemsl. ex Forbes et Hemsl. ●

387058　Tetradium fraxinifolium（Hook.）T. G. Hartley ＝Evodia fraxinifolia（D. Don）Hook. f. ●

387059　Tetradium glabrifolium（Champ. ex Benth.）T. G. Hartley ＝Evodia glabrifolia（Champ. ex Benth.）C. C. Huang ●

387060　Tetradium glabrifolium（Champ. ex Benth.）T. G. Hartley var. glaucum（Miq.）T. Yamaz. ＝Evodia glauca Miq. ●

387061　Tetradium ruticarpum（A. Juss.）T. G. Hartley ＝Evodia ruticarpa（Juss.）Benth. ●

387062　Tetradium ruticarpum（A. Juss.）T. G. Hartley var. officinale（Dode）T. G. Hartley ＝Evodia ruticarpa（Juss.）Benth. var. officinalis（Dode）C. C. Huang ●

387063　Tetradium trichotoma Lour. ＝Evodia trichotoma（Lour.）Pierre ●

387064　Tetradoa Pichon ＝Hunteria Roxb. ●

387065　Tetradoa hexaloba Pichon ＝Hunteria hexaloba（Pichon）Omino ●☆

387066　Tetradoa simii（Stapf）Pichon ＝Hunteria simii（Stapf）H. Huber ●☆

387067　Tetradoxa C. Y. Wu ＝Adoxa L. ■

387068　Tetradoxa C. Y. Wu（1981）;四福花属;Four Muskroot, Tetradoxa ■★

387069　Tetradoxa omeiensis（H. Hara）C. Y. Wu;四福花;Four Muskroot,Tetradoxa ■

387070　Tetradyas Danser ＝Cyne Danser ●☆

387071　Tetradymia DC. (1838);四蟹甲属;Horse Brush ●☆

387072　Tetradymia argyraea Munz et J. C. Roos;银叶四蟹甲●☆

387073　Tetradymia axillaris A. Nelson;腋花四蟹甲●☆

387074　Tetradymia axillaris A. Nelson var. longispina（M. E. Jones）Strother;长刺腋花四蟹甲●☆

387075　Tetradymia canescens DC. ;灰叶四蟹甲;Grey Felt Thorn, Grey Felt-thorn,Spineless Horsebrush,Stemless Horsebrush ●☆

387076　Tetradymia comosa A. Gray;白四蟹甲;White Felt Thorn,White Felt-thorn ●☆

387077　Tetradymia comosa A. Gray subsp. tetrameres S. F. Blake ＝

Tetradymia tetrameres (S. F. Blake) Strother ●☆

387078　Tetradymia filifolia Greene；线叶四蟹甲●☆

387079　Tetradymia glabrata Torr. et A. Gray；光叶四蟹甲；Bald-leaved Felt Thorn，Bald-leaved Felt-thorn，Coal Oil Brush，Littleleaf Horsebrush，Little-leaf Horsebrush，Spring Rabbitbrush ●☆

387080　Tetradymia inermis Nutt. = Tetradymia canescens DC. ●☆

387081　Tetradymia longispina (M. E. Jones) Rydb. = Tetradymia axillaris A. Nelson var. longispina (M. E. Jones) Strother ●☆

387082　Tetradymia nuttallii Torr. et A. Gray；纳托尔四蟹甲●☆

387083　Tetradymia ramosissima Torr. = Psathyrotes ramosissima (Torr.) A. Gray ■☆

387084　Tetradymia spinosa Hook. et Arn.；多刺四蟹甲；Spiny Horsebrush ●☆

387085　Tetradymia spinosa Hook. et Arn. var. longispina M. E. Jones；长刺四蟹甲；Cotton Thorn ●☆

387086　Tetradymia spinosa Hook. et Arn. var. longispina M. E. Jones = Tetradymia axillaris A. Nelson var. longispina (M. E. Jones) Strother ●☆

387087　Tetradymia stenolepis Greene；窄鳞四蟹甲；Narrow-scaled Felt-thorn，Narrow-sealed Felt Thorn ●☆

387088　Tetradymia tetrameres (S. F. Blake) Strother；四蟹甲●☆

387089　Tetradynansae Rchb. = Brassicaceae Burnett(保留科名)■●

387090　Tetradynansae Rchb. = Cruciferae Juss. (保留科名)■●

387091　Tetraedrocarpus O. Schwarz = Echiochilon Desf. ■☆

387092　Tetraedrocarpus arabicus O. Schwartz = Echiochilon arabicum (O. Schwartz) I. M. Johnst. ■☆

387093　Tetraena Maxim. (1889)；四合木属(油柴属)；Tetraena ●★

387094　Tetraena aegyptia (Hosny) Beier et Thulin = Zygophyllum aegyptium Hosny ●☆

387095　Tetraena alba (L. f.) Beier et Thulin = Zygophyllum album L. f. ■☆

387096　Tetraena applanata (Van Zyl) Beier et Thulin = Zygophyllum applanatum Van Zyl ●☆

387097　Tetraena chrysoptera (Retief) Beier et Thulin = Zygophyllum chrysopterum Retief ●☆

387098　Tetraena clavata (Schltr. et Diels) Beier et Thulin = Zygophyllum clavatum Schltr. et Diels ●☆

387099　Tetraena coccinea (L.) Beier et Thulin = Zygophyllum coccineum L. ●☆

387100　Tetraena cornuta (Coss.) Beier et Thulin = Zygophyllum album L. f. var. cornutum (Coss.) Murb. ●☆

387101　Tetraena cornuta (Coss.) Beier et Thulin = Zygophyllum cornutum Coss. ●☆

387102　Tetraena cylindrifolia (Schinz) Beier et Thulin = Zygophyllum cylindrifolium Schinz ●☆

387103　Tetraena decumbens (Delile) Beier et Thulin = Zygophyllum decumbens Delile ●☆

387104　Tetraena dumosa (Boiss.) Beier et Thulin = Zygophyllum dumosum Boiss. ●☆

387105　Tetraena gaetula (Emb. et Maire) Beier et Thulin = Zygophyllum gaetulum Emb. et Maire ●☆

387106　Tetraena gaetula (Emb. et Maire) Beier et Thulin subsp. waterlotii (Maire) Beier et Thulin = Zygophyllum waterlotii Maire ●☆

387107　Tetraena geslinii (Coss.) Beier et Thulin = Zygophyllum geslinii Coss. ●☆

387108　Tetraena giessii (Merxm. ex A. Schreib.) Beier et Thulin = Zygophyllum geslinii Coss. ●☆

387109　Tetraena hamiensis (Schweinf.) Beier et Thulin = Zygophyllum

hamiense Schweinf. ●☆

387110　Tetraena longicapsularis (Schinz) Beier et Thulin = Zygophyllum longicapsulare Schinz ●☆

387111　Tetraena longistipulata (Schinz) Beier et Thulin = Zygophyllum longistipulatum Schinz ●☆

387112　Tetraena madagascariensis (Baill.) Beier et Thulin = Zygophyllum madagascariense (Baill.) Stauffer ●☆

387113　Tetraena madecassa (H. Perrier) Beier et Thulin = Zygophyllum madecassum H. Perrier ●☆

387114　Tetraena microcarpa (Licht. ex Cham.) Beier et Thulin；大果四合木■☆

387115　Tetraena migiurtinora (Chiov.) Beier et Thulin = Zygophyllum migiurtinorum Chiov. ●☆

387116　Tetraena mongolica Maxim.；四合木(油柴)；Mongolian Tetraena，Tetraena ●◇

387117　Tetraena prismatica (Chiov.) Beier et Thulin = Zygophyllum prismaticum Chiov. ●☆

387118　Tetraena prismatocarpa (E. Mey. ex Sond.) Beier et Thulin = Zygophyllum prismatocarpum E. Mey. ex Sond. ●☆

387119　Tetraena pterocaulis (Van Zyl) Beier et Thulin = Zygophyllum pterocaule Van Zyl ●☆

387120　Tetraena retrofracta (Thunb.) Beier et Thulin = Zygophyllum retrofractum Thunb. ●☆

387121　Tetraena rigida (Schinz) Beier et Thulin = Zygophyllum rigidum Schinz ●☆

387122　Tetraena simplex (L.) Beier et Thulin = Zygophyllum simplex L. ●☆

387123　Tetraena somalensis (Hadidi) Beier et Thulin = Zygophyllum somalense Hadidi ●☆

387124　Tetraena stapfii (Schinz) Beier et Thulin = Zygophyllum stapffii Schinz ●☆

387125　Tetraeugenia Merr. = Syzygium R. Br. ex Gaertn. (保留属名)●☆

387126　Tetragamestus Rchb. f. (1854)；肖碗唇兰属■☆

387127　Tetragamestus Rchb. f. = Scaphyglottis Poepp. et Endl. (保留属名)■☆

387128　Tetragamestus aureus Rchb. f.；肖碗唇兰■☆

387129　Tetragastris Gaertn. (1790)；四囊榄属；Catuaba Herbal ●☆

387130　Tetragastris balsamifera Kuntze；香脂四囊榄；Pig Balsam ●☆

387131　Tetragastris panamensis (Engl.) Kuntze；巴拿马四囊榄●☆

387132　Tetraglochidion K. Schum. = Glochidion J. R. Forst. et G. Forst. (保留属名)●

387133　Tetraglochidium Bremek. (1944)；长苞蓝属(四锚属)●■

387134　Tetraglochidium Bremek. = Strobilanthes Blume ●■

387135　Tetraglochidium gigantodes (Lindau) C. Y. Wu et C. C. Hu；大苞蓝(大苞马蓝)●

387136　Tetraglochidium jugorum (Benoist) Bremek.；长苞蓝(长苞马蓝，垂序马蓝，红泽兰，马兰，日本马蓝)；Japan Conehead，Japanese Conehead ■

387137　Tetraglochidium jugorum (Benoist) Bremek. = Strobilanthes japonica (Thunb.) Miq. ■

387138　Tetraglochin Kuntze = Margyricarpus Ruiz et Pav. ●☆

387139　Tetraglochin Kuntze = Tetraglochin Kuntze ex Poepp. ●☆

387140　Tetraglochin Kuntze ex Poepp. (1833)；四尖蔷薇属●☆

387141　Tetraglochin Poepp. = Tetraglochin Kuntze ex Poepp. ●☆

387142　Tetraglochin acanthocarpa Speg.；刺果四尖蔷薇●☆

387143　Tetraglochin alata (Gillies ex Hook. et Arn.) Kuntze；翅四尖蔷薇●☆

387144　Tetraglochin longifolia Hauman；长叶四尖蔷薇●☆

387145　Tetraglochin microphylla Phil.；小叶四尖蔷薇●☆

387146　Tetraglossa Bedd. = Cleidion Blume ●

387147　Tetragocyanis Thouars = Epidendrum L.（保留属名）■☆

387148　Tetragocyanis Thouars = Phaius Lour. ■

387149　Tetragoga Bremek.（1944）；四苞蓝属（四苞爵床属）；Tetragoga ■

387150　Tetragoga Bremek. = Strobilanthes Blume ●■

387151　Tetragoga esquirolii（H. Lév.）E. Hossain；四苞蓝；Esquirol Tetragoga ●■

387152　Tetragoga nagaensis Bremek.；墨脱四苞蓝●■

387153　Tetragompha Bremek. = Strobilanthes Blume ●■

387154　Tetragonanthus S. G. Gmel. = Halenia Borkh.（保留属名）■

387155　Tetragonella Miq. = Tetragonia L. ●■

387156　Tetragonia L.（1753）；坚果番杏属（番杏属）；New Zealand Spinach，Tetragonia ●■

387157　Tetragonia acanthocarpa Adamson；刺果坚果番杏■☆

387158　Tetragonia arbuscula Fenzl；小乔木坚果番杏■☆

387159　Tetragonia borealis Batt. et Trab. = Tetragonia tetragonioides（Pall.）Kuntze ●■

387160　Tetragonia caesia Adamson；淡蓝坚果番杏■☆

387161　Tetragonia calycina Fenzl；萼状坚果番杏■☆

387162　Tetragonia chenopodioides Eckl. et Zeyh.；藜叶坚果番杏■☆

387163　Tetragonia chisimajensis Chiov. = Trianthema portulacastrum L. ■

387164　Tetragonia crystallina L'Hér.；结晶番杏●■☆

387165　Tetragonia cynocrambe Dinter = Tetragonia microptera Fenzl ex Harv. et Sond. ■☆

387166　Tetragonia decumbens Mill.；外倾坚果番杏■☆

387167　Tetragonia dimorphantha Pax = Tribulocarpus dimorphanthus（Pax）S. Moore ●☆

387168　Tetragonia distorta Fenzl；缠扭坚果番杏■☆

387169　Tetragonia echinata Aiton；具刺坚果番杏■☆

387170　Tetragonia erecta Adamson；直立坚果番杏■☆

387171　Tetragonia expansa Murray = Tetragonia tetragonioides（Pall.）Kuntze ●■

387172　Tetragonia fruticosa L.；灌丛坚果番杏■☆

387173　Tetragonia galenioides Fenzl ex Sond. = Aizoanthemum galenioides（Fenzl ex Sond.）Friedrich ■☆

387174　Tetragonia glauca Fenzl；灰绿坚果番杏■☆

387175　Tetragonia halimoides Fenzl ex Sond.；哈利木坚果番杏■☆

387176　Tetragonia haworthii Fenzl；霍沃斯坚果番杏■☆

387177　Tetragonia herbacea L.；草本坚果番杏■☆

387178　Tetragonia heterophylla Eckl. et Zeyh. = Tetragonia nigrescens Eckl. et Zeyh. ■☆

387179　Tetragonia hirsuta L. f.；粗毛坚果番杏■☆

387180　Tetragonia karasmontana Dinter ex Adamson = Tetragonia calycina Fenzl ■☆

387181　Tetragonia lanceolata Burm. f.；披针形坚果番杏■☆

387182　Tetragonia lasiantha Adamson；毛花坚果番杏■☆

387183　Tetragonia linearis Haw. = Tetragonia fruticosa L. ■☆

387184　Tetragonia macroptera Pax；大翅坚果番杏■☆

387185　Tetragonia macrostylis Schltr. = Tetragonia chenopodioides Eckl. et Zeyh. ■☆

387186　Tetragonia microptera Fenzl = Tetragonia microptera Fenzl ex Harv. et Sond. ■☆

387187　Tetragonia microptera Fenzl ex Harv. et Sond.；小翅坚果番杏■☆

387188　Tetragonia microptera Fenzl ex Harv. et Sond. var. monosperma ? = Tetragonia microptera Fenzl ex Harv. et Sond. ■☆

387189　Tetragonia microptera Fenzl ex Harv. et Sond. var. platycarpa Adamson = Tetragonia microptera Fenzl ex Harv. et Sond. ■☆

387190　Tetragonia microptera Fenzl ex Harv. et Sond. var. trisperma ? = Tetragonia microptera Fenzl ex Harv. et Sond. ■☆

387191　Tetragonia namaquensis Schltr.；纳马夸坚果番杏■☆

387192　Tetragonia nigrescens Eckl. et Zeyh.；黑坚果番杏■☆

387193　Tetragonia nigrescens Fenzl et Zeyh. var. hirsuta ? = Tetragonia nigrescens Eckl. et Zeyh. ■☆

387194　Tetragonia nigrescens Fenzl et Zeyh. var. hirta ? = Tetragonia nigrescens Eckl. et Zeyh. ■☆

387195　Tetragonia nigrescens Fenzl et Zeyh. var. maritima Sond. = Tetragonia nigrescens Eckl. et Zeyh. ■☆

387196　Tetragonia nigrescens Fenzl et Zeyh. var. pruinosa ? = Tetragonia nigrescens Eckl. et Zeyh. ■☆

387197　Tetragonia obovata Haw. = Tetragonia decumbens Mill. ■☆

387198　Tetragonia pentandra Balf. f.；五蕊坚果番杏■☆

387199　Tetragonia perfoliata（L. f.）Druce = Tetragonia decumbens Mill. ■☆

387200　Tetragonia pillansii Adamson；皮朗斯坚果番杏■☆

387201　Tetragonia portulacoides Fenzl；马齿苋坚果番杏■☆

387202　Tetragonia psiloptera Fenzl = Tetragonia robusta Fenzl var. psiloptera（Fenzl）Adamson ■☆

387203　Tetragonia quadricornis Stokes = Tetragonia tetragonioides（Pall.）Kuntze ●■

387204　Tetragonia rangeana Engl.；朗格坚果番杏■☆

387205　Tetragonia reduplicata Welw. ex Oliv.；二重坚果番杏■☆

387206　Tetragonia retusa Thulin；微凹坚果番杏■☆

387207　Tetragonia robusta Fenzl；粗壮坚果番杏■☆

387208　Tetragonia robusta Fenzl var. psiloptera（Fenzl）Adamson；光翅粗壮坚果番杏■☆

387209　Tetragonia rosea Schltr.；粉红坚果番杏■☆

387210　Tetragonia saligna Fenzl；柳坚果番杏■☆

387211　Tetragonia saligna Fenzl var. extrusa Adamson = Tetragonia saligna Fenzl ■☆

387212　Tetragonia saligna Fenzl var. latifolia Adamson = Tetragonia saligna Fenzl ■☆

387213　Tetragonia sarcophylla Fenzl；肉叶坚果番杏■☆

387214　Tetragonia sarcophylla Fenzl var. saxatilis（E. Phillips）Adamson；岩栖肉叶坚果番杏■☆

387215　Tetragonia saxatilis E. Phillips = Tetragonia sarcophylla Fenzl var. saxatilis（E. Phillips）Adamson ■☆

387216　Tetragonia schenckii Schinz；申克坚果番杏■☆

387217　Tetragonia somalensis Engl. = Tribulocarpus dimorphanthus（Pax）S. Moore ●☆

387218　Tetragonia sphaerocarpa Adamson；球状坚果番杏■☆

387219　Tetragonia spicata L. f.；穗状坚果番杏■☆

387220　Tetragonia spicata L. f. var. laxa Adamson；疏松穗状坚果番杏■☆

387221　Tetragonia tetragonioides（Pall.）Kuntze；番杏（白番苋，白番杏，白红菜，滨莴苣，法国菠菜，新西兰菠菜）；New Zealand Iceplant，New Zealand Spinach，Tetragonia，Warrigal，Warrigal Cabbage ●■

387222　Tetragonia tetrapteris Haw. = Tetragonia decumbens Mill. ■☆

387223　Tetragonia verrucosa Fenzl；多疣坚果番杏■☆

387224　Tetragonia virgata Schltr.；条纹坚果番杏■☆

387225　Tetragonia zeyheri Fenzl = Tetragonia decumbens Mill. ■☆

387226　Tetragoniaceae Lindl.（1836）（保留科名）；坚果番杏科（番杏

科）●■☆

387227　Tetragoniaceae Link　= Aizoaceae Martinov（保留科名）●■

387228　Tetragoniaceae Link　= Tetragoniaceae Lindl.（保留科名）●■☆

387229　Tetragoniaceae Nakai　= Aizoaceae Martinov（保留科名）●■

387230　Tetragoniaceae Nakai　= Tetragoniaceae Lindl.（保留科名）●■☆

387231　Tetragonobolus Scop.　= Lotus L. ■

387232　Tetragonobolus Scop.　= Tetragonolobus Scop.（保留属名）■☆

387233　Tetragonocalamus Nakai　= Bambusa Schreb.（保留属名）●

387234　Tetragonocalamus angulatus（Munro）Nakai　= Bambusa tuldoides Munro ●

387235　etragonocalamus quadrangularis（Fenzl）Nakai　= Chimonobambusa quadrangularis（Fenzl）Makino ex Nakai ●

387236　Tetragonocalamus quadrangularis Nakai　= Chimonobambusa quadrangularis（Fenzl）Makino ex Nakai ●

387237　Tetragonocarpos Mill.　= Tetragonia L. ●■

387238　Tetragonocarpus Hassk.　= Marsdenia R. Br.（保留属名）●

387239　Tetragonolobus Scop.（1772）（保留属名）；翅荚豌豆属；Bird's-foot Trefoil，Deervetch，Wing Pod Pea ■☆

387240　Tetragonolobus Scop.（保留属名）= Lotus L. ■

387241　Tetragonolobus biflorus（Desr.）DC.；翅荚豌豆■☆

387242　Tetragonolobus biflorus（Desr.）DC. var. leiolobus Maire　= Tetragonolobus biflorus（Desr.）DC. ■☆

387243　Tetragonolobus conjugatus（L.）Link；成对翅荚豌豆■☆

387244　Tetragonolobus conjugatus（L.）Link subsp. guttatus Pomel　= Tetragonolobus conjugatus（L.）Link subsp. requienii（Sanguin.）E. Domínguez et Galiano ●☆

387245　Tetragonolobus conjugatus（L.）Link subsp. requienii（Sanguin.）E. Domínguez et Galiano；北非荚豌豆■☆

387246　Tetragonolobus edulis Link　= Tetragonolobus purpureus Moench ■☆

387247　Tetragonolobus edulis Link.　= Lotus tetragonolobus L. ●

387248　Tetragonolobus gussonei A. L. P. Huet　= Tetragonolobus conjugatus（L.）Link ■☆

387249　Tetragonolobus guttatus Pomel　= Tetragonolobus conjugatus（L.）Link subsp. requienii（Sanguin.）E. Domínguez et Galiano ■☆

387250　Tetragonolobus maritimus（L.）Roth；滨海翅荚豌豆■☆

387251　Tetragonolobus maritimus Roth；海滨翅荚豌豆；Codded Trefoil，Dragon's Teeth，Horned Clover，Wild Tom Thumb，Winged Pea ■☆

387252　Tetragonolobus prostratus Moench　= Tetragonolobus siliquosus Roth ■☆

387253　Tetragonolobus purpureus Moench；紫翅荚豌豆●☆

387254　Tetragonolobus purpureus Moench　= Lotus tetragonolobus L. ■

387255　Tetragonolobus requienii（Sanguin.）Sanguin.　= Tetragonolobus conjugatus（L.）Link subsp. requienii（Sanguin.）E. Domínguez et Galiano ■☆

387256　Tetragonolobus siliquosus Roth；多果翅荚豌豆■☆

387257　Tetragonolobus siliquosus Roth　= Tetragonolobus maritimus（L.）Roth ■☆

387258　Tetragonolobus siliquosus Roth var. aureus Maire　= Tetragonolobus maritimus（L.）Roth ■☆

387259　Tetragonolobus siliquosus Roth var. bicolor Maire　= Tetragonolobus maritimus（L.）Roth ■☆

387260　Tetragonolobus tauricus Bunge ex Nyman　= Tetragonolobus siliquosus Roth ■☆

387261　Tetragonosperma Scheele　= Tetragonotheca L. ■☆

387262　Tetragonotheca Dill. ex L.　= Tetragonotheca L. ■☆

387263　Tetragonotheca L.（1753）；四角菊属■☆

387264　Tetragonotheca helianthoides L.；四角菊■☆

387265　Tetragonotheca ludoviciana（Torr. et A. Gray）A. Gray ex E. Hall；卢氏四角菊■☆

387266　Tetragonotheca repanda（Buckley）Small；残波四角菊■☆

387267　Tetragonotheca texana A. Gray et Engelm.；得州四角菊■☆

387268　Tetragyne Miq.　= Microdesmis Hook. f. ex Hook. ●

387269　Tetragyne acuminata Miq.　= Microdesmis casearifolia Planch. ●

387270　Tetrahit Adans.　= Sideritis L. ■●

387271　Tetrahit Gerard　= Stachys L. ●■

387272　Tetrahit Moench　= Galeopsis L. ■

387273　Tetrahitum Hoffmanns. et Link　= Tetrahit Gerard ●■

387274　Tetraith Bubani　= Galeopsis L. ■

387275　Tetraith Bubani　= Tetrahit Gerard ●■

387276　Tetralepis Steud.　= Cyathochaeta Nees ■☆

387277　Tetralix Griseb.（1866）（保留属名）；四蕊椴属●☆

387278　Tetralix Hill　= Cirsium Mill. ■

387279　Tetralix Hill　= Cnicus L.（保留属名）■●

387280　Tetralix Zinn（废弃属名）= Erica L. ●☆

387281　Tetralix Zinn.（废弃属名）= Tetralix Griseb.（保留属名）●☆

387282　Tetralix brachypetalus Griseb. et Urb.；四蕊椴●☆

387283　Tetralobus A. DC.　= Polypompholyx Lehm.（保留属名）■

387284　Tetralobus A. DC.　= Utricularia L. ■

387285　Tetralocularia O' Donell（1960）；四室旋花属■☆

387286　Tetralocularia pennellii O'Donell；四室旋花■☆

387287　Tetralopha Hook. f.　= Gynochthodes Blume ●☆

387288　Tetralyx Hill　= Cirsium Mill. ■

387289　Tetralyx Hill　= Tetralix Hill ■

387290　Tetramelaceae Airy Shaw　= Datiscaceae Dumort.（保留科名）■●

387291　Tetramelaceae Airy Shaw（1965）；四数木科●

387292　Tetramelaceae（Warb.）Airy Shaw　= Datiscaceae Dumort.（保留科名）■●

387293　Tetrameles R. Br.（1826）；四数木属；Tetrameles ●

387294　Tetrameles grahamiana（Nimmo）Wight　= Tetrameles nudiflora R. Br. ●◇

387295　Tetrameles grahamiana Wight　= Tetrameles nudiflora R. Br. ●◇

387296　Tetrameles grahamiana Wight var. ceylanica A. DC.　= Tetrameles nudiflora R. Br. ●◇

387297　Tetrameles nudiflora R. Br.；四数木（裸花四数木）；Bare-flowered Tetrameles，Kapong，Naked Flower Tetrameles，Nakedflower Tetrameles，Sompong ●◇

387298　Tetrameles rufinervis Miq.　= Tetrameles nudiflora R. Br. ●◇

387299　Tetrameranthus R. E. Fr.（1939）；四数花属●☆

387300　Tetrameranthus duckei R. E. Fr.；四数花●☆

387301　Tetrameranthus macrocarpus R. E. Fr.；大果四数花●☆

387302　Tetrameris Naudin　= Comolia DC. ●☆

387303　Tetramerista Miq.（1861）；四籽树属●☆

387304　Tetramerista crassifolia Hallier f.；厚叶四籽树●☆

387305　Tetramerista glabra Miq.；光四籽树；Punah ●☆

387306　Tetrameristaceae Hutch.（1959）；四籽树科●☆

387307　Tetramerium C F. Gaertn.（废弃属名）= Tetramerium Nees（保留属名）●☆

387308　Tetramerium C. F. Gaertn.（废弃属名）= Faramea Aubl. ●☆

387309　Tetramerium Nees（1846）（保留属名）；四分爵床属●☆

387310　Tetramerium angustius（Nees）T. F. Daniel；窄四分爵床●☆

387311　Tetramerium aureum Rose；黄四分爵床●☆

387312　Tetramerium diffusum Rose；铺散四分爵床●☆

387313　Tetramerium flavum Eastw.；鲜黄四分爵床●☆

387314　Tetramerium leptocaule Happ；细茎四分爵床●☆

387315 Tetramerium macrostachyum Happ;大穗四分爵床●☆

387316 Tetramerium nervosum Nees;多脉四分爵床●☆

387317 Tetramerium obovatum T. F. Daniel;倒卵四分爵床●☆

387318 Tetramerium polystachyum Nees;多穗四分爵床●☆

387319 Tetramerium rubrum Hopp;红四分爵床●☆

387320 Tetramerium tenuissimum Rose;纤细四分爵床●☆

387321 Tetramicra Lindl. (1831);四粉兰属（四隔兰属）■☆

387322 Tetramicra bicolor Rolfe;二色四粉兰■☆

387323 Tetramicra elegans Cogn. ;雅致四粉兰■☆

387324 Tetramicra minuta Rolfe;小四粉兰■☆

387325 Tetramicra montana Griseb. ;山地四粉兰■☆

387326 Tetramicra parviflora Lindl. ex Griseb. ;小花四粉兰■☆

387327 Tetramicra platyphylla Griseb. ;宽叶四粉兰■☆

387328 Tetramicra rigida Lindl. ;硬四粉兰■☆

387329 Tetramicra simplex Ames;简单四粉兰■☆

387330 Tetramixis Gagnep. = Spondias L. ●

387331 Tetramolopium Nees(1832);层菀木属●☆

387332 Tetramolopium bicolor Koster;二色层菀木●☆

387333 Tetramolopium fasciculatum Koster;簇生层菀木●☆

387334 Tetramolopium flaccidum Mattf. ;柔软层菀木●☆

387335 Tetramolopium gracile Koster;纤细层菀木●☆

387336 Tetramolopium polyphyllum Sherff;多叶层菀木●☆

387337 Tetramolopium tenue Koster;细层菀木●☆

387338 Tetramorphaea DC. = Centaurea L. (保留属名)●■

387339 Tetramorphandra Baill. = Hibbertia Andréws ●☆

387340 Tetramyxis Gagnep. = Spondias L. ●

387341 Tetrandra (Dc.) Miq. = Tournefortia L. ●■

387342 Tetrandra Miq. = Tournefortia L. ●■

387343 Tetranema Benth. (1843)(保留属名);四蕊花属■☆

387344 Tetranema Sweet = Desmodium Desv. (保留属名)●■

387345 Tetranema mexicanum Benth. = Tetranema roseum (Mart. et Galeotti) Standl. et Steyerm. ■☆

387346 Tetranema roseum (Mart. et Galeotti) Standl. et Steyerm. ;四蕊花（墨西哥毛地黄）;Mexican Foxglove,Mexican Violet ■☆

387347 Tetraneuris Greene = Hymenoxys Cass. ■☆

387348 Tetraneuris Greene(1898);四脉菊属(Bitterweed ■☆

387349 Tetraneuris acaulis (Pursh) Greene;无茎四脉菊;Angelita Daisy ■☆

387350 Tetraneuris acaulis (Pursh) Greene var. arizonica (Greene) K. L. Parker;亚利桑那四脉菊■☆

387351 Tetraneuris acaulis (Pursh) Greene var. caespitosa A. Nelson;丛生四脉菊■☆

387352 Tetraneuris acaulis (Pursh) Greene var. epunctata (A. Nelson) Kartesz et Gandhi;无斑无茎四脉菊■☆

387353 Tetraneuris angustata Greene = Tetraneuris scaposa (DC.) Greene ■☆

387354 Tetraneuris angustifolia Rydb. = Tetraneuris scaposa (DC.) Greene ■☆

387355 Tetraneuris argentea (A. Gray) Greene;银白四脉菊;Angelita Daisy ,Perky Sue ■☆

387356 Tetraneuris arizonica Greene = Tetraneuris acaulis (Pursh) Greene var. arizonica (Greene) K. L. Parker ■☆

387357 Tetraneuris brandegeei (Porter ex A. Gray) K. L. Parker = Hymenoxys brandegeei (Porter ex A. Gray) K. L. Parker ■☆

387358 Tetraneuris brevifolia Greene = Hymenoxys acaulis (Pursh) K. L. Parker var. caespitosa (A. Nelson) K. L. Parker ■☆

387359 Tetraneuris depressa (Torr. et A. Gray) Greene = Tetraneuris torreyana (Nutt.) Greene ■☆

387360 Tetraneuris eradiata A. Nelson = Tetraneuris acaulis (Pursh) Greene ■☆

387361 Tetraneuris fastigiata Greene = Tetraneuris scaposa (DC.) Greene ■☆

387362 Tetraneuris formosa Greene ex Wooton et Standl. = Tetraneuris argentea (A. Gray) Greene ■☆

387363 Tetraneuris glabra (Nutt.) Greene = Tetraneuris scaposa (DC.) Greene ■☆

387364 Tetraneuris grandiflora (Torr. et A. Gray) K. L. Parker = Hymenoxys grandiflora (Torr. et A. Gray) K. L. Parker ■☆

387365 Tetraneuris incana A. Nelson;灰白四脉菊■☆

387366 Tetraneuris intermedia Greene = Tetraneuris ivesiana Greene ■☆

387367 Tetraneuris ivesiana Greene;艾夫斯四脉菊■☆

387368 Tetraneuris lanata Greene = Hymenoxys acaulis (Pursh) K. L. Parker var. caespitosa (A. Nelson) K. L. Parker ■☆

387369 Tetraneuris lanigera Daniels = Hymenoxys acaulis (Pursh) K. L. Parker var. caespitosa (A. Nelson) K. L. Parker ■☆

387370 Tetraneuris leptoclada (A. Gray) Greene = Tetraneuris argentea (A. Gray) Greene ■☆

387371 Tetraneuris linearifolia (Hook.) Greene;线叶四脉菊■☆

387372 Tetraneuris linearifolia (Hook.) Greene subsp. dodgei Cockerell = Tetraneuris linearifolia (Hook.) Greene ■☆

387373 Tetraneuris linearifolia (Hook.) Greene var. arenicola Bierner;沙丘线叶四脉菊■☆

387374 Tetraneuris linearifolia (Hook.) Greene var. latior Cockerell = Tetraneuris linearifolia (Hook.) Greene ■☆

387375 Tetraneuris linearis (Nutt.) Greene = Tetraneuris scaposa (DC.) Greene ■☆

387376 Tetraneuris mancosensis A. Nelson = Tetraneuris ivesiana Greene ■☆

387377 Tetraneuris oblongifolia Greene = Tetraneuris linearifolia (Hook.) Greene ■☆

387378 Tetraneuris pilosa Greene = Tetraneuris ivesiana Greene ■☆

387379 Tetraneuris pygmaea (A. Gray) Wooton et Standl. = Tetraneuris incana A. Nelson ■☆

387380 Tetraneuris scaposa (DC.) Greene;粗糙四脉菊;Four-nerve Daisy ■☆

387381 Tetraneuris scaposa (DC.) Greene var. argyrocaulon (K. L. Parker) K. L. Parker;银茎粗糙四脉菊■☆

387382 Tetraneuris scaposa (DC.) Greene var. linearis (Nutt.) K. L. Parker = Tetraneuris scaposa (DC.) Greene ■☆

387383 Tetraneuris scaposa (DC.) Greene var. villosa (Shinners) Shinners = Tetraneuris scaposa (DC.) Greene ■☆

387384 Tetraneuris septentrionalis Rydb. = Tetraneuris acaulis (Pursh) Greene ■☆

387385 Tetraneuris simplex A. Nelson = Tetraneuris acaulis (Pursh) Greene ■☆

387386 Tetraneuris stenophylla Rydb. = Tetraneuris scaposa (DC.) Greene ■☆

387387 Tetraneuris torreyana (Nutt.) Greene;托里四脉菊■☆

387388 Tetraneuris trinervata Greene = Tetraneuris argentea (A. Gray) Greene ■☆

387389 Tetraneuris turneri (K. L. Parker) K. L. Parker;特纳四脉菊■☆

387390 Tetrantha Poit. = Riencourtia Cass. ■☆

387391 Tetrantha Poit. ex DC. = Riencourtia Cass. ■☆

387392 Tetranthera Jacq. (1797);四药樟属●

387393 Tetranthera Jacq. = Litsea Lam. (保留属名)●

387394 Tetranthera amara (Blume) Nees = Litsea umbellata (Lour.) Merr. ●

387395 Tetranthera amara Nees = Litsea umbellata (Lour.) Merr. ●

387396 Tetranthera bifaria Wall. = Lindera nacusua (D. Don) Merr. ●

387397 Tetranthera californica Hook. et Arn. = Umbellularia californica (Hook. et Arn.) Nutt. ●☆

387398 Tetranthera calophylla Kurz = Litsea martabanica (Kurz) Hook. f. ●

387399 Tetranthera caudatum Wall. = Lindera caudata (Nees) Hook. f. ●

387400 Tetranthera citrata Nees = Litsea cubeba (Lour.) Pers. ●

387401 Tetranthera cubeba Meisn. = Litsea cubeba (Lour.) Pers. ●

387402 Tetranthera cuipala D. Don = Neolitsea cuipala (D. Don) Kosterm. ●☆

387403 Tetranthera ferntginea R. Br. = Litsea umbellata (Lour.) Merr. ●

387404 Tetranthera japonica (Thunb.) Spreng. = Litsea japonica (Thunb.) Juss. ●☆

387405 Tetranthera lancifolia Roxb. ex Nees = Litsea lancifolia (Roxb. ex Nees) Benth. et Hook. f. ex Fern. -Vill. ●

387406 Tetranthera lanuginosa Wall. ex Nees = Neolitsea cuipala (D. Don) Kosterm. ●☆

387407 Tetranthera laurifolia Jacq.;四药樟●☆

387408 Tetranthera martabanica Kurz = Litsea martabanica (Kurz) Hook. f. ●

387409 Tetranthera monopetala Roxb. = Litsea monopetala (Roxb.) Pers. ●

387410 Tetranthera obovata Buch. -Ham. ex Wall. = Actinodaphne obovata (Nees) Blume ●

387411 Tetranthera pallens D. Don = Neolitsea pallens (D. Don) Momiy. et H. Hara ●

387412 Tetranthera panamanja Buch. -Ham. ex Nees = Litsea panamonja (Nees) Hook. f. ●

387413 Tetranthera panamonja Buch. -Ham. = Litsea panamonja (Nees) Hook. f. ●

387414 Tetranthera panamonja Nees = Litsea panamonja (Nees) Hook. f. ●

387415 Tetranthera pilosa (Lour.) Spreng. = Actinodaphne pilosa (Lour.) Merr. ●

387416 Tetranthera polyantha Wall. = Litsea cubeba (Lour.) Pers. ●

387417 Tetranthera pulcherrima Wall. = Lindera pulcherrima (Nees) Hook. f. ●

387418 Tetranthera pulcherrima Wall. = Lindera thomsonii C. K. Allen ●

387419 Tetranthera rotundifolia Wall. ex Nees = Litsea rotundifolia (Nees) Hemsl. ●

387420 Tetranthera salicifolia Roxb. ex Nees = Litsea salicifolia (Roxb. ex Nees) Hook. f. ●

387421 Tetranthera semecarpifolia Wall. ex Nees = Litsea semecarpifolia (Wall. ex Nees) Hook. f. ●

387422 Tetranthera sericea Nees = Litsea sericea (Nees) Hook. f. ●

387423 Tetranthera sericea Wall. ex Nees = Litsea sericea (Nees) Hook. f. ●

387424 Tetranthera tomentosa Roxb. ex Wall. ex Wight = Litsea deccanensis Gamble ●☆

387425 Tetranthus Sw. (1788);四花菊属●☆

387426 Tetranthus littoralis Sw.;四花菊■☆

387427 Tetraotis Reinw. = Enydra Lour. ■

387428 Tetrapanax (K. Koch) K. Koch(1859);通脱木属;Rice-paper Plant, Ricepaperplant, Ricepaper-plant ●★

387429 Tetrapanax Harms = Hoplopanax Post et Kuntze ●

387430 Tetrapanax Harms = Oplopanax (Torr. et A. Gray) Miq. ●

387431 Tetrapanax K. Koch = Tetrapanax (K. Koch) K. Koch ●★

387432 Tetrapanax papyrifer (Hook.) K. Koch;通脱木(白龙须,白通草,草片,草条,草枝,葱草,大木通,大通草,大通塔,大叶五加皮,方通,方通草,花草,活莌,活脱,空心通草,寇通,寇莌,寇脱,宽草,宽肠,离南,木通,木通树,泡通,片通,丝通草,天麻子,通草,通大海,通花,通花五加,蓪,蓪草,莌,万丈深,五加风,五角加皮,倚商,朱通草,紫金莲);Pith Paper, Pith Paper Plant, Rice Paper, Rice Paper Plant, Ricepaper Plant, Rice-paper Plant, Ricepaperplant, Ricepaper-plant ●

387433 Tetrapanax papyrifer (Hook.) K. Koch 'Variegata';斑叶通脱木●

387434 Tetrapanax papyrifer K. Koch = Tetrapanax papyrifer (Hook.) K. Koch ●

387435 Tetrapanax tibetanus G. Hoo;西藏通脱木;Tibet Ricepaperplant, Xizang Ricepaperplant ●

387436 Tetrapanax tibetanus G. Hoo = Merrilliopanax alpinus (C. B. Clarke) C. B. Shang ●

387437 Tetrapasma G. Don = Discaria Hook. ●☆

387438 Tetrapathaea (DC.) Rchb. = Passiflora L. ●■

387439 Tetrapathaea Rchb. = Passiflora L. ●■

387440 Tetrapeltis Lindl. = Otochilus Lindl. ■

387441 Tetrapeltis Wall. ex Lindl. = Otochilus Lindl. ■

387442 Tetrapeltis fragrans Wall. ex Lindl. = Otochilus porrectus Lindl. ■

387443 Tetraperone Urb. (1901);四带菊属■☆

387444 Tetraperone bellioides Urb.;四带菊■☆

387445 Tetrapetalum Miq. (1865);四瓣花属●☆

387446 Tetrapetalum borneense Merr.;四瓣花●☆

387447 Tetraphyla Rchb. = Sedum L. ●■

387448 Tetraphyla Rchb. = Tetraphyle Eckl. et Zeyh. ●■☆

387449 Tetraphylax (G. Don) de Vriese = Goodenia Sm. ●■☆

387450 Tetraphylax de Vriese = Goodenia Sm. ●■☆

387451 Tetraphyle Eckl. et Zeyh. = Crassula L. ●■☆

387452 Tetraphyle furcata Eckl. et Zeyh. = Crassula ericoides Haw. ●☆

387453 Tetraphyle lanceolata Eckl. et Zeyh. = Crassula lanceolata (Eckl. et Zeyh.) Endl. ex Walp. ●☆

387454 Tetraphyle littoralis Eckl. et Zeyh. = Crassula muscosa L. ●☆

387455 Tetraphyle lycopodioides (Lam.) Eckl. et Zeyh. = Crassula muscosa L. ●☆

387456 Tetraphyle muscosa (L.) Eckl. et Zeyh. = Crassula muscosa L. ●☆

387457 Tetraphyle polpodacea Eckl. et Zeyh. = Crassula muscosa L. var. polpodacea (Eckl. et Zeyh.) G. D. Rowley ●☆

387458 Tetraphyle propinqua Eckl. et Zeyh. = Crassula muscosa L. var. obtusifolia (Harv.) G. D. Rowley ●☆

387459 Tetraphyle pyramidalis (Thunb.) Eckl. et Zeyh. = Crassula pyramidalis Thunb. ●☆

387460 Tetraphyle quadrangula Eckl. et Zeyh. = Crassula pyramidalis Thunb. ●☆

387461 Tetraphyllaster Gilg(1897);四叶星属●☆

387462 Tetraphyllaster rosaceum Gilg;四叶星●☆

387463 Tetraphyllaster rosaceum Gilg = Tristemma leiocalyx Cogn. ●☆

387464 Tetraphyllum C. B. Clarke(1883);四叶苣苔属●☆

387465 Tetraphyllum Griff. = Tetraphyllum C. B. Clarke ■☆

387466 Tetraphyllum Griff. ex C. B. Clarke = Tetraphyllum C. B. Clarke ■☆

387467 Tetraphyllum bengalense C. B. Clarke;四叶苣苔●☆

387468 Tetraphyllum roseum Stapf;粉红四叶苣苔●☆

387469　Tetraphysa Schltr. (1906)；四室萝藦属■☆

387470　Tetraphysa lehmannii Schltr.；莱曼四室萝藦■☆

387471　Tetrapilus Lour. = Olea L. ●

387472　Tetrapilus brachiatus Lour. = Olea brachiata （Lour.） Merr. ex G. W. Groff et H. H. Groff ●

387473　Tetrapilus hainanensis （H. L. Li） L. Johnson = Olea hainanensis H. L. Li ●

387474　Tetraplacus Radlk. = Otacanthus Lindl. ■●☆

387475　Tetraplandra Baill. (1858)；四雄大戟属☆

387476　Tetraplandra grandifolia Glaz.；大叶四雄大戟☆

387477　Tetraplandra leandrii Baill.；四雄大戟☆

387478　Tetraplasandra A. Gray(1854)；四雄五加属●☆

387479　Tetraplasandra hawaiensis A. Gray；四雄五加●☆

387480　Tetraplasandra micrantha Sherff；小花四雄五加●☆

387481　Tetraplasandra paucidens Miq.；寡齿四雄五加●☆

387482　Tetraplasandra philippinensis Merr.；菲律宾四雄五加●☆

387483　Tetraplasia Rehder = Damnacanthus C. F. Gaertn. ●

387484　Tetraplasia Rehder(1920)；岛虎刺属(台虎刺属)；Tetraplasia ●

387485　Tetraplasia angustifolia （Hayata） Koidz. = Damnacanthus angustifolius Hayata ●

387486　Tetraplasia biflora Rehder = Damnacanthus biflorus （Rehder） Masam. ●☆

387487　Tetraplasia stenophylla Koidz.；岛虎刺（台虎刺）；Common Tetraplasia ●

387488　Tetraplasia stenophylla Koidz. = Damnacanthus angustifolius Hayata ●

387489　Tetraplasium Kuntze = Tetilla DC. ■☆

387490　Tetrapleura Benth. (1841)；四肋豆属■☆

387491　Tetrapleura Parl. = Tornabenea Parl. ex Webb ■☆

387492　Tetrapleura Parl. ex Webb = Tornabenea Parl. ex Webb ■☆

387493　Tetrapleura andongensis Welw. ex Oliv. = Amblygonocarpus andongensis （Welw. ex Oliv.） Exell et Torre ●☆

387494　Tetrapleura andongensis Welw. ex Oliv. var. schweinfurthii （Harms） Aubrév. = Amblygonocarpus andongensis （Welw. ex Oliv.） Exell et Torre ●☆

387495　Tetrapleura chevalieri （Harms） Baker f.；舍瓦利耶四肋豆●☆

387496　Tetrapleura insularis Webb = Tornabenea insularis （Webb） Webb ■☆

387497　Tetrapleura obtusangula Welw. ex Oliv. = Amblygonocarpus andongensis （Welw. ex Oliv.） Exell et Torre ●☆

387498　Tetrapleura tetraptera （Schumach. et Thonn.） Taub.；四肋豆（四肋草）■☆

387499　Tetrapleura tetraptera Taub. = Tetrapleura tetraptera （Schumach. et Thonn.） Taub. ■☆

387500　Tetrapleura thonningii Benth. = Tetrapleura tetraptera （Schumach. et Thonn.） Taub. ■☆

387501　Tetrapodenia Gleason = Burdachia Mart. ex A. Juss. ●☆

387502　Tetrapogon Desf. (1799)；四须草属■☆

387503　Tetrapogon bidentatus Pilg.；双齿四须草■☆

387504　Tetrapogon cenchriformis （A. Rich.） Clayton；蒺藜四须草■☆

387505　Tetrapogon cymbiferus Peter = Tetrapogon cenchriformis （A. Rich.） Clayton ■☆

387506　Tetrapogon ferrugineus （Renvoize） S. M. Phillips；锈色四须草■☆

387507　Tetrapogon flabellatus Hack. = Chloris flabellata （Hack.） Launert ■☆

387508　Tetrapogon monostachyus Peter = Tetrapogon tenellus （J. König ex Roxb.） Chiov. ■☆

387509　Tetrapogon mossambicensis （K. Schum.） Chippend. ex B. S. Fisher = Chloris mossambicensis K. Schum. ■☆

387510　Tetrapogon spathaceus （Hochst. ex Steud.） Hack. ex T. Durand et Schinz = Tetrapogon cenchriformis （A. Rich.） Clayton ■☆

387511　Tetrapogon tenellus （J. König ex Roxb.） Chiov.；细四须草■☆

387512　Tetrapogon villosus Desf.；四须草；Feathery Rhodes-grass ■☆

387513　Tetrapogon villosus Desf. f. distachyus Maire et Weiller = Tetrapogon villosus Desf. ■☆

387514　Tetrapogon villosus Desf. f. monostachyus （Batt. et Trab.） Maire et Weiller = Tetrapogon villosus Desf. ■☆

387515　Tetrapogon villosus Desf. f. monostachyus （Batt. et Trab.） Quézel = Tetrapogon villosus Desf. ■☆

387516　Tetrapogon villosus Desf. var. monostachyus Batt. et Trab. = Tetrapogon villosus Desf. ■☆

387517　Tetrapogon villosus Desf. var. monostachyus Trab. = Tetrapogon villosus Desf. ■☆

387518　Tetrapogon villosus Desf. var. sinaicus （Decne.） Täckh. = Tetrapogon villosus Desf. ■☆

387519　Tetrapogon villosus Desf. var. tibesticus Quézel = Tetrapogon villosus Desf. ■☆

387520　Tetrapollinia Maguire et B. M. Boom(1835)；四数龙胆属■☆

387521　Tetrapollinia caerulescens （Aubl.） Maguire et B. M. Boom；天蓝四数龙胆■☆

387522　Tetrapoma Turcz. = Rorippa Scop. ■

387523　Tetrapoma Turcz. ex Fisch. et C. A. Mey. = Rorippa Scop. ■

387524　Tetrapoma barbareifolium （DC.） Turcz. ex Fisch. et C. A. Mey. = Rorippa barbareifolia （DC.） Kitag. ■

387525　Tetrapoma barbarifolium Turcz. = Rorippa barbareifolia （DC.） Kitag. ■

387526　Tetrapoma kruhsianum Fisch. et C. A. Mey. = Rorippa barbareifolia （DC.） Kitag. ■

387527　Tetrapoma pyriforme Seem. = Rorippa barbareifolia （DC.） Kitag. ■

387528　Tetrapora Schauer = Baeckea L. ●

387529　Tetraptera Miers = Burmannia L. ■

387530　Tetraptera Miers ex Lindl. = Burmannia L. ■

387531　Tetraptera Phil. = Gaya Kunth ■●☆

387532　Tetrapteris Cav. = Tetrapterys Cav. （保留属名）●☆

387533　Tetrapteris Garden = Halesia J. Ellis ex L. （保留属名）●

387534　Tetrapteris acuminata Raeusch. ex A. Juss. = Tetrapterys acuminata Raeusch. ex A. Juss. ●☆

387535　Tetrapteris acutifolia Cav. = Tetrapterys acutifolia Cav. ●☆

387536　Tetrapteris affinis A. Juss. = Tetrapterys affinis A. Juss. ●☆

387537　Tetrapteris boliviensis Nied. = Tetrapterys boliviensis Nied. ●☆

387538　Tetrapteris brachyptera Mart. ex A. Juss. = Tetrapterys brachyptera Mart. ex A. Juss. ●☆

387539　Tetrapteris calophylla A. Juss. = Tetrapterys calophylla A. Juss. ●☆

387540　Tetrapteris chloroptera Cuatrec. = Tetrapterys chloroptera Cuatrec. ●☆

387541　Tetrapteris citrifolia Sw. = Tetrapterys citrifolia Sw. ●☆

387542　Tetrapteris cubensis Nied. = Tetrapterys cubensis Nied. ●☆

387543　Tetrapteris elliptica Rusby = Tetrapterys elliptica Rusby ●☆

387544　Tetrapteris fraxinifolia A. Juss. = Tetrapterys fraxinifolia A. Juss. ●☆

387545　Tetrapteris glabra Griseb. = Tetrapterys glabra Griseb. ●☆

387546　Tetrapteris glandulosa Griseb. = Tetrapterys glandulosa Griseb. ●☆

387547　Tetrapteris gracilis Walp. = Tetrapterys gracilis Walp. ●☆

387548　Tetrapteris lancifolia A. Juss. = Tetrapterys lancifolia A. Juss. ●☆

387549　Tetrapteris lasiocarpa A. Juss. = Tetrapterys lasiocarpa A. Juss. ●☆

387550　Tetrapteris laurifolia Griseb. = Tetrapterys laurifolia Griseb. ●☆

387551　Tetrapteris leucosepala A. Juss. = Tetrapterys leucosepala A. Juss. ●☆

387552　Tetrapteris longibracteata A. Juss. = Tetrapterys longibracteata A. Juss. ●☆

387553　Tetrapteris lucida A. Juss. = Tetrapterys lucida A. Juss. ●☆

387554　Tetrapteris macrocarpa I. M. Johnst. = Tetrapterys macrocarpa I. M. Johnst. ●☆

387555　Tetrapteris macrophylla Poepp. ex Griseb. = Tetrapterys macrophylla Poepp. ex Griseb. ●☆

387556　Tetrapteris mexicana Hook. et Arn. = Tetrapterys mexicana Hook. et Arn. ●☆

387557　Tetrapteris mollis Griseb. = Tetrapterys mollis Griseb. ●☆

387558　Tetrapteris mucronata Cav. = Tetrapterys mucronata Cav. ●☆

387559　Tetrapteris multiglandulosa A. Juss. = Tetrapterys multiglandulosa A. Juss. ●☆

387560　Tetrapteris nitida A. Juss. = Tetrapterys nitida A. Juss. ●☆

387561　Tetrapteris obovata Poepp. ex A. Juss. = Tetrapterys obovata Poepp. ex A. Juss. ●☆

387562　Tetrapteris ovalifolia Griseb. = Tetrapterys ovalifolia Griseb. ●☆

387563　Tetrapteris pallida Cuatrec. = Tetrapterys pallida Cuatrec. ●☆

387564　Tetrapteris paniculata Bello = Tetrapterys paniculata Bello ●☆

387565　Tetrapteris parvifolia Glaz. = Tetrapterys parvifolia Glaz. ●☆

387566　Tetrapteris pauciflora DC. = Tetrapterys pauciflora DC. ●☆

387567　Tetrapteris punctulata A. Juss. = Tetrapterys punctulata A. Juss. ●☆

387568　Tetrapteris reticulata Small = Tetrapterys reticulata Small ●☆

387569　Tetrapteris rotundifolia A. Juss. = Tetrapterys rotundifolia A. Juss. ●☆

387570　Tetrapteris salicifolia Nied. = Tetrapterys salicifolia Nied. ●☆

387571　Tetrapteris silvatica Cuatrec. = Tetrapterys silvatica Cuatrec. ●☆

387572　Tetrapteris styloptera A. Juss. = Tetrapterys styloptera A. Juss. ●☆

387573　Tetrapteris tenuistachys Rusby = Tetrapterys tenuistachys Rusby ●☆

387574　Tetrapteris trichocalyx Diels = Tetrapterys trichocalyx Diels ●☆

387575　Tetrapterocarpon Humbert(1939);四翅苏木属●☆

387576　Tetrapterocarpon geayi Humbert;四翅苏木●☆

387577　Tetrapterocarpon septentrionalis Du Puy et R. Rabev.;北方四翅苏木●☆

387578　Tetrapteron (Munz) W. L. Wagner et Hoch = Oenothera L. ●■

387579　Tetrapteron (Munz) W. L. Wagner et Hoch(2007);北美月见草属■☆

387580　Tetrapteron graciliflorum (Hook. et Arn.) W. L. Wagner et Hoch;纤花北美月见草■☆

387581　Tetrapteron palmeri (S. Watson) W. L. Wagner et Hoch;北美月见草■☆

387582　Tetrapterygium Fisch. et C. A. Mey. = Sameraria Desv. ■☆

387583　Tetrapterys A. Juss. = Tetrapterys Cav. (保留属名)●☆

387584　Tetrapterys Cav. (1790)('Tetrapteris')(保留属名);四翼木属(四翅金虎尾属)●☆

387585　Tetrapterys acuminata Raeusch. ex A. Juss. ;渐尖四翼木●☆

387586　Tetrapterys acutifolia Cav. ;尖叶四翼木●☆

387587　Tetrapterys affinis A. Juss. ;近缘四翼木●☆

387588　Tetrapterys boliviensis Nied. ;玻利维亚四翼木●☆

387589　Tetrapterys brachyptera Mart. ex A. Juss. ;短翅四翼木●☆

387590　Tetrapterys calophylla A. Juss. ;美叶四翼木●☆

387591　Tetrapterys chloroptera Cuatrec. ;绿翅四翼木●☆

387592　Tetrapterys citrifolia Sw. ;橘叶四翼木●☆

387593　Tetrapterys cubensis Nied. ;古巴四翼木●☆

387594　Tetrapterys elliptica Rusby;椭圆四翼木●☆

387595　Tetrapterys fraxinifolia A. Juss. ;白腊叶四翼木●☆

387596　Tetrapterys glabra Griseb. ;光四翼木●☆

387597　Tetrapterys glandulosa Griseb. ;腺点四翼木●☆

387598　Tetrapterys gracilis Walp. ;细四翼木●☆

387599　Tetrapterys lancifolia A. Juss. ;披针叶四翼木●☆

387600　Tetrapterys lasiocarpa A. Juss. ;毛果四翼木●☆

387601　Tetrapterys laurifolia Griseb. ;桂叶四翼木●☆

387602　Tetrapterys leucosepala A. Juss. ;白萼四翼木●☆

387603　Tetrapterys longibracteata A. Juss. ;长苞四翼木●☆

387604　Tetrapterys lucida A. Juss. ;亮四翼木●☆

387605　Tetrapterys macrocarpa I. M. Johnst. ;大果四翼木●☆

387606　Tetrapterys macrophylla Poepp. ex Griseb. ;大叶四翼木●☆

387607　Tetrapterys mexicana Hook. et Arn. ;墨西哥四翼木●☆

387608　Tetrapterys mollis Griseb. ;柔软四翼木●☆

387609　Tetrapterys mucronata Cav. ;钝尖四翼木●☆

387610　Tetrapterys multiglandulosa A. Juss. ;多腺四翼木●☆

387611　Tetrapterys nitida A. Juss. ;光亮四翼木●☆

387612　Tetrapterys obovata Poepp. ex A. Juss. ;倒卵四翼木●☆

387613　Tetrapterys ovalifolia Griseb. ;卵叶四翼木●☆

387614　Tetrapterys pallida Cuatrec. ;苍白四翼木●☆

387615　Tetrapterys paniculata Bello;圆锥四翼木●☆

387616　Tetrapterys parviflora Glaz. ;小花四翼木●☆

387617　Tetrapterys parvifolia Glaz. ;小叶四翼木●☆

387618　Tetrapterys pauciflora DC. ;少花四翼木●☆

387619　Tetrapterys punctulata A. Juss. ;斑点四翼木●☆

387620　Tetrapterys reticulata Small;网脉四翼木●☆

387621　Tetrapterys rotundifolia A. Juss. ;圆叶四翼木●☆

387622　Tetrapterys salicifolia Nied. ;柳叶四翼木●☆

387623　Tetrapterys silvatica Cuatrec. ;林地四翼木☆

387624　Tetrapterys styloptera A. Juss. ;柱翅四翼木●☆

387625　Tetrapterys tenuistachys Rusby;细穗四翼木●☆

387626　Tetrapterys trichocalyx Diels;毛萼四翼木●☆

387627　Tetrapteryx Dalla Torre et Harms = Tetrapterys Cav. (保留属名)●☆

387628　Tetraracus Klotzsch ex Engl. = Tapirira Aubl. ●☆

387629　Tetrardisia Mez(1902);四数紫金牛属●☆

387630　Tetrardisia denticulata Mez;四数紫金牛☆

387631　Tetrardisia fruticosa B. C. Stone;灌木四数紫金牛●☆

387632　Tetrarhaphis Miers = Oxytheca Nutt. ■☆

387633　Tetraria P. Beauv. (1816);四数莎草属■☆

387634　Tetraria angustifolia (Hochst.) Sweet = Tetraria bromoides (Lam.) Pfeiff. ■☆

387635　Tetraria aristata (Boeck.) C. B. Clarke = Tetraria flexuosa (Thunb.) C. B. Clarke ■☆

387636　Tetraria autumnalis Levyns = Tetraria ligulata (Boeck.) C. B. Clarke ■☆

387637　Tetraria bachmannii Kük. = Cyathocoma bachmannii (Kük.) C. Archer ■☆

387638　Tetraria bolusii C. B. Clarke;博卢斯四数莎草■☆

387639　Tetraria brachyphylla Levyns;短叶四数莎草■☆

387640　Tetraria brevicaulis C. B. Clarke = Capeobolus brevicaulis (C. B. Clarke) Browning ■☆

387641　Tetraria bromoides (Lam.) Pfeiff. ;燕麦莎草■☆

387642　Tetraria bromoides (Lam.) Pfeiff. var. angustifolia (Hochst.) Kük. = Tetraria bromoides (Lam.) Pfeiff. ■☆

387643　Tetraria burmannii (Vahl) C. B. Clarke;布尔曼四数莎草■☆

387644　Tetraria capillacea（Thunb.）C. B. Clarke；细毛四数莎草■☆

387645　Tetraria capitata Kük. = Schoenus nigricans L.■☆

387646　Tetraria circinalis（Schrad.）C. B. Clarke = Tetraria microstachys（Vahl）Pfeiff.■☆

387647　Tetraria circinalis（Schrad.）C. B. Clarke var. transiens Kük. = Tetraria pubescens Schönland et Turrill■☆

387648　Tetraria circinalis（Schrad.）C. B. Clarke var. usambarensis（K. Schum.）Kük. = Tetraria usambarensis K. Schum.■☆

387649　Tetraria compacta Levyns；紧密四数莎草■☆

387650　Tetraria compar（L.）T. Lestib.；伴侣四数莎草■☆

387651　Tetraria compar（L.）T. Lestib. var. minor（Boeck.）Kük. = Tetraria compar（L.）T. Lestib.■☆

387652　Tetraria compressa Turrill = Tetraria robusta（Kunth）C. B. Clarke■☆

387653　Tetraria crassa Levyns；粗四数莎草■☆

387654　Tetraria crinifolia（Nees）C. B. Clarke；丝叶莎草■☆

387655　Tetraria cuspidata（Rottb.）C. B. Clarke；骤尖四数莎草■☆

387656　Tetraria cuspidata（Rottb.）C. B. Clarke f. gracilis（Nees）Boeck. = Tetraria exilis Levyns■☆

387657　Tetraria cuspidata（Rottb.）C. B. Clarke f. robustior Kük. = Tetraria crassa Levyns■☆

387658　Tetraria dregeana（Boeck.）C. B. Clarke = Epischoenus dregeanus（Boeck.）Levyns■☆

387659　Tetraria exilis Levyns；瘦小四数莎草■☆

387660　Tetraria eximia C. B. Clarke；优异莎草■☆

387661　Tetraria fasciata（Rottb.）C. B. Clarke；带状四数莎草■☆

387662　Tetraria fasciata（Rottb.）C. B. Clarke var. maculata（Schönland et Turrill）Kük = Tetraria maculata Schönland et Turrill■☆

387663　Tetraria ferruginea C. B. Clarke；锈色四数莎草■☆

387664　Tetraria fimbriolata（Nees）C. B. Clarke；线四数莎草■☆

387665　Tetraria flexuosa（Thunb.）C. B. Clarke；曲折四数莎草■☆

387666　Tetraria fourcadei Turrill et Schönland；富尔四数莎草■☆

387667　Tetraria galpinii Schönland et Turrill；盖尔四数莎草■☆

387668　Tetraria gracilis Turrill = Epischoenus adnatus Levyns■☆

387669　Tetraria graminifolia Levyns；禾叶四数莎草■☆

387670　Tetraria involucrata（Rottb.）C. B. Clarke；总苞四数莎草■☆

387671　Tetraria ligulata（Boeck.）C. B. Clarke；舌状四数莎草■☆

387672　Tetraria lucida C. B. Clarke = Epischoenus lucidus（C. B. Clarke）Levyns■☆

387673　Tetraria macowaniana B. L. Burtt = Tetraria triangularis（Boeck.）C. B. Clarke■☆

387674　Tetraria macowanii C. B. Clarke = Tetraria triangularis（Boeck.）C. B. Clarke■☆

387675　Tetraria maculata Schönland et Turrill；斑点四数莎草■☆

387676　Tetraria microstachys（Vahl）H. Pfeiff. var. usambarensis（K. Schum.）Kük. = Tetraria usambarensis K. Schum.■☆

387677　Tetraria microstachys（Vahl）Pfeiff.；小穗四数莎草■☆

387678　Tetraria mlanjensis J. Raynal；姆兰杰四数莎草■☆

387679　Tetraria natalensis（C. B. Clarke）T. Koyama = Costularia natalensis C. B. Clarke■☆

387680　Tetraria nigrovaginata（Nees）C. B. Clarke；黑鞘四数莎草■☆

387681　Tetraria paludosa Levyns；沼泽四数莎草■☆

387682　Tetraria picta（Boeck.）C. B. Clarke；着色四数莎草■☆

387683　Tetraria pillansii Levyns；皮朗斯四数莎草■☆

387684　Tetraria pleosticha C. B. Clarke = Tetraria fasciata（Rottb.）C. B. Clarke■☆

387685　Tetraria pubescens Schönland et Turrill；短柔毛四数莎草■☆

387686　Tetraria punctoria（Vahl）C. B. Clarke = Neesenbeckia punctoria（Vahl）Levyns■☆

387687　Tetraria pygmaea Levyns；矮小四数莎草■☆

387688　Tetraria robusta（Kunth）C. B. Clarke；粗壮四数莎草■☆

387689　Tetraria rottboellii（Schrad.）C. B. Clarke = Tetraria bromoides（Lam.）Pfeiff.■☆

387690　Tetraria rottboellioides C. B. Clarke = Tetraria bromoides（Lam.）Pfeiff.■☆

387691　Tetraria scariosa Kük.；干膜质四数莎草■☆

387692　Tetraria spiralis（Hochst.）C. B. Clarke = Tetraria involucrata（Rottb.）C. B. Clarke■☆

387693　Tetraria sylvatica（Nees）C. B. Clarke；林地四数莎草■☆

387694　Tetraria sylvatica（Nees）C. B. Clarke var. pseudocuspidata Kük. = Tetraria brachyphylla Levyns■☆

387695　Tetraria sylvatica（Nees）C. B. Clarke var. triflora Kük.；三花林地四数莎草■☆

387696　Tetraria thermalis（L.）C. B. Clarke；温泉四数莎草■☆

387697　Tetraria thermalis（L.）C. B. Clarke var. eximia（C. B. Clarke）Kük. = Tetraria eximia C. B. Clarke■☆

387698　Tetraria thuarii P. Beauv. = Tetraria compar（L.）T. Lestib.■☆

387699　Tetraria triangularis（Boeck.）C. B. Clarke；三角四数莎草■☆

387700　Tetraria usambarensis K. Schum.；乌桑巴拉四数莎草■☆

387701　Tetraria ustulata（L.）C. B. Clarke；泡状四数莎草■☆

387702　Tetraria vaginata Schönland et Turrill；具鞘四数莎草■☆

387703　Tetraria variabilis Levyns；易变四数莎草■☆

387704　Tetraria wallichiana C. B. Clarke；沃里克四数莎草■☆

387705　Tetrariopsis C. B. Clarke = Tetraria P. Beauv.■☆

387706　Tetrarnorphaea DC. = Centaurea L.（保留属名）●■

387707　Tetrarrhena R. Br.（1810）；四雄禾属■☆

387708　Tetrarrhena R. Br. = Ehrharta Thunb.（保留属名）■☆

387709　Tetrarrhena acuminata R. Br.；渐尖四雄禾■☆

387710　Tetrarrhena laevis R. Br.；平滑四雄禾■☆

387711　Tetraselago Junell（1961）；四数玄参属■☆

387712　Tetraselago longituba（Rolfe）Hilliard et B. L. Burtt；长管四数玄参■☆

387713　Tetraselago natalensis（Rolfe）Junell；纳塔尔四数玄参■☆

387714　Tetraselago nelsonii（Rolfe）Hilliard et B. L. Burtt；纳尔逊四数玄参●☆

387715　Tetraselago wilmsii（Rolfe）Hilliard et B. L. Burtt；维尔姆斯四数玄参●☆

387716　Tetrasida Ulbr.（1916）；四稔属●☆

387717　Tetrasida polyantha Ulbr.；四稔●☆

387718　Tetrasiphon Urb.（1904）；四管卫矛属●☆

387719　Tetrasiphon jamaicensis Urb.；四管卫矛●☆

387720　Tetrasperma Steud. = Discaria Hook.●☆

387721　Tetrasperma Steud. = Tetrapasma G. Don●☆

387722　Tetraspidium Baker（1884）；四被列当属（四被玄参属）■☆

387723　Tetraspidium laxiflorum Baker；疏花四被列当■☆

387724　Tetraspis Chiov. = Kirkia Oliv.●☆

387725　Tetraspis ruspoliana Chiov. = Kirkia tenuifolia Engl.●☆

387726　Tetraspora Miq. = Baeckea L.●

387727　Tetrastemma Diels = Tetrastemon Hook. et Arn.●☆

387728　Tetrastemma Diels = Uvariopsis Engl. ex Engl. et Diels●☆

387729　Tetrastemma Diels ex H. Winkl. = Uvariopsis Engl. ex Engl. et Diels●☆

387730　Tetrastemma H. Winkl. = Uvariopsis Engl. ex Engl. et Diels●☆

387731　Tetrastemma bakerianum Hutch. et Dalziel = Uvariopsis bakeriana

（Hutch. et Dalziel）Robyns et Ghesq. ●☆

387732　Tetrastemma dioicum Diels ＝Uvariopsis dioica（Diels）et Ghesq. ●☆

387733　Tetrastemma pedunculosum Diels ＝Uvariopsis dioica（Diels）Robyns et Ghesq. ●☆

387734　Tetrastemma sessiliflorum Mildbr. et Diels ＝Uvariopsis sessiliflora（Mildbr. et Diels）Robyns et Ghesq. ●☆

387735　Tetrastemma solheidii De Wild. ＝Uvariopsis solheidii（De Wild.）Robyns et Ghesq. ●☆

387736　Tetrastemon Hook. et Arn.（1833）;四冠木属●☆

387737　Tetrastemon Hook. et Arn. ＝Myrrhinium Schott ●☆

387738　Tetrastemon loranthoides Hook. et Arn. ;四冠木●☆

387739　Tetrastichella Pichon ＝Arrabidaea DC. ●☆

387740　Tetrastigma（Miq.）Planch.（1887）;崖爬藤属（扁担藤属,崖藤属,岩藤属）;Javan Grape,Rockvine ●■

387741　Tetrastigma K. Schum. ＝Schumanniophyton Harms ●☆

387742　Tetrastigma Planch. ＝Tetrastigma（Miq.）Planch. ●■

387743　Tetrastigma alatum H. L. Li;翼柄崖爬藤（翅梗崖爬藤）;Winged Rockvine ●■

387744　Tetrastigma alatum H. L. Li ＝Tetrastigma hemsleyanum Diels et Gilg ●■

387745　Tetrastigma albiflorum C. Y. Wu;白花崖藤（白花崖爬藤,白花岩藤）;Whiteflower Rockvine ●

387746　Tetrastigma apiculatum Gagnep. ;草崖藤;Apiculate Rockvine ●

387747　Tetrastigma apiculatum Gagnep. var. pubescens C. L. Li;柔毛草崖藤;Pubescent Apiculate Rockvine ●

387748　Tetrastigma bioritsense（Hayata）Hsu et Kuoh ＝Tetrastigma hemsleyanum Diels et Gilg ●■

387749　Tetrastigma burmanicum Momiy. ＝Tetrastigma obtectum（Wall.）Planch. ●■

387750　Tetrastigma cambodianum Gagnep. ＝Tetrastigma cambodianum Pierre ex Gagnep. ●

387751　Tetrastigma cambodianum Gagnep. ＝Tetrastigma jinghongensis C. L. Li ●

387752　Tetrastigma cambodianum Pierre ex Gagnep. ;柬埔寨崖爬藤;Cambodia Rockvine,Cambodian Rockvine ●

387753　Tetrastigma campylocarpum（Kurz）Planch. ;多花崖爬藤;Manyflower Rockvine ●

387754　Tetrastigma caudatum Merr. et Chun;尾叶崖爬藤;Caudateleaf Rockvine,Caudate-leaved Rockvine ●

387755　Tetrastigma cauliflorum Merr. ;茎花崖爬藤;Cauliflory Rockvine,Cauli-flowered Rockvine ●

387756　Tetrastigma ceratopetalum C. Y. Wu;角花崖爬藤;Ceratopetal Rockvine ●

387757　Tetrastigma cerrulatum（Roxb.）Planch. ＝Tetrastigma napaulense（DC.）C. L. Li ●■

387758　Tetrastigma chapaense Merr. ＝Tetrastigma apiculatum Gagnep. ●

387759　Tetrastigma chingsiense Chun et F. C. How;靖西崖爬藤;Jingxi Rockvine ●

387760　Tetrastigma crassipes Planch. ＝Tetrastigma lincangense C. L. Li ●

387761　Tetrastigma crassipes Planch. var. strumarum Planch. ＝Tetrastigma pachyphyllum（Hemsl.）Chun ●

387762　Tetrastigma cruciatum Craib et Gagnep. ;十字崖爬藤;Cruciate Rockvine,Cruciform Rockvine ●

387763　Tetrastigma delavayi Gagnep. ;七小叶崖爬藤（七叶崖爬藤,乌蔹莓,一把笠）;Delavay Rockvine ●

387764　Tetrastigma delavayi Gagnep. f. majus W. T. Wang;福贡崖爬藤;Fugong Rockvine ●

387765　Tetrastigma delavayi Gagnep. f. majus W. T. Wang ＝Tetrastigma delavayi Gagnep. ●

387766　Tetrastigma dentatum（Hayata）H. L. Li;三脚鳖草;Dentate Rockvine ■

387767　Tetrastigma dentatum（Hayata）H. L. Li ＝Tetrastigma formosanum（Hemsl.）Gagnep. ●

387768　Tetrastigma dentatum（Hayata）H. L. Li ＝Tetrastigma hemsleyanum Diels et Gilg ●■

387769　Tetrastigma erubescens Planch. ;红枝崖爬藤（单籽崖爬藤,红崖爬藤）;Red-branch Rockvine,Single-seeded Rockvine ●

387770　Tetrastigma erubescens Planch. var. monophyllum Gagnep. ;单叶红枝崖爬藤●

387771　Tetrastigma erubescens Planch. var. monophyllum Gagnep. ＝Tetrastigma monophyllum（Gagnep.）C. Y. Wu ex W. T. Wang ●

387772　Tetrastigma erubescens Planch. var. monospermum Gagnep. ;单籽红枝崖爬藤（单籽崖爬藤）;One-seed Rockvine ●

387773　Tetrastigma erubescens Planch. var. monospermum Gagnep. ＝Tetrastigma erubescens Planch. ●

387774　Tetrastigma formosanum（Hemsl.）Gagnep. ;台湾崖爬藤（三叶崖爬藤）;Taiwan Rockvine ●

387775　Tetrastigma funingense C. L. Li;富宁崖爬藤;Funing Rockvine ●

387776　Tetrastigma godefroyanum Planch. ;柄果崖爬藤（过山青藤）;Stalkedfruit Rockvine ●

387777　Tetrastigma hainanense Chun et F. C. How ＝Tetrastigma papillatum（Hance）C. Y. Wu ●

387778　Tetrastigma harmandii Planch. ＝Tetrastigma pseudocruciatum C. L. Li ●

387779　Tetrastigma hemsleyanum Diels et Gilg;三叶崖爬藤（金线吊葫芦,拦山虎,雷胆子,三叶扁藤,三叶对,三叶青,三叶青根,蛇附子,石抱子,石猴子,石老鼠,丝线吊金钟,土经丸,小扁藤,有角乌蔹莓根）;Hemsley Rockvine ●■

387780　Tetrastigma henryi Gagnep. ;蒙自崖爬藤（滇琼崖爬藤）;Henry Rockvine ●

387781　Tetrastigma henryi Gagnep. var. mollifolium W. T. Wang;柔毛蒙自崖爬藤（柔毛崖爬藤）;Hairy-leaf Henry Rockvine ●

387782　Tetrastigma henryi Gagnep. var. mollifolium W. T. Wang ＝Tetrastigma henryi Gagnep. ●

387783　Tetrastigma hernyi Gagnep. ＝Tetrastigma erubescens Planch. ●

387784　Tetrastigma hypoglaucum Planch. ＝Tetrastigma hypoglaucum Planch. ex Franch. ●■

387785　Tetrastigma hypoglaucum Planch. ex Franch. ;叉须崖爬藤（白背崖爬藤,灯笼草,红葡萄,乌蔹莓,五虎下山,五爪金龙,五爪龙,狭叶崖爬藤,小红藤,小五爪金龙,雪里高）;Narrowleaf Rockvine,Narrow-leaved Rockvine ●■

387786　Tetrastigma hypoglaucum Planch. ex Franch. ＝Tetrastigma serrulatum（Roxb.）Planch. ●■

387787　Tetrastigma hypoglaucum Planch. ex Franch. var. puberulum W. T. Wang et Z. Y. Cao;毛狭叶崖爬藤;Hairy Narrowleaf Rockvine ●

387788　Tetrastigma hypoglaucum Planch. var. puberulum W. T. Wang ＝Tetrastigma serrulatum（Roxb.）Planch. ●■

387789　Tetrastigma hypoglaucum Planch. var. puberulum W. T. Wang et Z. Y. Cao ＝Tetrastigma serrulatum（Roxb.）Planch. var. puberulum W. T. Wang ●■

387790　Tetrastigma hypoglaucum Planch. var. puberulum W. T. Wang et Z. Y. Cao ＝Tetrastigma hypoglaucum Planch. ex Franch. var. puberulum W. T. Wang et Z. Y. Cao ●

387791 Tetrastigma indicum Maulik = Tetrastigma serrulatum（Roxb.）Planch. ●■

387792 Tetrastigma jingdongense C. L. Li；景东崖爬藤；Jingdong Rockvine ●

387793 Tetrastigma jinxiuense C. L. Li；金秀崖爬藤；Jinxiu Rockvine ●

387794 Tetrastigma kwangsiense C. L. Li；广西崖爬藤；Guangxi Rockvine ●

387795 Tetrastigma laevigatum Gagnep.；平滑崖爬藤 ●☆

387796 Tetrastigma lanyuense C. E. Chang；兰屿崖爬藤（兰屿粉藤）；Lanyu Rockvine ●

387797 Tetrastigma lanyuense C. E. Chang = Cissus lanyuensis（C. E. Chang）F. Y. Lu ●

387798 Tetrastigma lenticellatum C. Y. Wu ex W. T. Wang；显孔崖爬藤（大五爪龙）；Lenticellate Rockvine ●

387799 Tetrastigma lenticellatum C. Y. Wu ex W. T. Wang = Tetrastigma xishuangbannaense C. L. Li ●

387800 Tetrastigma lincangense C. L. Li；临沧崖爬藤；Lincang Rockvine ●

387801 Tetrastigma lineare W. T. Wang ex C. Y. Wu et C. L. Li；条叶崖爬藤；Lineleaf Rockvine ●

387802 Tetrastigma liukiuense T. Yamaz.；琉球崖爬藤 ●☆

387803 Tetrastigma longipedunculatum C. L. Li；长梗崖爬藤；Longpeicel Rockvine ●

387804 Tetrastigma lunglingense C. Y. Wu et W. T. Wang；龙陵崖爬藤；Longling Rockvine ●

387805 Tetrastigma lunglingense C. Y. Wu et W. T. Wang = Tetrastigma henryi Gagnep. ●

387806 Tetrastigma macrocorymbum Gagnep.；伞花崖爬藤；Bigumbel Rockvine ●

387807 Tetrastigma magnificum K. Schum. = Schumanniophyton magnificum（K. Schum.）Harms ●☆

387808 Tetrastigma megalocarpum W. T. Wang；大果崖爬藤；Bigfruit Rockvine，Big-fruited Rockvine ●

387809 Tetrastigma megalocarpum W. T. Wang = Tetrastigma jinghongensis C. L. Li ●

387810 Tetrastigma membranaceum C. Y. Wu ex W. T. Wang；膜叶崖爬藤；Membranous Rockvine，Membranousleaf Rockvine ●

387811 Tetrastigma membranaceum C. Y. Wu ex W. T. Wang = Tetrastigma cauliflorum Merr. ●

387812 Tetrastigma monophyllum（Gagnep.）C. Y. Wu ex W. T. Wang；单叶崖爬藤；Monophyllous Rockvine，Oneleaf Rockvine，Single-leaved Rockvine ●

387813 Tetrastigma monophyllum（Gagnep.）C. Y. Wu ex W. T. Wang = Tetrastigma erubescens Planch. var. monophyllum Gagnep. ●

387814 Tetrastigma myanmaricum Momiy. = Tetrastigma obtectum（Wall.）Planch. ●■

387815 Tetrastigma napaulense（DC.）C. L. Li；细齿崖爬藤；Serrulate Rockvine ●■

387816 Tetrastigma napaulense（DC.）C. L. Li = Tetrastigma serrulatum（Roxb.）Planch. ●■

387817 Tetrastigma napaulense（DC.）C. L. Li var. puberulum（W. T. Wang）C. L. Li；毛细齿崖爬藤（毛狭叶崖爬藤，毛枝细齿崖爬藤）；Hairy Serrulate Rockvine ●■

387818 Tetrastigma napaulense（Roxb.）Planch. var. puberulum（W. T. Wang）C. L. Li = Tetrastigma serrulatum（Roxb.）Planch. var. puberulum W. T. Wang ●■

387819 Tetrastigma obovatum（M. A. Lawson）Gagnep.；毛枝崖爬藤（大血藤，红五加，五爪龙）；Hairybranch Rockvine，Hairy-branched Rockvine ●

387820 Tetrastigma obovatum Gagnep. = Cayratia geniculata（Blume）Gagnep. ●

387821 Tetrastigma obtectum（Wall. ex M. A. Lawson）Planch. ex Franch. subsp. dichotomum W. T. Wang = Tetrastigma hypoglaucum Planch. ex Franch. ●■

387822 Tetrastigma obtectum（Wall.）Planch.；崖爬藤（红五加，毛五加，爬山虎，上树蜈蚣，蛇吴巴，蛇蜈巴，瘀五加，藤五加，藤五甲，铜丝绊，五加皮，五叶崖爬藤，五爪金龙，小红藤，小红药，小五爪金龙，小走游草，岩爬藤，岩五加，走游草）；Common Rockvine，Protected Rockvine ●■

387823 Tetrastigma obtectum（Wall.）Planch. subsp. dichotomum W. T. Wang = Tetrastigma hypoglaucum Planch. ●■

387824 Tetrastigma obtectum（Wall.）Planch. var. dichotomum W. T. Wang = Tetrastigma hypoglaucum Planch. ●■

387825 Tetrastigma obtectum（Wall.）Planch. var. glabrum（H. Lév. et Vaniot）Gagnep.；无毛崖爬藤（钝叶崖爬藤，光叶崖爬藤，毛五加，台湾崖爬藤，藤五加，铜丝绊，铜丝线，五爪金龙，小红藤，小红药，小九节铃，小五爪龙）；Glabrous Protected Rockvine，Hairlesss Rockvine ●■

387826 Tetrastigma obtectum（Wall.）Planch. var. pilosum Gagnep.；毛叶崖爬藤（毛崖爬藤）；Hairylraf Rockvine，Pilose Protected Rockvine ●■

387827 Tetrastigma obtectum（Wall.）Planch. var. pilosum Gagnep. = Tetrastigma obtectum（Wall.）Planch. ●■

387828 Tetrastigma obtectum（Wall.）Planch. var. potentilla（H. Lév. et Vaniot）Gagnep. = Tetrastigma obtectum（Wall.）Planch. ●■

387829 Tetrastigma obtectum（Wall.）Planch. var. potentilla（H. Lév. et Vaniot）Gagnep. = Tetrastigma obtectum（Wall.）Planch. var. pilosum Gagnep. ●■

387830 Tetrastigma obtectum（Wall.）Planch. var. potentilla（H. Lév. et Vaniot）Gagnep.；钝头崖爬藤 ●■

387831 Tetrastigma obtectum（Wall.）Planch. var. trichocarpum Gagnep.；巧家崖爬藤；Qiaojia Protected Rockvine ●■

387832 Tetrastigma obtectum（Wall.）Planch. var. trichocarpum Gagnep. = Tetrastigma obtectum（Wall.）Planch. ●■

387833 Tetrastigma obtectum（Wall.）Planch. var. trichocarpum Gagnep. = Tetrastigma obtectum（Wall.）Planch. var. pilosum Gagnep. ●■

387834 Tetrastigma oliviforme Planch.；橄榄形崖爬藤；Oliveform Rockvine ●☆

387835 Tetrastigma pachyphyllum（Hemsl.）Chun；厚叶崖爬藤（勾办，过山龙）；Thickleaf Rockvine，Thick-leaved Rockvine ●

387836 Tetrastigma papillatum（Hance）C. Y. Wu；海南崖爬藤（海南乌蔹莓）；Hainan Cayratia，Hainan Rockvine ●

387837 Tetrastigma planicaule（Hook. f.）Gagnep.；扁担藤（扁骨风，扁茎崖爬藤，扁藤，大芦藤，过江扁龙，过江扁藤，脚白藤，茎藤，铁带藤，羊带风，腰带藤，玉带藤）；Carrying Rockvine，Flatstem Rockvine，Flat-stemmed Rockvine ●

387838 Tetrastigma pseudocruciatum C. L. Li；过山崖爬藤（尖尾崖爬藤，小尾崖爬藤）；False Crusiate Rockvine，Harmand Rockvine ●

387839 Tetrastigma pubinerve Merr. et Chun；毛脉崖爬藤；Hairynerve Rockvine，Hairy-nerved Rockvine ●

387840 Tetrastigma retinervium Planch.；网脉崖爬藤；Net-veined Rockvine ●

387841 Tetrastigma retinervium Planch. var. pubescens C. L. Li；柔毛网

脉崖爬藤;Pubescent Rockvine ●

387842　Tetrastigma rumicispermum (Lawson) Planch.;喜马拉雅崖爬藤;Himalayan Rockvine ●

387843　Tetrastigma rumicispermum (Lawson) Planch. var. lasiogynum (W. T. Wang) C. L. Li;锈毛喜马拉雅崖爬藤;Rusty-hair Himalayan Rockvine ●

387844　Tetrastigma rupestre Planch.;石生崖爬藤;Rupestrine Rockvine ●

387845　Tetrastigma serrulatum (Roxb.) Planch.;狭叶崖爬藤(细齿崖爬藤,小五爪金龙);Serrulate Rockvine ●■

387846　Tetrastigma serrulatum (Roxb.) Planch. = Tetrastigma napaulense (DC.) C. L. Li ●■

387847　Tetrastigma serrulatum (Roxb.) Planch. var. lasiogynum W. T. Wang;毛蕊狭叶崖爬藤(毛蕊细齿崖爬藤);Hairy-stamen Rockvine ●■

387848　Tetrastigma serrulatum (Roxb.) Planch. var. lasiogynum W. T. Wang = Tetrastigma rumicispermum (Lawson) Planch. var. lasiogynum (W. T. Wang) C. L. Li ●

387849　Tetrastigma serrulatum (Roxb.) Planch. var. puberulum (W. T. Wang et Z. Y. Cao) C. L. Li = Tetrastigma serrulatum (Roxb.) Planch. var. puberulum W. T. Wang ●■

387850　Tetrastigma serrulatum (Roxb.) Planch. var. puberulum W. T. Wang = Tetrastigma napaulense (DC.) C. L. Li var. puberulum (W. T. Wang) C. L. Li ●■

387851　Tetrastigma serrulatum (Roxb.) Planch. var. pubinervium C. L. Li = Tetrastigma serrulatum (Roxb.) Planch. var. puberulum W. T. Wang ●■

387852　Tetrastigma shanglinense W. T. Wang;上林崖爬藤;Shanglin Rockvine ●

387853　Tetrastigma sichouense C. L. Li;西畴崖爬藤(大叶崖藤,老挝崖爬藤);Chestnut Vine,Xichou Rockvine ●

387854　Tetrastigma sichouense C. L. Li var. megalocarpum C. L. Li;大果西畴崖爬藤;Bigfruit Xichou Rockvine ●

387855　Tetrastigma sinodichotomum W. T. Wang = Tetrastigma hypoglaucum Planch. ex Franch. ●■

387856　Tetrastigma strumarum (Planch.) Gagnep. = Tetrastigma pachyphyllum (Hemsl.) Chun ●

387857　Tetrastigma strumarum Gagnep. = Tetrastigma pachyphyllum (Hemsl.) Chun ●

387858　Tetrastigma subtetragonum C. L. Li;红花崖爬藤;Redflower Rockvine ●

387859　Tetrastigma tenue Craib = Tetrastigma henryi Gagnep. ●

387860　Tetrastigma tonkinense Gagnep.;越南崖爬藤;Tonkin Rockvine,Viatnam ●

387861　Tetrastigma triphyllum (Gagnep.) W. T. Wang;菱叶崖爬藤;Rhombicleaf Rockvine,Rhombic-leaved Rockvine ●■

387862　Tetrastigma triphyllum (Gagnep.) W. T. Wang var. hirtum (Gagnep.) W. T. Wang;毛菱叶崖爬藤;Hairy Rhombicleaf Rockvine ●■

387863　Tetrastigma tsaianum C. Y. Wu;蔡氏崖爬藤;H. T. Tsai Rockvine ●

387864　Tetrastigma umbellatum (Hemsl.) Nakai = Tetrastigma obtectum (Wall.) Planch. var. glabrum (H. Lév. et Vaniot) Gagnep. ●■

387865　Tetrastigma venulosum C. Y. Wu;马关崖爬藤;Venulose Rockvine ●

387866　Tetrastigma voinierianum (Ball.) Pierre ex Gagnep. = Tetrastigma sichouense C. L. Li ●

387867　Tetrastigma voinierianum Pierre ex Gagnep. = Tetrastigma sichouense C. L. Li ●

387868　Tetrastigma xishuangbannaense C. L. Li;西双版纳崖爬藤(大五爪金龙,显孔崖爬藤);Xishuangbanna Rockvine ●

387869　Tetrastigma xizangense C. L. Li;西藏崖爬藤;Xizang Rockvine ●

387870　Tetrastigma yiwuense C. L. Li;易武崖爬藤;Yiwu Rockvine ●

387871　Tetrastigma yunnanense Gagnep.;云南崖爬藤(滇崖爬藤,飞蜈蚣,马头龙叶,爬树龙,软三角枫,三角枫,三爪龙,石葡萄,五爪金龙);Yunnan Rockvine ●■

387872　Tetrastigma yunnanense Gagnep. f. hirtum Gagnep. = Tetrastigma triphyllum (Gagnep.) W. T. Wang var. hirtum (Gagnep.) W. T. Wang ●■

387873　Tetrastigma yunnanense Gagnep. var. glabrum W. T. Wang;无毛云南崖爬藤;Glabrous Yunnan Rockvine ●■

387874　Tetrastigma yunnanense Gagnep. var. mollissimum C. Y. Wu;贡山崖爬藤;Gongshan Rockvine ●■

387875　Tetrastigma yunnanense Gagnep. var. pubipes W. T. Wang;毛柄云南崖爬藤;Hairy-stalk Yunnan Rockvine ●■

387876　Tetrastigma yunnanense Gagnep. var. pubipes W. T. Wang = Tetrastigma yunnanense Gagnep. ●■

387877　Tetrastigma yunnanense Gagnep. var. triphyllum Gagnep. = Tetrastigma papillatum (Hance) C. Y. Wu ●

387878　Tetrastigma yunnanense Gagnep. var. triphyllum Gagnep. = Tetrastigma triphyllum (Gagnep.) W. T. Wang ●■

387879　Tetrastigma yunnanense Gagnep. var. triphyllum Gagnep. f. glabrum Gagnep. = Tetrastigma triphyllum (Gagnep.) W. T. Wang ●■

387880　Tetrastigma yunnanense Gagnep. var. triphyllum Gagnep. f. hirtum Gagnep. = Tetrastigma triphyllum (Gagnep.) W. T. Wang var. hirtum (Gagnep.) W. T. Wang ●■

387881　Tetrastylidiaceae Calest. = Oleaceae Hoffmanns. et Link(保留科名)●

387882　Tetrastylidiaceae Tiegh. = Erythropalaceae Planch. ex Miq.(保留科名)●

387883　Tetrastylidiaceae Tiegh. = Oleaceae Hoffmanns. et Link(保留科名)●

387884　Tetrastylidium Engl. (1872);四柱木属●☆

387885　Tetrastylidium brasiliense Engl.;巴西四柱木●☆

387886　Tetrastylidium grandifolium Sleumer;大叶四柱木●☆

387887　Tetrastylidium peruvianum Sleumer;秘鲁四柱木●☆

387888　Tetrastylis Barb. Rodr. = Passiflora L. ●■

387889　Tetrasynandra Perkins(1898);聚药桂属●☆

387890　Tetrasynandra laxiflora (Benth.) Perkins;疏花聚药桂●☆

387891　Tetrasynandra longipes (Benth.) Perkins;长梗聚药桂●☆

387892　Tetrasynandra pubescens (Benth.) Perkins;毛聚药桂●☆

387893　Tetrataenium (DC.) Manden. (1959);四带芹属;Fourtapeparsley ■

387894　Tetrataenium (DC.) Manden. = Heracleum L. ■

387895　Tetrataenium candicans (Wall. ex DC.) Manden. = Heracleum candicans Wall. ex DC. ■

387896　Tetrataenium nepalense (D. Don) Manden.;尼泊尔四带芹(尼泊尔独活);Nepal Fourtapeparsley ■

387897　Tetrataenium nepalense (D. Don) Manden. = Heracleum nepalense D. Don ■

387898　Tetrataenium obtusifolium (Wall. ex DC.) Manden. = Heracleum candicans Wall. ex DC. var. obtusifolium (Wall. ex DC.) F. T. Pu et M. F. Watson ■

387899　Tetrataenium olgae (DC.) Manden.;大叶四带芹;Bigleaf Fourtapeparsley ■

387900　Tetrataenium olgae（Regel et Schmalh.）Manden. = Heracleum olgae Regel et Schmalh. ex Regel ■

387901　Tetrataenium rigens（Wall.）Manden.；硬尔四带芹■☆

387902　Tetrataxis Hook. f.（1867）；毛里求斯千屈菜属●☆

387903　Tetrataxis salicifolia（Tul.）Baker；毛里求斯千屈菜●☆

387904　Tetrateleia Arw. = Cleome L. ●■

387905　Tetrateleia Arw. = Tetratelia Sond. ●■

387906　Tetrateleia Sond. = Cleome L. ●■

387907　Tetratelia Sond. = Cleome L. ●■

387908　Tetratelia maculata（Sond.）Sond. = Cleome maculata（Sond.）Szyszyl. ■☆

387909　Tetratelia nationae（Burtt Davy）Pax et K. Hoffm. = Cleome macrophylla（Klotzsch）Briq. var. maculatiflora（Merxm.）Wild ■☆

387910　Tetratelia tenuifolia（Klotzsch）Arw. var. maculatiflora Merxm. = Cleome macrophylla（Klotzsch）Briq. var. maculatiflora（Merxm.）Wild ■☆

387911　Tetrathalamus Lauterb.（1905）；四室林仙属●☆

387912　Tetrathalamus Lauterb. = Zygogynum Baill. ●☆

387913　Tetrathalamus montanus Lauterb.；四室林仙●☆

387914　Tetratheca Sm.（1793）；四室木属（泰特雷瑟属）；Pink-Eye ●☆

387915　Tetratheca affinis Endl.；近缘四室木●☆

387916　Tetratheca aphylla F. Muell.；无叶四室木●☆

387917　Tetratheca ciliata Lindl.；睫毛四室木●☆

387918　Tetratheca elongata Schuch.；矩圆四室木●☆

387919　Tetratheca filiformis Benth.；线形四室木●☆

387920　Tetratheca gracilis Steetz；细四室木●☆

387921　Tetratheca hirsuta Lindl.；粗毛四室木●☆

387922　Tetratheca micrantha Schuch.；小花四室木●☆

387923　Tetratheca mollis E. Pritz.；软四室木●☆

387924　Tetratheca nuda Lindl.；裸四室木●☆

387925　Tetratheca parvifolia J. Thompson；小叶四室木●☆

387926　Tetratheca paucifolia J. Thompson；寡叶四室木●☆

387927　Tetratheca thymifolia Sm.；百里香叶四室木●☆

387928　Tetrathecaceae R. Br. = Elaeocarpaceae Juss.（保留科名）●

387929　Tetrathecaceae R. Br. = Tremandraceae R. Br. ex DC.（保留科名）●☆

387930　Tetrathylacium Poepp.（1841）；四囊木属●☆

387931　Tetrathylacium Poepp. et Endl. = Tetrathylacium Poepp. ●☆

387932　Tetrathylacium costaricense Standl.；四囊木；Costa Pica ●☆

387933　Tetrathyrium Benth.（1861）；四药门花属；Tetrathyrium ●★

387934　Tetrathyrium Benth. = Loropetalum R. Br. ex Rchb. ●

387935　Tetrathyrium simaoense Y. Y. Qian；思茅四药门花；Simao Tetrathyrium ●

387936　Tetrathyrium simaoense Y. Y. Qian = Loropetalum chinense（R. Br.）Oliv. ●

387937　Tetrathyrium subcordatum Benth.；四药门花（香港檵木）；Hongkong Tetrathyrium ●◇

387938　Tetrathyrium subcordatum Benth. = Loropetalum subcordatum（Benth.）Oliv. ●◇

387939　Tetratome Poepp. et Endl. = Mollinedia Ruiz et Pav. ●☆

387940　Tetratrichia avasmontana Dinter ex Merxm. = Pentatrichia avasmontana Merxm. ■☆

387941　Tetraulacium Turcz.（1843）；四沟玄参属■☆

387942　Tetraulacium veroniciforme Turcz.；四沟玄参■☆

387943　Tetrazygia Rich. = Tetrazygia Rich. ex DC. ●☆

387944　Tetrazygia Rich. ex DC.（1828）；四轭野牡丹属●☆

387945　Tetrazygia adenophora Griseb.；腺梗四轭野牡丹●☆

387946　Tetrazygia albicans Triana；灰白四轭野牡丹●☆

387947　Tetrazygia angustifolia DC.；窄叶四轭野牡丹●☆

387948　Tetrazygia argyrophylla（A. Rich.）Millsp.；银叶四轭野牡丹●☆

387949　Tetrazygia aurea R. A. Howard et W. R. Briggs；黄叶四轭野牡丹●☆

387950　Tetrazygia bicolor（Mill.）Cogn.；二色四轭野牡丹●☆

387951　Tetrazygia biflora（Cogn.）Urb.；双花四轭野牡丹●☆

387952　Tetrazygia brachycentra C. Wright；短距四轭野牡丹●☆

387953　Tetrazygia elegans Urb.；雅致四轭野牡丹●☆

387954　Tetrazygia lanceolata Urb.；披针叶四轭野牡丹●☆

387955　Tetrazygia laxiflora Naudin；疏花四轭野牡丹●☆

387956　Tetrazygia minor Urb.；小四轭野牡丹●☆

387957　Tetrazygia tetrandra（Sw.）DC.；四蕊四轭野牡丹●☆

387958　Tetrazygiopsis Borhidi = Tetrazygia Rich. ex DC. ●☆

387959　Tetrazygos Rich. ex DC. = Charianthus D. Don ●☆

387960　Tetreilema Turcz. = Frankenia L. ●■

387961　Tetrixus Tiegh. ex Lecomte = Viscum L. ●

387962　Tetrodea Raf. = Amphicarpaea Elliott ex Nutt.（保留属名）■

387963　Tetrodes Raf. = Tetrodea Raf. ■

387964　Tetrodon（Kraenzl.）M. A. Clem. et D. L. Jones = Eria Lindl.（保留属名）■

387965　Tetrodon（Kraenzl.）M. A. Clem. et D. L. Jones（1998）；四齿兰属■☆

387966　Tetrodus（Cass.）Cass. = Helenium L. ■

387967　Tetrodus（Cass.）Cass. = Mesodetra Raf. ■

387968　Tetrodus Cass. = Helenium L. ■

387969　Tetroncium Willd.（1808）；四疣麦冬属■☆

387970　Tetroncium magellanicum Willd.；四疣麦冬■☆

387971　Tetrorchidiopsis Rauschert = Tetrorchidium Poepp. ●☆

387972　Tetrorchidium Poepp.（1841）；四丸大戟属●☆

387973　Tetrorchidium Poepp. et Endl. = Tetrorchidium Poepp. ●☆

387974　Tetrorchidium congolense J. Léonard；刚果四丸大戟●☆

387975　Tetrorchidium didymostemon（Baill.）Pax et K. Hoffm.；双冠四丸大戟●☆

387976　Tetrorchidium gabonense Breteler；加蓬四丸大戟●☆

387977　Tetrorchidium minus（Prain）Pax et K. Hoffm. = Tetrorchidium didymostemon（Baill.）Pax et K. Hoffm. ●☆

387978　Tetrorchidium oppositifolium（Pax）Pax et K. Hoffm.；对叶四丸大戟●☆

387979　Tetrorchidium tenuifolium（Pax et K. Hoffm.）Pax et K. Hoffm. = Tetrorchidium oppositifolium（Pax）Pax et K. Hoffm. ●☆

387980　Tetrorchidium ulugurense Verdc.；乌卢古尔四丸大戟●☆

387981　Tetrorhiza Raf. = Tetrorhiza Raf. ex Jacks. ■

387982　Tetrorhiza Raf. ex Jacks. = Gentiana L. ■

387983　Tetrorhiza Raf. ex Jacks. = Tretorhiza Adans. ■

387984　Tetrorum Rose = Sedum L. ●■

387985　Tetrouratea Tiegh. = Ouratea Aubl.（保留属名）●

387986　Teucridium Hook. f.（1853）；小石蚕属●☆

387987　Teucridium parvifolium Hook. f.；小石蚕●☆

387988　Teucrion St. -Lag. = Teucrium L. ●■

387989　Teucrium L.（1753）；香科科属（石蚕属，香科属）；Germander，Wood Sage ●■

387990　Teucrium L. = Kinostemon Kudo ■★

387991　Teucrium abutiloides L'Hér.；苘麻香科科■☆

387992　Teucrium abyssinicum Hochst. ex Benth. = Teucrium scordium L. ■

387993　Teucrium africanum Thunb.；非洲香科科；Padda Foot ■☆

387994　Teucrium afrum（Emb. et Maire）Pau et Font Quer；热非香科■☆

387995　Teucrium afrum（Emb. et Maire）Pau et Font Quer subsp. rhiphaeum（Font Quer et Pau）Castrov. et Bayon；山地热非香科■☆

387996　Teucrium afrum（Emb. et Maire）Pau et Font Quer subsp. rubriflorum（Font Quer et Pau）Castrov. et Bayon；红花热非香科■☆

387997　Teucrium albidum Munby；白香科■☆

387998　Teucrium alborubrum Hemsl. = Kinostemon alborubrum（Hemsl.）C. Y. Wu et S. Chow■

387999　Teucrium alopecurus Noë；看麦娘香科■☆

388000　Teucrium anlungense C. Y. Wu et S. Chow；安龙香科科（安龙香科，野苏子）；Anlong Germander, Anlung Germander■

388001　Teucrium antiatlanticum（Maire）Sauvage et Vindt；安蒂香科■☆

388002　Teucrium aroanium Orph.；希腊香科●☆

388003　Teucrium atratum Pomel；黑香科■☆

388004　Teucrium aureocandidum Andr.；黄白香科■☆

388005　Teucrium aureum Schreb. = Teucrium luteum（Mill.）Degen■☆

388006　Teucrium aureum Schreb. subsp. flavovirens（Batt.）Puech = Teucrium luteum（Mill.）Degen subsp. flavovirens（Batt.）Greuter et Burdet■☆

388007　Teucrium aureum Schreb. subsp. gabesianum S. Puech = Teucrium luteum（Mill.）Degen subsp. gabesianum（S. Puech）Greuter et Burdet■☆

388008　Teucrium baeticum Boiss. et Reut.；伯蒂卡香科■☆

388009　Teucrium barbarum Jahand. et Maire；外来香科■☆

388010　Teucrium barbeyanum Asch. et Taub.；巴比香科■☆

388011　Teucrium barbeyanum Asch. et Taub. var. purpureum Pamp. = Teucrium barbeyanum Asch. et Taub.■☆

388012　Teucrium bidentatum Hemsl.；二齿香科科（白花石蚕，二齿香科，细沙虫草）；Bident Germander, Twodentate Germander■

388013　Teucrium bidentatum Hemsl. var. purpureum Diels = Teucrium bidentatum Hemsl.■

388014　Teucrium boreale E. P. Bicknell = Teucrium canadense L. var. occidentale（A. Gray）E. M. McClint. et Epling■☆

388015　Teucrium botrys L.；总状花石蚕（总状花序香科）；Cutleaf Germander, Cut-leaf Germander, Cut-leaved Germander, Oak of Jerusalem, Racemose Germander■☆

388016　Teucrium bovei Benth.；博韦香科■☆

388017　Teucrium bracteatum Desf.；具苞香科■☆

388018　Teucrium bracteatum Desf. var. acuticalyx Emb. = Teucrium bracteatum Desf.■☆

388019　Teucrium bracteatum Desf. var. riatarum（Humbert et Maire）Maire = Teucrium bracteatum Desf.■☆

388020　Teucrium bracteatum Desf. var. virescens Pau = Teucrium bracteatum Desf.■☆

388021　Teucrium brevifolium Schreb.；短叶香科■☆

388022　Teucrium bullatum Coss. et Balansa；泡状香科■☆

388023　Teucrium buxifolium Schreb. subsp. albidum（Munby）Greuter et Burdet = Teucrium albidum Munby■☆

388024　Teucrium buxifolium Schreb. var. albidum（Munby）Maire = Teucrium albidum Munby■☆

388025　Teucrium campanulatum L.；风铃草状石蚕■☆

388026　Teucrium canadense L.；加拿大石蚕；American Germander, Canadian Germander, Germander, Wood Sage■☆

388027　Teucrium canadense L. subsp. occidentale（A. Gray）W. A. Weber = Teucrium canadense L. var. occidentale（A. Gray）E. M. McClint. et Epling■☆

388028　Teucrium canadense L. var. angustatum A. Gray = Teucrium canadense L.■☆

388029　Teucrium canadense L. var. boreale（E. P. Bicknell）Shinners = Teucrium canadense L. var. occidentale（A. Gray）E. M. McClint. et Epling■☆

388030　Teucrium canadense L. var. boreale（E. P. Bicknell）Shinners = Teucrium canadense L.■☆

388031　Teucrium canadense L. var. littorale（E. P. Bicknell）Fernald = Teucrium canadense L.■☆

388032　Teucrium canadense L. var. occidentale（A. Gray）E. M. McClint. et Epling；西方石蚕；American Germander, Western Germander, Wild Germander, Wood Sage☆

388033　Teucrium canadense L. var. virginicum（L.）Eaton；美国香科；American Germander, Virginia Germander, Wood Sage■☆

388034　Teucrium canadense L. var. virginicum（L.）Eaton = Teucrium canadense L.■☆

388035　Teucrium canum Fisch. et C. A. Mey.；灰色香科；Grey Germander■☆

388036　Teucrium capense Thunb. = Teucrium trifidum Retz.■☆

388037　Teucrium capitatum L.；头状香科■☆

388038　Teucrium capitatum L. f. soloitanum Maire = Teucrium polium L.■☆

388039　Teucrium capitatum L. var. chamaedryfolium Pau et Font Quer = Teucrium capitatum L.■☆

388040　Teucrium capitatum L. var. max-onnoi Sennen = Teucrium polium L.■☆

388041　Teucrium cephalotes Pomel = Teucrium polium L.■☆

388042　Teucrium chamaedrys L.；紫石蚕（石蚕香科）；Chamaedrys Germander, Creeping Germander, English Treacle, Germander, Ground Oak, Herteclowre, Horsechire, Wall Germander■☆

388043　Teucrium chamaedrys L. subsp. algeriense Rech. f.；阿尔及利亚香科■☆

388044　Teucrium chamaedrys L. subsp. gracile（Batt.）Rech. f.；纤细紫石蚕■☆

388045　Teucrium chamaedrys L. subsp. maroccanum Rech. f.；摩洛哥香科■☆

388046　Teucrium charidemi Sandwith；查里香科■☆

388047　Teucrium chlorostachyum Pau et Font Quer；绿穗香科■☆

388048　Teucrium chlorostachyum Pau et Font Quer subsp. melillense（Maire et Sennen）El Oualidi et T. Navarro；梅利利香科■☆

388049　Teucrium cincinnatum Maire；卷毛香科■☆

388050　Teucrium cincinnatum Maire var. ghikae（Emb.）Maire = Teucrium cincinnatum Maire■☆

388051　Teucrium clementiae Ryding；克莱门特香科■☆

388052　Teucrium collincola Greuter et Burdet = Teucrium demnatense Batt.■☆

388053　Teucrium collinum Coss. et Balansa = Teucrium demnatense Batt.■☆

388054　Teucrium compactum Lag.；紧密香科■☆

388055　Teucrium compositum Pomel = Teucrium polium L.■☆

388056　Teucrium corymbiferum Desf.；伞房香科■☆

388057　Teucrium corymbiferum Desf. var. dissimile Sennen = Teucrium corymbiferum Desf.■☆

388058　Teucrium corymbosum R. Br.；伞房花香科●☆

388059　Teucrium cossonii D. Wood；科森香科■☆

388060　Teucrium crispum Pomel = Teucrium pseudoscorodonia Desf.■☆

388061　Teucrium cubense L.；古巴香科■☆

388062　Teucrium cylindraceum Greuter et Burdet = Teucrium aureocandidum Andr.■☆

388063　Teucrium cylindricum（Batt.）Sauvage et Vindt = Teucrium cylindraceum Greuter et Burdet ■☆

388064　Teucrium cylindricum（Batt.）Sauvage et Vindt var. decalvans（Maire）Sauvage et Vindt = Teucrium cylindraceum Greuter et Burdet ☆

388065　Teucrium cyrenaicum（Maire et Weiller）Brullo et Furnari;昔兰尼香科■☆

388066　Teucrium decaisnei C. Presl;德凯纳香科■☆

388067　Teucrium decipiens Coss. et Balansa;迷惑香科■☆

388068　Teucrium demnatense Batt.;代姆纳特香科■☆

388069　Teucrium depressum Small;矮香科;Low Germander ■☆

388070　Teucrium divaricatum Heldr.;叉开香科■☆

388071　Teucrium doumerguei Sennen;杜梅格香科■☆

388072　Teucrium ducellieri Batt.;迪塞利耶香科■☆

388073　Teucrium eburneum Thulin;象牙白香科■☆

388074　Teucrium embergeri（Sauvage et Vindt）El Oualidi, T. Navarro et A. Martin;恩贝格尔香科■☆

388075　Teucrium esquirolii H. Lév. = Isodon ternifolius（D. Don）Kudo ●■

388076　Teucrium excelsum Juz.;高香科;Tall Germander ■☆

388077　Teucrium faurei Maire;福雷香科■☆

388078　Teucrium fischeri Juz.;菲舍尔香科;Fischer Germander ■☆

388079　Teucrium flavum L.;黄香科■☆

388080　Teucrium flavum L. subsp. glaucum（Jord. et Fourr.）Ronniger;灰绿香科■☆

388081　Teucrium flavum L. var. leiophyllum Celak. = Teucrium flavum L. subsp. glaucum（Jord. et Fourr.）Ronniger ■☆

388082　Teucrium flavum L. var. lilacinum Pamp. = Teucrium flavum L. ■☆

388083　Teucrium foliosum Pomel = Teucrium polium L. ■☆

388084　Teucrium fortunei Benth. = Teucrium quadrifarium Buch.-Ham. ●

388085　Teucrium fruticans L.;密毛香科(灌木香科);Bush Germander, Shrubby Germander, Tree Germander, Tull Germander ●☆

388086　Teucrium fruticans L.'Azurea';蓝花密毛香科(深蓝灌木香科);Blue-flowered Germander ●☆

388087　Teucrium fruticans L. subsp. prostratum Gatt. et Maire;平卧香科■☆

388088　Teucrium fruticans L. var. lancifolium Debeaux = Teucrium fruticans L. ●☆

388089　Teucrium fruticans L. var. latifolium Rouy = Teucrium fruticans L. ●☆

388090　Teucrium fruticans L. var. linearifolium Clary = Teucrium fruticans L. ●☆

388091　Teucrium fruticans L. var. pallidum（Maire）Sauvage = Teucrium fruticans L. ●☆

388092　Teucrium fulvoaureum H. Lév. = Teucrium quadrifarium Buch.-Ham. ●

388093　Teucrium fulvum Hance = Teucrium quadrifarium Buch.-Ham. ●

388094　Teucrium gattefossei Emb.;加特福塞香科■☆

388095　Teucrium glomeratum Cav.;团集香科■☆

388096　Teucrium goetzei Gürke;格兹香科■☆

388097　Teucrium granatense Boiss. et Reut. = Teucrium rotundifolium Schreb. ■☆

388098　Teucrium granatense Boiss. et Reut. var. atlanticum Ball = Teucrium rotundifolium Schreb. ■☆

388099　Teucrium grosii Pau;格罗斯香科■☆

388100　Teucrium gussonei Nyman;古索内香科;Fruity Teucrium, Pineapple Germander ■☆

388101　Teucrium gypsophilum Emb. et Maire;喜钙香科■☆

388102　Teucrium helichrysoides（Diels）Greuter et Burdet;蜡菊香科■☆

388103　Teucrium heterophyllum L'Hér.;互叶香科■☆

388104　Teucrium hircanicum L.;西尔加香科;Hyrcania Germander ■☆

388105　Teucrium holocheilum W. E. Evans = Holocheila longipedunculata S. Chow ■

388106　Teucrium holocheilum W. E. Evans ex Kudo = Holocheila longipedunculata S. Chow ■

388107　Teucrium huoshanense S. W. Su et J. Q. He;霍山香科科;Huoshan Germander ■

388108　Teucrium huoshanense S. W. Su et J. Q. He = Teucrium pernyi Franch. ■

388109　Teucrium huotii Emb. et Maire var. grosii Font Quer = Teucrium grosii Pau ■☆

388110　Teucrium integrifolium C. Y. Wu et S. Chow;全叶香科科;Entireleaf Germander ■

388111　Teucrium iva L. = Ajuga iva（L.）Schreb. ■☆

388112　Teucrium jailae Juz.;扎意尔香科;Jail Germander ■☆

388113　Teucrium japonicum Houtt. var. continentale Kitag. = Teucrium ussuriense Kom. ■

388114　Teucrium japonicum Willd.;穗花香科科(毛秀才,石蚕,水藿香,穗花香科,野藿香);Japanese Germander, Spikeflower Germander ■

388115　Teucrium japonicum Willd. f. lanatum Y. Z. Sun ex C. H. Hu;绵毛穗花香科科;Hairyspike Japanese Germander ■

388116　Teucrium japonicum Willd. f. lanatum Y. Z. Sun ex C. H. Hu = Teucrium pilosum（Pamp.）C. Y. Wu et S. Chow ■

388117　Teucrium japonicum Willd. var. continentale Kitag. = Teucrium ussuriense Kom. ■

388118　Teucrium japonicum Willd. var. microphyllum C. Y. Wu et S. Chow;小叶穗花香科科(假荆芥);Littleleaf Japanese Germander, Smallleaf Japanese Germander ■

388119　Teucrium japonicum Willd. var. pilosum Pamp. = Teucrium pilosum（Pamp.）C. Y. Wu et S. Chow ■

388120　Teucrium japonicum Willd. var. tsungmingense C. Y. Wu et S. Chow;崇明穗花香科科■

388121　Teucrium jolyi Mathez et Sauvage;若利香科■☆

388122　Teucrium junii-victoris Sennen et Mauricio = Teucrium gypsophilum Emb. et Maire ■☆

388123　Teucrium kabylicum Batt.;卡比利亚香科■☆

388124　Teucrium kouytchouense H. Lév. = Teucrium quadrifarium Buch.-Ham. ●

388125　Teucrium kraussii Codd;克劳斯香科■☆

388126　Teucrium krymense Juz.;柔毛香科;Pubescent Germander ■☆

388127　Teucrium labiosum C. Y. Wu et S. Chow;大唇香科科(大唇香科,山苏麻,野薄荷);Biglip Germander, Largelip Germander ■

388128　Teucrium laciniatum Torr.;撕裂香科科;Cut-leaf Germander, Lacy Germander ■☆

388129　Teucrium latifolium L. = Teucrium fruticans L. ●☆

388130　Teucrium laxum D. Don;稀疏香科;Lax Germander ■☆

388131　Teucrium leucocladum Boiss.;白枝香科■☆

388132　Teucrium liouvillei Sennen;利乌维尔香科■☆

388133　Teucrium littorale E. P. Bicknell = Teucrium canadense L. ■☆

388134　Teucrium lucidum L.;光亮香科科■☆

388135　Teucrium luteum（Mill.）Degen;黄色香科■☆

388136　Teucrium luteum（Mill.）Degen subsp. flavovirens（Batt.）Greuter et Burdet;黄绿香科■☆

388137　Teucrium luteum（Mill.）Degen subsp. gabesianum（S. Puech）

Greuter et Burdet;加贝斯香科■☆

388138 Teucrium luteum (Mill.) Degen subsp. mesanidum (Litard. et Maire) Greuter et Burdet = Teucrium mesanidum (Litard. et Maire) Navarro et Rosua ■☆

388139 Teucrium luteum (Mill.) Degen subsp. xanthostachyum (Maire et Wilczek) Greuter et Burdet = Teucrium cylindraceum Greuter et Burdet ■☆

388140 Teucrium macrostachyum Wall. ex Benth. = Leucosceptrum canum Sm. ●

388141 Teucrium mairei Sennen = Teucrium doumerguei Sennen ■☆

388142 Teucrium mairei Sennen subsp. mairei = Teucrium doumerguei Sennen ■☆

388143 Teucrium mairei Sennen var. embergeri Sauvage et Vindt = Teucrium embergeri (Sauvage et Vindt) El Oualidi,T. Navarro et A. Martin ☆

388144 Teucrium manghuaense Y. Z. Sun ex S. Chow;巍山香科科(蒙化石蚕,蒙化香科科,巍山香科科);Menghua Germander, Weishan Germander ■

388145 Teucrium manghuaense Y. Z. Sun ex S. Chow var. angustum C. Y. Wu et S. Chow;狭苞巍山香科科■

388146 Teucrium maroccanum Sennen et Mauricio = Teucrium rifanum (Maire et Sennen) T. Navarro et El Oualidi ■☆

388147 Teucrium marum L. ;苦香科(猫香科);Cat Germander, Cat Thyme ●☆

388148 Teucrium massiliense L. = Teucrium viscidum Blume ■

388149 Teucrium maximowiczii Prob. = Teucrium viscidum Blume var. miquelianum (Maxim.) H. Hara ■

388150 Teucrium melillense Maire et Sennen = Teucrium chlorostachyum Pau et Font Quer subsp. melillense (Maire et Sennen) El Oualidi et T. Navarro ■☆

388151 Teucrium melillense Maire et Sennen var. embergeri Maire = Teucrium chlorostachyum Pau et Font Quer subsp. melillense (Maire et Sennen) El Oualidi et T. Navarro ■☆

388152 Teucrium mesanidum (Litard. et Maire) Navarro et Rosua;中间香科■☆

388153 Teucrium miquelianum (Maxim.) Kudo = Teucrium viscidum Blume var. miquelianum (Maxim.) H. Hara ■

388154 Teucrium montanum L. ;山香科(山石蚕);Montane Germander, Mountain Germander ■☆

388155 Teucrium multinodum (Bordz.) Juz. ;多节香科;Manynode Germander ■☆

388156 Teucrium murcicum Sennen subsp. hieronymi (Sennen) Navarro et Rosua;希罗香科■☆

388157 Teucrium musimonum Humbert;摩弗伦香科■☆

388158 Teucrium musimonum Humbert var. matris-filiae Emb. et Maire = Teucrium musimonum Humbert ■☆

388159 Teucrium musimonum Humbert var. mesatlanticum Emb. et Maire = Teucrium musimonum Humbert ■☆

388160 Teucrium musimonum Humbert var. virescens Maire = Teucrium musimonum Humbert ■☆

388161 Teucrium nanum C. Y. Wu et S. Chow;矮生香科科(矮生香科);Dwarf Germander ■

388162 Teucrium nepetifolium Benth. = Caryopteris nepetifolia (Benth.) Maxim.

388163 Teucrium nepetoides H. Lév. = Teucrium viscidum Blume var. nepetoides (H. Lév.) C. Y. Wu et S. Chow ■

388164 Teucrium ningpoense Hemsl. = Teucrium pernyi Franch.

388165 Teucrium nuchense K. Koch;努陈香科;Nuchen Germander ■☆

388166 Teucrium occidentale A. Gray;西方香科科;Hairy Germander, Wood Sage ■☆

388167 Teucrium occidentale A. Gray = Teucrium canadense L. var. occidentale (A. Gray) E. M. McClint. et Epling ■☆

388168 Teucrium occidentale A. Gray subsp. viscidum Piper = Teucrium canadense L. var. occidentale (A. Gray) E. M. McClint. et Epling ■☆

388169 Teucrium occidentale A. Gray var. boreale (E. P. Bicknell) Fernald = Teucrium canadense L. var. occidentale (A. Gray) E. M. McClint. et Epling ■☆

388170 Teucrium omeiense Y. Z. Sun ex S. Chow;峨眉香科科(峨眉石蚕,峨眉香科,野苏);Emei Germander, Omei Germander ■

388171 Teucrium omeiense Y. Z. Sun ex S. Chow var. cyanophyllum C. Y. Wu et S. Chow;蓝叶峨眉香科科■

388172 Teucrium orientale L. ;东方石蚕 (东方香科);Oriental Germander ■☆

388173 Teucrium ornatum Hemsl. = Kinostemon ornatum (Hemsl.) Kudo ■

388174 Teucrium oxylepis Font Quer = Teucrium afrum (Emb. et Maire) Pau et Font Quer ■☆

388175 Teucrium oxylepis Font Quer var. rhiphaeum Font Quer et Pau = Teucrium afrum (Emb. et Maire) Pau et Font Quer subsp. rhiphaeum (Font Quer et Pau) Castrov. et Bayon ■☆

388176 Teucrium palmatum Benth. ex Hook. f. = Rubiteucris palmata (Benth. ex Hook. f.) Kudo ■

388177 Teucrium pannonicum J. Kern. ;山地香科;Montane Germander ■☆

388178 Teucrium parviflorum Schreb. ;满山香科 (小花香科);Littleflower Germander ■☆

388179 Teucrium pernyi Franch. ;庐山香科科(白花石蚕,庐山香科,四齿莩草,细沙虫草);Lushan Germander, Perny Germander ■

388180 Teucrium philippinense Merr. = Teucrium viscidum Blume ■

388181 Teucrium pilosum (Pamp.) C. Y. Wu et S. Chow;长毛香科(长毛香科,铁马鞭);Longhair Germander, Pilose Germander, Poly Germander ■

388182 Teucrium pilosum L. var. macrophyllum C. Y. Wu et S. Chow;大叶长毛香科科;Bigleaf Pilose Germander ■

388183 Teucrium polioides Ryding;拟灰色香科科■☆

388184 Teucrium polium L. ;狭叶香科(灰白香科,灰石蚕,灰香科);Cat Thyme, Hulwort, Mediterranean Germanier, Narrowleaf Germander,Poleye,Polion ☆

388185 Teucrium polium L. subsp. antiatlanticum Maire = Teucrium antiatlanticum (Maire) Sauvage et Vindt ■☆

388186 Teucrium polium L. subsp. aurasiacum (Maire) Greuter et Burdet;奥拉斯香科科■☆

388187 Teucrium polium L. subsp. aureum (Schreb.) Arcang. = Teucrium luteum (Mill.) Degen ■☆

388188 Teucrium polium L. subsp. capitatum (L.) Briq. = Teucrium capitatum L. ■☆

388189 Teucrium polium L. subsp. chevalieri Maire;舍瓦利耶香科科■☆

388190 Teucrium polium L. subsp. cylindricum Maire = Teucrium cylindraceum Greuter et Burdet ■☆

388191 Teucrium polium L. subsp. cyrenaicum Maire et Weiller = Teucrium cyrenaicum (Maire et Weiller) Brullo et Furnari ■☆

388192 Teucrium polium L. subsp. flavovirens Batt. = Teucrium luteum (Mill.) Degen subsp. flavovirens (Batt.) Greuter et Burdet ■☆

388193 Teucrium polium L. subsp. gabesianum Le Houér. = Teucrium luteum (Mill.) Degen subsp. gabesianum (S. Puech) Greuter et

Burdet ■☆

388194　Teucrium polium L. subsp. geyrii Maire ＝Teucrium helichrysoides（Diels）Greuter et Burdet ■☆

388195　Teucrium polium L. subsp. ghikae Emb. ＝Teucrium cincinnatum Maire ■☆

388196　Teucrium polium L. subsp. helichrysoides（Diels）Maire ＝Teucrium helichrysoides（Diels）Greuter et Burdet ■☆

388197　Teucrium polium L. subsp. luteum（Mill.）Briq. ＝Teucrium luteum（Mill.）Degen ■☆

388198　Teucrium polium L. subsp. mairei（Sennen）Maire ＝Teucrium doumerguei Sennen ■☆

388199　Teucrium polium L. subsp. mesanidum Litard. et Maire ＝Teucrium mesanidum（Litard. et Maire）Navarro et Rosua ■☆

388200　Teucrium polium L. subsp. seuratianum Maire ＝Teucrium helichrysoides（Diels）Greuter et Burdet ■☆

388201　Teucrium polium L. subsp. thymoides（Pomel）Batt. ＝Teucrium thymoides Pomel ■☆

388202　Teucrium polium L. subsp. xanthostachyum Maire et Wilczek ＝Teucrium cylindraceum Greuter et Burdet ■☆

388203　Teucrium polium L. subsp. zanonii（Pamp.）Maire et Weiller ＝Teucrium zanonii Pamp. ■☆

388204　Teucrium polium L. var. adeliae Maire ＝Teucrium polium L. ■☆

388205　Teucrium polium L. var. albescens Sauvage ＝Teucrium cylindraceum Greuter et Burdet ■☆

388206　Teucrium polium L. var. angustifolium Benth. ＝Teucrium polium L. ■☆

388207　Teucrium polium L. var. atlanticum Ball ＝Teucrium polium L. ■☆

388208　Teucrium polium L. var. aurasiacum Maire ＝Teucrium polium L. ■☆

388209　Teucrium polium L. var. aureum（Schreb.）Boiss. ＝Teucrium polium L. ■☆

388210　Teucrium polium L. var. chamaedryfolium Pau et Font Quer ＝Teucrium polium L. ■☆

388211　Teucrium polium L. var. corymbiferum（Desf.）Maire ＝Teucrium corymbiferum Desf. ■☆

388212　Teucrium polium L. var. cylindrostachyum Maire ＝Teucrium mesanidum（Litard. et Maire）Navarro et Rosua ■☆

388213　Teucrium polium L. var. dissimile（Sennen）Jahand. et Maire ＝Teucrium corymbiferum Desf. ■☆

388214　Teucrium polium L. var. doumeri Sennen ＝Teucrium polium L. ■☆

388215　Teucrium polium L. var. faureanum Maire ＝Teucrium polium L. ■☆

388216　Teucrium polium L. var. flavidum Emb. et Maire ＝Teucrium doumerguei Sennen ■☆

388217　Teucrium polium L. var. gnaphalodess Ball;鼠曲香科■☆

388218　Teucrium polium L. var. helichrysoides Diels ＝Teucrium helichrysoides（Diels）Greuter et Burdet ■☆

388219　Teucrium polium L. var. hidumense Sennen ＝Teucrium doumerguei Sennen ■☆

388220　Teucrium polium L. var. humbertii Maire et Sennen ＝Teucrium capitatum L. ■☆

388221　Teucrium polium L. var. laxum Maire ＝Teucrium helichrysoides（Diels）Greuter et Burdet ■☆

388222　Teucrium polium L. var. longivillum Faure et Maire ＝Teucrium polium L. ■☆

388223　Teucrium polium L. var. majoricum（Rouy）Willk. ＝Teucrium polium L. ■☆

388224　Teucrium polium L. var. maximiliani Sennen et Mauricio ＝

Teucrium doumerguei Sennen ■☆

388225　Teucrium polium L. var. max-onnoi Sennen ＝Teucrium polium L. ■☆

388226　Teucrium polium L. var. mesatlanticum（Maire）Sauvage et Vindt ＝Teucrium luteum（Mill.）Degen subsp. flavovirens（Batt.）Greuter et Burdet ■☆

388227　Teucrium polium L. var. mesatlanticum Maire ＝Teucrium polium L. ■☆

388228　Teucrium polium L. var. pampaninianum Maire et Weiller ＝Teucrium zanonii Pamp. ■☆

388229　Teucrium polium L. var. pilosum Decne. ＝Teucrium polium L. ■☆

388230　Teucrium polium L. var. polycephalum（Pomel）Briq. ＝Teucrium polium L. ■☆

388231　Teucrium polium L. var. pseudohyssopus（Schreb.）Halácsy ＝Teucrium polium L. ■☆

388232　Teucrium polium L. var. rifanum Maire et Sennen ＝Teucrium rifanum（Maire et Sennen）T. Navarro et El Oualidi ■☆

388233　Teucrium polium L. var. scoparium Faure et Maire ＝Teucrium polium L. ■☆

388234　Teucrium polium L. var. soloitanum（Maire）Sauvage et Vindt ＝Teucrium polium L. ■☆

388235　Teucrium polium L. var. syrticum Maire et Weiller ＝Teucrium zanonii Pamp. ■☆

388236　Teucrium polium L. var. tetuanense Pau ＝Teucrium polium L. ■☆

388237　Teucrium polium L. var. tricolor Maire ＝Teucrium mesanidum（Litard. et Maire）Navarro et Rosua ■☆

388238　Teucrium polium L. var. tripolitanum Pamp. ＝Teucrium polium L. ■☆

388239　Teucrium polium L. var. valentinum（Schreb.）Maire ＝Teucrium capitatum L. ■☆

388240　Teucrium polium L. var. vulgare Benth. ＝Teucrium polium L. ■☆

388241　Teucrium polycephalum Pomel ＝Teucrium polium L. ■☆

388242　Teucrium praemontanum Klokov;美丽香科;Premontane Germander ■☆

388243　Teucrium pseudoscorodonia Desf. ;假蒜香科■☆

388244　Teucrium pseudoscorodonia Desf. var. baeticum（Boiss. et Reut.）Ball ＝Teucrium pseudoscorodonia Desf. ■☆

388245　Teucrium pseudoscorodonia Desf. var. crispum（Pomel）Batt. ＝Teucrium pseudoscorodonia Desf. ☆

388246　Teucrium pulchrius Juz. ;柔夷尔香科;Showy Germander ■☆

388247　Teucrium quadrifarium Buch. -Ham. ;铁轴草（凤凰草,黑头草,红杆一枝蒿, 红毛将军, 牛尾草, 绣球防风）;Fourfile Germander,Ironaxle Germander ●

388248　Teucrium radicans Bonnet et Barratte;具根香科■☆

388249　Teucrium ramosissimum Desf. ;多枝香科■☆

388250　Teucrium ramosissimum Desf. var. getulum（Maire）Quézel et Santa ＝Teucrium ramosissimum Desf. ■☆

388251　Teucrium resupinatum Desf. ;倒置香科科■☆

388252　Teucrium resupinatum Desf. var. xauense Pau et Font Quer ＝Teucrium resupinatum Desf. ■☆

388253　Teucrium riatarum Humbert et Maire ＝Teucrium bracteatum Desf. ■☆

388254　Teucrium rifanum（Maire et Sennen）T. Navarro et El Oualidi;里夫香科科■☆

388255　Teucrium rifeum Maire et Sennen ＝Teucrium rifanum（Maire et Sennen）T. Navarro et El Oualidi ■☆

388256　Teucrium riparium Hochst. ＝Teucrium kraussii Codd ■☆

388257 Teucrium rotundifolium Schreb. ;圆叶香科科■☆

388258 Teucrium rotundifolium Schreb. subsp. faurei（Maire）Sauvage et Vindt = Teucrium faurei Maire ■☆

388259 Teucrium rotundifolium Schreb. subsp. granatense（Boiss. et Reut.）Emb. = Teucrium rotundifolium Schreb. ■☆

388260 Teucrium rotundifolium Schreb. var. atlanticum（Ball）Maire = Teucrium rotundifolium Schreb. ■☆

388261 Teucrium rotundifolium Schreb. var. purpurascens Maire = Teucrium rotundifolium Schreb. ■☆

388262 Teucrium royleanum Wall. ;匍匐香科;Royle Germander ■☆

388263 Teucrium rupestre Coss. et Balansa ;岩生香科■☆

388264 Teucrium rupestre Coss. et Balansa var. villosum Maire = Teucrium rupestre Coss. et Balansa ■☆

388265 Teucrium salviastrum Schreb. subsp. afrum Emb. et Maire = Teucrium afrum（Emb. et Maire）Pau et Font Quer ■☆

388266 Teucrium salviastrum Schreb. subsp. rubriflorum（Font Quer et Pau）Maire = Teucrium afrum（Emb. et Maire）Pau et Font Quer subsp. rubriflorum（Font Quer et Pau）Castrov. et Bayon ■☆

388267 Teucrium sauvagei Le Houér. ;索瓦热香科■☆

388268 Teucrium saxatile Lam. var. getulum Maire = Teucrium ramosissimum Desf. ■☆

388269 Teucrium schoenenbergeri Nabli ;舍恩香科■☆

388270 Teucrium scordioides Schreb. ;沼泽香科科（拟蒜头石蚕,沼泽香科）;Marshy Germander ■

388271 Teucrium scordioides Schreb. = Teucrium scordium L. subsp. scordioides（Schreb.）Arcang. ■☆

388272 Teucrium scordium L. ;蒜味香科科（蒜头石蚕,蒜头香科,蒜味香科, 蒜叶香科科）;English Treacle, Garlic Germander, Garlicleaf Germander, Garlicsmell Germander, Garliek Germander, Marsh Germander,Scordion,Water Germander ■

388273 Teucrium scordium L. subsp. scordioides（Schreb.）Arcang. ; 蒜状香科■☆

388274 Teucrium scordium L. var. microphyllum A. Rich. = Teucrium scordium L. ■

388275 Teucrium scorodonia L. ;林石蚕;Ambrose, Catmint, Dog Mint, English Treacle, Garlic Germander, Garlic Sage, Garliek Germander, Gulseck-girse, Gypsy's Bacca, Gypsy's Sage, Large-leaved Germander,Poor Man's Treacle,Rock Mint,Sage Garlick,Sage-leaved Germander, Scordion, Wild Sage, Wood Germander, Wood Sage, Woodland Germander ■☆

388276 Teucrium scorodonia L. subsp. baeticum（Boiss. et Reut.）Tutin = Teucrium pseudoscorodonia Desf. ■☆

388277 Teucrium serpylloides Maire et Weiller;百里香叶香科■☆

388278 Teucrium sibiricum L. = Nepeta ucranica L. ■

388279 Teucrium simplex Vaniot ;香科科（荆芥,香科）;Germander, Simplex Germander ■

388280 Teucrium sinaicum Boiss. = Teucrium decaisnei C. Presl ■☆

388281 Teucrium soloitanum（Maire）T. Navarro et Rosua = Teucrium polium L. ■☆

388282 Teucrium somalense Ryding;索马里香科■☆

388283 Teucrium spicastrum Hedge et A. G. Mill. = Teucrium yemense Deflers ■☆

388284 Teucrium spinosum L. ;具刺香科■☆

388285 Teucrium staechadifolium Pomel = Teucrium polium L. ■☆

388286 Teucrium stocksianum Boiss. ;斯得香科■☆

388287 Teucrium stoloniferum Roxb. = Teucrium viscidum Blume ■

388288 Teucrium syspirense K. Koch;灰香科（泰勒香科）;Grey Germander ■☆

388289 Teucrium taiwanianum T. H. Hsieh et T. C. Huang;台湾香科科■

388290 Teucrium tananicum Maire;泰南香科■☆

388291 Teucrium taylori Boiss. ;泰勒香科（绒毛香科）;Taylor Germander ■☆

388292 Teucrium thymoides Pomel;百里香香科■☆

388293 Teucrium tochauense Kudo = Heterolamium debile（Hemsl.）C. Y. Wu var. tochauense（Kudo）C. Y. Wu ■

388294 Teucrium tomentosum K. Heyne;绒毛香科（香科）;Tomentose Germander ■☆

388295 Teucrium trifidum Retz. ;三裂香科■☆

388296 Teucrium tsinlingense C. Y. Wu et S. Chow;秦岭香科科（秦岭香科）;Qinling Germander,Tsinling Germander ■

388297 Teucrium tsinlingense C. Y. Wu et S. Chow var. porphyreum C. Y. Wu et S. Chow;紫萼秦岭香科科■

388298 Teucrium ussuriense Kom. ;黑龙江香科科（东北石蚕,黑龙江香科）;Heilongjiang Germander, Ussuri Germander ■

388299 Teucrium veronicoides Maxim. ;裂苞香科科（裂苞香科）;Splitbract Germander,Splitedbract Germander ■

388300 Teucrium veronicoides Maxim. var. brachytrichum Ohwi;短毛裂苞香科科■☆

388301 Teucrium virescens Pomel = Teucrium polium L. ■☆

388302 Teucrium virginicum L. = Teucrium japonicum Willd. ■

388303 Teucrium viscidum Blume;血见愁（布地锦,冲天泡,方骨苦草,方枝苦草,肺形草,黑星草,灰香科,假香菜,假紫苏,苦药菜,苦药草,蔓苦草,仁沙草,山黄荆,山藿香,蛇药,水苏麻,四方枝苦草,四棱香,消炎草,小苦草,血芙蓉,野薄荷,野石蚕,野苏麻,野芝麻,贼子草,粘毛石蚕,皱面草,皱面风,皱面苦草）;Stoloniferous Germander, Viscid Germander ■

388304 Teucrium viscidum Blume var. leiocalyx C. Y. Wu et S. Chow;光萼血见愁;Glabrouscalyx Viscid Germander ■

388305 Teucrium viscidum Blume var. longibracteatum C. Y. Wu et S. Chow;长苞血见愁;Longbract Viscid Germander ■

388306 Teucrium viscidum Blume var. macrosphanum C. Y. Wu et S. Chow;大唇血见愁;Biglip Viscid Germander ■

388307 Teucrium viscidum Blume var. miquelianum（Maxim.）H. Hara;蔓藿香■

388308 Teucrium viscidum Blume var. nepetoides（H. Lév.）C. Y. Wu et S. Chow;微毛血见愁;Hair Viscid Germander ■

388309 Teucrium viscidum Blume var. yingguanum Z. Y. Zhu;瘿冠血见愁■

388310 Teucrium wallichianum Benth. = Achyrospermum densiflorum Perkins ■☆

388311 Teucrium wallichianum Benth. = Achyrospermum wallichianum Benth. ex Hook. f. ■

388312 Teucrium werneri Emb. ;维尔纳香科■☆

388313 Teucrium wightii Hook. f. ;威特香科■☆

388314 Teucrium yemense Deflers;也门香科■☆

388315 Teucrium zaianum Emb. et Maire;宰哈奈香科■☆

388316 Teucrium zanonii Pamp. ;扎农香科■☆

388317 Teuscheria Garay(1958);托氏兰属（杜氏兰属）■☆

388318 Teuscheria cornucopia Garay;托氏兰■☆

388319 Teuscheria elegans Garay;雅致托氏兰■☆

388320 Teutliopsis（Dumort.）Čelak. = Atriplex L. ■●

388321 Texiera Jaub. et Spach = Glastaria Boiss. ■☆

388322 Textoria Miq. = Dendropanax Decne. et Planch. ●

388323 Textoria dentigera（Harms ex Diels）Nakai = Dendropanax dentiger（Harms ex Diels）Merr. ●

388324　Textoria dentigera（Harms）Nakai = Dendropanax dentiger（Harms ex Diels）Merr. ●

388325　Textoria hainanensis（Merr. et Chun）Nakai = Dendropanax hainanensis（Merr. et Chun）Chun ●

388326　Textoria japonica（Jungh.）Miq. = Dendropanax trifidus（Thunb.）Makino ex H. Hara ●

388327　Textoria japonica Miq. = Dendropanax trifidus（Thunb.）Makino ex H. Hara ●

388328　Textoria parviflora（Champ. ex Benth.）Nakai = Dendropanax proteus（Champ. ex Benth.）Benth. ●

388329　Textoria pellucidopunctata（Hayata）Kaneh. et Sasaki = Dendropanax dentiger（Harms ex Diels）Merr. ●

388330　Textoria pellucidopunctata（Hayata）Kaneh. et Sasaki = Dendropanax pellucidopunctatus（Hayata）Kaneh. ●

388331　Textoria protea（Champ. ex Benth.）Nakai = Dendropanax proteus（Champ. ex Benth.）Benth. ●

388332　Textoria sinensis（Nakai）Nakai = Dendropanax dentiger（Harms ex Diels）Merr. ●

388333　Textoria trifida（Thunb. ex Murray）Nakai ex Honda = Dendropanax trifidus（Thunb.）Makino ex H. Hara ●

388334　Textoria trifida（Thunb.）Nakai ex Honda = Dendropanax trifidus（Thunb.）Makino ex H. Hara ●

388335　Teyleria Backer（1939）;琼豆属;Jadebean,Teyleria ■

388336　Teyleria koordersii（Backer）Backer;琼豆;Hainan Teyleria,Jadebean ■

388337　Teysmannia Miq.（1857）= Pottsia Hook. et Arn. ●

388338　Teysmannia Miq.（1859）= Johannesteijsmannia H. E. Moore ●☆

388339　Teysmannia Miq.（1859）= Teyssmannia Rchb. f. et Zoll. ●

388340　Teysmannia Rchb. et Zoll. = Johannesteijsmannia H. E. Moore ●☆

388341　Teysmannia Rchb. f. et Zoll. = Johannesteijsmannia H. E. Moore ●☆

388342　Teysmannia Rchb. f. et Zoll. = Teysmannia Rchb. et Zoll. ●☆

388343　Teysmanniodendron Koord. = Teijsmanniodendron Koord. ●☆

388344　Teyssmannia Rchb. f. et Zoll. = Johannesteijsmannia H. E. Moore ●☆

388345　Thacla Spach = Caltha L. ■

388346　Thacla natans（Pall.）Deyl et Soják. = Caltha natans Pall. ■

388347　Thaeombauia Seem. = Flacourtia Comm. ex L'Her. ●

388348　Thaia Seidenf.（1975）;泰国兰属■☆

388349　Thaia saprophytica Seidenf.;泰国兰■☆

388350　Thailentadopsis Kosterm. = Havardia Small ●☆

388351　Thalamia Spreng. = Phyllocladus Rich. ex Mirb.（保留属名）●☆

388352　Thalamia Spreng. = Podocarpus Pers.（保留属名）●

388353　Thalasium Spreng. = Panicum L. ■

388354　Thalassia Banker et Sol. ex K. D. König = Thalassia Banks ex K. D. König ■

388355　Thalassia Banks = Thalassia Banks ex K. D. König ■

388356　Thalassia Banks et K. D. König = Thalassia Banks ex K. D. König ■

388357　Thalassia Banks ex K. D. König（1805）;海生草属（长喙藻属,泰来藻属）;Thalassia,Turtle Grass,Turtle-grass ■

388358　Thalassia Sol. ex K. D. König = Thalassia Banks ex K. D. König ■

388359　Thalassia ciliata（Forssk.）König = Thalassodendron ciliatum（Forssk.）Hartog ■☆

388360　Thalassia hemprichii（Ehrenb. ex Solms）Asch.;亨普海生草（泰来藻）;Thalassia,Turtle Grass,Turtle-grass ■

388361　Thalassia hemprichii（Ehrenb.）Asch. = Thalassia hemprichii

（Ehrenb. ex Solms）Asch. ■

388362　Thalassia testudinum Banks ex K. D. König;海生草;Thalassia,Turtle Grass,Turtle-grass ■☆

388363　Thalassiaceae Nakai = Hydrocharitaceae Juss.（保留科名）■

388364　Thalassiaceae Nakai = Thalictraceae Raf. ■☆

388365　Thalassiaceae Nakai(1943);海生草科■

388366　Thalassiophila Denizot = Thalassia Banks ex K. D. König ■

388367　Thalassodendron Hartog(1970);海丛藻属■☆

388368　Thalassodendron ciliatum（Forssk.）Hartog;缘毛海丛藻■☆

388369　Thaleropia Peter G. Wilson(1993);茂灌金娘属●☆

388370　Thaleropia iteophylla（Diels）Peter G. Wilson;茂灌金娘●☆

388371　Thalesia Mart. ex Pfeiff. = Sweetia Spreng.（保留属名）●☆

388372　Thalesia Raf. = Aphyllon Mitch. ■

388373　Thalesia Raf. = Orobanche L. ■

388374　Thalesia Raf. ex Britton = Aphyllon Mitch. ■

388375　Thalesia fasciculata（Nutt.）Britton = Orobanche fasciculata Nutt. ■☆

388376　Thalesia lutea（Parry）Rydb. = Orobanche fasciculata Nutt. ■☆

388377　Thalesia purpurea A. Heller = Orobanche uniflora L. ■☆

388378　Thalesia uniflora（L.）Britton = Orobanche uniflora L. ■☆

388379　Thalestris Rizzini = Justicia L. ●■

388380　Thalia L.（1753）;水竹芋属（塔里亚属,再力花属）;Thalia ■☆

388381　Thalia angustifolia Peters;狭叶水竹芋（狭叶塔里亚）■☆

388382　Thalia barbata Small = Thalia dealbata Fraser ■☆

388383　Thalia caerulea Ridl. = Thalia geniculata L. ■☆

388384　Thalia canniformis G. Forst. = Donax canniformis（G. Forst.）K. Schum. ■

388385　Thalia dealbata Fraser;水竹芋（白粉塔里亚）;Hardy Canna,Powdered Thalia,Powdery Thalia,Thalia,Water Canna ■☆

388386　Thalia divaricata Chapm. = Thalia geniculata L. ■☆

388387　Thalia geniculata L.;节花水竹芋（水竹芋）;Arrowroot,Fire Flag,Fire Flags,Fire-flag ■☆

388388　Thalia trichocalyx Gagnep. = Thalia geniculata L. ■☆

388389　Thalia welwitschii Ridl. = Thalia geniculata L. ■☆

388390　Thalianthus Klotzsch = Myrosma L. f. ■☆

388391　Thalianthus Klotzsch ex Körn. = Myrosma L. f. ■☆

388392　Thalictraceae Á. Löve et D. Löve = Ranunculaceae Juss.（保留科名）●■

388393　Thalictraceae Raf. = Ranunculaceae Juss.（保留科名）●■

388394　Thalictrella A. Rich. = Isopyrum L.（保留属名）■

388395　Thalictrodes Kuntze = Cimicifuga Wernisch. ●■

388396　Thalictrodes simplex（DC.）Kuntze = Cimicifuga simplex Wormsk. ex DC. ■

388397　Thalictrum L.（1753）;唐松草属（白莲草属,白蓬草属）;Meadow Rue,Meadowrue,Meadow-rue,Muskrat Wort ■

388398　Thalictrum Tourn. ex L. = Thalictrum L. ■

388399　Thalictrum actaeifolium Siebold et Zucc.;类叶升麻叶唐松草（紫银莲）■☆

388400　Thalictrum actaeifolium Siebold et Zucc. = Thalictrum minus L. var. kemense（Fr.）Trel. ■

388401　Thalictrum acutifolium（Hand.-Mazz.）B. Boivin;尖叶唐松草（石笋还阳）;Sharpleaf Meadowrue ■

388402　Thalictrum aduncum B. Boivin = Thalictrum rhynchocarpum Quart.-Dill. et A. Rich. ex A. Rich. ■☆

388403　Thalictrum affine Ledeb. = Thalictrum simplex L. var. affine（Ledeb.）Regel ■

388404　Thalictrum alpinum L.;高山唐松草（高山白蓬草,马尾黄

连）；Alpine Meadow Rue, Alpine Meadowrue, Alpine Meadow-rue, Arctic Meadow-rue, Dwarf Meadow-rue ■

388405 Thalictrum alpinum L. var. acutilobum H. Hara = Thalictrum alpinum L. var. elatum Ulbr. ■

388406 Thalictrum alpinum L. var. elatum Ulbr.；直梗高山唐松草（草岩连，大疖药，复叶披麻草，惊风草，亮星草，亮叶子，亮叶子草，披麻草，喜花草，小青草，岩连，直梗唐松草，紫金丹）；Straightstalk Alpine Meadowrue ■

388407 Thalictrum alpinum L. var. elatum Ulbr. f. puberulum W. T. Wang et S. H. Wang = Thalictrum alpinum L. var. elatum Ulbr. ■

388408 Thalictrum alpinum L. var. elatum Ulbr. f. puberulum W. T. Wang et S. H. Wang；毛叶高山唐松草■

388409 Thalictrum alpinum L. var. hebetum B. Boivin = Thalictrum alpinum L. ■

388410 Thalictrum alpinum L. var. microphyllum （Royle）Hand.-Mazz.；柄果高山唐松草（八仙草，白水脐疖药，亮叶子，三月参，喜花草，细花草，小棕树，眼疖药）；Littleleaf Alpine Meadowrue ■

388411 Thalictrum alpinum L. var. setulosinerve （H. Hara）W. T. Wang = Thalictrum alpinum L. var. elatum Ulbr. ■

388412 Thalictrum alpinum L. var. stipitatum Y. Yabe；短茎高山唐松草■☆

388413 Thalictrum altissimum Thomas ex De Massas = Thalictrum flavum L. ■

388414 Thalictrum amphibolum Greene = Thalictrum revolutum DC. ■

388415 Thalictrum amplissimum H. Lév. et Vaniot = Thalictrum minus L. var. hypoleucum （Siebold et Zucc.）Miq. ■

388416 Thalictrum amurense Maxim.；阿穆尔唐松草■☆

388417 Thalictrum anemonoides Michx.；银莲花状唐松草；Rue-anemone ■☆

388418 Thalictrum anemonoides Michx. = Anemonella thalictroides （L.）Spach ■☆

388419 Thalictrum anemonoides Michx. = Thalictrum thalictroides （L.）A. J. Eames et B. Boivin ■☆

388420 Thalictrum angustatum Weinm. ex Lecoy. = Thalictrum flavum L. ■

388421 Thalictrum angustifolium L.；狭叶唐松草；Narrow-leaved Meadow Rue, Narrow-leaved Meadow-rue ■☆

388422 Thalictrum anonymum Wall. ex Lecoy. = Thalictrum flavum L. ■

388423 Thalictrum aquilegiifolium L.；欧洲唐松草（糠斗菜叶唐松草）；Columbine Meadow Rue, Columbine Meadow-rue, Columbine-leaved Meadow Rue, Feathered Columbine, French Meadow-rue, Great Spanish Bastard Rhubarb, Greater Meadow Rue, Meadow Rue, Meadow-rue, Spanish Tuft, Tufted Columbine ■☆

388424 Thalictrum aquilegiifolium L. 'White Cloud'；白云欧洲唐松草（白云糠斗菜叶唐松草）■☆

388425 Thalictrum aquilegiifolium L. subsp. asiaticum （Nakai）Kitag. = Thalictrum aquilegiifolium L. var. sibiricum Regel et Tiling ■

388426 Thalictrum aquilegiifolium L. var. asiaticum Nakai = Thalictrum aquilegiifolium L. var. sibiricum Regel et Tiling ■

388427 Thalictrum aquilegiifolium L. var. daisenense （Nakai）Emura = Thalictrum aquilegiifolium L. var. sibiricum Regel et Tiling ■

388428 Thalictrum aquilegiifolium L. var. sibiricum Regel et Tiling；唐松草（草黄连，翅果唐松草，高山野花草，黑汉子腿，红莲，马尾连，猫爪子，土黄连，翼果白蓬草，翼果唐松草，紫花顿，西伯利亚唐松草）；Siberia Columbine Meadowrue, Siberian Golumbine Meadowrue ■

388429 Thalictrum argyi H. Lév. = Thalictrum javanicum Blume ■

388430 Thalictrum argyi H. Lév. et Vaniot = Thalictrum javanicum Blume ■

388431 Thalictrum atriplex Finet et Gagnep.；狭序唐松草（水黄连）；Narrowraceme Meadowrue ■

388432 Thalictrum baicalense Turcz.；贝加尔唐松草（草黄连，金丝莲，马尾黄连，马尾连，球果白蓬草，球果唐松草）；Baikal Meadowrue ■

388433 Thalictrum baicalense Turcz. f. levicarpum Tamura；光果贝加尔唐松草■☆

388434 Thalictrum baicalense Turcz. var. japonicum Boiss. = Thalictrum baicalense Turcz. ■

388435 Thalictrum baicalense Turcz. var. megalostigma Boivin；长柱贝加尔唐松草；Longstigma Baikal Meadowrue ■

388436 Thalictrum belgicum Jord. = Thalictrum flavum L. ■

388437 Thalictrum boivinianum Staner et J. Léonard = Thalictrum rhynchocarpum Quart.-Dill. et A. Rich. ex A. Rich. subsp. ruwenzoriense Lye ■☆

388438 Thalictrum bracteatum Roxb. = Clematis cadmia Buch.-Ham. ex Hook. f. et Thomson ■

388439 Thalictrum brevisericeum W. T. Wang et S. H. Wang；绢毛唐松草；Sericeous Meadowrue ■

388440 Thalictrum caffrum Eckl. et Zeyh. = Thalictrum minus L. subsp. maxwellii （Royle）Hand ■☆

388441 Thalictrum capitatum Jord. = Thalictrum flavum L. ■

388442 Thalictrum carolinianum Bosc ex DC. var. subpubescens DC. = Thalictrum pubescens Pursh ■☆

388443 Thalictrum caulophylloides Small = Thalictrum coriaceum （Britton）Small ■☆

388444 Thalictrum chaotungense W. T. Wang et S. H. Wang = Thalictrum glandulosissimum （Finet et Gagnep.）W. T. Wang et S. H. Wang var. chaotungense W. T. Wang et S. H. Wang ■

388445 Thalictrum chapinii B. Boivin = Thalictrum rhynchocarpum Quart.-Dill. et A. Rich. ex A. Rich. subsp. ruwenzoriense Lye ■☆

388446 Thalictrum chayuense W. T. Wang；察隅唐松草■

388447 Thalictrum chelidonii DC.；珠芽唐松草；Bulbule Meadowrue, Celandine Meadowrue ■

388448 Thalictrum chiaonis B. Boivin = Thalictrum acutifolium （Hand.-Mazz.）B. Boivin ■

388449 Thalictrum cirrhosum H. Lév.；星毛唐松草（淡色马尾连）；Starhair Meadowrue, Tendrilled Meadowrue ■

388450 Thalictrum clavatum DC.；山地光茎唐松草；Mountain Meadow-rue ■☆

388451 Thalictrum clavatum DC. var. acutifolium Hand.-Mazz. = Thalictrum acutifolium （Hand.-Mazz.）B. Boivin ■

388452 Thalictrum clavatum DC. var. filamentosum （Maxim.）Finet et Gagnep. = Thalictrum filamentosum Maxim. ■

388453 Thalictrum clavatum Hook. = Thalictrum sparsiflorum Turcz. ex Fisch. et C. A. Mey. ■

388454 Thalictrum clavatum Hook. var. acutifolium Hand.-Mazz. = Thalictrum acutifolium （Hand.-Mazz.）B. Boivin ■

388455 Thalictrum clavatum Hook. var. cavaleriei H. Lév. = Thalictrum acutifolium （Hand.-Mazz.）B. Boivin ■

388456 Thalictrum clavatum Hook. var. filamentosum （Maxim.）Finet et Gagnep. = Thalictrum filamentosum Maxim. ■

388457 Thalictrum clematidifolium Franch. = Thalictrum robustum Maxim. ■

388458 Thalictrum collinum Wallr. = Thalictrum minus L. ■

388459　Thalictrum confine Fernald;北美唐松草;Critical Meadow-rue,Northern Meadow-rue ■☆

388460　Thalictrum contortum L. ;盘旋唐松草;Contorted Meadowrue ■

388461　Thalictrum contortum L. = Thalictrum aquilegiifolium L. var. sibiricum Regel et Tiling ■

388462　Thalictrum cooleyi H. E. Ahles;库利唐松草;Cooley's Meadow-rue ■☆

388463　Thalictrum coreanum H. Lév. ;朝鲜唐松草■☆

388464　Thalictrum coriaceum (Britton) Small;革质唐松草■☆

388465　Thalictrum cultratum Wall. ;高原唐松草(草黄连,马尾黄连,马尾连,蚊子花);Highland Meadowrue,Plateau Meadowrue ■

388466　Thalictrum cultratum Wall. var. tsangense Brühl = Thalictrum squamiferum Lecoy. ■

388467　Thalictrum daisenense Nakai = Thalictrum aquilegiifolium L. var. sibiricum Regel et Tiling ■

388468　Thalictrum dalingo Buch. -Ham. ex DC. = Thalictrum foliolosum DC. ■

388469　Thalictrum dasycarpum Fisch. ,C. A. Mey. et Avé-Lall. ;粗果唐松草(粗果白蓬草,厚果唐松草,紫白蓬草,紫唐松草);Purple Meadow Rue,Purple Meadow-rue,Tall Meadow Rue,Tall Meadow-rue ■☆

388470　Thalictrum dasycarpum Fisch. , C. A. Mey. et Avé-Lall. f. hypoglaucum (Rydb.) Steyerm. = Thalictrum dasycarpum Fisch. , C. A. Mey. et Avé-Lall. var. hypoglaucum (Rydb.) B. Boivin ■☆

388471　Thalictrum dasycarpum Fisch. , C. A. Mey. et Avé-Lall. var. hypoglaucum (Rydb.) B. Boivin = Thalictrum dasycarpum Fisch. , C. A. Mey. et Avé-Lall. ■☆

388472　Thalictrum dasycarpum Fisch. , C. A. Mey. et Avé-Lall. var. hypoglaucum (Rydb.) B. Boivin;灰背粗果唐松草(灰绿粗果唐松草)■☆

388473　Thalictrum debile Buckley;柔软唐松草■☆

388474　Thalictrum debile Buckley var. texanum A. Gray = Thalictrum texanum (A. Gray) Small ■☆

388475　Thalictrum deciternatum B. Boivin = Thalictrum cultratum Wall. ■

388476　Thalictrum declinatum B. Boivin = Thalictrum acutifolium (Hand. -Mazz.) B. Boivin ■

388477　Thalictrum delavayi Franch. ;偏翅唐松草(大马尾连,马尾黄连,南马尾连,水黄连,土黄连);Chinese Meadow Rue, Chinese Meadow-rue,Delavay Meadowrue ■

388478　Thalictrum delavayi Franch. ' Hewitte's Double';休氏重瓣偏翅唐松草■☆

388479　Thalictrum delavayi Franch. f. appendiculatum W. T. Wang;德昌偏翅唐松草;Dechang Delavay Meadowrue ■

388480　Thalictrum delavayi Franch. var. acuminatum Franch. ;渐尖偏翅唐松草(具瓣南马尾黄连);Acuminate Delavay Meadowrue ■

388481　Thalictrum delavayi Franch. var. decorum Franch. ;宽萼偏翅唐松草(大花南马尾黄连,唐松草);Broadsepal Delavay Meadowrue ■

388482　Thalictrum delavayi Franch. var. mucronatum (Finet et Gagnep.) W. T. Wang et S. H. Wang;角药偏翅唐松草(锐尖南马尾黄连);Mucronate Delavay Meadowrue ■

388483　Thalictrum delavayi Franch. var. parviflorum Franch. = Thalictrum delavayi Franch. ■

388484　Thalictrum dichotomum Steud. = Thalictrum squarrosum Stephan ex Willd. ■

388485　Thalictrum diffusiflorum C. Marquand et Airy Shaw;堇花唐松草;Diffuseflower Meadowrue,Laxflower Meadowrue ■

388486　Thalictrum dioicum L. ;异株唐松草(异株白蓬草);Early Meadow Rue,Early Meadow-rue,Quicksilver Weed,Quicksilver-weed ■☆

388487　Thalictrum dioicum L. var. coriaceum Britton = Thalictrum coriaceum (Britton) Small ■☆

388488　Thalictrum dipterocarpum Franch. ;双翼果唐松草■☆

388489　Thalictrum dipterocarpum Franch. var. mucronatum Finet et Gagnep. = Thalictrum delavayi Franch. var. mucronatum (Finet et Gagnep.) W. T. Wang et S. H. Wang ■

388490　Thalictrum dipteroearpum Franch. = Thalictrum delavayi Franch. ■

388491　Thalictrum duclouxii H. Lév. = Thalictrum delavayi Franch. ■

388492　Thalictrum elegans Wall. ;小叶唐松草;Elegant Meadowrue, Smallleaf Meadowrue ■

388493　Thalictrum englerianum Ulbr. = Thalictrum virgatum Hook. f. et Thomson ■

388494　Thalictrum esquirolii H. Lév. et Vaniot = Thalictrum alpinum L. var. elatum Ulbr. ■

388495　Thalictrum faberi Ulbr. ;大叶唐松草(大叶马尾黄连,大叶马尾连,兰蓬草,蓝蓬草,马尾连);Bigleaf Meadowrue, Faber Meadowrue ■

388496　Thalictrum falcatum Pamp. = Thalictrum robustum Maxim. ■

388497　Thalictrum fargesii Franch. ex Finet et Gagnep. ;西南唐松草;Farges Meadowrue ■

388498　Thalictrum fauriei Hayata = Thalictrum urbainii Hayata ■

388499　Thalictrum fendleri Engelm. ex A. Gray;芬氏唐松草(芬氏白蓬草);Fendler Meadowrue,Fendler's Meadow-rue, Maid-of-the-mist ■

388500　Thalictrum fendleri Engelm. ex A. Gray var. platycarpum Trel. = Thalictrum fendleri Engelm. ex A. Gray ■

388501　Thalictrum fendleri Engelm. ex A. Gray var. polycarpum Torr. = Thalictrum polycarpum (Torr.) S. Watson ■☆

388502　Thalictrum fendleri Engelm. ex A. Gray var. wrightii (A. Gray) Trel. = Thalictrum fendleri Engelm. ex A. Gray ■

388503　Thalictrum filamentosum Maxim. ;花唐松草;Filamentary Meadowrue ■

388504　Thalictrum filipes Torr. et A. Gray = Thalictrum clavatum DC. ■☆

388505　Thalictrum finetii B. Boivin;滇川唐松草(滇川白蓬草,千里马);Finet Meadowrue ■

388506　Thalictrum flavum L. ;黄花唐松草(黄白蓬草,黄唐松草);Bastard Rhubarb,Common Meadow-rue,English Rhubarb,Fen Rue,Maidenhair Meadowrue,Meadow Rue,Meadow-rue,Paise Rhubarb,Poor Man's Rhubarb,Rue-weed,Small Meadowrue,Tassel Flower,Yellow Meadow Rue,Yellow Meadow-rue,Yellowflower Meadowrue ■

388507　Thalictrum flavum L. ' Illuminator';亮色黄花唐松草■☆

388508　Thalictrum flavum L. subsp. glaucum (DC.) Batt. = Thalictrum speciosissimum L. ■☆

388509　Thalictrum foeniculaceum Bunge;丝叶唐松草;Silk Meadowrue,Threadyleaf Meadowrue ■

388510　Thalictrum foetidum L. ;腺毛唐松草(臭唐松草,香毛唐松草,香唐松草);Glandularhairy Meadowrue ■

388511　Thalictrum foetidum L. var. glabrescens Takeda;扁果唐松草;Flatfruit Meadowrue ■

388512　Thalictrum foetidum L. var. glandulosissimum Finet et Gagnep. = Thalictrum glandulosissimum (Finet et Gagnep.) W. T. Wang et S. H. Wang ■

388513　Thalictrum foliolosum DC. ;多叶唐松草(贝加尔唐松草,草黄连,多叶白蓬草,金丝黄连,马尾黄连,马尾连,筛子花,水黄连,唐松草,土黄连,香唐松草,昭通唐松草);Manyleaf Meadowrue ■

388514 Thalictrum fortunei S. Moore;华东唐松草;Fortune Meadowrue ■

388515 Thalictrum fortunei S. Moore var. bulbiliferum B. Chen et X. J. Tian;珠芽华东唐松草■

388516 Thalictrum friesii Rupr. ;弗里斯唐松草■☆

388517 Thalictrum fusiforme W. T. Wang;纺锤唐松草■

388518 Thalictrum giraldii Ulbr. = Thalictrum baicalense Turcz. ■

388519 Thalictrum glandulosissimum (Finet et Gagnep.) W. T. Wang et S. H. Wang;金丝马尾连(多腺唐松草,马尾连);Goldenthread Meadowrue ■

388520 Thalictrum glandulosissimum (Finet et Gagnep.) W. T. Wang et S. H. Wang var. chaotungense W. T. Wang et S. H. Wang;昭通唐松草(昭通马尾连);Chaotum Meadowrue, Zhatong Meadowrue ■

388521 Thalictrum glareosum Hand. -Mazz. = Thalictrum squamiferum Lecoy. ■

388522 Thalictrum glaucum DC. = Thalictrum speciosissimum L. ■☆

388523 Thalictrum glaucum Desf. ;绿唐松草■☆

388524 Thalictrum glyphocarpum Wight et Arn. = Thalictrum javanicum Blume ■

388525 Thalictrum grandidentatum W. T. Wang et S. H. Wang;巨齿唐松草;Largedentate Meadowrue ■

388526 Thalictrum grandiflorum Maxim. ;大花唐松草;Largeflower Meadowrue ■

388527 Thalictrum gueguenii B. Boivin = Thalictrum umbricola Ulbr. ■

388528 Thalictrum hamatum Maxim. = Thalictrum uncatum Maxim. ■

388529 Thalictrum hayatanum Koidz. = Thalictrum urbainii Hayata ■

388530 Thalictrum heliophilum Wilken et DeMott;喜光唐松草■☆

388531 Thalictrum hepaticum Greene = Thalictrum revolutum DC. ■

388532 Thalictrum hernandeii Tausch. ;鹤氏唐松草■☆

388533 Thalictrum honanense W. T. Wang et S. H. Wang;河南唐松草(河南白蓬草);Henan Meadowrue, Honan Meadowrue ■

388534 Thalictrum hypoglaucum Rydb. = Thalictrum dasycarpum Fisch. ,C. A. Mey. et Avé-Lall. ■☆

388535 Thalictrum hypolecum Siebold et Zucc. = Thalictrum minus L. var. hypoleucum (Siebold et Zucc.) Miq. ■

388536 Thalictrum ichangense Lecoy. ex Oliv. ;盾叶唐松草(倒地拎,倒地掐,盾叶白蓬草,连钱草,龙眼草,石蒜还阳,水香草,小淫羊藿,岩扫把,羊耳,宜昌唐松草);Peltateleaf Meadowrue ■

388537 Thalictrum impexum B. Boivin = Thalictrum rhynchocarpum Quart. -Dill. et A. Rich. ex A. Rich. ■☆

388538 Thalictrum innitens B. Boivin = Thalictrum rhynchocarpum Quart. -Dill. et A. Rich. ex A. Rich. ■☆

388539 Thalictrum isopyroides C. A. Mey. ;紫堇叶唐松草(艾陶氏白蓬草,疏花唐松草);Corydalisleaf Meadowrue ■

388540 Thalictrum japonica Thunb. = Coptis japonica Makino ■☆

388541 Thalictrum javanicum Blume;爪哇唐松草(鬼见退,羊不食,玉山唐松草);Java Meadowrue,Javan Meadowrue ■

388542 Thalictrum javanicum Blume var. puberulum W. T. Wang;微毛爪哇唐松草;Puberulous Javan Meadowrue ■

388543 Thalictrum kemense Fr. = Thalictrum minus L. var. kemense (Fr.) Trel. ■

388544 Thalictrum kemense Fr. var. stipellatum C. A. Mey. = Thalictrum minus L. var. stipellatum (C. A. Mey.) Tamura ■

388545 Thalictrum kemense Fr. var. stipellatum C. A. Mey. ex Maxim. = Thalictrum minus L. var. kemense (Fr.) Trel. ■

388546 Thalictrum kiusianum Nakai;九州唐松草■☆

388547 Thalictrum lancangense Y. Y. Qian;澜沧唐松草;Lancang Meadowrue ■

388548 Thalictrum laxum Ulbr. ;疏序唐松草;Lax Meadowrue, Laxinflorescence Meadowrue ■

388549 Thalictrum lecoyeri Franch. ;微毛唐松草■

388550 Thalictrum lecoyeri Franch. = Thalictrum javanicum Blume ■

388551 Thalictrum leptophyllum F. Nyl. ;细叶唐松草■☆

388552 Thalictrum leuconotum Franch. ;白茎唐松草(长柱唐松草);Chinese Longstigma Meadowrue,Whitestem Meadowrue ■

388553 Thalictrum leve (Franch.) W. T. Wang;鹤庆唐松草;Heqing Meadowrue,Smooth Scabrousleaf Meadowrue ■

388554 Thalictrum longipedunculatum Hochst. ;长梗唐松草(长花梗白蓬草);Longstalk Meadowrue ■☆

388555 Thalictrum longistylum DC. ;长柱唐松草(长花柱白蓬草,长茎唐松草);Longstyle Meadowrue ■☆

388556 Thalictrum lucidum L. ;光亮唐松草(光泽白蓬草,狭叶唐松草);Shining Meadowrue ■☆

388557 Thalictrum macrocarpum Gren. ;大果唐松草■☆

388558 Thalictrum macrophyllum Migo = Thalictrum faberi Ulbr. ■

388559 Thalictrum macrorhynchum Franch. ;长喙唐松草(马尾莲,天星);Longbeak Meadowrue ■

388560 Thalictrum macrostigma Finet et Gagnep. = Thalictrum leuconotum Franch. ■

388561 Thalictrum macrostigma Lecoy. = Thalictrum virgatum Hook. f. et Thomson ■

388562 Thalictrum macrostylum Small et A. Heller;大柱唐松草■☆

388563 Thalictrum mairei H. Lév. = Thalictrum leuconotum Franch. ■

388564 Thalictrum majus Crantz;大唐松草;Greater Meadow Rue ■☆

388565 Thalictrum mannii Hutch. = Thalictrum rhynchocarpum Quart. -Dill. et A. Rich. ex A. Rich. ■☆

388566 Thalictrum maxwellii Royle = Thalictrum minus L. subsp. maxwellii (Royle) Hand ■☆

388567 Thalictrum megalostigma (B. Boivin) W. T. Wang = Thalictrum baicalense Turcz. var. megalostigma Boivin ■

388568 Thalictrum menthosma Stocks ex Lecoy. = Thalictrum reniforme Wall. ■

388569 Thalictrum microgynum Lecoy. ex Oliv. ;小果唐松草(飞蛾七,狗尾升麻,老虎香,石黄草,石笋一枝花,铁大乌草,铁线还阳,血经草,岩风七,雨点草);Smallfruit Meadowrue ■

388570 Thalictrum microphyllum Royle = Thalictrum alpinum L. var. microphyllum (Royle) Hand. -Mazz. ■

388571 Thalictrum microphyllum Royle = Thalictrum alpinum L. ■

388572 Thalictrum minus L. ;亚欧唐松草(欧亚唐松草,腾唐松草,小白蓬草,小金氏,小唐松草,新疆唐松草);Lesser Meadow Rue,Lesser Meadow-rue, Low Meadow Rue, Low Meadowrue, Meadow Rue,Small Meadow Rue ■

388573 Thalictrum minus L. 'Adiantifolium';矮小亚欧唐松草(铁线蕨叶唐松草);Dwarf Meadow Rue,Maidenhair ■☆

388574 Thalictrum minus L. subsp. elatum (Jacq.) Kerguélen;秋唐松草■☆

388575 Thalictrum minus L. subsp. kemense (Fr.) Cajander = Thalictrum minus L. var. kemense (Fr.) Trel. ■

388576 Thalictrum minus L. subsp. kemense (Fr.) Hultén = Thalictrum minus L. ■

388577 Thalictrum minus L. subsp. maxwellii (Royle) Hand;马氏亚欧唐松草■☆

388578 Thalictrum minus L. subsp. pubescens Arcang. ;毛亚欧唐松草■☆

388579 Thalictrum minus L. subsp. saxatile Hook. f. ;岩地亚欧唐松草■☆

388580 Thalictrum minus L. var. adiantifolium ? = Thalictrum minus L.

'Adiantifolium' ■☆

388581　Thalictrum minus L. var. amplissimum（H. Lév. et Vaniot）H. Lév. = Thalictrum minus L. var. hypoleucum（Siebold et Zucc.）Miq. ■

388582　Thalictrum minus L. var. elatum Lecoy. = Thalictrum minus L. var. hypoleucum（Siebold et Zucc.）Miq. ■

388583　Thalictrum minus L. var. hypoleucum（Siebold et Zucc.）Miq. ;东亚唐松草(佛爷指甲,金鸡脚下黄,马尾连,穷汉子腿,秋唐松草,腾唐松草,小果白蓬草,小金花,匈牙利小白蓬草,烟锅草); East-Asia Low Meadowrue ■

388584　Thalictrum minus L. var. kemense（Fr.）Trel. ;长梗亚欧唐松草(长梗欧亚唐松草);Longstalk Low Meadowrue ■

388585　Thalictrum minus L. var. majus Miq. = Thalictrum minus L. var. hypoleucum（Siebold et Zucc.）Miq. ■

388586　Thalictrum minus L. var. scabrivena Oliv. = Thalictrum minus L. subsp. maxwellii（Royle）Hand ■☆

388587　Thalictrum minus L. var. schimperianum（Hochst. ex Schweinf.）Chiov. = Thalictrum minus L. subsp. maxwellii（Royle）Hand ■☆

388588　Thalictrum minus L. var. stipellatum（C. A. Mey. ex Maxim.）Tamura = Thalictrum minus L. var. kemense（Fr.）Trel. ■

388589　Thalictrum minus L. var. stipellatum（C. A. Mey. ex Maxim.）Tamura = Thalictrum minus L. ■

388590　Thalictrum minus L. var. stipellatum（C. A. Mey.）Tamura = Thalictrum minus L. var. kemense（Fr.）Trel. ■

388591　Thalictrum minus L. var. stipellatum（C. A. Mey.）Tamura = Thalictrum minus L. ■

388592　Thalictrum minus L. var. thunbergii（DC.）Vorosch. = Thalictrum minus L. var. hypoleucum（Siebold et Zucc.）Miq. ■

388593　Thalictrum mirabile Small;奇异唐松草■☆

388594　Thalictrum morii Hayata;楔叶唐松草(森氏唐松草); Cuneateleaf Meadowrue ■

388595　Thalictrum moseleyi Greene = Thalictrum revolutum DC. ■

388596　Thalictrum multipeltatum（Pamp.）Pamp. = Thalictrum ichangense Lecoy. ex Oliv. ■

388597　Thalictrum multipeltatum Pamp. = Thalictrum ichangense Lecoy. ex Oliv. ■

388598　Thalictrum myriophyllum Ohwi;密叶唐松草(小叶唐松草); Denseleaf Meadowrue ■

388599　Thalictrum neurocarpum Royle = Thalictrum reniforme Wall. ■

388600　Thalictrum nudicaule Schwein. ex Torr. et A. Gray = Thalictrum clavatum DC. ■☆

388601　Thalictrum nudum H. Lév. et Vaniot ex Hand.-Mazz. = Thalictrum alpinum L. var. elatum Ulbr. ■

388602　Thalictrum occidentale A. Gray;西方唐松草;Western Meadow-rue ■☆

388603　Thalictrum occidentale A. Gray var. macounii B. Boivin = Thalictrum occidentale A. Gray ■☆

388604　Thalictrum occidentale A. Gray var. palousense H. St. John = Thalictrum occidentale A. Gray ■☆

388605　Thalictrum oligandrum Maxim. ;稀蕊唐松草;Fewstamen Meadowrue ■

388606　Thalictrum oligospermum Fisch. ex Sweet = Thalictrum squarrosum Stephan ex Willd. ■

388607　Thalictrum omeiense W. T. Wang et S. H. Wang;峨眉唐松草(倒水莲,黄芩,金鸡尾,野海棠);Emei Meadowrue, Omei Meadowrue ■

388608　Thalictrum orientale Boiss. ;东方唐松草■☆

388609　Thalictrum oshimae Masam. ;大岛唐松草(大岛氏铁大乌,大唐松草);Oshima Meadowrue ■

388610　Thalictrum osmundifolium Finet et Gagnep. ;川鄂唐松草; Royalfernleaf Meadowrue ■

388611　Thalictrum pallidum Franch. = Thalictrum fargesii Franch. ex Finet et Gagnep. ■

388612　Thalictrum pallidum Franch. = Thalictrum xingshanicum G. F. Tao ■

388613　Thalictrum pedunculatum Edgew. ;花梗唐松草(总花梗白蓬草);Penduculatum Meadowrue ■☆

388614　Thalictrum petaloideum L. ;瓣蕊唐松草(花唐松草,马尾黄连,肾叶唐松草);Petalformed Meadowrue ■

388615　Thalictrum petaloideum L. var. latifoliolatum Kitag. ;宽叶瓣蕊唐松草■

388616　Thalictrum petaloideum L. var. latifoliolatum Kitag. = Thalictrum petaloideum L. ■

388617　Thalictrum petaloideum L. var. supradecompositum（Nakai）Kitag. ;狭裂瓣蕊唐松草(卷叶唐松草,蒙古唐松草);Narrowcleft Petalformed Meadowrue ■

388618　Thalictrum philippinense C. B. Rob. ;菲律宾唐松草;Philippine Meadowrue ■

388619　Thalictrum podocarpum Kunth = Thalictrum polycarpum（Torr.）S. Watson ■☆

388620　Thalictrum polycarpum（Torr.）S. Watson;多果唐松草(柄果白蓬草,柄果唐松草);Tall Western Meadow-rue ■☆

388621　Thalictrum polygamum Muhl. ;杂性唐松草(高白蓬草,高唐松草,杂性白蓬草);Fall Meadow Rue, Fall Meadow-rue, King of the Meadow, Tall Meadow-rue ■

388622　Thalictrum polygamum Muhl. ex Spreng. = Thalictrum polygamum Muhl. ■

388623　Thalictrum polygamum Muhl. ex Spreng. = Thalictrum pubescens Pursh ■☆

388624　Thalictrum polygamum Muhl. ex Spreng. var. hebecarpum Fernald = Thalictrum pubescens Pursh ■☆

388625　Thalictrum polygamum Muhl. ex Spreng. var. intermedium B. Boivin = Thalictrum pubescens Pursh ■☆

388626　Thalictrum polygamum Muhl. ex Spreng. var. pubescens（Pursh）K. C. Davis = Thalictrum pubescens Pursh ■☆

388627　Thalictrum przewalskii Maxim. ;长柄唐松草(长柄花唐松草,莪正,甘青唐松草,拟散花唐松草,散花唐松草,直梗唐松草); Longstalk Meadowrue, Przewalski ■

388628　Thalictrum pubescens Pursh;高大唐松草;King-of-the-meadow, Late Meadow-rue, Meadow-weed, Muskrat-weed, Tall Meadow-rue ■☆

388629　Thalictrum pubescens Pursh var. hebecarpum（Fernald）B. Boivin = Thalictrum pubescens Pursh ■☆

388630　Thalictrum purdomii Clark = Thalictrum minus L. var. hypoleucum（Siebold et Zucc.）Miq. ■

388631　Thalictrum radiatum Royle = Thalictrum saniculiforme DC. ■

388632　Thalictrum ramosum B. Boivin;多枝唐松草(马尾连,软杆子,软杆子水黄连,软水黄连,水黄连,土黄连);Manybranch Meadowrue ■

388633　Thalictrum rariflorum Fr. ;稀花唐松草■☆

388634　Thalictrum reniforme Wall. ;美丽唐松草(鹅整);Beautiful Meadowrue, Shortglandular Meadowrue ■

388635　Thalictrum repens Schrad. = Thalictrum squarrosum Stephan ex Willd. ■

388636   Thalictrum reticulatum Franch.；网脉唐松草（草黄连，马尾黄连）；Reticulate Meadowrue ■

388637   Thalictrum reticulatum Franch. var. hirtellum W. T. Wang et S. H. Wang；毛叶网脉唐松草；Hairyleaf Reticulate Meadowrue ■

388638   Thalictrum revolutum DC.；外卷唐松草（外卷白蓬草）；Meadow Rue, Purple Meadow-rue, Revolute Meadowrue, Skunk Meadow-rue, Wax-leaf Meadow-rue, Wax-leaved Meadow Rue, Wax-leaved Meadow-rue, Waxy Meadow Rue, Waxy Meadow-rue ■

388639   Thalictrum revolutum DC. var. glandulosior B. Boivin = Thalictrum revolutum DC. ■

388640   Thalictrum rhynchocarpum Quart.-Dill. et A. Rich. = Thalictrum rhynchocarpum Quart.-Dill. et A. Rich. ex A. Rich. ■☆

388641   Thalictrum rhynchocarpum Quart.-Dill. et A. Rich. ex A. Rich.；喙果唐松草（喙果白蓬草）■☆

388642   Thalictrum rhynchocarpum Quart.-Dill. et A. Rich. ex A. Rich. subsp. ruwenzoriense Lye；鲁文佐里唐松草■☆

388643   Thalictrum richardsonii A. Gray = Thalictrum sparsiflorum Turcz. ex Fisch. et C. A. Mey. ■

388644   Thalictrum robustum Maxim.；粗壮唐松草；Strong Meadowrue ■

388645   Thalictrum rochebrunianum Franch. et Sav.；罗氏唐松草（罗蔡白蓬草，日本唐松草，紫金唐松，紫锦唐松草）■☆

388646   Thalictrum rockii Boivin = Thalictrum przewalskii Maxim. ■

388647   Thalictrum rostellatum Hook. f. et Thomson；小喙唐松草；Smallbeak Meadowrue ■

388648   Thalictrum rotundifolium DC.；圆叶唐松草；Roundleaf Meadowrue ■

388649   Thalictrum rubellum Siebold et Zucc. = Thalictrum aquilegiifolium L. var. sibiricum Regel et Tiling ■

388650   Thalictrum rubescens Ohwi；淡红唐松草（南湖唐松草，南湖铁大乌）；Pink Meadowrue, Redish Meadowrue ■

388651   Thalictrum rugosum Aiton；绉唐松草（皱纹白蓬草）；Rough Meadowrue ■

388652   Thalictrum rupestre Madden ex Lecoy. = Thalictrum saniculiforme DC. ■

388653   Thalictrum rutifolium Hook. f. et Thomson；芸香叶唐松草；Rueleaf Meadowrue ■

388654   Thalictrum sachalinense Lecoy.；库页岛唐松草（库页岛白蓬草）；Sachalin Meadowrue ■☆

388655   Thalictrum samariferum B. Boivin = Thalictrum elegans Wall. ■

388656   Thalictrum saniculiforme DC.；叉枝唐松草（黄连，马尾黄连）；Forkshoot Meadowrue ■

388657   Thalictrum scabrifolium Franch.；糙叶唐松草；Scabrifolious Meadowrue, Scabrousleaf Meadowrue ■

388658   Thalictrum scabrifolium Franch. var. leve Franch. = Thalictrum leve (Franch.) W. T. Wang ■

388659   Thalictrum scaposum W. E. Evans = Thalictrum microgynum Lecoy. ex Oliv. ■

388660   Thalictrum schimperianum Hochst. ex Schweinf. = Thalictrum minus L. subsp. maxwellii (Royle) Hand ■☆

388661   Thalictrum semiscandens W. W. Sm.；攀缘唐松草（水风藤）■

388662   Thalictrum sessile Hayata = Thalictrum javanicum Blume ■

388663   Thalictrum setulosinerve H. Hara = Thalictrum alpinum L. var. elatum Ulbr. ■

388664   Thalictrum shensiense W. T. Wang et S. H. Wang；陕西唐松草；Shaanxi Meadowrue, Shensi Meadowrue ■

388665   Thalictrum sibiricum Ledeb. = Thalictrum squarrosum Stephan ex Willd. ■

388666   Thalictrum simaoense W. T. Wang et G. Zhu；思茅唐松草；Simao Meadowrue ■

388667   Thalictrum simplex L.；箭头唐松草（单枝白蓬草，黄脚鸡，黄绿唐松草，黄唐松草，箭头白蓬草，金鸡脚下黄，水黄连，硬杆水黄连，硬水黄连）；Arrowhead Meadowrue, Slim-top Meadow Rue, Slimtop Meadowrue, Slim-top Meadow-rue ■

388668   Thalictrum simplex L. var. affine (Ledeb.) Regel；锐裂箭头唐松草（箭头白蓬草）；Sharpsplit Slimtop Meadowrue ■

388669   Thalictrum simplex L. var. brevipes Hara；短梗箭头唐松草（黄脚鸡，箭头唐松草，金鸡脚下黄，水黄连，硬秆水黄连，硬秆子水黄连，硬水黄连）；Shortstalk Slimtop Meadowrue ■

388670   Thalictrum simplex L. var. glandulosum W. T. Wang；腺毛箭头唐松草；Glandular Slimtop Meadowrue ■

388671   Thalictrum sinomacrostigma W. T. Wang = Thalictrum leuconotum Franch. ■

388672   Thalictrum smithii B. Boivin；鞭柱唐松草（水黄连）；Smith Meadowrue, Whipstyle Meadowrue ■

388673   Thalictrum sparsiflorum Turcz. ex Fisch. et C. A. Mey.；散花唐松草（散花白蓬草）；Few-flowered Meadow-rue, Flat-fruited Meadow Rue, Looseflower Meadowrue, Mountain Meadow-rue ■

388674   Thalictrum sparsiflorum Turcz. ex Fisch. et C. A. Mey. subsp. richardsonii (A. Gray) Cody = Thalictrum sparsiflorum Turcz. ex Fisch. et C. A. Mey. ■

388675   Thalictrum sparsiflorum Turcz. ex Fisch. et C. A. Mey. var. nevadense B. Boivin = Thalictrum sparsiflorum Turcz. ex Fisch. et C. A. Mey. ■

388676   Thalictrum sparsiflorum Turcz. ex Fisch. et C. A. Mey. var. richardsonii (A. Gray) B. Boivin = Thalictrum sparsiflorum Turcz. ex Fisch. et C. A. Mey. ■

388677   Thalictrum sparsiflorum Turcz. ex Fisch. et C. A. Mey. var. saximontanum B. Boivin = Thalictrum sparsiflorum Turcz. ex Fisch. et C. A. Mey. ■

388678   Thalictrum speciosissimum L.；极美唐松草■☆

388679   Thalictrum squamiferum Lecoy.；石砾唐松草（展枝唐松草）；Gravelly Meadowrue ■

388680   Thalictrum squarrosum Stephan ex Willd.；展枝唐松草（叉枝唐松草，坚唐松草，猫爪子，歧序唐松草，展枝白蓬草）；Nodding Meadowrue ■

388681   Thalictrum steeleanum B. Boivin = Thalictrum coriaceum (Britton) Small ■☆

388682   Thalictrum stolzii Ulbr.；斯托尔兹唐松草■☆

388683   Thalictrum strictum Ledeb.；直立唐松草（直立白蓬草）；Erect Meadowrue ■☆

388684   Thalictrum subrotundum B. Boivin = Thalictrum macrostylum Small et A. Heller ■☆

388685   Thalictrum sultanabadense Stapf；波斯唐松草■☆

388686   Thalictrum supradecompositum Nakai = Thalictrum petaloideum L. var. supradecompositum (Nakai) Kitag. ■

388687   Thalictrum tenii H. Lév. = Thalictrum trichopus Franch. ■

388688   Thalictrum tenue Franch.；细唐松草；Tender Meadowrue ■

388689   Thalictrum tenuisubulatum W. T. Wang；钻柱唐松草■

388690   Thalictrum texanum (A. Gray) Small；得州唐松草■☆

388691   Thalictrum thalictroides (L.) A. J. Eames et B. Boivin = Anemonella thalictroides (L.) Spach ■☆

388692   Thalictrum thubergii DC. var. majus (Miq.) Nakai = Thalictrum minus L. var. hypoleucum (Siebold et Zucc.) Miq. ■

388693   Thalictrum thunbergii DC. = Thalictrum minus L. var.

hypoleucum（Siebold et Zucc.）Miq. ■

388694　Thalictrum thunbergii DC. = Thalictrum minus L. var. kemense（Fr.）Trel. ■

388695　Thalictrum thunbergii DC. var. majus（Miq.）Nakai = Thalictrum minus L. var. hypoleucum（Siebold et Zucc.）Miq. ■

388696　Thalictrum tofieldioides Diels = Thalictrum alpinum L. var. elatum Ulbr. ■

388697　Thalictrum trichopus Franch.；毛发唐松草（多发唐松草，马尾黄连，水黄连，珍珠莲）；Slenderpedicel Meadowrue ■

388698　Thalictrum trigynum Fisch. ex DC. = Thalictrum squarrosum Stephan ex Willd. ■

388699　Thalictrum trigynum Fisch. ex Trevir. = Thalictrum squarrosum Stephan ex Willd. ■

388700　Thalictrum tripeltatum Maxim. = Thalictrum ichangense Lecoy. ex Oliv. ■

388701　Thalictrum triternatum Rupr.；三出唐松草■☆

388702　Thalictrum tsawarungense W. T. Wang et S. H. Wang；察瓦龙唐松草；Chawalong Meadowrue，Tsawa Meadowrue ■

388703　Thalictrum tuberiferum Maxim.；深山唐松草（深山白蓬草）；Montane Meadowrue，Remote Mountains Meadowrue ■

388704　Thalictrum turneri B. Boivin = Thalictrum confine Fernald ■☆

388705　Thalictrum umbricola Ulbr.；阴地唐松草；Shade Meadowrue ■

388706　Thalictrum uncatum Maxim.；钩柱唐松草（弩箭药）；Hookedstyle Meadowrue ■

388707　Thalictrum uncatum Maxim. var. angustialatum W. T. Wang；狭翅钩柱唐松草；Narrowwing Hookedstyle Meadowrue ■

388708　Thalictrum uncinulatum Franch. ex Lecoy.；弯柱唐松草；Curvedstyle Meadowrue ■

388709　Thalictrum unguiculatum B. Boivin = Thalictrum acutifolium（Hand.-Mazz.）B. Boivin ■

388710　Thalictrum urbainii Hayata；台湾唐松草（傅氏唐松草，台湾白蓬草，台湾铁大乌）；Taiwan Meadowrue ■

388711　Thalictrum urbainii Hayata var. majus T. Shimizu；大花台湾唐松草（大花傅氏唐松草）；Big Taiwan Meadowrue ■

388712　Thalictrum venulosum Trel.；北部唐松草；Early Meadow-rue，Northern Meadow-rue，Veined Meadow-rue，Veiny meadow-rue ■☆

388713　Thalictrum venulosum Trel. var. confine（Fernald）B. Boivin = Thalictrum confine Fernald ■☆

388714　Thalictrum venulosum Trel. var. lunellii（Greene）B. Boivin = Thalictrum venulosum Trel. ■☆

388715　Thalictrum verticillatum H. Lév. = Thalictrum virgatum Hook. f. et Thomson ■

388716　Thalictrum virgatum Hook. f. et Thomson；帚枝唐松草（惊风草，阴阳和，帚状唐松草）；Broom Meadowrue，Virgate Meadowrue ■

388717　Thalictrum virgatum Hook. f. et Thomson var. stipitatum Franch. = Thalictrum virgatum Hook. f. et Thomson ■

388718　Thalictrum viscosum W. T. Wang et S. H. Wang；黏唐松草；Viscous Meadowrue ■

388719　Thalictrum wangii B. Boivin；丽江唐松草；Lijiang Meadowrue，Wang Meadowrue ■

388720　Thalictrum wuyishanicum W. T. Wang et S. H. Wang；武夷唐松草；Wuyi Mountain Meadowrue，Wuyishan Meadowrue ■

388721　Thalictrum xingshanicum G. F. Tao；兴山唐松草；Xingshan Meadowrue ■

388722　Thalictrum yezoense Nakai；北海道唐松草；Yezo Meadowrue ■☆

388723　Thalictrum yui B. Boivin = Thalictrum cultratum Wall. ■

388724　Thalictrum yunnanense W. T. Wang；云南唐松草；Yunnan Meadowrue ■

388725　Thalictrum yunnanense W. T. Wang var. austro-yunnanense Y. Y. Qian；滇南唐松草；S. Yunnan Meadowrue ■

388726　Thalictrum zernyi Ulbr.；策尼唐松草■☆

388727　Thalliana Steud. = Colubrina Rich. ex Brongn.（保留属名）●

388728　Thalliana Steud. = Tralliana Lour. ●

388729　Thalysia Kuntze = Zea L. ●

388730　Thaminophyllum Harv.（1865）；帚粉菊属■☆

388731　Thaminophyllum latifolium Bond；宽叶帚粉菊■☆

388732　Thaminophyllum multiflorum Harv.；多花帚粉菊■☆

388733　Thaminophyllum mundii Harv.；帚粉菊■☆

388734　Thamnea Sol. ex Brongn.（1826）（保留属名）；鳞叶灌属●☆

388735　Thamnea Sol. ex R. Br. = Thamnea Sol. ex Brongn.（保留属名）●☆

388736　Thamnea depressa Oliv.；凹陷鳞叶灌●☆

388737　Thamnea diosmoides Oliv.；逸香鳞叶灌●☆

388738　Thamnea gracilis Oliv.；纤细鳞叶灌●☆

388739　Thamnea hirtella Oliv.；多毛鳞叶灌●☆

388740　Thamnea massoniana Dümmer；马森鳞叶灌●☆

388741　Thamnea thesioides Dümmer；百蕊鳞叶灌●☆

388742　Thamnea uniflora Sol. ex Brongn.；单花鳞叶灌●☆

388743　Thamnia P. Browne（废弃属名）= Laetia Loefl. ex L.（保留属名）●☆

388744　Thamnia P. Browne（废弃属名）= Thamnea Sol. ex Brongn.（保留属名）●☆

388745　Thamnium Klotzsch = Scyphogyne Brongn. ●☆

388746　Thamnocalamus Munro = Fargesia Franch. ●

388747　Thamnocalamus Munro（1868）；筱竹属（法氏竹属，华橘竹属）；Shrubbamboo，Umbrella Bamboo，Umbrella-bamboo ●

388748　Thamnocalamus aristatus（Gamble）E. G. Camus；有芒筱竹（筱竹）；Aristate Shrubbamboo，Aristate Umbrella Bamboo，Awned Umbrella-bamboo ●

388749　Thamnocalamus aristatus（Gamble）E. G. Camus = Thamnocalamus spathiflorus（Trin.）Munro ●

388750　Thamnocalamus collaris（T. P. Yi）T. P. Yi = Himalayacalamus collaris（T. P. Yi）Ohrnb. ●

388751　Thamnocalamus collaris T. P. Yi = Fargesia collaris T. P. Yi ●

388752　Thamnocalamus cuspidatus（Keng）P. C. Keng = Fargesia cuspidata（Keng）Z. P. Wang et G. H. Ye ●

388753　Thamnocalamus falconeri Munro = Himalayacalamus falconeri（Munro）P. C. Keng ●

388754　Thamnocalamus hindsii（Munro）E. G. Camus = Pseudosasa hindsii（Munro）S. L. Chen et G. Y. Sheng ex T. G. Liang ●

388755　Thamnocalamus hindsii（Munro）E. G. Camus var. graminea E. G. Camus = Pleioblastus gramineus（Bean）Nakai ●

388756　Thamnocalamus murielae（Gamble）Demoly = Fargesia murielae（Gamble）T. P. Yi ●

388757　Thamnocalamus prainii（Gamble）E. G. Camus = Neomicrocalamus prainii（Gamble）P. C. Keng ●

388758　Thamnocalamus prainii（Gamble）E. G. Camus = Racemobambos prainii（Gamble）P. C. Keng et T. H. Wen ●

388759　Thamnocalamus sparsiflorus（Rendle）P. C. Keng = Fargesia murielae（Gamble）T. P. Yi ●

388760　Thamnocalamus spathaceus（Franch.）Soderstr. = Fargesia spathacea Franch. ●

388761　Thamnocalamus spathiflorus（Trin.）Munro；筱竹；Muriel Bamboo，Umbrella Bamboo ●

388762　Thamnocalamus spathiflorus（Trin.）Munro var. crassinodus

（T. P. Yi）Stapleton;粗节筱竹●

388763　Thamnocalamus tessellatus（Nees）Soderstr. et R. P. Ellis;方格筱竹●☆

388764　Thamnocalamus unispiculatus T. P. Yi et J. Y. Shi;单穗筱竹●

388765　Thamnocharis W. T. Wang（1981）;辐花苣苔属（幅花苣苔属）;Thamnocharis■★

388766　Thamnocharis esquirolii（H. Lév.）W. T. Wang;辐花苣苔;Thamnocharis■

388767　Thamnochordus Kuntze = Thamnochortus P. J. Bergius■☆

388768　Thamnochortus P. J. Bergius（1767）;灌木帚灯草属●☆

388769　Thamnochortus acuminatus Pillans;渐尖灌木帚灯草●☆

388770　Thamnochortus aemulus Kunth = Staberoha aemula（Kunth）Pillans■☆

388771　Thamnochortus amoena H. P. Linder;秀丽灌木帚灯草●☆

388772　Thamnochortus arenarius Esterh.;沙地灌木帚灯草●☆

388773　Thamnochortus argenteus（Thunb.）Kunth = Hypodiscus argenteus（Thunb.）Mast.■☆

388774　Thamnochortus argenteus Pillans = Thamnochortus cinereus H. P. Linder●☆

388775　Thamnochortus bachmannii Mast.;巴克曼灌木帚灯草●☆

388776　Thamnochortus bromoides Kunth = Thamnochortus lucens（Poir.）H. P. Linder●☆

388777　Thamnochortus burchellii Mast. = Thamnochortus erectus（Thunb.）Mast.●☆

388778　Thamnochortus canescens Mast. = Thamnochortus dumosus Mast.●☆

388779　Thamnochortus caricinus Mast. = Thamnochortus erectus（Thunb.）Mast.●☆

388780　Thamnochortus cernuus（L. f.）Kunth = Staberoha cernua（L. f.）T. Durand et Schinz■☆

388781　Thamnochortus cinereus H. P. Linder;灰色灌木帚灯草●☆

388782　Thamnochortus comptonii Pillans = Thamnochortus platypteris Kunth●☆

388783　Thamnochortus consanguineus Kunth = Thamnochortus nutans（Thunb.）Pillans●☆

388784　Thamnochortus dichotomus（L.）Spreng. = Ischyrolepis capensis（L.）H. P. Linder■☆

388785　Thamnochortus dichotomus Mast. = Thamnochortus lucens（Poir.）H. P. Linder●☆

388786　Thamnochortus dichotomus Mast. var. hyalinus Pillans = Thamnochortus lucens（Poir.）H. P. Linder●☆

388787　Thamnochortus distichus（Rottb.）Mast. = Restio distichus Rottb.■☆

388788　Thamnochortus dumosus Mast.;灌丛帚灯草●☆

388789　Thamnochortus ecklonianus Kunth = Thamnochortus lucens（Poir.）H. P. Linder●☆

388790　Thamnochortus ellipticus Pillans;椭圆灌木帚灯草●☆

388791　Thamnochortus elongatus（Thunb.）Mast. = Thamnochortus erectus（Thunb.）Mast.●☆

388792　Thamnochortus erectus（Thunb.）Mast.;直立灌木帚灯草●☆

388793　Thamnochortus floribundus Kunth = Thamnochortus erectus（Thunb.）Mast.●☆

388794　Thamnochortus fraternus Pillans;兄弟帚灯草●☆

388795　Thamnochortus fruticosus P. J. Bergius;灌木帚灯草●☆

388796　Thamnochortus fruticosus P. J. Bergius var. glaber Mast. = Thamnochortus glaber（Mast.）Pillans●☆

388797　Thamnochortus giganteus Kunth = Rhodocoma gigantea（Kunth）H. P. Linder■☆

388798　Thamnochortus glaber（Mast.）Pillans;光滑灌木帚灯草●☆

388799　Thamnochortus gracilis Mast.;纤细灌木帚灯草●☆

388800　Thamnochortus guthrieae Pillans;格斯里灌木帚灯草●☆

388801　Thamnochortus imbricatus（Thunb.）Mast. = Staberoha distachyos（Rottb.）Kunth■☆

388802　Thamnochortus imbricatus（Thunb.）Mast. var. stenopterus（Kunth）Mast. = Staberoha aemula（Kunth）Pillans■☆

388803　Thamnochortus insignis Mast.;显著灌木帚灯草●☆

388804　Thamnochortus karooica H. P. Linder;卡鲁灌木帚灯草●☆

388805　Thamnochortus levynsiae Pillans;勒温斯灌木帚灯草●☆

388806　Thamnochortus lewisiae Pillans = Thamnochortus guthrieae Pillans●☆

388807　Thamnochortus lucens（Poir.）H. P. Linder;光亮灌木帚灯草●☆

388808　Thamnochortus maximus Kuntze = Thamnochortus spicigerus（Thunb.）Spreng.●☆

388809　Thamnochortus membranaceus Mast. = Restio distichus Rottb.■☆

388810　Thamnochortus micans（Nees）Kunth = Restio micans Nees■☆

388811　Thamnochortus modestus Kunth = Rhodocoma fruticosa（Thunb.）H. P. Linder■☆

388812　Thamnochortus muirii Pillans;缪里灌木帚灯草●☆

388813　Thamnochortus muticus Pillans = Thamnochortus sporadicus Pillans●☆

388814　Thamnochortus nervosus Pillans = Thamnochortus guthrieae Pillans●☆

388815　Thamnochortus nutans（Thunb.）Pillans;俯垂灌木帚灯草●☆

388816　Thamnochortus obtusus Pillans;钝灌木帚灯草●☆

388817　Thamnochortus occultus Mast. = Restio occultus（Mast.）Pillans■☆

388818　Thamnochortus paniculatus Mast.;圆锥灌木帚灯草●☆

388819　Thamnochortus papillosus Pillans = Thamnochortus lucens（Poir.）H. P. Linder●☆

388820　Thamnochortus papyraceus Pillans;纸质灌木帚灯草●☆

388821　Thamnochortus pellucidus Pillans;透明灌木帚灯草●☆

388822　Thamnochortus piketbergensis Pillans = Thamnochortus sporadicus Pillans●☆

388823　Thamnochortus platypteris Kunth;阔翅灌木帚灯草●☆

388824　Thamnochortus plumosus Pillans = Thamnochortus guthrieae Pillans●☆

388825　Thamnochortus pluristachyus Mast.;多穗灌木帚灯草●☆

388826　Thamnochortus pulcher Pillans;美丽灌木帚灯草●☆

388827　Thamnochortus punctatus Pillans;斑点灌木帚灯草●☆

388828　Thamnochortus rigidus Esterh.;硬灌木帚灯草●☆

388829　Thamnochortus robustus Kunth = Cannomois virgata（Rottb.）Steud.■☆

388830　Thamnochortus scabridus Pillans;微糙灌木帚灯草●☆

388831　Thamnochortus scariosus（Thunb.）Spreng. = Thamnochortus fruticosus P. J. Bergius●☆

388832　Thamnochortus schlechteri Pillans;施莱灌木帚灯草●☆

388833　Thamnochortus scirpiformis Mast. = Thamnochortus erectus（Thunb.）Mast.●☆

388834　Thamnochortus scirpoides Kunth = Cannomois scirpoides（Kunth）Mast.■☆

388835　Thamnochortus similis Pillans = Thamnochortus sporadicus Pillans●☆

388836　Thamnochortus spicigerus（Thunb.）Spreng.;穗花灌木帚灯草●☆

388837　Thamnochortus sporadicus Pillans;散布灌木帚灯草●☆

388838　Thamnochortus stokoei Pillans;斯托克灌木帚灯草●☆

388839　Thamnochortus striatus Hochst. = Thamnochortus spicigerus（Thunb.）Spreng. ●☆

388840　Thamnochortus strictus Kunth = Cannomois parviflora（Thunb.）Pillans ■☆

388841　Thamnochortus sulcatus Mast. = Thamnochortus bachmannii Mast. ●☆

388842　Thamnochortus umbellatus（Thunb.）Kunth = Staberoha distachyos（Rottb.）Kunth ■☆

388843　Thamnochortus virgatus（Rottb.）Kunth = Cannomois virgata（Rottb.）Steud. ■☆

388844　Thamnojusticia Mildbr.（1933）;灌木爵床属●■

388845　Thamnojusticia Mildbr. = Justicia L. ●■

388846　Thamnojusticia amabilis Mildbr. ;灌木爵床●■

388847　Thamnojusticia amabilis Mildbr. = Justicia asystasioides（Lindau）M. E. Steiner ■☆

388848　Thamnojusticia grandiflora Mildbr. = Justicia asystasioides（Lindau）M. E. Steiner ■☆

388849　Thamnosciadium Hartvig（1985）;灌伞芹属●☆

388850　Thamnosciadium junceum（Sm.）Hartvig,灌伞芹●☆

388851　Thamnoseris F. Phil.（1875）;肉苣木属●☆

388852　Thamnoseris lacera（Phil.）Johnst. ;肉苣木●☆

388853　Thamnosma Torr. et Frém.（1845）;香芸灌属●☆

388854　Thamnosma africana Engl. ;非洲香芸灌●☆

388855　Thamnosma africana Engl. var. rhodesica Baker f. = Thamnosma rhodesica（Baker f.）Mendonça ●☆

388856　Thamnosma hirschii Schweinf. ;希尔施香芸灌●☆

388857　Thamnosma montana Torr. et Frem. ;山地香芸灌;Turpentine Broom ●☆

388858　Thamnosma rhodesica（Baker f.）Mendonça;罗得西亚香芸灌●☆

388859　Thamnosma somalensis Thulin;索马里香芸灌●☆

388860　Thamnus Klotzsch = Blaeria L. ●☆

388861　Thamnus Klotzsch = Erica L. ●☆

388862　Thamnus Klotzsch（1838）;灌木杜鹃属●☆

388863　Thamnus L. = Tamus L. ■☆

388864　Thamnus multiflorus Klotzsch = Erica thamnoides E. G. H. Oliv. ●☆

388865　Thanatophorus Walp. = Danatophorus Blume ●

388866　Thanatophorus Walp. = Harpullia Roxb. ●

388867　Thapsandra Griseb. = Celsia L. ■☆

388868　Thapsia L.（1753）;毒胡萝卜属;Deadly Carrot ■☆

388869　Thapsia cinerea A. Pujadas;灰色毒胡萝卜■☆

388870　Thapsia garganica L. ;毒胡萝卜;Spanish Türpeth Root ■☆

388871　Thapsia garganica L. var. angusta Faure et Maire = Thapsia garganica L. ■☆

388872　Thapsia garganica L. var. platycarpa（Pomel）Batt. = Thapsia garganica L. ■☆

388873　Thapsia garganica L. var. sylphium（Viv.）Asch. = Thapsia garganica L. ■☆

388874　Thapsia lineariloba Pomel = Thapsia garganica L. ■☆

388875　Thapsia microcarpa Pomel = Thapsia villosa L. ■☆

388876　Thapsia platycarpa Pomel;宽果毒胡萝卜■☆

388877　Thapsia polygama Desf. = Rouya polygama（Desf.）Coincy ■☆

388878　Thapsia polygama Desf. var. glabra Maire = Rouya polygama（Desf.）Coincy ■☆

388879　Thapsia polygama Desf. var. hispida Maire = Rouya polygama（Desf.）Coincy ■☆

388880　Thapsia polygama Desf. var. tenuisecta Maire = Rouya polygama（Desf.）Coincy ■☆

388881　Thapsia silphia St. -Lag. ;西非毒胡萝卜■☆

388882　Thapsia stenocarpa Pomel = Thapsia garganica L. ■☆

388883　Thapsia tenuifolia Lag. ;细叶毒胡萝卜■☆

388884　Thapsia trifoliata L. ;三小叶毒胡萝卜■☆

388885　Thapsia villosa L. ;毛毒胡萝卜■☆

388886　Thapsia villosa L. var. dissecta Boiss. = Thapsia villosa L. ■☆

388887　Thapsia villosa L. var. microcarpa（Pomel）Batt. = Thapsia villosa L. ■☆

388888　Thapsia villosa L. var. stenoptera（Pomel）Batt. = Thapsia villosa L. ■☆

388889　Thapsium Walp. = Thaspium Nutt. ■☆

388890　Thapsus Raf. = Verbascum L. ■●

388891　Tharpia Britton et Rose = Cassia L.（保留属名）●■

388892　Tharpia Britton et Rose = Senna Mill. ●■

388893　Thaspium Nutt.（1818）;草地防风属（萨斯珀属）■☆

388894　Thaspium aureum（L.）Nutt. = Thaspium trifoliatum（L.）A. Gray var. flavum S. F. Blake ■☆

388895　Thaspium barbinode（Michx.）Nutt. ;草地防风;Meadow Parsnip ■☆

388896　Thaspium barbinode（Michx.）Nutt. var. angustifolium J. M. Coult. et Rose = Thaspium barbinode（Michx.）Nutt. ■☆

388897　Thaspium barbinode（Michx.）Nutt. var. chapmanii J. M. Coult. et Rose = Thaspium chapmanii（J. M. Coult. et Rose）Small ■☆

388898　Thaspium chapmanii（J. M. Coult. et Rose）Small;查普曼草地防风;Hairy Meadow-parsnip, Hairy-jointed Meadow-parsnip ■☆

388899　Thaspium trifoliatum（L.）A. Gray;三叶草地防风;Meadow Parsnip, Purple Meadow-parsnip, Smooth Meadow-parsnip ■☆

388900　Thaspium trifoliatum（L.）A. Gray var. apterum A. Gray = Zizia aptera（A. Gray）Fernald ■☆

388901　Thaspium trifoliatum（L.）A. Gray var. aureum（L.）Britton = Thaspium trifoliatum（L.）A. Gray var. flavum S. F. Blake ■☆

388902　Thaspium trifoliatum（L.）A. Gray var. flavum S. F. Blake;黄三叶草地防风;Purple Meadow-parsnip, Smooth Meadow-parsnip ■☆

388903　Thaumasianthes Danser（1933）;奇寄生属●☆

388904　Thaumasianthes amplifolia（Merr.）Danser;奇寄生●☆

388905　Thaumasianthes ovatibractea（Merr.）Danser;卵苞奇寄生●☆

388906　Thaumastochloa C. E. Hubb.（1936）;澳奇禾属（澳禾草属,假淡竹叶属,假蛇尾草属）;Thaumastochloa ■☆

388907　Thaumastochloa chenii C. C. Hsu = Mnesithea laevis（Retz.）Kunth var. chenii（C. C. Hsu）de Koning et Sosef ■

388908　Thaumastochloa chenii Y. C. Hsu = Heteropholis cochinchinensis（Lour.）Clayton var. chenii（Y. C. Hsu）Sosef et de Koning ■

388909　Thaumastochloa cochichinensis（Lour.）C. E. Hubb. ;澳奇禾（假淡竹叶,假蛇尾草）;Cochinchina Thaumastochloa ■☆

388910　Thaumastochloa cochinchinensis（Lour.）C. E. Hubb. = Heteropholis cochinchinensis（Lour.）Clayton ■

388911　Thaumastochloa cochinchinensis（Lour.）C. E. Hubb. = Mnesithea laevis（Retz.）Kunth ■

388912　Thaumastochloa cochinchinensis C. E. Hubb. = Heteropholis cochinchinensis（Lour.）Clayton ■

388913　Thaumastochloa cochinchinensis C. E. Hubb. f. shimadana Ohwi = Heteropholis cochinchinensis（Lour.）Clayton var. chenii（Y. C. Hsu）Sosef et de Koning ■

388914　Thaumastochloa pubescens（Domin）C. E. Hubb. ;柔毛澳奇禾（柔毛假蛇尾草）;Pubescent Thaumastochloa ■☆

388915　Thaumastochloa shimadana（Ohwi et Odash.）Ohwi et Odash. = Mnesithea laevis（Retz.）Kunth var. chenii（C. C. Hsu）de

Koning et Sosef ■

388916　Thaumatocaryon Baill. (1890);奇果紫草属■☆

388917　Thaumatocaryon dasyanthum I. M. Johnst. ;毛花奇果紫草■☆

388918　Thaumatocaryon hilarii Baill. ;奇果紫草■☆

388919　Thaumatocaryum Baill. = Thaumatocaryon Baill. ■☆

388920　Thaumatococcus Benth. (1883);奇果竹芋属(奇果属)■☆

388921　Thaumatococcus danielli (Bennet) Benth. ;奇果竹芋(奇异果,西非竹芋);Miraculous Fruit ■☆

388922　Thaumatococcus daniellii (Bennet) Benth. var. puberulifolius Dhetchuvi et Diafouka;毛叶奇果竹芋■☆

388923　Thaumatococcus daniellii Benth. = Thaumatococcus daniellii (Bennet) Benth. ■☆

388924　Thaumatococcus daniellii Benth. var. puberulifolius Dhetchuvi et Diafouka = Thaumatococcus daniellii (Bennet) Benth. var. puberulifolius Dhetchuvi et Diafouka ■☆

388925　Thaumatophyllum Schott = Philodendron Schott(保留属名)■●

388926　Thaumaza Salisb. = Eriospermum Jacq. ex Willd. ■☆

388927　Thaumuria Gaudich. = Parietaria L. ■

388928　Thawatchaia M. Kato, Koi et Y. Kita(1753);三裂川苔草属■☆

388929　Thawatchaia trilobata M. Kato, Koi et Y. Kita;三裂川苔草■☆

388930　Thea L. = Camellia L. ●

388931　Thea assamica J. W. Mast. = Camellia sinensis (L.) Kuntze var. assamica (Mast.) Kitam. ●◇

388932　Thea assamica J. W. Mast. = Camellia sinensis (L.) Kuntze ●

388933　Thea assimilis (Champ.) Seem. = Camellia assimilis Champ. ex Benth. ●

388934　Thea assimilis (Champ.) Seem. = Camellia caudata Wall. ●

388935　Thea bachamaensis Gagnep. = Camellia kissi Wall. ●

388936　Thea biflora Hayata = Camellia oleifera Abel ●

388937　Thea bohea L. = Camellia sinensis (L.) Kuntze ●

388938　Thea bolovensis Gagnep. = Camellia furfuracea (Merr.) Cohen-Stuart ●

388939　Thea brachystemon Gagnep. = Camellia kissi Wall. ●

388940　Thea brevistyla Hayata = Camellia brevistyla (Hayata) Cohen-Stuart ●

388941　Thea buisanensis (Sasaki) F. P. Metcalf = Pyrenaria microcarpa (Dunn) H. Keng var. ovalifolia (H. L. Li) T. L. Ming et S. X. Yang ●

388942　Thea camellia Hoffmanns. = Camellia japonica L. ●

388943　Thea camellia Hoffmanns. var. lucidissima H. Lév. = Camellia saluenensis Stapf ex Bean ●

388944　Thea cantonensis Lour. = Camellia sinensis (L.) Kuntze ●

388945　Thea caudata (Wall.) Seem. = Camellia caudata Wall. ●

388946　Thea caudata (Wall.) Seem. var. faberi Kochs = Camellia elongata (Rehder et E. H. Wilson) Rehder ●

388947　Thea caudata Wall. = Camellia caudata Wall. ●

388948　Thea caudata Wall. var. faberi Kochs = Camellia elongata (Rehder et E. H. Wilson) Rehder ●

388949　Thea cavaleriana H. Lév. = Camellia pitardii Cohen-Stuart ●

388950　Thea chinensis Sims = Camellia sinensis (L.) Kuntze ●

388951　Thea chinensis Sims var. androxantha H. Lév. = Camellia synaptica Sealy ●

388952　Thea chinensis Sims var. assamica (J. W. Mast.) Pierre = Camellia sinensis (L.) Kuntze var. assamica (Mast.) Kitam. ●◇

388953　Thea cochinchinensis Lour. = Camellia sinensis (L.) Kuntze ●

388954　Thea confusa Craib = Camellia kissii Wall. var. confusa (Craib) T. L. Ming ●

388955　Thea cordifolia F. P. Metcalf = Camellia cordifolia (F. P. Metcalf) Nakai ●

388956　Thea costei (H. Lév.) Rehder = Camellia costei H. Lév. ●

388957　Thea crapnelliana (Tutcher) Rehder = Camellia crapnelliana Tutcher ●◇

388958　Thea cuspidata Kochs = Camellia cuspidata (Kochs) Wright ex Garden ●

388959　Thea drupifera (Lour.) Pierre = Camellia drupifera Lour. ●

388960　Thea edithae (Hance) Kuntze = Camellia edithae Hance ●

388961　Thea elongata Rehder et E. H. Wilson = Camellia elongata (Rehder et E. H. Wilson) Rehder ●

388962　Thea euryoides (Lindl.) Booth = Camellia euryoides Lindl. ●

388963　Thea fluviatilis (Hand.-Mazz.) Merr. = Camellia fluviatilis Hand.-Mazz. ●

388964　Thea forrestii Diels = Camellia forrestii (Diels) Cohen-Stuart ●

388965　Thea fraterna (Hance) Kuntze = Camellia fraterna Hance ●

388966　Thea fraterna Kuntze = Camellia handelii Sealy ●

388967　Thea furfuracea Merr. = Camellia furfuracea (Merr.) Cohen-Stuart ●

388968　Thea fusiger Gagnep. = Camellia tsaii Hu ●

388969　Thea gaudichaudii Gagnep. = Camellia gaudichaudii (Gagnep.) Sealy ●

388970　Thea gilbertii A. Chev. = Camellia gilbertii (A. Chev.) Sealy ●

388971　Thea gnaphalocarpa Hayata = Camellia brevistyla (Hayata) Cohen-Stuart ●

388972　Thea gracilis (Hemsl.) Hayata = Camellia caudata Wall. var. gracilis (Hemsl.) Yamam. ex H. Keng ●

388973　Thea gracilis (Hemsl.) Hayata = Camellia caudata Wall. ●

388974　Thea grandifolia Salisb. = Camellia sinensis (L.) Kuntze ●

388975　Thea henryana (Cohen-Stuart) Rehder = Camellia henryana Cohen-Stuart ●

388976　Thea henryana (Cohen-Stuart) Rehder = Camellia yunnanensis (Pit. ex Diels) Cohen-Stuart ●

388977　Thea hongkongensis (Seem.) Pierre = Camellia hongkongensis Seem. ●

388978　Thea hozanensis Hayata = Camellia japonica L. ●

388979　Thea indochinensis (Merr.) Gagnep. = Camellia indochinensis Merr. ●

388980　Thea iniquicarpa (Clarke) Kochs = Camellia kissi Wall. ●

388981　Thea japonica (L.) Baill. = Camellia japonica L. ●

388982　Thea japonica (L.) Baill. var. hortensis Makino = Camellia japonica L. ●

388983　Thea japonica (L.) Baill. var. spontanea Makino = Camellia japonica L. ●

388984　Thea japonica Nois. = Camellia japonica L. ●

388985　Thea mairei H. Lév. = Camellia mairei (H. Lév.) Melch. ●

388986　Thea maliflora (Lindl.) Seem. = Camellia maliflora Lindl. ●

388987　Thea melliana (Hand.-Mazz.) Merr. = Camellia melliana Hand.-Mazz. ●

388988　Thea microphylla Merr. = Camellia brevistyla (Hayata) Cohen-Stuart var. microphylla (Merr.) T. L. Ming ●

388989　Thea microphylla Merr. = Camellia microphylla (Merr.) S. S. Chien ●

388990　Thea nakaii Hayata = Camellia japonica L. ●

388991　Thea nokoensis (Hayata) Makino et Nemoto = Camellia euryoides Lindl. var. nokoensis (Hayata) T. L. Ming ●

388992　Thea nokoensis (Hayata) Makino et Nemoto = Camellia nokoensis Hayata ●

388993   Thea olearia Lour. ex Gomes = Camellia sinensis (L.) Kuntze ●

388994   Thea oleifera (Abel) Rehder et E. H. Wilson = Camellia oleifera Abel ●

388995   Thea oleifera (C. Abel) Rehder et E. H. Wilson = Camellia oleifera Abel ●

388996   Thea oleosa Lour. = Camellia sinensis (L.) Kuntze ●

388997   Thea oxyanthera Gagnep. = Camellia kissii Wall. var. confusa (Craib) T. L. Ming ●

388998   Thea parvifolia Hayata = Camellia transarisanensis (Hayata) Cohen-Stuart ●

388999   Thea parvifolia Salisb. = Camellia sinensis (L.) Kuntze ●

389000   Thea paucipunctata Merr. et Chun = Camellia paucipunctata (Merr. et Chun) Chun ●

389001   Thea petelotii Merr. = Camellia petelotii (Merr.) Sealy ●

389002   Thea pitardii (Cohen-Stuart) Rehder = Camellia pitardii Cohen-Stuart ●

389003   Thea pitardii (Cohen-Stuart) Rehder = Camellia saluenensis Stapf ex Bean ●

389004   Thea pitardii (Cohen-Stuart) Rehder var. lucidissima (H. Lév.) Rehder = Camellia saluenensis Stapf ex Bean ●

389005   Thea podogyna H. Lév. = Camellia oleifera Abel ●

389006   Thea polygama (Hu) Nakai = Camellia forrestii (Diels) Cohen-Stuart ●

389007   Thea polygama Hu = Camellia forrestii (Diels) Cohen-Stuart ●

389008   Thea punctata Kochs = Camellia punctata (Kochs) Cohen-Stuart ●

389009   Thea reticulata (Lindl.) Kochs = Camellia reticulata Lindl. ●◇

389010   Thea reticulata (Lindl.) Kochs var. rosea Makino = Camellia uraku Kitam. ●

389011   Thea rosaeflora (Hook.) Kuntze = Camellia rosaeflora Hook. ●

389012   Thea rosaeflora (Hook.) Kuntze var. glabra Kochs = Camellia cuspidata (Kochs) Wright ex Garden ●

389013   Thea rosaeflora (Hook.) Kuntze var. pilosa Kochs = Camellia fraterna Hance ●

389014   Thea rosiflora (Hook.) Kuntze var. glabra Kochs. = Camellia cuspidata (Kochs) Wright ex Garden ●

389015   Thea rosthorniana (Hand.-Mazz.) Hand.-Mazz. = Camellia rosthorniana Hand.-Mazz. ●

389016   Thea salicifolia (Champ.) Seem. = Camellia salicifolia Champ. ex Benth. ●

389017   Thea salicifolia (Champ.) Seem. var. warburgii Kochs = Camellia salicifolia Champ. ex Benth. ●

389018   Thea sasanqua (Champ.) Seem. var. kissi (Wall.) Pierre = Camellia kissii Wall. ●

389019   Thea sasanqua (Thunb.) Cels var. kissii (Wall.) Pierre = Camellia kissii Wall. ●

389020   Thea sasanqua (Thunb.) Cels var. loureiroi Pierre = Camellia oleifera Abel ●

389021   Thea shinkoensis Hayata = Pyrenaria microcarpa (Dunn) H. Keng ●

389022   Thea shinkoensis Hayata = Tutcheria shinkoensis (Hayata) Nakai ●

389023   Thea sinensis L. = Camellia sinensis (L.) Kuntze ●

389024   Thea sinensis L. f. macrophylla Siebold ex Franch. et Sav. = Camellia sinensis (L.) Kuntze f. macrophylla (Siebold ex Miq.) Kitam. ●

389025   Thea sinensis L. f. rosea (Makino) Ohwi = Camellia sinensis (L.) Kuntze f. rosea (Makino) Kitam. ●☆

389026   Thea sinensis L. var. assamica (J. W. Mast.) Pierre = Camellia sinensis (L.) Kuntze var. assamica (Mast.) Kitam. ●◇

389027   Thea sinensis L. var. macrophylla Siebold = Camellia sinensis (L.) Kuntze ●

389028   Thea sinensis L. var. parvifolia Miq. = Camellia sinensis (L.) Kuntze ●

389029   Thea speciosa Kochs = Gordonia szechuanensis Hung T. Chang ●

389030   Thea speciosa Pit. ex Diels = Camellia pitardii Cohen-Stuart ●

389031   Thea spectabilis (Champ.) Kochs = Pyrenaria spectabilis (Champ.) C. Y. Wu et S. X. Yang ●

389032   Thea taliensis W. W. Sm. = Camellia taliensis (W. W. Sm.) Melch. ●

389033   Thea tenuiflora Hayata = Camellia brevistyla (Hayata) Cohen-Stuart ●

389034   Thea theiformis (Hance) Kuntze = Camellia euryoides Lindl. ●

389035   Thea transarisanensis Hayata = Camellia transarisanensis (Hayata) Cohen-Stuart ●

389036   Thea transnokoensis (Hayata) Makino et Nemoto = Camellia lutchuensis Ito ex Ito et Matsum. ●

389037   Thea transnokoensis (Hayata) Makino et Nemoto = Camellia transarisanensis (Hayata) Cohen-Stuart ●

389038   Thea trichoclada Rehder = Camellia trichoclada (Rehder) S. S. Chien ●

389039   Thea tsaii (Hu) Gagnep. = Camellia tsaii Hu ●

389040   Thea virgata Koidz. = Pyrenaria microcarpa (Dunn) H. Keng ●

389041   Thea viridis L. = Camellia sinensis (L.) Kuntze ●

389042   Thea viridis L. var. assamica (J. W. Mast.) Choisy = Camellia assamica (J. W. Mast.) Hung T. Chang ●

389043   Thea viridis L. var. assamica (J. W. Mast.) Choisy = Camellia sinensis (L.) Kuntze var. assamica (Mast.) Kitam. ●◇

389044   Thea yunnanensis Pit. ex Diels = Camellia yunnanensis (Pit. ex Diels) Cohen-Stuart ●

389045   Theaceae D. Don = Theaceae Mirb.(保留科名)●

389046   Theaceae Mirb.(1816)(保留科名);山茶科(茶科);Tea Family ●

389047   Theaceae Mirb.(保留科名) = Asteropeiaceae Takht. ●

389048   Theaceae Mirb. ex Ker Gaw. = Theaceae Mirb.(保留科名)●

389049   Theaphyla Raf. = Camellia L. ●

389050   Theaphyla Raf. = Theaphylla Raf. ●

389051   Theaphylla Raf. = Camellia L. ●

389052   Theaphylla Raf. = Thea L. ●

389053   Theaphylla cantonensis (Lour.) Raf. = Camellia sinensis (L.) Kuntze ●

389054   Theaphyllum Nutt. ex Turcz.(1863);茶叶卫矛属●☆

389055   Theaphyllum Nutt. ex Turcz. = Perrottetia Kunth ●

389056   Theaphyllum celastrinum Nutt. ex Turcz. = Perrottetia sandwicensis A. Gray ●☆

389057   Thebesia Neck. = Knowltonia Salisb. ■☆

389058   Theca Juss. = Tectona L. f.(保留属名)●

389059   Theca Juss. = Theka Adans.(废弃属名)●

389060   Thecacoris A. Juss.(1824);囊大戟属●☆

389061   Thecacoris annobonae Pax et K. Hoffm. ;安诺本囊大戟●☆

389062   Thecacoris batesii Hutch. ;贝茨囊大戟●☆

389063   Thecacoris bussei (Pax) Radcl.-Sm. = Thecacoris spathulifolia (Pax) Léandri ●☆

389064   Thecacoris chevalieri Beille = Thecacoris stenopetala (Müll.

Arg. ）Müll. Arg. ●☆

389065 Thecacoris glabrata Hutch. = Maesobotrya glabrata （Hutch.）Exell ●☆

389066 Thecacoris glabroglandulosa （J. Léonard）J. Léonard；光腺囊大戟●☆

389067 Thecacoris grandifolia （Pax et K. Hoffm.）Govaerts；大叶囊大戟●☆

389068 Thecacoris gymnogyne Pax = Thecacoris leptobotrya （Müll. Arg.）Brenan ●☆

389069 Thecacoris gymnogyne Pax var. glabroglandulosa J. Léonard = Thecacoris glabroglandulosa （J. Léonard）J. Léonard ●☆

389070 Thecacoris lancifolia Pax et K. Hoffm.；剑叶囊大戟●☆

389071 Thecacoris latistipula J. Léonard；宽托叶囊大戟●☆

389072 Thecacoris lenifolia J. Léonard = Thecacoris trichogyne Müll. Arg. ●☆

389073 Thecacoris leptobotrya （Müll. Arg.）Brenan；细穗囊大戟●☆

389074 Thecacoris lucida （Pax）Hutch.；光亮囊大戟●☆

389075 Thecacoris manniana （Müll. Arg.）Müll. Arg.；曼氏囊大戟●☆

389076 Thecacoris membranacea Pax；膜质囊大戟●☆

389077 Thecacoris obanensis Hutch. = Thecacoris leptobotrya （Müll. Arg.）Brenan ●☆

389078 Thecacoris reticulata Pax = Thecacoris batesii Hutch. ●☆

389079 Thecacoris spathulifolia （Pax）Léandri；匙叶囊大戟●☆

389080 Thecacoris stenopetala （Müll. Arg.）Müll. Arg.；窄瓣囊大戟●☆

389081 Thecacoris talbotae Hutch. = Thecacoris leptobotrya （Müll. Arg.）Brenan ●☆

389082 Thecacoris trichogyne Müll. Arg.；毛蕊囊大戟●☆

389083 Thecacoris trilliesii Beille = Spondianthus preussii Engl. subsp. glaber （Engl.）J. Léonard et Nkounkou ●☆

389084 Thecacoris usambarensis Verdc. = Cyathogyne usambarensis （Verdc.）J. Léonard ●☆

389085 Thecacoris viridis （Müll. Arg.）G. L. Webster = Cyathogyne viridis Müll. Arg. ●☆

389086 Thecagonum Babu = Oldenlandia L. ●■

389087 Thecagonum ovatifolium （Cav.）Babu = Hedyotis ovatifolia Cav. ■

389088 Thecanisia Raf. = Filipendula Mill. ■

389089 Thecanthes Wikstr. （1818）；囊花瑞香属●☆

389090 Thecanthes Wikstr. = Pimelea Banks ex Gaertn. （保留属名）●☆

389091 Thecanthes cornucopiae （M. Vahl）Wikstr.；囊花瑞香●☆

389092 Thecanthes filifolia Rye；线叶囊花瑞香●☆

389093 Thecanthes sanguinea （F. Muell.）Rye；血红囊花瑞香●☆

389094 Thecocarpus Boiss. （1844）；套果草属■☆

389095 Thecocarpus meifolius Boiss.；套果草■☆

389096 Thecophyllum André = Guzmania Ruiz et Pav. ■☆

389097 Thecopus Seidenf. （1984）；盒足兰属■☆

389098 Thecopus maingayi （Hook. f.）Seidenf.；盒足兰■☆

389099 Thecorchus Bremek. （1952）；箱果茜属●■☆

389100 Thecorchus Bremek. = Oldenlandia L. ●■

389101 Thecorchus wauensis （Hiern）Bremek.；箱果茜●■☆

389102 Thecorchus wauensis （Hiern）Bremek. var. scabrida Bremek. = Thecorchus wauensis （Hiern）Bremek. ●■☆

389103 Thecostele Rchb. f. （1857）；盒柱兰属■☆

389104 Thecostele alata E. C. Parish et Rchb. f.；翅盒柱兰■☆

389105 Thecostele zollingeri Rchb. f.；盒柱兰■☆

389106 Thedachloa S. W. L. Jacobs（2004）；西澳禾属●☆

389107 Theileamea Baill. = Chlamydacanthus Lindau ■☆

389108 Theileamia Willis = Theileamea Baill. ■☆

389109 Theilera E. Phillips（1926）；柱冠桔梗属☆

389110 Theilera capensis D. Y. Hong；好望角柱冠桔梗●☆

389111 Theilera guthriei （L. Bolus）E. Phillips；柱冠桔梗●☆

389112 Theis Salisb. ex DC. = Rhododendron L. ●

389113 Theka Adans. （废弃属名）= Tectona L. f. （保留属名）●

389114 Theka grandis （L. f.）Lam. = Tectona grandis L. f. ●

389115 Thela Lour. = Plumbago L. ●■

389116 Thela alba Lour. = Plumbago zeylanica L. ●

389117 Thela coccinea Lour. = Plumbago indica L. ●■

389118 Thelaia Alef. = Pyrola L. ●■

389119 Thelaia chlorantha （Sw.）Alef. = Pyrola chlorantha Sw. ●

389120 Thelaia media （Sw.）Alef. = Pyrola media Sw. ●

389121 Thelaia rotundifolia （L.）Alef. = Pyrola rotundifolia L. ●

389122 Thelasis Blume（1825）；矮柱兰属（八粉兰属）；Thelasis ■

389123 Thelasis carinata Blume；龙骨矮柱兰；Keeled Thelasis ■☆

389124 Thelasis clausa Fukuy. = Thelasis pygmaea （Griff.）Blume ■

389125 Thelasis elongata Blume = Thelasis pygmaea （Griff.）Blume ■

389126 Thelasis hongkongensis Rolfe = Thelasis pygmaea （Griff.）Blume ■

389127 Thelasis khasiana Hook. f.；滇南矮柱兰；Khas Thelasis ■

389128 Thelasis pygmaea （Griff.）Blume；矮柱兰（闭花八粉兰）；Dwarf Thelasis, Threawing Thelasis ■

389129 Thelasis pygmaea （Griff.）Blume var. khasiana （Hook. f.）Schltr. = Thelasis khasiana Hook. f. ■

389130 Thelasis pygmaea （Griff.）Blume var. multiflora Hook. f. = Thelasis pygmaea （Griff.）Blume ■

389131 Thelasis taiwaniana （Fukuy.）S. S. Ying = Phreatia taiwaniana Fukuy. ■

389132 Thelasis triptera Rchb. f. = Thelasis pygmaea （Griff.）Blume ■

389133 Thelecarpus Tiegh. = Phragmanthera Tiegh. ●☆

389134 Thelecarpus Tiegh. = Tapinanthus （Blume）Rchb. （保留属名）●☆

389135 Thelecarpus batangae （Engl.）Tiegh. = Phragmanthera batangae （Engl.）Balle ●☆

389136 Thelecarpus hexasepalus Tiegh. = Phragmanthera capitata （Spreng.）Balle ●☆

389137 Thelecarpus soyauxii （Engl.）Tiegh. = Phragmanthera capitata （Spreng.）Balle ●☆

389138 Thelecarpus thollonii Tiegh. = Phragmanthera capitata （Spreng.）Balle ●☆

389139 Thelechitonia Cuatrec. （1954）；乳甲菊属■☆

389140 Thelechitonia Cuatrec. = Wedelia Jacq. （保留属名）●●

389141 Thelechitonia gracilis （Rich.）H. Rob. et Cuatrec.；细乳甲菊■☆

389142 Thelechitonia muricata Cuatrec.；乳甲菊■☆

389143 Thelechitonia trilobata （L.）H. Rob. et Cuatrec. = Sphagneticola trilobata （L.）Pruski ■☆

389144 Theleophyton （Hook. f.）Moq. （1849）；乳头藜属●☆

389145 Theleophyton Moq. = Theleophyton （Hook. f.）Moq. ●☆

389146 Theleophyton billardierei （Moq.）Moq.；乳头藜●☆

389147 Thelepaepale Bremek. = Strobilanthes Blume ●■

389148 Thelepodium A. Nelson = Thelypodium Endl. ■☆

389149 Thelepogon Roth = Thelepogon Roth ex Roem. et Schult. ■☆

389150 Thelepogon Roth ex Roem. et Schult. （1817）；乳须草属■☆

389151 Thelepogon elegans Roth = Thelepogon elegans Roth ex Roem. et Schult. ☆

389152 Thelepogon elegans Roth ex Roem. et Schult.；乳须草■☆

389153　Thelesperma Less.（1831）；乳籽菊属（绿线菊属）；Green Thread ■●☆

389154　Thelesperma ambiguum A. Gray；科罗拉多乳籽菊；Colorado Green Thread，Showy Navajo Tea ■☆

389155　Thelesperma burridgeanum（Regel）S. F. Blake；布氏乳籽菊■☆

389156　Thelesperma caespitosum Dorn ＝ Thelesperma subnudum A. Gray ■☆

389157　Thelesperma curvicarpum Melchert ＝ Thelesperma simplicifolium（A. Gray）A. Gray ■☆

389158　Thelesperma filifolium（Hook.）A. Gray；线叶乳籽菊■☆

389159　Thelesperma filifolium（Hook.）A. Gray var. flavodiscum Shinners ＝ Thelesperma flavodiscum（Shinners）B. L. Turner ■☆

389160　Thelesperma filifolium（Hook.）A. Gray var. intermedium（Rydb.）Shinners ＝ Thelesperma filifolium（Hook.）A. Gray ■☆

389161　Thelesperma flavodiscum（Shinners）B. L. Turner；黄盘乳籽菊■☆

389162　Thelesperma fraternum Shinners ＝ Thelesperma ambiguum A. Gray ■☆

389163　Thelesperma gracile（Torr.）A. Gray ＝ Thelesperma longipes A. Gray ■☆

389164　Thelesperma intermedium Rydb.；间型乳籽菊■☆

389165　Thelesperma intermedium Rydb. ＝ Thelesperma filifolium（Hook.）A. Gray ■☆

389166　Thelesperma longipes A. Gray；乳籽菊（野茶）■☆

389167　Thelesperma marginatum Rydb. ＝ Thelesperma subnudum A. Gray ■☆

389168　Thelesperma megapotamicum（Spreng.）Kuntze；旱地乳籽菊■☆

389169　Thelesperma megapotamicum（Spreng.）Kuntze var. ambiguum（A. Gray）Shinners ＝ Thelesperma ambiguum A. Gray ■☆

389170　Thelesperma pubescens Dorn ＝ Thelesperma subnudum A. Gray ■☆

389171　Thelesperma pubescens Dorn var. caespitosum（Dorn）C. J. Hansen ＝ Thelesperma subnudum A. Gray ■☆

389172　Thelesperma simplicifolium（A. Gray）A. Gray；单叶乳籽菊■☆

389173　Thelesperma subnudum A. Gray；近裸乳籽菊■☆

389174　Thelesperma subnudum Dorn var. alpinum S. L. Welsh ＝ Thelesperma subnudum A. Gray ■☆

389175　Thelesperma subnudum Dorn var. marginatum（Rydb.）Cronquist ＝ Thelesperma subnudum A. Gray ■☆

389176　Thelesperma subnudum Dorn var. pubescens（Dorn）S. L. Welsh ＝ Thelesperma subnudum A. Gray ■☆

389177　Thelesperma trifidum（Poir.）Britton ＝ Thelesperma filifolium（Hook.）A. Gray ■☆

389178　Thelesperma windhamii C. J. Hansen ＝ Thelesperma subnudum A. Gray ■☆

389179　Thelethylax C. Cusset（1973）；瘤囊苔草属■☆

389180　Thelethylax minutiflora（Tul.）C. Cusset；瘤囊苔草■☆

389181　Theligonaceae Dumort.（1829）（保留科名）；假繁缕科（假牛繁缕科，牛繁缕科，纤花草科）；Cynocramba Family，Theligon Family，Theligona Family ■

389182　Theligonaceae Dumort.（保留科名）＝ Rubiaceae Juss.（保留科名）●■

389183　Theligonum L.（1753）；假繁缕属（假牛繁缕属，纤花草属）；Theligon，Theligonum，Thelygonum ■

389184　Theligonum cynocrambe L.；非洲假繁缕；Dog's Cabbage ■☆

389185　Theligonum formosanum（Ohwi）Ohwi et Tang S. Liu；台湾假繁缕（台湾假牛繁缕，台湾纤花草）；Taiwan Theligon，Taiwan Theligonum ■

389186　Theligonum japonicum Okubo et Makino；日本假繁缕；Japan Theligon，Japanese Theligonum ■

389187　Theligonum macranthum Franch.；假繁缕（假牛繁缕）；Largeflower Theligon，Largeflower Theligonum ■

389188　Thelionema R. J. F. Hend.（1985）；丝灵麻属■☆

389189　Thelionema caespitosum（R. Br.）R. J. F. Hend.；丝灵麻■☆

389190　Thelionema grande（C. T. White）R. J. F. Hend.；细丝灵麻■☆

389191　Thelira Thouars ＝ Parinari Aubl. ●☆

389192　Thellungia Prost ＝ Eragrostis Wolf ■

389193　Thellungia Stapf ＝ Eragrostis Wolf ■

389194　Thellungia Stapf ex Probst ＝ Eragrostis Wolf ■

389195　Thellungia advena Stapf ＝ Eragrostis advena（Stapf）S. M. Phillips ■☆

389196　Thellungiella O. E. Schulz ＝ Arabidopsis Heynh.（保留属名）■

389197　Thellungiella O. E. Schulz（1924）；盐芥属；Saltcress，Thellungiella ■

389198　Thellungiella halophila（C. A. Mey.）O. E. Schulz；小盐芥；Small Saltcress，Small Thellungiella ■

389199　Thellungiella parvula（Schrenk）Al-Shehbaz et O'Kane；条叶盐芥（小大蒜芥）■

389200　Thellungiella salsuginea（Pall.）O. E. Schulz；盐芥；Common Thellungiella，Saltcress ■

389201　Thelluntophace Godr. ＝ Lemna L. ■

389202　Thelluntophace Godr. ＝ Telmatophace Schleid. ■

389203　Thelocactus（K. Schum.）Britton et Rose（1922）；瘤玉属（瘤球属）；Thelocactus ●

389204　Thelocactus Britton et Rose ＝ Thelocactus（K. Schum.）Britton et Rose ●

389205　Thelocactus bicolor（Galeotti ex Pfeiff.）Britton et Rose；大统领（赤眼玉，两色玉，丽容丸，乳头仙人球）；Glory-of-texas，Texas Pride ■

389206　Thelocactus bicolor（Galeotti ex Pfeiff.）Britton et Rose var. bolaensis（Runge）F. M. Knuth；白针球（白刺大统领）■☆

389207　Thelocactus bicolor（Galeotti ex Pfeiff.）Britton et Rose var. flavidispinus Backeb.；黄刺大统领■☆

389208　Thelocactus bicolor（Galeotti ex Pfeiff.）Britton et Rose var. schottii（Engelm.）Krainz ＝ Thelocactus bicolor（Galeotti ex Sweet）Britton et Rose var. tricolor（K. Schum.）F. M. Knuth ■☆

389209　Thelocactus bicolor（Galeotti ex Pfeiff.）Britton et Rose var. tricolor（K. Schum.）F. M. Knuth ＝ Thelocactus bicolor（Galeotti ex Sweet）Britton et Rose var. tricolor（K. Schum.）F. M. Knuth ■☆

389210　Thelocactus bicolor（Galeotti ex Pfeiff.）Britton et Rose var. tricolor（K. Schum.）F. M. Knuth；花刺大统领（五色大统领）■☆

389211　Thelocactus bicolor（Galeotti ex Sweet）Britton et Rose ＝ Thelocactus bicolor（Galeotti ex Pfeiff.）Britton et Rose ■

389212　Thelocactus bicolor（Galeotti ex Sweet）Britton et Rose var. tricolor（K. Schum.）F. M. Knuth ＝ Thelocactus bicolor（Galeotti ex Pfeiff.）Britton et Rose var. tricolor（K. Schum.）F. M. Knuth ■☆

389213　Thelocactus ehrenbergii（Pfeiff.）F. M. Knuth；名山球（名山丸）■☆

389214　Thelocactus hastifer（Werderm. et Boed.）F. M. Knuth；赤岭球（赤岭丸，鹤美丸，银岭丸）■☆

389215　Thelocactus heterochromus（F. A. C. Weber）Oosten；多色玉■☆

389216　Thelocactus hexaedrophorus（Lem.）Britton et Rose；天晃（六角仙人球）■☆

389217　Thelocactus hexaedrophorus（Lem.）Britton et Rose var. major（Quehl）Y. Ito；绯冠龙■☆

389218　Thelocactus leucanthus Britton et Rose；白刺玉（白刺瘤玉）■☆

389219　Thelocactus lophothela（Salm-Dyck）Britton et Rose；狮子头；Lionhead Thelocactus ■

389220　Thelocactus macdowellii（Rebut ex Quehl）T. Marshall；大白球（大白丸，马氏瘤玉，月光球）●☆

389221　Thelocactus nidulans（Quehl）Britton et Rose；鹤巢球（鹤巢丸，鹤巢仙人球）；Nidulant Thelocactus ■

389222　Thelocactus phymatothelos Britton et Rose；眠狮子■☆

389223　Thelocactus rinconensis（Poselg.）Britton et Rose；拟美洲茶瘤玉■☆

389224　Thelocactus schwarzii Backeb.；春雨玉（白岭冠，黄铜丸，文天祥）■☆

389225　Thelocactus setispinus（Engelm.）E. F. Anderson = Hamatocactus bicolor（Terán et Berland.）I. M. Johnst.■☆

389226　Thelocactus tulensis Britton et Rose；长久球（布氏瘤玉，长久丸）■☆

389227　Thelocactus tulensis Britton et Rose var. longispinus Y. Ito；剑鬼球（剑鬼丸）■☆

389228　Thelocactus uncinatus（Galeotti）T. Marshall；罗纱锦■☆

389229　Thelocarpus Post et Kuntze = Tapinanthus（Blume）Rchb.（保留属名）●☆

389230　Thelocarpus Post et Kuntze = Thelecarpus Tiegh. ●☆

389231　Thelocephala Y. Ito = Neoporteria Britton et Rose ●■

389232　Thelocephala Y. Ito = Pyrrhocactus（A. Berger）Backeb. et F. M. Knuth ●■

389233　Thelomastus Frič = Thelocactus（K. Schum.）Britton et Rose + Echinornastus Britton et Rose ■●

389234　Thelomastus Frič = Thelocactus（K. Schum.）Britton et Rose ●

389235　Thelophytum Post et Kuntze = Theleophyton（Hook.f.）Moq. ●☆

389236　Thelopogon Post et Kuntze = Thelepogon Roth ex Roem. et Schult. ●☆

389237　Thelosperma Post et Kuntze = Thelesperma Less. ■●☆

389238　Thelychiton Endl. = Dendrobium Sw.（保留属名）■

389239　Thelycrania（Dumort.）Fourr.（1868）；肖楝木属●☆

389240　Thelycrania（Dumort.）Fourr. = Cornus L. ●

389241　Thelycrania（Dumort.）Fourr. = Swida Opiz ●

389242　Thelycrania Fourr. = Thelycrania（Dumort.）Fourr. ●☆

389243　Thelycrania alba（L.）Pojark. = Cornus alba L. ●

389244　Thelycrania alba（L.）Pojark. = Swida alba（L.）Opiz ●

389245　Thelycrania australis（C. A. Mey.）Sanadze = Cornus australis C. A. Mey. ●☆

389246　Thelycrania brachypoda（C. A. Mey.）Pojark. = Cornus brachypoda C. A. Mey. ●

389247　Thelycrania brachypoda（C. A. Mey.）Pojark. = Cornus macrophylla Wall. ●

389248　Thelycrania brachypoda（C. A. Mey.）Pojark. = Swida macrophylla（Wall.）Soják ●

389249　Thelycrania darvasica Pojark. = Swida darvasica（Pojark.）Soják ●☆

389250　Thelycrania iberica（Woronow）Pojark. = Cornus iberica Woronow ●☆

389251　Thelycrania meyeri Pojark.；迈尔肖楝木●☆

389252　Thelycrania sanguinea Fourr. = Cornus sanguinea L. ●

389253　Thelycrania sanguinea Fourr. = Swida sanguinea（L.）Opiz ●

389254　Thelycrania sericea（L.）Dandy = Cornus stolonifera Michx. ●☆

389255　Thelycrania stolonifera（Michx.）Pojark. = Cornus stolonifera Michx. ●☆

389256　Thelygonaceae Dumort. = Theligonaceae Dumort.（保留科名）■

389257　Thelygonum Schreb. = Theligonum L. ■

389258　Thelymitra J. R. Forst. et G. Forst.（1775）；柱帽兰属；Thelymitra ■☆

389259　Thelymitra antennifera Hook. f；柱帽兰；Rabbit-ears ■☆

389260　Thelymitra ixioides Sw.；鸟娇花柱帽兰；Spotted Sun Orchid ■☆

389261　Thelymitra malintana Blanco = Habenaria malintana（Blanco）Merr. ■

389262　Thelymitra stenopetala Hook. f. = Waireia stenopetala（Hook. f.）D. L. Jones, M. A. Clem. et Molloy ■☆

389263　Thelypetalum Gagnep. = Leptopus Decne. ●

389264　Thelypodiopsis Rydb.（1907）；类雌足芥属■☆

389265　Thelypodiopsis alpina（Standl. et Steyerm.）Rollins；高山类雌足芥■☆

389266　Thelypodiopsis aurea Rydb.；黄类雌足芥■☆

389267　Thelypodiopsis elegans Rydb.；雅致类雌足芥■☆

389268　Thelypodiopsis leptostachya O. E. Schulz；细穗类雌足芥■☆

389269　Thelypodiopsis linearifolia（Gray）Al-Shehbaz；线叶类雌足芥■☆

389270　Thelypodiopsis paniculata（A. Nelson）O. E. Schulz；圆锥类雌足芥■☆

389271　Thelypodiopsis stenopetala O. E. Schulz；狭瓣类雌足芥■☆

389272　Thelypodium Endl.（1839）；雌足芥属（女足芥属）■☆

389273　Thelypodium affine Greene；近缘雌足芥■☆

389274　Thelypodium aureum Eastw.；黄雌足芥■☆

389275　Thelypodium auriculatum S. Watson；小耳雌足芥■☆

389276　Thelypodium australe Brandegee；墨西哥雌足芥■☆

389277　Thelypodium brachycarpum Torr.；短果雌足芥■☆

389278　Thelypodium elegans M. E. Jones；雅致雌足芥■☆

389279　Thelypodium flavescens（Hook.）Jeps.；浅黄雌足芥■☆

389280　Thelypodium integrifolium Endl.；全缘叶雌足芥■☆

389281　Thelypodium lasiophyllum（Hook. et Arn.）Greene；毛叶雌足芥■☆

389282　Thelypodium leptosepalum Rydb.；细萼雌足芥■☆

389283　Thelypodium linearifolium S. Watson；线叶雌足芥■☆

389284　Thelypodium longifolium（Benth.）S. Watson；长叶雌足芥■☆

389285　Thelypodium longipes（Roll.-Germ.）Rollins；长梗雌足芥■☆

389286　Thelypodium macropetalum Rydb.；大瓣雌足芥■☆

389287　Thelypodium macrorrhizum Muschl.；大根雌足芥■☆

389288　Thelypodium ovalifolium Rydb.；卵叶雌足芥■☆

389289　Thelypodium pallidum Rose；苍白雌足芥■☆

389290　Thelypodium paniculatum A. Nelson；雌足芥■☆

389291　Thelypodium pinnatifidum（Michx.）S. Watson = Iodanthus pinnatifidus（Michx.）Steud. ■☆

389292　Thelypodium rigidum Greene；硬雌足芥■☆

389293　Thelypodium stenopetalum S. Watson；狭瓣雌足芥■☆

389294　Thelypogon Mutis ex Spreng. = Telipogon Mutis ex Kunth ■☆

389295　Thelypogon Spreng. = Telipogon Kunth ■☆

389296　Thelypotzium Gagnep. = Andrachne L. ●☆

389297　Thelyra DC. = Hirtella L. ●☆

389298　Thelyra Thouars = Hirtella L. ●☆

389299　Thelyschista Garay(1982)；分柱兰属■☆

389300　Thelyschista ghillanyi（Pabst）Garay；分柱兰■☆

389301　Thelysia Salisb. = Iris L. ■

389302　Thelysia Salisb. ex Parl. = Iris L. ■

389303　Thelysia alata（Poir.）Salisb. = Iris planifolia（Mill.）Fiori et Paol. ■☆

389304　Thelysia alata（Poir.）Salisb. var. micrantha Batt. = Iris planifolia（Mill.）Fiori et Paol. ■☆

389305　Thelythamnos A. Spreng. = Ursinia Gaertn.（保留属名）●■☆

389306　Thelythamnos filiformis A. Spreng. = Ursinia pinnata（Thunb.）Prassler ■☆

389307　Themeda Forssk.（1775）；菅属（菅草属）；Kangaroo Grass，Kangaroo-grass，Themeda ■

389308　Themeda acaulis B. S. Sun et S. Wang = Themeda helferi Munro ex Hack. ■

389309　Themeda anathera（Nees ex Steud.）Hack.；瘤菅；Tumor Themeda ■

389310　Themeda anathera（Nees ex Steud.）Hack. var. glabrescens（Andersson）Hack. = Themeda anathera（Nees ex Steud.）Hack. ■

389311　Themeda anathera（Nees ex Steud.）Hack. var. hirsuta（Andersson）Hack. = Themeda anathera（Nees ex Steud.）Hack. ■

389312　Themeda arguens（L.）Hack.；圣诞菅草；Christmas Grass ■☆

389313　Themeda arundinacea（Roxb.）A. Camus = Themeda arundinacea（Roxb.）Ridl. ■

389314　Themeda arundinacea（Roxb.）Ridl.；苇菅（韦菅）；Reed Themeda ■

389315　Themeda australis（R. Br.）Stapf；澳洲菅；Kangaroo Grass，Kangaroo-grass ■☆

389316　Themeda australis（R. Br.）Stapf = Themeda triandra Forssk. ■

389317　Themeda barbinodis B. S. Sun et S. Wang = Themeda triandra Forssk. ■

389318　Themeda caudata（Nees）A. Camus；苞子草（老虎须）；Bractlet grass，Caudate Themeda，Tail Themeda ■

389319　Themeda caudata（Nees）A. Camus var. matsudai Honda；松田苞子草（松田氏苞子草）；Matsuda Tail Themeda ■

389320　Themeda caudata（Nees）Dur. et Jack. = Themeda caudata（Nees）A. Camus ■

389321　Themeda caudata（Nees）Honda = Themeda caudata（Nees）A. Camus ■

389322　Themeda chinensis（A. Camus）S. L. Chen et T. D. Zhuang = Themeda quadrivalvis（L.）Kuntze ■

389323　Themeda ciliata（L. f.）Hack. subsp. chinensis A. Camus = Themeda chinensis（A. Camus）S. L. Chen et T. D. Zhuang ■

389324　Themeda ciliata（L. f.）Hack. subsp. chinensis A. Camus = Themeda quadrivalvis（L.）Kuntze ■

389325　Themeda ciliata（L. f.）Hack. subsp. helferi（Munro ex Hack.）A. Camus = Themeda helferi Munro ex Hack. ■

389326　Themeda echinata A. Camus ex Keng = Themeda chinensis（A. Camus）S. L. Chen et T. D. Zhuang ■

389327　Themeda echinata A. Camus ex Keng = Themeda quadrivalvis（L.）Kuntze ■

389328　Themeda echinata Keng = Themeda chinensis（A. Camus）S. L. Chen et T. D. Zhuang ■

389329　Themeda echinata Keng = Themeda quadrivalvis（L.）Kuntze ■

389330　Themeda foliosa（Kunth）Balansa = Hyparrhenia bracteata（Humb. et Bonpl. ex Willd.）Stapf ■

389331　Themeda forskalii Hack. var. glauca？ = Themeda triandra Forssk. ■

389332　Themeda forskalii Hook. = Themeda triandra Forssk. ■

389333　Themeda gigantea（Cav.）Hack. ex Duthie；巨菅（白华，苞子草，宝子草，大菅，高大菅，菅，菅茅，接骨草，苓草，蚂蚱草，野菅）；Giant Themeda，Ulla Grass ■☆

389334　Themeda gigantea（Cav.）Hack. ex Duthie subsp. arundinacea（Roxb.）Hack. = Themeda arundinacea（Roxb.）Ridl. ■

389335　Themeda gigantea（Cav.）Hack. ex Duthie subsp. caudata（Nees）Hack. = Themeda caudata（Nees）A. Camus ■

389336　Themeda gigantea（Cav.）Hack. ex Duthie subsp. caudata（Nees）Hack. = Themeda hookeri（Griseb.）A. Camus ■

389337　Themeda gigantea（Cav.）Hack. ex Duthie subsp. villosa（Poir.）Hack. = Themeda villosa（Poir.）A. Camus ■

389338　Themeda gigantea（Cav.）Hack. ex Duthie var. caudata（Nees）Keng = Themeda caudata（Nees）A. Camus ■

389339　Themeda gigantea（Cav.）Hack. ex Duthie var. villosa（Poir.）Keng；菅（大响铃草，接骨草，蚂蚱草）；Silky Kangaroo Grass，Villose Themeda ■

389340　Themeda gigantea（Cav.）Hack. ex Duthie var. villosa（Poir.）Keng = Themeda villosa（Poir.）A. Camus ■

389341　Themeda gigantea（Cav.）Hack. ex Duthie var. villosa Hack. = Themeda villosa（Poir.）A. Camus ■

389342　Themeda gigantea（Cav.）Hack. subsp. arundinacea（Roxb.）Hack. = Themeda arundinacea（Roxb.）Ridl. ■

389343　Themeda gigantea（Cav.）Hack. subsp. caudata（Nees）Hack. = Themeda caudata（Nees）A. Camus ■

389344　Themeda gigantea（Cav.）Hack. subsp. intermedia Hack. = Themeda intermedia（Hack.）Bor ■

389345　Themeda gigantea（Cav.）Hack. subsp. villosa（Poir.）Hack. = Themeda villosa（Poir.）A. Camus ■

389346　Themeda gigantea（Cav.）Hack. var. subsericans（Nees ex Steud.）Hack. = Themeda arundinacea（Roxb.）Ridl. ■

389347　Themeda gigantea（Cav.）Hack. var. villosa（Poir.）Hack. = Themeda villosa（Poir.）A. Camus ■

389348　Themeda glauca（Desf.）Hack. = Themeda triandra Forssk. ■

389349　Themeda glauca（Desf.）Hack. subsp. brachyantha（Boiss.）Trab. = Themeda triandra Forssk. ■

389350　Themeda gossweileri（Stapf）Roberty = Elymandra gossweileri（Stapf）Clayton ■☆

389351　Themeda helferi Munro ex Hack.；无茎菅 ■

389352　Themeda hookeri（Griseb.）A. Camus；西南菅草（虎氏菅草，小菅草）；Hooker Themeda ■

389353　Themeda imberbis（Retz.）Cooke = Themeda triandra Forssk. ■

389354　Themeda intermedia（Hack.）Bor；居中菅 ■

389355　Themeda japonica（Willd.）C. Tanaka；黄背菅草（草糖，黄背草，黄背茅，黄茅，黄米草，菅草，近肌草，进肌草，屈针草，日本苞子草，山红草）；Japan Themeda，Japanese Themeda，Yellowback Themeda ■

389356　Themeda japonica（Willd.）Tanaka = Themeda triandra Forssk. ■

389357　Themeda minor L. Liou；小菅草；Mini Themeda ■

389358　Themeda quadrivalvis（L.）Kuntze；中华菅；China Themeda，Chinese Themeda，Kangaroo Grass ■

389359　Themeda quadrivalvis（L.）Kuntze var. helferi（Munro ex Hack.）Bor = Themeda helferi Munro ex Hack. ■

389360　Themeda subsericans（Nees ex Steud.）Ridl. = Themeda arundinacea（Roxb.）Ridl. ■

389361　Themeda triandra Forssk.；黄背草（阿拉伯黄背草）；Arabia Themeda，Arabia Yellowback Grass，Rooigras，Triander Themeda ■

389362　Themeda triandra Forssk. subsp. japonica（Willd.）T. Koyama = Themeda japonica（Willd.）C. Tanaka ■

389363　Themeda triandra Forssk. subsp. japonica（Willd.）T. Koyama = Themeda triandra Forssk. var. japonica（Willd.）Makino ■

389364　Themeda triandra Forssk. var. brachyantha（Boiss.）Hack. = Themeda triandra Forssk. ■

389365　Themeda triandra Forssk. var. bracteosa Peter ＝ Themeda triandra Forssk. ■

389366　Themeda triandra Forssk. var. burchellii (Hack.) Stapf ＝ Themeda triandra Forssk. ■

389367　Themeda triandra Forssk. var. glauca (Desf.) Hack. ＝ Themeda triandra Forssk. ■

389368　Themeda triandra Forssk. var. glauca (Hack.) Thell. ＝ Themeda triandra Forssk. ■

389369　Themeda triandra Forssk. var. hispida Stapf ＝ Themeda triandra Forssk. ■

389370　Themeda triandra Forssk. var. imberbis (Retz.) A. Camus ＝ Themeda triandra Forssk. ■

389371　Themeda triandra Forssk. var. japonica (Willd.) Makino ＝ Themeda japonica (Willd.) C. Tanaka ■

389372　Themeda triandra Forssk. var. japonica (Willd.) Makino ＝ Themeda triandra Forssk. ■

389373　Themeda triandra Forssk. var. punctata (Hochst. ex A. Rich.) Stapf ＝ Themeda triandra Forssk. ■

389374　Themeda triandra Forssk. var. sublaevigata Chiov. ＝ Themeda triandra Forssk. ■

389375　Themeda triandra Forssk. var. trachyspathea Gooss. ＝ Themeda triandra Forssk. ■

389376　Themeda trichiata S. L. Chen et T. D. Zhuang;毛菅;Hair Themeda ■

389377　Themeda unica S. L. Chen et T. D. Zhuang;浙皖菅(浙江菅);Alone Themeda ■

389378　Themeda villosa (Poir.) A. Camus ＝ Themeda gigantea (Cav.) Hack. ex Duthie var. villosa (Poir.) Keng ■

389379　Themeda villosa (Poir.) Durieu et Jacks. ＝ Themeda villosa (Poir.) A. Camus ■

389380　Themeda yuanmounensis S. L. Chen et T. D. Zhuang;元谋菅;Yuanmou Themeda ■

389381　Themeda yuanmounensis S. L. Chen et T. D. Zhuang ＝ Themeda quadrivalvis (L.) Kuntze ■

389382　Themeda yunnanensis S. L. Chen et T. D. Zhuang;云南菅;Yunnan Themeda ■

389383　Themidaceae Salisb. (1866);菅科(菅草科,紫灯花科)■

389384　Themidaceae Salisb. ＝ Alliaceae Borkh. (保留科名)■

389385　Themis Salisb. (1866);无味葱属■☆

389386　Themis Salisb. ＝ Brodiaea Sm. (保留属名)■☆

389387　Themis Salisb. ＝ Triteleia Douglas ex Lindl. ■☆

389388　Themis ixioides (W. T. Aiton) Salisb. ＝ Triteleia ixioides (W. T. Aiton) Greene ■☆

389389　Themistoclesia Klotzsch(1851);南杞莓属●☆

389390　Themistoclesia alata Luteyn;翅南杞莓●☆

389391　Themistoclesia buxifolia Klotzsch;南杞莓●☆

389392　Themistoclesia caudata Sleumer;尾状南杞莓●☆

389393　Themistoclesia orientalis Luteyn;东方南杞莓●☆

389394　Themistoclesia pterocarpa Donn. Sm.;翅果南杞莓●☆

389395　Then L. ＝ Camellia L. ●

389396　Thenardia Kunth (1819);掌花夹竹桃属（特纳花属）;Thenardia ■☆

389397　Thenardia Moc. et Sessé ex DC. ＝ Rhynchanthera DC. (保留属名)●☆

389398　Thenardia floribunda Kunth;掌花夹竹桃（多花特纳花）;Manyflower Thenardia ■☆

389399　Theobroma L. (1753);可可属(可可树属);Cacao, Cacaotree, Chocolate Tree, Chocolatetree, Chocolate-tree ●

389400　Theobroma albiflora (Goudot) De Wild.;白花可可●☆

389401　Theobroma angustifolia DC.;狭叶可可;Monkey Cocoa ●☆

389402　Theobroma augusta L. ＝ Abroma augusta (L.) L. f. ●

389403　Theobroma bicolor Humb. et Bonpl.;二色可可树;Cupuacu, Patashte, Pataxte, Tiger Cacao, Tiger Cocoa ●☆

389404　Theobroma cacao L.;可可(可可树);Cacao, Cacao Tree, Cacaotree, Cacao-tree, Chocolate, Chocolate Bean, Chocolate Nut, Chocolate Tree, Chocolate-tree, Cocoa, Cocoa Tree ●

389405　Theobroma grandiflorum (Willd. ex Spreng.) Schum.;大花可可;Cupuacu ●☆

389406　Theobroma guazuma L. ＝ Guazuma ulmifolia Lam. ●☆

389407　Theobroma guianensis J. F. Gmel.;圭亚那可可●☆

389408　Theobroma ovatifolia ex DC.;卵叶可可●☆

389409　Theobroma speciosa Willd. ex Mart.;美丽可可●☆

389410　Theobromaceae J. Agardh ＝ Sterculiaceae Vent. (保留科名)●■

389411　Theobromaceae J. Agardh;可可科●

389412　Theobromataceae J. Agardh ＝ Malvaceae Juss. (保留科名)●■

389413　Theobromataceae J. Agardh ＝ Sterculiaceae Vent. (保留科名)●■

389414　Theobromodes Kuntze ＝ Theobroma L. ●

389415　Theodora Medik. (废弃属名) ＝ Schotia Jacq. (保留属名)●☆

389416　Theodora capitata (Bolle) Taub. ＝ Schotia capitata Bolle ●☆

389417　Theodora fischeri Taub. ＝ Scorodophloeus fischeri (Taub.) J. Léonard ●☆

389418　Theodora gnaphalodes (Royle) Kuntze ＝ Saussurea gnaphaloides (Royle) Sch. Bip. ■

389419　Theodora latifolia (Jacq.) Taub. ＝ Schotia latifolia Jacq. ●☆

389420　Theodora speciosa Medik. ＝ Schotia afra (L.) Thunb. ●☆

389421　Theodorea (Cass.) Cass. ＝ Saussurea DC. (保留属名)●■

389422　Theodorea Barb. Rodr. ＝ Rodrigueziella Kuntze ■☆

389423　Theodorea Cass. ＝ Saussurea DC. (保留属名)●■

389424　Theodorea auriculata Kuntze ＝ Saussurea auriculata (DC.) Sch. Bip. ■

389425　Theodorea baicalensis (Adams) Kuntze ＝ Saussurea baicalensis (Adams) B. L. Rob. ■

389426　Theodorea chinnampoensis (H. Lév. et Vaniot) Soják. ＝ Saussurea chinnampoensis H. Lév. et Vaniot ■

389427　Theodorea costus (Falc.) Kuntze ＝ Saussurea costus (Falc.) Lipsch. ■

389428　Theodorea costus Kuntze ＝ Aucklandia lappa Decne. ■

389429　Theodorea costus Kuntze ＝ Saussurea costus (Falc.) Lipsch. ■

389430　Theodorea deltoides Kuntze ＝ Saussurea deltoidea (DC.) Sch. Bip. ■

389431　Theodorea glomerata (Poir.) Soják. ＝ Saussurea amara (L.) DC. ■

389432　Theodorea gnaphaloides (Schrenk) Kuntze ＝ Saussurea gnaphaloides (Royle) Sch. Bip. ■

389433　Theodorea larionowii (C. Winkl.) Soják ＝ Saussurea larionowii C. Winkl. ■

389434　Theodorea pulchella (Fisch.) Cass. ＝ Saussurea pulchella (Fisch. ex Hornem.) Fisch. ■

389435　Theodorea taraxacifolia Kuntze ＝ Saussurea nepalensis Spreng. ■

389436　Theodoria Neck. ＝ Sterculia L. ●

389437　Theodorovia Kolak. ＝ Campanula L. ■●

389438　Theodorovia Kolak. ＝ Theodorovia Kolak. ex Ogan. ■☆

389439　Theodorovia Kolak. ex Ogan. (1991);特奥桔梗属■☆

389440　Theodorovia Kolak. ex Ogan. ＝ Campanula L. ■●

389441 Theodorovia karakuschensis（Grossh.）Kolak. ex Ogan.；特奥桔梗■☆

389442 Theophrasta L.（1753）；假轮叶属●☆

389443 Theophrasta americana L.；假轮叶●☆

389444 Theophrastaceae D. Don（1835）（保留科名）；假轮叶科（狄氏木科，拟棕科）●☆

389445 Theophrastaceae Link ＝Theophrastaceae D. Don（保留科名）●☆

389446 Theophroseris Andrae ＝Senecio L. ■●

389447 Theophroseris Andrae ＝Tephroseris（Rchb.）Rchb. ■

389448 Theopsis（Cohen-Stuart）Nakai ＝Camellia L. ●

389449 Theopsis Nakai ＝Camellia L. ●

389450 Theopsis amplexifolia（Merr. et Chun）Hu ＝Camellia amplexifolia Merr. et Chun ●

389451 Theopsis caudata（Wall.）Hu ＝Camellia caudata Wall. ●

389452 Theopsis chrysantha Hu ＝Camellia nitidissima C. W. Chi ●

389453 Theopsis chrysantha Hu ＝Camellia petelotii（Merr.）Sealy ●

389454 Theopsis elongata（Rehder et E. H. Wilson）Nakai ＝Camellia elongata（Rehder et E. H. Wilson）Rehder ●

389455 Theopsis euonymifolia Hu ＝Camellia kissi Wall. ●

389456 Theopsis forrestii（Diels）Nakai ＝Camellia forrestii（Diels）Cohen-Stuart ●

389457 Theopsis fraterna（Hance）Nakai ＝Camellia fraterna Hance ●

389458 Theopsis furfuracea（Merr.）Nakai ＝Camellia furfuracea（Merr.）Cohen-Stuart ●

389459 Theopsis longipedicellata Hu ＝Camellia longipedicellata（Hu）Hung T. Chang et D. Fang ●

389460 Theopsis lungyaiensis Hu ＝Camellia brevistyla（Hayata）Cohen-Stuart ●

389461 Theopsis maliflora（Lindl.）Nakai ＝Camellia maliflora Lindl. ●

389462 Theopsis microphylla（Merr.）Nakai ＝Camellia brevistyla（Hayata）Cohen-Stuart var. microphylla（Merr.）T. L. Ming ●

389463 Theopsis microphylla（Merr.）Nakai ＝Camellia microphylla（Merr.）S. S. Chien ●

389464 Theopsis nokoensis（Hayata）Nakai ＝Camellia euryoides Lindl. var. nokoensis（Hayata）T. L. Ming ●

389465 Theopsis nokoensis（Hayata）Nakai ＝Camellia nokoensis Hayata ●

389466 Theopsis parvilimba（Merr. et F. P. Metcalf）Nakai ＝Camellia euryoides Lindl. ●

389467 Theopsis parvilimba（Merr. et F. P. Metcalf）Nakai ＝Camellia parvilimba Merr. et F. P. Metcalf ●

389468 Theopsis polygama（Hu）Nakai ＝Camellia forrestii（Diels）Cohen-Stuart ●

389469 Theopsis transarisanensis（Hayata）Nakai ＝Camellia transarisanensis（Hayata）Cohen-Stuart ●

389470 Theopsis transnokoensis（Hayata）Nakai ＝Camellia lutchuensis Ito ex Ito et Matsum. ●

389471 Theopsis transnokoensis（Hayata）Nakai ＝Camellia transnokoensis Hayata ●

389472 Theopsis trichoclada（Rehder）Nakai ＝Camellia trichoclada（Rehder）S. S. Chien ●

389473 Theopyxis Griseb. ＝Lysimachia L. ●■

389474 Thepparatia Phuph.（2006）；泰国锦葵属☆

389475 Therebina Noronha ＝Pilea Lindl.（保留属名）■

389476 Therefon Raf. ＝Boykinia Nutt.（保留属名）●■☆

389477 Therefon Raf. ＝Therofon Raf. ●■☆

389478 Thereianthus G. J. Lewis（1941）；兽花鸢尾属■☆

389479 Thereianthus bracteolatus（Lam.）G. J. Lewis；小苞片兽花鸢尾■☆

389480 Thereianthus ixioides G. J. Lewis；鸟娇花兽花鸢尾■☆

389481 Thereianthus juncifolius（Baker）G. J. Lewis；灯心草叶兽花鸢尾■☆

389482 Thereianthus lapeyrousioides（Baker）G. J. Lewis；短丝花兽花鸢尾■☆

389483 Thereianthus lapeyrousioides（Baker）G. J. Lewis var. elatior G. J. Lewis ＝Thereianthus minutus（Klatt）G. J. Lewis ■☆

389484 Thereianthus lapeyrousioides（Baker）G. J. Lewis var. lapeyrousioides ＝Thereianthus minutus（Klatt）G. J. Lewis ■☆

389485 Thereianthus longicollis（Schltr.）G. J. Lewis；长颈花兽花鸢尾■☆

389486 Thereianthus minutus（Klatt）G. J. Lewis；微小兽花鸢尾■☆

389487 Thereianthus montanus J. C. Manning et Goldblatt；山地兽花鸢尾■☆

389488 Thereianthus racemosus（Klatt）G. J. Lewis；总状兽花鸢尾■☆

389489 Thereianthus spicatus（L.）G. J. Lewis；长穗兽花鸢尾■☆

389490 Thereianthus spicatus（L.）G. J. Lewis var. linearifolius G. J. Lewis；线叶长穗兽花鸢尾■☆

389491 Theresa Clos ＝Perilomia Kunth ●■

389492 Theresa Clos ＝Scutellaria L. ●■

389493 Theresia C. Koch ＝Fritillaria L. ■

389494 Theresia K. Koch ＝Fritillaria L. ■

389495 Theriophonum Blume（1837）；兽南星属■☆

389496 Theriophonum crenatum Blume；兽南星■☆

389497 Theriophonum indicum Engl.；印度兽南星■☆

389498 Theriophonum minutum Engl.；小兽南星■☆

389499 Thermia Nutt. ＝Thermopsis R. Br. ex W. T. Aiton ■

389500 Thermia rhombifolia Nutt. ex Pursh ＝Thermopsis rhombifolia（Nutt. ex Pursh）Nutt. ex Rich. ■☆

389501 Therminthos St. -Lag. ＝Pistacia L. ●

389502 Therminthos St. -Lag. ＝Terminthos St. -Lag. ●

389503 Thermophila Miers ＝Salacia L.（保留属名）●

389504 Thermopsis R. Br. ＝Thermopsis R. Br. ex W. T. Aiton ■

389505 Thermopsis R. Br. ex W. T. Aiton（1811）；野决明属（黄华属，霍州油菜属）；Bush Pea, False Lupin, False Lupine, Thermopsis, Wildsenna ■

389506 Thermopsis alpestris Czefr. ＝Thermopsis alpina（Pall.）Ledeb. ■

389507 Thermopsis alpina（Pall.）Ledeb.；高山野决明（高山黄华，紫花黄华）；Alp Wildsenna, Alpine Thermopsis ■

389508 Thermopsis alpina（Pall.）Ledeb. var. humilis Czefr. ＝Thermopsis smithiana E. Peter ■

389509 Thermopsis alpina（Pall.）Ledeb. var. yunnanensis Franch. ＝Thermopsis alpina（Pall.）Ledeb. ■

389510 Thermopsis alterniflora Regel et Schmalh. ex Regel；互生叶野决明（互花黄华，互生野决明）■☆

389511 Thermopsis annulocarpa A. Nelson ＝Thermopsis rhombifolia（Nutt. ex Pursh）Nutt. ex Rich. ■☆

389512 Thermopsis arenosa A. Nelson ＝Thermopsis rhombifolia（Nutt. ex Pursh）Nutt. ex Rich. ■☆

389513 Thermopsis atrata Czefr. ＝Thermopsis barbata Benth. ■

389514 Thermopsis barbata Benth.；紫花野决明（紫花黄华）；Purple Wildsenna, Purpleflower Thermopsis ■

389515 Thermopsis barbata Benth. f. chrysantha P. S. Li ＝Thermopsis gyirongensis S. Q. Wei ■

389516 Thermopsis barbata Royle ＝Thermopsis barbata Benth. ■

389517 Thermopsis caroliniana M. A. Curtis ＝Thermopsis villosa

(Walter) Fernald et B. G. Schub. ■☆

389518 Thermopsis chinensis Benth. ex S. Moore;霍州油菜(高脚豪猪脚,高脚猪猪豆,小叶黄华,小叶野决明,野决明);China Wildsenna,Chinese Thermopsis,Huozhou Bird Rape ■

389519 Thermopsis dahurica Czefr. = Thermopsis lanceolata R. Br. ■

389520 Thermopsis dolichocarpa V. V. Nikitin ex Vassilcz. ;长果野决明;Longfruit Thermopsis ■

389521 Thermopsis fabacea (L.) DC. = Thermopsis lupinoides (L.) Link ●■

389522 Thermopsis fabacea (Pall.) DC. = Thermopsis lupinoides (L.) Link ●■

389523 Thermopsis glabra Czefr. = Thermopsis lanceolata R. Br. var. glabra (Czefr.) Yakovlev ■

389524 Thermopsis grubovii Czefr. = Thermopsis mongolica Czefr. ■

389525 Thermopsis gyirongensis S. Q. Wei;吉隆野决明(吉隆黄华);Gyirong Thermopsis,Jilong Thermopsis,Jilong Wildsenna ■

389526 Thermopsis hirsutissima Czefr. = Thermopsis mongolica Czefr. ■

389527 Thermopsis inflata Cambess. ;轮生叶野决明(轮生叶野决明黄华,胀果黄华);Inflatfruit Thermopsis,Verticillate Wildsenna ■

389528 Thermopsis inflata Cambess. = Thermopsis smithiana E. Peter ■

389529 Thermopsis kaxgarica Chang Y. Yang = Thermopsis turkestanica Gand. ■

389530 Thermopsis kuenlunica Czefr. = Thermopsis przewalskii Czefr. ■

389531 Thermopsis laburnifolia D. Don = Piptanthus nepalensis (Hook.) Sweet ●

389532 Thermopsis ladyginii Czefr. = Thermopsis przewalskii Czefr. ■

389533 Thermopsis lanceolata R. Br. ;披针叶野决明(黄花苦豆,绞蛆爬,苦豆,苦豆子,拉豆,面人眼睛,牧马豆,披针叶黄华,西那,野决明);Lanceleaf Thermopsis,Lanceleaf Wildsenna ■

389534 Thermopsis lanceolata R. Br. = Thermopsis lupinoides (L.) Link ●■

389535 Thermopsis lanceolata R. Br. subsp. turkestanica (Gand.) Gubanov = Thermopsis turkestanica Gand. ■

389536 Thermopsis lanceolata R. Br. var. glabra (Czefr.) Yakovlev;东方野决明■

389537 Thermopsis laygyginii Czefr. = Thermopsis przewalskii Czefr. ■

389538 Thermopsis licentiana E. Peter;光叶黄华■☆

389539 Thermopsis licentiana E. Peter = Thermopsis alpina (Pall.) Ledeb. ■

389540 Thermopsis lupinoides (L.) Link;野决明(花豆秧,黄花苦豆子,黄华,霍州油菜,苦豆,牧马豆,披针叶黄花,披针叶黄华,枪叶野决明, 土马豆);Bean Thermopsis, Bush Pea, Luping-like Thermopsis,Wildsenna ●■

389541 Thermopsis lupinoides (L.) Link = Thermopsis lanceolata R. Br. ■

389542 Thermopsis mollis Curtis;软毛野决明;Soft Thermopsis ■☆

389543 Thermopsis mongolica Czefr. ;蒙古野决明(萨乌尔黄华);Mongol Wildsenna,Mongolian Thermopsis,Sawuer Thermopsis ■

389544 Thermopsis montana Nutt. ex Torr. et Gray = Thermopsis rhombifolia Rich. ■☆

389545 Thermopsis orientalis Czefr. = Thermopsis lanceolata R. Br. var. glabra (Czefr.) Yakovlev ■

389546 Thermopsis przewalskii Czefr. ;青海野决明;Qinghai Thermopsis, Qinghai Wildsenna ■

389547 Thermopsis rhombifolia (Nutt. ex Pursh) Nutt. ex Rich. ;菱叶野决明;Buffalo Bean, False Lupin, False-lupine, Golden Banner, Golden Pea,Golden-bean,Prairie Thermopsis,Yellow Bean ■☆

389548 Thermopsis rhombifolia (Nutt. ex Pursh) Nutt. ex Richardson var. annulocarpa (A. Nelson) L. O. Williams = Thermopsis rhombifolia (Nutt. ex Pursh) Nutt. ex Rich. ■☆

389549 Thermopsis rhombifolia (Nutt. ex Pursh) Nutt. ex Richardson var. arenosa (A. Nelson) Larisey = Thermopsis rhombifolia (Nutt. ex Pursh) Nutt. ex Rich. ■☆

389550 Thermopsis rhombifolia Rich. = Thermopsis rhombifolia (Nutt. ex Pursh) Nutt. ex Rich. ■☆

389551 Thermopsis saurensis Chang Y. Yang = Thermopsis mongolica Czefr. ■

389552 Thermopsis schischkinii Czefr. = Thermopsis mongolica Czefr. ■

389553 Thermopsis sibirica Czefr. = Thermopsis inflata Cambess. ■

389554 Thermopsis sibirica Czefr. = Thermopsis lanceolata R. Br. ■

389555 Thermopsis smithiana E. Peter;矮生野决明;Dwarf Thermopsis, Dwarf Wildsenna ■

389556 Thermopsis tibetica Czefr. = Thermopsis przewalskii Czefr. ■

389557 Thermopsis turkestanica Gand. ;新疆野决明(喀什黄华,新疆黄华); Sinkiang Thermopsis, Turkestan Thermopsis, Turkestan Wildsenna ■

389558 Thermopsis villosa (Walter) Fernald et B. G. Schub. ;美洲野决明(长柔毛黄华,美洲决明);Aaron's Rod, Blue-ridge Buckbean, Bush Pea, Carolina Bush Pea, Carolina Lupine, Carolina Thermopsis, False Lupine,Thermopsis ■☆

389559 Thermopsis yushuensis S. Q. Wei;玉树黄华(玉树野决明); Yushu Thermopsis,Yushu Wildsenna ■

389560 Therocistus Holub = Tuberaria (Dunal) Spach(保留属名)■●

389561 Therofon Raf. = Boykinia Nutt. (保留属名)●■☆

389562 Therogeron DC. = Minuria DC. ●●☆

389563 Therolepta Raf. = Marshallia Schreb. (保留属名)■☆

389564 Therophon Rydb. = Boykinia Nutt. (保留属名)●■☆

389565 Therophon Rydb. = Therofon Raf. ●■☆

389566 Therophonum Post et Kuntze = Therophon Rydb. ●■☆

389567 Theropogon Maxim. (1871);夏须草属;Theropogon ■

389568 Theropogon pallidus Maxim. ;夏须草;Common Theropogon, Theropogon ■

389569 Therorhodion (Maxim.) Small = Rhododendron L. ●

389570 Therorhodion (Maxim.) Small(1914);云间杜鹃属(弯柱杜鹃属);Therorhodion ●

389571 Therorhodion Small = Therorhodion (Maxim.) Small ●

389572 Therorhodion camtschaticum (Pall.) Small;云间杜鹃(勘察加杜鹃,勘察加云间杜鹃);Kamtschatka Azalea, Kamtschatka Rhododendron ●☆

389573 Therorhodion camtschaticum (Pall.) Small f. albiflorum Nakai; 白花云间杜鹃●☆

389574 Therorhodion camtschaticum (Pall.) Small var. barbatum Nakai = Therorhodion camtschaticum (Pall.) Small var. pumilum (E. A. Busch) T. Yamaz. ●☆

389575 Therorhodion camtschaticum (Pall.) Small var. pumilum (E. A. Busch) T. Yamaz. ;小云间杜鹃●☆

389576 Therorhodion glandulosum Standl. ex Small = Therorhodion camtschaticum (Pall.) Small var. pumilum (E. A. Busch) T. Yamaz. ●☆

389577 Therorhodion redowskianum (Maxim.) Hutch. ;苞叶云间杜鹃(苞叶杜鹃,弯柱杜鹃,叶状苞杜鹃,有苞杜鹃);Leaflikebract Azalea, Redowsk Therorhodion, Redowsk's Rhododendron, Summer Red Azalea ●◇

389578 Therorhodion redowskianum (Maxim.) Hutch. = Rhododendron redowskianum Maxim. ●◇

389579　Thesiaceae Vest ＝Santalaceae R. Br.（保留科名）●■

389580　Thesidium Sond.（1857）；小百蕊草属■☆

389581　Thesidium exocarpaeoides Sond. ＝Thesidium fragile（Thunb.）Sond. ■☆

389582　Thesidium fragile（Thunb.）Sond. ；纤细小百蕊草■☆

389583　Thesidium fruticulosum A. W. Hill；灌木状小百蕊草■☆

389584　Thesidium hirtum Sond. ；多毛小百蕊草■☆

389585　Thesidium leptostachyum（A. DC.）A. DC. ；细穗小百蕊草■☆

389586　Thesidium longifolium A. W. Hill ＝Thesidium fruticulosum A. W. Hill ■☆

389587　Thesidium microcarpum（A. DC.）A. DC. ＝Thesidium fragile（Thunb.）Sond. ■☆

389588　Thesidium minus A. W. Hill ＝Thesidium fruticulosum A. W. Hill ■☆

389589　Thesidium podocarpum（A. DC.）A. DC. ＝Thesidium fragile（Thunb.）Sond. ■☆

389590　Thesidium thunbergii Sond. ＝Thesidium fragile（Thunb.）Sond. ■☆

389591　Thesion St. -Lag. ＝Thesium L. ■

389592　Thesium L. （1753）；百蕊草属；Bastard Toadflax，Bastardtoadflax，Hundredstamen ■

389593　Thesium acuminatum A. W. Hill；渐尖百蕊草■☆

389594　Thesium acutissimum A. DC. ；尖百蕊草■☆

389595　Thesium affine Schltr. ；近缘百蕊草■☆

389596　Thesium afghanicum Hendrych ＝Thesium hookeri Hendrych ■☆

389597　Thesium alatavicum Kar. et Kir. ；阿拉套百蕊草■

389598　Thesium alatum Hilliard et B. L. Burtt；具翅百蕊草■☆

389599　Thesium albomontanum Compton；山地白百蕊草■☆

389600　Thesium alpinum L. ；高山百蕊草；Alpine Bastardtoadflax，Alpine Hundredstamen ■☆

389601　Thesium amicorum Lawalrée；可爱百蕊草■☆

389602　Thesium amplexicaule L. ＝Thesium euphorbioides P. J. Bergius ■☆

389603　Thesium andongense Hiern；安东百蕊草■☆

389604　Thesium angolense Pilg. ；安哥拉百蕊草■☆

389605　Thesium angulosum DC. ；棱角百蕊草☆

389606　Thesium annulatum A. W. Hill；环状百蕊草■☆

389607　Thesium annuum Lawalrée；一年百蕊草■☆

389608　Thesium apiculatum Sond. ＝Thesium acutissimum A. DC. ■☆

389609　Thesium archeri Compton；阿谢尔百蕊草■☆

389610　Thesium aristatum Schltr. ；具芒百蕊草■☆

389611　Thesium arvense Horv. ；田野百蕊草；Field Bastardtoadflax，Field Hundredstamen ■

389612　Thesium asperifolium A. W. Hill；糙叶百蕊草■☆

389613　Thesium assimile Sond. ＝Thesium carinatum A. DC. ■☆

389614　Thesium asterias A. W. Hill；星百蕊草■☆

389615　Thesium bangweolense R. E. Fr. ；班韦百蕊草■☆

389616　Thesium bequaertii Robyns et Lawalrée；贝卡尔百蕊草■☆

389617　Thesium bimalense Royle；露柱百蕊草■☆

389618　Thesium boissierianum A. DC. ；布瓦西耶百蕊草■☆

389619　Thesium bomiense C. Y. Wu et D. D. Tao；波密百蕊草；Bomi Bastardtoadflax，Bomi Hundredstamen ■

389620　Thesium brachyanthum Baker；短花百蕊草■☆

389621　Thesium brachycephalum Sond. ＝Thesium capitellatum A. DC. ■☆

389622　Thesium brachygyne Schltr. ；短蕊百蕊草■☆

389623　Thesium brachyphyllum Boiss. ；短叶百蕊草■☆

389624　Thesium brevibarbatum Pilg. ；短髯毛百蕊草■☆

389625　Thesium brevibracteatum P. C. Tam；短苞百蕊草；Shortbract Bastardtoadflax，Shortbract Hundredstamen ■

389626　Thesium brevifolium A. DC. ＝Thesium leptocaule Sond. ■☆

389627　Thesium breyeri N. E. Br. ；布鲁尔百蕊草■☆

389628　Thesium burchellii A. W. Hill；伯切尔百蕊草■☆

389629　Thesium burkei A. W. Hill；伯克百蕊草■☆

389630　Thesium caespitosum Robyns et Lawalrée；簇生百蕊草■☆

389631　Thesium capitatum L. ；头状百蕊草■☆

389632　Thesium capitellatum A. DC. ；短头百蕊草■☆

389633　Thesium capituliflorum Sond. ；杯花百蕊草■☆

389634　Thesium carinatum A. DC. ；龙骨状百蕊草■☆

389635　Thesium carinatum A. DC. var. pallidum A. W. Hill；苍白龙骨状百蕊草■☆

389636　Thesium cathaicum Hendrych；华北百蕊草；N. China Bastardtoadflax，N. China Hundredstamen ■

389637　Thesium chimanimaniense Brenan；奇马尼马尼百蕊草■☆

389638　Thesium chinense Turcz. ；百蕊草（白风草，百菜子，百乳草，草檀，打食草，地石榴，肚蹄草，凤芽蒿，疖积草，黄花蛇舌草，积药草，青龙草，青天白，珊瑚草，石菜子，松毛参，细须草，小草，一棵松，珍珠草）；China Hundredstamen，Chinese Bastardtoadflax ■

389639　Thesium chinense Turcz. var. longipedunculatum Y. C. Chu；长梗百蕊草；Longstalk Chinese Bastardtoadflax ■

389640　Thesium cinereum A. W. Hill；灰色百蕊草■☆

389641　Thesium classovianum Fisch. ex Trautv. ＝Thesium longifolium Turcz. ■

389642　Thesium colpoon L. f. ＝Osyris compressa（P. J. Bergius）A. DC. ●☆

389643　Thesium commutatum Sond. ；变异百蕊草■☆

389644　Thesium confine Sond. ；邻近百蕊草■☆

389645　Thesium congestum R. A. Dyer；团集百蕊草■☆

389646　Thesium conostylum Schltr. ；合柱百蕊草■☆

389647　Thesium cordatum A. W. Hill；心形百蕊草■☆

389648　Thesium coriarium A. W. Hill；革质百蕊草■☆

389649　Thesium cornigerum A. W. Hill；角状百蕊草■☆

389650　Thesium corymbuligerum Sond. ＝Thesium virgatum Lam. ■☆

389651　Thesium costatum A. W. Hill；单脉百蕊草■☆

389652　Thesium costatum A. W. Hill var. juniperinum ?；刺柏状单脉百蕊草■☆

389653　Thesium costatum A. W. Hill var. paniculatum N. E. Br. ；圆锥单脉百蕊草■☆

389654　Thesium crassipes Robyns et Lawalrée；粗梗百蕊草■☆

389655　Thesium cruciatum A. W. Hill ＝Thesium lacinulatum A. W. Hill ■☆

389656　Thesium cupressoides A. W. Hill；铜色百蕊草■☆

389657　Thesium cuspidatum A. W. Hill ＝Thesium capituliflorum Sond. ■☆

389658　Thesium cymosum A. W. Hill；聚伞百蕊草■☆

389659　Thesium cytisoides A. W. Hill；金雀花百蕊草■☆

389660　Thesium deceptum N. E. Br. ；十数百蕊草■☆

389661　Thesium decipiens Hilliard et B. L. Burtt；迷惑百蕊草■☆

389662　Thesium decurrens Blume ex A. DC. ＝Thesium chinense Turcz. ■

389663　Thesium densiflorum A. DC. ；密花百蕊草■☆

389664　Thesium densum N. E. Br. ；密集百蕊草■☆

389665　Thesium dinteri A. W. Hill ＝Thesium megalocarpum A. W. Hill ■☆

389666　Thesium disciflorum A. W. Hill；盘花百蕊草■☆

389667　Thesium disparile N. E. Br. ；异型百蕊草■☆

389668　Thesium dissitiflorum Schltr. ；疏花百蕊草■☆

389669　Thesium divaricatum Mert. et Koch；叉开百蕊草■☆

389670　Thesium divaricatum Mert. et Koch var. oranense Faure ＝ Thesium divaricatum Mert. et Koch ■☆

389671　Thesium diversifolium Sond. ;异叶百蕊草■☆

389672　Thesium dokerlaense C. Y. Wu ex D. D. Tao;德钦百蕊草; Deqin Bastardtoadflax, Deqin Hundredstamen ■

389673　Thesium dokerlaense C. Y. Wu ex D. D. Tao ＝ Thesium emodii Hendrych ■

389674　Thesium dolichomeres Brenan;长百蕊草■☆

389675　Thesium doloense Pilg. ;多罗百蕊草■☆

389676　Thesium durum Hilliard et B. L. Burtt;硬百蕊草■☆

389677　Thesium ebracteatum Hayne; 无苞百蕊草; Bractless Bastardtoadflax ■☆

389678　Thesium ecklonianum Sond. ;埃氏百蕊草☆

389679　Thesium elatius Sond. ;高百蕊草■☆

389680　Thesium emodii Hendrych;藏南百蕊草(短梗百蕊草);S. Xizang Bastardtoadflax, S. Xizang Hundredstamen ■

389681　Thesium ephedroides A. W. Hill ＝ Thesium lineatum L. f. ■☆

389682　Thesium equisetoides Welw. ex Hiern;木贼百蕊草■☆

389683　Thesium ericifolium A. DC. ;毛叶百蕊草☆

389684　Thesium erythronicum Pamp. ;浅红百蕊草■☆

389685　Thesium euphorbioides P. J. Bergius;大戟百蕊草■☆

389686　Thesium euphrasioides A. DC. ;小米草状百蕊草■☆

389687　Thesium exile N. E. Br. ;瘦小百蕊草■☆

389688　Thesium fallax Schltr. ;含糊百蕊草■☆

389689　Thesium fanshawei Hilliard;范肖百蕊草■☆

389690　Thesium fastigiatum A. W. Hill;束生百蕊草■☆

389691　Thesium ferganeuse Bobrov;费尔干百蕊草■☆

389692　Thesium filipes A. W. Hill;丝梗百蕊草■☆

389693　Thesium fimbriatum A. W. Hill;流苏百蕊草■☆

389694　Thesium flexuosum A. DC. ;曲折百蕊草■☆

389695　Thesium floribundum A. W. Hill ＝ Thesium pallidum A. DC. ☆

389696　Thesium foliosum A. DC. ;密叶百蕊草■☆

389697　Thesium foveolatum Schltr. ＝ Thesium capitellatum A. DC. ■☆

389698　Thesium fragile Link ex A. DC. ;脆百蕊草■☆

389699　Thesium frisea L. var. thunbergii A. DC. ;通氏百蕊草■☆

389700　Thesium fruticosum A. W. Hill;灌丛百蕊草■☆

389701　Thesium fulvum A. W. Hill;黄褐百蕊草■☆

389702　Thesium fuscum A. W. Hill;棕色百蕊草■☆

389703　Thesium galioides A. DC. ;拉拉藤百蕊草■☆

389704　Thesium germainii Robyns et Lawalrée;杰曼百蕊草■☆

389705　Thesium glaucescens A. W. Hill;灰绿百蕊草■☆

389706　Thesium glomeratum A. W. Hill;团聚百蕊草■☆

389707　Thesium glomeruliflorum Sond. ;团花百蕊草■☆

389708　Thesium gnidiaceum A. DC. ;格尼瑞香百蕊草■☆

389709　Thesium gnidiaceum A. DC. var. zeyheri (Sond.) A. DC. ;泽赫格尼瑞香百蕊草■☆

389710　Thesium goetzeanum Engl. ;格兹百蕊草■☆

389711　Thesium gontscharovii Bobrov;高恩恰洛夫百蕊草■☆

389712　Thesium gracile A. W. Hill;纤细百蕊草■☆

389713　Thesium griseum Sond. ;灰百蕊草■☆

389714　Thesium gypsophiloides A. W. Hill;喜钙百蕊草■☆

389715　Thesium hararensis A. G. Mill. ;哈拉雷百蕊草■☆

389716　Thesium helichrysoides A. W. Hill;蜡菊百蕊草■☆

389717　Thesium himalense Royle ＝ Thesium himalense Royle ex Edgew. ■

389718　Thesium himalense Royle ex Edgew. ;喜马拉雅百蕊草(九仙草,露珠百蕊草,露柱百蕊草,绿珊瑚,西域百蕊草);Himalayan Bastardtoadflax, Himalayas Hundredstamen ■

389719　Thesium himalense Royle var. pachyrhizum Hook. f. ＝ Thesium longiflorum Hand. -Mazz. ■

389720　Thesium hirsutum A. W. Hill;粗毛百蕊草■☆

389721　Thesium hispidulum Lam. ex Sond. var. subglabrum A. W. Hill;近光百蕊草■☆

389722　Thesium hockii Robyns et Lawalrée;霍克百蕊草■☆

389723　Thesium hollandii Compton;霍兰百蕊草■☆

389724　Thesium hookeri Hendrych;虎克百蕊草■☆

389725　Thesium hookeri Hendrych var. jarmilae (Hendrych) Y. J. Nasir ＝ Thesium jarmilae Hendrych ■

389726　Thesium horridum Pilg. ;多刺百蕊草■☆

389727　Thesium humifusum DC. ;叉枝百蕊草;Bastard Toadflax ■☆

389728　Thesium humifusum DC. subsp. divaricatum (Mert. et Koch) Maire ＝ Thesium divaricatum Mert. et Koch ■☆

389729　Thesium humifusum DC. var. longibracteatum Willk. ＝ Thesium divaricatum Mert. et Koch ■☆

389730　Thesium humifusum DC. var. oranense Faure ＝ Thesium divaricatum Mert. et Koch ■☆

389731　Thesium humifusum DC. var. transiens Maire ＝ Thesium divaricatum Mert. et Koch ■☆

389732　Thesium humile Vahl;矮小百蕊草■☆

389733　Thesium humile Vahl f. maritima N. D. Simpson ＝ Thesium humile Vahl var. maritima (N. D. Simpson) Sa'ad ■☆

389734　Thesium humile Vahl var. maritima (N. D. Simpson) Sa'ad ＝ Thesium humile Vahl ■☆

389735　Thesium hystricoides A. W. Hill;拟豪猪百蕊草■☆

389736　Thesium hystrix A. W. Hill;豪猪百蕊草■☆

389737　Thesium imbricatum Thunb. ;覆瓦百蕊草■☆

389738　Thesium imbricatum Thunb. var. zeyheri Sond. ＝ Thesium gnidiaceum A. DC. var. zeyheri (Sond.) A. DC. ■☆

389739　Thesium impeditum A. W. Hill;赘百蕊草■☆

389740　Thesium indicum Hendrych ＝ Thesium himalense Royle ex Edgew. ■

389741　Thesium inhambanense Hilliard;伊尼扬巴内百蕊草■☆

389742　Thesium inversum N. E. Br. ;俯垂百蕊草■☆

389743　Thesium jarmilae Hendrych;珠峰百蕊草(大果百蕊草); Bigfruit Bastardtoadflax, Bigfruit Hundredstamen ■

389744　Thesium jarmilae Hendrych ＝ Thesium hookeri Hendrych var. jarmilae (Hendrych) Y. J. Nasir ■

389745　Thesium jeaniae Brenan;琼氏百蕊草■☆

389746　Thesium junceum Bernh. ex Krauss;灯心草状百蕊草■☆

389747　Thesium junceum Bernh. ex Krauss var. mammosum A. W. Hill;多乳突百蕊草■☆

389748　Thesium juncifolium A. DC. ;灯心草叶百蕊草■☆

389749　Thesium junodii A. W. Hill;朱诺德百蕊草■☆

389750　Thesium karooicum Compton;卡鲁百蕊草■☆

389751　Thesium katangense Robyns et Lawalrée;加丹加百蕊草■☆

389752　Thesium kilimandscharicum Engl. ;基利百蕊草■☆

389753　Thesium kotschyanum Boiss. ;考奇百蕊草■☆

389754　Thesium lacinulatum A. W. Hill;条裂百蕊草■☆

389755　Thesium laetum Robyns et Lawalrée;愉悦百蕊草■☆

389756　Thesium leptocaule Sond. ;细茎百蕊草■☆

389757　Thesium leptostachyum A. DC. ＝ Thesidium leptostachyum (A. DC.) A. DC. ■☆

389758　Thesium lesliei N. E. Br. ;莱斯利百蕊草■☆

389759　Thesium leucanthum Gilg;白花百蕊草■☆

389760　Thesium lewallei Lawalrée;勒瓦莱百蕊草■☆

389761　Thesium libericum Hepper et Keay;利比里亚百蕊草■☆

389762　Thesium lineatum L. f. ;线纹百蕊草■☆

389763　Thesium linifolium Schrank;线叶百蕊草■☆

389764　Thesium linophyllum L. = Thesium divaricatum Mert. et Koch ■☆

389765　Thesium lisowskii Lawalrée;利索百蕊草■☆

389766　Thesium litoreum Brenan;滨海百蕊草■☆

389767　Thesium lobelioides A. DC. ;拟洛贝尔百蕊草■☆

389768　Thesium longiflorum Hand. -Mazz. ;长花百蕊草(长花绿珊瑚，长叶百蕊草，九龙草，酒草，绿珊瑚，撒花一棵针，山柏枝，西域百蕊草，细须草，小星宿草，一棵松，珍珠草);Longflower Bastardtoadflax, Longflower Hundredstamen ■

389769　Thesium longifolium Turcz. ;长叶百蕊草(九龙草，九仙草，酒草，酒龙草，酒仙草，绿珊瑚，茅草细辛，撒花一棵针，山柏枝，铁刷把，西域百蕊草，细须草，小星宿草，一棵松，珍珠草);Longleaf Bastardtoadflax, Longleaf Hundredstamen ■

389770　Thesium longifolium Turcz. var. vlassovianum A. DC. = Thesium longifolium Turcz. ■

389771　Thesium longirostre Schltr. = Thesium zeyheri A. DC. ■☆

389772　Thesium luembense Robyns et Lawalrée;卢恩贝百蕊草■☆

389773　Thesium lycopodioides Gilg;石松百蕊草■☆

389774　Thesium lynesii Robyns et Lawalrée;莱恩斯百蕊草■☆

389775　Thesium macrogyne A. W. Hill;大柱百蕊草■☆

389776　Thesium macrostachyum A. DC. ;大穗百蕊草■☆

389777　Thesium magnifructum Hilliard;大果百蕊草■☆

389778　Thesium malaissei Lawalrée;马莱泽百蕊草■☆

389779　Thesium manikense Robyns et Lawalrée;马尼科百蕊草■☆

389780　Thesium maritimum C. A. Mey. ;沼岸百蕊草■☆

389781　Thesium marlothii Schltr. ;马洛斯百蕊草■☆

389782　Thesium masukense Baker;马苏克百蕊草■☆

389783　Thesium matteii Chiov. ;马特百蕊草■☆

389784　Thesium mauritanicum Batt. ;毛里塔尼亚百蕊草■☆

389785　Thesium maximiliani Schltr. ;马克西米利亚诺百蕊草■☆

389786　Thesium megalocarpum A. W. Hill;热非大果百蕊草■☆

389787　Thesium microcarpum A. DC. = Thesidium fragile (Thunb.) Sond. ■☆

389788　Thesium microcephalum A. W. Hill;小头百蕊草■☆

389789　Thesium micromeria A. DC. ;小巧百蕊草■☆

389790　Thesium microphyllum Robyns et Lawalrée;小叶百蕊草■☆

389791　Thesium micropogon A. DC. ;小毛百蕊草■☆

389792　Thesium minkwitzianum B. Fedtsch. ;闵克百蕊草■☆

389793　Thesium mossii N. E. Br. ;莫西百蕊草■☆

389794　Thesium mukense A. W. Hill;穆卡百蕊草■☆

389795　Thesium multicaule Ledeb. ;多茎百蕊草■☆

389796　Thesium multiramulosum Pilg. ;多小枝百蕊草■☆

389797　Thesium myriocladum Baker;密枝百蕊草■☆

389798　Thesium namaquense Schltr. ;纳马夸百蕊草■☆

389799　Thesium natalense Sond. ;纳塔尔百蕊草■☆

389800　Thesium nigricans Rendle;变黑百蕊草■☆

389801　Thesium nigrum A. W. Hill;黑百蕊草■☆

389802　Thesium nudicaule A. W. Hill;裸茎百蕊草■☆

389803　Thesium nutans Robyns et Lawalrée;悬垂百蕊草■☆

389804　Thesium occidentale A. W. Hill;西方百蕊草■☆

389805　Thesium orgadophyum P. C. Tam;草地百蕊草;Lawn Bastardtoadflax, Lawn Hundredstamen ■

389806　Thesium orientale A. W. Hill;东部百蕊草■☆

389807　Thesium pallidum A. DC. ;苍白百蕊草■☆

389808　Thesium palliolatum A. W. Hill　= Thesium gracile A. W. Hill ■☆

389809　Thesium paniculatum L. ;圆锥百蕊草■☆

389810　Thesium paronychioides Sond. ;指甲草状百蕊草■☆

389811　Thesium passerinoides Robyns et Lawalrée;雀状百蕊草■☆

389812　Thesium patersonae A. W. Hill;帕特森百蕊草■☆

389813　Thesium patulum A. W. Hill;张开百蕊草■☆

389814　Thesium pawlowskianum Lawalrée;帕沃夫斯基百蕊草■☆

389815　Thesium penicillatum A. W. Hill;笔状百蕊草■☆

389816　Thesium phyllostachyum Sond. ;叶穗百蕊草■☆

389817　Thesium pilosum A. W. Hill;疏毛百蕊草■☆

389818　Thesium pinifolium A. DC. ;松叶百蕊草■☆

389819　Thesium pleuroloma A. W. Hill;侧边百蕊草■☆

389820　Thesium podocarpum A. DC. = Thesidium fragile (Thunb.) Sond. ■☆

389821　Thesium polycephalum Schltr. ;多头百蕊草■☆

389822　Thesium polygaloides A. W. Hill;远志百蕊草■☆

389823　Thesium pottiae N. E. Br. ;波蒂百蕊草■☆

389824　Thesium procerum N. E. Br. ;高大百蕊草■☆

389825　Thesium procumbens C. A. Mey. ;平铺百蕊草;Procumbent Bastardtoadflax ■☆

389826　Thesium prostratum A. W. Hill;平卧百蕊草■☆

389827　Thesium pseudovirgatum Levyns;假枝百蕊草■☆

389828　Thesium psilotocladum Svent. = Kunkeliella psilotoclada (Svent.) Stearn ●☆

389829　Thesium psilotoides Hance;白云百蕊草;Psilotum-like Bastardtoadflax, Whitecloud Hundredstamen ■

389830　Thesium pubescens A. DC. ;短柔毛百蕊草■☆

389831　Thesium pungens A. W. Hill;刚毛百蕊草■☆

389832　Thesium pycnanthum Schltr. ;西方密花百蕊草■☆

389833　Thesium pygmaeum Hilliard;矮生百蕊草■☆

389834　Thesium quarrei Robyns et Lawalrée;卡雷百蕊草■☆

389835　Thesium quinqueflorum Sond. ;五花百蕊草■☆

389836　Thesium racemosum Bernh. ex Krauss;总花百蕊草■☆

389837　Thesium radicans Hochst. ex A. Rich. ;辐射百蕊草■☆

389838　Thesium ramosissimum Bobrov;多枝百蕊草■☆

389839　Thesium ramosoides Hendrych;滇西百蕊草(绿珊瑚);W. Yunnan Bastardtoadflax, W. Yunnan Hundredstamen ■

389840　Thesium ramosum Hayne;分枝百蕊草;Branchy Bastardtoadflax ■☆

389841　Thesium rariflorum Sond. ;稀花百蕊草■☆

389842　Thesium rectangulum Welw. ex Hiern;直角百蕊草■☆

389843　Thesium reekmansii Lawalrée;里克曼斯百蕊草■☆

389844　Thesium refractum C. A. Mey. ;急折百蕊草(反折百蕊草，狗牙草，九龙草，九仙草，六天草，松毛参，西伯利亚百蕊草);Refracte Bastardtoadflax, Refracte Hundredstamen ■

389845　Thesium refractum C. A. Mey. var. hirtulum Krylov;毛急折百蕊草■

389846　Thesium remotebracteatum C. Y. Wu et D. D. Tao;远苞百蕊草(疏苞百蕊草);Laxbract Bastardtoadflax ■

389847　Thesium repandum A. W. Hill;浅波状百蕊草■☆

389848　Thesium repens Ledeb. ;匍匐百蕊草(匍生百蕊草);Creeping Bastardtoadflax ■

389849　Thesium resedoides A. W. Hill;木犀草百蕊草■☆

389850　Thesium resinifolium N. E. Br. ;胶叶百蕊草■☆

389851　Thesium rigidum Sond. = Thesium lineatum L. f. ■☆

389852　Thesium robynsii Lawalrée;罗宾斯百蕊草■☆

389853　Thesium rogersii A. W. Hill;罗杰斯百蕊草■☆

389854　Thesium rufescens A. W. Hill;焦黄百蕊草■☆

389855 Thesium rugulosum Bunge ex A. DC. = Thesium chinense Turcz. ■

389856 Thesium rupestre Ledeb. ;岩地百蕊草■☆

389857 Thesium saxatile Turcz. ex DC. ;岩生百蕊草■☆

389858 Thesium scabridulum A. W. Hill;微糙百蕊草■☆

389859 Thesium scabrum L. ;粗糙百蕊草■☆

389860 Thesium scandens E. Mey. ex A. DC. ;攀缘百蕊草■☆

389861 Thesium schimperianum Hochst. ex A. Rich. = Osyridicarpos schimperianus（Hochst. ex A. Rich.）A. DC. ●☆

389862 Thesium schlechteri A. W. Hill = Thesium zeyheri A. DC. ■☆

389863 Thesium schliebenii Pilg. ;施利本百蕊草■☆

389864 Thesium schmitzii Robyns et Lawalrée;施密茨百蕊草■☆

389865 Thesium schumannianum Schltr. ;舒曼百蕊草■☆

389866 Thesium schweinfurthii Engl. ;施韦百蕊草■☆

389867 Thesium scirpioides A. W. Hill;藨草状百蕊草■☆

389868 Thesium scoparium Peter;帚状百蕊草■☆

389869 Thesium sedifolium A. DC. ex Levyns;景天叶百蕊草■☆

389870 Thesium selagineum A. DC. ;石松状百蕊草■☆

389871 Thesium semotum N. E. Br. ;分离百蕊草■☆

389872 Thesium setulosum Robyns et Lawalrée;小刚毛百蕊草■☆

389873 Thesium shabense Lawalrée;沙巴百蕊草■☆

389874 Thesium singulare Hilliard;单一百蕊草■☆

389875 Thesium sonderianum Schltr. ;森诺百蕊草■☆

389876 Thesium sparteum R. Br. = Thesium lineatum L. f. ■☆

389877 Thesium spartioides A. W. Hill;绳索百蕊草■☆

389878 Thesium spathulatum Blume. = Dendrotrophe umbellata（Blume）Miq. ●

389879 Thesium sphaerocarpum Robyns et Lawalrée;球果百蕊草■☆

389880 Thesium spicatum L. ;穗状百蕊草■☆

389881 Thesium spinosum L. f. ;具刺百蕊草■☆

389882 Thesium spinulosum A. DC. ;细刺百蕊草■☆

389883 Thesium squarrosum L. f. ;粗鳞百蕊草■☆

389884 Thesium strictum P. J. Bergius;刚直百蕊草■☆

389885 Thesium strigulosum Welw. ex Hiern;硬毛百蕊草■☆

389886 Thesium stuhlmannii Engl. ;斯图尔曼百蕊草■☆

389887 Thesium subaphyllum Engl. ;亚无叶百蕊草■☆

389888 Thesium subnudum Sond. ;亚裸百蕊草■☆

389889 Thesium subnudum Sond. var. foliosum A. W. Hill;多叶百蕊草■☆

389890 Thesium subsimile N. E. Br. ;相似百蕊草■☆

389891 Thesium susannae A. W. Hill;苏珊娜百蕊草■☆

389892 Thesium symoensii Lawalrée;西莫百蕊草■☆

389893 Thesium szovitsii DC. ;绍氏百蕊草■☆

389894 Thesium tenuissimum Hook. f. ;极细百蕊草■☆

389895 Thesium teretifolium R. Br. = Thesium spinosum L. f. ■☆

389896 Thesium tetragonum A. W. Hill;四角百蕊草■☆

389897 Thesium thomsonii Hendrych = Thesium himalense Royle ex Edgew. ■

389898 Thesium thunbergianum A. DC. ;通贝里百蕊草■☆

389899 Thesium tongolicum Hendrych;藏东百蕊草（东俄洛百蕊草）;Tongol Bastardtoadflax,Tongol Hundredstamen ■

389900 Thesium transgariepinum Sond. = Thesium zeyheri A. DC. ■☆

389901 Thesium translucens A. W. Hill;半透明百蕊草■☆

389902 Thesium transvaalense Schltr. ;德兰士瓦百蕊草■☆

389903 Thesium triflorum Thunb. ex L. f. ;三花百蕊草●☆

389904 Thesium triste A. W. Hill;暗淡百蕊草■☆

389905 Thesium ulugurense Engl. ;乌卢古尔百蕊草■☆

389906 Thesium umbellatum L. = Comandra umbellata（L.）Nutt. ●☆

389907 Thesium umbelliferum A. W. Hill;伞花百蕊草■☆

389908 Thesium urceolatum A. W. Hill;坛状百蕊草■☆

389909 Thesium urundiense Robyns et Lawalrée;乌隆迪百蕊草■☆

389910 Thesium ussanguense Engl. ;乌桑古百蕊草■☆

389911 Thesium utile A. W. Hill;有用百蕊草■☆

389912 Thesium vimineum Robyns et Lawalrée;软枝百蕊草■☆

389913 Thesium virens E. Mey. ex A. DC. ;绿百蕊草■☆

389914 Thesium virgatum Lam. ;条纹百蕊草■☆

389915 Thesium viride A. W. Hill;绿花百蕊草■☆

389916 Thesium viridifolium Levyns;绿叶百蕊草■☆

389917 Thesium vlassovianum（A. DC.）Trautv. = Thesium longifolium Turcz. ■

389918 Thesium welwitschii Hiern;韦尔百蕊草■☆

389919 Thesium whitehillense Compton;怀特山百蕊草■☆

389920 Thesium whyteanum Rendle;怀特百蕊草■☆

389921 Thesium wightianum Wall. var. radicans DC. = Thesium radicans Hochst. ex A. Rich. ■☆

389922 Thesium wilczekianum Lawalrée;维尔切克百蕊草■☆

389923 Thesium wittei De Wild. et Staner;维特百蕊草■☆

389924 Thesium xerophyticum A. W. Hill;旱生百蕊草■☆

389925 Thesium zeyheri A. DC. ;泽赫百蕊草■☆

389926 Thesmophora Rourke(1993);好望角密穗木属●☆

389927 Thesmophora scopulosa Rourke;好望角密穗木●☆

389928 Thespesia Sol. ex Corrêa(1807)(保留属名);桐棉属（大萼葵属,截萼黄槿属,伞杨属,肖槿属）;Portia Tree,Portiatree ●

389929 Thespesia acutiloba（Baker f.）Exell et Mendonça;尖裂桐棉●☆

389930 Thespesia banalo Blanco = Thespesia populneoides（Roxb.）Kostel. ●

389931 Thespesia danis Oliv. ;丹尼斯桐棉●☆

389932 Thespesia danis Oliv. var. grandibracteata Chiov. = Thespesia danis Oliv. ●☆

389933 Thespesia danis Oliv. var. somalica Chiov. = Thespesia danis Oliv. ●☆

389934 Thespesia debeerstii De Wild. et T. Durand = Thespesia garckeana F. Hoffm. ●☆

389935 Thespesia garckeana F. Hoffm. ;罗得西亚桐棉（加尔开桐棉）;Quarters,Rhodesian Tree Hibiscus ●☆

389936 Thespesia grandiflora DC. ;大花桐棉●

389937 Thespesia hockii De Wild. = Thespesia garckeana F. Hoffm. ●☆

389938 Thespesia howii S. Y. Hu;长梗桐棉（长梗肖槿）;How Portia Tree,How Portiatree,Longstalk Portiatree ●

389939 Thespesia howii S. Y. Hu = Thespesia populnea（L.）Sol. ex Corrêa ●

389940 Thespesia lampas（Cav.）Dalzell et A. Gibson = Thespesia lampas（Cav.）Dalzell ex Dalzell et A. Gibson ●

389941 Thespesia lampas（Cav.）Dalzell ex Dalzell et A. Gibson;白脚桐棉（白脚桐,山棉花,桐棉,肖槿）;Chinese Portia Tree,Chinese Portiatree,Whitefoot Portiatree ●

389942 Thespesia macrophylla Blume = Thespesia lampas（Cav.）Dalzell et A. Gibson ●

389943 Thespesia macrophylla Blume = Thespesia populnea（L.）Sol. ex Corrêa ●

389944 Thespesia mossambicensis（Exell et Hillc.）Fryxell;类桐棉●☆

389945 Thespesia populnea（L.）Corrêa = Thespesia populnea（L.）Sol. ex Corrêa ●

389946 Thespesia populnea（L.）Sol. ex Corrêa;桐棉（恒春黄槿,截萼黄槿,伞杨,杨叶肖槿）;Bendy Tree,Bhendi Tree,Corktree,False

Rosewood,Mahoe,Portia Oil Nut,Portia Oilnut,Portia Tree,Portiatree, Rose Wood,Sandalwood,Seaside Mahoe,Tulip Tree ●

389947　Thespesia populnea(L.)Sol. ex Corrêa var. acutiloba Baker f. = Thespesia acutiloba(Baker f.)Exell et Mendonça ●☆

389948　Thespesia populnea Sol. ex Corrêa = Thespesia howii S. Y. Hu ●

389949　Thespesia populneoides(L.)Sol. ex Corrêa;长梗肖槿;Long-peduncled Portia Tree ●

389950　Thespesia populneoides(Roxb.)Kostel. = Thespesia populnea(L.)Sol. ex Corrêa ●

389951　Thespesia rehmannii Szyszyl. = Cienfuegosia gerrardii(Harv.)Hochr. ●☆

389952　Thespesia rogersii S. Moore = Thespesia garckeana F. Hoffm. ●☆

389953　Thespesia trilobata Baker f.;三裂桐棉●☆

389954　Thespesiopsis Exell et Hillc. = Thespesia Sol. ex Corrêa(保留属名)●

389955　Thespesiopsis mossambicensis Exell et Hillc. = Thespesia mossambicensis(Exell et Hillc.)Fryxell ●☆

389956　Thespesocarpus Pierre = Diospyros L. ●

389957　Thespesocarpus tiliaceus Pierre = Diospyros soyauxii Gürke et K. Schum. ●☆

389958　Thespidium F. Muell. = Thespidium F. Muell. ex Benth. ■☆

389959　Thespidium F. Muell. ex Benth.(1867);腋基菊属■☆

389960　Thespidium basiflorum F. Muell.;腋基菊■☆

389961　Thespis DC.(1833);歧伞菊属(鳞冠菊属);Thespis ■

389962　Thespis divaricata DC.;歧伞菊;Divaricate Thespis ■

389963　Thevenotia DC.(1833);毛叶刺苞菊属■☆

389964　Thevenotia persica DC.;毛叶刺苞菊■☆

389965　Thevetia(L.)Juss. ex Endl. = Thevetia L.(保留属名)●

389966　Thevetia Adans. = Cerbera L. ●

389967　Thevetia L.(1758)(保留属名);黄花夹竹桃属(黄夹竹桃属);Lucky Bean,Thevetia ●

389968　Thevetia Vell. = Thevetia L.(保留属名)●

389969　Thevetia Vell. = Thevetiana Kuntze ●

389970　Thevetia ahouai(L.)A. DC.;阔叶黄花夹竹桃(阔叶夹竹桃,阔叶竹桃);Broadleaf Thevetia,Broad-leaved Thevetia ●

389971　Thevetia ahouai(L.)Juss. ex Endl. = Thevetia ahouai(L.)A. DC. ●

389972　Thevetia linearis A. DC. = Thevetia peruviana(Pers.)K. Schum. ●

389973　Thevetia linearis Raf. = Cascabela thevetia(L.)Lippold ●

389974　Thevetia neriifolia Juss. ex Steud. = Cascabela thevetia(L.)Lippold ●

389975　Thevetia neriifolia Juss. ex Steud. = Thevetia peruviana(Pers.)K. Schum. ●

389976　Thevetia nitida DC.;光亮黄花夹竹桃●☆

389977　Thevetia ovata A. DC.;卵叶黄花夹竹桃(卵形夹竹桃)●☆

389978　Thevetia peruviana(Pers.)K. Schum.;黄花夹竹桃(大飞酸,吊钟花,番仔桃,黄花状元竹,夹竹桃,酒杯花,菱角树,柳木子,树都拉,台湾柳,铁石榴,相等子,竹桃);Bastard Oleander,Be-still Tree,Exile Tree,French Willow,Good Luck Tree,Lucky Nut,Luckynut Thevetia,Lucky-nut Thevetia,Trumpet-flower,Yellow Oleander,Yellow Pleander ●

389979　Thevetia peruviana(Pers.)K. Schum.' Aurantiaca';红酒杯花;Orange Luckynut Thevetia ●

389980　Thevetia peruviana(Pers.)K. Schum. = Cascabela thevetia(L.)Lippold ●

389981　Thevetia thevetia(L.)Millsp. = Thevetia peruviana(Pers.)

K. Schum. ●

389982　Thevetia thevetia Millsp. = Thevetia peruviana(Pers.)K. Schum. ●

389983　Thevetia thevetioides K. Schum.;大黄花夹竹桃(墨西哥夹竹桃);Be-still Tree,Giant Thevetia,Large-flowered Yellow Oleander ●☆

389984　Thevetia yccotli A. DC.;墨西哥黄花夹竹桃;Mexican Oleander ●☆

389985　Thevetiana Kuntze = Thevetia L.(保留属名)●

389986　Theyga Molina = Laurelia Juss.(保留属名)●☆

389987　Theyga Molina = Thiga Molina ●☆

389988　Theyodis A. Rich. = Oldenlandia L. ●■

389989　Theyodis octodon A. Rich. = Oldenlandia capensis L. f. var. pleiosepala Bremek. ●☆

389990　Thezera Raf. = ? Rhus L. ●

389991　Thibaudia Ruiz et Pav.(1805);赤宝花属●☆

389992　Thibaudia Ruiz et Pav. ex J. St. -Hil. = Thibaudia Ruiz et Pav. ●☆

389993　Thibaudia acuminata Hook. f.;渐尖赤宝花●☆

389994　Thibaudia affinis Griff.;近缘赤宝花●☆

389995　Thibaudia albiflora A. C. Sm.;白花赤宝花●☆

389996　Thibaudia axillaris Rusby;腋生赤宝花●☆

389997　Thibaudia biflora Hoerold;双花赤宝花●☆

389998　Thibaudia boliviensis(Kuntze)Hoerold;玻利维亚赤宝花●☆

389999　Thibaudia crassifolia Benth.;厚叶赤宝花●☆

390000　Thibaudia densiflora(Herzog)A. C. Sm.;密花赤宝花●☆

390001　Thibaudia diphylla Dunal;二叶赤宝花●☆

390002　Thibaudia elliptica Ruiz et Pav.;椭圆赤宝花●☆

390003　Thibaudia falcata Kunth;镰赤宝花●☆

390004　Thibaudia flava Nutt. ex Hook. f.;黄赤宝花●☆

390005　Thibaudia floribunda Blume;繁花赤宝花●☆

390006　Thibaudia gaultheriifolia Griff. = Vaccinium gaultherifolium(Griff.)Hook. f. ●

390007　Thibaudia glabra Griff.;光赤宝花●☆

390008　Thibaudia lateriflora A. C. Sm.;侧花赤宝花●☆

390009　Thibaudia laurifolia Blume;桂叶赤宝花●☆

390010　Thibaudia leucostoma Sleumer;白口赤宝花●☆

390011　Thibaudia longifolia Kunth;长叶赤宝花●☆

390012　Thibaudia multiflora Klotzsch;多花赤宝花●☆

390013　Thibaudia myrtifolia Griff. = Agapetes serpens(Wight)Sleumer ●

390014　Thibaudia myrtifolia Griff. = Pentapterygium serpens(Wight)Klotzsch ●

390015　Thibaudia nitida Kunth;光亮赤宝花●☆

390016　Thibaudia oblongifolia Remy;矩圆叶赤宝花●☆

390017　Thibaudia obovata(Wight)Griff. = Agapetes obovata(Wight)Benth. et Hook. f. ●

390018　Thibaudia octandra(Sleumer)J. F. Macbr.;八蕊赤宝花●☆

390019　Thibaudia ovata Hoerold;卵形赤宝花●☆

390020　Thibaudia pachyantha A. C. Sm.;粗花赤宝花●☆

390021　Thibaudia pachypoda A. C. Sm.;大足赤宝花●☆

390022　Thibaudia paniculata A. C. Sm.;圆锥赤宝花●☆

390023　Thibaudia parvifolia Hoerold;小叶赤宝花●☆

390024　Thibaudia punctata Ruiz et Pav.;斑点赤宝花●☆

390025　Thibaudia retusa Griff. = Vaccinium retusum(Griff.)Hook. f. ex C. B. Clarke ●

390026　Thibaudia revoluta Griff. = Vaccinium dunalianum Wight ●

390027　Thibaudia serrata Wall. = Vaccinium vacciniaceum(Roxb.)Sleumer ●

390028　Thibaudia sessiliflora(A. C. Sm.)Luteyn;无梗花赤宝花●☆

390029　Thibaudia setigera(D. Don ex G. Don)Endl.;印度树萝卜●☆

390030　Thibaudia truncata A. C. Sm. ;平截赤宝花●☆

390031　Thibaudia uniflora A. C. Sm. ;单花赤宝花●☆

390032　Thibaudia viridiflora (Kuntze) K. Schum. ;绿花赤宝花●☆

390033　Thibaudia viridiflora (Kuntze) Schum. = Cavendishia martii (Meisn.) A. C. Sm. ●☆

390034　Thicuania Raf. = Dendrobium Sw. (保留属名)■

390035　Thicuania moschata (Buch. -Ham.) Raf. = Dendrobium moschatum (Buch. -Ham.) Sw. ■

390036　Thiebautia Colla = Bletia Ruiz et Pav. ■☆

390037　Thiebautia nervosa Colla = Bletia purpurea (Lam.) DC. ■☆

390038　Thieleodoxa Cham. = Alibertia A. Rich. ex DC. ☆

390039　Thiersia Baill. = Faramea Aubl. ●☆

390040　Thiga Molina = Laurelia Juss. (保留属名)●☆

390041　Thilachium Lour. (1790);合萼山柑属●☆

390042　Thilachium africanum Lour. ;非洲合萼山柑●☆

390043　Thilachium alboviolaceum Gilg;白堇色合萼山柑●☆

390044　Thilachium angustifolium Bojer;窄叶合萼山柑●☆

390045　Thilachium densiflorum Gilg et Gilg-Ben. ;密花合萼山柑●☆

390046　Thilachium humbertii Hadj-Moust. ;亨伯特合萼山柑●☆

390047　Thilachium laurifolium Baker;月桂叶合萼山柑●☆

390048　Thilachium macrophyllum Gilg;大叶合萼山柑●☆

390049　Thilachium mildbraedii Gilg = Thilachium thomasii Gilg ●☆

390050　Thilachium monophyllum Hadj-Moust. ;单叶合萼山柑●☆

390051　Thilachium ovalifolium Juss. = Thilachium africanum Lour. ●☆

390052　Thilachium papillososcabrum Chiov. = Thilachium thomasii Gilg ●☆

390053　Thilachium paradoxum Gilg;奇异合萼山柑●☆

390054　Thilachium querimbense Klotzsch = Thilachium africanum Lour. ●☆

390055　Thilachium roseomaculatum Y. B. Harv. et Vollesen;粉红斑合萼山柑●☆

390056　Thilachium seyrigii Hadj-Moust. ;塞里格合萼山柑●☆

390057　Thilachium thomasii Gilg;托马斯合萼山柑●☆

390058　Thilachium verrucosum Klotzsch = Thilachium africanum Lour. ●☆

390059　Thilakium Lour. = Thilachium Lour. ●☆

390060　Thilcum Molina = Fuchsia L. ●■

390061　Thilcum Molina = Tilco Adans. ●■

390062　Thillaea Sang. = Crassula L. ●■☆

390063　Thillaea Sang. = Tillaea L. ■

390064　Thiloa Eichler = Combretum Loefl. (保留属名)●

390065　Thiloa Eichler(1866);肖风车子属●☆

390066　Thiloa gracilis Eichler;细肖风车子●☆

390067　Thiloa nitida Eichler;光亮肖风车子●☆

390068　Thimus Neck. = Thymus L. ●

390069　Thinicola J. H. Ross = Templetonia R. Br. ex W. T. Aiton ●☆

390070　Thinicola J. H. Ross(2001);滨海傲慢木属●☆

390071　Thinobia Phil. = Nardophyllum (Hook. et Arn.) Hook. et Arn. ●☆

390072　Thinogeton Benth. = Cacabus Bernh. ■☆

390073　Thinogeton Benth. = Exodeconus Raf. ☆

390074　Thinopyrum Á. Löve = Elymus L. ■

390075　Thinopyrum Á. Löve = Elytrigia Desv. ■

390076　Thinopyrum Á. Löve(1980);岸边披碱草属■☆

390077　Thinopyrum bessarabicum (Savul. et Rayss) Á. Löve;岸边披碱草■☆

390078　Thinopyrum distichum (Thunb.) Á. Löve;二列岸边披碱草■☆

390079　Thinopyrum intermedium (Host) Barkworth et D. R. Dewey = Elytrigia intermedia (Host) Nevski ■

390080　Thinopyrum ponticum (Podp.) D. R. Dewey = Elymus

elongatus (Host) Runemark ■☆

390081　Thinopyrum pycnanthum (Godr.) Barkworth;密花岸边披碱草;Tick Quackgrass ■☆

390082　Thinouia Triana et Planch. (1862);温美无患子属●☆

390083　Thinouia mucronata Radlk. ;钝尖温美无患子●☆

390084　Thinouia myriantha Triana et Planch. ;温美无患子●☆

390085　Thinouia tomocarpa Standl. ;毛果温美无患子●☆

390086　Thiodia Beun. = Laetia Loefl. ex L. (保留属名)●☆

390087　Thiodia Beun. = Lightfootia Sw. ●☆

390088　Thiodia Griseb. = Zuelania A. Rich. ●☆

390089　Thiollierea Montrouz. = Bikkia Reinw. (保留属名)●☆

390090　Thiollierea Montrouz. = Randia L. ●

390091　Thisantha Eckl. et Zeyh. = Crassula L. ●■☆

390092　Thisantha Eckl. et Zeyh. = Tillaea L. ■

390093　Thisbe Falc. = Herminium L. ■

390094　Thiseltonia Hemsl. (1905);柔鼠麴属■☆

390095　Thiseltonia gracillima (F. Muell. et Tate) Paul G. Wilson;柔鼠麴■☆

390096　Thismia Griff. (1845);水玉杯属(腐杯草属,肉质腐生草属)■

390097　Thismia abei (Akasawa) Hatus. ;阿拜水玉杯(阿拜腐生草)■☆

390098　Thismia americana N. Sweet;北美水玉杯(北美肉质腐生草)■☆

390099　Thismia taiwanensis S. Z. Yang, R. M. K. Saunders et C. J. Hsu;台湾水玉杯■

390100　Thismia tentaculata K. Larsen et Aver. ;三丝水玉杯■

390101　Thismia tuberculata Hatus. ;多疣水玉杯(多疣肉质腐生草)■☆

390102　Thismia winkleri Engl. = Afrothismia winkleri (Engl.) Schltr. ■☆

390103　Thismiaceae J. Agardh(1858)(保留科名);水玉杯科(腐杯草科,肉质腐生草科)■

390104　Thismiaceae J. Agardh(保留科名) = Burmanniaceae Blume(保留科名)■

390105　Thium Steud. = Astragalus L. ●■

390106　Thium Steud. = Tium Medik. ●■

390107　Thladiantha Bunge(1833);赤瓟属(赤雹属,赤瓟儿属,青牛胆属);Tuber Gourd, Tubergourd, Tuber-gourd ■

390108　Thladiantha africana C. Jeffrey;非洲赤瓟■☆

390109　Thladiantha borneensis Merr. = Siraitia borneensis (Merr.) C. Jeffrey ex A. M. Lu et Zhi Y. Zhang ■

390110　Thladiantha calcarata (Wall.) C. B. Clarke = Thladiantha cordifolia (Blume) Cogn. ■

390111　Thladiantha calcarata (Wall.) C. B. Clarke var. tonkinensis Cogn. = Thladiantha cordifolia (Blume) Cogn. var. tonkinensis (Cogn.) A. M. Lu et Zhi Y. Zhang ■

390112　Thladiantha calcarata C. B. Clarke = Thladiantha cordifolia (Blume) Cogn. ■

390113　Thladiantha calcarata C. B. Clarke var. subglabra Cogn. = Thladiantha cordifolia (Blume) Cogn. ■

390114　Thladiantha calcarata C. B. Clarke var. tonkinensis Cogn. = Thladiantha cordifolia (Blume) Cogn. ■

390115　Thladiantha capitata Cogn. ;头花赤瓟;Capitate Tubergourd ■

390116　Thladiantha cinerascens C. Y. Wu ex A. M. Lu et Zhi Y. Zhang;灰赤瓟;Grey Tubergourd ■

390117　Thladiantha cinerascens C. Y. Wu ex A. M. Lu et Zhi Y. Zhang = Thladiantha lijiangensis A. M. Lu et Zhi Y. Zhang ■

390118　Thladiantha clentata Cogn. = Thladiantha dentata Cogn. ■

390119　Thladiantha cordifolia (Blume) Cogn. ;大苞赤瓟(心叶赤瓟);Cordateleaf Tubergourd, Heartleaf Tubergourd, Largebract Tubergourd ■

390120　Thladiantha cordifolia (Blume) Cogn. var. tonkinensis (Cogn.)

A. M. Lu et Zhi Y. Zhang = Thladiantha cordifolia（Blume）Cogn. ■

390121   Thladiantha cordifolia（Blume）Cogn. var. tonkinensis（Cogn.）A. M. Lu et Zhi Y. Zhang；茸毛赤瓟（越南赤瓟）；Tomentose Tubergourd，Tonkin Tubergourd ■

390122   Thladiantha davidii Franch.；川赤瓟；David Tubergourd，Sichuan Tubergourd ■

390123   Thladiantha dentata Cogn.；齿叶赤瓟（龙须尖，猫儿爪）；Dentateleaf Tubergourd，Toothleaf Tubergourd ■

390124   Thladiantha dictyocarpa Hand.-Mazz. = Thladiantha henryi Hemsl. ■

390125   Thladiantha digitata H. Lév. = Thladiantha hookeri C. B. Clarke var. heptadactyla（Cogn.）A. M. Lu et Zhi Y. Zhang ■

390126   Thladiantha digitata H. Lév. = Thladiantha hookeri C. B. Clarke ■

390127   Thladiantha dimorphantha Hand.-Mazz.；山西赤瓟；Shansi Tubergourd，Twoformflower Tubergourd ■

390128   Thladiantha dubia Bunge；赤瓟（赤包，赤雹，赤雹子，公公须，老鸦瓜，马瓟瓜，气包，山屎瓜，山土豆，师姑草，屎包子，屎瓜，土瓜，菟瓜，王瓜，野丝瓜，野甜瓜）；Manchu Tuber Gourd，Manchu Tuber-gourd，Manchur Tubergourd，Manchurian Tubergourd ■

390129   Thladiantha formosana Hayata = Thladiantha nudiflora Hemsl. ex Forbes et Hemsl. ■

390130   Thladiantha glabra Cogn. = Thladiantha oliveri Cogn. ex Mottet ■

390131   Thladiantha glabra Cogn. ex Oliv. = Thladiantha oliveri Cogn. ex Mottet ■

390132   Thladiantha globicarpa A. M. Lu et Zhi Y. Zhang；球果赤瓟（野苦瓜）；Ballfruit Tubergourd，Globularfruit Tubergourd ■

390133   Thladiantha globicarpa A. M. Lu et Zhi Y. Zhang = Thladiantha cordifolia（Blume）Cogn. ■

390134   Thladiantha grandisepala A. M. Lu et Zhi Y. Zhang；大萼赤瓟；Bigsepal Tubergourd，Largesepal Tubergourd ■

390135   Thladiantha grosvenorii（Swingle）C. Jeffrey = Siraitia grosvenorii（Swingle）C. Jeffrey ex A. M. Lu et Zhi Y. Zhang ■

390136   Thladiantha harmsii Cogn. = Thladiantha nudiflora Hemsl. ex Forbes et Hemsl. ■

390137   Thladiantha henryi Cogn. var. subtomentosa Hand.-Mazz. = Thladiantha lijiangensis A. M. Lu et Zhi Y. Zhang ■

390138   Thladiantha henryi Hemsl.；皱果赤瓟（瓜缕藤，苦瓜，苦瓜蒌，癫瓜，米来瓜，南葛）；Henry Tubergourd ■

390139   Thladiantha henryi Hemsl. var. subtomentosa Hand.-Mazz. = Thladiantha lijiangensis A. M. Lu et Zhi Y. Zhang ■

390140   Thladiantha henryi Hemsl. var. verrucosa（Cogn.）A. M. Lu et Zhi Y. Zhang = Thladiantha henryi Hemsl. ■

390141   Thladiantha henryi Hemsl. var. verucosa（Cogn.）A. M. Lu et Z. Y. Zhang；喙赤瓟（老蛇头）；Beaked Tubergourd ■

390142   Thladiantha heptadactyla Cogn. = Thladiantha hookeri C. B. Clarke var. heptadactyla（Cogn.）A. M. Lu et Zhi Y. Zhang ■

390143   Thladiantha heptadactyla Cogn. = Thladiantha hookeri C. B. Clarke ■

390144   Thladiantha hookeri C. B. Clarke；异叶赤瓟（粗茎罗锅底，裂叶赤瓟）；Hooker Tubergourd ■

390145   Thladiantha hookeri C. B. Clarke f. quinquefoliata Chakr. = Thladiantha hookeri C. B. Clarke ■

390146   Thladiantha hookeri C. B. Clarke f. trifoliolata（Cogn.）Chakr. = Thladiantha hookeri C. B. Clarke ■

390147   Thladiantha hookeri C. B. Clarke var. heptadactyla（Cogn.）A. M. Lu et Zhi Y. Zhang = Thladiantha hookeri C. B. Clarke ■

390148   Thladiantha hookeri C. B. Clarke var. heptadactyla（Cogn.）A.

M. Lu et Zhi Y. Zhang；七叶赤瓟（山土瓜）；Sevenleaf Tubergourd，Sevenleaves Tubergourd ■

390149   Thladiantha hookeri C. B. Clarke var. palmatifolia Chakrav.；三叶赤瓟（老妈妈背捎果，天花粉，竹叶藤）；Threeleaf Tubergourd，Threeleaves Tubergourd ■

390150   Thladiantha hookeri C. B. Clarke var. palmatifolia Chakrav. = Thladiantha hookeri C. B. Clarke ■

390151   Thladiantha hookeri C. B. Clarke var. palmatifolia Chakrav. f. quinquefoliata Chakr. = Thladiantha hookeri C. B. Clarke var. heptadactyla（Cogn.）A. M. Lu et Zhi Y. Zhang ■

390152   Thladiantha hookeri C. B. Clarke var. pentadactyla（Cogn.）A. M. Lu et Zhi Y. Zhang = Thladiantha hookeri C. B. Clarke ■

390153   Thladiantha hookeri C. B. Clarke var. pentadactyla（Cogn.）A. M. Lu et Zhi Y. Zhang；五叶赤瓟（山土瓜，天花粉）；Fiveleaf Tubergourd，Fiveleaves Tubergourd ■

390154   Thladiantha indochinensis Merr. = Thladiantha nudiflora Hemsl. ex Forbes et Hemsl. ■

390155   Thladiantha legendrei Gagnep. = Thladiantha davidii Franch. ■

390156   Thladiantha lijiangensis A. M. Lu et Zhi Y. Zhang；丽江赤瓟；Lijiang Tubergourd ■

390157   Thladiantha lijiangensis A. M. Lu et Zhi Y. Zhang var. latisepala A. M. Lu et Zhi Y. Zhang = Thladiantha lijiangensis A. M. Lu et Zhi Y. Zhang ■

390158   Thladiantha lijiangensis A. M. Lu et Zhi Y. Zhang var. latisepala A. M. Lu et Zhi Y. Zhang；木里赤瓟（阔萼丽江赤瓟）；Broadsepal Tubergourd ■

390159   Thladiantha longifolia Cogn. ex Oliv.；长叶赤瓟（尖瓜，尖叶瓜，小苦瓜）；Longleaf Tubergourd ■

390160   Thladiantha longisepala C. Y. Wu ex A. M. Lu et Zhi Y. Zhang；长萼赤瓟；Longsepal Tubergourd ■

390161   Thladiantha maculata Cogn.；斑赤瓟（青牛胆）；Spotted Tubergourd ■

390162   Thladiantha medogensis A. M. Lu et J. Q. Li；墨脱赤瓟；Motuo Tubergourd ■

390163   Thladiantha montana Cogn.；山地赤瓟；Mountain Tubergourd，Mountainous Tubergourd ■

390164   Thladiantha nudiflora Hemsl. ex Forbes et Hemsl.；南赤瓟（赤爬儿，赤爬儿，地黄瓜，老蛇头，裸花赤瓟，毛瓜，毛苦瓜，南丝瓜，秦岭赤瓟，青牛胆，球子莲，丝瓜南，土瓜蒌，王瓜，野冬瓜，野丝瓜）；Nakedflower Tubergourd ■

390165   Thladiantha nudiflora Hemsl. ex Forbes et Hemsl. var. bracteata A. M. Lu et Zhi Y. Zhang；西固赤瓟（苞叶赤瓟）；Bracts Tubergourd ■

390166   Thladiantha nudiflora Hemsl. ex Forbes et Hemsl. var. macrocarpa Z. Zhang = Thladiantha nudiflora Hemsl. ex Forbes et Hemsl. ■

390167   Thladiantha nudiflora Hemsl. ex Forbes et Hemsl. var. macrocarpa Z. Zhang；大果赤瓟；Bigfruit Nakedflower Tubergourd ■

390168   Thladiantha nudiflora Hemsl. ex Forbes et Hemsl. var. membranacea Z. Zhang = Thladiantha nudiflora Hemsl. ex Forbes et Hemsl. ■

390169   Thladiantha nudiflora Hemsl. ex Forbes et Hemsl. var. membranacea Z. Zhang；绵赤瓟 ■

390170   Thladiantha oliveri Cogn. ex Mottet；鄂赤瓟（光赤瓟，苦瓜蒌，青牛胆，水葡萄，野瓜，野苦瓜藤）；Glabrous Tubergourd，Hubei Tubergourd，Oliver Tubergourd，Tall Tubergourd ■

390171   Thladiantha oliveri Cogn. ex Mottet = Thladiantha dentata Cogn. ■

390172   Thladiantha palmatipartita A. M. Lu et C. Jeffrey；掌叶赤瓟 ■

390173   Thladiantha pentadactyla Cogn. = Thladiantha hookeri C. B.

Clarke var. pentadactyla（Cogn.）A. M. Lu et Zhi Y. Zhang ■

390174   Thladiantha pentadactyla Cogn. = Thladiantha hookeri C. B. Clarke ■

390175   Thladiantha punctata Hayata；台湾赤瓟（斑花青牛胆）；Taiwan Tubergourd ■

390176   Thladiantha pustulata（H. Lév.）C. Jeffrey ex A. M. Lu et Zhi Y. Zhang var. jinfushanensis A. M. Lu et J. Q. Li；金佛山赤瓟；Jinfoshan Tubergourd ■

390177   Thladiantha pustulata（H. Lév.）C. Jeffrey ex A. M. Lu et Zhi Y. Zhang；云南赤瓟；Yunnan Tubergourd ■

390178   Thladiantha sessilifolia Hand.-Mazz.；短柄赤瓟（野黄瓜）；Shortstalk Tubergourd ■

390179   Thladiantha sessilifolia Hand.-Mazz. var. longipes A. M. Lu et Zhi Y. Zhang；沧源赤瓟；Longstalk Tubergourd ■

390180   Thladiantha setispina A. M. Lu et Zhi Y. Zhang；刚毛赤瓟（王瓜，西藏赤瓟）；Bristle Tubergourd ■

390181   Thladiantha siamensis Craib = Siraitia siamensis（Craib）C. Jeffrey ex S. Q. Zhong et D. Fang ■

390182   Thladiantha siamensis Craib = Thladiantha cordifolia（Blume）Cogn. ■

390183   Thladiantha taiwaniana Hayata = Sinobaijiania taiwaniana（Hayata）C. Jeffrey et W. J. de Wilde ■

390184   Thladiantha taiwaniana Hayata = Siraitia taiwaniana（Hayata）C. Jeffrey ex A. M. Lu et Zhi Y. Zhang ■

390185   Thladiantha tonkinensis Gagnep. = Thladiantha cordifolia（Blume）Cogn. ■

390186   Thladiantha trifoliolata（Cogn.）Merr. = Thladiantha hookeri C. B. Clarke var. palmatifolia Chakrav. ■

390187   Thladiantha trifoliolata（Cogn.）Merr. = Thladiantha hookeri C. B. Clarke ■

390188   Thladiantha verrucosa Cogn. = Thladiantha henryi Hemsl. ■

390189   Thladiantha villosula Cogn. = Thladiantha villosula Cogn. ex Diels ■

390190   Thladiantha villosula Cogn. ex Diels；长毛赤瓟（白斑王瓜，倒挂山余瓜，罗锅底，毛赤瓟，山墩，山余瓜，五叶赤瓟）；Villose Tubergourd ■

390191   Thladiantha villosula Cogn. ex Diels var. nigrita A. M. Lu et Zhi Y. Zhang；黑子赤瓟；Blackseed Tubergourd ■

390192   Thladiantha yunnanensis Gagnep. = Thladiantha pustulata（H. Lév.）C. Jeffrey ex A. M. Lu et Zhi Y. Zhang ■

390193   Thladianthopsis Cogn. = Thladiantha Bunge ■

390194   Thladianthopsis Cogn. ex Oliv.（1892）；类赤瓟属■☆

390195   Thladianthopsis Cogn. ex Oliv. = Thladiantha Bunge ■

390196   Thladianthopsis montana Cogn. ex Gagnep.；类赤瓟■☆

390197   Thlasidia Raf. = Scabiosa L. ●■

390198   Thlaspeocarpa C. A. Sm.（1931）；南非遏蓝菜属■☆

390199   Thlaspeocarpa C. A. Sm. = Heliophila Burm. f. ex L. ●■☆

390200   Thlaspeocarpa capensis（Sond.）C. A. Sm. = Heliophila suborbicularis Al-Shehbaz et Mummenhoff ■☆

390201   Thlaspeocarpa namaquensis Marais = Heliophila namaquensis（Marais）Al-Shehbaz et Mummenhoff ■☆

390202   Thlaspi L.（1753）；菥蓂属（遏蓝菜属）；Bastard Cress，Besomweed，Besomweed Pennycress，Pennycress，Penny-cress，Pennyeress ■

390203   Thlaspi affine Schott et Kotschy ex Boiss. = Thlaspi perfoliatum L. ■

390204   Thlaspi africanum Burm. f. = Lepidium africanum（Burm. f.）

DC. ■☆

390205   Thlaspi alliaceum L.；路旁菥蓂；Garlic Penny-cress，Roadside Pennycress ■☆

390206   Thlaspi alpestre L. = Thlaspi cochleariforme DC. ■

390207   Thlaspi alpestre L. = Thlaspi griffithianum（Boiss.）Boiss. ■☆

390208   Thlaspi alpinum Griseb.；高山菥蓂；Alpine Penny-cress ■☆

390209   Thlaspi amarum Crantz var. hesperidifolia ?；金星菥蓂■☆

390210   Thlaspi andersonii（Hook. f. et Thomson）O. E. Schulz；西藏菥蓂；Tibet Bastard Cress，Xizang Pennycress ■

390211   Thlaspi annuum K. Koch；一年菥蓂■☆

390212   Thlaspi armenum N. Busch；亚美尼亚菥蓂■☆

390213   Thlaspi arvense L.；菥蓂（败酱草，臭虫草，大蕺，大荠，遏蓝菜，葛菜，瓜子草，花叶荠，苦菜，苦稽，苦莩苈，老鼓草，老荠，犁头菜，犁头草，马驹，马辛，麦蓝菜，蔑菥，薽菥，南败酱大荠，荣目，水荠，析目，小山菠菜，洋辣罐，野榆钱）；Bastard Cress，Boor's Mustard，Bowyer's Mustard，Broomwort，Dish Mustard，Fanweed，Field Cress，Field Penny Cress，Field Pennycress，Field Penny-cress，Field Thlaspi，Frenchweed，Mithridate Mustard，Mithridatum，Money Plant，Money Tree，Penny Cress，Stinkweed，Treacle Mustard，Wild Cress ■

390214   Thlaspi arvense L. var. sinuatum H. Lév. = Thlaspi arvense L. ■

390215   Thlaspi atlanticum Batt.；大西洋菥蓂■☆

390216   Thlaspi bulbosum Griseb.；块茎菥蓂■☆

390217   Thlaspi bursa-pastoris L. = Capsella bursa-pastoris（L.）Medik. ■

390218   Thlaspi caerulescens J. Presl et C. Presl；蓝菥蓂；Alpine Penny Cress，Alpine Penny-cress ■☆

390219   Thlaspi campestre L. = Lepidium campestre（L.）Brot. ex Nyman ■

390220   Thlaspi cardiocarpum Hook. f. et Thomson = Thlaspi kotschyanum Boiss. et Hohen. ■☆

390221   Thlaspi carneum Banks et Sol. = Aethionema carneum（Banks et Sol.）B. Fedtsch. ■☆

390222   Thlaspi cartilagineum J. Mayer = Lepidium cartilagineum（J. Mayer）Thell. ■

390223   Thlaspi cochleariforme DC.；杓形菥蓂（山菥蓂）■

390224   Thlaspi cochleariforme DC. subsp. griffithianum（Boiss.）Jafri = Thlaspi griffithianum（Boiss.）Boiss. ■☆

390225   Thlaspi coclearioides Hook. f. et Thomson；假杓形菥蓂■☆

390226   Thlaspi collinum M. Bieb. = Thlaspi arvense L. ■

390227   Thlaspi cordatum Desf. = Aethionema cordatum（Desf.）Boiss. ■☆

390228   Thlaspi coronopifolium J. P. Bergeret ex Steud. = Teesdalia coronopifolia（Steud.）Thell. ■☆

390229   Thlaspi erraticum Jord. = Thlaspi perfoliatum L. ■

390230   Thlaspi exauriculatum Kom. = Thlaspi cochleariforme DC. ■

390231   Thlaspi ferganense N. Busch；新疆菥蓂（新疆遏蓝菜）；Sinkiang Bastard Cress，Xinjiang Bastard Cress，Xinjiang Pennycress ■

390232   Thlaspi flagelliferum O. E. Schulz；四川菥蓂；Sichuan Bastard Cress，Sichuan Pennycress，Szechuan Bastard Cress ■

390233   Thlaspi freynii N. Busch；弗雷菥蓂■☆

390234   Thlaspi griffithianum（Boiss.）Boiss.；格氏菥蓂■☆

390235   Thlaspi hirtum L. = Lepidium hirtum（L.）Sm. ■☆

390236   Thlaspi huetii Boiss.；休氏菥蓂■☆

390237   Thlaspi japonicum H. Boissieu；日本菥蓂；Japanese Bastard Cress ■☆

390238   Thlaspi japonicum H. Boissieu var. sagittatum Miyabe et Tatew.；箭头菥蓂■☆

390239   Thlaspi kotschyanum Boiss. et Hohen.；考奇菥蓂■☆

390240　Thlaspi luteum Biv. = Bivonaea lutea（Biv.）DC. ■☆

390241　Thlaspi macranthum N. Busch；大花菥蓂■☆

390242　Thlaspi montanum L.；山地菥蓂；Alpine Pennycress, Alpine Penny-cress ■☆

390243　Thlaspi multifidum Poir.；多裂菥蓂■☆

390244　Thlaspi neglectum Crép. = Thlaspi perfoliatum L. ■

390245　Thlaspi obtusatum Pomel = Thlaspi perfoliatum L. subsp. tinei（Nyman）Maire ■☆

390246　Thlaspi oliveri Engl. = Thlaspi alliaceum L. ■☆

390247　Thlaspi orbiculatum Stev.；圆形菥蓂■☆

390248　Thlaspi parviflorum A. Nelson；小花菥蓂■☆

390249　Thlaspi perfoliatum L.；全叶菥蓂（贯叶遏蓝菜）；Claspleaf Pennycress, Cotswoid Penny Cress, Penny Cress, Perfoliate Penny Cress, Perfoliate Penny-cress, Thoroughwort Pennycress ■☆

390250　Thlaspi perfoliatum L. subsp. tinei（Nyman）Maire；蒂内菥蓂■☆

390251　Thlaspi perfoliatum L. var. erraticum（Jord.）Gren. = Thlaspi perfoliatum L. ■

390252　Thlaspi perfoliatum L. var. neglectum（Crép.）Durand = Thlaspi perfoliatum L. ■

390253　Thlaspi perfoliatum L. var. rotundifolium Ball = Thlaspi perfoliatum L. subsp. tinei（Nyman）Maire ■

390254　Thlaspi pinnatum Beck；羽状菥蓂■☆

390255　Thlaspi platycarpum Fisch. et C. A. Mey.；宽果菥蓂■☆

390256　Thlaspi praecox Wulfen；早菥蓂■☆

390257　Thlaspi prolongi Boiss. = Jonopsidium prolongoi（Boiss.）Batt. ☆

390258　Thlaspi pumilum Ledeb.；矮菥蓂■☆

390259　Thlaspi roseolum N. Busch；粉红菥蓂■☆

390260　Thlaspi rostratum N. Busch；喙状菥蓂■☆

390261　Thlaspi rotundifolium Tineo；圆叶菥蓂■☆

390262　Thlaspi saxatile L. = Aethionema saxatile（L.）R. Br. ■☆

390263　Thlaspi septigerum（Bunge）Jafri；隔菥蓂■☆

390264　Thlaspi szovitsianum Boiss.；绍氏菥蓂■☆

390265　Thlaspi thlaspidioides（Pall.）Kitag.；山菥蓂（山遏蓝菜）；Mountain Bastard Cress, Wild Pennycress ■

390266　Thlaspi thlaspidioides（Pall.）Kitag. = Thlaspi cochleariforme DC. ■

390267　Thlaspi tineoi Nyman = Thlaspi perfoliatum L. subsp. tinei（Nyman）Maire ■☆

390268　Thlaspi umbellatum Stev.；散枝菥蓂■☆

390269　Thlaspi verna Lam. = Arabis alpina L. subsp. caucasica（Willd.）Briq. ■☆

390270　Thlaspi wendelboi Rech. f. = Thlaspi griffithianum（Boiss.）Boiss. ■☆

390271　Thlaspi yunnanense Franch.；云南菥蓂；Yunnan Bastard Cress, Yunnan Pennycress ■

390272　Thlaspi yunnanense Franch. var. dentata Diels；齿叶菥蓂；Dentate Yunnan Bastard Cress, Toothleaf Yunnan Pennycress ■

390273　Thlaspiaceae Martinov = Brassicaceae Burnett（保留科名）■●

390274　Thlaspiaceae Martinov = Cruciferae Juss.（保留科名）■●

390275　Thlaspiaceae Martinov；菥蓂科■

390276　Thlaspiceras F. K. Mey. = Thlaspi L. ■

390277　Thlaspidea Opiz = Thlaspi L. ■

390278　Thlaspidium Bubani = Thlaspi L. ■

390279　Thlaspidium Mill. = Biscutella L.（保留属名）■☆

390280　Thlaspidium Spach = Lepidium L. ■

390281　Thlaspius St. -Lag. = Thlaspi L. ■

390282　Thlipsocarpus Kuntze = Hyoseris L. ■☆

390283　Thlocephala Y. Ito = Neoporteria Britton et Rose ●■

390284　Thoa Aubl. = Gnetum L. ●

390285　Thoaceae Agardh = Gnetaceae Blume（保留科名）●

390286　Thoaceae Kuntze = Gnetaceae Blume（保留科名）●

390287　Thodaya Compton = Euryops（Cass.）Cass. ●■☆

390288　Thodaya elongata Compton = Euryops subcarnosus DC. ●☆

390289　Thodaya microphylla Compton = Euryops microphyllus（Compton）B. Nord. ●☆

390290　Thogsennia Aiello（1979）；陶格茜属●☆

390291　Thogsennia lindeniana（A. Rich.）Aiello；陶格茜●☆

390292　Thollonia Baill. = Icacina A. Juss. ●☆

390293　Thomandersia Baill.（1891）；托曼木属●☆

390294　Thomandersia anachoreta Heine；非洲托曼木●☆

390295　Thomandersia butayei De Wild.；布塔托曼木●☆

390296　Thomandersia congolana De Wild. et T. Durand；托曼木●☆

390297　Thomandersia hensii De Wild. et T. Durand；汉斯托曼木●☆

390298　Thomandersia laurentii De Wild.；洛朗托曼木●☆

390299　Thomandersia laurifolia（T. Anderson ex Benth.）Baill.；月桂叶托曼木●☆

390300　Thomandersia laurifolia Baill. = Thomandersia laurifolia（T. Anderson ex Benth.）Baill. ●☆

390301　Thomandersiaceae Sreemadh.（1977）；托曼木科●

390302　Thomandersiaceae Sreemadh. = Acanthaceae Juss.（保留科名）●■

390303　Thomasia J. Gay（1821）；中脉梧桐属（托玛斯木属）●☆

390304　Thomasia angustifolia Steud.；窄叶中脉梧桐●☆

390305　Thomasia diffusa G. Don；松散中脉梧桐●☆

390306　Thomasia glabrata Steud.；光中脉梧桐●☆

390307　Thomasia macrocalyx Steud.；大萼中脉梧桐●☆

390308　Thomasia macrocarpa Endl.；大果中脉梧桐●☆

390309　Thomasia montana Steud.；山地中脉梧桐●☆

390310　Thomasia pauciflora Lindl.；少花中脉梧桐●☆

390311　Thomasia petalocalyx F. Muell.；瓣萼中脉梧桐（托玛斯木）；Paperflower ●☆

390312　Thomasia purpurea（W. T. Aiton）J. Gay；紫中脉梧桐●☆

390313　Thomasia purpurea J. Gay = Thomasia purpurea（W. T. Aiton）J. Gay ●☆

390314　Thomasia rugosa Turcz.；皱中脉梧桐●☆

390315　Thomasia tenuivestita F. Muell.；细中脉梧桐●☆

390316　Thomasia triflora Walp.；三花中脉梧桐●☆

390317　Thomasia triloba Turcz.；三裂中脉梧桐●☆

390318　Thomasia triphylla（Labill.）J. Gay；三叶中脉梧桐●☆

390319　Thomasia triphylla J. Gay = Thomasia triphylla（Labill.）J. Gay ●☆

390320　Thomasia undulata Steetz；波叶中脉梧桐●☆

390321　Thomasia viridis Steud.；绿中脉梧桐●☆

390322　Thomassetia Hemsl. = Brexia Noronha ex Thouars（保留属名）●☆

390323　Thommasinia Steud. = Peucedanum L. ■

390324　Thommasinia Steud. = Tommasinia Bertol. ☆

390325　Thompsonella Britton et Rose（1909）；汤普森景天属■☆

390326　Thompsonella minutiflora（Rose）Britton et Rose；小花汤普森景天■☆

390327　Thompsonella platyphylla Rose；宽叶汤普森景天■☆

390328　Thompsonia R. Br. = Deidamia E. A. Noronha ex Thouars ■☆

390329　Thomsonia Wall.（废弃属名）= Amorphophallus Blume ex Decne.（保留属名）■●

390330　Thonandia H. P. Linder = Danthonia DC.（保留属名）■

390331　Thonandia H. P. Linder（1996）；托南德扁芒草属■☆

390332　Thonandia gracilis（Hook. f.）H. P. Linder；纤细托南德扁芒

草■☆

390333 Thonandia longifolia（R. Br.）H. P. Linder；长叶托南德扁芒草☆

390334 Thonnera De Wild. = Uvariopsis Engl. ex Engl. et Diels ●☆

390335 Thonnera congolana De Wild. = Uvariopsis congolana（De Wild.）R. E. Fr. ●☆

390336 Thonningia Vahl（1810）；非洲蛇菰属（特宁草属）■☆

390337 Thonningia angolensis Hemsl. = Thonningia sanguinea Vahl ■☆

390338 Thonningia dubia Hemsl. = Thonningia sanguinea Vahl ■☆

390339 Thonningia elegans Hemsl. = Thonningia sanguinea Vahl ■☆

390340 Thonningia malagasica Fawc. = Langsdorffia malagasica（Fawc.）B. Hansen ■☆

390341 Thonningia sanguinea Vahl；非洲蛇菰（特宁草）■☆

390342 Thonningia sessilis Lecomte = Thonningia sanguinea Vahl ■☆

390343 Thonningia ugandensis Hemsl. = Thonningia sanguinea Vahl ■☆

390344 Thora Fourr. = Ranunculus L. ■

390345 Thora Hill = Ranunculus L. ■

390346 Thoracocarpus Harling（1958）；甲果巴拿马草属■☆

390347 Thoracocarpus bissectus（Vell.）Harling；甲果巴拿马草■☆

390348 Thoracosperma Klotzsch = Eremia D. Don ●☆

390349 Thoracosperma Klotzsch = Erica L. ●☆

390350 Thoracosperma Klotzsch（1834）；甲籽杜鹃属●☆

390351 Thoracosperma articulatum（L.）Kuntze = Erica similis（N. E. Br.）E. G. H. Oliv. ●☆

390352 Thoracosperma bondiae Compton = Erica rosacea（L. Guthrie）E. G. H. Oliv. ●☆

390353 Thoracosperma ciliatum（Benth.）Kuntze = Erica paucifolia（J. C. Wendl.）E. G. H. Oliv. subsp. ciliata（Klotzsch）E. G. H. Oliv. ●☆

390354 Thoracosperma discolor（Klotzsch）Kuntze = Erica anguliger（N. E. Br.）E. G. H. Oliv. ●☆

390355 Thoracosperma fourcadei Compton = Erica rosacea（L. Guthrie）E. G. H. Oliv. ●☆

390356 Thoracosperma galpinii N. E. Br. = Erica rosacea（L. Guthrie）E. G. H. Oliv. ●☆

390357 Thoracosperma glaucum（Klotzsch）Kuntze = Erica similis（N. E. Br.）E. G. H. Oliv. ●☆

390358 Thoracosperma hirsutum（Benth.）Kuntze = Erica glabella Thunb. subsp. laevis E. G. H. Oliv. ●☆

390359 Thoracosperma hirtum（Klotzsch）Kuntze = Erica glabella Thunb. subsp. laevis E. G. H. Oliv. ●☆

390360 Thoracosperma interruptum N. E. Br. = Erica interrupta（N. E. Br.）E. G. H. Oliv. ●☆

390361 Thoracosperma marlothii N. E. Br. = Erica rosacea（L. Guthrie）E. G. H. Oliv. subsp. glabrata E. G. H. Oliv. ●☆

390362 Thoracosperma muirii L. Guthrie = Erica radicans（L. Guthrie）E. G. H. Oliv. ●☆

390363 Thoracosperma nanum N. E. Br. = Erica bolusanthus E. G. H. Oliv. ●☆

390364 Thoracosperma paniculatum（Thunb.）Klotzsch = Erica quadrifida（Benth.）E. G. H. Oliv. ●☆

390365 Thoracosperma parviflorum（Klotzsch）Kuntze = Erica anguliger（N. E. Br.）E. G. H. Oliv. ●☆

390366 Thoracosperma paucifolium（J. C. Wendl.）Kuntze = Erica paucifolia（J. C. Wendl.）E. G. H. Oliv. ●☆

390367 Thoracosperma puberulum（Klotzsch）N. E. Br. = Erica puberuliflora E. G. H. Oliv. ●☆

390368 Thoracosperma radicans L. Guthrie = Erica radicans（L. Guthrie）E. G. H. Oliv. ●☆

390369 Thoracosperma rosaceum L. Guthrie = Erica rosacea（L. Guthrie）E. G. H. Oliv. ●☆

390370 Thoracosperma scabrum（Thunb.）Kuntze = Erica glabella Thunb. ●☆

390371 Thoracosperma squarrosum（Benth.）Kuntze = Erica paucifolia（J. C. Wendl.）E. G. H. Oliv. subsp. squarrosa（Benth.）E. G. H. Oliv. ●☆

390372 Thoracosperma tenue（Benth.）Kuntze = Erica anguliger（N. E. Br.）E. G. H. Oliv. ●☆

390373 Thoracosperma viscidum L. Guthrie = Erica interrupta（N. E. Br.）E. G. H. Oliv. ●☆

390374 Thoracostachys Kurz = Thoracostachyum Kurz ■

390375 Thoracostachyum Kurz = Mapania Aubl. ■

390376 Thoracostachyum Kurz（1869）；野长蒲属；Thoracostachyum ■

390377 Thoracostachyum bancanum Kurz；邦卡野长蒲■☆

390378 Thoracostachyum pandanophyllum（F. Muell.）Domin；露兜树叶野长蒲（野长蒲）；Common Thoracostachyum, Screwpineleaf Thoracostachyum ■

390379 Thoraea Gand. = Pimpinella L. ■

390380 Thorea Briq. = Caropsis（Rouy et Camus）Rauschert ■☆

390381 Thorea Briq. = Thorella Briq. ■☆

390382 Thorea Rouy = Arrhenatherum P. Beauv. ■

390383 Thorea Rouy = Pseudarrhenatherum Rouy ■☆

390384 Thoreauea J. K. Williams（2002）；墨西哥夹竹桃属●☆

390385 Thoreauea paneroi J. K. Williams；墨西哥夹竹桃●☆

390386 Thoreldora Pierre = Glycosmis Corrêa（保留属名）●

390387 Thorelia Gagnep.（1920）（保留属名）；托雷尔菊属■☆

390388 Thorelia Gagnep.（保留属名）= Camchaya Gagnep. ■

390389 Thorelia Gagnep.（保留属名）= Thoreliella C. Y. Wu ■

390390 Thorelia Hance（废弃属名）= Thorelia Gagnep.（保留属名）■☆

390391 Thorelia Hance（废弃属名）= Tristania R. Br. ■

390392 Thorelia montana Gagnep.；托雷尔菊■☆

390393 Thorelia montana Gagnep. = Camchaya loloana Kerr ■

390394 Thoreliella C. Y. Wu = Camchaya Gagnep. ■

390395 Thoreliella C. Y. Wu = Thorelia Gagnep.（保留属名）■☆

390396 Thoreliella montana（Gagnep.）C. Y. Wu = Camchaya loloana Kerr ■

390397 Thorella Briq. = Caropsis（Rouy et Camus）Rauschert ■☆

390398 Thorelrella C. Y. Wu = Camchaya Gagnep. ■

390399 Thoreochloa Holub = Arrhenatherum P. Beauv. ■

390400 Thoreochloa Holub = Pseudarrhenatherum Rouy ■☆

390401 Thornbera Rydb. = Dalea L.（保留属名）●■☆

390402 Thorncroftia N. E. Br.（1912）；托恩草属■☆

390403 Thorncroftia longiflora N. E. Br.；长叶托恩草■☆

390404 Thorncroftia media Codd；中间托恩草■☆

390405 Thorncroftia succulenta（R. A. Dyer et E. A. Bruce）Codd；托恩草■☆

390406 Thorncroftia thorncroftii（S. Moore）Codd；非洲托恩草■☆

390407 Thornea Breedlove et E. M. McClint.（1976）；托纳藤属●☆

390408 Thornea calcicola（Standl. et Steyerm.）Breedlove et E. M. McClint.；盐碱托纳藤●☆

390409 Thornea matudae（Lundell）Breedlove et E. M. McClint.；托纳藤●☆

390410 Thorntonia Rchb.（废弃属名）= Kosteletzkya C. Presl（保留属名）■●☆

390411　Thorntonia Rchb.（废弃属名）= Pavonia Cav.（保留属名）●■☆

390412　Thorvaldsenia Liebm. = Chysis Lindl. ■☆

390413　Thottea Rottb.（1783）;线果兜铃属;Alpam Root,Thottea ●

390414　Thottea hainanensis（Merr. et Chun）Ding Hou;海南线果兜铃（阿柏麻）;Hainan Thottea ●

390415　Thottea siliquosa（Lam.）Ding Hou;长角线果兜铃●☆

390416　Thouarea Kunth = Thuarea Pers. ■

390417　Thouarsia Kuntze = Thuarea Pers. ■

390418　Thouarsia Post et Kuntze = Thouarsia Vent. ex DC. ●☆

390419　Thouarsia Vent. ex DC. = Psiadia Jacq. ●☆

390420　Thouarsiora Homolle ex Arènes = Ixora L. ●

390421　Thouinia Comm. ex Planch. = Vitis L. ●

390422　Thouinia Dombey ex DC. = Lardizabala Ruiz et Pav. ●☆

390423　Thouinia L. f. = Linociera Sw. ex Schreb.（保留属名）●

390424　Thouinia Poit.（1804）（保留属名）;索英木属;Thouinia ●☆

390425　Thouinia Sm. = Humbertia Comm. ex Lam. ●☆

390426　Thouinia Thunb. ex L. f.（废弃属名）= Thouinia Poit.（保留属名）●☆

390427　Thouinia acuminata S. Watson;渐尖索英木●☆

390428　Thouinia discolor Griseb.;异色索英木●☆

390429　Thouinia paucidentata Radlk.;少齿索英木●☆

390430　Thouinia spectabilis Sm. = Humbertia madagascariensis Lam. ●☆

390431　Thouinidium Radlk.（1878）;索英无患子属●☆

390432　Thouinidium decandrum Radlk.;索英无患子●☆

390433　Thouinidium hexandrum Radlk.;六蕊索英无患子●☆

390434　Thouinidium oblongum Radlk.;矩圆索英无患子●☆

390435　Thouvenotia Danguy = Beilschmiedia Nees ●

390436　Thouvenotia madagascariensis Danguy = Beilschmiedia velutina（Kosterm.）Kosterm. ●☆

390437　Thozetia F. Muell. ex Benth.（1868）;托兹萝藦属☆

390438　Thozetia racemosa F. Muell. ex Benth. ;托兹萝藦☆

390439　Thrasia Kunth = Thrasya Kunth ■☆

390440　Thrasya Kunth（1816）;勇夫草属■☆

390441　Thrasya axillaris（Swallen）A. G. Burm. ex Judz.;腋生勇夫草■☆

390442　Thrasya gracilis Swallen;细勇夫草■☆

390443　Thrasyopsis L. Parodi（1946）;拟勇夫草属■☆

390444　Thrasyopsis rawitscheri Parodi;拟勇夫草■☆

390445　Thraulococcus Radlk. = Lepisanthes Blume ●

390446　Threlkeldia R. Br.（1810）;肉被澳藜属●☆

390447　Threlkeldia diffusa R. Br. ;肉被澳藜●☆

390448　Thrica Gray = Leontodon L.（保留属名）■☆

390449　Thrica Gray = Thrincia Roth ■☆

390450　Thrinax L. f. = Thrinax L. f. ex Sw. ●☆

390451　Thrinax L. f. ex Sw.（1788）;白果棕（白桐属,白棕榈属,豆棕属,扇葵属,屋顶棕属,细叶风竹属）;Thatch Palm ●☆

390452　Thrinax Sw. = Thrinax L. f. ex Sw. ●☆

390453　Thrinax argentea Lodd. ex Schult. et Schult. f. var. garberi（Chapm.）Chapm. = Coccothrinax argentata（Jacq.）L. H. Bailey ●☆

390454　Thrinax bahamensis O. F. Cook = Thrinax morrisi H. Wendl. ●☆

390455　Thrinax excelsa Lodd. ex Mart.;豆棕●☆

390456　Thrinax floridana Sarg. = Thrinax radiata Lodd. ex Schult. et Schult. f. ●☆

390457　Thrinax garberi Chapm. = Coccothrinax argentata（Jacq.）L. H. Bailey ●☆

390458　Thrinax keyensis Sarg. = Thrinax morrisi H. Wendl. ●☆

390459　Thrinax martii Griseb. et H. Wendl. ex Griseb. = Thrinax radiata Lodd. ex Schult. et Schult. f. ●☆

390460　Thrinax microcarpa Sarg. ;小果屋顶棕;Brittle Thatch Palm,Key Palm ●☆

390461　Thrinax microcarpa Sarg. = Thrinax morrisi H. Wendl. ●☆

390462　Thrinax morrisi H. Wendl. ;矮屋顶棕（摩里扇葵,莫氏豆棕）;Brittle-thatch,Thatch Palm ●☆

390463　Thrinax parviflora Sw. ;牙买加（屋顶棕,五叶豆棕,小花白果棕）;Florida Thatch Palm,Jamaica Thatch Palm,Jamaican Thatch Palm,Mountain Thatch Palm,Thatch Palm ●☆

390464　Thrinax ponceana O. F. Cook = Thrinax morrisi H. Wendl. ●☆

390465　Thrinax praeceps O. F. Cook = Thrinax morrisi H. Wendl. ●☆

390466　Thrinax radiata Lodd. ex Desf. = Thrinax radiata Lodd. ex Schult. et Schult. f. ●☆

390467　Thrinax radiata Lodd. ex Schult. et Schult. f. ;佛罗里达白果棕（团叶豆棕）;Florida Thatch Palm,Thatch Palm ●☆

390468　Thrinax wendlandiana Becc. = Thrinax radiata Lodd. ex Schult. et Schult. f. ●☆

390469　Thrincia Roth = Leontodon L.（保留属名）■☆

390470　Thrincia hirta Roth = Leontodon saxatilis Lam. ■☆

390471　Thrincia hispida Roth = Leontodon saxatilis Lam. subsp. longirostris（Finch et P. D. Sell）P. Silva ■☆

390472　Thrincia maroccana Pers. = Leontodon maroccanus（Pers.）Ball ■☆

390473　Thrincia tingitana Boiss. et Reut. = Leontodon tingitanus（Boiss. et Reut.）Ball ■☆

390474　Thrincia tuberosa（L.）DC. = Leontodon tuberosus L. ■☆

390475　Thrincia tuberosa（L.）DC. var. tripolitana（Sch. Bip.）Durand et Barratte = Leontodon tuberosus L. ■☆

390476　Thrincoma O. F. Cook = Coccothrinax Sarg. ●☆

390477　Thringis O. F. Cook = Coccothrinax Sarg. ●☆

390478　Thrixa Dulac = Thrincia Roth ■☆

390479　Thrixanthocereus Backeb.（1937）;银衣柱属●☆

390480　Thrixanthocereus Backeb. = Espostoa Britton et Rose ●

390481　Thrixanthocereus blossfeldiorum（Werderm.）Backeb. ;银衣柱●☆

390482　Thrixanthocereus senilis F. Ritter;清凉殿●☆

390483　Thrixgyne Keng = Duthiea Hack. ■

390484　Thrixgyne dura Keng = Duthiea brachypodia（P. Candargy）Keng et P. C. Keng ■

390485　Thrixia Dulac = Leontodon L.（保留属名）■☆

390486　Thrixspermum Lour.（1790）;白点兰属（风兰属）;Thrixsperm,Thrixspermum ■

390487　Thrixspermum acuminatissimum（Blume）Rchb. f. ;尖叶白点兰;Acuminate Thrixspermum ■☆

390488　Thrixspermum album（Ridl.）Schltr. ;白花白点兰;Whiteflower Thrixspermum ■☆

390489　Thrixspermum amplexicaule（Blume）Rchb. f. ;抱茎白点兰;Amplectant Thrixsperm,Amplexicaul Thrixspermum ■

390490　Thrixspermum annamense（Guillaumin）Garay;海台白点兰（白毛风兰,海南白点兰,湖南白点兰）;Annam Thrixsperm,Annam Thrixspermum,Hunan Thrixsperm,Hunan Thrixspermum ■

390491　Thrixspermum arachnites（Blume）Rchb. f. ;蜘蛛白点兰;Spider Thrixspermum ■☆

390492　Thrixspermum arachnites（Blume）Rchb. f. = Thrixspermum centipeda Lour. ■

390493　Thrixspermum auriferum（Lindl.）Rchb. f. = Thrixspermum centipeda Lour. ■

390494　Thrixspermum austrosinense Ts. Tang et F. T. Wang = Thrixspermum annamense（Guillaumin）Garay ■

390495　Thrixspermum bravibracteatum J. J. Sm.；短苞白点兰；Shortbract Thrixspermum ■☆

390496　Thrixspermum calceolus（Blume）Rchb. f.；拖鞋白点兰；Calceolate Thrixspermum ■▲☆

390497　Thrixspermum carinatifolium（Ridl.）Schltr.；龙骨叶白点兰；Keeled-leaf Thrixspermum ■☆

390498　Thrixspermum centipeda Lour.；白点兰；Common Thrixsperm，Common Thrixspermum ■

390499　Thrixspermum devolium T. P. Lin et C. C. Hsu = Thrixspermum annamense（Guillaumin）Garay ■

390500　Thrixspermum eximium L. O. Williams；异色白点兰（异色瓣，异色瓣白娥兰，异色风兰）；Differcolor Thrixsperm，Differcolor Thrixspermum ■

390501　Thrixspermum falcilobum Schltr. = Thrixspermum subulatum Rchb. f. ■

390502　Thrixspermum fantasticum L. O. Williams；金唇白点兰（金唇风兰，金唇风铃兰，小风兰）；Goldenlip Thrixsperm，Goldenlip Thrixspermum ■

390503　Thrixspermum formosanum（Hayata）Schltr.；台湾白点兰（台湾风兰，台湾风铃兰）；Taiwan Thrixsperm，Taiwan Thrixspermum ■

390504　Thrixspermum hainanense（Rolfe）Schltr.；海南白点兰；Hainan Thrixspermum ■

390505　Thrixspermum hainanense（Rolfe）Schltr. = Thrixspermum centipeda Lour. ■

390506　Thrixspermum japonicum（Miq.）Rchb. f.；小叶白点兰（飞天草）；Japan Thrixsperm，Japanese Thrixspermum ■

390507　Thrixspermum kusukusense（Hayata）Schltr.；高士佛风兰 ■

390508　Thrixspermum kusukusense（Hayata）Schltr. = Thrixspermum merguense（Hook. f.）Kuntze ■

390509　Thrixspermum laurisilvaticum（Fukuy.）Garay；黄花白点兰（黄蛾兰）■

390510　Thrixspermum laurisilvaticum（Fukuy.）Garay = Thrixspermum saruwatarii（Hayata）Schltr. ■

390511　Thrixspermum laurisilvaticum（Fukuy.）S. S. Ying = Thrixspermum laurisilvaticum（Fukuy.）Garay ■

390512　Thrixspermum laurisilvaticum（Fukuy.）S. S. Ying = Thrixspermum saruwatarii（Hayata）Schltr. ■

390513　Thrixspermum laurisilvaticum（Fukuy.）S. Y. Hu = Thrixspermum laurisilvaticum（Fukuy.）Garay ■

390514　Thrixspermum laurisilvaticum（Fukuy.）S. Y. Hu = Thrixspermum saruwatarii（Hayata）Schltr. ■

390515　Thrixspermum laurisilvaticum（Fukuy.）Tang S. Liu et H. J. Su = Thrixspermum saruwatarii（Hayata）Schltr. ■

390516　Thrixspermum leopardinum E. C. Parish et Rchb. f. = Pteroceras leopardinum（Parl. et Rchb. f.）Seidenf. et Sm. ■

390517　Thrixspermum leopardium Par. et Rchb. f. = Pteroceras leopardinum（Parl. et Rchb. f.）Seidenf. et Sm. ■

390518　Thrixspermum merguense（Hook. f.）Kuntze；三毛白点兰（高士佛风兰，高士佛风铃兰）；Threehair Thrixsperm，Threehair Thrixspermum ■

390519　Thrixspermum neglectum Fukuy. = Thrixspermum fantasticum L. O. Williams ■

390520　Thrixspermum neglectum Fukuy. = Thrixspermum japonicum（Miq.）Rchb. f. ■

390521　Thrixspermum pardale（Ridl.）Schltr.；豹斑白点兰；Leopard-spot Thrixspermum ■☆

390522　Thrixspermum pendulicaule（Hayata）Schltr.；垂枝白点兰（倒垂风兰,悬垂风铃兰）；Nutant Thrixsperm，Nutant Thrixspermum ■

390523　Thrixspermum pendulicaule（Hayata）Schltr. = Thrixspermum pensile Schltr. ■

390524　Thrixspermum pensile Schltr.；倒垂风兰 ■

390525　Thrixspermum pricei（Rolfe）Schltr.；溪头风兰 ■

390526　Thrixspermum pricei（Rolfe）Schltr. = Thrixspermum formosanum（Hayata）Schltr. ■

390527　Thrixspermum saruwatarii（Hayata）Schltr.；长轴白点兰（黄娥兰,小白蛾兰）；Saruwatar Thrixsperm，Saruwatar Thrixspermum ■

390528　Thrixspermum sasaoi Masam.；埔里风兰 ■

390529　Thrixspermum sasaoi Masam. = Thrixspermum formosanum（Hayata）Schltr. ■

390530　Thrixspermum segawai Masam. = Chiloschista segawai（Masam.）Masam. et Fukuy. ■

390531　Thrixspermum subulatum Rchb. f.；厚叶白点兰（肥垂兰,厚叶风兰）；Thickleaf Thrixsperm，Thickleaf Thrixspermum ■

390532　Thrixspermum trichoglottis（Hook. f.）Kuntze；同色白点兰；Samecolor Thrixsperm，Samecolor Thrixspermum ■

390533　Thrixspermum tsii W. H. Chen et Y. M. Shui；吉氏白点兰 ■

390534　Thrixspermum xanthanthum Tuyama = Thrixspermum laurisilvaticum（Fukuy.）Garay ■

390535　Thrixspermum xanthanthum Tuyama = Thrixspermum saruwatarii（Hayata）Schltr. ■

390536　Thryallis L.（废弃属名）= Galphimia Cav. ●

390537　Thryallis L.（废弃属名）= Thryallis Mart.（保留属名）●

390538　Thryallis Mart.（1829）（保留属名）；金英属；Thryallis ●

390539　Thryallis brasiliensis Kuntze；巴西金英 ●☆

390540　Thryallis glauca（Cav.）Kuntze = Thryallis gracilis Kuntze ●

390541　Thryallis glauca Kuntze = Thryallis gracilis Kuntze ●

390542　Thryallis gracilis Kuntze；金英（小金英）；Gold Shower，Golden Thryallis，Goldshower Thryallis，Slender Goldshower，Slender Thryallis ●

390543　Thrycocephalum Steud. = Thryocephalon J. R. Forst. et G. Forst. ■

390544　Thryocephalon J. R. Forst. et G. Forst. = Kyllinga Rottb.（保留属名）■

390545　Thryocephalon nemorale J. R. Forst. et G. Forst. = Kyllinga nemoralis（J. R. Forst. et G. Forst.）Hutch. et Dalziel ■

390546　Thryothamnus Phil.（废弃属名）= Junellia Moldenke（保留属名）●☆

390547　Thryothamnus Phil.（废弃属名）= Verbena L. ■●

390548　Thryptomene Endl.（1839）（保留属名）；异岗松属；Thryptomene ●☆

390549　Thryptomene calycina（Lindl.）Stapf；格兰扁区异岗松；Grampians Thryptomene ●☆

390550　Thryptomene saxicola（Hook.）Schauer；岩生异岗松；Rock Thryptomene ●☆

390551　Thuarea Pers.（1805）；蒭雷草属（刍雷草属，卷轴草属，沙丘草属，砂滨草属）；Kuroiwa Grass，Thuarea

390552　Thuarea involuta（G. Forst.）R. Br. ex Roem. et Schult. = Thuarea involuta（G. Forst.）R. Br. ex Sm. ■

390553　Thuarea involuta（G. Forst.）R. Br. ex Sm.；刍雷草（常宫草）；Involute Thuarea，Kuroiwa Grass ■

390554　Thuarea latifolia R. Br. = Thuarea involuta（G. Forst.）R. Br. ex Sm. ■

390555　Thuarea media R. Br. = Thuarea involuta（G. Forst.）R. Br. ex Sm. ■

390556　Thuarea sarmentosa Pers. = Thuarea involuta（G. Forst.）R. Br. ex Sm. ■

390557 Thuessinkia Korth. ex Miq. = Caryota L. ●

390558 Thuia Scop. = Thuja L. ●

390559 Thuiacarpus Benth. et Hook. f. = Thuiaecarpus Trautv. ●

390560 Thuiaecarpus Trautv. = Juniperus L. ●

390561 Thuinia Raf. = Chionanthus L. ●

390562 Thuinia Raf. = Linociera Sw. ex Schreb. (保留属名) ●

390563 Thuinia Raf. = Thouinia Poit. (保留属名) ●☆

390564 Thuiopsis Endl. = Thuja L. ●

390565 Thuiopsis Endl. = Thujopsis Siebold et Zucc. ex Endl. (保留属名) ●

390566 Thuiopsis Siebold et Zucc. = Thujopsis Siebold et Zucc. ex Endl. (保留属名) ●

390567 Thuja L. (1753); 崖柏属 (侧柏属, 金钟柏属); Arborvitae, Arbor-vitae, Cedars, Red Cedar, Red-cedar, Thuja, Thuya, White Cedar ●

390568 Thuja aphylla L. = Tamarix aphylla (L.) H. Karst. ●

390569 Thuja articulata Desf. ; 节状崖柏 ●☆

390570 Thuja articulata Vahl = Tetraclinis articulata (Vahl) Mast. ●☆

390571 Thuja beverleyensis Rehder = Platycladus orientalis (L.) Franco 'Beverleyensis' ●

390572 Thuja chengii Borderes et Gaussen = Platycladus orientalis (L.) Franco ●

390573 Thuja dolabrata L. f. = Thujopsis dolabrata (L. f.) Siebold et Zucc. ●

390574 Thuja dolabrata Thunb. ex L. f. = Thujopsis dolabrata (Thunb. ex L. f.) Siebold et Zucc. ●

390575 Thuja gigantea Nutt. = Thuja plicata Donn ex D. Don ●

390576 Thuja gigantea Nutt. var. japonica (Maxim.) Franch. et Sav. = Thuja japonica Maxim. ●

390577 Thuja gigantea Nutt. var. japonica (Maxim.) Franch. et Sav. = Thuja standishii (Gordon) Carrière ●

390578 Thuja japonica Maxim. = Thuja standishii (Gordon) Carrière ●

390579 Thuja koraiensis Nakai; 朝鲜崖柏 (长白侧柏, 朝鲜柏); Korea Arborvitae, Korean Arborvitae, Korean Arbor-vitae, Korean Thuja ●◇

390580 Thuja koraiensis Nakai 'Glauca Prostrata'; 矮灰朝鲜崖柏 ●☆

390581 Thuja lineata Poir. = Taxodium ascendens Brongn. 'Nutans' ●

390582 Thuja lineata Poir. = Taxodium distichum (L.) Rich. var. imbricatum (Nutt.) Croom ●

390583 Thuja lobbii Gordon = Thuja plicata Donn ex D. Don ●

390584 Thuja macrolepis (Kurz) Voss = Calocedrus macrolepis Kurz ●◇

390585 Thuja obtusa (Siebold et Zucc.) Mast. = Chamaecyparis obtusa (Siebold et Zucc.) Siebold et Zucc. ex Endl. ●

390586 Thuja obtusa Moench = Thuja occidentalis L. ●

390587 Thuja occidentalis L.; 北美香柏 (北美侧柏, 北美崖柏, 黄心柏木, 金钟柏, 美国侧柏, 美国崖柏, 香柏); American Arborvitae, American Arbor-vitae, Arbor-vitae, Cedar, Chinese Arborvitae, Common Arbor-vitae, Eastern Arborvitae, Eastern Arbor-vitae, Eastern Thuja, Eastern White Cedar, Eastern White-cedar, Emerald Green Arborvitae, False White Cedar, Northern White Cedar, Northern White-cedar, Western White Cedar, White Cedar, Yellow Cedar ●

390588 Thuja occidentalis L. 'Aurea Nana'; 矮黄北美香柏 ●☆

390589 Thuja occidentalis L. 'Aurea'; 金黄北美香柏; Golden Eastern Arborvitae ●

390590 Thuja occidentalis L. 'Booth'; 布什北美香柏; Booth Eastern Arborvitae ●

390591 Thuja occidentalis L. 'Caespitosa'; 丛生北美香柏 (簇生北美香柏) ●☆

390592 Thuja occidentalis L. 'Colombia'; 白斑叶北美香柏; White-variegated Eastern Arborvitae ●

390593 Thuja occidentalis L. 'Compacta'; 密柱北美香柏; Compact Eastern Arborvitae ●

390594 Thuja occidentalis L. 'Conica'; 圆锥北美香柏; Conical Eastern Arborvitae, Danica Arborvitae ●

390595 Thuja occidentalis L. 'Douglasii Aurea'; 青铜色北美香柏; Bronzy-yellow Eastern Arborvitae ●

390596 Thuja occidentalis L. 'Douglasii Pyramidalis'; 密塔形北美香柏; Dense-pyramid Eastern Arborvitae ●

390597 Thuja occidentalis L. 'Ellwangeriana'; 爱塔北美香柏 (矮塔北美香柏); Dwarf-pyramid Eastern Arborvitae ●

390598 Thuja occidentalis L. 'Ericoides'; 灌丛北美香柏; Bushy Eastern Arborvitae ●

390599 Thuja occidentalis L. 'Europe Gold'; 欧洲金香柏 ●☆

390600 Thuja occidentalis L. 'Fastigiata'; 帚形北美香柏 ●☆

390601 Thuja occidentalis L. 'Filiformis'; 丝线北美香柏 (鞭枝北美香柏); Threadleaf Arborvitae ●☆

390602 Thuja occidentalis L. 'Globosa'; 球形北美香柏; Globe Eastern Arborvitae ●

390603 Thuja occidentalis L. 'Golden Globe'; 金球北美香柏 ●☆

390604 Thuja occidentalis L. 'Hetz Midget'; 袖珍北美香柏; Hetz's Midget Eastern Arborvitae ●☆

390605 Thuja occidentalis L. 'Holmstrup'; 霍姆斯特拉普北美香柏; Holmstrup Arborvitae ●☆

390606 Thuja occidentalis L. 'Hoveyi'; 矮球形北美香柏; Dwarf-globe Eastern Arborvitae ●

390607 Thuja occidentalis L. 'Little Champion'; 小冠军北美香柏 ●☆

390608 Thuja occidentalis L. 'Little Gem'; 小宝石北美香柏; Little Gem Arborvitae ●☆

390609 Thuja occidentalis L. 'Lutea Nana'; 矮金北美香柏 ●☆

390610 Thuja occidentalis L. 'Lutea'; 鲜黄北美香柏; Bright-yellow Eastern Arborvitae, Yellow Arborvitae ●

390611 Thuja occidentalis L. 'Master Lutea'; 鲜黄塔柏; Yellow-pyramid Eastern Arborvitae ●

390612 Thuja occidentalis L. 'Minima'; 小北美香柏; Minima Arborvitae ●☆

390613 Thuja occidentalis L. 'Nigra'; 深绿北美香柏 (黑人北美香柏); Dark-green Eastern Arborvitae ●

390614 Thuja occidentalis L. 'Ohlendorffii'; 奥利多菲北美香柏; Ohlendorf Eastern Arborvitae ●

390615 Thuja occidentalis L. 'Pumila'; 偃香柏 (小北美香柏); Minor Eastern Arborvitae ●

390616 Thuja occidentalis L. 'Pyramidalis'; 尖塔北美香柏 ●☆

390617 Thuja occidentalis L. 'Recurvar Nana'; 扭曲枝北美香柏; Eastern Arborvitae ●

390618 Thuja occidentalis L. 'Rheingold'; 雷宁古德北美香柏 (莱因金北美香柏); Gold Arborvitae, Rheingold Arborvitae ●☆

390619 Thuja occidentalis L. 'Robusta'; 绿塔北美香柏; Green-pyramid Eastern Arborvitae ●

390620 Thuja occidentalis L. 'Silver Queen'; 银色女王北美香柏 ●☆

390621 Thuja occidentalis L. 'Smaragd'; 祖母绿北美香柏 (亮绿北美香柏) ●☆

390622 Thuja occidentalis L. 'Spiralis'; 螺旋北美香柏 ●☆

390623 Thuja occidentalis L. 'Tiny Tim'; 小蒂姆北美香柏 ●☆

390624 Thuja occidentalis L. 'Umbraculifera'; 伞形北美香柏; Umbrella Eastern Arborvitae ●

390625 Thuja occidentalis L. 'Wintergreen'; 冬绿北美香柏 ●☆

390626　Thuja occidentalis L. 'Woodwardii';阔球形北美香柏(伍德沃德北美香柏);Wide-globe Eastern Arborvitae ●

390627　Thuja occidentalis L. pyramidalis ? = Thuja occidentalis L. 'Pyramidalis' ●☆

390628　Thuja orientalis L. 'Flagelliformis';下垂北美香柏●☆

390629　Thuja orientalis L. = Platycladus orientalis (L.) Franco ●

390630　Thuja orientalis L. f. beverleyensis (Rehder) Rehder = Platycladus orientalis (L.) Franco 'Beverleyensis' ●

390631　Thuja orientalis L. f. semperaurescens (Gordon) C. K. Schneid. = Platycladus orientalis (L.) Franco 'Semperaurescens' ●

390632　Thuja orientalis L. f. sieboldii (Endl.) Rehder = Platycladus orientalis (L.) Franco 'Sieboldii' ●

390633　Thuja orientalis L. var. argyi H. Lév. et Lemée = Platycladus orientalis (L.) Franco ●

390634　Thuja orientalis L. var. argyi Lemée et H. Lév. = Platycladus orientalis (L.) Franco ●

390635　Thuja orientalis L. var. aurea ? = Thuja occidentalis L. 'Aurea' ●

390636　Thuja orientalis L. var. beverleyensis Rehder = Platycladus orientalis (L.) Franco 'Beverleyensis' ●

390637　Thuja orientalis L. var. nana C. K. Schneid. = Platycladus orientalis (L.) Franco 'Sieboldii' ●

390638　Thuja orientalis L. var. pendula (Thunb.) Parl. = Thuja orientalis L. 'Flagelliformis' ●☆

390639　Thuja orientalis L. var. semperaurescens (Gordon) Nicholson = Platycladus orientalis (L.) Franco 'Semperaurescens' ●

390640　Thuja orientalis L. var. sieboldii (Endl.) Lawson = Platycladus orientalis (L.) Franco 'Sieboldii' ●

390641　Thuja pendula (Thunb.) Siebold et Zucc. = Thuja orientalis L. 'Flagelliformis' ●☆

390642　Thuja pendula (Thunb.) Siebold et Zucc. = Thuja orientalis L. var. pendula (Thunb.) Parl ●☆

390643　Thuja pensilis Staunton ex D. Don = Glyptostrobus pensilis (Staunton ex D. Don) K. Koch ●◇

390644　Thuja pisifera (Siebold et Zucc.) Mast. = Chamaecyparis pisifera (Siebold et Zucc.) Siebold et Zucc. ex Endl.

390645　Thuja plicata Donn ex D. Don;北美乔柏(北美香柏,红柏,巨侧柏,美国侧柏);American Arbor-vitae, Arbor-vitae, Canoe Cedar, Canoe-cedar, Cone Redcedar, Giant Arborvitae, Giant Arbor-vitae, Giant Cedar, Giant Thuya, Nootka Sound, Pacific Red Cedar, Pacific Redcedar, Red Cedar, Shinglewood, Western Arborvitae, Western Arbor-vitae, Western Red Ceder, Western Redcedar, Western Thuja, Western White Cedar ●

390646　Thuja plicata Donn ex D. Don 'Atrovirens';墨绿北美乔柏(深绿北美乔柏)●☆

390647　Thuja plicata Donn ex D. Don 'Aurea';金黄北美乔柏(金叶北美乔柏);Golden Western Arborvitae ●

390648　Thuja plicata Donn ex D. Don 'Collyer's Gold';绿金北美乔柏●☆

390649　Thuja plicata Donn ex D. Don 'Cuprea';铜色北美乔柏●☆

390650　Thuja plicata Donn ex D. Don 'Excelsa';高大北美乔柏●☆

390651　Thuja plicata Donn ex D. Don 'Fastigiata';帚状北美乔柏;Hogan Western Arborvitae ●

390652　Thuja plicata Donn ex D. Don 'George Washington';乔治·华盛顿北美乔柏●☆

390653　Thuja plicata Donn ex D. Don 'Hillieri';希里北美乔柏(希莱尔北美乔柏)●☆

390654　Thuja plicata Donn ex D. Don 'Stoneham Gold';嫩叶金黄北美乔柏(矮金北美乔柏)●☆

390655　Thuja plicata Donn ex D. Don 'Striblingii';密柱形北美乔柏;Dense-column Cedar ●

390656　Thuja plicata Donn ex D. Don 'Sunshin';日光北美乔柏●☆

390657　Thuja plicata Donn ex D. Don 'Virescens';淡绿北美乔柏●☆

390658　Thuja plicata Donn ex D. Don 'Zebrina';斑马北美乔柏(黄纹北美乔柏)●☆

390659　Thuja procera Salisb. = Thuja occidentalis L. ●

390660　Thuja standishii (Gordon) Carrière;日本香柏(日本侧柏);Japan Arborvitae, Japanese Arborvitae, Japanese Arbor-vitae, Standish Arbor-vitae ●

390661　Thuja sutchuanensis Franch.;崖柏(四川侧柏,崖柏树);Arborvitae, Sichuan Arborvitae, Szechwan Arborvitae, Szechwan Arbor-vitae ●◇

390662　Thuja theophrasti C. Bauhin ex Nieuwl. = Thuja occidentalis L. ●

390663　Thuja theophrasti Nieuwl. = Thuja occidentalis L. ●

390664　Thujaceae Burnett = Cupressaceae Gray(保留科名)●

390665　Thujaceae Burnett;崖柏科●

390666　Thujaecarpus Trautv. = Juniperus L. ●

390667　Thujaecarpus Trautv. = Thuiaecarpus Trautv. ●

390668　Thujocarpus Post et Kuntze = Juniperus L. ●

390669　Thujocarpus Post et Kuntze = Thujaecarpus Trautv. ●

390670　Thujopsidaceae Bessey = Cupressaceae Gray(保留科名)●

390671　Thujopsis Post et Kuntze = Tafalla D. Don ●☆

390672　Thujopsis Siebold et Zucc. = Thujopsis Siebold et Zucc. ex Endl.(保留属名)●

390673　Thujopsis Siebold et Zucc. ex Endl.(1842)(保留属名);罗汉柏属; Broadleaf Arborvitae, Broad-leaved Arbor-vitae, False Arborvitae, False Arbor-vitae, Hiba Arbor-vitae, Thujopsis ●

390674　Thujopsis dolabrata (L. f.) Siebold et Zucc. 'Nana' = Thujopsis dolabrata (Thunb. ex L. f.) Siebold et Zucc. 'Nana' ●

390675　Thujopsis dolabrata (L. f.) Siebold et Zucc. 'Variegata' = Thujopsis dolabrata (Thunb. ex L. f.) Siebold et Zucc. 'Variegata' ●

390676　Thujopsis dolabrata (L. f.) Siebold et Zucc. = Thujopsis dolabrata (Thunb. ex L. f.) Siebold et Zucc. ●

390677　Thujopsis dolabrata (L. f.) Siebold et Zucc. var. australis Henry = Thujopsis dolabrata (Thunb. ex L. f.) Siebold et Zucc. ●

390678　Thujopsis dolabrata (L. f.) Siebold et Zucc. var. hondae Makino = Thujopsis dolabrata (Thunb. ex L. f.) Siebold et Zucc. var. hondae Makino ●☆

390679　Thujopsis dolabrata (L. f.) Siebold et Zucc. var. nana Carrière = Thujopsis dolabrata (Thunb. ex L. f.) Siebold et Zucc. 'Nana' ●

390680　Thujopsis dolabrata (Thunb. ex L. f.) Siebold et Zucc.;罗汉柏(斧松,日本罗汉柏,蜈蚣柏);Broadleaf Arborvitae Hiba, Deerhorn Cedar, False Arborvitae, Hiba, Hiba Arbor-vitae, Hiba Cedar, Hiba Cedarvitae, Hiba False Arborvitae, Hiba False Arbor-vitae, Hiba False-arborvitae, Japanese Thuja ●

390681　Thujopsis dolabrata (Thunb. ex L. f.) Siebold et Zucc. 'Nana';矮生罗汉柏;Dwarf Broadleaf Arborvitae ●

390682　Thujopsis dolabrata (Thunb. ex L. f.) Siebold et Zucc. 'Variegata';金尖罗汉柏(白斑罗汉柏);Dwarf Variegata Thujopsis, Golden-tip Broadleaf Arborvitae ●

390683　Thujopsis dolabrata (Thunb. ex L. f.) Siebold et Zucc. var. australis A. Henry = Thujopsis dolabrata (Thunb. ex L. f.) Siebold et Zucc. ●

390684　Thujopsis dolabrata (Thunb. ex L. f.) Siebold et Zucc. var. hondae Makino;本田氏罗汉柏;Honda Hiba False Arborvitae ●☆

390685　Thujopsis standishii Gordon ＝ Thuja standishii (Gordon) Carrière ●

390686　Thulinia P. J. Cribb(1985);图林兰属■☆

390687　Thulinia alboltea P. J. Cribb;图林兰■☆

390688　Thumbergia Poit. ＝ Thunbergia Retz. (保留属名)●■

390689　Thumung J. König ＝ Zingiber Mill. (保留属名)■

390690　Thunbergia Montin (废弃属名) ＝ Gardenia Ellis(保留属名) ●

390691　Thunbergia Montin (废弃属名) ＝ Thunbergia Retz. (保留属名)●■

390692　Thunbergia Retz. (1773)(保留属名);老鸦嘴属(邓伯花属,老鸦咀属,山牵牛属,月桂藤属);Black-eyed Susan Vine,Clock Vine,Clockvine,Clock-vine,Dock Vine ●■

390693　Thunbergia abyssinica Turrill;阿比西尼亚老鸦嘴■☆

390694　Thunbergia acutibracteata De Wild.;尖苞老鸦嘴■☆

390695　Thunbergia adenocalyx Radlk.;腺萼老鸦嘴■☆

390696　Thunbergia adenophora W. W. Sm.;单腺山牵牛;Glandular Clockvine ■

390697　Thunbergia adenophora W. W. Sm. ＝ Thunbergia lacei Gamble ■

390698　Thunbergia adjumaensis De Wild.;阿朱马老鸦嘴■☆

390699　Thunbergia affinis S. Moore;邓伯花■☆

390700　Thunbergia affinis S. Moore var. pulvinata？＝ Thunbergia holstii Lindau ■☆

390701　Thunbergia alata Bojer ex Sims;翼叶山牵牛(黑眼花,翼叶老鸦嘴,翼叶老鸦嘴);Black-eyed Clockvine,Black-eyed Susan,Blackeyed Susan Vine,Black-eyed Susan Vine,Clock Vine ■

390702　Thunbergia alata Bojer ex Sims var. alba Paxton;白色翼叶山牵牛(白色邓伯花)■☆

390703　Thunbergia alata Bojer ex Sims var. aurantica Kuntze;橙色翼叶山牵牛(橙色邓伯花)■☆

390704　Thunbergia alata Bojer ex Sims var. minor S. Moore ＝ Thunbergia alata Bojer ex Sims ■

390705　Thunbergia alata Bojer ex Sims var. reticulata (Hochst. ex Nees) Burkill ＝ Thunbergia alata Bojer ex Sims ■

390706　Thunbergia alata Bojer ex Sims var. vixalata Burkill ＝ Thunbergia alata Bojer ex Sims ■

390707　Thunbergia alba S. Moore;白老鸦嘴■☆

390708　Thunbergia amanensis Lindau;阿曼老鸦嘴■☆

390709　Thunbergia amoena C. B. Clarke;秀丽老鸦嘴■☆

390710　Thunbergia angolensis S. Moore;安哥拉老鸦嘴■☆

390711　Thunbergia angulata Hils. et Bojer ex Hook.;棱角老鸦嘴■☆

390712　Thunbergia annua Hochst. ex Nees;一年老鸦嘴■☆

390713　Thunbergia annua Hochst. ex Nees var. ruspolii (Lindau) Burkill;鲁斯波利老鸦嘴■☆

390714　Thunbergia argentea Lindau;银白老鸦嘴■☆

390715　Thunbergia aspera Nees ＝ Thunbergia atriplicifolia E. Mey. ex Nees ■☆

390716　Thunbergia atacorensis Akoègn. et Lisowski et Sinsin;阿塔科老鸦嘴■☆

390717　Thunbergia atriplicifolia E. Mey. ＝ Thunbergia atriplicifolia E. Mey. ex Nees ■☆

390718　Thunbergia atriplicifolia E. Mey. ex Nees;纳塔尔翼叶山牵牛(纳塔尔邓伯花);Natal Primrose ■☆

390719　Thunbergia attenuata Benoist ＝ Anomacanthus congolanus (De Wild. et T. Durand) Brummitt ●☆

390720　Thunbergia aurea N. E. Br.;黄老鸦嘴■☆

390721　Thunbergia aureosetosa Mildbr.;黄刚毛老鸦嘴■☆

390722　Thunbergia bachmannii Lindau ＝ Thunbergia atriplicifolia E. Mey. ex Nees ■☆

390723　Thunbergia battiscombei Turrill;巴蒂老鸦嘴■☆

390724　Thunbergia bauri Lindau ＝ Thunbergia neglecta Sond. ■☆

390725　Thunbergia beniensis De Wild.;贝尼老鸦嘴■☆

390726　Thunbergia beninensioides De Wild.;拟贝尼老鸦嘴■☆

390727　Thunbergia bequaertii De Wild.;贝卡尔老鸦嘴■☆

390728　Thunbergia bianoensis De Wild. et Ledoux;比亚诺老鸦嘴■☆

390729　Thunbergia bikimaensis De Wild.;比基马老鸦嘴■☆

390730　Thunbergia bodinieri H. Lév. ＝ Thunbergia fragrans Roxb. ■

390731　Thunbergia brewerioides Schweinf. ex Lindau;伯纳旋花老鸦嘴■☆

390732　Thunbergia capensis Retz.;好望角老鸦嘴■☆

390733　Thunbergia castellaneana Buscal. et Muschl. ＝ Thunbergia erythreae Schweinf. ex Lindau ■☆

390734　Thunbergia cerinthoides Radlk.;蜡黄老鸦嘴■☆

390735　Thunbergia chinensis Merr. ＝ Thunbergia grandiflora (Roxb. ex Rottler) Roxb. ●

390736　Thunbergia chiovendae Fiori ＝ Thunbergia guerkeana Lindau ■☆

390737　Thunbergia chrysochlamys Baker ＝ Pseudocalyx saccatus Radlk. ■☆

390738　Thunbergia chrysops Hook.;金黄老鸦嘴■☆

390739　Thunbergia ciliata De Wild.;缘毛老鸦嘴■☆

390740　Thunbergia claessensii De Wild.;克莱森斯老鸦嘴■☆

390741　Thunbergia coccinea Wall. ＝ Thunbergia coccinia Wall. ex D. Don ●

390742　Thunbergia coccinea Wall. ex D. Don;红花山牵牛;Scarlet Clockvine ●

390743　Thunbergia collina S. Moore;山丘老鸦嘴■☆

390744　Thunbergia combretoides A. Chev. ＝ Mendoncia combretoides (A. Chev.) Benoist ●☆

390745　Thunbergia convolvulifolia Baker ＝ Thunbergia fragrans Roxb. ■

390746　Thunbergia cordata Lindau;心叶老鸦嘴■☆

390747　Thunbergia cordibracteolata C. B. Clarke ＝ Thunbergia atriplicifolia E. Mey. ex Nees ■☆

390748　Thunbergia cordifolia Nees ＝ Thunbergia grandiflora (Roxb. ex Willd.) Roxb. ●

390749　Thunbergia crispa Burkill;皱波老鸦嘴■☆

390750　Thunbergia cuanzensis S. Moore;宽扎老鸦嘴■☆

390751　Thunbergia cyanea Nees;蓝色老鸦嘴■☆

390752　Thunbergia cynanchifolia Benth.;鹅绒藤叶老鸦嘴■☆

390753　Thunbergia deflexiflora Baker ＝ Pseudocalyx saccatus Radlk. ■☆

390754　Thunbergia delamerei S. Moore;德拉米尔老鸦嘴■☆

390755　Thunbergia dregeana Nees;德雷老鸦嘴■☆

390756　Thunbergia eberhardtii Benoist;二色老鸦嘴(二色老鸦咀,二色山牵牛)■

390757　Thunbergia elliotii S. Moore;埃利奥特老鸦嘴■☆

390758　Thunbergia elskensii De Wild.;埃尔斯克老鸦嘴■☆

390759　Thunbergia erecta (Benth.) T. Anderson;直立山牵牛(金鱼木,蓝吊钟,立鹤花,硬枝老鸦嘴);Buch Clockvine,King's Mantle ●

390760　Thunbergia erecta (Benth.) T. Anderson var. alba Hort.;白色直立山牵牛(白色立鹤花)●☆

390761　Thunbergia erythreae Schweinf. ex Lindau;浅红老鸦嘴■☆

390762　Thunbergia exasperata Lindau;粗糙老鸦嘴■☆

390763　Thunbergia fasciculata De Wild. ＝ Thunbergia katentaniensis De Wild. ■☆

390764　Thunbergia fasciculata Lindau;簇生老鸦嘴■☆

390765　Thunbergia fischeri Engl.;菲舍尔老鸦嘴■☆

390766　Thunbergia flavohirta Lindau ＝ Thunbergia atriplicifolia E. Mey. ex Nees ■☆

390767　Thunbergia fragrans Roxb.；碗花草（老鸦嘴，山牵牛，铁贯藤）；Sweet Clockvine，Whitelady ■

390768　Thunbergia fragrans Roxb. = Thunbergia convolvulifolia Baker ■

390769　Thunbergia fragrans Roxb. subsp. hainanensis（C. Y. Wu et H. S. Lo）H. P. Tsui = Thunbergia hainanensis C. Y. Wu et H. S. Lo ■

390770　Thunbergia fragrans Roxb. subsp. lanceolata H. P. Tsui；滇南山牵牛（公鸡藤）■

390771　Thunbergia fragrans Roxb. var. laevis（Nees）C. B. Clarke = Thunbergia laevis Nees ■☆

390772　Thunbergia fragrans subsp. hainanensis（C. Y. Wu et H. S. Lo）H. P. Tsui；海南山牵牛 ■

390773　Thunbergia friesii Lindau；弗里斯老鸦嘴 ■☆

390774　Thunbergia fuscata T. Anderson ex Lindau = Thunbergia alata Bojer ex Sims ■

390775　Thunbergia galpinii Lindau = Thunbergia atriplicifolia E. Mey. ex Nees ■☆

390776　Thunbergia gentianoides Radlk.；龙胆老鸦嘴 ■☆

390777　Thunbergia geraniifolia Benth. = Thunbergia chrysops Hook. ■☆

390778　Thunbergia gibsonii S. Moore；吉布森老鸦嘴 ■☆

390779　Thunbergia gigantea Lindau = Thunbergia guerkeana Lindau ■☆

390780　Thunbergia glaberrima Lindau；无毛老鸦嘴 ■☆

390781　Thunbergia glandulifera Lindau；腺点老鸦嘴 ■☆

390782　Thunbergia glaucina S. Moore；灰绿老鸦嘴 ■☆

390783　Thunbergia gondarensis Chiov.；贡达尔老鸦嘴 ■☆

390784　Thunbergia gossweileri S. Moore；戈斯老鸦嘴 ■☆

390785　Thunbergia gracilis Benoist；纤细山牵牛 ■☆

390786　Thunbergia graminifolia De Wild.；禾叶老鸦嘴 ■☆

390787　Thunbergia grandiflora（Roxb. ex Rottler）Roxb.；大花山牵牛（白狗肠，大邓伯花，大花老鸦嘴，大花老鸦嘴，大青，老鼠黄瓜，老鸦杓，老鸦嘴，老鸦嘴，山牵牛，通骨消）；Bengal Clock Vine，Bengal Clockvine，Bengal Clock-vine，Bengal Trumpet，Blue Trumpet Vine，Blue Trumpet-vine，Blue-sky Vine，Clock-vine，Sky Vine，Sky-vine ●

390788　Thunbergia grandiflora（Roxb. ex Rottler）Roxb. = Thunbergia grandiflora（Roxb. ex Willd.）Roxb. ●

390789　Thunbergia grandiflora（Roxb. ex Willd.）Roxb = Thunbergia grandiflora（Roxb. ex Willd.）Roxb. ●

390790　Thunbergia grandiflora（Roxb. ex Willd.）Roxb. var. laurifolia（Lindl.）Benoist = Thunbergia laurifolia Lindl. ■

390791　Thunbergia gregorii S. Moore；橘黄山牵牛；Orange Clock Vine，Orange Clockvine ■☆

390792　Thunbergia guerkeana Lindau；盖尔克老鸦嘴 ■☆

390793　Thunbergia hainanensis C. Y. Wu et H. S. Lo；海南老鸦嘴（海南老鸦咀，海南山牵牛）；Hainan Clockvine ■

390794　Thunbergia hainanensis C. Y. Wu et H. S. Lo = Thunbergia fragrans Roxb. subsp. hainanensis（C. Y. Wu et H. S. Lo）H. P. Tsui ■

390795　Thunbergia hamata Lindau；顶钩老鸦嘴 ■☆

390796　Thunbergia hanningtonii Burkill；汉宁顿老鸦嘴 ■☆

390797　Thunbergia harrisii Hook. = Thunbergia laurifolia Lindl. ■

390798　Thunbergia hirsuta T. Anderson；粗毛老鸦嘴 ■☆

390799　Thunbergia hirtistyla C. B. Clarke = Thunbergia atriplicifolia E. Mey. ex Nees ■☆

390800　Thunbergia hispida Solms；硬毛老鸦嘴 ■☆

390801　Thunbergia hockii De Wild.；霍克老鸦嘴 ■☆

390802　Thunbergia holstii Lindau；霍尔老鸦嘴 ■☆

390803　Thunbergia homblei De Wild.；洪布勒老鸦嘴 ■☆

390804　Thunbergia hookeriana Lindau；胡克老鸦嘴 ■☆

390805　Thunbergia huillensis S. Moore；威拉老鸦嘴 ■☆

390806　Thunbergia humbertii Mildbr.；亨伯特老鸦嘴 ■☆

390807　Thunbergia hyalina S. Moore；透明老鸦嘴 ■☆

390808　Thunbergia jayii S. Moore；热氏老鸦嘴 ■☆

390809　Thunbergia kamatembica Mildbr.；卡马老鸦嘴 ■☆

390810　Thunbergia kamerunensis Lindau = Thunbergia vogeliana Benth. ■☆

390811　Thunbergia kassneri S. Moore；卡斯纳老鸦嘴 ■☆

390812　Thunbergia katangensis De Wild.；加丹加老鸦嘴 ■☆

390813　Thunbergia katentaniensis De Wild.；卡滕塔尼亚老鸦嘴 ■☆

390814　Thunbergia kirkiana T. Anderson；柯克老鸦嘴 ■☆

390815　Thunbergia kirkii Hook. f. = Thunbergia hookeriana Lindau ■☆

390816　Thunbergia lacei Gamble；黄毛山牵牛（长黄毛山牵牛，巢腺山牵牛，刚毛山牵牛）；Yellowhair Clockvine ■

390817　Thunbergia laevis Nees；平滑老鸦嘴 ■☆

390818　Thunbergia lamellata Hiern；片状老鸦嘴 ■☆

390819　Thunbergia lancifolia T. Anderson；热非老鸦嘴 ■☆

390820　Thunbergia laricifolia Lindl. = Thunbergia lancifolia T. Anderson ■☆

390821　Thunbergia lathyroides Burkill；山鬣豆老鸦嘴 ■☆

390822　Thunbergia laurifolia Lindl.；桂叶山牵牛（桂叶老鸦嘴，月桂藤，月桂叶山牵牛，樟叶邓伯花，樟叶老鸦嘴）；Blue Trumpetcreeper，Laurel Clockvine，Laurel-leaved Thunbergia，Skyflower ■

390823　Thunbergia leucorhiza Benoist；白根山牵牛 ■☆

390824　Thunbergia liebrechtsiana De Wild. et T. Durand；利布老鸦嘴 ■☆

390825　Thunbergia longepedunculata De Wild.；长梗老鸦嘴 ■☆

390826　Thunbergia longifolia Lindau；长叶老鸦嘴 ■☆

390827　Thunbergia longisepala Rendle；长萼老鸦嘴 ■☆

390828　Thunbergia lutea（Lindl.）T. Anderson；黄花山牵牛（羽麦山牵牛，羽脉山牵牛）；Yellowflower Clockvine ■

390829　Thunbergia lutea T. Anderson = Thunbergia lutea（Lindl.）T. Anderson ■

390830　Thunbergia malangana Lindau；马兰加老鸦嘴 ■☆

390831　Thunbergia manganjensis T. Anderson ex Lindau = Thunbergia alata Bojer ex Sims ■

390832　Thunbergia manikensis De Wild.；马尼科老鸦嘴 ■☆

390833　Thunbergia masisiensis De Wild.；马西西老鸦嘴 ■☆

390834　Thunbergia mechowii Lindau；梅休老鸦嘴 ■☆

390835　Thunbergia mellinocaulis Burkill；蜜色茎老鸦嘴 ■☆

390836　Thunbergia mestdaghi De Wild.；梅斯特老鸦嘴 ■☆

390837　Thunbergia micheliana De Wild.；米歇尔老鸦嘴 ■☆

390838　Thunbergia microchlamys S. Moore；小被老鸦嘴 ■☆

390839　Thunbergia mildbraediana Lebrun et Touss.；米尔德老鸦嘴 ■☆

390840　Thunbergia mollis Lindau；柔软老鸦嘴 ■☆

390841　Thunbergia monroi S. Moore；门罗老鸦嘴 ■☆

390842　Thunbergia mysorensis T. Anderson ex Bedd.；舌唇山牵牛；Clock Vine ■☆

390843　Thunbergia natalensis Hook.；纳塔尔山牵牛 ■☆

390844　Thunbergia neglecta Sond.；疏忽山牵牛 ■☆

390845　Thunbergia nidulans Lindau = Thunbergia fasciculata Lindau ■☆

390846　Thunbergia nymphaeifolia Lindau；睡莲叶老鸦嘴 ■☆

390847　Thunbergia oblongifolia Oliv.；矩圆叶老鸦嘴 ■☆

390848　Thunbergia oculata S. Moore；小眼老鸦嘴 ■☆

390849　Thunbergia oubanguiensis Benoist；乌班吉老鸦嘴 ■☆

390850　Thunbergia parvifolia Lindau；小叶老鸦嘴 ■☆

390851　Thunbergia paulitschkeana Beck var. lanceolata Chiov.；披针形老鸦嘴 ■☆

390852 Thunbergia petersiana Lindau;彼得斯老鸦嘴■☆

390853 Thunbergia pondoensis Lindau;庞多老鸦嘴■☆

390854 Thunbergia pratensis Lindau;草原老鸦嘴■☆

390855 Thunbergia primulina Hemsl.;报春老鸦嘴■☆

390856 Thunbergia prostrata Turrill;平卧老鸦嘴■☆

390857 Thunbergia proxima De Wild.;近基老鸦嘴■☆

390858 Thunbergia proximoides De Wild.;拟近基老鸦嘴■☆

390859 Thunbergia puberula Lindau;微毛老鸦嘴■☆

390860 Thunbergia purpurata Harv. ex C. B. Clarke;浅紫老鸦嘴■☆

390861 Thunbergia pynaertii De Wild.;皮那老鸦嘴■☆

390862 Thunbergia quadrialata Lindau;四翅老鸦嘴■☆

390863 Thunbergia randii S. Moore;兰德老鸦嘴■☆

390864 Thunbergia reticulata Hochst. ex Nees = Thunbergia alata Bojer ex Sims ■

390865 Thunbergia roberti Mildbr.;罗伯特老鸦嘴■☆

390866 Thunbergia rogersii Turrill;罗杰斯老鸦嘴■☆

390867 Thunbergia rufescens Lindau;焦黄老鸦嘴■☆

390868 Thunbergia rumicifolia Lindau;酸模老鸦嘴■☆

390869 Thunbergia ruspolii Lindau = Thunbergia annua Hochst. ex Nees var. ruspolii (Lindau) Burkill ■☆

390870 Thunbergia saltiana Steud. = Thunbergia alata Bojer ex Sims ■

390871 Thunbergia salweenensis W. W. Sm.;羽脉山牵牛;Salween Clockvine ■

390872 Thunbergia salwenensis W. W. Sm. = Thunbergia lutea (Lindl.) T. Anderson ■

390873 Thunbergia schimbensis S. Moore;欣贝老鸦嘴■☆

390874 Thunbergia schweinfurthii S. Moore;施韦老鸦嘴■☆

390875 Thunbergia sericea Burkill;绢毛老鸦嘴■☆

390876 Thunbergia sessilis Lindau;无柄老鸦嘴■☆

390877 Thunbergia spinulosa Chiov. = Thunbergia holstii Lindau ■☆

390878 Thunbergia squamuligera Lindau;鳞片老鸦嘴■☆

390879 Thunbergia stellarioides Burkill;星状老鸦嘴■☆

390880 Thunbergia stuhlmanniana Lindau;斯图尔曼老鸦嘴■☆

390881 Thunbergia subalata Lindau;翅老鸦嘴■☆

390882 Thunbergia subcordatifolia De Wild.;近心叶老鸦嘴■☆

390883 Thunbergia subfulva S. Moore;黄褐老鸦嘴■☆

390884 Thunbergia subnymphaeifolia Lindau = Thunbergia chrysops Hook. ■☆

390885 Thunbergia swynnertonii S. Moore;斯温纳顿老鸦嘴■☆

390886 Thunbergia talbotiae S. Moore = Thunbergia grandiflora (Roxb. ex Willd.) Roxb. ●

390887 Thunbergia thonneri De Wild. et T. Durand;托内老鸦嘴■☆

390888 Thunbergia togoensis Lindau;多哥老鸦嘴■☆

390889 Thunbergia torrei Benoist;托雷老鸦嘴■☆

390890 Thunbergia trinervis S. Moore;三脉老鸦嘴■☆

390891 Thunbergia usambarica Lindau;乌桑巴拉老鸦嘴■☆

390892 Thunbergia valida S. Moore;刚直老鸦嘴■☆

390893 Thunbergia variabilis De Wild.;易变老鸦嘴■☆

390894 Thunbergia venosa C. B. Clarke;多脉老鸦嘴■☆

390895 Thunbergia verdickii De Wild.;韦尔老鸦嘴■☆

390896 Thunbergia vincoides Benoist;蔓长春花老鸦嘴■☆

390897 Thunbergia vogeliana Benth.;沃格尔老鸦嘴■☆

390898 Thunbergia volubilis Pers. = Thunbergia fragrans Roxb. ■

390899 Thunbergia vossiana De Wild.;沃斯老鸦嘴■☆

390900 Thunbergia woodii Gand.;伍得老鸦嘴■☆

390901 Thunbergia xanthotricha Lindau = Thunbergia atriplicifolia E. Mey. ex Nees ■☆

390902 Thunbergia zernyi Mildbr.;策尼老鸦嘴■☆

390903 Thunbergiaceae Bremek. = Acanthaceae Juss.(保留科名)●■

390904 Thunbergiaceae Lilja = Acanthaceae Juss.(保留科名)●■

390905 Thunbergiaceae Tiegh.;老鸦嘴科(山牵牛科)●●

390906 Thunbergiaceae Tiegh. = Acanthaceae Juss.(保留科名)●■

390907 Thunbergianthus Engl.(1897);通氏列当属(桑氏花属)●☆

390908 Thunbergianthus quintasii Engl.;非洲通氏列当●☆

390909 Thunbergianthus ruwenzoriensis Good;通氏列当●☆

390910 Thunbergiella H. Wolff = Itasina Raf. ■☆

390911 Thunbergiella H. Wolff(1922);小老鸦嘴属(小山牵牛属)●☆

390912 Thunbergiella filiformis (Welw. ex Baker) Hiern = Itasina filifolia (Thunb.) Raf. ■☆

390913 Thunbergiella filiformis H. Wolff;小老鸦嘴(小山牵牛)■☆

390914 Thunbergiopsis Engl. = Thunbergianthus Engl. ●☆

390915 Thunbgeria Montin = Gardenia Ellis(保留属名)●

390916 Thunia Rchb. f.(1852);笋兰属(岩笋属,岩竹属);Thunia ■

390917 Thunia alba (Lindl.) Rchb. f.;笋兰(风兰,接骨丹,石笋,石竹子,通兰,岩角,岩笋,岩竹);Common Thunia,Thunia ■

390918 Thunia bensoniae Rchb. f.;本氏笋兰■☆

390919 Thunia marshalliana Rchb. f. = Thunia alba (Lindl.) Rchb. f. ■

390920 Thunia marshallina Rchb. f.;马氏笋兰■☆

390921 Thunia venosa Rolfe = Thunia alba (Lindl.) Rchb. f. ■

390922 Thuranthos C. H. Wright(1916);门花风信子属■☆

390923 Thuranthos aurantiacum (H. Lindb.) Speta;橘黄门花风信子■☆

390924 Thuranthos basuticum (E. Phillips) Oberm. = Drimia angustifolia Baker ■☆

390925 Thuranthos indicum (Roxb.) Speta;印度门花风信子■☆

390926 Thuranthos macranthum (Baker) C. H. Wright = Drimia macrantha (Baker) Baker ■☆

390927 Thuranthos noctiflorum (Batt. et Trab.) Speta;门花风信子■☆

390928 Thuranthos nocturnale R. A. Dyer = Drimia macrantha (Baker) Baker ■☆

390929 Thuranthos zambesiacum (Baker) Kativu;非洲门花风信子■☆

390930 Thuraria Nutt. = Grindelia Willd. ●■☆

390931 Thurberia A. Gray = Gossypium L. ●■

390932 Thurberia Benth. = Limnodea L. H. Dewey ■☆

390933 Thurnhausera Pohl ex G. Don = Curtia Cham. et Schltdl. ■☆

390934 Thurnheyssera Mart. ex Meisn. = Symmeria Benth. ●☆

390935 Thurnia Hook. f.(1883);圭亚那草属■☆

390936 Thurnia jenmanii Hook. f.;圭亚那草■☆

390937 Thurniaceae Engl.(1907)(保留科名);圭亚那草科(梭子草科)■☆

390938 Thurovia Rose = Gutierrezia Lag. ■●☆

390939 Thurovia Rose(1895);三花蛇黄花属■☆

390940 Thurovia triflora Rose;三花蛇黄花;Three-flower snakeweed ■☆

390941 Thurya Boiss. et Balansa(1856);刺缀属■☆

390942 Thurya capitata Boiss. et Balansa;刺缀■☆

390943 Thuspeinanta T. Durand(1888);总序旱草属■☆

390944 Thuspeinanta persica (Boiss.) Briq.;总序旱草■☆

390945 Thuya Adans. = Thuja L. ●

390946 Thuya L. = Thuja L. ●

390947 Thuya Thourn. ex L. = Thuja L. ●

390948 Thuya aphylla L. = Tamarix aphylla (L.) H. Karst. ●

390949 Thuyopsis Parl. = Thujopsis Siebold et Zucc. ex Endl.(保留属名)●

390950 Thya Adans. = Thuja L. ●

390951 Thyana Ham. = Thouinia Poit.(保留属名)●☆

390952 Thyarea Benth. = Thuarea Pers. ■

390953 Thyella Raf. = Jacquinia L. (保留属名)●☆

390954 Thyia Asch. = Thuja L. ●

390955 Thylacantha Nees et Mart. = Angelonia Bonpl. ■●☆

390956 Thylacanthus Tul. (1844); 囊花豆属■☆

390957 Thylacanthus ferrugineus Tul. ; 囊花豆■☆

390958 Thylachium DC. = Thilachium Lour. ●☆

390959 Thylacis Gagnep. = Thrixspermum Lour. ■

390960 Thylacitis Adans. = Gentiana L. ■

390961 Thylacitis Raf. = Centaurium Hill ■

390962 Thylacitis Ref. ex Adans. = Centaurium Hill ■

390963 Thylacium Spreng. = Thilachium Lour. ●☆

390964 Thylacodraba (Nábelek) O. E. Schulz = Draba L. ■

390965 Thylacophora Ridl. = Riedelia Oliv. (保留属名)■☆

390966 Thylacospennum ruprifragum Schrenk = Thylacospermum caespitosum (Cambess.) Schischk. ■

390967 Thylacospermum Fenzl(1840); 囊种草属(柔籽草属, 柔子草属); Sacseed ■

390968 Thylacospermum caespitosum (Cambess.) Schischk. ; 囊种草(簇生囊种草, 簇生柔子草); Clustered Sacseed ■

390969 Thylacospermum rupifragum (Kar. et Kir.) Schrenk. = Thylacospermum caespitosum (Cambess.) Schischk. ■

390970 Thylactitis Steud. = Gentiana L. ■

390971 Thylactitis Steud. = Thylacitis Adans. ■

390972 Thylax Raf. = Zanthoxylum L. ●

390973 Thylaxus Raf. = Thylax Raf ●

390974 Thylcis Gagnep. = Thrixspermum Lour. ■

390975 Thyloceras Steud. = Styloceras A. Juss. ●☆

390976 Thylocodraba O. E. Schulz = Draba L. ■

390977 Thyloglossa Nees = Tyloglossa Hochst. ●■

390978 Thylophora cordifolia (Link., Klotz. et Otto) Benth. et Hook. f. ex Kuntze = Belostemma cordifolium (Link, Klotzsch et Otto) M. G. Gilbert et P. T. Li ●

390979 Thylophora cordifolia Thwaites = Belostemma cordifolium (Link, Klotzsch et Otto) M. G. Gilbert et P. T. Li ●

390980 Thylophora macrantha Hance = Dregea volubilis (L. f.) Benth. ex Hook. f. ●

390981 Thylostemon Kunkel = Beilschmiedia Nees ●

390982 Thylostemon Kunkel = Tylostemon Engl. ●☆

390983 Thymalis Post et Kuntze = Euphorbia L. ●■

390984 Thymalis Post et Kuntze = Tumalis Raf. ●■

390985 Thymbra L. (1753); 香薄荷属●☆

390986 Thymbra Mill. = Satureja L. ●■

390987 Thymbra capitata (L.) Cav. ; 西班牙香薄荷; Conehead Thyme, Persian Hyssop, Spanish Oregano ■

390988 Thymbra capitata (L.) Cav. = Coridothymus capitatus (L.) Rchb. f. ■☆

390989 Thymbra capitata Griseb. = Thymus capitatus Hoffmanns. et Link ●☆

390990 Thymbra ciliata Desf. = Thymus munbyanus Boiss. et Reut. subsp. ciliatus (Desf.) Greuter et Burdet ●☆

390991 Thymbra spicata L. ; 穗花香薄荷■☆

390992 Thymelaea Adans. = Daphne L. ●

390993 Thymelaea All. = Thymelaea Mill. (保留属名)●■

390994 Thymelaea Mill. (1754)(保留属名); 欧瑞香属; Sparrowwort ●■

390995 Thymelaea Mill. (保留属名) = Daphne L. ●

390996 Thymelaea Tourn. ex Adans. = Thymelaea Mill. (保留属名)●■

390997 Thymelaea Tourn. ex Scop. = Thymelaea Mill. (保留属名)●■

390998 Thymelaea algeriensis (Chabert) Murb. = Thymelaea gussonei Boreau ■☆

390999 Thymelaea antiatlantica Maire; 安蒂欧瑞香●☆

391000 Thymelaea argentata (Lam.) Pau; 银白欧瑞香●☆

391001 Thymelaea arvensis Lam. = Thymelaea passerina (L.) Coss. et Germ. ■

391002 Thymelaea canescens (Schousb.) Endl. = Thymelaea lanuginosa (Lam.) Ceballos et Vicioso ●☆

391003 Thymelaea cneorum Scop. = Daphne cneorum L. ●☆

391004 Thymelaea gattefossei Kit Tan; 加特福塞欧瑞香●☆

391005 Thymelaea gussonei Boreau; 古索内欧瑞香■☆

391006 Thymelaea hirsuta (L.) Endl. ; 毛欧瑞香(梯莫莱瑞香); Mitnan ■☆

391007 Thymelaea hirsuta (L.) Endl. var. angustifolia Meisn. = Thymelaea hirsuta (L.) Endl. ■☆

391008 Thymelaea hirsuta Endl. = Thymelaea hirsuta (L.) Endl. ■☆

391009 Thymelaea lanuginosa (Lam.) Ceballos et Vicioso; 多毛瑞香●☆

391010 Thymelaea linifolia Andr. = Thymelaea antiatlantica Maire ●☆

391011 Thymelaea lythroides Barratte et Murb. ; 千屈菜欧瑞香●☆

391012 Thymelaea microphylla Coss. et Durieu; 小叶欧瑞香■☆

391013 Thymelaea myrtifolia (Poir.) D. A. Webb = Thymelaea velutina (Cambess.) Endl. ■☆

391014 Thymelaea nitida (Vahl) Endl. = Thymelaea argentata (Lam.) Pau ●☆

391015 Thymelaea passerina (L.) Coss. et Germ. ; 帕瑟欧瑞香(欧瑞香); Mezereon ■

391016 Thymelaea passerina (L.) Lange var. pubescens (Guss.) Maire = Thymelaea gussonei Boreau ■☆

391017 Thymelaea pubescens (L.) Meisn. var. virgata (Desf.) Pau = Thymelaea virgata (Desf.) Endl. ■☆

391018 Thymelaea putorioides Emb. et Maire; 臭茜欧瑞香●☆

391019 Thymelaea putorioides Emb. et Maire var. chlorantha ? = Thymelaea putorioides Emb. et Maire ●☆

391020 Thymelaea putorioides Emb. et Maire var. rhodantha ? = Thymelaea putorioides Emb. et Maire ●☆

391021 Thymelaea sempervirens Murb. ; 常绿欧瑞香■☆

391022 Thymelaea velutina (Cambess.) Endl. ; 绒毛欧瑞香■☆

391023 Thymelaea villosa (L.) Endl. ; 多毛欧瑞香■☆

391024 Thymelaea virescens Coss. et Durieu var. glaberrima (Batt.) Kit Tan = Thymelaea virescens Meisn. ■☆

391025 Thymelaea virescens Meisn. ; 浅绿欧瑞香■☆

391026 Thymelaea virgata (Desf.) Endl. ; 条纹欧瑞香■☆

391027 Thymelaea virgata (Desf.) Endl. subsp. broussonetii (Ball) Kit Tan; 布鲁索内欧瑞香■☆

391028 Thymelaea virgata (Desf.) Endl. var. broussonetii Ball = Thymelaea virgata (Desf.) Endl. subsp. broussonetii (Ball) Kit Tan ■☆

391029 Thymelaeaceae Adans. = Thymelaea Mill. (保留属名)●■

391030 Thymelaeaceae Juss. (1789)(保留科名); 瑞香科; Mezereum Family ●■

391031 Thymelina Hoffmanns. = Gnidia L. ●☆

391032 Thymium Post et Kuntze = Torreya Arn. ●

391033 Thymium Post et Kuntze = Tumion Raf. ●

391034 Thymocarpus Nicolson, Steyerm. et Sivad. = Calathea G. Mey. ●

391035 Thymophylla Lag. (1816); 丝叶菊属●■☆

391036 Thymophylla acerosa (DC.) Strother; 刺丝叶菊; Pricklyleaf Dogweed ●☆

391037　Thymophylla aurea（A. Gray）Greene;黄丝叶菊■☆

391038　Thymophylla aurea（A. Gray）Greene var. polychaeta（A. Gray）Strother;多毛黄丝叶菊■☆

391039　Thymophylla concinna（A. Gray）Strother;雅致丝叶菊■☆

391040　Thymophylla greggii A. Gray = Thymophylla setifolia Lag. var. greggii（A. Gray）Strother ■☆

391041　Thymophylla micropoides（DC.）Strother;小梗丝叶菊■☆

391042　Thymophylla pentachaeta（DC.）Small;五毛丝叶菊;Common Dogweed, Dogweed, Five-needled Fetid Marigold, Golden Dyssodia, Golden Fleece, Parralena ●■☆

391043　Thymophylla pentachaeta（DC.）Small var. belenidium（DC.）Strother;美洲五毛丝叶菊■☆

391044　Thymophylla pentachaeta（DC.）Small var. hartwegii（A. Gray）Strother;哈特韦格丝叶菊■☆

391045　Thymophylla pentachaeta（DC.）Small var. puberula（Rydb.）Strother;柔毛丝叶菊■☆

391046　Thymophylla polychaeta（A. Gray）Small = Thymophylla aurea（A. Gray）Greene var. polychaeta（A. Gray）Strother ■☆

391047　Thymophylla puberula Rydb. = Thymophylla pentachaeta（DC.）Small var. puberula（Rydb.）Strother ■☆

391048　Thymophylla setifolia Lag.;刚毛丝叶菊■☆

391049　Thymophylla setifolia Lag. var. greggii（A. Gray）Strother;刺叶刚毛丝叶菊■☆

391050　Thymophylla tenuiloba（DC.）Small;细裂丝叶菊;Dahlberg Daisy, Golden Fleece, Golden Fleece Daisy ■☆

391051　Thymophylla tenuiloba（DC.）Small var. texana（Cory）Strother;得州丝叶菊■☆

391052　Thymophylla tenuiloba（DC.）Small var. treculii（A. Gray）Strother;特雷屈尔丝叶菊■☆

391053　Thymophylla tenuiloba（DC.）Small var. wrightii（A. Gray）Strother;赖特丝叶菊■☆

391054　Thymophylla tephroleuca（S. F. Blake）Strother;灰白丝叶菊■☆

391055　Thymophyllum Benth. et Hook. f. = Thymophylla Lag. ●■☆

391056　Thymopsis Benth.（1873）（保留属名）;百香菊属（拟百里香属）■☆

391057　Thymopsis Jaub. et Spach（废弃属名）= Hypericum L. ■●

391058　Thymopsis Jaub. et Spach（废弃属名）= Thymopsis Benth.（保留属名）■☆

391059　Thymopsis wrightii Benth.;百香菊■☆

391060　Thymos St. -Lag. = Thymus L. ●

391061　Thymus L.（1753）;百里香属（地椒属）;Thyme ●

391062　Thymus acinos L. = Calamintha arvensis Lam. ■☆

391063　Thymus afer（Pau et Font Quer）Villar = Thymus willldenowii Boiss. ●☆

391064　Thymus alatauensis（Klokov et Des. -Shost.）Klokov;阿拉图百里香●☆

391065　Thymus albiflorus（Batt.）May = Thymus pallidus Batt. ●☆

391066　Thymus algeriensis Boiss. et Reut.;阿尔及利亚百里香●☆

391067　Thymus algeriensis Boiss. et Reut. var. abyleus（Font Quer et Maire）May = Thymus algeriensis Boiss. et Reut. ●☆

391068　Thymus algeriensis Boiss. et Reut. var. antiatlanticus（Emb. et Maire）May = Thymus willldenowii Boiss. ●☆

391069　Thymus algeriensis Boiss. et Reut. var. battandieri（Sennen）Maire = Thymus algeriensis Boiss. et Reut. ●☆

391070　Thymus algeriensis Boiss. et Reut. var. cinerascens（Murb.）Maire = Thymus algeriensis Boiss. et Reut. ●☆

391071　Thymus algeriensis Boiss. et Reut. var. masculensis Maire = Thymus algeriensis Boiss. et Reut. ●☆

391072　Thymus algeriensis Boiss. et Reut. var. pomelii Maire = Thymus algeriensis Boiss. et Reut. ●☆

391073　Thymus algeriensis Boiss. et Reut. var. villicaulis Maire = Thymus algeriensis Boiss. et Reut. ●☆

391074　Thymus altaicus Klokov et Des. -Shost. = Hyssopus altaicus Klokov et Des. -Shost. ●

391075　Thymus altaicus Serg. = Thymus altaicus Klokov et Des. -Shost. ●

391076　Thymus alternans Klokov;互生百里香●☆

391077　Thymus amictus Klokov;包被百里香●☆

391078　Thymus amurensis Klokov; 黑龙江百里香; Amur Thyme, Heilongjiang Thyme ●

391079　Thymus ararati-minoris Klokov et Des. -Shost. ;亚拉腊百里香●☆

391080　Thymus arcticus（Durand ex Kane）Ronniger;北极百里香;Creeping Thyme, Mother-of-thyme ●☆

391081　Thymus arcticus（Durand）Ronniger = Thymus praecox Opiz subsp. arcticus（Durand ex Kane）Jalas ●☆

391082　Thymus arenicola Sennen et Mauricio;沙生百里香●☆

391083　Thymus armeniacus Klokov et Des. -Shost. ;亚美尼亚百里香●☆

391084　Thymus arsenijevii Klokov;阿尔森百里香●☆

391085　Thymus aschurbajevii Klokov;阿舒百里香●☆

391086　Thymus asiaticus（Kitag.）Kitag. = Thymus quinquecostatus Celak. var. asiaticus（Kitag.）C. Y. Wu et Y. C. Huang ●

391087　Thymus asiaticus Kitag. = Thymus quinquecostatus Celak. var. asiaticus（Kitag.）C. Y. Wu et Y. C. Huang ●

391088　Thymus asiaticus Serg. = Thymus serpyllum L. var. asiaticus Kitag. ●

391089　Thymus atlanticus（Ball）Roussine;大西洋百里香●☆

391090　Thymus atlanticus（Ball）Roussine subsp. ayachicus（Humbert）Greuter et Burdet;阿亚希百里香●☆

391091　Thymus atlanticus（Ball）Roussine subsp. subayachicus（Emb. et Maire）Greuter et Burdet = Thymus atlanticus（Ball）Roussine ●☆

391092　Thymus atlanticus（Ball）Roussine var. leiodontus（Emb. et Maire）Dobignard = Thymus atlanticus（Ball）Roussine ●☆

391093　Thymus atlanticus（Ball）Roussine var. stenophyllus（Emb. et Maire）Dobignard = Thymus atlanticus（Ball）Roussine ●☆

391094　Thymus atlanticus（Ball）Roussine var. subayachicus（Emb. et Maire）Dobignard = Thymus atlanticus（Ball）Roussine ●☆

391095　Thymus azoricus Lodd. = Thymus caespititius Brot. ●☆

391096　Thymus baeticus Lacaita;伯蒂卡百里香●☆

391097　Thymus baeticus Lacaita var. capitatus Boiss. = Thymus baeticus Lacaita ●☆

391098　Thymus bashkiriensis Klokov et Des. -Shost. ;巴什基里亚百里香●☆

391099　Thymus biflorus Buch. -Ham. ex D. Don = Micromeria biflora（Buch. -Ham. ex D. Don）Benth. ●■

391100　Thymus biflorus Buch. -Ham. ex D. Don = Micromeria imbricata（Forssk.）C. Chr. ■☆

391101　Thymus binervulatus Klokov et Des. -Shost. ;二脉百里香●☆

391102　Thymus bituminosus Klokov;沥青百里香●☆

391103　Thymus bleicherianus Pomel;布莱谢百里香●☆

391104　Thymus bleicherianus Pomel var. humbertii Maire = Thymus bleicherianus Pomel ●☆

391105　Thymus bleicherianus Pomel var. pseudothymbroides Maire = Thymus bleicherianus Pomel ●☆

391106　Thymus borysthenicus Klokov et Des. -Shost. ;第聂伯百里香●☆

391107　Thymus bovei Benth. ;博韦百里香●☆

391108 Thymus brevidens（Maire et Weiller）Roussine = Thymus atlanticus（Ball）Roussine ●☆

391109 Thymus broussonetii Boiss.；布鲁索内百里香●☆

391110 Thymus broussonetii Boiss. var. hannonis（Maire）Maire = Thymus broussonetii Boiss. ●☆

391111 Thymus bucharicus Klokov；布哈尔百里香●☆

391112 Thymus bulgaricus（Domin et Podp.）Ronniger；布尔嘎尔百里香●☆

391113 Thymus buschianus Klokov et Des.-Shost.；布什百里香●☆

391114 Thymus caespititius Brot.；群生百里香●☆

391115 Thymus calcareus Klokov et Des.-Shost.；钙生百里香●☆

391116 Thymus callieri Halácsy ex Litv.；卡赖氏百里香●☆

391117 Thymus candidissimus Batt. = Thymus munbyanus Boiss. et Reut. subsp. coloratus（Boiss. et Reut.）Greuter et Burdet ●☆

391118 Thymus capitatus（L.）Hoffmanns. et Link = Coridothymus capitatus（L.）Rchb. f. ■☆

391119 Thymus capitatus Hoffmanns. et Link；头状百里香●☆

391120 Thymus carnosus Boiss.；肉叶百里香（葡萄牙百里香）●☆

391121 Thymus caucasicus Willd.；高加索百里香●☆

391122 Thymus cavaleriei H. Lév. = Micromeria biflora（Buch.-Ham. ex D. Don）Benth. ●■

391123 Thymus ciliatus（Desf.）Benth. = Thymus munbyanus Boiss. et Reut. subsp. ciliatus（Desf.）Greuter et Burdet ●☆

391124 Thymus ciliatus（Desf.）Benth. = Thymus munbyanus Boiss. et Reut. ●☆

391125 Thymus ciliatus（Desf.）Benth. subsp. abylaeus Font Quer et Maire = Thymus algeriensis Boiss. et Reut. ●☆

391126 Thymus ciliatus（Desf.）Benth. subsp. albiflorus Batt. = Thymus pallidus Batt. ●☆

391127 Thymus ciliatus（Desf.）Benth. subsp. algeriensis（Boiss. et Reut.）Batt. = Thymus algeriensis Boiss. et Reut. ●☆

391128 Thymus ciliatus（Desf.）Benth. subsp. coloratus（Boiss. et Reut.）Batt. = Thymus munbyanus Boiss. et Reut. subsp. coloratus（Boiss. et Reut.）Greuter et Burdet ●☆

391129 Thymus ciliatus（Desf.）Benth. subsp. munbyanus（Boiss. et Reut.）Batt. = Thymus munbyanus Boiss. et Reut. ●☆

391130 Thymus ciliatus（Desf.）Benth. var. angustatus Faure et Maire = Thymus munbyanus Boiss. et Reut. ●☆

391131 Thymus ciliatus（Desf.）Benth. var. angustibracteatus Maire = Thymus munbyanus Boiss. et Reut. ●☆

391132 Thymus ciliatus（Desf.）Benth. var. arenicola Maire = Thymus munbyanus Boiss. et Reut. ●☆

391133 Thymus ciliatus（Desf.）Benth. var. breviflorus（Batt.）Maire = Thymus munbyanus Boiss. et Reut. ●☆

391134 Thymus ciliatus（Desf.）Benth. var. cinerascens Faure et Maire = Thymus munbyanus Boiss. et Reut. ●☆

391135 Thymus ciliatus（Desf.）Benth. var. comosus Maire = Thymus munbyanus Boiss. et Reut. ●☆

391136 Thymus ciliatus（Desf.）Benth. var. elongatus Faure et Maire = Thymus munbyanus Boiss. et Reut. ●☆

391137 Thymus ciliatus（Desf.）Benth. var. gaetulus Humbert et Maire = Thymus munbyanus Boiss. et Reut. ●☆

391138 Thymus ciliatus（Desf.）Benth. var. intermedius Batt. = Thymus munbyanus Boiss. et Reut. ●☆

391139 Thymus ciliatus（Desf.）Benth. var. longespicatus Sennen et Mauricio = Thymus munbyanus Boiss. et Reut. ●☆

391140 Thymus ciliatus（Desf.）Benth. var. major Batt. = Thymus munbyanus Boiss. et Reut. ●☆

391141 Thymus ciliatus（Desf.）Benth. var. maurusius Maire et Sennen = Thymus munbyanus Boiss. et Reut. ●☆

391142 Thymus ciliatus（Desf.）Benth. var. oblongus Faure et Maire = Thymus munbyanus Boiss. et Reut. ●☆

391143 Thymus ciliatus（Desf.）Benth. var. reesei Maire = Thymus munbyanus Boiss. et Reut. ●☆

391144 Thymus ciliatus（Desf.）Benth. var. sublobatus（Pomel）Batt. = Thymus munbyanus Boiss. et Reut. ●☆

391145 Thymus ciliatus（Desf.）Benth. var. tetuanensis Pau = Thymus munbyanus Boiss. et Reut. ●☆

391146 Thymus ciliatus（Desf.）Benth. var. thymbroides（Pomel）Batt. = Thymus munbyanus Boiss. et Reut. subsp. coloratus（Boiss. et Reut.）Greuter et Burdet ●☆

391147 Thymus ciliatus（Desf.）Benth. var. transiens Maire = Thymus munbyanus Boiss. et Reut. ●☆

391148 Thymus ciliatus（Desf.）Benth. var. zattarellus（Pomel）Batt. = Thymus algeriensis Boiss. et Reut. ●☆

391149 Thymus ciliatus（Desf.）Benth. var. zygifolius Maire = Thymus munbyanus Boiss. et Reut. ●☆

391150 Thymus ciliatus Lam.；缘毛百里香●☆

391151 Thymus circumcinctus Klokov；围绕百里香●☆

391152 Thymus citriodorus Schreb.；橙味百里香；Lemon Thyme, Lemon-scented Thyme ●☆

391153 Thymus citriodorus Schreb. 'Aureus'；黄叶橙味百里香●☆

391154 Thymus citriodorus Schreb. 'Silver Queen'；银后橙味百里香●☆

391155 Thymus collinus M. Bieb.；山丘百里香●☆

391156 Thymus coloratus Boiss. et Reut. = Thymus munbyanus Boiss. et Reut. subsp. coloratus（Boiss. et Reut.）Greuter et Burdet ●☆

391157 Thymus commutatus（Batt.）Batt.；变异百里香●☆

391158 Thymus coriifolius Ronniger；革叶百里香●☆

391159 Thymus crebrifolius Klokov；密叶百里香●☆

391160 Thymus crenulatus Klokov；细圆齿百里香●☆

391161 Thymus cretaceus Klokov et Des.-Shost.；白垩百里香●☆

391162 Thymus creticola（Klokov et Des.-Shost.）Stank.；斯大林格勒百里香●☆

391163 Thymus cuneatus Klokov；楔形百里香●☆

391164 Thymus curtus Klokov；短毛百里香；Shorthair Thyme ●☆

391165 Thymus czernajevi Klokov et Des.-Shost.；捷氏百里香●

391166 Thymus dagestanicus Klokov et Des.-Shost.；达吉斯坦百里香●☆

391167 Thymus dauricus Serg.；兴安百里香（百里香，达呼里地椒）；Dahurian Thyme ●

391168 Thymus dauricus Serg. f. albiflora C. Y. Li；白花兴安百里香；Whiteflower Dahurian Thyme ●

391169 Thymus debilis Bunge = Calamintha debilis（Bunge）Benth. ■

391170 Thymus diminutus Klokov；缩小百里香●☆

391171 Thymus dimorphus Klokov et Des.-Shost.；二型百里香●

391172 Thymus disjunctus Klokov；长齿百里香；Longtooth Thyme ●

391173 Thymus diversifolius Klokov；异叶百里香●☆

391174 Thymus dreatensis Batt.；德雷特百里香●☆

391175 Thymus drucei Ronniger = Thymus serpyllum L. ●☆

391176 Thymus dubjanskyi Klokov et Des.-Shost.；杜张氏百里香；Dubjansky Thyme ●☆

391177 Thymus dzevanovskyi Klokov et Des.-Shost.；德氏百里香；Dzevanovsky Thyme ●☆

391178 Thymus elisabethae Klokov et Des.-Shost.；爱丽萨百里香●☆

391179 Thymus eltonicus Klokov et Des.-Shost.；埃尔唐百里香●☆

391180　Thymus enervius Klokov;无脉百里香●☆

391181　Thymus eravinensis Serg.;埃拉温百里香●☆

391182　Thymus eremita Klokov;埃雷米特百里香●☆

391183　Thymus eriophorus Ronniger;毛梗百里香●☆

391184　Thymus eubajcalensis Klokov;奥巴百里香●☆

391185　Thymus eupatoriensis Klokov et Des.-Shost.;耶夫帕拖里亚百里香●☆

391186　Thymus fedtschenkoi Ronniger;范德百里香●☆

391187　Thymus flexilis Klokov;弯曲百里香●☆

391188　Thymus fominii Klokov et Des.-Shost.;福明百里香●☆

391189　Thymus fontanesii Boiss. et Reut. = Thymus pallidus Batt. ●☆

391190　Thymus fontanesii Boiss. et Reut. var. heterophyllus Batt. = Thymus pallidus Batt. ●☆

391191　Thymus fragrantissimus ?;柠檬百里香;Orange Thyme ●☆

391192　Thymus gadorensis (Pau) Villar = Thymus willdenowii Boiss. ●☆

391193　Thymus glacialis Klokov;冰雪百里香●☆

391194　Thymus glandulosus Lag. var. maroccanus Pau = Thymus algeriensis Boiss. et Reut. ●☆

391195　Thymus graniticus Klokov et Des.-Shost.;岩百里香●☆

391196　Thymus grossheimii Ronniger;格罗百里香●☆

391197　Thymus guberlinensis Iljin;古必林百里香●☆

391198　Thymus guyonii Noë;居永百里香●☆

391199　Thymus hadzhievii Grossh.;哈德百里香●☆

391200　Thymus hasachstanicus Klokov et Des.-Shost.;哈萨克百里香●☆

391201　Thymus herba-barona Loisel.;高加索小百里香;Caraway Thyme,Corsican Thyme,Seed-cake Thyme,Seed-eake Thyme ●☆

391202　Thymus hesperidum Maire = Thymus maroccanus Ball subsp. rhombicus Villar ●☆

391203　Thymus hirsutus M. Bieb.;硬毛百里香●☆

391204　Thymus hirticaulis Klokov;毛茎百里香●☆

391205　Thymus hirtus Willd. = Thymus willdenowii Boiss. ●☆

391206　Thymus hirtus Willd. subsp. algeriensis (Boiss. et Reut.) Murb. = Thymus algeriensis Boiss. et Reut. ●☆

391207　Thymus hirtus Willd. var. albiflorus (Batt.) Briq. = Thymus willdenowii Boiss. ●☆

391208　Thymus hirtus Willd. var. antiatlanticus Emb. et Maire = Thymus willdenowii Boiss. ●☆

391209　Thymus hirtus Willd. var. battandieri Sennen = Thymus algeriensis Boiss. et Reut. ●☆

391210　Thymus hirtus Willd. var. cinerescens Murb. = Thymus algeriensis Boiss. et Reut. ●☆

391211　Thymus hirtus Willd. var. gadorensis (Pau) Maire = Thymus willdenowii Boiss. ●☆

391212　Thymus hirtus Willd. var. ketamarum Maire et Sennen = Thymus willdenowii Boiss. ●☆

391213　Thymus hirtus Willd. var. legitimus Boiss. = Thymus willdenowii Boiss. ●☆

391214　Thymus hirtus Willd. var. rifanus Emb. et Maire = Thymus willdenowii Boiss. ●☆

391215　Thymus hirtus Willd. var. saharae (Pomel) Batt. = Thymus zygis L. ●☆

391216　Thymus hirtus Willd. var. wallii Maire = Thymus willdenowii Boiss. ●☆

391217　Thymus hyemalis Lange;冬百里香●☆

391218　Thymus iljinii Klokov et Des.-Shost.;伊尔金百里香●☆

391219　Thymus imbricatus Forssk. = Micromeria imbricata (Forssk.) C. Chr. ■☆

391220　Thymus inaequalis Klokov;斜叶百里香;Obliqueleaf Thyme ●☆

391221　Thymus incertus Klokov;可疑百里香●☆

391222　Thymus inodorus Desf. = Micromeria inodora (Desf.) Benth. ■☆

391223　Thymus irtyschensis Klokov;额尔齐百里香●☆

391224　Thymus jajlae (Klokov et Des.-Shost.) Stank.;杰里百里香●☆

391225　Thymus japonicus (H. Hara) Kitag.;日本百里香;Japanese Thyme ●☆

391226　Thymus japonicus (H. Hara) Kitag. = Thymus quinquecostatus Celak. ●

391227　Thymus jenisseensis Iljin;热尼斯百里香●☆

391228　Thymus kabylicus (Batt.) Batt. = Thymus pallescens Noë ●☆

391229　Thymus kalmiussicus Klokov et Des.-Shost.;卡里密百里香●☆

391230　Thymus karamarianicus Klokov et Des.-Shost.;卡拉马百里香●☆

391231　Thymus karatavicus Dmitrieva;天山百里香;Tianshan Thyme ●☆

391232　Thymus karjaginli Grossh.;卡里亚百里香●☆

391233　Thymus kasakstanicus Klokov et Des.-Shost.;哈萨克斯坦百里香●☆

391234　Thymus kirgisorum Dubyansky;吉尔吉斯百里香●☆

391235　Thymus kitagawianus Tschern.;沙地百里香●

391236　Thymus kitagawianus Tschern. = Thymus serpyllum L. var. asiaticus Kitag. ●

391237　Thymus klokovii (Ronniger) Des.-Shost.;科罗考夫百里香●☆

391238　Thymus komarovii Serg.;科马罗夫百里香●☆

391239　Thymus kotschyanus Boiss. et Hohen.;考奇百里香●☆

391240　Thymus ladjanuricus Kem.-Nath.;拉德百里香●☆

391241　Thymus lanceolatus Desf.;披针百里香●☆

391242　Thymus lanceolatus Desf. subsp. kabylicus Batt. = Thymus pallescens Noë ●☆

391243　Thymus lanceolatus Desf. subsp. numidicus (Poir.) Batt. = Thymus numidicus Poir. ●☆

391244　Thymus lanceolatus Desf. var. heterophyllus Batt. = Thymus numidicus Poir. ●☆

391245　Thymus lanuginosus Mill.;多毛百里香●☆

391246　Thymus lanulosus Klokov et Des.-Shost.;绵毛百里香●☆

391247　Thymus latifolius (Besser) Andrz.;宽叶百里香;Broadleaf Thyme ●☆

391248　Thymus latifolius Noë = Thymus pallescens Noë ●☆

391249　Thymus leiodontus (Emb. et Maire) Roussine = Thymus atlanticus (Ball) Roussine subsp. ayachicus (Humbert) Greuter et Burdet ☆

391250　Thymus leptobotrys Murb. = Thymus maroccanus Ball subsp. leptobotrys (Murb.) Dobignard ●☆

391251　Thymus leptobotrys Murb. var. latifolius Maire = Thymus maroccanus Ball subsp. rhombicus Villar ●☆

391252　Thymus leucostegius Briq. = Thymus munbyanus Boiss. et Reut. subsp. ciliatus (Desf.) Greuter et Burdet ●☆

391253　Thymus leucotrichus Halácsy;白毛百里香;Juniper Thyme ●☆

391254　Thymus lipskyi Klokov et Des.-Shost.;利普斯基百里香●☆

391255　Thymus litoralis Klokov et Des.-Shost.;海滨百里香●☆

391256　Thymus loevyanus Opiz;廖氏百里香●☆

391257　Thymus longicaulis C. Presl;长茎百里香●☆

391258　Thymus longiflorus Boiss.;长花百里香●☆

391259　Thymus lovyanus Opiz;捷克百里香●☆

391260　Thymus lusitanicus Boiss. = Thymus villosus L. subsp. lusitanicus (Boiss.) Cout. ●☆

391261　Thymus lusitanicus Boiss. var. puberulus Villar et Maire = Thymus villosus L. subsp. lusitanicus (Boiss.) Cout. ●☆

391262 Thymus lythroides Murb. = Thymus maroccanus Ball subsp. lythroides (Murb.) Dobignard ●☆

391263 Thymus majkopensis Klokov et Des. -Shost. ;马伊科普百里香●☆

391264 Thymus mandschuricus Ronniger;短节百里香;Manchurian Thyme,Shortnode Thyme ●

391265 Thymus markhotensis Maleev;马尔浩特百里香●☆

391266 Thymus maroccanus Ball;摩洛哥百里香●☆

391267 Thymus maroccanus Ball subsp. leptobotrys (Murb.) Dobignard;细穗摩洛哥百里香●☆

391268 Thymus maroccanus Ball subsp. lythroides (Murb.) Dobignard;千屈菜百里香●☆

391269 Thymus maroccanus Ball subsp. rhombicus Villar;菱形摩洛哥百里香●☆

391270 Thymus maroccanus Ball var. capitantianus Emb. = Thymus maroccanus Ball ●☆

391271 Thymus maroccanus Ball var. lythroides (Murb.) Maire = Thymus maroccanus Ball subsp. lythroides (Murb.) Dobignard ●☆

391272 Thymus maroccanus Ball var. pycnostachys Maire et Weiller = Thymus maroccanus Ball ●☆

391273 Thymus maroccanus Ball var. rhombicus (Villar) Maire = Thymus maroccanus Ball subsp. rhombicus Villar ●☆

391274 Thymus marschallianus Willd. ;异株百里香;Marschall Thyme ●

391275 Thymus mastichenus L. ;乳胶百里香;Mastick Thyme ●☆

391276 Thymus mentagensis Batt. = Thymus willdenowii Boiss. ●☆

391277 Thymus micans Lowe = Thymus caespititius Brot. ●☆

391278 Thymus migricus Klokov et Des. -Shost. ;米格里百里香●☆

391279 Thymus minussinensis Serg. ;米努辛百里香●☆

391280 Thymus mohamedii Tahiri et Rejdali = Thymus willdenowii Boiss. ●☆

391281 Thymus mohamedii Tahiri et Rejdali subsp. rosatii Tahiri et Rejdali = Thymus willdenowii Boiss. ●☆

391282 Thymus moldavicus Klokov et Des. -Shost. ;摩尔达维亚百里香●☆

391283 Thymus monardii Noë = Thymus pallescens Noë ●

391284 Thymus mongolicus (Ronniger) Ronniger;百里香(地姜,地椒,地椒花,地椒叶,地角花,蒙古百里香,千里香,野百里香);Chinese Thyme,Mongolian Thyme ●

391285 Thymus mongolicus Klokov = Thymus mongolicus (Ronniger) Ronniger ●

391286 Thymus mongolicus Klokov = Thymus serpyllum L. var. asiaticus Kitag. ●

391287 Thymus mongolicus Ronniger var. leucanthus H. L. Liu et D. Z. Ma;白花蒙百里香;Whiteflower Mongolian Thyme ●

391288 Thymus mongolicus RRonniger = Thymus serpyllum L. var. mongolicus Ronniger ●

391289 Thymus mugodzharicus Klokov et Des. -Shost. ;穆戈札雷百里香●☆

391290 Thymus munbyanus Boiss. et Reut. ;芒比百里香●☆

391291 Thymus munbyanus Boiss. et Reut. subsp. abylaeus (Font Quer et Maire) Greuter et Burdet = Thymus algeriensis Boiss. et Reut. ●☆

391292 Thymus munbyanus Boiss. et Reut. subsp. ciliatus (Desf.) Greuter et Burdet;岩毛芒比百里香●☆

391293 Thymus munbyanus Boiss. et Reut. subsp. coloratus (Boiss. et Reut.) Greuter et Burdet. ;着色百里香●☆

391294 Thymus munbyanus Boiss. et Reut. var. benoistii Sennen et Mauricio = Thymus munbyanus Boiss. et Reut. ●☆

391295 Thymus munbyanus Boiss. et Reut. var. comosus (Maire) Sennen et Mauricio = Thymus munbyanus Boiss. et Reut. ●☆

391296 Thymus munbyanus Boiss. et Reut. var. mairei Sennen = Thymus munbyanus Boiss. et Reut. ●☆

391297 Thymus munbyanus Boiss. et Reut. var. mauritianus Sennen = Thymus munbyanus Boiss. et Reut. ●☆

391298 Thymus munbyanus Boiss. et Reut. var. maurusius (Maire et Sennen) Sennen et Mauricio = Thymus munbyanus Boiss. et Reut. ●☆

391299 Thymus mutisii Caball. = Micromeria inodora (Desf.) Benth. ■☆

391300 Thymus narymensis Serg. ;纳雷姆百里香●☆

391301 Thymus nerczensis Klokov;涅尔钦百里香●☆

391302 Thymus nervulosus Klokov;显脉百里香;Distinctvein Thyme,Distinct-veined Thyme ●

391303 Thymus nitidus Guss. ;光亮百里香●☆

391304 Thymus numidicus Poir. ;努米底亚百里香●☆

391305 Thymus numidicus Poir. var. kabylicus (Batt.) Dubuis = Thymus numidicus Poir. ●☆

391306 Thymus nummularius M. Bieb. ;铜钱百里香●☆

391307 Thymus ochotensis Klokov;鄂霍次克百里香●☆

391308 Thymus origanoides Webb et Berthel. ;牛至百里香●☆

391309 Thymus ovatus Mill. = Thymus pulegioides L. ●☆

391310 Thymus oxyodontus Klokov;尖齿百里香●☆

391311 Thymus pallasianus Hans Braun;帕拉氏百里香;Pallas Thyme ●☆

391312 Thymus pallescens Noë;变白百里香●☆

391313 Thymus pallidus Batt. ;苍白百里香●☆

391314 Thymus pallidus Batt. subsp. cossonianus Maire = Thymus pallidus Batt. ●☆

391315 Thymus pallidus Batt. subsp. eriodontus (Maire) Maire = Thymus pallidus Batt. ●☆

391316 Thymus pallidus Batt. var. camphoratus Maire = Thymus pallidus Batt. ●☆

391317 Thymus pallidus Batt. var. eriodontus Maire = Thymus pallidus Batt. ●☆

391318 Thymus pallidus Batt. var. hirsutissimus Maire = Thymus pallidus Batt. ●☆

391319 Thymus pallidus Batt. var. lindbergii Maire = Thymus pallidus Batt. ●☆

391320 Thymus pallidus Batt. var. vulcanicus Maire et Weiller = Thymus pallidus Batt. ●☆

391321 Thymus pannonicus All. ;潘城百里香●☆

391322 Thymus pastoralis Iljin;牧场百里香●☆

391323 Thymus paucifolius Klokov;寡叶百里香●☆

391324 Thymus petraeus Serg. ;岩生百里香●

391325 Thymus phyllopodus Klokov;叶梗百里香●☆

391326 Thymus podolicus Klokov et Des. -Shost. ;波多尔斯克百里香●☆

391327 Thymus polytrichus Kern. ex Borbás;密毛百里香;Serpolet Oil ●☆

391328 Thymus polytrichus Kern. ex Borbás = Thymus serpyllum L. ●☆

391329 Thymus praecox Opiz;早花百里香;Mother-of-thyme ●☆

391330 Thymus praecox Opiz = Thymus serpyllum L. ●☆

391331 Thymus praecox Opiz subsp. arcticus (Durand ex Kane) Jalas = Thymus arcticus (Durand ex Kane) Ronniger ●☆

391332 Thymus praecox Opiz subsp. arcticus (Durand) Jalas = Thymus arcticus (Durand ex Kane) Ronniger ●☆

391333 Thymus praecox Opiz var. arcticus (Durand) Karlsson = Thymus serpyllum L. ●☆

391334 Thymus proximus Serg. ;拟百里香;False Thyme,Similar Thyme ●

391335 Thymus przewalskii (Kom.) Nakai = Thymus quinquecostatus Celak. var. przewalskii (Kom.) Ronniger ●

391336 Thymus przewalskii Kom. ex Klokov = Thymus quinquecostatus

Celak. var. przewalskii（Kom.）Ronniger ●

391337   Thymus przewalskii Nakai ＝ Thymus quinquecostatus Celak. var. przewalskii（Kom.）Ronniger ●

391338   Thymus pseudograniticus Klokov et Des. -Shost. ;假岩百里香●☆

391339   Thymus pseudohumillimus Klokov et Des. -Shost. ;假矮百里香●☆

391340   Thymus pseudolanuginosus Ronniger;假多毛百里香;Woolly Thyme ●☆

391341   Thymus pseudomicromeria Font Quer ＝ Thymus maroccanus Ball subsp. rhombicus Villar ●☆

391342   Thymus pseudonummularius Klokov et Des. -Shost. ;假铜钱百里香●☆

391343   Thymus pseudopallidus H. Lindb. ＝ Thymus pallidus Batt. ●☆

391344   Thymus pseudopulegioides Klokov et Des. -Shost. ;假蚤草百里香●☆

391345   Thymus pulchellus C. A. Mey. ;美丽百里香●☆

391346   Thymus pulegioides L. ;薄荷百里香（卵叶百里香）;Large Thyme,Large Wild Thyme,Lemon Thyme,Ovate Thyme,Pennyroyal Thyme,Wild Thyme ●☆

391347   Thymus quinquecostatus Celak. ;地椒（百里香,五肋百里香,五肋地椒,五脉百里香,五脉地椒,烟台百里香）;Fiveribbed Thyme,Five-ribbed Thyme,Ground Chilli Thyme ●

391348   Thymus quinquecostatus Celak. f. albiflorus H. Hara;白花地椒●☆

391349   Thymus quinquecostatus Celak. f. maritimus H. Hara;海滨地椒●☆

391350   Thymus quinquecostatus Celak. var. asiaticus（Kitag.）C. Y. Wu et Y. C. Huang ＝ Thymus serpyllum L. var. asiaticus Kitag. ●

391351   Thymus quinquecostatus Celak. var. asiaticus（Kitag.）C. Y. Wu et Y. C. Huang;亚洲百里香（地椒,亚洲地椒）;Asian Thyme ●

391352   Thymus quinquecostatus Celak. var. canescens（C. A. Mey.）H. Hara;灰地椒●☆

391353   Thymus quinquecostatus Celak. var. ibukesis（Kudo）H. Hara;日本地椒（日本百里香）●☆

391354   Thymus quinquecostatus Celak. var. ibukiensis（Kudo）H. Hara ＝ Thymus quinquecostatus Celak. ●

391355   Thymus quinquecostatus Celak. var. japonicus H. Hara ＝ Thymus quinquecostatus Celak. ●

391356   Thymus quinquecostatus Celak. var. przewalskii（Kom.）Ronniger;展毛地椒（地椒,米凯百里香）●

391357   Thymus rariflorus K. Koch;稀花百里香●☆

391358   Thymus repens Buch. -Ham. ex D. Don ＝ Clinopodium repens（D. Don）Vell. ■

391359   Thymus repens D. Don ＝ Clinopodium repens（D. Don）Vell. ■

391360   Thymus roseus Schipcz. ;粉红百里香●☆

391361   Thymus saturejoides Coss. ;香草百里香●☆

391362   Thymus saturejoides Coss. subsp. commutatus Batt. ＝ Thymus commutatus（Batt.）Batt. ●☆

391363   Thymus schischkinii Serg. ;希施百里香●☆

391364   Thymus semiglaber Klokov;半光百里香●☆

391365   Thymus serpyllum L. ;欧百里香（百里香,地花椒,地椒,地椒草,铺地香,山胡椒,山椒,麝香草,野百里香）;Bank Thyme,Breckland Thyme,Breckland Wild Thyme,Brotherwort,Creeping Red Thyme,Creeping Thyme,English Thyme,Mother of Thyme,Mother-of-thyme,Motherthyme,Mountain Thyme,Puliall-mountain,Serpolet,Sheep's Thyme,Shepherd's Thyme,Thyme,Wild Thyme,Woolly Thyme ●☆

391366   Thymus serpyllum L. subsp. citriodorus Pers. ;柠檬铺地香;German Thyme,Lemon Thyme,Winter Thyme ●☆

391367   Thymus serpyllum L. subsp. quinquecostatus（Celak.）Kitam.

＝ Thymus quinquecostatus Celak. ●

391368   Thymus serpyllum L. subsp. quinquecostatus（Celak.）Kitam. var. canescens C. A. Mey. ＝ Thymus quinquecostatus Celak. var. canescens（C. A. Mey.）H. Hara ●☆

391369   Thymus serpyllum L. var. angustifolius Pers. ;狭叶欧百里香●☆

391370   Thymus serpyllum L. var. arcticus Durand ＝ Thymus arcticus（Durand ex Kane）Ronniger ●☆

391371   Thymus serpyllum L. var. asiaticus Kitag. ＝ Thymus quinquecostatus Celak. var. asiaticus（Kitag.）C. Y. Wu et Y. C. Huang ●

391372   Thymus serpyllum L. var. aureus ?;黄铺地香;Golden Thyme ●☆

391373   Thymus serpyllum L. var. coccineum ?;西方铺地香;Crimson Thyme ●☆

391374   Thymus serpyllum L. var. ibukiensis Kudo ＝ Thymus quinquecostatus Celak. ●

391375   Thymus serpyllum L. var. mongolicus Ronniger ＝ Thymus mongolicus（Ronniger）Ronniger ●

391376   Thymus serpyllum L. var. przewalskii Kom. ＝ Thymus quinquecostatus Celak. var. przewalskii（Kom.）Ronniger ●

391377   Thymus serpyllum L. var. villosus ? ＝ Thymus villosus L. ●☆

391378   Thymus sibiricus（Serg.）Klokov et Des. -Shost. ;西伯利亚百里香;Siberia Thyme ●☆

391379   Thymus sokolovii Klokov;索科罗夫百里香●☆

391380   Thymus sosnowskyi Grossh. ;索斯百里香●☆

391381   Thymus stepposus Klokov et Des. -Shost. ;草原百里香●☆

391382   Thymus subalpestris Klokov;亚高山百里香●☆

391383   Thymus subarcticus Klokov et Des. -Shost. ;亚北极百里香●☆

391384   Thymus talievii Klokov et Des. -Shost. ;台氏百里香●☆

391385   Thymus tauricus Klokov et Des. -Shost. ;克里木百里香●☆

391386   Thymus tianschanicus Dmitrieva ＝ Thymus karatavicus Dmitrieva ●☆

391387   Thymus tiflisiensis Klokov et Des. -Shost. ;梯弗里斯百里香●☆

391388   Thymus transcaspicus Klokov et Des. -Shost. ;里海百里香●☆

391389   Thymus transcaucasicus Ronniger;外高加索百里香●☆

391390   Thymus trautvetteri Klokov et Des. -Shost. ;特劳特百里香●☆

391391   Thymus turczaninovii Serg. ;图尔百里香●

391392   Thymus ucrainicus（Klokov et Des. -Shost.）Klokov;乌克兰百里香●☆

391393   Thymus ussuriensis Klokov;乌苏里百里香●☆

391394   Thymus villosus L. ;长柔毛百里香;Downy Thyme ●☆

391395   Thymus villosus L. subsp. lusitanicus（Boiss.）Cout. ;卢西塔尼亚毛百里香●☆

391396   Thymus virginicus L. ＝ Hyptis rhomboides M. Martens et Galeotti ■

391397   Thymus vulgaris L. ;麝香草（百里香）;Bank Thyme,Black Thyme,Common Garden Thyme,Common Thyme,Creeping Thyme,English Thyme,Garden Thyme,Horse Thyme,Mother-of-thyme,Motherwort,Motlrer Thyme,Pell-a-mountain,Penny Mountain,Piliol,Puliall Mountain,Running Thyme,Tea Grass,Tea-girse,Thyme,Wild Thyme ●

391398   Thymus willdenowii Boiss. ;威尔百里香●☆

391399   Thymus zelenetzkyi Klokov et Des. -Shost. ;切林氏百里香Zelenetzky Thyrocarpos ●☆

391400   Thymus zheguliensis Klokov et Des. -Shost. ;日古里百里香●☆

391401   Thymus ziaratinus Klokov et Serg. ;济亚拉蒂百里香●☆

391402   Thymus zygis L. ;葡百里香;French Thyme,Spanish Thyme ●☆

391403   Thymus zygis L. subsp. gracilis（Boiss.）R. Morales;纤细百里香●☆

391404　Thynninorchis D. L. Jones et M. A. Clem. (2002);鮪兰属■☆

391405　Thynninorchis huntiana (F. Muell.) D. L. Jones et M. A. Clem.;鮪兰■☆

391406　Thypha Costa = Typha L. ■

391407　Thyrasperma N. E. Br. = Apatesia N. E. Br. ■☆

391408　Thyrasperma N. E. Br. = Hymenogyne Haw. ■☆

391409　Thyridachne C. E. Hubb. (1949);盾草属■☆

391410　Thyridachne tisserantii C. E. Hubb.;盾草■☆

391411　Thyridiaceae Dulac = Orchidaceae Juss. (保留科名)■

391412　Thyridocalyx Bremek. (1956);盾萼茜属■☆

391413　Thyridocalyx ampandrandavae Bremek.;盾萼茜■☆

391414　Thyridolepis S. T. Blake(1972);窗草属■☆

391415　Thyridolepis mitchelliana (Nees) S. T. Blake;窗草■☆

391416　Thyridolepis xerophila (Domin) S. T. Blake;沙窗草■☆

391417　Thyridostachyum Nees = Mnesithea Kunth ■

391418　Thyridostachyum laeve (Retz.) Nees ex Steud. = Mnesithea laevis (Retz.) Kunth ■

391419　Thyrocarpus Hance(1862);盾果草属;Shieldfruit,Thyrocarpos ■★

391420　Thyrocarpus glochidiatus Maxim.;弯齿盾果草(弯果盾果草); Curvedtooth Shieldfruit,Curvedtooth Thyrocarpos ■

391421　Thyrocarpus sampsonii Hance;盾果草(盾形草,黑骨风,黑果风,铺墙草,森氏盾果草);Sampson Thyrocarpos,Shieldfruit ■

391422　Thyroma Miers = Aspidosperma Mart. et Zucc. (保留属名)●☆

391423　Thyrophora Neck. = Gentiana L. ■

391424　Thyrsacanthus Moric. = Thyrsacanthus Nees ●■☆

391425　Thyrsacanthus Nees = Odontonema Nees(保留属名)●■☆

391426　Thyrsanthella (Baill.) Pichon(1948);小杖花属●☆

391427　Thyrsanthella Pichon = Thyrsanthella (Baill.) Pichon ●☆

391428　Thyrsanthella difformis (Walter) Pichon;小杖花●☆

391429　Thyrsanthema Neck. = Chaptalia Vent. (保留属名)■☆

391430　Thyrsanthemum Pichon(1946);锥花草属■☆

391431　Thyrsanthemum floribundum (M. Martens et Galeotti) Pichon;多花锥花草■☆

391432　Thyrsanthemum macrophyllum (Greenm.) Rohweder;大叶锥花草■☆

391433　Thyrsanthera Pierre ex Gagnep. (1925);锥花大戟属☆

391434　Thyrsanthera suborbicularis Pierre ex Gagnep.;锥花大戟☆

391435　Thyrsanthus Benth. = Forsteronia G. Mey. ●☆

391436　Thyrsanthus Elliott = Wisteria Nutt. (保留属名)●

391437　Thyrsanthus Schrank = Lysimachia L. ●■

391438　Thyrsanthus Schrank = Naumburgia Moench ●■

391439　Thyrsia Stapf = Phacelurus Griseb. ■

391440　Thyrsia Stapf(1917);锥茅属;Awlquitch,Thyrsia ■

391441　Thyrsia thyrsoidea (Hack.) A. Camus = Phacelurus zea (C. B. Clarke) Clayton ■

391442　Thyrsia thyrsoidea (Hack.) A. Camus = Thyrsia zea (C. B. Clarke) Stapf ■

391443　Thyrsia undulatifolia (Chiov.) Robyns;波叶锥茅;Undulate-leaved Thyrsia ■☆

391444　Thyrsia viridula Ohwi = Phacelurus speciosus (Steud.) C. E. Hubb. ■☆

391445　Thyrsia zea (C. B. Clarke) Stapf;锥茅(华锥茅);Awlquitch, Common Thyrsia,Zea Thyrsia ■

391446　Thyrsia zea (C. B. Clarke) Stapf = Phacelurus zea (C. B. Clarke) Clayton ■

391447　Thyrsine Gled. = Cytinus L. (保留属名)■☆

391448　Thyrsodium Salzm. ex Benth. (1852);杖漆属●☆

391449　Thyrsodium africanum Engl. ;非洲杖漆●☆

391450　Thyrsodium bolivianum J. D. Mitchell et Daly;玻利维亚杖漆●☆

391451　Thyrsodium dasytrichum Sandwith;粗毛杖漆●☆

391452　Thyrsodium giganteum Engl. ;巨杖漆●☆

391453　Thyrsosalacia Loes. (1940);杖卫矛属●☆

391454　Thyrsosalacia racemosa (Loes. ex Harms) N. Hallé;杖卫矛●☆

391455　Thyrsosma Raf. = Viburnum L. ●

391456　Thyrsosma chinensis Raf. = Viburnum odoratissimum Ker Gawl. ●

391457　Thyrsostachys Gamble(1896);泰竹属(复穗竹属,廉序竹属,条竹属);Taibamboo,Thyrsostachys,Umbrella Bamboo ●

391458　Thyrsostachys oliveri Gamble;大泰竹;Oliver Taibamboo,Oliver Thyrsostachys ●

391459　Thyrsostachys regia (Thomson ex Munro) Bennet = Thyrsostachys siamensis (Kurz ex Munro) Gamble ●

391460　Thyrsostachys siamensis (Kurz ex Munro) Gamble;泰竹(柳竹,条竹,暹逻竹);Siam Thyrsostachys, Taibamboo, Umbrella Bamboo ●

391461　Thyrsostachys siamensis Gamble = Thyrsostachys siamensis (Kurz ex Munro) Gamble ●

391462　Thysamus Rchb. = Cnestis Juss. ●

391463　Thysamus Rchb. = Thysanus Lour. ●

391464　Thysanachne C. Presl = Arundinella Raddi ■

391465　Thysanella A. Gray = Polygonella Michx. ■☆

391466　Thysanella A. Gray ex Engelm. et A. Gray = Polygonella Michx. ■☆

391467　Thysanella Salisb. = Thysanotus R. Br. (保留属名)■

391468　Thysanella fimbriata (Elliott) A. Gray = Polygonella fimbriata (Elliott) Horton ■☆

391469　Thysanella robusta Small = Polygonella robusta (Small) G. L. Nesom et V. M. Bates ■☆

391470　Thysanocarex Börner = Carex L. ■

391471　Thysanocarpus Hook. (1830);缨果荠属(流苏果属)■☆

391472　Thysanocarpus curvipes Hook.;缨果荠(流苏果);Lacepod ■☆

391473　Thysanochilus Falc. = Eulophia R. Br. (保留属名)■

391474　Thysanoglossa Porto = Thysanoglossa Porto et Brade ■☆

391475　Thysanoglossa Porto et Brade(1940);缨舌兰属■☆

391476　Thysanoglossa jordanensis Porto et Brade;缨舌兰■☆

391477　Thysanolaena Nees (1835);棕叶芦属(粽叶芦属);Tiger Grass,Tigergrass,Tiger-grass ■

391478　Thysanolaena acarifera (Trin.) Wight et Arn. ex Nees = Thysanolaena latifolia (Roxb. ex Hornem.) Honda ■

391479　Thysanolaena acarifera (Trin.) Wight et Arn. ex Nees = Thysanolaena maxima (Roxb.) Kuntze ■

391480　Thysanolaena agrostis Nees = Thysanolaena latifolia (Roxb. ex Hornem.) Honda ■

391481　Thysanolaena agrostis Nees = Thysanolaena maxima (Roxb.) Kuntze ■

391482　Thysanolaena assamensis Gand. = Thysanolaena latifolia (Roxb. ex Hornem.) Honda ■

391483　Thysanolaena assamensis Gand. = Thysanolaena maxima (Roxb.) Kuntze ■

391484　Thysanolaena birmanica Gand. = Thysanolaena latifolia (Roxb. ex Hornem.) Honda ■

391485　Thysanolaena birmanica Gand. = Thysanolaena maxima (Roxb.) Kuntze ■

391486　Thysanolaena latifolia (Hornem.) Honda = Thysanolaena latifolia (Roxb. ex Hornem.) Honda ■

391487　Thysanolaena latifolia（Roxb. ex Hornem.）Honda;棕叶芦（莽草,扫把草,棕叶草,棕叶芦）; American Grass, Tiger Grass, Tigergrass ■

391488　Thysanolaena latifolia（Roxb.）Honda ＝ Thysanolaena latifolia（Roxb. ex Hornem.）Honda ■

391489　Thysanolaena maxima（Roxb.）Kuntze ＝ Thysanolaena latifolia（Roxb. ex Hornem.）Honda ■

391490　Thysanolaena mezii Janowski ＝ Neyraudia reynaudiana（Kunth）Keng ex Hitchc. ■

391491　Thysanolaena sikkimensis Gand. ＝ Thysanolaena latifolia（Roxb. ex Hornem.）Honda ■

391492　Thysanolaena sikkimensis Gand. ＝ Thysanolaena maxima（Roxb.）Kuntze ■

391493　Thysanospermum Champ. ex Benth. ＝ Coptosapelta Korth.（保留属名）●

391494　Thysanospermum diffusum Champ. ex Benth. ＝ Coptosapelta diffusa（Champ. ex Benth.）Steenis ●

391495　Thysanospermum diffusum Champ. var. longitubum Ohwi ＝ Coptosapelta diffusa（Champ. ex Benth.）Steenis ●

391496　Thysanostemon Maguire（1964）;缨蕊藤黄属●☆

391497　Thysanostemon fanshawei Maguire;缨蕊藤黄●☆

391498　Thysanostigma Imlay（1939）;缨柱爵床属■☆

391499　Thysanostigma siamense Imlay;缨柱爵床■☆

391500　Thysanotus R. Br.（1810）（保留属名）;异蕊草属; Fringed Lily, Fringed-lily, Fringelily ■

391501　Thysanotus chinensis Benth.;异蕊草（化骨龙）; Chinese Fringelily, Fringelily ■

391502　Thysanotus chrysantherus F. Muell. ＝ Thysanotus chinensis Benth. ■

391503　Thysantha Hook. ＝ Crassula L. ●■☆

391504　Thysantha Hook. ＝ Thisantha Eckl. et Zeyh. ●■☆

391505　Thysantha Hook. ＝ Tillaea L. ●■☆

391506　Thysanurus O. Hoffm. ＝ Geigeria Griess. ■●☆

391507　Thysanus Lour. ＝ Cnestis Juss. ●

391508　Thysanus palala Lour. ＝ Cnestis palala（Lour.）Merr. ●

391509　Thyselium Raf.（1840）;北亚草属■☆

391510　Thyselium Raf. ＝ Peucedanum L. ■

391511　Thyselium Raf. ＝ Thysselinum Hoffm. ■

391512　Thyselium palustre（L.）Raf.;北亚草■☆

391513　Thysselinum Adans. ＝ Selinum L.（保留属名）■

391514　Thysselinum Hoffm. ＝ Peucedanum L. ■

391515　Thysselinum Moench ＝ Pleurospermum Hoffm. ■

391516　Tianschaniella B. Fedtsch. ＝ Tianschaniella B. Fedtsch. ex Popov ■☆

391517　Tianschaniella B. Fedtsch. ex Popov（1951）;中亚紫草属■☆

391518　Tianschaniella umbellifera B. Fedtsch. ex Popov;中亚紫草■☆

391519　Tiaranthus Herb. ＝ Pancratium L. ■

391520　Tiarella L.（1753）;黄水枝属; False Mitrewort, Foam Flower, Foamflower ■

391521　Tiarella bodinieri H. Lév. ＝ Tiarella polyphylla D. Don ■

391522　Tiarella cordifolia L.;心叶黄水枝（泡沫花）; Allegany Foamflower, Coolwort, False Miterwort, Foam Flower, Foam-flowe, Foamflower, Heartleaf Foamflower, Heart-leaf Foam-flower, Miterwort ■☆

391523　Tiarella cordifolia L. subsp. collina Wherry ＝ Tiarella wherryi Lakela ■☆

391524　Tiarella polyphylla D. Don;黄水枝（博落,防风七,高脚铜告

碑,水前胡,紫背金钱）; Himalayan Foamflower, Himalayas Foamflower ■

391525　Tiarella unifoliata Hook.;单叶黄水枝; Coolwort Foamflower, False Mitrenort ■☆

391526　Tiarella wherryi Lakela;魏氏黄水枝（斐丽黄水枝）; Wherry Foam Flower ■☆

391527　Tiaridium Lehm. ＝ Heliotropium L. ●■

391528　Tiarocarpus Rech. f.（1972）;巾果菊属■☆

391529　Tiarocarpus Rech. f. ＝ Cousinia Cass. ●■

391530　Tiarocarpus neubaueri（Rech. f.）Rech. f.;巾果菊■☆

391531　Tiarrhena（Maxim.）Nakai ＝ Miscanthus Andersson ■

391532　Tiarrhena sacchariflora（Maxim.）Nakai ＝ Miscanthus sacchariflorus（Maxim.）Benth. ■

391533　Tibestina Maire ＝ Dicoma Cass. ●☆

391534　Tibetia（Ali）H. P. Tsui ＝ Gueldenstaedtia Fisch. ■

391535　Tibetia（Ali）H. P. Tsui（1979）;高山豆属; Alpbean, Tibetia ■

391536　Tibetia coelestis（Diels）H. P. Tsui ＝ Gueldenstaedtia coelestis Ulbr. ■

391537　Tibetia coelestis（Diels）H. P. Tsui ＝ Tibetia yunnanensis（Franch.）H. P. Tsui var. coelestis（Diels）X. Y. Zhu ■

391538　Tibetia forrestii（Ali）P. C. Li;中甸高山豆■

391539　Tibetia himalaica（Baker）H. P. Tsui;高山豆（单花米口袋,喜马拉雅米口袋,异叶米口袋）; Alpbean, Himalayan Gueldenstaedtia, Himalayan Tibetia, Singleflower Gueldenstaedtia ■

391540　Tibetia himalaica（Baker）H. P. Tsui ＝ Gueldenstaedtia himalaica Baker ■

391541　Tibetia himalaica（Baker）H. P. Tsui f. alba X. Y. Zhu ＝ Tibetia himalaica（Baker）H. P. Tsui ■

391542　Tibetia liangshanensis P. C. Li ＝ Tibetia forrestii（Ali）P. C. Li ■

391543　Tibetia tongolensis（Ulbr.）H. P. Tsui;黄花高山豆（黄花米口袋）; Yellow Alpbean, Yellowflower Tibetia ■

391544　Tibetia tongolensis（Ulbr.）H. P. Tsui f. coelestis（Diels）P. C. Li ＝ Tibetia yunnanensis（Franch.）H. P. Tsui var. coelestis（Diels）X. Y. Zhu ■

391545　Tibetia tongolensis（Ulbr.）H. P. Tsui var. coelestis（Diels）H. P. Tsui ＝ Tibetia yunnanensis（Franch.）H. P. Tsui var. coelestis（Diels）X. Y. Zhu ■

391546　Tibetia tongolensis（Ulbr.）H. P. Tsui var. coelestis（Diels）H. P. Tsui ＝ Tibetia coelestis（Diels）H. P. Tsui ■

391547　Tibetia yadongensis H. P. Tsui;亚东高山豆; Yadong Alpbean, Yadong Tibetia ■

391548　Tibetia yunnanensis（Franch.）H. P. Tsui;云南高山豆（云南米口袋）; Tibetia, Yunnan Alpbean, Yunnan Gueldenstaedtia, Yunnan Ricebag ■

391549　Tibetia yunnanensis（Franch.）H. P. Tsui var. coelestis（Diels）X. Y. Zhu;蓝花高山豆; Blue Alpbean, Blue Flower Tibetia ■

391550　Tibetoseris Sennikov（2007）;藏菊属■☆

391551　Tibetoseris angustifolia Tzvelev ＝ Pseudoyoungia angustifolia（Tzvelev）D. Maity et Maiti ■

391552　Tibetoseris conjunctiva（Babc. et Stebbins）Sennikov ＝ Youngia conjunctiva Babc. et Stebbins ■

391553　Tibetoseris cristata（C. Shih et C. Q. Cai）Sennikov ＝ Youngia cristata C. Shih et C. Q. Cai ■

391554　Tibetoseris depressa（Hook. f. et Thomson）Sennikov ＝ Crepis depressa Hook. f. et Thomson ■

391555　Tibetoseris depressa（Hook. f. et Thomson）Sennikov ＝ Soroseris glomerata（Decne.）Stebbins ■

391556　Tibetoseris depressa (Hook. f. et Thomson) Sennikov = Youngia depressa (Hook. f. et Thomson) Babc. et Stebbins ■

391557　Tibetoseris gracilipes (Hook. f.) Sennikov = Crepis gracilipes Hook. f. ■

391558　Tibetoseris gracilipes (Hook. f.) Sennikov = Youngia gracilipes (Hook. f.) Babc. et Stebbins ■

391559　Tibetoseris ladyginii Tzvelev;拉迪藏菊■

391560　Tibetoseris parva (Babc. et Stebbins) Sennikov = Crepis parva (Babc. et Stebbins) Hand. -Mazz. ■

391561　Tibetoseris parva (Babc. et Stebbins) Sennikov = Youngia parva Babc. et Stebbins ■

391562　Tibetoseris sericeus (C. Shih) Sennikov = Youngia sericea C. Shih ■

391563　Tibetoseris simulatrix (Babc.) Sennikov = Crepis simulatrix Babc. ■

391564　Tibetoseris simulatrix (Babc.) Sennikov = Youngia simulatrix (Babc.) Babc. et Stebbins ■

391565　Tibetoseris tianshanica (C. Shih) Tzvelev = Crepis tianshanica C. Shih ■

391566　Tibouchina Aubl. (1775);荣耀木属(蒂牡丹属,丽蓝木属); Glory Bush, Lasiandra ●■☆

391567　Tibouchina aspera Aubl. ;毛丹●☆

391568　Tibouchina aspera Aubl. var. asperrima Cogn. ;艮毛丹●☆

391569　Tibouchina granulosa (Desr.) Cogn. ;多色荣耀木; Brazilian Glorytree, Glory Bush ●☆

391570　Tibouchina herbacea (DC.) Cogn. ;草本荣耀木; Cane Tibouchina, Herbaceous Glorytree ■☆

391571　Tibouchina heteromalla Cogn. ;阔叶荣耀木;Glory Bush ●☆

391572　Tibouchina laxa Cogn. ;疏枝荣耀木●☆

391573　Tibouchina lepidota Baill. ;荣耀木;Glory Bush ●☆

391574　Tibouchina longifolia Millsp. ;长叶荣耀木;Longleaf Glorytree ●☆

391575　Tibouchina paratropica (Cogn.) Cogn. ;热带红花蕊●☆

391576　Tibouchina semidecandra (DC.) Cogn. = Tibouchina urvilleana (DC.) Cogn ●☆

391577　Tibouchina urvilleana (DC.) Cogn. ;五蕊荣耀木(翠蓝木,紫绀野牡丹); Brazilian Glorybush, Glory Bush, Glory Flower, Glorybush, Princess Flower, Princess-flower ●☆

391578　Tibouchinopsis Markgr. (1927);类荣耀木属●☆

391579　Tibouchinopsis glutinosa Markgr. ;类荣耀木●☆

391580　Tibouchinopsis mirabilis Brade et Markgr. ;奇异类荣耀木●☆

391581　Tibuchina Raf. = Tibouchina Aubl. ●■☆

391582　Ticanto Adans. = Caesalpinia L. ●

391583　Ticodendraceae Gómez-Laur. et L. D. Gómez(1991);太果木科 (核果桦科)●☆

391584　Ticodendron Gómez-Laur. et L. D. Gómez(1989);太果木属(核果桦属)●☆

391585　Ticodendron incognitum Gómez-Laur. et L. D. Gómez;太果木●☆

391586　Ticoglossum Lucas Rodr. ex Halb. (1983);第果兰属■☆

391587　Ticoglossum krameri (Rchb. f.) Lucas Rodr. ex Halb. ;第果兰■☆

391588　Ticorea A. St. -Hil. = Galipea Aubl. ●☆

391589　Ticorea Aubl. (1775);蒂克芸香属●☆

391590　Ticorea diandra Kallunki;双蕊蒂克芸香●☆

391591　Ticorea foetida Aubl. ;蒂克芸香●☆

391592　Ticorea tubiflora (A. C. Sm.) Gereau;管花蒂克芸香●☆

391593　Tidestromia Standl. (1916);星毛苋属●☆

391594　Tidestromia carnosa (Steyerm.) I. M. Johnst. ;肉质星毛苋■☆

391595　Tidestromia gemmata I. M. Johnst. = Tidestromia suffruticosa

391596　Tidestromia lanuginosa (Nutt.) Standl. ;绵毛星毛苋■☆

391597　Tidestromia lanuginosa (Nutt.) Standl. subsp. eliassoniana Sánch. Pino et Flores Olv. = Tidestromia lanuginosa (Nutt.) Standl. ■☆

391598　Tidestromia lanuginosa (Nutt.) Standl. var. carnosa (Steyerm.) Cory = Tidestromia carnosa (Steyerm.) I. M. Johnst. ■☆

391599　Tidestromia oblongifolia (S. Watson) Standl. = Tidestromia suffruticosa (Torr.) Standl. var. oblongifolia (S. Watson) Sánch. Pino et Flores Olv. ●☆

391600　Tidestromia oblongifolia (S. Watson) Standl. subsp. cryptantha (S. Watson) Wiggins = Tidestromia suffruticosa (Torr.) Standl. var. oblongifolia (S. Watson) Sánch. Pino et Flores Olv. ●☆

391601　Tidestromia suffruticosa (Torr.) Standl. ;灌木星毛苋●☆

391602　Tidestromia suffruticosa (Torr.) Standl. var. coahuilana I. M. Johnst. = Tidestromia suffruticosa (Torr.) Standl. ●☆

391603　Tidestromia suffruticosa (Torr.) Standl. var. oblongifolia (S. Watson) Sánch. Pino et Flores Olv. ;椭圆叶灌木星毛苋; Honeysweet ●☆

391604　Tiedemannia DC. = Oxypolis Raf. ■☆

391605　Tiedmannia Torr. et A. Gray = Tiedemannia DC. ■☆

391606　Tieghemella Pierre(1890);蒂氏山榄属●☆

391607　Tieghemella africana Pierre;非洲蒂氏山榄●☆

391608　Tieghemella heckelii (A. Chev.) Pierre ex Dubard;蒂氏山榄; African Cherry, Baku, Cherry Mahogany ●☆

391609　Tieghemella heckelii Pierre ex A. Chev. = Tieghemella heckelii (A. Chev.) Pierre ex Dubard ●☆

391610　Tieghemia Balle = Oncocalyx Tiegh. ●☆

391611　Tieghemopanax R. Vig. = Polyscias J. R. Forst. et G. Forst. ●

391612　Tieghemopanax cussonioides (Drake) R. Vig. = Polyscias cussonioides (Drake) Bernardi ●

391613　Tienmuia Hu = Phacellanthus Siebold et Zucc. ■

391614　Tienmuia triandra Hu = Phacellanthus tubiflorus Siebold et Zucc. ■

391615　Tietkensia P. S. Short(1990);长序金绒草属■☆

391616　Tietkensia corrickiae P. S. SHort. ;长序金绒草■☆

391617　Tigarea Aubl. = Doliocarpus Rol. + Tetracera L. ●

391618　Tigarea Pursh = Purshia DC. ex Poir. ●☆

391619　Tigivesta Luer = Pleurothallis R. Br. ■☆

391620　Tigivesta Luer(2007);哥伦比亚肋枝兰属■☆

391621　Tiglium Klotzsch = Croton L. ●

391622　Tigridia Juss. (1789);虎皮花属(虎菖蒲属,老虎花属,老虎莲属); Mexican Shell Flowers, Tiger Flower, Tigerflower, Tiger-flower, Tiger-iris ■

391623　Tigridia buccifera S. Watson = Alophia drummondii (Graham) R. C. Foster ■☆

391624　Tigridia grandiflora Diels;大花虎皮花; Mexican Shell Flower, Peacock Flower ■☆

391625　Tigridia grandiflora Diels = Sisyrinchium grandiflorum Cav. ■☆

391626　Tigridia grandiflora Salisb. = Tigridia pavonia (L. f.) Ker Gawl. ■

391627　Tigridia pavonia (L. f.) Ker Gawl. ;虎皮花(虎菖蒲,虎皮百合,老虎百合,老虎花); Cacomite, Common Tiger Flower, Common Tigerflower, Mexican Tiger-lily, Peach Tiger Flower, Peacock Flower, Peacock Tiger-flower, Peacockflower, Peacock-tigerflower, Shell Flower, Shell-flower, Tiger Flower, Tigerflower, Tiger-flower ■

391628　Tigridia pavonia (L. f.) Ker Gawl. ' Alba Grandiflora';大花白虎皮花(大花白)■

391629　Tigridia pavonia（L. f.）Ker Gawl.'Red Giant';大花红■

391630　Tigridia purpurea（Herb.）Shinners = Alophia drummondii（Graham）R. C. Foster■☆

391631　Tigridiopalma C. Chen（1979）;虎颜花属;Airplant,Tigridiopalma■★

391632　Tigridiopalma magnifica C. Chen;虎颜花（大莲蓬,熊掌）;Magnific Tigridiopalma■

391633　Tikalia Lundell = Blomia Miranda●☆

391634　Tikalia Lundell(1961);梯氏无患子属●☆

391635　Tikalia prisca（Standl.）Lundell;梯氏无患子●☆

391636　Tikusta Raf. = Tupistra Ker Gawl.■

391637　Tilco Adans. = Fuchsia L.●■

391638　Tilcusta Raf. = Campylandra Baker■

391639　Tildenia Miq. = Peperomia Ruiz et Pav.■

391640　Tilecarpus K. Schum. = Medusanthera Seem.●☆

391641　Tilecarpus K. Schum. = Tylecarpus Engl.●☆

391642　Tilesia G. Mey.（1818）;菱果菊属■☆

391643　Tilesia G. Mey. = Wulffia Neck. ex Cass.■☆

391644　Tilesia Thunb. ex Steud. = Gladiolus L.■

391645　Tilesia capitata G. Mey.;菱果菊■☆

391646　Tilia L.（1753）;椴树属（椴属）;Basswood,Lime,Lime Tree,Linden,Whitewood●

391647　Tilia 'Petiolaris';垂枝美洲椴;Pendent Sliver-lime,Silver Pendent Linden,Weeping Silver-lime,Weeping White Linden●☆

391648　Tilia alba Aiton = Tilia tomentosa Moench●☆

391649　Tilia americana L.;美洲椴（美国椴树,美洲白木,菩提木）;American Basswood,American Lime,American Linden,American White Wood,Bass Wood,Basswood,Bee-tree,Linden,Whitewood●☆

391650　Tilia americana L.'Ampelophylla';裂叶美洲椴;Lobe-leaved American Linden●

391651　Tilia americana L.'Fastigiata';塔形美洲椴;Pyramidal American Linden●

391652　Tilia americana L.'Macrophylla';大叶美洲椴;Large-leaved American Linden●

391653　Tilia americana L.'Pendula' = Tilia 'Petiolaris'●☆

391654　Tilia americana L.'Redmond';雷蒙德美洲椴●☆

391655　Tilia americana L. f. ampelophylla V. Engl. = Tilia americana L.'Ampelophylla'●

391656　Tilia americana L. f. dentata（Kirchn.）Rehder;齿叶美洲椴;Dentate-leaved American Linden●

391657　Tilia americana L. f. fastigiata（Slavin）Rehder = Tilia americana L.'Fastigiata'●

391658　Tilia americana L. f. macrophylla（Bayer）V. Engl. = Tilia americana L.'Macrophylla'●

391659　Tilia americana L. var. caroliniana（Mill.）E. Murray = Tilia caroliniana Mill.●

391660　Tilia americana L. var. neglecta（Spach）Fosberg = Tilia americana L.●☆

391661　Tilia amurensis Rupr.;紫椴（阿穆尔椴,白椴,椴树,籽椴）;Amur Lime,Amur Linden●

391662　Tilia amurensis Rupr. subsp. taquetii（C. K. Schneid.）Pigott = Tilia amurensis Rupr. var. taquetii（C. K. Schneid.）Liou et C. Y. Li●

391663　Tilia amurensis Rupr. var. araneosa Z. Wang et S. D. Zhao;毛紫椴;Cobwebby Amur Linden●

391664　Tilia amurensis Rupr. var. koreana（Nakai）Vorosch.;朝鲜紫椴●

391665　Tilia amurensis Rupr. var. sibirica（Fisch. ex Bayer）Y. C. Zhu;西伯利亚紫椴●

391666　Tilia amurensis Rupr. var. taquetii（C. K. Schneid.）Liou et C. Y. Li;小叶紫椴;Little-leaved Amur Linden●

391667　Tilia amurensis Rupr. var. tricuspidata Liou et C. Y. Li;裂叶紫椴（裂果紫椴）;Lobed Amur Linden●

391668　Tilia amurensis Rupr. var. tricuspidata Liou et C. Y. Li = Tilia amurensis Rupr.●

391669　Tilia angustibracteata Hung T. Chang;窄苞椴;Angustibracted Linden●

391670　Tilia angustibracteata Hung T. Chang = Tilia tuan Szyszyl.●

391671　Tilia argentea Desf.;银白椴;White Lime●☆

391672　Tilia argentea Desf. = Tilia tomentosa Moench●☆

391673　Tilia austroyunnanenica Hung T. Chang = Tilia mesembrinos Merr.●

391674　Tilia barbata Stokes;髯毛椴.●

391675　Tilia baroniana Diels = Tilia chinensis Maxim.●

391676　Tilia baroniana Diels var. investita V. Engl. = Tilia chinensis Maxim. var. investita（V. Engl.）Rehder●

391677　Tilia begonifolia Stev. = Tilia rubra DC.●☆

391678　Tilia begoniifolia Chun et H. D. Wong = Tilia endochrysea Hand. -Mazz.●

391679　Tilia breviradiata（Rehder）Hu et W. C. Cheng = Tilia chingiana Hu et W. C. Cheng●

391680　Tilia callidonta Hung T. Chang;美齿椴;Beaty-serrated Lime,Prettytooth Linden,Serrulate Linden●

391681　Tilia caroliniana Mill.;卡罗林椴（卡罗来纳椴,卡罗里纳椴）;Bee-tree,Carolina Basswood,Linden●

391682　Tilia caucasica Rupr.;高加索椴;Caucasian Lime,Caucasian Linden●☆

391683　Tilia chenmoui W. C. Cheng = Tilia tuan Szyszyl. var. chenmoui（W. C. Cheng）Y. Tang●◇

391684　Tilia chinensis Hu et W. C. Cheng = Tilia breviradiata（Rehder）Hu et W. C. Cheng●

391685　Tilia chinensis Maxim.;华椴（中华椴）;China Linden,Chinese Lime,Chinese Linden●

391686　Tilia chinensis Maxim. var. intonsa（E. H. Wilson）Y. C. Hsu et Zhuge;多毛椴●

391687　Tilia chinensis Maxim. var. investita（V. Engl.）Rehder;秃华椴;Glabrate Chinese Linden●

391688　Tilia chinensis Schneid. = Tilia tuan Szyszyl. var. chinensis Rehder et E. H. Wilson●

391689　Tilia chingiana Hu et W. C. Cheng;短毛椴（椴皮树,庐山椴）;Shorthair Linden,Short-haired Linden●

391690　Tilia chugokuensis Hatus. = Tilia mandshurica Rupr. et Maxim.●

391691　Tilia cordata Mill.;欧椴小叶椴（欧椴小叶椴,小叶欧椴,心叶椴）;Basswood,Bast,European Linden,European Little-leaved Linden,Littleleaf Linden,Little-leaf Linden,Little-leaved Lime,Red Lime,Small Leaved Lime,Small-leafed Lime,Smallleaved European Linden,Small-leaved European Linden,Small-leaved Lime,Small-leaved Limetree●☆

391692　Tilia cordata Mill.'Corinthian';科林斯欧洲小叶椴●☆

391693　Tilia cordata Mill.'Fairview';平等欧洲小叶椴●☆

391694　Tilia cordata Mill.'Greenspire';绿旋欧洲小叶椴（绿塔欧洲小叶椴）●☆

391695　Tilia cordata Mill.'June Bride';琼新娘欧洲小叶椴●☆

391696　Tilia cordata Mill.'Rancho';牧场欧洲小叶椴（牧场小屋欧

洲小叶椴）●☆

391697 Tilia cordata Mill. 'Swedish Upright'；瑞典直立欧洲小叶椴；Little-leaf Linden ●☆

391698 Tilia cordata Mill. var. japonica Miq. = Tilia japonica（Miq.）Simonk. ●

391699 Tilia cordifolia Besser；心叶椴●☆

391700 Tilia croizatii Chun et H. D. Wong = Tilia endochrysea Hand. -Mazz. ●

391701 Tilia dasystyla Steven；毛柱椴(毛茎椴)；Shaggy-style Linden ●☆

391702 Tilia dasystyla Steven = Tilia caucasica Rupr. ●☆

391703 Tilia dictyoneura V. Engl. ex C. K. Schneid. = Tilia paucicostata Maxim. var. dictyoneura（V. Engl. ex C. K. Schneid.）Hung T. Chang et E. W. Miao ●

391704 Tilia endochrysea Hand. -Mazz.；白毛椴(白毛椴树，火索树，湘椴)；Hunan Lime, Hunan Linden ●

391705 Tilia euchlora K. Koch；克里米亚椴；Baste-tree, Bast-tree, Caucasian Lime, Crimean Lime, Crimean Linden, Lime ●☆

391706 Tilia europaea L.；欧洲椴（欧椴，菩提木）；Carver's Tree, Common Lime, Common Linden, Europe Linden, European Lime, European Linden, Lenten, Lime, Lime Tree, Lin, Lind, Linden, Line, Linn, Locks-and-keys, Lynd, Pollard-flowers, Teil, Teyl, Tile, Tilet-tree, Tillet, Tillettree, Whitewood ●

391707 Tilia europaea L. 'Pallida'；灰欧洲椴；Pale-leaved Common Linden ●

391708 Tilia europaea L. 'Wratislaviensis'；侮莱迪斯拉维欧洲椴；Golden Linden ●☆

391709 Tilia europaea L. = Tilia cordata Mill. ●☆

391710 Tilia europaea L. = Tilia vulgaris Hayne ●

391711 Tilia europaea L. f. peduta Rehder；垂枝欧洲椴；Weeping Common Linden ●

391712 Tilia europaea L. var. pallida Rchb. = Tilia europaea L. 'Pallida' ●

391713 Tilia eurosinica Croizat = Tilia japonica（Miq.）Simonk. ●

391714 Tilia flaccida Host ex Bayer；美大椴；American-largeleaved Linden ●☆

391715 Tilia floridana Small；弗罗里达椴；Florida Linden ●☆

391716 Tilia flvescens A. Braun；美小椴；American-littleleaved Linden ●☆

391717 Tilia franchetiana C. K. Schneid.；日光椴；Franchet Linden ●☆

391718 Tilia franchetiana C. K. Schneid. = Tilia miqueliana Maxim. ●

391719 Tilia fulvosa Hung T. Chang；黄毛椴；Yellowhair Linden ●

391720 Tilia fulvosa Hung T. Chang = Tilia chinensis Maxim. var. intonsa（E. H. Wilson）Y. C. Hsu et Zhuge ●

391721 Tilia glabra Vent. = Tilia americana L. ●☆

391722 Tilia gracilis Hung T. Chang；纤椴；Thin Linden ●

391723 Tilia gracilis Hung T. Chang = Tilia tuan Szyszyl. ●

391724 Tilia grandifolia Ehrh. = Tilia platyphylla Scop. ●☆

391725 Tilia henryana Szyszyl.；糯米椴（椴树，粉椴，亨利椴树，毛糯米椴，糯米树）；Henry Linden ●

391726 Tilia henryana Szyszyl. var. subglabra V. Engl.；光叶糯米椴（糯米椴，秃糯米椴）；Glabrate Henry Linden ●

391727 Tilia heterophylla Vent.；异叶椴（白椴，白毛椴）；Bee-tree, Beetree Linden, Linden, White Basswood, White Linden ●☆

391728 Tilia heterophylla Vent. = Tilia americana L. ●☆

391729 Tilia heterophylla Vent. var. michauxii Sarg.；红枝白椴（红枝白毛椴）；Red-branched White Linden ●☆

391730 Tilia hupehensis W. C. Cheng ex Hung T. Chang；湖北毛椴（湖北椴，湖北糯米椴）；Hubei Lime, Hubei Linden ●

391731 Tilia hupehensis W. C. Cheng ex Hung T. Chang = Tilia tuan Szyszyl. ●

391732 Tilia hypoglauca Rehder = Tilia endochrysea Hand. -Mazz. ●

391733 Tilia insularis Nakai；岛生椴；Island Linden ●☆

391734 Tilia integerrima Hung T. Chang；全缘椴；Daoxian Linden, Entire Linden, Entire-leaved Linden ●

391735 Tilia integerrima Hung T. Chang = Tilia tuan Szyszyl. ●

391736 Tilia intermedia DC. = Tilia vulgaris Hayne ●

391737 Tilia intonsa E. H. Wilson = Tilia chinensis Maxim. var. intonsa（E. H. Wilson）Y. C. Hsu et Zhuge ●

391738 Tilia intonsa E. H. Wilson ex Rehder et E. H. Wilson；西蜀椴（多毛椴）；Barbed Lime, Barbed Linden, Hairy Linden, W. Sichuan Linden ●

391739 Tilia japonica（Miq.）Simonk.；华东椴（级木，日本椴）；Japan Lime, Japan Linden, Japanese Lime, Japanese Linden ●

391740 Tilia japonica（Miq.）Simonk. f. inaensis（Kusaka）H. Hara；伊那椴●☆

391741 Tilia japonica（Miq.）Simonk. f. pygmaea Satomi；矮小华东椴●☆

391742 Tilia japonica（Miq.）Simonk. f. stenoglossa（Honda）H. Hara；窄舌华东椴●☆

391743 Tilia japonica（Miq.）Simonk. var. leiocarpa Nakai；四国椴；Sikoku Linden ●☆

391744 Tilia japonica（Miq.）Simonk. var. magna H. Hara = Tilia japonica（Miq.）Simonk. ●

391745 Tilia japonica Simonk. = Tilia japonica（Miq.）Simonk. ●

391746 Tilia jiaodongensis S. B. Liang；胶东椴；Jiaodong Linden ●

391747 Tilia juranyana Simonk.；匈牙利椴；Hungarian Linden ●☆

391748 Tilia kinashii H. Lév. et Vaniot = Tilia miqueliana Maxim. ●

391749 Tilia kiusiana Makino et Shiras. ex Shiras.；小花椴（九州椴）；Kyushu Linden, Little-flower Linden ●☆

391750 Tilia koreana Nakai = Tilia amurensis Rupr. var. taquetii（C. K. Schneid.）Liou et C. Y. Li ●

391751 Tilia kueichouensis Hu；黔椴（贵州椴）；Guizhou Linden, Kweichou Lime, Kweichou Linden ●

391752 Tilia kwangtungensis Chun et H. D. Wong = Tilia miqueliana Maxim. ●

391753 Tilia laetevirens Rehder et E. H. Wilson；亮绿椴（长柄椴，亮绿叶椴）；Shiny-green Lime, Shinygreen Linden, Shiny-green Linden ●

391754 Tilia laetevirens Rehder et E. H. Wilson = Tilia chinensis Maxim. ●

391755 Tilia lepidota Rehder；鳞毛椴；Scalyhair Linden, Scaly-haired Lime, Scaly-haired Linden ●

391756 Tilia lepidota Rehder = Tilia endochrysea Hand. -Mazz. ●

391757 Tilia leptocarya Rehder = Tilia endochrysea Hand. -Mazz. ●

391758 Tilia leptocarya Rehder var. triloba Rehder = Tilia endochrysea Hand. -Mazz. ●

391759 Tilia likiangensis Hung T. Chang；丽江椴；Lijiang Lime, Lijiang Linden ●

391760 Tilia mandshurica Rupr. et Maxim.；糠椴（成道树，赤杨，大叶椴，椴，椴木，椴树，辽椴，杻，菩提纱，菩提树，山羊树，杝)；Liao Linden, Manchurian Lime, Manchurian Linden ●

391761 Tilia mandshurica Rupr. et Maxim. f. chugokuensis（Hatus.）T. Yamaz. = Tilia mandshurica Rupr. et Maxim. ●

391762 Tilia mandshurica Rupr. et Maxim. var. megaphylla（Nakai）Liou et C. Y. Li；棱果糠椴（棱果辽椴）；Prismatic Manchurian Linden, Ribfruit Linden ●

391763 Tilia mandshurica Rupr. et Maxim. var. ovalis（Nakai）Liou et

C. Y. Li;卵果糠椴(卵果辽椴);Ovate Manchurian Linden ●

391764 Tilia mandshurica Rupr. et Maxim. var. rufovillosa (Hatus.) Kitam. ;红毛糠椴 ●☆

391765 Tilia mandshurica Rupr. et Maxim. var. tuberculata Liou et C. Y. Li;瘤果糠椴(瘤果辽椴);Tuberculate Manchurian Linden ●

391766 Tilia maximowicziana Shiras. ;马氏椴(马克西莫维奇椴,日本大叶椴);Maximowicz Linden ●☆

391767 Tilia maximowicziana Shiras. f. yesoana (Nakai) H. Hara = Tilia maximowicziana Shiras. var. yesoana (Nakai) Tatew. ●☆

391768 Tilia maximowicziana Shiras. var. yesoana (Nakai) Tatew. ;北海道椴;Yezo Linden ●☆

391769 Tilia megaphylla Nakai = Tilia mandshurica Rupr. et Maxim. var. megaphylla (Nakai) Liou et C. Y. Li ●

391770 Tilia membranacea Hung T. Chang;膜叶椴;Membranousleaf Linden,Membranous-leaved Lime,Membranous-leaved Linden ●

391771 Tilia mesembrinos Merr. ;滇南椴;S. Yunnan Linden,South Yunnan Lime,South Yunnan Linden ●

391772 Tilia mesembrinos Merr. = Tilia tuan Szyszyl. ●

391773 Tilia miqueliana Maxim. ;南京椴(椴树,弥格椴,密克椴树,菩提树,山桑皮);Japanese Linden,Miquel Lime,Miquel Linden,Nanjing Linden ●

391774 Tilia miqueliana Maxim. var. chinensis Szyszyl. = Tilia tuan Szyszyl. var. chinensis Rehder et E. H. Wilson ●

391775 Tilia miqueliana Maxim. var. chinensis Diels = Tilia paucicostata Maxim. ●

391776 Tilia miqueliana Maxim. var. longipes P. C. Chiu;长柄南京椴;Long-pediceled Miquel Linden ●

391777 Tilia miqueliana Maxim. var. longipes P. C. Chiu = Tilia miqueliana Maxim. ●

391778 Tilia mofungensis Chun et H. D. Wong;帽峰椴(磨峰椴树);Maofeng Lime,Maofeng Linden ●

391779 Tilia mofungensis Chun et H. D. Wong = Tilia tuan Szyszyl. ●

391780 Tilia moltkei Spüth ex C. K. Schneid. ;莫特基椴;Moltke Linden ●☆

391781 Tilia mongolica Maxim. ;蒙椴(白皮椴,蒙古椴,米椴,小叶椴);Mongol Linden,Mongolian Lime,Mongolian Linden ●

391782 Tilia monticola Maxim. ;山生椴;Mountain Linden ●☆

391783 Tilia multiflora Ledeb. ;繁花椴 ●☆

391784 Tilia nanchuanensis Hung T. Chang;南川椴;Nanchuan Lime,Nanchuan Linden,S. Sichuan Linden ●

391785 Tilia nanchuanensis Hung T. Chang = Tilia kueichouensis Hu ●

391786 Tilia neglecta Spach;魁北克椴(簇毛椴,疏忽椴);Quebec Linden,Tufted Hair Linden ●☆

391787 Tilia neglecta Spach = Tilia americana L. ●☆

391788 Tilia nobilis Rehder et E. H. Wilson;大叶椴(大椴);Big-leaved Lime,Big-leaved Linden,Largeleaf Linden ●

391789 Tilia oblongifolia Rehder;矩圆叶椴(长圆叶椴,黄山椴树);Oblongifoliate Lime,Oblongifoliate Linden,Oblongleaf Linden ●

391790 Tilia oblongifolia Rehder = Tilia tuan Szyszyl. ●

391791 Tilia oblongifolia Rehder var. sangzhiensis B. R. Liao et W. X. Wang;桑植椴;Sangzhi Oblongleaf Linden ●

391792 Tilia oblongifolia Rehder var. sangzhiensis B. R. Liao et W. X. Wang = Tilia tuan Szyszyl. var. chinensis Rehder et E. H. Wilson ●

391793 Tilia obscura Hand. -Mazz. ;云山椴;Yunshan Lime,Yunshan Linden ●

391794 Tilia obscura Hand. -Mazz. = Tilia tuan Szyszyl. ●

391795 Tilia oliveri Szyszyl. ;粉椴(椴树,鄂椴);Chinese Lime,Oliver

Linden,Oliver's Lime,Powder Linden ●

391796 Tilia oliveri Szyszyl. var. cinerascens Rehder et E. H. Wilson;灰背椴(灰鄂椴);Hairy Oliver Linden ●

391797 Tilia omeiensis W. P. Fang;峨眉椴(白郎花);Emei Linden,Emei Tilia,Omei Lime ●

391798 Tilia omeiensis W. P. Fang = Tilia tuan Szyszyl. ●

391799 Tilia orbicularis Jouin;短柄椴;Petiole-shorted Linden ●☆

391800 Tilia orocryptica Croizat = Tilia breviradiata (Rehder) Hu et W. C. Cheng ●

391801 Tilia orocryptica Croizat = Tilia chingiana Hu et W. C. Cheng ●

391802 Tilia ovalis Nakai = Tilia mandshurica Rupr. et Maxim. var. megaphylla (Nakai) Liou et C. Y. Li ●

391803 Tilia ovalis Nakai = Tilia mandshurica Rupr. et Maxim. var. ovalis (Nakai) Liou et C. Y. Li ●

391804 Tilia palmeri F. C. Gates = Tilia americana L. ●☆

391805 Tilia parviflora Boiss. = Tilia parvifolia Ehrh. ●☆

391806 Tilia parvifolia Ehrh. = Tilia cordata Mill. ●☆

391807 Tilia paucicostata Maxim. ;少脉椴(简果椴,小叶椴);Fewnerve Linden,Few-nerved Lime,Few-nerved Linden ●

391808 Tilia paucicostata Maxim. var. dictyoneura (V. Engl. ex C. K. Schneid.) Hung T. Chang et E. W. Miao;红皮椴(网脉椴树);Redbark Fewnerve Linden ●

391809 Tilia paucicostata Maxim. var. firma V. Engl. = Tilia paucicostata Maxim. ●

391810 Tilia paucicostata Maxim. var. ningshanensis P. H. Yang;宁陕少脉椴;Ningshan Linden ●

391811 Tilia paucicostata Maxim. var. ningshanensis P. H. Yang = Tilia paucicostata Maxim. var. yunnanensis Diels ●

391812 Tilia paucicostata Maxim. var. tenuis V. Engl. = Tilia paucicostata Maxim. ●

391813 Tilia paucicostata Maxim. var. yunnanensis Diels;少脉毛椴;Yunnan Fewnerve Linden ●

391814 Tilia pekingensis Rupr. ex Maxim. = Tilia mandshurica Rupr. et Maxim. ●

391815 Tilia pendula V. Engl. ex Schneid. = Tilia oliveri Szyszyl. ●

391816 Tilia petiolaris DC. ;垂枝银毛椴;Pendent Silver Linden,Silver Pendent Linden ●☆

391817 Tilia platyphylla Scop. ;阔叶椴(大叶椴,夏菩提树);Bigleaf Linden,Big-leaf Linden,Broadleaf Lime,Broad-leaved Lime,Largeleaf Linden,Large-leafed Lime,Largeleaved Lime,Large-leaved Lime,Large-leaved Linden,Red Lime,Red-twigged Lime ●☆

391818 Tilia platyphylla Scop. ' Corallina ' = Tilia platyphylla Scop. var. rubra (West.) Rehder ●☆

391819 Tilia platyphylla Scop. 'Prince's Street';君主街阔叶椴 ●☆

391820 Tilia platyphylla Scop. ' Rubra ' = Tilia platyphylla Scop. var. rubra (West.) Rehder ●☆

391821 Tilia platyphylla Scop. f. fastigiata Rehder;塔形阔叶椴;Pyramidal Largeleaved Linden ●☆

391822 Tilia platyphylla Scop. var. aurea (Loudon) Kirchn. ;黄枝阔叶椴;Yellow-branched Largeleaved Linden ●☆

391823 Tilia platyphylla Scop. var. laciniata (Loudon) K. Koch;深裂叶阔叶椴;Deep-lobed Largeleaved Linden ●☆

391824 Tilia platyphylla Scop. var. rubra (West.) Rehder;红枝阔叶椴;Red-branched Largeleaved Linden,Red-twigged Lime ●☆

391825 Tilia platyphylla Scop. var. vitifolia (Host) Simonk. ;浅裂叶阔叶椴;Slight-lobed Largeleaved Linden ●☆

391826 Tilia populifolia Hung T. Chang;杨叶椴;Poplarleaf Linden,

Poplar-leaved Lime，Poplar-leaved Linden ●

391827 Tilia populifolia Hung T. Chang ＝ Tilia oliveri Szyszyl. var. cinerascens Rehder et E. H. Wilson ●

391828 Tilia pubescens Aiton；南方椴●☆

391829 Tilia rubra DC. ；红椴；Red Lime，Red Linden ●☆

391830 Tilia rufovillosa Hatus. ＝ Tilia mandshurica Rupr. et Maxim. var. rufovillosa（Hatus. ）Kitam. ●☆

391831 Tilia scalenophylla Y. Ling ＝ Tilia endochrysea Hand. -Mazz. ●

391832 Tilia sibirica Fisch. ；西伯利亚椴；Siberian Linden ●☆

391833 Tilia sylvestris Desf. ＝ Tilia cordata Mill. ●☆

391834 Tilia taishanensis S. B. Liang；泰山椴；Taishan Linden ●

391835 Tilia taquetii C. K. Schneid. ＝ Tilia amurensis Rupr. var. taquetii（C. K. Schneid. ）Liou et C. Y. Li ●

391836 Tilia tomentosa Moench；银毛椴（白背椴，毛椴，绒毛椴，银白椴，银椴，银叶椴）；European White Lime，Silver Lime，Silver Linden，White Linden ●☆

391837 Tilia tomentosa Moench 'Brabant'；布拉班特银毛椴●☆

391838 Tilia tomentosa Moench 'Nijmegen'；乃梅岁银毛椴●☆

391839 Tilia tomentosa Moench 'Sterling Silver'；纯银制品银毛椴●☆

391840 Tilia tristis Chun ex Hung T. Chang；淡灰椴；Dull-coloured Lime，Lightgrey Linden，Qunaxina Linden ●

391841 Tilia tristis Chun ex Hung T. Chang ＝ Tilia tuan Szyszyl. ●

391842 Tilia truncata Spach ＝ Tilia americana L. ●☆

391843 Tilia tuan Szyszyl. ；椴树（椴，滚筒树，家鹤儿，金桐力树，千层皮，青科椰，叶上果）；Tuan Lime，Tuan Linden ●

391844 Tilia tuan Szyszyl. f. divaricata V. Engl. ＝ Tilia tuan Szyszyl. ●

391845 Tilia tuan Szyszyl. var. breviradiata Rehder ＝ Tilia breviradiata （Rehder）Hu et W. C. Cheng ●

391846 Tilia tuan Szyszyl. var. breviradiata Rehder ＝ Tilia chingiana Hu et W. C. Cheng ●

391847 Tilia tuan Szyszyl. var. cavaleriei V. Engl. et H. Lév. ＝ Tilia tuan Szyszyl. ●

391848 Tilia tuan Szyszyl. var. cavaleriei V. Engl. et H. Lév. f. divaricata V. Engl. ＝ Tilia tuan Szyszyl. ●

391849 Tilia tuan Szyszyl. var. chenmoui（W. C. Cheng）Y. Tang；长苞椴；Binchuan Linden，Longbract Linden，Long-bracted Linden ●◇

391850 Tilia tuan Szyszyl. var. chinensis Rehder et E. H. Wilson；毛芽椴（矩圆叶椴，毛枝椴）；Chinese Tuan Linden，Woollytwig Tuan Linden ●

391851 Tilia tuan Szyszyl. var. pruinosa V. Engl. ＝ Tilia tuan Szyszyl. ●

391852 Tilia ulmifolia Scop. var. japonica（Miq. ）O. H. Sarg. ex Mayr ＝ Tilia japonica（Miq. ）Simonk. ●

391853 Tilia venulosa Sarg. ＝ Tilia americana L. ●☆

391854 Tilia vitifolia Hu et F. H. Chen ＝ Tilia endochrysea Hand. -Mazz. ●

391855 Tilia vulgaris Hayne ＝ Tilia europaea L. ●

391856 Tilia yunnanensis Hu；云南椴；Yunnan Lime，Yunnan Linden ●

391857 Tilia yunnanensis Hu ＝ Tilia chinensis Maxim. ●

391858 Tiliaceae Adans. ＝ Malvaceae Juss. （保留科名）●■

391859 Tiliaceae Adans. ＝ Tiliaceae Juss. （保留科名）●■

391860 Tiliaceae Juss. （1789）（保留科名）；椴树科（椴科，田麻科）；Lime Family，Linden Family ●■

391861 Tiliacora Colebr. （1821）（保留属名）；香料藤属（铁立藤属）●☆

391862 Tiliacora acuminata Miers ＝ Tiliacora racemosa Colebr. ●☆

391863 Tiliacora bequaertii De Wild. ＝ Tiliacora laurentii De Wild. var. bequaertii（De Wild. ）Troupin ●☆

391864 Tiliacora cabindensis Exell et Mendonça ＝ Beirnaertia

cabindensis（Exell et Mendonça）Troupin ●☆

391865 Tiliacora chrysobotrya Welw. ex Ficalho；金穗香料藤●☆

391866 Tiliacora dielsiana Hutch. et Dalziel；迪尔斯香料藤●☆

391867 Tiliacora dinklagei Engl. ；丁克香料藤●☆

391868 Tiliacora ealaensis Troupin；埃阿拉香料藤●☆

391869 Tiliacora funifera（Miers）Oliv. ；绳状香料藤（绳状铁立藤，梯里亚可拉）●☆

391870 Tiliacora funifera Oliv. ＝ Tiliacora funifera（Miers）Oliv. ●☆

391871 Tiliacora gabonensis Troupin；加蓬香料藤●☆

391872 Tiliacora gilletii De Wild. ＝ Triclisia dictyophylla Diels ●☆

391873 Tiliacora glycosmantha Diels ＝ Tiliacora funifera（Miers）Oliv. ●☆

391874 Tiliacora gossweileri Exell；戈斯香料藤●☆

391875 Tiliacora innularis Louis ex Troupin ＝ Tiliacora chrysobotrya Welw. ex Ficalho ●☆

391876 Tiliacora johannis Exell ＝ Tiliacora funifera（Miers）Oliv. ●☆

391877 Tiliacora kenyensis Troupin；肯尼亚香料藤●☆

391878 Tiliacora klaineana（Pierre）Diels；克莱恩香料藤●☆

391879 Tiliacora latifolia Troupin；宽叶香料藤●☆

391880 Tiliacora laurentii De Wild. ；洛朗香料藤●☆

391881 Tiliacora laurentii De Wild. var. bequaertii（De Wild. ）Troupin；贝卡尔香料藤●☆

391882 Tiliacora lehmbachii Engl. ；莱姆香料藤●☆

391883 Tiliacora leonardii Troupin；莱奥香料藤●☆

391884 Tiliacora leonensis（Scott-Elliot）Diels；莱昂香料藤●☆

391885 Tiliacora louisii Troupin；路易斯香料藤●☆

391886 Tiliacora macrophylla（Pierre）Diels；大叶香料藤●☆

391887 Tiliacora mayumbensis Exell ＝ Tiliacora chrysobotrya Welw. ex Ficalho ●☆

391888 Tiliacora nigerica Troupin；尼日利亚香料藤●☆

391889 Tiliacora odorata Engl. ；芳香香料藤●☆

391890 Tiliacora ovalis Diels；卵形香料藤●☆

391891 Tiliacora polygyna Diels ex Mildbr. ＝ Tiliacora louisii Troupin ●☆

391892 Tiliacora pynaertii De Wild. ＝ Tiliacora funifera（Miers）Oliv. ●☆

391893 Tiliacora racemosa Colebr. ；香料藤（铁立藤）●☆

391894 Tiliacora soyauxii Engl. ；索亚香料藤●☆

391895 Tiliacora tisserantii A. Chev. ＝ Tiliacora chrysobotrya Welw. ex Ficalho ●☆

391896 Tiliacora triandra（Colebr. ）Diels；三蕊香料藤（三蕊铁立藤）●☆

391897 Tiliacora triandra Diels ＝ Tiliacora triandra（Colebr. ）Diels ●☆

391898 Tiliacora trichantha Diels ＝ Triclisia dictyophylla Diels ●☆

391899 Tiliacora troupinii Cufod. ＝ Tiliacora funifera（Miers）Oliv. ●☆

391900 Tiliacora warneckei Engl. et Diels ＝ Tiliacora funifera（Miers）Oliv. ●☆

391901 Tilingia Regel et Tiling ＝ Ligusticum L. ■

391902 Tilingia Regel et Tiling ＝ Tilingia Regel ■

391903 Tilingia Regel（1859）；第岭芹属（第苓芹属，岩茴香属）■

391904 Tilingia ajanensis Regel et Tiling；第岭芹（黑水岩茴香）；Ajan Ligusticum ■

391905 Tilingia ajanensis Regel et Tiling ＝ Ligusticum ajanense（Regel et Tiling）Koso-Pol. ■

391906 Tilingia ajanensis Regel et Tiling f. latisecta（Takeda）Kitag. ；宽裂第岭芹■☆

391907 Tilingia ajanensis Regel et Tiling f. pectinata（Koidz. ）Kitag. ；篦状第岭芹■☆

391908 Tilingia ajanensis Regel et Tiling var. angustissima（Nakai ex H. Hara）Kitag. ；极窄第岭芹■☆

391909 Tilingia filisecta（Nakai et Kitag. ）Nakai et Kitag. ＝

Ligusticum tachiroei（Franch. et Sav.）M. Hiroe et Constance ■

391910 Tilingia filisecta（Nakai et Kitag.）Nakai et Kitag. = Rupiphila tachiroei（Franch. et Sav.）Pimenov et Lavrova ■

391911 Tilingia holopetala（Maxim.）Kitag. ;全瓣第岭芹■☆

391912 Tilingia holopetala（Maxim.）Kitag. = Ligusticum holopetalum（Maxim.）Hiroe et Constance ■☆

391913 Tilingia jeholensis（Nakai et Kitag.）Leute = Ligusticum jeholense（Nakai et Kitag.）Nakai et Kitag. ■

391914 Tilingia tachiroei（Franch. et Sav.）Kitag. = Ligusticum tachiroei（Franch. et Sav.）M. Hiroe et Constance ■

391915 Tilingia tachiroei（Franch. et Sav.）Kitag. = Rupiphila tachiroei（Franch. et Sav.）Pimenov et Lavrova ■

391916 Tilingia tsusimensis（Y. Yabe）Kitag. ;津岛第岭芹■☆

391917 Tilioides Medik. = Tilia L. ●

391918 Tillaea L.（1753）;东爪草属;Pygmyweed, Tillaea ■

391919 Tillaea L. = Crassula L. ●■☆

391920 Tillaea alata Viv. ;云南东爪草;Yunnan Pygmyweed, Yunnan Tillaea ■

391921 Tillaea alata Viv. = Crassula alata（Viv.）A. Berger ■

391922 Tillaea alsinoides Hook. f. = Crassula pellucida L. subsp. alsinoides（Hook. f.）Toelken ■☆

391923 Tillaea aquatica L. ;东 爪 草;Aquatic Tillaea, Common Pygmyweed, Pygmyweed ■

391924 Tillaea aquatica L. = Crassula aquatica（L.）Schönland ■

391925 Tillaea brevifolia（Eckl. et Zeyh.）Walp. = Crassula decumbens Thunb. var. brachyphylla（Adamson）Toelken ●☆

391926 Tillaea capensis L. f. = Crassula natans Thunb. ■☆

391927 Tillaea decumbens Willd. = Crassula thunbergiana Schult. ■☆

391928 Tillaea ecklonis Walp. = Crassula natans Thunb. var. minus（Eckl. et Zeyh.）G. D. Rowley ■☆

391929 Tillaea elatinoides（Eckl. et Zeyh.）Walp. = Crassula elatinoides（Eckl. et Zeyh.）Friedrich ●☆

391930 Tillaea erecta Hook. et Arn. ;直立东爪草;Pygmy-weed ■☆

391931 Tillaea filiformis（Eckl. et Zeyh.）Endl. ex Walp. = Crassula natans Thunb. ■☆

391932 Tillaea filiformis（Eckl. et Zeyh.）Endl. ex Walp. var. parvula ? = Crassula natans Thunb. ■☆

391933 Tillaea fluitans（Eckl. et Zeyh.）Endl. ex Walp. = Crassula natans Thunb. ■☆

391934 Tillaea fluitans（Eckl. et Zeyh.）Endl. ex Walp. var. intermedia ? = Crassula natans Thunb. ■☆

391935 Tillaea fluitans（Eckl. et Zeyh.）Endl. ex Walp. var. obovata ? = Crassula natans Thunb. ■☆

391936 Tillaea inanis（Thunb.）Steud. = Crassula inanis Thunb. ■☆

391937 Tillaea likiangensis H. Chuang;丽江东爪草;Lijiang Tillaea ■

391938 Tillaea macrantha Hook. f. = Crassula decumbens Thunb. ●☆

391939 Tillaea mongolica（Franch.）S. H. Fu;承德东爪草;Mongol Pygmyweed, Mongolian Tillaea ■

391940 Tillaea muscosa L. = Crassula tillaea Lest. -Garl. ■☆

391941 Tillaea muscosa L. var. trichopoda Post = Crassula alata（Viv.）A. Berger ■

391942 Tillaea pentandra Royle ex Edgew. = Crassula schimperi Fisch. et C. A. Mey. ■☆

391943 Tillaea pentandra Royle ex Edgew. = Tillaea schimperi（C. A. Mey.）M. G. Gilbert, H. Ohba et K. T. Fu ■

391944 Tillaea perfoliata L. f. = Crassula inanis Thunb. ■☆

391945 Tillaea perfoliata L. f. var. latifolium（Eckl. et Zeyh.）Walp. =

391946 Tillaea pharnaceoides Hochst. ex Britten = Tillaea alata Viv. ■

391947 Tillaea pharnaceoides Hochst. ex Steud. = Crassula alata（Viv.）A. Berger subsp. pharnaceoides（Fisch. et C. A. Mey.）Wickens et Bywater ■☆

391948 Tillaea recurva（Hook. f.）Hook. f. = Crassula helmsii（Kirk）Cockayne ●☆

391949 Tillaea recurva Hook. f. = Crassula helmsii（Kirk）Cockayne ●☆

391950 Tillaea reflexa（Eckl. et Zeyh.）Endl. ex Walp. = Crassula natans Thunb. ■☆

391951 Tillaea repens Peter = Crassula granvikii Mildbr. ■☆

391952 Tillaea rivularis Peter = Crassula granvikii Mildbr. ■☆

391953 Tillaea schimperi（C. A. Mey.）M. G. Gilbert, H. Ohba et K. T. Fu;五蕊东爪草;Fivestamen Pygmyweed, Fivestamen Tillaea ■

391954 Tillaea subulata（Hook. f.）Britten var. illecebroides Welw. ex Hiern = Crassula lanceolata（Eckl. et Zeyh.）Endl. ex Walp. ●☆

391955 Tillaea trichopoda Fenzl = Crassula alata（Viv.）A. Berger ■

391956 Tillaea trichopoda Fenzl ex Boiss = Tillaea alata Viv. ■

391957 Tillaea trichotoma（Eckl. et Zeyh.）Walp. = Crassula decumbens Thunb. ●☆

391958 Tillaea umbellata（Thunb.）Willd. = Crassula umbellata Thunb. ■☆

391959 Tillaea vaillantii Willd. ;威兰氏东爪草;Vaillant Tillaea ■☆

391960 Tillaea vaillantii Willd. = Crassula vaillantii（Willd.）Roth ●☆

391961 Tillaea yunnanensis S. H. Fu = Tillaea alata Viv. ■

391962 Tillaeaceae Martinov = Crassulaceae J. St. -Hil.（保留科名）●■

391963 Tillaeastrum Britton = Crassula L. ●■☆

391964 Tillaeastrum Britton = Tillaea L. ■

391965 Tillaeastrum vaillantii（Willd.）Britton = Crassula vaillantii（Willd.）Roth ●☆

391966 Tillandsia L.（1753）;花凤梨属（第伦丝属,第伦斯属,空气凤梨属,木柄凤梨属,俤兰德细亚属,铁兰属,紫凤梨属,紫花凤梨属）;Air Plant, Tillandsia ■☆

391967 Tillandsia × emilie Hort. ;大紫花凤梨（大紫花菠萝）■☆

391968 Tillandsia adpressa André;林生花凤梨（林生铁兰）■☆

391969 Tillandsia albida Mez et Purpus;白花花凤梨（白花铁兰）■☆

391970 Tillandsia aloifolia Hook. = Tillandsia flexuosa Sw. ■☆

391971 Tillandsia anceps Lodd. ;二棱花凤梨（二棱铁兰）■☆

391972 Tillandsia araujei Mez;阿氏花凤梨（阿氏铁兰）■☆

391973 Tillandsia arequitae André ex Mez;银叶花凤梨（银叶铁兰）■☆

391974 Tillandsia argentea Griseb. ;银白花凤梨（银叶铁兰）■☆

391975 Tillandsia atroviridipetala Matuda;绿瓣花凤梨（绿瓣铁兰）■☆

391976 Tillandsia baileyi Rose ex Small;贝利花凤梨（贝利铁兰）■☆

391977 Tillandsia balbisiana Schult. f. ;巴尔花凤梨（巴尔铁兰）■☆

391978 Tillandsia bartramii Elliott;巴特拉姆花凤梨（巴特拉姆铁兰）;Bartram's Airplant ■☆

391979 Tillandsia benthamiana Klotzsch;本瑟姆花凤梨（本瑟姆铁兰）■☆

391980 Tillandsia bergeri Mez;拜氏花凤梨（拜氏铁兰）■☆

391981 Tillandsia berteroniana Schult. f. = Catopsis berteroniana（Schult. f.）Mez ■☆

391982 Tillandsia brachycaulos Schltdl. ;短茎花凤梨（短茎铁兰）■☆

391983 Tillandsia bracteata Chapm. = Tillandsia fasciculata Sw. var. clavispica Mez ■☆

391984 Tillandsia bulbosa Hook. ;鳞茎花凤梨（鳞茎铁兰）;Knollige Tillandsie ☆

391985 Tillandsia butzii Urb;布氏花凤梨（布氏铁兰）■☆

391986 Tillandsia capillaris Ruiz et Pav. ;发状花凤梨（发状铁兰）■☆

391987 Tillandsia capitata Griseb.;头花花凤梨(头花铁兰)■☆

391988 Tillandsia caput-medusae E. Morren;蛇仙花凤梨(蛇仙铁兰)■☆

391989 Tillandsia circinnata Schltdl.;旋卷花凤梨(旋卷铁兰)■☆

391990 Tillandsia compacta Griseb.;密丛花凤梨(密丛铁兰)■☆

391991 Tillandsia complanata Benth.;扁平花凤梨(扁平铁兰)■☆

391992 Tillandsia cyanea (A. Dietr.) E. Morren = Tillandsia guatemalensis L. B. Sm.■☆

391993 Tillandsia cyanea Linden ex K. Koch;紫花凤梨(紫花菠萝)■

391994 Tillandsia dasyliriifolia Baker;毛沟花凤梨(毛沟铁兰)■☆

391995 Tillandsia decomposita Baker;重叠花凤梨(重叠铁兰)■☆

391996 Tillandsia didisticha (E. Morren) Baker;复二列花凤梨(复二列铁兰)■☆

391997 Tillandsia disticha Kunth;二列花凤梨(二列铁兰)■☆

391998 Tillandsia dura Baker;硬花凤梨(硬铁兰)■☆

391999 Tillandsia duratii Vis.;杜氏花凤梨(杜氏铁兰)■☆

392000 Tillandsia dyeriana André;垂花花凤梨(垂花铁兰)■☆

392001 Tillandsia fasciculata Sw.;束叶花凤梨(束花凤梨,束叶铁兰);Cardinal Airplant,Giant Airplant,Wild Pine,Wild Pineapple■☆

392002 Tillandsia fasciculata Sw. var. clavispica Mez;棒穗束花凤梨■☆

392003 Tillandsia fasciculata Sw. var. densispica Mez;密穗束花凤梨■☆

392004 Tillandsia fasciculata Sw. var. floridana L. B. Sm. = Tillandsia floridana (L. B. Sm.) H. Luther■☆

392005 Tillandsia filifolia Schltdl. et Charm.;线叶花凤梨(线叶铁兰)■☆

392006 Tillandsia flabellata Baker;歧花凤梨(歧花菠萝)■☆

392007 Tillandsia flexuosa Sw.;曲花凤梨(曲铁兰)■☆

392008 Tillandsia floridana (L. B. Sm.) H. Luther;佛罗里达花凤梨(佛罗里达铁兰)■☆

392009 Tillandsia fragrans André;芳香花凤梨(芳香铁兰)■☆

392010 Tillandsia fragrans André 'Variegata';斑叶香花凤梨(斑叶香铁兰)■☆

392011 Tillandsia funkiana Baker;方氏花凤梨(方氏铁兰)■☆

392012 Tillandsia geminiflora Brongn.;对花凤梨(对花铁兰)■☆

392013 Tillandsia graebneri Mez;格氏花凤梨(格氏铁兰)■☆

392014 Tillandsia grandis Mez;大花凤梨(大铁兰)■☆

392015 Tillandsia guatemalensis L. B. Sm.;蓝紫凤梨(花凤梨,蓝紫铁兰,紫花菠萝,紫花铁兰);Air Plant■☆

392016 Tillandsia harrisii Ehlers;哈氏花凤梨(哈氏老人须)■☆

392017 Tillandsia houzeavii Chapm. = Tillandsia variabilis Schltdl.■☆

392018 Tillandsia hystricina Small = Tillandsia fasciculata Sw. var. densispica Mez■☆

392019 Tillandsia incurva Griseb.;内曲花凤梨(内曲铁兰)■☆

392020 Tillandsia ionantha Planch.;淡紫花凤梨(淡紫铁兰,章鱼凤梨);Air Plant,Sky Plant■☆

392021 Tillandsia kammii Rauh;卡姆花凤梨(卡氏老人须)■☆

392022 Tillandsia karwinskyana Schult. f.;卡氏花凤梨(卡氏铁兰)■☆

392023 Tillandsia kautskyi E. Pereira;考氏花凤梨(考氏老人须)■☆

392024 Tillandsia leiboldiana Schltdl.;雷氏花凤梨(雷氏铁兰)■☆

392025 Tillandsia lepidosepala L. B. Sm.;鳞萼花凤梨(鳞萼铁兰)■☆

392026 Tillandsia lindeniana Regel;长苞花凤梨(长苞铁兰,铁兰);Blue-flowered Torch■☆

392027 Tillandsia lindenii (Lem.) Regel;长苞菠萝(长苞花凤梨);Blue-flowered Torch■☆

392028 Tillandsia mallemontii Glaz. et Mez;马氏花凤梨(马氏铁兰)■☆

392029 Tillandsia mauryana L. B. Sm.;莫氏花凤梨(莫氏老人须)■☆

392030 Tillandsia multicaulis Steud.;多茎花凤梨(多茎铁兰)■☆

392031 Tillandsia nutans Sw. = Catopsis nutans (Sw.) Griseb.■☆

392032 Tillandsia paucifolia Baker;少叶花凤梨(少叶铁兰)■☆

392033 Tillandsia plumosa Baker;羽状花凤梨(羽状铁兰)■☆

392034 Tillandsia polystachia (L.) L.;多穗花凤梨(多穗铁兰)■☆

392035 Tillandsia prodigiosa Baker;巨大花凤梨(巨大铁兰)■☆

392036 Tillandsia pruinosa Sw.;白粉花凤梨(白粉铁兰)■☆

392037 Tillandsia pulchella Hook.;美丽花凤梨(美丽铁兰)■☆

392038 Tillandsia punctulata Schltdl. et Charm.;细点花凤梨(细点铁兰)■☆

392039 Tillandsia recurvata (L.) L.;反曲花凤梨(反曲铁兰);Ball Moss,Ball-moss■☆

392040 Tillandsia rubra Ruiz et Pav. = Vriesea rubra (Ruiz et Pav.) Beer■☆

392041 Tillandsia schiedeana Steud.;施氏花凤梨(施氏铁兰)■☆

392042 Tillandsia setacea Sw.;多刚毛花凤梨(多刚毛铁兰);Southern Needleleaf■☆

392043 Tillandsia smalliana H. Luther;斯莫尔花凤梨(花凤梨)■☆

392044 Tillandsia sprengeliana Klotzsch ex Beer;斯氏花凤梨(斯氏老人须)■☆

392045 Tillandsia streptophylla Scheidw.;卷曲花凤梨(卷曲铁兰);Curly-locks■☆

392046 Tillandsia stricta Lindl.;刚直花凤梨(刚直铁兰)■☆

392047 Tillandsia sucrei E. Pereira;苏氏花凤梨(花凤梨)■☆

392048 Tillandsia tenuifolia L.;细叶花凤梨(细叶铁兰)■☆

392049 Tillandsia tricholepis Baker;丝鳞花凤梨(丝鳞铁兰)■☆

392050 Tillandsia tricolor Cham. et Schltdl.;三色花凤梨(三色铁兰);Tricolor Tillandsia■☆

392051 Tillandsia unca Griseb.;沟状花凤梨(沟状铁兰)■☆

392052 Tillandsia usuneoides (L.) L.;松萝花凤梨(老人须,松萝铁兰,苔花凤梨);Black-moss,Florida Moss,Long Moss,Old Man's Beard,Spanish Moss,Spanish-moss,Treebeard Tillandsia■☆

392053 Tillandsia utriculata L.;膨胀花凤梨(膨胀铁兰);Giant Wild-pine■☆

392054 Tillandsia valenzuelana A. Rich. = Tillandsia variabilis Schltdl.■☆

392055 Tillandsia variabilis Schltdl.;变叶花凤梨(变叶铁兰)■☆

392056 Tillandsia viridiflora (Regel) Wittm. ex Mez;绿花花凤梨(绿花铁兰)■☆

392057 Tillandsia wagneriana L. B. Sm.;粉苞花凤梨(粉苞铁兰)■☆

392058 Tillandsia walteri Mez;瓦氏花凤梨(瓦氏铁兰)■☆

392059 Tillandsia wilsonii S. Watson = Tillandsia fasciculata Sw. var. densispica Mez■☆

392060 Tillandsia wurdackii L. B. Sm.;吴氏花凤梨(吴氏铁兰)■☆

392061 Tillandsiaceae A. Juss.;花凤梨科■

392062 Tillandsiaceae A. Juss. = Bromeliaceae Juss.(保留科名)■

392063 Tillandsiaceae Wilbr. = Bromeliaceae Juss.(保留科名)■

392064 Tillea L. = Crassula L.●■☆

392065 Tillea L. = Tillaea L.■

392066 Tillia St.-Lag. = Tillea L.■

392067 Tilliandsia Michx. = Tillandsia L.■☆

392068 Tillospermum Griff. = Kunzea Rchb.(保留属名)●☆

392069 Tillospermum Salisb.(废弃属名) = Kunzea Rchb.(保留属名)●☆

392070 Tilmia O. F. Cook = Aiphanes Willd.●☆

392071 Tilocarpus Engl. = Lasianthera P. Beauv.●☆

392072 Timaeosia Klotzsch = Gypsophila L.■●

392073 Timaeosia cerastioides (D. Don) Klotzsch = Gypsophila cerastioides D. Don■

392074 Timandra Klotzsch = Croton L.●

392075 Timanthea Salisb. = Baltimora L.(保留属名)■☆

392076 Timbalia Clos = Pyracantha M. Roem. ●

392077 Timbuleta Steud. = Anarrhinum Desf.（保留属名）■●☆

392078 Timbuleta Steud. = Simbuleta Forssk.（废弃属名）■●☆

392079 Timeroya Benth. = Pisonia L. ●

392080 Timeroya Benth. et Hook. f. = Calpidia Thouars ●

392081 Timeroya Benth. et Hook. f. = Pisonia L. ●

392082 Timeroya Benth. et Hook. f. = Timeroyea Montrouz. ●

392083 Timeroyea Montrouz. = Calpidia Thouars ●

392084 Timeroyea Montrouz. = Pisonia L. ●

392085 Timmia J. F. Gmel. = Cyrtanthus Aiton（保留属名）■☆

392086 Timonius DC.（1830）（保留属名）；海茜树属（贝木属，海满树属，梯木属）；Timonius ●

392087 Timonius Rumph. = Timonius DC.（保留属名）●

392088 Timonius arboreus Elmer；海茜树（贝木，梯木）；Arborous Timonius, Lanyu Timonius, Tree Timonius ●

392089 Timoron Raf. = Capnophyllum Gaertn. ■☆

392090 Timotocia Moldenke = Casselia Nees et Mart.（保留属名）■●☆

392091 Timouria Roshev.（1916）；钝基草属（沙丘草属,帖木儿草属）；Dunegrass ■

392092 Timouria Roshev. = Stipa L. ■

392093 Timouria aurita Hitchc. = Trikeraia hookeri（Stapf）Bor ■

392094 Timouria mongolica（Hitchc.）Roshev. = Psammochloa villosa（Trin.）Bor ■

392095 Timouria saposhnikowii Roshev.；钝基草（帖木儿草）；Asia Dunegrass ■

392096 Timouria villosa（Trin.）Hand. -Mazz. = Psammochloa villosa（Trin.）Bor ■

392097 Tina Blume = Harpullia Roxb. ●

392098 Tina Roem. et Schult. = Ratonia DC. ●☆

392099 Tina Roem. et Schult. = Tina Schult. ●☆

392100 Tina Schult.（1819）；马岛无患子属●☆

392101 Tina Schult. = Tina Roem. et Schult. ●☆

392102 Tina alata Danguy et Choux = Tina chapelieriana（Cambess.）Kalkman ●☆

392103 Tina bongolavensis Choux = Tina isaloensis Drake ●☆

392104 Tina chapelieriana（Cambess.）Kalkman；沙普马岛无患子●☆

392105 Tina cupanioides Baill. = Tina chapelieriana（Cambess.）Kalkman ●☆

392106 Tina dasycarpa Radlk.；毛果马岛无患子●☆

392107 Tina fulvinervis Radlk.；黄褐脉马岛无患子●☆

392108 Tina isaloensis Drake；伊萨卢马岛无患子●☆

392109 Tina leptophylla Radlk. = Neotina isoneura（Radlk.）Capuron ●☆

392110 Tina madagascariensis（DC.）Radlk. = Tina chapelieriana（Cambess.）Kalkman ●☆

392111 Tina multifoveolata Choux = Tina isaloensis Drake ●☆

392112 Tina polyphylla Baker = Tina fulvinervis Radlk. ●☆

392113 Tina striata Radlk.；条纹马岛无患子●☆

392114 Tina thouarsiana（Cambess.）Capuron；图氏马岛无患子●☆

392115 Tina tsaratanensis Choux = Neotina isoneura（Radlk.）Capuron ●☆

392116 Tina unifoveolata Radlk. = Neotina isoneura（Radlk.）Capuron ●☆

392117 Tina velutina Baker = Tina fulvinervis Radlk. ●☆

392118 Tinaceae Martinov = Adoxaceae E. Mey.（保留科名）●■

392119 Tinadendron Achille = Guettarda L. ●

392120 Tinadendron Achille（2006）；新海岸桐属●☆

392121 Tinadendron kajewskii（Guillaumin）Achille；新海岸桐●☆

392122 Tinadendron kajewskii（Guillaumin）Achille = Guettarda kajewskii Guillaumin ●☆

392123 Tinadendron noumeanum（Baill.）Achille；赫布里底新海岸桐●☆

392124 Tinadendron noumeanum（Baill.）Achille = Guettarda noumeana Baill. ●☆

392125 Tinaea Boiss. = Neotinea Rchb. f. ■☆

392126 Tinaea Boiss. = Tinea Biv. ■☆

392127 Tinaea Garzia = Lamarckia Moench（保留属名）■☆

392128 Tinaea Garziaex Parl. = Lamarckia Moench（保留属名）■☆

392129 Tinantia Dumort.（废弃属名）= Cypella Herb. ■☆

392130 Tinantia Dumort.（废弃属名）= Tinantia Scheidw.（保留属名）■☆

392131 Tinantia M. Martens et Galeotti = Cyphomeris Standl. ■☆

392132 Tinantia Scheidw.（1839）（保留属名）；媚泪花属（蒂南草属）■☆

392133 Tinantia anomala（Torr.）C. B. Clarke；异常媚泪花（杜兰格蒂南草）；False Dayflower, Widow's-tears ■☆

392134 Tinantia erecta（Jacq.）Fenzl；直立媚泪花（直立蒂南草）■☆

392135 Tinantia fugax Scheidw.；媚泪花☆

392136 Tinda Rchb. = Streblus Lour. ●

392137 Tinea Biv. = Neotinea Rchb. f. ■☆

392138 Tinea Spreng. = Prockia P. Browne ex L. ●☆

392139 Tinea cylindrica Biv. = Neotinea maculata（Desf.）Stearn ■☆

392140 Tinea intacta Boiss. = Neotinea maculata（Desf.）Stearn ■☆

392141 Tineoa Post et Kuntze = Lamarckia Moench（保留属名）■☆

392142 Tineoa Post et Kuntze = Neotinea Rchb. f. ■☆

392143 Tineoa Post et Kuntze = Prockia P. Browne ex L. ●☆

392144 Tineoa Post et Kuntze = Tinaea Garzia ■☆

392145 Tineoa Post et Kuntze = Tinea Biv. ■☆

392146 Tineoa Post et Kuntze = Tinea Spreng. ●☆

392147 Tinguarra Parl.（1843）；加那利草属■☆

392148 Tinguarra cervariifolia（DC.）Parl.；非洲加那利草■☆

392149 Tinguarra montana（Christ）A. Hansen et Sunding；加那利草■☆

392150 Tinguarra sicula（L.）Benth. et Hook. f. = Athamanta sicula L. ■☆

392151 Tingulonga Kuntze = Protium Burm. f.（保留属名）●

392152 Tingulonga Kuntze = Tingulonga Rumph. ●

392153 Tingulonga Rumph. = Protium Burm. f.（保留属名）●

392154 Tingulonga Rumph. = Tingulonga Rumph. ex Kuntze ●

392155 Tingulonga Rumph. ex Kuntze = Protium Burm. f.（保留属名）●

392156 Tiniaria（Meisn.）Rchb. = Fallopia Adans. ●■

392157 Tiniaria Rchb. = Fallopia Adans. ●■

392158 Tiniaria aubertii（L. Henry）Hedberg ex Janch. = Fallopia aubertii（L. Henry）Holub ●

392159 Tiniaria cilinodis（Michx.）Small = Fallopia cilinodis（Michx.）Holub ■☆

392160 Tiniaria convolvulus（L.）Webb et Miq. = Fallopia convolvulus（L.）Á. Löve ■

392161 Tiniaria convolvulus（L.）Webb et Moq. = Fallopia convolvulus（L.）Á. Löve ■

392162 Tiniaria convolvulus（L.）Webb et Moq. = Polygonum convolvulus L. ■

392163 Tiniaria convolvulus（L.）Webb et Moq. ex Webb et Berthelot = Fallopia convolvulus（L.）Á. Löve ■

392164 Tiniaria cristata（Engelm. et A. Gray）Small = Fallopia scandens（L.）Holub ■☆

392165 Tiniaria cristata（Engelm. et A. Gray）Small = Polygonum scandens L. var. cristatum（Engelm. et A. Gray）Gleason ■☆

392166 Tiniaria dumetorum（L.）Opiz = Fallopia dumetora（L.）Holub ■

392167 Tiniaria japonica（Houtt.）Hedberg = Fallopia japonica

（Houtt.）Ronse Decr. ■

392168 Tiniaria japonica（Houtt.）Hedberg = Reynoutria japonica Houtt. ■

392169 Tiniaria pauciflora（Maxim.）Nakai ex T. Mori = Fallopia dumetora（L.）Holub var. pauciflora（Maxim.）A. J. Li ■

392170 Tiniaria sachalinensis（F. Schmidt）Janch. = Fallopia sachalinensis（F. Schmidt）Ronse Decr. ■☆

392171 Tiniaria sachalinensis（F. W. Schmidt ex Maxim.）Janch. = Polygonum sachalinense F. Schmidt ex Maxim. ■☆

392172 Tiniaria scandens（L.）Nakai var. dentatoalata（F. Schmidt）Nakai ex T. Mori = Fallopia dentatoalata（F. Schmidt ex Maxim.）Holub ■

392173 Tiniaria scandens（L.）Small = Fallopia scandens（L.）Holub ■☆

392174 Tinnea Kotschy et Peyr. = Tinnea Kotschy ex Hook. f. ●■☆

392175 Tinnea Kotschy ex Hook. f.（1867）；毒鱼草属●■☆

392176 Tinnea Vatke = Cyclocheilon Oliv. ●☆

392177 Tinnea aethiopica Kotschy ex Hook. f.；非洲毒鱼草●☆

392178 Tinnea aethiopica Kotschy ex Hook. f. subsp. litoralis Vollesen = Tinnea aethiopica Kotschy ex Hook. f. ●☆

392179 Tinnea aethiopica Kotschy ex Hook. f. var. dentata Hook. f. = Tinnea aethiopica Kotschy ex Hook. f. ●☆

392180 Tinnea apiculata Robyns et Lebrun；细尖毒鱼草●☆

392181 Tinnea arabica Baker = Asepalum eriantherum（Vatke）Marais ●☆

392182 Tinnea barbata Vollesen；髯毛毒鱼草●☆

392183 Tinnea barteri Gürke；巴特毒鱼草●☆

392184 Tinnea benguellensis Gürke；本格拉毒鱼草●☆

392185 Tinnea bequaertii De Wild. = Tinnea coerulea Gürke ●☆

392186 Tinnea caudata Taylor = Tinnea apiculata Robyns et Lebrun ●☆

392187 Tinnea coerulea Gürke；蓝毒鱼草●☆

392188 Tinnea coerulea Gürke var. linearifolia（Bamps）Vollesen；线叶蓝毒鱼草■☆

392189 Tinnea coerulea Gürke var. obovata（Robyns et Lebrun）Vollesen；卵叶蓝毒鱼草■☆

392190 Tinnea dinteri Gürke ex Dinter = Tinnea rhodesiana S. Moore ■☆

392191 Tinnea erianthera Vatke = Asepalum eriantherum（Vatke）Marais ●☆

392192 Tinnea eriocalyx Welw.；毛萼毒鱼草■☆

392193 Tinnea filipes Baker = Tinnea gracilis Gürke ■☆

392194 Tinnea fischeri Gürke = Tinnea aethiopica Kotschy ex Hook. f. ●☆

392195 Tinnea fusco-luteola Gürke = Tinnea vestita Baker ■☆

392196 Tinnea galpinii Briq.；盖尔毒鱼草■☆

392197 Tinnea gossweileri Robyns et Lebrun；戈斯毒鱼草■☆

392198 Tinnea gracilipedicellata Robyns et Lebrun = Tinnea somalensis Gürke ex Chiov. ■☆

392199 Tinnea gracilis Gürke；纤细毒鱼草■☆

392200 Tinnea heterotypica S. Moore = Renschia heteroptypica（S. Moore）Vatke ●☆

392201 Tinnea juttae Dinter = Tinnea rhodesiana S. Moore ■☆

392202 Tinnea lanuginosa Robyns et Lebrun = Tinnea eriocalyx Welw. ■☆

392203 Tinnea linarifolia Bamps = Tinnea coerulea Gürke var. linearifolia（Bamps）Vollesen ■☆

392204 Tinnea mirabilis（Bullock）Vollesen；奇异毒鱼草●☆

392205 Tinnea obovata Robyns et Lebrun = Tinnea coerulea Gürke var. obovata（Robyns et Lebrun）Vollesen ■☆

392206 Tinnea ochracea R. D. Good = Tinnea eriocalyx Welw. ■☆

392207 Tinnea pearsonii Robyns et Lebrun = Tinnea rhodesiana S. Moore ■☆

392208 Tinnea physalis Bruce；囊状毒鱼草■☆

392209 Tinnea physaloides Baker = Tinnea vesiculosa Gürke ■☆

392210 Tinnea platyphylla Briq.；宽叶毒鱼草■☆

392211 Tinnea rehmannii Schinz = Tinnea rhodesiana S. Moore ■☆

392212 Tinnea rhodesiana S. Moore；罗得西亚毒鱼草■☆

392213 Tinnea rogersii Robyns et Lebrun = Tinnea rhodesiana S. Moore ■☆

392214 Tinnea sacleuxii De Wild. = Tinnea aethiopica Kotschy ex Hook. f. ●☆

392215 Tinnea somalensis Gürke ex Chiov.；索马里毒鱼草■☆

392216 Tinnea stolzii Robyns et Lebrun = Tinnea aethiopica Kotschy ex Hook. f. ●☆

392217 Tinnea vesiculosa Gürke；多疱毒鱼草■☆

392218 Tinnea vestita Baker；包被毒鱼草■☆

392219 Tinnea zambesiaca Baker；赞比西毒鱼草■☆

392220 Tinnia Noronha = Leea D. Royen ex L.（保留属名）●■

392221 Tinomiscium Miers = Tinomiscium Miers ex Hook. f. et Thomson ●

392222 Tinomiscium Miers ex Hook. f. et Thomson（1855）；大叶藤属；Bigleafvine, Tinomiscium ●

392223 Tinomiscium petiolare Hook. f. et Thomson；大叶藤（黄藤,黄藤子,假黄藤,奶汁藤,藤黄连,土防己,土黄连）；Bigleafvine, Tonkin Tinomiscium ●

392224 Tinomiscium philippinense Diels；菲律宾大叶藤；Philippine Tinomiscium ●☆

392225 Tinomiscium phytocrenoides Kurz ex Teijsm. et Binn.；广西大叶藤；Guangxi Bigleafvine, Guangxi Tinomiscium ●

392226 Tinomiscium tonkinense Gagnep. = Tinomiscium petiolare Hook. f. et Thomson ●

392227 Tinopsis Radlk.（1887）；拟马岛无患子属●☆

392228 Tinopsis Radlk. = Tina Schult. ●☆

392229 Tinopsis antongiliensis Capuron；安通吉尔拟马岛无患●☆

392230 Tinopsis apiculata Radlk.；细尖拟马岛无患●☆

392231 Tinopsis chrysophylla Capuron；金叶拟马岛无患●☆

392232 Tinopsis conjugata（Thouars ex Radlk.）Capuron；成对拟马岛无患子●☆

392233 Tinopsis dissitiflora（Baker）Capuron；疏花拟马岛无患子●☆

392234 Tinopsis isoneura（Radlk.）Choux = Neotina isoneura（Radlk.）Capuron ●☆

392235 Tinopsis macrocarpa Capuron；大果拟马岛无患子●☆

392236 Tinopsis phellocarpa Capuron；栓果马岛无患子●☆

392237 Tinopsis tamatavensis Capuron；塔马塔夫马岛无患子●☆

392238 Tinopsis tampolensis Capuron；坦波尔马岛无患子●☆

392239 Tinopsis urschii Capuron；乌尔施马岛无患子●☆

392240 Tinopsis vadonii Capuron；瓦顿马岛无患子●☆

392241 Tinosolen Post et Kuntze = Heterophyllaea Hook. f. ●☆

392242 Tinosolen Post et Kuntze = Teinosolen Hook. f. ●☆

392243 Tinospora Miers（1851）（保留属名）；青牛胆属；Tinospora, Tinospore ●■

392244 Tinospora bakis（A. Rich.）Miers；巴基青牛胆●☆

392245 Tinospora bakis Miers = Tinospora bakis（A. Rich.）Miers ●☆

392246 Tinospora buchholzii Engl. = Platytinospora buchholzii（Engl.）Diels ●☆

392247 Tinospora caffra（Miers）Troupin；开菲尔青牛胆●☆

392248 Tinospora capillipes Gagnep.；纤梗青牛胆（地胆,地苦胆,金古榄,金果榄,金苦榄,金楛榄,金榄,金牛胆,金牛子,金狮藤,金线吊葫芦,金银袋,九连珠,九莲子,九龙胆,九牛胆,九牛子,破石珠,山茨菰,山慈姑,铜秤锤,雪里开,圆角金果榄）；Hairystalk Tinospora, Hairystalk Tinospore, Hairy-stalked Tinospore ●■

392249　Tinospora capillipes Gagnep. = Tinospora sagittata（Oliv.）Gagnep. ■

392250　Tinospora cordifolia Miers;心叶青牛胆（心叶宽筋藤）●■

392251　Tinospora craveniana S. Y. Hu = Tinospora sagittata（Oliv.）Gagnep. var. craveniana（S. Y. Hu）H. C. Lo ■

392252　Tinospora crispa（L.）Hook. f. et Thomson;波叶青牛胆（发冷藤,隔夜找娘,金鸡纳藤,克塞麻奈,癫浆包藤,绿包藤,千里找根,小赖藤,小轻藤,绉波青牛胆,皱波青牛胆）;Chilly Tinospore, Curled Tinospore, Thorel Tinospora, Waveleaf Tinospore, Waveleaf Tinospore ■

392253　Tinospora dentata Diels;台湾青牛胆（齿叶青牛胆,恒春青牛胆）;Taiwan Taiwan Tinospore,Taiwan Tinospora,Toothed Tinospore ●

392254　Tinospora fragosa（I. Verd.）I. Verd. et Troupin;脆青牛胆●☆

392255　Tinospora gibbericaulis Hand.-Mazz. = Tinospora crispa（L.）Hook. f. et Thomson ■

392256　Tinospora guangxiensis H. S. Lo;广西青牛胆（广西青牛藤）;Guangxi Tinospora,Guangxi Tinospore,Kwangsi Tinospore ●

392257　Tinospora hainanensis H. S. Lo et Z. X. Li;海南青牛胆;Hainan Tinospora,Hainan Tinospore ●■

392258　Tinospora hastata Elmer;戟形青牛胆（金黄果）●☆

392259　Tinospora imbricata S. Y. Hu = Tinospora sagittata（Oliv.）Gagnep. ■

392260　Tinospora intermedia S. Y. Hu = Tinospora sagittata（Oliv.）Gagnep. var. craveniana（S. Y. Hu）H. C. Lo ■

392261　Tinospora malabarica（Lam.）Hook. f. et Thomson = Tinospora sinensis（Lour.）Merr. ■

392262　Tinospora malabarica Miers;马拉巴尔青牛胆●☆

392263　Tinospora mastersii Diels = Tinospora crispa（L.）Hook. f. et Thomson ■

392264　Tinospora mossambicensis Engl.;莫桑比克青牛胆●☆

392265　Tinospora oblongifolia（Engl.）Troupin;矩圆叶青牛胆●☆

392266　Tinospora orophila Troupin;喜山青牛胆●☆

392267　Tinospora penninervifolia（Troupin）Troupin;羽脉青牛胆●☆

392268　Tinospora rumphii Boerl. = Tinospora crispa（L.）Hook. f. et Thomson ■

392269　Tinospora sagittata（Oliv.）Gagnep.;青牛胆（地胆,地蛋,地苦胆,叠基青牛胆,覆瓦叶金果榄,黄金古,箭叶青牛胆,金古榄,金果榄,金苦榄,金楷榄,金榄,金牛胆,金牛子,金狮藤,金线吊葫芦,金银袋,九连珠,九莲子,九龙胆,九牛胆,九牛子,破石珠,青鱼胆,山茨菇,山茨菰,天鹅蛋,铜秤锤,雪里开）;Arrowshaped Tinospora,Arrow-shaped Tinospore,Sagittate Tinospore,Tinospore ■

392270　Tinospora sagittata（Oliv.）Gagnep. var. craveniana（S. Y. Hu）H. C. Lo;峨眉青牛胆（江西青牛胆,中型青牛胆）;Emei Tinospora,Emei Tinospore ●

392271　Tinospora sagittata（Oliv.）Gagnep. var. leucocarpa Y. Wan et C. Z. Gao = Tinospora sagittata（Oliv.）Gagnep. ■

392272　Tinospora sagittata（Oliv.）Gagnep. var. leucocarpa Y. Wan ex C. Z. Gao;白果青牛胆■

392273　Tinospora sagittata（Oliv.）Gagnep. var. yunnanensis（S. Y. Hu）H. S. Lo;云南青牛胆（尖叶金果榄,金果榄,苦地胆）;Yunnan Tinospora,Yunnan Tinospore ●

392274　Tinospora sinensis（Lour.）Merr.;中华青牛胆（打不死,大接筋藤,大松身,吼筋藤,砍不死,宽筋藤,牛挣藤,青宽藤,软筋藤,赛筋藤,伸筋藤,舒筋藤,松根藤,松筋藤,无地生根,无地生须）;China Tinospore,Chinese Tinospora ■

392275　Tinospora smilacina Benth.;菝葜青牛胆●☆

392276　Tinospora stuhlmannii Engl. = Tinospora tenera Miers ●☆

392277　Tinospora szechuanensis S. Y. Hu;四川青牛胆●

392278　Tinospora szechuanensis S. Y. Hu = Tinospora crispa（L.）Hook. f. et Thomson ■

392279　Tinospora szechuanensis S. Y. Hu = Tinospora sagittata（Oliv.）Gagnep. ■

392280　Tinospora tenera Miers;柔弱青牛胆●☆

392281　Tinospora thorelii Gagnep.;发冷藤●

392282　Tinospora thorelii Gagnep. = Tinospora crispa（L.）Hook. f. et Thomson ■

392283　Tinospora tomentosa（Colebr.）Hook. f. et Thomson = Tinospora sinensis（Lour.）Merr. ■

392284　Tinospora tuberculata Beumee ex K. Heyne;块茎青牛胆●

392285　Tinospora uviforme（Baill.）Troupin;葡萄青牛胆■☆

392286　Tinospora yunnanensis S. Y. Hu = Tinospora sagittata（Oliv.）Gagnep. var. yunnanensis（S. Y. Hu）H. C. Lo ■

392287　Tinostachyum Post et Kuntze = Schizostachyum Nees ●

392288　Tinostachyum Post et Kuntze = Teinostachyum Munro ●☆

392289　Tintinabulum Rydb. = Gilia Ruiz et Pav. ■●☆

392290　Tintinnabularia Woodson(1936);铃竹桃属●☆

392291　Tintinnabularia mortonii Woodson;铃竹桃●☆

392292　Tinus Burm. = Ardisia Sw.（保留属名）●■

392293　Tinus Kuntze = Ardisia Sw.（保留属名）●■

392294　Tinus L.（1754）= Gillena Adans. ●

392295　Tinus L.（1754）= Premna L.（保留属名）●■

392296　Tinus L.（1754）= Volkameria P. Browne ●

392297　Tinus L.（1759）= Clethra Gronov. ex L. ●

392298　Tinus L.（1759）= Gilibertia J. F. Gmel. ●

392299　Tinus Mill. = Viburnum L. ●

392300　Tinus affinis（Hemsl.）Kuntze = Ardisia sinoaustralis C. Chen ●

392301　Tinus affinis Kuntze = Ardisia affinis Hemsl. ●

392302　Tinus bakeriana Kuntze = Oncostemum laurifolium（Bojer ex A. DC.）Mez ●☆

392303　Tinus bracteata（Baker）Kuntze = Ardisia bracteata Baker ●☆

392304　Tinus caudata（Hemsl.）Kuntze = Ardisia caudata Hemsl. ●

392305　Tinus caudata Kuntze = Ardisia caudata Hemsl. ●

392306　Tinus chinensis（Benth.）Kuntze = Ardisia chinensis Benth. ●

392307　Tinus chinensis Kuntze = Ardisia chinensis Benth. ●

392308　Tinus crispa（Thunb.）Kuntze = Ardisia crispa（Thunb.）A. DC. ●

392309　Tinus crispa Kuntze = Ardisia crispa（Thunb.）A. DC. ●

392310　Tinus depressa（C. B. Clarke）Kuntze = Ardisia thyrsiflora D. Don ●

392311　Tinus depressa Kuntze = Ardisia depressa C. B. Clarke ●

392312　Tinus faberi（Hemsl.）Kuntze = Ardisia faberi Hemsl. ●

392313　Tinus faberi Kuntze = Ardisia faberi Hemsl. ●

392314　Tinus fusco-pilosa Kuntze = Oncostemum leprosum Mez ●☆

392315　Tinus henryi（Hemsl.）Kuntze = Ardisia crispa（Thunb.）A. DC. ●

392316　Tinus henryi Kuntze = Ardisia crispa（Thunb.）A. DC. ●

392317　Tinus humilis（Vahl）Kuntze = Ardisia humilis Vahl ●

392318　Tinus japonica（Thunb.）Kuntze = Ardisia japonica（Thunb.）Blume ●

392319　Tinus japonica Kuntze = Ardisia japonica（Thunb.）Blume ●

392320　Tinus mamillata（Hance）Kuntze = Ardisia mamillata Hance ●

392321　Tinus mamillata Kuntze = Ardisia mamillata Hance ●

392322　Tinus nitidula（Baker）Kuntze = Oncostemum nitidulum（Baker）Mez ●☆

392323　Tinus primulifolia（Gardner et Champ.）Kuntze ＝ Ardisia primulifolia Gardner et Champ. ●

392324　Tinus primulifolia Kuntze ＝ Ardisia primulifolia Gardner et Champ. ●

392325　Tinus punctata（Lindl.）Kuntze ＝ Ardisia lindleyana D. Dietr. ●

392326　Tinus punctata Kuntze ＝ Ardisia lindleyana D. Dietr. ●

392327　Tinus pyrifolia（Willd. ex Roem. et Schult.）Kuntze ＝ Embelia pyrifolia（Willd. ex Roem. et Schult.）Mez ●☆

392328　Tinus sieboldii（Miq.）Kuntze ＝ Ardisia sieboldii Miq. ●

392329　Tinus sieboldii Kuntze ＝ Ardisia sieboldii Miq. ●

392330　Tinus squamulosa（C. Presl）Kuntze ＝ Ardisia elliptica Thunb. ●

392331　Tinus squamulosa Kuntze ＝ Ardisia elliptica Thunb. ●

392332　Tinus squamulosa Kuntze ＝ Ardisia squamulosa C. Presl ●

392333　Tinus thyrsiflora（D. Don）Kuntze ＝ Ardisia thyrsiflora D. Don ●

392334　Tinus triflora（Hemsl.）Kuntze ＝ Ardisia chinensis Benth. ●

392335　Tinus triflora Kuntze ＝ Ardisia chinensis Benth. ●

392336　Tinus virens（Kurz）Kuntze ＝ Ardisia virens Kurz ●

392337　Tinus virens Kuntze ＝ Ardisia virens Kurz ●

392338　Tipalia Dennst. ＝ Zanthoxylum L. ●

392339　Tipha Neck. ＝ Typha L. ■

392340　Tiphogeton Ehrh. ＝ Isnardia L. ●■

392341　Tiphogeton Ehrh. ＝ Ludwigia L. ●■

392342　Tipuana（Benth.）Benth.（1860）;迪普木属（梯普木属）;Pride of Bolivia,Tipu Tree ●☆

392343　Tipuana Benth. ＝ Tipuana（Benth.）Benth. ●☆

392344　Tipuana sericea Ducke ＝ Tipuana tipu（Benth.）Kuntze ●☆

392345　Tipuana tipu（Benth.）Kuntze;迪普木（地板檀,梯普木）;Palo Mortero,Pride of Bolivia,Pride-of-Bolivia,Pride-of-India,Tipa,Tipa Tree,Tipu,Tipu Tree ●☆

392346　Tipularia Nutt.（1818）;筒距兰属（蝇兰属）;Crane-fly Orchis,Tipularia ■

392347　Tipularia cunninghamii（King et Prain）S. C. Chen;软叶筒距兰（迪迪兰,细花软叶兰）;Cunningham Didiciea ■

392348　Tipularia discolor（Pursh）Nutt.;二色筒距兰;Cranefly Orchid,Discolor Tipularia ■☆

392349　Tipularia japonica Matsum.;日本筒距兰（日本花凤梨）;Japanese Tipularia ■☆

392350　Tipularia japonica Matsum. var. harae F. Maek.;日本迪迪兰;Japonese Didiciea ■☆

392351　Tipularia josephii Rchb. f. ex Lindl.;短柄筒距兰（约氏筒距兰）;Joseph Tipularia,Shortstalk Tipularia ■

392352　Tipularia odorata Fukuy.;台湾筒距兰（南湖蝇兰）;Taiwan Tipularia ■

392353　Tipularia szechuanica Schltr.;筒距兰;Sichuan Tipularia,Szechuan Tipularia ■

392354　Tipularia tipuloides（Willemet）Druce ＝ Calanthe triplicata（Willemet）Ames ■

392355　Tiputinia P. E. Berry et C. L. Woodw.（2007）;南美水玉簪属■☆

392356　Tiquilia Pers.（1805）;蒂基花属■☆

392357　Tiquilia Pers. ＝ Coldenia L. ■

392358　Tiquilia canescens（DC.）A. T. Richardson;灰蒂基花;Shrubby Coldenia,Shrubby Tiquilia ■☆

392359　Tiquilia greggii（Torr. et A. Gray）A. T. Richardson;格雷蒂基花;Plume Tiquilia ■☆

392360　Tiquilia hispidissima（Torr. et A. Gray）A. T. Richardson;蒂基花;Hairy Crinklemat ■☆

392361　Tiquilia nuttallii（Benth. ex Hook.）A. T. Richardson;纳托尔蒂基花■☆

392362　Tiquiliopsis A. Heller ＝ Tiquilia Pers. ■☆

392363　Tirania Pierre（1887）;六瓣山柑属●☆

392364　Tirania purpurea Pierre;六瓣山柑●☆

392365　Tirasekia G. Don ＝ Anagallis L. ■

392366　Tirasekia G. Don ＝ Jirasekia F. W. Schmidt ■

392367　Tiricta Raf. ＝ Daucus L. ■

392368　Tirpitzia Haller f.（1921）;青篱柴属;Hedgebavin,Tirpitzia ●

392369　Tirpitzia candida Hand. -Mazz. ＝ Tirpitzia sinensis（Hemsl.）Hallier f. ●

392370　Tirpitzia ovoidea Chun et F. C. How ex W. L. Sha;米念芭;Ovate Hedgebavin,Ovate Tirpitzia ●

392371　Tirpitzia sinensis（Hemsl.）Hallier f.;青篱柴（白豆瓣菜,白豆瓣草,白花柴,柄瓣木,米念巴,青皮柴,石海椒）;Chinese Tirpitzia,Hedgebavin ●

392372　Tirtalia Raf. ＝ Ipomoea L.（保留属名）●■

392373　Tirucalia Raf. ＝ Euphorbia L. ●■

392374　Tirucalia goetzei（Pax）P. V. Heath ＝ Euphorbia goetzei Pax ●☆

392375　Tirucalia gossypina（Pax）P. V. Heath ＝ Euphorbia gossypina Pax ■☆

392376　Tirucalia gossypina Pax var. coccinea（Pax）P. V. Heath ＝ Euphorbia gossypina Pax var. coccinea Pax ■☆

392377　Tirucalia tirucalli（L.）P. V. Heath ＝ Euphorbia tirucalli L. ●

392378　Tirucalla Raf. ＝ Euphorbia L. ●■

392379　Tischleria Schwantes ＝ Carruanthus（Schwantes）Schwantes ■☆

392380　Tischleria Schwantes（1951）;银桥属●☆

392381　Tischleria peersii Schwantes;银桥●☆

392382　Tisonia Baill.（1886）;蒂松木属●☆

392383　Tisonia ficulnea Baill.;蒂松木●☆

392384　Tisonia glabrata Baill.;光蒂松木●☆

392385　Tissa Adans.（废弃属名）＝ Spergularia（Pers.）J. Presl et C. Presl（保留属名）■

392386　Tissa leucantha Greene ＝ Spergularia macrotheca（Hornem. ex Cham. et Schltdl.）Heynh. var. leucantha（Greene）B. L. Rob. ■☆

392387　Tissa marina（L.）Britton ＝ Spergularia marina（L.）Griseb. ■

392388　Tissa marina（L.）Britton ＝ Spergularia salina J. Presl et C. Presl ■☆

392389　Tissa rubra（L.）Britton ＝ Spergularia rubra（L.）J. Presl et C. Presl ■

392390　Tissa rubra（L.）Britton var. perennans（Kindb.）Greene ＝ Spergularia rubra（L.）J. Presl et C. Presl ■

392391　Tissa sparsiflora Greene ＝ Spergularia marina（L.）Griseb. ■☆

392392　Tisserantia Humbert ＝ Sphaeranthus L. ■

392393　Tisserantia africana Humbert ＝ Sphaeranthus kirkii Oliv. et Hiern ■☆

392394　Tisserantiella Mimeur ＝ Thyridachne C. E. Hubb. ■☆

392395　Tisserantiella oubanguiensis Mimeur ＝ Thyridachne tisserantii C. E. Hubb. ■☆

392396　Tisserantiodoxa Aubrév. et Pellegr. ＝ Englerophytum K. Krause ●☆

392397　Tisserantiodoxa oubanguiensis Aubrév. et Pellegr. ＝ Englerophytum oubanguiense（Aubrév. et Pellegr.）Aubrév. et Pellegr. ●☆

392398　Tisserantodendron Sillans ＝ Fernandoa Welw. ex Seem. ●

392399　Tisserantodendron chevalieri Sillans ＝ Fernandoa adolfi-friderici（Gilg et Mildbr.）Heine ●☆

392400　Tisserantodendron walkeri Sillans ＝ Fernandoa ferdinandi（Welw.）K. Schum. ●☆

392401　Tita Scop. = Cassipourea Aubl. ●☆

392402　Titania Endl. = Oberonia Lindl.（保留属名）■

392403　Titanodendron A. V. Bobrov et Melikyan = Araucaria Juss. ●

392404　Titanodendron A. V. Bobrov et Melikyan（2006）；新几内亚杉属●☆

392405　Titanopsis Schwantes（1926）；宝玉草属（天女属）■☆

392406　Titanopsis calcarea（Marloth）Schwantes；宝玉草（天女）■☆

392407　Titanopsis crassipes（Marloth）N. E. Br. = Aloinopsis spathulata（Thunb.）L. Bolus ■☆

392408　Titanopsis fulleri Tischer；天女簪■☆

392409　Titanopsis fulleri Tischer = Titanopsis calcarea（Marloth）Schwantes ■☆

392410　Titanopsis hugo-schlechteri（Tischer）Dinter et Schwantes；天女扇■☆

392411　Titanopsis hugo-schlechteri（Tischer）Dinter et Schwantes var. alboviridis Dinter = Titanopsis hugo-schlechteri（Tischer）Dinter et Schwantes ■☆

392412　Titanopsis hugo-schlechteri（Tischer）Dinter et Schwantes var. alboviridis Dinter；白天女扇■☆

392413　Titanopsis luckhoffii L. Bolus = Aloinopsis luckhoffii（L. Bolus）L. Bolus ■☆

392414　Titanopsis luederitzii Tischer；天女盃■☆

392415　Titanopsis luederitzii Tischer = Titanopsis schwantesii（Dinter ex Schwantes）Schwantes ■☆

392416　Titanopsis primosii L. Bolus；天女影■☆

392417　Titanopsis primosii L. Bolus = Titanopsis schwantesii（Dinter ex Schwantes）Schwantes ■☆

392418　Titanopsis schwantesii（Dinter ex Schwantes）Schwantes；天女冠■☆

392419　Titanopsis schwantesii（Dinter）Schwantes = Titanopsis schwantesii（Dinter ex Schwantes）Schwantes ■☆

392420　Titanopsis schwantesii Schwantes = Titanopsis schwantesii（Dinter ex Schwantes）Schwantes ■☆

392421　Titanopsis setifera L. Bolus = Aloinopsis luckhoffii（L. Bolus）L. Bolus ■☆

392422　Titanopsis spathulata（Thunb.）Schwantes = Aloinopsis spathulata（Thunb.）L. Bolus ■☆

392423　Titanotrichum Soler.（1909）；台闽苣苔属（俄氏草属，台地黄属）；Titanotrichum ■

392424　Titanotrichum oldhamii（Hemsl.）Soler.；台闽苣苔（俄氏草，拉狸甲，拉狸莲，龙鳞草，台地黄，土毛地黄，鱼鳞甲）；Oldham Titanotrichum ■

392425　Titelbachia Klotzsch = Begonia L. ●■

392426　Titelbachia Klotzsch = Tittelbachia Klotzsch ●■

392427　Tithonia Desf. = Tithonia Desf. ex Juss. ●■

392428　Tithonia Desf. ex Juss.（1789）；肿柄菊属；Mexican Sunflower, Sunflowerweed, Tithonia ●■

392429　Tithonia Kuntze = Rivina L. ●

392430　Tithonia Raeusch. = Rudbeckia L. ■

392431　Tithonia argophylla D. C. Eaton = Enceliopsis argophylla（D. C. Eaton）A. Nelson ■☆

392432　Tithonia diversifolia（Hemsl.）A. Gray；肿柄菊（假向日葵，王爷葵，异叶肿柄菊）；Mexican Sun Flower, Mexican Sunflower, Mexican sunflowerweed, Tree Marigold, Yucatan Tithonia ●

392433　Tithonia palmeri Rose = Tithonia thurberi A. Gray ■☆

392434　Tithonia rotundifolia（Mill.）S. F. Blake；墨西哥肿柄菊（墨西哥向日葵，圆叶肿柄菊）；Mexican Sunflower ■☆

392435　Tithonia speciosa Hook. ex Griseb. = Tithonia rotundifolia（Mill.）S. F. Blake ■☆

392436　Tithonia tagetiflora Desf.；万寿花肿柄菊（圆叶肿柄菊）；Mexican Sunflower ●■☆

392437　Tithonia tagetiflora Lam. = Tithonia rotundifolia（Mill.）S. F. Blake ■☆

392438　Tithonia thurberi A. Gray；瑟伯肿柄菊■☆

392439　Tithymalaceae Vent. = Euphorbiaceae Juss.（保留科名）●■

392440　Tithymalis Raf. = Euphorbia L. ●■

392441　Tithymalis Raf. = Tithymalus Gaertn.（保留属名）●☆

392442　Tithymalis Raf. = Tithymalus Segnier = ●☆

392443　Tithymalodes Kuntze = Pedilanthus Neck. ex Poit.（保留属名）●

392444　Tithymalodes Kuntze = Tithymaloides Ortega ●

392445　Tithymalodes Ludw. ex Kuntze = Pedilanthus Neck. ex Poit.（保留属名）●

392446　Tithymaloides Ortega（废弃属名）= Pedilanthus Neck. ex Poit.（保留属名）●

392447　Tithymalopsis Klotzsch et Garcke = Agaloma Raf. ●■

392448　Tithymalopsis Klotzsch et Garcke = Euphorbia L. ●■

392449　Tithymalopsis corollata（L.）Klotzsch et Garcke = Euphorbia corollata L. ■☆

392450　Tithymalopsis marilandica（Greene）Small = Euphorbia corollata L. ■☆

392451　Tithymalopsis olivacea Small = Euphorbia corollata L. ■☆

392452　Tithymalus Gaertn.（1790）（保留属名）；原大戟属●☆

392453　Tithymalus Gaertn.（保留属名）= Pedilanthus Neck. ex Poit.（保留属名）●

392454　Tithymalus Hill = Euphorbia L. ●■

392455　Tithymalus Mill.（废弃属名）= Pedilanthus Neck. ex Poit.（保留属名）●

392456　Tithymalus Mill.（废弃属名）= Tithymalus Gaertn.（保留属名）●☆

392457　Tithymalus Scop. = Euphorbia L. ●■

392458　Tithymalus Ség. = Euphorbia L. ●■

392459　Tithymalus arkansanus（Engelm. et A. Gray）Klotzsch et Garcke = Euphorbia spathulata Lam. ■☆

392460　Tithymalus braunii Schweinf. = Euphorbia longetuberculosa Hochst. ex Boiss. ■☆

392461　Tithymalus commutatus（Engelm.）Klotzsch et Garcke = Euphorbia commutata Engelm. ■☆

392462　Tithymalus crispus Haw. = Euphorbia crispa（Haw.）Sweet ■☆

392463　Tithymalus cyparissias（L.）Hill = Euphorbia cyparissias L. ●■☆

392464　Tithymalus ecklonii Klotzsch et Garcke = Euphorbia ecklonii（Klotzsch et Garcke）A. Hässl. ■●☆

392465　Tithymalus esula（L.）Hill = Euphorbia esula L. ■

392466　Tithymalus esula（L.）Hill var. hondoensis ? = Euphorbia octoradiata H. Lév. et Vaniot ■☆

392467　Tithymalus helioscopius（L.）Hill = Euphorbia helioscopia L. ■

392468　Tithymalus himalayensis Klotzsch ex Klostzsch et Garcke = Euphorbia stracheyi Boiss. ■

392469　Tithymalus involucratus Klotzsch et Garcke = Euphorbia epicyparissias E. Mey. ex Boiss. ■☆

392470　Tithymalus longifolius（D. Don）Hurus. et Yas. Tanaka = Euphorbia donii Oudejans ■

392471　Tithymalus missouriensis（Norton）Small = Euphorbia spathulata Lam. ■☆

392472　Tithymalus nakaianus（H. Lév.）Hara = Euphorbia octoradiata H. Lév. et Vaniot ■☆

392473　Tithymalus obtusatus（Pursh）Klotzsch et Garcke ＝ Euphorbia obtusata Pursh ■☆

392474　Tithymalus ovatus E. Mey. ex Klotzsch et Garcke ＝ Euphorbia ovata（E. Mey. ex Klotzsch et Garcke）Boiss. ■☆

392475　Tithymalus pekinensis（Rupr.）Hara ＝ Euphorbia pekinensis Rupr. ■

392476　Tithymalus pekinensis（Rupr.）Hara subsp. barbellatus（Hurus.）Hurus. ＝ Euphorbia pekinensis Rupr. ■

392477　Tithymalus pekinensis（Rupr.）Hara subsp. lanceolatus（Liou）Hurus. ＝ Euphorbia pekinensis Rupr. ■

392478　Tithymalus peplus（L.）Gaertn.；原大戟●☆

392479　Tithymalus peplus（L.）Hill ＝ Euphorbia peplus L. ■

392480　Tithymalus sikkimensis（Boiss.）Hurus. et Yas. Tanaka ＝ Euphorbia sikkimensis Boiss. ■

392481　Tithymalus silenifolius Haw. ＝ Euphorbia silenifolia（Haw.）Sweet ■☆

392482　Tithymalus spathulatus（Lam.）W. A. Weber ＝ Euphorbia spathulata Lam. ■☆

392483　Tithymalus tchen-ngoi Soják ＝ Euphorbia pekinensis Rupr. ■

392484　Titragyne Salisb. ＝ Rohdea Roth ■

392485　Tittelbachia Klotzsch ＝ Begonia L. ●■

392486　Tittmannia Brongn.（1826）（保留属名）；蒂特曼木属●☆

392487　Tittmannia Rchb.（废弃属名）＝ Lindernia All. ■

392488　Tittmannia Rchb.（废弃属名）＝ Tittmannia Brongn.（保留属名）●☆

392489　Tittmannia esterhuyseniae Powrie；埃斯特特曼木●☆

392490　Tittmannia hispida Pillans；硬毛蒂特曼木●☆

392491　Tittmannia laevis Pillans；平滑蒂特曼木●☆

392492　Tittmannia lateriflora Brongn. ＝ Tittmannia laxa（Thunb.）C. Presl ●☆

392493　Tittmannia laxa（Thunb.）C. Presl；疏松蒂特曼木●☆

392494　Tittmannia laxa（Thunb.）C. Presl var. langebergensis Pillans；朗厄山蒂特曼木●☆

392495　Tittmannia obovata Bunge ＝ Mazus pumilus（Burm. f.）Steenis ■

392496　Tittmannia oliveri Dümmer ＝ Tittmannia laxa（Thunb.）C. Presl ●☆

392497　Tittmannia pruinosa Dümmer ＝ Tittmannia laxa（Thunb.）C. Presl ●☆

392498　Tittmannia stachydifolia Turcz. ＝ Mazus stachydifolius（Turcz.）Maxim. ■

392499　Tityrus Salisb. ＝ Narcissus L. ■

392500　Tium Medik. ＝ Astragalus L. ●■

392501　Tjongina Adans. ＝ Baeckea L. ●

392502　Tjutsjau Rumph. ＝ Salvia L. ●■

392503　Toanabo DC. ＝ Taonabo Aubl.（废弃属名）●

392504　Toanabo DC. ＝ Ternstroemia L. f. ●

392505　Toanabo DC. ＝ Ternstroemia Mutis ex L. f.（保留属名）●

392506　Tobagoa Urb.（1916）；托巴茜属☆

392507　Tobagoa maleolens Urb.；托巴茜☆

392508　Tobaphes Phil. ＝ Jobaphes Phil. ●☆

392509　Tobaphes Phil. ＝ Plazia Ruiz et Pav. ●☆

392510　Tobinia Desv. ＝ Fagara L.（保留属名）●

392511　Tobinia Desv. ex Ham. ＝ Fagara L.（保留属名）●

392512　Tobion Raf. ＝ Pimpinella L. ■

392513　Tobira Adans.（废弃属名）＝ Pittosporum Banks ex Gaertn.（保留属名）●

392514　Tobium Raf. ＝ Poterium L. ■☆

392515　Tocantinia Ravenna（2000）；巴西石蒜属■☆

392516　Tococa Aubl.（1775）；托考野牡丹属●☆

392517　Tococa acuminata Benth.；渐尖托考野牡丹●☆

392518　Tococa bolivarensis Gleason；玻利维亚托考野牡丹●☆

392519　Tococa discolor Pilg.；异色托考野牡丹●☆

392520　Tococa erioneura（Cogn.）Wurdack；毛脉托考野牡丹●☆

392521　Tococa erythrophylla（Ule）Wurdack；红叶托考野牡丹●☆

392522　Tococa ferruginea G. Nicholson；锈色托考野牡丹●☆

392523　Tococa filiformis（Gleason）Wurdack；线形托考野牡丹●☆

392524　Tococa grandifolia Standl.；大叶托考野牡丹●☆

392525　Tococa guianensis Aubl.；托考野牡丹●☆

392526　Tococa heterophylla D. Don；异叶托考野牡丹●☆

392527　Tococa lancifolia Spruce ex Triana；剑叶托考野牡丹●☆

392528　Tococa longisepala Cogn.；长萼托考野牡丹●☆

392529　Tococa macroptera Naudin；大翅托考野牡丹●☆

392530　Tococa macrosperma Mart.；大籽托考野牡丹●☆

392531　Tococa montana Gleason；山地托考野牡丹●☆

392532　Tococa obovata Gleason；倒卵托考野牡丹●☆

392533　Tococa oligantha Gleason；寡花托考野牡丹●☆

392534　Tococa pachystachya Wurdack；粗穗托考野牡丹●☆

392535　Tococa parviflora Spruce ex Triana；小花托考野牡丹●☆

392536　Tococa platphylla Benth.；宽叶托考野牡丹●☆

392537　Tococa pubescens Benth. ex Triana；毛托考野牡丹●☆

392538　Tococa sanguinea D. Don；血红托考野牡丹●☆

392539　Tocoyena Aubl.（1775）；托克茜属●☆

392540　Tocoyena acutiflora Mart.；渐尖托克茜●☆

392541　Tocoyena amazonica Standl.；亚马逊托克茜●☆

392542　Tocoyena brasiliensis Mart.；巴西托克茜●☆

392543　Tocoyena brevifolia Steyerm.；短叶托克茜●☆

392544　Tocoyena formosa K. Schum.；美丽托克茜●☆

392545　Tocoyena guianensis K. Schum.；圭亚那托克茜●☆

392546　Tocoyena latifolia Lam.；宽叶托克茜●☆

392547　Tocoyena longiflora K. Schum.；长花托克茜●☆

392548　Tocoyena macrophylla Kunth；大叶托克茜●☆

392549　Tocoyena microdon Mart.；小齿托克茜●☆

392550　Tocoyena mollis K. Krause；软托克茜●☆

392551　Todaroa A. Rich. et Galeotti ＝ Campylocentrum Benth. ■☆

392552　Todaroa Parl.（1843）；托达罗草属■☆

392553　Todaroa aurea Parl.；托达罗草■☆

392554　Todaroa aurea Parl. subsp. suaveolens Pérez；芳香托达罗草■☆

392555　Todaroa montana Webb ex Christ ＝ Athamanta montana（Christ）Spalik et Wojew. et Downie ■☆

392556　Toddalia Juss.（1789）（保留属名）；飞龙掌血属（黄肉树属）；Toddalia ●

392557　Toddalia aculeata Pers. ＝ Toddalia asiatica（L.）Lam. ●

392558　Toddalia angustifolia Lam. ＝ Toddalia asiatica（L.）Lam. ●

392559　Toddalia asiatica（L.）Lam.；飞龙掌血（八大王，抽皮筋，刺米通，大救驾，飞龙斩血，画眉跳，黄大金，黄椒，黄肉树，黄树根藤，鸡爪筋，见血飞，见血散，筋钩，猫爪筋，牛丹子，牛麻筋，爬山虎，入山虎，三百棒，三文藤，散血丹，散血飞，山桔，烧酒钩，通城虎，溪椒，细叶黄肉刺，下山虎，小金藤，血见愁，血见飞，血莲肠，油婆筋）；Asia Toddalia, Asian Toddalia, Asiatic Toddalia, Lopez Root ●

392560　Toddalia asiatica（L.）Lam. var. parva Z. M. Tan；小飞龙掌血；Small Asian Toddalia ●

392561　Toddalia crenulata Engl. ＝ Teclea crenulata（Engl.）Engl. ●☆

392562　Toddalia effusa Turcz. ＝ Toddalia asiatica（L.）Lam. ●

392563 Toddalia elliotii Radlk. = Vepris elliotii（Radlk.）I. Verd. ●☆

392564 Toddalia fischeri Engl. = Teclea fischeri（Engl.）Engl. ●☆

392565 Toddalia floribunda Wall. = Toddalia asiatica（L.）Lam. ●

392566 Toddalia glomerata F. Hoffm. = Vepris glomerata（F. Hoffm.）Engl. ●☆

392567 Toddalia lanceolata Lam. = Vepris lanceolata（Lam.）G. Don ●☆

392568 Toddalia macrophylla Baker = Vepris macrophylla（Baker）I. Verd. ●☆

392569 Toddalia macrophylla Baker var. macrocarpa Danguy = Vepris macrophylla（Baker）I. Verd. ●☆

392570 Toddalia natalensis Sond. = Teclea natalensis（Sond.）Engl. ●☆

392571 Toddalia nitida Lam. = Toddalia asiatica（L.）Lam. ●

392572 Toddalia pilosa Baker = Vepris pilosa（Baker）I. Verd. ●☆

392573 Toddalia pilosa Engl. = Teclea pilosa（Engl.）I. Verd. ●☆

392574 Toddalia polymorpha Danguy ex Lecomte = Vepris polymorpha（Danguy ex Lecomte）H. Perrier ●☆

392575 Toddalia rubricaulis Roem. et Schult. = Toddalia asiatica（L.）Lam. ●

392576 Toddalia sansibarensis Engl. = Vepris sansibarensis（Engl.）Mziray ●☆

392577 Toddalia schmidelioides Baker = Vepris schmidelioides（Baker）I. Verd. ●☆

392578 Toddalia simplicifolia Engl. = Vepris simplicifolia（Engl.）Mziray ●☆

392579 Toddalia simplicifolia Engl. var. eugeniifolia？ = Vepris eugeniifolia（Engl.）I. Verd. ●☆

392580 Toddalia tonkinensis Guillaumin = Toddalia asiatica（L.）Lam. ●

392581 Toddalia trichocarpa Engl. = Vepris trichocarpa（Engl.）Mziray ●☆

392582 Toddaliaceae Baum. -Bod. 飞龙掌血科●

392583 Toddaliaceae Baum. -Bod. = Rutaceae Juss.（保留科名）●■

392584 Toddaliopsis Engl.（1895）;拟飞龙掌血属●☆

392585 Toddaliopsis bremekampii I. Verd. ;拟飞龙掌血●☆

392586 Toddaliopsis ebolowensis Engl. ;埃博洛瓦拟飞龙掌血●☆

392587 Toddaliopsis heterophylla（Engl.）Engl. = Vepris heterophylla（Engl.）Letouzey ●☆

392588 Toddaliopsis sansibarensis（Engl.）Engl. = Vepris sansibarensis（Engl.）Mziray ●☆

392589 Todda-pana Adans. = Cycas L. ●

392590 Toddavaddia Kuntze = Biophytum DC. ■●

392591 Toechima Radlk.（1879）;特喜无患子属●☆

392592 Toechima hirsutum Radlk. ;特喜无患子●☆

392593 Toechima lanceolatum C. T. White;披针叶特喜无患子●☆

392594 Toechima monticola S. T. Reynolds;山地特喜无患子●☆

392595 Toelkenia P. V. Heath = Crassula L. ●■☆

392596 Toffieldoa Schrank = Tofieldia Huds. ■

392597 Tofielda Pers. = Tofieldia Huds. ■

392598 Tofieldia Huds.（1778）;岩菖蒲属;False Asphodel, Scottish Asphodel, Tofieldia ■

392599 Tofieldia Huds. = Narthecium Gerard（废弃属名）■

392600 Tofieldia brevistyla Franch. = Tofieldia divergens Bureau et Franch. ■

392601 Tofieldia calyculata Wahlenb. ;副萼岩菖蒲;German Asphodel ■☆

392602 Tofieldia cernua Sm. ;俯垂岩菖蒲■☆

392603 Tofieldia coccinea Rich. ;长白岩菖蒲（长岩菖蒲）;Northern False Asphodel, Scarlet Tofieldia ■

392604 Tofieldia divergens Bureau et Franch. ;叉柱岩菖蒲（扁竹参, 扁竹兰, 草灵芝, 复生草, 九节莲, 云南岩菖蒲）;Forkstyle

Tofieldia ■

392605 Tofieldia esquirolii H. Lév. = Tofieldia divergens Bureau et Franch. ■

392606 Tofieldia fauriei H. Lév. et Vaniot = Tofieldia coccinea Rich. ■

392607 Tofieldia glabra Nutt. ;白岩菖蒲;White False Asphodel, White Featherling ■☆

392608 Tofieldia glutinosa（Michx.）Pers. ;北美岩菖蒲;False Asphodel, Gluten Tofieldia, Glutinous False Asphodel, Sticky False Asphodel ■☆

392609 Tofieldia glutinosa（Michx.）Pers. = Triantha glutinosa（Michx.）Baker ■☆

392610 Tofieldia glutinosa（Michx.）Pers. subsp. absona C. L. Hitchc. = Triantha occidentalis（S. Watson）R. R. Gates subsp. brevistyla（C. L. Hitchc.）Packer ■☆

392611 Tofieldia glutinosa（Michx.）Pers. subsp. brevistyla C. L. Hitchc. = Triantha occidentalis（S. Watson）R. R. Gates subsp. brevistyla（C. L. Hitchc.）Packer ■☆

392612 Tofieldia glutinosa（Michx.）Pers. subsp. montana C. L. Hitchc. ;山地西方岩菖蒲■☆

392613 Tofieldia glutinosa（Michx.）Pers. subsp. montana C. L. Hitchc. = Triantha occidentalis（S. Watson）R. R. Gates subsp. montana（C. L. Hitchc.）Packer ■☆

392614 Tofieldia glutinosa（Michx.）Pers. var. absona（C. L. Hitchc.）R. J. Davis = Triantha occidentalis（S. Watson）R. R. Gates subsp. brevistyla（C. L. Hitchc.）Packer ■☆

392615 Tofieldia glutinosa（Michx.）Pers. var. brevistyla（C. L. Hitchc.）C. L. Hitchc. = Triantha occidentalis（S. Watson）R. R. Gates subsp. brevistyla（C. L. Hitchc.）Packer ■☆

392616 Tofieldia glutinosa（Michx.）Pers. var. brevistyla（C. L. Hitchc.）C. L. Hitchc. ;短柱西方岩菖蒲■☆

392617 Tofieldia glutinosa（Michx.）Pers. var. montana（C. L. Hitchc.）R. J. Davis = Tofieldia glutinosa（Michx.）Pers. subsp. montana C. L. Hitchc. ■☆

392618 Tofieldia glutinosa（Michx.）Pers. var. montana（C. L. Hitchc.）R. J. Davis = Triantha occidentalis（S. Watson）R. R. Gates subsp. montana（C. L. Hitchc.）Packer ■☆

392619 Tofieldia glutinosa（Michx.）Pers. var. occidentalis（S. Watson）C. L. Hitchc. = Tofieldia occidentalis S. Watson ■☆

392620 Tofieldia glutinosa Pers. = Triantha glutinosa（Michx.）Baker ■☆

392621 Tofieldia iridacea Franch. = Tofieldia thibetica Franch. ■

392622 Tofieldia japonica Miq. ;日本岩菖蒲;Japanese Tofieldia ■☆

392623 Tofieldia labordei H. Lév. et Vaniot = Tofieldia divergens Bureau et Franch. ■

392624 Tofieldia macilenta Franch. = Tofieldia thibetica Franch. ■

392625 Tofieldia nepalensis Wall. = Aletris pauciflora（G. Klotz）Franch. ■

392626 Tofieldia nutans Willd. = Tofieldia coccinea Rich. ■

392627 Tofieldia nutaus Willd. ex Schult. = Tofieldia coccinea Rich. ■

392628 Tofieldia occidentalis S. Watson;西方岩菖蒲■☆

392629 Tofieldia palustris Huds. ;沼泽岩菖蒲;Marshy Tofieldia, Scotch Asphodel, Scottish Asphodel ■☆

392630 Tofieldia pusilla（Michx.）Pers. ;苏格兰岩菖蒲;False Asphodel, Scotch False Asphodel, Scottis Asphodel, Scottish Asphodel ■☆

392631 Tofieldia pusilla Pers. = Tofieldia pusilla（Michx.）Pers. ■☆

392632 Tofieldia racemosa（Walter）Britton, Sterns et Poggenb. ;沿海岩菖蒲;Coastal False Asphodel ■☆

392633 Tofieldia racemosa（Walter）Britton, Sterns et Poggenb. =

Triantha racemosa（Walter）Small ■☆

392634　Tofieldia racemosa（Walter）Britton, Sterns et Poggenb. var. glutinosa（Michx.）H. E. Ahles = Triantha glutinosa（Michx.）Baker ■☆

392635　Tofieldia racemosa（Walter）Britton, Sterns et Poggenb. var. glutinosa（Michx.）H. E. Ahles = Tofieldia glutinosa Pers. ■☆

392636　Tofieldia setchuenensis Franch. = Tofieldia thibetica Franch. ■

392637　Tofieldia taquetii H. Lév. et Vaniot = Tofieldia coccinea Rich. ■

392638　Tofieldia tenella Hand.-Mazz. = Tofieldia divergens Bureau et Franch. ■

392639　Tofieldia tenuifolia（Michx.）Utech = Pleea tenuifolia Michx. ■☆

392640　Tofieldia thibetica Franch. ;岩菖蒲（岩飘子）;Tibet Tofieldia, Xizang Tofieldia ■

392641　Tofieldia yunnanensis Franch. = Tofieldia divergens Bureau et Franch. ■

392642　Tofieldiaceae Takht.（1995）;岩菖蒲科■

392643　Tofieldiaceae Takht. = Liliaceae Juss.（保留科名）■●

392644　Toisusu Kimura = Salix L.（保留属名）●

392645　Toisusu cardiophylla（Trautv. et C. A. Mey.）Kimura = Salix cardiophylla Trautv. et C. A. Mey. ●☆

392646　Toisusu cardiophylla（Trautv. et C. A. Mey.）Kimura var. maximowiczii（Kom.）Kimura. = Salix maximowiczii Kom. ●

392647　Toisusu cardiophylla（Trautv. et C. A. Mey.）Kimura var. maximowiczii Kimura = Salix maximowiczii Kom. ●

392648　Toisusu cardiophylla（Trautv. et C. A. Mey.）Kimura var. schneideri（Miyabe et Kudo）Kimura = Salix cardiophylla Trautv. et C. A. Mey. ●☆

392649　Toisusu cardiophylla（Trautv. et C. A. Mey.）Kimura var. urbaniana（Seemen）Kimura = Salix cardiophylla Trautv. et C. A. Mey. var. urbaniana（Seemen）Kudo ●☆

392650　Toisusu urbaniana（Seemen）Kimura = Salix cardiophylla Trautv. et C. A. Mey. var. urbaniana（Seemen）Kudo ●☆

392651　Toisusu urbaniana（Seemen）Kimura var. schneideri（Miyabe et Kudo）Kimura = Salix cardiophylla Trautv. et C. A. Mey. ●☆

392652　Toiyabea R. P. Roberts, Urbatsch et Neubig（2005）;山蛇菊属; Alpine Serpentweed ■☆

392653　Toiyabea alpina（L. C. Anderson et Goodrich）R. P. Roberts, Urbatsch et Neubig;山蛇菊;Alpine Serpentweed ■☆

392654　Tokoyena S. Rich. ex Steud. = Tocoyena Aubl. ●☆

392655　Tola Wedd. ex Benth. et Hook. f. = Lepidophyllum Cass. ●☆

392656　Tolbonia Kuntze = Calotis R. Br. ●

392657　Tolbonia Kuntze（1891）;陶尔菊属■☆

392658　Tolbonia anamitica Kuntze;陶尔菊■☆

392659　Toliara Judz.（2009）;托里禾属■☆

392660　Tollatia Endl. = Layia Hook. et Arn. ex DC.（保留属名）■☆

392661　Tollatia Endl. = Oxyura DC. ■☆

392662　Tolmachevia Á. Löve et D. Löve = Rhodiola L. ■

392663　Tolmiaea H. Buck = Tolmiea Hook.（废弃属名）■☆

392664　Tolmiea Hook.（废弃属名）= Cladothamnus Bong. ●☆

392665　Tolmiea Hook.（废弃属名）= Elliottia Muhl. ex Elliott ●☆

392666　Tolmiea Hook.（废弃属名）= Tolmiea Torr. et A. Gray（保留属名）■☆

392667　Tolmiea Torr. et A. Gray（1840）（保留属名）;千母草属（负儿草属）;Pig-a-back Plant, Piggyback Plant, Tolmiea ■☆

392668　Tolmiea memziesii（Pursh）Torr. et Gray;千母草（负儿草, 驮子草）;Memzies Tolmiea, Mother-of-thousands, Pickaback Plant, Pick-a-back Plant, Pig-a-back Plant, Piggyback Plant, Piggy-back

Plant, Thousand Mothers, Youth On Age, Youth-on-age ■☆

392669　Tolpis Adans.（1763）;糙缨苣属●■☆

392670　Tolpis abyssinica A. Rich. = Tolpis virgata（Desf.）Bertol. ■☆

392671　Tolpis altissima（Balb.）Pers. = Tolpis virgata（Desf.）Bertol. ■☆

392672　Tolpis barbata（L.）Gaertn. ;毛糙缨苣;European Umbrella Milkwort, Yellow Garden-hawkweed, Yellow Hawkweed ■☆

392673　Tolpis barbata（L.）Gaertn. subsp. liouvillei（Braun-Blanq. et Maire）H. Lindb. = Tolpis liouvillei Braun-Blanq. et Maire ■☆

392674　Tolpis barbata（L.）Gaertn. subsp. umbellata（Bertol.）Maire = Tolpis umbellata Bertol. ■☆

392675　Tolpis barbata（L.）Gaertn. var. grandiflora Ball = Tolpis nemoralis Font Quer ■☆

392676　Tolpis barbata（L.）Gaertn. var. macrantha Maire = Tolpis barbata（L.）Gaertn. ■☆

392677　Tolpis capensis（L.）Sch. Bip. ;好望角糙缨苣■☆

392678　Tolpis coronopifolia（Desf.）Biv. ;鸟足叶糙缨苣●☆

392679　Tolpis crassiuscula Svent. ;糙缨苣●☆

392680　Tolpis ephemera（Hiern）R. E. Fr. = Tolpis capensis（L.）Sch. Bip. ■☆

392681　Tolpis farinulosa（Webb）Walp. ;多粉糙缨苣■☆

392682　Tolpis glabrescens Kämmer;渐光糙缨苣●☆

392683　Tolpis glandulifera Bolle = Tolpis farinulosa（Webb）Walp. ■☆

392684　Tolpis laciniata Webb et Berthel. ;撕裂糙缨苣■☆

392685　Tolpis lagopoda C. Sm. ;兔足糙缨苣■☆

392686　Tolpis liouvillei Braun-Blanq. et Maire;利乌维尔糙缨苣■☆

392687　Tolpis macrorhiza（Lowe）Lowe;大根糙缨苣■☆

392688　Tolpis madagascariensis（DC. ex Froel.）Sch. Bip. = Tolpis capensis（L.）Sch. Bip. ■☆

392689　Tolpis mbalensis G. V. Pope;姆巴莱糙缨苣■☆

392690　Tolpis microcephala Pomel = Tolpis umbellata Bertol. ■☆

392691　Tolpis nemoralis Font Quer;森林糙缨苣■☆

392692　Tolpis somalensis R. E. Fr. = Prenanthes somaliensis C. Jeffrey ■☆

392693　Tolpis succulenta（Dryand.）Lowe;多汁糙缨苣■☆

392694　Tolpis umbellata Bertol. ;伞花糙缨苣■☆

392695　Tolpis umbellata Bertol. = Tolpis barbata（L.）Gaertn. ■☆

392696　Tolpis umbellata Bertol. var. minor Lange = Tolpis umbellata Bertol. ■☆

392697　Tolpis virgata（Desf.）Bertol. ;非洲糙缨苣■☆

392698　Tolpis webbii Webb et Berthel. ;韦布糙缨苣■☆

392699　Toludendron Ehrh. = Toluifera L.（废弃属名）●

392700　Toluifera L.（废弃属名）= Myroxylon L. f.（保留属名）●

392701　Toluifera Lour. = Glycosmis Corrêa（保留属名）●

392702　Toluifera Lour. = Loureira Meisn. ●

392703　Toluifera cochinchinensis Lour. = Glycosmis cochinchinensis（Lour.）Pierre ex Engl. ●

392704　Tolumnia Raf.（1837）;托卢兰属■☆

392705　Tolumnia Raf. = Oncidium Sw.（保留属名）■☆

392706　Tolumnia bahamensis（Nash ex Britton et Millsp.）Braem = Oncidium variegatum（Sw.）Sw. ■☆

392707　Tolumnia pulchella（Hook.）Raf. ;托卢兰■☆

392708　Tolumnia pulchella Raf. = Tolumnia pulchella（Hook.）Raf. ■☆

392709　Tolypanthus（Blume）Blume = Tolypanthus（Blume）Rchb. ●

392710　Tolypanthus（Blume）Rchb.（1841）;大苞寄生属;Tolypanthus ●

392711　Tolypanthus（Blume）Tiegh. = Tolypanthus（Blume）Rchb. ●

392712　Tolypanthus Blume = Tolypanthus（Blume）Rchb. ●

392713　Tolypanthus esquirolii（H. Lév.）Lauener;黔桂大苞寄生（榔榆寄生）;Esquiral Tolypanthus, Qiangui Maclure Tolypanthus, Qian-

Gui Tolypanthus ●

392714 Tolypanthus involucratus（Roxb.）Tiegh.；印度大苞寄生☆

392715 Tolypanthus maclurei（Merr.）Danser；大苞寄生（大苞桑寄生）；Big Bract Loranthus，Maclure Tolypanthus，Tolypanthus ●

392716 Tolypanthus maclurei Danser = Tolypanthus esquirolii（H. Lév.）Lauener ●

392717 Tolypeuma E. Mey. = Nesaea Comm. ex Kunth（保留属名）■●☆

392718 Tomabenea Parl.；托马草属 ☆ 392719 Tomantea Steud. = Tomanthea DC. ■☆

392720 Tomanthea DC.（1838）；汤姆菊属■☆

392721 Tomanthea DC. = Centaurea L.（保留属名）●■

392722 Tomanthea aucheri DC.；奥氏汤姆菊■☆

392723 Tomanthea daralaghezica（Fomin）Takht.；汤姆菊■☆

392724 Tomanthea phaeopappa（DC.）Takht. et Czerep.；褐冠毛汤姆菊■☆

392725 Tomanthea spectabilis（Fisch. et C. A. Mey.）Takht.；壮观汤姆菊■☆

392726 Tomanthera Raf.（废弃名属）= Agalinis Raf.（保留属名）■☆

392727 Tomanthera auriculata（Michx.）Raf. = Agalinis auriculata（Michx.）S. F. Blake ■☆

392728 Tomaris Raf. = Rumex L. ■●

392729 Tombea Brongn. et Gris = Chiratia Montrouz. ●

392730 Tombea Brongn. et Gris = Sonneratia L. f.（保留属名）●

392731 Tomentaurum G. L. Nesom（1991）；银毛菀属■☆

392732 Tomentaurum niveum（S. Watson）G. L. Nesom；雪白银毛菀■☆

392733 Tomentaurum vandevendcrorum G. L. Nesom；银毛菀■☆

392734 Tomex Forssk. = Dobera Juss. ●☆

392735 Tomex L. = Callicarpa L. ●

392736 Tomex Thunb. = Litsea Lam.（保留属名）●

392737 Tomex glabra Forssk. = Dobera glabra（Forssk.）Poir. ●☆

392738 Tomiephyllum Fourr. = Scrophularia L. ■●

392739 Tomilix Raf. = Macranthera Nutt. ex Benth. ■☆

392740 Tomiophyllum Fourr. = Scrophularia L. ■●

392741 Tomista Raf. = Florestina Cass. ■☆

392742 Tommasinia Bertol.（1838）；托氏草属■☆

392743 Tommasinia verticillaris（L.）Bertol.；托氏草☆

392744 Tomodon Raf. = Hymenocallis Salisb. ●

392745 Tomostima Raf. = Draba L. ■

392746 Tomostina Willis = Draba L. ■

392747 Tomostina Willis = Tomostima Raf. ■

392748 Tomostoma Merr. = Draba L. ■

392749 Tomostoma Merr. = Tomostima Raf. ■

392750 Tomostylis Montrouz. = Crossostyles Benth. et Hook. f. ●☆

392751 Tomostylis Montrouz. = Crossostylis J. R. Forst. et G. Forst. ●☆

392752 Tomotris Raf. = Corymborkis Thouars + Corymborkis Thouars ■

392753 Tomotris Raf. = Corymborkis Thouars ■

392754 Tomotris polystachya（Sw.）Raf. = Tropidia polystachya（Sw.）Ames ■☆

392755 Tomoxis Raf. = Ornithogalum L. ■

392756 Tomzanonia Nir = Dilomilis Raf. ■☆

392757 Tonabea Juss. = Taonabo Aubl.（废弃名属）●

392758 Tonabea Juss. = Ternstroemia Mutis ex L. f.（保留属名）●

392759 Tonalanthus Brandegee = Calea L. ●■☆

392760 Tonca Rich. = Bertholletia Bonpl. ●☆

392761 Tondin Vitman = Paullinia L. ●☆

392762 Tonduzia Boeck. ex Tonduz = Durandia Boeck. ■

392763 Tonduzia Boeck. ex Tonduz = Scleria P. J. Bergius ■

392764 Tonduzia Pittier = Alstonia R. Br.（保留属名）●

392765 Tonduzia Pittier（1908）；通杜木属●☆

392766 Tonduzia parvifolia Pittier；通杜木●☆

392767 Tonella Nutt. ex A. Gray（1868）；托尼婆婆纳属●☆

392768 Tonella collinsioides Nutt. ex A. Gray；托尼婆婆纳●☆

392769 Tonella floribunda A. Gray；多花托尼婆婆纳●☆

392770 Tonestus A. Nelson = Haplopappus Cass.（保留属名）■●☆

392771 Tonestus A. Nelson（1904）；蛇菊属；Serpentweed ■☆

392772 Tonestus aberrans（A. Nelson）G. L. Nesom et D. R. Morgan = Trinieurybia aberrans（A. Nelson）Brouillet，Urbatsch et R. P. Roberts ■☆

392773 Tonestus alpinus（L. C. Anderson et Goodrich）G. L. Nesom et D. R. Morgan = Toiyabea alpina（L. C. Anderson et Goodrich）R. P. Roberts，Urbatsch et Neubig ■☆

392774 Tonestus eximius（H. M. Hall）A. Nelson et J. F. Macbr.；湖畔蛇菊；Lake Tahoe Serpentweed ■☆

392775 Tonestus graniticus（Tiehm et L. M. Schultz）G. L. Nesom et D. R. Morgan；蛇菊；Lone Mountain Serpentweed ■☆

392776 Tonestus kingii（D. C. Eaton）G. L. Nesom = Herrickia kingii（D. C. Eaton）Brouillet ■☆

392777 Tonestus kingii（D. C. Eaton）G. L. Nesom var. barnebyana（S. L. Welsh et Goodrich）G. L. Nesom = Herrickia kingii（D. C. Eaton）Brouillet，Urbatsch et R. P. Roberts var. barnebyana（S. L. Welsh et Goodrich）Brouillet ■☆

392778 Tonestus lyallii（A. Gray）A. Nelson；莱尔蛇菊；Lyall's Serpentweed ■☆

392779 Tonestus microcephalus（Cronquist）G. L. Nesom et D. R. Morgan = Lorandersonia microcephala（Cronquist）Urbatsch，R. P. Roberts et Neubig ●☆

392780 Tonestus peirsonii（D. D. Keck）G. L. Nesom et D. R. Morgan = Lorandersonia peirsonii（D. D. Keck）Urbatsch，R. P. Roberts et Neubig ●☆

392781 Tonestus pygmaeus（Torr. et A. Gray）A. Nelson；矮蛇菊；Pygmy Serpentweed ■☆

392782 Tongoloa H. Wolff（1925）；东俄芹属（东谷芹属）；Tongoloa ■★

392783 Tongoloa achilleifolia（DC.）Pimenov et Kljukov = Meeboldia achilleifolia（DC.）P. K. Mukh. et Constance ■

392784 Tongoloa cnidiifolia K. T. Fu；蛇床东俄芹■

392785 Tongoloa cnidiifolia K. T. Fu. = Tongoloa elata H. Wolff ■

392786 Tongoloa dunnii（H. Boissieu）H. Wolff；宜昌东俄芹（红花芹，太白三七）；Dunn Tongoloa ■

392787 Tongoloa elata H. Wolff；大东俄芹；Large Tongoloa ■

392788 Tongoloa filicaudicis K. T. Fu；细颈东俄芹■

392789 Tongoloa fortunatii（H. Boissieu）Pimenov et Kljukov = Tongoloa silaifolia（H. Boissieu）H. Wolff ■

392790 Tongoloa gracilis H. Wolff；纤细东俄芹；Slender Tongoloa ■

392791 Tongoloa loloensis（H. Boissieu）H. Wolff；云南东俄芹（细裂东谷芹）；Yunnan Tongoloa ■

392792 Tongoloa napifera（H. Wolff）C. Norman；裂苞东俄芹■

392793 Tongoloa pauciradiata H. Wolff；少辐东俄芹■

392794 Tongoloa peucedanifolia（H. Boissieu）H. Wolff. = Tongoloa silaifolia（H. Boissieu）H. Wolff ■

392795 Tongoloa rockii H. Wolff；滇西东俄芹（丽江东谷芹）；Rock Tongoloa ■

392796 Tongoloa rubronervis S. L. Liou；红脉东俄芹；Redvein Tongoloa ■

392797 Tongoloa silaifolia（H. Boissieu）H. Wolff；城口东俄芹（太白三七，甜三七）；Chengkou Tongoloa ■

392798 Tongoloa smithii H. Wolff；短鞘东俄芹■

392799 Tongoloa stewardii H. Wolff；牯岭东俄芹（山蛇床）；Steward Tongoloa ■

392800 Tongoloa taeniophylla（H. Boissieu）H. Wolff；条叶东俄芹（带叶东谷芹）；Beltleaf Tongoloa ■

392801 Tongoloa tenuifolia H. Wolff；细叶东俄芹（细叶东谷芹）；Slenderleaf Tongoloa ■

392802 Tongoloa wolffiana H. Wolff ex Fedde ＝ Vicatia wolffiana（H. Wolff ex Fedde）C. Norman ■☆

392803 Tongoloa zhongdianensis S. L. Liou；中甸东俄芹；Zhongdian Tongoloa ■

392804 Tonguea Endl. ＝ Sisymbrium L. ■

392805 Tonina Aubl.（1775）；托尼谷精草属■☆

392806 Tonina fluviatilis Aubl.；托尼谷精草■☆

392807 Tonningia Juss. ＝ Cyanotis D. Don（保留属名）■

392808 Tonningia Neck. ＝ Cyanotis D. Don（保留属名）■

392809 Tonningia Neck. ex A. Juss. ＝ Cyanotis D. Don（保留属名）■

392810 Tonsella Schreb. ＝ Tontelea Miers（保留属名）●☆

392811 Tonsella africana Willd. ＝ Loeseneriella africana（Willd.）N. Hallé ●☆

392812 Tonsella pyriformis Sabine ＝ Salacia pyriformis（Sabine）Steud. ●☆

392813 Tonshia Buch.-Ham. ex D. Don ＝ Saurauia Willd.（保留属名）●

392814 Tontalea Aubl.（废弃属名）＝ Coccocypselum P. Browne（保留属名）●☆

392815 Tontanea Aubl. ＝ Coccocypselum P. Browne（保留属名）●☆

392816 Tontelea Aubl.（1775）（废弃属名）＝ Elachyptera A. C. Sm. ●☆

392817 Tontelea Aubl.（1775）（废弃属名）＝ Salacia L.（保留属名）●

392818 Tontelea Aubl.（1775）（废弃属名）＝ Tontelea Miers（1872）（保留属名）●☆

392819 Tontelea Miers（1872）（保留属名）；通特卫矛属●☆

392820 Tontelea attenuata Miers；通特卫矛●☆

392821 Tontelea prinoides Willd. ＝ Salacia prinoides（Willd.）DC. ●

392822 Toona（Endl.）M. Roem.（1846）；香椿属（椿属）；Chinese Toon，Toona ●

392823 Toona M. Roem. ＝ Cedrela P. Browne ●

392824 Toona M. Roem. ＝ Toona（Endl.）M. Roem. ●

392825 Toona australis（F. Muell.）Harms ＝ Toona ciliata M. Roem. ●◇

392826 Toona australis Harms；大洋洲香椿；Australian Redcedar，Australian Toona，Red-cedar ●

392827 Toona calantas Merr. et Rolfe；卡兰特香椿（菲律宾香椿，红棟子）；Philippine Toona ●

392828 Toona ciliata F. Muell. var. australis ＝ Toona ciliata M. Roem. ●◇

392829 Toona ciliata M. Roem.；红椿（赤昨工，红椿树，红棟子，缅甸椿，双翅香椿，缘毛椿）；Australian Cedar，Australian Red Cedar，Australian Redcedar，Bastard Mahogany，Burma Cedar，Burma Toon，Ciliate Toona，Moulmein Cedar，Red Cedar，Red Toona，Singapore Cedar，Suren Toona，Toon ●◇

392830 Toona ciliata M. Roem. ＝ Cedrela toona Roxb. ex Rottler et Willd. ●◇

392831 Toona ciliata M. Roem. var. australis（F. Muell.）Bahadur ＝ Toona australis Harms ●

392832 Toona ciliata M. Roem. var. henryi（C. DC.）C. Y. Wu；思茅红椿；Henry Toona ●

392833 Toona ciliata M. Roem. var. pubescens（Franch.）Hand.-Mazz.；毛红椿；Pubescent Toona ●

392834 Toona ciliata M. Roem. var. sublaxiflora（C. DC.）C. Y. Wu；疏花红椿；Looseflower Toona ●

392835 Toona ciliata M. Roem. var. yunnanensis（C. DC.）C. Y. Wu；滇红椿；Yunnan Toona ●

392836 Toona febrifuga M. Roem.；药用香椿；Medicinal Toona ●☆

392837 Toona microcarpa（C. DC.）Harms；紫椿（红椿树，小果香椿）；Small-fruit Toona，Small-fruited Toona ●

392838 Toona microcarpa（C. DC.）Harms ＝ Toona ciliata M. Roem. ●◇

392839 Toona rubriflora C. J. Tseng；红花香椿；Foreign Toona，Redflower Toona ●

392840 Toona serrata（Royle）M. Roem. ＝ Toona sinensis（A. Juss.）M. Roem. ●◇

392841 Toona sinensis（A. Juss.）M. Roem.；香椿（白椿，杶，春菜树，春尖，春甜树，春阳树，椿，椿榪，椿颠，椿木，椿树，椿芽木，椿芽树，橁，橒，大红椿树，红椿，红椿树，毛椿，树，香椿树，香树，猪椿）；China Toona，Chinese Cedar，Chinese Cedrela，Chinese Mahogany，Chinese Toon，Chinese Toona ●◇

392842 Toona sinensis（A. Juss.）M. Roem. 'Flamingo'；红鹤香椿●☆

392843 Toona sinensis（A. Juss.）M. Roem. ＝ Cedrela toona Roxb. ex Rottler et Willd. ●◇

392844 Toona sinensis（A. Juss.）M. Roem. var. grandis Pamp. ＝ Toona sinensis（A. Juss.）M. Roem. ●◇

392845 Toona sinensis（A. Juss.）M. Roem. var. hupehana（C. DC.）P. Y. Chen；湖北香椿；Hubei Toona ●

392846 Toona sinensis（A. Juss.）M. Roem. var. schensiana（C. DC.）P. Y. Chen；陕西香椿（毛椿）；Shaanxi Toona ●

392847 Toona sureni（Blume）Merr.；红棟子（赤昨工，红椿，红椿子）；Suren Toona ●☆

392848 Toona sureni（Blume）Merr. var. pubescens（Franch.）Chun ＝ Toona ciliata M. Roem. var. pubescens（Franch.）Hand.-Mazz. ●

392849 Toona sureni（Blume）Roem. ＝ Toona ciliata M. Roem. ●◇

392850 Topeinostemon C. Muell. ＝ Tapeinostemon Benth. ■☆

392851 Topiaris Raf. ＝ Cordia L.（保留属名）●

392852 Topiaris Raf. ＝ Cordiopsis Desv. ex Ham. ●

392853 Topobea Aubl.（1775）；托波野牡丹属■☆

392854 Topobea parasitica Aubl.；托波野牡丹■☆

392855 Toquera Raf. ＝ Cordia L.（保留属名）●

392856 Torcula Noronha ＝ Pithecellobium Mart.（保留属名）●

392857 Tordylioides Wall. ex DC. ＝ Heracleum L. ■

392858 Tordylioides brunonis Wall. ex DC. ＝ Tordyliopsis brunonis（Wall.）DC. ■

392859 Tordyliopsis DC.（1830）；阔翅芹属■

392860 Tordyliopsis DC. ＝ Tordylioides Wall. ex DC. ■

392861 Tordyliopsis brunonis（Wall.）DC.；珠峰阔翅芹■

392862 Tordyliopsis komarovii（Manden.）Manden. ＝ Semenovia dasycarpa（Regel et Schmalh.）Korovin ex Pimenov et V. N. Tikhom. ■

392863 Tordylium L.（1753）；环翅芹属（阔翅芹属）；Hartwort ■☆

392864 Tordylium L. ＝ Torilis Adans. ■

392865 Tordylium Tourn. ex L. ＝ Tordylium L. ■☆

392866 Tordylium absinthifolium Pers. ＝ Zosima absinthifolia（Vent.）Link ■☆

392867 Tordylium aegyptiacum（L.）Poir.；埃及环翅芹■☆

392868 Tordylium anthriscus L. ＝ Torilis japonica（Houtt.）DC. ■

392869 Tordylium apulum L.；地中海环翅芹（阿普环翅芹）；Ivory-fruited Hartwort，Mediterranean Hartwort ■☆

392870 Tordylium grandiflorum Moench；大花环翅芹■☆

392871 Tordylium humile Desf.；小环翅芹■☆

392872　Tordylium humile Desf. = Tordylium apulum L. ■☆

392873　Tordylium komarovii Manden. ;科马罗夫环翅芹■☆

392874　Tordylium latifolium L. = Turgenia latifolia（L.）Hoffm. ■

392875　Tordylium maximum L. ;大环翅芹;Hartwort ■☆

392876　Tordylium nodosum L. = Torilis nodosa（L.）Gaertn. ■☆

392877　Tordylium officinale L. ;药用环翅芹■☆

392878　Toreala B. D. Jacks. = Pithecellobium Mart.（保留属名）●

392879　Toreala B. D. Jacks. = Torcula Noronha ●

392880　Torenia L.（1753）;蝴蝶草属（倒地蜈蚣属，蓝猪耳属）;Butterflygrass,Torenia ■

392881　Torenia angolensis V. Naray. = Lindernia angolensis（V. Naray.）Eb. Fisch. ■☆

392882　Torenia arisanensis Sasaki;阿里山倒地蜈蚣■

392883　Torenia arisanensis Sasaki = Torenia concolor Lindl. ■

392884　Torenia asiatica L. ;光叶蝴蝶草（长叶蝴蝶草，倒胆草，光叶翼萼，苦生叶，蓝花草，蓝猪儿，老蛇药，老蛀药，水韩信草，水远志，翼萼）;Glabrous Butterflygrass, Glabrous Torenia, Wishbone Flower ■

392885　Torenia atropurpurea Ridl. ;紫色蝴蝶草■☆

392886　Torenia benthamiana Hance;毛叶蝴蝶草（粗毛蝴蝶草，地粘儿，黄蝴蝶草，毛叶蓝猪耳，毛叶翼萼）;Bentham Butterflygrass, Bentham Torenia ■

392887　Torenia bentharniana Hance = Torenia asiatica L. ■

392888　Torenia bentharniana Hance = Torenia glabra Osbeck ■

392889　Torenia bicolor Dalzell;二色蝴蝶草■☆

392890　Torenia bicolor Dalzell = Torenia asiatica L. ■

392891　Torenia biniflora T. L. Chin et D. Y. Hong;二花蝴蝶草;Twoflower Butterflygrass,Twoflower Torenia ■

392892　Torenia brevifolia Engl. et Pilg. = Crepidorhopalon goetzei（Engl.）Eb. Fisch. ■☆

392893　Torenia ciliata Hook. f. ;缘毛蝴蝶草;Ciliate Torenia ■☆

392894　Torenia ciliata Sm. = Torenia ciliata Hook. f. ■☆

392895　Torenia concolor Lindl. ;单色蝴蝶草（阿里山倒地蜈蚣，蚌壳草，单色翼萼，倒胞草，倒地蜈蚣，地蜈蚣，钉地蜈蚣，蝴蝶花，蓝猪草，蓝猪耳，老蛇药，散胆草，四角铜锣）;Concolorous Torenia, Purecolor Butterflygrass ■

392896　Torenia concolor Lindl. var. formosana T. Yamaz. = Torenia concolor Lindl. ■

392897　Torenia cordata（Griff.）N. M. Dutta;心叶蝴蝶草（长叶蝴蝶草）;Pansy Butterflygrass,Pansy Torenia ■

392898　Torenia cordata（Griff.）N. M. Dutta = Torenia asiatica L. ■

392899　Torenia cordifolia Roxb. ;西南蝴蝶草;Cordateleaf Butterflygrass,Cordateleaf Torenia ■

392900　Torenia crustacea（L.）Cham. et Schltdl. = Lindernia crustacea（L.）F. Muell. ■

392901　Torenia dinklagei Engl. ;丁克蝴蝶草■☆

392902　Torenia exappendiculata Regel = Torenia violacea（Azaola ex Blanco）Pennell ■

392903　Torenia flava Buch. -Ham. ex Benth. ;黄花蝴蝶草（黄蝴蝶草，黄花蓝猪耳，黄花翼萼，黄色蝴蝶草，黄色翼萼，母丁香，泡卜儿）;Yellowflower Butterflygrass,Yellowflower Torenia ■

392904　Torenia fordii Hook. f. ;紫斑蝴蝶草;Ford Torenia, Purplespot Butterflygrass ■

392905　Torenia fournieri Linden ex E. Fourn. ;蓝猪耳（兰猪耳，蓝翅蝴蝶草，夏堇，越南倒地蜈蚣）;Blue Butterflygrass, Blue Torenia, Blue Wings,Bluewings,Wishbone Flower ■

392906　Torenia fournieri Linden ex E. Fourn. 'Summer Wave';垂蔓夏堇■☆

392907　Torenia glabra Osbeck = Torenia asiatica L. ■

392908　Torenia goetzei（Engl.）Hepper = Crepidorhopalon goetzei（Engl.）Eb. Fisch. ■☆

392909　Torenia hirsuta Lam. = Torenia benthamiana Hance ■

392910　Torenia hirta Cham. et Schltdl. = Lindernia pusilla（Willd.）Bold. ■

392911　Torenia hirtella Hook. f. ;硬毛蝴蝶草;Hirsute Torenia ■☆

392912　Torenia hokutensis Hayata = Torenia flava Buch. -Ham. ex Benth. ■

392913　Torenia inaequalifolia Engl. = Crepidorhopalon spicatus（Engl.）Eb. Fisch. ■☆

392914　Torenia involucrata Philcox = Crepidorhopalon involucratus（Philcox）Eb. Fisch. ■☆

392915　Torenia kiusiana Ohwi = Torenia asiatica L. ■

392916　Torenia latibracteata（V. Naray.）Hepper = Crepidorhopalon latibracteatus（V. Naray.）Eb. Fisch. ■☆

392917　Torenia latibracteata（V. Naray.）Hepper subsp. parviflora Philcox = Crepidorhopalon parviflorus（Philcox）Eb. Fisch. ■☆

392918　Torenia ledermannii Hepper;纤细蝴蝶草■☆

392919　Torenia mannii V. Naray. ;曼氏蝴蝶草■☆

392920　Torenia monroi（S. Moore）Philcox = Lindernia monroi（S. Moore）Eb. Fisch. ■☆

392921　Torenia mucronulata Benth. ;短尖蝴蝶草;Mucronate Torenia ■☆

392922　Torenia nana Benth. = Craterostigma plantagineum Hochst. ■☆

392923　Torenia nantoensis Hayata;南投倒地蜈蚣;Nantou Butterflygrass ■

392924　Torenia nantoensis Hayata = Torenia benthamiana Hance ■

392925　Torenia oblonga（Benth.）Hance = Lindernia oblonga（Benth.）Merr. ■

392926　Torenia parviflora Buch. -Ham. ex Benth. = Torenia thouarsii（Cham. et Schltdl.）Kuntze ■☆

392927　Torenia parviflora Buch. -Ham. ex Wall. ;小花蝴蝶草;Smallflower Butterflygrass,Smallflower Torenia ■

392928　Torenia parviflora Ham. = Torenia flava Buch. -Ham. ex Benth. ■

392929　Torenia peduncularis Benth. = Torenia violacea（Azaola ex Blanco）Pennell ■

392930　Torenia peduncularis Benth. ex Hook. f. = Torenia violacea（Azaola ex Blanco）Pennell ■

392931　Torenia plantaginea（Hochst.）Benth. = Craterostigma plantagineum Hochst. ■☆

392932　Torenia polygonoides Benth. = Legazpia polygonoides（Benth.）T. Yamaz. ■

392933　Torenia pterocalyx Mildbr. = Torenia silvicola A. Raynal ■☆

392934　Torenia pubescens Peter;短柔毛蝴蝶草■☆

392935　Torenia pumila（Hochst.）Benth. = Craterostigma pumilum Hochst. ■☆

392936　Torenia radicans Vaniot = Torenia concolor Lindl. ■

392937　Torenia ramosissima Vatke = Torenia thouarsii（Cham. et Schltdl.）Kuntze ■☆

392938　Torenia rubens Benth. = Torenia concolor Lindl. ■

392939　Torenia schweinfurthii Oliv. = Crepidorhopalon schweinfurthii（Oliv.）Eb. Fisch. ■☆

392940　Torenia setulosa Maxim. = Lindernia setulosa（Maxim.）Tuyama ex H. Hara ■

392941　Torenia silvicola A. Raynal;森林蝴蝶草■☆

392942　Torenia spicata Engl. = Crepidorhopalon spicatus（Engl.）Eb.

Fisch. ■☆

392943 Torenia tenuifolia Philcox ＝ Crepidorhopalon tenuifolius (Philcox) Eb. Fisch. ■☆

392944 Torenia thouarsii (Cham. et Schltdl.) Kuntze；图氏蝴蝶草■☆

392945 Torenia vagans Roxb.；散播蝴蝶草(蝴蝶花,苦生叶,癫头草,蓝猪儿)；Inconstant Torenia ■☆

392946 Torenia vagans Roxb. ＝ Torenia asiatica L. ■

392947 Torenia vagans Roxb. ＝ Torenia glabra Osbeck ■

392948 Torenia violacea (Azaola ex Blanco) Pennell；紫萼蝴蝶草(长梗花蜈蚣,方形草,序柄蝴蝶草,紫萼翼萼)；Pendulate Torenia, Purplecalyx Butterflygrass, Purplecalyx Torenia ■

392949 Torenia violacea (Azaola ex Blanco) Pennell var. chinensis T. Yamaz. ＝ Torenia violacea (Azaola ex Blanco) Pennell ■

392950 Toresia Pers. ＝ Hierochloe R. Br. (保留属名)■

392951 Toresia Pers. ＝ Torresia Ruiz et Pav. ■

392952 Torfasadis Raf. ＝ Euphorbia L. ●■

392953 Torfosidis B. D. Jacks. ＝ Torfasadis Raf. ●■

392954 Torgesia Bornm. ＝ Crypsis Aiton(保留属名)■

392955 Toricellia DC. ＝ Torricellia DC. ●

392956 Toricellia intermedia Harms ex Diels ＝ Torricellia angulata Oliv. var. intermedia (Harms) Hu ●

392957 Toricelliaceae Hu ＝ Cornaceae Bercht. et J. Presl(保留科名)●■

392958 Toricelliaceae Hu ＝ Torricelliaceae Hu ●

392959 Torilis Adans. (1763)；窃衣属；Hedge Parsley, Hedgeparsley, Hedge-parsley ■

392960 Torilis africana (Thunb.) Spreng.；非洲窃衣■☆

392961 Torilis africana (Thunb.) Spreng. ＝ Torilis arvensis (Huds.) Link var. purpurea (Ten.) Thell. ■☆

392962 Torilis anthrisca (L.) C. C. Gmel. ＝ Torilis japonica (Houtt.) DC. ■

392963 Torilis anthrisca St. -Lag. ＝ Torilis japonica (Houtt.) DC. ■

392964 Torilis anthrisca St. -Lag. var. japonica (Houtt.) H. Boissieu ＝ Torilis japonica (Houtt.) DC. ■

392965 Torilis anthrisca St. -Lag. var. japonica Houtt. ＝ Torilis japonica (Houtt.) DC. ■

392966 Torilis anthriscus (L.) C. C. Gmel. ＝ Torilis japonica (Houtt.) DC. ■

392967 Torilis anthriscus C. C. Gmel. ＝ Torilis japonica (Houtt.) DC. ■

392968 Torilis arvensis (Huds.) Link；田野窃衣；Corn Bur Parsley, Field Bur Parsley, Field Hedgeparsley, Field Hedge-parsley, Hemlock Chervil, Spreading Bur Parsley, Spreading Hedge Parsley, Spreading Hedgeparsley, Spreading Hedge-parsley ■☆

392969 Torilis arvensis (Huds.) Link subsp. elongata (Hoffmanns. et Link) Cannon ＝ Torilis elongata (Hoffmanns. et Link) Samp. ■☆

392970 Torilis arvensis (Huds.) Link subsp. heterophylla (Guss.) Thell. ＝ Torilis arvensis (Huds.) Link subsp. purpurea (Ten.) Hayek ■☆

392971 Torilis arvensis (Huds.) Link subsp. neglecta (Spreng.) Thell. ＝ Torilis neglecta Roem. et Schultz ■☆

392972 Torilis arvensis (Huds.) Link subsp. purpurea (Ten.) Hayek；紫田野窃衣■☆

392973 Torilis arvensis (Huds.) Link subsp. recta Jury；直立野窃衣■☆

392974 Torilis arvensis (Huds.) Link var. biformis K. Malý ＝ Torilis arvensis (Huds.) Link ■☆

392975 Torilis arvensis (Huds.) Link var. elatior (Gaudin) Thell.；较高田野窃衣■☆

392976 Torilis arvensis (Huds.) Link var. heterocarpa (Batt.) Maire ＝ Torilis arvensis (Huds.) Link ■☆

392977 Torilis arvensis (Huds.) Link var. heterophylla (Guss.) Thell.；异叶田野窃衣；Spreading Hedgeparsley ■☆

392978 Torilis arvensis (Huds.) Link var. homomorpha (Chabert) Maire ＝ Torilis arvensis (Huds.) Link ■☆

392979 Torilis arvensis (Huds.) Link var. purpurea (Ten.) Thell. ＝ Torilis arvensis (Huds.) Link subsp. purpurea (Ten.) Hayek ■☆

392980 Torilis bifrons (Pomel) Jafri ＝ Torilis elongata (Hoffmanns. et Link) Samp. ■☆

392981 Torilis elongata (Hoffmanns. et Link) Samp.；伸长窃衣■☆

392982 Torilis gracilis Engl. ＝ Agrocharis incognita (C. Norman) Heywood et Jury ■☆

392983 Torilis gracilis Engl. f. umbrosa？ ＝ Agrocharis incognita (C. Norman) Heywood et Jury ■☆

392984 Torilis henryi C. Norman ＝ Torilis scabra (Thunb.) DC. ■

392985 Torilis heterophylla Guss.；异叶窃衣；Diverseleaf Hedgeparsley ■☆

392986 Torilis heterophylla Guss. ＝ Torilis arvensis (Huds.) Link var. heterophylla (Guss.) Thell. ■☆

392987 Torilis heterophylla Guss. var. helvetica (Jacq.) C. C. Gmel. ＝ Torilis arvensis (Huds.) Link subsp. recta Jury ■☆

392988 Torilis heterophylla Guss. var. homomorpha Chabert ＝ Torilis arvensis (Huds.) Link subsp. purpurea (Ten.) Hayek ■☆

392989 Torilis infesta (L.) Clairv.；创伤窃衣■☆

392990 Torilis infesta (L.) Clairv. ＝ Torilis arvensis (Huds.) Link ■☆

392991 Torilis infesta (L.) Clairv. var. neglecta (Spreng.) Batt. ＝ Torilis arvensis (Huds.) Link ■☆

392992 Torilis japonica (Houtt.) DC.；小窃衣(大叶山胡萝卜,华南鹤虱,破子草,窃衣)；Bur Parsley, Devil's Nightcap, Dill, Erect Hedgeparsley, Erect Hedge-parsley, Hedge Parsley, Hemlock Chervil, Hemlock Chervill, Hogweed, Honiton Lace, Japanese Hedgeparsley, Japanese Hedge-parsley, Lace-flower, Lady's Lace, Lady's Needlework, Mother-dee, Pig's Parsley, Queen Anne's Lace Handkerchief, Red Head, Red Kex, Rough Chervil, Rough Clcely, Scabby Heads, Upright Hedge Parsley, Upright Hedge-parsley ■

392993 Torilis leptophylla (L.) Rchb. f.；细叶窃衣；Bristlefruit Hedgeparsley ■☆

392994 Torilis melanantha (Hochst.) Vatke ＝ Agrocharis melanantha Hochst. ■☆

392995 Torilis neglecta Roem. et Schultz；暗窃衣■☆

392996 Torilis nodosa (L.) Gaertn.；有节窃衣(节窃衣)；Knotted Bur Parsley, Knotted Hedge Parsley, Knotted Hedge-parsley, Nodose Hedgeparsley ■☆

392997 Torilis nodosa (L.) Gaertn. var. bracteosa (Bianca) Murb. ＝ Torilis webbii Jury ■☆

392998 Torilis nodosa (L.) Gaertn. var. peduncularis Ten. ＝ Torilis nodosa (L.) Gaertn. ■☆

392999 Torilis praetermissa Hance ＝ Torilis japonica (Houtt.) DC. ■

393000 Torilis purpurea (Ten.) Guss. ＝ Torilis arvensis (Huds.) Link var. purpurea (Ten.) Thell. ■☆

393001 Torilis radiata Moench ＝ Torilis arvensis (Huds.) Link subsp. neglecta (Spreng.) Thell. ■☆

393002 Torilis radiata Moench.；铺散窃衣■☆

393003 Torilis rubella Moench ＝ Torilis japonica (Houtt.) DC. ■

393004 Torilis scabra (Thunb.) DC.；窃衣(繁花窃衣,紫花窃衣)；Common Hedgeparsley, Rough Hedgeparsley ■

393005 Torilis setifolia Boiss. ＝ Cuminum setifolium (Boiss.) Koso-Pol. ■☆

393006　Torilis tenella（Delile）Rchb. f. ;细小窃衣■☆

393007　Torilis ucrainica Spreng. ;乌克兰窃衣;Ukraine Hedgeparsley ■☆

393008　Torilis webbii Jury;韦布窃衣■☆

393009　Torilis xanthotricha（Stev.）Schischk. ;黄毛窃衣■☆

393010　Tormentilla L. = Potentilla L. ■●

393011　Tormentilla reptans L. = Potentilla reptans L. ■

393012　Tormentillaceae Martinov = Rosaceae Juss.（保留科名）●■

393013　Torminalis Medik. = Sorbus L. ●

393014　Torminaria（DC.）M. Roem. = Sorbus L. ●

393015　Torminaria（DC.）M. Roem. = Torminalis Medik. ●

393016　Torminaria M. Roem. = Sorbus L. ●

393017　Torminaria M. Roem. = Torminalis Medik. ●

393018　Torminaria Opiz = Sorbus L. ●

393019　Torminaria Opiz = Torminalis Medik. ●

393020　Tornabenea Parl. = Tornabenea Parl. ex Webb ■☆

393021　Tornabenea Parl. ex Webb（1850）;托尔纳草属■☆

393022　Tornabenea annua Bég. ;一年托尔纳草■☆

393023　Tornabenea bischoffii J. A. Schmidt = Tornabenea insularis（Webb）Webb ■☆

393024　Tornabenea hirta J. A. Schmidt = Tornabenea insularis（Webb）Webb ■☆

393025　Tornabenea humilis Lobin et K. H. Schmidt;低矮托尔纳草■☆

393026　Tornabenea insularis（Webb）Webb;海岛托尔纳草■☆

393027　Tornabenea tenuissima（A. Chev.）A. Hansen et Sunding;纤细托尔纳草■☆

393028　Tornabenia Benth. et Hook. f. = Tornabenea Parl. ex Webb ■☆

393029　Tornelia Gutierrez ex Schltdl. = Monstera Adans.（保留属名）●■

393030　Toronia L. Johnson et B. G. Briggs（1975）;新西兰龙眼属●☆

393031　Toronia toru（A. Cunn.）L. Johnson et B. G. Briggs;托龙眼●☆

393032　Torpesia（Endl.）M. Roem. = Trichilia P. Browne（保留属名）●

393033　Torpesia M. Roem. = Trichilia P. Browne（保留属名）●

393034　Torralbasia Krug et Urb.（1900）;托拉尔卫矛属●☆

393035　Torralbasia cuneifolia（C. Wright ex A. Gray）Krug et Urb. ;托拉尔卫矛●☆

393036　Torrentia Vell. = Ichthyothere Mart. ■●☆

393037　Torrenticola Domin ex Steenis = Cladopus H. Möller ■

393038　Torrenticola Domin（1928）;急流苔草属■☆

393039　Torrenticola queenslandica（Domin）Domin;急流苔草■☆

393040　Torresea Allemão = Amburana Schwacke et Taub. ●☆

393041　Torresea Allemão（1862）;伪香豆属●☆

393042　Torresea acreana Ducke = Amburana acreana（Ducke）A. C. Sm. ●☆

393043　Torresea cearensis Allemão;伪香豆●☆

393044　Torresea cearensis Allemão = Amburana cearensis（Allemão）A. C. Sm. ●☆

393045　Torresia Ruiz et Pav. = Hierochloe R. Br.（保留属名）■

393046　Torresia Willis = Torresea Allemão ●☆

393047　Torreya Arn.（1838）;榧树属（榧属）;California Nutmeg, Foetid Yew,Nutmeg-tree,Stiking Yew,Stinking-Cedar,Torreya ●

393048　Torreya Croom ex Meisn. = Croomia Torr. ■

393049　Torreya Eaton = Mentzelia L. ●■☆

393050　Torreya Eaton = Nuttallia Raf. ●■☆

393051　Torreya Raf.（1818）= Synandra Nutt. ■☆

393052　Torreya Raf.（1819）= Cyperus L. ■

393053　Torreya Raf.（1819）= Pycreus P. Beauv. ■

393054　Torreya Spreng.（1820）= Clerodendrum L. ●■

393055　Torreya Spreng.（1820）= Patulix Raf. ●■

393056　Torreya ascendens Nakai ex Uyeki = Typha minima Funck. ex Hoppe ■

393057　Torreya californica Torr. ;加州榧树;California Nutmeg, California Nutmeg Yew, California Torreya, Californian Nutmeg, California-nutmeg,Stinking Cedar,Stinking Yew ●☆

393058　Torreya fargesii Franch. ;巴山榧树（篦子杉,球果榧,铁头枞, 铁头枞,崖杉,紫柏）;Bashan Torreya,Farges Torreya ●◇

393059　Torreya fargesii Franch. var. yunnanensis（W. C. Cheng et L. K. Fu）N. Kang = Torreya yunnanensis W. C. Cheng et L. K. Fu ●◇

393060　Torreya grandis Fortune = Torreya grandis Fortune ex Lindl. ●◇

393061　Torreya grandis Fortune ex Lindl. ;榧树（凹叶榧,彼子,椒子, 赤果,大圆榧,钝叶榧树,榧,榧子,了木榧,栾泡榧,米榧,木榧, 细圆榧,香榧,小果榧,小果榧树,药榧,野杉,玉榧,玉山果,圆 榧,芝麻榧）;China Torreya, Chinese Nutmeg, Chinese Torreya, Grand Torreya,Tall Torreya ●◇

393062　Torreya grandis Fortune ex Lindl. 'Merrillii';香榧（细榧,羊角 榧）;Delicious Chinese Torreya ●

393063　Torreya grandis Fortune ex Lindl. f. major Hu = Torreya grandis Fortune ex Lindl. ●◇

393064　Torreya grandis Fortune ex Lindl. f. nonapiculata Hu = Torreya grandis Fortune ex Lindl. ●◇

393065　Torreya grandis Fortune ex Lindl. f. sargentii（Hu）W. C. Cheng et L. K. Fu;萨金特香榧;Sargent Grand Torreya ●☆

393066　Torreya grandis Fortune ex Lindl. var. chingii Hu = Torreya grandis Fortune ex Lindl. ●◇

393067　Torreya grandis Fortune ex Lindl. var. dielsii Hu = Torreya grandis Fortune ex Lindl. ●◇

393068　Torreya grandis Fortune ex Lindl. var. fargesii（Franch.）Silba = Torreya fargesii Franch. ●◇

393069　Torreya grandis Fortune ex Lindl. var. jiulongshanensis Zhi Y. Li,Z. C. Tang et N. Kang;九龙山榧;Jiulongshan Grand Torreya ●

393070　Torreya grandis Fortune ex Lindl. var. merrillii Hu = Torreya grandis Fortune ex Lindl. ●◇

393071　Torreya grandis Fortune ex Lindl. var. sargentii Hu = Torreya grandis Fortune ex Lindl. ●◇

393072　Torreya grandis Fortune ex Lindl. var. yunnanensis（W. C. Cheng et L. K. Fu）Silba = Torreya yunnanensis W. C. Cheng et L. K. Fu ●◇

393073　Torreya grandis Fortune ex Lindl. var. yunnanensis（W. C. Cheng et L. K. Fu）Silba = Torreya fargesii Franch. var. yunnanensis（W. C. Cheng et L. K. Fu）N. Kang ●◇

393074　Torreya jackii Chun;长叶榧树（长叶榧,浙榧）;Jack Torreya, Longleaf Torreya ●◇

393075　Torreya macrosperma Miyoshi ex Morik. ;大籽榧树;Big-seed Torreya ●☆

393076　Torreya myristica Hook. = Torreya californica Torr. ●☆

393077　Torreya nucifera（L.）Siebold et Zucc. ;日本榧树（柀,某,菜 子,榧,榧实,榧子,牛尾杉,排华,日榧,杉,文木,野杉）;Japan Torreya, Japanese Nutmeg, Japanese Nutmeg Yew, Japanese Plum- yew, Japanese Torreya, Kaya, Kaya Nut, Kaya Torreya, Nut-bearing Torreya,Plum-yew ●

393078　Torreya nucifera（L.）Siebold et Zucc. var. grandis（Fortune ex Lindl.）Pilg. = Torreya grandis Fortune ex Lindl. ●◇

393079　Torreya nucifera Siebold et Zucc. f. igaensis（Doi et Morik.）Kitam. ;伊贺榧树●☆

393080　Torreya nucifera Siebold et Zucc. f. macrosperma（Miyoshi）Kusaka;大籽日本榧树●☆

393081　Torreya nucifera Siebold et Zucc. f. nuda（Miyoshi）Kusaka；裸日本榧树●☆

393082　Torreya nucifera Siebold et Zucc. var. grandis（Fortune ex Lindl.）Pilg. = Torreya grandis Fortune ex Lindl. ●◇

393083　Torreya nucifera Siebold et Zucc. var. grandis（Fortune）Pilg. = Torreya grandis Fortune ex Lindl. ●◇

393084　Torreya nucifera Siebold et Zucc. var. igaensis（Doi et Morik.）Ohwi = Torreya nucifera Siebold et Zucc. f. igaensis（Doi et Morik.）Kitam. ●☆

393085　Torreya nucifera Siebold et Zucc. var. macrosperma（Miyoshi）Koidz. = Torreya nucifera Siebold et Zucc. f. macrosperma（Miyoshi）Kusaka ●☆

393086　Torreya nucifera Siebold et Zucc. var. nuda（Miyoshi）Makino = Torreya nucifera Siebold et Zucc. f. nuda（Miyoshi）Kusaka ●☆

393087　Torreya nucifera Siebold et Zucc. var. radicans Nakai；具根日本榧树●☆

393088　Torreya parvifolia T. P. Yi, Lin Yang et T. L. Long；小叶榧树●

393089　Torreya taxifolia Arn. ；佛罗里达榧树；Florida Yew, Foetid Yew, Gopherwood, Stinking Cedar, Stinking Yew, Stinking-cedar, Yew-leaved Torreya ●☆

393090　Torreya yunnanensis W. C. Cheng et L. K. Fu；云南榧树（滇榧子,沙松果,杉松果,云南榧子）；Yunnan Nutmeg Yew, Yunnan Torreya ●◇

393091　Torreya yunnanensis W. C. Cheng et L. K. Fu = Torreya fargesii Franch. var. yunnanensis（W. C. Cheng et L. K. Fu）N. Kang ●◇

393092　Torreyaceae Nakai = Taxaceae Gray（保留科名）●

393093　Torreyaceae Nakai；榧树科●

393094　Torreycactus Doweld = Echinocactus Link et Otto ●

393095　Torreycactus Doweld（1998）；托里球属●☆

393096　Torreycactus conothele（Regel et E. Klein bis）Doweld；托里球●☆

393097　Torreyochloa G. L. Church = Glyceria R. Br.（保留属名）■

393098　Torreyochloa G. L. Church（1949）；托里碱茅属■☆

393099　Torreyochloa fernaldii（Hitchc.）Church = Puccinellia fernaldii（Hitchc.）E. G. Voss ■☆

393100　Torreyochloa natans（Kom.）Church；扩散托里碱茅■☆

393101　Torreyochloa pallida（Torr.）G. L. Church；苍白托里碱茅；Weak Manna Grass ■☆

393102　Torreyochloa pallida（Torr.）G. L. Church = Puccinellia pallida（Torr.）R. T. Clausen ■☆

393103　Torreyochloa pallida（Torr.）G. L. Church subsp. natans（Kom.）T. Koyama et Kawano = Torreyochloa natans（Kom.）Church ■☆

393104　Torreyochloa pallida（Torr.）G. L. Church var. fernaldii（Hitchc.）Dore ex T. Koyama et Kawano = Puccinellia fernaldii（Hitchc.）E. G. Voss ■☆

393105　Torreyochloa viridis（Honda）Church = Glyceria viridis Honda ■☆

393106　Torricellia DC.（1830）；鞘柄木属（叩里木属,烂泥树属）；Torricellia ●

393107　Torricellia Harms ex Diels = Torricellia DC. ●

393108　Torricellia angulata Oliv. ；角叶鞘柄木（大接骨丹,接骨草树,蓝茶叶,烂泥树,裂叶鞘柄木,清明花,水冬瓜,水五加）；Angulate Torricellia ●

393109　Torricellia angulata Oliv. var. intermedia（Harms）Hu；齿叶鞘柄木（齿叶叩里木,齿叶叼里木,齿叶烂泥树,大接骨丹,叩里木,接骨草树,接骨丹,清明花,水东瓜,水冬瓜,水五加,有齿鞘柄木）；Intermediate Angulate Torricellia ●

393110　Torricellia tiliifolia（Wall.）DC. ；鞘柄木（大葫芦叶,大接骨,

大接骨丹,叩里木,椴叶叩里木,椴叶烂泥树,接骨丹,象耳朵）；Lindenleaf Torricellia, Linden-leaved Torricellia ●

393111　Torricelliaceae（Wangerin）Hu = Cornaceae Bercht. et J. Presl（保留科名）●■

393112　Torricelliaceae Hu = Cornaceae Bercht. et J. Presl（保留科名）●■

393113　Torricelliaceae Hu（1934）；鞘柄木科（烂泥树科）；Torricellia Family ●

393114　Torrubia Vell. = Guapira Aubl. ●☆

393115　Torrukia Vell. = Pisonia L. ●

393116　Tortipes Small = Streptopus Michx. ■

393117　Tortipes Small = Uvularia L. ■☆

393118　Tortipes amplexifolius（L.）Small = Streptopus amplexifolius（L.）DC. ■☆

393119　Tortuella Urb.（1927）；托尔图茜属☆

393120　Tortuella abietifolia Urb. et Ekman；托尔图茜☆

393121　Tortula Roxb. ex Willd. = Priva Adans. ■☆

393122　Torularia（Coss.）O. E. Schulz = Neotorularia Hedge et J. Léonard ■

393123　Torularia（Coss.）O. E. Schulz（1924）；肖念珠芥属■☆

393124　Torularia O. E. Schulz = Neotorularia Hedge et J. Léonard ■

393125　Torularia aculeolata（Boiss.）O. E. Schulz；尖念珠芥■☆

393126　Torularia aculeolata（Boiss.）O. E. Schulz var. grandifora Gilli = Torularia afghanica（Gilli）Hedge ■☆

393127　Torularia adpressa（Trautv.）O. E. Schulz = Torularia dentata（Freyn et Sint.）Kitam. ■☆

393128　Torularia afghanica（Gilli）Hedge；阿富汗肖念珠芥■☆

393129　Torularia brachycarpa Vassilcz. = Neotorularia brachycarpa（Vassilcz.）Hedge et J. Léonard ■

393130　Torularia bracteata S. L. Yang = Neotorularia brachycarpa（Vassilcz.）Hedge et J. Léonard ■

393131　Torularia bracteata S. L. Yang = Neotorularia bracteata（S. L. Yang）C. H. An ■

393132　Torularia brevipes（Kar. et Kir.）O. E. Schulz = Neotorularia brevipes（Kar. et Kir.）Hedge et J. Léonard ■

393133　Torularia brevipes（Kar. et Kir.）O. E. Schulz var. leiocarpa O. E. Schulz = Neotorularia brevipes（Kar. et Kir.）Hedge et J. Léonard ■

393134　Torularia brevipes O. E. Schulz = Torularia brevipes（Kar. et Kir.）O. E. Schulz ■

393135　Torularia conferta R. F. Huang = Neotorularia brachycarpa（Vassilcz.）Hedge et J. Léonard ■

393136　Torularia contortuplicata（Stephan）O. E. Schulz；卷褶马康草■☆

393137　Torularia dentata（Freyn et Sint.）Kitam. ；齿叶肖念珠芥■☆

393138　Torularia glandulosa（Kar. et Kir.）Vassilcz. = Arabis glandulosa Kar. et Kir. ■

393139　Torularia glandulosa（Kar. et Kir.）Vassilcz. = Dontostemon glandulosus（Kar. et Kir.）O. E. Schulz ■

393140　Torularia glandulosa（Kar. et Kir.）Vassilcz. var. pamirica Vassilcz. = Dontostemon glandulosus（Kar. etKir.）O. E. Schulz ■

393141　Torularia humilis（C. A. Mey.）O. E. Schulz = Neotorularia humilis（C. A. Mey.）Hedge et J. Léonard ■

393142　Torularia humilis（C. A. Mey.）O. E. Schulz f. angustifolia C. H. An = Neotorularia humilis（C. A. Mey.）Hedge et J. Léonard ■

393143　Torularia humilis（C. A. Mey.）O. E. Schulz f. angustifolia C. H. An = Neotorularia humilis（C. A. Mey.）Hedge et J. Léonard f. angustifolia（C. H. An）Ma ■

393144　Torularia humilis（C. A. Mey.）O. E. Schulz f. glabrata C. H.

An = Neotorularia humilis (C. A. Mey.) Hedge et J. Léonard ■

393145　Torularia humilis (C. A. Mey.) O. E. Schulz f. glabrata C. H. An = Neotorularia humilis (C. A. Mey.) Hedge et J. Léonard f. glabrata (C. H. An) Ma ■

393146　Torularia humilis (C. A. Mey.) O. E. Schulz f. grandiflora O. E. Schulz = Neotorularia humilis (C. A. Mey.) Hedge et J. Léonard ■

393147　Torularia humilis (C. A. Mey.) O. E. Schulz f. grandiflora O. E. Schulz = Neotorularia humilis (C. A. Mey.) Hedge et J. Léonard f. grandiflora (O. E. Schulz) Ma ■

393148　Torularia humilis (C. A. Mey.) O. E. Schulz f. hygrophila (E. Fourn.) O. E. Schulz = Neotorularia humilis (C. A. Mey.) O. E. Schulz f. hygrophila (E. Fourn.) C. H. An ■

393149　Torularia humilis (C. A. Mey.) O. E. Schulz f. hygrophila (E. Fourn.) O. E. Schulz = Neotorularia humilis (C. A. Mey.) Hedge et J. Léonard ■

393150　Torularia humilis (C. A. Mey.) O. E. Schulz var. maximoweiczii (Botsch.) H. L. Yang = Torularia humilis (C. A. Mey.) O. E. Schulz ■

393151　Torularia humilis (C. A. Mey.) O. E. Schulz var. maximowiczii (Botsch.) H. L. Yang = Neotorularia humilis (C. A. Mey.) Hedge et J. Léonard ■

393152　Torularia humilis (C. A. Mey.) O. E. Schulz var. piasezkii (Maxim.) H. L. Yang = Torularia humilis (C. A. Mey.) O. E. Schulz ■

393153　Torularia humilis (C. A. Mey.) O. E. Schulz var. piasezkii (Maxim.) Jafri = Neotorularia humilis (C. A. Mey.) Hedge et J. Léonard ■

393154　Torularia humilis (C. A. Mey.) O. E. Schulz var. venusta O. E. Schulz = Torularia humilis (C. A. Mey.) O. E. Schulz f. grandiflora O. E. Schulz ■

393155　Torularia korolkovii (Regel et Schmalh.) O. E. Schulz = Neotorularia korolkowii (Regel et Schmalh.) Hedge et J. Léonard ■

393156　Torularia korolkovii (Regel et Schmalh.) O. E. Schulz var. longicarpa C. H. An = Neotorularia korolkowii (Regel et Schmalh.) O. E. Schulz var. longicarpa (C. H. An) C. H. An ■

393157　Torularia korolkovii (Regel et Schmalh.) O. E. Schulz var. longicarpa C. H. An = Neotorularia korolkowii (Regel et Schmalh.) Hedge et J. Léonard ■

393158　Torularia korolkovii (Regel et Schmalh.) O. E. Schulz var. longistyla Vassilcz. = Torularia korolkovii (Regel et Schmalh.) O. E. Schulz ■

393159　Torularia korolkovii (Regel et Schmalh.) O. E. Schulz var. longistyla Vassilcz. = Neotorularia korolkowii (Regel et Schmalh.) Hedge et J. Léonard ■

393160　Torularia maximowiczii Botsch. = Neotorularia humilis (C. A. Mey.) Hedge et J. Léonard ■

393161　Torularia mollipila (Maxim.) O. E. Schulz = Neotorularia mollipila (Maxim.) C. H. An ■

393162　Torularia mollipila (Maxim.) O. E. Schulz = Sisymbriopsis mollipila (Maxim.) Botsch. ■

393163　Torularia parvia C. H. An = Neotorularia brachycarpa (Vassilcz.) Hedge et J. Léonard ■

393164　Torularia parvia C. H. An = Neotorularia parvia (C. H. An) C. H. An ■

393165　Torularia pectinata (DC.) Ovcz. et Junussov = Dontostemon pinnatifidus (Willd.) Al-Shehbaz et H. Ohba ■

393166　Torularia piasezkii (Maxim.) Botsch. = Neotorularia humilis

(C. A. Mey.) Hedge et J. Léonard ■

393167　Torularia rossica O. E. Schulz = Neotorularia rossica (O. E. Schulz) Hedge et J. Léonard ■☆

393168　Torularia rosulifolia K. C. Kuan et C. H. An = Lepidostemon rosularis (K. C. Kuan et C. H. An) Al-Shehbaz ■

393169　Torularia rosulifolia K. C. Kuan et C. H. An = Neotorularia korolkowii (Regel et Schmalh.) Hedge et J. Léonard ■

393170　Torularia rosulifolia K. C. Kuan et C. H. An = Neotorularia rosulifolia (K. C. Kuan et C. H. An) C. H. An ■

393171　Torularia sergievskiana Polozhij = Dimorphostemon glandulosus (Kar. et Kir.) Golubk. ■

393172　Torularia sergievskiana Polozhij = Dontostemon glandulosus (Kar. et Kir.) O. E. Schulz ■

393173　Torularia shuanghuica K. C. Kuan et C. H. An = Sisymbriopsis shuanghuica (K. C. Kuan et C. H. An) Al-Shehbaz et al. ■

393174　Torularia sulphurea (Korsh.) O. E. Schulz = Neotorularia korolkowii (Regel et Schmalh.) Hedge et J. Léonard ■

393175　Torularia sumbarensis (Lipsky) O. E. Schulz;桑巴尔肖念珠芥■☆

393176　Torularia tibetica C. H. An = Neotorularia brachycarpa (Vassilcz.) Hedge et J. Léonard ■

393177　Torularia tibetica C. H. An = Neotorularia tibetica (C. H. An) C. H. An ■

393178　Torularia torulosa (Desf.) O. E. Schulz = Neotorularia torulosa (Desf.) Hedge et J. Léonard ■

393179　Torularia torulosa (Desf.) O. E. Schulz var. hispida Maire = Neotorularia torulosa (Desf.) Hedge et J. Léonard ■

393180　Torularia torulosa (Desf.) O. E. Schulz var. scorpiuroides (Boiss.) Halácsy = Neotorularia torulosa (Desf.) Hedge et J. Léonard ■

393181　Torularia torulosa (Desf.) O. E. Schulz var. scorpiuroides (Boiss.) O. E. Schulz = Torularia torulosa (Desf.) O. E. Schulz ■

393182　Torularia torulosa (Desf.) O. E. Schulz var. scorpiuroides (Boiss.) O. E. Schulz = Neotorularia torulosa (Desf.) Hedge et J. Léonard ■

393183　Torulinium Desv. = Cyperus L. ■

393184　Torulinium Desv. = Torulinium Desv. ex Ham. ■

393185　Torulinium Desv. ex Ham. (1825);断节莎属;Torulinium ■

393186　Torulinium Desv. ex Ham. = Cyperus L. ■

393187　Torulinium angolense Turrill;安哥拉断节莎■☆

393188　Torulinium caucasicum Palla;勘察加断节莎■☆

393189　Torulinium confertum Desv. ex Ham. = Torulinium odoratum (L.) S. S. Hooper ■

393190　Torulinium confertum Ham. = Torulinium ferax (Rich.) Buch. -Ham. ■

393191　Torulinium confertum Ham. = Torulinium odoratum (L.) S. S. Hooper ■

393192　Torulinium eggersii (Boeck.) C. B. Clarke = Cyperus odoratus L. ■

393193　Torulinium eggersii (Boeck.) C. B. Clarke = Torulinium odoratum (L.) S. S. Hooper ■

393194　Torulinium ferax (Rich.) Buch. -Ham. ; 断节莎; Common Torulinium , Torulinium ■

393195　Torulinium ferax (Rich.) Buch. -Ham. = Cyperus odoratus L. ■

393196　Torulinium ferax (Rich.) Ham. = Cyperus odoratus L. ■

393197　Torulinium ferax (Rich.) Urb. = Cyperus odoratus L. ■

393198　Torulinium ferax (Rich.) Urb. = Torulinium ferax (Rich.) Buch. -Ham. ■

393199　Torulinium ferax（Rich.）Urb. = Torulinium odoratum（L.）S. S. Hooper ■

393200　Torulinium filiforme（Sw.）C. B. Clarke = Cyperus filiformis Sw. ■☆

393201　Torulinium macrocephalum（Liebm.）T. Koyama = Cyperus odoratus L. ■

393202　Torulinium macrocephalum（Liebm.）T. Koyama = Torulinium odoratum（L.）S. S. Hooper ■

393203　Torulinium michauxianum（Schult.）C. B. Clarke = Cyperus odoratus L. ■

393204　Torulinium michauxianum（Schult.）C. B. Clarke = Torulinium odoratum（L.）S. S. Hooper ■

393205　Torulinium odoratum（L.）S. S. Hooper;香断节莎（断节莎）;Flat Sedge, Flatsedge, Fragrant Cyperus, Nut Sedge, Rusty Flat Sedge, Rusty Flatsedge ■

393206　Torulinium odoratum（L.）S. S. Hooper = Cyperus odoratus L. ■

393207　Torymenes Salisb. = Amomum Roxb.（保留属名）■

393208　Tosagris P. Beauv. = Muhlenbergia Schreb. ■

393209　Tostimontia S. Díaz(2001);哥伦比亚山菊属 ☆

393210　Tostimontia gunnerifolia S. Díaz;哥伦比亚山菊 ☆

393211　Toubaouate Airy Shaw = Didelotia Baill. ●☆

393212　Toubaouate Aubrév. et Pellegr. = Didelotia Baill. ●☆

393213　Toubaouate Kunkel = Toubaouate Aubrév. et Pellegr. ●☆

393214　Toubaouate brevipaniculata（J. Léonard）Aubrév. et Pellegr. = Didelotia brevipaniculata J. Léonard ●☆

393215　Touchardia Gaudich.（1847—1848）;鱼线麻属 ●☆

393216　Touchardia latifolia Gaudich.;鱼线麻;Olona ■☆

393217　Touchiroa Aubl.（废弃属名）= Crudia Schreb.（保留属名）●☆

393218　Toulichiba Adans.（废弃属名）= Ormosia Jacks.（保留属名）●

393219　Toulicia Aubl.（1775）;图里无患子属 ●☆

393220　Toulicia acuminata Radlk.;渐尖图里无患子 ●☆

393221　Toulicia brachyphylla Radlk.;短叶图里无患子 ●☆

393222　Toulicia brasiliensis Casar.;巴西图里无患子 ●☆

393223　Toulicia crassifolia Radlk.;厚叶图里无患子 ●☆

393224　Toulicia elliptica Radlk.;椭圆图里无患子 ●☆

393225　Toulicia eriocarpa Radlk.;毛果图里无患子 ●☆

393226　Toulicia guianensis Aubl.;圭亚那图里无患子 ●☆

393227　Toulicia reticulata Radlk.;网脉图里无患子 ●☆

393228　Touloucouna M. Roem. = Carapa Aubl. ●☆

393229　Toumboa Naudin = Tumboa Welw. ■☆

393230　Toumboa Naudin = Welwitschia Hook. f.（保留属名）■☆

393231　Toumeya Britton et Rose = Pediocactus Britton et Rose ●☆

393232　Toumeya Britton et Rose = Sclerocactus Britton et Rose ●☆

393233　Toumeya Britton et Rose(1922);月童子属（月之童子属）■☆

393234　Toumeya bradyi（L. D. Benson）W. H. Earle = Pediocactus bradyi L. D. Benson ●☆

393235　Toumeya papyracantha（Engelm.）Britton et Rose;月童子（月之童子）;Pediocactus ■☆

393236　Toumeya papyracantha（Engelm.）Britton et Rose = Sclerocactus papyracanthus（Engelm.）N. P. Taylor ●☆

393237　Tounatea Aubl.（废弃属名）= Swartzia Schreb.（保留属名）●☆

393238　Tounatea madagascariensis（Desv.）Baill. = Swartzia madagascariensis Desv. ●☆

393239　Tournaya A. Schmitz = Bauhinia L. ●

393240　Tournefortia L.（1753）;紫丹属（清饭藤属,砂引草属）;Basket With, Tournefortia, Waftwort ●■

393241　Tournefortia arborea Blanco = Tournefortia argentea L. f. ●

393242　Tournefortia argentea L. f.;银毛树（白水草,白水木,山埔姜,水草）;Silverhair Messerschmidia, Silvery Messerschmidia, Silvery Tournefortia, Velvetleaf Soldierbush ●

393243　Tournefortia argentea L. f. = Argusia argentea（L. f.）Heine ●

393244　Tournefortia argentea L. f. = Heliotropium foertherianum Diane et Hilger ●

393245　Tournefortia argentea L. f. = Messerschmidia argentea（L. f.）I. M. Johnst. ●

393246　Tournefortia arguzia L. f. var. latifolia DC. = Tournefortia sibirica L. ●■

393247　Tournefortia arguzia Roem. et Schult. = Tournefortia sibirica L. ●■

393248　Tournefortia arguzia Roem. et Schult. var. angustior DC. = Tournefortia sibirica L. var. angustior（DC.）G. L. Chu et M. G. Gibert ●■

393249　Tournefortia arguzia Roem. et Schult. var. latifolia DC. = Tournefortia sibirica L. ●■

393250　Tournefortia boniana Gagnep. = Tournefortia montana Lour. ●

393251　Tournefortia brachyantha Merr. et Chun = Tournefortia montana Lour. ●

393252　Tournefortia fruticosa（L. f.）Ortega;灌木紫丹 ●☆

393253　Tournefortia gaudichandii Gagnep. = Tournefortia montana Lour. ●

393254　Tournefortia gnaphalodes（L.）Roem. et Schult. = Mallotonia gnaphalodes（L.）Britton ●☆

393255　Tournefortia hirsutissima L.;多毛紫丹;Trinidad Tournefortia ●☆

393256　Tournefortia intonsa Kerr.;具毛紫丹 ●☆

393257　Tournefortia kirkii（I. M. Johnst.）J. S. Mill.;柯克紫丹 ●☆

393258　Tournefortia linearis E. Mey. = Heliotropium lineare（A. DC.）Gürke ■☆

393259　Tournefortia micranthos DC. = Heliotropium micranthum（Pall.）Bunge ■

393260　Tournefortia mocquerysii Aug. DC. = Tournefortia puberula Baker ●☆

393261　Tournefortia montana Lour.;紫丹（长管滨紫,茨姆,卵形紫丹）;Montane Tournefortia ●

393262　Tournefortia ovata Wall. = Tournefortia montana Lour. ●

393263　Tournefortia ovata Wall. et G. Don = Tournefortia montana Lour. ●

393264　Tournefortia puberula Baker;微毛紫丹 ●☆

393265　Tournefortia puberula Baker var. kirkii I. M. Johnst. = Tournefortia kirkii（I. M. Johnst.）J. S. Mill. ●☆

393266　Tournefortia sampsonii Hance = Tournefortia montana Lour. ●

393267　Tournefortia sarmentosa Lam.;台湾紫丹（冷饭藤,清饭藤）;Bearing Runners Tournefortia, Taiwan Tournefortia ●

393268　Tournefortia sarmentosa Lam. subsp. usambarensis Verdc. = Tournefortia usambarensis（Verdc.）Verdc. ●☆

393269　Tournefortia sibirica L.;西伯利亚紫丹（地血,砂引草,西伯利亚砂引草,紫草,紫丹）;Siberia Tournefortia, Siberian Jellygrass, Siberian Messerschmidia ●■

393270　Tournefortia sibirica L. = Argusia sibirica（L.）Dandy ●■

393271　Tournefortia sibirica L. = Heliotropium japonicum A. Gray ■☆

393272　Tournefortia sibirica L. f. var. grandiflora H. Winkl. = Tournefortia sibirica L. ●■

393273　Tournefortia sibirica L. var. angustior（DC.）G. L. Chu et M. G. Gibert;细叶砂引草（挠挠糖,砂引草,狭叶砂引草,紫丹草）;Narrow Siberian Messerschmidia ●■

393274　Tournefortia sibirica L. var. grandiflora H. Winkl. = Tournefortia sibirica L. ●■

393275　Tournefortia sibirica L. var. rosmarinifolia Turcz. = Tournefortia sibirica L. var. angustior（DC.）G. L. Chu et M. G. Gibert ●■

393276 Tournefortia sosdiana（Bunge）Popov;索斯紫丹●☆

393277 Tournefortia stenoraca Klotzsch ＝ Heliotropium zeylanicum（Burm. f.）Lam. ■☆

393278 Tournefortia subulata Hochst. ex A. DC. ＝ Heliotropium zeylanicum（Burm. f.）Lam. ■☆

393279 Tournefortia tuberculosa Cham. ＝Heliotropium ciliatum Kaplan ■☆

393280 Tournefortia usambarensis（Verdc.）Verdc.;乌桑巴拉紫丹●☆

393281 Tournefortiopsis Rusby ＝ Guettarda L. ●

393282 Tournefortiopsis Rusby（1907）;拟紫丹属●■☆

393283 Tournefortiopsis reticulata Rusby;拟紫丹●☆

393284 Tournefortiopsis reticulata Rusby ＝ Guettarda tournefortiopsis Standl. ●☆

393285 Tournesol Adans.（废弃属名）＝ Chrozophora A. Juss.（保留属名）●

393286 Tournesolia Nissol. ex Scop. ＝ Chrozophora A. Juss.（保留属名）●

393287 Tournesolia Nissol. ex Scop. ＝ Tournesol Adans.（废弃属名）●

393288 Tournesolia Scop. ＝ Chrozophora A. Juss.（保留属名）●

393289 Tournesolia Scop. ＝ Tournesol Adans.（废弃属名）●

393290 Tourneuxia Coss.（1859）;双翅苣属■☆

393291 Tourneuxia variifolia Coss.;双翅苣■☆

393292 Tourneuxia variifolia Coss. var. bellidifolia Sauvage ＝ Tourneuxia variifolia Coss. ■☆

393293 Tournonia Moq.（1849）;柄落葵属■☆

393294 Tournonia hookeriana Moq.;柄落葵■☆

393295 Tourolia Stokes ＝ Touroulia Aubl. ●☆

393296 Touroubea Steud. ＝ Souroubea Aubl. ●☆

393297 Touroulia Aubl.（1775）;南美绒子树属●☆

393298 Touroulia guianensis Aubl. ;南美绒子树●☆

393299 Tourretia Foug. ＝ Tourrettia Foug.（保留属名）■☆

393300 Tourretia Juss. ＝ Tourrettia Foug.（保留属名）■☆

393301 Tourrettia DC. ＝ Tourrettia Foug.（保留属名）■☆

393302 Tourrettia Foug.（1787）（保留属名）;图紫葳属■☆

393303 Tourrettia lappocea（L'Her.）Willd. ;图紫葳■☆

393304 Toussaintia Boutique（1951）;图森木属（陶萨木属）●☆

393305 Toussaintia congolensis Boutique;刚果图森木（陶萨木）●☆

393306 Toussaintia hallei Le Thomas;图森木●☆

393307 Toussaintia orientalis Verdc. ;东方图森木（东方陶萨木）●☆

393308 Toussaintia patriciae Q. Luke et Deroin;非洲图森木●☆

393309 Touterea Eaton et Wright ＝ Mentzelia L. ●■☆

393310 Touteria Willis ＝ Touterea Eaton et Wright ●■☆

393311 Tovara Adans.（废弃属名）＝ Antenoron Raf. ■

393312 Tovara Adans.（废弃属名）＝ Persicaria（L.）Mill. ■

393313 Tovara Adans.（废弃属名）＝ Tovaria Ruiz et Pav.（保留属名）●■

393314 Tovara filiformis（Thunb.）Nakai ＝ Antenoron filiforme（Thunb.）Rob. et Vautier ■

393315 Tovara filiformis（Thunb.）Nakai var. kachina（Nieuwl.）H. L. Li ＝ Antenoron filiforme（Thunb.）Rob. et Vautier var. kachinum（Nieuwl.）H. Hara ■

393316 Tovara filiformis（Thunb.）Nakai var. neofiliforme（Nakai）Makino ＝ Antenoron filiforme（Thunb.）Rob. et Vautier var. neofiliforme（Nakai）A. J. Li ■

393317 Tovara neofiliformis（Nakai）Nakai ＝ Antenoron filiforme（Thunb.）Rob. et Vautier var. neofiliforme（Nakai）A. J. Li ■

393318 Tovara ryukyuensis Masam. ＝ Antenoron filiforme（Thunb.）Rob. et Vautier ■

393319 Tovara virginiana（L.）Raf. ＝ Persicaria virginiana（L.）Gaertn. ■☆

393320 Tovara virginiana（L.）Raf. ＝ Polygonum virginianum L. ■☆

393321 Tovara virginiana（L.）Raf. var. filiformis（Thunb.）Steward. ＝ Antenoron filiforme（Thunb.）Rob. et Vautier ■

393322 Tovara virginiana（L.）Raf. var. glaberrima Fernald ＝ Polygonum virginianum L. ■☆

393323 Tovara virginiana（L.）Raf. var. kachina Nieuwl. ＝ Antenoron filiforme（Thunb.）Rob. et Vautier var. kachinum（Nieuwl.）H. Hara ■

393324 Tovaria Baker ＝ Maianthemum F. H. Wigg.（保留属名）■

393325 Tovaria Neck. ＝ Maianthemum F. H. Wigg.（保留属名）■

393326 Tovaria Neck. ＝ Smilacina Desf.（保留属名）■

393327 Tovaria Neck. ex Baker ＝ Maianthemum F. H. Wigg.（保留属名）■

393328 Tovaria Ruiz et Pav.（1794）（保留属名）;烈味三叶草属（鹿药属）●■

393329 Tovaria atropurpurea Franch. ＝ Maianthemum atropurpureum（Franch.）LaFrankie ■

393330 Tovaria bodinieri H. Lév. et Vaniot ＝ Disporum bodinieri（H. Lév. et Vaniot）F. T. Wang et Ts. Tang ■

393331 Tovaria dahurica（Turcz. ex Fisch. et C. A. Mey.）Baker ＝ Maianthemum dahuricum（Turcz. ex Fisch. et C. A. Mey.）LaFrankie ■

393332 Tovaria dahurica Baker ＝ Maianthemum dahuricum（Turcz. ex Fisch. et C. A. Mey.）LaFrankie ■

393333 Tovaria delavayi Franch. ＝ Maianthemum paniculatum（Martens et Galeotti）LaFrankie ■

393334 Tovaria delavayi Franch. ＝ Maianthemum tatsienense（Franch.）LaFrankie ■

393335 Tovaria esquirolii H. Lév. ＝ Disporum trabeculatum Gagnep. ■

393336 Tovaria fargesii Franch. ＝ Maianthemum tubiferum（Batalin）LaFrankie ■

393337 Tovaria finitima W. W. Sm. ＝ Maianthemum fuscum（Wall.）LaFrankie ■

393338 Tovaria foresstii W. W. Sm. ＝ Maianthemum forrestii（W. W. Sm.）LaFrankie ■

393339 Tovaria fusca（Wall.）Baker ＝ Maianthemum fuscum（Wall.）LaFrankie ■

393340 Tovaria fusca Baker ＝Maianthemum fuscum（Wall.）LaFrankie ■

393341 Tovaria japonica（A. Gray）Baker ＝ Maianthemum japonicum（A. Gray）LaFrankie ■

393342 Tovaria japonica Baker ＝ Maianthemum japonicum（A. Gray）LaFrankie ■

393343 Tovaria lichiangensis W. W. Sm. ＝ Maianthemum lichiangense（W. W. Sm.）LaFrankie ■

393344 Tovaria longistyla H. Lév. et Vaniot ＝ Disporum longistylum（H. Lév. et Vaniot）H. Hara ■

393345 Tovaria miranda H. Lév. ＝ Pollia miranda（H. Lév.）H. Hara ■

393346 Tovaria oleracea Baker ＝ Maianthemum oleraceum（Baker）LaFrankie ■

393347 Tovaria oligophylla Baker ＝ Maianthemum purpureum（Wall.）LaFrankie ■

393348 Tovaria pallida（Royle）Baker ＝ Maianthemum purpureum（Wall.）LaFrankie ■

393349 Tovaria pallida Baker ＝ Maianthemum purpureum（Wall.）LaFrankie ■

393350 Tovaria pendula Ruiz et Pav. ;烈味三叶草●■☆

393351 Tovaria prattii Franch. = Maianthemum atropurpureum (Franch.) LaFrankie ■

393352 Tovaria prattii Franch. var. quadrifolia Franch. = Maianthemum tubiferum (Batalin) LaFrankie ■

393353 Tovaria prattii Franch. var. robusta Franch. = Maianthemum atropurpureum (Franch.) LaFrankie ■

393354 Tovaria purpurea (Wall.) Baker = Maianthemum purpureum (Wall.) LaFrankie ■

393355 Tovaria purpurea Baker = Maianthemum purpureum (Wall.) LaFrankie ■

393356 Tovaria rossii Baker = Maianthemum japonicum (A. Gray) LaFrankie ■

393357 Tovaria souliei Franch. = Maianthemum tubiferum (Batalin) LaFrankie ■

393358 Tovaria stenoloba Franch. = Maianthemum stenolobum (Franch.) S. C. Chen et Kawano ■

393359 Tovaria tatsienensis Franch. = Maianthemum tatsienense (Franch.) LaFrankie ■

393360 Tovaria trifolia (L.) Neck. ex Baker = Maianthemum trifolium (L.) Slobada ■

393361 Tovaria trifolia Neck. = Maianthemum trifolium (L.) Slobada ■

393362 Tovaria tubifera (Batalin) C. H. Wright = Maianthemum tubiferum (Batalin) LaFrankie ■

393363 Tovaria virginiana ?;弗吉尼亚烈味三叶草;Jumpsecd, Virginia Knotweed ■☆

393364 Tovaria wardii W. W. Sm. = Maianthemum atropurpureum (Franch.) LaFrankie ■

393365 Tovaria yunnanensis Franch. = Maianthemum tatsienense (Franch.) LaFrankie ■

393366 Tovaria yunnanensis Franch. var. rigida Franch. = Maianthemum tatsienense (Franch.) LaFrankie ■

393367 Tovariaceae Pax(1891)(保留科名);烈味三叶草科(多籽果科,鲜芹味科)●■☆

393368 Tovarochloa T. D. Macfarl. et But(1982);秘鲁托瓦草属■☆

393369 Tovarochloa peruviana Macfarl. et But;秘鲁托瓦草■☆

393370 Tovomia Pers. = Tovomita Aubl. ●☆

393371 Tovomita Aubl. (1775);托福木属●☆

393372 Tovomita albiflora A. C. Sm. ;托福木●☆

393373 Tovomita grandifolia L. O. Williams;大花托福木●☆

393374 Tovomita longifolia Hochr. ;长叶托福木●☆

393375 Tovomita micrantha A. C. Sm. ;小花托福木●☆

393376 Tovomita obovata Engl. ;倒卵托福木●☆

393377 Tovomita uniflora G. Don;单花托福木●☆

393378 Tovomitidium Ducke(1935);小托福木属●☆

393379 Tovomitidium clusiiflorum Ducke;小托福木●☆

393380 Tovomitidium speciosum (Ducke) Ducke;美丽小托福木●☆

393381 Tovomitopsis Planch. et Triana(1860);拟托福木属●☆

393382 Tovomitopsis allenii Maguire;阿伦拟托福木●☆

393383 Tovomitopsis angustifolia Maguire;窄叶拟托福木●☆

393384 Tovomitopsis colombiana Cuatrec. ;哥伦比亚拟托福木●☆

393385 Tovomitopsis membranifolia (Standl.) D'Arcy;膜叶拟托福木●☆

393386 Tovomitopsis micrantha (Engl.) D'Arcy;小花拟托福木●☆

393387 Townsendia Hook. (1834);孤菀属■☆

393388 Townsendia alpigena Piper = Townsendia montana M. E. Jones ■☆

393389 Townsendia alpigena Piper var. caelilinensis (S. L. Welsh) Kartesz et Gandhi = Townsendia montana M. E. Jones ■☆

393390 Townsendia alpigena Piper var. minima (Eastw.) Dorn = Townsendia minima Eastw. ■☆

393391 Townsendia aprica S. L. Welsh et Reveal;天窗孤菀■☆

393392 Townsendia arizonica A. Gray = Townsendia incana Nutt. ■☆

393393 Townsendia condensata Parry;紧缩孤菀■☆

393394 Townsendia condensata Parry var. anomala (Heiser) Dorn = Townsendia condensata Parry ■☆

393395 Townsendia eximia A. Gray;卓越孤菀■☆

393396 Townsendia exscapa Porter;无茎孤菀;Easter Daisy ■☆

393397 Townsendia fendleri A. Gray;芬德勒孤菀■☆

393398 Townsendia florifer (Hook.) A. Gray;多花孤菀■☆

393399 Townsendia florifer (Hook.) A. Gray var. watsonii (A. Gray) Cronquist = Townsendia florifer (Hook.) A. Gray ■☆

393400 Townsendia formosa Greene;落基山孤菀;Rocky Mountain Daisy ■☆

393401 Townsendia glabella A. Gray;近光孤菀■☆

393402 Townsendia grandiflora Nutt. ;大花孤菀■☆

393403 Townsendia gypsophila Lowrey et P. J. Knight;喜钙孤菀■☆

393404 Townsendia hookeri Beaman;虎克孤菀■☆

393405 Townsendia incana Nutt. ;亚利桑那孤菀■☆

393406 Townsendia jonesii (Beaman) Reveal;琼斯孤菀;Great Basin Naturalist ■☆

393407 Townsendia jonesii (Beaman) Reveal var. lutea S. L. Welsh = Townsendia aprica S. L. Welsh et Reveal ■☆

393408 Townsendia leptotes (A. Gray) Osterh. ;细孤菀■☆

393409 Townsendia mensana M. E. Jones var. jonesii Beaman = Townsendia jonesii (Beaman) Reveal ■☆

393410 Townsendia microcephala Dorn;小头孤菀■☆

393411 Townsendia minima Eastw. ;小孤菀■☆

393412 Townsendia montana M. E. Jones;山地孤菀■☆

393413 Townsendia montana M. E. Jones var. caelilinensis S. L. Welsh = Townsendia montana M. E. Jones ■☆

393414 Townsendia nuttallii Dorn = Townsendia hookeri Beaman ■☆

393415 Townsendia parryi D. C. Eaton;巴丽孤菀■☆

393416 Townsendia rothrockii A. Gray ex Rothr. ;罗思罗克孤菀■☆

393417 Townsendia scapigera D. C. Eaton;花葶孤菀;Ground Daisy ■☆

393418 Townsendia sericea Hook. = Townsendia exscapa Porter ■☆

393419 Townsendia sericea Hook. var. leptotes A. Gray = Townsendia leptotes (A. Gray) Osterh. ■☆

393420 Townsendia smithii L. M. Shultz et A. H. Holmgren;史密斯孤菀■☆

393421 Townsendia spathulata Nutt. ;匙叶孤菀■☆

393422 Townsendia strigosa Nutt. var. prolixa (M. E. Jones) S. L. Welsh = Townsendia spathulata Nutt. ■☆

393423 Townsendia texensis Larsen;得州孤菀■☆

393424 Townsonia Cheeseman(1906);汤森兰属■☆

393425 Townsonia deflexa Cheeseman;汤森兰■☆

393426 Townsonia viridis (Hook. f.) Schltr. ;绿汤森兰■☆

393427 Toxanthera Endl. ex Grtining = Monotaxis Brongn. ■☆

393428 Toxanthera Hook. f. = Kedrostis Medik. ■☆

393429 Toxanthera kwebensis N. E. Br. = Kedrostis hirtella (Naudin) Cogn. ■☆

393430 Toxanthera lugardae N. E. Br. = Kedrostis hirtella (Naudin) Cogn. ■☆

393431 Toxanthera natalensis Hook. f. = Kedrostis hirtella (Naudin) Cogn. ■☆

393432 Toxanthes Turcz. (1851);弓花鼠麹草属(腺叶鼠麹草属)■☆

393433 Toxanthes Turcz. = Millotia Cass. ■☆

393434　Toxanthes major Turcz. ;大弓花鼠麹草■☆

393435　Toxanthes perpusilla Turcz. ;弓花鼠麹草■☆

393436　Toxanthus Benth. = Toxanthes Turcz. ■☆

393437　Toxeumia L. Nutt. ex Scribn. et Merr. = Calamagrostis Adans. ■

393438　Toxicaria Aepnel ex Steud. = Antiaris Lesch.（保留属名）●

393439　Toxicaria Schreb. = Rouhamon Aubl. ●

393440　Toxicaria Schreb. = Strychnos L. ●

393441　Toxicodendron Mill.（1754）;漆树属（毒漆属,漆属,野葛属）;Lacquer Tree,Lacquertree,Lacquer-tree ●

393442　Toxicodendron Mill. = Rhus L. ●

393443　Toxicodendron acuminatum（DC.）C. Y. Wu et T. L. Ming;尖叶漆（尖叶野漆,尾叶漆）;Calcicole Lacquer-tree,Sharpleaf Lacquertree,Sharp-leaved Lacquer-tree ●

393444　Toxicodendron africanum Kuntze = Rhus lucida L. ●☆

393445　Toxicodendron altissimum Mill. = Ailanthus altissima（Mill.）Swingle ●

393446　Toxicodendron calcicola C. Y. Wu;石山漆;Calcicole Lacquertree,Rockliving Lacquer-tree ●

393447　Toxicodendron caudatum C. C. Huang ex T. L. Ming = Toxicodendron acuminatum（DC.）C. Y. Wu et T. L. Ming ●

393448　Toxicodendron crenatum Mill. = Rhus aromatica Aiton ●☆

393449　Toxicodendron delavayi（Franch.）F. A. Barkley;小漆树（漆树,山漆树,铁象杆,野漆树）;Delavay Lacquertree,Delavay Lacquer-tree ●

393450　Toxicodendron delavayi（Franch.）F. A. Barkley var. angustifolium C. Y. Wu;狭叶小漆树;Narrowleaf Delavay Lacquertree ●

393451　Toxicodendron delavayi（Franch.）F. A. Barkley var. quinquejugum（Rehder et E. H. Wilson）C. Y. Wu et T. L. Ming;多叶小漆树;Multifolia Lacquertree ●

393452　Toxicodendron desertorum Lunell = Toxicodendron rydbergii（Small ex Rydb.）Greene ●☆

393453　Toxicodendron diversilobum（Torr. et A. Gray）Greene;异裂毒漆;Poison Oak ●☆

393454　Toxicodendron fulvum（Craib）G. Y. Wu et T. L. Ming;黄毛漆;Yellowhair Lacquertree,Yellow-hairy Lacquer-tree ●

393455　Toxicodendron grandiflorum C. Y. Wu et T. L. Ming;大花漆;Big-flowered Lacquer-tree,Largeflower Lacquertree ●

393456　Toxicodendron grandiflorum C. Y. Wu et T. L. Ming var. longipes（Franch.）C. Y. Wu et T. L. Ming;长梗大花漆（长柄大花漆）;Longpediole Lacquertree ●

393457　Toxicodendron griffithii（Hook. f.）Kuntze;裂果漆;Griffith Lacquertree,Griffith Lacquer-tree ●

393458　Toxicodendron griffithii（Hook. f.）Kuntze var. barbatum C. Y. Wu et T. L. Ming;镇康裂果漆;Barbate Griffith Lacquertree,Zhenkang Griffith Lacquertree ●

393459　Toxicodendron griffithii（Hook. f.）Kuntze var. microcarpum C. Y. Wu et T. L. Ming;小果裂果漆;Littlefruit Griffith Lacquertree ●

393460　Toxicodendron hirtellum C. Y. Wu;硬毛漆;Hardhair Lacquertree,Hardhairy Lacquer-tree ●

393461　Toxicodendron hookeri（Sahni et Bahadur）C. Y. Wu et T. L. Ming;大叶漆;Big-leaved Lacquer-tree,Hooker Lacquertree,Hooker Lacquer-tree,Largeleaf Lacquertree ●

393462　Toxicodendron hookeri（Sahni et Bahadur）C. Y. Wu et T. L. Ming var. microcarpum（C. C. Huang ex T. L. Ming）C. Y. Wu et T. L. Ming;小果大叶漆;Littlefruit Hooker Lacquertree ●

393463　Toxicodendron insigne（Hook. f.）Kuntze = Toxicodendron

hookeri（Sahni et Bahadur）C. Y. Wu et T. L. Ming ●

393464　Toxicodendron insigne Kuntze var. microcarpum C. C. Huang ex T. L. Ming = Toxicodendron hookeri（Sahni et Bahadur）C. Y. Wu et T. L. Ming var. microcarpum（C. C. Huang ex T. L. Ming）C. Y. Wu et T. L. Ming ●

393465　Toxicodendron negundo Greene = Toxicodendron radicans（L.）Kuntze subsp. hispidum（Engl.）Gillis ●

393466　Toxicodendron orientale Greene;野葛（两似盐肤木）;Oriental Poison-ivy,Oriental Poison-oak ●

393467　Toxicodendron orientale Greene = Rhus ambigua H. Lév. ex Dippel ●

393468　Toxicodendron pubescens Mill. ;毛漆;Eastern Poison Oak,Poison Oak ●☆

393469　Toxicodendron quinquefoliolatum Q. H. Chen;五叶漆;Five-leaved Lacquer-tree ●

393470　Toxicodendron radicans（L.）Kuntze;毒漆藤（毒漆,野葛）;Common Eastern Poison-Ivy,Common Poison Ivy,Common Poison-ivy,Cow Itch,Markweed,Page Oak,Poison Ivy,Poison Oak,Poison Vine,Poison-ivy,Poisonous Lacquer-tree ●

393471　Toxicodendron radicans（L.）Kuntze subsp. hispidum（Engl.）Gillis;刺果毒漆藤（野葛,冶葛）;Common Eastern Poison-Ivy,Hispidfruit Poison-ivy,Poison-ivy ●

393472　Toxicodendron radicans（L.）Kuntze subsp. hispidum（Engl.）Gillis = Toxicodendron orientale Greene ●

393473　Toxicodendron radicans（L.）Kuntze subsp. negundo（Greene）Gillis = Toxicodendron radicans（L.）Kuntze subsp. hispidum（Engl.）Gillis ●

393474　Toxicodendron radicans（L.）Kuntze var. eximia（Greene）Berk. ;得克萨斯毒漆藤;Texas Poison-ivy ●☆

393475　Toxicodendron radicans（L.）Kuntze var. negundo（Greene）Reveal = Toxicodendron radicans（L.）Kuntze subsp. hispidum（Engl.）Gillis ●

393476　Toxicodendron radicans（L.）Kuntze var. rydbergii（Small ex Rydb.）Erskine = Toxicodendron rydbergii（Small ex Rydb.）Greene ●☆

393477　Toxicodendron rostratum T. L. Ming et Z. F. Chen;喙果漆;Rostrate Poison-ivy ●

393478　Toxicodendron rydbergii（Small ex Rydb.）Greene;雷氏漆;Rydberg's Poison-Ivy,Western Poison-Ivy ●☆

393479　Toxicodendron succedaneum（L.）Kuntze;野漆树（檫子漆,臭毛漆树,大木漆,红包树,林背子,栌,木腊树,木蜡树,漆,漆木,日本野漆,山漆,山漆树,山贼子,痒漆树,野漆）;Field Lacquertree,Field Lacquer-tree,Japanese Tallow,Japanese Wax Tree,Japanese Wild Sumac,Poison Sumac,Rhus Tree,Wax Tree,Wax-tree ●

393480　Toxicodendron succedaneum（L.）Kuntze = Rhus succedanea L. ●

393481　Toxicodendron succedaneum（L.）Kuntze var. acuminatum（Hook. f.）C. Y. Wu et T. L. Ming = Toxicodendron succedaneum（L.）Kuntze ●

393482　Toxicodendron succedaneum（L.）Kuntze var. acuminatum（Hook. f.）C. Y. Wu et T. L. Ming;尖叶野漆;Acuminate Lacquertree ●

393483　Toxicodendron succedaneum（L.）Kuntze var. kiangsiense C. Y. Wu;江西野漆;Jiangxi Field Lacquertree ●

393484　Toxicodendron succedaneum（L.）Kuntze var. microphyllum C. Y. Wu et T. L. Ming;小叶野漆;Littleleaf Field Lacquertree ●

393485　Toxicodendron sylvestre (Siebold et Zucc.) Kuntze;木蜡树(柴漆,木腊树,七月倍,染山红,山漆,山漆树,野毛漆,野漆疮树,野漆树);Oriental Poison Sumac, Woods Lacquertree, Woody Lacquertree ●

393486　Toxicodendron sylvestre (Siebold et Zucc.) Kuntze = Rhus sylvestris Siebold et Zucc. ●

393487　Toxicodendron toxicarium (Salisb.) Gillis = Toxicodendron pubescens Mill. ●☆

393488　Toxicodendron trichocarpum (Miq.) Kuntze;毛漆树(臭毛漆树);Hairyfruit Lacquer-tree, Hairy-fruited Lacquer-tree, Peelberry Poison Sumac ●

393489　Toxicodendron trichocarpum (Miq.) Kuntze = Rhus trichocarpa Miq. ●

393490　Toxicodendron vernicifera (DC.) F. A. Barkley = Toxicodendron vernicifluum (Stokes) F. A. Barkley ●

393491　Toxicodendron vernicifluum (Stokes) F. A. Barkley;漆树(大木漆,干漆,枦苗,美国毒漆,七,漆,山漆,黍,瞎妮子,小木漆,楂苫);Chinese Chi, Chinese Lacquer Tree, Japanese Lacquer, Japanese Lacquer Tree, Japanese Tree, Lacquer Tree, Lacquertree, Poison Ash, Poison Dogwood, Poison Elder, Poison Sumac, Poison Sumach, Swamp Sumac, True Lacquertree, True Lacquer-tree, Varnish Tree, Varnish-tree ●

393492　Toxicodendron vernicifluum (Stokes) F. A. Barkley var. shaanxiense J. Z. Zhang et Z. Y. Shang;陕西漆树(干漆,黑干漆,漆底,漆花,漆脚,漆渣,漆淬,续命筒);Shaanxi Lacquertree ●

393493　Toxicodendron vernix (L.) Kuntze = Toxicodendron vernicifluum (Stokes) F. A. Barkley ●

393494　Toxicodendron vernix Kuntze = Toxicodendron vernicifluum (Stokes) F. A. Barkley ●

393495　Toxicodendron wallichii (Hook. f.) Kuntze;绒毛漆;Chinese Lacquer-tree, Wallich Lacquertree, Wallich Lacquer-tree ●

393496　Toxicodendron wallichii (Hook. f.) Kuntze var. microcarpum C. C. Huang ex T. L. Ming;小果绒毛漆;Littlefruit Wallich Lacquertree ●

393497　Toxicodendron yunnanense C. Y. Wu;云南漆;Yunnan Lacquertree, Yunnan Lacquer-tree ●

393498　Toxicodendron yunnanense C. Y. Wu var. longipaniculatum C. Y. Wu et T. L. Ming;长序云南漆;Longpaniculate Yunnan Lacquertree ●

393499　Toxicodendrum Gaertn. = Allophylus L. ●

393500　Toxicodendrum Thunb. = Hyaenanche Lamb. ●☆

393501　Toxicodendrum acutifolium Benth. = Xymalos monospora (Harv.) Baill. ●☆

393502　Toxicodendrum capense Thunb. = Hyaenanche globosa (Gaertn.) Lamb. et Vahl ●☆

393503　Toxicodendrum globosum (Gaertn.) Pax et K. Schum. = Hyaenanche globosa (Gaertn.) Lamb. et Vahl ●☆

393504　Toxicophlaea Harv. = Acokanthera G. Don ●☆

393505　Toxicophlaea spectabilis Sond. = Acokanthera oblongifolia (Hochst.) Codd ●☆

393506　Toxicophloea Lindl. = Toxicophlaea Harv. ●☆

393507　Toxicoscordion Rydb. = Zigadenus Michx. ■

393508　Toxicoscordion acutum (Rydb.) Rydb. = Zigadenus venenosus S. Watson var. gramineus (Rydb.) O. S. Walsh ex C. L. Hitchc. ■☆

393509　Toxicoscordion arenicola A. Heller = Zigadenus venenosus S. Watson ■☆

393510　Toxicoscordion brevibracteatus (M. E. Jones) R. R. Gates = Zigadenus brevibracteatus (M. E. Jones) H. M. Hall ■☆

393511　Toxicoscordion exaltatum (Eastw.) A. Heller = Zigadenus exaltatus Eastw. ■☆

393512　Toxicoscordion falcatum (Rydb.) Rydb. = Zigadenus venenosus S. Watson var. gramineus (Rydb.) O. S. Walsh ex C. L. Hitchc. ■☆

393513　Toxicoscordion fremontii (Torr.) Rydb. = Zigadenus fremontii (Torr.) Torr. ex S. Watson ■☆

393514　Toxicoscordion fremontii (Torr.) Rydb. var. minor R. R. Gates = Zigadenus fremontii (Torr.) Torr. ex S. Watson ■☆

393515　Toxicoscordion gramineum (Rydb.) Rydb. = Zigadenus venenosus S. Watson var. gramineus (Rydb.) O. S. Walsh ex C. L. Hitchc. ■☆

393516　Toxicoscordion intermedium (Rydb.) Rydb. = Zigadenus venenosus S. Watson var. gramineus (Rydb.) O. S. Walsh ex C. L. Hitchc. ■☆

393517　Toxicoscordion micranthum (Eastw.) A. Heller = Zigadenus micranthus Eastw. ■☆

393518　Toxicoscordion nuttallii (A. Gray) Rydb. = Zigadenus nuttallii (A. Gray) S. Watson ■☆

393519　Toxicoscordion paniculatum (Nutt.) Rydb. = Zigadenus paniculatus (Nutt.) S. Watson ■☆

393520　Toxicoscordion salinum (A. Nelson) R. R. Gates = Zigadenus venenosus S. Watson ■☆

393521　Toxicoscordion texense Rydb. = Zigadenus nuttallii (A. Gray) S. Watson ■☆

393522　Toxicoscordion venenosum (S. Watson) Rydb. = Zigadenus venenosus S. Watson ■☆

393523　Toxicoscordum nuttallii (A. Gray ex S. Watson) Rydb. = Zigadenus nuttallii (A. Gray) S. Watson ■☆

393524　Toxina Noronha = Allophylus L. ●

393525　Toxocarpus Wight et Arn. (1834);弓果藤属;Bowfruitvine, Toxocarpus ●

393526　Toxocarpus africanus Oliv. = Secamone africana (Oliv.) Bullock ●☆

393527　Toxocarpus aurantiacus C. Y. Wu ex Tsiang et P. T. Li;云南弓果藤;Yunnan Bowfruitvine, Yunnan Toxocarpus ●

393528　Toxocarpus brevipes (Benth.) N. E. Br. = Secamone brevipes (Benth.) Klack. ●☆

393529　Toxocarpus caudiclavus Choux = Calyptranthera caudiclava (Choux) Klack. ●☆

393530　Toxocarpus fuscus Tsiang;锈毛弓果藤;Rustyhair Bowfruitvine, Rustyhair Toxocarpus, Rusty-haired Toxocarpus ●

393531　Toxocarpus hainanensis Tsiang;海南弓果藤;Hainan Bowfruitvine, Hainan Toxocarpus ●

393532　Toxocarpus himalensis Falc. ex Hook. f. ;西藏弓果藤(化肉藤);Himalayan Toxocarpus, Himalayas Bowfruitvine ●

393533　Toxocarpus himalensis Falc. ex Hook. f. = Toxocarpus fuscus Tsiang ●

393534　Toxocarpus laevigatus Tsiang;平滑弓果藤;Laevigate Bowfruitvine, Laevigate Lacquertree, Laevigate Toxocarpus, Smooth Toxocarpus ●

393535　Toxocarpus leonensis Scott-Elliot = Secamone leonensis (Scott-Elliot) N. E. Br. ●☆

393536　Toxocarpus letouzeyanus H. Huber = Secamone letouzeana (H. Huber) Klack. ●☆

393537　Toxocarpus lujaei (De Wild. et T. Durand) De Wild. = Secamone brevipes (Benth.) Klack. ●☆

393538　Toxocarpus ovalifolius Tsiang;圆叶弓果藤;Ovalleaf

Bowfruitvine，Ovalleaf Toxocarpus ●

393539　Toxocarpus ovalifolius Tsiang ＝ Toxocarpus wightianus Hook. et Arn. ●

393540　Toxocarpus parviflorus（Benth.）N. E. Br. ＝ Secamone brevipes（Benth.）Klack. ●☆

393541　Toxocarpus patens Tsiang；广花弓果藤；Patent Toxocarpus，Spread Bowfruitvine，Spreading Toxocarpus ●

393542　Toxocarpus paucinervis Tsiang；凌云弓果藤；Fewnerve Bowfruitvine，Fewnerve Toxocarpus，Few-nerved Toxocarpus，Lingyun Toxocarpus ●

393543　Toxocarpus racemosus（Benth.）N. E. Br. ＝ Secamone racemosa（Benth.）Klack. ●☆

393544　Toxocarpus schimperianus Hemsl.；斯帕弓果藤●☆

393545　Toxocarpus villosus（Blume）Decne.；毛弓果藤（大叶化肉藤）；Villous Bowfruitvine，Villous Toxocarpus ●

393546　Toxocarpus villosus（Blume）Decne. var. brevistylis Costantin；短柱弓果藤；Shortstyle Bowfruitvine，Shortstyle Villous Toxocarpus ●

393547　Toxocarpus villosus（Blume）Decne. var. thorelii Costantin；小叶弓果藤；Thorel Villous Bowfruitvine，Thorel Villous Toxocarpus ●

393548　Toxocarpus wangianus Tsiang；澜沧弓果藤；Lancangjian Toxocarpus，Wang Bowfruitvine，Wang Toxocarpus ●

393549　Toxocarpus wightianus Hook. et Arn.；弓果藤（牛茶藤，牛角藤，小羊角拗，小羊角藤）；Bowfruitvine，Wight Toxocarpus ●

393550　Toxophoenix Schott ＝ Astrocaryum G. Mey.（保留属名）●☆

393551　Toxopus Raf. ＝ Macranthera Nutt. ex Benth. ■☆

393552　Toxosiphon Baill.（1872）；弓管芸香属●☆

393553　Toxosiphon Baill. ＝ Erythrochiton Nees et Mart. ●☆

393554　Toxosiphon lindenii Baill.；弓管芸香●☆

393555　Toxostigma A. Rich.（1851）；弓柱紫草属☆

393556　Toxostigma A. Rich. ＝ Arnebia Forssk. ■●

393557　Toxostigma luteum A. Rich. ＝ Arnebia hispidissima（Sieber ex Lehm.）DC. ■☆

393558　Toxostigma purpurascens A. Rich. ＝ Arnebia purpurascens（A. Rich.）Baker ■☆

393559　Toxotrophis Planch. ＝ Taxotrophis Blume ●☆

393560　Toxotropis Turcz. ＝ Corynella DC. ■☆

393561　Toxylon Raf. ＝ Maclura Nutt.（保留属名）●

393562　Toxylon pomiferum Raf. ＝ Maclura pomifera（Raf.）C. K. Schneid. ●

393563　Toxylon pomiferum Raf. ex Sarg. ＝ Maclura pomifera（Raf.）C. K. Schneid. ●

393564　Toxylus Raf. ＝ Toxylon Raf. ●

393565　Tozzcttta Parl. ＝ Fritillaria L. ■

393566　Tozzcttta Parl. ＝ Theresia K. Koch ■

393567　Tozzettia Savi ＝ Alopecurus L. ■

393568　Tozzia L.（1753）；托齐列当属（阿尔卑斯玄参属）；Tozzia ■☆

393569　Tozzia alpina L.；阿尔卑斯托齐列当 ■☆

393570　Tozzia carpathica Wol.；托齐列当 ■☆

393571　Tracanthelium Kit. ex Schur ＝ Phyteuma L. ■☆

393572　Tracaulon Raf. ＝ Polygonum L.（保留属名）■●

393573　Tracaulon arifolium（L.）Raf. ＝ Persicaria arifolia（L.）Haraldson ■☆

393574　Tracaulon arifolium（L.）Raf. ＝ Polygonum arifolium L. ■

393575　Tracaulon maackianum（Regel）Greene ＝ Persicaria maackiana（Regel）Nakai ■

393576　Tracaulon maackianum（Regel）Greene ＝ Polygonum maackianum Regel ■

393577　Tracaulon muricatum（Meisn.）Greene ＝ Persicaria muricata（Meisn.）Nemoto ■

393578　Tracaulon muricatum（Meisn.）Greene ＝ Polygonum muricatum Meisn. ■

393579　Tracaulon pedunculare（Wall. ex Meisn.）Greene ＝ Polygonum dichotomum Blume ■

393580　Tracaulon perfoliatum（L.）Greene ＝ Persicaria perfoliata（L.）H. Gross ■

393581　Tracaulon perfoliatum（L.）Greene ＝ Polygonum perfoliatum L. ■

393582　Tracaulon praetermissum（Hook. f.）Greene ＝ Persicaria praetermissa（Hook. f.）H. Hara ■

393583　Tracaulon praetermissum（Hook. f.）Greene ＝ Polygonum praetermissum Hook. f. ■

393584　Tracaulon sagittatum（L.）Small ＝ Persicaria sagittata（L.）H. Gross ■

393585　Tracaulon sagittatum（L.）Small ＝ Polygonum sagittatum L. ■

393586　Tracaulon sagittatum（L.）Small var. gracilentum（Fernald）C. F. Reed ＝ Polygonum sagittatum L. ■

393587　Tracaulon sibiricum（Meisn.）Greene ＝ Polygonum sagittatum L. ■

393588　Tracaulon sieboldii（Meisn.）Greene ＝ Polygonum sagittatum L. ■

393589　Tracaulon strigosum（R. Br.）Greene ＝ Polygonum strigosum R. Br. ■

393590　Tracaulon tetragonum（Blume）Greene ＝ Polygonum dichotomum Blume ■

393591　Tracaulon thunbergii（Siebold et Zucc.）Greene ＝ Polygonum thunbergii Siebold et Zucc. ■

393592　Trachanthelium Schur ＝ Asyneuma Griseb. et Schenk ■

393593　Trachelanthus Klotzsch ＝ Begonia L. ●■

393594　Trachelanthus Klotzsch ＝ Trachelocarpus C. Muell. ●■

393595　Trachelanthus Kunze（1850）；喉花紫草属☆

393596　Trachelanthus hissaricus Lipsky；喉花紫草☆

393597　Trachelanthus korolkovii Lipsky；考氏喉花紫草☆

393598　Trachelioides Opiz ＝ Campanula L. ■●

393599　Tracheliopsis Buser ＝ Campanula L. ■●

393600　Tracheliopsis Opiz ＝ Campanula L. ■●

393601　Tracheliopsis Opiz（1852）；类疗喉草属●☆

393602　Tracheliopsis vulgaris Opiz；类疗喉草●☆

393603　Trachelium Hill ＝ Campanula L. ■●

393604　Trachelium L.（1753）；疗喉草属（喉草属）；Throatwort ■☆

393605　Trachelium Tourn. ex L. ＝ Trachelium L. ■☆

393606　Trachelium angustifolium Schousb. ＝ Feeria angustifolia（Schousb.）Buser ●☆

393607　Trachelium asperuloides Boiss. et Orph.；车叶草状疗喉草■☆

393608　Trachelium caeruleum L.；喉管花（喉草，疗喉草）；Blue Throatwort，Throatwort，Trachelium ■☆

393609　Trachelium diffusum L. f. ＝ Prismatocarpus diffusus（L. f.）A. DC. ●☆

393610　Trachelium tenuifolium L. f. ＝ Merciera tenuifolia（L. f.）A. DC. ●☆

393611　Trachelocarpns C. Muell. ＝ Begonia L. ●■

393612　Trachelosiphon Schltr. ＝ Eurystyles Wawra ■☆

393613　Trachelospermum Lem.（1851）；络石属；Chinese Ivy，Chinese Jasmine，Star Jasmine，Starjasmine，Star-jasmine ●

393614　Trachelospermum adnascens Hance ＝ Trachelospermum jasminoides（Lindl.）Lem. ●

393615 Trachelospermum anceps Dunn et Williams = Kibatalia macrophylla (Pierre ex Hua) Woodson ●

393616 Trachelospermum asiaticum (Siebold et Zucc.) Nakai;亚洲络石(白花藤,湖北络石,兰屿络石,络石,络石藤,日本络石,台湾络石,细梗络石);Asia Starjasmine, Asiatic Jasmine, Asiatic Starjasmine, Hubei Slenderstalk Starjasmine, Japanese Star Jasmine, Lanyu Starjasmine, Slender Stem Starjasmine, Slenderstalk Starjasmine, Taiwan Starjasmine, Yellow Star Jasmine ●

393617 Trachelospermum asiaticum (Siebold et Zucc.) Nakai var. brevisepalum (C. K. Schneid.) Tsiang = Trachelospermum asiaticum (Siebold et Zucc.) Nakai ●

393618 Trachelospermum asiaticum (Siebold et Zucc.) Nakai var. glabrum Nakai = Trachelospermum asiaticum (Siebold et Zucc.) Nakai ●

393619 Trachelospermum asiaticum (Siebold et Zucc.) Nakai var. glabrum Nakai;无毛日本络石;Smooth Asiatic Star-jasmine ●☆

393620 Trachelospermum asiaticum (Siebold et Zucc.) Nakai var. intermedium Nakai = Trachelospermum asiaticum (Siebold et Zucc.) Nakai ●

393621 Trachelospermum asiaticum (Siebold et Zucc.) Nakai var. intermedium Nakai;中型亚洲络石(丁络石,日本络石,亚洲络石);Intermediate Asian Starjasmine ●

393622 Trachelospermum asiaticum (Siebold et Zucc.) Nakai var. liukiuense (Hatus.) Hatus. = Trachelospermum gracilipes Hook. f. var. liukiuense (Hatus.) Kitam. ●

393623 Trachelospermum asiaticum (Siebold et Zucc.) Nakai var. majus (Nakai) Ohwi;大亚洲络石●☆

393624 Trachelospermum auritum C. K. Schneid. = Epigynum auritum (C. K. Schneid.) Tsiang et P. T. Li ●

393625 Trachelospermum axillare Hook. f.;紫花络石(车藤,辫辫果,络石,牛藤,藤杜仲,藤序络石,羊角果,脓花络石);Purpleflower Starjasmine, Purple-flowered Star-jasmine ●

393626 Trachelospermum bodinieri (H. Lév.) Woodson ex Rehder;贵州络石(长花络石,鸡屎藤,乳儿绳,台湾络石,温州络石,五根树,云南络石);Bodinier Starjasmine, Bodinier Star-jasmine, Cathaya Starjasmine, Cathayan Starjasmine, Formosan Trachelospermum, Long-flower Starjasmine, Yunnan Starjasmine ●

393627 Trachelospermum bodinieri Schneid. = Trachelospermum brevistylum Hand. -Mazz. ●

393628 Trachelospermum bowrighii Hemsl. = Gymnanthera oblonga (Burm. f.) P. S. Green ●

393629 Trachelospermum bowringii (Hance) Hemsl. = Gymnanthera oblonga (Burm. f.) P. S. Green ●

393630 Trachelospermum brevistylum Hand. -Mazz.;短柱络石;Shortstyle Starjasmine, Short-style Star-jasmine ●

393631 Trachelospermum cathayanum C. K. Schneid.;乳儿绳 ●

393632 Trachelospermum cathayanum C. K. Schneid. = Trachelospermum bodinieri (H. Lév.) Woodson ex Rehder ●

393633 Trachelospermum cathayanum C. K. Schneid. var. longipedicellatum Lingelsh. = Trachelospermum bodinieri (H. Lév.) Woodson ex Rehder ●

393634 Trachelospermum cathayanum C. K. Schneid. var. tetanocarpum (C. K. Schneid.) Tsiang et P. T. Li = Trachelospermum bodinieri (H. Lév.) Woodson ex Rehder ●

393635 Trachelospermum cathayanum C. K. Schneid. var. tetanocarpum (C. K. Schneid.) Tsiang et P. T. Li;长花络石 ●

393636 Trachelospermum cavaleriei H. Lév. = Cryptolepis buchananii Roem. et Schult. ●

393637 Trachelospermum cuneatum Tsiang = Trachelospermum brevistylum Hand. -Mazz. ●

393638 Trachelospermum difforme (Walter) A. Gray;异形络石;Climbing Dogbane ●☆

393639 Trachelospermum difforme A. Gray = Trachelospermum difforme (Walter) A. Gray ●☆

393640 Trachelospermum divaricatum K. Schum.;络石藤(白花藤,络石);Divaricate Starjasmine ●

393641 Trachelospermum divaricatum K. Schum. var. brevisepalum C. K. Schneid. = Trachelospermum asiaticum (Siebold et Zucc.) Nakai ●

393642 Trachelospermum dunnii (H. Lév.) H. Lév.;锈毛络石(大黑骨头,六角藤,韧皮络石,无腺络石,橡胶藤);Dunn Starjasmine, Firmbark Starjasmine, Glandless Star-jasmine, Glanduleless Starjasmine, Toughbark Star-jasmine ●

393643 Trachelospermum eglandulatum D. Fang = Trachelospermum dunnii (H. Lév.) H. Lév. ●

393644 Trachelospermum esquirolii H. Lév. = Melodinus hemsleyanus Diels ●

393645 Trachelospermum foetidum (Matsum. et Nakai) Nakai = Trachelospermum asiaticum (Siebold et Zucc.) Nakai ●

393646 Trachelospermum foetidum (Matsum. et Nakai) Nakai = Trachelospermum gracilipes Hook. f. var. liukiuense (Hatus.) Kitam. ●

393647 Trachelospermum formosanum Y. C. Liu et C. H. Ou;台湾络石 ●

393648 Trachelospermum formosanum Y. C. Liu et C. H. Ou = Trachelospermum bodinieri (H. Lév.) Woodson ex Rehder ●

393649 Trachelospermum fragrans Hook. f. = Trachelospermum lucidum (D. Don) Schum. ●☆

393650 Trachelospermum gracilipes Hook. f.;细梗络石 ●

393651 Trachelospermum gracilipes Hook. f. = Trachelospermum asiaticum (Siebold et Zucc.) Nakai ●

393652 Trachelospermum gracilipes Hook. f. = Trachelospermum gracilipes Hook. f. var. liukiuense (Hatus.) Kitam. ●

393653 Trachelospermum gracilipes Hook. f. var. cavaleriei (H. Lév.) Tsiang = Trachelospermum asiaticum (Siebold et Zucc.) Nakai ●

393654 Trachelospermum gracilipes Hook. f. var. hupehense Tsiang et P. T. Li;湖北络石 ●

393655 Trachelospermum gracilipes Hook. f. var. hupehense Tsiang et P. T. Li = Trachelospermum asiaticum (Siebold et Zucc.) Nakai ●

393656 Trachelospermum gracilipes Hook. f. var. liukiuense (Hatus.) Kitam.;琉球络石;Liuqiu Starjasmine ●

393657 Trachelospermum jasminoides (Lindl.) Lem.;络石(扒墙虎,白花藤,对叶肾,风藤,感冒藤,膏链,鬼系腰,过墙风,过桥风,合掌藤,交脚风,九庆藤,鲮石,领石,鹿角草,绿刺,略石,络石草,络石藤,明石,茉莉藤,耐冬,爬山虎,骑墙虎,墙络藤,乳风绳,软筋藤,石邦藤,石薜荔,石蹉,石鲮,石龙藤,石盘藤,石气柑,石血,双合草,酸树垣,台湾白花藤,薜络,剃头草,铁栏杆,铁线草,铁信,万字金银,万字茉莉,悬石,沿壁藤,羊角藤,云丹,云花,云英,云珠,折骨草);African Jasmine, China Starjasmine, Chinese Jasmine, Chinese Star Jasmine, Chinese Starjasmine, Chinese Star-jasmine, Confederate Jasmine, Confederate-jasmine, Diversifolious China Starjasmine, Diversifolious Chinese Starjasmine, Jasmine Confederate, Malayan Jasmine, Star Jasmine, Star-jasmine, Stoneblood ●

393658 Trachelospermum jasminoides (Lindl.) Lem. 'Tricolor';三色络石;Variegata Star Jasmine ●☆

393659 Trachelospermum jasminoides (Lindl.) Lem. 'Variegatum';变色络石(斑叶络石);Variegated Confederate-jasmine, Variegated

Star Jasmine ●

393660　Trachelospermum jasminoides（Lindl.）Lem. subsp. foetidum Matsu-mura et Nakai＝Trachelospermum asiaticum（Siebold et Zucc.）Nakai ●

393661　Trachelospermum jasminoides（Lindl.）Lem. subsp. foetidum Matsum. et Nakai＝Trachelospermum gracilipes Hook. f. var. liukiuense（Hatus.）Kitam. ●

393662　Trachelospermum jasminoides（Lindl.）Lem. subsp. foetidum Matsum. et Nakai＝Trachelospermum asiaticum（Siebold et Zucc.）Nakai ●

393663　Trachelospermum jasminoides（Lindl.）Lem. var. heterophyllum Tsiang＝Trachelospermum jasminoides（Lindl.）Lem. ●

393664　Trachelospermum jasminoides（Lindl.）Lem. var. pubescens Makino；毛脉络石；Hairyvein Starjasmine ●☆

393665　Trachelospermum jasminoides（Lindl.）Lem. var. pubescens Makino＝Trachelospermum jasminoides（Lindl.）Lem. ●

393666　Trachelospermum jasminoides（Lindl.）Lem. var. pubescens Makino f. variegatum Hatus. et S. Toyama；斑叶毛脉络石●☆

393667　Trachelospermum jasminoides（Lindl.）Lem. var. variegatum W. T. Mill.＝Trachelospermum jasminoides（Lindl.）Lem. 'Variegatum' ●

393668　Trachelospermum jasminoides（Lindl.）Lem. var. variegatum W. T. Mill.＝Trachelospermum jasminoides（Lindl.）Lem. ●

393669　Trachelospermum kuraruense Masam. ；屏东络石（台岛络石）；Pingdong Starjasmine ●

393670　Trachelospermum lanyuense C. E. Chang；兰屿络石●

393671　Trachelospermum lanyuense C. E. Chang＝Trachelospermum asiaticum（Siebold et Zucc.）Nakai ●

393672　Trachelospermum lanyuense C. E. Chang＝Trachelospermum gracilipes Hook. f. var. liukiuense（Hatus.）Kitam. ●

393673　Trachelospermum liukiuense Hatus. ＝Trachelospermum gracilipes Hook. f. var. liukiuense（Hatus.）Kitam. ●

393674　Trachelospermum longipedicellatum（Lingelsh.）Woodson＝Trachelospermum bodinieri（H. Lév.）Woodson ex Rehder ●

393675　Trachelospermum longipedicellatum Woodon＝Trachelospermum bodinieri（H. Lév.）Woodson ex Rehder ●

393676　Trachelospermum lucidum（D. Don）Schum. ；亮络石●☆

393677　Trachelospermum majus Nakai；大络石（巨络石）●

393678　Trachelospermum majus Nakai＝Trachelospermum asiaticum（Siebold et Zucc.）Nakai ●

393679　Trachelospermum navaillei H. Lév. ＝Aganosma schlechteriana H. Lév. ●

393680　Trachelospermum rubrinerve H. Lév. ＝Trachelospermum dunnii（H. Lév.）H. Lév. ●

393681　Trachelospermum siamense Craib＝Trachelospermum asiaticum（Siebold et Zucc.）Nakai ●

393682　Trachelospermum suaveolens Chun＝Trachelospermum brevistylum Hand. -Mazz. ●

393683　Trachelospermum tenax Tsiang；韧皮络石●

393684　Trachelospermum tenax Tsiang＝Trachelospermum dunnii（H. Lév.）H. Lév. ●

393685　Trachelospermum tetanocarpum C. K. Schneid. ＝Trachelospermum bodinieri（H. Lév.）Woodson ex Rehder ●

393686　Trachelospermum verrucosum（Blume）Boerl. ＝Chonemorpha verrucosa（Blume）D. J. Middleton ●

393687　Trachelospermum wenchowense Tsiang＝Trachelospermum bodinieri（H. Lév.）Woodson ex Rehder ●

393688　Trachelospermum wenchowense Tsiang＝Trachelospermum cathayanum C. K. Schneid. ●

393689　Trachelospermum yunnanense Tsiang et P. T. Li；云南络石●

393690　Trachelospermum yunnanense Tsiang et P. T. Li＝Trachelospermum bodinieri（H. Lév.）Woodson ex Rehder ●

393691　Trachinema Raf. ＝Anthericum L. ■☆

393692　Trachodes D. Don＝Sonchus L. ■

393693　Trachoma Garay＝Tuberolabium Yamam. ■

393694　Trachomitum Woodson＝Apocynum L. ●■

393695　Trachomitum Woodson（1930）；茶叶花属●☆

393696　Trachomitum armenum（Pobed.）Pobed. ＝Apocynum armenum Pobed. ●☆

393697　Trachomitum lancifolium（Russanov）Pobed. ＝Apocynum lancifolium Russanov ●☆

393698　Trachomitum lancifolium（Russanov）Pobed. ＝Apocynum venetum L. ●

393699　Trachomitum russanovii（Pobed.）Pobed. ＝Apocynum rusaanovii Pobed. ●☆

393700　Trachomitum sarmatiense Woodson＝Apocynum sarmatiense（Woodson）O. D. Wissjul. ●☆

393701　Trachomitum scabrum（Russanov）Pobed. ＝Apocynum scabrum Russanov ●☆

393702　Trachomitum scabrum（Russanov）Pobed. ＝Trachomitum venetum（L.）Woodson subsp. scabrum（Russanov）Rech. f. ●☆

393703　Trachomitum tauricum（Pobed.）Pobed. ＝Apocynum tauricum Pobed. ●☆

393704　Trachomitum venetum（L.）Woodson＝Apocynum venetum L. ●

393705　Trachomitum venetum（L.）Woodson subsp. armenum（Pobed.）Rech. f. ＝Apocynum armenum Pobed. ●☆

393706　Trachomitum venetum（L.）Woodson subsp. scabrum（Russanov）Rech. f. ＝Apocynum scabrum Russanov ●☆

393707　Trachomitum venetum（L.）Woodson subsp. tauricum（Pobed.）Greuter et Burdet＝Apocynum tauricum Pobed. ●☆

393708　Trachomitum venetum（L.）Woodson var. basikurumon（H. Hara）H. Hara＝Apocynum venetum L. var. basikurumon（H. Hara）H. Hara ●☆

393709　Trachomitum venetum（L.）Woodson var. basikurumon H. Hara＝Apocynum venetum L. ●

393710　Trachomitum venetum（L.）Woodson var. ellipticifolium（Bég. et Bél.）Woodson＝Apocynum venetum L. ●

393711　Trachopyron J. Gerard ex Raf. ＝Fagopyrum Mill. （保留属名）●■

393712　Trachopyron Raf. ＝Fagopyrum Mill. （保留属名）●■

393713　Trachyandra Kunth（1843）（保留属名）；糙蕊阿福花属■☆

393714　Trachyandra Kunth（保留属名）＝Anthericum L. ■☆

393715　Trachyandra acocksii Oberm. ；阿氏糙蕊阿福花■☆

393716　Trachyandra adamsonii（Compton）Oberm. ；亚当森糙蕊阿福花■☆

393717　Trachyandra affinis Kunth；近缘糙蕊阿福花■☆

393718　Trachyandra arenicola J. C. Manning et Goldblatt；沙生糙蕊阿福花■☆

393719　Trachyandra arvensis（Schinz）Oberm. ；田野阿福花■☆

393720　Trachyandra asperata Kunth；糙蕊阿福花■☆

393721　Trachyandra asperata Kunth var. basutoensis（Poelln.）Oberm. ；巴苏托阿福花■☆

393722　Trachyandra asperata Kunth var. carolinensis Oberm. ；卡罗来纳糙蕊阿福花■☆

393723　Trachyandra asperata Kunth var. macowanii（Baker）Oberm. ；

麦克欧文糙蕊阿福花■☆

393724　Trachyandra asperata Kunth var. nataglencoensis（Kuntze）Oberm.；纳塔尔糙蕊阿福花■☆

393725　Trachyandra asperata Kunth var. stenophylla（Baker）Oberm.；窄叶糙蕊阿福花■☆

393726　Trachyandra blepharophora（Roem. et Schult.）Kunth = Trachyandra ciliata（L. f.）Kunth ■☆

393727　Trachyandra brachypoda（Baker）Oberm.；短足糙蕊阿福花☆

393728　Trachyandra bracteosa Kunth = Trachyandra ciliata（L. f.）Kunth ■☆

393729　Trachyandra brehmeana（Schult. et Schult. f.）Kunth = Chlorophytum triflorum（Aiton）Kunth ■☆

393730　Trachyandra bulbinifolia（Dinter）Oberm.；球百合糙蕊阿福花■☆

393731　Trachyandra burkei（Baker）Oberm.；伯克阿福花■☆

393732　Trachyandra canaliculata（Aiton）Kunth = Trachyandra ciliata（L. f.）Kunth ■☆

393733　Trachyandra capillata（Poelln.）Oberm.；细毛糙蕊阿福花☆

393734　Trachyandra chlamydophylla（Baker）Oberm.；斗篷叶糙蕊阿福花■☆

393735　Trachyandra ciliata（L. f.）Kunth；睫毛糙蕊阿福花☆

393736　Trachyandra corymbosa Kunth = Trachyandra hirsuta（Thunb.）Kunth ■☆

393737　Trachyandra dissecta Oberm.；深裂糙蕊阿福花■☆

393738　Trachyandra divaricata（Jacq.）Kunth；叉开糙蕊阿福花■☆

393739　Trachyandra elongata（Willd.）Kunth = Trachyandra revoluta（L.）Kunth ■☆

393740　Trachyandra ensifolia（Sölch）Rössler；密叶糙蕊阿福花■☆

393741　Trachyandra erythrorrhiza（Conrath）Oberm.；红根糙蕊阿福花■☆

393742　Trachyandra esterhuyseniae Oberm.；埃斯特糙蕊阿福花■☆

393743　Trachyandra falcata（L. f.）Kunth；镰形糙蕊阿福花☆

393744　Trachyandra filiformis（Aiton）Oberm.；丝状糙蕊阿福花■☆

393745　Trachyandra fimbriata（Thunb.）Kunth = Trachyandra muricata（L. f.）Kunth ■☆

393746　Trachyandra flexifolia（L. f.）Kunth；弯叶糙蕊阿福花■☆

393747　Trachyandra gerrardii（Baker）Oberm.；杰勒德糙蕊阿福花■☆

393748　Trachyandra giffenii（F. M. Leight.）Oberm.；吉芬糙蕊阿福花■☆

393749　Trachyandra glandulosa（Dinter）Oberm.；具腺糙蕊阿福花■☆

393750　Trachyandra gracilenta Oberm.；细黏糙蕊阿福花☆

393751　Trachyandra hirsuta（Thunb.）Kunth；粗毛糙蕊阿福花☆

393752　Trachyandra hirsutiflora（Adamson）Oberm.；粗毛花糙蕊阿福花■☆

393753　Trachyandra hispida（L.）Kunth；硬毛糙蕊阿福花■☆

393754　Trachyandra humilis Kunth = Trachyandra asperata Kunth ■☆

393755　Trachyandra involucrata（Baker）Oberm.；总苞糙蕊阿福花■☆

393756　Trachyandra jacquiniana（Roem. et Schult.）Oberm.；雅坎糙蕊阿福花☆

393757　Trachyandra jacquinii Kunth = Trachyandra jacquiniana（Roem. et Schult.）Oberm.■☆

393758　Trachyandra karrooica Oberm.；卡卢糙蕊阿福花■☆

393759　Trachyandra lanata（Dinter）Oberm.；绵毛糙蕊阿福花■☆

393760　Trachyandra laxa（N. E. Br.）Oberm.；疏松阿福花■☆

393761　Trachyandra laxa（N. E. Br.）Oberm. var. erratica（Oberm.）Oberm. = Trachyandra laxa（N. E. Br.）Oberm. var. rigida（Suess.）Rössler ■☆

393762　Trachyandra laxa（N. E. Br.）Oberm. var. rigida（Suess.）Rössler；硬疏松阿福花■☆

393763　Trachyandra longepedunculata（Steud. ex Roem. et Schult.）Kunth = Trachyandra filiformis（Aiton）Oberm.■☆

393764　Trachyandra longifolia（Jacq.）Kunth = Trachyandra ciliata（L. f.）Kunth ■☆

393765　Trachyandra mandrarensis（H. Perrier）Marais et Reilly；曼德拉糙蕊阿福花■☆

393766　Trachyandra margaretae Oberm.；马格丽特糙蕊阿福花■☆

393767　Trachyandra montana J. C. Manning et Goldblatt；山地糙蕊阿福花■☆

393768　Trachyandra muricata（L. f.）Kunth；硬尖糙蕊阿福花■☆

393769　Trachyandra oligotricha（Baker）Oberm.；寡毛糙蕊阿福花■☆

393770　Trachyandra paniculata Oberm.；圆锥糙蕊阿福花■☆

393771　Trachyandra paradoxa（Schult. et Schult. f.）Kunth = Trachyandra hispida（L.）Kunth ■☆

393772　Trachyandra patens Oberm.；铺展糙蕊阿福花■☆

393773　Trachyandra pauciflora（Thunb.）Kunth = Chlorophytum triflorum（Aiton）Kunth ■☆

393774　Trachyandra peculiaris（Dinter）Oberm.；特殊糙蕊阿福花■☆

393775　Trachyandra prolifera P. L. Perry；多育糙蕊阿福花■☆

393776　Trachyandra pyrenicarpa（Welw. ex Baker）Oberm.；核果糙蕊阿福花■☆

393777　Trachyandra reflexipilosa（Kuntze）Oberm.；折毛糙蕊阿福花■☆

393778　Trachyandra revoluta（L.）Kunth；外卷糙蕊阿福花■☆

393779　Trachyandra sabulosa（Adamson）Oberm.；砂地阿福花■☆

393780　Trachyandra saltii（Baker）Oberm.；萨尔阿福花■☆

393781　Trachyandra saltii（Baker）Oberm. var. secunda（K. Krause et Dinter）Oberm. = Trachyandra saltii（Baker）Oberm.■☆

393782　Trachyandra scabra（L. f.）Kunth；粗糙糙蕊阿福花■☆

393783　Trachyandra schultesii Kunth = Chlorophytum triflorum（Aiton）Kunth ■☆

393784　Trachyandra tabularis（Baker）Oberm.；扁平糙蕊阿福花■☆

393785　Trachyandra thyrsoidea（Baker）Oberm.；聚伞糙蕊阿福花■☆

393786　Trachyandra tortilis（Baker）Oberm.；螺旋状糙蕊阿福花■☆

393787　Trachyandra triquetra Thulin；三棱糙蕊阿福花■☆

393788　Trachyandra undulata（Thunb.）Kunth = Trachyandra hispida（L.）Kunth ■☆

393789　Trachyandra vespertina（Jacq.）Kunth = Trachyandra ciliata（L. f.）Kunth ■☆

393790　Trachyandra zebrina（Schltr. ex Poelln.）Oberm.；条斑糙蕊阿福花■☆

393791　Trachycalymma（K. Schum.）Bullock（1953）；糙被萝藦属■☆

393792　Trachycalymma Bullock = Trachycalymma（K. Schum.）Bullock ■☆

393793　Trachycalymma amoenum（K. Schum.）Goyder；秀丽糙被萝藦■☆

393794　Trachycalymma buchwaldii（Schltr. et K. Schum.）Goyder；布赫糙被萝藦■☆

393795　Trachycalymma cristatum（Decne.）Bullock；冠状糙被萝藦●☆

393796　Trachycalymma cucullatum（Schltr.）Bullock = Asclepias cucullata（Schltr.）Schltr.■☆

393797　Trachycalymma fimbriatum（Weim.）Bullock；流苏糙被萝藦■☆

393798　Trachycalymma foliosum（K. Schum.）Goyder；多叶糙被萝藦■☆

393799　Trachycalymma graminifolium（Wild）Goyder；禾叶糙被萝藦●☆

393800　Trachycalymma minutiflorum Goyder；微花糙被萝藦■☆

393801　Trachycalymma pulchellum（Decne.）Bullock；美丽糙被萝藦■☆

393802　Trachycalymma shabaense Goyder；沙巴糙被萝藦■☆

393803　Trachycarpus H. Wendl.（1863）；棕榈属；Chusan Palm, Fan Palm, Hemp Palm, Palm, Windmill Palm, Windmillpalm, Windmillpalm ●

393804 Trachycarpus argyratus S. K. Lee et F. N. Wei = Guihaia argyrata（S. K. Lee et F. N. Wei）S. K. Lee, F. N. Wei et J. Dransf. ●

393805 Trachycarpus argyratus S. K. Lee et F. N. Wei = Trachycarpus nanus Becc. ●◇

393806 Trachycarpus caespitosus Roster；丛簇棕榈；Bush Windmill-palm ●☆

393807 Trachycarpus dracocephalus Ching et Y. C. Hsu = Trachycarpus nanus Becc. ●◇

393808 Trachycarpus excelsus H. Wendl. = Trachycarpus fortunei（Hook.）H. Wendl. ●

393809 Trachycarpus fortunei（Hook.）H. Wendl.；棕榈（垂叶棕榈，大崖棕，山棕，棕树）；China Coir Palm, Chinese Coir Palm, Chinese Fan Palm, Chinese Windmill Palm, Chusan Palm, Coir Palm, Fan Palm, Fortune, Fortune Palm, Fortune's Windmill Palm, Hemp Palm, Palm, Windmill Palm ●

393810 Trachycarpus fortunei（Hook.）H. Wendl. var. surculosa Henry；吸枝棕榈；Sucker Windmillpalm ●

393811 Trachycarpus khasianus H. Wendl. = Trachycarpus martianus（Wall.）H. Wendl. ●◇

393812 Trachycarpus latisectus Spanner, Noltie et Gibbons；宽齿山棕榈；Windamere Palm ●☆

393813 Trachycarpus martianus（Wall.）H. Wendl.；山棕榈；Himalayan Fan Palm, Mart Hemp Palm, Martius' Chusan Palm, Martius' Windmillpalm, Wild Palm ●◇

393814 Trachycarpus nanus Becc.；龙棕（龙棷）；Dwarf Windmillpalm, Gragon Palm ●◇

393815 Trachycarpus takil Becc.；塔基棕榈；Takil Palm, Takil Windmill-palm ●☆

393816 Trachycarpus wagnerianus Becc.；瓦氏棕榈（棕榈）；Wagner Windmillpalm ●

393817 Trachycaryon Klotzsch = Adriana Gaudich. ●☆

393818 Trachydium Lindl.（1835）；瘤果芹属（粗子芹属）；Trachydium ■

393819 Trachydium abyssinicum（Hochst.）Hiern = Haplosciadium abyssinicum Hochst. ■☆

393820 Trachydium abyssinicum（Hochst.）Hiern var. fischeri Engl. = Haplosciadium abyssinicum Hochst. ■☆

393821 Trachydium abyssinicum（Hochst.）Hiern var. kilimandschari Engl. = Haplosciadium abyssinicum Hochst. ■☆

393822 Trachydium abyssinicum（Hochst.）Hiern var. lindblomii H. Wolff = Haplosciadium abyssinicum Hochst. ■☆

393823 Trachydium affine W. WSm. = Chamaesium viridiflorum（Franch.）H. Wolff ex R. H. Shan ■

393824 Trachydium astrantioideum H. Boissieu = Pleurospermum astrantioideum（H. Boissieu）K. T. Fu et Y. C. Ho ■

393825 Trachydium chinense M. Hiroe = Ligusticum hispidum（Franch.）H. Wolff ■

393826 Trachydium chinense M. Hiroe = Ligusticum likiangense（H. Wolff）F. T. Pu et M. F. Watson ■

393827 Trachydium chloroleucum Diels. = Pleurospermum hookeri C. B. Clarke var. thomsonii C. B. Clarke ■

393828 Trachydium commutatum（Regel et Schmalh.）M. Hiroe = Pleurospermum simplex（Rupr.）Benth. et Hook. f. ex Drude ■

393829 Trachydium daucoides Franch. = Ligusticum daucoides（Franch.）Franch. ■

393830 Trachydium delavayi Franch. = Chamaesium delavayi（Franch.）R. H. Shan et S. L. Liou ■

393831 Trachydium depressum Boiss. subsp. chitralicum Rech. f. et Riedl；吉德拉尔瘤果芹■☆

393832 Trachydium dichotomum Korovin；二歧瘤果芹■☆

393833 Trachydium dissectum C. B. Clarke = Schulzia dissecta（C. B. Clarke）C. Norman ■

393834 Trachydium forrestii Diels = Physospermopsis forrestii（Diels）C. Norman ■

393835 Trachydium forrestii Diels = Physospermopsis shaniana C. Y. Wu et F. T. Pu ■

393836 Trachydium fuscopurpureum Hand.-Mazz. = Pleurospermum heterosciadium H. Wolff ■

393837 Trachydium hirsutulum C. B. Clarke = Physospermopsis obtusiuscula（Wall. ex DC.）C. Norman ■

393838 Trachydium hispidum Franch. = Ligusticum hispidum（Franch.）H. Wolff ■

393839 Trachydium hispidum H. Wolff = Ligusticum likiangense（H. Wolff）F. T. Pu et M. F. Watson ■

393840 Trachydium involucellatum R. H. Shan et F. T. Pu；裂苞瘤果芹；Splitbract Trachydium, Splitbract Tumorfruitparsley ■

393841 Trachydium kingdon-wardii H. Wolff；云南瘤果芹（阿墩粗子芹）；Kingdon-ward Trachydium, Yunnan Tumorfruitparsley ■

393842 Trachydium kopetdashense Korovin；科佩特瘤果芹■☆

393843 Trachydium lichiangense C. Y. Wu；刚毛粗子芹；Hispid Trachydium ■

393844 Trachydium lichiangense C. Y. Wu = Ligusticum likiangense（H. Wolff）F. T. Pu et M. F. Watson ■

393845 Trachydium loloense（Franch.）M. Hiroe = Tongoloa loloensis（H. Boissieu）H. Wolff ■

393846 Trachydium markgrafianum Fedde ex H. Wolff. = Chamaesium viridiflorum（Franch.）H. Wolff ex R. H. Shan ■

393847 Trachydium napiferum H. Wolff = Tongoloa napifera（H. Wolff）C. Norman ■

393848 Trachydium novemjugum C. B. Clarke var. tongolense H. Boissieu = Chamaesium novem-jugum（C. B. Clarke）C. Norman ■

393849 Trachydium obtusiusculum（Wall. ex DC.）C. B. Clarke = Physospermopsis obtusiuscula（C. B. Clarke）C. Norman ■

393850 Trachydium obtusiusculum C. B. Clarke = Physospermopsis obtusiuscula（C. B. Clarke）C. Norman ■

393851 Trachydium obtusiusculum C. B. Clarke var. strictum C. B. Clarke = Physospermopsis obtusiuscula（Wall. ex DC.）C. Norman ■

393852 Trachydium paradoxum（H. Wolff）M. Hiroe = Chamaesium paradoxum H. Wolff ■

393853 Trachydium purpurascens Franch. = Pleurospermum nanum Franch. ■

393854 Trachydium rockii H. Wolff；丽江粗子芹；Rock Trachydium ■

393855 Trachydium rockii H. Wolff = Ligusticum hispidum（Franch.）H. Wolff ■

393856 Trachydium roylei Lindl.；瘤果芹（粗子草）；Royle Trachydium, Tumorfruitparsley ■

393857 Trachydium rubrinerve Franch.；红脉粗子芹；Redvein Trachydium ■

393858 Trachydium rubrinerve Franch. = Physospermopsis rubrinervis（Franch.）C. Norman ■

393859 Trachydium simplicifolium W. W. Sm.；单叶瘤果芹（单叶粗子芹）；Simpleleaf Trachydium, Singleleaf Tumorfruitparsley ■

393860 Trachydium spatuliferum W. W. Sm. = Chamaesium novem-jugum（C. B. Clarke）C. Norman ■

393861 Trachydium spatuliferum W. W. Sm. = Chamaesium spatuliferum (W. W. Sm.) C. Norman ■

393862 Trachydium subnudum C. B. Clarke ex H. Wolff;密瘤瘤果芹;Dense Tumorfruitparsley,Warty Trachydium ■

393863 Trachydium thalictrifolium (H. Wolff) M. Hiroe = Chamaesium thalictrifolium H. Wolff ■

393864 Trachydium tianschanicum Korovin;天山瘤果芹;Tianshan Trachydium,Tianshan Tumorfruitparsley ■

393865 Trachydium tibetanicum H. Wolff;西藏瘤果芹(滇藏粗子芹);Tibet Trachydium,Xizang Trachydium,Xizang Tumorfruitparsley ■

393866 Trachydium trifoliatum H. Wolff;三叶瘤果芹(三叶粗子芹);Threeleves Trachydium ■

393867 Trachydium verrucosum R. H. Shan et F. T. Pu. = Trachydium subnudum C. B. Clarke ex H. Wolff ■

393868 Trachydium viridiflorum Franch. = Chamaesium viridiflorum (Franch.) H. Wolff ex R. H. Shan ■

393869 Trachydium wolffianum Fedde ex H. Wolff;白马山粗子芹;Wolff Trachydium ■

393870 Trachydium yunnanense M. Hiroe = Chamaesium wolffianum Fedde ex H. Wolff ■

393871 Trachylobium Hayne = Hymenaea L. ●

393872 Trachylobium Hayne(1827);粗裂豆属;Rasp Pod,Rasp-pod ■☆

393873 Trachylobium floribundum G. Don;多花粗裂豆 ●☆

393874 Trachylobium homemannianum Hayne = Hymenaea verrucosa Gaertn. ●

393875 Trachylobium mossambicense Klotzsch;莫桑比克粗裂豆■☆

393876 Trachylobium verrucosum (Gaertn.) Oliv. = Hymenaea verrucosa Gaertn. ●

393877 Trachyloma Pfeiff. = Scleria P. J. Bergius ■

393878 Trachyloma Pfeiff. = Trachylomia Nees ■

393879 Trachylomia Nees = Scleria P. J. Bergius ■

393880 Trachymarathrum Tausch = Hippomarathrum Link ■☆

393881 Trachymene DC. = Platysace Bunge ■☆

393882 Trachymene Rudge(1811);翠珠花属(蓝饰带花属)■☆

393883 Trachymene coerulea Graham;翠珠花;Blue Lace Flower,Blue Lace-flower,Hemp,Queen Anne's Lace,Rottnest Daisy ■☆

393884 Trachynia Link = Brachypodium P. Beauv. ■

393885 Trachynia Link(1827);特拉禾属■☆

393886 Trachynia distachya (L.) Link = Brachypodium distachyon (L.) P. Beauv. ■☆

393887 Trachynia distachya (L.) Link var. hispida (Pamp.) Bor = Brachypodium distachyon (L.) P. Beauv. ■

393888 Trachynia platystachya (Coss.) H. Scholz;宽穗特拉禾■☆

393889 Trachynotia Michx. = Spartina Schreb. ex J. F. Gmel. ■

393890 Trachynotia alterniflora (Loisel.) DC. = Spartina alterniflora Loisel. ■

393891 Trachyozus Rchb. = Trachys Pers. ■☆

393892 Trachypetalum Szlach. et Sawicka = Habenaria Willd. ■

393893 Trachypetalum Szlach. et Sawicka(2003);糙瓣兰属■☆

393894 Trachyphrynium Benth. (1883);糙柊叶属■☆

393895 Trachyphrynium K. Schum. = Hypselodelphys (K. Schum.) Milne-Redh. ■☆

393896 Trachyphrynium braunianum (K. Schum.) Baker;布劳恩糙柊叶■☆

393897 Trachyphrynium hirsutum Loes. = Hypselodelphys hirsuta (Loes.) Koechlin ■☆

393898 Trachyphrynium liebrechtsianum De Wild. et T. Durand = Haumania liebrechtsiana (De Wild. et T. Durand) J. Léonard ■☆

393899 Trachyphrynium poggeanum K. Schum. = Hypselodelphys poggeana (K. Schum.) Milne-Redh. ■☆

393900 Trachyphrynium violaceum Ridl. = Hypselodelphys violacea (Ridl.) Milne-Redh. ■☆

393901 Trachyphrynium zenkerianum K. Schum. = Hypselodelphys zenkeriana (K. Schum.) Milne-Redh. ■☆

393902 Trachyphytum Nutt. = Mentzelia L. ●■☆

393903 Trachyphytum Nutt. ex Torr. et A. Gray = Mentzelia L. ●■☆

393904 Trachypleurum Rchb. = Bupleurum L. ●■

393905 Trachypoa Bubani = Dactylis L. ■

393906 Trachypogon Nees(1829);糙须禾属■

393907 Trachypogon capensis (Thunb.) Trin. = Trachypogon spicatus (L. f.) Kuntze ■☆

393908 Trachypogon chevalieri (Stapf) Jacq. -Fél.;舍瓦利耶糙须禾■☆

393909 Trachypogon durus Stapf = Trachypogon spicatus (L. f.) Kuntze ■☆

393910 Trachypogon glaucescens Pilg. = Trachypogon spicatus (L. f.) Kuntze ■☆

393911 Trachypogon hirtus (L.) Nees = Hyparrhenia hirta (L.) Stapf ■☆

393912 Trachypogon involutus Pilg. = Trachypogon spicatus (L. f.) Kuntze ■☆

393913 Trachypogon ledermannii Pilg. = Trachypogon chevalieri (Stapf) Jacq. -Fél. ■☆

393914 Trachypogon planifolius Stapf = Trachypogon spicatus (L. f.) Kuntze ■☆

393915 Trachypogon plumosus (Humb. et Bonpl. ex Willd.) Nees = Trachypogon spicatus (L. f.) Kuntze ■☆

393916 Trachypogon polymorphus Hack. var. capensis ? = Trachypogon spicatus (L. f.) Kuntze ■☆

393917 Trachypogon polymorphus Hack. var. thollonii Franch. = Trachypogon spicatus (L. f.) Kuntze ■☆

393918 Trachypogon rufus Nees = Hyparrhenia rufa (Nees) Stapf ■☆

393919 Trachypogon spicatus (L. f.) Kuntze;穗状糙须禾■☆

393920 Trachypogon thollonii (Franch.) Stapf = Trachypogon spicatus (L. f.) Kuntze ■☆

393921 Trachypyrum Post et Kuntze = Fagopyrum Mill. (保留属名)●■

393922 Trachypyrum Post et Kuntze = Trachopyron Raf. ●■

393923 Trachyrhizum (Schltr.) Brieger = Dendrobium Sw. (保留属名)■

393924 Trachyrhynchium Nees = Cladium P. Browne ■

393925 Trachyrhyngium Kunth = Cladium P. Browne ■

393926 Trachyrhyngium Kunth = Trachyrhynchium Nees ■

393927 Trachyrhyngium Nees ex Kunth = Cladium P. Browne ■

393928 Trachys Pers. (1805);单翼草属■☆

393929 Trachys muricata (L.) Pers. ;单翼草■☆

393930 Trachysciadium (DC.) Eckl. et Zeyh. = Chamarea Eckl. et Zeyh. ■☆

393931 Trachysciadium (DC.) Eckl. et Zeyh. = Pimpinella L. ■

393932 Trachysciadium Eckl. et Zeyh. = Chamarea Eckl. et Zeyh. ■☆

393933 Trachysciadium Eckl. et Zeyh. = Pimpinella L. ■

393934 Trachysciadium capense Eckl. et Zeyh. = Ezosciadium capense (Eckl. et Zeyh.) B. L. Burtt ☆

393935 Trachysperma Raf. = Limnanthemum S. G. Gmel. ■

393936 Trachysperma Raf. = Nymphoides Ség. ■

393937 Trachyspermum Link(1821)(保留属名);糙果芹属(粗果芹属,蔓芹属);Ajowan,Roughfruitparsley ■

393938 Trachyspermum aethusifolium Chiov. = Trachyspermum pimpinelloides（Balf. f.）H. Wolff ■☆

393939 Trachyspermum aethusifolium Chiov. var. verruculosum C. C. Towns. = Trachyspermum pimpinelloides（Balf. f.）H. Wolff ■☆

393940 Trachyspermum ammi（L.）Sprague；细叶糙果芹（阿米糙果芹，阿米粗果芹，阿育魏，阿育魏实，糙果芹，粗糙芹，粗果芹）；Ajowan，Ajowan Caraway，Bombay Mix ■

393941 Trachyspermum clavatum Nasir；棍棒糙果芹 ■☆

393942 Trachyspermum copticum（L.）Link var. maritimum Chiov. = Trachyspermum pimpinelloides（Balf. f.）H. Wolff ■☆

393943 Trachyspermum copticum（L.）Link. = Trachyspermum ammi（L.）Sprague ■

393944 Trachyspermum involucratum（Roxb.）H. Wolff = Trachyspermum roxburghianum（DC.）H. Wolff ■

393945 Trachyspermum involucratum Maire = Stoibrax involucratum（Braun-Blanq. et Maire）B. L. Burtt ■☆

393946 Trachyspermum pimpinelloides（Balf. f.）H. Wolff；茴芹状糙果芹 ■☆

393947 Trachyspermum pomelianum Maire = Stoibrax pomelianum（Maire）B. L. Burtt ■☆

393948 Trachyspermum roxburghianum（DC.）Craib = Trachyspermum roxburghianum（DC.）H. Wolff ■

393949 Trachyspermum roxburghianum（DC.）H. Wolff；具苞糙果芹（滇南糙果芹，具苞蔓芹，罗氏粗子芹）；Roxburgh Roughfruitparsley ■

393950 Trachyspermum scaberulum（Franch.）H. Wolff = Trachyspermum scaberulum（Franch.）H. Wolff ex Hand.-Mazz. ■

393951 Trachyspermum scaberulum（Franch.）H. Wolff ex Hand.-Mazz.；糙果芹（粗子芹，微粗蔓芹）；Scabrous Roughfruitparsley ■

393952 Trachyspermum scaberulum（Franch.）H. Wolff ex Hand.-Mazz. var. ambrosiifolium（Franch.）R. H. Shan；豚草叶糙果芹（丽江粗蔓芹）；Lijiang Roughfruitparsley ■

393953 Trachyspermum stictocarpum（C. B. Clarke）H. Wolff. = Trachyspermum roxburghianum（DC.）H. Wolff ■

393954 Trachyspermum triradiatum H. Wolff；马尔康糙果芹；Malkang Roughfruitparsley ■

393955 Trachystachys A. Dietr. = Trachys Pers. ■☆

393956 Trachystella Steud. = Tetracera L. ●

393957 Trachystella Steud. = Trachytella DC. ●

393958 Trachystemma Meisn. = Trachystemon D. Don ●☆

393959 Trachystemon D. Don（1832）；糙蕊紫草属；Abraham-isaac-Jacob ●☆

393960 Trachystemon creticum D. Don ex G. Don；糙蕊紫草 ●☆

393961 Trachystemon orientalis（L.）G. Don；东方糙蕊紫草；Abraham-isaac-and-jacob，Eastern Borage ●☆

393962 Trachystigma C. B. Clarke（1883）；糙柱苣苔属 ■☆

393963 Trachystigma mannii C. B. Clarke；糙柱苣苔 ■☆

393964 Trachystoma O. E. Schulz（1916）；糙嘴芥属 ■☆

393965 Trachystoma aphanoneurum（Maire et Weiller）Maire；非洲糙嘴芥 ■☆

393966 Trachystoma ballii O. E. Schulz；巴尔糙嘴芥 ■☆

393967 Trachystoma labasii Maire；糙嘴芥 ■☆

393968 Trachystylis S. T. Blake（1937）；糙柱莎属 ■☆

393969 Trachystylis foliosa S. T. Blake；糙柱莎 ■☆

393970 Trachytella DC. = Tetracera L. ●

393971 Trachytheca Nutt. ex Benth. = Eriogonum Michx. ●■☆

393972 Trachythece Pierre；赤道山榄属 ●☆

393973 Tractema Raf.（1837）；欧非风信子属 ■☆

393974 Tractema ramburei（Boiss.）Speta；朗比尔欧非风信子 ■☆

393975 Tractema tingitana（Schousb.）Speta；丹吉尔欧非风信子 ■☆

393976 Tractocopevodia Raizada et K. Naray.（1946）；缅甸芸香属 ●☆

393977 Tractocopevodia burmahica Raizada et K. Naray.；缅甸芸香 ●☆

393978 Tracyanthus Small = Zigadenus Michx. ■

393979 Tracyanthus angustifolius（Michx.）Small = Zigadenus densus（Desr.）Fernald ■☆

393980 Tracyanthus angustifolius（Michx.）Small var. texanus Bush = Zigadenus densus（Desr.）Fernald ■☆

393981 Tracyanthus texanus（Bush）Small = Zigadenus densus（Desr.）Fernald ■☆

393982 Tracyina S. F. Blake（1937）；喙实菀属 ■☆

393983 Tracyina rostrata S. F. Blake；喙实菀 ■☆

393984 Tradescantella Small = Callisia Loefl. ■☆

393985 Tradescantella Small = Phyodina Raf. ■☆

393986 Tradescantella floridana（S. Watson）Small = Callisia cordifolia（Sw.）E. S. Anderson et Woodson ■☆

393987 Tradescantia L.（1753）；水竹草属（吊竹兰属，吊竹梅属，重扇属，紫背万年青属，紫露草属，紫万年青属）；Spider-lily，Spiderwort，Tradescantia，Virginia Spiderwort，Wandering Jew，Wanderingjew，Wandering-jew，Zebrina ■

393988 Tradescantia Rupp. ex L. = Tradescantia L. ■

393989 Tradescantia albiflora Kunth 'Aureovittata'；黄斑水竹草 ■☆

393990 Tradescantia albiflora Kunth 'Leakenensis Rainbow'；银线水竹草 ■☆

393991 Tradescantia albiflora Kunth 'Tricolor Minima'；彩虹水竹草 ■☆

393992 Tradescantia albiflora Kunth = Tradescantia fluminensis Vell. ■☆

393993 Tradescantia andersoniana W. Ludow et Rohweder = Tradescantia virginiana L. ■

393994 Tradescantia anomala Torr. = Tinantia anomala（Torr.）C. B. Clarke ■☆

393995 Tradescantia axillaris（L.）L. = Amischophacelus axillaris（L.）R. S. Rao et Kammathy ■☆

393996 Tradescantia axillaris（L.）L. = Cyanotis axillaris（L.）Sweet ■

393997 Tradescantia axillaris L. = Cyanotis axillaris（L.）D. Don ex Sweet ■

393998 Tradescantia barbata（D. Don）Spreng. = Cyanotis barbata D. Don ■

393999 Tradescantia blossfeldiana Mildbr. 'Variegata'；彩纹绿锦草（白纹绿锦草）■☆

394000 Tradescantia blossfeldiana Mildbr. = Tradescantia cerinthoides Kunth ■☆

394001 Tradescantia bracteata Small；具苞紫露草；Bracted Spiderwort，Long-bracted Spiderwort，Sticky Spiderwort ■☆

394002 Tradescantia brevicaulis Raf. = Tradescantia virginiana L. ■

394003 Tradescantia brevifolia（Torr.）Rose；短叶紫露草 ■☆

394004 Tradescantia buckleyi（I. M. Johnst.）D. R. Hunt；巴克利紫露草 ■☆

394005 Tradescantia canaliculata Raf. = Tradescantia ohiensis Raf. ■☆

394006 Tradescantia capitata Blume = Belosynapsis ciliata（Blume）R. S. Rao ■

394007 Tradescantia cerinthoides Kunth；绿锦草 ■☆

394008 Tradescantia ciliata Blume = Belosynapsis ciliata（Blume）R. S. Rao ■

394009 Tradescantia congesta（Moench）D. Don = Tradescantia virginiana L. ■

394010 Tradescantia cordifolia Griff. = Streptolirion volubile Edgew. ■

394011 Tradescantia cordifolia Sw. = Callisia cordifolia (Sw.) E. S. Anderson et Woodson ■☆

394012 Tradescantia crassifolia Cav.;厚叶紫露草;Wandering Jew ■☆

394013 Tradescantia crassula Link et Otto;肉质紫露草;Succulent Spiderwort ■☆

394014 Tradescantia cristata (L.) Jacq. = Cyanotis cristata (L.) Schult. f. ■

394015 Tradescantia discolor L'Hér. = Rhoeo discolor (L'Her.) Hance ■

394016 Tradescantia discolor L'Her. = Rhoeo spathacea (Sw.) Stearn ■

394017 Tradescantia discolor L'Her. = Tradescantia spathacea Sw. ■

394018 Tradescantia edwardsiana Tharp;埃德紫露草■☆

394019 Tradescantia ernestiana E. S. Anderson et Woodson;埃内斯特紫露草;Spiderwort ■☆

394020 Tradescantia floridana S. Watson = Callisia cordifolia (Sw.) E. S. Anderson et Woodson ■☆

394021 Tradescantia fluminensis Vell.;白苞紫露草(白花水竹草,白花紫露草,吊竹草,紫叶水竹草);Green Wandering Jew, Inch Plant, Small-leaf Spiderwort, Speedy Jenny, Wandering Jew, White Spider Wort, White-flowered Wandering Jew ■☆

394022 Tradescantia fluminensis Vell. 'Albovittata';银纹白花紫露草■☆

394023 Tradescantia fluminensis Vell. 'Tricolor';三色水竹草■☆

394024 Tradescantia fluminensis Vell. 'Variegata';花叶水竹草(黄白纹白花紫露草,黄斑水竹草)■☆

394025 Tradescantia fluminensis Vell. 'Viridia';绿叶水竹草■☆

394026 Tradescantia fluviatilis ?;河岸紫露草;Wandering Jew ■☆

394027 Tradescantia foliosa Small = Tradescantia ohiensis Raf. ■☆

394028 Tradescantia geniculata Lour. = Cyanotis loureiriana (Roem. et Schult.) Merr. ■

394029 Tradescantia gigantea Rose;巨紫露草■☆

394030 Tradescantia glomerata Willd. ex Schult. et Schult. f. = Floscopa glomerata (Willd. ex Schult. et Schult. f.) Hassk. ■☆

394031 Tradescantia hirsuticaulis Small;毛茎紫露草■☆

394032 Tradescantia hirsutiflora Bush;毛叶紫露草■☆

394033 Tradescantia humilis Rose;矮紫露草■☆

394034 Tradescantia imbricata Roxb. = Cyanotis cristata (L.) Schult. f. ■

394035 Tradescantia incarnata Small = Tradescantia ohiensis Raf. ■☆

394036 Tradescantia leiandra Torr.;光蕊紫露草■☆

394037 Tradescantia leiandra Torr. var. brevifolia Torr. = Setcreasea brevifolia (Torr.) K. Schum. et Syd. ■☆

394038 Tradescantia leiandra Torr. var. brevifolia Torr. = Tradescantia brevifolia (Torr.) Rose ■☆

394039 Tradescantia longifolia Sessé et Moc.;长叶紫露草■☆

394040 Tradescantia longipes E. S. Anderson et Woodson;长梗紫露草;Dwarf Spiderwort, Wild Crocus ■☆

394041 Tradescantia loureiriana Roem. et Schult. = Cyanotis loureiriana (Roem. et Schult.) Merr. ■

394042 Tradescantia loureiroana Schult. et Schult. f. = Cyanotis loureiriana (Roem. et Schult.) Merr. ■

394043 Tradescantia malabarica L. = Murdannia nudiflora (L.) Brenan ■

394044 Tradescantia micrantha Torr. = Callisia micrantha (Torr.) D. R. Hunt ■☆

394045 Tradescantia montana Shuttlew. = Tradescantia virginiana L. ■

394046 Tradescantia montana Shuttlew. ex Small et Vail = Tradescantia virginiana L. ■

394047 Tradescantia navicularis Ortgies;迭叶草(矮鸭舌草);Chain Plant ■☆

394048 Tradescantia navicularis Ortgies = Callisia navicularis (Ortgies) D. R. Hunt ■☆

394049 Tradescantia nodiflora Lam. = Cyanotis speciosa (L. f.) Hassk. ■☆

394050 Tradescantia occidentalis (Britton) Smyth;西方紫露草;Prairie Spiderwort, Western Spiderwort ■☆

394051 Tradescantia occidentalis (Britton) Smyth var. scopulorum (Rose) E. S. Anderson et Woodson;峭壁紫露草■☆

394052 Tradescantia occidentalis (Britton) Smyth var. typica E. S. Anderson et Woodson = Tradescantia occidentalis (Britton) Smyth ■☆

394053 Tradescantia ohiensis Raf.;北美紫露草(奥西紫露草,卷苞水竹草);Blue Jacket, Blue-jacket, Common Spiderwort, Ohi Spiderwort, Reflexed Spiderwort, Smooth Spiderwort, Spiderwort ■☆

394054 Tradescantia ohiensis Raf. var. foliosa (Small) MacRoberts = Tradescantia ohiensis Raf. ■☆

394055 Tradescantia ohiensis Raf. var. paludosa (E. S. Anderson et Woodson) MacRoberts = Tradescantia paludosa E. S. Anderson et Woodson ■☆

394056 Tradescantia ozarkana E. S. Anderson et Woodson;奥扎克紫露草;Ozark Spiderwort ■☆

394057 Tradescantia pallida (Rose) D. R. Hunt;苍白水竹草(苍白紫露草);Purple Heart, Purple Queen Tradescantia, Wandering Jew ■☆

394058 Tradescantia paludosa E. S. Anderson et Woodson;沼泽紫露草■☆

394059 Tradescantia pedicellata Celarier;小梗紫露草■☆

394060 Tradescantia pellucida M. Martens et Galeotti = Gibasis pellucida (M. Martens et Galeotti) D. R. Hunt ■☆

394061 Tradescantia pilosa Lehm. = Tradescantia subaspera Ker Gawl. ■☆

394062 Tradescantia pinetorum Greene;松林紫露草■☆

394063 Tradescantia radicans Royle = Cyanotis barbata D. Don ■

394064 Tradescantia reflexa Raf. = Tradescantia ohiensis Raf. ■☆

394065 Tradescantia reflexa Raf. f. albiflora Slavin et Nieuwl. = Tradescantia ohiensis Raf. ■☆

394066 Tradescantia reflexa Raf. f. lesteri Standl. = Tradescantia ohiensis Raf. ■☆

394067 Tradescantia reginae L. Linden et Rodigas = Dichorisandra reginae (L. Linden et Rodigas) W. Mill. ■☆

394068 Tradescantia reverchonii Bush;勒韦雄紫露草■☆

394069 Tradescantia rosea Vent.;红紫露草;Red Spiderwort ■☆

394070 Tradescantia rosea Vent. = Callisia rosea (Vent.) D. R. Hunt ■☆

394071 Tradescantia rosea Vent. var. graminea (Small) E. S. Anderson et Woodson = Callisia graminea (Small) G. C. Tucker ■☆

394072 Tradescantia rosea Vent. var. ornata (Small) E. S. Anderson et Woodson = Callisia ornata (Small) G. C. Tucker ■☆

394073 Tradescantia rupestris Raf. = Tradescantia virginiana L. ■

394074 Tradescantia scopulorum Rose = Tradescantia occidentalis (Britton) Smyth var. scopulorum (Rose) E. S. Anderson et Woodson ■☆

394075 Tradescantia sillamontana Matuda;白绢草(短绒水竹草,雪绢,紫霞草);White Velvet ■☆

394076 Tradescantia spathacea Sw.;紫背万年青(矮蚌花,蚌花,蚌壳叶,蚌兰,蚌兰花,蚌兰叶,佛焰紫万年青,蚶花,荷包花,荷包兰,红蚌兰,红蚌兰花,红川七草,红面将军,红竹叶,菱角花,青红兰,血见愁,舟百合,紫锦红,紫锦兰,紫兰,紫万年青,紫菹);Boat Lily, Boat-lily, Moses-in-the-boat, Moses-in-the-cradle, Muses-in-the-cradle, Oyster Plant, Oyster-plant, Purple-leaved Spiderwort, Three-men-in-a-boat, Tradescantia ■

394077 Tradescantia spathacea Sw. 'Vittata' = Rhoeo spathacea (Sw.) Stearn 'Vittata' ■☆

394078　Tradescantia spathacea Sw. = Rhoeo spathacea（Sw.）Stearn ■

394079　Tradescantia speciosa L. f. = Cyanotis speciosa（L. f.）Hassk. ■☆

394080　Tradescantia speciosa Salisb. = Tradescantia virginiana L. ■

394081　Tradescantia subacaulis Bush；近无茎紫露草●☆

394082　Tradescantia subaspera Ker Gawl.；粗糙紫露草（宽叶紫露草）；Scabrous Spiderwort，Spiderwort，Wide-leaved Spiderwort ■☆

394083　Tradescantia tharpii E. S. Anderson et Woodson；矮小紫露草；Dwarf Spiderwort ■☆

394084　Tradescantia thyrsiflora Blume = Pollia thyrsiflora（Blume）Endl. ex Hassk. ■

394085　Tradescantia tricolor C. B. Clarke = Tradescantia fluminensis Vell. ■☆

394086　Tradescantia vaga Lour. = Cyanotis vaga（Lour.）Roem. et Schult. ■

394087　Tradescantia velutina Kunth et Bouché = Tradescantia sillamontana Matuda ■☆

394088　Tradescantia virginiana L.；紫露草（安氏紫露草，本山金线，本山金线连，露水草，山地紫露草，血见愁，鸭舌草，鸭舌黄，紫鸭跖，紫鸭跖草）；Common Spiderwort，Day Spiderwort，Devil-in-the-pulpit，Flower of a Day，Flower-of-a-day，Life-of-man，Moses-and-the-bulrushes，Moses-in-the-bulrushes，One-day Flower，Spider Lily，SpiderWort，Trinity，Trinity Flower，Virginia Spiderwort，Wandering Jew，Widow's Tears，Widow's-tears ■

394089　Tradescantia virginiana L. var. alba Hook. ex Raf. = Tradescantia virginiana L. ■

394090　Tradescantia virginiana L. var. barbata Raf. = Tradescantia virginiana L. ■

394091　Tradescantia virginiana L. var. occidentalis Britton = Tradescantia occidentalis（Britton）Smyth ■☆

394092　Tradescantia wrightii Rose et Bush；赖氏紫露草●☆

394093　Tradescantia zebrina Bosse；水竹草（白带草，带毒散，吊竹菜，吊竹草，吊竹梅，二打不死，红苞鸭跖草，红骨竹仔菜，红莲，红舌草，红鸭跖草，红竹仔草，花叶水竹草，花叶竹夹菜，鸡舌话，金发草，进瓢梗，条纹紫露草，血见愁，鸭舌红，紫背金牛，紫背鸭跖草）；Inchplant，Silver Inch Plant，Silver Inch-plant，Wandering Jew，Wanderingjew，Wandering-jew，Wanderingjew Zebrina ■

394094　Tradescantia zebrina Loudon = Zebrina pendula Schnizl. ■

394095　Tradescantiaceae Salisb. = Commelinaceae Mirb.（保留科名）●■

394096　Traevia Neck. = Trewia L. ●

394097　Tragacantha Mill. = Astragalus L. ●■

394098　Traganopsis Maire et Wilczek（1936）；簇花蓬属●☆

394099　Traganopsis glomerata Maire et Wilczek；簇花蓬■☆

394100　Tragantha Endl. = Eupatorium L. ■●

394101　Tragantha Endl. = Traganthes Walk. ■●

394102　Tragantha Wallr. ex Endl. = Eupatorium L. ■●

394103　Tragantha Wallr. ex Endl. = Traganthes Walk. ■●

394104　Traganthes Walk. = Eupatorium L. ■●

394105　Traganthus Klotzsch = Adelia L.（保留属名）●☆

394106　Traganthus Klotzsch = Bernardia Mill.（废弃属名）●☆

394107　Traganum Delile（1813）；单花蓬属●☆

394108　Traganum acuminatum Maire et Weiller = Traganum nudatum Delile ●☆

394109　Traganum moquinii Webb ex Moq.；单花蓬●☆

394110　Traganum nudatum Delile；裸单花蓬●☆

394111　Traganum nudatum Delile var. acuminatum（Maire et Weiller）Maire = Traganum nudatum Delile ●☆

394112　Traganum nudatum Delile var. microphyllum Maire = Traganum nudatum Delile ●☆

394113　Traganum nudatum Delile var. obtusatum Maire et Weiller = Traganum nudatum Delile ●☆

394114　Tragia L.（1753）；刺痒藤属；Noseburn ●☆

394115　Tragia Plum. ex L. = Tragia L. ●☆

394116　Tragia abortiva M. G. Gilbert；败育刺痒藤●☆

394117　Tragia acalyphoides Radcl. -Sm.；铁苋菜刺痒藤●☆

394118　Tragia adenanthera Baill.；腺药刺痒藤●☆

394119　Tragia affinis Müll. Arg. ex Prain = Tragia wahlbergiana Prain ●☆

394120　Tragia akwapimensis Prain；阿夸平刺痒藤●☆

394121　Tragia ambigua S. Moore = Tragiella natalensis（Sond.）Pax et K. Hoffm. ●☆

394122　Tragia angolensis Müll. Arg.；安哥拉刺痒藤●☆

394123　Tragia angustifolia Benth. = Tragia vogelii Keay ●☆

394124　Tragia anisosepala Merr. et Chun = Cnesmone tonkinensis（Gagnep.）Croizat ●

394125　Tragia anomala Prain = Tragiella anomala（Prain）Pax et K. Hoffm. ●☆

394126　Tragia arabica Baill. = Tragia pungens（Forssk.）Müll. Arg. ●☆

394127　Tragia ballyi Radcl. -Sm.；博利刺痒藤●☆

394128　Tragia benthamii Baker；本瑟姆刺痒藤●☆

394129　Tragia betonicifolia Nutt.；水苏叶刺痒藤；Noseburn ●☆

394130　Tragia bolusii Kuntze = Tragia meyeriana Müll. Arg. ●☆

394131　Tragia brevipes Pax；短梗刺痒藤●☆

394132　Tragia buettneri Pax = Dalechampia ipomoeifolia Benth. ●☆

394133　Tragia calvescens Pax = Tragia tenuifolia Benth. ●☆

394134　Tragia cannabina L. f. = Tragia plukenetii Radcl. -Sm. ●☆

394135　Tragia cannabina L. f. var. intermedia（Müll. Arg.）Prain = Tragia hildebrandtii Müll. Arg. var. intermedia（Müll. Arg.）Cufod. ●☆

394136　Tragia capensis Thunb. = Ctenomeria capensis（Thunb.）Harv. ex Sond. ●☆

394137　Tragia chamaelea L. = Microstachys chamaelea（L.）Müll. Arg. ●☆

394138　Tragia chevalieri Beille；舍瓦利耶刺痒藤●☆

394139　Tragia chmnaelea L. = Sebastiania chamaelea（L.）F. Muell. ●

394140　Tragia cinerea（Pax）M. G. Gilbert et Radcl. -Sm.；灰色刺痒藤●☆

394141　Tragia collina Prain；山丘刺痒藤●☆

394142　Tragia cordata Michx.；心叶刺痒藤；Noseburn ●☆

394143　Tragia cordifolia Vahl = Tragia pungens（Forssk.）Müll. Arg. ●☆

394144　Tragia cordifolia Vahl var. cinerea（Pax）Prain = Tragia cinerea（Pax）M. G. Gilbert et Radcl. -Sm. ●☆

394145　Tragia crenata M. G. Gilbert；圆齿刺痒藤●☆

394146　Tragia descampsii De Wild.；德康刺痒藤●☆

394147　Tragia dinteri Pax；丁特刺痒藤●☆

394148　Tragia dioica Sond.；异株刺痒藤●☆

394149　Tragia durbanensis Kuntze = Tragia glabrata（Müll. Arg.）Pax et K. Hoffm. ●☆

394150　Tragia fasciculata Beille；簇生刺痒藤●☆

394151　Tragia friesiana Prain = Tragiella friesiana（Prain）Pax et K. Hoffm. ●☆

394152　Tragia gallabatensis Prain = Tragia plukenetii Radcl. -Sm. ●☆

394153　Tragia gardneri Prain；加德纳刺痒藤●☆

394154　Tragia glabrata（Müll. Arg.）Pax et K. Hoffm.；光滑刺痒藤●☆

394155　Tragia glabrata（Müll. Arg.）Pax et K. Hoffm. var. hispida Radcl. -Sm.；硬毛刺痒藤●☆

394156　Tragia glabrescens Pax；渐光刺痒藤●☆

394157　Tragia hildebrandtii Müll. Arg.；希尔德刺痒藤●☆

394158 Tragia hildebrandtii Müll. Arg. subsp. glaucescens Pax = Tragia hildebrandtii Müll. Arg. var. intermedia (Müll. Arg.) Cufod. ●☆

394159 Tragia hildebrandtii Müll. Arg. var. intermedia (Müll. Arg.) Cufod. ;间型希尔德刺痒藤●☆

394160 Tragia impedita Prain;累赘刺痒藤●☆

394161 Tragia incisifolia Prain;锐裂叶刺痒藤●☆

394162 Tragia insuavis Prain;芳香刺痒藤●☆

394163 Tragia involucrata L. = Cnesmone tonkinensis (Gagnep.) Croizat ●

394164 Tragia involucrata L. var. intermedia Müll. Arg. = Cnesmone mairei (H. Lév.) Croizat ●

394165 Tragia involucrata L. var. intermedia Müll. Arg. = Tragia hildebrandtii Müll. Arg. var. intermedia (Müll. Arg.) Cufod. ●☆

394166 Tragia involurata L. = Cnesmone mairei (H. Lév.) Croizat ●☆

394167 Tragia kassiliensis Beille = Tragia benthamii Baker ●☆

394168 Tragia keniensis Rendle;肯尼亚刺痒藤●☆

394169 Tragia kirkiana Müll. Arg. ;柯克刺痒藤●☆

394170 Tragia klingii Pax = Tragia tenuifolia Benth. ●☆

394171 Tragia lancifolia Dinter ex Pax et K. Hoffm. ;披针叶刺痒藤●☆

394172 Tragia lasiophylla Pax et K. Hoffm. ;毛叶刺痒藤●●☆

394173 Tragia lukafuensis De Wild. ;卢卡夫刺痒藤●●☆

394174 Tragia mairei (H. Lév.) Rehder = Cnesmone mairei (H. Lév.) Croizat ●

394175 Tragia manniana Müll. Arg. = Tragia tenuifolia Benth. ●☆

394176 Tragia mazoensis Radcl. -Sm. ;马索刺痒藤●☆

394177 Tragia meyeriana Müll. Arg. ;迈尔刺痒藤●☆

394178 Tragia meyeriana Müll. Arg. var. glabrata ? = Tragia glabrata (Müll. Arg.) Pax et K. Hoffm. ●☆

394179 Tragia micromeres Radcl. -Sm. ;小刺痒藤●☆

394180 Tragia mildbraediana Pax et K. Hoffm. ;米尔德刺痒藤●☆

394181 Tragia minor Sond. ;较小刺痒藤●☆

394182 Tragia mitis Hochst. ex A. Rich. ;柔软刺痒藤●☆

394183 Tragia mitis Hochst. ex Müll. Arg. var. cinerea Pax = Tragia cinerea (Pax) M. G. Gilbert et Radcl. -Sm. ●☆

394184 Tragia mixta M. G. Gilbert;混杂刺痒藤●☆

394185 Tragia monadelpha Schumach. et Thonn. ;单体雄蕊刺痒藤●☆

394186 Tragia natalensis Sond. = Tragiella natalensis (Sond.) Pax et K. Hoffm. ●☆

394187 Tragia negeliensis M. G. Gilbert;内盖里刺痒藤●☆

394188 Tragia parvifolia Pax;小叶刺痒藤●☆

394189 Tragia petiolaris Radcl. -Sm. ;柄叶刺痒藤●☆

394190 Tragia physocarpa Prain;囊果刺痒藤●☆

394191 Tragia platycalyx Radcl. -Sm. ;阔萼刺痒藤●☆

394192 Tragia plukenetii Radcl. -Sm. ;普拉克内特刺痒藤●☆

394193 Tragia pogostemonoides Radcl. -Sm. ;刺蕊草刺痒藤●☆

394194 Tragia polygonoides Prain;多节刺痒藤●☆

394195 Tragia preussii Pax;普罗伊斯刺痒藤●☆

394196 Tragia prionoides Radcl. -Sm. ;南非灯心草刺痒藤●☆

394197 Tragia prostrata Radcl. -Sm. ;平卧刺痒藤●☆

394198 Tragia pungens (Forssk.) Müll. Arg. ;多刺刺痒藤●☆

394199 Tragia pungens (Forssk.) Müll. Arg. var. cinerea (Pax) Pax = Tragia cinerea (Pax) M. G. Gilbert et Radcl. -Sm. ●☆

394200 Tragia ramosa Torr. ;多枝刺痒藤;Noseburn ●☆

394201 Tragia rhodesiae Pax;罗得西亚刺痒藤●☆

394202 Tragia rogersii Prain;罗杰斯刺痒藤●☆

394203 Tragia rupestris Sond. ;岩生刺痒藤●☆

394204 Tragia rupestris Sond. var. minor (Sond.) Müll. Arg. = Tragia minor Sond. ●☆

394205 Tragia scandens L. = Tetracera scandens (L.) Merr. ●

394206 Tragia schinzii Pax = Tragia dioica Sond. ●☆

394207 Tragia schlechteri Pax = Ctenomeria capensis (Thunb.) Harv. ex Sond. ●☆

394208 Tragia schultzeana Dinter ex Pax et K. Hoffm. = Plukenetia africana Sond. ●☆

394209 Tragia schweinfurthii Baker;施韦刺痒藤●☆

394210 Tragia senegalensis Müll. Arg. ;塞内加尔刺痒藤●☆

394211 Tragia shirensis Prain;希尔刺痒藤●☆

394212 Tragia shirensis Prain var. glabriuscula Radcl. -Sm. ;光刺痒藤●☆

394213 Tragia sonderi Prain;森诺刺痒藤●☆

394214 Tragia spathulata Benth. ;匙形刺痒藤●☆

394215 Tragia stipularis Radcl. -Sm. ;托叶刺痒藤●☆

394216 Tragia stolziana Pax et K. Hoffm. = Tragia kirkiana Müll. Arg. ●☆

394217 Tragia subsessilis Pax;近无柄刺痒藤●☆

394218 Tragia tenuifolia Benth. ;细叶刺痒藤●☆

394219 Tragia triandra Müll. Arg. = Seidelia triandra (E. Mey.) Pax ■☆

394220 Tragia tripartita Beille = Tragia plukenetii Radcl. -Sm. ●☆

394221 Tragia tripartita Schweinf. ;三深裂刺痒藤●☆

394222 Tragia triumfettoides M. G. Gilbert;刺蒴麻刺痒藤●☆

394223 Tragia ukambensis Pax;乌卡刺痒藤●☆

394224 Tragia ukambensis Pax var. ugandensis Radcl. -Sm. ;乌干达刺痒藤●☆

394225 Tragia uncinata M. G. Gilbert;具钩刺痒藤●☆

394226 Tragia velutina Pax = Tragia brevipes Pax ●☆

394227 Tragia villosa Thunb. = Acalypha capensis (L. f.) Prain et Hutch. ■☆

394228 Tragia vogelii Keay;沃格尔刺痒藤●☆

394229 Tragia volkensii Pax = Tragia brevipes Pax ●☆

394230 Tragia volubilis L. ;缠绕刺痒藤●☆

394231 Tragia wahlbergiana Prain;瓦尔贝里刺痒藤●☆

394232 Tragia wildemanii Beille;怀尔德曼刺痒藤●☆

394233 Tragia winkleri Pax = Tragia preussii Pax ●☆

394234 Tragia zenkeri Pax = Tragia tenuifolia Benth. ●☆

394235 Tragiaceae Raf. = Euphorbiaceae Juss. (保留科名)●■

394236 Tragiella Pax et K. Hoffm. (1919);特拉大戟属●☆

394237 Tragiella Pax et K. Hoffm. = Sphaerostylis Baill. ●☆

394238 Tragiella anomala (Prain) Pax et K. Hoffm. ;异常特拉大戟●☆

394239 Tragiella friesiana (Prain) Pax et K. Hoffm. ;特拉大戟●☆

394240 Tragiella natalensis (Sond.) Pax et K. Hoffm. ;纳塔尔特拉大戟●☆

394241 Tragiella pavoniifolia Chiov. = Dalechampia pavoniifolia (Chiov.) M. G. Gilbert ●☆

394242 Tragiella pyxostigma Radcl. -Sm. ;柱头特拉大戟●☆

394243 Tragiola Small et Pennell = Gratiola L. ■

394244 Tragiopsis H. Karst. = Sebastiania Spreng. ●

394245 Tragiopsis Pomel = Stoibrax Raf. ■☆

394246 Tragiopsis dichotoma Pomel = Stoibrax pomelianum (Maire) B. L. Burtt ■☆

394247 Tragiopsis hanotei Braun-Blanq. et Maire = Stoibrax hanotei (Maire) B. L. Burtt ■☆

394248 Tragiopsis pomeliana (Maire) Maire = Stoibrax pomelianum (Maire) B. L. Burtt ■☆

394249 Tragiopsis pomeliana (Maire) Maire var. vegeta (Pau et Font Quer) Maire = Stoibrax pomelianum (Maire) B. L. Burtt ■☆

394250 Tragiopsis scabriuscula Pomel = Stoibrax dichotomum (L.)

Raf. ■☆

394251 Tragium Spreng. = Pimpinella L. ■

394252 Tragium anisum (L.) Link. = Pimpinella anisum L. ■

394253 Tragoceras Spreng. = Tragoceros Kunth ●■

394254 Tragoceros Kunth = Zinnia L. (保留属名) ●■

394255 Tragolinum Raf. = Pimpinella L. ■

394256 Tragolium Raf. = Pimpinella L. ■

394257 Tragopogon L. (1753); 婆罗门参属; Goat's-Beard, Goatsbeard, Salsify ■

394258 Tragopogon × mirabilis Rouy; 奇异婆罗门参; Ontario Goatsbeard ■☆

394259 Tragopogon acanthocarpus Boiss.; 刺果婆罗门参■☆

394260 Tragopogon altaicus S. A. Nikitin et Schischk.; 阿尔泰婆罗门参 (阿山婆罗门参); Altai Salsify ■☆

394261 Tragopogon ameniacus Kuth.; 亚美尼亚婆罗门参■☆

394262 Tragopogon badachschanicus Boriss.; 巴达婆罗门参■☆

394263 Tragopogon bjelorussicus Artemczuk; 白俄罗斯婆罗门参■☆

394264 Tragopogon borysthenicus Artemczuk; 第聂伯婆罗门参■☆

394265 Tragopogon brevirostris DC.; 短喙婆罗门参; Short-beaked Salsify ■☆

394266 Tragopogon buphthalmoides (DC.) Boiss.; 牛眼婆罗门参■☆

394267 Tragopogon capensis Jacq. = Urospermum picrioides (L.) Scop. ex F. W. Schmidt ■☆

394268 Tragopogon capitatus S. A. Nikitin; 头状婆罗门参 (胀梗婆罗门参); Capitate Salsify ■

394269 Tragopogon charadzae Kuth.; 哈氏婆罗门参■☆

394270 Tragopogon colchicus Albov ex Grossh.; 黑海婆罗门参■☆

394271 Tragopogon collinus DC.; 山丘婆罗门参■☆

394272 Tragopogon coloratus C. A. Mey.; 着色婆罗门参■☆

394273 Tragopogon conduplicatus S. A. Nikitin; 对折婆罗门参■☆

394274 Tragopogon cretaceus S. A. Nikitin; 白垩婆罗门参■☆

394275 Tragopogon crocifolius L. subsp. samaritanii (Heldr. et Sart.) I. Richardson; 萨马利顿婆罗门参■☆

394276 Tragopogon daghestanicus (Artemczuk) Kuth.; 达赫斯坦婆罗门参■☆

394277 Tragopogon dalechampii L. = Urospermum dalechampii (L.) F. W. Schmidt ■☆

394278 Tragopogon dasyrhynchus Artemczuk; 粗喙婆罗门参■☆

394279 Tragopogon donetzicus Artemczuk; 顿涅茨山婆罗门参■☆

394280 Tragopogon dubianskyi Krasch. et S. A. Nikitin; 杜氏婆罗门参■☆

394281 Tragopogon dubius Scop.; 可疑婆罗门参 (大婆罗门参, 拟婆罗门参, 霜毛婆罗门参); Fistulous Goat's-beard, Goat's Beard, Greater Sand Goat's-beard, Western Salsify, Yellow Goat's Beard, Yellow Goatsbeard, Yellow Salsify ■

394282 Tragopogon dubius Scop. subsp. major (Jacq.) O. Bolòs et Vigo = Tragopogon dubius Scop. ■

394283 Tragopogon elatior Steven; 高婆罗门参■☆

394284 Tragopogon elongatus S. A. Nikitin; 长茎婆罗门参; Longstem Salsify ■

394285 Tragopogon filifolius Rehmann ex Boiss.; 丝叶婆罗门参■☆

394286 Tragopogon floccosus Waldst. et Kit.; 绵花婆罗门参; Woolly Goatsbeard ■☆

394287 Tragopogon gaudanicus Boriss.; 戈丹婆罗门参■☆

394288 Tragopogon glaber (L.) DC. = Geropogon hybridus (L.) Sch. Bip. ■☆

394289 Tragopogon gonocarpus (S. A. Nikitin) Stank. = Tragopogon marginifolius Pavlov ■

394290 Tragopogon gonocarpus S. A. Nikitin = Tragopogon marginifolius Pavlov ■

394291 Tragopogon gorskianus Rchb. f.; 高氏婆罗门参■☆

394292 Tragopogon gracilis D. Don; 纤细婆罗门参■

394293 Tragopogon graminifolius DC.; 禾叶婆罗门参■☆

394294 Tragopogon heteropappus C. H. An; 长苞婆罗门参; Longbract Salsify ■

394295 Tragopogon heterospermus Schweigg.; 异籽婆罗门参■☆

394296 Tragopogon hybridus L.; 杂种婆罗门参; Moscow Salsify, Ownbey's Goatsbeard, Pasture Goatsbeard, Slender Salsify ■☆

394297 Tragopogon idae Kuth.; 伊达婆罗门参■☆

394298 Tragopogon karelinii S. A. S. A. Nikitin; 喀里婆罗门参■☆

394299 Tragopogon karjaginii Kuth.; 卡尔婆罗门参■☆

394300 Tragopogon kasahstanicus S. A. Nikitin; 中亚婆罗门参; Central Asia Salsify ■

394301 Tragopogon kashmirianus G. Singh; 克什米尔婆罗门参■☆

394302 Tragopogon kemulariae Kuth.; 凯木婆罗门参■☆

394303 Tragopogon ketzkhovelii Kuth.; 凯兹婆罗门参■☆

394304 Tragopogon kopetdaghensis Boriss.; 科佩特婆罗门参■☆

394305 Tragopogon krascheninnikovii S. A. Nikitin; 克拉森婆罗门参■☆

394306 Tragopogon kultiassovii Popov; 库尔婆罗门参■☆

394307 Tragopogon lamottei Rouy; 拉氏婆罗门参; Jack-go-to-bed-at-noon ■☆

394308 Tragopogon latifolius Boiss.; 宽叶婆罗门参■☆

394309 Tragopogon lithuanicus (DC.) Boriss.; 立陶宛婆罗门参■☆

394310 Tragopogon macrocephalus Pomel = Tragopogon porrifolius L. subsp. macrocephalus (Pomel) Batt. ■☆

394311 Tragopogon macropogon C. A. Mey.; 大芒婆罗门参■☆

394312 Tragopogon major Jacq.; 大婆罗门参; Big Salsify ■☆

394313 Tragopogon major Jacq. = Tragopogon dubius Scop. ■

394314 Tragopogon makaschwilii Kuth.; 马卡什夫婆罗门参■☆

394315 Tragopogon marginatus Boiss. et Buhse; 具边婆罗门参■☆

394316 Tragopogon marginifolius Pavlov; 膜缘婆罗门参 (多面果婆罗门参, 近缘婆罗门参, 宽棱婆罗门参) ■☆

394317 Tragopogon maturatus Boriss.; 完美婆罗门参■☆

394318 Tragopogon meskheticus Kuth.; 迈斯亥特婆罗门参■☆

394319 Tragopogon mirus Ownbey; 华盛顿婆罗门参; Remarkable Goatsbeard ■☆

394320 Tragopogon montanus S. A. Nikitin; 山地婆罗门参 (山婆罗门参) ■☆

394321 Tragopogon orientale L.; 黄花婆罗门参 (东方婆罗门参); Oriental Salsify ■

394322 Tragopogon orientalis L. var. latifolius C. H. An; 宽叶黄花婆罗门参 (宽叶婆罗门参); Broadleaf Oriental Salsify ■

394323 Tragopogon paradoxus S. A. Nikitin; 中亚奇异婆罗门参■☆

394324 Tragopogon picroides L. = Urospermum picrioides (L.) Scop. ex F. W. Schmidt ■☆

394325 Tragopogon plantagineus Boiss. et Huet; 车前婆罗门参■☆

394326 Tragopogon podolicus Besser ex DC.; 波城婆罗门参■☆

394327 Tragopogon porrifolius L.; 蒜叶婆罗门参 (绿茇, 婆罗门参, 土泡参, 土洋参, 西洋牛蒡, 洋参); Chards, Garlicleaf Salsify, Goat's Beard, Leek-leaved Salsify, Lock-leaved Salsify, Mock Oyster, Nap-at-noon, Oyster, Oyster Plant, Oyster-plant, Purple Goat's Beard, Purple Goatsbeard, Purple Salsify, Salsify, Vegetable Oyster, Vegetable Oyster Salsify, Vegetable-oyster, Vegetable-oyster Salsify ■

394328 Tragopogon porrifolius L. subsp. australis (Jord.) Nyman; 南方蒜叶婆罗门参■☆

394329　Tragopogon porrifolius L. subsp. macrocephalus（Pomel）Batt.；大头蒜叶婆罗门参■☆

394330　Tragopogon porrifolius L. var. australis（Jord.）Nyman = Tragopogon porrifolius L. subsp. australis（Jord.）Nyman ■☆

394331　Tragopogon porrifolius L. var. cupani（Guss.）Fiori = Tragopogon porrifolius L. ■

394332　Tragopogon praecox S. A. Nikitin；早熟婆罗门参■☆

394333　Tragopogon pratensis L.；婆罗门参（草地婆罗门参，草原稿婆罗门参）；Flora's Clock，Goat's Beard，Goat's Foot，Goat's-beard，Go-to-bed-at-noon，Jack-abed-at-noon，Jack-by-the-hedge，Jack-go-to-bed-at-noon，John-go-to-bed-at-noon，Johnny-go-to-bed，John-that-goes-to-bed-at-noon，Joseph's Flower，Lesser Goat's-beard，Meadow Salsify，Nap-at-noon，Noon-day Flower，Noon-flower，Noontide，One-o'clock，Oyster Plant，Paintbrush，Salsify，Shepherd's Clock，Showy Goat's-beard，Sleep-at-noon，Sleepyhead，Star of Bethlehem，Star of Jerusalem，Star-of-jerusalem，Twelve-o'clock，Wild Salsify，Yellow Goatsbeard，Yellow Goat's-beard Buck's Beard，Yellow Salsify ■

394334　Tragopogon pseudomajor S. A. Nikitin；北疆婆罗门参（粗脖婆罗门参，赛大婆罗门参，小果婆罗门参）■

394335　Tragopogon pusillus M. Bieb.；微小婆罗门参■☆

394336　Tragopogon reticulatus Boiss. et Huet；网状婆罗门参■

394337　Tragopogon ruber S. T. Gmel.；红花婆罗门参（红婆罗门参，紫婆罗门参）；Redflower Salsify ■

394338　Tragopogon ruber S. T. Gmel. var. leucocarpus C. H. An；白果婆罗门参；Whitefruited Redflower Salsify ■

394339　Tragopogon ruthenicus Besser ex Claus；俄罗斯婆罗门参；Russia Salsify ■☆

394340　Tragopogon sabulosus Krasch. et S. A. Nikitin；沙地婆罗门参（沙婆罗门参，沙生婆罗门参）■

394341　Tragopogon samaritanii Heldr. et Sart. = Tragopogon crocifolius L. subsp. samaritanii（Heldr. et Sart.）I. Richardson ■☆

394342　Tragopogon scoparius S. A. Nikitin；帚状婆罗门参■

394343　Tragopogon segetus Kuth.；谷地婆罗门参■☆

394344　Tragopogon serotinus Sosn.；晚熟婆罗门参■☆

394345　Tragopogon sibiricus Ganesch.；西伯利亚婆罗门参；Siberia Salsify ■

394346　Tragopogon songoricus S. A. Nikitin；准噶尔婆罗门参；Dzungar Salsify ■

394347　Tragopogon sosnowskyi Kuth.；锁斯诺夫斯基婆罗门参■☆

394348　Tragopogon stepposus（S. A. Nikitin）Stankov；草原婆罗门参■☆

394349　Tragopogon subalpinus S. A. Nikitin；沙婆罗门参（北方婆罗门参，高山婆罗门参，中山婆罗门参）■

394350　Tragopogon tanaiticus Artemczuk；顿河婆罗门参■☆

394351　Tragopogon tomentosulus Boriss.；绒毛婆罗门参■☆

394352　Tragopogon trachycarpus S. A. Nikitin；糙果婆罗门参■☆

394353　Tragopogon tuberosus K. Koch；块状婆罗门参■☆

394354　Tragopogon turkestanicus S. A. Nikitin；土耳其斯坦婆罗门参；Turkestan Salsify ■☆

394355　Tragopogon ucrainicus Artemczuk；乌克兰婆罗门参；Ukraine Salsify ■☆

394356　Tragopogon verrucosobracteatus C. H. An；瘤苞婆罗门参■

394357　Tragopogon virginicus L. = Krigia biflora（Walter）S. F. Blake ■☆

394358　Tragopogon volgensis（S. A. Nikitin）S. A. Nikitin；伏尔加婆罗门参；Volga Salsify ■☆

394359　Tragopogon vvedenskyi Popov；韦坚斯基婆罗门参■☆

394360　Tragopogonodes Kuntze = Urospermum Scop. ●☆

394361　Tragopyrum M. Bieb. = Atraphaxis L. ●

394362　Tragopyrum laetevirens Ledeb. = Atraphaxis laetevirens（Ledeb.）Jaub. et Spach ●

394363　Tragopyrum lanceolatum M. Bieb. = Atraphaxis frutescens（L.）Eversm. ●

394364　Tragopyrum pungens M. Bieb. = Atraphaxis pungens（M. Bieb.）Jaub. et Spach ●

394365　Tragoriganum Gronov. = Satureja L. ●■

394366　Tragoselinum Haller = Pimpinella L. ■

394367　Tragoselinum Mill. = Pimpinella L. ■

394368　Tragoselinum Tourn. ex Haller = Pimpinella L. ■

394369　Tragoselinum luteum（Desf.）Pomel = Pimpinella lutea Desf. ■☆

394370　Tragosma C. A. Mey. ex Ledeb. = Cymbocarpum DC. ex C. A. Mey. ■☆

394371　Tragularia Koch. ex Roxb. = Pisonia L. ●

394372　Tragus Haller（1768）（保留属名）；锋芒草属（虱子草属）；Bur Grass，Burgrass，Bur-grass ■

394373　Tragus Panz. = Brachypodium P. Beauv. ■

394374　Tragus Panz. = Festuca L. ■

394375　Tragus alienus Schult. = Tragus berteronianus Schult. et Schult. f. ■

394376　Tragus alienus Schult. var. brevispinus Henrard = Tragus berteronianus Schult. et Schult. f. ■

394377　Tragus arenarius Bremek. et Oberm. = Tragus racemosus（L.）All. ■

394378　Tragus australianus S. T. Blake；澳洲锋芒草；Australian Burgrass，Australian Burr Grass ■☆

394379　Tragus berteronianus Schult. = Tragus berteronianus Schult. et Schult. f. ■

394380　Tragus berteronianus Schult. et Schult. f.；虱子草；African Burgrass，Spike Burgrass ■

394381　Tragus biflorus（Roxb.）Schult.；双花虱子草；Twoflower Burgrass ■

394382　Tragus biflorus sensu Schult. = Tragus roxburghii Panigrahi ■☆

394383　Tragus heptaneuron Clayton；肯尼亚锋芒草；Kenya Burr Grass ■☆

394384　Tragus koelerioides Asch.；大锋芒草■☆

394385　Tragus major（Hack.）Stapf = Tragus koelerioides Asch. ■☆

394386　Tragus mongolorum Ohwi = Tragus berteronianus Schult. et Schult. f. ■

394387　Tragus occidentalis Nees = Tragus berteronianus Schult. et Schult. f. ■

394388　Tragus paucispina Hack. = Tragus racemosus（L.）All. ■

394389　Tragus pedunculatus Pilg.；梗花锋芒草■☆

394390　Tragus racemosus（L.）All.；锋芒草（大虱子草）；Burgrass，European Bur-grass，Stalked Bur Grass，Stalked Burgrass，Stalked Bur-grass ■

394391　Tragus racemosus（L.）All. var. berteronianus（Schult. et Schult. f.）Hack. = Tragus berteronianus Schult. et Schult. f. ■

394392　Tragus racemosus（L.）All. var. berteronianus（Schult.）Hack. = Tragus berteronianus Schult. et Schult. f. ■

394393　Tragus racemosus（L.）All. var. decipiens（Fig. et De Not.）Maire = Tragus racemosus（L.）All. ■

394394　Tragus racemosus（L.）All. var. major Hack. = Tragus koelerioides Asch. ■☆

394395　Tragus racemosus（L.）All. var. paucispina（Hack.）Maire = Tragus racemosus（L.）All. ■

394396　Tragus roxburghii Panigrahi；罗氏锋芒草■☆

394397　Tragus roxburghii Panigrahi = Tragus mongolorum Ohwi ■

394398 Tragus tcheliensis Debeaux = Tragus berteronianus Schult. et Schult. f. ■

394399 Traillia Lindl. ex Endl. = Schimpera Steud. et Hochst. ex Endl. ■☆

394400 Trailliaedoxa W. W. Sm. et Forrest（1917）；丁茜属；Trailliaedoxa ●★

394401 Trailliaedoxa gracilis W. W. Sm. et Forrest；丁茜；Slender Trailliaedoxa ●◇

394402 Trallesia Zumag. = Matricaria L. ■

394403 Trallesia Zumagl. = Tripleurospermum Sch. Bip. ■

394404 Tralliana Lour. = Colubrina Rich. ex Brongn.（保留属名）●

394405 Trambis Raf. = Phlomis L. ●■

394406 Trambis Raf. = Phlomoides Moench ●■

394407 Tramoia Schwacke et Taub. ex Glaz.（1913）；巴西乳荨麻属☆

394408 Trankenia Thunb. = Frankenia L. ●■

394409 Transberingia Al-Shehbaz et O'Kane = Beringia R. A. Price, Al-Shehbaz et O'Kane ■☆

394410 Transberingia Al-Shehbaz et O'Kane（2003）；新白令芥属■☆

394411 Transcaucasia M. Hiroe = Astrantia L. ■☆

394412 Trapa L.（1753）；菱属；Water Chestnut, Waterchestnut ■

394413 Trapa acicularis V. N. Vassil.；针形菱■☆

394414 Trapa acornis Nakano；无角菱（南湖菱）；Nanhu Waterchestnut ■

394415 Trapa acornis Nakano = Trapa natans L. ■

394416 Trapa amurensis Flerow = Trapa natans L. ■

394417 Trapa amurensis Nakai；黑水菱；Amur Waterchestnut ■☆

394418 Trapa amurensis Nakai var. komarovi Skvortsov = Trapa manshurica Flerow ■

394419 Trapa amurensis Nakai var. komarovii Skvortsov = Trapa natans L. ■

394420 Trapa antennifera H. Lév. = Trapella sinensis Oliv. ■

394421 Trapa arcuata S. H. Li et Y. L. Chang；弓角菱；Arcuate Waterchestnut ■

394422 Trapa arcuata S. H. Li et Y. L. Chang = Trapa natans L. ■

394423 Trapa astrachanica（Flerow）Winter；阿斯特拉哈菱■☆

394424 Trapa austroafricana V. N. Vassil. = Trapa natans L. var. bispinosa（Roxb.）Makino ■

394425 Trapa bicornis L. f. = Trapa bicornis Osbeck ■

394426 Trapa bicornis Osbeck；乌菱（扒菱，大头菱，大湾角菱，风菱，红菱，菱，菱角，水菱角）；Ling Nut, Water Chestnut, Wu Waterchestnut ■

394427 Trapa bicornis Osbeck = Trapa natans L. ■

394428 Trapa bicornis Osbeck var. acornis（Nakano）Z. T. Xiong = Trapa natans L. ■

394429 Trapa bicornis Osbeck var. bispinosa（Roxb.）Nakano = Trapa natans L. ■

394430 Trapa bicornis Osbeck var. cochinchinensis（Lour.）Glück；越南菱（耳菱，菱）；Viatnam Waterchestnut ■

394431 Trapa bicornis Osbeck var. cochinchinensis（Lour.）Steenis = Trapa natans L. ■

394432 Trapa bicornis Osbeck var. quadrispinosa（Roxb.）Z. T. Xiong = Trapa natans L. ■

394433 Trapa bicornis Osbeck var. taiwanensis（Nakai）Z. T. Xiong；台湾菱；Taiwan Waterchestnut ■

394434 Trapa bicornis Osbeck var. taiwanensis（Nakai）Z. T. Xiong = Trapa natans L. ■

394435 Trapa bispinosa Osbeck var. iinumae Nakano = Trapa natans L. ■

394436 Trapa bispinosa Roxb.；菱（二角菱，风菱，芰，家菱，蕨攗，莲角，菱角，薐，菠，三角菱，沙角，双角菱，水栗，水菱，水菱角，乌菱）；Jesuit's-nut, Singhara Nut, Singharanut, Water Caltrops, Water Chestnut, Waterchestnut ■

394437 Trapa bispinosa Roxb. = Trapa japonica Flerow ■

394438 Trapa bispinosa Roxb. = Trapa natans L. var. bispinosa（Roxb.）Makino ■

394439 Trapa bispinosa Roxb. = Trapa natans L. ■

394440 Trapa bispinosa Roxb. var. culta Hu；水刺菱；Bigspine Waterchestnut ■☆

394441 Trapa bispinosa Roxb. var. iinumai Makino = Trapa bispinosa Roxb. ■

394442 Trapa bispinosa Roxb. var. iinumai Makino = Trapa japonica Flerow ■

394443 Trapa bispinosa Roxb. var. incisa（Siebold et Zucc.）Franch. et Sav. = Trapa incisa Siebold et Zucc. ■

394444 Trapa bispinosa Roxb. var. incisa Franch. et Sav. = Trapa incisa Siebold et Zucc. ■

394445 Trapa bispinosa Roxb. var. iwasakii？= Trapa japonica Flerow ■

394446 Trapa bispinosa Roxb. var. makinoi Nakano = Trapa japonica Flerow var. tuberculifera（V. N. Vassil.）Tzvelev ■

394447 Trapa carinthiaca（Beck）Vassiliev；克伦地亚野菱■☆

394448 Trapa chinensis Lour. = Trapa bicornis Osbeck var. cochinchinensis（Lour.）Glück ■

394449 Trapa chinensis Lour. = Trapa natans L. ■

394450 Trapa cochinchinensis Lour. = Trapa bicornis Osbeck var. cochinchinensis（Lour.）Glück ■

394451 Trapa cochinchinensis Lour. = Trapa natans L. ■

394452 Trapa congolensis V. N. Vassil. = Trapa natans L. var. bispinosa（Roxb.）Makino ■

394453 Trapa conocarpa？；束果野菱■☆

394454 Trapa cruciata（Glück）V. N. Vassil.；十字野菱■☆

394455 Trapa dimorphocarpa Z. S. Diao = Trapa bicornis Osbeck var. taiwanensis（Nakai）Z. T. Xiong ■

394456 Trapa dimorphocarpa Z. S. Diao = Trapa natans L. ■

394457 Trapa europaea Flerow；欧洲野菱■☆

394458 Trapa hyrcana Woronow；希尔康野菱■☆

394459 Trapa incisa Siebold et Zucc.；细果野菱（刺菱，四角刻叶菱，小果菱，野菱）；Incisa Waterchestnut ■

394460 Trapa incisa Siebold et Zucc. var. quadricaudata Glück；野菱（刺菱，四尾野菱）■

394461 Trapa incisa Siebold et Zucc. var. quadricaudata Glück = Trapa incisa Siebold et Zucc. ■

394462 Trapa insperata V. N. Vassil. = Trapa natans L. var. bispinosa（Roxb.）Makino ■

394463 Trapa japonica Flerow；日本菱（菱，菱角，丘角菱，沙角，无冠菱）；Japanese Waterchestnut ■

394464 Trapa japonica Flerow = Trapa natans L. ■

394465 Trapa japonica Flerow var. jeholensis（Nakai）Kitag. = Trapa natans L. ■

394466 Trapa japonica Flerow var. longicollum Z. T. Xiong = Trapa natans L. ■

394467 Trapa japonica Flerow var. magnicorona Z. T. Xiong = Trapa natans L. ■

394468 Trapa japonica Flerow var. makinoi（Nakano）Ohba = Trapa japonica Flerow var. tuberculifera（V. N. Vassil.）Tzvelev ■

394469 Trapa japonica Flerow var. pumila（Nakano ex Verdc.）Ohba = Trapa natans L. var. pumila Nakano ex Verdc. ■

394470 Trapa japonica Flerow var. rubeola（Makino）Ohwi = Trapa natans L. var. quadrispinosa（Roxb.）Makino ■

394471　Trapa japonica Flerow var. tuberculifera（V. N. Vassil.）Tzvelev；块菱■

394472　Trapa japonica Flerow var. tuberculifera（V. N. Vassil.）Tzvelev = Trapa natans L. ■

394473　Trapa jeholensis Nakai = Trapa natans L. ■

394474　Trapa komarovii V. N. Vassil. = Trapa japonica Flerow ■

394475　Trapa komarovii V. N. Vassil. = Trapa pseudoincisa Nakai ■

394476　Trapa komarovii V. N. Vassil. var. sungariensis A. I. Baranov et Skvortsov；松江格菱；Songjiang Waterchestnut ■

394477　Trapa komarovii V. N. Vassil. var. tetracorna A. I. Baranov et Skvortsov；四角格菱；Tetragonal Waterchestnut ■

394478　Trapa korshinskyi V. N. Vassil.；无冠菱；Crownless Waterchestnut ■☆

394479　Trapa korshinskyi V. N. Vassil. = Trapa natans L. ■

394480　Trapa litwinowii V. N. Vassil.；冠菱；Crown Waterchestnut ■

394481　Trapa litwinowii V. N. Vassil. = Trapa natans L. ■

394482　Trapa litwinowii V. N. Vassil. var. chihunensis S. F. Guan et Q. Lang = Trapa natans L. ■

394483　Trapa litwinowii V. N. Vassil. var. chihunensis S. F. Guan et Q. Lang = Trapa macropoda Miki var. bispinosa（Flerow）W. H. Wan ■

394484　Trapa longicornis V. N. Vassil.；长角菱■☆

394485　Trapa macropoda Miki；四角大柄菱（东北菱）；Bigstalk Waterchestnut ■

394486　Trapa macropoda Miki var. bispinosa（Flerow）W. H. Wan；二角大柄菱；Bispinose Bigstalk Waterchestnut ■

394487　Trapa maleevii V. N. Vassil.；马利菱■☆

394488　Trapa mammillifera Miki；四瘤菱■

394489　Trapa manshurica Flerow；东北菱（短颈东北菱）；Northern Waterchestnut ■

394490　Trapa manshurica Flerow = Trapa macropoda Miki ■

394491　Trapa manshurica Flerow = Trapa natans L. ■

394492　Trapa manshurica Flerow f. komarovi（Skvortsov）S. H. Li et Y. L. Chang = Trapa manshurica Flerow ■

394493　Trapa manshurica Flerow f. komarovii（Skvortsov）S. H. Li et Y. L. Chang = Trapa natans L. ■

394494　Trapa manshurica Flerow var. bispinosa Flerow = Trapa macropoda Miki var. bispinosa（Flerow）W. H. Wan ■

394495　Trapa manshurica Flerow var. bispinosa Flerow = Trapa natans L. ■

394496　Trapa maximowiczii Korsh.；细角野菱（刺菱，鬼菱，菱角，四角马氏菱，细果野菱，小果菱，野菱）；Maximowicz Waterchestnut ■

394497　Trapa maximowiczii Korsh. = Trapa incisa Siebold et Zucc. ■

394498　Trapa maximowiczii Korsh. = Trapa natans L. var. pumila Nakano ex Verdc. ■

394499　Trapa maximowiczii Korsh. var. tonkinensis Gag.？ = Trapa incisa Siebold et Zucc. ■

394500　Trapa natans L.；欧菱（芰，浮水菱，格菱，菩，芰，菱，菱角，菠，菠角，漂浮菱，水栗，四角菱，铁菱角，薢菩，野菱，野菱角）；Caltrop，Horn Nut，Horn-nut，Jesuit's Nut，Saligot，Saligot Nut，Spanish Chestnut，Water Caltrop，Water Caltrops，Water Chestnut ■

394501　Trapa natans L. = Trapa quadrispinosa Roxb. ■

394502　Trapa natans L. f. quadrispinosa（Roxb.）Makino = Trapa natans L. var. japonica Nakai ■

394503　Trapa natans L. f. quadrispinosa（Roxb.）Makino = Trapa natans L. ■

394504　Trapa natans L. var. africana Brenan；非洲欧菱■☆

394505　Trapa natans L. var. amurensis（Flerow）Kom. = Trapa natans L. ■

394506　Trapa natans L. var. bicornis Makino = Trapa taiwanensis Nakai ■

394507　Trapa natans L. var. bispinosa（Roxb.）Makino = Trapa bispinosa Roxb. ■

394508　Trapa natans L. var. bispinosa（Roxb.）Makino = Trapa japonica Flerow ■

394509　Trapa natans L. var. bispinosa（Roxb.）Makino = Trapa natans L. ■

394510　Trapa natans L. var. incisa（Siebold et Zucc.）Makino；裂叶菱■

394511　Trapa natans L. var. incisa（Siebold et Zucc.）Makino = Trapa incisa Siebold et Zucc. ■

394512　Trapa natans L. var. incisa Makino = Trapa incisa Siebold et Zucc. ■

394513　Trapa natans L. var. japonica Nakai；鬼菱（四角菱）■

394514　Trapa natans L. var. japonica Nakai = Trapa natans L. var. quadrispinosa（Roxb.）Makino ■

394515　Trapa natans L. var. japonica Nakai = Trapa natans L. ■

394516　Trapa natans L. var. pumila Nakano ex Verdc.；四角矮菱■

394517　Trapa natans L. var. pumila Nakano ex Verdc. = Trapa natans L. ■

394518　Trapa natans L. var. quadrispinosa（Roxb.）Makino = Trapa natans L. ■

394519　Trapa natans L. var. quadrispinosa（Roxb.）Makino = Trapa natans L. var. japonica Nakai ■

394520　Trapa natans L. var. rubeola Makino = Trapa natans L. var. quadrispinosa（Roxb.）Makino ■

394521　Trapa natans L. var. rubeola Makino f. viridis Sugim. = Trapa natans L. var. quadrispinosa（Roxb.）Makino ■

394522　Trapa octotuberculata Miki；八瘤菱（八角菱）■

394523　Trapa pectinata Vassiliev；篦菱■☆

394524　Trapa potaninii V. N. Vassil. = Trapa bicornis Osbeck var. cochinchinensis（Lour.）Glück ■

394525　Trapa potaninii V. N. Vassil. = Trapa natans L. ■

394526　Trapa pseudoincisa Nakai；格菱（假野菱）；Komarov Waterchestnut ■

394527　Trapa pseudoincisa Nakai = Trapa natans L. ■

394528　Trapa pseudoincisa Nakai var. aspinosa Z. T. Xiong = Trapa natans L. ■

394529　Trapa pseudoincisa Nakai var. aspinta Z. T. Xiong；无刺格菱；Spineless Waterchestnut ■

394530　Trapa pseudoincisa Nakai var. complana Z. T. Xiong；扁角格菱■

394531　Trapa pseudoincisa Nakai var. complana Z. T. Xiong = Trapa natans L. ■

394532　Trapa pseudoincisa Nakai var. nanchangensis W. H. Wan；南昌格菱；Nanchang Waterchestnut ■

394533　Trapa pseudoincisa Nakai var. nanchangensis W. H. Wan = Trapa natans L. ■

394534　Trapa pseudoincisa Nakai var. potaninii（V. N. Vassil.）Tzvelev = Trapa natans L. ■

394535　Trapa quadrispinosa Roxb.；四角菱（四角野菱，野菱）；Tetragonal Waterchestnut ■

394536　Trapa quadrispinosa Roxb. = Trapa natans L. ■

394537　Trapa quadrispinosa Roxb. var. yongxiuensis W. H. Wan；短四角菱■

394538　Trapa quadrispinosa Roxb. var. yongxiuensis W. H. Wan = Trapa natans L. ■

394539　Trapa rossica V. N. Vassil.；俄罗斯菱■☆

394540　Trapa saissanica（Flerow）V. N. Vassil. = Trapa natans L. ■

394541　Trapa sajanensis V. N. Vassil. ；萨因菱■☆

394542　Trapa sibirica Flerow；西伯利亚菱；Siberia Waterchestnut ■☆

394543　Trapa sibirica Flerow = Trapa natans L. ■

394544　Trapa sibirica Flerow var. saissanica Flerow = Trapa natans L. ■

394545　Trapa sibirica Flerow var. ussuriensis Flerow = Trapa natans L. ■

394546　Trapa sibirica Flerow var. ussuriensis V. N. Vassil. ；锚菱；Anchor Waterchestnut ■☆

394547　Trapa spryginii V. N. Vassil. ；斯普菱■☆

394548　Trapa taiwanensis Nakai = Trapa bicornis Osbeck var. taiwanensis（Nakai）Z. T. Xiong ■

394549　Trapa taiwanensis Nakai = Trapa natans L. ■

394550　Trapa transzchelii V. N. Vassil. = Trapa natans L. ■

394551　Trapa tranzschellii V. N. Vassil. ；川雪菱；Tranzschell Waterchestnut ■☆

394552　Trapa tuberculifera V. N. Vassil. ；瘤角菱；Tubercle Waterchestnut ■☆

394553　Trapa tuberculifera V. N. Vassil. = Trapa natans L. ■

394554　Trapaceae Dumort.（1829）（保留科名）；菱科；Water Chestnut Family，Waterchestnut Family ■

394555　Trapaceae Dumort.（保留科名）= Lythraceae J. St. -Hil.（保留科名）■●

394556　Trapaceae Dumort.（保留科名）= Trapellaceae Honda et Sakis. ■

394557　Trapaulos Rchb. = Hydrangea L. ●

394558　Trapaulos Rchb. = Traupalos Raf. ●

394559　Trapella Oliv.（1887）；茶菱属（茶菱角属，铁菱角属）；Trapella ■

394560　Trapella sinensis Oliv. ；茶菱（茅卡，荞米，三叉草，铁菱角，藻心）；China Trapella，Chinese Trapella ■

394561　Trapella sinensis Oliv. f. antennifera（H. Lév.）Kitag. = Trapella sinensis Oliv. ■

394562　Trapella sinensis Oliv. var. antennifera（H. Lév.）H. Hara = Trapella sinensis Oliv. ■

394563　Trapellaceae Honda et Sakis.（1930）；茶菱科■

394564　Trapellaceae Honda et Sakis. = Pedaliaceae R. Br.（保留科名）●■

394565　Trapellaceae Honda et Sakis. = Plantaginaceae Juss.（保留科名）■

394566　Trapellaceae Honda et Sakis. = Tremandraceae R. Br. ex DC.（保留科名）●☆

394567　Trarchydium rubrinerve Franch. = Physospermopsis rubrinervis（Franch.）C. Norman ■

394568　Trasera Raf. = Frasera Walter ■☆

394569　Trasi Lestib. = Trasis P. Beanv. ■

394570　Trasis P. Beanv. = Cladium P. Browne ■

394571　Trasus S. F. Gray = Carex L. ■

394572　Trattenikia Pers. = Marshallia Schreb.（保留属名）■☆

394573　Trattinickya A. Juss. = Trattinnickia Willd. ●☆

394574　Trattinnickia Willd.（1806）；特拉橄榄属●☆

394575　Traubia Moldenke（1963）；智利特石蒜属■☆

394576　Traubia chilensis（F. Phil.）Moldenke；智利特石蒜■☆

394577　Traunia K. Schum. = Toxocarpus Wight et Arn. ●

394578　Traunia albiflora K. Schum. = Dregea schimperi（Decne.）Bullock ●☆

394579　Traunsteinera Rchb.（1842）；特劳兰属（吐氏兰属）■☆

394580　Traunsteinera globosa（L.）Rchb. f. ；特劳兰；Globe Orchid ■☆

394581　Traunsteinera sphaerica（M. Bieb.）Schltdl. ；球形特劳兰■☆

394582　Traupalos Raf. = Hydrangea L. ●

394583　Trautvetteria Fisch. et C. A. Mey.（1835）；无瓣毛茛属（槭叶升麻属）；False Bugbane，Tassel-rue ■☆

394584　Trautvetteria caroliniensis（Walter）Vail；卡罗来纳无瓣毛茛（卡罗林无瓣毛茛）；False Bugbane，Tassel Rue，Tassel-rue ■☆

394585　Trautvetteria caroliniensis Fisch. et C. A. Mey. var. borealis（Hara）T. Shimizu = Trautvetteria caroliniensis（Walter）Vail ■☆

394586　Trautvetteria caroliniensis Fisch. et C. A. Mey. var. occidentalis（A. Gray）C. L. Hitchc. = Trautvetteria caroliniensis（Walter）Vail ■☆

394587　Trautvetteria grandis Nutt. = Trautvetteria caroliniensis（Walter）Vail ■☆

394588　Trautvetteria japonica Sieboldet Zucc. ；日本无瓣毛茛■☆

394589　Trautvetteria palmata（Michx.）Fisch. et C. A. Mey. ；无瓣毛茛（槭叶升麻）■☆

394590　Trautvetteria palmata（Michx.）Fisch. et C. A. Mey. = Trautvetteria caroliniensis（Walter）Vail ●☆

394591　Traversia Hook. f.（1864）；黏菊木属●☆

394592　Traversia Hook. f. = Senecio L. ■●

394593　Traversia baccharoides Hook. f. ；黏菊木●☆

394594　Traxara Raf. = Lobostemon Lehm. ■☆

394595　Traxilisa Raf. = Tetracera L. ●

394596　Traxilum Raf. = Ehretia P. Browne ●

394597　Trdeasea Rose = Tradescantia L. ■

394598　Trecacoris Pritz. = Thecacoris A. Juss. ●☆

394599　Trechonaetes Miers = Jaborosa Juss. ●☆

394600　Treculia Decne. = Treculia Decne. ex Trécul ●☆

394601　Treculia Decne. ex Trécul（1847）；非洲面包桑属（特里桑属，特丘林属）；Treculia ●☆

394602　Treculia acuminata Baill. ；尖非洲面包桑●☆

394603　Treculia affona N. E. Br. = Treculia africana Decne. ex Trécul ●☆

394604　Treculia africana Decne. = Treculia africana Decne. ex Trécul ●☆

394605　Treculia africana Decne. ex Trécul；非洲面包桑（非洲特里桑）；African Breadfruit，African Breadfruit-tree ●☆

394606　Treculia africana Decne. ex Trécul var. inversa Okafor；倒垂非洲面包桑●☆

394607　Treculia africana Decne. ex Trécul var. mollis（Engl.）J. Léonard；柔软非洲面包桑●☆

394608　Treculia africana Decne. var. inversa Okafor = Treculia africana Decne. ex Trécul var. inversa Okafor ●☆

394609　Treculia africana Decne. var. mollis（Engl.）J. Léonard = Treculia africana Decne. ex Trécul var. mollis（Engl.）J. Léonard ●☆

394610　Treculia brieyi De Wild. = Treculia obovoidea N. E. Br. ●☆

394611　Treculia centralis A. Chev. = Treculia africana Decne. ●☆

394612　Treculia dewevrei De Wild. et T. Durand = Treculia africana Decne. ●☆

394613　Treculia erinacea A. Chev. = Treculia africana Decne. ex Trécul ●☆

394614　Treculia mollis Engl. = Treculia africana Decne. var. mollis（Engl.）J. Léonard ●☆

394615　Treculia obovoidea N. E. Br. ；椭圆非洲面包桑●☆

394616　Treculia parva Engl. = Treculia acuminata Baill. ●☆

394617　Treculia staudtii Engl. = Treculia obovoidea N. E. Br. ●☆

394618　Treculia zenkeri Engl. = Treculia acuminata Baill. ●☆

394619　Treichelia Vatke（1874）；特雷桔梗属●☆

394620　Treichelia longebracteata（H. Buek）Vatke；长苞特雷桔梗●☆

394621　Treisia Haw. = Euphorbia L. ●■

394622　Treisteria Griff. = Curanga Juss. ■☆

394623　Treleasea Rose = Setcreasea K. Schum. et Syd. ■☆

394624　Treleasea brevifolia（Torr.）Rose = Setcreasea brevifolia

（Torr.）K. Schum. et Syd. ■☆

394625　Treleasea pallida Rose ＝Setcreasea pallida（Rose）K. Schum. et Syd. ■☆

394626　Trema Lour.（1790）；山黄麻属（山麻黄属）；Nettletree, Trema, Wildjute ●

394627　Trema africana（Planch.）Blume ＝Trema orientalis（L.）Blume ●

394628　Trema amboinensis（Willd.）Blume；安倍那山黄麻（山油麻）；Amboin Trema, Amboin Wildjute ●

394629　Trema amboinensis（Willd.）Blume ＝Trema cannabina Lour. ●

394630　Trema amboinensis（Willd.）Blume ＝Trema tomentosa（Roxb.）H. Hara ●

394631　Trema angustifolia（Planch.）Blume；狭叶山黄麻（麻脚树，狭叶山油麻，小叶山黄麻，野婆树）；Narrowleaf Trema, Narrowleaf Wildjute, Narrow-leaved Trema ●

394632　Trema angustifolia Blume ＝Trema angustifolia（Planch.）Blume ●

394633　Trema bracteolata（Hochst.）Blume ＝Trema orientalis（L.）Blume ●

394634　Trema calcicola S. X. Ren；钙土山黄麻●

394635　Trema cannabina Lour.；光叶山黄麻（果连丹，滑朗树，尖尾斧头树，尖尾叶谷木树，仁丹树，锐叶山黄麻，山海麻，山油麻，山榆麻，细叶山黄麻，野谷麻，野山麻，硬壳朗）；Glabrousleaf Trema, Glabrous-leaved Trema, Nakedleaf Wildjute, Shan-you-ma ●

394636　Trema cannabina Lour. var. dielsiana（Hand.-Mazz.）C. J. Chen ＝Trema dielsiana Hand.-Mazz. ●

394637　Trema dielsiana Hand.-Mazz.；山油麻（椰木，椰树，山脚麻，山水麻，山野麻，山油桐，羊角杯，野丝棉）；Diels Trema, Diels Wildjute ●

394638　Trema dielsiana Hand.-Mazz. ＝Trema cannabina Lour. var. dielsiana（Hand.-Mazz.）C. J. Chen ●

394639　Trema duaniana H. Lév. ＝Trema tomentosa（Roxb.）H. Hara ●

394640　Trema floridana Britton ＝Trema micrantha（L.）Blume ●☆

394641　Trema glomerata（Hochst.）Blume ＝Trema orientalis（L.）Blume ●

394642　Trema guineense（Schumach. et Thonn.）Ficalho；几内亚山黄麻；Guinean Trema ●☆

394643　Trema guineensis（Schumach. et Thonn.）Ficalho ＝Trema orientalis（L.）Blume ●

394644　Trema guineensis（Schumach. et Thonn.）Ficalho var. asperata（Solms）Cufod. ＝Trema orientalis（L.）Blume ●

394645　Trema guineensis（Schumach. et Thonn.）Ficalho var. hochstetteri（Planch.）Engl. ＝Trema orientalis（L.）Blume ●

394646　Trema guineensis（Schumach. et Thonn.）Ficalho var. parvifolia（Schumach. et Thonn.）Engl. ＝Trema orientalis（L.）Blume ●

394647　Trema guineensis（Schumach. et Thonn.）Ficalho var. paucinervia Hauman ＝Trema orientalis（L.）Blume ●

394648　Trema lamarckiana（Roem. et Schult.）Blume；西印度山黄麻；West Indian trema ●☆

394649　Trema lanceolata Merr. ＝Trema angustifolia（Planch.）Blume ●

394650　Trema levigata Hand.-Mazz.；羽脉山黄麻（光叶山黄麻，麻椰树，麻柳树，麻椰树，羽叶山黄麻）；Pinnatenerve Trema, Pinnatenerve Wildjute, Pinnate-nerved Trema ●

394651　Trema melinona Blume ＝Trema micrantha（L.）Blume ●☆

394652　Trema micrantha（L.）Blume；小花山黄麻；Florida Trema, Guacimilla ●☆

394653　Trema micrantha（L.）Blume var. floridana（Britton）Standl. et

Steyerm. ＝Trema micrantha（L.）Blume ●☆

394654　Trema nitens（Planch.）Blume ＝Trema orientalis（L.）Blume ●

394655　Trema nitida C. J. Chen；银毛叶山黄麻；Silverhairleaf Wildjute, Silverhairy Trema, Silver-hairy Trema ●

394656　Trema orientalis（L.）Blume；异色山黄麻（九层麻，麻桐树，山黄麻，山角麻，山麻黄，山麻木，山王麻）；Charcoal Tree, Differcolor Wildjute, Gunpowder Tree, India Charcoal-trema, Indian-charcoal Trema, Oriental Trema ●

394657　Trema orientalis（L.）Blume ＝Trema tomentosa（Roxb.）H. Hara ●

394658　Trema politoria（Planch.）Blume；亮山黄麻●☆

394659　Trema polygama Z. M. Wu et J. Y. Lin ＝Trema orientalis（L.）Blume ●

394660　Trema sampsonii（Hance）Merr. et Chun ＝Trema angustifolia（Planch.）Blume ●

394661　Trema strigosa（Planch.）Blume ＝Trema orientalis（L.）Blume ●

394662　Trema timorensis Blume ＝Trema cannabina Lour. ●

394663　Trema timorensis Hemsl. ＝Trema cannabina Lour. ●

394664　Trema tomentosa（Roxb.）H. Hara；山黄麻（九层麻，麻布树，麻络木，麻桐树，蚂蚁树，母子麻，母子树，山麻，山油麻，异色山黄麻）；India Charcoal Trema, Indian-charcoal Trema, Tomentose Trema, Wildjute ●

394665　Trema velutina（Planch.）Blume ＝Trema tomentosa（Roxb.）H. Hara ●

394666　Trema virgata（Roxb. ex Wall.）Blume ＝Trema cannabina Lour. ●

394667　Trema virgata（Roxb.）Blume ＝Trema cannabina Lour. ●

394668　Trema virgata Blume ＝Trema cannabina Lour. ●

394669　Trema virgata Blume ＝Trema levigata Hand.-Mazz. ●

394670　Tremacanthus S. Moore（1904）；穴刺爵床属■☆

394671　Tremacanthus roberti S. Moore；穴刺爵床■☆

394672　Tremacron Craib（1918）；短檐苣苔属（斜管苣苔属）；Tremacron ■★

394673　Tremacron aurantiacum K. Y. Pan；橙黄短檐苣苔；Orange Tremacron ■

394674　Tremacron begoniifolium H. W. Li；景东短檐苣苔；Begonialeaf Tremacron ■

394675　Tremacron forrestii Craib；短檐苣苔；Forrest Tremacron ■

394676　Tremacron mairei Craib；东川短檐苣苔；Maire Tremacron ■

394677　Tremacron obliquifolium K. Y. Pan；狭叶短檐苣苔；Narrowleaf Tremacron ■

394678　Tremacron rubrum Hand.-Mazz.；红短檐苣苔；Red Tremacron ■

394679　Tremacron urceolatum K. Y. Pan；木里短檐苣苔；Muli Tremacron ■

394680　Tremandra R. Br. ＝Tremandra R. Br. ex DC. ●☆

394681　Tremandra R. Br. ex DC.（1824）；孔药木属（孔药花属）●☆

394682　Tremandra diffusa R. Br. ex DC.；孔药木●☆

394683　Tremandraceae DC. ＝Elaeocarpaceae Juss.（保留科名）●

394684　Tremandraceae DC. ＝Tremandraceae R. Br. ex DC.（保留科名）●☆

394685　Tremandraceae R. Br. ex DC.（1824）（保留科名）；孔药木科（独勃门多拉科，假石南科，孔药花科）●☆

394686　Tremandraceae R. Br. ex DC.（保留科名）＝Elaeocarpaceae Juss.（保留科名）●

394687　Tremanthera Post et Kuntze ＝Saurauia Willd.（保留属名）●

394688　Tremanthera Post et Kuntze ＝Tremantanthera F. Muell. ●

394689　Tremanthus Pers. = Strigilia Cav. ●

394690　Tremanthus Pers. = Styrax L. ●

394691　Tremasperma Raf. = Calonyction Choisy ■

394692　Tremastelma Raf. = Lomelosia Raf. ■☆

394693　Tremastelma Raf. = Scabiosa L. ●■

394694　Trematanthera F. Muell. = Saurauia Willd.（保留属名）●

394695　Trematocarpus Zahlbr. = Trematolobelia Zahlbr. ex Rock ●☆

394696　Trematolobelia Zahlbr. ex Rock = Trematocarpus Zahlbr. ●☆

394697　Trematolobelia Zahlbr. ex Rock(1913)；凹半边莲属●☆

394698　Trematolobelia macrostachys Zahlbr. ex Rock；凹半边莲●☆

394699　Trematosperma Urb. = Pyrenacantha Hook. ex Wight ●

394700　Trematosperma Urb. = Pyrenacantha Wight（保留属名）●

394701　Trematosperma cordatum Urb. = Pyrenacantha malvifolia Engl. ●☆

394702　Trembleya DC.(1828)；特伦野牡丹属●☆

394703　Trembleya parviflora Cogn.；小花特伦野牡丹；Island Glorybush ●☆

394704　Trembleya phlogiformis DC.；特伦野牡丹；Island Glorybush ●☆

394705　Tremolsia Gand. = Atractylis L. ■☆

394706　Tremotis Raf. = Ficus L. ●

394707　Tremula Dumort. = Populus L. ●

394708　Tremularia Fabr. = Briza L. ■

394709　Tremularia Heist. = Briza L. ■

394710　Tremularia Heist. ex Fabr. = Briza L. ■

394711　Tremulina B. G. Briggs et L. A. S. Johnson(1998)；二室帚灯草属■☆

394712　Tremulina tremula（R. Br.）B. G. Briggs et L. A. S. Johnson；二室帚灯草■☆

394713　Trendelenburgia Klotzsch = Begonia L. ●■

394714　Trentepohlia Boeck. = Cyperus L. ■

394715　Trentepohlia Roth = Heliophila Burm. f. ex L. ●■☆

394716　Trentepohlia Roth(1800)；肖喜阳花属■☆

394717　Trentepohlia integrifolia Roth；全缘肖喜阳花■☆

394718　Trentepohlia iolithus ?；肖喜阳花；Rock-violet ■☆

394719　Trepadonia H. Rob.(1994)；藤状斑鸠菊属■☆

394720　Trepadonia mexiae（H. Rob.）H. Rob.；藤状斑鸠菊■☆

394721　Trepadonia oppositifolia H. Rob. et H. Beltrán；对叶藤状斑鸠菊■☆

394722　Trepnanthus Steud. = Spiranthera A. St. -Hil. ■☆

394723　Trepnanthus Steud. = Terpnanthus Nees et Mart. ■☆

394724　Trepocarpus Nutt. = Trepocarpus Nutt. ex DC. ■☆

394725　Trepocarpus Nutt. ex DC.(1829)；转果草属■☆

394726　Trepocarpus aethusae Nutt. ex DC.；转果草■☆

394727　Trepodandra Durand = Tripodandra Baill. ●☆

394728　Tresanthera H. Karst.(1859)；钻药茜属●☆

394729　Tresanthera condamineoides H. Karst.；钻药茜●☆

394730　Tresanthera pauciflora K. Schum. ex Soler.；少花钻药茜●☆

394731　Tresteira Hook. f. = Curanga Juss. ■☆

394732　Tresteira Hook. f. = Treisteria Griff. ■☆

394733　Tretocarya Maxim. = Microula Benth. ■

394734　Tretocarya pratensis Maxim. = Microula tibetica Benth. ex Maxim. var. pratensis（Maxim.）W. T. Wang ■

394735　Tretocarya pratensis Maxim. = Microula tibetica Benth. ex Maxim. ■

394736　Tretocarya sikkimensis（C. B. Clarke）Oliv. = Microula sikkimensis（C. B. Clarke）Hemsl. ■

394737　Tretocarya vaillantii Danguy = Microula diffusa（Maxim.）I. M. Johnst. ■

394738　Tretorhiza Adans. = Gentiana L. ■

394739　Tretorrhiza Renealm. ex Delarbre = Gentiana L. ■

394740　Treubania Tiegh. = Amylotheca Tiegh. ●☆

394741　Treubania Tiegh. = Decaisnina Tiegh. ●☆

394742　Treubaniaceae Tiegh. = Loranthaceae Juss.（保留科名）●

394743　Treubella Pierre = Palaquium Blanco ●

394744　Treubella Tiegh. = Amylotheca Tiegh. ●☆

394745　Treubella Tiegh. = Decaisnina Tiegh. ●☆

394746　Treubella Tiegh. = Treubania Tiegh. ●☆

394747　Treubellaceae Tiegh. = Treubaniaceae Tiegh. ●

394748　Treubia Pierre = Lophopyxis Hook. f. ●☆

394749　Treubia Pierre ex Boerl. = Lophopyxis Hook. f. ●☆

394750　Treuia Stokes = Trewia L. ●

394751　Treutlera Hook. f.(1883)；东喜马萝藦属☆

394752　Treutlera insignis Hook. f.；东喜马萝藦☆

394753　Trevauxia Steud. = Luffa Mill. ■

394754　Trevauxia Steud. = Trevouxia Scop. ■

394755　Trevesia Vis.(1842)；刺通草属(广叶参属,枞树属)；Trevesia ●

394756　Trevesia cavaleriei（H. Lév.）Grushv. et Skvortsova = Trevesia palmata（Roxb. ex Lindl.）Vis. ●

394757　Trevesia palmata（Roxb. ex Lindl.）Vis.；刺通草(党楠挡凹,广叶参,广叶参树,广叶蓧,脱萝,枞树,枞木)；Himalayan Trevesia ●

394758　Trevesia palmata（Roxb. ex Lindl.）Vis. var. costata H. L. Li = Trevesia palmata（Roxb. ex Lindl.）Vis. ●

394759　Trevesia palmata（Roxb.）Vis. = Trevesia palmata（Roxb. ex Lindl.）Vis. ●

394760　Trevesia palmata（Roxb.）Vis. var. costata H. L. Li；棱果刺通草(刺通草,肋果刺通草)；Ribfruit Himalayan Trevesia ●

394761　Trevesia pleiosperma Benth. et Hook. f.；少子刺通草●☆

394762　Trevesia sundaica Miq.；異他刺通草●☆

394763　Trevesia tomentella Craib；小毛刺通草●☆

394764　Trevia L. = Trewia L. ●

394765　Treviaceae Bullock = Trewiaceae Lindl. ●

394766　Treviaceae Lindl. = Euphorbiaceae Juss.（保留科名）●■

394767　Treviaceae Lindl. = Trewiaceae Lindl. ●

394768　Trevirana Willd. = Achimenes Pers.（保留属名）■☆

394769　Trevirania Heynh. = Psychotria L.（保留属名）●

394770　Trevirania Roth = Lindernia All. ■

394771　Trevirania Spreng. = Achimenes Pers.（保留属名）■☆

394772　Trevirania Spreng. = Trevirana Willd. ■☆

394773　Trevoa Miers = Trevoa Miers ex Hook. ●☆

394774　Trevoa Miers ex Hook.(1829)；四室鼠李属●☆

394775　Trevoa trinervia Miers ex Hook.；四室鼠李●☆

394776　Trevoria F. Lehm.(1897)；特雷兰属(特丽兰属)■☆

394777　Trevoria chloris F. Lehm.；特雷兰■☆

394778　Trevouxia Scop. = Luffa Mill. ■

394779　Trevouxia Scop. = Turia Forssk. ■

394780　Trewia L.(1753)；滑桃树属；Trewia ●

394781　Trewia Willd. = Mallotus Lour. ●

394782　Trewia africana Baill. = Erythrococca africana（Baill.）Prain ●☆

394783　Trewia nudiflora L.；滑桃树(苦皮树)；Bare-flowered Trewia, Naked Flower Trewia, Nakedflower Trewia ●

394784　Trewia nudifolia Hance = Mallotus repandus（Willd.）Müll. Arg. ●

394785　Trewiaceae Lindl.；滑桃树科●

394786　Trewiaceae Lindl. = Euphorbiaceae Juss.（保留科名）●■

394787　Triachne Cass. = Nassauvia Comm. ex Juss. ●☆

394788　Triachyrium Benth. = Sporobolus R. Br. ■

394789　Triachyrium Benth. = Triachyrum Hochst. ■

394790　Triachyrum Hochst. = Sporobolus R. Br. ■

394791　Triachyrum Hochst. ex A. Braun = Sporobolus R. Br. ■

394792　Triachyrum Hochst. ex Steud. = Sporobolus R. Br. ■

394793　Triachyrum adoense Hochst. ex A. Br. = Sporobolus discosporus Nees ■☆

394794　Triachyrum cordofanum Hochst. ex Steud. = Sporobolus cordofanus (Hochst. ex Steud.) Henriq. ex Coss. ■☆

394795　Triachyrum discosporum (Nees) Steud. = Sporobolus discosporus Nees ■☆

394796　Triachyrum longifolium Hochst. ex Steud. = Sporobolus panicoides A. Rich. ■☆

394797　Triachyrum micranthum Steud. = Sporobolus micranthus (Steud.) T. Durand et Schinz ■☆

394798　Triachyrum nilagiricum Steud. = Sporobolus pilifer (Trin.) Kunth ■

394799　Triacis Griseb. = Turnera L. ●■☆

394800　Triacma Van Hass. ex Miq. = Hoya R. Br. ●

394801　Triactina Hook. f. et Thomson = Sedum L. ●■

394802　Triactina verticillata Hook. f. et Thomson = Sedum triactina Berger ■

394803　Triadenia Miq. = Trachelospermum Lem. ●

394804　Triadenia Spach = Elodes Adans. ■●

394805　Triadenia Spach = Hypericum L. ■●

394806　Triadenia aegyptiaca (L.) Boiss. = Hypericum aegyptiacum L. ●☆

394807　Triadenum Raf. (1837) ;红花金丝桃属(三腺金丝桃属); Triadenum ●

394808　Triadenum asiaticum (Maxim.) Kom. = Triadenum japonicum (Blume) Makino ■

394809　Triadenum breviflorum (Wall. ex Dyer) Y. Kimura;三腺金丝桃;Shortflower Triadenum,White Flower Triadenum ■●

394810　Triadenum fraseri (Spach) Gleason;弗雷泽红花金丝桃;Bog St. John's-wort,Fraser's Marsh St. John's-wort ■☆

394811　Triadenum japonicum (Blume) Makino;红花金丝桃(地耳草,田基黄);Japan Triadenum,Japanese Triadenum ■

394812　Triadenum japonicum (Blume) Makino f. asiaticum (Maxim.) Y. Kimura = Triadenum japonicum (Blume) Makino ■

394813　Triadenum tubulosum (Walter) Gleason;沼泽金丝桃;Marsh St. John's-wort ■☆

394814　Triadenum virginicum (L.) Raf.;美洲金丝桃;Virginia Marsh St. John's-wort ■☆

394815　Triadenum virginicum (L.) Raf. subsp. fraseri (Spach) J. M. Gillett = Triadenum fraseri (Spach) Gleason ■☆

394816　Triadenum virginicum (L.) Raf. var. fraseri (Spach) Cooperr. = Triadenum fraseri (Spach) Gleason ■☆

394817　Triadenum walteri (J. F. Gmel.) Gleason;瓦氏金丝桃;Marsh St. John's-wort ■☆

394818　Triadica Lour. = Sapium Jacq. (保留属名) ●

394819　Triadica japonica (Siebold et Zucc.) Baill. = Neoshirakia japonica (Siebold et Zucc.) Esser ●☆

394820　Triadica japonica (Siebold et Zucc.) Baill. = Sapium japonicum (Siebold et Zucc.) Pax et K. Hoffm. ●

394821　Triadica japonica (Siebold et Zucc.) Baill. f. macrophylla ? = Sapium japonicum (Siebold et Zucc.) Pax et K. Hoffm. ●

394822　Triadica sebifera (L.) Small = Sapium sebiferum (L.) Roxb. ●

394823　Triadica sinensis Lour. = Sapium sebiferum (L.) Roxb. ●

394824　Triadodaphne Kosterm. (1974) ;异被土楠属●☆

394825　Triadodaphne Kosterm. = Endiandra R. Br. ●

394826　Triadodaphne myristicoides Kosterm. ;异被土楠●☆

394827　Triaena Kunth = Bouteloua Lag. (保留属名) ■

394828　Triaenacanthus Nees = Strobilanthes Blume ●■

394829　Triaenacanthus flexicaulis (Hayata) C. F. Hsieh et T. C. Huang = Parachampionella flexicaulis (Hayata) C. F. Hsieh et T. C. Huang ●■

394830　Triaenacanthus flexicaulis (Hayata) C. F. Hsieh et T. C. Huang = Strobilanthes flexicaulis Hayata ●■

394831　Triaenacanthus glandulifera (Hatus.) C. F. Hsieh et T. C. Huang = Strobilanthes glandulifera Hatus. ●■

394832　Triaenacanthus viginianum Raf. ;北美红花金丝桃●■

394833　Triaenanthus Nees = Strobilanthes Blume ●■

394834　Triaenanthus Nees = Triaenacanthus Nees ●■

394835　Triaenophora (Hook. f.) Soler. (1909) ;崖白菜属(呆白菜属,岩白菜属);Rockycabbage,Tridenwort ■★

394836　Triaenophora Soler. = Triaenophora (Hook. f.) Soler. ■★

394837　Triaenophora integra (H. L. Li) Ivanina;全缘叶崖白菜(全缘叶呆白菜);Entireleaf Tridenwort,EntireRockycabbage ■

394838　Triaenophora rupestris (Hemsl.) Soler. ;崖白菜(巴东岩白菜,呆白菜,岩白菜);Common Rockycabbage,Common Tridenworth ■

394839　Triaenophora shennongjiaensis X. D. Li, Y. Y. Zan et J. Q. Li;神农架崖白菜■

394840　Triaina Kunth = Botelua Lag. ■

394841　Triaina Kunth = Triaena Kunth ■

394842　Triainolepis Hook. f. (1873) ;三尖鳞茜草属■☆

394843　Triainolepis africana Hook. f. ;非洲三尖鳞茜草●■☆

394844　Triainolepis africana Hook. f. subsp. hildebrandtii (Vatke) Verdc. ;希尔非洲三尖鳞茜草●☆

394845　Triainolepis fryeri (Hemsl.) Bremek. = Triainolepis africana Hook. f. subsp. hildebrandtii (Vatke) Verdc. ■☆

394846　Triainolepis fryeri (Hemsl.) Bremek. var. latifolia Bremek. = Triainolepis africana Hook. f. subsp. hildebrandtii (Vatke) Verdc. ■☆

394847　Triainolepis hildebrandtii Vatke = Triainolepis africana Hook. f. subsp. hildebrandtii (Vatke) Verdc. ■☆

394848　Triainolepis sancta Verdc. ;三尖鳞茜草■☆

394849　Triallosia Raf. = Lachenalia J. Jacq. ex Murray ■☆

394850　Trianaea Planch. et Linden(1853) ;三尖茄属●☆

394851　Trianaea brevipes S. Knapp;短梗三尖茄●☆

394852　Trianaea nobilis Planch. et Linden;三尖茄●☆

394853　Trianaeopiper Trel. (1928) ;矮胡椒属●☆

394854　Trianaeopiper bullatum Cuatrec. ;泡矮胡椒●☆

394855　Trianaeopiper timbiquinum Trel. ;矮胡椒●☆

394856　Triandrophora O. Schwarz = Cleome L. ●■

394857　Trianea Karat. = Limnobium Rich. ■☆

394858　Triangia Thouars = Angraecum Bory ■

394859　Trianoptiles Fenzl = Trianoptiles Fenzl ex Endl. ■☆

394860　Trianoptiles Fenzl ex Endl. (1836) ;三尖莎属■☆

394861　Trianoptiles capensis (Steud.) Harv. ;好望角三尖莎■☆

394862　Trianoptiles stipitata Levyns;具柄三尖莎■☆

394863　Trianosperma (Torr. et A. Gray) Mart. = Cayaponia Silva Manso (保留属名) ■☆

394864　Trianosperma Mart. = Cayaponia Silva Manso(保留属名)■☆

394865　Trianosperma africana Hook. f. = Cayaponia africana (Hook. f.) Exell ■☆

394866　Triantha (Nutt.) Baker = Tofieldia Huds. ■

394867　Triantha (Nutt.) Baker(1879) ;三花岩菖蒲属■☆

394868　Triantha Baker = Tofieldia Huds. ■

394869　Triantha glutinosa（Michx.）Baker ＝Tofieldia glutinosa Pers. ■☆

394870　Triantha japonica Baker ＝Tofieldia japonica Miq. ■☆

394871　Triantha occidentalis（S. Watson）R. R. Gates ＝Tofieldia occidentalis S. Watson ■☆

394872　Triantha occidentalis（S. Watson）R. R. Gates subsp. brevistyla（C. L. Hitchc.）Packer ＝Tofieldia glutinosa（Michx.）Pers. var. brevistyla（C. L. Hitchc.）C. L. Hitchc. ■☆

394873　Triantha occidentalis（S. Watson）R. R. Gates subsp. montana（C. L. Hitchc.）Packer ＝Tofieldia glutinosa（Michx.）Pers. subsp. montana C. L. Hitchc. ■☆

394874　Triantha racemosa（Walter）Small ＝Tofieldia racemosa（Walter）Britton，Sterns et Poggenb. ■☆

394875　Trianthaea（DC.）Spach ＝Vernonia Schreb.（保留属名）●■

394876　Trianthaea Spach ＝Vernonia Schreb.（保留属名）●■

394877　Trianthella House ＝Tofieldia Huds. ■

394878　Trianthella House ＝Triantha（Nutt.）Baker ■☆

394879　Trianthema L.（1753）；假海马齿属（假海马齿属，三花草属，肖海马齿属）；Falseseapurslane，Horse-purslane ■

394880　Trianthema Spreng. ex Turcz. ＝Adenocline Turcz. ■☆

394881　Trianthema anceps Thunb. ＝Acrosanthes anceps（Thunb.）Sond. ■☆

394882　Trianthema camillei Cordem. ＝Zaleya camillei（Cordem.）H. E. K. Hartmann ■☆

394883　Trianthema ceratosepala Volkens et Irmsch.；喙萼假海马齿■☆

394884　Trianthema crystallina（Forssk.）Vahl；水晶假海马齿■☆

394885　Trianthema crystallina（Forssk.）Vahl var. oblongifolia Gamble ＝Trianthema salsoloides Fenzl ex Oliv. ■☆

394886　Trianthema crystallina（Forssk.）Vahl var. rubens Sond. ＝Trianthema parvifolia E. Mey. ex Sond. var. rubens（Sond.）Adamson ■☆

394887　Trianthema decandra L. ＝Zaleya decandra（L.）Burm. f. ■☆

394888　Trianthema fruticosa Vahl ＝Gymnocarpos decander Forssk. ●☆

394889　Trianthema glandulosa Peter；腺点假海马齿■☆

394890　Trianthema govindia Buch. -Ham. ex G. Don ＝Zaleya pentandra（L.）C. Jeffrey ■☆

394891　Trianthema hereroensis Schinz；赫雷罗假海马齿■☆

394892　Trianthema humifusa Thunb. ＝Acrosanthes humifusa（Thunb.）Sond. ■☆

394893　Trianthema lydaspica Edgew. ＝Sesuvium sesuvioides（Fenzl）Verdc. ■☆

394894　Trianthema monogyna L. ＝Trianthema portulacastrum L. ■

394895　Trianthema multiflora Peter；多花假海马齿■☆

394896　Trianthema nigricans Peter；黑假海马齿■☆

394897　Trianthema nyasica Baker ＝Sesuvium nyasicum（Baker）Gonc. ■☆

394898　Trianthema obcordata Roxb. ＝Trianthema portulacastrum L. ■

394899　Trianthema parvifolia E. Mey. ex Sond.；小花假海马齿■☆

394900　Trianthema parvifolia E. Mey. ex Sond. var. rubens（Sond.）Adamson；红小花假海马齿■☆

394901　Trianthema pentandra L. ＝Zaleya pentandra（L.）C. Jeffrey ■☆

394902　Trianthema pentandra L. var. hirtulum Batt. et Trab. ＝Zaleya pentandra（L.）C. Jeffrey ■☆

394903　Trianthema portulacastrum L.；假海马齿；Desert Horse-purslane，Falseseapurslane，Horse Purslane，Sea Purslane ■

394904　Trianthema procumbens Mill. ＝Trianthema portulacastrum L. ■

394905　Trianthema redimita Melville ＝Zaleya redimita（Melville）H. E. K. Hartmann ■☆

394906　Trianthema salsoloides Fenzl ex Oliv.；猪毛菜假海马齿■☆

394907　Trianthema salsoloides Fenzl ex Oliv. var. stenophylla Adamson；窄叶假海马齿■☆

394908　Trianthema salsoloides Fenzl ex Oliv. var. transvaalensis（Schinz）Adamson；德兰士瓦假海马齿■☆

394909　Trianthema sanguinea Volkens et Irmsch.；血红假海马齿■☆

394910　Trianthema sedifolia Vis. ＝Trianthema triquetra Willd. ex Spreng. ■☆

394911　Trianthema sennii Chiov. ＝Zaleya sennii（Chiov.）C. Jeffrey ■☆

394912　Trianthema transvaalensis Schinz ＝Trianthema salsoloides Fenzl ex Oliv. var. transvaalensis（Schinz）Adamson ■☆

394913　Trianthema triquetra Rottler ex Willd. subsp. parvifolia（E. Mey. ex Sond.）Jeffrey ＝Trianthema parvifolia E. Mey. ex Sond. ■☆

394914　Trianthema triquetra Willd. ex Spreng.；三棱假海马齿■☆

394915　Trianthema triquetra Willd. ex Spreng. var. sanguinea（Volkens et Irmsch.）C. Jeffrey ＝Trianthema sanguinea Volkens et Irmsch. ■☆

394916　Trianthera Wettst. ＝Stemotria Wettst. et Harms ex Engl. ●☆

394917　Trianthium Desv. ＝Chrysopogon Trin.（保留属名）■

394918　Trianthus Hook. f. ＝Nassauvia Comm. ex Juss. ●☆

394919　Triaristella（Rchb. f.）Brieger ＝Trisetella Luer ■☆

394920　Triaristella（Rchb. f.）Brieger ex Luer ＝Trisetella Luer ■☆

394921　Triaristella Brieger ＝Trisetella Luer ■☆

394922　Triaristella Luer ＝Trisetella Luer ■☆

394923　Triaristellina Rauschert ＝Triaristella Luer ■☆

394924　Triaristellina Rauschert ＝Trisetella Luer ■☆

394925　Triarrhena（Maxim.）Nakai ＝Miscanthus Andersson ■

394926　Triarrhena（Maxim.）Nakai（1950）；荻属；Silverreed ■

394927　Triarrhena lutarioriparia L. Liou ＝Miscanthus lutarioriparius L. Liou ex Renvoize et S. L. Chen ■

394928　Triarrhena lutarioriparia L. Liou var. elevatinodis L. Liou et P. F. Chen；秃节荻；Elevatinode Riparial Silverreed ■

394929　Triarrhena lutarioriparia L. Liou var. gongchai L. Liou；岗柴；Gangchai Riparial Silverreed ■

394930　Triarrhena lutarioriparia L. Liou var. gongchai L. Liou f. altissima L. Liou；一丈青■

394931　Triarrhena lutarioriparia L. Liou var. gongchai L. Liou f. chuiyeqing L. Liou ＝Triarrhena lutarioriparia L. Liou var. gongchai L. Liou f. pendulifolia L. Liou ■

394932　Triarrhena lutarioriparia L. Liou var. gongchai L. Liou f. coccinea L. Liou；胭脂红■

394933　Triarrhena lutarioriparia L. Liou var. gongchai L. Liou f. pendulifolia L. Liou；垂叶箐■

394934　Triarrhena lutarioriparia L. Liou var. gongchai L. Liou f. purpureorosa L. Liou；铁秆柴■

394935　Triarrhena lutarioriparia L. Liou var. gongchai L. Liou f. tiegangang L. Liou ＝Triarrhena lutarioriparia L. Liou var. gongchai L. Liou f. purpureorosa L. Liou ■

394936　Triarrhena lutarioriparia L. Liou var. gracilior L. Liou；茅荻；Thin Riparial Silverreed ■

394937　Triarrhena lutarioriparia L. Liou var. humilior L. Liou；细荻；Slender Riparial Silverreed ■

394938　Triarrhena lutarioriparia L. Liou var. junshanensis L. Liou；君山荻；Junshan Riparial Silverreed ■

394939　Triarrhena lutarioriparia L. Liou var. planiodis L. Liou；平节荻；Flatnode Riparial Silverreed ■

394940　Triarrhena lutarioriparia L. Liou var. shachai L. Liou；刹柴；Shachai Riparial Silverreed ■

394941　Triarrhena lutarioriparia L. Liou var. shachai L. Liou f. qingsha L. Liou;青刹■

394942　Triarrhena lutarioriparia L. Liou var. shachai L. Liou f. zisha L. Liou;紫刹■

394943　Triarrhena sacchariflora（Maxim.）Nakai;荻(阿穆尔芒,巴茅根,红柴,野苇子）;Amur Silver Grass, Amur Silvergrass, Amur Silver-grass, Japanese Plume Grass, Plume Grass, Sugarcaneflower Silverreed,Sweetcaneflower Silvergrass ■

394944　Triarrhena sacchariflora（Maxim.）Nakai = Miscanthus sacchariflorus（Maxim.）Benth. ex Hook. f. ■

394945　Triarrhena sacchariflora（Maxim.）Nakai = Miscanthus sacchariflorus（Maxim.）Benth. ■

394946　Triarthron Baill. = Phthirusa Mart. ●☆

394947　Trias Lindl.（1830）;三兰属■☆

394948　Trias crassifolia（Thwaites ex Trimen）C. S. Kumar;厚叶三兰■☆

394949　Trias mollis Seidenf. ;软三兰■☆

394950　Trias nana Seidenf. ;矮三兰■☆

394951　Trias oblonga Lindl. ;矩圆三兰■☆

394952　Trias ovata Lindl. ;卵三兰■☆

394953　Trias rosea（Ridl.）Seidenf. ;玫瑰三兰■☆

394954　Triascidium Benth. et Hook. f. = Huanaca Cav. ■☆

394955　Triascidium Benth. et Hook. f. = Trisciadium Phil. ■☆

394956　Triasekia G. Don = Anagallis L. ■

394957　Triasekia G. Don = Jirasekia F. W. Schmidt ■

394958　Triaspis Burch.（1824）;三盾草属●☆

394959　Triaspis acuminata Engl. = Flabellariopsis acuminata（Engl.）R. Wilczek ●☆

394960　Triaspis angolensis Nied. = Triaspis lateriflora Oliv. ●☆

394961　Triaspis aurea Nied. = Triaspis odorata（Willd.）A. Juss. ●☆

394962　Triaspis auriculata Radlk. = Caucanthus auriculatus（Radlk.）Nied. ●☆

394963　Triaspis bukombensis De Wild. = Triaspis mooreana Exell et Mendonça ●☆

394964　Triaspis canescens Engl. = Triaspis hypericoides（DC.）Burch. subsp. canescens（Engl.）Immelman ●☆

394965　Triaspis dumeticola Launert;灌丛三盾草●☆

394966　Triaspis emarginata De Wild. ;微缺三盾草●☆

394967　Triaspis emarginata De Wild. var. discolor R. Wilczek;异色三盾草●☆

394968　Triaspis erlangeri Engl. ;厄兰格三盾草●☆

394969　Triaspis flabellaria A. Juss. = Flabellaria paniculata Cav. ●☆

394970　Triaspis gilletii De Wild. = Triaspis macropteron Welw. ex Oliv. ●☆

394971　Triaspis glaucophylla Engl. ;灰绿三盾草●☆

394972　Triaspis hypericoides（DC.）Burch. ;金丝桃三盾草●☆

394973　Triaspis hypericoides（DC.）Burch. subsp. canescens（Engl.）Immelman;灰白三盾草●☆

394974　Triaspis hypericoides（DC.）Burch. subsp. nelsonii（Oliv.）Immelman;纳尔逊三盾草●☆

394975　Triaspis lateriflora Oliv. ;侧花三盾草●☆

394976　Triaspis leendertziae Burtt Davy = Triaspis glaucophylla Engl. ●☆

394977　Triaspis letestuana Launert;莱泰斯图三盾草●☆

394978　Triaspis macropteron Welw. ex Oliv. ;大翅三盾草●☆

394979　Triaspis macropteron Welw. ex Oliv. subsp. massaiensis（Nied.）Launert;马萨三盾草●☆

394980　Triaspis mooreana Exell et Mendonça;穆尔三盾草●☆

394981　Triaspis mozambica A. Juss. ;莫桑比克三盾草●☆

394982　Triaspis nelsonii Oliv. = Triaspis hypericoides（DC.）Burch.

subsp. nelsonii（Oliv.）Immelman ●☆

394983　Triaspis nelsonii Oliv. subsp. canescens（Engl.）Launert = Triaspis hypericoides（DC.）Burch. subsp. canescens（Engl.）Immelman ●☆

394984　Triaspis nelsonii Oliv. var. austro-occidentalis Schinz = Triaspis hypericoides（DC.）Burch. subsp. nelsonii（Oliv.）Immelman ●☆

394985　Triaspis niedenzuiana Engl. ;尼登楚三盾草●☆

394986　Triaspis odorata（Willd.）A. Juss. ;芳香三盾草●☆

394987　Triaspis ovata Bremek. = Triaspis hypericoides（DC.）Burch. subsp. nelsonii（Oliv.）Immelman ●☆

394988　Triaspis rogersii Burtt Davy = Triaspis hypericoides（DC.）Burch. subsp. nelsonii（Oliv.）Immelman ●☆

394989　Triaspis sapinii De Wild. ;萨潘三盾草●☆

394990　Triaspis schliebenii Ernst;施利本三盾草●☆

394991　Triaspis speciosa Nied. = Triaspis macropteron Welw. ex Oliv. subsp. massaiensis（Nied.）Launert ●☆

394992　Triaspis stipulata Oliv. ;托叶三盾草●☆

394993　Triaspis suffulta Launert;支柱三盾草●☆

394994　Triaspis ternata Greenway et Burtt Davy = Triaspis hypericoides（DC.）Burch. subsp. nelsonii（Oliv.）Immelman ●☆

394995　Triaspis thorncroftii Burtt Davy = Triaspis hypericoides（DC.）Burch. subsp. nelsonii（Oliv.）Immelman ●☆

394996　Triaspis transvalica Kuntze = Sphedamnocarpus pruriens（A. Juss.）Szyszyl. subsp. galphimiifolius（A. Juss.）P. D. de Villiers et D. J. Botha ●☆

394997　Triathera Desv. = Bouteloua Lag.（保留属名）■

394998　Triathera Roth ex Roem. et Schult. = Tripogon Roem. et Schult. ■

394999　Triathera bromoides Roth = Tripogon bromoides Roth ex Roem. et Schult. ■

395000　Triatherus Raf. = Ctenium Panz.（保留属名）■☆

395001　Triatherus Raf. = Monocera Elliott ■☆

395002　Triavenopsis P. Candargy = Duthiea Hack. ■

395003　Triavenopsis brachypodia P. Candargy = Duthiea brachypodia（P. Candargy）Keng et P. C. Keng ■

395004　Triavenopsis brachypodium P. Candargy = Duthiea brachypodia（P. Candargy）Keng et P. C. Keng ■

395005　Tribelaceae（Engl.）Airy Shaw = Escalloniaceae R. Br. ex Dumort.（保留科名）●

395006　Tribelaceae（Engl.）Airy Shaw = Grossulariaceae DC.（保留科名）●

395007　Tribelaceae Airy Shaw = Escalloniaceae R. Br. ex Dumort.（保留科名）●

395008　Tribelaceae Airy Shaw = Grossulariaceae DC.（保留科名）●

395009　Tribelaceae Airy Shaw;三齿叶科（三刺木科,智利木科）●☆

395010　Tribeles Phil.（1863）;三齿叶属（三刺木属,智利木属）●☆

395011　Tribeles australis Phil. ;三齿叶（三刺木）●☆

395012　Triblemma R. Br. ex DC. = Bertolonia Raddi(保留属名）■☆

395013　Triblemma R. Br. ex Spreng. = Bertolonia Raddi(保留属名）■☆

395014　Tribolacis Griseb. = Turnera L. ●■☆

395015　Tribolium Desv.（1831）;三尖草属■☆

395016　Tribolium Desv. = Lasiochloa Kunth ■☆

395017　Tribolium acutiflorum（Nees）Renvoize;尖花三尖草■☆

395018　Tribolium alternans（Nees）Renvoize = Plagiochloa uniolae（L. f.）Adamson et Sprague ■☆

395019　Tribolium amplexum Renvoize = Plagiochloa uniolae（L. f.）Adamson et Sprague ■☆

395020　Tribolium brachystachyum（Nees）Renvoize;短穗三尖草■☆

395021　Tribolium ciliare（Stapf）Renvoize；缘毛三尖草■☆

395022　Tribolium echinatum（Thunb.）Renvoize；刺三尖草■☆

395023　Tribolium hispidum（Thunb.）Desv.；毛三尖草■☆

395024　Tribolium obliterum（Hemsl.）Renvoize；三尖草；Capetown Grass■☆

395025　Tribolium obtusifolium（Nees）Renvoize；钝叶三尖草■☆

395026　Tribolium pusillum（Nees）H. P. Linder et Davidse；微小三尖草■☆

395027　Tribolium uniolae（L. f.）Renvoize = Plagiochloa uniolae（L. f.）Adamson et Sprague■☆

395028　Tribolium uniolae（L. f.）Renvoize = Tribolium alternans（Nees）Renvoize■☆

395029　Tribolium uniolae（L. f.）Renvoize = Tribolium amplexum Renvoize■☆

395030　Tribolium utriculosum（Nees）Renvoize；囊果三尖草■☆

395031　Tribonanthes Endl.（1839）；斗篷花属■☆

395032　Tribonanthes australis Endl.；长瓣斗篷花■☆

395033　Tribonanthes brachypetala Lindl.；短瓣斗篷花■☆

395034　Tribonanthes longipetala Lindl.；澳大利亚斗篷花■☆

395035　Tribrachia Lindl. = Bulbophyllum Thouars（保留属名）■

395036　Tribrachia hirta（Sm.）Lindl. = Bulbophyllum hirtum（Sm.）Lindl.■

395037　Tribrachia racemosa（Sm.）Lindl. = Sunipia scariosa Lindl.■

395038　Tribrachia reptans Lindl. = Bulbophyllum reptans（Lindl.）Lindl.■

395039　Tribrachium Benth. et Hook. f. = Bulbophyllum Thouars（保留属名）■

395040　Tribrachium Benth. et Hook. f. = Tribrachia Lindl.●■

395041　Tribrachya Korth. = Rennellia Korth.●☆

395042　Tribrachys Champ. ex Thw. = Thismia Griff.■

395043　Tribroma O. F. Cook = Theobroma L.●

395044　Tribroma bicolor O. F. Cook = Theobroma bicolor Humb. et Bonpl.●☆

395045　Tribula Hill = Seseli L.■

395046　Tribulaceae（Engl.）Hadidi = Zygophyllaceae R. Br.（保留科名）●■

395047　Tribulaceae Hadidi = Zygophyllaceae R. Br.（保留科名）●■

395048　Tribulaceae Trautv. = Zygophyllaceae R. Br.（保留科名）●■

395049　Tribulago Luer = Epidendrum L.（保留属名）■☆

395050　Tribulago Luer（2004）；三尖兰属■☆

395051　Tribulastrum B. Juss. ex Pfeiff. = Neurada L.■☆

395052　Tribulocarpus S. Moore（1921）；刺果番杏属●☆

395053　Tribulocarpus dimorphanthus（Pax）S. Moore；刺果番杏●☆

395054　Tribuloides Ség. = Trapa L.■

395055　Tribulopis R. Br.（1849）；拟蒺藜属■☆

395056　Tribulopis R. Br. = Tribulus L.■

395057　Tribulopis affinis（W. Fitzg.）H. Eichler；近缘拟蒺藜☆

395058　Tribulopis angustifolia R. Br.；窄叶拟蒺藜☆

395059　Tribulopis bicolor F. Muell.；二色拟蒺藜☆

395060　Tribulopis curvicarpa（W. Fitzg.）Keighery；弯果拟蒺藜■☆

395061　Tribulopis sessilis（Domin）H. Eichler；无梗拟蒺藜■☆

395062　Tribulopsis F. Muell. = Tribulopis R. Br.■☆

395063　Tribulopsis R. Br. = Tribulopis R. Br.☆

395064　Tribulus L.（1753）；蒺藜属；Bur Nut，Caltrap，Caltrop■

395065　Tribulus alatus Delile = Tribulus pentandrus Forssk.■☆

395066　Tribulus alatus Delile subsp. macropterus（Boiss.）Maire = Tribulus macropterus Boiss.■☆

395067　Tribulus alatus Delile var. micropteris Kralik = Tribulus pentandrus Forssk. var. micropteris（Kralik）Hosni■☆

395068　Tribulus alatus Delile var. monodii Maire = Tribulus macropterus Boiss.■☆

395069　Tribulus alatus Delile var. odontopteris Kralik = Tribulus pentandrus Forssk.■☆

395070　Tribulus alatus Delile var. vespertilio Maire = Tribulus pentandrus Forssk.■☆

395071　Tribulus albescens Schltr. ex Dinter = Tribulus terrestris L.■

395072　Tribulus bicornutus Fisch. et C. A. Mey. = Tribulus terrestris L. var. bicornutus（Fisch. et C. A. Mey.）Hadidi■

395073　Tribulus bicornutus Fisch. et C. A. Mey. = Tribulus terrestris L.■

395074　Tribulus bimucronatus Kralik = Tribulus terrestris L.■

395075　Tribulus bimucronatus Viv.；双尖蒺藜■☆

395076　Tribulus bimucronatus Viv. subsp. inermis（Kralik）Hosni = Tribulus bimucronatus Viv. var. inermis（Kralik）Hosni■☆

395077　Tribulus bimucronatus Viv. var. bispinulosus（Kralik）Hosni；双刺蒺藜■☆

395078　Tribulus bimucronatus Viv. var. inermis（Kralik）Hosni；无刺蒺藜■☆

395079　Tribulus bimucronatus Viv. var. tomentosus Batt. = Tribulus bimucronatus Viv.■☆

395080　Tribulus bispinulosus Kralik = Tribulus bimucronatus Viv. var. bispinulosus（Kralik）Hosni■☆

395081　Tribulus bispinulosus Kralik = Tribulus terrestris L.■

395082　Tribulus cistoides L.；大花蒺藜；Burr-nut，Jamaican Feverplant，Largeflower Caltrop，Puncture Vine■

395083　Tribulus cistoides L. var. medius（Engl.）Cufod.；药用大花蒺藜■☆

395084　Tribulus cristatus C. Presl；冠状蒺藜■☆

395085　Tribulus echinops Kers；刺蒺藜☆

395086　Tribulus ehrenbergii Asch. = Tribulus macropterus Boiss.☆

395087　Tribulus erectus Engl. = Tribulus zeyheri Sond.■☆

395088　Tribulus hispidus C. Presl = Tribulus terrestris L. var. bicornutus（Fisch. et C. A. Mey.）Hadidi■

395089　Tribulus inermis Engl. = Tribulus zeyheri Sond.■☆

395090　Tribulus inermis Kralik = Tribulus bimucronatus Viv. var. inermis（Kralik）Hosni■☆

395091　Tribulus intermedius Kralik = Tribulus parvispinus C. Presl var. intermedius（Kralik）Hosni■☆

395092　Tribulus intermedius Kralik = Tribulus terrestris L.■

395093　Tribulus lanuginosus L. = Tribulus terrestris L. var. bicornutus（Fisch. et C. A. Mey.）Hadidi■

395094　Tribulus lanuginosus L. var. microcarpus Chiov. = Tribulus terrestris L. var. bicornutus（Fisch. et C. A. Mey.）Hadidi■

395095　Tribulus lanuginosus L. var. orientalis（A. Kern.）M. P. Nayar et G. S. Giri = Tribulus terrestris L. var. orientalis（A. Kern.）Beck■☆

395096　Tribulus longipetalus Viv. = Tribulus pentandrus Forssk.■☆

395097　Tribulus longipetalus Viv. subsp. alatus（Delile）Ozenda et Quézel = Tribulus pentandrus Forssk.■☆

395098　Tribulus longipetalus Viv. subsp. macropterus（Boiss.）Maire = Tribulus macropterus Boiss.■☆

395099　Tribulus longipetalus Viv. subsp. macropterus（Boiss.）Maire ex Ozenda et Quézel = Tribulus macropterus Boiss.■☆

395100　Tribulus longipetalus Viv. var. anisopterus（Maire）Ozenda et Quézel = Tribulus pentandrus Forssk.■☆

395101　Tribulus longipetalus Viv. var. brevistylus（Maire）Ozenda et

Quézel = Tribulus pentandrus Forssk. ■☆

395102　Tribulus longipetalus Viv. var. macropterus（Boiss.）Zohary = Tribulus macropterus Boiss. ■☆

395103　Tribulus longipetalus Viv. var. mollis（Ehrenb. ex Schweinf.）Zohary = Tribulus mollis Ehrenb. ex Schweinf. ■☆

395104　Tribulus longipetalus Viv. var. monodii（Maire）Ozenda et Quézel = Tribulus macropterus Boiss. ■☆

395105　Tribulus longipetalus Viv. var. tibestiensis A. Chev. = Tribulus pentandrus Forssk. ■☆

395106　Tribulus longipetalus Viv. var. vespertilio（Maire）Ozenda et Quézel = Tribulus pentandrus Forssk. ■☆

395107　Tribulus macranthus Hassk. = Tribulus zeyheri Sond. subsp. macranthus（Hassk.）Hadidi ■☆

395108　Tribulus macropterus Boiss. ;大翅蒺藜■☆

395109　Tribulus macropterus Boiss. = Tribulus longipetalus Viv. subsp. macropterus（Boiss.）Maire ex Ozenda et Quézel ■☆

395110　Tribulus macropterus Boiss. var. anisopterus Maire = Tribulus macropterus Boiss. ■☆

395111　Tribulus macropterus Boiss. var. brevistylus Maire = Tribulus macropterus Boiss. ■☆

395112　Tribulus macropterus Boiss. var. collenetteae Hosni = Tribulus macropterus Boiss. ■☆

395113　Tribulus macropterus Boiss. var. ochroleucus Maire = Tribulus mollis Ehrenb. ex Schweinf. ■☆

395114　Tribulus macropterus Boiss. var. ochroleucus Maire = Tribulus ochroleucus（Maire）Ozenda et Quézel ■☆

395115　Tribulus macropterus Boiss. var. serolei Maire = Tribulus mollis Ehrenb. ex Schweinf. ■☆

395116　Tribulus megistopterus Kralik subsp. pterocarpus（Ehrenb. ex C. Müll.）Hosni = Tribulus pterophorus C. Presl ■☆

395117　Tribulus mollis Ehrenb. = Tribulus longipetalus Viv. ■☆

395118　Tribulus mollis Ehrenb. ex Schweinf. ;柔软蒺藜■☆

395119　Tribulus murex Schltr. ex Dinter = Tribulus terrestris L. ■

395120　Tribulus ochroleucus（Maire）Ozenda et Quézel;淡黄白蒺藜■☆

395121　Tribulus ochroleucus（Maire）Ozenda et Quézel = Tribulus mollis Ehrenb. ex Schweinf. ■☆

395122　Tribulus ochroleucus（Maire）Ozenda et Quézel var. perplexans = Tribulus mollis Ehrenb. ex Schweinf. ■☆

395123　Tribulus ochroleucus（Maire）Ozenda et Quézel var. serolei = Tribulus mollis Ehrenb. ex Schweinf. ■☆

395124　Tribulus orientalis A. Kern. = Tribulus terrestris L. var. orientalis（A. Kern.）Beck ■☆

395125　Tribulus orientalis A. Kern. = Tribulus terrestris L. ■

395126　Tribulus parvispinus C. Presl;小刺蒺藜☆

395127　Tribulus parvispinus C. Presl = Tribulus pentandrus Forssk. ■☆

395128　Tribulus parvispinus C. Presl var. intermedius（Kralik）Hosni;间型小刺蒺藜■☆

395129　Tribulus pechuelii Kuntze = Tribulus zeyheri Sond. ■☆

395130　Tribulus pentandrus Forssk. ;五蕊蒺藜■☆

395131　Tribulus pentandrus Forssk. var. macropterus（Boiss.）P. Singh et V. Singh = Tribulus macropterus Boiss. ■☆

395132　Tribulus pentandrus Forssk. var. micropteris（Kralik）Hosni;小翅五蕊蒺藜■☆

395133　Tribulus persicus Kralik = Tribulus longipetalus Viv. subsp. macropterus（Boiss.）Maire ex Ozenda et Quézel ■☆

395134　Tribulus persicus Kralik = Tribulus macropterus Boiss. ■☆

395135　Tribulus pterocarpus Ehrenb. ex C. Müll. = Tribulus

megistopterus Kralik subsp. pterocarpus（Ehrenb. ex C. Müll.）Hosni ■☆

395136　Tribulus pterophorus C. Presl;具翅蒺藜■☆

395137　Tribulus pterophorus C. Presl = Tribulus longipetalus Viv. subsp. pterophorus（C. Presl）Hadidi ■☆

395138　Tribulus pubescens G. Don = Kallstroemia pubescens（G. Don）Dandy ■☆

395139　Tribulus rajasthanensis Bhandari et V. S. Sharma;拉贾斯坦蒺藜■☆

395140　Tribulus revoilii Franch. = Kelleronia revoilii（Franch.）Chiov. ●☆

395141　Tribulus robustus Boiss. et Noë = Tribulus terrestris L. ■

395142　Tribulus saharae A. Chev. = Tribulus terrestris L. ■

395143　Tribulus securidocarpus Engl. = Tribulus pterophorus C. Presl ■☆

395144　Tribulus spurius Kralik;可疑蒺藜■☆

395145　Tribulus taiwanensis T. C. Huang et T. H. Hsieh;台湾蒺藜■

395146　Tribulus terrestris L. ;蒺藜（八角刺,白蒺藜,蒡通,犲羽,犲羽,炒蒺藜,茨,蕢,刺蒺藜,地菱,地菱儿,杜蒺藜,古冬非居塞,旱草,黄果刺,吉藜,吉利草,即藜,即藜,即藤,蒺骨子,蒺藜,蒺藜狗,蒺藜狗子,蒺藜菁葜,蒺藜角,蒺藜拉子,蒺藜子,腊居塞,旁通,七厘,秦尖,屈人,三角刺,三角蒺藜,三脚虎,升推,土蒺藜,休羽,盐蒺藜,野菱角,硬蒺藜,止行）;Bur Nut, Burnut, Bur-nut, Calthrop, Caltrap, Caltrop, Caltrops, Cat's Head, Devil Thorn, Devil's Thorn, Malta Cross, Maltese Cross, Puncture Vine, Puncturevine, Puncture-vine, Puncturevine Caltrop, Small Caltrops ■

395147　Tribulus terrestris L. subvar. medius Engl. = Tribulus cistoides L. var. medius（Engl.）Cufod. ■☆

395148　Tribulus terrestris L. var. bicornutus（Fisch. et C. A. Mey.）Hadidi = Tribulus terrestris L. ■

395149　Tribulus terrestris L. var. brachyceras Batt. et Trab. = Tribulus terrestris L. ■

395150　Tribulus terrestris L. var. desertorum Sond. = Tribulus parvispinus C. Presl ■☆

395151　Tribulus terrestris L. var. hispidissimus Sond. = Tribulus terrestris L. ■

395152　Tribulus terrestris L. var. inermis（Kralik）Chiov. = Tribulus bimucronatus Viv. var. inermis（Kralik）Hosni ■☆

395153　Tribulus terrestris L. var. inermis Boiss. = Tribulus terrestris L. ■

395154　Tribulus terrestris L. var. nogalensis Chiov. = Tribulus terrestris L. ■

395155　Tribulus terrestris L. var. orientalis（A. Kern.）Beck;东方蒺藜■☆

395156　Tribulus terrestris L. var. robustus（Boiss. et Noë）Boiss. = Tribulus terrestris L. ■

395157　Tribulus terrestris L. var. tomentosa Batt. et Trab. = Tribulus terrestris L. ■

395158　Tribulus zeyheri Sond. ;蔡氏蒺藜■☆

395159　Tribulus zeyheri Sond. subsp. macranthus（Hassk.）Hadidi;大花蔡氏蒺藜■☆

395160　Tricalistra Ridl.（1909）;隐柱铃兰属■☆

395161　Tricalistra ochracea Ridl. ;隐柱铃兰■☆

395162　Tricalysia A. Rich. = Diplospora DC. ●

395163　Tricalysia A. Rich. = Tricalysia A. Rich. ex DC. ●

395164　Tricalysia A. Rich. ex DC.（1830）;三萼木属(狗骨柴属,三萼草属,原狗骨柴属);Dogbonbavin, Tricalysia ●

395165　Tricalysia A. Rich. ex DC. = Diplospora DC. ●

395166　Tricalysia acidophylla Robbr. ;尖叶三萼木●☆

395167　Tricalysia acocantheroides K. Schum. ;尖药木三萼木●☆

395168　Tricalysia aequatoria Robbr. ;平三萼木●☆

395169　Tricalysia africana（Sim）Robbr. ;非洲三萼木●☆

395170　Tricalysia allenii（Stapf）Brenan;阿伦三萼木●☆

395171　Tricalysia allenii（Stapf）Brenan var. australis（Schweick.）Brenan = Tricalysia junodii（Schinz）Brenan ●☆

395172　Tricalysia allenii（Stapf）Brenan var. kirkii（Hook. f.）Brenan = Tricalysia allenii（Stapf）Brenan ●☆

395173　Tricalysia ambrensis Ranariv. et De Block;昂布尔三萼木●☆

395174　Tricalysia amplexicaulis Robbr. ;抱茎三萼木●☆

395175　Tricalysia analamazaotrensis Homolle ex Randrian. et De Block;阿纳拉马三萼木●☆

395176　Tricalysia andongensis Hiern = Sericanthe andongensis（Hiern）Robbr. ●☆

395177　Tricalysia angolensis A. Rich. ex DC. ;安哥拉三萼木●☆

395178　Tricalysia anomala E. A. Bruce;异常三萼木●☆

395179　Tricalysia anomala E. A. Bruce var. guineensis Robbr. ;几内亚异常三萼木●☆

395180　Tricalysia anomala E. A. Bruce var. montana Robbr. ;山地异常三萼木●☆

395181　Tricalysia anomalura N. Hallé = Tricalysia lasiodelphys（K. Schum. et K. Krause）A. Chev. subsp. anomalura（N. Hallé）Robbr. ●

395182　Tricalysia attenuata（Voigt）Hutch. ;三萼木（乌口树）●

395183　Tricalysia aurantiodora De Wild. = Tricalysia oligoneura K. Schum. ●☆

395184　Tricalysia auriculata Keay = Sericanthe auriculata（Keay）Robbr. ●☆

395185　Tricalysia bagshawei S. Moore;巴格肖三萼木●☆

395186　Tricalysia barumbuensis De Wild. = Tricalysia biafrana Hiern ●☆

395187　Tricalysia batesii A. Chev. = Tricalysia elliotii（K. Schum.）Hutch. et Dalziel var. centrafricana Robbr. ●☆

395188　Tricalysia benguellensis Welw. ex Hiern = Tricalysia griseiflora K. Schum. var. benguellensis（Welw. ex Hiern）Robbr. ●☆

395189　Tricalysia bequaertii De Wild. ;贝卡尔三萼木●☆

395190　Tricalysia biafrana Hiern;热非三萼木●☆

395191　Tricalysia bifida De Wild. ;双裂三萼木●☆

395192　Tricalysia boiviniana（Baill.）Ranariv. et De Block;博伊文三萼木●☆

395193　Tricalysia bracteata Hiern;显苞三萼木●☆

395194　Tricalysia breteleri Robbr. ;布勒泰尔三萼木●☆

395195　Tricalysia bussei K. Schum. = Tricalysia microphylla Hiern ●☆

395196　Tricalysia buxifolia Hiern;黄杨叶三萼木●☆

395197　Tricalysia cacondensis Hiern;卡孔达三萼木●☆

395198　Tricalysia capensis（Meisn. ex Hochst.）Sim;好望角三萼木●☆

395199　Tricalysia capensis（Meisn. ex Hochst.）Sim var. galpinii（Schinz）Robbr. ;盖尔三萼木●☆

395200　Tricalysia capensis（Meisn. ex Hochst.）Sim var. transvaalensis Robbr. ;德兰士瓦三萼木●☆

395201　Tricalysia chevalieri Aubrév. = Tricalysia faranahensis Aubrév. et Pellegr. ●☆

395202　Tricalysia chevalieri K. Krause = Sericanthe chevalieri（K. Krause）Robbr. ●☆

395203　Tricalysia coffeoides（A. Chev.）Hutch. et Dalziel = Sericanthe chevalieri（K. Krause）Robbr. var. coffeoides（A. Chev.）Robbr. ●☆

395204　Tricalysia coffeoides R. D. Good = Tricalysia elliotii（K. Schum.）Hutch. et Dalziel ●☆

395205　Tricalysia concolor N. Hallé;同色三萼木●☆

395206　Tricalysia congesta（Oliv.）Hiern;密集三萼木●☆

395207　Tricalysia congesta（Oliv.）Hiern subsp. chasei Bridson;蔡斯三萼木●☆

395208　Tricalysia corbisieri De Wild. = Tricalysia longituba De Wild. ●☆

395209　Tricalysia coriacea（Benth.）Hiern;革质三萼木●☆

395210　Tricalysia coriacea（Benth.）Hiern subsp. angustifolia（J. G. Garcia）Robbr. ;窄叶三萼木●☆

395211　Tricalysia coriacea（Benth.）Hiern subsp. nyassae（Hiern）Bridson;尼亚萨三萼木●☆

395212　Tricalysia coriacea K. Schum. = Tricalysia sonderiana Hiern ●☆

395213　Tricalysia coriaceoides De Wild. = Tricalysia coriacea（Benth.）Hiern ●☆

395214　Tricalysia cryptocalyx Baker;隐萼三萼木●☆

395215　Tricalysia cuneifolia Baker = Tricalysia ovalifolia Hiern ●☆

395216　Tricalysia davyi S. Moore = Tricalysia coriacea（Benth.）Hiern subsp. nyassae（Hiern）Bridson ●☆

395217　Tricalysia deightonii Brenan;戴顿三萼木●☆

395218　Tricalysia delagoensis Schinz;迪拉果三萼木●☆

395219　Tricalysia depauperata Hutch. = Tarenna depauperata Hutch. ●

395220　Tricalysia dewevrei De Wild. et T. Durand = Tricalysia glabra K. Schum. ●☆

395221　Tricalysia discolor Brenan;异色三萼木●☆

395222　Tricalysia djumaensis De Wild. = Tricalysia glabra K. Schum. ●☆

395223　Tricalysia djurensis Hiern = Tricalysia niamniamensis Schweinf. ex Hiern var. djurensis（Hiern）Robbr. ●☆

395224　Tricalysia dualensis A. Chev. = Tricalysia biafrana Hiern ●☆

395225　Tricalysia dubia（Lindl.）Ohwi = Diplospora dubia（Lindl.）Masam. ●

395226　Tricalysia dundensis Cavaco;敦达三萼木●☆

395227　Tricalysia dundusanensis De Wild. = Tricalysia oligoneura K. Schum. ●☆

395228　Tricalysia ealensis Wernham = Tricalysia pynaertii De Wild. ●☆

395229　Tricalysia elegans Robbr. ;雅致三萼木●☆

395230　Tricalysia elliotii（K. Schum.）Hutch. et Dalziel;埃利奥特三萼木●☆

395231　Tricalysia elliotii（K. Schum.）Hutch. et Dalziel var. centrafricana Robbr. ;中非三萼木●☆

395232　Tricalysia engleri（K. Krause）A. Chev. = Sericanthe andongensis（Hiern）Robbr. subsp. engleri（K. Krause）Bridson ●☆

395233　Tricalysia faranahensis Aubrév. et Pellegr. ;法拉纳三萼木●☆

395234　Tricalysia filiformi-stipulata（De Wild.）Brenan;线托叶三萼木●☆

395235　Tricalysia fililoba K. Krause = Psydrax gilletii（De Wild.）Bridson ●☆

395236　Tricalysia fruticosa（Hemsl.）K. Schum. = Diplospora fruticosa Hemsl. ●

395237　Tricalysia gabonica Hiern = Tricalysia pallens Hiern ●☆

395238　Tricalysia galpinii Schinz = Tricalysia capensis（Meisn. ex Hochst.）Sim var. galpinii（Schinz）Robbr. ●☆

395239　Tricalysia gilletii De Wild. = Tricalysia hensii De Wild. ●☆

395240　Tricalysia glabra K. Schum. ;光三萼木●☆

395241　Tricalysia glaucifolia A. Chev. ;灰蓝叶三萼木●☆

395242　Tricalysia gossweileri S. Moore;戈斯三萼木●☆

395243　Tricalysia grahamii Dale = Tricalysia ovalifolia Hiern var. taylorii（S. Moore）Brenan ☆

395244　Tricalysia griseiflora K. Schum. ;灰花三萼木●☆

395245　Tricalysia griseiflora K. Schum. var. barotseana Robbr. ;巴罗策三萼木●☆

395246　Tricalysia griseiflora K. Schum. var. benguellensis（Welw. ex Hiern）Robbr. ;本格拉三萼木●☆

395247　Tricalysia griseiflora K. Schum. var. longistipulata De Wild. et T. Durand　=Tricalysia pallens Hiern ●☆

395248　Tricalysia hensii De Wild.；亨斯三萼木●☆

395249　Tricalysia hookeri Hutch. et Dalziel　=Tricalysia biafrana Hiern ●☆

395250　Tricalysia humbertii Randriamb. et De Block；亨伯特三萼木●☆

395251　Tricalysia jasminiflora（Klotzsch）Benth. et Hook. f. ex Hiern；茉莉花三萼木●☆

395252　Tricalysia jasminiflora（Klotzsch）Benth. et Hook. f. ex Hiern var. hypotephros Brenan；灰背三萼木●☆

395253　Tricalysia jasminiflora（Klotzsch）Benth. et Hook. f. ex Hiern var. kirkii（Hiern）Robbr.　=Tricalysia jasminiflora（Klotzsch）Benth. et Hook. f. ex Hiern ●☆

395254　Tricalysia junodii（Schinz）Brenan；朱诺德三萼木●☆

395255　Tricalysia junodii（Schinz）Brenan var. kirkii（Hook. f.）Robbr.　=Tricalysia allenii（Stapf）Brenan ●☆

395256　Tricalysia katangensis De Wild.　=Tricalysia coriacea（Benth.）Hiern subsp. nyassae（Hiern）Bridson ●☆

395257　Tricalysia kirkii Hiern　=Tricalysia jasminiflora（Klotzsch）Benth. et Hook. f. ex Hiern var. kirkii（Hiern）Robbr. ●☆

395258　Tricalysia kirkii Hiern var. simonis J. G. Garcia　=Tricalysia jasminiflora（Klotzsch）Benth. et Hook. f. ex Hiern ●☆

395259　Tricalysia kivuensis Robbr.；基伍三萼木●☆

395260　Tricalysia lanceolata（Sond.）Burtt Davy；披针形三萼木●☆

395261　Tricalysia landanensis Good；刚果三萼木●☆

395262　Tricalysia lasiodelphys（K. Schum. et K. Krause）A. Chev.；毛三萼木●☆

395263　Tricalysia lasiodelphys（K. Schum. et K. Krause）A. Chev. subsp. anomalura（N. Hallé）Robbr.；异常乌口树●

395264　Tricalysia lastii K. Schum.　=Tricalysia coriacea（Benth.）Hiern subsp. nyassae（Hiern）Bridson ●☆

395265　Tricalysia laurentii De Wild.　=Tricalysia pynaertii De Wild. ●☆

395266　Tricalysia lecomteana Pierre ex Pellegr.　=Tricalysia elliotii（K. Schum.）Hutch. et Dalziel ●☆

395267　Tricalysia ledermannii K. Krause；莱德三萼木●☆

395268　Tricalysia legatii Hutch.　=Sericanthe andongensis（Hiern）Robbr. ●☆

395269　Tricalysia lejolyana Sonké et Cheek；勒若利三萼木●☆

395270　Tricalysia leucocarpa（Baill.）Randrian. et De Block；白果三萼木●☆

395271　Tricalysia ligustrina S. Moore　=Tricalysia delagoensis Schinz ●☆

395272　Tricalysia lineariloba Hutch.；线裂片三萼木●☆

395273　Tricalysia longistipulata（De Wild. et T. Durand）De Wild. et T. Durand　=Tricalysia pallens Hiern ●☆

395274　Tricalysia longituba De Wild.；长管三萼木●☆

395275　Tricalysia longituba De Wild. subsp. richardsiae Bridson；理查兹三萼木●☆

395276　Tricalysia longituba De Wild. var. velutina Robbr.；绒毛长管三萼木●☆

395277　Tricalysia lutea Hand. -Mazz.；黄花狗骨柴；Yellow-flowered Tricalysia ●

395278　Tricalysia lutea Hand. -Mazz.　=Diplospora dubia（Lindl.）Masam. ●

395279　Tricalysia macrochlamys（K. Schum.）A. Chev.　=Calycosiphonia macrochlamys（K. Schum.）Robbr. ●☆

395280　Tricalysia macrophylla K. Schum.；大叶三萼木●☆

395281　Tricalysia madagascariensis（Drake ex Dubard）A. Chev.；马岛三萼木●☆

395282　Tricalysia majungensis Ranariv. et De Block；马任加三萼木●☆

395283　Tricalysia maputensis Bridson et A. E. van Wyk；莫桑比克三萼木●☆

395284　Tricalysia mechowiana K. Schum.　=Tricalysia glabra K. Schum. ●☆

395285　Tricalysia micrantha Hiern；小花三萼木●☆

395286　Tricalysia microphylla Hiern；小叶三萼木●☆

395287　Tricalysia milanjiensis S. Moore　=Tricalysia acocantheroides K. Schum. ●☆

395288　Tricalysia mildbraedii Keay　=Tricalysia discolor Brenan ●☆

395289　Tricalysia mollissima（Hutch.）Hu　=Diplospora mollissima Hutch. ●

395290　Tricalysia mortehanii De Wild.　=Tricalysia oligoneura K. Schum. ●☆

395291　Tricalysia mucronulata K. Schum.　=Tricalysia acocantheroides K. Schum. ●☆

395292　Tricalysia myrtifolia S. Moore　=Tricalysia pallens Hiern ●☆

395293　Tricalysia ngalaensis Robbr.；恩加拉三萼木●☆

395294　Tricalysia niamniamensis Schweinf. ex Hiern；尼亚三萼木●☆

395295　Tricalysia niamniamensis Schweinf. ex Hiern subsp. nodosa（Robbr.）Bridson；多节三萼木●☆

395296　Tricalysia niamniamensis Schweinf. ex Hiern var. djurensis（Hiern）Robbr.；于拉三萼木●☆

395297　Tricalysia niamniamensis Schweinf. ex Hiern var. nodosa Robbr.　=Tricalysia niamniamensis Schweinf. ex Hiern subsp. nodosa（Robbr.）Bridson ●☆

395298　Tricalysia nyassae Hiern　=Tricalysia coriacea（Benth.）Hiern subsp. nyassae（Hiern）Bridson ●☆

395299　Tricalysia nyassae Hiern var. angustifolia J. G. Garcia　=Tricalysia coriacea（Benth.）Hiern subsp. angustifolia（J. G. Garcia）Robbr. ●☆

395300　Tricalysia obanensis Keay；奥班三萼木●☆

395301　Tricalysia obanensis Keay subsp. kwangoensis Robbr.；宽果河三萼木●☆

395302　Tricalysia odoratissima K. Schum.　=Sericanthe odoratissima（K. Schum.）Robbr. ●☆

395303　Tricalysia okelensis Hiern；奥凯尔三萼木●☆

395304　Tricalysia okelensis Hiern var. oblanceolata（Hutch. et Dalziel）Keay　=Tricalysia okelensis Hiern ●☆

395305　Tricalysia okelensis Hiern var. pubescens Aubrév. et Pellegr. ex Keay；短柔毛三萼木●☆

395306　Tricalysia oligoneura K. Schum.；寡脉三萼木●☆

395307　Tricalysia orientalis Randriamb. et De Block；东方三萼木●☆

395308　Tricalysia ovalifolia Hiern；卵叶三萼木●☆

395309　Tricalysia ovalifolia Hiern var. acutifolia Brenan　=Tricalysia ovalifolia Hiern ●☆

395310　Tricalysia ovalifolia Hiern var. glabrata（Oliv.）Brenan；光卵叶三萼木●☆

395311　Tricalysia ovalifolia Hiern var. taylorii（S. Moore）Brenan；泰勒三萼木●☆

395312　Tricalysia pachystigma K. Schum.　=Sericanthe andongensis（Hiern）Robbr. ●☆

395313　Tricalysia pachystigma K. Schum. var. praecox　=Sericanthe suffruticosa（Hutch.）Robbr. ●☆

395314　Tricalysia pallens Hiern；苍白三萼木●☆

395315　Tricalysia pallens Hiern var. dundensis（Cavaco）N. Hallé　=Tricalysia dundensis Cavaco ●☆

395316　Tricalysia pallens Hiern var. gabonica（Hiern）N. Hallé　=

Tricalysia pallens Hiern ●☆

395317 Tricalysia pangaensis De Wild. = Tricalysia oligoneura K. Schum. ●☆

395318 Tricalysia paroissei Aubrév. et Pellegr. = Sericanthe triloculanis (Scott-Elliot) Robbr. subsp. paroissei (Aubrév. et Pellegr.) Robbr. ●☆

395319 Tricalysia parva Keay；小三萼木●☆

395320 Tricalysia patentipilis K. Krause；展毛三萼木●☆

395321 Tricalysia pedicellata Robbr. ；梗花三萼木●☆

395322 Tricalysia pedunculosa (N. Hallé) Robbr. ；小梗三萼木●☆

395323 Tricalysia perrieri Homolle ex Randriamb. et De Block；佩里耶三萼木●☆

395324 Tricalysia pleiomera Hutch. = Tricalysia bifida De Wild. ●☆

395325 Tricalysia pluriovulata K. Schum. ex Hoyle = Tricalysia macrophylla K. Schum. ●☆

395326 Tricalysia pobeguinii Hutch. et Dalziel var. pubescens Aubrév. et Pellegr. = Tricalysia okelensis Hiern var. pubescens Aubrév. et Pellegr. ex Keay ●☆

395327 Tricalysia pseudoreticulata Aubrév. et Pellegr. = Tricalysia reticulata (Benth.) Hiern ●☆

395328 Tricalysia pynaertii De Wild. ；皮那三萼木●☆

395329 Tricalysia ramosissima De Wild. = Tricalysia pallens Hiern ●☆

395330 Tricalysia reflexa Hutch. ；反折三萼木●☆

395331 Tricalysia reflexa Hutch. var. ivorensis Robbr. ；伊沃里三萼木●☆

395332 Tricalysia repens Robbr. ；匍匐三萼木●☆

395333 Tricalysia reticulata (Benth.) Hiern；网脉三萼木●☆

395334 Tricalysia revoluta Hutch. ；外卷三萼木●☆

395335 Tricalysia roseoides De Wild. et T. Durand = Sericanthe roseoides (De Wild. et T. Durand) Robbr. ●☆

395336 Tricalysia ruandensis Bremek. = Tricalysia congesta (Oliv.) Hiern ●☆

395337 Tricalysia sapinii De Wild. = Tricalysia pallens Hiern ●☆

395338 Tricalysia schliebenii Robbr. ；施利本三萼木●☆

395339 Tricalysia semidecidua Bridson；半脱落三萼木●☆

395340 Tricalysia seretii De Wild. = Tricalysia niamniamensis Schweinf. ex Hiern ●☆

395341 Tricalysia somaliensis Robbr. ；索马里三萼木●☆

395342 Tricalysia sonderiana Hiern = Tricalysia ovalifolia Hiern ●☆

395343 Tricalysia soyauxii K. Schum. ；索亚三萼木●☆

395344 Tricalysia soyauxii K. Schum. var. pedunculosa N. Hallé = Tricalysia pedunculosa (N. Hallé) Robbr. ●☆

395345 Tricalysia soyauxii K. Schum. var. yangambiensis N. Hallé = Tricalysia yangambiensis (N. Hallé) Robbr. ●☆

395346 Tricalysia spathicalyx (K. Schum.) A. Chev. = Calycosiphonia spathicalyx (K. Schum.) Robbr. ●☆

395347 Tricalysia subsessilis K. Schum. ；近无柄三萼木●☆

395348 Tricalysia sudanica A. Chev. = Sericanthe chevalieri (K. Krause) Robbr. var. coffeoides (A. Chev.) Robbr. ●☆

395349 Tricalysia suffruticosa Hutch. = Sericanthe suffruticosa (Hutch.) Robbr. ●☆

395350 Tricalysia syrmanthera Hiern = Tricalysia bracteata Hiern ●☆

395351 Tricalysia talbotii (Wernham) Keay；塔尔博特三萼木●☆

395352 Tricalysia trachycarpa Robbr. ；糙果三萼木●☆

395353 Tricalysia trilocularis (Scott-Elliot) Hutch. et Dalziel = Sericanthe trilocularis (Scott-Elliot) Robbr. ●☆

395354 Tricalysia vadensis Robbr. ；瓦达三萼木●☆

395355 Tricalysia vanroechoudtii (Lebrun ex Van Roech.) Robbr. ；范罗三萼木●☆

395356 Tricalysia velutina Robbr. ；绒毛三萼木●☆

395357 Tricalysia verdcourtiana Robbr. ；韦尔德三萼木●☆

395358 Tricalysia vignei Aubrév. et Pellegr. = Tricalysia coriacea (Benth.) Hiern ●☆

395359 Tricalysia viridiflora (DC.) Matsum. = Diplospora dubia (Lindl.) Masam. ●

395360 Tricalysia viridiflora (DC.) Matsum. var. buisanensis (Hayata) Yamam. = Diplospora dubia (Lindl.) Masam. ●

395361 Tricalysia viridiflora (DC.) Matsum. var. tanakai (Hayata) Yamam. = Diplospora dubia (Lindl.) Masam. ●

395362 Tricalysia welwitschii K. Schum. ；韦尔三萼木●☆

395363 Tricalysia wernhamiana (Hutch. et Dalziel) Keay；沃纳姆三萼木●☆

395364 Tricalysia yangambiensis (N. Hallé) Robbr. ；扬甘比三萼木●☆

395365 Tricalysia zambesiaca Robbr. ；赞比西三萼木●☆

395366 Tricardia Torr. (1871)；三心田基麻属☆

395367 Tricardia Torr. ex S. Watson = Tricardia Torr. ☆

395368 Tricardia watsonii Torr. ；三心田基麻☆

395369 Tricarium Lour. = Phyllanthus L. ●■

395370 Tricarpelema J. K. Morton (1966)；三瓣果属；Tricarpelema, Tricarpweed ■

395371 Tricarpelema africanum Faden；非洲三瓣果■☆

395372 Tricarpelema chinense D. Y. Hong；三瓣果；China Tricarpweed, Chinese Tricarpelema ■

395373 Tricarpelema xizangense D. Y. Hong；西藏三瓣果；Xizang Tricarpelema, Xizang Tricarpweed ■

395374 Tricarpha Longpre = Sabazia Cass. ●■☆

395375 Tricaryum Spreng. = Phyllanthus L. ●■

395376 Tricaryum Spreng. = Tricarium Lour. ●■

395377 Tricatus Pritz. = Abronia Juss. ■☆

395378 Tricatus Pritz. = Tricratus L' Hér. ■☆

395379 Tricentrum DC. = Comolia DC. ●☆

395380 Tricera Schreb. = Buxus L. ●

395381 Triceraia Willd. ex Roem. et Schult. = Turpinia Vent. (保留属名)●

395382 Triceras Andrz. = Matthiola W. T. Aiton (保留属名)■●

395383 Triceras Andrz. ex Rchb. = Matthiola W. T. Aiton (保留属名)■●

395384 Triceras Post et Kuntze = Gomphogyne Griff. ■

395385 Triceras Post et Kuntze = Triceros Griff. ●

395386 Triceras Post et Kuntze = Turpinia Vent. (保留属名)●

395387 Triceras Wittst. = Staphylea L. ●

395388 Triceras Wittst. = Triceros Lour. (废弃属名)●

395389 Triceras Wittst. = Turpinia Vent. (保留属名)●

395390 Triceras fruticulosum (L.) Maire = Matthiola fruticulosa (L.) Maire ●☆

395391 Triceras fruticulosum (L.) Maire var. stenocarpum (Pau et Font Quer) Maire = Matthiola fruticulosa (L.) Maire ●☆

395392 Triceras fruticulosum (L.) Maire var. telum (Pomel) Maire = Matthiola fruticulosa (L.) Maire ●☆

395393 Triceras maroccanum (Coss.) Maire = Matthiola maroccana Coss. ■☆

395394 Triceras tricuspidatum (L.) Maire = Matthiola tricuspidata (L.) R. Br. ■☆

395395 Triceras tricuspidatum (L.) Maire var. glandulosus Faure et Maire = Matthiola tricuspidata (L.) R. Br. ■☆

395396 Tricerastes C. Presl = Datisca L. ●■

395397 Triceratella Brenan(1961)；黄剑茅属■☆

395398 Triceratella drummondii Brenan；黄剑茅■☆

395399 Triceratia A. Rich. = Sicydium Schltdl. ■☆

395400　Triceratorhynchus Summerh. (1951)；三角喙兰属■☆

395401　Triceratorhynchus viridiflorus Summerh. ；三角喙兰■☆

395402　Triceratostris（Szlach.）Szlach. et R. González = Deiregyne Schltr. ■☆

395403　Triceratostris（Szlach.）Szlach. et R. González(1996)；墨西哥颈柱兰属■☆

395404　Tricercandra A. Gray = Chloranthus Sw. ■●

395405　Tricercandra fortunei A. Gray = Chloranthus fortunei（A. Gray）Solms ■

395406　Tricercandra japonica（Siebold）Nakai = Chloranthus japonicus Siebold ■

395407　Tricercandra quadrifolia A. Gray = Chloranthus japonicus Siebold ■

395408　Tricerma Liebm. = Maytenus Molina ●

395409　Triceros Griff. = Gomphogyne Griff. ■

395410　Triceros Lour.（废弃属名）= Staphylea L. ●

395411　Triceros Lour.（废弃属名）= Turpinia Vent.（保留属名）●

395412　Triceros cochinchinensis Lour. = Turpinia cochinchinensis（Lour.）Merr. ●

395413　Trichacanthus Zoll. et Moritzi = Blepharis Juss. ●■

395414　Trichachne Nees = Digitaria Haller(保留属名)■

395415　Trichachne Nees = Panicum L. ■

395416　Trichachne Nees(1829)；酸草属■☆

395417　Trichachne insularis（L.）Nees = Digitaria insularis（L.）Mez ex Ekman ■☆

395418　Trichachne recalva Nees；酸草■☆

395419　Trichadenia Thwaites(1855)；毛腺木属●☆

395420　Trichadenia philippinensis Merr. ；菲律宾毛腺木●☆

395421　Trichadenia zeylanica Thwaites；毛腺木●☆

395422　Trichaeta P. Beauv. = Trisetaria Forssk. ■☆

395423　Trichaetolepis Rydb. = Adenophyllum Pers. ■●☆

395424　Trichandrum Neck. = Helichrysum Mill.（保留属名）●■

395425　Trichantha H. Karst. et Triana = Bonamia Thouars(保留属名)●■☆

395426　Trichantha H. Karst. et Triana = Breweria R. Br. ●☆

395427　Trichantha Hook.（1844）；毛花苣苔属；Trichantha ●☆

395428　Trichantha Hook. = Columnea L. ●■

395429　Trichantha Hook. = Trichantha Hook. f. ●☆

395430　Trichantha Hook. f. = Columnea L. ●■

395431　Trichantha Triana = Trichantha H. Karst. et Triana ●☆

395432　Trichantha minor Hook. ；毛花苣苔●☆

395433　Trichanthemis Regel et Schmalh.（1877）；毛春黄菊属■●☆

395434　Trichanthemis karataviensis Regel et Schmalh. ；毛春黄菊■●☆

395435　Trichanthera Ehrenb. = Eurynema Endl. ●☆

395436　Trichanthera Ehrenb. = Hermannia L. ●☆

395437　Trichanthera Kunth(1818)；毛药爵床属■☆

395438　Trichanthera gigantea（Humb. et Bonpl.）Nees；毛药爵床■☆

395439　Trichanthera modesta Ehrenb. = Hermannia modesta（Ehrenb.）Planch. ●☆

395440　Trichanthodium Sond. et F. Muell.（1853）；骨苞鼠麹草属■☆

395441　Trichanthodium Sond. et F. Muell. = Gnephosis Cass. ■☆

395442　Trichanthodium skeriophorum Sond. et F. Muell. ；骨苞鼠麹草■☆

395443　Trichanthus Phil. = Jaborosa Juss. ●■☆

395444　Trichapium Gilli = Clibadium F. Allam. ex L. ●■

395445　Tricharis Salisb. = Dipcadi Medik. ■☆

395446　Trichasma Walp. = Argyrolobium Eckl. et Zeyh.（保留属名）●☆

395447　Trichasma ciliatum Walp. = Argyrolobium tomentosum（Andréws）Druce ●☆

395448　Trichasterophyllum Willd. ex Link = Crocanthemum Spach ●☆

395449　Trichaulax Vollesen(1992)；中非爵床属☆

395450　Trichaurus Arn. = Tamarix L. ●

395451　Trichaurus ericoides（Rottler）Arn. = Tamarix ericoides Rottler et Willd. ●☆

395452　Trichaurus vaginata（Desv.）Walp. = Tamarix ericoides Rottler et Willd. ●☆

395453　Trichelostylis P. Beauv. ex T. Lestib. = Fimbristylis Vahl（保留属名）■

395454　Trichelostylis T. Lestib. = Fimbristylis Vahl（保留属名）■

395455　Trichelostylis contexta Nees = Abildgaardia contexta（Nees）Lye ■☆

395456　Trichelostylis geminata T. Lestib. et Nees = Fimbristylis autumnalis（L.）Roem. et Schult. ■

395457　Trichelostylis ludwigii（Steud.）Nees = Isolepis ludwigii（Steud.）Kunth ■☆

395458　Trichelostylis miliacea（L.）Nees = Fimbristylis miliacea（L.）Vahl ■

395459　Trichelostylis mucronulata（Michx.）Torr. = Fimbristylis autumnalis（L.）Roem. et Schult. ■

395460　Trichelostylis obtusifolia（Lam.）Nees = Fimbristylis obtusifolia（Lam.）Kunth ■

395461　Trichelostylis salbundia Nees = Fimbristylis salbundia（Nees）Kunth ■☆

395462　Trichera Schrad. = Knautia L. ■☆

395463　Tricherostigma Boiss. = Euphorbia L. ●■

395464　Tricherostigma Boiss. = Trichosterigma Klotasch et Garcke ●■

395465　Trichesmis pruniflora Kurz = Cratoxylum formosum Benth. et Hook. f. ex Dyer subsp. pruniflorum（Kurz）Gogelein ●

395466　Trichilia P. Browne(1756)（保留属名）；鹧鸪花属（老虎楝属，三唇属）；Bitterwood, Hittefwood, Partridgeflower, Trichilia ●

395467　Trichilia acutifoliata A. Chev. = Trichilia monadelpha（Thonn.）J. J. de Wilde ●☆

395468　Trichilia alata N. E. Br. = Ekebergia pterophylla（C. DC.）Hofmeyr ●☆

395469　Trichilia batesii C. DC. = Trichilia rubescens Oliv. ●☆

395470　Trichilia bilocularis Pax；双室鹧鸪花●☆

395471　Trichilia bipindeana C. DC. = Trichilia gilgiana Harms ●☆

395472　Trichilia buchananii C. DC. = Lepidotrichilia volkensii（Gürke）Leroy ●☆

395473　Trichilia caloneura Pierre ex Pellegr. = Trichilia welwitschii C. DC. ●☆

395474　Trichilia candollei A. Chev. = Trichilia monadelpha（Thonn.）J. J. de Wilde ●☆

395475　Trichilia capitata Klotzsch；头状鹧鸪花●☆

395476　Trichilia cedrata A. Chev. = Guarea cedrata（A. Chev.）Pellegr. ●☆

395477　Trichilia chirindensis Swynn. = Trichilia dregeana Sond. ●☆

395478　Trichilia connaroides（Wight et Arn. t）Bentv. = Heynea trijuga Roxb. ●

395479　Trichilia connaroides（Wight et Arn. t）Bentv. var. microcarpa（Pierre）Bentv. = Heynea trijuga Roxb. ●

395480　Trichilia connaroides（Wight et Arn.）Bentv. ；鹧鸪花（海木，假黄皮，老虎楝，小黄伞）；Common Trichilia, Connarus-like Trichilia, Partridgeflower ●

395481　Trichilia connaroides（Wight et Arn.）Bentv. f. glabra Bentv. ；光鹧鸪花（秃鹧鸪花，无毛鹧鸪花，鹧鸪花）；Glabrous Common Trichilia ●

395482　Trichilia connaroides（Wight et Arn.）Bentv. var. microcarpa（Pierre）Bentv. ；小果鹧鸪花（小果海木）；Littlefruit Trichilia ●

395483　Trichilia derumieri De Wild. = Trichilia rubescens Oliv. ●☆

395484　Trichilia djalonis A. Chev. ;贾隆鹧鸪花●☆

395485　Trichilia dregeana Sond. ;闪光鹧鸪花●☆

395486　Trichilia dregei E. Mey. ex C. DC. = Trichilia dregeana Sond. ●☆

395487　Trichilia ekebergia E. Mey. ex Sond. = Ekebergia capensis Sparrm. ●☆

395488　Trichilia emetica Vahl;催吐鹧鸪花（白鹧鸪花）;Cape Mahogany,Mafura Tallow,Mafurra,Mahogany,White Mahogany ●☆

395489　Trichilia emetica Vahl subsp. suberosa J. J. de Wilde;木栓质鹧鸪花●☆

395490　Trichilia gilgiana Harms;吉尔格鹧鸪花●☆

395491　Trichilia gilletii De Wild. ;吉勒特鹧鸪花●☆

395492　Trichilia grandifolia Oliv. ;大叶鹧鸪花●☆

395493　Trichilia guentheri Harms = Guarea laurentii De Wild. ●☆

395494　Trichilia havanensis Jacq. ;哈文鹧鸪花●☆

395495　Trichilia heterophylla Willd. ;异叶鹧鸪花●☆

395496　Trichilia heudelotii Planch. ex Oliv. = Trichilia monadelpha (Thonn.) J. J. de Wilde ●☆

395497　Trichilia heudelotii Planch. ex Oliv. var. zenkeri (Harms) Aubrév. = Trichilia ornithothera J. J. de Wilde ●☆

395498　Trichilia hirta L. ;硬毛鹧鸪花(毛鹧鸪花);Hairy Trichilia ●☆

395499　Trichilia hylobia Harms = Trichilia gilgiana Harms ●☆

395500　Trichilia integrifilamenta C. DC. = Trichilia monadelpha (Thonn.) J. J. de Wilde ●☆

395501　Trichilia johannis Harms = Trichilia monadelpha (Thonn.) J. J. de Wilde ●☆

395502　Trichilia kisoko De Wild. = Trichilia welwitschii C. DC. ☆

395503　Trichilia lanata A. Chev. = Trichilia tessmannii Harms ●☆

395504　Trichilia lancei Vermoesen = Trichilia tessmannii Harms ●☆

395505　Trichilia laurentii De Wild. = Trichilia rubescens Oliv. ●☆

395506　Trichilia ledermannii Harms = Trichilia dregeana Sond. ●☆

395507　Trichilia le-testui Pellegr. = Trichilia tessmannii Harms ●☆

395508　Trichilia lovettii Cheek ;洛维特鹧鸪花●☆

395509　Trichilia macrocarpa Spreng. ;大果鹧鸪花●☆

395510　Trichilia macrophylla Benth. ;小叶鹧鸪花●☆

395511　Trichilia megalantha Harms ;大花鹧鸪花●☆

395512　Trichilia mildbraedii Harms = Trichilia tessmannii Harms ●☆

395513　Trichilia minutiflora Standl. ;小花鹧鸪花●☆

395514　Trichilia monadelpha (Thonn.) J. J. de Wilde;单体雄蕊鹧鸪花●☆

395515　Trichilia montchali De Wild. = Trichilia tessmannii Harms ●☆

395516　Trichilia moschata Sw. ;麝香鹧鸪花;Pameruon Bark ●☆

395517　Trichilia oddoni De Wild. = Trichilia welwitschii C. DC. ●☆

395518　Trichilia oerstediana C. DC. ;欧氏鹧鸪花;Oersted Trichilia ●☆

395519　Trichilia ornithothera J. J. de Wilde;鸟状鹧鸪花●☆

395520　Trichilia papillosa Pierre ex A. Chev. = Trichilia rubescens Oliv. ●☆

395521　Trichilia prieuriana A. Juss. ;普里鹧鸪花●☆

395522　Trichilia prieuriana A. Juss. subsp. orientalis J. J. de Wilde;非洲普里鹧鸪花●☆

395523　Trichilia prieuriana A. Juss. subsp. vermoesenii J. J. de Wilde;东方普里鹧鸪花●☆

395524　Trichilia prieuriana A. Juss. var. vermoesenii Pellegr. = Trichilia prieuriana A. Juss. subsp. vermoesenii J. J. de Wilde ●☆

395525　Trichilia pterophylla C. DC. = Ekebergia pterophylla (C. DC.) Hofmeyr ☆

395526　Trichilia pynaertii De Wild. = Trichilia welwitschii C. DC. ●☆

395527　Trichilia quadrivalvis C. DC. ;五果片鹧鸪花●☆

395528　Trichilia redacta Burtt Davy et Bolton = Trichilia dregeana Sond. ●☆

395529　Trichilia retusa Oliv. ;微凹鹧鸪花●☆

395530　Trichilia reygaertii De Wild. = Guarea laurentii De Wild. ●☆

395531　Trichilia roka (Forssk.) Chiov. = Trichilia connaroides (Wight et Arn.) Bentv. ●

395532　Trichilia roka Chiov. = Trichilia emetica Vahl ●☆

395533　Trichilia rubescens Oliv. ;变红鹧鸪花●☆

395534　Trichilia rueppelliana Fresen. = Ekebergia capensis Sparrm. ☆

395535　Trichilia schliebenii Harms = Trichilia dregeana Sond. ●☆

395536　Trichilia senegalensis C. DC. = Trichilia prieuriana A. Juss. ●☆

395537　Trichilia siderotricha Chiov. = Brucea antidysenterica J. F. Mill. ●☆

395538　Trichilia sinensis Bentv. ;中华鹧鸪花(白骨走马,绒果海木,茸果鹧鸪花,中国鹧鸪花);Chinese Trichilia,Flossfruit Partridgeflower ●

395539　Trichilia sinensis Bentv. = Heynea velutina F. C. How et T. C. Chen ●

395540　Trichilia somalensis Chiov. = Trichilia emetica Vahl ●☆

395541　Trichilia splendida A. Chev. = Trichilia dregeana Sond. ●☆

395542　Trichilia strigulosa Welw. ex C. DC. = Trichilia dregeana Sond. ●☆

395543　Trichilia stuhlmannii Harms = Trichilia dregeana Sond. ●☆

395544　Trichilia suavis (Baill.) Harms;芳香鹧鸪花●☆

395545　Trichilia subcordata Gürke ;近心形鹧鸪花●☆

395546　Trichilia tessmannii Harms;泰斯曼鹧鸪花●☆

395547　Trichilia tomentosa A. Chev. = Trichilia dregeana Sond. ●☆

395548　Trichilia tripetala Blanco = Aphanamixis tripetala (Blanco) Merr. ●

395549　Trichilia umbrifera Swynn. et Baker f. = Trichilia emetica Vahl ●☆

395550　Trichilia umbrosa Vermoesen = Trichilia dregeana Sond. ●☆

395551　Trichilia velutina A. Chev. = Trichilia ornithothera J. J. de Wilde ●☆

395552　Trichilia vestita C. DC. = Trichilia dregeana Sond. ●☆

395553　Trichilia volkensii Gürke = Lepidotrichilia volkensii (Gürke) Leroy ●☆

395554　Trichilia volkensii Gürke var. genuina Pic. Serm. = Lepidotrichilia volkensii (Gürke) Leroy ●☆

395555　Trichilia welwitschii C. DC. ;韦尔鹧鸪花●☆

395556　Trichilia welwitschii C. DC. var. grandiflora ? = Trichilia dregeana Sond. ●☆

395557　Trichilia zenkeri Harms = Trichilia welwitschii C. DC. ●☆

395558　Trichinium R. Br. = Ptilotus R. Br. ■●☆

395559　Trichinium chrysurus Meisn. ex Moq. = Kyphocarpa trichinoides (Fenzl) Lopr. ■☆

395560　Trichinium remotiflorum Hook. = Sericorema remotiflora (Hook.) Lopr. ☆

395561　Trichinium zeyheri Moq. = Sericocoma avolans Fenzl ■☆

395562　Trichlisperma Raf. = Polygala L. ●■

395563　Trichlora Baker(1877);秘鲁葱属■☆

395564　Trichlora peruviana Baker;秘鲁葱■☆

395565　Trichloris E. Fourn. = Chloris Sw. ●■

395566　Trichloris E. Fourn. ex Benth. (1881);三花禾属(三花草属)■☆

395567　Trichloris crinita (Lag.) L. Parodi;长软毛三花禾■☆

395568　Trichloris pluriflora E. Fourn. ;三花禾■☆

395569　Trichoa Pars. = Abuta Aubl. ●☆

395570　Trichoballia C. Presl = Tetraria P. Beauv. ■☆

395571　Trichobasis Turcz. = Conothamnus Lindl. ●☆

395572　Trichocalyx Balf. f. (1884)(保留属名);毛萼爵床属☆

395573　Trichocalyx Schauer = Calytrix Labill. ●☆

395574　Trichocalyx obovatus Balf. f. ;倒卵毛萼爵床☆

395575　Trichocalyx orbiculatus Balf. f. ;圆毛萼爵床☆

395576　Trichocarpus Neck. = Persica Mill. ●

395577 Trichocarpus Schreb. = Ablania Aubl. ●

395578 Trichocarpus Schreb. = Sloanea L. ●

395579 Trichocarya Miq. = Angelesia Korth. + Diemenia Korth. ●☆

395580 Trichocarya Miq. = Licania Aubl. ●☆

395581 Trichocaulon N. E. Br. (1878);亚罗汉属■☆

395582 Trichocaulon alstonii N. E. Br. = Hoodia alstonii (N. E. Br.) Plowes ■☆

395583 Trichocaulon annulatum N. E. Br. = Hoodia pilifera (L. f.) Plowes subsp. annulata (N. E. Br.) Bruyns ■☆

395584 Trichocaulon cactiforme (Hook.) N. E. Br. = Larryleachia cactiformis (Hook.) Plowes ●☆

395585 Trichocaulon cinereum Pillans = Larryleachia perlata (Dinter) Plowes ●☆

395586 Trichocaulon clavatum (Willd.) H. Huber;棍棒亚罗汉■☆

395587 Trichocaulon columnare Nel = Richtersveldia columnaris (Nel) Meve et Liede ●☆

395588 Trichocaulon delaetianum Dinter = Hoodia officinalis (N. E. Br.) Plowes subsp. delaetiana (Dinter) Bruyns ■☆

395589 Trichocaulon dinteri A. Berger = Larryleachia marlothii (N. E. Br.) Plowes ●☆

395590 Trichocaulon engleri Dinter = Larryleachia picta (N. E. Br.) Plowes ●☆

395591 Trichocaulon felinum D. T. Cole = Larryleachia cactiformis (Hook.) Plowes var. felina (D. T. Cole) Bruyns ●☆

395592 Trichocaulon flavum N. E. Br. = Hoodia flava (N. E. Br.) Plowes ■☆

395593 Trichocaulon grande N. E. Br. = Hoodia grandis (N. E. Br.) Plowes ■☆

395594 Trichocaulon halenbergense Dinter = Hoodia alstonii (N. E. Br.) Plowes ■☆

395595 Trichocaulon karasmontanum Dinter;卡拉斯山亚罗汉■☆

395596 Trichocaulon keetmanshoopense Dinter = Larryleachia marlothii (N. E. Br.) Plowes ●☆

395597 Trichocaulon kubusense Nel = Larryleachia perlata (Dinter) Plowes ●☆

395598 Trichocaulon marlothii N. E. Br. = Larryleachia marlothii (N. E. Br.) Plowes ●☆

395599 Trichocaulon meloforme Marloth = Larryleachia picta (N. E. Br.) Plowes ●☆

395600 Trichocaulon mossamedense L. C. Leach = Hoodia mossamedensis (L. C. Leach) Plowes ■☆

395601 Trichocaulon officinale N. E. Br. = Hoodia officinalis (N. E. Br.) Plowes ■☆

395602 Trichocaulon pedicellatum Schinz = Hoodia pedicellata (Schinz) Plowes ■☆

395603 Trichocaulon perlatum Dinter = Larryleachia perlata (Dinter) Plowes ●☆

395604 Trichocaulon pictum N. E. Br. = Larryleachia picta (N. E. Br.) Plowes ●☆

395605 Trichocaulon piliferum (L. f.) N. E. Br. = Hoodia pilifera (L. f.) Plowes ■☆

395606 Trichocaulon pillansii N. E. Br. = Hoodia grandis (N. E. Br.) Plowes ■☆

395607 Trichocaulon pillansii N. E. Br. var. major ? = Hoodia grandis (N. E. Br.) Plowes ■☆

395608 Trichocaulon pubiflorum Dinter = Hoodia officinalis (N. E. Br.) Plowes ■☆

395609 Trichocaulon rusticum N. E. Br. = Hoodia officinalis (N. E. Br.) Plowes ■☆

395610 Trichocaulon simile N. E. Br. = Larryleachia cactiformis (Hook.) Plowes ●☆

395611 Trichocaulon sinus-luederitzii Dinter = Larryleachia marlothii (N. E. Br.) Plowes ●☆

395612 Trichocaulon sociarum A. C. White et B. Sloane = Larryleachia sociarum (A. C. White et B. Sloane) Plowes ●☆

395613 Trichocaulon somaliense Guillaumin = Echidnopsis planiflora P. R. O. Bally ■☆

395614 Trichocaulon triebneri Nel = Hoodia triebneri (Nel) Bruyns ■☆

395615 Trichocaulon truncatum Pillans = Larryleachia perlata (Dinter) Plowes ●☆

395616 Trichocentrum Poepp. et Endl. (1836);毛距兰属(骡耳兰属);Mule-ear Oncidium ■☆

395617 Trichocentrum undulatum (Sw.) Ackerman et M. W. Chase;波状毛距兰(波状骡耳兰)■☆

395618 Trichocephalum Schur = Virga Hill ■

395619 Trichocephalus Brongn. (1826);头毛鼠李属●☆

395620 Trichocephalus Brongn. = Phylica L. ●☆

395621 Trichocephalus comosus Eckl. et Zeyh. = Phylica comosa Sond. ●☆

395622 Trichocephalus gracilis Eckl. et Zeyh. = Phylica gracilis (Eckl. et Zeyh.) D. Dietr. ●☆

395623 Trichocephalus harveyi Arn. = Phylica harveyi (Arn.) Pillans ●☆

395624 Trichocephalus litoralis Eckl. et Zeyh. = Phylica litoralis (Eckl. et Zeyh.) D. Dietr. ●☆

395625 Trichocephalus stipularis (L.) Brongn. ;头毛鼠李●☆

395626 Trichocephalus trachyphyllus Eckl. et Zeyh. = Phylica trachyphylla (Eckl. et Zeyh.) D. Dietr. ●☆

395627 Trichocephalus verticillatus Eckl. et Zeyh. = Kogelbergia verticillata (Eckl. et Zeyh.) Rourke ●☆

395628 Trichocephalus virgatus Eckl. et Zeyh. = Phylica virgata (Eckl. et Zeyh.) D. Dietr. ●☆

395629 Trichoceras Spreng. = Trichoceros Kunth ■☆

395630 Trichocereus (A. Berger) Riccob. (1909);毛花柱属(大棱柱属);Trichocereus, Yellow Cactus ●

395631 Trichocereus (A. Berger) Riccob. = Echinopsis Zucc. ●

395632 Trichocereus Britton et Rose = Trichocereus (A. Berger) Riccob. ●

395633 Trichocereus Riccob. = Echinopsis Zucc. ●

395634 Trichocereus bridgesii Britton et Rose;天主阁●☆

395635 Trichocereus bruchii (Britton et Rose) F. Ritter;湘阳球●☆

395636 Trichocereus camarguensis Cardenas;金光龙●☆

395637 Trichocereus candicans (Salm-Dyck) Britton et Rose;金城球(金城,金城丸)●☆

395638 Trichocereus cephalomacrostibas (Werderm. et Backeb.) Backeb. ;毛冠柱●☆

395639 Trichocereus chilensis (Colla) Britton et Rose;锦龙柱(锦鸡龙)●☆

395640 Trichocereus deserticola (Werderm.) Looser;沙漠柱●☆

395641 Trichocereus formosus (Pfeiff.) F. Ritter;美丽毛花柱●☆

395642 Trichocereus grandiflorus Backeb. ;巨丽球(大花毛花柱,巨丽丸)●☆

395643 Trichocereus huascha Britton et Rose;湘南球(湘南丸)●☆

395644 Trichocereus litoralis (Johow) Looser;利升龙●☆

395645 Trichocereus macrogonus (Salm-Dyck) Riccob. ;钝角毛花柱(大棱柱);Largeangular Trichocereus ●

395646 Trichocereus macrogonus Riccob. ;大棱柱●☆

395647 Trichocereus neolamprochlorus Backeb. ;光绿柱●☆

395648 Trichocereus nigripilis（Phil.）Backeb.;虎锦柱●☆

395649 Trichocereus pachanoi Britton et Rose;毛花柱(巴查大棱柱,毛仙人掌);Pachan Trichocereus ●

395650 Trichocereus pasacana Britton et Rose;黄鹰柱(黄鹰)●☆

395651 Trichocereus poco Backeb.;金闪柱(金闪)●☆

395652 Trichocereus purpureopilosus Weing;紫红毛花柱●☆

395653 Trichocereus schickendantzii Britton et Rose;金棱柱(白晃丸,金棱)●☆

395654 Trichocereus spachianus（Lem.）Riccob. = Echinopsis spachiana（Lem.）H. Friedrich et G. D. Rowley ■☆

395655 Trichocereus strigosus（Salm-Dyck）Britton et Rose;硬毛柱●☆

395656 Trichocereus tacaquirensis（Vaupel）Cardenas ex Backeb.;鹰翔阁●☆

395657 Trichocereus tarijensis（Vaupel）Werderm.;塔丽亚毛花柱●☆

395658 Trichocereus terscheckii（Parm.）Britton et Rose;北斗阁●☆

395659 Trichocereus thelegonoides（Speg.）Britton et Rose;拟黑凤●☆

395660 Trichocereus thelegonus（F. A. C. Weber）Britton et Rose;黑凤●☆

395661 Trichocereus werdermannianus Backeb.;勇烈龙●☆

395662 Trichoceros Kunth(1816);毛角兰属■☆

395663 Trichoceros antennifer Kunth;毛角兰■☆

395664 Trichochaeta Steud. = Rhynchospora Vahl(保留属名)■

395665 Trichochilus Ames = Dipodium R. Br.■☆

395666 Trichochiton Kom.（1896);肖隐籽芥属■☆

395667 Trichochiton Kom. = Cryptospora Kar. et Kir.■

395668 Trichochiton inconspicum Kom.;肖隐籽芥■☆

395669 Trichochlaena Kuntze = Tricholaena Schrad. ex Schult. et Schult. f. ■☆

395670 Trichochlaena Post et Kuntze = Tricholaena Schrad. ex Schult. et Schult. f. ■☆

395671 Trichochloa DC. = Muhlenbergia Schreb.■

395672 Trichochloa P. Beauv. = Muhlenbergia Schreb.■

395673 Trichocladus Pers.（1807);毛枝梅属■☆

395674 Trichocladus crinitus（Thunb.）Pers.;长软毛毛枝梅●☆

395675 Trichocladus dentatus Hutch. = Trichocladus goetzei Engl.●☆

395676 Trichocladus ellipticus Eckl. et Zeyh.;椭圆毛枝梅●☆

395677 Trichocladus goetzei Engl.;格策毛枝梅●☆

395678 Trichocladus grandiflorus Oliv.;大花毛枝梅●☆

395679 Trichocladus peltatus Meisn. = Trichocladus crinitus（Thunb.）Pers.●☆

395680 Trichocladus verticillatus Eckl. et Zeyh. = Bowkeria verticillata（Eckl. et Zeyh.）Schinz ●☆

395681 Trichocladus vittatus Meisn. = Trichocladus crinitus（Thunb.）Pers.●☆

395682 Trichocline Cass.（1817);毛床菊属(毛丁草属)■☆

395683 Trichocline argentea Griseb.;阿根廷毛床菊■☆

395684 Trichocline incana Cass.;灰白毛床菊■☆

395685 Trichocoronis A. Gray(1849);虫泽兰属(下田菊属);Bugheal ■☆

395686 Trichocoronis riparia（Greene）Greene = Trichocoronis wrightii（Torr. et A. Gray）A. Gray ■☆

395687 Trichocoronis rivularis A. Gray = Shinnersia rivularis（A. Gray）R. M. King et H. Rob.■☆

395688 Trichocoronis wrightii（Torr. et A. Gray）A. Gray;赖特虫泽兰;Limestone Bugheal ■☆

395689 Trichocoryne S. F. Blake(1924);毛盔菊属■☆

395690 Trichocoryne connata S. F. Blake;毛盔菊■☆

395691 Trichocrepis Vis. = Crepis L.■

395692 Trichocrepis Vis. = Pterotheca Cass.■

395693 Trichocyamos Yakovlev = Ormosia Jacks.（保留属名)●

395694 Trichocyamos Yakovlev(1972);南红豆属■☆

395695 Trichocyamos inflata（Merr. et Chun）Yakovlev = Ormosia inflata Merr. et Chun ex H. Y. Chen ●

395696 Trichocyamos merrilliana（L. Chen）Yakovlev = Ormosia merrilliana H. Y. Chen ●

395697 Trichocyamos pachycarpa（Champ. ex Benth.）Yakovlev = Ormosia pachycarpa Champ. ex Benth.●

395698 Trichocyamos sericeolucida（L. Chen）Yakovlev = Ormosia sericeolucida H. Y. Chen ●

395699 Trichocyclus N. E. Br. = Brownanthus Schwantes ●☆

395700 Trichocyclus buchubergensis Dinter = Brownanthus pubescens（N. E. Br. ex C. A. Maass）Bullock ●☆

395701 Trichocyclus ciliatus（Aiton）N. E. Br. = Brownanthus vaginatus（Lam.）Chess. et M. Pignal ●☆

395702 Trichocyclus marlothii（Pax）N. E. Br. = Brownanthus marlothii（Pax）Schwantes ●☆

395703 Trichocyclus namibensis（Marloth）N. E. Br. ex C. A. Maass = Brownanthus namibensis（Marloth）Bullock ●☆

395704 Trichocyclus pillansii L. Bolus = Brownanthus pubescens（N. E. Br. ex C. A. Maass）Bullock ●☆

395705 Trichocyclus pubescens N. E. Br. ex C. A. Maass = Brownanthus pubescens（N. E. Br. ex C. A. Maass）Bullock ●☆

395706 Trichocyclus schenckii（Schinz）Dinter et Schwantes ex Range = Brownanthus vaginatus（Lam.）Chess. et M. Pignal subsp. schenckii（Schinz）Chess. et M. Pignal ●☆

395707 Trichocyclus simplex N. E. Br. ex C. A. Maass = Brownanthus vaginatus（Lam.）Chess. et M. Pignal subsp. schenckii（Schinz）Chess. et M. Pignal ●☆

395708 Trichodesma R. Br.（1810)（保留属名);毛束草属(碧果草属);Trichodesma ●■

395709 Trichodesma africanum（L.）Lehm. subsp. gracile（Batt.）Le Houér. = Trichodesma africanum（L.）Sm.■☆

395710 Trichodesma africanum（L.）Lehm. var. ehrenbergii（Schweinf. ex Boiss.）Post = Trichodesma ehrenbergii Schweinf. ex Boiss.■☆

395711 Trichodesma africanum（L.）R. Br. = Trichodesma africanum（L.）Sm.■☆

395712 Trichodesma africanum（L.）Sm.;非洲毛束草■☆

395713 Trichodesma africanum（L.）Sm. var. abyssinicum Brand = Trichodesma africanum（L.）Sm.■☆

395714 Trichodesma africanum（L.）Sm. var. heterotrichum Bornm. et Kneuck. = Trichodesma africanum（L.）Sm.■☆

395715 Trichodesma africanum（L.）Sm. var. homotrichum Bornm. et Kneuck. = Trichodesma africanum（L.）Sm.■☆

395716 Trichodesma africanum（L.）Sm. var. subcalcaratum Maire = Trichodesma africanum（L.）Sm.■☆

395717 Trichodesma ambacense Welw.;安巴萨毛束草■☆

395718 Trichodesma ambacense Welw. subsp. hockii（De Wild.）Brummitt;霍克毛束草■☆

395719 Trichodesma amplexicaule Roth = Trichodesma indicum（L.）R. Br.■☆

395720 Trichodesma angolense Brand = Trichodesma ambacense Welw.■☆

395721 Trichodesma angustifolium Harv.;窄叶毛束草■☆

395722 Trichodesma angustifolium Harv. subsp. argenteum Retief et A. E. van Wyk;银窄叶毛束草■☆

395723 Trichodesma arenicola Gürke;沙生毛束草■☆

395724 Trichodesma arenicola Gürke = Trichodesma ambacense Welw.

subsp. hockii（De Wild.）Brummitt ■☆

395725 Trichodesma arenicola Gürke subsp. concinnum Brummitt；整洁毛束草■☆

395726 Trichodesma barbatum Vaupel = Cystostemon barbatus（Vaupel）A. G. Mill. et Riedl ■☆

395727 Trichodesma baumii Gürke；鲍姆毛束草■☆

395728 Trichodesma bentii Baker et C. H. Wright = Trichodesma ehrenbergii Schweinf. ex Boiss. ■☆

395729 Trichodesma bequaertii De Wild.；贝卡尔毛束草■☆

395730 Trichodesma calathiforme Hochst. = Trichodesma trichodesmoides（Bunge）Gürke ■☆

395731 Trichodesma calathiforme Hochst. var. schimperi（Baker）Brand = Trichodesma trichodesmoides（Bunge）Gürke ■☆

395732 Trichodesma calcaratum Coss. ex Batt.；距毛束草■☆

395733 Trichodesma calcaratum Coss. ex Batt. var. homotrichum Guinet et Sauvage = Trichodesma calcaratum Coss. ex Batt. ■☆

395734 Trichodesma calcaratum Coss. ex Batt. var. monotrichum Guinet et Sauvage = Trichodesma calcaratum Coss. ex Batt. ■☆

395735 Trichodesma calcareum Craib = Trichodesma calycosum Collett et Hemsl. ●

395736 Trichodesma calycosum Collett et Hemsl.；毛束草（碧果草，假酸浆，毛束树，台湾毛束草）；Largecalyx Trichodesma，Taiwan Trichodesma ●

395737 Trichodesma calycosum Collett et Hemsl. var. formosanum（Matsum.）I. M. Johnst. = Trichodesma calycosum Collett et Hemsl. ●

395738 Trichodesma calycosum Collett et Hemsl. var. formosanum（Matsum.）I. M. Johnst.；台湾毛束草●

395739 Trichodesma cardiosepalum Oliv. = Trichodesma hildebrandtii Gürke ●☆

395740 Trichodesma dekindtianum Gürke = Trichodesma ambacense Welw. ■☆

395741 Trichodesma droogmansianum De Wild. et T. Durand = Trichodesma physaloides（Fenzl）A. DC. ■☆

395742 Trichodesma droogmansianum De Wild. et T. Durand var. glabrescens（Gürke）Brand = Trichodesma physaloides（Fenzl）A. DC. ■☆

395743 Trichodesma ehrenbergii Schweinf. ex Boiss.；爱伦堡毛束草■☆

395744 Trichodesma formosanum Matsum. = Trichodesma calycosum Collett et Hemsl. ●

395745 Trichodesma formsoanum Matsum. = Trichodesma calycosum Collett et Hemsl. var. formosanum（Matsum.）I. M. Johnst. ●

395746 Trichodesma frutescens K. Schum. = Trichodesma physaloides（Fenzl）A. DC. ■☆

395747 Trichodesma fruticosum Maire = Trichodesma africanum（L.）Sm. ■☆

395748 Trichodesma giganteum Quézel = Trichodesma africanum（L.）Sm. ■☆

395749 Trichodesma glabrescens Gürke = Trichodesma physaloides（Fenzl）A. DC. ■☆

395750 Trichodesma gracile Batt. et Trab. = Trichodesma africanum（L.）Sm. ■☆

395751 Trichodesma grandifolium Baker = Trichodesma trichodesmoides（Bunge）Gürke ■☆

395752 Trichodesma heliocharis S. Moore = Cystostemon heliocharis（S. Moore）A. G. Mill. et Riedl ■☆

395753 Trichodesma hemsleyanum H. Lév. = Trichodesma calycosum Collett et Hemsl. ●

395754 Trichodesma hildebrandtii Gürke；希尔德毛束草●☆

395755 Trichodesma hirsutum Edgew. = Trichodesma indicum（L.）R. Br. ■☆

395756 Trichodesma hispidum Baker et C. H. Wright = Cystostemon hispidus（Baker et C. H. Wright）A. G. Mill. et Riedl ■☆

395757 Trichodesma hockii De Wild. = Trichodesma ambacense Welw. subsp. hockii（De Wild.）Brummitt ■☆

395758 Trichodesma inaequale Edgew.；不等毛束草■☆

395759 Trichodesma incanum（Bunge）A. DC.；灰毛毛束草■☆

395760 Trichodesma indicum（L.）Lehm. = Trichodesma indicum（L.）R. Br. ■☆

395761 Trichodesma indicum（L.）Lehm. var. amplexicaule（Roth）T. Cooke = Trichodesma indicum（L.）Lehm. ■☆

395762 Trichodesma indicum（L.）R. Br.；印度毛束草■☆

395763 Trichodesma indicum L. var. amplexicaule（Roth）T. Cooke = Trichodesma indicum（L.）R. Br. ■☆

395764 Trichodesma khasianum C. B. Clarke；喀西毛束草（碧果草，假酸浆）；Khasi Trichodesma ■

395765 Trichodesma khasianum C. B. Clarke = Trichodesma calycosum Collett et Hemsl. var. formosanum（Matsum.）I. M. Johnst. ■

395766 Trichodesma lanceolatum Schinz = Trichodesma angustifolium Harv. ■☆

395767 Trichodesma ledermannii Vaupel = Trichodesma ambacense Welw. subsp. hockii（De Wild.）Brummitt ■☆

395768 Trichodesma longipedicellanum Rech. f. et Riedl = Trichodesma stocksii Boiss. ■☆

395769 Trichodesma macrantherum Gürke = Cystostemon macranthera（Gürke）A. G. Mill. et Riedl ■☆

395770 Trichodesma marsabiticum Brummitt；马萨比特毛束草■☆

395771 Trichodesma mechowii Vaupel = Cystostemon mechowii（Vaupel）A. G. Mill. et Riedl ■☆

395772 Trichodesma medusa Baker = Cystostemon medusa（Baker）A. G. Mill. et Riedl ■☆

395773 Trichodesma pauciflorum Baker = Trichodesma trichodesmoides（Bunge）Gürke ■☆

395774 Trichodesma physaloides（Fenzl）A. DC.；囊状毛束草■☆

395775 Trichodesma ringoetii De Wild. = Trichodesma physaloides（Fenzl）A. DC. ■☆

395776 Trichodesma schimperi Baker = Trichodesma trichodesmoides（Bunge）Gürke ■☆

395777 Trichodesma sedgwickianum S. P. Banerjee = Trichodesma inaequale Edgew. ■☆

395778 Trichodesma sinicum Brand = Trichodesma calycosum Collett et Hemsl. ●

395779 Trichodesma stenosepalum Baker = Cystostemon heliocharis（S. Moore）A. G. Mill. et Riedl ■☆

395780 Trichodesma stocksii Boiss.；斯托克斯毛束草■☆

395781 Trichodesma strictum Aitch. et Hemsl. = Trichodesma incanum（Bunge）A. DC. ■☆

395782 Trichodesma tinctorium Brand = Trichodesma ambacense Welw. subsp. hockii（De Wild.）Brummitt ■☆

395783 Trichodesma trichodesmoides（Bunge）Gürke；异叶毛束草■☆

395784 Trichodesma uniflorum Brand = Trichodesma inaequale Edgew. ■☆

395785 Trichodesma van-meelii Taton = Trichodesma bequaertii De Wild. ■☆

395786 Trichodesma verdickii Brand = Trichodesma ambacense Welw. subsp. hockii（De Wild.）Brummitt ■☆

395787 Trichodesma welwitschii Brand = Trichodesma ambacense Welw. ■☆

395788 Trichodesma zeylanicum（Burm. f.）R. Br.；锡兰毛束草；Camel Bush ■☆

395789 Trichodesma zeylanicum R. Br. = Trichodesma zeylanicum（Burm. f.）R. Br. ■☆

395790 Trichodia Griff. = Paropsia Noronha ex Thouars ●☆

395791 Trichodiadema Schwantes（1926）；仙花属（仙宝属）●☆

395792 Trichodiadema barbatum（L.）Schwantes；髯毛仙花●☆

395793 Trichodiadema bulbosum（Haw.）Schwantes = Trichodiadema intonsum（Haw.）Schwantes ●☆

395794 Trichodiadema burgeri L. Bolus；伯格仙花●☆

395795 Trichodiadema calvatum L. Bolus；光秃仙花●☆

395796 Trichodiadema concinnum L. Bolus = Trichodiadema intonsum（Haw.）Schwantes ●☆

395797 Trichodiadema decorum（N. E. Br.）Stearn ex H. Jacobsen；装饰仙花●☆

395798 Trichodiadema densum（Haw.）Schwantes；仙花●☆

395799 Trichodiadema densum Schwantes = Trichodiadema densum（Haw.）Schwantes ●☆

395800 Trichodiadema echinatum（Lam.）L. Bolus = Delosperma echinatum（Lam.）Schwantes ■☆

395801 Trichodiadema emarginatum L. Bolus；微缺仙花●☆

395802 Trichodiadema fergusoniae L. Bolus；费格森仙花●☆

395803 Trichodiadema fourcadei L. Bolus；富尔卡德仙花●☆

395804 Trichodiadema gracile L. Bolus；纤细仙花●☆

395805 Trichodiadema gracile L. Bolus var. piliferum ? = Trichodiadema gracile L. Bolus ●☆

395806 Trichodiadema gracile L. Bolus var. setiferum ? = Trichodiadema gracile L. Bolus ●☆

395807 Trichodiadema hallii L. Bolus；霍尔仙花●☆

395808 Trichodiadema hirsutum（Haw.）Stearn；粗毛仙花●☆

395809 Trichodiadema inornatum L. Bolus = Drosanthemum inornatum（L. Bolus）L. Bolus ●☆

395810 Trichodiadema intonsum（Haw.）Schwantes；须毛仙花●☆

395811 Trichodiadema introrsum（Haw. ex Hook. f.）Niesler；内向仙花■☆

395812 Trichodiadema littlewoodii L. Bolus；利特尔伍德仙花■☆

395813 Trichodiadema littlewoodii L. Bolus. f. alba ? = Trichodiadema littlewoodii L. Bolus ■☆

395814 Trichodiadema marlothii L. Bolus；马洛斯仙花■☆

395815 Trichodiadema mirabile（N. E. Br.）Schwantes；奇异仙花■☆

395816 Trichodiadema mirabile（N. E. Br.）Schwantes var. leptum L. Bolus = Trichodiadema mirabile（N. E. Br.）Schwantes ■☆

395817 Trichodiadema mirabile Schwantes = Trichodiadema mirabile（N. E. Br.）Schwantes ■☆

395818 Trichodiadema obliquum L. Bolus；偏斜仙花■☆

395819 Trichodiadema occidentale L. Bolus；西方仙花■☆

395820 Trichodiadema olivaceum L. Bolus；橄榄绿仙花■☆

395821 Trichodiadema orientale L. Bolus；东方仙花■☆

395822 Trichodiadema peersii L. Bolus；皮尔斯仙花■☆

395823 Trichodiadema pomeridianum L. Bolus；午后花仙花■☆

395824 Trichodiadema pygmaeum L. Bolus；矮小仙花■☆

395825 Trichodiadema rogersiae L. Bolus；罗杰斯仙花■☆

395826 Trichodiadema rupicola L. Bolus；岩生仙花■☆

395827 Trichodiadema ryderae L. Bolus；吕德仙花●☆

395828 Trichodiadema setuliferum（N. E. Br.）Schwantes；小刚毛仙花■☆

395829 Trichodiadema setuliferum（N. E. Br.）Schwantes var. niveum L. Bolus = Trichodiadema setuliferum（N. E. Br.）Schwantes ■☆

395830 Trichodiadema stayneri L. Bolus；斯泰纳仙花■☆

395831 Trichodiadema stellatum（Mill.）Schwantes；星状仙花；Karee Moer ■☆

395832 Trichodiadema stellatum（Mill.）Schwantes = Trichodiadema barbatum（L.）Schwantes ●☆

395833 Trichodiadema stelligerum（Haw.）Schwantes = Trichodiadema barbatum（L.）Schwantes ●☆

395834 Trichodiadema strumosum（Haw.）L. Bolus；多疣仙花■☆

395835 Trichodiclida Cerv.（1870）；重毛禾属■☆

395836 Trichodiclida Cerv. = Blepharidachne Hack. ■☆

395837 Trichodiclida linearis Cerv.；重毛禾■☆

395838 Trichodiclida prolifera Cerv.；多育重毛禾■☆

395839 Trichodium Michx. = Agrostis L.（保留属名）■

395840 Trichodium salmanticum Lag. = Agrostis salmantica（Lag.）Kunth ■

395841 Trichodon Benth. = Phragmites Adans. ■

395842 Trichodon Benth. = Trichoon Roth ■

395843 Trichodoum P. Beauv. ex Taub. = Dioclea Kunth ■☆

395844 Trichodrymonia Oerst.（1858）；毛林苣苔属（毛锥莫尼亚属）■☆

395845 Trichodrymonia Oerst. = Episcia Mart. ■☆

395846 Trichodrymonia Oerst. = Paradrymonia Hanst. ■●☆

395847 Trichodrymonia congesta Oerst.；毛林苣苔（毛锥莫尼亚）■☆

395848 Trichodypsis Baill. = Dypsis Noronha ex Mart. ●☆

395849 Trichodypsis hildebrandtii Baill. = Dypsis hildebrandtii（Baill.）Becc. ●☆

395850 Trichodypsis mocquerysiana Becc. = Dypsis mocquerysiana（Becc.）Becc. ●☆

395851 Trichogalium（DC.）Fourr. = Galium L. ■●

395852 Trichogalium Fourr. = Galium L. ■●

395853 Trichogamia Boehm. = Trichogamila P. Browne ●

395854 Trichogamila P. Browne = Styrax L. ●

395855 Trichoglottis Blume（1825）；毛舌兰属；Hairliporchis ■

395856 Trichoglottis breviracema（Hayata）Schltr. = Trichoglottis rosea（Lindl.）Ames ■

395857 Trichoglottis breviracema（Hayata）Schltr. = Trichoglottis rosea Ames var. breviracema（Hayata）Tang S. Liu et H. J. Su ■

395858 Trichoglottis dawsoniana（Rchb. f.）Rchb. f. = Staurochilus dawsonianus（Rchb. f.）Schltr. ■

395859 Trichoglottis ionosma（Lindl.）J. J. Sm. var. luchuensis（Rolfe）S. S. Ying = Staurochilus luchuensis（Rolfe）Fukuy. ■

395860 Trichoglottis ionosma Hayata = Staurochilus luchuensis（Rolfe）Fukuy. ■

395861 Trichoglottis ionosma J. J. Sm. var. luchuensis（Rolfe）S. S. Ying = Staurochilus luchuensis（Rolfe）Fukuy. ■

395862 Trichoglottis ionosma Lindl.；屈子花■

395863 Trichoglottis ionosma Lindl. var. luchuensis（Rolfe）S. S. Ying = Staurochilus luchuensis（Rolfe）Fukuy. ■

395864 Trichoglottis luchuensis（Rolfe）Garay et H. R. Sweet = Staurochilus luchuensis（Rolfe）Fukuy. ■

395865 Trichoglottis lutchuensis（Rolfe）Garay et H. R. Sweet；琉球毛舌兰（琉球万代兰）■☆

395866 Trichoglottis lutchuensis（Rolfe）Garay et H. R. Sweet var. grossemaculata（Koidz.）Hatus. = Trichoglottis lutchuensis（Rolfe）Garay et H. R. Sweet ■☆

395867 Trichoglottis oblongisepala（Hayata）Schltr. = Trichoglottis rosea（Lindl.）Ames ■

395868 Trichoglottis rosea（Lindl.）Ames；短穗毛舌兰（凤尾兰）；Shortspike Hairliporchis ■

395869 Trichoglottis rosea（Lindl.）Ames var. breviracema（Hayata）

Tang S. Liu et H. J. Su = Trichoglottis rosea (Lindl.) Ames ■

395870　Trichoglottis rosea Ames var. breviracema (Hayata) Tang S. Liu et H. J. Su = Trichoglottis rosea (Lindl.) Ames ■

395871　Trichoglottis triflora (Guillaumin) Garay et Seidenf.；毛舌兰（小毛舌兰）；Hairliporchis ■

395872　Trichogonia (DC.) Gardner(1846)；毛瓣柄泽兰属■●☆

395873　Trichogonia Gardner = Trichogonia (DC.) Gardner ■●☆

395874　Trichogonia alba V. M. Badillo；白毛瓣柄泽兰●☆

395875　Trichogonia multiflora Gardner；多花毛瓣柄泽兰●☆

395876　Trichogoniopsis R. M. King et H. Rob. (1972)；腺瓣柄泽兰属■●☆

395877　Trichogoniopsis adenantha (DC.) R. M. King et H. Rob.；腺花腺瓣柄泽兰■●☆

395878　Trichogoniopsis macrolepis (Baker) R. M. King et H. Rob.；大鳞腺瓣柄泽兰■●☆

395879　Trichogyne Less. (1831)；毛柱帚鼠麴属■☆

395880　Trichogyne Less. = Ifloga Cass. ■☆

395881　Trichogyne ambigua (L.) Druce；可疑毛柱帚鼠麴■☆

395882　Trichogyne candida (Hilliard) Anderb.；白毛柱帚鼠麴■☆

395883　Trichogyne cauliflora (Desf.) DC. = Ifloga spicata (Forssk.) Sch. Bip. ■☆

395884　Trichogyne decumbens (Thunb.) Less.；外倾毛柱帚鼠麴■☆

395885　Trichogyne glomerata Harv. = Ifloga glomerata (Harv.) Schltr. ■☆

395886　Trichogyne laricifolia (Lam.) Less. = Trichogyne ambigua (L.) Druce ■☆

395887　Trichogyne lerouxiae Beyers；勒鲁丹毛柱帚鼠麴■☆

395888　Trichogyne paronychioides DC.；指甲草毛柱帚鼠麴■☆

395889　Trichogyne pilulifera (Schltr.) Anderb.；小球毛柱帚鼠麴■☆

395890　Trichogyne polycnemoides (Fenzl) Anderb.；多节草毛柱帚鼠麴■☆

395891　Trichogyne radicans DC. = Trichogyne repens (L.) Anderb. ■☆

395892　Trichogyne reflexa (L. f.) Less. = Trichogyne repens (L.) Anderb. ■☆

395893　Trichogyne repens (L.) Anderb.；匍匐毛柱帚鼠麴■☆

395894　Trichogyne seriphioides Less. = Trichogyne ambigua (L.) Druce ■☆

395895　Trichogyne verticillata (L. f.) Less.；轮生毛柱帚鼠麴■☆

395896　Tricholaena Schrad. = Tricholaena Schrad. ex Schult. et Schult. f. ■☆

395897　Tricholaena Schrad. ex Schult. et Schult. f. (1824)；线衣草属■☆

395898　Tricholaena Schult. et Schult. f. = Tricholaena Schrad. ex Schult. et Schult. f. ■☆

395899　Tricholaena amethystea Franch. = Melinis amethystea (Franch.) Zizka ■☆

395900　Tricholaena arenaria Nees = Tricholaena capensis (Licht. ex Roem. et Schult.) Nees subsp. arenaria (Nees) Zizka ■☆

395901　Tricholaena arenaria Nees var. glauca (Hack.) Stapf = Tricholaena capensis (Licht. ex Roem. et Schult.) Nees subsp. arenaria (Nees) Zizka ■☆

395902　Tricholaena arenaria Nees var. semiglabra Hack. = Tricholaena monachne (Trin.) Stapf et C. E. Hubb. ■☆

395903　Tricholaena bellespicata Rendle = Melinis longiseta (A. Rich.) Zizka subsp. bellespicata (Rendle) Zizka ■☆

395904　Tricholaena bicolor (Schumach.) C. E. Hubb. = Tricholaena monachne (Trin.) Stapf et C. E. Hubb. ■☆

395905　Tricholaena brevipila Hack. = Melinis repens (Willd.) Zizka subsp. grandiflora (Hochst.) Zizka ■☆

395906　Tricholaena busseana (Mez) Peter = Melinis nerviglumis (Franch.) Zizka ■☆

395907　Tricholaena capensis (Licht. ex Roem. et Schult.) Nees；好望角毛柱帚鼠麴■☆

395908　Tricholaena capensis (Licht. ex Roem. et Schult.) Nees subsp. arenaria (Nees) Zizka；沙地好望角毛柱帚鼠麴■☆

395909　Tricholaena chinensis (Retz.) Domin = Eriachne pallescens R. Br.

395910　Tricholaena congoensis Franch. = Melinis nerviglumis (Franch.) Zizka ■☆

395911　Tricholaena delicatula Stapf et C. E. Hubb. = Tricholaena monachne (Trin.) Stapf et C. E. Hubb. ■☆

395912　Tricholaena dregeana (Nees) T. Durand et Schinz = Melinis repens (Willd.) Zizka ■

395913　Tricholaena eichingeri (Mez) Stapf et C. E. Hubb. = Tricholaena teneriffae (L. f.) Link ■☆

395914　Tricholaena filifolia Franch. = Melinis nerviglumis (Franch.) Zizka ■☆

395915　Tricholaena fragilis A. Braun = Melinis repens (Willd.) Zizka ■

395916　Tricholaena gillettii C. E. Hubb. = Tricholaena teneriffae (L. f.) Link ■☆

395917　Tricholaena glabra Stapf = Tricholaena monachne (Trin.) Stapf et C. E. Hubb. ■☆

395918　Tricholaena grandiflora Hochst. ex A. Rich. = Melinis repens (Willd.) Zizka subsp. grandiflora (Hochst.) Zizka ■☆

395919　Tricholaena grandiflora Hochst. ex A. Rich. var. collina Rendle = Melinis repens (Willd.) Zizka ■

395920　Tricholaena leucantha (A. Rich.) Stapf et C. E. Hubb. = Tricholaena teneriffae (L. f.) Link ■☆

395921　Tricholaena longiseta A. Rich. = Melinis longiseta (A. Rich.) Zizka ■☆

395922　Tricholaena madagascariense (Spreng.) Mez ex Pilg. = Tricholaena monachne (Trin.) Stapf et C. E. Hubb. ■☆

395923　Tricholaena maroccana Maire et Sam.；摩洛哥毛柱帚鼠麴■☆

395924　Tricholaena melinioides Stent = Melinis longiseta (A. Rich.) Zizka ■☆

395925　Tricholaena microstachya (Balf. f.) T. Durand et Schinz = Melinis repens (Willd.) Zizka subsp. grandiflora (Hochst.) Zizka ■☆

395926　Tricholaena minutiflora Rendle = Melinis longiseta (A. Rich.) Zizka ■☆

395927　Tricholaena monachne (Trin.) Stapf et C. E. Hubb.；单泡毛柱帚鼠麴■☆

395928　Tricholaena monachne (Trin.) Stapf et C. E. Hubb. var. annua J. G. Anderson = Tricholaena monachne (Trin.) Stapf et C. E. Hubb. ■☆

395929　Tricholaena monachyron Oliv. = Melinis repens (Willd.) Zizka subsp. grandiflora (Hochst.) Zizka ■☆

395930　Tricholaena nerviglumis Franch. = Melinis nerviglumis (Franch.) Zizka ■☆

395931　Tricholaena repens (Willd.) Hitchc. = Melinis repens (Willd.) Zizka ■

395932　Tricholaena repens (Willd.) Hitchc. = Rhynchelytrum repens (Willd.) C. E. Hubb. ■

395933　Tricholaena rhodesiana Rendle = Melinis nerviglumis (Franch.) Zizka ■☆

395934　Tricholaena rosea Nees = Melinis repens (Willd.) Zizka ■

395935　Tricholaena rosea Nees = Rhynchelytrum repens (Willd.) C. E. Hubb. ■

395936　Tricholaena rosea Nees var. ruderalis Vanderyst = Melinis repens (Willd.) Zizka ■

395937　Tricholaena rosea Nees var. setosa Peters = Melinis repens

（Willd.）Zizka subsp. grandiflora（Hochst.）Zizka ■☆

395938 Tricholaena rosea Nees var. sphacelata A. Chev. = Melinis repens（Willd.）Zizka ■

395939 Tricholaena rosea Nees var. vanheei Vanderyst = Melinis nerviglumis（Franch.）Zizka ■☆

395940 Tricholaena rosea Nees var. veminalis Vanderyst = Melinis amethystea（Franch.）Zizka ■☆

395941 Tricholaena ruficoma（Hochst. ex Steud.）Hack. = Rhynchelytrum ruficomum Hochst. ex Steud. ■☆

395942 Tricholaena rupicola Rendle = Melinis rupicola（Rendle）Zizka ■☆

395943 Tricholaena scabrida K. Schum. = Melinis scabrida（K. Schum.）Hack. ■☆

395944 Tricholaena setacea C. E. Hubb. = Tricholaena teneriffae（L. f.）Link ■☆

395945 Tricholaena setifolia Stapf = Melinis nerviglumis（Franch.）Zizka ■☆

395946 Tricholaena sphacelata Benth. = Melinis repens（Willd.）Zizka ■

395947 Tricholaena tanatricha Rendle = Melinis tanatricha（Rendle）Zizka ■☆

395948 Tricholaena teneriffae（L. f.）Link;特氏线衣草■☆

395949 Tricholaena teneriffae（L. f.）Link var. genuina Maire = Tricholaena teneriffae（L. f.）Link ■☆

395950 Tricholaena teneriffae（L. f.）Link var. sericea Maire = Tricholaena teneriffae（L. f.）Link ■☆

395951 Tricholaena teneriffae（L. f.）Link var. tibestica Maire = Tricholaena teneriffae（L. f.）Link ■☆

395952 Tricholaena tonsa Nees = Melinis repens（Willd.）Zizka ■

395953 Tricholaena tonsa Nees var. submutica Schweinf. = Melinis repens（Willd.）Zizka ■

395954 Tricholaena tuberculosa Hack. ex Hook. f. = Rhynchelytrum repens（Willd.）C. E. Hubb. ■

395955 Tricholaena uniglumis T. Durand et Schinz = Melinis repens（Willd.）Zizka subsp. grandiflora（Hochst.）Zizka ■☆

395956 Tricholaena vestita（Balf. f.）Stapf et C. E. Hubb. ;包被毛柱帚鼠麹■☆

395957 Tricholaena villosa（Parl.）T. Durand et Schinz = Rhynchelytrum repens（Willd.）C. E. Hubb. ■

395958 Tricholaena wightii Nees et Arn. ex Steud. = Rhynchelytrum repens（Willd.）C. E. Hubb. ■

395959 Tricholaser Gilli = Ducrosia Boiss. ■☆

395960 Tricholaser Gilli（1959）;毛石草属■☆

395961 Tricholaser afghanicum Gilli = Ducrosia ovatiloba Dunn et Williams ■☆

395962 Tricholemma（Röser）Röser = Avena L. ■

395963 Tricholemma（Röser）Röser（2009）;毛壳燕麦属■☆

395964 Tricholepis DC.（1833）;针苞菊属;Needlebractdaisy, Tricholepis ■

395965 Tricholepis furcata DC. ;针苞菊;Needlebractdaisy ■

395966 Tricholepis karensium Kurz ex C. B. Clarke;云南针苞菊（缅甸针苞菊）■

395967 Tricholepis tibetica Hook. f. et Thomson ex C. B. Clarke;红花针苞菊（西藏针苞菊）;Tibet Tricholepis, Xizang Needlebractdaisy ■

395968 Tricholepis trichocephala Lincz. ;毛头针苞菊■☆

395969 Tricholeptus Gand. = Daucus L. ■

395970 Tricholobos Turcz. = Sisymbrium L. ■

395971 Tricholobus Blume = Connarus L. ●

395972 Tricholobus africanus（Lam.）Heckel = Connarus africanus Lam. ●☆

395973 Tricholoma Benth. = Glossostigma Wight et Arn.（保留属名）■☆

395974 Tricholophus Spach = Polygala L. ●■

395975 Trichomaria Steud. = Tricomaria Gillies ex Hook. et Arn. ●☆

395976 Trichomema Gray = Romulea Maratti（保留属名）■☆

395977 Trichomema Gray = Trichonema Ker Gawl. ■☆

395978 Trichonema Ker Gawl. = Romulea Maratti（保留属名）■☆

395979 Trichonema coelestina（W. Bartram）Sweet = Calydorea coelestina（W. Bartram）Goldblatt et Henrich ■☆

395980 Trichonema dichotomum Klatt = Romulea dichotoma（Thunb.）Baker ■☆

395981 Trichonema longifolium Salisb. = Romulea rosea（L.）Eckl. var. australis（Ewart）M. P. de Vos ■☆

395982 Trichonema longitubum Klatt = Syringodea longituba（Klatt）Kuntze ■☆

395983 Trichonema monadelphum Sweet = Romulea monadelpha（Sweet）Baker ■☆

395984 Trichonema nivale Boiss. et Kotschy = Romulea bulbocodia（L.）Sebast. et Mauri ■☆

395985 Trichonema ornithogaloides（Licht. ex Roem. et Schult.）A. Dietr. = Geissorhiza ornithogaloides Klatt ■☆

395986 Trichonema pylium Herb. = Romulea bulbocodia（L.）Sebast. et Mauri ■☆

395987 Trichonema spirale Burch. = Geissorhiza spiralis（Burch.）M. P. de Vos ex Goldblatt ■☆

395988 Trichonema subpalustre Herb. = Romulea bulbocodia（L.）Sebast. et Mauri ■☆

395989 Trichoneura Andersson（1855）;毛肋茅属■☆

395990 Trichoneura ciliata（Peter）S. M. Phillips;缘毛毛肋茅■☆

395991 Trichoneura eleusinoides（Rendle）Ekman;穆毛肋茅■☆

395992 Trichoneura grandiglumis（Nees）Ekman;大颖毛肋茅■☆

395993 Trichoneura grandiglumis（Nees）Ekman var. minor（Rendle）Chippind. = Trichoneura grandiglumis（Nees）Ekman ■☆

395994 Trichoneura hirtella Napper = Trichoneura ciliata（Peter）S. M. Phillips ■☆

395995 Trichoneura mollis（Kunth）Ekman;绢毛毛肋茅■☆

395996 Trichoon Roth = Phragmites Adans. ■

395997 Trichoon phragmites（L.）Rendle = Phragmites australis（Cav.）Trin. ex Steud. ■

395998 Trichoon roxburghii（Kunth）Wight = Phragmites karka（Retz.）Trin. ex Steud. ■

395999 Trichopetalon Raf. = Trichopetalum Lindl. ■☆

396000 Trichopetalum Lindl.（1832）;毛瓣兰属■☆

396001 Trichopetalum Lindl. = Bottionea Colla ■☆

396002 Trichopetalum plumosum（Ruiz et Pav.）J. F. Macbr. ;羽状毛瓣兰■☆

396003 Trichophorum Pers.（1805）（保留属名）;毛莎草属（刚毛藨草属,芒莎草属,针蔺属）;Club-rush, Deer-grass, Trichophore ■

396004 Trichophorum Pers.（保留属名）= Scirpus L.（保留属名）■

396005 Trichophorum alpinum（L.）Pers. ;高山毛莎草（鳞苞针蔺）;Alpine Club-rush, Alpine Cotton-grass, Cotton Deer-grass, Hudsonian Club-rush ■☆

396006 Trichophorum alpinum（L.）Pers. var. hudsonianum（Michx.）Pers. = Trichophorum alpinum（L.）Pers. ■☆

396007 Trichophorum caespitosum（L.）Hartm. = Kreczetoviczia caespitosa（L.）Tzvelev ■☆

396008 Trichophorum cespitosum（L.）Hartm. = Trichophorum cespitosum（L.）Schur ■☆

396009　Trichophorum cespitosum (L.) Pers. = Scirpus cespitosus L. ■☆

396010　Trichophorum cespitosum (L.) Schur;簇生毛莎草;Tufted Bulrush,Tufted Club-rush,Tussock Bulrush ■☆

396011　Trichophorum clementis (M. E. Jones) S. G. Sm.;圣克利门蒂毛莎草■☆

396012　Trichophorum clintonii (A. Gray) S. G. Sm.;克林顿毛莎草;Clinton's Bulrush ■☆

396013　Trichophorum distigmaticum (Degl.) Egorova;双柱头针蔺■

396014　Trichophorum distigmaticum (Kuk.) Egorova = Trichophorum distigmaticum (Degl.) Egorova ■

396015　Trichophorum emergens Norman = Scirpus pumilus Vahl ■

396016　Trichophorum emergens Norman = Trichophorum pumilum (Vahl) Schinz et Thell. ■

396017　Trichophorum mattfeldianum (Kük.) S. Yun Liang;三棱针蔺■

396018　Trichophorum morrisonense (Hayata) Ohwi = Scirpus subcapitatus Thwaites et Hook. var. morrisonensis (Hayata) Ohwi ■

396019　Trichophorum planifolium (Spreng.) Palla;宽叶毛莎草;Bashful Bulrush ■☆

396020　Trichophorum pumilum (Vahl) Schinz et Thell.;矮针蔺(矮藨草);Low Bulrush ■

396021　Trichophorum pumilum (Vahl) Schinz et Thell. = Scirpus pumilus Vahl ■

396022　Trichophorum pumilum (Vahl) Schinz et Thell. var. rollandii (Fernald) Hultén = Scirpus pumilus Vahl ■

396023　Trichophorum pumilum (Vahl) Schinz et Thell. var. rollandii (Fernald) Hultén = Trichophorum pumilum (Vahl) Schinz et Thell. ■

396024　Trichophorum schansiense Hand.-Mazz.;太行山针蔺■

396025　Trichophorum subcapitatum (Thwaites et Hook.) D. A. Simpson;玉山针蔺(灯心草,类头序藨草,类头状花序藨草,龙须草,龙须莞,马胡须,青丝还阳,台湾藨草,野席草,针蔺);Headlike Bulrush,Taiwan Bulrush ■

396026　Trichophorum subcapitatum (Thwaites et Hook.) D. A. Simpson = Scirpus subcapitatus Thwaites et Hook. ■

396027　Trichophyllum Ehrh. = Eleocharis R. Br. ■

396028　Trichophyllum Ehrh. = Scirpus L.(保留属名)■

396029　Trichophyllum Ehrh. ex House = Eleocharis R. Br. ■

396030　Trichophyllum Ehrh. ex House = Scirpus L.(保留属名)■

396031　Trichophyllum House = Eleocharis R. Br. ■

396032　Trichophyllum Nutt. = Bahia Lag. ■☆

396033　Trichophyllum integrifolium Hook. = Eriophyllum lanatum (Pursh) J. Forbes var. integrifolium (Hook.) Smiley ■☆

396034　Trichophyllum oppositifolium Nutt. = Picradeniopsis oppositifolia (Nutt.) Rydb. ■☆

396035　Trichophyllum palustre (L.) Farw. var. calvum (Torr.) House = Eleocharis erythropoda Steud. ■☆

396036　Trichopilia Lindl. (1836);毛足兰属■☆

396037　Trichopodaceae Hutch. (1934)(保留科名);毛柄花科(发柄花科,毛柄科,毛脚科,毛脚薯科)■☆

396038　Trichopodaceae Hutch. (保留科名) = Dioscoreaceae R. Br. (保留科名)●■

396039　Trichopodium C. Presl = Dalea L. (保留属名)●■

396040　Trichopodium C. Presl = Trichopus Gaertn. ■☆

396041　Trichopodium Lindl. = Trichopus Gaertn. ■☆

396042　Trichopodium glandulosum C. Presl = Trichopus zeylanicus Gaertn. ■☆

396043　Trichopteria Lindl. = Trichopteryx Nees ■☆

396044　Trichopteria Nees = Trichopteryx Nees ■☆

396045　Trichopteris Neck. = Scabiosa L. ●■

396046　Trichopterya Lindl. = Trichopteryx Nees ■☆

396047　Trichopteryx Nees(1841);翼毛草属■☆

396048　Trichopteryx acuminata Stapf = Loudetia flavida (Stapf) C. E. Hubb. ■☆

396049　Trichopteryx ambiens K. Schum. = Loudetiopsis ambiens (K. Schum.) Conert ■☆

396050　Trichopteryx annua Stapf = Loudetia annua (Stapf) C. E. Hubb. ■☆

396051　Trichopteryx anthoxanthoides Stapf ex Vanderyst = Loudetia vanderystii (De Wild.) C. E. Hubb. ■☆

396052　Trichopteryx arundinacea (Hochst. ex A. Rich.) Hack. ex Engl. = Loudetia arundinacea (Hochst. ex A. Rich.) Steud. ■☆

396053　Trichopteryx arundinacea (Hochst. ex A. Rich.) Hack. ex Engl. var. trichantha Peter = Loudetia arundinacea (Hochst. ex A. Rich.) Steud. ■☆

396054　Trichopteryx barbata (Nees) Hack. ex T. Durand et Schinz = Danthoniopsis barbata (Nees) C. E. Hubb. ■☆

396055　Trichopteryx bequaertii De Wild. = Tristachya hubbardiana Conert ■☆

396056　Trichopteryx camerunensis Stapf = Loudetia simplex (Nees) C. E. Hubb. ■☆

396057　Trichopteryx catangensis Chiov. = Trichopteryx fruticulosa Chiov. ■☆

396058　Trichopteryx cerata Stapf = Loudetia annua (Stapf) C. E. Hubb. var. cerata (Stapf) Jacq.-Fél. ■☆

396059　Trichopteryx convoluta De Wild. = Loudetia arundinacea (Hochst. ex A. Rich.) Steud. ■☆

396060　Trichopteryx cuspidata Gilli = Loudetia flavida (Stapf) C. E. Hubb. ■☆

396061　Trichopteryx decumbens C. E. Hubb. = Trichopteryx marungensis Chiov. ■☆

396062　Trichopteryx delicatissima J. B. Phipps = Trichopteryx elegantula (Hook. f.) Stapf ■☆

396063　Trichopteryx demeusei De Wild. = Loudetia demeusei (De Wild.) C. E. Hubb. ■☆

396064　Trichopteryx densispica Rendle = Loudetia densispica (Rendle) C. E. Hubb. ■☆

396065　Trichopteryx dinteri Pilg. = Danthoniopsis dinteri (Pilg.) C. E. Hubb. ■☆

396066　Trichopteryx dobbelaerei De Wild. = Loudetia arundinacea (Hochst. ex A. Rich.) Steud. ■☆

396067　Trichopteryx dregeana Nees;德雷翼毛草■☆

396068　Trichopteryx dregeana Nees var. congoensis Franch. = Trichopteryx fruticulosa Chiov. ■☆

396069　Trichopteryx elegans (Hochst. ex A. Braun) Hack. ex Engl. = Loudetia simplex (Nees) C. E. Hubb. ■☆

396070　Trichopteryx elegans (Hochst. ex A. Braun) Hack. ex Engl. var. hensii De Wild. = Loudetia arundinacea (Hochst. ex A. Rich.) Steud. ■☆

396071　Trichopteryx elegantula (Hook. f.) Stapf;雅致翼毛草■☆

396072　Trichopteryx elegantula (Hook. f.) Stapf subsp. stolziana (Henrard) J. B. Phipps = Trichopteryx stolziana Henrard ■☆

396073　Trichopteryx elegantula (Hook. f.) Stapf var. katangensis Chippind. = Trichopteryx elegantula (Hook. f.) Stapf ■☆

396074　Trichopteryx elegantula Pilg. ex Peter = Trichopteryx elegantula (Hook. f.) Stapf ■☆

396075　Trichopteryxelisabethvilleana De Wild. = Loudetia simplex (Nees) C. E. Hubb. ■☆

396076 Trichopteryx figarii Chiov. = Loudetia togoensis（Pilg.）C. E. Hubb. ■☆

396077 Trichopteryx flavida Stapf = Loudetia flavida（Stapf）C. E. Hubb. ■☆

396078 Trichopteryx fruticulosa Chiov.；灌木状翼毛草■☆

396079 Trichopteryx fruticulosa Chiov. var. perlaxa（Pilg.）C. E. Hubb. = Trichopteryx fruticulosa Chiov. ■☆

396080 Trichopteryx fruticulosa Chiov. var. whytei C. E. Hubb. = Trichopteryx fruticulosa Chiov. ■☆

396081 Trichopteryx ganaensis Vanderyst = Loudetia vanderystii（De Wild.）C. E. Hubb. ■☆

396082 Trichopteryx gigantea Stapf = Tristachya superba（De Not.）Schweinf. et Asch. ■☆

396083 Trichopteryx gigantea Stapf var. gracilis Rendle = Tristachya superba（De Not.）Schweinf. et Asch. ■☆

396084 Trichopteryx gigantea Stapf var. phalacrotes Peter = Tristachya superba（De Not.）Schweinf. et Asch. ■☆

396085 Trichopteryx gigantea Stapf var. spiciformis Pilg. = Tristachya superba（De Not.）Schweinf. et Asch. ■☆

396086 Trichopteryx glabra Hack. = Loudetia flavida（Stapf）C. E. Hubb. ■☆

396087 Trichopteryx glabrata K. Schum. = Loudetiopsis glabrata（K. Schum.）Conert ■☆

396088 Trichopteryx glanvillei C. E. Hubb. = Trichopteryx elegantula（Hook. f.）Stapf ■☆

396089 Trichopteryx gracilis Peter = Loudetia simplex（Nees）C. E. Hubb. ■☆

396090 Trichopteryx gracillima C. E. Hubb. = Trichopteryx marungensis Chiov. ■☆

396091 Trichopteryx grisea K. Schum. = Loudetia arundinacea（Hochst. ex A. Rich.）Steud. ■☆

396092 Trichopteryx hockii De Wild. = Tristachya superba（De Not.）Schweinf. et Asch. ■☆

396093 Trichopteryx homblei De Wild. = Tristachya superba（De Not.）Schweinf. et Asch. ■☆

396094 Trichopteryx hordeiformis Stapf = Loudetia hordeiformis（Stapf）C. E. Hubb. ■☆

396095 Trichopteryx incompta Franch. = Loudetia simplex（Nees）C. E. Hubb. ■☆

396096 Trichopteryx kagerensis K. Schum. = Loudetia kagerensis（K. Schum.）C. E. Hubb. ex Hutch. ■☆

396097 Trichopteryx kapiriensis De Wild. = Loudetia simplex（Nees）C. E. Hubb. ■☆

396098 Trichopteryx katangensis De Wild. = Trichopteryx fruticulosa Chiov. ■☆

396099 Trichopteryx kerstingii Pilg. = Loudetiopsis kerstingii（Pilg.）Conert ■☆

396100 Trichopteryx lanata Stent et J. M. Rattray = Loudetia lanata（Stent et J. M. Rattray）C. E. Hubb. ■☆

396101 Trichopteryx lembaensis Vanderyst = Loudetia demeusei（De Wild.）C. E. Hubb. ■☆

396102 Trichopteryx lualabaensis De Wild. = Tristachya lualabaensis（De Wild.）J. B. Phipps ■☆

396103 Trichopteryx marungensis Chiov.；马龙加翼毛草■☆

396104 Trichopteryx migiurtina Chiov. = Loudetia migiurtina（Chiov.）C. E. Hubb. ■☆

396105 Trichopteryx mukuluensis De Wild. = Trichopteryx dregeana Nees ■☆

396106 Trichopteryx nigritiana Stapf = Loudetia simplex（Nees）C. E. Hubb. ■☆

396107 Trichopteryx parviflora Hack. = Trichopteryx dregeana Nees ■☆

396108 Trichopteryx pennata Chiov. = Loudetia flavida（Stapf）C. E. Hubb. ■☆

396109 Trichopteryx perlaxa Pilg. = Trichopteryx fruticulosa Chiov. ■☆

396110 Trichopteryx phragmitoides Peter = Loudetia phragmitoides（Peter）C. E. Hubb. ■☆

396111 Trichopteryx ramosa Stapf = Danthoniopsis ramosa（Stapf）Clayton ■☆

396112 Trichopteryx reflexa Pilg. = Loudetia kagerensis（K. Schum.）C. E. Hubb. ex Hutch. ■☆

396113 Trichopteryx sandaensis Vanderyst = Trichopteryx fruticulosa Chiov. ■☆

396114 Trichopteryx simplex（Nees）Benth. = Loudetia simplex（Nees）C. E. Hubb. ■☆

396115 Trichopteryx somalensis（Franch.）Engl. = Danthoniopsis barbata（Nees）C. E. Hubb. ●☆

396116 Trichopteryx spirathera K. Schum. = Loudetia kagerensis（K. Schum.）C. E. Hubb. ex Hutch. ■☆

396117 Trichopteryx stolziana Henrard；斯托尔兹翼毛草■☆

396118 Trichopteryx superba（De Not.）Chiov. = Tristachya superba（De Not.）Schweinf. et Asch. ■☆

396119 Trichopteryx ternata Stapf = Loudetiopsis ambiens（K. Schum.）Conert ■☆

396120 Trichopteryx togoensis Pilg. = Loudetia togoensis（Pilg.）C. E. Hubb. ■☆

396121 Trichopteryx vanderystii De Wild. = Loudetia vanderystii（De Wild.）C. E. Hubb. ■☆

396122 Trichopteryx verticillata De Wild. = Loudetia arundinacea（Hochst. ex A. Rich.）Steud. ■☆

396123 Trichopteryx viridis Rendle = Danthoniopsis viridis（Rendle）C. E. Hubb. ■☆

396124 Trichoptilium A. Gray（1859）；毛翼菊属（毛羽菊属）；Yellowhead ■☆

396125 Trichoptilium incisum（A. Gray）A. Gray；毛翼菊（毛羽菊）；Yellowhead ■☆

396126 Trichopus Gaertn.（1788）；毛柄花属（发柄花属，毛柄属）■☆

396127 Trichopus sempervirens（H. Perrier）Caddick et Wilkin；马岛毛柄花●☆

396128 Trichopus zeylanicus Gaertn.；毛柄花（发柄花）■☆

396129 Trichopyrum Á. Löve = Elymus L. ■

396130 Trichorhiza Lindl. ex Steud. = Luisia Gaudich. ■

396131 Trichoryne F. Muell. = Tricoryne R. Br. ■☆

396132 Trichosacme Zucc.（1846）；毛尖萝藦属●☆

396133 Trichosacme lanata Zucc.；毛尖萝藦●☆

396134 Trichosalpinx Luer(1983)；号角毛兰属■☆

396135 Trichosalpinx atropurpurea Luer et Hirtz；暗紫号角毛兰■☆

396136 Trichosalpinx bicolor（Barb. Rodr.）Luer；二色号角毛兰■☆

396137 Trichosalpinx ciliaris（Lindl.）Luer；睫毛号角毛兰■☆

396138 Trichosalpinx cryptantha（Barb. Rodr.）Luer；隐花号角毛兰■☆

396139 Trichosalpinx echinata Luer et Hirtz；刺号角毛兰■☆

396140 Trichosalpinx glabra D. E. Benn. et Christenson；光号角毛兰■☆

396141 Trichosalpinx montana（Barb. Rodr.）Luer；山地号角毛兰■☆

396142 Trichosalpinx orbicularis（Lindl.）Luer；圆号角毛兰■☆

396143 Trichosalpinx tenuis（C. Schweinf.）Luer；细号角毛兰■☆

396144　Trichosanchezia Mildbr. (1926);金毛爵床属☆

396145　Trichosanchezia chrysothrix Mildbr.;金毛爵床☆

396146　Trichosandra Decne. (1844);毛蕊萝藦属■☆

396147　Trichosandra borbonica Decne;毛蕊萝藦■☆

396148　Trichosantha Steud. = Stipa L. ■

396149　Trichosanthe microsiphon Kurz = Trichosanthes cordata Roxb. ■

396150　Trichosanthes L. (1753);栝楼属;Serpentagourd, Snake Gourd, Snakegourd ■

396151　Trichosanthes anguina L.;蛇瓜(豆角黄瓜,果蠃,蛇豆,丝瓜);Club Gourd, Edible Snake Gourd, Edible Snakegourd, Serpent Gourd, Serpentgourd, Snake Gourd, Viper Gourd ■

396152　Trichosanthes anguina L. = Trichosanthes cucumerina L. ■

396153　Trichosanthes anguina L. = Trichosanthes ovigera Blume ■

396154　Trichosanthes ascendens C. Y. Cheng et C. H. Yueh = Trichosanthes cucumeroides (Ser.) Maxim. ex Franch. et Sav. var. dicoelosperma (C. B. Clarke) S. K. Chen ■

396155　Trichosanthes baviensis Gagnep.;短序栝楼(假老鼠藤,野苦瓜);Shortinflorescence Snakegourd ■

396156　Trichosanthes boninensis Nakai ex Tuyama = Trichosanthes ovigera Blume var. boninensis (Nakai ex Tuyama) H. Ohba ■☆

396157　Trichosanthes bracteata (Lam.) Voigt = Trichosanthes laceribractea Hayata ■

396158　Trichosanthes bracteata (Lam.) Voigt = Trichosanthes tricuspidata Lour. ■

396159　Trichosanthes brevibracteata Kundu = Trichosanthes cucumerina L. ■

396160　Trichosanthes chingiana Hand. -Mazz. = Trichosanthes ovigera Blume ■

396161　Trichosanthes cochinchinensis M. Roem.;印支栝楼■☆

396162　Trichosanthes cordata Roxb.;心叶栝楼(心形栝楼);Cordateleaf Snakegourd, Heartleaf Snakegourd ■

396163　Trichosanthes crenulata C. Y. Cheng et C. H. Yueh = Trichosanthes rosthornii Harms ■

396164　Trichosanthes crispisepala C. Y. Wu ex S. K. Chen;皱萼栝楼;Curledsepal Snakegourd, Wrinklesepal Snakegourd ■

396165　Trichosanthes crispisepala C. Y. Wu ex S. K. Chen = Trichosanthes truncata C. B. Clarke ■

396166　Trichosanthes cucumerina L.;瓜叶栝楼(土瓜,王瓜);Melonleaf Snakegourd, Snake Gourd ■

396167　Trichosanthes cucumerina L. var. anguina (L.) Haines = Trichosanthes anguina L. ■

396168　Trichosanthes cucumerina L. var. anguina (L.) Haines = Trichosanthes cucumerina L. ■

396169　Trichosanthes cucumeroides (Ser.) Maxim. = Trichosanthes cucumeroides (Ser.) Maxim. ex Franch. et Sav. ■

396170　Trichosanthes cucumeroides (Ser.) Maxim. ex Franch. et Sav.;王瓜(雹瓜,长猫瓜,赤雹,吊瓜,杜瓜,鸽蛋瓜,公公须,钩,钩瓢,苦瓜莲,苦王瓜,藤姑,老鸦瓜,马雹儿,马剥儿,马廒儿,山冬瓜,师古草,水瓜,台湾师古草,天花粉,土瓜,土王瓜,菟瓜,狭果师古草,小苦兜,野甜瓜,野王瓜,野西瓜);Japanese Snakegourd, King Snakegourd, Narrowfruit Snakegourd ■

396171　Trichosanthes cucumeroides (Ser.) Maxim. ex Franch. et Sav. var. dicoelosperma (C. B. Clarke) S. K. Chen;波叶栝楼(翘子栝楼);Crispateleaf Snakegourd ■

396172　Trichosanthes cucumeroides (Ser.) Maxim. ex Franch. et Sav. var. formosana (Hayata) Kitam. ex Kitam. et Yoshida = Trichosanthes cucumeroides (Ser.) Maxim. ex Franch. et Sav. ■

396173　Trichosanthes cucumeroides (Ser.) Maxim. ex Franch. et Sav.

396173 (cont.)　var. formosana (Hayata) Kitam. = Trichosanthes cucumeroides (Ser.) Maxim. ■

396174　Trichosanthes cucumeroides (Ser.) Maxim. ex Franch. et Sav. var. globosa Honda = Trichosanthes cucumeroides (Ser.) Maxim. ex Franch. et Sav. ■

396175　Trichosanthes cucumeroides (Ser.) Maxim. ex Franch. et Sav. var. hainanensis (Hayata) S. K. Chen;海南栝楼(白瓜,老鸦瓜);Hainan Snakegourd ■

396176　Trichosanthes cucumeroides (Ser.) Maxim. ex Franch. et Sav. var. stenocarpa Honda = Trichosanthes cucumeroides (Ser.) Maxim. ex Franch. et Sav. ■

396177　Trichosanthes cucumeroides (Ser.) Maxim. ex Franch. et Sav. var. stenocarpa Honda;狭果师古草(狭果草栝楼)■

396178　Trichosanthes dafangensis N. G. Ye et S. J. Li;大方油栝楼;Dafang Snakegourd ■

396179　Trichosanthes damiaoshanensis C. Y. Cheng et C. H. Yueh;南方栝楼;Damiaoshan Snakegourd ■

396180　Trichosanthes damiaoshanensis C. Y. Cheng et C. H. Yueh = Trichosanthes rosthornii Harms var. multicirrata (C. Y. Cheng et C. H. Yueh) S. K. Chen ■

396181　Trichosanthes dicoelosperma C. B. Clarke = Trichosanthes cucumeroides (Ser.) Maxim. ex Franch. et Sav. var. dicoelosperma (C. B. Clarke) S. K. Chen ■

396182　Trichosanthes dioica Roxb.;异株栝楼(蒌皮)■☆

396183　Trichosanthes dunniana H. Lév.;糙点栝楼(邓氏栝楼,红花栝楼);Dunn Snakegourd ■

396184　Trichosanthes fissibracteata C. Y. Wu ex C. Y. Cheng et C. H. Yueh;裂苞栝楼(长房子栝楼);Splitbract Snakegourd ■

396185　Trichosanthes foetidissima Jacq. = Kedrostis foetidissima (Jacq.) Cogn. ■☆

396186　Trichosanthes formosana Hayata = Trichosanthes cucumeroides (Ser.) Maxim. ex Franch. et Sav. ■

396187　Trichosanthes grandibracteata Kurz = Trichosanthes wallichiana (Ser.) Wright ■

396188　Trichosanthes guizhouensis C. Y. Cheng et C. H. Yueh;贵州栝楼;Guizhou Snakegourd ■

396189　Trichosanthes guizhouensis C. Y. Cheng et C. H. Yueh = Trichosanthes rosthornii Harms ■

396190　Trichosanthes hainanensis Hayata = Trichosanthes cucumeroides (Ser.) Maxim. ex Franch. et Sav. var. hainanensis (Hayata) S. K. Chen ■

396191　Trichosanthes heteroclita Roxb. = Hodgsonia heteroclita (Roxb.) Hook. f. et Thomson ■

396192　Trichosanthes heteroclita Roxb. = Hodgsonia macrocarpa (Blume) Cogn. ■

396193　Trichosanthes himalensis C. B. Clarke;喜马拉雅栝楼(佛顶珠,实葫芦);Himalayan Snakegourd ■

396194　Trichosanthes himalensis C. B. Clarke = Trichosanthes ovigera Blume ■

396195　Trichosanthes himalensis C. B. Clarke var. indivisa Chakr. = Trichosanthes ovigera Blume ■

396196　Trichosanthes himalensis C. B. Clarke var. sikkimensis ? = Trichosanthes ovigera Blume ■

396197　Trichosanthes homophylla Hayata;芋叶栝楼(等叶栝楼,毛果栝楼,全叶栝楼);Sameleaf Snakegourd ■

396198　Trichosanthes homophylla Hayata var. ishigakiensis (E. Walker) T. Yamaz.;石垣栝楼■☆

396199 Trichosanthes hupehensis C. Y. Cheng et C. H. Yueh = Trichosanthes laceribractea Hayata ■

396200 Trichosanthes hylonoma Hand. -Mazz. ;湘桂栝楼(栝楼,雷山栝楼,圆子栝楼);Roundseed Snakegourd ■

396201 Trichosanthes integrifolia (Roxb.) Kurz = Gymnopetalum integrifolium (Roxb.) Kurz ■

396202 Trichosanthes ishigakiensis E. Walker = Trichosanthes homophylla Hayata var. ishigakiensis (E. Walker) T. Yamaz. ■☆

396203 Trichosanthes japonica (Miq.) Regel;日本栝楼(地白,栝楼);Japan Snakegourd,Yellowfruit Snakegourd ■☆

396204 Trichosanthes japonica (Miq.) Regel = Trichosanthes kirilowii Maxim. var. japonica (Miq.) Kitam. ■☆

396205 Trichosanthes javanica (Miq.) Regel = Thladiantha cordifolia (Blume) Cogn. ■

396206 Trichosanthes javanica Miq. = Thladiantha cordifolia (Blume) Cogn. ■

396207 Trichosanthes jinggangshanica C. H. Yueh;井冈栝楼(井冈山栝楼);Jinggangshan Snakegourd ■

396208 Trichosanthes kerrii Craib;长果栝楼;Kerr Snakegourd ■

396209 Trichosanthes khasiana Kundu = Trichosanthes truncata C. B. Clarke ■

396210 Trichosanthes kirilowii Maxim. ;栝楼(白药,大肚瓜,地瓜,地楼,杜瓜,钝裂栝楼,蒉,瓜蒌,瓜楼,果秾,果裸,果赢,花苦瓜,黄瓜,金丝莲,栝楼仁,山秋瓜,柿瓜,天瓜,天圆子,王瓜,王菩,鸭苦瓜,鸭屎瓜,药瓜,泽姑,泽巨,泽冶);Mongolian Snakegourd, Obtuselobe Snakegourd,Obtuselobed Snakegourd,Snakegourd ■

396211 Trichosanthes kirilowii Maxim. var. japonica (Miq.) Kitam. = Trichosanthes japonica (Miq.) Regel ■☆

396212 Trichosanthes koshunensis Hayata = Trichosanthes laceribractea Hayata ■

396213 Trichosanthes laceribractea Hayata;长萼栝楼(湖北栝楼,苦花粉,栝楼,裂苞栝楼,槭叶栝楼);Longcalyx Snakegourd,Longsepal Snakegourd ■

396214 Trichosanthes laceribracteata Hayata;槭叶括楼■

396215 Trichosanthes leishanensis C. Y. Cheng et C. H. Yueh;雷山栝楼;Leishan Snakegourd ■

396216 Trichosanthes leishanensis C. Y. Cheng et C. H. Yueh = Trichosanthes hylonoma Hand. -Mazz. ■

396217 Trichosanthes lepiniana (Naudin) Cogn. ;马干铃栝楼(马干铃);Lepin Snakegourd ■

396218 Trichosanthes macrocarpa Blume = Hodgsonia heteroclita (Roxb.) Hook. f. et Thomson ■

396219 Trichosanthes majuscula (C. B. Clarke) Kundu = Trichosanthes dunniana H. Lév. ■

396220 Trichosanthes majuscula (C. B. Clarke) Kundu = Trichosanthes rubriflos Thorel ex Cayla ■

396221 Trichosanthes matsudae Hayata = Trichosanthes cucumeroides (Ser.) Maxim. ex Franch. et Sav. var. stenocarpa Honda ■

396222 Trichosanthes mianyangensis C. H. Yueh et R. G. Liao;绵阳栝楼;Mianyang Snakegourd ■

396223 Trichosanthes microsiphon Kurz = Trichosanthes cordata Roxb. ■

396224 Trichosanthes miyagii Hayata;宫木括楼■☆

396225 Trichosanthes multicirrata C. Y. Cheng et C. H. Yueh = Trichosanthes rosthornii Harms var. multicirrata (C. Y. Cheng et C. H. Yueh) S. K. Chen ■

396226 Trichosanthes multiloba Miq. ;多裂栝楼(瓜蒌,花苦瓜,栝楼,裂叶栝楼);Manylobed Snakegourd ■

396227 Trichosanthes multiloba Miq. = Trichosanthes wallichiana (Ser.) Wright ■

396228 Trichosanthes multiloba Miq. var. majuscula C. B. Clarke = Trichosanthes rubriflos Thorel ex Cayla ■

396229 Trichosanthes multiloba Miq. var. majuscula C. B. Clarke = Trichosanthes dunniana H. Lév. ■

396230 Trichosanthes mushanensis Hayata = Trichosanthes homophylla Hayata ■

396231 Trichosanthes mushanensis Hayata = Trichosanthes reticulinervis C. Y. Wu ex S. K. Chen ■

396232 Trichosanthes obtusiloba C. Y. Wu ex C. Y. Cheng et C. H. Yueh;钝裂栝楼(金丝莲);Obtuselobe Snakegourd, Obtuselobed Snakegourd ■

396233 Trichosanthes obtusiloba C. Y. Wu ex C. Y. Cheng et C. H. Yueh = Trichosanthes kirilowii Maxim. ■

396234 Trichosanthes okamotoi Kitam. ;冈本氏括搂■

396235 Trichosanthes okamotoi Kitam. = Trichosanthes ovigera Blume ■

396236 Trichosanthes ovata Cogn. ;卵叶栝楼(皱叶栝楼);Ovateleaf Snakegourd ■

396237 Trichosanthes ovata Cogn. = Trichosanthes truncata C. B. Clarke ■

396238 Trichosanthes ovigera Blume;全缘栝楼(佛顶珠,冈本氏栝楼,喙果栝楼,假栝楼,实葫芦,喜马拉雅栝楼,喜马山栝楼,鸭屎瓜,野王瓜);Entire Snakegourd,Entireleaf Snakegourd ■

396239 Trichosanthes ovigera Blume subsp. cucumeroides (Ser.) C. Jeffrey = Trichosanthes cucumeroides (Ser.) Maxim. ex Franch. et Sav. ■

396240 Trichosanthes ovigera Blume var. boninensis (Nakai ex Tuyama) H. Ohba;小笠原栝楼■☆

396241 Trichosanthes ovigera Blume var. sikkimensis Kundu = Trichosanthes ovigera Blume ■

396242 Trichosanthes pachyrrhackis Kundu = Trichosanthes cucumerina L. ■

396243 Trichosanthes palmata Roxb. = Trichosanthes tricuspidata Lour. ■

396244 Trichosanthes palmata Roxb. var. scotanus C. B. Clarke = Trichosanthes wallichiana (Ser.) Wright ■

396245 Trichosanthes parviflora C. Y. Wu ex S. K. Chen;小花栝楼;Smallflower Snakegourd ■

396246 Trichosanthes parviflora C. Y. Wu ex S. K. Chen = Trichosanthes hylonoma Hand. -Mazz. ■

396247 Trichosanthes pedata Merr. et Chun;趾叶栝楼(叉指叶栝楼,瓜蒌,石蟾蜍,石蛤蟆);Pedateleaf Snakegourd ■

396248 Trichosanthes pedata Merr. et Chun var. yunnanensis C. Y. Cheng et C. H. Yueh = Trichosanthes pedata Merr. et Chun ■

396249 Trichosanthes pedata Merr. et Chun var. yunnanensis C. Y. Cheng et C. H. Yueh;云南栝楼;Yunnan Snakegourd ■

396250 Trichosanthes porphyrochlamys C. Y. Wu;紫背栝楼;Purpleback Snakegourd ■

396251 Trichosanthes prazeri Kundu = Trichosanthes dunniana H. Lév. ■

396252 Trichosanthes punctata Hayata = Trichosanthes laceribractea Hayata ■

396253 Trichosanthes quadricirrha Miq. = Trichosanthes japonica (Miq.) Regel ■☆

396254 Trichosanthes quinquangulata A. Gray;五角栝楼(兰屿栝楼);Fiveangle Snakegourd ■

396255 Trichosanthes quinquefolia C. Y. Wu ex C. Y. Cheng et C. H. Yueh;木基栝楼 (五叶栝楼);Fiveleaf Snakegourd, Fiveleaves Snakegourd ■

396256 Trichosanthes reticulinervis C. Y. Wu ex S. K. Chen;两广栝楼 (毛瓜蒌);Netvein Snakegourd,Reticulate Snakegourd ■

396257 Trichosanthes rosthornii Harms;华中栝楼(川贵栝楼,大肚瓜,地瓜,杜瓜,蕡,黄瓜,尖果栝楼,栝楼,日本栝楼,山秋瓜,柿瓜,双边栝楼,天瓜,天圆子,王瓜,王菩,鸭苦瓜,鸭屎瓜,泽巨,泽冶,中华栝楼);Oneflower Snakegourd,Rosthorn Snakegourd ∎

396258 Trichosanthes rosthornii Harms var. huangshanensis S. K. Chen;黄山栝楼;Huangshan Snakegourd ∎

396259 Trichosanthes rosthornii Harms var. multicirrata (C. Y. Cheng et C. H. Yueh) S. K. Chen;多卷须栝楼(瓜蒌,南方栝楼);Manytendril Snakegourd ∎

396260 Trichosanthes rosthornii Harms var. scabrella (C. H. Yueh et D. F. Gao) S. K. Chen;糙籽栝楼;Scabrouseed Snakegourd ∎

396261 Trichosanthes rostrata Kitam. = Trichosanthes ovigera Blume ∎

396262 Trichosanthes rubriflos Thorel ex Cayla;红花栝楼(大苞栝楼);Redflower Snakegourd ∎

396263 Trichosanthes rubriflos Thorel ex Cayla = Trichosanthes dunniana H. Lév. ∎

396264 Trichosanthes rubriflos Thorel ex Cayla f. macrosperma C. Y. Cheng et C. H. Yueh = Trichosanthes dunniana H. Lév. ∎

396265 Trichosanthes rubriflos Thorel ex Cayla f. macrosperma C. Y. Cheng et C. H. Yueh;大子红花栝楼;Bigseed Redflower Snakegourd ∎

396266 Trichosanthes rugatisemina C. Y. Cheng et C. H. Yueh;皱籽栝楼(皱子栝楼);Wrinkleseed Snakegourd ∎

396267 Trichosanthes scabra Lour. = Gymnopetalum integrifolium (Roxb.) Kurz ∎

396268 Trichosanthes scabrella C. H. Yueh et D. F. Gao = Trichosanthes rosthornii Harms var. scabrella (C. H. Yueh et D. F. Gao) S. K. Chen ∎

396269 Trichosanthes schizostroma Hayata = Trichosanthes laceribractea Hayata ∎

396270 Trichosanthes sericeifolia C. Y. Cheng et C. H. Yueh;丝毛栝楼;Sericeousleaf Snakegourd ∎

396271 Trichosanthes shikokiana Makino = Trichosanthes laceribractea Hayata ∎

396272 Trichosanthes sinopunctata C. Y. Cheng ex C. H. Yueh = Trichosanthes laceribractea Hayata ∎

396273 Trichosanthes smilacifolia C. Y. Wu ex C. H. Yueh et C. Y. Cheng;菝葜叶栝楼;Greenbrierleaf Snakegourd, Smilaxleaf Snakegourd ∎

396274 Trichosanthes stylopodifera C. Y. Cheng et C. H. Yueh;尖果栝楼;Sharpfruit Snakegourd ∎

396275 Trichosanthes stylopodifera C. Y. Cheng et C. H. Yueh = Trichosanthes rosthornii Harms ∎

396276 Trichosanthes subrosae C. Y. Cheng et C. H. Yueh;粉花栝楼;Paleroseflower Snakegourd,Roseflower Snakegourd ∎

396277 Trichosanthes tetragonosperma C. Y. Cheng et C. H. Yueh;方籽栝楼;Squareseed Snakegourd,Tetragonousseed Snakegourd ∎

396278 Trichosanthes tomentosa Chakr. = Trichosanthes kerrii Craib ∎

396279 Trichosanthes trichocarpa C. Y. Wu ex C. Y. Cheng et C. H. Yueh;杏籽栝楼(顶毛栝楼);Hairfruit Snakegourd, Hairyfruit Snakegourd,Redball Snakegourd ∎

396280 Trichosanthes tricuspidata Lour.;三尖栝楼(大苞栝楼,槭叶栝楼);Threecuspidate Snakegourd,Tricuspidate Snakegourd ∎

396281 Trichosanthes tricuspidata Lour. = Trichosanthes laceribractea Hayata ∎

396282 Trichosanthes tridentata C. Y. Cheng et C. H. Yueh;绿子栝楼;Tridentate Snakegourd ∎

396283 Trichosanthes tridentata C. Y. Cheng et C. H. Yueh = Trichosanthes dunniana H. Lév. ∎

396284 Trichosanthes truncata C. B. Clarke;截叶栝楼(大瓜蒌,大子栝楼);Truncate Snakegourd ∎

396285 Trichosanthes uniflora K. S. Hao = Trichosanthes rosthornii Harms ∎

396286 Trichosanthes villosa Blume;密毛栝楼;Villose Snakegourd ∎

396287 Trichosanthes wallichiana (Ser.) Wright;薄叶栝楼(全缘萼栝楼);Thinleaf Snakegourd, Wallich Snakegourd ∎

396288 Trichosanthes yixingense Z. P. Wang et Q. Z. Xie;喙果藤;Yixing Snakegourd ∎

396289 Trichosathera Ehrh. = Stipa L. ∎

396290 Trichoschoenus J. Raynal(1968);毛赤箭莎属∎☆

396291 Trichoschoenus bosseri Raynal;毛赤箭莎∎☆

396292 Trichoscypha Hook. f. (1862);毛杯漆属●☆

396293 Trichoscypha abut Engl. et Brehmer = Trichoscypha oddonii De Wild. ●☆

396294 Trichoscypha acuminata Engl.;渐尖毛杯漆●☆

396295 Trichoscypha africana Lecomte = Trichoscypha rubicunda Lecomte ●☆

396296 Trichoscypha albiflora Engl. = Trichoscypha lucens Oliv. ●☆

396297 Trichoscypha altescandens Van der Veken = Trichoscypha reygaertii De Wild. ●☆

396298 Trichoscypha arborea (A. Chev.) A. Chev.;树状杯漆●☆

396299 Trichoscypha arborescens Van der Veken = Trichoscypha reygaertii De Wild. ●☆

396300 Trichoscypha atropurpurea Engl. = Trichoscypha mannii Hook. f. ●☆

396301 Trichoscypha baldwinii Keay;鲍德温毛杯漆●☆

396302 Trichoscypha barbata Breteler;髯毛杯漆●☆

396303 Trichoscypha beguei Aubrév. et Pellegr. = Trichoscypha bijuga Engl. ●☆

396304 Trichoscypha bijuga Engl.;双轭杯漆●☆

396305 Trichoscypha bipindensis Engl. = Trichoscypha oliveri Engl. ●☆

396306 Trichoscypha bracteata Breteler;苞毛杯漆●☆

396307 Trichoscypha braunii Engl. = Trichoscypha acuminata Engl. ●☆

396308 Trichoscypha brieyi De Wild. = Trichoscypha oddonii De Wild. ●☆

396309 Trichoscypha buettneri Engl. = Trichoscypha acuminata Engl. ●☆

396310 Trichoscypha cabindensis Exell et Mendonça = Trichoscypha oddonii De Wild. ●☆

396311 Trichoscypha camerunensis Engl. = Trichoscypha laxiflora Engl. ●☆

396312 Trichoscypha cavalliensis Aubrév. et Pellegr.;卡瓦利杯漆●☆

396313 Trichoscypha chevalieri Aubrév. et Pellegr. = Trichoscypha lucens Oliv. ●☆

396314 Trichoscypha congensis Engl. = Trichoscypha acuminata Engl. ●☆

396315 Trichoscypha coriacea Engl. et Brehmer = Trichoscypha lucens Oliv. ●☆

396316 Trichoscypha dinklagei Engl. = Trichoscypha bijuga Engl. ●☆

396317 Trichoscypha diversifoliolata Van der Veken = Trichoscypha lucens Oliv. ●☆

396318 Trichoscypha ealensis Van der Veken = Trichoscypha lucens Oliv. ●☆

396319 Trichoscypha ejui Engl. et Brehmer = Trichoscypha oddonii De Wild. ●☆

396320 Trichoscypha engong Engl. et Brehmer;特斯曼杯漆●☆

396321 Trichoscypha escherichii Engl. = Trichoscypha rubicunda Lecomte ●☆

396322 Trichoscypha ferruginea Engl. = Trichoscypha acuminata Engl. ●☆

396323 Trichoscypha flamignii De Wild. = Trichoscypha acuminata Engl. ●☆

396324　Trichoscypha fusca Lecomte = Trichoscypha rubicunda Lecomte ●☆

396325　Trichoscypha gabonensis Lecomte = Trichoscypha oliveri Engl. ●☆

396326　Trichoscypha gambana Jongkind = Trichoscypha mannii Hook. f. ●☆

396327　Trichoscypha gossweileri Exell et Mendonça = Trichoscypha bijuga Engl. ●☆

396328　Trichoscypha hallei Breteler;哈勒毛杯漆●☆

396329　Trichoscypha heterophylla Engl. et Brehmer = Trichoscypha laxiflora Engl. ●☆

396330　Trichoscypha imbricata Engl.;覆瓦毛杯漆●☆

396331　Trichoscypha klainei Lecomte = Trichoscypha rubicunda Lecomte ●☆

396332　Trichoscypha kwangoensis Van der Veken = Trichoscypha lucens Oliv. ●☆

396333　Trichoscypha laurentii De Wild. = Trichoscypha acuminata Engl. ●☆

396334　Trichoscypha laxiflora Engl.;疏花毛杯漆●☆

396335　Trichoscypha laxissima Breteler;极疏毛杯漆●☆

396336　Trichoscypha ledermannii Engl. et Brehmer = Trichoscypha lucens Oliv. ●☆

396337　Trichoscypha lescrauwaetii De Wild. = Trichoscypha reygaertii De Wild. ●☆

396338　Trichoscypha le-testui Lecomte = Trichoscypha oddonii De Wild. ●☆

396339　Trichoscypha liberica Engl.;利比里亚毛杯漆●☆

396340　Trichoscypha liketensis Van der Veken = Trichoscypha lucens Oliv. ●☆

396341　Trichoscypha linderi Breteler;林德毛杯漆●☆

396342　Trichoscypha longifolia (Hook. f.) Engl.;长叶毛杯漆●☆

396343　Trichoscypha longipetala Baker f. = Trichoscypha mannii Hook. f. ●☆

396344　Trichoscypha lucens Oliv.;光亮毛杯漆●☆

396345　Trichoscypha macrophylla Lecomte = Trichoscypha oliveri Engl. ●☆

396346　Trichoscypha mannii Hook. f.;曼氏毛杯漆●☆

396347　Trichoscypha nigra Lecomte = Trichoscypha imbricata Engl. ●☆

396348　Trichoscypha nyangensis Pellegr.;尼扬加杯漆●☆

396349　Trichoscypha oba Aubrév. et Pellegr. = Trichoscypha lucens Oliv. ●☆

396350　Trichoscypha oddonii De Wild.;奥顿杯漆●☆

396351　Trichoscypha oliveri Engl.;奥里弗杯漆●☆

396352　Trichoscypha pallidiflora Engl. et Brehmer = Trichoscypha lucens Oliv. ●☆

396353　Trichoscypha panniculata Engl. = Trichoscypha patens (Oliv.) Engl. ●☆

396354　Trichoscypha parviflora Engl. = Trichoscypha oliveri Engl. ●☆

396355　Trichoscypha parvifloroides Pellegr. = Trichoscypha oliveri Engl. ●☆

396356　Trichoscypha parvifoliolata Van der Veken = Trichoscypha lucens Oliv. ●☆

396357　Trichoscypha patens (Oliv.) Engl.;铺展杯漆●☆

396358　Trichoscypha pauciflora Van der Veken;少花杯漆●☆

396359　Trichoscypha platycarpa Van der Veken = Trichoscypha oliveri Engl. ●☆

396360　Trichoscypha preussii Engl. = Trichoscypha bijuga Engl. ●☆

396361　Trichoscypha redingii De Wild. = Trichoscypha acuminata Engl. ●☆

396362　Trichoscypha reticulata Engl. = Trichoscypha bijuga Engl. ●☆

396363　Trichoscypha reygaertii De Wild.;赖氏杯漆●☆

396364　Trichoscypha rhoifolia Engl. et Brehmer = Trichoscypha lucens Oliv. ●☆

396365　Trichoscypha rubicunda Lecomte;红毛杯漆●☆

396366　Trichoscypha rubriflora Engl. et Brehmer = Trichoscypha arborea (A. Chev.) A. Chev. ●☆

396367　Trichoscypha scandens Van der Veken = Trichoscypha reygaertii De Wild. ●☆

396368　Trichoscypha silveirana Exell et Mendonça = Trichoscypha lucens Oliv. ●☆

396369　Trichoscypha smeathmannii Keay = Trichoscypha smythei Hutch. et Dalziel ●☆

396370　Trichoscypha smythei Hutch. et Dalziel;斯迈思杯漆●☆

396371　Trichoscypha soyauxii Engl. et Brehmer = Trichoscypha mannii Hook. f. ●☆

396372　Trichoscypha submontana Van der Veken = Trichoscypha lucens Oliv. ●☆

396373　Trichoscypha subretusa Engl. et Brehmer = Trichoscypha mannii Hook. f. ●☆

396374　Trichoscypha talbotii Baker f. = Trichoscypha laxiflora Engl. ●☆

396375　Trichoscypha tessmannii Engl. et Brehmer = Trichoscypha engong Engl. et Brehmer ●☆

396376　Trichoscypha turbinata Lecomte = Trichoscypha mannii Hook. f. ●☆

396377　Trichoscypha ulugurensis Mildbr.;乌卢古尔杯漆●☆

396378　Trichoscypha ulugurensis Mildbr. = Trichoscypha lucens Oliv. ●☆

396379　Trichoscypha ulugurensis Mildbr. subsp. submontana (Van der Veken) Kokwaro = Trichoscypha lucens Oliv. ●☆

396380　Trichoscypha victoriae Engl. = Trichoscypha patens (Oliv.) Engl. ●☆

396381　Trichoscypha volubilis Van der Veken = Trichoscypha lucens Oliv. ●☆

396382　Trichoscypha yapoensis Aubrév. et Pellegr. = Trichoscypha lucens Oliv. ●☆

396383　Trichoseris Poepp. et Endl. = Macrorhynchus Less. ■☆

396384　Trichoseris Poepp. et Endl. = Trochoseris Poepp. et Endl. ■☆

396385　Trichoseris Poepp. et Endl. = Troximon Gaertn. ■☆

396386　Trichoseris Vis. = Crepis L. ■

396387　Trichoseris Vis. = Pterotheca Cass. ■

396388　Trichosia Blume = Eria Lindl. (保留属名)■

396389　Trichosiphon Schott et Endl. = Sterculia L. ●

396390　Trichosiphon australe Schott et Endl. = Brachychiton australis (Schott et Endl.) A. Terracc. ●☆

396391　Trichosma Lindl. = Eria Lindl. (保留属名)■☆

396392　Trichosma coronaria (Lindl.) Kuntze = Eria coronaria (Lindl.) Rchb. f. ■

396393　Trichosma cylindropoda Griff. = Eria coronaria (Lindl.) Rchb. f. ■

396394　Trichosma simondii Gagnep. = Eria gagnepainii A. D. Hawkes et A. H. Heller ■

396395　Trichosma suavis Lindl. = Eria coronaria (Lindl.) Rchb. f. ■

396396　Trichospermum Blume(1825);毛籽椴属■☆

396397　Trichospermum P. Beauv. ex Cass. = Parthenium L. ■

396398　Trichospermum australe (Little) Kosterm.;南美毛籽椴●☆

396399　Trichospermum javanicum Blume;毛籽椴●

396400　Trichospira Kunth(1818);鬼角草属■☆

396401　Trichospira verticillata (L.) S. F. Blake;鬼角草■☆

396402　Trichosporum D. Don(废弃属名) = Aeschynanthus Jack(保留属名)●■

396403　Trichosporum acuminatum (Wall. ex A. DC.) Kuntze = Aeschynanthus acuminatus Wall. ex A. DC. ●

396404　Trichosporum lobianum (Hook.) Kuntze = Aeschynanthus lobianus Hook. ●☆

396405　Trichosporum macranthum Merr. = Aeschynanthus macranthus (Merr.) Pellegr. ●

396406 Trichosporum moningeriae Merr. = Aeschynanthus moningerae (Merr.) Chun ●

396407 Trichosporum speciosum Kuntze = Aeschynanthus speciosus Hook. ●

396408 Trichostachys Hook. f. (1873)(保留属名);毛穗茜属●☆

396409 Trichostachys Welw. (废弃属名) = Faurea Harv. ●

396410 Trichostachys Welw. (废弃属名) = Trichostachys Hook. f. (保留属名)●☆

396411 Trichostachys aurea Hiern;黄毛穗茜●☆

396412 Trichostachys ciliata Hiern = Psychotria globosa Hiern var. ciliata (Hiern) E. M. Petit ●☆

396413 Trichostachys hedraeocephala Bremek. = Psychotria globosa Hiern var. ciliata (Hiern) E. M. Petit ●☆

396414 Trichostachys interrupta K. Schum. ;间断毛穗茜●☆

396415 Trichostachys krauseana Wernham = Trichostachys interrupta K. Schum. ●☆

396416 Trichostachys laurentii De Wild. ;洛朗毛穗茜●☆

396417 Trichostachys lehmbachii K. Schum. ;莱姆毛穗茜●☆

396418 Trichostachys letestui Pellegr. ;莱泰斯图毛穗茜●☆

396419 Trichostachys longifolia Hiern;长叶毛穗茜●☆

396420 Trichostachys mayumbense De Wild. ;马永巴毛穗茜●☆

396421 Trichostachys microcarpa K. Schum. ;小果毛穗茜●☆

396422 Trichostachys petiolata Hiern;柄叶毛穗茜●☆

396423 Trichostachys soyauxii K. Schum. ;索亚毛穗茜●☆

396424 Trichostachys stenostachys K. Schum. ;狭穗毛穗茜●☆

396425 Trichostachys talbotii Wernham = Psychotria globosa Hiern var. ciliata (Hiern) E. M. Petit ●☆

396426 Trichostachys thollonii De Wild. ;托伦毛穗茜●☆

396427 Trichostachys vaginalis Hiern = Psychotria globosa Hiern var. ciliata (Hiern) E. M. Petit ●☆

396428 Trichostachys zenkeri De Wild. = Trichostachys aurea Hiern ●☆

396429 Trichostegia Turcz. = Athrixia Ker Gawl. ●■☆

396430 Trichostelma Baill. (1890);毛冠萝藦属●☆

396431 Trichostelma ciliatum Baill. ;毛冠萝藦●☆

396432 Trichostelma oblongifolium Donn. Sm. ;矩圆叶毛冠萝藦●☆

396433 Trichostema Gronov. ex L. = Trichostema L. ●■

396434 Trichostema L. (1753);蓝卷木属(丝蕊属);Blue Curls ●■

396435 Trichostema arizonicum A. Gray;亚利桑那蓝卷木; Arizona Blue Curls ●☆

396436 Trichostema brachiatum L. ;交互对生蓝卷木;False Pennyroyal, False-pennyroyal, Fluxweed ●☆

396437 Trichostema dichotomum L. ;北美蓝卷木;Blue Curls, Blue-curls ●☆

396438 Trichostema dichotomum L. var. puberulum Fernald et Griscom = Trichostema dichotomum L. ●☆

396439 Trichostema lanatum Benth. ;蓝卷木(披针丝蕊);Bastard Pennyroyal, Blue Curls, Woolly Blue Curls ●☆

396440 Trichostema setaceum Houtt. ;刚毛蓝卷木;Blue Curls ●☆

396441 Trichostemma Cass. = Trichostephium Cass. ■●

396442 Trichostemma Cass. = Wedelia Jacq. (保留属名)■●

396443 Trichostemma L. = Trichostema L. ●■☆

396444 Trichostemma R. Br. = Vernonia Schreb. (保留属名)●■

396445 Trichostemum Raf. = Trichostemma L. ●■

396446 Trichostephania Tardieu = Ellipanthus Hook. f. ●

396447 Trichostephanus Gilg(1908);毛冠木属●☆

396448 Trichostephanus acuminatus Gilg;非洲毛冠木●☆

396449 Trichostephanus gabonensis Breteler;毛冠木●☆

396450 Trichostephium Cass. = Trichostephus Cass. ■●

396451 Trichostephium Cass. = Wedelia Jacq. (保留属名)■●

396452 Trichostephus Cass. = Wedelia Jacq. (保留属名)■●

396453 Trichosterigma Klotzsch et Garcke = Euphorbia L. ●■

396454 Trichostigma A. Rich. (1845);毛柱属(毛柱珊瑚属);Haired Stygma ●☆

396455 Trichostigma octandrum (L.) H. Walter;八蕊毛柱;Hoopvine ●☆

396456 Trichostigma peruvianum (Moq.) H. Walter;秘鲁毛柱;Peru Haired Stygma ●☆

396457 Trichostomanthemum Domin = Melodinus J. R. Forst. et G. Forst. ●

396458 Trichostomanthemum Domin = Wrightia R. Br. ●

396459 Trichotaenia T. Yamaz. = Lindernia All. ■

396460 Trichothalamus Spreng. = Potentilla L. ■●

396461 Trichotheca (Nied.) Willis = Byrsonima Rich. ex Juss. ●☆

396462 Trichotolinum O. E. Schulz(1933);巴塔哥尼亚芥属■☆

396463 Trichotolinum deserticola (Speg.) O. E. Schulz;巴塔哥尼亚芥■☆

396464 Trichotosia Blume(1825);多毛兰属■

396465 Trichotosia annulata Blume;小环多毛兰■☆

396466 Trichotosia atroferruginea (Schltr.) P. F. Hunt;黑锈多毛兰■☆

396467 Trichotosia aurea (Ridl.) Carr;黄多毛兰■☆

396468 Trichotosia biflora Griff. ;双花多毛兰■☆

396469 Trichotosia brachybotrya (Schltr.) W. Kittr. ;短穗多毛兰■☆

396470 Trichotosia caespitosa (Rolfe) Kraenzl. = Ceratostylis hainanensis Z. H. Tsi ■

396471 Trichotosia ciliata Teijsm. et Binn. ;睫毛多毛兰■☆

396472 Trichotosia dasyphylla (Parl. et Rchb. f.) Kraenzl. ;瓜子毛鞘兰(瓜子毛兰);Densifolious Eria, Melonseed Hairorchis ■

396473 Trichotosia dasyphylla (Parl. et Rchb. f.) Kraenzl. = Eria dasyphylla Parl. et Rchb. f. ■

396474 Trichotosia dongfangensis X. H. Jin et L. P. Siu;东方毛鞘兰■

396475 Trichotosia latifolia (Blume) Seidenf. ;宽叶多毛兰■☆

396476 Trichotosia microphylla Blume;小叶毛鞘兰(毛叶毛兰,小叶毛兰);Littleleaf Hairorchis, Smallleaf Eria ■

396477 Trichotosia microphylla Blume = Eria microphylla (Blume) Blume ■

396478 Trichotosia mucronata Kraenzl. ;钝尖多毛兰■☆

396479 Trichotosia pulvinata (Lindl.) Kraenzl. ;高茎毛鞘兰(高茎毛兰);Tall Eria, Tall Hairorchis ■

396480 Trichotosia rufinula (Hook. f.) Kuntze = Trichotosia pulvinata (Lindl.) Kraenzl. ■

396481 Trichotosia rufinula (Rchb. f.) Kraenzl. = Eria pulvinata Lindl. ■

396482 Trichotosia rufinula (Rchb. f.) Kraenzl. = Trichotosia pulvinata (Lindl.) Kraenzl. ■

396483 Trichotosia uniflora (J. J. Wood) Schuit. et J. J. Wood;单花多毛兰■☆

396484 Trichouratea Tiegl. = Ouratea Aubl. (保留属名)●

396485 Trichovaselia Tiegh. = Elvasia DC. ●☆

396486 Trichroa Raf. = Rhamnus L. ●■

396487 Trichrysus Raf. = Helleborus L. ■

396488 Trichuriella Bennet(1985);小针叶苋属■

396489 Trichuriella monsoniae (L. f.) Bennet;小针叶苋■

396490 Trichurus C. C. Towns. = Trichuriella Bennet ■

396491 Trichurus Clem. (1896);针叶苋属;Trichurus ■☆

396492 Trichurus cylindricus Clem. et Shear;针叶苋;Common Trichurus, Trichurus ■☆

396493 Trichurus monsoniae (L. f.) C. C. Towns. = Trichuriella monsoniae (L. f.) Bennet ■

396494 Trichymenia Rydb. = Hymenothrix A. Gray ■☆

396495 Triclanthera Raf. = Crateva L. ●

396496　Tricliceras Thonn. ex DC.（1826）；三距时钟花属■☆

396497　Tricliceras auriculatum（A. Fern. et R. Fern.）R. Fern.；耳形三距时钟花■☆

396498　Tricliceras bivinianum（Tul.）R. Fern.；马达加斯加三距时钟花■☆

396499　Tricliceras brevicaule（Urb.）R. Fern.；短茎三距时钟花■☆

396500　Tricliceras brevicaule（Urb.）R. Fern. var. rosulatum（Urb.）R. Fern.；莲座三距时钟花■☆

396501　Tricliceras elatum（A. Fern. et R. Fern.）R. Fern.；高三距时钟花■☆

396502　Tricliceras glanduliferum（Klotzsch）R. Fern.；腺点三距时钟花■☆

396503　Tricliceras hirsutum（A. Fern. et R. Fern.）R. Fern.；粗毛三距时钟花■☆

396504　Tricliceras laceratum（Oberm.）Oberm. = Tricliceras schinzii（Urb.）R. Fern. subsp. laceratum（Oberm.）R. Fern. ■☆

396505　Tricliceras lanceolatum（A. Fern. et R. Fern.）R. Fern.；披针形三距时钟花■☆

396506　Tricliceras lobatum（Urb.）R. Fern.；浅裂三距时钟花■☆

396507　Tricliceras longepedunculatum（Mast.）R. Fern.；长梗时钟花■☆

396508　Tricliceras mossambicense（A. Fern. et R. Fern.）R. Fern.；莫桑比克三距时钟花■☆

396509　Tricliceras pilosum（Willd.）R. Fern.；疏毛三距时钟花■☆

396510　Tricliceras prittwitzii（Urb.）R. Fern.；普里特时钟花■☆

396511　Tricliceras schinzii（Urb.）R. Fern.；欣兹时钟花■☆

396512　Tricliceras schinzii（Urb.）R. Fern. subsp. laceratum（Oberm.）R. Fern.；撕裂三距时钟花■☆

396513　Tricliceras tanacetifolium（Klotzsch）R. Fern.；艾菊叶时钟花■☆

396514　Tricliceras xylorhizum Verdc.；木根三距时钟花■☆

396515　Triclicerus Thonn. ex DC. = Tricliceras Thonn. ex DC. ■☆

396516　Triclicerus Thonn. ex DC. = Wormskioldia Schumach. et Thonn. ■☆

396517　Triclinium Raf. = Sanicula L. ■

396518　Triclinium odoratum Raf. = Sanicula gregaria E. P. Bicknell ■☆

396519　Triclis Haller = Mollugo L. ■

396520　Triclisia Benth.（1862）；三被藤属●☆

396521　Triclisia angolensis Exell = Triclisia subcordata Oliv. ●☆

396522　Triclisia angustifolia Diels；窄叶三被藤●☆

396523　Triclisia coriacea Oliv.；革质三被藤●☆

396524　Triclisia dictyophylla Diels；网叶三被藤●☆

396525　Triclisia dielsii Hutch. et Dalziel = Triclisia subcordata Oliv. ●☆

396526　Triclisia flava Exell = Triclisia dictyophylla Diels ●☆

396527　Triclisia fragosa（I. Verd.）I. Verd. et Troupin = Tinospora fragosa（I. Verd.）I. Verd. et Troupin ●☆

396528　Triclisia gilletii（De Wild.）Staner = Triclisia dictyophylla Diels ●☆

396529　Triclisia hypochrysea Diels = Triclisia subcordata Oliv. ●☆

396530　Triclisia jumelliana Diels；朱迈尔三被藤●☆

396531　Triclisia lanceolata Troupin；剑叶三被藤●☆

396532　Triclisia loucoubensis Baill.；卢库巴三被藤●☆

396533　Triclisia louisii Troupin；路易斯三被藤●☆

396534　Triclisia macrocarpa（Baill.）Diels；大果三被藤●☆

396535　Triclisia macrophylla Oliv.；大叶三被藤●☆

396536　Triclisia patens Oliv.；铺展三被藤●☆

396537　Triclisia riparia Troupin；河岸三被藤●☆

396538　Triclisia sacleuxii（Pierre）Diels；萨克勒三被藤●☆

396539　Triclisia sacleuxii（Pierre）Diels var. ovalifolia Troupin；卵叶三被藤●☆

396540　Triclisia semnophylla Diels = Triclisia dictyophylla Diels ●☆

396541　Triclisia subcordata Oliv.；亚心形三被藤●☆

396542　Triclisia viridiflora（DC.）Matsum. var. buisanensis（Hayata）Yamam. = Diplospora dubia（Lindl.）Masam. ●

396543　Triclisia welwitschii Hiern = Syrrheonema welwitschii（Hiern）Diels ●☆

396544　Triclisperma Raf. = Polygala L. ●■

396545　Triclisperma paucifolia（Willd.）Nieuwl. = Polygala paucifolia Willd. ■☆

396546　Triclissa Salisb. = Kniphofia Moench（保留属名）■☆

396547　Triclissa Salisb. = Tritoma Ker Gawl. ■☆

396548　Triclissa uvaria（L.）Salisb. = Kniphofia uvaria（L.）Oken ■☆

396549　Tricloeladus Hutch. = Trichocladus Pers. ●☆

396550　Tricoceae Batsch = Euphorbiaceae Juss.（保留科名）●■

396551　Tricochilus Ames = Dipodium R. Br. ■☆

396552　Tricoilendus Raf. = Indigofera L. ●■

396553　Tricoilendus Raf. = Oustropis G. Don ●■

396554　Tricomaria Gillies = Tricomaria Gillies ex Hook. et Arn. ●☆

396555　Tricomaria Gillies ex Hook. = Tricomaria Gillies ex Hook. et Arn. ●☆

396556　Tricomaria Gillies ex Hook. et Arn.（1833）；三毛金虎尾属●☆

396557　Tricomaria usillo Hook. et Arn.；三毛金虎尾●☆

396558　Tricomariopsis Dubard = Sphedamnocarpus Planch. ex Benth. ●☆

396559　Tricondylus Knight = Lomatia R. Br.（保留属名）●☆

396560　Tricondylus Salisb. = Lomatia R. Br.（保留属名）●☆

396561　Tricondylus Salisb. ex Knight.（废弃属名）= Lomatia R. Br.（保留属名）●☆

396562　Tricoryne R. Br.（1810）；三棒吊兰属■☆

396563　Tricoryne acaulis D. Dietr.；无茎三棒吊兰■☆

396564　Tricoryne platyptera Rchb. f.；宽翅三棒吊兰■☆

396565　Tricoscypha Engl. = Trichoscypha Hook. f. ●☆

396566　Tricostularia Nees ex Lehm. = Tricostularia Nees ■

396567　Tricostularia Nees（1844）；三肋果莎属（三肋莎属）■

396568　Tricostularia undulate（Thwaites）Kern；三肋果莎■

396569　Tricratus L' Hér. = Abronia Juss. ■☆

396570　Tricratus L' Hér. ex Willd. = Abronia Juss. ■☆

396571　Tricuspidaria Ruiz et Pav. = Crinodendron Molina ●☆

396572　Tricuspidaria lanceolata Miq. = Crinodendron hookerianum Gay ●☆

396573　Tricuspis P. Beauv. = Tridens Roem. et Schult. ■☆

396574　Tricuspis Pers. = Crinodendron Molina ●☆

396575　Tricuspis Pers. = Tricuspidaria Ruiz et Pav. ●☆

396576　Tricycla Cav. = Bougainvillea Comm. ex Juss.（保留属名）●

396577　Tricyclandra Keraudren = Tricyclandra Rabenant. ●☆

396578　Tricyclandra Keraudren（1966）；三圆蕊属●☆

396579　Tricyclandra leandrii Rabenant.；三圆蕊●☆

396580　Tricyrtidaceae Takht.（1997）（保留科名）；油点草科■

396581　Tricyrtidaceae Takht.（保留科名）= Calochortaceae Dumort. ■

396582　Tricyrtidaceae Takht.（保留科名）= Liliaceae Juss.（保留科名）■●

396583　Tricyrtis Wall.（1826）（保留属名）；油点草属；Toad Lily, Toad-lilies, Toadlily, Toad-lily ■

396584　Tricyrtis affinis Makino；近缘油点草■☆

396585　Tricyrtis bakerii Koidz. = Tricyrtis latifolia Maxim. ■

396586　Tricyrtis bakerii Koidz. = Tricyrtis maculata（D. Don）J. F. Macbr. ■

396587　Tricyrtis chinensis Hir. Takah. bis；中国油点草■

396588　Tricyrtis formosana Baker；台湾油点草（小型油点草，小油点草，紫花油点草）；Taiwan Toadlily, Toad Lily ■

396589　Tricyrtis formosana Baker f. glandulosa T. Shimizu = Tricyrtis formosana Baker var. glandosa (Simizu) Tang S. Liu et S. S. Ying ■

396590　Tricyrtis formosana Baker f. glandulosa T. Shimizu = Tricyrtis formosana Baker ■

396591　Tricyrtis formosana Baker var. amethystina Masam. = Tricyrtis formosana Baker ■

396592　Tricyrtis formosana Baker var. glandosa (Simizu) Tang S. Liu et S. S. Ying;小型油点草(小油点草)■

396593　Tricyrtis formosana Baker var. glandulosa (T. Shimizu) Tang S. Liu et S. S. Ying = Tricyrtis formosana Baker ■

396594　Tricyrtis formosana Baker var. grandiflora S. S. Ying;大花油点草;Bigflower Taiwan Toadlily ■

396595　Tricyrtis formosana Baker var. lasiocarpa (Matsum.) Masam. = Tricyrtis lasiocarpa Matsum. ■

396596　Tricyrtis formosana Baker var. ovatifolia (S. S. Ying) S. S. Ying;卵叶油点草;Ovateleaf Taiwan Toadlily ■

396597　Tricyrtis formosana Baker var. stolonifera (Matsum.) Masam. = Tricyrtis stolonifera Matsum. ■

396598　Tricyrtis hirta Hook.;毛油点草(油点草);Hairy Toadlily, Japan Toad-lily,Toad Lily,Toadlily ■☆

396599　Tricyrtis lasiocarpa Matsum.;毛果油点草;Hairfruited Taiwan Toadlily ■

396600　Tricyrtis lasiocarpa Matsum. = Tricyrtis formosana Baker var. lasiocarpa (Matsum.) Masam. ■

396601　Tricyrtis latifolia Maxim.;宽叶油点草;Broadleaf Toadlily ■

396602　Tricyrtis macrantha Maxim.;日本大花油点草■☆

396603　Tricyrtis macranthopsis Masam.;拟大花油点草■☆

396604　Tricyrtis macropoda Miq.;油点草(白七,粗柄油点草,粗轴,粗轴油点草,红酸七,牛尾参,牛尾参油点草,水杨罗,竹叶七);Spikeled Toad Lily,Spikeled Toadlily,Toadlily ■

396605　Tricyrtis macrotpoda Miq. = Tricyrtis pilosa Wall. ■

396606　Tricyrtis maculata (D. Don) J. F. Macbr. = Tricyrtis pilosa Wall. ■

396607　Tricyrtis nana Yatabe;小油点草■☆

396608　Tricyrtis ohsumiensis Masam.;大隅油点草■☆

396609　Tricyrtis ovatifolia S. S. Ying = Tricyrtis formosana Baker var. ovatifolia (S. S. Ying) S. S. Ying ■

396610　Tricyrtis perfoliata Masam.;抱茎油点草■☆

396611　Tricyrtis pilosa Wall.;黄花油点草(粗柄油点草,粗轴油点草,大黄瓜香,瓜米菜,黑点草,黄瓜菜,黄瓜香,立竹根,柔毛油点草,山黄瓜,疏毛油点草);Yellowflower Toadlily ■

396612　Tricyrtis puberula Nakai et Kitag. = Tricyrtis latifolia Maxim. ■

396613　Tricyrtis ravenii C. I. Peng et C. L. Tiang;雷文油点草■

396614　Tricyrtis stolonifera Matsum.;山油点草(紫花油点草);Purpleflower Taiwan Toadlily ■

396615　Tricyrtis stolonifera Matsum. = Tricyrtis formosana Baker var. stolonifera (Matsum.) Masam. ■

396616　Tricyrtis stolonifera Matsum. = Tricyrtis formosana Baker ■

396617　Tricyrtis suzukii Masam.;侧花油点草(铃木氏油点草);Suzuki Toadlily ■

396618　Tricyrtis viridula Hir. Takah.;绿花油点草■

396619　Tridachne Liebm. ex Lindl. et Paxton = Notylia Lindl. ■☆

396620　Tridactyle Schltr. (1914);三指兰属■☆

396621　Tridactyle acutomarginata (De Wild.) Schltr. = Tridactyle scottellii (Rendle) Schltr. var. stipulata (De Wild.) Geerinck ■☆

396622　Tridactyle anthomaniaca (Rchb. f.) Summerh.;热非三指兰■☆

396623　Tridactyle anthomaniaca (Rchb. f.) Summerh. subsp. nana P. J. Cribb et Stévart;矮小三指兰■☆

396624　Tridactyle armeniaca (Lindl.) Schltr.;亚美尼亚三指兰■☆

396625　Tridactyle aurantiopunctata P. J. Cribb et Stévart;橘斑三指兰■☆

396626　Tridactyle bicaudata (Lindl.) Schltr.;双尾三指兰■☆

396627　Tridactyle bicaudata (Lindl.) Schltr. subsp. rupestris H. P. Linder;岩生双尾三指兰■☆

396628　Tridactyle bolusii (Rolfe) Schltr. = Tridactyle tridentata (Harv.) Schltr. ■☆

396629　Tridactyle brevicalcarata Summerh.;短距三指兰■☆

396630　Tridactyle brevifolia Mansf.;短叶三指兰■☆

396631　Tridactyle citrina P. J. Cribb;柠檬三指兰■☆

396632　Tridactyle crassifolia Summerh.;厚叶三指兰■☆

396633　Tridactyle eggelingii Summerh.;埃格林三指兰■☆

396634　Tridactyle exellii P. J. Cribb et Stévart;埃克塞尔三指兰■☆

396635　Tridactyle filifolia (Schltr.) Schltr.;线叶三指兰■☆

396636　Tridactyle fimbriata (Rendle) Schltr. = Tridactyle bicaudata (Lindl.) Schltr. ■☆

396637　Tridactyle fimbriatipetala (De Wild.) Schltr.;流苏瓣三指兰■☆

396638　Tridactyle flabellata P. J. Cribb;扇状三指兰■☆

396639　Tridactyle fragrans G. Will. = Tridactyle tricuspis (Bolus) Schltr. ■☆

396640　Tridactyle frommiana (Kraenzl.) Schltr. = Tridactyle tricuspis (Bolus) Schltr. ■☆

396641　Tridactyle furcistipes Summerh.;叉三指兰■☆

396642　Tridactyle gentilii (De Wild.) Schltr.;让蒂三指兰■☆

396643　Tridactyle goetzeana (Kraenzl.) Schltr. = Tridactyle tridentata (Harv.) Schltr. ■☆

396644　Tridactyle inaequilonga (De Wild.) Schltr. = Tridactyle scottellii (Rendle) Schltr. var. stipulata (De Wild.) Geerinck ■☆

396645　Tridactyle inflata Summerh.;膨胀三指兰■☆

396646　Tridactyle kindtiana (De Wild.) Schltr. = Tridactyle tridactylites (Rolfe) Schltr. ■☆

396647　Tridactyle lagosensis (Rolfe) Schltr.;拉各斯三指兰■☆

396648　Tridactyle latifolia Summerh.;宽叶三指兰■☆

396649　Tridactyle laurentii (De Wild.) Schltr.;洛朗三指兰■☆

396650　Tridactyle laurentii (De Wild.) Schltr. var. kabareensis Geerinck;卡巴雷三指兰■☆

396651　Tridactyle ledermanniana (Kraenzl.) Schltr. = Tridactyle tridactylites (Rolfe) Schltr. ■☆

396652　Tridactyle lepidota (Rolfe) Schltr. = Tridactyle anthomaniaca (Rchb. f.) Summerh. ■☆

396653　Tridactyle linearifolia (De Wild.) Schltr. = Tridactyle filifolia (Schltr.) Schltr. ■☆

396654　Tridactyle lisowskii (Szlach.) Szlach. et Olszewski;利索三指兰■☆

396655　Tridactyle minuta P. J. Cribb;微小三指兰■☆

396656　Tridactyle muriculata (Rendle) Schltr.;粗糙三指兰■☆

396657　Tridactyle nalaensis (De Wild.) Schltr.;纳拉三指兰■☆

396658　Tridactyle nigrescens Summerh.;变黑三指兰■☆

396659　Tridactyle nyassana Schltr. = Tridactyle tricuspis (Bolus) Schltr. ■☆

396660　Tridactyle oblongifolia Summerh. = Tridactyle anthomaniaca (Rchb. f.) Summerh. ■☆

396661　Tridactyle pentalobata P. J. Cribb et Stévart;五裂三指兰■☆

396662　Tridactyle phaeocephala Summerh.;褐头三指兰■☆

396663　Tridactyle polyschista Mansf. = Tridactyle bicaudata (Lindl.) Schltr. ■☆

396664　Tridactyle pulchella Schltr. = Tridactyle bicaudata (Lindl.) Schltr. ■☆

396665　Tridactyle rhodesiana（Rendle）Schltr. = Tridactyle tricuspis（Bolus）Schltr. ■☆

396666　Tridactyle sarcodantha Mansf.；肉花三指兰■☆

396667　Tridactyle scottellii（Rendle）Schltr.；斯科三指兰■☆

396668　Tridactyle scottellii（Rendle）Schltr. subsp. lisowskii Szlach. = Tridactyle lisowskii（Szlach.）Szlach. et Olszewski ■☆

396669　Tridactyle scottellii（Rendle）Schltr. var. stipulata（De Wild.）Geerinck；托叶三指兰■☆

396670　Tridactyle stevartiana Geerinck；斯特三指兰■☆

396671　Tridactyle stipulata（De Wild.）Schltr. = Tridactyle scottellii（Rendle）Schltr. var. stipulata（De Wild.）Geerinck ■☆

396672　Tridactyle tanneri P. J. Cribb；坦纳三指兰■☆

396673　Tridactyle teretifolia Schltr. = Tridactyle tridentata（Harv.）Schltr. ■☆

396674　Tridactyle thomensis P. J. Cribb et Stévart；爱岛三指兰■☆

396675　Tridactyle trachyrhiza（Schltr.）Schltr. = Tridactyle anthomaniaca（Rchb. f.）Summerh. ■☆

396676　Tridactyle translucens Summerh.；半透明三指兰■☆

396677　Tridactyle tricuspis（Bolus）Schltr.；三尖三指兰■☆

396678　Tridactyle tridactylites（Rolfe）Schltr.；三指兰■☆

396679　Tridactyle tridentata（Harv.）Schltr.；三齿三指兰■☆

396680　Tridactyle tridentata（Harv.）Schltr. var. subulifolia Summerh. = Tridactyle filifolia（Schltr.）Schltr. ■☆

396681　Tridactyle truncatiloba Summerh.；截形三指兰■☆

396682　Tridactyle unguiculata Mansf.；爪状三指兰■☆

396683　Tridactyle vanderlaaniana Geerinck；范德三指兰■☆

396684　Tridactyle verrucosa P. J. Cribb；多疣三指兰■☆

396685　Tridactyle virginea P. J. Cribb et la Croix；纯白三指兰■☆

396686　Tridactyle virgula（Kraenzl.）Schltr.；条纹三指兰■☆

396687　Tridactyle wakefieldii（Rolfe）Summerh. = Solenangis wakefieldii（Rolfe）P. J. Cribb et J. Stewart ■☆

396688　Tridactyle whitfieldii（Rendle）Schltr. = Tridactyle armeniaca（Lindl.）Schltr. ■☆

396689　Tridactylina（DC.）Sch. Bip.（1844）；三指菊属■☆

396690　Tridactylina Sch. Bip. = Tridactylina（DC.）Sch. Bip. ■☆

396691　Tridactylina kirilowii（Turcz. ex DC.）Sch. Bip.；三指菊■☆

396692　Tridactylites Haw. = Saxifraga L. ■

396693　Tridalia Noronha = Abroma Jacq. ●

396694　Tridaps Comm. ex Endl. = Artocarpus J. R. Forst. et G. Forst.（保留属名）●

396695　Tridax L.（1753）；羽芒菊属（长柄菊属，顶天草属）；Tridax ■●

396696　Tridax gaillardioides Hook. et Arn. = Layia gaillardioides（Hook. et Arn.）DC. ■☆

396697　Tridax procumbens L.；羽芒菊（长柄菊）；Coatbuttons, Procumbent Tridax ■

396698　Tridax trilobata Hemsl.；三裂羽芒菊；Trilobe Toadlily ■

396699　Triddenum Raf. = Triadenum Raf. ●

396700　Tridelta Luer = Cryptophoranthus Barb. Rodr. ■☆

396701　Tridelta Luer（2006）；三角窗兰属■☆

396702　Tridens Roem. et Schult.（1817）；短种脐草属■☆

396703　Tridens capensis Nees = Leptochloa fusca（L.）Kunth ■

396704　Tridens chapmanii（Small）Chase = Tridens flavus（L.）Hitchc. ■☆

396705　Tridens elongatus（Buckley）Nash = Tridens muticus（Torr.）Nash ■☆

396706　Tridens flavus（L.）Hitchc.；黄短种脐草；Purpletop, Purple-top, Tall Redtop ■☆

396707　Tridens flavus（L.）Hitchc. f. cupreus（Jacq.）Fosberg = Tridens flavus（L.）Hitchc. ■☆

396708　Tridens muticus（Torr.）Nash；纤细短种脐草；Slim Tridens ■☆

396709　Tridens oklahomensis（Feath.）Feath. ex Chase；俄地短种脐草■☆

396710　Tridens strictus（Nutt.）Nash；长穗短种脐草；Longspike Tridens ■☆

396711　Tridentea Haw.（1812）；三齿萝藦属■☆

396712　Tridentea Haw. = Stapelia L.（保留属名）■

396713　Tridentea aperta（Masson）L. C. Leach = Tromotriche aperta（Masson）Bruyns ■☆

396714　Tridentea baylissii（L. C. Leach）L. C. Leach = Tromotriche baylissii（L. C. Leach）Bruyns ■☆

396715　Tridentea baylissii（L. C. Leach）L. C. Leach var. ciliata L. C. Leach = Tromotriche baylissii（L. C. Leach）Bruyns ■☆

396716　Tridentea choanantha（Lavranos et H. Hall）L. C. Leach = Tromotriche choanantha（Lavranos et H. Hall）Bruyns ■☆

396717　Tridentea dwequensis（C. A. Lückh.）L. C. Leach；杜夸三齿萝藦■☆

396718　Tridentea gemmiflora（Masson）Haw.；芽花三齿萝藦■☆

396719　Tridentea herrei（Nel）L. C. Leach = Tromotriche herrei（Nel）Bruyns ■☆

396720　Tridentea jucunda（N. E. Br.）L. C. Leach；愉悦三齿萝藦■☆

396721　Tridentea jucunda（N. E. Br.）L. C. Leach var. cincta（Marloth）L. C. Leach = Tridentea jucunda（N. E. Br.）L. C. Leach ■☆

396722　Tridentea jucunda（N. E. Br.）L. C. Leach var. dinter（A. Berger）L. C. Leach = Tridentea jucunda（N. E. Br.）L. C. Leach ■☆

396723　Tridentea longii（C. A. Lückh.）L. C. Leach = Orbea longii（C. A. Lückh.）Bruyns ■☆

396724　Tridentea longipes（C. A. Lückh.）L. C. Leach = Tromotriche pedunculata（Masson）Bruyns subsp. longipes（C. A. Lückh.）Bruyns ■☆

396725　Tridentea marientalensis（Nel）L. C. Leach；马林塔尔三齿萝藦■☆

396726　Tridentea marientalensis（Nel）L. C. Leach subsp. albipilosa（Giess）L. C. Leach；白毛三齿萝藦■☆

396727　Tridentea moschata Haw. = Tridentea gemmiflora（Masson）Haw. ■☆

396728　Tridentea pachyrrhiza（Dinter）L. C. Leach；粗根三齿萝藦■☆

396729　Tridentea parvipuncta（N. E. Br.）L. C. Leach；小斑三齿萝藦■☆

396730　Tridentea parvipuncta（N. E. Br.）L. C. Leach subsp. truncata（C. A. Lückh.）Bruyns；平截小斑三齿萝藦■☆

396731　Tridentea parvipuncta（N. E. Br.）L. C. Leach var. truncata（C. A. Lückh.）L. C. Leach = Tridentea parvipuncta（N. E. Br.）L. C. Leach subsp. truncata（C. A. Lückh.）Bruyns ■☆

396732　Tridentea peculiaris（C. A. Lückh.）L. C. Leach；特殊三齿萝藦■☆

396733　Tridentea pedunculata（Masson）L. C. Leach = Tromotriche pedunculata（Masson）Bruyns ■☆

396734　Tridentea pusilla Frandsen = Tridentea parvipuncta（N. E. Br.）L. C. Leach subsp. truncata（C. A. Lückh.）Bruyns ■☆

396735　Tridentea ruschiana（Dinter）L. C. Leach = Tromotriche ruschiana（Dinter）Bruyns ■☆

396736　Tridentea simsii Haw. = Stapelia hirsuta L. var. vetula（Masson）Bruyns ■☆

396737　Tridentea stygia Haw. = Tridentea gemmiflora（Masson）Haw. ■☆

396738　Tridentea umdausensis（Nel）L. C. Leach = Tromotriche umdausensis（Nel）Bruyns ■☆

396739 Tridentea virescens (N. E. Br.) L. C. Leach;浅绿三齿萝藦■☆

396740 Tridermia Raf. = Grewia L. ●

396741 Tridesmis Lour. = Croton L. ●

396742 Tridesmis Spach = Cratoxylum Blume ●

396743 Tridesmis pruniflora Kurz = Cratoxylum formosum (Jack) Benth. et Hook. f. ex Dyer subsp. pruniflorum (Kurz) Gogelein ●

396744 Tridesmis tomentosa Lour. = Croton crassifolium Geiseler ●

396745 Tridesmostemon Engl. (1905);三链蕊属●☆

396746 Tridesmostemon Engl. et Pellegr. = Tridesmostemon Engl. ●☆

396747 Tridesmostemon claessenii De Wild.;克莱森斯三链蕊●☆

396748 Tridesmostemon congoense (Pierre ex A. Chev.) Aubrév. et Pellegr.;刚果三链蕊●☆

396749 Tridesmostemon mortehanii De Wild.;非洲三链蕊●☆

396750 Tridesmostemon omphalocarpoides Engl.;三链蕊●☆

396751 Tridesmus Steud. = Croton L. ●

396752 Tridesmus Steud. = Tridesmis Lour. ●

396753 Tridia Korth. = Hypericum L. ■●

396754 Tridianisia Baill. = Cassinopsis Sond. ●☆

396755 Tridimeris Baill. (1869);三双木属●☆

396756 Tridimeris baillonii G. E. Schatz ex Maas, E. A. Mennega et Westra;三双木●●☆

396757 Tridophyllum Neck. = Potentilla L. ■●

396758 Tridophyllum Neck. ex Greene = Potentilla L. ■●

396759 Tridynamia Gagnep. (1950);三翅藤属●

396760 Tridynamia Gagnep. = Porana Burm. f. ●■☆

396761 Tridynamia megalantha (Merr.) Staples;大花三翅藤(大花飞蛾藤,大锥序飞蛾藤,美飞蛾藤,锥序飞蛾藤);Beautiful Porana, Bigflower Porana, Bigthyrsiferous Porana ●

396762 Tridynamia sinensis (Hemsl.) Staples = Porana sinensis Hemsl. ●

396763 Tridynamia sinensis (Hemsl.) Staples = Poranopsis sinensis (Hemsl.) Staples ●

396764 Tridynamia sinensis (Hemsl.) Staples var. delavayi (Gagnep. et Courchet) Staples = Porana sinensis Hemsl. var. delavayi (Gagnep. et Courchet) Rehder ●

396765 Tridynamia sinensis (Hemsl.) Staples var. delavayi (Gagnep. et Courchet) Staples;密叶飞蛾藤(近无毛三翅藤);Denseleaved Porana, Denselraf Porana ●

396766 Tridynia Raf. = Lysimachia L. ●■

396767 Tridynia Raf. ex Steud. = Lysimachia L. ●■

396768 Tridyra Steud. = Tridynia Raf. ●■

396769 Trieenea Hilliard(1989);波籽玄参属■●☆

396770 Trieenea elsiae Hilliard;埃尔西亚波籽玄参■☆

396771 Trieenea frigida Hilliard;耐寒波籽玄参■●☆

396772 Trieenea glutinosa (Schltr.) Hilliard;黏性波籽玄参■●☆

396773 Trieenea lanciloba Hilliard;波籽玄参■●☆

396774 Trieenea lasiocephala Hilliard;毛头波籽玄参■☆

396775 Trieenea laxiflora Hilliard;疏花波籽玄参■☆

396776 Trieenea longipedicellata Hilliard;长梗波籽玄参■☆

396777 Trieenea schlechteri (Hiern) Hilliard;施莱波籽玄参■●☆

396778 Trieenea taylorii Hilliard;泰勒波籽玄参■●☆

396779 Triendilix Raf. = Glycine Willd. (保留属名)■

396780 Trientalis L. (1753);七瓣莲属(七叶莲属);Chickweed-wintergreen, Star Flower, Starflower, Star-flower ■

396781 Trientalis Rupp. ex L. = Trientalis L. ■

396782 Trientalis americana Pursh;北美七瓣莲■☆

396783 Trientalis americana Pursh = Trientalis borealis Raf. ■●☆

396784 Trientalis arctica Fisch. ex Hook. = Trientalis europaea L. var. arctica (Fisch. ex Hook.) Ledeb. ■☆

396785 Trientalis borealis Raf.;美洲七瓣莲;American Starflower, Chickweed Wintergreen, Starflower, Star-flower, Trientalis ■☆

396786 Trientalis europaea L.;七瓣莲;Chickweed Wintergreen, Europe Star Flower, Europe Starflower, European Starflower ■

396787 Trientalis europaea L. f. ramosa Iljinski = Trientalis europaea L. ■

396788 Trientalis europaea L. subsp. arctica (Fisch. ex Hook.) Hultén = Trientalis europaea L. var. arctica (Fisch. ex Hook.) Ledeb. ■☆

396789 Trientalis europaea L. var. arctica (Fisch. ex Hook.) Ledeb.;北极七瓣莲■☆

396790 Trientalis europaea L. var. latifolia (Hook.) Torr.;宽叶七瓣莲;Indian Potato, Star-flower, Western Starflower ■☆

396791 Trientalis latifolia Hook. = Trientalis europaea L. var. latifolia (Hook.) Torr. ■☆

396792 Triexastima Raf. = Heteranthera Ruiz et Pav. (保留属名)■☆

396793 Trifax Noronha = Reissantia N. Hallé ●

396794 Trifidacanthus Merr. (1917);三叉刺属;Threefork ●

396795 Trifidacanthus unifoliolatus Merr.;三叉刺;Single Leaflet Threefork, Singleleaf Threefork, Singleleaflet Threefork, Single-leaved Threefork ●

396796 Trifillium Medik. = Medicago L. (保留属名)●■

396797 Trifillium Medik. = Triphyllum Medik. ●■

396798 Triflorensia S. T. Reynolds = Diplospora DC. ●

396799 Triflorensia S. T. Reynolds(2005);澳狗骨柴属●☆

396800 Trifoliaceae Bercht. et J. Presl = Fabaceae Lindl. (保留科名)●■

396801 Trifoliaceae Bercht. et J. Presl = Leguminosae Juss. (保留科名)●■

396802 Trifoliada Rojas = Acosta Adans. ●■

396803 Trifoliada Rojas = Centaurea L. (保留属名)●■

396804 Trifoliastrum Moench = Trigonella L. ■

396805 Trifolium (Celak.) Gibelli et Belli = Micrantheum Desf. ●☆

396806 Trifolium L. (1753);车轴草属(三叶草属,三叶豆属,菽草属);Clover, Clover Trefoil, Hop-clover, Trefoil, Trifolium, Yellow Clover ■

396807 Trifolium abyssinicum D. Heller = Trifolium calocephalum Fresen. ■☆

396808 Trifolium acaule Steud. ex A. Rich.;无茎车轴草■☆

396809 Trifolium acaule Steud. ex A. Rich. var. emarginatum Chiov. ex Fiori = Trifolium acaule Steud. ex A. Rich. ■☆

396810 Trifolium acutiflorum Murb.;尖花车轴草■☆

396811 Trifolium africanum Ser.;非洲车轴草;Cape Clover, Red African Clover, Wild Pink Clover ■☆

396812 Trifolium africanum Ser. var. glabellum (E. Mey.) Harv. = Trifolium africanum Ser. ■☆

396813 Trifolium africanum Ser. var. lydenburgense J. B. Gillett;莱登堡车轴草■☆

396814 Trifolium agrarium L. = Trifolium aureum Pollich ■

396815 Trifolium agrarium L. = Trifolium campestre Schreb. ■

396816 Trifolium agrarium L. = Trifolium strepens Crantz ■

396817 Trifolium alexandrinum L.;亚历山大车轴草(埃及车轴草);Beerseem Clover, Berseem, Egypt Trefoil, Egyptian Clover, Egyptian Trefoil ■

396818 Trifolium alpestre L.;阿尔卑斯车轴草(阿尔卑斯三叶草);Alpine Clover, Purple-globe Clover ■☆

396819 Trifolium alpinum L.;高山车轴草(高山三叶草);Alpine Clover ■☆

396820 Trifolium amabile Kunth;秀丽车轴草;Aztec Clover ■☆

396821 Trifolium ambiguum M. Bieb.;可疑车轴草(可疑三叶草);Kura Clover ■☆

396822   Trifolium angulatum Waldst. et Kit. ;窄车轴草■☆

396823   Trifolium angustifolium L. ;窄叶车轴草(窄叶三叶草);Fine-leaved Clover,Narrow Clover,Narrowleaf Crimson Clover,Narrowleaf Trefoil ■☆

396824   Trifolium angustifolium L. var. acrogymnum Maire = Trifolium angustifolium L. ■☆

396825   Trifolium angustifolium L. var. acrolophum Maire = Trifolium angustifolium L. ■☆

396826   Trifolium ankaratrense Bosser;安卡拉特拉车轴草■☆

396827   Trifolium argutum Banks et Sol. ;亮车轴草■☆

396828   Trifolium arvense L. ;野车轴草(兔足三叶草,野三叶草);Dogs-and-cats,Harefoot,Hare's Foot Clover,Hare's Foot Trefoil,Hare's-foot,Hare's-foot Clover,Hare's-foot Trefoil,Old Field Clover,Rabbitfoot Clover,Rabbit-foot Clover,Rabbit's-foot Clover,Rough Trefoil,Stone Clover ■☆

396829   Trifolium arvense L. f. albiflorum Sylven;白花野车轴草(白花野三叶草)■☆

396830   Trifolium arvense L. var. australe Ten. = Trifolium arvense L. ■☆

396831   Trifolium arvense L. var. ballii Murb. = Trifolium arvense L. ■☆

396832   Trifolium arvense L. var. cyrenaicum Pamp. = Trifolium arvense L. ■☆

396833   Trifolium arvense L. var. lagopina Rouy = Trifolium arvense L. ■☆

396834   Trifolium arvense L. var. longisetum (Boiss. et Balansa) Maire = Trifolium arvense L. ■☆

396835   Trifolium arvense L. var. preslianum (Boiss.) Ball = Trifolium arvense L. ■☆

396836   Trifolium atlanticum Ball = Trifolium gemellum Willd. subsp. atlanticum (Ball) Dobignard ■☆

396837   Trifolium aureum Pollich = Trifolium strepens Crantz ■

396838   Trifolium baccarinii Chiov. ;巴卡林车轴草■☆

396839   Trifolium basileianum Chiov. = Trifolium burchellianum Ser. subsp. johnstonii (Oliv.) Cufod. ex J. B. Gillett ■☆

396840   Trifolium bellianum Chiov. = Trifolium steudneri Schweinf. ☆

396841   Trifolium bicorne Forssk. = Trifolium resupinatum L. ■☆

396842   Trifolium bilineatum Fresen. ;双线车轴草■☆

396843   Trifolium bocconei Savi;双头车轴草;Twin-headed Clover ■☆

396844   Trifolium bonanni C. Presl;双花车轴草;Twin-flowered Clover ■☆

396845   Trifolium bordsilovskyi Grossh. ;鲍氏车轴草■☆

396846   Trifolium bracteatum Willd. = Trifolium pratense L. ■

396847   Trifolium brunellii Chiov. ex Fiori = Trifolium semipilosum Fresen. var. brunellii Thulin ■☆

396848   Trifolium bullatum Boiss. = Trifolium tomentosum L. ■☆

396849   Trifolium burchellianum Ser. ;伯切尔车轴草■☆

396850   Trifolium burchellianum Ser. subsp. johnstonii (Oliv.) Cufod. ex J. B. Gillett;约翰斯顿车轴草;Uganda Clover ■☆

396851   Trifolium burchellianum Ser. var. oblongum J. B. Gillett;矩圆车轴草■☆

396852   Trifolium caeruleum L. = Trigonella coerulea (L.) Ser. ■

396853   Trifolium calocephalum Fresen. ;美头车轴草■☆

396854   Trifolium campestre Schreb. ;草原车轴草;Bishop Clove,Craid,Field Clover,Hop Clover,Hop Trefoil,Humble,Large Hop Clover,Large Hop-clover,Lawn Trefoil,Love-and-tangle,Low Hop Clover,Low Hop-clover,Mustard-tips,Pinnate Hop Clover,Tom Thumb,Yellow Clover ■

396855   Trifolium campestre Schreb. var. minus (Koch) Gremli = Trifolium campestre Schreb. ■

396856   Trifolium campestre Schreb. var. pandoi Font Quer = Trifolium campestre Schreb. ■

396857   Trifolium campestre Schreb. var. subsessile (Boiss.) Hayek = Trifolium campestre Schreb. ■

396858   Trifolium canescens Willd. ;灰白车轴草(灰白三叶草)■☆

396859   Trifolium carolinianum Michx. ;卡罗来纳车轴草(卡罗来纳三叶草);Carolina Clover ■☆

396860   Trifolium caucasicum Tausch;高加索车轴草■☆

396861   Trifolium cernuum Brot. ;俯垂车轴草;Nodding Clover ■☆

396862   Trifolium cheranganiense J. B. Gillett;切兰加尼车轴草■☆

396863   Trifolium cherleri L. ;米努草状车轴草■☆

396864   Trifolium chilaloense Thulin;奇拉罗车轴草■☆

396865   Trifolium ciliatum Nutt. ;缘毛三叶草;Tree Clover ■☆

396866   Trifolium clausonis Pomel = Trifolium physodes M. Bieb. ■☆

396867   Trifolium clusii Godr. et Gren. ;克鲁斯车轴草■☆

396868   Trifolium clusii Godr. et Gren. var. kahiricum Zohary = Trifolium clusii Godr. et Gren. ■☆

396869   Trifolium clypeatum L. ;圆盾车轴草(圆盾三叶草);Shield Clover ■☆

396870   Trifolium congestum Guss. ;密集车轴草■☆

396871   Trifolium cryptopodium Steud. ex A. Rich. ;隐足车轴草(隐足三叶草)■☆

396872   Trifolium cryptopodium Steud. ex A. Rich. var. kilimandscharicum (Taub.) J. B. Gillett = Trifolium cryptopodium Steud. ex A. Rich. ■☆

396873   Trifolium cyrenaicum (Pamp.) Maire et Weiller = Trifolium arvense L. ■☆

396874   Trifolium cytisoides Pall. = Lespedeza juncea (L. f.) Pers. ●

396875   Trifolium dahuricum Laxm. = Lespedeza dahurica (Laxm.) Schindl. ●

396876   Trifolium dalmaticum Vis. ;达儿马提亚车轴草;Dalmatian Clover ■☆

396877   Trifolium dasyurum C. Presl;毛尾车轴草(毛尾三叶草)■☆

396878   Trifolium dauricum Laxm. = Lespedeza daurica (Laxm.) Schindl. ●

396879   Trifolium decorum Chiov. ;装饰车轴草(装饰三叶草)■☆

396880   Trifolium dentatum Waldst. et Kit. = Melilotus dentatus (Waldst. et Kit.) Pers. ■

396881   Trifolium diffusum Ehrh. ;分散车轴草(分散三叶草)■☆

396882   Trifolium dubium Sibth. ;钝叶车轴草(黄菽草);Common Yellow Trefoil,Least Hop Clover,Lesser Hop Trefoil,Lesser Trefoil,Lesser Yellow Trefoil,Little Hop Clover,Obtuseleaf Trefoil,Shamrock,Small Hop Clover,Small Yellow Trefoil,Strawberry Clover,Suckling Clover,Yellow Clover ■

396883   Trifolium dubium Sibth. var. atlanticum Maire = Trifolium dubium Sibth. ■

396884   Trifolium durandoi Pomel = Trifolium physodes M. Bieb. ■☆

396885   Trifolium echinatum M. Bieb. ;刺车轴草;Hedgehog Clover,Prickly Clover ■☆

396886   Trifolium elegans Savi = Trifolium hybridum L. subsp. elegans (Savi) Asch. et Graebn. ■

396887   Trifolium elegans Savi = Trifolium hybridum L. ■

396888   Trifolium elegans Savi ex Gibelli et Belli = Trifolium hybridum L. ■

396889   Trifolium elgonense J. B. Gillett;埃尔贡车轴草■☆

396890   Trifolium elizabethae Grossh. ;伊丽莎白车轴草■☆

396891   Trifolium eriocephalum Nutt. ;西方毛头车轴草;Woolly-headed Clover ■☆

396892   Trifolium erubescens Fenzl;变红车轴草■☆

396893   Trifolium eximium Stephan ex DC. ;大花车轴草;Largeflower Trefoil ■

396894   Trifolium eximium Stephan ex Ser. = Trifolium eximium Stephan ex DC. ■

396895 Trifolium expansum Waldst. et Kit. ;扩大三叶草■☆

396896 Trifolium filiforme L. ;细茎三叶草;Thread Clover ■☆

396897 Trifolium filiforme L. = Trifolium dubium Sibth.

396898 Trifolium filiforme L. subsp. dubium (Sibth.) Maire = Trifolium dubium Sibth. ■

396899 Trifolium filiforme L. subsp. micranthum (Viv.) Bonnier et Layens = Trifolium micranthum Viv. ■☆

396900 Trifolium fimbriatum Lindl. ;流苏车轴草;Beach Clover ■☆

396901 Trifolium fistulosum Gilib. = Trifolium hybridum L. ■

396902 Trifolium flexuosum Jacq. = Trifolium medium L. ■

396903 Trifolium fontanum Bobrov;泉边车轴草■☆

396904 Trifolium formosum d'Urv. = Trifolium dasyurum C. Presl ■☆

396905 Trifolium fragiferum L. ;草莓车轴草(草莓三叶草);Strawberry Clover,Strawberry Trefoil ■

396906 Trifolium fragiferum L. var. orthodon Zohary;直齿车轴草■☆

396907 Trifolium fruticans L. = Otholobium fruticans (L.) C. H. Stirt. ●☆

396908 Trifolium fucatum Lindl. ;着色车轴草;Puff Clover ■☆

396909 Trifolium gemellum Willd. ;西班牙车轴草;Spanish Clover ■☆

396910 Trifolium gemellum Willd. subsp. atlanticum (Ball) Dobignard;北非车轴草■☆

396911 Trifolium gemellum Willd. var. atlanticum (Ball) Maire = Trifolium gemellum Willd. subsp. atlanticum (Ball) Dobignard ■☆

396912 Trifolium gemellum Willd. var. pedunculatum Maire, Weiller et Wilczek = Trifolium gemellum Willd. ■☆

396913 Trifolium gemellum Willd. var. phyllocephalum Zohary = Trifolium gemellum Willd. ■☆

396914 Trifolium gemellum Willd. var. tastetii (Pau) Devesa = Trifolium gemellum Willd. subsp. atlanticum (Ball) Dobignard ■☆

396915 Trifolium gillettianum Jacq. -Fél. ;吉莱特车轴草■☆

396916 Trifolium glabrescens J. B. Gillett = Trifolium semipilosum Fresen. var. brunellii Thulin ■☆

396917 Trifolium glomeratum L. ;圆头车轴草;Cluster Clover,Clustered Clover,Round-headed Trefoil ■☆

396918 Trifolium glomeratum L. var. condensatum Ball = Trifolium glomeratum L. ■☆

396919 Trifolium goetzenii Taub. = Trifolium tembense Fresen. ■☆

396920 Trifolium gordejevii (Kom.) Z. Wei;延边车轴草(宿瓣胡卢巴,延边苜蓿);Gordeiev Medic,Gordejev Clover,Yanbian Trefoil ■☆

396921 Trifolium grandiflorum Ledeb. = Trifolium eximium Stephan ex Ser. ■

396922 Trifolium guianense Aubl. = Stylosanthes guianensis (Aubl.) Sw. ■●

396923 Trifolium hedysaroides Pall. = Lespedeza juncea (L. f.) Pers. ●

396924 Trifolium hirsutum E. Mey. = Trifolium africanum Ser. ■☆

396925 Trifolium hirsutum E. Mey. var. glabellum ? = Trifolium africanum Ser. ■☆

396926 Trifolium hirtum All. ;短硬毛三叶草;Pin-pointed Clover,Rose Clover ■☆

396927 Trifolium hispidum Desf. = Trifolium hirtum All. ■☆

396928 Trifolium humboldtianum A. Braun et Asch. ;高加索三叶草■☆

396929 Trifolium humile Ball;矮小车轴草■☆

396930 Trifolium hybridum L. ;杂种车轴草(爱沙苜蓿,金花草,雅杂三叶草,亚尔西翘摇,杂三叶,杂三叶草);Alsatian Clover,Alsike Clover,Bastard Clover,Hybrid Trefoil,Showy Clover,Swedish Clover ■

396931 Trifolium hybridum L. subsp. elegans (Savi) Asch. et Graebn. = Trifolium hybridum L. ■

396932 Trifolium hybridum L. var. elegans (Savi) Boiss. = Trifolium hybridum L. ■

396933 Trifolium hybridum L. var. pratense Rabenh. = Trifolium hybridum L. ■

396934 Trifolium incarnatum L. ;绛车轴草(地中海三叶草,绛三叶,绛三叶草,猩红苜蓿,紫车轴草);Bloody Triumph,Crimson Clover,Crimson Trefoil, Italian Clover, Napoleon, Red Fingers, Scarlet Clover,Soldiers ■

396935 Trifolium incarnatum L. var. elatius Gibelli et Belli = Trifolium incarnatum L. ■

396936 Trifolium indicum L. = Melilotus indicus (L.) All. ■

396937 Trifolium infamia-ponertii Greuter = Trifolium angustifolium L. ■☆

396938 Trifolium intermedium Guss. = Trifolium angustifolium L. ■☆

396939 Trifolium involucratum Ortega;多花车轴草;Cow Clover, Multiflowered Clover ■☆

396940 Trifolium isodon Murb. ;同齿车轴草■☆

396941 Trifolium isodon Murb. var. maroccanum Humbert et Maire = Trifolium isodon Murb. ■☆

396942 Trifolium isodon Murb. var. miegeanum (Maire) Maire = Trifolium isodon Murb. ■☆

396943 Trifolium isthmocarpum Brot. ;颈果车轴草■☆

396944 Trifolium isthmocarpum Brot. var. jaminianum (Boiss.) Gibelli et Belli = Trifolium isthmocarpum Brot. ■☆

396945 Trifolium jaminianum Boiss. = Trifolium isthmocarpum Brot. ■☆

396946 Trifolium johnstonii Oliv. = Trifolium burchellianum Ser. subsp. johnstonii (Oliv.) Cufod. ex J. B. Gillett ■☆

396947 Trifolium juliani Batt. ;朱利安车轴草■☆

396948 Trifolium kilimandscharicum Taub. = Trifolium cryptopodium Steud. ex A. Rich. ■☆

396949 Trifolium lagopus Willd. = Trifolium sylvaticum Loisel. ■☆

396950 Trifolium lanceolatum (J. B. Gillett) J. B. Gillett;披针形车轴草■☆

396951 Trifolium lappaceum L. ;芒刺三叶草;Bur Clover, Burdock Clover ■☆

396952 Trifolium lappaceum L. var. carteiense (Coincy) Pau = Trifolium lappaceum L. ■☆

396953 Trifolium leucanthum M. Bieb. ;白花三叶草■☆

396954 Trifolium ligusticum Loisel. ;利古里亚车轴草■☆

396955 Trifolium lugardii Bullock;卢格德车轴草■☆

396956 Trifolium lupinaster L. ;野火球(也火秋,野车轴草,野火荻);Bastard Lupine,Lupinaster,Lupine Clover,Wild Clover,Wildfireball Trefoil ■

396957 Trifolium lupinaster L. f. albiflorum (Ser.) P. Y. Fu et Y. A. Chen = Trifolium lupinaster L. var. albiflorum Ser. ■

396958 Trifolium lupinaster L. subvar. obtusifolium Gibelli et Belli = Trifolium pacificum Bobrov ■

396959 Trifolium lupinaster L. var. albiflorum Ser. ;白花野火球;Whiteflower Clover ■

396960 Trifolium marginatum (Baker) Cufod. = Trifolium baccarinii Chiov. ■☆

396961 Trifolium maritimum Huds. ;海岸三叶草■☆

396962 Trifolium maritimum Huds. = Trifolium squarrosum L. ■☆

396963 Trifolium masaiense J. B. Gillett;吗西车轴草■☆

396964 Trifolium mattirolianum Chiov. ;马蒂奥里车轴草■☆

396965 Trifolium mauginianum Fiori = Trifolium polystachyum Fresen. ■☆

396966 Trifolium mauritanicum Ball = Trifolium isthmocarpum Brot. ■☆

396967 Trifolium medium L. ;中间车轴草;Cow Clover, Cow Grass,

Mammoth Clover, Marl-grass, Meadow Clover, Middle Trefoil, Zigzag Clover, Zig-zag Clover ■

396968　Trifolium melilotus-indica L. = Melilotus indicus（L.）All. ■

396969　Trifolium melilotus-officinalis L. = Melilotus officinalis（L.）Pall. ■

396970　Trifolium melilotus-ornithopodioides L. = Trifolium ornithopodioides L. ■☆

396971　Trifolium melilotus-ornithopodioides L. subsp. ornithopodioides（Sm.）Maire = Trifolium ornithopodioides L. ■☆

396972　Trifolium melilotus-ornithopodioides L. subsp. uniflorum（Munby）Maire = Trifolium ornithopodioides L. ■☆

396973　Trifolium melilotus-ornithopodioides L. var. meliloteum Mall. = Trifolium ornithopodioides L. ■☆

396974　Trifolium messanense L. = Melilotus messanensis（L.）All. ■☆

396975　Trifolium michelianum Savi；米氏车轴草；Bigflower Clover ■☆

396976　Trifolium michelianum Savi subsp. balansae（Boiss）Ponert；巴氏车轴草；Balansa Clover ■☆

396977　Trifolium micranthum Viv.；小花车轴草；Least Yellow Trefoil, Slender Trefoil, Slender Yellow Trefoil ■☆

396978　Trifolium micropetalum E. Mey. = Trifolium stipulaceum Thunb. ■☆

396979　Trifolium minus Sm. = Trifolium dubium Sibth. ■

396980　Trifolium molinerii Hornem.；长序车轴草 ■☆

396981　Trifolium montanum L.；山地三叶草；Mountain Clover ■☆

396982　Trifolium montanum L. = Trifolium spadiceum L. ■☆

396983　Trifolium multinerve A. Rich.；多脉车轴草 ■☆

396984　Trifolium neglectum Fisch. ,C. A. Mey. et Avé-Lall.；疏忽车轴草 ■☆

396985　Trifolium nigrescens Viv.；小黑三叶草；Small White Clover ■☆

396986　Trifolium obscurum Savi；隐匿车轴草 ■☆

396987　Trifolium obscurum Savi subsp. aequidentatum（Pérez Lara）Vicioso = Trifolium isodon Murb. ■☆

396988　Trifolium occidentale Coombe；西方车轴草（西方三叶草）；Western Clover ■☆

396989　Trifolium ochoroleucum L.；淡黄车轴草（淡黄三叶草）；Sulphur Clover, Yellow Clover ■☆

396990　Trifolium ochroleucon Huds.；淡黄白车轴草 ■☆

396991　Trifolium ochroleucon Huds. var. abbreviatum Jahand. et al. = Trifolium ochroleucon Huds. ■☆

396992　Trifolium ochroleucon Huds. var. pallidulum（Jord.）Asch. et Graebn. = Trifolium ochroleucon Huds. ■☆

396993　Trifolium officinalis L. = Melilotus officinalis（L.）Pall. ■

396994　Trifolium ornithopodioides L.；鸟爪车轴草；Bird's-thot Clover ■☆

396995　Trifolium oxaloides Bunge ex Nyman = Trifolium subterraneum L. ■☆

396996　Trifolium pacificum Bobrov；太平洋车轴草（阔叶野火球）■☆

396997　Trifolium pacificum Bobrov = Trifolium lupinaster L. ■☆

396998　Trifolium pallidum Waldst. et Kit.；苍白车轴草（苍白三叶草）■☆

396999　Trifolium pannonicum Jacq.；匈牙利三叶草；Hungarian Clover ■☆

397000　Trifolium panormitanum C. Presl；巴勒摩三叶草 ■☆

397001　Trifolium panormitanum C. Presl = Trifolium squarrosum L. ■☆

397002　Trifolium parviflorum Ehrh.；小花三叶草；Teasel Clover ■☆

397003　Trifolium parviflorum Ehrh. = Trifolium retusum L. ■☆

397004　Trifolium patens Schreb.；铺展车轴草 ■☆

397005　Trifolium petitianum A. Rich.；佩蒂蒂车轴草 ■☆

397006　Trifolium phleoides Pourr. ex Willd.；梯牧草车轴草 ■☆

397007　Trifolium phleoides Willd. = Trifolium phleoides Pourr. ex Willd. ■☆

397008　Trifolium phleoides Willd. subsp. audigieri Foucaud = Trifolium phleoides Willd. ■☆

397009　Trifolium phleoides Willd. subsp. gemellum（Willd.）Gibelli et Belli = Trifolium gemellum Willd. ■☆

397010　Trifolium phleoides Willd. subsp. willkommii（Chabert）Munoz Rodr.；维尔考姆车轴草 ■☆

397011　Trifolium phleoides Willd. var. minae（Ces.）Fiori = Trifolium gemellum Willd. ■☆

397012　Trifolium physodes Eichw. = Trifolium physodes M. Bieb. ■☆

397013　Trifolium physodes M. Bieb.；囊状车轴草 ■☆

397014　Trifolium physodes M. Bieb. var. durandoi（Pomel）Gibelli et Belli = Trifolium physodes M. Bieb. ■☆

397015　Trifolium polymorphum Poir.；多型车轴草；Peanut Clover ■☆

397016　Trifolium polyphyllum C. A. Mey.；多叶车轴草 ■☆

397017　Trifolium polystachyum Fresen.；多穗车轴草 ■☆

397018　Trifolium polystachyum Fresen. var. psoraleoides Welw. ex Hiern；补骨脂车轴草 ■☆

397019　Trifolium pratense L.；红车轴草（红荷兰翘摇，红花车轴草，红花车子，红花苜蓿，红芥兰翘摇，红三叶，红三叶草，红菽草，金花菜，三叶草）；Bee Bread, Broad Clover, Broad-grass, Clatter Malloch, Claver, Clover Rose, Cock's Head, Common Red Clover, Cow Clover, Cow Grass, Cow-grass, Dutch Clover, Fairy Pops, Honey-flower, Honeystalk, Honeysuck, Honeysuckle, Honeysuckle Trefoil, King's Crown, Knap, Lady's Posy, Lamb's Sucklings, Little Honeysuckle, Love-and-tangle, Marl-grass, Meadow Clover, Meadow Honeysuckle, Meadow Trefoil, Pea-vine Clover, Pinkies, Plyvens, Purple Clover, Red Clover, Red Cushion, Red Honeysuckle, Red Trefoil, Shamrock, Sleeping Maggie, Smere, Soukie Clover, Souks Soukie Clover, Sowkie, Sowkie Soó, Suck Bottle, Suck-bottle, Sucklers, Suckles, Sucklings, Sucks, Sugar Plum, Sugar-bosses, Sugar-busses, Sumark, Three-leaved Grass, Water Boats ■

397020　Trifolium pratense L. f. albiflorum Alef.；白花红车轴草 ■☆

397021　Trifolium pratense L. f. albiflorum Puskal = Trifolium pratense L. ■

397022　Trifolium pratense L. f. leucochraceum Asch. et Prantl = Trifolium pratense L. ■

397023　Trifolium pratense L. f. semipurpureum（Strobl）Asch. et Graebn. = Trifolium pratense L. ■

397024　Trifolium pratense L. subsp. baeticum（Boiss.）Vicioso；伯蒂卡车轴草 ■☆

397025　Trifolium pratense L. subsp. sativum（Schreb.）Schübl. et M. Martens = Trifolium pratense L. ■

397026　Trifolium pratense L. var. hirsutum Boiss. = Trifolium pratense L. ■

397027　Trifolium pratense L. var. mesatlanticum H. Lindb. = Trifolium pratense L. ■

397028　Trifolium pratense L. var. sativum（Mill.）Schreb. = Trifolium pratense L. ■

397029　Trifolium pratense L. var. sativum（Mill.）Schreb. f. flavicans（Vis.）Hayek = Trifolium pratense L. ■

397030　Trifolium pratense L. var. sativum Schreb. = Trifolium pratense L. ■

397031　Trifolium pratense L. var. sericeum Nègre = Trifolium pratense L. ■

397032　Trifolium pratense L. var. spontaneum Willk. = Trifolium pratense L. ■

397033　Trifolium pratense L. var. villosum Wahlenb. = Trifolium pratense L. ■

397034　Trifolium preussii Baker f. = Trifolium rueppellianum Fresen. ■☆

397035　Trifolium procumbens L. = Trifolium campestre Schreb. ■

397036　Trifolium pseudocryptopodium Chiov. ex Fiori = Trifolium usambarense Taub. ■☆

397037　Trifolium purpureum Loisel.；紫车轴草；Purple Clover ■☆

397038　Trifolium purseglovei J. B. Gillett；帕斯格洛夫车轴草 ■☆

397039　Trifolium quartinianum A. Rich. ;夸尔廷车轴草■☆

397040　Trifolium raddeanum Trautv. ;拉德车轴草■☆

397041　Trifolium reflexum L. ;布法罗车轴草;Buffalo Clover ■☆

397042　Trifolium reflexum L. var. glabrum Lojac. = Trifolium reflexum L. ■☆

397043　Trifolium repens L. ;白车轴草（白车苜蓿，白花三叶草，白三叶，白三球草，荷兰翘摇，兰翅摇，螃蟹花，三消草，菽草，紫云英）; Baa-lambs, Black Leaved Clover, Bobby Rose, Broad-grass, Claver, Cluber, Creeping Trefoil, Curldoddy, Dutch, Dutch Clover, Dutch White Clover, Honeystalk, Honeysuckle, Kentish Clover, Ladino Clover, Lamb's Sucklings, Mull, Mutton Rose, Persian Clover, Pussy Foot, Quillet, Shamrock, Sheep's Gowan, Smara, Smoora, Sucklers, Sucklings, Trefoy, White Clover, White Curldoddy, White Dutch Clover, White Honeysuckle, White Trefoil, Wild White Clover ■

397044　Trifolium repens L. ‘Purpurascens’;紫叶白车轴草（淡紫叶白车轴草）;Purple-leaf Clover ■☆

397045　Trifolium repens L. f. giganteum Lagr. -Foss. ;巨白车轴草■☆

397046　Trifolium repens L. f. giganteum Lagr. -Foss. = Trifolium repens L. ■

397047　Trifolium repens L. f. roseum Peterm. ;粉白车轴草■☆

397048　Trifolium repens L. var. giganteum Lagr. -Foss. = Trifolium repens L. ■

397049　Trifolium repens Thunb. = Trifolium burchellianum Ser. ■☆

397050　Trifolium resupinatum L. ;反曲三叶草;Persian Clover, Reversed Clover, Shabdar, Shaftal Clover, Stawbery Clover ■☆

397051　Trifolium resupinatum L. var. clusii (Gren. et Godr.) Murb. = Trifolium resupinatum L. ■☆

397052　Trifolium resupinatum L. var. majus Boiss. ;大反曲三叶草■☆

397053　Trifolium resupinatum L. var. microcephalum Zohary;小头反曲三叶草■☆

397054　Trifolium resupinatum L. var. minus Boiss. = Trifolium resupinatum L. ■☆

397055　Trifolium resupinatum L. var. robustum Rouy = Trifolium resupinatum L. ■☆

397056　Trifolium retusum L. ;微凹车轴草;Teasel Clover ■☆

397057　Trifolium rubens Aubry = Trifolium incarnatum L. ■

397058　Trifolium rubens Desc. = Trifolium alpestre L. ■☆

397059　Trifolium rubens L. ;西方红车轴草;Foxtail Clover ■☆

397060　Trifolium rueppellianum Fresen. ;鲁佩尔车轴草■☆

397061　Trifolium rueppellianum Fresen. f. minor A. Chev. = Trifolium baccarinii Chiov. ■☆

397062　Trifolium rueppellianum Fresen. var. lanceolatum J. B. Gillett = Trifolium lanceolatum (J. B. Gillett) J. B. Gillett ■☆

397063　Trifolium rueppellianum Fresen. var. minimiflorum J. B. Gillett;微花鲁佩尔车轴草■☆

397064　Trifolium rueppellianum Fresen. var. preussii (Baker f.) J. B. Gillett = Trifolium rueppellianum Fresen. ■☆

397065　Trifolium sativum Crome ex Boenn. ;栽培车轴草■☆

397066　Trifolium sativum Crome ex Boenn. subsp. praecox Bobrov;早熟车轴草;Early Red Clover ■☆

397067　Trifolium sativum Crome ex Boenn. subsp. serotium Bobrov;晚熟车轴草■☆

397068　Trifolium scabrum L. ;粗三叶草;Rough Clover, Rough Trefoil ■☆

397069　Trifolium scabrum L. var. glabrum Pamp. = Trifolium scabrum L. ■☆

397070　Trifolium scabrum L. var. hirsuticaulis H. Lindb. = Trifolium scabrum L. ■☆

397071　Trifolium schimperi A. Rich. ;欣珀车轴草■☆

397072　Trifolium scutatum Boiss. ;长圆盾形车轴草■☆

397073　Trifolium sebastianii Savi;塞氏车轴草■☆

397074　Trifolium semipilosum Fresen. ;半毛车轴草■☆

397075　Trifolium semipilosum Fresen. var. brunellii Thulin;布鲁内利车轴草■☆

397076　Trifolium semipilosum Fresen. var. glabrescens J. B. Gillett;渐光半毛车轴草■☆

397077　Trifolium semipilosum Fresen. var. intermedium Thulin;间型半毛车轴草■☆

397078　Trifolium semipilosum Fresen. var. kilimanjaricum Baker f. = Trifolium semipilosum Fresen. ■☆

397079　Trifolium seravschanicum Ovcz. ;塞拉夫车轴草☆

397080　Trifolium simense Fresen. ;锡米车轴草■☆

397081　Trifolium simense Fresen. f. albiflora Piovano = Trifolium simense Fresen. ■☆

397082　Trifolium somalense Taub. ex Harms;索马里车轴草■☆

397083　Trifolium spadiceum L. ;栗色三叶草■☆

397084　Trifolium speciosum Willd. ;美丽三叶草;Beautiful Clover ■☆

397085　Trifolium sphaerocephalum Desf. = Trifolium cherleri L. ■☆

397086　Trifolium spumosum L. ;海绵三叶草;Mediterranean Clover ■☆

397087　Trifolium squarrosum L. ;叉开三叶草;Sea Clover ■☆

397088　Trifolium squarrosum L. subsp. panorminatum (C. Presl) Pott. -Alap. = Trifolium squarrosum L. ■☆

397089　Trifolium stellatum L. ;星状车轴草;Sea Clover, Starry Clover ■☆

397090　Trifolium steudneri Schweinf. ;斯托德车轴草■☆

397091　Trifolium stipitatum Boiss. et Balansa ex Boiss. ;具柄车轴草■☆

397092　Trifolium stipulaceum Thunb. ;托叶状车轴草■☆

397093　Trifolium stoloniferum Muhl. ex Eaton;匍匐车轴草;Running Buffalo Clover ■☆

397094　Trifolium stolzii Harms;斯托尔兹车轴草■☆

397095　Trifolium stratum L. ;沟纹三叶草;Knotted Clover, Star Clover ■☆

397096　Trifolium strepens Crantz;黄车轴草（大车轴草，田生三叶草）; Golden Clover, Hop Clover, Hop-clover, Large Hop Trefoil, Large Hoptrefoil, Large Trefoil, Palmate Hop Clover, Yellow Clover, Yellow Hop Clover, Yellow Trefoil ■

397097　Trifolium striatum L. ;直立车轴草;Clover, Knotted Clover, Soft Clover, Soft Knotted Clover, Upright Clover ■☆

397098　Trifolium striatum L. var. atlanticum Pau et Font Quer = Trifolium striatum L. ■☆

397099　Trifolium striatum L. var. basalticum Nègre = Trifolium striatum L. ■☆

397100　Trifolium striatum L. var. lasiocalyx Batt. = Trifolium striatum L. ■☆

397101　Trifolium striatum L. var. nanum Rouy = Trifolium striatum L. ■☆

397102　Trifolium striatum L. var. spinescens Lange = Trifolium striatum L. ■☆

397103　Trifoliumstriatum L. var. tastetii Pau = Trifolium gemellum Willd. subsp. atlanticum (Ball) Dobignard ■☆

397104　Trifolium suaveolens Willd. = Trifolium resupinatum L. ■☆

397105　Trifolium subrotundum A. Rich. = Trifolium rueppellianum Fresen. ■☆

397106　Trifolium subterraneum L. ;地下车轴草（地下三叶草）;Burrowing Clover, Subclover, Subterranean Clover ■☆

397107　Trifolium subterraneum L. subsp. brachycalycinum Katzn. et Morley = Trifolium subterraneum L. subsp. oxaloides Nyman ■☆

397108　Trifolium subterraneum L. subsp. oxaloides Nyman;酢浆车轴草■☆

397109　Trifolium subterraneum L. var. brachycladum Gibelli et Belli;短枝车轴草■☆

397110　Trifolium subterraneum L. var. flagelliforme Guss. = Trifolium subterraneum L. ■☆

397111　Trifolium subterraneum L. var. longipes J. Gay = Trifolium subterraneum L. ■☆

397112 Trifolium suffocatum L. ;地中海车轴草■☆

397113 Trifolium sylvaticum Loisel. ;林地车轴草■☆

397114 Trifolium tastetii（Pau）Font Quer = Trifolium gemellum Willd. subsp. atlanticum（Ball）Dobignard ■☆

397115 Trifolium tembense Fresen. ;藤贝车轴草■☆

397116 Trifolium tomentosum L. ;毛车轴草；Woolly Trefoil ■☆

397117 Trifolium tomentosum L. var. chthonocephalum Bornm. = Trifolium tomentosum L. ■☆

397118 Trifolium tomentosum L. var. minus Rouy = Trifolium tomentosum L. ■☆

397119 Trifolium tomentosum L. var. orientale Bornm. = Trifolium tomentosum L. ■☆

397120 Trifolium trichocephalum M. Bieb. ;毛头车轴草■☆

397121 Trifolium tridentatum Lindl. ;三齿车轴草；Tomcat Clover, Woolly Clover ■☆

397122 Trifolium tumens Stev. ex M. Bieb. ;图马车轴草■☆

397123 Trifolium tunetanum Murb. ;图内特车轴草■☆

397124 Trifolium ukingense Harms ;热非车轴草■☆

397125 Trifolium umbellulatum A. Rich. = Trifolium tembense Fresen. ■☆

397126 Trifolium uniflorum L. ;单花车轴草■☆

397127 Trifolium uniflorum L. var. varians Vierh. = Trifolium uniflorum L. ■☆

397128 Trifolium unifolium Forssk = Cullen corylifolium（L.）Medik. ■☆

397129 Trifolium usambarense Taub. ;乌桑巴拉车轴草■☆

397130 Trifolium vesciculosum Savi;泡囊三叶草；Arrowleaf Clover ■☆

397131 Trifolium viciosoanum Pau = Trifolium pannonicum Jacq. ■☆

397132 Trifolium wentzelianum Harms;文策尔车轴草■☆

397133 Trifolium wentzelianum Harms var. stolzii（Harms）J. B. Gillett = Trifolium stolzii Harms ■☆

397134 Trifurcaria Endl. = Trifurcia Herb. ■☆

397135 Trifurcia Herb. = Herbertia Sweet ■☆

397136 Trifurcia caerulea Herb. = Herbertia caerulea（Herb.）Herb. ■☆

397137 Trifurcia caerulea Herb. = Herbertia lahue（Molina）Goldblatt ■☆

397138 Trifurcia lahue（Molina）Goldblatt = Herbertia lahue（Molina）Goldblatt ■☆

397139 Trifurcia lahue（Molina）Goldblatt subsp. caerulea（Herb.）Goldblatt = Herbertia lahue（Molina）Goldblatt ■☆

397140 Trigastrotheca F. Muell. = Mollugo L. ■

397141 Trigella Salisb. = Cyanella L. ■☆

397142 Triglochin L.（1753）;水麦冬属；Arrowgrass, Arrow-grass, Pod Grass, Podgrass, Pod-grass, Podograss, Squaw-root ■

397143 Triglochin ani K. Koch = Triglochin maritima L. ■☆

397144 Triglochin asiatica（Kitag.）Á. Löve et D. Löve;亚洲水麦冬■☆

397145 Triglochin barrelieri Loisel. ;巴尔水麦冬■☆

397146 Triglochin barrelieri Loisel. var. maurum Pau = Triglochin barrelieri Loisel. ■☆

397147 Triglochin bulbosa L. ;鳞茎水麦冬■☆

397148 Triglochin bulbosa L. subsp. laxiflora（Guss.）Rouy = Triglochin laxiflora Guss. ■☆

397149 Triglochin bulbosa L. var. maura（Pau）Maire = Triglochin barrelieri Loisel. ■☆

397150 Triglochin compacta Adamson = Triglochin bulbosa L. ■☆

397151 Triglochin concinna Burtt Davy = Triglochin maritima L. ■

397152 Triglochin concinna Burtt Davy var. debilis（M. E. Jones）J. T. Howell = Triglochin maritima L. ■

397153 Triglochin debilis（M. E. Jones）Á. Löve et D. Löve = Triglochin maritima L. ■

397154 Triglochin elata Nutt. = Triglochin maritima L. ■

397155 Triglochin elongata Buchenau = Triglochin bulbosa L. ■☆

397156 Triglochin gaspensis Lieth et D. Löve;加斯佩水麦冬■☆

397157 Triglochin himalensis Royle = Triglochin palustris L. ■

397158 Triglochin laxiflora Guss. ;疏花海韭菜■☆

397159 Triglochin maritima L. ;海韭菜（那冷门,三尖草,圆果水麦冬）;Arrow Grass, Arrow-grass, Common Bog Arrow-grass, Sea Arrow-grass, Sea Leek, Seaside Arrow-grass, Shore Pod Grass, Shore Pod-grass, Shore Podograss ■

397160 Triglochin maritima L. subsp. asiatica Kitag. = Triglochin asiatica（Kitag.）Á. Löve et D. Löve ■☆

397161 Triglochin maritima L. var. asiatica（Kitag.）Ohwi = Triglochin asiatica（Kitag.）Á. Löve et D. Löve ■☆

397162 Triglochin maritima L. var. elata（Nutt.）A. Gray = Triglochin maritima L. ■

397163 Triglochin milnei Horn;米尔恩水麦冬■☆

397164 Triglochin natalensis Gand. = Triglochin striata Ruiz et Pav. ■☆

397165 Triglochin palustris L. ;水麦冬（长果水麦冬,牛毛墩,小麦冬）;Arrow Pod Grass, Arrow Pod-grass, Arrow-grass, Marsh Arrow Grass, Marsh Arrow-grass, Podgrass, Slender Bog Arrow-grass ■

397166 Triglochin procera R. Br. ;澳洲水麦冬;Water-ribbons ■☆

397167 Triglochin roegneri K. Koch = Triglochin maritima L. ■

397168 Triglochin striata Ruiz et Pav. ;条纹水麦冬;Ridged Podograss ■☆

397169 Triglochin tenuifolia Adamson = Triglochin bulbosa L. ■☆

397170 Triglochin transcaucasica Bordz. = Triglochin maritima L. ■

397171 Triglochinaceae Bercht. et J. Presl = Juncaginaceae Rich.（保留科名）■

397172 Triglochinaceae Dumort. = Juncaginaceae Rich.（保留科名）■

397173 Triglossum Fisch. = Arundinaria Michx. ●

397174 Triglossum Roem. et Schult. = Arundinaria Michx. ●

397175 Trigonachras Radlk.（1879）;三籽木属●☆

397176 Trigonachras brachycarpa Radlk. ;三籽木●☆

397177 Trigonachras membranacea Radlk. ;膜三籽木●☆

397178 Trigonachras rigida Radlk. ;硬三籽木●☆

397179 Trigonanthe（Schltr.）Brieger = Dryadella Luer ■☆

397180 Trigonanthera André = Peperomia Ruiz et Pav. ■

397181 Trigonanthera André（1870）;三角药胡椒属■☆

397182 Trigonanthera resediflora（Linden et André）André = Peperomia resedaeflora Lindl. et André ■☆

397183 Trigonanthus Korth. ex Hook. f. = Ceratostylis Blume ■

397184 Trigonea Parl. = Allium L. ■

397185 Trigonea Parl. = Nectaroscordum Lindl. ■☆

397186 Trigonella L.（1753）;胡卢巴属;Fenugreek, Trigonella ■

397187 Trigonella adscendens Afan. et Gontsch. ;上举胡卢巴■☆

397188 Trigonella americana Nutt. = Lotus unifoliolata（Hook.）Benth. ☆

397189 Trigonella arabica Delile;阿拉伯胡卢巴■☆

397190 Trigonella arborea（L.）Vassilcz. = Medicago arborea L. ●

397191 Trigonella archiducis-nicolai（Sirj.）Vassilcz. = Medicago archiducis-nicolai Sirj. ■

397192 Trigonella arcuata C. A. Mey. ;弯果胡卢巴;Arcuate Trigonella, Curvefruit Fenugreek ■

397193 Trigonella armata Thunb. = Melolobium microphyllum（L. f.）Eckl. et Zeyh. ■☆

397194 Trigonella aschersoniana Urb. = Medicago hypogaea E. Small ■☆

397195 Trigonella aurantiaca Boiss. ;橙黄胡卢巴■☆

397196 Trigonella azurea C. A. Mey. ;蔚蓝胡卢巴■☆

397197 Trigonella bactrina Vassilcz. = Trigonella cachemiriana Cambess. ■

397198　Trigonella badachschanica Afan. ;巴达胡卢巴■☆

397199　Trigonella balansae Boiss. et Reut. ;巴兰萨胡卢巴■☆

397200　Trigonella besseriana Ser. ;白氏胡卢巴■☆

397201　Trigonella besseriana Ser. = Trigonella procumbens Rchb. ■☆

397202　Trigonella biflora Griseb. ;双花胡卢巴■☆

397203　Trigonella brachycarpa (Fisch.) Moris;短果胡卢巴■☆

397204　Trigonella cachemiriana Cambess. ;克什米尔胡卢巴;Kashmir Fenugreek ■

397205　Trigonella caerulea (Desr. ex Lam.) Ser. = Trigonella coerulea (L.) Ser. ■

397206　Trigonella cancellata Desf. ;网脉胡卢巴(网纹胡卢巴,纤细胡卢巴);Cancellate Trigonella, Netvein Fenugreek ■

397207　Trigonella cancellata Desf. var. arcuata (C. A. Mey.) Sirj. = Trigonella arcuata C. A. Mey. ■

397208　Trigonella coerulea (L.) Ser. ;蓝花胡卢巴(蓝胡卢巴,灵凌香,零陵香,卢豆,天蓝胡卢巴,盐湖胡卢巴);Blue Fenugreek, Blue Flowered Fenugreek, Blue Melilot, Blue Trigonella, Blue-white Trigoneila, Saltlake Fenugreek ■

397209　Trigonella coerulescens (M. Bieb.) Halácsy;浅蓝胡卢巴■☆

397210　Trigonella communis ?;普通胡卢巴■☆

397211　Trigonella corniculata L. ;小角胡卢巴;Cultivated Fenugreek, Sickle-fruited Fenugreek ■☆

397212　Trigonella cretacea (M. Bieb.) Grossh. ;白垩胡卢巴■☆

397213　Trigonella cretica (L.) Boiss. = Melilotus creticus (L.) Desr. ☆

397214　Trigonella cretica Boiss. ;克里特胡卢巴■☆

397215　Trigonella cylindracea Desv. ;柱形胡卢巴☆

397216　Trigonella emodii Benth. ;喜马拉雅胡卢巴(齿黄胡卢巴);Himalayan Trigonella, Himalayas Fenugreek ■

397217　Trigonella emodii Benth. = Medicago ruthenica (L.) Trautv. ■

397218　Trigonella emodii Benth. subsp. fimbriata (Royle ex Benth.) Ohashi = Trigonella fimbriata Royle ex Benth. ■

397219　Trigonella emodii Benth. var. fimbriata (Royle ex Benth.) ? = Trigonella fimbriata Royle ex Benth. ■

397220　Trigonella emodii Benth. var. firnbriata (Royle ex Benth.) Sirj. = Trigonella fimbriata Royle ex Benth. ■

397221　Trigonella emodii Benth. var. himalaica Sirj. = Trigonella emodii Benth. ■

397222　Trigonella emodii Benth. var. medicaginoides Sirj. = Trigonella cachemiriana Cambess. ■

397223　Trigonella esculenta Willd. ;食用胡卢巴■☆

397224　Trigonella falcata Balf. f. ;镰形胡卢巴■☆

397225　Trigonella fasciculata Bertol. ;簇生胡卢巴■☆

397226　Trigonella fimbriata Royle ex Benth. ;重齿胡卢巴;Fimbriate Fenugreek, Fimbriate Trigonella ■

397227　Trigonella fischeriana Ser. ;菲舍尔胡卢巴■☆

397228　Trigonella fischeriana Ser. = Medicago fischeriana (Ser.) Trautv. ■☆

397229　Trigonella foenum-graecum L. ;胡卢巴(胡巴,胡芦巴,葫芦巴,季豆,苦草,苦豆,苦朵菜,卢巴子,芦巴,芦肥,肾曹都护,肾胃都尉,香草,香草子,香豆,香豆子,香苜蓿,小木夏,芸香,芸香草);Classical Fenugreek, Fenugreek, Fenugreek Trigonella, Sicklefruit Fenugreek ■

397230　Trigonella geminiflora Bunge = Trigonella monantha C. A. Mey. ■

397231　Trigonella glabra Thunb. = Trigonella hamosa L. ■

397232　Trigonella gladiata M. Bieb. = Trigonella gladiata Steven ■☆

397233　Trigonella gladiata Steven;剑状胡卢巴■☆

397234　Trigonella gordejevii (Kom.) Grossh. = Trifolium gordejevii (Kom.) Z. Wei ■

397235　Trigonella gracilis Benth. ;纤细胡芦巴■

397236　Trigonella grandiflora Bunge;大花胡卢巴■☆

397237　Trigonella grossheimii Vassilcz. = Trigonella cachemiriana Cambess. ■

397238　Trigonella hamosa L. ;弯果胡卢巴■

397239　Trigonella hamosa L. = Trigonella glabra Thunb. ■

397240　Trigonella hirsuta Thunb. = Melolobium aethiopicum (L.) Druce ■☆

397241　Trigonella incisa Royle ex Benth. var. geminiflora (Bunge) Boiss. = Trigonella monantha C. A. Mey. ■

397242　Trigonella indica L. = Rothia indica (L.) Druce ■

397243　Trigonella karkarensis Semen. ex Vassilcz. = Medicago platycarpos (L.) Trautv. ■

397244　Trigonella korshinskii Grossh. = Medicago ruthenica (L.) Trautv. ■

397245　Trigonella laciniata L. ;撕裂胡卢巴■☆

397246　Trigonella laxiflora Aitch. et Baker;疏花胡卢巴■☆

397247　Trigonella lipskyi Sirj. ;利普斯基胡卢巴■☆

397248　Trigonella macroanguina Andr. = Trigonella anguina Delile ■☆

397249　Trigonella marginata Baker = Trifolium baccarinii Chiov. ☆

397250　Trigonella maritima Poir. ;滨海胡芦巴■☆

397251　Trigonella maritima Poir. var. leiosperma Maire = Trigonella maritima Poir. ■☆

397252　Trigonella maritima Poir. var. trachysperma Maire = Trigonella maritima Poir. ■☆

397253　Trigonella media Delile;中间胡卢巴■☆

397254　Trigonella monantha C. A. Mey. ;单花胡卢巴(双花胡卢巴);Monoflower Fenugreek, Oneflower Trigonella, Twoflowers Trigonella ■

397255　Trigonella monantha C. A. Mey. subsp. noeana (Boiss.) Hub. - Mor. ;诺氏胡卢巴■☆

397256　Trigonella noeana Boiss. = Trigonella monantha C. A. Mey. subsp. noeana (Boiss.) Hub. -Mor. ■☆

397257　Trigonella noeana Boiss. = Trigonella monantha C. A. Mey. ■

397258　Trigonella occulta Ser. ;隐蔽胡卢巴■☆

397259　Trigonella ornithopodioides (L.) DC. ;鸟脚胡卢巴;Bird's Foot Trigonella, Bird's Foot Clover, Bird's Foot Fenugreek, Bird's-foot Clover, Bird's-foot Fenugreek, Bird's-foot Trefoil, Fenugreek, Finnigig, Greek Hayseed, Trigonel ■☆

397260　Trigonella ornithopodioides (L.) DC. = Trifolium ornithopodioides L. ■☆

397261　Trigonella orthoceras Kar. et Kir. ;直果胡卢巴;Standingfruit Fenugreek ■

397262　Trigonella ovalis Boiss. ;椭圆胡卢巴■☆

397263　Trigonella pamirica Boriss. ;帕米尔胡卢巴;Pamir Trigonella ■☆

397264　Trigonella platycarpos L. = Medicago platycarpos (L.) Trautv. ■

397265　Trigonella polyceratia L. ;多角胡卢巴■☆

397266　Trigonella polyceratia L. var. atlantica Ball = Trigonella polyceratia L. ■☆

397267　Trigonella polyceratia L. var. glabrata (Sirj.) Maire = Trigonella polyceratia L. ■☆

397268　Trigonella polyceratia L. var. melanosperma Maire = Trigonella polyceratia L. ■☆

397269　Trigonella polyceratia L. var. pinnatifida (Cav.) Willk. = Trigonella polyceratia L. ■☆

397270　Trigonella polyceratia L. var. sericea Emb. = Trigonella polyceratia L. ■☆

397271　Trigonella popovii Korovin;波氏胡卢巴■☆

397272　Trigonella procumbens Rchb. ;伏卧胡卢巴;Trailing Fenugreek ■☆

397273　Trigonella pubescens Edgew. ex Baker = Medicago edgeworthii Sirj. ex Hand. -Mazz. ■

397274　Trigonella radiata (L.) Boiss. ;辐射胡卢巴■☆

397275　Trigonella ramosa ?;分枝胡卢巴;Branched Fenugreek ■☆

397276　Trigonella ruthenica L. = Medicago ruthenica（L.）Trautv. ■

397277　Trigonella schischkinii Vassilcz. = Medicago ruthenica（L.）Trautv. ■

397278　Trigonella spicata Sibth. et Sm.;穗胡卢巴■☆

397279　Trigonella stellata Forssk.;星状胡卢巴■☆

397280　Trigonella strangulata Boiss.;收缩胡卢巴■☆

397281　Trigonella striata L. f.;纤弱胡卢巴■☆

397282　Trigonella tenuis Fisch. = Trigonella cancellata Desf. ■

397283　Trigonella tenuis Fisch. ex M. Bieb. = Trigonella cancellata Desf. ■

397284　Trigonella tibetana（Alef.）Vassilcz. = Trigonella foenum-graecum L. ■

397285　Trigonella tibetica Vassilcz. = Medicago edgeworthii Sirj. ex Hand.-Mazz. ■

397286　Trigonella tibetica Vassilcz. = Trigonella pubescens Edgew. ex Baker ■

397287　Trigonella tomentosa Thunb. = Cullen tomentosum（Thunb.）J. W. Grimes ■☆

397288　Trigonella torulosa Griseb.;结节胡卢巴■☆

397289　Trigonella turkmena Popov;土库曼胡卢巴■☆

397290　Trigonella uncata Boiss. et Noë;钩状胡卢巴■☆

397291　Trigonella uniflora Munby = Trifolium ornithopodioides L. ■☆

397292　Trigonella verae Sirj.;薇拉胡卢巴■☆

397293　Trigonia Aubl.（1775）;三角果属（三棱果属,三数木属）●☆

397294　Trigonia bicolor Suess. et Overk.;二色三角果●☆

397295　Trigonia boliviana Warm.;玻利维亚三角果●☆

397296　Trigonia brasiliensis Jacob-Makoy;巴西三角果●☆

397297　Trigonia crassiflora A. C. Sm.;粗花三角果●☆

397298　Trigonia floribunda Oerst.;多花三角果●☆

397299　Trigonia laevis Aubl.;平滑三角果●☆

397300　Trigonia macrocarpa Benth.;大果三角果●☆

397301　Trigonia macrostachya Klotzsch;大穗三角果●☆

397302　Trigonia membranacea A. C. Sm.;膜质三角果●☆

397303　Trigonia micrantha Mart.;小花三角果●☆

397304　Trigonia microcarpa Sagot ex Warm.;小果三角果●☆

397305　Trigonia nivea Cambess.;雪白三角果●☆

397306　Trigonia reticulata E. Lleras;网脉三角果●☆

397307　Trigonia rigida Oerst.;硬三角果●☆

397308　Trigonia rotundifolia E. Lleras;圆叶三角果●☆

397309　Trigonia sericea Kunth;绢毛三角果●☆

397310　Trigoniaceae A. Juss.（1849）(保留科名);三角果科（三棱果科,三数木科）●☆

397311　Trigoniaceae Endl. = Trigoniaceae A. Juss.（保留科名）●☆

397312　Trigoniastrum Miq.（1861）(保留属名);小三角果属●☆

397313　Trigoniastrum hypoleucum Miq.;小三角果●☆

397314　Trigonidium Lindl.（1837）;美洲三角兰属■☆

397315　Trigonidium acuminatum Bateman;渐尖美洲三角兰■☆

397316　Trigonidium brachyglossum Schltr.;短舌美洲三角兰■☆

397317　Trigonidium macranthum Barb. Rodr.;大花美洲三角兰■☆

397318　Trigonidium monophyllum Griseb.;单叶美洲三角兰■☆

397319　Trigonidium peruvianum Schltr.;秘鲁美洲三角兰■☆

397320　Trigonidium tenue Lodd.;细美洲三角兰■☆

397321　Trigoniodendron E. F. Guim. et Miguel（1987）;三角木属●☆

397322　Trigoniodendron spiritusanctense E. F. Guim. et Miguel;三角木●☆

397323　Trigonis Jacq. = Cupania L. ●☆

397324　Trigonobalanus Forman = Formanodendron Nixon et Crepet ●

397325　Trigonobalanus Forman（1962）;三角栎属（三棱栎属）●

397326　Trigonobalanus doichangensis（A. Camus）Forman = Formanodendron

doichangensis（A. Camus）Nixon et Crepet ●◇

397327　Trigonobalanus excelsa Lozano,Hern. Cam. et Henao;南美三棱栎（高大三棱栎）●☆

397328　Trigonobalanus excelsa Lozano,Hern. Cam. et Henao = Colombobalanus excelsa（Lozano,Hern. Cam. et Henao）Nixon et Crepet ●☆

397329　Trigonobalanus verticillata Forman;轮叶三棱栎●☆

397330　Trigonocapnos Schltr.（1899）;三棱烟堇属■☆

397331　Trigonocapnos curvipes Schltr. = Trigonocapnos lichtensteinii（Cham. et Schltdl.）Lidén ■☆

397332　Trigonocapnos lichtensteinii（Cham. et Schltdl.）Lidén;三棱烟堇■☆

397333　Trigonocarpaea Steud. = Kokoona Thwaites ●☆

397334　Trigonocarpaea Steud. = Trigonocarpus Wall. ●☆

397335　Trigonocarpus Bert. ex Steud. = Chorizanthe R. Br. ex Benth. ■●☆

397336　Trigonocarpus Vell. = Cupania L. ●☆

397337　Trigonocarpus Wall. = Kokoona Thwaites ●☆

397338　Trigonocaryum Trautv.（1875）;角果五加属●☆

397339　Trigonocaryum involucratum（Steven）Kusn.;角果五加●☆

397340　Trigonochilum Königer et Schildh.（1994）;三棱兰属■☆

397341　Trigonochilum Königer et Schildh. = Cyrtochilum Kunth ■☆

397342　Trigonochilum Königer et Schildh. = Oncidium Sw.（保留属名）■☆

397343　Trigonochlamys Hook. f. = Santiria Blume ●

397344　Trigonopleura Hook. f.（1887）;棱脉大戟属●☆

397345　Trigonopleura malayana Hook. f.;棱脉大戟●☆

397346　Trigonopterum Steetz ex Andersson = Lipochaeta DC. ■☆

397347　Trigonopterum Steetz ex Andersson = Peperomia Ruiz et Pav. ■

397348　Trigonopterum Steetz ex Andersson(1854);角翅菊属●☆

397349　Trigonopterum laricifolium（Hook. f.）W. L. Wagner et H. Rob.;角翅菊■☆

397350　Trigonopyren Bremek. = Psychotria L.（保留属名）●

397351　Trigonosciadium Boiss.（1844）;三角伞芹属■☆

397352　Trigonosciadium intermedium Freyn et Sint. ex Sint.;间型三角伞芹■☆

397353　Trigonosciadium komarovii（Mandcnova）Tamamsch.;科氏三角伞芹■☆

397354　Trigonosciadium tuberosum Boiss.;三角伞芹■☆

397355　Trigonospermum Less.（1832）;三棱子菊属（角子菊属）■☆

397356　Trigonospermum floribundum Greenm.;多花三棱子菊●☆

397357　Trigonospermum tomentosum B. L. Rob. et Greenm.;毛三棱子菊■☆

397358　Trigonostemon Blume(1826)（'Trigostemon'）(保留属名);三宝木属;Trigonostemon,Triratna Tree ●

397359　Trigonostemon chinensis Merr.;三宝木;Chinese Trigonostemon,Triratna Tree ●

397360　Trigonostemon chinensis Merr. f. fengii（Merr.）Y. T. Chang;冯钦三宝木;Feng's Trigonostemon,Feng's Triratna Tree ●

397361　Trigonostemon filipes Y. T. Chang et X. L. Mo;丝梗三宝木;Silkstalk Triratna Tree,Thin-stalk Trigonostemon,Thread-stalked Trigonostemon ●

397362　Trigonostemon fungii Merr. = Trigonostemon chinensis Merr. f. fengii（Merr.）Y. T. Chang ●

397363　Trigonostemon heterophyllus Merr.;异叶三宝木;Differleaf Triratna Tree,Diverseleaf Trigonostemon,Heterophyllous Trigonostemon ●

397364　Trigonostemon howii Merr. et Chun;长序三宝木;F. C. How Trigonostemon,How Triratna Tree,How Trigonostemon ●

397365　Trigonostemon huangmosu Y. T. Chang;黄木树;Huangmushu Trigonostemon,Huangmushu Triratna Tree,Yellow-wood Trigonostemon ●

397366 Trigonostemon kwangsiensis Hand. -Mazz. ; 广西三宝木; Guangxi Triratna Tree ●

397367 Trigonostemon kwangsiensis Hand. -Mazz. var. viriculis H. S. Kiu; 细序三宝木; Viriculis Guangxi Triratna Tree ●

397368 Trigonostemon leucanthus Airy Shaw; 白花三宝木; White Triratna Tree, White-flower Trigonostemon ●

397369 Trigonostemon leucanthus Airy Shaw var. hainanensis H. S. Kiu; 海南三宝木; Hainan White-flower Trigonostemon ●

397370 Trigonostemon lii Y. T. Chang; 勐仑三宝木; Li Trigonostemon, Menglun Triratna Tree ●

397371 Trigonostemon lutescens Y. T. Chang et J. Y. Liang; 黄花三宝木(红花三宝木); Yellow Triratna Tree, Yellow-flower Trigonostemon ●◇

397372 Trigonostemon serratus Blume; 短柄三宝木; Serrate Triratna Tree ●☆

397373 Trigonostemon thyrsoideus Stapf; 长梗三宝木(花梗三宝木, 普柔树, 普黍树); Bunch-like Trigonostemon, Long-stalk Trigonostemon, Thyrselike Triratna Tree ●

397374 Trigonostemon xyphophylloides (Croizat) L. K. Dai et T. L. Wu; 剑叶三宝木; Lanceleaf Trigonostemon, Swordleaf Triratna Tree, Sword-leaved Trigonostemon ●

397375 Trigonotheca Hochst. = Catha Forssk. ●

397376 Trigonotheca Sch. Bip. = Melanthera Rohr ■●☆

397377 Trigonotheca natalensis Sch. Bip. = Melanthera scandens (Schumach. et Thonn.) Roberty subsp. dregei (DC.) Wild ■☆

397378 Trigonotheca serrata Hochst. = Catha edulis (Vahl) Forssk. ex Endl. ●

397379 Trigonotis Steven = Endogonia (Turcz.) Lindl. ■

397380 Trigonotis Steven(1851); 附地菜属(附地草属); Trigonotis ■

397381 Trigonotis amblyosepala Nakai et Kitag. = Trigonotis peduncularis (Trevis.) Benth. ex S. Moore et Baker var. amblyosepala (Nakai et Kitag.) W. T. Wang ■

397382 Trigonotis barkamensis Ching J. Wang; 金川附地菜(马尔康附地菜); Barkam Trigonotis ■

397383 Trigonotis bodinieri (H. Lév.) H. Lév. = Mitreola pedicellata Benth. ■

397384 Trigonotis bracteata Ching J. Wang; 全苞附地菜; Bracteate Trigonotis ■

397385 Trigonotis brevipes (Maxim.) Maxim. = Trigonotis brevipes (Maxim.) Maxim. ex Hemsl. ■☆

397386 Trigonotis brevipes (Maxim.) Maxim. ex Hemsl. ; 短梗附地菜■☆

397387 Trigonotis brevipes (Maxim.) Maxim. ex Hemsl. var. coronata (Ohwi) Ohwi; 具冠短梗附地菜■☆

397388 Trigonotis brevipes (Maxim.) Maxim. f. coronata (Ohwi) T. Yamaz. = Trigonotis brevipes (Maxim.) Maxim. var. coronata (Ohwi) Ohwi ■☆

397389 Trigonotis brevipes (Maxim.) Maxim. var. coronata (Ohwi) Ohwi = Trigonotis brevipes (Maxim.) Maxim. ex Hemsl. var. coronata (Ohwi) Ohwi ■☆

397390 Trigonotis brevipes Maxim. = Trigonotis cavaleriei (H. Lév.) Hand. -Mazz. ■

397391 Trigonotis cavaleriei (H. Lév.) Hand. -Mazz. ; 西南附地菜; Cavalerie Trigonotis ■

397392 Trigonotis cavaleriei (H. Lév.) Hand. -Mazz. var. angustifolia Ching J. Wang; 窄叶西南附地菜(窄叶附地菜); Narrowleaf Trigonotis ■

397393 Trigonotis chengkouensis W. T. Wang; 城口附地菜; Chengkou Trigonotis ■

397394 Trigonotis chuxiongensis H. Chuang; 楚雄附地菜; Chuxiong

397395 Trigonotis chuxiongensis H. Chuang = Trigonotis heliotropifolia Hand. -Mazz. ■

397396 Trigonotis cinereifolia Ching J. Wang; 灰叶附地菜; Greyleaf Trigonotis ■

397397 Trigonotis clavata Stev. = Trigonotis peduncularis (Trevis.) Benth. ex S. Moore et Baker ■

397398 Trigonotis compressa I. M. Johnst. ; 狭叶附地菜; Narrowleaf Trigonotis ■

397399 Trigonotis contortipes I. M. Johnst. = Trigonotis delicatula Hand. -Mazz. ■

397400 Trigonotis coreana Nakai = Trigonotis radicans (Turcz.) Steven subsp. sericea (Maxim.) Riedl ■

397401 Trigonotis coreana Nakai = Trigonotis radicans (Turcz.) Steven ■

397402 Trigonotis corispermoides Ching J. Wang; 虫实附地菜; Corispermumlike Trigonotis, Tickseedlike Trigonotis ■

397403 Trigonotis corispermoides Ching J. Wang var. sessilis W. T. Wang; 无柄虫实附地菜■

397404 Trigonotis coronata Ohwi = Trigonotis brevipes (Maxim.) Maxim. var. coronata (Ohwi) Ohwi ■☆

397405 Trigonotis cupulifera I. M. Johnst. = Omphalotrigonotis cupulifera (I. M. Johnst.) W. T. Wang ■

397406 Trigonotis delicatula Hand. -Mazz. ; 扭梗附地菜(扭柄附地菜); Delicate Trigonotis, Softstalk Trigonotis ■

397407 Trigonotis elevatovenosa Hayata = Trigonotis formosana Hayata var. elevatovenosa (Hayata) S. D. Shen et J. C. Wang ■

397408 Trigonotis faberi Hand. -Mazz. = Trigonotis cavaleriei (H. Lév.) Hand. -Mazz. ■

397409 Trigonotis floribunda I. M. Johnst. ; 多花附地菜; Manyflower Trigonotis ■

397410 Trigonotis formosana Hayata; 台湾附地菜(台湾附地草); Taiwan Trigonotis ■

397411 Trigonotis formosana Hayata var. elevatovenosa (Hayata) S. D. Shen et J. C. Wang; 凸脉附地菜(台北附地草); Convexvein Trigonotis ■

397412 Trigonotis funingensis H. Chuang; 富宁附地菜; Funing Trigonotis ■

397413 Trigonotis gamocalyx Hand. -Mazz. = Myosotis caespitosa Schultz ■

397414 Trigonotis gamocalyx Hand. -Mazz. = Myosotis laxa Lehm. ■

397415 Trigonotis giraldii Brand; 秦岭附地菜; Girald Trigonotis, Qinling Trigonotis ■

397416 Trigonotis gracilipes I. M. Johnst. ; 细梗附地菜(细柄附地菜); Slenderstalk Trigonotis, Thinstalk Trigonotis ■

397417 Trigonotis guilielmii (A. Gray) A. Gray ex Gürke = Trigonotis radicans (Turcz.) Steven ■

397418 Trigonotis harrysmithii R. R. Mill; 松潘附地菜(短梗附地菜); Smith Trigonotis ■

397419 Trigonotis heliotropifolia Hand. -Mazz. ; 毛花附地菜(毛脉附地菜, 天芥叶, 艳肠草); Hairflower Trigonotis, Hellotropeleaf Trigonotis ■

397420 Trigonotis iinumae (Maxim.) Makino; 饭沼附地菜■☆

397421 Trigonotis jinfoshanica W. T. Wang; 金佛山附地菜■

397422 Trigonotis koreana Nakai; 朝鲜附地菜■

397423 Trigonotis laxa I. M. Johnst. ; 南川附地菜(野甜菜); Nanchuan Trigonotis ■

397424 Trigonotis laxa I. M. Johnst. var. hirsuta W. T. Wang ex Ching J. Wang; 硬毛附地菜; Hirsute Nanchuan Trigonotis ■

397425 Trigonotis laxa I. M. Johnst. var. xichouensis (H. Huang) Ching J. Wang; 西畴附地菜; Xichou Trigonotis ■

397426 Trigonotis leucanta W. T. Wang; 白色附地菜(白花附地菜);

Whiteflower Trigonotis ■

397427 Trigonotis leyeensis W. T. Wang;乐业附地菜（乐叶附地菜）; Leye Trigonotis ■

397428 Trigonotis longipes W. T. Wang;长梗附地菜;Longstalk Trigonotis ■

397429 Trigonotis longiramosa W. T. Wang;长枝附地菜;Longbranch Trigonotis ■

397430 Trigonotis macrophylla Vaniot;大叶附地菜;Largeleaf Trigonotis ■

397431 Trigonotis macrophylla Vaniot var. trichocarpa Hand. -Mazz.;毛果附地菜;Hairfruit Largeleaf Trigonotis ■

397432 Trigonotis macrophylla Vaniot var. verrucosa Johnst.;瘤果附地菜;Errucose Largeleaf Trigonotis ■

397433 Trigonotis mairei（H. Lév.）I. M. Johnst.;迈氏附地菜（长梗附地菜）;Longstalk Trigonotis,Maire Trigonotis ■

397434 Trigonotis microcarpa（A. DC.）Benth.;毛脉附地菜;Hairvein Trigonotis,Littlefruit Trigonotis ■

397435 Trigonotis microcarpa（A. DC.）Benth. ex C. B. Clarke = Trigonotis heliotropifolia Hand. -Mazz. ■

397436 Trigonotis mollis Hemsl.;湖北附地菜;Hubei Trigonotis ■

397437 Trigonotis moupinense（Franch.）I. M. Johnst. = Sinojohnstonia moupinensis（Franch.）W. T. Wang ■

397438 Trigonotis moupinense（Franch.）Johnst. = Sinojohnstonia moupinensis（Franch.）W. T. Wang ex Z. Y. Zhang ■

397439 Trigonotis muliense W. T. Wang;木里附地菜;Muli Trigonotis ■

397440 Trigonotis muriculata I. M. Johnst. = Trigonotis mairei（H. Lév.）I. M. Johnst. ■

397441 Trigonotis myosotidea（Maxim.）Maxim.;水甸附地菜（勿忘草状附地菜）;Forgetmenot-like Trigonotis ■

397442 Trigonotis nakaii H. Hara = Trigonotis radicans（Turcz.）Steven subsp. sericea（Maxim.）Riedl ■

397443 Trigonotis nakaii H. Hara = Trigonotis radicans（Turcz.）Steven var. sericea（Maxim.）H. Hara ■

397444 Trigonotis nakaii H. Hara = Trigonotis radicans（Turcz.）Steven ■

397445 Trigonotis nandanensis Ching J. Wang;南丹附地菜;Nandan Trigonotis ■

397446 Trigonotis naokotaizanensis（Sasaki）Masam. et Ohwi ex Masam.;白花附地菜（南湖附地草）;White Trigonotis ■

397447 Trigonotis omeiensis Matsuda;峨眉附地菜;Emei Trigonotis,Omei Mountain Trigonotis ■

397448 Trigonotis orbiculariffolia Ching J. Wang;厚叶附地菜（圆叶附地菜）;Thickleaf Trigonotis ■

397449 Trigonotis peduncularis（Trevir.）Benth. ex Baker et Moore var. microcarpa（DC.）Brand = Trigonotis microcarpa（A. DC.）Benth. ■

397450 Trigonotis peduncularis（Trevis.）Benth. ex Baker et S. Moore var. vestita Hemsl. = Trigonotis vestita（Hemsl.）I. M. Johnst. ■

397451 Trigonotis peduncularis（Trevis.）Benth. ex S. Moore et Baker;附地菜（地胡椒,伏地菜,鸡肠,鸡肠草,野苜蓿）;Pedunculate Trigonotis ■

397452 Trigonotis peduncularis（Trevis.）Benth. ex S. Moore et Baker var. amblyosepala（Nakai et Kitag.）W. T. Wang;钝萼附地菜;Obtusesepal Trigonotis ■

397453 Trigonotis peduncularis（Trevis.）Benth. ex S. Moore et Baker var. macrophylla（Vaniot）H. Lév. = Trigonotis macrophylla Vaniot ■

397454 Trigonotis peduncularis（Trevis.）Benth. ex S. Moore et Baker var. macrantha W. T. Wang;大花附地菜■

397455 Trigonotis peduncularis（Trevis.）Benth. ex S. Moore et Baker var. microcarpa（A. DC.）Brand = Trigonotis microcarpa（A. DC.）Benth. ■

397456 Trigonotis peduncularis（Trevis.）Benth. ex S. Moore et Baker var. vestita Hemsl. = Trigonotis vestita（Hemsl.）I. M. Johnst. ■ ■

397457 Trigonotis pedunculata var. macrophylla（Vaniot）H. Lév. = Trigonotis macrophylla Vaniot ■

397458 Trigonotis petiolaris Maxim.;祁连山附地菜;Petiolate Trigonotis ■

397459 Trigonotis radicans（Turcz.）Steven;北附地菜;Rooted Trigonotis ■

397460 Trigonotis radicans（Turcz.）Steven subsp. sericea（Maxim.）H. Hara = Trigonotis radicans（Turcz.）Steven var. sericea（Maxim.）H. Hara ■

397461 Trigonotis radicans（Turcz.）Steven subsp. sericea（Maxim.）Riedl = Trigonotis radicans（Turcz.）Steven var. sericea（Maxim.）H. Hara ■

397462 Trigonotis radicans（Turcz.）Steven var. sericea（Maxim.）H. Hara;森林附地菜（北附地菜,朝鲜附地菜）;Korea Trigonotis,Korean Trigonotis ■

397463 Trigonotis radicans（Turcz.）Steven var. sericea（Maxim.）H. Hara = Trigonotis radicans（Turcz.）Steven ■

397464 Trigonotis rockii I. M. Johnst.;高山附地菜（高地附地菜）;Alpine Trigonotis,High Mountain Trigonotis ■

397465 Trigonotis rotundata I. M. Johnst.;圆叶附地菜;Rotundate Trigonotis,Roundleaf Trigonotis ■

397466 Trigonotis rotundifolia（Wall.）Benth. = Trigonotis rotundata I. M. Johnst. ■

397467 Trigonotis sericea（Maxim.）I. M. Johnst. = Trigonotis radicans（Turcz.）Steven var. sericea（Maxim.）H. Hara ■

397468 Trigonotis sericea（Maxim.）I. M. Johnst. = Trigonotis radicans（Turcz.）Steven ■

397469 Trigonotis sericea（Maxim.）Ohwi = Trigonotis radicans（Turcz.）Steven subsp. sericea（Maxim.）H. Hara ■

397470 Trigonotis sericea Ohwi = Trigonotis radicans（Turcz.）Steven ■

397471 Trigonotis smithii W. T. Wang = Trigonotis harrysmithii R. R. Mill ■

397472 Trigonotis tenera I. M. Johnst.;蒙山附地菜;Weak Trigonotis ■

397473 Trigonotis tibetica（C. B. Clarke）I. M. Johnst.;西藏附地菜（藏紫草）;Tibet Trigonotis,Xizang Trigonotis ■

397474 Trigonotis vestita（Hemsl.）I. M. Johnst.;灰毛附地菜;Greyhair Trigonotis ■

397475 Trigonotis xichougensis H. Chuang = Trigonotis laxa I. M. Johnst. var. xichouensis（H. Huang）Ching J. Wang ■

397476 Trigostemon Blume = Trigonostemon Blume(保留属名)●

397477 Triguera Cav.（1785）= Hibiscus L. ●■

397478 Triguera Cav.（1785）= Triguera Cav.（1786）(保留属名)■☆

397479 Triguera Cav.（1786）(保留属名);西班牙茄属■☆

397480 Triguera Cav.（1786）(保留属名)= Hibiscus L.(保留属名)●■

397481 Triguera ambrosiaca Cav.;西班牙茄■☆

397482 Triguera ambrosiaca Cav. = Solanum herculeum Bohs ■☆

397483 Triguera osbeckii（L.）Willk. = Solanum herculeum Bohs ■☆

397484 Trigula Noronha = Clematis L. ●■

397485 Trigynaea Schltdl.（1834）;三丽花属●☆

397486 Trigynaea angustifolia Benth.;窄叶三丽花●☆

397487 Trigynaea lanceipetala D. M. Johnson et N. A. Murray;披针叶三丽花●☆

397488 Trigynaea oblongifolia Schltdl.;矩圆叶三丽花●☆

397489 Trigyneia Rchb. = Trigynaea Schltdl. ●☆

397490 Trigynia Jacq. -Fél. = Leandra Raddi ●■☆

397491 Trigynia africana Jacq. -Fél. = Leandra quinquedentata（DC.）Cogn. ●☆

397492　Trigynia parviflora Jacq. -Fél. = Leandra quinquedentata（DC.）Cogn. ●☆

397493　Trihaloragis M. L. Moody et Les = Haloragis J. R. Forst. et G. Forst. ■●

397494　Trihaloragis M. L. Moody et Les（2007）；澳洲黄花小二仙草属■☆

397495　Trihaloragis hexandra（F. Muell.）M. L. Moody et Les；澳洲黄花小二仙草■☆

397496　Trihaloragis hexandra（F. Muell.）M. L. Moody et Les = Haloragis hexandra F. Muell. ■☆

397497　Trihesperus Herb. = Anthericum L. ■☆

397498　Trihesperus Herb. = Echeandia Ortega ■☆

397499　Trihexastigma Post et Kuntze = Heteranthera Ruiz et Pav.（保留属名）■☆

397500　Trihexastigma Post et Kuntze = Triexastima Raf. ■☆

397501　Trikadis Raf. = Suaeda Forssk. ex J. F. Gmel.（保留属名）●■

397502　Trikalis Raf.（1837）；三花藜属■☆

397503　Trikalis triflora Raf. ；三花藜■☆

397504　Trikalis trigyna Moq. ；三蕊三花藜■☆

397505　Trikeraia Bor（1955）；三角草属（三角草花属，三角颖属）；Trianglegrass，Trikeraia ■

397506　Trikeraia hookeri（Stapf）Bor；三角草；Hooker Speargrass，Hooker Trikeraia，Trianglegrass ■

397507　Trikeraia hookeri（Stapf）Bor var. ramosa Bor；展穗三角草；Ramose Trikeraia ■

397508　Trikeraia oreophila Cope；山地三角草（天山三角草）；Tianshan Trianglegrass，Tianshan Trikeraia ■

397509　Trikeraia pappiformis（Keng）P. C. Kuo et S. L. Lu；假冠毛草（假毛冠草）；False Pappograss，Papusform Speargrass ■

397510　Trikeraia tianshanica S. L. Lu et X. F. Lu = Trikeraia oreophila Cope ■

397511　Trilepidea Tiegh.（1895）；三鳞寄生属●☆

397512　Trilepidea adamsii（Cheeseman）Tiegh. ；三鳞寄生 ●☆

397513　Trilepis Nees（1834）；三鳞莎草属■☆

397514　Trilepis ciliatifolia T. Koyama；缘毛三鳞莎草■☆

397515　Trilepis lhotzkiana Nees；三鳞莎草■☆

397516　Trilepis pilosa Boeck. = Afrotrilepis pilosa（Boeck.）J. Raynal ■☆

397517　Trilepis royleana Nees = Kobresia royleana（Nees）Boeck. ■

397518　Trilepisium Thouars（1806）；三鳞桑属（鳞桑属）●☆

397519　Trilepisium madagascariense DC. ；三鳞桑●☆

397520　Triliena Raf. = Acnistus Schott ●☆

397521　Trilisa（Cass.）Cass.（1820）；毛鞭菊属（鹿舌菊属，拟蛇鞭菊属）■☆

397522　Trilisa（Cass.）Cass. = Carphephorus Cass. ■☆

397523　Trilisa Cass. = Carphephorus Cass. ■☆

397524　Trilisa Cass. = Trilisa（Cass.）Cass. ■☆

397525　Trilisa carnosa（Small）B. L. Rob. = Litrisa carnosa Small ■☆

397526　Trilisa odoratissima（J. F. Gmel.）Cass. ；香毛鞭菊（鹿舌菊，拟蛇鞭菊）■☆

397527　Trilisa odoratissima（J. F. Gmel.）Cass. = Carphephorus odoratissimus（J. F. Gmel.）H. J. -C. Hebert ■☆

397528　Trilisa odoratissima Cass. = Trilisa odoratissima（J. F. Gmel.）Cass. ■☆

397529　Trilisa paniculata（J. F. Gmel.）Cass. ；毛鞭菊；Hairy Chaffhead ■☆

397530　Trilisa paniculata（J. F. Gmel.）Cass. = Carphephorus paniculatus（J. F. Gmel.）H. J. -C. Hebert ■☆

397531　Trilix L. = Prockia P. Browne ex L. ●☆

397532　Trillesanthus Pierre ex A. Chev. = Marquesia Gilg ●☆

397533　Trilliaceae Chevall.（1827）（保留科名）；延龄草科（重楼科）；Trillium Family ■

397534　Trilliaceae Chevall.（保留科名）= Melanthiaceae Batsch ex Borkh.（保留科名）■

397535　Trilliaceae Chevall.（保留科名）= Trimeniaceae Gibbs（保留科名）●☆

397536　Trilliaceae Lindl. = Trilliaceae Chevall.（保留科名）■

397537　Trillidium Kunth = Trillium L. ■

397538　Trillidium govanianum（Royle）Kunth = Trillium govanianum Wall. ex Royle ■

397539　Trillidium govanianum（Wall. ex Royle）Kunth = Trillium govanianum Wall. ex Royle ■

397540　Trillium L.（1753）；延龄草属（头顶一颗珠属）；American Wood-lily，Birthroot，Ground Lily，Trillium，Trinity Flower，Wake Robin，Wakerobin，Wake-robin，White Wood Lily，Wood Lily ■

397541　Trillium affine Rendle = Trillium catesbaei Elliott ■☆

397542　Trillium albidum J. D. Freeman；白延龄草；White Toadshade ■☆

397543　Trillium angustipetalum（Torr.）J. D. Freeman；窄瓣延龄草；Narrow-petaled Trillium ■☆

397544　Trillium camschatcense Ker Gawl. ；白花延龄草（吉林延龄草）■

397545　Trillium camtschaticum Pall. ex Pursh = Trillium camschatcense Ker Gawl. ■

397546　Trillium catesbaei Elliott；卡氏延龄草；Bashful Trillium，Catesby's Trillium ■☆

397547　Trillium cernuum L. ；下垂延龄草（垂头延龄草）；Nodding Trillium，Whip-poor-will Flower ■☆

397548　Trillium cernuum L. var. macranthum A. J. Eames et Wiegand = Trillium cernuum L. ■☆

397549　Trillium chloropetalum（Torr.）Howell；加州延龄草（绿瓣延龄草）；California Wakerobin，Giant Trillium ■☆

397550　Trillium chloropetalum（Torr.）Howell var. giganteum（Hook. et Arn.）Munz；绿瓣延龄草■☆

397551　Trillium cuneatum Raf. ；休氏延龄草；Bloody Butcher，Cuneate Trillium，Hugher's Trillium，Large Toadshade，Little Sweet Betsy，Purple Toadshade，Sweet Betsy，Whip-poor-will Flower ■☆

397552　Trillium decipiens J. D. Freeman；迷惑延龄草；Deceiving Trillium ■☆

397553　Trillium declinatum Gleason = Trillium flexipes Raf. ■☆

397554　Trillium decumbens Harb. ；匍匐延龄草；Decumbent Trillium ■☆

397555　Trillium discolor T. Wray ex Hook. ；异色延龄草；Colorless Trillium，Pale Yellow Trillium，Small Yellow Toadshade ■☆

397556　Trillium erectum L. ；直立延龄草（褐花延龄草）；Bethroot，Birthroot，Brownflower Trillium，Ground Lily，Ill-scented Trillium，Indian Balm，Indian Shamrock，Lamb's Quarters，Nosebleed，Purple Trillium，Red Trillium，Squaw-root，Stinking Benjamin，Stinking Willie，Stinking-benjamin，Three-leaved Nightshade，Trillium，Trinity Flower，Wake Robin，Wakerobin，Wake-robin ■☆

397557　Trillium erectum L. f. albiflorum R. Hoffm. ；白花直立延龄草；White Erect Trillium，White Wakerobin ■☆

397558　Trillium erectum L. var. album（Michx.）Pursh；白直立延龄草■☆

397559　Trillium erectum L. var. blandum Jennison = Trillium flexipes Raf. ■☆

397560　Trillium erectum L. var. declinatum A. Gray = Trillium flexipes Raf. ■☆

397561　Trillium erectum L. var. vaseyi（Harb.）H. E. Ahles = Trillium vaseyi Harb. ■☆

397562 Trillium erythrocarpum Curtis = Trillium grandiflorum (Michx.) Salisb. ■☆

397563 Trillium flexipes Raf.;弯曲延龄草;Bent Trillium, Declined Trillium, Drooping Trillium, Nodding Wake-robin, White Trillium, White Wake Robin ■☆

397564 Trillium foetidissimum J. D. Freeman;臭延龄草;Fetid Trillium, Stinking Trillium ■☆

397565 Trillium giganteum (Hook. et Arn.) A. Heller var. angustipetalum (Torr.) R. R. Gates = Trillium angustipetalum (Torr.) J. D. Freeman ■☆

397566 Trillium giganteum (Hook. et Arn.) A. Heller var. chloropetalum (Torr.) R. R. Gates = Trillium chloropetalum (Torr.) Howell ☆

397567 Trillium gleasonii Fernald = Trillium flexipes Raf. ■☆

397568 Trillium govanianum Wall. ex Royle;西藏延龄草(藏延龄草, 头顶一颗珠,头顶珠,延龄草);Tibet Trillium, Xizang Wakerobin ■

397569 Trillium gracile J. D. Freeman;纤细延龄草;Graceful Trillium, Slender Trillium ■☆

397570 Trillium grandiflorum (Michx.) Salisb.;大花延龄草;Bethroot, Big White Trillium, Birthroot, Great White Trillium, Large White Trillium, Large-flowered Trillium, Mooseflower, Snow Trillium, Trinity Flower, Wake Robin, Wake-robin, White Trillium, White Wake-robin, Wood Lily ■☆

397571 Trillium grandiflorum Salisb. 'Flore Pleno';重瓣大花延龄草■☆

397572 Trillium japonicum (Franch. et Sav.) Matsum. = Paris japonica (Franch. et Sav.) Franch. ■

397573 Trillium kamtschaticum Pall. ex Miyabe = Trillium camschatcense Ker Gawl. ■

397574 Trillium kamtschaticum Pall. ex Pursh;吉林延龄草(白花延龄草);Jilin Wakerobin, Kamchatka Trillium ■

397575 Trillium kurabayashii J. D. Freeman;库拉延龄草■☆

397576 Trillium lancifolium Raf.;矛叶延龄草;Lance-leaved Trillium ■☆

397577 Trillium ludovicianum Harb.;路易斯延龄草;Louisiana Trillium ■☆

397578 Trillium luteum (Muhl.) Harb.;黄延龄草;Wax Trillium, Yellow Toadshade, Yellow Trillium ■☆

397579 Trillium luteum Harb.;黄花延龄草;Yellow Trillium, Yellowflower Trillium ■☆

397580 Trillium maculatum Raf.;斑点黄延龄草;Spotted Trillium ■☆

397581 Trillium morii Hayata = Trillium tschonoskii Maxim. ■

397582 Trillium nervosum Elliott;具脉延龄草;Nerved Trillium ■☆

397583 Trillium nervosum Elliott = Trillium catesbaei Elliott ☆

397584 Trillium nivale Riddell;矮延龄草(雪地矮延龄草);Dwarf Trillium, Dwarf White Trillium, Dwarf White Wake-robin, Dwarf White Wood Lily, Dwarf White Wood-lily, Snow Trillium ■☆

397585 Trillium ovatum Pursh;卵状延龄草;Western White Trillium, White Trillium ■☆

397586 Trillium ovatum Pursh subsp. oettingeri Munz et Thorne = Trillium ovatum Pursh var. oettingeri (Munz et Thorne) Case ■☆

397587 Trillium ovatum Pursh var. oettingeri (Munz et Thorne) Case; 萨尔蒙延龄草;Salmon Mountains Wake-robin ■☆

397588 Trillium ovatum Pursh var. stenosepalum R. R. Gates = Trillium ovatum Pursh ■☆

397589 Trillium ozarkanum E. J. Palmer et Steyerm. = Trillium pusillum Michx. ■☆

397590 Trillium parviflorum V. G. Soukup;小花延龄草;Small-flowered Trillium ■☆

397591 Trillium persistens W. H. Duncan;常绿延龄草■☆

397592 Trillium petiolatum Pursh;长梗延龄草;Long-petioled Trillium, Purple Trillium, Round-leaved Trillium ■☆

397593 Trillium pumilum Pursh = Trillium pusillum Michx. ■☆

397594 Trillium pusillum Michx.;矮小延龄草;Dwarf Trillium, Dwarf Wakerobin, Dwarf White Flowering Trillium, Least Trillium, Ozark Wake Robin ■☆

397595 Trillium pusillum Michx. var. monticulum Bodkin et Reveal = Trillium pusillum Michx. ■☆

397596 Trillium pusillum Michx. var. ozarkanum (E. J. Palmer et Steyerm.) Steyerm. = Trillium pusillum Michx. ■☆

397597 Trillium pusillum Michx. var. texanum (Buckley) Reveal et C. R. Broome = Trillium pusillum Michx. ■☆

397598 Trillium pusillum Michx. var. virginianum Fernald;弗吉尼亚延龄草(维尔吉亚延龄草)■☆

397599 Trillium recurvatum L. C. Becker;卷瓣延龄草(下弯延龄草); Bloody Butcher, Bloody Noses, Piairie Trillium, Purple Trillium, Purple Wake Robin, Red Trillium, Reflexed Sepal Trillium, Reflexed Trillium ■☆

397600 Trillium recurvatum L. C. Becker f. foliosum Steyerm. = Trillium recurvatum L. C. Becker ■☆

397601 Trillium recurvatum L. C. Becker f. petaloideum Steyerm. = Trillium recurvatum L. C. Becker ■☆

397602 Trillium recurvatum L. C. Becker f. shayi E. J. Palmer et Steyerm. = Trillium recurvatum L. C. Becker ■☆

397603 Trillium reliquum J. D. Freeman;残余延龄草;Relict Trillium ■☆

397604 Trillium rhombifolium Kom.;菱叶延龄草■☆

397605 Trillium rhomboideum Michx. var. album Michx. = Trillium erectum L. var. album (Michx.) Pursh ☆

397606 Trillium rhomboideum Michx. var. grandiflorum Michx. = Trillium grandiflorum (Michx.) Salisb. ■☆

397607 Trillium rivale S. Watson;溪畔延龄草(溪岸延龄草);Brook Wake-robin, Oregon Trillium ■☆

397608 Trillium rugelii Rendle;卢格延龄草;Southern Nodding Trillium ■☆

397609 Trillium sessile L.;无柄延龄草;Red Trillium, Sessile Trillium, Toad Lily, Toad Trillium, Toadshade Trillium, Wake Robin, Wake-robin ■☆

397610 Trillium sessile L. f. viridiflorum Beyer = Trillium sessile L. ■☆

397611 Trillium sessile L. var. angustipetalum Torr. = Trillium angustipetalum (Torr.) J. D. Freeman ■☆

397612 Trillium sessile L. var. chloropetalum Torr. = Trillium chloropetalum (Torr.) Howell ■☆

397613 Trillium sessile L. var. giganteum Hook. et Arn. = Trillium chloropetalum (Torr.) Howell var. giganteum (Hook. et Arn.) Munz ■☆

397614 Trillium sessile L. var. luteum Muhl. = Trillium luteum (Muhl.) Harb. ☆

397615 Trillium simile Gleason;甜白延龄草;Confusing Trillium, Sweet White Trillium ■☆

397616 Trillium smallii Maxim.;日本延龄草■☆

397617 Trillium stamineum Harb.;扭曲延龄草;Blue Ridge Wakerobin, Propellor Trillium, Twisted Trillium ■☆

397618 Trillium stylosum Nutt. = Trillium catesbaei Elliott ☆

397619 Trillium sulcatum T. S. Patrick;南部红延龄草;Barksdale Trillium, Southern Red Trillium ■☆

397620 Trillium taiwanense S. S. Ying;台湾延龄草;Taiwan Trillium ■

397621 Trillium texanum Buckley = Trillium pusillum Michx. ■☆

397622 Trillium tschonoskii Maxim.;延龄草(地珠,佛手七,华延龄草,黄花三七,三角七,尸儿七,狮儿七,头顶一颗珠,头顶珠,鱼儿七,玉儿七,芋儿七,直立延龄草);Tschonosk Trillium, Wakerobin ■

397623 Trillium tschonoskii Maxim. f. morii (Hayata) Yamam. = Trillium tschonoskii Maxim. ■

397624　Trillium tschonoskii Maxim. var. himalaicum H. Hara = Trillium tschonoskii Maxim. ■

397625　Trillium tschonoskii Maxim. var. morii（Hayata）Masam. = Trillium tschonoskii Maxim. ■

397626　Trillium underwoodii Small；安德伍德延龄草；Underwood's Trillium ■☆

397627　Trillium undulatum Willd.；波瓣延龄草（波叶延龄草）；Painted Lily, Painted Trillium, Painted Wood Lily, Painted Wood-lily ■☆

397628　Trillium vaseyi Harb.；瓦齐延龄草；Sweet Beth, Sweet Trillium, Vasey's Trillium ■☆

397629　Trillium vaseyi Harb. var. simile（Gleason）Barksdale = Trillium simile Gleason ■☆

397630　Trillium virginianum（Fernald）C. F. Reed = Trillium pusillum Michx. var. virginianum Fernald ■☆

397631　Trillium viride L. C. Beck；绿延龄草；Green Trillium ■☆

397632　Trillium viride L. C. Beck var. luteum（Muhl.）Gleason = Trillium luteum（Muhl.）Harb. ■☆

397633　Trillium viridescens Nutt.；奥扎克延龄草；Green Trillium, Ozark Green Trillium, Ozark Trillium ■☆

397634　Trilobachne Schenck ex Henrard（1931）；三裂果属 ■☆

397635　Trilobachne cookei（Stapf）Schenck ex Henrard；三裂果 ■☆

397636　TrilobuIina Raf. = Utricularia L. ■

397637　Trilocularia Schltr. = Balanops Baill. ●☆

397638　Trilomisa Raf. = Begonia L. ●■

397639　Trilophus Fisch. = Menispermum L. ●■

397640　Trilophus Lestib. = Kaempferia L. ■

397641　Trilopus Adans. = Hamamelis L. ●

397642　Trilopus Mitch. = Hamamelis L. ●

397643　Trilopus dentata Raf. = Hamamelis virginiana L. ●☆

397644　Trilopus estivalis Raf. = Hamamelis virginiana L. ●☆

397645　Trilopus nigra Raf. = Hamamelis virginiana L. ●☆

397646　Trilopus nigra Raf. var. catesbiana Raf. = Hamamelis virginiana L. ●☆

397647　Trilopus parvifolia（Nutt.）Raf. = Hamamelis virginiana L. ●☆

397648　Trilopus rotundifolia Raf. = Hamamelis virginiana L. ●☆

397649　Trilopus virginica（L.）Raf. = Hamamelis virginiana L. ●☆

397650　Trima Noronha = Mycetia Reinw. ●

397651　Trimelandra Raf. = Lonchostoma Wikstr.（保留属名）●☆

397652　Trimelopter Raf. = Ornithogalum L. ■

397653　Trimenia Seem.（1873）（保留属名）；早落瓣属（腺齿木属）●☆

397654　Trimenia weinmanniifolia Seem.；早落瓣 ●☆

397655　Trimeniaceae（Perkins et Gilg）Gibbs = Trimeniaceae Gibbs（保留科名）●☆

397656　Trimeniaceae Gibbs（1917）（保留科名）；早落瓣科（腺齿木科）●☆

397657　Trimeniaceae Perk. et Gilg = Trimeniaceae Gibbs（保留科名）●☆

397658　Trimeranthes（Cass.）Cass. = Sckuhria Moench（废弃属名）■☆

397659　Trimeranthes（Cass.）Cass. = Sigesbeckia L. ■

397660　Trimeranthes Cass. = Sigesbeckia L. ■

397661　Trimeranthus H. Karst. = Chaetolepis（DC.）Miq. ●☆

397662　Trimeria Harv.（1838）；三数木属 ●☆

397663　Trimeria alnifolia（Hook.）Harv. = Trimeria grandifolia（Hochst.）Warb. ●☆

397664　Trimeria bakeri Gilg = Trimeria grandifolia（Hochst.）Warb. subsp. tropica（Burkill）Sleumer ●☆

397665　Trimeria grandifolia（Hochst.）Warb.；大花三数木 ●☆

397666　Trimeria grandifolia（Hochst.）Warb. subsp. tropica（Burkill）Sleumer；热带大花三数木 ●☆

397667　Trimeria macrophylla Baker f. = Trimeria grandifolia（Hochst.）Warb. ●☆

397668　Trimeria rotundifolia（Hochst.）Gilg = Trimeria grandifolia（Hochst.）Warb. ●☆

397669　Trimeria trinervis Harv.；三脉三数木 ●☆

397670　Trimeria tropica Burkill = Trimeria grandifolia（Hochst.）Warb. subsp. tropica（Burkill）Sleumer ●☆

397671　Trimeris C. Presl = Lobelia L. ●■

397672　Trimeris C. Presl（1836）；赫勒纳桔梗属 ■☆

397673　Trimeris oblongifolia C. Presl；赫勒纳桔梗 ■☆

397674　Trimerisma C. Presl = Platylophus D. Don（保留属名）●☆

397675　Trimeriza Lindl. = Apama Lam. ●

397676　Trimeriza Lindl. = Thottea Rottb. ●

397677　Trimerocalyx（Murb.）Murb. = Linaria Mill. ■

397678　Trimerocalyx paradoxus Murb. = Linaria paradoxa Murb. ■☆

397679　Trimetra Moq. ex DC. = Borrichia Adans. ●■☆

397680　Trimeza Salisb. = Trimezia Salisb. ex Herb. ■☆

397681　Trimezia Salisb. ex Herb.（1844）；枝端花属 ■☆

397682　Trimezia bicolor Ravenna；二色枝端花 ■☆

397683　Trimezia caerulea（Ker Gawl.）Ravenna；蓝枝端花 ■☆

397684　Trimezia candida（Hassl.）Ravenna；白枝端花 ■☆

397685　Trimezia martinicensis（Jacq.）Herb.；枝端花；Martinique Trimezia ■☆

397686　Trimezia sylvestris ex Vell.；林地枝端花 ■☆

397687　Trimezia truncata Ravenna；平截枝端花 ■☆

397688　Trimezia violacea（Klatt）Ravenna；堇色枝端花 ■☆

397689　Trimista Raf. = Mirabilis L. ●

397690　Trimorpha Cass.（1817）；三形菊属 ■☆

397691　Trimorpha Cass. = Erigeron L. ■●

397692　Trimorpha acris（L.）Gray = Erigeron acris L. ■

397693　Trimorpha acris（L.）Gray var. debilis（A. Gray）G. L. Nesom = Erigeron nivalis Nutt. ■☆

397694　Trimorpha acris（L.）Gray var. kamtschatica（DC.）G. L. Nesom = Erigeron acer L. var. kamtschaticus（DC.）Herder ■

397695　Trimorpha acris Vierh. = Erigeron acer L. ■

397696　Trimorpha angulosa Vierh. = Erigeron elongatus Ledeb. ■

397697　Trimorpha armeriifolia（Turcz. ex DC.）Vierh. = Erigeron lonchophyllus Hook. ■

397698　Trimorpha armeriifolia Vierh. = Erigeron lonchophyllus Hook. ■

397699　Trimorpha elata（Hook.）G. L. Nesom = Erigeron elatus（Hook.）Greene ■☆

397700　Trimorpha elongata（Ledeb.）Vierh. = Erigeron elongatus Ledeb. ■

397701　Trimorpha lonchophylla（Hook.）G. L. Nesom = Erigeron lonchophyllus Hook. ■

397702　Trimorpha polita Vierh. = Erigeron elongatus Ledeb. ■

397703　Trimorpha vulgaris Cass. = Erigeron acer L. ■

397704　Trimorphaea Cass. = Trimorpha Cass. ■☆

397705　Trimorphandra Brongn. et Gris = Hibbertia Andréws ●☆

397706　Trimorphoea Benth. et Hook. f. = Erigeron L. ●

397707　Trimorphoea Benth. et Hook. f. = Trimorphaea Cass. ■☆

397708　Trimorphopetalum Baker = Impatiens L. ■

397709　Trinacte Gaertn. = Jungia L. f.（保留属名）■●☆

397710　Trinax D. Dietr. = Thrinax L. f. ex Sw. ●☆

397711　Trinchinettia Endl. = Geissopappus Benth. ●■☆

397712　Trinciatella Adans. = Hyoseris L. ■☆

397713　Trineuria C. Presl = Aspalathus L. ●☆

397714　Trineuria appendiculata（E. Mey.）C. Presl = Aspalathus ciliaris

L. ●☆

397715　Trineuria chenopoda（L.）C. Presl = Aspalathus chenopoda L. ●☆

397716　Trineuria ciliaris（L.）C. Presl = Aspalathus ciliaris L. ●☆

397717　Trineuria cochleariformis C. Presl = Aspalathus cymbiformis DC. ●☆

397718　Trineuria comosa（Thunb.）C. Presl = Aspalathus parviflora P. J. Bergius ●☆

397719　Trineuria cymbiformis（DC.）C. Presl = Aspalathus cymbiformis DC. ●☆

397720　Trineuria deciduifolia（Eckl. et Zeyh.）C. Presl = Aspalathus nigra L. ●☆

397721　Trineuria fuscescens C. Presl = Aspalathus nigra L. ●☆

397722　Trineuria linearifolia（DC.）C. Presl = Aspalathus linearifolia DC. ●☆

397723　Trineuria marginalis（Eckl. et Zeyh.）C. Presl = Aspalathus marginalis Eckl. et Zeyh. ●☆

397724　Trineuria papillosa（Eckl. et Zeyh.）C. Presl = Aspalathus ciliaris L. ●☆

397725　Trineuria rigescens（E. Mey.）C. Presl = Aspalathus setacea Eckl. et Zeyh. ●☆

397726　Trineuron Hook. f. = Abrotanella Cass. ■☆

397727　Tringa Roxb. = Hypolytrum Rich. ex Pers. ■

397728　Tringa Roxb. = Tunga Roxb. ■

397729　Trinia Hoffm.（1814）（保留属名）；特林芹属；Honewort,Trinia ■☆

397730　Trinia dahurica Turcz. ex Besser = Saposhnikovia divaricata（Turcz.）Schischk. ■

397731　Trinia glauca Hoffm. ex Steud.；灰特林芹；Corn Parsley, Honewort ■☆

397732　Trinia hispida Hoffm.；毛特林芹■☆

397733　Trinia kitaibelii M. Bieb.；基陶特林芹■☆

397734　Trinia leiogona（C. A. Mey.）B. Fedtsch.；光果特林芹■☆

397735　Trinia multicaulis（Poir.）Schischk.；多茎特林芹■☆

397736　Trinia muricata Godet；粗糙特林芹■☆

397737　Trinia polyclada Schischk.；多枝特林芹■☆

397738　Trinia stankovii Schischk.；斯坦考夫特林芹■☆

397739　Trinia swellendamensis Eckl. et Zeyh. = Anginon swellendamense（Eckl. et Zeyh.）B. L. Burtt ■☆

397740　Trinia ucrainica Schischk.；乌克兰特林芹■☆

397741　Trinia uitenhagensis Eckl. et Zeyh. = Anginon rugosum（Thunb.）Raf. ■☆

397742　Triniella Calest. = Trinia Hoffm.（保留属名）■☆

397743　Triniochloa Hitchc.（1913）；三重草属■☆

397744　Triniochloa alpestris（Kunth）Pittier；高山三重草■☆

397745　Triniochloa laxa Hitchc.；疏松三重草■☆

397746　Triniochloa micrantha Hitchc.；小花三重草■☆

397747　Triniteurybia Brouillet,Urbatsch et R. P. Roberts（2004）；三叉湖绿顶菊属■☆

397748　Triniteurybia aberrans（A. Nelson）Brouillet, Urbatsch et R. P. Roberts；三叉湖绿顶菊；Idaho Goldenweed ■☆

397749　Triniusa Steud. = Bromus L.（保留属名）■

397750　Triniusa danthoniae（Trin.）Steud. = Bromus danthoniae Trin. ex C. A. Mey. ■

397751　Trinogeton Walp. = Cacabus Bernh. ■☆

397752　Trinogeton Walp. = Thinogeton Benth. ■☆

397753　Triocles Salisb. = Kniphofia Moench（保留属名）■☆

397754　Triodanis Raf.（1838）；异檐花属■

397755　Triodanis biflora（Ruiz et Pav.）Greene；卵叶异檐花（异檐花）；Venus' Looking Glass ■

397756　Triodanis holzingeri McVaugh；豪尔异檐花；Venus' Looking Glass ■☆

397757　Triodanis lamprosperma McVaugh；维纳斯异檐花；Venus' Looking Glass ■☆

397758　Triodanis leptocarpa（Nutt.）Nieuwl.；细果异檐花；Slender-leaved Venus' Looking Glass ■☆

397759　Triodanis perfoliata（L.）Nieuwl.；穿叶异檐花；Clasping Venus'-looking-glass,Round-leaved Triodanis,Venus' Looking Glass ■

397760　Triodanis perfoliata（L.）Nieuwl. f. alba Voigt；白穿叶异檐花■☆

397761　Triodanis perfoliata（L.）Nieuwl. var. biflora（Ruiz et Pav.）T. R. Bradley = Triodanis biflora（Ruiz et Pav.）Greene ■

397762　Triodia R. Br.（1810）；三齿稃草属（三齿稃属）；Spinifex, Triodia ■☆

397763　Triodia capensis（Nees）T. Durand et Schinz = Leptochloa fusca（L.）Kunth ■

397764　Triodia ciliata（Steud.）T. Durand et Schinz = Trichoneura mollis（Kunth）Ekman ■☆

397765　Triodia decumbens（L.）P. Beauv. = Danthonia decumbens（L.）DC. ■☆

397766　Triodia flava（L.）Smyth = Tridens flavus（L.）Hitchc. ■☆

397767　Triodia flava（L.）Smyth f. cuprea（Jacq.）Fosberg = Tridens flavus（L.）Hitchc. ■☆

397768　Triodia livida（Nees）T. Durand et Schinz = Leptochloa fusca（L.）Kunth ■

397769　Triodia longiceps J. M. Black；长头三齿稃草；Buck Spinifex ■☆

397770　Triodia mollis（Kunth）T. Durand et Schinz = Trichoneura mollis（Kunth）Ekman ■☆

397771　Triodia plumosa（Andersson）Benth. = Leptocarydion vulpiastrum（De Not.）Stapf ■☆

397772　Triodia pulchella Kunth；美丽三齿稃草；Fluff Grass ■☆

397773　Triodica Steud. = Sapium Jacq.（保留属名）●

397774　Triodica Steud. = Triadica Lour. ●

397775　Triodoglossum Bullock（1962）；三齿舌萝藦属■☆

397776　Triodoglossum abyssinicum（Chiov.）Bullock；三齿舌萝藦■☆

397777　Triodon Baumg. = Danthonia DC.（保留属名）■

397778　Triodon Baumg. = Sieglingia Bernh.（废弃属名）■

397779　Triodon Baumg. = Triodia R. Br. ■☆

397780　Triodon DC. = Diodia L. ■

397781　Triodon DC. = Ebelia Rchb. ■

397782　Triodon Rich. = Rhynchospora Vahl（保留属名）■

397783　Triodon albus（L.）Farw. = Rhynchospora alba（L.）Vahl ■

397784　Triodon albus Farw. var. macer（C. B. Clarke ex Britton）Farw. = Rhynchospora macra（C. B. Clarke ex Britton）Small ■☆

397785　Triodon capillaceus（Torr.）Farw. = Rhynchospora capillacea Torr. ■☆

397786　Triodon glomeratus（L.）Farw. = Rhynchospora glomerata（L.）Vahl ■☆

397787　Triodris Thouars = Disperis Sw. ■

397788　Triodris Thouars = Dryopeia Thouars ■

397789　Triodus Raf. = Carex L. ■

397790　Triolaena T. Durand et Jacks. = Triolena Naudin ☆

397791　Triolena Naudin（1851）；三臂野牡丹属☆

397792　Triolena amazonica（Pilg.）Wurdack；亚马孙三臂野牡丹☆

397793　Triolena pumila Umaña et Almeda；偃伏三臂野牡丹☆

397794　Triolena spicata（Triana）L. O. Williams；三臂野牡丹☆

397795　Triolena stenophylla（Standl. et Steyerm.）Standl. et L. O. Williams；窄叶三臂野牡丹☆

397796　Triomma Hook. f.（1860）；三孔橄榄属●☆

397797　Triomma malaccensis Hook. f.；三孔橄榄●☆

397798   Trionaea Medik. = Hibiscus L. (保留属名) ●■

397799   Trioncinia (F. Muell.) Veldkamp(1991);折芒菊属■☆

397800   Trioncinia retroflexa (F. Muell.) Veldkamp;折芒菊■☆

397801   Trionfettaria Post et Kuntze = Triumfetta Plum. ex L. ●■

397802   Trionfettaria Post et Kuntze = Triumfettaria Rchb. ●■

397803   Trionfettia Post et Kuntze = Triumfetta Plum. ex L. ●■

397804   Trionum L. = Hibiscus L. (保留属名) ●■

397805   Trionum L. ex Schaeff. = Hibiscus L. (保留属名) ●■

397806   Trionum Schaeff. = Hibiscus L. (保留属名) ●■

397807   Trionum annuum Medik. = Hibiscus trionum L. ■

397808   Trionum trionum (L.) Wooton et Standl. = Hibiscus trionum L. ■

397809   Triopteris L. = Triopteris L. (保留属名) ●☆

397810   Triopteris ovata Cav. = Triopterys ovata Cav. ●

397811   Triopterys A. Juss. = Triopterys L. (保留属名) ●☆

397812   Triopterys L. (1753) ('Triopteris') (保留属名);三翅金虎尾属(三翅藤属) ●☆

397813   Triopterys ovata Cav.;三翅金虎尾(三翅藤,土利奥布得利藤) ●

397814   Triopteryx Dalla Torre et Harms = Triopterys L. (保留属名) ●☆

397815   Trioptolemea Benth. = Dalbergia L. f. (保留属名) ●

397816   Trioptolemea Benth. = Triptolemea Mart. ●

397817   Trioptolemea Mart. ex Benth. = Trioptolemea Benth. ●

397818   Triorchis Agosti = Spiranthes Rich. (保留属名) ■

397819   Triorchis Nieuwl. = Spiranthes Rich. (保留属名) ■

397820   Triorchos Small et Nash = Pteroglossaspis Rchb. f. ■☆

397821   Triorchos Small et Nash ex Small = Pteroglossaspis Rchb. f. ■☆

397822   Triorchos ecristatus (Fernald) Small = Pteroglossaspis ecristata (Fernald) Rolfe ■☆

397823   Triorchos strictus (Griseb.) Acuna = Pteroglossaspis ecristata (Fernald) Rolfe ■☆

397824   Triostemon Benth. et Hook. f. = Triosteum L. (保留属名) ■

397825   Triosteon Adans. = Triosteum L. (保留属名) ■

397826   Triosteon Dill. ex Adans. = Triosteum L. (保留属名) ■

397827   Triosteospermum Mill. = Triosteum L. (保留属名) ■

397828   Triosteum L. (1753) (保留属名);莛子藨属;Feverwort, Horsegentian, Horse-gentian, Horse Gentian ■

397829   Triosteum angustifolium L.;窄叶莛子藨;Yellow-flowered Horse Gentian ■☆

397830   Triosteum aurantiacum E. P. Bicknell;橘果莛子藨;Early Horse-gentian, Orange-fruit Horse-gentian, Orange-fruited Horse-gentian, Scarlet-fruited Horse-gentian ■☆

397831   Triosteum aurantiacum E. P. Bicknell f. glaucescens (Wiegand) F. C. Lane = Triosteum aurantiacum E. P. Bicknell ■☆

397832   Triosteum aurantiacum E. P. Bicknell var. glaucescens Wiegand = Triosteum aurantiacum E. P. Bicknell ■☆

397833   Triosteum aurantiacum E. P. Bicknell var. illinoense (Wiegand) E. J. Palmer et Steyerm.;伊利诺莛子藨;Horse Gentian, Illinois Horse-gentian, Orange-fruit Horse-gentian ■☆

397834   Triosteum erythrocarpum Harry Sm. = Triosteum himalayanum Wall. ■

397835   Triosteum fargesii Franch. = Triosteum himalayanum Wall. ■

397836   Triosteum himalayanum Wall.;穿心莛子藨(包谷陀子,穿心莛藨,大对月草,通天七,五转七,阴阳扇,钻子七);Himalayan Horsegentian, Himalayas Horsegentian ■

397837   Triosteum himalayanum Wall. var. chinense Diels et Graebn;中华莛子藨■

397838   Triosteum hirsutum Roxb. = Lasianthus hirsutus (Roxb.) Merr. ●

397839   Triosteum hirsutum Wall. = Triosteum himalayanum Wall. ■

397840   Triosteum illinoense (Wiegand) Rydb. = Triosteum aurantiacum E.

397841   P. Bicknell var. illinoense (Wiegand) E. J. Palmer et Steyerm. ■☆

397841   Triosteum illinoense (Wiegand) Rydb. f. glabrescens F. C. Lane = Triosteum aurantiacum E. P. Bicknell var. illinoense (Wiegand) E. J. Palmer et Steyerm. ■☆

397842   Triosteum intermedium Diels et Graebn. = Triosteum pinnatifidum Maxim. ■

397843   Triosteum maruyamae Ohwi = Triosteum pinnatifidum Maxim. ■

397844   Triosteum perfoliatum L.;穿叶莛子藨(普通莛子藨);Common Horse Gentian, Fever Wort, Feverwort, Late Horse-gentian, Tinker's Weed, Tinker's-weed, Wild Coffee, Wild-coffee ■☆

397845   Triosteum perfoliatum L. var. aurantiacum (E. P. Bicknell) Wiegand = Triosteum aurantiacum E. P. Bicknell ■☆

397846   Triosteum perfoliatum L. var. illinoense Wiegand = Triosteum aurantiacum E. P. Bicknell var. illinoense (Wiegand) E. J. Palmer et Steyerm. ■☆

397847   Triosteum pinnatifidum Maxim.;莛子藨(白果七,鸡爪七,四大天王,天王七,五转七,羽裂莛子藨,羽裂叶莛子藨);Featherycleft Horsegentian ■

397848   Triosteum rosthornii Diels et Graebn. = Triosteum pinnatifidum Maxim. ■

397849   Triosteum sinuatum Maxim.;腋花莛子藨;Sinuate Horsegentian ■

397850   Triotosiphon Schltr. ex Luer = Masdevallia Ruiz et Pav. ■☆

397851   Triotosiphon Schltr. ex Luer(2006);三数细瓣兰属■☆

397852   Tripagandra Raf. = Tripogandra Raf. ■☆

397853   Tripentas Casp. = Elodes Adans. ■●

397854   Tripentas Casp. = Hypericum L. ■●

397855   Tripetalanthus A. Chev. = Plagiosiphon Harms ■☆

397856   Tripetalanthus emarginatus (Hutch. et Dalziel) A. Chev. = Plagiosiphon emarginatus (Hutch. et Dalziel) J. Léonard ■☆

397857   Tripetalanthus gabonensis A. Chev. = Plagiosiphon gabonensis (A. Chev.) J. Léonard ■☆

397858   Tripetaleia Siebold et Zucc. (1843);三瓣木属(和鹃花属,切帕泰勒属) ●☆

397859   Tripetaleia Siebold et Zucc. = Elliottia Muhl. ex Elliott ●☆

397860   Tripetaleia bracteata Maxim.;深山三瓣木(深山切帕泰勒) ●☆

397861   Tripetaleia bracteata Maxim. = Cladothamnus bracteatus (Maxim.) T. Yamaz. ●☆

397862   Tripetaleia paniculata Siebold et Zucc.;三瓣木(和鹃花,穗踯躅,圆锥花切帕泰勒) ●☆

397863   Tripetaleia paniculata Siebold et Zucc. = Elliottia paniculata (Siebold et Zucc.) Hook. f. ●☆

397864   Tripetaleia paniculata Siebold et Zucc. f. albiflora Y. Kimura;白花三瓣木●☆

397865   Tripetaleia paniculata Siebold et Zucc. f. latifolia (Maxim.) Kitam. = Elliottia paniculata (Siebold et Zucc.) Hook. f. ●☆

397866   Tripetaleia paniculata Siebold et Zucc. var. latifolia Maxim. = Elliottia paniculata (Siebold et Zucc.) Hook. f. ●☆

397867   Tripetaleia paniculata Siebold et Zucc. var. yakushimensis Siebold et Zucc. = Elliottia paniculata (Siebold et Zucc.) Hook. f. ●☆

397868   Tripetaleia yakushimensis DC. = Elliottia paniculata (Siebold et Zucc.) Hook. f. ●☆

397869   Tripetalum K. Schum. = Garcinia L. ●

397870   Tripetalum Post et Kuntze = Sambucus L. ●■

397871   Tripetalum Post et Kuntze = Tripetelus Lindl. ●■

397872   Tripetelus Lindl. = Sambucus L. ●■

397873   Tripha Noronha = Mischocarpus Blume(保留属名) ●

397874   Triphaca Lour. = Sterculia L. ●

397875　Triphalia Banks et Sol. ex Hook. f. = Aristotelia L' Hér. (保留属名)●☆

397876　Triphasia Lour. (1790);酸橙果属(臭橘属,三囊属);Myrtle Lime,Three-birds Orchid ●☆

397877　Triphasia aurantiola Lour.;酸橙果;Nodding Pogonia,Three-birds Orchid ●

397878　Triphasia trifolia (Burm. f.) P. Wilson;三叶酸橙果;Chinese Lime,Lime Berry,Limeberry,Myrtle Lime ●☆

397879　Triphasia trifolia P. Wilson = Triphasia trifolia (Burm. f.) P. Wilson ●☆

397880　Triphasia trifoliata DC. = Poncirus trifoliata (L.) Raf. ●

397881　Triphasia trifoliata DC. = Triphasia aurantiola Lour. ●

397882　Triphelia R. Br. ex Endl. = Actinodium Schauer ●☆

397883　Triphlebia Stapf = Eragrostis Wolf ■

397884　Triphlebia Stapf = Stiburus Stapf ■☆

397885　Triphora Nutt. (1818);三褶兰属;Three Birds, Three-birds Orchid ■☆

397886　Triphora amazonica Schltr.;宽叶三褶兰■☆

397887　Triphora craigheadii G. M. Luer;克氏三褶兰■☆

397888　Triphora cubensis (Rchb. f.) Ames = Triphora gentianoides (Sw.) Ames et Schltr. ■☆

397889　Triphora gentianoides (Sw.) Ames et Schltr.;龙胆三褶兰■☆

397890　Triphora latifolia G. M. Luer = Triphora amazonica Schltr. ■☆

397891　Triphora rickettii G. M. Luer = Triphora yucatanensis Ames ■☆

397892　Triphora trianthophora (Sw.) Rydb.;三褶兰;Nodding Pogonia,Three Bird Orchid,Three Birds,Three-birds Orchid ■☆

397893　Triphora trianthophora (Sw.) Rydb. var. schaffneri Camp. = Triphora trianthophora (Sw.) Rydb. ■☆

397894　Triphora yucatanensis Ames;尤卡坦三褶兰■☆

397895　Triphylleion Suess. = Niphogeton Schltdl. ■☆

397896　Triphyllocynis Thouars = Cynorkis Thouars ■☆

397897　Triphylloides Moench = Trifolium L. ■

397898　Triphyllum Medik. = Medicago L. (保留属名)●■

397899　Triphyophyllaceae Emberger = Dioncophyllaceae Airy Shaw (保留科名)●☆

397900　Triphyophyllum Airy Shaw(1952);三叶木属(穗叶藤属)●☆

397901　Triphyophyllum peltatum (Hutch. et Dalziel) Airy Shaw;三叶木●☆

397902　Triphyophyllum peltatum Airy Shaw = Triphyophyllum peltatum (Hutch. et Dalziel) Airy Shaw ●☆

397903　Triphysaria Fisch. et C. A. Mey. (1836);直果草属(三叶参属)■

397904　Triphysaria Fisch. et C. A. Mey. = Orthocarpus Nutt. ■

397905　Triphysaria chinensis (D. Y. Hong) D. Y. Hong;直果草;China Orthocarpus,Chinese Orthocarpus ■

397906　Triphysaria floribunda (Benth.) T. I. Chuang et Heckard;多花直果草■☆

397907　Triphysaria hispida Rydb.;毛直果草■☆

397908　Triphysaria micrantha (Greene ex A. Heller) T. I. Chuang et Heckard;小花直果草■☆

397909　Tripidium H. Scholz = Ripidium Trin. ■

397910　Tripidium ravennae (L.) H. Scholz = Erianthus ravennae (L.) P. Beauv. ■

397911　Tripidium ravennae (L.) H. Scholz subsp. parviflorum (Pilg.) H. Scholz = Saccharum ravennae (L.) L. subsp. parviflorum (Pilg.) Scholz ■

397912　Tripinna Lour. = Colea Bojer ex Meisn. (保留属名)●☆

397913　Tripinna Lour. = Vitex L. ●

397914　Tripinna tripinnata Lour. = Colea tripinnata Seem. ●

397915　Tripinna tripinnata Lour. = Vitex tripinnata (Lour.) Merr. ●

397916　Tripinnaria Pers. = Colea Bojer ex Meisn. (保留属名)●☆

397917　Tripinnaria Pers. = Tripinna Lour. ●☆

397918　Tripinnaria Pers. = Vitex L. ●

397919　Tripinnaria africana Spreng. = Kigelia africana (Lam.) Benth. ●☆

397920　Triplachne Link(1833);光穗三芒草属■☆

397921　Triplachne nitens (Guss.) Link = Gastridium nitens (Guss.) Coss. et Durieu ■☆

397922　Triplachne nitens Link;光亮三芒草■☆

397923　Tripladenia D. Don = Kreysigia Rchb. ■☆

397924　Tripladenia D. Don = Schelhammera R. Br. (保留属名)■☆

397925　Tripladenia D. Don(1839);三腺兰属■☆

397926　Tripladenia cunninghamii D. Don;三腺兰■☆

397927　Triplandra Raf. = Croton L. ●

397928　Triplandron Benth. = Clusia L. ●☆

397929　Triplarina Raf. = Baeckea L. ●

397930　Triplaris Loefl. = Triplaris Loefl. ex L. ●

397931　Triplaris Loefl. ex L. (1758);蓼树属;Hormigo,Knotweedtree,Knotweed-tree,Long Jack,Volador ●

397932　Triplaris americana L.;蓼树(树蓼);Common Knotweedtree,Common Knotweed-tree,Knotweedtree ●

397933　Triplaris cumingiana Fisch. et C. A. Mey. ex C. A. Mey.;卡氏蓼树;Ant Tree ●☆

397934　Triplaris surinamensis Cham.;三角蓼树;Long John ●☆

397935　Triplasandra Seem. = Tetraplasandra A. Gray ●☆

397936　Triplasis P. Beauv. (1812);三重茅属■☆

397937　Triplasis americana P. Beauv.;美洲三重茅■☆

397938　Triplasis intermedia Nash = Triplasis purpurea (Walter) Chapm. ■☆

397939　Triplasis purpurea (Walter) Chapm.;紫三重茅;Purple Sand Grass,Sand Grass ■☆

397940　Triplateia Bartl. = Hymenella Moc. et Sessé ■

397941　Triplateia Bartl. = Minuartia L. ■

397942　Triplathera (Endl.) Lindl. = Bouteloua Lag. (保留属名)■

397943　Triplathera Endl. = Botelua Lag. ■

397944　Triplectrum D. Don ex Wight et Arn. = Medinilla Gaudich. ex DC. ■

397945　Triplectrum Wight et Arn. = Medinilla Gaudich. ex DC. ■

397946　Tripleura Lindl. = Zeuxine Lindl. (保留属名)■

397947　Tripleura pallida Lindl. = Pecteilis susannae (L.) Raf. ■

397948　Tripleurospermum Sch. Bip. (1844);三肋果属;Mayweed,Threebibachene,Threeribfruit ■

397949　Tripleurospermum ambiguum (Ledeb.) Franch. et Sav.;褐苞三肋果;Brownbract Threebibachene,Brownbract Threeribfruit ■☆

397950　Tripleurospermum ambiguum Franch. et Sav. = Tripleurospermum tetragonospermum (F. Schmidt) Pobed. ■

397951　Tripleurospermum auriculatum (Boiss.) Rech. f.;耳形三肋果■☆

397952　Tripleurospermum breviradiatum (Ledeb.) Pobed.;短线三肋果■☆

397953　Tripleurospermum caucasicum Hayek;高加索三肋果■☆

397954　Tripleurospermum colchicum (Manden.) Pobed.;黑海三肋果■☆

397955　Tripleurospermum decipiens (Fisch. et C. A. Mey.) Bornm.;迷惑三肋果■☆

397956　Tripleurospermum disciforme (C. A. Mey.) Sch. Bip.;盘状三肋果■☆

397957　Tripleurospermum disciforme Sch. Bip. = Tripleurospermum disciforme (C. A. Mey.) Sch. Bip. ■☆

397958　Tripleurospermum elongatum (Fisch. et C. A. Mey.) Bornm.;伸长三肋果■☆

397959　Tripleurospermum fuscatum (Desf.) Sch. Bip. = Heteromera fuscata（Desf.）Pomel ■☆

397960 Tripleurospermum grossheimii（Fed.）Pobed.；格氏三肋果■☆

397961 Tripleurospermum homogamum G. X. Fu = Tripleurospermum homogamum G. X. Fu ex Y. Ling et C. Shih ■

397962 Tripleurospermum homogamum G. X. Fu ex Y. Ling et C. Shih；无舌三肋果；Tongueless Threebibachene, Tongueless Threeribfruit ■

397963 Tripleurospermum hookeri Sch. Bip. = Tripleurospermum maritimum（L.）W. D. J. Koch subsp. phaeocephalum（Rupr.）Hämet-Ahti ■☆

397964 Tripleurospermum inodorum（L.）Sch. Bip.；新疆三肋果（不母菊,淡甘菊,无味母菊）；Common Threebibachene, False Chamomile, False Mayweed, Perforated Matricary, Perforated Mayweed, Scendess Mayweed, Scentless Chamomile, Scentless False Chamomile, Scentless Mayweed, Tassel Flower, Triplospermum, Xinjiang Threeribfruit ■

397965 Tripleurospermum inodorum（L.）Sch. Bip. = Chrysanthemum inodorum L. ■

397966 Tripleurospermum inodorum（L.）Sch. Bip. = Matricaria maritima L. ■☆

397967 Tripleurospermum inodorum（L.）Sch. Bip. = Tripleurospermum maritimum（L.）W. D. J. Koch ■☆

397968 Tripleurospermum karjaginii（Manden. et Sofijeva）Pobed.；卡氏三肋果■☆

397969 Tripleurospermum limosum（Maxim.）Pobed.；三肋果（幼母菊）；Marshy Threeribfruit ■

397970 Tripleurospermum maritimum（L.）W. D. J. Koch；海滨三肋果（海岸母菊,海滨母菊）；Camomile Goldins, Corn Feverfew, Corn Mayweed, Dog Daisy, Dog Gowan, Dog's Camomile, Dog's Camovyne, False Camomile, False Mayweed, Fern-leaved Daisy, Gypsy Daisy, Gypsy Flower, Harvest Daisy, Horse Daisy, Madarin, Scentless Camomile, Scentless Chamomile, Scentless Feverfew, Scentless Matricary, Scentless Maydweed, Sea Mayweed, Seashore Mayweed ■☆

397971 Tripleurospermum maritimum（L.）W. D. J. Koch = Matricaria maritima L. ■☆

397972 Tripleurospermum maritimum（L.）W. D. J. Koch subsp. inodorum（L.）Applequist = Tripleurospermum inodorum（L.）Sch. Bip. ■

397973 Tripleurospermum maritimum（L.）W. D. J. Koch subsp. phaeocephalum（Rupr.）Hämet-ahti；暗头海滨三肋果；Seashore Chamomile, Wild Chamomile ■☆

397974 Tripleurospermum monticola（Boiss. et Huet）Bornm.；山地三肋果■☆

397975 Tripleurospermum parviflorum（Willd.）Pobed.；小花三肋果■☆

397976 Tripleurospermum perforatum（Mérat）M. Lainz；穿孔三肋果；Scentless False Mayweed ■☆

397977 Tripleurospermum perforatum（Mérat）M. Lainz = Tripleurospermum inodorum（L.）Sch. Bip. ■

397978 ripleurospermum phaeocephalum（Rupr.）Pobed. = Tripleurospermum maritimum（L.）W. D. J. Koch subsp. phaeocephalum（Rupr.）Hämet-Ahti ■☆

397979 Tripleurospermum philaenorum（Maire et Weiller）Alavi = Heteromera philaenorum Maire et Weiller ■☆

397980 Tripleurospermum pulchrum（Ledeb.）Rupr. = Pyrethrum pulchrum Ledeb. ■

397981 Tripleurospermum rupestre（Sommier et H. Lév.）Pobed.；沼泽三肋果■☆

397982 Tripleurospermum sevanense（Manden.）Pobed.；塞凡三肋果■☆

397983 Tripleurospermum szowitzii（DC.）Pobed.；索氏三肋果■☆

397984 Tripleurospermum tchihatchewii（Boiss.）Bornm.；齐氏三肋果■☆

397985 Tripleurospermum tetragonospermum（Eastw.）Pobed. =

Tripleurospermum tetragonospermum（F. Schmidt）Pobed. ■

397986 Tripleurospermum tetragonospermum（F. Schmidt）Pobed.；东北三肋果（褐苞三肋果）；Fourangle Threeribfruit, Fourangleseed Threebibachene ■

397987 Tripleurospermum transcaucasicum（Manden.）Pobed.；外高加索三肋果■☆

397988 Triplima Raf. = Carex L. ■

397989 Triplisomeris（Baill.）Aubrév. et Pellegr.（1958）；肖仿花苏木属●☆

397990 Triplisomeris（Baill.）Aubrév. et Pellegr. = Anthonotha P. Beauv. ●☆

397991 Triplisomeris Aubrév. et Pellegrin = Anthonotha P. Beauv. ●☆

397992 Triplisomeris ernae（Dinkl.）Aubrév. et Pellegr.；肖仿花苏木●☆

397993 Triplisomeris explicans（Baill.）Aubrév. et Pellegr.；明显肖仿花苏木●☆

397994 Triplisomeris pellegrinii Aubrév.；佩尔格兰肖仿花苏木●☆

397995 Triplisomeris triplisomeris（Pellegr.）Aubrév. et Pellegr. = Triplisomeris pellegrinii Aubrév. ●☆

397996 Triplobaceae Raf = Malvaceae Juss.（保留科名）●■

397997 Triplobus Raf. = Sterculia L. ●

397998 Triplobus Raf. = Triphaca Lour. ●

397999 Triplocentron Cass. = Centaurea L.（保留属名）●■

398000 Triplocephalum O. Hoffm.（1894）；三头菊属●☆

398001 Triplocephalum holstii O. Hoffm.；三头菊●☆

398002 Triplochiton Alef.（废弃属名）= Triplochiton K. Schum.（保留属名）●☆

398003 Triplochiton K. Schum.（1900）（保留属名）；非洲梧桐属（切普劳奇属）●☆

398004 Triplochiton scleroxylon K. Schum.；非洲梧桐（切普劳奇）；Arere, Ayous, Obeche, Wawa, White Wood ●☆

398005 Triplochiton utile Sprague = Tarrietia utilis（Sprague）Sprague ●☆

398006 Triplochiton zambesiacus Milne-Redh.；赞比亚梧桐（赞比亚切普劳奇）●☆

398007 Triplochitonaceae K. Schum. = Malvaceae Juss.（保留科名）●■

398008 Triplochitonaceae K. Schum. = Sterculiaceae Vent.（保留科名）●■

398009 Triplochlamys Ulbr.（1915）；三罩锦葵属●☆

398010 Triplochlamys conferta Ulbr.；三罩锦葵●☆

398011 Triplochlamys longifolia Ulbr.；长叶三罩锦葵●☆

398012 Triplochlamys multiflora Ulbr.；多花三罩锦葵●☆

398013 Triplolepis Turcz. = Streptocaulon Wight et Arn. ■

398014 Triplomeia Raf. = Licaria Aubl. ●☆

398015 Triplopetalum Nyar. = Alyssum L. ■●

398016 Triplopogon Bor（1954）；三须颖草属■☆

398017 Triplopogon ramosissimus（Hack.）Bor；三须颖草■☆

398018 Triplorhiza Ehrh. = Leucorchis E. Mey. ■☆

398019 Triplorhiza Ehrh. = Pseudorchis Ség. ■☆

398020 Triplorhiza Ehrh. = Satyrium Sw.（保留属名）■

398021 Triplosperma G. Don = Ceropegia L. ■

398022 Triplostegia Wall. ex DC.（1830）；双参属（囊苞花属,小缬草属）；Triplostegia, Twinginseng ■

398023 Triplostegia delavayi Franch. ex Diels = Triplostegia grandiflora Gagnep. ■

398024 Triplostegia glandulifera Wall. ex DC.；双参（白都拉,都拉,都拉参,肚拉,对对参,萝卜参,萝卜都拉,三萼花草,土败酱,土洋参,西南囊苞花,一支蒿）；Common Triplostegia, Twinginseng ■

398025 Triplostegia grandiflora Gagnep.；大花双参（大花囊苞花,对对参,合合参,萝卜参,青羊参,山苦参,双参,童子参,羊蹄参,子母参）；Largeflower Triplostegia, Largeflower Twinginseng ■

398026 Triplostegia mairei H. Lév. = Chrysosplenium macrophyllum Oliv. ■

398027　Triplostegia pinifolia H. Lév. = Rhodiola fastigiata（Hook. f. et Thomson）S. H. Fu ■

398028　Triplostegia repens Hemsl. = Triplostegia glandulifera Wall. ex DC. ■

398029　Triplostegiaceae（Höck）A. E. Bobrov ex Airy Shaw（1965）；双参科■

398030　Triplostegiaceae A. E. Bobrov ex Airy Shaw = Dipsacaceae Juss.（保留科名）■●

398031　Triplostegiaceae A. E. Bobrov ex Airy Shaw = Triplostegiaceae（Höck）A. E. Bobrov ex Airy Shaw ■

398032　Triplostegiaceae A. E. Bobrov ex Airy Shaw = Valerianaceae Batsch（保留科名）●■

398033　Triplostegiaeeae（Höck）Airy Shaw = Triplostegiaceae（Höck）A. E. Bobrov ex Airy Shaw ■

398034　Triplostegiaeeae（Höck）Airy Shaw = Valerianaceae Batsch（保留科名）●■

398035　Triplotaxis Hutch. = Vernonia Schreb.（保留属名）●■

398036　Triplotaxis lundiensis Hutch. = Vernonia lundiensis（Hutch.）Wild et G. V. Pope ■☆

398037　Triplotaxis somalensis（O. Hoffm.）Hutch. = Gutenbergia somalensis（O. Hoffm.）M. G. Gilbert ■☆

398038　Triplotaxis stellulifera（Benth.）Hutch. = Vernonia stellulifera（Benth.）C. Jeffrey ■☆

398039　Tripodandra Baill. = Rhaptonema Miers ●☆

398040　Tripodandra thouarsiana Baill. = Rhaptonema thouarsiana（Baill.）Diels ●☆

398041　Tripodanthera M. Roem. = Gymnopetalum Arn. ■

398042　Tripodanthera cochinchinensis（Lour.）M. Roem. = Gymnopetalum chinense（Lour.）Merr. ■

398043　Tripodanthus（Eichler）Tiegh.（1895）；三足花属●☆

398044　Tripodanthus Tiegh. = Loranthus Jacq.（保留属名）●

398045　Tripodanthus Tiegh. = Tripodanthus（Eichler）Tiegh. ●☆

398046　Tripodanthus acutifolius Tiegh. ；尖叶三足花●☆

398047　Tripodanthus destructor Tiegh. ；三足花●☆

398048　Tripodanthus ligustrinus Tiegh. ；女贞三足花●☆

398049　Tripodanthus suaveolens Tiegh. ；香三足花●☆

398050　Tripodion Medik.（1787）；三足豆属■☆

398051　Tripodion Medik. = Anthyllis L. ■☆

398052　Tripodion kremerianum（Coss.）Lassen；非洲三足豆■☆

398053　Tripodion tetraphyllum（L.）Fourr. ；三足豆■☆

398054　Tripodium Medik. = Anthyllis L. ■☆

398055　Tripogandra Raf.（1837）；三芒蕊属■☆

398056　Tripogandra angustifolia（B. L. Rob.）Woodson；狭叶三芒蕊■☆

398057　Tripogandra cordifolia（Sw.）Aristeg. ；心叶三芒蕊■☆

398058　Tripogandra multiflora（Sw.）Steyerm. ；多花三芒蕊■☆

398059　Tripogandra parviflora（Ruiz et Pav.）Steyerm. ；小花三芒蕊■☆

398060　Tripogandra purpurascens（S. Schauer）Handlos；紫三芒蕊■☆

398061　Tripogandra rosea（Vent.）Woodson；粉红三芒蕊■☆

398062　Tripogandra stenophylla（Brandegee）Matuda；窄叶三芒蕊■☆

398063　Tripogon Bor = Tripogon Roem. et Schult. ■

398064　Tripogon Roem. et Schult.（1817）；草沙蚕属；Herbclamworm，Tripogon ■

398065　Tripogon Roth = Tripogon Roem. et Schult. ■

398066　Tripogon abyssinicus Nees = Tripogon purpurascens Duthie ■

398067　Tripogon abyssinicus Nees ex Steud. = Tripogon leptophyllus（A. Rich.）Cufod. ■☆

398068　Tripogon bromoides Roth ex Roem. et Schult. ；草沙蚕；Bromus-like Tripogon，Herbclamworm ■

398069　Tripogon bromoides Roth ex Roem. et Schult. var. yunnanensis（Keng ex J. L. Yang）S. L. Chen et X. L. Yang = Tripogon yunnanensis Keng ex J. L. Yang ■

398070　Tripogon calcicola A. Camus = Tripogon minimus（A. Rich.）Steud. ■☆

398071　Tripogon chinensis（Franch.）Hack. ；中华草沙蚕；China Herbclamworm，Chinese Tripogon ■

398072　Tripogon chinensis（Franch.）Hack. subsp. coreensis（Hack.）T. Koyama = Tripogon chinensis（Franch.）Hack. ■

398073　Tripogon chinensis（Franch.）Hack. var. coreensis Hack. = Tripogon chinensis（Franch.）Hack. ■

398074　Tripogon coreensis（Hack.）Ohwi = Tripogon chinensis（Franch.）Hack. ■

398075　Tripogon coreensis（Hack.）Ohwi var. longe-aristatus（Nakai）Hack. ex Mori = Tripogon longe-aristatus Honda ■

398076　Tripogon debilis L. B. Cai；柔弱草沙蚕■

398077　Tripogon festucoides Jaub. et Spach = Tripogon bromoides Roth ex Roem. et Schult. ■

398078　Tripogon filiformis Nees ex Steud. ；线形草沙蚕（小草沙蚕）；Thready Herbclamworm，Thready Tripogon ■

398079　Tripogon filiformis Nees ex Steud. var. tenuispicus Hook. f. = Tripogon filiformis Nees ex Steud. ■

398080　Tripogon hookeranus Bor ex Sultan et Stewart = Tripogon purpurascens Duthie ■

398081　Tripogon humbertianus A. Camus = Tripogon minimus（A. Rich.）Hochst. ex Steud. ■☆

398082　Tripogon humilis H. L. Yang；矮草沙蚕；Dwarf Herbclamworm，Small Tripogon ■

398083　Tripogon jacquemontii Stapf var. submuticus Hook. f. = Tripogon purpurascens Duthie ■

398084　Tripogon jaegerianus A. Camus = Tripogon major Hook. f. ■☆

398085　Tripogon japonicus（Honda）Ohwi = Tripogon longe-aristatus Hack. ex Honda var. japonica Honda ■

398086　Tripogon japonicus（Honda）Ohwi = Tripogon longe-aristatus Hack. ex Honda ■

398087　Tripogon leptophyllus（A. Rich.）Cufod. ；细叶草沙蚕■☆

398088　Tripogon liebenbergii C. E. Hubb. = Tripogon major Hook. f. ■☆

398089　Tripogon liouae S. M. Phillips et S. L. Chen；丽藕草沙蚕■

398090　Tripogon longe-aristatus（Nakai）Honda = Tripogon longe-aristatus Hack. ex Honda ■

398091　Tripogon longe-aristatus Hack. ex Honda；长芒草沙蚕；Longawn Herbclamworm，Longawn Tripogon ■

398092　Tripogon longe-aristatus Hack. ex Honda subsp. japonicus（Honda）T. Koyama = Tripogon longe-aristatus Hack. ex Honda ■

398093　Tripogon longe-aristatus Hack. ex Honda var. japonica Honda；日本草沙蚕■

398094　Tripogon longe-aristatus Hack. ex Honda var. japonicus Honda = Tripogon longe-aristatus Hack. ex Honda ■

398095　Tripogon longe-aristatus Honda = Tripogon longe-aristatus Hack. ex Honda ■

398096　Tripogon longe-aristatus Honda subsp. japonicus（Honda）T. Koyama = Tripogon longe-aristatus Hack. ex Honda ■

398097　Tripogon longe-aristatus Honda var. japonicus Honda = Tripogon longe-aristatus Hack. ex Honda ■

398098　Tripogon longe-aristatus Nakai = Tripogon longe-aristatus Hack. ex Honda ■

398099　Tripogon major Hook. f. ；大草沙蚕■☆

398100　Tripogon major Hook. f. subsp. deflexus Gledhill = Tripogon major Hook. f. ■☆

398101　Tripogon major Hook. f. subsp. jaegeranus Gledhill = Tripogon major Hook. f. ■☆

398102　Tripogon mandarensis A. Camus = Tripogon minimus（A. Rich.）Hochst. ex Steud. ■☆

398103　Tripogon minimus（A. Rich.）Hochst. ex Steud.；矮小草沙蚕■☆

398104　Tripogon minimus（A. Rich.）Steud. = Tripogon minimus（A. Rich.）Hochst. ex Steud. ■☆

398105　Tripogon modestus S. M. Phillips et Launert；适度草沙蚕■☆

398106　Tripogon montanus Chiov.；山地草沙蚕■☆

398107　Tripogon multiflorus Miré et H. Gillet；多花草沙蚕■☆

398108　Tripogon nanus Keng ex P. C. Keng et L. Liou；小草沙蚕；Dwarf Tripogon, Small Herbclamworm ■

398109　Tripogon nanus Keng ex P. C. Keng et L. Liou = Tripogon filiformis Nees ex Steud. ■

398110　Tripogon panxianensis H. Peng；盘县草沙蚕；Panxian Herbclamworm, Tripogon ■

398111　Tripogon panxianensis H. Peng = Tripogon longe-aristatus Hack. ex Honda ■

398112　Tripogon purpurascens Duthie；玫瑰紫草沙蚕；Rosepurple Herbclamworm, Tripogon ■

398113　Tripogon rupestris S. M. Phillips et S. L. Chen；岩生草沙蚕■

398114　Tripogon sichuanicus S. M. Phillips et S. L. Chen；四川草沙蚕■

398115　Tripogon snowdenii C. E. Hubb. = Tripogon major Hook. f. ■☆

398116　Tripogon subtilissimus Chiov.；纤细草沙蚕■☆

398117　Tripogon tibesticus Miré, H. Gillet et Quézel = Tripogon multiflorus Miré et H. Gillet ■☆

398118　Tripogon trifidus Munro ex Hook. f.；三裂草沙蚕■

398119　Tripogon unidentatus Nees ex Steud. = Tripogon filiformis Nees ex Steud. ■

398120　Tripogon unisetus Pilg. = Tripogon major Hook. f. ■☆

398121　Tripogon yunnanensis J. L. Yang ex S. M. Phillips et S. L. Chen；云南草沙蚕；Yunnan Herbclamworm, Yunnan Tripogon ■

398122　Tripogon yunnanensis Keng ex J. L. Yang = Tripogon bromoides Roth ex Roem. et Schult. var. yunnanensis（Keng ex J. L. Yang）S. L. Chen et X. L. Yang ■

398123　Tripogon yunnanensis Keng ex J. L. Yang = Tripogon yunnanensis J. L. Yang ex S. M. Phillips et S. L. Chen ■

398124　Tripolion Raf. = Tripolium Nees ■

398125　Tripolium Nees = Aster L. ●■

398126　Tripolium Nees(1832)；碱菀属（碱紫菀属）；Seastarwort ■

398127　Tripolium angustum Lindl. = Brachyactis ciliata（Ledeb.）Ledeb. subsp. angusta（Lindl.）A. G. Jones ■☆

398128　Tripolium frondosum Nutt. = Symphyotrichum frondosum（Nutt.）G. L. Nesom ■☆

398129　Tripolium pannonicum（Jacq.）Dobrocz. = Aster pannonicus Jacq. ■☆

398130　Tripolium pannonicum（Jacq.）Dobrocz. = Aster tripolium L. ■☆

398131　Tripolium pannonicum Schur = Aster tripolium L. ■☆

398132　Tripolium pannonicum Schur = Tripolium vulgare Nees ■

398133　Tripolium subulatum（Michx.）DC. var. parviflorum Nees = Symphyotrichum subulatum（Michx.）G. L. Nesom var. parviflorum（Nees）S. D. Sundb. ■☆

398134　Tripolium vulgare Nees；碱菀（灯笼花,金盏菜,麝香紫菀,铁秆蒿,竹叶菊）；Alkali Aster, Alkaliaster, Blue Camomile, Blue Daisy, Hog's Bean, Hog's Beans, Michaelmas Daisy, Musk Button, Musk Buttons, Purple Camomile, Sea Aster, Sea Starwort, Seastarwort, Serapia's Turbith, Sharewort, Starwort, Summer's Farewell, Toadwort, Tripoli Aster ■

398135　Tripolium vulgare Nees = Aster tripolium L. ■☆

398136　Tripora P. D. Cantino(1999)；三孔草属■☆

398137　Tripora divaricata（Maxim.）P. D. Cantino；三孔草■☆

398138　Tripsacum L.（1759）；磨擦禾属（磨擦草属）；Rubgrass, Gamagrass, Gama Grass, Gama-grass ■

398139　Tripsacum dactyloides（L.）L.；指状磨擦禾（鸭茅状磨擦禾,鸭足状磨擦草）；Eastern Gama Grass, Eastern Gamagrass, Gama Grass, Gama-grass ■☆

398140　Tripsacum fasciculatum Trin. ex Asch.；磨擦草（瓜地马拉草）；Guatemala Grass, Guatemala-grass, Lax Rubgrass ●■

398141　Tripsacum laxum Nash = Tripsacum fasciculatum Trin. ex Asch. ●■

398142　Tripsilina Raf. = Passiflora L. ●■

398143　Tripsilina foetida（L.）Raf. = Passiflora foetida L. ■

398144　Tripterachaenium Kuntze = Tripteris Less.（保留属名）■●☆

398145　Tripterachaenium humile Kuntze = Tripteris aghillana DC. ●☆

398146　Tripterachaenium scariosum（DC.）Kuntze = Tripteris aghillana DC. ●☆

398147　Tripteranthus Wall. ex Miers = Tripterella Michx. ■

398148　Tripterella Michx. = Burmannia L. ■

398149　Tripterella Michx. = Vogelia J. F. Gmel. ■

398150　Tripterellaceae Dumort. = Burmanniaceae Blume(保留科名)■

398151　Tripterigiaceae Dumort. = Celastraceae R. Br.（保留科名）●

398152　Tripteris Less.（1831）（保留属名）；三翅菊属■●☆

398153　Tripteris Less. = Osteospermum L. ●■

398154　Tripteris Thunb. = Triopterys L.（保留属名）●☆

398155　Tripteris afromontana（Norl.）B. Nord.；非洲山生三翅菊■☆

398156　Tripteris aghillana DC.；纳塔尔三翅菊●☆

398157　Tripteris aghillana DC. var. integrifolia Harv.；全叶纳塔尔三翅菊●☆

398158　Tripteris amplectens Harv. = Norlindhia amplectens（Harv.）B. Nord. ■☆

398159　Tripteris amplexicaulis（Thunb.）Less.；抱茎三翅菊●☆

398160　Tripteris amplexicaulis（Thunb.）Less. var. microtis DC. = Tripteris microcarpa Harv. ●☆

398161　Tripteris amplexicaulis DC. = Norlindhia amplectens（Harv.）B. Nord. ■☆

398162　Tripteris angolensis（Norl.）B. Nord.；安哥拉三翅菊●☆

398163　Tripteris angustissima S. Moore = Tripteris vaillantii Decne. ●☆

398164　Tripteris arborescens（Jacq.）Less. = Tripteris dentata（Burm. f.）Harv. ●☆

398165　Tripteris arborescens Harv. = Osteospermum grandiflorum DC. ●☆

398166　Tripteris atropurpurea Turcz. = Monoculus monstruosus（Burm. f.）B. Nord. ■☆

398167　Tripteris auriculata S. Moore；耳形三翅菊●☆

398168　Tripteris bolusii Compton = Osteospermum bolusii（Compton）Norl. ■☆

398169　Tripteris breviradiata（Norl.）B. Nord. = Norlindhia breviradiata（Norl.）B. Nord. ■☆

398170　Tripteris burchellii Hook. f. = Oligocarpus burchellii（Hook. f.）B. Nord. ■☆

398171　Tripteris cheiranthifolia Sch. Bip. = Tripteris vaillantii Decne. ●☆

398172　Tripteris clandestina Less. = Monoculus monstruosus（Burm. f.）B. Nord. ■☆

398173　Tripteris concordiae Schltr. = Osteospermum spinescens Thunb. ■☆

398174　Tripteris confusa Bolus = Norlindhia amplectens（Harv.）B. Nord.■☆

398175　Tripteris crassifolia O. Hoffm.；厚叶三翅菊●☆

398176　Tripteris dentata（Burm. f.）Harv.；齿叶三翅菊●☆

398177　Tripteris flexuosa Harv. = Tripteris aghillana DC.●☆

398178　Tripteris fruticosa Muschl. ex Dinter = Osteospermum karrooicum（Bolus）Norl.■☆

398179　Tripteris fruticosa Muschl. ex Engl. = Tripteris crassifolia O. Hoffm.●☆

398180　Tripteris glabra C. A. Sm. = Osteospermum glabrum N. E. Br.■☆

398181　Tripteris glabrata（Thunb.）Harv. = Tripteris oppositifolia（Aiton）B. Nord.●☆

398182　Tripteris glandulosa Muschl. = Tripteris microcarpa Harv.●☆

398183　Tripteris glandulosa Turcz. = Tripteris aghillana DC. var. integrifolia Harv.●☆

398184　Tripteris glandulosa Turcz. var. dentata Harv. = Tripteris aghillana DC.●☆

398185　Tripteris goetzei O. Hoffm. = Tripteris monocephala Oliv. et Hiern●☆

398186　Tripteris gossweileri Mattf. = Tripteris monocephala Oliv. et Hiern●☆

398187　Tripteris gracilis Hutch. = Norlindhia amplectens（Harv.）B. Nord.■☆

398188　Tripteris gweloensis Mattf. = Tripteris monocephala Oliv. et Hiern●☆

398189　Tripteris herbacea E. Mey. ex DC. = Monoculus monstruosus（Burm. f.）B. Nord.■☆

398190　Tripteris humilis Turcz. = Tripteris aghillana DC.●☆

398191　Tripteris hyoseroides DC. = Monoculus hyoseroides（DC.）B. Nord.■☆

398192　Tripteris hyoseroides DC. var. echinocarpa ? = Monoculus monstruosus（Burm. f.）B. Nord.■☆

398193　Tripteris incana Harv. = Tripteris oppositifolia（Aiton）B. Nord.●☆

398194　Tripteris integrifolia Dinter = Osteospermum karrooicum（Bolus）Norl.■☆

398195　Tripteris karrooica Bolus = Osteospermum karrooicum（Bolus）Norl.■☆

398196　Tripteris kraussii Sch. Bip. = Tripteris aghillana DC. var. integrifolia Harv.●☆

398197　Tripteris leptoloba Harv. = Osteospermum leptolobum（Harv.）Norl.●☆

398198　Tripteris limonifolia DC. = Tripteris polycephala DC.●☆

398199　Tripteris linearis Harv. = Tripteris sinuata DC. var. linearis（Harv.）B. Nord.●☆

398200　Tripteris lordii Oliv. et Hiern = Tripteris vaillantii Decne.●☆

398201　Tripteris macroptera DC. = Tripteris oppositifolia（Aiton）B. Nord.●☆

398202　Tripteris microcarpa Harv.；小果三翅菊●☆

398203　Tripteris microcarpa Harv. subsp. septentrionalis（Norl.）B. Nord.；北方三翅菊●☆

398204　Tripteris microtis（DC.）Hutch. = Tripteris microcarpa Harv.●☆

398205　Tripteris monocephala Oliv. et Hiern；单头三翅菊●☆

398206　Tripteris muschlerianus Dinter = Osteospermum spinescens Thunb.■☆

398207　Tripteris namaquensis Schltr. = Tripteris microcarpa Harv.●☆

398208　Tripteris natalensis Harv. = Tripteris aghillana DC.●☆

398209　Tripteris nervosa Hutch.；多脉三翅菊●☆

398210　Tripteris nyikensis（Norl.）B. Nord.；尼卡三翅菊●☆

398211　Tripteris oppositifolia（Aiton）B. Nord.；对叶三翅菊●☆

398212　Tripteris pachypteris（DC.）Harv. = Osteospermum spinescens Thunb.■☆

398213　Tripteris pallescens DC. = Tripteris oppositifolia（Aiton）B. Nord.●☆

398214　Tripteris petiolata DC. = Osteospermum grandiflorum DC.●☆

398215　Tripteris pinnatilobata（Norl.）B. Nord.；羽裂三翅菊●☆

398216　Tripteris polycephala DC.；多头三翅菊●☆

398217　Tripteris rhodesica R. E. Fr. = Tripteris monocephala Oliv. et Hiern●☆

398218　Tripteris rigida Harv. = Tripteris dentata（Burm. f.）Harv.●☆

398219　Tripteris rogersii S. Moore = Osteospermum rigidum Aiton■☆

398220　Tripteris rosulata（Norl.）B. Nord.；莲座三翅菊●☆

398221　Tripteris scariosa DC. = Monoculus monstruosus（Burm. f.）B. Nord.■☆

398222　Tripteris setifera DC. = Tripteris aghillana DC. var. integrifolia Harv.●☆

398223　Tripteris sinuata DC.；深波三翅菊●☆

398224　Tripteris sinuata DC. var. linearis（Harv.）B. Nord.；线叶深波三翅菊●☆

398225　Tripteris spathulata DC.；匙形三翅菊●☆

398226　Tripteris spinescens（Thunb.）Harv. = Osteospermum spinescens Thunb.■☆

398227　Tripteris spinigera Norl.；刺三翅菊●☆

398228　Tripteris thomii Harv. = Tripteris oppositifolia（Aiton）B. Nord.●☆

398229　Tripteris tomentosa（L. f.）Less. = Inuloides tomentosa（L. f.）B. Nord.■☆

398230　Tripteris vaillantii Decne.；瓦扬三翅菊●☆

398231　Tripteris volkensii O. Hoffm.；福尔三翅菊●☆

398232　Tripterium（DC.）Bercht. et J. Presl = Thalictrum L.■

398233　Tripterium Bercht. et J. Presl = Thalictrum L.■

398234　Tripterocalyx（Torr.）Hook.（1853）；沙烟花属（三翅萼属）；Sand-puffs■☆

398235　Tripterocalyx（Torr.）Hook. = Abronia Juss.■☆

398236　Tripterocalyx Hook. = Abronia Juss.■☆

398237　Tripterocalyx Hook. = Tripterocalyx（Torr.）Hook.■☆

398238　Tripterocalyx Hook. ex Standl. = Abronia Juss.■☆

398239　Tripterocalyx Torr. = Tripterocalyx（Torr.）Hook.■☆

398240　Tripterocalyx carneus（Greene）L. A. Galloway var. wootonii（Standl.）L. A. Galloway = Tripterocalyx wootonii Standl.■☆

398241　Tripterocalyx crux-maltae（Kellogg）Standl.；沙地沙烟花（沙地三翅萼）■☆

398242　Tripterocalyx micranthus（Torr.）Hook.；小花沙烟花（小花三翅萼）■☆

398243　Tripterocalyx wootonii Standl.；伍顿沙烟花（伍顿三翅萼）■☆

398244　Tripterocarpus Meisn. = Bridgesia Bertero ex Cambess.（保留属名）●☆

398245　Tripterococcus Endl.（1837）；三翅异雄蕊属■☆

398246　Tripterococcus brunonis Endl.；三翅异雄蕊☆

398247　Tripterodendron Radlk.（1890）；三翅木属☆

398248　Tripterodendron filicifolium Radlk；三翅木●☆

398249　Tripterospermum Blume = Crawfurdia Wall.■

398250　Tripterospermum Blume = Gentiana L.■

398251　Tripterospermum Blume（1826）；双蝴蝶属（肺形草属）；Dualbutterfly，Tripterospermum■

398252　Tripterospermum affine（Wall.）Harry Sm. = Tripterospermum nienkui（C. Marquand）C. J. Wu■

398253　Tripterospermum alutaceofolium（Tang S. Liu et Chin C. Kuo）J. Murata；台北肺形草（台北双蝴蝶，狭叶台湾肺形草）；Taibei

Dualbutterfly ■

398254　Tripterospermum angustatum （C. B. Clarke） Raizada = Crawfurdia angustata C. B. Clarke ■

398255　Tripterospermum australe J. Murata；南方双蝴蝶■

398256　Tripterospermum brevilobum D. Fang；短裂双蝴蝶；Shortlobed Dualbutterfly ■

398257　Tripterospermum bulleyanum （Forrest） Raizada = Crawfurdia campanulacea Wall. et Griff. ex C. B. Clarke ■

398258　Tripterospermum campanulacea （Wall. et Griff. ex C. B. Clarke） Raizada = Crawfurdia campanulacea Wall. et Griff. ex C. B. Clarke ■

398259　Tripterospermum canceolatum ?；方格双蝴蝶■☆

398260　Tripterospermum carlesii Harry Sm. = Tripterospermum chinense （Migo） Harry Sm. ■

398261　Tripterospermum caudatum （Marquand） Harry Sm. = Tripterospermum filicaule （Hemsl.） Harry Sm. ■

398262　Tripterospermum chinense （Migo） Harry Sm.；双蝴蝶（肺形草，华肺形草，黄金线，金交杯，金丝蝴蝶，蔓龙胆，山蝴蝶，石板青，四脚喜，铁板青，铁交杯，铜交杯）；Chinese Tripterospermum, Dualbutterfly ■

398263　Tripterospermum coeruleum （Hand.-Mazz. ex Harry Sm.） Harry Sm.；盐源双蝴蝶（小黄鳝藤，小筋骨藤）；Azure Dualbutterfly, Azure Tripterospermum ■

398264　Tripterospermum cordatum （C. Marquand） Harry Sm.；峨眉双蝴蝶；Emei Dualbutterfly, Common Tripterospermum ■

398265　Tripterospermum cordifolioides J. Murata；心叶双蝴蝶■

398266　Tripterospermum cordifolium （Yamam.） Satake；高山肺形草（缠竹黄）；Alpine Dualbutterfly, Alpine Tripterospermum ■

398267　Tripterospermum discoideum （C. Marquand） Harry Sm.；湖北双蝴蝶；Hubei Dualbutterfly, Hubei Tripterospermum ■

398268　Tripterospermum distylum J. Murata et Yahara；二柱双蝴蝶■☆

398269　Tripterospermum fasciculatum （Wall.） Chater；簇生双蝴蝶（双蝴蝶）■

398270　Tripterospermum filicaule （Hemsl.） Harry Sm.；细茎双蝴蝶；Thinstem Dualbutterfly, Thinstem Tripterospermum ■

398271　Tripterospermum hirticalyx C. Y. Wu ex C. J. Wu；毛萼双蝴蝶；Haircalyx Dualbutterfly, Haircalyx Tripterospermum ■

398272　Tripterospermum involubile N. Yonez. = Tripterospermum trinervium （Thunb.） H. Ohashi et H. Nakai var. involubile （N. Yonez.） H. Ohashi et H. Nakai ■☆

398273　Tripterospermum japonicum （Siebold et Zucc.） Maxim.；日本双蝴蝶（抽筋草，胆草，对叶林，蔓龙胆，青鱼胆草，日本蔓龙胆）；Japan Dualbutterfly, Japanese Tripterospermum, Japanese Vinegentin ■

398274　Tripterospermum japonicum （Siebold et Zucc.） Maxim. = Tripterospermum trinervium （Thunb.） H. Ohashi et H. Nakai ■☆

398275　Tripterospermum japonicum （Siebold et Zucc.） Maxim. f. albiflorum Honda = Tripterospermum japonicum （Siebold et Zucc.） Maxim. ■

398276　Tripterospermum japonicum （Siebold et Zucc.） Maxim. f. albiflorum Honda；白花日本双蝴蝶■☆

398277　Tripterospermum japonicum （Siebold et Zucc.） Maxim. f. leucanthum Honda = Tripterospermum japonicum （Siebold et Zucc.） Maxim. ■

398278　Tripterospermum japonicum （Siebold et Zucc.） Maxim. f. leucocarpum Honda = Tripterospermum japonicum （Siebold et Zucc.） Maxim. f. albiflorum Honda ■☆

398279　Tripterospermum japonicum （Siebold et Zucc.） Maxim. var. albiflorum Y. N. Lee = Tripterospermum japonicum （Siebold et Zucc.） Maxim. ■

398280　Tripterospermum japonicum （Siebold et Zucc.） Maxim. var. involubile （N. Yonez.） J. Murata f. albiflorum T. Shimizu = Tripterospermum japonicum （Siebold et Zucc.） Maxim. var. involubile （N. Yonez.） J. Murata f. album T. Shimizu ■☆

398281　Tripterospermum japonicum （Siebold et Zucc.） Maxim. var. involubile （N. Yonez.） J. Murata = Tripterospermum trinervium （Thunb.） H. Ohashi et H. Nakai var. involubile （N. Yonez.） H. Ohashi et H. Nakai ■☆

398282　Tripterospermum japonicum （Siebold et Zucc.） Maxim. var. involubile （N. Yonez.） J. Murata f. album T. Shimizu；白日本双蝴蝶■☆

398283　Tripterospermum japonicum （Siebold et Zucc.） Maxim. var. tenue （Masam.） Honda = Tripterospermum trinervium （Thunb.） H. Ohashi et H. Nakai ■☆

398284　Tripterospermum japonicum （Siebold et Zucc.） Maxim. var. tenue （Masam.） Honda = Tripterospermum japonicum （Siebold et Zucc.） Maxim. ■

398285　Tripterospermum lanceolatum （Hayata） Hara et Satake；玉山双蝴蝶（披针叶肺形草，玉山肺形草）；Lanceolate Dualbutterfly, Lanceolate Tripterospermum ■

398286　Tripterospermum luzonense （Vidal） Eggl. = Tripterospermum luzonense （Vidal） J. Murata ■

398287　Tripterospermum luzonense （Vidal） J. Murata；吕宋肺形草（高山双蝴蝶，黄花台湾肺形草）；Luzon Dualbutterfly ■

398288　Tripterospermum membranaceum （C. Marquand） Harry Sm.；膜叶双蝴蝶（腊叶双蝴蝶）■

398289　Tripterospermum microphyllum Harry Sm.；小叶双蝴蝶（西茎龙胆，细叶龙胆，小叶肺形草）；Smallleaf Dualbutterfly, Smallleaf Tripterospermum ■

398290　Tripterospermum nienkui （C. Marquand） C. J. Wu；香港双蝴蝶；Hongkong Dualbutterfly, Hongkong Tripterospermum ■

398291　Tripterospermum pallidum Harry Sm.；白花双蝴蝶；Pale Dualbutterfly, Pale Tripterospermum ■

398292　Tripterospermum pinbianense C. Y. Wu et C. J. Wu；屏边双蝴蝶；Pingbian Dualbutterfly, Pingbian Tripterospermum ■

398293　Tripterospermum puberulum （C. B. Clarke） Raizada = Crawfurdia puberula C. B. Clarke ■

398294　Tripterospermum speciosum （Wall.） Raizada = Crawfurdia speciosa Wall. ■

398295　Tripterospermum taiwanense （Masam.） Satake；台湾肺形草（高山双蝴蝶，黄花台湾肺形草，台湾双蝴蝶）；Taiwan Dualbutterfly, Taiwan Tripterospermum ■

398296　Tripterospermum taiwanense （Masam.） Satake f. alutaceifolium Tang S. Liu et Chin C. Kuo = Tripterospermum alutaceifolium （Tang S. Liu et Chin C. Kuo） J. Murata ■

398297　Tripterospermum taiwanense （Masam.） Satake f. alutaceofolium Tang S. Liu et Chin C. Kuo = Tripterospermum taiwanense （Masam.） Satake ■

398298　Tripterospermum taiwanense （Masam.） Satake var. alpinum Satake；狭叶台湾肺形草■

398299　Tripterospermum taiwanense （Masam.） Satake var. alpinum Satake = Tripterospermum luzonense （Vidal） J. Murata ■

398300　Tripterospermum taiwanense （Masam.） Satake var. alpinum Satake = Tripterospermum japonicum （Siebold et Zucc.） Maxim. ■

398301　Tripterospermum taiwanense （Masam.） Satake var. alutaceifolium Tang S. Liu et Chin C. Kuo = Tripterospermum alutaceifolium （Tang S. Liu et Chin C. Kuo） J. Murata ■

398302　Tripterospermum taiwanense （Masam.） Satake var. alutaceifolium

Tang S. Liu et Chin C. Kuo;黄花台湾肺形草■

398303　Tripterospermum taiwanense（Masam.）Satake var. alutaceofolium Tang S. Liu et Chin C. Kuo = Tripterospermum taiwanense（Masam.）Satake■

398304　Tripterospermum traillianum（Forrest）Raizada = Crawfurdia angustata C. B. Clarke ■

398305　Tripterospermum trinerve Blume = Crawfurdia blumei G. Don ■☆

398306　Tripterospermum trinervium（Thunb.）H. Ohashi et H. Nakai;三脉双蝴蝶;Threenerved Dualbutterfly,Threenerved Tripterospermum ■■☆

398307　Tripterospermum trinervium（Thunb.）H. Ohashi et H. Nakai var. involubile（N. Yonez.）H. Ohashi et H. Nakai;内卷三脉双蝴蝶■☆

398308　Tripterospermum volubile（D. Don）H. Hara;尼泊尔双蝴蝶（黄绿双蝴蝶,绕双蝴蝶,蛇药,小筋骨藤）;Nepal Dualbutterfly, Nepal Tripterospermum ■

398309　Tripterygiaceae Huber = Celastraceae R. Br.（保留科名）●

398310　Tripterygium Hook. f.（1862）;雷公藤属;Three-winged Nut, Threewingnut,Thundergodvine,Tripterygium ●

398311　Tripterygium bullockii Hance = Tripterygium wilfordii Hook. f. ●

398312　Tripterygium doianum Ohwi;道氏雷公藤●☆

398313　Tripterygium exesum（Sprague et Takeda）Rehder et Tsiang = Tripterygium forrestii Loes. ●

398314　Tripterygium forrestii A. C. Sm. = Tripterygium hypoglaucum（H. Lév.）Hutch. ●

398315　Tripterygium forrestii A. C. Sm. var. execum（Sprague et Takeda）C. H. Wang = Tripterygium hypoglaucum（H. Lév.）Hutch. ●

398316　Tripterygium forrestii Loes.;大理雷公藤（雷公藤）;Forrest Threewingnut ●

398317　Tripterygium hypoglaucum（H. Lév.）Hutch.;昆明雷公藤（大荞子,大叶黄藤,掉毛草,红毛山藤,黄藤根,火把花,火荞子,金刚藤,昆明山海棠,雷公藤,六方藤,胖关藤,山砒霜,紫金皮,紫金藤）;Glaucousback Threewingnut, Pale Three-winged Nut, Pale Threewingnut,Whiteback Thundergodvine ●

398318　Tripterygium regelii Sprague et Takeda;东北雷公藤（黑蔓）;Regel Three-winged Nut,Regel Threewingnut,Regel Thundergodvine ●

398319　Tripterygium regelii Sprague et Takeda f. hypoleucum（Hayashi）H. Hara;里白东北雷公藤●☆

398320　Tripterygium regelii Sprague et Takeda var. doianum（Ohwi）Masam. = Tripterygium doianum Ohwi ●☆

398321　Tripterygium regelii Sprague et Takeda var. occidentale T. Yamaz. = Tripterygium regelii Sprague et Takeda ●

398322　Tripterygium wilfordii Hook. f.;雷公藤（布洛克雷公藤,菜虫药,断肠草,红药,红紫根,黄腊藤,黄蜡藤,黄藤,黄藤根,黄藤木,黄药,莽草,南蛇根,三棱花,山砒霜,水莽草,水莽藤,水莽子,旱禾花）;Bullock Threewingnut, Common Threewingnut, Thundergodvine,Wilford Three-winged Nut,Wilford Three-wing-nut ●

398323　Tripterygium wilfordii Hook. f. = Tripterygium regelii Sprague et Takeda ●

398324　Tripterygium wilfordii Hook. f. var. bullockii（Hance）Matsuda = Tripterygium wilfordii Hook. f. ●

398325　Tripterygium wilfordii Hook. f. var. bullockii Matsuda = Tripterygium wilfordii Hook. f. ●

398326　Tripterygium wilfordii Hook. f. var. execum Sprague et Takeda = Tripterygium hypoglaucum（H. Lév.）Hutch. ●

398327　Tripterygium wilfordii Hook. f. var. exesum Sprague et Takeda = Tripterygium forrestii Loes. ●

398328　Triptilion Ruiz et Pav.（1794）;白蓝钝柱菊属■●☆

398329　Triptilion axillare Lag. ex Lindl.;腋生白蓝钝柱菊●☆

398330　Triptilion diffusum D. Don;铺散白蓝钝柱菊●☆

398331　Triptilion laxum Phil.;松散白蓝钝柱菊●☆

398332　Triptilion tenuifolium Phil.;细叶白蓝钝柱菊●☆

398333　Triptilodiscus Turcz.（1851）;锥托金绒草属●☆

398334　Triptilodiscus Turcz. = Helipterum DC. ex Lindl. ■☆

398335　Triptilodiscus pygmaeus Turcz.;锥托金绒草■☆

398336　Triptllium DC. = Triptilion Ruiz et Pav. ■●☆

398337　Triptolemaea Walp. = Triptolemea Mart. ●

398338　Triptolemea Mart. = Dalbergia L. f.（保留属名）●

398339　Triptorella Ritgen = Burmannia L. ●

398340　Triptorella Ritgen = Tripterella Michx. ●

398341　Tripudianthes（Seidenf.）Szlach. et Kras(2007);三鹿花属■☆

398342　Tripudianthes wallichii（Lindl.）Szlach. et Kras = Bulbophyllum wallichii Rchb. f. ■

398343　Triquetra Medik. = Astragalus L. ●■

398344　Triquiliopsis A. Heller ex Rydb. = Coldenia L. ■

398345　Triquiliopsis A. Heller ex Rydb. = Tiquiliopsis A. Heller ■☆

398346　Triquillopsis Rydb. = Coldenia L. ■

398347　Triquillopsis Rydb. = Tiquiliopsis A. Heller ■☆

398348　Triraphis Nees = Pentaschistis（Nees）Spach ■☆

398349　Triraphis R. Br.（1810）;三针草属（三芒针草属）■☆

398350　Triraphis abyssinica（Hochst. ex A. Rich.）Nees ex Engl. = Crinipes abyssinicus（Hochst. ex A. Rich.）Hochst. ■☆

398351　Triraphis andropogonoides（Steud.）E. Phillips;毛蕊三针草■☆

398352　Triraphis capensis Nees = Pentaschistis capensis（Nees）Stapf ■☆

398353　Triraphis compacta Cope;紧密三针草■☆

398354　Triraphis elliotii Rendle = Triraphis ramosissima Hack. ■☆

398355　Triraphis fleckii Hack. = Triraphis purpurea Hack. ■☆

398356　Triraphis glomerata A. Camus = Triraphis pumilio R. Br. ■☆

398357　Triraphis longipes Stapf et C. E. Hubb. = Nematopoa longipes（Stapf et C. E. Hubb.）C. E. Hubb. ■☆

398358　Triraphis mollis R. Br.;毛三芒针草;Purple Needlegrass ■☆

398359　Triraphis nana（Nees）Hack. = Triraphis pumilio R. Br. ■☆

398360　Triraphis nana Hack. = Triraphis pumilio R. Br. ■☆

398361　Triraphis nana Hack. var. conspica ? = Triraphis pumilio R. Br. ■☆

398362　Triraphis pumilio R. Br.;矮小三针草■☆

398363　Triraphis purpurea Hack.;紫三针草■☆

398364　Triraphis ramosissima Hack.;多枝三针草■☆

398365　Triraphis schinzii Hack.;欣兹三针草■☆

398366　Triraphis schlechteri Pilg. ex Stent = Triraphis schinzii Hack. ■☆

398367　Triraphis weberae Peter = Triraphis schinzii Hack. ■☆

398368　Triraphis welwitschii Rendle = Triraphis purpurea Hack. ■☆

398369　Trirhaphis Spreng. = Triraphis R. Br. ■☆

398370　Trirostellum Z. P. Wang et Q. Z. Xie = Gynostemma Blume ■

398371　Trirostellum cardiospermum（Cogn. ex Oliv.）Z. P. Wang et Q. Z. Xie = Gynostemma cardiospermum Cogn. ex Oliv. ■

398372　Trirostellum yixingense Z. P. Wang et Q. Z. Xie var. trichocarpum J. N. Ding = Gynostemma yixingense（Z. P. Wang et Q. Z. Xie）C. Y. Wu et S. K. Chen ■

398373　Trirostellum yizingense Z. P. Wang et Q. Z. Xie = Gynostemma yixingense（Z. P. Wang et Q. Z. Xie）C. Y. Wu et S. K. Chen ■

398374　Trisacarpis Raf. = Hippeastrum Herb.（保留属名）■

398375　Trisanthus Lour. = Centella L. ■

398376　Triscaphis Gagnep. = Picrasma Blume ●

398377　Triscenia Griseb.（1862）;古巴黍属■☆

398378　Triscenia ovina Griseb.;古巴黍■☆

398379　Trischidium Tul. = Swartzia Schreb.（保留属名）●☆

398380　Trisciadia Hook. f. = Coelospermum Blume ●

398381　Trisciadium Phil. = Huanaca Cav. ■☆

398382　Triscyphus Taub. = Thismia Griff. ■

398383　Triscyphus Taub. ex Warm. = Thismia Griff. ■

398384　Trisema Hook. f. = Hibbertia Andréws ●☆

398385　Trisepalum C. B. Clarke（1883）；唇萼苣苔属（斯里兰卡苣苔属）；Trisepalum ■

398386　Trisepalum birmanicum（Craib）B. L. Burtt；唇萼苣苔；Trisepalum ■

398387　Trisetaria Forssk.（1775）；三毛燕麦属■☆

398388　Trisetaria canariensis（Webb et Berthel.）Pignatti；加那利三毛燕麦■☆

398389　Trisetaria cavanillesii（Trin.）Maire = Trisetaria loeflingiana（L.）Paunero ■☆

398390　Trisetaria cristata（L.）Kerguélen = Rostraria cristata（L.）Tzvelev ■☆

398391　Trisetaria flavescens（L.）Paunero = Trisetum flavescens（L.）P. Beauv. ■☆

398392　Trisetaria flavescens（L.）Paunero subsp. pratensis（Pers.）Beck = Trisetum flavescens（L.）P. Beauv. ■☆

398393　Trisetaria flavescens（L.）Paunero var. africana（H. Lindb.）Maire = Trisetum flavescens（L.）P. Beauv. subsp. africanum（H. Lindb.）Dobignard ■☆

398394　Trisetaria flavescens（L.）Paunero var. clausonii Maire = Trisetum flavescens（L.）P. Beauv. ■☆

398395　Trisetaria flavescens（L.）Paunero var. dimorphantha Maire et Weiller = Trisetum flavescens（L.）P. Beauv. ■☆

398396　Trisetaria flavescens（L.）Paunero var. glabrata Asch. = Trisetum flavescens（L.）P. Beauv. ■☆

398397　Trisetaria flavescens（L.）Paunero var. griseovirens（H. Lindb.）Maire = Trisetum flavescens（L.）P. Beauv. subsp. griseovirens（Murb.）Dobignard ■☆

398398　Trisetaria flavescens（L.）Paunero var. macrathera（Maire et Trab.）Maire et Weiller = Trisetum flavescens（L.）P. Beauv. subsp. macratherum（Maire et Trab.）Dobignard ■☆

398399　Trisetaria flavescens（L.）Paunero var. villosa（Celak.）Maire et Weiller = Trisetum flavescens（L.）P. Beauv. ■☆

398400　Trisetaria fuscescens（Pomel）Maire = Rostraria festucoides（Link）Romero Zarco ■☆

398401　Trisetaria glumacea（Boiss.）Maire；颖毛燕麦■☆

398402　Trisetaria linearis Forssk.；线三毛燕麦■☆

398403　Trisetaria loeflingiana（L.）Paunero；卡瓦三毛草■☆

398404　Trisetaria loeflingiana（L.）Paunero var. cavanillesii（Trin.）Paunero = Trisetaria loeflingiana（L.）Paunero ■☆

398405　Trisetaria macrochaeta（Boiss.）Maire；大芒三毛燕麦■☆

398406　Trisetaria macrochaeta（Boiss.）Paunero var. boissieri Maire et Weiller = Trisetaria macrochaeta（Boiss.）Maire ■☆

398407　Trisetaria macrochaeta（Boiss.）Paunero var. pilosa（Cavara）Maire = Trisetaria macrochaeta（Boiss.）Maire ■☆

398408　Trisetaria michelii（Savi）D. Heller = Rostraria festucoides（Link）Romero Zarco ■☆

398409　Trisetaria nitida（Desf.）Maire；亮三毛燕麦■☆

398410　Trisetaria panicea（Lam.）Maire；食用三毛燕麦■☆

398411　Trisetaria panicea（Lam.）Paunero var. antiatlantica Weiller = Trisetaria panicea（Lam.）Maire ■☆

398412　Trisetaria panicea（Lam.）Paunero var. brachymera（Coss. et Durieu）Maire = Trisetaria panicea（Lam.）Maire ■☆

398413　Trisetaria panicea（Lam.）Paunero var. breviseta（Coss. et Durieu）Maire = Trisetaria panicea（Lam.）Maire ■☆

398414　Trisetaria panicea（Lam.）Paunero var. canariensis（Parl.）Maire et Weiller = Trisetaria canariensis（Webb et Berthel.）Pignatti ■☆

398415　Trisetaria panicea（Lam.）Paunero var. ciliata Willk. = Trisetaria panicea（Lam.）Maire ■☆

398416　Trisetaria panicea（Lam.）Paunero var. constricta Maire et Trab. = Trisetaria panicea（Lam.）Maire ■☆

398417　Trisetaria panicea（Lam.）Paunero var. includens（Domin）Maire et Weiller = Trisetaria panicea（Lam.）Maire ■☆

398418　Trisetaria panicea（Lam.）Paunero var. multiflora Trab. = Trisetaria panicea（Lam.）Maire ■☆

398419　Trisetaria panicea（Lam.）Paunero var. obtusata Maire et Trab. = Trisetaria panicea（Lam.）Maire ■☆

398420　Trisetaria panicea（Lam.）Paunero var. villiculmis Maire = Trisetaria panicea（Lam.）Maire ■☆

398421　Trisetaria parviflora（Desf.）Maire；小花三毛燕麦■☆

398422　Trisetaria pumila（Desf.）Maire = Lophochloa pumila（Desf.）Bor ■☆

398423　Trisetaria pumila（Desf.）Maire = Rostraria pumila（Desf.）Tzvelev ■☆

398424　Trisetaria pumila（Desf.）Maire subsp. fuscescens（Pomel）Maire et Weiller = Rostraria fuscescens（Pomel）Holub ■☆

398425　Trisetaria pumila（Desf.）Maire var. kilianii Maire = Rostraria pumila（Desf.）Tzvelev ■☆

398426　Trisetaria pumila（Desf.）Maire var. pomeliana Maire = Rostraria pumila（Desf.）Tzvelev ■☆

398427　Trisetaria quinqueseta（Steud.）Hochst. = Agrostis quinqueseta（Steud.）Hochst. ■☆

398428　Trisetaria vaccariana（Maire et Weiller）Maire；麦蓝菜三毛燕麦■☆

398429　Trisetaria vaccariana（Maire et Weiller）Maire var. glabriglumis？ = Trisetaria vaccariana（Maire et Weiller）Maire ■☆

398430　Trisetaria vaccariana（Maire et Weiller）Maire var. villiglumis？ = Trisetaria vaccariana（Maire et Weiller）Maire ■☆

398431　Trisetaria velutina（Boiss.）Paunero = Trisetaria loeflingiana（L.）Paunero ■☆

398432　Trisetarium Poir. = Trisetum Pers. ■

398433　Trisetella Luer（1980）；三尾兰属■☆

398434　Trisetella abbreviata Luer；三尾兰■☆

398435　Trisetella tenuissima（C. Schweinf.）Luer；细三尾兰■☆

398436　Trisetobromus Nevski = Bromus L.（保留属名）■

398437　Trisetum Pers.（1805）；三毛草属（蟹钓草属）；False Oat，Falseoat，Hair Grass，Trisetum，Yellow Oat-grass ■

398438　Trisetum aeneum（Hook. f.）R. R. Stewart；印巴三毛草■☆

398439　Trisetum alaskanum Nash = Trisetum spicatum（L.）K. Richt. subsp. alaskanum（Nash）Hultén ■

398440　Trisetum alpestre P. Beauv.；高山三毛草■☆

398441　Trisetum alpestre P. Beauv. subsp. glabrescens（Schur）Tzvelev；渐光高山三毛草■☆

398442　Trisetum altaicum（Stephan）Roshev.；高原三毛草（高山三毛草）；Altai Falseoat，Altai Trisetum ■

398443　Trisetum antarcticum Nees = Helictotrichon turgidulum（Stapf）Schweick. ☆

398444　Trisetum aureum Nees ex Steud. = Trisetum aeneum（Hook. f.）R. R. Stewart ■☆

398445　Trisetum aureum Ten.；金黄三毛草；Golden Oatgrass ■☆

398446　Trisetum balansae（Coss. et Durieu）Batt. = Rostraria balansae

(Coss. et Durieu) Holub ■☆

398447 Trisetum barbatum Nees = Helictotrichon barbatum (Nees) Schweick. ■☆

398448 Trisetum bifidum (Thunb.) Ohwi;三毛草;Twofork Falseoat ■

398449 Trisetum bifidum (Thunb.) Ohwi f. biaristatum (Nakai) M. Mizush.;双芒三毛草■☆

398450 Trisetum bifidum (Thunb.) Ohwi subsp. sibiricum (Rupr.) T. Koyama = Trisetum sibiricum Rupr. ■

398451 Trisetum biflorum Hochst. = Helictotrichon umbrosum (Hochst. ex Steud.) C. E. Hubb. ■☆

398452 Trisetum cavanillesii Trin. = Trisetaria loeflingiana (L.) Paunero ■☆

398453 Trisetum clarkei (Hook. f.) R. R. Stewart;长穗三毛草;C. B. Clarke Trisetum, Longspike Falseoat ■

398454 Trisetum clarkei (Hook. f.) R. R. Stewart var. kangdingense Z. L. Wu;康定三毛草;Kangding Falseoat, Kangding Trisetum ■

398455 Trisetum cristatum (L.) Potztal = Rostraria cristata (L.) Tzvelev ■☆

398456 Trisetum debile Chrtek;柔弱三毛草■

398457 Trisetum distichophyllum (Vill.) P. Beauv.;二列叶三毛草■☆

398458 Trisetum dregeanum Steud. = Helictotrichon namaquense Schweick. ■☆

398459 Trisetum faurei Sennen et Mauricio = Rostraria festucoides (Link) Romero Zarco ■☆

398460 Trisetum flavescens (L.) P. Beauv.;淡黄三毛草;Yellow Oat Grass ■☆

398461 Trisetum flavescens (L.) P. Beauv. = Trisetum pratense Pers. ■☆

398462 Trisetum flavescens (L.) P. Beauv. subsp. africanum (H. Lindb.) Dobignard;非洲淡黄三毛草■☆

398463 Trisetum flavescens (L.) P. Beauv. subsp. griseovirens (Murb.) Dobignard;灰绿三毛草■☆

398464 Trisetum flavescens (L.) P. Beauv. subsp. macratherum (Maire et Trab.) Dobignard;大花淡黄三毛草■☆

398465 Trisetum flavescens (L.) P. Beauv. subsp. sibiricum (Rupr.) T. Koyama = Trisetum sibiricum Rupr. ■

398466 Trisetum flavescens (L.) P. Beauv. var. africanum H. Lindb. = Trisetum flavescens (L.) P. Beauv. subsp. africanum (H. Lindb.) Dobignard ■☆

398467 Trisetum flavescens (L.) P. Beauv. var. bifidum (Thunb.) Makino = Trisetum bifidum (Thunb.) Ohwi ■

398468 Trisetum flavescens (L.) P. Beauv. var. bifidum Makino = Trisetum bifidum (Thunb.) Ohwi ■

398469 Trisetum flavescens (L.) P. Beauv. var. glabratum Asch. et Graebn. = Trisetum flavescens (L.) P. Beauv. ■☆

398470 Trisetum flavescens (L.) P. Beauv. var. griseovirens (H. Lindb.) Maire = Trisetum flavescens (L.) P. Beauv. subsp. griseovirens (Murb.) Dobignard ■☆

398471 Trisetum flavescens (L.) P. Beauv. var. macranthum Hack. = Trisetum bifidum (Thunb.) Ohwi ■

398472 Trisetum flavescens (L.) P. Beauv. var. nodosum Chabert = Trisetum flavescens (L.) P. Beauv. ■☆

398473 Trisetum flavescens (L.) P. Beauv. var. papilosum Hack. = Trisetum bifidum (Thunb.) Ohwi ■

398474 Trisetum flavescens (L.) P. Beauv. var. sibiricum (Rupr.) Ohwi = Trisetum sibiricum Rupr. ■

398475 Trisetum flavescens (L.) P. Beauv. var. villosum Asch. et Graebn. = Trisetum flavescens (L.) P. Beauv. ■☆

398476 Trisetum formosanum Honda = Trisetum spicatum (L.) K. Richt. subsp. alaskanum (Nash) Hultén ■

398477 Trisetum formosanum Honda = Trisetum spicatum (L.) K.

Richt. var. alascanum (Nash) Malte ex Louis-Marie ■

398478 Trisetum fuscescens Pomel = Rostraria fuscescens (Pomel) Holub ■☆

398479 Trisetum gaudinianum Boiss. = Trisetaria loeflingiana (L.) Paunero ■☆

398480 Trisetum glumaceum Boiss. = Trisetaria glumacea (Boiss.) Maire ■☆

398481 Trisetum griseovirens H. Lindb. = Trisetum flavescens (L.) P. Beauv. subsp. griseovirens (Murb.) Dobignard ■☆

398482 Trisetum henryi Rendle;湖北三毛草(亨利三毛草);Henry Falseoat ■

398483 Trisetum hirtulum Steud. = Helictotrichon hirtulum (Steud.) Schweick. ■☆

398484 Trisetum hispanicum Pers. = Trisetaria loeflingiana (L.) Paunero ■☆

398485 Trisetum imberbe Nees = Helictotrichon turgidulum (Stapf) Schweick. ■☆

398486 Trisetum koidzumianum Ohwi;小泉三毛草■☆

398487 Trisetum lachnanthum Hochst. ex A. Rich. = Helictotrichon lachnanthum (Hochst. ex A. Rich.) C. E. Hubb. ■☆

398488 Trisetum lineare (Forssk.) Boiss.;线形三毛草■☆

398489 Trisetum lineare (Forssk.) Boiss. = Trisetaria linearis Forssk. ■☆

398490 Trisetum litvinowii (Domin) Nevski = Koeleria litvinowii Domin ■

398491 Trisetum litvinowii Domin subsp. argenteum (Griseb.) Tzvelev = Koeleria litvinowii Domin subsp. argentea (Griseb.) S. M. Phillips et Z. L. Wu ■

398492 Trisetum litvinowii Domin var. argenteum (Griseb.) Tzvelev = Koeleria litvinowii Domin subsp. argentea (Griseb.) S. M. Phillips et Z. L. Wu ■

398493 Trisetum loeflingianum (L.) P. Beauv. = Trisetaria loeflingiana (L.) Paunero ■☆

398494 Trisetum longiaristum A. Rich. = Streblochaete longiarista (A. Rich.) Pilg. ■☆

398495 Trisetum longifolium Nees = Helictotrichon longifolium (Nees) Schweick. ■☆

398496 Trisetum macratherum Maire et Trab. = Trisetum flavescens (L.) P. Beauv. subsp. macratherum (Maire et Trab.) Dobignard ■☆

398497 Trisetum macrochaetum Boiss. = Trisetaria macrochaeta (Boiss.) Maire ■☆

398498 Trisetum macrochaetum Boiss. var. pilosum Cavara = Trisetaria macrochaeta (Boiss.) Maire ■☆

398499 Trisetum melicoides (Michx.) Vasey ex Scribn.;黍三毛草;Melic-oats, Purple False Oats ■☆

398500 Trisetum melicoides (Michx.) Vasey ex Scribn. var. majus (A. Gray) Hitchc. = Trisetum melicoides (Michx.) Vasey ex Scribn. ■☆

398501 Trisetum molle Kunth = Trisetum spicatum (L.) K. Richt. subsp. molle (Kunth) Hultén ■☆

398502 Trisetum molle Kunth subsp. alascanum (Nash) Rupr. = Trisetum spicatum (L.) K. Richt. var. alascanum (Nash) Malte ex Louis-Marie ■

398503 Trisetum neesii Steud. = Helictotrichon elongatum (Hochst. ex A. Rich.) C. E. Hubb. ■☆

398504 Trisetum nitidum (Desf.) Pers. = Trisetaria nitida (Desf.) Maire ■☆

398505 Trisetum paniceum (Lam.) Pers. = Trisetaria panicea (Lam.) Maire ■☆

398506 Trisetum paniceum (Lam.) Pers. var. brachymerum Coss. et Durieu = Trisetaria panicea (Lam.) Maire ■☆

398507 Trisetum paniceum (Lam.) Pers. var. canariense (Parl.) Webb et Berthel. = Trisetaria panicea (Lam.) Maire ■☆

398508　Trisetum paniceum（Lam.）Pers. var. ciliatum Willk. = Trisetaria panicea（Lam.）Maire ■☆

398509　Trisetum paniceum（Lam.）Pers. var. constrictum Maire et Trab. = Trisetaria panicea（Lam.）Maire ■☆

398510　Trisetum paniceum（Lam.）Pers. var. ducellieri Maire et Trab. = Trisetaria panicea（Lam.）Maire ■☆

398511　Trisetum paniceum（Lam.）Pers. var. obtusatum Maire et Trab. = Trisetaria panicea（Lam.）Maire ■☆

398512　Trisetum parviflorum（Desf.）Pers. = Trisetaria parviflora（Desf.）Maire ■☆

398513　Trisetum pauciflorum Keng = Trisetum pauciflorum Keng ex P. C. Kuo ■

398514　Trisetum pauciflorum Keng ex P. C. Kuo;贫花三毛草;Fewflower Trisetum,Poorflower Falseoat ■

398515　Trisetum pensylvanicum（L.）P. Beauv. =Sphenopholis pensylvanica（L.）Hitchc. ■☆

398516　Trisetum phleoides（Vill.）Trin. = Lophochloa phleoides（Vill.）Rchb. ■☆

398517　Trisetum pilosum Roem. et Schult. = Avena eriantha Durieu ■

398518　Trisetum pratense Pers.;黄三毛草（草原三毛草,黄燕麦）;Golden Oat Grass,Golden Oat-grass,Yellow Oat Grass,Yellow Oat-grass,Yellow Trisetum ■☆

398519　Trisetum pratense Pers. = Avena flavescens L. ■☆

398520　Trisetum pubiflorum Hack. = Trisetum spicatum（L.）K. Richt. ■

398521　Trisetum pumilum（Desf.）Kunth = Lophochloa pumila（Desf.）Bor ■☆

398522　Trisetum pumilum（Desf.）Kunth = Rostraria pumila（Desf.）Tzvelev ■☆

398523　Trisetum purpurascens Torr. = Schizachne purpurascens（Torr.）Swallen ■☆

398524　Trisetum rigidum Roem. et Schultz;硬三毛草■☆

398525　Trisetum rohlfsii Asch. = Lophochloa rohlfsii（Asch.）H. Scholz ■☆

398526　Trisetum scitulum Bor;优雅三毛草（锡金三毛草）;Pretty Falseoat,Pretty Trisetum ■

398527　Trisetum seravschanicum Roshev. = Trisetum spicatum（L.）K. Richt. subsp. virescens（Regel）Tzvelev ■

398528　Trisetum sibiricum Rupr.;西伯利亚三毛草（北蟹钓草）;Siberia Falseoat,Siberian Falseoat ■

398529　Trisetum sibiricum Rupr. subsp. umbratile（Kitag.）Tzvelev = Trisetum umbratile（Kitag.）Kitag. ■

398530　Trisetum sibiricum Rupr. var. umbratile Kitag. = Trisetum umbratile（Kitag.）Kitag. ■

398531　Trisetum spicatum（L.）K. Richt.;穗三毛草（穗三毛,亚高山三毛草）;False Oat,Narrow False Oats,Spike Falseoat,Spike Trisetum ■

398532　Trisetum spicatum（L.）K. Richt. subsp. alascanum（Nash）Hultén = Trisetum spicatum（L.）K. Richt. var. alascanum（Nash）Malte ex Louis-Marie ■

398533　Trisetum spicatum（L.）K. Richt. subsp. alaskanum（Nash）Hultén = Trisetum spicatum（L.）K. Richt. ■

398534　Trisetum spicatum（L.）K. Richt. subsp. alaskanum（Nash）Hultén = Trisetum spicatum（L.）K. Richt. var. alascanum（Nash）Malte ex Louis-Marie ■

398535　Trisetum spicatum（L.）K. Richt. subsp. congdonii（Scribn. et Merr.）Hultén = Trisetum spicatum（L.）K. Richt. ■

398536　Trisetum spicatum（L.）K. Richt. subsp. formosanum（Honda）Veldkamp = Trisetum spicatum（L.）K. Richt. subsp. alaskanum（Nash）Hultén ■

398537　Trisetum spicatum（L.）K. Richt. subsp. himalaicum Hultén = Trisetum spicatum（L.）K. Richt. var. himalaicum（Hultén）P. C. Kuo et Z. L. Wu ■

398538　Trisetum spicatum（L.）K. Richt. subsp. himalaicum Hultén ex Veldkamp = Trisetum spicatum（L.）K. Richt. subsp. virescens（Regel）Tzvelev ■

398539　Trisetum spicatum（L.）K. Richt. subsp. majus（Rydb.）Hultén = Trisetum spicatum（L.）K. Richt. ■

398540　Trisetum spicatum（L.）K. Richt. subsp. molle（Kunth）Hultén;柔软穗三毛■☆

398541　Trisetum spicatum（L.）K. Richt. subsp. molle（Michx.）Hultén = Trisetum spicatum（L.）K. Richt. ■

398542　Trisetum spicatum（L.）K. Richt. subsp. mongolicum Hultén = Trisetum spicatum（L.）K. Richt. var. mongolicum（Hultén）P. C. Kuo et Z. L. Wu ■

398543　Trisetum spicatum（L.）K. Richt. subsp. pilosiglume（Fernald）Hultén = Trisetum spicatum（L.）K. Richt. ■

398544　Trisetum spicatum（L.）K. Richt. subsp. tibeticum（P. C. Kuo et Z. L. Wu）Dickoré = Trisetum tibeticum P. C. Kuo et Z. L. Wu ■

398545　Trisetum spicatum（L.）K. Richt. subsp. virescens（Regel）Tzvelev;喜马拉雅穗三毛（绿变穗三毛）;Himalayan Trisetum, Himalayas Falseoat ■

398546　Trisetum spicatum（L.）K. Richt. var. alascanum（Nash）Malte ex Louis-Marie = Trisetum spicatum（L.）K. Richt. subsp. alascanum（Nash）Hultén ■

398547　Trisetum spicatum（L.）K. Richt. var. alascanum（Nash）Malte ex Louis-marie;台湾三毛草（大花穗三毛）;Largeflowered Trisetum ■

398548　Trisetum spicatum（L.）K. Richt. var. alaskanum（Nash）Malte ex Louis-Marie = Trisetum spicatum（L.）K. Richt. ■

398549　Trisetum spicatum（L.）K. Richt. var. alaskanum（Nash）Malte ex Louis-Marie = Trisetum spicatum（L.）K. Richt. subsp. alaskanum（Nash）Hultén ■

398550　Trisetum spicatum（L.）K. Richt. var. congdonii（Scribn. et Merr.）Hitchc. = Trisetum spicatum（L.）K. Richt. ■

398551　Trisetum spicatum（L.）K. Richt. var. formosanum（Honda）Ohwi = Trisetum spicatum（L.）K. Richt. subsp. alaskanum（Nash）Hultén ■

398552　Trisetum spicatum（L.）K. Richt. var. formosanum（Honda）Ohwi = Trisetum spicatum（L.）K. Richt. var. alascanum（Nash）Malte ex Louis-Marie ■

398553　Trisetum spicatum（L.）K. Richt. var. himalaicum（Hultén）P. C. Kuo et Z. L. Wu = Trisetum spicatum（L.）K. Richt. subsp. virescens（Regel）Tzvelev ■

398554　Trisetum spicatum（L.）K. Richt. var. kitadakense（Honda）Ohwi = Trisetum spicatum（L.）K. Richt. subsp. molle（Kunth）Hultén ■☆

398555　Trisetum spicatum（L.）K. Richt. var. maidenii（Gand.）Fernald = Trisetum spicatum（L.）K. Richt. ■

398556　Trisetum spicatum（L.）K. Richt. var. majus（Rydb.）Farw. = Trisetum spicatum（L.）K. Richt. ■

398557　Trisetum spicatum（L.）K. Richt. var. molle（Kunth）Beal = Trisetum spicatum（L.）K. Richt. subsp. molle（Kunth）Hultén ■☆

398558　Trisetum spicatum（L.）K. Richt. var. molle（Michx.）Beal = Trisetum spicatum（L.）K. Richt. ■

398559　Trisetum spicatum（L.）K. Richt. var. mongolicum（Hultén）P. C. Kuo et Z. L. Wu;蒙古穗三毛草（蒙古三毛）;Mongil Falseoat, Mongilian Trisetum ■

398560　Trisetum spicatum（L.）K. Richt. var. pilosiglume Fernald = Trisetum spicatum（L.）K. Richt. ■

398561　Trisetum spicatum（L.）K. Richt. var. villosissimum（Lange）Louis-Marie = Trisetum spicatum（L.）K. Richt. ■

398562　Trisetum steudelii Nees = Helictotrichon quinquesetum（Steud.）Schweick. ■☆

398563　Trisetum subalpestre Neuman = Trisetum spicatum（L.）K. Richt. ■

398564　Trisetum subspicatum（L.）P. Beauv. = Trisetum spicatum（L.）K. Richt. ■

398565　Trisetum tibeticum P. C. Kuo et Z. L. Wu；西藏三毛草；Xizang Falseoat，Xizang Trisetum ■

398566　Trisetum umbratile（Kitag.）Kitag.；绿穗三毛草；Green Spike Trisetum，Greenspike Falseoat ■

398567　Trisetum umbrosum Hochst. ex Steud. = Helictotrichon umbrosum（Hochst. ex Steud.）C. E. Hubb. ■☆

398568　Trisetum vaccarianum Maire et Weiller = Trisetaria vaccaria（Maire et Weiller）Maire ■☆

398569　Trisetum vaccarianum Maire et Weiller var. glabriglumis？ = Trisetaria vaccaria（Maire et Weiller）Maire ■☆

398570　Trisetum vaccarianum Maire et Weiller var. villiglumis？ = Trisetaria vaccaria（Maire et Weiller）Maire ■☆

398571　Trisetum viciosorum Sennen et Mauricio = Rostraria festucoides（Link）Romero Zarco ■☆

398572　Trisetum villosissimum（Lange）Louis-Marie = Trisetum spicatum（L.）K. Richt. ■

398573　Trisetum virescens（Regel）Roshev. = Trisetum spicatum（L.）K. Richt. subsp. virescens（Regel）Tzvelev ■

398574　Trisetum virescens Nees ex Steud. = Helictotrichon junghuhnii（Büse）Henrard. ■

398575　Trisetum virescens Nees ex Steud. = Helictotrichon virescens（Nees ex Steud.）Henrard ■

398576　Trisetum yunnanense Chrtek；云南三毛草■

398577　Trisiola Raf. = Distichlis Raf. ■☆

398578　Trisiola Raf. = Uniola L. ■☆

398579　Tristachya Nees（1829）；三联穗草属■☆

398580　Tristachya arundinacea Hochst. ex A. Rich. = Loudetia arundinacea（Hochst. ex A. Rich.）Steud. ■☆

398581　Tristachya atricha Peter = Tristachya nodiglumis K. Schum. ■☆

398582　Tristachya augusta J. B. Phipps = Tristachya superba（De Not.）Schweinf. et Asch. ■☆

398583　Tristachya aurea Chiov. = Zonotriche inamoena（K. Schum.）Clayton ■☆

398584　Tristachya auronitens P. A. Duvign.；黄光三联穗草■☆

398585　Tristachya barbata Nees = Danthoniopsis barbata（Nees）C. E. Hubb. ■☆

398586　Tristachya barbata of Hook. f. = Danthoniopsis stocksii（Boiss.）C. E. Hubb. ■☆

398587　Tristachya bequaertii De Wild.；贝卡尔三联穗草■☆

398588　Tristachya bequaertii De Wild. var. vanderystii？ = Tristachya nodiglumis K. Schum. ■☆

398589　Tristachya bicrinita（J. B. Phipps）Clayton；双毛三联穗草■☆

398590　Tristachya bricchettiana Chiov. = Danthoniopsis barbata（Nees）C. E. Hubb. ■☆

398591　Tristachya butuluensis Vanderyst = Tristachya nodiglumis K. Schum. ■☆

398592　Tristachya chevalieri Stapf = Loudetiopsis chevalieri（Stapf）Conert ■☆

398593　Tristachya chrysothrix Nees = Loudetiopsis chrysothrix（Nees）Conert ■☆

398594　Tristachya coarctata A. Camus = Loudetia coarctata（A. Camus）C. E. Hubb. ■☆

398595　Tristachya decora Stapf = Zonotriche decora（Stapf）J. B. Phipps ■☆

398596　Tristachya elegans（Hochst. ex A. Braun）A. Rich. = Loudetia simplex（Nees）C. E. Hubb. ■☆

398597　Tristachya elymoides Chiov. = Tristachya thollonii Franch. ■☆

398598　Tristachya elymoides Chiov. var. laevis？ = Tristachya thollonii Franch. ■☆

398599　Tristachya esculenta（C. E. Hubb.）Conert；食用三联穗草■☆

398600　Tristachya eylesii Stent et J. M. Rattray = Tristachya nodiglumis K. Schum. ■☆

398601　Tristachya fulva C. E. Hubb. = Loudetiopsis chrysothrix（Nees）Conert ■☆

398602　Tristachya glabra Stapf = Tristachya rehmannii Hack. ■☆

398603　Tristachya glabrinodis C. E. Hubb. = Loudetiopsis kerstingii（Pilg.）Conert ■☆

398604　Tristachya granulosa Chiov. = Tristachya leucothrix Trin. ex Nees ■☆

398605　Tristachya helenae Buscal. et Muschl. = Tristachya rehmannii Hack. ■☆

398606　Tristachya hispida（L. f.）K. Schum. = Tristachya leucothrix Trin. ex Nees ■☆

398607　Tristachya hitchcockii（C. E. Hubb.）Conert = Tristachya lualabaensis（De Wild.）J. B. Phipps ■☆

398608　Tristachya hockii De Wild. = Tristachya rehmannii Hack. ■☆

398609　Tristachya homblei De Wild. = Tristachya thollonii Franch. ■☆

398610　Tristachya hubbardiana Conert；哈伯德三联穗草■☆

398611　Tristachya huillensis Rendle；威拉三联穗草■☆

398612　Tristachya inamoena K. Schum. = Zonotriche inamoena（K. Schum.）Clayton ■☆

398613　Tristachya kagerensis（K. Schum.）A. Chev. = Loudetia kagerensis（K. Schum.）C. E. Hubb. ex Hutch. ■☆

398614　Tristachya kerstingii（Pilg.）C. E. Hubb. = Loudetiopsis kerstingii（Pilg.）Conert ■☆

398615　Tristachya leucothrix Trin. ex Nees；白毛三联穗草■☆

398616　Tristachya leucothrix Trin. ex Nees var. bolusii De Wild. = Tristachya leucothrix Trin. ex Nees ■☆

398617　Tristachya leucothrix Trin. ex Nees var. longiaristata De Wild. = Tristachya leucothrix Trin. ex Nees ■☆

398618　Tristachya leucothrix Trin. ex Nees var. sapinii De Wild. = Tristachya leucothrix Trin. ex Nees ■☆

398619　Tristachya longispiculata C. E. Hubb. = Tristachya nodiglumis K. Schum. ■☆

398620　Tristachya lualabaensis（De Wild.）J. B. Phipps；卢阿拉巴三联穗草■☆

398621　Tristachya microstachya Nees ex Steud. = Loudetiopsis tristachyoides（Trin.）Conert ■☆

398622　Tristachya minuta A. Chev. = Loudetiopsis capillipes（C. E. Hubb.）Conert ■☆

398623　Tristachya monocephala Hochst. = Loudetia simplex（Nees）C. E. Hubb. ■☆

398624　Tristachya mukuluensis Vanderyst = Tristachya leucothrix Trin. ex Nees ■☆

398625　Tristachya multinodis C. E. Hubb. = Loudetiopsis tristachyoides（Trin.）Conert ■☆

398626　Tristachya nodiglumis K. Schum.；节颖三联穗草■☆

398627　Tristachya nodiglumis K. Schum. var. laeviglumis = Tristachya nodiglumis K. Schum. ■☆

398628　Tristachya oligostachya Conert;寡穗三联穗草■☆

398629　Tristachya pallida Stent;苍白三联穗草■☆

398630　Tristachya parviflora Hack. = Loudetiopsis ambiens (K.
Schum.) Conert ■☆

398631　Tristachya pedicellata Stent = Loudetia pedicellata (Stent)
Chippind. ■☆

398632　Tristachya pilgeriana Buscal. et Muschl. = Zonotriche decora
(Stapf) J. B. Phipps ■☆

398633　Tristachya pseudoligulata J. B. Phipps = Tristachya hubbardiana
Conert ■☆

398634　Tristachya purpurea C. E. Hubb. = Loudetiopsis tristachyoides
(Trin.) Conert ■☆

398635　Tristachya rehmannii Hack.;拉赫曼三联穗草■☆

398636　Tristachya rehmannii Hack. subsp. mosambicensis J. B. Phipps;
莫桑比克三联穗草■☆

398637　Tristachya rehmannii Hack. var. helenae (Buscal. et Muschl.)
C. E. Hubb. = Tristachya rehmannii Hack. ■☆

398638　Tristachya rehmannii Hack. var. pilosa C. E. Hubb. = Tristachya
rehmannii Hack. ■☆

398639　Tristachya ringoetii De Wild. = Zonotriche inamoena (K.
Schum.) Clayton ■☆

398640　Tristachya simplex Nees = Loudetia simplex (Nees) C. E. Hubb. ■☆

398641　Tristachya somalensis Franch. = Danthoniopsis barbata (Nees)
C. E. Hubb. ■☆

398642　Tristachya spicata Pilg. ex Peter = Tristachya thollonii Franch. ☆

398643　Tristachya spiculata Peter = Tristachya thollonii Franch. ■☆

398644　Tristachya stocksii Boiss. = Danthoniopsis stocksii (Boiss.) C.
E. Hubb. ■☆

398645　Tristachya superba (De Not.) Schweinf. et Asch. ;华美三联穗
草■☆

398646　Tristachya superbiens Pilg. = Tristachya nodiglumis K. Schum. ■☆

398647　Tristachya thollonii Franch. ;托伦三联穗草■☆

398648　Tristachya tristachyoides (Trin.) C. E. Hubb. = Loudetiopsis
tristachyoides (Trin.) Conert ■☆

398649　Tristachya triticoides A. Camus et C. E. Hubb. ex A. Chev. =
Loudetia coarctata (A. Camus) C. E. Hubb. ■☆

398650　Tristachya tuberculata Stapf = Loudetiopsis tristachyoides (Trin.)
Conert ■☆

398651　Tristachya vanderystii De Wild. = Tristachya nodiglumis K.
Schum. ■☆

398652　Tristachya viridearistata (J. B. Phipps) Clayton;绿芒三联穗草■☆

398653　Tristachya welwitschii Rendle = Tristachya nodiglumis K. Schum. ■☆

398654　Tristachya welwitschii Rendle var. atricha Chiov. = Tristachya
nodiglumis K. Schum. ■☆

398655　Tristachya welwitschii Rendle var. major C. E. Hubb. =
Tristachya nodiglumis K. Schum. ■☆

398656　Tristachya welwitschii Rendle var. superbiens (Pilg.) C. E.
Hubb. = Tristachya nodiglumis K. Schum. ■☆

398657　Tristachya welwitschii Rendle var. trichophora Chiov. =
Tristachya nodiglumis K. Schum. ■☆

398658　Tristagma Poepp. (1833);三柱莲属(三滴葱属);Starflower ■☆

398659　Tristagma brevipes (Kunze) Traub;短梗三滴葱■☆

398660　Tristagma graminifolium (Phil.) Ravenna;禾叶三滴葱■☆

398661　Tristagma nivale Poepp. ;三滴葱■☆

398662　Tristagma pulchellum Speg. ;美丽三滴葱■☆

398663　Tristagma uniflorum (Lindl.) Traub;单花三滴葱■☆

398664　Tristagma uniflorum (Lindl.) Traub = Ipheion uniflorus (Lindl.)
Raf. ■☆

398665　Tristania Poir. = Spartina Schreb. ex J. F. Gmel. ■

398666　Tristania R. Br. (1812);红胶木属(三胶木属);Brush Box,
Brushbox,Tristania ●

398667　Tristania R. Br. ex Aiton = Tristania R. Br. ●

398668　Tristania conferta R. Br. = Lophostemon confertus (R. Br.)
Peter G. Wilson et J. T. Waterh. ●

398669　Tristania nereifolia R. Br. ;夹竹桃叶红胶木;Dwarf Water
Gum,Water Gum ●

398670　Tristania suaveolens Sol. ex Gaertn. = Lophostemon suaveolens
(Sol. ex Gaertn.) Peter G. Wilson et J. T. Waterh. ●☆

398671　Tristania whiteana Griff. ;怀特红胶木●☆

398672　Tristaniopsis Brongn. et Gris = Tristania R. Br. ●

398673　Tristaniopsis Brongn. et Gris(1863);异红胶木属●☆

398674　Tristaniopsis exiliflora (F. Muell.) Peter G. Wilson et J. T.
Waterh. ;北方异红胶木;Northern Water Gum ●☆

398675　Tristaniopsis laurina (Sm.) Peter G. Wilson et J. T. Waterh. ;
月桂异红胶木;Kanooka,Kanuka Box,Water Gum,Water-gum ●☆

398676　Tristeginaceae Link = Gramineae Juss. (保留科名)■●

398677　Tristeginaceae Link = Poaceae Barnhart(保留科名)■●

398678　Tristegis Nees = Melinis P. Beauv. ■

398679　Tristegis Nees = Suardia Schrank ■

398680　Tristegis glutinosa Nees = Melinis minutiflora P. Beauv. ■

398681　Tristellateia Thouars(1806);三星果属(三星果藤属,星果藤
属);Threestarfruit,Tristellateia ●

398682　Tristellateia africana S. Moore;非洲三星果●☆

398683　Tristellateia africana S. Moore var. somalensis (Chiov.) Arènes
= Caucanthus edulis Forssk. ●☆

398684　Tristellateia australasiae A. Rich. ;三星果(澳洲三星果,搭肉刺,
庚中藤,三星果藤,星果藤);Australian Tristellateia,Threestarfruit,
Tristellateia ●

398685　Tristellateia cynanchoides Chiov. = Triaspis niedenzuiana Engl. ●☆

398686　Tristellateia somalensis Chiov. = Caucanthus edulis Forssk. ●☆

398687　Tristemma Juss. (1789);三冠野牡丹属●☆

398688　Tristemma acuminatum A. Fern. et R. Fern. = Tristemma mauritianum
J. F. Gmel. ●☆

398689　Tristemma akeassii Jacq. -Fél. ;阿克斯三冠野牡丹●☆

398690　Tristemma albiflorum (G. Don) Benth. ;白花三冠野牡丹●☆

398691　Tristemma angolense Gilg = Tristemma mauritianum J. F. Gmel. ●☆

398692　Tristemma camerunense Jacq. -Fél. ;喀麦隆三冠野牡丹●☆

398693　Tristemma controversum A. Chev. et Jacq. -Fél. = Melastomastrum
theifolium (G. Don) A. Fern. et R. Fern. var. controversum (A. Chev. et
Jacq. -Fél.) Jacq. -Fél. ●☆

398694　Tristemma cornifolium Triana = Melastomastrum cornifolium
(Benth.) Jacq. -Fél. ●☆

398695　Tristemma coronatum Benth. ;顶饰三冠野牡丹●☆

398696　Tristemma demeusei De Wild. ;迪米三冠野牡丹●☆

398697　Tristemma erectum Guillaumin et Perr. = Melastomastrum capitatum
(Vahl) A. Fern. et R. Fern. ●☆

398698　Tristemma fruticulosum Gilg = Tristemma mauritianum J. F. Gmel. ●☆

398699　Tristemma grandiflorum (Cogn.) Gilg = Tristemma mauritianum
J. F. Gmel. ●☆

398700　Tristemma grandifolium (Cogn.) Gilg = Tristemma mauritianum
J. F. Gmel. ●☆

398701　Tristemma grandifolium (Cogn.) Gilg var. congolanum De
Wild. = Tristemma mauritianum J. F. Gmel. ●☆

398702　Tristemma hirtum P. Beauv. ;多毛三冠野牡丹●☆

398703　Tristemma hirtum Vent. ex DC. ＝ Tristemma hirtum P. Beauv. ●☆

398704　Tristemma incompletum R. Br. ＝ Tristemma mauritianum J. F. Gmel. ●☆

398705　Tristemma incompletum R. Br. var. grandifolium (Cogn.) Hiern ＝ Tristemma mauritianum J. F. Gmel. ●☆

398706　Tristemma involucratum (D. Don) Benth. ;总苞三冠野牡丹●☆

398707　Tristemma kassnerianum Kraenzl. ＝ Tristemma mauritianum J. F. Gmel. ●☆

398708　Tristemma leiocalyx Cogn. ;光萼三冠野牡丹●☆

398709　Tristemma leiocalyx Cogn. var. pierlotii Jacq. -Fél. ;皮氏三冠野牡丹●☆

398710　Tristemma leucanthum Gilg ex Engl. ＝ Tristemma demeusei De Wild. ●☆

398711　Tristemma littorale Benth. ;滨海三冠野牡丹●☆

398712　Tristemma mauritianum J. F. Gmel. ;毛里求斯三冠野牡丹●☆

398713　Tristemma mauritianum J. F. Gmel. var. mildbraedii (Gilg) Jacq. -Fél. ;米尔德三冠野牡丹●☆

398714　Tristemma mauritianum J. F. Gmel. var. thomense (J. H. Ferreira) Jacq. -Fél. ;爱岛三冠野牡丹●☆

398715　Tristemma mildbraedii Gilg ＝ Tristemma mauritianum J. F. Gmel. var. mildbraedii (Gilg) Jacq. -Fél. ●☆

398716　Tristemma monanthum Gilg ex Engl. ＝ Tristemma demeusei De Wild. ●☆

398717　Tristemma neglectum Naudin ＝ Melastomastrum cornifolium (Benth.) Jacq. -Fél. ●☆

398718　Tristemma oreophilum Gilg;喜山三冠野牡丹●☆

398719　Tristemma ovalifolium Engl. ＝ Melastomastrum cornifolium (Benth.) Jacq. -Fél. ●☆

398720　Tristemma quadriannulatum De Wild. ＝ Tristemma mauritianum J. F. Gmel. ●☆

398721　Tristemma radicans Gilg ＝ Tristemma demeusei De Wild. ●☆

398722　Tristemma roseum Gilg ＝ Tristemma leiocalyx Cogn. ●☆

398723　Tristemma rubens A. Fern. et R. Fern. ;变红三冠野牡丹●☆

398724　Tristemma schellenbergianum Gilg ex Engl. ＝ Tristemma oreophilum Gilg ●☆

398725　Tristemma schlechteri Gilg ＝ Melastomastrum segregatum (Benth.) A. Fern. et R. Fern. ●☆

398726　Tristemma schliebenii Markgr. ;施利本三冠野牡丹●☆

398727　Tristemma schumacheri Guillaumin et Perr. ＝ Tristemma albiflorum (G. Don) Benth. ●☆

398728　Tristemma segregatum (Benth.) Triana ＝ Melastomastrum segregatum (Benth.) A. Fern. et R. Fern. ●☆

398729　Tristemma theifolia (G. Don) Triana ＝ Melastomastrum theifolium (G. Don) A. Fern. et R. Fern. ●☆

398730　Tristemma thomense J. H. Ferreira ＝ Tristemma mauritianum J. F. Gmel. var. thomense (J. H. Ferreira) Jacq. -Fél. ●☆

398731　Tristemma verdickii De Wild. ＝ Heterotis canescens (E. Mey. ex R. A. Graham) Jacq. -Fél. ●☆

398732　Tristemma vestitum Jacq. -Fél. ;包被三冠野牡丹●☆

398733　Tristemma vincoides Gilg ＝ Tristemma leiocalyx Cogn. ●☆

398734　Tristemma virusamum Juss. ＝ Tristemma mauritianum J. F. Gmel. ●☆

398735　Tristemon Klotzsch ＝ Scyphogyne Brongn. ●☆

398736　Tristemon Klotzsch ＝ Scyphogyne Decne. ●☆

398737　Tristemon Raf. (1838) ＝ Juncus L. ■

398738　Tristemon Raf. (1919) ＝ Triglochin L. ■

398739　Tristemon Scheele ＝ Cucurbita L. ■

398740　Tristemon puberula Klotzsch ＝ Erica urceolata (Klotzsch) E. G. H. Oliv. ●☆

398741　Tristemon urceolatus Klotzsch ＝ Erica urceolata (Klotzsch) E. G. H. Oliv. ●☆

398742　Tristemonanthus Loes. (1940);三冠卫矛属●☆

398743　Tristemonanthus mildbraedianus Loes. ;非洲三冠卫矛●☆

398744　Tristemonanthus nigrisilvae (N. Hallé) N. Hallé;三冠卫矛●☆

398745　Tristeria Hook. f. ＝ Curanga Juss. ■☆

398746　Tristeria Hook. f. ＝ Treisteria Griff. ■☆

398747　Tristerix Mart. ＝ Macrosolen (Blume) Rchb. ●

398748　Tristicha Thouars(1806);三列苔草属■☆

398749　Tristicha alternifolia (Willd.) Spreng. f. sambesiaca Engl. ＝ Tristicha trifaria (Bory ex Willd.) Spreng. ■☆

398750　Tristicha dregeana Tul. ＝ Tristicha trifaria (Bory ex Willd.) Spreng. ■☆

398751　Tristicha hypnoides (A. Rich.) Spreng. var. pulchella Wedd. ＝ Tristicha trifaria (Bory ex Willd.) Spreng. subsp. pulchella (Wedd.) C. Cusset et G. Cusset ■☆

398752　Tristicha hypnoides Spreng. ＝ Tristicha trifaria (Bory ex Willd.) Spreng. ■☆

398753　Tristicha trifaria (Bory ex Willd.) Spreng. ;三列苔草■☆

398754　Tristicha trifaria (Bory ex Willd.) Spreng. subsp. pulchella (Wedd.) C. Cusset et G. Cusset;美丽三列苔草■☆

398755　Tristichaceae Willis ＝ Podostemaceae Rich. ex Kunth(保留科名)■

398756　Tristichaceae Willis;三列苔草科■

398757　Tristichocalyx F. Muell. ＝ Pachygone Miers ●

398758　Tristichocalyx Miers ＝ Legnephora Miers ●☆

398759　Tristichopsis A. Chev. ＝ Tristicha Thouars ■☆

398760　Tristichopsis riccioides A. Chev. ＝ Tristicha trifaria (Bory ex Willd.) Spreng. subsp. pulchella (Wedd.) C. Cusset et G. Cusset ■☆

398761　Tristira Radlk. (1879);菲律宾无患子属●☆

398762　Tristira harpullioides Radlk. ;菲律宾无患子●☆

398763　Tristira pubescens Merr. ;毛菲律宾无患子●☆

398764　Tristira triptera Radlk. ;三翅菲律宾无患子●☆

398765　Tristiropsis Radlk. (1887);拟菲律宾无患子属●☆

398766　Tristiropsis acutangula Radlk. ;拟菲律宾无患子●☆

398767　Tristylea Jord. et Fourr. ＝ Saxifraga L. ■

398768　Tristylium Turcz. ＝ Cleyera Thunb. (保留属名)●

398769　Tristylium ochanceum (DC.) Merr. ＝ Cleyera japonica Thunb. ●

398770　Tristylium ochnaceum (DC.) Merr. var. morii (Yamam.) Sasaki ＝ Cleyera japonica Thunb. ●

398771　Tristylium ochnaceum (DC.) Merr. var. morii Sasaki ＝ Cleyera japonica Thunb. var. morii (Yamam.) Masam. ●☆

398772　Tristylopsis Kaneh. et Hatus. ＝ Tristiropsis Radlk. ●☆

398773　Trisynsyne Baill. ＝ Nothofagus Blume(保留属名)●☆

398774　Tritaenicum Turcz. ＝ Asteriscium Cham. et Schltdl. ■☆

398775　Tritaxis Baill. ＝ Trigonostemon Blume(保留属名)●

398776　Tritelandra Raf. ＝ Epidendrum L. (保留属名)■☆

398777　Triteleia Douglas ex Lindl. (1830);美韭属; Pretty Face, Starflower, Triplet-lily, Triteleia, Wild Hyacinth ■☆

398778　Triteleia angustiflora A. Heller ＝ Triteleia laxa Benth. ■☆

398779　Triteleia anilina (Greene) Hoover ＝ Triteleia ixioides (W. T. Aiton) Greene subsp. anilina (Greene) L. W. Lenz ■☆

398780　Triteleia bicolor (Suksd.) A. Heller ＝ Triteleia grandiflora Lindl. ■☆

398781　Triteleia bridgesii (S. Watson) Greene;布里奇斯美韭(紫灯花);Bridges' Triteleia ■☆

398782　Triteleia bridgesii (S. Watson) Greene ＝ Brodiaea bridgesii S.

Watson ■☆

398783　Triteleia candida Greene = Triteleia laxa Benth. ■☆

398784　Triteleia clementina Hoover;圣克利门蒂美韭;San Clemente Island Triteleia ■☆

398785　Triteleia crocea（A. W. Wood）Greene;镉黄美韭;Yellow Triteleia ■☆

398786　Triteleia crocea（A. W. Wood）Greene var. modesta（H. M. Hall）Hoover = Triteleia crocea（A. W. Wood）Greene ■☆

398787　Triteleia dudleyi Hoover;达德利美韭;Dudley's Triteleia ■☆

398788　Triteleia gracilis（S. Watson）Greene = Triteleia montana Hoover ■☆

398789　Triteleia grandiflora Lindl.;大花美韭;Blue Umber-lily,Blue-lily,Large-flowered Triteleia,Wild Hyacinth ■☆

398790　Triteleia grandiflora Lindl. var. howellii（S. Watson）Hoover = Triteleia grandiflora Lindl. ■☆

398791　Triteleia hendersonii Greene;亨德森美韭;Henderson's Triteleia,Yellow Tiger-lily ■☆

398792　Triteleia hendersonii Greene var. leachiae（M. Peck）Hoover = Triteleia hendersonii Greene ■☆

398793　Triteleia howellii（S. Watson）Greene = Triteleia grandiflora Lindl. ■☆

398794　Triteleia hyacinthina（Lindl.）Greene;风信子美韭（乳白布罗地）;Fool's Onion,Hyacinth Brodiaea,White Brodiaea,Wild Hyacinth ■☆

398795　Triteleia hyacinthina（Lindl.）Greene var. greenei Hoover = Triteleia lilacina Greene ■☆

398796　Triteleia ixioides（W. T. Aiton）Greene;金美韭;Golden Brodiaea,Yellow-brodiaea ■☆

398797　Triteleia ixioides（W. T. Aiton）Greene subsp. anilina（Greene）L. W. Lenz;美丽金美韭;Pretty Face,Sierra-brodiaea ■☆

398798　Triteleia ixioides（W. T. Aiton）Greene subsp. cookii（Hoover）L. W. Lenz;库克美韭;Cook's Triteleia ■☆

398799　Triteleia ixioides（W. T. Aiton）Greene subsp. scabra（Greene）L. W. Lenz;粗糙金美韭 ■☆

398800　Triteleia ixioides（W. T. Aiton）Greene subsp. unifolia L. W. Lenz;单花金美韭■☆

398801　Triteleia ixioides（W. T. Aiton）Greene var. cookii Hoover = Triteleia ixioides（W. T. Aiton）Greene subsp. cookii（Hoover）L. W. Lenz ■☆

398802　Triteleia ixioides Greene = Triteleia ixioides（W. T. Aiton）Greene ■☆

398803　Triteleia ixioides Greene var. anilina（Greene）Hoover = Triteleia ixioides（W. T. Aiton）Greene subsp. anilina（Greene）L. W. Lenz ■☆

398804　Triteleia ixioides Greene var. scabra（Greene）Hoover = Triteleia ixioides（W. T. Aiton）Greene subsp. scabra（Greene）L. W. Lenz ■☆

398805　Triteleia lactea（Lindl.）Davidson et Moxley = Triteleia hyacinthina（Lindl.）Greene ■☆

398806　Triteleia laxa Benth.;疏花美韭（普通美韭,疏生布罗地）;Brodiaea,Californian Hyacinth,Common Triteleia,Grass Nut,Grass-nut,Ithuriel's Spear,Long Rayed Triteleia,Pretty Face,Triplet Lily,Triplet-lily ■☆

398807　Triteleia lemmoniae（S. Watson）Greene;莱蒙美韭;Lemmon's Triteleia,Oak Creek Triteleia ■☆

398808　Triteleia lilacina Greene;丁香美韭;Lilac-flowered Wild Hyacinth ■☆

398809　Triteleia lugens Greene;沿海美韭;Coast Range Triteleia ■☆

398810　Triteleia modesta（H. M. Hall）Abrams = Triteleia crocea（A.

W. Wood）Greene ■☆

398811　Triteleia montana Hoover;山地美韭;Mountain Triteleia ■☆

398812　Triteleia palmeri（S. Watson）Greene = Triteleiopsis palmeri（S. Watson）Hoover ■☆

398813　Triteleia peduncularis Lindl.;长梗美韭;Long-rayed Brodiaea ■☆

398814　Triteleia scabra（Greene）Hoover = Triteleia ixioides（W. T. Aiton）Greene subsp. scabra（Greene）L. W. Lenz ■☆

398815　Triteleia uniflora Lindl. = Tristagma uniflorum（Lindl.）Traub ■☆

398816　Triteleiopsis Hoover(1941);类美韭属;Baja-lily ■☆

398817　Triteleiopsis palmeri（S. Watson）Hoover;类美韭■☆

398818　Triteleya Phil. = Triteleia Douglas ex Lindl. ■☆

398819　Tritheca（Wight et Arn.）Miq. = Ammannia L. ■

398820　Tritheca Miq. = Ammannia L. ■

398821　Trithecanthera Tiegh.（1894）;三室寄生属●☆

398822　Trithecanthera flava Kosterm.;黄三室寄生●☆

398823　Trithecanthera xiphostachya Tiegh.;三室寄生●☆

398824　Trithrinax Mart.（1837）;三扇棕属（长刺棕属,刺鞘棕属,南美扇棕属,网刺棕属）;Trithrinax Palm ●☆

398825　Trithrinax acanthocoma Drude;簇生三扇棕（长刺棕,簇生刺鞘棕）;Caranda,Spiny Fiber Palm,Spiny Fibre-palm ●☆

398826　Trithrinax brasiliensis Mart.;三扇棕（巴西长刺棕,巴西刺鞘棕）;Brazil Trithrinax Palm,Carandai Palm,Saho,Saro,Trithrinax Palm ●☆

398827　Trithrinax campestris Drude et Griseb.;野生三扇棕;Campestre Palm ●☆

398828　Trithuria Hook. f.（1858）;三孔独蕊草属（三孔草属）■☆

398829　Trithuria occidentalis Benth.;西方三孔独蕊草（西方三孔草）■☆

398830　Trithuria submersa Hook. f.;三孔独蕊草（三孔草）■☆

398831　Trithyrocarpus Hassk. = Commelina L. ■

398832　Triticaceae Link = Gramineae Juss.（保留科名）■●

398833　Triticaceae Link = Poaceae Barnhart（保留科名）■●

398834　Triticum L.（1753）;小麦属;Triticum,Wheat ■

398835　Triticum abyssinicum（Vavilov）Flaksb. = Triticum aethiopicum Jakubz. ■☆

398836　Triticum abyssinicum Steud.;阿比西尼亚小麦;Ethiopian Wheat ■☆

398837　Triticum abyssinicum Steud. = Triticum polonicum L. ■

398838　Triticum aegilopoides Turcz. ex Griseb.;野生小麦■☆

398839　Triticum aestivum L.;小麦（二枝麦,浮麦,浮水麦,来,穌,麦,普通小麦）;Branched Wheat,Branchy-eared Wheat,Bread Wheat,Common Wheat,Wheat Vodka,White Beer ■

398840　Triticum aestivum L. subsp. durum（Desf.）Thell. = Triticum turgidum L. subsp. durum（Desf.）Husn. ■

398841　Triticum aestivum L. subsp. durum（Desf.）Trab. = Triticum turgidum L. subsp. durum（Desf.）Husn. ■

398842　Triticum aestivum L. subsp. tibeticum J. Z. Shao;西藏小麦■

398843　Triticum aestivum L. subsp. turgidum（L.）Domin = Triticum turgidum L. ■

398844　Triticum aestivum L. subsp. vulgare（Vill.）Trab. = Triticum aestivum L. ■

398845　Triticum aestivum L. subsp. yunnanense King ex S. L. Chen;云南小麦■

398846　Triticum aestivum L. var. durum（Desf.）Fiori = Triticum turgidum L. subsp. durum（Desf.）Husn. ■

398847　Triticum aestivum L. var. hybernum（L.）Fiori = Triticum aestivum L. ■

398848　Triticum aestivum L. var. monococcum（L.）L. H. Bailey = Triticum monococcum L. ■

398849　Triticum aestivum L. var. polonicum（L.）L. H. Bailey =

Triticum turgidum L. subsp. polonicum（L.）Thell. ■

398850　Triticum aestivum L. var. polonicum L. H. Bailey = Triticum polonicum L. ■

398851　Triticum aestivum L. var. polonicum L. H. Bailey = Triticum turgidum L. var. polonicum（L.）M. S. Yan ex P. C. Kuo ■

398852　Triticum aestivum L. var. turgidum（L.）Fiori = Triticum turgidum L. ■

398853　Triticum aethiopicum Jakubz.；非洲小麦；African Wheat ■☆

398854　Triticum angustum（Trin.）F. Herm. = Leymus angustus（Trin.）Pilg. ■

398855　Triticum aralense（Regel）F. Herm. = Leymus multicaulis（Kar. et Kir.）Tzvelev ■

398856　Triticum arktasianum F. Herm. = Elymus sibiricus L. ■

398857　Triticum armeniacum（Stolet.）Nevski；亚美尼亚小麦 ■☆

398858　Triticum arras Hochst. = Triticum dicoccon Schrank ■

398859　Triticum batalinii Krasn. = Kengyilia batalinii（Krasn.）J. L. Yang, C. Yen et B. R. Baum ■

398860　Triticum bicorne Forssk. = Aegilops bicornis（Forssk.）Jaub. et Spach ■☆

398861　Triticum bonaepartis Spreng. = Eremopyrum bonaepartis（Spreng.）Nevski ■

398862　Triticum boreale Turcz. = Elymus kronokensis（Kom.）Tzvelev ■

398863　Triticum brevisetum DC. = Brachypodium phoenicoides（L.）Roem. et Schult. ■☆

398864　Triticum buonaepartis Spreng. = Eremopyrum bonaepartis（Spreng.）Nevski ■

398865　Triticum caninum L. = Elymus caninus（L.）L. ■

398866　Triticum caninum L. = Roegneria canina（L.）Nevski ■

398867　Triticum caninum L. f. amurense Korsh. = Elymus pendulinus（Nevski）Tzvelev ■

398868　Triticum caninum L. var. gmelinii Ledeb. = Elymus gmelinii（Ledeb.）Tzvelev ■

398869　Triticum carthlicum Nevski；波斯小麦；Persian Wheat ■

398870　Triticum carthlicum Nevski = Triticum turgidum L. var. carthlicum（Nevski）M. S. Yan ex P. C. Kuo ■

398871　Triticum cereale（L.）Salisb. = Secale cereale L. ■

398872　Triticum cereale Salisb. = Secale cereale L. ■

398873　Triticum chinense Trin. ex Bunge = Leymus chinensis（Trin. ex Bunge）Tzvelev ■

398874　Triticum ciliare Trin. = Roegneria ciliaris（Trin.）Nevski ■

398875　Triticum ciliare Trin. ex Bunge = Elymus ciliaris（Trin. ex Bunge）Tzvelev ■

398876　Triticum ciliare Trin. f. pilosum Korsh. = Elymus ciliaris（Trin. ex Bunge）Tzvelev var. amurensis（Drobow）S. L. Chen ■

398877　Triticum compactum Host；密穗小麦；Club Wheat ■

398878　Triticum compactum Host = Triticum aestivum L. ■

398879　Triticum compactum Host var. compressum Körn. = Triticum aethiopicum Jakubz. ■☆

398880　Triticum compactum Host var. copticum Körn. = Triticum aethiopicum Jakubz. ■☆

398881　Triticum compositum L. = Triticum aestivum L. ■

398882　Triticum compositum L. = Triticum turgidum L. ■

398883　Triticum cylindricum（Host）Ces. = Aegilops cylindrica Host ■

398884　Triticum cylindricum（Host）Ces., Pass. et Gibelli = Aegilops cylindrica Host ■

398885　Triticum desertorum Fisch. ex Link = Agropyron desertorum（Fisch. ex Link）Schult. ■

398886　Triticum dicoccoides（Körn.）Aarons.；野生二粒小麦；Wild Dicoccum Wheat, Wild Wheat ■

398887　Triticum dicoccoides Körn. = Triticum turgidum L. var. dicoccoides（Körn.）Bowden ■

398888　Triticum dicoccon Schrank = Triticum dicoccum（Schrank）Schübeler ■

398889　Triticum dicoccon Schrank var. ajar Percival = Triticum dicoccon Schrank ■

398890　Triticum dicoccum（Schrank）Schübeler；二粒小麦（双粒小麦）；Dinkel Wheat, Emmer Wheat, Two-grain Wheat ■

398891　Triticum dicoccum（Schrank）Schübeler var. timopheevi Zhuk. = Triticum timopheevii Zhuk. ■

398892　Triticum dicoccum（Schrank）Schübler = Triticum turgidum L. subsp. dicoccum（Schrank）Thell. ■☆

398893　Triticum dicoccum L. var. pseudouncinatum Percival = Triticum aethiopicum Jakubz. ■☆

398894　Triticum dicoccum L. var. rufescens Percival = Triticum aethiopicum Jakubz. ■☆

398895　Triticum dicoccum L. var. schimperi（Körn.）Percival = Triticum aethiopicum Jakubz. ■☆

398896　Triticum dicoccum L. var. tomentosum Percival = Triticum aethiopicum Jakubz. ■☆

398897　Triticum dicoccum L. var. uncinatum Percival = Triticum aethiopicum Jakubz. ■☆

398898　Triticum distichum Thunb. = Elymus distichus（Thunb.）Melderis ■☆

398899　Triticum diversiflorum Steud. = Triticum aethiopicum Jakubz. ■☆

398900　Triticum durum Desf.；硬粒小麦（坚硬小麦）；Crack Wheat, Durum Wheat, Flint Wheat, Hard Wheat, Macaroni, Macaroni Wheat, Semolina, Spaghetti, Vermicelli ■

398901　Triticum durum Desf. = Triticum turgidum L. subsp. durum（Desf.）Husn. ■

398902　Triticum durum Desf. = Triticum turgidum L. var. durum（Desf.）M. S. Yan ex P. C. Kuo ■

398903　Triticum durum Desf. subsp. turanicum（Jakubz.）L. B. Cai = Triticum turanicum Jakubz. ■

398904　Triticum durum Desf. var. intermedium Ducell. = Triticum turgidum L. subsp. durum（Desf.）Husn. ■

398905　Triticum durum L. subsp. abyssinicum Vavilov = Triticum aethiopicum Jakubz. ■☆

398906　Triticum durum L. var. arraseita Hochst. ex Körn. = Triticum aethiopicum Jakubz. ■☆

398907　Triticum durum L. var. schimperi Körn. = Triticum aethiopicum Jakubz. ■☆

398908　Triticum elongatum Host = Elytrigia elongata（Host ex P. Beauv.）Nevski ■

398909　Triticum elymoides Hochst. ex A. Rich. = Elymus africanus Á. Löve ■☆

398910　Triticum farctum Viv. = Elytrigia juncea（L.）Nevski ■

398911　Triticum farrum Bayle-Bar. = Triticum dicoccon Schrank ■

398912　Triticum gmelinii Trin. = Elytrigia gmelinii（Trin.）Nevski ■

398913　Triticum hordeaceum Coss. et Durieu = Dasypyrum breviaristatum（H. Lindb.）Fred. ■☆

398914　Triticum hybernum L. = Triticum aestivum L. ■

398915　Triticum infestum Salisb. = Elytrigia repens（L.）Desv. ex B. D. Jacks. ■

398916　Triticum intermedium Host = Elytrigia intermedia（Host）Nevski ■

398917　Triticum junceum L. = Elytrigia juncea (L.) Nevski ■

398918　Triticum lasianthum (Boiss.) Steud. = Eremopyrum distans (K. Koch) Nevski ■

398919　Triticum lolium (Balansa) Batt. et Trab. = Agropyropsis lolium (Balansa) A. Camus ■☆

398920　Triticum macha Dekapr. et Menabde；莫迦小麦■☆

398921　Triticum macha Dekapr. et Menabde = Triticum aestivum L. ■

398922　Triticum maritimum L. = Cutandia maritima (L.) Barbey ■☆

398923　Triticum molle (Trin.) F. Herm. = Leymus mollis (Trin. ex Spreng.) Pilg. ■

398924　Triticum monococcum L.；一粒小麦（单粒小麦）；Eincorn，One-grain Wheat，One-grained Wheat ■

398925　Triticum monococcum L. var. cereale Asch. et Graebn. = Triticum monococcum L. ■

398926　Triticum orientale (L.) M. Bieb. = Eremopyrum orientale (L.) Jaub. et Spach ■

398927　Triticum orientale Flaksb. = Triticum turanicum Jakubz. ■

398928　Triticum orientale Percival = Triticum turanicum Jakubz. ■

398929　Triticum ovatum (L.) Raspail = Aegilops ovata L. ■

398930　Triticum ovatum L. subsp. violaceum Braun-Blanq. et Wilczek = Aegilops geniculata Roth ■☆

398931　Triticum patulum Willd. = Eremopyrum bonaepartis (Spreng.) Nevski ■

398932　Triticum pauciflorum Schwein. = Roegneria pauciflora (Schwein.) Hyl. ■

398933　Triticum percivalianum Parodi = Triticum turanicum Jakubz. ■

398934　Triticum peregrinum Hack. = Aegilops peregrina (Hack.) Maire et Weiller ■☆

398935　Triticum persicum Aitch. et Hemsl. = Triticum turgidum L. var. carthlicum (Nevski) M. S. Yan ex P. C. Kuo ■

398936　Triticum persicum Vavilov ex Zhuk. = Triticum turgidum L. var. carthlicum (Nevski) M. S. Yan ex P. C. Kuo ■

398937　Triticum petropavlovskyi Udachin et Migush.；新疆小麦■

398938　Triticum petropavlovskyi Udachin et Migush. = Triticum turgidum L. subsp. polonicum (L.) Thell. ■

398939　Triticum polonicum L.；波兰小麦；Giant Rye, Jerusalem Rye, Polish Wheat ■

398940　Triticum polonicum L. = Triticum turgidum L. subsp. polonicum (L.) Á. Löve et D. Löve ■

398941　Triticum polonicum L. = Triticum turgidum L. subsp. polonicum (L.) Thell. ■

398942　Triticum polonicum L. subsp. abyssinicum (Steud.) Vavilov = Triticum turgidum L. subsp. polonicum (L.) Á. Löve et D. Löve ■

398943　Triticum polonicum L. var. abissinicum Körn. = Triticum turgidum L. subsp. polonicum (L.) Á. Löve et D. Löve ■

398944　Triticum polonicum L. var. tibeticum Udachin = Triticum turgidum L. subsp. polonicum (L.) Thell. ■

398945　Triticum prostratum (Pall.) L. f. = Eremopyrum triticeum (Gaertn.) Nevski ■

398946　Triticum pyramidale Percival = Triticum durum Desf. ■

398947　Triticum pyramidale Percival = Triticum turgidum L. subsp. durum (Desf.) Husn. ■

398948　Triticum pyramidale Percival = Triticum turgidum L. var. durum (Desf.) M. S. Yan ex P. C. Kuo ■

398949　Triticum ramosum Trin. = Leymus ramosus Tzvelev ■

398950　Triticum recognitum Steud. = Triticum aethiopicum Jakubz. ■☆

398951　Triticum repens L. = Elytrigia repens (L.) Desv. ex B. D. Jacks. ■

398952　Triticum repens L. var. glaucum (Desf.) Coss. et Durieu = Elytrigia intermedia (Host) Nevski ■

398953　Triticum sativum Lam. = Triticum aestivum L. ■

398954　Triticum sativum Lam. subsp. durum (Desf.) K. Richt. = Triticum turgidum L. subsp. durum (Desf.) Husn. ■

398955　Triticum sativum Lam. subsp. durum (Desf.) Trab. = Triticum turgidum L. subsp. durum (Desf.) Husn. ■

398956　Triticum sativum Lam. subsp. vulgare (Vill.) Trab. = Triticum aestivum L. ■

398957　Triticum sativum Lam. var. aestivum (L.) A. Wood = Triticum aestivum L. ■

398958　Triticum sativum Lam. var. compositum (L.) A. W. Wood = Triticum turgidum L. ■

398959　Triticum sativum Lam. var. monococcum (L.) Vilm. = Triticum monococcum L. ■

398960　Triticum sativum Lam. var. polonicum (L.) Trab. = Triticum turgidum L. subsp. polonicum (L.) Á. Löve et D. Löve ■

398961　Triticum sativum Lam. var. turgidum (L.) Hack. = Triticum turgidum L. ■

398962　Triticum sativum Lam. var. vulgare Hack. = Triticum aestivum L. ■

398963　Triticum schimperi Hochst. ex A. Rich. = Brachypodium schimperi (Hochst. ex A. Rich.) Chiov. ■☆

398964　Triticum schrenkianum Fisch. et C. A. Mey. = Elymus schrenkianus (Fisch. et C. A. Mey.) Tzvelev ■

398965　Triticum schrenkianum Fisch. et C. A. Mey. = Roegneria schrenkiana (Fisch. et C. A. Mey.) Nevski ■

398966　Triticum secale Link = Secale cereale L. ■

398967　Triticum secalinum Georgi = Leymus secalinus (Georgi) Tzvelev ■

398968　Triticum segetale Salisb. = Triticum aestivum L. ■

398969　Triticum sibiricum Willd. = Agropyron sibiricum (Willd.) P. Beauv. ■

398970　Triticum spelta L.；斯佩尔特小麦（士卑尔脱小麦）；Spelt, Spelt Wheat ■☆

398971　Triticum spelta L. var. saharae Ducell. = Triticum spelta L. ■☆

398972　Triticum sphaerococcum Percival；印度矮生小麦（球粒小麦）；Sphere-seed Wheat ■☆

398973　Triticum sphaerococcum Percival = Triticum aestivum L. ■

398974　Triticum squarrosum Roth = Eremopyrum bonaepartis (Spreng.) Nevski ■

398975　Triticum subsecundum Link = Elymus trachycaulus (Link) Gould ex Shinners subsp. subsecundus (Link) Á. Löve et D. Löve ■☆

398976　Triticum subulatum Banks et Sol. = Loliolum subulatum (Banks et Sol.) Eig ■☆

398977　Triticum sylvaticum (Huds.) Moench. = Brachypodium sylvaticum (Huds.) P. Beauv. ■

398978　Triticum tauschii (Coss.) Schmalh. = Aegilops tauschii Coss. ■

398979　Triticum teres H. R. Jiang et X. X. Kong；圆柱小麦■

398980　Triticum thaoudar Reut. ex Boiss.；陶氏小麦■☆

398981　Triticum timopheevii Zhuk.；提莫非维小麦；Timopheev Wheat ■

398982　Triticum trachycaulum Link = Elymus trachycaulus (Link) Gould ex Shinners ■☆

398983　Triticum trichophorum Link = Elytrigia trichophora (Link) Nevski ■

398984　Triticum triunciale (L.) Raspail = Aegilops triuncialis L. ■

398985　Triticum turanicum Jakubz.；东方小麦（杂生小麦）■

398986　Triticum turgidum L.；圆锥小麦（圆柱小麦、肿胀小麦）；Cone Wheat, Poulard Wheat, Rivet Wheat, Swelling Wheat ■

398987 Triticum turgidum L. convar. durum（Desf.）Bowden = Triticum turgidum L. subsp. durum（Desf.）Husn. ■

398988 Triticum turgidum L. convar. polonicum（L.）Mackey = Triticum turgidum L. subsp. polonicum（L.）Thell. ■

398989 Triticum turgidum L. convar. turanicum（Jakubz.）Mackey = Triticum turanicum Jakubz. ■

398990 Triticum turgidum L. subsp. abyssinicum Vavass. = Triticum aethiopicum Jakubz. ■☆

398991 Triticum turgidum L. subsp. dicoccum（Schrank）Thell. = Triticum dicoccum（Schrank）Schübeler ■

398992 Triticum turgidum L. subsp. durum（Desf.）Husn. = Triticum durum Desf. ■

398993 Triticum turgidum L. subsp. polonicum（L.）Á. Löve et D. Löve = Triticum polonicum L. ■

398994 Triticum turgidum L. subsp. polonicum（L.）Thell. = Triticum polonicum L. ■

398995 Triticum turgidum L. subsp. turanicum（Jakubz.）Á. Löve et D. Löve = Triticum turanicum Jakubz. ■

398996 Triticum turgidum L. var. carthlicum（Nevski）M. S. Yan ex P. C. Kuo = Triticum carthlicum Nevski ■

398997 Triticum turgidum L. var. compositum（L.）Gaudin = Triticum turgidum L. ■

398998 Triticum turgidum L. var. dicoccoides（Körn.）Bowden = Triticum dicoccoides Körn. ■

398999 Triticum turgidum L. var. durum（Desf.）M. S. Yan ex P. C. Kuo = Triticum durum Desf. ■

399000 Triticum turgidum L. var. polonicum（L.）M. S. Yan ex P. C. Kuo = Triticum polonicum L. ■

399001 Triticum turgidum L. var. polonicum（L.）M. S. Yan ex P. C. Kuo = Triticum turgidum L. subsp. polonicum（L.）Thell. ■

399002 Triticum unilaterale L. = Vulpia unilateralis（L.）Stace ■☆

399003 Triticum vavilovianum ?；瓦韦罗夫小麦；Branched Wheat, Branchy-eared Wheat ■☆

399004 Triticum vavilovii（Thunb.）Jakubz.；瓦氏小麦；Vavilov Wheat ■☆

399005 Triticum ventricosum Ces. = Aegilops ventricosa Tausch ■

399006 Triticum volgense（Flaksb.）Nevski；伏尔加小麦■☆

399007 Triticum vulgare（L.）Salisb. = Hordeum vulgare L. ■

399008 Triticum vulgare（L.）Salisb. var. durum（Desf.）Alef. = Triticum turgidum L. subsp. durum（Desf.）Husn. ■

399009 Triticum vulgare Vill. = Triticum aestivum L. ■

399010 Triticum vulgare Vill. var. aestivum（L.）Spenn. = Triticum aestivum L. ■

399011 Triticum vulgare Vill. var. bidens Alef. = Triticum monococcum L. ■

399012 Triticum vulgare Vill. var. oasicola Ducell. = Triticum aestivum L. ■

399013 Triticum vulgare Vill. var. saharae Ducell. = Triticum aestivum L. ■

399014 Triticum vulgare Vill. var. turgidum（L.）Alef. = Triticum turgidum L. ■

399015 Triticum vulgare Vill. var. turgidum Alef. = Triticum turgidum L. ■

399016 Tritieum polonicum L. = Triticum turgidum L. var. polonicum（L.）M. S. Yan ex P. C. Kuo ■

399017 Tritillaria Raf. = Fritillaria L. ■

399018 Tritoma Ker Gawl. = Kniphofia Moench（保留属名）■☆

399019 Tritoma burchellii Lindl. = Kniphofia uvaria（L.）Oken ■☆

399020 Tritoma caulescens（Baker）Carrière = Kniphofia caulescens Baker ex Hook. f. ■☆

399021 Tritoma macowanii（Baker）Carrière = Kniphofia triangularis Kunth ■☆

399022 Tritoma media（Donn）Ker Gawl. = Kniphofia sarmentosa（Andréws）Kunth ■☆

399023 Tritoma rooperi T. Moore = Kniphofia rooperi（T. Moore）Lem. ■☆

399024 Tritoma uvaria（L.）Ker Gawl. = Kniphofia uvaria（L.）Oken ■☆

399025 Tritomanthe Link = Tritoma Ker Gawl. ■☆

399026 Tritomanthe burchellii（Lindl.）Steud. = Kniphofia uvaria（L.）Oken ■☆

399027 Tritomanthe media（Donn）Link = Kniphofia sarmentosa（Andréws）Kunth ■☆

399028 Tritomanthe uvaria（L.）Link = Kniphofia uvaria（L.）Oken ■☆

399029 Tritomium Link = Kniphofia Moench（保留属名）■☆

399030 Tritomium uvaria（L.）Link = Kniphofia uvaria（L.）Oken ■☆

399031 Tritomodon Turcz.（1848）；肖吊钟花属●☆

399032 Tritomodon Turcz. = Enkianthus Lour. ●

399033 Tritomodon campanulatus（Miq.）F. Maek. ex Okuyama = Enkianthus campanulatus（Miq.）G. Nicholson ●☆

399034 Tritomodon campanulatus（Miq.）F. Maek. ex Okuyama var. longilobus（Nakai）Okuyama = Enkianthus campanulatus（Miq.）G. Nicholson var. longilobus（Nakai）Makino ●☆

399035 Tritomodon campanulatus（Miq.）F. Maek. ex Okuyama var. rubicundus（Matsum. et Nakai）Okuyama = Enkianthus campanulatus（Miq.）G. Nicholson var. palibinii（Craib）Bean ●☆

399036 Tritomodon campanulatus（Miq.）F. Maek. ex Okuyama var. sikokianus（Palib.）Okuyama = Enkianthus sikokianus（Palib.）Ohwi ●☆

399037 Tritomodon cernuus（Siebold et Zucc.）Honda = Enkianthus cernuus（Siebold et Zucc.）Makino ●☆

399038 Tritomodon japonicus Turcz.；三齿肖吊钟花●☆

399039 Tritomodon subsessilis（Miq.）F. Maek. ex Okuyama = Enkianthus subsessilis（Miq.）Makino ●☆

399040 Tritomodon subsessillis（Miq.）F. Maek. var. nudipes（Honda）Okuyama = Enkianthus nudipes（Honda）Ohwi ●☆

399041 Tritomopterys（A. Juss. ex Endl.）Nied. = Gaudichaudia Kunth ●☆

399042 Tritomopterys Nied. = Gaudichaudia Kunth ●☆

399043 Tritonia Ker Gawl.（1802）；观音兰属（火焰兰属,鸢尾兰属）；Montbretia, Tritonia ■

399044 Tritonia abyssinica（R. Br. ex A. Rich.）Walp. = Lapeirousia abyssinica（R. Br. ex A. Rich.）Baker ■☆

399045 Tritonia acroloba Harms = Radinosiphon leptostachyis（Baker）N. E. Br. ■☆

399046 Tritonia atrorubens（N. E. Br.）L. Bolus；深红观音兰■☆

399047 Tritonia aurea Pappe ex Hook. = Crocosmia aurea（Pappe ex Hook.）Planch. ■

399048 Tritonia bakeri Klatt；贝克观音兰■☆

399049 Tritonia bakeri Klatt = Tritonia laxifolia（Klatt）Benth. et Hook. f. ■☆

399050 Tritonia bakeri Klatt subsp. lilacina（L. Bolus）M. P. de Vos；紫丁香色观音兰■☆

399051 Tritonia bongensis Pax = Zygotritonia bongensis（Pax）Mildbr. ■☆

399052 Tritonia bracteata Worsley = Tritonia laxifolia（Klatt）Benth. et Hook. f. ■☆

399053 Tritonia capensis（Houtt.）Ker Gawl. var. major Ker Gawl. = Tritonia flabellifolia（D. Delaroche）G. J. Lewis var. major（Ker Gawl.）M. P. de Vos ■☆

399054 Tritonia chrysantha Fourc.；金花观音兰■☆

399055 Tritonia cinnabarina Pax = Crocosmia aurea（Pappe ex Hook.）Planch. subsp. pauciflora（Milne-Redh.）Goldblatt ■☆

399056 Tritonia clusiana Worsley = Tritonia laxifolia（Klatt）Benth. et

Hook. f. ■☆

399057 Tritonia cooperi (Baker) Klatt;库珀观音兰■☆

399058 Tritonia cooperi (Baker) Klatt subsp. quadrialata M. P. de Vos;四翅观音兰■☆

399059 Tritonia crispa (L. f.) Ker Gawl. = Tritonia undulata (Burm. f.) Baker ■☆

399060 Tritonia crispa (L. f.) Ker Gawl. var. parviflora Baker = Tritonia undulata (Burm. f.) Baker ■☆

399061 Tritonia crocata (L.) Ker Gawl. = Tritonia crocata (Thunb.) Ker Gawl. ■

399062 Tritonia crocata (Thunb.) Ker Gawl.;观音兰(番红花观音兰,蒙蒂兰); Crocus Tritonia, Saffron Tritonia, Saffron-yellow Tritonia, Tritonia ■

399063 Tritonia crocata (Thunb.) Ker Gawl. var. miniata Baker;红观音兰;Red Tritonia ■☆

399064 Tritonia crocosmiflora Lemoine = Crocosmia crocosmiflora (Nicholson) N. E. Br. ■

399065 Tritonia crocosmiiflora (Lemoine) G. Nicholson = Crocosmia crocosmiflora (Nicholson) N. E. Br. ■

399066 Tritonia delpierrei M. P. de Vos = Tritonia marlothii M. P. de Vos subsp. delpierrei (M. P. de Vos) M. P. de Vos ■☆

399067 Tritonia deusta (Aiton) Ker Gawl.;黑斑观音兰;Blackblotch Tritonia ■☆

399068 Tritonia deusta (Aiton) Ker Gawl. subsp. miniata (Jacq.) M. P. de Vos;小黑斑观音兰■☆

399069 Tritonia deusta Ker Gawl. = Tritonia deusta (Aiton) Ker Gawl. ■☆

399070 Tritonia disticha (Klatt) Baker;二列观音兰■☆

399071 Tritonia disticha (Klatt) Baker subsp. rubrolucens (R. C. Foster) M. P. de Vos;红光二列观音兰■☆

399072 Tritonia disticha Baker subsp. rubrolucens (R. C. Foster) M. P. de Vos;南非观音兰■☆

399073 Tritonia drakensbergensis M. P. de Vos;德拉肯斯观音兰■☆

399074 Tritonia dubia Eckl. ex Klatt;可疑观音兰■☆

399075 Tritonia fenestrata (Jacq.) Ker Gawl. = Tritonia crocata (L.) Ker Gawl. ■

399076 Tritonia flabellifolia (D. Delaroche) G. J. Lewis;扇叶观音兰■☆

399077 Tritonia flabellifolia (D. Delaroche) G. J. Lewis var. major (Ker Gawl.) M. P. de Vos;大扇叶观音兰■☆

399078 Tritonia flabellifolia (D. Delaroche) G. J. Lewis var. thomasiae M. P. de Vos;毛扇叶观音兰■☆

399079 Tritonia flanaganii F. Bolus = Tritonia atrorubens (N. E. Br.) L. Bolus ■☆

399080 Tritonia flava (Aiton) Ker Gawl. = Tritonia flabellifolia (D. Delaroche) G. J. Lewis ■☆

399081 Tritonia flavida Schltr. = Tritonia lineata (Salisb.) Ker Gawl. ■☆

399082 Tritonia florentiae (Marloth) Goldblatt;弗洛朗蒂亚观音兰■☆

399083 Tritonia fucata Herb. = Crocosmia fucata (Herb.) M. P. de Vos ■☆

399084 Tritonia gladiolaris (Lam.) Goldblatt et J. C. Manning = Tritonia lineata (Salisb.) Ker Gawl. ■☆

399085 Tritonia graminifolia Baker = Freesia grandiflora (Baker) Klatt ■☆

399086 Tritonia hyalina (L. f.) Baker = Tritonia crocata (L.) Ker Gawl. ■

399087 Tritonia kamisbergensis Klatt;卡米斯伯格观音兰■☆

399088 Tritonia karooica M. P. de Vos;卡鲁观音兰■☆

399089 Tritonia kraussii Baker = Tritonia lineata (Salisb.) Ker Gawl. ■☆

399090 Tritonia lacerata (Burm. f.) Klatt;撕裂观音兰■☆

399091 Tritonia lancea (Thunb.) N. E. Br.;绵毛观音兰;Lanose Tritonia ■☆

399092 Tritonia laxifolia (Klatt) Benth. et Hook. f.;疏叶观音兰■☆

399093 Tritonia lilacina L. Bolus = Tritonia bakeri Klatt subsp. lilacina (L. Bolus) M. P. de Vos ■☆

399094 Tritonia lineata (Salisb.) Ker Gawl.;线叶观音兰■☆

399095 Tritonia lineata (Salisb.) Ker Gawl. = Tritonia gladiolaris (Lam.) Goldblatt et J. C. Manning ■☆

399096 Tritonia lineata (Salisb.) Ker Gawl. var. parvifolia M. P. de Vos = Tritonia gladiolaris (Lam.) Goldblatt et J. C. Manning ■☆

399097 Tritonia lineata Ker Gawl. = Tritonia lineata (Salisb.) Ker Gawl. ■☆

399098 Tritonia marlothii M. P. de Vos;马洛斯观音兰■☆

399099 Tritonia marlothii M. P. de Vos subsp. delpierrei (M. P. de Vos) M. P. de Vos;戴尔皮埃尔观音兰■☆

399100 Tritonia masonorum L. Bolus = Crocosmia masonorum (L. Bolus) N. E. Br. ■☆

399101 Tritonia mathewsiana L. Bolus = Crocosmia mathewsiana (L. Bolus) Goldblatt ■☆

399102 Tritonia mensensis Schweinf. = Gladiolus mensensis (Schweinf.) Goldblatt ■☆

399103 Tritonia miniata (Jacq.) Ker Gawl. = Tritonia deusta (Aiton) Ker Gawl. subsp. miniata (Jacq.) M. P. de Vos ■☆

399104 Tritonia moggii Oberm.;莫格观音兰■☆

399105 Tritonia nelsonii Baker;纳尔逊观音兰■☆

399106 Tritonia nervosa (Baker) Klatt = Tritoniopsis nervosa (Baker) G. J. Lewis ■☆

399107 Tritonia nervosa Baker = Tritoniopsis ramosa (Eckl. ex Klatt) G. J. Lewis var. unguiculata (Baker) G. J. Lewis ■☆

399108 Tritonia odorata Lodd. = Freesia refracta (Jacq.) Klatt ■

399109 Tritonia pallida Ker Gawl.;苍白观音兰■☆

399110 Tritonia pallida Ker Gawl. subsp. taylorae (L. Bolus) M. P. de Vos;泰勒观音兰■☆

399111 Tritonia parviflora Baker = Tritonia strictifolia (Klatt) Benth. ex Klatt ■☆

399112 Tritonia parvula N. E. Br.;较小观音兰■☆

399113 Tritonia pauciflora (Baker) Klatt = Gladiolus floribundus Jacq. ■☆

399114 Tritonia pauciflora Baker = Hesperantha pauciflora (Baker) G. J. Lewis ■☆

399115 Tritonia petrophila Baker = Tritonia nelsonii Baker ■☆

399116 Tritonia pottsii (Baker) Baker;英兰■☆

399117 Tritonia pottsii (Baker) Benth. et Hook. f. = Tritonia pottsii (Baker) Baker ■☆

399118 Tritonia pottsii (Macnab ex Baker) Baker = Crocosmia pottsii (Macnab ex Baker) N. E. Br. ■

399119 Tritonia rosea (Jacq.) Aiton = Tritonia flabellifolia (D. Delaroche) G. J. Lewis var. major (Ker Gawl.) M. P. de Vos ■☆

399120 Tritonia rosea Dryand. = Tritonia disticha Baker subsp. rubrolucens (R. C. Foster) M. P. de Vos ■☆

399121 Tritonia rosea Klatt = Tritonia disticha (Klatt) Baker subsp. rubrolucens (R. C. Foster) M. P. de Vos ■☆

399122 Tritonia rubrolucens R. C. Foster = Tritonia disticha (Klatt) Baker subsp. rubrolucens (R. C. Foster) M. P. de Vos ■☆

399123 Tritonia sanguinea Eckl. = Tritonia crocata (L.) Ker Gawl. ■

399124 Tritonia schimperi Asch. et Klatt = Lapeirousia schimperi (Asch. et Klatt) Milne-Redh. ■☆

399125 Tritonia schlechteri Baker = Ixia orientalis L. Bolus ■☆

399126 Tritonia scillaris (L.) Backer = Ixia scillaris L. ■☆

399127 Tritonia secunda (Eckl.) Steud. = Tritonia flabellifolia (D. Delaroche) G. J. Lewis var. major (Ker Gawl.) M. P. de Vos ■☆

399128　Tritonia securigera（Aiton）Ker Gawl. ;斧形观音兰■☆

399129　Tritonia securigera（Aiton）Ker Gawl. subsp. watermeyeri（L. Bolus）J. C. Manning et Goldblatt;瓦特斧形观音兰■☆

399130　Tritonia striata（Jacq.）Ker Gawl. = Babiana striata（Jacq.）G. J. Lewis ■☆

399131　Tritonia strictifolia（Klatt）Benth. ex Klatt;硬叶观音兰■☆

399132　Tritonia taylorae L. Bolus = Tritonia pallida Ker Gawl. subsp. taylorae（L. Bolus）M. P. de Vos ■☆

399133　Tritonia templemannii Baker = Pillansia templemannii（Baker）L. Bolus ■☆

399134　Tritonia teretifolia Baker = Gladiolus inandensis Baker ■☆

399135　Tritonia tigrina Pax = Gladiolus benguellensis Baker ■☆

399136　Tritonia trinervata Baker = Ixia trinervata（Baker）G. J. Lewis ■☆

399137　Tritonia tugwelliae L. Bolus;特格观音兰■☆

399138　Tritonia undulata（Burm. f.）Baker;波状观音兰■☆

399139　Tritonia unguiculata Baker = Tritoniopsis ramosa（Eckl. ex Klatt）G. J. Lewis var. unguiculata（Baker）G. J. Lewis ■☆

399140　Tritonia ventricosa Baker = Gladiolus brevitubus G. J. Lewis ■☆

399141　Tritonia watermeyeri L. Bolus = Tritonia securigera（Aiton）Ker Gawl. subsp. watermeyeri（L. Bolus）J. C. Manning et Goldblatt ■☆

399142　Tritonia watsonioides Baker = Watsonia watsonioides（Baker）Oberm. ■☆

399143　Tritonia wilsonii Baker = Gladiolus wilsonii（Baker）Goldblatt et J. C. Manning ■☆

399144　Tritonia xanthospila（DC.）Ker Gawl. ex Spreng. = Freesia xanthospila（DC.）Klatt ■☆

399145　Tritoniopsis L. Bolus（1929）;肖观音兰属■☆

399146　Tritoniopsis antholyza（Poir.）Goldblatt;花肖观音兰■☆

399147　Tritoniopsis apiculata（F. Bolus）G. J. Lewis;细尖肖观音兰■☆

399148　Tritoniopsis apiculata（F. Bolus）G. J. Lewis = Tritoniopsis revoluta（Burm. f.）Goldblatt ■☆

399149　Tritoniopsis apiculata（F. Bolus）G. J. Lewis var. minor G. J. Lewis = Tritoniopsis revoluta（Burm. f.）Goldblatt ■☆

399150　Tritoniopsis bicolor J. C. Manning et Goldblatt;二色肖观音兰■☆

399151　Tritoniopsis burchellii（N. E. Br.）Goldblatt;伯切尔肖观音兰■☆

399152　Tritoniopsis caffra（Ker Gawl. ex Baker）Goldblatt;开菲尔肖观音兰■☆

399153　Tritoniopsis caledonensis（R. C. Foster）G. J. Lewis;卡利登肖观音兰■☆

399154　Tritoniopsis dodii（G. J. Lewis）G. J. Lewis;多德肖观音兰■☆

399155　Tritoniopsis elongata（L. Bolus）G. J. Lewis;伸长肖观音兰■☆

399156　Tritoniopsis flava J. C. Manning et Goldblatt;黄肖观音兰■☆

399157　Tritoniopsis flexuosa（L. f.）G. J. Lewis;曲折肖观音兰■☆

399158　Tritoniopsis intermedia（Baker）Goldblatt;间型肖观音兰■☆

399159　Tritoniopsis lata（L. Bolus）G. J. Lewis;澳非肖观音兰■☆

399160　Tritoniopsis lata（L. Bolus）G. J. Lewis var. longibracteata ?;长苞肖观音兰■☆

399161　Tritoniopsis latifolia G. J. Lewis;宽叶肖观音兰■☆

399162　Tritoniopsis lesliei L. Bolus;莱斯利肖观音兰■☆

399163　Tritoniopsis longituba（Fourc.）Goldblatt = Tritoniopsis antholyza（Poir.）Goldblatt ■☆

399164　Tritoniopsis nemorosa（Klatt）G. J. Lewis;森林肖观音兰■☆

399165　Tritoniopsis nervosa（Baker）G. J. Lewis;多脉肖观音兰■☆

399166　Tritoniopsis parviflora（Jacq.）G. J. Lewis;小花肖观音兰■☆

399167　Tritoniopsis parviflora（Jacq.）G. J. Lewis var. angusta（L. Bolus）G. J. Lewis;狭小花肖观音兰■☆

399168　Tritoniopsis pulchella G. J. Lewis;美丽肖观音兰■☆

399169　Tritoniopsis pulchella G. J. Lewis var. alpina ?;高山美丽肖观音兰■☆

399170　Tritoniopsis pulchra（Baker）Goldblatt;雅丽肖观音兰■☆

399171　Tritoniopsis ramosa（Eckl. ex Klatt）G. J. Lewis;分枝肖观音兰■☆

399172　Tritoniopsis ramosa（Eckl. ex Klatt）G. J. Lewis var. robusta G. J. Lewis;粗壮分枝肖观音兰■☆

399173　Tritoniopsis ramosa（Eckl. ex Klatt）G. J. Lewis var. unguiculata（Baker）G. J. Lewis;爪状分枝肖观音兰■☆

399174　Tritoniopsis revoluta（Burm. f.）Goldblatt;外卷肖观音兰■☆

399175　Tritoniopsis triticea（Burm. f.）Goldblatt;小麦肖观音兰■☆

399176　Tritoniopsis unguicularis（Lam.）G. J. Lewis;爪状肖观音兰■☆

399177　Tritoniopsis williamsiana Goldblatt;威廉斯肖观音兰■☆

399178　Tritonixia Klatt = Tritonia Ker Gawl. ■☆

399179　Tritonixia coccinea Eckl. = Tritonia deusta（Aiton）Ker Gawl. ■☆

399180　Tritonixia crocata（L.）Klatt = Tritonia crocata（L.）Ker Gawl. ■

399181　Tritonixia deusta（Aiton）Klatt = Tritonia deusta（Aiton）Ker Gawl. ■☆

399182　Tritonixia disticha Klatt = Tritonia disticha（Klatt）Baker ■☆

399183　Tritonixia hyalina（L. f.）Klatt = Tritonia crocata（L.）Ker Gawl. ■

399184　Tritonixia lineata（Salisb.）Klatt = Tritonia gladiolaris（Lam.）Goldblatt et J. C. Manning ■☆

399185　Tritonixia miniata（Jacq.）Klatt = Tritonia deusta（Aiton）Ker Gawl. subsp. miniata（Jacq.）M. P. de Vos ■☆

399186　Tritonixia undulata（Burm. f.）Klatt = Tritonia undulata（Burm. f.）Baker ■☆

399187　Tritophus T. Lestib. = Kaempferia L. ■

399188　Tritophus T. Lestib. = Trilophus Lestib. ■

399189　Tritriela Raf. = Ornithogalum L. ■

399190　Triumfetta L.（1753）;刺蒴麻属（垂桉草属）;Triumfetta ●■

399191　Triumfetta Plum. ex L. = Triumfetta L. ●■

399192　Triumfetta abyssinica K. Schum. = Triumfetta pilosa Roth ●■

399193　Triumfetta actinocarpa S. Moore;星果刺蒴麻■☆

399194　Triumfetta angolensis Sprague et Hutch. ;安哥拉刺蒴麻■☆

399195　Triumfetta angulata Lam. = Triumfetta bartramia L. ■

399196　Triumfetta angulata Lam. = Triumfetta rhomboidea Jacq. ●■

399197　Triumfetta annua L. ;单毛刺蒴麻（小刺蒴麻,野卷单,粘人草）;Puny Triumfetta ■

399198　Triumfetta annua L. = Triumfetta japonica Makino ■☆

399199　Triumfetta annua L. f. piligera Sprague et Hutch. = Triumfetta japonica Makino ■☆

399200　Triumfetta antunesii Sprague et Hutch. ;安图内思刺蒴麻■☆

399201　Triumfetta arussorum Chiov. = Triumfetta heterocarpa Sprague et Hutch. ■☆

399202　Triumfetta bartramia L. ;垂桉草■

399203　Triumfetta bartramia L. = Triumfetta rhomboidea Jacq. ●■

399204　Triumfetta benguelensis Wawra ex Wawra et Peyr. ;本格拉刺蒴麻■☆

399205　Triumfetta bequaertii De Wild. ;贝卡尔刺蒴麻■☆

399206　Triumfetta bogotensis DC. ;波哥大刺蒴麻,Parquet Burr ■☆

399207　Triumfetta brachyceras K. Schum. ;短角刺蒴麻■☆

399208　Triumfetta brachyceras K. Schum. var. macrophylla Cufod. = Triumfetta brachyceras K. Schum. ■☆

399209　Triumfetta brachyceras K. Schum. var. rothii（K. Schum.）Cufod. = Triumfetta brachyceras K. Schum. ■☆

399210　Triumfetta brevipetiolata De Wild. ;短梗刺蒴麻■☆

399211　Triumfetta buettneriacea K. Schum. ;比特纳刺蒴麻■☆

399212　Triumfetta cana Blume;毛刺蒴麻（痴头婆,臭垂桉草,黄花痴

头婆,蓬绒木,绒毛螺旋草,山黄麻,细黄花,小铧叶,粘巴头);
Hair Triumfetta,Hairy Triumfetta ●■

399213 Triumfetta claessensii De Wild. ;克莱森斯刺蒴麻■☆

399214 Triumfetta claudinae J. -G. Adam = Triumfetta jaegeri Hochr. ■☆

399215 Triumfetta cordifolia A. Rich. ;心叶刺蒴麻■☆

399216 Triumfetta cordifolia A. Rich. var. pubescens R. Wilczek;短柔毛刺蒴麻■☆

399217 Triumfetta cordifolia A. Rich. var. tomentosa Sprague;绒毛刺蒴麻■☆

399218 Triumfetta cuneata A. Rich. = Triumfetta pentandra A. Rich. ■☆

399219 Triumfetta cuneata Hochst. ex A. Rich. = Triumfetta pentandra A. Rich. ■☆

399220 Triumfetta cupricola De Wild. = Triumfetta digitata (Oliv.) Sprague et Hutch. ■☆

399221 Triumfetta dekindtiana Engl. ;德金刺蒴麻■☆

399222 Triumfetta delicatula Sprague et Hutch. ;姣美刺蒴麻■☆

399223 Triumfetta dembianensis Chiov. = Triumfetta rhomboidea Jacq. ●■

399224 Triumfetta descampsii De Wild. et T. Durand = Triumfetta welwitschii Mast. ■☆

399225 Triumfetta digitata (Oliv.) Sprague et Hutch. ;指裂刺蒴麻■☆

399226 Triumfetta dilungensis Adamska et Lisowski;迪龙刺蒴麻■☆

399227 Triumfetta diversifolia E. Mey. ex Eyles = Triumfetta rhomboidea Jacq. ●■

399228 Triumfetta dubia De Wild. = Triumfetta setulosa Mast. ■☆

399229 Triumfetta dunalis Kuntze = Triumfetta grandidens Hance ■

399230 Triumfetta effusa E. Mey. ex Harv. = Triumfetta pilosa Roth var. effusa (E. Mey. ex Harv.) Wild ■☆

399231 Triumfetta elskensii De Wild. = Triumfetta brachyceras K. Schum. ■☆

399232 Triumfetta eriophlebia Hook. f. ;毛脉刺蒴麻■☆

399233 Triumfetta flabellato-pilosa R. Wilczek = Triumfetta digitata (Oliv.) Sprague et Hutch. ■☆

399234 Triumfetta flavescens Hochst. ex A. Rich. ;浅黄刺蒴麻■☆

399235 Triumfetta flavescens Hochst. ex A. Rich. var. benadiriana Fiori;贝纳迪尔刺蒴麻■☆

399236 Triumfetta geoides Welw. ex Mast. var. lanceolata R. Wilczek;披针形刺蒴麻■☆

399237 Triumfetta geoides Welw. ex Mast. var. rugosa Sprague et Hutch. ;褶皱刺蒴麻■☆

399238 Triumfetta gilletii De Wild. = Triumfetta setulosa Mast. ■☆

399239 Triumfetta glabrior (Sprague et Hutch.) Cheek;无毛刺蒴麻■☆

399240 Triumfetta glandulosa Lam. = Triumfetta rhomboidea Jacq. ●■

399241 Triumfetta glechomoides Welw. ex Mast. ;活血丹刺蒴麻■☆

399242 Triumfetta glechomoides Welw. ex Mast. var. glabra De Wild. = Triumfetta glechomoides Welw. ex Mast. ■☆

399243 Triumfetta gossweileri Exell et Mendonça;戈斯刺蒴麻■☆

399244 Triumfetta grandidens Hance;粗齿刺蒴麻;Broadtooth Triumfetta ■

399245 Triumfetta grandidens Hance var. glabra R. H. Miao ex Hung T. Chang;秃刺蒴麻■

399246 Triumfetta grandistipulata Wild;大托叶刺蒴麻■

399247 Triumfetta guazumifolia Bojer = Triumfetta pilosa Roth ●■

399248 Triumfetta heliocarpa K. Schum. = Triumfetta welwitschii Mast. ■☆

399249 Triumfetta hensii De Wild. et T. Durand = Triumfetta setulosa Mast. ■☆

399250 Triumfetta heptaphylla Exell;七叶刺蒴麻■☆

399251 Triumfetta heterocarpa Sprague et Hutch. ;七果刺蒴麻■☆

399252 Triumfetta heterocarpa Sprague et Hutch. var. glabrior ? = Triumfetta glabrior (Sprague et Hutch.) Cheek ■☆

399253 Triumfetta heudelotii Planch. ex Mast. ;厄德刺蒴麻■

399254 Triumfetta hirsuta Sprague et Hutch. = Triumfetta welwitschii Mast. var. hirsuta (Sprague et Hutch.) Wild ■☆

399255 Triumfetta holosericea Schinz = Triumfetta tomentosa Bojer ■

399256 Triumfetta humilis N. E. Br. = Triumfetta welwitschii Mast. var. lanata Brummitt et Seyani ■☆

399257 Triumfetta hundtii Exell et Mendonça;洪特刺蒴麻■☆

399258 Triumfetta indica Lam. = Triumfetta rhomboidea Jacq. ●■

399259 Triumfetta intermedia De Wild. = Triumfetta setulosa Mast. ■☆

399260 Triumfetta jaegeri Hochr. ;耶格刺蒴麻■☆

399261 Triumfetta japonica Makino;日本刺蒴麻; Japan Triumfetta, Japanese Triumfetta ■☆

399262 Triumfetta junodii Schinz = Corchorus junodii (Schinz) N. E. Br. ■☆

399263 Triumfetta katangensis R. Wilczek;加丹加刺蒴麻■☆

399264 Triumfetta keniensis R. E. Fr. ;肯尼亚刺蒴麻■☆

399265 Triumfetta kirkii Mast. ;柯克刺蒴麻■

399266 Triumfetta kundelungensis Adamska et Lisowski;昆德伦加刺蒴麻■☆

399267 Triumfetta lebrunii R. Wilczek;勒布伦刺蒴麻■☆

399268 Triumfetta lepidota K. Schum. ;小鳞刺蒴麻■☆

399269 Triumfetta likasiensis De Wild. ;利卡西刺蒴麻■☆

399270 Triumfetta longicornuta Hutch. et M. B. Moss;长角刺蒴麻■

399271 Triumfetta macrophylla K. Schum. = Triumfetta brachyceras K. Schum. ■☆

399272 Triumfetta macrophylla K. Schum. var. rothii Sprague et Hutch. = Triumfetta brachyceras K. Schum. ■☆

399273 Triumfetta macrophylla K. Schum. var. ruwenzoriensis (Sprague) Sprague et Hutch. = Triumfetta brachyceras K. Schum. ■☆

399274 Triumfetta marunguensis R. Wilczek;马龙古刺蒴麻■

399275 Triumfetta mastersii Baker f. var. heliocarpa (K. Schum.) Sprague et Hutch. = Triumfetta welwitschii Mast. ■☆

399276 Triumfetta melanocarpa Suess. = Triumfetta pilosa Roth var. effusa (E. Mey. ex Harv.) Wild ■☆

399277 Triumfetta micrantha K. Schum. ;小花刺蒴麻■☆

399278 Triumfetta morrumbalana De Wild. = Triumfetta pilosa Roth ●■

399279 Triumfetta nana Bojer = Triumfetta annua L. ■

399280 Triumfetta neghelliensis Lanza = Triumfetta flavescens Hochst. ex A. Rich. ■☆

399281 Triumfetta neglecta Wight et Arn. = Triumfetta pentandra A. Rich. ■☆

399282 Triumfetta oblonga Hornem. ex Schrank = Triumfetta pilosa Roth ●■

399283 Triumfetta oblongata Link. = Triumfetta pilosa Roth ●■

399284 Triumfetta obtusicornis Sprague et Hutch. ;钝角刺蒴麻■

399285 Triumfetta orbiculata König = Triumfetta rotundifolia Lam. ■☆

399286 Triumfetta orthacantha T. Durand et Schinz = Triumfetta setulosa Mast. ■☆

399287 Triumfetta orthacantha Welw. ex Mast. ;直刺刺蒴麻■

399288 Triumfetta palmatiloba Dunkley = Triumfetta trifida Sprague et Hutch. ■☆

399289 Triumfetta paradoxa (Welw. ex Hiern) Sprague et Hutch. ;奇异刺蒴麻■

399290 Triumfetta patulopilosa De Wild. = Triumfetta setulosa Mast. ■☆

399291 Triumfetta pedunculata De Wild. ;梗花刺蒴麻■☆

399292 Triumfetta pedunculata De Wild. var. stellato-pilosa ? = Triumfetta setulosa Mast. ■☆

399293 Triumfetta pedunculata De Wild. var. subglabra ? = Triumfetta pedunculata De Wild. ■☆

399294 Triumfetta pedunculata De Wild. var. tomentosa ? = Triumfetta pedunculata De Wild. ■☆

399295 Triumfetta pentandra A. Rich. ;五蕊刺蒴麻;Burweed ■☆

399296 Triumfetta pilosa Roth;长钩刺蒴麻(长毛螺旋草,长叶垂枝草,梗麻,金纳香,牛虱子,细山马栗);Longhook Triumfetta, Pilose Triumfetta ●■

399297 Triumfetta pilosa Roth var. effusa（E. Mey. ex Harv.）Wild;开展长钩刺蒴麻■☆

399298 Triumfetta pilosa Roth var. leiocarpa Fiori = Triumfetta brachyceras K. Schum. ■☆

399299 Triumfetta pilosa Roth var. nyasana Sprague et Hutch. = Triumfetta pilosa Roth ●■

399300 Triumfetta pilosa Roth var. tomentosa Szyszyl. ex Sprague et Hutch. ;绒毛长钩刺蒴麻■☆

399301 Triumfetta pilosa Roth. f. tricuspidata Sprague et Hutch. = Triumfetta tomentosa Bojer ■

399302 Triumfetta pleiacantha Sprague et Hutch. = Triumfetta actinocarpa S. Moore ■☆

399303 Triumfetta procumbens G. Forst. ;铺地刺蒴麻;Procumbent Triumfetta ●

399304 Triumfetta procumbens G. Forst. var. glaberrima Hatus. ;光铺地刺蒴麻●☆

399305 Triumfetta procumbens G. Forst. var. repens（Blume）Hatus. = Triumfetta repens（Blume）Merr. et Rolfe ■☆

399306 Triumfetta pseudocana Sprague et Craib = Triumfetta cana Blume ●■

399307 Triumfetta radicans Bojer = Triumfetta repens（Blume）Merr. et Rolfe ■☆

399308 Triumfetta rehmannii Szyszyl. = Triumfetta welwitschii Mast. ■☆

399309 Triumfetta repens（Blume）Merr. et Rolfe;匍匐刺蒴麻■☆

399310 Triumfetta reticulata Wild;网状刺蒴麻■☆

399311 Triumfetta rhodoneura K. Schum. ;红脉刺蒴麻■☆

399312 Triumfetta rhomboidea Jacq. ;刺蒴麻(苍耳叶狗核桃,黐头婆,垂枝草,黄花地桃花,黄花虱麻头,菱叶黐头婆,毛刺蒴麻,密麻椿,密马青,密马专,虱麻头,细叶黏头搔);Common Triumfetta, Rhomboid-leaved Triumphweed ●■

399313 Triumfetta rhomboidea Jacq. var. angulata（Lam.）Baker = Triumfetta rhomboidea Jacq. ●■

399314 Triumfetta rhomboidea Jacq. var. glandulosa Baker = Triumfetta rhomboidea Jacq. ●■

399315 Triumfetta riparia Hochst. = Triumfetta rhomboidea Jacq. ●■

399316 Triumfetta robynsii De Wild. = Triumfetta dekindtiana Engl. ■☆

399317 Triumfetta rogersii N. E. Br. = Triumfetta welwitschii Mast. var. rogersii（N. E. Br.）Brummitt et Seyani ■☆

399318 Triumfetta rotundifolia Lam. ;圆叶刺蒴麻■☆

399319 Triumfetta ruhengerensis De Wild. = Triumfetta tomentosa Bojer ■

399320 Triumfetta ruwenzoriensis Sprague = Triumfetta brachyceras K. Schum. ■☆

399321 Triumfetta sapinii De Wild. = Triumfetta dekindtiana Engl. ■☆

399322 Triumfetta scandens K. Schum. ;攀缘刺蒴麻■☆

399323 Triumfetta schimperi A. Rich. = Triumfetta annua L. ■

399324 Triumfetta schimperi Hochst. ex A. Rich. = Triumfetta annua L. ■

399325 Triumfetta semitriloba Jacq. ;菲岛刺蒴麻(菲岛垂枝草);Burbrush ●■

399326 Triumfetta setulosa Mast. ;细刚毛刺蒴麻■☆

399327 Triumfetta shinyangensis Verdc. ;希尼安加刺蒴麻■☆

399328 Triumfetta sonderi Ficalho et Hiern;森诺刺蒴麻■☆

399329 Triumfetta sonderiana Bolus = Triumfetta sonderi Ficalho et Hiern ■☆

399330 Triumfetta suborbiculata DC. = Triumfetta rotundifolia Lam. ■☆

399331 Triumfetta suffruticosa sensu Merr. = Triumfetta annua L. ■

399332 Triumfetta telekii Schweinf. = Triumfetta brachyceras K. Schum. ■☆

399333 Triumfetta tenuipedunculata Wild;细梗花刺蒴麻■☆

399334 Triumfetta tomentosa Baker;臭垂枝草（毛刺蒴麻）■

399335 Triumfetta tomentosa Bojer = Triumfetta cana Blume ●■

399336 Triumfetta tomentosa Bojer var. calvescens Franch. = Triumfetta cana Blume ●■

399337 Triumfetta tomentosa Bojer var. macrocerata Chiov. = Triumfetta tomentosa Bojer ■

399338 Triumfetta trachystema K. Schum. ;糙茎刺蒴麻■☆

399339 Triumfetta trichocarpa Hochst. ex A. Rich. ;毛果刺蒴麻■☆

399340 Triumfetta trichocarpa Hochst. ex A. Rich. var. heteracantha Sprague et Hutch. = Triumfetta trichocarpa Hochst. ex A. Rich. ■☆

399341 Triumfetta trichocarpa Sond. = Triumfetta sonderi Ficalho et Hiern ■☆

399342 Triumfetta trichoclada DC. = Triumfetta annua L. ■

399343 Triumfetta triclada Link = Triumfetta annua L. ■

399344 Triumfetta triclada Link = Triumfetta japonica Makino ■☆

399345 Triumfetta trifida Sprague et Hutch. ;三裂刺蒴麻■☆

399346 Triumfetta trigona Sprague et Hutch. ;三角刺蒴麻■☆

399347 Triumfetta trilocularis Roxb. = Triumfetta rhomboidea Jacq. ●■

399348 Triumfetta velutina Vahl = Triumfetta rhomboidea Jacq. ●■

399349 Triumfetta vermoesenii De Wild. = Triumfetta setulosa Mast. ■☆

399350 Triumfetta welwitschii Mast. ;韦尔刺蒴麻■☆

399351 Triumfetta welwitschii Mast. var. descampsii（De Wild. et Durieu）Brenan = Triumfetta welwitschii Mast. ■☆

399352 Triumfetta welwitschii Mast. var. hirsuta（Sprague et Hutch.）Wild;粗毛刺蒴麻■☆

399353 Triumfetta welwitschii Mast. var. lanata Brummitt et Seyani;绵毛刺蒴麻■☆

399354 Triumfetta welwitschii Mast. var. rogersii（N. E. Br.）Brummitt et Seyani;罗杰斯刺蒴麻■☆

399355 Triumfetta wittei De Wild. = Triumfetta digitata（Oliv.）Sprague et Hutch. ■☆

399356 Triumfetta youngii Exell et Mendonça;扬氏刺蒴麻■☆

399357 Triumfettaria Rchb. = Triumfetta Plum. ex L. ●■

399358 Triumfettoides Rauschert = Triumfetta Plum. ex L. ●■

399359 Triunia L. A. S. Johnson et B. G. Briggs(1975);厚被山龙眼属●☆

399360 Triunia montana（C. T. White）Foreman;山地厚被山龙眼●☆

399361 Triunia robusta（C. T. White）Foreman;粗壮厚被山龙眼●☆

399362 Triunia youngiana（F. Muell.）L. Johnson et B. G. Briggs;厚被山龙眼●☆

399363 Triunila Raf. = Uniola L. ■☆

399364 Triuranthera Backer = Driessenia Korth. ●■☆

399365 Triuridaceae Gardner(1843)（保留科名）;霉草科;Triuris Family ■

399366 Triuridopsis H. Maas et Maas(1994);秘鲁霉草属■☆

399367 Triuridopsis intermedia T. Franke,Beenken et C. Hahn;间型秘鲁霉草■☆

399368 Triuridopsis peruviana H. Maas et Maas;秘鲁霉草■☆

399369 Triuris L. = Triuris Miers ■☆

399370 Triuris Miers(1841);霉草属;Threefold ■☆

399371 Triuris hyalina Miers;霉草■☆

399372 Triurocodon Schltr. = Thismia Griff. ■

399373 Trivalvaria（Miq.）Miq. (1865);短梗玉盘属●☆

399374 Trivalvaria Miq. = Trivalvaria（Miq.）Miq. ●☆

399375 Trivalvaria argentea（Hook. f. et Thomson）J. Sinclair;银白短梗玉盘●☆

399376　Trivalvaria longirostris Becc. ex Scheff.；长喙短梗玉盘●☆

399377　Trivalvaria macrophylla Miq.；大叶短梗玉盘●☆

399378　Trivalvaria macrophylla Miq. var. sumatrana Heusden；苏门答腊短梗玉盘●☆

399379　Trivalvaria nervosa（Hook. f. et Thomson）J. Sinclair；多脉短梗玉盘●☆

399380　Trivalvaria pumila（King）J. Sinclair；小短梗玉盘●☆

399381　Trivolvulus Moc. et Sessé ex Choisy = Ipomoea L.（保留属名）●■

399382　Trixago Haller = Stachys L. ●■

399383　Trixago Raf. = Teucrium L. ●■

399384　Trixago Steven = Bellardia All. ■☆

399385　Trixago apula Steven = Bartsia trixago L. ■☆

399386　Trixago apula Steven var. versicolor（Lam.）Willk. et Lange = Bartsia trixago L. ■☆

399387　Trixago trixago（L.）All. = Bellardia trixago（L.）All. ■☆

399388　Trixago versicolor（Lam.）Webb et Berthel. = Bellardia trixago（L.）All. ■☆

399389　Trixanthera Raf. = Trichanthera Kunth ■☆

399390　Trixapias Raf. = Utricularia L. ■

399391　Trixella Fourr. = Stachys L. ●■

399392　Trixella Fourr. = Trixago Haller ●■

399393　Trixis Adans. = Proserpinaca L. ■☆

399394　Trixis P. Browne（1756）；三齿钝柱菊属；Threefold ■●☆

399395　Trixis Sw. = Baillieria Aubl. ●■☆

399396　Trixis Sw. = Clibadium F. Allam. ex L. ●■☆

399397　Trixis californica Kellogg；加州三齿钝柱菊；American Threefold，California Threefold，Trixis ●☆

399398　Trixis inula Crantz；墨西哥三齿钝柱菊；Mexican Trixis，Tropical Threefold ●☆

399399　Trixostis Raf. = Aristida L. ■

399400　Trizeuxis Lindl.（1821）；三轭兰属■☆

399401　Trizeuxis falcata Lindl.；三轭兰■☆

399402　Trocdaris Raf. = Carum L. ■

399403　Trochera Rich.（废弃属名）= Ehrharta Thunb.（保留属名）■☆

399404　Trochetia DC.（1823）；梭萼梧桐属●☆

399405　Trochetia grandiflora Bojer；大花梭萼梧桐●☆

399406　Trochetia laurifolia Benth.；桂叶梭萼梧桐●☆

399407　Trochetia triflora DC.；三花梭萼梧桐●☆

399408　Trochetia uniflora DC.；单花梭萼梧桐●☆

399409　Trochetiopsis Marais（1981）；拟梭萼梧桐属●☆

399410　Trochetiopsis erythroxylon（G. Forst.）Marais；红木拟梭萼梧桐●☆

399411　Trochetiopsis melanoxylon（R. Br.）Marais；黑木拟梭萼梧桐●☆

399412　Trochilocactus Linding. = Disocactus Lindl. ●☆

399413　Trochisandra Bedd. = Bhesa Buch.-Ham. ex Arn. ●

399414　Trochisandra Bedd. = Kurrimia Wall. ex Thwaites ●

399415　Trochiscanthes W. D. J. Koch（1824）；轮花草属■☆

399416　Trochiscanthes nodiflora W. D. J. Koch；轮花草☆

399417　Trochiscanthos St.-Lag. = Trochiscanthes W. D. J. Koch ■☆

399418　Trochiscus O. E. Schulz = Rorippa Scop. ■

399419　Trochiscus O. E. Schulz（1933）；滑轮芥属■☆

399420　Trochiscus cochlearioides（Roth）O. E. Schulz；滑轮芥■☆

399421　Trochiscus macrocarpus Gilli；大果滑轮芥■☆

399422　Trochocarpa R. Br.（1810）；轮果石南属（车轮果属）●☆

399423　Trochocarpa laurina R. Br.；轮果石南●☆

399424　Trochocarpa parviflora（Stschegl.）Benth.；小花轮果石南●☆

399425　Trochocarpa thymifolia（R. Br.）Spreng.；百里香轮果石南（百里香叶车轮果）●☆

399426　Trochocephalus（Mert. et W. D. J. Koch）Opiz = Scabiosa L. ●■

399427　Trochocephalus Opiz ex Bercht. = Scabiosa L. ●■

399428　Trochocodon P. Candargy（1897）；轮钟桔梗属■☆

399429　Trochocodon spicatus P. Candargy；轮钟桔梗■☆

399430　Trochodendraceae Eichler（1865）（保留科名）；昆栏树科●

399431　Trochodendraceae Prantl = Trochodendraceae Eichler（保留科名）●

399432　Trochodendron Siebold et Zucc.（1839）；昆栏树属（云叶属）；Wheel Tree，Wheelstamen Tree，Wheel-stamen Tree，Wheelstamentree ●

399433　Trochodendron aralioides Siebold et Zucc.；昆栏树（山车，云叶）；Bird-lime Tree，Wheel Tree，Wheelstamen Tree，Wheel-stamen Tree，Wheelstamentree ●

399434　Trochodendron aralioides Siebold et Zucc. f. longifolium（Maxim.）Ohwi = Trochodendron aralioides Siebold et Zucc. var. longifolium Maxim. ●☆

399435　Trochodendron aralioides Siebold et Zucc. var. longifolium Maxim.；长叶昆栏树；Longleaf Wheelstamentree ●☆

399436　Trochomeria Hook. f.（1867）；篏瓜属■☆

399437　Trochomeria atacorensis A. Chev. = Trochomeria macrocarpa（Sond.）Hook. f. ■☆

399438　Trochomeria baumiana Gilg；鲍姆篏瓜■☆

399439　Trochomeria brachypetala R. E. Fr. var. foliata R. Fern. = Trochomeria macrocarpa（Sond.）Hook. f. ■☆

399440　Trochomeria bussei Gilg = Trochomeria macrocarpa（Sond.）Hook. f. ■☆

399441　Trochomeria bussei Gilg = Trochomeria polymorpha（Welw.）Cogn. ■☆

399442　Trochomeria bussei Gilg var. tripartita R. Fern. = Trochomeria macrocarpa（Sond.）Hook. f. ■☆

399443　Trochomeria dalzielii Hutch. = Trochomeria macrocarpa（Sond.）Hook. f. ■☆

399444　Trochomeria debilis（Sond.）Hook. f.；弱小篏瓜■☆

399445　Trochomeria dentata Cogn. ex Harms = Trochomeria macrocarpa（Sond.）Hook. f. ■☆

399446　Trochomeria djurensis Schweinf. et Gilg = Trochomeria macrocarpa（Sond.）Hook. f. ■☆

399447　Trochomeria garcinii Hook. f. = Trochomeria hookeri Harv. ■☆

399448　Trochomeria harmsiana Bullock = Trochomeria macrocarpa（Sond.）Hook. f. ■☆

399449　Trochomeria hookeri Harv.；虎克篏瓜■☆

399450　Trochomeria longipetala A. Zimm. = Trochomeria macrocarpa（Sond.）Hook. f. ■☆

399451　Trochomeria macrocarpa（Sond.）Hook. f.；大篏瓜■☆

399452　Trochomeria macrocarpa（Sond.）Hook. f. subsp. vitifolia（Hook. f.）R. Fern. et A. Fern.；葡萄叶篏瓜■☆

399453　Trochomeria macroura Hook. f. = Trochomeria macrocarpa（Sond.）Hook. f. ■☆

399454　Trochomeria multiflora R. Fern. = Trochomeria macrocarpa（Sond.）Hook. f. ■☆

399455　Trochomeria nudiflora Burtt Davy = Trochomeria macrocarpa（Sond.）Hook. f. ■☆

399456　Trochomeria pectinata（Sond.）Cogn. = Trochomeria hookeri Harv. ■☆

399457　Trochomeria pectinata（Sond.）Cogn. var. subintegrifolia Cogn. = Trochomeria hookeri Harv. ■☆

399458　Trochomeria polymorpha（Welw.）Cogn.；多形篏瓜■☆

399459　Trochomeria polymorpha（Welw.）Cogn. var. stenoloba（Welw.）R. Fern. et A. Fern. = Trochomeria macrocarpa（Sond.）Hook. f. ■☆

399460　Trochomeriarotundata Burtt Davy = Trochomeria hookeri Harv. ■☆

399461　Trochomeria sagittata（Harv. ex Sond.）Cogn. ;箭头籇瓜■☆

399462　Trochomeria stefaninii（Chiov.）C. Jeffrey = Dactyliandra stefaninii（Chiov.）C. Jeffrey ■☆

399463　Trochomeria stenoloba（Welw.）Cogn. = Trochomeria macrocarpa（Sond.）Hook. f. ■☆

399464　Trochomeria stenoloba（Welw.）Cogn. = Trochomeria polymorpha（Welw.）Cogn. ■☆

399465　Trochomeria subglabra C. Jeffrey;光籇瓜■☆

399466　Trochomeria teixeirae R. Fern. et A. Fern. = Trochomeria macrocarpa（Sond.）Hook. f. ■☆

399467　Trochomeria teixeirae R. Fern. et A. Fern. = Trochomeria polymorpha（Welw.）Cogn. ■☆

399468　Trochomeria verdickii De Wild. = Trochomeria macrocarpa（Sond.）Hook. f. ■☆

399469　Trochomeria vitifolia Hook. f. = Trochomeria macrocarpa（Sond.）Hook. f. subsp. vitifolia（Hook. f.）R. Fern. et A. Fern. ■☆

399470　Trochomeria wyleyana（Sond.）Cogn. = Trochomeria debilis（Sond.）Hook. f. ■☆

399471　Trochomeriopsis Cogn.（1996）;拟籇瓜属■☆

399472　Trochomeriopsis diversifolia Cogn. ;拟籇瓜■☆

399473　Trochoseris Poepp. et Endl. = Troximon Gaertn. ■☆

399474　Trochoseris Poepp. et Endl. ex Endl. = Macrorhynchus Less. ■☆

399475　Trochostigma Siebold et Zucc. = Actinidia Lindl. ●

399476　Trochostigma arguta Siebold et Zucc. = Actinidia arguta（Siebold et Zucc.）Planch. ex Miq. ●

399477　Troglophyton Hilliard et B. L. Burtt（1981）;纸苞紫绒草属■☆

399478　Troglophyton acocksianum Hilliard;纸苞紫绒草■☆

399479　Troglophyton capillaceum（Thunb.）Hilliard et B. L. Burtt;细毛纸苞紫绒草■☆

399480　Troglophyton capillaceum（Thunb.）Hilliard et B. L. Burtt subsp. diffusum（DC.）Hilliard;铺散细毛纸苞紫绒草■☆

399481　Troglophyton elsiae Hilliard;埃尔西亚紫绒草■☆

399482　Troglophyton parvulum（Harv.）Hilliard et B. L. Burtt;小纸苞紫绒草■☆

399483　Troglophyton tenellum Hilliard;柔软紫绒草■☆

399484　Trollius L.（1753）;金莲花属; Globe Flower, Globeflower, Trollius ■

399485　Trollius × cultorum Bergmans;杂种金莲花; Globeflower ■☆

399486　Trollius × cultorum Bergmans 'Albaster';雪花石膏杂种金莲花■☆

399487　Trollius × cultorum Bergmans 'Earliest of All';早开杂种金莲花■☆

399488　Trollius × cultorum Bergmans 'Goldquelle';金红杂种金莲花■☆

399489　Trollius × cultorum Bergmans 'Orange Princess';橙黄公主杂种金莲花■☆

399490　Trollius albiflorus（A. Gray）Rydb. ;白花金莲花; White Globe-flower ■☆

399491　Trollius altaicus C. A. Mey. ;阿尔泰金莲花（金莲花,宽瓣金莲花）; Altai Globeflower ■

399492　Trollius americanus DC. = Trollius laxus Salisb. ■☆

399493　Trollius anemonifolius（Brühl）Stapf = Trollius yunnanensis（Franch.）Ulbr. var. anemonifolius（Brühl）W. T. Wang ■

399494　Trollius asiaticus L. ;宽瓣金莲花（金梅草,亚洲金莲花,重瓣金莲花）; Asiatic Globe Flower, Orange Globe Flower, Siberia Globeflower, Siberian Globeflower ■☆

399495　Trollius asiaticus L. var. chinensis（Bunge）Maxim. = Trollius chinensis Bunge ■

399496　Trollius asiaticus L. var. ledebourii Maxim. = Trollius hondoensis Nakai ■☆

399497　Trollius buddae Schipcz. ;川陕金莲花（骆驼七）; Sichuan-Shaanxi Globeflower, Szechwan-shensi Globeflower ■

399498　Trollius caucasicus Steven = Trollius ranunculinus（Sm.）Steam ■☆

399499　Trollius chartosepalus Schipcz. ;纸萼金莲花; Paperysepal Globeflower ■

399500　Trollius chinensis Bunge;金莲花（旱地莲,旱金莲,金芙蓉,金疙瘩,金梅草）; China Globeflower, Chinese Bog Trollius, Chinese Globe Flower, Chinese Globeflower ■

399501　Trollius chinensis Bunge 'Golden Queen';皇后金莲花; Globe Flower ■☆

399502　Trollius chinensis Bunge subsp. macropetalus（Regel）Luferov = Trollius macropetalus F. Schmidt ■

399503　Trollius dschungaricus Regel;准噶尔金莲花; Dzungar Globeflower, Turkestan Globeflower ■

399504　Trollius europaeus L. ;欧洲金莲花; Ballflower, Bolts, Bull Jumpling, Bummalkyte, Butter Basket, Butter Blob, Butter Bump, Butter-blob, Butterbumps, Cabbage Daisy, Common Globe Flower, Common Globeflower, Europe Globeflower, European Globe Flower, European Globeflower, Globe Crowfoot, Globe Crow's Foot, Globe Flower, Globe Ranunculus, Globeflower, Globes, Goldilocks, Golland, Kingcup, Lapper Gowan, Lapper-gowan, Lockan Gowan, Locken Gowan, Locken Gowlan, Locker Goulon, Lockergoulon, Lockin-magowan, Lockrea Gowlan, Lockren Gowlan, Lockyer-golden, Lopper Gowan, Lopper-gowan, Luckan Gowan, Lucken Gowan, Luckie Gowan, Lucky Gowan, Lukin Gowan, May Blob, May-blob, Stock, Troll-flower, Witch Gowan ■

399505　Trollius europaeus L. 'Canary Bird';金丝雀欧洲金莲花■☆

399506　Trollius europaeus L. var. songoricus Regel = Trollius dschungaricus Regel ■

399507　Trollius farreri Stapf;矮金莲花（五金草,一枝花）; Dwarf Globeflower, Farrer Globeflower ■

399508　Trollius farreri Stapf var. major W. T. Wang;大叶矮金莲花（大叶金莲花）; Largeleaf Globeflower ■

399509　Trollius hondoensis Nakai;本州金莲花（金梅草）; Ledebour Globe Flower ■☆

399510　Trollius japonicus Miq. ;长白金莲花（山地金莲花）; Japan Globeflower, Japanese Globeflower ■

399511　Trollius kansuensis（Brühl）Mukerjee = Trollius farreri Stapf ■

399512　Trollius laxus Salisb. ;美洲金莲花; American Globe Flower, American Globe-flower, Spreading Globeflower, Spreading Globe-flower ■☆

399513　Trollius laxus Salisb. var. albiflorus A. Gray = Trollius albiflorus（A. Gray）Rydb. ■☆

399514　Trollius ledebouri Rchb. var. macropetalus Regel = Trollius macropetalus F. Schmidt ■

399515　Trollius ledebourii Rchb. ;短瓣金莲花（金莲花）; Ledebour Globeflower, Shortpetal Globeflower ■

399516　Trollius ledebourii Rchb. f. plena（Kitag.）Y. C. Chu;短重瓣金莲花■

399517　Trollius ledebourii Rchb. var. macropetalus Regel = Trollius macropetalus F. Schmidt ■

399518　Trollius lilacinus Bunge;淡紫金莲花; Lilac Globeflower ■

399519　Trollius macropetalus F. Schmidt;长瓣金莲花; Longpetal Globeflower ■

399520　Trollius micranthus Hand. -Mazz. ;小花金莲花; Littleflower Globeflower ■

399521　Trollius papavereus Schipcz. = Trollius yunnanensis（Franch.）

Ulbr. ■

399522　Trollius patulus Salisb. ;开展金莲花;Spreading Globeflower ■

399523　Trollius pulcher Makino;美丽金莲花■☆

399524　Trollius pumilus D. Don;小金莲花(矮金莲花);Dwarf Globe Flower,Dwarf Globeflower,Small Globeflower ■

399525　Trollius pumilus D. Don subsp. anemonifolius Brühl = Trollius yunnanensis (Franch.) Ulbr. var. anemonifolius (Brühl) W. T. Wang ■

399526　Trollius pumilus D. Don subsp. normalis Brühl = Trollius pumilus D. Don ■

399527　Trollius pumilus D. Don subsp. normalis Brühl var. kansuensis Brühl = Trollius farreri Stapf ■

399528　Trollius pumilus D. Don subsp. normalis Brühl var. ranunculoides Brühl = Trollius ranunculoides Hemsl. ■

399529　Trollius pumilus D. Don subsp. normalis Brühl var. sikkimensis Brühl = Trollius pumilus D. Don ■

399530　Trollius pumilus D. Don subsp. normalis Brühl var. yunnanensis Brühl = Trollius yunnanensis (Franch.) Ulbr. ■

399531　Trollius pumilus D. Don var. alpinus Ulbr. = Trollius pumilus D. Don var. tanguticus Brühl ■

399532　Trollius pumilus D. Don var. foliosus (W. T. Wang) W. T. Wang;显叶金莲花;Manyleaf Dwarf Globeflower ■

399533　Trollius pumilus D. Don var. lobatus Schipcz. = Trollius farreri Stapf ■

399534　Trollius pumilus D. Don var. semifissus Schipcz. = Trollius pumilus D. Don ■

399535　Trollius pumilus D. Don var. tanguticus Brühl;青藏金莲花;Tangut Dwarf Globeflower,Tangut Globeflower ■

399536　Trollius pumilus D. Don var. tehkehensis (W. T. Wang) W. T. Wang;德格金莲花;Tehkeh Dwarf Globeflower ■

399537　Trollius pumilus D. Don var. yunnanensis Franch. = Trollius yunnanensis (Franch.) Ulbr. ■

399538　Trollius ranunculinus (Sm.) Steam;黄金莲花■☆

399539　Trollius ranunculoides Hemsl. ;毛茛状金莲花(西藏鸡爪草);Buttercup-like Globeflower ■

399540　Trollius riederianus Fisch. et C. A. Mey. ;里氏金莲花■☆

399541　Trollius sibiricus Schipcz. ;西伯利亚金莲花■☆

399542　Trollius stenopetalus Stapf = Trollius buddae Schipcz. ■

399543　Trollius taihasenzanensis Masam. ;台湾金莲花(金梅草);Taiwan Globeflower ■

399544　Trollius tanguticus (Brühl) W. T. Wang = Trollius pumilus D. Don var. tanguticus Brühl ■

399545　Trollius tanguticus (Brühl) W. T. Wang var. foliosus W. T. Wang = Trollius pumilus D. Don var. foliosus (W. T. Wang) W. T. Wang ■

399546　Trollius tehkehensis W. T. Wang = Trollius pumilus D. Don var. tehkehensis (W. T. Wang) W. T. Wang ■

399547　Trollius uralensis Gorodkov;乌拉尔金莲花;Ural Globeflower ■☆

399548　Trollius vaginatus Hand. -Mazz. ;鞘柄金莲花;Sheathstipe Globeflower ■

399549　Trollius yunnanensis (Franch.) Ulbr. ;云南金莲花(鸡爪草,金莲花);Yunnan Globeflower ■

399550　Trollius yunnanensis (Franch.) Ulbr. f. eupetala Stapf = Trollius yunnanensis (Franch.) Ulbr. var. eupetalus (Stapf) W. T. Wang ■

399551　Trollius yunnanensis (Franch.) Ulbr. f. ubera Stapf = Trollius yunnanensis (Franch.) Ulbr. ■

399552　Trollius yunnanensis (Franch.) Ulbr. subsp. anemonifolius (Brühl) Dorosz. = Trollius yunnanensis (Franch.) Ulbr. var. anemonifolius (Brühl) W. T. Wang ■

399553　Trollius yunnanensis (Franch.) Ulbr. var. anemonifolius (Brühl) W. T. Wang;覆裂云南金莲花(复裂云南金莲花)■

399554　Trollius yunnanensis (Franch.) Ulbr. var. eupetalus (Stapf) W. T. Wang;长瓣云南金莲花;Longpetal Yunnan Globeflower ■

399555　Trollius yunnanensis (Franch.) Ulbr. var. peltatus W. T. Wang;盾叶云南金莲花;Peltate Globeflower ■

399556　Trommsdorffia Bernh. (1800);特罗菊属■☆

399557　Trommsdorffia Bernh. = Hypochaeris L. ■

399558　Trommsdorffia Mart. = Hebanthe Mart. ■☆

399559　Trommsdorffia Mart. = Iresine P. Browne(保留属名)●■

399560　Trommsdorffia Mart. = Pedersenia Holub ■☆

399561　Trommsdorffia Mart. = Pfaffia Mart. ■☆

399562　Trommsdorffia ciliata (Thunb.) Soják = Hypochaeris ciliata (Thunb.) Makino ■

399563　Trommsdorffia crepidioides (Miyabe et Kudo) Soják = Hypochaeris crepidioides (Miyabe et Kudo) Tatew. et Kitam. ■☆

399564　Trommsdorffia maculata Bernh. ;特罗菊■☆

399565　Tromotriche Haw. (1812);颤毛萝藦属■☆

399566　Tromotriche Haw. = Stapelia L. (保留属名)■

399567　Tromotriche aperta (Masson) Bruyns;开放颤毛萝藦■☆

399568　Tromotriche baylissii (L. C. Leach) Bruyns;贝利斯颤毛萝藦■☆

399569　Tromotriche choanantha (Lavranos et H. Hall) Bruyns;漏斗花颤毛萝藦■☆

399570　Tromotriche engleriana (Schltr.) L. C. Leach = Stapelia engleriana Schltr. ■☆

399571　Tromotriche herrei (Nel) Bruyns;赫勒颤毛萝藦■☆

399572　Tromotriche longii (C. A. Lückh.) Bruyns;朗氏颤毛萝藦■☆

399573　Tromotriche longipes (C. A. Lückh.) Bruyns = Tromotriche pedunculata (Masson) Bruyns subsp. longipes (C. A. Lückh.) Bruyns ■☆

399574　Tromotriche pedunculata (Masson) Bruyns;梗花颤毛萝藦■☆

399575　Tromotriche pedunculata (Masson) Bruyns subsp. longipes (C. A. Lückh.) Bruyns;长梗颤毛萝藦■☆

399576　Tromotriche pruinosa (Masson) Haw. = Quaqua pruinosa (Masson) Bruyns ■☆

399577　Tromotriche revoluta (Masson) Haw. ;外卷颤毛萝藦■☆

399578　Tromotriche ruschiana (Dinter) Bruyns;鲁施颤毛萝藦■☆

399579　Tromotriche thudichumii (Pillans) L. C. Leach;图迪休姆颤毛萝藦■☆

399580　Tromotriche umdausensis (Nel) Bruyns;翁达颤毛萝藦■☆

399581　Tromsdorffia Benth. et Hook. f. = Iresine P. Browne(保留属名)●■

399582　Tromsdorffia Benth. et Hook. f. = Trommsdorffia Mart. ■

399583　Tromsdorffia Blume = Chirita Buch. -Ham. ex D. Don ●■

399584　Tromsdorffia Blume = Liebigia Endl. ●■

399585　Tromsdorffia Blume = Morstdorffia Steud. ●■

399586　Tromsdorffia R. Br. = Dichrotrichum Reinw. ■☆

399587　Tronicena Steud. = Aeginetia L. ■

399588　Troniceus Miq. = Aeginetia L. ■

399589　Troniceus Miq. = Tronicena Steud. ■

399590　Troostwyckia Benth. et Hook. f. = Troostwykia Miq. ●

399591　Troostwykia Miq. = Agelaea Sol. ex Planch. ●

399592　Troostwykia Miq. = Castanola Llanos ●

399593　Tropaeastrum Mabb. = Trophaeastrum Sparre ■☆

399594　Tropaeolaceae Bercht. et J. Presl = Tropaeolaceae Juss. ex DC. (保留科名)■

399595　Tropaeolaceae DC. = Tropaeolaceae Juss. ex DC. (保留科名)■

399596　Tropaeolaceae Juss. ex DC. (1824)(保留科名);旱金莲科;Nasturtium Family ■

399597　Tropaeolum L.（1753）；旱金莲属（金莲花属）；Great Indian Cress，Nasturtium，Tropaeolum ■

399598　Tropaeolum azureum Hook.；天蓝旱金莲；☆

399599　Tropaeolum canariense Lindl. et Moore = Tropaeolum peregrinum L. ■☆

399600　Tropaeolum lobbianum Veitch = Tropaeolum peltophorum Benth. ●■

399601　Tropaeolum majus L.；旱金莲（大红鸟，旱莲花，荷叶莲，金莲花，吐血丹）；Climbing Nasturtium，Common Nasturtium，Funny Face，Funny-face，Garden Nasturtium，Gramophone，Greater Indian Cress，Indian Cress，Jesuit's Cress，Jesuits' Cress，Large Indian Cress，Monkey Hat，Nasturtium，Nose Ticklers，Nose-smart，Nose-tickler，Nosetwitcher，Running Jacob，Storshin，Storshiner，Trumpet，Yellow Lamb's Head，Yellow Lark's Heel ■

399602　Tropaeolum majus L. f. nanum？；汤姆旱金莲；Tom Thumb ■☆

399603　Tropaeolum majus L. f. variegatum Hort.；变叶旱金莲■☆

399604　Tropaeolum minus L.；小旱金莲；Bush Nasturtium，Dwarf Indian Cress，Dwarf Nasturtium，Mini Nasturtium，Minor Nasturtium ■☆

399605　Tropaeolum peltophorum Benth.；盾叶金莲花；Lobb's Climbing Nasturtium，Nasturtium Lobbianum ●■☆

399606　Tropaeolum peregrinum L.；加那利旱金莲（裂叶金莲花，五裂叶旱金莲）；American Creeper，Canary Bird Flower，Canary Creeper，Canary Nasturtium，Canarybird Flower ■☆

399607　Tropaeolum polyphyllum Cav.；多叶旱金莲■☆

399608　Tropaeolum speciosum Poepp. et Endl.；六裂叶旱金莲（美丽金莲花）；Fire Screen，Fire-screen，Flame Climber，Flame Creeper，Flame Nasturtium，Flower Flames，Scotch Flame Flower ■☆

399609　Tropaeolum tricolor Lindl. = Tropaeolum trilorum Turcz. ■☆

399610　Tropaeolum trilorum Turcz.；三色金莲花☆

399611　Tropaeolum tuberosum Ruiz et Pav.；食用金莲花（块茎旱金莲）；Anu，Isanu，Mashua，Tuber Nasturtium，Tuberous Nasturtium ■☆

399612　Tropalanthe S. Moore = Pycnandra Benth. ●☆

399613　Tropentis Raf. = Seseli L. ■

399614　Tropexa Raf. = Aristolochia L. ■

399615　Tropexa Raf. = Howardia Klotzsch ●☆

399616　Trophaeastrum Sparre = Tropaeolum L. ■

399617　Trophaeastrum Sparre（1991）；南美旱金莲属■☆

399618　Trophaeastrum patagonicum（Speg.）Sparre；南美旱金莲■☆

399619　Trophaeum Kuntze = Tropaeolum L. ■

399620　Trophianthus Scheidw. = Aspasia Lindl. ■☆

399621　Trophis P. Browne（1756）（保留属名）；牛筋树属（牛盘藤属）●☆

399622　Trophis americana L.；牛筋树●☆

399623　Trophis brasiliensis Peckolt；巴西牛筋树●☆

399624　Trophis glabrata Liebm.；光牛筋树●☆

399625　Trophis macrostachya Donn. Sm.；大穗牛筋树●☆

399626　Trophis mexicana（Liebm.）Bureau；墨西哥牛筋树●☆

399627　Trophis scandens（Lour.）Hook. et Arn. = Malaisia scandens（Lour.）Planch. ●

399628　Trophis spinosa Roxb. ex Willd. = Maclura cochinchinensis（Lour.）Corner ●

399629　Trophis taxoides Heyne = Streblus taxoides（K. Heyne）Kurz ●

399630　Trophis taxoides Roth = Streblus taxoides（K. Heyne）Kurz ●

399631　Trophisomia Rojas Acosta = Sorocea A. St. -Hil. ●☆

399632　Trophospermum Walp. = Taphrospermum C. A. Mey. ■

399633　Tropidia Lindl.（1833）；竹茎兰属（摺唇兰属）；Tropidia ■

399634　Tropidia angulosa（Lindl.）Blume；阔叶竹茎兰（东亚摺唇兰，相马氏摺唇兰）；Broadleaf Tropidia ■

399635　Tropidia angulosa（Lindl.）Blume var. nipponica（Masam.）S. Ying = Tropidia nipponica Masam. ■

399636　Tropidia angustifolia C. L. Yeh et C. S. Leou；狭叶竹茎兰■

399637　Tropidia assamica Blume = Tropidia curculigoides Lindl. ■

399638　Tropidia barbeyana Schltr. = Tropidia angulosa（Lindl.）Blume ■

399639　Tropidia bellii Blatt. et McCann = Tropidia angulosa（Lindl.）Blume ■

399640　Tropidia calcarata Ames = Tropidia angulosa（Lindl.）Blume ■

399641　Tropidia curculigoides Lindl.；短穗竹茎兰（仙茅摺唇兰）；Shortspike Tropidia ■

399642　Tropidia eatonii Ames = Tropidia polystachya（Sw.）Ames ■☆

399643　Tropidia emeishanica K. Y. Lang；峨眉竹茎兰；Emeishan Tropidia ■

399644　Tropidia formosana Rolfe = Tropidia curculigoides Lindl. ■

399645　Tropidia govindovii Blume = Tropidia angulosa（Lindl.）Blume ■

399646　Tropidia graminea Blume = Tropidia curculigoides Lindl. ■

399647　Tropidia grandis Hance = Geodorum densiflorum（Lam.）Schltr. ■

399648　Tropidia hongkongensis Rolfe = Tropidia curculigoides Lindl. ■

399649　Tropidia nanhuae W. M. Lin，Kuo Huang et T. P. Lin；南华竹茎兰■

399650　Tropidia nipponica Masam.；竹茎兰（日本摺唇兰）；Japan Tropidia ■

399651　Tropidia nipponica Masam. var. hachijoensis F. Maek.；八丈岛竹茎兰■☆

399652　Tropidia pedunculata Blume；具梗竹茎兰■☆

399653　Tropidia polystachya（Sw.）Ames；多穗竹茎兰；Manyspike Tropidia，Young Palm Orchid ■☆

399654　Tropidia semilibera（Lindl.）Blume = Tropidia angulosa（Lindl.）Blume ■

399655　Tropidia somae Hayata；台湾竹茎兰■

399656　Tropidia somae Hayata = Tropidia angulosa（Lindl.）Blume ■

399657　Tropidia squamata Blume = Tropidia curculigoides Lindl. ■

399658　Tropidocarpum Hook.（1836）；龙骨果芥属■☆

399659　Tropidocarpum gracile Hook.；龙骨果芥■☆

399660　Tropidocarpum macrocarpum Hook. et Harv. ex Greene；大龙骨果芥■☆

399661　Tropidococcus Krapov. = Malva L. ■

399662　Tropidolepis Tausch = Chiliotrichum Cass. ●☆

399663　Tropidolepts Tausch = Chiliotrichum Cass. ●☆

399664　Tropidopetalum Turcz. = Bouea Meisn. ●

399665　Tropilis Raf. = Dendrobium Sw.（保留属名）■

399666　Tropitia Pichon = Tradescantia L. ●

399667　Tropitia Pichon = Tropitria Raf. ●

399668　Tropitoma Raf. = Desmodium Desv.（保留属名）●■

399669　Tropitria Raf. = Tradescantia L. ●

399670　Tropocarpa D. Don ex Meisn. = Orites R. Br. ●☆

399671　Tros Haw. = Narcissus L. ■

399672　Tros Haw. = Queltia Salisb. ■

399673　Troschelia Klotzsch et Schomb. = Schiekia Meisn. ■☆

399674　Trotula Comm. ex DC. = Nesaea Comm. ex Kunth（保留属名）●■☆

399675　Trouettea Pierre ex Aubrév. = Trouettia Pierre ex Baill. ■

399676　Trouettia Pierre ex Baill. = Apostasia Blume ■

399677　Trouettia Pierre ex Baill. = Chrysophyllum L. ■

399678　Trouettia Pierre ex Baill. = Niemeyera F. Muell.（保留属名）●☆

399679　Troxilanthes Raf. = Polygonatum Mill. ■

399680　Troximon Gaertn. = Krigia Schreb.（保留属名）■☆

399681　Troximon Gaertn. = Krigia Schreb. + Scorzonera L. ■

399682　Troximon Nutt. = Agoseris Raf. ■☆

399683　Troximon alpestre A. Gray = Nothocalais alpestris（A. Gray）K. L. Chambers ■☆

399684　Troximon apargioides Less. = Agoseris apargioides（Less.）

Greene ■☆

399685　Troximon aurantiacum Hook. = Agoseris aurantiaca Greene ■☆

399686　Troximon cuspidata Pursh = Microseris cuspidata (Pursh) Sch. Bip. ■☆

399687　Troximon cuspidatum Pursh = Nothocalais cuspidata (Pursh) Greene ■☆

399688　Troximon glaucum Pursh = Agoseris glauca Raf. ■☆

399689　Troximon glaucum Pursh var. dasycephalum Torr. et A. Gray = Agoseris glauca (Pursh) Raf. var. dasycephala (Torr. et A. Gray) Jeps. ■☆

399690　Troximon grandiflorum A. Gray var. laciniatum (Nutt.) A. Gray = Agoseris elata (Nutt.) Greene ■☆

399691　Troximon grandiflorum A. Gray var. tenuifolium A. Gray = Agoseris elata (Nutt.) Greene ■☆

399692　Troximon heterophyllum (Nutt.) Greene var. cryptopleuroides Suksd. = Agoseris heterophylla (Nutt.) Greene ■☆

399693　Troximon nuttallii A. Gray = Agoseris elata (Nutt.) Greene ■☆

399694　Troximon parviflorum Nutt. = Agoseris parviflora (Nutt.) D. Dietr. ■☆

399695　Troxirum Raf. = Peperomia Ruiz et Pav. ■

399696　Troxistemon Raf. = Hymenocallis Salisb. ■

399697　Trozelia Raf. (废弃属名) = Acnistus Schott ●☆

399698　Trozelia Raf. (废弃属名) = Iochroma Benth. (保留属名) ●☆

399699　Trudelia Garay(1986);特鲁兰属(图德兰属) ■☆

399700　Trudelia alpina (Lindl.) Garay;特鲁兰(垂头万代兰,高山万带兰);Alpine Vanda,Nutant Vanda ■☆

399701　Trudelia alpina (Lindl.) Garay = Vanda alpina Lindl. ■

399702　Trudelia pumila (Hook. f.) Senghas = Vanda pumila Hook. f. ■

399703　Truellum Houtt. = Polygonum L. (保留属名) ■●

399704　Truellum arifolium (L.) Soják = Persicaria arifolia (L.) Haraldson ■☆

399705　Truellum biconvexum (Hayata) Soják = Polygonum biconvexum Hayata ■

399706　Truellum darrisii (H. Lév.) Soják = Polygonum darrisii H. Lév. ■

399707　Truellum dichotomum (Blume) Soják = Persicaria dichotoma (Blume) Masam. ■

399708　Truellum dichotomum (Blume) Soják = Polygonum dichotomum Blume ■

399709　Truellum dissitiflorum (Hemsl.) Tzvelev = Persicaria dissitiflora (Hemsl.) H. Gross ex T. Mori ■

399710　Truellum dissitiflorum (Hemsl.) Tzvelev = Polygonum dissitiflorum Hemsl. ■

399711　Truellum glomeratum (Dammer) Soják = Persicaria glomerata (Dammer) S. Ortiz et Paiva ■☆

399712　Truellum hastato-auriculatum (Makino ex Nakai) Soják = Polygonum praetermissum Hook. f. ■

399713　Truellum hastato-sagittatum (Makino) Soják = Polygonum hastatosagittatum Makino ■

399714　Truellum japonicum Houtt. = Persicaria senticosa (Meisn.) H. Gross ■

399715　Truellum japonicum Houtt. = Polygonum senticosum (Meisn.) Franch. et Sav. ■

399716　Truellum korshinskianum (Nakai) Soják = Polygonum hastatosagittatum Makino ■

399717　Truellum maackianum (Regel) Soják = Persicaria maackiana (Regel) Nakai ■

399718　Truellum maackianum (Regel) Soják = Polygonum maackianum

Regel ■

399719　Truellum muricatum (Meisn.) Soják = Polygonum muricatum Meisn. ■

399720　Truellum nipponense (Makino) Soják = Persicaria muricata (Meisn.) Nemoto ■

399721　Truellum nipponense (Makino) Soják = Polygonum muricatum Meisn. ■

399722　Truellum oreophilum (Makino) Soják = Persicaria oreophila (Makino) Hiyama ■☆

399723　Truellum perfoliatum (L.) Soják = Persicaria perfoliata (L.) H. Gross ■

399724　Truellum perfoliatum (L.) Soják = Polygonum perfoliatum L. ■

399725　Truellum praetermissum (Hook. f.) Soják = Persicaria praetermissa (Hook. f.) H. Hara ■

399726　Truellum praetermissum (Hook. f.) Soják = Polygonum praetermissum Hook. f. ■

399727　Truellum sagittatum (L.) Soják = Persicaria sagittata (L.) H. Gross ■

399728　Truellum sagittatum (L.) Soják = Polygonum sagittatum L. ■

399729　Truellum sibiricum (Meisn.) Soják = Polygonum sagittatum L. ■

399730　Truellum sieboldii (Meisn.) Soják = Persicaria sagittata (L.) H. Gross var. sibirica (Meisn.) Miyabe ■

399731　Truellum strigosum (R. Br.) Soják = Persicaria strigosa (R. Br.) Nakai ■

399732　Truellum strigosum (R. Br.) Soják = Polygonum strigosum R. Br. ■

399733　Truellum thunbergii (Siebold et Zucc.) Soják = Persicaria thunbergii (Siebold et Zucc.) H. Gross ■

399734　Truellum thunbergii (Siebold et Zucc.) Soják = Polygonum thunbergii Siebold et Zucc. ■

399735　Trujanoa La Llave = Rhus L. ●

399736　Trukia Kaneh. (1935);特鲁茜属 ●☆

399737　Trukia Kaneh. = Randia L. ●

399738　Trukia megacarpa (Kaneh.) Kaneh. ;特鲁茜 ●☆

399739　Truncaria DC. = Miconia Ruiz et Pav. (保留属名) ●☆

399740　Trungboa Rauschert(1982);黄毛灌属 ●☆

399741　Trungboa poilanei (Gagnep.) Rauschert;黄毛灌 ●☆

399742　Tryallis C. Muell. = Thryallis Mart. (保留属名) ●

399743　Trybliocalyx Lindau = Chileranthemum Oerst. ☆

399744　Trychinolepis B. L. Rob. = Ophryosporus Meyen ■●☆

399745　Tryginia Jacq. -Fél. = Leandra Raddi ■☆

399746　Tryginia Jacq. -Fél. = Trigynia Jacq. -Fél. ●■☆

399747　Trygonanthus Endl. ex Steud. = Loranthus Jacq. (保留属名) ●

399748　Trymalium Fenzl(1837);杂分果鼠李属 ●☆

399749　Trymalium albicans Reissek;浅白杂分果鼠李 ●☆

399750　Trymalium brevifolium Reissek;短叶杂分果鼠李 ●☆

399751　Trymalium densiflorum Rye;密花杂分果鼠李 ●☆

399752　Trymalium parvifolium (Hook.) Reissek;小叶杂分果鼠李 ●☆

399753　Trymalium polycephalum Turcz. ;多头杂分果鼠李 ●☆

399754　Trymalium stenophyllum Reissek;窄叶杂分果鼠李 ●☆

399755　Trymatococcus Poepp. et Endl. (1838);曲药桑属 ●☆

399756　Trymatococcus africanus Baill. = Dorstenia africana (Baill.) C. C. Berg ●☆

399757　Trymatococcus amazonicus Poepp. et Endl. ;曲药桑 ●☆

399758　Trymatococcus conrauanus Engl. = Dorstenia africana (Baill.) C. C. Berg ●☆

399759　Trymatococcus dorstenioides Engl. = Dorstenia dorstenioides (Engl.) Hijman et C. C. Berg ■☆

399760　Trymatococcus gilletii De Wild. = Dorstenia kameruniana Engl. ■☆

399761　Trymatococcus kamerunianus (Engl.) Engl. = Dorstenia kameruniana Engl. ■☆

399762　Trymatococcus kamerunianus (Engl.) Engl. var. welwitschii Engl. = Dorstenia kameruniana Engl. ■☆

399763　Trymatococcus oligogynus Pellegr. = Dorstenia oligogyna (Pellegr.) Berg ●☆

399764　Trymatococcus parvifolius Engl. = Bosqueiopsis gilletii De Wild. et T. Durand ●☆

399765　Trymatococcus usambarensis Engl. = Dorstenia kameruniana Engl. ■☆

399766　Tryocephalum Endl. = Kyllinga Rottb. (保留属名)■

399767　Tryocephalum Endl. = Thryocephalon J. R. Forst. et G. Forst. ■

399768　Tryothamnus Willis = Thryothamnus Phil. (废弃属名)●☆

399769　Tryphane (Fenzl) Rchb. = Minuartia L. ■

399770　Tryphane Rchb. = Arenaria L. ■

399771　Tryphane Rchb. = Minuartia L. ■

399772　Tryphane rubella (Wahlenb.) Rchb. = Minuartia rubella (Wahlenb.) Hiern ■☆

399773　Tryphera Blume = Mollugo L. ■

399774　Tryphia Lindl. (废弃属名) = Holothrix Rich. ex Lindl. (保留属名)■☆

399775　Tryphia major Sond. = Holothrix secunda (Thunb.) Rchb. f. ■☆

399776　Tryphia orthoceras Harv. = Holothrix orthoceras (Harv.) Rchb. f. ■☆

399777　Tryphia parviflora Lindl. = Holothrix parviflora (Lindl.) Rchb. f. ■☆

399778　Tryphia secunda Lindl. = Holothrix parviflora (Lindl.) Rchb. f. ■☆

399779　Tryphostemma Harv. = Basananthe Peyr. ■●☆

399780　Tryphostemma alatopetiolatum Harms = Basananthe lanceolata (Engl.) W. J. de Wilde ■☆

399781　Tryphostemma apetalum Baker f. = Basananthe apetala (Baker f.) W. J. de Wilde ■☆

399782　Tryphostemma apetalum Baker f. var. serratum ? = Basananthe apetala (Baker f.) W. J. de Wilde ■☆

399783　Tryphostemma arenophilum Pott = Basananthe pedata (Baker f.) W. J. de Wilde ■☆

399784　Tryphostemma baumii Harms = Basananthe baumii (Harms) W. J. de Wilde ■☆

399785　Tryphostemma caerulescens A. Fern. et R. Fern. = Basananthe baumii (Harms) W. J. de Wilde var. caerulescens (A. Fern. et R. Fern.) W. J. de Wilde ■☆

399786　Tryphostemma cuneatum Engl. = Basananthe lanceolata (Engl.) W. J. de Wilde ■☆

399787　Tryphostemma foetidum Lebrun et Taton = Basananthe hanningtoniana (Mast.) W. J. de Wilde ■☆

399788　Tryphostemma friesii Norl. = Basananthe sandersonii (Harv.) W. J. de Wilde ■☆

399789　Tryphostemma gossweileri Hutch. et K. Pearce = Basananthe gossweileri (Hutch. et K. Pearce) W. J. de Wilde ■☆

399790　Tryphostemma hanningtonianum Mast. = Basananthe hanningtoniana (Mast.) W. J. de Wilde ■☆

399791　Tryphostemma harmsianum Dinter = Basananthe pedata (Baker f.) W. J. de Wilde ■☆

399792　Tryphostemma heterophyllum (Schinz) Engl. = Basananthe heterophylla Schinz ■☆

399793　Tryphostemma humile Dandy = Basananthe parvifolia (Baker f.) W. J. de Wilde ■☆

399794　Tryphostemma lanceolatum Engl. = Basananthe lanceolata (Engl.) W. J. de Wilde ■☆

399795　Tryphostemma littorale (Peyr.) Engl. = Basananthe littoralis Peyr. ■☆

399796　Tryphostemma longifolium Harms = Basananthe sandersonii (Harv.) W. J. de Wilde ■☆

399797　Tryphostemma mendesii A. Fern. et R. Fern. = Basananthe baumii (Harms) W. J. de Wilde ■☆

399798　Tryphostemma natalense Mast. = Basananthe sandersonii (Harv.) W. J. de Wilde ■☆

399799　Tryphostemma nummularium (Welw.) Engl. = Basananthe nummularia Welw. ■☆

399800　Tryphostemma papillosum A. Fern. et R. Fern. = Basananthe papillosa (A. Fern. et R. Fern.) W. J. de Wilde ■☆

399801　Tryphostemma parvifolium Baker f. = Basananthe parvifolia (Baker f.) W. J. de Wilde ■☆

399802　Tryphostemma pedatum Baker f. = Basananthe pedata (Baker f.) W. J. de Wilde ■☆

399803　Tryphostemma pilosum Harms = Basananthe zanzibarica (Mast.) W. J. de Wilde ■☆

399804　Tryphostemma polygaloides Hutch. et K. Pearce = Basananthe polygaloides (Hutch. et K. Pearce) W. J. de Wilde ■☆

399805　Tryphostemma reticulatum Baker f. = Basananthe reticulata (Baker f.) W. J. de Wilde ■☆

399806　Tryphostemma sagittatum Hutch. et K. Pearce = Basananthe triloba (Bolus) W. J. de Wilde ■☆

399807　Tryphostemma sandersonii Harv. = Basananthe sandersonii (Harv.) W. J. de Wilde ■☆

399808　Tryphostemma scabrifolium Dandy = Basananthe scabrifolia (Dandy) W. J. de Wilde ■☆

399809　Tryphostemma schinzianum Harms = Basananthe triloba (Bolus) W. J. de Wilde ■☆

399810　Tryphostemma schlechteri Schinz = Basananthe pedata (Baker f.) W. J. de Wilde ■☆

399811　Tryphostemma stuhlmannii Harms = Basananthe zanzibarica (Mast.) W. J. de Wilde ■☆

399812　Tryphostemma trilobum Bolus = Basananthe triloba (Bolus) W. J. de Wilde ■☆

399813　Tryphostemma viride Hutch. et K. Pearce = Basananthe sandersonii (Harv.) W. J. de Wilde ■☆

399814　Tryphostemma zanzibaricum Mast. = Basananthe zanzibarica (Mast.) W. J. de Wilde ■☆

399815　Tryptomene F. Muell. = Thryptomene Endl. (保留属名)●☆

399816　Tryptomene Walp. = Thryptomene Endl. (保留属名)●☆

399817　Tryssophyton Wurdack (1964);精美野牡丹属 ☆

399818　Tryssophyton merumense Wurdack;精美野牡丹 ☆

399819　Tsaiorchis Ts. Tang et F. T. Wang(1936);长喙兰属(假兜唇兰属);Longbeakorchis ■

399820　Tsaiorchis neottianthoides Ts. Tang et F. T. Wang;长喙兰;Longbeakorchis ■

399821　Tsavo Jarm. = Populus L. ●

399822　Tsavo ilicifolia (Engl.) Jarm. = Populus ilicifolia (Engl.) Rouleau ●☆

399823　Tschompskia Asch. et Graebn. = Arundinaria Michx. ●

399824　Tschudya DC. = Leandra Raddi ■●☆

399825　Tsebona Capuron(1962);蔡山榄属●☆

399826　Tsebona macrantha Capuron;蔡山榄●☆

399827　Tsia Adans. = Camellia L. ●

399828　Tsia Adans. = Thea L. ●

399829　Tsiana J. F. Gmel. = Costus L. ■

399830　Tsiana J. F. Gmel. = Hellenia Retz. ■

399831　Tsiangia But = Tsiangia But, H. H. Hsue et P. T. Li ●

399832　Tsiangia But, H. H. Hsue et P. T. Li(1986);蒋英木属;Tsiangia ●

399833　Tsiangia hongkongensis (Seem.) But, H. H. Hsue et P. T. Li;蒋英木;Hongkong Gaerthera, Hongkong Tsiangia, Tsiangia ●

399834　Tsiangia hongkongensis (Seem.) P. But = Tsiangia hongkongensis (Seem.) But, H. H. Hsue et P. T. Li ●

399835　Tsiemtani Adans. = Rumphia L. ●

399836　Tsiem-tani Adans. = Rumphia L. ●

399837　Tsilaitra R. Baron = Mascarenhasia A. DC. ●☆

399838　Tsilaitra micrantha (Baker) Baron = Mascarenhasia arborescens A. DC. ●☆

399839　Tsimatimia Jum. et H. Perrier = Garcinia L. ●

399840　Tsingya Capuron(1969);蒋英无患子属●☆

399841　Tsingya bemarana Capuron;蒋英无患子●☆

399842　Tsjeracanarinum T. Durand et Jacks. = Cansjera Juss.(保留属名)●

399843　Tsjeracanarinum T. Durand et Jacks. = Tsjeru-caniram Adans.(废弃属名)●

399844　Tsjeru-caniram Adans.(废弃属名)= Cansjera Juss.(保留属名)●

399845　Tsjerucaniram Adans. = Cansjera Juss.(保留属名)●

399846　Tsjinkia Adans. = Lagerstroemia L. ●

399847　Tsjinkia B. D. Jacks. = Tsjinkia Adans. ●

399848　Tsjuilang Rumph. = Aglaia Lour.(保留属名)●

399849　Tsoala Bosser et D'Arcy(1992);管花茄属●☆

399850　Tsoala tubiflora Bosser et D'Arcy;管花茄●☆

399851　Tsoongia Merr.(1923);假紫珠属(似荆属,钟萼木属,钟君木属,钟木属);False Beautyberry, Falsebeautyberry ●

399852　Tsoongia axillariflora Merr.;假紫珠(似荆,钟萼木,钟君木,钟木);False Beautyberry, Falsebeautyberry ●

399853　Tsoongia axillariflora Merr. var. trifoliolata H. L. Li = Tsoongia axillariflora Merr. ●

399854　Tsoongiodendron Chun = Michelia L. ●

399855　Tsoongiodendron Chun(1963);观光木属(宿萼木兰属,宿轴木兰属);Tsoong's Tree, Guanguangtree ●

399856　Tsoongiodendron odorum Chun;观光木(观光木兰,宿轴木兰,香花木);Chinese Tsoong's Tree, Guanguangtree, Tsoong's Tree ●

399857　Tsotria Raf. = Isotria Raf. ■☆

399858　Tsubaki Adans. = Camellia L. ●

399859　Tsubuki Kaempf. ex Adans. = Camellia L. ●

399860　Tsuga (Endl.) Carrière(1855);铁杉属(铁油杉属);Hemlock, Hemlock Spruce, Hemlock-spruce ●

399861　Tsuga Carrière = Tsuga (Endl.) Carrière ●

399862　Tsuga albertiana Sénécl. = Tsuga heterophylla (Raf.) Sarg. ●☆

399863　Tsuga argyrophylla (Chun et Kuang) de Laub. et Silba = Cathaya argyrophylla Chun et Kuang ●◇

399864　Tsuga brunoniana (Wall.) Carrière = Tsuga dumosa (D. Don) Eichler ●

399865　Tsuga brunoniana (Wall.) Carrière var. chinensis (Franch.) Mast. = Tsuga chinensis (Franch.) Pritz. ex Diels ●

399866　Tsuga calcarea Downie = Tsuga dumosa (D. Don) Eichler ●

399867　Tsuga canadensis (L.) Carrière;加拿大铁杉;Canada Hemlock, Canadian Hemlock, Common Hemlock, Eastern Hemlock, Eastern Hemlock-spruce, Hemlock, Hemlock Spruce, New England Hemlock, Northern Hemlock, Spruce Pine, White Hemlock ●☆

399868　Tsuga canadensis Carrière 'Aurea';金叶加拿大铁杉●☆

399869　Tsuga canadensis Carrière 'Bennette';贝内特加拿大铁杉(矮生加拿大铁杉)●☆

399870　Tsuga canadensis Carrière 'Cole's Prostrate';平卧加拿大铁杉●☆

399871　Tsuga canadensis Carrière 'Gracilis';纤细加拿大铁杉●☆

399872　Tsuga canadensis Carrière 'Jacqueline Verkade';杰奎林·瓦尔卡德加拿大铁杉●☆

399873　Tsuga canadensis Carrière 'Pendula';垂枝加拿大铁杉;Weeping Hemlock ●☆

399874　Tsuga caroliniana Engelm.;卡罗林铁杉(加罗林铁杉,卡罗来纳铁杉);Carolina Hemlock, Southern Hemlock ●☆

399875　Tsuga caroliniana Engelm. 'Le Bar Weeping';垂枝卡罗林铁杉(垂枝加罗林铁杉)●☆

399876　Tsuga chinensis (Franch.) E. Pritz. subsp. wardii (Downie) E. Murray = Tsuga dumosa (D. Don) Eichler ●

399877　Tsuga chinensis (Franch.) E. Pritz. var. formosana (Hayata) H. L. Li et H. Keng = Tsuga formosana Hayata ●

399878　Tsuga chinensis (Franch.) E. Pritz. var. oblongisquamata W. C. Cheng et L. K. Fu = Tsuga oblongisquamata (W. C. Cheng et L. K. Fu) W. C. Cheng et L. K. Fu ●

399879　Tsuga chinensis (Franch.) Pritz. ex Diels;铁杉(刺柏,假花板,尖头枞,铁林刺,仙柏,油杉);China Hemlock, Chinese Hemlock ●

399880　Tsuga chinensis (Franch.) Pritz. ex Diels subsp. patens (Downie) E. Murray = Tsuga chinensis (Franch.) Pritz. ex Diels var. patens (Downie) L. K. Fu et Nan Li ●

399881　Tsuga chinensis (Franch.) Pritz. ex Diels subsp. wardii (Downie) E. Murray = Tsuga dumosa (D. Don) Eichler ●

399882　Tsuga chinensis (Franch.) Pritz. ex Diels var. daibuensis S. S. Ying;大武铁杉;Dawu Hemlock ●

399883　Tsuga chinensis (Franch.) Pritz. ex Diels var. daibuensis S. S. Ying = Tsuga formosana Hayata ●

399884　Tsuga chinensis (Franch.) Pritz. ex Diels var. daibuensis S. S. Ying = Tsuga chinensis (Franch.) Pritz. ex Diels var. formosana (Hayata) H. L. Li et H. Keng ●

399885　Tsuga chinensis (Franch.) Pritz. ex Diels var. formosana (Hayata) H. L. Li et H. Keng = Tsuga formosana Hayata ●

399886　Tsuga chinensis (Franch.) Pritz. ex Diels var. forrestii (Downie) Silba = Tsuga forrestii Downie ●

399887　Tsuga chinensis (Franch.) Pritz. ex Diels var. oblongisquamata W. C. Cheng = Tsuga oblongisquamata (W. C. Cheng et L. K. Fu) W. C. Cheng et L. K. Fu ●

399888　Tsuga chinensis (Franch.) Pritz. ex Diels var. patens (Downie) L. K. Fu et Nan Li;长阳铁杉●

399889　Tsuga chinensis (Franch.) Pritz. ex Diels var. robusta W. C. Cheng et L. K. Fu;大果铁杉;Bigfruit Chinese Hemlock ●

399890　Tsuga chinensis (Franch.) Pritz. ex Diels var. tchekiangensis (Flous) W. C. Cheng et L. K. Fu = Tsuga tchekiangensis Flous ●

399891　Tsuga chinensis (Franch.) Pritz. ex Diels var. tchekiangensis (Flous) W. C. Cheng et L. K. Fu = Tsuga chinensis (Franch.) Pritz. ex Diels ●

399892　Tsuga crassifolia Flous;厚叶铁杉;Thickleaf Hemlock ●☆

399893　Tsuga crassifolia Flous = Tsuga mertensiana (Bong.) Carrière ●☆

399894　Tsuga cuneiformis W. C. Cheng et L. K. Fu = Tsuga chinensis (Franch.) Pritz. ex Diels ●

399895　Tsuga diversiflora (Maxim.) Mast.;北海道铁杉(日本铁杉,日本小叶铁杉);Hokkaiolo Hemlock, Japanese Hemlock, Northern Japanese Hemlock ●☆

399896　Tsuga dumosa (D. Don) Eichler;云南铁杉(喜马拉雅铁杉);

Himalayan Hemlock，Yunnan Hemlock ●

399897　Tsuga dumosa（D. Don）Eichler var. chinensis（Franch.）E. Pritz. = Tsuga dumosa（D. Don）Eichler ●

399898　Tsuga dumosa（D. Don）Eichler var. chinensis（Franch.）E. Pritz. = Tsuga chinensis（Franch.）Pritz. ex Diels ●

399899　Tsuga dumosa（D. Don）Eichler var. yunnanensis（Franch.）Silba = Tsuga dumosa（D. Don）Eichler ●

399900　Tsuga dura Downie = Tsuga dumosa（D. Don）Eichler ●

399901　Tsuga formosana Hayata；台湾铁杉（油松）；Taiwan Hemlock ●

399902　Tsuga formosana Hayata = Tsuga chinensis（Franch.）E. Pritz. var. formosana（Hayata）H. L. Li et H. Keng ●

399903　Tsuga forrestii Downie；丽江铁杉；Forrest Hemlock，Lijiang Hemlock ●◇

399904　Tsuga forrestii Downie = Tsuga chinensis（Franch.）Pritz. ex Diels var. forrestii（Downie）Siba ●◇

399905　Tsuga heterophylla（Raf.）Sarg.；异叶铁杉；Alaska Pine，Hemlock Fir，Hemlock Spruce，Hemlock-fir，Mountain Hemlock，Pacific Hemlock，Prince Albert's Fir，Unequalleaf Hemlock，West Coast Hemlock，Western Hemlock，Western Hemlock-spruce ●☆

399906　Tsuga heterophylla（Raf.）Sarg.'Argenteovariegata'；银斑异叶铁杉●☆

399907　Tsuga heterophylla（Raf.）Sarg.'Dumasa'；灌丛异叶铁杉●☆

399908　Tsuga heterophylla（Raf.）Sarg.'Laursen's Column'；柱状异叶铁杉●☆

399909　Tsuga hookeriana（A. Murray bis）Carrière = Tsuga mertensiana（Bong.）Carrière ●☆

399910　Tsuga intermedia Hand.-Mazz. = Tsuga dumosa（D. Don）Eichler ●

399911　Tsuga japonica Shiras. = Pseudotsuga japonica（Shiras.）Beissn. ●☆

399912　Tsuga jefferi Henry；哥伦比亚铁杉；Colombia Hemlock，Jeffer Hemlock ●☆

399913　Tsuga leptophylla Hand.-Mazz. = Tsuga dumosa（D. Don）Eichler ●

399914　Tsuga longibracteata W. C. Cheng；长苞铁杉（贵州杉，铁油杉）；Long-bracted Hemlock ●

399915　Tsuga mairei Lemée et H. Lév. = Taxus wallichiana Zucc. var. mairei（Lemée et H. Lév.）L. K. Fu et Nan Li ●◇

399916　Tsuga mertensiana（Bong.）Carrière；高山铁杉（大果铁杉，山地铁杉，山铁杉，西美山铁杉）；Alpine Hemlock，Alpine Spruce，Bigfruir Hemlock，Black Hemlock，Mountain Hemlock，Williamson's Spruce ●☆

399917　Tsuga mertensiana（Bong.）Carrière'Blue Star'；蓝星西美山铁杉●☆

399918　Tsuga mertensiana（Bong.）Carrière'Cascade'；瀑布西美山铁杉●☆

399919　Tsuga mertensiana（Bong.）Carrière'Glauca'；灰叶西美山铁杉（粉叶山地铁杉）●☆

399920　Tsuga oblongisquamata（W. C. Cheng et L. K. Fu）W. C. Cheng et L. K. Fu；矩鳞铁杉；Oblong-squamate Hemlock ●

399921　Tsuga patens Downie = Tsuga chinensis（Franch.）Pritz. ex Diels var. patens（Downie）L. K. Fu et Nan Li ●

399922　Tsuga pattoniana Downie var. hookeriana（A. Murray bis）Lemmon = Tsuga mertensiana（Bong.）Carrière ●☆

399923　Tsuga sieboldii Carrière；日本铁杉（铁杉，希氏铁杉，席氏铁杉）；Japan Hemlock，Japanese Hemlock，Japanese Hemlock Fir，Siebold Hemlock，Southern Japanese Hemlock ●☆

399924　Tsuga sieboldii Carrière = Tsuga chinensis（Franch.）Pritz. ex Diels ●

399925　Tsuga tchekiangensis Flous；南方铁杉（异萝松，浙江铁杉）；

Zhejiang Hemlock ●

399926　Tsuga tchekiangensis Flous = Tsuga chinensis（Franch.）Pritz. ex Diels ●

399927　Tsuga wardii Downie = Tsuga dumosa（D. Don）Eichler ●

399928　Tsuga yunnanensis（Franch.）E. Pritz. = Tsuga dumosa（D. Don）Eichler ●

399929　Tsuga yunnanensis（Franch.）E. Pritz. subsp. dura（Downie）E. Murray = Tsuga dumosa（D. Don）Eichler ●

399930　Tsusiophyllum Maxim.（1870）；杉叶鹃属●☆

399931　Tsusiophyllum Maxim. = Rhododendron L. ●

399932　Tsusiophyllum tanakae Maxim.；杉叶鹃●☆

399933　Tsusiophyllum tanakae Maxim. = Rhododendron tsusiophyllum Sugim. ●☆

399934　Tsutsusi Adans. = Azalea L.（废弃属名）●☆

399935　Tsutsusi Adans. = Loiseleuria Desv.（保留属名）●☆

399936　Tuamina Alef. = Vicia L. ■

399937　Tuba Spach = Nolana L. ex L. f. ■☆

399938　Tuba Spach = Tula Adans. ■☆

399939　Tubanthera Comm. ex DC. = Colubrina Rich. ex Brongn.（保留属名）●

399940　Tuberaria（Dunal）Spach（1836）（保留属名）；莲座半日花属（图贝花属）；Spotted Rockrose ■☆

399941　Tuberaria Spach = Tuberaria（Dunal）Spach（保留属名）■☆

399942　Tuberaria acuminata（Viv.）Grosser；渐尖莲座半日花■☆

399943　Tuberaria brevipes Willk.；短梗莲座半日花■☆

399944　Tuberaria commutata Gallego；变异莲座半日花■☆

399945　Tuberaria echioides（Lam.）Willk.；紫草莲座半日花■☆

399946　Tuberaria glomerata Willk. = Tuberaria guttata（L.）Fourr. ■☆

399947　Tuberaria guttata（L.）Fourr. = Tuberaria guttata Gross. ■☆

399948　Tuberaria guttata（L.）Fourr. subsp. acuminata（Viv.）Quézel et Santa = Tuberaria acuminata（Viv.）Grosser ■☆

399949　Tuberaria guttata（L.）Fourr. subsp. discolor（Pomel）Quézel et Santa = Tuberaria guttata（L.）Fourr. ■☆

399950　Tuberaria guttata（L.）Fourr. subsp. inconspicua（Pers.）Briq. = Tuberaria inconspicua（Pers.）Willk. ■☆

399951　Tuberaria guttata（L.）Fourr. subsp. lipopetala（Murb.）Quézel et Santa = Tuberaria lipopetala（Murb.）Greuter et Burdet ■☆

399952　Tuberaria guttata（L.）Fourr. subsp. macrosepala（Boiss.）Quézel et Santa = Tuberaria macrosepala（Boiss.）Willk. ■☆

399953　Tuberaria guttata（L.）Fourr. subsp. milleri（Rouy et Foucaud）Maire = Tuberaria guttata（L.）Fourr. ■☆

399954　Tuberaria guttata（L.）Fourr. subsp. praecox（Grosser）Quézel et Santa = Tuberaria praecox Grosser ■☆

399955　Tuberaria guttata（L.）Fourr. subsp. variabilis（Willk.）Litard. = Tuberaria guttata（L.）Fourr. ■☆

399956　Tuberaria guttata（L.）Fourr. subsp. villosissima（Pomel）Quézel et Santa = Tuberaria guttata（L.）Fourr. ■☆

399957　Tuberaria guttata Gross.；莲座半日花；Annual Rock Rose，Annual Rockrose，European Frostweed，Frostweed，Spotted Rock Rose，Spotted Rockrose ■☆

399958　Tuberaria inconspicua（Pers.）Willk.；显著莲座半日花■☆

399959　Tuberaria lignosa（Sweet）Samp.；木质莲座半日花(木质图贝花)■☆

399960　Tuberaria lipopetala（Murb.）Greuter et Burdet；唇瓣半日花■☆

399961　Tuberaria macrocephala Willk.；大头莲座半日花■☆

399962　Tuberaria macrosepala（Boiss.）Willk.；大萼莲座半日花■☆

399963　Tuberaria plantaginea（Willd.）Gallego = Tuberaria inconspicua（Pers.）Willk. ■☆

399964　Tuberaria praecox Grosser；早半日花■☆

399965　Tuberaria variabilis Willk. = Tuberaria guttata（L.）Fourr. ■☆

399966　Tuberaria variabilis Willk. subsp. masguindalii Sennen et Mauricio = Tuberaria guttata（L.）Fourr. ■☆

399967　Tuberaria villosissima（Pomel）Grosser = Tuberaria guttata（L.）Fourr. ■☆

399968　Tuberaria vulgaris Willk. = Tuberaria lignosa（Sweet）Samp. ■☆

399969　Tuberaria vulgaris Willk. var. lanatum ? = Tuberaria lignosa（Sweet）Samp. ■☆

399970　Tuberaria vulgaris Willk. var. trivialis Grosser = Tuberaria lignosa（Sweet）Samp. ■☆

399971　Tuberculocarpus Pruski(1996)；瘤果菊属■●☆

399972　Tuberculocarpus rubra（Aristeg.）Pruski；瘤果菊■☆

399973　Tuberogastris Thouars = Limodorum Boehm.（保留属名）■☆

399974　Tuberolabium Yamam.（1924）；管唇兰属；Tuberolabium ■

399975　Tuberolabium elobe（Seidenf.）Seidenf. = Parapteroceras elobe（Seidenf.）Aver. ■

399976　Tuberolabium kotoense Yamam.；管唇兰（袋状兰,红头兰,红头囊唇兰,兰屿管唇兰,兰屿囊唇兰,台湾红头兰）；Taiwan Tubelolabium ■

399977　Tuberosa Fabr. = Polianthes L. ■

399978　Tuberosa Heist. = Polianthes L. ■

399979　Tuberosa Heist. ex Fabr. = Polianthes L. ■

399980　Tuberostyles Benth. = Tuberostylis Steetz ●■☆

399981　Tuberostyles Benth. et Hook. f. = Tuberostylis Steetz ■●☆

399982　Tuberostylis Steetz(1853)；隆柱菊属■●☆

399983　Tuberostylis axillaris S. F. Blake；腋生隆柱菊■●☆

399984　Tuberostylis rhizophorae Steetz；隆柱菊■●☆

399985　Tubifilaceae Dulac = Malvaceae Juss.（保留科名）●■

399986　Tubiflora J. F. Gmel.（废弃属名）= Elytraria Michx.（保留属名）●☆

399987　Tubiflora acaulis（L. f.）Kuntze = Elytraria marginata Vahl ●☆

399988　Tubiflora acaulis Kuntze = Elytraria nodosa E. Hossain ●☆

399989　Tubilabium J. J. Sm.（1928）；筒兰属■

399990　Tubilabium J. J. Sm. = Myrmechis（Lindl.）Blume ■

399991　Tubilabium aureum J. J. Sm.；黄筒兰■☆

399992　Tubilabium bilobuliferum J. J. Sm.；筒兰■☆

399993　Tubilium Cass. = Pulicaria Gaertn. ●

399994　Tubocapsicum（Wettst.）Makino = Capsicum L. ●■

399995　Tubocapsicum（Wettst.）Makino（1908）；龙珠属；Dragonpearl, Tubocapsicum ■

399996　Tubocapsicum Makino = Capsicum L. ●■

399997　Tubocapsicum Makino = Tubocapsicum（Wettst.）Makino ■

399998　Tubocapsicum anomalum（Franch. et Sav.）Makino；龙珠（赤珠,大毛秀才,红珠草）；Dragonpearl, Japanese Tubocapsicum ■

399999　Tubocapsicum anomalum（Franch. et Sav.）Makino var. obtusum Makino = Tubocapsicum obtusum（Makino）Kitam. ■

400000　Tubocapsicum boninense（Koidz.）Koidz. ex H. Hara；小笠原龙珠■☆

400001　Tubocapsicum obtusum（Makino）Kitam.；日本龙珠■

400002　Tubocapsicum obtusum（Makino）Kitam. = Tubocapsicum anomalum（Franch. et Sav.）Makino ■

400003　Tubocytisus（DC.）Fourr. = Cytisus Desf.（保留属名）●

400004　Tubocytisus（DC.）Fourr. = Cytisus L. ●

400005　Tubocytisus（DC.）Fourr. = Viborgia Moench（废弃属名）●

400006　Tubocytisus Fourr. = Cytisus Desf.（保留属名）●

400007　Tubopadus Pomel = Cerasus Mill. ●

400008　Tubopadus prostrata（Labill.）Pomel = Prunus prostrata Labill. ●☆

400009　Tubutubu Rumph. = Tapeinochilos Miq.（保留属名）■☆

400010　Tubutubua T. Post et Kuntze = Tubutubu Rumph. ■☆

400011　Tuchiroa Kuntze = Crudia Schreb.（保留属名）●☆

400012　Tuchiroa Kuntze = Touchiroa Aubl.（废弃属名）●☆

400013　Tuchsia Raf. = Fuchsia L. ●■

400014　Tuckermania Klotzsch = Oakesia Tuck. ●☆

400015　Tuckermannia Klotzsch = Corema D. Don ●☆

400016　Tuckermannia Nutt. = Coreopsis L. ●■

400017　Tuckermannia Nutt. = Leptosyne DC. ●■

400018　Tuckermannia maritima Nutt. = Coreopsis maritima（Nutt.）Hook. f. ●■☆

400019　Tuckeya Gaudich. = Pandanus Parkinson ex Du Roi ●■

400020　Tucma Ravenna = Ennealophus N. E. Br. ■☆

400021　Tucnexia DC. = Morettia DC. ■☆

400022　Tucnexia DC. = Nectouxia DC. ■☆

400023　Tuctoria Reeder(1982)；春池草属■☆

400024　Tuctoria fragilis（Swallen）Reeder；春池草■☆

400025　Tuctoria greenei（Vasey）Reeder；格林春池草■☆

400026　Tuctoria mucronata（Crampton）Reeder；钝尖春池草■☆

400027　Tuerckheimia Dammer = Tuerckheimia Dammer ex Donn. Sm. ●☆

400028　Tuerckheimia Dammer ex Donn. Sm.（1905）；危地马拉棕属●☆

400029　Tuerckheimia ascendens Dammer；危地马拉棕●☆

400030　Tuerckheimocharis Urb.（1912）；西印度玄参属■☆

400031　Tuerckheimocharis Urb. = Scrophularia L. ■●

400032　Tuerckheimocharis domingensis Urb.；西印度玄参■☆

400033　Tuerckheimta Dammer ex Dorm. Sm. = Chamaedorea Willd.（保留属名）●☆

400034　Tuerckheimta Dammer ex Dorm. Sm. = Kinetostigma Dammer ●☆

400035　Tugarinovia Iljin(1928)；革苞菊属；Leatherybractdaisy, Tugarinovia ■

400036　Tugarinovia mongolica Iljin；革苞菊；Leatherybractdaisy, Mongolian Tugarinovia ■

400037　Tugarinovia mongolica Iljin var. ovatifolia Y. Ling et Ma；卵叶革苞菊；Ovateleaf Leatherybractdaisy, Ovateleaf Tugarinovia ■

400038　Tula Adans. = Nolana L. ex L. f. ■☆

400039　Tulakenia Raf. = Jurinea Cass. ●■

400040　Tulasnea Naudin = Siphanthera Pohl ■☆

400041　Tulasnea Wight = Dalzellia Wight ■

400042　Tulasnea Wight = Lawia Tul. ■

400043　Tulasnea Wight = Terniola Tul. ■

400044　Tulasneantha P. Royen(1951)；图氏川苔草属■☆

400045　Tulasneantha monadelpha（Bong.）P. Royen；图氏川苔草■☆

400046　Tulbachia D. Dietr. = Tulbaghia L.（保留属名）■☆

400047　Tulbaghia Fabr. = Agapanthus L' Hér.（保留属名）■☆

400048　Tulbaghia Heist.（废弃属名）= Tulbaghia L.（保留属名）●■☆

400049　Tulbaghia Heist. ex Kuntze = Tulbaghia L.（保留属名）■☆

400050　Tulbaghia L.（1771）（' Tulbagia'）（保留属名）；紫瓣花属(臭根葱莲属,土巴夫属,紫娇花属)；Tuibaghia ■☆

400051　Tulbaghia acutiloba Harv.；尖裂紫瓣花■☆

400052　Tulbaghia aequinoctialis Welw. ex Baker；昼夜紫瓣花■☆

400053　Tulbaghia aequinoctialis Welw. ex Baker subsp. monantha（Engl. et Gilg）R. B. Burb.；单花昼夜紫瓣花■☆

400054　Tulbaghia affinis Link；近缘紫瓣花■☆

400055　Tulbaghia alliacea L. f.；葱状紫瓣花■☆

400056　Tulbaghia brachystemma Kunth = Tulbaghia alliacea L. f. ■☆

400057　Tulbaghia calcarea Engl. et K. Krause；钙生紫瓣花■☆

400058　Tulbaghia cameronii Baker;卡梅伦紫瓣花■☆

400059　Tulbaghia campanulata N. E. Br. = Tulbaghia cernua Avé-Lall. ■☆

400060　Tulbaghia capensis L. ;好望角紫瓣花■☆

400061　Tulbaghia cepacea L. f. = Tulbaghia violacea Harv. ■☆

400062　Tulbaghia cepacea L. f. var. maritima Vosa;滨海紫瓣花■☆

400063　Tulbaghia cernua Avé-Lall. = Tulbaghia alliacea L. f. ■☆

400064　Tulbaghia coddii Vosa et Burb. ;科德紫瓣花■☆

400065　Tulbaghia cominsii Vosa;野紫瓣花;Wild Garlic ■☆

400066　Tulbaghia daviesii Grey = Tulbaghia simmleri P. Beauv. ■☆

400067　Tulbaghia dieterlenii E. Phillips = Tulbaghia leucantha Baker ■☆

400068　Tulbaghia dregeana Kunth;德雷紫瓣花■☆

400069　Tulbaghia fragrans I. Verd. = Tulbaghia simmleri P. Beauv. ■☆

400070　Tulbaghia friesii Suess. ;弗里斯紫瓣花■☆

400071　Tulbaghia galpinii Schltr. ;盖尔紫瓣花■☆

400072　Tulbaghia heisteri Fabric. = Agapanthus africanus (L.) Hoffmanns. ■☆

400073　Tulbaghia hockii De Wild. = Tulbaghia cameronii Baker ■☆

400074　Tulbaghia inodora Gaertn. = Tulbaghia alliacea L. f. ■☆

400075　Tulbaghia karasbergensis P. E. Glover = Tulbaghia tenuior K. Krause et Dinter ■☆

400076　Tulbaghia leucantha Baker;白花紫瓣花■☆

400077　Tulbaghia ludwigiana Harv. ;路德维格紫瓣花■☆

400078　Tulbaghia luebbertiana Engl. et K. Krause;吕贝特紫瓣花■☆

400079　Tulbaghia macrocarpa Vosa;大果紫瓣花■☆

400080　Tulbaghia monantha Engl. et Gilg = Tulbaghia aequinoctialis Welw. ex Baker subsp. monantha (Engl. et Gilg) R. B. Burb. ■☆

400081　Tulbaghia montana Vosa;山地紫瓣花■☆

400082　Tulbaghia narcissiflora Salisb. = Tulbaghia alliacea L. f. ■☆

400083　Tulbaghia natalensis Baker;纳塔尔紫瓣花■☆

400084　Tulbaghia nutans Vosa;俯垂紫瓣花■☆

400085　Tulbaghia pauciflora Baker;少花紫瓣花■☆

400086　Tulbaghia poetica Burb. = Tulbaghia coddii Vosa et Burb. ■☆

400087　Tulbaghia pulchella Avé-Lall. = Tulbaghia capensis L. ■☆

400088　Tulbaghia pulchella P. E. Barnes =Tulbaghia simmleri P. Beauv. ■☆

400089　Tulbaghia rhodesica R. E. Fr. ;罗得西亚紫瓣花■☆

400090　Tulbaghia rhodesica Weim. = Tulbaghia friesii Suess. ■☆

400091　Tulbaghia simmleri P. Beauv. ;西氏紫瓣花■☆

400092　Tulbaghia tenuior K. Krause et Dinter;瘦紫瓣花■☆

400093　Tulbaghia transvaalensis Vosa;德兰士瓦紫瓣花■☆

400094　Tulbaghia verdoornia Vosa et Burb. ;韦尔紫瓣花■☆

400095　Tulbaghia violacea Harv. ;堇色紫瓣花（紫娇花）;Society Garlic, Sweet Garlic, Wild Garlic ■☆

400096　Tulbaghiaceae Salisb. ;紫瓣花科■

400097　Tulbaghiaceae Salisb. = Alliaceae Borkh. (保留科名)■

400098　Tulbagia L. = Tulbaghia L. (保留属名)■☆

400099　Tulestea Aubrév. et Pellegr. (1961) ;肖神秘果属●☆

400100　Tulestea Aubrév. et Pellegr. = Synsepalum (A. DC.) Daniell ●☆

400101　Tulestea gabonensis Aubrév. et Pellegr. = Synsepalum gabonense (Aubrév. et Pellegr.) T. D. Penn. ●☆

400102　Tulestea kaessneri (Engl.) Aubrév. = Synsepalum kaessneri (Engl.) T. D. Penn. ●☆

400103　Tulestea koulamoutouensis Aubrév. et Pellegr. ;库拉穆图肖神秘果●☆

400104　Tulestea seretii (De Wild.) Aubrév. et Pellegr. = Synsepalum seretii (De Wild.) T. D. Penn. ●☆

400105　Tulestea tomentosa Aubrév. et Pellegr. ;绒毛肖神秘果●☆

400106　Tulexis Raf. = Brassavola R. Br. (保留属名)■☆

400107　Tulichiba Post et Kuntze = Ormosia Jacks. (保留属名)●

400108　Tulichiba Post et Kuntze = Toulichiba Adans. (废弃属名)●

400109　Tulicia Post et Kuntze = Toulicia Aubl. ●☆

400110　Tulipa L. (1753) ;郁金香属;Tulip ■

400111　Tulipa ‘ Cornuta ’ = Tulipa acuminata Vahl ex Hornem. ■☆

400112　Tulipa acuminata Vahl ex Hornem. ;土耳其郁金香（长尖郁金香,尖瓣郁金香）;Horned Tulip, Turkish Tulip ■☆

400113　Tulipa agenensis DC. ;埃及郁金香■☆

400114　Tulipa agenensis DC. subsp. boissieri ?;布瓦西耶郁金香;Rose of Sharon ■☆

400115　Tulipa aitchisonii A. D. Hall = Tulipa clusiana DC. ■☆

400116　Tulipa altaica Pall. ex Spreng. ;阿尔泰郁金香;Altai Mountain Tulip, Altai Tulip ■☆

400117　Tulipa anhuiensis X. S. Sheng;安徽郁金香;Anhui Tulip ■

400118　Tulipa anisophylla Vved. ;欧洲异叶郁金香■☆

400119　Tulipa aristata Regel = Tulipa kolpakowskiana Regel ■

400120　Tulipa aucheriana Baker;奥克郁金香■☆

400121　Tulipa australis Link = Tulipa sylvestris L. subsp. australis (Link) Pamp. ■☆

400122　Tulipa australis Link = Tulipa sylvestris L. ■☆

400123　Tulipa australis Link var. fragrans (Munby) Pau = Tulipa sylvestris L. subsp. australis (Link) Pamp. ■☆

400124　Tulipa bakeri A. D. Hall = Tulipa saxatilis Sieber ex Spreng. ■☆

400125　Tulipa batalinii Regel;巴塔林郁金香（乌兹别克郁金香）;Batalin Tulip ■☆

400126　Tulipa batalinii Regel ‘ Apricot Jewel ’;杏黄宝石巴塔林郁金香■☆

400127　Tulipa batalinii Regel ‘ Bright Gen ’;辉宝石巴塔林郁金香■☆

400128　Tulipa batalinii Regel ‘ Bronze Charm ’;铜魔巴塔林郁金香■☆

400129　Tulipa batalinii Regel var. orangecharm Hort. ;橙花郁金香;Orange Batalin Tulip ■☆

400130　Tulipa behmiana Regel;拜氏郁金香■☆

400131　Tulipa biebersteiniana Schult. f. ;毕伯氏郁金香;Bieberstein Tulip ■☆

400132　Tulipa biflora Pall. ;二花郁金香(柔毛郁金香,双花郁金香）;Pubescent Tulip, Twoflower Tulip ■

400133　Tulipa biflora Pall. var. turkestanica Hort. ;多花郁金香;Manyflower Tulip ■☆

400134　Tulipa bifloriformis Vved. ;二型花郁金香;Twoflower-type Tulip ■☆

400135　Tulipa borszczowii Regel;鲍尔郁金香■☆

400136　Tulipa brachystemon Regel;短丝郁金香;Shortfilament Tulip ■☆

400137　Tulipa buhseana Boiss. = Tulipa biflora Pall. ■

400138　Tulipa callieri Halácsy et H. Lév. ;卡赖氏郁金香;Callier Tulip ■☆

400139　Tulipa carinata Vved. ;龙骨状郁金香■☆

400140　Tulipa caucasica Lipsky;高加索郁金香■☆

400141　Tulipa celsiana DC. = Tulipa sylvestris L. subsp. australis (Link) Pamp. ■☆

400142　Tulipa celsiana DC. var. montana Batt. et Trab. = Tulipa sylvestris L. subsp. australis (Link) Pamp. ■☆

400143　Tulipa celsiana Henning;西尔斯郁金香;Cels Tulip ■☆

400144　Tulipa chrysantha Boiss. ex Baker = Tulipa clusiana DC. var. chrysantha Sealy ■☆

400145　Tulipa clusiana DC. ;克鲁斯氏郁金香（克鲁西郁金香,淑女郁金香）;Lady Tulip ■☆

400146　Tulipa clusiana DC. var. chrysantha Sealy;印度郁金香(黄花郁金香,黄郁金香）;India Tulip, Yellow Tulip ■☆

400147　Tulipa clusiana Shepherd ex Schult. f. f. stellata (Hook.) S.

Dasgupta et Deb;黄白郁金香■☆

400148 Tulipa dasystemon (Regel) Regel;毛蕊郁金香;Hairystamen Tulip ■

400149 Tulipa dasystemonoides Vved.;假毛蕊郁金香■☆

400150 Tulipa didieri Jord. var. mauriana Baker;艳红郁金香;Crimson Tulip ■☆

400151 Tulipa dubia Vved.;可疑郁金香■☆

400152 Tulipa edulis (Miq.) Baker = Amana edulis (Miq.) Honda ■☆

400153 Tulipa edulis (Miq.) Baker var. latifolia Makino = Tulipa erythronioides Baker ■

400154 Tulipa eichleri Regel;埃赫郁金香;Eichler Tulip ■☆

400155 Tulipa eichleri Regel = Tulipa undulata Jacq. ex Baker ■☆

400156 Tulipa erythronioides Baker;二叶郁金香(高山山慈姑,宽叶郁金香,阔叶老鸦瓣);Twoleaf Tulip ■

400157 Tulipa ferganica Vved.;费尔干郁金香■☆

400158 Tulipa florenskyi Woronow;福罗郁金香■☆

400159 Tulipa fosteriana Hoog. ex B. Fedtsch.;福斯特郁金香;Foster Tulip ■☆

400160 Tulipa fosteriana Hoog. ex B. Fedtsch. var. alba Hort.;白福斯特郁金香;White Foster Tulip ■☆

400161 Tulipa fragrans Munby = Tulipa sylvestris L. subsp. australis (Link) Pamp. ■☆

400162 Tulipa gesneriana L.;郁金香(红蓝花,洋荷花,郁香,紫述香);Common Garden Tulip, Common Tulip, Didier's Tulip, Late Tulip, Tulip ■

400163 Tulipa gesneriana L. 'Black Beauty';黑美人郁金香■

400164 Tulipa gesneriana L. 'Golden Emperor';皇金郁金香■

400165 Tulipa gesneriana L. 'Merry Widow';风流寡妇郁金香■

400166 Tulipa gesneriana L. 'Orange Emperor';皇橙郁金香■

400167 Tulipa gesneriana L. 'Red Emperor';皇红郁金香■

400168 Tulipa gesneriana L. 'White Emperor';皇白郁金香■

400169 Tulipa graminifolia Baker ex Moore = Amana edulis (Miq.) Honda ■☆

400170 Tulipa graminifolia Baker ex Moore = Tulipa edulis (Miq.) Baker ■☆

400171 Tulipa greigii Regel;格里克郁金香;Greig Tulip ■☆

400172 Tulipa hageri Heldr.;哈格尔郁金香(哈格郁金香);Hager Tulip ■☆

400173 Tulipa heteropetala Ledeb.;异瓣郁金香;Heteropetalous Tulip ■

400174 Tulipa heterophylla (Regel) Baker;异叶郁金香;Diversifolious Tulip ■

400175 Tulipa hissarica Popov et Vved.;希萨尔郁金香■☆

400176 Tulipa hoogiana B. Fedtsch.;胡氏郁金香;Hoog Tulip ■☆

400177 Tulipa humilis Herb.;矮生郁金香■☆

400178 Tulipa iliensis Regel;伊犁郁金香(光慈姑,伊犁山慈姑);Ili Tulip, Yili Tulip ■

400179 Tulipa ingens Hoog;巨大郁金香■☆

400180 Tulipa julia K. Koch;尤林郁金香■☆

400181 Tulipa kaufmanniana Regel;考夫曼郁金香(土耳其斯坦郁金香);Water Lily Tulip, Waterlily Tulip, Water-lily Tulip ■☆

400182 Tulipa koktebelica Junge;科克郁金香■☆

400183 Tulipa kolpakowskiana Regel;迟花郁金香;Late Tulip, Lateflower Tulip ■

400184 Tulipa korolkovii Regel;科罗克夫郁金香;Korolkov Tulip ■☆

400185 Tulipa korshinskyi Vved.;考尔施郁金香■☆

400186 Tulipa krauseana Regel;克鲁斯郁金香■☆

400187 Tulipa kuschkensis B. Fedtsch.;库什金郁金香■☆

400188 Tulipa latifolia (Makino) Makino = Tulipa erythronioides Baker ■

400189 Tulipa latifolia Makino = Tulipa erythronioides Baker ■

400190 Tulipa lehmanniana Merckl. ex Bunge;莱曼郁金香;Lehmann Tulip ■☆

400191 Tulipa linifolia Regel;线叶郁金香;Linearleaf Tulip ■☆

400192 Tulipa marjolettii Perrier et Songeon;马乔郁金香(中欧郁金香);Marjolett Tulip ■☆

400193 Tulipa maximowiczii Regel;马氏郁金香■☆

400194 Tulipa micheliana Hoog.;米歇尔郁金香■☆

400195 Tulipa mogoltavica Popov et Vved.;莫戈尔塔夫郁金香■☆

400196 Tulipa mongolica Y. Z. Zhao;蒙古郁金香■☆

400197 Tulipa montana Lindl.;山地郁金香;Mountain Tulip ■☆

400198 Tulipa nitida Hoog;光亮郁金香■☆

400199 Tulipa nutans (Trautv.) B. Fedtsch. = Tulipa uniflora (L.) Besser ex Baker ■

400200 Tulipa oculus-solis St.-Amans.;红光郁金香;Eye-of-the-sun Tulip ■☆

400201 Tulipa orphanidea Boiss. ex Heldr.;矮小郁金香(孤挺郁金香);Low Slender Tulip ■☆

400202 Tulipa ostrovskiana Regel;奥氏郁金香;Ostrovsk Tulip ■☆

400203 Tulipa patens Agardh ex Schult. f.;垂蕾郁金香(垂郁金香);Patent Tulip ■

400204 Tulipa platystemon Vved.;宽蕊郁金香■☆

400205 Tulipa polychroma Stapf = Tulipa biflora Pall. ■

400206 Tulipa praecox Ten.;早生郁金香;Clarimond Tulip, Early Tulip ■☆

400207 Tulipa praestans Hoog;尖被郁金香■☆

400208 Tulipa praestans Hoog 'Fusilier';枪手尖被郁金香■☆

400209 Tulipa praestans Hoog 'Unicum';独特尖被郁金香■☆

400210 Tulipa praestans Hoog 'Van Tubergen's Variety';范蒂伯根尖被郁金香■☆

400211 Tulipa praestans Hoog var. tubergenii Hort.;多花尖被郁金香;Manyflower Cuspidated Tulip ■☆

400212 Tulipa primulina Baker = Tulipa sylvestris L. subsp. primulina (Baker) Maire et Weiller ■☆

400213 Tulipa pulchella Boiss. ex Baker;美丽郁金香;Crocus Tulip ■☆

400214 Tulipa pulchella Boiss. ex Baker var. pallida Hort.;爱白郁金香;White Dwarf Tulip ■☆

400215 Tulipa pulchella Boiss. ex Baker var. persian pearl Hort.;波斯珍珠郁金香;Persian Pearl Tulip ■☆

400216 Tulipa pulchella Boiss. ex Baker var. rosea Hort.;矮粉郁金香;Rose Dwarf Tulip ■☆

400217 Tulipa pulchella Boiss. ex Baker var. violacea Hort.;矮堇郁金香;Violet Dwarf Tulip ■☆

400218 Tulipa regelii Krasn.;雷格尔郁金香(莱氏郁金香)■☆

400219 Tulipa rosen Vved.;粉郁金香■☆

400220 Tulipa saxatilis Sieber ex Spreng.;克里特郁金香(岩生郁金香);Cretan Tulip, Crete Tulip ■☆

400221 Tulipa schmidtii Fomin;史密氏郁金香■☆

400222 Tulipa schrenkii Regel;准噶尔郁金香(施运氏郁金香);Dzungar Tulip, Schrenk Tulip ■

400223 Tulipa sharonensis Dinsm.;莎伦郁金香;Sharon Tulip ■☆

400224 Tulipa sinkiangensis Z. M. Mao;新疆郁金香;Sinkiang Tulip, Xinjiang Tulip ■

400225 Tulipa sprengeri Baker;窄尖叶郁金香(窄小叶郁金香);Narrowleaf Tulip ■☆

400226 Tulipa stellata Hook.;浅白黄郁金香;Pale-yellow Tulip ■☆

400227 Tulipa suaveolens Roth;早开郁金香(香花郁金香);Duc Van Thol Tulip ■☆

400228　Tulipa subpraestans Vved. ;亚尖被郁金香■☆

400229　Tulipa sylvestris L. ;钟花郁金香(林生郁金香,星花郁金香); Australian Tulip,Bellflower Tulip,Florentine Tulip,Wild Tulip ■☆

400230　Tulipa sylvestris L. subsp. australis（Link）Pamp. ;南部钟花郁金香(星花郁金香)■☆

400231　Tulipa sylvestris L. subsp. cuspidata（Regel）Maire et Weiller;骤尖钟花郁金香■☆

400232　Tulipa sylvestris L. subsp. mauritii Sennen = Tulipa sylvestris L. ■☆

400233　Tulipa sylvestris L. var. celsiana（DC.）Levier = Tulipa sylvestris L. ■☆

400234　Tulipa sylvestris L. var. herbetei Sennen et Mauricio = Tulipa sylvestris L. ■☆

400235　Tulipa sylvestris L. var. mauritii Valdés = Tulipa sylvestris L. ■☆

400236　Tulipa sylvestris L. var. mediterranea Pamp. = Tulipa sylvestris L. ■☆

400237　Tulipa sylvestris L. var. melillensis Sennen et Mauricio = Tulipa sylvestris L. ■☆

400238　Tulipa sylvestris L. var. primulina（Baker）Maire = Tulipa sylvestris L. ■☆

400239　Tulipa tarbagataica D. Y. Tan et X. Wei;塔城郁金香;Tacheng Tulip ■☆

400240　Tulipa tarda Stapf;晚花郁金香(迟花郁金香);Latebloom Tulip ■☆

400241　Tulipa tetraphylla Regel;四叶郁金香;Fourleaf Tulip ■☆

400242　Tulipa tianschanica Regel;天山郁金香;Tianshan Mountain Tulip ,Tianshan Tulip ■

400243　Tulipa tianschanica Regel var. sailimuensis X. Wei et D. Y. Tan;赛里木湖郁金香;Salimu Tulip ■

400244　Tulipa tubergeniana Hoog;图氏郁金香■☆

400245　Tulipa turcomanica B. Fedtsch. ;土库曼郁金香■☆

400246　Tulipa turkestanica Regel;土耳其斯坦郁金香(土耳其郁金香);Turkestan Tulip ■☆

400247　Tulipa undulata Jacq. ex Baker;波叶郁金香(爱克勒郁金香);Eichler Tulip ■☆

400248　Tulipa uniflora（L.）Besser ex Baker;单花郁金香;Singleflower Tulip ■☆

400249　Tulipa urumiensis Stapf;乌鲁姆郁金香■☆

400250　Tulipa violacea Boiss. et Buhse;紫罗兰郁金香■☆

400251　Tulipa viridiflora Hort. ;绿花郁金香;Greenflower Tulip ☆

400252　Tulipa whittallii（Dykes）Elwes ex Newton;惠托尔郁金香(安欧郁金香);Wittall Tulip ☆

400253　Tulipa wilsoniana Hoog;威尔逊郁金香;E. H. Wilson Tulip ☆

400254　Tulipa zenaidae Vved. ;蔡氏郁金香■☆

400255　Tulipaceae Batsch ex Borkh. = Liliaceae Juss. （保留科名）■●

400256　Tulipaceae Horan. = Liliaceae Juss. （保留科名）■●

400257　Tulipastrum Spach = Magnolia L. ●

400258　Tulipastrum acuminatum（L.）Small = Magnolia acuminata（L.）L. ●☆

400259　Tulipastrum acuminatum（L.）Small var. aureum Ashe = Magnolia acuminata（L.）L. ●☆

400260　Tulipastrum acuminatum（L.）Small var. flavum Small = Magnolia acuminata（L.）L. ●☆

400261　Tulipastrum acuminatum（L.）Small var. ludovicianum（Sarg.）Ashe = Magnolia acuminata（L.）L. ●☆

400262　Tulipastrum acuminatum（L.）Small var. ozarkense（Ashe）Ashe = Magnolia acuminata（L.）L. ●☆

400263　Tulipastrum americanum Spach = Magnolia acuminata（L.）L. ●☆

400264　Tulipastrum americanum Spach var. subcordatum Spach = Magnolia acuminata（L.）L. ●☆

400265　Tulipastrum cordatum（Michx.）Small = Magnolia acuminata（L.）L. ●☆

400266　Tulipifera Herm. ex Mill. = Liriodendron L. ●

400267　Tulipifera Mill. = Liriodendron L. ●

400268　Tulipifera liriodendron Mill. = Liriodendron tulipifera L. ●

400269　Tulipiferae Vent. = Magnoliaceae Juss. （保留科名）●

400270　Tulisma Raf. （废弃属名）= Corytholoma（Benth.）Decne. ■☆

400271　Tulisma Raf. （废弃属名）= Rechsteineria Regel（保留属名）■☆

400272　Tulisma Raf. （废弃属名）= Sinningia Nees ■■☆

400273　Tulista Raf. = Haworthia Duval（保留属名）■☆

400274　Tullia Leavenw. = Pycnanthemum Michx. （保留属名）■☆

400275　Tullya Raf. = Pycnanthemum Michx. （保留属名）■☆

400276　Tulocarpus Hook. et Arn. = Guardiola Cerv. ex Bonpl. ■☆

400277　Tuloclinia Raf. = Metalasia R. Br. ●☆

400278　Tulophos Raf. = Triteleia Douglas ex Lindl. ■☆

400279　Tulorima Raf. = Saxifraga L. ■

400280　Tulotis Raf. （1833）;蜻蜓兰属;Dragonflyoechis ,Tulotis ■

400281　Tulotis Raf. = Platanthera Rich. （保留属名）■

400282　Tulotis asiatica H. Hara = Platanthera fuscescens（L.）Kraenzl. ■

400283　Tulotis asiatica H. Hara = Platanthera souliei Kraenzl. ■

400284　Tulotis asiatica H. Hara = Tulotis fuscescens Raf. ■

400285　Tulotis devolii T. P. Lin et T. W. Hu = Platanthera devolii（T. P. Lin et T. W. Hu）T. P. Lin et K. Inoue ■

400286　Tulotis fuscescens（L.）De Moor = Platanthera fuscescens（L.）Kraenzl. ■

400287　Tulotis fuscescens（L.）Raf. = Platanthera souliei Kraenzl. ■

400288　Tulotis fuscescens Raf. = Platanthera souliei Kraenzl. ■

400289　Tulotis iinumae（Makino）H. Hara = Platanthera iinumae（Makino）Makino ■☆

400290　Tulotis longicalcalata（Hayata）S. S. Ying = Platanthera longicalcarata Hayata ■

400291　Tulotis longicalcalata（Hayata）Tang S. Liu et H. J. Su = Platanthera longicalcarata Hayata ■

400292　Tulotis shensiana（Kraenzl.）H. Hara = Platanthera ussuriensis（Regel et Maack）Maxim. ■

400293　Tulotis shensiana（Kraenzl.）H. Hara = Tulotis ussuriensis（Regel et Maack）H. Hara ■

400294　Tulotis sonoharae（Masam.）Nackej. = Platanthera sonoharae Masam. ■☆

400295　Tulotis souliei（Kraenzl.）H. Hara = Platanthera souliei Kraenzl. ■

400296　Tulotis souliei（Kraenzl.）H. Hara = Tulotis fuscescens Raf. ■

400297　Tulotis taiwanensis S. S. Ying = Platanthera taiwaniana S. S. Ying ■

400298　Tulotis taiwaniana S. S. Ying = Platanthera taiwaniana S. S. Ying ■

400299　Tulotis transnokoensis（Ohwi et Fukuy.）S. S. Ying = Platanthera sachalinensis F. Schmidt ■

400300　Tulotis ussuriensis（Regel et Maack）H. Hara = Platanthera ussuriensis（Regel et Maack）Maxim. ■

400301　Tulotis ussuriensis（Regel et Maack）H. Hara var. transnokoensis（Ohwi et Fukuy.）Tang S. Liu et H. J. Su = Platanthera sachalinensis F. Schmidt ■

400302　Tulotis ussuriensis（Regel et Maack）H. Hara var. transnokoensis（Ohwi et Fukuy.）Tang S. Liu et H. J. Su;能高蜻蛉兰 ■

400303　Tulotis ussuriensis（Regel）H. Hara = Platanthera ussuriensis（Regel et Maack）Maxim. ■

400304　Tulotis ussuriensis（Regel）H. Hara var. transnokoensis（Ohwi et Fukuy.）Tang S. Liu et H. J. Su = Platanthera sachalinensis F. Schmidt ■

400305　Tulotis whangshanensis（S. S. Chien）H. Hara = Calanthe triplicata（Willemet）Ames ■

400306　Tulotis whangshanensis（S. S. Chien）H. Hara = Platanthera tipuloides（L. f.）Lindl. ■

400307　Tulucuna Post et Kuntze = Carapa Aubl. ●☆

400308　Tulucuna Post et Kuntze = Touloucouna M. Room. ●☆

400309　Tumalis Raf. = Euphorbia L. ●■

400310　Tumamoca Rose（1912）;图马瓜属■☆

400311　Tumamoca macdougalii Rose;图马瓜■☆

400312　Tumboa Welw. = Welwitschia Hook. f.（保留属名）■☆

400313　Tumboaceae Wettst. = Welwitschiaceae Caruel（保留科名）■☆

400314　Tumelaia Raf. = Daphne L. ●

400315　Tumelaia Raf. = Thymelaea Mill.（保留属名）●■

400316　Tumidinodus H. W. Li = Anna Pellegr. ■

400317　Tumidinodus purpureoruber H. W. Li = Anna submontana Pellegr. ■

400318　Tumion Raf. = Torreya Arn. ●

400319　Tumion Raf. ex Greene = Torreya Arn. ●

400320　Tumion californicum（Torr.）Greene = Torreya californica Torr. ●☆

400321　Tumion fargesii（Franch.）Skeels = Torreya fargesii Franch. ●◇

400322　Tumion grande（Fortune ex Lindl.）Greene = Torreya grandis Fortune ex Lindl. ●◇

400323　Tumion grandis Greene = Torreya grandis Fortune ex Lindl. ●◇

400324　Tumion taxifolium（Arn.）Greene = Torreya taxifolia Arn. ●☆

400325　Tumionella Greene = Haplopappus Cass.（保留属名）■●☆

400326　Tunaria Kuntze = Cantua Juss. ex Lam. ●☆

400327　Tunas Lunell = Opuntia Mill. ●

400328　Tunas polyacantha（Haw.）Nieuwl. et Lunell = Opuntia polyacantha Haw. ●☆

400329　Tunatea Kuntze = Swartzia Schreb.（保留属名）●☆

400330　Tunatea Kuntze = Tounatea Aubl.（废弃属名）●☆

400331　Tunga Roxb. = Hypolytrum Rich. ex Pers. ■

400332　Tunica（Hallier）Scop. = Petrorhagia（Ser. ex DC.）Link ■

400333　Tunica Haller = Petrorhagia（Ser. ex DC.）Link ■

400334　Tunica Ludw. = Dianthus L. ■

400335　Tunica Ludw. = Petrorhagia（Ser. ex DC.）Link ■

400336　Tunica Mert. et W. D. J. Koch = Petrorhagia（Ser. ex DC.）Link ■

400337　Tunica davaeana Coss. = Petrorhagia illyrica（Ard.）P. W. Ball et Heywood subsp. angustifolia（Poir.）P. W. Ball et Heywood ■☆

400338　Tunica illyrica（Ard.）Fisher et C. A. Mey. = Petrorhagia illyrica（Ard.）P. W. Ball et Heywood ■☆

400339　Tunica illyrica（Ard.）Fisher et C. A. Mey. subsp. angustifolia（Poir.）Maire = Petrorhagia illyrica（Ard.）P. W. Ball et Heywood subsp. angustifolia（Poir.）P. W. Ball et Heywood ■☆

400340　Tunica illyrica（Ard.）Fisher et C. A. Mey. var. australis（Batt.）Maire = Petrorhagia illyrica（Ard.）P. W. Ball et Heywood ■☆

400341　Tunica illyrica（Ard.）Fisher et C. A. Mey. var. laevis（Pamp.）Maire = Petrorhagia illyrica（Ard.）P. W. Ball et Heywood ■☆

400342　Tunica morrissii（Hance）Walp. = Dianthus caryophyllus L. ■

400343　Tunica pachygona Fisch. et C. A. Mey.;粗蕊膜萼花■☆

400344　Tunica prolifera（L.）Scop. = Petrorhagia prolifera（L.）P. W. Ball et Heywood ■☆

400345　Tunica prolifera（L.）Scop. subsp. nanteuilii（Burnat）Asch. et Graebn. = Petrorhagia nanteuilii（Burnat）P. W. Ball et Heywood ■☆

400346　Tunica prolifera（L.）Scop. subsp. velutina（Guss.）Briq. = Petrorhagia dubia（Raf.）G. López et Romo ☆

400347　Tunica prolifera（L.）Scop. var. glabricaulis Maire = Petrorhagia nanteuilii（Burnat）P. W. Ball et Heywood ■☆

400348　Tunica prolifera（L.）Scop. var. glandulifera Maire = Petrorhagia dubia（Raf.）G. López et Romo ■☆

400349　Tunica prolifera（L.）Scop. var. laevicaulis Rouy et Foucaud = Petrorhagia dubia（Raf.）G. López et Romo ■☆

400350　Tunica prolifera（L.）Scop. var. pubescens Rouy et Foucaud = Petrorhagia nanteuilii（Burnat）P. W. Ball et Heywood ■☆

400351　Tunica prolifera（L.）Scop. var. scabricaulis Maire = Petrorhagia prolifera（L.）P. W. Ball et Heywood ■☆

400352　Tunica prolifera（L.）Scop. var. scabrifolia Clavaud = Petrorhagia prolifera（L.）P. W. Ball et Heywood ■☆

400353　Tunica prolifera L. = Petrorhagia prolifera（L.）P. W. Ball et Heywood ■☆

400354　Tunica rhiphaea Pau et Font Quer = Petrorhagia rhiphaea（Pau et Font Quer）P. W. Ball et Heywood ■☆

400355　Tunica saxifraga（L.）Scop. = Petrorhagia saxifraga（L.）Link ■

400356　Tunica saxifraga（L.）Scop. var. alba Hort.;白花洋石竹;White Saxifrage Tunicflower ■☆

400357　Tunica saxifraga（L.）Scop. var. florepieno Hort.;重瓣洋石竹;Double Saxifrage Tunicflower ■☆

400358　Tunica saxifraga（L.）Scop. var. rosea Hort.;红花洋石竹;Pink Saxifrage Tunicflower ■☆

400359　Tunica scoparia Pamp. = Petrorhagia illyrica（Ard.）P. W. Ball et Heywood subsp. angustifolia（Poir.）P. W. Ball et Heywood ■☆

400360　Tunica stricta（Bunge）Fisch. et C. A. Mey. = Petrorhagia alpina（Hablitz）P. W. Ball et Heywood ■

400361　Tunica thessala Boiss. var. cyrenaica E. A. Durand et Barratte = Petrorhagia cyrenaica（E. A. Durand et Barratte）P. W. Ball ■☆

400362　Tunilla D. R. Hunt et Iliff = Opuntia Mill. ●

400363　Tunilla D. R. Hunt et Iliff（2000）;南美掌属●☆

400364　Tupa G. Don = Lobelia L. ●■

400365　Tupa deckenii Asch. = Lobelia deckenii（Asch.）Hemsl. ■☆

400366　Tupa montana（Fresen.）Vatke = Lobelia rhynchopetalum（Hochst. ex A. Rich.）Hemsl. ■☆

400367　Tupa rhynchopetalum Hochst. ex A. Rich. = Lobelia rhynchopetalum（Hochst. ex A. Rich.）Hemsl. ■☆

400368　Tupa schimperi Hochst. ex A. Rich. = Lobelia giberroa Hemsl. ■☆

400369　Tupeia Blume = Dendrotrophe Miq. ●

400370　Tupeia Blume = Henslowia Blume ●

400371　Tupeia Cham. et Schltdl.（1828）;新喀桑寄生属●☆

400372　Tupeia antarctica Cham. et Schltdl.;新喀桑寄生●☆

400373　Tupeianthus Takht. = Tepuianthus Maguire et Steyerm. ●☆

400374　Tupelo Adans. = Nyssa L. ●

400375　Tupidanthus Hook. f. et Thomson = Schefflera J. R. Forst. et G. Forst.（保留属名）●

400376　Tupidanthus Hook. f. et Thomson（1856）;多蕊木属（多蕊藤属,多蕊属,脱辟木属）;Anthrywood, Tupidanthus ●

400377　Tupidanthus calyptratus Hook. f. et Thomson;多蕊木（大七叶莲,单厄,老虎鞭,龙爪树,龙爪叶,七叶莲,脱辟木,珠砂莲）;Anthrywood, Common Tupidanthus, Mallet Flower ●

400378　Tupidanthus pueckleri K. Koch = Tupidanthus calyptratus Hook. f. et Thomson ●

400379　Tupistra Ker Gawl.（1814）;长柱开口箭属（长柱七叶,开口箭属）;Tupistra ●

400380　Tupistra annulata H. Li et J. L. Huang = Campylandra annulata（H. Li et J. L. Huang）M. N. Tamura, S. Yun Liang et N. J. Turland ●

400381　Tupistra aurantiaca Wall. ex Baker = Campylandra aurantiaca Wall. ex Baker ■

400382 Tupistra cavaleriei H. Lév. = Amischotolype hispida（A. Rich.）D. Y. Hong ■

400383 Tupistra chinensis Baker = Campylandra chinensis（Baker）M. N. Tamura,S. Yun Liang et N. J. Turland ■

400384 Tupistra delavayi Franch. = Campylandra delavayi（Franch.）M. N. Tamura,S. Yun Liang et N. J. Turland ■

400385 Tupistra emeiensis Z. Y. Zhu = Campylandra emeiensis（Z. Y. Zhu）M. N. Tamura,S. Yun Liang et N. J. Turland ■

400386 Tupistra ensifolia F. T. Wang et Ts. Tang = Campylandra ensifolia（F. T. Wang et Ts. Tang）M. N. Tamura,S. Yun Liang et N. J. Turland ■

400387 Tupistra esquirolii H. Lév. et Vaniot = Curculigo capitulata（Lour.）Kuntze ■

400388 Tupistra esquirolii H. Lév. et Vaniot = Molineria capitulata（Lour.）Herb. ■

400389 Tupistra fargesii Baill. = Campylandra chinensis（Baker）M. N. Tamura,S. Yun Liang et N. J. Turland ■

400390 Tupistra fimbriata Hand. -Mazz. = Campylandra fimbriata（Hand. -Mazz.）M. N. Tamura,S. Yun Liang et N. J. Turland ■

400391 Tupistra fimbriata Hand. -Mazz. = Tupistra ensifolia F. T. Wang et Ts. Tang ■

400392 Tupistra fimbriata Hand. -Mazz. var. breviloba H. Li et J. L. Huang = Campylandra fimbriata（Hand. -Mazz.）M. N. Tamura,S. Yun Liang et N. J. Turland ■

400393 Tupistra fimbriata Hand. -Mazz. var. breviloba H. Li et J. L. Huang = Tupistra ensifolia F. T. Wang et Ts. Tang ■

400394 Tupistra fungilliformis F. T. Wang et S. Yun Liang;伞柱开口箭;Fungiform Tupistra, Umbelstyle Tupistra ■

400395 Tupistra grandistigma F. T. Wang et S. Yun Liang;长柱开口箭;Longstyle Tupistra ■

400396 Tupistra heensis Y. Wan et X. H. Lu = Campylandra chinensis（Baker）M. N. Tamura,S. Yun Liang et N. J. Turland ■

400397 Tupistra japonica（Roth）F. T. Wang et Ts. Tang;日本开口箭（开喉箭,万年青）;Japan Tupistra,Japanese Tupistra ■

400398 Tupistra jinshanensis Z. L. Yang et X. G. Luo = Campylandra jinshanensis（Z. L. Yang et X. G. Luo）M. N. Tamura,S. Yun Liang et N. J. Turland ■

400399 Tupistra kwangtungensis S. S. Ying = Campylandra chinensis（Baker）M. N. Tamura,S. Yun Liang et N. J. Turland ■

400400 Tupistra liangshanensis Z. Y. Zhu = Campylandra liangshanensis（Z. Y. Zhu）M. N. Tamura,S. Yun Liang et N. J. Turland ■

400401 Tupistra lichuanensis Y. K. Yang, J. K. Wu et D. T. Peng = Campylandra lichuanensis（Y. K. Yang,J. K. Wu et D. T. Peng）M. N. Tamura,S. Yun Liang et N. J. Turland ■

400402 Tupistra longipedunculata F. T. Wang et S. Yun Liang = Campylandra longipedunculata（F. T. Wang et S. Yun Liang）M. N. Tamura,S. Yun Liang et N. J. Turland ■

400403 Tupistra longispica Y. T. Wan et X. H. Lu = Campylandra longispica Y. T. Wan et X. H. Lu ■

400404 Tupistra longispica Y. Wan et X. H. Lu;长穗开口箭;Longspike Tupistra ■

400405 Tupistra lorifolia Franch. = Campylandra chinensis（Baker）M. N. Tamura,S. Yun Liang et N. J. Turland ■

400406 Tupistra pingbianensis J. L. Huang et X. Z. Liu;屏边开口箭;Pingbian Tupistra ■

400407 Tupistra singapureana Baker = Neuwiedia singapureana（Baker）Rolfe ■

400408 Tupistra singapureana Wall. ex Baker = Neuwiedia singapureana（Baker）Rolfe ■

400409 Tupistra sparsiflora S. C. Chen et Y. T. Ma = Campylandra chinensis（Baker）M. N. Tamura,S. Yun Liang et N. J. Turland ■

400410 Tupistra tonkinensis Baill. = Campylandra wattii C. B. Clarke ■

400411 Tupistra tui（F. T. Wang et Ts. Tang）F. T. Wang et S. Yun Liang = Campylandra tui（F. T. Wang et Ts. Tang）M. N. Tamura,S. Yun Liang et N. J. Turland ■

400412 Tupistra urotepala（Hand. -Mazz.）F. T. Wang et Ts. Tang = Campylandra urotepala（Hand. -Mazz.）M. N. Tamura,S. Yun Liang et N. J. Turland ■

400413 Tupistra verruculosa Q. H. Chen = Campylandra verruculosa（Q. H. Chen）M. N. Tamura,S. Yun Liang et N. J. Turland ■

400414 Tupistra viridiflora Franch. = Campylandra chinensis（Baker）M. N. Tamura,S. Yun Liang et N. J. Turland ■

400415 Tupistra watanabei（Hayata）F. T. Wang et C. F. Liang = Campylandra chinensis（Baker）M. N. Tamura,S. Yun Liang et N. J. Turland ■

400416 Tupistra watanabei（Hayata）F. T. Wang et C. F. Liang = Campylandra watanabei（Hayata）Dandy ■

400417 Tupistra wattii（C. B. Clarke）Hook. f. = Campylandra wattii C. B. Clarke ■

400418 Tupistra yunnanensis F. T. Wang et S. Yun Liang = Campylandra yunnanensis（F. T. Wang et S. Yun Liang）M. N. Tamura,S. Yun Liang et N. J. Turland ■

400419 Tupistraceae Schnizl. = Ruscaceae M. Roem.（保留科名）●

400420 Turanga（Bunge）Kimura = Populus L. ●

400421 Turanga euphratica（Oliv.）Kimura = Populus euphratica Oliv. ●

400422 Turanga pruinosa（Schrenk）Kimura = Populus pruinosa Schrenk ●

400423 Turania Akhani et Roalson = Salsola L. ●■

400424 Turania Akhani et Roalson（2007）;图兰猪毛菜属●■☆

400425 Turania androssowii（Litv.）Akhani;安氏图兰猪毛菜■☆

400426 Turania aperta（Paulsen）Akhani;图兰猪毛菜■☆

400427 Turania deserticola（Iljin）Akhani;沙地图兰猪毛菜■☆

400428 Turaniphytum Poljakov（1961）;图兰蒿属■☆

400429 Turaniphytum eranthemum（Bunge）Poljakov;图兰蒿■☆

400430 Turbina Raf.（1838）;陀螺花属●■☆

400431 Turbina bracteata Deroin;具苞陀螺花■☆

400432 Turbina corymbosa（L.）Raf.;伞花陀螺花■☆

400433 Turbina curtoi（Rendle）A. Meeuse = Paralepistemon curtoi（Rendle）Lejoly et Lisowski ●☆

400434 Turbina holubii（Baker）A. Meeuse = Turbina holubii Baker ■☆

400435 Turbina holubii Baker;霍勒布陀螺花■☆

400436 Turbina longiflora Verdc.;长花陀螺花■☆

400437 Turbina mexicana（Hemsl.）Roberty;墨西哥陀螺花■☆

400438 Turbina oblongata（E. Mey. ex Choisy）A. Meeuse = Ipomoea oblongata E. Mey. ex Choisy ■☆

400439 Turbina oenotheroides（L. f.）A. Meeuse = Ipomoea oenotheroides（L. f.）Raf. ex Hallier f. ☆

400440 Turbina perbella Verdc.;极美陀螺花■☆

400441 Turbina pyramidalis（Hallier f.）A. Meeuse;圆锥陀螺花■☆

400442 Turbina rhodesiana Rendle = Turbina holubii（Baker）A. Meeuse ■☆

400443 Turbina robertsiana（Rendle）A. Meeuse = Ipomoea robertsiana Rendle ■☆

400444 Turbina shirensis（Oliv.）A. Meeuse = Paralepistemon shirensis（Oliv.）Lejoly et Lisowski ●☆

400445 Turbina stenosiphon（Hallier f.）A. Meeuse；细管陀旋花■☆

400446 Turbina stenosiphon（Hallier f.）A. Meeuse var. pubescens Verdc. = Turbina stenosiphon（Hallier f.）A. Meeuse■☆

400447 Turbina suffruticosa（Burch.）A. Meeuse = Ipomoea suffruticosa Burch.●☆

400448 Turbina wrightii（House）Alain；赖氏陀旋花■☆

400449 Turbinaceae Dulac = Oleaceae Hoffmanns. et Link（保留科名）●

400450 Turbine Willis = Turbina Raf.●■☆

400451 Turbinicarpus（Backeb.）Buxb. et Backeb.（1937）；姣丽球属■☆

400452 Turbinicarpus（Backeb.）Buxb. et Backeb. = Neolloydia Britton et Rose●☆

400453 Turbinicarpus Buxb. et Backeb. = Turbinicarpus（Backeb.）Buxb. et Backeb.■☆

400454 Turbinicarpus gielsdorfianus（Werderm.）V. John et Ríha = Gymnocactus gielsdorfianus（Werderm.）Backeb.●☆

400455 Turbinicarpus knuthianus（Bod.）V. John et Ríha = Gymnocactus knuthianus（Boed.）Backeb.●☆

400456 Turbinicarpus laui Glass et R. A. Foster；劳氏姣丽球●☆

400457 Turbinicarpus lophophoroides（Werderm.）Buxb. et Backeb.；姣丽球●☆

400458 Turbinicarpus macrochele（Werderm.）Buxb. et Backeb.；牙城球（牙城丸）■☆

400459 Turbinicarpus pseudomacrochele（Backeb.）Buxb. et Backeb.；长城球（长城丸）■☆

400460 Turbinicarpus saueri（Böd.）V. John et Ríha；绍尔姣丽球●☆

400461 Turbinicarpus schmiedickeanus（Boed.）Buxb. et Backeb.；羿龙球（升龙球，羿龙丸）■☆

400462 Turbinicarpus valdezianus（H. Moeller）Glass et R. A. Foster；蔷薇姣丽球●☆

400463 Turbinicarpus viereckii（Werderm.）V. John et Ríha；韦氏姣丽球●☆

400464 Turbith Tausch = Athamanta L.■☆

400465 Turbitha Raf. = Tribula Hill■

400466 Turczaninovia DC.（1836）；女菀属；Ladydaisy，Turczaninovia■

400467 Turczaninovia fastigiata（Fisch.）DC.；女菀（白菀，苆，女肠，女宛，羊须草，织女菀）；Common Turczaninovia，Ladydaisy■

400468 Turczaninovia fastigiata（Fisch.）DC. = Aster fastigiatus Fisch.■

400469 Turczaninoviella Koso-Pol.（1924）；图尔克草属■☆

400470 Turetta Vell. = ? Lauro-Cerasus Duhamel●

400471 Turgenia Hoffm.（1814）；刺果芹属；Turgenia■

400472 Turgenia heterocarpa DC. = Lisaea heterocarpa（DC.）Boiss.■☆

400473 Turgenia latifolia（L.）Hoffm.；刺果芹（宽叶高卡利）；Broad Caucalis，Broad Caucatis，Broadleaf False Carrot，Broadleaf Turgenia，Broad-leaved Hedgehog Parsley，Greater Bur Parsley，Greater Bur-parsley■

400474 Turgenia latifolia（L.）Hoffm. = Caucalis latifolia L.■

400475 Turgeniopsis Boiss.（1844）；类刺果芹属■☆

400476 Turgeniopsis Boiss. = Glochidotheca Fenzl■☆

400477 Turgeniopsis foeniculacea（Fenzl）Boiss.；类刺果芹■☆

400478 Turgosea Haw. = Crassula L.●■☆

400479 Turia Forssk. = Luffa Mill.■

400480 Turia Forssk. ex J. F. Gmel. = Luffa Mill.■

400481 Turia gijef J. F. Gmel. = Kedrostis gijef（J. F. Gmel.）C. Jeffrey■☆

400482 Turibana（Nakai）Nakai = Euonymus L.（保留属名）●

400483 Turibana Nakai = Euonymus L.（保留属名）●

400484 Turibana oxyphylla（Miq.）Nakai = Euonymus oxyphyllus Miq.●

400485 Turinia A. Juss. = Turpinia Vent.（保留属名）●

400486 Turnera L.（1753）；时钟花属（穗柱榆属，特纳草属，特纳，窝籽属）；Turnera●■☆

400487 Turnera Plum. ex L. = Turnera L.●■☆

400488 Turnera aphrodisiaca Ward = Turnera diffusa Willd. ex Schult.●☆

400489 Turnera diffusa Willd. ex Schult.；扩展时钟花（匍匐特纳树，特纳树）；Damiana，Rosecampi●☆

400490 Turnera oculata Story；小眼时钟花■☆

400491 Turnera oculata Story var. paucipilosa Oberm.；疏毛时钟花■☆

400492 Turnera thomasii（Urb.）Story；汤氏时钟花■☆

400493 Turnera ulmifolia L.；榆叶时钟花（时钟花）；Cat's Tongue，Sage Rose，Turnera，West Indian Holly■☆

400494 Turnera ulmifolia L. var. thomasii Urb. = Turnera thomasii（Urb.）Story■☆

400495 Turneraceae DC. = Turneraceae Kunth ex DC.（保留科名）●■☆

400496 Turneraceae Kunth ex DC.（1828）（保留科名）；时钟花科（穗柱榆科，窝籽科，有叶花科）●■☆

400497 Turpenia Wight = Turpinia Vent.（保留属名）●

400498 Turpethum Raf. = Merremia Dennst. ex Endl.（保留属名）●■

400499 Turpethum Raf. = Operculina Silva Manso（废弃属名）●■

400500 Turpinia Bonpl.（废弃属名）= Turpinia Vent.（保留属名）●

400501 Turpinia Cass. = Poiretia Vent.（保留属名）●■☆

400502 Turpinia Humb. et Bonpl. = Barnadesia Mutis ex L. f.●☆

400503 Turpinia Humb. et Bonpl. = Turpinia Vent.（保留属名）●

400504 Turpinia La Llave et Lex. = Turpinia Vent.（保留属名）●

400505 Turpinia La Llave et Lex. = Vernonia Schreb.（保留属名）●■

400506 Turpinia Lex. = Turpinia Vent.（保留属名）●

400507 Turpinia Lex. = Vernonia Schreb.（保留属名）●■

400508 Turpinia Lex. ex La Llave et Lex. = Critoniopsis Sch. Bip.●■☆

400509 Turpinia Lex. ex La Llave et Lex. = Turpinia Vent.（保留属名）●

400510 Turpinia Pers. = Poiretia Vent.（保留属名）●■☆

400511 Turpinia Raf. = Lobadium Raf.●

400512 Turpinia Raf. = Rhus L.●

400513 Turpinia Raf. = Schmaltzia Desv.●

400514 Turpinia Raf. = Turpinia Vent.（保留属名）●

400515 Turpinia Vent.（1807）（保留属名）；山香圆属；Fieldcitron，Turpinia●

400516 Turpinia Vent.（保留属名）= Staphylea L.●

400517 Turpinia affinis Merr. et L. M. Perry；硬毛山香圆（大果山香圆）；Hispid Fieldcitron，Hispid Turpinia●

400518 Turpinia arguta（Lindl.）Seem.；锐尖山香圆（黄柿，尖树，两指剑，七寸丁，千打锤，山香圆，五寸铁树）；Acute Fieldcitron，Acute Turpinia●

400519 Turpinia arguta（Lindl.）Seem. var. pubescens T. Z. Hsu；绒毛锐尖山香圆（假木棉，九节茶，梁山伯树，五寸刀）；Pubescent Acute Fieldcitron，Pubescent Acute Turpinia●

400520 Turpinia arguta sensu Kanehir = Turpinia formosana Nakai●

400521 Turpinia cochinchinensis（Lour.）Merr.；越南山香圆；Cochinchina Turpinia，Cochin-China Turpinia，Vietnam Fieldcitron，Vietnamese Turpinia●

400522 Turpinia formosana Nakai；台湾山香圆（山桂花，山香圆）；Formosan Turpinia，Taiwan Fieldcitron，Taiwan Turpinia●

400523 Turpinia glaberrima Merr. = Turpinia montana（Blume）Kurz var. glaberrima（Merr.）T. Z. Hsu●

400524 Turpinia glaberrima Merr. var. stenophylla Merr. et L. M. Perry = Turpinia montana（Blume）Kurz var. stenophylla（Merr. et L. M. Perry）T. Z. Hsu●

400525 Turpinia gracilis Nakai = Turpinia montana（Blume）Kurz●

400526 Turpinia indochinensis Merr.；疏脉山香圆；Indochina Fieldcitron，

Indochinese Turpinia ●

400527　Turpinia lucida Nakai = Turpinia ovalifolia Elmer ●

400528　Turpinia macrosperma C. C. Huang；大籽山香圆；Bigseed Fieldcitron，Bigseed Turpinia，Big-seeded Turpinia ●

400529　Turpinia microcarpa Wight et Arn. = Turpinia cochinchinensis（Lour.）Merr. ●

400530　Turpinia montana（Blume）Kurz；山香圆；Fieldcitron，Montane Turpinia，Mountanain Turpinia ●

400531　Turpinia montana（Blume）Kurz var. glaberrima（Merr.）T. Z. Hsu；光山香圆；Glabrous Fieldcitron，Glabrous Turpinia ●

400532　Turpinia montana（Blume）Kurz var. stenophylla（Merr. et L. M. Perry）T. Z. Hsu；狭叶山香圆；Narrowleaf Fieldcitron，Narrowleaf Turpinia ●

400533　Turpinia nepalensis Wall. ；尼泊尔山香圆；Nepal Fieldcitron，Nepal Turpinia ●

400534　Turpinia nepalensis Wall. = Turpinia cochinchinensis（Lour.）Merr. ●

400535　Turpinia nepalensis Wall. = Turpinia pomifea（Roxb.）DC. ●

400536　Turpinia ovalifolia Elmer；卵叶山香圆（羽叶山香圆，圆叶山香圆）；Ovateleaf Fieldcitron，Ovateleaf Turpinia，Ovate-leaved Turpinia ●

400537　Turpinia pachyphylla Merr. = Turpinia ovalifolia Elmer ●

400538　Turpinia parva Koord. et Valeton = Turpinia montana（Blume）Kurz ●

400539　Turpinia pomifea（Roxb.）DC. ；大果山香圆（山香圆）；Bigfruit Fieldcitron，Bigfruit Turpinia，Big-fruited Turpinia ●

400540　Turpinia pomifea（Roxb.）DC. var. minor C. C. Huang；山麻风树（小香圆）；Minor Fieldcitron，Minor Turpinia ●

400541　Turpinia pomifera（Roxb.）DC. = Turpinia ternata Nakai ●

400542　Turpinia robusta Craib；粗壮山香圆；Robust Fieldcitron，Robust Turpinia ●

400543　Turpinia simplicifolia Merr. ；亮叶山香圆；Lucidleaf Fieldcitron，Lucidleaf Turpinia，Lucid-leaved Turpinia ●

400544　Turpinia simplicifolia Merr. var. longipes C. Y. Wu；长柄亮叶山香圆；Longpes Turpinia，Longstipe Simpleleaf Turpinia ●

400545　Turpinia subsessilifolia C. Y. Wu；心叶山香圆；Heartleaf Fieldcitron，Subsessileleaf Turpinia，Subsessile-leaved Turpinia ●

400546　Turpinia ternata Nakai；三叶山香圆；Threeleaf Fieldcitron，Threeleaf Turpinia，Threeleaved Turpinia，Trileaved Turpinia ●

400547　Turpinia unifoliata Merr. et Chun = Turpinia simplicifolia Merr. ●

400548　Turpinis Miq. = Turpinia Vent. （保留属名）●

400549　Turpinium Baill. = Turpinia Vent. （保留属名）●

400550　Turpinium Baill. = Vernonia Schreb. （保留属名）●■

400551　Turpithum B. D. Jacks. = Merremia Dennst. ex Endl. （保留属名）●■

400552　Turpithum B. D. Jacks. = Operculina Silva Manso（废弃属名）●■

400553　Turpithum B. D. Jacks. = Turpethum Raf. ●■

400554　Turraea L. （1771）；杜楝属（金银楝属）；Starbush，Star-bush ●

400555　Turraea abyssinica Hochst. ex A. Rich. ；阿比西尼亚杜楝●☆

400556　Turraea abyssinica Hochst. ex A. Rich. var. longipedicellata Oliv. = Turraea holstii Gürke ●☆

400557　Turraea africana（Welw.）Cheek；非洲杜楝●☆

400558　Turraea angolensis Exell = Turraea cabrae De Wild. et T. Durand ●☆

400559　Turraea barbata Styles et F. White；髯毛杜楝●☆

400560　Turraea breviracemosa C. DC. = Turraea wakefieldii Oliv. ●☆

400561　Turraea cabrae De Wild. et T. Durand；安哥拉杜楝●☆

400562　Turraea cornucopia Styles et F. White；角状杜楝●☆

400563　Turraea cuneata Gürke = Turraea mombassana Hiern ex C. DC. subsp. cuneata（Gürke）Styles et F. White ●☆

400564　Turraea cylindrica Sim = Turraea wakefieldii Oliv. ●☆

400565　Turraea elephantina Styles et F. White；雅致杜楝●☆

400566　Turraea eylesii Baker f. = Turraea fischeri Gürke subsp. eylesii（Baker f. ）Styles et F. White ●☆

400567　Turraea fischeri Gürke；菲舍尔杜楝●☆

400568　Turraea fischeri Gürke subsp. eylesii（Baker f. ）Styles et F. White；艾尔斯杜楝●☆

400569　Turraea floribunda Hochst. ；多花杜楝●☆

400570　Turraea ghanensis J. B. Hall；加纳杜楝●☆

400571　Turraea glomeruliflora Harms；团花杜楝●☆

400572　Turraea goetzei Harms = Turraea robusta Gürke ●☆

400573　Turraea gracilis A. Chev. = Turraea heterophylla Sm. ●☆

400574　Turraea heterophylla Sm. ；互叶杜楝●☆

400575　Turraea holstii Gürke；霍尔杜楝●☆

400576　Turraea junodii Schinz = Turraea wakefieldii Oliv. ●☆

400577　Turraea kaessneri Baker f. = Turraea wakefieldii Oliv. ●☆

400578　Turraea kilimandscharica Gürke = Turraea abyssinica Hochst. ex A. Rich. ●☆

400579　Turraea kimbozensis Cheek；金博杜楝●☆

400580　Turraea kirkii Baker f. = Turraea wakefieldii Oliv. ●☆

400581　Turraea laurentii De Wild. ；洛朗杜楝●☆

400582　Turraea laxiflora C. DC. = Turraea holstii Gürke ●☆

400583　Turraea leonensis Keay；莱昂杜楝●☆

400584　Turraea lycioides Baker = Turraea parvifolia Deflers ●☆

400585　Turraea macrophylla A. Chev. ；大叶杜楝●☆

400586　Turraea mombassana Hiern ex C. DC. ；蒙巴萨杜楝●☆

400587　Turraea mombassana Hiern ex C. DC. subsp. cuneata（Gürke）Styles et F. White；楔形杜楝●☆

400588　Turraea mombassana Hiern ex C. DC. subsp. schliebenii（Harms）Styles et F. White；施楝●☆

400589　Turraea nilotica Kotschy et Peyr. ；小杜楝（尼罗杜楝）；Small Mahogany ●☆

400590　Turraea nilotica Kotschy et Peyr. var. glabrata Fiori = Turraea nilotica Kotschy et Peyr. ●☆

400591　Turraea oblancifolia Bremek. = Turraea obtusifolia Hochst. ●☆

400592　Turraea obtusifolia Hochst. ；钝叶杜楝；South African Honeysuckle，Star Bush ●☆

400593　Turraea obtusifolia Hochst. var. microphylla C. DC. = Turraea obtusifolia Hochst. ●☆

400594　Turraea parvifolia Deflers；小花杜楝●☆

400595　Turraea pellegriniana Keay；佩尔格兰杜楝●☆

400596　Turraea pevelingii Cheek = Turraea retusa Styles et F. White ●☆

400597　Turraea pinnata Wall. = Munronia pinnata（Wall. ）W. Theob. ●☆

400598　Turraea pubescens Hell. ；杜楝（金银楝）；Pubescet Starbush，Pubescet Star-bush，Starbush ●

400599　Turraea pulchella（Harms）T. D. Penn. ；美丽杜楝●☆

400600　Turraea randii Baker f. = Turraea nilotica Kotschy et Peyr. ●☆

400601　Turraea retusa Styles et F. White；微凹杜楝●☆

400602　Turraea robusta Gürke；粗壮杜楝●☆

400603　Turraea sacleuxii C. DC. = Turraea robusta Gürke ●☆

400604　Turraea schlechteri Harms = Turraea wakefieldii Oliv. ●☆

400605　Turraea schliebenii Harms = Turraea mombassana Hiern ex C. DC. subsp. schliebenii（Harms）Styles et F. White ●☆

400606　Turraea somaliensis P. T. Li et X. M. Chen = Turraea parvifolia Deflers ●☆

400607　Turraea squamulifera C. DC. = Turraea robusta Gürke ●☆

400608　Turraea stolzii Harms；斯托尔兹杜楝●☆

400609　Turraea streyi F. White et Styles;施特赖杜楝●☆

400610　Turraea thollonii Pellegr.;托伦杜楝●☆

400611　Turraea tisseranti Pellegr. = Turraea pellegriniana Keay ●☆

400612　Turraea tubulifera C. DC. = Turraea nilotica Kotschy et Peyr. ☆

400613　Turraea usambarensis Gürke = Turraea holstii Gürke ●☆

400614　Turraea vogelii Hook. f. ex Benth.;沃格尔杜楝●☆

400615　Turraea vogelioides Bagsh. et Baker f.;拟沃格尔杜楝●☆

400616　Turraea volkensii Gürke = Turraea robusta Gürke ●☆

400617　Turraea wakefieldii Oliv.;韦克杜楝●☆

400618　Turraea zambesica Styles et F. White;赞比西杜楝●☆

400619　Turraea zenkeri C. DC.;岑克尔杜楝●☆

400620　Turraeanthus Baill. (1874);肖杜楝属●

400621　Turraeanthus africanus (C. DC.) Pellegr. = Turraeanthus africanus (Welw. ex C. DC.) Pellegr. ●☆

400622　Turraeanthus africanus (Welw. ex C. DC.) Pellegr.;非洲肖杜楝;African Avodire ●☆

400623　Turraeanthus bracteolatus Harms = Turraeanthus longipes Baill. ●☆

400624　Turraeanthus longipes Baill.;长梗肖杜楝●☆

400625　Turraeanthus malchairi De Wild. = Turraeanthus africanus (Welw. ex C. DC.) Pellegr. ●☆

400626　Turraeanthus mannii Baill.;曼氏肖杜楝●☆

400627　Turraeanthus vignei Hutch. et Dalziel = Turraeanthus africanus (Welw. ex C. DC.) Pellegr. ●☆

400628　Turraeanthus zenkeri Harms = Turraeanthus africanus (Welw. ex C. DC.) Pellegr. ●☆

400629　Turraya Wall. = Leersia Sw. (保留属名)■

400630　Turretia DC. = Turrettia Poir. ■☆

400631　Turrettia Poir. = Tourrettia Foug. (保留属名)■☆

400632　Turricula J. F. Macbr. (1917);小塔草属●☆

400633　Turricula parryi J. F. Macbr.;小塔草●☆

400634　Turrigera Decne. = Tweedia Hook. et Arn. ●☆

400635　Turrillia A. C. Sm. = Bleasdalea F. Muell. ex Domin ●

400636　Turrita Wallr. = Arabis L. ●■

400637　Turritis Adans. = Arabis L. ●■

400638　Turritis L. (1753);旗杆芥属(赛南芥属,塔儿芥);Rock Cress,Towercress ■

400639　Turritis L. = Arabis L. ●■

400640　Turritis Tourn. ex L. = Turritis L. ■

400641　Turritis brachycarpa Torr. et A. Gray = Arabis divaricarpa A. Nelson ■☆

400642　Turritis dregeana Sond. = Turritis glabra L. ■

400643　Turritis drummondii (A. Gray) Lunell = Arabis drummondii A. Gray ■☆

400644　Turritis glabra L.;旗杆芥(光筷子芥,南芥,南芥菜);Glabrous Arabis, Glabrous Rockcress, Glabrous Towercress, Smooth Tower Mustard, Tower Cress, Tower Mustard, Tower Mustard Rockcress, Tower Rock-cress, Tower-mustard Rock Cross, Towers Mustard, Tower's Treacle ■

400645　Turritis glabra L. = Arabis glabra (L.) Bernh. ☆

400646　Turritis glabra L. var. lilacina O. E. Schulz = Turritis glabra L. ■

400647　Turritis hirsuta L. = Arabis hirsuta (L.) Scop. ■

400648　Turritis laevigata Muhl. ex Willd. = Arabis laevigata (Muhl. ex Willd.) Poir. ■☆

400649　Turritis pseudoturritis (Boiss. et Huldr.) Velen. = Turritis glabra L. ■

400650　Turritis pubescens Desf. = Arabis pubescens (Desf.) Poir. ■☆

400651　Turritis sagittata Bertol. = Arabis sagittata (Bertol.) DC. ■

400652　Tursenia Cass. = Baccharis L. (保留属名)●■☆

400653　Tursitis Raf. = Kickxia Dumort. ●☆

400654　Turukhania Vassilcz. = Medicago L. (保留属名)●■

400655　Turukhania Vassilcz. = Trifillium Medik. ●■

400656　Turukhania karkarensis (Semen. ex Vassilcz.) Vassilcz. = Medicago platycarpos (L.) Trautv. ■

400657　Turukhania platycarpos (L.) Vassilcz. = Medicago platycarpos (L.) Trautv. ■

400658　Turulia Post et Kuntze = Touroulia Aubl. ●☆

400659　Tussaca Raf. = Goodyera R. Br. ■

400660　Tussaca Rchb. = Chrysothemis Decne. ■☆

400661　Tussaca Rchb. = Tussacia Rchb. ■☆

400662　Tussacia Beer = Catopsis Griseb. ■☆

400663　Tussacia Benth. = Chrysothemis Decne. ■☆

400664　Tussacia Klotzsch ex Beer = Catopsis Griseb. ■☆

400665　Tussacia Raf. ex Desv. = Spiranthes Rich. (保留属名)■

400666　Tussacia Rchb. = Chrysothemis Decne. ■☆

400667　Tussacia Willd. ex Beer = Catopsis Griseb. ■☆

400668　Tussacia Willd. ex Schult. et Schult. f. = ? Catopsis Griseb. ■☆

400669　Tussilagaceae Bercht. et J. Presl = Asteraceae Bercht. et J. Presl (保留科名)●■

400670　Tussilagaceae Bercht. et J. Presl = Compositae Giseke(保留科名)●■

400671　Tussilago L. (1753);款冬属;Colts Foot,Coltsfoot ■

400672　Tussilago albicans Sw. = Chaptalia albicans (Sw.) Vent. ex B. D. Jacks. ■☆

400673　Tussilago anandria L. = Gerbera anandria (L.) Sch. Bip. ■

400674　Tussilago anandria L. = Leibnitzia anandria (L.) Turcz. ■

400675　Tussilago farfara L.;款冬(艾冬花,八角乌,氏冬,冬花,蜂斗花,虎须,九苨草,九九花,看灯花,颗冬,颗冻,苦萃,款冬花,款冻,款花,密灸冬,救肺侯,石薹,氏冬,水斗花,菟奚,菟蒵,橐吾,钻冬,钻冻);Asa's Foot, Baccy-plant, Bullfoot, Bull's Foot, Calf's Foot, Calf's-foot, Clatterclogs, Clayweed, Claywort, Cleat, Colt-herb, Colts Foot, Colt's Foot, Coltsfoot, Colts-foot, Common Coltsfoot, Coosil, Cowheave, Cow-heave, Crow's Foot, Dishalaga, Dishilago, Dishlago, Dishylagie, Donnhove, Dove Dock, Dove-dock, Dunny Nettle, Dunny-leaf, Dunny-weed, Fieldhove, Foalfoot, Foal-foot, Foal's Foot, Foalsfoot, Foal's-foot, Foalswort, Foile Foot, Foile-foot, Foolwort, Hogweed, Hoofs, Hooves, Horse Hoof, Horsefoot, One-o' clock, Poor Man's Baccy, Poor Man's Tobacco, Ram's Claws, Son-before-the-father, Sow Foot, Sweep's Brush, Tun-hoof, Tushy-lueky Gowan, Wild Rhubarb, Yellow Star, Yellow Trumpet ■

400676　Tussilago japonica L. = Farfugium japonicum (L.) Kitam. ■

400677　Tussilago palmata Aiton = Petasites frigidus (L.) Fr. var. palmatus (Aiton) Cronquist ■☆

400678　Tussilago palmata Aiton = Petasites frigidus (L.) Fr. ■☆

400679　Tussilago petasites Thunb. = Petasites japonicus (Siebold et Zucc.) Maxim. ■

400680　Tussilago rubella J. F. Gmel. = Petasites rubellus (J. F. Gmel.) J. Toman ■

400681　Tussilago sagittata Banks ex Pursh = Petasites frigidus (L.) Fr. var. sagittatus (Banks ex Pursh) Chern. ■☆

400682　Tussilago saxatilis Turcz. ex DC. = Petasites rubellus (J. F. Gmel.) J. Toman ■

400683　Tutcheria Dunn = Pyrenaria Blume ●

400684　Tutcheria Dunn(1908);石笔木属(楉捷木属);Slatepentree, Tutcheria ●

400685　Tutcheria acutiserrata Hung T. Chang;尖齿石笔木;Acuteserrated Tutcheria, Acutiserrate Tutcheria, Sharptooth Slatepentree ●

400686　Tutcheria acutiserrata Hung T. Chang = Pyrenaria sophiae（Hu）S. X. Yang et T. L. Ming ●

400687　Tutcheria austrosinica Hung T. Chang;华南石笔木;S. China Slatepentree, South China Tutcheria, South-China Tutcheria ●

400688　Tutcheria austrosinica Hung T. Chang = Pyrenaria spectabilis（Champ.）C. Y. Wu et S. X. Yang var. greeniae（Chun）S. X. Yang ●

400689　Tutcheria brachycarpa Hung T. Chang;短果石笔木;Shortfruit Slatepentree, Shortfruit Tutcheria, Short-fruited Tutcheria ●

400690　Tutcheria brachycarpa Hung T. Chang = Pyrenaria spectabilis（Champ.）C. Y. Wu et S. X. Yang ●

400691　Tutcheria championii Nakai = Pyrenaria spectabilis（Champ.）C. Y. Wu et S. X. Yang ●

400692　Tutcheria championii Nakai = Tutcheria spectabilis Dunn ●

400693　Tutcheria greeniae Chun = Pyrenaria spectabilis（Champ.）C. Y. Wu et S. X. Yang var. greeniae（Chun）S. X. Yang ●

400694　Tutcheria hexalocularia Hu et S. Ye Liang = Tutcheria hexalocularia Hu et S. Ye Liang ex Hung T. Chang ●◇

400695　Tutcheria hexalocularia Hu et S. Ye Liang ex Hung T. Chang;六片石笔木(六瓣石笔木);Sixloculed Tutcheria, Sixroom Slatepentree ●◇

400696　Tutcheria hexalocularia Hu et S. Ye Liang ex Hung T. Chang = Pyrenaria spectabilis（Champ.）C. Y. Wu et S. X. Yang ●

400697　Tutcheria hirta（Hand. -Mazz.）H. L. Li = Pyrenaria hirta（Hand. -Mazz.）H. Keng ●

400698　Tutcheria hirta（Hand. -Mazz.）H. L. Li var. cordatula H. L. Li = Pyrenaria hirta（Hand. -Mazz.）H. Keng var. cordatula（H. L. Li）S. X. Yang et T. L. Ming ●

400699　Tutcheria hirta（Hand. -Mazz.）H. L. Li var. grandiflora（Y. C. Wu）H. L. Li = Pyrenaria hirta（Hand. -Mazz.）H. Keng ●

400700　Tutcheria kwangsiensis（Hung T. Chang）Hung T. Chang et C. X. Ye = Pyrenaria kwangsiensis Hung T. Chang ●

400701　Tutcheria kweichouensis Hung T. Chang et Y. K. Li;贵州石笔木;Guizhou Slatepentree, Guizhou Tutcheria ●

400702　Tutcheria kweichouensis Hung T. Chang et Y. K. Li = Pyrenaria pingpienensis（Hung T. Chang）S. X. Yang et T. L. Ming ●

400703　Tutcheria longisepala Hung T. Chang;长萼石笔木;Long-sepaled Tutcheria ●

400704　Tutcheria maculatoclada Y. K. Li = Pyrenaria maculatoclada（Y. K. Li）S. X. Yang ●

400705　Tutcheria maculatoclada Y. K. Li = Tutcheria greeniae Chun ●

400706　Tutcheria microcarpa Dunn = Pyrenaria microcarpa（Dunn）H. Keng ●

400707　Tutcheria multisepala Merr. et Chun = Parapyrenaria multisepala（Merr. et Chun）Hung T. Chang ●◇

400708　Tutcheria multisepala Merr. et Chun = Pyrenaria jonquieriana Pierre ex Lanessan subsp. multisepala（Merr. et Chun）S. X. Yang ●

400709　Tutcheria obtusifolia Hung T. Chang;钝叶石笔木(石笔木);Obtuse-leaved Tutcheria ●

400710　Tutcheria ovalifolia H. L. Li = Pyrenaria microcarpa（Dunn）H. Keng var. ovalifolia（H. L. Li）T. L. Ming et S. X. Yang ●

400711　Tutcheria paraspectabilis Hung T. Chang;肖石笔木;False Slatepentree, False Tutcheria ●

400712　Tutcheria pingpienensis Hung T. Chang = Pyrenaria pingpienensis（Hung T. Chang）S. X. Yang et T. L. Ming ●

400713　Tutcheria pubicostata Hung T. Chang;毛肋石笔木;Pubicostate Tutcheria ●

400714　Tutcheria pubicostata Hung T. Chang = Pyrenaria sophiae（Hu）S. X. Yang et T. L. Ming ●

400715　Tutcheria pubifolia Merr. ex Hung T. Chang = Pyrenaria hirta（Hand. -Mazz.）H. Keng ●

400716　Tutcheria pubifolia Merr. ex Hung T. Chang = Tutcheria hirta（Hand. -Mazz.）H. L. Li ●

400717　Tutcheria rostrata Hung T. Chang;尖喙石笔木;Beak-shaped Tutcheria, Rostrate Tutcheria, Sharpbeak Slatepentree ●

400718　Tutcheria rostrata Hung T. Chang = Pyrenaria spectabilis（Champ. ex Benth.）C. Y. Wu et S. X. Yang var. greeniae（Chun）S. X. Yang ●

400719　Tutcheria sessiliflora Hung T. Chang;无梗石笔木(无柄石笔木);Sessile-flowered Tutcheria ●

400720　Tutcheria shinkoensis（Hayata）Nakai = Pyrenaria microcarpa（Dunn）H. Keng ●

400721　Tutcheria sophiae（Hu）Hung T. Chang = Pyrenaria sophiae（Hu）S. X. Yang et T. L. Ming ●

400722　Tutcheria spectabilis（Champ. ex Benth.）Dunn;石笔木(石笔,楣捷木);Champion Slatepentree, Common Tutcheria, Tutcheria ●

400723　Tutcheria spectabilis（Champ.）Dunn = Pyrenaria spectabilis（Champ.）C. Y. Wu et S. X. Yang ●

400724　Tutcheria spectabilis（Champ.）Dunn = Tutcheria spectabilis（Champ. ex Benth.）Dunn ●

400725　Tutcheria spectabilis Dunn = Tutcheria spectabilis（Champ. ex Benth.）Dunn ●

400726　Tutcheria subsessiliflora Hung T. Chang;短梗石笔木(无柄石笔木);Sessile Tutcheria, Sessilflower Slatepentree, Stalkless Tutcheria ●

400727　Tutcheria subsessiliflora Hung T. Chang = Pyrenaria hirta（Hand. -Mazz.）H. Keng ●

400728　Tutcheria symplocifolia Merr. et Chun;锥果石笔木;Conicalfruit Slatepentree, Conicalfruit Tutcheria, Conical-fruited Tutcheria ●

400729　Tutcheria symplocifolia Merr. et F. P. Metcalf = Pyrenaria microcarpa（Dunn）H. Keng var. ovalifolia（H. L. Li）T. L. Ming et S. X. Yang ●

400730　Tutcheria taiwanica Hung T. Chang et S. X. Ren;台湾石笔木;Taiwan Slatepentree, Taiwan Tutcheria ●

400731　Tutcheria taiwanica Hung T. Chang et S. X. Ren = Pyrenaria microcarpa（Dunn）H. Keng var. ovalifolia（H. L. Li）T. L. Ming et S. X. Yang ●

400732　Tutcheria tenuifolia Hung T. Chang;薄叶石笔木;Thinleaf Slatepentree, Thinleaf Tutcheria, Thin-leaved Tutcheria ●

400733　Tutcheria tenuifolia Hung T. Chang = Pyrenaria microcarpa（Dunn）H. Keng ●

400734　Tutcheria vietnamensis Hung T. Chang = Pyrenaria hirta（Hand. -Mazz.）H. Keng var. cordatula（H. L. Li）S. X. Yang et T. L. Ming ●

400735　Tutcheria villosa Y. C. Wu = Pyrenaria hirta（Hand. -Mazz.）H. Keng ●

400736　Tutcheria villosa Y. C. Wu = Tutcheria hirta（Hand. -Mazz.）H. L. Li ●

400737　Tutcheria villosa Y. C. Wu var. grandiflora Y. C. Wu = Pyrenaria hirta（Hand. -Mazz.）H. Keng ●

400738　Tutcheria virgata（Koidz.）Nakai = Pyrenaria microcarpa（Dunn）H. Keng ●

400739　Tutcheria wuana Hung T. Chang = Pyrenaria wuana（Hung T. Chang）S. X. Yang ●

400740　Tutuca Molina(1810);智利杜鹃花属 ●☆

400741　Tutuca chilensis Molina;智利杜鹃花 ●☆

400742　Tutuca fistulosa Molina;管智利杜鹃花●☆

400743　Tuxtla Villasenor et Strother = Zexmenia La Llave ●■☆

400744　Tuxtla Villasenor et Strother(1989);微弯菊●☆

400745　Tuxtla pittieri (Greenm.) Villaseñor et Strother;微弯菊●☆

400746　Tuxtla pittieri (Greenm.) Villaseñor et Strother = Zexmenia pittieri Greenm. ●☆

400747　Tuyamaea T. Yamaz. = Lindernia All. ■

400748　Tweedia Hook. et Arn. (1834);尖瓣藤属●☆

400749　Tweedia caerulea D. Don = Tweedia caerulea D. Don ex Sweet ●☆

400750　Tweedia caerulea D. Don ex Sweet;尖瓣藤(天蓝尖瓣木);Blue-flowered Milkweed,Southern Star ●☆

400751　Twisselmannia Al-Shehbaz = Tropidocarpum Hook. ■☆

400752　Tydaea Decne. = Kohleria Regel ●■☆

400753　Tydea C. Muell. = Tydaea Decne. ●■☆

400754　Tylacantha Endl. = Angelonia Bonpl. ●■☆

400755　Tylacantha Endl. = Thylacantha Nees et Mart. ●■☆

400756　Tylachenia Post et Kuntze = Jurinea Cass. ●■

400757　Tylachenia Post et Kuntze = Tulakenia Raf. ●■

400758　Tylachium Grig = Thilachium Lour. ●☆

400759　Tylanthera C. Hansen(1990);胀药野牡丹属●☆

400760　Tylanthera cordata C. Hansen;泰国胀药野牡丹●☆

400761　Tylanthera tuberosa C. Hansen;胀药野牡丹●☆

400762　Tylanthus Reissek = Phylica L. ●☆

400763　Tylanthus emirnensis Tul. = Phylica emirnensis (Tul.) Pillans ●☆

400764　Tylecarpus Engl. = Medusanthera Seem. ●☆

400765　Tylecodon Toelken(1979);棒毛萼属(奇峰锦属)●☆

400766　Tylecodon albiflorus Bruyns;棒毛萼●☆

400767　Tylecodon aurusbergensis G. Will. et Van Jaarsv. ;阿乌棒毛萼●☆

400768　Tylecodon bayeri Van Jaarsv. ;巴耶尔棒毛萼●☆

400769　Tylecodon buchholzianus (Schuldt et Stephens) Toelken;布赫棒毛萼●☆

400770　Tylecodon buchholzianus (Schuldt et Stephens) Toelken subsp. fasciculatus G. Will. ;簇生棒毛萼●☆

400771　Tylecodon cacalioides (L. f.) Toelken;蟹甲草棒毛萼●☆

400772　Tylecodon cordiformis G. Will.;心形棒毛萼●☆

400773　Tylecodon cremnophilus Bruyns = Tylecodon ellaphieae Van Jaarsv. ●☆

400774　Tylecodon decipiens Toelken;迷惑棒毛萼●☆

400775　Tylecodon ellaphieae Van Jaarsv.;南非棒毛萼●☆

400776　Tylecodon fragilis (R. A. Dyer) Toelken;脆棒毛萼●☆

400777　Tylecodon grandiflorus (Burm. f.) Toelken;大花棒毛萼●☆

400778　Tylecodon hallii (Toelken) Toelken;霍尔棒毛萼●■☆

400779　Tylecodon hirtifolius (W. F. Barker) Toelken;毛叶棒毛萼●☆

400780　Tylecodon jarmilae Halda = Tylecodon ventricosus (Burm. f.) Toelken ●☆

400781　Tylecodon leucothrix (C. A. Sm.) Toelken;白毛棒毛萼●☆

400782　Tylecodon longipes Van Jaarsv. et G. Will.;长梗棒毛萼●☆

400783　Tylecodon mallei G. Will. ;马勒棒毛萼●☆

400784　Tylecodon nigricaulis G. Will. et Van Jaarsv.;黑茎棒毛萼●☆

400785　Tylecodon occultans (Toelken) Toelken;隐蔽棒毛萼●☆

400786　Tylecodon paniculatus (L. f.) Toelken;阿房宫;Botterboom, Butter Tree ●☆

400787　Tylecodon papillaris (L.) G. D. Rowley;乳突棒毛萼■☆

400788　Tylecodon papillaris (L.) G. D. Rowley subsp. wallichii (Harv.) G. D. Rowley;奇峰锦●☆

400789　Tylecodon pearsonii (Schönland) Toelken;皮尔逊棒毛萼●☆

400790　Tylecodon peculiaris Van Jaarsv.;特殊棒毛萼●☆

400791　Tylecodon pusillus Bruyns;微小棒毛萼●☆

400792　Tylecodon pygmaeus (W. F. Barker) Toelken;矮小棒毛萼●☆

400793　Tylecodon pygmaeus (W. F. Barker) Toelken var. tenuis (Toelken) Toelken = Tylecodon tenuis (Toelken) Bruyns ●☆

400794　Tylecodon racemosus (Harv.) Toelken;总花棒毛萼●☆

400795　Tylecodon reticulatus (L. f.) Toelken;万物相(网脉长筒莲);Barbed-wire Plant ●☆

400796　Tylecodon reticulatus (L. f.) Toelken subsp. phyllopodium Toelken;柄叶万物相●☆

400797　Tylecodon rubrovenosus (Dinter) Toelken;红脉棒毛萼●☆

400798　Tylecodon scandens Van Jaarsv. = Tylecodon tenuis (Toelken) Bruyns ●☆

400799　Tylecodon schaeferianus (Dinter) Toelken;谢弗棒毛萼●☆

400800　Tylecodon similis (Toelken) Toelken;相似棒毛萼●☆

400801　Tylecodon singularis (R. A. Dyer) Toelken;单一棒毛萼●☆

400802　Tylecodon stenocaulis Bruyns;窄茎棒毛萼●☆

400803　Tylecodon striatus (Hutchison) Toelken;条纹棒毛萼●☆

400804　Tylecodon suffultus Bruyns ex Toelken;支柱棒毛萼●☆

400805　Tylecodon sulphureus (Toelken) Toelken;硫色棒毛萼●☆

400806　Tylecodon tenuis (Toelken) Bruyns;细棒毛萼●☆

400807　Tylecodon torulosus Toelken;结节棒毛萼●☆

400808　Tylecodon tuberosus Toelken;块状棒毛萼●☆

400809　Tylecodon ventricosus (Burm. f.) Toelken;偏肿棒毛萼●☆

400810　Tylecodon viridiflorus (Toelken) Toelken;绿花棒毛萼●☆

400811　Tylecodon wallichii (Harv.) Toelken = Tylecodon papillaris (L.) G. D. Rowley subsp. wallichii (Harv.) G. D. Rowley ●☆

400812　Tylecodon wallichii (Harv.) Toelken subsp. ecklonianus (Harv.) Toelken;埃克棒毛萼●☆

400813　Tyleria Gleason(1931);泰勒木属●☆

400814　Tyleria floribunda Gleason;多花泰勒木●☆

400815　Tyleria grandiflora Gleason;大花泰勒木●☆

400816　Tyleropappus Greenm. (1931);泰勒菊属■☆

400817　Tyleropappus dichotomus Greenm. 泰勒菊■☆

400818　Tylexis Post et Kuntze = Brassavola R. Br. (保留属名)●☆

400819　Tylexis Post et Kuntze = Tulexis Raf. ■☆

400820　Tylisma Post et Kuntze = Corytholoma (Benth.) Decne. ■☆

400821　Tylisma Post et Kuntze = Tulisma Raf. (废弃属名)●☆

400822　Tylista Post et Kuntze = Haworthia Duval(保留属名)■☆

400823　Tylista Post et Kuntze = Tulista Raf. ■☆

400824　Tylloma D. Don = Chaetanthera Ruiz et Pav. ■☆

400825　Tylocarpus Post et Kuntze = Guardiola Cerv. ex Bonpl. ■☆

400826　Tylocarpus Post et Kuntze = Medusanthera Seem. ●☆

400827　Tylocarpus Post et Kuntze = Tulocarpus Hook. et Arn. ■☆

400828　Tylocarpus Post et Kuntze = Tylecarpus Engl. ●☆

400829　Tylocarya Nelmes = Fimbristylis Vahl(保留属名)■

400830　Tylochilus Nees = Cyrtopodium R. Br. ■☆

400831　Tyloclinta Post et Kuntze = Metalasia R. Br. ●☆

400832　Tyloclinta Post et Kuntze = Tuloclinia Raf. ●☆

400833　Tyloderma Miers = Hylenaea Miers ●☆

400834　Tylodontia Griseb. = Cynanchum L. ●■

400835　Tyloglossa Hochst. = Justicia L. ●■

400836　Tyloglossa acuminata (Nees) Hochst. = Justicia palustris (Hochst.) T. Anderson ■☆

400837　Tyloglossa kotschyi Hochst. = Justicia ladanoides Lam. ■☆

400838　Tyloglossa major Hochst. = Justicia flava (Vahl) Vahl ■☆

400839　Tyloglossa palustris Hochst. = Justicia palustris (Hochst.) T. Anderson ■☆

400840 Tyloglossa pubescens Hochst. = Dicliptera capensis Nees ●☆

400841 Tyloglossa schimperi Hochst. = Justicia ladanoides Lam. ■☆

400842 Tylomium C. Presl = Lobelia L. ●■

400843 Tylopetalum Barneby et Krukoff = Sciadotenia Miers ●☆

400844 Tylophora R. Br. (1810);娃儿藤属(欧蔓属);Blue-Flowered Vine,Childvine,Tylophora ●■

400845 Tylophora adalinae K. Schum. = Tylophora sylvatica Decne. ●☆

400846 Tylophora anfracta N. E. Br. = Tylophora oblonga N. E. Br. ●☆

400847 Tylophora angustiniana (Hemsl.) Craib;宜昌娃儿藤;Yichang Childvine,Yichang Tylophora ●

400848 Tylophora anomala N. E. Br. ;异常娃儿藤●☆

400849 Tylophora anthopotamica (Hand.-Mazz.) Tsiang et H. T. Zhang;花溪娃儿藤(飞菜);Flower-river Tylophora,Huaxi Childvine,Huaxi Tylophora ●

400850 Tylophora apiculata K. Schum. ;细尖娃儿藤●☆

400851 Tylophora arachnoidea Goyder;蛛网娃儿藤●☆

400852 Tylophora arenicola Merr. ;虎须娃儿(老虎须,老虎须藤,沙地娃儿藤);Sandhill Tylophora,Tigerbeard ●

400853 Tylophora aristolochioides Miq. ;马兜铃娃儿藤●☆

400854 Tylophora astephanoides Tsiang et P. T. Li;阔叶娃儿藤(阔叶车前);Broadleaf Childvine ■

400855 Tylophora asthmatica Wight et Arn. ;印度娃儿藤(治喘娃儿藤);India Childvine ●

400856 Tylophora atrofolliculata F. P. Metcalf;毛果娃儿藤(黎针,三分丹,蛇花藤);Brackfollicle Tylophora,Sanfendan Childvine,Snakeflower Tylophora ●

400857 Tylophora atrofolliculata F. P. Metcalf = Tylophora ovata (Lindl.) Hook. ex Steud. ●

400858 Tylophora badia (E. Mey.) Schltr. ;栗色娃儿藤●☆

400859 Tylophora badia (E. Mey.) Schltr. var. latifolia N. E. Br. ;宽叶栗色娃儿藤●☆

400860 Tylophora balansae Costantin = Tylophora kerrii Craib ●

400861 Tylophora belostemma Benth. = Belostemma hirsutum Wall. ex Wight ●

400862 Tylophora brevipes (Turcz.) Fern. -Vill. ;短柄娃儿藤(短柄鸥蔓)●

400863 Tylophora brownii Hayata;光叶娃儿藤●

400864 Tylophora brownii Hayata = Tylophora ovata (Lindl.) Hook. ex Steud. ●

400865 Tylophora caffra Meisn. = Sphaerocodon caffer (Meisn.) Schltr. ●☆

400866 Tylophora cameroonica N. E. Br. ;喀麦隆娃儿藤●☆

400867 Tylophora carnosa Wall. ex Wight = Tylophora flexuosa R. Br. ■

400868 Tylophora chingtungensis Tsiang et P. T. Li;景东娃儿藤(景东车前,显脉娃儿藤);Jingdong Childvine,Jingdong Tylophora ●

400869 Tylophora chungii Merr. ex F. P. Metcalf = Tylophora floribunda Miq. ●■

400870 Tylophora cirrosa Asch. = Pentatropis nivalis (J. F. Gmel.) D. V. Field et J. R. I. Wood ■☆

400871 Tylophora coddii Bullock;科德娃儿藤●☆

400872 Tylophora congoensis Schltr. ;刚果娃儿藤●☆

400873 Tylophora conspicua N. E. Br. ;显著娃儿藤●☆

400874 Tylophora cordata (Thunb.) Druce;心形娃儿藤●☆

400875 Tylophora cordifolia (Link,Klotzsch et Otto) Benth. et Hook. f. ex Kuntze = Belostemma cordifolium (Link,Klotzsch et Otto) M. G. Gilbert et P. T. Li ●

400876 Tylophora cordifolia Thwaites = Belostemma cordifolium (Link,Klotzsch et Otto) M. G. Gilbert et P. T. Li ●

400877 Tylophora corollae Bullock ex Meve et Liede = Tylophora lugardae Bullock ●☆

400878 Tylophora crebriflora S. T. Blake;密花娃儿藤●☆

400879 Tylophora cycleoides Tsiang;轮环娃儿藤;Annular Tylophora,Cycle Childvine,Cyclelike Tylophora ●

400880 Tylophora dahomensis K. Schum. ;达荷姆娃儿藤●☆

400881 Tylophora deightonii Hutch. = Tylophora oculata N. E. Br. ■☆

400882 Tylophora dielsii (H. Lév.) Hu = Tylophora flexuosa R. Br. ■

400883 Tylophora erecta F. Muell. ex Benth. ;直立娃儿藤●☆

400884 Tylophora erubescens (Liede et Meve) Liede;变红娃儿藤●☆

400885 Tylophora flanaganii Schltr. ;弗拉纳根娃儿藤●☆

400886 Tylophora fleckii (Schltr.) N. E. Br. ;弗莱克娃儿藤●☆

400887 Tylophora flexuosa R. Br. ;小叶娃儿藤(平伐车前,平伐娃儿藤);Diels Childvine,Littleleaf Childvine,Small-leaf Childvine,Small-leaf Tylophora ■

400888 Tylophora floribunda Miq. ;多花娃儿藤(白龙须,黄茅细辛,老君须,毛管细辛,七层楼,三十六荡,三十六根,双飞蝴蝶,藤老君须,土细辛,娃儿藤,小尾伸根,须参,一见香);Manyflower Childvine,Manyflower Tylophora ●■

400889 Tylophora forrestii M. G. Gilbert et P. T. Li;大花娃儿藤;Big-flower Childvine,Big-flower Tylophora ●

400890 Tylophora gilletii De Wild. ;吉勒特娃儿藤●☆

400891 Tylophora glabra Costantin;长梗娃儿藤;Glabrous Childvine,Glabrous Tylophora,Long-pediceled Tylophora ●

400892 Tylophora glauca Bullock;灰蓝娃儿藤●☆

400893 Tylophora gracilenta Tsiang et P. T. Li;天峨娃儿藤;Slender Childvine,Slender Tylophora,Tian'e Tylophora ●

400894 Tylophora gracilis De Wild. ;纤细娃儿藤●☆

400895 Tylophora gracillima Markgr. ;细长娃儿藤●☆

400896 Tylophora hainanensis Tsiang;海南娃儿藤;Hainan Childvine,Hainan Tylophora ●

400897 Tylophora hainanensis Tsiang = Lygisma inflexum (Costantin) Kerr ■

400898 Tylophora henryi Warb. ;紫花娃儿藤(紫花车前);Purpleflower Childvine ■

400899 Tylophora heterophylla A. Rich. ;互叶娃儿藤●☆

400900 Tylophora hirsuta (Wall.) Wight = Tylophora ovata (Lindl.) Hook. ex Steud. ●

400901 Tylophora hispida Decne. = Tylophora ovata (Lindl.) Hook. ex Steud. ●

400902 Tylophora hispida Decne. var. browni Hayata = Tylophora ovata (Lindl.) Hook. ex Steud. ●

400903 Tylophora hispida Decne. var. tanakae (Maxim.) Hatus. = Tylophora tanakae Maxim. ex Franch. et Sav. ●☆

400904 Tylophora hoyopsis H. Lév. = Tylophora flexuosa R. Br. ■

400905 Tylophora hui Tsiang;建水娃儿藤;Hu Childvine,Hu Tylophora ●

400906 Tylophora incana Brunner = Leptadenia hastata (Pers.) Decne. ●☆

400907 Tylophora indica Merr. = Tylophora asthmatica Wight et Arn. ●

400908 Tylophora inhambanensis Schltr. ;伊尼扬巴内娃儿藤●☆

400909 Tylophora insulana Tsiang et P. T. Li;台湾娃儿藤●

400910 Tylophora insulana Tsiang et P. T. Li = Tylophora ovata (Lindl.) Hook. ex Steud. ●

400911 Tylophora jacquemontii Decne. = Tylophora hirsuta (Wall.) Wight ●

400912 Tylophora japonica Miq. = Vincetoxicum sublanceolatum (Miq.) Maxim. var. albiflorum (Franch. et Sav.) Kitag. ●☆

400913 Tylophora kerrii Craib;人参娃儿藤(山豆根,土牛七,土牛膝,土人参);Gingseng Childvine,Kerr Tylophora ●

400914 Tylophora koi Merr. ;通天莲(白花鸡屎藤,红药,乳汁藤,双飞蝴蝶,通天连);Ko Childvine, Ko Tylophora ●

400915 Tylophora lanyuensis Y. C. Liu et F. Y. Lu;兰屿欧蔓●

400916 Tylophora lanyuensis Y. C. Liu et F. Y. Lu = Tylophora ovata (Lindl.) Hook. ex Steud. ●

400917 Tylophora leptantha Tsiang;广花娃儿藤;Broadinflorescence Childvine, Broadinflorescence Tylophora, Broad-inflorescence Tylophora ●

400918 Tylophora leveilleana Schltr. ex H. Lév. = Tylophora kerrii Craib

400919 Tylophora liberica N. E. Br. = Tylophora oblonga N. E. Br. ●☆

400920 Tylophora longepedunculata Schltr. = Tylophora heterophylla A. Rich. ●☆

400921 Tylophora longifolia Wight;长叶娃儿藤;Long-leaf Childvine, Long-leaf Tylophora ●

400922 Tylophora longipedicellata Tsiang et P. T. Li;斑胶藤;Longpedicel Childvine, Longpedicel Tylophora ●

400923 Tylophora longipedicellata Tsiang et P. T. Li = Tylophora glabra Costantin ●

400924 Tylophora lugardae Bullock;卢格德娃儿藤●☆

400925 Tylophora lycioides (E. Mey.) Decne. ;枸杞状娃儿藤■☆

400926 Tylophora macrantha Hance = Dregea volubilis (L. f.) Benth. ex Hook. f. ●

400927 Tylophora matsumurae (T. Yamaz.) T. Yamash. et Tateishi;松村氏娃儿藤■☆

400928 Tylophora maximowicziana Warb. = Cynanchum sublanceolatum (Miq.) Matsum. ■

400929 Tylophora membranacea Tsiang et P. T. Li;膜叶娃儿藤(膜叶车前);Filmleaf Childvine, Membranousleaf Childvine ■

400930 Tylophora micrantha Decne. = Secamone elliptica R. Br. ●■

400931 Tylophora minutiflora A. Chev. = Tylophora dahomensis K. Schum. ●☆

400932 Tylophora minutiflora Woodson = Secamone minutiflora (Woodson) Tsiang ●■

400933 Tylophora mollissima Wall. ex Wight = Tylophora ovata (Lindl.) Hook. ex Steud. ●

400934 Tylophora mollissima Wight;绵毛娃儿藤(大白前,绵毛车前,三白根,通脉丹);Woolly Childvine ■

400935 Tylophora mollissima Wight = Tylophora ovata (Lindl.) Hook. ex Steud. ●

400936 Tylophora nana C. K. Schneid. ;汶川娃儿藤;Dwarf Childvine, Dwarf Tylophora ●

400937 Tylophora nana C. K. Schneid. var. gansuensis L. C. Wang et X. G. Sun;甘肃娃儿藤;Gansu Dwarf Childvine, Gansu Dwarf Tylophora ●

400938 Tylophora nana C. K. Schneid. var. gansuensis L. C. Wang et X. G. Sun = Tylophora nana C. K. Schneid. ●

400939 Tylophora nikoensis (Franch. et Sav.) Matsum. = Tylophora floribunda Miq. ●■

400940 Tylophora oblonga N. E. Br. ;矩圆娃儿藤●☆

400941 Tylophora oculata N. E. Br. ;小眼娃儿藤■☆

400942 Tylophora oligophylla (Tsiang) M. G. Gilbert, W. D. Stevens et P. T. Li;滑藤(稀叶藤);Childvine, Fewleaf Absolmsia ●

400943 Tylophora orthocaulis K. Schum. = Tylophora congolana (Baill.) Bullock ●☆

400944 Tylophora oshimae Hayata;少花娃儿藤(少花车前,疏花欧蔓,狭叶欧蔓);Fewflower Childvine, Oshima Childvine ■

400945 Tylophora ovata (Lindl.) Hook. ex Steud. ;娃儿藤(白龙须,缠竹消,关腰草,光叶娃儿藤,黄芽细辛,鸡母香,金钱吊丝馅,兰屿欧蔓,老虎须,老君须,黎针,卵叶娃儿藤,落地金瓜,落地蜘蛛,落土香,芒尾蛇,毛管细辛,毛果娃儿藤,欧蔓,婆婆针线包,三十六荡,三十六根,山辣子,蛇花藤,双飞蝴蝶,台湾娃儿藤,藤霸王,藤细辛,藤叶细辛,土细辛,虾箱须,小霸王,哮喘草);Brown Ovate Tylophora, Brown Tylophora, Insula Childvine, Lanyu Tylophora, Lucidleaf Childvine, Ovate Childvine, Ovate Tylophora, Ovate-leaf Tylophora, Taiwan Tylophora ●

400946 Tylophora ovata (Lindl.) Hook. ex Steud. var. brownii (Hayata) Tsiang et P. T. Li = Tylophora ovata (Lindl.) Hook. ex Steud. ●

400947 Tylophora ovata (Lindl.) Hook. ex Steud. var. brownii (Hayata) Tsiang et P. T. Li = Tylophora brownii Hayata ●

400948 Tylophora panzhutenga Z. Y. Zhu;攀竹藤;Panzhuteng Childvine ■

400949 Tylophora panzhutenga Z. Y. Zhu = Tylophora brownii Hayata ●

400950 Tylophora panzhutenga Z. Y. Zhu = Tylophora ovata (Lindl.) Hook. ex Steud. ●

400951 Tylophora parvifolia Robyns et Lebrun;非洲小叶娃儿藤●☆

400952 Tylophora picta Tsiang;紫叶娃儿藤;Purpleleaf Tylophora, Purple-leaved Tylophora, Violetleaf Childvine ●

400953 Tylophora pseudotenerrima Costantin = Tylophora kerrii Craib ●

400954 Tylophora renchangii Tsiang;扒地蜈蚣(假白前,扛棺回);Renchang Childvine, Renchang Tylophora ●

400955 Tylophora renchangii Tsiang = Tylophora glabra Costantin ●

400956 Tylophora rockii M. G. Gilbert et P. T. Li;山娃儿藤;Rock Childvine ●

400957 Tylophora rotundifolia Buch. -Ham. ex Wight;圆叶娃儿藤;Rotundifolious Tylophora, Roundleaf Childvine ●

400958 Tylophora secamonoides Tsiang;蛇胆草(戏班须);Secamone-like Tylophora, Snakegoll Childvine ●

400959 Tylophora shikokiana Matsum. ex Nakai = Tylophora floribunda Miq. ●■

400960 Tylophora silvestrii (Pamp.) Tsiang et P. T. Li;湖北娃儿藤(贵州娃儿藤,湖北车前);Hubei Childvine ■

400961 Tylophora silvestris Tsiang;贵州娃儿藤;Guizhou Childvine, Woods Tylophora, Woody Tylophora ●

400962 Tylophora smilacina S. Moore = Tylophora gilletii De Wild. ●☆

400963 Tylophora sootepensis Craib = Tylophora angustiniana (Hemsl.) Craib ●

400964 Tylophora sootepensis Craib = Tylophora koi Merr. ●

400965 Tylophora stenoloba (K. Schum.) N. E. Br. ;窄裂片娃儿藤●☆

400966 Tylophora sublanceolata Miq. = Cynanchum sublanceolatum (Miq.) Matsum. ■

400967 Tylophora sylvatica Decne. ;林地娃儿藤●☆

400968 Tylophora syringifolia E. Mey. = Tylophora cordata (Thunb.) Druce ●☆

400969 Tylophora taiwanensis Hatus. ;台湾欧蔓●

400970 Tylophora taiwanensis Hatus. = Tylophora koi Merr. ●

400971 Tylophora tanakae Maxim. = Tylophora tanakae Maxim. ex Franch. et Sav. ■☆

400972 Tylophora tanakae Maxim. ex Franch. et Sav. ;蔓毛林花;Tanaka Tylophora ■☆

400973 Tylophora tanakae Maxim. var. glabrescens Hatus. ex T. Yamaz. ;光蔓毛林花■☆

400974 Tylophora tenerrima Wight;极细娃儿藤●☆

400975 Tylophora tengii Tsiang;普定娃儿藤;Puding Childvine, Teng Tylophora ●

400976 Tylophora tenuipedunculata K. Schum. ;梗花娃儿藤●☆

400977 Tylophora tenuis Blume = Tylophora flexuosa R. Br. ■

400978 Tylophora tenuissima (Roxb.) Wight et Arn. = Tylophora flexuosa

R. Br. ■

400979 Tylophora tetrapetala（Dennst.）Suresh. = Tylophora flexuosa R. Br. ■

400980 Tylophora trichophylla Tsiang = Tylophora rotundifolia Buch. - Ham. ex Wight ●

400981 Tylophora tsiangii（P. T. Li）M. G. Gilbert，W. D. Stevens et P. T. Li；曲序娃儿藤；Tsiang's Childvine，Tsiang's Tylophora ●

400982 Tylophora tuberculata M. G. Gilbert et P. T. Li；个旧娃儿藤；Tuberculate Childvine，Tuberculate Tylophora ●

400983 Tylophora umbellata Schltr.；伞花娃儿藤●☆

400984 Tylophora uncinata M. G. Gilbert et P. T. Li；钩毛娃儿藤；Uncinate Childvine，Uncinate Tylophora ●

400985 Tylophora urceolata Meve；坛状娃儿藤●☆

400986 Tylophora yemensis Deflers = Tylophora heterophylla A. Rich. ● ☆

400987 Tylophora yunnanensis Schltr.；云南娃儿藤（白龙须，白藤，白薇，金线包，老妈妈针线包，山辣子，蛇辣子，水辣子，娃儿藤，小白薇，野辣椒，野辣子）；Yunnan Childvine，Yunnan Tylophora ●

400988 Tylophora zenkeri Schltr.；岑克尔娃儿藤●☆

400989 Tylophoropsis N. E. Br.（1894）；类娃儿藤属●☆

400990 Tylophoropsis erubescens Liede et Meve；变红类娃儿藤●☆

400991 Tylophoropsis fleckii Schltr. = Tylophora fleckii（Schltr.）N. E. Br. ●☆

400992 Tylophoropsis heterophylla（A. Rich.）N. E. Br. = Tylophora heterophylla A. Rich. ●☆

400993 Tylophoropsis nyeriana Sambo；涅里类娃儿藤●☆

400994 Tylophoropsis yemense（Deflers）N. E. Br. = Tylophora heterophylla A. Rich. ●☆

400995 Tylophus Post et Kuntze = Triteleia Douglas ex Lindl. ■☆

400996 Tylophus Post et Kuntze = Tulophos Raf. ■☆

400997 Tylopsacas Leeuwenb.（1960）；肿粒苣苔属■☆

400998 Tylopsacas cuneata（Gleason）Leeuwenb.；肿粒苣苔■☆

400999 Tylorima Post et Kuntze = Saxifraga L. ■

401000 Tylorima Post et Kuntze = Tulorima Raf. ■

401001 Tylosema（Schweinf.）Torre et Hillc.（1955）；热非羊蹄甲属●☆

401002 Tylosema（Schweinf.）Torre et Hillc. = Bauhinia L. ●

401003 Tylosema angolensis P. Silveira et S. Castro；安哥拉热非羊蹄甲●☆

401004 Tylosema argentea（Chiov.）Brenan；银白热非羊蹄甲●☆

401005 Tylosema esculenta（Burch.）A. Schreib.；食用热非羊蹄甲●☆

401006 Tylosema fassoglensis（Schweinf.）Torre et Hillc.；热非羊蹄甲●☆

401007 Tylosema humifusa（Pic. Serm. et Roti Mich.）Brenan；平伏热非羊蹄甲●☆

401008 Tylosepalum Kurz ex Teijsm. et Binn. = Trigonostemon Blume（保留属名）●

401009 Tylosperma Botsch. = Potentilla L. ■

401010 Tylosperma Leeuwenb. = Tylopsacas Leeuwenb. ■☆

401011 Tylostemon Engl.（1899）；疣蕊樟属●☆

401012 Tylostemon Engl. = Beilschmiedia Nees ●

401013 Tylostemon acutifolius Engl. et K. Krause = Beilschmiedia acuta Kosterm. ●☆

401014 Tylostemon anacardioides Engl. et K. Krause = Beilschmiedia anacardioides（Engl. et K. Krause）Robyns et R. Wilczek ●☆

401015 Tylostemon asteranthus Mildbr.；星花疣蕊樟●☆

401016 Tylostemon barensis Engl. et K. Krause = Beilschmiedia barensis（Engl. et K. Krause）Robyns et R. Wilczek ●☆

401017 Tylostemon batangensis Engl. = Beilschmiedia batangensis（Engl.）Robyns et R. Wilczek ●☆

401018 Tylostemon caudatus（Stapf）Stapf = Beilschmiedia caudata（Stapf）A. Chev. ●☆

401019 Tylostemon cinnamomeus Stapf = Beilschmiedia cinnamomea（Stapf）Robyns et R. Wilczek ●☆

401020 Tylostemon confertus S. Moore = Beilschmiedia gaboonensis（Meisn.）Benth. et Hook. f. ●☆

401021 Tylostemon congestiflorus Engl. et K. Krause = Beilschmiedia congestiflora（Engl. et K. Krause）Robyns et R. Wilczek ●☆

401022 Tylostemon corbisieri Robyns = Beilschmiedia corbisieri（Robyns）Robyns et R. Wilczek ●☆

401023 Tylostemon crassifolius Engl. = Beilschmiedia crassifolia（Engl.）Robyns et R. Wilczek ●☆

401024 Tylostemon crassipes Engl. et K. Krause = Beilschmiedia crassipes（Engl. et K. Krause）Robyns et R. Wilczek ●☆

401025 Tylostemon cuspidatus Krause = Beilschmiedia cuspidata（Krause）Robyns et R. Wilczek ●☆

401026 Tylostemon dinklagei Engl. = Beilschmiedia dinklagei（Engl.）Robyns et R. Wilczek ●☆

401027 Tylostemon euryneura（Stapf）Stapf = Beilschmiedia caudata（Stapf）A. Chev. ●☆

401028 Tylostemon foliosus S. Moore = Beilschmiedia foliosa（S. Moore）Robyns et R. Wilczek ●☆

401029 Tylostemon fruticosus（Engl.）Stapf = Beilschmiedia fruticosa Engl. ●☆

401030 Tylostemon gaboonensis（Meisn.）Stapf = Beilschmiedia gaboonensis（Meisn.）Benth. et Hook. f. ●☆

401031 Tylostemon grandifolius Stapf = Beilschmiedia grandifolia（Stapf）Robyns et R. Wilczek ●☆

401032 Tylostemon jabassensis Engl. et K. Krause = Beilschmiedia jabassensis（Engl. et K. Krause）Robyns et R. Wilczek ●☆

401033 Tylostemon kamerunensis Engl. et K. Krause = Beilschmiedia mannii（Meisn.）Benth. et Hook. f. ●☆

401034 Tylostemon kenyensis Chiov. = Ocotea kenyensis（Chiov.）Robyns et R. Wilczek ●☆

401035 Tylostemon lancifolius Engl. et K. Krause = Beilschmiedia lancilimba Kosterm. ●☆

401036 Tylostemon ledermannii Engl. et K. Krause = Beilschmiedia robynsiana Kosterm. ●☆

401037 Tylostemon letestui Pellegr. = Beilschmiedia cinnamomea（Stapf）Robyns et R. Wilczek ●☆

401038 Tylostemon longipes Stapf = Beilschmiedia mannii（Meisn.）Benth. et Hook. f. ●☆

401039 Tylostemon macrophyllus Hutch. et Dalziel = Beilschmiedia hutchinsoniana Robyns et R. Wilczek ●☆

401040 Tylostemon mannii（Meisn.）Stapf = Beilschmiedia mannii（Meisn.）Benth. et Hook. f. ●☆

401041 Tylostemon mannii Stapf；曼氏疣蕊樟●☆

401042 Tylostemon membranaceus Stapf = Beilschmiedia membranifolia Kosterm. ●☆

401043 Tylostemon minutiflorus（Meisn.）Stapf = Beilschmiedia minutiflora（Meisn.）Benth. et Hook. f. ●☆

401044 Tylostemon myrciifolius S. Moore = Beilschmiedia myrciifolia（S. Moore）Robyns et R. Wilczek ●☆

401045 Tylostemon ndongensis Engl. et K. Krause = Beilschmiedia ndongensis（Engl. et K. Krause）Robyns et R. Wilczek ●☆

401046 Tylostemon ngriki A. Chev. = Beilschmiedia jacques-felixii Robyns et R. Wilczek ●☆

401047 Tylostemon obscurus Stapf = Beilschmiedia obscura（Stapf）

Engl. ex A. Chev. ●☆

401048 Tylostemon papyraceus Stapf = Beilschmiedia papyracea (Stapf) Robyns et R. Wilczek ●☆

401049 Tylostemon preussii (Engl.) Stapf = Beilschmiedia preussii Engl. ●☆

401050 Tylostemon sessilifolius Stapf = Beilschmiedia sessilifolia (Stapf) Engl. ex Fouilloy ●☆

401051 Tylostemon staudtii (Engl.) Stapf = Beilschmiedia staudtii Engl. ●☆

401052 Tylostemon talbotiae S. Moore = Beilschmiedia talbotiae (S. Moore) Robyns et R. Wilczek ●☆

401053 Tylostemon ugandensis (Rendle) Stapf = Beilschmiedia ugandensis Rendle ●☆

401054 Tylostemon zahnii Krause = Beilschmiedia zahnii (Krause) Robyns et R. Wilczek ●☆

401055 Tylostigma Schltr. (1916);膨头兰属■☆

401056 Tylostigma filiforme H. Perrier;线形膨头兰■☆

401057 Tylostigma foliosum Schltr.;多叶膨头兰■☆

401058 Tylostigma hildebrandtii (Ridl.) Schltr.;希氏膨头兰■☆

401059 Tylostigma madagascariense Schltr.;马岛膨头兰■☆

401060 Tylostigma nigrescens Schltr.;黑膨头兰■☆

401061 Tylostigma perrieri Schltr.;佩里耶膨头兰■☆

401062 Tylostigma tenellum Schltr.;细膨头兰■☆

401063 Tylostylis Blume = Callostylis Blume ■

401064 Tylostylis Blume = Eria Lindl. (保留属名)■

401065 Tylostylis discolor (Lindl.) Hook. f. = Callostylis rigida Blume ■

401066 Tylostylis rigida (Blume) Blume = Callostylis rigida Blume ■

401067 Tylothrasya Döll = Panicum L. ■

401068 Tylothrasya Döll = Thrasya Kunth ■☆

401069 Tylothrasya Döll(1877);肖勇夫草属■☆

401070 Tylothrasya petrosa (Trin.) Döll;肖勇夫草■☆

401071 Tylotis Post et Kuntze = Tulotis Raf. ■

401072 Tympananthe Hassk. = Dictyanthus Decne. ●

401073 Tympananthe Hassk. = Matelea Aubl. ●☆

401074 Tynanthus Miers(1863);丁花属●☆

401075 Tynanthus elongatus Miers;丁花●☆

401076 Tynanthus laxiflorus Miers;疏花丁花●☆

401077 Tynanthus macranthus L. O. Williams;大花丁花●☆

401078 Tynanthus micranthus Corr. Mello ex K. Schum.;小花丁花●☆

401079 Tynanthus polyanthus (Bureau) Sandwith;多花丁花●☆

401080 Tynanthus pubescens A. H. Gentry;毛丁花●☆

401081 Tynnanthus K. Schum. = Tynanthus Miers ●☆

401082 Tynnanthus Miers = Tynanthus Miers ●☆

401083 Tynus J. Presl = Tinus Mill. ●

401084 Tynus J. Presl = Viburnum L. ●

401085 Typha L. (1753);香蒲属;Bulrush, Cat-o'-nine-tails, Cat's Tail,Cattail,Cat-tail,Cat-tail Flag,Cumbungi,Cumbungi Reed,Reed Mace,Reedmace,Reed-mace ■

401086 Typha ×glauca Godr.;灰绿香蒲;Hybrid Cat-tail,White Cat-tail ■☆

401087 Typha × suwensis T. Shimizu;取访香蒲■☆

401088 Typha aequalis Schnizl. = Typha domingensis (Pers.) Steud. ■

401089 Typha aethiopica (Rohrb.) Kronfeldt = Typha domingensis (Pers.) Steud. ■

401090 Typha angusta Bory et Chaub. = Typha angustifolia L. ■

401091 Typha angustata Bory et Chaub. = Typha domingensis (Pers.) Steud. ■

401092 Typha angustata Bory et Chaub. var. abyssinica Graebn. = Typha

domingensis (Pers.) Steud. ■

401093 Typha angustata Bory et Chaub. var. aethiopica Rohrb. = Typha domingensis (Pers.) Steud. ■

401094 Typha angustifolia L.;水烛香蒲(板枝,苞香蒲,草蒲黄,蕭董,甘蒲,鬼蜡烛,醮,醮石,金簪草,藾,芦油烛,芦烛,毛蜡烛,蒲,蒲棒,蒲包草,蒲草,蒲草黄,蒲桙,蒲槌,蒲儿根,蒲黄,蒲黄草,蒲笋,莎草,水蜡烛,水烛,睢,睢蒲,随手香,狭叶香蒲,香蒲);Lesser Bulrush, Lesser Reedmace, Maori Bulrush, Narrow Leaved Cattail, Narrowleaf Cattail, Narrow-leaved Cattail, Narrow-leaved Cat-tail, Narrow-leaved Reedmace, Raupo, Smaller Cat's Tail, Smaller Cat's Tails, Water Cattail ■

401095 Typha angustifolia L. subsp. angustata (Bory et Chaub.) Briq. = Typha domingensis (Pers.) Steud. ■

401096 Typha angustifolia L. subsp. australis (Schum. et Thonn.) Graebn. = Typha domingensis (Pers.) Steud. ■

401097 Typha angustifolia L. var. calumetensis Peattie = Typha angustifolia L. ■

401098 Typha angustifolia L. var. elongata (Dudley) Wiegand = Typha angustifolia L. ■

401099 Typha australis Schumach. = Typha domingensis (Pers.) Steud. ■

401100 Typha australis Schumach. et Thonn. = Typha domingensis Pers. ■

401101 Typha capensis (Rohrb.) N. E. Br.;好望角香蒲■☆

401102 Typha changbaiensis M. J. Wu et Y. T. Zhao;长白香蒲■

401103 Typha davidiana (Kronf.) Hand.-Mazz.;达香蒲■

401104 Typha davidiana (Kronf.) Hand.-Mazz. = Typha laxmannii Lepech. ■

401105 Typha domingensis (Pers.) Steud.;长苞香蒲(多明香蒲,蒲草,水蜡,水蜡烛,水烛);Longbract Cattail, Narrow-leaf Cumbungi, Southern Cattail, Southern Cat-tail ■

401106 Typha domingensis Pers. = Typha domingensis (Pers.) Steud. ■

401107 Typha elephantina Roxb.;象蒲(草芽);Elephant Cattail ■

401108 Typha glauca Godr.;灰蓝香蒲;Hybrid Cattail, White Cat-tail ■☆

401109 Typha gracilis Jord.;短序香蒲;Shortspike Cattail ■

401110 Typha japonica Miq. = Typha orientalis C. Presl ■

401111 Typha lanceolata ?;剑叶香蒲;Black Boys ■☆

401112 Typha latifolia L.;宽叶香蒲(蒲黄,香蒲);Baccobolts,Baccybolts, Black Poker, Black Pudding, Black Sticks, Blackamore, Blackcap, Blackie Toppers, Blacky-more, Broad-leaf Cat Tail, Broadleaf Cattai, Broad-leaved Cat-tail, Bull Seg, Bulrush, Cat O'nine Tails, Cat Tail, Cat Tall Flag, Cat's Tail, Cat's Tails, Cat's Tall, Cat's-tail, Cat-tail, Cat-tail Flag, Chimney Sweeper, Chimney Sweeper's Brush, Common Cat Tail, Common Cattail, Common Cat-tail, Cossack Asparagus, Devil's Poker, Dod, Dodder, Fairy Wives' Distaff, False Bulrush, Flags, Flaxtail, Flue Brush, Flue Brushes, Goss, Great Cat-tail, Great Grass, Great Reedmace, Greater Cat's Tail, Greater Cat's Tails, Holy Poker, Levver, Lyver, Marish Beetle, Marsh Beetle, Marsh Pestle, Mat Reed, Poker, Poker-plant, Pull Poker, Pull-poker, Pussies, Pussy's Tail, Pussy's Tails, Reedmace, Serge, Whitehead ■

401113 Typha latifolia L. 'Variegata';斑点宽叶香蒲(花叶宽叶香蒲)■☆

401114 Typha latifolia L. f. ambigua (Sond.) Kronf. = Typha latifolia L. ■

401115 Typha latifolia L. subsp. capensis Rohrb. = Typha capensis (Rohrb.) N. E. Br. ■☆

401116 Typha latifolia L. subsp. maresii (Batt.) Trab. = Typha elephantina Roxb. ■

401117 Typha latifolia L. subsp. orientalis C. Presl = Typha orientalis C.

Presl ■

401118　Typha latifolia L. var. orientalis（C. Presl）Rohrb. = Typha orientalis C. Presl ■

401119　Typha laxmanni sensu Ledeb. = Typha minima Funck ex Hoppe ■

401120　Typha laxmannii Lepech.；无苞香蒲（达香蒲，大卫香蒲，拉氏香蒲，毛腊黄，蒙古香蒲，缺苞香蒲，线叶香蒲，香蒲）；Bractless Cattail，David Cattail ■

401121　Typha laxmannii Lepech. var. davidiana（Kronf.）C. F. Fang = Typha davidiana（Kronf.）Hand. -Mazz. ■

401122　Typha laxmannii Lepech. var. planifolia Kronf. = Typha laxmannii Lepech. ■

401123　Typha maresii Batt. = Typha elephantina Roxb. ■

401124　Typha martini H. Lév. et Vaniot var. davidiana Kronf. = Typha davidiana（Kronf.）Hand. -Mazz. ■

401125　Typha martini H. Lév. et Vaniot var. davidiana Kronf. = Typha laxmannii Lepech. ■

401126　Typha martini Thomson ex Aitch. = Typha laxmannii Lepech. ■

401127　Typha minima Funck = Typha minima Funck. ex Hoppe ■

401128　Typha minima Funck. ex Hoppe；小香蒲（水香蒲，细叶香蒲）；Dwarf Reedmace，Little Cattail，Mini Cattail ■

401129　Typha orientalis C. Presl；香蒲（苞香蒲，草蒲黄，东方香蒲，东香蒲，毛蜡烛，蒙古香蒲，蒲棒，蒲包草，蒲草黄，蒲棰，蒲槌，蒲儿根，蒲笋，水蜡烛，水烛）；Broad-leaf Cumbungi，Oriental Cattail ■

401130　Typha orientalis C. Presl var. brunnea Skvortsov = Typha orientalis C. Presl ■

401131　Typha pallida Pobed.；球序香蒲；Ballspike Cattail ■

401132　Typha przewalskii Skvortsov；普香蒲；Przewalsk Cattail ■

401133　Typha schimperi Rohrb. = Typha domingensis（Pers.）Steud. ■

401134　Typha shuttleworthii Koch et Sond.；毛蜡烛 ■☆

401135　Typha shuttleworthii Koch et Sond. = Typha orientalis C. Presl ■

401136　Typha shuttleworthii Koch et Sond. subsp. orientalis（C. Presl）Graebn. = Typha orientalis C. Presl ■

401137　Typha shuttleworthii Koch et Sond. var. orientalis Rohrb. = Typha orientalis C. Presl ■

401138　Typha stenophylla Fisch. et C. A. Mey. = Typha laxmannii Lepech. ■

401139　Typha veresczagini Krylov et Schischk.；韦氏香蒲 ■☆

401140　Typhaceae Juss.（1789）（保留科名）；香蒲科；Cattail Family，Cat-tail Family ■

401141　Typhalea（DC.）C. Presl = Pavonia Cav.（保留属名）●■☆

401142　Typhalea Neck. = Pavonia Cav.（保留属名）●■☆

401143　Typhodes Post et Kuntze = Phalaris L. ■

401144　Typhodes Post et Kuntze = Typhoides Moench ■

401145　Typhoides Moench = Phalaris L. ■

401146　Typhoides arundinacea（L.）Moench = Phalaris arundinacea L. ■

401147　Typhonium Schott ex Endl. = Typhonium Schott ■

401148　Typhonium Schott（1829）；犁头尖属（独角莲属，独脚莲属，犁头草属，土半夏属）；Ploughpoint，Typhonium ■

401149　Typhonium albidinervum C. Z. Tang et H. Li；白脉犁头尖；Whitevein Ploughpoint，Whitevein Typhonium ■

401150　Typhonium alpinum C. Y. Wu ex H. Li，Y. Shiao et S. L. Tseng；高山犁头尖（贝母）；Alpine Ploughpoint，Alpine Typhonium ■

401151　Typhonium anstrotibeticum H. Li；藏南犁头尖；S. Xizang Ploughpoint，S. Xizang Typhonium ■

401152　Typhonium blumei Nicolson et Sivad.；土半夏（半夏，犁头尖，青半夏，生半夏，瓮半夏，瓮菜黄）■

401153　Typhonium brevipes Hook. f. = Sauromatum brevipes（Hook. f.）N. E. Br. ■

401154　Typhonium calcicola C. Y. Wu ex H. Li，Y. Shiao et S. L. Tseng；单籽犁头尖（岩生犁头尖）；Rocky Typhonium，Singleseed Ploughpoint ■

401155　Typhonium cuspidatum（Blume）Decne. = Typhonium flagelliforme（Lodd.）Blume ■

401156　Typhonium divaricatum（L.）Decne.；犁头尖（白附子，百步还原，半夏，充半夏，茨菇七，打麻刺，大叶半夏，地金莲，独角莲，狗半夏，耗子尾巴，金半夏，老鼠尾，犁半夏，犁头草，犁头七，坡芋，青半夏，三步镖，三角青，三角蛇，山半夏，山茨菇，生半夏，鼠尾巴，田间半夏，土巴豆，土半夏，瓮半夏，瓮菜黄，小独角莲，小独脚莲，小野芋，药狗丹，野慈姑，野附子，芋头草，芋头七，芋叶半夏）；Divaricate Typhonium，Ploughpoint ■

401157　Typhonium divaricatum（L.）Decne. = Typhonium blumei Nicolson et Sivad. ■

401158　Typhonium diversifolium Wall.；高原犁头尖；Highland Ploughpoint，Highland Typhonium ■

401159　Typhonium diversifolium Wall. var. huegelianum（Schott）Engl. = Typhonium diversifolium Wall. ■

401160　Typhonium flagelliforme（Lodd.）Blume；鞭檐犁头尖（白苞犁头尖，半夏，疯狗薯，戟叶犁头尖，水半夏，田三七，土半夏）；Flageliforme Typhonium，Whipformed Ploughpoint ■

401161　Typhonium giganteum Engl.；独角莲（白附子，白禹，滴水参，疔毒豆，独脚莲，副本一粒红，红南星，鸡心白附，剪刀草，雷振子，犁头尖，麻芋子，麦夫子，牛奶白附，天南星，野半夏，野慈姑，野芋，禹白附，玉如意，芋叶半夏）；Giant Typhonium，Gigantic Typhonium，Onehorn Ploughpoint ■

401162　Typhonium giganteum Engl. = Arisaema fargesii Buchet ■

401163　Typhonium giganteum Engl. var. giraldii Baroni = Typhonium giganteum Engl. ■

401164　Typhonium giraldii（Baroni）Engl. = Typhonium giganteum Engl. ■

401165　Typhonium hongyanense Z. Y. Zhu；洪雅犁头尖（白休籽）；Hongya Ploughpoint，Hongya Typhonium ■

401166　Typhonium huegelianum Schott = Typhonium diversifolium Wall. ■

401167　Typhonium hunanense H. Li et Z. Q. Liu；湖南犁头尖（地花生，地金莲，夏无影）；Hunan Ploughpoint，Hunan Typhonium ■

401168　Typhonium kungmingense H. Li；昆明犁头尖；Kunming Ploughpoint，Kunming Typhonium ■

401169　Typhonium mangkamgense H. Li；芒康犁头尖；Mangkang Ploughpoint，Mangkang Typhonium ■

401170　Typhonium omeiense H. Li；西南犁头尖（半夏，红南星，鸡包谷，野水芋）；Emei Ploughpoint，Emei Typhonium ■

401171　Typhonium orixense Schott = Typhonium trilobatum（L.）Schott ■

401172　Typhonium roxburghii Schott；金慈姑（金慈菇）；Golden Ploughpoint，Three-lobed Arum，Typhonium ■

401173　Typhonium siamense Engl. = Typhonium trilobatum（L.）Schott ■

401174　Typhonium stoliczkae Engl. = Typhonium giganteum Engl. ■

401175　Typhonium trifoliatum F. T. Wang et H. S. Lo ex H. Li，Y. Shiao et S. L. Tseng；三叶犁头尖（代半夏，范半夏，花半夏，三裂叶犁头尖）；Threeleaf Ploughpoint，Trilobe Typhonium ■

401176　Typhonium trilobatum（L.）Schott；马蹄犁头尖（裂叶犁头尖，马蹄跌打，三裂犁头尖，三裂叶犁头尖，山半夏，小黑牛，野半夏）；Hoof Ploughpoint，Hoof Typhonium ■

401177　Typhonium venosum（Dryand. ex Aiton）Hett. et P. C. Boyce；多脉犁头尖 ■☆

401178　Typhonium venosum（Dryand. ex Aiton）Hett. et P. C. Boyce = Sauromatum venosum（Dryand. ex Aiton）Kunth ■

401179　Typhonodorum Schott（1857）；旋囊南星属 ■☆

401180　Typhonodorum lindleyanum Schott；旋囊南星 ■☆

401181 Tyrbastes B. G. Briggs et L. A. S. Johnson(1998);寡颖帚灯草属■☆

401182 Tyrbastes glaucescens B. G. Briggs et L. A. S. Johnson;寡颖帚灯草■☆

401183 Tyria Klotzsch = Macleania Hook. ●☆

401184 Tyria Klotzsch ex Endl. = Adelia L. (保留属名)●☆

401185 Tyria Klotzsch ex Endl. = Bernardia Mill. (废弃属名)●☆

401186 Tyrimnus (Cass.) Cass. (1826);髭苣蓟属■☆

401187 Tyrimnus Cass. = Tyrimnus (Cass.) Cass. ■☆

401188 Tyrimnus leucographus (L.) Cass. ;髭苣蓟■☆

401189 Tyrimnus leucographus Cass. = Tyrimnus leucographus (L.) Cass. ■☆

401190 Tysonia Bolus = Afrotysonia Rauschert ●☆

401191 Tysonia F. Muell. = Neotysonia Dalla Torre et Harms ■☆

401192 Tysonia F. Muell. = Swinburnia Ewart ■☆

401193 Tysonia africana Bolus = Afrotysonia africana (Bolus) Rauschert ●☆

401194 Tysonia glochidiata R. R. Mill = Afrotysonia glochidiata (R. R. Mill) R. R. Mill ●☆

401195 Tyssacia Steud. = Chrysothemis Decne. ■☆

401196 Tyssacia Steud. = Tussacia Rchb. ■☆

401197 Tytonia G. Don = Hydrocera Blume ex Wight et Arn.(保留属名)■

401198 Tytonia natans (Willd.) G. Don = Hydrocera triflora (L.) Wight et Arn. ■

401199 Tytonia triflora (L.) C. E. Wood = Hydrocera triflora (L.) Wight et Arn. ■

401200 Tytthostemma Nevski = Stellaria L. ■

401201 Tzellemtinia Chiov. = Bridelia Willd.(保留属名)●

401202 Tzellemtinia nervosa Chiov. = Bridelia scleroneura Müll. Arg. ●☆

401203 Tzeltalia E. Estrada et M. Martínez(1998);采尔茄属●☆

401204 Tzeltalia amphitricha (Bitter) E. Estrada et M. Martínez;周毛采尔茄●☆

401205 Tzeltalia calidaria (Standl. et Steyerm.) E. Estrada et M. Martínez;采尔茄●☆

401206 Tzvelevia E. B. Alexeev = Festuca L. ■

401207 Tzvelevopyrethrum Kamelin = Chrysanthemum L.(保留属名)■

401208 Tzvelevopyrethrum Kamelin(1993);土耳其蒿属■☆

401209 Tzvelevopyrethrum walteri (C. Winkl.) Kamelin;土耳其蒿■☆

401210 Uapaca Baill. (1858);瓦帕大戟属;Uapaca ■☆

401211 Uapaca acuminata (Hutch.) Pax et K. Hoffm. ;渐尖瓦帕大戟■☆

401212 Uapaca albida De Wild. = Uapaca kirkiana Müll. Arg. ■☆

401213 Uapaca angolensis Hutch. ex Pax et K. Hoffm. = Uapaca kirkiana Müll. Arg. ■☆

401214 Uapaca angustipyrena De Wild. ;细核瓦帕大戟■☆

401215 Uapaca benguelensis Müll. Arg. = Uapaca gossweileri Hutch. ■☆

401216 Uapaca benguelensis Müll. Arg. f. glabra P. A. Duvign. = Uapaca kirkiana Müll. Arg. ■☆

401217 Uapaca benguelensis Müll. Arg. f. pilosa P. A. Duvign. = Uapaca kirkiana Müll. Arg. ■☆

401218 Uapaca benguelensis Müll. Arg. var. pedunculata P. A. Duvign. = Uapaca kirkiana Müll. Arg. ■☆

401219 Uapaca bingervillensis Beille = Uapaca guineensis Müll. Arg. ■☆

401220 Uapaca brevipedunculata De Wild. ;短梗瓦帕大戟■☆

401221 Uapaca brieyii De Wild. ;布里瓦帕大戟■☆

401222 Uapaca chevalieri Beille;舍氏瓦帕大戟■☆

401223 Uapaca corbisieri De Wild. ;科比西尔瓦帕大戟■☆

401224 Uapaca dubia De Wild. = Uapaca kirkiana Müll. Arg. ■☆

401225 Uapaca ealaensis De Wild. ;埃阿拉大戟■☆

401226 Uapaca esculenta A. Chev. ex Aubrév. et Léandri;食用瓦帕大戟■☆

401227 Uapaca goetzei Pax = Uapaca kirkiana Müll. Arg. ■☆

401228 Uapaca goossensii De Wild. ;古森斯瓦帕大戟■☆

401229 Uapaca gossweileri Hutch. ;戈斯瓦帕大戟■☆

401230 Uapaca greenwayi Suess. = Uapaca kirkiana Müll. Arg. ■☆

401231 Uapaca guignardii A. Chev. ex Beille = Uapaca togoensis Pax ■☆

401232 Uapaca guignardii A. Chev. ex Beille var. sudanica Beille = Uapaca guineensis Müll. Arg. ■☆

401233 Uapaca guineensis Müll. Arg. ;几内亚瓦帕大戟;Red Cedar, Sugar Plum ■☆

401234 Uapaca heudelotii Baill. ;厄德瓦帕大戟■☆

401235 Uapaca heudelotii Baill. var. acuminata Hutch. = Uapaca acuminata (Hutch.) Pax et K. Hoffm. ■☆

401236 Uapaca homblei De Wild. = Uapaca kirkiana Müll. Arg. ■☆

401237 Uapaca katentaniensis De Wild. ;卡滕塔尼亚大戟■☆

401238 Uapaca kirkiana Müll. Arg. ;柯克瓦帕大戟;Wild Loquat ■☆

401239 Uapaca kirkiana Müll. Arg. var. dubia (De Wild.) P. A. Duvign. 可疑柯克瓦帕大戟■☆

401240 Uapaca kirkiana Müll. Arg. var. goetzei (Pax) Pax;格兹瓦帕大戟■☆

401241 Uapaca kirkiana Müll. Arg. var. kwangoensis P. A. Duvign. = Uapaca kirkiana Müll. Arg. ■☆

401242 Uapaca kirkiana Müll. Arg. var. sessilifolia P. A. Duvign. ;无柄柯克大戟■☆

401243 Uapaca lebruni De Wild. ;勒布伦瓦帕大戟■☆

401244 Uapaca letestuana A. Chev. ;勒泰斯蒂瓦帕大戟■☆

401245 Uapaca lissopyrena Radcl. -Sm. ;光果瓦帕大戟■☆

401246 Uapaca macrocephala Pax et K. Hoffm. = Uapaca pilosa Hutch. f. petiolata P. A. Duvign. ■☆

401247 Uapaca macrostipulata De Wild. ;大托叶瓦帕大戟■☆

401248 Uapaca marquesii Pax = Uapaca heudelotii Baill. ■☆

401249 Uapaca masuku De Wild. = Uapaca pilosa Hutch. ■☆

401250 Uapaca microphylla Pax = Uapaca nitida Müll. Arg. ■☆

401251 Uapaca microphylla Pax var. hendrickxii De Wild. = Uapaca nitida Müll. Arg. ■☆

401252 Uapaca mole Pax = Uapaca guineensis Müll. Arg. ■☆

401253 Uapaca multinervata De Wild. ;多脉瓦帕大戟■☆

401254 Uapaca munamensis De Wild. = Uapaca kirkiana Müll. Arg. var. dubia (De Wild.) P. A. Duvign. ■☆

401255 Uapaca neomasuku De Wild. = Uapaca kirkiana Müll. Arg. ■☆

401256 Uapaca nitida Müll. Arg. ;光亮瓦帕大戟■☆

401257 Uapaca nitida Müll. Arg. f. acuta P. A. Duvign. = Uapaca nitida Müll. Arg. ■☆

401258 Uapaca nitida Müll. Arg. f. bianoensis P. A. Duvign. = Uapaca nitida Müll. Arg. ■☆

401259 Uapaca nitida Müll. Arg. f. latiuscula P. A. Duvign. = Uapaca nitida Müll. Arg. ■☆

401260 Uapaca nitida Müll. Arg. f. longifolia P. A. Duvign. = Uapaca nitida Müll. Arg. var. longifolia (P. A. Duvign.) Radcl. -Sm. ■☆

401261 Uapaca nitida Müll. Arg. f. scalarinervosa P. A. Duvign. = Uapaca nitida Müll. Arg. ■☆

401262 Uapaca nitida Müll. Arg. f. sokolobe P. A. Duvign. = Uapaca nitida Müll. Arg. ■☆

401263 Uapaca nitida Müll. Arg. var. longifolia (P. A. Duvign.) Radcl. -Sm. ;长叶瓦帕大戟■☆

401264 Uapaca nitida Müll. Arg. var. mulengo P. A. Duvign. = Uapaca nitida Müll. Arg. ■☆

401265 Uapaca nitida Müll. Arg. var. nsambi P. A. Duvign. = Uapaca nitida Müll. Arg. ■☆

401266 Uapaca nitida Müll. Arg. var. rufopilosa De Wild. = Uapaca rufopilosa (De Wild.) P. A. Duvign. ■☆

401267 Uapaca nitida Müll. Arg. var. suffrutescens P. A. Duvign. = Uapaca nitida Müll. Arg. ■☆

401268 Uapaca nymphaeantha Pax et K. Hoffm.;睡莲大戟■☆

401269 Uapaca paludosa Aubrév. et Léandri;沼泽瓦帕大戟■☆

401270 Uapaca perrotii Beille = Uapaca guineensis Müll. Arg. ■☆

401271 Uapaca pilosa Hutch.;疏毛瓦帕大戟■☆

401272 Uapaca pilosa Hutch. f. hirsuta P. A. Duvign. = Uapaca pilosa Hutch. f. petiolata P. A. Duvign. ■☆

401273 Uapaca pilosa Hutch. f. petiolata P. A. Duvign.;柄叶疏毛瓦帕大戟■☆

401274 Uapaca pilosa Hutch. f. subglabra P. A. Duvign. = Uapaca pilosa Hutch. f. petiolata P. A. Duvign. ■☆

401275 Uapaca pynaertii De Wild.;皮那瓦帕大戟■☆

401276 Uapaca robynsii De Wild.;罗宾斯瓦帕大戟■☆

401277 Uapaca rufopilosa (De Wild.) P. A. Duvign.;红毛瓦帕大戟■☆

401278 Uapaca sansibarica Pax;桑给巴尔瓦帕大戟■☆

401279 Uapaca similis Pax et K. Hoffm. = Uapaca nitida Müll. Arg. ■☆

401280 Uapaca somon Aubrév. et Léandri = Uapaca togoensis Pax ■☆

401281 Uapaca staudtii Pax;施陶瓦帕大戟■☆

401282 Uapaca stipularis Pax et K. Hoffm.;托叶瓦帕大戟■☆

401283 Uapaca teusczii Pax = Uapaca kirkiana Müll. Arg. ■☆

401284 Uapaca togoensis Pax;多哥瓦帕大戟■☆

401285 Uapaca vanderystii De Wild.;范德瓦帕大戟■☆

401286 Uapaca vanhouttei De Wild.;瓦努特瓦帕大戟;Vanhout Spirea ■☆

401287 Uapaca verruculosa De Wild.;小疣瓦帕大戟■☆

401288 Uapacaceae (Müll. Arg.) Airy Shaw = Euphorbiaceae Juss. (保留科名)●■

401289 Uapacaceae (Müll. Arg.) Airy Shaw = Phyllanthaceae J. Agardh ●■

401290 Uapacaceae Airy Shaw = Euphorbiaceae Juss. (保留科名)●■

401291 Uapacaceae Airy Shaw = Phyllanthaceae J. Agardh ●■

401292 Ubiaea J. Gay = Landtia Less. ■☆

401293 Ubiaea J. Gay ex A. Rich. = Landtia Less. ■☆

401294 Ubidium Raf. = Ubium J. F. Gmel. ■

401295 Ubium J. F. Gmel. = Dioscorea L. (保留属名)■

401296 Ubochea Baill. (1891);好望角马鞭草属●☆

401297 Ubochea Baill. = Stachytarpheta Vahl(保留属名)■●

401298 Ubochea dichotoma Baill.;好望角马鞭草■☆

401299 Ubochea dichotoma Baill. = Stachytarpheta fallax A. E. Gonc. ●☆

401300 Ucacea Cass. = Blainvillea Cass. ■●

401301 Ucacea Cass. = Synedrella Gaertn. (保留属名)■

401302 Ucacea Cass. = Ucacou Adans. (废弃属名)■

401303 Ucacou Adans. (废弃属名) = Synedrella Gaertn. (保留属名)■

401304 Uchi Post et Kuntze = Sacoglottis Mart. ●☆

401305 Ucnopsolen A. W. Hill = Ucnopsolon Raf. ■

401306 Ucnopsolen Raf. = Lindernia All. ■

401307 Ucnopsolon Raf. = Lindernia All. ■

401308 Ucria Pfeiff. = Ambrosinia L. ■☆

401309 Ucria Targ. ex Pfeiff. = Ambrosinia L. ■☆

401310 Ucriana Spreng. = Augusta Pohl + Tocoyena Aubl. ●☆

401311 Ucriana Willd. = Tocoyena Aubl. ●☆

401312 Ucriana racemosa Schumach. et Thonn. = Oxyanthus racemosus (Schumach. et Thonn.) Keay ■☆

401313 Udani Adans. = Quisqualis L. ●

401314 Udora Nutt. = Anacharis Rich. ■☆

401315 Udora Nutt. = Elodea Michx. ■☆

401316 Udora cordofana Hochst. = Lagarosiphon cordofana Casp. ■☆

401317 Udora verticillata (L. f.) Spreng. var. minor Engelm. ex Casp. = Elodea nuttallii (Planch.) H. St. John ■☆

401318 Udoza Raf. = Udora Nutt. ■☆

401319 Udrastina Raf. = Laportea Gaudich. (保留属名)●■

401320 Uebelinia Hochst. (1841);林仙翁属■☆

401321 Uebelinia abyssinica Hochst.;阿比西尼亚林仙翁■☆

401322 Uebelinia crassifolia T. C. E. Fr.;厚叶林仙翁■☆

401323 Uebelinia hispida Pax = Uebelinia abyssinica Hochst. ■☆

401324 Uebelinia kigesiensis R. D. Good;基盖西林仙翁■☆

401325 Uebelinia kiwuensis T. C. E. Fr.;基乌林仙翁■☆

401326 Uebelinia kiwuensis T. C. E. Fr. subsp. erlangeriana (Engl.) Ousted;厄兰格林仙翁■☆

401327 Uebelinia nigerica Turrill = Uebelinia abyssinica Hochst. ■☆

401328 Uebelinia rotundifolia Oliv.;圆叶林仙翁■☆

401329 Uebelinia rotundifolia Oliv. var. erlangeriana Engl. = Uebelinia kiwuensis T. C. E. Fr. subsp. erlangeriana (Engl.) Ousted ■☆

401330 Uebelinia scottii Turrill;司科特林仙翁■☆

401331 Uebelinia spathulifolia Hochst. ex T. C. E. Fr. = Uebelinia abyssinica Hochst. ■☆

401332 Uebelmannia Buining (1967);尤伯球属(假般若属);Uebelmannia ●☆

401333 Uebelmannia buiningii Donald;布氏尤伯球■☆

401334 Uebelmannia gummifera (Backeb. et Voll) Buining;产胶尤伯球(橡胶尤伯球)■☆

401335 Uebelmannia meninensis Buining;拟小尤伯球■☆

401336 Uebelmannia pectinifera Buining;尤伯球■☆

401337 Uebelmannia pectinifera Buining var. pseudopectinifera Buining;假栉齿尤伯球■☆

401338 Uechtritzia Freyn(1892);粉丁草属■☆

401339 Uechtritzia armena Freyn;粉丁草■☆

401340 Uechtritzia kokanica (Regel et Schmalh.) Pobed.;浩罕粉丁草■☆

401341 Uffenbachia Fabr. = Uvularia L. ■☆

401342 Uffenbachia Heist. ex Fabr. = Uvularia L. ■☆

401343 Ugamia Pavlov(1950);垂甘菊属●☆

401344 Ugamia angrenica (Krasch.) Tzvelev;垂甘菊■☆

401345 Ugni Turcz. (1848);异香桃木属(小红果属)●☆

401346 Ugni molinae Turcz.;智利异香桃木(莫里异香桃木,小红果);Chilean Cranberry,Chilean Guava,Strawberry Myrtle ●☆

401347 Ugona Adans. = Hugonia L. ●☆

401348 Uhdea Kunth = Montanoa Cerv. ●■☆

401349 Uhdea bipinnatifida Kunth = Montanoa bipinnatifida (Kunth) K. Koch ●☆

401350 Uitenia Noronha = Erioglossum Blume ●

401351 Uittienia Steenis = Dialium L. ●☆

401352 Uladendron Marc. -Berti(1971);尤拉木属●☆

401353 Uladendron codesuri Marc. -Berti;尤拉木●☆

401354 Ulantha Hook. = Chloraea Lindl. ■☆

401355 Ulanthia Raf. = Ulantha Hook. ■☆

401356 Ulbrichia Urb. = Thespesia Sol. ex Corrêa(保留属名)●

401357 Uldinia J. M. Black(1922);澳中草属■☆

401358 Uldinia ceratocarpa (W. Fitzg.) N. T. Burb.;角果澳中草■☆

401359 Uldinia mercurialis J. M. Black;澳中草■☆

401360 Ulea C. B. Clarke ex H. Pfeiff. = Exochogyne C. B. Clarke ■☆

401361　Ulea-flos A. W. Hill = Exochogyne C. B. Clarke ■☆

401362　Ulea-flos C. B. Clarke ex H. Pfeiff. = Lagenocarpus Nees ■☆

401363　Uleanthus Harms(1905);荆花豆属■☆

401364　Uleanthus erythrinoides Harms;荆花豆■☆

401365　Ulearum Engl. (1905);荆南星属■☆

401366　Ulearum sagittatum Engl.;荆南星■☆

401367　Uleiorchis Hoehne(1944);荆兰属■☆

401368　Uleiorehis ulaei (Cogn.) Handro;荆兰■☆

401369　Uleodendron Rauschert = Naucleopsis Miq. ●☆

401370　Uleophytum Hieron. (1907);腋序亮泽兰属■☆

401371　Uleophytum scandens Hieron. ;腋序亮泽兰■☆

401372　Uleopsis Fedde = Chamaeanthus Schltr. ex J. J. Sm. ■

401373　Ulex L. (1753);荆豆属;Furze,Gorse,Whin ●

401374　Ulex africanus Webb = Ulex parviflorus Pourr. subsp. africanus (Webb) Greuter ●☆

401375　Ulex africanus Webb var. delestrei = Ulex parviflorus Pourr. subsp. africanus (Webb) Greuter ●☆

401376　Ulex africanus Webb var. discolor Maire et Sennen = Ulex parviflorus Pourr. subsp. africanus (Webb) Greuter ●☆

401377　Ulex baeticus Boiss. ;伯蒂卡荆豆●☆

401378　Ulex baeticus Boiss. subsp. scaber (Kunze) Cubas;粗糙荆豆●☆

401379　Ulex boivinii Webb = Stauracanthus boivinii (Webb) Samp. ●☆

401380　Ulex boivinii Webb var. cossonii (Webb) Maire = Stauracanthus boivinii (Webb) Samp. ●☆

401381　Ulex boivinii Webb var. megalorites (Webb) Ball = Stauracanthus boivinii (Webb) Samp. ●☆

401382　Ulex boivinii Webb var. narcissii Sennen et Mauricio = Ulex boivinii Webb ●☆

401383　Ulex boivinii Webb var. pilosulus (Sennen) Maire = Stauracanthus boivinii (Webb) Samp. ●☆

401384　Ulex boivinii Webb var. salzmannii (Webb) Ball = Stauracanthus boivinii (Webb) Samp. ●☆

401385　Ulex boivinii Webb var. tazensis (Braun-Blanq. et Maire) Maire = Stauracanthus boivinii (Webb) Samp. ●☆

401386　Ulex boivinii Webb var. webbianus (Coss.) Maire = Stauracanthus boivinii (Webb) Samp. ●☆

401387　Ulex congestus (Webb) Pau = Ulex baeticus Boiss. subsp. scaber (Kunze) Cubas ●☆

401388　Ulex europaeus L. ; 荆豆; Bunch of Keys, Bunch-of-keys, Common Gorse, Crannock, Creggans, Eithin, European Gorse, Fingers-and-thumbs, Fray, French Fuzz, Frez, Friz, Froz, Furra, Furze, Furzen, Fuzz, Fuzzen, Fyrrys, Golden Gorse, Gorse, Gorst, Goss, Gost, Great Furze, Hauth, Hawth, Honeybottle, Hoth, King's Bush, Ling, Needles-and-pins, Pins-and-needles, Prickly Broom, Prickly Ghost, Rocky Mountain Whortleberry Whun, Ruffet, Thorn Broom, Thumbs-and-fingers, Turr, Vuz, Vuzz, Vuzzen, Whin ●

401389　Ulex europaeus L. 'Flore Pleno';重瓣荆豆(重瓣花荆豆); Double-flowered Gorse ●☆

401390　Ulex gallii Planch. ;加利荆豆(高尔荆豆); Bed Furze, Bed-furze, Cat Whin, Cornish Fuzz, Dwarf Furze, Dwarf Gorse, Tam Fuzz, Tame, Western Gorse ●☆

401391　Ulex genistoides Brot. subsp. spectabilis (Webb) Cout. = Stauracanthus spectabilis Webb ●☆

401392　Ulex mauritii Sennen = Ulex parviflorus Pourr. subsp. africanus (Webb) Greuter ●☆

401393　Ulex maximilianii Sennen et Mauricio = Ulex parviflorus Pourr. ●☆

401394　Ulex microclada Sennen = Ulex parviflorus Pourr. subsp. africanus (Webb) Greuter ●☆

401395　Ulex minor Roth; 小荆豆; Bed Furze, Bed-furze, Cat Whin, Dwarf Autumnal Furze, Dwarf Furze, Dwarf Gorse, Genista Anglica, Lesser Furze, Lesser Gorse, Manx Gorse, Petty Whin, Petty Whir, Small Furze, Tam Furze, Taxi Fuzz, Whin ●☆

401396　Ulex nanus Lees;矮荆豆;Dwarf Gorse ●☆

401397　Ulex narcissi Sennen = Stauracanthus boivinii (Webb) Samp. ●☆

401398　Ulex parviflorus Pourr. ;小花荆豆●☆

401399　Ulex parviflorus Pourr. subsp. africanus (Webb) Greuter;非洲小花荆豆●☆

401400　Ulex parviflorus Pourr. subsp. funkii (Webb) Guinea = Ulex parviflorus Pourr. subsp. africanus (Webb) Greuter ●☆

401401　Ulex parviflorus Pourr. var. funkii (Webb) Rothm. = Ulex parviflorus Pourr. ●☆

401402　Ulex scaber Kuntze var. congestus Webb = Ulex baeticus Boiss. ●☆

401403　Ulex spectabilis (Webb) Willk. = Stauracanthus spectabilis Webb ●☆

401404　Ulex tazensis (Braun-Blanq. et Maire) Pau et Font Quer = Stauracanthus boivinii (Webb) Samp. ●☆

401405　Ulex webbianus Coss. = Stauracanthus boivinii (Webb) Samp. ●☆

401406　Ulex webbianus Coss. var. tazensis Braun-Blanq. et Maire = Stauracanthus boivinii (Webb) Samp. ●☆

401407　Ulina Opiz = Inula L. ●■

401408　Ulleria Bremek. = Ruellia L. ■●

401409　Ulloa Pers. = Juanulloa Ruiz et Pav. ●☆

401410　Ullucaceae Nakai = Basellaceae Raf. (保留科名)■

401411　Ullucus Caldas(1809);块根落葵属(皱果薯属);Ullucus ■☆

401412　Ullucus tuberosus Caldas;块根落葵(安第斯薯);Oca Quina, Oca Quirts, Olluco, Tuber Ullucu, Tuber Ullucus, Ulluco, Ullucu ☆

401413　Ullucus tuberosus Caldas f. albiflorus Kuntze;白花块根落葵;Whiteflower Tuber Ullucus ■☆

401414　Ullucus tuberosus Caldas f. rubriflorus Kuntze;红花块根落葵;Redflower Tuber Ullucus ■☆

401415　Ulmaceae Mirb. (1815)(保留科名);榆科;Elm Family ●

401416　Ulmaria (Tourn.) Hill. = Filipendula Mill. ■

401417　Ulmaria Hill. = Spiraea L. ●

401418　Ulmaria Mill. = Filipendula Mill. ■

401419　Ulmaria aruncus (L.) Hill = Aruncus sylvester Kostel. ex Maxim. ■

401420　Ulmaria aruncus Hill = Aruncus sylvester Kostel. ex Maxim. ■

401421　Ulmaria rubra Hill = Filipendula rubra (Hill) B. L. Rob. ■☆

401422　Ulmariaceae Gray = Rosaceae Juss. (保留科名)●■

401423　Ulmarronia Friesen = Cordia L. (保留属名)●

401424　Ulmarronia Friesen = Varronia P. Browne ●☆

401425　Ulmus L. (1753);榆属;Elm ●

401426　Ulmus 'Dicksonii' = Ulmus carpinifolia Gled. 'Dicksonii'●☆

401427　Ulmus 'Wheatleyi Aurea' = Ulmus 'Dicksonii'●☆

401428　Ulmus 'Wheatleyi Aurea' = Ulmus carpinifolia Gled. 'Dicksonii'●☆

401429　Ulmus 'Wheatleyi Aurea' = Ulmus 'Dicksonii'●☆

401430　Ulmus alata Michx.;翼枝长序榆(栓翅榆,翼枝行序榆);Cork Elm, Red Elm, Wahoo, Wahoo Elm, Winged Elm ●

401431　Ulmus americana L.;美洲榆(灰榆,美国白榆,美国榆,美榆,软榆,水榆);America Elm, American Elm, American White Elm, Gray Elm, Orhamwood, Soft Elm, Swamp Elm, Water Elm, White Elm ●

401432　Ulmus americana L. 'Ascendens';上升美洲榆●☆

401433　Ulmus americana L. 'Augustine';雄伟美洲榆●☆

401434　Ulmus americana L. 'Aurea';金叶美洲榆●☆

401435　Ulmus americana L. 'Columnaris';柱状美洲榆●☆

401436　Ulmus americana L. 'Delaware';特拉华美洲榆●☆

401437　Ulmus americana L. 'Incisa';深裂美洲榆●☆

401438　Ulmus americana L. 'Nigricans';黑色美洲榆●☆

401439　Ulmus americana L. f. alba (Aiton) Fernald = Ulmus americana L. ●

401440　Ulmus americana L. f. intercedens Fernald = Ulmus americana L. ●

401441　Ulmus americana L. f. laevior Fernald = Ulmus americana L. ●

401442　Ulmus americana L. f. pendula (Aiton) Fernald = Ulmus americana L. ●

401443　Ulmus americana L. var. alba Aiton = Ulmus americana L. ●

401444　Ulmus americana L. var. aspera Chapm. = Ulmus americana L. ●

401445　Ulmus americana L. var. floridana (Chapm.) Little = Ulmus americana L. ●

401446　Ulmus americana L. var. pendula Aiton = Ulmus americana L. ●

401447　Ulmus amoena W. C. Cheng;喜悦榆;Agreeable Elm ●

401448　Ulmus androssowii Litv.;安氏榆●☆

401449　Ulmus androssowii Litv. var. subhirsuta (C. K. Schneid.) P. H. Huang,F. Y. Gao et L. H. Zhuo;毛枝榆(大树皮,红榔木,榔榆,毛白榆,棉榔树,榆树);Hairtwig Elm,Hairyshoot Elm,Hairy-shooted Elm ●

401450　Ulmus androssowii Litv. var. virgata (Planch.) Grudz. = Ulmus androssowii Litv. var. subhirsuta (C. K. Schneid.) P. H. Huang,F. Y. Gao et L. H. Zhuo ●

401451　Ulmus androssowii Litv. var. virgata (Wall. ex Planch.) Grudz. = Ulmus androssowii Litv. var. subhirsuta (C. K. Schneid.) P. H. Huang,F. Y. Gao et L. H. Zhuo ●

401452　Ulmus angustifolia Moench = Ulmus minor Mill. subsp. angustifolia (Weston) Stace ●☆

401453　Ulmus angustifolia Moench var. cornubiensis ? = Ulmus cornubiensis K. Koch ●☆

401454　Ulmus bergmanniana C. K. Schneid.;兴山榆;Bergmann Elm ●

401455　Ulmus bergmanniana C. K. Schneid. var. lasiophylla C. K. Schneid.;蜀榆(西蜀榆,叶上珠,一点红);Hairyleaf Elm,Sichuan Elm ●

401456　Ulmus brandisiana C. K. Schneid. = Ulmus wallichiana Planch. ●☆

401457　Ulmus brandisiana Melville et Heybroek = Ulmus wallichiana Planch. ●☆

401458　Ulmus campestris H. Lév. = Ulmus procera Salisb. ●☆

401459　Ulmus campestris H. Lév. var. chinensis Loudon = Ulmus parvifolia Jacq. ●

401460　Ulmus campestris H. Lév. var. japonica Rehder = Ulmus davidiana Planch. var. japonica (Rehder) Nakai ●

401461　Ulmus campestris H. Lév. var. parvifolia f. pendula Kirchn. = Ulmus pumila L. 'Tenue' ●

401462　Ulmus campestris L. subsp. procera (Salisb.) Maire = Ulmus minor Mill. ●☆

401463　Ulmus campestris L. var. chinensis Loudon = Ulmus parvifolia Jacq. ●

401464　Ulmus campestris L. var. japonica Rehder = Ulmus davidiana Planch. var. japonica (Rehder) Nakai ●

401465　Ulmus campestris L. var. pumila (L.) Maxim. = Ulmus pumila L. ●

401466　Ulmus canescens Melville;灰毛榆;Hairy Elm ●☆

401467　Ulmus carpinifolia Borkh. = Ulmus minor Mill. ●☆

401468　Ulmus carpinifolia Gled.欧洲光叶榆(鹅耳枥叶榆,鹅耳枥榆,光叶榆);English Elm,Europe Smoothleaf Elm,Field Elm,Hornbeamleaf Elm,Smoothleaf Elm,Smooth-leaf Elm,Smooth-leaved Elm ●

401469　Ulmus carpinifolia Gled. 'Dicksonii';迪克森欧洲光叶榆;Cornish Golden Elm,Dickson's Golden Elm ●☆

401470　Ulmus carpinifolia Gled. 'Purpurea';紫叶欧洲光叶榆●☆

401471　Ulmus carpinifolia Gled. 'Sarniensis Aurea' = Ulmus carpinifolia

Gled. 'Dicksonii'●☆

401472　Ulmus carpinifolia Gled. 'Silvery Gem';银色宝石欧洲光叶榆●☆

401473　Ulmus carpinifolia Gled. 'Variegata';斑叶欧洲光叶榆●☆

401474　Ulmus carpinifolia Gled. var. cornubiensis ? = Ulmus cornubiensis K. Koch ●☆

401475　Ulmus carpinifolia Gled. var. suberosa ?;木栓质欧洲光叶榆;Cork Elm,Cork-barked Smooth-leaved Elm ●☆

401476　Ulmus castaneifolia Hemsl.;多脉榆(栗叶榆,铁锈榆,锈毛榆);Ferruginous Elm,Many-vein Elm,Multi-veined Elm,Veiny Elm ●

401477　Ulmus cavaleriei H. Lév. = Pteroceltis tartarinowii Maxim. ●

401478　Ulmus celtidea (Rogow.) Litv.;朴叶欧洲白榆●☆

401479　Ulmus changii W. C. Cheng;杭州榆(江南榆);Chang's Elm ●

401480　Ulmus changii W. C. Cheng var. kunmingensis (W. C. Cheng) W. C. Cheng et L. K. Fu;昆明榆;Kunming Elm ●

401481　Ulmus chemnoui W. C. Cheng;琅琊榆;Langya Elm,Langya Mountain Elm ●◇

401482　Ulmus chinensis Pers. = Ulmus parvifolia Jacq. ●

401483　Ulmus chumlia Melville et Heybroek = Ulmus androssowii Litv. var. subhirsuta (C. K. Schneid.) P. H. Huang,F. Y. Gao et L. H. Zhuo ●

401484　Ulmus coreana Nakai = Ulmus parvifolia Jacq. ●

401485　Ulmus cornubiensis K. Koch;科努比榆;Cornish Elm,Cornish Smooth-leaved Elm ●☆

401486　Ulmus crassifolia Nutt.;厚叶榆;Basket Elm,Cedar Elm,Southern Rock Elm ●

401487　Ulmus crispa Willd. = Ulmus rubra Muhl. ●☆

401488　Ulmus davidiana L. var. levigata (C. K. Schneid.) Nakai = Ulmus davidiana Planch. var. japonica (Rehder) Nakai ●

401489　Ulmus davidiana Planch.;黑榆(白榆,东北黑榆,枥榆,热河榆,山毛榆);Black Elm,David Elm ●

401490　Ulmus davidiana Planch. var. japonica (Rehder) Nakai;春榆(白皮榆,光叶春榆,红榆,蜡条榆,日本榆,沙榆,山榆,栓皮春榆);Japanese Elm,Spring Elm ●

401491　Ulmus davidiana Planch. var. japonica (Rehder) Nakai f. laevis Miyabe = Ulmus davidiana Planch. var. japonica (Rehder) Nakai f. levigata (C. K. Schneid.) W. T. Lee ●☆

401492　Ulmus davidiana Planch. var. japonica (Rehder) Nakai f. levigata (C. K. Schneid.) W. T. Lee;光滑春榆●☆

401493　Ulmus davidiana Planch. var. japonica (Rehder) Nakai f. suberosa (Turcz.) Nakai = Ulmus davidiana Planch. var. japonica (Rehder) Nakai ●

401494　Ulmus davidiana Planch. var. japonica (Rehder) Nakai f. suberosa Nakai = Ulmus davidiana Planch. var. japonica (Rehder) Nakai ●

401495　Ulmus davidiana Planch. var. levigata (C. K. Schneid.) Nakai = Ulmus davidiana Planch. var. japonica (Rehder) Nakai ●

401496　Ulmus davidiana Planch. var. mandshurica Skvortsov;东北黑榆;Manchrian Elm ●

401497　Ulmus davidiana Planch. var. mandshurica Skvortsov = Ulmus davidiana Planch. ●

401498　Ulmus davidiana Planch. var. pubescens Skvortsov = Ulmus davidiana Planch. ●

401499　Ulmus densa Litv.;圆冠榆;Round-crown Elm,Roundedcrown Elm ●

401500　Ulmus diversifolia Melville;东盎格鲁榆;East Anglian Elm,Lock Elm,Smali-leaved Elm ●☆

401501　Ulmus effusa Sibth.;铺散榆;Spreading Elm ●☆

401502　Ulmus elegantissima Horw.;雅致榆;Midland Elm ●☆

401503　Ulmus elliptica K. Koch;椭榆●☆

401504　Ulmus elongata L. K. Fu et C. S. Ding；长序榆（榔柏，牛皮筋，野榔皮，野榆，叶榔）；Longraceme Elm，Long-raceme Elm ●

401505　Ulmus erythrocarpa W. C. Cheng = Ulmus szechuanica W. P. Fang ●

401506　Ulmus ferruginea W. C. Cheng = Ulmus castaneifolia Hemsl. ●

401507　Ulmus floridana Chapm. = Ulmus americana L. ●

401508　Ulmus foliacea Gilib. ；叶榆；English Elm，Field Elm，Leafy Elm，Smooth-leaved Elm ●☆

401509　Ulmus fulva Michx. ；糙枝榆；Indian Elm，Moose Elm，Red Elm，Rock Elm，Slippery Elm，Sweet Elm ●☆

401510　Ulmus fulva Michx. = Ulmus rubra Muhl. ●☆

401511　Ulmus gaussenii W. C. Cheng；毛榆（醉翁榆）；Gaussen Elm，Hairy Elm ●◇

401512　Ulmus glabra Huds. ；山榆（糙榆，光榆，山地榆）；Bough Elm，Broad-lcaved Elm，Camperdown Elm，Chair Elm，Chewbark，Drunken Elm，Elm-wych，Halse，Helm，Holme，Horn Birch，Hornbeam，Mountain Elm，Ome-tree，Quicken-tree，Scotch Elm，Scots Elm，Switch Elm，Witan Elm，Witch，Witch Elm，Witch Halse，Witch Hazel，Witch Tree，Witchwood，Wych Elm，Wych Halse，Wych Hazel，Wych Tree，Wychwood ●

401513　Ulmus glabra Huds. 'Camperdownii'；龙爪山榆（龙爪光榆）；Camperdown Elm ●☆

401514　Ulmus glabra Huds. 'Exoniensis'；埃克赛特山榆（埃克赛特光榆）；Exeter Elm，Ford's Elm ●☆

401515　Ulmus glabra Huds. 'Horizontalis'；桌山榆；Tabletop Scotch Elm ●☆

401516　Ulmus glabra Huds. 'Pendula'；垂枝山榆；Weeping Elm，Weeping Wych-elm ●☆

401517　Ulmus glabra Huds. var. camperdownii ？ = Ulmus glabra Huds. 'Camperdownii' ●☆

401518　Ulmus glabra Huds. var. exoniensis ？ = Ulmus glabra Huds. 'Exoniensis' ●☆

401519　Ulmus glaucescens Franch. ；旱榆（粉榆，灰榆，山榆，崖榆）；Dryland Elm，Glaucescent Elm ●

401520　Ulmus glaucescens Franch. var. lasiocarpa Rehder；毛果旱榆；Hairyfruit Elm，Hairyfruit Glaucescent Elm ●

401521　Ulmus harbinensis S. Q. Nie et G. Q. Huang；哈尔滨榆；Harbin Elm ●

401522　Ulmus hollandica Mill. ；荷兰榆；Dutch Elm，Huntingdon Elm ●☆

401523　Ulmus hollandica Mill. 'Belgica'；比利时荷兰榆●☆

401524　Ulmus hollandica Mill. 'Groenveldt'；格鲁维尔特荷兰榆●☆

401525　Ulmus hollandica Mill. 'Hillier'；希里荷兰榆●☆

401526　Ulmus hollandica Mill. 'Jaqueline Hillier'；杰奎琳·希里荷兰榆（杰奎琳·希莱尔荷兰榆）●☆

401527　Ulmus hollandica Mill. 'Major'；壮美荷兰榆●☆

401528　Ulmus hollandica Mill. 'Modolina'；莫道里娜荷兰榆●☆

401529　Ulmus hollandica Mill. 'Vegeta'；瓦卡塔荷兰榆（亨廷顿榆）；Chichester Elm，Huntingdon Elm ●☆

401530　Ulmus hollandica Mill. var. vegeta ？ = Ulmus hollandica Mill. 'Vegeta' ●☆

401531　Ulmus japonica (Rehder) Sarg. = Ulmus davidiana Planch. var. japonica (Rehder) Nakai ●

401532　Ulmus japonica (Rehder) Sarg. f. suberosa (Turcz.) Kitag. = Ulmus davidiana Planch. var. japonica (Rehder) Nakai f. suberosa (Turcz.) Nakai ●

401533　Ulmus japonica (Rehder) Sarg. var. levigata C. K. Schneid. = Ulmus davidiana Planch. var. japonica (Rehder) Nakai f. levigata (C. K. Schneid.) W. T. Lee ●☆

401534　Ulmus japonica (Rehder) Sarg. var. levigata C. K. Schneid. = Ulmus davidiana Planch. var. japonica (Rehder) Nakai ●

401535　Ulmus japonica Sarg. = Ulmus davidiana Planch. var. japonica (Rehder) Nakai ●

401536　Ulmus japonica Siebold = Ulmus parvifolia Jacq. ●

401537　Ulmus keaki Siebold = Zelkova serrata (Thunb.) Makino ●

401538　Ulmus koopmanni Starcs；库普曼榆●☆

401539　Ulmus kunmingensis W. C. Cheng = Ulmus changii W. C. Cheng var. kunmingensis (W. C. Cheng) W. C. Cheng et L. K. Fu ●

401540　Ulmus kunmingensis W. C. Cheng var. qingchengshanensis T. P. Yi = Ulmus changii W. C. Cheng var. kunmingensis (W. C. Cheng) W. C. Cheng et L. K. Fu ●

401541　Ulmus kunmingensis W. C. Cheng var. qingchengshanensis T. P. Yi = Ulmus changii W. C. Cheng ●

401542　Ulmus laciniata (Trautv.) Mayr；裂叶榆（大青榆，大叶榆，尖尖榆，麻榆，青榆，粘榆）；Japanese Wych Elm，Laciniate Elm，Lobedleaf Elm，Manchurian Elm ●

401543　Ulmus laciniata (Trautv.) Mayr f. holophylla Nakai = Ulmus laciniata (Trautv.) Mayr ●

401544　Ulmus laciniata (Trautv.) Mayr var. laevigata Inokuma = Ulmus laciniata (Trautv.) Mayr ●

401545　Ulmus laciniata (Trautv.) Mayr var. nikkoensis Rehder；名古屋裂叶榆；Nagoya Elm，Nagoya Lobedleaf Elm ●☆

401546　Ulmus laciniata (Trautv.) Mayr var. nikkoensis Rehder f. holophylla Nakai = Ulmus laciniata (Trautv.) Mayr ●

401547　Ulmus laciniata (Trautv.) Mayr var. nikkoensis Rehder f. laevigata (Inokuma) = Ulmus laciniata (Trautv.) Mayr ●

401548　Ulmus laevigata Royle = Ulmus villosa Brandis ex Gamble ●☆

401549　Ulmus laevis Pall. ；欧洲白榆（大叶榆，欧洲大叶榆，新疆大叶榆）；Cork Elm，Europe White Elm，European White Elm，Fluttering Elm，Rock Elm，Russia Elm，Russian Elm，Spreading Elm，White Elm ●

401550　Ulmus lamellosa Z. Wang et S. L. Chang；脱皮榆（沙包榆，太行榆）；Deciduosbark Elm，Decidous-bark Elm，Molt Elm ●◇

401551　Ulmus lanceifolia Roxb. = Ulmus lanceifolia Roxb. ex Wall. ●

401552　Ulmus lanceifolia Roxb. ex Wall. ；常绿榆（常绿滇榆，大树皮，滇榆，火绳树，榔榆，榆树，越南榆）；Evergreen Elm，Tonkin Elm，Vietnam Elm ●

401553　Ulmus lasiophylla (C. K. Schneid.) W. C. Cheng；西蜀榆；Hairyleaf Elm，Sichuan Elm ●

401554　Ulmus lasiophylla (C. K. Schneid.) W. C. Cheng = Ulmus bergmanniana C. K. Schneid. var. lasiophylla C. K. Schneid. ●

401555　Ulmus macrocaphylla Nakai = Ulmus macrocarpa Hance ●

401556　Ulmus macrocarpa Hance；大果榆（矮形大果榆，矮形黄榆，白贵子，白芜荑，迸榆，扁榆，翅枝黄榆，臭芜荑，大芜荑，倒卵果黄榆，姑榆，广卵果黄榆，黄榆，柳榆，毛榆，蒙古大果榆，蒙古黄榆，山扁榆，山粉榆，山松榆，山榆，无姑，无夷，芜姑，芜荑，蓝荑，榆）；Bigfruit Elm，Big-fruited Elm，Low Bigfruit Elm，Mongolian Bigfruit Elm ●

401557　Ulmus macrocarpa Hance var. glabra S. Q. Nie et G. Q. Huang；光果黄榆（光秃大果榆，光秃大黄榆）；Glabrous Bigfruit Elm ●

401558　Ulmus macrocarpa Hance var. mandshurica Skvortsov = Ulmus macrocarpa Hance ●

401559　Ulmus macrocarpa Hance var. mongolica Liou et C. Y. Li = Ulmus macrocarpa Hance ●

401560　Ulmus macrocarpa Hance var. nana Liou et C. Y. Li = Ulmus macrocarpa Hance ●

401561　Ulmus macrocarpa Hance var. suberosa Skvortsov = Ulmus macrocarpa Hance ●

401562 Ulmus macrocarpa W. C. Cheng = Ulmus gaussenii W. C. Cheng ●◇

401563 Ulmus macrophylla Nakai;大叶榆●☆

401564 Ulmus macrophylla Nakai = Ulmus macrocarpa Hance ●

401565 Ulmus major Hohen. var. heterophylla Maxim. = Ulmus laciniata (Trautv.) Mayr ●

401566 Ulmus manshurica Nakai = Ulmus pumila L. ●

401567 Ulmus mexicana (Liebm.) Planch. ;墨西哥长序榆●☆

401568 Ulmus mianzhuensis T. P. Yi et L. Yang;绵竹榆●

401569 Ulmus microcarpa L. K. Fu;小果榆;Smallfruit Elm, Small-fruited Elm ●

401570 Ulmus microphylla Pers. = Ulmus pumila L. ●

401571 Ulmus minor Mill. ' Cornubiensis' = Ulmus minor Mill. subsp. angustifolia (Weston) Stace ●☆

401572 Ulmus minor Mill. = Ulmus carpinifolia Gled. ●

401573 Ulmus minor Mill. subsp. angustifolia (Weston) Stace;狭叶榆;Cornish Elm,Goodyer's Elm ●☆

401574 Ulmus minor Mill. var. lockii (Druce) Richens = Ulmus plotii Druce ●☆

401575 Ulmus minor Mill. var. sarniensis (Loudon) Richens;萨尼亚榆;Jersey Elm,Wheatley Elm ●☆

401576 Ulmus montana Stokes = Ulmus glabra Huds. ●

401577 Ulmus montana Stokes var. laciniata Trautv. = Ulmus laciniata (Trautv.) Mayr ●

401578 Ulmus montana With. var. laciniata Trautv. = Ulmus laciniata (Trautv.) Mayr ●

401579 Ulmus multinervis W. C. Cheng = Ulmus castaneifolia Hemsl. ●

401580 Ulmus nitens Moench = Ulmus carpinifolia Gled. ●

401581 Ulmus parvifolia Jacq. ;榔榆(豹皮榆,豺皮榆,大树皮,掉皮榆,枸丝榆,构树榆,红鸡油,榔,朗榆,檆,檆木,挠皮榆,牛筋树,排钱树,秋榆,松心木,田柳榆,田柳榆,脱皮榆,蚊子树,细叶榆,小叶榆,榆栈树);Chinese Elm,Lace Bark Elm,Langyu Elm ●☆

401582 Ulmus parvifolia Jacq. ' Catlin';卡特林榔榆●☆

401583 Ulmus parvifolia Jacq. ' Evergreen';常绿榔榆;Chinese Elm ●☆

401584 Ulmus parvifolia Jacq. ' Frosty';银白榔榆●☆

401585 Ulmus parvifolia Jacq. ' Hansen';翰森榔榆●☆

401586 Ulmus parvifolia Jacq. ' King's Choice';圣旨榔榆●☆

401587 Ulmus parvifolia Jacq. ' Pendens';垂枝榔榆●☆

401588 Ulmus parvifolia Jacq. ' True Green';翠绿榔榆(常绿榔榆)●☆

401589 Ulmus pedunculata Foug. = Ulmus laevis Pall. ●

401590 Ulmus pendula Willd. = Ulmus rubra Muhl. ●☆

401591 Ulmus pinnato-ramosa Dieck ex Koehne;羽枝榆●☆

401592 Ulmus plotii Druce;普氏榆;Plot's Elm ●☆

401593 Ulmus procera Salisb. ;英国榆(白榆,粉,家榆,榆);Allom-tree, Aum, Aum-tree, Carpathian Burl Elm, Common Elm, Elem, Ellem, Ellum, Elm, Elmin Elmen, Elven, Emmel Emmal, English Elm,European Elm, Field Elm, Holm, Horse May, May, Nave Elm, Ome-tree,Owm, Red Elm, Smooth-leaved Elm, Ullum, Warwickshire Weed,Wiltshire Weed ●☆

401594 Ulmus procera Salisb. ' Argenteovariegata';银斑英国榆●☆

401595 Ulmus procera Salisb. ' Louis van Houtte';路易斯·范·胡特英国榆●☆

401596 Ulmus procera Salisb. ' Purpurea';紫叶英国榆●☆

401597 Ulmus procera Salisb. ' Viminalis Aurea';金叶英国榆;Gold-leaved English Elm ●☆

401598 Ulmus propinqua Koidz. = Ulmus davidiana Planch. var. japonica (Rehder) Nakai ●

401599 Ulmus prunifolia W. C. Cheng et L. K. Fu;李叶榆;Plumleaf Elm ,Plum-leaved Elm ●

401600 Ulmus pseudopropinqua F. T. Wang et H. L. Li;假春榆;False Japanese Elm, False Spring Elm ●

401601 Ulmus pubescens Walter = Ulmus rubra Muhl. ●☆

401602 Ulmus pumila L. ;榆树(白粉,白榆,长叶家榆,粉,粉榆,红榔木,黄药家榆,家榆,零榆,毛白榆,棉榔树,钱榆,西伯利亚榆,榆,榆钱树,钻天榆);Dwarf Asiatic Elm, Dwarf Elm, Elm, Littleleaf Elm, Siberia Elm, Siberian Elm ●

401603 Ulmus pumila L. ' Leptodermis';细皮榆;Thin-bark Siberian Elm ●

401604 Ulmus pumila L. ' Parvifolia';小叶榆;Small-leaf Siberian Elm ●

401605 Ulmus pumila L. ' Pendula' = Ulmus pumila L. ' Tenue' ●

401606 Ulmus pumila L. ' Pseudopropinqua' = Ulmus pseudopropinqua F. T. Wang et H. L. Li ●

401607 Ulmus pumila L. ' Pyramidalis';钻天榆(河南钻天榆);Pyramidal Siberian Elm ●

401608 Ulmus pumila L. ' Tenue';垂枝榆(倒榆);Pendentbranch Siberian Elm ●

401609 Ulmus pumila L. ' Tortuosa';龙爪榆;Dragonclaw Elm ●

401610 Ulmus pumila L. f. pendula (G. Kirchn.) Rehder = Ulmus pumila L. ' Tortuosa' ●

401611 Ulmus pumila L. f. tenue S. Y. Wang = Ulmus pumila L. ' Tenue' ●

401612 Ulmus pumila L. var. genuina Skvortsov = Ulmus pumila L. ●

401613 Ulmus pumila L. var. gracila S. Y. Wang;细枝榆;Thin-branch Elm ●

401614 Ulmus pumila L. var. microphylla Pers. = Ulmus pumila L. ●

401615 Ulmus pumila L. var. pendula Rehder = Ulmus pumila L. ' Tortuosa' ●

401616 Ulmus pumila L. var. pilosa Rehder = Ulmus androssowii Litv. var. subhirsuta (C. K. Schneid.) P. H. Huang,F. Y. Gao et L. H. Zhuo ●

401617 Ulmus pumila L. var. sabulosa J. H. Guo,Y. S. Li et J. H. Li;锡盟沙地榆●

401618 Ulmus racemosa D. Thomas;栓皮榆(宽果长序榆,山胡桃榆,石榆,索马榆,硬榆);Cliff Elm, Cork Elm, Hard Elm, Hickory Elm, Rock Elm, Spreading Elm ●

401619 Ulmus racemosa D. Thomas = Ulmus thomasii Sarg. ●☆

401620 Ulmus rubra Muhl. ;红榆(灰榆,软榆,驼鹿榆);Gray Elm, Moose Elm, Red Elm, Slippery Elm, Soft Elm ●☆

401621 Ulmus rubra Muhl. = Ulmus fulva Michx. ●☆

401622 Ulmus rubrocarpa W. C. Cheng = Ulmus szechuanica W. P. Fang ●

401623 Ulmus sarniensis (C. K. Schneid.) Melville;萨尔尼安榆;Jersey Elm ●☆

401624 Ulmus sarniensis (C. K. Schneid.) Melville ' Aurea' = Ulmus ' Dicksonii' ●☆

401625 Ulmus scabra Mill. = Ulmus glabra Huds. ●

401626 Ulmus serotina Sarg. ;秋榆(美国红榆);Autumn Elm, Red Elm,September Elm ●

401627 Ulmus shirasawana Daveau = Ulmus parvifolia Jacq. ●

401628 Ulmus sibirica H. Lév. = Ulmus pumila L. ●

401629 Ulmus sieboldii Daveau = Ulmus parvifolia Jacq. ●

401630 Ulmus sieboldii Daveau f. shirasawana Nakai = Ulmus parvifolia Jacq. ●

401631 Ulmus stricta Lindl. ;康沃尔榆;Cornish Elm ●☆

401632 Ulmus stricta Lindl. var. sarniensis ?;格恩榆;Guernsey Elm, Jersey Elm,Wheatley Elm ●☆

401633 Ulmus suberosa Moench;栓榆;Cork Elm, Cork-bark Smooth-leaved Elm, Suberose Elm ●☆

401634　Ulmus szechuanica W. P. Fang；红果榆（明陵榆，蓉榆）；Redfruit Elm，Red-fruited Elm，Sichuan Elm ●

401635　Ulmus taihangshanensis S. Y. Wang；太行榆；Taihangshan Elm ●

401636　Ulmus taihangshanensis S. Y. Wang = Ulmus lamellosa Z. Wang et S. L. Chang ●◇

401637　Ulmus taihangshanensis S. Y. Wang = Ulmus macrocarpa Hance ●

401638　Ulmus thomasii Sarg.；托马斯榆；Cork Elm，Hickory Elm，Rock Elm ●☆

401639　Ulmus thomasii Sarg. = Ulmus racemosa D. Thomas ●

401640　Ulmus tonkinensis Gagnep. = Ulmus lanceifolia Roxb. ex Wall. ●

401641　Ulmus turkestanica Req. = Ulmus pumila L. ●

401642　Ulmus uyematsui Hayata；阿里山榆（台湾榆）；Alishan Elm，Arishan Elm，Taiwan Elm ●

401643　Ulmus villosa Brandis ex Gamble；西喜马拉雅榆；West Himalayan Elm ●☆

401644　Ulmus virgata Wall. ex Planch. = Ulmus androssowii Litv. var. subhirsuta（C. K. Schneid.）P. H. Huang，F. Y. Gao et L. H. Zhuo ●

401645　Ulmus wallichiana Planch.；沃利克榆●☆

401646　Ulmus wallichiana Planch. subsp. xanthoderma Melville et Heybroek = Ulmus wallichiana Planch. ●☆

401647　Ulmus wilsoniana C. K. Schneid.；威尔逊榆；E. H. Wilson Elm ●

401648　Ulmus wilsoniana C. K. Schneid. = Ulmus davidiana Planch. var. japonica（Rehder）Nakai ●

401649　Ulmus wilsoniana C. K. Schneid. var. psilophylla Schneid. = Ulmus davidiana Planch. var. japonica（Rehder）Nakai ●

401650　Ulmus wilsoniana C. K. Schneid. var. subhirsuta C. K. Schneid. = Ulmus androssowii Litv. var. subhirsuta（C. K. Schneid.）P. H. Huang，F. Y. Gao et L. H. Zhuo ●

401651　Uloma Raf.（废弃属名）= Colea Bojer ex Meisn.（保留属名）●☆

401652　Uloma Raf.（废弃属名）= Rhodocolea Baill. ●☆

401653　Uloptera Fenzl = Ferula L. ■

401654　Ulospermum Link = Capnophyllum Gaertn. ■☆

401655　Ulostoma D. Don = Gentiana L. ■

401656　Ulostoma D. Don ex G. Don = Gentiana L. ■

401657　Ulostoma G. Don = Gentiana L. ■

401658　Ulricia Jacq. ex Steud. = Lepechinia Willd. ●■☆

401659　Ulticona Raf.（废弃属名）= Hebecladus Miers（保留属名）●☆

401660　Ultragossypium Roberty = Gossypium L. ●■

401661　Ulugbekia Zakirov = Arnebia Forssk. ●■

401662　Ulugbekia tschimganica（B. Fedtsch.）Zakirov = Arnebia tschimganica（B. Fedtsch.）G. L. Zhu ■

401663　Uluxia Juss. = Columellia Ruiz et Pav.（保留属名）●☆

401664　Ulva Adans. = Carex L. ■

401665　Umari Adans. = Geoffroea Jacq. ●☆

401666　Umbellifera Honigb. = ? Ligusticum L. ■

401667　Umbelliferae Juss.（1789）（保留科名）；伞形花科（伞形科）■●

401668　Umbelliferae Juss.（保留科名）= Apiaceae Lindl.（保留科名）●■

401669　Umbellulanthus S. Moore = Triaspis Burch. ●☆

401670　Umbellularia（Nees）Nutt.（1842）（保留属名）；加州桂属（北美木姜子属，加州月桂属，伞桂属）；Californian Bay，Headache Tree，Oregon Myrtle ●☆

401671　Umbellularia Nutt. = Umbellularia（Nees）Nutt.（保留属名）●☆

401672　Umbellularia californica（Hook. et Arn.）Nutt.；加州桂（俄勒冈香桃木，海湾桂木，椒木，香桃木）；California Laurel，Californian Bay，Californian Laurel，Californian Olive，Californian Sassafras，Headache Tree，Laurel，Myrtle，Myrtle-wood，Oregon Myrtle，Oregon-myrtle，Pepperwood，Spice Tree ●☆

401673　Umbellularia californica Nutt. = Umbellularia californica（Hook. et Arn.）Nutt. ●☆

401674　Umbellularia parvifolia Hemsl.；小叶加州桂●☆

401675　Umbilicaria Fabr. = Omphalodes Mill. ■☆

401676　Umbilicaria Heist. ex Fabr. = Omphalodes Mill. ■☆

401677　Umbilicaria Pers. = Cotyledon L. ●■☆

401678　Umbilicus DC.（1801）；脐景天属；Navelwort ■☆

401679　Umbilicus affinis Schrenk = Pseudosedum affine（Schrenk）A. Berger ●■

401680　Umbilicus alpestris Kar. et Kir. = Rosularia alpestris（Kar. et Kir.）Boriss. ■

401681　Umbilicus botryoides Hochst. ex A. Rich.；葡萄脐景天■☆

401682　Umbilicus deflexus Pomel = Umbilicus rupestris（Salisb.）Dandy ■☆

401683　Umbilicus erectus DC.；脐景天■☆

401684　Umbilicus erubescens Maxim. = Orostachys spinosa（L.）A. Berger ■

401685　Umbilicus erubescens Maxim. = Orostachys spinosa（L.）Sweet ■

401686　Umbilicus fimbriatus（Turcz.）Turcz. = Orostachys fimbriata（Turcz.）A. Berger ■

401687　Umbilicus gaditanus Boiss.；加迪特脐景天■☆

401688　Umbilicus gaditanus Boiss. var. giganteus Batt. = Umbilicus gaditanus Boiss. ■☆

401689　Umbilicus giganteus Batt. = Umbilicus gaditanus Boiss. ■☆

401690　Umbilicus heylandianus Webb et Berthel.；海兰德脐景天●☆

401691　Umbilicus hispidus DC. = Sedum mucizonia（Ortega）Raym. - Hamet ■☆

401692　Umbilicus horizontalis（Guss.）DC.；平展脐景天■☆

401693　Umbilicus horizontalis（Guss.）DC. var. intermedius（Boiss.）D. F. Chamb. = Umbilicus intermedius Boiss. ■☆

401694　Umbilicus intermedius Boiss.；间型脐景天■☆

401695　Umbilicus leucanthus（Ledeb.）Ledeb. = Orostachys thyrsiflora Fisch. ■

401696　Umbilicus linearifolius Franch. = Rhodiola semenovii（Regel et Herder）Boriss. ■

401697　Umbilicus linifolius Ost. -Sack. et Rupr. = Rhodiola semenovii（Regel et Herder）Boriss. ■

401698　Umbilicus malacophyllus（Pall.）DC. = Orostachys malacophylla（Pall.）Fisch. ■

401699　Umbilicus maroccanus Gand. = Umbilicus horizontalis（Guss.）DC. ■☆

401700　Umbilicus micranthus Pomel = Umbilicus patens Pomel ■☆

401701　Umbilicus mirus（Pamp.）Greuter；奇异脐景天■☆

401702　Umbilicus oppositifolius Ledeb.；对叶脐景天■☆

401703　Umbilicus oreades Decne. = Sedum oreades（Decne.）Raym. - Hamet ■

401704　Umbilicus paniculiformis Wickens；锥形脐景天■☆

401705　Umbilicus patens Pomel；铺展脐景天■☆

401706　Umbilicus patulus Pomel = Umbilicus rupestris（Salisb.）Dandy ■☆

401707　Umbilicus pendulinus DC.；下垂脐景天■☆

401708　Umbilicus pendulinus DC. = Umbilicus rupestris（Salisb.）Dandy ■☆

401709　Umbilicus platyphyllus Schrenk = Rosularia platyphylla（Schrenk）A. Berger ■

401710　Umbilicus ramosissimus Maxim. = Orostachys fimbriata（Turcz.）A. Berger ■

401711　Umbilicus rupestris（Salisb.）Dandy；岩生脐景天；Bachelor's Buttons，Corn-leaf，Cows，Cup-and-saucer，Cut-finger，Dimplewort，Great Stonecrop，Halfpennies-and-pennies，Hap Pennies-and-pennies, Happenies-and-pennies, Hipwort, Ice-plant, Jack-in-the-bush, Kidneyweed, Kidneywort, Lady's Navel, Lover's Links, Lovers' Links, Lucky Moon, Maid-in-the-mist, Milk the Cows, Money Penny, Navel of the Earth, Navelwort, Nipplewort, Pancake, Penny Cake, Penny Cakes, Penny Cap, Penny Cod, Penny Cods, Penny Flower, Penny Hat, Penny Hats, Penny Leaves, Penny Pie, Penny Plate, Penny Plates, Pennygrass, Pennywall, Pennywort, Prince's Feathers, Royal Penny, Sibthorpia Europaea, Sunshade, Umbrella, Venns'-navelwort, Venus' Navelwort, Wall Pennyroyal, Wall Pennywort, Wallwort, Wushleen ■☆

401712 Umbilicus schmidtii Bolle；施密特脐景天■☆

401713 Umbilicus semenovii Regel et Herder = Rhodiola semenovii (Regel et Herder) Boriss. ■

401714 Umbilicus semiensis A. Rich. = Rosularia semiensis (J. Gay ex A. Rich.) H. Ohba ■☆

401715 Umbilicus thyrsiflorus (Fisch.) DC. = Orostachys thyrsiflora Fisch. ■

401716 Umbilicus thyrsiflorus DC. = Orostachys thyrsiflora Fisch. ■

401717 Umbilicus tropaeolifolius Boiss.；旱金莲叶脐景天■☆

401718 Umbilicus turkestanicus Regel et Winkler = Rosularia turkestanica (Regel et Winkl.) A. Berger ■

401719 Umbraculum Kuntze = Aegiceras Gaertn. ●

401720 Umbraculum Rumph. = Aegiceras Gaertn. ●

401721 Umbraculum Rumph. ex Kuntze = Aegiceras Gaertn. ●

401722 Umbraculum corniculatum Kuntze = Aegiceras corniculatum (L.) Blanco ●

401723 Umsema Raf. = Pontederia L. ■☆

401724 Umsema Raf. = Unisema Raf. ■☆

401725 Umtiza Sim(1907)；乌巴树属●☆

401726 Umtiza listeriana Sim；乌巴树●☆

401727 Unamia Greene = Aster L. ●■

401728 Unamia alba (Nutt.) Rydb. = Solidago ptarmicoides (Nees) B. Boivin ■☆

401729 Unamia ptarmicoides (Nees) Greene = Solidago ptarmicoides (Nees) B. Boivin ■☆

401730 Unannea Steud. = Stemodia L. (保留属名)■☆

401731 Unannea Steud. = Unanuea Ruiz et Pav. ex Pennell ■☆

401732 Unanuea Ruiz et Pav. = Stemodia L. (保留属名)■☆

401733 Unanuea Ruiz et Pav. ex Benth. = Stemodia L. (保留属名)■☆

401734 Unanuea Ruiz et Pav. ex Pennell = Stemodia L. (保留属名)■☆

401735 Unanuea Ruiz, Pav. et Pennell = Stemodia L. (保留属名)■☆

401736 Uncaria Burch. = Harpagophytum DC. ex Meisn. ■☆

401737 Uncaria Schreb. (1789)(保留属名)；钩藤属；Gambir Plant, Gambirplant, Gambir-plant, Hookvine ●

401738 Uncaria africana G. Don；非洲钩藤●☆

401739 Uncaria africana G. Don subsp. angolensis (Havil.) Ridsdale；安哥拉钩藤●☆

401740 Uncaria africana G. Don subsp. lacus-victoriae Verdc.；维多利亚钩藤●☆

401741 Uncaria africana G. Don var. angolensis Havil. = Uncaria africana G. Don subsp. angolensis (Havil.) Ridsdale ●☆

401742 Uncaria africana G. Don var. bequaertii De Wild. = Uncaria africana G. Don var. angolensis Havil. ●☆

401743 Uncaria africana G. Don var. hydrophila E. M. Petit；喜水非洲钩藤●☆

401744 Uncaria africana G. Don var. myrmecophyta De Wild.；蚂蚁非洲钩藤●☆

401745 Uncaria africana G. Don var. xerophila E. M. Petit；喜沙非洲钩

藤●☆

401746 Uncaria angolensis (Havil.) Welw. ex Hutch. et Dalziel = Uncaria africana G. Don var. angolensis Havil. ●☆

401747 Uncaria attenuata Korth.；狭钩藤；Narrow Gambirplant ●☆

401748 Uncaria bernaysii F. Muell.；博纳钩藤●☆

401749 Uncaria burchellii Kuntze；布氏钩藤●☆

401750 Uncaria callophylla Korth.；马来西亚钩藤●☆

401751 Uncaria donisii E. M. Petit；多尼斯钩藤●☆

401752 Uncaria elliptica R. Br. et G. Don；椭圆钩藤；Elliptic Gambirplant ●☆

401753 Uncaria florida Vidal；多花钩藤；Florida Gambirplant ●☆

401754 Uncaria formosana (Matsum.) Hayata；台湾钩藤；Taiwan Gambirplant, Taiwan Gambir-plant ●

401755 Uncaria formosana (Matsum.) Hayata = Uncaria hirsuta Havil. ●

401756 Uncaria gambier Roxb. = Uncaria gambir (Hunter) Roxb. ●

401757 Uncaria gambir (Hunter) Roxb.；儿茶钩藤(阿仙药, 槟榔膏, 黑儿茶, 马来钩藤, 孟加拉钩藤)；Bengal Gambir-plant, Gambir, Gambier Cutch, Gambier Plant, Gambier-plant, Gambier, Pale Catechu, White Cutch ●☆

401758 Uncaria grandidieri Kuntze；格氏钩藤●☆

401759 Uncaria guianensis (Aubl.) J. F. Gmel.；圭亚那钩藤●☆

401760 Uncaria hernagsia F. Muell.；海尔钩藤●☆

401761 Uncaria hirsuta Havil.；毛钩藤(川上钩藤, 倒吊风藤, 倒挂刺, 倒挂金钩, 吊风根, 吊藤, 吊藤钩, 钓钩藤, 钓藤, 勾丁, 钩丁, 钩耳, 钩藤, 金钩草, 金钩藤, 老鹰爪, 嫩钩钩, 双钩, 双钩藤, 台湾风藤, 台湾钩藤, 莺爪风, 鹰爪风)；Hirsute Gambirplant, Hookvine, Kawakami's Gambirplant, Taiwan Gambirplant ●

401762 Uncaria homomalla Miq.；越南钩藤(北越钩藤, 东京钩藤, 四楞通, 印支钩藤)；Homomallous Gambir-plant, N. Vietnam Hookvine, North Viet Nam Gambirplant ●

401763 Uncaria inermis Willd. = Mitragyna inermis (Willd.) K. Schum. ●☆

401764 Uncaria kawakamii Hayata = Uncaria hirsuta Havil. ●

401765 Uncaria laevigata Wall. et G. Don；双钩藤(倒挂金钩, 钩藤, 光钩藤, 平滑钩藤)；Smooth Gambirplant, Smooth Gambir-plant, Velvet Hookvine ●

401766 Uncaria lancifolia Hutch.；披针叶钩藤(倒挂金钩, 钩藤)；Inverse Hookvine, Lance-leved Gambirplant ●

401767 Uncaria lanosa Wall. f. setiloba (Benth.) Ridsdale = Uncaria lanosa Wall. var. appendiculata (Benth.) Ridsdale ●

401768 Uncaria lanosa Wall. var. appendiculata (Benth.) Ridsdale；恒春钩藤(线萼钩藤)；Hengchun Gambirplant, Hengchun Hookvine, Linearsepal Gambirplant, Linearsepal Hookvine, Linear-sepaled Gambir-plant ●

401769 Uncaria leptocarpa Kuntze；小果钩藤●☆

401770 Uncaria macrophylla Wall.；大叶钩藤(大钩丁, 倒挂刺, 倒挂金钩, 吊风根, 吊藤, 吊藤钩, 钓钩藤, 钓藤, 勾丁, 钩丁, 钩耳, 钩藤, 金钩草, 金钩藤, 老鹰爪, 嫩钩钩, 双钩, 双钩藤, 莺爪风, 鹰爪风)；Largeleaf Gambirplant, Largeleaf Hookvine, Macrophyllous Gambir-plant ●

401771 Uncaria membranifolia F. C. How = Uncaria sinensis (Oliv.) Havil. ●

401772 Uncaria procumbens Burch. = Harpagophytum procumbens (Burch.) DC. ex Meisn. ■☆

401773 Uncaria rhynchophylla (Miq.) B. D. Jacks.；钩藤(大叶钩藤, 倒挂刺, 倒挂金钩, 吊风根, 吊藤, 吊藤钩, 钓钩藤, 钓藤, 勾丁, 勾勾, 勾藤, 勾田, 勾苻, 钩丁, 钩耳, 钩藤钩子, 挂钩藤, 孩儿茶, 华钩藤, 金钩草, 金钩藤, 老鹰爪, 类钩藤, 毛钩藤, 嫩钩钩, 嫩双钩, 攀茎钩藤, 披针叶钩藤, 双丁, 双勾, 双钩, 双钩藤, 天吊藤, 无柄果钩藤,莺爪风,鹰爪风,嘴叶钩藤);Gambir Plant, Sharpleaf Gambirplant, Sharpleaf Hookvine, Sharp-leaved Gambir-plant ●

401774 Uncaria rhynchophylla (Miq.) B. D. Jacks. var. koutong T. Yamaz. = Uncaria rhynchophylla (Miq.) B. D. Jacks. ●

401775 Uncaria rhynchophylla (Miq.) Miq. ex Havil. = Uncaria rhynchophylla (Miq.) B. D. Jacks. ●

401776 Uncaria rhynchophylla (Miq.) Miq. var. hirsuta Kitam.;硬毛钩藤●☆

401777 Uncaria rhynchophylloides F. C. How;假钩藤(方枝钩藤,侯钩藤,侯氏钩藤,类钩藤);Beakedleaf Hookvine, False-sharpleaf Gambirplant ●

401778 Uncaria scandens (Sm.) Hutch.;攀枝钩藤(倒钩风,钩藤,攀茎钩藤,鹰爪风);Climbing Gambirplant, Climbing Gambir-plant, Scandent Hookvine ●

401779 Uncaria sessilifructus Roxb.;白钩藤(倒挂刺,倒挂金钩,吊风根,吊藤,吊藤钩,钓钩藤,钓藤,勾丁,钩丁,钩耳,钩藤,金钩草,金钩藤,老鹰爪,嫩钩钩,双钩,双钩藤,无柄钩滕,无柄果钩藤,莺爪风,鹰爪风);White Gambirplant, White Gambir-plant, White Hookvine ●

401780 Uncaria setiloba Benth. = Uncaria lanosa Wall. f. setiloba (Benth.) Ridsdale ●

401781 Uncaria sinensis (Oliv.) Havil.;华钩藤(倒挂刺,倒挂金钩,吊风根,吊藤,吊藤钩,钓钩藤,钓藤,勾丁,钩丁,钩耳,钩藤,金钩草,金钩藤,老鹰爪,米钩,膜叶钩藤,嫩钩钩,双钩,双钩藤,莺爪风,鹰爪风,中华钩藤);China Hookvine, Chinese Gambirplant, Chinese Gambir-plant ●

401782 Uncaria talbotii Wernham;塔尔博特钩藤●☆

401783 Uncaria tomentosa (Willd.) DC.;绒毛钩藤(猫爪,茸毛钩藤)●☆

401784 Uncaria tonkinensis Havil. = Uncaria homomalla Miq. ●

401785 Uncaria uraiensis Hayata = Uncaria hirsuta Havil. ●

401786 Uncaria wangii F. C. How;鹰爪风(王氏钩藤);Eegleclaw, Wang Gambirplant ●

401787 Uncaria wangii F. C. How = Uncaria scandens (Sm.) Hutch. ●

401788 Uncaria yunnanensis K. C. Hsia;云南钩藤;Yunnan Gambirplant, Yunnan Gambir-plant, Yunnan Hookvine ●

401789 Uncarina (Baill.) Stapf(1895);黄花胡麻属●☆

401790 Uncarina Stapf = Uncarina (Baill.) Stapf ●☆

401791 Uncarina grandidieri (Baill.) Stapf;黄花胡麻●☆

401792 Uncariopsis H. Karst. = Salvia L. ●■

401793 Uncariopsis H. Karst. = Schradera Vahl(保留属名)■☆

401794 Uncasia Greene = Eupatorium L. ■●

401795 Uncifera Lindl. (1858);叉喙兰属;Uncifera ■

401796 Uncifera acuminata Lindl.;叉喙兰;Common Uncifera ■

401797 Uncifera buccosa (Rchb. f.) Finet ex Guillaumin = Robiquetia succisa (Lindl.) Seidenf. et Garay ■

401798 Uncifera tenuicaulis (Hook. f.) Holttum;细茎叉喙兰;Finestem Uncifera ■☆

401799 Uncifera thailandica Seidenf. et Smitinand;中泰叉喙兰■

401800 Unciferia (Luer) Luer = Pleurothallis R. Br. ■☆

401801 Uncina C. A. Mey. = Uncinia Pers. ■☆

401802 Uncinaria Rchb. = Uncaria Schreb.(保留属名)●

401803 Uncinia Pers. (1807);钩莎属■☆

401804 Uncinia hamata (Sw.) Urb.;顶钩莎;Birdcatching Sedge ■☆

401805 Uncinia hamata Urb. = Uncinia hamata (Sw.) Urb. ■☆

401806 Uncinia lehmannii Nees = Schoenoxiphium lehmannii (Nees) Steud. ■☆

401807 Uncinia microglochin (Wahlenb.) Spreng. = Carex microglochin Wahlenb. ■

401808 Uncinia nepalensis Nees = Kobresia nepalensis (Nees) Kük. ■

401809 Uncinia spartea (Wahlenb.) Spreng. = Schoenoxiphium sparteum (Wahlenb.) C. B. Clarke ■☆

401810 Uncinia sprengelii Nees = Schoenoxiphium sparteum (Wahlenb.) C. B. Clarke ■☆

401811 Uncinus Raeusch. = Melodinus J. R. Forst. et G. Forst. ●

401812 Uncinus Raeusch. = Oncinus Lour. ●

401813 Unedo Hoffmanns. et Link = Arbutus L. ●☆

401814 Ungeria Nees ex C. B. Clarke = Cyperus L. ■

401815 Ungeria Schott et Endl. (1832);繁花梧桐属●☆

401816 Ungeria floribunda Schott et Endl.;繁花梧桐●☆

401817 Ungernia Bunge(1875);波斯石蒜属■☆

401818 Ungernia ferganica Vved.;费尔干波斯石蒜■☆

401819 Ungernia flava Boiss. et Hausskn. ex Boiss.;黄波斯石蒜■☆

401820 Ungernia minor Vved.;小波斯石蒜■☆

401821 Ungernia oldhamii Maxim.;奥氏波斯石蒜■☆

401822 Ungernia sewerzowii (Regel) B. Fedtsch.;西氏波斯石蒜■☆

401823 Ungernia spiralis ?;螺纹波斯石蒜■☆

401824 Ungernia tadshikorum Vved.;塔什克波斯石蒜■☆

401825 Ungernia trisphaera Bunge;波斯石蒜■☆

401826 Ungernia victoria Vved. = Ungernia victoria Vved. ex Artjush. ■☆

401827 Ungernia victoria Vved. ex Artjush.;三球波斯石蒜■☆

401828 Ungnadia Endl. (1835);翁格木属(翁格那木属)●

401829 Ungnadia speciosa Endl.;美丽翁格木(美丽翁格那木);Mexican Buck Eye, Mexican Buckeye, Monilla ●☆

401830 Unguacha Hochst. = Strychnos L. ●

401831 Unguella Luer = Pleurothallis R. Br. ■☆

401832 Unguiculabia Mytnik et Szlach. (2008);尤兰属■☆

401833 Unguiculabia Mytnik et Szlach. = Polystachya Hook.(保留属名)■

401834 Ungula Barlow = Amyema Tiegh. ●☆

401835 Ungulipetalum Moldenke(1938);蹄瓣藤属●☆

401836 Ungulipetalum filipendulum (C. Mart.) Moldenke;蹄瓣藤●☆

401837 Unifolium Boehm. = Maianthemum F. H. Wigg.(保留属名)■

401838 Unifolium Haller = Maianthemum F. H. Wigg.(保留属名)■

401839 Unifolium Ludw. = Maianthemum F. H. Wigg.(保留属名)■

401840 Unifolium Zinn = Maianthemum F. H. Wigg.(保留属名)■

401841 Unifolium amplexicaule (Nutt.) Greene = Maianthemum racemosum (L.) Link subsp. amplexicaule (Nutt.) LaFrankie ■☆

401842 Unifolium canadense (Desf.) Greene = Maianthemum canadense Desf. ■☆

401843 Unifolium dilatatum (A. Wood) Greene = Maianthemum dilatatum (A. W. Wood) A. Nelson et J. F. Macbr. ■☆

401844 Unifolium kamtschaticum (J. F. Gmel.) Gorman = Maianthemum dilatatum (A. W. Wood) A. Nelson et J. F. Macbr. ■☆

401845 Unifolium liliaceum Greene = Maianthemum stellatum (L.) Link ■☆

401846 Unifolium racemosum (L.) Britton = Maianthemum racemosum (L.) Link ■☆

401847 Unifolium sessilifolium (Nutt. ex Baker) Greene = Maianthemum stellatum (L.) Link ■☆

401848 Unifolium stellatum (L.) Greene = Maianthemum stellatum (L.) Link ■☆

401849 Unifolium trifolium (L.) Greene = Maianthemum trifolium (L.) Slobada ■

401850 Unigenes E. Wimm. (1948);单生桔梗属■☆

401851 Unigenes humifusa A. DC.;单生桔梗■☆

401852   Uniola L. (1753);牧场草属■☆

401853   Uniola bipinnata (L.) L. = Desmostachya bipinnata (L.) Stapf ■

401854   Uniola jardinii Steud. = Eragrostis superba Peyr. ■☆

401855   Uniola lappacea (L.) Trin. = Centotheca lappacea (L.) Desv. ■

401856   Uniola latifolia Michx. – Chasmanthium latifolium (Michx.) H. O. Yates ■☆

401857   Uniola laxa (L.) Britton,Sterns et Poggenb. = Chasmanthium laxum (L.) H. O. Yates ■☆

401858   Uniola mucronata L. = Halopyrum mucronatum (L.) Stapf ■☆

401859   Uniola paniculata L. ;圆锥牧场草;Sea Oats ■☆

401860   Uniola sessiliflora Poir. = Chasmanthium laxum (L.) H. O. Yates ■☆

401861   Uniola spicata L. = Distichlis spicata (L.) Greene ■☆

401862   Unisema Raf. = Pontederia L. ■☆

401863   Unisema cordata (L.) Farw. = Pontederia cordata L. ■☆

401864   Unisemataceae Raf. = Pontederiaceae Kunth(保留科名)■

401865   Univiscidiatus (Kores) Szlach. = Acianthopsis Szlach. ■☆

401866   Univiscidiatus (Kores) Szlach. = Acianthus R. Br. ■☆

401867   Unjala Blume = Schefflera J. R. Forst. et G. Forst. (保留属名)●

401868   Unjala Reinw. ex Blume = Schefflera J. R. Forst. et G. Forst. (保留属名)●

401869   Unona Hook. f. et Thomson = Desmos Lour. + Dasymaschalon (Hook. f. et Thomson) Dalla Torre et Harms ●

401870   Unona L. f. = Desmos Lour. ●

401871   Unona L. f. = Xylopia L. (保留属名)●

401872   Unona acutiflora Dunal = Xylopia acutiflora (Dunal) A. Rich. ●☆

401873   Unona aethiopica Dunal = Xylopia aethiopica (Dunal) A. Rich. ●☆

401874   Unona albida Engl. = Friesodielsia gracilipes (Benth.) Steenis ●☆

401875   Unona buchananii Engl. = Monanthotaxis buchananii (Engl.) Verdc. ●☆

401876   Unona chinensis DC. = Desmos chinensis Lour. ●

401877   Unona confinis Pierre = Duguetia confinis (Engl. et Diels) Chatrou ●☆

401878   Unona congensis Engl. et Diels = Monanthotaxis laurentii (De Wild.) Verdc. ●☆

401879   Unona desmos Dunal var. grandifolia Finet et Gagnep. = Desmos grandifolius (Finet et Gagnep.) C. Y. Wu ex P. T. Li ●

401880   Unona desmos Raeusch. var. grandifolia Finet et Gagnep. = Desmos grandifolius (Finet et Gagnep.) C. Y. Wu ex P. T. Li ●

401881   Unona dielsiana Engl. = Friesodielsia dielsiana (Engl.) Steenis ●☆

401882   Unona discolor Vahl = Desmos chinensis Lour. ●

401883   Unona dumosa Roxb. = Desmos dumosus (Roxb.) Saff. ●

401884   Unona elegans (Engl. et Diels) Thwaites = Monanthotaxis elegans (Engl. et Diels) Verdc. ●☆

401885   Unona eminii Engl. = Monanthotaxis ferruginea (Oliv.) Verdc. ●☆

401886   Unona ferruginea Oliv. = Monanthotaxis ferruginea (Oliv.) Verdc. ●☆

401887   Unona glauca Engl. et Diels = Monanthotaxis oligandra Exell ●☆

401888   Unona grandiflora DC. = Uvaria grandiflora Roxb. ●

401889   Unona grandiflora Leschen. ex DC. = Uvaria grandiflora Roxb. ●

401890   Unona hirsuta Benth. = Friesodielsia hirsuta (Benth.) Steenis ●☆

401891   Unona latifolia Dunal = Fissistigma latifolium (Dunn) Merr. ●

401892   Unona lepidota Oliv. = Meiocarpidium lepidotum (Oliv.) Engl. et Diels ●☆

401893   Unona longifolia Dunal = Polyalthia longifolia (Sonn.) Thwaites ●

401894   Unona lucidula Oliv. = Monanthotaxis lucidula (Oliv.) Verdc. ●☆

401895   Unona macrocarpa DC. = Uvaria chamae P. Beauv. ●☆

401896   Unona millenii Engl. et Diels = Friesodielsia gracilis (Hook. f.) Steenis ●☆

401897   Unona montana Engl. et Diels = Friesodielsia montana (Engl. et Diels) Steenis ●☆

401898   Unona obanensis Baker f. = Friesodielsia enghiana (Diels) Verdc. ●☆

401899   Unona obovata Benth. = Friesodielsia obovata (Benth.) Verdc. ●☆

401900   Unona ovata DC. = Uvaria ovata (DC.) A. DC. ●☆

401901   Unona ovata DC. var. afzeliana DC. = Uvaria ovata (DC.) A. DC. subsp. afzeliana (DC.) Keay ●☆

401902   Unona oxypetala DC. = Xylopia acutiflora (Dunal) A. Rich. ●☆

401903   Unona parvifolia Oliv. = Monanthotaxis parvifolia (Oliv.) Verdc. ●☆

401904   Unona parvifolia Oliv. var. petersii Engl. = Cleistochlamys kirkii (Benth.) Oliv. ●☆

401905   Unona polycarpa DC. = Annickia polycarpa (DC.) Setten et Maas ●☆

401906   Unona simiarum Baill. ex Pierre = Polyalthia simiarum (Ham. ex Hook. f. et Thomson) Benth. ex Hook. f. et Thomson ●

401907   Unona stuhlmannii Engl. = Polyalthia stuhlmannii (Engl.) Verdc. ●☆

401908   Unona uncinata (Lam.) Dunal = Artabotrys hexapetalus (L. f.) Bhandari ●

401909   Unonopsis R. E. Fr. (1900);类番鹰爪属●☆

401910   Unonopsis veneficiorum (C. Mart.) R. E. Fr. ;类番鹰爪●☆

401911   Unxia Bert. ex Colla = Blennosperma Less. ■☆

401912   Unxia Kunth = Villanova Lag. (保留属名)■☆

401913   Unxia L. f. (废弃属名) = Villanova Lag. (保留属名)■☆

401914   Upata Adans. = Avicennia L. ●

401915   Upoda Adans. = Hypoxis L. ■

401916   Upopion Raf. = Thaspium Nutt. ■☆

401917   Upoxis Adans. = Gagea Salisb. ■

401918   Upudalia Raf. = Eranthemum L. ●■

401919   Upuna Symington(1941);加岛香属●☆

401920   Upuna borneensis Symington;加岛香■☆

401921   Upuntia Raf. = Opuntia Mill. ●

401922   Urachne Trin. = Oryzopsis Michx. ■

401923   Urachne Trin. = Piptatherum P. Beauv. ■

401924   Urachne acutigluma Steud. = Garnotia acutigluma (Steud.) Ohwi ■

401925   Urachne lanata Trin. et Rupr. = Oryzopsis hymenoides (Roem. et Schult.) Ricker et Piper ■

401926   Urachne songarica Trin. et Rupr. = Oryzopsis songarica (Trin. et Rupr.) B. Fedtsch. ■

401927   Urachne songarica Trin. et Rupr. = Piptatherum songaricum (Trin. et Rupr.) Roshev. ■

401928   Uragoga Baill. (1879);肖九节属●☆

401929   Uragoga Baill. = Cephaelis Sw. (保留属名)●

401930   Uragoga Baill. = Psychotria L. (保留属名)●

401931   Uragoga acuta De Wild. ;尖肖九节●☆

401932   Uragoga afzelii (Hiern) Kuntze = Chassalia afzelii (Hiern) K. Schum. ■☆

401933   Uragoga anetoclada (Hiern) Kuntze = Psychotria subobliqua Hiern ●☆

401934   Uragoga ansellii (Hiern) Kuntze = Chassalia laxiflora Benth. ■☆

401935   Uragoga articulata (Hiern) Kuntze = Psychotria articulata (Hiern) E. M. Petit ●☆

401936   Uragoga atenensis De Wild. = Psychotria peduncularis (Salisb.) Steyerm. ●☆

401937   Uragoga barteri Kuntze = Psychotria calva Hiern ●☆

401938   Uragoga bayakaensis De Wild. = Psychotria peduncularis (Salisb.) Steyerm. ●☆

401939   Uragoga benthamiana (Hiern) Kuntze = Chassalia kolly (Schumach.) Hepper ■☆

401940   Uragoga bequaerti De Wild.;贝卡尔肖九节●☆

401941   Uragoga biaurita Hutch. et Dalziel = Psychotria biaurita (Hutch. et Dalziel) Verdc. ●☆

401942   Uragoga bidentata (Thunb. ex Roem. et Schult.) Kuntze = Psychotria bidentata (Thunb. ex Roem. et Schult.) Hiern ●☆

401943   Uragoga bieleri De Wild. = Psychotria peduncularis (Salisb.) Steyerm. ●☆

401944   Uragoga bifaria (Hiern) Kuntze = Psychotria bifaria Hiern ●☆

401945   Uragoga boa De Wild. = Psychotria peduncularis (Salisb.) Steyerm. ●☆

401946   Uragoga brachyantha (Hiern) Kuntze = Psychotria brachyantha Hiern ●☆

401947   Uragoga brachypus K. Schum. et K. Krause;短足肖九节●☆

401948   Uragoga bracteosa (Hiern) Kuntze = Peripeplus bracteosus (Hiern) E. M. Petit ●☆

401949   Uragoga brassii (Hiern) Kuntze = Psychotria brassii Hiern ●☆

401950   Uragoga brazzai De Wild.;布拉扎肖九节●☆

401951   Uragoga brunnea (Schweinf. ex Hiern) Kuntze = Psychotria brunnea Schweinf. ex Hiern ●☆

401952   Uragoga butaensis De Wild. = Psychotria peduncularis (Salisb.) Steyerm. ●☆

401953   Uragoga calva (Hiern) Kuntze = Psychotria calva Hiern ●☆

401954   Uragoga ceratoloba K. Schum.;角裂肖九节●☆

401955   Uragoga ciliato-stipulata De Wild. = Psychotria peduncularis (Salisb.) Steyerm. var. ciliato-stipulata Verdc. ●☆

401956   Uragoga congensis K. Schum.;刚果肖九节●☆

401957   Uragoga crispa (Hiern) Kuntze = Psychotria latistipula Benth. ●☆

401958   Uragoga cristata (Hiern) Kuntze = Chassalia cristata (Hiern) Bremek. ■☆

401959   Uragoga cyanocarpa K. Krause = Psychotria peduncularis (Salisb.) Steyerm. var. suaveolens (Hiern) Verdc. ●☆

401960   Uragoga debeauxii De Wild. = Psychotria peduncularis (Salisb.) Steyerm. ●☆

401961   Uragoga densifolia De Wild. = Psychotria peduncularis (Salisb.) Steyerm. ●☆

401962   Uragoga dewevrei De Wild.;德韦肖九节●☆

401963   Uragoga doniana (Benth.) Kuntze = Chassalia doniana (Benth.) G. Taylor ■☆

401964   Uragoga eminiana Kuntze = Psychotria eminiana (Kuntze) E. M. Petit ●☆

401965   Uragoga foliosa (Hiern) Kuntze = Psychotria foliosa Hiern ●☆

401966   Uragoga gabonensis De Wild.;加蓬肖九节●☆

401967   Uragoga gabonica (Hiern) Kuntze = Psychotria gabonica Hiern ●☆

401968   Uragoga giorgii De Wild.;乔治肖九节●☆

401969   Uragoga globosa (Hiern) Kuntze = Psychotria globosa Hiern ●☆

401970   Uragoga globoso-capitata De Wild. = Psychotria peduncularis (Salisb.) Steyerm. ●☆

401971   Uragoga goossensii De Wild. = Psychotria peduncularis (Salisb.) Steyerm. ●☆

401972   Uragoga grandifolia De Wild. = Psychotria peduncularis (Salisb.) Steyerm. ●☆

401973   Uragoga guerzeensis Schnell = Psychotria peduncularis (Salisb.) Steyerm. ●☆

401974   Uragoga guerzeensis Schnell f. puberula？ = Psychotria peduncularis (Salisb.) Steyerm. ●☆

401975   Uragoga guerzeensis Schnell f. saouroana？ = Psychotria peduncularis (Salisb.) Steyerm. ●☆

401976   Uragoga guineensis (Schnell) Schnell = Psychotria peduncularis (Salisb.) Steyerm. var. guineensis (Schnell) Verdc. ●☆

401977   Uragoga guineensis (Schnell) Schnell var. bindelyensis Schnell = Psychotria peduncularis (Salisb.) Steyerm. var. guineensis (Schnell) Verdc. ●☆

401978   Uragoga gumola De Wild. = Psychotria peduncularis (Salisb.) Steyerm. ●☆

401979   Uragoga hexamera K. Schum. = Psychotria globosa Hiern var. ciliata (Hiern) E. M. Petit ●☆

401980   Uragoga hiernii Kuntze = Chassalia hiernii (Kuntze) G. Taylor ●☆

401981   Uragoga hirsuticalyx R. D. Good;粗毛萼肖九节●☆

401982   Uragoga homblei De Wild.;洪布勒肖九节●☆

401983   Uragoga humilis (Hiern) Kuntze = Psychotria humilis Hiern ●☆

401984   Uragoga hydrophila K. Krause;喜水肖九节●☆

401985   Uragoga ibaliensis De Wild. = Psychotria peduncularis (Salisb.) Steyerm. ●☆

401986   Uragoga ikengaensis De Wild. = Psychotria peduncularis (Salisb.) Steyerm. ●☆

401987   Uragoga infundibularis (Hiern) Kuntze = Psychotria infundibularis Hiern ●☆

401988   Uragoga isimbi De Wild. = Psychotria peduncularis (Salisb.) Steyerm. ●☆

401989   Uragoga ituriensis De Wild. = Psychotria ituriensis De Wild. ex E. M. Petit ●☆

401990   Uragoga ivorensis Schnell = Psychotria peduncularis (Salisb.) Steyerm. var. ivorensis (Schnell) Verdc. ●☆

401991   Uragoga kirkii (Hiern) Kuntze = Psychotria kirkii Hiern ●☆

401992   Uragoga klainei De Wild.;克莱恩肖九节●☆

401993   Uragoga kolly (Schumach.) Kuntze = Chassalia kolly (Schumach.) Hepper ■☆

401994   Uragoga konguensis (Hiern) Kuntze = Psychotria konguensis Hiern ●☆

401995   Uragoga korrowalensis K. Krause;科罗瓦尔肖九节●☆

401996   Uragoga lateralis K. Schum.;侧生肖九节●☆

401997   Uragoga latistipula (Benth.) Kuntze = Psychotria latistipula Benth. ●☆

401998   Uragoga lebruni De Wild. = Psychotria peduncularis (Salisb.) Steyerm. ●☆

401999   Uragoga lecomtei De Wild. = Psychotria peduncularis (Salisb.) Steyerm. ●☆

402000   Uragoga lecomtei De Wild. var. nimbana Schnell = Psychotria peduncularis (Salisb.) Steyerm. var. suaveolens (Hiern) Verdc. ●☆

402001   Uragoga ledermannii K. Krause;莱德肖九节●☆

402002   Uragoga lemairei De Wild. = Psychotria peduncularis (Salisb.) Steyerm. ●☆

402003   Uragoga leptophylla (Hiern) Kuntze = Psychotria leptophylla Hiern ●☆

402004   Uragoga letestui De Wild.;莱泰斯图肖九节●☆

402005   Uragoga librevillensis De Wild.;利伯维尔肖九节●☆

402006   Uragoga lonkasa De Wild. = Psychotria peduncularis (Salisb.) Steyerm. ●☆

402007   Uragoga lophoclada (Hiern) Kuntze = Chazaliella lophoclada

（Hiern）E. M. Petit et Verdc. ●☆

402008　Uragoga lubutuensis De Wild. ;卢布图肖九节●☆

402009　Uragoga lucens（Hiern）Kuntze = Psychotria lucens Hiern ●☆

402010　Uragoga macrophylla K. Krause = Psychotria tanganyicensis Verdc. ●☆

402011　Uragoga malchairi De Wild. ;马尔谢里肖九节●☆

402012　Uragoga mannii（Hiern）Kuntze = Psychotria mannii Hiern ●☆

402013　Uragoga mannii（Hook. f.）Hutch. et Dalziel = Psychotria camptopus Verdc. ●☆

402014　Uragoga mayumbensis De Wild. ;马永巴肖九节●☆

402015　Uragoga mayumbensis R. D. Good = Uragoga mayumbensis De Wild. ●☆

402016　Uragoga melanochlora K. Schum. ;墨绿肖九节●☆

402017　Uragoga membranifolia Mildbr. ;膜叶肖九节●☆

402018　Uragoga mildbraedii K. Krause;米尔德肖九节●☆

402019　Uragoga monticola（Hiern）Kuntze = Psychotria nubicola G. Taylor ●☆

402020　Uragoga mortehani De Wild. ;莫特汉肖九节●☆

402021　Uragoga mucronata（Hiern）Kuntze = Psychotria kirkii Hiern var. mucronata（Hiern）Verdc. ●☆

402022　Uragoga multinervata De Wild. = Psychotria peduncularis（Salisb.）Steyerm. ●☆

402023　Uragoga nigropunctata（Hiern）Kuntze = Psychotria nigropunctata Hiern ●☆

402024　Uragoga nimbana（Schnell）Schnell = Psychotria peduncularis（Salisb.）Steyerm. var. suaveolens（Hiern）Verdc. ●☆

402025　Uragoga nubica（Delile）Kuntze = Psychotria nubica Delile ●☆

402026　Uragoga nutans K. Krause;俯垂肖九节●☆

402027　Uragoga nyassana K. Krause = Psychotria peduncularis（Salisb.）Steyerm. var. nyassana（K. Krause）Verdc. ●☆

402028　Uragoga oblanceolata R. D. Good = Psychotria oblanceolata（R. D. Good）Ruhsam ●☆

402029　Uragoga ombrophila Schnell = Psychotria ombrophila（Schnell）Verdc. ●☆

402030　Uragoga owariensis（P. Beauv.）Kuntze = Psychotria owariensis（P. Beauv.）Hiern ●☆

402031　Uragoga pachyphylla K. Krause;厚叶肖九节●☆

402032　Uragoga pauridiantha（Hiern）Kuntze = Psychotria bifaria Hiern var. pauridiantha（Hiern）E. M. Petit ☆

402033　Uragoga peduncularis（Salisb.）K. Schum. = Psychotria peduncularis（Salisb.）Steyerm. ●☆

402034　Uragoga peduncularis（Salisb.）K. Schum. var. guineensis Schnell = Psychotria peduncularis（Salisb.）Steyerm. var. guineensis（Schnell）Verdc. ●☆

402035　Uragoga pobeguini De Wild. = Psychotria peduncularis（Salisb.）Steyerm. ●☆

402036　Uragoga prantliana Kuntze = Psychotria recurva Hiern ●☆

402037　Uragoga psychotrioides Schnell = Psychotria rufipilis De Wild. ●☆

402038　Uragoga psychotrodes（DC.）Kuntze = Psychotria psychotrioides（DC.）Roberty ●☆

402039　Uragoga pumila（Hiern）Kuntze = Psychotria pumila Hiern ●☆

402040　Uragoga punctata（Vatke）Kuntze = Psychotria punctata Vatke ●☆

402041　Uragoga recurva（Hiern）Kuntze = Psychotria recurva Hiern ●☆

402042　Uragoga repens De Wild. = Psychotria peduncularis（Salisb.）Steyerm. ●☆

402043　Uragoga reptans（Benth.）Kuntze = Psychotria reptans Benth. ●☆

402044　Uragoga reygaerti De Wild. = Psychotria peduncularis（Salisb.）Steyerm. ●☆

402045　Uragoga sangalkamensis Schnell = Psychotria bidentata（Thunb. ex Roem. et Schult.）Hiern ●☆

402046　Uragoga sapini De Wild. ;萨潘肖九节●☆

402047　Uragoga scaphus K. Schum. = Hymenocoleus scaphus（K. Schum.）Robbr. ■☆

402048　Uragoga schweinfurthii（Hiern）Kuntze = Psychotria schweinfurthii Hiern ●☆

402049　Uragoga semlikiensis De Wild. = Psychotria peduncularis（Salisb.）Steyerm. var. semlikiensis Verdc. ●☆

402050　Uragoga sereti De Wild. = Psychotria peduncularis（Salisb.）Steyerm. ●☆

402051　Uragoga setacea（Hiern）Kuntze = Psychotria leptophylla Hiern ●☆

402052　Uragoga setistipulata R. D. Good = Psychotria setistipulata（R. D. Good）E. M. Petit ●☆

402053　Uragoga soyauxii（Hiern）Kuntze = Psychotria schweinfurthii Hiern ●☆

402054　Uragoga spathacea（Hiern）Hutch. et Dalziel = Psychotria spathacea（Hiern）Verdc. ●☆

402055　Uragoga sphaerocarpa（Hiern）Kuntze = Psychotria fernandopoensis E. M. Petit ●☆

402056　Uragoga staudtii De Wild. = Psychotria peduncularis（Salisb.）Steyerm. ●☆

402057　Uragoga suaveolens（Hiern）K. Schum. = Psychotria peduncularis（Salisb.）Steyerm. var. suaveolens（Hiern）Verdc. ●☆

402058　Uragoga subherbacea（Hiern）Kuntze = Chassalia subherbacea（Hiern）Hepper ■☆

402059　Uragoga subipecacuanha K. Schum. = Hymenocoleus subipecacuanha（K. Schum.）Robbr. ■☆

402060　Uragoga subnuda（Hiern）Kuntze = Chassalia subnuda（Hiern）Hepper ■☆

402061　Uragoga subobliqua（Hiern）Kuntze = Psychotria subobliqua Hiern ●☆

402062　Uragoga subpunctata（Hiern）Kuntze = Psychotria subpunctata Hiern ●☆

402063　Uragoga subsessilifolia K. Schum. ;近无柄叶肖九节●☆

402064　Uragoga subsessilis De Wild. ;近无柄肖九节●☆

402065　Uragoga succulenta（Hiern）Kuntze = Psychotria succulenta（Schweinf. ex Hiern）E. M. Petit ●☆

402066　Uragoga thollonii De Wild. = Hymenocoleus scaphus（K. Schum.）Robbr. ■☆

402067　Uragoga thonneri De Wild. et T. Durand;托内肖九节●☆

402068　Uragoga thonningii Kuntze;通宁肖九节●☆

402069　Uragoga tumbaensis De Wild. = Psychotria psychotrioides（DC.）Roberty ●☆

402070　Uragoga umbraticola（Vatke）Kuntze = Chassalia umbraticola Vatke ■☆

402071　Uragoga venosa（Hiern）Kuntze = Psychotria venosa（Hiern）E. M. Petit ●☆

402072　Uragoga verschuerenii De Wild. ;费许伦肖九节●☆

402073　Uragoga vogeliana（Benth.）Kuntze = Psychotria vogeliana Benth. ●☆

402074　Uragoga wellensii De Wild. ;韦伦斯肖九节●☆

402075　Uragoga wendjiensis De Wild. = Psychotria peduncularis（Salisb.）Steyerm. ●☆

402076　Uragoga yapoensis Schnell = Psychotria yapoensis（Schnell）Verdc. ●☆

Steyerm. ●☆

402077　Uragoga zambesiana（Hiern）Kuntze = Psychotria capensis（Eckl.）Vatke ●☆

402078　Uragoga zanguebarica（Hiern）Kuntze = Chassalia umbraticola Vatke ■☆

402079　Uragoga zenkeri De Wild. = Psychotria peduncularis（Salisb.）Steyerm. ●☆

402080　Uragoga zombamontana Kuntze = Psychotria zombamontana（Kuntze）E. M. Petit ●☆

402081　Uralepis Nutt. = Triplasis P. Beauv. ■☆

402082　Uralepis Raf. = Triplasis P. Beauv. ■☆

402083　Uralepis Raf. = Uralepsis Nutt. ■☆

402084　Uralepis arenaria Hochst. et Steud. ex Steud. = Trichoneura mollis（Kunth）Ekman ■☆

402085　Uralepis capensis（Nees）Steud. = Leptochloa fusca（L.）Kunth ■

402086　Uralepis ciliata Steud. = Trichoneura mollis（Kunth）Ekman ■☆

402087　Uralepsis Nutt. = Triplasis P. Beauv. ■☆

402088　Urananthus（Griseb.）Benth. = Eustoma Salisb. ■☆

402089　Urananthus Benth. = Eustoma Salisb. ■☆

402090　Urandra Thwaites = Stemonurus Blume ●

402091　Urania DC. = Uraria Desv. ●■

402092　Urania Schreb. = Ravenala Adans. ●■

402093　Uranodactylus Gilli = Winklera Regel ■☆

402094　Uranodactylus afghanicus Gilli = Heldreichia silaifolia Hook. f. et Thomson ■☆

402095　Uranodactylus patrinoides（Regel）Gilli = Winklera patrinoides Regel ■☆

402096　Uranodactylus silaifolius（Hook. f. et Thomson）Jafri = Heldreichia silaifolia Hook. f. et Thomson ■☆

402097　Uranostachys Fourr. = Veronica L. ■

402098　Uranthera Naudin = Acisanthera P. Browne ●■☆

402099　Uranthera Pax et K. Hoffm.（1911）;尾药大戟属●☆

402100　Uranthera Pax et K. Hoffm. = Phyllanthodendron Hemsl. ●

402101　Uranthera Pax et K. Hoffm. = Phyllanthus L. ●■

402102　Uranthera Raf. = Justicia L. ●■

402103　Uranthera siamensis Pax et K. Hoffm.；尾药大戟●☆

402104　Uranthoecium Stapf（1916）;扁轴草属■☆

402105　Uranthoecium truncatum（Maiden et Betche）Stapf;扁轴草☆

402106　Uraria Desv.（1813）;狸尾豆属（兔尾草属,猪腰豆属）;Uraria ●■

402107　Uraria Wall. = Urariopsis C. K. Schneid. ●

402108　Uraria aequilobata Hosok.；圆叶兔尾草■

402109　Uraria aequilobata Hosok. = Uraria lagopodioides（L.）Desv. ex DC. ■

402110　Uraria aequilobata Hosok. = Uraria neglecta Prain ●■

402111　Uraria campanulata Wall. = Christia campanulata（Wall.）Thoth. ●

402112　Uraria clarkei（C. B. Clarke）Gagnep.;野番豆;C. B. Clarke Uraria ●

402113　Uraria clarkei Gagnep. = Uraria lacei Craib ●■

402114　Uraria cochinchinensis Schindl. = Urariopsis brevissima Yen C. Yang et P. H. Huang ●

402115　Uraria comosa（Vahl）DC. = Uraria crinita（L.）Desv. ex DC. ●■

402116　Uraria cordifolia Wall. = Urariopsis cordifolia（Wall.）Schindl. ●

402117　Uraria cornosa Span. = Uraria crinita（L.）Desv. ex DC. ●■

402118　Uraria crinita（L.）Desv. ex DC.；玉树野决明（布狗尾,长穗狸尾草,大本山菁,防虫草,狗尾射,古钱窗草,狐狸草,虎尾轮,猫公树,猫上树,猫尾草,猫尾豆,猫尾射,牛春花,七狗尾,铁金铜,统天草,土狗尾,兔狗尾,兔尾草）;Cattail Uraria,Hairy Uraria ●■

402119　Uraria crinita（L.）Desv. ex DC. var. macrostachya Wall.；长穗猫尾豆（布狗尾,长穗狸猫尾草,长穗猫尾草,防虫草,狐狸尾,虎尾轮,狼尾草,猫公树,猫公尾,猫上树,猫尾草,土狗尾,兔狗尾）●■

402120　Uraria crinita（L.）Desv. ex DC. var. macrostachya Wall. = Uraria crinita（L.）Desv. ex DC. ●■

402121　Uraria formosana（Hayata）Hayata = Christia campanulata（Benth.）Thoth. ●

402122　Uraria formosana Hayata = Christia campanulata（Wall.）Thoth. ●

402123　Uraria fujianensis Yen C. Yang et P. H. Huang = Uraria neglecta Prain ■

402124　Uraria gossweileri Baker f.；戈斯狸尾豆■☆

402125　Uraria hamosa（Sweet）Wall. ex Wight et Arn. = Uraria rufescens（DC.）Schindl. ●

402126　Uraria hamosa（Sweet）Wall. ex Wight et Arn. var. formosana Matsum. = Christia campanulata（Wall.）Thoth. ●

402127　Uraria hamosa（Sweet）Wall. ex Wight et Arn. var. sinensis Hemsl. = Uraria sinensis（Hemsl.）Franch. ●

402128　Uraria hamosa Sweet ex Arn. = Uraria rufescens（DC.）Schindl. ●

402129　Uraria henryi Schindl. = Desmodium hispidum Franch. ●

402130　Uraria lacei Craib;滇南狸尾豆（滇岩黄耆）;S. Yunnan Uraria,South Yunnan Uraria ●■

402131　Uraria lagopodioides（L.）Desv. = Uraria lagopodioides（L.）Desv. ex DC. ■

402132　Uraria lagopodioides（L.）Desv. ex DC.；狸尾豆（大叶兔尾草,狐狸尾,狸尾草,猫尾草,兔尾草,圆叶兔尾草,猪屎豆）;Haretail Uraria,Uraria ■

402133　Uraria lagopoides DC. = Uraria lagopodioides（L.）Desv. ex DC. ■

402134　Uraria latisepala Hayata = Christia campanulata（Benth.）Thoth. ●

402135　Uraria leucantha Zipp. ex Span = Uraria picta（Jacq.）Desv. ex DC. ●

402136　Uraria linearis Hassk. = Uraria picta（Jacq.）Desv. ex DC. ●

402137　Uraria longibracteata Yen C. Yang et P. H. Huang；长苞狸尾豆;Longbract Uraria,Long-bracted Uraria ●

402138　Uraria macrostachya（Wall.）Prain = Uraria crinita（L.）Desv. ex DC. ●■

402139　Uraria macrostachya（Wall.）Schindl. = Uraria crinita（L.）Desv. ex DC. ●■

402140　Uraria macrostachya（Wall.）Schneid. = Uraria crinita（L.）Desv. ex DC. var. macrostachya Wall. ●■

402141　Uraria neglecta Prain；福建狸尾豆（长苞狸尾豆）●■

402142　Uraria paniculata Hassk. = Uraria clarkei（C. B. Clarke）Gagnep. ●

402143　Uraria picta（Jacq.）DC. = Uraria picta（Jacq.）Desv. ex DC. ●

402144　Uraria picta（Jacq.）Desv. ex DC.；美花狸尾豆（翅果槐,叠果豆,狐狸尾,美花兔尾草,美花兔尾木,美丽兔尾草,密马,退蛆草,蜈蚣草,羽叶兔尾草）;Beautifulflower Uraria, Beautiful-flowered Uraria,Spiffyflower Uraria ●

402145　Uraria retroflexa Drake = Desmodium styracifolium（Osbeck）Merr. ●■

402146　Uraria rufescens（DC.）Schindl.；钩柄狸尾豆（钩柄狸尾草,银合欢）;Hookstalk Uraria,Reddish Uraria ●

402147　Uraria sinensis（Hemsl.）Franch.；中华狸尾豆（华南兔尾木,猫尾草,中华兔尾草）;China Uraria,Chinese Uraria ●

402148　Uraria sinensis Franch. = Uraria rufescens（DC.）Schindl. ●

402149　Urariopsis C. K. Schneid.（1916）;算珠豆属;Abacubeadbean,Urariopsis ●

402150　Urariopsis C. K. Schneid. = Uraria Desv. ●■

402151　Urariopsis brevissima Yen C. Yang et P. H. Huang;短序算珠豆;Shortinflorescence Urariopsis, Shortraceme Abacubeadbean ●

402152　Urariopsis brevissima Yen C. Yang et P. H. Huang = Uraria cochinchinensis Schindl. ●

402153　Urariopsis cordifolia（Wall.）Schindl.；算珠豆（大金钱草）；Abacubeadbean，Heartleaf Urariopsis，Heart-leaved Urariopsis ●

402154　Urariopsis cordifolia（Wall.）Schindl. = Uraria cordifolia Wall. ●

402155　Uraspermum Nutt.（废弃属名）= Osmorhiza Raf.（保留属名）■

402156　Uraspermum aristatum（Thunb.）Kuntze = Osmorhiza aristata（Thunb.）Makino et Y. Yabe ■

402157　Uraspermum aristatum Kuntze = Osmorhiza aristata（Thunb.）Makino et Y. Yabe ■

402158　Uratea J. F. Gmel. = Ouratea Aubl.（保留属名）●

402159　Uratella Post et Kuntze = Ouratea Aubl.（保留属名）●

402160　Uratella Post et Kuntze = Ouratella Tiegh. ●

402161　Urbananthus R. M. King et H. Rob.（1971）；光果亮泽兰属 ●☆

402162　Urbananthus critoniformis（Urb.）R. M. King et H. Rob.；光果亮泽兰 ●☆

402163　Urbanella Pierre = Lucuma Molina ●

402164　Urbanella Pierre = Pouteria Aubl. ●

402165　Urbania Phil.（1891）（保留属名）；乌尔班草属 ●☆

402166　Urbania Vatke（废弃属名）= Lyperia Benth. ■☆

402167　Urbania Vatke（废弃属名）= Urbania Phil.（保留属名）●☆

402168　Urbania lyperiiflora Vatke = Camptoloma lyperiiflorum（Vatke）Hilliard ■●☆

402169　Urbania pappigera Phil.；乌尔班草 ●☆

402170　Urbaniella Dusén ex Melch. = Haplolophium Cham.（保留属名）●☆

402171　Urbaniella Dusén ex Melch. = Urbanolophium Melch. ●☆

402172　Urbaniella Melch. = Haplolophium Cham.（保留属名）●☆

402173　Urbanisol Kuntze = Tithonia Desf. ex Juss. ●■

402174　Urbanodendron Mez（1889）；疣楠属 ●☆

402175　Urbanodendron verrucosum Mez；疣楠 ●☆

402176　Urbanodoxa Muschl.（1908）；乌尔班芥属（乌尔巴诺芥属）■☆

402177　Urbanodoxa Muschl. = Cremolobus DC. ■☆

402178　Urbanodoxa rhomboidea Muschl.；乌尔班芥 ■☆

402179　Urbanoguarea Harms = Guarea F. Allam.（保留属名）●☆

402180　Urbanolophium Melch. = Haplolophium Cham.（保留属名）●☆

402181　Urbanosdadium H. Wolff = Niphogeton Schltdl. ■☆

402182　Urbinella Greenm.（1903）；小秀菊属 ■☆

402183　Urbinella palmeri Greenm.；小秀菊 ■☆

402184　Urbinia Rose = Echeveria DC. ●■☆

402185　Urceodiscus W. J. de Wilde et Duyfjes = Melothria L. ■

402186　Urceodiscus W. J. de Wilde et Duyfjes（2006）；壶瓜属 ■☆

402187　Urceola Roxb.（1799）（保留属名）；水壶藤属（乐东藤属，小壶藤属）；Urceola ●

402188　Urceola Vand.（废弃属名）= Urceola Roxb.（保留属名）●

402189　Urceola esculenta Benth. ex Hook. f.；水壶藤；Kyetpaung ●☆

402190　Urceola huaitingii（Chun et Tsiang）D. J. Middleton；毛杜仲藤（鸡头藤，力酱梗，藤杜仲，续断，银花藤，引汁藤）；Hair Parabarium，Hairy Parabarium，Hairy Urceola ●

402191　Urceola huaitingii（Chun et Tsiang）D. J. Middleton = Parabarium huaitingii Chun et Tsiang ●

402192　Urceola linearicarpa（Pierre）D. J. Middleton；线果水壶藤（杜仲藤，牛角藤，养当杜）；Linear-fruit Parabarium，Linear-fruited Urceola，Oxhorn Parabarium ●

402193　Urceola linearicarpa（Pierre）D. J. Middleton = Parabarium linearicarpum（Pierre）Pichon ●

402194　Urceola micrantha（Wall. ex G. Don）D. J. Middleton；杜仲藤（奥蓼，白杜仲，白喉崩，白胶藤，大种笔须藤，杜浓，红杜仲，红及藤，花皮胶藤，鸡嘴藤，假杜仲，结衣藤，九年藤，老鸦嘴藤，牛腿子藤，乳藤，软羔藤，松筋藤，藤杜仲，藤仲，土杜仲，小赛格多，英蓼,中赛格多）；Parabarium，Small-flower Parabarium，Spire Parabarium，Spire Urceola ●

402195　Urceola micrantha（Wall. ex G. Don）D. J. Middleton = Ecdysanthera Utilis Hayata et Kawak. ●

402196　Urceola micrantha（Wall. ex G. Don）D. J. Middleton = Parabarium micranthum（Wall. ex G. Don）Pierre ●

402197　Urceola napeensis（Quintaret）D. J. Middleton；华南水壶藤（华南小壶藤）；Nape Urceola ●

402198　Urceola quintaretii（Pierre）D. J. Middleton；华南杜仲藤；South-China Urceola ●

402199　Urceola quintaretii（Pierre）D. J. Middleton = Parabarium chunianum Tsiang ●

402200　Urceola quintaretii（Pierre）D. J. Middleton = Parabarium hainanense Tsiang ●

402201　Urceola rosea（Hook. et Arn.）D. J. Middleton；酸叶胶藤（斑鸠藤，风藤，黑风藤，红背酸藤，厚皮藤，花皮胶藤，牛卷藤，乳藤，三酸藤，伞风藤，十八症，石酸藤，酸藤，酸藤木，酸叶藤，头林心，头淋沁，细叶榕藤）；Pinkflower Ecdysanthera，Pink-flowered Urceola，Sour Creeper ●

402202　Urceola rosea（Hook. et Arn.）D. J. Middleton = Ecdysanthera rosea Hook. et Arn. ●

402203　Urceola tournieri（Pierre）D. J. Middleton；云南水壶藤（大赛格多，赫马结）；Tournier Parabarium ●

402204　Urceola tournieri（Pierre）D. J. Middleton = Parabarium tournieri（Pierre）Pierre ex Spire ●

402205　Urceola xylinabariopsoides（Tsiang）D. J. Middleton = Chunechites xylinabariopsoides Tsiang ●

402206　Urceolaria F. Dietr. = Utricularia L. ●

402207　Urceolaria Herb. = Urceolina Rchb.（保留属名）■☆

402208　Urceolaria Huth = Sarmienta Ruiz et Pav.（保留属名）■●☆

402209　Urceolaria Molina = Sarmienta Ruiz et Pav.（保留属名）■●☆

402210　Urceolaria Molina ex J. D. Brandis（废弃属名）= Sarmienta Ruiz et Pav.（保留属名）■●☆

402211　Urceolaria Willd. = Schradera Vahl（保留属名）■☆

402212　Urceolaria Willd. ex Cothen. = Schradera Vahl（保留属名）■☆

402213　Urceolina Rchb.（1829）（保留属名）；耳蓝石蒜属 ■☆

402214　Urceolina fulva Herb.；黄耳壶石蒜 ■☆

402215　Urechites Müll. Arg.（1860）；蛇尾蔓属（黄花葵属，金香藤属，蛇尾曼属）；Vipertail ■☆

402216　Urechites Müll. Arg. = Pentalinon Voigt ●☆

402217　Urechites suberectus Müll. Arg.；蛇尾蔓（黄花葵）■☆

402218　Urelytrum Hack.（1887）；圆叶舌茅属 ■☆

402219　Urelytrum agropyroides（Hack.）Hack.；冰草状圆叶舌茅 ■☆

402220　Urelytrum annuum Stapf；一年生圆叶舌茅 ■☆

402221　Urelytrum auriculatum C. E. Hubb.；耳状圆叶舌茅 ■☆

402222　Urelytrum coronulatum Stapf = Urelytrum digitatum K. Schum. ■☆

402223　Urelytrum digitatum K. Schum.；掌状圆叶舌茅 ■☆

402224　Urelytrum fasciculatum Stapf ex C. E. Hubb. = Urelytrum digitatum K. Schum. ■☆

402225　Urelytrum giganteum Pilg.；大圆叶舌茅 ■☆

402226　Urelytrum gracilius C. E. Hubb. = Urelytrum agropyroides（Hack.）Hack. ■☆

402227　Urelytrum henrardii Chippind.；昂拉尔圆叶舌茅 ■☆

402228　Urelytrum monostachyum Peter = Urelytrum agropyroides（Hack.）Hack. ■☆

402229　Urelytrum muricatum C. E. Hubb.；短尖圆叶舌茅 ■☆

402230　Urelytrum pallidum C. E. Hubb. = Urelytrum agropyroides

（Hack.）Hack. ■☆

402231　Urelytrum semispirale Clayton = Urelytrum agropyroides（Hack.）Hack. ■☆

402232　Urelytrum setaceosubulatum P. A. Duvign. = Urelytrum agropyroides（Hack.）Hack. ■☆

402233　Urelytrum squarrosum Hack. = Urelytrum agropyroides（Hack.）Hack. ■☆

402234　Urelytrum stapfianum C. E. Hubb. = Urelytrum digitatum K. Schum. ■☆

402235　Urelytrum strigosum Gledhill = Loxodera strigosa（Gledhill）Clayton ■☆

402236　Urelytrum thyrsoides Stapf = Urelytrum giganteum Pilg. ■☆

402237　Urelytrum vanderystii Robyns = Urelytrum agropyroides（Hack.）Hack. ■☆

402238　Urena L.（1753）;梵天花属（地桃花属，野棉花属）;Buddhamallow, Indian Mallow ●■

402239　Urena capitata（L.）M. Gómez = Malachra capitata（L.）L. ■☆

402240　Urena chinensis Osbeck = Urena lobata L. var. chinensis（Osbeck）S. Y. Hu ■

402241　Urena diversifolia Schum. = Urena lobata L. ■

402242　Urena diversifolia Schumach. = Urena lobata L. ■

402243　Urena glabra R. Br. = Pavonia flavoferruginea（Forssk.）Hepper et J. R. I. Wood ●☆

402244　Urena lappago Sm. var. glauca Blume = Urena lobata L. var. glauca（Blume）Borss. Waalk. ■

402245　Urena lobata L.;地桃花（八卦草，八卦拦路虎，痴头婆，刺头婆，大梅花树，大迷马桩棵，大叶马松子，刀伤药，梵天花，狗脚迹，红孩儿，红花地桃花，红花虱母头，厚皮草，假桃花，毛桐子，迷马桩，牛毛七，千下锤，千下槌，三角风，山棋菜，虱麻头，虱母草，虱母子，石松毛，桃子草，天下锤，头婆，土杜仲，土口芪，小朝阳，肖梵天花，羊带归，野鸡花，野梅花，野棉花，野茄子，野桃花，野桐乔，油玲花，黏油子）;Aramina Fibre, Buddhamallow, Cadillo, Caesar Weed, Caesar's Weed, Caesarweed, Congo Jute, Guaxima, Lobate Wildcotfon, Rose Mallow ■

402246　Urena lobata L. subsp. sinuata（L.）Borss. Waalk. = Urena procumbens L. ●

402247　Urena lobata L. subsp. sinuata（L.）Borss. Waalk. = Urena sinuata L. ●

402248　Urena lobata L. var. alba X. Q. Li;白花肖梵天花（白花地桃花）;Whiteflower Lobate Wildcotfon ■

402249　Urena lobata L. var. chinensis（Osbeck）S. Y. Hu;中华地桃花（糙脉梵天花）;China Buddhamallow, Chinese Cadillo, Chinese Rose Mallow ■

402250　Urena lobata L. var. glauca（Blume）Borss. Waalk.;粗叶地桃花（粗糙野棉花，粗叶梵天花，狗扯尾，千金垂草，田芙蓉，消风草）;Scabrousleaf Buddhamallow, Scabrousleaf Cadillo, Scabrousleaf Rose Mallow ■

402251　Urena lobata L. var. henryi S. Y. Hu;湖北地桃花■

402252　Urena lobata L. var. reticulata（Cav.）Gurke;网状地桃花■☆

402253　Urena lobata L. var. scabriuscula（DC.）Walp. = Urena lobata L. var. glauca（Blume）Borss. Waalk. ■

402254　Urena lobata L. var. scabriuscula（DC.）Walp. = Urena lobata L. ■

402255　Urena lobata L. var. sinuata（L.）Borss. Waalk. = Urena sinuata L. ●

402256　Urena lobata L. var. sinuata（L.）Hochr. = Urena sinuata L. ●

402257　Urena lobata L. var. tomentosa（Blume）Walp. = Urena lobata L. ■

402258　Urena lobata L. var. viminea（Cav.）Gürke;柳枝地桃花■☆

402259　Urena lobata L. var. xichangensis Z. Y. Zhu;西昌地桃花■

402260　Urena lobata L. var. yunnanensis S. Y. Hu;云南地桃花;Yunnan Buddhamallow, Yunnan Cadillo, Yunnan Rose Mallow ■

402261　Urena mollis R. Br. = Pavonia burchellii（DC.）R. A. Dyer ●☆

402262　Urena monopetala Lour. = Urena lobata L. ■

402263　Urena morifolia DC. = Urena procumbens L. ●

402264　Urena ovalifolia Forssk. = Hibiscus ovalifolius（Forssk.）Vahl ●☆

402265　Urena polyflora Lour. = Triumfetta bartramia L. ■

402266　Urena polyflora Lour. = Triumfetta rhomboidea Jacq. ●■

402267　Urena procumbens L.;梵天花（八大锤，波叶棉花，痴头婆，黐头婆，地棉花，狗脚迹，狗脚跡，红野棉花，花蝴蝶，假棉花，假肉花，卷耳草，棉花肾，破布勒，七姐妹，犬跤迹，犬跤跡，犬跤爪，犬咬爪，三合枫，三角枫，山棉花，深裂梵天花，虱麻头，铁包金，乌云盖雪，五龙会，野棉花，野棉桃，野木棉，野茄，野桃花，黏花衣）;Buddhamallow, Procumbent Indian Mallow ●

402268　Urena procumbens L. = Urena lobata L. subsp. sinuata（L.）Borss. Waalk. ●

402269　Urena procumbens L. var. micraphylla K. M. Feng;小叶梵天花（白野棉花）;Littleleaf Procumbent Mallow ●

402270　Urena procumbens sensu Lour. = Triumfetta grandidens Hance ■

402271　Urena repanda Roxb.;波叶梵天花;Repand Indian Mallow, Waveleaf Buddhamallow ●

402272　Urena ricinocarpa Eckl. et Zeyh. = Sparrmannia ricinocarpa（Eckl. et Zeyh.）Kuntze ●☆

402273　Urena scabriuscula DC. = Urena lobata L. var. glauca（Blume）Borss. Waalk. ●

402274　Urena scabriuscula DC. = Urena lobata L. ■

402275　Urena scabriuscum DC. = Urena lobata L. var. scabriuscula（DC.）Walp. ■

402276　Urena sinuata L. = Urena lobata L. var. sinuata（L.）Borss. Waalk. ●

402277　Urena sinuata L. = Urena procumbens L. ●

402278　Urena speciosa Wall. = Urena repanda Roxb. ●

402279　Urena tomentosa Blume = Urena lobata L. ■

402280　Urena viminea Cav. = Urena lobata L. var. viminea（Cav.）Gürke ■☆

402281　Urera Gaudich.（1830）;肉果荨麻属（烧麻属）;Scrntchbush ●☆

402282　Urera baccifera（L.）Gaudin;肉果荨麻（浆果烧麻）■☆

402283　Urera batesii Rendle;贝茨肉果荨麻●☆

402284　Urera benthamii Wedd. = Urera thonneri De Wild. et T. Durand ■☆

402285　Urera bequaertii De Wild. ex Hauman;贝卡尔肉果荨麻●☆

402286　Urera braunii Engl. = Urera sansibarica Engl. ■☆

402287　Urera cameroonensis Wedd. = Urera trinervis（Hochst.）Friis et Immelman ■☆

402288　Urera cordifolia Engl.;心叶肉果荨麻（心叶烧麻）■☆

402289　Urera cuneata Rendle;楔形肉果荨麻（楔形烧麻）■☆

402290　Urera elliotii Rendle = Urera rigida（Benth.）Keay ■☆

402291　Urera engleriana Dinter = Obetia carruthersiana（Hiern）Rendle ●☆

402292　Urera fischeri Engl. = Urera sansibarica Engl. ■☆

402293　Urera flamigniana Lambinon;弗拉米尼肉果荨麻■☆

402294　Urera henriquesii Engl.;亨利克斯肉果荨麻■☆

402295　Urera keayi Letouzey;凯伊肉果荨麻■☆

402296　Urera mannii（Wedd.）Benth. et Hook. f. ex Rendle;曼氏肉果荨麻（曼氏烧麻）■☆

402297　Urera oblongifolia Benth.;矩圆叶肉果荨麻（矩圆叶烧麻）■☆

402298　Urera obovata Benth.;卵叶肉果荨麻（卵叶烧麻）■☆

402299　Urera radula Baker = Obetia radula（Baker）B. D. Jacks. ●☆

402300　Urera repens（Wedd.）Rendle;匍匐肉果荨麻（匍匐烧麻）■☆

402301　Urera rigida（Benth.）Keay;硬肉果荨麻（硬烧麻）■☆

402302　Urera robusta A. Chev. ;粗壮肉果荨麻(粗壮烧麻)■☆

402303　Urera sansibarica Engl. ;布劳恩肉果荨麻(布劳恩烧麻)■☆

402304　Urera talbotii Rendle;塔尔博特肉果荨麻(塔尔博特烧麻)■☆

402305　Urera tenax N. E. Br. = Obetia tenax (N. E. Br.) Friis ●☆

402306　Urera thonneri De Wild. et T. Durand;托内肉果荨麻(托内烧麻)■☆

402307　Urera trinervis (Hochst.) Friis et Immelman;三脉肉果荨麻(三脉烧麻)■☆

402308　Urera woodii N. E. Br. = Urera trinervis (Hochst.) Friis et Immelman ■☆

402309　Ureskinnera Post et Kuntze = Uroskinnera Lindl. ●☆

402310　Uretia Kuntze = Aerva Forssk. (保留属名)●■

402311　Uretia Post et Kuntze = Aerva Forssk. (保留属名)●■

402312　Uretia Post et Kuntze = Ouret Adans. (废弃属名)●■

402313　Uretia Raf. = Aerva Forssk. (保留属名) + Digera Forssk. ■☆

402314　Urginea Steinh. (1834);海葱属(仙葱属);Sea Onion,Sea-onion ■☆

402315　Urginea Steinh. = Drimia Jacq. ex Willd. ■☆

402316　Urginea acinacifolia Schinz = Albuca abyssinica Jacq. ■☆

402317　Urginea alooides Bolus = Aloe alooides (Bolus) Druten ■☆

402318　Urginea altissima (L. f.) Baker;高海葱■☆

402319　Urginea altissima (L. f.) Baker = Drimia altissima (L. f.) Ker Gawl. ■☆

402320　Urginea altissima Baker = Urginea altissima (L. f.) Baker ■☆

402321　Urginea amboensis Baker = Drimia indica (Roxb.) Jessop ■☆

402322　Urginea angolensis Baker;安哥拉海葱■☆

402323　Urginea angustisepala Engl. ;窄瓣海葱■☆

402324　Urginea anthericoides (Poir.) Steinh. var. secundiflora Maire = Charybdis numidica (Jord. et Fourr.) Speta ■☆

402325　Urginea arenosa Adamson = Drimia calcarata (Baker) Stedje ■☆

402326　Urginea aurantiaca H. Lindb. = Thuranthos aurantiacum (H. Lindb.) Speta ■☆

402327　Urginea autumnalis (L.) El Gadi = Prospero autumnale (L.) Speta ■☆

402328　Urginea bakeri Chiov. = Ornithogalum tenuifolium F. Delaroche ■☆

402329　Urginea basutica E. Phillips = Drimia angustifolia Baker ■☆

402330　Urginea beccarii Baker = Albuca abyssinica Jacq. ■☆

402331　Urginea bequaertii De Wild. ;贝卡尔海葱■☆

402332　Urginea brachystachys Baker = Drimia brachystachys (Baker) Stedje ■☆

402333　Urginea bragae Engl. ;布拉加海葱■☆

402334　Urginea brevifolia Baker;短叶海葱■☆

402335　Urginea brevipes Baker;短梗海葱■☆

402336　Urginea burkei Baker;布奇海葱■☆

402337　Urginea burkei Baker = Drimia sanguinea (Schinz) Jessop ■☆

402338　Urginea calcarata (Baker) Hilliard et B. L. Burtt = Drimia calcarata (Baker) Stedje ■☆

402339　Urginea capitata (Hook. f.) Baker = Drimia depressa (Baker) Jessop ■☆

402340　Urginea chlorantha Welw. ex Baker = Urginea angolensis Baker ■☆

402341　Urginea ciliata (L. f.) Baker = Drimia ciliata (L. f.) J. C. Manning et Goldblatt ■☆

402342　Urginea comosa Welw. ex Baker = Ornithogalum pulchrum Schinz ■☆

402343　Urginea corradii Chiov. = Ledebouria kirkii (Baker) Stedje et Thulin ■☆

402344　Urginea delagoensis Baker = Drimia delagoensis (Baker) Jessop ■☆

402345　Urginea depressa Baker = Drimia depressa (Baker) Jessop ■☆

402346　Urginea dimorphantha Baker = Ornithogalum pulchrum Schinz ■☆

402347　Urginea dregei Baker = Drimia dregei (Baker) J. C. Manning et Goldblatt ■☆

402348　Urginea duthieae Adamson = Drimia duthieae (Adamson) Jessop ■☆

402349　Urginea echinostachys Baker = Drimia macrocentra (Baker) Jessop ■☆

402350　Urginea ecklonii Baker;埃氏海葱■☆

402351　Urginea ecklonii Duthie = Drimia duthieae (Adamson) Jessop ■☆

402352　Urginea ensifolia (Thonn.) Hepper = Drimia glaucescens (Engl. et K. Krause) Scholz ■☆

402353　Urginea epigea R. A. Dyer = Drimia altissima (L. f.) Ker Gawl. ■☆

402354　Urginea eriospermoides Baker = Drimia anomala (Baker) Benth. ■☆

402355　Urginea exilis Adamson = Drimia minor (A. V. Duthie) Jessop ■☆

402356　Urginea exuviata (Jacq.) Steinh. = Drimia exuviata (Jacq.) Jessop ■☆

402357　Urginea filifolia (Jacq.) Steinh. = Drimia filifolia (Jacq.) J. C. Manning et Goldblatt ■☆

402358　Urginea fischeri Baker;菲舍尔海葱■☆

402359　Urginea flavovirens (Baker) Weim. ;黄绿海葱■☆

402360　Urginea flexuosa Adamson = Drimia exuviata (Jacq.) Jessop ■☆

402361　Urginea forsteri Baker = Drimia capensis (Burm. f.) Wijnands ■☆

402362　Urginea fragrans (Jacq.) Steinh. = Drimia fragrans (Jacq.) J. C. Manning et Goldblatt ■☆

402363　Urginea fugax (Moris) Steinh. ;早萎海葱■☆

402364　Urginea fugax (Moris) Steinh. var. major Litard. et Maire = Urginea fugax (Moris) Steinh. ■☆

402365　Urginea garuensis Engl. et K. Krause;加鲁海葱■☆

402366　Urginea gigantea (Jacq.) Oyewole;大海葱■☆

402367　Urginea glaucescens Engl. et K. Krause = Drimia glaucescens (Engl. et K. Krause) Scholz ■☆

402368　Urginea gracilis Duthie = Drimia calcarata (Baker) Stedje ■☆

402369　Urginea grandiflora Baker;大花海葱■☆

402370　Urginea hildebrandtii Baker;希氏海葱■☆

402371　Urginea indica (Roxb.) Kunth;印度海葱■☆

402372　Urginea indica (Roxb.) Kunth = Drimia indica (Roxb.) Jessop ■☆

402373　Urginea indica Kunth = Urginea indica (Roxb.) Kunth ■☆

402374　Urginea insignis Engl. et K. Krause;显著海葱■☆

402375　Urginea johnstonii Baker;约翰斯顿海葱■☆

402376　Urginea kniphofioides Baker = Drimia kniphofioides (Baker) J. C. Manning et Goldblatt ■☆

402377　Urginea langii Bremek. = Ornithogalum seineri (Engl. et K. Krause) Oberm. ■☆

402378　Urginea lilacina Baker = Drimia macrocentra (Baker) Jessop ■☆

402379　Urginea lydenburgensis R. A. Dyer = Drimia delagoensis (Baker) Jessop ■☆

402380　Urginea macrantha (Baker) E. Phillips = Drimia macrantha (Baker) Baker ■☆

402381　Urginea macrocentra Baker = Drimia macrocentra (Baker) Jessop ■☆

402382　Urginea mandalensis Baker = Drimia calcarata (Baker) Stedje ■☆

402383　Urginea mankonensis (A. Chev.) Hutch. = Albuca sudanica A. Chev. ■☆

402384　Urginea marginata (Thunb.) Baker = Drimia marginata (Thunb.) Jessop ■☆

402385　Urginea maritima (L.) Baker = Charybdis maritima (L.) Speta ■☆

402386　Urginea maritima (L.) Baker = Scilla maritima (L.) Baker ■☆

402387　Urginea maritima (L.) Baker subsp. maura (Maire) Maire = Charybdis maura (Maire) Speta ■☆

402388　Urginea maritima（L.）Baker var. angustifolia Maire ＝ Charybdis maritima（L.）Speta ■☆

402389　Urginea maritima（L.）Baker var. anthericoides（Poir.）Maire et Weiller ＝ Charybdis anthericoides（Poir.）Dobignard et Vela ■☆

402390　Urginea maritima（L.）Baker var. numidica（Jord. et Fourr.）Maire et Weiller ＝ Charybdis numidica（Jord. et Fourr.）Speta ■☆

402391　Urginea maritima（L.）Baker var. sphaeroidea（Jord. et Fourr.）Maire et Weiller ＝ Charybdis maritima（L.）Speta ■☆

402392　Urginea maritima（L.）Baker var. stenophylla Maire ＝ Charybdis maura（Maire）Speta ■☆

402393　Urginea maritima（L.）Baker var. tadlaensis Nègre ＝ Charybdis maritima（L.）Speta ■☆

402394　Urginea maura Maire ＝ Charybdis maura（Maire）Speta ■☆

402395　Urginea micrantha（A. Rich.）Solms ＝ Drimia altissima（L. f.）Ker Gawl. ■☆

402396　Urginea minor A. V. Duthie ＝ Drimia minor（A. V. Duthie）Jessop ■☆

402397　Urginea modesta Baker ＝ Drimia calcarata（Baker）Stedje ■☆

402398　Urginea mouretii Batt. et Trab. ＝ Ornithogalum caudatum Aiton ■☆

402399　Urginea muirii N. E. Br. ;缪里海葱■☆

402400　Urginea multifolia G. J. Lewis ＝ Drimia multifolia（G. J. Lewis）Jessop ■☆

402401　Urginea multisetosa Baker ＝ Drimia multisetosa（Baker）Jessop ■☆

402402　Urginea narcissifolia（A. Chev.）Hutch. ＝ Albuca sudanica A. Chev. ■☆

402403　Urginea natalensis Baker ＝ Drimia calcarata（Baker）Stedje ■☆

402404　Urginea nigritana Baker ＝ Albuca nigritana（Baker）Troupin ■☆

402405　Urginea noctiflora Batt. et Trab. ＝ Thuranthos noctiflorum（Batt. et Trab.）Speta ■☆

402406　Urginea noctiflora Batt. et Trab. subsp. aurantiaca（H. Lindb.）Maire ＝ Thuranthos aurantiacum（H. Lindb.）Speta ■☆

402407　Urginea noctiflora Batt. et Trab. subsp. helicophylla Maire ＝ Thuranthos noctiflorum（Batt. et Trab.）Speta ■☆

402408　Urginea numidica Jord. et Fourr. ＝ Charybdis numidica（Jord. et Fourr.）Speta ■☆

402409　Urginea nyasae Rendle;尼亚萨海葱■☆

402410　Urginea ollivieri Maire;奥氏海葱■☆

402411　Urginea paludosa Engl. et K. Krause;沼泽海葱■☆

402412　Urginea patersoniae Schönl. ＝ Urginea patersoniae Schönl ■☆

402413　Urginea pauciflora Baker ＝ Ebertia pauciflora（Baker）Speta ●☆

402414　Urginea pedunculata Adamson ＝ Drimia calcarata（Baker）Stedje ■☆

402415　Urginea petitiana（A. Rich.）Solms ＝ Albuca abyssinica Jacq. ■☆

402416　Urginea petrophila A. Chev. ;喜岩海葱■☆

402417　Urginea physodes（Jacq.）Baker ＝ Drimia physodes（Jacq.）Jessop ■☆

402418　Urginea pilosula Engl. ＝ Trachyandra saltii（Baker）Oberm. ■☆

402419　Urginea porphyrantha Bullock ＝ Drimia porphyrantha（Bullock）Stedje ■☆

402420　Urginea pretoriensis Baker ＝ Drimia calcarata（Baker）Stedje ■☆

402421　Urginea psilostachya Welw. ex Baker;裸穗海葱■☆

402422　Urginea pulchra（Schinz）Sölch ＝ Ornithogalum pulchrum Schinz ■☆

402423　Urginea pusilla（Jacq.）Baker ＝ Drimia physodes（Jacq.）Jessop ■☆

402424　Urginea pygmaea Duthie ＝ Drimia minor（A. V. Duthie）Jessop ■☆

402425　Urginea quartiniana（A. Rich.）Solms ＝ Albuca abyssinica Jacq. ■☆

402426　Urginea rautanenii Baker ＝ Drimia sanguinea（Schinz）Jessop ■☆

402427　Urginea revoluta A. V. Duthie ＝ Drimia hesperantha J. C. Manning et Goldblatt ■☆

402428　Urginea rigidifolia Baker ＝ Drimia sclerophylla J. C. Manning et Goldblatt ■☆

402429　Urginea riparia Baker ＝ Drimia calcarata（Baker）Stedje ■☆

402430　Urginea rubella Baker ＝ Drimia calcarata（Baker）Stedje ■☆

402431　Urginea salmonea Berhaut;鲑色海葱■☆

402432　Urginea salteri Compton ＝ Drimia salteri（Compton）J. C. Manning et Goldblatt ■☆

402433　Urginea sanguinea Schinz ＝ Drimia sanguinea（Schinz）Jessop ■☆

402434　Urginea saniensis Hilliard et B. L. Burtt ＝ Drimia saniensis（Hilliard et B. L. Burtt）J. C. Manning et Goldblatt ■☆

402435　Urginea schlechteri Baker ＝ Drimia macrocentra（Baker）Jessop ■☆

402436　Urginea scilla Steinh. ＝ Charybdis maritima（L.）Speta ■☆

402437　Urginea scilla Steinh. ＝ Scilla maritima（L.）Baker ■☆

402438　Urginea sebirei Berhaut ＝ Drimia indica（Roxb.）Jessop ■☆

402439　Urginea secunda Baker ＝ Ornithogalum secundum Jacq. ■☆

402440　Urginea segetalis A. Chev. ;谷地海葱■☆

402441　Urginea simensis（Hochst. ex A. Rich.）Schweinf. ＝ Charybdis maritima（L.）Speta ■☆

402442　Urginea somalensis Chiov. ＝ Ornithogalum donaldsonii（Rendle）Greenway ■☆

402443　Urginea tayloriana Rendle ＝ Ornithogalum tenuifolium F. Delaroche ■☆

402444　Urginea tazensis（Batt. et Maire）Maire ＝ Charybdis tazensis（Maire）Speta ■☆

402445　Urginea tenella Baker ＝ Drimia calcarata（Baker）Stedje ■☆

402446　Urginea ubomboensis R. A. Dyer ＝ Drimia delagoensis（Baker）Jessop ■☆

402447　Urginea umgeniensis Poelln. ;乌姆加尼海葱■☆

402448　Urginea undulata（Desf.）Steinh. ＝ Charybdis undulata（Desf.）Speta ■☆

402449　Urginea undulata（Desf.）Steinh. subsp. tazensis（Batt. et Maire）Maire et Weiller ＝ Charybdis tazensis（Maire）Speta ■☆

402450　Urginea undulata（Desf.）Steinh. var. major Gatt. et Weiller ＝ Charybdis undulata（Desf.）Speta ■☆

402451　Urginea undulatifolia Steinh. ＝ Charybdis undulata（Desf.）Speta ■☆

402452　Urginea unifolia Duthie ＝ Drimia exuviata（Jacq.）Jessop ■☆

402453　Urginea virens Schltr. ＝ Drimia virens（Schltr.）J. C. Manning et Goldblatt ■☆

402454　Urginea viridula Baker;浅绿海葱■☆

402455　Urginea volubilis H. Perrier ＝ Igidia volubilis（H. Perrier）Speta ■☆

402456　Urginea zambesiaca Baker ＝ Thuranthos zambesiacum（Baker）Kativu ■☆

402457　Urgineopsis Compton ＝ Drimia Jacq. ex Willd. ■☆

402458　Urgineopsis Compton（1930）;类海葱属■☆

402459　Urgineopsis salteri Compton;类海葱■☆

402460　Urgneopsis Compton ＝ Urginea Steinh. ■☆

402461　Uribea Dugand et Romero（1962）;假罗望子属■☆

402462　Uribea tamarindoides Dugand et Romero;假罗望子■☆

402463　Uricola Boerl. ＝ Urceola Roxb.（保留属名）●

402464　Urinaria Medik. ＝ Phyllanthus L. ●■

402465　Urmenetea Phil.（1860）;锈冠菊属■☆

402466　Urmenetia Phil. ＝ Urmenetea Phil. ■☆

402467　Urmenetia atacamensis Phil. ;锈冠菊■☆

402468　Urnectis Raf. ＝ Salix L.（保留属名）●

402469　Urnularia P. Karst.（废弃属名）＝ Urnularia Stapf ●☆

402470　Urnularia Stapf = Willughbeia Roxb. (保留属名) ●☆

402471　Urobotrya Stapf (1905) (保留属名) ; 尾球木属 (鳞尾木属) ; Urobotrya ●

402472　Urobotrya afzelii (Engl.) Stapf ex Hutch. et Dalziel = Urobotrya congolana (Baill.) Hiepko subsp. afzelii (Engl.) Hiepko ●☆

402473　Urobotrya angustifolia Stapf = Urobotrya congolana (Baill.) Hiepko ●☆

402474　Urobotrya congolana (Baill.) Hiepko ; 康哥尔尾球木●☆

402475　Urobotrya congolana (Baill.) Hiepko subsp. afzelii (Engl.) Hiepko ; 阿氏尾球木●☆

402476　Urobotrya latifolia Stapf = Urobotrya congolana (Baill.) Hiepko subsp. afzelii (Engl.) Hiepko ●

402477　Urobotrya latisquama (Gagnep.) Hiepko ; 尾球木 ; Broadscale Urobotrya , Broad-scaled Urobotrya , Urobotrya ●

402478　Urobotrya latisquama (Gagnep.) Stapf = Urobotrya latisquama (Gagnep.) Hiepko ●

402479　Urobotrya minutiflora Stapf = Urobotrya congolana (Baill.) Hiepko ●☆

402480　Urobotrya sparsiflora (Engl.) Hiepko ; 疏花尾球木●☆

402481　Urobotrya sparsiflora (Engl.) Hiepko subsp. bruneelii (De Wild.) Hiepko ; 布氏尾球木●☆

402482　Urobotyra trinervia Stapf = Urobotrya congolana (Baill.) Hiepko subsp. afzelii (Engl.) Hiepko ●☆

402483　Urobotrya stapfiana Hutch. et Dalziel = Urobotrya congolana (Baill.) Hiepko ●☆

402484　Urocarpidium Ulbr. (1916) ; 尾果锦葵属■☆

402485　Urocarpidium chilense (A. Braun et C. D. Bouché) Krapov. ; 智利尾果锦葵■☆

402486　Urocarpidium echinatum (C. Presl) Krapov. ; 刺尾果锦葵■☆

402487　Urocarpidium leptocalyx Krapov. ; 细萼尾果锦葵■☆

402488　Urocarpidium macrocarpum Krapov. ; 大果尾果锦葵■☆

402489　Urocarpus J. L. Drumm. ex Harv. = Asterolasia F. Muell. ●☆

402490　Urochilus D. L. Jones et M. A. Clem. (2002) ; 尾唇兰属■☆

402491　Urochilus D. L. Jones et M. A. Clem. = Pterostylis R. Br. (保留属名) ■☆

402492　Urochilus vittatus (Lindl.) D. L. Jones et M. A. Clem. ; 尾唇兰■☆

402493　Urochlaena Nees = Tribolium Desv. ■☆

402494　Urochlaena Nees (1841) ; 风滚尾属■☆

402495　Urochlaena major Rendle = Tribolium pusillum (Nees) H. P. Linder et Davidse ■☆

402496　Urochlaena pusilla Nees = Tribolium pusillum (Nees) H. P. Linder et Davidse ■☆

402497　Urochloa P. Beauv. (1812) ; 尾稃草属 ; Signal-grass , Tailgrass , Urochloa ■

402498　Urochloa ambigua (Trin.) Pilg. = Urochloa paspaloides J. Presl ex C. Presl ■

402499　Urochloa arrecta (Hack. ex T. Durand et Schinz) Morrone et Zuloaga ; 非洲尾稃草 ; African Signalgrass ■☆

402500　Urochloa bifalcigera (Stapf) Stapf = Brachiaria platynota (K. Schum.) Robyns ■☆

402501　Urochloa bolbodes (Steud.) Stapf = Urochloa oligotricha (Fig. et De Not.) Henrard ■☆

402502　Urochloa brachyphylla Gilli = Urochloa trichopus (Hochst.) Stapf ■☆

402503　Urochloa brachyura (Hack.) Stapf ; 短尾尾稃草■☆

402504　Urochloa cimicina (L.) Kunth = Alloteropsis cimicina (L.) Stapf ■

402505　Urochloa coccosperma (Steud.) Stapf ex Reeder = Brachiaria villosa (Lam.) A. Camus ■

402506　Urochloa cordata Keng ex S. L. Chen et Y. X. Jin ; 心叶尾稃草 ; Cordate Urochloa , Heartleaf Tailgrass ■

402507　Urochloa cordata Keng ex S. L. Chen et Y. X. Jin = Urochloa setigera (Retz.) Stapf ■

402508　Urochloa cruciata Boiss. ; 十字形尾稃草■☆

402509　Urochloa decidua Morrone et Zuloaga = Rupichloa decidua (Morrone et Zuloaga) Salariato et Morrone ■☆

402510　Urochloa decumbens (Stapf) R. D. Webster = Brachiaria decumbens Stapf ■☆

402511　Urochloa dictyoneura (Fig. et De Not.) Veldkamp = Brachiaria dictyoneura (Fig. et De Not.) Stapf ■☆

402512　Urochloa distachya (L.) T. Q. Nguyen ; 热带尾稃草 ; Tropical Signalgrass ■☆

402513　Urochloa distachya (L.) T. Q. Nguyen = Brachiaria distachya (L.) Stapf ■☆

402514　Urochloa echinolaenoides Stapf ; 刺被尾稃草■☆

402515　Urochloa engleri Pilg. = Urochloa trichopus (Hochst.) Stapf ■☆

402516　Urochloa fusiformis (Reeder) Veldkamp. = Brachiaria fusiformis Reeder ■☆

402517　Urochloa geniculata C. E. Hubb. = Urochloa brachyura (Hack.) Stapf ■☆

402518　Urochloa gorinii Chiov. = Urochloa rudis Stapf ■☆

402519　Urochloa helopus (Trin.) Stapf = Urochloa panicoides P. Beauv. ■

402520　Urochloa insculpta (Steud.) Stapf = Brachiaria lata (Schumach.) C. E. Hubb. ■☆

402521　Urochloa jinshaicola B. S. Sun et Z. H. Hu ; 金沙尾稃草 ; Jinsha Tailgrass , Jinsha Urochloa ■

402522　Urochloa jinshaicola B. S. Sun et Z. H. Hu = Urochloa panicoides P. Beauv. ■

402523　Urochloa kurzii (Hook. f.) T. Q. Nguyen = Brachiaria kurzii (Hook. f.) A. Camus ■

402524　Urochloa lata (Schumach.) C. E. Hubb. = Brachiaria lata (Schumach.) C. E. Hubb. ■☆

402525　Urochloa longifolia B. S. Sun et Z. H. Hu ; 长叶尾稃草 ; Longleaf Tailgrass , Longleaf Urochloa ■

402526　Urochloa longifolia B. S. Sun et Z. H. Hu = Urochloa panicoides P. Beauv. ■

402527　Urochloa longifolia B. S. Sun et Z. H. Hu var. yuanmouensis (B. S. Sun et Z. H. Hu) S. L. Chen et Y. X. Jin ; 元谋尾稃草 ; Yuanmou Tailgrass , Yuanmou Urochloa ■

402528　Urochloa longifolia B. S. Sun et Z. H. Hu var. yuanmuensis (B. S. Sun et Z. H. Hu) S. L. Chen et Y. X. Jin = Urochloa panicoides P. Beauv. ■

402529　Urochloa marathensis Henrard = Urochloa panicoides P. Beauv. ■

402530　Urochloa marathensis Henrard var. velutina Henrard = Urochloa panicoides P. Beauv. ■

402531　Urochloa maxima (Jacq.) R. D. Webster ; 几内亚尾稃草 ; Green Panic Grass , Guinea Grass , Guineagrass ■☆

402532　Urochloa maxima (Jacq.) R. D. Webster = Panicum maximum Jacq. ■

402533　Urochloa mosambicensis (Hack.) Dandy ; 沙拉草 ; African Liverseed Grass , Sabigrass ■☆

402534　Urochloa mutica (Forssk.) T. Q. Nguyen = Brachiaria mutica (Forssk.) Stapf ■

402535　Urochloa novemnervia C. E. Hubb. = Urochloa brachyura (Hack.)

Stapf ■☆

402536　Urochloa oligobrachiata（Pilg.）Kartesz；寡枝尾稃草；Weak Signalgrass ■☆

402537　Urochloa oligotricha（Fig. et De Not.）Henrard；少毛尾稃草■☆

402538　Urochloa panicoides P. Beauv.；类黍尾稃草；Panic Liverseed Grass,Panicgrasslike Tailgrass,Panicgrasslike Urochloa ■

402539　Urochloa panicoides P. Beauv. var. marathensis（Henrard）Bor. = Urochloa panicoides P. Beauv. ■

402540　Urochloa panicoides P. Beauv. var. pubescens（Kunth）Bor = Urochloa panicoides P. Beauv. ■

402541　Urochloa panicoides P. Beauv. var. velutina（Henrard）Bor = Urochloa panicoides P. Beauv. ■

402542　Urochloa paniculata Benth. = Alloteropsis paniculata（Benth.）Stapf ■☆

402543　Urochloa paspaloides J. Presl ex C. Presl；雀稗尾稃草；Paspalum Tailgrass,Paspalum Urochloa,Sharp-flowered Signal-grass ■

402544　Urochloa plantaginea（Link）R. D. Webster；车前尾稃草；Creeping Signal Grass,Plantain Signalgrass ■☆

402545　Urochloa platynota（K. Schum.）Pilg. = Brachiaria platynota（K. Schum.）Robyns ■☆

402546　Urochloa platyphylla（Griseb.）R. D. Webster；宽叶尾稃草；Broad-leaved Signal-grass ■☆

402547　Urochloa platyphylla（Munro ex C. Wright）R. D. Webster = Urochloa platyphylla（Griseb.）R. D. Webster ■☆

402548　Urochloa platyrrhachis C. E. Hubb.；阔轴尾稃草■☆

402549　Urochloa pubescens Kunth = Urochloa panicoides P. Beauv. ■

402550　Urochloa pullulans Stapf = Urochloa mosambicensis（Hack.）Dandy ■☆

402551　Urochloa quintasii Mez = Alloteropsis cimicina（L.）Stapf ■

402552　Urochloa ramosa（L.）T. Q. Nguyen = Brachiaria ramosa（L.）Stapf ■

402553　Urochloa repans（L.）Stapf；尾稃草（匍匐臂形草）；Creeping Signalgrass,Creeping Tailgrass,Creeping Urochloa ■

402554　Urochloa repans（L.）Stapf var. glabra S. L. Chen et Y. X. Jin；光尾稃草；Glabrous Creeping Tailgrass,Glabrous Creeping Urochloa ■

402555　Urochloa reptans（L.）Stapf = Brachiaria reptans（L.）C. A. Gardner et C. E. Hubb. ■

402556　Urochloa rhodesiensis Stent = Urochloa mosambicensis（Hack.）Dandy ■☆

402557　Urochloa rudis Stapf；粗糙尾稃草■☆

402558　Urochloa ruschii Pilg. = Urochloa panicoides P. Beauv. ■

402559　Urochloa ruziziensis（R. Germ. et C. M. Evrard）Crins；鲁济济尾稃草■☆

402560　Urochloa sclerochlaena Chiov.；硬被尾稃草■☆

402561　Urochloa sclerochlaena Chiov. var. commelinoides？= Urochloa sclerochlaena Chiov. ■☆

402562　Urochloa semialata（R. Br.）Kunth = Alloteropsis semialata（R. Br.）Hitchc. ■

402563　Urochloa semialatus Desv. = Alloteropsis semialata（R. Br.）Hitchc. ■

402564　Urochloa setigera（Retz.）Stapf；刺毛尾稃草■

402565　Urochloa stolonifera（Gooss.）Chippind.；匍匐尾稃草■☆

402566　Urochloa subquadripara（Trin.）R. D. Webster = Brachiaria subquadripara（Trin.）Hitchc. ■

402567　Urochloa texana（Buckley）R. D. Webster；得州尾稃草；Browntop Signalgrass,Texas Millet,Texas Panic Grass ■☆

402568　Urochloa trichopus（Hochst.）Stapf；毛梗尾稃草■☆

402569　Urochloa villosa（Lam.）T. Q. Nguyen = Brachiaria villosa（Lam.）A. Camus ■

402570　Urochloa yuanmouensis B. S. Sun et Z. H. Hu = Urochloa longifolia B. S. Sun et Z. H. Hu var. yuanmouensis（B. S. Sun et Z. H. Hu）S. L. Chen et Y. X. Jin ■

402571　Urochloa yuanmuensis B. S. Sun et Z. H. Hu = Urochloa panicoides P. Beauv. ■

402572　Urochondra C. E. Hubb.（1947）；毛子房草属■☆

402573　Urochondra setulosa（Trin.）C. E. Hubb.；毛子房草■☆

402574　Urodesmium Naudin = Pachyloma DC. ●☆

402575　Urodon Turcz. = Pultenaea Sm. ●☆

402576　Urogentias Gilg et Gilg-Ben.（1933）；尾龙胆属■☆

402577　Urogentias ulugurensis Gilg et Gilg-Ben.；尾龙胆■☆

402578　Urolepis（DC.）R. M. King et H. Rob.（1971）；尾鳞菊属■●☆

402579　Urolepis hecatantha（DC.）R. M. King et H. Rob.；尾鳞菊■☆

402580　Uromorus Bureau = Streblus Lour. ●

402581　Uromyrtus Burret（1941）；尾香木属●☆

402582　Uromyrtus australis A. J. Scott；尾香木●☆

402583　Uromyrtus rostrata（Lauterb.）N. Snow et Guymer；喙尾香木●☆

402584　Uropappus Nutt.（1841）；尾毛菊属；Silver-puffs ■☆

402585　Uropappus Nutt. = Microseris D. Don ■☆

402586　Uropappus heterocarpus Nutt. = Stebbinsoseris heterocarpa（Nutt.）K. L. Chambers ■☆

402587　Uropappus lindleyi（DC.）Nutt.；尾毛菊；Lindley's Silver Puff ■☆

402588　Uropappus linearifolius Nutt. = Uropappus lindleyi（DC.）Nutt. ■☆

402589　Uropedilum Pfitzer = Phragmipedium Rolfe（保留属名）■☆

402590　Uropedilum Pfitzer = Uropedium Lindl.（废弃属名）■☆

402591　Uropedium Lindl.（废弃属名）= Phragmipedium Rolfe（保留属名）■☆

402592　Uropetalon Burch. ex Ker Gawl. = Zuccangnia Thunb.（废弃属名）●☆

402593　Uropetalon Ker Gawl. = Dipcadi Medik. ■☆

402594　Uropetalum Burch. = Dipcadi Medik. ■☆

402595　Uropetalum Burch. = Uropetalon Ker Gawl. ■☆

402596　Uropetalum ciliare Eckl. et Zeyh. ex Harv. = Ornithogalum cirrhulosum J. C. Manning et Goldblatt ■☆

402597　Uropetalum glaucum Burch. ex Ker Gawl. = Dipcadi glaucum（Burch. ex Ker Gawl.）Baker ■☆

402598　Uropetalum longifolium Lindl. = Dipcadi longifolium（Lindl.）Baker ■☆

402599　Uropetalum minimum Steud. ex A. Rich. = Ornithogalum viride（L.）J. C. Manning et Goldblatt ■☆

402600　Uropetalum serotinum（L.）Ker Gawl. = Dipcadi serotinum（L.）Medik. ■☆

402601　Uropetalum tacazzeanum A. Rich. = Ornithogalum viride（L.）J. C. Manning et Goldblatt ■☆

402602　Uropetalum umbonatum Baker = Ornithogalum viride（L.）J. C. Manning et Goldblatt ■☆

402603　Urophyllon Salisb. = Ornithogalum L. ■

402604　Urophyllum C. Koch = Urospatha Schott ■☆

402605　Urophyllum Jack ex Wall.（1824）；尖叶木属（尾叶树属）；Tailleaftree,Urophyllum ●

402606　Urophyllum K. Koch = Urospatha Schott ■☆

402607　Urophyllum Wall. = Urophyllum Jack ex Wall. ●

402608　Urophyllum afzelii Hiern = Pauridiantha afzelii（Hiern）Bremek. ●☆

402609 Urophyllum annobonense Mildbr. = Pauridiantha insularis（Hiern）Bremek. ●☆

402610 Urophyllum bequaertii De Wild. = Pauridiantha bequaertii（De Wild.）Bremek. ●☆

402611 Urophyllum butaguense De Wild. = Pauridiantha paucinervis（Hiern）Bremek. subsp. butaguensis（De Wild.）Verdc. ●☆

402612 Urophyllum butaguense De Wild. var. exsertostylosum ? = Pauridiantha paucinervis（Hiern）Bremek. subsp. butaguensis（De Wild.）Verdc. ☆

402613 Urophyllum callicarpoides Hiern = Pauridiantha callicarpoides（Hiern）Bremek. ●☆

402614 Urophyllum camposii G. Taylor = Pauridiantha camposii（G. Taylor）Exell ●☆

402615 Urophyllum cauliflorum R. D. Good = Stelechantha cauliflora（R. D. Good）Bremek. ■☆

402616 Urophyllum chinense Merr. et Chun;尖叶木;China Tailleaftree, Chinese Urophyllum ●

402617 Urophyllum chloranthum K. Schum. = Rhipidantha chlorantha（K. Schum.）Bremek. ●☆

402618 Urophyllum dewevrei De Wild. et T. Durand = Pauridiantha dewevrei（De Wild. et T. Durand）Bremek. ●☆

402619 Urophyllum divaricatum K. Schum. = Pauridiantha divaricata（K. Schum.）Bremek. ●☆

402620 Urophyllum eketense Wernham;埃凯特尖叶木 ●☆

402621 Urophyllum filiformi-stipulatum De Wild. = Tricalysia filiformi-stipulata（De Wild.）Brenan ●☆

402622 Urophyllum floribundum K. Schum. et K. Krause = Pauridiantha floribunda（K. Schum. et K. Krause）Bremek. ●☆

402623 Urophyllum gilletii De Wild. et T. Durand = Pauridiantha viridiflora（Hiern）Hepper ●☆

402624 Urophyllum guangxiensis H. C. Lo;广西尖叶木（小花尖叶木）;Guangxi Tailleaftree, Guangxi Urophyllum ●

402625 Urophyllum hirtellum Benth. = Pauridiantha hirtella（Benth.）Bremek. ●☆

402626 Urophyllum holstii K. Schum. = Pauridiantha paucinervis（Hiern）Bremek. subsp. holstii（K. Schum.）Verdc. ●☆

402627 Urophyllum insculptum Hutch. et Dalziel = Pauridiantha insculpta（Hutch. et Dalziel）Bremek. ●☆

402628 Urophyllum insulare Hiern = Pauridiantha insularis（Hiern）Bremek. ●☆

402629 Urophyllum kassaiense K. Schum. ex Mildbr. = Pauridiantha dewevrei（De Wild. et T. Durand）Bremek. ●☆

402630 Urophyllum lanuriense De Wild. = Pauridiantha paucinervis（Hiern）Bremek. subsp. butaguensis（De Wild.）Verdc. ●☆

402631 Urophyllum liebrechtsiana De Wild. et T. Durand = Pauridiantha liebrechtsiana（De Wild. et T. Durand）Ntore et Dessein ●☆

402632 Urophyllum linderi Hutch. et Dalziel = Schizocolea linderi（Hutch. et Dalziel）Bremek. ●☆

402633 Urophyllum mayumbense R. D. Good = Pauridiantha mayumbensis（R. D. Good）Bremek. ●☆

402634 Urophyllum micranthum Hiern = Pauridiantha micrantha（Hiern）Bremek. ●☆

402635 Urophyllum parviflorum F. C. How ex H. S. Lo;小花尖叶木;Littleflower Urophyllum, Smallflower Tailleaftree ●

402636 Urophyllum paucinerve Hiern = Pauridiantha paucinervis（Hiern）Bremek. ●☆

402637 Urophyllum pyramidatum K. Krause = Pauridiantha pyramidata（K. Krause）Bremek. ●☆

402638 Urophyllum rubens Benth. = Pauridiantha rubens（Benth.）Bremek. ●☆

402639 Urophyllum setiflorum R. D. Good = Poecilocalyx setiflorus（R. D. Good）Bremek. ■☆

402640 Urophyllum stelechanthum Mildbr. = Stelechantha cauliflora（R. D. Good）Bremek. ■☆

402641 Urophyllum stenophyllum K. Krause = Pauridiantha multiflora K. Schum. ●☆

402642 Urophyllum stipulosum Hutch. et Dalziel = Poecilocalyx stipulosa（Hutch. et Dalziel）N. Hallé ■☆

402643 Urophyllum sylvicola Hutch. et Dalziel = Pauridiantha sylvicola（Hutch. et Dalziel）Bremek. ●☆

402644 Urophyllum symplocoides S. Moore = Pauridiantha symplocoides（S. Moore）Bremek. ●☆

402645 Urophyllum talbotii Wernham;塔尔博特尖叶木 ●☆

402646 Urophyllum tsaianum F. C. How ex H. S. Lo;滇南尖叶木;H. T. Tsai Urophyllum, Tsai Tailleaftree ●

402647 Urophyllum verticillatum De Wild. et T. Durand = Pauridiantha verticillata（De Wild. et T. Durand）N. Hallé ●☆

402648 Urophyllum viridiflorum Hiern = Pauridiantha viridiflora（Hiern）Hepper ●☆

402649 Urophyllum xanthorrhoeum K. Schum. = Danais xanthorrhoea（K. Schum.）Bremek. ■☆

402650 Urophyllum ziamaeanum Jacq. -Fél. = Stelechantha ziamaeana（Jacq. -Fél.）N. Hallé ■☆

402651 Urophysa Ulbr.（1929）;尾囊草属;Tailsacgrass, Urophysa ■

402652 Urophysa henryi（Oliv.）Ulbr.;尾囊草（尾囊果,岩蝴蝶）;Henry Tailsacgrass, Henry Urophysa ■

402653 Urophysa rockii Ulbr.;距瓣尾囊草;Rock Tailsacgrass, Spurredpetal Urophysa ■

402654 Uroskinnera Lindl.（1857）;尾婆纳属 ●☆

402655 Uroskinnera spectabilis Lindl.;尾婆纳 ●☆

402656 Urospatha Schott（1853）;尾苞南星属 ■☆

402657 Urospatha affinis Schott;近缘尾苞南星 ■☆

402658 Urospatha angusta K. Krause;窄尾苞南星 ■☆

402659 Urospathella G. S. Bunting = Urospatha Schott ■☆

402660 Urospermum Scop.（1777）;尾籽菊属（喙果苣属,尾子菊属）;Sheep's-beard ■☆

402661 Urospermum capense（Jacq.）Spreng. = Urospermum picrioides（L.）Scop. ex F. W. Schmidt ■☆

402662 Urospermum dalechampii（L.）F. W. Schmidt;达氏尾籽菊 ■☆

402663 Urospermum picrioides（L.）Scop. ex F. W. Schmidt;尾子菊;Prickly Goldenfleece ■☆

402664 Urospermum picroides（L.）F. W. Schmidt = Urospermum picrioides（L.）Scop. ex F. W. Schmidt ■☆

402665 Urospermum picroides（L.）Scop. var. laeve Maire = Urospermum picrioides（L.）Scop. ex F. W. Schmidt ■☆

402666 Urostachya（Lindl.）Brieger = Eria Lindl.（保留属名）■

402667 Urostachya（Lindl.）Brieger = Pinalia Lindl. ■

402668 Urostachya（Lindl.）Brieger（1981）;尾花兰属 ■

402669 Urostachya chrysantha（Schltr.）Brieger;金尾花兰 ■☆

402670 Urostachya floribunda（Lindl.）Brieger;多花尾花兰 ■☆

402671 Urostachya pachystachya（Lindl.）Rauschert;粗穗尾花兰 ■☆

402672 Urostelma Bunge = Metaplexis R. Br. ●■

402673 Urostelma chinensis Bunge = Metaplexis japonica（Thunb.）Makino ●■

402674　Urostemon B. Nord. (1978);澳洲尾药菊属■☆

402675　Urostemon B. Nord. = Brachyglottis J. R. Forst. et G. Forst. ●■☆

402676　Urostemon kirkii (Hook. f. ex Kirk) B. Nord. ;澳洲尾药菊■☆

402677　Urostephanus B. L. Rob. et Greenm. (1895);尾冠萝藦属●☆

402678　Urostephanus gonoloboides Rob. et Greenm. ;尾冠萝藦●☆

402679　Urostigma Gasp. = Ficus L. ●

402680　Urostigma Gasp. = Mastosuke Raf. ●

402681　Urostigma abutilifolium Miq. = Ficus abutilifolia (Miq.) Miq. ●☆

402682　Urostigma amblyphyllum Miq. = Ficus microcarpa L. f. ●

402683　Urostigma annulatum (Blume) Miq. = Ficus annulata Blume ●

402684　Urostigma benjamina (L.) Miq. = Ficus benjamina L. ●

402685　Urostigma benjamina Miq. = Ficus benjamina L. ●

402686　Urostigma benjaminum (L.) Miq. = Ficus benjamina L. ●

402687　Urostigma benjaminum Miq. var. nudum (Miq.) Miq. = Ficus benjamina L. var. nuda (Miq.) Barrett ●

402688　Urostigma benjaminum Miq. var. nudum Miq. = Ficus benjamina L. var. nuda (Miq.) Barrett ●

402689　Urostigma burkei Miq. = Ficus burkei (Miq.) Miq. ●☆

402690　Urostigma caulocarpum Miq. = Ficus caulocarpa (Miq.) Miq. ●

402691　Urostigma circumscissum Miq. = Ficus elastica Roxb. ●

402692　Urostigma concinnum Miq. = Ficus concinna (Miq.) Miq. ●

402693　Urostigma cordifolium (Roxb.) Miq. = Ficus rumphii Blume ●

402694　Urostigma dasycarpum Miq. = Ficus drupacea Thunb. var. pubescens (Roth) Corner ●

402695　Urostigma dekdekena Miq. = Ficus thonningii Blume ●☆

402696　Urostigma elasticum (Roxb. ex Hornem.) Miq. = Ficus elastica Roxb. ●

402697　Urostigma elasticum (Roxb.) Miq. = Ficus elastica Roxb. ●

402698　Urostigma elasticum Miq. = Ficus elastica Roxb. ●

402699　Urostigma elegans Miq. = Ficus artocarpoides Warb. ●☆

402700　Urostigma fazokelense Miq. = Ficus glumosa Delile ●☆

402701　Urostigma flavescens (Blume) Miq. = Ficus annulata Blume ●☆

402702　Urostigma fraseri Miq. = Ficus virens Aiton ●

402703　Urostigma glaberrimum (Blume) Miq. = Ficus glaberrima Blume ●

402704　Urostigma haematocarpum (Blume ex Decne.) Miq. = Ficus benjamina L. ●

402705　Urostigma hasseltii Miq. = Ficus pubilimba Merr. ●

402706　Urostigma hochstetteri Miq. = Ficus thonningii Blume ●☆

402707　Urostigma ilicinum Sond. = Ficus ilicina (Sond.) Miq. ●☆

402708　Urostigma infectorium Miq. = Ficus virens Aiton ●

402709　Urostigma ingens Miq. = Ficus ingens (Miq.) Miq. ●☆

402710　Urostigma karet Miq. = Ficus elastica Roxb. ●

402711　Urostigma kotschyanum Miq. = Ficus platyphylla Delile ●☆

402712　Urostigma microcarpa (L. f.) Miq. = Ficus microcarpa L. f. ●

402713　Urostigma modestum Miq. = Ficus nervosa K. Heyne ex Roth ●

402714　Urostigma mysorense (Roth ex Roem. et Schult.) Miq. = Ficus drupacea Thunb. var. pubescens (Roth) Corner ●

402715　Urostigma nervosum (B. Heyne ex Roth) Miq. = Ficus nervosa K. Heyne ex Roth ●

402716　Urostigma nervosum (Heyne) Miq. = Ficus nervosa K. Heyne ex Roth ●

402717　Urostigma nudum Miq. = Ficus benjamina L. var. nuda (Miq.) Barrett ●

402718　Urostigma obtusifolium (Roxb.) Miq. = Ficus curtipes Corner ●

402719　Urostigma odoratum Miq. = Ficus elastica Roxb. ●

402720　Urostigma ottoniifolium Miq. = Ficus ottoniifolia (Miq.) Miq. ●☆

402721　Urostigma parvifolium Miq. = Ficus concinna (Miq.) Miq. ●

402722　Urostigma pisocarpum (Blume) Miq. = Ficus pisocarpa Blume ●

402723　Urostigma religiosum (L.) Gasp. = Ficus religiosa L. ●

402724　Urostigma rhododendrifolium Miq. = Ficus maclellandi King ●

402725　Urostigma rubicundum Miq. = Ficus glumosa Delile ●☆

402726　Urostigma rumphii (Blume) Miq. = Ficus rumphii Blume ●

402727　Urostigma rumphii Miq. = Ficus rumphii Blume ●

402728　Urostigma schimperi Miq. = Ficus thonningii Blume ●☆

402729　Urostigma stipulosum Miq. = Ficus caulocarpa (Miq.) Miq. ●

402730　Urostigma strictum Miq. = Ficus stricta Miq. ●

402731　Urostigma superbum Miq. = Ficus superba (Miq.) Miq. ●

402732　Urostigma thonningii Miq. = Ficus sur Forssk. ●☆

402733　Urostigma trichocarpum (Blume) Miq. = Ficus trichocarpa Blume ●

402734　Urostigma vogelii Miq. = Ficus lutea Vahl ●☆

402735　Urostigma wightianum Miq. = Ficus virens Aiton ●

402736　Urostylis Meisn. = Layia Hook. et Arn. ex DC. (保留属名)■☆

402737　Urotheca Gilg = Gravesia Naudin ●☆

402738　Urotheca hylophila Gilg = Gravesia hylophila (Gilg) A. Fern. et R. Fern. ●☆

402739　Ursia Vassilcz. = Trifolium L. ■

402740　Ursia gordejevii (Kom.) Vassilcz. = Trifolium gordejevii (Kom.) Z. Wei ■

402741　Ursifolium Doweld = Trifolium L. ■

402742　Ursinea Willis = Ursinia Gaertn. (保留属名)●■☆

402743　Ursinia Gaertn. (1791) (保留属名);熊菊属(乌寝花属); Ursinia ●■☆

402744　Ursinia abrotanifolia (R. Br.) Spreng. ;美叶熊菊■☆

402745　Ursinia abyssinica Sch. Bip. ex Walp. = Ursinia nana DC. ■☆

402746　Ursinia affinis Harv. = Ursinia nana DC. ■☆

402747　Ursinia alpina N. E. Br. ;高山熊菊■☆

402748　Ursinia anethifolia (Less.) N. E. Br. = Ursinia paleacea (L.) Moench ●☆

402749　Ursinia anethoides (DC.) N. E. Br. ;莳萝熊菊■☆

402750　Ursinia annua Less. = Ursinia nana DC. ■☆

402751　Ursinia annua Less. ex Harv. = Ursinia nana DC. ■☆

402752　Ursinia annua Less. ex Harv. var. indecora (DC.) Harv. = Ursinia nana DC. ■☆

402753　Ursinia annua Less. ex Harv. var. nana Harv. = Ursinia nana DC. ■☆

402754　Ursinia anthemoides (L.) Poir. ;春黄熊菊(春黄菊状熊菊); African Daisy ■☆

402755　Ursinia anthemoides (L.) Poir. subsp. versicolor (DC.) Prassler;变色春黄熊菊■☆

402756　Ursinia apiculata DC. = Ursinia montana DC. subsp. apiculata (DC.) Prassler ■☆

402757　Ursinia argentea Compton = Ursinia sericea (Thunb.) N. E. Br. ●☆

402758　Ursinia bolusii Thell. = Ursinia scariosa (Aiton) Poir. ●☆

402759　Ursinia brachyloba (Kunze) N. E. Br. ;短裂春黄熊菊■☆

402760　Ursinia brachypoda (Harv.) N. E. Br. = Ursinia hispida (DC.) N. E. Br. ●☆

402761　Ursinia brevicaulis J. M. Wood et M. S. Evans = Ursinia montana DC. ■☆

402762　Ursinia caledonica (E. Phillips) Prassler;卡利登熊菊●☆

402763　Ursinia chrysanthemoides (Less.) Harv. ;蒿菊状熊菊■☆

402764　Ursinia chrysanthemoides (Less.) Harv. var. geyeri (L. Bolus et Harry Hall) Prassler = Ursinia chrysanthemoides (Less.) Harv. ■☆

402765　Ursinia ciliaris (DC.) N. E. Br. = Ursinia tenuifolia (L.)

Poir. subsp. ciliaris（DC.）Prassler ●☆

402766 Ursinia concolor（Harv.）N. E. Br. = Ursinia punctata（Thunb.）N. E. Br. ●☆

402767 Ursinia coronopifolia（Less.）N. E. Br.；鸟足叶熊菊■☆

402768 Ursinia crithmifolia（R. Br.）Spreng. = Ursinia paleacea（L.）Moench ●☆

402769 Ursinia crithmifolia（R. Br.）Spreng. var. grandiflora（Harv.）N. E. Br. = Ursinia paleacea（L.）Moench ●☆

402770 Ursinia crithmoides（P. J. Bergius）Poir. = Ursinia paleacea（L.）Moench ●☆

402771 Ursinia dentata（L.）Poir.；齿熊菊■☆

402772 Ursinia dentata（L.）Poir. var. setigera（DC.）N. E. Br. = Ursinia dentata（L.）Poir.■☆

402773 Ursinia discolor（Less.）N. E. Br.；异色熊菊■☆

402774 Ursinia dregeana（DC.）N. E. Br.；德雷熊菊●☆

402775 Ursinia eckloniana（Sond.）N. E. Br.；埃氏熊菊■☆

402776 Ursinia engleriana Muschl. = Ursinia nana DC.■☆

402777 Ursinia filiformis（A. Spreng.）Griess. = Ursinia pinnata（Thunb.）Prassler ■☆

402778 Ursinia filipes（E. Mey. ex DC.）N. E. Br.；丝梗熊菊●☆

402779 Ursinia foeniculacea（Jacq.）N. E. Br. = Ursinia anthemoides（L.）Poir.■☆

402780 Ursinia foeticulacea Poir.；茴香叶熊菊●☆

402781 Ursinia frutescens Dinter；灌木熊菊●☆

402782 Ursinia geyeri L. Bolus et Harry Hall = Ursinia chrysanthemoides（Less.）Harv.■☆

402783 Ursinia heterodonta（DC.）N. E. Br.；异齿熊菊●☆

402784 Ursinia hispida（DC.）N. E. Br.；硬毛熊菊●☆

402785 Ursinia incisa（DC.）N. E. Br. = Ursinia serrata（L. f.）Poir. ●☆

402786 Ursinia indecora DC. = Ursinia nana DC.■☆

402787 Ursinia jacottetiana Thell. = Ursinia montana DC.■☆

402788 Ursinia longiscapa E. Phillips = Ursinia paleacea（L.）Moench ●☆

402789 Ursinia macropoda（DC.）N. E. Br.；大足熊菊●☆

402790 Ursinia matricariifolia Dinter = Ursinia nana DC.■☆

402791 Ursinia merxmuelleri Prassler；梅尔熊菊●☆

402792 Ursinia montana DC.；山地熊菊■☆

402793 Ursinia montana DC. subsp. apiculata（DC.）Prassler；细尖熊菊■☆

402794 Ursinia montana DC. subsp. tenuiloba（DC.）Prassler = Ursinia tenuiloba DC.■☆

402795 Ursinia nana DC.；矮熊菊■☆

402796 Ursinia nana DC. subsp. leptophylla Prassler；细叶矮熊菊■☆

402797 Ursinia natalensis（Sch. Bip.）N. E. Br. = Ursinia tenuiloba DC.■☆

402798 Ursinia nudicaulis（Thunb.）N. E. Br.；裸茎熊菊●☆

402799 Ursinia paleacea（L.）Moench；糠熊菊●☆

402800 Ursinia pauciloba（DC.）N. E. Br. = Ursinia punctata（Thunb.）N. E. Br. ●☆

402801 Ursinia pilifera（P. J. Bergius）Poir.；纤毛熊菊■☆

402802 Ursinia pinnata（Thunb.）Prassler；双羽熊菊■☆

402803 Ursinia pulchra N. E. Br.；美丽熊菊；Orange Ursinia ■☆

402804 Ursinia pulchra N. E. Br. = Ursinia anthemoides（L.）Poir.■☆

402805 Ursinia punctata（Thunb.）N. E. Br.；斑点熊菊●☆

402806 Ursinia pusilla（DC.）N. E. Br. = Ursinia anthemoides（L.）Poir.■☆

402807 Ursinia pygmaea DC.；矮小熊菊●☆

402808 Ursinia quinquepartita（DC.）N. E. Br.；五深裂熊菊●☆

402809 Ursinia rigidula（DC.）N. E. Br.；稍硬熊菊●☆

402810 Ursinia saxatilis N. E. Br.；岩栖熊菊●☆

402811 Ursinia scariosa（Aiton）Poir.；干膜质熊菊■☆

402812 Ursinia scariosa（Aiton）Poir. subsp. subhirsuta（DC.）Prassler；粗毛干膜质熊菊●☆

402813 Ursinia schinzii Dinter = Ursinia nana DC.■☆

402814 Ursinia sericea（Thunb.）N. E. Br.；绢毛熊菊●☆

402815 Ursinia sericea N. E. Br. = Ursinia sericea（Thunb.）N. E. Br. ●☆

402816 Ursinia serrata（L. f.）Poir.；具齿熊菊●☆

402817 Ursinia speciosa DC.；美花熊菊●☆

402818 Ursinia subflosculosa（DC.）Prassler；多小花熊菊●☆

402819 Ursinia subhirsuta（DC.）N. E. Br. = Ursinia scariosa（Aiton）Poir. subsp. subhirsuta（DC.）Prassler ●☆

402820 Ursinia subintegrifolia Bolus = Ursinia montana DC.■☆

402821 Ursinia tenuifolia（L.）Poir.；细叶熊菊●☆

402822 Ursinia tenuifolia（L.）Poir. subsp. ciliaris（DC.）Prassler；缘毛细叶熊菊●☆

402823 Ursinia tenuiloba DC.；细裂熊菊●☆

402824 Ursinia trifida（Thunb.）N. E. Br.；三裂熊菊■☆

402825 Ursinia trifida（Thunb.）N. E. Br. f. calva Prassler；光秃熊菊●☆

402826 Ursinia trifurca（Harv.）N. E. Br. = Ursinia tenuifolia（L.）Poir. subsp. ciliaris（DC.）Prassler ●☆

402827 Ursinia tysoniana E. Phillips = Ursinia tenuiloba DC. ●☆

402828 Ursinia versicolor（DC.）N. E. Br. = Ursinia anthemoides（L.）Poir. subsp. versicolor（DC.）Prassler ■☆

402829 Ursiniopsis E. Phillips = Ursinia Gaertn.（保留属名）●■☆

402830 Ursiniopsis E. Phillips（1951）；类熊菊属●☆

402831 Ursiniopsis caledonica E. Phillips；类熊菊●☆

402832 Ursiniopsis caledonica E. Phillips = Ursinia caledonica（E. Phillips）Prassler ●☆

402833 Ursiniopsis eckloniana（Sond.）E. Phillips = Ursinia eckloniana（Sond.）N. E. Br.■☆

402834 Ursiniopsis quinquepartita（DC.）E. Phillips = Ursinia quinquepartita（DC.）N. E. Br. ●☆

402835 Ursopuntia P. V. Heath = Opuntia Mill. ●

402836 Ursopuntia P. V. Heath（1994）；阿根廷熊掌属●☆

402837 Ursopuntia textoris P. V. Heath；阿根廷熊掌●☆

402838 Ursulaea Read et Baensch（1944）；柄矛光萼荷属■☆

402839 Ursulaea macvaughii（L. B. Sm.）Read et Baensch；柄矛光萼荷■☆

402840 Urtica L.（1753）；荨麻属；Nettle ■

402841 Urtica acuminata Roxb. = Oreocnide integrifolia（Gaudich.）Miq. ●

402842 Urtica adoensis Steud. = Girardinia diversifolia（Link）Friis ■

402843 Urtica aestuans L. = Laportea aestuans（L.）Chew ■

402844 Urtica alienata L. = Pouzolzia zeylanica（L.）Benn. et R. Br.■

402845 Urtica angulata Blume = Pilea angulata（Blume）Blume ■

402846 Urtica angustata Blume = Debregeasia longifolia（Burm. f.）Wedd. ●

402847 Urtica angustifolia Fisch. ex Hornem.；狭叶荨麻（白活麻，哈拉海，蠚草，蕁麻，活麻草，火麻草，宽叶荨麻，荨麻，螫麻，螫麻子，细荨麻，小荨麻，蝎子草）；Narrowleaf Nettle ■

402848 Urtica angustifolia Fisch. ex Hornem. var. sikokiana（Makino）Ohwi；日光荨麻（长叶荨麻）■☆

402849 Urtica anisophylla Wall. = Pilea anisophylla Wedd. ■

402850 Urtica appendiculata Wall. = Oreocnide integrifolia（Gaudich.）Miq. ●

402851 Urtica arborea L. f. = Gesnouinia arborea（L. f.）Gaudich. ●☆

402852　Urtica arborescens Link = Pipturus arborescens（Link）C. B. Rob. ●

402853　Urtica ardens Blume = Dendrocnide sinuata（Blume）Chew ●

402854　Urtica ardens Link；喜马拉雅荨麻（小花荨麻，须弥荨麻）；Himalayas Nettle，Littleflower Nettle ■

402855　Urtica atrichocaulis（Hand. -Mazz.）C. J. Chen；小果荨麻（秃茎荨麻，无刺茎荨麻，细荨麻，小苎麻）；Smallfruit Nettle ■

402856　Urtica balearica L. = Urtica pilulifera L. ●☆

402857　Urtica bicolar Roxb. = Debregeasia salicifolia（D. Don）Rendle ●

402858　Urtica bicolor Roxb. = Debregeasia saeneb（Forssk.）Hepper et J. R. I. Wood ●

402859　Urtica bicolor Wall. ex Roxb. = Debregeasia saeneb（Forssk.）Hepper et J. R. I. Wood ●

402860　Urtica breweri S. Watson = Urtica dioica L. subsp. holosericea（Nutt.）Thorne ■☆

402861　Urtica bulbifera Siebold et Zucc. = Laportea bulbifera（Siebold et Zucc.）Wedd. ■

402862　Urtica bullosa Steud. = Girardinia bullosa（Steud.）Wedd. ■☆

402863　Urtica buraei H. Lév. = Girardinia diversifolia（Link）Friis ■

402864　Urtica burchellii N. E. Br. = Urtica lobulata Blume ■☆

402865　Urtica caffra Thunb. = Didymodoxa caffra（Thunb.）Friis et Wilmot-Dear ■☆

402866　Urtica californica Greene = Urtica dioica L. subsp. gracilis（Aiton）Selander ■☆

402867　Urtica canadensis L. = Laportea canadensis（L.）Wedd. ■☆

402868　Urtica candicans Burm. f. = Callicarpa candicans（Burm. f.）Hochr. ●

402869　Urtica cannabina L.；麻叶荨麻（赤麻子，哈拉海，火麻，火麻草，宽叶荨麻，裂叶荨麻，荨麻，螫麻子，蝎子草，焮麻，焖麻）；Coast Nettle，Hempleaf Nettle ■

402870　Urtica cannabina L. f. angustiloba Chu = Urtica cannabina L. ■

402871　Urtica capensis L. f. = Didymodoxa capensis（L. f.）Friis et Wilmot-Dear ■☆

402872　Urtica capitata L. = Urtica thunbergiana Siebold et Zucc. ■

402873　Urtica cardiophylla Rydb. = Urtica dioica L. subsp. gracilis（Aiton）Selander ■☆

402874　Urtica caudata Vahl = Urtica membranacea Poir. ■☆

402875　Urtica chamaedryoides Pursh；西方麻叶荨麻；Nettle，Weak Nettle ■☆

402876　Urtica chamaedryoides Pursh var. runyonii Correll = Urtica chamaedryoides Pursh ■☆

402877　Urtica condensata Steud. = Girardinia diversifolia（Link）Friis ■

402878　Urtica crenulata Roxb. = Dendrocnide sinuata（Blume）Chew ●

402879　Urtica cyanescens Kom. = Urtica laetevirens Maxim. subsp. cyanescens（Kom. ex Jarm.）C. J. Chen ■

402880　Urtica cyanescens Kom. ex Jarm. = Urtica laetevirens Maxim. subsp. cyanescens（Kom. ex Jarm.）C. J. Chen ■

402881　Urtica cylindrica L. = Boehmeria cylindrica（L.）Sw. ●☆

402882　Urtica dentata Hand. -Mazz. = Urtica laetevirens Maxim. subsp. dentata（Hand. -Mazz.）C. J. Chen ■

402883　Urtica dentata Hand. -Mazz. = Urtica laetevirens Maxim. ■

402884　Urtica dentata Hand. -Mazz. var. atrichocaulis Hand. -Mazz.；无刺茎荨麻（秃茎荨麻，小荨麻）；Spineless-stem Nettle ■☆

402885　Urtica dichotoma Blume = Debregeasia longifolia（Burm. f.）Wedd. ●

402886　Urtica dioica L.；异株荨麻（大荨麻，单性荨麻，宽叶荨麻，欧荨麻，钱麻，三洲荨麻，线麻，小荨麻）；Big Nettle，Bigsting Nettle，Big-sting Nettle，Common Nettle，Common Stinging Nettle，Devil's Apron，Devil's Leaf，Devil's Plaything，Dioecism Nettle，Ettley，Female Nettle，Great Nettle，Heg-beg，Hettle，Hidgy-pidgy，Hoky-poky，Jaggy Nettle，Jenny Nettle，Naughty Man's Plaything，Nettle，Nettle Fibre，Niddle，Scaddie，Sting Nettle，Stinger，Stinging Nettle，Stingy Nettle，Swedish Hemp，Tall Nettle，Tanging Nettle ■

402887　Urtica dioica L. = Urtica thunbergiana Siebold et Zucc. ■

402888　Urtica dioica L. subsp. afghanica Chrtek；尾尖异株荨麻■

402889　Urtica dioica L. subsp. gansuensis C. J. Chen；甘肃异株荨麻（甘肃荨麻）；Gansu Nettle ■

402890　Urtica dioica L. subsp. gracilis（Aiton）Selander；纤细异株荨麻；Stinging Nettle ■☆

402891　Urtica dioica L. subsp. holosericea（Nutt.）Thorne；布鲁尔异株荨麻■☆

402892　Urtica dioica L. subsp. xingjiangensis C. J. Chen = Urtica dioica L. ■

402893　Urtica dioica L. subsp. xinjiangensis C. J. Chen；新疆异株荨麻；Xingjiang Nettle ■

402894　Urtica dioica L. var. angustifolia（Fisch. ex Hornem.）Ledeb. = Urtica angustifolia Fisch. ex Hornem. ■

402895　Urtica dioica L. var. angustifolia H. Lév. = Urtica atrichocaulis（Hand. -Mazz.）C. J. Chen ■

402896　Urtica dioica L. var. angustifolia Ledeb. = Urtica angustifolia Fisch. ex Hornem. ■

402897　Urtica dioica L. var. angustifolia Schltdl. = Urtica dioica L. subsp. gracilis（Aiton）Selander ■☆

402898　Urtica dioica L. var. angustifolia Schltr. = Urtica dioica L. ■

402899　Urtica dioica L. var. atrichocaulis Hand. -Mazz. = Urtica atrichocaulis（Hand. -Mazz.）C. J. Chen ■

402900　Urtica dioica L. var. californica（Greene）C. L. Hitchc. = Urtica dioica L. subsp. gracilis（Aiton）Selander ■☆

402901　Urtica dioica L. var. capensis Wedd. = Urtica dioica L. ■

402902　Urtica dioica L. var. gracilis（Aiton）R. L. Taylor et MacBryde = Urtica dioica L. subsp. gracilis（Aiton）Selander ■☆

402903　Urtica dioica L. var. holosericea（Nutt.）C. L. Hitchc. = Urtica dioica L. subsp. holosericea（Nutt.）Thorne ■☆

402904　Urtica dioica L. var. lyallii（S. Watson）C. L. Hitchc. = Urtica dioica L. subsp. gracilis（Aiton）Selander ■☆

402905　Urtica dioica L. var. occidentalis S. Watson = Urtica dioica L. subsp. holosericea（Nutt.）Thorne ■☆

402906　Urtica dioica L. var. procera（Muhl. ex Willd.）Wedd. = Urtica dioica L. subsp. gracilis（Aiton）Selander ■☆

402907　Urtica dioica L. var. procera（Muhl. ex Willd.）Wedd. = Urtica dioica L. ■

402908　Urtica dioica L. var. vulgaris Wedd. = Urtica dioica L. ■

402909　Urtica divaricata L. = Laportea canadensis（L.）Wedd. ■☆

402910　Urtica diversifolia Link = Girardinia diversifolia（Link）Friis ■

402911　Urtica dubia Forssk. = Urtica membranacea Poir. ■☆

402912　Urtica eckloniana Blume = Urtica dioica L. ■

402913　Urtica eckloniana Blume var. flavovirens？ = Urtica dioica L. ■

402914　Urtica ferox G. Forst.；树荨麻；Tree Nettle ●☆

402915　Urtica fissa E. Pritz. = Urtica fissa E. Pritz. ex Diels ■

402916　Urtica fissa E. Pritz. ex Diels；荨麻（白活麻，白蛇麻，活麻草，火麻，火麻草，裂叶荨麻，蛇麻草，台湾咬人猫，透骨风）；Lobedleaf Nettle，Nettle ■

402917　Urtica foliosa Blume = Urtica angustifolia Fisch. ex Hornem. ■

402918　Urtica frutescens Thunb. = Oreocnide frutescens（Thunb.）Miq. ●

402919　Urtica frutescens Thunb. = Villebrunea frutescens（Thunb.）Blume ●

402920　Urtica gemina Lour. = Acalypha australis L. ■

402921　Urtica glaberrima Blume = Pilea glaberrima（Blume）Blume ●■

402922　Urtica gracilenta Greene；细黏荨麻■☆

402923　Urtica gracilis Aiton；纤细荨麻；Slender Nettle ■☆

402924　Urtica gracilis Aiton = Urtica dioica L. subsp. gracilis（Aiton）Selander ■☆

402925　Urtica gracilis Aiton subsp. holosericea（Nutt.）W. A. Weber = Urtica dioica L. subsp. holosericea（Nutt.）Thorne ■☆

402926　Urtica gracilis Aiton var. greenei（Jeps.）Jeps. = Urtica dioica L. subsp. holosericea（Nutt.）Thorne ■☆

402927　Urtica gracilis Aiton var. holosericea（Nutt.）Jeps. = Urtica dioica L. subsp. holosericea（Nutt.）Thorne ■☆

402928　Urtica hamiltoniana Wall. = Boehmeria hamiltoniana Wedd. ●

402929　Urtica haussknechtii Boiss.；豪斯荨麻■☆

402930　Urtica helanshanica W. Z. Di et W. B. Liao；贺兰山荨麻；Helanshan Nettle ■

402931　Urtica heterophylla D. Don = Girardinia diversifolia（Link）Friis ■

402932　Urtica heterophylla Vahl = Girardinia palmata（Forssk.）Gaudich. ■

402933　Urtica heterophylla Wahl = Girardinia diversifolia（Link）Friis ■

402934　Urtica himalayensis Kunth et Bouché = Urtica ardens Link ■

402935　Urtica hirta Blume = Gonostegia hirta（Blume ex Hassk.）Miq. ■

402936　Urtica holosericea Nutt.；绢毛荨麻■☆

402937　Urtica holosericea Nutt. = Urtica dioica L. subsp. holosericea（Nutt.）Thorne ■☆

402938　Urtica hyperborea Jacq. ex Wedd.；高原荨麻（北方荨麻）；Plateau Nettle ■

402939　Urtica iners Forssk. = Droguetia iners（Forssk.）Schweinf. ■☆

402940　Urtica interrupta L. = Laportea interrupta（L.）Chew ■

402941　Urtica japonica L. f. = Boehmeria japonica（L. f.）Miq. ●■

402942　Urtica japonica Thunb. = Fatoua villosa（Thunb.）Nakai ■

402943　Urtica kioviensis Rogow.；基辅荨麻■☆

402944　Urtica kunlunshanica Chang Y. Yang；昆仑荨麻；Kunlunshan Nettle ■

402945　Urtica kunlunshanica Chang Y. Yang = Urtica hyperborea Jacq. ex Wedd. ■

402946　Urtica laetevirens Maxim.；宽叶荨麻（哈拉海，虎麻草，林生荨麻，荨麻，螫麻，螫麻子，蝎子草，痒痒草）；Broadleaf Nettle, Forest Nettle ■

402947　Urtica laetevirens Maxim. subsp. cyanescens（Kom. ex Jarm.）C. J. Chen；乌苏里荨麻（哈拉海）■

402948　Urtica laetevirens Maxim. subsp. cyanescens（Kom.）C. J. Chen = Urtica laetevirens Maxim. subsp. cyanescens（Kom. ex Jarm.）C. J. Chen ■

402949　Urtica laetevirens Maxim. subsp. dentata（Hand. -Mazz.）C. J. Chen；齿叶荨麻（湖北红活麻）；Dentate Nettle, Toothed Broadleaf Nettle ■

402950　Urtica laetevirens Maxim. subsp. dentata（Hand. -Mazz.）C. J. Chen = Urtica laetevirens Maxim. ■

402951　Urtica lobata Blume = Urtica lobulata Blume ■☆

402952　Urtica lobatifolia S. S. Ying；裂叶荨麻；Lobe Nettle ■

402953　Urtica lobotifolia S. S. Ying = Girardinia diversifolia（Link）Friis ■

402954　Urtica lobulata Blume；浅裂荨麻■☆

402955　Urtica longifolia Burm. f. = Debregeasia longifolia（Burm. f.）Wedd. ●

402956　Urtica lyallii S. Watson = Urtica dioica L. subsp. gracilis（Aiton）Selander ■☆

402957　Urtica lyallii S. Watson var. californica（Greene）Jeps. = Urtica dioica L. subsp. gracilis（Aiton）Selander ■☆

402958　Urtica macrorrhiza Hand. -Mazz.；粗根荨麻（白活麻，活麻，火麻，巨根荨麻，老虎麻，荨麻，青活麻）；Thickroot Nettle ■

402959　Urtica macrorrhiza Hand. -Mazz. = Urtica thunbergiana Siebold et Zucc. ■

402960　Urtica mairei H. Lév.；滇藏荨麻（大荨麻，云南荨麻）；Maire Nettle, Yunnan Nettle ■

402961　Urtica mairei H. Lév. var. oblongifolia C. J. Chen；长圆叶荨麻；Oblong Maire Nettle ■

402962　Urtica mairei H. Lév. var. oblongifolia C. J. Chen = Urtica ardens Link ■

402963　Urtica major H. P. Fuchs = Urtica dioica L. subsp. gracilis（Aiton）Selander ■☆

402964　Urtica malabarica Wall. = Boehmeria malabarica Wedd. ●

402965　Urtica massaica Mildbr.；马萨荨麻■☆

402966　Urtica melastomoides Poir. = Pilea melastomoides（Poir.）Wedd. ●■

402967　Urtica membranacea Poir.；膜质荨麻■☆

402968　Urtica membranifolia C. J. Chen；膜叶荨麻；Filmleaf Nettle ■

402969　Urtica meyeniana Walp. = Dendrocnide meyeniana（Walp.）Chew ●

402970　Urtica meyeri Wedd. = Urtica lobulata Blume ■☆

402971　Urtica microphylla Swart = Pilea microphylla（L.）Liebm. ■

402972　Urtica mitis Hochst. = Laportea peduncularis（Wedd.）Chew ●☆

402973　Urtica moluccana Blume = Cypholophus moluccanus（Blume）Miq. ●

402974　Urtica morifolia Poir.；桑叶荨麻●☆

402975　Urtica muralis Vahl = Pouzolzia parasitica（Forssk.）Schweinf. ●☆

402976　Urtica naucleiflora Roxb. = Poikilospermum naucleiflorum（Roxb. ex Lindl.）Chew ●

402977　Urtica nivea L. = Boehmeria nivea（L.）Gaudich. ●

402978　Urtica nummularifolia Sw. = Pilea nummulariifolia（Sw.）Wedd. ■☆

402979　Urtica obesa Wall. = Pilea umbrosa Blume var. obesa Wedd. ■

402980　Urtica pachyrrhachis Hand. -Mazz. = Urtica laetevirens Maxim. ■

402981　Urtica palmata Forssk. = Girardinia diversifolia（Link）Friil. ■

402982　Urtica palmata Forssk. = Girardinia palmata（Forssk.）Gaudich. ■

402983　Urtica parasitica Forssk. = Pouzolzia parasitica（Forssk.）Schweinf. ●☆

402984　Urtica parviflora Roxb.；圆果荨麻●

402985　Urtica parviflora Roxb. = Urtica ardens Link ■

402986　Urtica parviflora Wedd. = Urtica membranifolia C. J. Chen ■

402987　Urtica pauciflora Steud. = Droguetia iners（Forssk.）Schweinf. ■☆

402988　Urtica pentandra Roxb. = Gonostegia pentandra（Roxb.）Miq. ●

402989　Urtica pentandra Roxb. = Pouzolzia pentandra（Roxb.）Benn. ●☆

402990　Urtica petiolaris Siebold et Zucc. = Pilea angulata（Blume）Blume subsp. petiolaris（Siebold et Zucc.）C. J. Chen ■

402991　Urtica pilosiuscula Blume = Boehmeria pilosiuscula（Blume）Hassk. ●■

402992　Urtica pilulifera L.；小球荨麻（蝎子草）；Greek Nettle, Male Nettle, Nettles Burning, Pill Nettle, Roman Nettle, Wild Nettle ●☆

402993　Urtica pilulifera L. var. balearica（L.）Willk. et Lange = Urtica pilulifera L. ●☆

402994　Urtica pilulifera L. var. dodartii ?；西班牙荨麻；Spanish Marjoram ■☆

402995　Urtica pinfaensis H. Lév. et Vaniot = Urtica fissa E. Pritz. ex Diels ■

402996　Urtica platyphylla Wedd.；北海道荨麻；Yezo Nettle ■☆

402997　Urtica platyphylla Wedd. = Urtica laetevirens Maxim. subsp. cyanescens（Kom. ex Jarm.）C. J. Chen ■

402998　Urtica polystachya Wall. = Boehmeria polystachya Wedd. ●■

402999 Urtica procera Muhl. ex Willd. ；高荨麻；Stingdng Nettle, Tall Nettle ■☆

403000 Urtica procera Muhl. ex Willd. = Urtica dioica L. subsp. gracilis （Aiton）Selander ■☆

403001 Urtica pubesceos Ledeb. ；毛荨麻■☆

403002 Urtica pulcherrima Roxb. = Sarcochlamys pulcherrima Gaudich. ●

403003 Urtica pumila L. = Pilea pumila （L.）A. Gray ■

403004 Urtica puya Buch. -Ham. et Wall. = Maoutia puya （Wall. ex Wedd.）Wedd. ●

403005 Urtica rubescens Blume = Oreocnide rubescens （Blume）Miq. ●

403006 Urtica sanguinea Blume = Pouzolzia sanguinea （Blume）Merr. ●

403007 Urtica scabra Blume = Oreocnide rubescens （Blume）Miq. ●

403008 Urtica scabrella Roxb. = Boehmeria macrophylla D. Don var. scabrella （Roxb.）D. G. Long ●

403009 Urtica scorpioides ?；蝎子荨麻（蝎子麻）■☆

403010 Urtica scripta Buch. -Ham. ex D. Don = Pilea scripta （Buch. - Ham. ex D. Don）Wedd. ■☆

403011 Urtica sessiliflora Sw. = Urtica thunbergiana Siebold et Zucc. ■

403012 Urtica sikokiana （Makino）Makino = Urtica angustifolia Fisch. ex Hornem. var. sikokiana （Makino）Ohwi ■☆

403013 Urtica silvatica Hand. -Mazz. = Urtica laetevirens Maxim. ■

403014 Urtica simensis Hochst. ex A. Rich. ；锡米荨麻■☆

403015 Urtica sinuata Blume = Dendrocnide sinuata （Blume）Chew ●

403016 Urtica sondenii （Simmons）Avrorin；松顿荨麻■☆

403017 Urtica spicata Thunb. = Boehmeria gracilis C. H. Wright ■

403018 Urtica spicata Thunb. = Boehmeria spicata （Thunb.）Thunb. ■

403019 Urtica squamigera Wall. = Chamabainia cuspidata Wight ■

403020 Urtica stachyoides Webb et Berthel. ；穗荨麻■☆

403021 Urtica stimulans L. f. = Dendrocnide stimulans （L. f.）Chew ●

403022 Urtica stipulosa Miq. = Pilea angulata （Blume）Blume ■

403023 Urtica strigosissima Rydb. = Urtica dioica L. subsp. gracilis （Aiton）Selander ■☆

403024 Urtica subincisa Benth. ；锐裂荨麻■☆

403025 Urtica sylvatica Blume = Oreocnide rubescens （Blume）Miq. ●

403026 Urtica taiwaniana S. S. Ying；台湾荨麻；Taiwan Nettle ■

403027 Urtica tetraphylla Steud. = Pilea tetraphylla （Steud.）Blume ■☆

403028 Urtica thunbergiana Siebold et Zucc. ；咬人荨麻（刺草，火麻，荨麻，蛇麻，蛇麻草，咬人猫）；Snappish Nettle ■

403029 Urtica thunbergiana Siebold et Zucc. = Urtica macrorrhiza Hand. -Mazz. ■

403030 Urtica tibetica W. T. Wang = Urtica dioica L. ■

403031 Urtica tibetica W. T. Wang ex C. J. Chen；西藏荨麻；Xizang Nettle ■

403032 Urtica triangularis Hand. -Mazz. ；三角叶荨麻（花叶活麻，火麻）；Deltleaf Nettle, Triangleleaf Nettle ■

403033 Urtica triangularis Hand. -Mazz. f. pinnatifida Hand. -Mazz. = Urtica triangularis Hand. -Mazz. subsp. pinnatifida （Hand. -Mazz.）C. J. Chen ■

403034 Urtica triangularis Hand. -Mazz. subsp. pinnatifida （Hand. - Mazz.）C. J. Chen；羽裂荨麻■

403035 Urtica triangularis Hand. -Mazz. subsp. trichocarpa C. J. Chen；毛果荨麻；Hairfruit Nettle ■

403036 Urtica trianthemoides Sw. = Pilea trianthemoides （Sw.）Lindl. ■☆

403037 Urtica trinervia Roxb. = Pilea melastomoides （Poir.）Wedd. ●■

403038 Urtica umbrosa Wall. = Pilea umbrosa Blume ■

403039 Urtica urens L. ；欧荨麻（小荨麻）；Annual Nettle, Burning Nettle, Dog Nettle, Dwarf Nettle, Europe Nettle, Nettle, Nettles Burning, Small Nettle, Stinging Nettle ■

403040 Urtica urens L. var. iners （Forssk.）Wedd. = Droguetia iners （Forssk.）Schweinf. ■☆

403041 Urtica urophylla Wall. = Oreocnide integrifolia （Gaudich.）Miq. ●

403042 Urtica utilis L. = Boehmeria nivea （L.）Gaudich. ●

403043 Urtica villosa Thunb. = Fatoua villosa （Thunb.）Nakai ■

403044 Urtica viminea Wall. = Pouzolzia sanguinea （Blume）Merr. ●

403045 Urtica viridis Rydb. = Urtica dioica L. subsp. gracilis （Aiton）Selander ■☆

403046 Urtica virulenta Wall. = Dendrocnide sinuata （Blume）Chew ●

403047 Urtica virulenta Wall. = Urtica ardens Blume ●

403048 Urtica zayuensis C. J. Chen；察隅荨麻；Chayu Nettle ■

403049 Urtica zayuensis C. J. Chen = Urtica ardens Link ■

403050 Urticaceae Juss. （1789）（保留科名）；荨麻科；Nettle Family ●■

403051 Urticastrum Fabr. = Laportea Gaudich. （保留属名）●■

403052 Urticastrum Heist. ex Fabr. （废弃属名）= Laportea Gaudich. （保留属名）●■

403053 Urticastrum carruthersianum Hiern = Obetia carruthersiana （Hiern）Rendle ●☆

403054 Urticastrum divaricatum （L.）Kuntze = Laportea canadensis （L.）Wedd. ■☆

403055 Urucu Adans. = Bixa L. ●

403056 Urumovia Stef. = Jasione L. ■☆

403057 Uruparia Raf. = Ourouparia Aubl. （废弃属名）●

403058 Uruparia Raf. = Uncaria Schreb. （保留属名）●

403059 Uruparia africana （G. Don）Kuntze = Uncaria africana G. Don ●☆

403060 Urvillaea DC. = Urvillea Kunth ●☆

403061 Urvillea Kunth（1821）；于维尔无患子属●☆

403062 Urvillea affinis Schltdl. ；近缘于维尔无患子●☆

403063 Urvillea ferruginea Lindl. ；锈色于维尔无患子●☆

403064 Urvillea glabra Benth. ；光于维尔无患子●☆

403065 Urvillea laevis Radlk. ；平滑于维尔无患子●☆

403066 Urvillea mexicana A. Gray；墨西哥于维尔无患子●☆

403067 Urvillea paucidentata Ferrucci；少齿于维尔无患子●☆

403068 Urvillea pubescens Klotzsch；毛于维尔无患子●☆

403069 Urvillea tridentata Miq. ；三齿于维尔无患子●☆

403070 Urvillea triphylla Radlk. ；三叶于维尔无患子●☆

403071 Usionis Raf. = Salix L. （保留属名）●

403072 Usoricum Lunell = Oenothera L. ●■

403073 Ussuria Tzvelev = Neoussuria Tzvelev ■☆

403074 Usteria Cav. = Maurandya Ortega ■☆

403075 Usteria Dennst. = Acalypha L. ●■

403076 Usteria Medik. = Endymion Dumort. ■☆

403077 Usteria Medik. = Hylomenes Salisb. ■☆

403078 Usteria Willd. （1790）；于斯马钱属●☆

403079 Usteria guineensis Willd. ；于斯马钱●☆

403080 Usteria volubilis Afzel. = Usteria guineensis Willd. ●☆

403081 Usubis Burm. f. = Allophylus L. ●

403082 Utahia Britton et Rose = Pediocactus Britton et Rose ●☆

403083 Utahia Britton et Rose（1922）；天狼属●☆

403084 Utahia peeblesianas （Croizat）Kladiwa = Pediocactus peeblesianus （Croizat）L. D. Benson ●☆

403085 Utahia sileri （Engelm. ex J. M. Coult.）Britton et Rose = Pediocactus sileri （Engelm. ex J. M. Coult.）L. D. Benson ●☆

403086 Utahia sileri （Engelm.）Britton et Rose = Pediocactus sileri （Engelm. ex J. M. Coult.）L. D. Benson ●☆

403087 Utania G. Don = Fagraea Thunb. ●

403088　Utea J. St. -Hil. = Macrolobium Schreb.（保留属名）●☆

403089　Uterveria Bertol. = Capparis L. ●

403090　Utleria Bedd. ex Benth. = Utleria Bedd. ex Benth. et Hook. f. ☆

403091　Utleria Bedd. ex Benth. et Hook. f.（1876）;柳叶萝藦属☆

403092　Utleria salicifolia Bedd. ex Hook. f.，柳叶萝藦☆

403093　Utleya Wilbur et Luteyn(1977);五敛莓属●☆

403094　Utleya costaricensi Wilbur et Luteyn;五敛莓●☆

403095　Utricularia L.（1753）;狸藻属（挖耳草属）;Bladder Wort，Bladderwort，Draws Ear Grass ■☆

403096　Utricularia aberrans Bosser = Utricularia welwitschii Oliv. ■☆

403097　Utricularia acicularis Sol. ex Stapf = Utricularia bisquamata Schrank ■☆

403098　Utricularia affinis Wight;紫花挖耳草（紫挖耳草）■

403099　Utricularia affinis Wight = Utricularia uliginosa Vahl ■

403100　Utricularia affinis Wight var. griffithii ? = Utricularia uliginosa Vahl ■

403101　Utricularia afromontana R. E. Fr. = Utricularia livida E. Mey. ■☆

403102　Utricularia alpina Jacq. ;高山狸藻■☆

403103　Utricularia ambigua A. DC. = Utricularia gibba L. ■

403104　Utricularia andongensis Welw. ex Hiern;安东狸藻■☆

403105　Utricularia angolensis Kamienski = Utricularia subulata L. ■☆

403106　Utricularia appendiculata E. A. Bruce;附属物狸藻■☆

403107　Utricularia arenaria A. DC. ;沙狸藻■☆

403108　Utricularia aurea Lour. ;黄花狸藻（黄花挖耳草，金鱼草，狸藻，水上一枝黄花）;Yellow Bladderwort，Yellowflower Bladderwort ■

403109　Utricularia aurea Lour. f. immaculata Tamura;无斑黄花狸藻■☆

403110　Utricularia australis R. Br. ;南方狸藻（狗尾巴草，台湾狸藻，野狸藻，幽狸藻，鱼刺草）;Austral Bladderwort，Bladderwort，Greater Bladderwort ■

403111　Utricularia australis R. Br. f. tenuicaulis（Miki）Komiya et C. Shibata = Utricularia australis R. Br. ■

403112　Utricularia bangweolensis R. E. Fr. = Utricularia reflexa Oliv. ■☆

403113　Utricularia baouleensis A. Chev. ;海南挖耳草;Hainan Bladderwort，Hainan Draws ear grass ■

403114　Utricularia baouleensis A. Chev. = Utricularia foveolata Edgew. ■

403115　Utricularia baumii Kamienski = Utricularia spiralis Sm. ■☆

403116　Utricularia baumii Kamienski var. leptocheilos Pellegr. = Utricularia spiralis Sm. ■☆

403117　Utricularia benjaminiana Oliv. ;本氏狸藻■☆

403118　Utricularia bifida Bojer ex A. DC. = Utricularia prehensilis E. Mey. ■☆

403119　Utricularia bifida L. ;挖耳草;Bifid Bladderwort，Bifid Draws Ear Grass ■

403120　Utricularia bifida L. = Utricularia limosa R. Br. ■

403121　Utricularia bifidocalcar R. D. Good = Utricularia gibba L. ■

403122　Utricularia biflora Hayata = Utricularia bifida L. ■

403123　Utricularia biflora Lam. = Utricularia gibba L. ■

403124　Utricularia biflora Roxb. = Utricularia exoleta R. Br. ■

403125　Utricularia bisquamata Schrank;双鳞狸藻■☆

403126　Utricularia brachiata Oliv. = Utricularia striatula J. Sm. ■

403127　Utricularia brachyceras Schltr. = Utricularia bisquamata Schrank ■☆

403128　Utricularia bracteata R. D. Good;具苞狸藻■☆

403129　Utricularia bremii Heer = Utricularia minor L. ■

403130　Utricularia bremii Heer ex Koelliker;波氏狸藻■☆

403131　Utricularia bryophila Ridl. = Utricularia mannii Oliv. ■☆

403132　Utricularia caerulea L. ;短梗挖耳草（长距挖耳草，密花狸藻，折苞挖耳草）;Blue Bladderwort，Shortstalk Draws Ear Grass ■

403133　Utricularia caerulea L. = Utricularia graminifolia Vahl ■

403134　Utricularia caerulea L. = Utricularia uliginosa Vahl ■

403135　Utricularia caerulea L. f. leucantha（Komiya）Komiya;白花短梗挖耳草■☆

403136　Utricularia caerulea L. var. affinis（Wight）Thwaites = Utricularia uliginosa Vahl ■

403137　Utricularia caerulea L. var. filicaulis（Wall. ex A. DC.）Haines = Utricularia caerulea L. ■

403138　Utricularia caerulea L. var. graminifolia（Vahl）Bhattach. = Utricularia graminifolia Vahl ■

403139　Utricularia capensis Spreng. = Utricularia bisquamata Schrank ■☆

403140　Utricularia capensis Spreng. var. brevicalcarata Oliv. = Utricularia bisquamata Schrank ■☆

403141　Utricularia capensis Spreng. var. elatior Kamienski = Utricularia bisquamata Schrank ■☆

403142　Utricularia cavalerii H. Lév. = Utricularia bifida L. ■

403143　Utricularia cavalerii Stapf = Utricularia caerulea L. ■

403144　Utricularia cervicornuta H. Perrier = Utricularia benjaminiana Oliv. ■☆

403145　Utricularia charoidea Stapf = Utricularia reflexa Oliv. ■☆

403146　Utricularia clandestina Nutt. = Utricularia geminiscapa Benj. ■☆

403147　Utricularia cornuta Michx. ;有距狸藻;Horned Bladderwort，Naked Bladderwort ■☆

403148　Utricularia cucullata Afzel. ex Kamienski = Utricularia micropetala Sm. ■☆

403149　Utricularia cymbantha Oliv. ;舟花狸藻■☆

403150　Utricularia deightonii F. E. Lloyd et G. Taylor = Utricularia pubescens Sm. ■☆

403151　Utricularia delicata Kamienski = Utricularia bisquamata Schrank ■☆

403152　Utricularia denticulata Benj. ;细齿狸藻■☆

403153　Utricularia diantha Roxb. ex Roem. et Schult. = Utricularia exoleta R. Br. ■

403154　Utricularia diantha Roxb. ex Roem. et Schult. = Utricularia gibba L. ■

403155　Utricularia dichotoma Labill. ;二歧狸藻■☆

403156　Utricularia dimorphantha Makino;异花狸藻■☆

403157　Utricularia diploglossa Welw. ex Oliv. = Utricularia reflexa Oliv. ■☆

403158　Utricularia dregei Kamienski = Utricularia livida E. Mey. ■☆

403159　Utricularia dregei Kamienski var. stricta ? = Utricularia livida E. Mey. ■☆

403160　Utricularia dusenii Sylven;杜氏狸藻■☆

403161　Utricularia eburnea R. E. Fr. = Utricularia livida E. Mey. ■☆

403162　Utricularia ecklonii Spreng. = Utricularia bisquamata Schrank ■☆

403163　Utricularia ecklonii Spreng. var. lutea H. Perrier = Utricularia firmula Welw. ex Oliv. ■☆

403164　Utricularia elevata Kamienski = Utricularia livida E. Mey. ■☆

403165　Utricularia elevata Kamienski var. macowanii ? = Utricularia livida E. Mey. ■☆

403166　Utricularia engleri Kamienski = Utricularia livida E. Mey. ■☆

403167　Utricularia exilis Oliv. = Utricularia arenaria A. DC. ■☆

403168　Utricularia exilis Oliv. var. arenaria（A. DC.）Kamienski = Utricularia bisquamata Schrank ■☆

403169　Utricularia exilis Oliv. var. bryoides Welw. ex Hiern = Utricularia arenaria A. DC. ■☆

403170　Utricularia exilis Oliv. var. ecklonii（Spreng.）Kamienski = Utricularia bisquamata Schrank ■☆

403171　Utricularia exilis Oliv. var. elatior Kamienski = Utricularia bisquamata

Schrank ■☆

403172 Utricularia exilis Oliv. var. hirsuta Kamienski = Utricularia arenaria A. DC. ■☆

403173 Utricularia exilis Oliv. var. minor Kamienski = Utricularia bisquamata Schrank ■☆

403174 Utricularia exilis Oliv. var. nematoscapa Welw. ex Hiern = Utricularia arenaria A. DC. ■☆

403175 Utricularia exoleta R. Br. = Utricularia gibba L. ■

403176 Utricularia exoleta R. Br. f. natans (Komiya) Komiya;漂浮狸藻■☆

403177 Utricularia extensa Hance = Utricularia aurea Lour. ■☆

403178 Utricularia falcata R. D. Good = Utricularia tortilis Welw. ex Oliv. ■☆

403179 Utricularia fernaldiana F. E. Lloyd et G. Taylor = Utricularia pubescens Sm. ■☆

403180 Utricularia filicaulis Wall. = Utricularia caerulea L. ■

403181 Utricularia filicaulis Wall. ex A. DC. = Utricularia caerulea L. ■

403182 Utricularia firmula Welw. ex Oliv.;坚硬狸藻■☆

403183 Utricularia flexuosa Benth. = Utricularia australis R. Br. ■

403184 Utricularia flexuosa Vahl = Utricularia aurea Lour. ■

403185 Utricularia flexuosa Vahl var. parviflora Kamienski = Utricularia stellaris L. f. ■☆

403186 Utricularia foliosa L.;多叶狸藻■☆

403187 Utricularia foliosa L. var. gracilis Kamienski = Utricularia foliosa L. ■☆

403188 Utricularia forrestii P. Taylor;长距狸藻■

403189 Utricularia foveolata Edgew.;海南狸藻■

403190 Utricularia geminiscapa Benj.;双茎狸藻;Hidden-fruited Bladderwort,Twin-stemmed Bladderwort ■☆

403191 Utricularia gibba L.;丝叶狸藻(环翅狸藻,少花狸藻);Creeping Bladderwort,Fewflower Bladderwort,Humped Bladderwort,Swollen-spurred Bladderwort ■

403192 Utricularia gibba L. = Utricularia exoleta R. Br. ■

403193 Utricularia gibba L. subsp. exoleta (R. Br.) P. Taylor = Utricularia exoleta R. Br. ■

403194 Utricularia gibba L. subsp. exoleta (R. Br.) P. Taylor = Utricularia gibba L. ■

403195 Utricularia gibba L. subsp. exoleta (R. Br.) P. Taylor f. natans Komiya = Utricularia exoleta R. Br. f. natans (Komiya) Komiya ■☆

403196 Utricularia gibbsiae Stapf = Utricularia scandens Benj. ■

403197 Utricularia gilletii De Wild. et T. Durand = Utricularia benjaminiana Oliv. ■☆

403198 Utricularia glochidiata Wight = Utricularia striata J. Sm. ■

403199 Utricularia graminifolia Vahl;禾叶挖耳草;Grassleaf Bladderwort,Grassleaf Draws Ear Grass ■

403200 Utricularia grandivesiculosa Czech = Utricularia reflexa Oliv. ■☆

403201 Utricularia graniticola A. Chev. et Pellegr. = Utricularia pubescens Sm. ■☆

403202 Utricularia griffithii Wight = Utricularia uliginosa Vahl ■

403203 Utricularia gyrans Suess. = Utricularia tortilis Welw. ex Oliv. ■☆

403204 Utricularia harlandii Oliv. ex Benth. = Utricularia striatula J. Sm. ■

403205 Utricularia hians A. DC. = Utricularia prehensilis E. Mey. ■☆

403206 Utricularia humbertiana H. Perrier = Utricularia livida E. Mey. ■☆

403207 Utricularia humbertiana H. Perrier var. andringitrensis？ = Utricularia livida E. Mey. ■☆

403208 Utricularia humilis E. Phillips = Utricularia livida E. Mey. ■☆

403209 Utricularia hydrocotyloides F. E. Lloyd et G. Taylor = Utricularia pubescens Sm. ■☆

403210 Utricularia ibarensis Baker = Utricularia livida E. Mey. ■☆

403211 Utricularia imerinensis H. Perrier = Utricularia reflexa Oliv. ■☆

403212 Utricularia incerta Kamienski = Utricularia australis R. Br. ■

403213 Utricularia inflata Afzel. ex Kamienski = Utricularia micropetala Sm. ■☆

403214 Utricularia inflata Walter;膀胱挖耳草;Inflated Bladderwort,Swollen Buttercup ■☆

403215 Utricularia inflexa Forssk.;内弯狸藻■☆

403216 Utricularia inflexa Forssk. var. major Kamienski = Utricularia inflexa Forssk. ■☆

403217 Utricularia inflexa Forssk. var. remota Kamienski = Utricularia inflexa Forssk. ■☆

403218 Utricularia inflexa Forssk. var. stellaris (L. f.) P. Taylor = Utricularia stellaris L. f. ■☆

403219 Utricularia inflexa Forssk. var. tenuifolia Kamienski = Utricularia inflexa Forssk. ■☆

403220 Utricularia intermedia Hayne;异枝狸藻(小狸藻);Differentshoot Bladderwort,Flat-leaved Bladderwort,Intermediate Bladderwort,Irish Bladderwort,Northern Bladderwort,Smaller Bladderwort ■

403221 Utricularia intermedia Heyne f. ochroleuca (R. Hartm.) Komiya = Utricularia ochroleuca R. Hartm. ■☆

403222 Utricularia japonica Makino;日本狸藻;Japanese Bladderwort ■

403223 Utricularia japonica Makino = Utricularia australis R. Br. ■

403224 Utricularia japonica Makino = Utricularia vulgaris L. var. japonica (Makino) Tamura ■

403225 Utricularia juncea Vahl;灯心草狸藻■☆

403226 Utricularia kalmaloensis A. Chev. = Utricularia gibba L. ■

403227 Utricularia kirkii Stapf = Utricularia arenaria A. DC. ■☆

403228 Utricularia kumaonensis Oliv.;毛籽狸藻■☆

403229 Utricularia lateriflora R. Br.;侧花狸藻■☆

403230 Utricularia lehmannii Benj. = Utricularia bisquamata Schrank ■☆

403231 Utricularia letestuii P. Taylor;莱泰斯图狸藻■☆

403232 Utricularia limosa R. Br.;长梗挖耳草;Longstalk Bladderwort,Longstalk Draws ear grass ■

403233 Utricularia linarioides Welw. ex Oliv. = Utricularia welwitschii Oliv. ■☆

403234 Utricularia lingulata Baker = Utricularia prehensilis E. Mey. ■☆

403235 Utricularia livida E. Mey.;肝色狸藻■☆

403236 Utricularia livida E. Mey. var. engleri (Kamienski) Stapf = Utricularia livida E. Mey. ■☆

403237 Utricularia livida E. Mey. var. micrantha Kamienski = Utricularia livida E. Mey. ■☆

403238 Utricularia longecalcarata Benj. = Utricularia livida E. Mey. ■☆

403239 Utricularia longifolia Garden;长叶狸藻■☆

403240 Utricularia macrocheilos (P. Taylor) P. Taylor;大唇狸藻■☆

403241 Utricularia macrophylla Masam. et Syozi = Utricularia uliginosa Vahl ■

403242 Utricularia macrorhiza J. Le Conte = Utricularia vulgaris L. subsp. macrorhiza (J. Le Leconte) R. T. Clausen ■

403243 Utricularia macrorhiza J. Le Conte = Utricularia vulgaris L. ■

403244 Utricularia madagascariensis A. DC. = Utricularia livida E. Mey. ■☆

403245 Utricularia magnavesica R. D. Good = Utricularia reflexa Oliv. ■☆

403246 Utricularia major Schmidel;大狸藻■☆

403247 Utricularia mannii Oliv.;曼氏狸藻■☆

403248 Utricularia mauroyae H. Perrier = Utricularia livida E. Mey. ■☆

403249 Utricularia microcalyx (P. Taylor) P. Taylor;小萼狸藻■☆

403250 Utricularia micropetala Sm.;小瓣狸藻■☆

403251 Utricularia micropetala Sm. var. macrocheilos P. Taylor =

Utricularia macrocheilos（P. Taylor）P. Taylor ■☆

403252 Utricularia minor L.；细叶狸藻（姬狸藻，小狸藻）；Lesser Bladderwort, Small Bladderwort, Smallleaf Bladderwort, Smallleaf Draws Ear Grass ■

403253 Utricularia minor L. f. natans Komiya = Utricularia minor L. ■

403254 Utricularia minor L. f. stricta Komiya = Utricularia minor L. ■

403255 Utricularia minor L. var. terrestris Glück = Utricularia minor L. ■

403256 Utricularia minor L. var. multispinosa Miki = Utricularia minor L. ■

403257 Utricularia minutissima Vahl；斜果挖耳草；Obliquefruit Bladderwort, Obliquefruit Draws Ear Grass ■

403258 Utricularia minutissima Vahl f. albiflora Komiya；白花斜果挖耳草 ■☆

403259 Utricularia monophylla Dinter = Utricularia arenaria A. DC. ■☆

403260 Utricularia multicaulis Oliv. = Utricularia striatula J. Sm. ■

403261 Utricularia multispinosa（Miki）Miki = Utricularia minor L. ■

403262 Utricularia nagurae Makino = Utricularia exoleta R. Br. ■

403263 Utricularia neglecta Lehm. = Utricularia australis R. Br. ■

403264 Utricularia nepalensis Kitam. = Utricularia minor L. ■

403265 Utricularia nipponica Makino = Utricularia minutissima Vahl ■

403266 Utricularia obtusa Sw. = Utricularia gibba L. ■

403267 Utricularia obtusiloba Benj. = Utricularia caerulea L. ■

403268 Utricularia ochroleuca R. Hartm.；苍白狸藻；Pale Bladderwort ■☆

403269 Utricularia odontosepala Stapf；齿萼狸藻 ■☆

403270 Utricularia odontosperma Stapf = Utricularia livida E. Mey. ■☆

403271 Utricularia oliveri Kamienski = Utricularia inflexa Forssk. ■☆

403272 Utricularia oliveri Kamienski var. fimbriata ？ = Utricularia inflexa Forssk. ■☆

403273 Utricularia oliveri Kamienski var. schweinfurthii ？ = Utricularia inflexa Forssk. ■☆

403274 Utricularia orbiculata Wall. = Utricularia striatula J. Sm. ■

403275 Utricularia orbiculata Wall. ex A. DC. = Utricularia striatula J. Sm. ■

403276 Utricularia papillosa Stapf = Utricularia pubescens Sm. ■☆

403277 Utricularia parkeri Baker = Utricularia bisquamata Schrank ■☆

403278 Utricularia pedicellata Wight = Utricularia graminifolia Vahl ■

403279 Utricularia peltatifolia A. Chev. et Pellegr. = Utricularia pubescens Sm. ■☆

403280 Utricularia pentadactyla P. Taylor；五指狸藻 ■☆

403281 Utricularia perpusilla A. DC. = Utricularia subulata L. ■☆

403282 Utricularia philetas R. D. Good = Utricularia striatula J. Sm. ■

403283 Utricularia pilifera A. Chev. = Utricularia reflexa Oliv. ■☆

403284 Utricularia pilosa（Makino）Makino = Utricularia aurea Lour. ■

403285 Utricularia platyptera Stapf = Utricularia reflexa Oliv. ■☆

403286 Utricularia pobeguinii Pellegr.；波别狸藻 ■☆

403287 Utricularia praelonga A. St. -Hil.；长狸藻 ■☆

403288 Utricularia prehensilis E. Mey.；蔓狸藻 ■☆

403289 Utricularia prehensilis E. Mey. var. hians（A. DC.）Kamienski = Utricularia prehensilis E. Mey. ■☆

403290 Utricularia prehensilis E. Mey. var. huillensis Welw. ex Kamienski = Utricularia prehensilis E. Mey. ■☆

403291 Utricularia prehensilis E. Mey. var. lingulata（Baker）Kamienski = Utricularia prehensilis E. Mey. ■☆

403292 Utricularia prehensilis E. Mey. var. parviflora Oliv. = Utricularia andongensis Welw. ex Hiern ■☆

403293 Utricularia pubescens Sm.；细毛狸藻 ■☆

403294 Utricularia pumila Walter = Utricularia gibba L. ■

403295 Utricularia punctata Wall. ex A. DC.；盾鳞狸藻；Punctate Bladderwort ■

403296 Utricularia purpurea Walter；紫狸藻；Eastern Purple Bladderwort, Purple Bladderwort, Spotted Bladderwort ■☆

403297 Utricularia quadricarinata Suess. = Utricularia prehensilis E. Mey. ■☆

403298 Utricularia racemosa Wall. = Utricularia caerulea L. ■

403299 Utricularia racemosa Wall. ex Walp.；密花狸藻（长距挖耳草）；Denseflower Bladderwort ■

403300 Utricularia racemosa Wall. ex Walp. = Utricularia caerulea L. ■

403301 Utricularia racemosa Wall. var. filicaulis（Wall. ex A. DC.）C. B. Clarke = Utricularia caerulea L. ■

403302 Utricularia radiata Small；辐射狸藻；Little Floating Bladderwort ■☆

403303 Utricularia ramosa Vahl var. filicaulis（Wall. ex A. DC.）Clarke = Utricularia caerulea L. ■

403304 Utricularia raynalii P. Taylor；雷纳尔狸藻 ■☆

403305 Utricularia recta P. Taylor = Utricularia scandens Benj. var. firmula（Oliv.）Z. Y. Li ■

403306 Utricularia reflexa Oliv.；反折狸藻 ■☆

403307 Utricularia reflexa Oliv. var. parviflora P. Taylor = Utricularia reflexa Oliv. ■☆

403308 Utricularia rehmannii Kamienski = Utricularia bisquamata Schrank ■☆

403309 Utricularia rendlei F. E. Lloyd = Utricularia subulata L. ■☆

403310 Utricularia reniformis A. St. -Hil.；肾叶狸藻 ■☆

403311 Utricularia resupinata B. D. Greene ex Bigelow；西北狸藻；Northeastern Bladderwort, Resupinate Bladderwort ■☆

403312 Utricularia reticulata J. Sm. = Utricularia graminifolia Vahl ■

403313 Utricularia reticulata J. Sm. var. uliginosa（Vahl）C. B. Clarke = Utricularia uliginosa Vahl ■

403314 Utricularia riccioides A. Chev. = Utricularia gibba L. ■

403315 Utricularia rigida Benj.；硬狸藻 ■☆

403316 Utricularia salwinensis Hand. -Mazz.；怒江挖耳草；Nujiang Bladderwort, Nujiang Draws Ear Grass ■

403317 Utricularia sandersonii Oliv.；桑氏狸藻 ■☆

403318 Utricularia sandersonii Oliv. var. treubii（Kamienski）Kamienski = Utricularia sandersonii Oliv. ■☆

403319 Utricularia sanguinea Oliv. = Utricularia livida E. Mey. ■☆

403320 Utricularia sanguinea Oliv. var. minor Kamienski = Utricularia livida E. Mey. ■☆

403321 Utricularia scandens Benj.；缠绕挖耳草；Slendent Bladderwort, Twine Draws Ear Grass ■

403322 Utricularia scandens Benj. subsp. firmula（Oliv.）Z. Y. Li = Utricularia scandens Benj. var. firmula（Oliv.）Z. Y. Li ■

403323 Utricularia scandens Benj. subsp. schweinfurthii（Baker ex Stapf）P. Taylor = Utricularia scandens Benj. ■

403324 Utricularia scandens Benj. var. firmula（Oliv.）Z. Y. Li；尖萼挖耳草；Sharpcalyx Bladderwort ■

403325 Utricularia scandens Oliv. = Utricularia baouleensis A. Chev. ■

403326 Utricularia scandens Oliv. = Utricularia foveolata Edgew. ■

403327 Utricularia scandens Oliv. var. firmula（Oliv.）C. Y. Wu = Utricularia scandens Benj. ■

403328 Utricularia scandens Oliv. var. firmula（Oliv.）Subr. et Banerjee = Utricularia scandens Benj. var. firmula（Oliv.）Z. Y. Li ■

403329 Utricularia schinzii Kamienski = Utricularia bisquamata Schrank ■☆

403330 Utricularia schweinfurthii Baker ex Stapf = Utricularia scandens Benj. ■

403331 Utricularia sematophora Stapf = Utricularia livida E. Mey. ■☆

403332 Utricularia siakujiiensis S. Nakaj. ex H. Hara = Utricularia australis

R. Br. ■

403333 Utricularia simulans Pilg. ;相似狸藻■☆

403334 Utricularia spartea Baker = Utricularia livida E. Mey. ■☆

403335 Utricularia spartea Baker var. marojejensis H. Perrier = Utricularia livida E. Mey. ■☆

403336 Utricularia spartea Baker var. subspicata H. Perrier = Utricularia livida E. Mey. ■☆

403337 Utricularia spartioides Scott-Elliot ex H. Perrier = Utricularia livida E. Mey. ■☆

403338 Utricularia spiralis Sm. ;螺旋狸藻■☆

403339 Utricularia spiralis Sm. var. pobeguinii (Pellegr.) P. Taylor = Utricularia pobeguinii Pellegr. ■☆

403340 Utricularia spiralis Sm. var. tortilis (Welw. ex Oliv.) P. Taylor = Utricularia tortilis Welw. ex Oliv. ■☆

403341 Utricularia sprengelii Kamienski = Utricularia bisquamata Schrank ■☆

403342 Utricularia sprengelii Kamienski var. acuticeras ? = Utricularia bisquamata Schrank ■☆

403343 Utricularia sprengelii Kamienski var. humilis ? = Utricularia livida E. Mey. ■☆

403344 Utricularia stanfieldii P. Taylor;斯氏狸藻■☆

403345 Utricularia stellaris L. = Utricularia intermedia Hayne ■

403346 Utricularia stellaris L. f. ;星狸藻■☆

403347 Utricularia stellaris L. f. var. breviscapa Kamienski = Utricularia stellaris L. f. ■☆

403348 Utricularia stellaris L. f. var. dilatata Kamienski = Utricularia stellaris L. f. ■☆

403349 Utricularia stellaris L. f. var. filiformis Kamienski = Utricularia stellaris L. f. ■☆

403350 Utricularia stellaris L. f. var. inflexa (Forssk.) C. B. Clarke = Utricularia inflexa Forssk. ■☆

403351 Utricularia striatula J. Sm. ;圆叶挖耳草(条纹挖耳草,圆叶狸藻);Roundleaf Bladderwort,Roundleaf Draws ear grass ■

403352 Utricularia strumosa Sol. ex Stapf = Utricularia bisquamata Schrank ■☆

403353 Utricularia stygia G. Thor;北极狸藻;Arctic Bladderwort,Nordic Bladderwort ■☆

403354 Utricularia suaveolens Afzel. ex Benj. = Utricularia rigida Benj. ■☆

403355 Utricularia subsessilis Schltr. ex Kamienski = Utricularia firmula Welw. ex Oliv. ■☆

403356 Utricularia subulata L. ;钻狸藻;Slender Bladderwort ■☆

403357 Utricularia subulata L. var. minuta Kamienski = Utricularia subulata L. ■☆

403358 Utricularia taikankoensis Yamam. = Utricularia striatula J. Sm. ■

403359 Utricularia tenerrima Merr. = Utricularia baouleensis A. Chev. ■

403360 Utricularia tenerrima Merr. = Utricularia foveolata Edgew. ■

403361 Utricularia tenuicaulis Miki = Utricularia australis R. Br. ■

403362 Utricularia tetraloba P. Taylor;四裂狸藻■☆

403363 Utricularia thomasii F. E. Lloyd et G. Taylor = Utricularia pubescens Sm. ■☆

403364 Utricularia thonningii Schumach. = Utricularia inflexa Forssk. ■☆

403365 Utricularia thonningii Schumach. var. laciniata Stapf = Utricularia inflexa Forssk. ■☆

403366 Utricularia tortilis Welw. ex Oliv. ;螺旋状狸藻■☆

403367 Utricularia tortilis Welw. ex Oliv. var. andongensis (Welw. ex Hiern) Kamienski = Utricularia andongensis Welw. ex Hiern ■☆

403368 Utricularia transrugosa Stapf = Utricularia livida E. Mey. ■☆

403369 Utricularia treubii Kamienski = Utricularia sandersonii Oliv. ■☆

403370 Utricularia tribracteata Hochst. ex A. Rich. = Utricularia arenaria A. DC. ■☆

403371 Utricularia trichoschiza Stapf = Utricularia stellaris L. f. ■☆

403372 Utricularia tricolor A. St. -Hil. ;三色狸藻■☆

403373 Utricularia tricrenata Baker ex Hiern = Utricularia gibba L. ■

403374 Utricularia triloba R. D. Good = Utricularia subulata L. ■☆

403375 Utricularia troupinii P. Taylor;特鲁皮尼狸藻■☆

403376 Utricularia uliginoides Wight = Utricularia graminifolia Vahl ■

403377 Utricularia uliginosa Vahl;齿萼挖耳草(紫花挖耳草);Swamp Bladderwort,Toothcalyx Draws Ear Grass ■

403378 Utricularia uliginosa Vahl = Utricularia graminifolia Vahl ■

403379 Utricularia uliginosa Vahl f. albida (Makino) Komiya et C. Shibata;白花齿萼挖耳草■☆

403380 Utricularia verticillata Benj. = Utricularia limosa R. Br. ■

403381 Utricularia villosula Stapf = Utricularia benjaminiana Oliv. ■☆

403382 Utricularia vulgaris L. ;狸藻(水豆儿,葳菜,闸草);Bladderwort, Common Bladderwort,Great Bladderwort,Greater Bladderwort,Hooded Milfoil ■

403383 Utricularia vulgaris L. = Utricularia macrorhiza J. Le Conte ■

403384 Utricularia vulgaris L. subsp. macrorhiza (J. Le Leconte) R. T. Clausen = Utricularia vulgaris L. ■

403385 Utricularia vulgaris L. subsp. macrorhiza (J. Le Leconte) R. T. Clausen;大根狸藻;Common Bladderwort ■

403386 Utricularia vulgaris L. var. americana A. Gray = Utricularia vulgaris L. ■

403387 Utricularia vulgaris L. var. americana A. Gray = Utricularia vulgaris L. subsp. macrorhiza (J. Le Leconte) R. T. Clausen ■

403388 Utricularia vulgaris L. var. formosana F. T. Kuo = Utricularia australis R. Br. ■

403389 Utricularia vulgaris L. var. japonica (Makino) S. Yamanaka = Utricularia australis R. Br. ■

403390 Utricularia vulgaris L. var. japonica (Makino) Tamura = Utricularia japonica Makino ■

403391 Utricularia vulgaris L. var. japonica (Makino) Tamura f. fixa Komiya = Utricularia japonica Makino ■

403392 Utricularia vulgaris L. var. japonica (Makino) Tamura f. tenuicaulis (Miki) Komiya = Utricularia australis R. Br. ■

403393 Utricularia vulgaris L. var. pilosa Makino = Utricularia aurea Lour. ■

403394 Utricularia vulgaris L. var. tenuicaulis (Miki) F. T. Kuo = Utricularia australis R. Br. ■

403395 Utricularia vulgaris L. var. tenuicaulis (Miki) F. T. Kuo ex J. Y. Hsiao = Utricularia australis R. Br. ■

403396 Utricularia wallichiana Benj. = Utricularia bifida L. ■

403397 Utricularia wallichiana Wight = Utricularia scandens Benj. ■

403398 Utricularia wallichiana Wight var. firmula Oliv. = Utricularia scandens Benj. subsp. firmula (Oliv.) Z. Y. Li ■

403399 Utricularia wallichiana Wight var. firmula Oliv. = Utricularia scandens Benj. var. firmula (Oliv.) Z. Y. Li ■

403400 Utricularia wallichii Wight = Utricularia scandens Benj. ■

403401 Utricularia warburgii K. I. Goebel;钩距挖耳草■☆

403402 Utricularia welwitschii Oliv. ;韦尔狸藻■☆

403403 Utricularia welwitschii Oliv. var. microcalyx P. Taylor = Utricularia microcalyx (P. Taylor) P. Taylor ■☆

403404 Utricularia welwitschii Oliv. var. odontosepala (Stapf) P. Taylor = Utricularia odontosepala Stapf ■☆

403405 Utricularia welwitschii Oliv. var. pusilla Suess. = Utricularia welwitschii Oliv. ■☆

403406　Utricularia yakusimensis Masam. = Utricularia uliginosa Vahl ■

403407　Utricularia yakusimensis Masam. f. albida（Makino）H. Hara = Utricularia uliginosa Vahl f. albida（Makino）Komiya et C. Shibata ■☆

403408　Utriculariaceae Hoffmanns. et Link = Lentibulariaceae Rich.（保留科名）■

403409　Utsetela Pellegr.（1928）；头序桑属●☆

403410　Utsetela gabonensis Pellegr.；头序桑●☆

403411　Utsetela neglecta Jongkind；疏忽头序桑●☆

403412　Uva Kuntze = Uvaria L. ●

403413　Uvaria L.（1753）；紫玉盘属；Uvaria ●

403414　Uvaria Torr. et A. Gray = Asimina Adans. ●☆

403415　Uvaria acuminata Oliv.；尖紫玉盘●☆

403416　Uvaria aethiopica（Dunal）A. Rich. = Xylopia aethiopica（Dunal）A. Rich. ●☆

403417　Uvaria afzelii Scott-Elliot；阿氏紫玉盘●☆

403418　Uvaria amuyon Blanco = Goniothalamus amuyon（Blanco）Merr. ●◇

403419　Uvaria angolensis Oliv. = Uvaria angolensis Welw. ex Oliv. ●☆

403420　Uvaria angolensis Oliv. subsp. guineensis Keay = Uvaria angolensis Welw. ex Oliv. ●☆

403421　Uvaria angolensis Welw. ex Oliv.；安哥拉紫玉盘●☆

403422　Uvaria angustifolia Engl. et Diels = Uvariodendron angustifolium（Engl. et Diels）R. E. Fr. ●☆

403423　Uvaria anonoides Baker f.；宽叶紫玉盘●☆

403424　Uvaria asterias S. Moore = Asteranthe asterias（S. Moore）Engl. et Diels ●☆

403425　Uvaria badiiflora Hance = Uvaria macrophylla Roxb. ●

403426　Uvaria badiiflora Hance = Uvaria microcarpa Champ. et Benth. ●

403427　Uvaria baumannii Engl. et Diels；鲍曼紫玉盘●☆

403428　Uvaria bequaertii De Wild. = Afroguatteria bequaertii（De Wild.）Boutique ●☆

403429　Uvaria bipindensis Engl.；比平迪紫玉盘●☆

403430　Uvaria boniana Finet et Gagnep.；光叶紫玉盘（挪藤）；Glabrousleaf Uvaria，Glabrous-leaved Uvaria ●

403431　Uvaria brevistipitata De Wild.；短柄紫玉盘●☆

403432　Uvaria buchholzii Engl. et Diels = Balonga buchholzii（Engl. et Diels）Le Thomas ●☆

403433　Uvaria bukobensis Engl. = Uvaria angolensis Welw. ex Oliv. ●☆

403434　Uvaria cabindensis Exell；卡宾达紫玉盘●☆

403435　Uvaria cabrae De Wild. et T. Durand；卡布拉紫玉盘●☆

403436　Uvaria caffra E. Mey. ex Sond.；开菲尔紫玉盘●☆

403437　Uvaria caillei A. Chev. ex Hutch. et Dalziel = Friesodielsia hirsuta（Benth.）Steenis ●☆

403438　Uvaria calamistrata Hance；刺果紫玉盘（毛荔枝藤，山香蕉，细叶酒饼木）；Spinyfruit Uvaria，Spiny-fruited Uvaria ●

403439　Uvaria cavaleriei H. Lév. = Fissistigma cavaleriei（H. Lév.）Rehder ●

403440　Uvaria cerasoides Roxb. = Polyalthia cerasoides（Roxb.）Benth. et Hook. f. ●

403441　Uvaria chamae P. Beauv.；矮紫玉盘；Dwarf Uvaria ●☆

403442　Uvaria chariensis A. Chev.；沙里紫玉盘●☆

403443　Uvaria clavata Pierre ex Engl. et Diels；棍棒紫玉盘●☆

403444　Uvaria comperei Le Thomas；孔佩尔紫玉盘●☆

403445　Uvaria connivens Benth. = Uvariodendron connivens（Benth.）R. E. Fr. ●☆

403446　Uvaria cordata（Dunal）Alston = Uvaria macrophylla Roxb. ●

403447　Uvaria cordata Schumach. et Thonn. = Uvaria ovata（DC.）A. DC. ●☆

403448　Uvaria cornuana Engl. et Diels；角状紫玉盘●☆

403449　Uvaria corynocarpa Diels = Uvaria scabrida Oliv. ●☆

403450　Uvaria crassipetala Engl. ex Engl. et Diels = Anonidium mannii（Oliv.）Engl. et Diels ●☆

403451　Uvaria cristata R. Br. = Uvaria chamae P. Beauv. ●☆

403452　Uvaria cuanzensis Paiva；宽扎紫玉盘●☆

403453　Uvaria cylindrica Schumach. et Thonn. = Uvaria chamae P. Beauv. ●☆

403454　Uvaria decidua Diels；脱落紫玉盘●☆

403455　Uvaria denhardtiana Engl. et Diels；小柱紫玉盘●☆

403456　Uvaria dependens Engl. et Diels；悬垂紫玉盘●☆

403457　Uvaria dielsii R. E. Fr. = Uvaria lucida Benth. ●☆

403458　Uvaria dinklagei Engl. et Diels；丁克紫玉盘●☆

403459　Uvaria doeringii Diels；多林紫玉盘●☆

403460　Uvaria dolichoclada Hayata = Uvaria macclurei Diels ●

403461　Uvaria dolichoclada Hayata = Uvaria macrophylla Roxb. ●

403462　Uvaria echinata A. Chev. = Uvaria chamae P. Beauv. ●☆

403463　Uvaria edulis N. Robson；可食紫玉盘●☆

403464　Uvaria elliotiana Engl. et Diels；埃氏紫玉盘●☆

403465　Uvaria elliotiana Engl. et Diels = Mischogyne elliotianum（Engl. et Diels）R. E. Fr. ●☆

403466　Uvaria engleriana Exell = Uvaria baumannii Engl. et Diels ●☆

403467　Uvaria esculenta Roxb. ex Rottler = Artabotrys hexapetalus（L. f.）Bhandari ●

403468　Uvaria faulknerae Verdc.；福氏紫玉盘●☆

403469　Uvaria fruticosa Engl. = Uvaria lucida Benth. ●☆

403470　Uvaria fujianensis Yen C. Yang et P. H. Huang = Uraria neglecta Prain ●■

403471　Uvaria fusca Benth. = Uvariodendron fuscum（Benth.）R. E. Fr. ●☆

403472　Uvaria gabonensis Engl. et Diels；加蓬紫玉盘●☆

403473　Uvaria gazensis Baker f. = Uvaria lucida Benth. subsp. virens（N. E. Br.）Verdc. ●☆

403474　Uvaria gigantea Engl. = Uvariodendron giganteum（Engl.）R. E. Fr. ●☆

403475　Uvaria globosa Hook. f. = Uvaria ovata（DC.）A. DC. ●☆

403476　Uvaria globosa Hook. f. var. warneckei Engl. = Uvaria ovata（DC.）A. DC. ●☆

403477　Uvaria glomerulata A. Chev. = Uvaria scabrida Oliv. ●☆

403478　Uvaria gossweileri Exell = Polyceratocarpus gossweileri（Exell）Paiva ●☆

403479　Uvaria gracilipes N. Robson；细梗紫玉盘●☆

403480　Uvaria gracilis Hook. f. = Friesodielsia gracilis（Hook. f.）Steenis ●☆

403481　Uvaria grandiflora Roxb.；大花紫玉盘（川血乌，各骆子藤，红肉梨，酒饼树，匍匐木，葡萄木，山芭蕉罗，山椒子，水香桃，细藤周公）；Largeflower Uvaria，Large-flowered Uvaria ●

403482　Uvaria guangxiensis W. L. Sha = Uraria lacei Craib ●■

403483　Uvaria hahniana Baill. = Tridimeris baillonii G. E. Schatz ex Maas, E. A. Mennega et Westra ●☆

403484　Uvaria hamiltonii Hook. f. et Thomson var. kurzii King = Uvaria kurzii（King）P. T. Li ●

403485　Uvaria hamiltonii Hook. f. var. kurzii King = Uvaria kurzii（King）P. T. Li ●

403486　Uvaria hamosa（Roxb.）Wall. ex Wight et Arn. var. formosana Matsum. = Uraria neglecta Prain ●■

403487　Uvaria heteroclita Roxb. = Kadsura heteroclita（Roxb.）Craib ●

403488　Uvaria heterotricha Pellegr.；异毛紫玉盘●☆

403489　Uvaria hexaloboides R. E. Fr. = Uvariastrum hexaloboides（R. E. Fr.）R. E. Fr. ●☆

403490　Uvaria hispida（Franch.）Schindl. = Desmodium hispidum Franch. ●

403491　Uvaria hispido-costata Pierre ex Engl. et Diels；硬毛紫玉盘●☆

403492　Uvaria holstii Engl. = Uvaria acuminata Oliv. ●☆

403493　Uvaria huillensis Engl. et Diels = Hexalobus monopetalus（A. Rich.）Engl. et Diels ●☆

403494　Uvaria insculpta Engl. et Diels = Uvariastrum insculptum（Engl. et Diels）Sprague et Hutch. ●☆

403495　Uvaria japonica L. = Kadsura japonica（L.）Dunal ●

403496　Uvaria johannis Exell；约翰紫玉盘●☆

403497　Uvaria kirkii Hook. f. = Uvaria kirkii Oliv. ex Hook. f. ●☆

403498　Uvaria kirkii Oliv. ex Hook. f. ；柯克紫玉盘●☆

403499　Uvaria klaineana Engl. et Diels；克莱恩紫玉盘●☆

403500　Uvaria klaineana Engl. et Diels var. chrysophylla Pellegr. = Uvaria klaineana Engl. et Diels ●☆

403501　Uvaria klainei Pierre ex Engl. et Diels；克氏紫玉盘●☆

403502　Uvaria kurzii（King）P. T. Li；黄花紫玉盘；Kurz Uvaria，Yellow Uvaria ●

403503　Uvaria kweichowensis P. T. Li；瘤果紫玉盘（贵州紫玉盘）；Guizhou Uvaria，Tumorfruit Uvaria ●

403504　Uvaria lagopus DC. var. neglecta（Prain）H. Ohashi = Uraria neglecta Prain ●■

403505　Uvaria lastoursvillensis Pellegr. ；拉斯图维尔紫玉盘●☆

403506　Uvaria latifolia（Scott-Elliot）Engl. et Diels = Uvaria anonoides Baker f. ●☆

403507　Uvaria laurentii De Wild. ；洛朗紫玉盘●☆

403508　Uvaria leonensis Engl. et Diels = Uvaria ovata（DC.）A. DC. subsp. afzeliana（DC.）Keay ●☆

403509　Uvaria leopoldvillensis De Wild. ；利奥波德维尔紫玉盘●☆

403510　Uvaria leptocladon Oliv. ；长枝紫玉盘●☆

403511　Uvaria leptocladon Oliv. subsp. septentrionalis Verdc. ；北方紫玉盘●☆

403512　Uvaria leptocladon Oliv. var. holstii（Engl.）Engl. et Diels = Uvaria acuminata Oliv. ●☆

403513　Uvaria letestui Pellegr. = Uvariodendron molundense（Engl. et Diels）R. E. Fr. ●☆

403514　Uvaria longibracteata Yen C. Yang et P. H. Huang = Uraria neglecta Prain ●■

403515　Uvaria longifolia Sonn. = Polyalthia longifolia（Sonn.）Thwaites ●

403516　Uvaria lucida Benth. ；光亮紫玉盘●☆

403517　Uvaria lucida Benth. subsp. virens（N. E. Br.）Verdc. ；绿花光亮紫玉盘●☆

403518　Uvaria lucida Bojer ex Sweet = Uvaria lucida Benth. ●☆

403519　Uvaria macclurei Diels；那大紫玉盘（石山紫玉盘）；Hayata Uvaria，Macclure Uvaria，Nada Uvaria ●

403520　Uvaria macclurei Diels = Uvaria macrophylla Roxb. ●

403521　Uvaria macrophylla Roxb. ；紫玉盘（大叶紫玉盘，蕉藤，酒饼，酒饼木，酒饼婆，酒饼叶，酒饼子，牛刀树，牛荃子，牛头罗，山巴豆，石龙叶，土枇杷，小十八风藤，油椎）；Bigleaf Uvaria，Littlefruit Uvaria，Uvaria ●

403522　Uvaria macrophylla Roxb. var. microcarpa（Champ. ex Benth.）Fient et Gagnep. = Uvaria macrophylla Roxb. ●

403523　Uvaria macrophylla Roxb. var. microcarpa（Champ.）Finet et Gagnep. = Uvaria microcarpa Champ. et Benth. ●

403524　Uvaria mannii Hutch. et Dalziel = Uvariodendron molundense（Engl. et Diels）R. E. Fr. ●☆

403525　Uvaria marginata Diels = Uvaria obanensis Baker f. ●☆

403526　Uvaria mayumbensis Exell = Uvariodendron molundense（Engl. et Diels）R. E. Fr. ●☆

403527　Uvaria mendesii Paiva；门代斯紫玉盘●☆

403528　Uvaria microcarpa Champ. ex Benth. = Uvaria macrophylla Roxb. ●

403529　Uvaria microphylla A. Chev. = Uvaria ovata（DC.）A. DC. subsp. afzeliana（DC.）Keay ●☆

403530　Uvaria microstyla Chiov. = Uvaria denhardtiana Engl. et Diels ●☆

403531　Uvaria microtricha Engl. et Diels = Polyceratocarpus microtrichus（Engl. et Diels）Ghesq. ex Pellegr. ●☆

403532　Uvaria mollis Engl. et Diels；柔软紫玉盘●☆

403533　Uvaria molundensis Engl. et Diels = Uvariodendron molundense（Engl. et Diels）R. E. Fr. ●☆

403534　Uvaria monopetala A. Rich. = Hexalobus monopetalus（A. Rich.）Engl. et Diels ●☆

403535　Uvaria muricata Pierre ex Engl. et Diels；短尖紫玉盘●☆

403536　Uvaria muricata Pierre ex Engl. et Diels var. suaveolens（Louis ex Boutique）Le Thomas；芳香短尖紫玉盘●☆

403537　Uvaria muricata Pierre ex Engl. et Diels var. yalingensis Tisser. ；亚林加紫玉盘●☆

403538　Uvaria narum Wall. ；那尔紫玉盘●☆

403539　Uvaria ngounyensis Pellegr. ；恩戈尼亚紫玉盘●☆

403540　Uvaria nigrescens Engl. et Diels = Uvaria chamae P. Beauv. ●☆

403541　Uvaria nyassensis Engl. et Diels = Uvaria lucida Benth. subsp. virens（N. E. Br.）Verdc. ●☆

403542　Uvaria obanensis Baker f. ；奥班紫玉盘●☆

403543　Uvaria oblanceolata W. T. Wang；狭叶紫玉盘；Narrowleaf Uvaria ●

403544　Uvaria oblanceolata W. T. Wang = Polyalthia petelotii Merr. ●

403545　Uvaria obovata（Willd.）Torr. et A. Gray = Asimina obovata（Willd.）Nash ●☆

403546　Uvaria obovatifolia Hayata = Uvaria macrophylla Roxb. ●

403547　Uvaria obovatifolia Hayata = Uvaria microcarpa Champ. et Benth. ●

403548　Uvaria odorata Lam. = Cananga odorata（Lam.）Hook. f. et Thomson ●

403549　Uvaria odoratissima Roxb. = Artabotrys hexapetalus（L. f.）Bhandari ●

403550　Uvaria osmantha Diels；木犀紫玉盘●☆

403551　Uvaria ovata（DC.）A. DC. ；卵紫玉盘●☆

403552　Uvaria ovata（DC.）A. DC. subsp. afzeliana（DC.）Keay；阿氏卵紫玉盘●☆

403553　Uvaria pandensis Verdc. ；潘德紫玉盘●☆

403554　Uvaria parviflora（Michx.）Torr. et A. Gray = Asimina parviflora（Michx.）Dunal ●☆

403555　Uvaria parviflora A. Rich. = Xylopia parviflora（A. Rich.）Benth. ●☆

403556　Uvaria pecoensis Exell = Uvaria smithii Engl. ●☆

403557　Uvaria platypetala Champ. ex Benth. = Uvaria grandiflora Roxb. ●

403558　Uvaria platyphylla Boutique = Uvaria anonoides Baker f. ●☆

403559　Uvaria poggei Engl. et Diels；波格紫玉盘●☆

403560　Uvaria poggei Engl. et Diels var. anisotricha Le Thomas；异毛波格紫玉盘●☆

403561　Uvaria polyantha Wall. = Fissistigma polyanthum（Hook. f. et Thomson）Merr. ●

403562　Uvaria puguensis D. M. Johnson；普古紫玉盘●☆

403563　Uvaria pulchra Louis ex Boutique；美丽紫玉盘●☆

403564　Uvaria purpurea Blume = Uvaria grandiflora Roxb. ●

403565　Uvaria pycnophylla Diels = Uvariodendron pycnophyllum

（Diels）R. E. Fr. ●☆

403566　Uvaria pygmaea（W. Bartram）Torr. et A. Gray = Asimina pygmaea（W. Bartram）Dunal ●☆

403567　Uvaria rhodantha Hance = Uvaria grandiflora Roxb. ●

403568　Uvaria rivularis Louis ex Boutique;溪边紫玉盘●☆

403569　Uvaria rufa Blume;小花紫玉盘;Littleflower Uvaria, Little-flowered Uvaria ●

403570　Uvaria sassandrensis Jongkind;萨桑德拉紫玉盘●☆

403571　Uvaria scaberrima Exell = Uvaria osmantha Diels ●☆

403572　Uvaria scabrida Oliv. ;微糙紫玉盘●☆

403573　Uvaria scabrida Oliv. var. parviflora Pellegr. = Uvaria scabrida Oliv. ●☆

403574　Uvaria scheffleri Diels;谢飞紫玉盘●☆

403575　Uvaria schweinfurthii Engl. et Diels;施韦紫玉盘●☆

403576　Uvaria smithii Engl. ;史密斯紫玉盘●☆

403577　Uvaria sofa Scott-Elliot;软紫玉盘●☆

403578　Uvaria spectabilis A. Chev. ex Hutch. et Dalziel = Uvariopsis guineensis Keay ●☆

403579　Uvaria staudtii Engl. et Diels = Duguetia staudtii（Engl. et Diels）Chatrou ●☆

403580　Uvaria stuhlmannii Engl. = Uvaria kirkii Hook. f. ●☆

403581　Uvaria suaveolens Louis ex Boutique = Uvaria muricata Pierre ex Engl. et Diels var. suaveolens（Louis ex Boutique）Le Thomas ●☆

403582　Uvaria suberosa Roxb. = Polyalthia suberosa（Roxb. ）Thwaites ●

403583　Uvaria tanzaniae Verdc. ;坦桑紫玉盘●☆

403584　Uvaria thomasii Sprague et Hutch. ;托马斯紫玉盘●☆

403585　Uvaria tonkinensis Finet et Gagnep. ;东京紫玉盘（扣匹，乌藤）;Tonkin Uvaria ●

403586　Uvaria tonkinensis Finet et Gagnep. var. subglabra Finet et Gagnep. = Uvaria tonkinensis Finet et Gagnep. ●

403587　Uvaria tonkinensis Finet et Gagnep. var. subglabra Finet et Gagnep. ;乌藤;Subglabrous Uvaria ●

403588　Uvaria tortilis A. Chev. ex Hutch. et Dalziel;螺旋状紫玉盘●☆

403589　Uvaria triloba（L. ）Torr. et A. Gray = Asimina triloba（L. ）Dunal ●☆

403590　Uvaria ugandensis（Bagsh. et Baker f. ）Exell = Uvaria schweinfurthii Engl. et Diels ●☆

403591　Uvaria uncata Lour. = Artabotrys hexapetalus（L. f. ）Bhandari ●

403592　Uvaria valvata De Wild. = Uvaria welwitschii（Hiern）Engl. et Diels ●☆

403593　Uvaria variabilis De Wild. = Uvaria angolensis Welw. ex Oliv. ●☆

403594　Uvaria velutina Dunal = Miliusa velutina（Dunal）Hook. f. et Thomson ●

403595　Uvaria verrucosa Engl. et Diels = Uvaria baumannii Engl. et Diels ●☆

403596　Uvaria versicolor Pierre ex Engl. et Diels;变色紫玉盘●☆

403597　Uvaria villosa Roxb. = Miliusa velutina（Dunal）Hook. f. et Thomson ●

403598　Uvaria virens N. E. Br. = Uvaria lucida Benth. subsp. virens（N. E. Br. ）Verdc. ●☆

403599　Uvaria vogelii Hook. f. = Monanthotaxis vogelii（Hook. f. ）Verdc. ●☆

403600　Uvaria welwitschii（Hiern）Engl. et Diels;韦尔紫玉盘●☆

403601　Uvaria zenkeri Engl. = Meiocarpidium lepidotum（Oliv. ）Engl. et Diels ●☆

403602　Uvariastrum Engl. （1901）;肖紫玉盘属（拟紫玉盘属）●☆

403603　Uvariastrum Engl. et Diels = Uvariastrum Engl. ●☆

403604　Uvariastrum dependens（Engl. et Diels）Engl. et Diels = Uvaria dependens Engl. et Diels ●☆

403605　Uvariastrum elliotianum（Engl. et Diels）Sprague et Hutch. = Mischogyne elliotianum（Engl. et Diels）R. E. Fr. ●☆

403606　Uvariastrum elliotianum（Engl. et Diels）Sprague et Hutch. var. gabonensis Pellegr. = Mischogyne elliotianum（Engl. et Diels）R. E. Fr. var. gabonensis Pellegr. ex Le Thomas ●☆

403607　Uvariastrum elliotianum（Engl. et Diels）Sprague et Hutch. var. glabrum Keay = Mischogyne elliotianum（Engl. et Diels）R. E. Fr. var. glabra（Keay）Evrard ●☆

403608　Uvariastrum elliotianum（Engl. et Diels）Sprague et Hutch. var. sericeum Keay = Mischogyne elliotianum（Engl. et Diels）R. E. Fr. ●☆

403609　Uvariastrum germainii Boutique;杰曼肖紫玉盘●☆

403610　Uvariastrum hexaloboides（R. E. Fr. ）R. E. Fr. ;六裂肖紫玉盘●☆

403611　Uvariastrum insculptum（Engl. et Diels）Sprague et Hutch. ;雕刻肖紫玉盘●☆

403612　Uvariastrum modestum Diels;适度肖紫玉盘●☆

403613　Uvariastrum neglectum Paiva;疏忽肖紫玉盘●☆

403614　Uvariastrum pierreanum Engl. ;皮埃尔肖紫玉盘●☆

403615　Uvariastrum pynaertii De Wild. ;皮那肖紫玉盘●☆

403616　Uvariastrum zenkeri Engl. et Diels;岑克尔肖紫玉盘●☆

403617　Uvariastrum zenkeri Engl. et Diels var. nigritianum Baker f. = Uvariastrum zenkeri Engl. et Diels ●☆

403618　Uvariella Ridl. = Uvaria L. ●

403619　Uvariodendron（Engl. et Diels）R. E. Fr. （1930）;玉盘木属●☆

403620　Uvariodendron angustifolium（Engl. et Diels）R. E. Fr. ;窄叶玉盘木●☆

403621　Uvariodendron anisatum Verdc. ;异型玉盘木●☆

403622　Uvariodendron calophyllum R. E. Fr. ;美叶玉盘木●☆

403623　Uvariodendron connivens（Benth. ）R. E. Fr. ;靠合玉盘木●☆

403624　Uvariodendron fuscum（Benth. ）R. E. Fr. ;棕色玉盘木●☆

403625　Uvariodendron giganteum（Engl. ）R. E. Fr. ;巨大玉盘木●☆

403626　Uvariodendron gossweileri（Exell）Exell et Mendonça = Polyceratocarpus gossweileri（Exell）Paiva ●☆

403627　Uvariodendron kirkii Verdc. ;柯克玉盘木●☆

403628　Uvariodendron letestui（Pellegr. ）R. E. Fr. = Uvariodendron molundense（Engl. et Diels）R. E. Fr. ●☆

403629　Uvariodendron magnificum Verdc. ;壮观肖紫玉盘●☆

403630　Uvariodendron mayumbense（Exell）R. E. Fr. = Uvariodendron molundense（Engl. et Diels）R. E. Fr. ●☆

403631　Uvariodendron mirabile R. E. Fr. ;奇异肖紫玉盘●☆

403632　Uvariodendron molundense（Engl. et Diels）R. E. Fr. ;莫卢肖紫玉盘●☆

403633　Uvariodendron occidentale Le Thomas;西方肖玉盘木●☆

403634　Uvariodendron oligocarpum Verdc. ;寡果肖玉盘木●☆

403635　Uvariodendron pycnophyllum（Diels）R. E. Fr. ;密叶肖玉盘木●☆

403636　Uvariodendron usambarense R. E. Fr. ;乌桑巴拉肖玉盘木●☆

403637　Uvariopsis Engl. = Uvariopsis Engl. ex Engl. et Diels ●☆

403638　Uvariopsis Engl. et Diels = Uvariopsis Engl. ex Engl. et Diels ●☆

403639　Uvariopsis Engl. ex Engl. et Diels（1899）;拟紫玉盘属●☆

403640　Uvariopsis bakeriana（Hutch. et Dalziel）Robyns et Ghesq. ;贝克拟紫玉盘●☆

403641　Uvariopsis batesii Robyns et Ghesq. = Uvariopsis solheidii（De Wild. ）Robyns et Ghesq. ●☆

403642　Uvariopsis bisexualis Verdc. ;双性拟紫玉盘●☆

403643　Uvariopsis chevalieri Robyns et Ghesq. = Mischogyne elliotianum（Engl. et Diels）R. E. Fr. ●☆

403644　Uvariopsis congensis Robyns et Ghesq. ;刚果拟紫玉盘●☆

403645　Uvariopsis congolana（De Wild. ）R. E. Fr. ;康戈尔拟紫玉盘●☆

403646 Uvariopsis dioica (Diels) Robyns et Ghesq. ;异株拟紫玉盘●☆

403647 Uvariopsis globiflora Keay;球花拟紫玉盘●☆

403648 Uvariopsis guineensis Keay;几内亚拟紫玉盘●☆

403649 Uvariopsis korupensis Gereau et Kenfack;科鲁普拟紫玉盘●☆

403650 Uvariopsis letestui Pellegr. ;莱泰斯图拟紫玉盘●☆

403651 Uvariopsis noldeae Exell et Mendonça;诺尔德拟紫玉盘●☆

403652 Uvariopsis pedunculosa (Diels) Robyns et Ghesq. = Uvariopsis dioica (Diels) Robyns et Ghesq. ●☆

403653 Uvariopsis sessiliflora (Mildbr. et Diels) Robyns et Ghesq. ;无花梗拟紫玉盘●☆

403654 Uvariopsis solheidii (De Wild.) Robyns et Ghesq. ;索尔海德拟紫玉盘●☆

403655 Uvariopsis submontana Kenfack et Gosline et Gereau;亚山生拟紫玉盘●☆

403656 Uvariopsis vanderystii Robyns et Ghesq. ;范德拟紫玉盘●☆

403657 Uvariopsis zenkeri Engl. ;岑克尔拟紫玉盘●☆

403658 Uva-ursi Duhamel(废弃属名) = Arctostaphylos Adans. (保留属名)●☆

403659 Uva-ursi uva-ursi (L.) Britton = Arctostaphylos uva-ursi (L.) Spreng. ●☆

403660 Uvedalia R. Br. = Mimulus L. ●■

403661 Uvifera Kuntze = Coccoloba P. Browne(保留属名)●

403662 Uvirandra J. St. -Hil. = Aponogeton L. f. (保留属名)■

403663 Uvirandra J. St. -Hil. = Ouvirandra Thouars ■

403664 Uvulana Raf. = Uvularia L. ■☆

403665 Uvularia L. (1753);细钟花属(垂铃儿属,颚花属);Bell Wort,Bellwort,Merry Bells,Merrybells,Wild Oats Bellwort ■☆

403666 Uvularia amplexifolia L. = Streptopus amplexifolius (L.) DC. ■☆

403667 Uvularia chinensis Ker Gawl. = Disporum cantoniense (Lour.) Merr. ■

403668 Uvularia cirrhosa Thunb. = Fritillaria thunbergii Miq. ■

403669 Uvularia floridana Chapm. ;佛罗里达细钟花;Florida Bellwort ■☆

403670 Uvularia grandiflora Sm. ;大细钟花(大颚花,大花垂铃儿);Bellwort, Cowbells, Large Bellwort, Large Merry-bell, Largeflower Bellwort,Large-flowered Bellwort,Merry Bells,Merry-bells ■☆

403671 Uvularia lanceolata Aiton = Streptopus lanceolatus (Aiton) Reveal ■☆

403672 Uvularia perfoliata L. ;穿叶细钟花(垂铃儿);Bellwort,Perfoliate Bellwort ■☆

403673 Uvularia puberula Michx. ;山地细钟花;Mountain Bellwort ■☆

403674 Uvularia pudica (Walter) Fernald = Uvularia puberula Michx. ■☆

403675 Uvularia sessilifolia L. ;无柄细钟花;Mountain Bellwort, Sessile Bellwort,Sessile-leaf Bellwort,Sessile-leaved Bellwort,Small Bellwort, Straw-lily,Wild Oats,Wild-oats ■☆

403676 Uvularia sessilis Thunb. = Disporum uniflorum Baker ex S. Moore ■

403677 Uvularia smithii Hook. = Prosartes smithii (Hook.) Utech ■☆

403678 Uvularia viridescens Maxim. = Disporum viridescens (Maxim.) Nakai ■

403679 Uvulariaceae A. Gray ex Kunth(1843)(保留科名);细钟花科(悬阶草科)■

403680 Uvulariaceae A. Gray ex Kunth(保留科名) = Colchicaceae DC. (保留科名)■

403681 Uvulariaceae A. Gray ex Kunth(保留科名) = Melanthiaceae Batsch ex Borkh. (保留科名)■

403682 Uvulariaceae Kunth = Uvulariaceae A. Gray ex Kunth(保留科名)■

403683 Uwarowia Bunge = Verbena L. ■●

403684 Uxi Almeida = Sacoglottis Mart. ●☆

403685 Vaccaria Medik. = Vaccaria Wolf ■

403686 Vaccaria Wolf(1776);麦蓝菜属(王不留行属);Cowherb, Cowherb Soapwort ■

403687 Vaccaria hispanica (Mill.) Rauschert;麦蓝菜(不留,不留行) 长豉草,大麦牛,大麦片,道灌草,豆篮子,对经草,莪蒿,孩儿,剪 金花,剪金子,角蒿,金剪刀草,金钱银台,金盏银台,禁宫花,妈 不流,马不留,麦加菜子,麦加子,麦兰菜状肥皂草,麦篮子,麦连 子,面条棵子,木蓝子,奶米,普通麦蓝菜,兔儿草子,王不流行, 王不留,王不留行,王留,王母牛,王母片,王牡牛);Colorado Corn Cockle,Cow Basil, Cow Cockle, Cow Herb, Cow Soapwort, Cowcockle, Cow-fat,Cowherb,Cow-herb,Cowherb Soapwort ■

403688 Vaccaria hispanica (Mill.) Rauschert subsp. grandiflora (Ser.) Holub = Vaccaria hispanica (Mill.) Rauschert ■

403689 Vaccaria hispanica (Mill.) Rauschert subsp. oxyodonta (Boiss.) Greuter et Burdet;尖齿麦蓝菜■☆

403690 Vaccaria hispanica (Mill.) Rauschert subsp. pyramidata (Medik.) Holub = Vaccaria hispanica (Mill.) Raeusch. ■

403691 Vaccaria hispanica (Mill.) Rauschert var. grandiflora (Ser.) J. Léonard = Vaccaria hispanica (Mill.) Rauschert ■

403692 Vaccaria hispanica (Mill.) Rauschert var. vaccaria (L.) Greuter = Vaccaria hispanica (Mill.) Rauschert ■

403693 Vaccaria officinalis L. ;药用麦蓝菜;Medicinal Cowherb ■☆

403694 Vaccaria parviflora Moench = Vaccaria hispanica (Mill.) Rauschert ■

403695 Vaccaria pyramidata Medik. = Vaccaria hispanica (Mill.) Raeusch. ■

403696 Vaccaria pyramidata Medik. = Vaccaria hispanica (Mill.) Rauschert var. vaccaria (L.) Greuter ■

403697 Vaccaria pyramidata Medik. var. grandiflora Ser. = Vaccaria hispanica (Mill.) Rauschert ■

403698 Vaccaria segetalis (Neck.) Garcke = Vaccaria hispanica (Mill.) Raeusch. ■

403699 Vaccaria segetalis (Neck.) Garcke ex Asch. = Vaccaria hispanica (Mill.) Raeusch. ■

403700 Vaccaria segetalis Garcke ex Asch. = Vaccaria hispanica (Mill.) Raeusch. ■

403701 Vaccaria vaccaria (L.) Britton = Vaccaria hispanica (Mill.) Raeusch. ■

403702 Vaccaria vulgaris Host. = Vaccaria hispanica (Mill.) Raeusch. ■

403703 Vacciniaceae Adans. = Ericaceae Juss. (保留科名)●

403704 Vacciniaceae Adans. = Vacciniaceae DC. ex Perleb(保留科名)●

403705 Vacciniaceae DC. ex Gray = Ericaceae Juss. (保留科名)●

403706 Vacciniaceae DC. ex Perleb(1818)(保留科名);越橘科(乌饭树科);Blueberry Family ●

403707 Vacciniaceae DC. ex Perleb(保留科名) = Ericaceae Juss. (保留科名)●

403708 Vacciniaceae Gray = Ericaceae Juss. (保留科名)●

403709 Vacciniopsis Rusby = Disterigma (Klotzsch) Nied. ●☆

403710 Vaccinium L. (1753);越橘属(乌饭树属,越桔属);Bilbery, Billberry, Blueberry, Bluet, Buck Berry, Cranberry, Huckleberry, Whortleberry ●

403711 Vaccinium albidens H. Lév. et Vaniot;白花越橘;White Blueberry,Whiteflower Blueberry,Whitish Blueberry ●

403712 Vaccinium albidens H. Lév. et Vaniot = Vaccinium iteophyllum Hance ●

403713 Vaccinium albiflorum Hook. = Vaccinium corymbosum L. ●☆

403714 Vaccinium albutoides C. B. Clarke;草莓树状越橘;Strawberry Bladderwort,Tree-like Blueberry ●

403715 Vaccinium amamianum Hatus. = Vaccinium emarginatum Hayata ●

403716　Vaccinium angustifolium Aiton；狭叶越橘（矮灌蓝莓，窄叶乌饭树）；Early Low Blueberry，Early Low-blueberry，Huckleberry，Late Low Blueberry，Low Sweet Blueberry，Lowbush Blueberry，Low-bush Blueberry，Lowsweet Blueberry，Narrowleaf Bladderwort，Sweet Hurts ●

403717　Vaccinium angustifolium Aiton var. hypolasium Fernald = Vaccinium angustifolium Aiton ●

403718　Vaccinium angustifolium Aiton var. laevifolium House；平滑狭叶越橘；Lowbush Blueberry，Low-bush Blueberry ● ☆

403719　Vaccinium angustifolium Aiton var. laevifolium House = Vaccinium angustifolium Aiton ●

403720　Vaccinium angustifolium Aiton var. myrtilloides （ Michx. ） House = Vaccinium myrtilloides Michx. ● ☆

403721　Vaccinium angustifolium Aiton var. nigrum （ A. W. Wood ） Dole；黑狭叶越橘；Black-fruited Blueberry ● ☆

403722　Vaccinium angustifolium Aiton var. nigrum （ A. W. Wood ） Dole = Vaccinium angustifolium Aiton ●

403723　Vaccinium anthonyi Merr. = Vaccinium fragile Franch. ●

403724　Vaccinium arboreum Marshall；树越橘（白莓）；Farkleberry，Farkleberry Tree，Sparkleberry，Tree Huckleberry ● ☆

403725　Vaccinium arboreum Marshall var. glaucescens （ Greene ） Sarg. = Vaccinium arboreum Marshall ● ☆

403726　Vaccinium arbuscula （ A. Gray ） Merriam = Vaccinium cespitosum Michx. ● ☆

403727　Vaccinium arbutoides C. B. Clarke；草莓树状越橘●

403728　Vaccinium arctostaphylos L. ；熊果越橘（高加索越橘）；Bearberry Vaccinium，Brousea Tea，Caucasian Whortleberry，Greek Strawberry Tree ● ☆

403729　Vaccinium ardisioides Hook. f. ex C. B. Clarke；紫梗越橘（大叶树萝卜，红梗越橘）；Ardisia-like Blueberry，Redstipe Bladderwort ●

403730　Vaccinium ashei Reade；兔眼越橘；Rabbit-eye Blueberry ● ☆

403731　Vaccinium atrococcum （ Grey ） A. Heller；湿地越橘（黑果高越橘）；Black Blueberry，Black Highbush Blueberry，Downy Swamp Blueberry ● ☆

403732　Vaccinium axillare Nakai = Vaccinium ovalifolium Sm. ●

403733　Vaccinium bancanum C. B. Clarke = Vaccinium exaristatum Kurz ●

403734　Vaccinium boninense Nakai；小笠原越橘● ☆

403735　Vaccinium brachyandrum C. Y. Wu et R. C. Fang；短蕊越橘；Short-stamen Blueberry，Short-stamened Blueberry ●

403736　Vaccinium brachybotrys （ Franch. ） Hand. -Mazz. ；短序越橘；Shortraceme Blueberry，Short-racemed Blueberry，Shortspike Bladderwort ●

403737　Vaccinium brachybotrys （ Franch. ） Hand. -Mazz. var. glaucocarpum C. Y. Wu. = Vaccinium brachybotrys （ Franch. ） Hand. -Mazz. ●

403738　Vaccinium bracteatum Thunb. ；南烛（苞越橘，草木之王，秤杆树，大禾子，饭筒树，黑饭草，猴菽，猴药，后卓，康菊紫，冷饭籽，零丁子，米饭花，米饭树，米碎子木，墨饭草，男续，南竺，南烛草木，牛筋，染椒，染菽，惟那木，乌草，乌饭草，乌饭树，乌饭叶，乌饭子，乌米饭树，羊爪子，珍珠花）；Bracted Races Blueberry，Oriental Blueberry，South Candle，Sweet Fruits Blueberry，Sweet-fruited Blueberry ●

403739　Vaccinium bracteatum Thunb. var. chinense （ Lodd. ） Chun ex Sleumer；小叶南烛（小叶乌饭树）；Chinese Oriental Blueberry ●

403740　Vaccinium bracteatum Thunb. var. formosanum （ Hayata ） S. S. Ying = Vaccinium wrightii A. Gray var. formosanum （ Hayata ） H. L. Li ●

403741　Vaccinium bracteatum Thunb. var. lanceolatum Nakai；披针叶南烛●

403742　Vaccinium bracteatum Thunb. var. lanceolatum Nakai = Vaccinium bracteatum Thunb. ●

403743　Vaccinium bracteatum Thunb. var. lanceolatum Nakai = Vaccinium randaiense Hayata ●

403744　Vaccinium bracteatum Thunb. var. longitubum Hayata = Vaccinium bracteatum Thunb. ●

403745　Vaccinium bracteatum Thunb. var. obovatum C. Y. Wu et R. C. Fang；倒卵叶南烛（倒卵叶乌饭树，石子陵木）；Obovateleaf Blueberry ●

403746　Vaccinium bracteatum Thunb. var. rubellum P. S. Hsu et al. ；淡红南烛（淡红乌饭树）；Reddish Oriental Blueberry ●

403747　Vaccinium bracteatum Thunb. var. wrightii （ Gray ） Rehder et E. H. Wilson = Vaccinium wrightii A. Gray ●

403748　Vaccinium brevipedicellatum C. Y. Wu ex W. P. Fang et Z. H. Pan；短梗越橘（短梗乌饭）；Shortpedicel Blueberry，Short-pediceled Blueberry，Shortstalk Bladderwort ●

403749　Vaccinium brittonii Porter ex E. P. Bicknell = Vaccinium angustifolium Aiton ●

403750　Vaccinium bullatum （ Dop ） Sleumer；泡泡叶越橘（山木薯）；Bullate Blueberry，Bullate-leaf Blueberry，Bullate-leaved Blueberry，Bullbleleaf Bladderwort ●

403751　Vaccinium bulleyanum （ Diels ） Sleumer；灯台越橘；Bulley Blueberry，Lampstand Bladderwort ●

403752　Vaccinium buxifolium （ H. Lév. et Vaniot ） H. Lév. = Vaccinium triflorum Rehder ●

403753　Vaccinium caesium Greene；鹿莓越橘；Deerberry，Squaw Huckleberry ● ☆

403754　Vaccinium caespitosum Michx. ；丛生越橘（矮越橘）；Dwarf Bilberry，Dwarf Blueberry，Fascicular Bladderwort ●

403755　Vaccinium calycinum Sm. ；长叶越橘● ☆

403756　Vaccinium camphorifolium Hand. -Mazz. = Vaccinium dunalianum Wight var. urophyllum Rehder et E. H. Wilson ●

403757　Vaccinium canadense Kalm ex A. Rich. = Vaccinium myrtilloides Michx. ● ☆

403758　Vaccinium canadense Kalm ex Rich. ；加拿大越橘；Canada Blueberry，Sour Top ● ☆

403759　Vaccinium candicans Michx. ；白亮越橘；Whiteshing Blueberry ● ☆

403760　Vaccinium carlesii Dunn；短尾越橘（小庆果，早禾子树，乌饭子）；Carles Blueberry，Shorttail Bladderwort ●

403761　Vaccinium carlesii Dunn var. longicaudatum （ Chun ex W. P. Fang et Z. H. Pan ） P. C. Tam = Vaccinium longicaudatum Chun ex W. P. Fang et Z. H. Pan ●

403762　Vaccinium caudatifolium Hayata = Vaccinium dunalianum Wight var. caudatifolium （ Hayata ） H. L. Li ●

403763　Vaccinium cavaleriei H. Lév. et Vaniot = Schoepfia jasminodora Siebold et Zucc. ●

403764　Vaccinium cavinerve C. Y. Wu；圆顶越橘；Cavate-nerved Blueberry，Roundtop Blueberry ●

403765　Vaccinium cespitosum Michx. ；簇生越橘；Dwarf Bilberry，Dwarf Huckleberry ● ☆

403766　Vaccinium cespitosum Michx. var. arbuscula A. Gray = Vaccinium cespitosum Michx. ● ☆

403767　Vaccinium chaetothrix Sleumer；团叶越橘（圆叶越橘）；Roundleaf Blueberry，Round-leaved Blueberry ●

403768　Vaccinium chamaebuxus C. Y. Wu；矮越橘；Dwarf Blueberry ●

403769　Vaccinium chapaense Merr. = Agapetes rubrobracteata R. C. Fang et S. H. Huang ●

403770　Vaccinium chengae W. P. Fang；四川越橘（诚君珍珠树）；Cheng Blueberry，Sichuan Bladderwort ●

403771　Vaccinium chengae W. P. Fang var. pilosum C. Y. Wu；毛萼四川越橘（毛萼珍珠树）；Pilose Cheng Blueberry ●

403772　Vaccinium chinense（Lodd.）Champ. = Vaccinium bracteatum Thunb. var. chinense（Lodd.）Chun ex Sleumer ●

403773　Vaccinium chinense（Lodd.）Champ. ex Benth. = Vaccinium bracteatum Thunb. var. chinense（Lodd.）Chun ex Sleumer ●

403774　Vaccinium chingii Sleumer = Vaccinium henryi Hemsl. var. chingii（Sleumer）C. Y. Wu et R. C. Fang ●

403775　Vaccinium chunii Merr. ex Sleumer；蓝果越橘；Bluefruit Bladderwort，Chun Blueberry ●

403776　Vaccinium ciliatum Thunb.；睫毛乌饭树；Ciliate Blueberry ●☆

403777　Vaccinium conchophyllum Rehder；贝叶越橘；Shell-leaf Blueberry，Shell-leaved Blueberry ●

403778　Vaccinium constablaei A. Gray = Vaccinium corymbosum L. ●☆

403779　Vaccinium corymbiferum Miq. = Sorbus corymbifera（Miq.）T. H. Nguyên et Yakovlev ●

403780　Vaccinium corymbosum L.；高大越橘（伞房花越橘，伞花越橘,湿地越橘,沼生越橘）；Blueberry, Corymb Bladderwort, High Bush Blueberry, Highbush Blueberry, High-bush Blueberry, Highbush Huckleberry, High-bush Huckleberry, Jubilee Blueberry, Swamp Blueberry, Tall Blueberry ●☆

403781　Vaccinium corymbosum L. 'Blue Ray'；蓝光高大越橘●☆

403782　Vaccinium corymbosum L. 'Earliblue'；早蓝高大越橘●☆

403783　Vaccinium corymbosum L. 'Pioneer'；先锋湿地越橘●☆

403784　Vaccinium corymbosum L. 'Tomahawk'；战斧高大越橘●☆

403785　Vaccinium corymbosum L. var. albiflorum（Hook.）Fernald = Vaccinium corymbosum L. ●☆

403786　Vaccinium corymbosum L. var. glabrum A. Gray = Vaccinium corymbosum L. ●☆

403787　Vaccinium craspedotum Sleumer；长萼越橘；Long-calyx Blueberry，Long-sepaled Blueberry ●

403788　Vaccinium craspedotum Sleumer var. brevipes C. Y. Wu；短梗长萼越橘；Short-stalk Blueberry ●

403789　Vaccinium craspedotum Sleumer var. brevipes C. Y. Wu. = Vaccinium craspedotum Sleumer ●

403790　Vaccinium crassifolium Andréws；匍匐越橘（厚叶越橘）；Creeping Blueberry ●☆

403791　Vaccinium crassivenium Sleumer；网脉越橘；Netvein Bladderwort，Net-veined Blueberry ●

403792　Vaccinium cuspidifolium C. Y. Wu et R. C. Fang；凸尖越橘；Cuspidateleaf Bladderwort, Cuspid-leaf Blueberry, Retinerved Blueberry，Sharp-leaved Blueberry ●

403793　Vaccinium cylindraceum Sm.；柱花越橘；Azores Blueberry, Tubular Flowers Blueberry ●☆

403794　Vaccinium darrowii Camp；常绿越橘；Evergreen Blueberry ●☆

403795　Vaccinium delavayi Franch.；苍山越橘（老鸦果，山梨儿，山檀，山檀香，土地瓜，岩檀香，野万年青）；Cangshan Bladderwort, Delavay Blueberry ●

403796　Vaccinium delavayi Franch. subsp. merrillianum（Hayata）R. C. Fang；台湾越橘（高山越橘，玉山岩桃）；Formosan Blue Berry, Taiwan Blueberry ●

403797　Vaccinium delavayi Franch. subsp. merrillianum（Hayata）R. C. Fang = Vaccinium merrillianum Hayata ●

403798　Vaccinium deliciosum Piper；美味越橘（甜越橘）；Sweet Blueberry ●☆

403799　Vaccinium dendrocharis Hand. -Mazz.；树生越橘；Epiphyte Blueberry，Living tree Bladderwort ●

403800　Vaccinium diaphanoloma Hand. -Mazz. = Vaccinium gaultherifolium（Griff.）Hook. f. ●

403801　Vaccinium donianum Wight = Vaccinium wrightii A. Gray ●

403802　Vaccinium donianum Wight var. austrosinense Hand. -Mazz. = Vaccinium mandarinorum Diels ●

403803　Vaccinium donianum Wight var. austrosinense Hand. -Mazz. = Vaccinium mandarinorum L. var. austrosinense（Hand. -Mazz.）F. P. Metcalf ●

403804　Vaccinium donianum Wight var. brachybotrys Franch. = Vaccinium brachybotrys（Franch.）Hand. -Mazz. ●

403805　Vaccinium donianum Wight var. hangchouense Matsuda = Vaccinium mandarinorum Diels ●

403806　Vaccinium donianum Wight var. hangchouense Matsuda = Vaccinium wrightii A. Gray ●

403807　Vaccinium donianum Wight var. laetum（Diels）Rehder et E. H. Wilson = Vaccinium mandarinorum Diels ●

403808　Vaccinium donianum Wight var. laetum（Diels）Rehder et E. H. Wilson = Vaccinium laetum Diels ●

403809　Vaccinium dopii H. F. Copel. = Vaccinium dunalianum Wight var. urophyllum Rehder et E. H. Wilson ●

403810　Vaccinium duclouxii（H. Lév.）Hand. -Mazz.；云南越橘；Ducloux Blueberry，Yunnan Blueberry ●

403811　Vaccinium duclouxii（H. Lév.）Hand. -Mazz. var. hirtellum C. Y. Wu et R. C. Fang；毛果云南越橘；Hairyfruit Yunnan Blueberry ●

403812　Vaccinium duclouxii（H. Lév.）Hand. -Mazz. var. hirticaule C. Y. Wu；毛茎云南越橘；Setose Blueberry ●

403813　Vaccinium duclouxii（H. Lév.）Hand. -Mazz. var. pubipes C. Y. Wu；柔毛云南越橘（柔毛越橘）；Hairy Yunnan Blueberry ●

403814　Vaccinium dunalianum Wight；樟叶越橘（长尾叶越橘，长尾越橘,饭米果,喜马越橘,珍珠花）；Cinnamonleaf Bladderwort，Dunal's Blueberry, Himalaya Blueberry ●

403815　Vaccinium dunalianum Wight var. calycinum Dop = Vaccinium dunalianum Wight var. urophyllum Rehder et E. H. Wilson ●

403816　Vaccinium dunalianum Wight var. caudatifolium（Hayata）H. L. Li；长尾叶越橘（大透骨草，樟叶越橘，珍珠花）；Caudate-leaf Blueberry ●

403817　Vaccinium dunalianum Wight var. megaphyllum Sleumer；大樟叶越橘；Big-leaf Caudate-leaf Blueberry ●

403818　Vaccinium dunalianum Wight var. urophyllum Rehder et E. H. Wilson；尾叶越橘（大透骨草）；Tail-leaf Blueberry ●

403819　Vaccinium dunnianum Sleumer；长穗越橘（白黄果）；Longspike Blueberry，Long-spiked Blueberry ●

403820　Vaccinium elliotii Chapm.；埃利越橘；Elliott's Blueberry, Elliott's Huckleberry ●☆

403821　Vaccinium emarginatum Hayata；凹顶越橘（凹叶岩桃，凹叶越橘，老鼠连枝）；Emarginate Blueberry，Tuber-bearing Blue Berry ●

403822　Vaccinium erytheocarpum Michx.；红果越橘；Dingleberry ●☆

403823　Vaccinium erythrocarpum Michx. subsp. japonicum（Miq.）Van der Kloet = Vaccinium japonicum Miq. ●

403824　Vaccinium exaristatum Kurz；隐距越橘；Awnles Blueberry, Exaristate Blueberry, Hiddenspur Bladderwort ●

403825　Vaccinium exaristatum Kurz var. pubescens Kurz = Vaccinium exaristatum Kurz ●

403826　Vaccinium fimbribracteatum C. Y. Wu；齿苞越橘；Fimbribracteate Blueberry, Fimbribriate-bracted Blueberry, Toothbract Bladderwort ●

403827　Vaccinium fimbricalyx Chun et W. P. Fang；流苏萼越橘（流苏萼乌饭树）；Fimbribriate-calyxed Blueberry, Fimbricalyx Blueberry, Tasselcalyx Bladderwort ●

403828　Vaccinium foetidissimum H. Lév. et Vaniot；臭越橘；Foetid

Blueberry，Stinkkest Bladderwort ●

403829　Vaccinium forbesii Hook. = Vaccinium madagascariense（Thouars ex Poir.）Sleumer ●☆

403830　Vaccinium formosanum Hayata = Vaccinium wrightii A. Gray var. formosanum（Hayata）H. L. Li ●

403831　Vaccinium forrestii Diels = Vaccinium duclouxii（H. Lév.）Hand. -Mazz. ●

403832　Vaccinium fragile Franch. ；乌鸦果(纯阳子,黑果叶,老鸦果,老鸦泡,冷饭果,毛叶乌饭树,米饭果,千年矮,沙汤果,土千年剑,土千年健,乌饭果,乌饭子)；Crowfruit，Edible Fruit Blueberry，Fragile Blueberry，Sichuan Bladderwort，Sichuan Blueberry，Szechwan Blueberry ●

403833　Vaccinium fragile Franch. var. crinitum Franch. = Vaccinium fragile Franch. ●

403834　Vaccinium fragile Franch. var. mekongense（W. W. Sm.）Sleumer；大叶乌鸦果(大理土千年健,救军粮,左黑果)；Bigleaf Fragile Blueberry ●

403835　Vaccinium fragile Franch. var. myrtifolium Franch. = Vaccinium fragile Franch. ●

403836　Vaccinium fuscatum Aiton；暗棕越橘；Dark-brown Blueberry ●☆

403837　Vaccinium gaultherifolium（Griff.）Hook. f. ；软骨边越橘(软骨叶越橘)；Caltilage Bladderwort，Thin Leathery Leaves Blueberry，Thin-leathered Leaves Blueberry，Winter-greenleaf Blueberry ●

403838　Vaccinium gaultherifolium（Griff.）Hook. f. var. glaucorubrum C. Y. Wu；粉花软骨边越橘；Glauco-reded Winter-greenleaf Blueberry ●

403839　Vaccinium glandulosissimum C. Y. Wu ex W. P. Fang et Z. H. Pan；腺毛乌饭树；Glandule-hair Blueberry ●

403840　Vaccinium glandulosissimum C. Y. Wu ex W. P. Fang et Z. H. Pan = Agapetes inopinata Airy Shaw ●

403841　Vaccinium glauco-album Hook. f. ex C. B. Clarke；粉白越橘(白背越橘,灰白越橘)；Grey-white Blueberry，Pale Bladderwort，White-greyleaf Blueberry ●

403842　Vaccinium glaucophyllum C. Y. Wu et R. C. Fang；灰叶乌饭；Greyleaf Blueberry，Grey-leaved Blueberry ●

403843　Vaccinium griffithianum C. Y. Wu et R. C. Fang var. glabratum Hook. f. et Thomson = Vaccinium bracteatum Thunb. ●

403844　Vaccinium guangdongense W. P. Fang et Z. H. Pan；广东乌饭；Guangdong Blueberry，Kwangtung Blueberry ●

403845　Vaccinium hainanense Sleumer；海南越橘；Hainan Blueberry ●

403846　Vaccinium haitangense Sleumer；海棠越橘；Haitang Blueberry ●

403847　Vaccinium hanceockiae Merr. = Vaccinium randaiense Hayata ●

403848　Vaccinium hancockiae Merr. = Vaccinium randaiense Hayata ●

403849　Vaccinium hangchouense（Matsuda）Komatsu = Vaccinium mandarinorum Diels ●

403850　Vaccinium harmandianum Dop；长冠越橘；Harmand's Blueberry，Longcorolla Bladderwort ●

403851　Vaccinium henryi Hemsl. ；无梗越橘；Henry Blueberry，Stalkless Bladderwort ●

403852　Vaccinium henryi Hemsl. var. chingii（Sleumer）C. Y. Wu et R. C. Fang；有梗越橘(秦氏越橘)；Ching Blueberry，Ching Henry Blueberry ●

403853　Vaccinium hirsutum Buckley；毛越橘(多毛越橘)；Hairy Huckleberry，Sessile-leaf Blueberry ●☆

403854　Vaccinium hirtum Thunb. ；多毛越橘(红果越橘,绒毛越橘,硬毛越橘)；Hairy Huckleberry，Redfruit Bladderwort，Rough-haired Blueberry ●

403855　Vaccinium hirtum Thunb. f. lasiocarpum（Koidz.）Ohwi =

Vaccinium hirtum Thunb. ●

403856　Vaccinium hirtum Thunb. f. lasiocarpum（Koidz.）Ohwi = Vaccinium lasiocarpum Nakai ●☆

403857　Vaccinium hirtum Thunb. var. kiusianum（Koidz.）H. Hara；九州越橘●☆

403858　Vaccinium hirtum Thunb. var. koreanum（Nakai）Kitam. = Vaccinium koreanum Nakai ●

403859　Vaccinium hirtum Thunb. var. pubescens（Koidz.）T. Yamaz. ；短柔毛越橘●☆

403860　Vaccinium hirtum Thunb. var. smallii（A. Gray）Maxim. = Vaccinium smallii A. Gray var. glabrum Koidz. ●☆

403861　Vaccinium impressinerve C. Y. Wu；凹脉越橘；Concavevein Bladderwort，Concave-veined Blueberry，Sunkenvein Blueberry ●

403862　Vaccinium inokumae Tatew. = Vaccinium hirtum Thunb. ●

403863　Vaccinium iteophyllum Hance；黄背越橘(鼠刺叶乌饭树)；Like Iteo Leaf Blueberry，Willowleaf Blueberry，Willow-leaved Blueberry ●

403864　Vaccinium iteophyllum Hance var. fragrans Rehder et E. H. Wilson = Vaccinium iteophyllum Hance ●

403865　Vaccinium iteophyllum Hance var. fragrans Rehder et E. H. Wilson. = Vaccinium iteophyllum Hance ●

403866　Vaccinium iteophyllum Hance var. glaudulosum C. Y. Wu et R. C. Fang；腺毛米饭花(腺毛越橘)；Glandulose Willowleaf Blueberry ●

403867　Vaccinium iteophyllum Hance var. hispidum Hand. -Mazz. = Vaccinium trichocladum Merr. et F. P. Metcalf ●

403868　Vaccinium japonicum Miq. ；日本扁枝越橘(扁枝越橘)；Japan Bladderwort，Japanese Blueberry，Japanese Cranberry ●

403869　Vaccinium japonicum Miq. f. incisum（F. Maek.）T. Yamaz. ；锐裂扁枝越橘●☆

403870　Vaccinium japonicum Miq. var. ciliare Matsum. ex Komatsu；缘毛日本扁枝越橘●☆

403871　Vaccinium japonicum Miq. var. lasiostemon Hayata；台湾扁枝越橘(毛蕊花)；Woollstamen Japanese Cranberry ●

403872　Vaccinium japonicum Miq. var. sinicum（Nakai）Rehder；扁枝越橘(山小蘖,扇木,深红越橘,小叶梨状越橘)；Chinese Japanese Cranberry ●

403873　Vaccinium jesoense Miq. = Vaccinium vitis-idaea L. ●

403874　Vaccinium kachinense Brandis；卡钦越橘；Kachin Blueberry，Kaqin Blueberry ●

403875　Vaccinium kengii C. E. Chang；鞍马山越橘；Keng Blueberry ●

403876　Vaccinium kingdon-wardii Sleumer；纸叶越橘；Kingdon-ward Blueberry，Paperleaf Bladderwort ●

403877　Vaccinium kiusianum Koidz. = Vaccinium hirtum Thunb. var. kiusianum（Koidz.）H. Hara ●☆

403878　Vaccinium koreanum Nakai；朝鲜越橘(红果越橘)；Korean Blueberry ●

403879　Vaccinium koreanum Nakai = Vaccinium hirtum Thunb. var. koreanum（Nakai）Kitam. ●

403880　Vaccinium laetum Diels；西南越橘(饱饭花,米饭花,乌饭子,小叶珍珠花)；Bright-coloured Blueberry，Cheerful Blueberry，SW. China Bladderwort ●

403881　Vaccinium laetum Diels = Vaccinium mandarinorum Diels ●

403882　Vaccinium laetum Diels var. undulatum Y. C. Yang = Vaccinium mandarinorum Diels ●

403883　Vaccinium lamarckii Camp = Vaccinium angustifolium Aiton ●

403884　Vaccinium lamprophyllum C. Y. Wu et R. C. Fang；亮叶越橘；Brightleaf Bladderwort，Shinyleaf Blueberry，Shiny-leaved Blueberry ●

403885 Vaccinium lanigerum Sleumer;羽毛越橘;Feather Bladderwort,Wool-bearing Blueberry,Woolly Blueberry ●

403886 Vaccinium lasiocarpum Nakai;毛果越橘;Hairfruit Blueberry ●☆

403887 Vaccinium leucobotrys (Nutt.) Nicholson;白果越橘;White-fruit Blueberry,White-fruited Blueberry ●

403888 Vaccinium lincangense W. P. Fang et Z. H. Pan;临沧乌饭;Lincang Blueberry ●

403889 Vaccinium lincangense W. P. Fang et Z. H. Pan. = Vaccinium dunalianum Wight var. megaphyllum Sleumer ●

403890 Vaccinium longeracemosum Franch. et Sav. = Vaccinium sieboldii Miq. ●☆

403891 Vaccinium longicaudatum Chun ex W. P. Fang et Z. H. Pan;长尾乌饭;Longcaudate Blueberry,Long-caudated Blueberry,Longtail Bladderwort ●

403892 Vaccinium longitubulosum J. J. Sm. = Vaccinium exaristatum Kurz ●

403893 Vaccinium loquihense Dop et Trochain - Marquès = Vaccinium dunalianum Wight var. urophyllum Rehder et E. H. Wilson ●

403894 Vaccinium macrocarpon Aiton;大果越橘(蔓越橘);American Cranberry,Commercial Cranberry,Cranberry,Large Cranberry,Large-fruit Cranberry,Small Cranberry ●☆

403895 Vaccinium madagascariense (Thouars ex Poir.) Sleumer;马岛越橘●☆

403896 Vaccinium mairei H. Lév. = Lyonia ovalifolia (Wall.) Drude var. lanceolata (Wall.) Hand. -Mazz. ●

403897 Vaccinium malaccense Wight = Vaccinium bracteatum Thunb. var. chinense (Lodd.) Chun ex Sleumer ●

403898 Vaccinium malaccense Wight = Vaccinium bracteatum Thunb. ●

403899 Vaccinium mandarinorum Diels;江南越橘(米饭花,糯米饭,乌饭,五桐子,夏菠,香白珠,小三条筋子树,羊豆饭,杨春花树,早禾酸,早禾子);China Bladderwort,Mandarin Blueberry,South China Blueberry ●

403900 Vaccinium mandarinorum L. var. austrosinense (Hand. -Mazz.) F. P. Metcalf;具苞江南越橘;South China Mandarin Blueberry ●

403901 Vaccinium mandarinorum L. var. austrosinense (Hand. -Mazz.) Metcalf = Vaccinium mandarinorum Diels ●

403902 Vaccinium mandarinorum L. var. laetum (Diels) Metcalf = Vaccinium mandarinorum Diels ●

403903 Vaccinium mekongense W. W. Sm. = Vaccinium fragile Franch. var. mekongense (W. W. Sm.) Sleumer ●

403904 Vaccinium membranaceum Douglas ex Hook.;薄叶越橘(膜质越橘);Blueberry,Tall Bilberry,Thinleaf Huckleberry,Thin-leaf Huckleberry ●☆

403905 Vaccinium merkii Fisch. ex Herder;迈吉越橘●☆

403906 Vaccinium merrillianum Hayata = Vaccinium delavayi Franch. subsp. merrillianum (Hayata) R. C. Fang ●

403907 Vaccinium microcarpos (Turcz. ex Rupr.) Schmalh. = Vaccinium oxycoccos L. ●

403908 Vaccinium microcarpum (Turcz. ex Rupr.) Schmalh. = Oxycoccus microcarpus Turcz. ex Rupr. ●

403909 Vaccinium miniatum (Griff.) Kurz. = Agapetes miniata (Griff.) Benth. et Hook. f. ●

403910 Vaccinium modestum W. W. Sm.;大苞越橘;Bigbract Blueberry,Big-bracted Blueberry ●

403911 Vaccinium mortinia Hook. f.;莫丁越橘;Mortinia,Mortinia Blueberry ●☆

403912 Vaccinium moupinense Franch.;宝兴越橘(穆坪越橘);Baoxing Blueberry,Paohsing Blueberry ●

403913 Vaccinium mucronatum L. = Ilex mucronata (L.) M. Powell,Savol. et S. Andréws ●☆

403914 Vaccinium myrsinites Lam.;常绿蓝果越橘;Evergreen Blueberry ●☆

403915 Vaccinium myrtilloides Michx.;天鹅绒叶越橘(酸梢越橘);Bilberry,Blaeberry,Canada Blueberry,Sour Top Blueberry,Sourtop Blueberry,Velvetleaf,Velvet-leaf Blueberry,Velvet-leaf Huckleberry,Whortleberry ●☆

403916 Vaccinium myrtillus L.;黑果越橘(蔓越橘,欧洲越橘);Arts,Bilberry,Black Hurs,Black Whort,Black Whortle,Blackberry,Blackfruit Bladderwort,Blackhearts,Blackwort,Blaeberry,Blayberry,Bleeberry,Blueberry,Brylocks,Bullberry,Coraseena,Cowberry,Croneberry,Crowberry,Fayberry,Fraghan,Fraughan,Fraughans,Frawn,Frawns,Frocken,Frughan,Fruog,Fuighans,Hartberry,Harts,Heartmint,Hearts,Heathberry,Heatherberry,Herts,Hortleberry,Horts,Huckleberry,Hurds,Hurtleberry,Hurts,Lingberry,Mossberry,Mulberry,Myrtle Blueberry,Myrtle Whortleberry,Sour-top Blueberry,Truckleberry,Urts,Whimberry,Whinberry,Whinberry Whortleberry,Whortleberry,Whortle-bilberry,Whorts,Whurt,Winberry,Windberry,Worts ●

403917 Vaccinium myrtillus L. var. yatabei (Makino) Matsum. et Komatsu;谷田越橘●☆

403918 Vaccinium nigrum (A. W. Wood) Britton = Vaccinium angustifolium Aiton ●

403919 Vaccinium nikkoense Nakai;日光越橘●☆

403920 Vaccinium nummularia Hook. f. et Thomson ex C. B. Clarke;抱石越橘(西藏越橘,圆叶越橘);Coin-shaped Blueberry,Orbicular-leaf Blueberry ●

403921 Vaccinium nummularia Hook. f. et Thomson ex C. B. Clarke var. oblongifolium R. C. Fang;长叶抱石越橘;Longleaf Coin-shaped Blueberry ●

403922 Vaccinium nummularia var. oblongifolium C. Y. Wu et R. C. Fang = Vaccinium nummularia Hook. f. et Thomson ex C. B. Clarke ●

403923 Vaccinium obovatum Wight = Agapetes obovata (Wight) Benth. et Hook. f. ●

403924 Vaccinium occidentalis A. Gray;西方越橘;Western Blueberry ●☆

403925 Vaccinium oldhamii Miq.;腺齿越橘;Oldham's Blueberry ●

403926 Vaccinium oldhamii Miq. f. glaucinum (Nakai) Kitam. = Vaccinium oldhamii Miq. f. glaucum (Koidz.) Hiyama ●☆

403927 Vaccinium oldhamii Miq. f. glaucum (Koidz.) Hiyama;灰蓝腺齿越橘●☆

403928 Vaccinium oldhamii Miq. var. glaucum Koidz. = Vaccinium oldhamii Miq. f. glaucum (Koidz.) Hiyama ●☆

403929 Vaccinium omeiense W. P. Fang;峨眉越橘(峨眉珍珠树);Emei Blueberry,Omei Mountain Blueberry ●

403930 Vaccinium oreophilum Rydb.;落基山越橘;Rocky Mountain Whortleberry ●☆

403931 Vaccinium oreotrephes W. W. Sm. = Vaccinium sikkimense C. B. Clarke ●

403932 Vaccinium ovalifolium Sm.;早越橘(卵叶越橘);Blue Huckleberry,Blue Whortleberry,Early Blueberry,Mathers,Oval-leaved Huckleberry ●

403933 Vaccinium ovalifolium Sm. f. angustifolium (Nakai) Honda;狭叶早越橘;Narrowleaf Early Blueberry ●☆

403934 Vaccinium ovalifolium Sm. f. angustifolium (Nakai) Ohwi = Vaccinium ovalifolium Sm. ●

403935 Vaccinium ovalifolium Sm. f. obovoideum (Takeda) H. Hara ex Ohwi = Vaccinium ovalifolium Sm. ●

403936 Vaccinium ovalifolium Sm. f. platyanthum (Nakai) H. Hara ex

Ohwi = Vaccinium ovalifolium Sm. ●

403937　Vaccinium ovalifolium Sm. f. platyantum（Nakai）Honda；壶花早越橘●☆

403938　Vaccinium ovalifolium Sm. f. villosum（H. Boissieu）T. Shimizu；长柔毛早越橘●☆

403939　Vaccinium ovalifolium Sm. var. alpinum（Tatew.）T. Yamaz.；好山早越橘●☆

403940　Vaccinium ovalifolium Sm. var. coriaceum H. Boissieu = Vaccinium ovalifolium Sm. ●

403941　Vaccinium ovalifolium Sm. var. membranaceum H. Boissieu = Vaccinium ovalifolium Sm. ●

403942　Vaccinium ovalifolium Sm. var. sachalinense T. Yamaz.；库页早越橘●☆

403943　Vaccinium ovalifolium Sm. var. shikokianum（Nakai）H. Hara = Vaccinium shikokianum Nakai ●☆

403944　Vaccinium ovatum Pursh；加州越橘（卵叶越橘）；Box Blueberry, California Huckleberry, Cranberry, Evergreen Huckleberry ●

403945　Vaccinium oxycoccos L. var. intermedium A. Gray = Vaccinium oxycoccos L. ●

403946　Vaccinium oxycoccos L. var. microphyllum（Lange）J. Rousseau et Raymond = Vaccinium oxycoccos L. ●

403947　Vaccinium oxycoccos L. var. ovalifolium Michx. = Vaccinium oxycoccos L. ●

403948　Vaccinium oxycoccus L. = Oxycoccus oxycoccus（L.）MacMill. ●

403949　Vaccinium oxycoccus L. subsp. microcarpum（Turcz. ex Rupr.）Kitam. = Vaccinium microcarpum（Turcz. ex Rupr.）Schmalh. ●

403950　Vaccinium oxycoccus L. var. microcarpum（Turcz. ex Rupr.）B. Fedtsch. et Flerow = Vaccinium microcarpum（Turcz. ex Rupr.）Schmalh. ●

403951　Vaccinium padifolium Sm.；稠李叶越橘；Madeira Whortleberry, Padue-leaf Blueberry ●☆

403952　Vaccinium pallidum Aiton；旱地越橘（灰果越橘）；Blue Ridge Blueberry, Dryland Blueberry, Dry-land Blueberry, Early Low-bush Blueberry, Glaucos Fruits Blueberry, Hillside Blueberry, Lowbush Blueberry ●☆

403953　Vaccinium papillatum P. F. Stevens；粉果越橘（粉红越橘，山羊叶）；Farinosefruit Bladderwort, Papillate Blueberry ●

403954　Vaccinium papulosum C. Y. Wu et R. C. Fang；瘤果越橘；Papustular Blueberry, Populose Blueberry, Tumorfruit Bladderwort ●

403955　Vaccinium parvibracteatum Hayata = Vaccinium mandarinorum Diels ●

403956　Vaccinium parvifolium Sm.；小叶越橘（红果越橘）；Littleleaf Blueberry, Red Bilberry, Red Huckleberry, Red Whortleberry, Small-leaf Blueberry ●☆

403957　Vaccinium pennsylvanicum Lam.；北美越橘；Blue Huckleberry, Blueberry, Early Blueberry, Low Blueberry, Sweet Blueberry ●☆

403958　Vaccinium pensilvanicum Lam. = Vaccinium angustifolium Aiton ●

403959　Vaccinium pensilvanicum Lam. var. angustifolium（Aiton）A. Gray = Vaccinium angustifolium Aiton ●

403960　Vaccinium pensilvanicum Lam. var. myrtilloides（Michx.）Fernald = Vaccinium myrtilloides Michx. ●☆

403961　Vaccinium pensilvanicum Lam. var. nigrum A. W. Wood = Vaccinium angustifolium Aiton ●

403962　Vaccinium petelotii Dop = Vaccinium dunalianum Wight var. urophyllum Rehder et E. H. Wilson ●

403963　Vaccinium petelotii Dop = Vaccinium dunalianum Wight ●

403964　Vaccinium petelotii Merr.；大叶越橘；Bigleaf Blueberry, Big-leaved Blueberry ●

403965　Vaccinium piliferum（Hook. f.）Sleumer = Agapetes pilifera Hook. f. ex C. B. Clarke ●

403966　Vaccinium podocarpoideum W. P. Fang et Z. H. Pan；罗汉松叶乌饭；Podocarpifolius Blueberry, Podocarpus-leaf Blueberry, Podocarpus-like Bladderwort ●

403967　Vaccinium praestans Lamb.；勘察加越橘（樱桃越橘）；Kamtschatka Blueberry ●☆

403968　Vaccinium pratense P. C. Tam ex C. Y. Wu et R. C. Fang；草地越橘；Grass-land Blueberry, Lawn Bladderwort, Meadow Blueberry ●

403969　Vaccinium pseudobullatum W. P. Fang et Z. H. Pan；拟泡泡叶乌饭；Fake Bubbleleaf Bladderwort, False Bullate Blueberry, False-bullated Blueberry ●

403970　Vaccinium pseudorobustum Sleumer；椭圆叶越橘（大首伞，假硬毛越橘）；Elliptic-leaf Blueberry, Elliptic-leaved Blueberry ●

403971　Vaccinium pseudospadiceum Dop；耳叶越橘；Auriculate-leaf Blueberry, Earleaf Bladderwort, Ear-leaved Blueberry ●

403972　Vaccinium pseudotonkinense Sleumer；腺萼越橘；Glandcalyx Bladderwort, Glandular-calyxed Blueberry ●

403973　Vaccinium pubicalyx Franch.；毛萼越橘（短毛萼越橘）；Hairycalyx Blueberry, Hairy-calyxed Blueberry ●

403974　Vaccinium pubicalyx Franch. var. anomalum J. Anthony；少毛毛萼越橘；Fewhairs Hairycalyx Blueberry ●

403975　Vaccinium pubicalyx Franch. var. leucocalyx（H. Lév.）Rehder；多毛毛萼越橘；Whitecalyx Hairycalyx Blueberry ●

403976　Vaccinium randaiense Hayata；峦大越橘；Luandashan Blueberry, Luantashan Blueberry, Randa Blueberry ●

403977　Vaccinium repens（H. Lév.）Rehder = Vaccinium fragile Franch. ●

403978　Vaccinium reticulatum Sm.；网状越橘；Ohelo Berry ●☆

403979　Vaccinium retusum（Griff.）Hook. f. ex C. B. Clarke；西藏越橘；Tibet Blueberry, Xizang Blueberry ●

403980　Vaccinium rugosum Hook. f. et Thomson ex Hook. = Agapetes incurvata（Griff.）Sleumer ●

403981　Vaccinium salweenense W. W. Sm. = Vaccinium fragile Franch. var. mekongense（W. W. Sm.）Sleumer ●

403982　Vaccinium sangtavanense Dop et = Vaccinium dunalianum Wight var. urophyllum Rehder et E. H. Wilson ●

403983　Vaccinium saxicola Chun ex Sleumer；石生越橘（石瓜子莲）；Rocky Bladderwort, Saxatile Blueberry, Saxicolous Blueberry ●

403984　Vaccinium scalarinervium C. Y. Wu et R. C. Fang；梯脉越橘；Ladder-veins Blueberry ●

403985　Vaccinium scalarinervium C. Y. Wu et R. C. Fang = Vaccinium subdissitifolium P. F. Stevens ●

403986　Vaccinium sciaphilum C. Y. Wu；林生越橘；Forest Blueberry, Shady Bladderwort, Sylvan Blueberry ●

403987　Vaccinium scoparium Leiberg；松鸡越橘（帚状越橘）；Broom-like Blueberry, Grouse Whortleberry, Grouseberry, Low Huckleberry, Red Tiny Whortleberry ●☆

403988　Vaccinium scopulorum W. W. Sm.；岩生越橘；Cliff Blueberry, Scopulous Bladderwort ●

403989　Vaccinium secundiflorum Hook. = Vaccinium madagascariense（Thouars ex Poir.）Sleumer ●☆

403990　Vaccinium serpens Wight = Agapetes serpens（Wight）Sleumer ●

403991　Vaccinium serratum（G. Don）Wight = Vaccinium vacciniaceum（Roxb.）Sleumer ●

403992　Vaccinium serratum（G. Don）Wight var. leucobotrys（Nutt.）C. B. Clarke = Vaccinium leucobotrys（Nutt.）Nicholson ●

403993　Vaccinium serrulatum W. P. Fang et Z. H. Pan；细齿乌饭；Serrulate Blueberry，Smalltooth Bladderwort ●

403994　Vaccinium setosum Anthony = Vaccinium fragile Franch. ●

403995　Vaccinium setosum C. H. Wright = Vaccinium fragile Franch. ●

403996　Vaccinium shikokianum Nakai；四国越橘；Sikoku Bladderwort ●☆

403997　Vaccinium siccum H. Lév. et Vaniot = Vaccinium japonicum Miq. var. sinicum（Nakai）Rehder ●

403998　Vaccinium sieboldii Miq.；西氏越橘（席氏越橘）；Siebold Blueberry ●☆

403999　Vaccinium sikangense Y. C. Yang = Vaccinium moupinense Franch. ●

404000　Vaccinium sikkimense C. B. Clarke；荚蒾叶越橘（锡金越橘）；Sikkim Blueberry ●

404001　Vaccinium sinicum Sleumer；广西越橘（路边针，中国越橘）；Guangxi Blueberry，Kwangsi Blueberry ●

404002　Vaccinium slaweenense W. W. Sm. = Vaccinium fragile Franch. var. mekongense（W. W. Sm.）Sleumer ●

404003　Vaccinium smallii A. Gray；斯莫尔氏越橘（大叶越橘）；Small Blueberry ●☆

404004　Vaccinium smallii A. Gray f. glabrescens（H. Hara）M. Mizush. et O. Mori = Vaccinium smallii A. Gray ●☆

404005　Vaccinium smallii A. Gray var. glabrum Koidz.；光越橘；Smooth Small Blueberry ●☆

404006　Vaccinium smallii A. Gray var. minus ？ = Vaccinium smallii A. Gray var. glabrum Koidz. ●☆

404007　Vaccinium smallii A. Gray var. versicolor（Koidz.）T. Yamaz.；变色斯莫尔氏越橘●☆

404008　Vaccinium spicatum（Lour.）Poir. = Vaccinium bracteatum Thunb. ●

404009　Vaccinium spicigerum W. W. Sm. = Vaccinium pubicalyx Franch. ●

404010　Vaccinium spiculatum C. Y. Wu et R. C. Fang；小尖叶越橘；Spiculate Blueberry ●

404011　Vaccinium sprengelii（G. Don）Sleumer；米饭花（米饭树，无柄越橘，夏波，夏菠，小三条筋子树，羊豆饭，珍珠花）；Sessile-leaf Blueberry，Sprengel Bladderwort ●

404012　Vaccinium stamineum L.；长蕊越橘（鹿莓越橘，鹿越橘）；Deerberry，Highbush Huckleberry，Southern Gooseberry，Squaw，Squaw Huckleberry ●☆

404013　Vaccinium stamineum L. var. interius（Ashe）E. J. Palmer et Steyerm. = Vaccinium stamineum L. ●☆

404014　Vaccinium stamineum L. var. melanocarpum C. Mohr = Vaccinium stamineum L. ●☆

404015　Vaccinium stamineum L. var. neglectum（Small）Deam = Vaccinium stamineum L. ●☆

404016　Vaccinium stanleyi Schweinf.；斯坦利越橘●☆

404017　Vaccinium stapfianum Sleumer；黑斑草叶越橘；Black Dotted Leaves Blueberry ●☆

404018　Vaccinium subdissitifolium P. F. Stevens；亚梯脉越橘（梯脉越橘）；Laddervein Bladderwort，Scalar-nerved Blueberry，Subladder-veins Blueberry ●

404019　Vaccinium subfalcatum Merr. ex Sleumer；镰叶越橘（白花乌板紫）；Falcate-leaved Blueberry，Sickleleaf Bladderwort，Subfalcate Blueberry ●

404020　Vaccinium supracostatum Hand. -Mazz.；凸脉越橘；Covexvein Bladderwort，Covexvein Blueberry，Covex-veined Blueberry，Supra-vein Leaves Blueberry ●

404021　Vaccinium taliense W. W. Sm. = Vaccinium fragile Franch. var. mekongense（W. W. Sm.）Sleumer ●

404022　Vaccinium tenuiflorum R. C. Fang；狭花越橘●

404023　Vaccinium trichocladum Merr. et F. P. Metcalf；刺毛越橘；Rigid Hair Branches Blueberry，Spinehairy Blueberry，Spiny-haired Blueberry ●

404024　Vaccinium trichocladum Merr. et F. P. Metcalf var. glabriracemosum C. Y. Wu；光序刺毛越橘；Glabrousraceme Spinehairy Blueberry ●

404025　Vaccinium triflorum Rehder；三花越橘；Threeflower Blueberry，Triflorous Blueberry ●

404026　Vaccinium truncatocalyx Chun ex W. P. Fang et Z. H. Pan；平萼乌饭；Truncate-calyx Blueberry，Truncate-calyxed Blueberry ●

404027　Vaccinium uliginosum L.；笃斯越橘（地果，甸果，笃斯，黑豆树，荒野越橘，龙果蛤塘果，湿生越橘，水越橘）；Bilberry，Blaeberry，Bog Bilberry，Bog Blueberry，Bog Whortleberry，Large Bilberry，Moorberry，Northern Bilberry ●

404028　Vaccinium uliginosum L. var. albium J. Y. Ma et Yue Zhang；白果笃斯越橘；Thitefruit Moorberry ●

404029　Vaccinium uliginosum L. var. album J. Y. Ma et Yue Zhang = Vaccinium uliginosum L. ●

404030　Vaccinium uliginosum L. var. alpinum Bigelow；高山笃斯●☆

404031　Vaccinium uliginosum L. var. japonicum T. Yamaz.；日本笃斯●☆

404032　Vaccinium uliginosum L. var. microphyllum Lange = Vaccinium oxycoccos L. ●

404033　Vaccinium urceolatum Hemsl.；红花越橘；Redflower Blueberry，Redflowered Blueberry，Red-flowered Blueberry，Urn-like Flowers Blueberry ●

404034　Vaccinium urceolatum Hemsl. var. pubescens C. Y. Wu et R. C. Fang；毛序红花越橘；Pubescent Redflower Blueberry ●

404035　Vaccinium usunoki Nakai = Vaccinium hirtum Thunb. var. pubescens（Koidz.）T. Yamaz. ●☆

404036　Vaccinium vacciniaceum（Roxb.）Sleumer；小轮叶越橘（轮生叶越橘）；Smallwhorlleaf Blueberry，Vaccinium-like Blueberry，Veined-leaf Blueberry，Whorlleaf Blueberry ●

404037　Vaccinium vacciniaceum（Roxb.）Sleumer subsp. glabritubum P. F. Stevens；秃冠小轮叶越橘；Hairless Whorlleaf Blueberry ●

404038　Vaccinium vacciniaceum（Roxb.）Sleumer var. hispidum（C. B. Clarke）Sleumer = Vaccinium subdissitifolium P. F. Stevens ●

404039　Vaccinium vacillans Kalm ex Torr. = Vaccinium pallidum Aiton ●☆

404040　Vaccinium vacillans Torr.；糖甜越橘；Dryland Blueberry，Late Low Blueberry，Late Low-blueberry ●☆

404041　Vaccinium vacillans Torr. = Vaccinium pallidum Aiton ●☆

404042　Vaccinium vacillans Torr. var. crinitum Fernald = Vaccinium pallidum Aiton ●☆

404043　Vaccinium vacillans Torr. var. missouriense Ashe = Vaccinium pallidum Aiton ●☆

404044　Vaccinium venosum Wight；轮生叶越橘；Veiny Blueberry，Whorl-leaved Blueberry ●

404045　Vaccinium venosum Wight var. hispidum C. B. Clarke = Vaccinium leucobotrys（Nutt.）Nicholson ●

404046　Vaccinium venosum Wight var. hispidum C. B. Clarke = Vaccinium subdissitifolium P. F. Stevens ●

404047　Vaccinium versicolor Nakai；变色越橘●☆

404048　Vaccinium versicolor Nakai = Vaccinium smallii A. Gray var. versicolor（Koidz.）T. Yamaz. ●☆

404049　Vaccinium viburnoides Rehder et E. H. Wilson = Vaccinium sikkimense C. B. Clarke ●

404050　Vaccinium virgatum Aiton；细枝越橘（兔眼越橘）；Rabbit-eye Blueberry，Slender-twigs Blueberry ●☆

404051　Vaccinium vitis-idaea L.；越橘（矮越橘，浜梨，甘露梅，红豆，温普，小越橘，熊果，牙疙瘩，牙疙疸）；Brawlins，Clusterberry，Cow

Berry，Cowberry，Cranberry，Cranberry Wire，Cranberry-wire，Crowberry，Flowering Box，Foxberry，Granberry，Grape-of-mount-Ida，Lingberry，Lingonberry，Moonogs，Mountain Cranberry，Munshook，Partridgeberry，Red Bilberry，Red Huckleberry，Red Whortleberry，Red Worts，Whimberry ●

404052　Vaccinium vitis-idaea L. subsp. minus（Lodd.）Hultén；小越橘；Lingonberry，Mountain Cranberry，Partridgeberry ●☆

404053　Vaccinium vitis-idaea L. subsp. minus（Lodd.）Hultén = Vaccinium vitis-idaea L. ●

404054　Vaccinium vitis-idaea L. var. genuinum Herder = Vaccinium vitis-idaea L. ●

404055　Vaccinium vitis-idaea L. var. merrillianum（Hayata）S. S. Ying = Vaccinium merrillianum Hayata ●

404056　Vaccinium vitis-idaea L. var. merrillianum（Hayata）S. S. Ying = Vaccinium delavayi Franch. subsp. merrillianum（Hayata）R. C. Fang ●

404057　Vaccinium vitis-idaea L. var. minus Lodd. = Vaccinium vitis-idaea L. subsp. minus（Lodd.）Hultén ●☆

404058　Vaccinium vitis-idaea L. var. punctatum Moench = Vaccinium vitis-idaea L. subsp. minus（Lodd.）Hultén ●☆

404059　Vaccinium wardii Adamson = Vaccinium fragile Franch. ●

404060　Vaccinium wrightii A. Gray；海岛越橘（大叶越橘，来特氏越橘）；Island Bladderwort，Lingonberry，Mountain Cranberry，Partridgeberry，Wright Blue Berry，Wright Blueberry ●

404061　Vaccinium wrightii A. Gray var. formosanum（Hayata）H. L. Li；长柄海岛越橘（台湾大叶越橘，台湾越橘，乌饭树）；Formosan Blueberry，Taiwan Blueberry，Taiwan Wright Blueberry ●

404062　Vaccinium yakushimense Makino；屋久岛越橘 ●☆

404063　Vaccinium yaoshanicum Sleumer；瑶山越橘；Yaoshan Blueberry ●

404064　Vaccinium yaoshanicum Sleumer var. megaphyllum C. Y. Wu et R. C. Fang；大叶瑶山越橘；Bigleaf Yaoshan Blueberry ●

404065　Vaccinium yaoshanicum Sleumer var. megaphyllum C. Y. Wu et R. C. Fang = Vaccinium yaoshanicum Sleumer ●

404066　Vaccinium yatabei Makino；八部越橘；Yatabe Blueberry ●☆

404067　Vaccinium yersinii A. Chev. = Vaccinium dunalianum Wight var. urophyllum Rehder et E. H. Wilson ●

404068　Vaccinium yunnanense Franch. = Gaultheria leucocarpa Blume var. crenulata（Kurz）T. Z. Hsu ●

404069　Vaccinium yunnanense Franch. = Gaultheria leucocarpa Blume var. yunnanensis（Franch.）T. Z. Hsu et R. C. Fang ●

404070　Vaccinium yunnanense Franch. var. franchetianum H. Lév. = Gaultheria leucocarpa Blume var. yunnanensis（Franch.）T. Z. Hsu et R. C. Fang ●

404071　Vaccinium yunnanense Franch. var. franchetianum H. Lév. = Gaultheria leucocarpa Blume var. crenulata（Kurz）T. Z. Hsu ●

404072　Vaccinium yunnanense Franch. var. pubipes C. Y. Wu = Vaccinium sprengelii（G. Don）Sleumer ●

404073　Vachellia Wight et Arn.（1834）；编条金合欢属 ●☆

404074　Vachellia Wight et Arn. = Acacia Mill.（保留属名）●■

404075　Vachellia farnesiana（L.）Wight et Arn. = Acacia farnesiana（L.）Willd. ●

404076　Vachendorfia Adans. = Wachendorfia Burm. ■☆

404077　Vacoparis Spangler = Sorghum Moench（保留属名）■

404078　Vada-Kodi Adans. = Justicia L. ●■

404079　Vadia O. F. Cook = Chamaedorea Willd.（保留属名）●☆

404080　Vadulia Plowes = Caralluma R. Br. ■

404081　Vadulia Plowes（2003）；南非水牛角属 ■☆

404082　Vadulia maughanii（R. A. Dyer）Plowes；南非水牛角 ■☆

404083　Vagaria Herb.（1837）；漫游石蒜属 ■☆

404084　Vagaria gattefossei Maire = Vagaria ollivieri Maire ■☆

404085　Vagaria legionariorum（Emb.）Maire = Vagaria ollivieri Maire ■☆

404086　Vagaria ollivieri Maire；漫游石蒜 ■☆

404087　Vagaria ollivieri Maire var. gattefossei（Maire）Maire et Weiller = Vagaria ollivieri Maire ■☆

404088　Vagaria ollivieri Maire var. legionariorum（Emb.）Maire et Weiller = Vagaria ollivieri Maire ■☆

404089　Vagaria parviflora Herb.；小花漫游石蒜 ■☆

404090　Vaginaria Kunth = Vagaria Herb. ■☆

404091　Vaginaria Pers. = Fuirena Rottb. ■

404092　Vaginaria richardii Pers. = Fuirena scirpoidea Michx. ■☆

404093　Vagnera Adans.（废弃属名）= Maianthemum F. H. Wigg.（保留属名）■

404094　Vagnera Adans.（废弃属名）= Smilacina Desf.（保留属名）■

404095　Vagnera amplexicaulis（Nutt.）Greene = Maianthemum racemosum（L.）Link subsp. amplexicaule（Nutt.）LaFrankie ■☆

404096　Vagnera australis Rydb. = Maianthemum racemosum（L.）Link ■☆

404097　Vagnera dahurica（Turcz. ex Fisch. et C. A. Mey.）Makino = Maianthemum dahuricum（Turcz. ex Fisch. et C. A. Mey.）LaFrankie ■

404098　Vagnera liliacea（Greene）Rydb. = Maianthemum stellatum（L.）Link ■☆

404099　Vagnera pallescens Greene = Maianthemum racemosum（L.）Link subsp. amplexicaule（Nutt.）LaFrankie ■☆

404100　Vagnera racemosa（L.）Morong = Maianthemum racemosum（L.）Link ■☆

404101　Vagnera retusa Raf. = Maianthemum racemosum（L.）Link ■☆

404102　Vagnera sessilifolia（Nutt. ex Baker）Greene = Maianthemum stellatum（L.）Link ■☆

404103　Vagnera stellata（L.）Morong = Maianthemum stellatum（L.）Link ■☆

404104　Vagnera trifolia（L.）Morong = Maianthemum trifolium（L.）Slobada ■

404105　Vahadenia Stapf（1902）；瓦腺木属 ●☆

404106　Vahadenia caillei（A. Chev.）Stapf ex Hutch. et Dalziel；卡耶瓦腺木 ●☆

404107　Vahadenia laurentii（De Wild.）Stapf；瓦腺木 ●☆

404108　Vahadenia laurentii（De Wild.）Stapf f. obtusifolia De Wild. = Vahadenia laurentii（De Wild.）Stapf ●☆

404109　Vahadenia laurentii（De Wild.）Stapf var. grandiflora（De Wild.）De Wild. = Vahadenia laurentii（De Wild.）Stapf ●☆

404110　Vahadenia talbotii Wernham = Vahadenia laurentii（De Wild.）Stapf ●☆

404111　Vahea Lam. = Landolphia P. Beauv.（保留属名）●☆

404112　Vahea comorensis Bojer = Saba comorensis（Bojer ex A. DC.）Pichon ●☆

404113　Vahea comorensis Bojer ex A. DC. = Saba comorensis（Bojer ex A. DC.）Pichon ●☆

404114　Vahea florida（Benth.）F. Muell. = Saba comorensis（Bojer ex A. DC.）Pichon ●☆

404115　Vahea gummifera Lam. = Landolphia gummifera（Lam.）K. Schum. ●☆

404116　Vahea heudelotii（A. DC.）F. Muell. = Landolphia heudelotii A. DC. ●☆

404117　Vahea kirkii（R. A. Dyer）Sadeb. = Landolphia kirkii R. A. Dyer ●☆

404118　Vahea owariensis (P. Beauv.) F. Muell. = Landolphia owariensis P. Beauv. ●☆

404119　Vahea senegalensis A. DC. = Saba senegalensis (A. DC.) Pichon ●☆

404120　Vahea senegambensis Hell. var. traunii (Sadeb.) Sadeb. = Landolphia heudelotii A. DC. ●☆

404121　Vahea tomentosa Lepr. et Perr. ex Baucher = Landolphia heudelotii A. DC. ●☆

404122　Vahea traunii Sadeb. = Landolphia heudelotii A. DC. ●☆

404123　Vahlbergella Blytt = Melandrium Röhl. ■

404124　Vahlbergella Blytt = Wahlbergella Fr. ■

404125　Vahlenbergella Pax et Hottm. = Vahlbergella Blytt ■

404126　Vahlia Dahl = Dombeya Cav. (保留属名) ●☆

404127　Vahlia Thunb. (1782) (保留属名) ；二歧草属■☆

404128　Vahlia Thunb. = Bistorta (L.) Adans. ■

404129　Vahlia capensis (L. f.) Thunb. ；好望角二歧草■☆

404130　Vahlia capensis (L. f.) Thunb. subsp. ellipticifolia Bridson；椭圆叶二歧草■☆

404131　Vahlia capensis (L. f.) Thunb. subsp. macrantha (Klotzsch) Bridson；大花二歧草■☆

404132　Vahlia capensis (L. f.) Thunb. var. latifolia Burtt Davy；宽叶二歧草■☆

404133　Vahlia capensis (L. f.) Thunb. var. linearis E. Mey. ex Bridson；线叶二歧草■☆

404134　Vahlia capensis (L. f.) Thunb. var. longifolia (Gand.) Bridson；长叶二歧草■☆

404135　Vahlia capensis (L. f.) Thunb. var. verbasciflora Oliv. ；毛蕊花二歧草■☆

404136　Vahlia capensis (L. f.) Thunb. var. vulgaris Bridson；普通二歧草■☆

404137　Vahlia capensis Thunb. = Vahlia capensis (L. f.) Thunb. ■☆

404138　Vahlia cynodonteti Dinter = Vahlia capensis (L. f.) Thunb. var. linearis E. Mey. ex Bridson ■☆

404139　Vahlia dichotoma (Murray) Kuntze；二歧草■☆

404140　Vahlia digyna (Retz.) Kuntze = Bistella digyna (Retz.) Bullock ☆

404141　Vahlia geminiflora (Delile) Bridson；对花二歧草■☆

404142　Vahlia glandulosa Schltr. ex Engl. = Vahlia capensis (L. f.) Thunb. ■☆

404143　Vahlia goddingii Bruce = Vahlia somalensis Chiov. subsp. goddingii (Bruce) Bridson ■☆

404144　Vahlia longifolia Gand. = Vahlia capensis (L. f.) Thunb. var. longifolia (Gand.) Bridson ■☆

404145　Vahlia macrantha Klotzsch = Vahlia capensis (L. f.) Thunb. subsp. macrantha (Klotzsch) Bridson ■☆

404146　Vahlia menyharthii Schinz = Vahlia digyna (Retz.) Kuntze ■☆

404147　Vahlia oldenlandoides Roxb. = Vahlia dichotoma (Murray) Kuntze ■☆

404148　Vahlia ramosissima DC. = Vahlia digyna (Retz.) Kuntze ■☆

404149　Vahlia sessiliflora DC. = Bistella digyna (Retz.) Bullock ☆

404150　Vahlia sessiliflora DC. = Vahlia digyna (Retz.) Kuntze ■☆

404151　Vahlia somalensis Chiov. ；索马里二歧草■☆

404152　Vahlia somalensis Chiov. subsp. goddingii (Bruce) Bridson；戈鼎二歧草■☆

404153　Vahlia verbasciflora (Oliv.) Mendes = Vahlia capensis (L. f.) Thunb. var. verbasciflora Oliv. ■☆

404154　Vahlia viscosa Roxb. = Bistella digyna (Retz.) Bullock ■☆

404155　Vahlia viscosa Roxb. = Vahlia digyna (Retz.) Kuntze ■☆

404156　Vahliaceae Dandy = Saxifragaceae Juss. (保留科名) ●■

404157　Vahliaceae Dandy (1959) ；二歧草科■☆

404158　Vahlodea Fr. (1842) ；沃尔禾属■☆

404159　Vahlodea Fr. = Deschampsia P. Beauv. ■

404160　Vahlodea atropurpurea (Wahlenb.) Fr. ；暗紫沃尔禾■☆

404161　Vahlodea atropurpurea (Wahrenb.) Fr. ex Hartm. = Deschampsia atropurpurea (Wahrenb.) Scheele subsp. paramushirensis (Kudo) T. Koyama ■☆

404162　Vahlodea atropurpurea (Wahrenb.) Fr. ex Hartm. subsp. paramushirensis (Kudo) Hultén = Deschampsia atropurpurea (Wahrenb.) Scheele subsp. paramushirensis (Kudo) T. Koyama ■☆

404163　Vahlodea flexuosa (Honda) Ohwi = Deschampsia atropurpurea (Wahrenb.) Scheele subsp. paramushirensis (Kudo) T. Koyama ■☆

404164　Vahlodea paramushirensis (Kudo) Roshev. ；幌筵岛沃尔禾■☆

404165　Vailia Rusby (1898) ；韦尔萝藦属■☆

404166　Vailia mucronata Rusby；韦尔萝藦■☆

404167　Vailia salicina (Decne.) Morillo；柳叶韦尔萝藦■☆

404168　Vaillanta Raf. = Vaillantia Neck. ex Hoffm. ■☆

404169　Vaillantia Hoffm. = Valantia L. ■☆

404170　Vaillantia Neck. ex Hoffm. = Valantia L. ■☆

404171　Vainilla Salisb. = Vanilla Plum. ex Mill. ■

404172　Valantia L. (1753) ；瓦朗茜属■☆

404173　Valantia articulata L. = Cruciata articulata (L.) Ehrend. ■☆

404174　Valantia columella (Ehrenb. ex Boiss.) Bald. ；小圆柱瓦朗茜■☆

404175　Valantia cucullaria L. = Callipeltis cucullaris (L.) Steven ■☆

404176　Valantia glabra L. = Cruciata glabra (L.) Ehrend. ■☆

404177　Valantia hispida L. ；硬毛瓦朗茜■☆

404178　Valantia incrassata Pomel = Valantia muralis L. ■☆

404179　Valantia lanata Delile ex Coss. = Valantia columella (Ehrenb. ex Boiss.) Bald. ■☆

404180　Valantia muralis (L.) DC. var. incrassata (Pomel) Batt. = Valantia muralis L. ■☆

404181　Valantia muralis L. ；厚壁瓦朗茜■☆

404182　Valantia pedemontana Bellardi = Cruciata pedemontana (Bellardi) Ehrend. ■☆

404183　Valarum Schur = Sisymbrium L. ■

404184　Valbomia Raf. = Tetracera L. ●

404185　Valbomia Raf. = Wahlbomia Thunb. ●

404186　Valdesia Ruiz et Pav. = Blakea P. Browne ■☆

404187　Valdia Boehm. = Clerodendrum L. ●■

404188　Valdia Boehm. = Ovieda L. ●■

404189　Valdimiria Iljin = Dolomiaea DC. ■

404190　Valdivia Gay ex J. Rémy (1848) ；瓦尔鼠刺属●☆

404191　Valdivia J. Rémy = Valdivia Gay ex J. Rémy ●☆

404192　Valdivia gayana J. Rémy；瓦尔鼠刺●☆

404193　Valentiana Raf. = Thunbergia Retz. (保留属名) ●■

404194　Valentiana Raf. ex DC. = Thunbergia Retz. (保留属名) ●■

404195　Valentiana volubilis Raf. = Thunbergia alata Bojer ex Sims ■

404196　Valentina Speg. = Heliotropium L. ●■

404197　Valentina Speg. = Valentiniella Speg. ●■

404198　Valentinia Fabr. = Maianthemum F. H. Wigg. (保留属名) ■

404199　Valentinia Heist. ex Fabr. = Maianthemum F. H. Wigg. (保留属名) ■

404200　Valentinia Neck. = Tachigalia Aubl. ●☆

404201　Valentinia Raeusch. = Xanthophyllum Roxb. (保留属名) ●

404202　Valentinia Sw. = Casearia Jacq. ●

404203　Valentiniella Speg. = Heliotropium L. ●■

404204　Valenzuela B. D. Jacks. = Picramnia Sw. (保留属名)●☆

404205　Valenzuela B. D. Jacks. = Valenzuelia Bertero ex Cambess. ●☆

404206　Valenzuelia Bertero = Valenzuelia Bertero ex Cambess. ●☆

404207　Valenzuelia Bertero ex Cambess. = Guindilia Gillies ex Hook. et Arn. ●☆

404208　Valenzuelia S. Mutis ex Caldas = Picramnia Sw. (保留属名)●☆

404209　Valeranda Neck. = Orphium E. Mey. (保留属名)●☆

404210　Valeranda Neck. ex Kuntze = Orphium E. Mey. (保留属名)●☆

404211　Valerandia T. Durand et Jacks. = Valeranda Neck. ●☆

404212　Valeria Minod = Stemodia L. (保留属名)■☆

404213　Valeriana L. (1753);缬草属;Valerian ●■

404214　Valeriana aberdarica T. C. E. Fr. = Swertia kilimandscharica Engl. ■☆

404215　Valeriana acuminata Royle = Valeriana hardwickii Wall. ■

404216　Valeriana acutiloba Rydb. ;尖叶缬草;Sharpleaf Valerian ■☆

404217　Valeriana ajanensis Kom. ;阿扬湾缬草■☆

404218　Valeriana alliariifolia Troitsky = Valeriana alliariifolia Vahl ■

404219　Valeriana alliariifolia Vahl;欧亚缬草■☆

404220　Valeriana alpestris Stev. ;山生缬草■☆

404221　Valeriana alternifolia Bunge = Valeriana officinalis L. ■

404222　Valeriana alternifolia Bunge f. angustifolia (Kom.) Kitag. = Valeriana officinalis L. ■

404223　Valeriana alternifolia Bunge f. verticillata (Kom.) S. X. Li = Valeriana officinalis L. ■

404224　Valeriana alternifolia Bunge var. angustifolia (Kom.) S. H. Li = Valeriana officinalis L. ■

404225　Valeriana amurensis C. C. Davis ex Kom. ;黑水缬草(拔地麻);Amur Valerian ■

404226　Valeriana amurensis C. C. Davis ex Kom. f. leiocarpa Hara = Valeriana amurensis C. C. Davis ex Kom. ■

404227　Valeriana angustifolia Mill. ;日本缬草■☆

404228　Valeriana arizonica A. Gray;阿里桑那缬草(亚美尼亚缬草);Arizona Valerian ■☆

404229　Valeriana armena P. A. Smirn. ;亚美尼亚缬草■☆

404230　Valeriana barbulata Diels;髯毛缬草(通经草,细须缬草);Beard Valerian ■

404231　Valeriana barbulata Diels var. gymnostorna Hand. -Mazz. = Valeriana barbulata Diels ■

404232　Valeriana briquetiana H. Lév. ;单叶缬草(滇北缬草)■

404233　Valeriana capensis Thunb. ;好望角缬草■☆

404234　Valeriana capensis Thunb. var. lanceolata N. E. Br. = Valeriana capensis Thunb. ■☆

404235　Valeriana capensis Thunb. var. nana B. L. Burtt;矮小缬草■☆

404236　Valeriana capitata Pall. ex Link;头状缬草■☆

404237　Valeriana carinata var. navicularis Krok = Valerianella plagiostephana Fisch. et C. A. Mey. ■

404238　Valeriana celtica L. ;西欧甘松香;Bouncing Bess, Celtic Nard, Celtic Spikenard, Celtic Valerian, Delicate Bess, French Spikenard ■☆

404239　Valeriana cephalantha Schltdl. ;头花缬草■☆

404240　Valeriana chinensis Kreyer ex Kom. = Valeriana officinalis L. ■

404241　Valeriana chinensis L. = Commicarpus chinensis (L.) Heimerl ●

404242　Valeriana chionophila Popov et Kult. ;喜雪缬草■☆

404243　Valeriana ciliata Torr. et A. Gray = Valeriana edulis Nutt. ex Torr. et A. Gray subsp. ciliata (Torr. et A. Gray) F. G. Mey. ■☆

404244　Valeriana colchica Utkin;黑海缬草■☆

404245　Valeriana coreana Briq. = Valeriana officinalis L. ■

404246　Valeriana coreana Briq. subsp. leiocarpa (Kitag.) Vorosch. = Valeriana officinalis L. ■

404247　Valeriana daghestanica Rupr. ex Boiss. ;达赫斯坦缬草■☆

404248　Valeriana daphniflora Hand. -Mazz. ;瑞香缬草;Daphneflower Valerian ■

404249　Valeriana delavayi Franch. = Valeriana daphniflora Hand. -Mazz. ■

404250　Valeriana delavayi Franch. ex Hand. -Mazz. = Valeriana daphniflora Hand. -Mazz. ■

404251　Valeriana dioica L. ;沼泽缬草;Cherry Pie, Cherry-pie, Gooseberry Pie, Lesser Valerian, Marsh Valerian, Poor Man's Pepper ■☆

404252　Valeriana dubia Bunge;土耳其斯坦缬草■☆

404253　Valeriana dubia Bunge = Valeriana officinalis L. ■

404254　Valeriana edulis Nutt. ex Torr. et A. Gray;可食缬草(食用缬草);Edible Valerian, Tap-rooted Valerian ■☆

404255　Valeriana edulis Nutt. ex Torr. et A. Gray subsp. ciliata (Torr. et A. Gray) F. G. Mey. ;睫毛可食缬草;Edible Valerian, Kooyah, Tap-rooted Valerian, Tobacco-root ■☆

404256　Valeriana edulis Nutt. ex Torr. et A. Gray var. ciliata (Torr. et A. Gray) Cronquist = Valeriana edulis Nutt. ex Torr. et A. Gray subsp. ciliata (Torr. et A. Gray) F. G. Mey. ■☆

404257　Valeriana elgonensis Mildbr. = Swertia kilimandscharica Engl. ■☆

404258　Valeriana exaltata J. C. Mikan ex Pohl;高缬草■☆

404259　Valeriana excelsa Poir. ;莲座缬草;Rosette Valerian ■☆

404260　Valeriana faberi Graebn. = Valeriana flaccidissima Maxim. ■

404261　Valeriana fauriei Briq. ;法氏缬草(吉草,缬草)■☆

404262　Valeriana fauriei Briq. = Valeriana officinalis L. ■

404263　Valeriana fauriei Briq. f. albiflora Akasawa;白花法氏缬草■☆

404264　Valeriana fauriei Briq. f. yezoensis (Kudo) H. Hara;北海道缬草■☆

404265　Valeriana fauriei Briq. f. yezoensis (Kudo) H. Hara = Valeriana fauriei Briq. ■☆

404266　Valeriana fauriei Briq. var. leiocarpa (Kitag.) Kitag. = Valeriana officinalis L. ■

404267　Valeriana fedtschenkoi Coincy;新疆缬草;Fedtschenko Valerian, Xinjiang Valerian ■

404268　Valeriana ficariifolia Boiss. ;芥叶缬草■

404269　Valeriana flaccidissima Maxim. ;柔垂缬草(臭草,嫩茎缬草,细臭灵丹,小蜘蛛香,岩边香);Flaccid Valerian ■

404270　Valeriana flagellifera Batalin;鞭枝缬草(秀丽缬草);Flagelibranch Valerian ■

404271　Valeriana grossheimii Vorosch. ;格氏缬草■

404272　Valeriana hardwickii Wall. ;长序缬草(阔叶缬草,老君须,蛇头细辛,通经草,西南缬草,小蜘蛛香,岩参);Hardwick Valerian, Longspike Valerian ■

404273　Valeriana hardwickii Wall. var. arnottiana Wight = Valeriana hardwickii Wall. ■

404274　Valeriana hardwickii Wall. var. hoffmeisteri Klotzsch = Valeriana hardwickii Wall. ■

404275　Valeriana hardwickii Wall. var. leiocarpa Miq. = Valeriana hardwickii Wall. ■

404276　Valeriana harmsii Graebn. = Valeriana jatamansi Jones ■

404277　Valeriana helictes Graebn. = Valeriana hardwickii Wall. ■

404278　Valeriana hengduanensis D. Y. Hong;横断山缬草;Hengduanshan Valerian ■

404279　Valeriana heterophylla Turcz. = Valeriana turczaninovii Grubov ■

404280　Valeriana hiemalis Graebn. ;全缘叶缬草(漏斗花缬草,全叶缬草);Entireleaf Valerian ■

404281　Valeriana hirticalyx Lien C. Chiu;毛果缬草;Hairyfruit Valerian ■

404282　Valeriana hygrobia Briq. = Valeriana jatamansi Jones ■

404283　Valeriana infundibulum Franch. = Valeriana daphniflora Hand. -Mazz. ■

404284　Valeriana japonica Miq.；连香草；Japan Valerian, Japanese Valerian ■☆

404285　Valeriana jatamansi Jones；蜘蛛香（臭狗药，臭药，大救驾，豆豉菜，豆豉草，狗臭药，鬼见愁，九转香，老虎七，老君须，老龙须，雷公七，连香草，马蹄香，猫儿屎，磨脚花，土细辛，乌参，香草，香草子，心叶缬草，养心莲，养血莲，印度缬草，珍珠香）；Jatamans Valerian, Spider Valerian ■

404286　Valeriana jatamansi Jones = Valeriana wallichii DC. ■

404287　Valeriana jatamansi Jones var. frondosa Hand. -Mazz. = Valeriana jatamansi Jones ■

404288　Valeriana jatamansi Jones var. glabra Merr. = Valeriana jatamansi Jones ■

404289　Valeriana jatamansi Jones var. hydrobia（Briq.）Hand. -Mazz.；湿生蜘蛛香（湿生马蹄香）■

404290　Valeriana jatamansi Jones var. hygrobia（Briq.）Hand. -Mazz. = Valeriana jatamansi Jones ■

404291　Valeriana jelenevskyi P. A. Smirn.；叶列涅夫斯基缬草 ■☆

404292　Valeriana kawakamii Hayata；高山缬草；Alpine Valerian ■

404293　Valeriana keniensis T. C. E. Fr. = Swertia kilimandscharica Engl. ■☆

404294　Valeriana kilimandscharica Engl.；基列满缬草 ■☆

404295　Valeriana kilimandscharica Engl. subsp. aberdarica（T. C. E. Fr.）Hedberg = Valeriana kilimandscharica Engl. ■☆

404296　Valeriana kilimandscharica Engl. subsp. elgonensis（Mildbr.）Hedberg = Valeriana kilimandscharica Engl. ■☆

404297　Valeriana lancifolia Hand. -Mazz.；披针叶缬草 ■

404298　Valeriana leiocarpa Kitag. = Valeriana officinalis L. ■

404299　Valeriana leschenaultii DC.；赖氏缬草 ■☆

404300　Valeriana leschenaultii DC. var. brononiana ？；印度缬草 ■☆

404301　Valeriana locusta L. = Valerianella locusta（L.）Laterr. ■☆

404302　Valeriana locusta L. var. coronata L. = Valerianella coronata（L.）DC. ■☆

404303　Valeriana locusta L. var. dentata L. = Valerianella dentata（L.）Pollich ■☆

404304　Valeriana locusta L. var. olitoria ？ = Valerianella locusta（L.）Laterr. ■☆

404305　Valeriana mairei Briq. = Valeriana jatamansi Jones ■

404306　Valeriana martjanowii Krylov；香毛草（小缬草）■

404307　Valeriana meonantha C. Y. Cheng et H. B. Chen；细花缬草；Thinflower Valerian ■

404308　Valeriana mexicana DC.；墨西哥缬草 ■☆

404309　Valeriana minutiflora Hand. -Mazz.；小花缬草；Littleflower Valerian ■

404310　Valeriana montana L.；山缬草；Mountain Valerian ■☆

404311　Valeriana muliensis S. K. Wu；木里缬草；Muli Valerian ■

404312　Valeriana muliensis S. K. Wu = Valeriana trichostoma Hand. -Mazz. ■

404313　Valeriana nipponica Nakai ex Kitag. = Valeriana officinalis L. ■

404314　Valeriana nitida Kreyer；光亮缬草（闪光缬草）；Shining Valerian ■☆

404315　Valeriana nokozanensis Yamam. = Valeriana flaccidissima Maxim. ■

404316　Valeriana officinalis L.；缬草（拔地麻，朝鲜缬草，臭草，穿心排草，大救驾，东北缬草，法氏缬草，甘松，黑水缬草，互生叶缬草，吉草，阔叶缬草，鹿子草，满坡香，满山香，猫食菜，毛节缬草，欧缬草，七里香，五里香，西南缬草，媳妇菜，香草，小救驾，珍珠香，蜘蛛七，蜘蛛香，抓地虎）；Allheal, Black Elder, Broadleaf Valerian, Capon's Feathers, Capon's Tail, Capon's Tails, Cat Trail, Cat's Valerian, Cat's Love, Cats' Love, Cat's Valerian, Cat-trail,

Cherry Pie, Cherry-pie, Common Valerian, Countryman's Treacle, Cutfinger, Cut-finger Leaf, Cut-heal, Drunken Slots, English Orris, Faurie Valerian, Filaera, Garden Heliotrope, Garden Valerian, Gardenheliotrope, Grandmother's Needle, Heal-all, Herb Bennett, Phu, Poor Man's Remedy, Setwall, St. Bennet's Herb, St. George's Herb, Valaeria, Valara, Valerian, Villera, Wild Valerian ■

404317　Valeriana officinalis L. ' Alba'；白花缬草 ■☆

404318　Valeriana officinalis L. 'Rubra'；红花缬草 ■☆

404319　Valeriana officinalis L. var. alternifolia（Bunge）Ledeb. = Valeriana alternifolia Bunge ■

404320　Valeriana officinalis L. var. alternifolia（Bunge）Ledeb. = Valeriana officinalis L. ■

404321　Valeriana officinalis L. var. alternifolia Ledeb. = Valeriana officinalis L. ■

404322　Valeriana officinalis L. var. angustifolia Miq. = Valeriana officinalis L. ■

404323　Valeriana officinalis L. var. incisa Nakai ex Mori = Valeriana amurensis C. C. Davis ex Kom. ■

404324　Valeriana officinalis L. var. latifolia Briq. = Valeriana officinalis L. ■

404325　Valeriana officinalis L. var. latifolia Miq.；宽叶缬草（广州拔地麻，阔叶缬草，墨香，七里香，随手香，缬草）；Broadleaf Common Valerian, Broadleaf Valerian ■

404326　Valeriana oligantha Boiss. et Heldr.；少花缬草 ■☆

404327　Valeriana palustris Garsault = Valeriana dioica L. ■☆

404328　Valeriana palustris Kreyer = Valeriana officinalis L. ■

404329　Valeriana pauciflora Michx.；疏花缬草 ■☆

404330　Valeriana petrophila Bunge；喜岩缬草 ■☆

404331　Valeriana phu L.；高加索缬草（大园缬草）；Cretan Spikenard ■☆

404332　Valeriana phu L. ' Aurea'；黄叶高加索缬草（黄枝芬缬草）■☆

404333　Valeriana pilosa Ruiz et Pav.；疏毛缬草 ■☆

404334　Valeriana pseudodioica Pax et K. Hoffm. = Valeriana flagellifera Batalin ■

404335　Valeriana pseudofficinalis C. Y. Cheng et H. B. Chen；拟缬草 ■

404336　Valeriana pseudofficinalis C. Y. Cheng et H. B. Chen = Valeriana officinalis L. ■

404337　Valeriana pyrenaica L.；比利牛斯缬草；Capon's Tails, Giant Valerian, Pyrenean Valerian, Setwall ■☆

404338　Valeriana pyrolifolia Decne.；鹿蹄草叶缬草；Pyrolaeleaf Valerian ■☆

404339　Valeriana rhodoleuca H. B. Chen et C. Y. Cheng；川滇缬草；Chuan-Dian Valerian ■

404340　Valeriana rhodoleuca H. B. Chen et C. Y. Cheng = Valeriana hardwickii Wall. ■

404341　Valeriana rosthornii Graebn. = Valeriana hardwickii Wall. ■

404342　Valeriana rubra L. = Centranthus ruber（L.）DC. ■☆

404343　Valeriana rupestris Pall. = Patrinia rupestris（Pall.）Juss. ■

404344　Valeriana ruthenica Willd. = Patrinia sibirica（L.）Juss. ■

404345　Valeriana salicariifolia Vahl；柳叶缬草 ■☆

404346　Valeriana sambucifolia Mikan；接骨木叶缬草 ■☆

404347　Valeriana sambucifolia Mikan = Valeriana officinalis L. ■

404348　Valeriana sambucifolia Mikan var. fauriei（Briq.）H. Hara = Valeriana fauriei Briq. ■☆

404349　Valeriana septentrionalis Rydb.；北方缬草 ■☆

404350　Valeriana sibirica L. = Patrinia sibirica（L.）Juss. ■

404351　Valeriana sichuanica D. Y. Hong；川缬草；Sichuan Valerian ■

404352　Valeriana simplicifolia Kab. ？；独叶缬草 ■☆

404353　Valeriana sisymbriifolia sensu Lien C. Chiu = Valeriana ficariifolia Boiss. ■

404354　Valerianasisymbriifolia Vahl;欧洲芥叶缬草;Mustard Valerian ■

404355　Valeriana sitchensis Bong.;西奇缬草■☆

404356　Valeriana sitchensis Bong. subsp. uliginosa（Torr. et A. Gray）F. G. Mey. = Valeriana uliginosa（Torr. et A. Gray）Rydb. ■☆

404357　Valeriana sitchensis Bong. var. uliginosa（Torr. et A. Gray）B. Boivin = Valeriana uliginosa（Torr. et A. Gray）Rydb. ■☆

404358　Valeriana stabendorfii Kreyer ex Kom. = Valeriana officinalis L. ■

404359　Valeriana stenoptera Diels;窄叶缬草（窄裂缬草）;Narrowwing Valerian ■

404360　Valeriana stenoptera Diels var. cardaminea Hand. -Mazz.;细花窄裂缬草;Smallflower Narrowwing Valerian ■

404361　Valeriana stolonifera Czern. = Valeriana officinalis L. ■

404362　Valeriana stubendorfii Kreyer ex Kom. = Valeriana officinalis L. ■

404363　Valeriana subbipinnatifolia A. I. Baranov;羽叶缬草■

404364　Valeriana subbipinnatifolia A. I. Baranov = Valeriana officinalis L. ■

404365　Valeriana supina L.;奥地利缬草;Austrian Valerian, Dwarf Valerian ■☆

404366　Valeriana sylvatica Banks ex Richardson var. uliginosa Torr. et A. Gray = Valeriana uliginosa（Torr. et A. Gray）Rydb. ■☆

404367　Valeriana tangutica Batalin;小缬草（唐古特缬草,西北缬草,香草仔,香毛草,小香草）;Tangut Valerian ■

404368　Valeriana tenera Wall. ex DC. = Valeriana hardwickii Wall. ■

404369　Valeriana tianschanica Kreyer ex Hand. -Mazz. = Valeriana officinalis L. ■

404370　Valeriana tiliifolia Troitsky;椴叶缬草■☆

404371　Valeriana toluccana DC.;托卢卡缬草■☆

404372　Valeriana trichostoma Hand. -Mazz.;毛口缬草（毛房缬草）■

404373　Valeriana tripteris L.;三翅缬草■☆

404374　Valeriana tripteroides Hand. -Mazz. = Valeriana flaccidissima Maxim. ■

404375　Valeriana truncata Betcke = Valerianella muricata（Steven ex Roem. et Schult.）W. H. Baxter ■☆

404376　Valeriana truncata Betcke var. muricata（Stev.）Boiss. = Valerianella muricata（Steven ex Roem. et Schult.）W. H. Baxter ■☆

404377　Valeriana tuberosa L.;块根缬草;Tuberose Valerian ■

404378　Valeriana tuberosa L. var. ateridoi Pau = Valeriana tuberosa L. ■

404379　Valeriana turczaninovii Grubov;北疆缬草;Turczaninov Valerian ■

404380　Valeriana turczaninovii sensu Lien C. Chiu = Valeriana fedtschenkoi Coincy ■

404381　Valeriana turkestanica Samn.？ = Valeriana dubia Bunge ■☆

404382　Valeriana udicola Briq. = Valeriana hardwickii Wall. ■

404383　Valeriana uliginosa（Torr. et A. Gray）Rydb.;湿地缬草;Bog Valerian, Marsh Valerian, Mountain Valerian, Swamp Valerian ■☆

404384　Valeriana venusta Lien C. Chiu;秀丽缬草;Grace Valerian ■

404385　Valeriana venusta Lien C. Chiu = Valeriana flagellifera Batalin ■

404386　Valeriana villosa Thunb. = Patrinia villosa（Thunb.）Juss. ■

404387　Valeriana volkensii Engl.;福尔缬草■☆

404388　Valeriana wallichii DC. = Valeriana jatamansi Jones ■

404389　Valeriana wallichii DC. var. violifolia Franch. = Valeriana jatamansi Jones ■

404390　Valeriana xiaheensis Lien C. Chiu;夏河缬草;Xiahe Valerian ■

404391　Valeriana xiaheensis Lien C. Chiu = Valeriana flagellifera Batalin ■

404392　Valerianaceae Batsch(1802)(保留科名);缬草科(败酱科);Valerian Family ●■

404393　Valerianaceae Lam. et DC. = Valerianaceae Batsch(保留科名)●■

404394　Valerianella Mill.(1754);歧缬草属(拟缬草属,新缬草属);Corn Salad, Cornsalad, Corn-salad ■

404395　Valerianella abyssinica Fresen. = Valerianella microcarpa Loisel. ■☆

404396　Valerianella amblyotis Fisch. et C. A. Mey. ex Hohen.;钝歧缬草■☆

404397　Valerianella anodon Lincz.;无节歧缬草■☆

404398　Valerianella aucheri Boiss. = Valerianella szovitsiana Fisch. et C. A. Mey. ■☆

404399　Valerianella auricula DC. = Valerianella rimosa Bastard ■☆

404400　Valerianella brachystephana Bertol.;短冠歧缬草■☆

404401　Valerianella bushii Dyal = Valerianella ozarkana Dyal ■☆

404402　Valerianella carinata Loisel.;龙骨状歧缬草;European Cornsalad, Keeled Corn Salad, Keeled Cornsalad, Keeled-fruited Cornsalad ■☆

404403　Valerianella caucasica Hoek;高加索歧缬草■☆

404404　Valerianella chenopodifolia（Pursh）DC.;藜叶歧缬草;Goosefoot Corn-salad, Great Lakes Corn-salad ■☆

404405　Valerianella chlorodonta Coss. et Durieu;绿齿歧缬草■☆

404406　Valerianella chlorostephana Pomel = Valerianella pomelii Batt. ■☆

404407　Valerianella conjungens Boiss. = Valerianella leiocarpa（K. Koch）Kuntze ■☆

404408　Valerianella coronata（L.）DC.;冠状歧缬草■☆

404409　Valerianella coronata（L.）DC. subsp. discoidea（L.）Loisel. = Valerianella discoidea（L.）Loisel. ■☆

404410　Valerianella costata Betcke;单脉歧缬草■☆

404411　Valerianella cymbicarpa C. A. Mey.;舟形歧缬草■

404412　Valerianella dactylophylla Boiss.;指叶歧缬草■☆

404413　Valerianella dentata（L.）Pollich;狭果歧缬草;Narrowfruit Cornsalad, Narrow-fruited Cornsalad ■☆

404414　Valerianella dichotoma Gilib.;二歧缬草;Lamb's-lettuce ■☆

404415　Valerianella diodon Boiss.;双齿歧缬草■☆

404416　Valerianella diodon Boiss. = Valerianella oxyrrhyncha Fisch. et C. A. Mey. ■☆

404417　Valerianella discoidea（L.）Loisel.;盘状歧缬草■☆

404418　Valerianella discoidea（L.）Loisel. var. multidentata（Loscos）Batt. = Valerianella discoidea（L.）Loisel. ■☆

404419　Valerianella dufresnia Bunge ex Boiss.;迪氏歧缬草■☆

404420　Valerianella dufresnia Bunge ex Boiss. = Valerianella leiocarpa（K. Koch）Kuntze ■☆

404421　Valerianella echinata（L.）DC.;刺歧缬草■☆

404422　Valerianella eriocarpa Derv. -Sok.;意大利歧缬草;Hairy-fruited Cornsalad, Italian Corn Salad, Italian Cornsalad, Italian Corn-salad ■☆

404423　Valerianella eriocarpa Desv.;红果歧缬草■☆

404424　Valerianella eriocarpa Desv. subsp. truncata（Betcke）Burnat = Valerianella muricata（Steven ex Roem. et Schult.）W. H. Baxter ■☆

404425　Valerianella eriocarpa Desv. var. macrocyathus（Pomel）Batt. = Valerianella muricata（Steven ex Roem. et Schult.）W. H. Baxter ■☆

404426　Valerianella eriocarpa Desv. var. plagiocyathus（Pomel）Batt. = Valerianella muricata（Steven ex Roem. et Schult.）W. H. Baxter ■☆

404427　Valerianella eriocarpa Desv. var. truncata（Betcke）Batt. = Valerianella muricata（Steven ex Roem. et Schult.）W. H. Baxter ■☆

404428　Valerianella fallax Batt.;迷惑歧缬草■☆

404429　Valerianella huetii Boiss. = Valerianella cymbicarpa C. A. Mey. ■

404430　Valerianella kotschyi Boiss.;考氏歧缬草■☆

404431　Valerianella kulabensis Lipsky ex Lincz.;库拉波歧缬草■☆

404432　Valerianella lasiocarpa（Steven）Betcke = Valerianella lasiocarpa Stev. ex Betcke ■☆

404433　Valerianella lasiocarpa Stev. ex Betcke;毛果歧缬草■☆

404434　Valerianella leiocarpa（K. Koch）Kuntze;光果歧缬草■☆

404435　Valerianella leiocarpa（K. Koch）Kuntze var. orientalis（DC.）Kuntze = Valerianella leiocarpa（K. Koch）Kuntze ■☆

404436　Valerianella leptocarpa Pomel；细果歧缬草■☆

404437　Valerianella lipskyi Lincz.；利普斯基歧缬草■☆

404438　Valerianella locusta（L.）Betcke = Valerianella locusta（L.）Laterr. ■☆

404439　Valerianella locusta（L.）Laterr.；小白缬草；Common Cornsalad, Corn Lettuce, Corn Salad, Cornsalad, European Corn Salad, Lamb Lettuce, Lamb's Lettuce, Lamb's-lettuce, Lewiston Cornsalad, Little Valerian, Loblolly, Milkgrass, White Potherb ■☆

404440　Valerianella macrocyatha Pomel = Valerianella eriocarpa Desv. ■☆

404441　Valerianella microcarpa Loisel.；小果歧缬草■☆

404442　Valerianella microcarpa Loisel. = Valerianella dentata（L.）Pollich ■☆

404443　Valerianella microcarpa Loisel. var. puberula（DC.）Batt. = Valerianella microcarpa Loisel. ■☆

404444　Valerianella microdonta Sennen et Mauricio = Valerianella coronata（L.）DC. ■☆

404445　Valerianella monodon K. Koch = Valerianella cymbicarpa C. A. Mey. ■

404446　Valerianella morisonii DC. = Valerianella dentata（L.）Pollich ■☆

404447　Valerianella morisonii DC. subsp. dentata（Pollich）P. Fourn. = Valerianella dentata（L.）Pollich ■☆

404448　Valerianella morisonii DC. subsp. microcarpa（Loisel.）P. Fourn. = Valerianella microcarpa Loisel. ■☆

404449　Valerianella multidentata Loscos = Valerianella discoidea（L.）Loisel. ■☆

404450　Valerianella muricata（M. Bieb.）J. W. Loudon = Valerianella muricata（Steven ex Roem. et Schult.）W. H. Baxter ■☆

404451　Valerianella muricata（Stev.）W. H. Baxter = Valerianella muricata（Steven ex Roem. et Schult.）W. H. Baxter ■☆

404452　Valerianella muricata（Steven ex Roem. et Schult.）W. H. Baxter；短尖歧缬草■☆

404453　Valerianella muricata M. Bieb. ex W. H. Baxter = Valerianella muricata（Steven ex Roem. et Schult.）W. H. Baxter ■☆

404454　Valerianella olitoria（L.）Pollich；歧缬草（野莴）；Cornsalad, Corn-salad, Lamb's Lettuce ■☆

404455　Valerianella olitoria（L.）Pollich = Valerianella locusta（L.）Betcke ■☆

404456　Valerianella otodontata Pomel = Valerianella dentata（L.）Pollich ■☆

404457　Valerianella oxyrhyncha Fisch. et C. A. Mey.；尖嘴歧缬草■☆

404458　Valerianella oxyrrhyncha Fisch. et C. A. Mey. = Valerianella oxyrhyncha Fisch. et C. A. Mey. ■☆

404459　Valerianella ozarkana Dyal；奥扎克歧缬草；Ozark Corn Salad ■☆

404460　Valerianella persica Boiss. = Valerianella szovitsiana Fisch. et C. A. Mey. ■☆

404461　Valerianella plagiocyata Pomel = Valerianella eriocarpa Desv. ■☆

404462　Valerianella plagiostephana Fisch. et C. A. Mey.；斜冠歧缬草■

404463　Valerianella platycarpa Trautv.；宽果歧缬草■☆

404464　Valerianella pomelii Batt.；波梅尔缬草■☆

404465　Valerianella pontica Velen.；蓬特缬草■☆

404466　Valerianella pumila（Willd.）DC. = Valerianella coronata（L.）DC. ■☆

404467　Valerianella pumila DC.；小歧缬草■☆

404468　Valerianella radiata（L.）Dufr.；辐射缬草；Corn Salad ■☆

404469　Valerianella radiata（L.）Dufr. f. parviflora（Dyal）Egg. Ware；小花辐射缬草■☆

404470　Valerianella radiata（L.）Dufr. var. fernaldii Dyal = Valerianella radiata（L.）Dufr. ■☆

404471　Valerianella radiata（L.）Dufr. var. missouriensis Dyal = Valerianella radiata（L.）Dufr. ■☆

404472　Valerianella rimosa Bastard；总状歧缬草；Broad-fruited Cornsalad, European Cornsalad ■☆

404473　Valerianella sclerocarpa Fisch. et C. A. Mey.；核果歧缬草■☆

404474　Valerianella stenocarpa（Engelm.）Krok var. parviflora Dyal = Valerianella radiata（L.）Dufr. ■☆

404475　Valerianella stocksii Boiss. = Valerianella oxyrrhyncha Fisch. et C. A. Mey. ■☆

404476　Valerianella szovitsiana Fisch. et C. A. Mey.；绍氏歧缬草■☆

404477　Valerianella truncata Betcke = Valerianella muricata（Steven ex Roem. et Schult.）W. H. Baxter ■☆

404478　Valerianella tuberculata Boiss.；多疣缬草■☆

404479　Valerianella turgida Betcke；膨胀缬草■☆

404480　Valerianella turkestanica Regel et Schmalh. ex Regel；土耳其歧缬草■☆

404481　Valerianella uncinata Dufr.；具钩缬草■☆

404482　Valerianella vesicaria Moench；水泡缬草■☆

404483　Valerianella vvedenskyi Lincz.；韦氏歧缬草■☆

404484　Valerianella woodsiana（Torr. et A. Gray）Walp.；伍茨歧缬草；Corn Salad ■☆

404485　Valerianodes Kuntze = Stachytarpheta Vahl（保留属名）■●

404486　Valerianodes Kuntze = Valerianoides Medik.（废弃属名）■●

404487　Valerianodes T. Durand et Jacks. = Stachytarpheta Vahl（保留属名）■●

404488　Valerianoides Medik.（废弃属名）= Stachytarpheta Vahl（保留属名）■●

404489　Valerianopsis C. A. Muell. = Valeriana L. ●■

404490　Valerioa Standl. et Steyerm. = Peltanthera Benth.（保留属名）●☆

404491　Valerioanthus Lundell = Ardisia Sw.（保留属名）●■

404492　Valerioanthus Lundell（1982）；缬花紫金牛属●☆

404493　Valerioanthus nevermannii（Standl.）Lundell；缬花紫金牛●☆

404494　Valetonia T. Durand = Pleurisanthes Baill. ●☆

404495　Valetonia T. Durand ex Engl. = Pleurisanthes Baill. ●☆

404496　Validallium Small = Allium L. ■

404497　Validallium tricoccum（Sol.）Small = Allium tricoccum Sol. ■☆

404498　Valiha S. Dransf.（1998）；马岛竹属●☆

404499　Valiha diffusa S. Dransf.；马岛竹●☆

404500　Valiha perrieri（A. Camus）S. Dransf.；佩里耶马岛竹●☆

404501　Valihaha Adans. = Memecylon L. ●

404502　Valisneria Scop. = Vallisneria L. ■

404503　Valkera Stokes = Ouratea Aubl.（保留属名）●

404504　Valkera Stokes = Walkera Schreb. ●

404505　Vallantia A. Dietr. = Valantia L. ■☆

404506　Vallariopsis Woodson（1936）；拟纽子花属●☆

404507　Vallariopsis lancifolta（Hook. f.）Woodson；拟纽子花●☆

404508　Vallaris Burm. f.（1768）；纽子花属；Vallaris, Buttonflower ●

404509　Vallaris Raf. = Euphorbia L. ●■

404510　Vallaris anceps（Dunn et Williams）C. E. C. Fisch. = Kibatalia macrophylla（Pierre ex Hua）Woodson ●

404511　Vallaris arborea C. E. C. Fisch. = Kibatalia macrophylla（Pierre ex Hua）Woodson ●

404512　Vallaris controversa Spreng. = Lepistemon binectarifer（Wall.）Kuntze ■

404513　Vallaris divaricata（Lour.）G. Don = Strophanthus divaricatus（Lour.）Hook. et Arn. ●

404514　Vallaris divaricata G. Don = Strophanthus divaricatus（Lour.）Hook. et Arn. ●

404515　Vallaris grandiflora Hemsl. et E. H. Wilson = Vallaris indecora（Baill.）Tsiang et P. T. Li ●

404516　Vallaris heynii Spreng. ;玉盏藤●

404517　Vallaris heynii Spreng. = Vallaris solanacea（Roth）Kuntze ●

404518　Vallaris indecora（Baill.）Tsiang et P. T. Li;大纽子花(纽子花);Giant Buttonflower,Giant Vallaris ●

404519　Vallaris laxiflora Blume = Pottsia laxiflora（Blume）Kuntze ●

404520　Vallaris sinensis（Lour.）G. Don = Cryptolepis sinensis（Lour.）Merr. ●

404521　Vallaris sinensis G. Don = Cryptolepis sinensis（Lour.）Merr. ●

404522　Vallaris solanacea（Roth）K. Schum. = Vallaris solanacea（Roth）Kuntze ●

404523　Vallaris solanacea（Roth）Kuntze;纽子花(海因纽子花,茄形纽子花,玉盏藤);Buttonflower,Nightshade-like Vallaris ●

404524　Vallea Mutis ex L. f.（1782）;瓦莱木属(瓦拉木属)●☆

404525　Vallea stipularis L. f. ;瓦莱木(瓦拉木)●☆

404526　Vallesia Ruiz et Pav.（1794）;河谷木属(瓦来斯木属,瓦来西亚属);Vallesia ●☆

404527　Vallesia antilana Woodson;河谷木(瓦来西亚)●☆

404528　Vallesia flexuosa Woodson;曲折河谷木(曲折瓦来斯木);Flexuose Vallesia ●☆

404529　Vallesia glabra（Cav.）Link;无毛河谷木;Pearlberry ●☆

404530　Vallesneriaceae Link = Hydrocharitaceae Juss.（保留科名）■

404531　Vallisneria L.（1753）;苦草属;Bittergrass,Eelgrass,Lapegrass,Tapegrass,Tape-grass,Wild Celery,Wild-celery ■

404532　Vallisneria Mich. ex L. = Vallisneria L. ■

404533　Vallisneria aethiopica Fenzl = Vallisneria spiralis L. ■

404534　Vallisneria alternifolia Roxb. = Nechamandra alternifolia（Roxb. ex Wight）Thwaites ■

404535　Vallisneria alternifolia Roxb. ex Wight = Nechamandra alternifolia（Roxb. ex Wight）Thwaites ■

404536　Vallisneria americana Michx. ;美洲苦草(大苦草);American Eelgrass,American Wild Celery,Eel-grass,Tape Grass,Water-celery ■☆

404537　Vallisneria anhuiensis X. S. Shen;安徽苦草;Anhui Bittergrass ■

404538　Vallisneria asiatica Michx. = Vallisneria americana Michx. ■☆

404539　Vallisneria asiatica Miki = Vallisneria natans（Lour.）H. Hara ■

404540　Vallisneria asiatica Miki var. biwaensis Miki = Vallisneria natans（Lour.）H. Hara var. biwaensis（Miki）H. Hara ■☆

404541　Vallisneria asiatica Miki var. higoensis Miki = Vallisneria natans（Lour.）H. Hara var. higoensis（Miki）H. Hara ■☆

404542　Vallisneria asiatica Miki var. higoensis Miki = Vallisneria natans（Lour.）H. Hara ■

404543　Vallisneria biwaensis（Miki）Ohwi = Vallisneria natans（Lour.）H. Hara var. biwaensis（Miki）H. Hara ■☆

404544　Vallisneria denseserrulata（Makino）Makino;密刺苦草;Spiny Bittergrass ■

404545　Vallisneria denseserrulata（Makino）Makino = Vallisneria spiralis L. f. aethiopica（Fenzl）T. Durand et Schinz ■☆

404546　Vallisneria gigantea Graebn. ;大苦草■

404547　Vallisneria gigantea Graebn. = Vallisneria americana Michx. ■☆

404548　Vallisneria gigantea Graebn. = Vallisneria natans（Lour.）H. Hara ■

404549　Vallisneria gigantea Graebn. var. biwaensis（Miki）Kitam. =

Vallisneria natans（Lour.）H. Hara var. biwaensis（Miki）H. Hara ■☆

404550　Vallisneria gigantea Graebn. var. higoensis（Miki）Kitam. = Vallisneria natans（Lour.）H. Hara var. higoensis（Miki）H. Hara ■☆

404551　Vallisneria gigantea Graebn. var. higoensis（Miki）Kitam. = Vallisneria natans（Lour.）H. Hara ■

404552　Vallisneria higoensis（Miki）Ohwi = Vallisneria natans（Lour.）H. Hara ■

404553　Vallisneria higoensis（Miki）Ohwi = Vallisneria natans（Lour.）H. Hara var. higoensis（Miki）H. Hara ■☆

404554　Vallisneria longipedunculata X. S. Shen;长梗苦草;Longpedicel Bittergrass ■

404555　Vallisneria natans（Lour.）H. Hara;苦草(篦藻,扁草,大苦草,带子孔,脚带小草,韭菜草,蓼萍草,密齿苦草,面条草,石菖藻,水苗,亚洲苦草);Asia Bittergrass,Bittergrass,Gigant Bittergrass ■

404556　Vallisneria natans（Lour.）H. Hara var. biwaensis（Miki）H. Hara;比瓦苦草■☆

404557　Vallisneria natans（Lour.）H. Hara var. higoensis（Miki）H. Hara;肥后苦草■☆

404558　Vallisneria natans（Lour.）H. Hara var. higoensis（Miki）H. Hara = Vallisneria natans（Lour.）H. Hara ■

404559　Vallisneria neotropicalis Vict. = Vallisneria americana Michx. ■☆

404560　Vallisneria numidica Pomel = Vallisneria spiralis L. f. aethiopica（Fenzl）T. Durand et Schinz ■☆

404561　Vallisneria octandra Roxb. = Blyxa octandra（Roxb.）Planch. ex Thwaites ■

404562　Vallisneria spinulosa S. Z. Yan;刺苦草(苦草);Spine Bittergrass ■

404563　Vallisneria spiralis L. ;螺旋苦草(苦草);Eelgrass,Spiral Wild Celery,Tape-grass ■

404564　Vallisneria spiralis L. = Vallisneria natans（Lour.）H. Hara ■

404565　Vallisneria spiralis L. f. aethiopica（Fenzl）T. Durand et Schinz;埃塞俄比亚苦草■☆

404566　Vallisneria spiralis L. var. asiatica（Michx.）Torr. = Vallisneria americana Michx. ■☆

404567　Vallisneria spiralis L. var. denseserrulata Makino = Vallisneria denseserrulata（Makino）Makino ■☆

404568　Vallisneria spiralis L. var. denseserrulata Makino = Vallisneria spiralis L. f. aethiopica（Fenzl）T. Durand et Schinz ■☆

404569　Vallisneria spiralis L. var. numidica（Pomel）Maire et Weiller = Vallisneria spiralis L. ■

404570　Vallisneria spiraloides Roxb. = Vallisneria natans（Lour.）H. Hara ■

404571　Vallisneria verticillata Roxb. = Hydrilla verticillata（L. f.）Royle ■

404572　Vallisneriaceae Dumort. ;苦草科■

404573　Vallisneriaceae Dumort. = Hydrocharitaceae Juss.（保留科名）■

404574　Vallota Herb. = Cyrtanthus Aiton（保留属名）■☆

404575　Vallota Salisb. ex Herb.（1821）（保留属名）;瓦氏石蒜属;George,George Lily,Lily,Scarborough Lily,Scarborough-lily ■☆

404576　Vallota Salisb. ex Herb.（保留属名）= Cyrtanthus Aiton（保留属名）■☆

404577　Vallota Steud. = Trichachne Nees ■☆

404578　Vallota miniata Lindl. = Clivia miniata（Lindl.）Regel ■

404579　Vallota purpurea（Aiton）Herb. = Cyrtanthus elatus（Jacq.）Traub ■☆

404580　Vallota purpurea Herb. = Vallota speciosa T. Durand et Schinz ■☆

404581　Vallota speciosa（L. f.）T. Durand et Schinz;瓦氏石蒜;George Lily,Jersey Lily,Knysna Lily,Scarborough Lily,Scarborough-lily ■☆

404582　Vallota speciosa（L. f.）T. Durand et Schinz = Cyrtanthus elatus（Jacq.）Traub ■☆

404583　Vallota speciosa T. Durand et Schinz = Vallota speciosa（L. f.）T. Durand et Schinz ■☆

404584　Vallota speciosa Voss = Vallota speciosa T. Durand et Schinz ■☆

404585　Vallotita Post et Kuntze = Vallota Steud. ■☆

404586　Valoradia Hochst. = Ceratostigma Bunge ●■

404587　Valoradia abyssinica Hochst. = Ceratostigma abyssinicum（Hochst.）Schweinf. et Asch. ●☆

404588　Valoradia plumbaginoides（Bunge）Boiss. = Ceratostigma plumbaginoides Bunge ●■

404589　Valota Adans.（废弃属名）= Digitaria Haller（保留属名）■

404590　Valota Adans.（废弃属名）= Trichachne Nees ■☆

404591　Valota Adans.（废弃属名）= Vallota Salisb. ex Herb.（保留属名）■☆

404592　Valota Dumort. = Vallota Salisb. ex Herb.（保留属名）■☆

404593　Valsonica Scop. = Watsonia Mill.（保留属名）■☆

404594　Valteta Raf.（废弃属名）= Iochroma Benth.（保留属名）●☆

404595　Valvanthera C. T. White = Hernandia L. ●

404596　Valvaria Ser. = Clematis L. ●■

404597　Valvinterlobus Dulac = Schultesia Roth ■●

404598　Valvinterlobus Dulac = Wahlenbergia Schrad. ex Roth（保留属名）■●

404599　Vanalphimia Lesch. ex DC. = Saurauia Willd.（保留属名）●

404600　Vanalpighmia Steud. = Vanalphimia Lesch. ex DC. ●

404601　Vananthes Willis = Vauanthes Haw. ●■☆

404602　Vanasta Raf. = Manettia Mutis ex L.（保留属名）●■☆

404603　Vanasta Raf. = Vanessa Raf. ●■☆

404604　Vanasushava P. K. Mukh. et Constance（1974）；南印度草属■☆

404605　Vanasushava pedata（Wight）P. K. Mukh. et Constance；南印度草■☆

404606　Vanclevea Greene = Chrysothamnus Nutt.（保留属名）■●☆

404607　Vanclevea Greene（1899）；芒黄花属●☆

404608　Vanclevea stylosa（Eastw.）Greene；芒黄花；Resinbush ●☆

404609　Vanclevea stylosa（Eastw.）Greene = Chrysothamnus stylosus（Eastw.）Urbatsch, R. P. Roberts et Neubig ●☆

404610　Vancouveria C. Moore et Decne.（1834）；折瓣花属（多萼草属，范库弗草属，弗草属）；Inside-out Flower, Vancouveria ■☆

404611　Vancouveria Decne. = Vancouveria C. Moore et Decne. ■☆

404612　Vancouveria chrysantha Greene；黄花折瓣花（黄花多萼草，黄花库弗草）；Golden Inside-out Flower, Siskiyou Inside-out Flower, Yellow Vancouveria ■☆

404613　Vancouveria chrysantha Greene var. parviflora Jeps. = Vancouveria planipetala Calloni ■☆

404614　Vancouveria concolor Greene；同色折瓣花（同色范库弗草）；Concolor Vancouveria ■☆

404615　Vancouveria hexandra（Hook.）C. Morren et Decne.；折瓣花（北范库弗草，多萼草）；Inside-out-flower, Northern Inside-out Flower, Northern Vancouveria ■☆

404616　Vancouveria hexandra C. Moore et Decne. = Vancouveria hexandra（Hook.）C. Morren et Decne. ■☆

404617　Vancouveria hexandra C. Moore et Decne. var. chrysantha Greene = Vancouveria chrysantha Greene ■☆

404618　Vancouveria parviflora Greene = Vancouveria planipetala Calloni ■☆

404619　Vancouveria planipetala Calloni；小花折瓣花（小花范库弗草）；Redwood Inside-out Flower, Redwood-ivy, Smallflowered Vancouveria ■☆

404620　Vanda Jones ex R. Br.（1820）；万代兰属（万带兰属）；Cowslip-scented Orchid, Vanda ■☆

404621　Vanda R. Br. = Vanda Jones ex R. Br. ■

404622　Vanda alpina Lindl. = Trudelia alpina（Lindl.）Garay ■

404623　Vanda amesiana Hook. f. = Holcoglossum amesianum（Rchb. f.）Christenson ■

404624　Vanda amiensis Masam. et Segawa = Vanda lamellata Lindl. ■

404625　Vanda bensoni Bateman；宾森万代兰（宾森万带兰）；Benson Vanda ■☆

404626　Vanda bicolor Griff.；双色万代兰（双色万带兰）；Bicolor Vanda ■☆

404627　Vanda brunnea Rchb. f.；白柱万代兰（白花万代兰，白花万带兰）；Whitestyle Vanda ■

404628　Vanda clarkei（Rchb. f.）N. E. Br. = Esmeralda clarkei Rchb. f. ■

404629　Vanda coerulea Griff. ex Lindl.；大花万代兰（白花万带兰，大花万带兰，万代兰）；Bigflower Vanda, Blue Orchid, Largeflower Vanda ■

404630　Vanda coerulescens Griff.；小蓝万代兰（蓝花万带兰，小花万代兰）；Blueflower Vanda, Skyblue Vanda ■

404631　Vanda coerulescens Griff. ex Lindl. = Vanda brunnea Rchb. f. ■

404632　Vanda concolor Blume；琴唇万代兰（琴唇万带兰，树兰，松兰）；Concolorous Vanda, Samecolor Vanda ■

404633　Vanda cristata Lindl.；叉唇万代兰（鸡冠万带兰）；Cristate Vanda ■

404634　Vanda dearei Lindl.；戴氏万代兰■☆

404635　Vanda denisoniana Benson et Rchb. f.；德尼森万代兰（白花万带兰，大花万带兰，德尼森万带兰）；Denison Vanda ■☆

404636　Vanda denisoniana Benson et Rchb. f. = Vanda brunnea Rchb. f. ■

404637　Vanda denisoniana Benson et Rchb. f. var. hebraica Rchb. f. = Vanda brunnea Rchb. f. ■

404638　Vanda densiflora Lindl. = Rhynchostylis gigantea（Lindl.）Ridl. ■

404639　Vanda esquirolei Schltr. = Vanda concolor Blume ■

404640　Vanda falcata（Thunb.）Beer = Neofinetia falcata（Thunb.）Hu ■

404641　Vanda foetida J. J. Sm.；臭味万代兰（臭味万带兰，臭叶万带兰）；Scented Vanda ■☆

404642　Vanda fuscoviridis Lindl.；广东万代兰（广东万带兰）；Guangdong Vanda ■

404643　Vanda gigantea Lindl. = Vandopsis gigantea（Lindl.）Pfitzer ■

404644　Vanda hainanensis Rolfe = Rhynchostylis gigantea（Lindl.）Ridl. ■

404645　Vanda henryi Schltr. = Vanda brunnea Rchb. f. ■

404646　Vanda hookeriana Rchb. f.；虎克万代兰（虎克万带兰）；Hooker Vanda ■☆

404647　Vanda hybrida Hort.；杂交万代兰■☆

404648　Vanda insignis Blume；美丽万代兰（美丽万带兰）；Beautiful Vanda ■☆

404649　Vanda kimballiana Rchb. f. = Holcoglossum kimballianum（Rchb. f.）Garay ■

404650　Vanda kwangtungensis S. J. Cheng et C. Z. Tang = Vanda concolor Blume ■

404651　Vanda kwangtungensis S. J. Cheng et C. Z. Tang = Vanda fuscoviridis Lindl. ■

404652　Vanda lamellata Lindl.；雅美万代兰（雅美万带兰）；Elegant Vanda ■

404653　Vanda longifolia Lindl. = Acampe rigida（Buch. -Ham. ex Sm.）P. F. Hunt ■

404654　Vanda lowii Lindl. = Dimorphorchis lowii（Lindl.）Rolfe ■☆

404655　Vanda luzonica Loher ex Rolfe；吕宋万代兰（吕宋万带兰）；Luzon Vanda ■☆

404656　Vanda multiflora Lindl. = Acampe rigida（Buch. -Ham. ex Sm.）P. F. Hunt ■

404657　Vandapaniculata（Ker Gawl.）R. Br. = Cleisostoma paniculatum（Ker Gawl.）Garay ■

404658　Vanda parishii Rchb. f. = Hygrochilus parishii（Veitch et Rchb. f.）Pfitzer ■

404659　Vanda parviflora Lindl.；小花万代兰（小花万带兰）；Smallflower Vanda ■☆

404660　Vanda parviflora Lindl. = Vanda coerulescens Griff. ■

404661　Vanda pumila Hook. f.；矮万代兰（矮生万带兰）；Dwarf Vanda ■

404662　Vanda rostrata Lodd. = Cleisostoma rostratum（Lodd.）Seidenf. ex Aver. ■

404663　Vanda roxburghii R. Br. = Vanda subconcolor Ts. Tang et F. T. Wang ■

404664　Vanda rupestris Hand. -Mazz. = Holcoglossum rupestre（Hand. -Mazz.）Garay ■

404665　Vanda sanderiana Rchb. f. = Euanthe sanderiana（Rchb. f.）Schltr. ■☆

404666　Vanda saprophytica Gagnep. = Holcoglossum kimballianum（Rchb. f.）Garay ■

404667　Vanda simondii Gagnep. = Cleisostoma simondii（Gagnep.）Seidenf. ■

404668　Vanda simondii Gagnep. = Vanda watsonii Rolfe ■☆

404669　Vanda subconcolor Ts. Tang et F. T. Wang；纯色万代兰（纯色万带兰，黑珊瑚，罗氏万代兰，罗氏万带兰，三色万带兰）；Purecolor Vanda ■

404670　Vanda subconcolor Ts. Tang et F. T. Wang var. distincha Ts. Tang et F. T. Wang = Vanda subconcolor Ts. Tang et F. T. Wang ■

404671　Vanda subconcolor Ts. Tang et F. T. Wang var. distincha Ts. Tang et F. T. Wang；密叶万代兰（密叶万带兰）；Denseleaf Vanda ■

404672　Vanda subulifolia Rchb. f. = Holcoglossum subulifolium（Rchb. f.）Christenson ■

404673　Vanda subulifolia Rchb. f. = Vanda watsonii Rolfe ■☆

404674　Vanda sumatrana Schltr.；苏门答腊万代兰（苏门答腊万带兰）；Sumatra Vanda ■☆

404675　Vanda taiwaniana S. S. Ying = Papilionanthe taiwaniana（S. S. Ying）Ormerod ■

404676　Vanda teres（Roxb.）Lindl.；棒叶万代兰（棒叶万带兰）■

404677　Vanda teres（Roxb.）Lindl. = Papilionanthe teres（Roxb.）Schltr. ■

404678　Vanda teretifolia Lindl. = Cleisostoma simondii（Gagnep.）Seidenf. ■

404679　Vanda tessellata（Roxb.）Hook. ex G. Don；方格纹万代兰（方格纹万带兰）；Chequer Vanda ■☆

404680　Vanda tricolor Lindl.；三色万代兰（纯色万带兰，三色万带兰）；Tricolor Vanda ■☆

404681　Vanda tricolor Lindl. var. suavis（Lindl.）Veitch；长序三色万带兰；Long-inflorescence Vanda ■☆

404682　Vanda undulata Lindl. = Vandopsis undulata（Lindl.）J. J. Sm. ■

404683　Vanda watsonii Rolfe；瓦氏万代兰 ■☆

404684　Vanda yamiensis Masam. et Segawa = Vanda lamellata Lindl. ■

404685　Vandalea（Fourr.）Fourr. = Sisymbrium L. ■

404686　Vandalea Fourr. = Sisymbrium L. ■

404687　Vandasia Domin = Vandasina Rauschert ■☆

404688　Vandasina Rauschert（1982）；铁扇三叶豆属 ■☆

404689　Vandasina retusa（Benth.）Rauschert；铁扇三叶豆 ■☆

404690　Vandea Griff. = Vanda Jones ex R. Br. ■

404691　Vandellia L. = Lindernia All. ■

404692　Vandellia L. = Vandellia P. Browne ex L. ■

404693　Vandellia P. Browne = Lindernia All. ■

404694　Vandellia P. Browne ex L.（1767）；旱田草属 ■

404695　Vandellia P. Browne ex L. = Lindernia All. ■

404696　Vandellia anagallis（Burm. f.）T. Yamaz.；定经草 ■☆

404697　Vandellia anagallis（Burm. f.）T. Yamaz. = Lindernia anagallis（Burm. f.）Pennell ■

404698　Vandellia anagallis（Burm. f.）T. Yamaz. = Lindernia antipoda（L.）Alston ■

404699　Vandellia anagallis（Burm. f.）T. Yamaz. = Ruellia anagallis Burm. f. ■

404700　Vandellia anagallis（Burm. f.）T. Yamaz. var. verbenifolia（Colsm.）T. Yamaz. = Lindernia anagallis（Burm. f.）Pennell ■

404701　Vandellia anagallis（Burm. f.）T. Yamaz. var. verbenifolia（Colsm.）T. Yamaz. = Lindernia antipoda（L.）Alston ■

404702　Vandellia angustifolia Benth. = Lindernia micrantha D. Don ■

404703　Vandellia angustifolia Benth. f. leucantha Hiyama；白花狭叶母草 ■☆

404704　Vandellia antipoda（L.）T. Yamaz.；旱田草（定经草）■

404705　Vandellia antipoda（L.）T. Yamaz. = Lindernia antipoda（L.）Alston ■

404706　Vandellia bodinieri H. Lév. = Lindernia crustacea（L.）F. Muell. ■

404707　Vandellia brevipedunculata（Migo）T. Yamaz. = Lindernia brevipedunculata Migo ■

404708　Vandellia callitrichifolia H. Lév. = Lindernia anagallis（Burm. f.）Pennell ■

404709　Vandellia cavaleriei H. Lév. = Lindernia setulosa（Maxim.）Tuyama ex H. Hara ■

404710　Vandellia chinensis T. Yamaz. = Lindernia nummulariifolia（D. Don）Wettst. ■

404711　Vandellia ciliata（Colsm.）T. Yamaz. = Lindernia ciliata（Colsm.）Pennell ■

404712　Vandellia cordifolia（Colsm.）G. Don；心叶母草 ■

404713　Vandellia cordifolia（Colsm.）G. Don = Lindernia anagallis（Burm. f.）Pennell ■

404714　Vandellia corymbosa Baker = Lindernia nummulariifolia（D. Don）Wettst. ■

404715　Vandellia crustacea（L.）Benth. = Lindernia crustacea（L.）F. Muell. ■

404716　Vandellia diffusa L. = Lindernia difusa（L.）Wettst. ■☆

404717　Vandellia diffusa L. var. pedunculata（Benth.）V. Naray. = Lindernia difusa（L.）Wettst. ■☆

404718　Vandellia diffusa L. var. pedunculata Benth. = Lindernia difusa（L.）Wettst. ■☆

404719　Vandellia elata Benth. = Lindernia elata（Benth.）Wettst. ■

404720　Vandellia erecta Benth. = Lindernia procumbens（Krock.）Philcox ■

404721　Vandellia hirta（Cham. et Schltdl.）T. Yamaz. = Lindernia pusilla（Willd.）Bold. ■

404722　Vandellia japonica Miq. = Mazus miquelii Makino ■

404723　Vandellia lobelioides Oliv. = Lindernia oliverana Dandy ■☆

404724　Vandellia mollis Benth. = Lindernia mollis（Benth.）Wettst. ■

404725　Vandellia montana（Blume）Benth. = Lindernia mollis（Benth.）Wettst. ■

404726　Vandellia nummularifolia D. Don = Lindernia nummularifolia（D. Don）Wettst. ■

404727　Vandellia oblonga Benth. = Lindernia oblonga（Benth.）Merr. ■

404728　Vandellia obovata Walp. = Mazus pumilus（Burm. f.）Steenis ■

404729　Vandellia pyxidaria（L.）Maxim. = Lindernia procumbens（Krock.）Philcox ■

404730　Vandellia scutellariiformis（T. Yamaz.）T. Yamaz. = Lindernia scutellariiformis T. Yamaz. ■

404731　Vandellia senegalensis Benth. = Lindernia senegalensis（Benth.）V. Naray. ■☆

404732　Vandellia sessilifolia Benth. = Lindernia nummularifolia（D. Don）Wettst. ■

404733　Vandellia setulosa（Maxim.）T. Yamaz. = Lindernia setulosa（Maxim.）Tuyama ex H. Hara ■

404734　Vandellia stachydifolia Walp. = Mazus stachydifolius（Turcz.）Maxim. ■

404735　Vandellia subcrenulata Miq. = Lindernia oblonga（Benth.）Merr. ■

404736　Vandellia tenuifolia（Colsm.）Hainens = Lindernia tenuifolia（Vahl）Alston ■

404737　Vandellia urticifolia Hance = Lindernia elata（Benth.）Wettst. ■

404738　Vandellia veronicifolia（Retz.）Haines = Lindernia antipoda（L.）Alston ■

404739　Vandellia viscosa（Hornem.）Merr. = Lindernia viscosa（Hornem.）Merr. ■

404740　Vandera Raf. = Croton L. ●

404741　Vanderystia De Wild. = Ituridendron De Wild. ●☆

404742　Vanderystia De Wild. = Omphalocarpum P. Beauv. ●☆

404743　Vanderystia congolensis De Wild. = Omphalocarpum pachysteloides Mildbr. ex Hutch. et Dalziel ●☆

404744　Vandesia Salisb. = Bomarea Mirb. ■☆

404745　Vandopsis Pfitzer（1889）；拟万代兰属（假万带兰属）；False Vanda, Vandopsis ■

404746　Vandopsis chinensis（Rolfe）Schltr. = Vandopsis gigantea（Lindl.）Pfitzer ■

404747　Vandopsis gigantea（Lindl.）Pfitzer；拟万代兰（大万朵兰,假万带兰）；Gigantic False Vanda, Vandopsis Gigantic ■

404748　Vandopsis lissochiloides（Gaudich.）Pfitzer；香万朵兰（菲律宾假万带兰）■☆

404749　Vandopsis luchuensis（Rolfe）Schltr.；琉球万代兰■☆

404750　Vandopsis luchuensis（Rolfe）Schltr. = Staurochilus luchuensis（Rolfe）Fukuy. ■

404751　Vandopsis lutchuensis（Rolfe）Schltr. = Trichoglottis lutchuensis（Rolfe）Garay et H. R. Sweet ■☆

404752　Vandopsis osmantha Fukuy. ex Masam. = Cleisostoma paniculatum（Ker Gawl.）Garay ■

404753　Vandopsis parishii（Rchb. f.）Schltr. = Hygrochilus parishii（Veitch et Rchb. f.）Pfitzer ■

404754　Vandopsis undulata（Lindl.）J. J. Sm.；白花拟万代兰（波状万带兰,船唇兰）；Undulate Vandopsis, White Vandopsis ■

404755　Vandopsis warocqueana（Rolfe）Schltr.；瓦洛氏假万带兰；Warocque Vandopsis ■☆

404756　Vanessa Raf. = Manettia Mutis ex L.（保留属名）●■☆

404757　Vangueria Comm. ex Juss.（1789）；瓦氏茜属（万格茜属）■☆

404758　Vangueria Juss. = Vangueria Comm. ex Juss. ■☆

404759　Vangueria abyssinica A. Rich. = Vangueria madagascariensis J. F. Gmel. var. abyssinica（A. Rich.）Puff ■☆

404760　Vangueria acuminatissima K. Schum. = Rytigynia acuminatissima（K. Schum.）Robyns ■☆

404761　Vangueria acutiloba Robyns = Vangueria madagascariensis J. F. Gmel. ■☆

404762　Vangueria adenodonta K. Schum. = Rytigynia adenodonta（K. Schum.）Robyns ■☆

404763　Vangueria apiculata K. Schum.；尖瓦氏茜（尖万格茜）■☆

404764　Vangueria argentea Wernham = Rytigynia argentea（Wernham）Robyns ■☆

404765　Vangueria armata K. Schum. = Plectroniella armata（K. Schum.）Robyns ●☆

404766　Vangueria bagshawei S. Moore = Rytigynia bagshawei（S. Moore）Robyns ■☆

404767　Vangueria barnimiana Schweinf. = Vangueria madagascariensis J. F. Gmel. var. abyssinica（A. Rich.）Puff ■☆

404768　Vangueria beniensis De Wild. = Rytigynia beniensis（De Wild.）Robyns ■☆

404769　Vangueria bequaertii De Wild. = Rytigynia bagshawei（S. Moore）Robyns ■☆

404770　Vangueria bicolor K. Schum.；二色瓦氏茜■☆

404771　Vangueria bicolor K. Schum. var. crassiramis ？= Vangueria volkensii K. Schum. ■☆

404772　Vangueria binata K. Schum. = Rytigynia binata（K. Schum.）Robyns ●☆

404773　Vangueria bomiliensis De Wild. = Rytigynia bomiliensis（De Wild.）Robyns ■☆

404774　Vangueria brachytricha K. Schum. = Fadogia triphylla Baker var. giorgii（De Wild.）Verdc. ●☆

404775　Vangueria butaguensis De Wild. = Rytigynia bugoyensis（K. Krause）Verdc. ■☆

404776　Vangueria caffra Sim = Pachystigma macrocalyx（Sond.）Robyns ●☆

404777　Vangueria campanulata Robyns = Vangueria infausta Burch. var. campanulata（Robyns）Verdc. ■☆

404778　Vangueria canthioides Benth.；鱼骨木瓦氏茜（鱼骨木万格茜）■☆

404779　Vangueria canthioides Benth. = Rytigynia canthioides（Benth.）Robyns ■☆

404780　Vangueria canthoides De Wild. = Rytigynia demeusei（De Wild.）Robyns ■☆

404781　Vangueria chariensis A. Chev. ex Robyns；沙里瓦氏茜■☆

404782　Vangueria chartacea Robyns = Vangueria randii S. Moore subsp. chartacea（Robyns）Verdc. ■☆

404783　Vangueria cinnamomea Dinter；朱红瓦氏茜■☆

404784　Vangueria cistifolia（Welw. ex Hiern）Lantz；岩蔷薇瓦氏茜■☆

404785　Vangueria cistifolia（Welw. ex Hiern）Lantz var. latifolia（Verdc.）Lantz；宽叶瓦氏茜■☆

404786　Vangueria claessensii De Wild. = Rytigynia claessensii（De Wild.）Robyns ■☆

404787　Vangueria concolor Hiern = Rytigynia umbellulata（Hiern）Robyns ■☆

404788　Vangueria congesta K. Krause = Rytigynia congesta（K. Krause）Robyns ■☆

404789　Vangueria crassa（Hiern）Schweinf. ex Hiern = Multidentia crassa（Hiern）Bridson et Verdc. ●☆

404790　Vangueria cyanescens Robyns；浅蓝瓦氏茜■☆

404791　Vangueria dalzielii Hutch. = Fadogia erythrophloea（K. Schum. et K. Krause）Hutch. et Dalziel ●☆

404792　Vangueria dasyothamnus K. Schum. = Rytigynia dasyothamnus（K. Schum.）Robyns ■☆

404793　Vangueria demeusei De Wild. = Rytigynia demeusei（De Wild.）

Robyns ■☆

404794　Vangueria dewevrei De Wild. = Rytigynia demeusei（De Wild.）Robyns ■☆

404795　Vangueria dewevrei De Wild. et T. Durand = Rytigynia dewevrei（De Wild. et T. Durand）Robyns ●☆

404796　Vangueria dryadum S. Moore = Lagynias dryadum（S. Moore）Robyns ■☆

404797　Vangueria edulis（Vahl）Vahl = Vangueria madagascariensis J. F. Gmel. ■☆

404798　Vangueria edulis Lam. = Vangueria madagascariensis J. F. Gmel. ■☆

404799　Vangueria edulis Lam. var. pubescens Fiori = Vangueria madagascariensis J. F. Gmel. var. abyssinica（A. Rich.）Puff ■☆

404800　Vangueria emirnensis Baker；艾米瓦氏茜（艾米万格茜）■☆

404801　Vangueria erythrophloea K. Schum. et K. Krause = Fadogia erythrophloea（K. Schum. et K. Krause）Hutch. et Dalziel ●☆

404802　Vangueria esculenta S. Moore；食用瓦氏茜■☆

404803　Vangueria esculenta S. Moore var. glabra ? = Vangueria esculenta S. Moore ■☆

404804　Vangueria euonymoides Schweinf. ex Hiern = Rytigynia lecomtei Robyns ■☆

404805　Vangueria floribunda Robyns = Vangueria madagascariensis J. F. Gmel. ■☆

404806　Vangueria fyffei Robyns = Vangueria volkensii K. Schum. var. fyffei（Robyns）Verdc. ■☆

404807　Vangueria glabra K. Schum. = Rytigynia celastroides（Baill.）Verdc. ●☆

404808　Vangueria glabrata K. Schum.；光滑瓦氏茜■☆

404809　Vangueria glabrifolia De Wild. = Rytigynia glabrifolia（De Wild.）Robyns ■☆

404810　Vangueria gracilipetiolata De Wild. = Rytigynia gracilipetiolata（De Wild.）Robyns ■☆

404811　Vangueria infausta Burch.；逊瓦氏茜（逊万格茜）■☆

404812　Vangueria infausta Burch. subsp. rotundata（Robyns）Verdc.；圆形瓦氏茜■☆

404813　Vangueria infausta Burch. var. campanulata（Robyns）Verdc.；风铃草状瓦氏茜■☆

404814　Vangueria infausta Burch. var. rotundata（Robyns）Verdc. = Vangueria infausta Burch. subsp. rotundata（Robyns）Verdc. ■☆

404815　Vangueria ituriensis De Wild. = Rytigynia umbellulata（Hiern）Robyns ■☆

404816　Vangueria junodii Schinz = Rytigynia umbellulata（Hiern）Robyns ■☆

404817　Vangueria kaessneri S. Moore = Tapiphyllum cinerascens（Hiern）Robyns var. inaequale（Robyns）Verdc. ■☆

404818　Vangueria katangensis K. Schum. = Fadogia elskensii De Wild. ●☆

404819　Vangueria kerstingii Robyns；克斯廷瓦氏茜■☆

404820　Vangueria kiwuensis K. Krause = Rytigynia kiwuensis（K. Krause）Robyns ■☆

404821　Vangueria lasiantha（Sond.）Sond. = Lagynias lasiantha（Sond.）Bullock ■☆

404822　Vangueria lasioclados K. Schum. = Vangueria proschii Briq. ■☆

404823　Vangueria lateritia Dinter = Vangueriopsis lanciflora（Hiern）Robyns ●☆

404824　Vangueria latifolia（Sond.）Sond. = Pachystigma latifolium Sond. ●☆

404825　Vangueria laurentii De Wild. = Rytigynia laurentii（De Wild.）

Robyns ■☆

404826　Vangueria leonensis K. Schum. = Rytigynia leonensis（K. Schum.）Robyns ■☆

404827　Vangueria linearisepala K. Schum. = Vangueria volkensii K. Schum. ■☆

404828　Vangueria longicalyx Robyns = Vangueria apiculata K. Schum. ■☆

404829　Vangueria longiflora Hutch. = Hutchinsonia barbata Robyns ●☆

404830　Vangueria longisepala K. Krause = Rytigynia decussata（K. Schum.）Robyns ■☆

404831　Vangueria loranthifolia K. Schum. = Pachystigma loranthifolium（K. Schum.）Verdc. ●☆

404832　Vangueria macrocalyx Sond. = Pachystigma macrocalyx（Sond.）Robyns ●☆

404833　Vangueria madagascariensis J. F. Gmel.；马岛瓦氏茜；Voa Vanga，Voavanga ■☆

404834　Vangueria madagascariensis J. F. Gmel. var. abyssinica（A. Rich.）Puff；阿比西尼亚瓦氏茜■☆

404835　Vangueria membranacea Hiern = Rytigynia membranacea（Hiern）Robyns ■☆

404836　Vangueria microphylla K. Schum. = Rytigynia celastroides（Baill.）Verdc. ●☆

404837　Vangueria monantha K. Schum. = Rytigynia monantha（K. Schum.）Robyns ■☆

404838　Vangueria munjiro S. Moore = Vangueria infausta Burch. ■☆

404839　Vangueria nana K. Schum. = Pachystigma pygmaeum（Schltr.）Robyns ●☆

404840　Vangueria neglecta K. Schum.；疏忽瓦氏茜（疏忽万格茜）■☆

404841　Vangueria neglecta K. Schum. = Rytigynia uhligii（K. Schum. et K. Krause）Verdc. ■☆

404842　Vangueria neglecta K. Schum. var. puberula ? = Rytigynia schumannii Robyns var. puberula（K. Schum.）Robyns ■☆

404843　Vangueria nigerica S. Moore = Rytigynia nigerica（S. Moore）Robyns ■☆

404844　Vangueria nigrescens Scott-Elliot ex Oliv. = Cuviera nigrescens（Scott-Elliot ex Oliv.）Wernham ■☆

404845　Vangueria nodulosa K. Schum. = Rytigynia nodulosa（K. Schum.）Robyns ●☆

404846　Vangueria oblanceolata Wernham = Pauridiantha liebrechtsiana（De Wild. et T. Durand）Ntore et Dessein ●☆

404847　Vangueria obtusifolia K. Schum. = Tapiphyllum obtusifolium（K. Schum.）Robyns ■☆

404848　Vangueria oligacantha K. Schum. = Rytigynia celastroides（Baill.）Verdc. ●☆

404849　Vangueria oxyantha K. Schum. = Rytigynia leonensis（K. Schum.）Robyns ■☆

404850　Vangueria parvifolia Sond.；小叶瓦氏茜■☆

404851　Vangueria pauciflora Schweinf. ex Hiern = Rytigynia pauciflora（Schweinf. ex Hiern）Robyns ■☆

404852　Vangueria praecox Verdc.；早瓦氏茜■☆

404853　Vangueria proschii Briq.；普罗施瓦氏茜■☆

404854　Vangueria pygmaea Schltr. = Pachystigma pygmaeum（Schltr.）Robyns ●☆

404855　Vangueria randii S. Moore；兰德瓦氏茜■☆

404856　Vangueria randii S. Moore subsp. acuminata Verdc.；渐尖瓦氏茜■☆

404857　Vangueria randii S. Moore subsp. chartacea（Robyns）Verdc.；纸质瓦氏茜■☆

404858　Vangueria reygaertii De Wild. = Vangueriella orthacantha（Mildbr.）Bridson et Verdc. ●☆

404859　Vangueria rhodesiana S. Moore = Pachystigma pygmaeum（Schltr.）Robyns ●☆

404860　Vangueria robynsii Tennant = Vangueria madagascariensis J. F. Gmel. ■☆

404861　Vangueria rotundata Robyns = Vangueria infausta Burch. subsp. rotundata（Robyns）Verdc. ■☆

404862　Vangueria rubiginosa K. Schum. = Rytigynia rubiginosa（K. Schum.）Robyns ■☆

404863　Vangueria rupicola Robyns = Vangueria proschii Briq. ■☆

404864　Vangueria ruwenzoriensis De Wild. = Rytigynia ruwenzoriensis（De Wild.）Robyns ■☆

404865　Vangueria ruwenzoriensis De Wild. var. breviflora？ = Rytigynia ruwenzoriensis（De Wild.）Robyns ■☆

404866　Vangueria sapini De Wild. = Vangueriella sapini（De Wild.）Verdc. ●☆

404867　Vangueria senegalensis（Blume）Hiern = Rytigynia senegalensis Blume ●■☆

404868　Vangueria setosa Conrath = Pachystigma pygmaeum（Schltr.）Robyns ●☆

404869　Vangueria soutpansbergensis N. Hahn;索特潘瓦氏茜■☆

404870　Vangueria sparsifolia S. Moore = Rytigynia umbellulata（Hiern）Robyns ■☆

404871　Vangueria spinosa Roxb.;多刺瓦氏茜（多刺万格茜）■☆

404872　Vangueria spinosa Roxb. var. mollis？ = Meyna pubescens Robyns ●☆

404873　Vangueria squamata De Wild. = Rytigynia squamata（De Wild.）Robyns ■☆

404874　Vangueria stenophylla K. Krause = Pygmaeothamnus zeyheri（Sond.）Robyns ●☆

404875　Vangueria subbiflora Mildbr. = Rytigynia subbiflora（Mildbr.）Robyns ■☆

404876　Vangueria tetraphylla Schweinf. ex Hiern = Meyna tetraphylla（Schweinf. ex Hiern）Robyns ●☆

404877　Vangueria tomentosa Hochst. = Vangueria infausta Burch. ■☆

404878　Vangueria tomentosa K. Schum. = Pachystigma schumannianum（Robyns）Bridson et Verdc. subsp. mucronulatum（Robyns）Bridson et Verdc. ●☆

404879　Vangueria tristis K. Schum. = Fadogia cienkowskii Schweinf. ●☆

404880　Vangueria uhligii K. Schum. et K. Krause = Rytigynia uhligii（K. Schum. et K. Krause）Verdc. ■☆

404881　Vangueria umbellulata Hiern = Rytigynia umbellulata（Hiern）Robyns ■☆

404882　Vangueria velutina Hiern = Tapiphyllum velutinum（Hiern）Robyns ■☆

404883　Vangueria velutina Hiern var. laevior K. Schum. = Tapiphyllum cinerascens（Hiern）Robyns var. laevius（K. Schum.）Verdc. ■☆

404884　Vangueria velutina Hook. = Vangueria infausta Burch. ■☆

404885　Vangueria venosa（Hochst.）Sond. = Pachystigma venosum Hochst. ●☆

404886　Vangueria venosa Robyns = Vangueria madagascariensis J. F. Gmel. ■☆

404887　Vangueria verdickii K. Schum. = Fadogia schumanniana Robyns ●☆

404888　Vangueria verruculosa K. Krause = Rytigynia verruculosa（K. Krause）Robyns ■☆

404889　Vangueria volkensii K. Schum.;福尔瓦氏茜■☆

404890　Vangueria volkensii K. Schum. var. fyffei（Robyns）Verdc.;法伊夫瓦氏茜■☆

404891　Vangueria zeyheri（Sond.）Sond. = Pygmaeothamnus zeyheri（Sond.）Robyns ●☆

404892　Vangueriella Verdc.（1987）;小瓦氏茜属●☆

404893　Vangueriella campylacantha（Mildbr.）Verdc.;弯刺小瓦氏茜●☆

404894　Vangueriella chlorantha（K. Schum.）Verdc.;绿花小瓦氏茜●☆

404895　Vangueriella discolor（Benth.）Verdc.;异色小瓦氏茜●☆

404896　Vangueriella georgesii Verdc.;乔治小瓦氏茜●☆

404897　Vangueriella glabrescens（Robyns）Verdc.;渐光小瓦氏茜●☆

404898　Vangueriella laxiflora（K. Schum.）Verdc.;疏花小瓦氏茜●☆

404899　Vangueriella letestui Verdc.;莱泰斯图瓦氏茜●☆

404900　Vangueriella nigerica（Robyns）Verdc.;尼日利亚小瓦氏茜●☆

404901　Vangueriella nigricans（Robyns）Verdc.;变黑小瓦氏茜●☆

404902　Vangueriella olacifolia（Robyns）Verdc.;铁青树小瓦氏茜●☆

404903　Vangueriella orthacantha（Mildbr.）Bridson et Verdc.;直刺小瓦氏茜●☆

404904　Vangueriella rhamnoides（Hiern）Verdc.;鼠李小瓦氏茜●☆

404905　Vangueriella rufa（Robyns）Verdc.;浅红小瓦氏茜●☆

404906　Vangueriella sapini（De Wild.）Verdc.;萨潘小瓦氏茜●☆

404907　Vangueriella soyauxii（K. Schum.）Verdc.;索亚小瓦氏茜●☆

404908　Vangueriella spinosa（Schumach. et Thonn.）Verdc.;具刺小瓦氏茜●☆

404909　Vangueriella spinosa（Schumach. et Thonn.）Verdc. = Canthium stenosepalum Lantz ●☆

404910　Vangueriella spinosa（Schumach. et Thonn.）Verdc. = Canthium thonningii Benth. ●☆

404911　Vangueriella vanguerioides（Hiern）Verdc.;普通小瓦氏茜●☆

404912　Vangueriella zenkeri Verdc.;大果小瓦氏茜●☆

404913　Vangueriopsis Robyns = Vangueriopsis Robyns ex R. D. Good ●■☆

404914　Vangueriopsis Robyns ex R. D. Good（1928）;拟瓦氏茜属●■☆

404915　Vangueriopsis calycophila（K. Schum.）Robyns = Vangueriella laxiflora（K. Schum.）Verdc. ●☆

404916　Vangueriopsis castaneae Robyns = Multidentia castaneae（Robyns）Bridson et Verdc. ●☆

404917　Vangueriopsis chlorantha（K. Schum.）Robyns = Vangueriella chlorantha（K. Schum.）Verdc. ●☆

404918　Vangueriopsis coriacea Robyns = Cuviera acutiflora DC. ■☆

404919　Vangueriopsis decidua（K. Schum.）Robyns = Vangueriella soyauxii（K. Schum.）Verdc. ●☆

404920　Vangueriopsis discolor（Benth.）Robyns = Vangueriella discolor（Benth.）Verdc. ●☆

404921　Vangueriopsis glabrescens Robyns = Vangueriella glabrescens（Robyns）Verdc. ●☆

404922　Vangueriopsis gossweileri Robyns;戈斯瓦氏茜●☆

404923　Vangueriopsis lanceolata Robyns = Vangueriella nigerica（Robyns）Verdc. ●☆

404924　Vangueriopsis lanciflora（Hiern）Robyns;剑叶拟瓦氏茜;Livelong,Never-die,Wild Medlar ●☆

404925　Vangueriopsis lancifolia Greenway = Vangueriopsis lanciflora（Hiern）Robyns ●☆

404926　Vangueriopsis leucodermis（K. Krause）Hutch. et Dalziel = Vangueriella spinosa（Schumach. et Thonn.）Verdc. ●☆

404927　Vangueriopsis longiflora Verdc.;长叶拟瓦氏茜●☆

404928　Vangueriopsis membranacea Robyns = Vangueriella spinosa（Schumach. et Thonn.）Verdc. ●☆

404929　Vangueriopsis nigerica Robyns = Vangueriella nigerica（Robyns）

Verdc. ●☆

404930　Vangueriopsis nigricans Robyns = Vangueriella nigricans（Robyns）Verdc. ●☆

404931　Vangueriopsis olacifolia Robyns = Vangueriella olacifolia（Robyns）Verdc. ●☆

404932　Vangueriopsis rubiginosa Robyns；锈红拟瓦氏茜●☆

404933　Vangueriopsis rufa Robyns = Vangueriella rufa（Robyns）Verdc. ●☆

404934　Vangueriopsis setosa Robyns = Vangueriella campylacantha（Mildbr.）Verdc. ●☆

404935　Vangueriopsis sillitoei Bullock = Multidentia dichrophylla（Mildbr.）Bridson ■☆

404936　Vangueriopsis spinosa（Schumach. et Thonn.）Hepper = Vangueriella spinosa（Schumach. et Thonn.）Verdc. ●☆

404937　Vangueriopsis subulata Robyns = Vangueriella campylacantha（Mildbr.）Verdc. ●☆

404938　Vangueriopsis vanguerioides（Hiern）Robyns = Vangueriella vanguerioides（Hiern）Verdc. ●☆

404939　Vangueriopsis violacea Robyns = Vangueriella spinosa（Schumach. et Thonn.）Verdc. ●☆

404940　Vanhallia Schult. et Schult. f. = Munnickia Blume ex Rchb. ●

404941　Vanhallia Schult. f. = Apama Lam. ●

404942　Vanheerdea L. Bolus ex H. E. K. Hartmann（1992）；胧玉属（黄龙幻属）●☆

404943　Vanheerdea angusta（L. Bolus）L. Bolus = Vanheerdea roodiae（N. E. Br.）L. Bolus ex H. E. K. Hartmann ●☆

404944　Vanheerdea divergens（L. Bolus）L. Bolus = Vanheerdea roodiae（N. E. Br.）L. Bolus ex H. E. K. Hartmann ●☆

404945　Vanheerdea primosii（L. Bolus）L. Bolus ex H. E. K. Hartmann；翠帐●☆

404946　Vanheerdea roodiae（N. E. Br.）L. Bolus ex H. E. K. Hartmann；胧玉●☆

404947　Vanheerdia L. Bolus = Vanheerdea L. Bolus ex H. E. K. Hartmann ●☆

404948　Vanheerdia divergens（L. Bolus）L. Bolus；秘色玉●☆

404949　Vanheerdia primosii（L. Bolus）L. Bolus = Vanheerdea primosii（L. Bolus）L. Bolus ex H. E. K. Hartmann ●☆

404950　Vanheerdia roodiae（N. E. Br.）L. Bolus = Vanheerdea roodiae（N. E. Br.）L. Bolus ex H. E. K. Hartmann ●☆

404951　Vanhouttea Lem.（1845）；豪特苣苔属●☆

404952　Vanhouttea calcarata Lem.；豪特苣苔●☆

404953　Vanhouttea mollis Fritsch；软豪特苣苔●☆

404954　Vania F. K. Mey. = Thlaspi L. ■

404955　Vaniera J. St. -Hil. = Vanieria Lour.（废弃属名）●

404956　Vanieria Lour.（废弃属名）= Cudrania Trécul（保留属名）●

404957　Vanieria Lour.（废弃属名）= Maclura Nutt.（保留属名）●

404958　Vanieria Montrouz. = Hibbertia Andréws ●☆

404959　Vanieria Montrouz. = Trisema Hook. f. ●☆

404960　Vanieria bodinieri（H. Lév.）Chun = Capparis cantoniensis Lour. ●

404961　Vanieria cochinchinensis Lour. = Maclura cochinchinensis（Lour.）Corner ●

404962　Vanieria cochinchinensis Lour. var. gerontogea（Siebold et Zucc.）Nakai = Maclura cochinchinensis（Lour.）Corner ●

404963　Vanieria fruticosa（Roxb.）Chun = Maclura fruticosa（Roxb.）Corner ●

404964　Vanieria fruticosa（Wight）Chun = Cudrania fruticosa（Roxb.）Wight ex Kurz ●

404965　Vanieria pubescens（Trécul）Chun = Maclura pubescens（Trécul）Z. K. Zhou et M. G. Gilbert ●

404966　Vanieria tricuspidata（Carrière）Hu = Cudrania tricuspidata（Carrière）Bureau ex Lavalleé ●

404967　Vanieria tricuspidata（Carrière）Hu = Maclura tricuspidata Carrière ●

404968　Vanieria triloba（Hance）Satake = Cudrania tricuspidata（Carrière）Bureau ex Lavaleé ●

404969　Vanieria triloba（Hance）Satake = Maclura tricuspidata Carrière ●

404970　Vanilla Mill. = Vanilla Plum. ex Mill. ■

404971　Vanilla Plum. ex Mill.（1754）；香荚兰属（凡尼属，香草属，梵尼兰属，香果兰属，香子兰属）；Vanilla ■

404972　Vanilla acuminata Rolfe；尖香荚兰■☆

404973　Vanilla africana Lindl.；非洲香荚兰■☆

404974　Vanilla africana Lindl. var. gilletii（De Wild.）Portères = Vanilla ramosa Rolfe ■☆

404975　Vanilla africana Lindl. var. laurentiana（De Wild.）Portères = Vanilla ramosa Rolfe ■☆

404976　Vanilla albida Blume = Vanilla somai Hayata ■

404977　Vanilla anaromatica Griseb. = Vanilla mexicana Mill. ■☆

404978　Vanilla annamica Gagnep.；南方香荚兰（西南香果兰，香果兰，香子兰，越南香荚兰）；Annam Vanilla ■

404979　Vanilla aphylla（Roxb.）Blume；无叶香荚兰（无叶香果兰）；Leafless Vanilla ■☆

404980　Vanilla aromatica Sw. = Vanilla mexicana Mill. ■☆

404981　Vanilla articulata Northr. = Vanilla barbellata Rchb. f. ■☆

404982　Vanilla barbellata Rchb. f.；短毛香荚兰（短毛香果兰）；Shorthair Vanilla，Worm Vine ■☆

404983　Vanilla beauchenei A. Chev. = Vanilla ramosa Rolfe ■☆

404984　Vanilla chalotii Finet；沙洛香荚兰■☆

404985　Vanilla coursii H. Perrier；库尔斯香荚兰■☆

404986　Vanilla crenulata Rolfe；细圆齿香荚兰■☆

404987　Vanilla cucullata Kraenzl. ex J. Braun et K. Schum.；僧帽状香荚兰■☆

404988　Vanilla decaryana H. Perrier；德卡里香荚兰■☆

404989　Vanilla dilloniana Correll；德隆氏香荚兰（德隆氏香果兰）；Dillon Vanilla，Leafless Vanilla，Mrs. Lott's Vanilla ■☆

404990　Vanilla fragrans（Salisb.）Ames；香荚兰（香果兰）；Common Vanilla ■

404991　Vanilla fragrans（Salisb.）Ames = Vanilla mexicana Mill. ■☆

404992　Vanilla fragrans（Salisb.）Ames = Vanilla planifolia Andréws ■

404993　Vanilla fragrans Ames = Vanilla planifolia Andréws ■

404994　Vanilla francoisii H. Perrier；法兰西斯香荚兰■☆

404995　Vanilla grandifolia Lindl.；大叶香荚兰■☆

404996　Vanilla grandifolia Lindl. var. lujae（De Wild.）Geerinck；卢亚香荚兰■☆

404997　Vanilla griffithii Rchb. f. = Vanilla somai Hayata ■

404998　Vanilla griffithii Rchb. f. var. formosana Ito = Vanilla somai Hayata ■

404999　Vanilla griffithii Rchb. f. var. ronoensis（Hayata）S. S. Ying = Vanilla somai Hayata ■

405000　Vanilla hallei Szlach. et Olszewski；哈勒香荚兰■☆

405001　Vanilla heterolopha Summerh.；异冠香荚兰■☆

405002　Vanilla humblotii Rchb. f.；洪布氏香荚兰（洪布氏香果兰）；Humblot Vanilla ■☆

405003　Vanilla imperialis Kraenzl.；壮丽香荚兰■☆

405004　Vanilla imperialis Kraenzl. var. congolensis De Wild. = Vanilla imperialis Kraenzl. ■☆

405005　Vanilla inodora Schiede = Vanilla mexicana Mill. ■☆

405006　Vanilla laurentiana De Wild. = Vanilla ramosa Rolfe ■☆

405007　Vanilla laurentiana De Wild. var. gilletii ？ = Vanilla ramosa Rolfe ■☆

405008　Vanilla lujae De Wild. = Vanilla grandifolia Lindl. var. lujae（De Wild.）Geerinck ■☆

405009　Vanilla madagascariensis Rolfe;马岛香荚兰■☆

405010　Vanilla mexicana Mill. ;墨西哥香荚兰（香荚兰）;Mexican Vanilla ■☆

405011　Vanilla nigerica Rendle;尼日利亚香荚兰■☆

405012　Vanilla ochyrae Szlach. et Olszewski;奥吉拉香荚兰■☆

405013　Vanilla ovalifolia Rolfe = Vanilla ramosa Rolfe ■☆

405014　Vanilla perrieri Schltr. ;佩里耶香荚兰■☆

405015　Vanilla phaeantha Rchb. f. ;褐花香果兰;Brownflower Vanilla, Leafy Vanilla,Oblong-leaved Vanilla ■☆

405016　Vanilla phalaenopsis Rchb. f. ex Van Houtte;蝴蝶香果兰; Butterfly Vanilla ■☆

405017　Vanilla planifolia Andréws;扁叶香果兰（上树蜈蚣,香草,香草兰,香荚兰,香子兰）;Commercial Vanilla, Common Vanilla, Flatleaf Vanilla,Mexican Vanilla,Vanilla,Vanilla Bran,Vanilla Orchid ■

405018　Vanilla planifolia Andréws = Vanilla mexicana Mill. ■☆

405019　Vanilla planifolia Jacks. = Vanilla planifolia Andréws ■

405020　Vanilla polylepis Summerh. ;多鳞香荚兰■☆

405021　Vanilla pompona Schiede;蓬蓬纳香草;West Indian Vanilla ■☆

405022　Vanilla ramosa Rolfe;多枝香荚兰■☆

405023　Vanilla ronoensis Hayata = Vanilla somai Hayata ■

405024　Vanilla roscheri Rchb. f. ;罗氏香荚兰■☆

405025　Vanilla seretii De Wild. ;赛雷香荚兰■☆

405026　Vanilla shenzhenica Z. J. Liu et S. C. Chen;深圳香荚兰■

405027　Vanilla siamensis Rolfe ex Downie;大香荚兰（大香果兰,石蚕,香果兰）;Large Vanilla ■

405028　Vanilla somai Hayata;台湾香荚兰（姬氏凡尼兰,姬氏梵尼兰,台湾凡尼兰,台湾梵尼兰）;Griffith Vanilla,Taiwan Vanilla ■

405029　Vanilla somai Hayata = Vanilla albida Blume ■

405030　Vanilla tahitensis J. W. Moore;塔西提香荚兰■☆

405031　Vanilla taiwaniana S. S. Ying;宝岛香荚兰■

405032　Vanilla tisserantii Portères;蒂斯朗特香荚兰■☆

405033　Vanilla vanilla（L.）Britton = Vanilla mexicana Mill. ■☆

405034　Vanilla zanzibarica Rolfe = Vanilla ramosa Rolfe ■☆

405035　Vanillaceae Lindl. ;香荚兰科■

405036　Vanillaceae Lindl. = Orchidaceae Juss.（保留科名）■

405037　Vanillophorum Neck. = Vanilla Plum. ex Mill. ■

405038　Vanillosma Spach = Piptocarpha R. Br. ●☆

405039　Vanillosma Spach(1847)（'Vannillosma'）;肖香荚兰属■☆

405040　Vanillosma acuminata Mart. ex Baker;尖肖香荚兰■☆

405041　Vanillosma albida Mart. ex Baker;白肖香荚兰■☆

405042　Vanillosma bicolor Mart. ex Baker;二色肖香荚兰■☆

405043　Vanillosma multiflora Mart. ex Baker;多花肖香荚兰■☆

405044　Vanillosma pyrifolia Mart. ex Sch. Bip. ;梨叶肖香荚兰■☆

405045　Vanillosmopsis Sch. Bip.（1861）;香兰菊属■☆

405046　Vanillosmopsis Sch. Bip. = Eremanthus Less. ●☆

405047　Vanillosmopsis erythropappa Sch. Bip. ;香兰菊（红冠毛香兰菊,香子兰菊）■☆

405048　Vaniotia H. Lév. = Petrocosmea Oliv. ■

405049　Vaniotia martinii H. Lév. = Petrocosmea martinii（H. Lév.）H. Lév. ■

405050　Vanoverberghia Merr.（1912）;法氏姜属■

405051　Vanoverberghia sasakiana Funk. et H. Ohashi;兰屿法氏姜■

405052　Van-royena Aubrév. = Pouteria Aubl. ●

405053　Vanroyenella Novelo et C. T. Philbrick(1993);小桃榄属●☆

405054　Vantanea Aubl.（1775）;文塔木属●☆

405055　Vantanea barbourii Standl. ;文塔木●☆

405056　Vanwykia Wiens(1979);万氏寄生属●☆

405057　Vanwykia remota（Baker et Sprague）Wiens;万氏寄生●☆

405058　Vanwykia rubella Polhill et Wiens;非洲万氏寄生●☆

405059　Vanzijlia L. Bolus(1927);白莲玉属■☆

405060　Vanzijlia angustipetala（L. Bolus）N. E. Br. = Vanzijlia annulata（A. Berger）L. Bolus ■☆

405061　Vanzijlia annulata（A. Berger）L. Bolus;白莲玉■☆

405062　Vanzijlia rostella L. Bolus = Cephalophyllum rostellum（L. Bolus）H. E. K. Hartmann ■☆

405063　Varangevillea Willis = Rhodocolea Baill. + Vitex L. ●

405064　Varangevillea Willis = Varengevillea Baill. ●

405065　Varasia Phil. = Gentiana L. ■

405066　Vareca Gaertn. = Casearia Jacq. ●

405067　Vareca Roxb. = Rinorea Aubl.（保留属名）●

405068　Varengevillea Baill. = Rhodocolea Baill. + Vitex L. ●

405069　Varengevillea hispidissima（Seem.）Baill. = Vitex congesta Oliv. ●☆

405070　Varennea DC. = Eysenhardtia Kunth(保留属名)●☆

405071　Varennea DC. = Viborquia Ortega(废弃属名)●☆

405072　Vargasia Bert. ex Spreng. = Thouinia Poit.（保留属名）●☆

405073　Vargasia DC. = Galinsoga Ruiz et Pav. ■●

405074　Vargasia DC. = Vasargia Stoud. ■●

405075　Vargasia Ernst = Caracasia Szyszyl. ●☆

405076　Vargasiella C. Schweinf.（1952）;瓦尔兰属■☆

405077　Vargasiella peruviana C. Schweinf. ;瓦尔兰■☆

405078　Varilla A. Gray(1849);棒菊属■☆

405079　Varilla texana A. Gray;得州棒菊●☆

405080　Varinga Raf. = Ficus L. ●

405081　Variphylis Thouars = Bulbophyllum Thouars(保留属名)■

405082　Varnera L. = Gardenia Ellis(保留属名)●

405083　Varnera augusta L. = Gardenia augusta（L.）Merr. ●

405084　Varonthe Juss. ex Rchb. = Physena Noronha ex Thouars ●☆

405085　Varronia P. Browne = Cordia L.（保留属名）●

405086　Varronia P. Browne(1756);肖破布木属●☆

405087　Varronia abyssinica（R. Br.）DC. = Cordia africana Lam. ●☆

405088　Varronia curassavica Jacq. ;肖破布木●☆

405089　Varronia sinensis Lour. = Cordia dichotoma G. Forst. ●

405090　Varroniopsis Friesen = Cordia L.（保留属名）●

405091　Varroniopsis Friesen = Varronia P. Browne ●☆

405092　Vartheimia Benth. et Hook. f. = Varthemia DC. ●■☆

405093　Varthemia DC.（1836）;分尾菊属●■☆

405094　Varthemia arabica Boiss. = Pluchea arabica（Boiss.）Qaiser et Lack ■●☆

405095　Varthemia candicans（Delile）Boiss. = Jasonia candicans（Delile）Botsch. ■☆

405096　Varthemia hesperia（Maire et Wilczek）Dobignard;西方分尾菊■☆

405097　Varthemia kotschyi Sch. Bip. ex Schweinf. et Asch. = Pentanema indicum（L.）Y. Ling ■

405098　Varthemia montana（Vahl）Boiss. = Jasonia montana（Delile）Botsch. ■☆

405099　Varthemia persica DC. ;分尾菊■☆

405100　Varthemia sericea（Batt. et Trab.）Diels;绢毛分尾菊■☆

405101　Varthemia sericea（Batt. et Trab.）Diels subsp. virescens Maire;浅绿分尾菊■☆

405102　Vasargia Stoud. = Galinsoga Ruiz et Pav. ■●

405103　Vasargia Stoud. = Vargasia DC. ■●

405104　Vascoa DC. (1824) = Mundia Kunth ■☆

405105　Vascoa DC. (1824) = Nylandtia Dumort. ■☆

405106　Vascoa DC. (1825) = Rafnia Thunb. ■☆

405107　Vascoa acuminata E. Mey. = Rafnia acuminata (E. Mey.) G. J. Campb. et B. -E. van Wyk ■☆

405108　Vascoa amplexicaulis (L.) DC. = Rafnia amplexicaulis (L.) Thunb. ■☆

405109　Vascoa perfoliata (Thunb.) DC. = Rafnia acuminata (E. Mey.) G. J. Campb. et B. -E. van Wyk ■☆

405110　Vascoa perfoliata (Thunb.) DC. var. acuminata (E. Mey.) Walp. = Rafnia acuminata (E. Mey.) G. J. Campb. et B. -E. van Wyk ■☆

405111　Vasconcellea A. St. -Hil. (1837);单干木瓜属●☆

405112　Vasconcellea A. St. -Hil. = Carica L. ●

405113　Vasconcellea cauliflora A. DC.;茎花单干木瓜●☆

405114　Vasconcellea chilensis Planch.;智利单干木瓜●☆

405115　Vasconcellea crassipetala (V. M. Badillo) V. M. Badillo;厚瓣单干木瓜●☆

405116　Vasconcellea glandulosa A. DC.;多腺单干木瓜●☆

405117　Vasconcellea heterophylla A. DC.;异叶单干木瓜●☆

405118　Vasconcellea lanceolata A. DC.;披针叶单干木瓜●☆

405119　Vasconcellea longiflora (V. M. Badillo) V. M. Badillo;长花单干木瓜●☆

405120　Vasconcellea microcarpa (Jacq.) A. DC.;小果单干木瓜●☆

405121　Vasconcellea parviflora A. DC.;小花单干木瓜●☆

405122　Vasconcellea quercifolia A. St. -Hil.;单干木瓜●☆

405123　Vasconcellea sphaerocarpa (García-Barr. et Hern. Cam.) V. M. Badillo;球果单干木瓜●☆

405124　Vasconcellia Mart. = Arrabidaea DC. ●☆

405125　Vasconcellosia Caruel = Carica L. ●

405126　Vasconella Regel = Vasconcellosia Caruel ●

405127　Vaselia Tiegh. = Elvasia DC. ●☆

405128　Vaseya Thurb. = Muhlenbergia Schreb. ■

405129　Vaseyanthus Cogn. (1891);瓦齐花属■☆

405130　Vaseyanthus rosei Cogn.;瓦齐花■☆

405131　Vaseyochloa Hitchc. (1933);多脉草属■☆

405132　Vaseyochloa multinervosa (Vasey) Hitchc.;多脉草■☆

405133　Vasivaea Baill. (1872);群蕊椴属●☆

405134　Vasivaea alchorneoides Baill.;群蕊椴●☆

405135　Vasivaea podocarpa Kuhlm.;柄果群蕊椴●☆

405136　Vasovulaceae Dulac = Aquifoliaceae Bercht. et J. Presl(保留科名)●

405137　Vasquezia Phil. = Villanova Lag. (保留属名)■☆

405138　Vasqueziella Dodson(1982);巴斯兰属●☆

405139　Vasqueziella boliviana Dodson;巴斯兰●☆

405140　Vassilczenkoa Lincz. (1979);大苞补血草属■☆

405141　Vassilczenkoa Lincz. = Chaetolimon (Bunge) Lincz. ■☆

405142　Vassilczenkoa sogdiana (Lincz.) Lincz.;大苞补血草●☆

405143　Vassobia Rusby(1907);瓦索茄属●☆

405144　Vassobia breviflora (Sendtn.) Hunz.;短花瓦索茄●☆

405145　Vassobia fasciculata (Miers) Hunz.;簇生瓦索茄●☆

405146　Vatairea Aubl. (1775);瓦泰豆属●☆

405147　Vatairea guianensis Aubl.;圭亚那瓦泰豆●☆

405148　Vataireopsis Ducke = Vatairea Aubl. ●☆

405149　Vataireopsis Ducke(1932);拟瓦泰豆属(瓦泰里属)●☆

405150　Vataireopsis araroba (Aguiar) Ducke;拟瓦泰豆●☆

405151　Vataireopsis araroba (Aguiar) Ducke = Andira araroba Aguiar ●☆

405152　Vataireopsis speciosa Ducke;美丽拟瓦泰豆●☆

405153　Vateria L. (1753);瓦特木属(达玛脂树属,瓦蒂香属,瓦泰特里亚属,印度胶脂树属);Vateria ●☆

405154　Vateria indica L.;印度瓦特木(白达玛脂树,达玛脂树);Dhupa Fat,Malabar Tallow,Piney Tree,Piney Varnish ●☆

405155　Vateria lanceifolia Roxb. = Vatica lanceifolia (Roxb.) Blume ●

405156　Vateriopsis F. Heim(1892);类瓦特木属(拟瓦蒂香属,拟印度胶脂树属)●☆

405157　Vateriopsis seychellarum Heim;类类瓦特木●☆

405158　Vatica L. (1771);青梅属;Vatica ●

405159　Vatica africana (A. DC.) Welw. = Monotes africanus A. DC. ●☆

405160　Vatica africana (A. DC.) Welw. var. hypoleuca Welw. = Monotes hypoleucus (Oliv.) Gilg ●☆

405161　Vatica apteranthera Blanco = Vatica mangachapoi Blanco ●◇

405162　Vatica astrotricha Hance = Vatica mangachapoi Blanco ●◇

405163　Vatica bancana Scheff.;班卡青梅(班卡油楠)●☆

405164　Vatica chinensis L.;东印青梅;E. India Vatica ●☆

405165　Vatica cordata Hu = Porana sinensis Hemsl. ●

405166　Vatica cordata Hu = Tridynamia sinensis (Hemsl.) Staples ●

405167　Vatica fleuryana Tardieu;拟版纳青梅●

405168　Vatica guangxiensis X. L. Mo;广西青梅;Guangxi Vatica, Kwangsi Vatica ●◇

405169　Vatica guangxiensis X. L. Mo subsp. xishuangbannaensis (G. D. Tao et J. H. Zhang) Y. K. Yang et J. K. Wu = Vatica guangxiensis X. L. Mo ●◇

405170　Vatica hainanensis H. T. Chang et L. C. Wang = Vatica mangachapoi Blanco ●◇

405171　Vatica hainanensis Hung T. Chang et L. Z. Wang = Vatica mangachapoi Blanco ●◇

405172　Vatica hainanensis Hung T. Chang et L. Z. Wang var. glandipetala L. C. Wang = Vatica mangachapoi Blanco ●◇

405173　Vatica hainanensis Hung T. Chang et L. Z. Wang var. parvifolia H. T. Chang = Vatica mangachapoi Blanco ●◇

405174　Vatica hypoleuca Welw. = Monotes hypoleucus (Oliv.) Gilg ●☆

405175　Vatica indica ?;印度青梅;White Dammar ●☆

405176　Vatica katangensis De Wild. = Monotes katangensis (De Wild.) De Wild. ●☆

405177　Vatica lanceifolia (Roxb.) Blume;西藏青梅●

405178　Vatica mangachapoi Blanco;青梅(海梅,海南青皮,苦香,青楣,青皮,油楠);Mangachapo Vatica,Stellatehair Vatica,Vatica ●◇

405179　Vatica mangachapoi Blanco subsp. hainanensis (H. T. Chang et L. C. Wang) Y. K. Yang et J. K. Wu = Vatica mangachapoi Blanco ●◇

405180　Vatica mangachapoi Blanco var. glandipetala (L. C. Wang) Y. K. Yang et J. K. Wu = Vatica mangachapoi Blanco ●◇

405181　Vatica mangachapoi Blanco var. parvifolia (H. T. Chang) Y. K. Yang et J. K. Wu = Vatica mangachapoi Blanco ●◇

405182　Vatica pachyphylla Merr.;厚叶青梅(厚叶油楠);Thickleaf Vatica ●☆

405183　Vatica shingkeng Dunn = Hopea shingkeng (Dunn) Bor ●

405184　Vatica xishuangbannaensis G. D. Tao et J. H. Zhang;版纳青梅;Banna Vatica,Xishuangbanna Vatica ●◇

405185　Vatica xishuangbannaensis G. D. Tao et J. H. Zhang = Vatica guangxiensis X. L. Mo ●◇

405186　Vatkea Hildeb. et O. Hoffm. = Martynia L. ■

405187　Vatkea O. Hoffm. = Martynia L. ■

405188　Vatovaea Chiov. (1951);瓦托豆属■☆

405189　Vatovaea biloba Chiov. = Vatovaea pseudolablab (Harms) J. B.

Gillett ■☆

405190　Vatovaea pseudolablab（Harms）J. B. Gillett;瓦托豆■☆

405191　Vatricania Backeb.（1950）;金装龙属●☆

405192　Vatricania Backeb. = Espostoa Britton et Rose ●

405193　Vatricania guentheri（Kupper）Backeb.;金装龙●☆

405194　Vauanthes Haw. = Crassula L. ●■☆

405195　Vauanthes chloraeflora Haw. = Crassula dichotoma L. ●☆

405196　Vauanthes dichotoma（L.）Kuntze = Crassula dichotoma L. ●☆

405197　Vaughania S. Moore = Indigofera L. ●■

405198　Vaughania S. Moore（1920）;沃恩木蓝属●☆

405199　Vaughania cloiselii（Drake）Du Puy, Labat et Schrire;克卢塞尔沃恩木蓝●☆

405200　Vaughania depauperata（Drake）Du Puy, Labat et Schrire;萎缩沃恩木蓝●☆

405201　Vaughania dionaeifolia S. Moore;捕蝇草沃恩木蓝●☆

405202　Vaughania humbertiana（M. Pelt.）Du Puy, Labat et Schrire;亨伯特沃恩木蓝●☆

405203　Vaughania interrupta Du Puy, Labat et Schrire;间断沃恩木蓝●☆

405204　Vaughania longidentata Du Puy, Labat et Schrire;长齿沃恩木蓝●☆

405205　Vaughania mahafalensis Du Puy, Labat et Schrire;马哈法尔沃恩木蓝●☆

405206　Vaughania perrieri（R. Vig.）Du Puy, Labat et Schrire;佩里耶沃恩木蓝●☆

405207　Vaughania xerophila（R. Vig.）Du Puy, Labat et Schrire;旱生沃恩木蓝●☆

405208　Vaupelia Brand（保留属名）= Cystostemon Balf. f. ■☆

405209　Vaupelia barbata（Vaupel）Brand = Cystostemon barbatus（Vaupel）A. G. Mill. et Riedl ■☆

405210　Vaupelia heliocharis（S. Moore）Brand = Cystostemon heliocharis（S. Moore）A. G. Mill. et Riedl ■☆

405211　Vaupelia hispida（Baker et C. H. Wright）Brand = Cystostemon hispidus（Baker et C. H. Wright）A. G. Mill. et Riedl ■☆

405212　Vaupelia hispidissima S. Moore = Cystostemon hispidissimus（S. Moore）A. G. Mill. et Riedl ■☆

405213　Vaupelia macranthera（Gürke）Brand = Cystostemon macranthera（Gürke）A. G. Mill. et Riedl ■☆

405214　Vaupelia mechowii（Vaupel）Brand = Cystostemon mechowii（Vaupel）A. G. Mill. et Riedl ■☆

405215　Vaupelia medusa（Baker）Brand = Cystostemon medusa（Baker）A. G. Mill. et Riedl ■☆

405216　Vaupellia Griseb.（废弃属名）= Pentarhaphia Lindl. ■☆

405217　Vaupellia Griseb.（废弃属名）= Vaupelia Brand（保留属名）■☆

405218　Vaupesia R. E. Schult.（1955）;沃佩大戟属 ☆

405219　Vaupesia cataractarum R. E. Schult.;沃佩大戟 ☆

405220　Vauquelinia Corrêa ex Bonpl.（1807）;西方红木属●☆

405221　Vauquelinia Corrêa ex Humb. et Bonpl. = Vauquelinia Corrêa ex Bonpl. ●☆

405222　Vauquelinia Humb. et Bonpl. = Vauquelinia Corrêa ex Bonpl. ●☆

405223　Vauquelinia californica（Torr.）Sarg.;西方红木; Arizona Rosewood, California Rosewood, Desert Rose ●☆

405224　Vauquelinia corymbosa Bonpl. var. angustifolia（Rydb.）W. J. Hess et Henrickson;细叶西方红木; Chisos Rosewood, Guayal, Palo Prieto ●☆

405225　Vauquelinia corymbosa Bonpl. var. heterodon（I. M. Johnst.）W. J. Hess et Henrickson;异齿西方红木; Slimleaf Rosewood ●☆

405226　Vausagesia Baill. = Sauvagesia L. ●

405227　Vausagesia africana Baill. = Sauvagesia africana（Baill.）Bamps ●☆

405228　Vausagesia bellidifolia Engl. et Gilg = Sauvagesia africana（Baill.）Bamps ●☆

405229　Vauthiera A. Rich. = Cladium P. Browne ■

405230　Vavaea Benth.（1843）;瓦楝属●☆

405231　Vavaea amicorum Benth.;瓦楝●☆

405232　Vavanga Rohr = Vangueria Comm. ex Juss. ■☆

405233　Vavanga chinensis Rohr = Vangueria madagascariensis J. F. Gmel. ■☆

405234　Vavanga edulis Vahl = Vangueria madagascariensis J. F. Gmel. ■☆

405235　Vavara Benoist（1962）;马岛瓦爵床属 ☆

405236　Vavara breviflora Benoist;马岛瓦爵床 ☆

405237　Vavilovia Fed.（1939）;美丽豌豆属■☆

405238　Vavilovia formosa（Steven）Fed.;美丽豌豆■☆

405239　Vazquezia Phil. = Vasquezia Phil. ■☆

405240　Vazquezia Phil. = Villanova Lag.（保留属名）■☆

405241　Veatchia A. Gray = Pachycormus Coville ex Standl. ●☆

405242　Veatchia Kellogg = Hesperoscordum Lindl. ■☆

405243　Veatchia crystallina Kellogg = Triteleia hyacinthina（Lindl.）Greene ■☆

405244　Veconcibea（Müll. Arg.）Pax et K. Hoffm. = Conceveiba Aubl. ●☆

405245　Veconcibea Pax et K. Hoffm. = Conceveiba Aubl. ●☆

405246　Vedela Adans.（废弃属名）= Ardisia Sw.（保留属名）●■

405247　Veeresia Monach. et Moldenke = Reevesia Lindl. ●

405248　Vegaea Urb.（1913）;维加木属●☆

405249　Vegaea pungens Urb.;维加木●☆

405250　Vegelia Neck. = Weigela Thunb. ●

405251　Veillonia H. E. Moore.（1978）;银椰属（维罗尼亚椰属,银桐属）●☆

405252　Veillonia alba H. E. Moore;银椰●☆

405253　Veitchia H. Wendl.（1868）（保留属名）;圣诞椰属（斐济椰子属,斐济棕属,圣诞椰子属,维契棕属）; Veech Palm, Veitchia ●☆

405254　Veitchia Lindl.（废弃属名）= Picea A. Dietr. ●

405255　Veitchia Lindl.（废弃属名）= Veitchia H. Wendl.（保留属名）●☆

405256　Veitchia arecina Becc.;槟榔圣诞椰●☆

405257　Veitchia joannis Wendl.;乔氏圣诞椰（斐济椰子）●☆

405258　Veitchia merrillii（Becc.）Moore et H. Wendl.;圣诞椰（马尼拉棕,马尼拉棕榈,维契棕）; Christmas Palm, Manila Palm, Merrill Palm, Merrill Veitchia ●☆

405259　Velaea D. Dietr. = Arracacia Bancr. ■☆

405260　Velaea DC. = Tauschia Schltdl.（保留属名）■☆

405261　Velaga Adans. = Pentapetes L. ■●

405262　Velaga Adans. = Pterospermum Schreb.（保留属名）●

405263　Velarum（DC.）Rchb. = Sisymbrium L. ■

405264　Velarum Rchb. = Sisymbrium L. ■

405265　Velascoa Calderón et Rzed.（1997）;斗花亮籽属●☆

405266　Velascoa recondita G. Calderón et Rzed.;斗花亮籽●☆

405267　Velasquezia Pritz. = Triplaris Loefl. ex L. ●

405268　Velasquezia Pritz. = Vellasquezia Bertol. ●

405269　Velezia L.（1753）;硬石竹属; Velezia ■☆

405270　Velezia lagrangei Coss. = Silene lagrangei（Coss.）Greuter et Burdet ■☆

405271　Velezia rigida L.;硬石竹; Velezia ■☆

405272　Velezia rigida L. var. glabrata Regel = Velezia rigida L. ■☆

405273　Velezia rigida L. var. sessiliflora Williams = Velezia rigida L. ■☆

405274　Velheimia Scop. = Veltheimia Gled. ■☆

405275　Vella DC. = Vella L. ●☆

405276　Vella L. (1753)；瓦拉木属（水堇菜属）●☆

405277　Vella L. = Vella DC. + Carrichtera Adans. (废弃属名)●☆

405278　Vella annua L. = Carrichtera annua (L.) DC. ■☆

405279　Vella glabrescens Coss. = Vella pseudocytisus L. subsp. glabrata Greuter ●☆

405280　Vella integrifolia Salisb. = Vella pseudocytisus L. ●☆

405281　Vella mairei Humbert；迈氏瓦拉木●☆

405282　Vella mairei Humbert var. eriocarpus Maire = Vella mairei Humbert ●●☆

405283　Vella mairei Humbert var. leiocarpus Maire = Vella mairei Humbert ●●☆

405284　Vella mairei Humbert var. macrantha Fern. Casas = Vella mairei Humbert ●●☆

405285　Vella pseudocytisus L.；非洲瓦拉木●☆

405286　Vella pseudocytisus L. subsp. anremerica Litard. et Maire = Matthiola scapifera Humbert ■☆

405287　Vella pseudocytisus L. subsp. glabrata Greuter；光非洲瓦拉木●☆

405288　Vella pseudocytisus L. subsp. glabrescens (Coss.) Litard. et Maire = Vella pseudocytisus L. subsp. glabrata Greuter ●☆

405289　Vella spinosa Boiss.；刺瓦拉木●☆

405290　Vella tenuissima Pall. = Litwinowia tenuissima (Pall.) Woronow ex Pavlov ■

405291　Vellasquezia Bertol. = Triplaris Loefl. ex L. ●

405292　Vellea D. Dietr. ex Steud. = Tauschia Schltdl. (保留属名)■☆

405293　Vellea D. Dietr. ex Steud. = Velaea DC. ■☆

405294　Velleia Sm. (1798)；翅籽草海桐属■☆

405295　Velleia foliosa K. Krause；多叶翅籽草海桐■☆

405296　Velleia glabrata Carolin；光翅籽草海桐■☆

405297　Velleia lanceolata Lindl.；披针叶翅籽草海桐■☆

405298　Velleia macrocalyx de Vriese；大萼翅籽草海桐■☆

405299　Velleia montana Hook. f.；山地翅籽草海桐■☆

405300　Velleia paradoxa R. Br.；奇异翅籽草海桐■☆

405301　Velleia parvisepta Carolin；小萼翅籽草海桐■☆

405302　Velleia rosea S. Moore；粉红翅籽草海桐■☆

405303　Velleia trinervis Labill.；三脉翅籽草海桐■☆

405304　Vellereophyton Hilliard et B. L. Burtt(1981)；白鼠麴属；White Cudweed ■☆

405305　Vellereophyton dealbatum (Thunb.) Hilliard et B. L. Burtt；白色白鼠麴■☆

405306　Vellereophyton gracillimum Hilliard；细长白鼠麴■☆

405307　Vellereophyton lasianthum (Schltr. et Moeser) Hilliard；毛花白鼠麴■☆

405308　Vellereophyton niveum Hilliard；白鼠麴■☆

405309　Vellereophyton pulvinatum Hilliard；叶枕白鼠麴■☆

405310　Vellereophyton vellereum (R. A. Dyer) Hilliard；羊毛白鼠麴●☆

405311　Velleruca Pomel = Eruca Mill. ■

405312　Velleruca longistyla Pomel = Eruca vesicaria (L.) Cav. ■☆

405313　Velleruca setulosa (Boiss. et Reut.) Pomel = Guenthera setulosa (Boiss. et Reut.) Gómez-Campo ■☆

405314　Velleruca vesicaria (L.) Pomel = Eruca vesicaria (L.) Cav. ■☆

405315　Velleya Roem. et Schult. = Velleia Sm. ■☆

405316　Velleya Walp. = Velleia Sm. ■☆

405317　Vellosia Spreng. = Vellozia Vand. ■☆

405318　Vellosiella Baill. (1888)；韦略列当属■●☆

405319　Vellosiella dracocephaloides Baill.；韦略列当■●☆

405320　Vellozia Vand. (1788)；翡若翠属（巴西蒜属，斐若翠属，尖叶棱枝草属，尖叶鳞枝属）■☆

405321　Vellozia Vand. = Xerophyta Juss. ●■☆

405322　Vellozia acuminata Baker = Xerophyta acuminata (Baker) N. L. Menezes ●☆

405323　Vellozia aequatorialis Rendle = Xerophyta spekei Baker ■☆

405324　Vellozia argentea Wild = Xerophyta argentea (Wild) L. B. Sm. et Ayensu ●☆

405325　Vellozia clavata (Baker) Baker = Xerophyta retinervis Baker ■☆

405326　Vellozia dasylirioides (Baker) Poiss. = Xerophyta dasylirioides Baker ■☆

405327　Vellozia elegans (Balf.) Oliv. ex Hook. f. = Talbotia elegans Balf. ■☆

405328　Vellozia elegans (Balf.) Oliv. ex Hook. f. var. minor Baker = Talbotia elegans Balf. ■☆

405329　Vellozia elegans Talbot ex Balf.；雅致翡若翠（雅致斐若翠）■☆

405330　Vellozia equisetoides (Baker) Baker = Xerophyta equisetoides Baker ●☆

405331　Vellozia equisetoides (Baker) Baker var. trichophylla Baker = Xerophyta equisetoides Baker var. trichophylla (Baker) L. B. Sm. et Ayensu ☆

405332　Vellozia eylesii Greves = Xerophyta eylesii (Greves) N. L. Menezes ■☆

405333　Vellozia hereroensis (Schinz) Baker = Xerophyta viscosa Baker ■☆

405334　Vellozia humilis Baker = Xerophyta humilis (Baker) T. Durand et Schinz ■☆

405335　Vellozia kirkii Hemsl. = Xerophyta kirkii (Hemsl.) L. B. Sm. et Ayensu ☆

405336　Vellozia minuta Baker = Xerophyta humilis (Baker) T. Durand et Schinz ■☆

405337　Vellozia monroi Greves = Xerophyta villosa (Baker) L. B. Sm. et Ayensu ☆

405338　Vellozia retinervis (Baker) Baker = Xerophyta retinervis Baker ■☆

405339　Vellozia rosea Baker = Xerophyta schlechteri (Baker) N. L. Menezes ■☆

405340　Vellozia schlechteri Baker = Xerophyta schlechteri (Baker) N. L. Menezes ■☆

405341　Vellozia schnizleinia (Hochst.) Baker var. occidentalis Milne-Redh. = Xerophyta schnizleinia (Hochst.) Baker ■☆

405342　Vellozia schnizleinia (Hochst.) Martelli var. somalensis A. Terracc. = Xerophyta schnizleinia (Hochst.) Baker var. somalensis (A. Terracc.) Lye ■☆

405343　Vellozia splendens Rendle = Xerophyta splendens (Rendle) N. L. Menezes ●☆

405344　Vellozia squarrosa (Baker) Baker = Xerophyta squarrosa Baker ●☆

405345　Vellozia suaveolens Greves = Xerophyta suaveolens (Greves) N. L. Menezes ●☆

405346　Vellozia talbotii Balf. = Talbotia elegans Balf. ■☆

405347　Vellozia villosa Baker = Xerophyta villosa (Baker) L. B. Sm. et Ayensu ■☆

405348　Vellozia violacea Baker = Xerophyta villosa (Baker) L. B. Sm. et Ayensu ■☆

405349　Vellozia viscosa (Baker) Baker = Xerophyta viscosa Baker ■☆

405350　Velloziaceae Endl. = Velloziaceae J. Agardh(保留科名)■

405351　Velloziaceae J. Agardh(1858)(保留科名)；翡若翠科（巴西蒜科,尖叶棱枝草科,尖叶鳞枝科）■

405352　Velloziella Baill. = Vellosiella Baill. ■●☆

405353　Vellozoa Lem. = Vellozia Vand. ■☆

405354　Velpeaulia Gaudich. = Dolia Lindl. ■☆

405355　Velpeaulia Gaudich. = Nolana L. ex L. f. ■☆

405356　Veltheimia Gled.（1771）；仙火花属（维西美属）；Veltheimia, Cape Lily ■☆

405357　Veltheimia abyssinica DC. = Kniphofia pumila（Aiton）Kunth ■☆

405358　Veltheimia bracteata Harv. ex Baker；显苞仙火花（维西美）；Forest Lily, Veltheimia ■☆

405359　Veltheimia capensis（L.）DC.；仙火花 ■☆

405360　Veltheimia capensis DC. = Veltheimia capensis（L.）DC. ■☆

405361　Veltheimia deasii P. E. Barnes = Veltheimia capensis（L.）DC. ■☆

405362　Veltheimia glauca（Aiton）Jacq. = Veltheimia capensis（L.）DC. ■☆

405363　Veltheimia glauca Jacq. = Veltheimia capensis DC. ■☆

405364　Veltheimia media Donn = Kniphofia sarmentosa（Andréws）Kunth ■☆

405365　Veltheimia repens Ker Gawl. = Kniphofia sarmentosa（Andréws）Kunth ■☆

405366　Veltheimia roodeae E. Phillips = Veltheimia capensis（L.）DC. ■☆

405367　Veltheimia sarmentosa（Andréws）Willd. = Kniphofia sarmentosa（Andréws）Kunth ■☆

405368　Veltheimia undulata Moench = Veltheimia bracteata Harv. ex Baker ■☆

405369　Veltheimia uvaria（L.）Willd. = Kniphofia uvaria（L.）Oken ■☆

405370　Veltheimia viridifolia Jacq. = Veltheimia bracteata Harv. ex Baker ■☆

405371　Veltis Adans. = ? Centaurea L.（保留属名）●■

405372　Velvetia Tiegh. = Loranthus Jacq.（保留属名）●

405373　Velvetia Tiegh. = Psittacanthus Mart. ●☆

405374　Velvitsia Hiern. = Melasma P. J. Bergius ■

405375　Velvitsia calycina Hiern = Melasma calycinum（Hiern）Hemsl. ■☆

405376　Vemonia Edgew. = Vernonia Schreb.（保留属名）●■

405377　Venana Lam.（废弃属名）= Brexia Noronha ex Thouars（保留属名）●☆

405378　Venana madagascariensis Lam. = Brexia madagascariensis（Lam.）Ker Gawl. ●☆

405379　Venatris Raf. = Aster L. ●■

405380　Vendredia Baill. = Rhetinodendron Meisn. ●☆

405381　Vendredia Baill. = Robinsonia DC.（保留属名）●☆

405382　Venegasia DC.（1838）；谷菊属；Canyon Sunflower ●☆

405383　Venegasia carpesioides DC.；谷菊；Canyon Sunflower ■☆

405384　Venegazia Benth. et Hook. f. = Venegasia DC. ●☆

405385　Venelia Comm. ex Bndi. = Erythroxylum P. Browne ●

405386　Venidium Less.（1831）；凉菊属（拟金盏菊属）■☆

405387　Venidium Less. = Arctotis L. ●■☆

405388　Venidium angustifolium DC.；窄叶凉菊 ■☆

405389　Venidium arctotoides（L. f.）Less. = Arctotis arctotoides（L. f.）O. Hoffm. ■☆

405390　Venidium aureum DC.；黄凉菊 ■☆

405391　Venidium aureum DC. = Arctotis fastuosa Jacq. ■☆

405392　Venidium bellidiastrum S. Moore = Arctotis bellidiastrum（S. Moore）Lewin ■☆

405393　Venidium bolusii S. Moore = Arctotis bolusii（S. Moore）Lewin ■☆

405394　Venidium calendulaceum Less. = Arctotheca calendula（L.）Levyns ■☆

405395　Venidium canescens DC.；灰凉菊 ■☆

405396　Venidium cinerarium DC. = Arctotis perfoliata（Less.）Beauverd ■☆

405397　Venidium decurrens Less. = Arctotheca calendula（L.）Levyns ■☆

405398　Venidium decurrens Less. = Arctotis arctotoides（L. f.）O. Hoffm. ■☆

405399　Venidium discolor Less.；异色凉菊 ■☆

405400　Venidium discolor Less. = Arctotis discolor（Less.）Beauverd ■☆

405401　Venidium erosum Harv. = Arctotis erosa（Harv.）Beauverd ■☆

405402　Venidium fastuosum（Jacq.）Stapf = Arctotis fastuosa Jacq. ■☆

405403　Venidium fastuosum（Jacq.）Stapf = Sanvitalia procumbens Lam. ■

405404　Venidium fugax Harv. = Arctotis hirsuta（Harv.）Beauverd ■☆

405405　Venidium hirsutum Harv. = Arctotis hirsuta（Harv.）Beauverd ■☆

405406　Venidium hispidulum DC. = Arctotis hirsuta（Harv.）Beauverd ■☆

405407　Venidium hispidulum Less. = Arctotis hispidula（Less.）Beauverd ■☆

405408　Venidium kraussii Sch. Bip. = Arctotis hirsuta（Harv.）Beauverd ■☆

405409　Venidium macrocephalum DC.；大头凉菊 ■☆

405410　Venidium macrocephalum DC. = Arctotis fastuosa Jacq. ■☆

405411　Venidium macrospermum DC. = Arctotis macrosperma（DC.）Beauverd ■☆

405412　Venidium microcephalum DC. = Arctotis microcephala（DC.）Beauverd ■☆

405413　Venidium perfoliatum Less. = Arctotis perfoliata（Less.）Beauverd ■☆

405414　Venidium puberulum DC. = Arctotis hispidula（Less.）Beauverd ■☆

405415　Venidium rogersii S. Moore = Arctotis rogersii（Benson）M. C. Johnst. ■☆

405416　Venidium semipapposum DC. = Arctotis semipapposa（DC.）Beauverd ■☆

405417　Venidium serpens S. Moore = Arctotis serpens（S. Moore）Lewin ■☆

405418　Venidium subacaule DC. = Arctotis hirsuta（Harv.）Beauverd ■☆

405419　Venidium wyleyi Harv. = Arctotis fastuosa Jacq. ■☆

405420　Veniera Salisb. = Narcissus L. ■

405421　Venilia（G. Don）Fourr. = Ceramanthe（Rchb.）Dumort. ■●

405422　Venilia（G. Don）Fourr. = Scrophularia L. ■●

405423　Venilia Fourr. = Ceramanthe（Rchb.）Dumort. ■●

405424　Venilia Fourr. = Scrophularia L. ■●

405425　Ventana J. F. Macbr. = Vantanea Aubl. ●☆

405426　Ventenata Koeler(1802)（保留属名）；风草属 ■☆

405427　Ventenata avenacea Koeler = Ventenata dubia（Leers）Coss. et Durieu ■☆

405428　Ventenata dubia（Leers）Coss. et Durieu = Ventenata dubia（Leers）F. W. Schultz ■☆

405429　Ventenata dubia（Leers）F. W. Schultz；北非风草（北非维西美）；North Africa Grass, Soft-bearded Oat Grass, Ventenata Grass ■☆

405430　Ventenatia Cav. = Astroloma R. Br. + Melichrus R. Br. ●☆

405431　Ventenatia Cav. = Vintenatia Cav.（废弃属名）■☆

405432　Ventenatia P. Beauv. = Caloncoba Gilg ●☆

405433　Ventenatia P. Beauv. = Oncoba Forssk. ●

405434　Ventenatia Sm. = Stylidium Sw. ex Willd.（保留属名）■

405435　Ventenatia Tratt. = Pedilanthus Neck. ex Poit.（保留属名）●

405436　Ventenatia glauca P. Beauv. = Oncoba glauca（P. Beauv.）Planch. ●☆

405437　Ventenatum Leschen. ex Rchb. = Diplolaena R. Br. ●☆

405438　Ventilago Gaertn.（1788）；翼核果属（翼核木属）；Ironweed, Ventilago, Wingdrupe ●☆

405439　Ventilago africana Exell；非洲翼核果 ●☆

405440　Ventilago bracteata Heyne = Ventilago maderaspatana Gaertn. ●

405441　Ventilago calyculata Tul.；毛果翼核果（副萼翼核果, 河边

茶）；Hairyfruit Ironweed，Hairyfruit Ventilago，Hairyfruit Wingdrupe，Hairy-fruited Ventilago ●

405442　Ventilago calyculata Tul. var. trichoclada Y. L. Chen et P. K. Chou；毛枝翼核果；Hairy-branched Ventilago，Hairy-branched Wingdrupe ●

405443　Ventilago cristata Pierre = Ventilago inaequilateralis Merr. et Chun ●

405444　Ventilago diffusa（G. Don）Exell；松散翼核果●☆

405445　Ventilago elegans Hemsl.；雅致翼核果（台湾翼核果，翼核果，翼核木）；Elagant Ironweed，Elagant Ventilago，Taiwan Wingdrupe ●

405446　Ventilago inaequilateralis Merr. et Chun；海南翼核果（扁果藤，翼核果藤）；Crested Ventilago，Crested Wingdrupe，Hainan Ironweed，Hainan Ventilago，Hainan Wingdrupe ●

405447　Ventilago leiocarpa Benth.；翼核果（扁果藤，穿破石，光果翼核果，光果翼核木，红蛇根，拉牛入石，牛参，青筋藤，青藤，铁牛入石，乌多年，血风根，血风藤，血枫藤）；Smoothfruit Ironweed，Smoothfruit Ventilago，Smooth-fruited Ventilago，Wingdrupe ●

405448　Ventilago leiocarpa Benth. var. pubescens Y. L. Chen et P. K. Chou；毛叶翼核果；Hairyleaf Ironweed，Hairyleaf Ventilago，Hairyleaf Wingdrupe ●

405449　Ventilago leptadenia Tul.；细腺翼核果●☆

405450　Ventilago maderaspatana Gaertn.；印度翼核果；Indian Ironweed，Indian Ventilago，IndiaWingdrupe ●

405451　Ventilago maderaspatana Gaertn. = Ventilago calyculata Tul. ●

405452　Ventilago oblongifolia Blume；矩叶翼核果；Oblongleaf Ironweed，Oblongleaf Ventilago，Oblongleaf Wingdrupe，Oblong-leaved Ventilago ●

405453　Ventilago viminalis Hook.；柳枝状翼核果●☆

405454　Ventraceae Dulac = Cucurbitaceae Juss.（保留科名）●■

405455　Ventricularia Garay（1972）；腹兰属■☆

405456　Ventricularia tenuicaulia（Hook. f.）Garay.；腹兰■☆

405457　Veprecella Naudin = Gravesia Naudin ●☆

405458　Veprecella Naudin（1851）；灌丛野牡丹属●☆

405459　Veprecella acuminata Cogn. = Gravesia macrophylla（Naudin）Baill. ■☆

405460　Veprecella apiculata Cogn. = Gravesia apiculata（Cogn.）H. Perrier ●☆

405461　Veprecella biformis Baker = Gravesia laxiflora（Naudin）Baill. ●☆

405462　Veprecella bullosa Cogn. = Gravesia bullosa（Cogn.）H. Perrier ●☆

405463　Veprecella foliosa Cogn. = Gravesia nigrescens（Naudin）Baill. ■☆

405464　Veprecella hispida Baker = Gravesia hispida（Baker）H. Perrier ●☆

405465　Veprecella lanceolata Cogn. = Gravesia lanceolata（Cogn.）H. Perrier ■☆

405466　Veprecella laxiflora Naudin = Gravesia laxiflora（Naudin）Baill. ●☆

405467　Veprecella lutea Naudin = Gravesia lutea（Naudin）H. Perrier ■☆

405468　Veprecella macrophylla Naudin = Gravesia macrophylla（Naudin）Baill. ■☆

405469　Veprecella microphylla Cogn. = Gravesia microphylla（Cogn.）H. Perrier ■☆

405470　Veprecella nigrescens Naudin = Gravesia nigrescens（Naudin）Baill. ■☆

405471　Veprecella oblongifolia Cogn. = Gravesia oblongifolia（Cogn.）H. Perrier ■☆

405472　Veprecella ovalifolia Cogn. = Gravesia macrophylla（Naudin）Baill. ■☆

405473　Veprecella parvifolia Aug. DC. = Gravesia parvifolia（Aug. DC.）H. Perrier ■☆

405474　Veprecella pilosula Cogn. = Gravesia pilosula（Cogn.）Baill. ■☆

405475　Veprecella riparia Cogn. = Gravesia macrophylla（Naudin）Baill. ■☆

405476　Veprecella rubra Jum. et H. Perrier = Gravesia rubra（Jum. et H. Perrier）H. Perrier ■☆

405477　Veprecella schizocarpa Baker = Gravesia macrophylla（Naudin）Baill. ■☆

405478　Veprecella tetraptera Cogn. = Gravesia tetraptera（Cogn.）H. Perrier ■☆

405479　Veprecella vestita Baker = Gravesia vestita（Baker）H. Perrier ■☆

405480　Veprecella violacea Jum. et H. Perrier = Gravesia violacea（Jum. et H. Perrier）H. Perrier ■☆

405481　Vepris A. Juss. = Vepris Comm. ex A. Juss. ●☆

405482　Vepris Comm. ex A. Juss.（1825）；刺橘属●☆

405483　Vepris afzelii（Engl.）Mziray；阿氏刺橘●☆

405484　Vepris allenii I. Verd.；阿伦刺橘●☆

405485　Vepris amaniensis（Engl.）Mziray；亚美尼亚刺橘●☆

405486　Vepris angolensis（Hiern）Mziray = Vepris hiernii Gereau ●☆

405487　Vepris angolensis Engl. = Fagaropsis angolensis（Engl.）Dale ●☆

405488　Vepris aralioides H. Perrier；槭木刺橘●☆

405489　Vepris arenicola H. Perrier；沙生刺橘●☆

405490　Vepris arushensis Kokwaro；阿鲁沙刺橘●☆

405491　Vepris bachmannii（Engl.）Mziray；巴克曼刺橘●☆

405492　Vepris bilocularis Engl.；二室刺橘●☆

405493　Vepris borenensis（M. G. Gilbert）Mziray；北方刺橘●☆

405494　Vepris calcicola H. Perrier；钙生刺橘●☆

405495　Vepris carringtoniana Mendonça = Teclea pilosa（Engl.）I. Verd. ●☆

405496　Vepris cauliflora H. Perrier；茎花刺橘●☆

405497　Vepris dainellii（Pic. Serm.）Kokwaro；戴尼尔刺橘●☆

405498　Vepris decaryana H. Perrier；德卡里刺橘●☆

405499　Vepris drummondii Mendonça；德拉蒙德刺橘●☆

405500　Vepris eggelingii（Kokwaro）Mziray；埃格林刺橘●☆

405501　Vepris elegantissima F. White et Pannell；雅致刺橘●☆

405502　Vepris elliotii（Radlk.）I. Verd.；埃氏刺橘●☆

405503　Vepris eugeniifolia（Engl.）I. Verd.；番樱桃叶刺橘●☆

405504　Vepris fadenii（Kokwaro）Mziray；法登刺橘●☆

405505　Vepris fanshawei Mendonça；范肖刺橘●☆

405506　Vepris felicis Breteler；多育刺橘●☆

405507　Vepris gabonensis（Pierre）Mziray；加蓬刺橘●☆

405508　Vepris glaberrima（Engl.）J. B. Hall ex D. J. Harris；无毛刺橘●☆

405509　Vepris glandulosa（Hoyle et Leakey）Kokwaro；腺刺橘●☆

405510　Vepris glomerata（F. Hoffm.）Engl.；团集刺橘●☆

405511　Vepris glomerata（F. Hoffm.）Engl. var. glabra Kokwaro；光滑团集刺橘●☆

405512　Vepris gossweileri（I. Verd.）Mziray；戈斯刺橘●☆

405513　Vepris grandifolia（Engl.）Mziray；大花刺橘●☆

405514　Vepris hanangensis（Kokwaro）Mziray；坦桑尼亚刺橘●☆

405515　Vepris heterophylla（Engl.）Letouzey；互叶刺橘●☆

405516　Vepris hiernii Gereau；希尔刺橘●☆

405517　Vepris humbertii H. Perrier；亨伯特刺橘●☆

405518　Vepris lanceolata（Lam.）G. Don；剑叶刺橘●☆

405519　Vepris leandriana H. Perrier；利安刺橘●☆

405520　Vepris lepidota Capuron；小鳞刺橘●☆

405521　Vepris louisii G. C. C. Gilbert；路易斯刺橘（路易刺橘）●☆

405522　Vepris louvelii H. Perrier；卢韦尔刺橘●☆

405523　Vepris macedoi（Exell et Mendonça）Mziray；马塞多刺橘●☆

405524　Vepris macrophylla（Baker）I. Verd.；大叶刺橘●☆

405525　Vepris madagascarica（Baill.）H. Perrier；马岛刺橘●☆

405526　Vepris madagascariensis（H. Perrier）Mziray；马岛大刺橘●☆

405527　Vepris mendoncana Mziray;门东萨刺橘●☆

405528　Vepris mildbraediana G. M. Schulze;米尔德刺橘●☆

405529　Vepris morogorensis (Kokwaro) Mziray;莫罗戈罗刺橘●☆

405530　Vepris myrei (Exell et Mendonça) Mziray;米雷刺橘●☆

405531　Vepris ngamensis I. Verd. ;恩加姆刺橘●☆

405532　Vepris nitida (Baker) I. Verd. ;光亮刺橘●☆

405533　Vepris nobilis (Delile) Mziray;名贵刺橘●☆

405534　Vepris noldeae (Exell et Mendonça) Mziray;诺尔德刺橘●☆

405535　Vepris pilosa (Baker) I. Verd. ;毛刺橘●☆

405536　Vepris pilosa Engl. = Teclea pilosa (Engl.) I. Verd. ●☆

405537　Vepris polymorpha (Danguy ex Lecomte) H. Perrier;多型刺橘●☆

405538　Vepris punctata (I. Verd.) Mziray;斑点刺橘●☆

405539　Vepris reflexa I. Verd. ;反折刺橘●☆

405540　Vepris reflexifolia H. Perrier;曲叶刺橘●☆

405541　Vepris renieri (G. C. C. Gilbert) Mziray;雷尼尔刺橘●☆

405542　Vepris rogersii (Mendonça) Mziray;罗杰斯刺橘●☆

405543　Vepris samburuensis Kokwaro;桑布鲁刺橘●☆

405544　Vepris sansibarensis (Engl.) Mziray;桑给巴尔刺橘●☆

405545　Vepris schliebenii Mildbr. ;施利本刺橘●☆

405546　Vepris schmidelioides (Baker) I. Verd. ;异木患刺橘●☆

405547　Vepris sclerophylla H. Perrier;硬叶刺橘●☆

405548　Vepris simplicifolia (Engl.) Mziray;单叶刺橘●☆

405549　Vepris soyauxii (Engl.) Mziray;索亚刺橘●☆

405550　Vepris stolzii I. Verd. ;斯托尔兹刺橘●☆

405551　Vepris suaveolens (Engl.) Mziray;芳香刺橘●☆

405552　Vepris sudanica (A. Chev.) Letouzey = Vepris heterophylla (Engl.) Letouzey ●☆

405553　Vepris tabouensis (Aubrév. et Pellegr.) Mziray;塔布刺橘●☆

405554　Vepris trichocarpa (Engl.) Mziray;毛果刺橘●☆

405555　Vepris trifoliolata (Engl.) Mziray;三小叶刺橘●☆

405556　Vepris undulata (Thunb.) I. Verd. et C. A. Sm. = Vepris lanceolata (Lam.) G. Don ●☆

405557　Vepris verdoorniana (Engl. et Mendonça) Mziray;喀麦隆刺橘●☆

405558　Vepris welwitschii (Hiern) Exell;韦尔刺橘●☆

405559　Vepris whitei Mendonça;怀特刺橘●☆

405560　Vepris zambesiaca S. Moore;赞比西刺橘●☆

405561　Veratraceae C. Agardh = Melanthiaceae Batsch ex Borkh. (保留科名)■

405562　Veratraceae Salisb. = Melanthiaceae Batsch ex Borkh. (保留科名)■

405563　Veratraceae Vest = Melanthiaceae Batsch ex Borkh. (保留科名)■

405564　Veratrilla (Baill.) Franch. (1899);黄秦艽属(滇黄芩属);Veratrilla ■

405565　Veratrilla Baill. ex Franch. = Veratrilla (Baill.) Franch. ■

405566　Veratrilla Franch. = Veratrilla (Baill.) Franch. ■

405567　Veratrilla baillonii Franch. ;黄秦艽(大苦参,滇黄芩,黄龙胆,金不换,丽江金不换);Baillon Veratrilla,Veratrilla ■

405568　Veratrilla burkilliana (W. W. Sm.) Harry Sm. ;短叶黄秦艽;Shortleaf Veratrilla ■

405569　Veratronia Miq. = Hanguana Blume ■☆

405570　Veratrum L. (1753);藜芦属;Corn-lily, False Hellebore, Falsehellebore, False-hellebore, Skunk-cabbage ■

405571　Veratrum album L. ;白花藜芦(白藜芦,藜芦,藜芦,蒜藜芦);European White Hellebore, Langwort, Lingwort, Madberry, Neesewort, Neesing Root, Sneezewort, Tunsing-wort, White False Hellebore, White False Helleborine, White Falsehellebore, White False-hellebore, White Hellebore, White Neesewort, Woodberry ■☆

405572　Veratrum album L. = Veratrum grandiflorum (Maxim.) O. Loes. ■

405573　Veratrum album L. var. dahuricum Turcz. = Veratrum dahuricum (Turcz.) O. Loes. ■

405574　Veratrum album L. var. grandiflorum Maxim. = Veratrum grandiflorum (Maxim.) O. Loes. ■

405575　Veratrum album L. var. lobelianum (Bernh.) Suess. = Veratrum lobelianum Bernh. ■

405576　Veratrum album L. var. oxysepalum (Turcz.) Miyabe et Kudo;尖萼白花藜芦■☆

405577　Veratrum album L. var. viride (Aiton) Baker = Veratrum oxysepalum Turcz. ■

405578　Veratrum atroviolaceum O. Loes. = Veratrum schindleri O. Loes. ■

405579　Veratrum bohnhofii O. Loes. = Veratrum maackii Regel ■

405580　Veratrum bracteatum Batalin = Veratrum nigrum L. ■

405581　Veratrum californicum Durand;加州藜芦;Californian False Hellebore, Corn Lily, False Hellebore, Skunk Cabbage ■☆

405582　Veratrum californicum Durand var. caudatum (A. Heller) C. L. Hitchc. ;尾状加州藜芦■☆

405583　Veratrum calyciflorum Kom. ;萼花藜芦■☆

405584　Veratrum caudatum A. Heller = Veratrum californicum Durand var. caudatum (A. Heller) C. L. Hitchc. ■☆

405585　Veratrum dahuricum (Turcz.) O. Loes. ;兴安藜芦;Dahur Falsehellebore, Dahuria Falsehellebore ■

405586　Veratrum dolichopetalum O. Loes. = Veratrum oxysepalum Turcz. ■

405587　Veratrum escholtzianum Loes. = Veratrum viride Aiton var. eschscholtzianum (Roem. et Schult.) Breitung ■☆

405588　Veratrum eschscholtzianum A. Gray = Veratrum viride Aiton var. eschscholtzianum (Roem. et Schult.) Breitung ■☆

405589　Veratrum eschscholtzianum Rydb. ex A. Heller = Veratrum viride Aiton var. eschscholtzianum (Roem. et Schult.) Breitung ■☆

405590　Veratrum eschscholtzii A. Gray;埃氏藜芦(埃希首氏藜芦)■☆

405591　Veratrum eschscholtzii A. Gray var. watsonii Baker = Veratrum californicum Durand ■☆

405592　Veratrum fimbriatum A. Gray;流苏藜芦;Fringed False Hellebore ■☆

405593　Veratrum formosanum O. Loes. ;台湾藜芦;Taiwan Falsehellebore ■

405594　Veratrum formosanum O. Loes. = Veratrum japonicum (Baker) O. Loes. ■

405595　Veratrum formosanum O. Loes. f. albiflorum (Masam.) Masam. = Veratrum formosanum O. Loes. ■

405596　Veratrum formosanum O. Loes. var. albiflorum Masam. = Veratrum formosanum O. Loes. ■

405597　Veratrum gongshanense S. Z. Chen et G. J. Xu;贡山藜芦;Gongshan Falsehellebore ■

405598　Veratrum grandiflorum (Maxim.) O. Loes. ;毛叶藜芦(大花白藜芦,人头发,蒜藜芦,算藜芦);Bigflower Falsehellebore, Largeflower Falsehellebore ■

405599　Veratrum insolitum Jeps. ;锡斯基尤藜芦;Siskiyou False Hellebore ■☆

405600　Veratrum intermedium Chapm. = Melanthium woodii (Robbins ex A. W. Wood) Bodkin ■☆

405601　Veratrum japonicum (Baker) O. Loes. ;黑紫藜芦(藜芦,日本藜芦);Japan Falsehellebore, Japanese Falsehellebore ■

405602　Veratrum jonesii A. Heller = Veratrum californicum Durand ■☆

405603　Veratrum kudoi Masam. = Veratrum formosanum O. Loes. ■

405604　Veratrum lobelianum Bernh. ;阿尔泰藜芦(半边莲状藜芦,藜芦,新疆藜芦);Altai Falsehellebore, Altaian Falsehellebore ■

405605　Veratrum lobelianum Bernh. var. eschscholtzianum Roem. et

Schult. = Veratrum viride Aiton var. eschscholtzianum（Roem. et Schult.）Breitung ■☆

405606 Veratrum maackii Regel；毛穗藜芦（马氏藜芦）；Hairspike Falsehellebore，Maack Falsehellebore ■

405607 Veratrum maackii Regel var. reymondianum（O. Loes.）Hara et Mizush.；雷氏藜芦■☆

405608 Veratrum mairei H. Lév. = Curculigo capitulata（Lour.）Kuntze ■

405609 Veratrum mairei H. Lév. = Molineria capitulata（Lour.）Herb. ■

405610 Veratrum mandschuricum O. Loes. = Veratrum maackii Regel ■

405611 Veratrum maximowiczii Baker；闽浙藜芦■

405612 Veratrum maximowiczii Baker = Veratrum schindleri O. Loes. ■

405613 Veratrum maximowiczii Baker var. hupehense Pamp. = Veratrum oblongum O. Loes. ■

405614 Veratrum mengtzeanum O. Loes.；蒙自藜芦（翻天印，黄龙须，批麻草，披麻草，细毒蒜，小藜芦，小天蒜，小棕包）；Mengtzu Falsehellebore，Mengzi Falsehellebore ■

405615 Veratrum micranthum F. T. Wang et Ts. Tang；小花藜芦；Smallflower Falsehellebore ■

405616 Veratrum misae（Sirj.）Loes.；北极藜芦■☆

405617 Veratrum nanchuanense S. Z. Chen et G. J. Xu；南川藜芦；Nanchuan Falsehellebore ■

405618 Veratrum nigrum L.；藜芦（葱白藜芦，葱葵，葱苒，葱葖，大叶藜芦，毒药草，丰芦，公苒，憨葱，旱葱，黑藜芦，蕙葵，老旱葱，梨卢，芦莲，鹿葱，七厘丹，人头发，山白菜，山苞米，山葱，山棕榈，药蝇子草，棕包头，棕色脚）；Black False Hellebore，Black Falsehellebore，Black False-hellebore，Falsehellebore ■

405619 Veratrum nigrum L. subsp. ussuriense（Loes.）Vorosch. = Veratrum nigrum L. ■

405620 Veratrum nigrum L. var. japonicum Baker = Veratrum japonicum（Baker）O. Loes. ■

405621 Veratrum nigrum L. var. maackii（Regel）Maxim. = Veratrum maackii Regel ■

405622 Veratrum nigrum L. var. microcarpum Loes. = Veratrum nigrum L. ■

405623 Veratrum nigrum L. var. ussuriense Loes. = Veratrum nigrum L. ■

405624 Veratrum oblongum O. Loes.；长梗藜芦；Longstalk Falsehellebore ■

405625 Veratrum oblongum O. Loes. = Veratrum maackii Regel ■

405626 Veratrum omeiensis F. T. Wang et Ts. Tang；峨眉藜芦；Emei Falsehellebore ■

405627 Veratrum oxysepalum Turcz.；尖被藜芦（光脉藜芦，毛脉藜芦）；Sharpsepal Falsehellebore ■

405628 Veratrum oxysepalum Turcz. = Veratrum album L. var. oxysepalum（Turcz.）Miyabe et Kudo ■☆

405629 Veratrum ozysepalum Turcz. = Veratrum oxysepalum Turcz. ■

405630 Veratrum parviflorum Michx. = Melanthium parviflorum（Michx.）S. Watson ■☆

405631 Veratrum patulum O. Loes. = Veratrum oxysepalum Turcz. ■

405632 Veratrum puberulum O. Loes. = Veratrum grandiflorum（Maxim.）O. Loes. ■

405633 Veratrum schindleri O. Loes.；牯岭藜芦（闽浙藜芦，天目藜芦，邢氏藜芦）；Schneidler Falsehellebore ■

405634 Veratrum shuehshanarum S. S. Ying；雪山藜芦■

405635 Veratrum speciosum Rydb. = Veratrum californicum Durand ■☆

405636 Veratrum stenophyllum Diels；狭叶藜芦；Narrowleaf Falsehellebore ■

405637 Veratrum stenophyllum Diels var. taronense O. Loes.；滇北藜芦（独龙藜芦）；Northern Yunnan Falsehellebore ■

405638 Veratrum taliense O. Loes.；大理藜芦（披麻草，七仙草）；Dali Falsehellebore，Tali Falsehellebore ■

405639 Veratrum tenuipetalum A. Heller = Veratrum californicum Durand ■☆

405640 Veratrum ussuriense（Loes.）Nakai = Veratrum nigrum L. ■

405641 Veratrum versicolor Nakai f. brunneum Nakai = Veratrum maackii Regel ■

405642 Veratrum virginicum（L.）Aiton = Melanthium virginicum L. ■☆

405643 Veratrum viride Aiton；绿藜芦（藜芦，美国白藜芦，美国藜芦）；American False Hellebore，American Falsehellebore，American False-hellebore，American Hellebore，American White Hellebore，False Hellebore，Green False Hellebore，Green Falsehellebore，Green False-hellebore，Green Hellebore，Hellebore，Indian Hellebore，Indian Poke，Itchweed，Poke，Swamp Hellebore，White Hellebore ■☆

405644 Veratrum viride Aiton subsp. eschscholtzii（A. Gray）Á. Löve et D. Löve = Veratrum viride Aiton var. eschscholtzianum（Roem. et Schult.）Breitung ■☆

405645 Veratrum viride Aiton var. escholtzianoides Loes. = Veratrum viride Aiton var. eschscholtzianum（Roem. et Schult.）Breitung ■☆

405646 Veratrum viride Aiton var. eschscholtzianum（Roem. et Schult.）Breitung；埃氏金美韭■☆

405647 Veratrum warburgii O. Loes. = Veratrum schindleri O. Loes. ■

405648 Veratrum wilsonii C. H. Wright ex O. Loes. = Veratrum mengtzeanum O. Loes. ■

405649 Veratrum woodii Robbins ex A. Wood；伍德藜芦；False Hellebore ■☆

405650 Veratrum woodii Robbins ex A. Wood = Melanthium woodii（Robbins ex A. W. Wood）Bodkin ■☆

405651 Veratrum yunnanense O. Loes.；云南藜芦■

405652 Veratrum yunnanense O. Loes. = Veratrum stenophyllum Diels ■

405653 Verbascaceae Bercht. et J. Presl = Verbascaceae Nees ■●

405654 Verbascaceae Bonnier = Scrophulariaceae Juss.（保留科名）●■

405655 Verbascaceae Nees = Scrophulariaceae Juss.（保留科名）●■

405656 Verbascaceae Nees；毛蕊花科■●

405657 Verbascum L.（1753）；毛蕊花属；Celsia，Mullein，Verbascum ■●

405658 Verbascum alpigenum K. Koch；高山毛蕊花■☆

405659 Verbascum arcturus L.；北极毛蕊花（黄花姬毛蕊花）；Cretan Bears-tail ■☆

405660 Verbascum arnaizii Sennen = Verbascum sinuatum L. ■☆

405661 Verbascum artvinense E. Wulff；阿尔特温毛蕊花■☆

405662 Verbascum atlanticum Batt.；大西洋毛蕊花■☆

405663 Verbascum ballii（Batt.）Hub. -Mor.；鲍尔毛蕊花■☆

405664 Verbascum battandieri（Murb.）Hub. -Mor.；巴坦毛蕊花■☆

405665 Verbascum benthamianum Hepper；丛毛毛蕊花■☆

405666 Verbascum betonicifolium（Desf.）Kuntze；药水苏叶毛蕊花■☆

405667 Verbascum blattaria L.；毛瓣毛蕊花（黄毛蕊花）；Moth Mullein ■

405668 Verbascum blattaria L. f. albiflora（G. Don）House = Verbascum blattaria L. f. erubescens Brügger ■☆

405669 Verbascum blattaria L. f. albiflorum（G. Don）House = Verbascum blattaria L. ■

405670 Verbascum blattaria L. f. erubescens Brügger；白花毛瓣毛蕊花■☆

405671 Verbascum blattaria L. f. erubescens Brügger = Verbascum blattaria L. ■

405672 Verbascum blattaria L. var. albiflorum G. Don = Verbascum blattaria L. ■

405673 Verbascum blattaria L. var. brevipedicellatum Halácsy = Verbascum blattaria L. ■

405674 Verbascum bombyciforme Boiss.；绢毛毛蕊花；Broussa Mullein ■☆

405675 Verbascum brevipedicellatum（Engl.）Hub. -Mor. = Rhabdotosperma brevipedicellata（Engl.）Hartl ●☆

405676　Verbascum caboverdeanum Sunding = Verbascum capitis-viridis Hub. -Mor. ■☆

405677　Verbascum calycinum Ball;萼状毛蕊花■☆

405678　Verbascum capitis-viridis Hub. -Mor. ;绿头毛蕊花■☆

405679　Verbascum celsioides Benth. ;真正毛蕊花;Celsia Mullein ■☆

405680　Verbascum chaixii Vill. ;麻叶毛蕊花;Nettle-leaf Mullein, Nettle-leaved Mullein ■☆

405681　Verbascum chaixii Vill. subsp. orientale (M. Bieb.) Hayek;东方毛蕊花;Oriental Mullein ■

405682　Verbascum chaixii Vill. subsp. orientale Hayek = Verbascum chaixii Vill. subsp. orientale (M. Bieb.) Hayek ■

405683　Verbascum cheiranthifolium Boiss. ;桂竹香叶毛蕊花■☆

405684　Verbascum chinense (L.) Druce = Verbascum chinense (L.) Santapau ■

405685　Verbascum chinense (L.) Santapau;琴叶毛蕊花(细穗玄参);Violineleaf Mullein ■

405686　Verbascum collinum Schrad. ;山毛蕊花■☆

405687　Verbascum commixtum (Murb.) Hub. -Mor. ;混合毛蕊花■☆

405688　Verbascum cordatum Desf. = Verbascum blattaria L. ■

405689　Verbascum coromandelianum (Vahl) Kuntze = Verbascum chinense (L.) Santapau ■

405690　Verbascum cossonianum Ball = Verbascum dentifolium Delile ■☆

405691　Verbascum creticum (L.) Cav. ;克里特毛蕊花(姬毛蕊花);Cretan Mullein ■☆

405692　Verbascum demnatense (Maire et Murb.) Dobignard et Lambinon;代姆纳特毛蕊花■☆

405693　Verbascum densiflorum Bertol. ;密花毛蕊花■☆

405694　Verbascum densifolium (Hook. f.) Hub. -Mor. ;黄毛蕊花(黄毛蕊,毡毛毛蕊);Denseflower Mullein, Dense-flower Mullein, Dense-flowered Mullein,Mullein,Wool Mullein ■☆

405695　Verbascum densifolium (Hook. f.) Hub. -Mor. = Rhabdotosperma densifolia (Hook. f.) Hartl ●☆

405696　Verbascum dentifolium Delile;齿叶毛蕊花■☆

405697　Verbascum dumulosum P. H. Davis et Hub. -Mor. ex Hub. -Mor. ;灌状毛蕊花■☆

405698　Verbascum erianthum Benth. = Verbascum sinaiticum Benth. ■☆

405699　Verbascum erivanicum E. Wulff;埃里温毛蕊花■☆

405700　Verbascum erosum Cav. ;啮蚀状毛蕊花■☆

405701　Verbascum faurei (Murb.) Hub. -Mor. ;福雷毛蕊花■☆

405702　Verbascum faurei (Murb.) Hub. -Mor. subsp. acanthifolium (Pau) Benedî et J. M. Monts. ;刺叶福雷毛蕊花■☆

405703　Verbascum flavidum (Boiss.) Freyn et Bornm. ;黄色毛蕊花■☆

405704　Verbascum flexuosum E. Wulff;反折毛蕊花■☆

405705　Verbascum floccosum (Benth.) Kuntze = Celsia floccosa Benth. ■☆

405706　Verbascum formosum Fisch. ;美丽毛蕊花■☆

405707　Verbascum gaetulum (Maire) Murb. = Verbascum simplex Hoffmanns. et Link ■☆

405708　Verbascum georgicum Benth. ;乔治亚毛蕊花■☆

405709　Verbascum glomeratum Boiss. ;团集毛蕊花■☆

405710　Verbascum gnaphalodes M. Bieb. ;鼠曲草毛蕊花■☆

405711　Verbascum gossypinum M. Bieb. ;棉毛蕊花■☆

405712　Verbascum haenseleri Boiss. = Verbascum rotundifolium Ten. subsp. haenseleri (Boiss.) Murb. ☆

405713　Verbascum hajastanicum Bordz. ;哈贾斯坦毛蕊花■☆

405714　Verbascum hookerianum Ball;胡克毛蕊花■☆

405715　Verbascum hookerianum Ball var. ballii Murb. = Verbascum hookerianum Ball ■☆

405716　Verbascum hookerianum Ball var. pseudocalycinum Maire et Murb. = Verbascum hookerianum Ball ■☆

405717　Verbascum hookerianum Ball var. tagadirtense (Murb.) Maire = Verbascum hookerianum Ball ■☆

405718　Verbascum interruptum (Fresen.) Kuntze = Celsia interrupta Fresen. ■☆

405719　Verbascum kabylianum Debeaux = Verbascum rotundifolium Ten. ■☆

405720　Verbascum keniense (Murb.) Hub. -Mor. = Rhabdotosperma keniensis (Murb.) Hartl ●☆

405721　Verbascum kerneri Fritsch;克纳毛蕊花;Kerner's Mullein ■☆

405722　Verbascum laciniatum (Poir.) Kuntze = Verbascum erosum Cav. ■☆

405723　Verbascum latesulcatifolium Sennen et Mauricio = Verbascum rotundifolium Ten. subsp. haenseleri (Boiss.) Murb. ■☆

405724　Verbascum laxum Fil. et Jav. ;疏花毛蕊花■☆

405725　Verbascum ledermannii (Murb.) Hub. -Mor. = Rhabdotosperma ledermannii (Murb.) Hartl ●☆

405726　Verbascum letourneuxii Asch. et Schweinf. ;勒图尔呐毛蕊花■☆

405727　Verbascum longifolium Ten. et Murb. ;长叶毛蕊花■☆

405728　Verbascum longirostre (Murb.) Hub. -Mor. ;长喙毛蕊花■☆

405729　Verbascum lychnitis L. ;剪秋毛蕊花(白毛蕊花);Adam's Flanne,White Mullein ■☆

405730　Verbascum lychnitis L. var. giganteum Maire = Verbascum lychnitis L. ■☆

405731　Verbascum macrocarpum Boiss. ;大果毛蕊花■☆

405732　Verbascum mairei (Murb.) Hub. -Mor. ;迈氏毛蕊花■☆

405733　Verbascum maroccanum (Ball) Hub. -Mor. ;摩洛哥毛蕊花■☆

405734　Verbascum masguindalii (Pau) Benedî et J. M. Monts. ;马斯毛蕊花■☆

405735　Verbascum maurum Maire et Murb. ;模糊毛蕊花■☆

405736　Verbascum megaphlomos (Boiss. et Heldr.) Halaesy;糙苏毛蕊花■☆

405737　Verbascum nigrum L. ;黑喉毛蕊花(黑毛蕊花);Black Mullein,Dark Mullein,Mack Mullein ■☆

405738　Verbascum nobile Velen. ;壮丽毛蕊花■☆

405739　Verbascum numidicum Pomel = Verbascum rotundifolium Ten. ■☆

405740　Verbascum olympicum Boiss. ;奥林匹克毛蕊花■☆

405741　Verbascum oranense (Murb.) Dobignard;奥兰毛蕊花■☆

405742　Verbascum oreophilum K. Koch;喜山毛蕊花■☆

405743　Verbascum orientale M. Bieb. = Verbascum chaixii Vill. subsp. orientale (M. Bieb.) Hayek ■

405744　Verbascum ovalifolium Donn;卵叶毛蕊花■☆

405745　Verbascum paniculatum E. Wulff;圆锥毛蕊花■☆

405746　Verbascum pedunculosum (Steud. et Hochst. ex Benth.) Kuntze = Celsia pedunculosa Steud. et Hochst. ex Benth. ■☆

405747　Verbascum phlomoides L. ;抱茎毛蕊花(绒毛蕊花);Clasping Mullein,Orange Mullein ■☆

405748　Verbascum phoeniceum L. ;紫毛蕊花(紫毛蕊);Purple Mullein, Purplered Mullein ■

405749　Verbascum pinnatifidum Vahl;羽裂毛蕊花;Pinnatifid Mullein ■☆

405750　Verbascum pseudocreticum Benedî et J. M. Monts. subsp. demnatense (Maire et Murb.) Ibn Tattou;代姆纳特羽裂毛蕊花■☆

405751　Verbascum pulverulentum Nicotra = Verbascum sinuatum L. ■☆

405752　Verbascum pulverulentum Salisb. = Verbascum lychnitis L. ■☆

405753　Verbascum pulverulentum Spreng. = Verbascum phlomoides L. ■☆

405754　Verbascum pulverulentum Vill. ;阔叶毛蕊花;Broad-leaf Mullein, Hoary Mullein ■☆

405755 Verbascum punalense Boiss. et Buhse;普纳尔毛蕊花■☆

405756 Verbascum pyramidatum M. Bieb.；塔形毛蕊花；Pyramidal Mullein ■☆

405757 Verbascum rhiphaeum（Murb.）Hub.-Mor.；山地毛蕊花■☆

405758 Verbascum rotundifolium Ten.；圆叶毛蕊花■☆

405759 Verbascum rotundifolium Ten. subsp. haenseleri（Boiss.）Murb.；黑泽勒毛蕊花■☆

405760 Verbascum rotundifolium Ten. var. castellorum Maire = Verbascum rotundifolium Ten. ■☆

405761 Verbascum rotundifolium Ten. var. kabylianum（Batt.）Murb. = Verbascum rotundifolium Ten. ■☆

405762 Verbascum rotundifolium Ten. var. numidicum（Pomel）Batt. et Trab. = Verbascum rotundifolium Ten. ■☆

405763 Verbascum rubiginosum Waldst. et Kit.；红棕毛蕊花■☆

405764 Verbascum saccatum K. Koch；囊状毛蕊花■☆

405765 Verbascum schimperi V. Naray. = Rhabdotosperma schimperi（V. Naray.）Hartl ●☆

405766 Verbascum scrophulariifolium（Hochst. ex A. Rich.）Hub.-Mor. = Rhabdotosperma scrophulariifolia（Hochst. ex A. Rich.）Hartl ●☆

405767 Verbascum sessiliflorum Murb.；无梗毛蕊花■☆

405768 Verbascum simplex Hoffmanns. et Link；简单毛蕊花■☆

405769 Verbascum simplex Hoffmanns. et Link var. dyris（Litard. et Maire）Murb. = Verbascum simplex Hoffmanns. et Link ■☆

405770 Verbascum simplex Hoffmanns. et Link var. gaetulum（Maire）Maire = Verbascum simplex Hoffmanns. et Link ■☆

405771 Verbascum simplex Hoffmanns. et Link var. valentinum（Burnat et Barbey）Murb. = Verbascum simplex Hoffmanns. et Link ■☆

405772 Verbascum sinaiticum Benth.；索马里毛蕊花■☆

405773 Verbascum sinense H. Lév. et Giraud = Verbascum chinense（L.）Santapau ■

405774 Verbascum sinuatum L.；波叶毛蕊花（深波毛蕊花）；Mediterranean Mullein, Wavyleaf Mullein ■☆

405775 Verbascum sinuatum L. var. adenosepalum Murb. = Verbascum sinuatum L. ■☆

405776 Verbascum somaliense Baker = Verbascum sinaiticum Benth. ■☆

405777 Verbascum songoricum Schrenk；准噶尔毛蕊花；Dzungar Mullein, Dzungarian Mullein ■

405778 Verbascum songoricum Schrenk ex Fisch. et C. A. Mey. = Verbascum songoricum Schrenk ■

405779 Verbascum speciosum Schrad.；艳丽毛蕊花；Beautiful Mullein, Hungarian Mullein, Showy Mullein ■☆

405780 Verbascum spectabile M. Bieb.；华丽毛蕊花■☆

405781 Verbascum stachydiforme Boiss. et Buhse；穗状毛蕊花■☆

405782 Verbascum sudanicum（Murb.）Hepper = Celsia sudanica（Murb.）Wickens ■☆

405783 Verbascum szovitsianum Boiss.；绍氏毛蕊花■☆

405784 Verbascum tagadirtense Murb. = Verbascum hookerianum Ball ■☆

405785 Verbascum ternacha Hochst. ex A. Rich. = Verbascum sinaiticum Benth. ■☆

405786 Verbascum tetrandrum Barratte et Murb.；四蕊毛蕊花■☆

405787 Verbascum thapsiforme Schrad. = Verbascum densifolium（Hook. f.）Hub.-Mor. ■☆

405788 Verbascum thapsus L.；毛蕊花（霸王鞭, 大毛蕊花, 大毛叶, 毒鱼草, 海绵蒲, 虎尾鞭, 毛蕊草, 牛耳草, 熊耳毛蕊花, 一炷香）；Aaron's Flannel, Aaron's Rod, Abraham's Blanket, Adam's Flannel, Ag-paper, Bear's Beard, Beggar's Blanket, Beggar's Stalk, Bird's Candles, Blanket, Blanket Leaf, Blanket Mullein, Blanket-leaf, Bullock's Lungwort, Bull's Ear, Bull's Ears, Bunny's Ear, Bunny's Ears, Candleweek Flower, Candlewick Flower, Close Sciences, Clote, Clown's Lungwort, Common Mullein, Coses Sciences, Cow's Ear, Cow's Lungwort, Cuddie's Lungs, Cuddy's Lugs, Cuddy's Lungs, Devil's Blanket, Donkey's Ear, Donkey's Ears, Duffle, Fairy's Wand, Feltwort, Flannel, Flannel Flower, Flannel Jacket, Flannel Leaf, Flannel Mullein, Flannel Petticoat, Flannel Petticoats, Flannel Plant, Flannel-leaved Mullein, Fluffweed, Foxglove, French Poppy, Giant Mullein, Golden Grain, Golden Rod, Goldilocks, Great Broad-leaved Mullein, Great Mullein, Hag Taper, Hag-leaf, Hag's Taper, Hag-taper, Hare's Beard, Hate's Beard, Hedge Poppy, Hedge Taper, Hedge-taper, Hig Taper, Higgis Taper, High Taper, High-taper, Hig-taper, Hill Poppy, Jacob's Staff, Jupiter's Staff, King's Taper, Lady's Candle, Lady's Candles, Lady's Flannel, Lady's Foxglove, Lady's Taper, Large-flowered Mullein, Longwort, Miller's Jerkin, Molayne, Moses' Blanket, Moth Mullein, Moth Plant, Mullein, Mullein Dock, Mullen, Old Man's Flannel, Our Lady's Candle, Our Lady's Candles, Our Lady's Flannel, Our Lord's Flannel, Our Saviour's Flannel, Our's Flannel, Peter's Staff, Poor Man's Blanket, Poor Man's Flannel, Rag Paper, Rag-paper, Sea Cabbage, Shepherd's Club, Shepherd's Staff, Snake's Flower, Snake's Head, Soldiers, Soldier's Tears, Sweethearts, Taper, Torch, Torch Plant, Torchwort, Tower of Babel, Velvet Dock, Velvet Leaf, Velvet Mullein, Velvet Plant, Velvet Poppy, Virgin Mary's Candle, Virgin Mary's Candles, White Mullein, Wild Ice-leaf, Wool Mullein, Woollen, Woolly Mullein ■

405789 Verbascum thapsus L. f. candicans House；白毛蕊花■☆

405790 Verbascum thapsus L. subsp. gaetulum Maire = Verbascum simplex Hoffmanns. et Link ■☆

405791 Verbascum tibesticum（Quézel）Hub.-Mor.；提贝斯提毛蕊花■☆

405792 Verbascum transcaucasicum E. Wulff；外高加索毛蕊花■☆

405793 Verbascum turcomanicum Murb.；土库曼毛蕊花■☆

405794 Verbascum turkestanicum Franch.；突厥斯坦毛蕊花■☆

405795 Verbascum varians Freyn et Sint.；易变毛蕊花■☆

405796 Verbascum virgatum Stokes；大花毛蕊花；Aaron's Rod, Large-flowered Mullein, Twiggy Mullein, Wand Mullein ■☆

405797 Verbascum warionis Batt. = Verbascum simplex Hoffmanns. et Link ■☆

405798 Verbascum wiedemannianum Fisch. et C. A. Mey.；威氏毛蕊花■☆

405799 Verbascum wilhelmsianum K. Koch；威尔毛蕊花■☆

405800 Verbascum zaianense（Murb.）Hub.-Mor.；宰哈奈毛蕊花■☆

405801 Verbena L.（1753）；马鞭草属；Verbena, Vervain ■●

405802 Verbena Rumph. = Aerva Forssk.（保留属名）●■

405803 Verbena × aubletia L. = Verbena canadensis（L.）Britton ■☆

405804 Verbena × incompta P. W. Michael；装饰马鞭草■☆

405805 Verbena ambrosiifolia Rydb. ex Small = Verbena bipinnatifida Nutt. ■☆

405806 Verbena angustifolia Michx. = Verbena simplex Lehm. ■☆

405807 Verbena angustifolia Mill. = Stachytarpheta indica（L.）Vahl ●

405808 Verbena aristigera S. Moore；芒马鞭草■☆

405809 Verbena bipinnatifida Nutt.；二羽裂马鞭草；Cut-leaved Verbena, Dakota Vervain ■☆

405810 Verbena bipinnatifida Nutt. = Glandularia bipinnatifida（Nutt.）Nutt. ■☆

405811 Verbena bipinnatifida Nutt. var. latiloba L. M. Perry = Verbena bipinnatifida Nutt. ■☆

405812 Verbena bonariensis L.；柳叶马鞭草；Argentine Vervain, Brazilian Verbena, Buenos Ayres Verbena, Garden Vervain, Purpletop，Purpletop Vervain，South American Vervain，Tall Vervain ■

405813　Verbena bonariensis L. var. conglomerata Briq.；团聚柳叶马鞭草；Purpletop Vervain ■☆

405814　Verbena bonariensis L. var. conglomerata Briq. = Verbena bonariensis L. ■

405815　Verbena bracteata Cav. ex Lag. et Rodr.；宽裂马鞭草；Creeping Vervain，Large-bracted Vervain，Prostrate Vervain ■

405816　Verbena bracteata Lag. et Rodr. = Verbena bracteata Cav. ex Lag. et Rodr. ■

405817　Verbena bracteosa Michx.；多苞马鞭草；Bracteose Verbena ■☆

405818　Verbena bracteosa Michx. = Verbena bracteata Cav. ex Lag. et Rodr. ■

405819　Verbena brasiliensis Vell.；巴西马鞭草；Brazilian Vervain，Vervain ■☆

405820　Verbena canadensis（L.）Britton；加拿大马鞭草；Canada Verbena，Canada Vervain，Clump Verbena，Rose Verbena，Rose Vervain ■☆

405821　Verbena canadensis（L.）Britton var. atroviolacea Dermen = Verbena canadensis（L.）Britton ■☆

405822　Verbena canadensis（L.）Britton var. compacta Dermen = Verbena canadensis（L.）Britton ■☆

405823　Verbena canadensis（L.）Britton var. drummondii（Lindl.）E. M. Baxter = Verbena canadensis（L.）Britton ■☆

405824　Verbena canadensis（L.）Britton var. grandiflora（Haage et Schmidt）Moldenke = Verbena canadensis（L.）Britton ■☆

405825　Verbena canadensis（L.）Britton var. lambertii（Sims）Thell. = Verbena canadensis（L.）Britton ■☆

405826　Verbena capensis Thunb. = Lippia alba（Mill.）N. E. Br. ex Britton et Rose ●☆

405827　Verbena caracasana Kunth = Verbena litoralis Kunth ■☆

405828　Verbena carolina L.；卡罗来纳马鞭草（加罗林马鞭草，卡罗林马鞭草）■☆

405829　Verbena cayennensis Rich. = Stachytarpheta cayennensis（Rich.）Vahl ●☆

405830　Verbena ciliata Benth. = Verbena bipinnatifida Nutt. ■☆

405831　Verbena ciliata Benth. var. longidentata L. M. Perry = Verbena bipinnatifida Nutt. ■☆

405832　Verbena ciliata Benth. var. pubera（Greene）L. M. Perry = Verbena bipinnatifida Nutt. ■☆

405833　Verbena citriodora（Paláu）Cav. = Aloysia citriodora Paláu ●☆

405834　Verbena corymbosa Ruiz et Pav.；伞房花马鞭草●☆

405835　Verbena deamii Moldenke；迪姆马鞭草；Deam's Verbena，Deam's Vervain，Vervain ■☆

405836　Verbena demareei Moldenke = Verbena bipinnatifida Nutt. ■☆

405837　Verbena dichotoma Ruiz et Pav. = Stachytarpheta cayennensis（Rich.）Vahl ●☆

405838　Verbena dodgei B. Boivin = Verbena deamii Moldenke ■☆

405839　Verbena elongata Salisb. = Verbena bonariensis L. ■

405840　Verbena engelmannii Moldenke；恩格尔曼马鞭草；Engelmann's Vervain，Vervain ■☆

405841　Verbena erinoides Lam.；狐地黄状马鞭草；Moss Verbena ■☆

405842　Verbena forskaolaei Vahl = Priva adhaerens（Forssk.）Chiov. ■☆

405843　Verbena halei Small = Verbena officinalis L. ■

405844　Verbena hastata L.；戟叶马鞭草（戟形马鞭草）；American Vervain，Blue Vervain，False Vervain，Hastate Verbena，Simpler's-joy，Swamp Verbena，Wild Hyssop，Wild Vervain ■☆

405845　Verbena hastata L. f. albiflora Moldenke = Verbena hastata L. ■☆

405846　Verbena hastata L. f. rosea Cheney = Verbena hastata L. ■☆

405847　Verbena hastata L. var. scabra Moldenke = Verbena hastata L. ■☆

405848　Verbena hybrida Groem. et Rümpler = Verbena hybrida Voss ex Groem. et Rümpler ■

405849　Verbena hybrida Voss 'Fern Verbena'；羊齿美女樱■

405850　Verbena hybrida Voss 'Sparkle'；喷彩美女樱■

405851　Verbena hybrida Voss = Verbena hybrida Voss ex Groem. et Rümpler ■

405852　Verbena hybrida Voss ex Groem. et Rümpler；美女樱（美人樱）；Common Garden Verbena，Common Garden Vervain，Garden Verbena，Garden Vervain，Hybrid Verbena，Verbena ■

405853　Verbena illicita Moldenke；缘毛马鞭草；Bastard Vervain，Illicit Verbena ■☆

405854　Verbena imbricata Wooton et Standl. = Verbena bracteata Cav. ex Lag. et Rodr. ■

405855　Verbena incisa Hook.；锐裂马鞭草■☆

405856　Verbena indica L. = Stachytarpheta indica（L.）Vahl ●

405857　Verbena jamaicensis L.；牙买加马鞭草；Jamaica Vervain ■☆

405858　Verbena jamaicensis L. = Stachytarpheta jamaicensis（L.）Vahl ■●

405859　Verbena javanica Burm. f. = Lippia javanica（Burm. f.）Spreng. ●■☆

405860　Verbena lambertii Sims = Verbena canadensis（L.）Britton ■☆

405861　Verbena lappulacea L.；多芒马鞭草；Burry Vervain ■☆

405862　Verbena lappulacea L. = Priva lappulacea（L.）Pers. ■☆

405863　Verbena lasiostachys Link；毛蕊马鞭草■☆

405864　Verbena lilacina Greene；雪松马鞭草；Cedros Island Verbena ■☆

405865　Verbena litoralis Kunth；海滨马鞭草；Ha'uowi，Oi，Owi，Seashore Vervain ■☆

405866　Verbena litoralis Kunth var. brasiliensis（Vell.）Briq. = Verbena brasiliensis Vell. ■☆

405867　Verbena litoralis Kunth var. caracasana（Kunth）Briq. = Verbena litoralis Kunth ■☆

405868　Verbena moechina Moldenke；莫氏马鞭草；Verbena ■☆

405869　Verbena montevidensis Spreng.；乌拉圭马鞭草；Uruguayan Vervain ■☆

405870　Verbena mutabilis Jacq. = Stachytarpheta mutabilis（Jacq.）Vahl ●☆

405871　Verbena nodiflora L. = Phyla nodiflora（L.）Greene ■

405872　Verbena officinalis L.；马鞭草（白马鞭，风须草，风颈草，蛤蟆棵，狗牙草，鹤膝风，红藤草，苦练草，马鞭梢，马鞭稍，马鞭子，疟马鞭，蜻蜓草，蜻蜓饭，散血草，顺捋草，田鸡草，铁麻鞭，铁马鞭，铁马莲，铁扫手，透骨草，土荆芥，土马鞭，兔子草，退血草，须捋草，燕尾草，野荆芥，粘身蓝被，紫顶龙芽）；Ashthroat，Berbine，Blood of Mercury，Columbine，Common Verbena，Common Vervain，Devil's Bane，Devil's Hate，Europe Vervain，European Verbena，European Vervain，Frogfoot，Herb of Grace，Herb of the Cross，Holy Herb，Holy Vervain，Iron-hard，Juno's Tears，La Campana，Mercury's Moist Blood，Nan Wade，Pigeon's Grass，Pigeon's Meat，Simpler's Joy，Varvine，Verbena，Vervain，Vevine ■

405873　Verbena officinalis L. subsp. africana R. Fern. et Verdc.；非洲马鞭草■☆

405874　Verbena officinalis L. var. natalensis Krauss = Verbena officinalis L. subsp. africana R. Fern. et Verdc. ■☆

405875　Verbena officinalis L. var. prostrata ?；平卧马鞭草；Prostrate Verbena ■☆

405876　Verbena officinalis L. var. ramosa H. Lév. = Verbena officinalis L. ■

405877　Verbena oklahomensis Moldenke = Verbena canadensis（L.）

Britton ■☆

405878　Verbena paniculatistricta Engelm. = Verbena rydbergii Moldenke ■☆

405879　Verbena patagonica Speg. = Verbena bonariensis L. ■

405880　Verbena perriana Moldenke; 佩尔美女樱; Perry's Vervain, Vervain ■☆

405881　Verbena peruviana (L.) Britton; 小花美女樱; Verbena ■☆

405882　Verbena phlogiflora Cham. = Verbena hybrida Voss ex Groem. et Rümpler ■

405883　Verbena pubera Greene = Verbena bipinnatifida Nutt. ■☆

405884　Verbena quadrangularis Vell. = Verbena brasiliensis Vell. ■☆

405885　Verbena rigida Spreng.; 糙叶美人樱 (坚挺美女樱); Coarse Verbena, Purple Verbena, Rigid Verbena, Sandpaper Verbena, Slender Vervain, Tuberous Vervain, Veined Vervain, Verbena ■☆

405886　Verbena rugosa D. Don = Verbena rigida Spreng. ☆

405887　Verbena rydbergii Moldenke; 雷氏马鞭草; Rydberg's Vervain, Vervain ■☆

405888　Verbena simplex Lehm.; 狭叶马鞭草; Narrow-leaved Vervain ■☆

405889　Verbena stricta Vent.; 绒毛马鞭草; Fever Weed, Hoary Verbena, Hoary Vervain, Upright Vervain ■☆

405890　Verbena stricta Vent. f. albiflora Wadmond; 白花绒毛马鞭草■☆

405891　Verbena stricta Vent. f. albiflora Wadmond = Verbena stricta Vent. ■☆

405892　Verbena stricta Vent. f. roseiflora Benke = Verbena stricta Vent. ■☆

405893　Verbena supina L.; 卧马鞭草 (仰卧马鞭草); Prostrate Verbena, Supine Vervain ■☆

405894　Verbena supina L. f. erecta Moldenke; 直立马鞭草■☆

405895　Verbena supina L. var. erecta Pau = Verbena supina L. f. erecta Moldenke ■☆

405896　Verbena tenera Spreng.; 细叶美女樱 (裂叶美人樱); Thinleaf Vervain ■

405897　Verbena tenuisecta Briq.; 细齿马鞭草; Moss Verbena ■☆

405898　Verbena tenuisecta Briq. = Verbena aristigera S. Moore ■☆

405899　Verbena tenuispicata Stapf = Verbena officinalis L. ■

405900　Verbena trichotoma Moench = Verbena bonariensis L. ■

405901　Verbena triphylla L'Hér. = Aloysia citriodora Paláu ●☆

405902　Verbena triphylla L'Hér. = Lippia citrodora (Paláu) Kunth ●☆

405903　Verbena urticifolia L.; 白马鞭草; Nettle-leaved Vervain, Nettle-leaved Vervein, White Vervain ■☆

405904　Verbena urticifolia L. var. leiocarpa L. M. Perry et Fernald; 光果白马鞭草; Nettle-leaved Vervain, White Vervain ■☆

405905　Verbena venosa Gillies et Hook. = Verbena rigida Spreng. ■☆

405906　Verbenaceae Adans. = Verbenaceae J. St. -Hil. (保留科名) ●■

405907　Verbenaceae J. St. -Hil. (1805) (保留科名); 马鞭草科; Verbena Family, Vervain Family ●■

405908　Verbenastrum Lippi ex Del. = Capraria L. ■☆

405909　Verbenoxylum Tronc. (1971); 巴西马鞭木属 ●☆

405910　Verbenoxylum reitzii (Moldenke) Tronc.; 巴西马鞭木 ●☆

405911　Verbesina L. (1753) (保留属名); 冠须菊属 (冠须菊, 马鞭菊属, 韦伯西菊属); Crownbeard, Crown Beard ●■☆

405912　Verbesina abyssinica (Sch. Bip. ex Walp.) A. Rich. = Bidens camporum (Hutch.) Mesfin ■☆

405913　Verbesina acmella L. = Blainvillea acmella (L.) Phillipson ■

405914　Verbesina alba L. = Eclipta prostrata (L.) L. ■

405915　Verbesina alternifolia (L.) Britton ex Kearney; 互叶冠须菊 (互叶奇瓣菊, 翼枝菊); Wingstem, Wing-stem, Yellow Ironweed ■☆

405916　Verbesina aristata (Elliott) A. Heller; 具芒冠须菊■☆

405917　Verbesina aschenborniana ?; 阿申冠须菊 (阿申斑鸠菊) ■☆

405918　Verbesina asteroides L. = Amellus asteroides (L.) Druce ■☆

405919　Verbesina barclayae H. Rob.; 巴克利冠须菊 ●☆

405920　Verbesina biblianensis Domke; 比夫利安冠须菊 ●☆

405921　Verbesina biflora L. = Melanthera biflora (L.) Wild ■☆

405922　Verbesina biflora L. = Wedelia biflora (L.) DC. ■

405923　Verbesina bosvallia L. f. = Glossocardia bosvallia (L. f.) DC. ☆

405924　Verbesina brachyopoda S. F. Blake; 短梗冠须菊 ●☆

405925　Verbesina calendulacea L. = Wedelia chinensis (Osbeck) Merr. ■

405926　Verbesina chapmanii J. R. Coleman; 查普曼冠须菊■☆

405927　Verbesina chinensis L. = Anisopappus chinensis (L.) Hook. et Arn. ■

405928　Verbesina ciliata Schumach. = Aspilia helianthoides (Schumach. et Thonn.) Oliv. et Hiern subsp. ciliata (Schumach.) C. D. Adams ■☆

405929　Verbesina crocata Less.; 镉黄冠须菊 (镉黄韦伯西菊) ■☆

405930　Verbesina dentata Kunth; 齿冠须菊 ●☆

405931　Verbesina dichotoma Murray = Blainvillea rhomboidea Cass. ■

405932　Verbesina dissita A. Gray; 分离冠须菊■☆

405933　Verbesina encelioides (Cav.) A. Gray = Verbesina encelioides (Cav.) Benth. et Hook. f. ex A. Gray ■☆

405934　Verbesina encelioides (Cav.) A. Gray var. exauriculata B. L. Rob. et Greenm.; 亚里桑冠须菊■☆

405935　Verbesina encelioides (Cav.) Benth. et Hook. f. = Verbesina encelioides (Cav.) Benth. et Hook. f. ex A. Gray ■☆

405936　Verbesina encelioides (Cav.) Benth. et Hook. f. ex A. Gray; 韦伯西菊; Butter Daisy, Golden Crownbeard, Golden Crown-beard, Toothache Plant ■☆

405937　Verbesina encelioides (Cav.) Benth. et Hook. f. ex A. Gray subsp. exauriculata (B. L. Rob. et Greenm.) J. R. Coleman = Verbesina encelioides (Cav.) Benth. et Hook. f. ex A. Gray ■☆

405938　Verbesina encelioides (Cav.) Benth. et Hook. f. var. exauriculata B. L. Rob. et Greenm. = Verbesina encelioides (Cav.) Benth. et Hook. f. ■☆

405939　Verbesina encelioides (Cav.) Benth. et Hook. var. exauriculata C. B. Rob. et Greenm. = Verbesina encelioides (Cav.) Benth. et Hook. f. ex A. Gray ■☆

405940　Verbesina hallii Hieron.; 豪尔冠须菊 ●☆

405941　Verbesina helianthoides Michx.; 向日葵冠须菊; Yellow Crownbeard ■☆

405942　Verbesina heterophylla (Chapm.) A. Gray; 异叶冠须菊■☆

405943　Verbesina humboldtii Spreng.; 洪堡冠须菊 ●☆

405944　Verbesina inuloides Hieron.; 旋覆花冠须菊 ●☆

405945　Verbesina involucrata (Sch. Bip. ex Walp.) A. Rich. = Bidens pachyloma (Oliv. et Hiern) Cufod. ■☆

405946　Verbesina kingii H. Rob.; 金氏冠须菊 ●☆

405947　Verbesina laciniata Nutt. = Verbesina virginica L. ■☆

405948　Verbesina laciniata Walter = Verbesina virginica L. ■☆

405949　Verbesina latisquamata S. F. Blake; 宽鳞冠须菊 ●☆

405950　Verbesina lavenia L. = Adenostemma lavenia (L.) Kuntze ■

405951　Verbesina lindheimeri B. L. Rob. et Greenm.; 里氏冠须菊■☆

405952　Verbesina lineata A. Rich. = Bidens setigera (Sch. Bip. ex Walp.) Sherff ■☆

405953　Verbesina longifolia (A. Gray) A. Gray; 长叶冠须菊■☆

405954　Verbesina macrantha (Sch. Bip.) A. Rich. = Bidens macroptera (Sch. Bip. ex Chiov.) Mesfin ■☆

405955　Verbesina microptera DC.; 小翅冠须菊■☆

405956　Verbesina monticola Hook. f. = Bidens mannii T. G. J. Rayner ■☆

405957　Verbesinanana (A. Gray) B. L. Rob. et Greenm.; 矮冠须菊■☆

405958　Verbesina nodiflora L. = Synedrella nodiflora（L.）Gaertn. ■

405959　Verbesina occidentalis（L.）Walter；西方须菊；Southern Flatseed Sunflower ■☆

405960　Verbesina oreophila Wooton et Standl. ；沙丘须菊■☆

405961　Verbesina pentantha S. F. Blake；五花须菊●☆

405962　Verbesina persicifolia DC. ；桃叶冠须菊（桃叶韦伯西菊）■☆

405963　Verbesina podocephala A. Gray = Lasianthaea podocephala（A. Gray）K. M. Becker ■☆

405964　Verbesina prostrata Hook. et Arn. = Wedelia prostrata（Hook. et Arn.）Hemsl. ■

405965　Verbesina prostrata L. = Eclipta prostrata（L.）L. ■

405966　Verbesina rothrockii B. L. Rob. et Greenm. ；罗思罗克冠须菊■☆

405967　Verbesina rueppellii（Sch. Bip. ex Walp.）A. Rich. = Bidens rueppellii（Sch. Bip. ex Walp.）Sherff ■☆

405968　Verbesina saloyensis Domke；萨洛亚冠须菊●☆

405969　Verbesina sativa Roxb. = Guizotia abyssinica（L. f.）Cass. ■☆

405970　Verbesina sodiroi Hieron. ；苏迪洛冠须菊●☆

405971　Verbesina veris A. Rich. = Bidens prestinaria（Sch. Bip. ex Walp.）Cufod. ■☆

405972　Verbesina villonacoensis H. Rob. ；比略纳克冠须菊●☆

405973　Verbesina virginica L. ；弗州冠须菊；Frost Weed, Tickweed, White Crownbeard ■☆

405974　Verbesina virginica L. var. laciniata A. Gray = Verbesina virginica L. ■☆

405975　Verbesina walteri Shinners；瓦尔特冠须菊■☆

405976　Verbiascum Fenzl = Verbascum L. ■●

405977　Verbvesina alba L. = Eclipta prostrata（L.）L. ■

405978　Verdcourtia R. Wilczek = Dipogon Liebm. ■☆

405979　Verdickia De Wild. = Chlorophytum Ker Gawl. ■

405980　Verea Willd. = Kalanchoe Adans. ●■

405981　Verea Willd. = Vereia Andréws ●■

405982　Verea crenata Andréws = Kalanchoe crenata（Andréws）Haw. ■☆

405983　Verea laciniata（L.）Willd. = Kalanchoe laciniata（L.）DC. ■

405984　Verea rotundifolia（Haw.）D. Dietr. = Kalanchoe rotundifolia（Haw.）Haw. ■☆

405985　Vereia Andréws = Kalanchoe Adans. ●■

405986　Vereia pinnata（Lam.）Andréws = Kalanchoe pinnata（Lam.）Pers. ■

405987　Vereia pinnata（Lam.）Willd. = Kalanchoe pinnata（Lam.）Pers. ■

405988　Verena Minod = Stemodia L. （保留属名）■☆

405989　Verhuellia Miq. （1843）；巴西草胡椒属■☆

405990　Verhuellia Miq. = Peperomia Ruiz et Pav. ■

405991　Verhuellia elegans Miq. ；巴西草胡椒■☆

405992　Verhuellia knoblocheriana（Schott）C. DC. = Peperomia pellucida（L.）Kunth ■

405993　Veriangia Neck. = Argania Roem. et Schult. （保留属名）●☆

405994　Verinea Merino = Melica L. ■

405995　Verinea Pomel = Asphodelus L. ■☆

405996　Verinea fistulosa（L.）Pomel = Asphodelus fistulosus L. ■☆

405997　Verinea tenuifolius（Cav.）Pomel = Asphodelus tenuifolius Cav. ■☆

405998　Verlangia Neck. = Argania Roem. et Schult. （保留属名）●☆

405999　Verlangia Neck. ex Raf. = Argania Roem. et Schult. （保留属名）●☆

406000　Verlotia E. Fourn. = Marsdenia R. Br. （保留属名）●

406001　Vermeulenia Á. Löve et D. Löve = Orchis L. ●

406002　Vermicularia Moench = Stachytarpheta Vahl（保留属名）■●

406003　Vermifrux J. B. Gillett = Lotus L. ■

406004　Vermifrux abyssinica（A. Rich.）J. B. Gillett = Dorycnopsis abyssinica（A. Rich.）Tikhom. et D. D. Sokoloff ■☆

406005　Vermifuga Ruiz et Pav. = Flaveria Juss. ■●

406006　Verminiaria Hon. = Viminaria Sm. ●☆

406007　Vermoneta Comm. ex Juss. = Homalium Jacq. ●

406008　Vernonia Edgew. = Vernonia Schreb. （保留属名）●■

406009　Vermontea Steud. = Homalium Jacq. ●

406010　Vermontea Steud. = Vermoneta Comm. ex Juss. ●

406011　Vernasolis Raf. = Coreopsis L. ●■

406012　Vernicaceae Link = Anacardiaceae R. Br. （保留科名）●

406013　Vernicaceae Schultz Sch. = Anacardiaceae R. Br. （保留科名）●

406014　Vernicia Lour. （1790）；油桐属；Tungoiltree, Oiltung, Tung-oil Tree, Tung-oil-tree ●

406015　Vernicia Lour. = Aleurites J. R. Forst. et G. Forst. ●

406016　Vernicia cordata（Thunb.）Airy Shaw；日本木油桐（荏桐,日本木油树,日本油桐,桐树,桐油,桐子,桐子树,小桐树,心叶油桐,罂子桐,油桐）；Japan Wood Oil Tree, Japan Wood-oil Tree, Japanese Wood Oil Tree, Japanese Wood-oil Tree ●☆

406017　Vernicia cordata（Thunb.）Airy Shaw = Aleurites cordata（Thunb.）R. Br. ex Steud. ●☆

406018　Vernicia fordii（Hemsl.）Airy Shaw；油桐（百年桐,光面桐,光桐,虎子桐,荏桐,三年桐,桐油树,桐子树,五年桐,罂子树,罂子桐）；China Wood-oil Tree, China-wood, Chinese Wood-oil Tree, Oiltung, Tung Nut, Tung Oil Tree, Tung Tree, Tungoil Tree, Tung-oil Tree, Tungoiltree, Tung-oil-tree, Wood Oil ●

406019　Vernicia montana Lour. ；木油桐（广东油桐,木油树,千年桐,乌龟桐,五爪桐,皱果桐,皱皮桐,皱桐）；Candle Nut Tree, Mu Oil Tree, Mu Oiltree, Mu Tree, Mu-oil, Mu-oil Tree, Muoiltree, Mu-oiltree, Muyou Oiltung, Tun Oil Tree, Wood Oil Tree, Wood Oil-tree, Wood-oil Tree ●

406020　Verniseckia Steud. = Houmiri Aubl. ●☆

406021　Verniseckia Steud. = Wernisekia Scop. ●☆

406022　Vernix Adans. = Rhus L. ●

406023　Vernix Adans. = Toxicodendron Mill. ●

406024　Vernonanthura H. Rob. （1992）；方晶斑鸠菊属●☆

406025　Vernonanthura angulata（H. Rob.）H. Rob. ；窄方晶斑鸠菊●☆

406026　Vernonanthura auriculata（Griseb.）H. Rob. ；耳方晶斑鸠菊●☆

406027　Vernonanthura cordata（Kunth）H. Rob. ；心形方晶斑鸠菊●☆

406028　Vernonanthura cuneifolia（Gardner）H. Rob. ；楔叶方晶斑鸠菊●☆

406029　Vernonanthura discolor（Less.）H. Rob. ；异色方晶斑鸠菊●☆

406030　Vernonanthura ferruginea（Less.）H. Rob. ；锈色方晶斑鸠菊●☆

406031　Vernonanthura laxa（Gardner）H. Rob. ；松散方晶斑鸠菊●☆

406032　Vernonanthura lucida（Less.）H. Rob. ；亮方晶斑鸠菊●☆

406033　Vernonanthura membranacea Gardner；膜质方晶斑鸠菊●☆

406034　Vernonanthura nudiflora（Less.）H. Rob. ；裸花方晶斑鸠菊●☆

406035　Vernonella Sond. （1850）；非洲小斑鸠菊属■☆

406036　Vernonella Sond. = Centrapalus Cass. ■☆

406037　Vernonella Sond. = Vernonia Schreb. （保留属名）●■

406038　Vernonella africana Sond. ；非洲小斑鸠菊■☆

406039　Vernonella africana Sond. = Vernonia africana（Sond.）Druce ■☆

406040　Vernonia Schreb. （1791）（保留属名）；斑鸠菊属；Bitterleaf, Cabbage Tree, Iron Weed, Ironweed ●■

406041　Vernonia abbreviata（Wall.）DC. = Vernonia cinera（L.）Less. ■

406042　Vernonia abyssinica Sch. Bip. ex Walp. = Vernonia schimperi DC. ■☆

406043　Vernonia acaulis（Walter）Gleason；无茎斑鸠菊■☆

406044　Vernonia acrocephala Klatt；尖头斑鸠菊■☆

406045　Vernonia acuminatissima S. Moore；渐尖斑鸠菊■☆

406046　Vernonia acuta De Wild. = Vernonia sunzuensis Wild ■☆

406047　Vernonia adenocephala S. Moore；腺头斑鸠菊■☆

406048　Vernonia adenosticta Fenzl ex Walp. = Vernonia amygdalina Delile ■☆

406049　Vernonia adoensis Sch. Bip. ex Walp. ；阿多斑鸠菊■☆

406050　Vernonia adoensis Sch. Bip. ex Walp. = Ascaricida richardii Steetz ■☆

406051　Vernonia adoensis Sch. Bip. ex Walp. var. kotschyana（Sch. Bip. ex Walp.）G. V. Pope = Baccharoides adoensis（Sch. Bip. ex Walp.）H. Rob. var. kotschyana（Sch. Bip. ex Walp.）Isawumi, El-Ghazaly et B. Nord. ■☆

406052　Vernonia adoensis Sch. Bip. ex Walp. var. mossambiquensis（Steetz）G. V. Pope = Baccharoides adoensis（Sch. Bip. ex Walp.）H. Rob. var. mossambiquensis（Steetz）Isawumi, El-Ghazaly et B. Nord. ●☆

406053　Vernonia adolfi-friderici Muschl. = Vernonia calvoana（Hook. f.）Hook. f. subsp. adolfi-friderici（Muschl.）C. Jeffrey ■☆

406054　Vernonia aemulans Vatke；匹敌斑鸠菊■☆

406055　Vernonia africana（Sond.）Druce = Vernonella africana Sond. ■☆

406056　Vernonia afromontana R. E. Fr. = Vernonia galamensis（Cass.）Less. var. afromontana（R. E. Fr.）M. G. Gilbert ■☆

406057　Vernonia agrianthoides C. Jeffrey；田花斑鸠菊●☆

406058　Vernonia agricola S. Moore = Vernonia subaphylla Baker ■☆

406059　Vernonia alata Heyne ex DC. = Laggera alata（D. Don）Sch. Bip. ex Oliv. ■

406060　Vernonia albocinerascens C. Jeffrey；灰白斑鸠菊■☆

406061　Vernonia albosquama Y. L. Chen；白苞斑鸠菊●

406062　Vernonia alboviolacea De Wild. = Vernonia ringoetii De Wild. ●☆

406063　Vernonia alboviolacea Muschl. ；浅堇色斑鸠菊■☆

406064　Vernonia aldabrensis Hemsl. = Vernonia colorata（Willd.）Drake subsp. grandis（DC.）C. Jeffrey ■☆

406065　Vernonia alleizettei Humbert；阿雷斑鸠菊●☆

406066　Vernonia alsodea Klatt = Vernonia diversifolia Bojer ex DC. ■☆

406067　Vernonia alticola G. V. Pope；高原斑鸠菊■☆

406068　Vernonia altissima Nutt. ；高斑鸠菊（紫霞菊）■☆

406069　Vernonia altissima Nutt. = Vernonia gigantea（Walter）Trel. ex Branner et Coville ■☆

406070　Vernonia altissima Nutt. f. alba Moldenke = Vernonia gigantea（Walter）Trel. ex Branner et Coville ■☆

406071　Vernonia altissima Nutt. var. taeniotricha S. F. Blake = Vernonia gigantea（Walter）Trel. ex Branner et Coville ■☆

406072　Vernonia amaniensis Muschl. ；阿马尼斑鸠菊■☆

406073　Vernonia ambigua Kotschy et Peyr. ；可疑斑鸠菊■☆

406074　Vernonia amblyolepis Baker；钝鳞斑鸠菊■☆

406075　Vernonia ambolensis Humbert；安布尔斑鸠菊●☆

406076　Vernonia ambongensis Humbert；安邦斑鸠菊●☆

406077　Vernonia ambrensis Humbert；昂布尔斑鸠菊●☆

406078　Vernonia amoena S. Moore；秀丽斑鸠菊●☆

406079　Vernonia ampandrandavensis Humbert；安潘斑鸠菊●☆

406080　Vernonia ampla O. Hoffm. = Vernonia myriantha Hook. f. ●☆

406081　Vernonia ampla Vaniot = Conyza leucantha（D. Don）Ludlow et Raven ■

406082　Vernonia amplexicaulis Baker = Vernonia uncinata Oliv. et Hiern ■☆

406083　Vernonia amygdalina Delile；扁桃斑鸠菊（扁桃状斑鸠菊，杏叶斑鸠菊）；Bitter Leaf, Bitterleaf ■☆

406084　Vernonia anandrioides S. Moore；无雄斑鸠菊●☆

406085　Vernonia andapensis Humbert；安达帕斑鸠菊●☆

406086　Vernonia andersonii C. B. Clarke；细脉斑鸠菊（毒根斑鸠菊，印度斑鸠菊）■

406087　Vernonia andersonii C. B. Clarke = Vernonia cumingiana Benth. ●■

406088　Vernonia andersonii C. B. Clarke var. albipappa Hayata = Vernonia gratiosa Hance ●■

406089　Vernonia andersonii Henry = Vernonia gratiosa Hance ●■

406090　Vernonia andersonii Henry var. albipappa Hayata = Vernonia gratiosa Hance ●■

406091　Vernonia andohii C. D. Adams = Vernonia doniana DC. ●☆

406092　Vernonia andrangovalensis Humbert；安德朗斑鸠菊●☆

406093　Vernonia angolensis（O. Hoffm.）N. E. Br. = Distephanus angolensis（O. Hoffm.）H. Rob. et B. Kahn ●☆

406094　Vernonia angulifolia DC. = Distephanus angulifolius（DC.）H. Rob. et B. Kahn ■☆

406095　Vernonia angustifolia Michx. ；狭叶斑鸠菊■☆

406096　Vernonia angustifolia Michx. subsp. mohrii（S. B. Jones）S. B. Jones et W. Z. Faust = Vernonia angustifolia Michx. ■☆

406097　Vernonia angustifolia Michx. subsp. scaberrima（Nutt.）S. B. Jones et W. Z. Faust = Vernonia angustifolia Michx. ■☆

406098　Vernonia angustifolia Michx. var. scaberrima（Nutt.）A. Gray = Vernonia angustifolia Michx. ■☆

406099　Vernonia anisochaetoides Sond. = Distephanus anisochaetoides（Sond.）H. Rob. et B. Kahn ■☆

406100　Vernonia antandroyi Humbert = Distephanus antandroy（Humbert）H. Rob. et B. Kahn ●☆

406101　Vernonia antanossi Scott-Elliot = Centauropsis antanossi（Scott-Elliot）Humbert ●☆

406102　Vernonia anthelmintica（L.）Willd. ；驱虫斑鸠菊（毒根斑鸠菊，发痧藤，过山龙，红脉斑鸠菊，虎三头，惊风红，藤牛七，藤三七，细脉斑鸠菊，夜牵牛，印度山茴香）；Anthelmintic Ironweed, Vermifuge Ironweed ■

406103　Vernonia anthelmintica（L.）Willd. = Baccharoides anthelmintica（L.）Moench ●

406104　Vernonia antinoriana Avetta = Vernonia urticifolia A. Rich. ■☆

406105　Vernonia antunesii O. Hoffm. ；安图内思斑鸠菊●☆

406106　Vernonia appendiculata Less. ；附属物斑鸠菊●☆

406107　Vernonia arabica Davies；阿拉伯斑鸠菊●☆

406108　Vernonia arbor H. Lév. = Vernonia esculenta Hemsl. ex Forbes et Hemsl. ●

406109　Vernonia arborea Buch. -Ham. ；树斑鸠菊（乔木斑鸠菊）；Arborous Ironweed, Tree Ironweed, Tree-like Ironweed ●

406110　Vernonia arborea Welw. ex O. Hoffm. = Vernonia conferta Benth. ●☆

406111　Vernonia arenicola S. Moore；沙生斑鸠菊●☆

406112　Vernonia argutidens Chiov. = Vernonia buchingeri（Steetz）Oliv. et Hiern ■☆

406113　Vernonia aristata（DC.）Sch. Bip. = Vernonia natalensis Sch. Bip. ex Walp. ■☆

406114　Vernonia arkansana DC. ；阿肯色斑鸠菊；Arkansas Ironweed, Ironweed, Ozark Ironweed ■☆

406115　Vernonia armerioides O. Hoffm. = Vernonia subaphylla Baker ■☆

406116　Vernonia aschersonii Sch. Bip. ；阿舍逊斑鸠菊■☆

406117　Vernonia aschersonii Sch. Bip. var. robusta Cufod. = Vernonia popeana C. Jeffrey ●☆

406118　Vernonia aschersonioides Chiov. = Vernonia aschersonii Sch. Bip. ■☆

406119　Vernonia aspera（Roxb.）Buch.-Ham.；糙叶斑鸠菊（糙叶咸虾花，黑升麻，六月雪）；Roughleaf Ironweed ■

406120　Vernonia assimilis S. Moore = Vernonia holstii O. Hoffm. ■☆

406121　Vernonia asterifolia Baker = Vernonia schweinfurthii Oliv. et Hiern ■☆

406122　Vernonia atriplicifolia Jaub. et Spach = Vernonia spatulata（Forssk.）Sch. Bip. ■☆

406123　Vernonia attenuata（Wall.）DC.；狭长斑鸠菊；Attenuate Ironweed，Long and Narrow Ironweed ■

406124　Vernonia aurantiaca（O. Hoffm.）N. E. Br. = Distephanus divaricatus（Steetz）H. Rob. et B. Kahn ●☆

406125　Vernonia auriculifera Hiern；具耳斑鸠菊 ■☆

406126　Vernonia baillonii Scott-Elliot；巴永斑鸠菊 ●☆

406127　Vernonia bainesii Oliv. et Hiern；贝恩斯斑鸠菊 ●☆

406128　Vernonia bainesii Oliv. et Hiern subsp. wildii（Merxm.）Wild；维尔德斑鸠菊 ■☆

406129　Vernonia bakeri Vatke；贝克斑鸠菊 ■☆

406130　Vernonia baldwinii Torr.；鲍尔温斑鸠菊；Ironweed ■☆

406131　Vernonia baldwinii Torr. subsp. interior（Small）W. Z. Faust = Vernonia baldwinii Torr. ■☆

406132　Vernonia baldwinii Torr. var. interior（Small）B. G. Schub. = Vernonia baldwinii Torr. ■☆

406133　Vernonia ballyi C. Jeffrey；博利斑鸠菊 ■☆

406134　Vernonia bambilorensis Berhaut；班比斑鸠菊 ■☆

406135　Vernonia bamendae C. D. Adams；巴门达斑鸠菊 ■☆

406136　Vernonia baoulensis A. Chev.；巴乌莱斑鸠菊 ■☆

406137　Vernonia barbosae Wild = Vernonia ugandensis S. Moore ■☆

406138　Vernonia baronii Baker；巴龙斑鸠菊 ●☆

406139　Vernonia bauchiensis Hutch. et Dalziel；包奇斑鸠菊 ■☆

406140　Vernonia baumii O. Hoffm. = Vernonia temnolepis O. Hoffm. ■☆

406141　Vernonia beforonensis Humbert = Vernonia humblotii Drake ■☆

406142　Vernonia bellinghamii S. Moore；贝林厄姆斑鸠菊 ●☆

406143　Vernonia benguellensis Hiern；本格拉斑鸠菊 ■☆

406144　Vernonia benthamiana Oliv. et Hiern = Kinghamia angustifolia（Benth.）C. Jeffrey ■☆

406145　Vernonia bequaertii De Wild. = Vernonia adoensis Sch. Bip. ex Walp. var. kotschyana（Sch. Bip. ex Walp.）G. V. Pope ■☆

406146　Vernonia betonicifolia Baker；药水苏叶斑鸠菊 ■☆

406147　Vernonia betsilensis Drake；贝齐尔斑鸠菊 ■☆

406148　Vernonia biafrae Oliv. et Hiern；比亚斑鸠菊 ■☆

406149　Vernonia biafrae Oliv. et Hiern var. tufnelliae Isawumi et al. = Vernonia biafrae Oliv. et Hiern ■☆

406150　Vernonia blanda（Wall.）DC.；喜斑鸠菊（蔓斑鸠菊）；Bland Ironweed，Bock Ironweed，Smooth Ironweed ●■

406151　Vernonia blodgettii Small；布洛杰特斑鸠菊 ■☆

406152　Vernonia blumeoides Hook. f.；艾纳香斑鸠菊 ●☆

406153　Vernonia bockiana Diels；南川斑鸠菊；Bock's Ironweed，Nanchuan Ironweed，S. Sichuan Ironweed ●

406154　Vernonia bojeri Less.；博耶尔斑鸠菊 ■☆

406155　Vernonia bothrioclinoides C. H. Wright = Vernonia karaguensis Oliv. ●☆

406156　Vernonia brachycalyx O. Hoffm.；短萼斑鸠菊 ■☆

406157　Vernonia brachycalyx O. Hoffm. var. megana Cufod. = Vernonia brachycalyx O. Hoffm. ■☆

406158　Vernonia brachylaenoides S. Moore = Vernonia suprafastigiata Klatt ●☆

406159　Vernonia brachyscypha Baker = Vernonia pachyclada Baker ■☆

406160　Vernonia brachytrichoides C. Jeffrey；短毛斑鸠菊 ■☆

406161　Vernonia bracteata Wall. = Vernonia nantcianensis（Pamp.）Hand.-Mazz. ■

406162　Vernonia bracteata Wall. var. nantcianensis Pamp. = Vernonia nantcianensis（Pamp.）Hand.-Mazz. ■

406163　Vernonia bracteosa O. Hoffm.；小苞斑鸠菊 ■☆

406164　Vernonia braunii Muschl. = Vernonia lasiopus O. Hoffm. var. iodocalyx（O. Hoffm.）C. Jeffrey ●☆

406165　Vernonia brazzavillensis Aubrév. ex Compère；布拉柴维尔斑鸠菊 ●☆

406166　Vernonia brideliifolia O. Hoffm.；土密树斑鸠菊 ●☆

406167　Vernonia britteniana Hiern；布里滕斑鸠菊 ■☆

406168　Vernonia brownii S. Moore = Vernonia lasiopus O. Hoffm. ●☆

406169　Vernonia bruceae C. Jeffrey；布鲁斯斑鸠菊 ■☆

406170　Vernonia bruceana Wild；布氏斑鸠菊 ●☆

406171　Vernonia buchananii Baker = Vernonia bainesii Oliv. et Hiern ●☆

406172　Vernonia buchingeri（Steetz）Oliv. et Hiern；布赫斑鸠菊 ■☆

406173　Vernonia bukamaensis De Wild. = Vernonia schweinfurthii Oliv. et Hiern var. bukamaensis（De Wild.）Kalanda et Lisowski ●☆

406174　Vernonia bullulata S. Moore；泡状斑鸠菊 ●☆

406175　Vernonia bulo-burtiensis Mesfin；布洛布尔提斑鸠菊 ●☆

406176　Vernonia burtonii Oliv. et Hiern = Vernonia karaguensis Oliv. ●☆

406177　Vernonia caboverdeana Lobin = Vernonia colorata（Willd.）Drake ■☆

406178　Vernonia calongensis Muschl. = Vernonia syringifolia O. Hoffm. ●☆

406179　Vernonia calulu Hiern var. carinata（Hutch. et B. L. Burtt）Kalanda；龙骨状斑鸠菊 ●☆

406180　Vernonia calvoana（Hook. f.）Hook. f.；苦斑鸠菊；Bitterleaf ●☆

406181　Vernonia calvoana（Hook. f.）Hook. f. subsp. adolfi-friderici（Muschl.）C. Jeffrey；弗里德里西斑鸠菊 ■☆

406182　Vernonia calvoana（Hook. f.）Hook. f. subsp. leucocalyx（O. Hoffm.）C. Jeffrey；白萼苦斑鸠菊 ■☆

406183　Vernonia calvoana（Hook. f.）Hook. f. subsp. meridionalis（Wild）C. Jeffrey；南方苦斑鸠菊 ■☆

406184　Vernonia calvoana（Hook. f.）Hook. f. subsp. oehleri（Muschl.）C. Jeffrey；奥勒苦斑鸠菊 ■☆

406185　Vernonia calvoana（Hook. f.）Hook. f. subsp. ruwenzoriensis C. Jeffrey；鲁文佐里苦斑鸠菊 ■☆

406186　Vernonia calvoana（Hook. f.）Hook. f. subsp. ulugurensis（O. Hoffm.）C. Jeffrey；乌卢古尔斑鸠菊 ■☆

406187　Vernonia calvoana（Hook. f.）Hook. f. subsp. usambarensis C. Jeffrey；乌桑巴拉苦斑鸠菊 ■☆

406188　Vernonia calvoana（Hook. f.）Hook. f. var. acuta（C. D. Adams）C. Jeffrey；尖锐斑鸠菊 ■☆

406189　Vernonia calvoana（Hook. f.）Hook. f. var. mesocephala C. D. Adams = Vernonia hymenolepis A. Rich. ■☆

406190　Vernonia calvoana（Hook. f.）Hook. f. var. microcephala C. D. Adams；小头苦斑鸠菊 ■☆

406191　Vernonia calvoana（Hook. f.）Hook. f. var. mokaensis（Mildbr. et Mattf.）C. Jeffrey；莫卡斑鸠菊 ■☆

406192　Vernonia calyculata S. Moore；小萼斑鸠菊 ■☆

406193　Vernonia campanea S. Moore = Vernonia karaguensis Oliv. ●☆

406194　Vernonia campenonii Drake = Vernonia baronii Baker ●☆

406195　Vernonia campicola S. Moore；平原斑鸠菊 ■☆

406196　Vernonia camporum A. Chev.；弯斑鸠菊 ■☆

406197　Vernonia candelabricephala Gilli = Vernonia usafuensis O. Hoffm. ■☆

406198　Vernonia candollei Vatke = Vernonia erythromarula Klatt ■☆

406199　Vernonia cannabina Muschl. = Vernonia nestor S. Moore ●☆

406200　Vernonia capensis（Houtt.）Druce；好望角斑鸠菊■☆

406201　Vernonia capuronii Humbert；凯普伦斑鸠菊■☆

406202　Vernonia caput-medusae S. Moore = Vernonia ugandensis S. Moore ■☆

406203　Vernonia cardiolepis O. Hoffm. = Vernonia guineensis Benth. ■☆

406204　Vernonia cardiophylla Gilli = Vernonia syringifolia O. Hoffm. ●☆

406205　Vernonia carinata Hutch. et B. L. Burtt = Vernonia calulu Hiern var. carinata（Hutch. et B. L. Burtt）Kalanda ●☆

406206　Vernonia carnea Hiern；肉色斑鸠菊■☆

406207　Vernonia carnea Hiern var. monocephala Mendonça = Vernonia carnea Hiern ■☆

406208　Vernonia carnotiana Humbert；卡尔诺斑鸠菊●☆

406209　Vernonia castellana S. Moore = Vernonia praemorsa Muschl. ■●☆

406210　Vernonia catumbensis Hiern；卡通贝斑鸠菊●☆

406211　Vernonia caudata Drake = Vernoniopsis caudata（Drake）Humbert ●☆

406212　Vernonia centaureoides Klatt；矢车菊状斑鸠菊●☆

406213　Vernonia cephalophora Oliv.；头状斑鸠菊●☆

406214　Vernonia cernua Bojer = Moquinia bojeri DC. ●☆

406215　Vernonia chapelieri Drake；沙普斑鸠菊●☆

406216　Vernonia chapmanii C. D. Adams；查普曼斑鸠菊●☆

406217　Vernonia chariensis O. Hoffm. = Vernonia nestor S. Moore ●☆

406218　Vernonia chevalieri O. Hoffm. = Vernonia guineensis Benth. ■☆

406219　Vernonia chiarugii Pic. Serm. = Vernonia myriantha Hook. f. ●☆

406220　Vernonia chiliocephala O. Hoffm.；千头斑鸠菊●☆

406221　Vernonia chinensis Less. = Vernonia patula（Dryand.）Merr. ■

406222　Vernonia chingiana Hand.-Mazz.；广西斑鸠菊（黑升麻，棠菊）；Ching's Ironweed，Guangxi Ironweed，Kwangsi Ironweed ●

406223　Vernonia chlorolepis S. Moore = Vernonia verrucata Klatt ●☆

406224　Vernonia chloropappa Baker；绿斑鸠菊●☆

406225　Vernonia chthonocephala O. Hoffm.；对叶鼠麴草斑鸠菊●☆

406226　Vernonia chunii C. C. Chang；少花斑鸠菊；Chun's Ironweed，Fewflower Ironweed ●

406227　Vernonia cinera（L.）Less. = Cyanthillium cinereum（L.）H. Rob. ■

406228　Vernonia cinera（L.）Less. var. parviflora（Reinw. ex Blume）DC.；小花夜香牛（小花斑鸠菊）；Smallflower Ashycoloured Ironweed ■

406229　Vernonia cinerascens Sch. Bip.；浅灰斑鸠菊●☆

406230　Vernonia cinerea（L.）Less. = Cyanthillium cinereum（L.）H. Rob. ■

406231　Vernonia cinerea（L.）Less. var. antoniensis Bolle = Vernonia cinerea（L.）Less. ■

406232　Vernonia cinerea（L.）Less. var. lentii（O. Hoffm.）C. Jeffrey；伦特夜香牛■☆

406233　Vernonia cinerea（L.）Less. var. parviflora（Reinw. ex Blume）DC. = Vernonia cinerea（L.）Less. var. parviflora（Reinw. ex Blume）DC. ■

406234　Vernonia cinerea（L.）Less. var. parviflora（Reinw.）DC. = Vernonia cinera（L.）Less. var. parviflora（Reinw. ex Blume）DC. ■

406235　Vernonia cinerea（L.）Less. var. ugandense C. Jeffrey；乌干达夜香牛■☆

406236　Vernonia cirrifera S. Moore = Vernonia colorata（Willd.）Drake subsp. oxyura（O. Hoffm.）C. Jeffrey ■☆

406237　Vernonia cistifolia O. Hoffm. = Vernonia karaguensis Oliv. ●☆

406238　Vernonia cistifolia O. Hoffm. var. bothrioclinoides（C. H.

Wright）Brenan = Vernonia karaguensis Oliv. ●☆

406239　Vernonia cistifolia O. Hoffm. var. rosea ？ = Vernonia karaguensis Oliv. ●☆

406240　Vernonia clarenceana Hook. f. = Conyza clarenceana（Hook. f.）Oliv. et Hiern ■☆

406241　Vernonia cleanthoides O. Hoffm.；蓝星花斑鸠菊■☆

406242　Vernonia clinopodioides O. Hoffm.；风轮菜斑鸠菊■☆

406243　Vernonia clivorum Hance；山岗斑鸠菊；Hillock Ironweed，Hilly Ironweed ■

406244　Vernonia cloiselii S. Moore = Distephanus cloiselii（S. Moore）H. Rob. et B. Kahn ●☆

406245　Vernonia collina Klatt = Pegolettia lanceolata Harv. ■☆

406246　Vernonia colorata（Willd.）Drake；有色斑鸠菊；Bitterleaf ■☆

406247　Vernonia colorata（Willd.）Drake subsp. grandis（DC.）C. Jeffrey；大有色斑鸠菊■☆

406248　Vernonia colorata（Willd.）Drake subsp. oxyura（O. Hoffm.）C. Jeffrey；尖尾色斑鸠菊■☆

406249　Vernonia colorata Drake = Vernonia colorata（Willd.）Drake ■☆

406250　Vernonia concinna S. Moore；整洁斑鸠菊■☆

406251　Vernonia condensata Baker；密集斑鸠菊■☆

406252　Vernonia conferta Benth.；稠密斑鸠菊；Cabbage Tree ●☆

406253　Vernonia conferta Benth. var. seretii De Wild.；团集斑鸠菊■☆

406254　Vernonia conferta Sch. Bip. ex Baker = Vernonia conferta Benth. ●☆

406255　Vernonia congesta Benth. = Duhaldea chinensis DC. ●■

406256　Vernonia congesta Benth. = Inula cappa（Buch.-Ham.）DC. ●■

406257　Vernonia congolensis De Wild. et Muschl.；康戈尔斑鸠菊■☆

406258　Vernonia congolensis De Wild. et Muschl. subsp. longiflora C. Jeffrey；长花康戈尔斑鸠菊■☆

406259　Vernonia congolensis De Wild. et Muschl. subsp. vernonioides（Sch. Bip. ex Walp.）C. Jeffrey；普通斑鸠菊■☆

406260　Vernonia conyzoides DC.；假蓬斑鸠菊■☆

406261　Vernonia conyzoides Hutch. et Dalziel = Vernonia klingii O. Hoffm. et Muschl. ●☆

406262　Vernonia corchoroides Muschl.；黄麻斑鸠菊■☆

406263　Vernonia corymbosa（L. f.）Less. = Vernonia tigna Klatt ●☆

406264　Vernonia coursii Humbert；库尔斯斑鸠菊●☆

406265　Vernonia crataegifolia Hutch.；山楂叶斑鸠菊●☆

406266　Vernonia crinita Raf.；长刚毛斑鸠菊；Bur Ironweed ■☆

406267　Vernonia crinita Raf. = Vernonia arkansana DC. ■☆

406268　Vernonia cruda Klatt = Vernonia thomsoniana Oliv. et Hiern ex Oliv. ■☆

406269　Vernonia cryptocephala Baker；隐头长刚毛斑鸠菊■☆

406270　Vernonia cryptocephala Baker var. concinna Airy Shaw = Vernonia cryptocephala Baker ■☆

406271　Vernonia cuanzensis Welw. = Inula cuanzensis（Welw.）Hiern ■☆

406272　Vernonia cumingiana Benth.；毒根斑鸠菊（大木菊，发痧藤，过山龙，虎三头，惊风红，蔓斑鸠菊，藤牛七，细脉斑鸠菊，夜牵牛）；Cuming Ironweed，Poisonous-root Ironweed ●■

406273　Vernonia cylindrica Sch. Bip. ex Walp.；柱蕊毒根斑鸠菊■☆

406274　Vernonia cylindriceps C. B. Clarke = Vernonia extensa（Wall.）DC. ●

406275　Vernonia daphnifolia O. Hoffm.；月桂叶斑鸠菊■☆

406276　Vernonia decaryana Humbert；德卡里斑鸠菊■☆

406277　Vernonia dekindtii O. Hoffm. = Vernonia hochstetteri Sch. Bip. ex Walp. var. dekindtii（O. Hoffm.）C. Jeffrey ●☆

406278　Vernonia demulans Vatke = Vernonia aemulans Vatke ■☆

406279　Vernonia densicapitulata De Wild. = Vernonia thomsoniana

Oliv. et Hiern ex Oliv. ■☆

406280　Vernonia denudata Hutch. et B. L. Burtt;裸露毒根斑鸠菊■☆

406281　Vernonia descampsii De Wild. = Vernonia longipedunculata De Wild. var. manikensis (De Wild.) G. V. Pope ■☆

406282　Vernonia devredii Kalanda;德夫雷斑鸠菊●☆

406283　Vernonia dewildemaniana Muschl.;德氏斑鸠菊■☆

406284　Vernonia divergens (DC.) Edgew.;叉枝毒根斑鸠菊(叉枝斑鸠菊);Divaricate Ironweed,Divergent Ironweed ■

406285　Vernonia diversifolia Bojer ex DC.;异叶斑鸠菊■☆

406286　Vernonia djalonensis A. Chev. ex Hutch. et Dalziel;贾隆斑鸠菊●☆

406287　Vernonia doniana DC.;多尼斑鸠菊●☆

406288　Vernonia dregeana Sch. Bip.;德雷斑鸠菊■☆

406289　Vernonia duemmeri S. Moore = Vernonia purpurea Sch. Bip. ex Walp. ■☆

406290　Vernonia dumicola S. Moore = Vernonia lasiopus O. Hoffm. ●☆

406291　Vernonia dupuisii Klatt = Vernonia undulata Oliv. et Hiern ■☆

406292　Vernonia duvigneaudii Kalanda;迪维尼奥斑鸠菊■☆

406293　Vernonia edulis Steud.;可食斑鸠菊■☆

406294　Vernonia elaeagnoides (DC.) Sch. Bip. = Vernonia oligocephala (DC.) Sch. Bip. ex Walp. ■☆

406295　Vernonia elegantissima Hutch. et Dalziel;雅致斑鸠菊●☆

406296　Vernonia elisabethvilleana De Wild. = Vernonia bellinghamii S. Moore ■☆

406297　Vernonia elliotii S. Moore = Vernonia karaguensis Oliv. ●☆

406298　Vernonia elliptica DC. = Tarlmounia elliptica (DC.) H. Rob., S. C. Keeley,Skvarla et R. Chan ■☆

406299　Vernonia entohylea Wild = Vernonia dewildemaniana Muschl. ■☆

406300　Vernonia erinacea Wild = Vernonia rhodanthoidea Muschl. ■☆

406301　Vernonia eriocephala Klatt = Vernonia petersii Oliv. et Hiern ex Oliv. ●☆

406302　Vernonia eriophylla Drake = Distephanus eriophyllus (Drake) H. Rob. et B. Kahn ●☆

406303　Vernonia eriosematoides Walp. = Duhaldea chinensis DC. ●■

406304　Vernonia eriosematoides Walp. = Inula cappa (Buch.-Ham.) DC. ●■

406305　Vernonia eritreana Klatt = Vernonia amygdalina Delile ■☆

406306　Vernonia erlangeriana O. Hoffm. ex Engl.;厄兰格斑鸠菊■☆

406307　Vernonia erubescens Hochst. = Vernonia aschersonii Sch. Bip. ■☆

406308　Vernonia erythromarula Klatt;浅红斑鸠菊■☆

406309　Vernonia esculenta Hemsl. = Vernonia bockiana Diels ●

406310　Vernonia esculenta Hemsl. = Vernonia esculenta Hemsl. ex Forbes et Hemsl. ●

406311　Vernonia esculenta Hemsl. ex Forbes et Hemsl.;斑鸠菊(大藤菊,豆腐渣树,火烧叶,火炭树,火炭叶,火烫叶,鸡菊花,聋耳朵树);Edible Ironweed,Iron Weed ●

406312　Vernonia esquirolii H. Lév. = Vernonia volkameriifolia (Wall.) DC. ●

406313　Vernonia esquirolii Vaniot = Cissampelopsis volubilis (Blume) Miq. ●

406314　Vernonia evrardiana Kalanda;埃夫拉尔斑鸠菊●☆

406315　Vernonia exasperata Wild = Vernonia kandtii Muschl. ■☆

406316　Vernonia exillis Miq. = Vernonia cinerea (L.) Less. ■

406317　Vernonia exserta Baker;伸展斑鸠菊●☆

406318　Vernonia exsertiflora Baker;展花斑鸠菊●☆

406319　Vernonia exsertiflora Baker var. tenuicalyx G. V. Pope;细萼展花斑鸠菊●☆

406320　Vernonia extensa (Wall.) DC.;展枝斑鸠菊(棒头斑鸠菊,扩张鸡菊花,茄叶一枝蒿,小黑升麻);Explanate Ironweed,Prone Ironweed,Spread Ironweed ●

406321　Vernonia eylesii S. Moore;艾尔斯斑鸠菊●☆

406322　Vernonia fandrarazanensis Humbert = Vernonia hispidula Drake ●☆

406323　Vernonia faradifani Scott-Elliot = Oliganthes triflora Cass. ●☆

406324　Vernonia fargesii Franch. = Synotis nagensis (C. B. Clarke) C. Jeffrey et Y. L. Chen ■

406325　Vernonia fasciculata Michx.;簇生斑鸠菊;Common Ironweed, Prairie Ironweed,Smooth Ironweed ■☆

406326　Vernonia fasciculata Michx. subsp. corymbosa (Schwein.) S. B. Jones = Vernonia fasciculata Michx. ■☆

406327　Vernonia fasciculata Michx. var. corymbosa (Schwein.) B. G. Schub. = Vernonia fasciculata Michx. ■☆

406328　Vernonia fastigiata Oliv. et Hiern;刷状斑鸠菊■☆

406329　Vernonia filigera Oliv. et Hiern;具丝斑鸠菊■☆

406330　Vernonia filipendula Hiern;垂丝斑鸠菊■☆

406331　Vernonia filisquama M. G. Gilbert = Vernonia galamensis (Cass.) Less. subsp. filisquama (M. G. Gilbert) C. Jeffrey ■☆

406332　Vernonia firma Oliv. et Hiern = Vernonia guineensis Benth. ■☆

406333　Vernonia fischeri O. Hoffm.;菲舍尔斑鸠菊■☆

406334　Vernonia flaccidifolia Small;软叶斑鸠菊■☆

406335　Vernonia flanaganii (E. Phillips) Hilliard;弗拉纳根斑鸠菊■☆

406336　Vernonia flexuosa Sims;曲折斑鸠菊■☆

406337　Vernonia fontinalis S. Moore = Vernonia ugandensis S. Moore ■☆

406338　Vernonia forrestii Anthony;滇西斑鸠菊;Forrest's Ironweed,W. Yunnan Ironweed ●◇

406339　Vernonia fortunei Sch. Bip. = Vernonia solanifolia Benth. ●

406340　Vernonia francavillana Oliv. et Hiern = Vernonia rueppellii Sch. Bip. ●☆

406341　Vernonia fraterna N. E. Br. = Vernonia bellinghamii S. Moore ■☆

406342　Vernonia friisii Mesfin;弗里斯斑鸠菊■☆

406343　Vernonia frondosa Oliv. et Hiern;多叶斑鸠菊■☆

406344　Vernonia fulgens Gilli = Vernonia lasiopus O. Hoffm. var. iodocalyx (O. Hoffm.) C. Jeffrey ●☆

406345　Vernonia fulviseta S. Moore = Baccharoides adoensis (Sch. Bip. ex Walp.) H. Rob. var. mossambiquensis (Steetz) Isawumi, El-Ghazaly et B. Nord. ●☆

406346　Vernonia galamensis (Cass.) Less.;加拉姆斑鸠菊■☆

406347　Vernonia galamensis (Cass.) Less. subsp. filisquama (M. G. Gilbert) C. Jeffrey;丝鳞斑鸠菊■☆

406348　Vernonia galamensis (Cass.) Less. subsp. gibbosa M. G. Gilbert = Vernonia galamensis (Cass.) Less. var. gibbosa (M. G. Gilbert) C. Jeffrey ■☆

406349　Vernonia galamensis (Cass.) Less. subsp. lushotoensis M. G. Gilbert = Vernonia galamensis (Cass.) Less. var. lushotoensis (M. G. Gilbert) C. Jeffrey ■☆

406350　Vernonia galamensis (Cass.) Less. subsp. mutomoensis M. G. Gilbert;穆托莫斑鸠菊■☆

406351　Vernonia galamensis (Cass.) Less. subsp. nairobiensis M. G. Gilbert;内罗比斑鸠菊■☆

406352　Vernonia galamensis (Cass.) Less. var. afromontana (R. E. Fr.) M. G. Gilbert;非洲山生斑鸠菊■☆

406353　Vernonia galamensis (Cass.) Less. var. australis M. G. Gilbert;南方非洲山生斑鸠菊■☆

406354　Vernonia galamensis (Cass.) Less. var. gibbosa (M. G. Gilbert) C. Jeffrey;浅囊斑鸠菊■☆

406355　Vernonia galamensis (Cass.) Less. var. lushotoensis (M. G.

Gilbert) C. Jeffrey;卢绍托斑鸠菊■☆

406356　Vernonia galamensis (Cass.) Less. var. petitiana (A. Rich.) M. G. Gilbert;佩蒂蒂斑鸠菊■☆

406357　Vernonia galpinii Klatt;盖尔斑鸠菊■☆

406358　Vernonia garambaensis Kalanda = Aedesia spectabilis Mattf. ■☆

406359　Vernonia garnieriana Klatt = Distephanus garnierianus (Klatt) H. Rob. et B. Kahn ●☆

406360　Vernonia gerberiformis Oliv. et Hiern;火石花斑鸠菊■☆

406361　Vernonia gerberiformis Oliv. et Hiern subsp. macrocyanus (O. Hoffm.) C. Jeffrey;大萼斑鸠菊■☆

406362　Vernonia gerberiformis Oliv. et Hiern var. hockii (De Wild. et Muschl.) G. V. Pope;霍克斑鸠菊■☆

406363　Vernonia gerrardii Harv.;杰勒德斑鸠菊■☆

406364　Vernonia gigantea (Walter) Branner et Coville = Vernonia gigantea (Walter) Trel. ex Branner et Coville ■☆

406365　Vernonia gigantea (Walter) Trel. ex Branner et Coville;巨斑鸠菊;Ironweed ■☆

406366　Vernonia gigantea (Walter) Trel. ex Branner et Coville subsp. ovalifolia (Torr. et A. Gray) Urbatsch = Vernonia gigantea (Walter) Trel. ex Branner et Coville ■☆

406367　Vernonia gilbertii Mesfin;吉尔伯特斑鸠菊■☆

406368　Vernonia giorgii De Wild. = Vernonia amygdalina Delile ■☆

406369　Vernonia glaberrima Welw. ex O. Hoffm.;无毛斑鸠菊■☆

406370　Vernonia glabra (Steetz) Vatke;光滑斑鸠菊■☆

406371　Vernonia glabra (Steetz) Vatke var. hillii (Hutch. et Dalziel) C. D. Adams = Vernonia ituriensis Muschl. ■☆

406372　Vernonia glabra (Steetz) Vatke var. laxa (Steetz) Brenan;松散斑鸠菊■☆

406373　Vernonia glabra (Steetz) Vatke var. occidentalis C. D. Adams = Vernonia ituriensis Muschl. var. occidentalis (C. D. Adams) C. Jeffrey ■☆

406374　Vernonia glabra (Steetz) Vatke var. ondongensis (Klatt) Merxm. = Vernonia glabra (Steetz) Vatke ■☆

406375　Vernonia glandulicincta Humbert = Distephanus glandulicinctus (Humbert) H. Rob. et B. Kahn ●☆

406376　Vernonia glandulosa DC. = Vernonia madagascariensis Less. ●☆

406377　Vernonia glauca (L.) Willd.;灰蓝斑鸠菊■☆

406378　Vernonia glutinosa DC. = Distephanus glutinosus (DC.) H. Rob. et B. Kahn ●☆

406379　Vernonia goetzeana O. Hoffm. = Vernonia bellinghamii S. Moore ■☆

406380　Vernonia goetzei Muschl. = Vernonia adoensis Sch. Bip. ex Walp. ■☆

406381　Vernonia goetzenii O. Hoffm.;格氏斑鸠菊■☆

406382　Vernonia gofensis O. Hoffm.;戈法斑鸠菊■☆

406383　Vernonia golungensis Welw. ex Mendonça;戈龙斑鸠菊■☆

406384　Vernonia gomphophylla Baker = Psiadia incana Oliv. et Hiern ●☆

406385　Vernonia gossweileri S. Moore;戈斯斑鸠菊☆

406386　Vernonia graciliflora De Wild. = Vernonia filipendula Hiern ■☆

406387　Vernonia gracilipes S. Moore = Vernonia wollastonii S. Moore ■☆

406388　Vernonia gracilipes S. Moore var. minor ? = Vernonia nepetifolia Wild ■☆

406389　Vernonia grandis (DC.) Humb. = Vernonia colorata (Willd.) Drake subsp. grandis (DC.) C. Jeffrey ■☆

406390　Vernonia graniticola G. V. Pope;花岗岩斑鸠菊■☆

406391　Vernonia grantii Oliv. = Vernonia adoensis Sch. Bip. ex Walp. ■☆

406392　Vernonia gratiosa Hance;台湾斑鸠菊(过山龙);Taiwan Ironweed ●■

406393　Vernonia grevei Drake;格雷弗斑鸠菊●☆

406394　Vernonia grisea Baker = Vernonia exserta Baker ●☆

406395　Vernonia griseopapposa G. V. Pope;灰冠毛斑鸠菊●☆

406396　Vernonia guineensis Benth.;几内亚斑鸠菊■☆

406397　Vernonia guineensis Benth. var. cameroonica C. D. Adams;喀麦隆斑鸠菊●☆

406398　Vernonia guineensis Benth. var. procera (O. Hoffm.) C. D. Adams = Vernonia procera O. Hoffm. ●☆

406399　Vernonia hamata Klatt;顶钩斑鸠菊●☆

406400　Vernonia harperi Gleason = Vernonia noveboracensis (L.) Michx. ☆

406401　Vernonia helenae Buscal. et Muschl.;海伦娜斑鸠菊●☆

406402　Vernonia henryi Dunn;黄花斑鸠菊;Henry's Ironweed, Yellow Ironweed ●

406403　Vernonia hensii Klatt = Vernonia glaberrima Welw. ex O. Hoffm. ■☆

406404　Vernonia heterocarpa Chiov. = Vernonia wollastonii S. Moore ■☆

406405　Vernonia hierniana S. Moore = Vernonia guineensis Benth. ■☆

406406　Vernonia hildebrandtii Baker = Vernonia rubicunda Klatt ●☆

406407　Vernonia hildebrandtii Vatke;希尔德斑鸠菊■☆

406408　Vernonia hillii Hutch. et Dalziel = Vernonia ituriensis Muschl. ■☆

406409　Vernonia hindei S. Moore = Vernonia glabra (Steetz) Vatke var. laxa (Steetz) Brenan ■☆

406410　Vernonia hirsuta (DC.) Sch. Bip. ex Walp.;粗毛斑鸠菊●☆

406411　Vernonia hirsuta (DC.) Sch. Bip. ex Walp. var. flanaganii E. Phillips = Vernonia flanaganii (E. Phillips) Hilliard ■☆

406412　Vernonia hirsuta (DC.) Sch. Bip. ex Walp. var. obtusifolia Harv. = Vernonia hirsuta (DC.) Sch. Bip. ex Walp. ●☆

406413　Vernonia hispidula Drake;硬毛斑鸠菊●☆

406414　Vernonia hochstetteri Sch. Bip. ex Walp.;霍赫斑鸠菊■☆

406415　Vernonia hochstetteri Sch. Bip. ex Walp. var. dekindtii (O. Hoffm.) C. Jeffrey;德金斑鸠菊●☆

406416　Vernonia hochstetteri Sch. Bip. ex Walp. var. kivuensis (Humb. et Staner) C. Jeffrey;基伍斑鸠菊●☆

406417　Vernonia hockii De Wild. et Muschl. = Vernonia gerberiformis Oliv. et Hiern var. hockii (De Wild. et Muschl.) G. V. Pope ■☆

406418　Vernonia hoffmanniana Hutch. et Dalziel = Vernonia musofensis S. Moore var. miamensis (S. Moore) G. V. Pope ●☆

406419　Vernonia hoffmanniana S. Moore = Vernonia brachycalyx O. Hoffm. ■☆

406420　Vernonia holstii O. Hoffm.;霍尔斑鸠菊■☆

406421　Vernonia homblei De Wild. = Vernonia polysphaera Baker ●☆

406422　Vernonia homilantha S. Moore;团花斑鸠菊■☆

406423　Vernonia homilocephala S. Moore = Vernonia hymenolepis A. Rich. ■☆

406424　Vernonia homolleae Humbert;奥莫勒斑鸠菊●☆

406425　Vernonia huillensis Hiern;威拉斑鸠菊■☆

406426　Vernonia humblotii Drake;洪布斑鸠菊☆

406427　Vernonia humilis C. H. Wright = Stomatanthes africanus (Oliv. et Hiern) R. M. King et H. Rob. ■☆

406428　Vernonia humillima Humbert;矮小斑鸠菊■☆

406429　Vernonia hymenolepis A. Rich.;膜鳞斑鸠菊(膜质斑鸠菊)■☆

406430　Vernonia hymenolepis A. Rich. subsp. leucocalyx (O. Hoffm.) Wild = Vernonia calvoana (Hook. f.) Hook. f. subsp. leucocalyx (O. Hoffm.) C. Jeffrey ■☆

406431　Vernonia hymenolepis A. Rich. subsp. meridionalis Wild = Vernonia calvoana (Hook. f.) Hook. f. subsp. meridionalis (Wild) C. Jeffrey ■☆

406432　Vernonia ibityensis Humbert;伊比提斑鸠菊●☆

406433   Vernonia ikongensis Humbert;伊孔古斑鸠菊●☆

406434   Vernonia inanis S. Moore;空斑鸠菊■☆

406435   Vernonia incompta S. Moore;装饰斑鸠菊●☆

406436   Vernonia infundibularis Oliv. et Hiern;漏斗状斑鸠菊■☆

406437   Vernonia inhacensis G. V. Pope;伊尼亚卡斑鸠菊■☆

406438   Vernonia integra S. Moore = Vernonia adoensis Sch. Bip. ex Walp. ■☆

406439   Vernonia interior Small = Vernonia baldwinii Torr. ■☆

406440   Vernonia inulifolia Baker = Vernonia bakeri Vatke ■☆

406441   Vernonia inulifolia Steud. ex Walp. = Vernonia purpurea Sch. Bip. ex Walp. ■☆

406442   Vernonia iodocalyx O. Hoffm. = Vernonia lasiopus O. Hoffm. var. iodocalyx (O. Hoffm.) C. Jeffrey ●☆

406443   Vernonia isalensis Humbert;伊萨卢斑鸠菊●☆

406444   Vernonia ischnophylla Muschl.;细长叶斑鸠菊■☆

406445   Vernonia isoetifolia Wild;水韭叶斑鸠菊■☆

406446   Vernonia ituriensis Muschl.;伊图里斑鸠菊■☆

406447   Vernonia ituriensis Muschl. var. occidentalis (C. D. Adams) C. Jeffrey;西方斑鸠菊■☆

406448   Vernonia jaceoides A. Rich. = Vernonia purpurea Sch. Bip. ex Walp. ■☆

406449   Vernonia jaegeri C. D. Adams;耶格斑鸠菊■☆

406450   Vernonia jelfiae S. Moore;杰尔夫斑鸠菊●☆

406451   Vernonia jelfiae S. Moore var. albida G. V. Pope;苍白耶格斑鸠菊■☆

406452   Vernonia jodopappa Chiov. = Vernonia brachycalyx O. Hoffm. ■☆

406453   Vernonia jodopapposa Chiov. ex Lanza = Vernonia brachycalyx O. Hoffm. ■☆

406454   Vernonia jodopapposa Chiov. ex Lanza. f. erecta？ = Vernonia brachycalyx O. Hoffm. ■☆

406455   Vernonia jugalis Oliv. et Hiern;非常斑鸠菊■☆

406456   Vernonia jugalis Oliv. et Hiern = Vernonia hochstetteri Sch. Bip. ex Walp. ■☆

406457   Vernonia kaessneri S. Moore = Vernonia lasiopus O. Hoffm. ●☆

406458   Vernonia kandtii Muschl.;坎德斑鸠菊■☆

406459   Vernonia kapirensis De Wild.;卡皮里斑鸠菊●☆

406460   Vernonia kapolowensis Kalanda;卡波斑鸠菊●☆

406461   Vernonia karaguensis Oliv.;卡拉古斑鸠菊●☆

406462   Vernonia karongensis Baker = Vernonia petersii Oliv. et Hiern ex Oliv. ●☆

406463   Vernonia kasaiensis Kalanda;开赛斑鸠菊■☆

406464   Vernonia kassneri De Wild. et Muschl. = Vernonia chloropappa Baker ●☆

406465   Vernonia katangensis O. Hoffm. = Vernonia schweinfurthii Oliv. et Hiern ■☆

406466   Vernonia kawakamii Hayata = Vernonia maritima Hayata ex Merr. ●■

406467   Vernonia kawoziensis Davies;卡沃兹斑鸠菊●☆

406468   Vernonia kayuniana G. V. Pope;卡尤尼斑鸠菊●☆

406469   Vernonia keniensis R. E. Fr. = Vernonia purpurea Sch. Bip. ex Walp. ■☆

406470   Vernonia kingii C. B. Clarke = Vernonia clivorum Hance ■

406471   Vernonia kirkii Oliv. et Hiern;柯克斑鸠菊●☆

406472   Vernonia kirungae R. E. Fr.;基龙加斑鸠菊●☆

406473   Vernonia kivuensis Humb. et Staner = Vernonia hochstetteri Sch. Bip. ex Walp. var. kivuensis (Humb. et Staner) C. Jeffrey ●☆

406474   Vernonia klingii O. Hoffm. et Muschl.;克林斑鸠菊●☆

406475   Vernonia kotschyana Sch. Bip.;科切斑鸠菊●☆

406476   Vernonia kotschyana Sch. Bip. ex Walp. = Vernonia adoensis Sch. Bip. ex Walp. var. kotschyana (Sch. Bip. ex Walp.) G. V. Pope ■☆

406477   Vernonia kraussii Sch. Bip. ex Walp. = Vernonia oligocephala (DC.) Sch. Bip. ex Walp. ■☆

406478   Vernonia kreismannii Welw. ex Hiern = Vernonia exsertiflora Baker ●☆

406479   Vernonia kroneana Miq. = Vernonia cinerea (L.) Less. ■

406480   Vernonia kuluina S. Moore = Vernonia praemorsa Muschl. ■●☆

406481   Vernonia kwangolana P. A. Duvign. et Hotyat = Vernonia bullulata S. Moore ●☆

406482   Vernonia lafukensis S. Moore;拉夫卡斑鸠菊●☆

406483   Vernonia lampropappa O. Hoffm.;亮冠毛斑鸠菊●☆

406484   Vernonia lanata Mesfin = Vernonia newbouldii Beentje et Mesfin ●☆

406485   Vernonia lancibracteata S. Moore = Vernonia filipendula Hiern ■☆

406486   Vernonia lantziana Drake = Vernonia exserta Baker ●☆

406487   Vernonia laosensis Gand. = Vernonia parishii Hook. f. ●

406488   Vernonia lappoides O. Hoffm. = Vernonia musofensis S. Moore var. miamensis (S. Moore) G. V. Pope ●☆

406489   Vernonia larseniae B. L. King et S. B. Jones;拉尔森斑鸠菊■☆

406490   Vernonia lasensis Gand. = Vernonia parishii Hook. f. ●

406491   Vernonia lasiolepis O. Hoffm. = Vernonia tenoreana Oliv. ■☆

406492   Vernonia lasiopus O. Hoffm.;毛梗斑鸠菊●☆

406493   Vernonia lasiopus O. Hoffm. var. acuta C. Jeffrey;尖毛斑鸠菊●☆

406494   Vernonia lasiopus O. Hoffm. var. caudata C. Jeffrey;尾状斑鸠菊●☆

406495   Vernonia lasiopus O. Hoffm. var. grandiceps C. Jeffrey;大头毛梗斑鸠菊●☆

406496   Vernonia lasiopus O. Hoffm. var. iodocalyx (O. Hoffm.) C. Jeffrey;落萼斑鸠菊●☆

406497   Vernonia lastellei Drake = Distephanus lastellei (Drake) H. Rob. et B. Kahn ●☆

406498   Vernonia lastellei Drake var. bernieri (Drake) Humbert = Distephanus lastellei (Drake) H. Rob. et B. Kahn ●☆

406499   Vernonia lastii Engl. = Vernonia cinerea (L.) Less. var. lentii (O. Hoffm.) C. Jeffrey ■☆

406500   Vernonia latisquama Mattf. = Vernonia adoensis Sch. Bip. ex Walp. ■☆

406501   Vernonia laurentii De Wild. = Vernonia auriculifera Hiern ■☆

406502   Vernonia lavandulifolia Muschl. ex De Wild.;薰衣草叶斑鸠菊■☆

406503   Vernonia lawalreeana Kalanda = Vernonia robinsonii Wild ●☆

406504   Vernonia leandrii Humbert;利安斑鸠菊■☆

406505   Vernonia leandrii subsp. sambiranensis Humbert = Vernonia sambiranensis (Humbert) Humbert ●☆

406506   Vernonia lebrunii Staner = Vernonia infundibularis Oliv. et Hiern ■☆

406507   Vernonia lecomtei Humbert = Oliganthes lecomtei (Humbert) Humbert ●☆

406508   Vernonia ledermannii Mattf.;莱德斑鸠菊●☆

406509   Vernonia ledocteana P. A. Duvign. et Van Bockstal;莱多斑鸠菊■☆

406510   Vernonia lejolyana Adamska et Lisowski = Ageratinastrum lejolyanum (Adamska et Lisowski) Kalanda ■☆

406511   Vernonia lemurica Humbert;莱穆拉斑鸠菊●☆

406512   Vernonia lentii Volkens et O. Hoffm. = Vernonia cinerea (L.) Less. var. lentii (O. Hoffm.) C. Jeffrey ■☆

406513   Vernonia leptantha Klatt;细花斑鸠菊■☆

406514   Vernonia leptolepis Baker = Baccharoides adoensis (Sch. Bip. ex Walp.) H. Rob. var. kotschyana (Sch. Bip. ex Walp.) Isawumi, El-Ghazaly et B. Nord. ■☆

406515   Vernonia leptolepis O. Hoffm. = Vernonia biafrae Oliv. et Hiern

■☆

406516　Vernonia lescrauwaetii De Wild. = Vernonia suprafastigiata Klatt ●☆

406517　Vernonia lettermannii Engelm. ex A. Gray；莱特曼斑鸠菊■☆

406518　Vernonia leucocalyx O. Hoffm. = Vernonia calvoana（Hook. f.）Hook. f. subsp. leucocalyx（O. Hoffm.）C. Jeffrey ■☆

406519　Vernonia leucocalyx O. Hoffm. var. acuta C. D. Adams = Vernonia calvoana（Hook. f.）Hook. f. var. acuta（C. D. Adams）C. Jeffrey ☆

406520　Vernonia leucophylla Baker = Psiadia leucophylla（Baker）Humbert ●☆

406521　Vernonia levelillei Fedde ex H. Lév. = Vernonia volkameriifolia（Wall.）DC. ●

406522　Vernonia limosa O. Hoffm. ；湿地斑鸠菊■☆

406523　Vernonia lindheimeri A. Gray et Engelm. ；林德斑鸠菊■☆

406524　Vernonia lindheimeri A. Gray et Engelm. var. leucophylla Larsen = Vernonia lindheimeri A. Gray et Engelm. ■☆

406525　Vernonia lisowskii Kalanda；利索斑鸠菊■☆

406526　Vernonia livingstoniana Oliv. et Hiern = Vernonia thomsoniana Oliv. et Hiern ex Oliv. ■☆

406527　Vernonia loandensis S. Moore；罗安达斑鸠菊■☆

406528　Vernonia loloana Dunn ex Kerr. = Camchaya loloana Kerr ■

406529　Vernonia longibracteata S. Ortiz et Rodr. Oubina；长苞斑鸠菊■☆

406530　Vernonia longipedunculata De Wild. ；长梗斑鸠菊■☆

406531　Vernonia longipedunculata De Wild. var. manikensis（De Wild.）G. V. Pope；马尼科斑鸠菊■☆

406532　Vernonia longipedunculata De Wild. var. retusa（R. E. Fr.）G. V. Pope；微凹斑鸠菊■☆

406533　Vernonia longipetiolata Muschl. = Vernonia colorata（Willd.）Drake subsp. oxyura（O. Hoffm.）C. Jeffrey ■☆

406534　Vernonia louvelii Humbert；卢韦尔斑鸠菊●☆

406535　Vernonia luaboensis Kalanda = Vernonia bruceana Wild ●☆

406536　Vernonia lualabaensis De Wild. ；卢阿拉巴斑鸠菊●☆

406537　Vernonia luederitziana O. Hoffm. = Vernonia cinerascens Sch. Bip. ●☆

406538　Vernonia luembensis De Wild. et Muschl. ；卢恩贝斑鸠菊■☆

406539　Vernonia luhomeroensis Q. Luke et Beentje；卢赫梅罗斑鸠菊●☆

406540　Vernonia lujae De Wild. = Vernonia myriantha Hook. f. ●☆

406541　Vernonia lumbilae Gilli = Gutenbergia cordifolia Benth. ex Oliv. ■☆

406542　Vernonia lundiensis（Hutch.）Wild et G. V. Pope；伦迪斑鸠菊■☆

406543　Vernonia lutea N. E. Br. = Distephanus angolensis（O. Hoffm.）H. Rob. et B. Kahn ●☆

406544　Vernonia luteoalbida De Wild. = Vernonia stenocephala Oliv. ■☆

406545　Vernonia lycioides Wild；枸杞状斑鸠菊●☆

406546　Vernonia macrocephala A. Rich. = Vernonia adoensis Sch. Bip. ex Walp. ■☆

406547　Vernonia macrocyanus O. Hoffm. = Vernonia gerberiformis Oliv. et Hiern subsp. macrocyanus（O. Hoffm.）C. Jeffrey ■☆

406548　Vernonia macrocyanus O. Hoffm. var. ambacensis Hiern = Vernonia gerberiformis Oliv. et Hiern ■☆

406549　Vernonia macrophylla Chiov. = Vernonia theophrastifolia Schweinf. ex Oliv. et Hiern ■☆

406550　Vernonia madagascariensis Less. ；马岛斑鸠菊●☆

406551　Vernonia mahafaly Humbert；马哈法里斑鸠菊●☆

406552　Vernonia mairei H. Lév. = Synotis erythropappa（Bureau et Franch.）C. Jeffrey et Y. L. Chen ■

406553　Vernonia majungensis Humbert = Distephanus lastellei（Drake）H. Rob. et B. Kahn ●☆

406554　Vernonia malacophyta Baker = Distephanus lastellei（Drake）H. Rob. et B. Kahn ●☆

406555　Vernonia malaissei Kalanda = Vernonia robinsonii Wild ●☆

406556　Vernonia malosana Baker = Stomatanthes africanus（Oliv. et Hiern）R. M. King et H. Rob. ■☆

406557　Vernonia manambolensis Humbert = Distephanus manambolensis（Humbert）H. Rob. et B. Kahn ●☆

406558　Vernonia mandrarensis Humbert；曼德拉斑鸠菊●☆

406559　Vernonia mangokensis Humbert = Distephanus mangokensis（Humbert）H. Rob. et B. Kahn ●☆

406560　Vernonia mangokensis Humbert var. subulata Humbert = Distephanus mangokensis（Humbert）H. Rob. et B. Kahn ☆

406561　Vernonia manikensis De Wild. = Vernonia longipedunculata De Wild. var. manikensis（De Wild.）G. V. Pope ■☆

406562　Vernonia mannii Hook. f. = Inula mannii（Hook. f.）Oliv. et Hiern ■☆

406563　Vernonia manongarivensis Humbert；马农加斑鸠菊●☆

406564　Vernonia marginata Oliv. et Hiern = Gutenbergia cordifolia Benth. ex Oliv. ■☆

406565　Vernonia maritima Hayata = Vernonia maritima Hayata ex Merr. ●■

406566　Vernonia maritima Hayata ex Merr. ；滨海斑鸠菊（滨斑鸠菊）；Ironweed，Seashore Ironweed ●■

406567　Vernonia maritima Merr. = Vernonia maritima Hayata ex Merr. ●■

406568　Vernonia marojejyensis Humbert；马罗斑鸠菊●☆

406569　Vernonia martinii Vaniot = Vernonia saligna（Wall.）DC. ■

406570　Vernonia masaiensis S. Moore = Vernonia holstii O. Hoffm. ■☆

406571　Vernonia mashonica N. E. Br. = Vernonia glaberrima Welw. ex O. Hoffm. ■☆

406572　Vernonia mbalensis G. V. Pope；姆巴莱斑鸠菊●☆

406573　Vernonia meiocalyx S. Moore = Vernonia brachycalyx O. Hoffm. ■☆

406574　Vernonia melanacrophylla Cufod. = Vernonia karaguensis Oliv. ●☆

406575　Vernonia melanacrophylla Cufod. var. hispida ？ = Vernonia karaguensis Oliv. ●☆

406576　Vernonia melanacrophylla Cufod. var. pseudoblumeoides ？ = Vernonia karaguensis Oliv. ●☆

406577　Vernonia melanocoma C. Jeffrey；黑边斑鸠菊●☆

406578　Vernonia melleri Oliv. et Hiern；梅勒斑鸠菊●☆

406579　Vernonia melleri Oliv. et Hiern var. superba（O. Hoffm.）C. Jeffrey；华美梅勒斑鸠菊■☆

406580　Vernonia mellifera Muschl. = Vernonia urticifolia A. Rich. ■☆

406581　Vernonia merenskiana Dinter ex Merxm. = Erlangea misera（Oliv. et Hiern）S. Moore ■☆

406582　Vernonia mespilifolia Less. var. subcanescens DC. = Vernonia crataegifolia Hutch. ●☆

406583　Vernonia mgetae Gilli = Vernonia ugandensis S. Moore ■☆

406584　Vernonia miamensis S. Moore = Vernonia musofensis S. Moore var. miamensis（S. Moore）G. V. Pope ●☆

406585　Vernonia migeodii S. Moore；米容德斑鸠菊■☆

406586　Vernonia mikumiensis C. Jeffrey；米库米斑鸠菊●☆

406587　Vernonia milanjiana S. Moore；米兰吉斑鸠菊●☆

406588　Vernonia mildbraedii Muschl. ；米尔德斑鸠菊●☆

406589　Vernonia milne-redheadii Wild = Vernonia mumpullensis Hiern ●☆

406590　Vernonia miombicola Wild；赞比亚斑鸠菊●☆

406591　Vernonia miombicoloides C. Jeffrey；拟赞比亚斑鸠菊●☆

406592　Vernonia misera Oliv. et Hiern = Erlangea misera（Oliv. et Hiern）S. Moore ■☆

406593　Vernonia missurica Raf. ；密苏里斑鸠菊；Drummond's Ironweed，

Drummond's Iron-weed, Ironweed ■☆

406594　Vernonia mogadoxensis Chiov. ;索马里斑鸠菊●☆

406595　Vernonia mokaensis Mildbr. et Mattf. = Vernonia calvoana (Hook. f.) Hook. f. var. mokaensis (Mildbr. et Mattf.) C. Jeffrey ■☆

406596　Vernonia monantha Humbert;单花斑鸠菊●☆

406597　Vernonia monocephala Harv. = Vernonia galpinii Klatt ■☆

406598　Vernonia monosis sensu Franch. = Vernonia esculenta Hemsl. ex Forbes et Hemsl. ●

406599　Vernonia moramballae Oliv. et Hiern = Bothriocline moramballae (Oliv. et Hiern) O. Hoffm. ■☆

406600　Vernonia mossambicensis Buscal. et Muschl. ;莫桑比克斑鸠菊●☆

406601　Vernonia mossambiquensis (Steetz) Oliv. et Hiern = Baccharoides adoensis (Sch. Bip. ex Walp.) H. Rob. var. mossambiquensis (Steetz) Isawumi, El-Ghazaly et B. Nord. ●☆

406602　Vernonia muelleri Wild;米勒斑鸠菊●☆

406603　Vernonia muelleri Wild subsp. brevicuspis G. V. Pope;短尖斑鸠菊●☆

406604　Vernonia muelleri Wild subsp. integra C. Jeffrey;全缘米勒斑鸠菊●☆

406605　Vernonia muhiensis Kalanda = Vernonia ituriensis Muschl. ■☆

406606　Vernonia multiflora De Wild. = Vernonia suprafastigiata Klatt ●☆

406607　Vernonia mumpullensis Hiern;蒙普尔斑鸠菊●☆

406608　Vernonia mushituensis Wild;穆希图斑鸠菊●☆

406609　Vernonia musofensis S. Moore;穆索富斑鸠菊●☆

406610　Vernonia musofensis S. Moore var. miamensis (S. Moore) G. V. Pope;迈阿密斑鸠菊●☆

406611　Vernonia myriantha Hook. f. ;多花斑鸠菊●☆

406612　Vernonia myriantha Hook. f. var. ampla (O. Hoffm.) Isawumi = Vernonia myriantha Hook. f. ●☆

406613　Vernonia myriantha Hook. f. var. subuligera (O. Hoffm.) Isawumi = Vernonia myriantha Hook. f. ●☆

406614　Vernonia myrianthoides Muschl. = Vernonia myriantha Hook. f. ●☆

406615　Vernonia myriocephala A. Rich. = Vernonia theophrastifolia Schweinf. ex Oliv. et Hiern ■☆

406616　Vernonia myriotricha Baker = Inula mannii (Hook. f.) Oliv. et Hiern ■☆

406617　Vernonia nandensis S. Moore = Vernonia gerberiformis Oliv. et Hiern ■☆

406618　Vernonia nantcianensis (Pamp.) Hand. -Mazz. ;南漳斑鸠菊 (狗仔草);Nanzhang Ironweed ■

406619　Vernonia natalensis Sch. Bip. ex Walp. ;纳塔尔斑鸠菊■☆

406620　Vernonia neocorymbosa Hilliard = Vernonia tigna Klatt ●☆

406621　Vernonia neocoursiana Humbert;新库尔斯斑鸠菊●☆

406622　Vernonia neoperrieriana Humbert;佩里耶斑鸠菊●☆

406623　Vernonia nepetifolia Wild;荆芥叶斑鸠菊■☆

406624　Vernonia nestor S. Moore;奈斯托尔斑鸠菊●☆

406625　Vernonia neumanniana O. Hoffm. ;纽曼斑鸠菊■☆

406626　Vernonia newbouldii Beentje et Mesfin;纽博尔德斑鸠菊●☆

406627　Vernonia nigritiana Oliv. et Hiern;尼格里塔斑鸠菊■☆

406628　Vernonia nimbaensis C. D. Adams;尼恩巴斑鸠菊■☆

406629　Vernonia noveboracensis (L.) Michx. ;哈珀斑鸠菊■☆

406630　Vernonia noveboracensis (L.) Michx. var. tomentosa Britton = Vernonia noveboracensis (L.) Michx. ■☆

406631　Vernonia noveboracensis (L.) Willd. ;新巴拉森斑鸠菊;Common Ironweed, Flat-tops, Ironweed, Iron-weed, New York Ironweed ■☆

406632　Vernonia nudicaulis Less. ;裸茎斑鸠菊■☆

406633　Vernonia nummulariifolia (Klatt) Klatt = Distephanus nummulariifolius (Klatt) H. Rob. et B. Kahn ●☆

406634　Vernonia nuxioides O. Hoffm. et Muschl. ;努西木斑鸠菊■☆

406635　Vernonia nyassae Oliv. ;尼亚萨斑鸠菊■☆

406636　Vernonia obbreviata (Wall.) DC. = Vernonia cinerea (L.) Less. ■

406637　Vernonia obconica Oliv. et Hiern;倒圆锥斑鸠菊■☆

406638　Vernonia obconica Oliv. et Hiern = Vernonia glabra (Steetz) Vatke ■☆

406639　Vernonia obionifolia O. Hoffm. ;滨藜叶斑鸠菊■☆

406640　Vernonia obionifolia O. Hoffm. subsp. dentata Merxm. ;尖齿斑鸠菊■☆

406641　Vernonia ochroleuca Baker = Distephanus ochroleucus (Baker) H. Rob. et B. Kahn ●☆

406642　Vernonia ochyrae Lisowski;奥吉拉斑鸠菊■☆

406643　Vernonia oehleri Muschl. = Vernonia calvoana (Hook. f.) Hook. f. subsp. oehleri (Muschl.) C. Jeffrey ■☆

406644　Vernonia oligocephala (DC.) Sch. Bip. ex Walp. ;少头斑鸠菊■☆

406645　Vernonia oligocephala Sch. Bip. = Vernonia oligocephala (DC.) Sch. Bip. ex Walp. ■☆

406646　Vernonia oliveriana Pic. Serm. = Vernonia myriantha Hook. f. ●☆

406647　Vernonia ondongensis Klatt = Vernonia glabra (Steetz) Vatke ■☆

406648　Vernonia oocephala Baker = Vernonia stenocephala Oliv. ■☆

406649　Vernonia oocephala Baker var. angustifolia S. Moore = Vernonia stenocephala Oliv. ■☆

406650　Vernonia orchidorrhiza Welw. ex Hiern;兰根斑鸠菊■☆

406651　Vernonia orgyalis S. Moore;银斑鸠菊■☆

406652　Vernonia ornata S. Moore;饰冠斑鸠菊■☆

406653　Vernonia otophora Mattf. ;耳斑鸠菊■☆

406654　Vernonia ovalifolia Torr. et A. Gray = Vernonia gigantea (Walter) Trel. ex Branner et Coville ■☆

406655　Vernonia oxyura O. Hoffm. = Vernonia colorata (Willd.) Drake subsp. oxyura (O. Hoffm.) C. Jeffrey ■☆

406656　Vernonia pachyclada Baker;粗枝斑鸠菊■☆

406657　Vernonia pandurata Link;琴形斑鸠菊■☆

406658　Vernonia papillosa Franch. = Vernonia esculenta Hemsl. ex Forbes et Hemsl. ●

406659　Vernonia papillosissima Chiov. ;乳头斑鸠菊■☆

406660　Vernonia paraemulans C. Jeffrey;拟匹敌斑鸠菊■☆

406661　Vernonia parapetersii C. Jeffrey;假彼得斯斑鸠菊●☆

406662　Vernonia parishii Hook. f. ;滇缅斑鸠菊(大发散,大红花远志,豆豉叶,野辣烟,镇心丸);Parish's Ironweed, Yunnan-Burma Ironweed ●

406663　Vernonia parviflora Reinw. = Vernonia cinerea (L.) Less. var. parviflora (Reinw. ex Blume) DC. ■

406664　Vernonia parviflora Reinw. ex Blume;小花斑鸠菊■

406665　Vernonia parviflora Reinw. ex Blume = Vernonia cinerea (L.) Less. var. parviflora (Reinw. ex Blume) DC. ■

406666　Vernonia pascuosa S. Moore = Vernonia purpurea Sch. Bip. ex Walp. ■☆

406667　Vernonia patula (Dryand.) Merr. ;咸虾花(大风艾,大叶咸虾花,狗狗木,狗仔菜,狗仔花,鬼点火,鲫鱼草,岭南野菊,牛鞭子草,蜻蜓饭,万重花,烟斗菜,展叶斑鸠菊);Halfspreading Ironweed, Saltshrimp Flower ■

406668　Vernonia pauciflora (Willd.) Less. = Vernonia galamensis (Cass.) Less. ■☆

406669　Vernonia pechuelii Kuntze = Blumea cafra (DC.) O. Hoffm. ■☆

406670　Vernonia pectoralis Baker;胸骨状斑鸠菊■☆

406671　Vernonia peculiaris Verdc. ;特殊斑鸠菊■☆

406672　Vernonia pellegrinii Humbert;佩尔格兰斑鸠菊●☆

406673　Vernonia perparva S. Moore = Vernonia chthonocephala O. Hoffm. ●☆

406674　Vernonia perrottetii Sch. Bip. ex Walp.;佩罗斑鸠菊●☆

406675　Vernonia petersii Oliv. et Hiern ex Oliv.;彼得斯斑鸠菊●☆

406676　Vernonia petitiana A. Rich. = Vernonia galamensis (Cass.) Less. var. petitiana (A. Rich.) M. G. Gilbert ■☆

406677　Vernonia philipsoniana Lawalrée = Vernonia musofensis S. Moore var. miamensis (S. Moore) G. V. Pope ●☆

406678　Vernonia phillipsiae S. Moore;菲利斑鸠菊●☆

406679　Vernonia phlomoides Muschl.;糙苏斑鸠菊●☆

406680　Vernonia pinifolia (Lam.) Less. = Vernonia capensis (Houtt.) Druce ■☆

406681　Vernonia piovanii Chiov. = Vernonia glabra (Steetz) Vatke var. laxa (Steetz) Brenan ■☆

406682　Vernonia platylepis Drake;宽鳞斑鸠菊●☆

406683　Vernonia platyseta S. Ortiz;阔刚毛斑鸠菊●☆

406684　Vernonia pleiotaxoides Hutch. et B. L. Burtt = Vernonia longipedunculata De Wild. var. retusa (R. E. Fr.) G. V. Pope ■☆

406685　Vernonia plumbaginifolia Fenzl ex Oliv. et Hiern;白花丹叶斑鸠菊●☆

406686　Vernonia plumbaginifolia Fenzl ex Oliv. et Hiern var. kenyensis C. Jeffrey;肯尼亚斑鸠菊●☆

406687　Vernonia plumosa (O. Hoffm.) J.-P. Lebrun et Stork = Distephanus plumosus (O. Hoffm.) Mesfin ●☆

406688　Vernonia pobeguinii Aubrév. = Vernonia thomsoniana Oliv. et Hiern ex Oliv. ■☆

406689　Vernonia podocoma Sch. Bip. ex Vatke = Vernonia myriantha Hook. f. ●☆

406690　Vernonia podocoma Sch. Bip. ex Vatke var. glabrata Fiori = Vernonia rueppellii Sch. Bip. ●☆

406691　Vernonia poggeana O. Hoffm.;波格斑鸠菊●☆

406692　Vernonia pogosperma Klatt = Vernonia glabra (Steetz) Vatke ■☆

406693　Vernonia poissonii Humbert;普瓦松斑鸠菊●☆

406694　Vernonia polygalifolia Less. = Distephanus polygalifolius (Less.) H. Rob. et B. Kahn ●☆

406695　Vernonia polymorpha Vatke = Vernonia adoensis Sch. Bip. ex Walp. ■☆

406696　Vernonia polymorpha Vatke var. accedens ? = Vernonia adoensis Sch. Bip. ex Walp. ■☆

406697　Vernonia polymorpha Vatke var. ambigua ? = Vernonia adoensis Sch. Bip. ex Walp. ■☆

406698　Vernonia polymorpha Vatke var. microcephala ? = Vernonia adoensis Sch. Bip. ex Walp. ■☆

406699　Vernonia polysphaera Baker;多球斑鸠菊●☆

406700　Vernonia polytricholepis Baker;多毛鳞斑鸠菊●☆

406701　Vernonia polyura O. Hoffm. = Vernonia colorata (Willd.) Drake subsp. oxyura (O. Hoffm.) C. Jeffrey ■☆

406702　Vernonia popeana C. Jeffrey;波普斑鸠菊●☆

406703　Vernonia porphyrolepis S. Moore = Vernonia karaguensis Oliv. ●☆

406704　Vernonia porta-taurinae Dinter ex Merxm. = Vernonia cinerascens Sch. Bip. ●☆

406705　Vernonia poskeana Vatke et Hildebrandt subsp. samfyana (G. V. Pope) G. V. Pope;萨姆菲亚斑鸠菊■☆

406706　Vernonia poskeana Vatke et Hildebrandt var. centaureoides (Klatt) Wild = Vernonia centaureoides Klatt ●☆

406707　Vernonia poskeana Vatke et Hildebrandt var. chlorolepis (Steetz) O. Hoffm. = Vernonia steetziana Oliv. et Hiern ●☆

406708　Vernonia poskeana Vatke et Hildebrandt var. elegantissima (Hutch. et Dalziel) C. D. Adams = Vernonia elegantissima Hutch. et Dalziel ●☆

406709　Vernonia poskeana Vatke et Hildebrandt var. vulgaris Hiern = Vernonia rhodanthoidea Muschl. ■☆

406710　Vernonia praecox Welw. ex O. Hoffm.;早生斑鸠菊■☆

406711　Vernonia praemorsa Muschl.;啮蚀斑鸠菊●■○☆

406712　Vernonia pratensis Hiern = Vernonia kandtii Muschl. ■☆

406713　Vernonia pratensis Klatt = Bothriocline madagascariensis (DC.) C. Jeffrey ■☆

406714　Vernonia praticola S. Moore;草原斑鸠菊■☆

406715　Vernonia primulina O. Hoffm. = Vernonia gerberiformis Oliv. et Hiern subsp. macrocyanus (O. Hoffm.) C. Jeffrey ■☆

406716　Vernonia printzioides Muschl.;尾药菀斑鸠菊■☆

406717　Vernonia printzioides Muschl. var. tomentosa Cufod.;绒毛斑鸠菊■☆

406718　Vernonia pristis Hutch. et B. L. Burtt = Vernonia gerberiformis Oliv. et Hiern var. hockii (De Wild. et Muschl.) G. V. Pope ■☆

406719　Vernonia procera O. Hoffm.;高大斑鸠菊●☆

406720　Vernonia proclivicola S. Moore = Vernonia ugandensis S. Moore ■☆

406721　Vernonia prolixa S. Moore;铺展斑鸠菊■☆

406722　Vernonia pseudoappendiculata Humbert;假附属物斑鸠菊●☆

406723　Vernonia pseudocorymbosa Thell. = Vernonia crataegifolia Hutch. ●☆

406724　Vernonia pseudonatalensis Wild = Vernonia natalensis Sch. Bip. ex Walp. ■☆

406725　Vernonia pteropoda Oliv. et Hiern;翅足斑鸠菊■☆

406726　Vernonia pulchella Small;美丽斑鸠菊■☆

406727　Vernonia pumila Kotschy et Peyr.;偃俯斑鸠菊●☆

406728　Vernonia punctulata De Wild. = Vernonia ugandensis S. Moore ■☆

406729　Vernonia purpurea Sch. Bip. ex Walp.;紫斑鸠菊■☆

406730　Vernonia purpurea Sch. Bip. ex Walp. var. schnellii C. D. Adams = Vernonia purpurea Sch. Bip. ex Walp. ■☆

406731　Vernonia quadriflora Baker = Vernonia secundifolia Bojer ex DC. ■☆

406732　Vernonia quangensis O. Hoffm.;热非斑鸠菊●☆

406733　Vernonia quangensis O. Hoffm. f. angustifolia ? = Vernonia quangensis O. Hoffm. ●☆

406734　Vernonia quartiniana A. Rich. = Vernonia congolensis De Wild. et Muschl. subsp. vernonioides (Sch. Bip. ex Walp.) C. Jeffrey ■☆

406735　Vernonia quartziticola Humbert;阔茨斑鸠菊●☆

406736　Vernonia randii S. Moore = Vernonia amygdalina Delile ■☆

406737　Vernonia retifolia S. Moore;网叶斑鸠菊■☆

406738　Vernonia retusa R. E. Fr. = Vernonia longipedunculata De Wild. var. retusa (R. E. Fr.) G. V. Pope ■☆

406739　Vernonia rhaponticoides Baker = Centauropsis rhaponticoides (Baker) Drake ●☆

406740　Vernonia rhodanthoidea Muschl.;粉红花斑鸠菊■☆

406741　Vernonia rhodanthoidea Muschl. var. densifolia Verdc. = Vernonia rhodanthoidea Muschl. ■☆

406742　Vernonia rhodanthoidea Muschl. var. psammophila (Muschl.) Verdc. = Vernonia rhodanthoidea Muschl. ■☆

406743　Vernonia rhodocalymma Chiov. = Vernonia biafrae Oliv. et Hiern ■☆

406744　Vernonia rhodolepis Baker;红鳞斑鸠菊■☆

406745　Vernonia rhodopappa Baker;红冠毛斑鸠菊●☆

406746　Vernonia rhodophylla O. Hoffm.;红叶斑鸠菊■☆

406747　Vernonia richardiana (Kuntze) Pic. Serm. = Vernonia

theophrastifolia Schweinf. ex Oliv. et Hiern ■☆

406748　Vernonia rigidifolia Hiern;硬叶斑鸠菊■☆

406749　Vernonia rigiophylla DC. = Vernonia squarrosa (D. Don) Less. ■

406750　Vernonia rigorata S. Moore = Vernonia purpurea Sch. Bip. ex Walp. ■☆

406751　Vernonia ringoetii De Wild. ;林戈斑鸠菊●☆

406752　Vernonia robecchiana Muschl. ;罗贝克斑鸠菊●☆

406753　Vernonia robinsonii Wild;鲁滨逊斑鸠菊●☆

406754　Vernonia rochonioides Humbert = Distephanus rochonioides (Humbert) H. Rob. et B. Kahn ●☆

406755　Vernonia rochonioides subsp. ambositrensis Humbert = Distephanus rochonioides (Humbert) H. Rob. et B. Kahn ●☆

406756　Vernonia rochonioides subsp. fiherenensis Humbert = Distephanus rochonioides (Humbert) H. Rob. et B. Kahn ●☆

406757　Vernonia rochonioides subsp. rochonioides = Distephanus rochonioides (Humbert) H. Rob. et B. Kahn ●☆

406758　Vernonia rogersii S. Moore = Vernonia acuminatissima S. Moore ■☆

406759　Vernonia rosenii R. E. Fr. ;罗森斑鸠菊■☆

406760　Vernonia roseopapposa Gilli = Vernonia glabra (Steetz) Vatke ☆

406761　Vernonia roseoviolacea De Wild. ;粉堇色斑鸠菊■☆

406762　Vernonia rothii Oliv. et Hiern = Vernonia hymenolepis A. Rich. ■☆

406763　Vernonia rotundisquamata S. Moore = Vernonia guineensis Benth. ■☆

406764　Vernonia roxburghii Less. = Vernonia aspera (Roxb.) Buch.-Ham. ■

406765　Vernonia rubens Wild;变淡红斑鸠菊●☆

406766　Vernonia rubicunda Klatt;稍红斑鸠菊●☆

406767　Vernonia rueppellii Sch. Bip. ;吕佩尔斑鸠菊●☆

406768　Vernonia rufuensis Muschl. ;赤褐斑鸠菊●☆

406769　Vernonia ruwenzoriensis S. Moore;鲁文佐里斑鸠菊■☆

406770　Vernonia sabulosa Beentje et Mesfin;砂地斑鸠菊■☆

406771　Vernonia sakalava Humbert;萨卡拉瓦斑鸠菊●☆

406772　Vernonia saligna (Wall.) DC. ;柳叶斑鸠菊(白龙须,白头升麻,牙金药);Willowleaf Ironweed ■

406773　Vernonia saligna DC. = Vernonia clivorum Hance ■

406774　Vernonia salutarii S. Moore = Vernonia lasiopus O. Hoffm. var. iodocalyx (O. Hoffm.) C. Jeffrey ●☆

406775　Vernonia sambiranensis (Humbert) Humbert;桑比朗斑鸠菊●☆

406776　Vernonia samfyana G. V. Pope = Vernonia poskeana Vatke et Hildebrandt subsp. samfyana (G. V. Pope) G. V. Pope ■☆

406777　Vernonia sapinii De Wild. ;萨潘斑鸠菊■☆

406778　Vernonia saussureoides Hutch. = Vernonia infundibularis Oliv. et Hiern ■☆

406779　Vernonia scaberrima Nutt. = Vernonia angustifolia Michx. ■☆

406780　Vernonia scabrida C. H. Wright = Vernonia purpurea Sch. Bip. ex Walp. ■☆

406781　Vernonia scabrifolia O. Hoffm. = Vernonia melleri Oliv. et Hiern var. superba (O. Hoffm.) C. Jeffrey ■☆

406782　Vernonia scabrifolia O. Hoffm. var. amplifolia ? = Vernonia melleri Oliv. et Hiern ●☆

406783　Vernonia scandens DC. = Vernonia blanda (Wall.) DC. ●■

406784　Vernonia scandens Merr. = Vernonia cumingiana Benth. ●■

406785　Vernonia scandens Sensu Merr. = Vernonia cumingiana Benth. ●■

406786　Vernonia scapiformis (DC.) Drake = Vernonia nudicaulis Less. ■☆

406787　Vernonia schimperi DC. ;阿比西尼亚斑鸠菊■☆

406788　Vernonia schimperi Sch. Bip. = Vernonia purpurea Sch. Bip. ex Walp. ■☆

406789　Vernonia schinzii O. Hoffm. = Vernonia fastigiata Oliv. et Hiern ■☆

406790　Vernonia schlechteri O. Hoffm. ;施莱斑鸠菊■☆

406791　Vernonia schliebenii Mattf. ;施利本斑鸠菊■☆

406792　Vernonia schoenfelderiana Dinter ex Merxm. = Pleiotaxis antunesii O. Hoffm. ■☆

406793　Vernonia schubotziana Muschl. ;舒博斑鸠菊■☆

406794　Vernonia schweinfurthii Oliv. et Hiern;施韦斑鸠菊■☆

406795　Vernonia schweinfurthii Oliv. et Hiern var. bukamaensis (De Wild.) Kalanda et Lisowski;布卡马斑鸠菊●☆

406796　Vernonia sciaphila S. Moore = Vernonia luembensis De Wild. et Muschl. ■☆

406797　Vernonia sclerophylla O. Hoffm. ;核叶斑鸠菊■☆

406798　Vernonia scoparia O. Hoffm. ;帚状斑鸠菊■☆

406799　Vernonia scorpioides (Lam.) Pers. ;蝎尾斑鸠菊■☆

406800　Vernonia sculptifolia Hiern = Vernonia teucrioides Welw. ex O. Hoffm. ■☆

406801　Vernonia secundifolia Bojer ex DC. ;单侧叶斑鸠菊■☆

406802　Vernonia seguini Vaniot = Vernonia saligna (Wall.) DC. ■

406803　Vernonia senecioides A. Chev. = Vernonia biafrae Oliv. et Hiern ■☆

406804　Vernonia senegalensis (Pers.) Less. = Vernonia colorata (Willd.) Drake ■☆

406805　Vernonia senegalensis (Pers.) Less. var. acuminata S. Moore = Vernonia colorata (Willd.) Drake subsp. grandis (DC.) C. Jeffrey ■☆

406806　Vernonia sengana S. Moore;森加斑鸠菊■☆

406807　Vernonia sennii Chiov. = Vernonia homilantha S. Moore ■☆

406808　Vernonia seretii De Wild. = Vernonia theophrastifolia Schweinf. ex Oliv. et Hiern ■☆

406809　Vernonia sericolepis O. Hoffm. ;绢毛鳞斑鸠菊■☆

406810　Vernonia seyrigii Humbert;塞里格斑鸠菊●☆

406811　Vernonia shabensis Kalanda;沙巴斑鸠菊■☆

406812　Vernonia shirensis Oliv. et Hiern = Baccharoides adoensis (Sch. Bip. ex Walp.) H. Rob. var. mossambiquensis (Steetz) Isawumi, El-Ghazaly et B. Nord. ●☆

406813　Vernonia sidamensis O. Hoffm. ;锡达莫斑鸠菊●☆

406814　Vernonia silhetensis var. nantcianensis (Pamp.) Hand.-Mazz. = Vernonia nantcianensis (Pamp.) Hand.-Mazz. ■

406815　Vernonia smaragdopappa S. Moore = Vernonia chloropappa Baker ●☆

406816　Vernonia smithiana Less. ;史密斯斑鸠菊■☆

406817　Vernonia solanifolia Benth. ;茄叶斑鸠菊(白沉沙,白花毛桃,斑鸠菊,斑鸠木,大过山龙,大消藤,空心癌麻,牛舌癀,茄叶咸虾花,咸虾花,硬骨过山龙,硬骨夜牵牛,月中风,肿风);Eggplant Ironweed, Nightshadeleaf Ironweed, Nightshade-leaved Ironweed ●

406818　Vernonia solidaginifolius Bojer ex DC. ;密叶斑鸠菊■☆

406819　Vernonia solweziensis Wild;索卢韦齐斑鸠菊●☆

406820　Vernonia somalensis Franch. = Vernonia aschersonii Sch. Bip. ■☆

406821　Vernonia spatulata (Forssk.) Sch. Bip. ;匙形斑鸠菊■☆

406822　Vernonia speiracephala Baker;马达加斯加斑鸠菊●☆

406823　Vernonia spelaeicola Vaniot = Cissampelopsis spelaeicola (Vaniot) C. Jeffrey et Y. L. Chen ●

406824　Vernonia sphacelata Klatt;毒斑鸠菊●☆

406825　Vernonia sphaerocalyx O. Hoffm. = Vernonia exsertiflora Baker ●☆

406826　Vernonia spiciforma Klatt;矛斑鸠菊■☆

406827　Vernonia spirei Gand. ;折苞斑鸠菊(金沙斑鸠菊,六月雪);Flexbract Ironweed, Spire's Ironweed ■

406828　Vernonia squarrosa (D. Don) Less. ;刺苞斑鸠菊(白脚威灵,白脚威灵仙,黑继参,剪刀草,剪子草,圆柱斑鸠菊,紫花地丁);Spinebract Ironweed, Squarrous Ironweed ■

406829　Vernonia squarrosa (D. Don) Less. var. orientalis Kitam. =

Vernonia squarrosa（D. Don）Less. ■

406830　Vernonia squarrosa Dinter ex Merxm. = Vernonia cinerascens Sch. Bip. ●☆

406831　Vernonia staehelinoides Harv.；卷翅菊状斑鸠菊●☆

406832　Vernonia steetziana Oliv. et Hiern；斯蒂兹斑鸠菊●☆

406833　Vernonia stellulifera（Benth.）C. Jeffrey；星状斑鸠菊■☆

406834　Vernonia stenocephala Oliv.；窄头斑鸠菊■☆

406835　Vernonia stenoclinoides Baker = Helichrysum stenoclinoides（Baker）Humbert ●☆

406836　Vernonia stenolepis Oliv. = Baccharoides anthelmintica（L.）Moench ●

406837　Vernonia stenostegia（Stapf）Hutch. et Dalziel；窄盖斑鸠菊■☆

406838　Vernonia stibaliae Hand. -Mazz. = Vernonia spirei Gand. ■

406839　Vernonia stipulacea Klatt = Vernonia myriantha Hook. f. ●☆

406840　Vernonia streptoclada Baker = Distephanus streptocladus（Baker）H. Rob. et B. Kahn ●☆

406841　Vernonia stuhlmannii O. Hoffm.；斯图尔曼斑鸠菊■☆

406842　Vernonia subaphylla Baker；亚无叶斑鸠菊■☆

406843　Vernonia subaphylla Muschl. = Vernonia subaphylla Baker ■☆

406844　Vernonia subarborea Vaniot = Vernonia extensa（Wall.）DC. ●

406845　Vernonia sublanata Drake = Oliganthes sublanata（Drake）Humbert ●☆

406846　Vernonia subscandens R. E. Fr.；亚攀缘斑鸠菊■☆

406847　Vernonia subuligera O. Hoffm. = Vernonia myriantha Hook. f. ●☆

406848　Vernonia sunzuensis Wild；孙祖斑鸠菊■☆

406849　Vernonia superba O. Hoffm. = Vernonia melleri Oliv. et Hiern var. superba（O. Hoffm.）C. Jeffrey ■☆

406850　Vernonia suprafastigiata Klatt；帚斑鸠菊●☆

406851　Vernonia sutherlandii Harv. = Hilliardiella sutherlandii（Harv.）H. Rob. ■☆

406852　Vernonia swinglei Humbert = Distephanus swinglei（Humbert）H. Rob. et B. Kahn ●☆

406853　Vernonia swynnertonii S. Moore = Vernonia kirkii Oliv. et Hiern ●☆

406854　Vernonia sylvatica Dunn；林生斑鸠菊（藤菊）；Forest Ironweed，Woody Ironweed ●

406855　Vernonia sylvatica sensu Merr. et Chun = Vernonia chunii C. C. Chang ●

406856　Vernonia sylvicola G. V. Pope；西尔维亚斑鸠菊●☆

406857　Vernonia syringifolia O. Hoffm.；丁香叶斑鸠菊●☆

406858　Vernonia tanalensis Baker；塔纳尔斑鸠菊●☆

406859　Vernonia tanganyikensis R. E. Fr.；坦噶尼喀斑鸠菊■☆

406860　Vernonia tayloriana（Isawumi）J. -P. Lebrun et Stork = Vernonia hymenolepis A. Rich. ■☆

406861　Vernonia taylorii S. Moore = Vernonia hildebrandtii Vatke ■☆

406862　Vernonia teitensis O. Hoffm.；泰塔斑鸠菊■☆

406863　Vernonia temnolepis O. Hoffm.；毛鳞斑鸠菊■☆

406864　Vernonia tenoreana Oliv.；泰诺雷斑鸠菊■☆

406865　Vernonia tephrodoides Chiov. = Vernonia cinerascens Sch. Bip. ●☆

406866　Vernonia teres sensu Merr. = Vernonia aspera（Roxb.）Buch. -Ham. ■

406867　Vernonia teres Wall. ex DC. = Vernonia squarrosa（D. Don）Less. ■

406868　Vernonia terniflora Less. = Oliganthes triflora Cass. ●☆

406869　Vernonia teucrioides Welw. ex O. Hoffm.；香科斑鸠菊■☆

406870　Vernonia teuszcii Klatt；托伊斑鸠菊■☆

406871　Vernonia tewoldei Mesfin；特沃斑鸠菊■☆

406872　Vernonia texana（A. Gray）Small；得州斑鸠菊■☆

406873　Vernonia theophrastifolia Schweinf. ex Oliv. et Hiern；假轮叶斑鸠菊■☆

406874　Vernonia theophrastifolia Schweinf. ex Oliv. et Hiern var. richardiana（Kuntze）Isawumi = Vernonia theophrastifolia Schweinf. ex Oliv. et Hiern ■☆

406875　Vernonia thomasii Hutch. = Vernonia klingii O. Hoffm. et Muschl. ●☆

406876　Vernonia thomsoniana Oliv. et Hiern ex Oliv.；托马森斑鸠菊■☆

406877　Vernonia thomsoniana Oliv. et Hiern ex Oliv. var. livingstoniana（Oliv. et Hiern）Pic. Serm. = Vernonia thomsoniana Oliv. et Hiern ex Oliv. ■☆

406878　Vernonia thulinii Mesfin；图林斑鸠菊■☆

406879　Vernonia tigna Klatt；蒂格纳斑鸠菊●☆

406880　Vernonia tigrensis Oliv. et Hiern = Vernonia adoensis Sch. Bip. ex Walp. ■☆

406881　Vernonia timpermaniana Kalanda；廷珀曼斑鸠菊■☆

406882　Vernonia tinctosetosa C. Jeffrey；色毛斑鸠菊■☆

406883　Vernonia titanophylla Brenan；叶斑鸠菊■☆

406884　Vernonia towaensis De Wild. = Vernonia gerberiformis Oliv. et Hiern subsp. macrocyanus（O. Hoffm.）C. Jeffrey ■☆

406885　Vernonia trachyphylla Muschl.；粗叶斑鸠菊■☆

406886　Vernonia transvaalensis Hutch. = Vernonia wollastonii S. Moore ■☆

406887　Vernonia trichocalyx Muschl. ex De Wild.；毛萼斑鸠菊■☆

406888　Vernonia trichodesma Baker = Vernonia exserta Baker ●☆

406889　Vernonia tricholoba C. Jeffrey；毛片斑鸠菊■☆

406890　Vernonia triflora Bremek. = Gymnanthemum triflorum（Bremek.）H. Rob. ■

406891　Vernonia trinervis（Bojer ex DC.）Drake = Distephanus trinervis Bojer ex DC. ●☆

406892　Vernonia trinervis（Bojer ex DC.）Drake f. capitata（Bojer ex DC.）Humbert = Distephanus trinervis Bojer ex DC. ●☆

406893　Vernonia tuberculata Hutch. et B. L. Burtt = Vernonia ringoetii De Wild. ●☆

406894　Vernonia tuberifera R. E. Fr.；块茎斑鸠菊■☆

406895　Vernonia tufnelliae S. Moore = Vernonia biafrae Oliv. et Hiern ■☆

406896　Vernonia turbinata Oliv. et Hiern；陀螺形斑鸠菊■☆

406897　Vernonia ugandensis S. Moore；乌干达斑鸠菊■☆

406898　Vernonia uhligii Muschl. = Vernonia myriantha Hook. f. ●☆

406899　Vernonia ulophylla O. Hoffm. = Vernonia guineensis Benth. ■☆

406900　Vernonia ulophylla O. Hoffm. var. hoffmanniana Hiern = Vernonia guineensis Benth. ■☆

406901　Vernonia ulugurensis O. Hoffm. = Vernonia calvoana（Hook. f.）Hook. f. subsp. ulugurensis（O. Hoffm.）C. Jeffrey ■☆

406902　Vernonia umbratica Oberm. = Vernonia wollastonii S. Moore ■☆

406903　Vernonia uncinata Oliv. et Hiern；具钩斑鸠菊■☆

406904　Vernonia undulata Hiern = Vernonia golungensis Welw. ex Mendonça ■☆

406905　Vernonia undulata Oliv. et Hiern；波状斑鸠菊■☆

406906　Vernonia uniflora Hutch. et Dalziel = Vernonia auriculifera Hiern ■☆

406907　Vernonia upembaensis Kalanda；乌彭巴斑鸠菊■☆

406908　Vernonia urophylla Muschl. = Vernonia pteropoda Oliv. et Hiern ■☆

406909　Vernonia urticifolia A. Rich.；荨麻叶斑鸠菊■☆

406910　Vernonia usafuensis O. Hoffm.；乌沙夫斑鸠菊■☆

406911　Vernonia usambarensis O. Hoffm.；乌桑巴拉斑鸠菊■☆

406912　Vernonia vaginata O. Hoffm.；具鞘斑鸠菊■☆

406913　Vernonia vallicola S. Moore；河谷斑鸠菊■☆

406914　Vernonia vanmeelii Lawalrée = Vernonia melleri Oliv. et Hiern var. superba（O. Hoffm.）C. Jeffrey ■☆

406915　Vernonia venosa S. Moore = Vernonia retifolia S. Moore ■☆

406916　Vernonia verdickii O. Hoffm. = Vernonia dewildemaniana Muschl. ■☆

406917　Vernonia verdickii O. Hoffm. et Muschl. = Vernonia dewildemaniana Muschl. ■☆

406918　Vernonia vernonella Harv. = Vernonia africana (Sond.) Druce ■☆

406919　Vernonia vernonioides (Sch. Bip. ex Walp.) Cufod. = Vernonia congolensis De Wild. et Muschl. subsp. vernonioides (Sch. Bip. ex Walp.) C. Jeffrey ■☆

406920　Vernonia verrucata Klatt;疣斑鸠菊●☆

406921　Vernonia verschuerenii De Wild. = Vernonia biafrae Oliv. et Hiern ■☆

406922　Vernonia vialis DC. = Vernonia cinerea (L.) Less. ■

406923　Vernonia viatorum S. Moore;扩散斑鸠菊■☆

406924　Vernonia vilersii Drake = Vernonia bakeri Vatke ■☆

406925　Vernonia violacea Oliv. et Hiern ex Oliv.;堇色斑鸠菊■☆

406926　Vernonia violaceopapposa De Wild. ;冠毛斑鸠菊■☆

406927　Vernonia violaceopapposa De Wild. subsp. nuttii G. V. Pope;纳特斑鸠菊■☆

406928　Vernonia vitellina N. E. Br. = Distephanus divaricatus (Steetz) H. Rob. et B. Kahn ●☆

406929　Vernonia vogeliana Benth. = Vernonia amygdalina Delile ■☆

406930　Vernonia vohemarensis Humbert;武海马尔斑鸠菊●☆

406931　Vernonia volkameriifolia (Wall.) DC. ;大叶斑鸠菊(大叶鸡菊花,当毫温);Largeleaf Ironweed, Large-leaved Ironweed ■☆

406932　Vernonia volkameriifolia (Wall.) DC. var. lanata S. Y. Hu = Vernonia parishii Hook. f. ●

406933　Vernonia vollesenii C. Jeffrey;福勒森斑鸠菊■☆

406934　Vernonia voluta Baker = Senecio volutus (Baker) Humbert ■☆

406935　Vernonia wakefieldii Oliv. ;韦克菲尔德斑鸠菊■☆

406936　Vernonia weisseana Muschl. = Vernonia amygdalina Delile ■☆

406937　Vernonia welwitschii O. Hoffm. ;韦尔斑鸠菊■☆

406938　Vernonia whyteana Britten = Baccharoides adoensis (Sch. Bip. ex Walp.) H. Rob. var. mossambiquensis (Steetz) Isawumi, El-Ghazaly et B. Nord. ●☆

406939　Vernonia wildii Merxm. = Vernonia bainesii Oliv. et Hiern subsp. wildii (Merxm.) Wild ■☆

406940　Vernonia wittei Hutch. et B. L. Burtt = Vernonia longipedunculata De Wild. ■☆

406941　Vernonia wollastonii S. Moore;沃韦尔斑鸠菊■☆

406942　Vernonia woodii O. Hoffm. = Baccharoides adoensis (Sch. Bip. ex Walp.) H. Rob. var. kotschyana (Sch. Bip. ex Walp.) Isawumi, El-Ghazaly et B. Nord. ■☆

406943　Vernonia yatesii S. Moore = Herderia truncata Cass. ■☆

406944　Vernonia zambesiaca S. Moore = Vernonia kirkii Oliv. et Hiern ●☆

406945　Vernonia zanzibarensis Less. ;桑给巴尔斑鸠菊■☆

406946　Vernonia zernyi Gilli;策尼斑鸠菊■☆

406947　Vernoniaceae Bessey = Asteraceae Bercht. et J. Presl(保留科名)●■

406948　Vernoniaceae Bessey = Compositae Giseke(保留科名)●■

406949　Vernoniaceae Bessey;斑鸠菊科(绿菊科)●

406950　Vernoniaceae Burmeist. = Asteraceae Bercht. et J. Presl(保留科名)●■

406951　Vernoniaceae Burmeist. = Compositae Giseke(保留科名)●■

406952　Vernoniastrum H. Rob. (1999);斑鸠瘦片菊属■☆

406953　Vernoniastrum H. Rob. = Vernonia Schreb. (保留属名)●■

406954　Vernoniastrum aemulans (Vatke) H. Rob. = Vernonia aemulans Vatke ■☆

406955　Vernoniastrum ambiguum (Kotschy et Peyr.) H. Rob. = Vernonia ambigua Kotschy et Peyr. ■☆

406956　Vernoniastrum latifolium (Steetz) H. Rob. = Vernonia petersii Oliv. et Hiern ex Oliv. ●☆

406957　Vernoniastrum musofense (S. Moore) H. Rob. = Vernonia musofensis S. Moore ●☆

406958　Vernoniastrum nestor (S. Moore) H. Rob. = Vernonia nestor S. Moore ●☆

406959　Vernoniastrum ugandense (S. Moore) H. Rob. = Vernonia ugandensis S. Moore ■☆

406960　Vernoniastrum uncinatum (Oliv. et Hiern) H. Rob. = Vernonia uncinata Oliv. et Hiern ■☆

406961　Vernoniastrum viatorum (S. Moore) H. Rob. = Vernonia viatorum S. Moore ■☆

406962　Vernoniopsis Dusén = Vernoniopsis Humbert ●■☆

406963　Vernoniopsis Humbert(1955);距药菊属●■☆

406964　Vernoniopsis caudata (Drake) Humbert;距药菊●☆

406965　Vernoniopsis crassipes Dusén;粗梗距药菊■☆

406966　Veronica L. (1753);婆婆纳属(锹形草属);Bird's Eye, Speedwell, Veronica ■

406967　Veronica × franciscana (Eastw.) J. T. Howell, P. H. Raven et P. Rubtzov = Hebe × franciscana (Eastw.) Souster ●☆

406968　Veronica × myriantha Tos. Tanaka;多花婆婆纳■☆

406969　Veronica aberdarica R. E. Fr. = Veronica glandulosa Hochst. ex Benth. ■☆

406970　Veronica abyssinica Fresen. ;阿比西尼亚婆婆纳■☆

406971　Veronica acinifolia L. ;葡萄叶婆婆纳■☆

406972　Veronica africana Hook. f. = Veronica abyssinica Fresen. ■☆

406973　Veronica agrestis L. ;耕地婆婆纳;Field Speedwell, Garden Speedwell, Germander Chickweed, Green Field Speedwell, Green Field-speedwell, Prncumbent Garden Speedwell, Winterweed ■☆

406974　Veronica agrestis L. = Veronica didyma Ten. ■

406975　Veronica agrestis L. = Veronica polita Fr. ■

406976　Veronica agrestis L. subsp. polita (Fr.) Rouy = Veronica polita Fr. ■

406977　Veronica alatavica Popov = Pseudolysimachion alatavicum (Popov) Holub ■

406978　Veronica albanica K. Koch;阿尔邦婆婆纳■☆

406979　Veronica alpina L. ;高山婆婆纳;Alpine Speedwell ■☆

406980　Veronica alpina L. 'Alba';白高山婆婆纳;White Alpine Speedwell ■☆

406981　Veronica alpina L. subsp. pumila (All.) Dostal;短花柱婆婆纳;Shortstyle Speedwell ■

406982　Veronica alpina L. var. australis Wahlenb. = Veronica alpina L. subsp. pumila (All.) Dostal ■

406983　Veronica americana (Raf.) Schwein. ex Benth. ;美洲婆婆纳;American Brooklime, American Speedwell ■☆

406984　Veronica americana (Raf.) Schwein. ex Benth. f. albiflora Tatew. ex H. Hara;白花美洲婆婆纳■☆

406985　Veronica americana Schwein. ex Benth. = Veronica americana (Raf.) Schwein. ex Benth. ■☆

406986　Veronica amoena Stev. ;秀丽婆婆纳■☆

406987　Veronica amygdalina ?;膀胱状婆婆纳;Bittedeaf ■☆

406988　Veronica anagallidiformis Boreau;腺鳞草婆婆纳■☆

406989　Veronica anagallis L. = Veronica anagallis-aquatica L. ■

406990　Veronica anagallis L. = Veronica undulata Wall. ex Jack ■

406991　Veronica anagallis L. var. punctata ? = Veronica undulata Wall. ex

Jack ■

406992　Veronica anagallis-aquatica L.;北水苦荬(半边山,虫虫草,大仙桃草,二代草,接骨桃,接骨仙桃草,蛤蛙草,芒种草,秋麻子,水波浪,水菠菜,水对叶莲,水接骨丹,水苦荬,水上浮萍,水莴苣,水窝窝,水仙桃草,水泽兰,蚊子草,仙人对座草,仙桃草,谢婆菜,鸭儿草,珍珠草);Blue Water-speedwell, Cress, Fat Grass, Water Pimpernel, Water Speedwell, Water-speedwell, Watery Speedwell ■

406993　Veronica anagallis-aquatica L. = Veronica undulata Wall. ex Jack ■

406994　Veronica anagallis-aquatica L. subsp. anagalloides (Guss.) Batt. = Veronica anagalloides Guss. ■

406995　Veronica anagallis-aquatica L. subsp. aquatica (Bernh.) Maire = Veronica anagallis-aquatica L. ■

406996　Veronica anagallis-aquatica L. subsp. divaricata Krösche;日本水苦荬■☆

406997　Veronica anagallis-aquatica L. subsp. oxycarpa (Boiss.) Elenevsky = Veronica oxycarpa Boiss. ■

406998　Veronica anagallis-aquatica L. subsp. undulata (Wall. ex Jack.) Elenevsky = Veronica undulata Wall. ex Jack ■

406999　Veronica anagallis-aquatica L. var. elata Hoffmanns. et Link = Veronica anagallis-aquatica L. ■

407000　Veronica anagallis-aquatica L. var. glandulosa Farw. = Veronica anagallis-aquatica L. ■

407001　Veronica anagallis-aquatica L. var. laevipes Celak. = Veronica anagallis-aquatica L. ■

407002　Veronica anagallis-aquatica L. var. savatieri Makino;大连水苦荬■

407003　Veronica anagalloides Guss.;长果水苦荬(拟水苦荬);Pimpernel-like Speedwell ■

407004　Veronica andersonii Lindl. et Paxton = Hebe andersonii Cockayne ●☆

407005　Veronica angustifolia Fisch. ex Link = Pseudolysimachion lineariifolium (Pall. ex Link) Holub ■

407006　Veronica angustifolia Fisch. ex Link var. dilatata Nakai et Tamg. = Pseudolysimachion linariifolium (Pall. ex Link) Holub subsp. dilatatum (Nakai et Kitag.) D. Y. Hong ■

407007　Veronica angustifolia Fisch. var. dilatata Nakai et Kitag. = Pseudolysimachion linariifolium (Pall. ex Link) Holub subsp. dilatatum (Nakai et Kitag.) D. Y. Hong ■

407008　Veronica aphylla L.;无叶婆婆纳■☆

407009　Veronica aquatica Bernh. = Veronica undulata Wall. ex Jack ■

407010　Veronica arenosa (Serg.) Boriss.;砂婆婆纳■☆

407011　Veronica argute-serrata Regel et Schmalh.;尖齿婆婆纳(博恩婆婆纳)■

407012　Veronica armena Boiss. et Huet;亚美尼亚婆婆纳■☆

407013　Veronica arotunda Nakai var. subintegra (Nakai) T. Yamaz.;东北婆婆纳;Northeast Speedwell ■

407014　Veronica arvensis L.;直立婆婆纳(脾寒草,平地婆婆纳);Common Speedwell, Corn Speedwell, Field Speedwell, Stand Speedwell, Wall Speedwell ■

407015　Veronica arvensis L. subsp. pseudoarvensis (Tineo) H. Lindb. = Veronica arvensis L. var. pseudoarvensis (Tineo) Fiori ■

407016　Veronica arvensis L. var. atlantica Batt. = Veronica arvensis L. ■

407017　Veronica arvensis L. var. pseudoarvensis (Tineo) Fiori = Veronica arvensis L. ■

407018　Veronica austriaca L.;奥地利婆婆纳;Broadleaf Speedwell, Broad-leaf Speedwell, Large Speedwell, Speedwell ■☆

407019　Veronica austriaca L. subsp. teucrium (L.) D. A. Webb;香科状婆婆纳(邓木址,卷毛婆婆纳);Broad-leaf Speedwell, Floccosous Speedwell, Wood-sage Speedwell ■

407020　Veronica austriaca L. subsp. teucrium (L.) D. A. Webb 'Royal Blue';品蓝香科状婆婆纳■☆

407021　Veronica austriaca L. var. maroccana Pau et Font Quer = Veronica rosea Desf. ■☆

407022　Veronica bachofenii Heuff.;巴氏婆婆纳(巴豪婆婆纳);Bachofen's Speedwell, Heart-leaf Speedwell ■☆

407023　Veronica balfouriana Hook. f.;巴尔氏婆婆纳●☆

407024　Veronica baranetzkii Bordz.;巴拉婆婆纳■☆

407025　Veronica barrelieri Schult.;巴列尔婆婆纳■☆

407026　Veronica bartsiifolia Boiss. ex Freyn. = Veronica argute-serrata Regel et Schmalh. ■

407027　Veronica battiscombei R. E. Fr. = Veronica glandulosa Hochst. ex Benth. ■☆

407028　Veronica baumgartenii Roem. et Schult.;鲍姆婆婆纳■☆

407029　Veronica beccabunga L.;茶苦荬(水茴芹婆婆纳,有柄水苦荬);Beccabunga Speedwell, Becky-leaves, Bekkabung, Biddy's Eyes, Bird's Eye, Brooklem, Brooklime, Burileek, Cow Cress, European Brooklime, European Speedwell, Horse Cress, Horse Well Cress, Horse Well-cress, Horse Wellgrass, Limewort, Limpwort, Stalked Speedwell, Wall Ink, Water Bird's Eye, Water Bird's Eyes, Water Pimpernel, Water Pumpy, Water Purple, Water Speedwell, Wellink ■

407030　Veronica beccabunga L. 'Grease';油脂水苦荬;Pig's Grease ■☆

407031　Veronica beccabunga L. f. minima Engl. = Veronica beccabunga L. ■

407032　Veronica beccabunga L. subsp. muscosa (Korsh.) Elenevsky;有柄水苦荬■

407033　Veronica beccabunga L. var. americana Raf. = Veronica americana (Raf.) Schwein. ex Benth. ■☆

407034　Veronica beccabunga L. var. limosa (Lej.) = Veronica beccabunga L. ■

407035　Veronica beccabunga L. var. muscosa Korsh. = Veronica beccabunga L. subsp. muscosa (Korsh.) Elenevsky ■

407036　Veronica beccabunga L. var. xauenensis Pau = Veronica beccabunga L. ■

407037　Veronica beccabungoides Bornm.;假柄婆婆纳(假有柄水苦荬)■☆

407038　Veronica bellidifolia Juz.;雅叶婆婆纳■☆

407039　Veronica bellidioides L.;雅致婆婆纳■☆

407040　Veronica biloba L.;两裂婆婆纳;Bilobed Speedwell, Twolobe Speedwell, Two-lobe Speedwell ■

407041　Veronica bobrovii Nevski;鲍勃罗夫婆婆纳■☆

407042　Veronica bornmuelleri Hausskn. = Veronica argute-serrata Regel et Schmalh. ■

407043　Veronica brachiata Benth. ex Oerst.;对枝婆婆纳■☆

407044　Veronica brasiliana (L.) Druce;巴西婆婆纳■☆

407045　Veronica bucharica B. Fedtsch.;布赫婆婆纳■☆

407046　Veronica buxbaumii Ten. = Veronica persica Poir. ■

407047　Veronica buxifolia Benth.;黄杨叶婆婆纳●☆

407048　Veronica callitrichoides Kom.;美毛婆婆纳■☆

407049　Veronica campylopoda Boiss.;弯果婆婆纳;Bent-foot Speedwell, Curvedfruit Speedwell ■

407050　Veronica cana Wall.;灰毛婆婆纳(青地蚕子);Greyhairy Speedwell ■

407051　Veronica cana Wall. ex Benth. = Veronica taiwanica T. Yamaz. ■

407052　Veronica cana Wall. ex Benth. subsp. henryi (T. Yamaz.) Elenevsky = Veronica henryi T. Yamaz. ■

407053　Veronica cana Wall. ex Benth. var. miqueliana (Nakai) Ohwi =

Veronica miqueliana Nakai ■☆

407054　Veronica cana Wall. ex Benth. var. takedana Makino;武田灰毛婆婆纳■☆

407055　Veronica cana Wall. subsp. henryi（T. Yamaz.）Elenevsky = Veronica henryi T. Yamaz. ■

407056　Veronica caninotesticulata Makino ex H. Hara = Veronica polita Fr. subsp. lilacina（T. Yamaz.）T. Yamaz. ■☆

407057　Veronica capensis Fenzl = Veronica anagallis-aquatica L. ■

407058　Veronica capillipes Nevski;细毛婆婆纳■☆

407059　Veronica capitata Royle ex Benth. ;头花婆婆纳（头花柱婆婆纳）;Capitateflower Speedwell ■

407060　Veronica capitata Royle ex Benth. var. sikkimensis Hook. f. = Veronica szechuanica Batalin subsp. sikkimensis（Hook. f.）D. Y. Hong ■

407061　Veronica capitata Royle var. sikkimensis Hook. f. = Veronica szechuanica Batalin subsp. sikkimensis（Hook. f.）D. Y. Hong ■

407062　Veronica capitata Royle var. sikkimensis Hook. f. = Veronica umbelliformis Pennell ■

407063　Veronica cardiocarpa（Kar. et Kir.）Walp. ;心果婆婆纳;Heartfruit Speedwell ■

407064　Veronica catenata Pennell;粉色水婆婆纳;Pink Water Speedwell, Pink Water-speedwell,Water Speedwell ■☆

407065　Veronica catenata Pennell = Veronica anagallis-aquatica L. ■

407066　Veronica catenata Pennell subsp. pseudocatenata（Chrtek）Osb. -Kos. ;假链粉色水婆婆纳■☆

407067　Veronica catenata Pennell var. glandulosa（Farw.）Pennell = Veronica anagallis-aquatica L. ■

407068　Veronica caucasica M. Bieb. ;高加索婆婆纳■☆

407069　Veronica cephaloides Pennell = Veronica ciliata Fisch. subsp. cephaloides（Pennell）D. Y. Hong ■

407070　Veronica cerassifolia Monjuschko;厚叶婆婆纳■☆

407071　Veronica ceratocarpa C. A. Mey. ;角果婆婆纳■☆

407072　Veronica chamaedrys L. ;石蚕叶婆婆纳（欧婆婆纳）;Angel's Eyes, Angel's Tears, Angel's-eyes, Base Vervain, Biddy's Eyes, Billy Bright Eyes, Billy Bright-eye, Bird's-eye, Bird's-eye Speedwell, Blewart,Blind Flower,Blue Bird's Eye,Blue Bird's Eyes,Blue Eyes, Blue Star, Bobberty, Bobby's Eye, Bobby's Eyes, Botherum, Break-basin,Bright Eye,Bright Eyes,Brighteye,Bullock's Eyes,Cat's Eye, Cat's Eyes, Cat's-eye, Cup-and-saucer, Devil's Eye, Devil's Eyes, Devil's Flower, Dotherum, Eye of Christ, Eyebright, Flat Vervain, Forget-me-not, Germander Chickweed, Germander Speedwell, Germanderleaf Speedwell, God's Eye, God's Eyes, Hawk-your-Mother's-eyes-out, Hawk-your-Mother's-eyes-out, Jerrymander, Lady's Thimble,Lady's Thimbles,Lark's Eye,Lark's Eyes,Little-and-pretty, Love-me-not, Mary's Rest, Milkmaid's Eye, Milkmaid's Eyes, Mother-breaks-her-heart, Nancy Pretty, Nancy-pretty, Pick-your-mother's-eyes-out,Poor Man's Tea, Remember-me, Shamrock, Sore Eyes, St. Joseph's Flower, Strike-fire, Teab-your-mother's-eyes-out, Thunderbolts, Wild Germander,Wish-me-well ■

407073　Veronica chamydryoides Engl. = Veronica javanica Blume ■

407074　Veronica chantavica Pavlov;哈恩塔夫婆婆纳■☆

407075　Veronica charadzeae Kem. -Nath. ;哈拉婆婆纳■☆

407076　Veronica chayuensis D. Y. Hong;察隅婆婆纳;Chayu Speedwell,Tsayu Speedwell ■

407077　Veronica chingii H. L. Li = Veronica ciliata Fisch. ■

407078　Veronica chinoalpina T. Yamaz. ;河北婆婆纳;Hebei Speedwell,Hopeh Speedwell ■

407079　Veronica ciliata Fisch. ;长果婆婆纳（八夏噶,青海婆婆纳,纤毛婆婆纳）;Longfruit Speedwell ■

407080　Veronica ciliata Fisch. subsp. cephaloides（Pennell）D. Y. Hong;拉萨长果婆婆纳(矮小长果婆婆纳）;Lasa Longfruit Speedwell ■

407081　Veronica ciliata Fisch. subsp. zhongdianensis D. Y. Hong;中甸长果婆婆纳;Zhongdian Longfruit Speedwell ■

407082　Veronica cinerea Raf. ;灰色婆婆纳■☆

407083　Veronica comosa A. G. Richt. = Veronica catenata Pennell ■☆

407084　Veronica comosa K. Richt. var. glaberrima（Pennell）B. Boivin = Veronica anagallis-aquatica L. ■

407085　Veronica comosa K. Richt. var. glandulosa（Farw.）B. Boivin = Veronica anagallis-aquatica L. ■

407086　Veronica condensata Baker;紧缩婆婆纳■☆

407087　Veronica conferta Boiss. = Veronica pusilla Hohen. et Boiss. ex Benth.

407088　Veronica connata Raf. subsp. glaberrima Pennell = Veronica anagallis-aquatica L. ■

407089　Veronica connata Raf. var. glaberrima（Pennell）Fassett = Veronica anagallis-aquatica L. ■

407090　Veronica connata Raf. var. typica Pennell = Veronica anagallis-aquatica L. ■

407091　Veronica coreana Nakai = Pseudolysimachion rotundum（Nakai）T. Yamaz. var. coreanum（Nakai）D. Y. Hong ■

407092　Veronica crista-galli Stev. ;鸡冠婆婆纳;Crested Field-speedwell ■☆

407093　Veronica cuneifolia G. Don subsp. atlantica Ball = Veronica rosea Desf. ■☆

407094　Veronica cuneifolia G. Don var. atlantica（Ball）Ball = Veronica rosea Desf. ■☆

407095　Veronica cupressoides Hook. f. = Hebe cupressoides Andersen ●☆

407096　Veronica cymbalaria Bodard;船形婆婆纳;Glandular Speedwell, Pale Speedwell ■☆

407097　Veronica cymbalaria Bodard subsp. panorminata（Guss.）Nyman = Veronica cymbalaria Bodard ■☆

407098　Veronica cymbalaria Bodard var. panorminata（Guss.）Murb. = Veronica cymbalaria Bodard ■☆

407099　Veronica daghestanica Trautv. ;达赫斯坦婆婆纳■☆

407100　Veronica dahurica Steven = Pseudolysimachion dahuricum（Steven）Holub ■

407101　Veronica daisenensis Makino = Pseudolysimachion schmidtianum（Regel）T. Yamaz. subsp. senanense（Maxim.）T. Yamaz. f. daisenense（Makino）T. Yamaz. ■☆

407102　Veronica daurica Steven = Pseudolysimachion dahuricum（Steven）Holub ■

407103　Veronica deltigera Wall. ;南亚长柄婆婆纳;Longstalk Speedwell ■

407104　Veronica denkichiana Honda = Pseudolysimachion ovatum（Nakai）T. Yamaz. subsp. maritimum（Nakai）T. Yamaz. ■☆

407105　Veronica denkichiana Honda var. canescens Satake = Pseudolysimachion ovatum（Nakai）T. Yamaz. subsp. maritimum（Nakai）T. Yamaz. f. canescens（Satake）T. Yamaz. ■☆

407106　Veronica densiflora Ledeb. ;密花婆婆纳;Denseflower Speedwell ■

407107　Veronica dentata F. Schmidt;具齿婆婆纳■☆

407108　Veronica denudata Albov;裸露婆婆纳■☆

407109　Veronica didyma Ten. ;婆婆纳（狗卵草,卵子草,脾寒草,桑肾子,石补丁,石补钉,双果草,双肾草,双铜锤,双珠草）;Geminate Speedwell,Speedwell ■

407110　Veronica didyma Ten. = Veronica polita Fr. ■

407111   Veronica didyma Ten. subsp. dilatata（Nakai et Kitag.）D. Y. Hong = Pseudolysimachion lineariifolium（Pall. ex Link）Holub subsp. dilatatum（Nakai et Kitag.）D. Y. Hong ■

407112   Veronica didyma Ten. var. lilacina T. Yamaz. = Veronica polita Fr. ■

407113   Veronica didyma Ten. var. lilacina T. Yamaz. = Veronica polita Fr. subsp. lilacina（T. Yamaz.）T. Yamaz. ■☆

407114   Veronica dillenii Crantz；迪伦婆婆纳（欧洲蚊母草）；Dillenius' Speedwell ■☆

407115   Veronica diosmifolia Knowles et Westc. = Hebe diosmifolia Andersen ●☆

407116   Veronica elliptica G. Forst. = Hebe elliptica（G. Forst.）Pennell ●☆

407117   Veronica eriogyne H. Winkl.；毛果婆婆纳（青海婆婆纳）；Hairyfruit Speedwell ■

407118   Veronica exaltata Maund = Veronica longifolia L. ■

407119   Veronica exortiva Kitag. = Pseudolysimachion longifolium（L.）Opiz ■

407120   Veronica fargesii Franch.；城口婆婆纳；Farges Speedwell ■

407121   Veronica fedtschenkoi Boriss.；费氏婆婆纳■☆

407122   Veronica ferganica Popov = Veronica rubrifolia Boiss. ■

407123   Veronica filifolia Lipsky；线叶婆婆纳■☆

407124   Veronica filiformis Sm.；线形婆婆纳；Creeping Speedwell，Round-leaved Speedwell，Slender Speedwell，Threadstalk Speedwell ■☆

407125   Veronica filipes P. C. Tsoong；丝梗婆婆纳；Filiformed Pedicel Speedwell，Silkstalk Speedwell ■

407126   Veronica formosana Masam. = Veronicastrum formosanum（Masam.）T. Yamaz. ■

407127   Veronica forrestii Diels；大理婆婆纳；Dali Speedwell ■

407128   Veronica fraterna N. E. Br.；兄弟婆婆纳■☆

407129   Veronica fruticans Jacq.；灌木婆婆纳；Rock Speedwell，Shrubby Speedwell ●☆

407130   Veronica fruticulosa L.；欧洲婆婆纳；Rock Speedwell ■☆

407131   Veronica galactites Hance = Pseudolysimachion linariifolium（Pall. ex Link）Holub subsp. dilatatum（Nakai et Kitag.）D. Y. Hong ■

407132   Veronica galactites Hance = Veronica linariifolia Pall. ex Link ■

407133   Veronica gentianoides Vahl；龙胆状婆婆纳；Gentian Speedwell ■☆

407134   Veronica glaberrima Boiss. et Balansa = Veronica pusilla Hohen. et Boiss. ex Benth. ■

407135   Veronica glabrifolia Boriss.；光叶婆婆纳■☆

407136   Veronica glabrifolia Kitag. = Pseudolysimachion kiusianum（Furumi）T. Yamaz. ■

407137   Veronica glabrifolia Kitag. = Veronica kiusiana Furumi ■

407138   Veronica glandifera Pennell = Veronica anagallis-aquatica L. ■

407139   Veronica glandulosa Hochst. ex Benth.；具腺婆婆纳■☆

407140   Veronica glandulosa Hochst. ex Benth. subsp. mannii（Hook. f.）Elenevsky = Veronica mannii Hook. f. ■☆

407141   Veronica gorbunovii Gontsch.；高尔布婆婆纳■☆

407142   Veronica grandiflora Gaertn.；大花婆婆纳■☆

407143   Veronica grandis Fisch. ex Spreng. = Pseudolysimachion dahuricum（Steven）Holub ■

407144   Veronica hederifolia L.；常春藤叶婆婆纳；Biddy's Eyes，Bird's Eye，Botherum，Corn Speedwell，Dotherum，Ivy Chickweed，Ivyleaf Speedwell，Ivy-leaved Speedwell，Lesser Henbit，Morgeline，Mother of Wheat，Mother-of-wheat，Winterweed ■☆

407145   Veronica hederifolia L. subsp. maura Murb.；模糊婆婆纳■☆

407146   Veronica hederifolia L. subsp. sibthorpioides（Debeaux et al.）Walters = Veronica sibthorpiodes Debeaux et Degen et Hervier ■

407147   Veronica hederifolia L. var. brevipes（Pomel）Maire = Veronica hederifolia L. ■☆

407148   Veronica hederifolia L. var. eriocalyx Batt. = Veronica hederifolia L. ■☆

407149   Veronica hederifolia L. var. sibthorpioides（Debeaux et al.）Maire = Veronica sibthorpiodes Debeaux et Degen et Hervier ■

407150   Veronica henryi T. Yamaz.；华中婆婆纳；Henry Speedwell ■

407151   Veronica himalensis D. Don；喜马拉雅婆婆纳（大花婆婆纳）；Himalayan Speedwell，Himalayas Speedwell ■

407152   Veronica himalensis D. Don subsp. yunnanensis（P. C. Tsoong）D. Y. Hong；多腺大花婆婆纳；Yunnan Himalayan Speedwell ■

407153   Veronica himalensis D. Don var. yunnanensis P. C. Tsoong = Veronica himalensis D. Don subsp. yunnanensis（P. C. Tsoong）D. Y. Hong ■

407154   Veronica hjuleri Paulsen = Veronica beccabunga L. subsp. muscosa（Korsh.）Elenevsky ■

407155   Veronica hololeuca Juz.；全白婆婆纳■☆

407156   Veronica humifusa Dicks. = Veronica serpyllifolia L. subsp. humifusa（Dicks.）Syme ex Sowerby ■☆

407157   Veronica humifusa Dicks. = Veronica serpyllifolia L. ■

407158   Veronica imagalis-aquatica？；水生婆婆纳■☆

407159   Veronica imeretica Kem.-Nath.；伊梅里特婆婆纳■☆

407160   Veronica incana L. = Pseudolysimachion incanum（L.）Holub ■

407161   Veronica intercedens Bornm.；中间婆婆纳■☆

407162   Veronica japonensis Makino；日本山地婆婆纳■☆

407163   Veronica japonensis Makino var. occidentalis Murata ex T. Yamaz. = Veronica muratae T. Yamaz. ■☆

407164   Veronica javanica Blume；爪哇婆婆纳（多枝婆婆纳，细号疳壳草，小败火草，爪哇水苦荬）；Java Speedwell ■

407165   Veronica jeholensis Nakai = Pseudolysimachion linariifolium（Pall. ex Link）Holub subsp. dilatatum（Nakai et Kitag.）D. Y. Hong ■

407166   Veronica karatavica Pavlov ex Nevski = Veronica argute-serrata Regel et Schmalh. ■

407167   Veronica kemulariae Kuth.；凯姆婆婆纳■☆

407168   Veronica keniensis R. E. Fr. = Veronica glandulosa Hochst. ex Benth. ■☆

407169   Veronica khoroasanica Czern.；霍罗森婆婆纳■☆

407170   Veronica kitamurae（Ohwi）Nemoto = Veronicastrum kitamurae（Ohwi）T. Yamaz. ■

407171   Veronica kiusiana Furumi = Pseudolysimachion kiusianum（Furumi）T. Yamaz. ■

407172   Veronica kiusiana Furumi = Pseudolysimachion ovatum（Nakai）T. Yamaz. subsp. kiusianum（Furumi）T. Yamaz. ■☆

407173   Veronica kiusiana Furumi subsp. maritima（Nakai）T. Yamaz. = Pseudolysimachion ovatum（Nakai）T. Yamaz. subsp. maritimum（Nakai）T. Yamaz. ■☆

407174   Veronica kiusiana Furumi subsp. miyabei（Nakai et Honda）T. Yamaz. = Pseudolysimachion ovatum（Nakai）T. Yamaz. subsp. miyabei（Nakai et Honda）T. Yamaz. ■☆

407175   Veronica kiusiana Furumi subsp. miyabei（Nakai et Honda）T. Yamaz. var. japonica（Miq.）T. Yamaz. = Pseudolysimachion ovatum（Nakai）T. Yamaz. subsp. miyabei（Nakai et Honda）T. Yamaz. var. japonicum（Miq.）T. Yamaz. ■☆

407176   Veronica kiusiana Furumi subsp. miyabei（Nakai et Honda）T. Yamaz. var. japonica（Miq.）T. Yamaz. f. yamamotoi T. Yamaz. = Pseudolysimachion ovatum（Nakai）T. Yamaz. subsp. miyabei

（Nakai et Honda）T. Yamaz. var. japonicum（Miq.）T. Yamaz. ■☆

407177 Veronica kiusiana Furumi var. canescens Satake = Pseudolysimachion ovatum（Nakai）T. Yamaz. subsp. maritimum（Nakai）T. Yamaz. f. canescens（Satake）T. Yamaz. ■☆

407178 Veronica kiusiana Furumi var. glabrifolia（Kitag.）Kitag. = Pseudolysimachion kiusianum（Furumi）T. Yamaz. ■

407179 Veronica kiusiana Furumi var. kitadakemontana T. Yamaz. = Pseudolysimachion ovatum（Nakai）T. Yamaz. subsp. kiusianum（Furumi）T. Yamaz. var. kitadakemontanum（T. Yamaz.）T. Yamaz. ■☆

407180 Veronica kiusiana Furumi var. maritima（Nakai）T. Yamaz. = Pseudolysimachion ovatum（Nakai）T. Yamaz. subsp. maritimum（Nakai）T. Yamaz. ■☆

407181 Veronica kiusiana Furumi var. miyabei（Nakai et Honda）T. Yamaz. = Pseudolysimachion ovatum（Nakai）T. Yamaz. subsp. miyabei（Nakai et Honda）T. Yamaz. ■☆

407182 Veronica kiusiana Furumi var. villosa（Furumi）T. Yamaz. = Pseudolysimachion ovatum（Nakai）T. Yamaz. subsp. miyabei（Nakai et Honda）T. Yamaz. var. villosum（Furumi）T. Yamaz. ■☆

407183 Veronica kojimae Ohwi = Veronica morrisonicola Hayata ■

407184 Veronica komarovii Monjuschko；科马罗夫婆婆纳■☆

407185 Veronica komarovii Monjuschko = Pseudolysimachion rotundum（Nakai）T. Yamaz. var. subintegrum（Nakai）D. Y. Hong ■

407186 Veronica komarovii Monjuschko var. petiolata Nakai = Pseudolysimachion rotundum（Nakai）Holub var. petiolatum（Nakai）T. Yamaz. ■☆

407187 Veronica komarovii Monjuschko var. petiolata Nakai f. albiflora（H. Hara）H. Hara = Saussurea pulchella（Fisch. ex Hornem.）Fisch. f. albiflora（Kitam.）Kitam. ■☆

407188 Veronica kopetdaghensis B. Fedtsch.；科佩特婆婆纳■☆

407189 Veronica krylovii Schischk. = Veronica austriaca L. subsp. teucrium（L.）D. A. Webb ■

407190 Veronica krylovii Schischk. = Veronica teucrium L. subsp. altaica Watzl ■

407191 Veronica krylovii Schischk. = Veronica teucrium L. ■

407192 Veronica kurdica Benth.；库尔得婆婆纳■☆

407193 Veronica lackschewitzii J. Keller = Veronica anagallis-aquatica L. ■

407194 Veronica laeta Kar. et Kir. = Pseudolysimachion pinnatum（L.）Holub ■

407195 Veronica lanosa Royle ex Benth.；长梗婆婆纳■

407196 Veronica lanuginosa Benth. ex Hook. f.；绵毛婆婆纳；Downy Speedwell ■

407197 Veronica lasiocarpa Pennell = Veronica alpina L. subsp. pumila（All.）Dostal ■

407198 Veronica latifolia L.；宽叶婆婆纳；Hungarian Speedwell, Saw-leaved Speedwell ■☆

407199 Veronica latifolia L. = Veronica austriaca L. subsp. teucrium（L.）D. A. Webb ■

407200 Veronica laxa Benth.；疏花婆婆纳；Laxflower Speedwell ■

407201 Veronica laxa Benth. = Veronica melissifolia Poir. ■

407202 Veronica laxissima D. Y. Hong；极疏花婆婆纳■

407203 Veronica linariifolia Pall. ex Link = Pseudolysimachion lineariifolium（Pall. ex Link）Holub ■

407204 Veronica linariifolia Pall. ex Link f. dilatata（Nakai et Kitag.）T. Yamaz. = Pseudolysimachion lineariifolium（Pall. ex Link）Holub subsp. dilatatum（Nakai et Kitag.）D. Y. Hong ■

407205 Veronica linariifolia Pall. ex Link subsp. dilatata（Nakai et Kitag.）D. Y. Hong = Pseudolysimachion lineariifolium（Pall. ex Link）Holub subsp. dilatatum（Nakai et Kitag.）D. Y. Hong ■

407206 Veronica linariifolia Pall. ex Link var. dilatata（Nakai et Kitag.）Nakai et Kitag. = Pseudolysimachion lineariifolium（Pall. ex Link）Holub subsp. dilatatum（Nakai et Kitag.）D. Y. Hong ■

407207 Veronica linariifolia Pall. ex Link var. dilatata（Nakai et Kitag.）D. Y. Hong = Pseudolysimachion lineariifolium（Pall. ex Link）Holub subsp. dilatatum（Nakai et Kitag.）D. Y. Hong ■

407208 Veronica linariifolia Pall. ex Link var. dilatata Nakai et Kitag. = Pseudolysimachion linariifolium（Pall. ex Link）Holub subsp. dilatatum（Nakai et Kitag.）D. Y. Hong ■

407209 Veronica linariifolia Pall. ex Link var. dilatata Nakai et Kitag. = Veronica linariifolia Pall. ex Link subsp. dilatata（Nakai et Kitag.）D. Y. Hong ■

407210 Veronica linariifolia Pall. ex Link var. jeholensis（Nakai）Kitag. = Pseudolysimachion linariifolium（Pall. ex Link）Holub subsp. dilatatum（Nakai et Kitag.）D. Y. Hong ■

407211 Veronica linnaeoides R. E. Fr. = Veronica glandulosa Hochst. ex Benth. ■☆

407212 Veronica longifolia L. = Pseudolysimachion longifolium（L.）Opiz ■

407213 Veronica longifolia L. var. exortiva（Kitag.）Kitag. = Pseudolysimachion longifolium（L.）Opiz ■

407214 Veronica longifolia L. var. grandis Regel = Pseudolysimachion dahuricum（Steven）Holub ■

407215 Veronica longipetiolata D. Y. Hong；长柄婆婆纳；Longpetiolate Speedwell, Longpetiole Speedwell ■

407216 Veronica lutkeana Rupr.；里德婆婆纳■☆

407217 Veronica lycica E. Lehm. = Cochlidiosperma lycica（E. Lehm.）D. Y. Hong et S. Nilsson ■☆

407218 Veronica lysimachioides Boiss.；珍珠菜婆婆纳■☆

407219 Veronica macrocarpa Vahl = Hebe macrocarpa Cockayne et Allan ●☆

407220 Veronica macrostemonoides Zakirov；普通婆婆纳■☆

407221 Veronica mannii Hook. f.；曼氏婆婆纳■☆

407222 Veronica maritima L. = Pseudolysimachion longifolium（L.）Opiz ■

407223 Veronica martini H. Lév. = Veronicastrum caulopterum（Hance）T. Yamaz. ■

407224 Veronica maxima Mill.；高大婆婆纳■☆

407225 Veronica maximowicziana Vorosch. = Veronica peregrina L. f. xalapensis（Humb.）Kitag. ■

407226 Veronica melissifolia Poir. = Veronica laxa Benth. ■

407227 Veronica michauxii Lam.；米氏婆婆纳■☆

407228 Veronica microcarpa Boiss.；小果婆婆纳■☆

407229 Veronica minima（K. Koch）K. Koch；弱小婆婆纳■☆

407230 Veronica minima K. Koch = Veronica minima（K. Koch）K. Koch ■☆

407231 Veronica minuta C. A. Mey.；微小婆婆纳■☆

407232 Veronica miqueliana Nakai；米克尔婆婆纳■☆

407233 Veronica miqueliana Nakai f. leucantha Nakai；白花婆婆纳■☆

407234 Veronica miqueliana Nakai f. takedana（Makino）T. Yamaz.；武田婆婆纳■☆

407235 Veronica miqueliana Nakai var. takedana（Makino）Nemoto = Veronica miqueliana Nakai f. takedana（Makino）T. Yamaz. ■☆

407236 Veronica miqueliana Nakai var. takedoi Makino = Veronica taiwanica T. Yamaz. ■

407237 Veronica miyabei Nakai et Honda = Pseudolysimachion ovatum

（Nakai）T. Yamaz. subsp. miyabei（Nakai et Honda）T. Yamaz. ■☆

407238　Veronica montana L.；山地婆婆纳；Mountain Speedwell，Wood Speedwell ■☆

407239　Veronica monticola Trautv.；山生婆婆纳 ■☆

407240　Veronica montioides Boiss.；拟山生婆婆纳 ■☆

407241　Veronica morrisonicola Hayata；匍茎婆婆纳（玉山水苦荬）；Creeping Speedwell ■

407242　Veronica morrisonicola Hayata f. kojimae（Ohwi）T. Yamaz. = Veronica morrisonicola Hayata ■

407243　Veronica morrisonicola Hayata var. kojimae Ohwi = Veronica morrisonicola Hayata ■

407244　Veronica morrisonicola Hayata var. tsugitakaensis（Masam.）H. L. Li = Veronica morrisonicola Hayata ■

407245　Veronica multifida L.；多裂婆婆纳 ■☆

407246　Veronica muratae T. Yamaz.；日本西部婆婆纳 ■☆

407247　Veronica murorum Maxim. = Veronica javanica Blume ■

407248　Veronica myrsinoides Oliv. = Veronica glandulosa Hochst. ex Benth. ■☆

407249　Veronica nana Pennell = Veronica ciliata Fisch. subsp. cephaloides（Pennell）D. Y. Hong ■

407250　Veronica nevskii Boriss.；聂氏婆婆纳 ■☆

407251　Veronica nigricans K. Koch；黑婆婆纳 ■☆

407252　Veronica nipponica Makino ex Furumi；日本婆婆纳；Japanese Speedwell ■☆

407253　Veronica nipponica Makino ex Furumi var. sinanoalpina H. Hara；锡南婆婆纳 ■☆

407254　Veronica noveboracensis ?；新博尔卡婆婆纳；Ironweed，New York Ironweed，Tall ■☆

407255　Veronica nudicaulis Kar. et Kir. = Veronica pusilla Hohen. et Boiss. ex Benth. ■

407256　Veronica officinalis L.；药用婆婆纳（药用水蔓菁）；Biddy's Eyes，Bird's Eye，Bird's-eye，Cancerwort，Common Gypsyweed，Common Gypsy-weed，Common Speedwell，Drug Speedwell，European Tea，Fluellen，Gipsywort，Ground-hale，Ground-heele，Gypsyweed，Heath Speedwell，Male Speedwell，Medicinal Speedwell，Paul's Betony，Speedwell，Wood Pennyroyal ■☆

407257　Veronica officinalis L. var. tournefortii ?；图内福尔婆婆纳；Common Gypsyweed ■☆

407258　Veronica olgensis Kom.；奥尔加婆婆纳 ■☆

407259　Veronica oligosperma Hayata；少籽婆婆纳（贫子水苦荬）；Fewseed Speedwell ■

407260　Veronica oltensis Woronow；奥尔特婆婆纳 ■☆

407261　Veronica onoei Franch. et Sav.；小野婆婆纳 ■☆

407262　Veronica opaca Fr.；暗色婆婆纳 ■☆

407263　Veronica orchidea Crantz；兰状婆婆纳 ■☆

407264　Veronica orientalis Mill.；东方婆婆纳 ■☆

407265　Veronica ornata Monjuschko；华美婆婆纳 ■☆

407266　Veronica ovata Nakai = Pseudolysimachion ovatum（Nakai）T. Yamaz. ■☆

407267　Veronica oxycarpa Boiss.；尖果水苦荬；Acutefruit Speedwell ■

407268　Veronica paniculata L. = Pseudolysimachion spurium（L.）Rauschert ■

407269　Veronica paniculata L. = Veronica rotunda Nakai var. subintegra（Nakai）T. Yamaz. ■

407270　Veronica panormitana Tineo ex Guss. = Cochlidiosperma panormitana（Tineo ex Guss.）D. Y. Hong et S. Nilsson ■☆

407271　Veronica pectinata L.；篦齿婆婆纳；Speedwell ■☆

407272　Veronica pectinata L. 'Rosea'；玫瑰篦齿婆婆纳 ■☆

407273　Veronica peduncularis M. Bieb.；梗花婆婆纳 ■☆

407274　Veronica pedunculata Labill.；梗序婆婆纳；Speedwell-creeping，Veronica ■☆

407275　Veronica peregrina L.；蚊母草（八卦仙桃，本地老鸦草，波丝婆婆纳，波丝水苦荬，灯笼草，夺命丹，活血丹，活血接骨丹，接骨草，接骨丹，接骨仙桃，接骨仙桃草，毛虫婆婆纳，蟠桃草，蟠仙桃，肾子草，水菨衣，水莴苣，水仙桃草，无水自动草，仙桃草，小红头，小头红，鸭儿草，蚊公草，英桃草，止血草）；American Speedwell，Necklace Weed，Neckweed，Neck-weed，Purslane Speedwell ■

407276　Veronica peregrina L. f. xalapensis（Humb.）Kitag. = Veronica peregrina L. var. xalapensis（Kunth）Pennell ■

407277　Veronica peregrina L. subsp. xalapensis（Kunth）Pennell = Veronica peregrina L. var. xalapensis（Kunth）Pennell ■

407278　Veronica peregrina L. subsp. xalapensis（Kunth）Pennell = Veronica peregrina L. ■

407279　Veronica peregrina L. var. pubescens Honda = Veronica peregrina L. ■

407280　Veronica peregrina L. var. pubescens Honda = Veronica peregrina L. f. xalapensis（Humb.）Kitag. ■

407281　Veronica peregrina L. var. typica Pennell = Veronica peregrina L. ■

407282　Veronica peregrina L. var. xalapensis（Humb.）Pennell = Veronica peregrina L. var. xalapensis（Kunth）Pennell ■

407283　Veronica peregrina L. var. xalapensis（Kunth）Pennell；毛虫婆婆纳；Hairy Purslane Speedwell，Purslane Speedwell ■

407284　Veronica peregrina L. var. xalapensis（Kunth）Pennell = Veronica peregrina L. ■

407285　Veronica perfoliata R. Br. = Parahebe perfoliata（R. Br.）B. G. Briggs et Ehrend. ●☆

407286　Veronica perpusilla Boiss. ex Benth. = Veronica pusilla Hohen. et Boiss. ex Benth. ■

407287　Veronica persica Poir.；阿拉伯婆婆纳（波斯婆婆纳，波斯水苦荬，灯笼草，灯笼婆婆纳，曲果草，肾子草，台北水苦荬）；Arab Speedwell，Birdeye Speedwell，Bird's Eye，Bird's-eye，Bird's-eye Speedwell，Blue Eyes，Bullock's Eyes，Buxbaum's Speedwell，Cat's Eyes，Common Field Speedwell，Common Field-speedwell，Field Speedwell，Iran Speedwell，Large Field Speedwell，Persian Speedwell，Pickpocket，Tournefort Speedwell，Tournefort's Speedwell ■

407288　Veronica petitiana A. Rich. = Veronica abyssinica Fresen. ■☆

407289　Veronica petraea（M. Bieb.）Stev.；岩生婆婆纳 ■☆

407290　Veronica pinnata L. = Pseudolysimachion pinnatum（L.）Holub ■

407291　Veronica piroliformis Franch.；鹿蹄草婆婆纳；Pyrolaformed Speedwell ■

407292　Veronica polita Fr.；蓝婆婆纳；Grey Field Speedwell，Grey Field-speedwell，Island Speedwell，Wayside Speedwell ■

407293　Veronica polita Fr. = Veronica didyma Ten. ■

407294　Veronica polita Fr. subsp. lilacina（T. Yamaz.）T. Yamaz.；紫丁香色婆婆纳 ■☆

407295　Veronica porphyriana Pavlov；波尔婆婆纳 ■☆

407296　Veronica porphyriana Pavlov = Pseudolysimachion spicatum（L.）Opiz ■

407297　Veronica praecox All.；早熟婆婆纳；Breck Speedwell，Breckland Speedwell ■☆

407298　Veronica propinqua Boriss.；邻近婆婆纳 ■☆

407299　Veronica prostrata L.；平卧婆婆纳；Harebell Hungarian Speedwell，Harewell Speedwell，Prostrate Speedwell ■

407300　Veronica prostrata L. 'Kapitan'；卡皮坦平卧婆婆纳 ■☆

407301 Veronica prostrata L. 'Spode Blue';黛蓝平卧婆婆纳■☆

407302 Veronica prostrata L. 'Trehane';特里汉平卧婆婆纳■☆

407303 Veronica pseudolongifolia Printz = Pseudolysoimachion longifolium (L.) Opiz ■☆

407304 Veronica pumila All. = Veronica alpina L. subsp. pumila (All.) Dostal ■

407305 Veronica pusilla Hohen. et Boiss. = Veronica pusilla Hohen. et Boiss. ex Benth. ■

407306 Veronica pusilla Hohen. et Boiss. ex Benth.;侏倭婆婆纳;Dwarf Speedwell ■

407307 Veronica qingheensis Y. Z. Zhao;清河婆婆纳■

407308 Veronica ramosissima Boriss.;密枝婆婆纳■☆

407309 Veronica repens DC.;匍匐婆婆纳;Corsican Speedwell,Creeping Speedwell,Speedwell ■☆

407310 Veronica repens DC. var. cyanea Litard. et Maire = Veronica serpyllifolia L. ■

407311 Veronica riae H. Winkl.;膜叶婆婆纳;Membraneousleaf Speedwell ■

407312 Veronica riae H. Winkl. = Veronica henryi T. Yamaz. ■

407313 Veronica rockii H. L. Li;光果婆婆纳(两股钗);Rock Speedwell ■

407314 Veronica rockii H. L. Li subsp. stenocarpa (H. L. Li) D. Y. Hong;尖果婆婆纳(尖果光果婆婆纳)■

407315 Veronica rosea Desf.;粉红婆婆纳■☆

407316 Veronica rosea Desf. subsp. atlantica (Ball) I. Soriano = Veronica rosea Desf. ■☆

407317 Veronica rosea Desf. subsp. virgata (Emb. et Maire) Dobignard et D. Jord.;枝条粉红婆婆纳■☆

407318 Veronica rosea Desf. var. atlantica (Ball) Murb. = Veronica rosea Desf. ■☆

407319 Veronica rosea Desf. var. glabrescens Emb. et Maire = Veronica rosea Desf. ■☆

407320 Veronica rosea Desf. var. lacera Alleiz. = Veronica rosea Desf. ■☆

407321 Veronica rosea Desf. var. macrantha Pau ex I. Soriano = Veronica rosea Desf. ■☆

407322 Veronica rosea Desf. var. maroccana (Pau et Font Quer) Maire = Veronica rosea Desf. ■☆

407323 Veronica rosea Desf. var. pallida Maire = Veronica rosea Desf. ■☆

407324 Veronica rosea Desf. var. virgata Emb. et Maire = Veronica rosea Desf. subsp. virgata (Emb. et Maire) Dobignard et D. Jord. ■☆

407325 Veronica rotunda Nakai = Pseudolysimachion rotundum (Nakai) T. Yamaz. ■

407326 Veronica rotunda Nakai var. coreana (Nakai) T. Yamaz. = Pseudolysimachion rotundum (Nakai) T. Yamaz. var. coreanum (Nakai) D. Y. Hong ■

407327 Veronica rotunda Nakai var. petiolata (Nakai) Nakai ex Ohwi = Pseudolysimachion rotundum (Nakai) Holub var. petiolatum (Nakai) T. Yamaz. ■☆

407328 Veronica rotunda Nakai var. subintegra (Nakai) T. Yamaz. = Pseudolysimachion rotundum (Nakai) T. Yamaz. var. subintegrum (Nakai) D. Y. Hong ■

407329 Veronica rubrifolia Boiss.;红叶婆婆纳;Redleaf Speedwell ■

407330 Veronica rupestris Aiton et Hemsl. = Veronica deltigera Wall. ■

407331 Veronica rupestris Aiton et Hemsl. = Veronica prostrata L. ■

407332 Veronica ruprechtii Lipsky;卢普婆婆纳■☆

407333 Veronica sachalinensis Boriss.;库页婆婆纳■☆

407334 Veronica sajanensis Printz;萨因婆婆纳■☆

407335 Veronica salicifolia G. Forst.;柳叶婆婆纳■☆

407336 Veronica salina Schur. = Veronica undulata Wall. ex Jack ■

407337 Veronica schmidtiana Regel;施密特婆婆纳■☆

407338 Veronica schmidtiana Regel subsp. bandaiana (Makino) Kitam. et Murata = Astilbe odontophylla Miq. var. bandaica (Honda) H. Hara ■☆

407339 Veronica schmidtiana Regel subsp. daisenensis (Makino) Kitam. et Murata = Pseudolysimachion schmidtianum (Regel) T. Yamaz. subsp. senanense (Maxim.) T. Yamaz. f. daisenense (Makino) T. Yamaz. ■☆

407340 Veronica schmidtiana Regel subsp. senanensis (Maxim.) Kitam. et Murata = Pseudolysimachion schmidtianum (Regel) T. Yamaz. subsp. senanense (Maxim.) T. Yamaz. ■☆

407341 Veronica schmidtiana Regel subsp. yesoalpina (Koidz. ex H. Hara) Kitam. et Murata f. exigua (Takeda) Kitam. et Murata = Pseudolysimachion schmidtianum (Regel) T. Yamaz. var. yezoalpinum (Koidz. ex H. Hara) T. Yamaz. f. exiguum (Takeda) T. Yamaz. ■☆

407342 Veronica schmidtiana Regel subsp. yesoalpina (Koidz. ex H. Hara) Kitam. et Murata = Pseudolysimachion schmidtianum (Regel) T. Yamaz. var. yezoalpinum (Koidz. ex H. Hara) T. Yamaz. ■☆

407343 Veronica schmidtiana Regel var. bandaiana Makino = Astilbe odontophylla Miq. var. bandaica (Honda) H. Hara ■☆

407344 Veronica schmidtiana Regel var. bandaiana Makino f. daisenensis (Makino) T. Yamaz. = Pseudolysimachion schmidtianum (Regel) T. Yamaz. subsp. senanense (Maxim.) T. Yamaz. f. daisenense (Makino) T. Yamaz. ■☆

407345 Veronica schmidtiana Regel var. bandaiana Makino f. senanensis (Maxim.) T. Yamaz. = Pseudolysimachion schmidtianum (Regel) T. Yamaz. subsp. senanense (Maxim.) T. Yamaz. ■☆

407346 Veronica schmidtiana Regel var. bandaiana Makino f. tomentosa T. Yamaz. = Pseudolysimachion schmidtianum (Regel) T. Yamaz. subsp. senanense (Maxim.) T. Yamaz. f. tomentosum (T. Yamaz.) T. Yamaz. ■☆

407347 Veronica schmidtiana Regel var. senanensis (Maxim.) Ohwi = Pseudolysimachion schmidtianum (Regel) T. Yamaz. subsp. senanense (Maxim.) T. Yamaz. ■☆

407348 Veronica schmidtiana Regel var. yesoalpina (Koidz. ex H. Hara) T. Yamaz. = Pseudolysimachion schmidtianum (Regel) T. Yamaz. var. yezoalpinum (Koidz. ex H. Hara) T. Yamaz. ■☆

407349 Veronica scutellata L.;矩盾状婆婆纳;Bog Speedwell, Marsh Speedwell,Narrow-leaved Speedwell,Skullcap Speedwell ■☆

407350 Veronica scutellata L. f. villosa (Schumach.) Pennell = Veronica scutellata L. ■☆

407351 Veronica scutellata L. var. villosa Schumach. = Veronica scutellata L. ■☆

407352 Veronica semiamplexicaulis D. Y. Hong;半抱茎婆婆纳(西藏婆婆纳);Semiamplectant Speedwell,Semiamplexicaul Speedwell ■

407353 Veronica semiamplexicaulis D. Y. Hong = Veronica deltigera Wall. ■

407354 Veronica senanensis Maxim. = Pseudolysimachion schmidtianum (Regel) T. Yamaz. subsp. senanense (Maxim.) T. Yamaz. ■☆

407355 Veronica senanensis Maxim. var. daisenensis (Makino) H. Hara = Pseudolysimachion schmidtianum (Regel) T. Yamaz. subsp. senanense (Maxim.) T. Yamaz. f. daisenense (Makino) T. Yamaz. ■☆

407356 Veronica senanensis Maxim. var. yesoalpina Koidz. ex H. Hara = Pseudolysimachion schmidtianum (Regel) T. Yamaz. var. yezoalpinum (Koidz. ex H. Hara) T. Yamaz. ■☆

407357 Veronica septentrionalis Boriss.;北方婆婆纳■☆

407358 Veronica serpyllifolia L.;小婆婆纳(白里香,百里香叶婆婆

纳,地涩涩,荞皮草,仙桃草,圆叶婆婆纳);Little Smooth Speedwell, Mountain Speedwell, Paul's Betony, Thymeleaf Speedwell, Thyme-leaved Speedwell ■

407359 Veronica serpyllifolia L. subsp. humifusa (Dicks.) Pennell = Veronica serpyllifolia L. ■

407360 Veronica serpyllifolia L. subsp. humifusa (Dicks.) Syme = Veronica serpyllifolia L. subsp. humifusa (Dicks.) Syme ex Sowerby ■☆

407361 Veronica serpyllifolia L. subsp. humifusa (Dicks.) Syme = Veronica serpyllifolia L. ■

407362 Veronica serpyllifolia L. subsp. humifusa (Dicks.) Syme ex Sowerby;铺散小婆婆纳;Bright-blue Speedwell, Humifuse Speedwell, Thyme-leaved Speedwell ■☆

407363 Veronica serpyllifolia L. subsp. humifusa (Dicks.) Syme ex Sowerby = Veronica serpyllifolia L. ■

407364 Veronica serpyllifolia L. var. borealis Laest. = Veronica serpyllifolia L. subsp. humifusa (Dicks.) Syme ex Sowerby ■☆

407365 Veronica serpyllifolia L. var. humifusa (Dicks.) Sm. = Veronica serpyllifolia L. ■

407366 Veronica serpyllifolia L. var. humifusa (Dicks.) Vahl = Veronica serpyllifolia L. subsp. humifusa (Dicks.) Syme ex Sowerby ■☆

407367 Veronica serpyllifolia L. var. humifusa (Dicks.) Vahl = Veronica serpyllifolia L. ■

407368 Veronica serpyllifolia L. var. nummularioides Lecoq et Lamotte = Veronica serpyllifolia L. ■

407369 Veronica serpylloides Regel;百里香婆婆纳■☆

407370 Veronica sherwoodii M. Peck = Veronica peregrina L. var. xalapensis (Kunth) Pennell ■

407371 Veronica sibirica L. ;西伯利亚婆婆纳■

407372 Veronica sibirica L. = Veronica virginica L. ■

407373 Veronica sibirica L. = Veronicastrum sibiricum (L.) Pennell ■

407374 Veronica sibirica L. var. glabra Nakai = Veronicastrum sibiricum (L.) Pennell ■

407375 Veronica sibthorpioides Debeaux ex Degen et Hervier = Cochlidiosperma sibthorpioides (Debeaux ex Degen et Herv.) D. Y. Hong et S. Nilsson ■☆

407376 Veronica sieboldiana Miq. = Pseudolysimachion sieboldianum (Miq.) Holub ■☆

407377 Veronica simensis Fresen. ;锡米婆婆纳■☆

407378 Veronica speciosa L. = Hebe speciosa Andersen ●☆

407379 Veronica spicata L. = Pseudolysimachion spicatum (L.) Opiz ■

407380 Veronica spicata L. subsp. incana (L.) Walters;灰白水苦荬■☆

407381 Veronica spicata L. subsp. porphyriana (Pavlov) Elenevsky = Pseudolysimachion spicatum (L.) Opiz ■

407382 Veronica spuria L. = Pseudolysimachion lineariifolium (Pall. ex Link) Holub ■

407383 Veronica spuria L. = Pseudolysimachion rotundum (Nakai) T. Yamaz. var. subintegrum (Nakai) D. Y. Hong ■

407384 Veronica spuria L. = Pseudolysimachion spurium (L.) Rauschert ■

407385 Veronica spuria L. var. subintegra Nakai = Pseudolysimachion rotundum (Nakai) T. Yamaz. var. subintegrum (Nakai) D. Y. Hong ■

407386 Veronica stamatiadae M. Fisch. et Greuter = Cochlidiosperma stamatiadae (M. Fisch. et Greuter) D. Y. Hong et S. Nilsson ■☆

407387 Veronica stelleri Pall. ex Link;长白婆婆纳;Longstyle Speedwell ■

407388 Veronica stelleri Pall. ex Link var. longistyla Kitag. ;长柱长白婆婆纳(长白婆婆纳)■☆

407389 Veronica stelleri Pall. subsp. nipponica (Makino ex Furumi)

Jelen. = Veronica nipponica Makino ex Furumi ■☆

407390 Veronica stenocarpa H. L. Li = Veronica rockii H. L. Li subsp. stenocarpa (H. L. Li) D. Y. Hong ■

407391 Veronica stenocarpa H. L. Li = Veronica rockii H. L. Li ■

407392 Veronica stewartii Pennell = Cochlidiosperma stewartii (Pennell) D. Y. Hong et S. Nilsson ■☆

407393 Veronica stylophora Popov;冠柱婆婆纳■☆

407394 Veronica subincanovelutina Koidz. = Pseudolysimachion ovatum (Nakai) T. Yamaz. subsp. miyabei (Nakai et Honda) T. Yamaz. var. villosum (Furumi) T. Yamaz. ■☆

407395 Veronica subincanovelutina Koidz. var. glabrescens (H. Hara) Satake = Pseudolysimachion ovatum (Nakai) T. Yamaz. subsp. miyabei (Nakai et Honda) T. Yamaz. var. japonicum (Miq.) T. Yamaz. ■☆

407396 Veronica sublobata M. Fisch. = Cochlidiosperma sublobata (M. Fisch.) D. Y. Hong et S. Nilsson ■☆

407397 Veronica subsessilis (Miq.) Carrière = Pseudolysimachion subsessile (Miq.) Holub ■☆

407398 Veronica sutchuensis Franch. ;川西婆婆纳;Sichuan Speedwell,W. Sichuan Speedwell ■

407399 Veronica syriaca Roem. et Schult. ;叙利亚婆婆纳■☆

407400 Veronica szechuanica Batalin;四川婆婆纳;Sichuan Speedwell, Szechwan Speedwell ■

407401 Veronica szechuanica Batalin subsp. sikkimensis (Hook. f.) D. Y. Hong;多毛四川婆婆纳(多毛婆婆纳);Manyhair Sichuan Speedwell ■

407402 Veronica taiwanalpine S. S. Ying = Veronica morrisonicola Hayata ■

407403 Veronica taiwanica T. Yamaz. ;台湾婆婆纳(台湾水苦荬); Taiwan Speedwell ■

407404 Veronica taurica Willd. ;克里木婆婆纳■☆

407405 Veronica teberdensis (Kem. -Nath.) Boriss. ;捷别尔达婆婆纳■☆

407406 Veronica telephiifolia Vahl;疣叶婆婆纳■☆

407407 Veronica tenella All. = Veronica serpyllifolia L. subsp. humifusa (Dicks.) Syme ex Sowerby ■☆

407408 Veronica tenella All. = Veronica serpyllifolia L. ■

407409 Veronica tenuissima Boriss. ;丝茎婆婆纳;Thinstem Speedwell ■

407410 Veronica tetraphylla Popov = Veronica tenuissima Boriss. ■

407411 Veronica teucrium L. = Veronica austriaca L. subsp. teucrium (L.) D. A. Webb ■

407412 Veronica teucrium L. prostrata ? = Veronica prostrata L. ■

407413 Veronica teucrium L. subsp. altaica Watzl;卷毛婆婆纳■

407414 Veronica thunbergii A. Gray = Veronica laxa Benth. ■

407415 Veronica tianachanica Lincz. ;天山婆婆纳■☆

407416 Veronica tibetica D. Y. Hong;西藏婆婆纳;Tibet Speedwell, Xizang Speedwell ■

407417 Veronica tournefortii C. C. Gmel. = Veronica persica Poir. ■

407418 Veronica trichadena Jord. et Fourr. = Cochlidiosperma trichadena (Jord. et Fourr.) D. Y. Hong et S. Nilsson ■☆

407419 Veronica triloba Opiz = Cochlidiosperma triloba (Opiz) D. Y. Hong et S. Nilsson ■☆

407420 Veronica tripartita Boriss. ;三深裂婆婆纳■☆

407421 Veronica triphyllos L. ;三叶婆婆纳;Finger Speedwell,Fingered Speedwell,Trifid Speedwell ■☆

407422 Veronica tsinglingensis D. Y. Hong;陕川婆婆纳;Chinling Mountains Speedwell,Fingered Speedwell,Qinling Speedwell ■

407423 Veronica tsugitakaensis Masam. = Veronica morrisonicola Hayata ■

407424 Veronica tubiflora Fisch. et C. A. Mey. ;管花婆婆纳■☆

407425　Veronica tubiflora Fisch. et C. A. Mey. = Veronicastrum tubiflorum（Fisch. et C. A. Mey.）H. Hara ■

407426　Veronica turkmenorum B. Fedtsch.；土库曼婆婆纳■

407427　Veronica umbelliformis Pennell = Veronica szechuanica Batalin subsp. sikkimensis（Hook. f.）D. Y. Hong ■

407428　Veronica uncinata Pennell = Veronica rubrifolia Boiss. ■

407429　Veronica undulata Wall. = Veronica undulata Wall. ex Jack ■

407430　Veronica undulata Wall. ex Jack；水苦荬（半边山，北水苦荬，虫虫草，大仙桃草，二代草，海绿婆婆纳，接骨银，蚧蛙草，芒种草，水波浪，水菠菜，水对叶莲，水接骨丹，水上浮萍，水莴苣，水窝窝，水仙桃草，水泽兰，兔子草，蚊子草，仙人对座草，仙桃草，谢婆菜，鸭儿草）；Undulate Speedwell ■

407431　Veronica vandellioides Maxim.；唐古拉婆婆纳；Tangula Range Speedwell，Tangula Speedwell ■

407432　Veronica verna L.；裂叶婆婆纳（春婆婆纳）；Lobedleaf Speedwell，Spring Speedwell，Vernal Speedwell ■

407433　Veronica violifolia Hochst. ex Benth.；堇叶婆婆纳■☆

407434　Veronica virginica L.；美婆婆纳（北美草本威灵仙，草本威灵仙，能消，威灵仙，西伯利亚婆婆纳）；Black Root，Blackfoot，Bowman Root，Brinton Root，Culver's Root ■

407435　Veronica virginica L. = Veronicastrum virginicum（L.）Farw. ■☆

407436　Veronica virginica L. var. japonica？ = Veronicastrum japonicum（Nakai）T. Yamaz. ■☆

407437　Veronica wogorensis Hochst. ex A. Rich. = Veronica javanica Blume ■

407438　Veronica xalapensis Kunth = Veronica peregrina L. var. xalapensis（Kunth）Pennell ■

407439　Veronica xilinensis Y. Z. Zhao；锡林婆婆纳；Xilinhaote Speedwell ■

407440　Veronica xilinensis Y. Z. Zhao = Pseudolysimachion incanum（L.）Holub ■

407441　Veronica yunnanensis D. Y. Hong；云南婆婆纳；Yunnan Speedwell ■

407442　Veronica yushanchienshanica S. S. Ying = Veronica morrisonicola Hayata ■

407443　Veronicaceae Cassel = Plantaginaceae Juss.（保留科名）■

407444　Veronicaceae Cassel = Scrophulariaceae Juss.（保留科名）●■

407445　Veronicaceae Horan.；婆婆纳科■

407446　Veronicaceae Horan. = Plantaginaceae Juss.（保留科名）■

407447　Veronicaceae Horan. = Scrophulariaceae Juss.（保留科名）●■

407448　Veronicaceae Raf. = Scrophulariaceae Juss.（保留科名）●■

407449　Veronicastrum Heist. ex Fabr.（1759）；腹水草属（草本威灵仙属，四方麻属）；Ascitesgrass，Culver's Root，Culver's-physic，Veronicastrum ■

407450　Veronicastrum Heist. ex Fabr. = Veronica L. ■

407451　Veronicastrum Moench = Veronicastrum Heist. ex Fabr. ■

407452　Veronicastrum Opiz = Veronica L. ■

407453　Veronicastrum axillare（Siebold et Zucc.）T. Yamaz.；爬岩红（串鱼草，钓鱼杆，钓鱼竿，多穗草，腹水草，见毒消，毛脉腹水草，小串鱼，小钓鱼竿，新竹腹水草，腋生腹水草，一串鱼）；Axillary Ascitesgrass，Axillary Veronicastrum ■

407454　Veronicastrum axillare（Siebold et Zucc.）T. Yamaz. subsp. venosum（Hemsl.）D. Y. Hong；毛脉腹水草■☆

407455　Veronicastrum axillare（Siebold et Zucc.）Yamaz. var. simadai（Masam.）H. Y. Liu；新竹腹水草■

407456　Veronicastrum blattaria L.；毛瓣腹水草（毛瓣毛蕊花）；Hairypetal Ascitesgrass，Hairypetal Veronicastrum ■☆

407457　Veronicastrum borissovae（De Moor）Soják = Veronicastrum sibiricum（L.）Pennell subsp. yezoense（H. Hara）T. Yamaz. ■☆

407458　Veronicastrum brunonianum（Benth.）D. Y. Hong；美穗草（高山四方麻，黑升麻，咳药，小寒药）；Beautifulspikegrass，Brunon Veronicastrum ■

407459　Veronicastrum brunonianum（Benth.）D. Y. Hong subsp. sutchuenense（Franch.）D. Y. Hong；川鄂美穗草■

407460　Veronicastrum caulopterum（Hance）T. Yamaz.；四方麻（狼尾拉花，青鱼胆，山练草，四方青，四方消，四角草，四棱草，四棱三百棒，鱼珠草）；Wingystem Ascitesgrass，Wingystem Veronicastrum ■

407461　Veronicastrum formosanum（Masam.）T. Yamaz.；台湾腹水草；Taiwan Ascitesgrass，Taiwan Veronicastrum ■

407462　Veronicastrum japonicum（Nakai）T. Yamaz.；日本腹水草■☆

407463　Veronicastrum japonicum（Nakai）T. Yamaz. f. album T. Yamaz.；白花日本腹水草■☆

407464　Veronicastrum japonicum（Nakai）T. Yamaz. f. humile（Nakai）T. Yamaz. = Veronicastrum japonicum（Nakai）T. Yamaz. var. humile（Nakai）T. Yamaz. ■☆

407465　Veronicastrum japonicum（Nakai）T. Yamaz. var. australe（T. Yamaz.）T. Yamaz.；南方日本腹水草■☆

407466　Veronicastrum japonicum（Nakai）T. Yamaz. var. australe（T. Yamaz.）T. Yamaz. f. albiflorum（Akasawa）T. Yamaz.；白花南方日本腹水草■☆

407467　Veronicastrum japonicum（Nakai）T. Yamaz. var. humile（Nakai）T. Yamaz.；小日本腹水草■☆

407468　Veronicastrum kitamurae（Ohwi）T. Yamaz.；直立腹水草（高山腹水草）；Erect Ascitesgrass，Kitamura Veronicastrum ■

407469　Veronicastrum kitamurae（Ohwi）T. Yamaz. = Veronicastrum formosanum（Masam.）T. Yamaz. ■

407470　Veronicastrum latifolium（Hemsl.）T. Yamaz.；宽叶腹水草（钓鱼竿）；Broadleaf Ascitesgrass，Broadleaf Veronicastrum ■

407471　Veronicastrum liukiuense（Ohwi）T. Yamaz.；琉球腹水草■

407472　Veronicastrum longispicatum（Merr.）T. Yamaz.；长穗腹水草（吊杆青筋，钓鱼竿）；Longspike Ascitesgrass，Longspike Veronicastrum ■

407473　Veronicastrum longispicatum（Merr.）T. Yamaz. subsp. nanchuanense T. L. Chin et D. Y. Hong；南川长穗腹水草（南川腹水草）■

407474　Veronicastrum longispicatum（Merr.）T. Yamaz. subsp. plukenetii（Yamaz.）D. Y. Hong = Veronicastrum stenostachyum（Hemsl.）T. Yamaz. subsp. plukenetii（T. Yamaz.）D. Y. Hong ■

407475　Veronicastrum lungtsuanense M. Cheng et Z. J. Feng；龙泉腹水草；Longquan Ascitesgrass，Longquan Veronicastrum ■

407476　Veronicastrum lungtsuanense M. Cheng et Z. J. Feng = Veronicastrum villosulum（Miq.）T. Yamaz. var. parviflorum T. L. Chin et D. Y. Hong ■

407477　Veronicastrum martini H. Lév. = Veronicastrum caulopterum（Hance）T. Yamaz. ■

407478　Veronicastrum plukenetii（T. Yamaz.）T. Yamaz. = Veronicastrum stenostachyum（Hemsl.）T. Yamaz. subsp. plukenetii（T. Yamaz.）D. Y. Hong ■

407479　Veronicastrum rhombifolium（Hand.-Mazz.）P. C. Tsoong；菱叶腹水草；Rhombicleaf Ascitesgrass，Rhomboidleaf Veronicastrum ■

407480　Veronicastrum robustum（Diels）D. Y. Hong；粗壮腹水草；Robust Ascitesgrass，Robust Veronicastrum ■

407481　Veronicastrum robustum（Diels）D. Y. Hong subsp. grandifolium T. L. Chin et D. Y. Hong；大叶腹水草（九拱桥）；Bigleaf Robust Veronicastrum ■

407482　Veronicastrum sachalinense T. Yamaz. = Veronicastrum sibiricum（L.）Pennell subsp. yezoense（H. Hara）T. Yamaz. ■☆

407483　Veronicastrum serpyllifolium（L.）Fourr. = Veronica serpyllifolia

L.■

407484 Veronicastrum serpyllifolium（L.）Fourr. subsp. humifusum（Dicks.）W. A. Weber = Veronica serpyllifolia L. subsp. humifusa（Dicks.）Syme ex Sowerby ■☆

407485 Veronicastrum sibiricum（L.）Pennell；草本威灵仙（草本灵仙,草灵仙,草玉梅,秤杆升麻,九节草,九龙草,九轮草,狼尾巴花,轮叶婆婆纳,山鞭草,威灵仙,斩龙剑）；Herb Weiling Ascitesgrass,Siberian Veronicastrum ■

407486 Veronicastrum sibiricum（L.）Pennell subsp. japonicum（Nakai）T. Yamaz. f. humile（Nakai）T. Yamaz. = Veronicastrum japonicum（Nakai）T. Yamaz. var. humile（Nakai）T. Yamaz. ■☆

407487 Veronicastrum sibiricum（L.）Pennell subsp. japonicum（Nakai）T. Yamaz. = Veronicastrum japonicum（Nakai）T. Yamaz. ■☆

407488 Veronicastrum sibiricum（L.）Pennell subsp. yezoense（H. Hara）T. Yamaz.；北海道草本威灵仙■☆

407489 Veronicastrum sibiricum（L.）Pennell var. australe T. Yamaz. = Veronicastrum japonicum（Nakai）T. Yamaz. var. australe（T. Yamaz.）T. Yamaz. ■☆

407490 Veronicastrum sibiricum（L.）Pennell var. japonicum（Nakai）H. Hara f. album Sugim. ex T. Shimizu = Veronicastrum japonicum（Nakai）T. Yamaz. f. album T. Yamaz. ■☆

407491 Veronicastrum sibiricum（L.）Pennell var. japonicum（Nakai）H. Hara = Veronicastrum japonicum（Nakai）T. Yamaz. ■☆

407492 Veronicastrum sibiricum（L.）Pennell var. japonicum H. Hara；日本威灵仙■

407493 Veronicastrum sibiricum（L.）Pennell var. yezoense H. Hara；北海道威灵仙■☆

407494 Veronicastrum sibiricum（L.）Pennell var. yezoense H. Hara = Veronicastrum sibiricum（L.）Pennell subsp. yezoense（H. Hara）T. Yamaz. ■☆

407495 Veronicastrum sibiricum（L.）Pennell var. yezoense H. Hara f. albiflorum Hideki Takah. = Veronicastrum sibiricum（L.）Pennell var. yezoense H. Hara f. candidum Yonek. ■☆

407496 Veronicastrum sibiricum（L.）Pennell var. yezoense H. Hara f. candidum Yonek. ；白北海道威灵仙■☆

407497 Veronicastrum simadae（Masam.）T. Yamaz. = Veronicastrum axillare（Siebold et Zucc.）T. Yamaz. ■

407498 Veronicastrum stenostachyum（Hemsl.）T. Yamaz.；腹水草（长穗腹水草,钓鱼竿,虎尾悬铃草,柔毛腹水草,细穗腹水草,腋生多穗草,腋生腹水草）；Ascitesgrass,Stenostachyous Veronicastrum ■

407499 Veronicastrum stenostachyum（Hemsl.）T. Yamaz. subsp. nanchuanense T. L. Chin et D. Y. Hong；南川腹水草；Nanchuan Veronicastrum ■

407500 Veronicastrum stenostachyum（Hemsl.）T. Yamaz. subsp. plukenetii（T. Yamaz.）D. Y. Hong；细穗腹水草■

407501 Veronicastrum tagawae（Ohwi）T. Yamaz.；田川腹水草■☆

407502 Veronicastrum tubiflorum（Fisch. et C. A. Mey.）H. Hara；管花腹水草（柳叶婆婆纳）；Tubeflower Ascitesgrass,Tubeflower Veronicastrum ■

407503 Veronicastrum villosulum（Miq.）T. Yamaz.；毛叶腹水草（穿山鞭,翠梅草,倒地龙,吊杆风,吊线风,钓竿藤,钓鱼草,钓鱼竿,疔疮草,多穗草,二头马兰,腹水草,腹水草藤,过山龙,过天桥,虎尾悬铃草,金鸡尾,金桑鸟草,惊天雷,两头绷,两头根,两头爬,两头蛇,两头生根,两头粘,两头镇,毛腹水草,毛叶仙桥,梅叶伸经,爬崖红,秋草,三节两梗,散血丹,双头粘,霜里红,汤生草,天桥草,万里云,仙株草,仙人搭桥,蟹珠草,悬铃草,叶下红,腋生多穗草,腋生腹水草,一条筋）；Villosulous Ascitesgrass,Villosulous Veronicastrum ■

407504 Veronicastrum villosulum（Miq.）T. Yamaz. var. glabrum T. L. Chin et D. Y. Hong；铁钓竿（两头忙）；Glabrous Ascitesgrass, Glabrous Veronicastrum,Ironhook pole ■

407505 Veronicastrum villosulum（Miq.）T. Yamaz. var. hirsutum T. L. Chin et D. Y. Hong；刚毛毛叶腹水草（刚毛腹水草）■

407506 Veronicastrum villosulum（Miq.）T. Yamaz. var. parviflorum T. L. Chin et D. Y. Hong；两头连■

407507 Veronicastrum virginicum（L.）Farw.；弗吉尼亚腹水草（维州腹水草）；Culver's Root,Culver's-physic,Culver's-root ■☆

407508 Veronicastrum virginicum（L.）Farw. f. villosum Pennell = Veronicastrum virginicum（L.）Farw. ■☆

407509 Veronicastrum yamatsutae（T. Yamaz.）T. Yamaz. = Veronicastrum stenostachyum（Hemsl.）T. Yamaz. ■

407510 Veronicastrum yunnanense（W. W. Sm.）T. Yamaz.；云南腹水草（金钩连）；Yunnan Ascitesgrass,Yunnan Veronicastrum ■

407511 Veronicella Fourr. = Veronica L. ■

407512 Verreauxia Benth.（1868）；韦罗草海桐属●■☆

407513 Verreauxia paniculata Benth. ；韦罗草海桐●■☆

407514 Verrucaria Medik. = Tournefortia L. ■

407515 Verrucifera N. E. Br. = Titanopsis Schwantes ■☆

407516 Verrucularia A. Juss.（1840）；疣虎尾属；Wartwort ●☆

407517 Verrucularia A. Juss. = Verrucularina Rauschert ●☆

407518 Verrucularia glaucophylla A. Juss. ；疣虎尾●☆

407519 Verrucularina Rauschert = Verrucularia A. Juss. ●☆

407520 Verschaffeltia H. Wendl.（1865）；根柱凤尾椰属（扶摇棕属,外沙佛棕属,韦嘉夫桐属,竹马椰子属）；Verschaffelt Palm,Verschaffeltia,Waftwort ●☆

407521 Verschaffeltia splendida H. Wendl. ；根柱凤尾椰（外沙佛棕）；Tall Verschaffeltia,Verschaffelt Palm ●☆

407522 Versteegia Valeton（1911）；巴布亚茜草属●☆

407523 Versteegia cauliflora Valeton；茎花巴布亚茜草●☆

407524 Versteegia grandifolia Valeton；大叶巴布亚茜草●☆

407525 Versteegia minor Valeton；小巴布亚茜草●☆

407526 Versteggia Willis = Versteegia Valeton ●☆

407527 Verticillaceae Dulac = Hippuridaceae Vest（保留科名）■

407528 Verticillaria Ruiz et Pav. = Rheedia L. ●☆

407529 Verticordia DC.（1828）（保留属名）；羽花木属；Feather-flower,Morrison ●☆

407530 Verticordia chrysantha Endl. ；黄花羽花木（黄羽花木）●☆

407531 Verticordia grandis J. L. Drumm. ；高大羽花木●☆

407532 Verticordia nitens（Lindl.）Endl. ；莫里森羽花木；Morrison Feather Flower ●☆

407533 Verticordia picta Endl. ；着色羽花木；Painted Feather-flower ●☆

407534 Verticordia plumosa（Desf.）Druce；羽花木●☆

407535 Verulamia DC. ex Poir. = Pavetta L. ●

407536 Verutina Cass. = Centaurea L.（保留属名）●■

407537 Verzinum Raf. = Thermopsis R. Br. ex W. T. Aiton ■

407538 Vesalea M. Martens et Galeotti = Abelia R. Br. ●

407539 Vescisepalum（J. J. Sm.）Garay, Hamer et Siegerist = Bulbophyllum Thouars（保留属名）■

407540 Veselskya Opiz = Pyramidium Boiss. ■☆

407541 Veselskya Opiz（1964）；韦塞尔芥属■☆

407542 Veselskya griffithiana（Boiss.）Opiz；韦塞尔芥■☆

407543 Veseyochloa J. B. Phipps = Tristachya Nees ■☆

407544 Veseyochloa viridiaristata J. B. Phipps = Tristachya viridearistata（J. B. Phipps）Clayton ■☆

407545 Vesicarex Steyerm. = Carex L. ■

407546 Vesicaria Adans. = Alyssoides Mill. ■●☆

407547 Vesicaria Adans. = Lesquerella S. Watson ■☆

407548 Vesicaria Adans. = Vesicaria Tourn. ex Adans. ■☆

407549 Vesicaria Tourn. ex Adans. (1763);膀胱芥属■☆

407550 Vesicaria graeca Reut. ;膀胱芥■☆

407551 Vesicaria leiocarpa (Trautv.) H. Buch;光果膀胱芥■☆

407552 Vesicaria microcarpa Vis. = Zilla spinosa (L.) Prantl ■☆

407553 Vesicarpa Rydb. = Artemisia L. ●■

407554 Vesicarpa Rydb. = Sphaeromeria Nutt. ■☆

407555 Vesicarpa potentilloides (A. Gray) Rydb. = Sphaeromeria potentilloides (A. Gray) A. Heller ■☆

407556 Vesicarpa potentilloides (A. Gray) Rydb. var. nitrophilum (Cronquist) Kartesz = Sphaeromeria potentilloides (A. Gray) A. Heller var. nitrophila (Cronquist) A. H. Holmgren, L. M. Shultz et Lowrey ■☆

407557 Vesicisepalum (J. J. Sm.) Garay, Hamer et Siegerist(1994);泡萼兰属■☆

407558 Vesicisepalum folliculiferum (J. J. Sm.) Garay, Hamer et Siegerist;泡萼兰■☆

407559 Vesiculina Raf. = Utricularia L. ■

407560 Vesiculina gibba (L.) Raf. = Utricularia gibba L. ■

407561 Vesiculina purpurea (Walter) Raf. = Utricularia purpurea Walter ■☆

407562 Vesiculina saccata Raf. = Utricularia purpurea Walter ■☆

407563 Veslingia Fabr. = Aizoon L. ■☆

407564 Veslingia Heist. ex Fabric. = Aizoon L. ■☆

407565 Veslingia Vis. = Guizotia Cass. (保留属名)■●

407566 Veslingia caulifloris Moench = Aizoon canariense L. ■☆

407567 Veslingia scabra Vis. = Guizotia scabra (Vis.) Chiov. ■☆

407568 Vespuccia Parl. = Hydrocleys Rich. ■☆

407569 Vesquella Heim = Stemonoporus Thwaites ●☆

407570 Vesselowskya Pamp. (1905);维赛木属(瓦萨罗斯卡属,瓦萨木属);Southern Marara ●☆

407571 Vesselowskya rubifolia (F. Muell.) Pamp. ;维赛木(瓦萨罗斯卡,瓦萨木);Southern Marara ●☆

407572 Vestia Willd. (1809);维斯特木属(垂管花属)●☆

407573 Vestia foetida (Ruiz et Pav.) Hoffmanns. ;维斯特木(垂管花)●☆

407574 Vestia foetida Hoffmanns. = Vestia foetida (Ruiz et Pav.) Hoffmanns. ●☆

407575 Vestia lycioides Willd. = Vestia foetida (Ruiz et Pav.) Hoffmanns. ●☆

407576 Vestigium Luer = Pleurothallis R. Br. ■☆

407577 Vetiveria Bory = Chrysopogon Trin. (保留属名)■

407578 Vetiveria Bory = Vetiveria Bory ex Lem. ■

407579 Vetiveria Bory ex Lem. (1822);香根草属(培地茅属);Aromaticroot, Oil Grass, Vetiver ■

407580 Vetiveria fulvibarbis (Trin.) Stapf;黄褐毛香根草■☆

407581 Vetiveria muricata (Retz.) Griseb. = Vetiveria zizanioides (L.) Nash ■

407582 Vetiveria nigritana (Benth.) Stapf = Chrysopogon nigritanus (Benth.) Veldkamp ■☆

407583 Vetiveria odoratissima Bory = Vetiveria zizanioides (L.) Nash ■

407584 Vetiveria odoratissima Lem. = Vetiveria zizanioides (L.) Nash ■

407585 Vetiveria zizanioides (L.) Nash;香根草(糙须芒草,培地茅,岩兰,岩兰草);Aromaticroot, Cuscus, Khus Khus, Khus-Khus, Sevendara, Sevendara Grass, Vetiver, Vetivergrass ■

407586 Vetiveria zizanioides (L.) Nash var. nigritana (Benth.) A. Camus = Chrysopogon nigritanus (Benth.) Veldkamp ■☆

407587 Vetiveria zizanioides (L.) Nash. = Chrysopogon zizanioides (L.) Roberty ■

407588 Vetrix Raf. = Salix L. (保留属名)●

407589 Vexatorella Rourke(1984);平叶山龙眼属●☆

407590 Vexatorella alpina (Salisb. ex Knight) Rourke;高山平叶山龙眼●☆

407591 Vexatorella amoena (Rourke) Rourke;非洲平叶山龙眼●☆

407592 Vexatorella latebrosa Rourke;平叶山龙眼●☆

407593 Vexatorella obtusata (Thunb.) Rourke;钝毛平叶山龙眼●☆

407594 Vexatorella obtusata (Thunb.) Rourke subsp. albomontana (Rourke) Rourke;山地白平叶山龙眼●☆

407595 Vexibia Raf. (1825);苦豆子属●■

407596 Vexibia Raf. = Sophora L. ●■

407597 Vexibia alopecuroides (L.) W. A. Weber = Sophora alopecuroides L. ■●

407598 Vexibia alopecuroides (L.) Yakovlev = Sophora alopecuroides L. ■●

407599 Vexibia alppecuroides (L.) W. A. Weber var. tomentosa (Boiss.) Yakovlev = Sophora alopecuroides L. var. tomentosa (Boiss.) Ponert ●

407600 Vexibia nuttalliana (Turner) Yakovlev = Sophora nuttalliana B. L. Turner ●☆

407601 Vexibia pachycarpa (C. A. Mey.) W. A. Weber = Sophora pachycarpa Schrenk ex C. A. Mey. ●

407602 Vexibia pachycarpa (Schrenk ex C. A. Mey.) Yakovlev = Sophora pachycarpa Schrenk ex C. A. Mey. ●

407603 Vexibia pachycarpa (Schrenk) Yakovlev = Sophora pachycarpa Schrenk ex C. A. Mey. ●

407604 Vexibia stenophylla (A. Gray) W. A. Weber = Sophora stenophylla A. Gray ●☆

407605 Vexillabium F. Maek. = Kuhlhasseltia J. J. Sm. ■

407606 Vexillabium fissum F. Maek. ;半裂旗唇兰■☆

407607 Vexillabium humilum (Fukuy.) S. S. Ying = Kuhlhasseltia yakushimensis (Yamam.) Ormerod ■

407608 Vexillabium humilum (Fukuy.) S. S. Ying = Vexillabium yakushimense (Yamam.) F. Maek. ■

407609 Vexillabium inamii Ohwi = Vexillabium fissum F. Maek. ■☆

407610 Vexillabium integrum (Fukuy.) S. S. Ying = Kuhlhasseltia yakushimensis (Yamam.) Ormerod ■

407611 Vexillabium integrum (Fukuy.) S. S. Ying = Vexillabium yakushimense (Yamam.) F. Maek. ■

407612 Vexillabium nakaianum F. Maek. ;中井氏旗唇兰■☆

407613 Vexillabium yakushimense (Gogelein) F. Maek. = Vexillabium yakushimense (Yamam.) F. Maek. ■

407614 Vexillabium yakushimense (Yamam.) F. Maek. = Kuhlhasseltia yakushimensis (Yamam.) Ormerod ■

407615 Vexillaria Benth. = Centrosema (DC.) Benth. (保留属名)●■☆

407616 Vexillaria Eaton = Clitoria L. ●

407617 Vexillaria Hoffmanns. = Centrosema (DC.) Benth. (保留属名)●■☆

407618 Vexillaria Hoffmanns. ex Benth. = Centrosema (DC.) Benth. (保留属名)●■☆

407619 Vexillaria Raf. = Clitoria L. ●

407620 Vexillifera Ducke = Dussia Krug et Urb. ex Taub. ■☆

407621 Veyretella Szlach. et Olszewski = Habenaria Willd. ■

407622 Veyretella Szlach. et Olszewski(1998);小巴拉圭绶草属■☆

407623 Veyretella hetaerioides (Summerh.) Szlach. et Olszewski;小巴

拉圭绶草■☆

407624 Veyretia Szlach. (1995);巴拉圭绶草属■☆

407625 Veyretia Szlach. = Spiranthes Rich. (保留属名)■

407626 Veyretia hassleri (Cogn.) Szlach.;巴拉圭绶草■☆

407627 Vialia Vis. = Melhania Forssk. ●■

407628 Vialia Vis. ex Schltdl. = Melhania Forssk. ●■

407629 Vibexia Raf. = Sophora L. ●■

407630 Vibexia Raf. = Vexibia Raf. ●■

407631 Vibo Medik. (废弃属名) = Emex Campd. (保留属名)■☆

407632 Vibones Raf. = Rumex L. ■●

407633 Viborgia Moench(废弃属名) = Cytisus Desf. (保留属名)●

407634 Viborgia Moench(废弃属名) = Tubocytisus (DC.) Fourr. ●

407635 Viborgia Moench(废弃属名) = Wiborgia Thunb. (保留属名)■☆

407636 Viborgia Spreng. = Galinsoga Ruiz et Pav. ■●

407637 Viborgia Spreng. = Wiborgia Thunb. (保留属名)■☆

407638 Viborquia Ortega(废弃属名) = Eysenhardtia Kunth (保留属名)●☆

407639 Viburnaceae Dumort. = Adoxaceae E. Mey. (保留科名)●■

407640 Viburnaceae Dumort. = Caprifoliaceae Juss. (保留科名)●■

407641 Viburnaceae Dumort. = Viburnum L. + Sambucus L. ●■

407642 Viburnaceae Dumort. = Violaceae Batsch(保留科名)●■

407643 Viburnaceae Raf. (1820);荚蒾科●

407644 Viburnaceae Raf. = Caprifoliaceae Juss. (保留科名)●■

407645 Viburnum L. (1753);荚蒾属(绣球花属);Arrowwood, Arrow-wood, Snowball, Viburnum ●

407646 Viburnum 'Cayuga';卡尤佳荚蒾;Fragrant Viburnum ●☆

407647 Viburnum 'Mohawk';莫霍克荚蒾;Fragrant Viburnum ●☆

407648 Viburnum × bodnantense Aberc. ex Steam;鲍德南特荚蒾(博德兰特荚蒾);Bodnant Viburnum ●☆

407649 Viburnum × bodnantense Aberc. ex Steam 'Charles Lamont';查尔斯·拉蒙特鲍德南特荚蒾●☆

407650 Viburnum × bodnantense Aberc. ex Steam 'Deben';德本博德兰特荚蒾●☆

407651 Viburnum × bodnantense Aberc. ex Steam 'Down';黎明博德兰特荚蒾(破晓鲍德南特荚蒾)●☆

407652 Viburnum × burkwoodii Burkwood et Skipwith;布克伍德荚蒾(伯氏荚蒾);Burkwood Viburnum, Burkwood's Viburnum ●☆

407653 Viburnum × burkwoodii Burkwood et Skipwith 'Anne Russell';安尼·鲁塞尔布克伍德荚蒾(拉塞尔伯氏荚蒾)●☆

407654 Viburnum × burkwoodii Burkwood et Skipwith 'Park Farm Hybrid';公园农场杂交种布克伍德荚蒾(园圃杂种伯氏荚蒾)●☆

407655 Viburnum × burwoodii Burkwood et Skipwith 'Mohawk';莫霍克布克伍德荚蒾;Mohawk Viburnum ●☆

407656 Viburnum × hizenense (Hatus.) Hatus. ex H. Hara;肥前荚蒾●☆

407657 Viburnum × juddii Rehder;朱地荚蒾(贾德荚蒾);Judd Viburnum ●☆

407658 Viburnum × kiusianum Hatus.;九州荚蒾●☆

407659 Viburnum × pragense ?;布拉格荚蒾;Pragense Viburnum, Prague Viburnum ●☆

407660 Viburnum × rhytidophylloides Suringar;皱叶苜苔荚蒾;Lantanaphyllum Viburnum ●☆

407661 Viburnum acerifolium Bobg. = Viburnum acerifoliun L. ●

407662 Viburnum acerifolium L.;槭叶荚蒾;Arrowwood, Arrow-wood, Dockmackie, Mapleleaf Arrowwood, Mapleleaf Viburnum, Maple-leaved Arrow-wood, Maple-leaved Viburnum, Possum-haw ●

407663 Viburnum acerifolium L. f. collinsii Rouleau = Viburnum acerifolium L. ●

407664 Viburnum acerifolium L. f. eburneum House = Viburnum acerifolium L. ●

407665 Viburnum acerifolium L. f. ovatum Rehder = Viburnum acerifolium L. ●

407666 Viburnum acuminatum Wall. ex DC. = Viburnum punctatum Buch. -Ham. ex D. Don ●

407667 Viburnum adenophorum W. W. Sm. = Viburnum betulifolium Batalin ●

407668 Viburnum adenophorum W. W. Sm. = Viburnum hupehense Rehder ●

407669 Viburnum affine Bush ex C. K. Schneid. = Viburnum rafinesquianum Schult. var. affine (Bush ex C. K. Schneid.) House ●☆

407670 Viburnum affine Bush ex C. K. Schneid. var. australe (C. V. Morton) McAtee = Viburnum rafinesquianum Schult. var. affine (Bush ex C. K. Schneid.) House ●☆

407671 Viburnum affine Bush ex C. K. Schneid. var. hypomalacum S. F. Blake = Viburnum rafinesquianum Schult. ●

407672 Viburnum affine Bush ex Rehder;近缘荚蒾;Affined Arrow-wood ●☆

407673 Viburnum ajugifolium H. Lév. = Viburnum foetidum Wall. var. ceanothoides (C. H. Wright) Hand. -Mazz. ●

407674 Viburnum alnifolium Marshall;桤叶荚蒾;American Wayfaring Tree, Devil's Shoestrings, Down-you-go, Hobble Bush, Hobblebush, Hobble-bush, Hobblebush Arrowwood, Hobblebush Viburnum, Mooseberry, Moosewood, Tangle-legs, Witch Hobble, Witchhobble, Witch-hobble ●

407675 Viburnum americanum Mill.;北美荚蒾;Cranberry-tree ●☆

407676 Viburnum amplifolium Rehder;宽叶荚蒾(广叶荚蒾);Broadleaf Arrowwood, Broadleaf Viburnum, Broad-leaved Viburnum ●

407677 Viburnum amurense Oett.;阿穆尔荚蒾;Amur Arrowwood ●☆

407678 Viburnum arborescens Hemsl. = Viburnum macrocephalum Fortune ●

407679 Viburnum arboricola Hayata;著生珊瑚树;Tree-inhabiting Viburnum ●

407680 Viburnum arboricola Hayata = Viburnum awabuki K. Koch ●

407681 Viburnum arboricola Hayata = Viburnum odoratissimum Ker Gawl. var. awabuki (K. Koch) Zabel ●

407682 Viburnum arboricola Hayata = Viburnum odoratissimum Ker Gawl. ●

407683 Viburnum arcuatum Kom. = Viburnum burejaeticum Regel et Herder ●

407684 Viburnum atrocyaneum C. B. Clarke = Viburnum atrocyaneum C. B. Clarke et Diels ●

407685 Viburnum atrocyaneum C. B. Clarke et Diels;蓝黑果荚蒾(光荚蒾,水红花,小黑果,小铁果,朱启树);Blueblack Arrowwood, Blueblack Viburnum, Dark-blue Viburnum ●

407686 Viburnum atrocyaneum C. B. Clarke et Diels f. harryanum (Rehder) P. S. Hsu = Viburnum atrocyaneum C. B. Clarke et Diels ●

407687 Viburnum atrocyaneum C. B. Clarke et Diels subsp. harryanum (Rehder) P. S. Hsu = Viburnum atrocyaneum C. B. Clarke et Diels ●

407688 Viburnum atrocyaneum C. B. Clarke et Diels subsp. harryanum (Rehder) P. S. Hsu = 毛枝荚蒾(哈利荚蒾);Arrowwood, Hairy-branch Blueblack Viburnum ●

407689 Viburnum atrocyaneum C. B. Clarke et Diels var. puberulum (C. K. Schneid.) P. S. Hsu = Viburnum atrocyaneum C. B. Clarke et Diels ●

407690 Viburnum atrocyaneum C. B. Clarke et Diels var. puberulum

(C. K. Schneid.) P. S. Hsu = Viburnum atrocyaneum C. B. Clarke et Diels subsp. harryanum (Rehder) P. S. Hsu ●

407691　Viburnum australe C. V. Morton = Viburnum rafinesquianum Schult. var. affine (Bush ex C. K. Schneid.) House ●☆

407692　Viburnum awabuki Hort. ex K. Koch = Viburnum odoratissimum Ker Gawl. ●

407693　Viburnum awabuki K. Koch = Viburnum odoratissimum Ker Gawl. var. awabuki (K. Koch) Zabel ●

407694　Viburnum awabuki K. Koch = Viburnum odoratissimum Ker Gawl. ●

407695　Viburnum barbigerum H. Lév. = Viburnum dilatatum Thunb. ●

407696　Viburnum betulifolium Batalin；桦叶荚蒾（对节子，红对节子，老婆子，卵叶荚蒾，山杞子，新高山荚蒾）；Birchleaf Arrowwood, Birchleaf Viburnum, Birch-leaved Viburnum, Morrison Arrowwood, Morrison Viburnum, Yushan Viburnum ●

407697　Viburnum betulifolium Batalin var. flocculosum (Rehder) P. S. Hsu；卷毛荚蒾●

407698　Viburnum betulifolium Batalin var. flocculosum (Rehder) P. S. Hsu = Viburnum betulifolium Batalin ●

407699　Viburnum bitchiuense Makino = Viburnum carlesii Hemsl. var. bitchiuense (Makino) Nakai ●

407700　Viburnum bockii Graebn. = Viburnum utile Hemsl. ex Forbes et Hemsl. ●

407701　Viburnum bodinieri H. Lév. = Viburnum setigerum Hance ●

407702　Viburnum boninsimense (Makino) Koidz. ex Nakai = Viburnum japonicum (Thunb.) Spreng. var. boninsimense Makino ●☆

407703　Viburnum botryoideum H. Lév. = Viburnum erubescens Wall. var. prattii (Graebn.) Rehder ●

407704　Viburnum botryoideum H. Lév. = Viburnum erubescens Wall. ●

407705　Viburnum brachyandrum Nakai；新岛荚蒾●☆

407706　Viburnum brachybotryum Hemsl. ex Forbes et Hemsl.；短序荚蒾（短球荚蒾，尖果荚蒾，球花荚蒾）；Shorttraceme Arrowwood, Shortraceme Viburnum, Short-racemed Arrowwood ●

407707　Viburnum brachybotryum Hemsl. ex Forbes et Hemsl. var. tengyuehense W. W. Sm. = Viburnum tengyuehense (W. W. Sm.) P. S. Hsu ●

407708　Viburnum brachybotryum Hemsl. var. tengyuehense W. W. Sm. = Viburnum tengyuehense (W. W. Sm.) P. S. Hsu ●

407709　Viburnum bracteatum Rehder；苞叶荚蒾；Bracted Viburnum ●☆

407710　Viburnum brevipes Rehder；短柄荚蒾；Shortstalk Arrowwood, Shortstalk Viburnum, Short-stalked Viburnum ●

407711　Viburnum brevipes Rehder = Viburnum dilatatum Thunb. ●

407712　Viburnum brevitubum (P. S. Hsu) P. S. Hsu；短筒荚蒾；Shorttube Arrowwood, Shorttube Viburnum, Short-tubed Viburnum ●

407713　Viburnum buddleifolium C. H. Wright；醉鱼草状荚蒾（醉鱼草荚蒾）；Buddlejaleaf Viburnum, Summerlilacleaf Arrowwood, Tutterflybuch-like Viburnum ●

407714　Viburnum buddleifolium C. H. Wright = Viburnum rhytidophyllum Hemsl. ex Forbes et Hemsl. ●

407715　Viburnum burejaeticum Regel et Herder；修枝荚蒾（河朔绣球，河朔绣球花，暖木条荚蒾，暖木条子，乌荚蒾）；Manchurian Arrowwood, Manchurian Viburnum ●

407716　Viburnum burejanum Herder = Viburnum burejaeticum Regel et Herder ●

407717　Viburnum burmanicum (Rehder) C. Y. Wu ex P. S. Hsu；滇缅荚蒾；Burman Arrowwood, Burman Viburnum ●

407718　Viburnum burmanicum (Rehder) C. Y. Wu ex P. S. Hsu = Viburnum erubescens Wall. ●

407719　Viburnum burmanicum (Rehder) C. Y. Wu ex P. S. Hsu var. motoense P. S. Hsu = Viburnum erubescens Wall. ●

407720　Viburnum burmanicum (Rehder) C. Y. Wu ex P. S. Hsu var. motoense P. S. Hsu；墨脱荚蒾；Motuo Arrowwood, Motuo Viburnum ●

407721　Viburnum bushii Ashe = Viburnum prunifolium L. ●

407722　Viburnum calvum Rehder = Viburnum atrocyaneum C. B. Clarke et Diels ●

407723　Viburnum calvum Rehder var. kwapiense Hand. -Mazz. = Viburnum atrocyaneum C. B. Clarke et Diels subsp. harryanum (Rehder) P. S. Hsu ●

407724　Viburnum calvum Rehder var. kwapiense Hand. -Mazz. = Viburnum atrocyaneum C. B. Clarke et Diels ●

407725　Viburnum calvum Rehder var. puberulum Schneid. = Viburnum atrocyaneum C. B. Clarke et Diels subsp. harryanum (Rehder) P. S. Hsu ●

407726　Viburnum calvum Rehder var. puberulum Schneid. = Viburnum atrocyaneum C. B. Clarke et Diels ●

407727　Viburnum carlesii Hemsl. ex Forbes et Hemsl.；红雷荚蒾（大丁字荚蒾）；Carles Arrowwood, Carles Viburnum, Korean Spice Viburnum, Korean Spice-viburnum, Koreanspice Viburnum ●☆

407728　Viburnum carlesii Hemsl. ex Forbes et Hemsl. ‘ Aurora ’；金黄红雷荚蒾；Korean Spice Viburnum, Koreanspice Viburnum ●☆

407729　Viburnum carlesii Hemsl. ex Forbes et Hemsl. ‘ Compactum ’；朝鲜红雷荚蒾；Korean Spice Viburnum, Korean Viburnum ●☆

407730　Viburnum carlesii Hemsl. ex Forbes et Hemsl. ‘ Diana ’；黛安娜红雷荚蒾（月亮神红雷荚蒾）●☆

407731　Viburnum carlesii Hemsl. ex Forbes et Hemsl. var. bitchiuense Nakai；冈山红雷荚蒾●☆

407732　Viburnum carlesii Hemsl. var. bitchiuense (Makino) Nakai；备中荚蒾（比奇荚蒾）●

407733　Viburnum carnosulum (W. W. Sm.) P. S. Hsu = Viburnum chingii P. S. Hsu var. carnosulum (W. W. Sm.) P. S. Hsu ●

407734　Viburnum carnosulum (W. W. Sm.) P. S. Hsu = Viburnum chingii P. S. Hsu ●

407735　Viburnum carnosulum (W. W. Sm.) P. S. Hsu var. impressinervium P. S. Hsu = Viburnum chingii P. S. Hsu var. impressinervium (P. S. Hsu) P. S. Hsu ●

407736　Viburnum carnosulum (W. W. Sm.) P. S. Hsu var. impressinervum P. S. Hsu = Viburnum brevitubum (P. S. Hsu) P. S. Hsu ●

407737　Viburnum cassinoides L.；决明状荚蒾；Appalachian Tea, Blue Haw, Mountain Ash, Northern Wild Raisin, Swamp Haw, Tea-berry, White-rod, Wild Raisin, Witherod, Withe-rod, Witherod Arrowwood, Witherod Viburnum ●

407738　Viburnum cassinoides L. var. harbisonii McAtee = Viburnum nudum L. var. cassinoides (L.) Torr. et A. Gray ●☆

407739　Viburnum cassinoides L. var. nitidum (Aiton) McAtee = Viburnum nudum L. var. cassinoides (L.) Torr. et A. Gray ●☆

407740　Viburnum cavaleriei H. Lév. = Viburnum chinshanense Graebn. ●

407741　Viburnum ceanothoides C. H. Wright = Viburnum foetidum Wall. var. ceanothoides (C. H. Wright) Hand. -Mazz. ●

407742　Viburnum chaffanjoni H. Lév. = Viburnum ternatum Rehder ●

407743　Viburnum chingii P. S. Hsu；漾濞荚蒾（秦氏荚蒾）；Ching Arrowwood, Ching Viburnum ●

407744　Viburnum chingii P. S. Hsu var. carnosulum (W. W. Sm.) P. S. Hsu；肉叶荚蒾；Flesh Ching Arrowwood ●

407745　Viburnum chingii P. S. Hsu var. carnosulum（W. W. Sm.）P. S. Hsu = Viburnum chingii P. S. Hsu ●

407746　Viburnum chingii P. S. Hsu var. impressinervium（P. S. Hsu）P. S. Hsu；凹脉肉叶荚蒾；Sunkenvein Ching Arrowwood，Sunkenvein Ching Viburnum ●

407747　Viburnum chingii P. S. Hsu var. impressinervium（P. S. Hsu）P. S. Hsu = Viburnum brevitubum（P. S. Hsu）P. S. Hsu ●

407748　Viburnum chingii P. S. Hsu var. patentiserratum P. S. Hsu = Viburnum chingii P. S. Hsu ●

407749　Viburnum chingii P. S. Hsu var. tenuipes P. S. Hsu；细梗漾濞荚蒾；Slenderstalk Arrowwood，Slenderstalk Viburnum ●

407750　Viburnum chingii P. S. Hsu var. tenuipes P. S. Hsu = Viburnum chingii P. S. Hsu ●

407751　Viburnum chinshanense Graebn.；金佛山荚蒾（贵州荚蒾，黑桃子，金山荚蒾，雀儿屎树）；Chinshan Viburnum，Jinfoshan Viburnum，Jinshan Arrowwood，Jinshan Viburnum ●

407752　Viburnum chunii P. S. Hsu；金腺荚蒾（陈氏荚蒾，酒子）；Chun Viburnum，Goldenglandular Arrowwood ●

407753　Viburnum chunii P. S. Hsu subsp. chengii P. S. Hsu = Viburnum chunii P. S. Hsu ●

407754　Viburnum chunii P. S. Hsu var. piliferum P. S. Hsu；毛枝金腺荚蒾；Hairy-branch Chun Arrowwood，Hairy-branch Chun Viburnum ●

407755　Viburnum chunii P. S. Hsu var. piliferum P. S. Hsu = Viburnum chunii P. S. Hsu ●

407756　Viburnum cinnamomifolium Rehder；樟叶荚蒾；Cinnamonleaf Arrowwood，Cinnamonleaf Viburnum，Cinnamon-leaved Viburnum，Evergreen Viburnum ●

407757　Viburnum colebrookeanum Wall. ex DC. = Viburnum lutescens Blume ●

407758　Viburnum congestum Rehder；密花荚蒾（密生荚蒾）；Dense Arrowwood，Dense Viburnum，Dense-flowered Viburnum ●

407759　Viburnum cordifolium Wall. ex DC.；心叶荚蒾；Heartleaf Arrowwood ●

407760　Viburnum cordifolium Wall. ex DC. = Viburnum nervosum D. Don ●

407761　Viburnum cordifolium Wall. ex DC. = Viburnum sympodiale Graebn. ●

407762　Viburnum cordifolium Wall. ex DC. var. hypsophilum Hand. -Mazz. = Viburnum nervosum D. Don ●

407763　Viburnum coriaceum Blume = Viburnum cylindricum Buch. -Ham. ex D. Don ●

407764　Viburnum corymbiflorum P. S. Hsu et S. C. Hsu；伞房荚蒾（雷公子）；Corymbose Viburnum，Corymbous Arrowwood ●

407765　Viburnum corymbiflorum P. S. Hsu et S. C. Hsu subsp. malifolium P. S. Hsu；苹果叶荚蒾；Appleleaf Arrowwood，Appleleaf Viburnum ●

407766　Viburnum corymbiflorum P. S. Hsu et S. C. Hsu var. longipedunculatum P. S. Hsu = Viburnum longipedunculatum（P. S. Hsu）P. S. Hsu ●

407767　Viburnum cotinifolium D. Don；黄栌叶荚蒾；Smoketreeleaf Arrowwood，Smoketreeleaf Viburnum，Smoke-tree-leaved Viburnum ●

407768　Viburnum crassifolium Rehder = Viburnum cylindricum Buch. -Ham. ex D. Don ●

407769　Viburnum cylindricum Buch. -Ham. ex D. Don；水红木（斑鸠柘，抽刀红，大路通，吊白叶，粉果叶，粉帕叶，粉桐叶，粉叶荚蒾，狗肋巴，红经果，灰包木，灰色树，揉白叶，山女贞，水红木荚蒾，睡眠果，四季青，羊脆骨）；Tubeflower Arrowwood，Tubeflower Viburnum，Tube-flowered Viburnum ●

407770　Viburnum cylindricum Buch. -Ham. ex D. Don var. crassifolium（Rehder）Schneid. = Viburnum cylindricum Buch. -Ham. ex D. Don ●

407771　Viburnum dalzielii W. W. Sm.；粤赣荚蒾；Delziel Arrowwood，Delziel Viburnum ●

407772　Viburnum dasyanthum Rehder；毛花荚蒾；Hairy-flower Arrowwood，Hairy-flower Viburnum，Hairy-flowered Viburnum ●

407773　Viburnum dasyanthum Rehder = Viburnum betulifolium Batalin ●

407774　Viburnum davidii Franch.；川西荚蒾；David Arrowwood，David Viburnum，David's Viburnum ●

407775　Viburnum davidii Franch. 'Longleaf'；长叶川西荚蒾；David's Viburnum ●

407776　Viburnum davuricum Maxim. = Viburnum burejaeticum Regel et Herder ●

407777　Viburnum davuricum Pall. = Viburnum mongolicum（Pall.）Rehder ●

407778　Viburnum dentatum L.；齿叶荚蒾（尖荚蒾）；Arrowwood，Arrow-wood，Arrowwood Viburnum，Southern Arrow Wood，Southern Arrowwood，Southern Arrow-wood ●

407779　Viburnum dentatum L. 'Morton'；莫顿齿叶荚蒾；Arrowwood，Northern Burgundy ●☆

407780　Viburnum dentatum L. 'Ralph Senior'；拉尔夫·西尼尔齿叶荚蒾●☆

407781　Viburnum dentatum L. 'Synnestvedt'；芝加哥齿叶荚蒾；Arrowwood，Chicago Lustre ●☆

407782　Viburnum dentatum L. var. lucidum Aiton；亮齿叶荚蒾；Southern Arrow-wood ●☆

407783　Viburnum dielsii Graebn. = Viburnum schensianum Maxim. ●

407784　Viburnum dielsii H. Lév. = Callicarpa rubella Lindl. ●

407785　Viburnum dilatatum Thunb.；荚蒾（孩儿拳头，红楂梅，火柴果，击迷，酒籽，苦柴子，弄先，山梨儿，酸梅子，酸汤杆，土兰条，乌酸木，野花绣球，弈先，羿先）；Arrowwood，Linden Arrowwood，Linden Viburnum ●

407786　Viburnum dilatatum Thunb. 'Catskil'；卡次启尔荚蒾（卡特斯基尔荚蒾）●☆

407787　Viburnum dilatatum Thunb. 'Erie'；伊利荚蒾；Erie Viburnum ●☆

407788　Viburnum dilatatum Thunb. 'Iroquois'；易洛魁荚蒾●☆

407789　Viburnum dilatatum Thunb. 'Oneida'；奥奈达荚蒾；Oneida Viburnum ●☆

407790　Viburnum dilatatum Thunb. f. heterophyllum（Nakai）Sugim.；异叶荚蒾●☆

407791　Viburnum dilatatum Thunb. f. microphyllum（Nakai）Sugim.；小叶酸梅子●☆

407792　Viburnum dilatatum Thunb. f. nikoense（Hiyama）H. Hara；日光荚蒾●☆

407793　Viburnum dilatatum Thunb. f. xanthocarpum Rehder；黄果荚蒾●☆

407794　Viburnum dilatatum Thunb. var. formosanum（Hance）Maxim. = Viburnum formosanum Hayata ●

407795　Viburnum dilatatum Thunb. var. fulvotomentosum（P. S. Hsu）P. S. Hsu；庐山荚蒾（黄褐绒毛荚蒾）；Arrowwood，Luahan Viburnum，Yellow-brown Arrowwood ●

407796　Viburnum dilatatum Thunb. var. fulvotomentosum（P. S. Hsu）P. S. Hsu = Viburnum dilatatum Thunb. ●

407797　Viburnum dilatatum Thunb. var. hizenense（Hatus.）Ohwi = Viburnum × hizenense（Hatus.）Hatus. ex H. Hara ●☆

407798　Viburnum dilatatum Thunb. var. macrophyllum P. S. Hsu = Viburnum dilatatum Thunb. ●

407799　Viburnum edule（Michx.）Raf.；可食荚蒾；Mooseberry，

Moose-berry Viburnum, Moosewood Viburnum, Squashberry, Squash-berry ●☆

407800 Viburnum erosum A. Gray = Viburnum dilatatum Thunb. ●

407801 Viburnum erosum Batalin = Viburnum erosum Thunb. ●

407802 Viburnum erosum Thunb.;宜昌荚蒾(对节木,对叶散花,苦索花,糯米条子,蚀齿荚蒾,松田氏荚蒾,小叶荚蒾,小鱼辣树,野绣球,猪婆子藤);Erose Viburnum, Japanese Arrowwood Viburnum, Littleleaf Viburnum, Yichang Arrowwood, Yichang Viburnum ●

407803 Viburnum erosum Thunb. f. aurantiacum Hayashi;黄宜昌荚蒾●☆

407804 Viburnum erosum Thunb. f. sikokianum (Koidz.) H. Hara;四国荚蒾;Sikoku Arrowwood ●☆

407805 Viburnum erosum Thunb. f. taquetii (H. Lév.) Sugim. = Viburnum erosum Thunb. var. taquetii (H. Lév.) Rehder ●

407806 Viburnum erosum Thunb. f. xanthocarpum (Sugim.) H. Hara;黄果宜昌荚蒾●☆

407807 Viburnum erosum Thunb. subsp. ichangense (Hemsl.) P. S. Hsu = Viburnum erosum Thunb. ●

407808 Viburnum erosum Thunb. subsp. ichangense (Hemsl.) P. S. Hsu var. taquetii (H. Lév.) P. S. Hsu = Viburnum erosum Thunb. var. taquetii (H. Lév.) Rehder ●

407809 Viburnum erosum Thunb. var. formosanum Hance = Viburnum formosanum Hayata ●

407810 Viburnum erosum Thunb. var. furcipila Franch. et Sav. = Viburnum erosum Thunb. var. taquetii (H. Lév.) Rehder ●

407811 Viburnum erosum Thunb. var. hirsutum Pamp. = Viburnum dilatatum Thunb. ●

407812 Viburnum erosum Thunb. var. ichangense Hemsl. = Viburnum erosum Thunb. ●

407813 Viburnum erosum Thunb. var. laevis Franch. et Sav. = Viburnum erosum Thunb. var. taquetii (H. Lév.) Rehder ●

407814 Viburnum erosum Thunb. var. punctataum Franch. et Sav. f. taquetii (H. Lév.) Sugim. = Viburnum erosum Thunb. var. taquetii (H. Lév.) Rehder ●

407815 Viburnum erosum Thunb. var. punctatum Franch. et Sav. = Viburnum erosum Thunb. ●

407816 Viburnum erosum Thunb. var. punctatum Franch. et Sav. = Viburnum erosum Thunb. var. taquetii (H. Lév.) Rehder ●

407817 Viburnum erosum Thunb. var. punctatum Franch. et Sav. f. taquetii (H. Lév.) Sugim. = Viburnum erosum Thunb. var. taquetii (H. Lév.) Rehder ●

407818 Viburnum erosum Thunb. var. setchuenense Graebn. = Viburnum erosum Thunb. ●

407819 Viburnum erosum Thunb. var. taquetii (H. Lév.) Rehder;裂叶宜昌荚蒾;Lobeleaf Yichang Arrowwood ●

407820 Viburnum erubescens Wall.;淡红荚蒾(红荚蒾);Reddish Arrowwood, Reddish Viburnum ●

407821 Viburnum erubescens Wall. var. brevitubum P. S. Hsu = Viburnum brevitubum (P. S. Hsu) P. S. Hsu ●

407822 Viburnum erubescens Wall. var. burmanicum Rehder = Viburnum burmanicum (Rehder) C. Y. Wu ex P. S. Hsu ●

407823 Viburnum erubescens Wall. var. burmanicum Rehder = Viburnum erubescens Wall. ●

407824 Viburnum erubescens Wall. var. carnosulum W. W. Sm. = Viburnum chingii P. S. Hsu ●

407825 Viburnum erubescens Wall. var. carnosulum W. W. Sm. = Viburnum chingii P. S. Hsu var. carnosulum (W. W. Sm.) P. S. Hsu ●

407826 Viburnum erubescens Wall. var. gracilipes Rehder;细梗淡红荚蒾(细梗红荚蒾,细梗荚蒾);Slender Reddish Arrowwood, Slender Reddish Viburnum ●

407827 Viburnum erubescens Wall. var. gracilipes Rehder = Viburnum erubescens Wall. ●

407828 Viburnum erubescens Wall. var. limitaneum W. W. Sm. = Viburnum subalpinum Hand. -Mazz. ●

407829 Viburnum erubescens Wall. var. limitaneum W. W. Sm. = Viburnum subalpinum Hand. -Mazz. var. limitaneum (W. W. Sm.) P. S. Hsu ●

407830 Viburnum erubescens Wall. var. neurophyllum Hand. -Mazz. = Viburnum chingii P. S. Hsu ●

407831 Viburnum erubescens Wall. var. parvum P. S. Hsu et S. C. Hsu;小淡红荚蒾(小红荚蒾);Small Reddish Arrowwood, Small Reddish Viburnum ●

407832 Viburnum erubescens Wall. var. parvum P. S. Hsu et S. C. Hsu = Viburnum erubescens Wall. ●

407833 Viburnum erubescens Wall. var. prattii (Graebn.) Rehder;紫药淡红荚蒾(臭叶子,油麻树,紫药红荚蒾);Pratt Reddish Arrowwood ●

407834 Viburnum erubescens Wall. var. prattii (Graebn.) Rehder = Viburnum erubescens Wall. ●

407835 Viburnum fallax Graebn. = Viburnum chinshanense Graebn. ●

407836 Viburnum fallax Graebn. = Viburnum utile Hemsl. ex Forbes et Hemsl. ●

407837 Viburnum farreri Stearn;香荚蒾(丹春,丁香花,探春,香探春,野绣球);Farrer Viburnum, Fragrant Arrowwood ●

407838 Viburnum farreri Stearn 'Candidissimum';纯白香荚蒾●☆

407839 Viburnum farreri Stearn var. stellipilum D. Z. Ma et H. L. Liu;毛香荚蒾;Hairy Fragrant Arrowwood ●

407840 Viburnum farreri Stearn var. stellipilum D. Z. Ma et H. L. Liu = Viburnum farreri Stearn ●

407841 Viburnum flavescens W. W. Sm.;川滇荚蒾;Yellowish Arrowwood, Yellowish Viburnum ●

407842 Viburnum flavescens W. W. Sm. = Viburnum betulifolium Batalin ●

407843 Viburnum foetens Decne. = Viburnum grandiflorum Wall. ex DC. ●

407844 Viburnum foetidum Wall.;臭荚蒾(冷饭果,糯米果);Foetid Viburnum, Stink Arrowwood ●

407845 Viburnum foetidum Wall. f. integrifolium (Hayata) Nakai = Viburnum integrifolium Hayata ●

407846 Viburnum foetidum Wall. var. ceanothoides (C. H. Wright) Hand. -Mazz.;珍珠荚蒾(老米酒,冷饭子,糯米果,山五味子,碎米团果,珍珠花);Pearl Arrowwood, Pearl Viburnum ●

407847 Viburnum foetidum Wall. var. integrifolium (Hayata) Kaneh. et Hatus. = Viburnum integrifolium Hayata ●

407848 Viburnum foetidum Wall. var. malacotrichum Hand. -Mazz. = Viburnum foetidum Wall. var. rectangulatum (Graebn.) Rehder ●

407849 Viburnum foetidum Wall. var. penninervium Hand. -Mazz. = Viburnum foetidum Wall. var. rectangulatum (Graebn.) Rehder ●

407850 Viburnum foetidum Wall. var. rectangulatum (Graebn.) Rehder;直角荚蒾(半牛尾藤,豆搭子,冷饭果,糯米果,山羊柿子,太平山荚蒾,狭叶荚蒾);Rightangle Arrowwood, Rightangle Viburnum, Taipingshan Viburnum ●

407851 Viburnum foochowense W. W. Sm. = Viburnum luzonicum Rolfe ●

407852 Viburnum fordiae Hance;南方荚蒾(东南荚蒾,火柴树,火斋,猫尿果,猫屎树,酸闷木,酸汤果,酸汤泡,小雷公子);Southern Arrowwood, Southern Viburnum ●

407853 Viburnum formosanum Hayata;台中荚蒾(红子荚蒾,净花荚

莲,台湾荚蒾);Formosa Viburnum,Taiwan Arrowwood,Taizhong Arrowwood,Taizhong Viburnum ●

407854 Viburnum formosanum Hayata f. morrisonense（Hayata）Nakai = Viburnum betulifolium Batalin ●

407855 Viburnum formosanum Hayata f. morrisonense（Hayata）Nakai = Viburnum morrisonense Hayata ●

407856 Viburnum formosanum Hayata f. mushanense Nakai = Viburnum luzonicum Rolfe ●

407857 Viburnum formosanum Hayata subsp. leiogynum P. S. Hsu;光萼荚蒾;Glabrous-calyx Taizhong Arrowwood,Glabrous-calyx Taizhong Viburnum ●

407858 Viburnum formosanum Hayata subsp. leiogynum P. S. Hsu var. pubigerum P. S. Hsu = Viburnum formosanum Hayata subsp. pubigerum（P. S. Hsu）P. S. Hsu ●

407859 Viburnum formosanum Hayata subsp. pubigerum（P. S. Hsu）P. S. Hsu;毛枝台中荚蒾;Hairy-branch Taizhong Arrowwood,Hairy-branch Taizhong Viburnum ●

407860 Viburnum formosanum Hayata var. mushanense Nakai = Viburnum luzonicum Rolfe ●

407861 Viburnum fragrans Bunge = Viburnum farreri Stearn ●

407862 Viburnum fulvotomentosum P. S. Hsu = Viburnum dilatatum Thunb. var. fulvotomentosum（P. S. Hsu）P. S. Hsu ●

407863 Viburnum fulvotomentosum P. S. Hsu = Viburnum dilatatum Thunb. ●

407864 Viburnum furcatum Blume = Viburnum sympodiale Graebn. ●

407865 Viburnum furcatum Blume ex Hook. f. et Thomson;假绣球（粉团团）;Round-leaved Viburnum ●

407866 Viburnum furcatum Blume ex Hook. f. et Thomson var. melanophyllum（Hayata）H. Hara = Viburnum sympodiale Graebn. ●

407867 Viburnum fusiforme Nakai;琉球荚蒾;Liuqiu Arrowwood ●☆

407868 Viburnum giraldii Graebn. = Viburnum schensianum Maxim. ●

407869 Viburnum glomeratum Maxim.;聚花荚蒾（丛花荚蒾,球花荚蒾）;Glomerate Arrowwood,Glomerate Viburnum ●

407870 Viburnum glomeratum Maxim. subsp. magnificum（P. S. Hsu）P. S. Hsu;壮大荚蒾;Big Glomerate Arrowwood,Big Glomerate Viburnum ●

407871 Viburnum glomeratum Maxim. subsp. magnificum（P. S. Hsu）P. S. Hsu = Viburnum glomeratum Maxim. ●

407872 Viburnum glomeratum Maxim. subsp. rotundifolium（P. S. Hsu）P. S. Hsu;圆叶荚蒾;Round-leaf Glomerate Arrowwood,Round-leaf Viburnum ●

407873 Viburnum glomeratum Maxim. subsp. rotundifolium（P. S. Hsu）P. S. Hsu = Viburnum glomeratum Maxim. ●

407874 Viburnum glomeratum Maxim. var. rockii Rehder;小叶聚花荚蒾●

407875 Viburnum glomeratum Maxim. var. rockii Rehder = Viburnum glomeratum Maxim. ●

407876 Viburnum grandiflorum Wall. ex DC.;大花荚蒾;Largeflower Arrowwood,Largeflower Viburnum,Large-flowered Viburnum ●

407877 Viburnum hainanense Merr. et Chun;海南荚蒾（牛瘦鞭,油炸木）;Hainan Arrowwood,Hainan Viburnum ●

407878 Viburnum hanceanum Maxim.;蝶花荚蒾（假沙梨）;Butterfluflower Arrowwood,Hance Viburnum ●

407879 Viburnum harryanum Rehder = Viburnum atrocyaneum C. B. Clarke et Diels subsp. harryanum（Rehder）P. S. Hsu ●

407880 Viburnum harryanum Rehder = Viburnum atrocyaneum C. B. Clarke et Diels ●

407881 Viburnum hengshanicum Tsiang ex P. S. Hsu;衡山荚蒾; Hengshan Arrowwood,Hengshan Viburnum ●

407882 Viburnum henryi Hemsl. ;巴东荚蒾;Badong Arrowwood,Henry Viburnum ●

407883 Viburnum henryi Hemsl. var. xerocarpa Graebn. = Viburnum henryi Hemsl. ●

407884 Viburnum hillieri W. T. Wang;常青荚蒾（希利荚蒾）●

407885 Viburnum hirtulum Rehder = Viburnum fordiae Hance ●

407886 Viburnum hupehense Rehder;湖北荚蒾;Hubei Arrowwood,Hubei Viburnum ●

407887 Viburnum hupehense Rehder = Viburnum betulifolium Batalin ●

407888 Viburnum hupehense Rehder subsp. septentrionale P. S. Hsu;北方荚蒾;North Hubei Arrowwood,North Hubei Viburnum ●

407889 Viburnum hupehense Rehder subsp. septentrionale P. S. Hsu = Viburnum betulifolium Batalin ●

407890 Viburnum hypoleucum Rehder = Viburnum chinshanense Graebn. ●

407891 Viburnum ichangense（Hemsl.）Rehder = Viburnum erosum Batalin ●

407892 Viburnum ichangense Rehder = Viburnum erosum Thunb. ●

407893 Viburnum ichangense Rehder var. atratocarpum（P. S. Hsu）T. R. Dudley et S. C. Sun = Viburnum erosum Thunb. ●

407894 Viburnum inopinatum Craib;厚绒荚蒾（毛叶荚蒾,特异荚蒾,猪脚杆树）;Casual Arrowwood,Unexpected Viburnum ●

407895 Viburnum integrifolium Hayata;全叶荚蒾（狭叶糯米树,玉山荚蒾,玉山糯米树）;Entire Viburnum,Entireleaf Arrowwood,Integrifolious Viburnum,Narrow-leaved Viburnum ●

407896 Viburnum involucratum Wall. ex DC. = Viburnum mullaha Buch. -Ham. ex D. Don ●

407897 Viburnum japonicum（Thunb.）Spreng. ;日本荚蒾（坚荚树）;Japan Arrowwood,Japanese Viburnum ●

407898 Viburnum japonicum（Thunb.）Spreng. var. boninsimense Makino;小笠原荚蒾●☆

407899 Viburnum japonicum（Thunb.）Spreng. var. fruticosum Nakai;灌木日本荚蒾●☆

407900 Viburnum kansuense Batalin et Hand. -Mazz. ;甘肃荚蒾（甘肃琼花）;Gansu Arrowwood,Gansu Viburnum,Kansu Viburnum ●

407901 Viburnum kerrii Geddes = Viburnum odoratissimum Ker Gawl. ●

407902 Viburnum keteleeri Carrière = Viburnum macrocephalum Fortune f. keteleeri（Carrière）Rehder ●

407903 Viburnum komarovii H. Lév. et Vaniot = Photinia komarovii（H. Lév. et Vaniot）L. T. Lu et C. L. Li ●

407904 Viburnum komarovii H. Lév. et Vaniot = Photinia parvifolia（Pritz.）C. K. Schneid. ●

407905 Viburnum koreanum Nakai;朝鲜荚蒾（朝鲜条子,少花荚蒾）;Few-flowered Cranberry-tree,Korean Arrowwood,Korean Viburnum ●

407906 Viburnum lancifolium P. S. Hsu;披针叶荚蒾（长叶荚蒾,六角藤,沙罗树,猪母柴）;Lanceleaf Viburnum,Lanceolateleaf Arrowwood,Lanceolate-leaved Viburnum ●

407907 Viburnum lantana L. ;绵毛荚蒾（黑果绣球,山地荚蒾）;Balm Tree,Coban Tree,Cobin Tree,Coven-tree,Dog Timber,Dogberry,Dogwood,Lithewort,Lithy-tree,Meal Tree,Mealy Guelder Rose,Mealy-tree,Red Royal Oak,Shepherd's Delight,Twistwood,Twist-wood,Wayfaring Tree,Wayfaring Viburnum,Wayfaringtree,Wayfaring-tree,Wayfaringtree Viburnum,Whipcrop,Whiptop,Whitewood,Whitney,Whittenbeam,Whitten-tree,Whltty-tree ●☆

407908 Viburnum lantana L. ' Aureum ';金黄绵毛荚蒾;Wayfaring Tree ●☆

407909 Viburnum lantana L. ' Mohican ';莫希干绵毛荚蒾;Wayfaring

Tree Viburnum ●☆

407910　Viburnum lantana L. 'Variegatum';斑叶绵毛荚蒾;Variegata Wayfaringtree, Variegated Wayfaring Tree ●☆

407911　Viburnum lantana L. 'Versicolor';异色绵毛荚蒾●☆

407912　Viburnum lantana L. var. glabratum Chabert = Viburnum lantana L. ●☆

407913　Viburnum lantanoides Michx. = Viburnum alnifolium Marshall ●

407914　Viburnum laterale Rehder;侧花荚蒾;Lateral Viburnum, Lateralleaf Arrowwood ●

407915　Viburnum leiocarpum P. S. Hsu;光果荚蒾;Glabrous-fruit Viburnum, Psilateleaf Arrowwood, Smooth-fruited Viburnum ●

407916　Viburnum leiocarpum P. S. Hsu var. punctatum P. S. Hsu;斑点光果荚蒾;Spot Glabrous-fruit Viburnum, Spot Psilateleaf Arrowwood ●

407917　Viburnum lentago L.;细枝荚蒾;Black Haw, Blackhaw, Granberry-bushes, Highbush-cranberry, Nannyberry, Nannyberry Sweet Viburnum, Nannyberry Viburnum, Sheepberry, Sweet Arrowwood, Sweet Viburnum, Wild Raisin ●☆

407918　Viburnum lepidotulum Merr. et Chun = Viburnum punctatum Buch. -Ham. ex D. Don subsp. lepidotulum (Merr. et Chun) P. S. Hsu ●

407919　Viburnum lobophyllum Graebn.;阔叶荚蒾(沛阳花);Broad-leaf Viburnum, Lobedleaf Arrowwood, Lobedleaf Viburnum ●

407920　Viburnum lobophyllum Graebn. = Viburnum betulifolium Batalin ●

407921　Viburnum lobophyllum Graebn. = Viburnum melanocarpum P. S. Hsu ●

407922　Viburnum lobophyllum Graebn. var. flocculosum Rehder = Viburnum betulifolium Batalin var. flocculosum (Rehder) P. S. Hsu ●

407923　Viburnum lobophyllum Graebn. var. flocculosum Rehder = Viburnum betulifolium Batalin ●

407924　Viburnum lobophyllum Graebn. var. silvestrii Pamp.;腺叶荚蒾●

407925　Viburnum lobophyllum Graebn. var. silvestrii Pamp. = Viburnum betulifolium Batalin ●

407926　Viburnum lobophyullum Graebn. var. flocculosum Rehder = Viburnum betulifolium Batalin var. flocculosum (Rehder) P. S. Hsu ●

407927　Viburnum longipedunculatum (P. S. Hsu) P. S. Hsu;长梗荚蒾;Long-pedicel Viburnum, Long-petioled Viburnum, Longstalk Arrowwood ●

407928　Viburnum longiradiatum P. S. Hsu et S. W. Fan;长伞梗荚蒾;Long-radiate Viburnum, Long-radiated Viburnum, Longzxle Arrowwood ●

407929　Viburnum lutescens Blume;淡黄荚蒾(黄荚蒾,早禾子树,猪肚木);Paleyellow Arrowwood, Paleyellow Viburnum, Pale-yellow Viburnum ●

407930　Viburnum luzonicum Rolfe;吕宋荚蒾(半伴木,淡黄荚蒾,福州荚蒾,红子仔,罗盖叶,细叶火柴枝树);Luzon Arrowwood, Luzon Viburnum ●

407931　Viburnum luzonicum Rolfe var. formosanum (Hance) Rehder = Viburnum formosanum Hayata ●

407932　Viburnum luzonicum Rolfe var. formosanum (Hance) Rehder f. mushanense Kaneh. et Sasaki ex Sasaki = Viburnum luzonicum Rolfe ●

407933　Viburnum luzonicum Rolfe var. formosanum (Hance) Rehder f. subglabrum (Hayata) Kaneh. et Sasaki ex Sasaki = Viburnum formosanum Hayata ●

407934　Viburnum luzonicum Rolfe var. morrisonense (Hayata) S. S. Ying;玉山荚蒾;Yushan Arrowwood ●

407935　Viburnum luzonicum Rolfe var. morrisonense (Hayata) S. S. Ying = Viburnum betulifolium Batalin ●

407936　Viburnum luzonicum Rolfe var. oblongum (Kaneh. et Sasaki) H. L. Li;长椭圆叶荚蒾;Oblongleaf Luzon Viburnum ●

407937　Viburnum luzonicum Rolfe var. tashiroi (Nakai) Hatus. = Viburnum tashiroi Nakai ●☆

407938　Viburnum luzonicum var. formosanum (Hance) Rehder var. mushanense Kaneh. et Sasaki = Viburnum luzonicum Rolfe ●

407939　Viburnum macrocephalum Fortune;绣球荚蒾(八仙花,大绣球,斗球,木球荚蒾,木绣球,琼花,绣球,绣球花,中国绣球花,中国绣球荚蒾,紫阳花);China Arrowwood, Chinese Snowball, Chinese Snowball Bush, Chinese Snowball Tree, Chinese Snowball Viburnum, Chinese Viburnum ●

407940　Viburnum macrocephalum Fortune 'Sterile';不育木绣球●☆

407941　Viburnum macrocephalum Fortune f. keteleeri (Carrière) Nicholson = Viburnum macrocephalum Fortune f. keteleeri (Carrière) Rehder ●

407942　Viburnum macrocephalum Fortune f. keteleeri (Carrière) Rehder;琼花(八仙花,蝴蝶花,木绣球);Wild Chinese Arrowwood, Wild Chinese Viburnum ●

407943　Viburnum macrocephalum Fortune var. indutum Hand. -Mazz.;多毛绣球荚蒾(多毛琼花,毛琼花);Arrowwood, Hairy Chinese Viburnum ●

407944　Viburnum macrocephalum Fortune var. indutum Hand. -Mazz. = Viburnum macrocephalum Fortune ●

407945　Viburnum macrocephalum Fortune var. sterile Dippel = Viburnum macrocephalum Fortune ●

407946　Viburnum mairei H. Lév. = Viburnum congestum Rehder ●

407947　Viburnum martinii H. Lév. = Viburnum sympodiale Graebn. ●

407948　Viburnum matsudae Hayata;松田氏荚蒾●

407949　Viburnum matsudae Hayata = Viburnum erosum Batalin ●

407950　Viburnum melanocarpum P. S. Hsu;黑果荚蒾;Blackfruit Arrowwood, Blackfruit Viburnum, Melanocarpous Viburnum ●

407951　Viburnum melanophyllum Hayata = Viburnum sympodiale Graebn. ●

407952　Viburnum meyer-waldeckii Loes. = Viburnum erosum Thunb. var. taquetii (H. Lév.) Rehder ●

407953　Viburnum molle Michx.;肯塔基荚蒾;Arrow Wood, Kentucky Viburnum ●☆

407954　Viburnum mongolicum (Pall.) Rehder;蒙古荚蒾(白暖条,蒙古条子,蒙古绣球花,土连树);MongolArrowwood, Mongolian Viburnum ●

407955　Viburnum monogynum Blume = Viburnum lutescens Blume ●

407956　Viburnum morrisonense Hayata = Viburnum betulifolium Batalin ●

407957　Viburnum morrisonensis Hayata = Viburnum formosanum Hayata ●

407958　Viburnum mullaha Buch. -Ham. ex D. Don;西域荚蒾(星芒荚蒾);Himalayan Viburnum, Western Arrowwood ●

407959　Viburnum mullaha Buch. -Ham. ex D. Don = Viburnum dalzielii W. W. Sm. ●

407960　Viburnum mullaha Buch. -Ham. ex D. Don var. glabrescens (C. B. Clarke) Kitam.;少毛西域荚蒾;Glabrescent Western Arrowwood ●

407961　Viburnum mullaha Buch. -Ham. ex D. Don var. tashiroi (Nakai) Hatus. = Viburnum tashiroi Nakai ●☆

407962　Viburnum multratum K. Koch = Viburnum cotinifolium D. Don ●

407963　Viburnum mushanense Hayata = Viburnum luzonicum Rolfe ●

407964　Viburnum nervosum D. Don;显脉荚蒾(黄柏,心叶荚蒾);Distinctvein Viburnum, Heartleaf Viburnum, Nervose Arrowwood, Veined Viburnum ●

407965　Viburnum nervosum D. Don = Viburnum grandiflorum Wall. ex DC. ●

407966   Viburnum nervosum D. Don var. hypsophilum（Hand. -Mazz.）H. W. Li = Viburnum nervosum D. Don ●

407967   Viburnum nervosum Hook. et Arn. = Viburnum sempervirens K. Koch ●

407968   Viburnum nervosum Hook. f. et Thomson = Viburnum grandiflorum Wall. ex DC. ●

407969   Viburnum nitidum Aiton = Viburnum nudum L. var. cassinoides（L.）Torr. et A. Gray ●☆

407970   Viburnum nudum L.；裸莢蒾（光泽叶莢蒾）；Naked Whiterod, Possum Haw, Possumhaw, Possum-haw, Possumhaw Viburnum, Possum-haw Viburnum, Smooth White-rod, Smooth Withered, Swamphaw ●☆

407971   Viburnum nudum L. var. cassinoides（L.）Torr. et A. Gray；北方裸莢蒾；Northern Wild-raisin, Possum-haw, Withe-rod ●☆

407972   Viburnum oblongum P. S. Hsu；长圆莢蒾；Oblong Arrowwood ●

407973   Viburnum oblongum P. S. Hsu = Viburnum tengyuehense（W. W. Sm.）P. S. Hsu ●

407974   Viburnum oblongum P. S. Hsu var. polyneurum P. S. Hsu = Viburnum tengyuehense（W. W. Sm.）P. S. Hsu var. polyneurum（P. S. Hsu）P. S. Hsu ●

407975   Viburnum oblongum P. S. Hsu var. tengyuehense（W. W. Sm.）P. S. Hsu = Viburnum tengyuehense（W. W. Sm.）P. S. Hsu ●

407976   Viburnum obovatum Walter；倒卵叶莢蒾 ●☆

407977   Viburnum odoratissimum Ker Gawl.；极香莢蒾（法国冬青，枫饭树，篱椿木，麻香油，麻油客，沙糖木，沙掌树，砂糖树，山猪肉，珊瑚树，四季青，香柄树，早禾树，猪耳木，着生珊瑚树）；Evergreen Tree Viburnum, Sweet Arrowwood, Sweet Viburnum ●

407978   Viburnum odoratissimum Ker Gawl. var. arboricola（Hayata）Yamam. = Viburnum odoratissimum Ker Gawl. var. awabuki（K. Koch）Zabel ●

407979   Viburnum odoratissimum Ker Gawl. var. arboricola（Hayata）Yamam. = Viburnum odoratissimum Ker Gawl. ●

407980   Viburnum odoratissimum Ker Gawl. var. awabuki（K. Koch）Zabel；日本珊瑚树（矮大叶珊瑚树，法国冬青，泡花莢蒾，日本极香莢蒾，山猪肉，珊瑚树）；Japan Arrowwood, Japanese Viburnum ●

407981   Viburnum odoratissimum Ker Gawl. var. awabuki（K. Koch）Zabel ex Rümpler = Viburnum awabuki K. Koch ●

407982   Viburnum odoratissimum Ker Gawl. var. awabuki（K. Koch）Zabel ex Rümpler = Viburnum odoratissimum Ker Gawl. ●

407983   Viburnum odoratissimum Ker Gawl. var. conspersum W. W. Sm. = Viburnum odoratissimum Ker Gawl. ●

407984   Viburnum odoratissimum Ker Gawl. var. officinale O. M. Hu；油麻香 ●

407985   Viburnum odoratissimum Ker Gawl. var. sessiliflorum（Geddes）Fukuoka = Viburnum odoratissimum Ker Gawl. ●

407986   Viburnum odoratissimum Ker Gawl. var. sessiliflorum（Geddes）Fukuoka；云南珊瑚树；Yunnan Coraltree, Yunnan Sweet Viburnum ●

407987   Viburnum oliganthum Batalin；少花莢蒾（野红枣）；Fewflower Arrowwood, Fewflower Viburnum, Oligoflorous Viburnum ●

407988   Viburnum omeiensis P. S. Hsu；峨眉莢蒾；Emei Arrowwood, Emei Viburnum, Omei Viburnum ●

407989   Viburnum opulus L.；欧洲莢蒾（白雪树，佛头花，欧木绣球，欧洲绣球，琼花莢蒾，析伤木，雪球莢蒾）；Cherry-wood, Club Bunches, Common Snowball, Cramp Bark, Crampbark, Cranberry Bush, Cranberry Tree, Cranberry Viburnum, Cranberrybush, Cranberry-bush Viburnum, Cranberry-tree, Dog Eller, Dog Rose, Dog Rowan, Dog Tree, Dogberry, Dogwood, Dwarf Plane, Elder Rose, Europe Arrowwood, European Cranberry, European Cranberry Bush, European Cranberry Viburnum, European Cranberrybush, European Cranberry-bush, European Cranberrybush Viburnum, European Highbush Cranberry, European Highbush-cranberry, European Snowball, Gadrise, Gaelder Rose, Gatten-tree, Gatteridge-tree, Gelders Rose, Gottridge, Guelder Rose, Guelder-rose, High-bush Cranberry, King's Crown, Marsh Elder, May Tassel, May Tossel, May Tossels, May Tosty, May-ball, Maypole, May-rose, May-tassels, Mugget, Mugget Rose, Opier, Ople-tree, Opple-tree, Parnell, Pincushion, Pincushion Tree, Pincushiontree, Queen's Cushion, Queen's Pincushion, Red Elder, Rose Elder, Silver Ball, Silver Bells, Snow Toss, Snowball, Snowball Bush, Snowball Rose, Snowball Tree, Snowball-bush, Snowbaw, Snowflake, Stink Tree, Tisty-tosty, Tossy Ball, Tossy Balls, Water Elder, White Dogwood, White Elder, Whitsun Ball, Whitsun Balls, Whitsun Boss, Whitsun Flower, Whitsun Tassel, Whitsun Tassels, Whitten-tree, Wintsun Rose ●

407990   Viburnum opulus L. 'Aureum'；黄枝欧洲莢蒾（金叶欧洲绣球）；Golden-leafed European Cranberry Bush, Golden-leaved European Viburnum ●

407991   Viburnum opulus L. 'Compactum'；密枝欧洲莢蒾（稠密欧洲莢蒾）；Compact Cranberry Bush ●

407992   Viburnum opulus L. 'Leonard's Dwarf'；莱奥欧洲莢蒾；Dwarf European Cranberry Bush ●☆

407993   Viburnum opulus L. 'Nanum'；矮生欧洲莢蒾（矮生欧洲绣球）●

407994   Viburnum opulus L. 'Notcutt's Variety'；欧洲绣球努特卡特变种 ●☆

407995   Viburnum opulus L. 'Roseum'；玫瑰欧洲绣球；Snowball Tree, Whitsunfide Boss ●

407996   Viburnum opulus L. 'Roseum' = Viburnum opulus L. 'Sterile' ●

407997   Viburnum opulus L. 'Sterile'；雪球欧洲莢蒾；Snowball Bush, Snowball Tree ●

407998   Viburnum opulus L. 'Xanthocarpum'；黄果欧洲莢蒾；Yellow-fruited Cranberry. Bush, Yellow-fruited European Cranberry Bush ●

407999   Viburnum opulus L. subsp. calvescens（Rehder）Sugim. = Viburnum opulus L. var. calvescens（Rehder）H. Hara ●

408000   Viburnum opulus L. subsp. calvescens（Rehder）Sugim. = Viburnum opulus L. var. sargentii（Koehne）Takeda ●

408001   Viburnum opulus L. subsp. trilobum（Marshall）R. T. Clausen；三裂欧洲莢蒾；American Cranberry-bush, Cranberry Viburnum, High-bush Cranberry ●☆

408002   Viburnum opulus L. var. americanum Aiton；美洲莢蒾；American Cranberrybush, Cramp-bark, Cranberry Tree, High Cranberry, Highbush Cranberry, Pembina ●☆

408003   Viburnum opulus L. var. americanum Aiton 'Red Wings'；红翼美洲莢蒾；American Cranberry Bush ●☆

408004   Viburnum opulus L. var. americanum Aiton = Viburnum opulus L. subsp. trilobum（Marshall）R. T. Clausen ●☆

408005   Viburnum opulus L. var. calvescens（Rehder）H. Hara = Viburnum opulus L. var. sargentii（Koehne）Takeda ●

408006   Viburnum opulus L. var. calvescens（Rehder）H. Hara f. puberulum（Kom.）Sugim. = Viburnum opulus L. var. sargentii（Koehne）Takeda f. puberulum（Kom.）Sugim. ●

408007   Viburnum opulus L. var. calvescens（Rehder）H. Hara f. puberulum（Kom.）Sugim. = Viburnum opulus L. var. sargentii（Koehne）Takeda ●

408008   Viburnum opulus L. var. roseum L. = Viburnum opulus L. ●

408009   Viburnum opulus L. var. sargentii（Koehne）Takeda；秃莢蒾

（春花子，佛头花，高山野黄肉，红枣鸭雀食，鸡树条，鸡树条荚蒾，鸡树条子，糯米条，沙氏荚蒾，山竹子，少毛鸡条树，双合草，双肾参，天目琼花）；Sargent Arrowwood, Sargent Crandberrybush, Sargent Viburnum, Smooth Cranberrybush, Tianmu Arrowwood, Tianmu Viburnum ●

408010 Viburnum opulus L. var. sargentii（Koehne）Takeda f. flavum（Rehder）H. Hara;黄秃荚蒾●

408011 Viburnum opulus L. var. sargentii（Koehne）Takeda f. hydrangeoides（Nakai）H. Hara;绣球秃荚蒾●☆

408012 Viburnum opulus L. var. sargentii（Koehne）Takeda f. puberulum（Kom.）Sugim.;毛叶鸡树条●

408013 Viburnum opulus L. var. sterilis？ = Viburnum opulus L. 'Sterile' ●

408014 Viburnum opulus L. var. trilobum（Marshall）McAtee = Viburnum opulus L. subsp. trilobum（Marshall）R. T. Clausen ●☆

408015 Viburnum orientale Pall. ;东方荚蒾●☆

408016 Viburnum ovatifolium Rehder;卵叶荚蒾（高粱花，红对节子，藤草）; Ovateleaf Arrowwood, Ovateleaf Viburnum, Ovate-leaved Viburnum ●

408017 Viburnum ovatifolium Rehder = Viburnum betulifolium Batalin ●

408018 Viburnum pallidum Franch. = Viburnum foetidum Wall. var. rectangulatum（Graebn.）Rehder ●

408019 Viburnum parvifolium Hayata;小叶荚蒾; Littleleaf Viburnum, Smallleaf Arrowwood,Smallleaved Viburnum,Small-leaved Viburnum ●

408020 Viburnum parvifolium W. W. Sm. = Viburnum luzonicum Rolfe ●

408021 Viburnum parvilimbum Merr. = Viburnum foetidum Wall. var. rectangulatum（Graebn.）Rehder ●

408022 Viburnum pauciflorum Bach. Pyl. ex Torr. et A. Gray = Viburnum edule（Michx.）Raf. ●☆

408023 Viburnum pauciflorum Raf. = Viburnum koreanum Nakai ●

408024 Viburnum phlebotrichum Siebold et Zucc. ;扶老杖荚蒾（基隆荚蒾）●

408025 Viburnum phlebotrichum Siebold et Zucc. f. xanthocarpum Hayashi;黄果扶老杖荚蒾●☆

408026 Viburnum pinfaense H. Lév. = Viburnum sempervirens K. Koch var. trichophorum Hand.-Mazz. ●

408027 Viburnum plicatum Thunb. ;雪球荚蒾（粉团，粉团花，粉团荚蒾,蝴蝶树,蝴蝶戏珠花）; Doublefile Viburnum, Japanese Snowball, Japanese Snow-ball,Japanese Snowball Tree,Snowball Arrowwood ●

408028 Viburnum plicatum Thunb. 'Mariesii';玛丽雪球荚蒾●☆

408029 Viburnum plicatum Thunb. 'Nanum Semperflorens';矮常花雪球荚蒾●☆

408030 Viburnum plicatum Thunb. 'Pink Beauty';艳粉雪球荚蒾●☆

408031 Viburnum plicatum Thunb. f. glabrum（Koidz. ex Nakai）Rehder;光雪球荚蒾●☆

408032 Viburnum plicatum Thunb. f. parvifolium（Miq.）Rehder = Viburulum plicatum Thunb. var. parvifolium Miq. ●☆

408033 Viburnum plicatum Thunb. f. tomentosum（Miq.）Rehder = Viburnum plicatum Thunb. var. tomentosum Miq. ●

408034 Viburnum plicatum Thunb. f. tomentosum（Thunb. ex Murray）Rehder;蝴蝶荚蒾（蝴蝶花,蝴蝶树,蝴蝶戏珠花,苦酸汤,糯米树,糯树,绣球花）; Butterfly Arrowwood, Doublefile Viburnum, Japanese Doublefile Viburnum, Japanese Snowball, Japanese Snowball Tree,Tomentose Japanese Viburnum ●

408035 Viburnum plicatum Thunb. f. tomentosum（Thunb. ex Murray）Rehder 'Mariesii';马利斯荚蒾; Doublefile Viburnum, Mariesi Viburnum ●☆

408036 Viburnum plicatum Thunb. f. tomentosum（Thunb. ex Murray）Rehder 'Shoshoni';肖肖恩荚蒾; Doublefile Viburnum ●☆

408037 Viburnum plicatum Thunb. f. tomentosum（Thunb.）Rehder = Viburnum plicatum Thunb. var. tomentosum（Thunb.）Miq. ●

408038 Viburnum plicatum Thunb. subsp. glabrum（Koidz. ex Nakai）Kitam. = Viburnum plicatum Thunb. f. glabrum（Koidz. ex Nakai）Rehder ●☆

408039 Viburnum plicatum Thunb. var. formosanum Y. C. Liu et C. H. Ou;台湾蝴蝶戏珠花●

408040 Viburnum plicatum Thunb. var. formosanum Y. C. Liu et C. H. Ou = Viburnum plicatum Thunb. ●

408041 Viburnum plicatum Thunb. var. glabrum（Koidz. ex Nakai）H. Hara = Viburnum plicatum Thunb. f. glabrum（Koidz. ex Nakai）Rehder ●☆

408042 Viburnum plicatum Thunb. var. parvifolium Miq. ;小叶雪球荚蒾●☆

408043 Viburnum plicatum Thunb. var. parvifolium Miq. = Viburnum plicatum Thunb. var. tomentosum（Thunb.）Miq. ●

408044 Viburnum plicatum Thunb. var. parvifolium Miq. f. plenum Honda;重瓣小叶雪球荚蒾●☆

408045 Viburnum plicatum Thunb. var. parvifolium Miq. f. watanabei Honda;渡边荚蒾●☆

408046 Viburnum plicatum Thunb. var. plenum Miq. = Viburnum plicatum Thunb. ●

408047 Viburnum plicatum Thunb. var. tomentosum（Thunb. ex Murray）Rehder 'Mariesii' = Viburnum plicatum Thunb. f. tomentosum（Thunb. ex Murray）Rehder 'Mariesii'●☆

408048 Viburnum plicatum Thunb. var. tomentosum（Thunb. ex Murray）Rehder f. lanceatum（Rehder）Rehder;披针状荚蒾●☆

408049 Viburnum plicatum Thunb. var. tomentosum（Thunb. ex Murray）Rehder = Viburnum plicatum Thunb. f. tomentosum（Thunb. ex Murray）Rehder ●

408050 Viburnum plicatum Thunb. var. tomentosum（Thunb.）Miq. = Viburnum plicatum Thunb. f. tomentosum（Thunb. ex Murray）Rehder ●

408051 Viburnum plicatum Thunb. var. tomentosum Miq. = Viburnum plicatum Thunb. f. tomentosum（Thunb. ex Murray）Rehder ●

408052 Viburnum polycarpum Wall. ex DC. = Viburnum cotinifolium D. Don ●

408053 Viburnum prattii Graebn. = Viburnum erubescens Wall. var. prattii（Graebn.）Rehder ●

408054 Viburnum prattii Graebn. = Viburnum erubescens Wall. ●

408055 Viburnum premnaceum Wall. ex DC. = Viburnum foetidum Wall. ●

408056 Viburnum propinquum Hemsl. ;球核荚蒾（臭药，高山荚蒾，淮角筋，鸡壳精，六股筋，水马蹄，兴山荚蒾，兴山绣球）; Ballpit Arrowwood, Evergreen Viburnum, Mountain Viburnum, Neighbouring Viburnum,Xingshan Viburnum ●

408057 Viburnum propinquum Hemsl. f. parvifolium（Graebn. ex Diels）Nakai = Viburnum propinquum Hemsl. ●

408058 Viburnum propinquum Hemsl. var. mairei W. W. Sm. ;狭叶球核荚蒾（滇南兴山荚蒾）; Maire Ballpit Arrowwood, Maire Xingshan Viburnum ●

408059 Viburnum propinquum Hemsl. var. parvifolium Graebn. ex Diels;小叶球核荚蒾●

408060 Viburnum propinquum Hemsl. var. parvifolium Graebn. ex Diels = Viburnum propinquum Hemsl. ●

408061 Viburnum prunifolium L. ;梨叶荚蒾（樱叶荚蒾）; American Sloe, Black Haw, Black Haw Viburnum, Blackhaw, Black-haw,

Blackhaw Viburnum, Black-haw Viburnum, Cherryleaf Arrowwood, Haw , Plum Leaf Viburnum , Smooth Black-haw , Stagbush , Sweethaw ●

408062　Viburnum prunifolium L. 'Early Red';早红樱叶荚蒾;Black Haw ●☆

408063　Viburnum prunifolium L. var. bushii（Ashe）E. J. Palmer et Steyerm. = Viburnum prunifolium L. ●

408064　Viburnum prunifolium L. var. globosum Nash ex C. K. Schneid. = Viburnum prunifolium L. ●

408065　Viburnum pubescens Pursh;柔毛荚蒾;Arrowwood, Downy Viburnum ●☆

408066　Viburnum pubigerum Wight et Arn. = Viburnum erubescens Wall. ●

408067　Viburnum pubinerve Blume ex Miq. = Viburnum opulus L. var. calvescens（Rehder）H. Hara ●

408068　Viburnum pubinerve Blume ex Miq. = Viburnum opulus L. var. sargentii（Koehne）Takeda ●

408069　Viburnum pubinerve Blume ex Miq. f. calvescens（Rehder）Nakai = Viburnum opulus L. var. calvescens（Rehder）H. Hara ●

408070　Viburnum pubinerve Blume ex Miq. f. calvescens（Rehder）Rehder = Viburnum opulus L. var. sargentii（Koehne）Takeda ●

408071　Viburnum pubinerve Blume ex Miq. f. puberulum（Kom.）Nakai = Viburnum opulus L. var. sargentii（Koehne）Takeda f. puberulum（Kom.）Sugim. ●

408072　Viburnum pubinerve Blume ex Miq. f. puberulum（Kom.）Nakai = Viburnum opulus L. var. calvescens（Rehder）H. Hara f. puberulum（Kom.）Sugim. ●

408073　Viburnum pubinerve Blume ex Miq. f. puberulum（Kom.）Nakai = Viburnum opulus L. var. sargentii（Koehne）Takeda ●

408074　Viburnum punctatum Buch. -Ham. ex D. Don;鳞斑荚蒾（大青藤,点叶荚蒾,小鳞荚蒾）;Plotchleaf Arrowwood, Variegatedleaf Viburnum , Variegated-leaved Viburnum ●

408075　Viburnum punctatum Buch. -Ham. ex D. Don subsp. lepidotulum（Merr. et Chun）P. S. Hsu;鳞毛荚蒾（大果鳞斑荚蒾,鳞秕荚蒾）;Hairy Viburnum , Smallscale Blotch Arrowwood ●

408076　Viburnum punctatum Buch. -Ham. ex D. Don var. lepidotulum（Merr. et Chun）P. S. Hsu = Viburnum punctatum Buch. -Ham. ex D. Don subsp. lepidotulum（Merr. et Chun）P. S. Hsu ●

408077　Viburnum pyramidatum Rehder;锥序荚蒾（尖锥荚蒾）;Paniculate Arrowwood , Paniculate Viburnum ●

408078　Viburnum pyrifolium Poir. = Viburnum prunifolium L. ●

408079　Viburnum rafinesquianum Schult.;灰棕枝荚蒾;Arrow-wood, Downy Arrow Wood, Downy Arrowwood, Downy Arrow-wood, Downy Arrowwood Viburnum , Downy Viburnum , Missouri Viburnum ●

408080　Viburnum rafinesquianum Schult. var. affine（Bush ex C. K. Schneid.）House;近缘灰棕枝荚蒾;Arrow-wood, Downy Arrow-wood ●☆

408081　Viburnum recognitum Fernald;平滑荚蒾;Arrow Wood, Arrow-wood , Smooth Arrowwood , Smooth Viburnum ●

408082　Viburnum recognitum Fernald = Viburnum dentatum L. var. lucidum Aiton ●☆

408083　Viburnum recognitum Fernald var. alabamense McAtee = Viburnum dentatum L. var. lucidum Aiton ●☆

408084　Viburnum rectangulare Graebn. ex Hayata = Viburnum foetidum Wall. var. rectangulatum（Graebn.）Rehder ●

408085　Viburnum rectangulatum Graebn. = Viburnum foetidum Wall. var. rectangulatum（Graebn.）Rehder ●

408086　Viburnum rectangulatum Graebn. ex Hayata = Viburnum foetidum Wall. var. rectangulatum（Graebn.）Rehder ●

408087　Viburnum rhytidophylloides Suringar;杂种绵毛荚蒾●☆

408088　Viburnum rhytidophyllum Hemsl. = Viburnum rhytidophyllum Hemsl. ex Forbes et Hemsl. ●

408089　Viburnum rhytidophyllum Hemsl. ex Forbes et Hemsl.;皱叶荚蒾（黑汉条子,毛羊尿树,枇杷叶荚蒾,山枇杷,野枇杷,皱皮荚蒾）;Evergreen Viburnum, Leatherleaf Arrowwood, Leatherleaf Viburnum, Leather-leaved Viburnum , Loquatleaf Arrowwood , Wrinkled Viburnum ●

408090　Viburnum rhytidophyllum Hemsl. ex Forbes et Hemsl. 'Aldenhamense';阿登哈曼斯枇杷叶荚蒾●☆

408091　Viburnum rhytidophyllum Hemsl. ex Forbes et Hemsl. 'Roseum';玫瑰红枇杷叶荚蒾●☆

408092　Viburnum rigidum Vent.;硬叶荚蒾;Canary Island Viburnum ●☆

408093　Viburnum rosthornii Graebn. = Viburnum chinshanense Graebn. ●

408094　Viburnum rosthornii Graebn. var. xerocarpa Graebn. = Viburnum henryi Hemsl. ●

408095　Viburnum rufidulum Raf.;锈红荚蒾（南方樱叶荚蒾）;Black Haw, Rusty Blackhaw, Southern Black Haw, Southern Black-Haw, Southern Blackhaw Viburnum , Wild Raisin ●☆

408096　Viburnum sambucinum Reinw. ex Blume var. tomentosum Hallier f.;毛叶荚蒾;Hairyleaf Arrowwood, Hairyleaf Viburnum, Tomentose Viburnum ●☆

408097　Viburnum sambucinum Reinw. ex Blume var. tomentosum Hallier f. = Viburnum inopinatum Craib ●

408098　Viburnum sargentii Koehne;鸡树条荚蒾（鸡树条子,天目琼花）;Sargent Viburnum ●

408099　Viburnum sargentii Koehne 'Flavum';黄果鸡树条荚蒾（黄果天目琼花）;Yellow-fruited Sargent Viburnum ●

408100　Viburnum sargentii Koehne 'Onondaga';奥内达加鸡树条荚蒾（奥农多加鸡树条荚蒾）;Onondaga Viburnum ●

408101　Viburnum sargentii Koehne 'Susquehanna';圆球鸡树条荚蒾;Sargent Viburnum , Susquehanna Viburnum ●☆

408102　Viburnum sargentii Koehne = Viburnum opulus L. var. calvescens（Rehder）H. Hara ●

408103　Viburnum sargentii Koehne = Viburnum opulus L. var. sargentii（Koehne）Takeda ●

408104　Viburnum sargentii Koehne f. calvescens（Rehder）Rehder = Viburnum opulus L. var. sargentii（Koehne）Takeda ●

408105　Viburnum sargentii Koehne f. calvescens（Rehder）Rehder = Viburnum opulus L. var. calvescens（Rehder）H. Hara ●

408106　Viburnum sargentii Koehne f. glabra Kom. = Viburnum opulus L. var. calvescens（Rehder）H. Hara ●

408107　Viburnum sargentii Koehne f. glabra Kom. = Viburnum opulus L. var. sargentii（Koehne）Takeda ●

408108　Viburnum sargentii Koehne f. puberulum Kom. = Viburnum opulus L. var. calvescens（Rehder）H. Hara f. puberulum（Kom.）Sugim. ●

408109　Viburnum sargentii Koehne f. puberulum Kom. = Viburnum opulus L. var. sargentii（Koehne）Takeda f. puberulum（Kom.）Sugim. ●

408110　Viburnum sargentii Koehne f. puberulum Kom. = Viburnum opulus L. var. sargentii（Koehne）Takeda ●

408111　Viburnum sargentii Koehne var. bracteatum Y. Q. Zhu;泰山琼花荚蒾;Taishan Sargent Arrowwood ●

408112　Viburnum sargentii Koehne var. calvescens Rehder = Viburnum opulus L. var. calvescens（Rehder）H. Hara ●

408113　Viburnum sargentii Koehne var. calvescens Rehder = Viburnum opulus L. var. sargentii（Koehne）Takeda ●

408114　Viburnum sargentii Koehne var. intermedium (Nakai) Kitag. = Viburnum opulus L. var. sargentii (Koehne) Takeda ●

408115　Viburnum sargentii Koehne var. puberulum (Kom.) Kitag. = Viburnum opulus L. var. calvescens (Rehder) H. Hara f. puberulum (Kom.) Sugim. ●

408116　Viburnum sargentii Koehne var. puberulum (Kom.) Kitag. = Viburnum opulus L. var. sargentii (Koehne) Takeda f. puberulum (Kom.) Sugim. ●

408117　Viburnum sargentii Koehne var. puberulum (Kom.) Kitag. = Viburnum opulus L. var. sargentii (Koehne) Takeda ●

408118　Viburnum schensianum Maxim. ;陕西荚蒾(冬兰条,冬栾条,鸡骨头,土兰条,土连材,土栾树,土栾条);Chinese Viburnum, Shaanxi Arrowwood,Shaanxi Viburnum ●

408119　Viburnum schensianum Maxim. subsp. chekiangense P. S. Hsu et P. L. Chiu = Viburnum schensianum Maxim. ●

408120　Viburnum schensianum Maxim. subsp. chekiangense P. S. Hsu et P. L. Chiu;浙江荚蒾;Zhejiang Arrowwood,Zhejiang Viburnum ●

408121　Viburnum schensianum Maxim. var. chekiangense (P. S. Hsu et P. L. Chiu) Y. Ren et W. Z. Di = Viburnum schensianum Maxim. ●

408122　Viburnum schneiderianum Hand. -Mazz. = Viburnum atrocyaneum C. B. Clarke et Diels ●

408123　Viburnum sempervirens K. Koch;常绿荚蒾(冬红果,坚荚树,苦柴枝,咸鱼汁树,猪妈柴);Evergreen Arrowwood, Evergreen Viburnum ●

408124　Viburnum sempervirens K. Koch var. trichophorum Hand. -Mazz. ;毛枝长绿荚蒾(具毛长绿荚蒾,毛坚荚蒾);Hairy-branch Evergreen Arrowwood,Hairy-branch Evergreen Viburnum ●

408125　Viburnum sessiliflorum Geddes = Viburnum odoratissimum Ker Gawl. var. sessiliflorum (Geddes) Fukuoka ●

408126　Viburnum sessiliflorum Geddes = Viburnum odoratissimum Ker Gawl. ●

408127　Viburnum setigerum Hance;茶荚蒾(垂果荚蒾,大雷公子,饭汤子,刚毛荚蒾,虎柴子,虎降子,鸡公柴,糯米树,糯树,跑路杆子,霜降子,水茶子,汤饭子,甜茶);Tea Arrowwood,Tea Viburnum ●

408128　Viburnum setigerum Hance var. sulcatum P. S. Hsu;沟核茶荚蒾(具沟刚毛荚蒾);Sulcate Tea Arrowwood,Sulcate Tea Viburnum ●

408129　Viburnum setigerum Hance var. sulcatum P. S. Hsu = Viburnum setigerum Hance ●

408130　Viburnum shweliense W. W. Sm. ;瑞丽荚蒾;Ruili Arrowwood, Ruili Viburnum ●

408131　Viburnum sieboldii Miq. ;西氏荚蒾(希博尔荚蒾,席氏荚蒾,沼生荚蒾);Siebold Arrowwood,Siebold Viburnum,Siebold's Arrowwood ●☆

408132　Viburnum sieboldii Miq. 'Seneca';塞内加西氏荚蒾(塞内加希博尔荚蒾);Siebold Viburnum ●☆

408133　Viburnum sieboldii Miq. subsp. obovatifolium (Yanagita) Kitam. = Viburnum sieboldii Miq. var. obovatifolium (Yanagita) Sugim. ●☆

408134　Viburnum sieboldii Miq. var. obovatifolium (Yanagita) Sugim. ;倒卵叶西氏荚蒾(倒卵叶席氏荚蒾);Obovateleaf Arrowwood ●☆

408135　Viburnum sikokianum Koidz. = Viburnum erosum Thunb. f. sikokianum (Koidz.) H. Hara ●☆

408136　Viburnum simonsii Hook. f. et Thomson = Viburnum odoratissimum Ker Gawl. ●

408137　Viburnum smithianum H. L. Li = Viburnum luzonicum Rolfe ●

408138　Viburnum smithii F. P. Metcalf = Viburnum luzonicum Rolfe ●

408139　Viburnum sphaerocarpum Y. C. Liu et C. H. Ou;球果荚蒾●

408140　Viburnum sphaerocarpum Y. C. Liu et C. H. Ou = Viburnum arboricola Hayata ●

408141　Viburnum squamulosum P. S. Hsu;瑶山荚蒾(细鳞荚蒾);Squamulose Viburnum,Yaoshan Arrowwood,Yaoshan Viburnum ●

408142　Viburnum stapfianum H. Lév. = Viburnum oliganthum Batalin ●

408143　Viburnum stellulatum Wall. = Viburnum mullaha Buch. -Ham. ex D. Don ●

408144　Viburnum stellulatum Wall. ex DC. var. glabrescens C. B. Clarke = Viburnum mullaha Buch. -Ham. ex D. Don var. glabrescens (C. B. Clarke) Kitam. ●

408145　Viburnum stellulatum Wall. var. glabrescens C. B. Clarke = Viburnum mullaha Buch. -Ham. ex D. Don var. glabrescens (C. B. Clarke) Kitam. ●

408146　Viburnum subalpinum Hand. -Mazz. ;亚高山荚蒾;Subalpine Arrowwood,Subalpine Viburnum ●

408147　Viburnum subalpinum Hand. -Mazz. var. limitaneum (W. W. Sm.) P. S. Hsu;边沿荚蒾;Limitaneous Subalpine Arrowwood, Limitaneous Subalpine Viburnum ●

408148　Viburnum subalpinum Hand. -Mazz. var. limitaneum (W. W. Sm.) P. S. Hsu = Viburnum subalpinum Hand. -Mazz. ●

408149　Viburnum subglabrum Hayata = Viburnum formosanum Hayata ●

408150　Viburnum sundaicum Miq. = Viburnum lutescens Blume ●

408151　Viburnum suspensum Lindl. ;悬垂荚蒾(比尤库荚蒾,长筒荚蒾);Ryukyu Viburnum,Sandankwa Viburnum ●☆

408152　Viburnum sympodiale Graebn. ;合轴荚蒾(假绣球);Sympodial Arrowwood,Sympodial Viburnum ●

408153　Viburnum taihasense Hayata = Viburnum betulifolium Batalin ●

408154　Viburnum taihasense Hayata = Viburnum morrisonense Hayata ●

408155　Viburnum taitoense Hayata;台东荚蒾;Taidong Arrowwood, Taidong Viburnum,Taitung Viburnum ●

408156　Viburnum taiwanianum Hayata;台湾荚蒾(台湾高山荚蒾);Taiwan Arrowwood,Taiwan Viburnum ●

408157　Viburnum taiwanianum Hayata = Viburnum urceolatum Siebold et Zucc. ●

408158　Viburnum taquetii H. Lév. = Viburnum erosum Thunb. var. taquetii (H. Lév.) Rehder ●

408159　Viburnum tashiroi Nakai;田代氏荚蒾●☆

408160　Viburnum tengyuehense (W. W. Sm.) P. S. Hsu;腾越荚蒾;Tengyue Arrowwood,Tengyue Viburnum ●

408161　Viburnum tengyuehense (W. W. Sm.) P. S. Hsu var. polyneurum (P. S. Hsu) P. S. Hsu;多脉腾越荚蒾;Manyvein Tengyue Arrowwood,Manyvein Tengyue Viburnum ●

408162　Viburnum ternatum Rehder;三叶荚蒾(三出叶荚蒾);Threeleaf Arrowwood,Threeleaf Viburnum,Trifoliate Viburnum ●

408163　Viburnum thaiyongense W. W. Sm. ;戴云山荚蒾(汕头荚蒾);Daiyunshan Arrowwood,Daiyunshan Viburnum ●

408164　Viburnum theiferum Rehder = Viburnum setigerum Hance ●

408165　Viburnum thibeticum C. Y. Wu et Y. F. Huang;西藏荚蒾;Tibet Viburnum,Xizang Arrowwood,Xizang Viburnum ●

408166　Viburnum thibeticum C. Y. Wu et Y. F. Huang = Viburnum erubescens Wall. ●

408167　Viburnum tinus L. ;地中海绵毛荚蒾(绵毛荚蒾,月桂荚蒾);Garden May, Laurustine, Laurustinus, Laurustinus Viburnum, May, Wild Bay ●☆

408168　Viburnum tinus L. 'Eve Price';伊夫·普里斯地中海绵毛荚蒾(普赖斯月桂荚蒾)●☆

408169　Viburnum tinus L. 'Gwenllian';格温里恩月桂荚蒾●☆

408170　Viburnum tinus L. 'Lucidum';光亮地中海绵毛荚蒾●☆

408171　Viburnum tinus L. 'Purpureum';紫叶地中海绵毛荚蒾●☆

408172 Viburnum tinus L. 'Spring Bouquet';春季花束地中海绵毛荚蒾;Spring Bouquet Viburnum ●☆

408173 Viburnum tinus L. 'Variegatum';斑叶地中海绵毛荚蒾●☆

408174 Viburnum tinus L. subsp. rigidum (Vent.) P. Silva;硬地中海绵毛荚蒾●☆

408175 Viburnum tinus L. var. mauritii Sennen = Viburnum tinus L. ●☆

408176 Viburnum tomentosum Thunb. = Viburnum hanceanum Maxim. ●

408177 Viburnum tomentosum Thunb. = Viburnum plicatum Thunb. var. tomentosum (Thunb.) Miq. ●

408178 Viburnum tomentosum Thunb. ex Murray 'Plicamm' = Viburnum plicatum Thunb. ●

408179 Viburnum tomentosum Thunb. ex Murray 'Sterile' = Viburnum plicatum Thunb. ●

408180 Viburnum tomentosum Thunb. ex Murray = Viburnum plicatum Thunb. f. tomentosum (Thunb. ex Murray) Rehder ●

408181 Viburnum tomentosum Thunb. ex Murray = Viburnum plicatum Thunb. ●

408182 Viburnum tomentosum Thunb. var. parvifolium (Miq.) Rehder = Viburnum plicatum Thunb. var. tomentosum (Thunb.) Miq. ●

408183 Viburnum tomentosum Thunb. var. plenum Rehder = Viburnum plicatum Thunb. ●

408184 Viburnum tomentosum Thunb. var. plicatum Maxim. = Viburnum plicatum Thunb. ●

408185 Viburnum tomentosum Thunb. var. sterile K. Koch = Viburnum plicatum Thunb. ●

408186 Viburnum touchanense H. Lév. = Viburnum foetidum Wall. var. rectangulatum (Graebn.) Rehder ●

408187 Viburnum trabeculosum C. Y. Wu ex P. S. Hsu;横脉荚蒾;Acrossvein Arrowwood,Trabeculose Viburnum ●

408188 Viburnum trilobum Marshall;三裂叶荚蒾;American Cranberry Bush, American Cranberry-bush, American Highbush Cranberry, Cranberry, Cranberry Bush, Cranberrybush Viburnum, High Bush-cranberry,High Cranberry,Highbush Cranberry ●☆

408189 Viburnum trilobum Marshall 'Bailey Compact';贝蕾紧凑三裂叶荚蒾●☆

408190 Viburnum trilobum Marshall 'Compactum';紧凑三裂叶荚蒾●☆

408191 Viburnum trilobum Marshall 'Wentworth';温特沃斯三裂叶荚蒾;American Cranberry Bush Viburnum ●☆

408192 Viburnum trilobum Marshall = Viburnum opulus L. subsp. trilobum (Marshall) R. T. Clausen ●☆

408193 Viburnum triplinerve Hand.-Mazz.;三脉叶荚蒾(三脉荚蒾);Threeveined Arrowwood,Trinerves Viburnum,Triplinerved Viburnum ●

408194 Viburnum tsangii Rehder = Viburnum hainanense Merr. et Chun ●

408195 Viburnum tsangii Rehder f. xanthocarpum Rehder = Viburnum hainanense Merr. et Chun ●

408196 Viburnum tubulosum P. S. Hsu;管花荚蒾;Tubeflower Arrowwood,Tube-flower Viburnum ●

408197 Viburnum tubulosum P. S. Hsu = Viburnum taitoense Hayata ●

408198 Viburnum urceolatum Siebold et Zucc.;壶花荚蒾;Asidiform Viburnum,Urceolar Arrowwood,Urceolate Viburnum ●

408199 Viburnum urceolatum Siebold et Zucc. f. procumbens (Nakai) H. Hara;平卧壶花荚蒾●☆

408200 Viburnum urceolatum Siebold et Zucc. var. procumbens Nakai = Viburnum urceolatum Siebold et Zucc. f. procumbens (Nakai) H. Hara ●☆

408201 Viburnum utile Hemsl. = Viburnum utile Hemsl. ex Forbes et Hemsl. ●

408202 Viburnum utile Hemsl. ex Forbes et Hemsl.;烟管荚蒾(黑擦子树花,黑汉条,灰猫条,灰毛条,灰猪藤,冷饭团,牛屎柴,羊奶根,羊尿子,羊舌条,羊石子,羊食子,羊屎柴,羊屎条,羊屎子,洋石子,有用荚蒾);Flue Arrowwood,Service Viburnum,Useful Viburnum ●

408203 Viburnum utile Hemsl. ex Forbes et Hemsl. 'Eskimo';爱斯基摩烟管荚蒾;Eskimo,Viburnum ●☆

408204 Viburnum utile Hemsl. ex Forbes et Hemsl. var. elaeagnifolium Rehder = Viburnum chinshanense Graebn. ●

408205 Viburnum utile Hemsl. ex Forbes et Hemsl. var. minor Pamp. = Viburnum utile Hemsl. ex Forbes et Hemsl. ●

408206 Viburnum utile Hemsl. var. elaeagnifolium Rehder = Viburnum chinshanense Graebn. ●

408207 Viburnum utile Hemsl. var. minor Pamp. = Viburnum utile Hemsl. ex Forbes et Hemsl. ●

408208 Viburnum utile Hemsl. var. ningqiangense Y. Ren et W. Z. Di = Viburnum utile Hemsl. ex Forbes et Hemsl. ●

408209 Viburnum valerianicum Elmer = Viburnum propinquum Hemsl. ●

408210 Viburnum veitchii C. H. Wright;维奇荚蒾;Chinese Wayfaring Tree,Flat-topped Viburnum,Veitch Viburnum ●☆

408211 Viburnum veitchii C. H. Wright = Viburnum glomeratum Maxim. ●

408212 Viburnum veitchii C. H. Wright subsp. magnificum P. S. Hsu = Viburnum glomeratum Maxim. ●

408213 Viburnum veitchii C. H. Wright subsp. magnificum P. S. Hsu = Viburnum glomeratum Maxim. subsp. magnificum (P. S. Hsu) P. S. Hsu ●

408214 Viburnum veitchii C. H. Wright subsp. rotundifolium P. S. Hsu = Viburnum glomeratum Maxim. ●

408215 Viburnum veitchii C. H. Wright subsp. rotundifolium P. S. Hsu = Viburnum glomeratum Maxim. subsp. rotundifolium (P. S. Hsu) P. S. Hsu ●

408216 Viburnum venulosum Benth. = Viburnum sempervirens K. Koch ●

408217 Viburnum vetteri Zabel = Viburnum lentago L. ●☆

408218 Viburnum villosifolium Hayata = Viburnum erosum Thunb. ●

408219 Viburnum wightianum Wall. = Viburnum erubescens Wall. ●

408220 Viburnum willeanum Graebn. = Viburnum betulifolium Batalin ●

408221 Viburnum wilsonii Rehder;西南荚蒾;E. H. Wilson Arrowwood, E. H. Wilson Viburnum ●

408222 Viburnum wilsonii Rehder = Viburnum hupehense Rehder ●

408223 Viburnum wilsonii Rehder var. adenophorum (W. W. Sm.) Hand.-Mazz. = Viburnum hupehense Rehder ●

408224 Viburnum wrightii Miq.;浙皖荚蒾(赖特氏荚蒾);Wright Arrowwood,Wright Viburnum ●

408225 Viburnum wrightii Miq. = Viburnum hengshanicum Tsiang ex P. S. Hsu ●

408226 Viburnum wrightii Miq. f. eglandulosum (Nakai) Hiyama = Viburnum wrightii Miq. var. stipellatum Nakai ●☆

408227 Viburnum wrightii Miq. f. minus (Nakai) Sugim.;小浙皖荚蒾●☆

408228 Viburnum wrightii Miq. f. nikoense Hiyama = Viburnum dilatatum Thunb. f. nikoense (Hiyama) H. Hara ●☆

408229 Viburnum wrightii Miq. f. sylvestre (Koidz.) Hiyama = Viburnum wrightii Miq. var. stipellatum Nakai ●☆

408230 Viburnum wrightii Miq. var. lucidum Hatus.;亮浙皖荚蒾●☆

408231 Viburnum wrightii Miq. var. stipellatum Nakai;小托叶荚蒾●☆

408232 Viburnum wrightii Miq. var. stipellatum Nakai f. kaiense Hiyama = Viburnum wrightii Miq. var. stipellatum Nakai ●☆

408233 Viburnum wrightii Miq. var. sylvestre Koidz. = Viburnum wrightii Miq. var. stipellatum Nakai ●☆

408234　Viburnum yamadae Bartlett et Yamam. = Viburnum parvifolium Hayata ●

408235　Viburnum yunnanense Rehder;云南荚蒾(黄香根);Yunnan Arrowwood,Yunnan Viburnum ●

408236　Vicarya Stocks = Myriopteron Griff. ●

408237　Vicarya Wall. ex Voigt = Myriopteron Griff. ●

408238　Vicatia DC. (1830);凹乳芹属;Vicatia ■

408239　Vicatia achilleifolia (DC.) P. K. Mukh. = Meeboldia achilleifolia (DC.) P. K. Mukh. et Constance ■

408240　Vicatia atrosanguinea (Kar. et Kir.) P. K. Mukh. et Pimenov = Carum atrosanguineum Kar. et Kir. ■

408241　Vicatia bipinnata R. H. Shan et F. T. Pu;少裂凹乳芹(土当归);Bipinnate Vicatia ■

408242　Vicatia coniifolia (Wall.) DC. = Vicatia coniifolia Wall. ex DC. ■

408243　Vicatia coniifolia DC. = Vicatia coniifolia Wall. ex DC. ■

408244　Vicatia coniifolia Wall. ex DC.;凹乳芹(沟果芹);Poisonhemlockleaf Vicatia,Vicatia ■

408245　Vicatia millefolia (Klotzsch) C. B. Clarke = Vicatia coniifolia Wall. ex DC. ■

408246　Vicatia millefolia C. B. Clarke = Vicatia coniifolia Wall. ex DC. ■

408247　Vicatia thibetica H. Boissieu;西藏凹乳芹(独脚当归,西康凹乳芹,野当归);Tibet Vicatia,Xizang Vicatia ■

408248　Vicatia wolffiana (Fedde ex H. Wolff) C. Norman;沃尔夫凹乳芹■☆

408249　Vicatia wolffiana (H. Wolff ex Fedde) C. Norman = Vicatia wolffiana (Fedde ex H. Wolff) C. Norman ■☆

408250　Vicentia Allemáo = Terminalia L. (保留属名)●

408251　Vicia L. (1753);野豌豆属(蚕豆属,巢菜属);Tare, Tufted Vetch,Vetch,Wild Pea ■

408252　Vicia abbreviata (S. H. Fu et F. H. Chen) Z. D. Xia;短序野豌豆(辽野豌豆);Short Inflorescens Vetch ■

408253　Vicia abyssinica Alef. = Vicia sativa L. var. angustifolia (L. ex Reichard) Wahlenb. ■☆

408254　Vicia albiflora Z. D. Xia;白野豌豆;Whiteflower Vetch ■

408255　Vicia alpestris Stev. ;高山野豌豆■☆

408256　Vicia altissima Desf. ;高大野豌豆■☆

408257　Vicia americana Muhl. ex Willd. ;美国野豌豆;American Tare, American Vetch,Chick Pea,Purple Vetch ■☆

408258　Vicia americana Muhl. ex Willd. subsp. oregana (Nutt.) Abrams = Vicia americana Muhl. ex Willd. ■☆

408259　Vicia americana Muhl. ex Willd. var. oregana (Nutt.) A. Nelson = Vicia americana Muhl. ex Willd. ■☆

408260　Vicia americana Muhl. ex Willd. var. truncata (Nutt.) W. H. Brewer = Vicia americana Muhl. ex Willd. ■☆

408261　Vicia americana Muhl. ex Willd. var. villosa (Kellogg) F. J. Herm. = Vicia americana Muhl. ex Willd. ■☆

408262　Vicia amoena Fisch. ex DC. ;山野豌豆(草藤,东北透骨草,豆豆苗,豆碗碗,涝豆秧,芦豆苗,落豆秧,山豆苗,山黑豆,山豌豆,宿根草藤,宿根巢菜,宿根苕子,透骨草);Broadleaf Vetch,Wild Vetch ■

408263　Vicia amoena Fisch. ex DC. f. alba Sugim. ;白色山野豌豆■☆

408264　Vicia amoena Fisch. ex DC. f. albiflora P. Y. Fu et Y. A. Chen;白花山野豌豆;Whiteflower Broadleaf Vetch ■

408265　Vicia amoena Fisch. ex DC. var. angusta Freyn. = Vicia amoena Fisch. ex DC. ■

408266　Vicia amoena Fisch. ex DC. var. macrophylla Litv. ;大野山野豌豆;Largeleaf Vetch ■

408267　Vicia amoena Fisch. ex DC. var. macrophylla Litv. = Vicia amoena Fisch. ex DC. ■

408268　Vicia amoena Fisch. ex DC. var. macrophylla Litv. ex B. Festsch. = Vicia amoena Fisch. ex DC. ■

408269　Vicia amoena Fisch. ex DC. var. oblongifolia Regel;狭叶山野豌豆(芦豆苗);Oblongleaf Vetch ■

408270　Vicia amoena Fisch. ex DC. var. oblongifolia Regel = Vicia amoena Fisch. ex DC. ■

408271　Vicia amoena Fisch. ex DC. var. pubescens Turcz. ;柔毛山野豌豆;Pubescent Vetch ■

408272　Vicia amoena Fisch. ex DC. var. sericea Kitag. ;绢毛山野豌豆(毛山野豌豆);Sericeous Vetch ■

408273　Vicia amoena Fisch. ex Ser. f. alba Sugim. = Vicia amoena Fisch. ex DC. f. alba Sugim. ■☆

408274　Vicia amphicarpa Dorthes;圆果巢菜■☆

408275　Vicia amuransis Oett. f. senheensis Y. Q. Jiang et S. M. Fu = Vicia amuransis Oett. var. alba H. Ohasi et Tateishi ■

408276　Vicia amuransis Oett. var. alba H. Ohasi et Tateishi;白花黑龙江野豌豆(三河野豌豆);Sanhe Vetch,White Flower Amur Vetch ■

408277　Vicia amuransis Oett. var. pratensi (Kom.) Hara = Vicia amuransis Oett. ■

408278　Vicia amurensis Oett. ;黑龙江野豌豆(大巢菜,透骨草,圆叶草藤);Amur Vetch,Heilongjiang Vetch ■

408279　Vicia amurensis Oett. f. alba H. Ohashi et Tateishi = Vicia amuransis Oett. var. alba H. Ohasi et Tateishi ■

408280　Vicia amurensis Oett. f. alborosea Sugim. ;粉白黑龙江野豌豆■

408281　Vicia amurensis Oett. f. sanheensis Y. Q. Jiang et S. M. Fu = Vicia amurensis Oett. f. alba H. Ohashi et Tateishi ■

408282　Vicia amurensis Oett. var. pallida (Trautv.) Kitag. = Vicia japonica A. Gray var. pallida (Trautv.) H. Hara ■☆

408283　Vicia angustifolia L. = Vicia sativa L. subsp. nigra (L.) Ehrh. ■☆

408284　Vicia angustifolia L. ex Reichard;窄叶野豌豆(大巢菜,大巢豆,大巢叶,救荒野豌豆,苦豆子,闹豆子,山豆子,铁豆秧,狭叶野豌豆,野绿豆,野豌豆,紫花苕子);Narrowleaf Vetch, Narrowleaved Vetch,Narrow-leaved Vetch,Summer Vetch ■

408285　Vicia angustifolia L. ex Reichard = Vicia sativa L. subsp. nigra (L.) Ehrh. ■☆

408286　Vicia angustifolia L. ex Reichard var. minor (Bertol.) Ohwi = Vicia sativa L. subsp. nigra (L.) Ehrh. var. minor (Bertol.) Gaudin ■☆

408287　Vicia angustifolia L. ex Reichard var. segetalis (Thuill.) K. Koch = Vicia sativa L. subsp. nigra (L.) Ehrh. ■☆

408288　Vicia angustifolia L. ex Reichard var. segetalis (Thuill.) Koch;箭头巢菜■☆

408289　Vicia angustifolia L. ex Reichard var. segetalis (Thuill.) Koch f. albiflora Honda;白花箭头巢菜■☆

408290　Vicia angustifolia L. ex Reichard var. segetalis (Thuill.) Koch f. normalis (Makino) Ohwi = Vicia sativa L. subsp. nigra (L.) Ehrh. f. normalis (Makino) Kitam. ■☆

408291　Vicia angustifolia L. var. segetalis (Thuill.) W. D. J. Koch = Vicia sativa L. subsp. nigra (L.) Ehrh. ■☆

408292　Vicia angustifolia L. var. uncinata (Desf. ex Nyman) Rouy = Vicia sativa L. subsp. nigra (L.) Ehrh. ■☆

408293　Vicia angustifolia Roth = Vicia angustifolia L. ex Reichard ■

408294　Vicia angustiunguiculata Z. D. Xia;窄爪野豌豆■

408295　Vicia antiqua Grossh. ;古老野豌豆■☆

408296　Vicia apoda (Maxim.) Z. D. Xia;短序歪头菜;Shartspike Pair Vetch,Shortinflorescens Vetch ■

408297　Vicia articulata Hornem. = Vicia monantha Retz. ■☆

408298　Vicia atlantica Pomel = Vicia ochroleuca Ten. subsp. atlantica（Pomel）Greuter et Burdet ■☆

408299　Vicia atlantica Pomel var. mesatlantica（Maire）Pau et Font Quer = Vicia tenuifolia Roth subsp. villosa（Batt.）Greuter ■

408300　Vicia atropurpurea Desf. = Vicia benghalensis L. ■☆

408301　Vicia baborensis Batt. et Trab. = Vicia ochroleuca Ten. subsp. baborensis（Batt. et Trab.）Greuter et Burdet ■☆

408302　Vicia baicalensis（Turcz.）B. Fedtsch.；老豆秧（贝加野豌豆，野崂豆）；Baivcale Vetch ■

408303　Vicia bakeri Ali；察隅野豌豆；Baker Vetch，Chayu Vetch ■

408304　Vicia balansae Boiss.；巴兰野豌豆■☆

408305　Vicia benghalensis L.；紫野豌豆；Purple Vetch，Reddish Tufted Vetch ■☆

408306　Vicia benghalensis L. subsp. aquitana（Clavaud）Quézel et Santa = Vicia benghalensis L. ■☆

408307　Vicia benghalensis L. subsp. atropurpurea（Desf.）Maire = Vicia benghalensis L. ■☆

408308　Vicia benghalensis L. subsp. heterocalyx（Maire et Weiller）Hormat；异萼紫野豌豆■☆

408309　Vicia benghalensis L. var. atropurpurea（Desf.）Maire = Vicia benghalensis L. ■☆

408310　Vicia benghalensis L. var. heterocalyx Maire et Weiller = Vicia benghalensis L. subsp. heterocalyx（Maire et Weiller）Hormat ■☆

408311　Vicia benghalensis L. var. perennis（DC.）Fiori = Vicia benghalensis L. ■☆

408312　Vicia bequaertii De Wild. = Vicia paucifolia Baker ■☆

408313　Vicia biebersteinii Besser ex M. Bieb.；毕伯氏巢菜■☆

408314　Vicia biennia L.；二年生野豌豆■☆

408315　Vicia biflora Desf. = Vicia monantha Retz. ■☆

408316　Vicia biflora Desf. subsp. calcarata（Desf.）Maire = Vicia monantha Retz. subsp. calcarata（Desf.）Romero Zarco ■☆

408317　Vicia biflora Desf. subsp. cinerea（Munby）Maire = Vicia monantha Retz. ■☆

408318　Vicia biflora Desf. var. marmorata Maire et Weiller = Vicia monantha Retz. ■☆

408319　Vicia biflora Desf. var. trichocarpa Maire = Vicia monantha Retz. ■☆

408320　Vicia bifolia Nakai = Vicia unijuga A. Braun ■

408321　Vicia bifurcata Z. D. Xia；苞叶野豌豆；Twice Forked Vetch ■

408322　Vicia bithynica（L.）L.；小亚细亚巢菜；Bithynian Vetch ■☆

408323　Vicia bithynica（L.）L. var. major Arcang. = Vicia bithynica（L.）L. ■☆

408324　Vicia bithynica L. = Vicia bithynica（L.）L. ■☆

408325　Vicia bobartii E. Forst.；包巴氏巢菜■☆

408326　Vicia boissieri Freyn；布瓦西耶野豌豆■☆

408327　Vicia bungei Ohwi；大花野豌豆（老豆蔓，毛苕子，三齿草藤，三齿萼野豌豆，三齿野豌豆，山黧豆，野豌豆）；Bigflower Vetch，Bunge Vetch ■

408328　Vicia calcarata Desf.；距野豌豆■☆

408329　Vicia calcarata Desf. = Vicia monantha Retz. ■☆

408330　Vicia calcarata Desf. var. biflora（Desf.）Batt. = Vicia monantha Retz. ■☆

408331　Vicia calcarata Desf. var. cossoniana Batt. = Vicia monantha Retz. ■☆

408332　Vicia calcarata Desf. var. marmarica Asch. et Schweinf. = Vicia monantha Retz. ■☆

408333　Vicia californica Greene = Vicia americana Muhl. ex Willd. ■☆

408334　Vicia californica Greene var. madrensis Jeps. = Vicia americana Muhl. ex Willd. ■☆

408335　Vicia capensis P. J. Bergius = Lessertia capensis（P. J. Bergius）Druce ■☆

408336　Vicia caroliniana Walter；卡罗来纳野豌豆（卡罗莱纳野豌豆）；Carolina Vetch，Pale Vetch，Wood Vetch ■☆

408337　Vicia cassubica L.；扭曲巢菜；Danzig Vetch ■☆

408338　Vicia chianschanensis（P. Y. Fu et Y. A. Chen）Z. D. Xia；千山野豌豆（山绿豆）；Chianshan Vetch，Qianshan Vetch ■

408339　Vicia chinensis Franch.；华野豌豆；China Vetch，Chinese Vetch ■

408340　Vicia chinensis Franch. var. angustifolia Z. D. Xia；窄叶华野豌豆；Narrowleaf Chinese Vetch ■

408341　Vicia chinensis Franch. var. longiracemosa Z. D. Xia；长序华野豌豆；Longracemose Vetch ■

408342　Vicia chosenensis Ohwi；朝鲜野豌豆■☆

408343　Vicia ciliatula Lipsky；缘毛野豌豆■☆

408344　Vicia cinerea M. Bieb. = Vicia monantha Retz. ■☆

408345　Vicia cirrhosa Webb et Berthel.；卷须野豌豆■☆

408346　Vicia claessensii De Wild. = Vicia paucifolia Baker ■☆

408347　Vicia cordata Hoppe = Vicia sativa L. subsp. cordata（Hoppe）Batt. ■☆

408348　Vicia cordata Wulfen ex Hoppe；心叶巢菜；Heartleaf Vetch ■☆

408349　Vicia cossoniana Batt. = Vicia monantha Retz. subsp. calcarata（Desf.）Romero Zarco ■☆

408350　Vicia costata Ledeb.；新疆野豌豆（白花野豌豆，肋脉野豌豆）；Ribbed Vetch，Xinjiang Vetch ■

408351　Vicia costata Ledeb. var. angusta Z. D. Xia；窄叶新疆野豌豆；Narrowleaf Ribbed Vetch ■

408352　Vicia cracca L.；广布野豌豆（草藤，多花野豌豆，肥田草，落豆秧，苕，细叶落豆秧）；Bird Vetch，Blue Girsf，Blue Tar Fitch，Blue Tar-fitch，Blue Vetch，Bush Vetch，Canada Pea，Cat Pea，Common Vetch，Cow Vetch，Crow Vetch，Fingers-and-thumbs，Fitchacks，Gerard Vetch，Goose-and-gander，Huggaback，Mouse Pea，Mouse's Pease，Tar Grass，Tare Fitch，Tare Vetch，Tare-fitch，Tare-grass，Tine，Tine-grass，Tine-weed，Titatch，Tufted Vetch，Twine-grass，Wild Fetch，Wild Tare，Wild Thetch-grass，Wild Vetch ■

408353　Vicia cracca L. f. albida Peterm. = Vicia cracca Ledeb. var. albiflora Trautv. ■☆

408354　Vicia cracca L. f. canescens Maxim. = Vicia cracca Ledeb. var. canescens Maxim. ex Franch. et Sav. ■☆

408355　Vicia cracca L. f. leucantha Nakai = Vicia cracca L. f. albida Peterm. ■☆

408356　Vicia cracca L. subsp. tenuifolia（Roth）Gaudin；细叶广布野豌豆；Bird Vetch，Cow Vetch ■☆

408357　Vicia cracca L. subsp. tenuifolia（Roth）Gaudin = Vicia tenuifolia Roth ■

408358　Vicia cracca L. var. albiflora Trautv.；白花广布野豌豆；Whiteflower Bird Vetch ■

408359　Vicia cracca L. var. canescens（Maxim.）Maxim. ex Franch. et Sav. = Scaevola taccada（Gaertn.）Roxb. f. moomomiana（O. Deg. et Greenwell）T. Yamaz. ■☆

408360　Vicia cracca L. var. canescens Maxim. ex Franch. et Sav.；灰野豌豆；Grey Bird Vetch ■

408361　Vicia cracca L. var. japonica Miq.；日本广布野豌豆；Japanese Bule Vetch ■

408362　Vicia cracca L. var. lilacina（Ledeb.）Krylov = Vicia lilacina

Ledeb. ■

408363 Vicia cracca L. var. tenuifolia（Roth）Beck ＝ Vicia cracca L. subsp. tenuifolia（Roth）Gaudin ■☆

408364 Vicia cracca L. var. tenuifolia Beck ＝ Vicia tenuifolia Roth ■

408365 Vicia cracca Ledeb. var. albiflora Trautv. ＝ Vicia cracca L. var. albiflora Trautv. ■

408366 Vicia cracca Ledeb. var. canescens Maxim. ex Franch. et Sav. ＝ Vicia cracca L. var. canescens Maxim. ex Franch. et Sav. ■

408367 Vicia cracca Ledeb. var. japonica Miq. ＝ Vicia cracca L. var. japonica Miq. ■

408368 Vicia cracca Poir. ＝ Vicia villosa Roth ■

408369 Vicia cuneata Guss. ＝ Vicia sativa L. subsp. nigra（L.）Ehrh. ■☆

408370 Vicia dalmatica J. Kern. ;达玛巢菜 ■☆

408371 Vicia dasycarpa Ten. ;毛果野豌豆（苕子）;Thickfruit Vetch, Woolly-pod Vetch ■

408372 Vicia dasycarpa Ten. ＝ Vicia villosa Roth ■

408373 Vicia dasycarpa Ten. var. glabrescens（Koch）Beck ＝ Vicia villosa Roth subsp. varia（Host）Corb. ■

408374 Vicia deflexa Nakai ;弯折巢菜（羽叶野豌豆）;Deflexed Vetch ■

408375 Vicia deflexa Nakai ＝ Vicia venosa（Willd. ex Link）Maxim. var. cuspidata Maxim. ■

408376 Vicia delmasii Emb. et Maire ;戴尔马野豌豆 ■☆

408377 Vicia dichroantha Diels ;二色野豌豆（高原苕子）;Bicolor Vetch, Coloured-flower Vetch ■

408378 Vicia disperma DC. ;异籽野豌豆 ;European Vetch ■☆

408379 Vicia disperma DC. var. suberviformis（Maire）Raynaud ＝ Vicia disperma DC. ■☆

408380 Vicia disperma DC. var. subuniflora Pau ＝ Vicia disperma DC. ■☆

408381 Vicia dumetorum L. ;灌丛巢菜 ;German Vetch ■☆

408382 Vicia durandii Boiss. ＝ Vicia altissima Desf. ■☆

408383 Vicia ecirrhosa Rupr. ex Boiss. ;无须野豌豆 ■☆

408384 Vicia edentata F. T. Wang et Ts. Tang ;无萼齿野豌豆 ;Toothless Vetch ■

408385 Vicia edentata F. T. Wang et Ts. Tang ＝ Vicia kulingiana L. H. Bailey ■

408386 Vicia edentata F. T. Wang et Ts. Tang f. minima L. L. Lou ;小叶无萼齿野豌豆 ;Little-leaf Toothless Vetch ■

408387 Vicia elegans Guss. ;雅致野豌豆 ■☆

408388 Vicia elegans Guss. ＝ Vicia tenuifolia Roth subsp. elegans（Guss.）Nyman ■☆

408389 Vicia embergeri Font Quer et Maire ＝ Vicia lecomtei Humbert et Maire subsp. embergeri（Font Quer et Maire）Maire ■☆

408390 Vicia eriocarpa（Hausskn.）Halácsy ＝ Vicia villosa Roth subsp. eriocarpa（Hausskn.）P. W. Ball ■☆

408391 Vicia erviformis Boiss. ＝ Vicia vicioides（Desf.）Cout. ■☆

408392 Vicia ervilia（L.）Willd. ;欧维野豌豆（苦野豌豆）;Bitter Vetch, Wild Bittervetch ■☆

408393 Vicia faba L. ;蚕豆（大豆,佛豆,寒豆,胡豆,柤豆,罗泛豆,罗汉豆,马齿豆,南豆,竖豆,田豆,湾豆,夏豆,仙豆）;Bean, Broad Bean, Broadbean, English Bean, Fava Bean, Field Bean, Horse Bean, Horsebean, House Bean, Household Bean, Mait-banes, Scotch Bean, Silkwormbean, Windsor Bean ■

408394 Vicia faba L. var. equina Pers. ;马蚕豆 ;Horse Bean, Pigeon Bean ■☆

408395 Vicia faba L. var. minor（Peterm.）;小蚕豆 ;Small Horse Bean ■☆

408396 Vicia faba L. var. minor（Peterm.）＝ Vicia faba L. ■

408397 Vicia faba L. var. pliniana Trab. ＝ Vicia faba L. ■

408398 Vicia fairchildiana Maire ;费尔豌豆 ■☆

408399 Vicia fauriei Franch. ;福氏野豌豆 ■☆

408400 Vicia fauriei Franch. var. unijuga Matsum. ＝ Vicia unijuga A. Braun ■

408401 Vicia fedtschenkoana V. V. Nikitin ;费氏野豌豆 ■☆

408402 Vicia ferreirensis Goyder ;费雷尔豌豆 ■☆

408403 Vicia filicaulis Webb et Berthel. ;线茎野豌豆 ■☆

408404 Vicia fulgens Batt. ;光亮豌豆 ■☆

408405 Vicia garbiensis Font Quer et Pau ＝ Vicia villosa Roth subsp. garbiensis（Font Quer et Pau）Maire ■☆

408406 Vicia geminiflora Trautv. ;索伦野豌豆（三尺草藤,双花野豌豆）;Suolun Vetch, Twoflowers Vetch ■

408407 Vicia gigantea Bunge ;大野豌豆（大巢菜,山木樨,薇,薇菜,薇山扁豆,野豌豆）;Giant Vetch, Sikta Vetch ■

408408 Vicia glauca C. Presl ;灰蓝野豌豆 ■☆

408409 Vicia glauca C. Presl subsp. giennensis（Cuatrec.）Blanca et F. Valle ;日安野豌豆 ■☆

408410 Vicia glauca C. Presl var. anremerica Maire ＝ Vicia glauca C. Presl ■☆

408411 Vicia glauca C. Presl var. aurasiaca Maire ＝ Vicia glauca C. Presl ■☆

408412 Vicia glauca C. Presl var. ayachica Emb. ＝ Vicia glauca C. Presl ■☆

408413 Vicia glauca C. Presl var. mesatlantica Emb. et Maire ＝ Vicia glauca C. Presl ■☆

408414 Vicia glauca C. Presl var. montisferrati Maire ＝ Vicia glauca C. Presl ■☆

408415 Vicia glauca C. Presl var. rerayensis Ball ＝ Vicia glauca C. Presl ■☆

408416 Vicia glauca C. Presl var. rifana Maire ＝ Vicia glauca C. Presl ■☆

408417 Vicia gracilior Popov ;纤细野豌豆 ■☆

408418 Vicia gracilis Loisel. ;美巢菜 ;Slender Vetch ■☆

408419 Vicia grandiflora Scop. ;大花巢菜 ;Bigflower Vetch, Large Yellow Vetch, Showy Vetch ■☆

408420 Vicia grossheimii Ekutim. ;格罗野豌豆 ■☆

408421 Vicia hajastana Grossh. ;哈贾斯坦野豌豆 ■☆

408422 Vicia heterophylla C. Presl ＝ Vicia sativa L. subsp. nigra（L.）Ehrh. ■☆

408423 Vicia hirsuta（L.）Gray ;小巢菜（白花苕菜,白翘摇,漂摇草,飘摇草,翘摇,翘摇车,雀野豆,雀野豌豆,苕,苕饶,苕子,薇,小巢豆,小野麻豌,小野麻豌豆,摇车,野蚕豆,野豌豆,硬毛果野豌豆,元修菜,柱夫）;Bindweed, Dill, Dother, Hairy Tare, Hairy Vetch, Hirsute Vetch, Lintels, Mouse Pea, Pigeon Vetch, Rough-podded Tine Tare, Rough-podded Tine-tare, Strangle Tare, Strangler Vetch, Tare, Tare Fitch, Tare Vetch, Tare-fitch, Tare-grass, Tare-Vetch, Tine, Tine Tare, Tine-grass, Tine-tare, Tine-weed, Tiny Vetch, Titters, Viciu Cracca, Wild Fitch, Wild Thetch-grass ■

408424 Vicia hirsuta（L.）Gray var. cyrenaica Maire et Weiller ＝ Vicia hirsuta（L.）Gray ■

408425 Vicia hirsuta（L.）Gray var. hefeiana J. Q. He ;合肥小巢菜 ;Hefei Vetch ■

408426 Vicia hirta DC. ＝ Vicia lutea L. ■☆

408427 Vicia hololasia Woronow ;全毛野豌豆 ■☆

408428 Vicia hugeri Small ＝ Vicia caroliniana Walter ■☆

408429 Vicia hybrida L. ;杂种巢菜 ;Hairy Yellow Vetch, Hairy Yellow-vetch ■☆

408430 Vicia hybrida L. var. cyrenaica Maire et Weiller ＝ Vicia hybrida L. ■☆

408431 Vicia hyrcanica Fisch. et C. A. Mey. ;希尔康野豌豆 ■☆

408432 Vicia iberica Grossh. ;伊比利亚野豌豆 ■☆

408433　Vicia incisa M. Bieb. ;锐裂巢菜■☆

408434　Vicia iranica Boiss. ;伊朗野豌豆■☆

408435　Vicia japonica A. Gray；东方野豌豆（日本野豌豆）；Beach Pea，Japan Vetch，Oriental Vetch，Wood Vetch ■

408436　Vicia japonica A. Gray f. albiflora Honda;白花东方野豌豆■☆

408437　Vicia japonica A. Gray subsp. amurensis（Oett.）Kitam. = Vicia amurensis Oett. ■

408438　Vicia japonica A. Gray subsp. pallida（Trautv.）Vorosch. = Vicia japonica A. Gray var. pallida（Trautv.）H. Hara■☆

408439　Vicia japonica A. Gray var. comosa H. Boissieu;簇毛东方野豌豆■☆

408440　Vicia japonica A. Gray var. laxiracemis Ohwi = Vicia japonica A. Gray var. pallida（Trautv.）H. Hara■☆

408441　Vicia japonica A. Gray var. laxiracemis Ohwi = Vicia japonica A. Gray ■

408442　Vicia japonica A. Gray var. pallida（Trautv.）H. Hara;苍白东方野豌豆■☆

408443　Vicia japonica A. Gray var. pratensis Kom. = Vicia amurensis Oett. ■

408444　Vicia japonica A. Gray var. silvatica Kom. = Vicia amurensis Oett. ■

408445　Vicia johannis Tamamsch. ;约翰野豌豆■☆

408446　Vicia kioshanica L. H. Bailey;确山野豌豆（确山巢菜,山豆根）;Kioshan Vetch,Queshan Vetch ■

408447　Vicia kitaibeliana Koch;基陶伊贝尔巢菜■☆

408448　Vicia kokanica Regel et Schmalh. ;浩罕野豌豆■☆

408449　Vicia kulingiana L. H. Bailey;牯岭野豌豆（红花豆,山蚕豆,山绿豆,四叶豆）;Guling Vetch ■

408450　Vicia laevigata Sm. = Vicia lutea L. ■☆

408451　Vicia lagopus Pomel = Vicia vicioides（Desf.）Cout. ■☆

408452　Vicia lathyroides L. ;山鼍豆巢菜;Chichling Vetch, Spring Vetch ■☆

408453　Vicia latibracteolata K. T. Fu;宽苞野豌豆（三昆草藤）;Broadbract Vetch ■

408454　Vicia latibracteolata K. T. Fu var. acerosa K. T. Fu;针苞野豌豆;Needlebract Vetch ■

408455　Vicia latibracteolata K. T. Fu var. acerosa K. T. Fu = Vicia latibracteolata K. T. Fu ■

408456　Vicia latiuniquiculata Z. D. Xia;宽爪野豌豆;Broad-claw Vetch ■

408457　Vicia latiuniquiculata Z. D. Xia f. albiflora Z. D. Xia;白花宽爪野豌豆;Whiteflower Broad-claw Vetch ■

408458　Vicia laxiflora Boiss. ;疏花巢菜;Slender Vetch ■☆

408459　Vicia laxiflora Brot. = Vicia parviflora Cav. ■☆

408460　Vicia lecomtei Humbert et Maire;勒孔特野豌豆■☆

408461　Vicia lecomtei Humbert et Maire subsp. embergeri（Font Quer et Maire）Maire;勒恩■☆

408462　Vicia lecomtei Humbert et Maire var. dolichocarpa Font Quer = Vicia lecomtei Humbert et Maire ■☆

408463　Vicia lecomtei Humbert et Maire var. embergeri（Font Quer et Maire）Font Quer = Vicia lecomtei Humbert et Maire subsp. embergeri（Font Quer et Maire）Maire ■☆

408464　Vicia lens L. ;金麦菜■☆

408465　Vicia leucantha Biv. ;白花野豌豆■☆

408466　Vicia lilacina Ledeb. ;阿尔泰野豌豆;Altai Vetch ■

408467　Vicia longicuspis Z. D. Xia;长齿野豌豆（阿坝野豌豆）;Long-cusp Vetch,Longtooth Vetch ■

408468　Vicia ludoviciana Nutt. ;鹿野豌豆;Deer Pea Vetch ■☆

408469　Vicia lutea L. ;黄巢菜;Rough-podded Yellow Vetch, Smooth Yellow Vetch,Yellow Vetch,Yellow-vetch ■☆

408470　Vicia lutea L. subsp. cavanillesii（Mart. Mart.）Romero Zarco;卡氏黄巢菜■☆

408471　Vicia lutea L. subsp. vestita（Boiss.）Rouy;包被黄巢菜■☆

408472　Vicia lutea L. var. hirta（DC.）Loisel. = Vicia lutea L. ■☆

408473　Vicia lutea L. var. laevigata（Sm.）Boiss. = Vicia lutea L. ■☆

408474　Vicia lutea L. var. muricata Ser. = Vicia lutea L. ■☆

408475　Vicia lutea L. var. nitida Ball = Vicia lutea L. ■☆

408476　Vicia lutea L. var. tuberculata Willk. = Vicia lutea L. subsp. cavanillesii（Mart. Mart.）Romero Zarco ■☆

408477　Vicia lutea L. var. vestita（Boiss.）Batt. = Vicia lutea L. ■☆

408478　Vicia macrocarpa Bertol. ;大果野豌豆;Big-pod Vetch ■☆

408479　Vicia macrophylla（Maxim.）B. Fedtsch. ;大叶野豌豆■☆

408480　Vicia mairei H. Lév. = Vicia dichroantha Diels ■

408481　Vicia malosana（Baker）Baker f. = Vicia paucifolia Baker subsp. malosana（Baker）Verdc. ■☆

408482　Vicia mauritanica Batt. = Vicia vicioides（Desf.）Cout. ■☆

408483　Vicia megalosperma M. Bieb. ;大籽野豌豆■☆

408484　Vicia megalotropis Ledeb. ;大龙骨野豌豆(小巢菜,窄叶大龙骨巢菜,窄叶大龙骨野豌豆）;Largekeel Vetch, Narrowleaf Large-keel Vetch ■

408485　Vicia megalotropis Ledeb. f. stenophylla Franch. = Vicia megalotropis Ledeb. ■

408486　Vicia megalotropis Ledeb. var. stenophylla（Franch.）F. T. Wang et Ts. Tang = Vicia megalotropis Ledeb. ■

408487　Vicia megalotropis Ledeb. var. stenophylla（Franch.）F. T. Wang et Ts. Tang ex Z. D. Xia = Vicia megalotropis Ledeb. ■

408488　Vicia melanops Sibth. et Sm. ;黑野豌豆;Black Vetch ■☆

408489　Vicia meyeri Boiss. ;梅氏巢菜■☆

408490　Vicia michauxii Spreng. ;米氏野豌豆■☆

408491　Vicia microphylla d'Urv. = Vicia villosa Roth subsp. microphylla（d'Urv.）P. W. Ball ■☆

408492　Vicia minutiflora F. Dietr. ;微花野豌豆;Pygmy-flowered Vetch ■☆

408493　Vicia monantha Retz. ;单花野豌豆;Barn Vetch, Oneflower Vetch,One-flower Vetch ■☆

408494　Vicia monantha Retz. subsp. biflora（Desf.）Maire = Vicia monantha Retz. subsp. calcarata（Desf.）Romero Zarco ■☆

408495　Vicia monantha Retz. subsp. calcarata（Desf.）Romero Zarco;距单花野豌豆■☆

408496　Vicia monantha Retz. subsp. cinerea（M. Bieb.）Maire = Vicia monantha Retz. ■☆

408497　Vicia monantha Retz. subsp. triflora（Ten.）Burtt et Lewis = Vicia monantha Retz. subsp. calcarata（Desf.）Romero Zarco ■☆

408498　Vicia monantha Retz. var. marmorata（Maire et Weiller）Maire = Vicia monantha Retz. ■☆

408499　Vicia monantha Retz. var. trichocarpa（Maire）Maire = Vicia monantha Retz. ■☆

408500　Vicia monanthos（L.）Desf. = Vicia articulata Hornem. ■☆

408501　Vicia monardii Boiss. et Reut. ;莫纳尔豌豆■☆

408502　Vicia multicaulis Ledeb. ;多茎野豌豆（豆豌豌,金豌豆,野毛耳）;Manystem Vetch,Stemmy Vetch ■

408503　Vicia multijuga Z. D. Xia;多叶野豌豆;Leafy Vetch ■

408504　Vicia murbeckii Maire;穆尔拜克豌豆■☆

408505　Vicia narbonensis L. ;纳博巢菜;Narbonne Vetch,Purple Broad Vetch ■☆

408506　Vicia narbonensis L. var. affinis Asch. et Schweinf. = Vicia narbonensis L. ■☆

408507 Vicia narbonensis L. var. serratifolia（Jacq.）Ser. = Vicia narbonensis L. ■☆

408508 Vicia nipponica Matsum.；日本野豌豆■☆

408509 Vicia nipponica Matsum. f. albiflora Sugim.；白花日本野豌豆■☆

408510 Vicia nipponica Matsum. f. normalis Hiyama；普通日本野豌豆■☆

408511 Vicia nipponica Matsum. var. ramosa H. Nakam.；分枝日本野豌豆■☆

408512 Vicia nummularia Hand.-Mazz.；西南野豌豆（黄花野苕子，黄花野豌豆）；South-western Vetch, SW. China Vetch ■

408513 Vicia ochroleuca Ten.；淡黄白野豌豆■☆

408514 Vicia ochroleuca Ten. subsp. atlantica（Pomel）Greuter et Burdet；亚特兰大野豌豆■☆

408515 Vicia ochroleuca Ten. subsp. baborensis（Batt. et Trab.）Greuter et Burdet；巴布尔野豌豆■☆

408516 Vicia ohwiana Hosok.；头序歪头菜（长齿歪头菜，弯老腰）；Longtooth Vetch, Ohw Vetch ■

408517 Vicia olbiensis Reut. ex Timb.-Lagr.；奥尔比亚野豌豆■☆

408518 Vicia onobrychioides L.；驴喜豆野豌豆；False Sainfoin ■☆

408519 Vicia onobrychioides L. subsp. alborosea Dobignard = Vicia onobrychioides L. ■☆

408520 Vicia onobrychioides L. var. alborosea（Dobignard）Dobignard = Vicia onobrychioides L. ■☆

408521 Vicia oregana Nutt. = Vicia americana Muhl. ex Willd. ■☆

408522 Vicia orobus DC.；苦野豌豆；Bitter Vetch, Bittervetch, Bitter-vetch, Horse Pease, Upright Vetch, Wood Bitter-vetch ■☆

408523 Vicia pallida H. W. Kung = Vicia latibracteolata K. T. Fu ■

408524 Vicia pallida Turcz. = Vicia bakeri Ali ■

408525 Vicia pallida Turcz. = Vicia japonica A. Gray ■

408526 Vicia pallida Turcz. var. pratensis（Kom.）Nakai = Vicia amurensis Oett. ■

408527 Vicia pannonica Crantz；褐毛野豌豆（多毛野豌豆，匈牙利野豌豆）；Brownhair Vetch, Hungarian Vetch, Many-hair Vetch ■

408528 Vicia pannonica Crantz subsp. striata（M. Bieb.）Nyman = Vicia pannonica Crantz ■

408529 Vicia pannonica Crantz var. pannonica ? = Vicia pannonica Crantz ■

408530 Vicia pannonica Crantz var. purpurascens Ser. = Vicia pannonica Crantz ■

408531 Vicia parviflora Cav.；小花野豌豆；Slender Tare ■☆

408532 Vicia paucifolia Baker；疏叶野豌豆■☆

408533 Vicia paucifolia Baker subsp. malosana（Baker）Verdc.；马洛斯野豌豆■☆

408534 Vicia paucifolia Baker var. malosana（Baker）Brenan = Vicia paucifolia Baker subsp. malosana（Baker）Verdc. ■☆

408535 Vicia paucijuga（Trautv.）B. Fedtsch.；少轭野豌豆■☆

408536 Vicia pectinata Lowe；篦状野豌豆■☆

408537 Vicia peregrina L.；奇巢菜；Wandering Vetch ■☆

408538 Vicia peregrina L. var. cyrenaea Maire et Weiller = Vicia peregrina L. ■☆

408539 Vicia perelegans K. T. Fu；精致野豌豆；Delicacy Vetch, Elegant Vetch ■

408540 Vicia picta Fisch. et C. A. Mey.；杂花巢菜（彩巢菜）■☆

408541 Vicia pilosa M. Bieb.；毛野豌豆（毛巢菜）；Pilose Vetch ■☆

408542 Vicia pilosa M. Bieb. var. albiflora Z. D. Xia；白花毛野豌豆；Whiteflower Pilose Vetch ■

408543 Vicia pisiformis L.；豌豆状巢菜；Pea Vetch ■☆

408544 Vicia polyphylla Desf. = Vicia tenuifolia Roth ■

408545 Vicia polyphylla Z. D. Xia；众叶野豌豆（多叶野豌豆）；Many Leaves Vetch ■

408546 Vicia portosanctana Meneses = Vicia ferreirensis Goyder ■☆

408547 Vicia pseudocracca Bertol. = Vicia villosa Roth subsp. pseudocracca（Bertol.）Rouy ■☆

408548 Vicia pseudocracca Bertol. var. ambigua（Guss.）Durand et Barratte = Vicia villosa Roth subsp. pseudocracca（Bertol.）Rouy ■☆

408549 Vicia pseudocracca Bertol. var. brevipes Willk. = Vicia villosa Roth subsp. pseudocracca（Bertol.）Rouy ■☆

408550 Vicia pseudocracca Bertol. var. wilczekii Maire = Vicia villosa Roth subsp. pseudocracca（Bertol.）Rouy ■☆

408551 Vicia pseudo-orobus Fisch. ex C. A. Mey.；假香野豌豆（大叶草藤，大叶香豌豆，大叶野豌豆，槐条花，芦豆苗）；False Robust Vetch, False Vetch, Largeleaf Vetch ■

408552 Vicia pseudo-orobus Fisch. ex C. A. Mey. f. albiflora（H. Koidz. ex Honda）Honda = Vicia pseudo-orobus Fisch. ex C. A. Mey. f. albiflora（Nakai）P. Y. Fu et Y. A. Chen ■

408553 Vicia pseudo-orobus Fisch. ex C. A. Mey. f. albiflora（Nakai）P. Y. Fu et Y. A. Chen；白花大野豌豆；Whiteflower Falserobust Vetch ■

408554 Vicia pseudo-orobus Fisch. ex C. A. Mey. f. alborosea Sugim.；粉白大叶野豌豆■☆

408555 Vicia pseudo-orobus Fisch. ex C. A. Mey. f. breviramea P. Y. Fu et Y. C. Teng；短序大野豌豆；Short Inflorescence Falserobust Vetch ■

408556 Vicia pseudo-orobus Fisch. ex C. A. Mey. f. rotundifolia Sugim.；圆大叶野豌豆■☆

408557 Vicia pseudo-orobus Fisch. ex C. A. Mey. var. albiflora Nakai = Vicia pseudo-orobus Fisch. ex C. A. Mey. f. albiflora（Nakai）P. Y. Fu et Y. A. Chen ■

408558 Vicia pseudo-orobus Fisch. ex C. A. Mey. var. tanakae Makino；四叶大野豌豆；Fourleaves Falserobust Vetch ■

408559 Vicia pubescens（DC.）Link；短柔毛巢菜■☆

408560 Vicia pubescens Link = Vicia pubescens（DC.）Link ■☆

408561 Vicia purpurascens DC. = Vicia pannonica Crantz ■

408562 Vicia quinquenervia Miq. = Lathyrus quinquenervius（Miq.）Litv. ■

408563 Vicia ramuliflora（Maxim.）Ohwi；北野豌豆；Branchlet Flower Vetch, Northern Vetch ■

408564 Vicia ramuliflora（Maxim.）Ohwi f. abbreviata P. Y. Fu et Y. A. Chen；辽野豌豆■

408565 Vicia ramuliflora（Maxim.）Ohwi f. baicalensis（Turcz.）P. Y. Fu et Y. A. Chen = Vicia ramuliflora（Maxim.）Ohwi ■

408566 Vicia ramuliflora（Maxim.）Ohwi f. baicalensis（Turcz.）P. Y. Fu et Y. A. Chen；贝加尔野豌豆；Baikal Vetch ■

408567 Vicia ramuliflora（Maxim.）Ohwi f. chianschanensis P. Y. Fu et Y. A. Chen = Vicia chianschanensis（P. Y. Fu et Y. A. Chen）Z. D. Xia ■

408568 Vicia raynaudii Coulot et Dobignard；雷纳德野豌豆■☆

408569 Vicia rerayensis（Ball）Murb. = Vicia glauca C. Presl ■☆

408570 Vicia rigidula Royle；坚挺野豌豆■☆

408571 Vicia sativa L.；救荒野豌豆（草藤，巢菜，垂水，春巢菜，大巢菜，大叶野豌豆，薔仔豆，肥田草，黄藤子，箭舌豌豆，箭舌野豌豆，箭豌豆，留豆，马豆，马豆草，普通苕子，雀雀豆，山扁豆，苕饶，苕子，豌豆，薇，薇菜，野豆，野菜豆，野麻豌，野豌豆子，野毛豆，野豌豆）；Chichelings, Cichlings, Common Tare, Common Vetch, Cultiva Vetch, Datch, Fatch, Fetch, Fitchacks, Fodder Vetch, Garden Vetch, Gore Thetch, Gore-thetch, Gypsy Pea, Gypsy Peas,

Lint, Lintin, Pebble Vetch, Pebble-vetch, Racers, Spring Vetch, Tar, Tar Vetch, Tare, Tar-vetch, Tere, Thetch, Titatch, Twadger, Urles, Vatch, Wild Fitch ■

408572　Vicia sativa L. subsp. angustifolia (L. ex Reichard) Gaudin var. segetalis (Thuill.) Ser. = Vicia sativa L. subsp. nigra (L.) Ehrh. ■☆

408573　Vicia sativa L. subsp. angustifolia (L.) Batt. = Vicia sativa L. subsp. nigra (L.) Ehrh. ■☆

408574　Vicia sativa L. subsp. cordata (Hoppe) Batt.;心叶救荒野豌豆■☆

408575　Vicia sativa L. subsp. cuneata (Guss.) Maire = Vicia sativa L. subsp. nigra (L.) Ehrh. ■☆

408576　Vicia sativa L. subsp. macrocarpa (Moris) Arcang.;大果救荒野豌豆■☆

408577　Vicia sativa L. subsp. maculata (C. Presl) Batt. = Vicia sativa L. subsp. nigra (L.) Ehrh. ■☆

408578　Vicia sativa L. subsp. nigra (L.) Ehrh.;黑救荒野豌豆(野豌豆); Common Vetch, Garden Vetch, Narrow-leaved Vetch, Spring Vetch ■☆

408579　Vicia sativa L. subsp. nigra (L.) Ehrh. f. normalis (Makino) Kitam.;正常黑救荒野豌豆■☆

408580　Vicia sativa L. subsp. nigra (L.) Ehrh. var. minor (Bertol.) Gaudin;小黑救荒野豌豆■☆

408581　Vicia sativa L. subsp. obovata (Ser.) Schinz et Thell. = Vicia sativa L. ■

408582　Vicia sativa L. var. abyssinica (Alef.) Baker = Vicia sativa L. var. angustifolia (L. ex Reichard) Wahlenb. ☆

408583　Vicia sativa L. var. amphicarpa (L.) Boiss. = Vicia sativa L. ■

408584　Vicia sativa L. var. angustifolia (L. ex Reichard) Wahlenb.;窄叶救荒野豌豆■☆

408585　Vicia sativa L. var. angustifolia (L. ex Reichard) Wahlenb. = Vicia sativa L. subsp. nigra (L.) Ehrh. ■☆

408586　Vicia sativa L. var. angustifolia (L.) Ser. = Vicia sativa L. subsp. nigra (L.) Ehrh. ■☆

408587　Vicia sativa L. var. angustifolia (L.) Wahlenb. = Vicia sativa L. ■

408588　Vicia sativa L. var. angustifolia Wahlb. = Vicia angustifolia L. ex Reichard ■

408589　Vicia sativa L. var. aristulata Chiov. = Vicia sativa L. var. angustifolia (L. ex Reichard) Wahlenb. ■☆

408590　Vicia sativa L. var. bobartii (E. Forst.) Burnat = Vicia sativa L. ■

408591　Vicia sativa L. var. cordata (Hoppe) Arcang. = Vicia sativa L. ■

408592　Vicia sativa L. var. cosentini (Guss.) Arcang. = Vicia sativa L. ■

408593　Vicia sativa L. var. ecirrhosa J. Q. He;无卷须巢菜■

408594　Vicia sativa L. var. heterophylla (C. Presl) Asch. et Graebn. = Vicia sativa L. subsp. nigra (L.) Ehrh. ■☆

408595　Vicia sativa L. var. linearis Lange = Vicia sativa L. ■

408596　Vicia sativa L. var. macrocarpa Moris = Vicia sativa L. subsp. macrocarpa (Moris) Arcang. ■☆

408597　Vicia sativa L. var. nemoralis Pers. = Vicia sativa L. ■

408598　Vicia sativa L. var. nigra L. = Vicia angustifolia L. ex Reichard ■

408599　Vicia sativa L. var. nigra L. = Vicia sativa L. subsp. nigra (L.) Ehrh. ☆

408600　Vicia sativa L. var. nigra L. = Vicia sativa L. ■

408601　Vicia sativa L. var. segetalis (Thuill.) Ser. = Vicia sativa L. subsp. nigra (L.) Ehrh. ■☆

408602　Vicia sativa L. var. segetalis (Thuill.) Ser. = Vicia sativa L. ■

408603　Vicia sativa L. var. segetalis Thuill. = Vicia sativa L. ■

408604　Vicia sativa L. var. villosa Maire, Weiller et Wilczek = Vicia sativa L. ■

408605　Vicia saxatilis (Vent.) Tropea;岩地野豌豆■☆

408606　Vicia scandens Murray;攀缘野豌豆■☆

408607　Vicia segetalis Thuill. = Vicia angustifolia L. ex Reichard var. segetalis (Thuill.) Koch ■☆

408608　Vicia segetalis Thuill. = Vicia sativa L. subsp. nigra (L.) Ehrh. ■☆

408609　Vicia semenovii (Regel et Herder) B. Fedtsch.;赛氏野豌豆■☆

408610　Vicia semiglabra Rupr. ex Boiss.;半光野豌豆■☆

408611　Vicia sepium L.;野豌豆(滇野豌豆); Buch Vetch, Bush Vetch, Crow Peas, Dill, Fingers-and-thumbs, Hedge Vetch, Tare, Thetch, Vetch, Wild Tare ■

408612　Vicia serratifolia Jacq.;齿叶野豌豆■☆

408613　Vicia sicula (Raf.) Guss.;西西里野豌豆■☆

408614　Vicia silvatica L.;林巢菜■☆

408615　Vicia sinkiangensis H. W. Kung = Vicia costata Ledeb. ■

408616　Vicia sosnovskyi Ekutim.;索斯野豌豆■☆

408617　Vicia sparsifolia Nutt. ex Torr. et A. Gray var. truncata (Nutt.) S. Watson = Vicia americana Muhl. ex Willd. ■☆

408618　Vicia striata M. Bieb.;条纹巢菜■☆

408619　Vicia subvillosa (Ledeb.) Trautv.;近无毛野豌豆■☆

408620　Vicia sylvatica L.;林地野豌豆; Culverkeys, Wood Vetch ■☆

408621　Vicia sylvatica L. var. tingitana Martínez = Vicia altissima Desf. ☆

408622　Vicia taipaica K. T. Fu;太白野豌豆; Taibai Vetch ■

408623　Vicia tenera Graham var. yunnanensis Franch. = Vicia dichroantha Diels ■

408624　Vicia tenuifolia Roth;细叶野豌豆(黑子野豌豆,三齿草藤,细叶巢菜); Bramble Vetch, Fine-leaved Vetch, Slenderleaf Vetch ■

408625　Vicia tenuifolia Roth = Vicia cracca L. subsp. tenuifolia (Roth) Gaudin ■☆

408626　Vicia tenuifolia Roth subsp. elegans (Guss.) Nyman;雅致细叶野豌豆■☆

408627　Vicia tenuifolia Roth subsp. villosa (Batt.) Greuter = Vicia tenuifolia Roth ■

408628　Vicia tenuifolia Roth var. mesatlantica Maire = Vicia tenuifolia Roth ■

408629　Vicia tenuifolia Roth var. rifana Emb. et Maire = Vicia tenuifolia Roth ■

408630　Vicia tenuifolia Roth var. stenophylla Boiss. = Vicia tenuifolia Roth ■

408631　Vicia tenuifolia Roth var. villosa (Batt.) Jahand. et Maire = Vicia tenuifolia Roth ■

408632　Vicia tenuifolia Z. D. Xia f. albiflora Z. D. Xia;白花细叶野豌豆; Whiteflower Fine-leaved Vetch ■

408633　Vicia tenuissima Schinz et Thell.;极细野豌豆; Slender Tare ■☆

408634　Vicia tenuissima Schinz et Thell. = Vicia parviflora Cav. ■☆

408635　Vicia ternata Z. D. Xia;三尖野豌豆; Threetine Vetch ■

408636　Vicia tetrantha H. W. Kung;四花野豌豆; Fourflower Vetch ■

408637　Vicia tetrasperma (L.) Moench = Vicia tetrasperma (L.) Schreb. ■

408638　Vicia tetrasperma (L.) Schreb.;四籽野豌豆(鸟喙豆,鸟啄豆,乔乔子,苕子,丝翘翘,四籽草藤,乌嘴豆,小乔菜,野扁豆,野苕子); Fourseed Vetch, Four-seed Vetch, Four-seeded Vetch, Lentil Vetch, Slender Tare, Slender Vetch, Smooth Tare, Smooth-podded Tine Tare, Smooth-podded Tine-tare, Sparrow Vetch ■

408639　Vicia tetrasperma (L.) Schreb. subsp. gracilis Hook. f. = Vicia parviflora Cav. ■☆

408640　Vicia tetrasperma (L.) Schreb. subsp. pubescens (DC.) Bonnier et Layens = Vicia pubescens (DC.) Link ■☆

408641　Vicia tibetica Prain ex C. E. C. Fisch.;西藏野豌豆; Tibet

Vetch，Xizang Vetch ■

408642　Vicia tricuspidata Steven；三尖巢菜■☆

408643　Vicia tridentata Bunge = Vicia bungei Ohwi ■

408644　Vicia tridentifolia Z. D. Xia；三齿野豌豆；Three Teethed-leaf Vetch ■

408645　Vicia truncata Nutt. = Vicia americana Muhl. ex Willd. ■☆

408646　Vicia truncatula Fisch. ；截形巢菜■☆

408647　Vicia uncinata Desf. ex Nyman = Vicia sativa L. subsp. nigra（L. ）Ehrh. ■☆

408648　Vicia unijuga A. Braun；歪头菜(草豆，豆菜，豆苗菜，豆叶菜，二叶野豌豆，两叶豆苗，偏头草，三铃子，山豌豆，水皂荚，鲜豆苗，野豌豆)；Askew Vetch，Pair Vetch，Two-leaved Vetch ■

408649　Vicia unijuga A. Braun f. angustifolia（Makino）Makino ex W. T. Lee；窄叶歪头菜■☆

408650　Vicia unijuga A. Braun f. minor Nakai；小歪头菜■☆

408651　Vicia unijuga A. Braun f. trifoliolata（Z. D. Xia）Y. Endo et H. Ohashi；三小叶歪头菜■☆

408652　Vicia unijuga A. Braun f. venusta（Nakai）H. Ohashi；雅致巢菜■☆

408653　Vicia unijuga A. Braun var. albiflora Kitag. ；白花歪头菜；Whiteflower Pair Vetch ■

408654　Vicia unijuga A. Braun var. angustifolia Nakai；狭叶歪头菜；Narrowleaf Pair Vetch ■

408655　Vicia unijuga A. Braun var. angustifolia Nakai = Vicia unijuga A. Braun ■

408656　Vicia unijuga A. Braun var. apoda Maxim. = Vicia apoda（Maxim. ）Z. D. Xia ■

408657　Vicia unijuga A. Braun var. apoda Maxim. = Vicia ohwiana Hosok. ■

408658　Vicia unijuga A. Braun var. austrohigoensis（Honda）Sugim. ；南肥后歪头菜■☆

408659　Vicia unijuga A. Braun var. bracteata Franch. et Sav. = Vicia bifolia Nakai ■

408660　Vicia unijuga A. Braun var. bracteata Franch. et Sav. = Vicia unijuga A. Braun ■

408661　Vicia unijuga A. Braun var. breviramea Nakai；复总花歪头菜■

408662　Vicia unijuga A. Braun var. breviramea Nakai = Vicia unijuga A. Braun ■

408663　Vicia unijuga A. Braun var. ohwiana（Hosok. ）Nakai = Vicia ohwiana Hosok. ■

408664　Vicia unijuga A. Braun var. trifoliolata Z. D. Xia；三叶歪头菜；Three-leaved Pair Vetch ■

408665　Vicia ussuriensis Oett. = Vicia amurensis Oett. ■

408666　Vicia varia Host；欧洲苕子；Europe Vetch，Winter Vetch ■

408667　Vicia varia Host = Vicia villosa Roth subsp. varia（Host）Corb. ■

408668　Vicia varia Host var. eriocarpa Hausskn. = Vicia villosa Roth subsp. eriocarpa（Hausskn. ）P. W. Ball ■☆

408669　Vicia varia Host var. villosa（Roth）Batt. = Vicia villosa Roth subsp. varia（Host）Corb. ■

408670　Vicia variabilis Freyn et Sint. ；可变野豌豆■☆

408671　Vicia variegata Willd. ；斑叶野豌豆■☆

408672　Vicia venosa（Willd. ex Link）Maxim. ；柳叶野豌豆(老豆秧，脉草藤，脉基巢菜，脉叶野豌豆，细脉巢菜)；Veined Vetch，Willowleaf Vetch ■

408673　Vicia venosa（Willd. ex Link）Maxim. subsp. cuspidata（Maxim. ）Y. Endo et H. Ohashi = Vicia deflexa Nakai ■

408674　Vicia venosa（Willd. ex Link）Maxim. subsp. cuspidata（Maxim. ）Y. Endo et H. Ohashi var. glabristyla Y. Endo et H. Ohashi；光柱野豌豆■☆

408675　Vicia venosa（Willd. ex Link）Maxim. subsp. stolonifera（Y. Endo et H. Ohashi）Y. Endo et H. Ohashi；匍匐柳叶野豌豆■☆

408676　Vicia venosa（Willd. ex Link）Maxim. subsp. yamanakae（Y. Endo et H. Ohashi）Y. Endo et H. Ohashi；山中柳叶野豌豆■☆

408677　Vicia venosa（Willd. ex Link）Maxim. var. albiflora Maxim. ；白花柳叶野豌豆■

408678　Vicia venosa（Willd. ex Link）Maxim. var. alpina Kitag. ；长白山柳叶野豌豆■

408679　Vicia venosa（Willd. ex Link）Maxim. var. cuspidata Maxim. = Vicia deflexa Nakai ■

408680　Vicia venosa（Willd. ex Link）Maxim. var. fauriei（Franch. ）Okuyama = Vicia fauriei Franch. ■☆

408681　Vicia venosa（Willd. ex Link）Maxim. var. stolonifera Y. Endo et H. Ohashi = Vicia venosa（Willd. ex Link）Maxim. subsp. stolonifera（Y. Endo et H. Ohashi）Y. Endo et H. Ohashi ■☆

408682　Vicia venosa（Willd. ex Link）Maxim. var. willdenowiana Maxim. = Vicia venosa（Willd. ex Link）Maxim. ■

408683　Vicia venosa（Willd. ex Link）Maxim. var. willdenowiana Miura = Lathyrus vaniotii H. Lév. ■

408684　Vicia venosa（Willd. ex Link）Maxim. var. yamanakae Y. Endo et H. Ohashi = Vicia venosa（Willd. ex Link）Maxim. subsp. yamanakae（Y. Endo et H. Ohashi）Y. Endo et H. Ohashi ■☆

408685　Vicia venosa（Willd. ）Maxim. = Vicia venosa（Willd. ex Link）Maxim. ■

408686　Vicia venulosa Boiss. et Hohen. ；细脉野豌豆■☆

408687　Vicia vexillata（L. ）A. Rich. = Vigna vexillata（L. ）A. Rich. ■

408688　Vicia vexillata（L. ）A. Rich. var. yunnanensis Franch. = Vigna vexillata（L. ）A. Rich. ■

408689　Vicia vicioides（Desf. ）Cout. ；卷曲野豌豆■☆

408690　Vicia vicioides（Desf. ）Cout. var. erviformis（Boiss. ）Maire = Vicia vicioides（Desf. ）Cout. ■☆

408691　Vicia vicioides（Desf. ）Cout. var. sericea（Batt. ）Maire = Vicia vicioides（Desf. ）Cout. ■☆

408692　Vicia vicioides（Desf. ）Cout. var. subcapitata（Pérez）Maire = Vicia vicioides（Desf. ）Cout. ■☆

408693　Vicia villosa Roth；长柔毛野豌豆(毛苕子，毛叶苕子，柔毛苕子，柔毛野豌豆)；Fodder Vetch，Hairy Vetch，Lesser Tttfted Vetch，Russia Vetch，Russian Vetch，Siberian Vetch，Villose Vetch，Winter Vetch，Woolly Vetch，Woolly-pod Vetch ■

408694　Vicia villosa Roth f. albiflora Z. D. Xia；白花柔毛野豌豆；Whiteflower Villose Vetch ■

408695　Vicia villosa Roth subsp. ambigua（Guss. ）Kerguélen = Vicia villosa Roth subsp. pseudocracca（Bertol. ）Rouy ■☆

408696　Vicia villosa Roth subsp. dasycarpa（Ten. ）Cavill. = Vicia villosa Roth subsp. varia（Host）Corb. ■

408697　Vicia villosa Roth subsp. eriocarpa（Hausskn. ）P. W. Ball；毛果长柔毛野豌豆(毛果野豌豆)■☆

408698　Vicia villosa Roth subsp. garbiensis（Font Quer et Pau）Maire；加尔比长柔毛野豌豆■☆

408699　Vicia villosa Roth subsp. microphylla（d' Urv. ）P. W. Ball；小叶长柔毛野豌豆■☆

408700　Vicia villosa Roth subsp. pseudocracca（Bertol. ）Rouy；假巢菜野豌豆；Winter Vetch ■☆

408701　Vicia villosa Roth subsp. simulans Maire；相似长柔毛野豌豆■☆

408702　Vicia villosa Roth subsp. varia（Host）Corb. = Vicia varia Host ■

408703 Vicia villosa Roth subsp. varia Corb. = Vicia varia Host ■

408704 Vicia villosa Roth var. brevipes Willk. = Vicia villosa Roth ■

408705 Vicia villosa Roth var. eriosolen Faure et Maire = Vicia villosa Roth subsp. eriocarpa (Hausskn.) P. W. Ball ■☆

408706 Vicia villosa Roth var. glabrescens W. D. J. Koch = Vicia villosa Roth ■

408707 Vicia villosa Roth var. leiosolen Faure et Maire = Vicia villosa Roth ■

408708 Vicia villosa Roth var. wilczekii Maire = Vicia villosa Roth ■

408709 Vicia woroschilovii N. S. Pavlova = Vicia japonica A. Gray var. pallida (Trautv.) H. Hara ■☆

408710 Vicia wushanica Z. D. Xia；武山野豌豆；Wushan Vetch ■

408711 Viciaceae Bercht. et J. Presl = Fabaceae Lindl. (保留科名)●■

408712 Viciaceae Bercht. et J. Presl = Leguminosae Juss. (保留科名)●■

408713 Viciaceae Dostal = Fabaceae Lindl. (保留科名)●■

408714 Viciaceae Dostal = Leguminosae Juss. (保留科名)●■

408715 Viciaceae Dostal；野豌豆科■

408716 Viciaceae Oken = Fabaceae Lindl. (保留科名)●■

408717 Viciaceae Oken = Leguminosae Juss. (保留科名)●■

408718 Vicilla Schur = Vicia L. ■

408719 Vicioides Moench = Vicia L. ■

408720 Vicoa Cass. = Pentanema Cass. ■●

408721 Vicoa appendiculata (Wall.) DC. = Pentanema indicum (L.) Y. Ling ■

408722 Vicoa appendiculata Wall. ex DC. = Pentanema indicum (L.) Y. Ling ■

408723 Vicoa auriculata Cass. = Pentanema indicum (L.) Y. Ling var. hypoleucum (Hand. -Mazz.) Y. Ling ■

408724 Vicoa auriculata Cass. = Pentanema indicum (L.) Y. Ling ■

408725 Vicoa aurita DC. = Pentanema indicum (L.) Y. Ling ■

408726 Vicoa cernua Dalzell = Pentanema cernuum (Dalzell et A. Gibson) Y. Ling ■

408727 Vicoa cernua Dalzell et A. Gibson = Pentanema cernuum (Dalzell et A. Gibson) Y. Ling ■

408728 Vicoa divaricata Oliv. et Hiern = Pulicaria petiolaris Jaub. et Spach ■☆

408729 Vicoa indica (L.) DC. = Pentanema indicum (L.) Y. Ling ■

408730 Vicoa leptoclada (Webb) Dandy = Pentanema indicum (L.) Y. Ling ■

408731 Vicoa vestita Benth. = Pentanema vestitum (Wall. ex DC.) Y. Ling ■

408732 Vicq-aziria Buc'hoz = Gurania (Schltdl.) Cogn. ■☆

408733 Victoria Lindl. (1837)；王莲属；Royal Water Lily, Victoria, Water Platter, Water-platter ■

408734 Victoria amazonica (Poepp.) J. C. Sowerby；王莲（大鬼莲）；Amazon Water Lily, Amazon Water-platter, Giant Water Lily, Royal Water Lily, Water Maize ■

408735 Victoria cruziana A. D. Orb.；克鲁兹王莲；Santa Cruz Water Lily, Santa Cruz Water-lily, Santa Cruz Water-platter ■☆

408736 Victoria regia Lindl. = Victoria amazonica (Poepp.) J. C. Sowerby ■

408737 Victorinia Léon = Cnidoscolus Pohl ●☆

408738 Victoriperrea Hombr. = Freycinetia Gaudich. ●

408739 Victoriperrea Hombr. et Jacquinot ex Decne. = Freycinetia Gaudich. ●

408740 Vidalasia Tirveng. (1998)；维达茜属●☆

408741 Vidalasia fusca (Craib) Tirveng.；褐维达茜●☆

408742 Vidalasia morindifolia (Elmer) Tirveng.；巴戟天叶维达茜●☆

408743 Vidalasia murina (Craib) Tirveng.；维达茜●☆

408744 Vidalasia pubescens (Tirveng. et Sastre) Tirveng.；毛维达茜●☆

408745 Vidalasia tonkinensis (Pit.) Tirveng.；越南维达茜●☆

408746 Vidalia Fern. -Vill. = Mesua L. ●

408747 Vidoricum Kuntze = Illipe Gras ●

408748 Vidoricum Kuntze = Madhuca Buch. -Ham. ex J. F. Gmel. ●

408749 Vidoricum Rumph. = Madhuca Buch. -Ham. ex J. F. Gmel. ●

408750 Vidoricum Rumph. ex Kuntze = Madhuca Buch. -Ham. ex J. F. Gmel. ●

408751 Vieillardia Brongn. et Gris = Calpidia Thouars ●

408752 Vieillardia Brongn. et Gris = Pisonia L. ●

408753 Vieillardia Brongn. et Gris = Timeroyea Montrouz. ●

408754 Vieillardia Montrouz. = Castanospermum A. Cunn. ex Hook. ●☆

408755 Vieillardorchis Kraenzl. = Goodyera R. Br. ■

408756 Viellardia Benth. et Hook. f. (1865) = Castanospermum A. Cunn. ex Hook. ●☆

408757 Viellardia Benth. et Hook. f. (1865) = Vieillardia Montrouz. ●☆

408758 Viellardia Benth. et Hook. f. (1880) = Calpidia Thouars ●

408759 Viellardia Benth. et Hook. f. (1880) = Vieillardia Brongn. et Gris ●

408760 Vieraea Sch. Bip. = Vieraea Webb ex Sch. Bip. ●☆

408761 Vieraea Webb et Berthel. = Vieraea Webb ex Sch. Bip. ●☆

408762 Vieraea Webb ex Sch. Bip. (1844)；光覆花属●☆

408763 Vieraea laevigata (Willd.) Webb = Vieraea laevigata (Willd.) Webb et Berthel. ■☆

408764 Vieraea laevigata (Willd.) Webb et Berthel.；光覆花☆

408765 Viereckia R. M. King et H. Rob. (1975)；三脉亮泽兰属●☆

408766 Viereckia tamaulipasensis R. M. King et H. Rob.；三脉亮泽兰●☆

408767 Viereya Stand. = Vireya Blume ●

408768 Viereya Steud. = Alloplectus Mart. (保留属名)●■

408769 Viereya Steud. = Rhododendron L. ●

408770 Viereya Steud. = Vireya Raf. (废弃属名)●■

408771 Vierhapperia Hand. -Mazz. = Nannoglottis Maxim. ■★

408772 Vierhapperia hieraciphylla Hand. -Mazz. = Nannoglottis hieraciphylla (Hand. -Mazz.) Y. Ling et Y. L. Chen ■

408773 Vieria Webb et Berthel. = Vieraea Webb ex Sch. Bip. ●☆

408774 Vieria Webb ex Sch. Bip. = Buphthalmum L. ■

408775 Vietnamia P. T. Li(1994)；越南萝藦属■☆

408776 Vietnamia inflexa P. T. Li；越南萝藦■☆

408777 Vietnamocalamus T. Q. Nguyen(1991)；越南竹属●☆

408778 Vietnamocalamus catbaensis T. Q. Nguyen；越南竹●☆

408779 Vietnamochloa Veldkamp et R. Nowack(1995)；越南禾属■☆

408780 Vietnamochloa aurea Veldkamp et R. Nowack；越南禾■☆

408781 Vietnamosasa T. Q. Nguyen(1990)；越南笹属■☆

408782 Vietnamosasa ciliata (A. Camus) T. Q. Nguyen；睫毛越南笹■☆

408783 Vietnamosasa darlacensis T. Q. Nguyen；越南笹■☆

408784 Vietorchis Aver. et Averyanova = Silvorchis J. J. Sm. ■☆

408785 Vietorchis Aver. et Averyanova(2003)；越南林兰属■☆

408786 Vietorchis aurea Aver. et Averyanova；越南林兰■☆

408787 Vietsenia C. Hansen(1984)；越南野牡丹属■☆

408788 Vietsenia laxiflora C. Hansen；睫毛越南野牡丹■☆

408789 Vietsenia poilanei C. Hansen；越南野牡丹■☆

408790 Vietsenia rotundifolia C. Hansen；圆叶越南野牡丹■☆

408791 Vieusseuxia D. Delaroche = Moraea Mill. (保留属名)■

408792 Vieusseuxia angustifolia Eckl. = Moraea gawleri Spreng. ■☆

408793 Vieusseuxia aristata D. Delaroche = Moraea aristata (D.

Delaroche）Asch. et Graebn. ■☆

408794　Vieusseuxia bellendenii Sweet = Moraea bellendenii（Sweet）N. E. Br. ■☆

408795　Vieusseuxia bituminosa（L. f.）Eckl. = Moraea bituminosa（L. f.）Ker Gawl. ■☆

408796　Vieusseuxia brehmii Eckl. = Moraea gawleri Spreng. ■☆

408797　Vieusseuxia edulis（L. f.）Link = Moraea fugax（D. Delaroche）Jacq. ■☆

408798　Vieusseuxia freuchenia（Eckl.）Steud. = Moraea ramosissima（L. f.）Druce ■☆

408799　Vieusseuxia fugax D. Delaroche = Moraea fugax（D. Delaroche）Jacq. ■☆

408800　Vieusseuxia geniculata Eckl. = Moraea lugubris（Salisb.）Goldblatt ■☆

408801　Vieusseuxia glaucopis DC. = Moraea aristata（D. Delaroche）Asch. et Graebn. ■☆

408802　Vieusseuxia graminifolia Eckl. = Moraea vegeta L. ■☆

408803　Vieusseuxia intermedia Eckl. = Moraea papilionacea（L. f.）Ker Gawl. ■☆

408804　Vieusseuxia lurida（Ker Gawl.）Sweet = Moraea lurida Ker Gawl. ■☆

408805　Vieusseuxia mutila C. H. Berg ex Eckl. = Moraea tripetala（L. f.）Ker Gawl. ■☆

408806　Vieusseuxia nervosa Eckl. = Moraea papilionacea（L. f.）Ker Gawl. ■☆

408807　Vieusseuxia pavonia（L. f.）DC. = Moraea tulbaghensis L. Bolus ■☆

408808　Vieusseuxia pulchra Eckl. = Moraea tripetala（L. f.）Ker Gawl. ■☆

408809　Vieusseuxia rivularis Eckl. = Moraea vegeta L. ■☆

408810　Vieusseuxia spiralis D. Delaroche = Moraea bellendenii（Sweet）N. E. Br. ■☆

408811　Vieusseuxia tenuis（Ker Gawl.）Roem. et Schult. = Moraea unguiculata Ker Gawl. ■☆

408812　Vieusseuxia tricuspis（Thunb.）Spreng. = Moraea tricuspidata（L. f.）G. J. Lewis ■☆

408813　Vieusseuxia tripetala（L. f.）Klatt = Moraea tripetala（L. f.）Ker Gawl. ■☆

408814　Vieusseuxia tripetaloides DC. = Moraea tripetala（L. f.）Ker Gawl. ■☆

408815　Vieusseuxia unguiculata（Ker Gawl.）Roem. et Schult. = Moraea unguiculata Ker Gawl. ■☆

408816　Vieusseuxia villosa（Ker Gawl.）Spreng. = Moraea villosa（Ker Gawl.）Ker Gawl. ■☆

408817　Vieusseuxia viscaria（L. f.）Eckl. = Moraea viscaria（L. f.）Ker Gawl. ■☆

408818　Vigethia W. A. Weber(1943)；墨腺菊属●☆

408819　Vigethia mexicana（S. Watson）W. A. Weber；墨腺菊■☆

408820　Vigia Vell. = Fragariopsis A. St. -Hil. ●☆

408821　Vigiera Benth. et Hook. f. = Vigieria Vell. ●☆

408822　Vigieria Vell. = Escallonia Mutis ex L. f. ●☆

408823　Vigineixia Pomel = Picris L. ■

408824　Vigineixia balansae（Coss. et Durieu）Pomel = Helminthotheca balansae（Coss. et Durieu）Lack ■☆

408825　Vigna Savi（1824）（保留属名）；豇豆属；Cowpea,Cow-pea, Mung-bean ■

408826　Vigna abyssinica Taub. = Vigna heterophylla A. Rich. ■☆

408827　Vigna abyssinica Taub. var. ugandensis Baker f. = Vigna heterophylla A. Rich. ■☆

408828　Vigna aconifolia（Jacq.）Maréchal；扇叶豇豆■☆

408829　Vigna aconitifolia（Jacq.）Maréchal；乌头叶豇豆（乌头叶菜豆）；Aconitum-leaf Cowpea, Blackheadleaf Cowpea, Dew Bean, Mat Bean, Moth Bean, Moth-bean, Tepary Bean ■

408830　Vigna aconitifolia（Jacq.）Maréchal = Vigna aconifolia（Jacq.）Maréchal ■☆

408831　Vigna acuminata Hayata；狭叶豇豆；Narrowleaf Cowpea ■

408832　Vigna adenantha（G. Mey.）Maréchal, Mascherpa et Stainier；腺药豇豆■

408833　Vigna afzelii Baker = Vigna gracilis（Guillaumin et Perr.）Hook. f. ■☆

408834　Vigna alba（G. Don）Baker f. = Vigna unguiculata（L.）Walp. subsp. alba（G. Don）Pasquet ■☆

408835　Vigna ambacensis Welw. ex Baker = Vigna heterophylla A. Rich. ■☆

408836　Vigna ambacensis Welw. ex Baker var. pubigera（Baker）Maréchal et al. ；短毛豇豆■☆

408837　Vigna andongensis Baker = Vigna reticulata Hook. f. ●☆

408838　Vigna angivensis Baker；马达加斯加豇豆■☆

408839　Vigna angularis（Willd.）Ohwi et H. Ohashi；赤豆(赤小豆,春湾豆,饭豆,腐婢,红豆,红饭豆,红小豆,虭稤豆,小豆,朱赤豆,朱小豆)；Adsuki Bean, Adzuki Bean, Red Bean, Red Cowpea ■

408840　Vigna angularis（Willd.）Ohwi et H. Ohashi var. nippoensis（Ohwi）Ohwi et Ohashi = Vigna angularis（Willd.）Ohwi et H. Ohashi ■

408841　Vigna angularis（Willd.）Ohwi et H. Ohashi var. nipponensis（Ohwi）Ohwi et H. Ohashi；日本赤豆(日本菜豆,野红豆)■

408842　Vigna angustifolia（Schumach. et Thonn.）Hook. f. = Vigna vexillata（L.）A. Rich. var. angustifolia（Schumach. et Thonn.）Baker ■☆

408843　Vigna angustifolia Benth. = Sphenostylis angustifolia Sond. ■☆

408844　Vigna angustifoliolata Verdc. = Vigna unguiculata（L.）Walp. subsp. stenophylla（Harv.）Maréchal et al. ■☆

408845　Vigna anomala Walp. = Vigna marina（Burm.）Merr. ■

408846　Vigna antunesii Harms；安图内思豇豆■☆

408847　Vigna aurea ?；金黄豇豆■☆

408848　Vigna baoulensis A. Chev. = Vigna unguiculata（L.）Walp. subsp. baoulensis（A. Chev.）Pasquet ■☆

408849　Vigna benthamii Vatke = Pseudovigna argentea（Willd.）Verdc. ■☆

408850　Vigna benuensis Pasquet et Maréchal；博努豇豆■☆

408851　Vigna bequaertii R. Wilczek；贝卡尔豇豆■☆

408852　Vigna bosseri Du Puy et Labat；博瑟豇豆●☆

408853　Vigna brachystachys Benth. = Vigna luteola（Jacq.）Benth. ■

408854　Vigna briartii De Wild. = Sphenostylis briartii（De Wild.）Baker f. ■☆

408855　Vigna buchneri Harms = Vigna frutescens A. Rich. var. buchneri（Harms）Verdc. ■☆

408856　Vigna bukobensis Harms = Vigna luteola（Jacq.）Benth. ■

408857　Vigna burchellii（DC.）Harv. = Otoptera burchellii DC. ●☆

408858　Vigna caesia Chiov. = Vigna membranacea A. Rich. subsp. caesia（Chiov.）Verdc. ■☆

408859　Vigna calcarata（Roxb.）Kurz = Vigna umbellata（Thunb.）Ohwi et H. Ohashi ■

408860　Vigna capensis（L.）Walp. = Vigna vexillata（L.）A. Rich. ■

408861　Vigna capitata De Wild. = Sphenostylis erecta（Baker f.）

Hutch. ex Baker f. ■☆

408862 Vigna caracalla（L.）Verdc.；蜗牛花；Snail Flower, Snail Plant, Snail Vine ■☆

408863 Vigna catjang（Burm. f.）Walp. = Vigna unguiculata（L.）Walp. subsp. cylindrica（L.）Verdc. ■

408864 Vigna catjang Walp. = Vigna unguiculata（L.）Walp. subsp. cylindrica（L.）Verdc. ■

408865 Vigna chiovendae Baker f. = Vigna heterophylla A. Rich. ■☆

408866 Vigna coerulea Baker = Vigna unguiculata（L.）Walp. var. tenuis（E. Mey.）Maréchal et Mascherpa et Stainier ■☆

408867 Vigna comosa Baker；簇毛豇豆■☆

408868 Vigna comosa Baker subsp. abercornensis Verdc.；阿伯康豇豆■☆

408869 Vigna congoensis Baker f. = Vigna heterophylla A. Rich. ■☆

408870 Vigna crinita A. Rich. = Vigna vexillata（L.）A. Rich. ■

408871 Vigna cylindrica（L.）Skeels = Vigna unguiculata（L.）Walp. subsp. cylindrica（L.）Verdc. ■

408872 Vigna dauciformis A. Chev. = Vigna stenophylla Harms ■☆

408873 Vigna davyi Bolus = Vigna vexillata（L.）A. Rich. var. davyi（Bolus）B. J. Pienaar ■☆

408874 Vigna debanensis Martelli = Vigna frutescens A. Rich. ■☆

408875 Vigna decipiens Harv. = Vigna frutescens A. Rich. ■☆

408876 Vigna dekindtiana Harms = Vigna unguiculata（L.）Walp. var. dekindtiana（Harms）Verdc. ■☆

408877 Vigna desmodioides R. Wilczek；束状豇豆■☆

408878 Vigna dinteri Harms = Vigna lobatifolia Baker ■☆

408879 Vigna dolichonema Harms = Vigna vexillata（L.）A. Rich. ■

408880 Vigna dolomitica R. Wilczek；多罗米蒂豇豆■☆

408881 Vigna donii Baker = Vigna racemosa（G. Don）Hutch. et Dalziel ■☆

408882 Vigna esculenta（De Wild.）De Wild. = Vigna frutescens A. Rich. ■☆

408883 Vigna filicaulis Hepper；线茎豇豆■☆

408884 Vigna filicaulis Hepper var. pseudovenulosa Maréchal et al.；细脉线茎豇豆■☆

408885 Vigna fischeri Harms = Vigna luteola（Jacq.）Benth. ■

408886 Vigna fragrans Baker f. = Vigna frutescens A. Rich. ■☆

408887 Vigna friesiorum Harms；弗里斯豇豆■☆

408888 Vigna friesiorum Harms var. angustifolia Verdc.；窄叶弗里斯豇豆■☆

408889 Vigna friesiorum Harms var. ulugurensis（Harms）Verdc.；乌卢古尔豇豆■☆

408890 Vigna frutescens A. Rich.；灌木状豇豆■☆

408891 Vigna frutescens A. Rich. subsp. incana（Taub.）Verdc.；灰毛灌木状豇豆■☆

408892 Vigna frutescens A. Rich. subsp. kotschyi（Schweinf.）Verdc.；科奇豇豆■☆

408893 Vigna frutescens A. Rich. var. buchneri（Harms）Verdc.；布赫纳豇豆■☆

408894 Vigna galpinii Burtt Davy = Vigna nervosa Markötter ■☆

408895 Vigna gazensis Baker f.；加兹扁豆☆

408896 Vigna glabra Savi；光扁豆■☆

408897 Vigna glabra Savi = Vigna luteola（Jacq.）Benth. ■

408898 Vigna glandulosa Chiov. = Vigna frutescens A. Rich. ■☆

408899 Vigna golungensis Baker = Vigna vexillata（L.）A. Rich. ■

408900 Vigna gracilicaulis（Ohwi）Ohwi et Ohashi；细茎豇豆；Slenderstem Cowpea ■

408901 Vigna gracilicaulis（Ohwi）Ohwi et Ohashi = Vigna minima（Roxb.）Ohwi et H. Ohashi ■

408902 Vigna gracilis（Guillaumin et Perr.）Hook. f.；纤细豇豆■☆

408903 Vigna gracilis（Guillaumin et Perr.）Hook. f. var. multiflora（Hook. f.）Maréchal et al.；多花纤细豇豆■☆

408904 Vigna grahamiana（Wight et Arn.）Verdc. = Wajira grahamiana（Wight et Arn.）Thulin et Lavin ■☆

408905 Vigna harmsiana Buscal. et Muschl. = Vigna frutescens A. Rich. ■☆

408906 Vigna harmsii R. Vig. = Vigna angivensis Baker ■☆

408907 Vigna hastifolia Baker = Nesphostylis holosericea（Baker）Verdc. ■☆

408908 Vigna haumaniana R. Wilczek；豪曼豇豆■☆

408909 Vigna haumaniana R. Wilczek var. pedunculata ？ = Vigna haumaniana R. Wilczek ■☆

408910 Vigna heterophylla A. Rich.；互叶豇豆■☆

408911 Vigna heterophylla A. Rich. var. lanceolata R. Wilczek = Vigna heterophylla A. Rich. ■☆

408912 Vigna hirta Hook. = Vigna vexillata（L.）A. Rich. ■

408913 Vigna hispida（E. Mey.）Walp. = Vigna unguiculata（L.）Walp. subsp. stenophylla（Harv.）Maréchal et al. ■☆

408914 Vigna holosericea Baker = Nesphostylis holosericea（Baker）Verdc. ■☆

408915 Vigna holstii Harms = Vigna luteola（Jacq.）Benth. ■

408916 Vigna homblei De Wild. = Sphenostylis briartii（De Wild.）Baker f. ■☆

408917 Vigna hosei（Craib）Backer；和氏豇豆；Sarawak Bean ■

408918 Vigna hosei（Craib）Backer ex K. Heyne = Vigna hosei（Craib）Backer ■

408919 Vigna hosei（Craib）Backer var. pubescens Maréchal et al.；毛和氏豇豆■☆

408920 Vigna huillensis Welw. ex Baker = Vigna unguiculata（L.）Walp. var. dekindtiana（Harms）Verdc. ■☆

408921 Vigna hundtii Rossberg = Vigna antunesii Harms ■☆

408922 Vigna hygrophila Harms = Vigna oblongifolia A. Rich. var. parviflora（Baker）Verdc. ■☆

408923 Vigna incana Taub. = Vigna frutescens A. Rich. subsp. incana（Taub.）Verdc. ■☆

408924 Vigna jaegeri Harms = Vigna luteola（Jacq.）Benth. ■

408925 Vigna juncea Milne-Redh.；灯心草豇豆■☆

408926 Vigna junodii Harms = Dolichos junodii（Harms）Verdc. ■☆

408927 Vigna juruana（Harms）Verdc.；热带豇豆；Tropical Cowpea ■☆

408928 Vigna kassneri R. Wilczek = Vigna radicans Welw. ex Baker ■☆

408929 Vigna katangensis（De Wild.）T. Durand et H. Durand = Vigna frutescens A. Rich. var. buchneri（Harms）Verdc. ■☆

408930 Vigna katangensis De Wild. = Sphenostylis stenocarpa（Hochst. ex A. Rich.）Harms ■☆

408931 Vigna keniensis Harms = Vigna frutescens A. Rich. ■☆

408932 Vigna keraudrenii Du Puy et Labat；克罗德朗豇豆●☆

408933 Vigna kirkii（Baker）J. B. Gillett；柯克豇豆■☆

408934 Vigna kotschyi Schweinf. = Vigna frutescens A. Rich. subsp. kotschyi（Schweinf.）Verdc. ■☆

408935 Vigna lanceolata Benth.；剑叶豇豆；Maloga Bean ■☆

408936 Vigna lancifolia A. Rich. = Vigna oblongifolia A. Rich. ■☆

408937 Vigna laurentii De Wild.；洛朗豇豆■☆

408938 Vigna lebrunii Baker f. = Vigna comosa Baker ■☆

408939 Vigna ledermannii Harms = Vigna frutescens A. Rich. ■☆

408940 Vigna leptodon Harms = Vigna membranacea A. Rich. ■☆

408941 Vigna linearifolia Hook. f. = Vigna reticulata Hook. f. ●☆

408942 Vigna linearifolia Hutch. = Vigna multinervis Hutch. et Dalziel ■☆

408943　Vigna lobatifolia Baker;裂叶豇豆■☆

408944　Vigna longepedunculata Taub. = Vigna schimperi Baker ■☆

408945　Vigna longifolia（Benth.）Verdc.;长叶豇豆■☆

408946　Vigna longiloba Burtt Davy = Vigna frutescens A. Rich. ■☆

408947　Vigna longissima Hutch. ;极长豇豆■☆

408948　Vigna lutea（Sw.）A. Gray = Vigna marina（Burm.）Merr. ■

408949　Vigna lutea（Sw.）A. Gray var. minor Matsum. = Vigna minima（Roxb.）Ohwi et H. Ohashi var. minor（Matsum.）Tateishi ■

408950　Vigna luteola（Jacq.）Benth. ;浅黄豇豆（长叶豇豆）;Longleaf Cowpea,Palrymple Bean ■

408951　Vigna macrantha Harms = Spathionema kilimandscharicum Taub. ■☆

408952　Vigna macrodon Robyns et Boutique = Vigna membranacea A. Rich. subsp. macrodon（Robyns et Boutique）Verdc. ■☆

408953　Vigna macrorhyncha（Harms）Milne-Redh. ;大喙豇豆■☆

408954　Vigna malosana Baker = Vigna unguiculata（L.）Walp. var. tenuis（E. Mey.）Maréchal et Mascherpa et Stainier ■☆

408955　Vigna maranguensis（Taub.）Harms = Vigna parkeri Baker subsp. maranguensis（Taub.）Verdc. ■☆

408956　Vigna marginata（E. Mey.）Benth. ;具边豇豆■☆

408957　Vigna marina（Burm.）Merr. ;滨豇豆;Seashore Cowpea ■

408958　Vigna matengoana Harms = Sphenostylis briartii（De Wild.）Baker f. ■☆

408959　Vigna membranacea A. Rich. ;膜质豇豆■☆

408960　Vigna membranacea A. Rich. subsp. caesia（Chiov.）Verdc. ;淡蓝豇豆■☆

408961　Vigna membranacea A. Rich. subsp. macrodon（Robyns et Boutique）Verdc. ;大足豇豆■☆

408962　Vigna membranaceoides Robyns et Boutique = Vigna membranacea A. Rich. ■☆

408963　Vigna mendesii Torre;门代斯豇豆■☆

408964　Vigna micrantha Chiov. = Vigna heterophylla A. Rich. ■☆

408965　Vigna micrantha Harms = Vigna comosa Baker ■☆

408966　Vigna micrantha Harms var. lebrunii（Baker f.）R. Wilczek = Vigna comosa Baker ■☆

408967　Vigna microsperma R. Vig. ;小籽豇豆●☆

408968　Vigna mildbraedii Harms = Vigna luteola（Jacq.）Benth. ■

408969　Vigna minima（Roxb.）Ohwi et H. Ohashi;贼小豆（山绿豆,台湾赤小豆,台湾小豇豆,细叶小豇豆,狭叶菜豆,小豇豆）;Mini Cowpea,Small Cowpea ■

408970　Vigna minima（Roxb.）Ohwi et H. Ohashi f. heterophylla（Hayata）Ohwi et H. Ohashi = Vigna minima（Roxb.）Ohwi et H. Ohashi var. dimorphophylla T. L. Wu ■

408971　Vigna minima（Roxb.）Ohwi et H. Ohashi f. heterophylla（Hosok.）Ohwi et H. Ohashi = Vigna minima（Roxb.）Ohwi et H. Ohashi ■

408972　Vigna minima（Roxb.）Ohwi et H. Ohashi f. linealis（Hosok.）T. C. Huang et H. Ohashi = Vigna minima（Roxb.）Ohwi et H. Ohashi ■

408973　Vigna minima（Roxb.）Ohwi et H. Ohashi var. dimorphophylla T. L. Wu;台湾小豇豆;Different Leaves Small Cowpea,Taiwan Cowpea ■

408974　Vigna minima（Roxb.）Ohwi et H. Ohashi var. linealis（Hosok.）T. C. Huang et H. Ohashi = Vigna minima（Roxb.）Ohwi et H. Ohashi ■

408975　Vigna minima（Roxb.）Ohwi et H. Ohashi var. linealis（Hosok.）T. C. Huang et H. Ohashi;细叶小豇豆（山绿豆,贼小豆）;Linearleaf Small Cowpea,Little Phaseolus,Wild Green Bean ■

408976　Vigna minima（Roxb.）Ohwi et H. Ohashi var. minor（Matsum.）Tateishi;小叶豇豆■

408977　Vigna monantha Thulin;山地豇豆■☆

408978　Vigna monophylla Taub. ;单叶豇豆■☆

408979　Vigna mudenia B. J. Pienaar;默登豇豆■☆

408980　Vigna multiflora Hook. f. = Vigna gracilis（Guillaumin et Perr.）Hook. f. var. multiflora（Hook. f.）Maréchal et al. ■☆

408981　Vigna multinervis Hutch. et Dalziel;多脉豇豆■☆

408982　Vigna mungo（L.）Hepper;小豆（吉豆,蒙戈豇豆）;Black Gram,Mung Bean,Mung-bean,Mungo Bean,Urd,Urd Bean,Woolly Pyrul ■

408983　Vigna mungo（L.）Hepper = Phaseolus mungo L. ■

408984　Vigna mungo（L.）Hepper = Vigna radiata（L.）R. Wilczek ■

408985　Vigna nakashimae（Ohwi）Ohwi et H. Ohashi = Vigna minima（Roxb.）Ohwi et H. Ohashi ■

408986　Vigna nervosa Markötter;密脉豇豆■☆

408987　Vigna neumannii Harms = Vigna frutescens A. Rich. subsp. kotschyi（Schweinf.）Verdc. ■☆

408988　Vigna nigerica A. Chev. = Vigna luteola（Jacq.）Benth. ■

408989　Vigna nilotica（Delile）Hook. f. = Vigna luteola（Jacq.）Benth. ■

408990　Vigna ntemensis Pellegr. = Vigna comosa Baker ■☆

408991　Vigna nuda N. E. Br. = Vigna antunesii Harms ■☆

408992　Vigna nyangensis Mithen;尼扬加豇豆■☆

408993　Vigna oblongifolia A. Rich. ;矩圆叶豇豆■☆

408994　Vigna oblongifolia A. Rich. var. parviflora（Baker）Verdc. ;小花矩圆叶豇豆■☆

408995　Vigna occidentalis Baker f. = Vigna gracilis（Guillaumin et Perr.）Hook. f. ■☆

408996　Vigna oligosperma Baker = Vigna hosei（Craib）Backer ex K. Heyne ■

408997　Vigna ornata Welw. ex Baker = Sphenostylis stenocarpa（Hochst. ex A. Rich.）Harms ■☆

408998　Vigna ostinii Chiov. = Vigna monophylla Taub. ■☆

408999　Vigna oubanguensis Pellegr. = Vigna radicans Welw. ex Baker ■☆

409000　Vigna paludosa Milne-Redh. = Vigna longifolia（Benth.）Verdc. ■☆

409001　Vigna parkeri Baker;帕克豇豆■☆

409002　Vigna parkeri Baker subsp. acutifolia Verdc. = Vigna parkeri Baker subsp. maranguensis（Taub.）Verdc. ■☆

409003　Vigna parkeri Baker subsp. maranguensis（Taub.）Verdc. ;马兰古豇豆■☆

409004　Vigna parviflora Baker = Vigna oblongifolia A. Rich. var. parviflora（Baker）Verdc. ■☆

409005　Vigna parvifolia Planch. ex Baker = Vigna gracilis（Guillaumin et Perr.）Hook. f. ■☆

409006　Vigna phaseoloides Baker = Vigna vexillata（L.）A. Rich. ■

409007　Vigna phoenix Brummitt;凤凰豇豆■☆

409008　Vigna pilosa（Klein ex Willd.）Baker ex K. Heyne;毛豇豆;Dolicho Vigna,Hair Cowpea,Hairy Cowpea,Pilose Sicklelobe ■

409009　Vigna platyloba Welw. ex Hiern;宽裂片豇豆■☆

409010　Vigna polytricha Baker = Vigna reticulata Hook. f. ●☆

409011　Vigna pongolensis Burtt Davy = Vigna frutescens A. Rich. ■☆

409012　Vigna praecox Verdc. = Wajira praecox（Verdc.）Thulin et Lavin ■☆

409013　Vigna proboscidella Chiov. = Wajira grahamiana（Wight et Arn.）Thulin et Lavin ■☆

409014　Vigna procera Welw. ex Hiern;高大豇豆■☆

409015　Vigna pseudolablab Harms = Vatovaea pseudolablab（Harms）J. B. Gillett ■☆

409016　Vigna pseudotriloba Harms = Vigna frutescens A. Rich. ■☆

409017　Vigna pubescens R. Wilczek = Vigna unguiculata (L.) Walp. subsp. protracta (E. Mey.) B. J. Pienaar ■☆

409018　Vigna pubigera Baker = Vigna ambacensis Welw. ex Baker var. pubigera (Baker) Maréchal et al. ■☆

409019　Vigna punctata Micheli = Adenodolichos punctatus (Micheli) Harms ■☆

409020　Vigna pusilla A. Chev. ;微小豇豆■☆

409021　Vigna pygmaea R. E. Fr. ;矮小豇豆■☆

409022　Vigna pygmaea R. E. Fr. var. grandiflora Verdc. = Vigna phoenix Brummitt ■☆

409023　Vigna racemosa (G. Don) Hutch. et Dalziel;总花豇豆■☆

409024　Vigna racemosa (G. Don) Hutch. et Dalziel. f. glabrescens Baker f. = Vigna racemosa (G. Don) Hutch. et Dalziel ■☆

409025　Vigna radiata (L.) R. Wilczek;绿豆(拔绿,官绿,荨豆,青小豆,文豆,油绿,摘绿,植豆);Bean Sprouts, Beans, Golden Gram, Green Bean, Green Gram, Greenbean, Mung, Mung Bean, Mung-hean ■

409026　Vigna radiata (L.) R. Wilczek var. sublobata (Roxb.) Verdc. ;三裂叶绿豆(三裂叶豇豆)■☆

409027　Vigna radicans Welw. ex Baker;辐射豇豆■☆

409028　Vigna reflexopilosa Hayata;卷毛豇豆(曲毛豇豆);Crinklehair Cowpea ■

409029　Vigna repens (L.) Kuntze = Vigna luteola (Jacq.) Benth. ■☆

409030　Vigna reticulata Hook. f. ;网状豇豆●☆

409031　Vigna reticulata Hook. f. var. linearifolia (Hook. f.) Baker = Vigna reticulata Hook. f. ●☆

409032　Vigna retusa (E. Mey.) Walp. = Vigna marina (Burm.) Merr. ■

409033　Vigna rhomboidea Burtt Davy = Vigna unguiculata (L.) Walp. subsp. protracta (E. Mey.) B. J. Pienaar ■☆

409034　Vigna richardsiae Verdc. ;理查兹豇豆■☆

409035　Vigna ringoetii (De Wild.) De Wild. = Vigna antunesii Harms ■☆

409036　Vigna riukiuensis (Ohwi) Ohwi et H. Ohashi;琉球豇豆;Liuqiu Cowpea, Riukiu Cowpea ■

409037　Vigna riukiuensis (Ohwi) Ohwi et H. Ohashi = Vigna minima (Roxb.) Ohwi et H. Ohashi var. minor (Matsum.) Tateishi ■

409038　Vigna scabra (De Wild.) T. Durand et H. Durand = Vigna unguiculata (L.) Walp. var. spontanea (Schweinf.) Pasquet ■☆

409039　Vigna scabrida Burtt Davy = Vigna unguiculata (L.) Walp. var. spontanea (Schweinf.) Pasquet ■☆

409040　Vigna schimperi Baker;欣珀豇豆■☆

409041　Vigna schlechteri Harms;施莱豇豆■☆

409042　Vigna schliebenii Harms = Vigna kirkii (Baker) J. B. Gillett ■☆

409043　Vigna senegalensis A. Chev. = Vigna vexillata (L.) A. Rich. ■

409044　Vigna sesquipedalis (L.) Fruwirth = Dolichos sesquipedalis L. ■

409045　Vigna sesquipedalis (L.) Fruwirth = Vigna unguiculata (L.) Walp. subsp. sesquipedalis (L.) Verdc. ■

409046　Vigna sesquipedalis (L.) Fruwirth var. purpurascens Nakai;紫花长豇豆■

409047　Vigna sinensis (L.) Endl. ex Hassk. = Vigna unguiculata (L.) Walp. ■

409048　Vigna sinensis (L.) Endl. ex Hassk. var. contorta Nakai;眼镜菜豆■

409049　Vigna sinensis (L.) Endl. ex Hassk. var. monachalis Nakai;鞍菜豆■

409050　Vigna sinensis (L.) Endl. ex Hassk. var. sanguinea Nakai;红菜豆■

409051　Vigna sinensis (L.) Hassk. var. spontanea Schweinf. = Vigna unguiculata (L.) Walp. var. spontanea (Schweinf.) Pasquet ■☆

409052　Vigna sinensis (L.) Savi = Vigna unguiculata (L.) Walp. ■

409053　Vigna sinensis (L.) Savi ex Hassk. = Vigna unguiculata (L.) Walp. ■

409054　Vigna sinensis (L.) Savi subsp. cylindrica (L.) Van Eselt. = Vigna unguiculata (L.) Walp. subsp. cylindrica (L.) Verdc. ■

409055　Vigna sinensis (L.) Savi subsp. sesquipedalis (L.) Van Eselt. = Vigna unguiculata (L.) Walp. subsp. sesquipedalis (L.) Verdc. ■

409056　Vigna sinensis (L.) Savi var. catjang (Burm. f.) Chiov. = Vigna unguiculata (L.) Walp. subsp. cylindrica (L.) Verdc. ■

409057　Vigna sinensis (L.) Savi var. sesquipedalis (L.) Asch. et Schweinf. = Vigna unguiculata (L.) Walp. subsp. sesquipedalis (L.) Verdc. ■

409058　Vigna somaliensis Baker f. ;索马里豇豆■☆

409059　Vigna spartioides Taub. = Vigna frutescens A. Rich. var. buchneri (Harms) Verdc. ■☆

409060　Vigna speciosa (Kunth) Verdc. ;美丽豇豆;Wondering Cowpea ■☆

409061　Vigna stenocarpa Engl. ;细果豇豆■☆

409062　Vigna stenodactyla Harms = Vigna radicans Welw. ex Baker ■☆

409063　Vigna stenophylla (Harv.) Burtt Davy = Vigna unguiculata (L.) Walp. subsp. stenophylla (Harv.) Maréchal et al. ■☆

409064　Vigna stenophylla Harms;窄叶豇豆■☆

409065　Vigna stipulata Hayata;黑种豇豆;Black Cowpea, Peltate-stipule Cowpea ■

409066　Vigna stipulata Hayata = Vigna radiata (L.) Wilczek var. sublobata (Roxb.) Verdc. ■

409067　Vigna strophiolata Piper = Vigna racemosa (G. Don) Hutch. et Dalziel ■☆

409068　Vigna stuhlmannii Harms = Vigna heterophylla A. Rich. ■☆

409069　Vigna subterranea (L.) Verdc. ;地下豇豆(非洲花生);African Peanut, Bambara Groundnut, Congo Goober ■☆

409070　Vigna subterranea (L.) Verdc. var. spontanea (Harms) Pasquet;野生地下豇豆■☆

409071　Vigna sudanica Baker f. = Vigna frutescens A. Rich. ■☆

409072　Vigna taubertii Volkens ex Harms = Vigna frutescens A. Rich. ■☆

409073　Vigna tenuis (E. Mey.) D. Dietr. = Vigna unguiculata (L.) Walp. var. tenuis (E. Mey.) Maréchal et Mascherpa et Stainier ■☆

409074　Vigna thonningii Hook. f. = Vigna vexillata (L.) A. Rich. ■

409075　Vigna tisserantiana Pellegr. = Vigna tisserantii Pellegr. ■☆

409076　Vigna tisserantii Pellegr. ;蒂斯朗特豇豆■☆

409077　Vigna triloba Walp. = Vigna unguiculata (L.) Walp. subsp. stenophylla (Harv.) Maréchal et al. ■☆

409078　Vigna triloba Walp. var. somalensis Chiov. = Vigna monantha Thulin ■☆

409079　Vigna triloba Walp. var. stenophylla Harv. = Vigna unguiculata (L.) Walp. subsp. stenophylla (Harv.) Maréchal et al. ■☆

409080　Vigna trilobata (L.) Verdc. ;三裂叶豇豆(三瓣菜豆,三裂叶菜豆);Trilobate Cowpea, Trilobate Leaf Cowpea ■

409081　Vigna triphylla (R. Wilczek) Verdc. ;三叶豇豆■☆

409082　Vigna tuberosa A. Rich. = Vigna vexillata (L.) A. Rich. ■

409083　Vigna uliginosa R. Vig. = Vigna oblongifolia A. Rich. ■☆

409084　Vigna ulugurensis Harms = Vigna friesiorum Harms var. ulugurensis (Harms) Verdc. ■☆

409085　Vigna umbellata (Thunb.) Ohwi et H. Ohashi;赤小豆(赤豆,春湾豆,蛋白豆,杜赤豆,饭豆,腐婢,红豆,红小豆,金红小豆,米豆,虱犆豆,朱赤豆,朱小豆,猪肝赤);Avecote, Chixiaodou Cowpea, Jack's Ladder, Multiflora Bean, Rice Bean, Scarlet Runner, Scarlet Runner Bean ■

409086　Vigna unguiculata（L.）Walp.；豇豆（八月角豆，长豆，长荚豆，豆角，饭豆，粉豆，浆豆，角豆，角豆角，戳豆，米豆，裙带豆，羊豆，羊角，腰豆，紫豇豆）；Asparagus Bean，Black-eyed Bean，Blackeyed Pea，Black-eyed Pea，Chinese Long Bean，Common Cowpea，Common Cow-pea，Cowpea，Cow-pea，Cuba Bean，Gubgub，Southern Pea，Yard-long Bean，Yawa ■

409087　Vigna unguiculata（L.）Walp. subsp. alba（G. Don）Pasquet；白豇豆■☆

409088　Vigna unguiculata（L.）Walp. subsp. baoulensis（A. Chev.）Pasquet；几内亚豇豆■☆

409089　Vigna unguiculata（L.）Walp. subsp. burundiensis Pasquet；布隆迪豇豆■☆

409090　Vigna unguiculata（L.）Walp. subsp. catjang（Burm. f.）A. Chev. = Vigna unguiculata（L.）Walp. subsp. cylindrica（L.）Verdc.■

409091　Vigna unguiculata（L.）Walp. subsp. cylindrica（L.）Van Eselt. = Vigna unguiculata（L.）Walp.■

409092　Vigna unguiculata（L.）Walp. subsp. cylindrica（L.）Verdc.；短豇豆（白豆，白目豆，菜豆，短荚豇豆，饭豆，饭豇豆，甘豆，眉豆，旗豇豆）；Browbean，Catiang Bean，Catjang Cowpea，Catjang Cow-pea，Hindu Cowpea，Marble Pea ■

409093　Vigna unguiculata（L.）Walp. subsp. letouzeyi Pasquet；勒图豇豆■☆

409094　Vigna unguiculata（L.）Walp. subsp. protracta（E. Mey.）B. J. Pienaar；伸长豇豆■☆

409095　Vigna unguiculata（L.）Walp. subsp. pubescens（R. Wilczek）Pasquet；短柔毛豇豆■☆

409096　Vigna unguiculata（L.）Walp. subsp. sesquipedalis（L.）Verdc.；长豇豆（菜豆，尺八豇，豆角，裙带豆，十八豇，十六豇豆）；Asparagus Bean，Asparagus Cowpea，Asparagus Pea，Chowlee，Cow Pea，Gubgub，Snake Bean，Yard Long Cowpea，Yard-long Bean，Yardlong Cowpea，Yard-long Cow-pea，Yawa ■

409097　Vigna unguiculata（L.）Walp. subsp. stenophylla（Harv.）Maréchal et al.；窄叶长豇豆■☆

409098　Vigna unguiculata（L.）Walp. var. catjang（Burm. f.）Bertoni = Vigna unguiculata（L.）Walp. subsp. cylindrica（L.）Verdc.■

409099　Vigna unguiculata（L.）Walp. var. cylindrica（L.）H. Ohashi = Vigna unguiculata（L.）Walp. subsp. cylindrica（L.）Verdc.■

409100　Vigna unguiculata（L.）Walp. var. dekindtiana（Harms）Verdc.；德金豇豆■☆

409101　Vigna unguiculata（L.）Walp. var. huillensis（Welw. ex Baker）B. J. Pienaar；威拉豇豆■☆

409102　Vigna unguiculata（L.）Walp. var. huillensis（Welw. ex Baker）Mithen = Vigna unguiculata（L.）Walp. var. dekindtiana（Harms）Verdc.■☆

409103　Vigna unguiculata（L.）Walp. var. ovata（E. Mey.）B. J. Pienaar；卵豇豆■☆

409104　Vigna unguiculata（L.）Walp. var. protracta（E. Mey.）Verdc. = Vigna unguiculata（L.）Walp. subsp. stenophylla（Harv.）Maréchal et al.■☆

409105　Vigna unguiculata（L.）Walp. var. pubescens（R. Wilczek）Maréchal et Mascherpa et Stainier = Vigna unguiculata（L.）Walp. subsp. pubescens（R. Wilczek）Pasquet■☆

409106　Vigna unguiculata（L.）Walp. var. sesquipedalis（L.）H. Ohashi = Vigna unguiculata（L.）Walp. subsp. sesquipedalis（L.）Verdc.■

409107　Vigna unguiculata（L.）Walp. var. spontanea（Schweinf.）Pasquet；野生豇豆■☆

409108　Vigna unguiculata（L.）Walp. var. stenophylla（Harv.）Mithen = Vigna unguiculata（L.）Walp. subsp. stenophylla（Harv.）Maréchal et al.■☆

409109　Vigna unguiculata（L.）Walp. var. tenuis（E. Mey.）Maréchal et Mascherpa et Stainier；细豇豆■☆

409110　Vigna unguiculata（L.）Walp. var. tenuis（E. Mey.）Mithen = Vigna unguiculata（L.）Walp. var. tenuis（E. Mey.）Maréchal et Mascherpa et Stainier■☆

409111　Vigna venulosa Baker；细脉豇豆■☆

409112　Vigna verticillata Engl.；轮生豇豆■☆

409113　Vigna vexillata（L.）A. Rich.；野豇豆（红参，假人参，山豆根，山马豆，山土瓜，土白参，土高丽参，野绿豆，野马豆，野蛮豆，云南山土瓜，云南野豇豆）；Field Cowpea ■

409114　Vigna vexillata（L.）A. Rich. var. angustifolia（Schumach. et Thonn.）Baker；窄叶野豇豆■☆

409115　Vigna vexillata（L.）A. Rich. var. davyi（Bolus）B. J. Pienaar；戴维豇豆■☆

409116　Vigna vexillata（L.）A. Rich. var. dolichonema（Harms）Verdc. = Vigna vexillata（L.）A. Rich.■

409117　Vigna vexillata（L.）A. Rich. var. hirta（Hook.）Baker f. = Vigna vexillata（L.）A. Rich.■

409118　Vigna vexillata（L.）A. Rich. var. lobatifolia（Baker）Pasquet；裂叶野豇豆■☆

409119　Vigna vexillata（L.）A. Rich. var. ovata（E. Mey.）B. J. Pienaar；卵叶野豇豆■☆

409120　Vigna vexillata（L.）A. Rich. var. puriflora Franch. = Vigna vexillata（L.）A. Rich.■

409121　Vigna vexillata（L.）A. Rich. var. thonningii（Hook. f.）Baker = Vigna vexillata（L.）A. Rich.■

409122　Vigna vexillata（L.）A. Rich. var. tsusimensis Matsum.；津岛野豇豆■

409123　Vigna vexillata（L.）A. Rich. var. tuberosa（A. Rich.）Chiov. = Vigna vexillata（L.）A. Rich.■

409124　Vigna vexillata（L.）A. Rich. var. yunnanensis Franch.；云南野豇豆（山豆根，山马豆，细活血，野马豆，野汤豆，云南山土瓜）；Yunnan Cowpea ■

409125　Vigna vexillata（L.）A. Rich. var. yunnanensis Franch. = Vigna vexillata（L.）A. Rich.■

409126　Vigna violacea Hutch. = Vigna frutescens A. Rich.■☆

409127　Vigna virescens Thulin；浅绿豇豆■☆

409128　Vigna wilmsii Burtt Davy = Vigna oblongifolia A. Rich.■☆

409129　Vigna wittei Baker f. = Vigna radicans Welw. ex Baker■☆

409130　Vignaldia A. Rich. = Pentas Benth.●■

409131　Vignaldia occidentalis Hook. f. = Pentas schimperiana（A. Rich.）Vatke subsp. occidentalis（Hook. f.）Verdc.●☆

409132　Vignaldia quartiniana A. Rich. = Pentas lanceolata（Forssk.）Deflers var. quartiniana（A. Rich.）Verdc.●☆

409133　Vignaldia quartiniana A. Rich. var. grandiflora Schweinf. = Pentas lanceolata（Forssk.）K. Schum.●

409134　Vignaldia schimperiana A. Rich. = Pentas schimperiana（A. Rich.）Vatke●☆

409135　Vignantha Schur = Carex L.■

409136　Vignaudia Schweinf. = Pentas Benth.●■

409137　Vignaudia Schweinf. = Vignaldia A. Rich.●■

409138　Vignaudia luteola（Delile）Schweinf. = Pseudomussaenda flava Verdc.■☆

409139　Vignea P. Beauv. = Carex L.■

409140　Vignea P. Beauv. ex T. Lestib. = Carex L. ■

409141　Vigneopsis De Wild. (1902);热非豇豆属■☆

409142　Vigneopsis De Wild. = Psophocarpus Neck. ex DC. (保留属名)■

409143　Vigneopsis lukafuensis De Wild. ;热非豇豆■☆

409144　Vignidula Börner = Carex L. ■

409145　Vignopsis lukafuensis De Wild. = Psophocarpus lukafuensis (De Wild.) R. Wilczek ■☆

409146　Vigolina Poir. = Galinsoga Ruiz et Pav. ■●

409147　Viguiera Kunth(1818);金目菊属(金眼菊属,维格菊属); Golden Eye,Goldeneye ●■☆

409148　Viguiera annua (Jones) S. F. Blake;一年金眼菊;Annual Golden Eye ■☆

409149　Viguiera annua (M. E. Jones) S. F. Blake = Heliomeris longifolia (B. L. Rob. et Greenm.) Cockerell var. annua (M. E. Jones) W. F. Yates ■☆

409150　Viguiera ciliata (B. L. Rob. et Greenm.) S. F. Blake = Heliomeris hispida (A. Gray) Cockerell ■☆

409151　Viguiera ciliata (B. L. Rob. et Greenm.) S. F. Blake var. hispida (A. Gray) S. F. Blake = Heliomeris hispida (A. Gray) Cockerell ■☆

409152　Viguiera cordifolia A. Gray;心叶金眼菊;Heartleaf Goldeneye ■☆

409153　Viguiera deltoidea A. Gray;金目菊(金眼菊);Golden Eye ■☆

409154　Viguiera deltoidea A. Gray var. parishii (Greene) Vasey et Rose = Bahiopsis parishii (Greene) E. E. Schill. et Panero ■☆

409155　Viguiera dentata (Cav.) Spreng. ;齿状金眼菊(齿状维格菊);Sunflower Goldeneye,Toothleaf ■☆

409156　Viguiera fusiformis S. F. Blake;纺锤金眼菊■☆

409157　Viguiera laciniata A. Gray = Bahiopsis laciniata (A. Gray) E. E. Schill. et Panero ■☆

409158　Viguiera lanceolata Britton;剑叶金眼菊●☆

409159　Viguiera longifolia (B. L. Rob. et Greenm.) S. F. Blake = Heliomeris longifolia (B. L. Rob. et Greenm.) Cockerell ■☆

409160　Viguiera macrorhiza Baker;大根金眼菊■☆

409161　Viguiera multiflora (Nutt.) S. F. Blake = Heliomeris multiflora Nutt. ■☆

409162　Viguiera oligodonta S. F. Blake = Viguiera tucumanensis (Hook. et Arn.) Griseb. ●☆

409163　Viguiera ovalis S. F. Blake = Heliomeris multiflora Nutt. var. brevifolia (Greene ex Wooton et Standl.) W. F. Yates ■☆

409164　Viguiera parishii Greene = Bahiopsis parishii (Greene) E. E. Schill. et Panero ■☆

409165　Viguiera pinnatilobata (Sch. Bip.) S. F. Blake;羽裂金眼菊(羽裂维格菊)■☆

409166　Viguiera porteri (A. Gray) S. F. Blake = Helianthus porteri (A. Gray) Pruski ■☆

409167　Viguiera reticulata S. Watson = Bahiopsis reticulata (S. Watson) E. E. Schill. et Panero ■☆

409168　Viguiera soliceps Barneby = Heliomeris soliceps (Barneby) W. F. Yates ■☆

409169　Viguiera stenoloba S. F. Blake;狭裂金眼菊;Skeletonleaf Goldeneye ■☆

409170　Viguiera stenophylla Griseb. ;狭叶金眼菊■☆

409171　Viguiera triloba (A. Gray) Johan-Olsen = Zaluzania grayana B. L. Rob. et Greenm. ■☆

409172　Viguiera tucumanensis (Hook. et Arn.) Griseb. ;土库曼金眼菊●☆

409173　Viguieranthus Villiers(2002);维吉豆属●☆

409174　Viguieranthus alternans (Benth.) Villiers;互生维吉豆●☆

409175　Viguieranthus ambongensis (R. Vig.) Villiers;安邦维吉豆●☆

409176　Viguieranthus brevipennatus Villiers;短羽状维吉豆●☆

409177　Viguieranthus cylindricostachys Villiers;柱穗维吉豆●☆

409178　Viguieranthus densinervus Villiers;密脉维吉豆●☆

409179　Viguieranthus glaber Villiers;光滑维吉豆●☆

409180　Viguieranthus glandulosus Villiers;具腺维吉豆●☆

409181　Viguieranthus longiracemosus Villiers;长序维吉豆●☆

409182　Viguieranthus megalophyllus (R. Vig.) Villiers;大叶维吉豆●☆

409183　Viguieranthus perrieri (R. Vig.) Villiers;佩里耶维吉豆●☆

409184　Viguieranthus pervillei (Drake) Villiers;佩尔维吉豆●☆

409185　Viguieranthus scottianus (R. Vig.) Villiers;司科特维吉豆●☆

409186　Viguieranthus simulans (R. Vig.) Villiers;相似维吉豆●☆

409187　Viguieranthus subauriculatus Villiers;耳形维吉豆●☆

409188　Viguieranthus umbilicus Villiers;脐维吉豆●☆

409189　Viguieranthus variabilis Villiers;易变维吉豆●☆

409190　Viguierella A. Camus(1926);维吉禾属(马岛旱禾属,维格尔禾属)■☆

409191　Viguierella madagascariensis A. Camus;维吉禾(马岛旱禾,维格尔禾)■☆

409192　Vilaria Guett. (废弃属名) = Berardia Vill. ■☆

409193　Vilaria Guett. (废弃属名) = Villaria Rolfe(保留属名)●☆

409194　Vilbouchevitchia A. Chev. = Alafia Thouars ■☆

409195　Vilbouchevitchia atropurpurea A. Chev. = Alafia whytei Stapf ●☆

409196　Vilfa Adans. = Agrostis L. (保留属名)■

409197　Vilfa P. Beauv. = Sporobolus R. Br. ■

409198　Vilfa alba (L.) P. Beauv. var. ramosa Gray. = Agrostis gigantea Roth ■

409199　Vilfa arabica (Boiss.) Steud. = Sporobolus arabicus Boiss. ■☆

409200　Vilfa arguta Nees = Sporobolus coromandelianus (Retz.) Kunth ■

409201　Vilfa brachystachys C. Presl = Crypsis alopecuroides (Piller et Mitterp.) Schrad. ■☆

409202　Vilfa capensis P. Beauv. = Sporobolus africanus (Poir.) Robyns et Tournay ■☆

409203　Vilfa centrifuga Trin. = Sporobolus centrifugus (Trin.) Nees ■☆

409204　Vilfa commutata Trin. = Sporobolus coromandelianus (Retz.) Kunth ■

409205　Vilfa confinis Steud. = Sporobolus confinis (Steud.) Chiov. ■☆

409206　Vilfa coromandeliana (Retz.) P. Beauv. = Sporobolus coromandelianus (Retz.) Kunth ■

409207　Vilfa diandra (Retz.) Trin. = Sporobolus diander (Retz.) P. Beauv. ■

409208　Vilfa fimbriata Trin. = Sporobolus fimbriatus (Trin.) Nees ■☆

409209　Vilfa geniculata Nees ex Steud. = Sporobolus tremulus (Willd.) Kunth ■☆

409210　Vilfa glaucifolia Hochst. ex Steud. = Sporobolus helvolus (Trin.) T. Durand et Schinz ■☆

409211　Vilfa glaucifolia Steud. = Sporobolus helvolus (Trin.) T. Durand et Schinz ■☆

409212　Vilfa helvola Trin. = Sporobolus helvolus (Trin.) T. Durand et Schinz ■☆

409213　Vilfa heterolepis A. Gray = Sporobolus heterolepis (A. Gray) A. Gray ■☆

409214　Vilfa ioclados Nees ex Trin. = Sporobolus ioclados (Nees ex Trin.) Nees ■☆

409215　Vilfa ioclados Trin. = Sporobolus ioclados (Trin.) Nees ■☆

409216　Vilfa marginata (Hochst. ex A. Rich.) Steud. = Sporobolus ioclados (Nees ex Trin.) Nees ■☆

409217　Vilfa mauritiana Steud. = Sporobolus mauritianus（Steud.）T. Durand et Schinz ■☆

409218　Vilfa minutiflora Trin. = Sporobolus tenuissimus（Mart. ex Schrank）Kuntze ■

409219　Vilfa natalensis Steud. = Sporobolus natalensis（Steud.）T. Durand et Schinz ■☆

409220　Vilfa nervosa（Hochst.）Schweinf. = Sporobolus nervosus Hochst. ■☆

409221　Vilfa pallida Nees ex Trin. = Sporobolus ioclados（Nees ex Trin.）Nees ■☆

409222　Vilfa paniculata Trin. = Sporobolus paniculatus（Trin.）T. Durand et Schinz ■☆

409223　Vilfa pilifera Trin. = Sporobolus pilifer（Trin.）Kunth ■

409224　Vilfa retzii Steud. = Sporobolus diander（Retz.）P. Beauv. ■

409225　Vilfa setulosa Trin. = Urochondra setulosa（Trin.）C. E. Hubb. ■☆

409226　Vilfa tremula（Willd.）Trin. = Sporobolus tremulus（Willd.）Kunth ■☆

409227　Vilfa virginica（L.）P. Beauv. = Sporobolus virginicus（L.）Kunth ■

409228　Vilfagrostis A. Br. et Asch. ex Döll = Eragrostis P. Beauv. ■

409229　Vilfagrostis Döll = Eragrostis Wolf ■

409230　Villadia Rose（1903）；塔莲属■☆

409231　Villadia acuta Moran et C. H. Uhl；尖塔莲■☆

409232　Villadia albiflora（Hemsl.）Rose；白花塔莲■☆

409233　Villadia alpina（Fröd.）H. Jacobsen；高山塔莲■☆

409234　Villadia diffusa Rose；铺散塔莲■☆

409235　Villadia fusca（Hemsl.）H. Jacobsen；褐塔莲■☆

409236　Villadia mexicana（Schltdl.）H. Jacobsen；墨西哥塔莲■☆

409237　Villadia minutiflora Rose；小花塔莲■☆

409238　Villadia platyphylla（Rose）E. Walther；宽叶塔莲■☆

409239　Villamilla（Moq.）Benth. et Hook. f. = Trichostigma A. Rich. ●☆

409240　Villamilla Ruiz et Pav. = Trichostigma A. Rich. ●☆

409241　Villamilla Ruiz et Pav. ex Moq. = Trichostigma A. Rich. ●☆

409242　Villamillia Lopez = Villamilla Ruiz et Pav. ●☆

409243　Villanova Lag.（1816）（保留属名）；扁角菊属■☆

409244　Villanova Ortega（废弃属名）= Parthenium L. ■●

409245　Villanova Ortega（废弃属名）= Villanova Lag.（保留属名）■☆

409246　Villanova Pourr. ex Cutanda = Colmeiroa Reut. ●☆

409247　Villanova Pourr. ex Cutanda = Flueggea Willd. ●

409248　Villanova Pourr. ex Cutanda = Securinega Comm. ex Juss.（保留属名）●☆

409249　Villanova alternifolia Lag.；互叶扁角菊■☆

409250　Villanova oppositifolia Lag.；对叶扁角菊■☆

409251　Villaresia Ruiz et Pav.（1794）；维拉木属；Villaresia ●☆

409252　Villaresia Ruiz et Pav.（1794）= Villanova Lag.（保留属名）■☆

409253　Villaresia Ruiz et Pav.（1802）= Citronella D. Don ●☆

409254　Villaresia adenophylla Domin；腺叶维拉木●☆

409255　Villaresia chilensis Stuntz；智利维拉木●☆

409256　Villaresia gonghona C. Müll.；维拉木●☆

409257　Villaresia grandifolia Fisch. et C. A. Mey. ex Engl.；大叶维拉木●☆

409258　Villaresia latifolia Merr.；宽叶维拉木●☆

409259　Villaresia macrocarpa Scheff.；大果维拉木●☆

409260　Villaresia philippinensis Merr.；菲律宾维拉木●☆

409261　Villaresiopsis Sleumer = Citronella D. Don ●☆

409262　Villarezia Rocm. et Schult. = Villaresia Ruiz et Pav. ●☆

409263　Villaria Bally = Villarsia Vent.（保留属名）■☆

409264　Villaria DC. = Berardia Vill. ■☆

409265　Villaria DC. = Villaria Rolfe（保留属名）●☆

409266　Villaria Rolfe（1884）（保留属名）；维勒茜属●☆

409267　Villaria Schreb. = Villaria Rolfe（保留属名）●☆

409268　Villaria V. Bally = Villarsia Vent.（保留属名）■☆

409269　Villaria acutifolia Merr.；尖叶维勒茜●☆

409270　Villaria philippinensis Rolfe；维勒茜●☆

409271　Villarsia J. F. Gmel.（废弃属名）= Nymphoides Ség. ■

409272　Villarsia J. F. Gmel.（废弃属名）= Villarsia Vent.（保留属名）■☆

409273　Villarsia Neck. = Cabomba Aubl. ■

409274　Villarsia Neck. = Villarsia Vent.（保留属名）■☆

409275　Villarsia Post et Kuntze = Berardia Vill. ■☆

409276　Villarsia Post et Kuntze = Villarsia Vent.（保留属名）■☆

409277　Villarsia Sm. = Villaria Schreb. ●☆

409278　Villarsia Sm. = Villarsia Vent.（保留属名）■☆

409279　Villarsia Vent.（1803）（保留属名）；维拉尔睡菜属■☆

409280　Villarsia capensis（Houtt.）Merr.；好望角维拉尔睡菜■☆

409281　Villarsia manningiana Ornduff；曼宁维拉尔睡菜■☆

409282　Villarsia ovata（L. f.）Vent. = Villarsia capensis（Houtt.）Merr. ■☆

409283　Villarsia senegalensis G. Don = Nymphoides indica（L.）Kuntze subsp. occidentalis A. Raynal ■☆

409284　Villasenoria B. L. Clark（1999）；大羽千里光属■☆

409285　Villasenoria orcuttii（Greenm.）B. L. Clark；大羽千里光■☆

409286　Villebrunea Gaudich. = Oreocnide Miq. ●

409287　Villebrunea Gaudich. ex Wedd. = Oreocnide Miq. ●

409288　Villebrunea appendiculata Wedd. = Oreocnide integrifolia（Gaudich.）Miq. ●

409289　Villebrunea boniana Gagnep. = Oreocnide boniana（Gagnep.）Hand. -Mazz. ●

409290　Villebrunea frutescens（Thunb.）Blume = Oreocnide frutescens（Thunb.）Miq. subsp. occidentalis C. J. Chen ●

409291　Villebrunea frutescens（Thunb.）Blume var. hirsuta Pamp. = Oreocnide frutescens（Thunb.）Miq. ●

409292　Villebrunea fruticosa（Gaudich.）Nakai = Oreocnide frutescens（Thunb.）Miq. ●

409293　Villebrunea fruticosa Nakai = Oreocnide pedunculata（Shirai）Masam. ●

409294　Villebrunea intefrifolia Gaudich. = Oreocnide integrifolia（Gaudich.）Miq. subsp. subglabra C. J. Chen ●

409295　Villebrunea integrifolia Gaudich. = Oreocnide integrifolia（Gaudich.）Miq. ●

409296　Villebrunea integrifolia Gaudich. var. sylvatica Hook. f. = Oreocnide rubescens（Blume）Miq. ●

409297　Villebrunea microcephala（Benth.）Nakai = Oreocnide frutescens（Thunb.）Miq. ●

409298　Villebrunea paradoxa Gagnep. = Oreocnide obovata（C. H. Wright）Merr. var. paradoxa（Gagnep.）C. J. Chen ●

409299　Villebrunea pedunculata Shirai = Oreocnide pedunculata（Shirai）Masam. ●

409300　Villebrunea petelotii Gagnep. = Oreocnide obovata（C. H. Wright）Merr. ●

409301　Villebrunea rubescens（Blume）Blume = Oreocnide rubescens（Blume）Miq. ●

409302　Villebrunea scabra（Blume）Wedd. = Oreocnide rubescens（Blume）Miq. ●

409303　Villebrunea sylvatica（Blume）Blume = Oreocnide rubescens（Blume）Miq. ●

409304　Villebrunea sylvatica（Blume）Blume var. integrifolia Wedd. =

Oreocnide integrifolia（Gaudich.）Miq. subsp. subglabra C. J. Chen ●

409305　Villebrunea sylvatica Blume var. integrifolia Wedd. = Oreocnide integrifolia（Gaudich.）Miq. ●

409306　Villebrunea tonkinensis Gagnep. = Oreocnide tonkinensis（Gagnep.）Merr. et Chun ●

409307　Villebrunea trinervis Wedd. = Oreocnide trinervis（Wedd.）Miq. ●

409308　Villebrunia Willis = Villebrunea Gaudich. ●

409309　Villemetia Moq. = Chenolea Thunb. ●☆

409310　Villemetia Moq. = Willemetia Maerkl. ■☆

409311　Villocuspis（A. DC.）Aubrév. et Pellegr. = Chrysophyllum L. ●

409312　Villosogastris Thouars = Limodorum Boehm.（保留属名）■☆

409313　Villosogastris Thouars = Phaius Lour. ■

409314　Villouratea Tiegh. = Ouratea Aubl.（保留属名）●

409315　Vilmorinia DC. = Poitea Vent. ●☆

409316　Vilobia Strother = Tagetes L. ■●

409317　Vilobia Strother（1968）;玻利维亚菊属■●☆

409318　Vilobia praetermissa Strother;玻利维亚菊■●☆

409319　Vimen P. Browne = Hyperbaena Miers ex Benth.（保留属名）●☆

409320　Vimen Raf. = Salix L.（保留属名）●

409321　Viminaria Sm.（1805）;澳洲豆树属（折枝扫帚属）●☆

409322　Viminaria denudata（Vent.）Sm. = Viminaria juncea（Schrad.）Hoffmanns. ●☆

409323　Viminaria denudata Sm. = Viminaria juncea（Schrad.）Hoffmanns. ●☆

409324　Viminaria juncea（Schrad.）Hoffmanns.;澳洲豆树;Australian Native Broom, Native Broom ●☆

409325　Vinca L.（1753）;蔓长春花属（长春花属）;Periwinkle ■

409326　Vinca L. = Catharanthus G. Don ●■

409327　Vinca difformis Pourr.;异形蔓长春花（异形长春花）;Diverse Periwinkle, Intermediate Periwinkle ■

409328　Vinca difformis Pourr. var. dubia（Batt.）Maire = Vinca difformis Pourr. ■

409329　Vinca erecta Regel et Schmalh.;直立长春花;Erect Periwinkle ■☆

409330　Vinca grandiflora Salisb. = Vinca major L. ■

409331　Vinca guilelmi-waldemarii Klotzsch = Catharanthus roseus（L.）G. Don ■

409332　Vinca herbacea Waldst. et Kit. = Vinca major L. ■

409333　Vinca lancea Bojer ex A. DC. = Catharanthus lanceus（Bojer ex A. DC.）Pichon ■☆

409334　Vinca libanonica Zucc.;黎巴嫩长春花;Libanon Periwinkle ■☆

409335　Vinca major L.;蔓长春花（草本长春花,草本蔓长春花,草质长春花,大长春花,大蔓长春花,攀缠长春花,藤本日日春）;Band Plant, Big Leaf Periwinkle, Big Periwinkle, Bigleaf Periwinkle, Big-leaf Periwinkle, Blue Betsy, Blue Buttons, Blue Fingers, Blue Jack, Blue-eyed Beauty, Cockle, Common Periwinkle, Cutfinger, Cut-finger, Dicky Dilver, Dwarf Periwinkle, Granter Periwinkle, Great Periwinkle, Greater Periwinkle, Herbaceous Periwinkle, Large Periwinkle, Myrtle, Penniwinkle, Periwinkle, Pucelage, Quater, Running Myrtle, Wreath Flower ■

409336　Vinca major L. ' Variegata';花叶蔓长春花;Variegatedleaf Periwinkle ■

409337　Vinca major L. var. variegata Loudon = Vinca major L. ■

409338　Vinca media Hoffmanns. et Link = Vinca difformis Pourr. ■

409339　Vinca minor L.;小蔓长春花（蔓长春花,小长春花）;Bachelor's Buttons, Bill Buttons, Billy Buttons, Blue Betsy, Blue Buttons, Blue Jack, Blue Myrtle, Blue Smock, Bluebell, Bluebottle, Cockle Shells, Cockle-shell, Common Periwinkle, Cut-finger, Dicky

Dilver, Dwarf Periwinkle, Dwinkle, Fairy's Paintbrush, Fairy's Paintbrushes, Flower of Death, Greater Periwinkle, Ground Ivy, Joy-of-the-ground, Lesser Periwinkle, Little Star, Myrtle, Old Woman's Eye, Old Woman's Eyes, Pennyrinkle, Pennywinkle, Periwinkle, Pervenkle, Pervinca, Pervinkle, Perwinkle, Pinpatch, Pucellage, Running Myrtle, Running-myrtle, Sengreen, Small Periwinkle, Sorcerer's Violet, St. Cundida's Eyes, Trailing Myrtle, Tutsan, Umbrella, Violet ■

409340　Vinca minor L. ' Alba Variegata';黄白边小蔓长春花■☆

409341　Vinca minor L. 'Bowles' Blue' = Vinca minor L. 'La Grave' ■☆

409342　Vinca minor L. 'Bowles' White';大白花小蔓长春花■☆

409343　Vinca minor L. ' Gertrude Jekyll';小白花小蔓长春花■☆

409344　Vinca minor L. ' La Grave';庄严花小蔓长春花■☆

409345　Vinca pubescens d'Urv.;柔毛长春花;Pubescent Periwinkle ■☆

409346　Vinca pusilla Murray;细小长春花■☆

409347　Vinca pusilla Murray = Catharanthus pusillus（Murray）G. Don ■☆

409348　Vinca rosea L. = Catharanthus roseus（L.）G. Don ■

409349　Vinca rosea L. var. alba（G. Don）Sweet = Catharanthus roseus（L.）G. Don ■

409350　Vinca rosea L. var. alba Sweet ex Hubb. = Catharanthus roseus（L.）G. Don ■

409351　Vinca speciosa Salisb. = Catharanthus roseus（L.）G. Don ■

409352　Vinca trichophylla Baker = Catharanthus trichophyllus（Baker）Pichon ■☆

409353　Vincaceae Gray = Apocynaceae Juss.（保留科名）●■

409354　Vincaceae Gray;蔓长春花科■

409355　Vincaceae Vest = Apocynaceae Juss.（保留科名）●■

409356　Vincentella Pierre = Synsepalum（A. DC.）Daniell ●☆

409357　Vincentella brenanii Heine = Synsepalum brenanii（Heine）T. D. Penn. ●☆

409358　Vincentella camerounensis Pierre ex Aubrév. et Pellegr. = Synsepalum revolutum（Baker）T. D. Penn. ●☆

409359　Vincentella densiflora（Baker）Pierre = Synsepalum revolutum（Baker）T. D. Penn. ●☆

409360　Vincentella kemoensis（Dubard）A. Chev. = Synsepalum kemoense（Dubard）Aubrév. ●☆

409361　Vincentella longistyla（Baker）Pierre = Synsepalum brevipes（Baker）T. D. Penn. ●☆

409362　Vincentella micrantha（A. Chev.）A. Chev. = Synsepalum afzelii（Engl.）T. D. Penn. ●☆

409363　Vincentella muelleri Kupicha = Synsepalum muelleri（Kupicha）T. D. Penn. ●☆

409364　Vincentella ovatostipulata（De Wild.）Aubrév. et Pellegr. = Pachystela ovatostipulata De Wild. ●☆

409365　Vincentella passargei（Engl.）Aubrév. = Synsepalum passargei（Engl.）T. D. Penn. ●☆

409366　Vincentella revoluta（Baker）Pierre = Synsepalum revolutum（Baker）T. D. Penn. ●☆

409367　Vincentella sapinii（De Wild.）Brenan = Synsepalum passargei（Engl.）T. D. Penn. ●☆

409368　Vincentia Bojer = Vinticena Steud. ●

409369　Vincentia Gaudich. = Machaerina Vahl ■

409370　Vincentia boehmiana（F. Hoffm.）Burret = Grewia boehmiana F. Hoffm. ●☆

409371　Vincentia caffra（Meisn.）Burret = Grewia caffra Meisn. ●☆

409372　Vincentia carpinifolia（Juss.）Burret = Grewia carpinifolia Juss. ●☆

409373　Vincentia falcistipula（K. Schum.）Burret = Grewia falcistipula K. Schum. ●☆

409374　Vincentia flavescens (Juss.) Burret = Grewia flavescens Juss. ●☆

409375　Vincentia forbesii (Harv. ex Mast.) Burret = Grewia forbesii Harv. ex Mast. ●☆

409376　Vincentia holstii (Burret) Burret = Grewia holstii Burret ●☆

409377　Vincentia kerstingii (Burret) Burret = Grewia lasiodiscus K. Schum. ●☆

409378　Vincentia lasiodiscus (K. Schum.) Burret = Grewia lasiodiscus K. Schum. ●☆

409379　Vincentia olukondae (Schinz) Burret = Grewia olukondae Schinz ●☆

409380　Vincentia platyclada (K. Schum.) Burret = Grewia flavescens Juss. ●☆

409381　Vincentia retinervis (Burret) Burret = Grewia retinervis Burret ●☆

409382　Vincentia schweinfurthii (Burret) Burret = Grewia schweinfurthii Burret ●☆

409383　Vincentia triflora Bojer = Grewia triflora (Bojer) Walp. ●☆

409384　Vincentia viscosa (Boivin ex Baill.) Burret = Grewia triflora (Bojer) Walp. ●☆

409385　Vincentia welwitschii (Burret) Burret = Grewia welwitschii Burret ●☆

409386　Vincetoxicitm lateriflorum (Hemsl.) Kitag. = Cynanchum mongolicum (Maxim.) Hemsl. ■

409387　Vincetoxicopsis Costantin(1912);类白前属■☆

409388　Vincetoxicopsis harmandii Costantin;类白前■☆

409389　Vincetoxicum Medik. = Cynanchum L. ●■

409390　Vincetoxicum Walter = Gonolobus Michx. ●☆

409391　Vincetoxicum Walter = Vincetoxicum Wolf ●■

409392　Vincetoxicum Wolf = Cynanchum L. ●■

409393　Vincetoxicum Wolf = Gonolobus Michx. ●☆

409394　Vincetoxicum Wolf(1781);白前属;Black Swallowwort ●■

409395　Vincetoxicum × purpurascens C. Morren et Decne.;变紫白前■☆

409396　Vincetoxicum acuminatum C. Morren et Decne. = Cynanchum acuminatifolium Hemsl. ■

409397　Vincetoxicum acuminatum Decne. = Cynanchum acuminatifolium Hemsl. ■

409398　Vincetoxicum affine (Hemsl.) Kuntze = Cynanchum mooreanum Hemsl. ■

409399　Vincetoxicum affine Kuntze = Cynanchum mooreanum Hemsl. ■

409400　Vincetoxicum alatum Kuntze = Cynanchum alatum Buch. -Ham. ex Wight et Arn. ■

409401　Vincetoxicum album Asch. = Cynanchum vincetoxicum (L.) Pers. ■

409402　Vincetoxicum ambiguum Maxim.;可疑白前■☆

409403　Vincetoxicum amplexicaule Siebold et Zucc. = Cynanchum amplexicaule (Siebold et Zucc.) Hemsl. ■

409404　Vincetoxicum amplexicaule Siebold et Zucc. f. castaneum (Makino) Kitag. = Cynanchum amplexicaule (Siebold et Zucc.) Hemsl. var. castaneum Makino ■

409405　Vincetoxicum amplexicaule Siebold et Zucc. var. castaneum (Makino) Kitag. = Cynanchum amplexicaule (Siebold et Zucc.) Hemsl. var. castaneum Makino ■

409406　Vincetoxicum amplexicaule Siebold et Zucc. var. castaneum (Makino) Kitag. = Cynanchum amplexicaule (Siebold et Zucc.) Hemsl. ■

409407　Vincetoxicum amplexicaule Siebold et Zucc. var. castaneum Kitag. = Cynanchum amplexicaule (Siebold et Zucc.) Hemsl. ■

409408　Vincetoxicum arnottianum (Wight) Wight;阿诺特白前■☆

409409　Vincetoxicum ascyrifolium Franch. et Sav. = Cynanchum acuminatifolium Hemsl. ■

409410　Vincetoxicum ascyrifolium Franch. et Sav. = Cynanchum ascyrifolium (Franch. et Sav.) Matsum. ■

409411　Vincetoxicum atratum (Bunge) C. Morren et Decne. f. viridescens (H. Hara) Sugim. = Cynanchum atratum Bunge f. viridescens H. Hara ■☆

409412　Vincetoxicum atratum (Bunge) E. Morren et Decne. = Cynanchum atratum Bunge ■

409413　Vincetoxicum atratum (Bunge) Morren et Decne. = Cynanchum atratum Bunge ■

409414　Vincetoxicum auriculatum (Royle ex Wight) Kuntze = Cynanchum auriculatum Royle ex Wight ●■

409415　Vincetoxicum auriculatum Kuntze = Cynanchum auriculatum Royle ex Wight ●■

409416　Vincetoxicum austrokiusianum (Koidz.) Kitag.;南九州白前■☆

409417　Vincetoxicum balfourianum (Schltr.) C. Y. Wu et D. Z. Li = Cynanchum balfourianum (Schltr.) Tsiang et H. D. Zhang ■

409418　Vincetoxicum balfourianum (Schltr.) C. Y. Wu et D. Z. Li = Cynanchum forrestii Schltr. ■

409419　Vincetoxicum biondioides (W. T. Wang ex Tsiang et P. T. Li) C. Y. Wu et D. Z. Li = Cynanchum biondioides W. T. Wang ex Tsiang et P. T. Li ■

409420　Vincetoxicum biondioides (W. T. Wang) C. Y. Wu et D. Z. Li = Cynanchum biondioides W. T. Wang ex Tsiang et P. T. Li ■

409421　Vincetoxicum bojerianum (Decne.) Kuntze = Cynanchum bojerianum (Decne.) Choux ●☆

409422　Vincetoxicum calcareum (H. Ohashi) Akasawa;石灰白前■☆

409423　Vincetoxicum callialatum (Buch. -Ham. ex Wight) Kuntze = Cynanchum callialatum Buch. -Ham. ex Wight ■

409424　Vincetoxicum callialatum Kuntze = Cynanchum callialatum Buch. -Ham. ex Wight ■

409425　Vincetoxicum canescens (Willd.) Decne. = Cynanchum canescens (Willd.) K. Schum. ■

409426　Vincetoxicum capense (L. f.) Kuntze = Pentatropis capensis (L. f.) Bullock ■☆

409427　Vincetoxicum capense (R. Br.) Kuntze = Cynanchum obtusifolium L. f. ●☆

409428　Vincetoxicum cardiostephanum (Rech. f.) Rech. f.;心冠白前■☆

409429　Vincetoxicum chekiangense (M. Cheng) C. Y. Wu et D. Z. Li = Cynanchum chekiangense M. Cheng ex Tsiang et P. T. Li ■

409430　Vincetoxicum chinense S. Moore = Cynanchum mooreanum Hemsl. ■

409431　Vincetoxicum darvasicum B. Fedtsch.;达尔瓦斯白前■☆

409432　Vincetoxicum deltoideum Kuntze = Cynanchum otophyllum C. K. Schneid. ■

409433　Vincetoxicum doianum (Koidz.) Kitag.;道氏白前■☆

409434　Vincetoxicum dregeanum (Decne.) Kuntze = Cynanchum obtusifolium L. f. ●☆

409435　Vincetoxicum eurychitoides K. Schum. = Cynanchum eurychitoides (K. Schum.) K. Schum. ●☆

409436　Vincetoxicum flexuosum (R. Br.) Kuntze = Tylophora flexuosa R. Br. ■

409437　Vincetoxicum floribundum (Miq.) Franch. et Sav. = Tylophora floribunda Miq. ●■

409438　Vincetoxicum fordii (Hemsl.) Kuntze = Cynanchum fordii Hemsl. ■

409439　Vincetoxicum fordii Kuntze = Cynanchum fordii Hemsl. ■

409440   Vincetoxicum formosanum（Maxim.）Kuntze = Cynanchum formosanum（Maxim.）Hemsl. ex Forbes et Hemsl. ●■

409441   Vincetoxicum formosanum Kuntze = Cynanchum formosanum（Maxim.）Hemsl. ex Forbes et Hemsl. ●■

409442   Vincetoxicum forrestii（Schltr.）C. Y. Wu et D. Z. Li = Cynanchum forrestii Schltr. ■

409443   Vincetoxicum forrestii（Schltr.）C. Y. Wu et D. Z. Li var. stenolobum（Tsiang et Zhang）C. Y. Wu et D. Z. Li = Cynanchum forrestii Schltr. var. stenolobum Tsiang et H. D. Zhang ■

409444   Vincetoxicum forrestii（Schltr.）C. Y. Wu et D. Z. Li var. stenolobum（Tsiang et Zhang）C. Y. Wu et D. Z. Li = Cynanchum forrestii Schltr. ■

409445   Vincetoxicum fradinii Pomel = Vincetoxicum hirundinaria Medik. ■☆

409446   Vincetoxicum fruticulosum（Decne.）Decne. = Blyttia fruticulosa（Decne.）D. V. Field ■☆

409447   Vincetoxicum glabrum（Nakai）Kitag. ;无毛白前■☆

409448   Vincetoxicum glabrum（Nakai）Kitag. f. viridescens（Murata）Sugim. ;绿花无毛白前■☆

409449   Vincetoxicum glabrum（Nakai）Kitag. var. rotundifolium（Honda）Sugim. ;圆叶无毛白前■☆

409450   Vincetoxicum glaucescens（Decne.）C. Y. Wu et D. Z. Li = Cynanchum glaucescens（Decne.）Hand. -Mazz. ●■

409451   Vincetoxicum glaucum（Wall. ex Wight）Rech. f. = Cynanchum canescens（Willd.）K. Schum. ■

409452   Vincetoxicum glaucum（Wall. ex Wight）Rech. f. = Cynanchum glaucum Wall. ex Wight ■

409453   Vincetoxicum glaucum（Wall.）Rech. f. = Vincetoxicum canescens（Willd.）Decne. ■

409454   Vincetoxicum gracilipes（Tsiang et H. D. Zhang）C. Y. Wu et D. Z. Li = Cynanchum taihangense Tsiang et H. D. Zhang ■

409455   Vincetoxicum gracilipes（Tsiang et H. D. Zhang）C. Y. Wu et D. Z. Li = Cynanchum gracilipes Tsiang et H. D. Zhang ■

409456   Vincetoxicum hancockianum（Maxim.）C. Y. Wu et D. Z. Li = Cynanchum hancockianum（Maxim.）Iljinski ■

409457   Vincetoxicum hancockianum（Maxim.）C. Y. Wu et D. Z. Li = Cynanchum mongolicum（Maxim.）Hemsl. ■

409458   Vincetoxicum hastatum（Bunge）Kuntze = Cynanchum bungei Decne. ●■

409459   Vincetoxicum hastatum Kuntze = Cynanchum bungei Decne. ●■

409460   Vincetoxicum heterophylla（A. Rich.）Vatke = Tylophora heterophylla A. Rich. ●☆

409461   Vincetoxicum heydei（Hook. f.）Kuntze = Cynanchum heydei Hook. f. ■

409462   Vincetoxicum heydei Kuntze = Cynanchum heydei Hook. f. ■

409463   Vincetoxicum hirundinaria Medik. ;燕白前■☆

409464   Vincetoxicum hirundinaria Medik. subsp. glaucum（Wall. ex Wight）H. Hara = Cynanchum canescens（Willd.）K. Schum. ■

409465   Vincetoxicum hirundinaria Medik. subsp. glaucum（Wall. ex Wight）Hara = Cynanchum canescens（Willd.）K. Schum. ■

409466   Vincetoxicum holstii K. Schum. = Cynanchum abyssinicum Decne. ●☆

409467   Vincetoxicum hoyoense T. Yamash. ;丰予白前■☆

409468   Vincetoxicum huteri Vis. et Asch. ;于泰白前■☆

409469   Vincetoxicum hybanthera Kuntze = Belostemma cordifolium（Link，Klotzsch et Otto）M. G. Gilbert et P. T. Li ●

409470   Vincetoxicum hydrophilum（Tsiang et H. D. Zhang）C. Y. Wu et D. Z. Li = Cynanchum hydrophilum Tsiang et H. D. Zhang ■

409471   Vincetoxicum inamoenum Maxim. = Cynanchum inamoenum（Maxim.）Loes. ex Gilg et Loes. ■

409472   Vincetoxicum insulanum（Hance）Kuntze = Cynanchum insulanum（Hance）Hemsl. ■

409473   Vincetoxicum insulanum Kuntze = Cynanchum insulanum（Hance）Hemsl. ■

409474   Vincetoxicum izuense T. Yamash. ;伊豆白前■☆

409475   Vincetoxicum jacquemontianum（Decne.）DC. = Cynanchum jacquemontianum Decne. ■☆

409476   Vincetoxicum japonicum C. Morren et Decne. = Cynanchum japonicum C. Morren et Decne. ■☆

409477   Vincetoxicum japonicum C. Morren et Decne. f. maritimum Sugim. ;滨海白前■☆

409478   Vincetoxicum japonicum C. Morren et Decne. f. puncticlatum（Koidz.）Kitag. = Cynanchum japonicum C. Morren et Decne. var. puncticulatum（Koidz.）H. Hara ■☆

409479   Vincetoxicum japonicum C. Morren et Decne. var. albiflorum（Franch. et Sav.）Kitag. = Cynanchum japonicum C. Morren et Decne. var. albiflorum（Franch. et Sav.）H. Hara ■☆

409480   Vincetoxicum katoi（Ohwi）Kitag. ;加藤白前■☆

409481   Vincetoxicum katoi（Ohwi）Kitag. f. albescens（H. Hara）Kitag. ;白花加藤白前■☆

409482   Vincetoxicum kitagawae Hiyama = Vincetoxicum macrophyllum Siebold et Zucc. var. nikoense Maxim. ■☆

409483   Vincetoxicum krameri Franch. et Sav. ;克雷默白前■☆

409484   Vincetoxicum lateriflorum（Hemsl.）Kitag. = Cynanchum mongolicum（Maxim.）Hemsl. ■

409485   Vincetoxicum leucanthum K. Schum. = Cynanchum leucanthum（K. Schum.）K. Schum. ■☆

409486   Vincetoxicum limprichtii（Schltr.）C. Y. Wu et D. Z. Li = Cynanchum forrestii Schltr. ■

409487   Vincetoxicum linearifolium（Hemsl.）Kuntze = Cynanchum stauntonii（Decne.）Schltr. ex H. Lév. ●■

409488   Vincetoxicum linearifolium Kuntze = Cynanchum stauntonii（Decne.）Schltr. ex H. Lév. ●■

409489   Vincetoxicum luridinum Stocks = Vincetoxicum stocksii Ali et Khatoon ■☆

409490   Vincetoxicum macrophyllum Siebold et Zucc. = Cynanchum grandifolium Hemsl. ■☆

409491   Vincetoxicum macrophyllum Siebold et Zucc. var. nikoense Maxim. = Cynanchum inamoenum（Maxim.）Loes. ex Gilg et Loes. ■

409492   Vincetoxicum macrophyllum Siebold et Zucc. var. nikoense Maxim. = Cynanchum grandifolium Hemsl. var. nikoense（Maxim.）Ohwi ■☆

409493   Vincetoxicum madagascariense K. Schum. = Cynanchum madagascariense K. Schum. ■☆

409494   Vincetoxicum magnificum（Nakai）Kitag. ;华丽白前■☆

409495   Vincetoxicum makinoi Honda = Cynanchum paniculatum（Bunge）Kitag. ex H. Hara var. latifolium（Makino）H. Hara ■☆

409496   Vincetoxicum makinoi Honda = Vincetoxicum pycnostelma Kitag. f. latifolium（Makino）Kitag. ■☆

409497   Vincetoxicum mandschuricum Hance = Cynanchum versicolor Bunge ●■

409498   Vincetoxicum mannii Scott-Elliot = Cynanchum adalinae（K. Schum.）K. Schum. subsp. mannii（Scott-Elliot）Bullock ■☆

409499   Vincetoxicum matsumurae（T. Yamaz.）H. Ohashi = Tylophora matsumurae（T. Yamaz.）T. Yamash. et Tateishi ■☆

409500　Vincetoxicum meyeri（Decne.）Benth. et Hook. f. = Cynanchum meyeri（Decne.）Schltr. ●☆

409501　Vincetoxicum mongolicum Maxim. = Cynanchum mongolicum（Maxim.）Hemsl. ■

409502　Vincetoxicum mongolicum Maxim. var. hancockianum Maxim. = Cynanchum mongolicum（Maxim.）Hemsl. ■

409503　Vincetoxicum muliense（Tsiang）C. Y. Wu et D. Z. Li = Cynanchum forrestii Schltr. ■

409504　Vincetoxicum muliense（Tsiang）C. Y. Wu et D. Z. Li = Cynanchum muliense Tsiang ■

409505　Vincetoxicum multinerve Franch. et Sav. = Cynanchum atratum Bunge ■

409506　Vincetoxicum nigrum（L.）Moench = Cynanchum louiseae Kartesz et Gandhi ■☆

409507　Vincetoxicum nikoense（Maxim.）Kitag. = Vincetoxicum macrophyllum Siebold et Zucc. var. nikoense Maxim. ■☆

409508　Vincetoxicum nipponicum（Matsum.）Kitag.；本州白前■☆

409509　Vincetoxicum nipponicum（Matsum.）Kitag. f. abukumense（Koidz.）Kitag.；阿武隈白前■☆

409510　Vincetoxicum obtusifolium（L. f.）Kuntze = Cynanchum obtusifolium L. f. ●☆

409511　Vincetoxicum officinale Moench = Cynanchum vincetoxicum（L.）Pers. ■

409512　Vincetoxicum officinale Moench = Vincetoxicum hirundinaria Medik. ■☆

409513　Vincetoxicum officinale Moench subsp. fradinii（Pomel）Batt. = Vincetoxicum hirundinaria Medik. ■☆

409514　Vincetoxicum officinale Moench var. acutatum Chabert = Vincetoxicum hirundinaria Medik. ■☆

409515　Vincetoxicum officinale Moench var. dentiferum Chabert = Vincetoxicum hirundinaria Medik. ■☆

409516　Vincetoxicum officinale Moench var. floribundum Chabert = Vincetoxicum hirundinaria Medik. ■☆

409517　Vincetoxicum paniculatum（Bunge）C. Y. Wu et D. Z. Li = Cynanchum paniculatum（Bunge）Kitag. ex H. Hara ■

409518　Vincetoxicum petrense（Hemsl. et Lace）Rech. f. = Cynanchum petrense Hemsl. et Lace ●☆

409519　Vincetoxicum pilosum（R. Br.）G. Nicholson = Cynanchum africanum（L.）Hoffmanns. ●☆

409520　Vincetoxicum polyanthum K. Schum. = Cynanchum polyanthum K. Schum. ■☆

409521　Vincetoxicum pubescens（Bunge）Kuntze = Cynanchum chinense R. Br. ■

409522　Vincetoxicum pubescens Kuntze = Cynanchum chinense R. Br. ■

409523　Vincetoxicum purpureum（Pall.）Kuntze = Cynanchum purpureum（Pall.）K. Schum. ■

409524　Vincetoxicum purpureum Kuntze = Cynanchum purpureum（Pall.）K. Schum. ■

409525　Vincetoxicum pycnostachys Kitag. = Cynanchum paniculatum（Bunge）Kitag. ex H. Hara ■

409526　Vincetoxicum pycnostelma Kitag. = Cynanchum paniculatum（Bunge）Kitag. ex H. Hara ■

409527　Vincetoxicum pycnostelma Kitag. f. latifolium（Makino）Kitag. = Cynanchum paniculatum（Bunge）Kitag. ex H. Hara var. latifolium（Makino）H. Hara ■☆

409528　Vincetoxicum repandum（Decne.）Kuntze = Cynanchum

repandum（Decne.）K. Schum. ●☆

409529　Vincetoxicum riparium（Tsiang et H. D. Zhang）C. Y. Wu et D. Z. Li = Cynanchum riparium Tsiang et H. D. Zhang ■

409530　Vincetoxicum rutenbergianum Vatke = Cynanchum repandum（Decne.）K. Schum. ●☆

409531　Vincetoxicum sakesarense Ali et Khatoon；瑟盖瑟尔白前■☆

409532　Vincetoxicum sarcostemmoides Schweinf. ex Penz. = Cynanchum gerrardii（Harv.）Liede ●☆

409533　Vincetoxicum sibiricum（L.）Decne. = Cynanchum thesioides（Freyn）K. Schum. ■

409534　Vincetoxicum sibiricum（L.）Decne. var. australe Maxim. = Cynanchum thesioides（Freyn）K. Schum. ■

409535　Vincetoxicum sibiricum（L.）Decne. var. boreale Maxim. = Cynanchum thesioides（Freyn）K. Schum. ■

409536　Vincetoxicum sibiricum Decne. = Cynanchum thesioides（Freyn）K. Schum. ■

409537　Vincetoxicum sibiricum Decne. var. australe Maxim. = Cynanchum thesioides（Freyn）K. Schum. var. australe（Maxim.）Tsiang et P. T. Li ■

409538　Vincetoxicum sibiricum Decne. var. australe Maxim. = Cynanchum thesioides（Freyn）K. Schum. ■

409539　Vincetoxicum sibiricum Decne. var. boreale Maxim. = Cynanchum thesioides（Freyn）K. Schum. ■

409540　Vincetoxicum stauntonii（Decne.）C. Y. Wu et D. Z. Li = Cynanchum stauntonii（Decne.）Schltr. ex H. Lév. ●■

409541　Vincetoxicum stenophyllum（Hemsl.）Kuntze = Cynanchum stenophyllum Hemsl. ■

409542　Vincetoxicum stenophyllum Standl. = Cynanchum stenophyllum Hemsl. ■

409543　Vincetoxicum steppicola（Hand.-Mazz.）C. Y. Wu et D. Z. Li = Cynanchum forrestii Schltr. ■

409544　Vincetoxicum steppicola（Hand.-Mazz.）C. Y. Wu et D. Z. Li = Cynanchum steppicola Hand.-Mazz. ■

409545　Vincetoxicum stocksii Ali et Khatoon；斯托克斯白前■☆

409546　Vincetoxicum sublanceolatum（Miq.）Maxim. = Cynanchum sublanceolatum（Miq.）Matsum. ■

409547　Vincetoxicum sublanceolatum（Miq.）Maxim. f. albiflorum（Franch. et Sav.）H. Ohashi = Cynanchum sublanceolatum（Miq.）Matsum. f. albiflorum（Franch. et Sav.）T. Yamaz. ●☆

409548　Vincetoxicum sublanceolatum（Miq.）Maxim. var. albiflorum（Franch. et Sav.）Kitag. = Cynanchum sublanceolatum（Miq.）Matsum. f. albiflorum（Franch. et Sav.）T. Yamaz. ●☆

409549　Vincetoxicum sublanceolatum（Miq.）Maxim. var. auriculatum Franch. et Sav. = Cynanchum sublanceolatum（Miq.）Matsum. var. auriculatum（Franch. et Sav.）Matsum. ■☆

409550　Vincetoxicum sublanceolatum（Miq.）Maxim. var. macranthum Maxim. = Cynanchum sublanceolatum（Miq.）Matsum. var. macranthum（Maxim.）Matsum. ■☆

409551　Vincetoxicum sublanceolatum（Miq.）Maxim. var. macranthum Maxim. f. yesoense（Nakai）Kitag.；北海道白前■☆

409552　Vincetoxicum sublanceolatum Maxim. = Cynanchum sublanceolatum（Miq.）Matsum. ■

409553　Vincetoxicum taihangense（Tsiang et H. D. Zhang）C. Y. Wu et D. Z. Li = Cynanchum taihangense Tsiang et H. D. Zhang ■

409554　Vincetoxicum thesioides Freyn = Cynanchum thesioides（Freyn）K. Schum. ■

409555　Vincetoxicum tsiangii（P. T. Li）P. T. Li = Cynanchum tsiangii

P. T. Li ■

409556　Vincetoxicum tsiangii（P. T. Li）P. T. Li = Tylophora tsiangii（P. T. Li）M. G. Gilbert，W. D. Stevens et P. T. Li ●

409557　Vincetoxicum vernyi Franch. et Sav. ;韦尔尼白前■☆

409558　Vincetoxicum versicolor（Bunge）Decne. = Cynanchum versicolor Bunge ●■

409559　Vincetoxicum versicolor Decne. = Cynanchum versicolor Bunge ●■

409560　Vincetoxicum verticillatum（Hemsl.）Kuntze = Cynanchum verticillatum Hemsl. ●■

409561　Vincetoxicum verticillatum Kuntze = Cynanchum verticillatum Hemsl. ●■

409562　Vincetoxicum verticillatum Kuntze var. arenicola（Tsiang et Zhang）C. Y. Wu et D. Z. Li = Cynanchum verticillatum Hemsl. var. arenicola Tsiang et H. D. Zhang ex Tsiang et P. T. Li ●■

409563　Vincetoxicum verticillatum Kuntze var. arenicola（Tsiang et Zhang）C. Y. Wu et D. Z. Li = Cynanchum verticillatum Hemsl. ●■

409564　Vincetoxicum verticillatum var. arenicola（Tsiang et H. D. Zhang）C. Y. Wu et D. Z. Li = Cynanchum verticillatum Hemsl. ●■

409565　Vincetoxicum virescens K. Schum. = Cynanchum eurychiton（Decne.）K. Schum. ●☆

409566　Vincetoxicum volubile Maxim. = Cynanchum volubile（Maxim.）Hemsl. ■

409567　Vincetoxicum wallichii（Wight）Kuntze = Cynanchum wallichii Wight ■

409568　Vincetoxicum wallichii Kuntze = Cynanchum wallichii Wight ■

409569　Vincetoxicum wangii（P. T. Li et W. Kittr.）Liede = Cynanchum wangii P. T. Li et W. Kittr. ■

409570　Vincetoxicum wilfordii（Maxim.）Franch. et Sav. = Cynanchum wilfordii（Maxim.）Hook. f. ■

409571　Vincetoxicum wilfordii Franch. et Sav. = Cynanchum wilfordii（Maxim.）Hemsl. ■

409572　Vincetoxicum yamanakae（Ohwi et H. Ohashi）H. Ohashi;山中白前■☆

409573　Vincetoxicum yonakuniense（Hatus.）T. Yamash. et Tateishi;与那国白前■☆

409574　Vinchia DC. = Alstonia R. Br.（保留属名）●

409575　Vindasia Benoist(1962);文达爵床属☆

409576　Vindasia virgata Benoist;文达爵床☆

409577　Vindicta Raf. = Aceranthus C. Morren et Decne. ■

409578　Vindicta Raf. = Epimedium L. ■

409579　Vinicia Dematt.（2007）;维尼菊属☆

409580　Vinicia tomentosa Dematt. ;维尼菊☆

409581　Vinkia Meijden = Myriophyllum L. ■

409582　Vinkia Meijden(1975);文克草属■☆

409583　Vinkia callitrichoides（Orchard）Meijden;文克草■☆

409584　Vinkiella R. Johns. ;小文克草属●☆

409585　Vinsonia Gaudich. = Pandanus Parkinson ex Du Roi ●■

409586　Vintenatia Cav.（废弃属名）= Astroloma R. Br. + Melichrus R. Br. ●☆

409587　Vintenatia Cav.（废弃属名）= Ventenata Koeler(保留属名)■☆

409588　Vintera Humb. et Bonpl. = Drimys J. R. Forst. et G. Forst.（保留属名）●☆

409589　Vintera Humb. et Bonpl. = Wintera Humb. et Bonpl. ●☆

409590　Vinticena Steud. = Grewia L. ●

409591　Vinticena boehmiana（F. Hoffm.）Burret = Grewia boehmiana F. Hoffm. ●☆

409592　Vinticena caffra（Meisn.）Burret = Grewia caffra Meisn. ●☆

409593　Vinticena carpinifolia（Juss.）Burret = Grewia carpinifolia Juss. ●☆

409594　Vinticena falcistipula（K. Schum.）Burret = Grewia falcistipula K. Schum. ●☆

409595　Vinticena flavescens（Juss.）Burret = Grewia flavescens Juss. ●☆

409596　Vinticena forbesii（Harv. ex Mast.）Burret = Grewia forbesii Harv. ex Mast. ●☆

409597　Vinticena holstii（Burret）Burret = Grewia holstii Burret ●☆

409598　Vinticena kerstingii（Burret）Burret = Grewia lasiodiscus K. Schum. ●☆

409599　Vinticena lasiodiscus（K. Schum.）Burret = Grewia lasiodiscus K. Schum. ●☆

409600　Vinticena macromischa Burret = Grewia triflora（Bojer）Walp. ●☆

409601　Vinticena platyclada（K. Schum.）Burret = Grewia flavescens Juss. ●☆

409602　Vinticena retinervis（Burret）Burret = Grewia retinervis Burret ●☆

409603　Vinticena rugosifolia（De Wild.）Robyns et Lawalrée = Grewia rugosifolia De Wild. ●☆

409604　Vinticena schweinfurthii（Burret）Burret = Grewia schweinfurthii Burret ●☆

409605　Vinticena triflora（Bojer）Steud. = Grewia triflora（Bojer）Walp. ●☆

409606　Vinticena viscosa（Boivin ex Baill.）Burret = Grewia triflora（Bojer）Walp. ●☆

409607　Vinticena welwitschii（Burret）Burret = Grewia welwitschii Burret ●☆

409608　Viola L.（1753）;菫菜属;Pansy，Violet ■●

409609　Viola × chinoi F. Maek. ;千布菫菜■☆

409610　Viola × ibukiana Makino;伊吹山菫菜■☆

409611　Viola × interposita Kitag. = Viola tenuicornis W. Becker ■

409612　Viola × jettmarii Hand. -Mazz. = Viola dissecta Ledeb. var. incisa（Turcz.）Y. S. Chen ■

409613　Viola × kisoana Nakai;木曽菫菜■☆

409614　Viola × lii Kitag. = Viola dissecta Ledeb. ■

409615　Viola × martinii F. Maek. ;马丁菫菜■☆

409616　Viola × ogawae Nakai;小川菫菜■☆

409617　Viola × polysecta Nakai = Viola × savatieri Makino ■☆

409618　Viola × savatieri Makino;萨瓦捷菫菜■☆

409619　Viola × savatieri Makino nothof. variegata E. Hama ex T. Shimizu;彩色萨瓦捷菫菜■☆

409620　Viola × takahashii（Nakai）Taken. = Viola takahashii（Nakai）Taken. ■

409621　Viola × tanakaeana Makino = Viola violacea Makino var. tanakaeana（Makino）T. Hashim. ■☆

409622　Viola × tokyoensis F. Maek. et T. Hashim. ;东京菫菜■☆

409623　Viola × williamsii ?;威廉斯菫菜;Bedding Viola ■☆

409624　Viola × wittrockiana Gams;杂种三色菫;Bedding Violet，Garden Pansy，Heartsease，Pansy ■☆

409625　Viola aberrans Greene = Viola conjugens Greene ■☆

409626　Viola abundans House = Viola sagittata Aiton ■☆

409627　Viola abyssinica Steud. ex Oliv. ;阿比西尼亚菫菜■☆

409628　Viola abyssinica Steud. ex Oliv. var. eminii Engl. = Viola eminii（Engl.）R. E. Fr. ■☆

409629　Viola abyssinica Steud. ex Oliv. var. ulugurensis Engl. = Viola eminii（Engl.）R. E. Fr. ■☆

409630　Viola acuminata Ledeb. ;鸡腿菫菜（红铧头菜，胡森菫菜，鸡脚菫菜，鸡腿菜，尖叶菫，尖叶菫菜，菫菜，锐叶菫菜，走边疆）;Acuminate Violet，Chickenleg Violet，Toothed-stipuled Wild Violet ■

409631　Viola acuminata Ledeb. f. alba Moriya;白鸡腿堇菜■☆

409632　Viola acuminata Ledeb. f. alba Moriya = Viola acuminata Ledeb. ■

409633　Viola acuminata Ledeb. f. glaberrima（H. Hara）Kitam.;光鸡腿堇菜■☆

409634　Viola acuminata Ledeb. f. glaberrima（H. Hara）Kitam. = Viola acuminata Ledeb. ■

409635　Viola acuminata Ledeb. f. shikokuensis（W. Becker）F. Maek. = Viola acuminata Ledeb. f. glaberrima（H. Hara）Kitam. ■☆

409636　Viola acuminata Ledeb. subsp. austroussuriensis W. Becker = Viola acuminata Ledeb. ■

409637　Viola acuminata Ledeb. var. austroussuriensis（W. Becker）Kitag. = Viola acuminata Ledeb. ■

409638　Viola acuminata Ledeb. var. brevistipulata（W. Becker）Kitag. = Viola acuminata Ledeb. ■

409639　Viola acuminata Ledeb. var. dentata W. Becker = Viola acuminata Ledeb. ■

409640　Viola acuminata Ledeb. var. glaberrima H. Hara = Viola acuminata Ledeb. f. glaberrima（H. Hara）Kitam. ■☆

409641　Viola acuminata Ledeb. var. intermedia Nakai = Viola acuminata Ledeb. ■

409642　Viola acuminata Ledeb. var. pilifera Ching J. Wang;毛花鸡腿堇菜;Hairflower ■

409643　Viola acutifolia（Kar. et Kir.）W. Becker;尖叶堇菜;Sharpleaved Violet ■

409644　Viola acutilabella Hayata = Viola nagasawae Makino et Hayata ■

409645　Viola adenothrix Hayata = Viola fargesii H. Boissieu ■

409646　Viola adenothrix Hayata var. tsugitakaensis（Masam.）Ching J. Wang et T. C. Huang = Viola tsugitakaensis Masam. ■

409647　Viola adenothrix Hayata var. tsugitakaensis（Masam.）Ching J. Wang et T. C. Huang = Viola fargesii H. Boissieu ■

409648　Viola adunca Sm.;钩堇;Blue Violet, Hooked Violet, Hooked-spur Violet, Hook-spur Violet, Sand Violet, Western Dog Violet ■☆

409649　Viola adunca Sm. f. albiflora Vict. et J. Rousseau = Viola adunca Sm. ■☆

409650　Viola adunca Sm. f. glabra（Brainerd）G. N. Jones;光钩堇;Hook-spur Violet, Sand Violet ■☆

409651　Viola adunca Sm. var. glabra Brainerd = Viola adunca Sm. f. glabra（Brainerd）G. N. Jones ■☆

409652　Viola adunca Sm. var. minor（Hook.）Fernald = Viola labradorica Schrank ■☆

409653　Viola aetolica Boiss. et Heldr.;埃托利亚堇菜■☆

409654　Viola affinis Leconte;近缘堇菜;Le Conte's Violet, Sand Violet ■☆

409655　Viola alaica Vved.;阿莱堇菜■☆

409656　Viola alata Burgersd.;白犁头草（堇菜,如意草）;Violet ■

409657　Viola alata Burgersd. = Viola arcuata Blume ■

409658　Viola alata Burgersd. = Viola hamiltoniana D. Don ■

409659　Viola alata Burgersd. subsp. verecunda（A. Gray）W. Becker = Viola arcuata Blume ■

409660　Viola alata Burgersd. subsp. verecunda（A. Gray）W. Becker = Viola verecunda A. Gray ■

409661　Viola alba Besser;白堇菜;Pale Meadow Violet, Pale Meadow-violet, Parma Violet, White Violet ■☆

409662　Viola alba Besser subsp. dehnhardtii（Ten.）W. Becker;德恩哈特堇菜■☆

409663　Viola albida Besser f. sieboldiana（Maxim.）F. Maek. = Viola chaerophylloides（Regel）W. Becker var. sieboldiana（Maxim.）Makino ■

409664　Viola albida Besser f. takahashii（Nakai）Kitag. = Viola takahashii（Nakai）Taken. ■

409665　Viola albida Palib.;朝鲜堇菜;Korean Violet, White Violet ■

409666　Viola albida Palib. f. takahashii（Nakai）Kitag. = Viola takahashii（Nakai）Taken. ■

409667　Viola albida Palib. var. chaerophylloides（Regel）F. Maek. = Viola chaerophylloides（Regel）W. Becker ■

409668　Viola albida Palib. var. chaerophylloides（Regel）F. Maek. f. sieboldiana（Maxim.）F. Maek. = Viola chaerophylloides（Regel）W. Becker var. sieboldiana（Maxim.）Makino ■

409669　Viola albida Palib. var. chaerophylloides F. Maek. = Viola takahashii（Nakai）Taken. ■

409670　Viola albida Palib. var. takahashii（Nakai）Nakai = Viola takahashii（Nakai）Taken. ■

409671　Viola albida Palib. var. takahaskii Nakai = Viola takahashii（Nakai）Taken. ■

409672　Viola alexandrowiana（W. Becker）Juz.;阿来堇菜■☆

409673　Viola alisoviana Kiss = Viola philippica Cav. ■

409674　Viola alisoviana Kiss f. candida（Kitag.）Taken. = Viola philippica Cav. ■

409675　Viola alisoviana Kiss f. intermedia（Kitag.）Taken. = Viola mandshurica W. Becker ■

409676　Viola alisoviana Kiss. = Viola philippica Cav. ■

409677　Viola alisoviana Kiss. = Viola yedoensis Makino ■☆

409678　Viola alisoviana Kiss. f. candida（Kitag.）Taken. = Viola philippica Cav. ■

409679　Viola alisoviana Kiss. f. intermedia（Kitag.）Taken. = Viola philippica Cav. ■

409680　Viola alliariifolia Nakai;葱芥叶堇菜■☆

409681　Viola alpestris W. Becker;高山堇菜;Alpine Violet ■☆

409682　Viola altaica Ker Gawl.;阿尔泰堇菜;Altai Violet, Tartarian Violet ■

409683　Viola altaica Ker Gawl. subsp. typica W. Becker = Viola altaica Ker Gawl. ■

409684　Viola altaica Ker Gawl. var. typica Kupffer = Viola altaica Ker Gawl. ■

409685　Viola amamiana Hatus.;奄美堇菜■☆

409686　Viola ambigua Waldst. et Kit.;多变堇菜■☆

409687　Viola amurica W. Becker;阿穆尔堇菜（额穆尔堇菜,黑龙江堇菜）;Amur Violet ■

409688　Viola amurica W. Becker = Viola arcuata Blume ■

409689　Viola angellae Pollard = Viola palmata L. ■☆

409690　Viola angustipulata C. C. Chang;狭托叶堇菜;Narrowstipule Violet ■

409691　Viola angustistipulata C. C. Chang = Viola tienschiensis W. Becker ■

409692　Viola arborescens L.;树状堇菜;Shrubby Violet ■☆

409693　Viola arborescens L. var. serratifolia DC. = Viola arborescens L. ■☆

409694　Viola arcuata Blume = Viola hamiltoniana D. Don ■

409695　Viola arcuata Blume f. radicans（Makino）Nakai = Viola arcuata Blume ■

409696　Viola arcuata Blume var. verecunda（A. Gray）Nakai = Viola arcuata Blume ■

409697　Viola arcuata Blume var. verecunda（A. Gray）Nakai = Viola verecunda A. Gray ■

409698　Viola arenaria DC.;沙生堇菜;Sand Violet ■☆

409699　Viola arenaria DC. = Viola rupestris F. W. Schmidt ■

409700 Viola arenicola Chabert = Viola reichenbachiana Boreau ■☆

409701 Viola arisanensis W. Becker = Viola fargesii H. Boissieu ■

409702 Viola arisanensis W. Becker = Viola formosana Hayata ■

409703 Viola arvensis Murray;野菫菜（耕地菫菜，田野菫菜，野生菫菜）;Biddy's Eye,Biddy's Eyes,Corn Pansy,European Field Pansy,Field Pansy,Field Violet,Heartsease,Pansy,Wild Pansy ■

409704 Viola arvensis Murray var. subatlantica Maire = Viola subatlantica（Maire）Ibn Tattou ■☆

409705 Viola arvensis Murray var. tezensis（Ball）Murb. = Viola arvensis Murray ■

409706 Viola atroviolacea W. Becker;暗紫菫菜■☆

409707 Viola aurasiaca Pomel = Viola arvensis Murray ■

409708 Viola austroussuriensis（W. Becker）Kom. = Viola acuminata Ledeb. ■

409709 Viola avatschensis W. Becker et Hultén;勘察加菫菜■☆

409710 Viola baicalensis W. Becker = Viola variegata Fisch. ex Link ■

409711 Viola bambusetorum Hand. -Mazz.;盐源菫菜;Bamboo Violet ■

409712 Viola bambusetorum Hand. -Mazz. = Viola striatella H. Boissieu ■

409713 Viola battandieri Becker = Viola munbyana Boiss. et Reut. ■☆

409714 Viola belophylla H. Boissieu;枪叶菫菜（多花菫菜，凤凰菫菜，维西菫菜）;Monbeig Violet ■☆

409715 Viola bernardii Greene;伯纳德菫菜;Bernard's Violet,Violet ■☆

409716 Viola betonicifolia Sm.;戟叶菫菜（铧头草，箭叶菫菜，菫菫菜,尼泊尔菫菜,青地黄瓜,小叶犁头草,野半夏）;Harberdleaf Violet,Mountain Violet ■

409717 Viola betonicifolia Sm. f. pubescens H. Hara = Viola trichopetala C. C. Chang ■

409718 Viola betonicifolia Sm. subsp. dielsiana W. Becker = Viola betonicifolia Sm. ■

409719 Viola betonicifolia Sm. subsp. jaunsariensis（W. Becker）H. Hara = Viola tienschiensis W. Becker ■

409720 Viola betonicifolia Sm. subsp. napalensis（Ging.）W. Becker;尼泊尔菫菜(箭叶菫菜,紫花地丁）;Napal Violet ■

409721 Viola betonicifolia Sm. subsp. nepalensis（Ging.）W. Becker = Viola betonicifolia Sm. ■

409722 Viola betonicifolia Sm. subsp. novaguineensis D. M. Moore = Viola inconspicua Blume ■

409723 Viola betonicifolia Sm. var. albescens（Nakai）F. Maek. et Hashim.;白花戟叶菫菜■☆

409724 Viola betonicifolia Sm. var. oblongosagittata（Nakai）F. Maek. et T. Hashim. = Viola inconspicua Blume ■

409725 Viola betonicifolia Sm. var. oblongosagittata（Nakai）F. Maek. et T. Hashim.;圆戟菫菜（短圆箭叶菫菜，戟叶菫菜）■☆

409726 Viola bhutanica H. Hara = Viola yunnanfuensis W. Becker ■

409727 Viola biacuta W. Becker = Viola yezoensis Maxim. ■

409728 Viola bicolor Pursh;双色菫菜;Field Pansy,Johnny-jump-up,Wild Pansy ■☆

409729 Viola biflora L.;双花菫菜（短距黄花菫菜，短距黄菫,谷穗补,阔叶菫菜,孪生菫菜,双花紫菫,双黄花菫菜,硬毛双花菫菜）;Broadleaf Twinflower Violet,Hirsute Twinflower Violet,Twin Flowered Violet,Twinflower Violet,Twin-flowered Violet,Yellow Wood Violet ■

409730 Viola biflora L. f. glabrifolia Hid. Takah.;光叶双花菫菜■☆

409731 Viola biflora L. var. acutifolia H. Boissieu = Viola szetschwanensis W. Becker et H. Boissieu ■

409732 Viola biflora L. var. acutifolia Kar. et Kir. = Viola acutifolia（Kar. et Kir.）W. Becker ■

409733 Viola biflora L. var. akaishiensis Hid. Takah. et Ohba;明石菫菜■☆

409734 Viola biflora L. var. ciliicalyx H. Boissieu = Viola szetschwanensis W. Becker et H. Boissieu ■

409735 Viola biflora L. var. hirsuta W. Becker = Viola biflora L. ■

409736 Viola biflora L. var. nudicaulis W. Becker = Viola biflora L. ■

409737 Viola biflora L. var. platyphylla Delavay ex Franch. = Viola biflora L. ■

409738 Viola biflora L. var. platyphylla Franch. = Viola biflora L. ■

409739 Viola biflora L. var. platyphylla Franch. = Viola confertifolia C. C. Chang ■

409740 Viola biflora L. var. rockiana（W. Becker）Y. S. Chen;圆叶双花菫菜（圆叶小菫菜）■

409741 Viola biflora L. var. typica H. Boissieu = Viola biflora L. ■

409742 Viola biflora L. var. valdepilosa Hand. -Mazz. = Viola biflora L. ■

409743 Viola binchuanensis S. H. Huang;宾川菫菜;Binchuan Violet ■

409744 Viola binchuanensis S. H. Huang = Viola grandisepala W. Becker ■

409745 Viola bissellii House;比斯尔菫菜;Bissell's Violet ■☆

409746 Viola bisseti Maxim.;比塞特菫菜■☆

409747 Viola bisseti Maxim. f. albiflora Nakai ex F. Maek.;白花比氏菫菜■☆

409748 Viola bissetii Maxim. f. variegata Nakai = Viola bissetii Maxim. var. kiusiana Terao ■☆

409749 Viola bissetii Maxim. var. kiusiana Terao;九州菫菜■☆

409750 Viola bissetii Maxim. var. kiusiana Terao = Viola bisseti Maxim. ■☆

409751 Viola blanda Willd.;白花芳香菫菜;Aweet White Violet,Sweet White Violet ■☆

409752 Viola blanda Willd. var. palustriformis A. Gray;拟沼泽菫菜;Sweet White Violet ■☆

409753 Viola blandaeformis Nakai;光滑菫菜■☆

409754 Viola blandaeformis Nakai var. pilosa H. Hara = Viola hultenii W. Becker ■☆

409755 Viola boissieuana Makino;布瓦菫菜■☆

409756 Viola boissieuana Makino var. iwagawae（Makino）Ohwi = Viola iwagawae Makino ■☆

409757 Viola boissieui H. Lév. = Viola delavayi Franch. ■

409758 Viola boissieui H. Lév. et Maire = Viola delavayi Franch. ■

409759 Viola borealis Weinm. = Viola selkirkii Pursh ex Goldie ■

409760 Viola brachycentra Hayata = Viola adenothrix Hayata ■

409761 Viola brachycentra Hayata = Viola fargesii H. Boissieu ■

409762 Viola brachyceras Turcz.;兴安圆叶菫菜;Shortspur Violet ■

409763 Viola brachysepala Maxim. = Viola mirabilis L. ■

409764 Viola brainerdii Greene = Viola renifolia A. Gray ■☆

409765 Viola brevistipulata（Franch. et Sav.）W. Becker f. acuminata（Nakai）S. Watan.;渐尖菫菜■☆

409766 Viola brevistipulata（Franch. et Sav.）W. Becker f. ciliata（M. Kikuchi）F. Maek. = Viola brevistipulata（Franch. et Sav.）W. Becker var. ciliata M. Kikuchi ■☆

409767 Viola brevistipulata（Franch. et Sav.）W. Becker f. laciniata（H. Boissieu）F. Maek. = Viola brevistipulata（Franch. et Sav.）W. Becker var. laciniata（H. Boissieu）W. Becker ■☆

409768 Viola brevistipulata（Franch. et Sav.）W. Becker f. pubescens（Nakai）F. Maek.;短柔毛小托叶菫菜■☆

409769 Viola brevistipulata（Franch. et Sav.）W. Becker subsp. hidakana（Nakai）S. Watan. f. incisa S. Watan. = Viola brevistipulata（Franch. et Sav.）W. Becker var. laciniata（H. Boissieu）W. Becker ■☆

409770　Viola brevistipulata（Franch. et Sav.）W. Becker subsp. hidakana（Nakai）S. Watan. ;日高菫菜■☆

409771　Viola brevistipulata（Franch. et Sav.）W. Becker subsp. hidakana（Nakai）S. Watan. var. incisa（S. Watan.）F. Maek. et T. Hashim. ;锐裂小托叶堇菜■☆

409772　Viola brevistipulata（Franch. et Sav.）W. Becker subsp. hidakana（Nakai）S. Watan. var. yezoana Toyok. ex H. Nakai, H. Igarashi et H. Ohashi;虾夷堇菜■☆

409773　Viola brevistipulata（Franch. et Sav.）W. Becker subsp. hidakana（Nakai）S. Watan. var. yezoana Toyok. ex H. Nakai, H. Igarashi et H. Ohashi f. parviflora S. Watan. ;小花小托叶堇菜■☆

409774　Viola brevistipulata（Franch. et Sav.）W. Becker subsp. hidakana（Nakai）S. Watan. var. yezoana Toyok. ex H. Nakai, H. Igarashi et H. Ohashi f. glabra S. Watan. ;光滑小托叶堇菜■☆

409775　Viola brevistipulata（Franch. et Sav.）W. Becker subsp. minor（Nakai）F. Maek. et T. Hashim. ;微小托叶堇菜■☆

409776　Viola brevistipulata（Franch. et Sav.）W. Becker var. acuminata Nakai = Viola brevistipulata（Franch. et Sav.）W. Becker f. acuminata（Nakai）S. Watan. ■☆

409777　Viola brevistipulata（Franch. et Sav.）W. Becker var. ciliata M. Kikuchi;睫毛小托叶堇菜■☆

409778　Viola brevistipulata（Franch. et Sav.）W. Becker var. crassifolia（Koidz.）F. Maek. ex Akiyama, H. Ohba et Tabuchi = Viola yubariana Nakai ■☆

409779　Viola brevistipulata（Franch. et Sav.）W. Becker var. hidakana（Nakai）S. Watan. ;日高小托叶堇菜■☆

409780　Viola brevistipulata（Franch. et Sav.）W. Becker var. kishidae（Nakai）F. Maek. et T. Hashim. ;岸田堇菜■☆

409781　Viola brevistipulata（Franch. et Sav.）W. Becker var. laciniata（H. Boissieu）W. Becker;撕裂小托叶堇菜■☆

409782　Viola brevistipulata（Franch. et Sav.）W. Becker var. minor Nakai = Viola brevistipulata（Franch. et Sav.）W. Becker subsp. minor（Nakai）F. Maek. et T. Hashim. ■☆

409783　Viola brevistipulata（Franch. et Sav.）W. Becker var. renifolia（Koidz.）F. Maek. = Viola alliariifolia Nakai ■☆

409784　Viola brevistipulata（Franch. et Sav.）W. H. Baker;小托叶堇菜■☆

409785　Viola brittoniana Pollard;海岸堇菜;Coast Violet ■☆

409786　Viola brunneostipulosa Hand. -Mazz. = Viola grandisepala W. Becker ■

409787　Viola buchaniana DC. ex D. Don = Viola pilosa Blume ■

409788　Viola bulbosa Maxim. ;鳞茎堇菜;Bulbous Violet ■

409789　Viola bulbosa Maxim. = Viola tuberifera Franch. ■

409790　Viola bulbosa Maxim. subsp. tuberifera（Franch.）W. Becker = Viola bulbosa Maxim. ■

409791　Viola bulbosa Maxim. subsp. tuberifera（Franch.）W. Becker = Viola tuberifera Franch. ■

409792　Viola bulbosa Maxim. var. brevipedicellata S. Y. Chen = Viola tuberifera Franch. ■

409793　Viola bulbosa Maxim. var. franchetii H. Boissieu = Viola bulbosa Maxim. ■

409794　Viola bulbosa Maxim. var. franchetii H. Boissieu = Viola tuberifera Franch. ■

409795　Viola caespitosa D. Don = Viola betonicifolia Sm. ■

409796　Viola calcarata L. ;距堇（显距堇菜）;Calcarate Violet, Long-spurred Pansy ■☆

409797　Viola cameleo H. Boissieu;阔紫叶堇菜;Broadpurpleleaf Violet ■

409798　Viola canadensis L. ;北美菫菜;Canada Violet, Canadian Violet, Canadian White Violet, North America Violet, Tall White Violet ■☆

409799　Viola canadensis L. var. rugulosa（Greene）C. L. Hitchc. ;皱纹北美堇菜;Canadian Violet, Creeping-root Violet ■☆

409800　Viola canescens Wall. ;灰堇菜;Grey Violet ■

409801　Viola canescens Wall. f. glabrescens W. Becker = Viola fargesii H. Boissieu ■

409802　Viola canescens Wall. subsp. lanuginosa W. Becker = Viola fargesii H. Boissieu ■

409803　Viola canescens Wall. subsp. lanuginosa W. Becker = Viola principis H. Boissieu ■

409804　Viola canina L. ;普通菫菜（贯头尖，犁头尖，犬堇菜）;Butter Pat, Butter Pats, Dog Violet, Dog-violet, Heath Dog Violet, Heath Dog-violet, Heath Violet, Little Bluebell ■☆

409805　Viola canina L. var. acuminata（Ledeb.）Regel = Viola acuminata Ledeb. ■

409806　Viola canina L. var. kamtschatica Ging. = Viola sacchalinensis H. Boissieu ■

409807　Viola canina L. var. rupestris（F. W. Schmidt）Regel = Viola rupestris F. W. Schmidt ■

409808　Viola canina L. var. rupestris Regel et Herder = Viola rupestris F. W. Schmidt ■

409809　Viola capensis Thunb. = Hybanthus capensis（Thunb.）Engl. ●■

409810　Viola capillaris Pers. ;白毛堇菜■☆

409811　Viola carnosula W. Becker;肉质堇菜■☆

409812　Viola cazorlensis Gand. ;卡索拉堇菜■☆

409813　Viola cenisia L. ;塞尼斯堇菜■☆

409814　Viola cerasifolia A. St. -Hil. ;樱叶堇■☆

409815　Viola cestrica House = Viola sagittata Aiton ■☆

409816　Viola chaerophylloides（Regel）W. Becker;南山菫菜（胡堇菜，胡堇草，泥鳅草，蜈蚣草，细芹叶堇）;Chervil-like Violet ■

409817　Viola chaerophylloides（Regel）W. Becker f. sieboldiana（Maxim.）F. Maek. et T. Hashim. = Viola chaerophylloides（Regel）W. Becker var. sieboldiana（Maxim.）Makino ■

409818　Viola chaerophylloides（Regel）W. Becker var. eizanensis（Makino）Ohwi = Viola eizanensis（Makino）Makino ■☆

409819　Viola chaerophylloides（Regel）W. Becker var. sieboldiana（Maxim.）Makino;细裂菫菜（席氏南山堇菜）■

409820　Viola chalcosperma Brainerd = Viola palmata L. ■☆

409821　Viola cheiranthifolia Kunth;桂竹香叶堇菜■☆

409822　Viola chinensis G. Don = Viola inconspicua Blume ■

409823　Viola chinensis G. Don = Viola prionantha Bunge ■

409824　Viola chinensis G. Don f. alboviolacea Skvortsov = Viola philippica Cav. ■

409825　Viola chinensis G. Don f. anomala Skvortsov = Viola philippica Cav. ■

409826　Viola chinensis G. Don f. communis Skvortsov = Viola philippica Cav. ■

409827　Viola chinensis G. Don f. dissecta Skvortsov = Viola philippica Cav. ■

409828　Viola chinensis G. Don f. glabra Skvortsov = Viola philippica Cav. ■

409829　Viola cinerea Boiss. ;灰色堇菜■☆

409830　Viola cinerea Boiss. var. erythraea Fiori;浅红堇菜■☆

409831　Viola cinerea Boiss. var. somalensis（Engl.）Chiov. = Viola cinerea Boiss. var. stocksii（Boiss.）W. Becker ■☆

409832　Viola cinerea Boiss. var. soyrae （Chiov.） Cufod. = Viola cinerea Boiss. var. erythraea Fiori ■☆

409833　Viola cinerea Boiss. var. stocksii （Boiss.） W. Becker;斯托克斯董菜■☆

409834　Viola collina Besser;球果董菜（白毛叶地丁草,白毛叶地丁子,匙头菜,地丁子,地核桃,花头菜,怀胎草,箭头草,毛果董菜,山核桃,银地匙,圆叶毛董菜）;Cornt Violet,Hairyfruit Violet ■

409835　Viola collina Besser var. intramongolica Ching J. Wang;光叶球果董菜■

409836　Viola concolor T. F. Forst. = Hybanthus concolor （T. F. Forst.） Spreng. ●☆

409837　Viola concordifolia Ching J. Wang;心叶董菜（毛董菜,紫花地丁）;Heartleaved Violet ■

409838　Viola concordifolia Ching J. Wang = Viola yunnanfuensis W. Becker ■

409839　Viola concordifolia Ching J. Wang var. hirtipedicellata Ching J. Wang = Viola japonica Langsd. ex Ging. ■

409840　Viola concordifolia Ching J. Wang var. hirtipedicellata Ching J. Wang;毛梗心叶董菜;Hairypedicel Heartleaved Violet, Heartleaved Violet ■

409841　Viola conferta （W. Becker） Nakai = Viola orientalis （Maxim.） W. Becker ■

409842　Viola confertifolia C. C. Chang;密叶董菜;Denseleaved Violet ■

409843　Viola confusa Champ. ex Benth. ;短毛董菜（毛董菜）;Hairy Violet ■

409844　Viola confusa Champ. ex Benth. = Viola inconspicua Blume ■

409845　Viola confusa Champ. ex Benth. = Viola pilosa Blume ■

409846　Viola confusa Champ. ex Benth. subsp. nagasakiensis （W. Becker） F. Maek. et T. Hashim. = Viola inconspicua Blume subsp. nagasakiensis （W. Becker） Ching J. Wang et T. C. Huang ■

409847　Viola confusa Champ. ex Benth. subsp. nagasakiensis （W. Becker） F. Maek. et T. Hashim. = Viola inconspicua Blume ■

409848　Viola confusa Champ. ex Benth. subsp. nagashakiensis （W. Becker） F. Maek. et T. Hashim. = Viola inconspicua Blume subsp. nagasakiensis （W. Becker） Ching J. Wang et T. C. Huang ■

409849　Viola conilii Franch. et Sav. = Viola phalacrocarpa Maxim. ■

409850　Viola conjugens Greene;接合董菜■☆

409851　Viola conspersa Rchb. ;丰花董菜（数花董菜）;American Dog Violet,American Dog-violet,Dog Violet,Multiflorous Violet ■☆

409852　Viola conspersa Rchb. = Viola labradorica Schrank ■☆

409853　Viola conspersa Rchb. f. masonii （Farw.） House = Viola striata Aiton ■☆

409854　Viola conspersa Rchb. var. masonii Farw. = Viola striata Aiton ■☆

409855　Viola conturbata House = Viola bissellii House ■☆

409856　Viola cordifolia W. Becker = Viola concordifolia Ching J. Wang ■

409857　Viola cordifolia W. Becker = Viola yunnanfuensis W. Becker ■

409858　Viola cornuta L. ;角董（有距董菜）;Bedding Pansy Violet,Black Pansy,Butterfly Viola,Horned Pansy,Horned Violet,Tufted Pansy ■☆

409859　Viola cornuta L. 'Alba';白角董菜;White Horned Violet ■☆

409860　Viola cornuta L. ' Atropurpyrea ';紫角董菜;Dark-purple Horned Violet ■☆

409861　Viola cornuta L. ' Lutea Splendens ';金黄角董菜;Golden-yellow Horned Violet ■☆

409862　Viola crassa Makino;厚叶董菜■☆

409863　Viola crassa Makino subsp. alpicola Hid. Takah. ;高山厚叶董菜■☆

409864　Viola crassa Makino subsp. borealis Hid. Takah. ;北方厚叶董菜■☆

409865　Viola crassa Makino var. kitamiana （Nakai） Toyok. = Viola kitamiana Nakai ■☆

409866　Viola crassicalcarata Ching J. Wang;粗距董菜■

409867　Viola crassicalcarata Ching J. Wang = Viola japonica Langsd. ex Ging. ■

409868　Viola crassicornis W. Becker et Hultén;粗角董菜■☆

409869　Viola cucullata Aiton = Viola obliqua Aiton ■☆

409870　Viola cucullata Aiton f. albiflora Britton = Viola cucullata Aiton ■☆

409871　Viola cucullata Aiton f. prionosepala （Greene） Brainerd = Viola cucullata Aiton ■☆

409872　Viola cucullata Aiton f. thurstonii （Twining） House = Viola cucullata Aiton ■☆

409873　Viola cucullata Aiton var. microtitis Brainerd = Viola cucullata Aiton ■☆

409874　Viola cucullata Aiton var. thurstonii Twining = Viola cucullata Aiton ■☆

409875　Viola cunninghamii Hook. f. ;坎宁安董菜;New Zealand Violet ■☆

409876　Viola cuspidifolia W. Becker;鄂西董菜（光叶董菜,锐尖董菜）;W. Hubei Violet ■

409877　Viola dacica Borbás;达西卡董菜■☆

409878　Viola dactyloides Roem. et Schult. ;掌叶董菜;Dactyloideus Violet,Palmleaf Violet ■

409879　Viola dactyloides Roem. et Schult. var. multipartita W. Becker = Viola dactyloides Roem. et Schult. ■

409880　Viola daiskei Kitag. ;代氏董菜;Phoenix Violet ■☆

409881　Viola davidii Franch. ;深圆齿董菜;Crenulate Violet ■

409882　Viola davidii Franch. var. paucicrenata W. Becker = Viola davidii Franch. ■

409883　Viola declinata Waldst. et Kit. ;下曲董菜■☆

409884　Viola decumbens L. f. ;匍匐野董菜;Wild Violet ■☆

409885　Viola dehnhardtii Ten. = Viola alba Besser subsp. dehnhardtii （Ten.） W. Becker ■☆

409886　Viola dehnhardtii Ten. var. atlantica Braun-Blanq. et Maire = Viola alba Besser subsp. dehnhardtii （Ten.） W. Becker ■☆

409887　Viola dehnhardtii Ten. var. gomarica Emb. et Maire = Viola alba Besser subsp. dehnhardtii （Ten.） W. Becker ■☆

409888　Viola delavayi Franch. ;灰叶董菜（黄花草,黄花地草果,黄花地丁,黄花董菜,黄花细辛,踏膀药,土细辛,小黄药）;Delavay Violet ■

409889　Viola delavayi Franch. f. depauperata Diels = Viola delavayi Franch. ■

409890　Viola delavayi Franch. var. depauperata Diels;少花灰叶董菜（少花董菜）■

409891　Viola delavayi Franch. var. villosa W. Becker = Viola delavayi Franch. ■

409892　Viola delphinantha Boiss. ;翠雀花董菜■☆

409893　Viola deltoidea Yatabe = Viola raddeana Regel ■

409894　Viola dentariifolia H. Boissieu = Viola chaerophylloides （Regel） W. Becker ■

409895　Viola diamantiaca Nakai;大叶董菜（寸节七,大铧头草,大叶紫董）;Bigleaf Violet ■

409896　Viola diamantiaca Nakai f. glabrior （Kitag.） Kitag. = Viola diamantiaca Nakai ■

409897　Viola diamantiaca Nakai var. glabrior Kitag. = Viola diamantiaca Nakai ■

409898 Viola diffusa Ging. ;七星莲(白菜仔,白地黄瓜,白花耳钩草, 白花散血草,茶匙黄,抽脓拔,地白菜,地白草,狗儿草,黄瓜菜, 黄瓜草,黄瓜香,黄花香,鸡疴粘草,鸡痾黏草,冷毒草,蔓茎堇, 蔓茎堇菜,毛毛藤,匍伏草,匍匐堇菜,葡伏堇,石白菜,天芥菜 草,王瓜草,细通草,雪里青,野白菜,银茶匙);Climbing Violet, Sevenstar Lotus ■

409899 Viola diffusa Ging. subsp. tenuis (Benth.) W. Becker = Viola diffusa Ging. ■

409900 Viola diffusa Ging. subsp. tenuis W. Becker = Viola diffusa Ging. ■

409901 Viola diffusa Ging. var. brevibarbata Ching J. Wang;短须毛七 星莲■

409902 Viola diffusa Ging. var. brevibarbata Ching J. Wang = Viola diffusa Ging. ■

409903 Viola diffusa Ging. var. brevisepala W. Becker = Viola diffusa Ging. ■

409904 Viola diffusa Ging. var. glabella H. Boissieu;光匍匐堇(九州堇 菜)■

409905 Viola diffusa Ging. var. glabella H. Boissieu = Viola diffusa Ging. ■

409906 Viola diffusa Ging. var. glabella H. Boissieu = Viola diffusoides Ching J. Wang ■

409907 Viola diffusa Ging. var. glaberrima W. Becker = Viola diffusa Ging. ■

409908 Viola diffusa Ging. var. tenuis (Benth.) W. Becker = Viola diffusa Ging. ■

409909 Viola diffusa Ging. var. tomentosa W. Becker = Viola diffusa Ging. ■

409910 Viola diffusa Vell. = Viola cerasifolia A. St. -Hil. ■☆

409911 Viola diffusoides Ching J. Wang;光蔓茎堇菜(抽脓草,光蔓堇 菜,光匍匐堇,光叶匍匐堇菜,九州堇菜);Nakespread Violet, Smooth-leaved Climbing Violet ■

409912 Viola diffusoides Ching J. Wang = Viola diffusa Ging. ■

409913 Viola dimorphophylla Y. S. Chen et Q. E. Yang;轮叶堇菜■

409914 Viola dissecta Ledeb. ;裂叶堇菜(疔毒草,深裂叶堇菜,亚尔 母堂);Dissected Violet,Splitleaf Violet ■

409915 Viola dissecta Ledeb. f. pubescens (Regel) Kitag. = Viola dissecta Ledeb. ■

409916 Viola dissecta Ledeb. f. sieboldiana (Maxim.) Makino = Viola chaerophylloides (Regel) W. Becker var. sieboldiana (Maxim.) Makino ■

409917 Viola dissecta Ledeb. subvar. albida (Palib.) Makino = Viola albida Palib. ■

409918 Viola dissecta Ledeb. var. albida (Palib.) Nakai = Viola albida Palib. ■

409919 Viola dissecta Ledeb. var. angustisecta W. Becker = Viola dissecta Ledeb. ■

409920 Viola dissecta Ledeb. var. chaerophylloides (Regel) Makino = Viola chaerophylloides (Regel) W. Becker ■

409921 Viola dissecta Ledeb. var. chaerophylloides (Regel) Makino f. eizanensis (Makino) E. Ito = Viola eizanensis (Makino) Makino ■☆

409922 Viola dissecta Ledeb. var. chaerophylloides (Regel) Makino f. simplicifolia (Makino) Ohwi = Viola eizanensis (Makino) Makino var. simplicifolia (Makino) Makino ■☆

409923 Viola dissecta Ledeb. var. chaerophylloides (Regel) Makino subvar. albida (Palib.) Makino = Viola albida Palib. ■

409924 Viola dissecta Ledeb. var. chaerophylloides (Regel) Makino

409925 Viola dissecta Ledeb. var. incisa (Turcz.) Y. S. Chen;总裂叶 堇菜;Splitleaf Violet,Split-leaved Violet ■

409926 Viola dissecta Ledeb. var. latisecta W. Becker = Viola dissecta Ledeb. ■

409927 Viola dissecta Ledeb. var. multifuta W. Becker = Viola dissecta Ledeb. ■

409928 Viola dissecta Ledeb. var. pubescens (Regel) Kitag. ;短毛裂叶 堇菜;Pubescent Dissected Violet ■

409929 Viola dissecta Ledeb. var. pubescens (Regel) Kitag. = Viola dissecta Ledeb. ■

409930 Viola dissecta Ledeb. var. sieboldiana (Maxim.) Nakai = Viola chaerophylloides (Regel) W. Becker var. sieboldiana (Maxim.) Makino ■

409931 Viola dissecta Ledeb. var. takahashii Nakai = Viola takahashii (Nakai) Taken. ■

409932 Viola distans Wall. = Viola arcuata Blume ■

409933 Viola distans Wall. = Viola hamiltoniana D. Don ■

409934 Viola dolichocentra Botsch. ;长刺堇菜■☆

409935 Viola dolichoceras Ching J. Wang = Viola pekinensis (Regel) W. Becker ■

409936 Viola domestica Pollard = Viola sororia Willd. ■☆

409937 Viola douglasii Steud. ;道格拉斯堇菜;Douglas Violet ■☆

409938 Viola dudouxii W. Becker;紫点堇菜(白花地丁,山白花)■

409939 Viola dyris Maire;荒地堇菜■☆

409940 Viola dyris Maire var. calcarea Litard. et Maire = Viola dyris Maire ■☆

409941 Viola dyris Maire var. orientalis Emb. = Viola dyris Maire ■☆

409942 Viola eizanensis (Makino) Makino;睿山堇(日本堇菜); Japanese Violet ■☆

409943 Viola eizanensis (Makino) Makino f. candida Hiyama;白睿山 堇■☆

409944 Viola eizanensis (Makino) Makino f. simplicifolia (Makino) F. Maek. = Viola eizanensis (Makino) Makino var. simplicifolia (Makino) Makino ■☆

409945 Viola eizanensis (Makino) Makino var. simplicifolia (Makino) Makino f. leucantha Hiyama;白花单叶山堇■☆

409946 Viola eizanensis (Makino) Makino var. simplicifolia (Makino) Makino;单叶山堇■☆

409947 Viola elatior Fr. ;高堇菜;Tall Violet ■

409948 Viola elatior Fr. = Viola montana L. ■

409949 Viola elegantula Schott;紫红堇菜;Rose-purple Violet ■☆

409950 Viola elisabethae Klokov;埃氏堇菜■☆

409951 Viola emarginata (Nutt.) Leconte = Viola sagittata Aiton ■☆

409952 Viola emarginata (Nutt.) Leconte var. acutiloba Brainerd = Viola sagittata Aiton ■☆

409953 Viola emarginata (Nutt.) Leconte var. subsinuata Greene = Viola subsinuata (Greene) Greene ■☆

409954 Viola emeiensis Ching J. Wang;峨眉堇菜;Emei Violet ■

409955 Viola emeiensis Ching J. Wang = Viola striatella H. Boissieu ■

409956 Viola eminii (Engl.) R. E. Fr. ;埃明堇菜■☆

409957 Viola enneasperma L. = Hybanthus enneaspermus (L.) F. Muell. ●

409958 Viola epipsila Ledeb. ;溪堇菜(土丘堇);Rivulet Violet,Stream Violet ■

409959 Viola epipsila Ledeb. subsp. palustroides W. Becker = Viola epipsila Ledeb. ■

409960　Viola epipsila Ledeb. subsp. repens（Turcz.）W. Becker = Viola epipsila Ledeb. ■

409961　Viola epipsila Ledeb. subsp. repens W. Becker = Viola epipsiloides Á. Löve et D. Löve ■

409962　Viola epipsiloides Á. Löve et D. Löve;假溪堇菜（溪堇菜）■

409963　Viola eriocarpa Schwein.;光果黄堇;Smoothish Yellow Violet ■☆

409964　Viola eriocarpa Schwein. = Viola pubescens Aiton var. scabriuscula Schwein. ex Torr. et A. Gray ■☆

409965　Viola eriocarpa Schwein. f. leiocarpa（Fernald et Wiegand）Deam = Viola pubescens Aiton var. scabriuscula Schwein. ex Torr. et A. Gray ■☆

409966　Viola eriocarpa Schwein. var. leiocarpa Fernald et Wiegand = Viola pubescens Aiton var. scabriuscula Schwein. ex Torr. et A. Gray ■☆

409967　Viola erratica House = Viola sagittata Aiton ■☆

409968　Viola erythraea（Fiori）Chiov. = Viola cinerea Boiss. var. erythraea Fiori ■☆

409969　Viola erythraea（Fiori）Chiov. var. soyrae Chiov. = Viola cinerea Boiss. var. erythraea Fiori ■☆

409970　Viola esculenta Elliott = Viola palmata L. ■☆

409971　Viola etbaica Schweinf. = Viola cinerea Boiss. var. stocksii（Boiss.）W. Becker ■☆

409972　Viola excisa Hance = Viola arcuata Blume ■

409973　Viola excisa Hance = Viola hamiltoniana D. Don ■

409974　Viola falconeri Hook. f. et Thomson;法尔堇菜■☆

409975　Viola fargesii H. Boissieu;柔毛堇菜（白阿飞,宝剑草,犁头草,马蹄嫩草,瓮菜癀,喜岩堇菜,岩生堇菜,紫叶堇菜）;Pubescent Violet, Rocky Violet ■

409976　Viola faurieana W. Becker;长梗紫花堇菜;Longstalk Violet ■

409977　Viola faurieana W. Becker f. albifaurieana E. Hama;白长梗紫花堇菜■☆

409978　Viola faurieana W. Becker var. rhizomata（Nakai）F. Maek. et T. Hashim. = Viola grypoceras A. Gray var. rhizomata（Nakai）Ohwi ■☆

409979　Viola fedtschenkoana W. Becker;费氏堇菜（范得堇菜）■☆

409980　Viola fergesii H. Boissieu;密毛堇菜（黄瓜香,野白菜）;Ferges Violet ■

409981　Viola fernaldii House = Viola conjugens Greene ■☆

409982　Viola filifera Kom. = Viola bulbosa Maxim. ■

409983　Viola fimbriatula Sm.;流苏堇菜;Ovate-leaved Violet ■☆

409984　Viola fimbriatula Sm. = Viola sagittata Aiton var. ovata（Nutt.）Torr. et A. Gray ■☆

409985　Viola fischeri Sweet = Viola gmeliniana Roem. et Schult. ■

409986　Viola fissifolia Kitag. = Viola dissecta Ledeb. var. incisa（Turcz.）Y. S. Chen ■

409987　Viola flavida Bureau et Franch. = Viola tienschiensis W. Becker ■

409988　Viola flaviflora Nakai = Viola brevistipulata（Franch. et Sav.）W. Becker f. acuminata（Nakai）S. Watan. ■☆

409989　Viola flettii Piper;岩堇（弗雷特堇菜）;Olympic Violet ■☆

409990　Viola florariensis Karrer;四季堇菜;Ever-blooming Horned Violet ■☆

409991　Viola floridana Brainerd = Viola affinis Leconte ■☆

409992　Viola formosa Vuk. = Viola reichenbachiana Boreau ■☆

409993　Viola formosana Hayata;台湾堇菜（塔山堇菜）;Taiwan Violet ■

409994　Viola formosana Hayata var. kawakamii（Hayata）Ching J. Wang = Viola formosana Hayata var. kawakamii（Hayata）Y. S. Chen et Q. E. Yang ■

409995　Viola formosana Hayata var. kawakamii（Hayata）Y. S. Chen et Q. E. Yang;长柄台湾堇菜（川上氏堇菜）;Longstalk Taiwan Violet ■

409996　Viola formosana Hayata var. stenopetala（Hayata）Ching J. Wang,T. C. Huang et T. Hashim. = Viola formosana Hayata var. kawakamii（Hayata）Y. S. Chen et Q. E. Yang ■

409997　Viola formosana Hayata var. stenopetala（Hayata）Ching J. Wang,T. C. Huang et T. Hashim.;川上氏堇菜■

409998　Viola formosana Hayata var. tozanensis（Hayata）C. F. Hsieh;塔山堇菜■

409999　Viola formosana Hayata var. tozanensis（Hayata）C. F. Hsieh = Viola formosana Hayata ■

410000　Viola formosana W. Becker var. tozanensis（Hayata）C. F. Hsieh = Viola formosana Hayata ■

410001　Viola forrestiana W. Becker;羽裂堇菜（昌都堇菜,门空堇菜）;Forrest Violet ■

410002　Viola franchetii H. Boissieu = Viola rossii Hemsl. ex Forbes et Hemsl. ■

410003　Viola fukienensis W. Becker = Viola kosanensis Hayata ■

410004　Viola funghuangensis P. Y. Fu et Y. C. Teng = Viola monbeigii W. Becker ■

410005　Viola funghuangensis P. Y. Fu et Y. C. Teng = Viola tokubuchiana Makino var. takedana（Makino）F. Maek. ■

410006　Viola fusiformis Sm. = Viola gmeliniana Roem. et Schult. ■

410007　Viola ganchouenensis W. Becker = Viola tienschiensis W. Becker ■

410008　Viola glabella Nutt.;西海岸堇菜（溪堇菜）;Stream Violet, West-coast Violet ■☆

410009　Viola glabella Nutt. = Viola diffusoides Ching J. Wang ■

410010　Viola gmeliniana Roem. et Schult.;兴安堇菜;Xing'an Violet ■

410011　Viola gmeliniana Roem. et Schult. var. albiflora W. Becker = Viola gmeliniana Roem. et Schult. ■

410012　Viola gmeliniana Roem. et Schult. var. glabra Ledeb. = Viola gmeliniana Roem. et Schult. ■

410013　Viola gmeliniana Roem. et Schult. var. hispida Ledeb. = Viola gmeliniana Roem. et Schult. ■

410014　Viola gracilis Sibth. et Sm.;细线堇菜（纤细堇菜）;Thread Violet ■☆

410015　Viola gracilis Sibth. et Sm. var. aurasiaca（Pomel）Batt. = Viola arvensis Murray ■

410016　Viola grandiflora L.;大花堇菜;Showy Vetch ■☆

410017　Viola grandisepala W. Becker;阔萼堇菜（长茎堇菜,大萼堇菜,峨眉堇菜,阔叶堇菜）;Broadsepal Violet, Brownstipule Violet, Longstem Violet ■

410018　Viola grayi Franch. et Sav.;长梗堇菜■☆

410019　Viola grayi Franch. et Sav. = Viola faurieana W. Becker ■

410020　Viola grayi Franch. et Sav. f. glabra（W. Becker）F. Maek. = Viola grayi Franch. et Sav. ■☆

410021　Viola grayi Franch. et Sav. var. candida H. Boissieu = Viola grypoceras A. Gray ■

410022　Viola grayi Franch. et Sav. var. erecta M. Nagas. = Viola grayi Franch. et Sav. ■☆

410023　Viola grayi Franch. et Sav. var. erecta M. Nagas. f. pubescens（Nakai）E. Hama = Viola grayi Franch. et Sav. ■☆

410024　Viola grayi Franch. et Sav. var. glabra ? = Viola grayi Franch. et Sav. ■☆

410025　Viola grayi Franch. et Sav. var. magnifica ? = Viola sachalinensis H. Boissieu ■

410026　Viola grayi Franch. et Sav. var. pubescens（Nakai）E. Hama = Viola grayi Franch. et Sav. ■☆

410027　Viola greenei House;格林堇菜;Violet ■☆

410028　Viola greenei House = Viola conjugens Greene ■☆

410029　Viola greenmanii House = Viola conjugens Greene ■☆

410030　Viola grypoceras A. Gray;紫花堇菜(白蒂黄瓜,地黄瓜,铧嘴菜,黄瓜香,曲角堇,肾气草,紫花高茎堇菜);Purpleflower Violet ■

410031　Viola grypoceras A. Gray f. albiflora Makino = Viola grypoceras A. Gray f. leucantha H. Hara ■☆

410032　Viola grypoceras A. Gray f. leucantha H. Hara;日本白花堇菜■☆

410033　Viola grypoceras A. Gray f. pubescens (Nakai) M. Mizush.;柔毛紫花堇菜;Pubescent Purpleflower Violet ■☆

410034　Viola grypoceras A. Gray f. purpurellocalcarata (Makino) Hiyama ex F. Maek.;紫距紫花堇菜■☆

410035　Viola grypoceras A. Gray f. rosipetala Hiyama;粉瓣紫花堇菜■☆

410036　Viola grypoceras A. Gray f. trifolia Nakai = Viola grypoceras A. Gray ■

410037　Viola grypoceras A. Gray f. variegata Nakai;斑叶紫花堇菜■☆

410038　Viola grypoceras A. Gray f. viridans Hiyama = Viola grypoceras A. Gray f. pubescens (Nakai) M. Mizush. ■☆

410039　Viola grypoceras A. Gray f. viridiflora Makino ex F. Maek. = Viola grypoceras A. Gray ■

410040　Viola grypoceras A. Gray f. yakusimensis (Masam.) Sugim. = Viola grypoceras A. Gray var. exilis (Miq.) Nakai ■☆

410041　Viola grypoceras A. Gray var. barbata W. Becker = Viola grypoceras A. Gray ■

410042　Viola grypoceras A. Gray var. exilis (Miq.) Nakai;柔弱堇菜■☆

410043　Viola grypoceras A. Gray var. exilis (Miq.) Nakai f. chionantha F. Maek.;雪花堇菜■☆

410044　Viola grypoceras A. Gray var. hichitoana (Nakai) F. Maek. f. kikuzatoi K. Nakaj.;白紫花堇菜■☆

410045　Viola grypoceras A. Gray var. imberbis (A. Gray) Ohwi = Viola grayi Franch. et Sav. ■☆

410046　Viola grypoceras A. Gray var. pubescens Nakai = Viola grypoceras A. Gray f. pubescens (Nakai) M. Mizush. ■☆

410047　Viola grypoceras A. Gray var. pubescens Nakai = Viola grypoceras A. Gray ■

410048　Viola grypoceras A. Gray var. rhizomata (Nakai) Ohwi;根茎紫花堇菜■☆

410049　Viola grypoceras A. Gray var. ripensis N. Yamada et M. Okamoto;河岸紫花堇菜■☆

410050　Viola grypoceras A. Gray var. yakusimensis Masam. = Viola grypoceras A. Gray var. exilis (Miq.) Nakai ■☆

410051　Viola hallii A. Gray;霍尔堇菜;Hall Violet ■☆

410052　Viola hamiltoniana D. Don;如意草(白犁头草,弧茎堇菜);Hamilton Violet ■

410053　Viola hancockii W. Becker;西山堇菜;Hancock Violet ■

410054　Viola hancockii W. Becker var. fangshanensis J. W. Wang = Viola hancockii W. Becker ■

410055　Viola harae Miyabe et Tatew. = Viola sachalinensis H. Boissieu ■

410056　Viola hastata Michx.;戟堇;Halberd-leaved Violet, Hastate Violet ■☆

410057　Viola hebeiensis J. W. Wang et T. G. Ma = Viola mongolica Franch. ■

410058　Viola hederacea Labill.;澳洲堇菜(常春藤叶堇菜);Australian Native Violet, Australian Violet, Ivy-leaf Violet, Ivy-leaved Violet ■☆

410059　Viola hediniana W. Becker;紫叶堇菜;Purpleleaf Violet ■

410060　Viola henryi H. Boissieu;巫山堇菜(紫叶堇菜);Purpleleaf Violet, Wushan Violet ■

410061　Viola henryi H. Boissieu = Viola hediniana W. Becker ■

410062　Viola henryi H. Boissieu var. cameleo (H. Boissieu) Chang = Viola cameleo H. Boissieu ■

410063　Viola hidakana Nakai = Viola brevistipulata (Franch. et Sav.) W. Becker var. hidakana (Nakai) S. Watan. ■☆

410064　Viola hirsutula Brainerd;粗毛堇菜;Southern Wood-violet ■☆

410065　Viola hirsutula Brainerd f. albicans L. K. Henry;白花粗毛堇菜■☆

410066　Viola hirta L.;硬毛堇菜(茸堇菜,硬毛香堇,紫花地丁);Hairy Violet, Hardhair Violet, Horse Violet ■

410067　Viola hirta L. = Viola collina Besser ■

410068　Viola hirta L. subsp. brevifimbriata W. Becker = Viola hirta L. ■

410069　Viola hirta L. var. collina (Besser) Regel = Viola collina Besser ■

410070　Viola hirta L. var. glabella Regel = Viola phalacrocarpa Maxim. ■

410071　Viola hirta L. var. japonica Maxim. = Viola hondoensis W. Becker et H. Boissieu ■

410072　Viola hirtipedoides W. Becker = Viola hirtipes S. Moore ■

410073　Viola hirtipes S. Moore;毛柄堇菜(大深山堇菜);Hairstalk Violet, Hairy Stalk Violet ■

410074　Viola hirtipes S. Moore f. glabra E. Hama;光毛柄堇菜■☆

410075　Viola hirtipes S. Moore f. grisea (Nakai) Hiyama;灰毛柄堇菜■☆

410076　Viola hirtipes S. Moore f. nudipes Hiyama;裸梗堇菜■☆

410077　Viola hirtipes S. Moore f. rhodovenia (Nakai) Hiyama ex F. Maek.;粉脉毛柄堇菜■☆

410078　Viola hirtipes S. Moore var. rhodovenia Nakai = Viola hirtipes S. Moore f. rhodovenia (Nakai) Hiyama ex F. Maek. ■☆

410079　Viola hispida Lam.;黑毛堇菜■☆

410080　Viola hissarica Juz.;希萨尔堇菜■☆

410081　Viola hondoensis W. Becker et H. Boissieu;日本球果堇菜(本州堇菜)■

410082　Viola hookeri Franch. = Viola bulbosa Maxim. ■

410083　Viola hookeri Thomson = Viola sikkimensis W. Becker ■

410084　Viola hookeri Thomson et W. Becker = Viola sikkimensis W. Becker ■

410085　Viola hopheinensis Ching J. Wang et T. G. Ma = Viola yezoensis Maxim. var. hopeiensis (Ching J. Wang et T. G. Ma) Ching J. Wang et J. Yang ■

410086　Viola hossei W. Becker = Viola sumatrana Miq. ■

410087　Viola hsinganensis N. L. Zhu;厚叶紫花地丁;Thickleaf Violet ■

410088　Viola hsinganensis N. L. Zhu = Viola mandshurica W. Becker ■

410089　Viola hsinganensis Taken. = Viola mandshurica W. Becker ■

410090　Viola hultenii W. Becker;胡尔滕堇菜■☆

410091　Viola hunanensis Hand.-Mazz.;湖南堇菜(拟长萼堇菜);Hunan Violet ■

410092　Viola hunanensis Hand.-Mazz. = Viola inconspicua Blume ■

410093　Viola hupeiana W. Becker = Viola arcuata Blume ■

410094　Viola hypoleuca Hayata = Viola formosana Hayata var. kawakamii (Hayata) Y. S. Chen et Q. E. Yang ■

410095　Viola impatiens H. Lév. = Viola delavayi Franch. ■

410096　Viola incisa Turcz. = Viola dissecta Ledeb. var. incisa (Turcz.) Y. S. Chen ■

410097　Viola incisa Turcz. var. acuminata Franch. et Sav. = Viola savatieri Makino ■

410098　Viola incognita Brainerd = Viola blanda Willd. var. palustriformis A. Gray ■☆

410099　Viola incognita Brainerd var. forbesii Brainerd = Viola blanda Willd. var. palustriformis A. Gray ■☆

410100　Viola inconspicua Blume;长萼堇菜(铧头草,犁头草,紫花地

丁）; Longsepal Violet ■

410101　Viola inconspicua Blume subsp. dielsiana W. Becker = Viola betonicifolia Sm. ■

410102　Viola inconspicua Blume subsp. dielsiana W. Becker = Viola inconspicua Blume ■

410103　Viola inconspicua Blume subsp. nagasakiensis （W. Becker） Ching J. Wang et T. C. Huang = Viola inconspicua Blume ■

410104　Viola inconspicua Blume subsp. nagasakiensis （W. Becker） Ching J. Wang et T. C. Huang f. albescens （Taken.） Yonek. ;白小菫菜■

410105　Viola inconspicua Blume subsp. nagasakiensis （W. Becker） Ching J. Wang et T. C. Huang;小菫菜(微菫菜)■

410106　Viola isopetala Juz. ;同瓣菫菜■☆

410107　Viola iwagawae Makino;岩川菫菜■☆

410108　Viola japonica Langsd. ex Ging. ;犁头草（瘩背草,地丁,地丁草,耳钩草,铧头草,箭头草,烙铁草,犁铧尖,犁头尖,犁嘴草,如意草,三角草,小甜水茄,玉如意,紫地丁,紫花地丁,紫金锁）; Share-like Violet ■

410109　Viola japonica Langsd. ex Ging. f. albida F. Maek. ;日本白犁头草■☆

410110　Viola japonica Langsd. ex Ging. f. barbata （Hiyama） Hiyama ex F. Maek. ;髯毛犁头草■☆

410111　Viola japonica Langsd. ex Ging. f. variegata （Hatus.） F. Maek. ;杂色犁头草■☆

410112　Viola japonica Langsd. ex Ging. var. stenopetala Franch. ex H. Boissieu = Viola japonica Langsd. ex Ging. ■

410113　Viola jettmari Hand. -Mazz. = Viola fissifolia Kitag. ■

410114　Viola jizushanensis S. H. Huang;鸡足山菫菜;Jizushan Violet ■

410115　Viola jizushanensis S. H. Huang = Viola biflora L. var. rockiana （W. Becker） Y. S. Chen ■

410116　Viola jooi Janka;乔菫;Joo Violet ■☆

410117　Viola jordanii Hanry;约尔丹菫菜■☆

410118　Viola jordanii Hanry var. falconeri ? = Viola falconeri Hook. f. et Thomson ■☆

410119　Viola kamtchadalorum W. Becker et Hultén = Viola langsdorffii Fisch. ex Ging. subsp. sachalinensis W. Becker var. parviflora （Regel） Nakai ■☆

410120　Viola kamtchadalorum W. Becker et Hultén var. parviflora （Regel） Tatew. = Viola langsdorffii Fisch. ex Ging. subsp. sachalinensis W. Becker var. parviflora （Regel） Nakai ■☆

410121　Viola kamtschatica Ging. = Viola selkirkii Pursh ex Goldie ■

410122　Viola kamtschatica Ging. var. pekinensis Regel = Viola pekinensis （Regel） W. Becker ■

410123　Viola kanoi Sasaki = Viola biflora L. ■

410124　Viola kansuensis W. Becker = Viola pendulicarpa W. Becker ■

410125　Viola kansuensis W. Becker var. oblonga W. Becker = Viola pendulicarpa W. Becker ■

410126　Viola karakalensis Klokov;卡拉卡利菫菜■☆

410127　Viola kawakamii Hayata = Viola formosana Hayata var. kawakamii （Hayata） Y. S. Chen et Q. E. Yang ■

410128　Viola kawakamii Hayata = Viola formosana Hayata var. kawakamii （Hayata） Ching J. Wang ■

410129　Viola kawakamii Hayata var. stenopetala Hayata = Viola formosana Hayata var. kawakamii （Hayata） Ching J. Wang, T. C. Huang et T. Hashim. ■

410130　Viola kawakamii Hayata var. stenopetala Hayata = Viola formosana Hayata var. kawakamii （Hayata） Y. S. Chen et Q. E. Yang ■

410131　Viola keiskei Miq. ;伊藤氏菫菜■☆

410132　Viola keiskei Miq. f. barbata Hiyama ex F. Maek. ;髯毛伊藤氏菫菜■☆

410133　Viola keiskei Miq. f. hirsutior W. Becker = Viola sphaerocarpa W. Becker ■

410134　Viola keiskei Miq. f. okuboi （Makino） F. Maek. = Viola keiskei Miq. ■☆

410135　Viola keiskei Miq. var. glabra （Makino） W. Becker = Viola keiskei Miq. ■☆

410136　Viola kiangsiensis W. Becker;江西菫菜;Jiangxi Violet, Kiangsi Violet ■

410137　Viola kiangsiensis W. Becker = Viola kosanensis Hayata ■

410138　Viola kishidae Nakai = Viola brevistipulata （Franch. et Sav.） W. Becker var. kishidae （Nakai） F. Maek. et T. Hashim. ■☆

410139　Viola kitaibeliana Roem. et Schult. var. rafinesquii （Greene） Fernald = Viola bicolor Pursh ■☆

410140　Viola kitaibeliana Roem. et Schultz;基陶伊贝尔菫菜;Dwarf Pansy, Field Pansy ■☆

410141　Viola kitaibeliana Schult. = Viola kitaibeliana Roem. et Schultz ■☆

410142　Viola kitaibeliana Schult. var. brevicalcarata Font Quer et Svent. = Viola kitaibeliana Roem. et Schultz ■☆

410143　Viola kitamiana Nakai;北村菫菜■☆

410144　Viola kiusiana Makino = Viola diffusa Ging. ■

410145　Viola komarovii W. Becker = Viola sachalinensis H. Boissieu ■

410146　Viola koraiensis Nakai = Viola sachalinensis H. Boissieu f. alpina （H. Hara） F. Maek. et T. Hashim. ■

410147　Viola kosanensis Hayata;福建菫菜■

410148　Viola krugiana W. Becker = Viola grypoceras A. Gray ■

410149　Viola kunawarensis Royle;西藏菫菜(藏东菫菜); Kunawar Violet ■

410150　Viola kunawarensis Royle f. longifolia Ching J. Wang = Viola kunawarensis Royle ■

410151　Viola kunawarensis Royle var. angustifolia W. Becker = Viola kunawarensis Royle ■

410152　Viola kupfferi Klokov;库泊菫菜■☆

410153　Viola kurilensis Nakai = Viola langsdorffii Fisch. ex Ging. subsp. sachalinensis W. Becker var. parviflora （Regel） Nakai ■☆

410154　Viola kusanoana Makino;草野氏菫菜■☆

410155　Viola kusanoana Makino f. alba Masam. ;白花草野氏菫菜■☆

410156　Viola kusanoana Makino f. brevicalcarata （H. Hara） F. Maek. ;短距草野氏菫菜■☆

410157　Viola kusanoana Makino f. pubescens （Nakai） M. Mizush. ;毛草野氏菫菜■☆

410158　Viola kusnezowiana W. Becker;库兹菫菜■☆

410159　Viola kwangtungensis Melch. ;广东菫菜■

410160　Viola kwangtungensis Melch. = Viola mucronulifera Hand. -Mazz. ■

410161　Viola labradorica Schrank;拉布拉多菫菜;Alpine Violet, Dog Violet, Labrador Violet ■☆

410162　Viola laciniata （H. Boissieu） Koidz. = Viola brevistipulata （Franch. et Sav.） W. Becker var. laciniata （H. Boissieu） W. Becker ■☆

410163　Viola laciniosa A. Gray = Viola acuminata Ledeb. ■

410164　Viola lactea Sm. ;乳白花菫菜;Pale Dog-violet, Pale Heath Violet ■☆

410165　Viola lactiflora Nakai;白花菫菜(宽叶白花菫菜);Milky-flowered Violet, White Violet ■☆

410166　Viola lanceolata L. ;披针叶菫菜;Bog White Violet, Lance-leaved Violet, Lanceolate Violet, Strap-leaved Violet ■☆

410167　Viola langloisii Greene = Viola affinis Leconte ■☆

410168　Viola langloisii Greene var. pedatiloba Brainerd = Viola palmata L. ■☆

410169　Viola langsdorffii Fisch. ex Ging. ;朗氏堇菜■☆

410170　Viola langsdorffii Fisch. ex Ging. subsp. sachalinensis W. Becker;库页朗氏堇菜■☆

410171　Viola langsdorffii Fisch. ex Ging. subsp. sachalinensis W. Becker f. parviflora（Regel）F. Maek. = Viola langsdorffii Fisch. ex Ging. subsp. sachalinensis W. Becker var. parviflora（Regel）Nakai ■☆

410172　Viola langsdorffii Fisch. ex Ging. subsp. sachalinensis W. Becker f. pubescens（Miyabe et Tatew.）F. Maek. ;柔毛库页朗氏堇菜■☆

410173　Viola langsdorffii Fisch. ex Ging. subsp. sachalinensis W. Becker var. parviflora（Regel）Nakai;小花库页朗氏堇菜■☆

410174　Viola lasiostipes Nakai = Viola muehldorfii Kiss ■

410175　Viola latiuscula Greene = Viola sororia Willd. ■☆

410176　Viola lavandulacea E. P. Bicknell = Viola conjugens Greene ■☆

410177　Viola leveillei H. Boissieu = Viola grypoceras A. Gray ■

410178　Viola lianhuashanensis Ching J. Wang et K. Sun;莲花山堇菜;Lianhuashan Violet ■

410179　Viola lianhuashanensis Ching J. Wang et K. Sun = Viola striatella H. Boissieu ■

410180　Viola liaosiensis P. Y. Fu et Y. C. Teng = Viola pekinensis（Regel）W. Becker ■

410181　Viola liaoxiensis P. Y. Fu et Y. C. Teng;辽西堇菜;Liaoxi Violet ■

410182　Viola limprichtiana W. Becker = Viola lactiflora Nakai ■

410183　Viola littoralis Spreng. ;滨海堇菜■☆

410184　Viola lobata Benth. ;长裂堇菜;Long-lobe Violet ■☆

410185　Viola longistipulata Hayata = Viola philippica Cav. var. pseudojaponica（Nakai）Y. S. Chen ■

410186　Viola lucens W. Becker;亮毛堇菜(亮叶堇);Brighthair Violet, Shining-leaved Violet ■

410187　Viola lutea Sessé et Moc. ;欧洲黄堇菜;Bonewort, European Yellow Violet, Great Yellow Pansy, Mountain Pansy, Shepherd's Pansy, Yellow Violet ■☆

410188　Viola macedonica Boiss. et Heldr. = Viola tricolor L. ■

410189　Viola macloskeyi F. E. Lloyd;小白堇菜;Northern White Violet, Small White Violet, Smooth White Violet, Sweet White Violet, Wild White Violet ■☆

410190　Viola macloskeyi F. E. Lloyd subsp. pallens（Banks ex Ging.）M. S. Baker;苍白小堇菜;Small White Violet, Wild White Violet ■☆

410191　Viola macloskeyi F. E. Lloyd var. pallens（Banks ex Ging.）C. L. Hitchc. = Viola macloskeyi F. E. Lloyd subsp. pallens（Banks ex Ging.）M. S. Baker ■☆

410192　Viola macroceras Bunge;大距堇菜;Bigspur Violet ■

410193　Viola maculata Cav. ;斑点堇菜■☆

410194　Viola maderensis Lowe = Viola odorata L. var. maderensis（Lowe）Webb ■

410195　Viola magnifica Ching J. Wang et X. D. Wang;犁头叶堇菜;Ploughshareleaf Violet ■

410196　Viola mainlingensis S. Y. Chen;米林堇菜;Milin Violet ■

410197　Viola mainlingensis S. Y. Chen = Viola szetschwanensis W. Becker et H. Boissieu ■

410198　Viola mairei H. Lév. = Viola moupinensis（Franch.）Franch. ■

410199　Viola makinoi H. Boissieu = Viola violacea Makino var. makinoi（H. Boissieu）Hiyama ex F. Maek. ■☆

410200　Viola manaslensis F. Maek. = Viola szetschwanensis W. Becker et H. Boissieu ■

410201　Viola mandshurica W. Becker;东北堇菜（紫花地丁）;Manchurian Violet, NE. China Violet ■

410202　Viola mandshurica W. Becker f. albiflora P. Y. Fu et Y. C. Teng;白花东北堇菜;Whiteflower NE. China Violet ■

410203　Viola mandshurica W. Becker f. albiflora P. Y. Fu et Y. C. Teng = Viola mandshurica W. Becker ■

410204　Viola mandshurica W. Becker f. albovariegata？ = Viola mandshurica W. Becker ■

410205　Viola mandshurica W. Becker f. ciliata（Nakai）F. Maek. = Viola mandshurica W. Becker ■

410206　Viola mandshurica W. Becker f. crassa（Tatew.）F. Maek. ;厚叶东北堇菜■☆

410207　Viola mandshurica W. Becker f. dispar Honda = Viola mandshurica W. Becker ■

410208　Viola mandshurica W. Becker f. glabra（Nakai）Hiyama = Viola mandshurica W. Becker ■

410209　Viola mandshurica W. Becker f. glabra（Nakai）Hiyama ex Maek. = Viola mandshurica W. Becker ■

410210　Viola mandshurica W. Becker f. glabripetala？ = Viola mandshurica W. Becker ■

410211　Viola mandshurica W. Becker f. hasegawae Hiyama = Viola mandshurica W. Becker ■

410212　Viola mandshurica W. Becker f. horomuiensis？ = Viola mandshurica W. Becker ■

410213　Viola mandshurica W. Becker f. macrantha（Maxim.）Nakai et Kitag. = Viola mandshurica W. Becker ■

410214　Viola mandshurica W. Becker f. macrantha（Maxim.）Nakai ex Kitag. ;大花东北堇菜■☆

410215　Viola mandshurica W. Becker f. plena F. Maek. ;重瓣东北堇菜■

410216　Viola mandshurica W. Becker f. plena F. Maek. = Viola mandshurica W. Becker ■

410217　Viola mandshurica W. Becker f. villosa？ = Viola mandshurica W. Becker ■

410218　Viola mandshurica W. Becker subsp. nagasakiensis W. Becker = Viola inconspicua Blume ■

410219　Viola mandshurica W. Becker subsp. nagasakiensis W. Becker = Viola inconspicua Blume subsp. nagasakiensis（W. Becker）Ching J. Wang et T. C. Huang ■

410220　Viola mandshurica W. Becker var. ciliata Nakai;堇色堇菜■

410221　Viola mandshurica W. Becker var. ciliata Nakai = Viola mandshurica W. Becker ■

410222　Viola mandshurica W. Becker var. ciliata Nakai f. aureoreticulata？ = Viola mandshurica W. Becker ■

410223　Viola mandshurica W. Becker var. crassa Tatew. = Viola mandshurica W. Becker ■

410224　Viola mandshurica W. Becker var. crassa Tatew. = Viola mandshurica W. Becker f. crassa（Tatew.）F. Maek. ■☆

410225　Viola mandshurica W. Becker var. media？ = Viola mandshurica W. Becker ■

410226　Viola mandshurica W. Becker var. triangularis（Franch. et Sav.）M. Mizush. ;三角东北堇菜■☆

410227　Viola mariae W. Becker = Viola sacchalinensis H. Boissieu ■

410228　Viola maroccana Maire;摩洛哥堇菜■☆

410229　Viola matsudae Hayata = Viola formosana Hayata var. kawakamii（Hayata）Y. S. Chen et Q. E. Yang ■

410230　Viola matsudai Hayata = Viola formosana Hayata var. kawakamii（Hayata）Ching J. Wang ■

410231　Viola matsumurae Makino = Viola rossii Hemsl. ex Forbes et Hemsl. ■

410232　Viola matutina Klokov;清晨堇菜■☆

410233　Viola mauritii Tepl.;茂丽堇菜;Maurit Violet ■

410234　Viola maximowicziana Makino;马氏堇菜■☆

410235　Viola maximowicziana Makino f. rubescens Makino;红色马氏堇菜■☆

410236　Viola mearnsii Standl. = Viola eminii (Engl.) R. E. Fr. ■☆

410237　Viola meyeriana (Rupr.) Klokov;梅耶尔堇菜■☆

410238　Viola micrantha Turcz. = Viola acuminata Ledeb. ■

410239　Viola microdonta C. C. Chang;细齿堇菜;Smalltoothed Violet ■

410240　Viola microdonta Chang = Viola collina Besser ■

410241　Viola minor (Makino) Makino;微堇菜■

410242　Viola minor (Makino) Makino = Viola inconspicua Blume subsp. nagashakiensis (W. Becker) Ching J. Wang et T. C. Huang ■

410243　Viola minor (Makino) Makino = Viola inconspicua Blume ■

410244　Viola minor (Makino) Makino f. albescens (Taken.) F. Maek. = Viola inconspicua Blume subsp. nagasakiensis (W. Becker) Ching J. Wang et T. C. Huang f. albescens (Taken.) Yonek. ■

410245　Viola minor (Makino) Makino f. pilosa F. Maek. = Viola inconspicua Blume subsp. nagasakiensis (W. Becker) Ching J. Wang et T. C. Huang ■

410246　Viola mirabilis L.;奇异堇菜(见肿消,伊吹堇菜);Eastern Dog-violet, Mir Violet, Strange Violet, Wonder Violet ■

410247　Viola mirabilis L. f. latisepala W. Becker = Viola mirabilis L. ■

410248　Viola mirabilis L. var. brachysepala (Maxim.) Regel = Viola mirabilis L. ■

410249　Viola mirabilis L. var. brachyspaia Regel = Viola mirabilis L. ■

410250　Viola mirabilis L. var. brevicalcarata Nakai = Viola mirabilis L. ■

410251　Viola mirabilis L. var. glaberrima W. Becker = Viola mirabilis L. ■

410252　Viola mirabilis L. var. platysepala Kitag. = Viola mirabilis L. ■

410253　Viola mirabilis L. var. subglabra f. latisepala W. Becker = Viola mirabilis L. ■

410254　Viola mirabilis L. var. subglabra f. strigosa W. Becker = Viola mirabilis L. ■

410255　Viola mirabilis L. var. subglabra Ledeb.;光奇异堇菜■☆

410256　Viola mirabilis L. var. subglabra Ledeb. = Viola mirabilis L. ■

410257　Viola mirabilis L. var. subglabra Ledeb. f. seiichii E. Hama;塞氏光奇异堇菜■☆

410258　Viola mirabilis L. var. vulgaris Ledeb. = Viola mirabilis L. ■

410259　Viola miranda W. Becker = Viola sacchalinensis H. Boissieu ■

410260　Viola missouriensis Greene;密苏里堇菜;Missouri Violet ■☆

410261　Viola missouriensis Greene = Viola affinis Leconte ■☆

410262　Viola mistassinica Greene = Viola renifolia A. Gray ■☆

410263　Viola miyabei Makino = Viola hirtipes S. Moore ■

410264　Viola miyakei Nakai = Viola sachalinensis H. Boissieu ■

410265　Viola modesta Ball = Viola parvula Tineo ■☆

410266　Viola modestula Klokov;适度堇菜■☆

410267　Viola mollicula House = Viola primulifolia L. ■☆

410268　Viola monbeigii W. Becker = Viola belophylla H. Boissieu ■

410269　Viola mongolica Franch.;蒙古堇菜(白花堇菜);Mongol Violet, Mongolian Violet ■

410270　Viola mongolica Franch. f. longisepala P. Y. Fu et Y. C. Teng;长萼蒙古堇菜;Longcalyx Violet, Longsepal Violet ■

410271　Viola mongolica Franch. f. longisepala P. Y. Fu et Y. C. Teng = Viola pekinensis (Regel) W. Becker ■

410272　Viola monochroa Klokov;单色堇菜■☆

410273　Viola montana L.;山地堇菜(高堇菜)■

410274　Viola montana L. var. elatior (Fr.) Regel = Viola montana L. ■

410275　Viola moupinensis (Franch.) Franch.;堇(白三百棒,宝兴堇菜,黄堇,鸡心七,筋骨七,乌蔗莲,萱);Moupin Violet, Muping Violet ■

410276　Viola moupinensis (Franch.) Franch. var. lijiangensis Ching J. Wang = Viola moupinensis (Franch.) Franch. ■

410277　Viola moupinensis Franch. var. lijiangensis Ching J. Wang;黄花萱■

410278　Viola mucronulifera Hand. -Mazz.;小尖堇菜;Apiculate Violet ■

410279　Viola muehldorfii Kiss;大黄花堇菜;Big Yellowflowered Violet, Muehldorf Violet ■

410280　Viola muliensis Y. S. Chen et Q. E. Yang;木里堇菜■

410281　Viola multistolonifera Ching J. Wang;多伏茎堇菜■

410282　Viola multistolonifera Ching J. Wang = Viola bulbosa Maxim. ■

410283　Viola munbyana Boiss. et Reut.;芒比堇菜■☆

410284　Viola munbyana Boiss. et Reut. var. battandieri (Becker) Batt. = Viola munbyana Boiss. et Reut. ■☆

410285　Viola munbyana Boiss. et Reut. var. kabylica Batt. = Viola munbyana Boiss. et Reut. ■☆

410286　Viola munbyana Boiss. et Reut. var. rifana Emb. et Maire;里夫堇菜■☆

410287　Viola mutsuensis W. Becker = Viola sacchalinensis H. Boissieu ■

410288　Viola nagamiana T. Hashim. = Viola kosanensis Hayata ■

410289　Viola nagasawae Makino et Hayata;台北堇菜;Nagasawa Violet ■

410290　Viola nagasawae Makino et Hayata var. acutilabella (Hayata) Nakai = Viola nagasawae Makino et Hayata ■

410291　Viola nagasawae Makino et Hayata var. pricei (W. Becker) Ching J. Wang;锐叶台北堇菜(普莱氏堇菜)■

410292　Viola nannae R. E. Fr.;南纳堇菜■☆

410293　Viola nantouensis S. S. Ying = Viola philippica Cav. var. pseudojaponica (Nakai) Y. S. Chen ■

410294　Viola napae House = Viola sororia Willd. ■☆

410295　Viola napellifolia Nakai = Viola chaerophylloides (Regel) W. Becker ■

410296　Viola napellifolia Nakai var. sieboldiana (Maxim.) Nakai = Viola chaerophylloides (Regel) W. Becker var. sieboldiana (Maxim.) Makino ■

410297　Viola nephrophylla Greene;北方沼地堇菜;Blue Prairie Violet, Northern Bog Violet, Northern Bog-violet ■☆

410298　Viola nephrophylla Greene f. albinea Farw. = Viola nephrophylla Greene ■☆

410299　Viola nikkoensis Nakai = Viola tokubuchiana Makino ■☆

410300　Viola nipponica Maxim.;葵堇■☆

410301　Viola novae-angliae House;新英格兰堇菜;New England Blue Violet ■☆

410302　Viola nubica Hutch. = Viola cinerea Boiss. var. stocksii (Boiss.) W. Becker ■☆

410303　Viola nuda W. Becker;裸堇菜(无毛堇菜);Nake Violet ■

410304　Viola nudicaulis (W. Becker) S. Y. Chen = Viola biflora L. ■

410305　Viola nuttallii Pursh;纽托尔堇菜;Nuttall Violet ■☆

410306　Viola obliqua Aiton;头巾堇菜(堇蓝花湿生菜);Blue Hooded Violet, Blue Marsh Violet, Hooded Violet, Marsh Blue Violet, Marsh Violet, Wood Violet ■☆

410307　Viola obliqua Hill = Viola cucullata Aiton ■☆

410308　Viola oblongosagittata Nakai = Viola betonicifolia Sm. ■

410309　Viola oblongosagittata Nakai = Viola inconspicua Blume ■

410310　Viola oblongosagittata Nakai f. ishizakii Yamam. = Viola mandshurica W. Becker ■

410311　Viola oblongosagittata Nakai var. violascens Nakai = Viola betonicifolia Sm. ■

410312　Viola obtusa (Makino) Makino;钝齿堇菜■☆

410313　Viola obtusa (Makino) Makino f. chibae (Makino) Hiyama ex F. Maek.;白花钝齿堇菜■☆

410314　Viola obtusa (Makino) Makino f. hemileuca Sugim.;半白钝齿堇菜■☆

410315　Viola obtusa (Makino) Makino var. tsuifengensis Hashim.;翠峰堇菜■

410316　Viola obtusa Makino = Viola obtusa (Makino) Makino ■☆

410317　Viola obtusa Makino f. nuda (Ohwi) F. Maek.;裸露钝齿堇菜■☆

410318　Viola occulta Lehm.;隐蔽堇菜■☆

410319　Viola ocellata Torr. et Gray;眼斑堇菜;Ocellate Violet ■☆

410320　Viola odorata L.;香堇菜(香堇);Common Violet, Devon Violet, English Violet, Fineleaf, Fire Lights, Florists Violet, Fragrant Violet, Garden Violet, Humpbacks, March Violet, Miss Modesty, Miss Scenty, Modest Maiden, Parma Violet, Sweet Violet, Vilip, Violet, Violet Tea ■

410321　Viola odorata L. f. albiflora Oborny = Viola odorata L. ■

410322　Viola odorata L. var. maderensis (Lowe) Webb = Viola odorata L. ■

410323　Viola oligoceps C. C. Chang;分蘖堇菜;Tiller Violet ■

410324　Viola oligoceps C. C. Chang = Viola tienschiensis W. Becker ■

410325　Viola oreades M. Bieb.;克里木堇菜■☆

410326　Viola orientalis (Maxim.) W. Becker;东方堇菜(朝鲜堇菜,黄花堇菜,小堇菜);Eastern Violet, Oriental Violet ■

410327　Viola orientalis (Maxim.) W. Becker f. laciniata (Taken.) F. Maek.;条裂东方堇菜■☆

410328　Viola orientalis (Maxim.) W. Becker var. conferta W. Becker = Viola orientalis (Maxim.) W. Becker ■

410329　Viola orthoceras Ledeb.;直角堇菜■☆

410330　Viola ovato-oblonga (Miq.) Makino;矩卵形堇菜■☆

410331　Viola ovato-oblonga (Miq.) Makino f. albiflora Honda;白花矩卵形堇菜■☆

410332　Viola ovato-oblonga (Miq.) Makino f. luteoviridiflora (Araki) F. Maek.;黄白花矩卵形堇菜■☆

410333　Viola ovato-oblonga (Miq.) Makino f. pubescens (Nakai) F. Maek.;毛矩卵形堇菜■☆

410334　Viola ovato-oblonga (Miq.) Makino f. variegata E. Hama;变色矩卵形堇菜■☆

410335　Viola ovato-oblonga (Miq.) Makino var. obtusa Makino = Viola obtusa Makino ■☆

410336　Viola oxycentra Juz.;新疆香堇(香堇)■

410337　Viola pallens (Banks ex DC.) Brainerd = Viola macloskeyi F. E. Lloyd ■☆

410338　Viola pallens (Banks ex Ging.) Brainerd = Viola macloskeyi F. E. Lloyd subsp. pallens (Banks ex Ging.) M. S. Baker ■☆

410339　Viola pallens (Banks ex Ging.) Brainerd var. subreptans J. Rousseau = Viola macloskeyi F. E. Lloyd subsp. pallens (Banks ex Ging.) M. S. Baker ■☆

410340　Viola palmata L.;掌裂堇菜;Cleft Violet, Early Blue Violet, Johnny-jump-up, Palmate Violet, Palmate-lobed Violet, Three-lobed Violet, Wood Violet ■☆

410341　Viola palmata L. var. angellae (Pollard) Stone = Viola palmata L. ■☆

410342　Viola palmata L. var. pedatifida (G. Don) Cronquist = Viola pedatifida G. Don ■☆

410343　Viola palmata L. var. sororia (Willd.) Pollard = Viola sororia Willd. ■☆

410344　Viola palmata L. var. triloba (Schwein.) Ging. ex DC. = Viola palmata L. ■☆

410345　Viola palmata Patrin ex Ging. = Viola dactyloides Roem. et Schult. ■

410346　Viola palmensis Webb et Berthel.;帕尔马堇菜■☆

410347　Viola palustris L.;沼泽堇菜(湿生堇菜);Alpine Marsh Violet, Bog Violet, Marsh Violet ■☆

410348　Viola palustris L. var. epipsila (Ledeb.) Maxim. = Viola epipsila Ledeb. ■

410349　Viola palustris L. var. moupinensis (Franch.) Franch. = Viola moupinensis (Franch.) Franch. ■

410350　Viola palustris L. var. moupinensis Franch. = Viola moupinensis (Franch.) Franch. ■

410351　Viola papilionacea Pursh;蝶形花堇菜;Blue Violet, Butterfly Violet, Common Blue Violet, Common Violet, Confederate Violet, Meadow Blue Violet, Meadow Violet, Papilionaceus Violet ■☆

410352　Viola papilionacea Pursh = Viola sororia Willd. ■☆

410353　Viola papilionacea Pursh f. albiflora Grover = Viola sororia Willd. ■☆

410354　Viola papilionacea Pursh var. priceana (Pollard) Alexander = Viola sororia Willd. ■☆

410355　Viola paradoxa Lowe;马岛奇异堇菜■☆

410356　Viola paravaginata H. Hara = Viola moupinensis (Franch.) Franch. ■

410357　Viola parviflora L. f. = Hybanthus parviflorus (L. f.) Baill. ■☆

410358　Viola parvula Tineo;较小堇菜■☆

410359　Viola parvula Tineo var. tenella (Webb) Pau = Viola parvula Tineo ■☆

410360　Viola patrinii DC. ex Ging.;白花地丁(白花堇菜,白犁头尖,宝剑草,报春叶堇菜,地黄瓜,贯头尖,铧头草,桦头草,堇,烙铁草,犁头草,犁头尖,米布袋,青地黄瓜,紫花地丁);China Violet, Primula-leaved Violet, Whiteflower Violet ■

410361　Viola patrinii DC. ex Ging. = Viola betonicifolia Sm. ■

410362　Viola patrinii DC. ex Ging. = Viola lactiflora Nakai ■

410363　Viola patrinii DC. ex Ging. f. angustifolia (Regel) F. Maek. = Viola patrinii DC. ex Ging. var. angustifolia Regel ■☆

410364　Viola patrinii DC. ex Ging. f. glabra (Nakai) F. Maek. = Viola mandshurica W. Becker ■

410365　Viola patrinii DC. ex Ging. f. glabra (Nakai) F. Maek. = Viola patrinii DC. ex Ging. ■

410366　Viola patrinii DC. ex Ging. f. hispida W. Becker = Viola patrinii DC. ex Ging. ■

410367　Viola patrinii DC. ex Ging. f. prunellifolia (Nakai) F. Maek. = Viola patrinii DC. ex Ging. ■

410368　Viola patrinii DC. ex Ging. f. toyokoroensis Koji Ito;东京白花地丁■☆

410369　Viola patrinii DC. ex Ging. var. acuminata (Franch. et Sav.) Makino = Viola savatieri Makino ■

410370　Viola patrinii DC. ex Ging. var. angustifolia Regel;狭叶白花地丁■☆

410371　Viola patrinii DC. ex Ging. var. brevicalcarata Skvortsov = Viola patrinii DC. ex Ging. ■

410372　Viola patrinii DC. ex Ging. var. caespitosa (D. Don) Ridl. =

Viola betonicifolia Sm. ■

410373 Viola patrinii DC. ex Ging. var. caespitosa Ridl. = Viola betonicifolia Sm. ■

410374 Viola patrinii DC. ex Ging. var. chinensis Ging. = Viola philippica Cav. ■

410375 Viola patrinii DC. ex Ging. var. laotiana H. Boissieu = Viola betonicifolia Sm. ■

410376 Viola patrinii DC. ex Ging. var. macrantha Maxim. = Viola mandshurica W. Becker ■

410377 Viola patrinii DC. ex Ging. var. minor Makino = Viola inconspicua Blume ■

410378 Viola patrinii DC. ex Ging. var. nepaulensis Ging. = Viola betonicifolia Sm. ■

410379 Viola patrinii DC. ex Ging. var. subsagitata Maxim. = Viola patrinii DC. ex Ging. ■

410380 Viola pedata L. ;鸟爪堇菜(鸟趾堇菜,鸟足堇菜);Bird's-foot Violet,Bird's Foot Violet,Crowfoot Violet,Cut-leaved Violet,Hans and Roosters,Pansy Violet ■☆

410381 Viola pedata L. f. alba（Thurb.）Britton = Viola pedata L. ■☆

410382 Viola pedata L. f. bicolor Britton;双色鸟趾堇菜■☆

410383 Viola pedata L. f. lineariloba（DC.）F. Seym. = Viola pedata L. f. rosea Sanders ■☆

410384 Viola pedata L. f. lineariloba（DC.）F. Seym. = Viola pedata L. ■☆

410385 Viola pedata L. f. rosea Sanders;粉鸟爪堇菜;Bird's-foot Violet ■☆

410386 Viola pedata L. var. concolor Holmgren;大花鸟爪堇菜;Bigflower Birds-foot Violet ■☆

410387 Viola pedata L. var. concolor Holmgren = Viola pedata L. f. rosea Sanders ■☆

410388 Viola pedata L. var. lineariloba DC. ;线裂鸟爪堇菜;Bird's-foot Violet ■☆

410389 Viola pedata L. var. lineariloba DC. = Viola pedata L. f. rosea Sanders ■☆

410390 Viola pedata L. var. lineariloba DC. = Viola pedata L. ■☆

410391 Viola pedata L. var. ranunculifolia DC. = Viola pedata L. ■☆

410392 Viola pedatifida G. Don;草原堇菜;Larkspur Violet,Prairie Violet ■☆

410393 Viola pedatifida G. Don = Viola subsinuata（Greene）Greene ■☆

410394 Viola pedunculata Torr. et Gray;加州堇菜;Californian Violet,Johnny-jump-up,Yeilow Pansy Violet,Yellow Pansy Violet ■☆

410395 Viola pekinensis（Regel）W. Becker;北京堇菜（拟弱距堇菜）;Beijing Violet,Peking Violet ■

410396 Viola pendulicarpa W. Becker;悬果堇菜（垂果堇菜）;Nutantfruit Violet ■

410397 Viola pensylvanica Michx. = Viola pubescens Aiton ■☆

410398 Viola pensylvanica Michx. var. leiocarpa（Fernald et Wiegand）Fernald = Viola pubescens Aiton ■☆

410399 Viola pensylvanica Michx. var. leiocarpa（Fernald et Wiegand）Fernald = Viola pubescens Aiton var. scabriuscula Schwein. ex Torr. et A. Gray ■☆

410400 Viola perpusilla H. Boissieu;极细堇菜■

410401 Viola persicifolia Roth;池塘堇菜（直立堇菜）;Fen Violet ■☆

410402 Viola persicifolia Schreb. = Viola persicifolia Roth ■☆

410403 Viola phalacrocarpa Maxim. ;茜堇菜（白果堇菜,秃果堇菜）;Whitefruit Violet ■

410404 Viola phalacrocarpa Maxim. f. candida Kitag. ;纯白花茜堇菜■☆

410405 Viola phalacrocarpa Maxim. f. chionantha Hiyama;雪花茜堇菜■☆

410406 Viola phalacrocarpa Maxim. f. glaberrima（W. Becker）F. Maek. ;光秃茜堇菜■☆

410407 Viola phalacrocarpa Maxim. f. leucantha Hiyama;白花茜堇菜■☆

410408 Viola phalacrocarpa Maxim. f. plena Okuhara ex T. Shimizu;重瓣茜堇菜■☆

410409 Viola phalacrocarpa Maxim. f. subpubescens Hiyama ex F. Maek. ;近无毛茜堇菜■☆

410410 Viola phalacrocarpa Maxim. var. glaberrima W. Becker = Viola phalacrocarpa Maxim. f. glaberrima（W. Becker）F. Maek. ■☆

410411 Viola phalacrocarpa Maxim. var. pallida Yatabe = Viola hirtipes S. Moore ■

410412 Viola philippica Cav. ;紫花地丁（宝剑草,地丁,菲律宾堇菜,光瓣堇菜,鸡舌草,戟叶紫花地丁,犁头草,辽堇菜,马蹄癀草,台湾如意草,瓮菜癀,野堇菜,紫花堇菜）;Arrow-leaved Violet,Neat Philippine Violet,Purpleflower Violet,Taiwan Violet ■

410413 Viola philippica Cav. f. candida（Kitag.）Kitag. = Viola philippica Cav. ■

410414 Viola philippica Cav. f. intermedia（Kitag.）Kitag. = Viola mandshurica W. Becker ■

410415 Viola philippica Cav. f. intermedia（Kitag.）Kitag. = Viola philippica Cav. ■

410416 Viola philippica Cav. subsp. malesica W. Becker;金盘银盏（拔疔草,地草果,地果草,铧头菜,剪刀菜,犁头草,犁嘴菜,紫花地丁）■

410417 Viola philippica Cav. subsp. malesica W. Becker = Viola inconspicua Blume ■

410418 Viola philippica Cav. subsp. malesica W. Becker = Viola philippica Cav. ■

410419 Viola philippica Cav. subsp. munda W. Becker = Viola philippica Cav. ■

410420 Viola philippica Cav. var. pseudojaponica（Nakai）Y. S. Chen;琉球堇菜■

410421 Viola philippica Cav. var. yunnanfuensis W. Becker = Viola trichopetala C. C. Chang ■

410422 Viola pilosa Blume;匍匐堇菜（地黄瓜,冷毒草）;Creeping Violet ■

410423 Viola pinnata L. ;羽叶堇菜■☆

410424 Viola pinnata L. subsp. multifida W. Becker = Viola dissecta Ledeb. ■

410425 Viola pinnata L. var. chaerophylloides Regel = Viola chaerophylloides（Regel）W. Becker ■

410426 Viola pinnata L. var. dissecta（Ledeb.）Regel = Viola dissecta Ledeb. ■

410427 Viola pinnata L. var. dissecta Turcz. = Viola dissecta Ledeb. ■

410428 Viola pinnata L. var. sibirica Ging. = Viola dissecta Ledeb. ■

410429 Viola pinnata L. var. sieboldiana Maxim. = Viola chaerophylloides（Regel）W. Becker var. sieboldiana（Maxim.）Makino ■

410430 Viola plantaginea H. Christ;车前堇菜■☆

410431 Viola pogonantha W. W. Sm. = Viola pilosa Blume ■

410432 Viola polymorpha C. C. Chang;多形堇菜;Multiple Violet ■

410433 Viola polymorpha C. C. Chang = Viola pendulicarpa W. Becker ■

410434 Viola populifolia Greene;杨叶堇菜;Violet ■☆

410435 Viola populifolia Greene = Viola palmata L. ■☆

410436 Viola praemorsa Douglas;啮蚀堇菜;Praemorse Violet ■☆

410437 Viola pranii W. Becker = Viola szetschwanensis W. Becker et H. Boissieu ■

410438　Viola pratincola Greene = Viola nephrophylla Greene ■☆

410439　Viola priceana Pollard;丛生堇菜;Confederate Violet ■☆

410440　Viola priceana Pollard = Viola sororia Willd. ■☆

410441　Viola pricei W. Becker = Viola nagasawae Makino et Hayata var. pricei (W. Becker) Ching J. Wang ■

410442　Viola primulifolia L. ;报春叶堇菜;Violet ■☆

410443　Viola primulifolia L. = Viola patrinii DC. ex Ging. ■

410444　Viola primulifolia L. subsp. villosa (Eaton) N. H. Russell = Viola primulifolia L. ■☆

410445　Viola primulifolia L. var. acuta (Bigelow) Torr. et A. Gray = Viola primulifolia L. ■☆

410446　Viola primulifolia L. var. glabra Nakai = Viola patrinii DC. ex Ging. ■

410447　Viola primulifolia L. var. villosa Eaton = Viola primulifolia L. ■☆

410448　Viola primulifolia Lour. = Viola inconspicua Blume ■

410449　Viola principis H. Boissieu = Viola fargesii H. Boissieu ■

410450　Viola principis H. Boissieu var. acutifolia Ching J. Wang;尖叶柔毛堇菜;Sharpleaf ■

410451　Viola principis H. Boissieu var. acutifolia Ching J. Wang = Viola fargesii H. Boissieu ■

410452　Viola prionantha Bunge;早开堇菜(光瓣堇菜,尖瓣堇菜,早花地丁,早开地丁,紫花地丁);Serrate Violet ■

410453　Viola prionantha Bunge subsp. jaunsariensis W. Becker = Viola tienschiensis W. Becker ■

410454　Viola prionantha Bunge var. incisa Kitag. = Viola albida Palib. ■

410455　Viola prionantha Bunge var. sylvatica Kitag. = Viola prionantha Bunge ■

410456　Viola prionantha Bunge var. trichantha Ching J. Wang;毛花早开堇菜;Hairflower Violet ■

410457　Viola prionantha Bunge var. trichantha Ching J. Wang = Viola prionantha Bunge ■

410458　Viola pseudoarcuata C. C. Chang;假如意草(拟弧茎堇菜)■☆

410459　Viola pseudoarcuata C. C. Chang = Viola pendulicarpa W. Becker ■

410460　Viola pseudobambusetorum C. C. Chang = Viola striatella H. Boissieu ■

410461　Viola pseudojaponica Nakai = Viola confusa Champ. ex Benth. ■

410462　Viola pseudojaponica Nakai = Viola philippica Cav. var. pseudojaponica (Nakai) Y. S. Chen ■

410463　Viola pseudojaponica Nakai = Viola yedoensis Makino var. pseudojaponica (Nakai) T. Hashim. ■

410464　Viola pseudomonbeigii C. C. Chang;多花堇菜(拟多花堇菜);Flowery Violet ■

410465　Viola pseudomonbeigii C. C. Chang = Viola inconspicua Blume ■

410466　Viola pubescens Aiton;北美柔毛堇菜;Downy Yellow Violet,Pubescent Violet,Smooth Yellow Violet,Yellow Forest Violet,Yellow Violet ■☆

410467　Viola pubescens Aiton var. eriocarpa (Schwein.) N. H. Russell = Viola pubescens Aiton var. scabriuscula Schwein. ex Torr. et A. Gray ■☆

410468　Viola pubescens Aiton var. eriocarpa (Schwein.) N. H. Russell = Viola pubescens Aiton ■☆

410469　Viola pubescens Aiton var. eriocarpon Nutt. = Viola pubescens Aiton ■☆

410470　Viola pubescens Aiton var. peckii House = Viola pubescens Aiton ■☆

410471　Viola pubescens Aiton var. scabriuscula Schwein. ex Torr. et A.

Gray f. leiocarpa (Fernald et Wiegand) Farw. = Viola pubescens Aiton var. scabriuscula Schwein. ex Torr. et A. Gray ■☆

410472　Viola pubescens Aiton var. scabriuscula Schwein. ex Torr. et A. Gray;黄柔毛堇菜;Smooth Yellow Violet, Yellow Forest Violet, Yellow Violet ■☆

410473　Viola pulla W. Becker = Viola fargesii H. Boissieu ■

410474　Viola pumila Chaix;矮堇菜;Meadow Violet ■☆

410475　Viola pumila W. Becker = Viola sieboldii Maxim. ■☆

410476　Viola pycnophylla Franch. et Sav. = Viola yezoensis Maxim. ■

410477　Viola pyrenaica DC. var. maroccana Maire = Viola maroccana Maire ■☆

410478　Viola raddeana Regel;立堇菜(直立堇菜);Radde Violet ■

410479　Viola raddeana Regel var. japonica Makino = Viola raddeana Regel ■

410480　Viola rafinesquii Greene;拉非堇菜;Field Pansy,Johnny-jump-well,Wild Pansy ■☆

410481　Viola rafinesquii Greene = Viola bicolor Pursh ■☆

410482　Viola reichenbachiana Boreau = Viola reichenbachiana Jord. Puyf. ex Boreau ■☆

410483　Viola reichenbachiana Jord. Puyf. ex Boreau;附生堇菜(雷氏堇菜, 林堇菜); Dog-violet, Early Dog-violet, Pale Wood Violet, Pertaining Violet, Snake Violet, Snake's Violet, Sylvan Violet, Wood Dog Violet, Wood Violet, Woodland Violet ■☆

410484　Viola renifolia A. Gray;肾叶堇菜;Kidney-leaved Violet, White Violet ■☆

410485　Viola renifolia A. Gray var. brainerdii (Greene) Fernald = Viola renifolia A. Gray ■☆

410486　Viola reniformis Wall. = Viola hederacea Labill. ■☆

410487　Viola reniformis Wall. = Viola wallichiana Ging. ■

410488　Viola repens Turcz. = Viola epipsila Ledeb. ■

410489　Viola repens Turcz. ex Trautv. et C. A. Mey. = Viola epipsiloides Á. Löve et D. Löve ■

410490　Viola retusa Greene = Viola nephrophylla Greene ■☆

410491　Viola rhodosepala Kitag. ;红萼堇菜;Red-calyx Violet ■☆

410492　Viola rhodosepala Kitag. = Viola mandshurica W. Becker ■

410493　Viola riviniana Rchb. ;里文堇菜(喜木堇菜);Blue Mice, Blue Violet, Common Dog-violet, Common Violet Cuckoo's Shoes, Cuckoo's Stockings, Dog Violet, Gypsy Violet, Hedge Violet, Horse Violet, Hypocrites, Pig Violet, Rivini Violet, Shoes-and-stockings, Snake Violet, Summer Violet, Wood Violet ■☆

410494　Viola riviniana Rchb. ' Purpurea';紫叶里文堇菜■☆

410495　Viola rockiana W. Becker;圆叶小堇菜(圆叶黄堇菜);Rock Violet ■

410496　Viola rockiana W. Becker = Viola biflora L. var. rockiana (W. Becker) Y. S. Chen ■

410497　Viola rosacea Brainerd = Viola affinis Leconte ■☆

410498　Viola rossii Hemsl. ex Forbes et Hemsl. ;辽宁堇菜(白铧头草,寸节七,大铧头草,庐山堇菜,洛氏堇菜,洛雪堇菜);Ross Violet ■

410499　Viola rossii Hemsl. ex Forbes et Hemsl. f. atropurpurea (Nakai) F. Maek. ;深紫辽宁堇菜■☆

410500　Viola rossii Hemsl. ex Forbes et Hemsl. f. lactiflora (Nakai) Hiyama ex F. Maek. ;大花辽宁堇菜■☆

410501　Viola rosthornii E. Pritz. = Viola moupinensis (Franch.) Franch. ■

410502　Viola rostrata Pursh;长距堇菜(喙状堇菜);Long-spurred Violet ■☆

410503　Viola rostrata Pursh f. albiflora Y. Ueno;白花长距堇菜■☆

410504　Viola rostrata Pursh f. alpina E. Hama;高山长距堇菜■☆

410505　Viola rostrata Pursh subsp. japonica W. Becker et H. Boissieu；日本长距堇菜■☆

410506　Viola rostrata Pursh var. elongata Farw. = Viola rostrata Pursh ■☆

410507　Viola rostrata Pursh var. japonica（W. Becker et H. Boissieu）Ohwi = Viola rostrata Pursh subsp. japonica W. Becker et H. Boissieu ■☆

410508　Viola rothomagensis Desf.；罗托堇菜；Rouen Pansy ■☆

410509　Viola rotundifolia Michx.；北美圆叶堇菜；Early Yellow Violet，Rotundifolious Violet，Round-leaved Violet ■☆

410510　Viola rotundifolia Michx. var. pallens Banks ex Ging. = Viola macloskeyi F. E. Lloyd subsp. pallens（Banks ex Ging.）M. S. Baker ■☆

410511　Viola rugolosa Greene；匍匐茎堇菜（匍匐堇菜）；Runner Violet ■☆

410512　Viola rugolosa Greene = Viola canadensis L. var. rugulosa（Greene）C. L. Hitchc. ■☆

410513　Viola rupestris F. W. Schmidt；石生堇菜；Heartsease，Herb Trinity，Johnny-jump-up，Lady's Delight，Lithophilous Violet，Love-in-idleness，Teesdale Violet ■

410514　Viola rupestris F. W. Schmidt var. licentii W. Becker；长托叶石生堇菜（长托叶堇菜）■

410515　Viola rupicola Elmer；喜岩堇菜■

410516　Viola ruppii All.；汝氏堇菜■☆

410517　Viola sacchalinensis H. Boissieu；库页堇菜（林堇菜）；Sachalin Violet ■

410518　Viola sacchalinensis H. Boissieu f. albialpina E. Hama = Viola sachalinensis H. Boissieu ■

410519　Viola sacchalinensis H. Boissieu f. alpina（H. Hara）F. Maek. et T. Hashim. = Viola sacchalinensis H. Boissieu ■

410520　Viola sacchalinensis H. Boissieu f. alpina（H. Hara）F. Maek. et T. Hashim.；高山库页堇菜（长白山堇菜）■

410521　Viola sacchalinensis H. Boissieu f. chionantha E. Hama = Viola sacchalinensis H. Boissieu ■

410522　Viola sacchalinensis H. Boissieu var. alpicola P. Y. Fu et Y. C. Teng；山生库页堇菜■

410523　Viola sacchalinensis H. Boissieu var. alpicola P. Y. Fu et Y. C. Teng = Viola sacchalinensis H. Boissieu f. alpina（H. Hara）F. Maek. et T. Hashim. ■

410524　Viola sacchalinensis H. Boissieu var. alpina H. Hara = Viola sacchalinensis H. Boissieu f. alpina（H. Hara）F. Maek. et T. Hashim. ■

410525　Viola sacchalinensis H. Boissieu var. miyakei（Nakai）Ohwi；三宅堇菜■☆

410526　Viola sachalinensis H. Boissieu f. alpina（H. Hara）F. Maek. et T. Hashim. = Viola sacchalinensis H. Boissieu f. alpina（H. Hara）F. Maek. et T. Hashim. ■

410527　Viola sagittata Aiton；箭堇；Arrowhead Violet，Arrow-leaved Violet，Sagittate Violet ■☆

410528　Viola sagittata Aiton var. ovata（Nutt.）Torr. et A. Gray；卵叶箭堇；Arrow-leaved Violet，Sand Violet ■☆

410529　Viola sagittata Aiton var. subsagittata（Greene）Pollard = Viola sagittata Aiton ■☆

410530　Viola sarmentosa M. Bieb.；林地堇菜；Wood Violet ■☆

410531　Viola savatieri Makino；辽东堇菜（萨氏堇菜，萨瓦特堇菜）；Liaodong Violet，Savatier Violet ■

410532　Viola savatieri Makino f. detonsa（Kitag.）Kitag. = Viola savatieri Makino ■

410533　Viola savatieri Makino f. detonsa（Kitag.）Kitag. = Viola takahashii（Nakai）Taken. ■

410534　Viola savatieri Makino var. detonsa Kitag. = Viola savatieri Makino ■

410535　Viola savatieri Makino var. detonsa Kitag. = Viola takahashii（Nakai）Taken. ■

410536　Viola saxatilis F. W. Schmidt；岩生堇菜（石岩生堇菜）；Rocky Violet ■☆

410537　Viola saxifraga Maire；虎耳草堇菜■☆

410538　Viola saxifraga Maire var. ciliata？ = Viola saxifraga Maire ■☆

410539　Viola schensiensis W. Becker；陕西堇菜；Shaanxi Violet ■

410540　Viola schensiensis W. Becker = Viola striatella H. Boissieu ■

410541　Viola schneideri W. Becker；浅圆齿堇菜；C. K. Schneid. Violet ■

410542　Viola schneideri W. Becker = Viola davidii Franch. ■

410543　Viola schulzeana W. Becker；黄花肾叶堇菜（黄花堇菜，肾叶堇菜）；Nephroidleaf Violet ■

410544　Viola scorpiuroides Coss.；蝎尾堇菜■☆

410545　Viola scorpiuroides Coss. var. inflata Pamp. = Viola scorpiuroides Coss. ■☆

410546　Viola segobricensis Pau = Viola suavis M. Bieb. ■☆

410547　Viola selkirkii Pursh ex Goldie；深山堇菜（一口血，阴生堇菜）；Great-spurred Violet，Northern Violet，Selkirk's Violet，Wilderness Violet ■

410548　Viola selkirkii Pursh ex Goldie f. alba Tatew.；白深山堇菜■☆

410549　Viola selkirkii Pursh ex Goldie f. albiflora（Nakai）F. Maek. = Viola selkirkii Pursh ex Goldie f. alba Tatew. ■☆

410550　Viola selkirkii Pursh ex Goldie f. subglabra（W. Becker）M. Mizush.；亚光深山堇菜■☆

410551　Viola selkirkii Pursh ex Goldie f. variegata（Nakai）F. Maek.；斑点深山堇菜■☆

410552　Viola selkirkii Pursh ex Goldie var. albiflora Nakai = Viola selkirkii Pursh ex Goldie ■

410553　Viola selkirkii Pursh ex Goldie var. angustistipulata W. Becker = Viola selkirkii Pursh ex Goldie ■

410554　Viola selkirkii Pursh ex Goldie var. brevicalcarata W. Becker = Viola selkirkii Pursh ex Goldie ■

410555　Viola selkirkii Pursh ex Goldie var. subbarbata W. Becker；须毛深山堇菜（须毛堇菜）■

410556　Viola selkirkii Pursh ex Goldie var. subbarbata W. Becker = Viola selkirkii Pursh ex Goldie ■

410557　Viola selkirkii Pursh ex Goldie var. variegata Nakai = Viola selkirkii Pursh ex Goldie ■

410558　Viola sempervirens Greene；常绿堇菜；Evergreen Violet ●☆

410559　Viola senzanensis Hayata；尖山堇菜■

410560　Viola septentrionalis Greene；北方堇菜；Northern Blue Violet ■☆

410561　Viola septentrionalis Greene = Viola sororia Willd. ■☆

410562　Viola septentrionalis Greene f. alba Vict. et M. Rousseau = Viola sororia Willd. ■☆

410563　Viola septentrionalis Greene var. grisea Fernald = Viola novae-angliae House ■☆

410564　Viola serpens Wall. = Viola pilosa Blume ■

410565　Viola serpens Wall. ex Ging. = Viola pilosa Blume ■

410566　Viola serpens Wall. ex Ging. var. confusa（Benth.）Hook. f. et Thomson = Viola inconspicua Blume ■

410567　Viola serpens Wall. subsp. gurhwalensis W. Becker = Viola pilosa Blume ■

410568　Viola serpens Wall. var. canescens（Wall.）Hook. f. et Thomson = Viola canescens Wall. ■

410569　Viola serpens Wall. var. pseudoscotophylla H. Boissieu = Viola pilosa Blume ■

410570　Viola serrula W. Becker；小齿堇菜；Smalltoothed Violet ■

410571 Viola shikokiana Makino;四国堇菜■☆

410572 Viola shinchikuensis Yamam. ;新竹堇菜■

410573 Viola shinchikuensis Yamam. = Viola kosanensis Hayata ■

410574 Viola sieboldiana（Maxim.）Makino = Viola chaerophylloides（Regel）W. Becker var. sieboldiana（Maxim.）Makino ■

410575 Viola sieboldiana（Maxim.）Makino var. chaerophylloides（Regel）Nakai = Viola chaerophylloides（Regel）W. Becker ■

410576 Viola sieboldii Maxim. ;西氏堇菜■☆

410577 Viola sieboldii Maxim. f. variegata（Nagas. ex F. Maek.）F. Maek. et T. Hashim. ex T. Shimizu;杂色西氏堇菜■☆

410578 Viola sieboldii Maxim. subsp. boissieuana（Makino）F. Maek. et T. Hashim. = Viola boissieuana Makino ■☆

410579 Viola sieheana W. Becker;喜氏堇菜■☆

410580 Viola sikkimensis W. Becker;锡金堇菜（锡京堇菜）;Sikkim Violet ■

410581 Viola sikkimensis W. Becker var. debilis W. Becker = Viola davidii Franch. ■

410582 Viola sikkimensis W. Becker var. debilis W. Becker = Viola kwangtungensis Melch. ■

410583 Viola silvestriformis W. Becker = Viola sacchalinensis H. Boissieu ■

410584 Viola silvestris Rchb. ;林堇菜■☆

410585 Viola smithiana W. Becker = Viola davidii Franch. ■

410586 Viola somalensis Engl. = Viola cinerea Boiss. var. stocksii（Boiss.）W. Becker ■☆

410587 Viola sororia Willd. ;美洲普通堇菜;Broad-leaved Wood Violet, Butterfly Violet, Common Blue Violet, Common Violet, Confederate Violet, Door-yard Violet, Hairy Wood Violet, Meadow Violet, Sister Violet, Speckled Violet, Woolly Blue Violet ■☆

410588 Viola sororia Willd. f. beckwithae House = Viola sororia Willd. ■☆

410589 Viola sororia Willd. f. priceana（Pollard）Cooperr. = Viola sororia Willd. ■☆

410590 Viola sororia Willd. subsp. affinis（Leconte）R. J. Little = Viola affinis Leconte ■☆

410591 Viola sororia Willd. var. affinis（Leconte）McKinney = Viola affinis Leconte ■☆

410592 Viola sororia Willd. var. missouriensis（Greene）McKinney = Viola missouriensis Greene ■☆

410593 Viola sororia Willd. var. novae-angliae（House）McKinney = Viola novae-angliae House ■☆

410594 Viola sphaerocarpa W. Becker;圆果堇菜;Roundfruit Violet ■

410595 Viola stagnina Kit. = Viola persicifolia Roth ■☆

410596 Viola stenocentra Hayata ex Nakai = Viola philippica Cav. var. pseudojaponica（Nakai）Y. S. Chen ■

410597 Viola stenocentra Hayata ex Nakai = Viola philippica Cav. ■

410598 Viola stewardiana W. Becker;庐山堇菜（拟蔓地草）;Lushan Violet ■

410599 Viola stocksii Boiss. = Viola cinerea Boiss. var. stocksii（Boiss.）W. Becker ■☆

410600 Viola stoloniflora Yokota et Higa;垂花堇菜■☆

410601 Viola stoneana House = Viola palmata L. ■☆

410602 Viola striata Aiton;具纹堇菜;Cream Violet, Creamy Violet, Pale Violet, Striate Violet, Striped Violet, Striped White Violet ■☆

410603 Viola striata Aiton f. albiflora Farw. = Viola striata Aiton ■☆

410604 Viola striatella H. Boissieu;圆叶堇菜;Roundleaf Violet ■

410605 Viola suavis M. Bieb. ;可人堇菜■☆

410606 Viola subatlantica（Maire）Ibn Tattou;亚特兰大堇菜■☆

410607 Viola subdelavayi S. H. Huang;淡黄堇菜■

410608 Viola subdelavayi S. H. Huang = Viola urophylla Franch. ■

410609 Viola suberosa Desf. = Viola arborescens L. ■☆

410610 Viola sublanceolata House = Viola primulifolia L. ■☆

410611 Viola subsagittata Greene = Viola sagittata Aiton ■☆

410612 Viola subsinuata（Greene）Greene;深波堇菜;Early Blue Violet, Lobed Violet ■☆

410613 Viola suffruticosa L. ;亚灌木堇菜■☆

410614 Viola suffruticosa L. = Hybanthus enneaspermus（L.）F. Muell. ●

410615 Viola sumatrana Miq. ;光叶堇菜;Glabrousleaf Violet ■

410616 Viola sylvatica Fr. ex Hartm. var. imberbis A. Gray = Viola grypoceras A. Gray ■

410617 Viola sylvestris Kitag. = Viola grypoceras A. Gray ■

410618 Viola sylvestris Lam. = Viola reichenbachiana Jord. Puyf. ex Boreau ■☆

410619 Viola sylvestris Lam. var. candida（H. Boissieu）H. Lév. = Viola grypoceras A. Gray ■

410620 Viola sylvestris Lam. var. grypoceras（A. Gray）Maxim. = Viola grypoceras A. Gray ■

410621 Viola sylvestris Ledeb. = Viola sacchalinensis H. Boissieu ■

410622 Viola szetschwanensis W. Becker et H. Boissieu;四川堇菜（川黄堇菜,米林堇菜）;Sichuan Violet ■

410623 Viola szetschwanensis W. Becker et H. Boissieu var. kangdienensis C. C. Chang = Viola szetschwanensis W. Becker et H. Boissieu ■

410624 Viola szetschwanensis W. Becker et H. Boissieu var. kangdingensis C. C. Chang;心叶四川堇菜（康滇堇菜,康定堇菜）■

410625 Viola szetschwanensis W. Becker et H. Boissieu var. nudicaulis W. Becker = Viola szetschwanensis W. Becker et H. Boissieu ■

410626 Viola taishanensis Ching J. Wang;泰山堇菜■

410627 Viola taishanensis Ching J. Wang = Viola prionantha Bunge ■

410628 Viola taiwanensis W. Becker = Viola adenothrix Hayata ■

410629 Viola taiwanensis W. Becker = Viola formosana Hayata ■

410630 Viola taiwaniana Nakai = Viola inconspicua Blume subsp. nagasakiensis（W. Becker）Ching J. Wang et T. C. Huang ■

410631 Viola taiwaniana Nakai = Viola mandshurica W. Becker ■

410632 Viola taiwaniana Nakai = Viola philippica Cav. var. pseudojaponica（Nakai）Y. S. Chen ■

410633 Viola takahashii（Nakai）Taken. ;菊叶堇菜（菊叶朝鲜堇菜）;Mumleaved Violet ■

410634 Viola takasagoensis Koidz. = Viola formosana Hayata var. kawakamii（Hayata）Y. S. Chen et Q. E. Yang ■

410635 Viola takasagoensis Koidz. = Viola formosana Hayata var. stenopetala（Hayata）Ching J. Wang,T. C. Huang et T. Hashim. ■

410636 Viola takedana Makino = Viola tokubuchiana Makino var. takedana（Makino）F. Maek. ■

410637 Viola takedana Makino f. variegata（Nakai）Makino = Viola tokubuchiana Makino var. takedana（Makino）F. Maek. f. variegata（Nakai）F. Maek. ■☆

410638 Viola takedana Makino var. austroyezoensis Kawano = Viola tokubuchiana Makino var. takedana（Makino）F. Maek. f. austroyezoensis（Kawano）F. Maek. et T. Hashim. ■☆

410639 Viola tanaitica Grosset;塔奈特堇菜■☆

410640 Viola tarbagataica Klokov;塔城堇菜■

410641 Viola tashiroi Makino;田代堇菜■☆

410642 Viola tashiroi Makino f. alboreticulata Nackej. ;白网田代氏堇菜■☆

410643 Viola tashiroi Makino f. takushii K. Nakaj.；卓志菫菜■☆

410644 Viola tashiroi Makino subsp. iwagawae（Makino）K. Nakaj. = Viola iwagawae Makino ■☆

410645 Viola tashiroi Makino var. tairae K. Nakaj.；平菫菜■☆

410646 Viola tayemonii Hayata = Viola biflora L. ■

410647 Viola tenuicornis W. Becker；细距菫菜（弱距菫菜）；Littlespur Violet，Smallspur Violet ■

410648 Viola tenuicornis W. Becker = Viola trichosepala（W. Becker）Juz. ■☆

410649 Viola tenuicornis W. Becker subsp. primorskajensis W. Becker = Viola variegata Fisch. ex Link ■

410650 Viola tenuicornis W. Becker subsp. trichosepala W. Becker；毛萼细距菫菜(毛萼菫菜)；Haircalyx Violet ■

410651 Viola tenuicornis W. Becker var. brachytricha W. Becker = Viola tenuicornis W. Becker ■

410652 Viola tenuis Benth.；心叶茶匙黄■

410653 Viola tenuis Benth. = Viola diffusa Ging. ■

410654 Viola tenuissima C. C. Chang；纤茎菫菜；Verythin Violet ■

410655 Viola teshioensis Miyabe et Tatew. = Viola collina Besser ■

410656 Viola teshioensis Miyabe et Tatew. f. albiflora Tatew.；白花山丘菫菜■☆

410657 Viola tezensis Ball = Viola parvula Tineo ■☆

410658 Viola thesiifolia Juss. ex Poir. = Hybanthus enneaspermus（L.）F. Muell. ●

410659 Viola thianschanica Maxim. = Viola kunawarensis Royle ■

410660 Viola thomsonii Oudem.；毛菫菜（贡山菫菜）；Thomson Violet ■

410661 Viola thrichopoda Hayata = Viola fargesii H. Boissieu ■

410662 Viola tianschanica Maxim.；天山菫菜；Tianshan Violet ■

410663 Viola tienschiensis W. Becker；滇西菫菜■

410664 Viola tokubuchiana Makino；德渊菫菜■☆

410665 Viola tokubuchiana Makino f. concolor E. Hama；同色德渊菫菜■☆

410666 Viola tokubuchiana Makino f. lactiflora E. Hama；乳白花德渊菫菜■☆

410667 Viola tokubuchiana Makino var. takedana（Makino）F. Maek.；凤凰菫菜■

410668 Viola tokubuchiana Makino var. takedana（Makino）F. Maek. f. albiflora Hayashi；白花凤凰菫菜■☆

410669 Viola tokubuchiana Makino var. takedana（Makino）F. Maek. f. austroyezoensis（Kawano）F. Maek. et T. Hashim.；南北海道菫菜■☆

410670 Viola tokubuchiana Makino var. takedana（Makino）F. Maek. f. variegata（Nakai）F. Maek.；花叶凤凰菫菜■☆

410671 Viola tozanensis Hayata = Viola formosana Hayata ■

410672 Viola triangulifolia W. Becker；三角叶菫菜（扣子兰，蔓地草，蔓地犁）；Threeanguledleaf Violet，Triangular Violet ■

410673 Viola trichopetala C. C. Chang；毛瓣菫菜；Hairpetal Violet ■

410674 Viola trichopoda Hayata = Viola adenothrix Hayata ■

410675 Viola trichosepala（W. Becker）Juz.；三瓣菫菜■☆

410676 Viola trichosepala（W. Becker）Juz. = Viola tenuicornis W. Becker subsp. trichosepala W. Becker ■

410677 Viola tricolor L.；三色菫菜（蝴蝶花，蝴蝶梅，三色菫）；Back-to-back，Beaty Eyes，Beedy's Eyes，Biddy's Eye，Biddy's Eyes，Bird's Eye，Bleeding Heart，Butterfly Flower，Buttery-entry，Caddie-me-to-you，Call-me-to-you，Cat's Face，Coach Horse，Coach-horses，Come-and-kiss-me，Corn Pansy，Cuddle-me，Cuddle-me-to-you，Cull-me-to-you，Cupid's Flower，Eyebright，Face-and-hood，Fairy Queen，Fancy，Field Pansy，Flame Flower，Flamy，Forget-me-not，Funny-face，Garden Pansy，Gentleman John，Gentleman Tailor，Godfathers And Godmothers，Granny's Face，Hartsease，Heart Pansy，Hearts at Ease，Hearts Ease，Heart's Ease，Heartsease，Heart's-ease，Heartsease Wild Pansy，Heartseed，Herb of the Trinity，Herb Trinity，Horse Violet，Jack-behind-the-garden-gate，Jack-jump-up-and-kiss-me，Johnny Jumpup，Johnny-jump-up，Johnny-jump-well，Johnny-run-the-street，Jump-up-and-kiss-me，Kiss-and-look-up，Kiss-at-the-garden-gate，Kisses，Kiss-me，Kiss-me-at-the-garden-gate，Kiss-me-behind-the-garden-gate，Kiss-me-ere-I-rise，Kiss-me-love，Kiss-me-love-at-the-garden-gate，Kiss-me-over-the-garden-gate，Kiss-me-quick，Kit-run-the-fields，Kitty-run-the-street，Kiss-me-John-at-the-garden-gate，Ladies' Delight，Ladies-and-gentlemen，Ladies' Flower，Lark's Eye，Lark's Eyes，Leap-up-and-kiss-me，Live-in-idleness，Living Idols，Look-vp-and-kiss-me，Love Idol，Love-and-idleness，Love-and-idols，Love-idol，Love-in-Idle，Love-in-idleness，Love-in-vain，Love-lies-bleeding，Lover's Thoughts，Lovers' Thoughts，Love-true，Loving Lidols，Loving Lydles，Meet-her-in-the-entry-kiss-her-in-the-buttery，Meet-me-love，Meet-me-love-behind-the-garden-door，Men's Face，Men's Faces，Monkey Face，Monkey Faces，Needles-and-pins，Nuffin Idols，Numman-idles，Old Maid's Flower，Old Maid's Last Friend，Old Man's Face，Pance，Pansy，Pansy Violet，Paunce，Paunsy，Pink of My John，Pinkeney John，Pink-eyed John，Pink-of-my-John，Pink-o'-my-John，Pussy Face，Shame Face，Shame-face，Shoes-and-stockings，Sister-in-law，Stepdaughters，Stepmother，Stepmother's Son，Summer Hat，Summer Hats，The Longer-the-dearer，Three Faces-under-a-hood，Three-coloured Violet，Three-faces-in-a-hood，Threefaces-under-a-hood，Tittle-my-fancy，Trinity Flower，Trinity Violet，Two-faces-in-a-hood，Two-faces-under-one-hat，Two-faces-under-the-sun，Wild Pansy ■

410678 Viola tricolor L. 'Bowles Black'；黑花三色菫菜■☆

410679 Viola tricolor L. subsp. arvensis Murray = Viola arvensis Murray ■

410680 Viola tricolor L. subsp. minima Gaudin = Viola kitaibeliana Schult. ■☆

410681 Viola tricolor L. subsp. parvula（Tineo）Rouy et Foucaud = Viola parvula Tineo ■☆

410682 Viola tricolor L. var. arvensis（Murr.）Boiss. = Viola arvensis Murray ■

410683 Viola tricolor L. var. curtisii ?；海滨三色菫菜；Sand Pansy，Seaside Pansy ■☆

410684 Viola tricolor L. var. hortensis DC. = Viola tricolor L. ■

410685 Viola tricolor L. var. parvula（Tineo）Batt. = Viola parvula Tineo ■☆

410686 Viola tricolor L. var. subatlantica Maire = Viola subatlantica（Maire）Ibn Tattou ■☆

410687 Viola triloba Schwein. = Viola palmata L. ■☆

410688 Viola triloba Schwein. f. albida Steyerm. = Viola palmata L. ■☆

410689 Viola triloba Schwein. var. dilatata（Elliott）Brainerd = Viola palmata L. ■☆

410690 Viola trinervata（Howell）Howell ex A. Gray；三脉菫菜；Trinervate Violet ■☆

410691 Viola tsugitakaensis Masam.；雪山菫菜；Snowmountain Violet ■

410692 Viola tsugitakaensis Masam. = Viola adenothrix Hayata ■

410693 Viola tsugitakaensis Masam. = Viola fargesii H. Boissieu ■

410694 Viola tuberifera Franch.；块茎菫菜；Tuberous Violet ■

410695 Viola tuberifera Franch. = Viola bulbosa Maxim. ■

410696 Viola tuberifera Franch. var. brevipedicellata S. Y. Chen；短梗块茎菫菜；Shortstalk Tuberous Violet ■

410697 Viola tuberifera Franch. var. brevipedicellata S. Y. Chen = Viola bulbosa Maxim. ■

410698   Viola tuberifera Franch. var. pseudopalustris H. Lév. = Viola bulbosa Maxim. ■

410699   Viola tuberifera Franch. var. pseudopalustris H. Lév. = Viola tuberifera Franch. ■

410700   Viola turczaninowii Juz.；图尔菫菜■☆

410701   Viola turczaninowii Juz. = Viola acuminata Ledeb. ■

410702   Viola uliginosa Besser；泥地菫菜（泥泞地菫菜）■☆

410703   Viola umbrosa Fr. = Viola selkirkii Pursh ex Goldie ■

410704   Viola uniflora L.；单花菫菜■☆

410705   Viola uniflora L. = Viola orientalis (Maxim.) W. Becker ■

410706   Viola uniflora L. var. kareliniana Maxim. = Viola acutifolia (Kar. et Kir.) W. Becker ■

410707   Viola uniflora L. var. orientalis Maxim. = Viola orientalis (Maxim.) W. Becker ■

410708   Viola urophylla Franch.；粗齿菫菜（尾叶黄菫菜）；Tailleaf Violet ■

410709   Viola urophylla Franch. var. densivillosa Ching J. Wang；密毛粗齿菫菜■

410710   Viola vaginata Maxim.；大叶菫（白三百棒，红三百棒，茛，鸡心七，菫，母犁头菜，鞘柄菫菜，如意草，山羊臭，乌蔗连，乌泡连，萱）；Bigleaf Violet, Vaginate Violet ■

410711   Viola vaginata Maxim. = Viola moupinensis (Franch.) Franch. ■

410712   Viola vaginata Maxim. f. albescens Sugim.；白斑大叶菫■☆

410713   Viola vaginata Maxim. f. albiflora Honda；白花大叶菫■☆

410714   Viola vaginata Maxim. subsp. alata W. Becker = Viola moupinensis (Franch.) Franch. ■

410715   Viola vaginata Maxim. var. sutchuensis Franch. ex H. Boissieu = Viola moupinensis (Franch.) Franch. ■

410716   Viola vakasagoensis Koidz. = Viola formosana Hayata var. kawakamii (Hayata) Ching J. Wang ■

410717   Viola variabilis Greene = Viola palmata L. ■☆

410718   Viola variegata Fisch. ex Ging. f. nipponica (Makino) F. Maek. = Viola variegata Fisch. ex Ging. var. nipponica Makino ■☆

410719   Viola variegata Fisch. ex Ging. var. nipponica Makino；日本斑叶菫菜■☆

410720   Viola variegata Fisch. ex Ging. var. nipponica Makino = Viola variegata Fisch. ex Link ■

410721   Viola variegata Fisch. ex Link；斑叶菫菜（贯头尖，犁头尖，天蹄）；Blotchleaf Violet, Variegatedleaf Violet ■

410722   Viola variegata Fisch. ex Link = Viola tenuicornis W. Becker ■

410723   Viola variegata Fisch. ex Link f. viridis (Kitag.) P. Y. Fu et Y. C. Teng = Viola variegata Fisch. ex Link ■

410724   Viola variegata Fisch. ex Link f. viridis (Kitag.) P. Y. Fu et Y. C. Teng；绿斑叶菫菜；Green Blotchleaf Violet ■

410725   Viola variegata Fisch. ex Link var. chinensis Bunge ex Regel = Viola tenuicornis W. Becker ■

410726   Viola variegata Fisch. ex Link var. typica Regel = Viola variegata Fisch. ex Link ■

410727   Viola variegata Fisch. ex Link var. viridis Kitag. = Viola tenuicornis W. Becker ■

410728   Viola variegata Fisch. ex Link var. viridis Kitag. = Viola variegata Fisch. ex Link ■

410729   Viola venusta Nakai = Viola sacchalinensis H. Boissieu ■

410730   Viola verecunda A. Gray；菫菜（白花蚶壳草，白老碗，地黄瓜，罐嘴菜，箭头草，菫菫菜，葡菫菜，如意草，三角金砖，水白地黄瓜，消毒药，小犁头草，玉如意）；Common Violet ■

410731   Viola verecunda A. Gray = Viola arcuata Blume ■

410732   Viola verecunda A. Gray f. candidissima M. Mizush. ex E. Hama et Nackej.；极白菫菜■☆

410733   Viola verecunda A. Gray f. hensoaensis Kudo et Sasaki；日月潭葡菫菜■

410734   Viola verecunda A. Gray f. hensoaensis Kudo et Sasaki = Viola arcuata Blume ■

410735   Viola verecunda A. Gray f. lilacina Sugaya；紫丁香色菫菜■☆

410736   Viola verecunda A. Gray f. radicans Makino；茶匙黄■

410737   Viola verecunda A. Gray f. radicans Makino = Viola arcuata Blume ■

410738   Viola verecunda A. Gray f. variegata Honda；斑驳菫菜■☆

410739   Viola verecunda A. Gray var. excisa (Hance) Maxim. = Viola verecunda A. Gray var. subaequiloba (Franch. et Sav.) F. Maek. ■☆

410740   Viola verecunda A. Gray var. fibrillosa (W. Becker) Ohwi；须毛菫菜■☆

410741   Viola verecunda A. Gray var. semilunaris Maxim.；新月菫菜■☆

410742   Viola verecunda A. Gray var. semilunaris Maxim. = Viola arcuata Blume ■

410743   Viola verecunda A. Gray var. subaequiloba (Franch. et Sav.) F. Maek.；近等裂菫菜■☆

410744   Viola verecunda A. Gray var. yakushimana (Nakai) Ohwi；屋久岛菫菜■☆

410745   Viola viarum Pollard；平原菫菜；Plains Violet ■☆

410746   Viola viarum Pollard = Viola palmata L. ■☆

410747   Viola viarum Pollard f. pilifera E. J. Palmer et Steyerm. = Viola viarum Pollard ■☆

410748   Viola villarii Sennen et Mauricio = Viola munbyana Boiss. et Reut. var. rifana Emb. et Maire ■☆

410749   Viola violacea Makino；紫背菫菜；Pansy, Viola, Violet ■

410750   Viola violacea Makino f. albida (Nakai) F. Maek.；白紫背菫菜■☆

410751   Viola violacea Makino f. concolor Nakash.；同色紫背菫菜■☆

410752   Viola violacea Makino f. pictifolia Honda；色叶紫背菫菜■☆

410753   Viola violacea Makino f. versicolor E. Hama；变色紫背菫菜■☆

410754   Viola violacea Makino var. makinoi (H. Boissieu) Hiyama ex F. Maek.；牧野氏菫菜■☆

410755   Viola violacea Makino var. makinoi (H. Boissieu) Hiyama ex F. Maek. f. variegata E. Hama；花叶牧野氏菫菜■☆

410756   Viola violacea Makino var. tanakaeana (Makino) T. Hashim.；田中氏菫菜■☆

410757   Viola wallichiana Ging.；西藏细距菫菜（细距菫菜）；Wallich Violet, Xizang Smallspar Violet ■

410758   Viola websteri Hemsl. ex Forbes et Hemsl.；蓼叶菫菜（朝鲜蓼叶菫菜）；Knotweed Violet, Knotweedleaf Violet ■

410759   Viola weixiensis Ching J. Wang = Viola pendulicarpa W. Becker ■

410760   Viola weixienss Ching J. Wang；维西菫菜（滇西菫菜）；Weixi Violet ■

410761   Viola wiedemanni Boiss. = Viola odorata L. ■

410762   Viola wilsonii W. Becker = Viola diffusa Ging. ■

410763   Viola xanthopetala Nakai；黄花菫菜；Yellow-flowered Violet ■

410764   Viola xanthopetala Nakai = Viola orientalis (Maxim.) W. Becker ■

410765   Viola xanthopetala Nakai f. laciniata？ = Viola orientalis (Maxim.) W. Becker ■

410766   Viola xanthopetala Nakai var. laciniata？ = Viola orientalis (Maxim.) W. Becker ■

410767   Viola yamatsutae Ishid. ex Kitag. = Viola mongolica Franch. ■

410768　Viola yatabei Makino = Viola yezoensis Maxim. ■

410769　Viola yazawana Makino;矢沢菫菜■☆

410770　Viola yedoensis Makino;日本紫花地丁（白花菫菜,白毛菫菜,宝剑草,北菫菜,地丁,地丁草,独行虎,光瓣地丁草,光瓣菫菜,铧头草,箭头草,金剪刀,堇菜,犁铧草,犁头菜,辽菫菜,如意草,羊角子,野菫菜,云南菫菜,紫色地丁）;Nakepetal Violet, Wild Violet ■☆

410771　Viola yedoensis Makino = Viola philippica Cav. ■

410772　Viola yedoensis Makino f. albescens Hiyama ex F. Maek. ;变白紫花地丁■☆

410773　Viola yedoensis Makino f. barbata Hiyama = Viola yedoensis Makino ■☆

410774　Viola yedoensis Makino f. candida Kitag. = Viola philippica Cav. ■

410775　Viola yedoensis Makino f. candida Kitag. = Viola yedoensis Makino f. albescens Hiyama ex F. Maek. ■☆

410776　Viola yedoensis Makino f. glaberrima F. Maek. = Viola yedoensis Makino ■☆

410777　Viola yedoensis Makino f. intermedia Kitag. = Viola mandshurica W. Becker ■

410778　Viola yedoensis Makino f. intermedia Kitag. = Viola philippica Cav. ■

410779　Viola yedoensis Makino var. codida Kitag. ;白瓣紫花地丁■

410780　Viola yedoensis Makino var. confusa（Champ.）Hashim. = Viola confusa Champ. ex Benth. ■

410781　Viola yedoensis Makino var. pseudojaponica（Nakai）T. Hashim. = Viola philippica Cav. var. pseudojaponica（Nakai）Y. S. Chen ■

410782　Viola yedoensis Makino var. pseudojaponica（Nakai）T. Hashim. ex E. Hama et K. M. Nakai = Viola philippica Cav. var. pseudojaponica（Nakai）Y. S. Chen ■

410783　Viola yedoensis Makino var. pseudojaponica（Nakai）T. Hashim. f. sonoharae E. Hama;白紫地丁■☆

410784　Viola yezoensis Maxim. ;阴地菫菜;Shady Violet ■

410785　Viola yezoensis Maxim. f. discolor（Nakai）Hiyama ex F. Maek. ;异色阴地菫菜■☆

410786　Viola yezoensis Maxim. var. hebeiensis（J. W. Wang et T. G. Ma）J. W. Wang et J. Yang = Viola mongolica Franch. ■

410787　Viola yezoensis Maxim. var. hopeiensis（Ching J. Wang et T. G. Ma）Ching J. Wang et J. Yang;河北菫菜■

410788　Viola yubariana Nakai;日本厚叶菫菜■☆

410789　Viola yunnanensis W. Becker et H. Boissieu;云南菫菜（滇菫菜,滇中菫菜,昆明菫菜,拟柔毛菫菜,紫罗兰）;Yunnan Violet ■

410790　Viola yunnanfuensis W. Becker;滇中菫菜(昆明菫菜,心叶菫菜,紫罗兰)■

410791　Viola zongia Tul. = Viola abyssinica Steud. ex Oliv. ■☆

410792　Violaceae Batsch(1802)(保留科名);菫菜科;Violet Fnmily ●■

410793　Violaceae Lam. et DC. = Violaceae Batsch(保留科名)●■

410794　Violaeoides Michx. ex DC. = Noisettia Kunth ■☆

410795　Violaria Post et Kuntze = Talauma Juss. ●

410796　Vionaea Neck. = Leucadendron R. Br. (保留属名)●

410797　Viorna（Pers.）Rchb. = Cheiropsis（DC.）Bercht. et J. Presl ■

410798　Viorna（Pers.）Rchb. = Clematis L. ●■

410799　Viorna Rchb. = Cheiropsis Bercht. et J. Presl ●■

410800　Viorna Rchb. = Clematis L. ●■

410801　Viorna addisonii（Britton）Small = Clematis addisonii Britton ■☆

410802　Viorna arizonica（A. Heller）A. Heller = Clematis hirsutissima Pursh ■☆

410803　Viorna bakeri（Greene）Rydb. = Clematis hirsutissima Pursh ■☆

410804　Viorna baldwinii（Torr. et A. Gray）Small = Clematis baldwinii Torr. et A. Gray ■☆

410805　Viorna beadlei Small = Clematis viorna L. ■☆

410806　Viorna coccinea（A. Gray）Small = Clematis texensis Buckley ●☆

410807　Viorna crispa（L.）Small = Clematis crispa L. ■☆

410808　Viorna eriophora Rydb. = Clematis hirsutissima Pursh ■☆

410809　Viorna flaccida（Small ex Rydb.）Small = Clematis viorna L. ■☆

410810　Viorna fremontii（S. Watson）A. Heller = Clematis fremontii S. Watson ■☆

410811　Viorna gattingeri Small = Clematis viorna L. ■☆

410812　Viorna obliqua Small = Clematis crispa L. ■☆

410813　Viorna ochroleuca（Aiton）Small = Clematis ochroleuca Aiton ●☆

410814　Viorna onesii（Kuntze）Rydb. = Clematis hirsutissima Pursh ■☆

410815　Viorna pitcheri（Torr. et A. Gray）Britton = Clematis pitscheri Torr. et Gray ■☆

410816　Viorna reticulata（Walter）Small = Clematis reticulata Walter ■☆

410817　Viorna subreticulata Harb. ex Small = Clematis reticulata Walter ■☆

410818　Viorna versicolor（Small ex Rydb.）Small = Clematis versicolor Small ex Rydb. ■☆

410819　Viorna viorna（L.）Small = Clematis viorna L. ■☆

410820　Viorna wyethii（Nutt.）Rydb. = Clematis hirsutissima Pursh ■☆

410821　Viposia Lundell = Plenckia Reissek(保留属名)●☆

410822　Viraea Vahl ex Benth. et Hook. f. = Leontodon L. (保留属名)■☆

410823　Viraea Vahl ex Benth. et Hook. f. = Virea Adans. ■☆

410824　Viraea asplenoides（L.）Batt. = Picris sinuata（Lam.）Lack ■☆

410825　Viraya Gaudich. = Waitzia J. C. Wendl. ■☆

410826　Virchowia Schenk = Ilysanthes Raf. ■

410827　Virchowia Schenk ex Urb. = Ilysanthes Raf. ■

410828　Virdika Adans. = Albuca L. ■☆

410829　Virea Adans. = Leontodon L. (保留属名)■☆

410830　Virecta Afzel. ex Sm. = Virectaria Bremek. ■☆

410831　Virecta L. f. = Sipanea Aubl. ●■

410832　Virecta Sm. = Virectaria Bremek. ■☆

410833　Virecta angustifolia Hiern = Virectaria angustifolia（Hiern）Bremek. ■☆

410834　Virecta heteromera K. Schum. = Virectaria angustifolia（Hiern）Bremek. ■☆

410835　Virecta kaessneri S. Moore = Virectaria major（K. Schum.）Verdc. ●☆

410836　Virecta lanceolata（Forssk.）Baill. = Pentas lanceolata（Forssk.）K. Schum. ●■

410837　Virecta major K. Schum. = Virectaria major（K. Schum.）Verdc. ●☆

410838　Virecta multiflora Sm. = Virectaria multiflora（Sm.）Bremek. ■☆

410839　Virecta obscura K. Schum. = Parapentas silvatica（K. Schum.）Bremek. ■☆

410840　Virecta petrophila Mildbr. = Virectaria herbacoursi N. Hallé var. petrophila ■☆

410841　Virecta procumbens Sm. = Virectaria procumbens（Sm.）Bremek. ■☆

410842　Virecta salicoides C. H. Wright = Virectaria salicoides（C. H. Wright）Bremek. ■☆

410843　Virecta setigera Hiern = Parapentas setigera（Hiern）Verdc. ■☆

410844　Virectaria Bremek. (1952);绿洲茜属■☆

410845　Virectaria angustifolia（Hiern）Bremek. ;窄叶绿洲茜■☆

410846　Virectaria angustifolia（Hiern）Bremek. var. schlechteri

Verdc. ;施莱绿洲茜■☆

410847　Virectaria herbacoursi N. Hallé var. petrophila ?;喜岩绿洲茜■☆

410848　Virectaria heteromera （K. Schum.） Bremek. = Virectaria angustifolia （Hiern） Bremek. ■☆

410849　Virectaria kaessneri （S. Moore） Bremek. = Virectaria major （K. Schum.） Verdc. ●☆

410850　Virectaria major （K. Schum.） Verdc. ;大绿洲茜●☆

410851　Virectaria major （K. Schum.） Verdc. subsp. decumbens Verdc. ;外倾绿洲茜■☆

410852　Virectaria major （K. Schum.） Verdc. subsp. spathulata （Verdc.） Dessein et Robbr. ;匙形绿洲茜●☆

410853　Virectaria major （K. Schum.） Verdc. var. spathulata Verdc. = Virectaria major （K. Schum.） Verdc. subsp. spathulata （Verdc.） Dessein et Robbr. ●☆

410854　Virectaria multiflora （Sm.） Bremek. ;多花绿洲茜■☆

410855　Virectaria petrophila Mildbr. = Virectaria herbacoursi N. Hallé var. petrophila ? ■☆

410856　Virectaria procumbens （Sm.） Bremek. ;平铺绿洲茜■☆

410857　Virectaria salicoides （C. H. Wright） Bremek. ;柳叶绿洲茜■☆

410858　Virectaria tenella J. B. Hall;柔弱绿洲茜■☆

410859　Vireya Blume = Rhododendron L. ●

410860　Vireya Post et Kuntze = Viraya Gaudich. ■☆

410861　Vireya Post et Kuntze = Waitzia J. C. Wendl. ■☆

410862　Vireya Raf. （废弃属名） = Alloplectus Mart. （保留属名）●■☆

410863　Vireya Raf. = Columnea L. ●■☆

410864　Virga Hill = Dipsacus L. ■

410865　Virga atrata （Hook. f. et Thomsonex C. B. Clarke） Holub = Dipsacus atratus Hook. f. et Thomson ex C. B. Clarke ■

410866　Virga inermis （Wall.） Holub. = Dipsacus inermis Wall. ■

410867　Virgilia L' Hér. （废弃属名） = Gaillardia Foug. ■

410868　Virgilia L' Hér. （废弃属名） = Virgilia Poir. （保留属名）●☆

410869　Virgilia Poir. （1808）（保留属名）;南非槐属（维吉尔豆属）;Virgilia ●☆

410870　Virgilia aurea （Aiton） Lam. = Calpurnia aurea （Aiton） Benth. ■☆

410871　Virgilia capensis （L.） Lam. = Virgilia oroboides （P. J. Bergius） T. M. Salter ●☆

410872　Virgilia capensis Lam. = Virgilia oroboides （P. J. Bergius） T. M. Salter ●☆

410873　Virgilia divaricata Adamson;维吉尔豆●☆

410874　Virgilia grandis E. Mey. = Millettia grandis （E. Mey.） Skeels ●☆

410875　Virgilia helioides L'Hér. = Gaillardia pulchella Foug. ■

410876　Virgilia intrusa R. Br. = Calpurnia intrusa （R. Br.） E. Mey. ■☆

410877　Virgilia lutea F. Michx. = Cladrastis lutea （F. Michx.） K. Koch ●☆

410878　Virgilia oroboides （P. J. Bergius） T. M. Salter;南非槐（相思子维吉尔豆）;Cape Lilac,Cape Virgilia,Tree-in-a-hurry ●☆

410879　Virgilia oroboides （P. J. Bergius） T. M. Salter subsp. ferruginea B. -E. van Wyk;锈色南非槐●☆

410880　Virgilia sylvatica （Burch.） DC. = Calpurnia aurea （Aiton） Benth. ■☆

410881　Virginia （DC.） Nicoli = Helichrysum Mill. （保留属名）●■

410882　Virgularia Ruiz et Pav. （废弃属名） = Agalinis Raf. （保留属名）■☆

410883　Virgularia Ruiz et Pav. （废弃属名） = Gerardia L. （废弃属名）■☆

410884　Virgulaster Semple = Aster L. ●■

410885　Virgulus Raf. = Aster L. ●■

410886　Virgulus adnatus （Nutt.） Reveal et Keener = Symphyotrichum adnatum （Nutt.） G. L. Nesom ■☆

410887　Virgulus amethystinus （Nutt.） Reveal et Keener = Aster amethystinus Nutt. ■☆

410888　Virgulus amethystinus （Nutt.） Reveal et Keener = Symphyotrichum amethystinum （Nutt.） G. L. Nesom ■☆

410889　Virgulus campestris （Nutt.） Reveal et Keener = Symphyotrichum campestre （Nutt.） G. L. Nesom ■☆

410890　Virgulus carolinianus （Walter） Reveal et Keener = Ampelaster carolinianus （Walter） G. L. Nesom ■☆

410891　Virgulus concolor （L.） Reveal et Keener = Symphyotrichum concolor （L.） G. L. Nesom ■☆

410892　Virgulus ericoides （L.） Reveal et Keener = Symphyotrichum ericoides （L.） G. L. Nesom ■☆

410893　Virgulus ericoides （L.） Reveal et Keener var. pansus （S. F. Blake） Reveal et Keener = Symphyotrichum ericoides （L.） G. L. Nesom var. pansum （S. F. Blake） G. L. Nesom ■☆

410894　Virgulus falcatus （Lindl.） Reveal et Keener = Symphyotrichum falcatum （Lindl.） G. L. Nesom ■☆

410895　Virgulus falcatus （Lindl.） Reveal et Keener subsp. commutatus （Torr. et A. Gray） Schaak = Symphyotrichum falcatum （Lindl.） G. L. Nesom var. commutatum （Torr. et A. Gray） G. L. Nesom ■☆

410896　Virgulus georgianus （Alexander） Semple = Symphyotrichum georgianum （Alexander） G. L. Nesom ■☆

410897　Virgulus grandiflorus （L.） Reveal et Keener = Symphyotrichum grandiflorum （L.） G. L. Nesom ■☆

410898　Virgulus novae-angliae （L.） Reveal et Keener = Aster novi-belgii L. ■

410899　Virgulus novae-angliae （L.） Reveal et Keener = Symphyotrichum novae-angliae （L.） G. L. Nesom ■☆

410900　Virgulus oblongifolius （Nutt.） Reveal et Keener = Aster oblongifolius Nutt. ■☆

410901　Virgulus oblongifolius （Nutt.） Reveal et Keener = Symphyotrichum oblongifolium （Nutt.） G. L. Nesom ■☆

410902　Virgulus patens （Aiton） Reveal et Keener = Symphyotrichum patens （Aiton） G. L. Nesom ■☆

410903　Virgulus patens （Aiton） Reveal et Keener var. georgianus （Alexander） Reveal et Keener = Symphyotrichum georgianum （Alexander） G. L. Nesom ■☆

410904　Virgulus patens （Aiton） Reveal et Keener var. gracilis （Hook.） Reveal et Keener = Symphyotrichum patens （Aiton） G. L. Nesom var. gracile （Hook.） G. L. Nesom ■☆

410905　Virgulus patens （Aiton） Reveal et Keener var. patentissimus （Lindl. ex DC.） Reveal et Keener = Symphyotrichum patens （Aiton） G. L. Nesom var. patentissimum （Lindl. ex DC.） G. L. Nesom ■☆

410906　Virgulus patens （Aiton） Reveal et Keener var. phlogifolius （Muhl. ex Willd.） Reveal et Keener = Symphyotrichum phlogifolium （Muhl. ex Willd.） G. L. Nesom ■☆

410907　Virgulus sericeus （Vent.） Reveal et Keener = Aster sericeus Vent. ■☆

410908　Virgulus sericeus （Vent.） Reveal et Keener = Symphyotrichum sericeum （Vent.） G. L. Nesom ■☆

410909　Virgulus walteri （Alexander） Reveal et Keener = Symphyotrichum walteri （Alexander） G. L. Nesom ■☆

410910　Virgulus yukonensis （Cronquist） Reveal et Keener = Symphyotrichum yukonense （Cronquist） G. L. Nesom ■☆

410911　Viridantha Espejo = Tillandsia L. ■☆

410912　Viridantha Espejo（2002）;绿花凤梨属■☆

410913　Viridivia J. H. Hemsl. et Verdc. （1956）;杯花西番莲属●☆

410914 Viridivia suberosa J. H. Hemsl. et Verdc. ;杯花西番莲●☆

410915 Virletia Sch. Bip. ex Benth. et Hook. f. = Bahia Lag. ■☆

410916 Virola Aubl. (1775);蔻木属(美洲肉豆蔻属,南美肉豆蔻属)●☆

410917 Virola albidiflora Ducke;白花蔻木(白花南美肉豆蔻)●☆

410918 Virola calophylla Warb. ;美叶蔻木(美叶肉豆蔻,美洲肉豆蔻)●☆

410919 Virola calophylla Warb. = Myristica calophylla Spruce ●☆

410920 Virola carinata Warb. ;龙骨蔻木(龙骨南美肉豆蔻)●☆

410921 Virola elongata (Benth.) Warb. ;伸长蔻木●☆

410922 Virola koschnyi Warb. ;科申蔻木(科申南美肉豆蔻);Banak ●☆

410923 Virola micrantha A. C. Sm. ;小花蔻木(小花南美肉豆蔻)●☆

410924 Virola multiflora (Standl.) A. C. Sm. ;多花蔻木(多花南美肉豆蔻)●☆

410925 Virola officinalis Warb. ;药用蔻木(药用南美肉豆蔻)●☆

410926 Virola polyneura W. A. Rodrigues;多脉蔻木(多脉南美肉豆蔻)●☆

410927 Virola sebifera Aubl. ;蜡质蔻木●☆

410928 Virola surinamensis (Rottb.) Warb. = Myristica surinamensis Roll. -Germ. ●☆

410929 Virola venezuelensis Warb. ;委内瑞拉蔻木(委内瑞拉南美肉豆蔻)●☆

410930 Virotia L. A. S. Johnson et B. G. Briggs(1975);维氏山龙眼属●☆

410931 Virotia angustifolia (Virot) P. H. Weston et A. R. Mast;窄叶维氏山龙眼●☆

410932 Virotia leptophylla (Guillaumin) L. A. S. Johnson et B. G. Briggs;细叶维氏山龙眼●☆

410933 Virotia neurophylla (Guillaumin) P. H. Weston et A. R. Mast;脉叶维氏山龙眼●☆

410934 Viscaceae Batsch(1802);槲寄生科;Mistletoe Family ●

410935 Viscaceae Miq. = Santalaceae R. Br. (保留科名)●■

410936 Viscaceae Miq. = Viscaceae Batsch ●

410937 Viscaceae Miq. = Vitaceae Juss. (保留科名)●■

410938 Viscago Zinn = Cucubalus L. + Silene L. (保留属名)■

410939 Viscago Zinn = Silene L. (保留属名)■

410940 Viscago wolgensis Hornem. = Silene wolgensis (Willd.) Besser ex Spreng. ■

410941 Viscainoa Greene(1888);黏蒺藜属●☆

410942 Viscainoa geniculata Greene;黏蒺藜●☆

410943 Viscaria Bernh. = Lychnis L. (废弃属名)■

410944 Viscaria Bernh. = Silene L. (保留属名)■

410945 Viscaria Comm. = Korthalsella Tiegh. ●

410946 Viscaria Comm. ex Danser = Korthalsella Tiegh. ●

410947 Viscaria Riv. ex Rupp. = Lychnis L. (废弃属名)■

410948 Viscaria Röhl. (1812);黏石竹属■☆

410949 Viscaria Röhl. = Steris Adans. (废弃属名)●

410950 Viscaria alpina (L.) G. Don;北极黏石竹(北极剪秋罗);Alpine Catchfly, Alpine Red Catchfly, Arctic Campion ■☆

410951 Viscaria alpina (L.) G. Don = Silene suecica (Lodd.) Greuter et Burdet ■☆

410952 Viscaria alpina (L.) G. Don subsp. americana (Fernald) Böcher = Silene suecica (Lodd.) Greuter et Burdet ■☆

410953 Viscaria viscosa (Scop.) Asch. ;黏石竹(黏剪秋罗);Clammy Campion ■☆

410954 Viscaria viscosa (Scop.) Asch. = Silene viscaria (L.) Jess. ■☆

410955 Viscaria viscosa Asch. = Viscaria viscosa (Scop.) Asch. ■☆

410956 Viscaria vulgaris Bernh. = Silene viscaria (L.) Jess. ■☆

410957 Viscaria vulgaris Röhl. ;普通黏石竹(洋剪秋罗);Clammy Campion , German Campion , German Catchfly , Red Catchfly , Red German Catchfly ,Sticky Catchfly ■☆

410958 Viscoides Jacq. = Psychotria L. (保留属名)●

410959 Viscum L. (1753);槲寄生属;Mistletoe ●

410960 Viscum album L. ;白果槲寄生(柏寄生,冬青,槲寄生,欧寄生,欧洲槲寄生,松萝);Birdlime Mistletoe, Churchman's Greeting, Common Mistletoe, European Mistletoe, Kiss-and-go, Masslin, Misceldin, Miscelto, Misle, Mislin-bush, Missel, Misselto, Misseltoe, Misseltow, Missletoe, Mistletoe, Mizzeltoe, White Mistletoe, Whitefruit Mistletoe ●

410961 Viscum album L. = Viscum coloratum (Kom.) Nakai ●

410962 Viscum album L. subsp. austriacum (Wiesb.) Vollm. ;澳洲白果槲寄生■☆

410963 Viscum album L. subsp. coloratum Kom. = Viscum coloratum (Kom.) Nakai ●

410964 Viscum album L. subsp. coloratum Kom. f. rubroaurantiacum (Makino) Ohwi = Viscum coloratum (Kom.) Nakai f. rubroaurantiacum (Makino) Kitag. ●

410965 Viscum album L. subsp. meridianum (Danser) D. G. Long = Viscum album L. var. meridianum Danser ●

410966 Viscum album L. var. coloratum (Kom.) Ohwi = Viscum album L. subsp. coloratum Kom. ●

410967 Viscum album L. var. coloratum (Kom.) Ohwi = Viscum coloratum (Kom.) Nakai ●

410968 Viscum album L. var. lutescens Makino = Viscum coloratum (Kom.) Nakai f. lutescens Kitag. ●

410969 Viscum album L. var. meridianum Danser;阔叶槲寄生(白果槲寄生,卵叶槲寄生);Ovateleaf Mistletoe ●

410970 Viscum album L. var. rubroaurantiacum Makino f. lutescens (Makino) H. Hara = Viscum album L. subsp. coloratum Kom. ●

410971 Viscum album Lecomte = Viscum coloratum (Kom.) Nakai ●

410972 Viscum alniformosanae Hayata;台湾槲寄生;Taiwan Mistletoe ●

410973 Viscum alniformosanae Hayata = Viscum coloratum (Kom.) Nakai ●

410974 Viscum anceps E. Mey. ex Sprague;二棱槲寄生■☆

410975 Viscum angulatum K. Heyne ex DC. = Viscum diospyrosicolum Hayata ●

410976 Viscum angulatum Merr. = Viscum diospyrosicolum Hayata ●

410977 Viscum apiculatum Lecomte;细尖槲寄生●☆

410978 Viscum apodum Baker = Viscum multicostatum Baker ●☆

410979 Viscum articulatum Burm. f. ;扁枝槲寄生(百子痰梗,扁枝寄生,赤柯寄生,椆栎柿寄生,风饭寄生,枫寄生,枫树寄生,枫香寄生,槲寄生,寄生包,栗寄生,路路通寄生,麻栎寄生,螃蟹夹,榕树寄生,柿寄生,粟寄生,桐树寄生,桐子寄生,虾蚶草,虾脚寄生,蟹爪寄生);Flatshoot Mistletoe, Flat-shooted Mistletoe, Oak-inhabiting Mistletoe ●

410980 Viscum articulatum Burm. f. var. dichotomum (D. Don) Kurz = Viscum articulatum Burm. f. ●

410981 Viscum articulatum Burm. f. var. liquidambaricola (Hayata) Sesh. Rao = Viscum liquidambaricola Hayata ●

410982 Viscum bagshawei Rendle;巴格肖槲寄生●☆

410983 Viscum bequaertii De Wild. = Viscum congolense De Wild. ●☆

410984 Viscum bivalve (Tiegh.) Engl. = Viscum obscurum Thunb. ●☆

410985 Viscum boivinii Tiegh. ;博伊文槲寄生●☆

410986 Viscum bongariense Hayata = Viscum liquidambaricola Hayata ●

410987 Viscum bosciae-foetidae Dinter = Viscum rotundifolium L. f. ●☆

410988 Viscum brevifolium (Harv.) Engl. = Viscum obscurum Thunb.

●☆

410989   Viscum calcaratum Balle;距槲寄生●☆

410990   Viscum camporum Engl. et K. Krause = Viscum tuberculatum A. Rich. ●☆

410991   Viscum capense L. f.;好望角槲寄生;Cape Mistletoe ●☆

410992   Viscum coloratum（Kom.）Nakai;槲寄生(北寄生,倒吊草,冬青,冬青条,冻青,黄寄生,寄生,寄生柴,寄生子,柳寄生,桑寄生,台湾赤杨寄生,台湾槲寄生,有色槲寄生);Colored Mistletoe, Formosan Alder Mistletoe ●

410993   Viscum coloratum（Kom.）Nakai = Viscum album L. subsp. coloratum Kom. ●

410994   Viscum coloratum（Kom.）Nakai f. lutescens Kitag.;黄果槲寄生(冬青,黄白槲寄生);Yellowish Colored Mistletoe ●

410995   Viscum coloratum （ Kom.） Nakai f. rubroaurantiacum（Makino）Kitag.;橙红槲寄生(红果槲寄生)●

410996   Viscum coloratum（Kom.）Nakai var. alniformosanae（Hayata）Iwata = Viscum coloratum（Kom.）Nakai ●

410997   Viscum coloratum （ Kom.） Nakai var. rubroaurantiacum（Makino）Miyabe = Viscum album L. subsp. coloratum Kom. f. rubroaurantiacum（Makino）Ohwi ●

410998   Viscum combreticola Engl.;风车子槲寄生●☆

410999   Viscum comorense Lecomte = Viscum triflorum DC. ●☆

411000   Viscum congolense De Wild.;刚果槲寄生●☆

411001   Viscum congolense De Wild. var. chevalieri Balle = Viscum congolense De Wild. ●☆

411002   Viscum continuum E. Mey. ex Sprague;连续槲寄生●☆

411003   Viscum costatum Gamble = Viscum album L. var. meridianum Danser ●

411004   Viscum coursii Balle;库尔斯槲寄生●☆

411005   Viscum crassulae Eckl. et Zeyh.;厚叶槲寄生●☆

411006   Viscum cruciatum Boiss. = Viscum cruciatum Sieber ex Spreng. ●☆

411007   Viscum cruciatum Sieber = Viscum cruciatum Sieber ex Spreng. ●☆

411008   Viscum cruciatum Sieber ex Spreng.;十字槲寄生●☆

411009   Viscum cuneifolium Baker;楔叶槲寄生●☆

411010   Viscum cylindricum Polhill et Wiens;柱形槲寄生●☆

411011   Viscum decaryi Lecomte;德卡里槲寄生●☆

411012   Viscum decurrens（Engl.）Baker et Sprague;下延槲寄生●☆

411013   Viscum dichotomum D. Don = Viscum articulatum Burm. f. ●

411014   Viscum dichotomum D. Don var. elegans Engl. = Viscum engleri Tiegh. ●☆

411015   Viscum dielsianum Dinter ex Neusser;迪尔斯槲寄生●☆

411016   Viscum diospyrosicolum Hayata;棱枝槲寄生(枫木寄生,梨寄生,青刚栎寄生,桑寄生,柿寄生,桐木寄生,万寿木寄生,樟木寄生);Ebonyshoot Mistletoe, Ebony-shooted Mistletoe, Persimmon-loving Mistletoe ●

411017   Viscum echinocarpum Baker;刺果槲寄生●☆

411018   Viscum elegans（Engl.）Engl. = Viscum engleri Tiegh. ●☆

411019   Viscum engleri Tiegh.;恩格勒槲寄生●☆

411020   Viscum eucleae Eckl. et Zeyh. = Viscum pauciflorum L. f. ●☆

411021   Viscum euphorbiae E. Mey. ex Drège = Viscum crassulae Eckl. et Zeyh. ●☆

411022   Viscum farafanganense Lecomte = Viscum multicostatum Baker ●☆

411023   Viscum fargesii Lecomte;线叶槲寄生(寄生);Farges Mistletoe ●

411024   Viscum fastigiatum Balle;帚状槲寄生●☆

411025   Viscum filipendulum Hayata = Viscum diospyrosicolum Hayata ●

411026   Viscum fischeri Engl.;菲舍尔槲寄生●☆

411027   Viscum galpinianum Schinz = Viscum subserratum Schltr. ●☆

411028   Viscum gilletii De Wild. = Viscum congolense De Wild. ●☆

411029   Viscum glaucum Eckl. et Zeyh. = Viscum rotundifolium L. f. ●☆

411030   Viscum goetzei Engl.;格兹槲寄生●☆

411031   Viscum gracile Polhill et Wiens;纤细槲寄生●☆

411032   Viscum grandidieri（Tiegh.）Lecomte = Viscum echinocarpum Baker ●☆

411033   Viscum grandifolium Engl. = Viscum congolense De Wild. ●☆

411034   Viscum griseum Polhill et Wiens;灰槲寄生●☆

411035   Viscum hainanense R. L. Han et D. X. Zhang;海南槲寄生●

411036   Viscum heteranthum Wall. ex DC. = Dendrotrophe platyphylla（Spreng.）N. H. Xia et M. G. Gilbert ●

411037   Viscum heterantum Wall. ex DC. = Dendrotrophe heteranta（Wall. ex DC.）A. N. Henry et B. Roy ●

411038   Viscum hexapterum Balle;六翅槲寄生●☆

411039   Viscum hildebrandtii Engl.;希尔德槲寄生●

411040   Viscum holstii Engl. = Viscum tuberculatum A. Rich. ●☆

411041   Viscum hoolei（Wiens）Polhill et Wiens;胡尔槲寄生●☆

411042   Viscum iringense Polhill et Wiens;伊林加槲寄生●☆

411043   Viscum japonicum Thunb. = Korthalsella japonica（Thunb.）Engl. ●

411044   Viscum japonicum Thunb. = Korthalsella opuntia（Thunb.）Merr. ●

411045   Viscum junodii（Tiegh.）Engl. = Viscum shirense Sprague ●☆

411046   Viscum kaempferi DC. = Taxillus kaempferi（DC.）Danser ●

411047   Viscum lenticellatum De Wild. et T. Durand = Agelanthus djurensis（Engl.）Polhill et Wiens ●☆

411048   Viscum liquidambaricola Hayata;枫香槲寄生(扁寄生,赤柯寄生,枫寄生,枫树寄生,枫香寄生,寄生草,螃蟹脚,桐树寄生,狭叶枫寄生);Sweetgumshoot Mistletoe, Sweet-gum-shooted Mistletoe ●

411049   Viscum littorum Polhill et Wiens;滨海槲寄生●☆

411050   Viscum longiarticulatum Engl.;长节槲寄生●☆

411051   Viscum longipetiolatum Balle;长柄槲寄生●☆

411052   Viscum loranthi Elmer;聚花槲寄生;Conferted Flower Mistletoe, Gregariousflower Mistletoe, Scurrula Mistletoe ●

411053   Viscum macowanii Engl. = Viscum rotundifolium L. f. ●☆

411054   Viscum macrofalcatum R. L. Han et D. X. Zhang;大镰叶槲寄生●☆

411055   Viscum matabelense Engl. = Viscum menyharthii Engl. et Schinz ●☆

411056   Viscum melanocarpum Peter = Viscum longiarticulatum Engl. ●☆

411057   Viscum menyharthii Engl. et Schinz;迈尼哈尔特槲寄生●☆

411058   Viscum minimum Harv.;极小槲寄生●☆

411059   Viscum minutiflorum Engl. et K. Krause = Viscum triflorum DC. ●☆

411060   Viscum moniliforme Wight et Arn. = Korthalsella japonica（Thunb.）Engl. ●

411061   Viscum monoicum Roxb.;五脉槲寄生;Fivenerve Mistletoe, Five-nerved Mistletoe ●

411062   Viscum multicostatum Baker;多脉槲寄生●☆

411063   Viscum multicostatum Baker var. laevibaccatum Balle = Viscum multicostatum Baker ●☆

411064   Viscum multiflorum Lecomte;多花槲寄生●☆

411065   Viscum multinerve（Hayata）Hayata;柄果槲寄生(刀叶槲寄生,寄生茶,相思叶寄生);Many-nerved Mistletoe, Multinerved Mistletoe, Stipefruit Mistletoe, Stipe-fruited Mistletoe ●

411066   Viscum myriophlebium Baker;脉槲寄生●☆

411067   Viscum nepalense Spreng. = Viscum articulatum Burm. f. ●

411068   Viscum nervosum Hochst. ex A. Rich. = Viscum triflorum DC. ●☆

411069   Viscum nervosum Hochst. ex A. Rich. var. angustifolium Sprague

= Viscum triflorum DC. ●☆

411070　Viscum nervosum Hochst. ex A. Rich. var. nyanzense（Rendle）Sprague = Viscum triflorum DC. ●☆

411071　Viscum nudum Danser；绿茎槲寄生；Glabrous Mistletoe, Greenstem Mistletoe, Naked Mistletoe ●

411072　Viscum nyanzense Rendle = Viscum triflorum DC. ●☆

411073　Viscum obovatum Harv.；倒卵槲寄生●☆

411074　Viscum obscurum Thunb.；隐匿槲寄生●☆

411075　Viscum obscurum Thunb. var. brevifolium Harv. = Viscum obscurum Thunb. ●☆

411076　Viscum obscurum Thunb. var. decurrens Engl. = Viscum decurrens（Engl.）Baker et Sprague ●☆

411077　Viscum obscurum Thunb. var. longiflorum Harv. = Viscum obscurum Thunb. ●☆

411078　Viscum opuntia Thunb. = Korthalsella japonica（Thunb.）Engl. ●

411079　Viscum opuntia Thunb. = Korthalsella opuntia（Thunb.）Merr. ●

411080　Viscum oreophilum Wiens；喜山槲寄生●☆

411081　Viscum orientale DC. = Viscum cruciatum Sieber ex Spreng. ●☆

411082　Viscum orientale Willd.；石榴槲寄生（东方槲寄生，瘤果槲寄生，禄柚寄生，柚寄生，柚子寄生）；Pomegranate Mistletoe ●☆

411083　Viscum orientale Willd. = Viscum ovalifolium DC. ●

411084　Viscum orientale Willd. var. multinerve Hayata = Viscum multinerve（Hayata）Hayata ●

411085　Viscum ovalifolium DC.；瘤果槲寄生（禄柚寄生，柚寄生，柚树寄生，柚子寄生）；Ovalleaf Mistletoe, Oval-leaved Mistletoe, Tumorfruit Mistletoe ●

411086　Viscum oxycedri DC. = Arceuthobium oxycedri（DC.）M. Bieb. ●

411087　Viscum pauciflorum L. f.；少花槲寄生●☆

411088　Viscum pauciflorum L. f. var. eucleae（Eckl. et Zeyh.）Harv. = Viscum pauciflorum L. f. ●☆

411089　Viscum pedicellatum Lecomte = Viscum boivinii Tiegh. ●☆

411090　Viscum pentanthum Baker；五花槲寄生●☆

411091　Viscum perrieri Lecomte；佩里耶槲寄生●☆

411092　Viscum petiolatum Polhill et Wiens；柄叶槲寄生●☆

411093　Viscum platyphyllum Spreng. = Dendrotrophe platyphylla（Spreng.）N. H. Xia et M. G. Gilbert ●

411094　Viscum pulchellum Sprague = Viscum obovatum Harv. ●☆

411095　Viscum querci-morii Hayata = Viscum liquidambaricola Hayata ●

411096　Viscum radula Baker；刮刀槲寄生●☆

411097　Viscum ramosissinum Hand. -Mazz. = Viscum diospyrosicolum Hayata ●

411098　Viscum rhipsaloides Baker；仙人棒槲寄生●☆

411099　Viscum rigidum Engl. et K. Krause = Viscum capense L. f. ●☆

411100　Viscum robustum Eckl. et Zeyh. = Viscum capense L. f. ●☆

411101　Viscum roncartii Balle；龙卡特槲寄生●☆

411102　Viscum rotundifolium L. f.；圆叶槲寄生●☆

411103　Viscum schimperi Engl.；欣珀槲寄生●☆

411104　Viscum semiplanum（Tiegh.）Engl. = Viscum schimperi Engl. ●☆

411105　Viscum shirense Sprague；热非槲寄生●☆

411106　Viscum spathulatum Lecomte = Viscum triflorum DC. ●☆

411107　Viscum spragueanum Burtt Davy = Viscum tuberculatum A. Rich. ●☆

411108　Viscum staudtii Engl. = Viscum congolense De Wild. ●☆

411109　Viscum stellatum Buch-Ham. ex D. Don = Viscum album L. ●

411110　Viscum stipitatum Lecomte = Viscum multinerve（Hayata）Hayata ●

411111　Viscum subcylindricum Weim. = Viscum menyharthii Engl. et Schinz ●☆

411112　Viscum subserratum Schltr.；具齿槲寄生●☆

411113　Viscum subverrucosum Polhill et Wiens；小疣槲寄生●☆

411114　Viscum tarchonanthum Welw. ex Tiegh. = Viscum tuberculatum A. Rich. ●☆

411115　Viscum tenue Engl.；瘦槲寄生●☆

411116　Viscum thymifolium C. Presl = Viscum rotundifolium L. f. ●☆

411117　Viscum trachycarpum Baker；糙果槲寄生●☆

411118　Viscum tricostatum E. Mey. ex Harv. = Viscum rotundifolium L. f. ●☆

411119　Viscum triflorum DC.；三花槲寄生●☆

411120　Viscum triflorum DC. subsp. nervosum（Hochst. ex A. Rich.）M. G. Gilbert = Viscum triflorum DC. ●☆

411121　Viscum tsaratananense Lecomte；察拉塔纳纳槲寄生●☆

411122　Viscum tuberculatum A. Rich.；多疣槲寄生●☆

411123　Viscum ugandense Sprague = Viscum combreticola Engl. ●☆

411124　Viscum umbellatum Blume = Dendrotrophe umbellata（Blume）Miq. ●

411125　Viscum venosum DC. var. lanceolatum DC. = Viscum triflorum DC. ●☆

411126　Viscum verrucosum Harv.；密疣槲寄生●☆

411127　Viscum yunnanense H. S. Kiu；云南槲寄生；Yunnan Mistletoe ●

411128　Viscum zenkeri Engl. = Viscum congolense De Wild. ●☆

411129　Viscum ziziphi-mucronati Dinter = Viscum rotundifolium L. f. ●☆

411130　Visena Schult. = Visenia Houtt. ●■

411131　Visenia Houtt. = Melochia L.（保留属名）●■

411132　Visenia corchorifolia（L.）Spreng. = Melochia corchorifolia L. ●■

411133　Visiania A. DC. = Ligustrum L. ●

411134　Visiania Gasp. = Ficus L. ●

411135　Visiania Gasp. = Macrophthalma Gasp. ●

411136　Visiania elastica（Roxb. ex Hornem.）Gasp. = Ficus elastica Roxb. ●

411137　Visiania paniculata（Roxb.）DC. = Ligustrum lucidum W. T. Aiton ●

411138　Visiania robusta（Roxb.）DC. = Ligustrum robustum（Roxb.）Blume ●

411139　Visinia Turcz. = Vismia Vand.（保留属名）●☆

411140　Vismia Vand.（1788）（保留属名）；维斯木属；Vismia ●☆

411141　Vismia affinis Oliv.；近缘维斯木●☆

411142　Vismia angusta Miq.；狭叶维斯木●☆

411143　Vismia cayennensis（Jacq.）Pers.；卡宴维斯木●☆

411144　Vismia frondosa Oliv. = Vismia affinis Oliv. ●☆

411145　Vismia guineensis（L.）Choisy；几内亚维斯木●☆

411146　Vismia japurensis Reichardt；血树●☆

411147　Vismia laurentii De Wild.；洛朗维斯木●☆

411148　Vismia laurentii De Wild. var. polyandra Hochr. = Vismia guineensis（L.）Choisy ●☆

411149　Vismia leonensis Hook. f. = Vismia guineensis（L.）Choisy ●☆

411150　Vismia orientalis Engl.；东方维斯木●☆

411151　Vismia pauciflora Milne-Redh.；少花维斯木●☆

411152　Vismia rubescens Oliv.；红维斯木●☆

411153　Vismia striatipetala Mildbr. ex Engl. = Vismia laurentii De Wild. ●☆

411154　Vismia torrei Mendes；托雷维斯木●☆

411155　Vismianthus Mildbr.（1935）；维斯花属●☆

411156　Vismianthus punctatus Mildbr.；斑点维斯花●☆

411157　Visnaga Gaertn. = Ammi L. ■

411158　Visnaga Mill. = Ammi L. ■

411159　Visnaga daucoides Gaertn. = Ammi visnaga (L.) Lam. ■

411160　Visnea L. f. (1782);长萼厚皮香属●☆

411161　Visnea Steud. ex Endl. = Barbacenia Vand. ■☆

411162　Visnea mocanera L. f. ;长萼厚皮香●☆

411163　Vissadali Adans. = Knoxia L. ●

411164　Vistnu Adans. = Evolvulus L. ●■

411165　Vitaceae Juss. (1789)(保留科名);葡萄科;Grape Family, Grape-vine Family ●■

411166　Vitaeda Börner = Ampelopsis Michx. ●

411167　Vitaliana Sesl. (废弃属名) = Androsace L. ■

411168　Vitaliana Sesl. (废弃属名) = Douglasia Lindl. (保留属名)■☆

411169　Vitekorchis Romowicz et Szlach. (2006);维特兰属■☆

411170　Vitekorchis Romowicz et Szlach. = Oncidium Sw. (保留属名)■☆

411171　Vitellaria C. F. Gaertn. (1807);蛋黄榄属●

411172　Vitellaria C. F. Gaertn. = Butyrospermum Kotschy ●

411173　Vitellaria paradoxa C. F. Gaertn. = Butyrospermum parkii (G. Don) Kotschy ●

411174　Vitellaria paradoxa C. F. Gaertn. subsp. nilotica (Kotschy) A. N. Henry, Chithra et N. C. Nair = Butyrospermum parkii (G. Don) Kotschy var. niloticum (Kotschy) Pierre ex Engl. ●☆

411175　Vitellariopsis (Baill.) Dubard = Vitellariopsis Baill. ex Dubard ●☆

411176　Vitellariopsis Baill. = Vitellariopsis Baill. ex Dubard ●☆

411177　Vitellariopsis Baill. ex Dubard(1915);拟蛋黄榄属●☆

411178　Vitellariopsis cuneata (Engl.) Aubrév.;楔形拟蛋黄榄●☆

411179　Vitellariopsis dispar (N. E. Br.) Aubrév.;异型拟蛋黄榄●☆

411180　Vitellariopsis ferruginea Kupicha;锈色拟蛋黄榄●☆

411181　Vitellariopsis kirkii (Baker) Dubard;柯克拟蛋黄榄●☆

411182　Vitellariopsis marginata (N. E. Br.) Aubrév.;具边拟蛋黄榄●☆

411183　Vitellariopsis sylvestris (S. Moore) Aubrév. = Vitellariopsis marginata (N. E. Br.) Aubrév. ●☆

411184　Vitenia Noronha = Erioglossum Blume ●

411185　Vitenia Noronha ex Cambess. = Erioglossum Blume ●

411186　Vitenia Noronha ex Cambess. = Vitenia Noronha ●

411187　Vitex L. (1753);牡荆属(黄荆属,荆条属);Chaste Tree, Chastetree, Chaste-tree ●

411188　Vitex aesculifolia Baker = Vitex congolensis De Wild. et T. Durand ●☆

411189　Vitex agnus-castus L. ;穗花牡荆(淡紫花牡荆,树牡荆,洋莺哥木);Abraham's Balm, Agnus Castus, Agnus-castus, Chaste Tree, Chaste Willow, Chaste-lamb Tree, Chastetree, Chaste-tree, Hemptree, Hemp-tree, Lilac Chaste Tree, Lilac Chastetree, Lilac Chaste-tree, Monk's Pepper, Monk's Pepper Tree, Monks' Pepper Tree, Small Pepper, Tree-of-chastity, Wild Pepper ●

411190　Vitex agnus-castus L. var. subtrisecta Kuntze = Vitex trifolia L. var. subtrisecta (Kuntze) Moldenke ●

411191　Vitex agnus-castus L. var. subtrisecta Kuntze = Vitex trifolia L. ●

411192　Vitex agraria A. Chev. = Vitex thyrsiflora Baker ●☆

411193　Vitex allasia Planch. = Vitex payos (Lour.) Merr. ●☆

411194　Vitex amboniensis Gürke = Vitex ferruginea Schumach. et Thonn. ●☆

411195　Vitex amboniensis Gürke var. schlechteri W. Piep. = Vitex ferruginea Schumach. et Thonn. var. amboniensis (Gürke) Verdc. ●☆

411196　Vitex andongensis Baker = Vitex fischeri Gürke ●☆

411197　Vitex annamensis Dop = Vitex tripinnata (Lour.) Merr. ●

411198　Vitex arborea Desf. = Vitex negundo L. ●

411199　Vitex arborea Fisch. = Vitex negundo L. ●

411200　Vitex aurata Endl. = Diplocyclos (Endl.) Post et Kuntze ■

411201　Vitex aurata Hils. et Bojer = Vitex chrysomallum Steud. ●☆

411202　Vitex aurea Moldenke;黄牡荆●☆

411203　Vitex balbi Chiov. = Vitex keniensis Turrill ●☆

411204　Vitex barbata Baker = Vitex madiensis Oliv. ●☆

411205　Vitex bequaertii De Wild. = Vitex fischeri Gürke ●☆

411206　Vitex beraviensis Vatke;贝拉瓦牡荆●☆

411207　Vitex betsiliensis Humbert;贝齐里牡荆●☆

411208　Vitex bicolor Willd. ;二色牡荆;Simpleleaf Chastetree ●

411209　Vitex bicolor Willd. = Vitex negundo L. ●

411210　Vitex bicolor Willd. = Vitex trifolia L. ●

411211　Vitex bipindensis Gürke = Vitex grandifolia Gürke ●☆

411212　Vitex bojeri Schau;博耶尔牡荆●☆

411213　Vitex bracteata Scott-Elliot;具苞牡荆●☆

411214　Vitex bracteosa Mildbr. = Vitex welwitschii Gürke var. laurentii (De Wild.) W. Piep. ●☆

411215　Vitex buchananii Gürke;布坎南牡荆●☆

411216　Vitex buchananii Gürke var. quadrangula (Gürke) W. Piep. = Vitex buchananii Gürke ●☆

411217　Vitex buchneri Gürke;布赫纳牡荆●☆

411218　Vitex bunguensis Moldenke = Vitex zanzibarensis Vatke ●☆

411219　Vitex burmensis Moldenke;长叶牡荆;Lanceleaf Chastetree, Lance-leaved Chaste-tree, Longleaf Chastetree ●

411220　Vitex caespitosa Exell;丛生牡荆●☆

411221　Vitex camporum Büttner = Vitex madiensis Oliv. ●☆

411222　Vitex canescens Kurz;灰毛牡荆(灰布荆,灰毛黄荆,灰毛荆,灰牡荆);Greyhair Chastetree, Grey-haired Chaste-tree ●

411223　Vitex cannabifolia Siebold et Zucc. = Vitex negundo L. var. cannabifolia (Siebold et Zucc.) Hand. -Mazz. ●

411224　Vitex cannabifolia Siebold et Zucc. = Vitex negundo L. ●

411225　Vitex carvalhi Gürke;卡瓦牡荆●☆

411226　Vitex cauliflora Moldenke;茎花牡荆●☆

411227　Vitex cestroides Baker;夜香树牡荆●☆

411228　Vitex chariensis A. Chev. ;沙里牡荆●☆

411229　Vitex chinensis Mill. = Vitex negundo L. var. heterophylla (Franch.) Rehder ●

411230　Vitex chinensis Mill. = Vitex negundo L. var. incisa (Lam.) C. B. Clarke ●

411231　Vitex chrysocarpa Planch. ex Benth. ;金果牡荆●☆

411232　Vitex chrysoclada Bojer = Premna chrysoclada (Bojer) Gürke ●☆

411233　Vitex chrysomallum Steud. ;金毛牡荆●☆

411234　Vitex cienkowskii Kotschy et Peyr. = Vitex doniana Sweet ●☆

411235　Vitex ciliata Pierre ex Pellegr. ;缘毛牡荆●☆

411236　Vitex cilio-foliolata A. Chev. = Vitex rivularis Gürke ●☆

411237　Vitex cofassa Reinw. ex Blume;新几内亚牡荆;New Guinea Teak, Vitex ●☆

411238　Vitex congensis A. Chev. ;刚果牡荆●☆

411239　Vitex congesta Oliv. ;密集牡荆●☆

411240　Vitex congolensis De Wild. et T. Durand;康戈尔牡荆●☆

411241　Vitex congolensis De Wild. et T. Durand var. gillettii (Gürke) W. Piep. ;秸莱特牡荆●☆

411242　Vitex cordata Aubrév. = Vitex madiensis Oliv. ●☆

411243　Vitex coriacea Schltr. ex Piep. = Vitex chrysomallum Steud. ●☆

411244　Vitex coursii Moldenke;库尔斯牡荆●☆

411245　Vitex crenata A. Chev. ;圆齿牡荆●☆

411246　Vitex cuneata Thonn. = Vitex doniana Sweet ●☆

411247　Vitex cuneata Thonn. var. parvifolia Engl. = Vitex doniana Sweet ●☆

411248　Vitex cuspidata Hiern;骤尖牡荆●☆

411249　Vitex cymoss Benth.;伞花牡荆●☆

411250　Vitex dekindtiana Gürke = Vitex grisea Baker var. dekindtiana（Gürke）W. Piep.●☆

411251　Vitex dentata Klotzsch;尖齿牡荆●☆

411252　Vitex dewevrei De Wild. et T. Durand = Vitex doniana Sweet●☆

411253　Vitex dinklagei Gürke;丁克牡荆●☆

411254　Vitex divaricata Baker = Vitex doniana Sweet●☆

411255　Vitex divaricata Sw.;叉开牡荆●☆

411256　Vitex diversifolia Baker = Vitex madiensis Oliv.●☆

411257　Vitex djumaensis De Wild.;朱马牡荆●☆

411258　Vitex doniana Sweet;西非牡荆;West African Plum●☆

411259　Vitex dryadum S. Moore;森林牡荆●☆

411260　Vitex duboisii Moldenke;杜氏牡荆●☆

411261　Vitex duclouxii Dop;金沙荆;Ducloux Chastetree, Ducloux Chaste-tree●

411262　Vitex erythrocarpa Gürke = Vitex mombassae Vatke●☆

411263　Vitex esquirolii H. Lév. = Buddleja asiatica Lour.●

411264　Vitex eylesii S. Moore = Vitex payos（Lour.）Merr.●☆

411265　Vitex ferruginea Bojer ex Schauer = Vitex bojeri Schau●☆

411266　Vitex ferruginea Schumach. et Thonn.;锈色牡荆●☆

411267　Vitex ferruginea Schumach. et Thonn. var. amaniensis（W. Piep.）Verdc. = Vitex ferruginea Schumach. et Thonn.●☆

411268　Vitex ferruginea Schumach. et Thonn. var. amboniensis（Gürke）Verdc. = Vitex ferruginea Schumach. et Thonn.●☆

411269　Vitex finlaysoniana Wall. = Vitex vestita Wall.●

411270　Vitex fischeri Gürke;菲舍尔牡荆●☆

411271　Vitex flavescens Rolfe = Vitex mombassae Vatke●☆

411272　Vitex flavescens Rolfe var. parviflora Gibbs = Vitex mombassae Vatke●☆

411273　Vitex fosteri Wright = Vitex ferruginea Schumach. et Thonn.●☆

411274　Vitex gabunensis Gürke;加蓬牡荆（加本牡荆）●☆

411275　Vitex geminata H. Pearson = Vitex harveyana H. Pearson●☆

411276　Vitex gillettii Gürke = Vitex congolensis De Wild. et T. Durand var. gillettii（Gürke）W. Piep.●☆

411277　Vitex gillettii Mildbr. = Vitex congolensis De Wild. et T. Durand●☆

411278　Vitex giorgii De Wild.;乔治牡荆●☆

411279　Vitex glabrata R. Br.;无毛牡荆;Smooth Chastetree●☆

411280　Vitex goetzei Gürke = Vitex mombassae Vatke●☆

411281　Vitex gomphophylla Baker = Cordia myxa L.●☆

411282　Vitex grandidiana W. Piep.;格兰牡荆●☆

411283　Vitex grandifolia Gürke;大叶牡荆●☆

411284　Vitex grandifolia Gürke var. bipindensis（Gürke）W. Piep. = Vitex grandifolia Gürke●☆

411285　Vitex grisea Baker;灰牡荆●☆

411286　Vitex grisea Baker var. dekindtiana（Gürke）W. Piep.;德金牡荆●☆

411287　Vitex guerkeana De Wild. = Vitex buchananii Gürke●☆

411288　Vitex guerkeana H. Pearson = Vitex pearsonii W. Piep.●☆

411289　Vitex harveyana H. Pearson;哈维牡荆●☆

411290　Vitex harveyana H. Pearson f. geminata（H. Pearson）Moldenke = Vitex harveyana H. Pearson●☆

411291　Vitex heterophylla Roxb. = Vitex quinata（Lour.）F. N. Williams●

411292　Vitex heterophylla Roxb. var. puberula H. J. Lam = Vitex quinata（Lour.）F. N. Williams var. puberula（H. J. Lam.）Moldenke●

411293　Vitex hildebrandtii Vatke = Vitex payos（Lour.）Merr.●☆

411294　Vitex hildebrandtii Vatke var. glabrescens W. Piep. = Vitex payos（Lour.）Merr. var. glabrescens（W. Piep.）Moldenke●☆

411295　Vitex hirsutissima Baker;粗毛牡荆●☆

411296　Vitex homblei De Wild. = Vitex doniana Sweet●☆

411297　Vitex huillensis Hiern = Vitex grisea Baker●☆

411298　Vitex humbertii Moldenke;亨伯特牡荆●☆

411299　Vitex impressinervia Mildbr. ex W. Piep.;陷脉牡荆●☆

411300　Vitex incisa Lam. = Vitex negundo L. var. heterophylla（Franch.）Rehder●

411301　Vitex incisa Lam. var. heterophylla Franch. = Vitex negundo L. var. heterophylla（Franch.）Rehder●

411302　Vitex involucrata C. Presl = Sphenodesme involucrata（C. Presl）B. L. Rob.●

411303　Vitex iringensis Gürke = Vitex payos（Lour.）Merr. var. glabrescens（W. Piep.）Moldenke●☆

411304　Vitex iriomotensis Ohwi = Vitex bicolor Willd.●

411305　Vitex isotjensis Gibbs;同位牡荆●☆

411306　Vitex kapirensis De Wild.;卡皮里牡荆●☆

411307　Vitex keniensis Turrill;肯尼亚牡荆;Meru Oak●☆

411308　Vitex kirkii Baker = Vitex petersiana Klotzsch●☆

411309　Vitex kwangchouensis C. P'ei = Vitex canescens Kurz●

411310　Vitex kwangsiensis C. P'ei;广西牡荆;Guangxi Chastetree, Guangxi Chaste-tree, Kwangsi Chastetree●

411311　Vitex kweichowensis C. P'ei = Vitex canescens Kurz●

411312　Vitex laevigata Baker = Vitex ferruginea Schumach. et Thonn. var. amboniensis（Gürke）Verdc.●☆

411313　Vitex lanceifolia S. C. Huang = Vitex burmensis Moldenke●

411314　Vitex lanceolata C. P'ei = Vitex lanceifolia S. C. Huang●

411315　Vitex lanceolata C. P'ei = Vitex burmensis Moldenke●

411316　Vitex lanigera Schauer;绵毛牡荆●☆

411317　Vitex lastellei Moldenke;拉斯泰勒牡荆●☆

411318　Vitex laurentii De Wild. = Vitex welwitschii Gürke var. laurentii（De Wild.）W. Piep.●☆

411319　Vitex leandrii Moldenke;利安牡荆●☆

411320　Vitex lebrunii Moldenke;勒布伦牡荆●☆

411321　Vitex lehmbachii Gürke;莱姆牡荆●☆

411322　Vitex littoralis A. Cunn.;滨海牡荆（滨牡荆）;New Zealand Teak●☆

411323　Vitex lobata Moldenke;浅裂牡荆●☆

411324　Vitex lokundjensis W. Piep.;洛昆牡荆●☆

411325　Vitex lomiensis Mildbr. = Vitex welwitschii Gürke●☆

411326　Vitex longeacuminata A. Chev. = Vitex micrantha Gürke●☆

411327　Vitex longipetiolata Gürke;长梗牡荆●☆

411328　Vitex loureiri Hook. et Arn. = Vitex quinata（Lour.）F. N. Williams●

411329　Vitex lucens Kirk;新西兰牡荆;Puriri●☆

411330　Vitex lukafuensis De Wild.;卢卡夫牡荆●☆

411331　Vitex lundensis Gürke;隆德牡荆●☆

411332　Vitex lutea A. Chev. = Vitex grandifolia Gürke●☆

411333　Vitex lutea Exell;黄色牡荆●☆

411334　Vitex madagascariensis Moldenke;马岛牡荆●☆

411335　Vitex madiensis Oliv.;马迪牡荆●☆

411336　Vitex madiensis Oliv. var. milanjiensis（Britten）F. White;米兰吉牡荆●☆

411337　Vitex madiensis Oliv. var. parvifolia Hiern = Vitex madiensis Oliv.●☆

411338　Vitex madiensis Oliv. var. schweinfurthii（Gürke）W. Piep. =

Vitex madiensis Oliv. ●☆

411339   Vitex masoalensis G. E. Schatz;马苏阿拉牡荆●☆

411340   Vitex mechowii Gürke = Vitex mombassae Vatke ●☆

411341   Vitex megapotamica (Spreng.) Moldenke;巴西牡荆;Brazil
Chastetree ●☆

411342   Vitex melleri Baker = Vitex chrysomallum Steud. ●☆

411343   Vitex micrantha Gürke;小花牡荆●☆

411344   Vitex microcalyx Baker = Holmskioldia microcalyx (Baker) W.
Piep. ●☆

411345   Vitex microphylla (Hand. -Mazz.) C. P'ei ex C. Y. Wu = Vitex
negundo L. var. microphylla Hand. -Mazz. ●

411346   Vitex microphylla (Hand. -Mazz.) C. P'ei ex C. Y. Wu = Vitex
negundo L. var. microphylla Hand. -Mazz. ●

411347   Vitex microphylla Moldenke;小叶牡荆●☆

411348   Vitex milanjiensis Britten = Vitex madiensis Oliv. var.
milanjiensis (Britten) F. White ●☆

411349   Vitex milnei W. Piep. = Vitex doniana Sweet ●☆

411350   Vitex mollis Kunth;软牡荆●☆

411351   Vitex mombassae Vatke;蒙巴萨牡荆●☆

411352   Vitex mombassae Vatke var. acuminata W. Piep. = Vitex
mombassae Vatke ●☆

411353   Vitex mombassae Vatke var. parviflora (Gibbs) W. Piep. =
Vitex mombassae Vatke ●☆

411354   Vitex monroviana W. Piep. = Vitex phaeotricha Mildbr. ex W.
Piep. ●☆

411355   Vitex mooiensis H. Pearson = Premna mooiensis (H. Pearson)
W. Piep. ■☆

411356   Vitex mooiensis H. Pearson var. rudolphii ? = Premna mooiensis
(H. Pearson) W. Piep. ■☆

411357   Vitex morogoroensis Walsingham et S. Atkins;莫罗戈罗牡荆●☆

411358   Vitex mossambicensis Gürke;莫桑比克牡荆●☆

411359   Vitex mossambicensis Gürke var. oligantha (Baker) W. Piep. =
Vitex mossambicensis Gürke ●☆

411360   Vitex mufutu De Wild. = Vitex mombassae Vatke ●☆

411361   Vitex myrmecophila Mildbr. = Vitex thyrsiflora Baker ●☆

411362   Vitex negundo L.;黄荆(白背叶,白毛黄荆,白毛荆,白叶黄
荆,不惊茶,布荆,布惊,黄金,黄金条,黄荆条,酱草,荆条,荆仔,
马藤,蔓荆,牡荆,埔姜,埔姜花,埔姜仔,埔荆茶,七叶黄荆,山黄
荆,山荆,山埔姜,土柴胡,土常山,蚊烟柴,蚊兹草,蚊子柴,五指
风,五指柑);Chaste Tree,Chastetree,Chinese Chaste Tree,Hemp
Tree,Negundo Chaste Tree,Negundo Chastetree,Negundo Chastetree,Vites,Whitehair Negundo Chastetree,Whitehairy Chastetree ●

411363   Vitex negundo L. f. alba C. P'ei;白毛黄荆●

411364   Vitex negundo L. f. alba C. P'ei = Vitex negundo L. ●

411365   Vitex negundo L. f. albiflora H. W. Jen et Y. J. Chang;白花荆
条;White-flowered Chaste-tree ●

411366   Vitex negundo L. f. intermedia C. P'ei = Vitex negundo L. var.
cannabifolia (Siebold et Zucc.) Hand. -Mazz. ●

411367   Vitex negundo L. f. intermedia C. P'ei = Vitex negundo L. ●

411368   Vitex negundo L. f. laxipaniculata C. P'ei;疏序黄荆;
Laxpaniculated Chastetree,Laxpaniculated Negundo Chastetree ●

411369   Vitex negundo L. f. laxipaniculata C. P'ei = Vitex negundo L. ●

411370   Vitex negundo L. var. bicolor (Willd.) H. J. Lam = Vitex trifolia L. ●

411371   Vitex negundo L. var. bicolor (Willd.) H. J. Lam. = Vitex
trifolia L. var. bicolor (Willd.) Moldenke ●

411372   Vitex negundo L. var. bicolor H. J. Lam = Vitex trifolia L. ●

411373   Vitex negundo L. var. cannabifolia (Siebold et Zucc.) Hand. -

Mazz.;牡荆(荆条,梦子,牡荆条,奶疸,五指柑,小荆);
Chastetree,Hempleaf Negundo Chastetree,Negundo Chastetree ●

411374   Vitex negundo L. var. heterophylla (Franch.) Rehder;荆条
(楚,黄荆条,荆棵,刻叶黄荆,刻叶荆条,山荆子);Heterophyllous
Chastetree,Heterophyllous Negundo Chastetree,Negundo Chastetree ●

411375   Vitex negundo L. var. incisa (Lam.) C. B. Clarke;刻叶黄荆
(刻叶荆条)●

411376   Vitex negundo L. var. incisa (Lam.) C. B. Clarke = Vitex
negundo L. var. heterophylla (Franch.) Rehder ●

411377   Vitex negundo L. var. intermedia ? = Vitex negundo L. var.
cannabifolia (Siebold et Zucc.) Hand. -Mazz. ●

411378   Vitex negundo L. var. lobata H. W. Jen;妙峰山荆条;Chaste-
tree,Miaofengshan Negundo ●

411379   Vitex negundo L. var. microphylla Hand. -Mazz.;小叶荆(小叶
黄荆);Littleleaf Negundo Chastetree ●

411380   Vitex negundo L. var. sichuanensis J. L. Liu;四川黄荆●

411381   Vitex negundo L. var. thyrsoides C. P'ei et S. L. Liou;拟黄荆;
Thyrselike Negundo Chastetree ●

411382   Vitex negundo L. var. typica Lam. = Vitex negundo L. var.
cannabifolia (Siebold et Zucc.) Hand. -Mazz. ●

411383   Vitex obanensis Wernham = Vitex thyrsiflora Baker ●☆

411384   Vitex obovata E. Mey.;倒卵牡荆●☆

411385   Vitex obovata E. Mey. subsp. wilmsii (Gürke) Bredenk. et D. J.
Botha;维尔姆斯牡荆●☆

411386   Vitex oligantha Baker = Vitex mossambicensis Gürke ●☆

411387   Vitex ovata Thunb. = Vitex rotundifolia L. f. ●

411388   Vitex ovata Thunb. = Vitex trifolia L. var. simplicifolia Champ. ●

411389   Vitex ovata Thunb. var. subtrisecta Kuntze = Vitex trifolia L. var.
subtrisecta (Kuntze) Moldenke ●

411390   Vitex oxycuspis Baker;尖凸牡荆●☆

411391   Vitex oxycuspis Baker var. mossambicensis Moldenke;莫桑比克
尖凸牡荆●☆

411392   Vitex pachyclada Baker;粗枝牡荆●☆

411393   Vitex pachyphylla Baker;厚叶牡荆●☆

411394   Vitex paludosa Vatke = Vitex doniana Sweet ●☆

411395   Vitex paniculata Lam. = Vitex negundo L. ●

411396   Vitex parviflora A. Juss.;马尼拉白埔姜;Smallflower
Chastetree,Smallflower Chaste-tree ●

411397   Vitex patula E. A. Bruce;张开牡荆●☆

411398   Vitex payos (Lour.) Merr.;巧克力色牡荆;Chocolate Berry,
Coffee Bean Tree ●☆

411399   Vitex payos (Lour.) Merr. var. glabrescens (W. Piep.) Moldenke;
光巧克力色牡荆●☆

411400   Vitex payos (Lour.) Merr. var. stipitata Moldenke = Vitex fischeri
Gürke ●☆

411401   Vitex pearsonii W. Piep.;皮尔逊牡荆●☆

411402   Vitex peduncularis Wall.;长序牡荆(长序荆);Longspike
Chastetree,Long-spiked Chaste-tree ●

411403   Vitex perrieri Danguy;佩里耶牡荆●☆

411404   Vitex pervillei Baker;佩尔牡荆●☆

411405   Vitex petersiana Klotzsch;彼得斯牡荆●☆

411406   Vitex petersiana Klotzsch var. tettensis (Klotzsch) W. Piep.;泰
特牡荆●☆

411407   Vitex phaeotricha Mildbr. ex W. Piep.;褐毛牡荆●☆

411408   Vitex phillyreifolia Baker;欧女贞叶牡荆●☆

411409   Vitex pierreana Dop;莺哥木;Parrot Chastetree,Pierre Chastetree,
Pierre Chaste-tree ●

411410 Vitex pinnata L. ;羽状牡荆●☆

411411 Vitex pobeguinii Aubrév. = Vitex madiensis Oliv. ●☆

411412 Vitex poggei Gürke;波格牡荆●☆

411413 Vitex polyantha Baker = Vitex ferruginea Schumach. et Thonn. var. amboniensis（Gürke）Verdc. ●☆

411414 Vitex pseudochrysocarpa W. Piep. = Vitex chrysocarpa Planch. ex Benth. ●☆

411415 Vitex pseudocuspidata Mildbr. ex W. Piep. ;假尖牡荆●☆

411416 Vitex puberula Baker;微毛牡荆●☆

411417 Vitex pubescens Vahl;短毛牡荆●☆

411418 Vitex pulchra Moldenke;美丽山牡荆●☆

411419 Vitex quadrangula Gürke = Vitex buchananii Gürke ●☆

411420 Vitex quinata（Lour.）F. N. Williams;山牡荆（布荆,布惊,荆条,麻仔,埔姜木,全叶牡荆,山布惊,山埔姜,山紫荆,乌甜树,五指风,五指柑,莺哥）;Cloth Alarm, Fiveleaf Chaste Tree, Fiveleaf Chastetree, Five-leaved Chaste-tree, Wild Chastetree ●

411421 Vitex quinata（Lour.）F. N. Williams f. lungchowensis S. L. Liou;龙州山牡荆（微毛布荆）;Longzhou Chastetree ●

411422 Vitex quinata（Lour.）F. N. Williams var. puberula（H. J. Lam.）Moldenke;微毛布荆（微毛布惊）;Puberulent Chastetree, Puberulent Fiveleaf Chastetree ●

411423 Vitex radula W. Piep. = Vitex buchananii Gürke ●☆

411424 Vitex reflexa H. Pearson = Vitex obovata E. Mey. ●☆

411425 Vitex rehmannii Gürke;拉赫曼牡荆●☆

411426 Vitex resinifera Moldenke;胶牡荆●☆

411427 Vitex ringoetii De Wild. = Vitex madiensis Oliv. var. milanjiensis（Britten）F. White ●☆

411428 Vitex rivularis Gürke;溪边牡荆●☆

411429 Vitex robynsii De Wild. ;罗宾斯牡荆●☆

411430 Vitex rotundifolia L. f. ;单叶蔓荆（白埔姜,海埔姜,蔓荆,蔓荆子,山埔姜）;Beech Vitex, Roundleaf Chastetree, Round-leaf Vitex, Simpleleaf Shrub Chastetree ●

411431 Vitex rotundifolia L. f. f. albescens Hiyama;白单叶蔓荆●☆

411432 Vitex rotundifolia L. f. f. heterophylla（Makino ex H. Hara）Kitam. = Vitex rotundifolia L. f. var. subtrisecta（Kuntze）Moldenke ●☆

411433 Vitex rotundifolia L. f. f. rosea Satomi;粉单叶蔓荆●☆

411434 Vitex rotundifolia L. f. f. uyekii Honda;植木蔓荆●☆

411435 Vitex rotundifolia L. f. var. heterophylla Makino ex H. Hara = Vitex rotundifolia L. f. var. subtrisecta（Kuntze）Moldenke ●☆

411436 Vitex rotundifolia L. f. var. subtrisecta（Kuntze）Moldenke;三裂单叶蔓荆●☆

411437 Vitex rubra Moldenke;红牡荆●☆

411438 Vitex rubro-aurantiaca De Wild. ;红黄牡荆●☆

411439 Vitex rufa A. Chev. ex Hutch. et Dalziel = Vitex phaeotricha Mildbr. ex W. Piep. ●☆

411440 Vitex rufescens Gürke = Vitex ferruginea Schumach. et Thonn. ●☆

411441 Vitex sampsonii Hance;广东牡荆;Sampson Chastetree, Sampson Chaste-tree ●

411442 Vitex schlechteri Gürke = Vitex harveyana H. Pearson ●☆

411443 Vitex schliebenii Moldenke;施利本牡荆●☆

411444 Vitex schweinfurthii Baker = Vitex madiensis Oliv. ●☆

411445 Vitex schweinfurthii Gürke = Vitex madiensis Oliv. ●☆

411446 Vitex seineri Gürke ex W. Piep. ;塞纳牡荆●☆

411447 Vitex seretii De Wild. ;赛雷牡荆●☆

411448 Vitex shirensis Baker = Vitex payos（Lour.）Merr. ●☆

411449 Vitex simplicifolia Oliv. = Vitex madiensis Oliv. ●☆

411450 Vitex simplicifolia Oliv. var. vogelii（Baker）W. Piep. = Vitex madiensis Oliv. ●☆

411451 Vitex staudtii Gürke = Vitex thyrsiflora Baker ●☆

411452 Vitex stellata Moldenke;星状牡荆●☆

411453 Vitex sulphurea Baker;硫色牡荆●☆

411454 Vitex swynnertonii S. Moore = Vitex ferruginea Schumach. et Thonn. var. amboniensis（Gürke）Verdc. ●☆

411455 Vitex syringifolia Baker = Cordia senegalensis Juss. ●☆

411456 Vitex taihangensis L. B. Guo et S. Q. Zhou = Vitex trifolia L. var. taihangensis（L. B. Guo et S. Q. Zhou）S. L. Chen ●

411457 Vitex tangensis Gürke = Vitex ferruginea Schumach. et Thonn. var. amboniensis（Gürke）Verdc. ●☆

411458 Vitex tettensis Klotzsch = Vitex petersiana Klotzsch var. tettensis（Klotzsch）W. Piep. ●☆

411459 Vitex thomasii De Wild. ;托马斯牡荆●☆

411460 Vitex thonneri De Wild. ;托内牡荆●☆

411461 Vitex thyrsiflora Baker;聚伞牡荆●☆

411462 Vitex trichantha Baker;毛花牡荆●☆

411463 Vitex trifolia Forbes et Hemsl. = Vitex rotundifolia L. f. ●

411464 Vitex trifolia L. ;蔓荆（白背草,白背风,白背木耳,白背杨,白布荆,白叶,大荆,单叶蔓荆,黄荆,京子,荆子,蔓荆子,蔓荆子,蔓青,蔓条子,三叶蔓荆,水稔,水稔子,万金子,万京,万京子,万荆,小刀豆藤,小蔓）;Shrub Chastetree, Shrub Chaste-tree, Simpleleaf Chaste Tree, Simpleleaf Chastetree, Threeleaf Chastetree ●

411465 Vitex trifolia L. subsp. litoralis Steenis = Vitex rotundifolia L. f. ●

411466 Vitex trifolia L. var. bicolor（Willd.）Moldenke = Vitex bicolor Willd. ●

411467 Vitex trifolia L. var. bicolor（Willd.）Moldenke = Vitex trifolia L. ●

411468 Vitex trifolia L. var. ovata（Thunb.）Makino = Vitex rotundifolia L. f. ●

411469 Vitex trifolia L. var. parviflora Benth. = Vitex trifolia L. ●

411470 Vitex trifolia L. var. simplicifolia Champ. = Vitex rotundifolia L. f. ●

411471 Vitex trifolia L. var. subtrisecta（Kuntze）Moldenke;异叶蔓荆;Simpleleaf Chastetree, Subtrisect Threeleaf Chastetree ●

411472 Vitex trifolia L. var. taihangensis（L. B. Guo et S. Q. Zhou）S. L. Chen;太行荆●

411473 Vitex trifolia L. var. trifoliolata Schauer = Vitex trifolia L. ●

411474 Vitex trifolia L. var. unifoliolata Schauer = Vitex rotundifolia L. f. ●

411475 Vitex trifolia L. var. unifoliolata Schauer = Vitex trifolia L. var. simplicifolia Champ. ●

411476 Vitex trifolia L. var. variegata Moldenke;斑叶蔓荆;Varigated Chastetree ●☆

411477 Vitex tripinnata（Lour.）Merr. ;越南牡荆;Tripinate Chastetree, Tripinate Chaste-tree, Vietnam Chastetree ●

411478 Vitex tristis Scott-Elliot;暗淡牡荆●☆

411479 Vitex ubanghensis A. Chev. ;乌班吉牡荆●☆

411480 Vitex ugogensis Verdc. ;热非牡荆●☆

411481 Vitex umbrosa G. Don ex Sabine = Vitex doniana Sweet ●☆

411482 Vitex uniflora Baker;单花牡荆●☆

411483 Vitex venulosa Moldenke;细脉牡荆●☆

411484 Vitex vermoesenii De Wild. ;韦尔蒙森牡荆●☆

411485 Vitex verticillata A. Chev. ;轮生牡荆●☆

411486 Vitex vestita Wall. ;黄毛牡荆;Yellowhair Chastetree, Yellow-haired Chaste-tree, Yellowhairy Chastetree ●

411487 Vitex vestita Wall. var. brevituba Z. Y. Huang et S. Y. Liu;短管黄毛牡荆●

411488 Vitex villosa Sim;长柔毛牡荆●

411489　Vitex vogelii Baker = Vitex madiensis Oliv. ●☆
411490　Vitex volkensii Gürke = Vitex buchananii Gürke ●☆
411491　Vitex waterlotii Danguy;瓦泰洛牡荆●☆
411492　Vitex wellensis De Wild. ;韦伦斯牡荆●☆
411493　Vitex welwitschii Gürke;韦氏牡荆●☆
411494　Vitex welwitschii Gürke var. laurentii (De Wild.) W. Piep. ;洛朗牡荆●☆
411495　Vitex wilmsii Gürke = Vitex obovata E. Mey. subsp. wilmsii (Gürke) Bredenk. et D. J. Botha ●☆
411496　Vitex wilmsii Gürke var. reflexa (H. Pearson) W. Piep. = Vitex obovata E. Mey. ●☆
411497　Vitex yaundensis Gürke;雅温德牡荆●☆
411498　Vitex yunnanensis W. W. Sm. ;滇牡荆; Yunnan Chastetree, Yunnan Chaste-tree ●
411499　Vitex zambesiaca Baker = Vitex payos (Lour.) Merr. var. glabrescens (W. Piep.) Moldenke ●☆
411500　Vitex zanzibarensis Vatke;桑给巴尔牡荆●☆
411501　Vitex zechii Gürke ex W. Piep. = Vitex chrysocarpa Planch. ex Benth. ●☆
411502　Vitex zenkeri Gürke;岑克尔牡荆●☆
411503　Vitex zeyheri Sond. ex Schauer;泽赫牡荆●☆
411504　Viticaceae Juss. ;牡荆科●■
411505　Viticaceae Juss. = Labiatae Juss. (保留科名)●
411506　Viticaceae Juss. = Lamiaceae Martinov(保留科名)●■
411507　Viticaceae Juss. = Verbenaceae J. St. -Hil. (保留科名)●■
411508　Viticastrum C. Presl = Sphenodesme Jack ●
411509　Viticella Dill. ex Moench = Clematis L. ●■
411510　Viticella Mitch. (废弃属名) = Galax L. (废弃属名)■☆
411511　Viticella Mitch. (废弃属名) = Nemophila Nutt. (保留属名)■☆
411512　Viticella Moench = Clematis L. ●■
411513　Viticella orientalis (L.) W. A. Weber = Clematis orientalis L. ●
411514　Viticella viticella (L.) Small = Clematis viticella L. ●☆
411515　Viticena Benth. = Vinticena Steud. ●
411516　Viticipremna H. J. Lam(1919);荆鞭木属●☆
411517　Viticipremna philippinensis H. J. Lam;荆鞭木●☆
411518　Vitidaceae Juss. = Vitaceae Juss. (保留科名)●■
411519　Vitiphoenix Becc. = Veitchia H. Wendl. (保留属名)●☆
411520　Vitis Adans. = Cissus L. ●
411521　Vitis L. (1753);葡萄属;Grape,Grape-vine, Vine ●
411522　Vitis × tsukubana (Makino) F. Maek. ex Hisauti = Vitis flexuosa Thunb. var. tsukubana Makino ●☆
411523　Vitis abyssinica Hochst. ex A. Rich. = Ampelocissus abyssinica (Hochst. ex A. Rich.) Planch. ●☆
411524　Vitis aconitifolia (Bunge) Hance = Ampelopsis aconitifolia Bunge ●
411525　Vitis aconitifolia Hance = Ampelopsis aconitifolia Bunge ●
411526　Vitis adenoclada Hand. -Mazz. ;腺枝葡萄;Glandular Grape ●
411527　Vitis adnata (Roxb.) Wall. = Cissus adnata Roxb. ●
411528　Vitis adstricta Hance = Vitis bryoniifolia Bunge var. adstricta (Hance) W. T. Wang ●
411529　Vitis adstricta Hance = Vitis bryoniifolia Bunge ●
411530　Vitis adstricta Hance var. ternata W. T. Wang = Vitis bryoniifolia Bunge var. ternata (W. T. Wang) C. L. Li ●
411531　Vitis aestivalis Michx. ;夏葡萄(花蕾葡萄);Bunch Grape, Pigeon Grape,Summer Grape ●
411532　Vitis aestivalis Michx. var. argentifolia (Munson) Fernald;银叶夏葡萄;Silver-leaf Grape,Summer Grape ●☆
411533　Vitis aestivalis Michx. var. argentifolia (Munson) Fernald =

411533(cont) Vitis argentifolia Munson ●☆
411534　Vitis aestivalis Michx. var. bicolor Deam = Vitis aestivalis Michx. var. argentifolia (Munson) Fernald ●☆
411535　Vitis aestivalis Michx. var. glauca L. H. Bailey = Vitis linsecomii Buckley var. glauca Munson ●☆
411536　Vitis aestivalis Michx. var. linsecumii Munson = Vitis linsecomii Buckley ●☆
411537　Vitis aestivalis Michx. var. monticola Engelm. = Vitis berlandieri Planch. ●☆
411538　Vitis afzelii Baker = Cissus diffusiflora (Baker) Planch. ●☆
411539　Vitis amplexa Baker = Cyphostemma bororense (Klotzsch) Desc. ex Wild et R. B. Drumm. ●☆
411540　Vitis amurensis Rupr. ;山葡萄(阿穆尔葡萄,黑龙江葡萄,黑水葡萄,山藤藤,山藤藤秧,乌苏里葡萄,野葡萄);Amur Grape, Amur Grape Vine,Amur Grapevine,Amurland Grape ●
411541　Vitis amurensis Rupr. f. hermaphrodita J. Y. Li;双锦山葡萄●
411542　Vitis amurensis Rupr. var. coignetiae (Pulliat ex Planch.) Nakai = Vitis coignetiae Pulliat ex Planch. ●☆
411543　Vitis amurensis Rupr. var. dissecta Skvortsov;深裂山葡萄; Deeplobed Grape ●
411544　Vitis amurensis Rupr. var. genuina Skvortsov = Vitis amurensis Rupr. ●
411545　Vitis amurensis Rupr. var. shiragai (Makino) Ohwi = Vitis amurensis Rupr. ●
411546　Vitis amurensis Rupr. var. yanshanensis D. Z. Lu et H. P. Liang; 燕山葡萄;Yanshan Grape ●
411547　Vitis amurensis Rupr. var. yanshanensis D. Z. Lu et H. P. Liang = Vitis amurensis Rupr. var. dissecta Skvortsov ●
411548　Vitis andongensis Welw. ex Baker = Cyphostemma chloroleucum (Welw. ex Baker) Desc. ex Wild et R. B. Drumm. ●☆
411549　Vitis angolensis Baker = Ampelocissus angolensis (Baker) Planch. ●☆
411550　Vitis aralioides Baker = Cissus aralioides (Baker) Planch. ●☆
411551　Vitis argentifolia Munson;蓝葡萄(银叶葡萄);Blue Grape, Silver-leaf Grape,Silverlraf Grape,Summer Grape ●☆
411552　Vitis argentifolia Munson = Vitis aestivalis Michx. var. argentifolia (Munson) Fernald ●☆
411553　Vitis arisanensis (Hayata) Hayata = Tetrastigma obtectum (Wall.) Planch. ●■
411554　Vitis arisanensis (Hayata) Hayata = Tetrastigma obtectum (Wall.) Planch. var. glabrum (H. Lév. et Vaniot) Gagnep. ●■
411555　Vitis arizonensis Parry = Vitis arizonica Engelm. ●☆
411556　Vitis arizonica Engelm. ;峡谷葡萄;Arizona Wild Grape,Canon Grape,Canyon Grape ●☆
411557　Vitis arizonica Engelm. var. glabra Munson;无毛峡谷葡萄; Glabrous Canon Grape ●☆
411558　Vitis armata Diels et Gilg = Vitis davidii (Rom. Caill.) Foëx ●
411559　Vitis armata Diels et Gilg var. cyanocarpa Gagnep. = Vitis davidii (Rom. Caill.) Foëx var. cyanocarpa (Gagnep.) Sarg. ●
411560　Vitis assamica M. A. Lawson = Cissus assamica (Lawson) Craib ●
411561　Vitis austrina Small = Vitis cinerea (Engelm.) Millardet var. floridana Munson ●☆
411562　Vitis austrokoreana Hatus. = Vitis ficifolia Bunge f. glabrata (Nakai) W. T. Lee ●
411563　Vitis baihensis P. C. He = Vitis piasezkii Maxim. ●
411564　Vitis baihuashanensis M. S. Kang et D. Z. Lu;百花山葡萄; Baihuashan Grape ●

411565　Vitis baihuashanensis M. S. Kang et D. Z. Lu = Vitis amurensis Rupr. ●

411566　Vitis baihuashanensis M. S. Kang et D. Z. Lu = Vitis amurensis Rupr. var. dissecta Skvortsov ●

411567　Vitis baileyana Munson;强葡萄;Possum Grape ●☆

411568　Vitis balanseana Planch. ;小果葡萄(补刀藤,穿过山,大血藤,光叶葡萄,假葡萄,葡萄血藤,山葫芦,蚊树,小果山葡萄,小果野葡萄,小葡萄,野葡萄);Littlefruit Grape,Small-fruited Grape ●

411569　Vitis balanseana Planch. var. ficifolioides (W. T. Wang) C. L. Li = Vitis ficifolioides W. T. Wang ●

411570　Vitis balanseana Planch. var. tomentosa C. L. Li;绒毛小果葡萄;Tomentose Littlefruit Grape ●

411571　Vitis barteri Baker = Cissus barteri (Baker) Planch. ●☆

411572　Vitis bashanica P. C. He;麦黄葡萄;Bashan Grape ●

411573　Vitis bellula (Rehder) W. T. Wang;美丽葡萄(小叶毛葡萄);Beautiful Grape ●

411574　Vitis bellula (Rehder) W. T. Wang var. pubigera C. L. Li;华南美丽葡萄;South-China Beautiful Grape ●

411575　Vitis berlandieri Planch. ;冬葡萄; Fall Grape, Inter Grape, Spanish Grape,Winter Grape ●☆

411576　Vitis betulifolia Diels et Gilg;桦叶葡萄;Birchleaf Grape,Birch-leaved Grape ●

411577　Vitis betulifolia Diels et Gilg = Vitis heyneana Roem. et Schult. ●

411578　Vitis bicolor Le Conte;二色葡萄(灰青色葡萄,两色葡萄);Blue Grape,Summer Grape ●☆

411579　Vitis bicolor Leconte = Vitis aestivalis Michx. var. argentifolia (Munson) Fernald ●☆

411580　Vitis bioritsensis Hayata = Tetrastigma hemsleyanum Diels et Gilg ●■

411581　Vitis bipinnata Torr. et Gray = Ampelopsis arborea (L.) Koehne ●☆

411582　Vitis biternata Chiov. = Cyphostemma ternato-multifidum (Chiov.) Desc. ●☆

411583　Vitis blandii W. R. Prince = Vitis labrusca L. ●

411584　Vitis bodinieri H. Lév. et Vaniot = Ampelopsis bodinieri (H. Lév. et Vaniot) Rehder ●

411585　Vitis bombycina Baker = Ampelocissus bombycina (Baker) Planch. ●☆

411586　Vitis brevipedunculata (Maxim.) Dippel = Ampelopsis brevipedunculata (Maxim.) Trautv. ●

411587　Vitis brevipedunculata (Maxim.) Dippel = Ampelopsis glandulosa (Wall.) Momiy. var. brevipedunculata (Maxim.) Momiy. ●

411588　Vitis brevipedunculata (Maxim.) Dippel = Ampelopsis heterophylla (Thunb.) Siebold et Zucc. var. brevipedunculata (Regel) C. L. Li ●

411589　Vitis bryoniifolia Bunge;华北葡萄(甘古藤,禾花子藤,禾黄藤,接骨藤,猫耳藤,猫眼睛,木龙,千岁蔂,山红羊,山苦瓜,山葡萄,山蒲桃,酸古藤,烟黑,燕薁,野葡萄,野葡萄藤,蔓舌,蔓薁,薁);N. China Grape,North China Grape ●

411590　Vitis bryoniifolia Bunge var. adstricta (Hance) W. T. Wang;蔓薁(扁担酸,多裂华北葡萄,甘古藤,禾花子藤,禾黄藤,接骨藤,猫耳藤,猫眼睛,木龙,平布藤,山红草,山红羊,山苦瓜,山葡萄,山蒲桃,酸古藤,烟黑,燕薁,野葡萄,野葡萄藤,蔓舌,薁);Manylobed North China Grape,Wild Grape,Ying'ao Grape ●

411591　Vitis bryoniifolia Bunge var. mairei (H. Lév.) W. T. Wang = Vitis bryoniifolia Bunge ●

411592　Vitis bryoniifolia Bunge var. multilobata S. Y. Wang et Y. H. Hu = Vitis bryoniifolia Bunge ●

411593　Vitis bryoniifolia Bunge var. ternata (W. T. Wang) C. L. Li;三出蔓薁●

411594　Vitis californica Benth. ;加州葡萄;California Wild Grape, Pacific Grape,W. Wild Grape ●☆

411595　Vitis campylocarpa Kurz = Tetrastigma campylocarpum (Kurz) Planch. ●

411596　Vitis candicans Engelm. ;白背叶葡萄;Mustang Grape ●☆

411597　Vitis cantoniensis Seem. = Ampelopsis cantoniensis (Hook. et Arn.) Planch. ●

411598　Vitis cantoniensis Seem. = Ampelopsis hypoglauca (Hance) C. L. Li ●

411599　Vitis capensis (Willd.) Thunb. = Rhoicissus tomentosa (Lam.) Wild et R. B. Drumm. ●☆

411600　Vitis capensis Thunb. ;常绿葡萄;Evergreen Grape ●☆

411601　Vitis capriolata D. Don = Tetrastigma serrulatum (Roxb.) Planch. ●■

411602　Vitis cardiospermoides Franch. = Cayratia cardiospermoides (Planch.) Gagnep. ●

411603　Vitis caribaea DC. ;西方野葡萄●☆

411604　Vitis carnosa (Lam.) Roxb. = Cissus carnosa (L.) Lam. ●

411605　Vitis carnosa (Lam.) Wall. ex M. A. Lawson = Cayratia trifolia (L.) Domin ●

411606　Vitis carnosa Wall. ex M. A. Lawson = Cayratia trifolia (L.) Domin ●

411607　Vitis cavaleriei H. Lév. = Vitis flexuosa Thunb. ●

411608　Vitis cavaleriei H. Lév. et Vaniot = Vitis flexuosa Thunb. ●

411609　Vitis cavicaulis Baker = Ampelocissus abyssinica (Hochst. ex A. Rich.) Planch. ●☆

411610　Vitis chaffanjoni H. Lév. et Vaniot = Ampelopsis chaffanjonii (H. Lév. et Vaniot) Rehder ●

411611　Vitis champinii Planch. ;山平氏葡萄(石灰葡萄);Champin Grape ●☆

411612　Vitis chloroleuca Welw. ex Baker = Cyphostemma chloroleucum (Welw. ex Baker) Desc. ex Wild et R. B. Drumm. ●☆

411613　Vitis chunganensis Hu;东南葡萄;SE. China Grape,Southeathern China Grape ●

411614　Vitis chungii F. P. Metcalf;闽赣葡萄(背带藤,红扁藤);Chung Grape ●

411615　Vitis cinerea (Engelm.) Engelm. ex Millardet;灰葡萄(灰背叶葡萄);Downy Grape, Grayback Grape, Gray-bark Grape, Pigeon Grape,Sweet Winter Grape ●☆

411616　Vitis cinerea (Engelm.) Millardet = Vitis cinerea (Engelm.) Engelm. ex Millardet ●☆

411617　Vitis cinerea (Engelm.) Millardet var. canescens L. H. Bailey;圆叶灰葡萄(圆叶灰背葡萄);Roundleaf Grayback Grape ●☆

411618　Vitis cinerea (Engelm.) Millardet var. floridana Munson;佛罗里达灰葡萄;Florida Grayback Grape,Florida Pigeon Grape ●☆

411619　Vitis cinerea Engelm. = Vitis cinerea (Engelm.) Engelm. ex Millardet ●☆

411620　Vitis cirrhosa Thunb. = Cyphostemma cirrhosum (Thunb.) Desc. ex Wild et R. B. Drumm. ●☆

411621　Vitis cirrhosa Thunb. var. transvaalensis Szyszyl. = Cyphostemma cirrhosum (Thunb.) Desc. ex Wild et R. B. Drumm. subsp. transvaalense (Szyszyl.) Wild et R. B. Drumm. ●☆

411622　Vitis coignetiae Pulliae ex Planch. = Vitis heyneana Roem. et Schult. ●

411623　Vitis coignetiae Pulliat ex Planch. ;紫葛葡萄(山葡萄,酸葡萄藤,紫葛);Crimson Glory Vine, Crimson Glory-vine, Gloryvine,

Japanese Crimson Glory-vine ●☆

411624 Vitis coignetiae Pulliat ex Planch. f. glabrescens (Nakai) H. Hara;光紫葛葡萄●☆

411625 Vitis coignetiae Pulliat ex Planch. var. glabrescens Nakai = Vitis coignetiae Pulliat ex Planch. ●☆

411626 Vitis coignetiae Pulliat ex Planch. var. glabrescens Nakai = Vitis coignetiae Pulliat ex Planch. f. glabrescens (Nakai) H. Hara ●☆

411627 Vitis concinna Baker = Ampelocissus concinna (Baker) Planch. ●☆

411628 Vitis congesta Baker = Cyphostemma congestum (Baker) Desc. ex Wild et R. B. Drumm. ●☆

411629 Vitis congoensis Hort. = Cyphostemma congoense Desc. ●☆

411630 Vitis constricta Baker = Cissus aralioides (Baker) Planch. ●☆

411631 Vitis cordifolia Michx. ;心叶葡萄(霜葡萄);Chicken Grape, Frost Grape,Raccoon,True Frost Grape,Winter Grape ●☆

411632 Vitis cordifolia Michx. = Vitis vulpina L. ●☆

411633 Vitis cordifolia Michx. var. foetida Engelm. ;香果心叶葡萄;Stinking Winter Grape ●☆

411634 Vitis cordifolia Michx. var. riparia Torr. et Gray = Vitis riparia Michx. ●☆

411635 Vitis cordifolia Michx. var. sempervirens Munson;光叶心叶葡萄;Sharpleaf Winter Grape ●☆

411636 Vitis coriacea Servett. ex K. Koch;革叶葡萄;Callous Grape, Leather-leaf Grape ●☆

411637 Vitis corniculata Benth. = Cayratia corniculata (Benth.) Gagnep. ●

411638 Vitis cornifolia Baker = Cissus cornifolia (Baker) Planch. ●☆

411639 Vitis crassifolia Baker = Cissus rotundifolia (Forssk.) Vahl ●☆

411640 Vitis crassiuscula Baker = Cyphostemma crassiusculum (Baker) Desc. ●☆

411641 Vitis curvipoda Baker = Cyphostemma curvipodum (Baker) Desc. ●☆

411642 Vitis cuspidata (Planch.) Palacky = Cissus floribunda (Baker) Planch. ●☆

411643 Vitis davidiana (Carrière) N. E. Br. = Ampelopsis heterophylla (Thunb.) Siebold et Zucc. ●

411644 Vitis davidiana Dippel = Vitis davidii (Rom. Caill.) Foëx ●

411645 Vitis davidiana G. Nicholson = Ampelopsis humulifolia Bunge ●

411646 Vitis davidii (Rom. Caill.) Foëx;刺葡萄;Brier Grape ●

411647 Vitis davidii (Rom. Caill.) Foëx var. brachytricha Merr. = Vitis pilosonerva F. P. Metcalf ●

411648 Vitis davidii (Rom. Caill.) Foëx var. cyanocarpa (Gagnep.) Sarg. ;蓝果刺葡萄(瘤枝葡萄,疣枝刺葡萄);Tuberculate-branched Grape ●

411649 Vitis davidii (Rom. Caill.) Foëx var. ferruginea Merr. et Chun;锈毛刺葡萄;Rustyhair Brier Grape ●

411650 Vitis davidii (Rom. Caill.) Foëx var. hispida X. D. Wang et S. C. Chen;顺昌刺葡萄;Hispid Brier Grape ●

411651 Vitis debilis Baker = Cayratia debilis (Baker) Suess. ●☆

411652 Vitis dentata Hayata = Tetrastigma hemsleyanum Diels et Gilg ●■

411653 Vitis diffusiflora Baker = Cissus diffusiflora (Baker) Planch. ●☆

411654 Vitis discolor (Blume) Dalzell = Cissus discolor Blume ●

411655 Vitis discolor (Blume) Dalzell = Cissus javana DC. ●

411656 Vitis dissecta Carrière = Ampelopsis aconitifolia Bunge ●

411657 Vitis diversifolia W. R. Prince = Vitis linsecomii Buckley ●☆

411658 Vitis doaniana Munson;斑点葡萄(斑点叶葡萄);Panhandle Grape ●

411659 Vitis elegans K. Koch = Ampelopsis heterophylla (Thunb.)

411660 Vitis elongata (Roxb.) Wall. = Cissus elongata Roxb. ●

411661 Vitis elongata (Roxb.) Wall. ex M. A. Lawson = Cissus elongata Roxb. ●

411662 Vitis embergeri Galet = Vitis tsoi Merr. ●

411663 Vitis erythrodes Fresen. = Rhoicissus tridentata (L. f.) Wild et R. B. Drumm. subsp. cuneifolia (Eckl. et Zeyh.) Urton ●☆

411664 Vitis erythrophylla W. T. Wang;红叶葡萄;Redleaf Grape, Red-leaved Grape ●

411665 Vitis esquirolii H. Lév. et Vaniot = Tetrastigma hemsleyanum Diels et Gilg ●■

411666 Vitis fagifolia Hu = Vitis hancockii Hance ●

411667 Vitis feddei H. Lév. = Parthenocissus feddei (H. Lév.) C. L. Li ●

411668 Vitis fengqinensis C. L. Li;凤庆葡萄;Fengqing Grape ●

411669 Vitis ficifolia Bunge;榕叶葡萄(桑叶葡萄,细本葡萄);Figleaf Grape,Fig-leaved Grape ●

411670 Vitis ficifolia Bunge = Vitis heyneana Roem. et Schult. subsp. ficifolia (Bunge) C. L. Li ●

411671 Vitis ficifolia Bunge f. glabrata (Nakai) W. T. Lee;光榕叶葡萄●

411672 Vitis ficifolia Bunge f. sinuata (Regel) Murata;深裂榕叶葡萄●☆

411673 Vitis ficifolia Bunge var. austrokoreana (Hatus.) Hatus. = Vitis ficifolia Bunge f. glabrata (Nakai) W. T. Lee ●

411674 Vitis ficifolia Bunge var. ganebu Hatus. = Vitis ficifolia Bunge ●

411675 Vitis ficifolia Bunge var. izuinsularis (Tuyama) H. Hara;伊豆岛榕叶葡萄●☆

411676 Vitis ficifolia Bunge var. lobata (Regel) Nakai = Vitis ficifolia Bunge ●

411677 Vitis ficifolia Bunge var. pentagona Pamp. = Vitis heyneana Roem. et Schult. ●

411678 Vitis ficifolia Bunge var. pentagona Pamp. = Vitis quinquangularis Rehder ●

411679 Vitis ficifolia Bunge var. sinuata (Regel) H. Hara = Vitis ficifolia Bunge f. sinuata (Regel) Murata ●☆

411680 Vitis ficifolia Bunge var. taiwaniana (F. Y. Lu) Yang ?;台湾小叶葡萄(小叶葡萄)●

411681 Vitis ficifolia Bunge var. thunbergii (Siebold et Zucc.) Nakai = Vitis ficifolia Bunge f. sinuata (Regel) Murata ●☆

411682 Vitis ficifolioides W. T. Wang;龙州葡萄;Fig-leaved-like Grape, Longzhou Grape ●

411683 Vitis flavicans Baker = Cyphostemma flavicans (Baker) Desc. ●☆

411684 Vitis flcifolioides W. T. Wang = Vitis balanseana Planch. var. ficifolioides (W. T. Wang) C. L. Li ●

411685 Vitis flcifolioides W. T. Wang = Vitis ficifolioides W. T. Wang ●

411686 Vitis flexuosa Thunb. ;葛藟葡萄(割舍镰藤,葛藟,光叶葡萄,葫芦藤,蓝,藥芜,蔓山葡萄,鸟娃子,千岁蓝,千岁藥,千岁木,山葡萄,菰藥,乌哇子,芜,小叶葛蓝,野葡萄,栽秧藤);Gelu Grape, Littleleaf Oriental Grape,Oriental Grape,Small-leaf Oriental Grape ●

411687 Vitis flexuosa Thunb. f. crassifolia ? = Vitis flexuosa Thunb. ●

411688 Vitis flexuosa Thunb. f. malayana Planch. = Vitis flexuosa Thunb. ●

411689 Vitis flexuosa Thunb. f. parvifolia (Roxb.) Planch. = Vitis flexuosa Thunb. ●

411690 Vitis flexuosa Thunb. f. parvifolia Planch. = Vitis flexuosa Thunb. ●

411691 Vitis flexuosa Thunb. f. typica Planch. = Vitis flexuosa Thunb. ●

411692 Vitis flexuosa Thunb. subsp. rufotomentosa (Makino) Murata = Vitis flexuosa Thunb. var. rufotomentosa Makino ●☆

411693 Vitis flexuosa Thunb. var. chinensis Veitch = Vitis flexuosa Thunb. ●

411694 Vitis flexuosa Thunb. var. gaudichaudii Planch. = Vitis balanseana

Planch. ●

411695　Vitis flexuosa Thunb. var. mairei H. Lév. = Vitis bryoniifolia Bunge ●

411696　Vitis flexuosa Thunb. var. malayana Planch.；大叶千岁藟（大叶岁藟）●

411697　Vitis flexuosa Thunb. var. parvifolia（Roxb.）Gagnep.；小叶葛藟（山葡萄）●

411698　Vitis flexuosa Thunb. var. parvifolia（Roxb.）Gagnep. = Vitis flexuosa Thunb. ●

411699　Vitis flexuosa Thunb. var. parvifolia（Roxb.）Planch. = Vitis parvifolia Roxb. ●

411700　Vitis flexuosa Thunb. var. rufotomentosa Makino；红毛葛藟葡萄●☆

411701　Vitis flexuosa Thunb. var. tsukubana Makino；筑波葛藟●☆

411702　Vitis flexuosa Thunb. var. wallichii（DC.）F. S. Wang = Vitis flexuosa Thunb. ●

411703　Vitis flexuosa Thunb. var. wilsonii Veitch = Vitis flexuosa Thunb. var. parvilolia（Roxb.）Gagnep. ●

411704　Vitis floribunda Baker = Cissus floribunda（Baker）Planch. ●☆

411705　Vitis foexeana Planch. = Vitis monticola Buckley ●☆

411706　Vitis formosana Hemsl. = Tetrastigma formosanum（Hemsl.）Gagnep. ●

411707　Vitis gentiliana H. Lév. et Vaniot = Ampelopsis delavayana（Franch.）Planch. var. gentiliana（H. Lév. et Vaniot）Hand. -Mazz. ●

411708　Vitis gentiliana H. Lév. et Vaniot. = Ampelopsis delavayana（Franch.）Planch. var. setulosa（Diels et Gilg）C. L. Li ●

411709　Vitis gilvotomentosa Makino et F. Maek. = Vitis flexuosa Thunb. var. rufotomentosa Makino ●☆

411710　Vitis girdiana Munson；谷地葡萄；Desert Grape, Southern California Grape, Valley Grape ●☆

411711　Vitis glabrata D. Don = Ampelocissus latifolia（Roxb.）Planch. ●

411712　Vitis glandulosa Wall. = Ampelopsis glandulosa（Wall.）Momiy. ●

411713　Vitis glandulosa Wall. = Ampelopsis heterophylla（Thunb.）Siebold et Zucc. var. vestita Rehder ●

411714　Vitis glossopetala Baker = Cissus glossopetala（Baker）Suess. ●☆

411715　Vitis goudoti（Planch.）Palacky = Cyphostemma microdiptera（Baker）Desc. ●☆

411716　Vitis grantii Baker = Ampelocissus africana（Lour.）Merr. ●☆

411717　Vitis grisea Baker = Cissus grisea（Baker）Planch. ●☆

411718　Vitis guerkeana Büttner = Cissus guerkeana（Büttner）T. Durand et Schinz ●☆

411719　Vitis hancockii Hance；菱叶葡萄（菱状叶葡萄，庐山葡萄，山毛榉叶葡萄）；Hancock Grape ●

411720　Vitis hederacea Ehrh. = Parthenocissus quinquefolius（L.）Planch. ●

411721　Vitis hekouensis C. L. Li；河口葡萄；Hekou Grape ●

411722　Vitis hekouensis C. L. Li = Vitis retordii Rom. Caill. ex Planch. ●

411723　Vitis helleri（L. H. Bailey）Small；北美圆叶葡萄（圆形叶葡萄）；Roundleaf Grape ●☆

411724　Vitis helleri Small = Vitis helleri（L. H. Bailey）Small ●☆

411725　Vitis henryana Hemsl. = Parthenocissus henryana（Hemsl.）Diels et Gilg ●

411726　Vitis heptaphylla L. = Schefflera heptaphylla（L.）Frodin ●

411727　Vitis heterophylla Thunb. = Ampelopsis glandulosa（Wall.）Momiy. var. heterophylla（Thunb.）Momiy. ●

411728　Vitis heterophylla Thunb. = Ampelopsis heterophylla（Thunb.）Siebold et Zucc. ●

411729　Vitis heterophylla Thunb. var. aconitifolia H. Lév. et Vaniot =

Ampelopsis aconitifolia Bunge ●

411730　Vitis heterophylla Thunb. var. humulifolia（Bunge）Hook. = Ampelopsis heterophylla（Thunb.）Siebold et Zucc. ●

411731　Vitis heterophylla Thunb. var. humulifolia Hook. f. = Ampelopsis heterophylla（Thunb.）Siebold et Zucc. ●

411732　Vitis heterophylla Thunb. var. maximowiczii Regel = Ampelopsis glandulosa（Wall.）Momiy. var. heterophylla（Thunb.）Momiy. ●

411733　Vitis heterophylla Thunb. var. maximowiczii Regel = Ampelopsis heterophylla（Thunb.）Siebold et Zucc. ●

411734　Vitis hexamera Gagnep. = Vitis betulifolia Diels et Gilg ●

411735　Vitis heyneana Roem. et Schult.；毛葡萄（大风藤，飞天白鹤，粉田子，蝴蝶艾，基隆葡萄，绿葡萄，毛叶葡萄，绵毛葡萄，木果藤，绒毛葡萄，五角葡萄，五角叶葡萄，橡根藤，野葡萄，野葡萄藤，止血藤）；Hairy Grape, Hairyleaf Grape, Heyne's Grape ●

411736　Vitis heyneana Roem. et Schult. subsp. ficifolia（Bunge）C. L. Li；桑叶葡萄（河南毛葡萄，毛葡萄，榕叶葡萄，野葡萄）；Mulberryleaf Grape ●

411737　Vitis himalayana（Royle）Brandis = Parthenocissus semicordata（Wall.）Planch. ●

411738　Vitis himalayana（Royle）Brandis var. semicordata（Wall. ex Roxb.）M. A. Lawson = Parthenocissus semicordata（Wall.）Planch. ●

411739　Vitis himalayana Brandis = Parthenocissus semicordata（Wall. ex Roxb.）Planch. var. roylei（King ex Parker）Nazim. et Qaiser ●☆

411740　Vitis himalayana Brandis = Parthenocissus semicordata（Wall.）Planch. ●

411741　Vitis hochstetteri Miq. = Cissus petiolata Hook. f. ●☆

411742　Vitis hui W. C. Cheng；庐山葡萄；Hu Grape, Lushan Grape ●

411743　Vitis humilis N. E. Br. = Cyphostemma humile（N. E. Br.）Desc. ex Wild et R. B. Drumm. ●☆

411744　Vitis humulifolia Hort. f. glabra ? = Ampelopsis heterophylla（Thunb.）Siebold et Zucc. ●

411745　Vitis hypoglauca（A. Gray）F. Muell.；粉背叶葡萄（灰背白粉藤）；Hypoglaucous Grape ●☆

411746　Vitis illex L. H. Bailey；海牛葡萄；Manatree Grape ●☆

411747　Vitis imerinensis Baker = Cayratia imerinensis（Baker）Desc. ●☆

411748　Vitis inconstans Miq. = Parthenocissus tricuspidatus（Siebold et Zucc.）Planch. ●

411749　Vitis indivisa Willd. = Ampelopsis cordata Michx. ●

411750　Vitis inserta A. Kern. = Parthenocissus vitacea（Knerr）Hitchc. ●☆

411751　Vitis integrifolia Baker = Cissus integrifolia（Baker）Planch. ●☆

411752　Vitis islandica Heard.；冰岛葡萄●☆

411753　Vitis jacquemontii Parker；雅克蒙葡萄●☆

411754　Vitis japonica Thunb. = Cayratia japonica（Thunb.）Gagnep. ●

411755　Vitis jatrophoides Baker = Cyphostemma junceum（Webb）Wild et R. B. Drumm. subsp. jatrophoides（Baker）Verdc. ●☆

411756　Vitis jinggangensis W. T. Wang；井岗葡萄（井冈葡萄）；Jinggang Grape ●

411757　Vitis jinzhainensis X. S. Shen；金寨山葡萄；Jinzhai Grape ●

411758　Vitis juncea（Webb）Baker = Cyphostemma junceum（Webb）Wild et R. B. Drumm. ●☆

411759　Vitis kaempferi K. Koch；克氏葡萄；Kaempfer Grape ●☆

411760　Vitis kelungensis Momiy.；基隆葡萄●

411761　Vitis kelungensis Momiy. = Vitis heyneana Roem. et Schult. ●

411762　Vitis kiusiana Momiy. = Vitis romanetii Rom. Caill. ●

411763　Vitis labordei H. Lév. et Vaniot = Tetrastigma hemsleyanum Diels et Gilg ●■

411764　Vitis labrusca L.；美洲葡萄（白肚，狐葡萄，酒葡萄，美国葡萄，鸟娃子,烟黑,野葡萄,蔓奥); Catawbarebe, Fox Grape, Fuchsrebe, Fuchstraube, Isabellarebe, Labruscan Vineyard Grape, Skunk Grape ●

411765 Vitislabrusca L. var. aestivalis (F. Michx.) Regel = Vitis aestivalis Michx. ●

411766 Vitis labrusca L. var. ficifolia (Bunge) Regel = Vitis heyneana Roem. et Schult. subsp. ficifolia (Bunge) C. L. Li ●

411767 Vitis labrusca L. var. ficifolia Regel = Vitis ficifolia Bunge ●

411768 Vitis labrusca L. var. ficifolia Regel = Vitis heyneana Roem. et Schult. subsp. ficifolia (Bunge) C. L. Li ●

411769 Vitis labrusca L. var. subedentata Fernald = Vitis labrusca L. ●

411770 Vitis labruscana L. H. Bailey = Vitis labrusca L. ●

411771 Vitis lanata Roxb. = Vitis heyneana Roem. et Schult. ●

411772 Vitis lanceolatifoliosa C. L. Li;鸡足葡萄(披针小叶葡萄,狭复叶葡萄); Lanceolate-foliolate Grape ●

411773 Vitis landuk Miq. = Parthenocissus dalzielii Gagnep. ●

411774 Vitis latifolia Buch. -Ham. ex Wall.;宽叶野葡萄(宽叶葡萄)●☆

411775 Vitis latifolia Roxb. = Ampelocissus latifolia (Roxb.) Planch. ●

411776 Vitis lecontiana House = Vitis aestivalis Michx. var. argentifolia (Munson) Fernald ●☆

411777 Vitis leeoides Maxim. = Ampelopsis cantoniensis (Hook. et Arn.) Planch. ●

411778 Vitis lenticellata Baker = Cissus floribunda (Baker) Planch. ●☆

411779 Vitis leucocarpa (Blume) Hayata = Cayratia japonica (Thunb.) Gagnep. ●

411780 Vitis lincecumii Buckley var. glauca Munson = Vitis aestivalis Michx. ●

411781 Vitis lincecumii Buckley var. lactea Small = Vitis aestivalis Michx. ●

411782 Vitis linsecomii Buckley;土耳其葡萄; Pine-wodd, Postoak Grape, Post-oak Grape, Turkey Grape ●☆

411783 Vitis linsecomii Buckley var. glauca Munson;光叶土耳其葡萄; Glaucous Turkey Grape ●☆

411784 Vitis longii W. R. Prince et Prince;野葡萄; Bush Grape ●☆

411785 Vitis longii W. R. Prince et Prince var. microsperma Bailey;小籽野葡萄;Small-seed Bush Grape ●☆

411786 Vitis longquanensis P. L. Chiu;龙泉葡萄;Longquan Grape ●

411787 Vitis luochengensis W. T. Wang;罗城葡萄;Luocheng Grape ●

411788 Vitis luochengensis W. T. Wang var. tomentoso-nerva C. L. Li;连山葡萄;Tomentose-nerve Grape ●

411789 Vitis lyjoannis H. Lév. = Ampelopsis bodinieri (H. Lév. et Vaniot) Rehder var. cinerea (Gagnep.) Rehder ●

411790 Vitis mannii Baker = Cyphostemma mannii (Baker) Desc. ●☆

411791 Vitis marchandii H. Lév. = Vitis wilsonae Veitch ●

411792 Vitis martini H. Lév. et Vaniot = Gynostemma pentaphyllum (Thunb.) Makino ■

411793 Vitis masukuensis Baker = Cyphostemma masukuense (Baker) Desc. ex Wild et R. B. Drumm. ●☆

411794 Vitis megalophylla H. Lév. = Ampelopsis chaffanjonii (H. Lév. et Vaniot) Rehder ●

411795 Vitis megalophylla Veitch = Ampelopsis megalophylla Diels et Gilg ●

411796 Vitis menghaiensis C. L. Li;勐海葡萄;Menghai Grape ●

411797 Vitis mengziensis C. L. Li;蒙自葡萄;Mengzi Grape ●

411798 Vitis micans (Rehder) Bean. = Ampelopsis bodinieri (H. Lév. et Vaniot) Rehder ●

411799 Vitis microdiptera Baker = Cyphostemma microdiptera (Baker) Desc. ●☆

411800 Vitis microdonta Baker = Cissus microdonta (Baker) Planch. ●☆

411801 Vitis mollis Wall. = Cayratia japonica (Thunb.) Gagnep. var. mollis (Wall.) C. L. Li ●

411802 Vitis mollis Wall. ex M. A. Lawson = Cayratia japonica (Thunb.) Gagnep. var. mollis (Wall. ex M. A. Lawson) Momiy. ●

411803 Vitis monosperma Michx. = Vitis rubra Michx. ●☆

411804 Vitis montana Buckley = Vitis berlandieri Planch. ●☆

411805 Vitis monticola Buckley;山地葡萄(甜山地葡萄); Mountain Grape, Sweet Mountain Grape ●☆

411806 Vitis monticola Mill. = Vitis berlandieri Planch. ●☆

411807 Vitis morifolia Baker = Cissus auricoma Desc. ●☆

411808 Vitis mossambicensis Klotzsch = Ampelocissus africana (Lour.) Merr. ●☆

411809 Vitis multijugata H. Lév. et Vaniot = Ampelopsis cantoniensis (Hook. et Arn.) Planch. ●

411810 Vitis multistriata Baker = Ampelocissus multistriata (Baker) Planch. ●☆

411811 Vitis munsoniana Simpson;乌葡萄; Everbearing Grape, Everlasting Grape, Little Muscadine Grape ●☆

411812 Vitis mustangensis Buckley = Vitis candicans Engelm. ●☆

411813 Vitis natalitia Szyszyl. = Cyphostemma natalitium (Szyszyl.) J. J. M. van der Merwe ●☆

411814 Vitis nortoni W. R. Prince = Vitis aestivalis Michx. ●

411815 Vitis novisinensis Vassilcz. = Vitis bryoniifolia Bunge ●

411816 Vitis nuevomexicana Lemmon = Vitis longii W. R. Prince et Prince ●☆

411817 Vitis nymphiifolia Welw. ex Baker = Cissus nymphiifolia (Welw. ex Baker) Planch. ●☆

411818 Vitis obovata M. A. Lawson = Tetrastigma obovatum (M. A. Lawson) Gagnep. ●

411819 Vitis obtecta Wall. = Tetrastigma obtectum (Wall.) Planch. ●■

411820 Vitis obtecta Wall. ex M. A. Lawson = Tetrastigma obtectum (Wall.) Planch. ●■

411821 Vitis obtecta Wall. ex M. A. Lawson var. potentilla f. pilosum (Planch.) H. Lév. = Tetrastigma obtectum (Wall.) Planch. var. pilosum Gagnep. ●■

411822 Vitis obtecta Wall. f. pilosa (Gagnep.) H. Lév. = Tetrastigma obtectum (Wall.) Planch. ●■

411823 Vitis obtusata Welw. ex Baker = Ampelocissus obtusata (Welw. ex Baker) Planch. ●☆

411824 Vitis odorata Hort. = Vitis riparia Michx. ●☆

411825 Vitis odoratissima Donn = Vitis riparia Michx. ●☆

411826 Vitis oligoarpa H. Lév. et Vaniot = Cayratia oligocarpa Gagnep. ●

411827 Vitis oligocarpa H. Lév. et Vaniot = Cayratia oligocarpa (H. Lév. et Vaniot) Gagnep. ●

411828 Vitis oxyodonta Baker = Cissus oxyodonta (Baker) Desc. ●☆

411829 Vitis oxyphylla A. Rich. = Cyphostemma oxyphyllum (A. Rich.) Vollesen ●☆

411830 Vitis pachyphylla Hemsl. = Tetrastigma pachyphyllum (Hemsl.) Chun ●

411831 Vitis pagnuccii Rom. Caill. = Vitis piasezkii Maxim. var. pagnuccii (Rom. Caill.) Rehder ●

411832 Vitis pagnuccii Rom. Caill. = Vitis piasezkii Maxim. ●

411833 Vitis pallida Wight et Arn. = Cissus repanda Vahl ●

411834 Vitis palmata Vahl;掌叶葡萄; Cat Grape, Catbird Grape, Missouri Grape, Red Grape ●☆

411835 Vitis palmatifida Baker = Cissus palmatifida (Baker) Planch. ●☆

411836　Vitis papillata Hance = Tetrastigma papillatum（Hance）C. Y. Wu ●

411837　Vitis parvifolia Roxb. = Vitis flexuosa Thunb. var. parvilolia（Roxb.）Gagnep. ●

411838　Vitis parvifolia Roxb. = Vitis flexuosa Thunb. ●

411839　Vitis pendula Welw. ex Baker = Cyphostemma pendulum（Welw. ex Baker）Desc. ●☆

411840　Vitis pentagona Diels et Gilg = Vitis heyneana Roem. et Schult. ●

411841　Vitis pentagona Diels et Gilg var. bellula Rehder = Vitis bellula（Rehder）W. T. Wang ●

411842　Vitis pentagona Diels et Gilg var. bellula Rehder = Vitis bellula（Rehder）W. T. Wang var. pubigera C. L. Li ●

411843　Vitis pentagona Diels et Gilg var. honanensis Rehder = Vitis ficifolia Bunge ●

411844　Vitis pentagona Diels et Gilg var. honanensis Rehder = Vitis heyneana Roem. et Schult. ●

411845　Vitis pentagona Diels et Gilg var. laotica Gagnep. = Vitis romanetii Rom. Caill. ●

411846　Vitis pentaphylla Guillaumin et Perr. = Ampelocissus multistriata（Baker）Planch. ●☆

411847　Vitis pentaphylla Thunb. = Gynostemma pentaphyllum（Thunb.）Makino ■

411848　Vitis persica Boiss. = Ampelopsis vitifolia（Boiss.）Planch. ●☆

411849　Vitis piasezkii Maxim.；变叶葡萄（刺葡萄，复叶葡萄，黑葡萄，麻羊藤，皮氏葡萄，甜茶，野葡萄）；Piasezky Grape ●

411850　Vitis piasezkii Maxim. var. angusta W. T. Wang = Vitis lanceolatifoliosa C. L. Li ●

411851　Vitis piasezkii Maxim. var. angustata W. T. Wang = Vitis lanceolatifoliosa C. L. Li ●

411852　Vitis piasezkii Maxim. var. baroniana Diels et Gilg = Vitis piasezkii Maxim. ●

411853　Vitis piasezkii Maxim. var. pagnuccii（Rom. Caill.）Rehder；少毛变叶葡萄（少毛复叶葡萄，少毛葡萄，无毛变叶葡萄）；Pagnucc Grape ●

411854　Vitis piasezkii Maxim. var. pagnuccii（Rom. Caill.）Rehder = Vitis piasezkii Maxim. ●

411855　Vitis pilosonerva F. P. Metcalf；毛脉葡萄；Pilose-nernes Grape ●

411856　Vitis planicaulis Hook. f. = Tetrastigma planicaule（Hook. f.）Gagnep. ●

411857　Vitis potaninii Kom. = Ampelopsis delavayana（Franch.）Planch. var. gentiliana（H. Lév. et Vaniot）Hand. -Mazz. ●

411858　Vitis potentilla H. Lév. et Vaniot = Tetrastigma obtectum（Wall.）Planch. ●■

411859　Vitis potentilla H. Lév. et Vaniot = Tetrastigma obtectum（Wall.）Planch. var. pilosum Gagnep. ●■

411860　Vitis potentilla H. Lév. et Vaniot var. glabra H. Lév. = Tetrastigma obtectum（Wall.）Planch. var. glabrum（H. Lév. et Vaniot）Gagnep. ●■

411861　Vitis prunisapida H. Lév. = Vitis davidii（Rom. Caill.）Foëx ●

411862　Vitis prunisapida H. Lév. et Vaniot = Vitis davidii（Rom. Caill.）Foëx ●

411863　Vitis pseudoreticulata W. W. Wang；华东葡萄；E. China Grape, Falsenetted Grape, False-netted Grape ●

411864　Vitis pteroclada（Hayata）Hayata = Cissus pteroclada Hayata ●

411865　Vitis pteroclada Hayata = Cissus hastata（Miq.）Planch. ●

411866　Vitis purani Buch. -Ham. ex D. Don = Vitis flexuosa Thunb. ●

411867　Vitis purani Don = Vitis flexuosa Thunb. ●

411868　Vitis quadrangularis（L.）Wall.；四棱葡萄（仙素莲）；

Cactuform Cissus, Winged Grape ●

411869　Vitis quadrangularis（L.）Wall. ex Wight et Arn. = Cissus quadrangularis L. ●

411870　Vitis quelpaertensis H. Lév. = Gynostemma pentaphyllum（Thunb.）Makino ■

411871　Vitis quinquangularis Rehder = Vitis heyneana Roem. et Schult. ●

411872　Vitis quinquangularis Rehder var. bellula（Rehder）Rehder = Vitis bellula（Rehder）W. T. Wang ●

411873　Vitis quinquefolia（L.）Lam. = Parthenocissus quinquefolius（L.）Planch. ●

411874　Vitis quinquefolia Lam. = Parthenocissus quinquefolius（L.）Planch. ●

411875　Vitis rapnanda Wight et Arn.；波叶葡萄 ●

411876　Vitis repanda（Vahl）Wight et Arn. = Cissus repanda Vahl ●

411877　Vitis repens（Lam.）Wight et Arn. = Cissus repens（Wight et Arn.）Lam. ●

411878　Vitis repens Veitch = Ampelopsis bodinieri（H. Lév. et Vaniot）Rehder ●

411879　Vitis repens Wight et Arn. = Cissus repens（Wight et Arn.）Lam. ●

411880　Vitis reticulata Gagnep. = Vitis wilsonae Veitch ●

411881　Vitis reticulata Pamp. = Vitis wilsonae Veitch ●

411882　Vitis retordii Rom. Caill. ex Planch.；绵毛葡萄（毛葡萄，绒毛葡萄）；Tomentose Grape ●

411883　Vitis retordii Rom. Caill. ex Planch. = Vitis hekouensis C. L. Li ●

411884　Vitis rhodotricha Baker = Cissus rhodotricha（Baker）Desc. ●☆

411885　Vitis rhombifolia Khakhlov；热美菱叶葡萄（菱叶白粉藤）；Grape Ivy, Grape Vine Ivy, Grape-vine Ivy, Natal Vine, Oak Leaf Ivy, Oak-leaf Ivy, Rhombicleaf Grape ●☆

411886　Vitis rigida H. Lév. et Vaniot = Ampelopsis delavayana（Franch.）Planch. ●

411887　Vitis riparia Michx.；河岸葡萄（狐色葡萄）；Chicken Grape, Forest Grape, Frost Grape, June Grape, River Bank Grape, River Grape, Riverbank Grape, Wild Grape, Winter Grape ●☆

411888　Vitis riparia Michx. var. praecox Engelm. ex L. H. Bailey = Vitis riparia Michx. ●☆

411889　Vitis riparia Michx. var. syrticola（Fernald et Wiegand）Fernald = Vitis riparia Michx. ●☆

411890　Vitis romanetii Rom. Caill.；秋葡萄（扁担藤，刺葡萄，黑葡萄，洛氏葡萄，山葡萄，腺葡萄，野葡萄，紫葡萄）；Autumn Grape, Romanet Grape ●

411891　Vitis romanetii Rom. Caill. var. arachnoidea Y. L. Cao et Y. H. He = Vitis romanetii Rom. Caill. ●

411892　Vitis romanetii Rom. Caill. var. tomentosa Y. L. Cao et Y. H. He；绒毛秋葡萄 ●

411893　Vitis rotundifolia Michx.；圆叶葡萄；Bull Grape, Bullace Grape, Bullet, Mascadine, Mascadine Grape, Muscadine Grape, Scuppernong, Southern Fox Grape ●☆

411894　Vitis rubifolia Wall. = Ampelopsis rubifolia（Wall.）Planch. ●

411895　Vitis rubiginosa Welw. ex Baker = Cissus rubiginosa（Welw. ex Baker）Planch. ●☆

411896　Vitis rubra Michx.；红葡萄；Cat Grape, Red Grape ●☆

411897　Vitis rubrifolia H. Lév. et Vaniot = Parthenocissus semicordata（Wall.）Planch. ●

411898　Vitis rufotomentosa Small = Vitis aestivalis Michx. ●

411899　Vitis rumicisperma M. A. Lawson = Tetrastigma rumicispermum（Lawson）Planch. ●

411900　Vitis rupestris E. Scheele; 沙地葡萄; Mountain Grape, Rock Grape, Sand Grape ●☆

411901　Vitis rutilans Carrière = Vitis romanetii Rom. Caill. ●

411902　Vitis ruyuanensis C. L. Li; 乳源葡萄; Ruyuan Grape ●

411903　Vitis saccharifera Makino ex Matsum. ; 糖葡萄 ●☆

411904　Vitis sarcocephala Schweinf. ex Oliv. = Ampelocissus sarcocephala (Schweinf. ex Oliv.) Planch. ●☆

411905　Vitis schimperiana Hochst. ex A. Rich. = Ampelocissus schimperiana (Hochst. ex A. Rich.) Planch. ●☆

411906　Vitis schuttleworthii House = Vitis coriacea Servett. ex K. Koch ●☆

411907　Vitis seguinii H. Lév. = Iodes balansae Gagnep. ●

411908　Vitis seguinii H. Lév. = Iodes seguinii (H. Lév.) Rehder ●

411909　Vitis semicordata Wall. = Parthenocissus semicordata (Wall.) Planch. ●

411910　Vitis semicordata Wall. ex Roxb. = Parthenocissus semicordata (Wall. ex Roxb.) Planch. ●

411911　Vitis semicordata Wall. var. himalayana (Royle) Kurz ex Hance = Parthenocissus semicordata (Wall.) Planch. ●

411912　Vitis semicordata Wall. var. roylei King. ex Parker = Parthenocissus semicordata (Wall. ex Roxb.) Planch. var. roylei (King ex Parker) Nazim. et Qaiser ●☆

411913　Vitis serianiifolia (Bunge) Maxim. = Ampelopsis japonica (Thunb.) Makino ●

411914　Vitis serjanifolia Franch. et Sav. = Ampelopsis japonica (Thunb.) Makino ●

411915　Vitis serjanifolia Koch = Ampelopsis japonica (Thunb.) Makino ●

411916　Vitis sessilifolia Baker; 无柄叶葡萄 ●☆

411917　Vitis shenxiensis C. L. Li; 陕西葡萄; Shaanxi Grape ●

411918　Vitis shifunensis Hayata; 三叶葡萄; Shifun Grape ●

411919　Vitis shimenensis W. T. Wang = Vitis betulifolia Diels et Gilg ●

411920　Vitis shiragai Makino = Vitis amurensis Rupr. ●

411921　Vitis sieboldii Hort. ex K. Koch = Vitis ficifolia Bunge ●

411922　Vitis sikkimensis M. A. Lawson = Ampelocissus sikkimensisi (Lawson) Planch. ●

411923　Vitis silvestrii Pamp. ; 湖北葡萄; Hubei Grape ●

411924　Vitis simpsonii Munson; 醋栗葡萄; Currant Grape ●☆

411925　Vitis simpsonii Munson = Vitis aestivalis Michx. ●

411926　Vitis sinica Miq. = Ampelopsis glandulosa (Wall.) Momiy. ●

411927　Vitis sinica Miq. = Ampelopsis heterophylla (Thunb.) Siebold et Zucc. var. vestita Rehder ●

411928　Vitis sinica Miq. = Ampelopsis sinica (Miq.) W. T. Wang ●

411929　Vitis sinocinerea W. T. Wang; 小叶葡萄; Small-leaf Grape, Small-leaved Grape ●

411930　Vitis smalliana L. H. Bailey; 大叶葡萄; Figleaf Grape ●☆

411931　Vitis smalliana L. H. Bailey = Vitis aestivalis Michx. ●

411932　Vitis smithiana Baker = Cissus smithiana (Baker) Planch. ●☆

411933　Vitis sola L. H. Bailey; 短穗葡萄; Curtiss Grape ●☆

411934　Vitis solonis Planch. = Vitis longii W. R. Prince et Prince ●☆

411935　Vitis solonis Planch. var. microsperma Munson = Vitis longii W. R. Prince et Prince ●☆

411936　Vitis stenoloba Welw. ex Baker = Cyphostemma stenolobum (Welw. ex Baker) Desc. ex Wild et R. B. Drumm. ●☆

411937　Vitis stipulacea Baker = Cyphostemma stipulaceum (Baker) Desc. ●☆

411938　Vitis striata Miq. = Cissus striata Ruiz et Pav. ●☆

411939　Vitis subciliata Baker = Cyphostemma subciliatum (Baker) Desc. ex Wild et R. B. Drumm. ●☆

411940　Vitis suberosa Baker = Cissus petiolata Hook. f. ●☆

411941　Vitis succulenta Galpin = Cissus cactiformis Gilg ●☆

411942　Vitis taquetii H. Lév. = Parthenocissus tricuspidata (Siebold et Zucc.) Planch. ●

411943　Vitis tenuifolia Wight et Arn. = Cayratia japonica (Thunb.) Gagnep. ●

411944　Vitis texana Munson = Vitis monticola Buckley ●☆

411945　Vitis thalictrifolia (Planch.) Palacky = Cayratia triternata (Baker) Desc. ●☆

411946　Vitis thomsonii M. A. Lawson = Yua thomsonii (M. A. Lawson) C. L. Li ●

411947　Vitis thonningii Baker = Cyphostemma cymosum (Schumach. et Thonn.) Desc. ●☆

411948　Vitis thunbergii (Siebold et Zucc.) Druce = Parthenocissus tricuspidatus (Siebold et Zucc.) Planch. ●

411949　Vitis thunbergii Siebold et Zucc. = Vitis adstricta Hance ●

411950　Vitis thunbergii Siebold et Zucc. = Vitis amurensis Rupr. ●

411951　Vitis thunbergii Siebold et Zucc. = Vitis bryoniifolia Bunge ●

411952　Vitis thunbergii Siebold et Zucc. = Vitis ficifolia Bunge ●

411953　Vitis thunbergii Siebold et Zucc. = Vitis heyneana Roem. et Schult. subsp. ficifolia (Bunge) C. L. Li ●

411954　Vitis thunbergii Siebold et Zucc. var. adsrricta (Hance) Gagnep. = Vitis bryoniifolia Bunge var. adstricta (Hance) W. T. Wang ●

411955　Vitis thunbergii Siebold et Zucc. var. adstricta (Hance) Gagnep. = Vitis sinocinerea W. T. Wang ●

411956　Vitis thunbergii Siebold et Zucc. var. adstricta (Hance) Gagnep. = Vitis bryoniifolia Bunge ●

411957　Vitis thunbergii Siebold et Zucc. var. cinerea Gagnep. = Vitis bryoniifolia Bunge var. adstricta (Hance) W. T. Wang ●

411958　Vitis thunbergii Siebold et Zucc. var. cinerea Gagnep. = Vitis sinocinerea W. T. Wang ●

411959　Vitis thunbergii Siebold et Zucc. var. izuinsularis Tuyama = Vitis ficifolia Bunge var. izuinsularis (Tuyama) H. Hara ●☆

411960　Vitis thunbergii Siebold et Zucc. var. mairei (H. Lév.) Lauener = Vitis bryoniifolia Bunge ●

411961　Vitis thunbergii Siebold et Zucc. var. sinuata (Regel) Rehder; 小叶蘡薁 ●☆

411962　Vitis thunbergii Siebold et Zucc. var. taiwaniana F. Y. Lu = Vitis sinocinerea W. T. Wang ●

411963　Vitis thunbergii Siebold et Zucc. var. yunnanensis Planch. = Vitis heyneana Roem. et Schult. ●

411964　Vitis thunbergii Siebold et Zucc. var. yunnanensis Planch. ex Franch. = Vitis heyneana Roem. et Schult. ●

411965　Vitis thyrsiflora Miq. ; 聚伞葡萄 ●☆

411966　Vitis tiubaensis X. L. Niu = Vitis piasezkii Maxim. ●

411967　Vitis treleasei Munson; 山谷葡萄; Gulch Grape ●☆

411968　Vitis trichoclada Diels et Gilg = Vitis betulifolia Diels et Gilg ●

411969　Vitis trifolia L. = Cayratia trifolia (L.) Domin ●

411970　Vitis trifolia L. = Cissus carnosa (L.) Lam. ●

411971　Vitis triphylla Hayata = Vitis shifunensis Hayata ●

411972　Vitis triternata Baker = Cayratia triternata (Baker) Desc. ●☆

411973　Vitis tsoi Merr. ; 狭叶葡萄; Narrowleaf Grape, Narrow-leaved Grape ●

411974　Vitis umbellata Hemsl. = Tetrastigma obtectum (Wall.) Planch. var. glabrum (H. Lév. et Vaniot) Gagnep. ●■

411975　Vitis umbellata Hemsl. var. arisanensis Hayata = Tetrastigma obtectum (Wall.) Planch. var. glabrum (H. Lév. et Vaniot) Gagnep. ●■

411976　Vitis unifoliata （Harv.） Kuntze ＝ Rhoicissus microphylla （Turcz.） Gilg et M. Brandt ●☆

411977　Vitis variifolia Baker ＝ Cyphostemma crotalarioides （Planch.） Desc. ex Wild et R. B. Drumm. ●☆

411978　Vitis verrucosa （Zoll.） Backer ＝ Vitis rotundifolia Michx. ●☆

411979　Vitis vinifera L.；葡萄（草龙珠，赐紫樱桃，马乳葡萄，欧洲葡萄，菩提子，蒲陶，山葫芦，水晶葡萄，索索葡萄，野葡萄，紫葡萄）；Brandy See，Chicken Grape，Common Grape，Common Vine，Currant，Europe Grape，European Grape，Frost Grape，Gouty Vine，Grape，Grape Vine，Grapevine，Muscatel，Raisin，Vine，Vine Grape，Wine Grape，Wine See ●

411980　Vitis vinifera L. 'Incana'；灰毛葡萄；Dusty-miller Grape ●☆

411981　Vitis vinifera L. 'Purpurea'；紫叶葡萄；Claret Vine，Purpleleaf European Grape，Teinturier，Teinturier Grape ●☆

411982　Vitis vinifera L. 'Variegata'；斑叶葡萄；Variegated Grape ●☆

411983　Vitis vinifera L. subsp. sylvestris （C. C. Gmel.） Hegi；森林葡萄；Forest European Grape，Woodland European Grape ●☆

411984　Vitis vinifera L. var. amurensis （Rupr.） Regel ＝ Vitis amurensis Rupr. ●

411985　Vitis vinifera L. var. apiifolia Loudon；细裂叶葡萄；Cearyleaf European Grape ●☆

411986　Vitis vinifera L. var. purpurea Bean ＝ Vitis vinifera L. 'Purpurea' ●☆

411987　Vitis vinifera L. var. sativa DC.；栽培葡萄；Caltivated European Grape ●☆

411988　Vitis vinifera L. var. sylvestis Willd. ＝ Vitis vinifera L. subsp. sylvestris （C. C. Gmel.） Hegi ●☆

411989　Vitis vitacea L. ＝ Parthenocissus insertus （A. Kern.） Fritsch ●☆

411990　Vitis voanonala Baker ＝ Cayratia triternata （Baker） Desc. ●☆

411991　Vitis vulpina L.；狐色葡萄；Frost Grape ●☆

411992　Vitis vulpina L. ＝ Vitis riparia Michx. ●☆

411993　Vitis vulpina L. subsp. riparia （Michx.） R. T. Clausen ＝ Vitis riparia Michx. ●☆

411994　Vitis vulpina L. var. amurensis Regel ＝ Vitis amurensis Rupr. ●

411995　Vitis vulpina L. var. parvifolia （Roxb.） Regel ＝ Vitis flexuosa Thunb. ●

411996　Vitis vulpina L. var. praecox （Engelm. ex L. H. Bailey） L. H. Bailey；六月山葡萄；June Grape ●☆

411997　Vitis vulpina L. var. praecox （Engelm. ex L. H. Bailey） L. H. Bailey ＝ Vitis riparia Michx. ●☆

411998　Vitis vulpina L. var. praecox L. H. Bailey ＝ Vitis vulpina L. var. praecox （Engelm. ex L. H. Bailey） L. H. Bailey ●☆

411999　Vitis vulpina L. var. syrticola Fernald et Wiegand ＝ Vitis riparia Michx. ●☆

412000　Vitis wallichii DC. ＝ Vitis flexuosa Thunb. ●

412001　Vitis welwitschii Baker ＝ Cissus welwitschii （Baker） Planch. ●☆

412002　Vitis wenchowensis C. Ling ex W. T. Wang；温州葡萄；Wenzhou Grape ●

412003　Vitis wentsaiana P. L. Chiu ＝ Vitis hancockii Hance ●

412004　Vitis wilsonae Veitch；网脉葡萄（川鄂葡萄，大叶山天萝，鸟葡萄，威氏葡萄，魏氏葡萄，野葡萄）；E. H. Wilson Grape，Netvein Grape，Wilson Grape ●

412005　Vitis wuhanensis C. L. Li；武汉葡萄；Wuhan Grape ●

412006　Vitis wuhanensis C. L. Li var. arachnoidea X. D. Wang et C. L. Li；毛叶武汉葡萄；Hairy-leaf Wuhan Grape ●

412007　Vitis wuhanensis C. L. Li var. arachnoidea X. D. Wang et C. L. Li ＝ Vitis wuhanensis C. L. Li ●

412008　Vitis yuenlingensis W. T. Wang；源陵葡萄；Yuanling Grape ●

412009　Vitis yunnanensis C. L. Li；云南葡萄；Yunnan Grape ●

412010　Vitis zhejiang-adstricta P. L. Chiu；浙江葡萄（浙江蘡薁）；Zhejiang Grape ●

412011　Vitis zombensis Baker ＝ Cyphostemma zombense （Baker） Desc. ex Wild et R. B. Drumm. ●☆

412012　Vitis-Idaea Ség. ＝ Vaccinium L. ●

412013　Vitis-idaea Tourn. ex Moench ＝ Vaccinium L. ●

412014　Vitmania Turr. ex Cav. ＝ Mirabilis L. ■

412015　Vitmannia Endl. ＝ Mirabilis L. ■

412016　Vitmannia Torr. ＝ Mirabilis L. ■

412017　Vitmannia Torr. ex Cav. ＝ Mirabilis L. ■

412018　Vitmannia Vahl ＝ Quassia L. ●☆

412019　Vitmannia Vahl ＝ Samadera Gaertn. （保留属名）●☆

412020　Vitmannia Wight et Arn. ＝ Noltea Rchb. ●☆

412021　Vittadenia Steud. ＝ Vittadinia A. Rich. ■☆

412022　Vittadinia A. Rich. （1832）；簇毛层菀属（维塔丁尼亚属，维太菊属）■☆

412023　Vittadinia australis A. Rich.；南方簇毛层菀（澳大利亚维太菊，南方维太菊）■☆

412024　Vittadinia triloba Hort. ＝ Erigeron mucronatus DC. ■☆

412025　Vittetia R. M. King et H. Rob. （1974）；点腺柄泽兰属●☆

412026　Vittetia orbiculata （DC.） R. M. King et H. Rob.；点腺柄泽兰●☆

412027　Vittmannia Endl. ＝ Mirabilis L. ■

412028　Vittmannia Endl. ＝ Vitmannia Turr. ex Cav. ■

412029　Viviana Cav. ＝ Viviania Cav. ■☆

412030　Viviana Colla ＝ Melanopsidium Colla ●☆

412031　Viviana Merr. ＝ Guettarda L. ●

412032　Viviana Merr. ＝ Viviania Raf. ●

412033　Viviana Raf. ＝ Guettarda L. ●

412034　Vivianaceae Klotzsch ＝ Geraniaceae Juss. （保留科名）●■

412035　Vivianaceae Klotzsch ＝ Vivianiaceae Klotzsch ■☆

412036　Viviania Cav. （1804）；青蛇胚属（曲胚属，韦韦苗属）■☆

412037　Viviania Colla ＝ Billiottia DC. ●☆

412038　Viviania Colla ＝ Melanopsidium Colla ●☆

412039　Viviania Raf. ＝ Guettarda L. ●

412040　Viviania Raf. ex DC. ＝ Guettarda L. ●

412041　Viviania Willd. ex Less. ＝ Liabum Adans. ■●☆

412042　Viviania marifolia Cav.；青蛇胚■☆

412043　Vivianiaceae Klotzsch ＝ Geraniaceae Juss. （保留科名）■●

412044　Vivianiaceae Klotzsch；青蛇胚科（曲胚科，韦韦苗科）■☆

412045　Vladimirea Iljin ＝ Dolomiaea DC. ■

412046　Vladimiria Iljin ＝ Dolomiaea DC. ■

412047　Vladimiria berardioides （Franch.） Y. Ling ＝ Dolomiaea berardioides （Franch.） C. Shih ■

412048　Vladimiria calophylla ？ ＝ Dolomiaea calophylla Y. Ling ■

412049　Vladimiria crispo-undulata （C. C. Chang） C. Shih et S. Y. Jin ＝ Dolomiaea crispo-undulata （C. C. Chang） Y. Ling ■

412050　Vladimiria denticulata Y. Ling ＝ Dolomiaea denticulata （Y. Ling） C. Shih ■

412051　Vladimiria edulis （Franch.） Y. Ling ＝ Dolomiaea edulis （Franch.） C. Shih ■

412052　Vladimiria edulis （Franch.） Y. Ling f. bracteata Y. Ling ＝ Dolomiaea edulis （Franch.） C. Shih ■

412053　Vladimiria edulis （Franch.） Y. Ling f. caulescens Y. Ling ＝ Dolomiaea edulis （Franch.） C. Shih ■

412054　Vladimiria forrestii （Diels） Y. Ling ＝ Dolomiaea forrestii

（Diels）C. Shih ■

412055 Vladimiria georgii（Anthony）Y. Ling = Dolomiaea georgii（J. Anthony）C. Shih ■

412056 Vladimiria muliensis（Hand. -Mazz.）Y. Ling = Dolomiaea souliei（Franch.）C. Shih var. mirabilis（J. Anthony）C. Shih ■

412057 Vladimiria platylepis（Hand. -Mazz.）Y. Ling = Dolomiaea platylepis（Hand. -Mazz.）C. Shih ■

412058 Vladimiria salwinensis（Hand. -Mazz.）Iljin = Dolomiaea salwinensis（Hand. -Mazz.）C. Shih ■

412059 Vladimiria scabrida C. Shih et S. Y. Jin = Dolomiaea scabrida（C. Shih et S. Y. Jin）C. Shih ■

412060 Vladimiria souliei（Franch.）Y. Ling = Dolomiaea souliei（Franch.）C. Shih ■

412061 Vladimiria souliei（Franch.）Y. Ling var. cinerea Y. Ling = Dolomiaea souliei（Franch.）C. Shih var. mirabilis（J. Anthony）C. Shih ■

412062 Vladimiria taraxacifolia（Anthony）Y. Ling = Dolomiaea souliei（Franch.）C. Shih var. mirabilis（J. Anthony）C. Shih ■

412063 Vladimiria trachyloma（Hand. -Mazz.）Y. Ling = Dolomiaea souliei（Franch.）C. Shih var. mirabilis（J. Anthony）C. Shih ■

412064 Vlamingia Buse ex de Vriese = Hybanthus Jacq.（保留属名）●■

412065 Vlamingia Buse ex de Vriese = Ionidium Vent. ●■

412066 Vlamingia de Vriese = Hybanthus Jacq.（保留属名）●■

412067 Vlamingia de Vriese = Ionidium Vent. ●■

412068 Vlechia Raf. = Vleckia Raf. ■

412069 Vleckia Raf. = Agastache J. Clayton ex Gronov. ■

412070 Vleckia Raf. = Lophanthus Adans. ■●

412071 Vleisia Toml. et Posl.（1976）；肖加利亚草属■☆

412072 Vleisia Toml. et Posl. = Pseudalthenia（Graebn.）Nakai ■☆

412073 Vleisia aschersoniana（Graebn.）Toml. et Posl. = Pseudalthenia aschersoniana（Graebn.）Hartog ■☆

412074 Vlokia S. A. Hammer（1994）；好望角番杏属☆

412075 Vlokia ater S. A. Hammer；好望角番杏☆

412076 Vlokia montana Klak；山地好望角番杏☆

412077 Vlrgaria Raf. ex DC. = Aster L. ●■

412078 Voacanga Thouars（1806）；马铃果属（伏康树属，老刺木属）●

412079 Voacanga africana Stapf；非洲马铃果（非洲伏康树，非洲沃坎加树，伏康树，沃坎加树）●

412080 Voacanga africana Stapf = Voacanga africana Stapf ex Scott-Elliot ●

412081 Voacanga africana Stapf ex Scott-Elliot = Voacanga africana Stapf ●

412082 Voacanga africana Stapf var. auriculata Pichon = Voacanga africana Stapf ●

412083 Voacanga africana Stapf var. glabra（K. Schum.）Pichon = Voacanga africana Stapf ●

412084 Voacanga africana Stapf var. lutescens（Stapf）Pichon = Voacanga africana Stapf ●

412085 Voacanga africana Stapf var. typica Pichon = Voacanga africana Stapf ●

412086 Voacanga angolensis Stapf ex Hiern = Voacanga africana Stapf ●

412087 Voacanga angustifolia K. Schum. = Voacanga africana Stapf ●

412088 Voacanga bequaertii De Wild. = Voacanga africana Stapf ●

412089 Voacanga boehmii K. Schum. = Voacanga africana Stapf ●

412090 Voacanga bracteata Stapf；苞片马铃果●☆

412091 Voacanga bracteata Stapf var. lanceolata？ = Voacanga bracteata Stapf ●☆

412092 Voacanga bracteata Stapf var. zenkeri（Stapf）H. Huber = Voacanga psilocalyx Pierre ex Stapf ●☆

412093 Voacanga caudiflora Stapf；干花马铃果●☆

412094 Voacanga chalotiana Pierre ex Stapf；马铃果●

412095 Voacanga densiflora K. Schum. ex Engl. ；密花马铃果●

412096 Voacanga dichotoma K. Schum. = Tabernaemontana pachysiphon Stapf ●☆

412097 Voacanga diplochlamys K. Schum. = Voacanga bracteata Stapf ●☆

412098 Voacanga dregei E. Mey. = Voacanga thouarsii Roem. et Schult. ●☆

412099 Voacanga eketensis Wernham = Voacanga africana Stapf ●

412100 Voacanga glaberrima Wernham = Voacanga africana Stapf ●

412101 Voacanga glabra K. Schum. = Voacanga africana Stapf ●

412102 Voacanga globosa Merr. ；球状伏康树（沃坎加树）●☆

412103 Voacanga klainei Pierre ex Stapf = Voacanga africana Stapf ●

412104 Voacanga lemosii Philipson = Voacanga africana Stapf ●

412105 Voacanga lutescens Stapf = Voacanga africana Stapf ●

412106 Voacanga magnifolia Wernham = Voacanga africana Stapf ●

412107 Voacanga micrantha Pichon = Voacanga bracteata Stapf ●☆

412108 Voacanga obanensis Wernham = Voacanga bracteata Stapf ●☆

412109 Voacanga obtusa K. Schum. = Voacanga thouarsii Roem. et Schult. ●☆

412110 Voacanga obtusata K. Schum. ex De Wild. et T. Durand = Voacanga thouarsii Roem. et Schult. ●☆

412111 Voacanga pachyceras Leeuwenb. ；粗角马铃果●☆

412112 Voacanga psilocalyx Pierre ex Stapf；光萼马铃果●☆

412113 Voacanga puberula K. Schum. = Voacanga africana Stapf ●

412114 Voacanga schweinfurthii Stapf = Voacanga africana Stapf ●

412115 Voacanga schweinfurthii Stapf var. puberula（K. Schum.）Pichon = Voacanga africana Stapf ●

412116 Voacanga spectabilis Stapf = Voacanga africana Stapf ●

412117 Voacanga talbotii Wernham = Voacanga bracteata Stapf ●☆

412118 Voacanga thouarsii Roem. et Schult. ；图氏马铃果●☆

412119 Voacanga thouarsii Roem. et Schult. var. dregei（E. Mey.）Pichon = Voacanga thouarsii Roem. et Schult. ●☆

412120 Voacanga thouarsii Roem. et Schult. var. obtusa（K. Schum.）Pichon = Voacanga thouarsii Roem. et Schult. ●☆

412121 Voacanga zenkeri Stapf = Voacanga psilocalyx Pierre ex Stapf ●☆

412122 Voandzeia Thouars（废弃属名）= Vigna Savi（保留属名）■

412123 Voandzeia poissonii A. Chev. = Macrotyloma geocarpum（Harms）Maréchal et Baudet ■☆

412124 Voandzeia subterranea（L.）DC. = Vigna subterranea（L.）Verdc. ■☆

412125 Voandzeia subterranea（L.）DC. f. spontanea Harms = Vigna subterranea（L.）Verdc. var. spontanea（Harms）Pasquet ■☆

412126 Voandzeia subterranea（L.）DC. var. spontanea（Harms）Hepper = Vigna subterranea（L.）Verdc. var. spontanea（Harms）Pasquet ■☆

412127 Voanioala J. Dransf.（1989）；多体椰属（森林椰子属）●☆

412128 Voanioala gerardii J. Dransf. ；多体椰●☆

412129 Voatamalo Capuron ex Bosser（1976）；沃大戟属☆

412130 Voatamalo capuronii Bosser；马岛沃大戟☆

412131 Voatamalo eugenioides Capuron ex Bosser；沃大戟☆

412132 Vochisia Juss. = Vochysia Aubl.（保留属名）●☆

412133 Vochy Aubl. = Vochysia Aubl.（保留属名）●☆

412134 Vochya Vell. ex Vand. = Vochysia Aubl.（保留属名）●☆

412135 Vochysia Aubl.（1775）（保留属名）独蕊属（囊萼花属）●☆

412136 Vochysia Poir. = Vochysia Aubl.（保留属名）●☆

412137 Vochysia acuminata Bong. ；渐尖独蕊●☆

412138 Vochysia allenii Standl. et L. O. Williams；阿伦独蕊●☆

412139 Vochysia alpestris Mart. ；高山独蕊●☆

412140 Vochysia alternifolia Glaz. ；互叶独蕊●☆

412141　Vochysia angustifolia Ducke;窄叶独蕊●☆

412142　Vochysia aurea Stafleu;黄独蕊●☆

412143　Vochysia crassifolia Warm. ;厚叶独蕊●☆

412144　Vochysia dasyantha Warm. ;毛花独蕊●☆

412145　Vochysia densiflora Spruce ex Warm. ;密花独蕊●☆

412146　Vochysia elegans Stafleu;雅致独蕊●☆

412147　Vochysia elliptica Mart. ;椭圆独蕊●☆

412148　Vochysia ferruginea Standl. ;锈色独蕊●☆

412149　Vochysia floribunda Mart. ;多花独蕊●☆

412150　Vochysia guianensis Aubl. ;圭亚那独蕊●☆

412151　Vochysia herbacea Pohl;草本独蕊■☆

412152　Vochysia lanceolata Stafleu;披针叶独蕊●☆

412153　Vochysia laurifolia Warm. ;桂叶独蕊●☆

412154　Vochysia macrophylla Stafleu;大叶独蕊●☆

412155　Vochysia maliformis Klotzsch ex Warm. ;苹果独蕊●☆

412156　Vochysia megalantha Stafleu;大花独蕊●☆

412157　Vochysia oblongifolia Warm. ;矩圆叶独蕊●☆

412158　Vochysia obovata Stafleu;倒卵独蕊●☆

412159　Vochysia pachyantha Ducke;厚花独蕊●☆

412160　Vochysia parviflora Villada;小花独蕊●☆

412161　Vochysia stenophylla Briq. ;巴西窄叶独蕊●☆

412162　Vochysia tetraphylla DC. ;四叶独蕊●☆

412163　Vochysia tomentosa DC. ;毛独蕊●☆

412164　Vochysiaceae A. St. -Hil. (1820)(保留科名);独蕊科(蜡烛树科,囊萼花科)●■☆

412165　Voelckeria Klotzsch et H. Karst. = Ternstroemia Mutis ex L. f. (保留属名)●

412166　Voelckeria Klotzsch et H. Karst. ex Endl. = Ternstroemia Mutis ex L. f. (保留属名)●

412167　Vogelia J. F. Gmel. = Burmannia L. ■

412168　Vogelia Lam. = Dyerophytum Kuntze ●☆

412169　Vogelia Medik. = Neslia Desv. (保留属名)■☆

412170　Vogelia africana Lam. = Dyerophytum africanum (Lam.) Kuntze ■☆

412171　Vogelia apiculata (Fisch. , Mey. et Avé-Lall.) Vierh. = Neslia apiculata Fisch. ,C. A. Mey. et Avé-Lall. ■☆

412172　Vogelia capitata Walter ex J. F. Gmel. = Burmannia capitata (Walter ex J. F. Gmel.) Mart. ■☆

412173　Vogelia paniculata (L.) Hornem. = Neslia paniculata (L.) Desv. ■

412174　Vogelocassia Bntton = Cassia L. (保留属名)●■

412175　Vogelocassia Britton = Senna Mill. ●■

412176　Voglera P. Gaertn. ,B. Mey. et Scherb. = Genista L. ●

412177　Voharanga Costantin et Bois = Cynanchum L. ●■

412178　Voharanga madagascariensis Const. et Bois = Cynanchum arenarium Jum. et H. Perrier ●☆

412179　Vohemaria Buchenau = Cynanchum L. ●■

412180　Vohemaria Buchenau(1889);武海马尔萝藦属■☆

412181　Vohemaria implicata (Jum. et H. Perrier) Jum. et H. Perrier = Cynanchum implicatum (Jum. et H. Perrier) Jum. et H. Perrier ■☆

412182　Vohemaria messeri Buchenau;武海马尔萝藦■☆

412183　Vohemaria messeri Buchenau = Cynanchum messeri (Buchenau) Jum. et H. Perrier ●☆

412184　Vohiria Juss. = Voyria Aubl. ■☆

412185　Voigtia Klotzsch = Bathysa C. Presl ■☆

412186　Voigtia Roth = Andryala L. ■☆

412187　Voigtia Spreng. = Barnadesia Mutis ex L. f. ●☆

412188　Voigtia Spreng. = Turpinia Vent. (保留属名)●

412189　Voladeria Benoist = Oreobolus R. Br. ■☆

412190　Volataceae Duhc = Aceraceae Juss. (保留科名)●

412191　Volcameria Heist. ex Fabr. = Cedronella Moench ●☆

412192　Volckameria Fabr. = Cedronella Moench ●☆

412193　Volhensiophyton Lindau = Lepidagathis Willd. ●■

412194　Volkamera Post et Kuntze = Capparis L. ●

412195　Volkamera Post et Kuntze = Volkameria Burm. f. ●

412196　Volkameria Burm. f. = Capparis L. ●

412197　Volkameria L. = Clerodendrum L. ●■

412198　Volkameria P. Browne = Gilibertia J. F. Gmel. ●

412199　Volkameria P. Browne = Gillena Adans. ●

412200　Volkameria acerbiana Vis. = Clerodendrum acerbianum (Vis.) Benth. ●☆

412201　Volkameria aculeata L. = Clerodendrum aculeatum (L.) Griseb. ●☆

412202　Volkameria alata (Thonn.) Kuntze = Sesamum alatum Thonn. ■☆

412203　Volkameria angulata Lour. = Clerodendrum paniculatum L. ●

412204　Volkameria antirrhinoides (Welw. ex Asch.) Kuntze = Sesamum schinzianum Asch. ■☆

412205　Volkameria bicolor Roxb. ex Hardw. = Caryopteris bicolor (Roxb. ex Hardw.) Mabb. ●

412206　Volkameria capitata Willd. = Clerodendrum capitatum (Willd.) Schumach. ●☆

412207　Volkameria cordifolia Hochst. = Clerodendrum umbellatum Poir. ●☆

412208　Volkameria fragrans Vent. = Clerodendrum chinense (Osbeck) Mabb. ●

412209　Volkameria fragrans Vent. = Clerodendrum philippinum Schauer ●

412210　Volkameria inermis L. = Clerodendrum inerme (L.) Gaertn. ●

412211　Volkameria japonica Thunb. = Clerodendrum chinense (Osbeck) Mabb. ●

412212　Volkameria japonica Thunb. = Clerodendrum japonicum (Thunb.) Sweet ●

412213　Volkameria kaemoferi Jacq. = Clerodendrum japonicum (Thunb.) Sweet ●

412214　Volkameria nereifolia Roxb. = Clerodendrum inerme (L.) Gaertn. ●

412215　Volkameria odorata Buch. -Ham. ex Roxb. = Caryopteris bicolor (Roxb. ex Hardw.) Mabb. ●

412216　Volkameria odorata Buch. -Ham. ex Roxb. = Caryopteris odorata (D. Don) B. L. Rob. ●

412217　Volkameria odorata D. Don = Caryopteris odorata (D. Don) B. L. Rob. ●

412218　Volkameria odorata Ham. ex Roxb. = Caryopteris odorata (Ham. ex Roxb.) Rob. ●

412219　Volkameria orientalis (L.) Kuntze = Sesamum indicum L. ■

412220　Volkameria pumila Lour. = Clerodendrum fortunatum L. ●

412221　Volkameria serrata L. = Clerodendrum serratum (L.) Moon ●

412222　Volkensia O. Hoffm. = Bothriocline Oliv. ex Benth. ■☆

412223　Volkensia argentea O. Hoffm. = Bothriocline argentea (O. Hoffm.) Wild et G. V. Pope ■☆

412224　Volkensia duemmeri (S. Moore) B. L. Burtt = Bothriocline bagshawei (S. Moore) C. Jeffrey ■☆

412225　Volkensia elliotii Muschl. = Bothriocline ruwenzoriensis (S. Moore) C. Jeffrey ■☆

412226　Volkensia glomerata O. Hoffm. et Muschl. = Bothriocline glomerata (O. Hoffm. et Muschl.) C. Jeffrey ■☆

412227　Volkensia moramballae (Oliv. et Hiern) B. L. Burtt = Bothriocline moramballae (Oliv. et Hiern) O. Hoffm. ■☆

412228　Volkensia syneilema Wech. = Bothriocline glomerata（O. Hoffm. et Muschl.）C. Jeffrey ■☆

412229　Volkensiella H. Wolff = Oenanthe L. ■

412230　Volkensiella procumbens H. Wolff = Oenanthe procumbens（H. Wolff）C. Norman ■☆

412231　Volkensinia Schinz（1912）；长柄苋属■☆

412232　Volkensinia grandiflora Suess. = Volkensinia prostrata（Volkens ex Gilg）Schinz ■☆

412233　Volkensinia prostrata（Volkens ex Gilg）Schinz；长柄苋■☆

412234　Volkensinia prostrata（Volkens ex Gilg）Schinz f. lanceolata Suess. = Volkensinia prostrata（Volkens ex Gilg）Schinz ■☆

412235　Volkensiophyton Lindau = Lepidagathis Willd. ●■

412236　Volkensiophyton neuracanthoides Lindau = Lepidagathis scariosa Nees ■☆

412237　Volkensteinia Tiegh. = Ouratea Aubl.（保留属名）●

412238　Volkensteinia Tiegh. = Wolkensteinia Regel ●

412239　Volkiella Merxm. et Czech（1953）；沃尔克莎属■☆

412240　Volkiella disticha Merxm. et Czech；沃尔克莎■☆

412241　Volkmannia Jacq. = Clerodendrum L. ●■

412242　Volubilis Catesby = Vanilla Plum. ex Mill. ■

412243　Volucrepis Thouars = Epidendrum L.（保留属名）■☆

412244　Volucrepis Thouars = Oeonia Lindl.（保留属名）■☆

412245　Voluterella Cass. = Amberboi Adans.（废弃属名）■

412246　Voluterella Cass. = Volutaria Cass. ■☆

412247　Voluterella crupinoides（Desf.）Ball = Volutaria crupinoides（Desf.）Cass. ex Maire ■☆

412248　Voluterella leucantha（Coss.）Ball = Volutaria sinaica（DC.）Wagenitz ■☆

412249　Voluterella lippii（L.）Cass. = Volutaria lippii（L.）Cass. ■☆

412250　Voluterella muricata（L.）Benth. et Hook. f. = Volutaria muricata（L.）Maire ■☆

412251　Voluterella omphalodes Benth. et Hook. f. = Stephanochilus omphalodes（Benth. et Hook. f.）Maire ■☆

412252　Volutaria Cass.（1816）；旋瓣菊属■☆

412253　Volutaria Cass. = Amberboa（Pers.）Less. ■

412254　Volutaria Cass. = Amberboi Adans.（废弃属名）■

412255　Volutaria abyssinica（A. Rich.）C. Jeffrey；阿比西尼亚旋瓣菊■☆

412256　Volutaria abyssinica（A. Rich.）C. Jeffrey subsp. aylmeri（Baker）Wagenitz；艾梅旋瓣菊■☆

412257　Volutaria abyssinica（A. Rich.）C. Jeffrey subsp. inornata Wagenitz；无饰阿比西尼亚旋瓣菊■☆

412258　Volutaria bollei（Bolle）A. Hansen et Sunding；博勒旋瓣菊■☆

412259　Volutaria boranensis（Cufod.）Wagenitz；加那利旋瓣菊■☆

412260　Volutaria canariensis Wagenitz = Volutaria boranensis（Cufod.）Wagenitz ■☆

412261　Volutaria crupinoides（Desf.）Cass. ex Maire；半毛菊状旋瓣菊■☆

412262　Volutaria crupinoides（Desf.）Cass. ex Maire var. libyca（Viv.）Maire et Weiller = Volutaria crupinoides（Desf.）Cass. ex Maire ■☆

412263　Volutaria leucantha（L. Chevall.）Maire = Volutaria sinaica（DC.）Wagenitz ■☆

412264　Volutaria lippii（L.）Cass.；里普旋瓣菊■☆

412265　Volutaria lippii（L.）Cass. subsp. medians（Maire）Wagenitz；中间旋瓣菊■☆

412266　Volutaria lippii（L.）Cass. subsp. tubuliflora（Murb.）Maire；管花里普旋瓣菊■☆

412267　Volutaria lippii（L.）Cass. var. atlantica（Pit.）Maire =

412268　Volutaria lippii（L.）Cass. var. medians（Maire）Maire = Volutaria lippii（L.）Cass. subsp. medians（Maire）Wagenitz ■☆

412269　Volutaria lippii（L.）Cass. var. microcephala Maire = Volutaria lippii（L.）Cass. ■☆

412270　Volutaria lippii（L.）Cass. var. ramosissima（Pit.）Maire = Volutaria lippii（L.）Cass. ■☆

412271　Volutaria maroccana（Barratte et Murb.）Maire；摩洛哥旋瓣菊■☆

412272　Volutaria muricata（L.）Maire；短尖旋瓣菊；Morocco knapweed ■☆

412273　Volutaria muricata（L.）Maire var. eradiata（Braun-Blanq. et Maire）Maire = Volutaria muricata（L.）Maire ■☆

412274　Volutaria muricata（L.）Maire var. micractis（Boiss.）Maire = Volutaria muricata（L.）Maire ■☆

412275　Volutaria omphalodes（Benth. et Hook. f.）Maire = Stephanochilus omphalodes（Benth. et Hook. f.）Maire ■☆

412276　Volutaria saharae（L. Chevall.）Wagenitz；左原旋瓣菊■☆

412277　Volutaria sinaica（DC.）Wagenitz；支那旋瓣菊■☆

412278　Volutaria somalensis（Oliv. et Hiern）C. Jeffrey = Volutaria abyssinica（A. Rich.）C. Jeffrey ■☆

412279　Volutaria tubuliflora（Murb.）Sennen = Volutaria lippii（L.）Cass. subsp. tubuliflora（Murb.）Maire ■☆

412280　Volutella Forssk. = Cassytha L. ■●

412281　Volvulopsis Roberty = Evolvulus L. ●■

412282　Volvulopsis nummularium（L.）Roberty = Evolvulus nummularius（L.）L. ■

412283　Volvulus Medik.（废弃属名）= Calystegia R. Br.（保留属名）■

412284　Volvulus hederaceus（Wall.）Kuntze = Calystegia hederacea Wall. ex Roxb. ■

412285　Volvulus japonicus（Thunb.）Farw. var. pubescens（Lindl.）Farw. = Calystegia pubescens Lindl. ■

412286　Vonitra Becc.（1906）；马岛椰属（碱椰子属，马岛棕属，王尼爪桐属，我你他棕属）●☆

412287　Vonitra Becc. = Dypsis Noronha ex Mart. ●☆

412288　Vonitra crinita Jum. et H. Perrier = Dypsis crinita（Jum. et H. Perrier）Beentje et J. Dransf. ●☆

412289　Vonitra fibrosa（C. H. Wright）Becc. = Dypsis fibrosa（C. H. Wright）Beentje et J. Dransf. ●☆

412290　Vonitra nossibensis（Becc.）H. Perrier = Dypsis nossibensis（Becc.）Beentje et J. Dransf. ●☆

412291　Vonitra thouarsiana（Baill.）Becc.；马岛椰●☆

412292　Vonitra thouarsiana（Baill.）Becc. = Dypsis fibrosa（C. H. Wright）Beentje et J. Dransf. ●☆

412293　Vonitra utilis Jum. = Dypsis utilis（Jum.）Beentje et J. Dransf. ●☆

412294　Vonroemeria J. J. Sm. = Octarrhena Thwaites ■☆

412295　Vormia Adans. = Selago L. ●☆

412296　Vorstia Adans. = Galphimia Cav. ●

412297　Vorstia Adans. = Thryallis L.（废弃属名）●

412298　Vorstia Adans. = Thryallis Mart.（保留属名）●

412299　Vosacan Adans. = Helianthus L. ■

412300　Vossia Adans.（废弃属名）= Glottiphyllum Haw. ex N. E. Br. ■☆

412301　Vossia Adans.（废弃属名）= Vossia Wall. et Griff.（保留属名）■☆

412302　Vossia Wall. et Griff.（1836）（保留属名）；河马草属■☆

412303　Vossia cuspidata（Roxb.）Griff.；河马草■☆

412304　Vossia cuspidata（Roxb.）Griff. var. polystachya Koechlin = Vossia cuspidata（Roxb.）Griff. ■☆

412305　Vossia procera Wall. et Griff. = Vossia cuspidata（Roxb.）Griff. ■☆

412306　Vossia speciosa（Steud.）Benth. = Phacelurus speciosus（Steud.）C. E. Hubb. ■☆

412307　Vossianthus Kuntze = Sparrmannia L. f.（保留属名）●☆

412308　Votomita Aubl.（1775）;沃套野牡丹属●☆

412309　Votomita guianensis Aubl. ;圭亚那沃套野牡丹●☆

412310　Votomita monantha（Urb.）Morley;山地沃套野牡丹●☆

412311　Votomita pubescens Morley;毛沃套野牡丹●☆

412312　Votschia B. Stahl（1993）;沃氏假轮叶属●☆

412313　Votschia nemophila（Pittier）B. Stahl;沃氏假轮叶●☆

412314　Vouacapoua Aubl.（废弃属名）= Andira Lam.（保留属名）●☆

412315　Vouapa Aubl.（废弃属名）= Macrolobium Schreb.（保留属名）●☆

412316　Vouapa coerulea Taub. = Paramacrolobium coeruleum（Taub.）J. Léonard ●☆

412317　Vouapa crassifolia Baill. = Anthonotha crassifolia（Baill.）J. Léonard ●☆

412318　Vouapa demonstrans Baill. = Gilbertiodendron demonstrans（Baill.）J. Léonard ●☆

412319　Vouapa explicans Baill. = Triplisomeris explicans（Baill.）Aubrév. et Pellegr. ●☆

412320　Vouapa limba（Scott-Elliot）Taub. = Gilbertiodendron limba（Scott-Elliot）J. Léonard ●☆

412321　Vouarana Aubl.（1775）;圭亚那无患子属●☆

412322　Vouarana guianensis Aubl. ;圭亚那无患子●☆

412323　Vouay Aubl. = Geonoma Willd. ●☆

412324　Voucapoua Steud. = Andira Lam.（保留属名）●☆

412325　Voucapoua Steud. = Vouacapoua Aubl.（废弃属名）●☆

412326　Voyara Aubl. = Capparis L. ●

412327　Voyria Aubl.（1775）;沃伊龙胆属■☆

412328　Voyria platypetala Baker = Voyria primuloides Baker ■☆

412329　Voyria primuloides Baker;沃伊龙胆■☆

412330　Voyriaceae Doweld = Gentianaceae Juss.（保留科名）●■

412331　Voyriella Miq.（1851）;小沃伊龙胆属■☆

412332　Voyriella parviflora Miq. ;小沃伊龙胆■☆

412333　Vrena Noronha = Urena L. ●■

412334　Vriesea Beer = Vriesea Lindl.（保留属名）■☆

412335　Vriesea Hassk.（废弃属名）= Lindernia All. ■

412336　Vriesea Hassk.（废弃属名）= Vriesea Lindl.（保留属名）■☆

412337　Vriesea Lindl.（1843）（保留属名）;丽穗凤梨属（斑氏凤梨属,弗里西属,虎尾凤梨属,花叶兰属,剑凤梨属,剑叶兰属,丽穗兰属,莺哥凤梨属,鹦哥凤梨属,鹦哥）;Vriesea ■☆

412338　Vriesea 'Marjan';黄宝剑■☆

412339　Vriesea 'Orange Marie';玛丽橙红剑丽穗凤梨(玛丽橙红剑)■☆

412340　Vriesea 'Rose Marie';玛丽红剑丽穗凤梨(玛丽红剑)■☆

412341　Vriesea barilleti E. Morren;巴丽丽穗凤梨■☆

412342　Vriesea bituminosa Wawra;褐斑丽穗凤梨■☆

412343　Vriesea carinata Wawra;背棱丽穗凤梨(莺歌菠萝);Crab-claw Plum, Lobster Claws ■☆

412344　Vriesea chrysostachys E. Morren;金色丽穗凤梨■☆

412345　Vriesea confusa L. B. Sm. ;铺散丽穗凤梨■☆

412346　Vriesea densiflora Mez;密花丽穗凤梨■☆

412347　Vriesea drepanocarpa（Baker）Mez;镰果丽穗凤梨■☆

412348　Vriesea fenestralis E. Morren;网纹丽穗凤梨(网纹菠萝,网纹凤梨)■☆

412349　Vriesea fenestralis E. Morren 'Variegata';花叶网纹丽穗凤梨■☆

412350　Vriesea flammea L. B. Sm. ;焰苞丽穗凤梨;Flame Vriesea ■☆

412351　Vriesea fosteriana L. B. Sm. ;福德丽穗凤梨(福氏丽穗凤梨);Foster Vriesea ■☆

412352　Vriesea fosteriana L. B. Sm. 'Red Chestnut';红栗丽穗凤梨■☆

412353　Vriesea fosteriana L. B. Sm. 'Seideliana';黄带丽穗凤梨■☆

412354　Vriesea gigantea Mez;巨大丽穗凤梨■☆

412355　Vriesea guttata André et Linden;豹纹丽穗凤梨;Dusted Vriesea ■☆

412356　Vriesea heliconioides（Kunth）Lindl. ;蝎尾蕉丽穗凤梨■☆

412357　Vriesea hieroglyphica E. Morren;纹叶凤梨(纹叶丽穗凤梨);King of the Bromeliads, King-of-bromeliads, King-of-the-bromeliads ■☆

412358　Vriesea imperialis Carrière;皇帝丽穗凤梨;Gjant Vriesea ■☆

412359　Vriesea incurva（Griseb.）Read;内折丽穗凤梨■☆

412360　Vriesea incurvata Gaudich. ;曲叶丽穗凤梨;Sidewinder Vriesea ■☆

412361　Vriesea longiscapa Ule;长花梗丽穗凤梨■☆

412362　Vriesea petropolitana L. B. Sm. ;贝多丽穗凤梨;Petropolit Vriesea ■☆

412363　Vriesea platynema Gaudin. ;紫尖凤梨■☆

412364　Vriesea procera Wittm. ;高株丽穗凤梨■☆

412365　Vriesea psittacina（Hook.）Lindl. ;鹦鹉凤梨(山莺菠萝)■☆

412366　Vriesea recurvata Gaudich. ;反卷丽穗凤梨■☆

412367　Vriesea rubra（Ruiz et Pav.）Beer;红丽穗凤梨■☆

412368　Vriesea saundersii（Carrière）Morren;丽斑菠萝■☆

412369　Vriesea simplex（Vell.）Beer;单茎丽穗凤梨;Simple Stem Vriesea ■☆

412370　Vriesea speciosa Hook. ;黄色花菠萝■☆

412371　Vriesea splendens（Brongn.）Lem. ;丽穗凤梨(虎纹菠萝,虎纹凤梨,剑凤梨);Flaming Sword ■☆

412372　Vriesea splendens Lem. 'Anderken Carl Wolf';纵带丽穗兰■☆

412373　Vriesea splendens Lem. 'Chantrieri';紫斑丽穗兰■☆

412374　Vriesea splendens Lem. 'Major';红剑丽穗凤梨■☆

412375　Vriesea splendens Lem. 'Variegata';花叶丽穗凤梨■☆

412376　Vrieseida Rojas Acosta（1897）;阿根廷凤梨属■☆

412377　Vrieseida foetida Rojas Acosta;阿根廷凤梨■☆

412378　Vriesia Lindl. = Vriesea Lindl.（保留属名）■☆

412379　Vrlesia Lindl. = Vrieseida Rojas Acosta ■☆

412380　Vroedea Bubani = Glaux Ehrh. ■

412381　Vroedea Bubani = Glaux L. ■

412382　Vrolicida Steud. = Vrolikia Spreng ■☆

412383　Vrolikia Spreng. = Heteranthia Nees et Mart. ■☆

412384　Vrydagzenia Benth. et Hook. f. = Vrydagzynea Blume ■

412385　Vrydagzynea Blume（1858）; 二尾兰属; Doubletail Orchis, Vrydagzynea ■

412386　Vrydagzynea albida Blume var. formosana（Hayata）T. Hashim. = Vrydagzynea nuda Blume ■

412387　Vrydagzynea formosana Hayata = Vrydagzynea nuda Blume ■

412388　Vrydagzynea nuda Blume;二尾兰(矮二尾兰,台湾二尾兰);Bare Vrydagzynea, Doubletail Orchis ■

412389　Vuacapua Kuntze = Andira Lam.（保留属名）●☆

412390　Vuacapua Kuntze = Vouacapoua Aubl.（废弃属名）●☆

412391　Vuapa Kuntze = Macrolobium Schreb.（保留属名）●☆

412392　Vuapa Kuntze = Vouapa Aubl.（废弃属名）●☆

412393　Vulneraria Mill. = Anthyllis L. ■☆

412394　Vulpia C. C. Gmel.（1805）;鼠茅属;Fescue ■

412395　Vulpia alopecuros（Schousb.）Link;狐狸鼠茅■☆

412396　Vulpia alopecuros（Schousb.）Link subsp. fibrosa H. Lindb. = Vulpia alopecuros（Schousb.）Link ■☆

412397　Vulpia alopecuros（Schousb.）Link subsp. schousboei H.

Lindb. = Vulpia alopecuros (Schousb.) Link ■☆

412398　Vulpia alopecuros (Schousb.) Link var. glabra H. Lindb. = Vulpia alopecuros (Schousb.) Link ■☆

412399　Vulpia alopecuros (Schousb.) Link var. glabrata Lange = Vulpia alopecuros (Schousb.) Link ■☆

412400　Vulpia alopecuros (Schousb.) Link var. lanata (Boiss.) H. Lindb. = Vulpia alopecuros (Schousb.) Link ■☆

412401　Vulpia alopecuros (Schousb.) Link var. lindbergii Maire et Weiller = Vulpia alopecuros (Schousb.) Link ■☆

412402　Vulpia alopecuros (Schousb.) Link var. oranensis (Trab.) Maire = Vulpia alopecuros (Schousb.) Link ■☆

412403　Vulpia alopecuros (Schousb.) Link var. sylvatica (Boiss.) H. Lindb. = Vulpia alopecuros (Schousb.) Link ■☆

412404　Vulpia alpina L.;高原鼠茅■

412405　Vulpia ambigua More;可疑短鼠茅;Bearded Fescue ■☆

412406　Vulpia brevis Boiss. et Kotschy;短鼠茅■☆

412407　Vulpia bromoides (L.) Gray;雀麦草鼠茅;Annual Fescue, Brome Fescue,Squirrel-taft Fescue ■☆

412408　Vulpia broteri Boiss. et Reut. = Vulpia myuros (L.) C. C. Gmel. subsp. sciuroides (Roth) Rouy ■☆

412409　Vulpia ciliata (Pers.) Link;缘毛鼠茅;Bearded Fescue, Fringed Fescue ■☆

412410　Vulpia ciliata Dumort. = Vulpia ciliata (Pers.) Link ■☆

412411　Vulpia ciliata Dumort. var. danthonii (Asch. et Graebn.) Maire et Weiller = Vulpia ciliata Dumort. ■☆

412412　Vulpia ciliata Dumort. var. imberbis (Vis.) Hayek = Vulpia ciliata Dumort. ■☆

412413　Vulpia ciliata Dumort. var. penicellata Murb. = Vulpia ciliata Dumort. ■☆

412414　Vulpia ciliata Dumort. var. subglabra Litard. et Sauvage = Vulpia ciliata Dumort. ■☆

412415　Vulpia ciliata Dumort. var. tripolitana (Pamp.) Maire et Weiller = Vulpia ciliata Dumort. ■☆

412416　Vulpia cynosuroides (Desf.) Parl. = Ctenopsis cynosuroides (Desf.) Paunero ex Romero García ■☆

412417　Vulpia danthonii (Asch. et Graebn.) Volkart = Vulpia ciliata Dumort. ■☆

412418　Vulpia danthonii (Asch. et Graebn.) Volkart var. tripolitana Pamp. = Vulpia ciliata Dumort. ■☆

412419　Vulpia delicatula (Lag.) Dumort. = Ctenopsis pectinella (Delile) De Not. ■☆

412420　Vulpia dertonensis (All.) Gola = Vulpia myuros (L.) C. C. Gmel. subsp. sciuroides (Roth) Rouy ■☆

412421　Vulpia elliotea (Raf.) Fernald;埃利鼠茅■☆

412422　Vulpia fasciculata (Forssk.) Samp.;簇生鼠茅;Dune Fescue ■☆

412423　Vulpia flavescens Sennen et Mauricio = Vulpia geniculata (L.) Link ■☆

412424　Vulpia geniculata (L.) Link;膝曲鼠茅■☆

412425　Vulpia geniculata (L.) Link subsp. attenuata (Parl.) Trab.;渐狭膝曲鼠茅■☆

412426　Vulpia geniculata (L.) Link subsp. breviglumis (Trab.) Murb.;短颖膝曲鼠茅■☆

412427　Vulpia geniculata (L.) Link subsp. monantha Maire;山地膝曲鼠茅■☆

412428　Vulpia geniculata (L.) Link subsp. pauana (Font Quer) Maire;波氏鼠茅■☆

412429　Vulpia geniculata (L.) Link var. ciliata Parl. = Vulpia

geniculata (L.) Link ■☆

412430　Vulpia geniculata (L.) Link var. dasyantha Henrard = Vulpia geniculata (L.) Link ■☆

412431　Vulpia geniculata (L.) Link var. dianthera Maire = Vulpia geniculata (L.) Link ■☆

412432　Vulpia geniculata (L.) Link var. eriantha Maire = Vulpia geniculata (L.) Link ■☆

412433　Vulpia geniculata (L.) Link var. glabriglumis Maire = Vulpia geniculata (L.) Link ■☆

412434　Vulpia geniculata (L.) Link var. hirsuta (H. Lindb.) Maire = Vulpia geniculata (L.) Link ■☆

412435　Vulpia geniculata (L.) Link var. hispida Batt. et Trab. = Vulpia geniculata (L.) Link ■☆

412436　Vulpia geniculata (L.) Link var. leiantha Maire = Vulpia geniculata (L.) Link ■☆

412437　Vulpia geniculata (L.) Link var. longiglumis Caball. = Vulpia geniculata (L.) Link ■☆

412438　Vulpia geniculata (L.) Link var. reesei Maire = Vulpia geniculata (L.) Link ■☆

412439　Vulpia gracilis Scholz;纤细鼠茅■☆

412440　Vulpia hispanica (Reichard) Kerguélen = Vulpia unilateralis (L.) Stace ■☆

412441　Vulpia hispanica (Reichard) Kerguélen subsp. montana (Boiss. et Reut.) Devesa = Vulpia unilateralis (L.) Stace subsp. montana (Boiss. et Reut.) Cabezudo et al. ■☆

412442　Vulpia hybrida (Brot.) Pau;杂种鼠茅■☆

412443　Vulpia incrassata (Loisel.) Parl. = Vulpiella stipoides (L.) Maire ■☆

412444　Vulpia inops (Delile) Hack. = Vulpia brevis Boiss. et Kotschy ■☆

412445　Vulpia inops (Delile) Hack. var. glabra Hack. = Vulpia inops (Delile) Hack. ■☆

412446　Vulpia inops (Delile) Hack. var. spiralis Asch. et Hack. = Vulpia inops (Delile) Hack. ■☆

412447　Vulpia inops (Delile) Hack. var. strigosa Hack. = Vulpia inops (Delile) Hack. ■☆

412448　Vulpia inops (Delile) Hack. var. subdisticha Asch. et Hack. = Vulpia inops (Delile) Hack. ■☆

412449　Vulpia ligustica (All.) Link;利古里亚鼠茅■☆

412450　Vulpia ligustica (All.) Link var. hispidula Parl. = Vulpia ligustica (All.) Link ■☆

412451　Vulpia ligustica (All.) Link var. intermedia Rouy = Vulpia ligustica (All.) Link ■☆

412452　Vulpia ligustica Link = Vulpia ligustica (All.) Link ■☆

412453　Vulpia litardiereana (Maire) A. Camus;利塔尔鼠茅■☆

412454　Vulpia litardiereana (Maire) A. Camus var. glabrivaginata Maire = Vulpia litardiereana (Maire) A. Camus ■☆

412455　Vulpia litardiereana (Maire) A. Camus var. pubivaginata Maire = Vulpia litardiereana (Maire) A. Camus ■☆

412456　Vulpia longiseta (Brot.) Hack.;长刚毛鼠茅■☆

412457　Vulpia megalura (Nutt.) Rydb. = Festuca myuros L. ■

412458　Vulpia megalura (Nutt.) Rydb. = Vulpia myuros (L.) C. C. Gmel. var. megalura (Nutt.) Rydb. ■☆

412459　Vulpia megalura (Nutt.) Rydb. = Vulpia myuros (L.) C. C. Gmel. ■

412460　Vulpia megastachya Nees = Festuca vulpioides Steud. ■☆

412461　Vulpia membranacea (L.) Dumort.;膜质鼠茅■☆

412462　Vulpia membranacea (L.) Dumort. var. longiseta (Brot.)

Maire et Weiller = Vulpia longiseta（Brot.）Hack.■☆

412463　Vulpia muralis（Kunth）Nees；厚壁鼠茅■☆

412464　Vulpia muralis（L.）C. C. Gmel. = Vulpia hybrida（Brot.）Pau ■☆

412465　Vulpia myuros（L.）C. C. Gmel.；鼠茅；Annual Fescue，Foxtail Fescue，Rat's-tail Fescue，Rattail Fescue，Rat-tail Fescue，Rattail Fescuegrass，Red-tail Fescue ■

412466　Vulpia myuros（L.）C. C. Gmel. = Festuca myuros L. ■

412467　Vulpia myuros（L.）C. C. Gmel. subsp. pseudomyuros（Soy. - Will.）Maire et Weiller = Vulpia myuros（L.）C. C. Gmel. ■

412468　Vulpia myuros（L.）C. C. Gmel. subsp. sciuroides（Roth）Rouy；松鼠尾鼠茅■☆

412469　Vulpia myuros（L.）C. C. Gmel. var. hirsuta Hack. = Vulpia myuros（L.）C. C. Gmel. ■

412470　Vulpia myuros（L.）C. C. Gmel. var. megalura（Nutt.）Rydb.；大尾鼠茅■☆

412471　Vulpia myuros（L.）C. C. Gmel. var. subuniglumis Hack. = Vulpia myuros（L.）C. C. Gmel. ■

412472　Vulpia myuros（L.）C. C. Gmel. var. tenella（Boiss.）Maire et Weiller = Vulpia hybrida（Brot.）Pau ■☆

412473　Vulpia obtusa Trab.；钝鼠茅■☆

412474　Vulpia octoflora（Walter）Rydb.；八花鼠茅；Eight-flowered Fescue-grass，Sixweeks Fescue，Six-weeks Fescue ■☆

412475　Vulpia octoflora（Walter）Rydb. var. glauca（Nutt.）Fernald；苍白鼠茅；Six-weeks Fescue ■☆

412476　Vulpia octoflora（Walter）Rydb. var. tenella（Willd.）Fernald = Vulpia octoflora（Walter）Rydb. var. glauca（Nutt.）Fernald ■☆

412477　Vulpia octoflora（Walter）Rydb. var. tenella（Willd.）Fernald = Vulpia octoflora（Walter）Rydb. ■☆

412478　Vulpia pectinella（Delile）Boiss. = Ctenopsis pectinella（Delile）De Not. ■☆

412479　Vulpia persica（Boiss. et Buhse）V. I. Krecz. et Bobrov；波斯鼠茅■☆

412480　Vulpia sciurea（Nutt.）Henrard = Vulpia elliotea（Raf.）Fernald ■☆

412481　Vulpia sciuroides（Roth）C. C. Gmel. = Vulpia myuros（L.）C. C. Gmel. subsp. sciuroides（Roth）Rouy ■☆

412482　Vulpia sciuroides（Roth）C. C. Gmel. var. broteroi（Boiss. et Reut.）Trab. = Vulpia bromoides（L.）Gray ■☆

412483　Vulpia setacea Parl. = Vulpia sicula（C. Presl）Link ■☆

412484　Vulpia sicula（C. Presl）Link；西西里鼠茅■☆

412485　Vulpia sicula（C. Presl）Link subsp. setacea（Parl.）Trab. = Vulpia sicula（C. Presl）Link ■☆

412486　Vulpia sicula（C. Presl）Link var. setacea（Parl.）Hack. = Vulpia sicula（C. Presl）Link ■☆

412487　Vulpia stipoides（L.）Dumort. = Vulpiella stipoides（L.）Maire ■☆

412488　Vulpia stipoides（L.）Dumort. subsp. letourneuxii（Asch.）H. Scholz = Vulpiella stipoides（L.）Maire ■☆

412489　Vulpia stipoides（L.）Dumort. subsp. tenuis（Tineo）Scholz = Vulpiella tenuis（Tineo）Kerguélen ■☆

412490　Vulpia subalata Sennen = Vulpia geniculata（L.）Link ■☆

412491　Vulpia tenuis（Tineo）Parl. = Vulpiella tenuis（Tineo）Kerguélen ■☆

412492　Vulpia uniglumis（Sol.）Dumort. = Vulpia fasciculata（Forssk.）Samp. ■☆

412493　Vulpia unilateralis（L.）Stace；中亚鼠茅；Mat-grass Fescue ■☆

412494　Vulpia unilateralis（L.）Stace subsp. montana（Boiss. et Reut.）Cabezudo et al.；山地中亚鼠茅■☆

412495　Vulpia villosa Maslenn.；长柔毛鼠茅■☆

412496　Vulpiella（Batt. et Trab.）Andrews（1927）；小鼠茅属■☆

412497　Vulpiella incrassata（Lam.）Andr. = Vulpiella stipoides（L.）Maire ■☆

412498　Vulpiella incrassata（Lam.）Andr. var. stipoides（Desf.）Andr. = Vulpiella stipoides（L.）Maire ■☆

412499　Vulpiella incrassata（Lam.）Andr. var. tenuis（Tineo）Andr. = Vulpiella tenuis（Tineo）Kerguélen ■☆

412500　Vulpiella incrassata（Lam.）Burollet = Vulpiella stipoides（L.）Maire ■☆

412501　Vulpiella stipoides（L.）Maire；小鼠茅■☆

412502　Vulpiella stipoides（L.）Maire subsp. tenuis（Tineo）Scholz = Vulpiella tenuis（Tineo）Kerguélen ■☆

412503　Vulpiella stipoides（L.）Maire var. letourneuxii（Asch.）Maire = Vulpiella stipoides（L.）Maire ■☆

412504　Vulpiella stipoides（L.）Maire var. multiflora Trotter = Vulpiella stipoides（L.）Maire ■☆

412505　Vulpiella stipoides（L.）Maire var. submutica Trotter = Vulpiella stipoides（L.）Maire ■☆

412506　Vulpiella stipoides（L.）Maire var. tenuis（Tineo）Maire = Vulpiella tenuis（Tineo）Kerguélen ■☆

412507　Vulpiella tenuis（Tineo）Kerguélen；细小鼠茅■☆

412508　Vulvaria Bubani = Chenopodium L. ■●

412509　Vvedenskya Korovin（1947）；韦坚草属☆

412510　Vvedenskya pinnatifolia Korovin；韦坚草☆

412511　Vvedenskyella Botsch. = Christolea Cambess. ■

412512　Vvedenskyella Botsch. = Phaeonychium O. E. Schulz ■

412513　Vvedenskyella kashgarica Botsch. = Phaeonychium kashgaricum（Botsch.）Al-Shehbaz ■

412514　Vvedenskyella pumila（Kurz）Botsch. = Christolea pumila（Kurz）Jafri ■

412515　Vvedenskyella pumila（Kurz）Botsch. = Desideria pumila（Kurz）Al-Shehbaz ■

412516　Vyenomus C. Presl = Euonymus L.（保留属名）●

412517　Vyenomus pendulus C. Presl = Euonymus obovatus Nutt. ●

412518　Wacchendorfia Burm. f. = Wachendorfia Burm. ■☆

412519　Wachendorfia Burm.（1757）；折扇草属；Wachendorfia ■☆

412520　Wachendorfia Burm. ex L. = Wachendorfia Burm. ■☆

412521　Wachendorfia Loefl. = Callisia Loefl. ■☆

412522　Wachendorfia brachyandra W. F. Barker；短蕊折扇草■☆

412523　Wachendorfia brevifolia Sol. ex Ker Gawl. = Wachendorfia paniculata Burm. ■☆

412524　Wachendorfia graminea Thunb. = Wachendorfia paniculata Burm. ■☆

412525　Wachendorfia graminifolia L. f. = Wachendorfia paniculata Burm. ■☆

412526　Wachendorfia herbertii Sweet = Wachendorfia paniculata Burm. ■☆

412527　Wachendorfia hirsuta Thunb. = Wachendorfia paniculata Burm. ■☆

412528　Wachendorfia multiflora（Klatt）J. C. Manning et Goldblatt；多花折扇草■☆

412529　Wachendorfia paniculata Burm.；圆锥折扇草■☆

412530　Wachendorfia parviflora W. F. Barker = Wachendorfia multiflora（Klatt）J. C. Manning et Goldblatt ■☆

412531　Wachendorfia tenella Thunb. = Wachendorfia paniculata Burm. ■☆

412532　Wachendorfia thyrsiflora Burm.；折扇草■☆

412533　Wachendorfia thyrsiflora L. = Wachendorfia thyrsiflora Burm. ■☆

412534　Wachendorfiaceae Herb. = Haemodoraceae R. Br. (保留科名) ■☆

412535　Wadapus Raf. = Gomphrena L. ●▥

412536　Waddingtonia Phil. = Nicotiana L. ●■

412537　Waddingtonia Phil. = Petunia Juss. (保留属名) ■

412538　Wadea Raf. = Cestrum L. ●

412539　Wagatea Dalzell = Moullava Adans. ■☆

412540　Wagatea spicata (Dalziel) Wight = Caesalpinia spicata Dalziel ●☆

412541　Wageneria Klotzsch = Begonia L. ●■

412542　Wagenitzia Dostál = Centaurea L. (保留属名) ●■

412543　Wagnera Post et Kuntze = Maianthemum F. H. Wigg. (保留属名) ■

412544　Wagnera Post et Kuntze = Smilacina Desf. (保留属名) ■

412545　Wagnera Post et Kuntze = Vagnera Adans. (废弃属名) ■

412546　Wagneria Klotzsch = Begonia L. ●■

412547　Wagneria Klotzsch = Wageneria Klotzsch ●■

412548　Wagneria Lem. = Diervilla Mill. ●☆

412549　Wahabia Fenzl = Barleria L. ●■

412550　Wahabia longiflora Fenzl = Barleria acanthoides Vahl ■☆

412551　Wahabia longiflora Fenzl ex Solms = Barleria acanthoides Vahl ■☆

412552　Wahlbergella Fr. = Gastrolychnis (Fenzl) Rchb. ■

412553　Wahlbergella Fr. = Melandrium Röhl. ■

412554　Wahlbergella Fr. = Silene L. (保留属名) ■

412555　Wahlbergella attenuata (Farr) Rydb. = Silene uralensis (Rupr.) Bocquet ■☆

412556　Wahlbergella drummondii (Hook.) Rydb. = Silene drummondii Hook. ■☆

412557　Wahlbergella kingii (S. Watson) Rydb. = Silene kingii (S. Watson) Bocquet ■☆

412558　Wahlbergella montana (S. Watson) Rydb. = Silene hitchguirei Bocquet ■☆

412559　Wahlbergella parryi (S. Watson) Rydb. = Silene parryi (S. Watson) C. L. Hitchc. et Maguire ■☆

412560　Wahlbergella striata (Rydb.) Rydb. = Silene drummondii Hook. subsp. striata (Rydb.) J. K. Morton ■☆

412561　Wahlbergella tayloriae (B. L. Rob.) Rydb. = Silene involucrata (Cham. et Schltdl.) Bocquet subsp. tenella (Tolm.) Bocquet ■☆

412562　Wahlbergella triflora (R. Br. ex Sommerf.) Fr. = Silene sorensenis (B. Boivin) Bocquet ■☆

412563　Wahlbomia Thunb. = Tetracera L. ●

412564　Wahlenbergia Blume = Tarenna Gaertn. ●

412565　Wahlenbergia R. Br. = Dichapetalum Thouars ●

412566　Wahlenbergia R. Br. ex Wall. = Dichapetalum Thouars ●

412567　Wahlenbergia Schrad. = Wahlenbergia Schrad. ex Roth (保留属名) ■●

412568　Wahlenbergia Schrad. ex Roth (1821) (保留属名); 蓝花参属 (兰花参属); Bellflower, Blue Bell, Harebell, Ivy-leaved Beltfower, Rockbell, Tufty Bells ■●

412569　Wahlenbergia Schumach. = Enydra Lour. ■

412570　Wahlenbergia aberdarica T. C. E. Fr. = Wahlenbergia capillacea (L. f.) A. DC. subsp. tenuior (Engl.) Thulin ■☆

412571　Wahlenbergia abyssinica (Hochst. ex A. Rich.) Thulin; 阿比西尼亚蓝花参 ■☆

412572　Wahlenbergia abyssinica (Hochst. ex A. Rich.) Thulin subsp. parvipetala Thulin; 小瓣阿比西尼亚蓝花参 ■☆

412573　Wahlenbergia acaulis E. Mey.; 无茎蓝花参 ■☆

412574　Wahlenbergia acicularis Brehmer; 针形蓝花参 ■☆

412575　Wahlenbergia acuminata Brehmer; 渐尖蓝花参 ■☆

412576　Wahlenbergia adamsonii Lammers; 亚当森蓝花参 ■☆

412577　Wahlenbergia adpressa (Thunb.) Sond.; 匍匐蓝花参 ■☆

412578　Wahlenbergia agrestis A. DC. = Wahlenbergia marginata (Thunb.) A. DC. ■

412579　Wahlenbergia albens (Spreng. ex A. DC.) Lammers; 白蓝花参 ■☆

412580　Wahlenbergia albicaulis (Sond.) Lammers; 白茎蓝花参 ■☆

412581　Wahlenbergia albomarginata Hook.; 新西兰蓝花参; New Zealand Bluebell ■☆

412582　Wahlenbergia androsacea A. DC.; 点地梅蓝花参 ■☆

412583　Wahlenbergia annua (A. DC.) Thulin; 一年生蓝花参 ■☆

412584　Wahlenbergia annularis A. DC.; 小环蓝花参 ■☆

412585　Wahlenbergia annuliformis Brehmer; 环状蓝花参 ■☆

412586　Wahlenbergia arabidifolia (Engl.) Brehmer = Wahlenbergia krebsii Cham. subsp. arguta (Hook. f.) Thulin ■☆

412587　Wahlenbergia arcta Thulin; 直蓝花参 ■☆

412588　Wahlenbergia arenaria A. DC. = Wahlenbergia androsacea A. DC. ■☆

412589　Wahlenbergia arguta Hook. f. = Wahlenbergia krebsii Cham. subsp. arguta (Hook. f.) Thulin ■☆

412590　Wahlenbergia arguta Hook. f. var. longifusiformis Brehmer = Wahlenbergia krebsii Cham. subsp. arguta (Hook. f.) Thulin ■☆

412591　Wahlenbergia arguta Hook. f. var. parvilocula Brehmer = Wahlenbergia krebsii Cham. subsp. arguta (Hook. f.) Thulin ■☆

412592　Wahlenbergia asparagoides (Adamson) Lammers; 天门冬蓝花参 ■☆

412593　Wahlenbergia asperifolia Brehmer; 糙叶蓝花参 ■☆

412594　Wahlenbergia axillaris (Sond.) Lammers; 腋花蓝花参 ■☆

412595　Wahlenbergia banksiana A. DC.; 班克斯蓝花参 ■☆

412596　Wahlenbergia basutica E. Phillips = Craterocapsa montana (A. DC.) Hilliard et B. L. Burtt ■☆

412597　Wahlenbergia bernardii Leredde = Wahlenbergia campanuloides (Delile) Vatke ■☆

412598　Wahlenbergia bojeri A. DC. = Wahlenbergia undulata (L. f.) A. DC. ■☆

412599　Wahlenbergia bolusiana Schltr. et Brehmer; 博卢斯蓝花参 ■☆

412600　Wahlenbergia bowkerae Sond.; 鲍克蓝花参 ■☆

412601　Wahlenbergia brachiata (Adamson) Lammers; 短蓝花参 ■☆

412602　Wahlenbergia brachycarpa Schltr.; 短果蓝花参 ■☆

412603　Wahlenbergia brachyphylla (Adamson) Lammers; 短叶蓝花参 ■☆

412604　Wahlenbergia brehmeri Lammers; 布雷默蓝花参 ■☆

412605　Wahlenbergia brevipes Hemsl. = Homocodon brevipes (Hemsl.) D. Y. Hong ■

412606　Wahlenbergia brevisquamifolia Brehmer; 短鳞蓝花参 ■☆

412607　Wahlenbergia caffra A. DC.; 开菲尔蓝花参 ■☆

412608　Wahlenbergia calcarea (Adamson) Lammers; 石灰蓝花参 ■☆

412609　Wahlenbergia caledonica Sond. = Wahlenbergia undulata (L. f.) A. DC. ☆

412610　Wahlenbergia caledonica Sond. var. cyanea (Engl. et Gilg) Brehmer = Wahlenbergia undulata (L. f.) A. DC. ■☆

412611　Wahlenbergia campanuloides (Delile) Vatke; 风铃草蓝花参 ■☆

412612　Wahlenbergia candolleana (Hiern) Thulin; 康多勒蓝花参 ■☆

412613　Wahlenbergia capensis (L.) A. DC.; 好望角蓝花参 ■☆

412614　Wahlenbergia capillacea (L. f.) A. DC.; 纤毛蓝花参 ■☆

412615　Wahlenbergia capillacea (L. f.) A. DC. subsp. tenuior (Engl.) Thulin; 小纤毛蓝花参 ■☆

412616　Wahlenbergia capillacea (L. f.) A. DC. var. tenuior Engl. = Wahlenbergia capillacea (L. f.) A. DC. subsp. tenuior (Engl.)

Thulin ■☆

412617　Wahlenbergia capillaris（H. Buek）Lammers = Wahlenbergia thulinii Lammers ■☆

412618　Wahlenbergia capillata Brehmer;发状蓝花参■☆

412619　Wahlenbergia capillifolia E. Mey. ex Brehmer;毛叶蓝花参■☆

412620　Wahlenbergia capitata（Baker）Thulin;头状蓝花参■☆

412621　Wahlenbergia cephalodina Thulin;小头蓝花参■☆

412622　Wahlenbergia cernua（Thunb.）A. DC.;俯垂蓝花参■☆

412623　Wahlenbergia cervicina A. DC. = Wahlenbergia campanuloides（Delile）Vatke ■☆

412624　Wahlenbergia ciliolata A. DC. = Wahlenbergia cernua（Thunb.）A. DC. ■☆

412625　Wahlenbergia cinerea（L. f.）Lammers;灰色蓝花参■☆

412626　Wahlenbergia clavata Brehmer;棍棒蓝花参■☆

412627　Wahlenbergia clavatula Brehmer;小棒蓝花参■☆

412628　Wahlenbergia claviculata E. Mey. = Wahlenbergia exilis A. DC. ■☆

412629　Wahlenbergia clematidea Schrenk = Codonopsis clematidea（Schrenk）C. B. Clarke ■

412630　Wahlenbergia coerulea H. J. P. Winkl. = Wahlenbergia krebsii Cham. subsp. arguta（Hook. f.）Thulin ■☆

412631　Wahlenbergia collomioides（A. DC.）Thulin;黏胶花蓝花参■☆

412632　Wahlenbergia communis Carolin;普通蓝花参;Bluebell ■☆

412633　Wahlenbergia compacta Brehmer;紧密蓝花参■☆

412634　Wahlenbergia congesta（Cheeseman）N. E. Br.;密集匍匐蓝花参■☆

412635　Wahlenbergia congesta Thulin = Wahlenbergia tenuiloba Thulin ■☆

412636　Wahlenbergia congestifolia Brehmer;密叶蓝花参■☆

412637　Wahlenbergia constricta Brehmer;缢缩蓝花参■☆

412638　Wahlenbergia cooperi Brehmer;库珀蓝花参■☆

412639　Wahlenbergia cordata（Adamson）Lammers;心形蓝花参■☆

412640　Wahlenbergia costata A. DC.;单脉蓝花参■☆

412641　Wahlenbergia cuspidata Brehmer;骤尖蓝花参■☆

412642　Wahlenbergia cyanea Engl. et Gilg = Wahlenbergia undulata（L. f.）A. DC. ■☆

412643　Wahlenbergia cylindrica Pax. et Hoffm. = Campanula aristata Wall. ■

412644　Wahlenbergia debilis H. Buek;瘦小蓝花参■☆

412645　Wahlenbergia decipiens A. DC.;迷惑蓝花参■☆

412646　Wahlenbergia dehiscens A. DC. = Wahlenbergia marginata（Thunb.）A. DC. ■

412647　Wahlenbergia densicaulis Brehmer;密茎蓝花参■☆

412648　Wahlenbergia dentata Brehmer;齿叶蓝花参■☆

412649　Wahlenbergia denticulata（Burch.）A. DC.;小齿蓝花参■☆

412650　Wahlenbergia denticulata（Burch.）A. DC. var. scabra A. DC. = Wahlenbergia denticulata（Burch.）A. DC. ■☆

412651　Wahlenbergia denticulata（Burch.）A. DC. var. transvaalensis（Adamson）Welman;德兰士瓦蓝花参■☆

412652　Wahlenbergia dentifera Brehmer = Wahlenbergia cuspidata Brehmer ■☆

412653　Wahlenbergia denudata A. DC.;裸露蓝花参■☆

412654　Wahlenbergia depressa J. M. Wood et M. S. Evans;凹陷蓝花参■☆

412655　Wahlenbergia depressa Wolley-Dod = Treichelia longebracteata（H. Buek）Vatke ■☆

412656　Wahlenbergia desmantha Lammers;束花蓝花参■☆

412657　Wahlenbergia dicentrifolia C. B. Clarke = Codonopsis dicentrifolia（C. B. Clarke）W. W. Sm. ■

412658　Wahlenbergia dichotoma A. DC.;二歧蓝花参■☆

412659　Wahlenbergia dieterlenii（E. Phillips）Lammers;迪特尔蓝花参☆

412660　Wahlenbergia dilatata Brehmer;膨大蓝花参■☆

412661　Wahlenbergia dinteri Brehmer = Wahlenbergia undulata（L. f.）A. DC. ■☆

412662　Wahlenbergia distincta Brehmer;离生蓝花参■☆

412663　Wahlenbergia divergens A. DC.;稍叉蓝花参■☆

412664　Wahlenbergia dregeana A. DC. = Wahlenbergia ecklonii H. Buek ■☆

412665　Wahlenbergia ecklonii H. Buek;埃氏蓝花参☆

412666　Wahlenbergia effusa（Adamson）Lammers;开展蓝花参☆

412667　Wahlenbergia emirnensis A. DC. = Gunillaea emirnensis（A. DC.）Thulin ■☆

412668　Wahlenbergia engleri Brehmer = Wahlenbergia undulata（L. f.）A. DC. ■☆

412669　Wahlenbergia erecta（Roth ex Roem. et Schult.）Tuyn;直立蓝花参☆

412670　Wahlenbergia ericoidella（P. A. Duvign. et Denaeyer）Thulin;欧石南蓝花参☆

412671　Wahlenbergia erophiloides Markgr.;绮春蓝花参■☆

412672　Wahlenbergia etbaica（Schweinf.）Vatke = Wahlenbergia lobelioides（L. f.）Link subsp. nutabunda（Guss.）Murb. ■☆

412673　Wahlenbergia exilis A. DC.;弱小蓝花参☆

412674　Wahlenbergia fasciculata Brehmer;簇生蓝花参☆

412675　Wahlenbergia filicaulis R. D. Good = Prismatocarpus sessilis Eckl. ex A. DC. ●☆

412676　Wahlenbergia filipes Brehmer;线梗蓝花参■☆

412677　Wahlenbergia fistulosa Brehmer;管蓝花参■☆

412678　Wahlenbergia flaccida A. DC.;柔软蓝花参☆

412679　Wahlenbergia flexuosa（Hook. f. et Thomson）Thulin;曲折蓝花参☆

412680　Wahlenbergia floribunda Schltr. et Brehmer;繁花蓝花参☆

412681　Wahlenbergia foliosa Brehmer = Wahlenbergia banksiana A. DC. ■☆

412682　Wahlenbergia fruticosa Brehmer;灌丛蓝花参■☆

412683　Wahlenbergia furcata Brehmer = Wahlenbergia cuspidata Brehmer ■☆

412684　Wahlenbergia galpiniae Schltr.;盖尔蓝花参■☆

412685　Wahlenbergia galpinii E. Phillips = Wahlenbergia lobulata Brehmer ■☆

412686　Wahlenbergia glandulifera Brehmer;腺体蓝花参■☆

412687　Wahlenbergia glandulosa Brehmer = Wahlenbergia androsacea A. DC. ■☆

412688　Wahlenbergia globularis Schumach. et Thonn. = Enydra radicans（Willd.）Lack ■☆

412689　Wahlenbergia gloriosa Lothian;华丽蓝花参;Royal Bluebell ☆

412690　Wahlenbergia gracilis（G. Forst.）Schrad. var. misera Hemsl. = Wahlenbergia marginata（Thunb.）A. DC. ■

412691　Wahlenbergia gracilis E. Mey.;纤细蓝花参;Australian Bluebell ■☆

412692　Wahlenbergia gracilis Schrad. = Wahlenbergia marginata（Thunb.）A. DC. ■

412693　Wahlenbergia gracillima S. Moore = Wahlenbergia banksiana A. DC. ■☆

412694　Wahlenbergia graminifolia A. DC.;禾叶蓝花参■☆

412695　Wahlenbergia grandiflora Brehmer;大花蓝花参■☆

412696　Wahlenbergia guthriei L. Bolus = Theilera guthriei（L. Bolus）E. Phillips ●☆

412697　Wahlenbergia hederacea Rchb.;常春藤状蓝花参;Ivy Campanula, Ivy-leaved Bellflower,Ivy-leaved Harebell,Witches' Thimbles ■☆

412698　Wahlenbergia hederifolia Bubani;常春藤叶蓝花参;Ivy Campanula, Witches' thimbles ■☆

412699　Wahlenbergia hilsenbergii A. DC. = Wahlenbergia madagascariensis A. DC. ■☆

412700　Wahlenbergia hirsuta（Edgew.）Tuyn;硬毛蓝花参■☆

412701　Wahlenbergia hispidula（Thunb.）A. DC. ;细毛蓝花参■☆

412702　Wahlenbergia hookeri（C. B. Clarke）Tuyn;星花草;Hooker Cephalostigma ■

412703　Wahlenbergia hookeri（C. B. Clarke）Tuyn = Cephalostigma hookeri C. B. Clarke ■

412704　Wahlenbergia huillana A. DC. = Gunillaea emirnensis（A. DC.）Thulin ■☆

412705　Wahlenbergia huillana A. DC. var. pusilla ? = Gunillaea emirnensis（A. DC.）Thulin ■☆

412706　Wahlenbergia humbertii Thulin;亨伯特蓝花参■☆

412707　Wahlenbergia humifusa Markgr. = Wahlenbergia campanuloides（Delile）Vatke ■☆

412708　Wahlenbergia humilis A. DC. = Wahlenbergia lobelioides（L. f.）Link subsp. riparia（A. DC.）Thulin ■☆

412709　Wahlenbergia humpatensis Brehmer = Wahlenbergia candolleana（Hiern）Thulin ■☆

412710　Wahlenbergia huttonii（Sond.）Thulin;赫顿蓝花参■☆

412711　Wahlenbergia inconspicua A. DC. ;显著蓝花参■☆

412712　Wahlenbergia inhambanensis Klotzsch = Wahlenbergia androsacea A. DC. ■☆

412713　Wahlenbergia intricatissima Dinter ex Range = Wahlenbergia patula A. DC. ■☆

412714　Wahlenbergia juncea（H. Buek）Lammers;灯心草蓝花参■☆

412715　Wahlenbergia kilimandscharica Engl. = Wahlenbergia capillacea（L. f.）A. DC. subsp. tenuior（Engl.）Thulin ■☆

412716　Wahlenbergia kilimandscharica Engl. var. intermedia Brehmer = Wahlenbergia capillacea（L. f.）A. DC. subsp. tenuior（Engl.）Thulin ■☆

412717　Wahlenbergia kowiensis R. A. Dyer;克温斯蓝花参■☆

412718　Wahlenbergia krebsii Cham. ;克雷布斯蓝花参■☆

412719　Wahlenbergia krebsii Cham. subsp. arguta（Hook. f.）Thulin;光亮克雷布斯蓝花参■☆

412720　Wahlenbergia lasiocarpa Schltr. et Brehmer;毛果蓝花参■☆

412721　Wahlenbergia lateralis Brehmer = Wahlenbergia ramosissima（Hemsl.）Thulin subsp. lateralis（Brehmer）Thulin ■☆

412722　Wahlenbergia lavandulifolia A. DC. = Wahlenbergia marginata（Thunb.）A. DC. ■

412723　Wahlenbergia laxiflora（Sond.）Lammers;疏花蓝花参■☆

412724　Wahlenbergia leucantha Engl. et Gilg = Wahlenbergia banksiana A. DC. ■☆

412725　Wahlenbergia linarioides Lam. ;柳穿鱼蓝花参;Tuffybells ■☆

412726　Wahlenbergia linearis A. DC. = Microcodon lineare（L. f.）H. Buek ■☆

412727　Wahlenbergia littoralis Schltr. et Brehmer = Wahlenbergia orae Lammers ■☆

412728　Wahlenbergia lobata Brehmer;浅裂蓝花参■☆

412729　Wahlenbergia lobelioides（L. f.）Link;小花蓝花参■☆

412730　Wahlenbergia lobelioides（L. f.）Link subsp. nutabunda（Guss.）Murb. ;摇摆蓝花参■☆

412731　Wahlenbergia lobelioides（L. f.）Link subsp. riparia（A. DC.）Thulin;河岸蓝花参■☆

412732　Wahlenbergia lobelioides（L. f.）Link var. gussonei Webb et Berthel. = Wahlenbergia lobelioides（L. f.）Link ■☆

412733　Wahlenbergia lobelioides（L. f.）Link var. linnaei Webb et Berthel. = Wahlenbergia lobelioides（L. f.）Link ■☆

412734　Wahlenbergia lobelioides（L. f.）Link var. macilenta Webb et Berthel. = Wahlenbergia lobelioides（L. f.）Link ■☆

412735　Wahlenbergia lobulata Brehmer;小裂片蓝花参■☆

412736　Wahlenbergia longifolia（A. DC.）Lammers;长叶蓝花参■☆

412737　Wahlenbergia longifolia（A. DC.）Lammers var. corymbosa（Adamson）Welman;伞花长叶蓝花参■☆

412738　Wahlenbergia longisepala Brehmer;长萼蓝花参■☆

412739　Wahlenbergia longisquamifolia Brehmer;长鳞叶蓝花参■☆

412740　Wahlenbergia lycopodioides Schltr. et Brehmer;石松蓝花参■☆

412741　Wahlenbergia macra Schltr. et Brehmer = Wahlenbergia ecklonii H. Buek ■☆

412742　Wahlenbergia macrostachys（A. DC.）Lammers;大穗蓝花参■☆

412743　Wahlenbergia maculata Brehmer;斑点蓝花参■☆

412744　Wahlenbergia madagascariensis A. DC. ;马岛蓝花参■☆

412745　Wahlenbergia mairei H. Lév. = Cyananthus flavus C. Marquand subsp. montanus（C. Y. Wu）D. Y. Hong et L. M. Ma ■

412746　Wahlenbergia malaissei Thulin;马莱泽蓝花参■☆

412747　Wahlenbergia mannii Vatke = Wahlenbergia silenoides Hochst. ex A. Rich. ■☆

412748　Wahlenbergia mannii Vatke var. intermedia Brehmer = Wahlenbergia silenoides Hochst. ex A. Rich. ■☆

412749　Wahlenbergia mannii Vatke var. virgulta Brehmer = Wahlenbergia silenoides Hochst. ex A. Rich. ■☆

412750　Wahlenbergia marginata（Thunb.）A. DC. ;蓝花参(霸王草, 鼓锤草, 拐棒参, 拐棍参, 鹌鹑草, 罐罐草, 寒草, 葫芦草, 金线草, 金线吊葫芦, 兰花参, 蓝花草, 毛鸡腿, 牛奶菜, 破石珠, 雀舌草, 乳浆草, 沙参草, 蛇须草, 天蓬草, 土参, 娃儿菜, 娃儿草, 玩儿草, 细蓝花参, 细叶蓝花参, 细叶沙参, 小绿细辛, 一窝鸡);Australian Hareball, Gentian Rockbell, Marginate Rockbell, Southern Rockbell ■

412751　Wahlenbergia marunguensis Thulin;马龙古蓝花参■☆

412752　Wahlenbergia mashonica N. E. Br. = Wahlenbergia banksiana A. DC. ■☆

412753　Wahlenbergia massonii A. DC. ;马森蓝花参■☆

412754　Wahlenbergia meyeri A. DC. ;迈尔蓝花参■☆

412755　Wahlenbergia miarie H. Lév. = Cyananthus montanus C. Y. Wu ■

412756　Wahlenbergia microphylla（Adamson）Lammers;澳非小叶蓝花参■☆

412757　Wahlenbergia minuta Brehmer;微小蓝花参■☆

412758　Wahlenbergia mollis Brehmer;绢毛蓝花参■☆

412759　Wahlenbergia monotropa Killick = Wahlenbergia lobulata Brehmer ■☆

412760　Wahlenbergia montana A. DC. ;山地蓝花参■☆

412761　Wahlenbergia montana A. DC. = Craterocapsa montana（A. DC.）Hilliard et B. L. Burtt ■☆

412762　Wahlenbergia multiflora Conrath = Wahlenbergia banksiana A. DC. ■☆

412763　Wahlenbergia namaquana Sond. ;纳马夸蓝花参■☆

412764　Wahlenbergia nana Brehmer;矮小蓝花参■☆

412765　Wahlenbergia napiformis（A. DC.）Thulin;芜菁形蓝花参■☆

412766　Wahlenbergia neorigida Lammers;坚挺蓝花参■☆

412767　Wahlenbergia neostricta Lammers;劲直蓝花参■☆

412768　Wahlenbergia nodosa（H. Buek）Lammers;多节蓝花参■☆

412769　Wahlenbergia nudicaulis A. DC. = Wahlenbergia androsacea A. DC. ■☆

412770　Wahlenbergia nutabunda（Guss.）A. DC. = Wahlenbergia lobelioides（L. f.）Link subsp. nutabunda（Guss.）Murb. ■☆

412771　Wahlenbergia nutabunda（Guss.）A. DC. var. erythreae Chiov. = Wahlenbergia lobelioides（L. f.）Link subsp. nutabunda（Guss.）Murb. ■☆

412772　Wahlenbergia oatesii Rolfe = Wahlenbergia undulata（L. f.）A. DC. ■☆

412773　Wahlenbergia obovata Brehmer;倒卵蓝花参■☆

412774　Wahlenbergia okavangensis N. E. Br. = Wahlenbergia banksiana A. DC. ■☆

412775　Wahlenbergia oligantha Lammers;寡花蓝花参■☆

412776　Wahlenbergia oligotricha Schltr. et Brehmer;寡毛蓝花参■☆

412777　Wahlenbergia oliveri Schweinf. ex Engl. = Wahlenbergia capillacea（L. f.）A. DC. subsp. tenuior（Engl.）Thulin ■☆

412778　Wahlenbergia oocarpa Sond. ;卵果蓝花参■☆

412779　Wahlenbergia oppositifolia A. DC. = Wahlenbergia madagascariensis A. DC. ■☆

412780　Wahlenbergia orae Lammers;奥拉蓝花参■☆

412781　Wahlenbergia ovalis Brehmer = Craterocapsa montana（A. DC.）Hilliard et B. L. Burtt ■☆

412782　Wahlenbergia oxyphylla A. DC. ;尖叶蓝花参■

412783　Wahlenbergia pallidiflora Hilliard et B. L. Burtt;苍白花蓝花参■

412784　Wahlenbergia paludicola Thulin;沼泽蓝花参■

412785　Wahlenbergia paniculata（Thunb.）A. DC. ;圆锥蓝花参■☆

412786　Wahlenbergia parviflora A. DC. = Wahlenbergia dichotoma A. DC. ■☆

412787　Wahlenbergia parvifolia（P. J. Bergius）Lammers;小叶蓝花参■☆

412788　Wahlenbergia patula A. DC. ;张开蓝花参■☆

412789　Wahlenbergia paucidentata Schinz;稀齿蓝花参■☆

412790　Wahlenbergia pauciflora A. DC. ;少花蓝花参■

412791　Wahlenbergia pavida Launert = Wahlenbergia erophiloides Markgr. ■☆

412792　Wahlenbergia perennis Brehmer = Wahlenbergia androsacea A. DC. ■☆

412793　Wahlenbergia perotifolia Wight et Arn. = Wahlenbergia erecta（Roth ex Roem. et Schult.）Tuyn ■☆

412794　Wahlenbergia perrieri Thulin;佩里耶蓝花参■☆

412795　Wahlenbergia perrottetii（A. DC.）Thulin;佩罗蓝花参■☆

412796　Wahlenbergia persimilis Thulin;相似蓝花参■☆

412797　Wahlenbergia petrae Thulin;岩生蓝花参■☆

412798　Wahlenbergia pilosa A. DC. = Wahlenbergia polyclada A. DC. ■☆

412799　Wahlenbergia pilosa H. Buek;疏毛蓝花参■☆

412800　Wahlenbergia pinifolia N. E. Br. ;松针蓝花参■☆

412801　Wahlenbergia pinnata Compton;羽状蓝花参■☆

412802　Wahlenbergia pinnata Compton var. simplicifolia？ = Wahlenbergia capillacea（L. f.）A. DC. ■☆

412803　Wahlenbergia polyantha Lammers;多花蓝花参■☆

412804　Wahlenbergia polycephala（Mildbr.）Thulin;多头蓝花参■☆

412805　Wahlenbergia polychotoma Brehmer = Wahlenbergia undulata（L. f.）A. DC. ■☆

412806　Wahlenbergia polyclada A. DC. ;多枝蓝花参■☆

412807　Wahlenbergia polyclada Hook. f. = Wahlenbergia silenoides Hochst. ex A. Rich. ■☆

412808　Wahlenbergia polyphylla Thulin;多叶蓝花参■☆

412809　Wahlenbergia polytrichifolia Schltr. ;多毛叶蓝花参■☆

412810　Wahlenbergia procumbens（Thunb.）A. DC. ;平铺蓝花参■☆

412811　Wahlenbergia prostrata A. DC. ;平卧蓝花参■☆

412812　Wahlenbergia psammophila Schltr. ;喜沙蓝花参■☆

412813　Wahlenbergia pseudoandrosacea Brehmer;假点地梅蓝花参■☆

412814　Wahlenbergia pseudoinhambanensis Brehmer;假伊尼扬巴内蓝花参■☆

412815　Wahlenbergia pseudonudicaulis Brehmer;假裸茎蓝花参■☆

412816　Wahlenbergia pulchella Thulin;美丽蓝花参■☆

412817　Wahlenbergia pulchella Thulin subsp. laurentii Thulin;洛朗蓝花参■☆

412818　Wahlenbergia pulchella Thulin subsp. mbalensis Thulin;姆巴莱蓝花参■☆

412819　Wahlenbergia pulchella Thulin subsp. michelii Thulin;米歇尔蓝花参■☆

412820　Wahlenbergia pulchella Thulin subsp. paradoxa Thulin;奇异美丽蓝花参■☆

412821　Wahlenbergia pulchella Thulin subsp. pedicellata Thulin;梗花美丽蓝花参■☆

412822　Wahlenbergia pulvillus-gigantis Hilliard et B. L. Burtt;巨大蓝花参■☆

412823　Wahlenbergia purpurea（Wall.）A. DC. = Codonopsis purpurea Wall. ■

412824　Wahlenbergia purpurea A. DC. = Codonopsis purpurea Wall. ■

412825　Wahlenbergia pusilla Hochst. ex A. Rich. ;瘦弱蓝花参■☆

412826　Wahlenbergia pyrophila Lammers;喜炎蓝花参■☆

412827　Wahlenbergia ramifera Brehmer;枝生蓝花参■☆

412828　Wahlenbergia ramosissima（Hemsl.）Thulin;密枝蓝花参■☆

412829　Wahlenbergia ramosissima（Hemsl.）Thulin subsp. centiflora Thulin;百花蓝花参■☆

412830　Wahlenbergia ramosissima（Hemsl.）Thulin subsp. lateralis（Brehmer）Thulin;侧生多枝蓝花参■☆

412831　Wahlenbergia ramosissima（Hemsl.）Thulin subsp. lobelioides Thulin;半边莲蓝花参■☆

412832　Wahlenbergia ramosissima（Hemsl.）Thulin subsp. richardsiae Thulin;理查兹蓝花参■☆

412833　Wahlenbergia ramosissima（Hemsl.）Thulin subsp. subcapitata Thulin;亚头多枝蓝花参■☆

412834　Wahlenbergia ramosissima（Hemsl.）Thulin subsp. zambiensis Thulin;赞比亚蓝花参■☆

412835　Wahlenbergia ramulosa E. Mey. ;分枝蓝花参■☆

412836　Wahlenbergia rara Schltr. et Brehmer;稀花蓝花参■☆

412837　Wahlenbergia recurvata Brehmer = Wahlenbergia virgata Engl. ■☆

412838　Wahlenbergia rhodesiana S. Moore = Wahlenbergia banksiana A. DC. ■☆

412839　Wahlenbergia rhytidosperma Thulin;皱籽蓝花参■☆

412840　Wahlenbergia rigida Bernh. = Wahlenbergia robusta Sond. ■☆

412841　Wahlenbergia riparia A. DC. = Wahlenbergia lobelioides（L. f.）Link subsp. riparia（A. DC.）Thulin ■☆

412842　Wahlenbergia riparia A. DC. var. clavata Brehmer = Wahlenbergia lobelioides（L. f.）Link subsp. riparia（A. DC.）Thulin ■☆

412843　Wahlenbergia riparia A. DC. var. etbaica Brehmer = Wahlenbergia lobelioides（L. f.）Link subsp. nutabunda（Guss.）Murb. ■☆

412844　Wahlenbergia riparia A. DC. var. segregata Brehmer = Wahlenbergia lobelioides（L. f.）Link subsp. riparia（A. DC.）Thulin ■☆

412845　Wahlenbergia riparia A. DC. var. virgulta Brehmer = Wahlenbergia lobelioides（L. f.）Link subsp. riparia（A. DC.）

Thulin ■☆

412846　Wahlenbergia riversdalensis Lammers;里弗斯代尔蓝花参■☆

412847　Wahlenbergia rivularis Diels;溪边蓝花参■☆

412848　Wahlenbergia robusta Sond. ;粗壮蓝花参■☆

412849　Wahlenbergia rosulata Brehmer = Wahlenbergia androsacea A. DC. ■☆

412850　Wahlenbergia rotundifolia Brehmer = Wahlenbergia brehmeri Lammers ■☆

412851　Wahlenbergia rubens（H. Buek）Lammers;淡红蓝花参■☆

412852　Wahlenbergia rubens（H. Buek）Lammers var. brachyphylla（Adamson）Welman;短叶变淡红蓝花参■☆

412853　Wahlenbergia rubioides（Banks ex A. DC.）Lammers;盖茜蓝花参■☆

412854　Wahlenbergia rubioides（Banks ex A. DC.）Lammers var. stokoei（Adamson）Welman;斯托科蓝花参■☆

412855　Wahlenbergia sabulosa Brehmer = Wahlenbergia densicaulis Brehmer ■☆

412856　Wahlenbergia saginoides S. Moore = Wahlenbergia banksiana A. DC. ■☆

412857　Wahlenbergia sarmentosa T. C. E. Fr. = Wahlenbergia krebsii Cham. subsp. arguta（Hook. f.）Thulin ■☆

412858　Wahlenbergia saxifragoides Brehmer;虎耳草状蓝花参■☆

412859　Wahlenbergia schimperi（Hochst. ex A. Rich.）Schweinf. et Asch. = Wahlenbergia erecta（Roth ex Roem. et Schult.）Tuyn ■☆

412860　Wahlenbergia schlechteri Brehmer;施莱蓝花参■☆

412861　Wahlenbergia scoparia Brehmer = Wahlenbergia undulata（L. f.）A. DC. ■☆

412862　Wahlenbergia scottii Thulin;司科特蓝花参■☆

412863　Wahlenbergia serpentina Brehmer;蛇形蓝花参■☆

412864　Wahlenbergia sessiliflora Brehmer;无花梗蓝花参■☆

412865　Wahlenbergia silenoides Hochst. ex A. Rich. ;千里光蓝花参■☆

412866　Wahlenbergia silenoides Hochst. ex A. Rich. var. elongata Brehmer = Wahlenbergia silenoides Hochst. ex A. Rich. ■☆

412867　Wahlenbergia sonderi Lammers;森诺蓝花参■☆

412868　Wahlenbergia sparticula Chiov. = Wahlenbergia virgata Engl. ■☆

412869　Wahlenbergia sphaerica Brehmer;球形蓝花参■☆

412870　Wahlenbergia spinulosa A. DC. ;细刺蓝花参■☆

412871　Wahlenbergia spinulosa Engl. = Wahlenbergia denticulata（Burch.）A. DC. ■☆

412872　Wahlenbergia squamifolia Brehmer;鳞叶蓝花参■☆

412873　Wahlenbergia squarrosa Brehmer;粗鳞蓝花参■☆

412874　Wahlenbergia steingroeveri Engl. = Wahlenbergia oxyphylla A. DC. ■

412875　Wahlenbergia stellarioides Cham. et Schltdl. ;星状蓝花参■☆

412876　Wahlenbergia stricta（R. Br.）Sweet;澳洲蓝花参;Australian Bluebell ■☆

412877　Wahlenbergia stricta Sweet = Wahlenbergia stricta（R. Br.）Sweet ■☆

412878　Wahlenbergia subaphylla（Baker）Thulin;近无叶蓝花参■☆

412879　Wahlenbergia subaphylla（Baker）Thulin subsp. scoparia（Wild）Thulin;帚状近无叶蓝花参■☆

412880　Wahlenbergia subaphylla（Baker）Thulin subsp. thesioides Thulin;百蕊草蓝花参■☆

412881　Wahlenbergia subfusiformis Brehmer;纺锤形蓝花参■☆

412882　Wahlenbergia subnuda Conrath = Wahlenbergia virgata Engl. ■☆

412883　Wahlenbergia subpilosa Brehmer;微毛蓝花参■☆

412884　Wahlenbergia subrosulata Brehmer;莲座蓝花参■☆

412885　Wahlenbergia subtilis Brehmer;细小蓝花参■☆

412886　Wahlenbergia subulata（L'Hér.）Lammers;钻形蓝花参■☆

412887　Wahlenbergia subulata（L'Hér.）Lammers var. congesta（Adamson）Welman;密集钻形蓝花参■☆

412888　Wahlenbergia subulata（L'Hér.）Lammers var. tenuifolia（Adamson）Welman;细叶钻形蓝花参■☆

412889　Wahlenbergia subumbellata Markgr. ;亚小伞蓝花参■☆

412890　Wahlenbergia swellendamensis H. Buek = Wahlenbergia ecklonii H. Buek ■☆

412891　Wahlenbergia tenella（L. f.）Lammers;细蓝花参■☆

412892　Wahlenbergia tenella（L. f.）Lammers var. palustris（Adamson）Welman;沼泽柔软蓝花参■☆

412893　Wahlenbergia tenella（L. f.）Lammers var. stokoei（Adamson）Welman;斯托克蓝花参■☆

412894　Wahlenbergia tenerrima（H. Buek）Lammers;极细蓝花参■☆

412895　Wahlenbergia tenerrima（H. Buek）Lammers var. montana（Adamson）Welman;山地极细蓝花参■☆

412896　Wahlenbergia tenuifolia A. DC. ;细叶蓝花参■☆

412897　Wahlenbergia tenuiloba Thulin;细裂蓝花参■☆

412898　Wahlenbergia tetramera Thulin;四蓝花参■☆

412899　Wahlenbergia thalictrifolia A. DC. = Codonopsis thalictrifolia Wall. ■

412900　Wahlenbergia thulinii Lammers;图林蓝花参■☆

412901　Wahlenbergia thunbergiana（H. Buek）Lammers;通贝里蓝花参■☆

412902　Wahlenbergia tibestica Quézel;提贝斯提蓝花参■☆

412903　Wahlenbergia tomentosula Brehmer;绒毛蓝花参■☆

412904　Wahlenbergia tortilis Brehmer;螺旋状蓝花参■☆

412905　Wahlenbergia transvaalensis Brehmer = Wahlenbergia denticulata（Burch.）A. DC. var. transvaalensis（Adamson）Welman ■☆

412906　Wahlenbergia tumida Brehmer;肿胀蓝花参■☆

412907　Wahlenbergia turbinata A. DC. = Wahlenbergia ecklonii H. Buek ■☆

412908　Wahlenbergia tysonii Zahlbr. = Wahlenbergia rivularis Diels ■☆

412909　Wahlenbergia uitenhagensis（H. Buek）Lammers;埃滕哈赫蓝花参■☆

412910　Wahlenbergia uitenhagensis（H. Buek）Lammers var. debilis（Sond.）Welman;弱小埃滕哈赫蓝花参■☆

412911　Wahlenbergia uitenhagensis（H. Buek）Lammers var. filifolia（Adamson）Welman;线叶埃滕哈赫蓝花参■☆

412912　Wahlenbergia umbellata（Adamson）Lammers;小伞蓝花参■☆

412913　Wahlenbergia undulata（L. f.）A. DC. ;尾状蓝花参;Melton bluebird ■☆

412914　Wahlenbergia undulata A. DC. = Wahlenbergia undulata（L. f.）A. DC. ■☆

412915　Wahlenbergia unidentata（L. f.）Lammers;单齿蓝花参■☆

412916　Wahlenbergia upembensis Thulin;乌彭贝蓝花参■☆

412917　Wahlenbergia variabilis E. Mey. = Wahlenbergia krebsii Cham. ■☆

412918　Wahlenbergia verbascoides Thulin;毛蕊花蓝花参■☆

412919　Wahlenbergia virgata Engl. ;条纹蓝花参■☆

412920　Wahlenbergia virgata Engl. var. longisepala Brehmer = Wahlenbergia virgata Engl. ■☆

412921　Wahlenbergia virgata Engl. var. tenuis Brehmer = Wahlenbergia virgata Engl. ■☆

412922　Wahlenbergia virgata Engl. var. valida Brehmer = Wahlenbergia virgata Engl. ■☆

412923　Wahlenbergia welwitschii（A. DC.）Thulin;韦尔蓝花参■☆

412924　Wahlenbergia wittei Thulin;维特蓝花参■☆

412925　Wahlenbergia zeyheri H. Buek = Wahlenbergia krebsii Cham. ■☆

412926　Wailesia Lindl. = Dipodium R. Br. ■☆

412927　Waireia D. L. Jones, Molloy et M. A. Clem. (1997);奥克兰柱帽兰属■☆

412928　Waireia D. L. Jones, Molloy et M. A. Clem. = Thelymitra J. R. Forst. et G. Forst. ■☆

412929　Waireia stenopetala (Hook. f.) D. L. Jones, M. A. Clem. et Molloy;奥克兰帽兰■☆

412930　Waitzia J. C. Wendl. (1808);尖柱鼠麹草属■☆

412931　Waitzia Rchb. = Tritonia Ker Gawl. ■

412932　Waitzia acuminata Steetz;尖柱鼠麹草■☆

412933　Waitzia aurea (Benth.) Steetz;黄尖柱鼠麹草■☆

412934　Waitzia dasycarpa Turcz. ;毛果尖柱鼠麹草■☆

412935　Waitzia discolor Turcz. ;杂色尖柱鼠麹草■☆

412936　Waitzia grandiflora Naudin;大花尖柱鼠麹草■☆

412937　Waitzia nitida (Lindl.) Paul G. Wilson;亮尖柱鼠麹草■☆

412938　Waitzia suaveolens Druce;香尖柱鼠麹草■☆

412939　Waitzia sulphurea Steetz;硫色尖柱鼠麹草■☆

412940　Waitzia xanthospila (DC.) Heynh. = Freesia xanthospila (DC.) Klatt ■☆

412941　Wajira Thulin(1982);肯尼亚豇豆属■☆

412942　Wajira albescens Thulin;白肯尼亚豇豆■☆

412943　Wajira grahamiana (Wight et Arn.) Thulin et Lavin;肯尼亚豇豆■☆

412944　Wajira praecox (Verdc.) Thulin et Lavin;早肯尼亚豇豆■☆

412945　Wajira virescens (Thulin) Thulin et Lavin;浅绿肯尼亚豇豆■☆

412946　Wakilia Gilli = Phaeonychium O. E. Schulz ■

412947　Walafrida E. Mey. (1838);瓦拉玄参属●☆

412948　Walafrida E. Mey. = Selago L. ●☆

412949　Walafrida albanensis (Schltr.) Rolfe = Selago recurva E. Mey. ●☆

412950　Walafrida alopecuroides (Rolfe) Rolfe = Selago alopecuroides Rolfe ●☆

412951　Walafrida angolensis (Rolfe) Rolfe = Selago angolensis Rolfe ●☆

412952　Walafrida apiculata (E. Mey.) Rolfe = Glumicalyx apiculatus (E. Mey.) Hilliard et B. L. Burtt ■☆

412953　Walafrida articulata (Thunb.) Rolfe = Selago articulata Thunb. ●☆

412954　Walafrida barabei Mielcarek = Selago barabei (Mielcarek) Hilliard ●☆

412955　Walafrida basutica E. Phillips = Selago saxatilis E. Mey. ●☆

412956　Walafrida chongweensis Rolfe = Selago angolensis Rolfe ●☆

412957　Walafrida ciliata (L. f.) Rolfe = Selago ciliata L. f. ●☆

412958　Walafrida cinerea (L. f.) Rolfe = Selago cinerea L. f. ●☆

412959　Walafrida congesta (Rolfe) Rolfe = Selago congesta Rolfe ●☆

412960　Walafrida crassifolia Rolfe = Selago crassifolia (Rolfe) Hilliard ●☆

412961　Walafrida decipiens (E. Mey.) Rolfe = Selago decipiens E. Mey. ●☆

412962　Walafrida densiflora (Rolfe) Rolfe = Selago densiflora Rolfe ●☆

412963　Walafrida diffusa Rolfe = Selago linearis Rolfe ●☆

412964　Walafrida dinteri (Rolfe) Rolfe = Selago dinteri Rolfe ●☆

412965　Walafrida distans (E. Mey.) Rolfe = Selago distans E. Mey. ●☆

412966　Walafrida fleckii Rolfe = Selago dinteri Rolfe ●☆

412967　Walafrida geniculata (L. f.) Rolfe = Selago geniculata L. f. ●☆

412968　Walafrida goetzei (Rolfe) Brenan = Selago goetzei Rolfe ●☆

412969　Walafrida goetzei (Rolfe) Brenan var. brevipila Brenan = Selago goetzei Rolfe subsp. ambigua Hilliard ●☆

412970　Walafrida goetzei (Rolfe) Brenan var. pubescentior Brenan = Selago goetzei Rolfe subsp. ambigua Hilliard ●☆

412971　Walafrida gracilis Rolfe = Selago gracilis (Rolfe) Hilliard ●☆

412972　Walafrida lacunosa (Klotzsch) Rolfe = Selago lacunosa Klotzsch ●☆

412973　Walafrida loganii Hutch. = Selago rigida Rolfe ●☆

412974　Walafrida macowani Rolfe = Selago decipiens E. Mey. ●☆

412975　Walafrida merxmuelleri Rössler = Selago divaricata L. f. ●☆

412976　Walafrida micrantha (Choisy) Rolfe = Selago paniculata Thunb. ●☆

412977　Walafrida minuta Rolfe = Cromidon minutum (Rolfe) Hilliard ■☆

412978　Walafrida muralis (Benth. et Hook.) Rolfe = Selago muralis Benth. et Hook. f. ●☆

412979　Walafrida myrtifolia (Rchb.) Rolfe = Selago myrtifolia Rchb. ●☆

412980　Walafrida nachtigalii (Rolfe) Rolfe = Selago nachtigalii Rolfe ●☆

412981　Walafrida nitida E. Mey. = Selago myrtifolia Rchb. ●☆

412982　Walafrida paniculata (Thunb.) Rolfe = Selago paniculata Thunb. ●☆

412983　Walafrida polycephala (Otto ex Walp.) Rolfe = Selago polycephala Otto ex Walp. ●☆

412984　Walafrida polystachya Rolfe = Selago densiflora Rolfe ●☆

412985　Walafrida pubescens Rolfe = Selago gracilis (Rolfe) Hilliard ●☆

412986　Walafrida pusilla Rössler = Cromidon pusillum (Rössler) Hilliard ■☆

412987　Walafrida recurva (E. Mey.) Rolfe = Selago recurva E. Mey. ●☆

412988　Walafrida rigida (Rolfe) Hutch. = Selago rigida Rolfe ●☆

412989　Walafrida rotundifolia (L. f.) Rolfe = Selago rotundifolia L. f. ●☆

412990　Walafrida saxatilis (E. Mey.) Rolfe = Selago saxatilis E. Mey. ●☆

412991　Walafrida schinzii Rolfe = Selago alopecuroides Rolfe ●☆

412992　Walafrida squarrosa Rolfe = Selago gracilis (Rolfe) Hilliard ●☆

412993　Walafrida swynnertonii S. Moore = Selago swynnertonii (S. Moore) Hilliard ●☆

412994　Walafrida swynnertonii S. Moore var. leiophylla Brenan = Selago swynnertonii (S. Moore) Hilliard var. leiophylla (Brenan) Hilliard ●☆

412995　Walafrida tenuifolia Rolfe = Selago tenuifolia (Rolfe) Hilliard ●☆

412996　Walafrida trimera Hochst. = Selago myrtifolia Rchb. ●☆

412997　Walafrida witbergensis (E. Mey.) Rolfe = Selago witbergensis E. Mey. ●☆

412998　Walafrida zeyheri (Choisy) Rolfe = Selago zeyheri Choisy ●☆

412999　Walafrida zuurbergensis Rolfe = Selago zeyheri Choisy ●☆

413000　Walberia Mill. ex Ehret = Nolana L. ex L. f. ■☆

413001　Walcottia F. Muell. = Lachnostachys Hook. ●☆

413002　Walcuffa J. F. Gmel. = Dombeya Cav. (保留属名)●☆

413003　Walcuffa torrida J. F. Gmel. = Dombeya torrida (J. F. Gmel.) Bamps ●☆

413004　Waldeckia Klotzsch = Hirtella L. ●☆

413005　Waldemaria Klotzsch = Rhododendron L. ●

413006　Waldemaria argentea (Hook. f.) Klotzsch = Rhododendron grande Wight ●

413007　Waldheimia Kar. et Kir. (1842);扁芒菊属;Flatawndaisy, Waldheimia ■

413008　Waldheimia Kar. et Kir. = Allardia Decne. ■

413009　Waldheimia glabra (Decne.) Regel;西藏扁芒菊;Tibet Flatawndaisy, Xizang Flatawndaisy, Xizang Waldheimia ■

413010　Waldheimia glabra (Decne.) Regel = Allardia tridactylites (Kar. et Kir.) Sch. Bip. ■

413011　Waldheimia huegelii (Sch. Bip.) Tzvelev;多毛扁芒菊;Huigel Waldheimia, Machhair Flatawndaisy ■

413012　Waldheimia huegelii (Sch. Bip.) Tzvelev = Allardia huegelii Sch. Bip. ■

413013　Waldheimia korolkowii Regel et Schmalh. = Waldheimia stoliczkae (C. B. Clarke) Ostenf. ■

413014　Waldheimia lasiocarpa G. X. Fu = Allardia lasiocarpa (G. X. Fu) Bremer et Humphries ■

413015　Waldheimia nivea (Hook. f. et Thomsonex C. B. Clarke) Regel = Allardia nivea Hook. et Roem. ex C. B. Clarke ■

413016　Waldheimia stoliczkae (C. B. Clarke) Ostenf. = Allardia stoliczkae C. B. Clarke ■

413017　Waldheimia stracheyana Regel = Waldheimia huegelii (Sch. Bip.) Tzvelev ■

413018　Waldheimia tomentosa (Decne.) Regel;羽叶扁芒菊;Tomentose Flatawndaisy,Tomentose Waldheimia ■

413019　Waldheimia tomentosa (Decne.) Regel = Allardia tomentosa Decne. ■

413020　Waldheimia transalaica Tzvelev;外阿拉扁芒菊■☆

413021　Waldheimia tridactylites Kar. et Kir.;扁芒菊(新疆扁芒菊);Cock's Foot Waldheimia,Flatawndaisy ■

413022　Waldheimia tridactylites Kar. et Kir. = Allardia tridactylites (Kar. et Kir.) Sch. Bip. ■

413023　Waldheimia vestita (Hook. f. et Thomson ex C. B. Clarke) Pamp.;厚毛扁芒菊;Thickhairy Waldheimia ■

413024　Waldheimia vestita (Hook. f. et Thomson ex C. B. Clarke) Pamp. = Allardia vestita Hook. f. et Thomson ex C. B. Clarke ■

413025　Waldschmidia F. H. Wigg. = Nymphoides Ség. ■

413026　Waldschmidia Weber = Nymphoides Ség. ■

413027　Waldschmidtia Bluff et Firgerh. = Waldschmidia Weber ■

413028　Waldschmidtia Scop. = Apalatoa Aubl. (废弃属名)●☆

413029　Waldschmidtia Scop. = Crudia Schreb. (保留属名)●☆

413030　Waldsteinia Willd. (1799);林石草属;Barren Strawberry,Waldsteinia ■

413031　Waldsteinia fragarioides (Michx.) Tratt.;草莓状林石草;Barren Strawberry,Barren-strawberry,Dry Strawberry,Waldsteinia ■☆

413032　Waldsteinia lobata Torr. et Gray;浅裂林石草■☆

413033　Waldsteinia sibirica Tratt. = Waldsteinia ternata (Stephan) Fritsch ■

413034　Waldsteinia ternata (Stephan) Fritsch;林石草(西伯利亚林石草,小金梅);Dry Strawberry,Siberia Waldsteinia,Ternate Waldsteinia,Yellow Strawberry ■

413035　Waldsteinia ternata (Stephan) Fritsch var. glabriuscula Te T. Yu et C. L. Li;光叶林石草;Glabrous Ternate Waldsteinia ■

413036　Waldsteinia trifolia Rochel ex Koch = Waldsteinia ternata (Stephan) Fritsch ■

413037　Walidda (A. DC.) Pichon = Wrightia R. Br. ●

413038　Walkera Schreb. = Campylospermum Tiegh. ●

413039　Walkera Schreb. = Meesia Gaertn. ●

413040　Walkera Schreb. = Ouratea Aubl. (保留属名)●

413041　Walkera serrata (Gaertn.) Willd. = Campylospermum serratum (Gaertn.) Bittrich et M. C. E. Amaral ●

413042　Walkeria A. Chev. = Lecomtedoxa (Pierre ex Engl.) Dubard ●☆

413043　Walkeria A. Chev. = Nogo Baehni ●☆

413044　Walkeria Mill. ex Ehret = Atropa L. ■

413045　Walkeria Mill. ex Ehret = Nolana L. ex L. f. ■☆

413046　Walkeria Mill. ex Ehret = Zwingera Hofer ■

413047　Walkeria heitziana A. Chev. = Lecomtedoxa nogo (A. Chev.) Aubrév. ●☆

413048　Walkeria nogo A. Chev. = Lecomtedoxa nogo (A. Chev.) Aubrév. ●☆

413049　Walkuffa Bruce ex Steud. = Dombeya Cav. (保留属名)●☆

413050　Walkuffa Bruce ex Steud. = Walcuffa J. F. Gmel. ●☆

413051　Wallacea Spruce = Wallacea Spruce ex Benth. et Hook. f. ●☆

413052　Wallacea Spruce ex Benth. et Hook. f. (1862);华莱士木属●☆

413053　Wallacea Spruce ex Hook. = Wallacea Spruce ex Benth. et Hook. f. ●☆

413054　Wallacea insignis Spruce ex Benth. et Hook. f.;华莱士木●☆

413055　Wallacea multiflora Ducke;多花华莱士木●☆

413056　Wallaceaceae Tiegh. = Ochnaceae DC. (保留科名)●■

413057　Wallaceodendron Koord. (1898);褐冠豆属●☆

413058　Wallaceodendron celebicum Koord.;褐冠豆●☆

413059　Wallenia Sw. (1788)(保留属名);沃伦紫金牛属;Wallenia ●☆

413060　Wallenia Sw. (保留属名) = Cybianthus Mart. (保留属名)●☆

413061　Wallenia angustifolia Nees et Mart.;窄叶沃伦紫金牛●☆

413062　Wallenia discolor Urb.;杂色沃伦紫金牛●☆

413063　Wallenia gracilis Alain;细沃伦紫金牛●☆

413064　Wallenia laxiflora Mart.;疏花沃伦紫金牛●☆

413065　Wallenia punctulata Urb.;斑点沃伦紫金牛●☆

413066　Wallenia purpurascens Mez;浅紫沃伦紫金牛●☆

413067　Wallenia sylvestris Urb.;森林沃伦紫金牛●☆

413068　Walleniella P. Wilson = Solonia Urb. ●☆

413069　Walleria J. Kirk(1864);肉根草属■☆

413070　Walleria armata Schltr. et K. Krause = Walleria gracilis (Salisb.) S. Carter ■☆

413071　Walleria gracilis (Salisb.) S. Carter;纤细肉根草■☆

413072　Walleria mackenzii J. Kirk;马氏肉根草■☆

413073　Walleria muricata N. E. Br. = Walleria nutans J. Kirk ■☆

413074　Walleria nutans J. Kirk;肉根草■☆

413075　Walleriaceae H. Huber ex Takht. (1995)(保留科名);肉根草科■☆

413076　Walleriaceae H. Huber ex Takht. (保留科名) = Liliaceae Juss. (保留科名)■●

413077　Walleriaceae H. Huber ex Takht. (保留科名) = Tecophilaeaceae Leyb. (保留科名)■☆

413078　Wallia Alef. = Juglans L. ●

413079　Wallia cinerea (L.) Alef. = Juglans cinerea L. ●☆

413080　Wallia nigra (L.) Alef. = Juglans nigra L. ●☆

413081　Wallichia DC. = Eriolaena DC. ●

413082　Wallichia DC. = Schillera Rchb. ●

413083　Wallichia Reinw. = Urophyllum Jack ex Wall. ●

413084　Wallichia Reinw. ex Blume = Urophyllum Jack ex Wall. ●

413085　Wallichia Roxb. (1820);瓦理棕属(华立加椰子属,华立氏椰子属,华羽棕属,琴叶椰属,娃利嘉桐属,瓦理椰属,小董棕属,羽毛椰子属);Wallich Palm,Wallichia,Wallichpalm ●

413086　Wallichia caryotoides Roxb.;琴叶瓦理棕;Fishtail Wallichpalm,Violin-leaved Wallichpalm ●

413087　Wallichia caryotoides Roxb. = Wallichia disticha T. Anderson ●

413088　Wallichia caudata (Lour.) Mart. = Arenga caudata (Lour.) H. E. Moore ●

413089　Wallichia caudata (Lour.) Mart. = Didymosperma caudatum (Lour.) H. Wendl. et Drude ●

413090　Wallichia chinensis Burret;瓦理棕(小董棕);Chinese Wallich Palm,Chinese Wallichia,Wallichpalm ●

413091　Wallichia densiflora Mart.;密花瓦理棕(密花小董棕);Denseflower Wallichia,Dense-flowered Wallichia,Flowery Wallichpalm,Takoru Palm ● u

413092　Wallichia densiflora Mart. = Wallichia caryotoides Roxb. ●

413093　Wallichia disticha T. Anderson；二列瓦理棕（二列小堇棕，华立氏椰子）；Distichous-leaved Wallichpalm, Straddleleaf Wallichpalm, Twolow Wallichpalm, Wallich Palm ●

413094　Wallichia mooreana S. K. Basu；云南瓦理棕（摩氏瓦理棕）；Moore Wallichpalm, Yunnan Wallichpalm, Yunnan Wallichpalm ●

413095　Wallichia oblongifolia Griff. = Wallichia densiflora Mart. ● u

413096　Wallichia porphyrocarpa Blume ex Mart. = Didymosperma porphyrocarpum（Blume ex Mart.）H. Wendl. et Drude ex Hook. f. ●☆

413097　Wallichia quinquelocularis（Wight et Arn.）Steud. = Eriolaena quinquelocularis（Wight et Arn.）Wight ●

413098　Wallichia quinquelocularis Steud. = Eriolaena quinquelocularis（Wight et Arn.）Wight ●

413099　Wallichia siamensis Becc.；泰国瓦理棕；Siam Wallichia, Tailand Wallichpalm ●

413100　Wallichia spectabilia DC. = Eriolaena spectabilis（DC.）Planch. ex Hook. f. ●

413101　Wallichia yomae Kurz = Wallichia disticha T. Anderson ●

413102　Wallinia Moq. = Lophiocarpus Turcz. ■☆

413103　Wallinia polystachya（Turcz.）Moq. = Lophiocarpus polystachyus Turcz. ■☆

413104　Wallisia（Regel）E. Morren = Tillandsia L. ■☆

413105　Wallisia E. Morren = Tillandsia L. ■☆

413106　Wallisia Regel = Lisianthus P. Browne ■☆

413107　Wallisia Regel = Schlimia Regel ■☆

413108　Wallnoeferia Szlach.（1994）；秘鲁喜湿兰属■☆

413109　Wallnoeferia peruviana Szlach.；秘鲁喜湿兰■☆

413110　Wallrothia Roth = Vitex L. ●

413111　Wallrothia Spreng. = Bunium L. ■☆

413112　Wallrothia Spreng. = Seseli L. ■

413113　Walnewa Hort. = Leochilus Knowles et Westc. ■☆

413114　Walnewa Hort. = Waluewa Regel ■☆

413115　Walpersia Harv.（1862）（保留属名）；瓦尔豆属●☆

413116　Walpersia Harv.（保留属名）= Phyllota（DC.）Benth. ●☆

413117　Walpersia Harv. et Send. = Walpersia Harv.（保留属名）●☆

413118　Walpersia Harv. et Sond. = Phyllota（DC.）Benth. ●☆

413119　Walpersia Meisn. ex Krauss = Rhynchosia Lour.（保留属名）●■

413120　Walpersia Reissek = Phylica L. ●☆

413121　Walpersia Reissek ex Endl.（废弃属名）= Trichocephalus Brongn. ●☆

413122　Walpersia Reissek ex Endl.（废弃属名）= Walpersia Harv.（保留属名）●☆

413123　Walpersia rigida C. Presl；硬瓦尔豆●☆

413124　Walsura Roxb.（1832）；割舌树属；Cuttonguetree, Walsura ●

413125　Walsura cochinchinensis（Baill.）Harms；越南割舌树（南方割舌树）；Cochinchina Walsura, Cochin-China Walsura, Vietnam Cuttonguetree, Vietnamese Cuttonguetree ●

413126　Walsura monophylla Elmer ex Merr.；单叶割舌树●☆

413127　Walsura piscidia Roxb.；毒鱼割舌树●☆

413128　Walsura pubescens Kurz = Trichilia connaroides（Wight et Arn.）Benth. ●

413129　Walsura robusta Roxb.；割舌树；Robust Walsura, Sturdy Cuttonguetree ●

413130　Walsura trijuga（Roxb.）Kurz = Heynea trijuga Roxb. ●

413131　Walsura trijuga（Roxb.）Kurz var. microcarpa（Pierre）S. Y. Hu = Heynea trijuga Roxb. ●

413132　Walsura trijuga Kurz = Trichilia connaroides（Wight et Arn.）Benty. ●

413133　Walsura trijuga Kurz var. microcarpa（Pierre）Hu = Trichilia connaroides（Wight et Arn.）Bentv. var. microcarpa（Pierre）Bentv. ●

413134　Walsura tubulata Hiern = Walsura yunnanensis C. Y. Wu ●

413135　Walsura xizangensis C. Y. Wu et H. Li = Glycosmis xizangensis（C. Y. Wu et H. Li）D. D. Tao ●

413136　Walsura yunnanensis C. Y. Wu；云南割舌树；Yunnan Cuttonguetree, Yunnan Walsura ●

413137　Walteranthus Keighery（1985）；澳洲环蕊木属●☆

413138　Walteranthus erectus Keighery.；澳洲环蕊木●☆

413139　Walteria A. St. -Hil. = Vateria L. ●☆

413140　Walteria Scop. = Waltheria L. ●■

413141　Walteriana Fraser ex Endl. = Cliftonia Banks ex C. F. Gaertn. ●☆

413142　Waltheria L.（1753）；蛇婆子属（草梧桐属）；Waltheria ●■

413143　Waltheria americana L.；草梧桐●■

413144　Waltheria americana L. = Waltheria indica L. ●■

413145　Waltheria americana L. var. indica（L.）K. Schum. = Waltheria indica L. ●■

413146　Waltheria americana L. var. sahelica Roberty = Waltheria indica L. ●■

413147　Waltheria americana L. var. subspicata K. Schum. = Waltheria indica L. ●■

413148　Waltheria elliptica Cav. = Waltheria indica L. ●■

413149　Waltheria indica L.；蛇婆子（草梧桐，倒地梅，合他草，满地毯，美洲蛇婆子，山胶浊，仙人撒网，印度蛇婆子）；Common Waltheria, Florida Waltheria, India Waltheria ●■

413150　Waltheria indica L. var. americana（L.）R. Br. ex Hosaka = Waltheria indica L. ●■

413151　Waltheria lanceolata R. Br. ex Mast.；剑叶蛇婆子■☆

413152　Waltheria laxa Thulin；松散蛇婆子■☆

413153　Waltheria makinoi Hayata = Waltheria indica L. ●■

413154　Waluewa Regel = Leochilus Knowles et Westc. ■☆

413155　Walwhalleya Wills et J. J. Bruhl（1991）；瓦尔草属■☆

413156　Walwhalleya proluta（F. Muell.）Wills et J. J. Bruhl；瓦尔草■☆

413157　Wamalchitamia Strother（1991）；棱果菊属●☆

413158　Wamalchitamia appressipila（S. F. Blake）Strother；棱果菊●☆

413159　Wangenheimia A. Dietr. = Dendropanax Decne. et Planch. ●

413160　Wangenheimia F. Dietr. = Gilibertia Ruiz et Pav. ●

413161　Wangenheimia Moench（1794）；万根鼠茅属■☆

413162　Wangenheimia demnatensis（Murb.）Stace；非洲万根鼠茅■☆

413163　Wangenheimia lima（L.）Trin.；万根鼠茅■☆

413164　Wangenheimia lima（L.）Trin. var. glabra Maire = Wangenheimia lima（L.）Trin. ■☆

413165　Wangenheimia lima（L.）Trin. var. villosula Maire = Wangenheimia lima（L.）Trin. ■☆

413166　Wangerinia E. Franz = Microphyes Phil. ■☆

413167　Warburgia Engl.（1895）（保留属名）；十数樟属●☆

413168　Warburgia breyeri Pott = Warburgia salutaris（G. Bertol.）Chiov. ●☆

413169　Warburgia elongata Verdc.；伸长十数樟●☆

413170　Warburgia salutaris（G. Bertol.）Chiov.；澳非十数樟●☆

413171　Warburgia stuhlmannii Engl.；斯图尔曼十数樟●☆

413172　Warburgia ugandensis Sprague subsp. longifolia Verdc.；长叶十数樟●☆

413173　Warburgina Eig（1927）；瓦尔茜属☆

413174　Warburgina factorovskyi Eig；瓦尔茜☆

413175　Warburtonia F. Muell. = Hibbertia Andréws ●☆

413176 Warczewiczella Rchb. f. = Warrea Lindl. ■☆
413177 Warczewitzia Skinner = Catasetum Rich. ex Kunth ■☆
413178 Wardaster J. Small = Aster L. ●■
413179 Wardaster J. Small(1926);华菀属■☆
413180 Wardaster lanuginosus J. Small;华菀■☆
413181 Wardaster lanuginosus J. Small = Aster lanuginosus (J. Small) Y. Ling ■
413182 Wardenia King = Brassaiopsis Decne. et Planch. ●
413183 Warea C. B. Clarke = Biswarea Cogn. ■
413184 Warea Nutt. (1834);韦尔芥属■☆
413185 Warea amplexifolia Nutt. ;韦尔芥■☆
413186 Warea tonglensis C. B. Clarke = Biswarea tonglensis (C. B. Clarke) Cogn. ■
413187 Waria Aubl. = Uvaria L. ●
413188 Warionia Benth. et Coss. (1872);沙菊木属●☆
413189 Warionia saharae Benth. et Coss. ;沙菊木●☆
413190 Warmingia Engl. (废弃属名) = Spondias L. ●
413191 Warmingia Engl. (废弃属名) = Warmingia Rchb. f. (保留属名)■☆
413192 Warmingia Rchb. f. (1881)(保留属名);瓦明兰属■☆
413193 Warmingia eugenii Rchb. f. ;瓦明兰■☆
413194 Warmingia pauciflora Engl. = Spondias purpurea L. ●☆
413195 Warneckea Gilg = Memecylon L. ●
413196 Warneckea Gilg(1904);沃内野牡丹属●☆
413197 Warneckea acutifolia (De Wild.) Jacq.-Fél.;尖叶沃内野牡丹●☆
413198 Warneckea amaniensis Gilg;阿马尼沃内野牡丹●☆
413199 Warneckea bebaiensis (Gilg ex Engl.) Jacq.-Fél.;贝贝沃内野牡丹●☆
413200 Warneckea bequaertii (De Wild.) Jacq.-Fél. = Lijndenia bequaertii (De Wild.) Borhidi ●☆
413201 Warneckea cauliflora Jacq.-Fél.;茎花沃内野牡丹●☆
413202 Warneckea cinnamomoides (G. Don) Jacq.-Fél.;肉桂色沃内野牡丹●☆
413203 Warneckea congolensis (A. Fern. et R. Fern.) Jacq.-Fél.;刚果沃内野牡丹●☆
413204 Warneckea erubescens (Gilg) Jacq.-Fél.;变红沃内野牡丹●☆
413205 Warneckea fascicularis (Planch. ex Benth.) Jacq.-Fél.;扁沃内野牡丹●☆
413206 Warneckea floribunda Jacq.-Fél.;繁花沃内野牡丹●☆
413207 Warneckea fosteri (Hutch. et Dalziel) Jacq.-Fél.;福斯特沃内野牡丹●☆
413208 Warneckea gilletii (De Wild.) Jacq.-Fél.;吉勒特沃内野牡丹●☆
413209 Warneckea golaensis (Baker f.) Jacq.-Fél.;戈拉沃内野牡丹●☆
413210 Warneckea guineensis (Keay) Jacq.-Fél.;几内亚沃内野牡丹●☆
413211 Warneckea hedbergiorum (Borhidi) Borhidi;赫德沃内野牡丹●☆
413212 Warneckea jasminoides (Gilg) Jacq.-Fél. = Lijndenia jasminoides (Gilg) Borhidi ●☆
413213 Warneckea lecomteana Jacq.-Fél.;勒孔特沃内野牡丹●☆
413214 Warneckea macrantha Jacq.-Fél.;大花沃内野牡丹●☆
413215 Warneckea masoalae R. D. Stone;马苏阿拉沃内野牡丹●☆
413216 Warneckea membranifolia (Hook. f.) Jacq.-Fél.;膜叶沃内野牡丹●☆
413217 Warneckea memecyloides (Benth.) Jacq.-Fél.;谷木沃内野牡丹●☆
413218 Warneckea microphylla (Gilg) Borhidi;小叶沃内野牡丹●☆

413219 Warneckea mouririfolia (Brenan) Borhidi;拟穆里野牡丹●☆
413220 Warneckea pulcherrima (Gilg) Jacq.-Fél.;美丽沃内野牡丹●☆
413221 Warneckea reygaertii (De Wild.) Jacq.-Fél.;赖氏沃内野牡丹●☆
413222 Warneckea sansibarica (Taub.) Jacq.-Fél.;桑给巴尔沃内野牡丹●☆
413223 Warneckea sansibarica (Taub.) Jacq.-Fél. var. buchananii (Gilg) A. Fern. et R. Fern.;布坎南沃内野牡丹●☆
413224 Warneckea sapinii (De Wild.) Jacq.-Fél.;安哥拉沃内野牡丹●☆
413225 Warneckea schliebenii (Markgr.) Jacq.-Fél.;施利本沃内野牡丹●☆
413226 Warneckea sessilicarpa (A. Fern. et R. Fern.) Jacq.-Fél.;无梗沃内野牡丹●☆
413227 Warneckea sousae (A. Fern. et R. Fern.) A. E. van Wyk;索萨沃内野牡丹●☆
413228 Warneckea superba (A. Fern. et R. Fern.) Jacq.-Fél.;华美沃内野牡丹●☆
413229 Warneckea walikalensis (A. Fern. et R. Fern.) Jacq.-Fél.;瓦利卡莱沃内野牡丹●☆
413230 Warneckea wildeana Jacq.-Fél.;维尔德沃内野牡丹●☆
413231 Warneckea yangambensis (A. Fern. et R. Fern.) Jacq.-Fél.;扬甘比沃内野牡丹●☆
413232 Warnera Mill. = Hydrastis Ellis ex L. ■☆
413233 Warnera Mill. = Warneria Mill. ●
413234 Warneria Ellis = Gardenia Ellis(保留属名)●
413235 Warneria Ellis = Varnera L. ●
413236 Warneria Mill. = Hydrastis Ellis ex L. ■☆
413237 Warneria Mill. ex L. = Watsonia Mill. (保留属名)■☆
413238 Warnockia M. W. Turner(1996);马岛瓦尔草属■☆
413239 Warnockia scuttelarioides (Engelm. et A. Gray) M. W. Turner;马岛瓦尔草■☆
413240 Warpuria Stapf = Podorungia Baill. ■☆
413241 Warpuria clandestina Stapf = Podorungia clandestina (Stapf) Benoist ■☆
413242 Warpuria serotina Benoist = Podorungia serotina (Benoist) Benoist ■☆
413243 Warrea Lindl. (1843);瓦利兰属;Varrea ■☆
413244 Warrea costaricensis Schltr. ;哥斯达黎加瓦利兰;Costarica Varrea ■☆
413245 Warrea discolor Lindl. = Chondrorhyncha discolor (Lindl.) P. H. Allen ■☆
413246 Warrea warreana (Lodd. ex Lindl.) C. Schweinf. ;瓦利兰;Common Varrea ■☆
413247 Warreella Schltr. (1914);小瓦利兰属■☆
413248 Warreella cyanea (Lindl.) Schltr. ;小瓦利兰■☆
413249 Warreopsis Garay(1973);类瓦利兰属■☆
413250 Warreopsis pardina (Rchb. f.) Garay;类瓦利兰■☆
413251 Warreopsis wightiana (Wall. ex Wight et Arn.) M. Roem. ;怀氏类瓦利兰■☆
413252 Warscaea Szlach. (1994);瓦尔绥草属(盘龙参属)■☆
413253 Warscaea Szlach. = Spiranthes Rich. (保留属名)■
413254 Warscewiczella Rchb. f. (1852);瓦氏兰属■☆
413255 Warscewiczella Rchb. f. = Chondrorhyncha Lindl. ■☆
413256 Warscewiczella Rchb. f. = Cochleanthes Raf. ■☆
413257 Warscewiczella Rchb. f. = Warczewiczella Rchb. f. ■☆
413258 Warscewiczella Rchb. f. = Warrea Lindl. ■☆

413259　Warscewiczella Rchb. f. = Zygopetalum Hook. ■☆

413260　Warscewiczella discolor Rchb. f.；异色瓦氏兰（二色喙柱兰，二色康多兰）■☆

413261　Warscewiczia Klotzsch(1853)；瓦氏芸香属（沃泽维奇属）■☆

413262　Warscewiczia Post et Kuntze = Catasetum Rich. ex Kunth ■☆

413263　Warscewiczia Post et Kuntze = Warczewitzia Skinner ■☆

413264　Warszewiczella Benth. et Hook. f. = Cochleanthes Raf. ■☆

413265　Warszewiczella Benth. et Hook. f. = Warscewiczella Rchb. f. ■☆

413266　Warszewiczella Rchb. f. = Warscewiczella Rchb. f. ■☆

413267　Warszewiczella discolor Rchb. f.；异色瓦氏芸香■☆

413268　Warszewiczia Klotzsch = Warscewiczia Klotzsch ■☆

413269　Warszewiczia coccinea（Vahl）Klotzsch；亮红瓦氏芸香（亮红沃泽维奇）；Chaconia，Wild Poinsettia ●☆

413270　Warthemia Boiss. = Iphiona Cass.（保留属名）●■☆

413271　Warthemia Boiss. = Varthemia DC. ●■☆

413272　Wartmannia Müll. Arg. = Homalanthus A. Juss.（保留属名）●

413273　Wasabia Matsum.（1899）；生鱼芥属■

413274　Wasabia Matsum. = Eutrema R. Br. ■

413275　Wasabia bracteata（S. Moore）Hisauchi = Eutrema tenue（Miq.）Makino ■

413276　Wasabia hederifolia（Franch. et Sav.）Matsum. = Eutrema tenuis Makino ■

413277　Wasabia japonica（Miq.）Matsum. = Eutrema wasabii（Siebold）Maxim. ■

413278　Wasabia japonica（Miq.）Matsum. var. sachalinensis（Miyabe et Miyake）Hisauti = Eutrema japonicum（Miq.）Koidz. var. sachalinense（Miyabe et Miyake）Nemoto ■☆

413279　Wasabia koreana Nakai = Eutrema wasabii（Siebold）Maxim. ■

413280　Wasabia pungens Matsum. = Eutrema wasabii（Siebold）Maxim. ■

413281　Wasabia tenuis（Miq.）Matsum. = Eutrema tenuis Makino ■

413282　Wasabia tenuis（Miq.）Matsum. var. okinosimensis（Taken.）Kitam. = Eutrema okinosimense Taken. ■

413283　Wasabia wasabi（Siebold）Makino = Eutrema wasabi（Siebold）Maxim. ■

413284　Wasabia wasabi Makino = Eutrema wasabi（Siebold）Maxim. ■

413285　Wasabia yunnanense（Franch.）Nakai = Eutrema yunnanense Franch. ■

413286　Wasatchia M. E. Jones = Festuca L. ■

413287　Wasatchia M. E. Jones = Hesperochloa Rydb. ■

413288　Washingtonia C. Winslow = Washingtonia H. Wendl.（保留属名）●

413289　Washingtonia H. Wendl.（1879）（保留属名）；丝葵属（华盛顿椰属，华盛顿棕子属，华盛顿棕属，加州葵属，加州蒲葵属，老人葵属，裙棕属，银丝棕属）；Fan Palm，Silkpalm，Washington Palm，Washingtonia ●

413290　Washingtonia Raf. = Osmorhiza Raf.（保留属名）■

413291　Washingtonia Winsl. = Sequoiadendron J. Buchholz ●

413292　Washingtonia claytonii（Michx.）Britton = Osmorhiza aristata（Thunb.）Makino et Y. Yabe ■

413293　Washingtonia claytonii（Michx.）Britton = Osmorhiza claytonii（Michx.）C. B. Clarke ■☆

413294　Washingtonia divaricata Britton = Osmorhiza berteroi DC. ■☆

413295　Washingtonia filamentosa（Franeeschi）Kuntze = Washingtonia filifera（Linden ex André）H. Wendl. ex de Bary ●

413296　Washingtonia filamentosa（H. Wendl. ex Franeeschi）Kuntze = Washingtonia filifera（Linden ex André）H. Wendl. ex de Bary ●

413297　Washingtonia filifera（Linden ex André）H. Wendl. = Washingtonia filifera（Linden ex André）H. Wendl. ex de Bary ●

413298　Washingtonia filifera（Linden ex André）H. Wendl. ex de Bary；丝葵（华盛顿椰子，华盛顿棕，华盛顿棕榈,加州蒲葵，老人葵）；California Fan Palm，California Palm，California Washington Palm，California Washingtonia，Cotton Palm，Desert Fan Palm，Fan Palm，Fanpalm，Gray Green Washingtonia，Petticoat Palm，Silkpalm，Washingtonia Palm ●

413299　Washingtonia filifera（Linden ex André）H. Wendl. ex de Bary var. arizonica（O. F. Cook ex Annon.？）M. E. Jones = Washingtonia filifera（Linden ex André）H. Wendl. ex de Bary ●

413300　Washingtonia filifera（Linden ex André）H. Wendl. ex de Bary var. robusta（H. Wendl.）Parish = Washingtonia robusta H. Wendl. ●

413301　Washingtonia filifera（Linden ex André）H. Wendl. ex de Bary var. sonorae（S. Watson）M. E. Jones = Washingtonia robusta H. Wendl. ●

413302　Washingtonia filifera（Linden ex André）H. Wendl. ex de Bary var. typica M. E. Jones = Washingtonia filifera（Linden ex André）H. Wendl. ex de Bary ●

413303　Washingtonia filifera（Linden）H. Wendl. = Washingtonia filifera（Linden ex André）H. Wendl. ex de Bary ●

413304　Washingtonia gracilis Parish = Washingtonia robusta H. Wendl. ●

413305　Washingtonia laxa（Royle）Koso-Pol. ex B. Fedtsch. = Osmorhiza aristata（Thunb.）Makino et Y. Yabe var. laxa（Royle）Constance et R. H. Shan ■

413306　Washingtonia longistylis（Torr.）Britton = Osmorhiza iongistylis（Torr.）DC. ■☆

413307　Washingtonia robusta H. Wendl.；大丝葵（光叶加州蒲葵,壮干棕榈）；Big Silkpalm，Brilliant-leaf Washingtonia，Cotton Palm，Desert Palm，Fan Palm，Mexican Fan Palm，Mexican Fanpalm，Mexican Washington Palm，Mexican Washingtonia，Thread Palm，Washington Fan Palm ●

413308　Washingtonia robusta H. Wendl. var. gracilis（Parish）Becc. = Washingtonia robusta H. Wendl. ●

413309　Washingtonia sonorae（S. Watson）Rose = Washingtonia robusta H. Wendl. ●

413310　Washingtonia sonorae S. Watson = Washingtonia robusta H Wendl. ●

413311　Waterhousea B. Hyland(1983)；沃特桃金娘属●☆

413312　Waterhousea floribunda（F. Muell.）B. Hyland；沃特桃金娘；Weeping Lillypilly，Weeping Satinash ●☆

413313　Watsonamra Kuntze = Pentagonia Benth.（保留属名）■☆

413314　Watsonia Boehm. = Byttneria Loefl.（保留属名）●

413315　Watsonia Mill.（1758）（保留属名）；沃森花属；Bugle Lily，Buglelily，Bugle-lily，Watson Flower ■☆

413316　Watsonia albertiniensis P. E. Glover = Watsonia laccata（Jacq.）Ker Gawl. ■☆

413317　Watsonia aletroides（Burm. f.）Ker Gawl.；粉条儿菜沃森花■☆

413318　Watsonia amabilis Goldblatt；秀丽沃森花■☆

413319　Watsonia angusta Ker Gawl.；狭沃森花■☆

413320　Watsonia archbelliae L. Bolus = Watsonia pillansii L. Bolus ■☆

413321　Watsonia ardenei Sander = Watsonia borbonica（Pourr.）Goldblatt ■☆

413322　Watsonia ardernei Sander = Watsonia borbonica（Pourr.）Goldblatt subsp. ardernei（Sander）Goldblatt ■☆

413323　Watsonia bachmannii L. Bolus；巴克曼沃森花■☆

413324　Watsonia baurii L. Bolus = Watsonia gladioloides Schltr. ■☆

413325　Watsonia beatricis J. W. Mathews et L. Bolus = Watsonia pillansii L. Bolus ■☆

413326 Watsonia beatricis J. W. Mathews et L. Bolus. ;短筒沃森花;
Shorttube Watson Flower ■☆

413327 Watsonia bella N. E. Br. ex Goldblatt;雅致沃森花■☆

413328 Watsonia borbonica（Pourr.）Goldblatt;锥穗沃森花;Bugle-lily,Cape Bugle-lily,Watsonia ■☆

413329 Watsonia borbonica（Pourr.）Goldblatt subsp. ardernei（Sander）Goldblatt;阿代纳沃森花■☆

413330 Watsonia brevifolia Ker Gawl. = Watsonia laccata（Jacq.）Ker Gawl. ■☆

413331 Watsonia bulbillifera J. W. Mathews et L. Bolus = Watsonia meriana（L.）Mill. ■☆

413332 Watsonia caledonica Baker = Watsonia laccata（Jacq.）Ker Gawl. ■☆

413333 Watsonia canaliculata Goldblatt;具沟沃森花■☆

413334 Watsonia coccinea Herb. ex Baker;绯红沃森花■☆

413335 Watsonia comptonii L. Bolus = Watsonia zeyheri L. Bolus ■☆

413336 Watsonia confusa Goldblatt;混乱沃森花■☆

413337 Watsonia cooperi（Baker）L. Bolus = Watsonia borbonica（Pourr.）Goldblatt ■☆

413338 Watsonia densiflora Baker;密花沃森花(喇叭鸢尾);Bugle Lily ■☆

413339 Watsonia densiflora Baker 'Alba';白色密花沃森花(白色喇叭鸢尾)■☆

413340 Watsonia desmidtii L. Bolus = Watsonia wilmaniae J. W. Mathews et L. Bolus ■☆

413341 Watsonia distans L. Bolus;远离沃森花■☆

413342 Watsonia dubia Eckl. ex Klatt;可疑沃森花■☆

413343 Watsonia elimensis L. Bolus = Watsonia zeyheri L. Bolus ■☆

413344 Watsonia elsiae Goldblatt;埃尔西亚沃森花■☆

413345 Watsonia fergusoniae L. Bolus;费格森沃森花■☆

413346 Watsonia flavida L. Bolus = Watsonia watsonioides（Baker）Oberm. ■☆

413347 Watsonia fourcadei J. W. Mathews et L. Bolus;长筒沃森花■☆

413348 Watsonia galpinii L. Bolus;盖尔沃森花■☆

413349 Watsonia gladioloides Schltr. ;唐菖蒲沃森花■☆

413350 Watsonia humilis Mill. ;低矮沃森花■☆

413351 Watsonia hutchinsonii L. Bolus = Watsonia pillansii L. Bolus ■☆

413352 Watsonia hyacinthoides Pers. = Watsonia laccata（Jacq.）Ker Gawl. ■☆

413353 Watsonia hysterantha J. W. Mathews et L. Bolus;宫花沃森花■☆

413354 Watsonia inclinata Goldblatt;下倾沃森花■☆

413355 Watsonia iridifolia（Jacq.）Ker Gawl. = Watsonia meriana（L.）Mill. ■☆

413356 Watsonia iridifolia（Jacq.）Ker Gawl. var. obrienii N. E. Br. = Watsonia borbonica（Pourr.）Goldblatt subsp. ardernei（Sander）Goldblatt ■☆

413357 Watsonia juncifolia（Baker）Baker = Thereianthus juncifolius（Baker）G. J. Lewis ■☆

413358 Watsonia knysnana L. Bolus;克尼斯纳沃森花■☆

413359 Watsonia laccata（Jacq.）Ker Gawl. ;撕裂沃森花■☆

413360 Watsonia lapeyrousioides Baker = Thereianthus minutus（Klatt）G. J. Lewis ■☆

413361 Watsonia latifolia N. E. Br. ex Oberm. ;宽叶沃森花■☆

413362 Watsonia leipoldtii L. Bolus = Watsonia meriana（L.）Mill. ■☆

413363 Watsonia lepida N. E. Br. ;小鳞沃森花■☆

413364 Watsonia longicollis Schltr. = Thereianthus longicollis（Schltr.）G. J. Lewis ■☆

413365 Watsonia lucidor Eckl. = Tritoniopsis triticea（Burm. f.）Goldblatt ■☆

413366 Watsonia marginata（L. f.）Ker Gawl. ;香沃森花;Fragrant Bugle-lily ■☆

413367 Watsonia marlothii L. Bolus;马洛斯沃森花■☆

413368 Watsonia masoniae L. Bolus = Watsonia pillansii L. Bolus ■☆

413369 Watsonia meriana（L.）Mill. ;玛丽沃森花;Bulbil Bugle-lily,Lakepypie,Marian Watson Flower ■☆

413370 Watsonia meriana（L.）Mill. var. bulbillifera（J. W. Mathews et L. Bolus）D. A. Cooke;球根沃森花■☆

413371 Watsonia meriana Mill. = Watsonia meriana（L.）Mill. ■☆

413372 Watsonia middlemostii L. Bolus = Watsonia rogersii L. Bolus ■☆

413373 Watsonia minima Goldblatt;极小沃森花■☆

413374 Watsonia minuta Klatt = Thereianthus minutus（Klatt）G. J. Lewis ■☆

413375 Watsonia muirii E. Phillips = Watsonia laccata（Jacq.）Ker Gawl. ■☆

413376 Watsonia natalensis Eckl. = Gladiolus dalenii Van Geel ■☆

413377 Watsonia neglecta N. E. Br. = Watsonia densiflora Baker ■☆

413378 Watsonia obrienii（N. E. Br.）Tubergen = Watsonia borbonica（Pourr.）Goldblatt subsp. ardernei（Sander）Goldblatt ■☆

413379 Watsonia occulta L. Bolus;隐蔽沃森花■☆

413380 Watsonia pauciflora L. Bolus = Watsonia distans L. Bolus ■☆

413381 Watsonia paucifolia Goldblatt;少叶沃森花■☆

413382 Watsonia pillansii L. Bolus;橙红沃森花■☆

413383 Watsonia pilosa Klatt = Gladiolus bonaspei Goldblatt et M. P. de Vos ■☆

413384 Watsonia plantii N. E. Br. = Watsonia densiflora Baker ■☆

413385 Watsonia pondoensis Goldblatt;庞多沃森花■☆

413386 Watsonia pottbergensis Eckl. = Watsonia stenosiphon L. Bolus ■☆

413387 Watsonia praecox（Andréws）Pers. = Gladiolus watsonius Thunb. ■☆

413388 Watsonia priorii L. Bolus = Watsonia pillansii L. Bolus ■☆

413389 Watsonia pulchra N. E. Br. ex Goldblatt;美丽沃森花■☆

413390 Watsonia punctata（Andréws）Ker Gawl. = Thereianthus spicatus（L.）G. J. Lewis ■☆

413391 Watsonia pyramidata（Andr.）Stapf;塔形沃森花;Pyramidal Watson Flower ■☆

413392 Watsonia pyramidata（Andréws）Klatt = Watsonia borbonica（Pourr.）Goldblatt ■☆

413393 Watsonia racemosa Klatt = Thereianthus racemosus（Klatt）G. J. Lewis ■☆

413394 Watsonia retusa Klatt = Ixia micrandra Baker var. minor G. J. Lewis ■☆

413395 Watsonia revoluta Pers. = Gladiolus watsonius Thunb. ■☆

413396 Watsonia rogersii L. Bolus;罗杰斯沃森花■☆

413397 Watsonia rosea Banks ex Ker = Watsonia borbonica（Pourr.）Goldblatt ■☆

413398 Watsonia rosea Ker Gawl. ;玫红沃森花;Rose Watson Flower ■☆

413399 Watsonia rosea Ker Gawl. var. adernei Mathews et L. Bolus;喇叭兰■☆

413400 Watsonia roseoalba Ker Gawl. = Watsonia humilis Mill. ■☆

413401 Watsonia rourkei Goldblatt;鲁尔克沃森花■☆

413402 Watsonia ryderae L. Bolus = Watsonia fourcadei J. W. Mathews et L. Bolus ■☆

413403 Watsonia schinzii L. Bolus = Watsonia schlechteri L. Bolus ■☆

413404 Watsonia schlechteri L. Bolus;施莱沃森花■☆

413405 Watsonia socium J. W. Mathews et L. Bolus = Watsonia pillansii

L. Bolus ■☆

413406　Watsonia spectabilis Schinz;壮观沃森花■☆

413407　Watsonia stanfordiae L. Bolus = Watsonia fourcadei J. W. Mathews et L. Bolus ■☆

413408　Watsonia starkeae L. Bolus = Watsonia wilmaniae J. W. Mathews et L. Bolus ■☆

413409　Watsonia stenosiphon L. Bolus;窄管沃森花■☆

413410　Watsonia stokoei L. Bolus;斯托克沃森花■

413411　Watsonia strictiflora Ker Gawl. ;刚直沃森花■☆

413412　Watsonia tabularis Eckl. = Watsonia tabularis J. W. Mathews et L. Bolus ■☆

413413　Watsonia tabularis J. W. Mathews et L. Bolus;扁平沃森花■☆

413414　Watsonia tigrina Eckl. = Tritoniopsis triticea (Burm. f.) Goldblatt ■☆

413415　Watsonia transvaalensis Baker;德兰士瓦沃森花■☆

413416　Watsonia tubulosa (Andréws) Pers. = Watsonia aletroides (Burm. f.) Ker Gawl. ■☆

413417　Watsonia tubulosa Eckl. = Watsonia aletroides (Burm. f.) Ker Gawl. ■☆

413418　Watsonia vanderspuyiae L. Bolus;范德沃森花■☆

413419　Watsonia vivipara J. W. Mathews et L. Bolus = Watsonia meriana (L.) Mill. ■☆

413420　Watsonia watsonioides (Baker) Oberm. ;沃森花■☆

413421　Watsonia wilmaniae J. W. Mathews et L. Bolus;维尔曼沃森花■☆

413422　Watsonia wilmsii L. Bolus;维尔姆斯沃森花■☆

413423　Watsonia wordsworthiana J. W. Mathews et L. Bolus = Watsonia borbonica (Pourr.) Goldblatt subsp. ardernei (Sander) Goldblatt ■☆

413424　Watsonia zeyheri L. Bolus;泽赫沃森花■☆

413425　Wattakaka (Decne.) Hassk. = Dregea E. Mey. (保留属名)●

413426　Wattakaka Hassk. = Dregea E. Mey. (保留属名)●

413427　Wattakaka corrugata (C. K. Schneid.) Stapf = Dregea sinensis Hemsl. var. corrugata (C. K. Schneid.) Tsiang et P. T. Li ●

413428　Wattakaka corrugata (Scheid.) Stapf = Dregea sinensis Hemsl. var. corrugata (C. K. Schneid.) Tsiang et P. T. Li ●

413429　Wattakaka sinensis (C. K. Schneid.) Stapf var. corrugata (C. K. Schneid.) Tsiang = Dregea sinensis Hemsl. var. corrugata (C. K. Schneid.) Tsiang et P. T. Li ●

413430　Wattakaka sinensis (Hemsl.) Stapf = Dregea sinensis Hemsl. ●

413431　Wattakaka sinensis (Hemsl.) Stapf var. corrugata (Schneid.) Tsiang = Dregea sinensis Hemsl. var. corrugata (C. K. Schneid.) Tsiang et P. T. Li ●

413432　Wattakaka volubilis (L. f.) Stapf = Dregea volubilis (L. f.) Benth. ex Hook. f. ●

413433　Wattakaka yunnanensis Tsiang = Dregea yunnanensis (Tsiang) Tsiang et P. T. Li ●

413434　Wattakaka yunnanensis Tsiang var. major Tsiang = Dregea yunnanensis (Tsiang) Tsiang et P. T. Li ●

413435　Webbia DC. = Vernonia Schreb. (保留属名)●■

413436　Webbia Ruiz et Pav. ex Engl. = Dictyoloma A. Juss. (保留属名)●☆

413437　Webbia Sch. Bip. = Conyza Less. (保留属名)■

413438　Webbia Spach = Huebneria Rchb. ■●

413439　Webbia Spach = Hypericum L. ■●

413440　Webbia aristata DC. = Vernonia natalensis Sch. Bip. ex Walp. ■☆

413441　Webbia elaeagnoides DC. = Vernonia oligocephala (DC.) Sch. Bip. ex Walp. ■☆

413442　Webbia hirsuta DC. = Vernonia hirsuta (DC.) Sch. Bip. ex Walp. ●☆

413443　Webbia kraussii Sch. Bip. = Conyza obscura DC. ■☆

413444　Webbia nudicaulis DC. = Vernonia dregeana Sch. Bip. ■☆

413445　Webbia oligocephala DC. = Vernonia oligocephala (DC.) Sch. Bip. ex Walp. ■☆

413446　Webbia pinifolia (Lam.) DC. = Vernonia capensis (Houtt.) Druce ■☆

413447　Webbia serratuloides DC. = Vernonia perrottetii Sch. Bip. ex Walp. ●☆

413448　Webera Cramer = Plectronia L. ●☆

413449　Webera J. F. Gmel. = Bellucia Neck. ex Raf. (保留属名)●☆

413450　Webera Schreb. = Chomelia L. (废弃属名)●☆

413451　Webera Schreb. = Tarenna Gaertn. ●

413452　Webera attenuata Hook. f. = Tarenna attenuata (Voigt) Hutch. ●

413453　Webera cavaleriei H. Lév. = Randia wallichii Hook. f. ●

413454　Webera cavaleriei H. Lév. = Tarennoidea wallichii (Hook. f.) Tirveng. ●

413455　Webera gracilis Stapf = Tarenna gracilis (Stapf) Keay ●☆

413456　Webera henryi H. Lév. = Randia wallichii Hook. f. ●

413457　Webera henryi H. Lév. = Tarennoidea wallichii (Hook. f.) Tirveng. ●

413458　Webera marchandii H. Lév. = Daphniphyllum macropodum Miq. ●

413459　Webera mollissima Benth. ex Hance = Tarenna mollissima (Hook. et Arn.) Rob. ●

413460　Webera pallida Franch. ex Brandis = Randia wallichii Hook. f. ●

413461　Webera pallida Franch. ex Brandis = Tarennoidea wallichii (Hook. f.) Tirveng. ●

413462　Webera pavettoides (Harv.) Benth. et Hook. f. = Tarenna pavettoides (Harv.) Sim ●☆

413463　Webera saxatilis Scott-Elliot = Paracephaelis saxatilis (Scott-Elliot) De Block ●☆

413464　Weberaster Á. Löve et D. Löve = Aster L. ●■

413465　Weberaster modestus (Lindl.) Á. Löve et D. Löve = Canadanthus modestus (Lindl.) G. L. Nesom ☆

413466　Weberaster radulinus (A. Gray) Á. Löve et D. Löve = Eurybia radulina (A. Gray) G. L. Nesom ☆

413467　Weberbauera Gilg et Muschl. (1909);韦伯芥属■☆

413468　Weberbauera densiflora (Muschl.) Gilg et Muschl. ;韦伯芥■☆

413469　Weberbauerella Ulbr. (1906);小韦豆属■☆

413470　Weberbauerella brongniartioides Ulbr. ;小韦豆■☆

413471　Weberbaueriella Ferreyra = Chucoa Cabrera ●☆

413472　Weberbauerocereus Backeb.(1942);韦伯柱属(韦伯掌属,魏氏仙人柱属)●☆

413473　Weberbauerocereus Backeb. = Haageocereus Backeb. ●☆

413474　Weberbauerocereus cephalomacrostibas (Werderm. et Backeb.) F. Ritter;大头韦伯柱(大头魏氏仙人柱)●☆

413475　Weberbauerocereus johnsonii F. Ritter;韦伯柱（韦伯掌）;Golden Column ●☆

413476　Weberbauerocereus rauhii Backeb. ;白乌龙●☆

413477　Weberiopuntia Frič = Opuntia Mill. ●

413478　Weberiopuntia Frič ex Kreuz. = Opuntia Mill. ●

413479　Weberocereus Britton et Rose(1909);瘤果鞭属●☆

413480　Weberocereus biolleyi (F. A. C. Weber) Britton et Rose;美玉恋■☆

413481　Weberocereus tonduzii (F. A. C. Weber) F. H. Brandt;舞女花;Ballerina Flower ■☆

413482　Websteria S. H. Wright(1887);韦氏莎草属■☆

413483 Websteria confervoides（Poir.）S. S. Hooper；韦氏莎草■☆

413484 Websteria limnophila S. H. Wright = Websteria confervoides（Poir.）S. S. Hooper ■☆

413485 Websteria submersa（C. Wright）Britton = Websteria confervoides（Poir.）S. S. Hooper ■☆

413486 Weddellina Tul.（1849）；韦德尔川苔草属■☆

413487 Weddellina squamulosa Tul.；韦德尔川苔草■☆

413488 Wedela Steud. = Ardisia Sw.（保留属名）●■

413489 Wedela Steud. = Vedela Adans.（废弃属名）●■

413490 Wedelia Jacq.（1760）（保留属名）；蟛蜞菊属；Crabdaisy，Wedelia ■●

413491 Wedelia Loefl.（废弃属名）= Allionia L.（保留属名）■☆

413492 Wedelia Loefl.（废弃属名）= Wedelia Jacq.（保留属名）■●

413493 Wedelia Post et Kuntze = Ardisia Sw.（保留属名）●■

413494 Wedelia Post et Kuntze = Vedela Adans.（废弃属名）●■

413495 Wedelia acapulcensis Kunth；阿地蟛蜞菊■☆

413496 Wedelia acapulcensis Kunth var. hispida（Kunth）Strother = Zexmenia hispida（Kunth）A. Gray ■☆

413497 Wedelia affinis De Wild. = Aspilia mossambicensis（Oliv.）Wild ■☆

413498 Wedelia africana P. Beauv. = Aspilia africana（P. Beauv.）C. D. Adams ■☆

413499 Wedelia africana P. Beauv. var. ambigua（C. D. Adams）Isawumi = Aspilia africana（P. Beauv.）C. D. Adams ■☆

413500 Wedelia africana P. Beauv. var. guineensis（C. D. Adams）Isawumi = Aspilia africana（P. Beauv.）C. D. Adams ■☆

413501 Wedelia africana P. Beauv. var. minor（C. D. Adams）Isawumi = Aspilia africana（P. Beauv.）C. D. Adams ■☆

413502 Wedelia albiflora Hiern = Aspilia angolensis（Klatt）Muschl. ■☆

413503 Wedelia angolensis Klatt = Aspilia angolensis（Klatt）Muschl. ■☆

413504 Wedelia angustifolia（Oliv. et Hiern）Isawumi = Aspilia angustifolia Oliv. et Hiern ■☆

413505 Wedelia asperrima（Decne.）Benth.；无子蟛蜞菊■☆

413506 Wedelia biflora（L.）DC.；孪花蟛蜞菊（大蟛蜞菊，黄泥菜，蟛蜞菊，双花蟛蜞菊）；Twinflower Crabdaisy，Twoflower Wedelia ■

413507 Wedelia biflora（L.）DC. var. ryukyensis H. Koyama；琉球蟛蜞菊（孪花蟛蜞菊，双花蟛蜞菊）；Liuqiu Wedelia ■

413508 Wedelia biflora（L.）Wight = Melanthera biflora（L.）Wild ■☆

413509 Wedelia bracteosa（C. D. Adams）Isawumi = Aspilia helianthoides（Schumach. et Thonn.）Oliv. et Hiern subsp. prieuriana（DC.）C. D. Adams ■☆

413510 Wedelia bussei（O. Hoffm. et Muschl.）Isawumi = Aspilia bussei O. Hoffm. et Muschl. ■☆

413511 Wedelia cachimboensis（H. Rob.）B. L. Turner = Wedelia chinensis（Osbeck）Merr. ■

413512 Wedelia calendulacea（L.）Less. = Wedelia chinensis（Osbeck）Merr. ■

413513 Wedelia chevalieri（O. Hoffm. et Muschl.）Isawumi = Aspilia chevalieri O. Hoffm. et Muschl. ■☆

413514 Wedelia chinensis（Osbeck）Merr.；蟛蜞菊（海砂菊，黄花草，黄花龙舌草，黄花墨菜，黄花蟛蜞草，黄花曲草，金盏蟛蜞菊，卤地菊，路边菊，马兰草，蟛蜞菊，水兰，田黄菊）；China Crabdaisy，Chinese Wedelia ■

413515 Wedelia chinensis（Osbeck）Merr. var. robusta（Makino）Masam.；山素英■

413516 Wedelia cryptocephala Peter = Synedrella nodiflora Gaertn. ■

413517 Wedelia diversipapposa S. Moore = Aspilia mossambicensis（Oliv.）Wild ■☆

413518 Wedelia elongata（DC.）Vatke；伸长蟛蜞菊■☆

413519 Wedelia glauca（Ortega）Hoffm. ex Hicken = Pascalia glauca Ortega ■

413520 Wedelia gossweileri S. Moore；戈斯蟛蜞菊●☆

413521 Wedelia helianthoides（Schumach. et Thonn.）Isawumi；向日葵蟛蜞菊●☆

413522 Wedelia helianthoides（Schumach. et Thonn.）Isawumi subsp. ciliata（Schumach.）Isawumi = Aspilia helianthoides（Schumach. et Thonn.）Oliv. et Hiern subsp. ciliata（Schumach.）C. D. Adams ■☆

413523 Wedelia helianthoides（Schumach. et Thonn.）Isawumi subsp. helianthoides = Aspilia helianthoides（Schumach. et Thonn.）Oliv. et Hiern ■☆

413524 Wedelia helianthoides（Schumach. et Thonn.）Isawumi subsp. papposa（O. Hoffm. et Muschl.）Isawumi = Aspilia helianthoides（Schumach. et Thonn.）Oliv. et Hiern subsp. prieuriana（DC.）C. D. Adams ■☆

413525 Wedelia helianthoides（Schumach. et Thonn.）Isawumi subsp. prieuriana（DC.）Isawumi = Aspilia helianthoides（Schumach. et Thonn.）Oliv. et Hiern subsp. prieuriana（DC.）C. D. Adams ■☆

413526 Wedelia hirtella Humbert；多毛蟛蜞菊●☆

413527 Wedelia hispida Kunth = Zexmenia hispida（Kunth）A. Gray ■☆

413528 Wedelia huillensis Hiern = Aspilia angolensis（Klatt）Muschl. ■☆

413529 Wedelia incarnata（L.）Kuntze subsp. villosa Standl. = Allionia incarnata L. var. villosa（Standl.）Munz ■☆

413530 Wedelia instar S. Moore = Aspilia mossambicensis（Oliv.）Wild ■☆

413531 Wedelia katangensis De Wild. = Aspilia natalensis（Sond.）Wild ■☆

413532 Wedelia kotschyi（Sch. Bip.）Soldano = Aspilia kotschyi（Sch. Bip.）Oliv. ■☆

413533 Wedelia kotschyi（Sch. Bip.）Soldano var. alba（Berhaut）Isawumi = Aspilia kotschyi（Sch. Bip.）Oliv. var. alba Berhaut ■☆

413534 Wedelia linearifolia（Oliv. et Hiern）Isawumi = Aspilia angustifolia Oliv. et Hiern ■☆

413535 Wedelia lundii DC.；伦迪蟛蜞菊■☆

413536 Wedelia macrorrhiza（Chiov.）Chiov. = Aspilia macrorrhiza Chiov. ■☆

413537 Wedelia madagascariensis Humbert = Wedelia elongata（DC.）Vatke ■☆

413538 Wedelia magnifica Chiov. = Aspilia africana（P. Beauv.）C. D. Adams subsp. magnifica（Chiov.）Wild ■☆

413539 Wedelia menotriche Oliv. et Hiern = Aspilia mossambicensis（Oliv.）Wild ■☆

413540 Wedelia montana（Blume）Boerl. var. wallichii（Less.）H. Koyama = Wedelia wallichii Less. ■

413541 Wedelia mortonii（C. D. Adams）Isawumi = Aspilia angustifolia Oliv. et Hiern ■☆

413542 Wedelia mossambicensis Oliv. = Aspilia mossambicensis（Oliv.）Wild ■☆

413543 Wedelia natalensis Sond. = Aspilia natalensis（Sond.）Wild ■☆

413544 Wedelia oblonga Hutch. = Guizotia scabra（Vis.）Chiov. ■☆

413545 Wedelia paludosa（Berhaut）Isawumi = Aspilia paludosa Berhaut ■☆

413546 Wedelia paludosa DC. = Sphagneticola trilobata（L.）Pruski ■☆

413547 Wedelia perrieri Humbert = Chrysogonum perrieri（Humbert）Humbert ■☆

413548 Wedelia pratensis Vatke；草原蟛蜞菊■☆

413549　Wedelia prostrata（Hook. et Arn.）Hemsl. ;卤地菊（单花蟛蜞菊，海滨蟛蜞菊，黄花冬菊，黄花龙舌草，黄花蜜菜，黄花蟛蜞草，黄野蒿，瘌草，尖刀草，龙舌草，龙舌三尖刀，三尖刀，山蟛蜞，天蓬草舅）;Prostrate Crabdaisy,Prostrate Wedelia ■

413550　Wedelia prostrata（Hook. et Arn.）Hemsl. var. robusta Makino;大天蓬草舅（背草，瘌草）■

413551　Wedelia radiosa Ker Gawl. ;射线蟛蜞菊■☆

413552　Wedelia ringoetii De Wild. = Aspilia mossambicensis（Oliv.）Wild ■☆

413553　Wedelia robusta（Makino）Kitam. = Wedelia prostrata（Hook. et Arn.）Hemsl. var. robusta Makino ■

413554　Wedelia rudis（Oliv. et Hiern）Isawumi;粗糙蟛蜞菊■☆

413555　Wedelia rudis（Oliv. et Hiern）Isawumi = Aspilia rudis Oliv. et Hiern ■☆

413556　Wedelia rudis（Oliv. et Hiern）Isawumi subsp. fontinaloides（C. D. Adams）Isawumi = Aspilia rudis Oliv. et Hiern subsp. fontinaloides C. D. Adams ■☆

413557　Wedelia spenceriana（Muschl.）Isawumi = Aspilia rudis Oliv. et Hiern ■☆

413558　Wedelia thouarsii（DC.）H. Rob. ;图氏蟛蜞菊■☆

413559　Wedelia trilobata（L.）Hitchc. ;南美蟛蜞菊（三裂蟛蜞菊）;Creeping Oxeye, Orange Zexmenia, Prostrate Wedelia, Rough Zexmenia,Singapore Daisy,Yellow Dots ■☆

413560　Wedelia trilobata（L.）Hitchc. = Sphagneticola trilobata（L.）Pruski ■☆

413561　Wedelia triseta Peter = Blainvillea gayana Cass. ■☆

413562　Wedelia triternata Klatt = Melanthera triternata（Klatt）Wild ■☆

413563　Wedelia urticifolia DC. ;麻叶蟛蜞菊（滴血根，小血藤，血参）;Nettleleaf Crabdaisy,Nettleleaf Wedelia ■

413564　Wedelia urticifolia DC. = Wedelia wallichii Less. ■

413565　Wedelia wallichii Less. ;山蟛蜞菊（麻叶蟛蜞菊，乳腺草，细针果，血参）;Wallich's Wedelia, Wild Crabdaisy ■

413566　Wedeliella Cockerell = Allionia L.（保留属名）■☆

413567　Wedeliella incarnta（L.）Cockerell = Allionia incarnata L. ■☆

413568　Wedeliopsis Planch. ex Benth. = Dissotis Benth.（保留属名）●☆

413569　Wehlia F. Muell. = Homalocalyx F. Muell. ●☆

413570　Weigela Thunb.（1780）;锦带花属;Brocadebeldflower, Cardinal Shrub,Japanese Honeysuckle,Weigela ●

413571　Weigela × fujisanensis（Makino）Nakai;富士山锦带花●☆

413572　Weigela × fujisanensis（Makino）Nakai f. cremea（Nakai）H. Hara;悬垂富士山锦带花●☆

413573　Weigela × hakonensis Nakai;箱根锦带花●☆

413574　Weigela amabilis（Carrière）Hook. = Weigela coraeensis Thunb. ●

413575　Weigela amagiensis Nakai = Weigela decora（Nakai）Nakai var. amagiensis（Nakai）H. Hara ●☆

413576　Weigela coraeensis Thunb. ;海仙花（门关柴）;Korean Weigela ●

413577　Weigela coraeensis Thunb. f. alba（Voss）Rehder;白海仙花●☆

413578　Weigela coraeensis Thunb. f. rubriflora Momiy. ;红海仙花●☆

413579　Weigela coraeensis Thunb. var. fragrans（Ohwi）H. Hara;香海仙花●☆

413580　Weigela decora（Nakai）Nakai;美丽锦带花●☆

413581　Weigela decora（Nakai）Nakai f. nivea Sugim. ;雪白美丽锦带花●☆

413582　Weigela decora（Nakai）Nakai f. unicolor（Nakai）H. Hara;单色美丽锦带花●☆

413583　Weigela decora（Nakai）Nakai var. amagiensis（Nakai）H. Hara;天城山锦带花●☆

413584　Weigela decora（Nakai）Nakai var. amagiensis（Nakai）H. Hara f. bicolor Sugim. ;二色天城山锦带花●☆

413585　Weigela decora（Nakai）Nakai var. amagiensis（Nakai）H. Hara f. viridiflava（Nakai）H. Hara;绿花天城山锦带花●☆

413586　Weigela decora（Nakai）Nakai var. rosea（Makino）H. Hara = Weigela × fujisanensis（Makino）Nakai ●☆

413587　Weigela decora（Nakai）Nakai var. rosea（Makino）H. Hara f. fujisanensis（Makino）H. Hara = Weigela × fujisanensis（Makino）Nakai ●☆

413588　Weigela floribunda（Siebold et Zucc.）K. Koch;路边花（美丽锦带花）;Rosy Weigela ●

413589　Weigela floribunda（Siebold et Zucc.）K. Koch f. leucantha Honda;白路边花●☆

413590　Weigela floribunda（Siebold et Zucc.）K. Koch var. nakaii（Makino）H. Hara;中井氏路边花●☆

413591　Weigela florida（Bunge）A. DC. ;锦带花（海仙，锦带，空枝子,连蕚锦带花,山脂麻,文官花）;Apple Shrub, Brocadebeldflower, Oldfashioned Weigela, Old-fashioned Weigela, Red Prince Weigela, Weigela ●

413592　Weigela florida（Bunge）A. DC. 'Alba';白花锦带花●☆

413593　Weigela florida（Bunge）A. DC. 'Alexandra';亚历山大锦带花●☆

413594　Weigela florida（Bunge）A. DC. 'Foliis Purpureis';福利斯紫锦带花（紫叶锦带花）●☆

413595　Weigela florida（Bunge）A. DC. 'Java Red';爪哇红锦带花●☆

413596　Weigela florida（Bunge）A. DC. 'Pink Princess';粉红公主锦带花;Old Fashioned Weigela ●☆

413597　Weigela florida（Bunge）A. DC. 'Purpurea';紫色锦带花●☆

413598　Weigela florida（Bunge）A. DC. 'Rosea';玫瑰红锦带花●☆

413599　Weigela florida（Bunge）A. DC. 'Variegata Nana' = Weigela florida（Bunge）A. DC. 'Variegata'●☆

413600　Weigela florida（Bunge）A. DC. 'Variegata';彩叶锦带花（白边锦带花）●☆

413601　Weigela florida（Bunge）A. DC. 'Victoria';维多利亚锦带花;Old Fashioned Weigela ●☆

413602　Weigela florida（Bunge）A. DC. 'Wine and Roses' = Weigela florida（Bunge）A. DC. 'Alexandra'●☆

413603　Weigela florida（Bunge）A. DC. f. albiflora Y. C. Chu = Weigela florida（Bunge）A. DC. ●

413604　Weigela florida（Bunge）A. DC. f. leucantha Nakai;白锦带花●☆

413605　Weigela florida（Bunge）A. DC. var. glabra Nakai = Weigela florida（Bunge）A. DC. ●

413606　Weigela florida（Bunge）A. DC. var. nakaii Hara;血红锦带花●☆

413607　Weigela florida（Bunge）A. DC. var. praecox（Lemoine）Y. C. Chu = Weigela florida（Bunge）A. DC. ●

413608　Weigela hortensis（Siebold et Zucc.）K. Koch;园圃锦带花●☆

413609　Weigela hortensis（Siebold et Zucc.）K. Koch f. albiflora（Siebold et Zucc.）Rehder;白花园圃锦带花●☆

413610　Weigela hybrida Hort. ;杂交锦带花;Weigela ●

413611　Weigela japonica Thunb. ;日本锦带花（杨庐耳,杨栌,杨栌耳）;Japan Brocadebeldflower,Japanese Weigela ●

413612　Weigela japonica Thunb. var. decora（Nakai）Okuyama = Weigela decora（Nakai）Nakai ●☆

413613　Weigela japonica Thunb. var. sinica（Rehder）Bailey;水马桑（白马桑,半边日,半边月,包头杆子,大号黄山掌,鸡骨柴,铃钟花,麻布柴,木绣球,水吞骨,杨栌）;China Brocadebeldflower, Chinese Weigela ●

413614　Weigela japonica Thunb. var. sinica Rehder = Weigela japonica Thunb. var. sinica (Rehder) Bailey ●

413615　Weigela kariyosensis Nakai；刈谷锦带花●☆

413616　Weigela maximowiczii (S. Moore) Rehder；马氏锦带花；Maximowicz Weigela ●☆

413617　Weigela middendorffiana (Carrière) K. Koch；米氏锦带花(米登氏锦带花,远东锦带花)●☆

413618　Weigela middendorffiana (Carrière) K. Koch = Macrodiervilla middendorffiana (Carrière) Nakai ●☆

413619　Weigela nikkoensis Makino；日光锦带花●☆

413620　Weigela pauciflora A. DC. = Weigela florida (Bunge) A. DC. ●

413621　Weigela praecox (Lemoine) L. H. Bailey；早锦带花(密枝锦带花,早开锦带花)；Precocious Brocadebeldflower,Prococious Weigela ●

413622　Weigela praecox (Lemoine) L. H. Bailey 'Variegata'；斑叶密枝锦带花(白缘叶锦带花)●☆

413623　Weigela praecox (Lemoine) L. H. Bailey var. pilosa Nakai；微毛早锦带花●

413624　Weigela sanguinea (Nakai) Nakai；绒毛锦带花●☆

413625　Weigela sanguinea (Nakai) Nakai f. leucantha (Nakai) H. Hara；白花绒毛锦带花●☆

413626　Weigela sanguinea (Nakai) Nakai var. nakai Makino；中井氏锦带花●☆

413627　Weigela sinica (Rehder) C. C. Chang = Weigela japonica Thunb. var. sinica (Rehder) Bailey ●

413628　Weigela suavis (Kom.) L. H. Bailey；芳香锦带花●☆

413629　Weigela suavis (Kom.) L. H. Bailey = Diervilla suavis Kom. ●☆

413630　Weigelastrum (Nakal) Nakai = Weigela Thunb. ●

413631　Weigelastrum maximowiczii (S. Moore) Nakai = Weigela maximowiczii (S. Moore) Rehder ●☆

413632　Weigelia Pers. = Weigela Thunb. ●

413633　Weigeltia A. DC. = Cybianthus Mart. (保留属名)●☆

413634　Weihea Eckl. = Geissorhiza Ker Gawl. ■☆

413635　Weihea Rchb. = Burtonia R. Br. (保留属名)●☆

413636　Weihea Spreng. (1825)(保留属名)；魏厄木属●☆

413637　Weihea Spreng. (保留属名) = Cassipourea Aubl. ●☆

413638　Weihea Spreng. ex Eichler = Phthirusa Mart. ●☆

413639　Weihea Spreng. ex Eichler = Weihea Spreng. (保留属名)●☆

413640　Weihea abyssinica Engl. = Cassipourea malosana (Baker) Alston ●☆

413641　Weihea africana (Benth.) Oliv. = Cassipourea congoensis DC. ●☆

413642　Weihea afzelii Oliv. = Cassipourea afzelii (Oliv.) Alston ●☆

413643　Weihea avettae Chiov. = Cassipourea malosana (Baker) Alston ●☆

413644　Weihea bequaertii De Wild. = Cassipourea ruwensorensis (Engl.) Alston ●☆

413645　Weihea boranensis Cufod. = Cassipourea malosana (Baker) Alston ●☆

413646　Weihea dinklagei (Engl.) Engl. = Cassipourea dinklagei (Engl.) Alston ●☆

413647　Weihea eickii Engl. = Cassipourea malosana (Baker) Alston ●☆

413648　Weihea elliottii Engl. = Cassipourea malosana (Baker) Alston ●☆

413649　Weihea flanaganii Schinz = Cassipourea flanaganii (Schinz) Alston ●☆

413650　Weihea gerrardii Schinz = Cassipourea malosana (Baker) Alston ●☆

413651　Weihea huillensis Engl. = Cassipourea huillensis (Engl.) Alston ●☆

413652　Weihea ilicifolia Brehmer = Cassipourea malosana (Baker) Alston ●☆

413653　Weihea kamerunensis (Engl.) Engl. = Cassipourea kamerunensis (Engl.) Alston ●☆

413654　Weihea madagascariensis Spreng. ；魏厄木●☆

413655　Weihea malosana Baker = Cassipourea malosana (Baker) Alston ●☆

413656　Weihea mawambensis Engl. = Cassipourea ruwensorensis (Engl.) Alston ●☆

413657　Weihea mildbraedii Engl. = Cassipourea ruwensorensis (Engl.) Alston ●☆

413658　Weihea mollis R. E. Fr. = Cassipourea mollis (R. E. Fr.) Alston ●☆

413659　Weihea mossambicensis Brehmer = Cassipourea mossambicensis (Brehmer) Alston ●☆

413660　Weihea plumosa Oliv. = Cassipourea plumosa (Oliv.) Alston ●☆

413661　Weihea rotundifolia Engl. = Cassipourea rotundifolia (Engl.) Alston ●☆

413662　Weihea ruwensoriensis Engl. = Cassipourea ruwensorensis (Engl.) Alston ●☆

413663　Weihea salvago-raggei Chiov. = Cassipourea malosana (Baker) Alston ●☆

413664　Weihea sericea (Engl.) Engl. = Cassipourea sericea (Engl.) Alston ●☆

413665　Weihea subpeltata Sim = Androstachys johnsonii Prain ●☆

413666　Weihea zenkeri Engl. = Cassipourea zenkeri (Engl.) Alston ●☆

413667　Weilbachia Klotzsch et Oerat. = Begonia L. ●■

413668　Weingaertneria Bernh. (废弃属名) = Corynephorus P. Beauv. (保留属名)■☆

413669　Weingaertneria articulata (Desf.) Asch. et Graebn. = Corynephorus divaricatus (Pourr.) Breistr. ■☆

413670　Weingaertneria articulata (Desf.) Asch. et Graebn. var. gracilis (Guss.) Font Quer = Corynephorus divaricatus (Pourr.) Breistr. ■☆

413671　Weingartia Werderm. (1937)；花笠球属(轮冠属)■☆

413672　Weingartia Werderm. = Gymnocalycium Pfeiff. ex Mittler ●

413673　Weingartia Werderm. = Rebutia K. Schum. ●

413674　Weingartia ambigua (Hildm. ex K. Schum.) Backeb. ；黄昏玉●☆

413675　Weingartia fidaiana (Backeb.) Werderm. ；花饰球(花饰玉)■☆

413676　Weingartia hediniana Backeb. ；白绵毛花笠球●☆

413677　Weingartia lanata F. Ritter；软毛花笠球■☆

413678　Weingartia neocumingii Backeb. ；花笠球(花笠丸)■☆

413679　Weingartia neumanniana (Backeb.) Werderm. ；花钿玉■☆

413680　Weingartneria Benth. = Corynephorus P. Beauv. (保留属名)■☆

413681　Weingartneria Benth. = Weingaertneria Bernh. (废弃属名)■☆

413682　Weinmannia L. (1759)(保留属名)；温曼木属(万恩曼属,万灵木属,维玛木属,魏曼树属)；Weinmannia ●☆

413683　Weinmannia arguta (Bernardi) J. Bradford；亮温曼木●☆

413684　Weinmannia bifida Poepp. ex Engl. ；二裂温曼木(二裂万灵木)●☆

413685　Weinmannia bojeriana Tul. ；博耶尔温曼木●☆

413686　Weinmannia commersonii Bernardi；科梅逊温曼木●☆

413687　Weinmannia decora Tul. ；装饰温曼木●☆

413688　Weinmannia henricorum Bernardi；昂里克温曼木●☆

413689　Weinmannia hepaticarum Bernardi；肝色温曼木●☆

413690　Weinmannia hildebrandtii var. arguta Bernardi = Weinmannia arguta (Bernardi) J. Bradford ●☆

413691　Weinmannia humbertiana Bernardi；亨伯特温曼木●☆

413692　Weinmannia humblotii Baill. ；洪布温曼木●☆

413693　Weinmannia integrifolia J. Bradford；全缘叶温曼木●☆

413694　Weinmannia louveliana Bernardi；卢韦尔温曼木●☆

413695　Weinmannia lowryana J. Bradford；劳里温曼木●☆

413696　Weinmannia lucens Baker；光亮温曼木●☆

413697　Weinmannia magnifica J. Bradford et Z. S. Rogers；壮观温曼木●☆

413698　Weinmannia mammea Bernardi；乳突温曼木●☆

413699　Weinmannia marojejyensis J. S. Mill. et J. Bradford；马罗温曼木●☆

413700　Weinmannia pauciflora J. Bradford；少花温曼木●☆

413701　Weinmannia pinnata L. ；西印温曼木（西印万灵木，羽叶维玛木）●☆

413702　Weinmannia racemosa L. f. ；单叶温曼木（单叶万灵木，单叶维玛木）；Kamahi，Towai Bark ●☆

413703　Weinmannia rutenbergii Engl. ；鲁滕贝格温曼木●☆

413704　Weinmannia sanguisugarum Bernardi；血红温曼木●☆

413705　Weinmannia stenostachya Baker；窄穗温曼木●☆

413706　Weinmannia trichosperma Cav. ；毛籽温曼木（丛生维玛木，毛籽万恩曼，毛籽万灵木）；Maden，Tineo ●☆

413707　Weinmannia trifoliata L. f. = Platylophus trifoliatus（L. f. ）D. Don ●☆

413708　Weinmannia venosa J. Bradford；多脉温曼木●☆

413709　Weinmannia venusta Bernardi；雅致温曼木（雅致万灵木）●☆

413710　Weinmanniaphyllum R. J. Carp. et A. M. Buchanan（1993）；澳洲火把树属●☆

413711　Weinmanniaphyllum bernardii R. J. Carp. et A. M. Buchanan；澳洲火把树●☆

413712　Weinreichia Rchb. = Echinodiscus（DC. ）Benth. ●

413713　Weinreichia Rchb. = Pterocarpus Jacq. （保留属名）●

413714　Weitenwebera Opiz = Campanula L. ■●

413715　Welchiodendron Peter G. Wilson et J. T. Waterh. （1982）；韦尔木属●☆

413716　Welchiodendron longivalve（F. Muell. ）Peter G. Wilson et J. T. Waterh. ；韦尔木●☆

413717　Weldena Pohl ex K. Schum. = Abutilon Mill. ●■

413718　Weldenia Rchb. = ? Hibbertia Andréws ●☆

413719　Weldenia Schult. f. （1829）；银瓣花属■☆

413720　Weldenia candida Schult. f. ；银瓣花■☆

413721　Welezia Neck. = Velezia L. ■☆

413722　Welfia H. Wendl. （1869）；羽叶椰属（外尔非桐属，维夫棕属，杏果椰属，羽叶棕属）●☆

413723　Welfia H. Wendl. ex Andre = Welfia H. Wendl. ●☆

413724　Welfia georgii H. Wendl. ex Burret；羽叶椰●☆

413725　Wellingtonia Lindl. = Sequoiadendron J. Buchholz ●

413726　Wellingtonia Meisn. = Meliosma Blume ●

413727　Wellingtonia gigantea Lindl. = Sequoiadendron giganteum（Lindl. ）Buchholz ●

413728　Wellingtoniaceae Meisn. = Meliosmaceae Endl. ●

413729　Wellingtoniaceae Meisn. = Millingtoniaceae Wight et Arn. ●

413730　Wellingtoniaceae Meisn. = Sabiaceae Blume（保留科名）●

413731　Wellstedia Balf. f. （1884）；四室果属■●☆

413732　Wellstedia dinteri Pilg. ；丁特四室果●☆

413733　Wellstedia dinteri Pilg. var. gracilior Hunt；纤细四室果●☆

413734　Wellstedia laciniata Thulin et A. Johanss. ；撕裂四室果●☆

413735　Wellstedia robusta Thulin；粗壮四室果●☆

413736　Wellstedia socotrana Balf. f. ；索科特拉四室果●■☆

413737　Wellstedia somalensis Thulin et A. Johanss. ；索马里四室果●■☆

413738　Wellstediaceae（Pilg. ）Novák = Boraginaceae Juss. （保留科名）■●

413739　Wellstediaceae Novák = Boraginaceae Juss. （保留科名）■●

413740　Wellstediaceae Novák；四室果科（番厚壳树科）●■☆

413741　Welwitschia Hook. f. （1862）（保留属名）；百岁兰属（千岁兰属）；Welwitschia ■☆

413742　Welwitschia Post et Kuntze = Melasma P. J. Bergius ■

413743　Welwitschia Post et Kuntze = Velvitsia Hiern. ■

413744　Welwitschia Rchb. （废弃属名）= Eriastrum Wooton et Standl. ■●☆

413745　Welwitschia Rchb. （废弃属名）= Gilia Ruiz et Pav. ■●☆

413746　Welwitschia Rchb. （废弃属名）= Welwitschia Hook. f. （保留属名）■☆

413747　Welwitschia bainesii Carrière = Welwitschia mirabilis Hook. f. ■☆

413748　Welwitschia mirabilis Hook. f. ；百岁兰（千岁兰）；Miracle Tree，Welwitschia ■☆

413749　Welwitschiaceae（Engl. ）Markgr. = Welwitschiaceae Caruel（保留科名）■☆

413750　Welwitschiaceae Caruel（1879）（保留科名）；百岁兰科■☆

413751　Welwitschiaceae Markgr. = Welwitschiaceae Caruel（保留科名）■☆

413752　Welwitschiella Engl. = Triclisia Benth. ●☆

413753　Welwitschiella Engl. = Welwitschiina Engl. ●☆

413754　Welwitschiella O. Hoffm. （1894）；无舌山黄菊属■☆

413755　Welwitschiella neriifolia O. Hoffm. ；无舌山黄菊■☆

413756　Welwitschiina Engl. = Triclisia Benth. ●☆

413757　Welwitschiina macrophylla（Hiern）Engl. = Chondodendron macrophyllum Hiern ■☆

413758　Wenchengia C. Y. Wu et S. Chow（1965）；保亭花属；Baotingflower，Wenchengia ●■

413759　Wenchengia alternifolia C. Y. Wu et S. Chow；保亭花（连丝果）；Chinese Baotingflower，Chinese Wenchengia ●■

413760　Wendelboa Soest = Taraxacum F. H. Wigg. （保留属名）■

413761　Wenderothia Schltdl. = Canavalia Adans. （保留属名）●■

413762　Wendia Hoffm. （废弃属名）= Heracleum L. ■

413763　Wendia Hoffm. （废弃属名）= Wendtia Meyen（保留属名）■☆

413764　Wendlandia Bartl. = Wendlandia Bartl. ex DC. （保留属名）●

413765　Wendlandia Bartl. ex DC. （1830）（保留属名）；水锦树属；Wendlandia ●

413766　Wendlandia DC. = Wendlandia Bartl. ex DC. （保留属名）●

413767　Wendlandia Willd. （废弃属名）= Androphylax J. C. Wendl. （废弃属名）●

413768　Wendlandia Willd. （废弃属名）= Cocculus DC. （保留属名）●

413769　Wendlandia Willd. （废弃属名）= Wendlandia Bartl. ex DC. （保留属名）●

413770　Wendlandia aberrans F. C. How；广西水锦树；Guangxi Wendlandia，Kwangsi Wendlandia ●

413771　Wendlandia angustinii Cowan；南水锦树（思茅水锦树）；Angustine Wendlandia，Simao Wendlandia ●

413772　Wendlandia arabica Deflers；阿拉伯水锦树●☆

413773　Wendlandia arabica Deflers subsp. aethiopica ?；埃塞俄比亚水锦树●☆

413774　Wendlandia bouvardioides Hutch. ；红花水锦树（薄叶水锦树）；Red-flower Wendlandia，Red-flowered Wendlandia，Thinleaf Wendlandia ●

413775　Wendlandia brevipaniculata W. C. Chen；吹树；Blowytree Wendlandia，Short-paniculate Wendlandia ●

413776　Wendlandia brevituba Chun et F. C. How ex W. C. Chen；短筒水锦树；Shorttube Wendlandia ●

413777　Wendlandia cavaleriei H. Lév. ；滇越水锦树（贵州水锦树）；Cavalerie Wendlandia，Guizhou Wendlandia ●

413778　Wendlandia chinensis Merr. = Wendlandia uvariifolia Hance subsp. chinensis（Merr. ）Cowan ●

413779　Wendlandia erythroxylon Cowan；红木水锦树；Red-wood

Wendlandia, Rosewood Wendlandia ●

413780　Wendlandiaerythroxylon Cowan = Wendlandia uvariifolia Hance ●

413781　Wendlandia feddei H. Lév. = Wendlandia cavaleriei H. Lév. ●

413782　Wendlandia floribunda Craib = Wendlandia tinctoria（Roxb.）DC. subsp. barbata Cowan ●

413783　Wendlandia floribunda Craib = Wendlandia tinctoria（Roxb.）DC. subsp. floribunda（Craib）Cowan ●

413784　Wendlandia formosana Cowan；水金京（红木，假鸡纳树，水魂仔，水金定，水金惊）；Formosa Wendlandia, Taiwan Wendlandia ●

413785　Wendlandia formosana Cowan subsp. breviflora F. C. How；短花水金京（短花台湾水锦树）；Short-flower Formosa Wendlandia ●

413786　Wendlandia glabrata DC. var. floribunda Craib = Wendlandia tinctoria（Roxb.）DC. subsp. floribunda（Craib）Cowan ●

413787　Wendlandia grandis（Hook. f.）Cowan；西藏水锦树；Tibet Wendlandia, Xizang Wendlandia ●

413788　Wendlandia guangdongensisi W. C. Chen；广东水锦树；Guangdong Wendlandia, Kwangtung Wendlandia ●

413789　Wendlandia handelii（Cowan）Cowan；蛮耗红皮（野麻栗树）；Hendel Wendlandia ●

413790　Wendlandia henryi Oliv. = Wendlandia longidens（Hance）Hutch. ●

413791　Wendlandia jingdongensis W. C. Chen；景东水锦树；Jingdong Wendlandia ●

413792　Wendlandia laxa S. K. Wu ex W. C. Chen；疏花水锦树；Loose-flower Wendlandia, Scatterflower Wendlandia ●

413793　Wendlandia ligustriana（Wall.）Wall.；小叶水锦树；Littleleaf Wendlandia, Small-leaf Wendlandia, Small-leaved Wendlandia ●

413794　Wendlandia litseifolia F. C. How；木姜子叶水锦树；Litseleaf Wendlandia ●

413795　Wendlandia longidens（Hance）Hutch.；水晶稞子（黑仔，绵柳，水丝条）；Crystal Wendlandia, Longtooth Wendlandia, Long-toothed Wendlandia ●

413796　Wendlandia longifolia（Wall.）DC. = Mycetia longifolia（Wall.）Kuntze ●

413797　Wendlandia longipedicellata F. C. How；长柄水锦树（长梗水锦树）；Longpedicel Wendlandia, Long-petioled Wendlandia ●

413798　Wendlandia luzoniensis DC.；吕宋水锦树；Luzon Wendlandia ●

413799　Wendlandia membranifolia Elmer = Wendlandia luzoniensis DC. ●

413800　Wendlandia merrilliana Cowan；海南水锦树；Merrill Wendlandia ●

413801　Wendlandia merrilliana Cowan var. parvifolia F. C. How；细叶海南水锦树（小叶水锦树）；Littleleaf Wendlandia, Small-leaf Merrill Wendlandia ●

413802　Wendlandia multiflora Bani. ex DC. = Wendlandia luzoniensis DC. ●

413803　Wendlandia myriantha F. C. How；密花水锦树；Denseflower Wendlandia, Dense-flowered Wendlandia ●

413804　Wendlandia oligantha W. C. Chen；龙州水锦树；Longzhou Wendlandia ●

413805　Wendlandia Paniculata（Roxb.）DC. subsp. scabra Cowan = Wendlandia scabra Kurz ●

413806　Wendlandia paniculata DC. = Wendlandia uvariifolia Hance ●

413807　Wendlandia paniculata DC. subsp. scabra Cowan = Wendlandia scabra Kurz ●

413808　Wendlandia parviflora W. C. Chen；小花水锦树；Littleflower Wendlandia, Smallflower Wendlandia ●

413809　Wendlandia pendula（Wall.）DC.；垂枝水锦树（垂序水锦树）；Drooping Wendlandia, Nutanttwig Wendlandia, Pendent Wendlandia ●

413810　Wendlandia pilosa G. Don = Bertiera spicata（C. F. Gaertn.）K. Schum. ■☆

413811　Wendlandia pingpienensis F. C. How；屏边水锦树（红木树）；Pingbian Wendlandia ●

413812　Wendlandia puberula DC. = Wendlandia sikkimensis Cowan ●☆

413813　Wendlandia pubigera W. C. Chen；大叶水锦树（大叶木莲红）；Largeleaf Wendlandia ●

413814　Wendlandia racemosa G. Don = Bertiera racemosa（G. Don）K. Schum. ■☆

413815　Wendlandia salicifolia Franch. = Wendlandia salicifolia Franch. ex Drake ●

413816　Wendlandia salicifolia Franch. ex Drake；柳叶水锦树；Willowleaf Wendlandia, Willow-leaved Wendlandia ●

413817　Wendlandia scabra Kurz；粗叶水锦树（黄皮花树，千里木，碎米花树，朱赤木）；Coarseleaf Wendlandia, Scabrous Wendlandia ●

413818　Wendlandia scabra Kurz var. dependens Cowan；悬花水锦树（下垂黄皮花树）；Dropin Scabrous Wendlandia ●

413819　Wendlandia scabra Kurz var. forresti Cowan；滇西水锦树；Forrest's Scabrous Wendlandia ●

413820　Wendlandia scabra Kurz var. pilife R. C. How ex W. C. Chen；毛粗叶水锦树；Hairy Scabrous Wendlandia ●

413821　Wendlandia scabra Kurz var. speciosa Cowan = Wendlandia speciosa Cowan ●

413822　Wendlandia sikkimensis Cowan = Wendlandia puberula DC. ●☆

413823　Wendlandia speciosa Cowan；美丽水锦树（美水锦树）；Beautiful Scabrous Wendlandia, Beautiful Wendlandia, Specious Wendlandia ●

413824　Wendlandia speciosa Cowan var. forrestii Cowan = Wendlandia speciosa Cowan ●

413825　Wendlandia subalpina W. W. Sm.；高山水锦树（虎跳涧水晶稞）；Alp Wendlandia, Alpine Wendlandia ●

413826　Wendlandia sulcata G. Don；纵沟水锦树●☆

413827　Wendlandia taiwaniana（Cowan）Cowan；台湾水锦树；Taiwan Wendlandia ●

413828　Wendlandia taiwaniana（Cowan）Cowan var. brevifolia F. C. How；虾须木；Shortflower Wendlandia, Shrimpfeelerswood ●

413829　Wendlandia tinctoria（Roxb.）DC.；染色水锦树（红皮水锦树）；Dye Wendlandia, Tinctorial Wendlandia ●

413830　Wendlandia tinctoria（Roxb.）DC. = Wendlandia grandis（Hook. f.）Cowan ●

413831　Wendlandia tinctoria（Roxb.）DC. subsp. affinis F. C. How ex W. C. Chen；毛冠水锦树；Affined Tinctorial Wendlandia ●

413832　Wendlandia tinctoria（Roxb.）DC. subsp. barbata Cowan；粗毛水锦树（须毛水锦树）；Bearded Tinctorial Wendlandia ●

413833　Wendlandia tinctoria（Roxb.）DC. subsp. callitricha（Cowan）W. C. Chen；厚毛水锦树；Thickhair Wendlandia ●

413834　Wendlandia tinctoria（Roxb.）DC. subsp. floribunda（Craib）Cowan；多花水锦树；Manyflower Wendlandia ●

413835　Wendlandia tinctoria（Roxb.）DC. subsp. handelii Cowan；麻栗水锦树（野麻栗树，野麻栗水锦树）；Handel Wendlandia ●

413836　Wendlandia tinctoria（Roxb.）DC. subsp. intermedia（F. C. How）W. C. Chen；红皮水锦树；Redbark Wendlandia, Tinctorial Wendlandia ●

413837　Wendlandia tinctoria（Roxb.）DC. subsp. orientalis Cowan；东方水锦树（沙牛木）；Oriental Wendlandia ●

413838　Wendlandia tinctoria（Roxb.）DC. var. callitricha Cowan = Wendlandia tinctoria（Roxb.）DC. subsp. callitricha（Cowan）W.

C. Chen ●

413839 Wendlandia tinctoria（Roxb.）DC. var. grandis Hook. f. = Wendlandia grandis（Hook. f.）Cowan ●

413840 Wendlandia tinctoria（Roxb.）DC. var. intermedia F. C. How = Wendlandia tinctoria（Roxb.）DC. subsp. intermedia（F. C. How）W. C. Chen ●

413841 Wendlandia tinctoria（Roxb.）DC. var. normalis ？ = Wendlandia tinctoria（Roxb.）DC. ●

413842 Wendlandia uvariifolia Hance；水锦树（大虫耳，饭汤木，红木，黄廊芽，黄紫茅，毛水锦树，牛伴木，双耳蛇，猪血木）；Common Wendlandia，Uvarialeaf Wendlandia，Uvaria-leaved Wendlandia ●

413843 Wendlandia uvariifolia Hance subsp. chinensis（Merr.）Cowan；中华水锦树（黄廊木，黄廊牙）；Chinese Elkwood，Chinese Wendlandia ●

413844 Wendlandia uvariifolia Hance subsp. dunniana（H. Lév.）Cowan = Wendlandia uvariifolia Hance ●

413845 Wendlandia uvariifolia Hance subsp. pilosa W. C. Chen；疏毛水锦树；Pilose Chinese Wendlandia ●

413846 Wendlandia uvariifolia Hance subsp. rotundifolia（Hand.-Mazz.）Cowan = Wendlandia uvariifolia Hance ●

413847 Wendlandia uvariifolia Hance subsp. rufula Cowan = Wendlandia uvariifolia Hance ●

413848 Wendlandia uvariifolia Hance subsp. yunnanensis Cowan = Wendlandia uvariifolia Hance ●

413849 Wendlandia uvariifolla Hance subsp. rotundifolia（Hand.-Mazz.）Cowan = Wendlandia uvariifolia Hance ●

413850 Wendlandia villosa W. C. Chen；毛叶水锦树；Hairleaf Wendlandia，Villose Wendlandia ●

413851 Wendlandia virgata G. Don = Pouchetia africana A. Rich. ex DC. ●☆

413852 Wendlandia zooi C. How = Wendlandia scabra Kurz ●

413853 Wendlandiella Dammer（1905）；单梗苞椰属（文兰代桐属，文氏椰属，文氏棕属）●☆

413854 Wendlandiella gracilis Dammer；单梗苞椰●☆

413855 Wendtia Ledeb. = Heracleum L. ■

413856 Wendtia Meyen = Balbisia Cav.（保留属名）●☆

413857 Wendtia Meyen（1834）（保留属名）；文氏草属■☆

413858 Wendtia argentea Griseb.；银白文氏草■☆

413859 Wendtia gracilis Meyen；文氏草■☆

413860 Wendtia miniata I. M. Johnst.；小文氏草■☆

413861 Wensea J. C. Wendl. = Pogostemon Desf. ●■

413862 Wentsaiboea D. Fang et D. H. Qin（2004）；文采苣苔属■

413863 Wentsaiboea renifolia D. Fang et W. T. Wang；文采苣苔；Westringia ■

413864 Wenzelia Merr.（1915）；文策尔芸香属●☆

413865 Wenzelia brevipes Merr.；文策尔芸香●☆

413866 Wenzelia grandiflora（Lauterb.）Swingle；大花文策尔芸香●☆

413867 Wenzelia tenuifolia Swingle；细叶文策尔芸香●☆

413868 Wepferia Fabr. = Aethusa L. ■☆

413869 Wepferia Heist. ex Fabr. = Aethusa L. ■☆

413870 Werauhia J. R. Grant（1995）；指纹瓣凤梨属■☆

413871 Werauhia acuminata（Mez et Wercklé）J. R. Grant；尖指纹瓣凤梨■☆

413872 Werauhia bicolor（L. B. Sm.）J. R. Grant；二色指纹瓣凤梨■☆

413873 Werauhia graminifolia（Mez et Wercklé）J. R. Grant；禾叶指纹瓣凤梨■☆

413874 Werauhia laxa（Mez et Wercklé）J. R. Grant；松散指纹瓣凤梨■☆

413875 Werauhia macrantha（Mez et Wercklé）J. R. Grant；大花指纹瓣凤梨■☆

413876 Werauhia pycnantha（L. B. Sm.）J. R. Grant；密花指纹瓣凤梨■☆

413877 Werauhia rubra（Mez et Wercklé）J. R. Grant；红指纹瓣凤梨■☆

413878 Werauhia rugosa（Mez et Wercklé）J. R. Grant；皱指纹瓣凤梨■☆

413879 Werauhia stenophylla（Mez et Wercklé）J. R. Grant；窄叶指纹瓣凤梨■☆

413880 Wercklea Pittier et Standl.（1916）；韦克锦葵属●■☆

413881 Wercklea insignis Pittier et Standl.；韦克锦葵●☆

413882 Werckleocereus Britton et Rose = Weberocereus Britton et Rose ●☆

413883 Werckleocereus Britton et Rose（1909）；刺萼三棱柱属●☆

413884 Werckleocereus glaber（Eichlam）Britton et Rose；无毛刺萼三棱柱●☆

413885 Werckleocereus tonduzii（F. A. C. Weber）Britton et Rose；刺萼三棱柱●☆

413886 Werdermannia O. E. Schulz（1928）；韦德曼芥属■☆

413887 Werdermannia macrostachya O. E. Schulz；韦德曼芥■☆

413888 Wernera Kuntze = Werneria Kunth ■☆

413889 Werneria Kunth；光莲菊属（沃纳菊属）■☆

413890 Werneria africana Oliv. et Hiern = Senecio nanus Sch. Bip. ex A. Rich. ■☆

413891 Werneria antinorii Avetta = Euryops antinorii（Avetta）S. Moore ●☆

413892 Werneria ellisii Hook. f. = Cremanthodium ellisii（Hook. f.）Kitam. ■

413893 Werneria glaberrima Phil.；光莲菊（光沃纳菊，极光沃纳菊）■☆

413894 Werneria nana（Decne.）Benth. = Cremanthodium nanum（Decne.）W. W. Sm. ■

413895 Wernhamia S. Moore（1922）；玻利维亚茜属■☆

413896 Wernhamia boliviensis S. Moore；玻利维亚茜■☆

413897 Wernisekia Scop. = Houmiri Aubl. ●☆

413898 Wernisekia Scop. = Humiria Aubl.（保留属名）●☆

413899 Werrinuwa Heyne = Guizotia Cass.（保留属名）■●

413900 Westeringia Dum. Cours. = Westringia Sm. ●☆

413901 Westia Vahl（废弃属名）= Berlinia Sol. ex Hook. f.（保留属名）+ Afzelia Sm.（保留属名）●

413902 Westia Vahl（废弃属名）= Berlinia Sol. ex Hook. f.（保留属名）●☆

413903 Westia grandiflora Vahl = Berlinia grandiflora（Vahl）Hutch. et Dalziel ●☆

413904 Westia parviflora Vahl = Afzelia parviflora（Vahl）Hepper ●☆

413905 Westia stipulacea（Benth.）J. F. Macbr. = Gilbertiodendron stipulaceum（Benth.）J. Léonard ●☆

413906 Westonia Spreng. = Rothia Pers.（保留属名）■

413907 Westoniella Cuatrec.（1977）；紫绒菀属■●☆

413908 Westoniella chirripoensis Cuatrec.；紫绒菀■●☆

413909 Westphalina A. Robyns et Bamps（1977）；威斯椴属●☆

413910 Westphalina macrocarpa A. Robyns et Bamps；威斯椴●☆

413911 Westringia Sm.（1797）；澳迷迭香属（维斯特灵属）；Westringia ●☆

413912 Westringia brevifolia Benth.；短叶澳迷迭香（短叶维斯特灵）；Short-leafed Westringia ●☆

413913 Westringia eremicola A. Cunn. ex Benth.；纤细澳迷迭香（纤细维斯特灵）；Slender Western Rosemary，Slender Westringia ●☆

413914 Westringia fruticosa Druce；澳迷迭香（海滨维斯特灵）；Australian Rosemary，Coast Rosemary，Coastal Rosemary，Morning Light，Native Rosemary，Westringia ●☆

413915 Westringia glabra R. Br.；栗斑澳迷迭香（栗斑维斯特灵）；Violet Westringia ●☆

413916 Westringia rosmariniformis Labill. ex Benth. = Westringia fruticosa Druce ●☆

413917 Wetria Baill. (1858);韦大戟属●☆

413918 Wetria trewioides Baill. ;韦大戟●☆

413919 Wetriaria (Müll. Arg.) Kuntze = Argomuellera Pax ●☆

413920 Wetriaria Kuntze = Argomuellera Pax ●☆

413921 Wetriaria Pax = Argomuellera Pax ●☆

413922 Wettinella O. F. Cook et Doyle = Wettinia Poepp. ex Endl. ●☆

413923 Wettinia Poepp. = Wettinia Poepp. ex Endl. ●☆

413924 Wettinia Poepp. ex Endl. (1837);韦廷棕属●☆

413925 Wettinia augusta Poepp. ex Endl. ;韦廷棕●☆

413926 Wettiniicarpus Burret = Wettinia Poepp. ex Endl. ●☆

413927 Wettsteinia Petr. = Carduus L. ■

413928 Wettsteinia Petr. = Olgaea Iljin ■

413929 Wettsteiniola Suess. (1935);阿根廷川苔草属■☆

413930 Wettsteiniola pinnata Suess. ;阿根廷川苔草■☆

413931 Whalleya Wills et J. J. Bruhl = Walwhalleya Wills et J. J. Bruhl ■☆

413932 Wheelerella G. B. Grant = Cryptantha Lehm. ex G. Don ■☆

413933 Wheelerella G. B. Grant = Greeneocharis Gürke et Harms ■☆

413934 Whipplea Torr. (1857);惠普属●☆

413935 Whipplea modesta Torr. ;惠普木●☆

413936 Whitefieldia Nees = Whitfieldia Hook. ■☆

413937 Whiteheadia Harv. (1868);怀特风信子属■☆

413938 Whiteheadia bifolia (Jacq.) Baker;双叶怀特风信子(双小叶怀特风信子,双叶凤梨百合)■☆

413939 Whiteheadia etesionamibensis U. Müll.-Doblies et D. Müll.-Doblies;怀特风信子■☆

413940 Whiteochloa C. E. Hubb. (1952);怀特黍属■☆

413941 Whiteochloa semitonsa (F. Muell. ex Benth.) C. E. Hubb. ;怀特黍■☆

413942 Whiteodendron Steenis(1952);加岛桃金娘属●☆

413943 Whiteodendron moultonianum (W. W. Sm.) Steenis;加岛桃金娘●☆

413944 White-Sloanea Chiov. (1937);索马里萝藦属■☆

413945 White-Sloanea crassa (N. E. Br.) Chiov. ;索马里萝藦■☆

413946 White-Sloanea migiurtina Chiov. = Pseudolithos migiurtinus (Chiov.) P. R. O. Bally ■☆

413947 Whitfieldia Hook. (1845);惠特爵床属■☆

413948 Whitfieldia arnoldiana De Wild. et T. Durand;阿诺德惠特爵床■☆

413949 Whitfieldia brazzae (Baill.) C. B. Clarke;布拉扎惠特爵床■☆

413950 Whitfieldia colorata C. B. Clarke ex Stapf;着色惠特爵床■☆

413951 Whitfieldia elongata (P. Beauv.) De Wild. et T. Durand;伸长惠特爵床;White Candles ■☆

413952 Whitfieldia elongata (P. Beauv.) De Wild. et T. Durand var. dewevrei De Wild. = Whitfieldia elongata (P. Beauv.) De Wild. et T. Durand ■☆

413953 Whitfieldia gilletii De Wild. = Whitfieldia thollonii (Baill.) Benoist ■☆

413954 Whitfieldia laurentii (Lindau) C. B. Clarke;洛朗惠特爵床■☆

413955 Whitfieldia letestui Benoist;莱泰斯图惠特爵床■☆

413956 Whitfieldia liebrechtsiana De Wild. et T. Durand;利布惠特爵床■☆

413957 Whitfieldia longiflora S. Moore = Whitfieldia elongata (P. Beauv.) De Wild. et T. Durand ■☆

413958 Whitfieldia longifolia T. Anderson;长叶惠特爵床■☆

413959 Whitfieldia longifolia T. Anderson var. perglabra (C. B. Clarke) Hutch. et Dalziel = Whitfieldia elongata (P. Beauv.) De Wild. et T. Durand ■☆

413960 Whitfieldia orientalis Vollesen;东方惠特爵床■☆

413961 Whitfieldia perglabra C. B. Clarke = Whitfieldia elongata (P. Beauv.) De Wild. et T. Durand ■☆

413962 Whitfieldia preussii (Lindau) C. B. Clarke;普罗伊斯惠特爵床■☆

413963 Whitfieldia purpurata (Benoist) Heine;紫惠特爵床■☆

413964 Whitfieldia rutilans Heine;橙红惠特爵床■☆

413965 Whitfieldia seretii De Wild. = Whitfieldia stuhlmannii (Lindau) C. B. Clarke ■☆

413966 Whitfieldia seretii De Wild. var. elliptica ? = Whitfieldia stuhlmannii (Lindau) C. B. Clarke ■☆

413967 Whitfieldia striata (S. Moore) Vollesen = Whitfieldia liebrechtsiana De Wild. et T. Durand ■☆

413968 Whitfieldia stuhlmannii (Lindau) C. B. Clarke;斯图惠特爵床■☆

413969 Whitfieldia subviridis C. B. Clarke = Whitfieldia elongata (P. Beauv.) De Wild. et T. Durand ■☆

413970 Whitfieldia sylvatica De Wild. = Whitfieldia brazzae (Baill.) C. B. Clarke ■☆

413971 Whitfieldia tanganyikensis C. B. Clarke = Whitfieldia elongata (P. Beauv.) De Wild. et T. Durand ■☆

413972 Whitfieldia thollonii (Baill.) Benoist;托伦惠特爵床■☆

413973 Whitfordia Elmer = Whitfordiodendron Elmer ●

413974 Whitfordiodendron Elmer = Callerya Endl. ●■

413975 Whitfordiodendron Elmer(1910);猪腰豆属(大荚藤属,猪腰子属);Whitfordiodendron,Porkkidneybean ●

413976 Whitfordiodendron filipes (Dunn) Dunn = Afgekia filipes (Dunn) R. Geesink ● u

413977 Whitfordiodendron filipes (Dunn) Dunn var. tomentosum Z. Wei = Afgekia filipes (Dunn) R. Geesink var. tomentosa (Z. Wei) Y. F. Deng et H. N. Qin ●

413978 Whitfordiodendron taiwanianum (Hayata) Ohwi = Millettia pachycarpa Benth. ●■

413979 Whitia Blume = Cyrtandra J. R. Forst. et G. Forst. ●■

413980 Whitlavia Harv. = Phacelia Juss. ■☆

413981 Whitleya D. Don = Anisodus Link ex Spreng. ■

413982 Whitleya D. Don = Scopolia Jacq. (保留属名)■

413983 Whitleya D. Don ex Sweet = Anisodus Link ex Spreng. ■

413984 Whitleya D. Don ex Sweet = Scopolia Jacq. (保留属名)■

413985 Whitleya Sweet = Anisodus Link ex Spreng. ■

413986 Whitleya Sweet = Scopolia Jacq. (保留属名)■

413987 Whitleya streamonifolia Sweet = Anisodus luridus Link et Otto ■

413988 Whitmorea Sleumer(1969);所罗门木属●☆

413989 Whitmorea grandiflora Sleumer;所罗门木●☆

413990 Whitneya A. Gray = Arnica L. ●■☆

413991 Whitneya A. Gray(1865);惠特尼菊属●☆

413992 Whitneya dealbata A. Gray = Arnica dealbata (A. Gray) B. G. Baldwin ●☆

413993 Whittonia Sandwith(1962);南美围盘树属●☆

413994 Whittonia guianensis Sandwith;南美围盘树●☆

413995 Whyanbeelia Airy Shaw et B. Hyland(1976);怀亚大戟属☆

413996 Whyanbeelia terrae-reginae Airy Shaw et B. Hyland;怀亚大戟☆

413997 Whytockia W. W. Sm. (1919);异叶苣苔属(玉玲花属);Whytockia ■★

413998 Whytockia bijieensis Yin Z. Wang et Z. Yu Li;毕节异叶苣苔;Bijie Whytockia ■

413999 Whytockia chiritiflora (Oliv.) W. W. Sm. ;异叶苣苔;

Diversifolious Whytockia, Whytockia ■

414000　Whytockia chiritiflora (Oliv.) W. W. Sm. var. minor W. W. Sm. = Whytockia hekouensis Yin Z. Wang var. minor (W. W. Sm.) Yin Z. Wang ■

414001　Whytockia chiritiflora (Oliv.) W. W. Sm. var. minor W. W. Sm. = Whytockia tsiangiana (Hand.-Mazz.) A. Weber var. minor (W. W. Sm.) A. Weber ■

414002　Whytockia gongshanensis Yin Z. Wang et H. Li;贡山异叶苣苔;Gongshan Whytockia ■

414003　Whytockia hekouensis Yin Z. Wang;河口异叶苣苔;Hekou Whytockia ■

414004　Whytockia hekouensis Yin Z. Wang var. minor (W. W. Sm.) Yin Z. Wang;屏边异叶苣苔;Pingbian Whytockia ■

414005　Whytockia purpurascens Yin Z. Wang;紫红异叶苣苔;Purpurascent Whytockia ■

414006　Whytockia sasakii (Hayata) B. L. Burtt;台湾异叶苣苔(玉玲花);Taiwan Whytockia ■

414007　Whytockia tsiangiana (Hand.-Mazz.) A. Weber;白花异叶苣苔;Whiteflower Whytockia ■

414008　Whytockia tsiangiana (Hand.-Mazz.) A. Weber var. minor (W. W. Sm.) A. Weber = Whytockia hekouensis Yin Z. Wang var. minor (W. W. Sm.) Yin Z. Wang ■

414009　Whytockia tsiangiana (Hand.-Mazz.) A. Weber var. wilsonii A. Weber;峨眉异叶苣苔;Emei Whytockia ■

414010　Wiasemskya Klotzsch(废弃属名) = Tammsia H. Karst.(保留属名)☆

414011　Wibelia P. Gaertn.,B. Mey. et Scherb. = Crepis L. ■

414012　Wibelia Pers. = Paypayrola Aubl. ■☆

414013　Wibelia Roehl. = Chondrilla L. ■

414014　Wiborgia Kuntze = Eysenhardtia Kunth(保留属名)●☆

414015　Wiborgia Kuntze = Viborquia Ortega(废弃属名)●☆

414016　Wiborgia Post et Kuntze = Cytisus Desf.(保留属名)●

414017　Wiborgia Post et Kuntze = Tubocytisus (DC.) Fourr. ●

414018　Wiborgia Post et Kuntze = Viborgia Moench(废弃属名)●

414019　Wiborgia Roth = Galinsoga Ruiz et Pav. ■●

414020　Wiborgia Roth = Vigolina Poir. ■●

414021　Wiborgia Thunb. (1800)(保留属名);维堡豆属■☆

414022　Wiborgia angustifolia Benth. = Wiborgia tenuifolia E. Mey. ■☆

414023　Wiborgia apterophora R. Dahlgren = Wiborgia humilis (Thunb.) R. Dahlgren ■☆

414024　Wiborgia armata (Thunb.) Harv. = Wiborgia mucronata (L. f.) Druce ■☆

414025　Wiborgia armata (Thunb.) Harv. var. puberula Harv. = Wiborgia monoptera E. Mey. ■☆

414026　Wiborgia cuspidata (E. Mey.) Benth. = Wiborgia incurvata E. Mey. ■☆

414027　Wiborgia flexuosa E. Mey. = Wiborgia fusca Thunb. ■☆

414028　Wiborgia floribunda Lodd. = Wiborgia obcordata (P. J. Bergius) Thunb. ■☆

414029　Wiborgia fusca Thunb. ;棕色维堡豆■☆

414030　Wiborgia fusca Thunb. subsp. macrocarpa R. Dahlgren;大果维堡豆■☆

414031　Wiborgia grandiflora E. Mey. = Lebeckia sessilifolia (Eckl. et Zeyh.) Benth. ■☆

414032　Wiborgia heteroclados E. Mey. = Wiborgia mucronata (L. f.) Druce ■☆

414033　Wiborgia humilis (Thunb.) R. Dahlgren;低矮维堡豆■☆

414034　Wiborgia incurvata E. Mey. ;内折维堡豆■☆

414035　Wiborgia lanceolata E. Mey. = Wiborgia sericea Thunb. ■☆

414036　Wiborgia leptoptera R. Dahlgren;窄翅维堡豆■☆

414037　Wiborgia monoptera E. Mey. ;单翅维堡豆■☆

414038　Wiborgia mucronata (L. f.) Druce;短尖维堡豆■☆

414039　Wiborgia obcordata (P. J. Bergius) Thunb. ;倒心形维堡豆■☆

414040　Wiborgia oblongata E. Mey. = Wiborgia fusca Thunb. ■☆

414041　Wiborgia oblongata E. Mey. var. cuspidata ? = Wiborgia incurvata E. Mey. ■☆

414042　Wiborgia parviflora Kunth = Galinsoga parviflora Cav. ■

414043　Wiborgia parvifolia E. Mey. = Wiborgia tenuifolia E. Mey. ■☆

414044　Wiborgia sericea Thunb. ;绢毛维堡豆■☆

414045　Wiborgia spinescens Eckl. et Zeyh. = Wiborgia mucronata (L. f.) Druce ■☆

414046　Wiborgia tenuifolia E. Mey. ;细叶维堡豆■☆

414047　Wiborgia tetraptera E. Mey. ;四翅维堡豆■☆

414048　Wiborgia urticifolia Kunth = Galinsoga ciliata (Raf.) S. F. Blake ■

414049　Wiborgiella Boatwr. et B. -E. van Wyk = Lebeckia Thunb. ■☆

414050　Wiborgiella Boatwr. et B. -E. van Wyk(2009);小维堡豆属■☆

414051　Wichuraea M. Roem. = Bomarea Mirb. ■☆

414052　Wichuraea Nees = Cryptandra Sm. ●☆

414053　Wichuraea Nees ex Reissek = Cryptandra Sm. ●☆

414054　Wichurea Benth. et Hook. f. = Cryptandra Sm. ●☆

414055　Wichurea Benth. et Hook. f. = Wichuraea Nees ●☆

414056　Wickstroemia Rchb. = Laplacea Kunth(保留属名)●☆

414057　Wickstroemia Rchb. = Wikstroemia Endl. (保留属名)●

414058　Widdringtonia Endl. (1842);维氏柏属(南非柏属,维林图柏属);African Cypress, Clanwilliam Cedar, Widdringtonia ●☆

414059　Widdringtonia cedarbergensis J. A. Marsh;维氏柏;Clanwilliam Cedar ●☆

414060　Widdringtonia cupressoides (L.) Endl. ;冰山维氏柏(冰山维林图柏);Berg Cypress, Cypress Pine ●☆

414061　Widdringtonia cupressoides Endl. = Widdringtonia cupressoides (L.) Endl. ●☆

414062　Widdringtonia juniperoides Endl. = Callitris arborea Schrad. ex E. Mey. ●☆

414063　Widdringtonia juniperoides Endl. = Widdringtonia cedarbergensis J. A. Marsh ●☆

414064　Widdringtonia nodiflora (L.) Powrie = Brunia nodiflora L. ●☆

414065　Widdringtonia schwarzii Mast. ;斯切沃维氏柏(斯切沃维林图柏);Schwarz Widdringtonia, Willowmore Cedar ●☆

414066　Widdringtonia whytei Rendle;瓦特维氏柏; Minnie Cedar, Mlanji Widdringtonia ●☆

414067　Widdringtonia whytei Rendle = Brunia nodiflora L. ●☆

414068　Widdringtonia whytei Rendle = Widdringtonia nodiflora (L.) Powrie ●☆

414069　Widdringtoniaceae Doweld = Cupressaceae Gray(保留科名)●

414070　Widgrenia Malme(1900);维德萝藦属■☆

414071　Widgrenia corymbosa Malme. ;维德萝藦■☆

414072　Wiedemannia Fisch. et C. A. Mey. (1838);威德曼草属(魏德曼草属)■☆

414073　Wiedemannia Fisch. et C. A. Mey. = Lamium L. ■

414074　Wiedemannia multifida (L.) Benth. ;多裂威德曼草(多裂魏德曼草)■☆

414075　Wiedemannia orientalis Fisch. et C. A. Mey. ;威德曼草(魏德曼草)■☆

414076　Wiegmannia Hochst. et Steud. ex Steud. = Maerua Forssk. ●☆

414077　Wiegmannia Meyen = Hedyotis L. (保留属名)●■

414078　Wiegmannia Meyen = Kadua Cham. et Schltdl. ●■

414079　Wielandia Baill. (1858);维兰德大戟属●☆

414080　Wielandia elegans Baill. ;维兰德大戟●☆

414081　Wielandia elegans Baill. var. perrieri Léandri = Wielandia elegans Baill. ●☆

414082　Wierzbickia Rchb. = Minuartia L. ■

414083　Wierzbickia macrocarpa (Pursh) Rchb. = Minuartia macrocarpa (Pursh) Ostenf. ■

414084　Wiesbauria Gand. = Viola L. ■●

414085　Wiesneria Micheli(1881);威森泻属■☆

414086　Wiesneria filifolia Hook. f. ;线叶威森泻■☆

414087　Wiesneria schweinfurthii Hook. f. ;非洲威森泻■☆

414088　Wiesneria sparganiifolia Graebn. = Wiesneria schweinfurthii Hook. f. ■☆

414089　Wiestia Boiss. = Boissiera Hochst. ex Steud. ■

414090　Wiestia Sch. Bip. = Lactuca L. ■

414091　Wiganda St. -Lag. = Wigandia Kunth(保留属名)●■☆

414092　Wigandia Kunth(1819)(保留属名);威根麻属(维甘木属,维康草属);Wigandla ●■☆

414093　Wigandia Neck. = Disparago Gaertn. (保留属名)●☆

414094　Wigandia Neck. ex Less. = Disparago Gaertn. (保留属名)●☆

414095　Wigandia caracasana Kunth;加拉加斯威根麻(卡拉卡萨维甘木,维康草)■☆

414096　Wigandia kunthii Choisy;库氏威根麻(库氏维甘木)●☆

414097　Wigandia kunthii Choisy. f. africana Brand = Wigandia urens (Ruiz et Pav.) Kunth f. africana (Brand) Verdc. ■☆

414098　Wigandia leucocephala (DC.) Sch. Bip. = Stoebe leucocephala DC. ●☆

414099　Wigandia urens (Ruiz et Pav.) Kunth;矮生威根麻(矮生维甘木);Caracus Wigandia ●☆

414100　Wigandia urens (Ruiz et Pav.) Kunth f. africana (Brand) Verdc. ;非洲威根麻■☆

414101　Wigandia urens (Ruiz et Pav.) Kunth f. caracasana (Kunth) Gibson = Wigandia caracasana Kunth ■☆

414102　Wigandia urens Choisy = Wigandia urens (Ruiz et Pav.) Kunth ●☆

414103　Wiggersia P. Gaertn. ,B. Mey. et Scherb. = Vicia L. ■

414104　Wigginsia D. M. Porter = Parodia Speg. (保留属名)●

414105　Wigginsia D. M. Porter(1964);金鹰仙人球属■☆

414106　Wigginsia sellowii (Link. et Otto) D. M. Porter;金鹰仙人球■☆

414107　Wigginsia vorwerkiana (Werderm.) D. M. Porter;软果金鹰仙人球(软果仙人球);Colombian Ball Cactus ■☆

414108　Wightia Spreng. ex DC. = Centratherum Cass. ■☆

414109　Wightia Wall. (1830);美丽桐属(岩梧桐属);Wightia ●

414110　Wightia alpinii Craib = Wightia speciosissima (D. Don) Merr. ●

414111　Wightia elliptica Merr. = Wightia speciosissima (D. Don) Merr. ●

414112　Wightia gigantea Wall. = Wightia speciosissima (D. Don) Merr. ●

414113　Wightia lacei Craib = Wightia speciosissima (D. Don) Merr. ●

414114　Wightia speciosissima (D. Don) Merr. ;美丽桐(大美丽桐,石牡丹,岩梧桐);Beautiful Wightia,Gigantic Wightia ●

414115　Wigmannia Walp. = Hedyotis L. (保留属名)●■

414116　Wigmannia Walp. = Kadua Cham. et Schltdl. ●■

414117　Wigmannia Walp. = Wiegmannia Meyen ●■

414118　Wikstroemia Endl. (1833)('Wickstroemia')(保留属名);荛花属(雁皮属);Stringbush,Wikstroemia ●

414119　Wikstroemia Schrad. (废弃属名) = Laplacea Kunth(保留属名)●☆

414120　Wikstroemia Schrad. (废弃属名) = Wikstroemia Endl. (保留属名)●

414121　Wikstroemia Spreng. = Wikstroemia Endl. (保留属名)●

414122　Wikstroemia alba Hand. -Mazz. = Wikstroemia trichotoma (Thunb.) Makino ●

414123　Wikstroemia albiflora Yatabe;白花荛花●☆

414124　Wikstroemia albiflora Yatabe = Diplomorpha albiflora (Yatabe) Nakai ●☆

414125　Wikstroemia alternifolia Batalin;互生叶荛花;Alternateleaf Stringbush, Alternate-leaved Stringbush ●

414126　Wikstroemia alternifolia Batalin var. multiflora Lecomte;多花互生叶荛花;Manyflower Alternateleaf Stringbush ●

414127　Wikstroemia alternifolia Batalin var. multiflora Lecomte = Wikstroemia alternifolia Batalin ●

414128　Wikstroemia androsaemifolia Hand. -Mazz. = Wikstroemia lamatsoensis Hamaya ●

414129　Wikstroemia angustifolia Hemsl. ;狭叶荛花(黄构皮,岩杉树);Angustifoliate Stringbush,Narrowleaf Stringbush ●

414130　Wikstroemia angustiloba (Rehder) Domke = Daphne angustiloba Rehder ●

414131　Wikstroemia angustiloba (Rehder) Domke = Daphne tenuiflora Bureau et Franch. ●

414132　Wikstroemia angustissima Merr. = Wikstroemia lanceolata Merr. ●

414133　Wikstroemia anhuiensis D. C. Zhang et X. P. Zhang;安徽荛花;Anhui Stringbush ●

414134　Wikstroemia aurantiaca (Diels) Domke = Daphne aurantiaca Diels ●

414135　Wikstroemia aurantiaca (Diels) Domke var. pulvinata Domke = Daphne aurantiaca Diels ●

414136　Wikstroemia axillaris Merr. et Chun = Daphne axillaris (Merr. et Chun) Chun et C. F. Wei ●

414137　Wikstroemia baimashanensis S. C. Huang;白马山荛花(白马荛花);Baimashan Stringbush ●

414138　Wikstroemia balansae Drake = Rhamnoneuron balansae (Drake) Gilg ●

414139　Wikstroemia bodinieri H. Lév. = Alyxia schlechteri H. Lév. ●

414140　Wikstroemia bodinieri H. Lév. = Daphne tangutica Maxim. ●

414141　Wikstroemia brevipaniculata Rehder = Wikstroemia micrantha Hemsl. ●

414142　Wikstroemia calcicola (W. W. Sm.) Domke = Daphne aurantiaca Diels ●

414143　Wikstroemia canescens (Wall.) Meisn. ;荛花(矮陀陀,长花荛花,黄荛花,黄荛花,灰白荛花,老虎麻,老龙树,土箭芪,一把香);Canescent Stringbush,Hoary Stringbush ●

414144　Wikstroemia canescens (Wall.) Meisn. = Wikstroemia dolichantha Diels ●

414145　Wikstroemia canescens Meisn. ;灰白荛花●☆

414146　Wikstroemia canescens Meisn. = Wikstroemia pilosa W. C. Cheng var. kulingensis (Domke) S. C. Huang ●

414147　Wikstroemia capitata Rehder;头序荛花(赶山尖,滑皮树,黄狗皮,木兰条,香叶子);Capitate Stringbush ●

414148　Wikstroemia capitatoracemosa S. C. Huang;短总序荛花;Capitate-racemose Stringbush,Shortrace Stringbush ●

414149　Wikstroemia capitellata (H. Hara) H. Hara = Daphnimorpha capitellata (H. Hara) Nakai ●☆

414150　Wikstroemia chamaedaphne Meisn. ;河朔荛花(矮雁皮,北荛花,番泻叶,甘遂,拐拐花,花鱼梢,黄闷头花,黄荛花,黄芫花,黄雁雁,叩皮花,老虎麻,芫,芫蒿,芫花,雁皮花,羊厌厌,羊燕花,羊冤冤,痒眼花,药鱼梢,野瑞香,野雁皮,岳彦花);Lowdaphne Stringbush,Low-daphne Stringbush ●

414151　Wikstroemia chamaedaphne Meisn. var. galioides Batalin ex Lecomte = Wikstroemia stenophylla E. Pritz. ex Diels ●

414152　Wikstroemia chamaejasme (L.) Domke = Stellera chamaejasme L. ■

414153　Wikstroemia chinensis Meisn. ;中国荛花(中华荛花)●

414154　Wikstroemia chuii Merr. ;琼岛荛花(窄叶荛花);Chu Stringbush ●

414155　Wikstroemia circinata (Lecomte) Domke = Wikstroemia dolichantha Diels ●

414156　Wikstroemia circinata (Lecomte) Domke var. divaricata (Lecomte) Domke = Wikstroemia dolichantha Diels ●

414157　Wikstroemia ciricnata (Lecomte) Domke var. divaricata (Lecomte) Domke = Wikstroemia dolichantha Diels ●

414158　Wikstroemia clivicola (Hand. -Mazz.) Domke = Daphne rosmarinifolia Rehder ●

414159　Wikstroemia cochlearifolia S. C. Huang;匙叶荛花;Spatulate-leaved Stringbush,Spoonleaf Stringbush ●

414160　Wikstroemia delavayi Lecomte;澜沧荛花(澜沧江荛花);Delavay Stringbush,Lancang Stringbush ●

414161　Wikstroemia diffusa (Lecomte) Domke = Daphne rosmarinifolia Rehder ●

414162　Wikstroemia dolichantha Diels;一把香荛花(矮陀陀,长花荛花,长毛花荛花,构皮荛花,光叶荛花,山棉花,山条一把香,山皮香,铁扁担,土箭七,土箭芫,香构,一把香,一柱香,竹腊皮);Handful Stringbush, Longflower Stringbush, Longflower Stringbush Bashful,Long-flowered Stringbush,Spreading Stringbush ●

414163　Wikstroemia dolichantha Diels var. effusa (Rehder) C. Y. Chang = Wikstroemia dolichantha Diels ●

414164　Wikstroemia dolichantha Diels var. pubescens Domke = Wikstroemia dolichantha Diels ●

414165　Wikstroemia domkeana H. L. Li = Daphne gracilis E. Pritz. ●

414166　Wikstroemia effusa Rehder = Wikstroemia dolichantha Diels ●

414167　Wikstroemia ellipsocarpa Maxim. = Diplomorpha trichotoma (Thunb.) Nakai ●

414168　Wikstroemia ellipsocarpa Maxim. = Wikstroemia trichotoma (Thunb.) Makino ●

414169　Wikstroemia ericifolia Domke = Wikstroemia micrantha Hemsl. ●

414170　Wikstroemia eriophylla H. Winkl. = Daphne holoserica (Diels) Hamaya var. thibetensis (Lecomte) Hamaya ●

414171　Wikstroemia eriophylla H. Winkl. = Pentathymelaea thibetensis Lecomte ●

414172　Wikstroemia fargesii (Lecomte) Domke;城口荛花;Farges Stringbush ●

414173　Wikstroemia flaviflora (H. Winkl.) Domke = Daphne penicillata Rehder ●

414174　Wikstroemia foetida Hillebr. var. oahuensis Gray;瓦胡臭荛花(南岭荛花);Oahu Stringbush ●☆

414175　Wikstroemia forsteri Decne. = Wikstroemia indica (L.) C. A. Mey. ●

414176　Wikstroemia fuminensis Y. D. Qi et Yin Z. Wang;富民荛花 ●

414177　Wikstroemia ganpi (Siebold et Zucc.) Maxim. ;小瑞香 ●

414178　Wikstroemia ganpi (Siebold et Zucc.) Maxim. = Diplomorpha ganpi (Siebold et Zucc.) Nakai ●

414179　Wikstroemia gemmata (E. Pritz.) Domke = Daphne gemmata E. Pritz. ex Diels ●

414180　Wikstroemia genkwa (Siebold et Zucc.) Domke = Daphne genkwa Siebold et Zucc. ●

414181　Wikstroemia glabra W. C. Cheng;光叶荛花(光洁荛花,山荆);Glabrous Stringbush ●

414182　Wikstroemia glabra W. C. Cheng f. purpurea (W. C. Cheng) S. C. Huang;紫背光叶荛花;Purple Glabrous Stringbush ●

414183　Wikstroemia glabra W. C. Cheng var. purpurea W. C. Cheng = Wikstroemia glabra W. C. Cheng f. purpurea (W. C. Cheng) S. C. Huang ●

414184　Wikstroemia gracilis (E. Pritz.) Domke = Daphne gracilis E. Pritz. ●

414185　Wikstroemia gracilis Hemsl. ;纤细荛花;Slender Stringbush ●

414186　Wikstroemia hainanensis Merr. ;海南荛花;Hainan Stringbush ●

414187　Wikstroemia hainanensis Merr. = Wikstroemia nutans Champ. ex Benth. ●

414188　Wikstroemia haoii Domke;武都荛花; Hao Stringbush, Wudu Stringbush ●

414189　Wikstroemia hemsleyana H. Lév. = Alstonia mairei H. Lév. ●

414190　Wikstroemia holosericea Diels = Daphne holoserica (Diels) Hamaya ●

414191　Wikstroemia huidongensis C. Yu Chang;会东荛花; Huidong Stringbush ●

414192　Wikstroemia inamoena Meisn. = Wikstroemia canescens (Wall.) Meisn. ●

414193　Wikstroemia indica (L.) C. A. Mey. ;了哥王(白棉儿,爆牙郎,别南根,布英,赤坡,大黄头树,地巴麻,地棉根,地棉麻树,地棉皮,毒鱼藤,哥春光,狗颈树,狗信药,红赤七,红灯笼,黄皮子,火索木,鸡杜头,鸡子麻,假黄皮,金腰带,九信菜,九信草,了哥麻,了王麻,南岭荛花,铺银草,蒲仑,蒲仑头,埔根,千年矮,雀儿麻,雀仔麻,雀子麻,三雁皮,山豆子,山黄皮,山六麻,山络麻,山麻皮,山棉皮,山埔仑,山埔银,山石榴,山雁皮,山之一,石谷皮,石棉皮,铁骨伞,铁骨散,铁乌散,桐皮子,乌子麻,消山药,小金腰带,小叶金腰带,野麻朴,野棉之,指皮麻);Indian Stringbush, Indian Wikstroemia,Mynaking ●

414194　Wikstroemia indica (L.) C. A. Mey. var. viridiflora Hook. f. = Wikstroemia indica (L.) C. A. Mey. ●

414195　Wikstroemia japonica (Siebold et Zucc.) Miq. = Wikstroemia trichotoma (Thunb.) Makino ●

414196　Wikstroemia japonica Miq. = Wikstroemia trichotoma (Thunb.) Makino ●

414197　Wikstroemia kudoi Makino = Daphnimorpha kudoi (Makino) Nakai ●☆

414198　Wikstroemia kulingensis Domke = Wikstroemia pilosa W. C. Cheng var. kulingensis (Domke) S. C. Huang ●

414199　Wikstroemia kulingensis Domke = Wikstroemia pilosa W. C. Cheng ●

414200　Wikstroemia lamatsoensis Hamaya;金丝桃叶荛花(金丝桃荛花);Rockjasmine Stringbush,Rockjasmineleaf Wikstroemia ●

414201　Wikstroemia lanceolata Merr. ;披针叶荛花;Lanceleaf Stringbush ●

414202　Wikstroemia lecomteana Domke = Daphne rosmarinifolia Rehder ●

414203　Wikstroemia leptophylla W. W. Sm. ;细叶荛花;Leptophyllous Stringbush,Thinleaf Stringbush ●

414204　Wikstroemia leptophylla W. W. Sm. var. atroviclacea Hand. -Mazz. ;黑紫荛花(黑紫细叶荛花);Dark Purple Thinleaf Stringbush ●

414205　Wikstroemia leuconeura (Rehder) Domke = Daphne esquirolii

H. Lév. ●

414206 Wikstroemia liangii Merr. et Chun;大叶荛花;Liang Stringbush ●

414207 Wikstroemia lichiangensis W. W. Sm.;丽江荛花(醉鱼草);Lijiang Stringbush ●

414208 Wikstroemia ligustrina Rehder;白蜡叶荛花(白腊叶荛花,羊眼子);Ashleaf Stringbush,Ash-leaved Stringbush ●

414209 Wikstroemia linearifolia H. F. Zhou = Wikstroemia linearifolia H. F. Zhou ex C. Yu Chang ●

414210 Wikstroemia linearifolia H. F. Zhou ex C. Yu Chang;线叶荛花;Stripedleaf Stringbush ●

414211 Wikstroemia linoides Hemsl. ex Forbes et Hemsl.;亚麻叶荛花(麻叶荛花,亚麻荛花);Flaxleaf Stringbush,Flax-leaved Stringbush ●

414212 Wikstroemia longipaniculata S. C. Huang;长锥序荛花;Longpanicle Stringbush,Long-peniculate Stringbush ●

414213 Wikstroemia lungtzeensis S. C. Huang;隆子荛花;Longzi Stringbush ●

414214 Wikstroemia mairei (Lecomte) Domke = Daphne esquirolii H. Lév. ●

414215 Wikstroemia mekongensis W. W. Sm. = Wikstroemia delavayi Lecomte ●

414216 Wikstroemia micrantha Hemsl.;小黄构(冬青,短锥序荛花,黄构,黄构皮,藤构,娃娃皮,香构,香叶,小黄狗皮,野棉皮);Littleflower Stringbush,Little-flowered Stringbush,Shortpaniculate Stringbush ●

414217 Wikstroemia micrantha Hemsl. var. paniculata (H. L. Li) S. C. Huang;圆锥荛花(耗子皮,两广荛花,小雀儿麻);Panicled Littleflower Stringbush,Paniculate Stringbush ●

414218 Wikstroemia micrantha Hemsl. var. paniculata (H. L. Li) S. C. Huang = Wikstroemia micrantha Hemsl. ●

414219 Wikstroemia modesta (Rehder) Domke = Daphne modesta Rehder ●

414220 Wikstroemia monnula Hance;北江荛花(地棉根,黄皮子,山谷麻,山谷皮,山花皮,山棉,山棉皮,土坝天);Lovely Stringbush ●

414221 Wikstroemia monnula Hance var. xiuningensis D. C. Zhang et J. Z. Shao;休宁荛花;Xiuning Stringbush ●

414222 Wikstroemia mononectraria Hayata;独鳞荛花(红荛花,乌来荛花);Uninectary Stringbush,Urai Wikstroemia ●

414223 Wikstroemia myrtilloides (Nitsche) Domke = Daphne myrtilloides Nitsche ●

414224 Wikstroemia nutans Champ. ex Benth.;细轴荛花(垂穗荛花,地麻棉,地棉麻,狗颈树,金腰带,山皮棉,石棉麻,野发麻,野棉花);Drooping Stringbush ●

414225 Wikstroemia nutans Champ. ex Benth. var. brevior Hand.-Mazz.;短细轴荛花;Short Drooping Stringbush ●

414226 Wikstroemia obovata Hemsl. = Wikstroemia retusa A. Gray ●

414227 Wikstroemia ohsumiensis Hatus.;高隈荛花●☆

414228 Wikstroemia ovalifolia Decne. = Wikstroemia indica (L.) C. A. Mey. ●

414229 Wikstroemia ovata Fernández-Villar = Wikstroemia indica (L.) C. A. Mey. ●

414230 Wikstroemia pachyrachis S. L. Tsai;粗轴荛花(厚轴荛花);Thickrachis Stringbush,Thick-rachised Stringbush ●

414231 Wikstroemia paimashanensis S. C. Huang = Wikstroemia baimashanensis S. C. Huang ●

414232 Wikstroemia pampaninii Rehder;鄂北荛花;N. Hubei Stringbush,Pampanin Stringbush ●

414233 Wikstroemia paniculata H. L. Li = Wikstroemia micrantha

Hemsl. ●

414234 Wikstroemia paniculata H. L. Li. = Wikstroemia micrantha Hemsl. var. paniculata (H. L. Li) S. C. Huang ●

414235 Wikstroemia parviflora S. C. Huang;小花荛花;Littleflower Stringbush,Small-flowered Stringbush ●

414236 Wikstroemia parviflora S. C. Huang = Wikstroemia sinoparviflora Yin Z. Wang et M. G. Gilbert ●

414237 Wikstroemia pauciflora (Franch. et Sav.) Franch. et Sav. ex Shirai;少花荛花●☆

414238 Wikstroemia pauciflora (Franch. et Sav.) Franch. et Sav. ex Shirai var. phymatoglossa (Koidz.) Hatus. = Diplomorpha phymatoglossa (Koidz.) Nakai ●☆

414239 Wikstroemia pauciflora (Franch. et Sav.) Franch. et Sav. ex Shirai var. yakusimensis Makino = Diplomorpha pauciflora (Franch. et Sav.) Nakai var. yakushimensis (Makino) T. Yamanaka ●☆

414240 Wikstroemia pauciflora (Franch. et Sav.) Franch. et Sav. ex Shirai var. yakusimensis Makino;屋久岛荛花●☆

414241 Wikstroemia pauciflora (Franch. et Sav.) Franch. et Sav. ex Shirai = Diplomorpha pauciflora (Franch. et Sav.) Nakai ●☆

414242 Wikstroemia paxiana H. Winkl.;懋公荛花(藏东荛花);Pax Stringbush ●

414243 Wikstroemia phymatoglossa Koidz.;瘤舌荛花●☆

414244 Wikstroemia pilosa W. C. Cheng;毛花荛花(地棉皮,多毛荛花,柔毛荛花,山棉皮,浙雁皮);Hairyflower Stringbush,Hairy-flowered Stringbush ●

414245 Wikstroemia pilosa W. C. Cheng var. kulingensis (Domke) S. C. Huang;绢毛荛花;Guling Stringbush ●

414246 Wikstroemia pilosa W. C. Cheng var. kulingensis S. C. Huang = Wikstroemia pilosa W. C. Cheng ●

414247 Wikstroemia pretiosa (Balf. f. et W. W. Sm. ex Farrer) Domke = Daphne myrtilloides Nitsche ●

414248 Wikstroemia pseudoretusa Koidz.;假倒卵叶荛花●☆

414249 Wikstroemia reginaldi-farreri (Halda) Yin Z. Wang et M. G. Gilbert;甘肃荛花●

414250 Wikstroemia retusa A. Gray;倒卵叶荛花(凹叶荛花);Obovateleaf Stringbush ●

414251 Wikstroemia rosmarinifolia (Rehder) Domke = Daphne rosmarinifolia Rehder ●

414252 Wikstroemia rosmarinifolia H. Winkl. = Wikstroemia stenophylla E. Pritz. ex Diels ●

414253 Wikstroemia salicina H. Lév. et Vaniot ex Rehder;柳状荛花(柳叶荛花);Willowleaf Stringbush ●

414254 Wikstroemia scytophylla Diels;革叶荛花(小构树);Coriaceousleaf Stringbush,Coriaceous-leaved Stringbush,Leatherleaf Stringbush ●

414255 Wikstroemia sericea Domke = Wikstroemia pilosa W. C. Cheng var. kulingensis (Domke) S. C. Huang ●

414256 Wikstroemia sericea Domke = Wikstroemia pilosa W. C. Cheng ●

414257 Wikstroemia sikokiana Franch. et Sav.;四国荛花(雁皮);Diplomorpha Stringbush,Gampi,Sikoku Stringbush ●☆

414258 Wikstroemia sikokiana Franch. et Sav. = Diplomorpha sikokiana (Franch. et Sav.) Honda ●☆

414259 Wikstroemia sinoparviflora Yin Z. Wang et M. G. Gilbert;中国小花荛花(小花荛花)●

414260 Wikstroemia stenantha Hemsl. = Wikstroemia monnula Hance ●

414261 Wikstroemia stenophylla E. Pritz. ex Diels;轮叶荛花(线叶荛花,窄叶荛花);Narrowleaf Stringbush,Stenophyllous Stringbush,

Whorlleaf Stringbush ●

414262 Wikstroemia stenophylla E. Pritz. ex Diels var. ziyangensis C. Y. Yu;岩山树;Ziyang Whorlleaf Stringbush ●

414263 Wikstroemia stenophylla E. Pritz. ex Diels var. ziyangensis C. Y. Yu = Wikstroemia stenophylla E. Pritz. ex Diels ●

414264 Wikstroemia subcyclolepidota L. P. Liu et Y. S. Lian;亚环鳞荛花●

414265 Wikstroemia taiwanensis S. C. Chang;台湾荛花;Taiwan Stringbush ●

414266 Wikstroemia techinensis S. C. Huang;德钦荛花;Deqin Stringbush ●

414267 Wikstroemia tenuiflora (Bureau et Franch.) Domke = Daphne tenuiflora Bureau et Franch. ●

414268 Wikstroemia tenuiflora (Bureau et Franch.) Domke var. legendrei (Lecomte) Domke = Daphne tenuiflora Bureau et Franch. var. legendrei (Lecomte) Hamaya ●

414269 Wikstroemia thibetensis (Lecomte) Domke var. thibetensis (Lecomte) Hamaya = Daphne holoserica (Diels) Hamaya var. thibetensis (Lecomte) Hamaya ●

414270 Wikstroemia thibetensis (Lecomte) Domke = Pentathymelaea thibetensis Lecomte ●

414271 Wikstroemia trichotoma (Thunb.) Makino;三歧荛花(白花荛花,荛花,日本荛花);Japanese Wikstroemia, White Stringbush, Whiteflower Stringbush,White-flowered Stringbush ●

414272 Wikstroemia trichotoma (Thunb.) Makino = Diplomorpha trichotoma (Thunb.) Nakai ●

414273 Wikstroemia trichotoma (Thunb.) Makino f. pilosa (Hamaya) Ohwi = Diplomorpha trichotoma (Thunb.) Nakai f. pilosa Hamaya ●

414274 Wikstroemia trichotoma (Thunb.) Makino f. pilosa (Hamaya) Ohwi = Wikstroemia trichotoma (Thunb.) Makino ●

414275 Wikstroemia trichotoma (Thunb.) Makino var. flavianthera S. Y. Liu;黄药三歧荛花(黄药白花荛花)●

414276 Wikstroemia uva-ursi A. Gray = Wikstroemia foetida Hillebr. var. oahuensis Gray ●☆

414277 Wikstroemia vaccinium (H. Lév.) Rehder;平伐荛花(凯里荛花,越橘荛花)●

414278 Wikstroemia valbrayi H. Lév. = Wikstroemia indica (L.) C. A. Mey. ●

414279 Wikstroemia viridiflora Meisn. = Wikstroemia indica (L.) C. A. Mey. ●

414280 Wikstroemia yakushimensis (Makino) Nakai ex Masam. = Diplomorpha pauciflora (Franch. et Sav.) Nakai var. yakushimensis (Makino) T. Yamanaka ●☆

414281 Wikstroemia yakushimensis (Makino) Nakai ex Masam. = Wikstroemia pauciflora (Franch. et Sav.) Franch. et Sav. ex Shirai var. yakusimensis Makino ●☆

414282 Wilberforcia Hook. f. ex Planch. = Bonamia Thouars(保留属名)●☆

414283 Wilbrandia Silva Manso(1836);威尔瓜属■☆

414284 Wilbrandia ebracteata Cogn.;无苞片威尔瓜■☆

414285 Wilckea Scop. = Vitex L. ●

414286 Wilckia Scop.(废弃属名) = Malcolmia W. T. Aiton(保留属名)■

414287 Wilckia africana (L.) F. Muell. = Malcolmia africana (L.) R. Br. ■

414288 Wilckia africana (L.) F. Muell. var. stenopetala (Bernh. ex Fisch. et C. A. Mey.) Grossh. = Malcolmia africana (L.) R. Br. ■

414289 Wilckia africana (L.) F. Muell. var. trichocarpa (Boiss. et Buhse) Grossh. = Malcolmia africana (L.) R. Br. ■

414290 Wilckia stenopetala (Bernh. ex Fisch. et C. A. Mey.) N. Busch = Malcolmia africana (L.) R. Br. ■

414291 Wilcoxia Britton et Rose = Echinocereus Engelm. ●

414292 Wilcoxia Britton et Rose = Peniocereus (A. Berger) Britton et Rose ●

414293 Wilcoxia Britton et Rose(1909);威尔掌属(威氏仙人掌属)■☆

414294 Wilcoxia albiflora Backeb.;白花威尔掌■☆

414295 Wilcoxia diguetii (F. A. C. Weber) Diguet et Guillaumin = Peniocereus striatus (Brandegee) Buxb. ●☆

414296 Wilcoxia poselgeri (Lem.) Britton et Rose;球根仙人鞭(银纽)■☆

414297 Wilcoxia poselgeri (Lem.) Britton et Rose = Echinocereus poselgeri Lem. ●☆

414298 Wilcoxia schmollii (Weing.) Knuth ex Backeb.;毛刺仙人鞭(珠毛柱)■

414299 Wilcoxia schmollii (Weing.) Knuth ex Backeb. = Echinocereus schmollii (Weing.) N. P. Taylor ●☆

414300 Wilcoxia striata (Brandegee) Britton et Rose = Peniocereus striatus (Brandegee) Buxb. ●☆

414301 Wilcoxia stricta (Brandegee) Britton et Rose;劲直威尔掌■☆

414302 Wildemaniodoxa Aubrév. et Pellegr. = Englerophytum K. Krause ●☆

414303 Wildenowia Thunb. = Willdenowia Thunb. ■☆

414304 Wildpretia U. Reifenb. et A. Reifenb. = Sonchus L. ■

414305 Wildpretina Kuntze = Ixanthus Griseb. ■

414306 Wildungenia Weuder. = Sinningia Nees ●■☆

414307 Wilhelminia Hochr. = Hibiscus L. (保留属名)●■

414308 Wilhelmsia C. Koch = Koeleria Pers. ■

414309 Wilhelmsia K. Koch = Koeleria Pers. ■

414310 Wilhelmsia K. Koch = Rostraria Trin. ■☆

414311 Wilhelmsia Rchb. (1829);极地蚤缀属;Merckia ■☆

414312 Wilhelmsia physodes (Fisch. ex Ser.) McNeill;极地蚤缀■☆

414313 Wilibalda Roth = Coleanthus Seidel(保留属名)■

414314 Wilibalda Sternb. = Coleanthus Seidel(保留属名)■

414315 Wilibalda Sternb. ex Roth = Coleanthus Seidel(保留属名)■

414316 Wilibalda Sternb. ex Roth = Schmidtia Tratt. ■

414317 Wilibalda subtilis (Tratt.) Roth = Coleanthus subtilis (Tratt.) Seidl ■

414318 Wilibald-schmidtia Conrad = Danthonia DC. (保留属名)■

414319 Wilibald-Schmidtia Seidel = Danthonia DC. (保留属名)■■

414320 Wilkea Post et Kuntze = Vitex L. ●

414321 Wilkea Post et Kuntze = Wilckea Scop. ●

414322 Wilkesia A. Gray(1852);多轮菊属●☆

414323 Wilkesia gymnoxiphium A. Gray;多轮菊;Iliau ■☆

414324 Wilkia F. Muell. = Malcolmia W. T. Aiton(保留属名)■

414325 Wilkia F. Muell. = Wilckia Scop. (废弃属名)■

414326 Wilkiea F. Muell. (1858);澳洲盖裂桂属●☆

414327 Wilkiea angustifolia (Bailey) Perkins;窄叶澳洲盖裂桂●☆

414328 Wilkiea calyptrocalyx F. Muell.;澳洲盖裂桂●☆

414329 Wilkiea rigidifolia (A. C. Sm.) Whiffin et Foreman;硬叶澳洲盖裂桂●☆

414330 Wilkstroemia Spreng. = Eupatorium L. ■●

414331 Willardia Rose(1891);墨矛果豆属●☆

414332 Willardia argyrotricha (Harms) F. J. Herm.;银毛墨矛果豆●☆

414333 Willardia mexicana Rose;墨矛果豆●☆

414334 Willardia obovata (Benth.) F. J. Herm.;倒卵墨矛果豆●☆

414335 Willardia parviflora Rose;小花墨矛果豆●☆

414336 Willbleibia Herter = Willkommia Hack. ex Schinz ■☆

414337 Willbleibia annua (Hack.) Herter = Willkommia annua Hack. ■☆

414338 Willbleibia newtonii (Hack.) Herter = Willkommia newtonii Hack. ■☆

414339 Willbleibia sarmentosa (Hack.) Herter = Willkommia sarmentosa Hack. ■☆

414340 Willdampia A. S. George = Clianthus Sol. ex Lindl. (保留属名)●

414341 Willdampia A. S. George = Donia G. Don et D. Don ex G. Don ●

414342 Willdenovia J. F. Gmel. = Cyrtanthus Aiton (保留属名)■☆

414343 Willdenovia J. F. Gmel. = Posoqueria Aubl. ●☆

414344 Willdenovia J. F. Gmel. = Rondeletia L. ●

414345 Willdenovia Thunb. = Willdenowia Thunb. ■☆

414346 Willdenowa Cav. = Adenophyllum Pers. ■●☆

414347 Willdenowa Cav. = Schlechtendalia Willd. (废弃属名)■●☆

414348 Willdenowia Steud. = Rondeletia L. ●

414349 Willdenowia Steud. = Willdenovia J. F. Gmel. ●

414350 Willdenowia Steud. = Willdenowa Cav. ■●☆

414351 Willdenowia Thunb. (1788) ('Wildenowia') ;威尔帚灯草属■☆

414352 Willdenowia affinis Pillans;近缘威尔帚灯草■☆

414353 Willdenowia argentea (Kunth) Hieron. = Ceratocaryum argenteum Kunth ■☆

414354 Willdenowia bolusii Pillans;博卢斯威尔帚灯草■☆

414355 Willdenowia compressa Thunb. = Cannomois parviflora (Thunb.) Pillans ■☆

414356 Willdenowia cuspidata Mast. = Willdenowia incurvata (Thunb.) H. P. Linder ■☆

414357 Willdenowia decipiens N. E. Br. = Ceratocaryum decipiens (N. E. Br.) H. P. Linder ■☆

414358 Willdenowia ecklonii (Nees) Kunth = Willdenowia incurvata (Thunb.) H. P. Linder ■☆

414359 Willdenowia ecklonii (Nees) T. Durand et Schinz = Anthochortus ecklonii Nees ■☆

414360 Willdenowia esterhuyseniae Pillans = Ceratocaryum fimbriatum (Kunth) H. P. Linder ■☆

414361 Willdenowia fimbriata Kunth = Ceratocaryum fimbriatum (Kunth) H. P. Linder ■☆

414362 Willdenowia fistulosa (Mast.) Pillans = Ceratocaryum fistulosum Mast. ■☆

414363 Willdenowia fraterna N. E. Br. = Willdenowia teres Thunb. ■☆

414364 Willdenowia galpinii N. E. Br. = Willdenowia teres Thunb. ■☆

414365 Willdenowia glomerata (Thunb.) H. P. Linder;团集威尔帚灯草■☆

414366 Willdenowia humilis Mast.;低矮威尔帚灯草■☆

414367 Willdenowia incurvata (Thunb.) H. P. Linder;内折威尔帚灯草■☆

414368 Willdenowia lucaeana Kunth = Willdenowia glomerata (Thunb.) H. P. Linder ■☆

414369 Willdenowia neglecta Steud. = Willdenowia incurvata (Thunb.) H. P. Linder ■☆

414370 Willdenowia peninsularis N. E. Br. = Willdenowia teres Thunb. ■☆

414371 Willdenowia purpurea Pillans;紫威尔帚灯草■☆

414372 Willdenowia rugosa Esterh.;皱褶威尔帚灯草■☆

414373 Willdenowia simplex N. E. Br. = Hypodiscus argenteus (Thunb.) Mast. ■☆

414374 Willdenowia stokoei Pillans;斯托克威尔帚灯草■☆

414375 Willdenowia striata Thunb. = Willdenowia incurvata (Thunb.) H. P. Linder ■☆

414376 Willdenowia sulcata Mast. ;纵沟威尔帚灯草■☆

414377 Willdenowia teres Thunb. ;圆柱威尔帚灯草■☆

414378 Willdenowia xerophila Pillans = Ceratocaryum xerophilum (Pillans) H. P. Linder ■☆

414379 Willemeta Cothen. = Koelreuteria Laxm. ●

414380 Willemetia Brongn. = Noltea Rchb. ●☆

414381 Willemetia Maerkl. = Bassia All. + Kochia Roth ●■

414382 Willemetia Maerkl. = Kochia Roth ●■

414383 Willemetia Neck. = Chondrilla L. ■

414384 Willemetia Neck. = Willemetia Neck. ex Cass. ■☆

414385 Willemetia Neck. ex Cass. (1777-1778) ;鳞果苣属■☆

414386 Willemetia Neck. ex Cass. = Calycocorsus F. W. Schmidt ■☆

414387 Willemetia scandens Eckl. et Zeyh. = Helinus integrifolius (Lam.) Kuntze ●☆

414388 Willemetia sedoides Moq. = Bassia sedoides (Schrad.) Asch. ■

414389 Willemetia tuberosa Fisch. et C. A. Mey. ex DC. ;鳞果苣■☆

414390 Williamia Baill. = Phyllanthus L. ●■

414391 Williamodendron Kubitzki et H. G. Richter (1987) ;威廉桂属●☆

414392 Williamodendron glaucophyllum (van der Werff) Kubitzki et H. G. Richter;灰蓝威廉桂●☆

414393 Williamodendron spectabile Kubitzki et H. G. Richter;威廉桂●☆

414394 Williamsia Merr. = Praravinia Korth. ●☆

414395 Willibalda Steud. = Coleanthus Seidel (保留属名)■

414396 Willibalda Steud. = Wilibalda Sternb. ■

414397 Willichia Mutis ex L. = Sibthorpia L. ■☆

414398 Willisellus Gray = Elatine L. ■

414399 Willisia Warm. (1901) ;威利斯川苔草属■☆

414400 Willisia selaginoides Warm. ex J. C. Willis;威利斯川苔草■☆

414401 Willkommia Hack. (1888) ;结脉草属■☆

414402 Willkommia Hack. ex Schinz = Willkommia Hack. ■☆

414403 Willkommia Sch. Bip. ex Nyman = Senecio L. ■●

414404 Willkommia annua Hack. ;一年结脉草■☆

414405 Willkommia newtonii Hack. ;纽敦结脉草■☆

414406 Willkommia sarmentosa Hack. ;结脉草■☆

414407 Willoughbeia Hook. f. = Willughbeia Roxb. (保留属名)●☆

414408 Willoughbya Kuntze = Mikania Willd. (保留属名)■

414409 Willoughbya Kuntze = Willugbaeya Neck. ■

414410 Willoughbya Neck. ex Kuntze = Mikania Willd. (保留属名)■

414411 Willrusselia feliciana A. Chev. = Pitcairnia feliciana (A. Chev.) Harms et Mildbr. ■☆

414412 Willrussellia A. Chev. = Pitcairnia L' Hér. (保留属名)■☆

414413 Willugbaeya Neck. = Mikania Willd. (保留属名)■

414414 Willughbeia Klotzsch = Landolphia P. Beauv. (保留属名)●☆

414415 Willughbeia Roxb. (1820) (保留属名) ;胶乳藤属(乳藤属,威乐比属)●☆

414416 Willughbeia Scop. = Ambelania Aubl. + Pacouria Aubl. (废弃属名)●☆

414417 Willughbeia cordata Klotzsch = Saba comorensis (Bojer ex A. DC.) Pichon ●☆

414418 Willughbeia coriacea Wall. ;胶乳藤;Borneo Rubber ●☆

414419 Willughbeia firma Blume = Willughbeia coriacea Wall. ●☆

414420 Willughbeia petersiana Klotzsch = Ancylobotrys petersiana (Klotzsch) Pierre ●☆

414421 Willughbeia senensis Klotzsch = Ancylobotrys petersiana (Klotzsch) Pierre ●☆

414422 Willughbeiaceae J. Agardh = Apocynaceae Juss. (保留科名)●■

414423 Willughbeiaceae J. Agardh ;胶乳藤科●

414424 Willughbeiopsis Rauschert = Urnularia Stapf ●☆

414425 Willughbeiopsis Rauschert = Willughbeia Roxb.(保留属名)●☆

414426 Willughbeja Scop. ex Schreb.（废弃属名）= Willughbeia Roxb.（保留属名）●☆

414427 Willwebera Á. Löve et D. Löve = Arenaria L. ■

414428 Wilmattea Britton et Rose = Hylocereus（A. Berger）Britton et Rose ●

414429 Wilmattea Britton et Rose(1920);姬花蔓柱属(威尔玛太属)●☆

414430 Wilmattea minutiflora Britton et Rose;小花姬花蔓柱●☆

414431 Wilmattea venezuelensis Croizat;姬花蔓柱●☆

414432 Wilmattia Willis = Wilmattea Britton et Rose ●☆

414433 Wilsonia Hook. = Dipyrena Hook. ●☆

414434 Wilsonia R. Br.（1810）;威尔逊旋花属■☆

414435 Wilsonia crassifolia F. Muell.;厚叶威尔逊旋花■☆

414436 Wilsonia humilis R. Br.;矮威尔逊旋花■☆

414437 Wilsonia ovalifolia Hallier f.;卵叶威尔逊旋花■☆

414438 Wimmera Post et Kuntze = Wimmeria Schltdl. et Cham. ●☆

414439 Wimmerella Serra, M. B. Crespo et Lammers(1999);维默桔梗属■☆

414440 Wimmerella arabidea（C. Presl）Serra, M. B. Crespo et Lammers;阿拉维默桔梗■☆

414441 Wimmerella bifida（Thunb.）Serra, M. B. Crespo et Lammers;双裂维默桔梗■☆

414442 Wimmerella frontidentata（E. Wimm.）Serra, M. B. Crespo et Lammers;齿叶维默桔梗■☆

414443 Wimmerella giftbergensis（E. Phillips）Serra, M. B. Crespo et Lammers = Wimmerella bifida（Thunb.）Serra, M. B. Crespo et Lammers ■☆

414444 Wimmerella hederacea（Sond.）Serra, M. B. Crespo et Lammers;常春藤维默桔梗■☆

414445 Wimmerella hedyotidea（Schltr.）Serra, M. B. Crespo et Lammers;耳草维默桔梗■☆

414446 Wimmerella longitubus（E. Wimm.）Serra, M. B. Crespo et Lammers;长管维默桔梗■☆

414447 Wimmerella mariae（E. Wimm.）Serra, M. B. Crespo et Lammers;玛利亚维默桔梗■☆

414448 Wimmerella pygmaea（Thunb.）Serra, M. B. Crespo et Lammers;矮小维默桔梗■☆

414449 Wimmerella secunda（L. f.）Serra, M. B. Crespo et Lammers;单侧维默桔梗■☆

414450 Wimmeria Nees ex Meisn. = Beilschmiedia Nees ●

414451 Wimmeria Schltdl. = Wimmeria Schltdl. et Cham. ●☆

414452 Wimmeria Schltdl. et Cham.（1831）;维默卫矛属■☆

414453 Wimmeria acuminata L. O. Williams;尖维默卫矛●☆

414454 Wimmeria mexicana（Moc. et Sessé ex DC.）Lundell;墨西哥维默卫矛●☆

414455 Wimmeria microphylla Radlk.;小叶维默卫矛●☆

414456 Wimmeria montana Lundell;山地维默卫矛●☆

414457 Wimmeria pubescens Radlk.;毛维默卫矛●☆

414458 Winchia A. DC.（1844）;盆架树属;Washstand Tree, Winchia. ●

414459 Winchia A. DC. = Alstonia R. Br.（保留属名）●

414460 Winchia calophylla A. DC. = Alstonia rostrata C. E. C. Fisch. ●

414461 Winchia glaucescens（Wall. ex G. Don）K. Schum. = Alstonia rostrata C. E. C. Fisch. ●

414462 Winchia glaucescens K. Schum. = Alstonia rostrata C. E. C. Fisch. ●

414463 Windmannia P. Browne（废弃属名）= Weinmannia L.（保留属名）●☆

414464 Windsoria Nutt. = Tridens Roem. et Schult. ■☆

414465 Windsoria pallida Torr. = Puccinellia pallida（Torr.）R. T. Clausen ■☆

414466 Windsorina Gleason(1923);小梗偏穗草属■☆

414467 Windsorina guianensis Gleason;小梗偏穗草■☆

414468 Winifredia L. A. S. Johnson et B. G. Briggs(1986);威尼帚灯草属■☆

414469 Winifredia sola L. A. S. Johnson et B. G. Briggs;威尼帚灯草■☆

414470 Winika M. A. Clem., D. L. Jones et Molloy = Dendrobium Sw.（保留属名）■

414471 Winklera Post et Kuntze = Mertensia Roth（保留属名）■

414472 Winklera Post et Kuntze = Winkleria Rchb.

414473 Winklera Regel = Uranodactylus Gilli ☆

414474 Winklera Regel(1886);温克勒芥属■☆

414475 Winklera patrinoides Regel;温克勒芥■☆

414476 Winklera silaifolia（Hook. f. et Thomson）Korsh. = Heldreichia silaifolia Hook. f. et Thomson ■☆

414477 Winklera silaifolia Korsh. = Heldreichia silaifolia Hook. f. et Thomson ■☆

414478 Winklerella Engl.（1905）;温克勒苔草属■☆

414479 Winklerella dichotoma Engl.;温克勒苔草■☆

414480 Winkleria Rchb. = Mertensia Roth（保留属名）■

414481 Wintera G. Forst. = Pseudowintera Dandy ●☆

414482 Wintera Humb. et Bonpl. = Drimys J. R. Forst. et G. Forst.（保留属名）●☆

414483 Wintera Humb. et Bonpl. = Wintera Murray ●☆

414484 Wintera Murray = Drimys J. R. Forst. et G. Forst.（保留属名）●☆

414485 Winteraceae Lindl. = Winteraceae R. Br. ex Lindl.（保留科名）●

414486 Winteraceae R. Br. ex Lindl.（1830）（保留科名）;林仙科（冬木科,假八角科,辛辣木科）●

414487 Winterana L. = Canella P. Browne（保留属名）●☆

414488 Winterana Sol. ex Meclik. = Drimys J. R. Forst. et G. Forst.（保留属名）●☆

414489 Winteranaceae Warb. = Canellaceae Mart.（保留科名）●☆

414490 Winterania L. = Canella P. Browne（保留属名）●☆

414491 Winterania L. = Winterana L. ●☆

414492 Winterania Post et Kuntze = Drimys J. R. Forst. et G. Forst.（保留属名）●☆

414493 Winterania Post et Kuntze = Winterana Sol. ex Meclik. ●☆

414494 Winteria F. Ritter = Cleistocactus Lem. ●☆

414495 Winteria F. Ritter = Hildewintera F. Ritter ■☆

414496 Winterlia Moench = Ilex L. ●

414497 Winterlia Spreng. = Ammannia L. ■

414498 Winterlia Spreng. = Sellowia Schult. ■

414499 Winterlia integra（Thunb.）K. Koch = Ilex integra Thunb. ●◇

414500 Winterocereus Backeb. = Cleistocactus Lem. ●☆

414501 Winterocereus Backeb. = Hildewintera F. Ritter ■☆

414502 Wirtgenia Doll = Paspalum L. ■

414503 Wirtgenia H. Andres = Andresia Sleumer ■

414504 Wirtgenia H. Andres = Cheilotheca Hook. f. ■

414505 Wirtgenia Jungh. ex Hassk. = Spondias L. + Lannea A. Rich.（保留属名）●

414506 Wirtgenia Nees ex Doell = Paspalum L. ■

414507 Wirtgenia Sch. Bip. = Aspilia Thouars ■☆

414508 Wirtgenia Sch. Bip. = Cheilotheca Hook. f. ■

414509　Wirtgenia abyssinica Sch. Bip. = Aspilia helianthoides (Schumach. et Thonn.) Oliv. et Hiern subsp. ciliata (Schumach.) C. D. Adams ■☆

414510　Wirtgenia multiflora Fenzl = Aspilia helianthoides (Schumach. et Thonn.) Oliv. et Hiern subsp. prieuriana (DC.) C. D. Adams ■☆

414511　Wirtgenia schimperi Sch. Bip. ex A. Rich. = Aspilia helianthoides (Schumach. et Thonn.) Oliv. et Hiern subsp. prieuriana (DC.) C. D. Adams ■☆

414512　Wirtgenia schimperi Sch. Bip. ex A. Rich. var. gracilis ? = Aspilia helianthoides (Schumach. et Thonn.) Oliv. et Hiern subsp. prieuriana (DC.) C. D. Adams ■☆

414513　Wirtgenia schimperi Sch. Bip. ex A. Rich. var. robusta Sch. Bip. = Aspilia helianthoides (Schumach. et Thonn.) Oliv. et Hiern subsp. prieuriana (DC.) C. D. Adams ■☆

414514　Wisconsin willow-herb = Epilobium wisconsinense Ugent ■☆

414515　Wisenia J. F. Gmel. = Melochia L. (保留属名) ●■

414516　Wisenia J. F. Gmel. = Visenia Houtt. ●■

414517　Wislizenia Engelm. (1848);维斯山柑属■☆

414518　Wislizenia refracta Engelm.;维斯山柑;Jackass Clover ■☆

414519　Wisneria Micheli = Wiesneria Micheli ■☆

414520　Wisneria filifolia Hook. f.;线叶威斯纳泽泻■☆

414521　Wisneria triandra Micheli;威斯纳泽泻■☆

414522　Wissadula Medik. (1787);隔蒴苘属;Wissadula ●■☆

414523　Wissadula amplissima (L.) R. E. Fr.;膨大隔蒴苘■☆

414524　Wissadula amplissima (L.) R. E. Fr. var. rostrata (Schumach.) R. E. Fr. = Wissadula rostrata (Schumach.) Hook. f. ●☆

414525　Wissadula contracta (Link) R. E. Fr.;紧缩隔蒴苘■☆

414526　Wissadula grandifolia Baker f. ex Rusby;大叶隔蒴苘●☆

414527　Wissadula periplocifolia (L.) C. Presl ex Thwaites;隔蒴苘;Silkvineleaf Wissadula,Silk-vine-leaved Wissadula ●

414528　Wissadula periplocifolia (L.) Thwaites = Wissadula periplocifolia (L.) C. Presl ex Thwaites ●

414529　Wissadula periplocifolia C. Presl = Wissadula periplocifolia (L.) C. Presl ex Thwaites ●

414530　Wissadula rostrata (Schumach.) Hook. f.;喙状叶隔蒴苘●☆

414531　Wissadula rostrata (Schumach.) Hook. f. var. zeylanica Mast. = Wissadula periplocifolia (L.) C. Presl ex Thwaites ●

414532　Wissadula rostrata Planch. = Wissadula periplocifolia (L.) C. Presl ex Thwaites ●

414533　Wissadula spicata (Cav.) K. Schum.;穗状隔蒴苘;Paco-paco ●☆

414534　Wissadula zeylanica Medik. = Wissadula periplocifolia (L.) C. Presl ex Thwaites ●

414535　Wissmania Burret = Livistona R. Br. ●

414536　Wissmannia Burret = Livistona R. Br. ●

414537　Wissmannia carinensis (Chiov.) Burret = Livistona carinensis (Chiov.) J. Dransf. et N. W. Uhl ●☆

414538　Wistaria Nutt. = Wisteria Nutt. (保留属名)●

414539　Wistaria Nutt. ex Spreng. = Wisteria Nutt. (保留属名)●

414540　Wistaria Spreng. = Wisteria Nutt. (保留属名)●

414541　Wisteria Nutt. (1818)(保留属名);紫藤属;Purplevine,Wistaria,Wisteria ●

414542　Wisteria alba Lindl. = Wisteria sinensis (Sims) Sweet ●

414543　Wisteria brachybotrys Hemsl. = Wisteria villosa Rehder ●■

414544　Wisteria brachybotrys Hemsl. f. alba (W. Willer) Hurus. = Wisteria venusta Rehder et E. H. Wilson ●■

414545　Wisteria brachybotrys Hemsl. var. alba W. Willer = Wisteria venusta Rehder et E. H. Wilson ●■

414546　Wisteria brachybotrys Siebold et Zucc.;短穗紫藤●

414547　Wisteria brachybotrys Siebold et Zucc. f. alba (Mill.) Ohwi = Wisteria sinensis (Sims) Sweet var. alba Lindl. ●

414548　Wisteria brachybotrys Siebold et Zucc. var. alba Mill. = Wisteria sinensis (Sims) Sweet var. alba Lindl. ●

414549　Wisteria brachybotrys Siebold et Zucc. var. alba Mill. = Wisteria venusta Rehder et E. H. Wilson ●■

414550　Wisteria brevidentata Rehder;短齿紫藤(短梗紫藤,山藤);Short-toothed Purplevine,Short-toothed Wisteria ●■

414551　Wisteria breviwitata Rehder;短梗紫藤;Short-pedicel Wisteria,Shortstalk Purplevine ●■

414552　Wisteria chinensis Bunge = Wisteria villosa Rehder ●■

414553　Wisteria chinensis DC. = Wisteria sinensis (Sims) Sweet ●

414554　Wisteria floribunda (Willd.) DC.;多花紫藤(女萝,日本紫藤,藤花菜,藤萝花,朱藤);Flowery Purplevine,Japanese Wistaria,Japanese Wisteria,Multiflorous Wisteria,Wistaria,Wisteria ●■

414555　Wisteria floribunda (Willd.) DC. 'Alba';白色多花紫藤(白花多花紫藤)●■

414556　Wisteria floribunda (Willd.) DC. 'Macrobotrys';长序多花紫藤;Japanese Wisteria,Long-cluster Japanese Wistaria,Wisteria ●■

414557　Wisteria floribunda (Willd.) DC. 'Multijuga' = Wisteria floribunda (Willd.) DC. 'Macrobotrys' ●■

414558　Wisteria floribunda (Willd.) DC. 'Royar Purple';重瓣多花紫藤●■

414559　Wisteria floribunda (Willd.) DC. 'Shiro Noda' = Wisteria floribunda (Willd.) DC. 'Alba' ●■

414560　Wisteria floribunda (Willd.) DC. f. alba (Lindl.) Rehder et E. H. Wilson = Wisteria floribunda (Willd.) DC. 'Alba' ●■

414561　Wisteria floribunda (Willd.) DC. f. alba (Lindl.) Rehder et E. H. Wilson = Wisteria sinensis (Sims) Sweet var. alba Lindl. ●

414562　Wisteria floribunda (Willd.) DC. f. alborosea (Makino) Okuyama;粉白多花紫藤■☆

414563　Wisteria floribunda (Willd.) DC. f. macrobotrys (Siebold ex Neubert) Verhaeghe = Wisteria floribunda (Willd.) DC. 'Macrobotrys' ●■

414564　Wisteria formosa Rehder;美丽藤萝■☆

414565　Wisteria frutescens (L.) Poir.;美洲紫藤;American Wisteria,Long-cluster Japanese Wisteria,Wisteria ●☆

414566　Wisteria frutescens (L.) Poir. var. macrostachya (Nutt. ex Torr. et A. Gray) Torr. et A. Gray = Wisteria frutescens (L.) Poir. ●☆

414567　Wisteria japonica Siebold et Zucc.;日本崖豆●☆

414568　Wisteria japonica Siebold et Zucc. = Millettia kiangsiensis Z. Wei ●

414569　Wisteria japonica Siebold et Zucc. f. alborosea (Sakata) Yonek.;粉白日本崖豆●☆

414570　Wisteria japonica Siebold et Zucc. f. microphylla (Makino) H. Ohashi;小叶日本崖豆●☆

414571　Wisteria macrostachya Nutt.;大穗紫藤;Kentucky Wisteria ●☆

414572　Wisteria multijuga Van Houtte;多裂紫藤;Long-cluster Japanese Wistaria ●☆

414573　Wisteria multijuga Van Houtte = Wisteria floribunda (Willd.) DC. ●■

414574　Wisteria multijuga Van Houtte = Wisteria sinensis (Sims) Sweet ●

414575　Wisteria praecox Hand. -Mazz. = Wisteria sinensis (Sims) Sweet ●

414576　Wisteria sinensis (Sims) Sweet;紫藤(豆藤,葛花,葛萝树,黄环,黄纤藤,绞藤,徽藤,轿藤,狼跋子,女萝,藤,藤花,藤花菜,藤萝,藤罗花,土木鳖,小黄藤,招豆藤,朱藤,紫金藤);American

Wisteria, Blue Wisteria, Chinese Kidney Bean, Chinese Wisteria, Purplevine, Wisteria ●

414577　Wisteria sinensis (Sims) Sweet 'Alba';白花紫藤(白花藤萝,银藤);White Chinese Wisteria, White Wisteria, Whiteflower Purplevine ●

414578　Wisteria sinensis (Sims) Sweet 'Prolific';繁花紫藤■☆

414579　Wisteria sinensis (Sims) Sweet var. alba Lindl. = Wisteria sinensis (Sims) Sweet 'Alba' ●

414580　Wisteria sinensis DC. = Wisteria floribunda (Willd.) DC. ●■

414581　Wisteria sinensis DC. f. alba (Lindl.) Rehder et E. H. Wilson = Wisteria sinensis (Sims) Sweet 'Alba' ●

414582　Wisteria sinensis DC. var. albiflora Lem. = Wisteria sinensis (Sims) Sweet 'Alba' ●

414583　Wisteria sinensis DC. var. albiflora Lem. = Wisteria sinensis (Sims) Sweet ●

414584　Wisteria venusta Rehder et E. H. Wilson;白花藤萝(白花藤,白龙藤,白藤,大发汗,大发汗藤,断肠叶,紫藤);Silky Wistaria, White Silky Wisteria, White-flower Purplevine, White-flower Wisteria, White-flowered Wisteria ●■

414585　Wisteria villosa Rehder;藤萝;Villous Purplevine, Villous Wisteria ●■

414586　Withania Pauquy(1825)(保留属名);睡茄属(醉茄属);Withania, Sleepingeeg ●■

414587　Withania adpressa Coss. ex Batt.;匍匐睡茄●☆

414588　Withania aristata (Aiton) Pauquy;具芒睡茄●☆

414589　Withania chevalieri A. E. Gonc.;舍瓦利耶睡茄●☆

414590　Withania coagulans (Stocks) Dunal;凝固睡茄(凝固醉茄,睡茄);Panirband, Vegetable Rennet ●■☆

414591　Withania coagulans Dunal = Withania coagulans (Stocks) Dunal ●■☆

414592　Withania flexuosa ?;曲折睡茄●☆

414593　Withania frutescens (L.) Pauquy;灌木睡茄●☆

414594　Withania grisea (Hepper et Boulos) Thulin;灰睡茄●☆

414595　Withania holstii Dammer = Discopodium penninervium Hochst. ■☆

414596　Withania kansuensis Kuang et A. M. Lu;睡茄●■

414597　Withania kansuensis Kuang et A. M. Lu = Withania somnifera (L.) Dunal ●■

414598　Withania microphysalis Suess. = Withania somnifera (L.) Dunal ●■

414599　Withania obtusifolia Täckh.;钝睡茄●☆

414600　Withania reichenbachii (Vatke) Bitter;赖兴巴赫茄■☆

414601　Withania somnifera (L.) Dunal;南非睡茄(催眠睡茄,甘肃睡茄,南非醉茄,睡茄,醉茄);Ashwagandha, Gansu Withania, S. Africa Withania, Sleepingeeg ●■

414602　Withania somnifera (L.) Dunal var. macrocalyx Chiov. = Withania somnifera (L.) Dunal ●■

414603　Withania somnifera (L.) Dunal var. somalensis Schinz = Withania sphaerocarpa Hepper et Boulos ●☆

414604　Withania somnifera Dunal = Withania somnifera (L.) Dunal ●■

414605　Withania sphaerocarpa Hepper et Boulos;球果睡茄●☆

414606　Withania sphaerocarpa Hepper et Boulos var. grisea Hepper et Boulos = Withania grisea (Hepper et Boulos) Thulin ●☆

414607　Witharia Rchb. = Withania Pauquy(保留属名)●■

414608　Witheringia L' Hér. (1789);威瑟属●☆

414609　Witheringia L' Hér. = Bassovia Aubl. ●■

414610　Witheringia L' Hér. = Solanum L. ●■

414611　Witheringia Miers = Athenaea Sendtn. (保留属名)●☆

414612　Witheringia affinis (C. V. Morton) Hunz.;近缘威瑟茄●☆

414613　Witheringia alata Miers;翅威瑟茄●☆

414614　Witheringia angustifolia Dunal;窄叶威瑟茄●☆

414615　Witheringia biflora Miers;双花威瑟茄●☆

414616　Witheringia ciliata Kunth;睫毛威瑟茄●☆

414617　Witheringia macrophylla Dunal;大叶威瑟茄●☆

414618　Witheringia maculata (C. V. Morton et Standl.) Hunz.;斑点威瑟茄●☆

414619　Witsenia Thunb. (1782) ('Witsena');威特鸢属●☆

414620　Witsenia capitata Klatt = Nivenia argentea Goldblatt ●☆

414621　Witsenia corymbosa Ker Gawl. = Nivenia corymbosa (Ker Gawl.) Baker ●☆

414622　Witsenia fruticosa (L. f.) Ker Gawl. = Nivenia fruticosa (L. f.) Baker ●☆

414623　Witsenia maura Thunb.;威特鸢尾●☆

414624　Witsenia ramosa Vahl = Nivenia fruticosa (L. f.) Baker ●☆

414625　Wittea Kunth = Downingia Torr. (保留属名)■☆

414626　Wittelsbachia Mart. = Cochlospermum Kunth(保留属名)●☆

414627　Wittelsbachia Mart. et Zucc. = Maximilianea Mart. (废弃属名)●

414628　Wittia K. Schum. = Disocactus Lindl. ●☆

414629　Wittia K. Schum. = Wittiocactus Rauschert ●☆

414630　Wittiocactus Rauschert = Disocactus Lindl. ●☆

414631　Wittmackanthus Kuntze = Pallasia Klotzsch ●☆

414632　Wittmackanthus Kuntze(1891);维特茜属●☆

414633　Wittmackanthus stanleyanus Kuntze;维特茜●☆

414634　Wittmackia Mez = Aechmea Ruiz et Pav. (保留属名)■☆

414635　Wittmannia Vahl = Vangueria Comm. ex Juss. ■☆

414636　Wittrockia Lindm. (1891);韦氏凤梨属(光尊凤梨属,辉勒草属)■☆

414637　Wittrockia amazonica (Baker) L. B. Sm.;亚马孙韦氏凤梨■☆

414638　Wittrockia echinata Leme;刺韦氏凤梨■☆

414639　Wittrockia superba Lindm.;韦氏凤梨■☆

414640　Wittsteinia F. Muell. (1861);澳洲假海桐属●☆

414641　Wittsteinia vacciniacea F. Muell.;澳洲假海桐●☆

414642　Wodyetia A. K. Irvine(1983);狐尾椰属(二枝棕属);Foxtail Palm ●☆

414643　Wodyetia bifurcata A. K. Irvine;狐尾椰(二枝棕);Foxtail Palm ●☆

414644　Woehleria Griseb. (1861);四被苋属■☆

414645　Woehleria serpyllifolia Griseb.;四被苋■☆

414646　Woikoia Baehni = Pouteria Aubl. ●

414647　Wokoia Baehni = Pouteria Aubl. ●

414648　Wokoia Baehni = Woikoia Baehni ●

414649　Wolffia Horkel ex Schleid. (1844)(保留属名);芜萍属(微萍属,无根萍属);Duckweed, Rootless Duckweed, Water Meal, Water-meal, Wolffia ■

414650　Wolffia Schleid. = Pseudowolffia Hartog et Plas ■☆

414651　Wolffia Schleid. = Wolffia Horkel ex Schleid. (保留属名)■

414652　Wolffia angusta Landolt;细芜萍■☆

414653　Wolffia arrhiza (L.) Horkel ex Wimm.;芜萍(卵萍,萍沙,水蚤萍,微萍,无根浮萍,无根萍);Least Duckweed, Rootless Duckweed, Rootless Wolffia ■

414654　Wolffia borealis (Engelm. ex Hegelm.) Landolt;北方芜萍;Northern Water-meal ■☆

414655　Wolffia borealis (Engelm.) Landolt = Wolffia borealis (Engelm. ex Hegelm.) Landolt ■☆

414656　Wolffia brasiliensis Wedd.;巴西芜萍;Brazilian Water-meal, Watermeal ■☆

414657　Wolffia brasiliensis Wedd. var. borealis Engelm. = Wolffia borealis (Engelm. ex Hegelm.) Landolt ■☆

414658　Wolffia brasiliensis Wedd. var. borealis Engelm. ex Hegelm. = Wolffia borealis (Engelm. ex Hegelm.) Landolt ■☆

414659　Wolffia columbiana H. Karst. ;普通芜萍;Common Water-meal, Water Meal, Watermeal ■☆

414660　Wolffia cylindracea Hegelm. ;柱形芜萍■☆

414661　Wolffia denticulata Hegelm. = Wolffiella denticulata (Hegelm.) Hegelm. ☆

414662　Wolffia gladiata Hegelm. = Wolffiella gladiata (Hegelm.) Hegelm. ■☆

414663　Wolffia gladiata Hegelm. var. floridana Donn. Sm. = Wolffiella gladiata (Hegelm.) Hegelm. ■☆

414664　Wolffia globosa (Roxb.) Hartog et Plas;球状芜萍☆

414665　Wolffia lingulata Hegelm. = Wolffiella lingulata (Hegelm.) Hegelm. ■☆

414666　Wolffia michelii Schleid. = Wolffia arrhiza (L.) Horkel. ex Wimm. ■

414667　Wolffia microscopica (Griff. ex Voigt) Kurz;小芜萍■☆

414668　Wolffia papulifera C. H. Thomps. = Wolffia brasiliensis Wedd. ■☆

414669　Wolffia punctata Griseb. = Wolffia borealis (Engelm. ex Hegelm.) Landolt ☆

414670　Wolffia punctata Griseb. = Wolffia brasiliensis Wedd. ■☆

414671　Wolffia repanda Hegelm. = Wolffiella repanda (Hegelm.) Monod ☆

414672　Wolffia schleideni Miq. = Wolffia globosa (Roxb.) Hartog et Plas ■☆

414673　Wolffia welwitschii Hegelm. = Wolffiella welwitschii (Hegelm.) Monod ■☆

414674　Wolffiaceae (Engl.) Nakai = Lemnaceae Martinov(保留科名)■

414675　Wolffiaceae Bubani = Lemnaceae Martinov(保留科名)■

414676　Wolffiaceae Nakai = Araceae Juss. (保留科名)■●

414677　Wolffiaceae Nakai = Lemnaceae Martinov(保留科名)■

414678　Wolffiaceae Nakai;芜萍科(微萍科)■

414679　Wolffiella (Hegelm.) Hegelm. (1895);小芜萍属(小微萍属)■☆

414680　Wolffiella Hegelm. = Wolffiella (Hegelm.) Hegelm. ■☆

414681　Wolffiella denticulata (Hegelm.) Hegelm. ;齿小芜萍■☆

414682　Wolffiella floridana (Donn. Sm.) C. H. Thomps. = Wolffiella gladiata (Hegelm.) Hegelm. ■☆

414683　Wolffiella floridana C. H. Thomps. = Wolffiella gladiata (Hegelm.) Hegelm. ■☆

414684　Wolffiella gladiata (Hegelm.) Hegelm. ;佛罗里达小芜萍; Bog-mat, Florida Mudmidget, Wolffiella ■☆

414685　Wolffiella hyalina (Delile) Monod;透明小芜萍■☆

414686　Wolffiella lingulata (Hegelm.) Hegelm. ;舌状小芜萍■☆

414687　Wolffiella monodii Ast = Wolffiella hyalina (Delile) Monod ■☆

414688　Wolffiella oblonga (Phil.) Hegelm. ;矩圆小芜萍■☆

414689　Wolffiella repanda (Hegelm.) Monod;匍匐小芜萍■☆

414690　Wolffiella rotunda Landolt;圆小芜萍■☆

414691　Wolffiella welwitschii (Hegelm.) Monod;韦氏小芜萍■☆

414692　Wolffiopsis (Hegelm.) Hartog et Plas(1970);类芜萍属(拟微萍属)■☆

414693　Wolffiopsis Hartog et Plas = Wolffiella (Hegelm.) Hegelm. ■☆

414694　Wolffiopsis welwitschii (Hegelm.) Hartog et Plas = Wolffiella welwitschii (Hegelm.) Monod ■☆

414695　Wolfia Dennst. = ? Renealmia L. f. (保留属名)☆

414696　Wolfia Kunth = Wolffia Horkel ex Schleid. (保留属名)■

414697　Wolfia Post et Kuntze = Orchidantha N. E. Br. ■

414698　Wolfia Schreb. (废弃属名) = Casearia Jacq. ●

414699　Wolfia Schreb. (废弃属名) = Wolffia Horkel ex Schleid. (保留属名)■

414700　Wolfia Spreng. (1824);沃氏兰花蕉属■☆

414701　Wolfia spectabilis Dennst. = Eulophia spectabilis (Dennst.) Suresh ■

414702　Wolkensteinia Regel = Ouratea Aubl. (保留属名)●

414703　Wollastonia DC. ex Decne. (1834);滨沙菊属■●☆

414704　Wollastonia DC. ex Decne. = Wedelia Jacq. (保留属名)■●

414705　Wollastonia biflora (L.) DC. = Melanthera biflora (L.) Wild ■☆

414706　Wollastonia biflora (L.) DC. = Wedelia biflora (L.) DC. ■

414707　Wollastonia biflora (L.) DC. var. canescens (Gaudich.) Fosberg = Wedelia biflora (L.) DC. ■

414708　Wollastonia prostrata Hook. et Arn. = Wedelia prostrata (Hook. et Arn.) Hemsl. ■

414709　Wollastonia scabriuscula DC. ex Decne. ;滨沙菊■●☆

414710　Wollastonia zanzibarensis DC. = Melanthera biflora (L.) Wild ■☆

414711　Wollastonia zanzibarensis DC. = Wedelia elongata (DC.) Vatke ■☆

414712　Wollemia W. G. Jones, K. D. Hill et J. M. Allen(1995);恶来杉属(沃勒米杉属);Wollemi Pine ●☆

414713　Wollemia nobilis W. G. Jones, K. D. Hill et J. M. Allen;恶来杉 (沃勒米杉);Wollemi Pine ●☆

414714　Woodburnia Prain(1904);缅甸五加属●☆

414715　Woodburnia penduliflora Prain;缅甸五加●☆

414716　Woodfordia Salisb. (1806);虾子花属(吴福花属);Shrimpflower, Woodfordia ●

414717　Woodfordia floribunda Salisb. = Woodfordia fruticosa (L.) Kurz ●

414718　Woodfordia fruticosa (L.) Kurz;虾子花(红蜂蜜花,红虾花, 破血草,破血药,沙花,吴福花,五福花,虾花,野红花);Shrubby Shrimpflower, Shrubby Woodfordia ●

414719　Woodfordia tomentosa (Roxb.) Bedd. = Woodfordia fruticosa (L.) Kurz ●

414720　Woodfordia uniflora (A. Rich.) Koehne;单花虾子花●

414721　Woodia Schltr. (1894);伍得萝藦属■☆

414722　Woodia marginata (E. Mey.) Schltr. = Woodia mucronata (Thunb.) N. E. Br. ■☆

414723　Woodia mucronata (Thunb.) N. E. Br. ;短尖伍得萝藦■☆

414724　Woodia mucronata (Thunb.) N. E. Br. var. trifurcata (Schltr.) N. E. Br. = Woodia mucronata (Thunb.) N. E. Br. ■☆

414725　Woodia singularis N. E. Br. ;单一伍得萝藦■☆

414726　Woodia trifurcata (Schltr.) Schltr. = Woodia mucronata (Thunb.) N. E. Br. ■☆

414727　Woodia trifurcata (Schltr.) Schltr. var. planifolia Schltr. = Woodia mucronata (Thunb.) N. E. Br. ■☆

414728　Woodia verruculosa Schltr. ;小疣伍得萝藦■☆

414729　Woodiella Merr. (1922);伍得番荔枝属●☆

414730　Woodiella Merr. = Woodiellantha Rauschert ●☆

414731　Woodiella sympetala Merr. ;伍得番荔枝●☆

414732　Woodiellantha Rauschert(1982);乌德花属●☆

414733　Woodiellantha sympetala (Merr.) Rauschert. ;乌德花●☆

414734　Woodier Roxb. ex Kostel. = Lannea A. Rich. (保留属名)●

414735　Woodier Roxb. ex Kostel. = Odina Roxb. ●

414736　Woodrowia Stapf = Dimeria R. Br. ■

414737　Woodsonia L. H. Bailey = Neonicholsonia Dammer ●☆

414738　Woodvillea DC. = Erigeron L. ●

414739　Wooleya L. Bolus(1960);粉玉树属■●☆

414740　Wooleya farinosa (L. Bolus) L. Bolus;粉玉树■☆

414741　Woollsia F. Muell. (1873);辣石南属●☆

414742　Woollsia·pungens F. Muell. ;辣石南●☆

414743　Woonyoungia Y. W. Law = Kmeria（Pierre）Dandy ●☆

414744　Woonyoungia Y. W. Law(1997);焕镛木属●

414745　Woonyoungia septeatrionalia（Dandy）Y. W. Law;焕镛木●

414746　Wootonella Standl. = Verbesina L. (保留属名)●■☆

414747　Wootonia Greene = Dicranocarpus A. Gray ■☆

414748　Worcesterianthus Merr. = Microdesmis Hook. f. ex Hook. ●

414749　Wormia Post et Kuntze = Selago L. ●☆

414750　Wormia Post et Kuntze = Vormia Adans. ●☆

414751　Wormia Rottb. = Dillenia Heist. ex Fabr. ■☆

414752　Wormia Rottb. = Dillenia L. ●

414753　Wormia Rottb. = Sherardia L. ■☆

414754　Wormia Vahl = Ancistrocladus Wall. (保留属名) ●

414755　Wormskioldia Schumach. et Thonn. = Tricliceras Thonn. ex DC. ■☆

414756　Wormskioldia Thonn. = Tricliceras Thonn. ex DC. ■☆

414757　Wormskioldia Thonn. = Wormskioldia Schumach. et Thonn. ■☆

414758　Wormskioldia auriculata A. Fern. et R. Fern. = Tricliceras auriculatum（A. Fern. et R. Fern.）R. Fern. ■☆

414759　Wormskioldia biviniana Tul. = Tricliceras bivinianum（Tul.）R. Fern. ■☆

414760　Wormskioldia brevicaulis Urb. = Tricliceras brevicaule（Urb.）R. Fern. ■☆

414761　Wormskioldia glandulifera Klotzsch = Tricliceras glanduliferum（Klotzsch）R. Fern. ■☆

414762　Wormskioldia lacerata Oberm. = Tricliceras laceratum（Oberm.）Oberm. ■☆

414763　Wormskioldia lanceolata A. Fern. et R. Fern. = Tricliceras lanceolatum（A. Fern. et R. Fern.）R. Fern. ■☆

414764　Wormskioldia lobata Urb. = Tricliceras lobatum（Urb.）R. Fern. ■☆

414765　Wormskioldia longepedunculata Mast. = Tricliceras longepedunculatum（Mast.）R. Fern. ■☆

414766　Wormskioldia mossambicensis A. Fern. et R. Fern. = Tricliceras mossambicense（A. Fern. et R. Fern.）R. Fern. ■☆

414767　Wormskioldia pilosa（Willd.）Schweinf. ex Urb. = Tricliceras pilosum（Willd.）R. Fern. ■☆

414768　Wormskioldia prittwitzii Urb. = Tricliceras prittwitzii（Urb.）R. Fern. ■☆

414769　Wormskioldia rostulata Urb. = Tricliceras brevicaule（Urb.）R. Fern. var. rosulatum（Urb.）R. Fern. ■☆

414770　Wormskioldia schinzii Urb. var. hirsuta A. Fern. et R. Fern. = Tricliceras hirsutum（A. Fern. et R. Fern.）R. Fern. ■☆

414771　Wormskioldia tanacetifolia Klotzsch = Tricliceras tanacetifolium（Klotzsch）R. Fern. ■☆

414772　Woronowia Juz.（1941）;沃氏蔷薇属●☆

414773　Woronowia Juz. = Geum L. ●

414774　Woronowia speciosa（Alb.）Juz. ;沃氏蔷薇●☆

414775　Wormia J. F. Gmel. = Dillenia L. ●

414776　Wormia J. F. Gmel. = Wormia Rottb. ●

414777　Worsleya（Traub）Traub = Hippeastrum Herb. (保留属名)■

414778　Worsleya（Traub）Traub = Worsleya（W. Watson ex Traub）Traub ■☆

414779　Worsleya（W. Watson ex Traub）Traub（1944）;孤莛蓝属;Blue Amaryllis ■☆

414780　Worsleya Traub = Worsleya（W. Watson ex Traub）Traub ■☆

414781　Worsleya W. Watson = Hippeastrum Herb. (保留属名)■

414782　Worsleya W. Watson = Worsleya（W. Watson ex Traub）Traub ■

414783　Worsleya procera（Duch.）Traub = Worsleya rayneri（Hook. f.）Traub et Moldenke ■☆

414784　Worsleya rayneri（Hook. f.）Traub et Moldenke;孤莛蓝;Blue Amaryllis ■☆

414785　Woytkowskia Woodson(1960);沃伊夹竹桃属●☆

414786　Woytkowskia spermatochorda Woodson;沃伊夹竹桃●☆

414787　Wredowia Eckl. = Aristea Sol. ex Aiton ■☆

414788　Wredowia pulchra Eckl. = Pillansia templemannii（Baker）L. Bolus ■☆

414789　Wrenciala A. Gray = Lawrencia Hook. ●☆

414790　Wrenciala A. Gray = Plagianthus J. R. Forst. et G. Forst. ●☆

414791　Wrightea Roxb. = Wallichia Roxb. ●

414792　Wrightea Tussac = Meriania Sw. (保留属名)●☆

414793　Wrightea caryotoides Roxb. = Wallichia caryotoides Roxb. ●

414794　Wrightia R. Br.（1810）;倒吊笔属;Wrightia ●

414795　Wrightia Sol. ex Naudin = Meriania Sw. (保留属名)●☆

414796　Wrightia afzelii K. Schum. = Pleioceras afzelii（K. Schum.）Stapf ●☆

414797　Wrightia annamensis Eberh. et Dubard = Wrightia pubescens R. Br. ●

414798　Wrightia antidysenterica R. Br. ;抗痢倒吊笔●☆

414799　Wrightia arborea（Dennst.）Mabb. ;毛倒吊笔（胭木，胭树）;Rougewood,Tomentose Wrightia ●

414800　Wrightia boranensis（Chiov.）Cufod. = Wrightia demartiniana Chiov. ●☆

414801　Wrightia coccinea（Roxb.）Sims;云南倒吊笔;Scarlet Wrightia ●

414802　Wrightia coccinea Sims = Wrightia coccinea（Roxb.）Sims ●

414803　Wrightia demartiniana Chiov. ;德马丁倒吊笔●☆

414804　Wrightia hainanensis Merr. = Wrightia laevis Hook. f. ●

414805　Wrightia hainanensis Merr. var. chingii Tsiang = Wrightia sikkimensis Gamble ●

414806　Wrightia hainanensis Merr. var. variabilis Tsiang = Wrightia laevis Hook. f. ●

414807　Wrightia kwangtungensis Tsiang;广东倒吊笔;Guangdong Wrightia,Kwangtung Wrightia ●

414808　Wrightia kwangtungensis Tsiang = Wrightia pubescens R. Br. ●

414809　Wrightia laevis Hook. f. ;蓝树（板蓝根，大蓝靛，大青叶，蓝木，岭刀把，米木，木靛，木蓝，七星树，山蓝树，羊角汁，滞良）;Baskettree,Smooth Wrightia ●

414810　Wrightia laniti（Blanco）Merr. = Wrightia pubescens R. Br. ●

414811　Wrightia natalensis Stapf;纳塔尔倒吊笔●☆

414812　Wrightia parviflora Stapf = Pleioceras barteri Baill. ●☆

414813　Wrightia pubescens R. Br. ;倒吊笔（常子，刀柄，倒吊蜡烛，九龙木，苦常，苦杨，蓝靛木，马凌，墨柱根，乳酱树，神仙蜡烛，土北芪，细姑木，章表，枝桐木，猪菜母，猪松木）;Common Wrightia,Wrightia ●

414814　Wrightia pubescens R. Br. subsp. laniti（Blanco）Ngan = Wrightia pubescens R. Br. ●

414815　Wrightia religiosa（Teijsm. et Binn.）Benth. ;无冠倒吊笔;Water Jasmine,Wild Water Plum ●

414816　Wrightia schlechteri H. Lév. = Wrightia sikkimensis Gamble ●

414817　Wrightia sikkimensis Gamble;锡金倒吊笔（个薄，个溥）;Sikkim Wrightia ●

414818　Wrightia stuhlmannii K. Schum. = Alafia lucida Stapf ●☆

414819　Wrightia tinctoria R. Br. ;染用倒吊笔●

414820　Wrightia tinctoria R. Br. var. laevis（Hook. f.）Pichon = Wrightia laevis Hook. f. ●

414821　Wrightia tomentosa（Roxb.）Roem. et Schult. = Wrightia arborea

（Dennst.）Mabb. ●

414822　Wrightia tomentosa Roem. et Schult. = Wrightia tomentosa（Roxb.）Roem. et Schult. ●

414823　Wrixonia F. Muell.（1876）；里克森草属；Feather Flower ●☆

414824　Wrixonia prostantheroides F. Muell.；里克森草 ●☆

414825　Wuerschmittia Sch. Bip. ex Hochst. = Melanthera Rohr ●■☆

414826　Wuerschmittia abyssinica Sch. Bip. ex A. Rich. = Melanthera abyssinica（Sch. Bip. ex A. Rich.）Vatke ■☆

414827　Wuerthia Regel = Ixia L.（保留属名）■☆

414828　Wuerthia elegans Regel = Ixia polystachya L. ■☆

414829　Wulfenia Jacq.（1781）；石墙花属（乌鲁芬草属）；Wulfenia ■☆

414830　Wulfenia amherstiana Benth. = Wulfeniopsis amherstiana（Benth.）D. Y. Hong ■☆

414831　Wulfenia amherstiana Benth. var. nepalensis（T. Yamaz.）T. Yamaz.；尼泊尔显脉石墙花 ■☆

414832　Wulfenia bullii（Eaton）Barnhart = Besseya bullii（Eaton）Rydb. ■☆

414833　Wulfenia carinthiaca Jacq.；石墙花 ☆

414834　Wulfenia obliqua Wall. = Rhynchoglossum obliquum Blume ■

414835　Wulfeniopsis D. Y. Hong（1980）；拟石墙花属 ■☆

414836　Wulfeniopsis amherstiana（Benth.）D. Y. Hong；拟石墙花（显脉石墙花）■☆

414837　Wulfeniopsis amherstiana（Benth.）D. Y. Hong = Wulfenia amherstiana Benth. ■☆

414838　Wulfeniopsis nepalensis（T. Yamaz.）D. Y. Hong = Wulfenia amherstiana Benth. var. nepalensis（T. Yamaz.）T. Yamaz. ■☆

414839　Wulffia Neck. = Wulffia Neck. ex Cass. ■☆

414840　Wulffia Neck. ex Cass.（1825）；伍尔夫菊属 ■☆

414841　Wulffia stenoglossa DC.；窄舌伍尔夫菊 ■☆

414842　Wulfhorstia C. DC.（1900）；热非楝属 ●☆

414843　Wulfhorstia C. E. C. Fisch. = Entandrophragma C. E. C. Fisch. ●☆

414844　Wulfhorstia ekebergioides Harms = Entandrophragma ekebergioides（Harms）Sprague ●☆

414845　Wulfhorstia spicata C. DC. = Entandrophragma spicatum（C. DC.）Sprague ●☆

414846　Wullschlaegelia Rchb. f.（1863）；伍尔兰属 ■☆

414847　Wullschlaegelia aphylla Rchb. f.；伍尔兰 ■☆

414848　Wunderlichia Riedel ex Benth. = Wunderlichia Riedel ex Benth. et Hook. f. ■☆

414849　Wunderlichia Riedel ex Benth. et Hook. f.（1873）；羽冠菊属 ■☆

414850　Wunderlichia tomentosa Glaz.；毛羽冠菊 ■☆

414851　Wunschmannia Urb. = Distictis Mart. ex Meisn. ●☆

414852　Wurdackanthus Maguire（1985）；沃达龙胆属 ☆

414853　Wurdackanthus argyreus Maguire；沃达龙胆 ■☆

414854　Wurdackia Moldenke = Paepalanthus Kunth（保留属名）■☆

414855　Wurdackia Moldenke = Rondonanthus Herzog ■☆

414856　Wurdastom B. Walln.（1996）；新美洲野牡丹属 ●☆

414857　Wurdastom bullata（Wurdack）B. Walln.；新美洲野牡丹 ●☆

414858　Wurfbaeinia Steud. = Amomum Roxb.（保留属名）■

414859　Wurfbainia Giseke（废弃属名）= Amomum Roxb.（保留属名）■

414860　Wurmbaea Steud. = Wurmbea Thunb. ☆

414861　Wurmbea Thunb.（1781）；伍尔秋水仙属 ■☆

414862　Wurmbea angustifolia B. Nord.；窄叶伍尔秋水仙 ■☆

414863　Wurmbea burttii B. Nord.；伯特伍尔秋水仙 ■☆

414864　Wurmbea campanulata Willd. = Wurmbea spicata（Burm. f.）T. Durand et Schinz ■☆

414865　Wurmbea campanulata Willd. var. latifolia Baker = Wurmbea

variabilis B. Nord. ■☆

414866　Wurmbea campanulata Willd. var. longiflora Baker = Wurmbea dolichantha B. Nord. ■☆

414867　Wurmbea campanulata Willd. var. marginata Schltdl. = Wurmbea spicata（Burm. f.）T. Durand et Schinz ■☆

414868　Wurmbea campanulata Willd. var. pumila Schult. et Schult. f. = Wurmbea capensis Thunb. ☆

414869　Wurmbea campanulata Willd. var. purpurea（Aiton）Schltdl. = Wurmbea marginata（Desr.）B. Nord. ■☆

414870　Wurmbea campanulata Willd. var. unicolor Schltdl. = Wurmbea dolichantha B. Nord. ■☆

414871　Wurmbea capensis Thunb.；好望角伍尔秋水仙 ☆

414872　Wurmbea capensis Thunb. var. latifolia（Baker）Baker = Wurmbea variabilis B. Nord. ■☆

414873　Wurmbea capensis Thunb. var. longiflora（Baker）Baker = Wurmbea dolichantha B. Nord. ■☆

414874　Wurmbea capensis Thunb. var. purpurea（Aiton）Baker = Wurmbea marginata（Desr.）B. Nord. ■☆

414875　Wurmbea capensis Thunb. var. truncata（Schltdl.）Baker = Wurmbea monopetala（L. f.）B. Nord. ☆

414876　Wurmbea compacta B. Nord.；紧密伍尔秋水仙 ■☆

414877　Wurmbea conferta N. E. Br. = Wurmbea spicata（Burm. f.）T. Durand et Schinz var. ustulata（B. Nord.）B. Nord. ☆

414878　Wurmbea dioica（R. Br.）F. Muell.；异株伍尔秋水仙；Early Nancy ■☆

414879　Wurmbea dolichantha B. Nord.；长花伍尔秋水仙 ■☆

414880　Wurmbea elatior B. Nord.；较高伍尔秋水仙 ■☆

414881　Wurmbea elongata B. Nord.；伸长伍尔秋水仙 ■☆

414882　Wurmbea goetzei Engl. = Wurmbea tenuis（Hook. f.）Baker subsp. goetzei（Engl.）B. Nord. ■☆

414883　Wurmbea hamiltonii Wendelbo = Wurmbea tenuis（Hook. f.）Baker subsp. hamiltonii（Wendelbo）B. Nord. ■☆

414884　Wurmbea hiemalis B. Nord.；冬伍尔秋水仙 ■☆

414885　Wurmbea homblei De Wild. = Wurmbea tenuis（Hook. f.）Baker subsp. goetzei（Engl.）B. Nord. ■☆

414886　Wurmbea kraussii Baker；克劳斯秋水仙 ■☆

414887　Wurmbea longiflora Willd. = Wurmbea dolichantha B. Nord. ■☆

414888　Wurmbea marginata（Desr.）B. Nord.；具边伍尔秋水仙 ■☆

414889　Wurmbea minima B. Nord.；极小秋水仙 ■☆

414890　Wurmbea monopetala（L. f.）B. Nord.；单瓣伍尔秋水仙 ■☆

414891　Wurmbea pumila Willd. = Wurmbea capensis Thunb. ☆

414892　Wurmbea purpurea Aiton = Wurmbea marginata（Desr.）B. Nord. ■☆

414893　Wurmbea pusilla E. Phillips；微小伍尔秋水仙 ■☆

414894　Wurmbea recurva B. Nord.；反折伍尔秋水仙 ■☆

414895　Wurmbea robusta B. Nord.；粗壮伍尔秋水仙 ■☆

414896　Wurmbea spicata（Burm. f.）T. Durand et Schinz；长穗伍尔秋水仙 ■☆

414897　Wurmbea spicata（Burm. f.）T. Durand et Schinz f. marginata（Desr.）T. Durand et Schinz = Wurmbea marginata（Desr.）B. Nord. ■☆

414898　Wurmbea spicata（Burm. f.）T. Durand et Schinz f. pumila T. Durand et Schinz = Wurmbea capensis Thunb. ■☆

414899　Wurmbea spicata（Burm. f.）T. Durand et Schinz f. purpurea（Aiton）T. Durand et Schinz = Wurmbea marginata（Desr.）B. Nord. ■☆

414900　Wurmbea spicata（Burm. f.）T. Durand et Schinz f. revoluta T.

Durand et Schinz = Wurmbea recurva B. Nord. ■☆

414901　Wurmbea spicata（Burm. f.）T. Durand et Schinz f. truncata（Schltdl.）T. Durand et Schinz = Wurmbea monopetala（L. f.）B. Nord. ■☆

414902　Wurmbea spicata（Burm. f.）T. Durand et Schinz var. latifolia（Baker）T. Durand et Schinz = Wurmbea variabilis B. Nord. ■☆

414903　Wurmbea spicata（Burm. f.）T. Durand et Schinz var. longiflora（Baker）T. Durand et Schinz = Wurmbea dolichantha B. Nord. ■☆

414904　Wurmbea spicata（Burm. f.）T. Durand et Schinz var. truncata（Schltdl.）Adamson = Wurmbea monopetala（L. f.）B. Nord. ■☆

414905　Wurmbea spicata（Burm. f.）T. Durand et Schinz var. ustulata（B. Nord.）B. Nord. ;焦秋水仙■☆

414906　Wurmbea tenuis（Hook. f.）Baker;细秋水仙■☆

414907　Wurmbea tenuis（Hook. f.）Baker subsp. australis B. Nord. ;南方细秋水仙■☆

414908　Wurmbea tenuis（Hook. f.）Baker subsp. goetzei（Engl.）B. Nord. ;格兹秋水仙■☆

414909　Wurmbea tenuis（Hook. f.）Baker subsp. hamiltonii（Wendelbo）B. Nord. ;汉密尔顿秋水仙■☆

414910　Wurmbea truncata Schltdl. = Wurmbea monopetala（L. f.）B. Nord. ■☆

414911　Wurmbea ustulata B. Nord. = Wurmbea spicata（Burm. f.）T. Durand et Schinz var. ustulata（B. Nord.）B. Nord. ■☆

414912　Wurmbea variabilis B. Nord. ;易变伍尔秋水仙■☆

414913　Wurmschnittia Benth. = Melanthera Rohr ■●☆

414914　Wurmschnittia Benth. = Wuerschmittia Sch. Bip. ex Hochst. ■●☆

414915　Wurtzia Baill. = Margaritaria L. f. ●

414916　Wurtzia Baill. = Phyllanthus L. ●■

414917　Wutongshania Z. J. Liu et J. N. Zhang = Cymbidium Sw. ■

414918　Wutongshania guangdongensis Z. J. Liu et J. N. Zhang = Cymbidium sinense（Jacks. ex Andréws）Willd. ■

414919　Wycliffea Ewart et A. H. K. Petrie = Glinus L. ■

414920　Wydlera Post et Kuntze = Apium L. ■

414921　Wydlera Post et Kuntze = Wydleria Fisch. et Trautv. ■

414922　Wydleria DC. = Carum L. ■

414923　Wydleria Fisch. et Trautv. = Apium L. ■

414924　Wyethia Nutt.（1834）;韦斯菊属（骡耳菊属）;Mule-ears, Mules-ears ■☆

414925　Wyethia amplexicaulis（Nutt.）Nutt. ;抱茎韦斯菊;Mule's-ears ■☆

414926　Wyethia amplexicaulis（Nutt.）Nutt. subsp. major Piper = Wyethia amplexicaulis（Nutt.）Nutt. ■☆

414927　Wyethia amplexicaulis（Nutt.）Nutt. subsp. subresinosa Piper = Wyethia amplexicaulis（Nutt.）Nutt. ■☆

414928　Wyethia angustifolia（DC.）Nutt. ;狭叶韦斯菊■☆

414929　Wyethia angustifolia（DC.）Nutt. var. foliosa（Congdon）H. M. Hall = Wyethia angustifolia（DC.）Nutt. ■☆

414930　Wyethia arizonica A. Gray;亚利桑那韦斯菊■☆

414931　Wyethia bolanderi（A. Gray）W. A. Weber = Agnorhiza bolanderi（A. Gray）W. A. Weber ■☆

414932　Wyethia coriacea A. Gray = Agnorhiza ovata（Torr. et Gray）W. A. Weber ■☆

414933　Wyethia elata H. M. Hall = Agnorhiza elata（H. M. Hall）W. A. Weber ■☆

414934　Wyethia glabra A. Gray;光韦斯菊■☆

414935　Wyethia helenioides（DC.）Nutt. ;堆心韦斯菊■☆

414936　Wyethia helenioides Nutt. = Wyethia helenioides（DC.）Nutt. ■☆

414937　Wyethia helianthoides Nutt. ;向日葵韦斯菊■☆

414938　Wyethia invenusta（Greene）W. A. Weber = Agnorhiza invenusta（Greene）W. A. Weber ■☆

414939　Wyethia lanceolata Howell = Wyethia amplexicaulis（Nutt.）Nutt. ■☆

414940　Wyethia longicaulis A. Gray;长茎韦斯菊■☆

414941　Wyethia mollis A. Gray;柔毛韦斯菊■☆

414942　Wyethia ovata Torr. et A. Gray = Agnorhiza ovata（Torr. et Gray）W. A. Weber ■☆

414943　Wyethia reticulata Greene = Agnorhiza reticulata（Greene）W. A. Weber ■☆

414944　Wyethia scabra Hook. = Scabrethia scabra（Hook.）W. A. Weber ■☆

414945　Wyethia scabra Hook. var. attenuata W. A. Weber = Scabrethia scabra（Hook.）W. A. Weber subsp. attenuata（W. A. Weber）W. A. Weber ■☆

414946　Wyethia scabra Hook. var. canescens W. A. Weber = Scabrethia scabra（Hook.）W. A. Weber subsp. canescens（W. A. Weber）W. A. Weber ■☆

414947　Wylia Hoffm. = Scandix L. ■

414948　Wyomingia A. Nelson = Erigeron L. ■●

414949　Xaathoxalis Small = Oxalis L. ■●

414950　Xaiasme Raf. = Stellera L. ■●

414951　Xalkitis Raf. = ? Aster L. ●■

414952　Xalkitis Raf. = Bindera Raf. ●■

414953　Xamacrista Raf. = Cassia L.（保留属名）●■

414954　Xamacrista Raf. = Chamaecrista Moench ■●

414955　Xamesike Raf. = Chamaesyce Gray ●■

414956　Xamesike Raf. = Euphorbia L. ●■

414957　Xamesuke Raf. = Xamesike Raf. ●■

414958　Xamilenis Raf. = Silene L.（保留属名）■

414959　Xamilensis acaulis（L.）Tzvelev = Silene acaulis L. ■☆

414960　Xananthes Raf. = Utricularia L. ■

414961　Xananthes minor（L.）Raf. = Utricularia minor L. ■

414962　Xanthaea Rchb. = Centaurium Hill ■

414963　Xanthanthos St. -Lag. = Anthoxanthum L. ■

414964　Xanthe Schreb. = Clusia L. ●☆

414965　Xanthe Schreb. = Quapoya Aubl. ●☆

414966　Xantheranthemum Lindau（1895）;黄可爱花属■☆

414967　Xantheranthemum igneum（Regel）Lindau;黄可爱花■☆

414968　Xanthiaceae Vest = Asteraceae Bercht. et J. Presl（保留科名）●■

414969　Xanthiaceae Vest = Compositae Giseke（保留科名）●■

414970　Xanthidium Delpino = Franseria Cav.（保留属名）●■☆

414971　Xanthisma DC.（1836）;眠雏菊属;Sleepy-daisy ●■☆

414972　Xanthisma blephariphyllum（A. Gray）D. R. Morgan et R. L. Hartm. ;缘毛叶眠雏菊●☆

414973　Xanthisma coloradoense（A. Gray）D. R. Morgan et R. L. Hartm. ;眠雏菊■☆

414974　Xanthisma glaberrimum（Rydb.）G. L. Nesom et O'Kennon = Xanthisma spinulosum（Pursh）D. R. Morgan et R. L. Hartm. var. glaberrimum（Rydb.）D. R. Morgan et R. L. Hartm. ■☆

414975　Xanthisma gracile（Nutt.）D. R. Morgan et R. L. Hartm. ;纤细眠雏菊;Slender Goldenweed ■☆

414976　Xanthisma grindelioides（Nutt.）D. R. Morgan et R. L. Hartm. ;胶草状眠雏菊;Goldenweed, Gumweed Aster ■☆

414977　Xanthisma grindelioides（Nutt.）D. R. Morgan et R. L. Hartm. var. depressum（Maguire）D. R. Morgan et R. L. Hartm. ;凹陷眠雏菊■☆

414978　Xanthisma gypsophilum（B. L. Turner）D. R. Morgan et R. L. Hartm.；喜钙眠雏菊●☆

414979　Xanthisma junceum（Greene）D. R. Morgan et R. L. Hartm.；圆滑眠雏菊●☆

414980　Xanthisma spinulosum（Pursh）D. R. Morgan et R. L. Hartm.；裂叶眠雏菊；Cut-leaf Ironplant，Spiny Goldenweed ●■☆

414981　Xanthisma spinulosum（Pursh）D. R. Morgan et R. L. Hartm. var. chihuahuanum（B. L. Turner et R. L. Hartm.）D. R. Morgan et R. L. Hartm.；奇瓦瓦眠雏菊■☆

414982　Xanthisma spinulosum（Pursh）D. R. Morgan et R. L. Hartm. var. glaberrimum（Rydb.）D. R. Morgan et R. L. Hartm.；光裂叶眠雏菊■☆

414983　Xanthisma spinulosum（Pursh）D. R. Morgan et R. L. Hartm. var. gooddingii（A. Nelson）D. R. Morgan et R. L. Hartm.；古丁眠雏菊●☆

414984　Xanthisma spinulosum（Pursh）D. R. Morgan et R. L. Hartm. var. paradoxum（B. L. Turner et R. L. Hartm.）D. R. Morgan et R. L. Hartm.；奇异裂叶眠雏菊■☆

414985　Xanthisma texanum DC.；得州眠雏菊；Star-of-texas，Texas Sleepy-daisy ■☆

414986　Xanthisma texanum DC. subsp. drummondii（Torr. et A. Gray）Semple = Xanthisma texanum DC. var. drummondii（Torr. et A. Gray）A. Gray ■☆

414987　Xanthisma texanum DC. var. drummondii（Torr. et A. Gray）A. Gray；德拉蒙德眠雏菊■☆

414988　Xanthisma texanum DC. var. orientale Semple；东方眠雏菊■☆

414989　Xanthisma viscidum（Wooton et Standl.）D. R. Morgan et R. L. Hartm.；黏眠雏菊■☆

414990　Xanthium L.（1753）；苍耳属；Bur Weed，Cocklebur，Noosoora Bur ■

414991　Xanthium abyssinicum Wallr. = Xanthium strumarium L. ■

414992　Xanthium acerosum Greene = Xanthium canadense Mill. ■☆

414993　Xanthium ambrosioides Hook. et Arn. = Xanthium spinosum L. ■

414994　Xanthium americanum Walter = Xanthium strumarium L. var. glabratum（DC.）Cronquist ■☆

414995　Xanthium americanum Walter = Xanthium strumarium L. ■

414996　Xanthium antiquorum Wallr. = Xanthium strumarium L. ■

414997　Xanthium brasilicum Vell.；巴西苍耳■☆

414998　Xanthium brevirostre Hochst. = Xanthium strumarium L. ■

414999　Xanthium californicum Greene；加州苍耳■☆

415000　Xanthium californicum Greene = Xanthium canadense Mill. ■☆

415001　Xanthium californicum Greene var. rotundifolium Widder = Xanthium canadense Mill. ■☆

415002　Xanthium campestre Greene = Xanthium canadense Mill. ■☆

415003　Xanthium canadense Mill.；加拿大苍耳；Burdock，Butterbur，Canada Cocklebur，Clot-burr，Ditch Bur ■☆

415004　Xanthium catharticum Kunth；泻苍耳■☆

415005　Xanthium cavanillesii Schouw = Xanthium canadense Mill. ■☆

415006　Xanthium cavanillesii Schouw = Xanthium italicum Moretti ■☆

415007　Xanthium cenchroides Millsp. et Sherff = Xanthium canadense Mill. ■☆

415008　Xanthium chasei Fernald = Xanthium strumarium L. ■

415009　Xanthium chinense Mill. = Xanthium strumarium L. var. glabratum（DC.）Cronquist ■☆

415010　Xanthium chinense Mill. = Xanthium strumarium L. ■

415011　Xanthium cloessplateaum D. Z. Ma；高原苍耳；Plereau Cocklebur ■

415012　Xanthium commune Britton = Xanthium canadense Mill. ■☆

415013　Xanthium curvescens Millsp. et Sherff = Xanthium strumarium L. ■

415014　Xanthium cylindricum Millsp. et Fernald = Xanthium strumarium L. ■

415015　Xanthium echinatum Murray = Xanthium canadense Mill. ■☆

415016　Xanthium echinatum Murray = Xanthium strumarium L. ■

415017　Xanthium echinellum Greene ex Rydb. = Xanthium strumarium L. ■

415018　Xanthium glanduliferum Greene = Xanthium canadense Mill. ■☆

415019　Xanthium globosum C. Shull = Xanthium strumarium L. ■

415020　Xanthium inaequilaterum DC.；偏基苍耳；Unequalbase Cocklebur ■

415021　Xanthium inaequilaterum DC. = Xanthium strumarium L. ■

415022　Xanthium inflexum Mack. et Bush = Xanthium strumarium L. ■

415023　Xanthium italicum Moretti；意大利苍耳■☆

415024　Xanthium italicum Moretti = Xanthium canadense Mill. ■☆

415025　Xanthium italicum Moretti = Xanthium strumarium L. ■

415026　Xanthium japonicum Widder = Xanthium sibiricum Patrin ex Widder ■

415027　Xanthium japonicum Widder = Xanthium strumarium L. ■

415028　Xanthium macounii Britton = Xanthium canadense Mill. ■☆

415029　Xanthium macrocarpum DC. var. glabratum DC. = Xanthium strumarium L. var. glabratum（DC.）Cronquist ■☆

415030　Xanthium macrocrpum DC. = Xanthium orientale L. ■

415031　Xanthium mongolicum Kitag.；蒙古苍耳（东北苍耳）；Mongol Cocklebur，Mongolian Cocklebur ■

415032　Xanthium natalense Widder = Xanthium strumarium L. ■

415033　Xanthium occidentale Bertol.；西方苍耳；Noogoora Burr ■☆

415034　Xanthium oligacanthum Piper = Xanthium canadense Mill. ■☆

415035　Xanthium orientale Blume = Xanthium inaequilaterum DC. ■

415036　Xanthium orientale L. = Xanthium strumarium L. ■

415037　Xanthium orientale Zenkert；东方苍耳；Clot-burr ■☆

415038　Xanthium oviforme Wallr. = Xanthium canadense Mill. ■☆

415039　Xanthium oviforme Wallr. = Xanthium strumarium L. ■

415040　Xanthium pennsylvanicum Wallr. = Xanthium canadense Mill. ■☆

415041　Xanthium pennsylvanicum Wallr. = Xanthium strumarium L. ■

415042　Xanthium pungens Wallr.；辛辣苍耳；Cocldebur ■☆

415043　Xanthium pungens Wallr. = Xanthium strumarium L. ■

415044　Xanthium riparium Itzigs. et Hertsch；河岸苍耳■☆

415045　Xanthium saccharatum Wallr. = Xanthium canadense Mill. ■☆

415046　Xanthium sibiricum Patrin ex Widder；苍耳（白痴头婆，白胡，白花虱母头，白猪母络，敝子，菜耳，苍耳，苍刺头，苍耳蒺藜，苍耳药，苍耳子，苍浪子，苍子，苍子棵，常思，常思菜，常枲，痴头猛，痴头婆，刺儿棵，刺儿苗，道人头，地葵，疔苍草，疔疮草，饿虱子，耳珰菜，耳珰草，佛耳，狗耳朵草，狗子耳，喝起草，胡苍耳，胡寝子，胡菱，胡枲，假矮瓜，缣丝草，进贡菜，进贤菜，卷耳，爵耳，老苍子，苓耳，毛苍子，绵苍浪子，棉蝗螂，牛虱子，羌子棵子，抢子，芩耳，青棘子，虱麻头，蒵，丝枲，枲耳，羊大归，羊带归，羊负来，野缣丝，野落苏，野茄，野茄猪耳，野茄子，野丝，野紫菜，黏头婆，粘粘葵，只刺，猪耳）；Siberia Cocklebur，Siberian Cocklebur ■

415047　Xanthium sibiricum Patrin ex Widder = Xanthium strumarium L. ■

415048　Xanthium sibiricum Patrin ex Widder var. jingyuanense H. G. Ho et Y. T. Lu = Xanthium sibiricum Patrin ex Widder ■

415049　Xanthium sibiricum Patrin ex Widder var. jingyuanense H. G. Ho et Y. T. Lu；一室苍耳■

415050　Xanthium sibiricum Patrin ex Widder var. subinerme（Winkl.）Widder；近无刺苍耳（稀刺苍耳）■

415051   Xanthium speciosum Kearney = Xanthium canadense Mill. ■☆

415052   Xanthium speciosum Kearney = Xanthium strumarium L. ■

415053   Xanthium spinosum L. ;刺苍耳; Argentine Cocklebur, Bastard Burr, Bathurst Bur, Burweed, Clotbur, Clot-burr, Dagger Cocklebur, Dagger Weed, Daggerweed, Spiny Burweed, Spiny Clotbur, Spiny Clot-burr, Spiny Cocklebur, Thorny Burweed ■

415054   Xanthium spinosum L. = Acanthoxanthium spinosum（L.）Fourr. ■

415055   Xanthium spinosum L. f. inerme（Bel）O. Bolòs et Vigo = Xanthium spinosum L. ■

415056   Xanthium spinosum L. var. inerme Bel = Xanthium spinosum L. ■

415057   Xanthium strumarium L. ;欧洲苍耳(苍耳,苍耳子,常思菜,常枲,道人头,地葵,耳璫草,喝起草,胡枲,回菜场子花,卷耳,蒼耳,爵耳,苓耳,美国苍耳,母猪癞,菠,葹,羊带来,羊负来,硬刺苍耳); Beach Clotbur, Beach Cocklebur, Bur Weed, Burweed, Clot-burr, Cocklebur, Cockle-bur, Common Cocklebur, Ditch Bur, Hardspiny Cocklebur, Rough Cocklebur, Small Burdock, Spiny Cocklebur ■

415058   Xanthium strumarium L. subsp. cavanillesii（Schouw）D. Löve et Dans. = Xanthium italicum Moretti ■☆

415059   Xanthium strumarium L. subsp. italicum（Moretti）D. Löve = Xanthium italicum Moretti ■☆

415060   Xanthium strumarium L. subsp. sibiricum（Widder）Greuter = Xanthium strumarium L. ■

415061   Xanthium strumarium L. var. brasilicum（Vell.）Fiori = Xanthium brasilicum Vell. ■☆

415062   Xanthium strumarium L. var. canadense（Mill.）Torr. et A. Gray = Xanthium canadense Mill. ■☆

415063   Xanthium strumarium L. var. glabratum（DC.）Cronquist;光加拿大苍耳;Burweed,Clotbur,Common Clotbur,Common Cocklebur ■☆

415064   Xanthium strumarium L. var. japonicum（Widder）H. Hara = Xanthium strumarium L. ■

415065   Xanthium strumarium L. var. oviforme（Wallr.）M. Peck = Xanthium canadense Mill. ■☆

415066   Xanthium strumarium L. var. pensylvanicum（Wallr.）M. Peck = Xanthium canadense Mill. ■☆

415067   Xanthium strumarium Lour. = Xanthium inaequilaterum DC. ■

415068   Xanthium strumarium Lour. var. canadense（Mill.）Torr. et A. Gray = Xanthium strumarium L. ■

415069   Xanthium strumarium Lour. var. glabratum（DC.）Cronquist = Xanthium strumarium L. ■

415070   Xanthium strumarium Lour. var. inaequilaterale C. B. Clarke = Xanthium inaequilaterum DC. ■

415071   Xanthium strumarium Lour. var. japonica（Widder）Hara = Xanthium strumarium L. ■

415072   Xanthium strumarium Lour. var. subinerme Winkl. = Xanthium sibiricum Patrin ex Widder var. subinerme（Winkl.）Widder ■

415073   Xanthium varians Greene = Xanthium canadense Mill. ■☆

415074   Xanthium varians Greene = Xanthium strumarium L. ■

415075   Xanthium wootonii Cockerell = Xanthium strumarium L. ■

415076   Xantho J. Rémy = Hologymne Bartl. ■☆

415077   Xantho J. Rémy = Lasthenia Cass. ■☆

415078   Xanthobrychis Galushko = Onobrychis Mill. ■

415079   Xanthocephalum Willd. (1807);黄头菊属;Snakeweed ■☆

415080   Xanthocephalum Willd. = Gutierrezia Lag. ■●☆

415081   Xanthocephalum amoenum Shinners = Amphiachyris amoena（Shinners）Solbrig ■☆

415082   Xanthocephalum amoenum Shinners var. intermedium Shinners = Amphiachyris dracunculoides（DC.）Nutt. ■☆

415083   Xanthocephalum arizonicum（A. Gray）Shinners = Gutierrezia arizonica（A. Gray）M. A. Lane ■☆

415084   Xanthocephalum californicum（DC.）Greene = Gutierrezia californica（DC.）Torr. et A. Gray ■☆

415085   Xanthocephalum gymnospermoides（A. Gray）Benth. et Hook. f. ;裸籽黄头菊;San Pedro Matchweed ■☆

415086   Xanthocephalum microcephalum（DC.）Shinners = Gutierrezia microcephaia（DC.）A. Gray ■☆

415087   Xanthocephalum petradoria S. L. Welsh et Goodrich = Gutierrezia petradoria（S. L. Welsh et Goodrich）S. L. Welsh ■☆

415088   Xanthocephalum sarothrae（Pursh）Shinners;黄头菊;Snakeweed ■☆

415089   Xanthocephalum sarothrae（Pursh）Shinners = Gutierrezia sarothrae（Pursh）Britton et Rusby ■☆

415090   Xanthocephalum sphaerocephalum（A. Gray）Shinners = Gutierrezia sphaerocephala A. Gray ■☆

415091   Xanthocephalum sphaerocephalum（A. Gray）Shinners var. eriocarpum（A. Gray）Shinners = Gutierrezia sphaerocephala A. Gray ■☆

415092   Xanthocephalum texanum（DC.）Shinners = Gutierrezia texana（DC.）Torr. et A. Gray ■☆

415093   Xanthocephalum wrightii（A. Gray）A. Gray = Gutierrezia wrightii A. Gray ■☆

415094   Xanthoceras Bunge（1833）;文冠果属;Yellow Horn, Yellowhorn, Yellow-horn ●

415095   Xanthoceras enkianthiflorum H. Lév. = Staphylea holocarpa Hemsl. ●

415096   Xanthoceras sorbifolium Bunge;文冠果(木瓜,土木瓜,温旦革子,文官果,文冠花,文冠木,文冠树,文光果,崖木瓜,珍珠梅); Chinese Flowering Chestnut, Chinese Xanthoceras, Shinleaf Yellowhorn, Shinyleaf Yellowhorn, Shiny-leaf Yellowhorn, Shiny-leaved Yellow-horn, Yellowhorn ●

415097   Xanthocercis Baill.（1870）;黄尾豆属●■☆

415098   Xanthocercis madagascariensis Baill. ;马岛黄尾豆●☆

415099   Xanthocercis rabiensis Maesen;拉比黄尾豆●☆

415100   Xanthocercis zambesiaca（Baker）Dumaz-le-Grand;黄尾豆;Nyala Tree ■☆

415101   Xanthochloa（Krivot.）Tzvelev(2006);黄草属■☆

415102   Xanthochloa griffithiana（St. -Yves）Tzvelev;黄草■☆

415103   Xanthochrysum Turcz. = Helichrysum Mill.（保留属名）●■

415104   Xanthochrysum Turcz. = Schoenia Steetz ■☆

415105   Xanthochymus Roxb. = Garcinia L. ●

415106   Xanthochymus pictorius Roxb. = Garcinia xanthochymus Hook. f. ex T. Anderson ●

415107   Xanthochymus tinctorius DC. = Garcinia xanthochymus Hook. f. ex T. Anderson ●

415108   Xanthocoma Kunth = Xanthocephalum Willd. ■☆

415109   Xanthocromyon H. Karst. = Trimeza Salisb. ■☆

415110   Xanthocyparis Farjon et T. H. Nguyên = Cupressus L. ●

415111   Xanthocyparis Farjon et T. H. Nguyên(2002);黄金柏属●

415112   Xanthocyparis nootkatensis（D. Don）Farjon et D. K. Harder = Chamaecyparis nootkatensis（D. Don）Spach ●

415113   Xanthocyparis nootkatensis（D. Don）Farjon et D. K. Harder = Cupressus nootkatensis D. Don ●

415114   Xanthocyparis vietnamensis Farjon et T. H. Nguyên;黄金柏●☆

415115   Xanthogalum Avé-Lall.（1842）;黄盔芹属■☆

415116   Xanthogalum Avé-Lall. = Angelica L. ■

415117　Xanthogalum purpurascens Avé-Lall. ;黄盔芹■☆

415118　Xanthogalum sachokianum Karyagin;萨氏黄盔芹■☆

415119　Xanthogalum tatianae (Bordz.) Schischk. ;塔氏黄盔芹■☆

415120　Xantholepis Willd. ex Less. = Cacosmia Kunth ●☆

415121　Xantholinum Rchb. = Linum L. ●■

415122　Xanthomyrtus Diels(1922);黄桃木属●☆

415123　Xanthomyrtus angustifolia A. J. Scott;窄叶黄桃木●☆

415124　Xanthomyrtus aurea Merr. ;金黄桃木●☆

415125　Xanthomyrtus grandiflora A. J. Scott;大花黄桃木●☆

415126　Xanthomyrtus ovata A. J. Scott;卵形黄桃木●☆

415127　Xanthomyrtus polyclada Diels;多枝黄桃木●☆

415128　Xanthomyrtus rostrata Merr. et L. M. Perry;喙黄桃木●☆

415129　Xanthonanthus St. -Lag. = Anthoxanthum L. ●

415130　Xanthopappus C. Winkl. (1893);黄缨菊属(黄冠菊属);
　　　　Xanthopappus ■★

415131　Xanthopappus multicephalus Y. Ling = Xanthopappus subacaulis
　　　　C. Winkl. ■

415132　Xanthopappus subacaulis C. Winkl. ;黄缨菊(黄冠菊,九头刺
　　　　盖,九头妖,马刺盖,魔);Common Xanthopappus ■

415133　Xanthophthalmum Sch. Bip. = Chrysanthemum L. (保留属名)■●

415134　Xanthophyllaceae (Chodat) Gagnep. = Polygalaceae Hoffmanns. et
　　　　Link(保留科名)■●

415135　Xanthophyllaceae (Chodat) Gagnep. = Xanthophyllaceae Gagnep.
　　　　ex Reveal et Hoogland ●

415136　Xanthophyllaceae Gagnep. = Polygalaceae Hoffmanns. et Link
　　　　(保留科名)■●

415137　Xanthophyllaceae Gagnep. = Xanthophyllaceae Gagnep. ex
　　　　Reveal et Hoogland ●

415138　Xanthophyllaceae Gagnep. ex Reveal et Hoogland = Polygalaceae
　　　　Hoffmanns. et Link(保留科名)■●

415139　Xanthophyllaceae Gagnep. ex Reveal et Hoogland;黄叶树科;
　　　　Xanthophyllum Family ●

415140　Xanthophyllaceae Reveal et Hoogland = Xanthophyllaceae
　　　　Gagnep. ex Reveal et Hoogland ●

415141　Xanthophyllon St. -Lag. = Xanthoxylon Spreng. ●

415142　Xanthophyllon St. -Lag. = Zanthoxylum L. ●

415143　Xanthophyllum Roxb. (1820)(保留属名);黄叶树属;
　　　　Xanthophyllum,Yellow Leaf Tree,Yellowleaftree,Yellow-leaved Tree ●

415144　Xanthophyllum hainanense Hu;黄叶树(海南黄叶树,黄肖,黄
　　　　枝木,青蓝);Hainan Xanthophyllum, Hainan Yellow Leaf Tree,
　　　　Hainan Yellowleaftree, Hainan Yellow-leaved Tree ●

415145　Xanthophyllum oliganthum C. Y. Wu;少花黄叶树;Fewflower
　　　　Yellowleaftree, Few-flowered Yellow-leaved Tree, Paucity Flower
　　　　Yellowleaftree,Poorflower Yellowleaftree ●

415146　Xanthophyllum racemosum Chodat = Xanthophyllum hainanense
　　　　Hu ●

415147　Xanthophyllum siamense Craib;泰国黄叶树(山龙眼);Tailand
　　　　Yellowleaftree,Tailand Yellow-leaved Tree ●

415148　Xanthophyllum yunnanense C. Y. Wu;云南黄叶树;Yunnan
　　　　Yellowleaftree,Yunnan Yellow-leaved Tree ●

415149　Xanthophytopsis Pit. (1922);拟岩黄树属(假树属,拟黄树属)●

415150　Xanthophytopsis Pit. = Xanthophytum Reinw. ex Blume ●

415151　Xanthophytopsis balansae Pit. = Xanthophytum balansae (Pierre
　　　　ex Pit.) H. C. Lo ●

415152　Xanthophytopsis balanse Pierre ex Pit. = Xanthophytum balansae
　　　　(Pierre ex Pit.) H. C. Lo ●

415153　Xanthophytopsis kwangtungensis Chun et F. C. How =

Xanthophytum kwangtungense (Chun et F. C. How) H. C. Lo ●

415154　Xanthophytum Reinw. ex Blume(1827);岩黄树属(黄树属,岩
　　　　果树属);Xanthophytum Rockyellowtree ●

415155　Xanthophytum attopevense (Pierre ex Pit.) H. S. Lo;琼岛岩黄
　　　　树;Attopev Xanthophytum,Hainan Rockyellowtree ●

415156　Xanthophytum balansae (Pierre ex Pit.) H. S. Lo;长梗岩黄树
　　　　(越南拟黄树);Balanse Xanthophytopsis,Longstalk Rockyellowtree ●

415157　Xanthophytum kwangtungense (Chun et F. C. How) H. S. Lo;岩黄
　　　　树(拟黄树);Guangdong Rockyellowtree, Guangdong Xanthophytopsis,
　　　　Kwangtung Xanthophytopsis ●

415158　Xanthophytum polyanthum Pit. ;多花岩黄树●

415159　Xanthophytum sinicum H. S. Lo = Lerchea sinica (H. C. Lo) H.
　　　　S. Lo ●

415160　Xanthopsis (DC.) K. Koch = Centaurea L. (保留属名)●■

415161　Xanthopsis C. Koch = Centaurea L. (保留属名)●■

415162　Xanthopsis K. Koch = Centaurea L. (保留属名)●■

415163　Xanthorhiza L' Hér. = Xanthorhiza Marshall ●☆

415164　Xanthorhiza Marshall(1785);木黄连属(黄根木属,黄根树
　　　　属,黄根);Shrub Yellow Root, Shrub Yellowroot, Shrub Yellow-
　　　　root,Yellowroot ●☆

415165　Xanthorhiza apiifolia L'Hér. = Xanthorhiza simplicissima
　　　　Marshall ●☆

415166　Xanthorhiza simplicissima Marshall;木黄连;Brook-feather,
　　　　Shrub Yellowroot,Yellowroot ●☆

415167　Xanthorhiza simplicissima Marshall = Zanthorhiza simplicissima
　　　　Marshall ●☆

415168　Xanthorhizaceae Bercht. et J. Presl = Ranunculaceae Juss. (保留
　　　　科名)●■

415169　Xanthorrhiza L' Hér. = Xanthorhiza Marshall ●☆

415170　Xanthorrhiza Marshall = Xanthorhiza Marshall ●☆

415171　Xanthorrhiza apifolia L'Hér. = Zanthorhiza simplicissima
　　　　Marshall ●☆

415172　Xanthorrhoea Sm. (1798);黄脂木属(草树胶属,草树属,刺
　　　　叶树属,禾木胶属,黄胶木属,黄万年青属,黄脂草属,木根旱生
　　　　草属,树草属);Acaroid Resin, Australian Grass-tree, Black Boy,
　　　　Blackboy,Grass Gum,Grass Tree,Yacca,Yellow Gum ●■☆

415173　Xanthorrhoea arborea R. Br. ;树黄脂木(黑孩子,乔木黄万年
　　　　青,树百合);Black Boy, Forest Grass Tree, Grass Tree ●☆

415174　Xanthorrhoea australis R. Br. ;南方黄脂木(澳洲黄胶木,澳洲
　　　　香树,黑子树,黄脂草,南方禾木胶);Blackboy, Grass Tree,
　　　　Southern Grass Tree ■☆

415175　Xanthorrhoea glauca D. J. Bedford;灰黄脂木(灰禾木胶);
　　　　Narrow-leafed Grass Tree ●☆

415176　Xanthorrhoea hastilis R. Br. ;矛叶黄脂木(矛叶草树)●☆

415177　Xanthorrhoea johnsonii A. T. Lee;昆士兰黄脂木(昆士兰禾木
　　　　胶);Queensland Grass Tree ●■☆

415178　Xanthorrhoea preissii Endl. ;西澳洲禾木胶;Black Boy,
　　　　Blackboy,Grass Gum,Grass Tree,Western Australian Grass Tree ●☆

415179　Xanthorrhoea pumilio R. Br. ;短叶黄脂木●☆

415180　Xanthorrhoea quadrangulata F. Muell. ;四棱黄脂木(草树);
　　　　Grass Tree ●☆

415181　Xanthorrhoea resinosa Pers. ;黄脂木;Acaroid Resin, Yellow
　　　　Gum,Yellow Resin ●☆

415182　Xanthorrhoeaceae Dumort. (1829)(保留科名);黄脂木科(草
　　　　树胶科,刺叶树科,禾木胶科,黄胶木科,黄万年青科,黄脂草科,
　　　　木根旱生草科)●■☆

415183　Xanthorrhoeaceae Dumort. (保留科名) = Dasypogonaceae

Dumort. ■☆

415184　Xanthoselinum Schur = Peucedanum L. ■

415185　Xanthoselinum Schur(1866);黄亮蛇床属■☆

415186　Xanthoselinum alsaticum (L.) Schur;黄亮蛇床■☆

415187　Xanthoselinum alsaticum (L.) Schur = Peucedanum alsaticum L. ■☆

415188　Xanthosia Rudge(1811);黄伞草属■☆

415189　Xanthosia rotundifolia DC. ;圆叶黄伞草;Southern Cross ■☆

415190　Xanthosoma Schott(1832);千年芋属(黄肉芋属,黄体芋属,角柱芋属,南美芋属);Malanga,Malango,Tanier,Yautia ■

415191　Xanthosoma atrovirens K. Koch et C. D. Bouché;浓绿黄肉芋;Yautia Amarilla ■☆

415192　Xanthosoma atrovirens K. Koch et C. D. Bouché = Xanthosoma sagittifolium (L.) Schott ■

415193　Xanthosoma atrovirens K. Koch et C. D. Bouché var. appendiculatum Engl. ;附体浓绿黄肉芋■☆

415194　Xanthosoma lindenii (André) Engl. = Caladium lindenii (André) Madison ■☆

415195　Xanthosoma mafaffa Schott = Xanthosoma sagittifolium (L.) Schott ■

415196　Xanthosoma mexicana Liebm. ;墨西哥黄肉芋■☆

415197　Xanthosoma nigrum (Vell.) Mansf. = Xanthosoma violaceum Schott ■

415198　Xanthosoma nigrum (Vell.) Stellfeld = Xanthosoma sagittifolium (L.) Schott ■

415199　Xanthosoma robustum Schott;粗黄肉芋(壮黄肉芋)■☆

415200　Xanthosoma roseum Schott;粉色黄肉芋;Rosy Malanga ■☆

415201　Xanthosoma sagittifolium (L.) Schott;千年芋(慈姑叶黄肉芋,黄肉芋,箭叶海芋,箭叶黄肉芋,箭叶疆南星,野姜);Cocayams, Elephant Ear, Malanga, New Cocoyam, Tania, Tanias, Tannia ■

415202　Xanthosoma violaceum Schott;紫柄千年芋(紫大芋,紫茎黄肉芋,紫脉黄肉芋);Blue Tannia, Blue Taro, Purplestem Taro ■

415203　Xanthosoma violaceum Schott = Xanthosoma sagittifolium (L.) Schott ■

415204　Xanthostachya Bremek. = Strobilanthes Blume ●■

415205　Xanthostemon F. Muell. (1857) (保留属名);黄蕊桃金娘属●☆

415206　Xanthostemon chrysanthus (F. Muell.) Benth. ;金花黄蕊桃金娘;Golden Penda ●☆

415207　Xanthostemon oppositifolius Bailey;对叶黄蕊桃金娘●☆

415208　Xanthoxalis Small = Oxalis L. ■●

415209　Xanthoxalis bushii (Small) Small = Oxalis stricta L. ■

415210　Xanthoxalis coloradensis (Rydb.) Rydb. = Oxalis stricta L. ■

415211　Xanthoxalis corniculata (L.) Small = Oxalis corniculata L. ■

415212　Xanthoxalis corniculata (L.) Small var. atropurpurea (Planch.) Moldenke = Oxalis corniculata L. ■

415213　Xanthoxalis corniculata (L.) Small var. repens (Thunb.) Nakai = Oxalis corniculata L. ■

415214　Xanthoxalis cymosa (Small) Small = Oxalis stricta L. ■

415215　Xanthoxalis dillenii (Jacq.) Holub = Oxalis dillenii Jacq. ■☆

415216　Xanthoxalis dillenii (Jacq.) Holub var. piletocarpa (Wiegand) Holub = Oxalis stricta L. ■

415217　Xanthoxalis europaea (Jord.) Moldenke = Oxalis stricta L. ■

415218　Xanthoxalis fontana (Bunge) Holub = Oxalis stricta L. ■

415219　Xanthoxalis interior Small = Oxalis stricta L. ■

415220　Xanthoxalis langloisii Small = Oxalis corniculata L. ■

415221　Xanthoxalis repens (Thunb.) Moldenke = Oxalis corniculata L. ■

415222　Xanthoxalis rufa (Small) Small = Oxalis stricta L. ■

415223　Xanthoxalis stricta (L.) Small = Oxalis stricta L. ■

415224　Xanthoxalis stricta (L.) Small var. piletocarpa (Wiegand) Moldenke = Oxalis stricta L. ■

415225　Xanthoxylaceae Nees et Mart. = Rutaceae Juss. (保留科名)●■

415226　Xanthoxylon Spreng. = Xanthoxylum J. F. Gmel. ●

415227　Xanthoxylon Spreng. = Zanthoxylum L. ●

415228　Xanthoxylum Engl. = Zanthoxylum L. ●

415229　Xanthoxylum J. F. Gmel. = Zanthoxylum L. ●

415230　Xanthoxylum Mill. = Zanthoxylum L. ●

415231　Xanthoxylum americanum Mill. = Zanthoxylum americanum Mill. ●☆

415232　Xanthoxylum mantschuricum Benn. = Zanthoxylum schinifolium Siebold et Zucc. ●

415233　Xantium Gilib. = Xanthium L. ■

415234　Xantolis Raf. (1838);刺榄属(荷包果属);Spine Olive, Xantolis ●

415235　Xantolis boniana (Dubard) Royen;越南刺榄;Viatnam, Viet Nam Xantolis ●

415236　Xantolis boniana (Dubard) Royen var. rostrata (Merr.) Royen;喙果刺榄;Beakedfruit Spine olive, Beak-fruited Xantolis, Rostrate Viet Nam Xantolis ●

415237　Xantolis boniana (Dubard) Royen var. rostrata (Merr.) Royen = Xantolis stenosepala (Hu) Royle ●

415238　Xantolis embeliifolia (Merr.) P. Royen = Xantolis longispinosa (Merr.) H. S. Lo ●

415239　Xantolis longispinosa (Merr.) H. S. Lo;琼刺榄(刺山榄,信筒子山榄);Long Spine olive, Longspine Xantolis, Long-spined Xantolis ●

415240　Xantolis shweliensis (W. W. Sm.) P. Royen;瑞丽刺榄(瑞丽荷包果);Ruili Spine olive, Ruili Xantolis, Showeli Xantolis ●◇

415241　Xantolis stenosepala (Hu) Royle;滇刺榄(鸡心果,漫板树,狭萼荷包果);Dian Spine olive, Narrowsepal Xantolis, Stenosepallous Xantolis ●

415242　Xantolis stenosepala (Hu) Royle var. brevistylis C. Y. Wu;短柱滇刺榄(短柱滇刺果,短柱荷包果);Short-style Narrowsepal Xantolis ●

415243　Xantonnea Pierre ex Pit. (1923);东南亚茜属●☆

415244　Xantonnea parvifolia (Kuntze) Craib;小叶东南亚茜●☆

415245　Xantonnea quocensis Pierre ex Pit. ;东南亚茜●☆

415246　Xantonneopsis Pit. (1923);拟东南亚茜属☆

415247　Xantonneopsis robinsonii Pit. ;拟东南亚茜●☆

415248　Xantophtalmum Sang. = Leucopoa Griseb. ■

415249　Xantophtalmum Sang. = Xanthophthalmum Sch. Bip. ■

415250　Xantophtalmum Sch. Bip. = Glebionis Cass. ■

415251　Xantophtalmum segetum (L.) Sch. Bip. = Chrysanthemum segetum L. ■

415252　Xantophtalmum segetum (L.) Sch. Bip. = Glebionis segetum (L.) Fourr. ■

415253　Xantorrhoea Diels = Xanthorrhoea Sm. ●■☆

415254　Xaritonia Raf. = Oncidium Sw. (保留属名)■☆

415255　Xaritonia Raf. = Tolumnia Raf. ■☆

415256　Xartthochymus Roxb. = Garcinia L. ●

415257　Xatardia Meisn. = Xatardia Meisn. et Zeyh. ■☆

415258　Xatardia Meisn. et Zeyh. (1838);法西草属■☆

415259　Xatardia scabra Meisn. = Xatardia scabra Meisn. et Zeyh. ■☆

415260　Xatardia scabra Meisn. et Zeyh. ;法西草■☆

415261　Xatartia St. -Lag. = Xatardia Meisn. et Zeyh. ■☆

415262 Xatatia Bubani = Xatardia Meisn. et Zeyh. ■☆

415263 Xaveria Endl. = Anemonopsis Siebold et Zucc. ■☆

415264 Xeilyathum Raf. = Oncidium Sw. (保留属名)■☆

415265 Xenacanthus Bremek. = Strobilanthes Blume ●■

415266 Xenia Gerbaulet(1992);阿根廷马齿苋属■☆

415267 Xenia vulcanensis (Anon) Gerbaulet;外来马齿苋■☆

415268 Xeniatrum Salisb. = Clintonia Raf. ■

415269 Xeniatrum umbellulatum (Michx.) Small = Clintonia umbellulata (Michx.) Morong ■☆

415270 Xenikophyton Garay(1974);西大洋兰属■☆

415271 Xenikophyton smeeanum (Rchb. f.) Garay;西大洋兰■☆

415272 Xenismia DC. = Dimorphotheca Vaill. (保留属名)■●☆

415273 Xenismia DC. = Oligocarpus Less. ☆

415274 Xenismia acanthosperma DC. = Osteospermum acanthospermum (DC.) Norl. ■☆

415275 Xenocarpus Cass. = Cineraria L. ■●☆

415276 Xenochloa Licht. = ? Danthonia DC. (保留属名)■

415277 Xenochloa Licht. = Phragmites Adans. ■

415278 Xenochloa Licht. ex Roem. et Schult. = Phragmites Adans. ■

415279 Xenochloa Roem. et Schult. = Phragmites Adans. ■

415280 Xenochloa arundinacea Licht. ex Roem. et Schult. = Phragmites xenochloa Trin. ex Steud. ■☆

415281 Xenodendron K. Schum. et Lauterb. = Acmena DC. ●☆

415282 Xenodendron K. Schum. et Lauterb. = Syzygium R. Br. ex Gaertn. (保留属名)●

415283 Xenophonta Benth. et Hook. f. = Xenophontia Vell. ●☆

415284 Xenophontia Vell. = Barnadesia Mutis ex L. f. ●☆

415285 Xenophya Schott = Alocasia (Schott) G. Don(保留属名)■

415286 Xenophyllum V. A. Funk(1997);变叶菊属■☆

415287 Xenophyllum dactylophyllum (Sch. Bip.) V. A. Funk;指叶变叶菊■☆

415288 Xenophyllum rigidum (Kunth) V. A. Funk;硬叶变叶菊■☆

415289 Xenophyllum roseum (Hieron.) V. A. Funk;粉红变叶菊■☆

415290 Xenopoma Willd. (废弃属名) = Clinopodium L. ■●

415291 Xenopoma Willd. (废弃属名) = Micromeria Benth. (保留属名)■●

415292 Xenoscapa (Goldblatt) Goldblatt et J. C. Manning(1995);西南非鸢尾属■☆

415293 Xenoscapa fistulosa (Spreng. ex Klatt) Goldblatt et J. C. Manning;簇生西南非鸢尾■☆

415294 Xenoscapa uliginosa Goldblatt et J. C. Manning;西南非鸢尾■☆

415295 Xenosia Luer = Pleurothallis R. Br. ■☆

415296 Xenosia Luer(2004);外来兰属■☆

415297 Xenostegia D. F. Austin et Staples(1981);地旋花属(戟叶菜栾藤属)■

415298 Xenostegia medium (L.) D. F. Austin et Staples = Merremia medium (L.) Hallier f. ■☆

415299 Xenostegia tridentata (L.) D. F. Austin et Staples;地旋花(飞洋草,过腰蛇,戟叶菜栾藤,尖尊山猪菜,尖尊鱼黄草,凉粉草,三齿菜栾藤,三齿鱼黄草,三莴萝藤,野通心菜);Hastate Merremia, Threeteeth Merremia ■

415300 Xenostegia tridentata (L.) D. F. Austin et Staples = Merremia tridentata (L.) Hallier f. ■

415301 Xenostegia tridentata (L.) D. F. Austin et Staples subsp. alatipes (Dammer) Lejoly et Lisowski = Merremia tridentata (L.) Hallier f. subsp. alatipes (Dammer) Verdc. ■☆

415302 Xenostegia tridentata (L.) D. F. Austin et Staples subsp. alatipes (Dammer) Lejoly et Lisowski;翼梗地旋花■☆

415303 Xenostegia tridentata (L.) D. F. Austin et Staples subsp. angustifolia (Jacq.) Lejoly et Lisowski = Merremia tridentata (L.) Hallier f. var. angustifolia (Jacq.) Ooststr. ■☆

415304 Xenostegia tridentata (L.) D. F. Austin et Staples subsp. angustifolia (Jacq.) Lejoly et Lisowski;窄叶地旋花(尖尊鱼黄草)■☆

415305 Xeodolon Salisb. = Scilla L. ■

415306 Xeracina Raf. = Adelobotrys DC. ●☆

415307 Xeractis Oliv. = Xerotia Oliv. ●☆

415308 Xeraea Kuntze = Gomphrena L. ●■

415309 Xeraenanthus Mart. ex Koehne = Pleurophora D. Don ■☆

415310 Xeralis Raf. = Charachera Forssk. ●

415311 Xeralsine Fourr. = Minuartia L. ■

415312 Xerandra Raf. = Iresine P. Browne(保留属名)●■

415313 Xeranthemaceae Döll = Asteraceae Bercht. et J. Presl(保留科名)●■

415314 Xeranthemaceae Döll = Compositae Giseke(保留科名)●■

415315 Xeranthemum L. (1753);旱花属(干花菊属,灰毛菊属);Immortelle ■☆

415316 Xeranthemum Tourn. ex L. = Xeranthemum L. ■☆

415317 Xeranthemum annuum L. ;旱花(干花菊,千年菊);Annual Dry-flower,Common Immortal,Common Immortelle,Immortelle ■☆

415318 Xeranthemum annuum L. = Xeranthemum inapertum (L.) Mill. ■☆

415319 Xeranthemum argenteum Thunb. = Syncarpha argentea (Thunb.) B. Nord. ■☆

415320 Xeranthemum australe Pomel = Xeranthemum inapertum (L.) Mill. ■☆

415321 Xeranthemum bracteatum Vent. = Bracteantha bracteata (Vent.) Anderb. et Haegi ■

415322 Xeranthemum bracteatum Vent. = Helichrysum bracteatum (Vent.) Andréws ■

415323 Xeranthemum bracteatum Vent. = Xerochrysum bracteatum (Vent.) Tzvelev ■

415324 Xeranthemum canescens L. = Syncarpha canescens (L.) B. Nord. ■☆

415325 Xeranthemum ciliatum L. = Polyarrhena reflexa (L.) Cass. ●☆

415326 Xeranthemum cylindraceum Sm. ;长筒旱花(长筒干花菊,灰毛菊);Cylinder Immortelle ■☆

415327 Xeranthemum erectum J. Presl et C. Presl = Xeranthemum inapertum (L.) Mill. ☆

415328 Xeranthemum ericoides Lam. = Dolichothrix ericoides (Lam.) Hilliard et B. L. Burtt ■☆

415329 Xeranthemum fasciculatum Andréws = Edmondia fasciculata (Andréws) Hilliard ●☆

415330 Xeranthemum ferrugineum Lam. = Syncarpha ferruginea (Lam.) B. Nord. ■☆

415331 Xeranthemum fulgidum L. f. = Helichrysum aureum (Houtt.) Merr. ●☆

415332 Xeranthemum herbaceum Andréws = Helichrysum herbaceum (Andréws) Sweet ■☆

415333 Xeranthemum humile Andréws = Edmondia pinifolia (Lam.) Hilliard ●☆

415334 Xeranthemum inapertum (L.) Mill. ;闭旱花(闭干花菊)■☆

415335 Xeranthemum inapertum (L.) Mill. var. australe (Pomel) Batt. et Trab. = Xeranthemum inapertum (L.) Mill. ■☆

415336 Xeranthemum inapertum (L.) Mill. var. reboudianum Verl. =

Xeranthemum inapertum（L.）Mill. ■☆

415337　Xeranthemum lancifolium Thunb. = Helichrysum lancifolium（Thunb.）Thunb. ●☆

415338　Xeranthemum modestum Ball = Xeranthemum inapertum（L.）Mill. ■☆

415339　Xeranthemum paniculatum L. = Syncarpha paniculata（L.）B. Nord. ■☆

415340　Xeranthemum pinifolium Lam. = Edmondia pinifolia（Lam.）Hilliard ●☆

415341　Xeranthemum proliferum L. = Phaenocoma prolifera（L.）D. Don ●☆

415342　Xeranthemum recurvatum L. f. = Syncarpha recurvata（L. f.）B. Nord. ■☆

415343　Xeranthemum retortum L. = Helichrysum retortum（L.）Willd. ●☆

415344　Xeranthemum sesamoides L. = Edmondia sesamoides（L.）Hilliard ●☆

415345　Xeranthemum speciosissimum L. = Syncarpha speciosissima（L.）B. Nord. ●☆

415346　Xeranthemum spinosum L. = Macledium spinosum（L.）S. Ortiz ●☆

415347　Xeranthemum squamosum Jacq. = Edmondia pinifolia（Lam.）Hilliard ●☆

415348　Xeranthemum staehelina L. = Syncarpha staehelina（L.）B. Nord. ■☆

415349　Xeranthemum stoloniferum L. f. = Helichrysum stoloniferum（L. f.）Willd. ●☆

415350　Xeranthemum striatum Thunb. = Syncarpha striata（Thunb.）B. Nord. ■☆

415351　Xeranthemum variegatum P. J. Bergius = Syncarpha variegata（P. J. Bergius）B. Nord. ■☆

415352　Xeranthemum vestitum L. = Syncarpha vestita（L.）B. Nord. ■☆

415353　Xeranthemum virgatum P. J. Bergius = Syncarpha virgata（P. J. Bergius）B. Nord. ■☆

415354　Xeranthium Lepech. = Xanthium L. ■

415355　Xeranthus Miers = Grahamia Gillies ex Hook. et Arn. ●☆

415356　Xeregathis Raf. = Baccharis L.（保留属名）●■☆

415357　Xeria C. Presl ex Rohrb. = Pycnophyllum J. Rémy ■☆

415358　Xeris Medik. = Iris L. ■

415359　Xeroaloysia Tronc.（1963）;旱鞭木属●☆

415360　Xeroaloysia ovatifolia（Moldenke）Tronc. ;旱鞭木●☆

415361　Xerobius Cass. = Egletes Cass. ■☆

415362　Xerobotrys Nutt. = Arctostaphylos Adans.（保留属名）●☆

415363　Xerocarpa（G. Don）Spach（废弃属名）= Xerocarpa H. J. Lam（保留属名）●☆

415364　Xerocarpa H. J. Lam（1919）;干果马鞭草属■☆

415365　Xerocarpa H. J. Lam.（保留属名）= Teijsmanniodendron Koord. ●☆

415366　Xerocarpa Spach = Scaevola L.（保留属名）●■

415367　Xerocarpus Guill. et Perr. = Rothia Pers.（保留属名）■

415368　Xerocarpus hirsutus Guillaumin et Perr. = Rothia hirsuta（Guillaumin et Perr.）Baker ■☆

415369　Xerocassia Britton et Rose = Cassia L.（保留属名）●■

415370　Xerocassia Britton et Rose = Senna Mill. ●■

415371　Xerochlamys Baker = Leptolaena Thouars ●☆

415372　Xerochlamys arenaria F. Gérard = Leptolaena arenaria（F. Gerard）Cavaco ●☆

415373　Xerochlamys bernieri（Baill.）H. Perrier = Leptolaena bernieri Baill. ●☆

415374　Xerochlamys bojeriana（Baill.）Baker = Leptolaena bojeriana（Baill.）Cavaco ●☆

415375　Xerochlamys elliptica F. Gérard = Leptolaena elliptica（F. Gérard）Hong-Wa ●☆

415376　Xerochlamys grandidieri（Baill.）Baker = Leptolaena bojeriana（Baill.）Cavaco ●☆

415377　Xerochlamys luteola H. Perrier = Leptolaena villosa（F. Gerard）G. E. Schatz et Lowry ●☆

415378　Xerochlamys pilosa Baker = Leptolaena bojeriana（Baill.）Cavaco ●☆

415379　Xerochlamys pubescens Baker = Leptolaena bojeriana（Baill.）Cavaco ●☆

415380　Xerochlamys tampoketsensis F. Gérard = Leptolaena tampoketsensis（F. Gérard）Hong-Wa ●☆

415381　Xerochlamys villosa F. Gérard = Leptolaena villosa（F. Gérard）G. E. Schatz et Lowry ●☆

415382　Xerochloa R. Br.（1810）;灯草旱禾属■☆

415383　Xerochloa barbata R. Br. ;灯草旱禾■☆

415384　Xerochloa latifolia Hassk. ;宽叶灯草旱禾■☆

415385　Xerochrysum Tzvelev（1990）;麦秆菊属（小蜡菊属）;Paper Daisy ■☆

415386　Xerochrysum bracteatum（Vent.）Tzvelev;麦秆菊(贝细工,脆菊,蜡菊,麦藁菊);Bracted Strawflower, Bush Strawflower, Cudweed, Everlasting, Everlasting Flower, Golden Everlasting, Golden Paper Daisy, Immortelle, Paper Daisy, Strawflower, Straw-flower, Waxdaisy ■

415387　Xerochrysum bracteatum（Vent.）Tzvelev = Helichrysum bracteatum（Vent.）Andréws ■

415388　Xerocladia Harv.（1862）;干枝豆属●☆

415389　Xerocladia viridiramis（Burch.）Taub. ;干枝豆●☆

415390　Xerocladia zeyheri Harv. = Xerocladia viridiramis（Burch.）Taub. ●☆

415391　Xerococcus Oerst. = Hoffmannia Sw. ●■☆

415392　Xerodanthia J. B. Phipps = Danthoniopsis Stapf ■☆

415393　Xerodanthia barbata（Nees）J. B. Phipps = Danthoniopsis barbata（Nees）C. E. Hubb. ■☆

415394　Xerodanthia stocksii（Boiss.）Phipps = Danthoniopsis stocksii（Boiss.）C. E. Hubb. ■☆

415395　Xerodenis Roberty = Ostryocarpus Hook. f. ■☆

415396　Xerodera Fourr. = Ranunculus L. ■

415397　Xeroderis Roberty = Xeroderris Roberty ●☆

415398　Xeroderris Roberty（1954）;干鱼藤属●☆

415399　Xeroderris stuhlmannii（Taub.）Mendonça et E. P. Sousa;斯图尔曼干鱼藤●☆

415400　Xerodraba Skottsb.（1916）;干葶苈属●■☆

415401　Xerodraba colobanthoides Skottsb. ;干葶苈■☆

415402　Xerodraba microphylla Skottsb. ;小叶干葶苈■☆

415403　Xerodraba monantha Skottsb. ;山地干葶苈■☆

415404　Xerogona Raf. = Passiflora L. ●■

415405　Xerolekia Anderb. = Buphthalmum L. ■

415406　Xerolirion A. S. George（1986）;旱百合属■☆

415407　Xerolirion divaricata A. S. George;旱百合■☆

415408　Xerololophus B. D. Jacks. = Xerolophus Dulac ■

415409　Xerololophus Dulac = Thesium L. ■

415410　Xeroloma Cass. = Xeranthemum L. ■☆

415411　Xerolophus Dulac = Thesium L. ■

415412　Xeromalon Raf. = Crataegus L. ●

415413　Xeromphis Raf. = Catunaregam Wolf ●

415414　Xeromphis Raf. = Randia L. ●

415415　Xeromphis keniensis Tennant = Tennantia sennii（Chiov.）Verdc. et Bridson ●☆

415416　Xeromphis nilotica（Stapf）Keay = Catunaregam nilotica（Stapf）Tirveng. ●☆

415417　Xeromphis obovata（Hochst.）Keay = Catunaregam obovata（Hochst.）A. E. Gonc. ●☆

415418　Xeromphis retzii Raf. = Catunaregam spinosa（Thunb.）Tirveng. ●

415419　Xeromphis rudis（E. Mey. ex Harv.）Codd = Coddia rudis（E. Mey. ex Harv.）Verdc. ●☆

415420　Xeromphis spinosa（Thunb.）Keay = Catunaregam spinosa（Thunb.）Tirveng. ●

415421　Xeronema Brongn. = Xeronema Brongn. et Gris ■☆

415422　Xeronema Brongn. et Gris（1865）；鸢尾麻属■☆

415423　Xeronema callistemon W. R. B. Oliv.；鸢尾麻；Poor Knighl's Lily ■☆

415424　Xeronemataceae M. W. Chase，Rudall et M. F. Fay（2000）；鸢尾麻科（血剑草科）■☆

415425　Xeropappus Wall. = Dicoma Cass. ●☆

415426　Xeropetalon Hook. = Viviania Cav. ■☆

415427　Xeropetalum Delile = Dombeya Cav.（保留属名）●☆

415428　Xeropetalum Rchb. = Dillwynia Sm. ●☆

415429　Xeropetalum brucei Hochst. = Dombeya torrida（J. F. Gmel.）Bamps ●☆

415430　Xeropetalum minus Endl. = Dombeya quinqueseta（Delile）Exell ●☆

415431　Xeropetalum multiflorum Endl. = Dombeya quinqueseta（Delile）Exell ●☆

415432　Xeropetalum quinquesetum Delile = Dombeya quinqueseta（Delile）Exell ●☆

415433　Xeropetalum rotundifolium Hochst. = Dombeya rotundifolia（Hochst.）Planch. ●☆

415434　Xeropetalum tiliaceum Endl. = Dombeya tiliacea（Endl.）Planch. ●☆

415435　Xerophyllaceae Takht = Melanthiaceae Batsch ex Borkh.（保留科名）■

415436　Xerophyllaceae Takht.（1994）；密花草科■☆

415437　Xerophyllaceae Takht. = Liliaceae Juss.（保留科名）■●

415438　Xerophyllaceae Takht. = Xyridaceae C. Agardh（保留科名）■

415439　Xerophyllum Michx.（1803）；密花草属（旱叶草属）■☆

415440　Xerophyllum asphodeloides（L.）Nutt.；土耳其密花草（土耳其旱叶草）；Mountain-asphodel，Turkey Beard，Turkey-beard ■☆

415441　Xerophyllum tenax（Pursh）Nutt.；密花草（旱叶草）；Bear Grass，Beargrass，Bear-grass，Elk-grass，Indian-basket-grass，Squaw-grass ■☆

415442　Xerophylum Raf. = Xerophyllum Michx. ■☆

415443　Xerophysa Steven = Astragalus L. ●■

415444　Xerophyta Juss.（1789）；干若翠属●■☆

415445　Xerophyta acuminata（Baker）N. L. Menezes；渐尖干若翠●☆

415446　Xerophyta aequatorialis（Rendle）N. L. Menezes = Xerophyta spekei Baker ■☆

415447　Xerophyta analavelonensis Phillipson et Lowry；阿纳拉干若翠●☆

415448　Xerophyta argentea（Wild）L. B. Sm. et Ayensu；银白干若翠●☆

415449　Xerophyta barbarae P. A. Duvign. et Dewit = Xerophyta equisetoides Baker var. trichophylla（Baker）L. B. Sm. et Ayensu ●☆

415450　Xerophyta barbarae P. A. Duvign. et Dewit subsp. cuprophila？

= Xerophyta equisetoides Baker var. trichophylla（Baker）L. B. Sm. et Ayensu ●☆

415451　Xerophyta capillaris Baker；发状干若翠●☆

415452　Xerophyta capillaris Baker var. occultans L. B. Sm. et Ayensu；隐蔽干若翠●☆

415453　Xerophyta clavata Baker = Xerophyta retinervis Baker ■☆

415454　Xerophyta concolor L. B. Sm. et Ayensu；同色干若翠●☆

415455　Xerophyta dasylirioides Baker；毛百合干若翠■☆

415456　Xerophyta eglandulosa H. Perrier；无腺干若翠●☆

415457　Xerophyta eglandulosa H. Perrier var. hirtocarpa H. Perrier = Xerophyta eglandulosa H. Perrier ●☆

415458　Xerophyta eglandulosa H. Perrier var. trichocarpa H. Perrier = Xerophyta eglandulosa H. Perrier ●☆

415459　Xerophyta elegans（Balf.）Baker = Talbotia elegans Balf. ■☆

415460　Xerophyta equisetoides Baker；木贼干若翠●☆

415461　Xerophyta equisetoides Baker var. pauciramosa L. B. Sm. et Ayensu；少枝干若翠●☆

415462　Xerophyta equisetoides Baker var. pubescens L. B. Sm. et Ayensu；短柔毛干若翠●☆

415463　Xerophyta equisetoides Baker var. setosa L. B. Sm. et Ayensu；刚毛干若翠●☆

415464　Xerophyta equisetoides Baker var. trichophylla（Baker）L. B. Sm. et Ayensu；毛叶干若翠●☆

415465　Xerophyta eylesii（Greves）N. L. Menezes；艾尔斯干若翠■☆

415466　Xerophyta goetzei（Harms）L. B. Sm. et Ayensu；格兹干若翠■☆

415467　Xerophyta hereroensis（Schinz）N. L. Menezes = Xerophyta viscosa Baker ■☆

415468　Xerophyta humilis（Baker）T. Durand et Schinz；低矮干若翠■☆

415469　Xerophyta kirkii（Hemsl.）L. B. Sm. et Ayensu；柯克干若翠■☆

415470　Xerophyta longicaulis Hilliard；长茎干若翠●☆

415471　Xerophyta madagascariensis J. F. Gmel. = Xerophyta pinifolia Lam. ●☆

415472　Xerophyta melleri Baker = Xerophyta equisetoides Baker ●☆

415473　Xerophyta minuta Baker = Talbotia elegans Balf. ■☆

415474　Xerophyta monroi（Greves）N. L. Menezes = Xerophyta villosa（Baker）L. B. Sm. et Ayensu ■☆

415475　Xerophyta nutans L. B. Sm. et Ayensu；俯垂干若翠●☆

415476　Xerophyta parviflora Phillipson et Lowry；小花干若翠●☆

415477　Xerophyta pectinata Baker；篦状干若翠●☆

415478　Xerophyta pinifolia Lam.；松叶干若翠●☆

415479　Xerophyta retinervis Baker；网脉干若翠●☆

415480　Xerophyta retinervis Baker var. equisetoides（Baker）Coetzee = Xerophyta equisetoides Baker ●☆

415481　Xerophyta retinervis Baker var. wentzeliana（Harms）Coetzee = Xerophyta equisetoides Baker var. pauciramosa L. B. Sm. et Ayensu ●☆

415482　Xerophyta rosea（Baker）N. L. Menezes = Xerophyta schlechteri（Baker）N. L. Menezes ■☆

415483　Xerophyta scabrida（Pax）T. Durand et Schinz；微糙干若翠■☆

415484　Xerophyta schlechteri（Baker）N. L. Menezes；施莱干若翠■☆

415485　Xerophyta schnizleinia（Hochst.）Baker；施尼干若翠■☆

415486　Xerophyta schnizleinia（Hochst.）Baker var. somalensis（A. Terracc.）Lye；索马里干若翠■☆

415487　Xerophyta somalensis（A. Terracc.）N. L. Menezes = Xerophyta schnizleinia（Hochst.）Baker var. somalensis（A. Terracc.）Lye ■☆

415488　Xerophyta spekei Baker；斯皮克干若翠■☆

415489　Xerophyta spinulosa Ridl. = Xerophyta pectinata Baker ●☆

415490　Xerophyta splendens（Rendle）N. L. Menezes；光亮干若翠●☆

415491　Xerophyta squarrosa Baker;粗鳞干若翠●☆

415492　Xerophyta suaveolens（Greves）N. L. Menezes;芳香干若翠●☆

415493　Xerophyta suaveolens（Greves）N. L. Menezes var. vestita L. B. Sm. et Ayensu;包被干若翠●☆

415494　Xerophyta trichophylla（Baker）N. L. Menezes ＝ Xerophyta equisetoides Baker var. trichophylla（Baker）L. B. Sm. et Ayensu ●☆

415495　Xerophyta tulearensis（H. Perrier）Phillipson et Lowry;图莱亚尔干若翠●☆

415496　Xerophyta velutina Welw. ex Baker;绒毛干若翠●☆

415497　Xerophyta villosa（Baker）L. B. Sm. et Ayensu;长柔毛干若翠■☆

415498　Xerophyta violacea（Baker）N. L. Menezes ＝ Xerophyta villosa（Baker）L. B. Sm. et Ayensu ■☆

415499　Xerophyta viscosa Baker;黏干若翠■☆

415500　Xerophyta wentzeliana（Harms）Sölch ＝ Xerophyta equisetoides Baker var. pauciramosa L. B. Sm. et Ayensu ●☆

415501　Xerophyta zambiana L. B. Sm. et Ayensu;赞比亚干若翠●☆

415502　Xeroplana Briq.（1895）;旱密穗属（干密穗草属）●☆

415503　Xeroplana zeyheri Briq. ;旱密穗●☆

415504　Xeroplana zeyheri Briq. ＝ Stilbe overbergensis Rourke ●☆

415505　Xerorchis Schltr.（1912）;旱兰属■☆

415506　Xerorchis amazonica Schltr. ;旱兰■☆

415507　Xerorchis trichorhiza（Kraenzl.）Garay;毛根旱兰■☆

415508　Xerosicyos Humbert(1939);沙葫芦属（碧雷鼓属）●■☆

415509　Xerosicyos danguyi Humbert;银沙葫芦;Silver Dollar Plant ●☆

415510　Xerosicyos decaryi Guillaumin et Rabenant.;德卡里沙葫芦●☆

415511　Xerosicyos perrieri Humbert;佩里耶沙葫芦●☆

415512　Xerosicyos pubescens Rabenant.;短柔毛沙葫芦●☆

415513　Xerosiphon Turcz.（1843）;旱苋属■☆

415514　Xerosiphon Turcz. ＝ Gomphrena L. ●■

415515　Xerosiphon angustiflorus（Mart.）Pedersen;狭花旱苋■☆

415516　Xerosiphon aphyllus（Pohl ex Moq.）Pedersen;无叶旱苋■☆

415517　Xerosiphon gracilis Turcz. ;旱苋■☆

415518　Xerosollya Turcz. ＝ Sollya Lindl. ●☆

415519　Xerospermum Blume（1849）;干果木属（假荔枝属）;Xerospermum ●

415520　Xerospermum bonii（Lecomte）Radlk. ;干果木（云南干果木）;Bon Xerospermum,Xerospermum ●◇

415521　Xerospermum topengii Merr. ＝ Nephelium topengii（Merr.）H. S. Lo ●

415522　Xerospermum yunnanense W. T. Wang ＝ Dimocarpus yunnanensis（W. T. Wang）C. Y. Wu et T. L. Ming ●

415523　Xerosphaera Soják ＝ Galearia C. Presl(废弃属名)■

415524　Xerosphaera Soják ＝ Trifolium L. ■

415525　Xerospiraea Henrickson(1986);旱绣线菊属●☆

415526　Xerospiraea hartwegiana（Rydb.）Henrickson;旱绣线菊●☆

415527　Xerotaceae Endl. ＝ Dasypogonaceae Dumort. ●☆

415528　Xerotaceae Endl. ＝ Laxmanniaceae Bubani ■

415529　Xerotaceae Endl. ＝ Xanthorrhoeaceae Dumort.（保留科名）●■

415530　Xerotaceae Hassk. ＝ Dasypogonaceae Dumort. ■☆

415531　Xerotaceae Hassk. ＝ Laxmanniaceae Bubani ■

415532　Xerotaceae Hassk. ＝ Lomandraceae Lotsy ●■

415533　Xerotaceae Hassk. ＝ Xanthorrhoeaceae Dumort.（保留科名）●■☆

415534　Xerotecoma J. C. Gomes ＝ Godmania Hemsl. ●☆

415535　Xerotes R. Br. ＝ Lomandra Labill. ●■☆

415536　Xerothamnella C. T. White(1944);旱灌爵床属●☆

415537　Xerothamnella parvifolia C. T. White;旱灌爵床●☆

415538　Xerothamnus DC. ＝ Osteospermum L. ●■☆

415539　Xerothamnus ecklonianus DC. ＝ Gibbaria scabra（Thunb.）Norl. ■☆

415540　Xerotia Oliv.（1895）;假麻黄属●☆

415541　Xerotia arabica Oliv. ;假麻黄●☆

415542　Xerotis Hoffmanns. ＝ Lomandra Labill. ■●☆

415543　Xerotis Hoffmanns. ＝ Xerotes R. Br. ●■☆

415544　Xerotium Bluff et Fingerh. ＝ Filago L.（保留属名）■

415545　Xerxes J. R. Grant(1994);无茎叉毛菊属■☆

415546　Xestaea Griseb. ＝ Schultesia Mart.（保留属名）■☆

415547　Xetola Raf. ＝ Cephalaria Schrad.（保留属名）■

415548　Xetoligus Raf. ＝ Stevia Cav. ■●☆

415549　Xilophia Ausier ＝ Xilopia Juss. ●

415550　Xilopia Juss. ＝ Xylopia L.（保留属名）●

415551　Ximenesia Cav. ＝ Verbesina L.（保留属名）●■☆

415552　Ximenesia encelioides Cav. ;无茎叉毛菊■☆

415553　Ximenesia encelioides Cav. ＝ Verbesina encelioides（Cav.）Benth. et Hook. f. ■☆

415554　Ximenesia encelioides Cav. var. nana A. Gray ＝ Verbesina nana（A. Gray）B. L. Rob. et Greenm. ■☆

415555　Ximenia L.（1753）;海檀木属;Tallowwood,Tallow-wood ●

415556　Ximenia Plum. ex L. ＝ Ximenia L. ●

415557　Ximenia aegyptiaca L. ＝ Balanites aegyptiaca（L.）Delile ●☆

415558　Ximenia americana L. ;海檀木（山梅树,西门木）;America Tallowwood, American Tallowwood, American Tallow-wood, Bastard, Hog Plum, Monkey Plum, Seaside Plum, Spiny Plum, Tallow Nut, Tallow-wood,Wild Lime,Wild Olive ●

415559　Ximenia americana L. var. microphylla Welw. ex Oliv. ;小叶海檀木●☆

415560　Ximenia caffra Sond. ;酸海檀木（卡夫拉海檀木）;Monkey Plum,Sour Plum ●☆

415561　Ximenia caffra Sond. var. natalensis ?;纳塔尔酸海檀木●☆

415562　Ximenia ferox Poir. ＝ Balanites aegyptiaca（L.）Delile var. ferox（Poir.）DC. ●☆

415563　Ximenia gabonensis Laness. ;加蓬海檀木●☆

415564　Ximenia inermis L. ＝ Ximenia americana L. ●

415565　Ximenia olacoides Wight et Arn. ＝ Olax imbricata Roxb. ●

415566　Ximenia rogersii Burtt Davy ＝ Ximenia americana L. var. microphylla Welw. ex Oliv. ●☆

415567　Ximeniaceae Horan. ＝ Olacaceae R. Br.（保留科名）●

415568　Ximeniaceae Martinet ＝ Olacaceae R. Br.（保留科名）●

415569　Ximeniaceae Tiegh. ;海檀木科●

415570　Ximeniaceae Tiegh. ＝ Olacaceae R. Br.（保留科名）●

415571　Ximeniopsis Alain ＝ Ximenia L. ●

415572　Ximeniopsis Alain(1980);类海檀木属●☆

415573　Ximeniopsis horridus（Urb. et Ekman）Alain;类海檀木●☆

415574　Xiphagrostis Coville ＝ Miscanthus Andersson ■

415575　Xiphidiaceae Dumort. ＝ Haemodoraceae R. Br.（保留科名）■☆

415576　Xiphidium Aubl.（1775）;剑草属■☆

415577　Xiphidium Loefl. ＝ Xiphidium Aubl. ■☆

415578　Xiphidium Loefl. ex Aubl. ＝ Xiphidium Aubl. ■☆

415579　Xiphidium coeruleum Aubl. ;蓝剑草■☆

415580　Xiphidium floribundum Sw. ;多花蓝剑草■☆

415581　Xiphidium xanthorrhizon Wright ex Griseb. ;黄根蓝剑草■☆

415582　Xiphion Mill. ＝ Iris L. ■

415583　Xiphion Tourn. ex Mill. ＝ Iris L. ■

415584　Xiphion Tourn. ex Mill. ＝ Xiphion Mill. ■

415585　Xiphium Mill. ＝ Iris L. ■

415586　Xiphium Mill. = Xiphion Mill. ■

415587　Xiphium kolpakowskiana（Regel）Baker = Iris kolpakowskiana Regel ■

415588　Xiphizusa Rchb. f. = Bulbophyllum Thouars（保留属名）■

415589　Xiphocarpus C. Presl = Tephrosia Pers.（保留属名）●■

415590　Xiphochaeta Poepp.（1843）;沼生斑鸠菊属■☆

415591　Xiphochaeta Poepp. et Endl. = Stilpnopappus Mart. ex DC. ●■

415592　Xiphochaeta aquatica Poepp.;沼生斑鸠菊■☆

415593　Xiphocoma Steven = Ranunculus L. ■

415594　Xiphodendron Raf. = Yucca L. ●■

415595　Xipholepis Steetz = Vernonia Schreb.（保留属名）●■

415596　Xiphophyllum Ehrh. = Cephalanthera Rich. ■

415597　Xiphophyllum Ehrh. = Serapias L.（保留属名）■☆

415598　Xiphosium Griff. = Cryptochilus Wall. ■

415599　Xiphosium Griff. = Eria Lindl.（保留属名）■

415600　Xiphosium roseum（Lindl.）Griff. = Cryptochilus roseus（Lindl.）S. C. Chen et J. J. Wood ■

415601　Xiphosium roseum（Lindl.）Griff. = Eria rosea Lindl. ■

415602　Xiphostylis Gasp. = Trigonella L. ■

415603　Xiphotheca Eckl. et Zeyh.（1836）;刀囊豆属■☆

415604　Xiphotheca Eckl. et Zeyh. = Priestleya DC. ■☆

415605　Xiphotheca canescens（Thunb.）A. L. Schutte et B. -E. van Wyk;灰白刀囊豆■☆

415606　Xiphotheca cordifolia A. L. Schutte et B. -E. van Wyk;心叶刀囊豆■☆

415607　Xiphotheca elliptica（DC.）A. L. Schutte et B. -E. van Wyk;椭圆刀囊豆■☆

415608　Xiphotheca fruticosa（L.）A. L. Schutte et B. -E. van Wyk;灌丛刀囊豆■☆

415609　Xiphotheca guthriei（L. Bolus）A. L. Schutte et B. -E. van Wyk;格斯里刀囊豆■☆

415610　Xiphotheca lanceolata（E. Mey.）Eckl. et Zeyh. ;披针形刀囊豆■☆

415611　Xiphotheca phylicoides A. L. Schutte et B. -E. van Wyk;菲利木刀囊豆■☆

415612　Xiphotheca polycarpa Eckl. et Zeyh. = Xiphotheca tecta（Thunb.）A. L. Schutte et B. -E. van Wyk ■☆

415613　Xiphotheca reflexa（Thunb.）A. L. Schutte et B. -E. van Wyk;反折刀囊豆■☆

415614　Xiphotheca rotundifolia（Eckl. et Zeyh.）Walp. = Xiphotheca tecta（Thunb.）A. L. Schutte et B. -E. van Wyk ■☆

415615　Xiphotheca rotundifolia Eckl. et Zeyh. = Xiphotheca tecta（Thunb.）A. L. Schutte et B. -E. van Wyk ■☆

415616　Xiphotheca tecta（Thunb.）A. L. Schutte et B. -E. van Wyk;屋顶刀囊豆■☆

415617　Xiria Raf. = Xyris L. ■

415618　Xizangia D. Y. Hong = Pterygiella Oliv. ●■★

415619　Xizangia D. Y. Hong（1986）;马松蒿属（藏草属）;Xizangia ■

415620　Xizangia bartschioides（Hand. -Mazz.）C. Y. Wu et D. D. Tao = Xizangia serrata D. Y. Hong ■

415621　Xizangia bartschioides（Hand. -Mazz.）D. Y. Hong;藏草（齿叶翅茎草, 马松蒿）;Toothedleaf Pterygiella, Toothleaf Pterygiella, Toothleaf Woodbetony ■

415622　Xizangia serrata D. Y. Hong;马松蒿（藏草）;Xizangia ■

415623　Xizangia serrata D. Y. Hong = Xizangia bartschioides（Hand. -Mazz.）C. Y. Wu et D. D. Tao ■

415624　Xolantha Raf.（废弃属名）= Helianthemum Mill. ●■

415625　Xolantha Raf.（废弃属名）= Tuberaria（Dunal）Spach（保留属名）■☆

415626　Xolantha commutata（Gallego）Gallego et Munoz Garm. = Tuberaria commutata Gallego ■☆

415627　Xolantha echioides（Lam.）Gallego et Munoz Garm. et Navarro = Tuberaria echioides（Lam.）Willk. ■☆

415628　Xolantha guttata（L.）Raf. = Tuberaria guttata（L.）Fourr. subsp. praecox（Grosser）Quézel et Santa ■☆

415629　Xolantha macrosepala（Boiss.）Gallego et Munoz Garm. = Tuberaria macrosepala（Boiss.）Willk. ■☆

415630　Xolantha plantaginea（Willd.）Gallego et Munoz Garm. et Navarro = Tuberaria inconspicua（Pers.）Willk. ■☆

415631　Xolantha praecox（Boiss. et Reut.）Gallego et Munoz Garm. = Tuberaria guttata（L.）Fourr. subsp. praecox（Grosser）Quézel et Santa ■☆

415632　Xolantha tuberaria（L.）Gallego et Munoz Garm. et C. Navarro = Tuberaria lignosa（Sweet）Samp. ■☆

415633　Xolanthes Raf. = Helianthemum Mill. ●■

415634　Xolanthes Raf. = Tuberaria（Dunal）Spach（保留属名）■☆

415635　Xolanthes Raf. = Xolantha Raf.（废弃属名）●■

415636　Xolemia Raf. = Gentiana L. ■

415637　Xolisma Raf. = Lyonia Nutt.（保留属名）●

415638　Xolisma compta（W. W. Sm. et Jeffrey）Rehder = Lyonia compta（W. W. Sm. et Jeffrey）Hand. -Mazz. ●

415639　Xolisma compta（W. W. Sm. et Jeffrey）Rehder = Lyonia ovalifolia（Wall.）Drude var. lanceolata（Wall.）Hand. -Mazz. ●

415640　Xolisma elliptica（Siebold et Zucc.）Nakai = Lyonia ovalifolia（Wall.）Drude var. elliptica（Siebold et Zucc.）Hand. -Mazz. ●

415641　Xolisma formosana（Komatsu）Nakai = Lyonia ovalifolia（Wall.）Drude var. elliptica（Siebold et Zucc.）Hand. -Mazz. ●

415642　Xolisma formosana（Komatsu）Nakai var. pilosa（Komatsu）Nakai = Lyonia ovalifolia（Wall.）Drude var. elliptica（Siebold et Zucc.）Hand. -Mazz. ●

415643　Xolisma formosana var. pilosa（Komatsu）Nakai = Lyonia ovalifolia（Wall.）Drude var. elliptica（Siebold et Zucc.）Hand. -Mazz. ●

415644　Xolisma ovalifolia（Wall.）Drude var. elliptica（Siebold et Zucc.）Rehder = Lyonia ovalifolia（Wall.）Drude var. elliptica（Siebold et Zucc.）Hand. -Mazz. ●

415645　Xolisma ovalifolia（Wall.）Rehder = Lyonia ovalifolia（Wall.）Drude ●

415646　Xolisma ovalifolia（Wall.）Rehder var. elliptica（Siebold et Zucc.）Rehder = Lyonia ovalifolia（Wall.）Drude var. elliptica（Siebold et Zucc.）Hand. -Mazz. ●

415647　Xolisma ovalifolia（Wall.）Rehder var. hebecarpa（Franch. ex Forbes et Hemsl.）F. P. Metcalf = Lyonia ovalifolia（Wall.）Drude var. hebecarpa（Franch. ex Forbes et Hemsl.）Chun ●

415648　Xolisma ovalifolia（Wall.）Rehder var. lanceolata（Wall.）Rehder = Lyonia ovalifolia（Wall.）Drude var. lanceolata（Wall.）Hand. -Mazz. ●

415649　Xolisma sphaerantha Hand. -Mazz. = Lyonia ovalifolia（Wall.）Drude var. sphaerantha（Hand. -Mazz.）Hand. -Mazz. ●

415650　Xolisma sphaerantha Hand. -Mazz. = Lyonia villosa（Hook. f. ex C. B. Clarke）Hand. -Mazz. var. sphaerantha（Hand. -Mazz.）Hand. -Mazz. ●

415651　Xolisma villosa（Wall. ex C. B. Clarke）Rehder = Lyonia villosa（Hook. f. ex C. B. Clarke）Hand. -Mazz. ●

415652　Xolisma villosa（Wall. ex C. B. Clarke）Rehder var. pubescens（Franch.）Rehder = Lyonia villosa（Hook. f. ex C. B. Clarke）Hand. -Mazz. ●

415653　Xolisma vollosa（Wall. ex C. B. Clarke）Rehder var. pubescenes（Franch.）Rehder = Lyonia villosa（Hook. f. ex C. B. Clarke）Hand. -Mazz. var. pubescens（Franch.）Judd ●

415654　Xolocotzia Miranda（1965）;墨西哥鞭木属●☆

415655　Xolocotzia asperifolia Miranda;墨西哥鞭木●☆

415656　Xoxylon Raf. = Maclura Nutt.（保留属名）●

415657　Xoxylon Raf. = Toxylon Raf. ●

415658　Xuaresia Pers. = Capraria L. ■☆

415659　Xuaresia Pers. = Xuarezia Ruiz et Pav. ■☆

415660　Xuarezia Ruiz et Pav. = Capraria L. ■☆

415661　Xuris Adans. = Iris L. ■

415662　Xuris Adans. = Xyris L. ■

415663　Xuris Raf. = Xyris L. ■

415664　Xyladenius Desv. = Banara Aubl. ●☆

415665　Xyladenius Desv. ex Ham. = Banara Aubl. ●☆

415666　Xylanche Beck = Boschniakia C. A. Mey. ex Bong. ■

415667　Xylanche Beck（1893）;丁座草属（千斤坠属）■

415668　Xylanche himalaica（Hook. f. et Thomson）Beck = Boschniakia himalaica Hook. f. et Thomson ■

415669　Xylanche himalaica Beck = Boschniakia himalaica Hook. f. et Thomson ■

415670　Xylanche kawakamii（Hayata）Beck = Boschniakia himalaica Hook. f. et Thomson ■

415671　Xylanthema Neck. = Cirsium Mill. ■

415672　Xylanthemum Tzvelev（1961）;木花菊属（木菊属）●☆

415673　Xylanthemum fischerae（Aitch. et Hemsl.）Tzvelev;菲舍尔木花菊■☆

415674　Xylanthemum macropodum（Hemsl. et Lace）K. Bremer et Humphries;大足木花菊■☆

415675　Xylanthemum pamiricum（Hoffm.）Tzvelev;帕米尔木花菊■☆

415676　Xylanthemum rupestre（Popov）Tzvelev;湿地木花菊■☆

415677　Xylanthemum tianschanicum（Krasch.）Muradyan;天山木花菊☆

415678　Xylia Benth.（1842）;木荚豆属●

415679　Xylia Benth. = Esclerona Raf. ●

415680　Xylia africana Harms;非洲木荚豆●☆

415681　Xylia dinklagei（Harms）Roberty = Calpocalyx dinklagei Harms ●☆

415682　Xylia dolabriformis Benth. = Xylia xylocarpa（Roxb.）W. Theob. ●

415683　Xylia evansii Hutch. ;埃文斯木荚豆●☆

415684　Xylia fraterna（Vatke）Baill. ;兄弟木荚豆●☆

415685　Xylia ghesquierei Robyns;盖斯基埃木荚豆●☆

415686　Xylia hildebrandtii Baill. = Xylia hoffmannii（Vatke）Drake ●☆

415687　Xylia hoffmannii（Vatke）Drake;豪夫曼木荚豆●☆

415688　Xylia longipes Baill. = Xylia fraterna（Vatke）Baill. ●☆

415689　Xylia mendoncae Torre;门东萨木荚豆●☆

415690　Xylia perrieri Drake = Xylia hoffmannii（Vatke）Drake ●☆

415691　Xylia schliebenii Harms;施利本木荚豆●☆

415692　Xylia torreana Brenan;托尔木荚豆●☆

415693　Xylia xylocarpa（Roxb.）W. Theob. ;木荚豆（柄木, 缅甸铁木）;Burma Pyingndo, Ironwood, Pyinkado ●

415694　Xylinabaria Pierre = Urceola Roxb.（保留属名）●

415695　Xylinabaria reynaudii Jum. = Urceola napeensis（Quintaret）D. J. Middleton ●

415696　Xylinabariopsis Lý = Urceola Roxb.（保留属名）●

415697　Xylinabariopsis Pit. = Ecdysanthera Hook. et Arn. ●

415698　Xylinabariopsis napeensis（Quint.）F. P. Metcalf = Urceola napeensis（Quintaret）D. J. Middleton ●

415699　Xylinabariopsis reynaudii（Jum.）Pit. = Urceola napeensis（Quintaret）D. J. Middleton ●

415700　Xylinabariopsis ventii Lý = Chunechites xylinabariopsoides Tsiang ●

415701　Xylinabariopsis ventii Lý = Urceola xylinabariopsoides（Tsiang）D. J. Middleton ●

415702　Xylinabariopsis xylinabariopsoides（Tsiang）Lý = Urceola xylinabariopsoides（Tsiang）D. J. Middleton ●

415703　Xylinabariupsis Pit. = Urceola Roxb.（保留属名）●

415704　Xylobium Lindl.（1825）;西劳兰属;Xylobium ■☆

415705　Xylobium decolor Nicholson;无色西劳兰■☆

415706　Xylobium elongatum（Lindl. et Paxton）Hemsl. ;长茎西劳兰;Longstem Xylobium ■☆

415707　Xylobium fovestum（Lindl.）Nicholson;落叶西劳兰;Deciduous Xylobium ■☆

415708　Xylobium palmifolium（Sw.）Fawc. ;棕叶西劳兰;Palmleaf Xylobium ■☆

415709　Xylobium squalens Lindl. ;黄色西劳兰■☆

415710　Xylocalyx Balf. f.（1883）;木萼列当属●☆

415711　Xylocalyx carterae Thulin;卡特拉木萼列当●☆

415712　Xylocalyx hispidus S. Carter;硬毛木萼列当●☆

415713　Xylocalyx recurvus S. Carter;反折木萼列当●☆

415714　Xylocarpus J. König（1784）;木果楝属;Xylocarpus ●

415715　Xylocarpus benadirensis Mattei = Xylocarpus granatum J. König ●

415716　Xylocarpus granatum J. König;木果楝（海柚）;Common Xylocarpus ●

415717　Xylocarpus moluccensis（Lam.）M. Roem. ;马六甲木果楝●☆

415718　Xylocarpus obovatus Juss. = Xylocarpus granatum J. König ●

415719　Xylochlaena Dalla Torre et Harms = Scleroolaena Baill. ●☆

415720　Xylochlaena Dalla Torre et Harms = Xyloolaena Baill. ●☆

415721　Xylochlamys Domin = Amyema Tiegh. ●☆

415722　Xylococcus Nutt. = Arctostaphylos Adans.（保留属名）●☆

415723　Xylococcus R. Br. ex Britten et S. Moore = Petalostigma F. Muell. ●☆

415724　Xylolaena Baill. = Scleroolaena Baill. ●☆

415725　Xylolaena Baill. = Xyloolaena Baill. ●☆

415726　Xylolobus Kuntze = Esclerona Raf. ●

415727　Xylolobus Kuntze = Xylia Benth. ●

415728　Xylomelum Sm.（1798）;木果山龙眼属;Woody Pear, Woody-pear ●☆

415729　Xylomelum occidentale R. Br. ;西方木果山龙眼;Western Woody-pear ●☆

415730　Xylomelum pyriforme（Gaertn.）Knight;木果山龙眼;Eastern Woody-pear, Woody Pear ●☆

415731　Xylon Kuntze = Bombax L.（保留属名）●

415732　Xylon L. = Ceiba Mill. ●

415733　Xylon Mill. = Gossypium L. ●■

415734　Xylon pentandrum（L.）Kuntze = Ceiba pentandra（L.）Gaertn. ●

415735　Xylonagra Donn. Sm. et Rose（1913）;加州月见草属■☆

415736　Xylonagra arborea（Kellogg）J. D. Sm. et Rose;加州月见草■☆

415737　Xylonymus Kalkman ex Ding Hou = Xylonymus Kalkman ●☆

415738　Xylonymus Kalkman（1963）;木果卫矛属●☆

415739　Xylonymus versteeghii Kalkman;木果卫矛●☆

415740　Xyloolaena Baill.（1886）;木苞杯花属●☆

415741　Xyloolaena humbertii Cavaco;亨伯特木苞杯花●☆

415742　Xyloolaena perrieri F. Gérard;佩里耶木苞杯花●☆

415743　Xyloolaena richardii（Baill.）Baill.;理查德木苞杯花☆

415744　Xyloolaena sambiranensis Lowry et G. E. Schatz;马岛木苞杯花●☆

415745　Xyloolaena speciosa Lowry et G. E. Schatz;美丽木苞杯花●☆

415746　Xylophacos Rydb. = Astragalus L. ●■

415747　Xylophacos Rydb. ex Small = Astragalus L. ●■

415748　Xylophragma Sprague（1903）;木栅紫葳属●☆

415749　Xylophragma myrianthum Sprague;木栅紫葳●☆

415750　Xylophragma xanthophyllum（Bur. et K. Schum.）J. F. Macbr. ;黄叶木栅紫葳●☆

415751　Xylophylla L.（废弃属名）= Exocarpos Labill.（保留属名）●☆

415752　Xylophylla L.（废弃属名）= Phyllanthus L. + Exocarpos Labill. ●■

415753　Xylophylla L.（废弃属名）= Phyllanthus L. ●■

415754　Xylophylla ensifolia Bojer ex Drake = Phylloxylon xylophylloides（Baker）Du Puy，Labat et Schrire ●☆

415755　Xylophylla obovata Willd. = Flueggea virosa（Roxb. ex Willd.）Voigt ●

415756　Xylophylla ramiflora Aiton = Flueggea suffruticosa（Pall.）Baill. ●

415757　Xylophyllos Kuntze = Exocarpos Labill.（保留属名）●☆

415758　Xylophyllos Rumph. = Exocarpos Labill.（保留属名）●☆

415759　Xylophyllos Rumph. ex Kuntze = Exocarpos Labill.（保留属名）●☆

415760　Xylopia L.（1759）（保留属名）;木瓣树属;Xylopia ●

415761　Xylopia acutiflora（Dunal）A. Rich.;尖花木瓣树●☆

415762　Xylopia aethiopica（Dunal）A. Rich.;埃塞俄比亚木瓣树;African Pepper, Ethiopian Pepper, Guinea Pepper, Guinea Pepper-tree，Negro Pepper ●☆

415763　Xylopia africana（Benth.）Oliv. ;非洲木瓣树●☆

415764　Xylopia antunesii Engl. et Diels = Xylopia odoratissima Welw. ex Oliv. ●☆

415765　Xylopia antunesii Engl. et Diels var. shirensis ? = Xylopia parviflora（A. Rich.）Benth. ●☆

415766　Xylopia arenaria Engl.;沙地木瓣树●☆

415767　Xylopia aurantiiodora De Wild. et T. Durand;橘齿木瓣树●☆

415768　Xylopia bequaertii De Wild. = Xylopia aurantiiodora De Wild. et T. Durand ●☆

415769　Xylopia bokoli De Wild. et T. Durand = Monanthotaxis bokoli（De Wild. et T. Durand）Verdc. ●☆

415770　Xylopia brieyi De Wild. = Xylopia hypolampra Mildbr. ●☆

415771　Xylopia butayei De Wild. = Xylopia rubescens Oliv. ●☆

415772　Xylopia chrysophylla Louis ex Boutique = Xylopia cupularis Mildbr. ●☆

415773　Xylopia collina Diels;考林木瓣树●☆

415774　Xylopia congolensis De Wild. ;刚果木瓣树●☆

415775　Xylopia cupularis Mildbr.;杯状木瓣树●☆

415776　Xylopia dekeyzeriana De Wild. = Xylopia aethiopica（Dunal）A. Rich. ●☆

415777　Xylopia dinklagei Engl. et Diels = Xylopia acutiflora（Dunal）A. Rich. ●☆

415778　Xylopia discreta（L. f.）Sprague et Hutch. ;分离木瓣树;Separated Xylopia ●☆

415779　Xylopia elliotii Engl. et Diels;埃利木瓣树●☆

415780　Xylopia eminii Engl. = Xylopia aethiopica（Dunal）A. Rich. ●☆

415781　Xylopia flamignii Boutique;弗拉米尼木瓣树●☆

415782　Xylopia gilbertii Boutique;吉尔伯特木瓣树●☆

415783　Xylopia gilletii De Wild. = Xylopia aethiopica（Dunal）A. Rich. ●☆

415784　Xylopia gilviflora Exell = Xylopia cupularis Mildbr. ●☆

415785　Xylopia gossweileri Exell = Xylopia rubescens Oliv. ●☆

415786　Xylopia holtzii Engl. = Xylopia parviflora（A. Rich.）Benth. ●☆

415787　Xylopia humilis Engl. et Diels = Xylopia rubescens Oliv. ●☆

415788　Xylopia hypolampra Mildbr. ;喀麦隆木瓣树●☆

415789　Xylopia katangensis De Wild. ;加丹加木瓣树●☆

415790　Xylopia katangensis De Wild. var. gillardinii Boutique = Xylopia katangensis De Wild. ●☆

415791　Xylopia klaineana Pierre ex Engl. et Diels = Xylopia rubescens Oliv. var. klaineana Pellegr. ●☆

415792　Xylopia lanepoolei Sprague et Hutch. = Xylopia quintasii Pierre ex Engl. et Diels ●☆

415793　Xylopia latipetala Verdc. ;宽瓣木瓣树●☆

415794　Xylopia latoursvillei Pellegr. = Xylopia mildbraedii Diels ●☆

415795　Xylopia letestui Pellegr. ;莱泰斯图木瓣树●☆

415796　Xylopia letestui Pellegr. var. longepilosa Le Thomas;长毛莱泰斯图木瓣树●☆

415797　Xylopia longipetala De Wild. et T. Durand = Xylopia parviflora（A. Rich.）Benth. ●☆

415798　Xylopia macrocarpa A. Chev. = Xylopia villosa Chipp ●☆

415799　Xylopia mayombensis De Wild. = Xylopia staudtii Engl. et Diels ●☆

415800　Xylopia mendoncae Exell = Xylopia tomentosa Exell ●☆

415801　Xylopia mildbraedii Diels;米尔德木瓣树●☆

415802　Xylopia odoratissima Welw. ex Oliv. ;苦木瓣树●☆

415803　Xylopia odoratissima Welw. ex Oliv. var. minor Engl. = Xylopia tomentosa Exell ●☆

415804　Xylopia otunga Exell = Polyalthia suaveolens Engl. et Diels ●☆

415805　Xylopia oxypetala（DC.）Oliv. = Xylopia acutiflora（Dunal）A. Rich. ●☆

415806　Xylopia paniculata Exell;圆锥木瓣树●☆

415807　Xylopia parviflora（A. Rich.）Benth. ;小花木瓣树●☆

415808　Xylopia polycarpa（DC.）Oliv. = Annickia polycarpa（DC.）Setten et Maas ●☆

415809　Xylopia polycarpa Oliv. ;多果木瓣树●☆

415810　Xylopia pynaertii De Wild. ;皮那木瓣树●☆

415811　Xylopia pyrifolia Engl. ;梨叶木瓣树●☆

415812　Xylopia quintasii Engl. et Diels = Xylopia quintasii Pierre ex Engl. et Diels ●☆

415813　Xylopia quintasii Pierre ex Engl. et Diels;昆塔木瓣树☆

415814　Xylopia rubescens Oliv. ;变红木瓣树☆

415815　Xylopia rubescens Oliv. var. klaineana Pellegr. ;克莱恩木瓣树●☆

415816　Xylopia seretii De Wild. = Xylopia acutiflora（Dunal）A. Rich. ●☆

415817　Xylopia staudtii Engl. et Diels;施陶木瓣树●☆

415818　Xylopia striata Engl. = Xylopia quintasii Pierre ex Engl. et Diels ●☆

415819　Xylopia talbotii Exell;塔尔博特木瓣树●☆

415820　Xylopia thomsonii Oliv. = Xylopia acutiflora（Dunal）A. Rich. ●☆

415821　Xylopia tomentosa Exell;绒毛木瓣树●☆

415822　Xylopia torrei N. Robson;托雷木瓣树●☆

415823　Xylopia toussaintii Boutique;图森特木瓣树●☆

415824　Xylopia undulata P. Beauv. = Xylopia aethiopica（Dunal）A. Rich. ●☆

415825　Xylopia vallotii Chipp ex Exell = Xylopia parviflora（A. Rich.）Benth. ●☆

415826　Xylopia vielana Pierre;木瓣树;Common Xylopia，Xylopia ●

415827　Xylopia villosa Chipp;长毛木瓣树●☆

415828　Xylopia wilwerthii De Wild. et T. Durand;维尔沃斯木瓣树●☆

415829　Xylopia wilwerthii De Wild. et T. Durand var. cuneata ? =

Xylopia wilwerthii De Wild. et T. Durand ●☆

415830　Xylopia zenkeri Engl. et Diels = Xylopia rubescens Oliv. ●☆

415831　Xylopiastrum Robert = Xylopia L. (保留属名)●

415832　Xylopiastrum Roberty = Uvaria L. ●

415833　Xylopiastrum villosum (Chipp) Aubrév. =Xylopia villosa Chipp ●☆

415834　Xylopicron Adans. = Xylopia L. (保留属名)●

415835　Xylopicrum P. Browne(废弃属名) = Xylopia L. (保留属名)●

415836　Xylopicrum aethiopicum (Dunal) Kuntze = Xylopia aethiopica (Dunal) A. Rich. ●☆

415837　Xylopicrum africanum Kuntze = Xylopia africana (Benth.) Oliv. ●☆

415838　Xylopicrum odoratissimum (Welw. ex Oliv.) Kuntze = Xylopia odoratissima Welw. ex Oliv. ●☆

415839　Xylopleurum Spach = Oenothera L. ●■

415840　Xylopodia Weigend(1997);秘鲁刺莲花属●☆

415841　Xylopodia klaprothioides Weigend;秘鲁刺莲花●☆

415842　Xylorhiza Nutt. (1840);木根菊属;Woody-aster ●■

415843　Xylorhiza Nutt. = Machaeranthera Nees ■☆

415844　Xylorhiza Salisb. = Allium L. ■

415845　Xylorhiza brandegeei Rydb. = Xanthisma coloradoense (A. Gray) D. R. Morgan and R. L. Hartm. ■☆

415846　Xylorhiza cognata (H. M. Hall) T. J. Watson;近缘木根菊;Mecca Woody-aster ●☆

415847　Xylorhiza confertifolia (Cronquist) T. J. Watson;针叶木根菊;Henrieville Woody-aster ●☆

415848　Xylorhiza cronquistii S. L. Welsh et N. D. Atwood;克朗木根菊;Cronquist's Woody-aster ●☆

415849　Xylorhiza glabriuscula Nutt. ;无毛木根菊;Smooth Woody-aster ●☆

415850　Xylorhiza glabriuscula Nutt. var. linearifolia T. J. Watson = Xylorhiza linearifolia (T. J. Watson) G. L. Nesom ●☆

415851　Xylorhiza glabriuscula Nutt. var. villosa (Nutt.) A. Nelson = Xylorhiza glabriuscula Nutt. ●☆

415852　Xylorhiza linearifolia (T. J. Watson) G. L. Nesom;摩押木根菊;Moab Woody-aster ●☆

415853　Xylorhiza orcuttii (Vasey et Rose) Greene;奥科特木根菊;Orcutt's Woody-aster ●☆

415854　Xylorhiza tortifolia (Torr. et A. Gray) Greene;莫哈韦木根菊;Mojave Woody-aster ●☆

415855　Xylorhiza tortifolia (Torr. et A. Gray) Greene var. imberbis (Cronquist) T. J. Watson;光滑莫哈韦木根菊;Smooth Mojave Woody-aster ●☆

415856　Xylorhiza venusta (M. E. Jones) A. Heller;娇媚木根菊;Charming Woody-aster ●■☆

415857　Xylorhiza villosa Nutt. = Xylorhiza glabriuscula Nutt. ●☆

415858　Xylorhiza wrightii (A. Gray) Greene;赖特木根菊;Big Bend Woody-aster ●■☆

415859　Xylosalsola Tzvelev = Salsola L. ●■

415860　Xylosalsola arbuscula (Pall.) Tzvelev = Salsola arbuscula Pall. ●

415861　Xyloselinum Pimenov et Kljuykov(2006);越南蛇床属■☆

415862　Xyloselinum vietnamense Pimenov et Kljuykov;越南蛇床■☆

415863　Xylosma G. Forst. (1786)(保留属名);柞木属;Manzanilla, Manzanillo,Xylosma ●

415864　Xylosma Harv. = Xymalos Baill. ●☆

415865　Xylosma J. R. Forst. et G. Forst. = Xylosma G. Forst. (保留属名)●

415866　Xylosma apactis Koidz. = Xylosma congesta (Lour.) Merr. ●

415867　Xylosma apactis Koidz. = Xylosma japonica (Thunb.) A. Gray ●

415868　Xylosma apactis Koidz. = Xylosma racemosa (Siebold et Zucc.) Miq. ●

415869　Xylosma congesta (Lour.) Merr. ;柞木(刺凿,刺柞,冬青,冻青树,孤奴,红蒙子根,红心刺,葫芦刺,蒙子树,鼠木,凿树,凿子木,凿子树,柞树);Congested Xylosma, Dense Logwood, Japanese Xylosma,Shiny Xylosma,Xylosma ●

415870　Xylosma congesta (Lour.) Merr. = Xylosma japonica (Thunb.) A. Gray ●

415871　Xylosma congesta (Lour.) Merr. = Xylosma racemosa (Siebold et Zucc.) Miq. ●

415872　Xylosma congesta (Lour.) Merr. var. caudata S. S. Lai = Xylosma congesta (Lour.) Merr. ●

415873　Xylosma congesta (Lour.) Merr. var. caudata S. S. Lai = Xylosma racemosa (Siebold et Zucc.) Miq. var. caudata (S. S. Lai) S. S. Lai ●

415874　Xylosma congesta (Lour.) Merr. var. kwangtungensis F. P. Metcalf = Xylosma longifolia Clos ●

415875　Xylosma congesta (Lour.) Merr. var. pubescens (Rehder et E. H. Wilson) Chun = Xylosma racemosa (Siebold et Zucc.) Miq. var. glaucescens Franch. ●

415876　Xylosma congesta (Lour.) Merr. var. pubescens (Rehder et E. H. Wilson) Chun = Xylosma congesta (Lour.) Merr. ●

415877　Xylosma controversa Clos;南岭柞木(岭南柞木);S. China Xylosma,South China Xylosma ●

415878　Xylosma controversa Clos var. glabra S. S. Lai;光叶柞木;Glabrous South China Xylosma ●

415879　Xylosma controversa Clos var. pubescens Q. E. Yang;毛叶南岭柞木●

415880　Xylosma elliptica Tul. = Flacourtia indica (Burm. f.) Merr. ●

415881　Xylosma fasciculiflora S. S. Lai;丛花柞木;Denseflower Xylosma ●

415882　Xylosma flanaganii Bolus = Scolopia flanaganii (Bolus) Sim ●☆

415883　Xylosma japonica (Thunb.) A. Gray = Xylosma racemosa (Siebold et Zucc.) Miq. ●

415884　Xylosma japonica (Walp.) A. Gray = Xylosma racemosa (Siebold et Zucc.) Miq. ●

415885　Xylosma japonica (Walp.) A. Gray var. pubescens (Rehder et E. H. Wilson) C. Yu Chang = Xylosma racemosa (Siebold et Zucc.) Miq. var. glaucescens Franch. ●

415886　Xylosma japonica A. Gray = Xylosma congesta (Lour.) Merr. ●

415887　Xylosma japonica A. Gray = Xylosma racemosa (Siebold et Zucc.) Miq. ●

415888　Xylosma japonica A. Gray var. pubescens (Rehder et E. H. Wilson) C. Y. Chang = Xylosma congesta (Lour.) Merr. ●

415889　Xylosma laxiflora Merr. et Chun = Xylosma controversa Clos var. glabra S. S. Lai ●

415890　Xylosma leprosipes Clos = Bennettiodendron leprosipis (Clos) Merr. ●

415891　Xylosma longifolia Clos;长叶柞木(长叶榨木,跌破笋,狗牙木,笋齿树,笋凿树,簕凿树,耙齿木,小角刺);Longleaf Xylosma, Long-leaved Xylosma ●

415892　Xylosma longipes Oliv. = Bennettiodendron leprosipis (Clos) Merr. ●

415893　Xylosma maximowiczii Rupr. = Lonicera maximowiczii (Rupr. ex Maxim.) Rupr. ex Maxim. ●

415894　Xylosma monospora Harv. = Xymalos monospora (Harv.) Baill. ●☆

415895　Xylosma racemosa (Siebold et Zucc.) Miq. = Xylosma congesta (Lour.) Merr. ●

415896　Xylosma racemosa（Siebold et Zucc.）Miq. = Xylosma japonica（Thunb.）A. Gray ●

415897　Xylosma racemosa（Siebold et Zucc.）Miq. var. caudata（S. S. Lai）S. S. Lai = Xylosma congesta（Lour.）Merr. ●

415898　Xylosma racemosa（Siebold et Zucc.）Miq. var. caudata（S. S. Lai）S. S. Lai;尾叶柞木;Caudate Japanese Xylosma ●

415899　Xylosma racemosa（Siebold et Zucc.）Miq. var. glaucescens Franch.;毛枝柞木（檬子刺根,檬子树,蒙子树,柔毛柞木,柞木）;Hairy-branch Japanese Xylosma ●

415900　Xylosma racemosa（Siebold et Zucc.）Miq. var. glaucescens Franch. = Xylosma congesta（Lour.）Merr. ●

415901　Xylosma racemosa（Siebold et Zucc.）Miq. var. kwangtungensis（F. P. Metcalf）Rehder = Xylosma longifolia Clos ●

415902　Xylosma racemosa（Siebold et Zucc.）Miq. var. pubescens Rehder et E. H. Wilson = Xylosma racemosa（Siebold et Zucc.）Miq. var. glaucescens Franch. ●

415903　Xylosma racemosa（Siebold et Zucc.）Miq. var. pubescens Rehder et E. H. Wilson = Xylosma congesta（Lour.）Merr. ●

415904　Xylosma senticosa Hance = Xylosma congesta（Lour.）Merr. ●

415905　Xylosteon Adans. = Lonicera L. ●■

415906　Xylosteon Mill. = Lonicera L. ●■

415907　Xylosteon Tourn. ex Adans. = Lonicera L. ●■

415908　Xylosteon ciliatum Pursh = Lonicera canadensis Marshall ex Roem. et Schult. ●

415909　Xylosteon sieversiana Rupr. = Lonicera microphylla Willd. ex Roem. et Schult. ●

415910　Xylosterculia Kosterm. = Sterculia L. ●

415911　Xylosteum Rupr. = Xylosteon Adans. ●■

415912　Xylosteum chrysanthum var. subtomentosum Rupr. = Lonicera ruprechtiana Regel ●

415913　Xylosteum involucratum Richardson = Lonicera involucrata（Rich.）Banks ex Spreng. ●☆

415914　Xylosteum karelini Kuntze = Lonicera heterophylla Decne. ●

415915　Xylosteum karelini Rupr. = Lonicera heterophylla Decne. ●

415916　Xylosteum ligustrinum（Wall.）D. Don = Lonicera ligustrina Wall. ●

415917　Xylosteum ligustrinum D. Don = Lonicera ligustrina Wall. ●

415918　Xylosteum maackii Rupr. = Lonicera maackii（Rupr.）Maxim. ●

415919　Xylosteum maximowiczii Rupr. = Lonicera maximowiczii（Rupr. ex Maxim.）Rupr. ex Maxim. ●

415920　Xylosteum sieversianum Rupr. = Lonicera microphylla Willd. ex Roem. et Schult. ●

415921　Xylosteum spinosum（Jacquem. ex Walp.）Decne. = Lonicera spinosa Jacquem. ex Walp. ●

415922　Xylosteum spinosum Decne. = Lonicera spinosa Jacquem. ex Walp. ●

415923　Xylosteum tataricum（L.）Medik. = Lonicera tatarica L. ●

415924　Xylothamia G. L. Nesom, Y. B. Suh, D. R. Morgan et B. B. Simpson = Gundlachia A. Gray ●☆

415925　Xylothamia palmeri（A. Gray）G. L. Nesom = Neonesomia palmeri（A. Gray）Urbatsch et R. P. Roberts ●☆

415926　Xylothamia triantha（S. F. Blake）G. L. Nesom = Haplopappus trianthus S. F. Blake ■☆

415927　Xylotheca Hochst.（1843）;木果大风子属●☆

415928　Xylotheca fissistyla（Warb.）Gilg = Oncoba tettensis（Klotzsch）Harv. var. fissistyla（Warb.）Hul et Breteler ●☆

415929　Xylotheca glutinosa Gilg = Oncoba tettensis（Klotzsch）Harv.

var. kirkii（Oliv.）Hul et Breteler ●☆

415930　Xylotheca holtzii Gilg = Oncoba tettensis（Klotzsch）Harv. var. fissistyla（Warb.）Hul et Breteler ●☆

415931　Xylotheca kirkii（Oliv.）Gilg = Oncoba tettensis（Klotzsch）Harv. var. kirkii（Oliv.）Hul et Breteler ●☆

415932　Xylotheca kotzei E. Phillips = Oncoba kraussiana（Hochst.）Planch. ●☆

415933　Xylotheca kraussiana Hochst. = Oncoba kraussiana（Hochst.）Planch. ●☆

415934　Xylotheca kraussiana Hochst. var. glabrifolia Wild = Oncoba kraussiana（Hochst.）Planch. ●☆

415935　Xylotheca lasiopetala Gilg = Oncoba kraussiana（Hochst.）Planch. ●☆

415936　Xylotheca longipes（Gilg）Gilg;长梗木果大风子●☆

415937　Xylotheca macrophylla（Klotzsch）Sleumer = Oncoba tettensis（Klotzsch）Harv. var. macrophylla（Klotzsch）Hul et Breteler ●☆

415938　Xylotheca stuhlmannii（Gürke）Gilg = Oncoba tettensis（Klotzsch）Harv. var. macrophylla（Klotzsch）Hul et Breteler ●☆

415939　Xylotheca sulcata Gilg = Oncoba tettensis（Klotzsch）Harv. var. fissistyla（Warb.）Hul et Breteler ●☆

415940　Xylotheca tettensis（Klotzsch）Gilg;泰特木果大风子●☆

415941　Xylotheca tettensis（Klotzsch）Gilg = Oncoba tettensis（Klotzsch）Harv. ●☆

415942　Xylotheca tettensis（Klotzsch）Gilg var. fissistyla（Warb.）Sleumer = Oncoba tettensis（Klotzsch）Harv. var. fissistyla（Warb.）Hul et Breteler ●☆

415943　Xylotheca tettensis（Klotzsch）Gilg var. kirkii（Oliv.）Wild = Oncoba tettensis（Klotzsch）Harv. var. kirkii（Oliv.）Hul et Breteler ●☆

415944　Xylotheca tettensis（Klotzsch）Gilg var. macrophylla（Klotzsch）Wild = Oncoba tettensis（Klotzsch）Harv. var. macrophylla（Klotzsch）Hul et Breteler ●☆

415945　Xylothermia Greene = Pickeringia Nutt.（保留属名）●☆

415946　Xylovirgata Urbatsch et R. P. Roberts = Haplopappus Cass.（保留属名）●■☆

415947　Xylovirgata Urbatsch et R. P. Roberts(2004);帚黄花属●☆

415948　Xylovirgata pseudobaccharis（S. F. Blake）Urbatsch et R. P. Roberts;帚黄花■☆

415949　Xylum Post et Kuntze = Ceiba Mill. ●

415950　Xymalobium Steud. = Xysmalobium R. Br. ■☆

415951　Xymalos Baill.（1887）;单心桂属●☆

415952　Xymalos Baill. et Warb. = Xymalos Baill. ●☆

415953　Xymalos monospora（Harv.）Baill.;单心桂●☆

415954　Xymalos monospora Baill. = Xymalos monospora（Harv.）Baill. ●☆

415955　Xymalos mossambicensis Cavaco;莫桑比克单心桂;Lemon-wood ●☆

415956　Xymalos ulugurensis（Engl.）Engl. = Xymalos monospora（Harv.）Baill. ●☆

415957　Xymalos usambarensis（Engl.）Engl. = Xymalos monospora（Harv.）Baill. ●☆

415958　Xynophylla Montrouz. = Exocarpos Labill.（保留属名）●☆

415959　Xynophylla Montrouz. = Xylophylla L.（废弃属名）●■

415960　Xyochlaena Stapf = Tricholaena Schrad. ex Schult. et Schult. f. ■☆

415961　Xyochlaena monachne（Trin.）Stapf = Tricholaena monachne（Trin.）Stapf et C. E. Hubb. ■☆

415962　Xyochlaena vestita（Balf. f.）Stapf = Tricholaena vestita（Balf. f.）Stapf et C. E. Hubb. ■☆

415963　Xyphanthus Raf. = Erythrina L. ●■

415964　Xypherus Raf. = Amphicarpaea Elliott ex Nutt.（保留属名）■

415965　Xyphidium Neck. = Xiphidium Aubl. ■☆

415966　Xyphidium Steud. = Iris L. ■

415967　Xyphidium Steud. = Xiphion Mill. ■

415968　Xyphion Medik. = Xyphidium Steud. ■

415969　Xyphion filifolium（Boiss.）Baker = Iris filifolia Boiss. ■☆

415970　Xyphion filifolium（Boiss.）Baker var. intermedium Baker = Iris filifolia Boiss. ■☆

415971　Xyphion foetidissimum（L.）Parl. = Iris foetidissima L. ■☆

415972　Xyphion pseudacorus（L.）Parl. = Iris pseudacorus L. ■

415973　Xyphion tingitanum（Boiss. et Reut.）Baker = Iris tingitana Boiss. et Reut. ■☆

415974　Xyphostylis Raf. = Canna L. ■

415975　Xyridaceae C. Agardh（1823）（保留科名）；黄眼草科（黄谷精科,芴草科）；Yellow-eyed-grass Family, Yelloweyegrass Family ■

415976　Xyridanthe Lindl. = Helipterum DC. ex Lindl. ■☆

415977　Xyridanthe Lindl. = Rhodanthe Lindl. ●■☆

415978　Xyridion（Tausch）Fourr. = Iris L. ■

415979　Xyridion Fourr. = Iris L. ■

415980　Xyridium Steud. = Xyridion Fourr. ■

415981　Xyridium Tausch ex Steud. = Xyridion Fourr. ■

415982　Xyridopsis B. Nord. = Psednotrichia Hiern ■☆

415983　Xyridopsis Welw. ex B. Nord.（1978）；鸢尾菊属■☆

415984　Xyridopsis Welw. ex B. Nord. = Emilia（Cass.）Cass. ■

415985　Xyridopsis Welw. ex O. Hoffm. = Oligothrix DC. ■☆

415986　Xyridopsis newtonii（O. Hoffm.）B. Nord. = Psednotrichia newtonii（O. Hoffm.）Anderb. et P. O. Karis ■☆

415987　Xyridopsis welwitschii B. Nord.；鸢尾菊■☆

415988　Xyridopsis welwitschii B. Nord. = Psednotrichia xyridopsis（O. Hoffm.）Anderb. et P. O. Karis ■☆

415989　Xyris Gronov. ex L. = Xyris L. ■

415990　Xyris L.（1753）；黄眼草属（黄谷精属,芴草属）；Morning Yellow-eyed-grass, Sword Plant, Yellow eye, Yellow-eyed Grass, Yellow-eyed-grasses, Yelloweyegrass ■

415991　Xyris aberdarica Malme；阿伯德尔黄眼草■☆

415992　Xyris affinis Welw. ex Rendle；近缘黄眼草■☆

415993　Xyris ambigua Beyr. ex Kunth；可疑黄眼草（两似眼草）；Morning Yellow-eyed-grass ■☆

415994　Xyris anceps Lam.；二棱黄眼草■☆

415995　Xyris anceps Lam. var. minima（Steud.）Lock；小二棱黄眼草■☆

415996　Xyris anceps Pers. = Xyris jupicai Rich. ■☆

415997　Xyris angularis N. E. Br.；棱角黄眼草■☆

415998　Xyris angustifolia De Wild. et T. Durand；窄叶黄眼草■☆

415999　Xyris anisophylla Welw. ex Rendle；异叶黄眼草■☆

416000　Xyris arenicola Miq. = Xyris jupicai Rich. ■☆

416001　Xyris arenicola Small = Xyris caroliniana Walter ■☆

416002　Xyris aristata N. E. Br.；具芒黄眼草■☆

416003　Xyris asterotricha Lock；星毛黄眼草■☆

416004　Xyris atrata Malme；黑黄眼草■☆

416005　Xyris bakeri L. A. Nilsson = Xyris congensis Büttner ■☆

416006　Xyris baldwiniana Schult.；鲍德温黄眼草■☆

416007　Xyris baldwiniana Schult. var. tenuifolia（Chapm.）Malme = Xyris baldwiniana Schult. ■☆

416008　Xyris bancana Miq.；中国黄眼草；China Yelloweyegrass ■

416009　Xyris baronii Baker ex Malme = Xyris congensis Büttner ■☆

416010　Xyris baronii Malme；巴龙黄眼草■☆

416011　Xyris barteri N. E. Br.；巴特黄眼草■☆

416012　Xyris batokana N. E. Br. = Xyris congensis Büttner ■☆

416013　Xyris baumii L. A. Nilsson = Xyris congensis Büttner ■☆

416014　Xyris bayardii Fernald = Xyris difformis Chapm. var. curtissii（Malme）Král ■☆

416015　Xyris bobartioides Dinter；博巴鸢尾黄眼草■☆

416016　Xyris brevifolia Michx.；短叶黄眼草■☆

416017　Xyris brunnea L. A. Nilsson = Xyris obscura N. E. Br. ■☆

416018　Xyris bulbosa Kunth = Xyris torta Sm. ■☆

416019　Xyris calocephala Miq. = Xyris indica L. ■

416020　Xyris capensis Thunb.；南非黄眼草（黄谷精）；S. Africa Yelloweyegrass ■

416021　Xyris capensis Thunb. f. schoenoides（Mart.）Nilsson = Xyris capensis Thunb. var. schoenoides（Mart.）L. A. Nilsson ■

416022　Xyris capensis Thunb. var. angolensis Malme = Xyris capensis Thunb. ■

416023　Xyris capensis Thunb. var. medullosa N. E. Br. = Xyris huillensis Rendle ■☆

416024　Xyris capensis Thunb. var. microcephala Malme = Xyris capensis Thunb. ■

416025　Xyris capensis Thunb. var. multicaulis L. A. Nilsson = Xyris capensis Thunb. ■

416026　Xyris capensis Thunb. var. nilagiriensis（Steud.）Engl. = Xyris capensis Thunb. ■

416027　Xyris capensis Thunb. var. pallescens Malme；苍白南非黄眼草■☆

416028　Xyris capensis Thunb. var. schoenoides（Mart.）L. A. Nilsson = Xyris capensis Thunb. ■

416029　Xyris capillaris Malme；发状黄眼草■☆

416030　Xyris capito Hance = Xyris indica L. ■

416031　Xyris capnoides Malme；延胡索黄眼草■☆

416032　Xyris caroliniana Walter；卡罗来纳黄眼草；Carolina Yellow-eyed-grass ■☆

416033　Xyris caroliniana Walter var. olneyi A. W. Wood = Xyris smalliana Nash ■☆

416034　Xyris chapmanii E. L. Bridges et Orzell；查普曼黄眼草■☆

416035　Xyris chinensis Malme = Xyris bancana Miq. ■

416036　Xyris communis Kunth = Xyris jupicai Rich. ■☆

416037　Xyris complanata R. Br.；硬叶黄眼草（硬叶葱草,硬叶芴草）；Flattened Yelloweyegrass, Hard Yelloweyegrass, Hawaii Yelloweyed Grass ■

416038　Xyris congdonii Small = Xyris smalliana Nash ■☆

416039　Xyris congensis Büttner；伸展黄眼草■☆

416040　Xyris conocephala Wright = Xyris caroliniana Walter ■☆

416041　Xyris curtissii Malme = Xyris difformis Chapm. var. curtissii（Malme）Král ■☆

416042　Xyris decipiens N. E. Br.；迷惑黄眼草■☆

416043　Xyris decipiens N. E. Br. var. vanderystii（Malme）Malme = Xyris angularis N. E. Br. ■☆

416044　Xyris densa Malme；密集黄眼草■☆

416045　Xyris difformis Chapm.；大黄眼草；Bog Yellow-eyed-grass, Tall Yellow-eyed-grass ■☆

416046　Xyris difformis Chapm. var. curtissii（Malme）Král；柯蒂斯黄眼草■☆

416047　Xyris difformis Chapm. var. floridana Král；佛罗里达黄眼草■☆

416048　Xyris dilungensis Brylska = Xyris aristata N. E. Br. ■☆

416049　Xyris dispar N. E. Br. = Xyris rehmannii L. A. Nilsson ■☆

416050　Xyris dissimilis Malme；不似黄眼草■☆

416051　Xyris drummondii Malme；德拉蒙德黄眼草■☆

416052　Xyris elata Chapm. = Xyris difformis Chapm. ■☆

416053　Xyris elegantula Malme;雅致黄眼草■☆

416054　Xyris elliottii Chapm. ;埃利奥特黄眼草;Elliott Yellow-eyed-grass ■☆

416055　Xyris elliottii Chapm. var. stenotera Malme = Xyris elliottii Chapm. ■☆

416056　Xyris elongata Rudge = Xyris complanata R. Br. ■

416057　Xyris erosa Lock;啮蚀状黄眼草■☆

416058　Xyris erubescens Rendle;变红黄眼草■☆

416059　Xyris exigua Malme;弱小黄眼草■☆

416060　Xyris extensa Malme = Xyris congensis Büttner ■☆

416061　Xyris filiformis Lam. ;线形黄眼草■☆

416062　Xyris fimbriata Elliott;缘叶黄眼草■☆

416063　Xyris flabelliformis Chapm. ;扇形黄眼草■☆

416064　Xyris flexuosa Muhl. ex Elliott = Xyris caroliniana Walter ■☆

416065　Xyris flexuosa Muhl. ex Elliott var. pallescens (Mohr) Barnhart = Xyris caroliniana Walter ■☆

416066　Xyris flexuosa Muhl. var. pusilla A. Gray = Xyris montana Ries ■☆

416067　Xyris flexuosa sensu Chapm. = Xyris torta Sm. ■☆

416068　Xyris foliolata L. A. Nilsson;托叶黄眼草■☆

416069　Xyris formosana Hayata;台湾黄眼草(黄壳精,黄芴,桃园草);Taiwan Yelloweyegrass ■

416070　Xyris friesii Malme;弗里斯黄眼草■☆

416071　Xyris fugaciflora Rendle;早萎黄眼草■☆

416072　Xyris gerrardii N. E. Br. ;杰勒德黄眼草■☆

416073　Xyris gossweileri Malme;戈斯黄眼草■☆

416074　Xyris gymnoptera Griseb. = Xyris jupicai Rich. ■☆

416075　Xyris hildebrandtii L. A. Nilsson = Xyris congensis Büttner ■☆

416076　Xyris hildebrandtii L. A. Nilsson var. angustifolia Malme = Xyris congensis Büttner ■☆

416077　Xyris huillensis Rendle;威拉黄眼草■☆

416078　Xyris humilis Kunth = Xyris anceps Lam. ■☆

416079　Xyris humilis Kunth f. minima (Steud.) L. A. Nilsson = Xyris anceps Lam. var. minima (Steud.) Lock ■☆

416080　Xyris humpatensis N. E. Br. ;洪帕塔黄眼草■☆

416081　Xyris humpatensis N. E. Br. var. rhodolepis Malme = Xyris rhodolepis (Malme) Malme ex Lock ■☆

416082　Xyris indica L. ;黄眼草;Indian Yelloweyegrass,Yelloweyegrass ■

416083　Xyris insularis Steud. = Xyris anceps Lam. ■☆

416084　Xyris intermedia Malme = Xyris brevifolia Michx. ■☆

416085　Xyris iridifolia Chapm. ;鸢尾叶黄眼草■☆

416086　Xyris juncea Baldwin ex Elliott = Xyris baldwiniana Schult. ■☆

416087　Xyris jupicae Michx. = Xyris jupicai Rich. ■☆

416088　Xyris jupicai Rich. ;尤皮卡黄眼草;Yellow-eyed Grass ■☆

416089　Xyris jupicai Rich. var. brachylepis Malme = Xyris jupicai Rich. ■☆

416090　Xyris kibaraensis Lisowski;基巴拉黄眼草■☆

416091　Xyris kornasiana Brylska et Lisowski;科纳斯黄眼草■☆

416092　Xyris kundelungensis Brylska;昆德龙黄眼草■☆

416093　Xyris kwangolana P. A. Duvign. et Homès;宽果河黄眼草■☆

416094　Xyris laciniata Hutch. = Xyris aristata N. E. Br. ■☆

416095　Xyris laniceps Lock;毛梗黄眼草■☆

416096　Xyris laxifolia Mart. ;疏叶黄眼草■☆

416097　Xyris laxifolia Mart. var. iridifolia (Chapm.) Král = Xyris iridifolia Chapm. ■☆

416098　Xyris ledermannii Malme = Xyris barteri N. E. Br. ■☆

416099　Xyris lejolyanus Lisowski;勒若利黄眼草■☆

416100　Xyris leonensis Hepper;莱昂黄眼草■☆

416101　Xyris leptophylla Malme = Xyris congensis Büttner ■☆

416102　Xyris longisepala Král;长萼黄眼草■☆

416103　Xyris louisianica E. L. Bridges et Orzell = Xyris stricta Chapm. var. obscura Král ■☆

416104　Xyris macrocephala Vahl f. minor (Mart.) M. Kuhlm. et Kuhn = Xyris jupicai Rich. ■☆

416105　Xyris madagascariensis Malme = Xyris congensis Büttner ■☆

416106　Xyris malaccensis Steud. = Xyris complanata R. Br. ■

416107　Xyris mallocephala Lock;毛头黄眼草■☆

416108　Xyris melanocephala Miq. = Xyris capensis Thunb. var. schoenoides (Mart.) L. A. Nilsson ■

416109　Xyris minima Steud. = Xyris anceps Lam. var. minima (Steud.) Lock ■☆

416110　Xyris montana Ries;山地黄眼草(北方黄眼草);Bog Yellow-eyed-grass,Mountain Xyris,Northern Yellow-eyed-grass ■☆

416111　Xyris multicaulis N. E. Br. = Xyris straminea L. A. Nilsson ■☆

416112　Xyris natalensis L. A. Nilsson;纳塔尔黄眼草■☆

416113　Xyris neglecta Small = Xyris difformis Chapm. var. curtissii (Malme) Král ■☆

416114　Xyris nilagiriensis Steud. = Xyris capensis Thunb. ■

416115　Xyris nitida L. A. Nilsson = Xyris congensis Büttner ■☆

416116　Xyris nitida Willd. ex A. Dietr. = Xyris anceps Lam. ■☆

416117　Xyris nivea Welw. ex Rendle;雪白黄眼草■☆

416118　Xyris novoguineensis Hatus. = Xyris capensis Thunb. var. schoenoides (Mart.) L. A. Nilsson ■

416119　Xyris obscura N. E. Br. ;隐匿黄眼草■☆

416120　Xyris operculata R. Br. ;具盖黄眼草■☆

416121　Xyris ornithoptera Lock;鸟翅黄眼草■☆

416122　Xyris pallescens Small = Xyris caroliniana Walter ■☆

416123　Xyris paludosa R. Br. = Xyris indica L. ■

416124　Xyris papillosa Fassett = Xyris montana Ries ■☆

416125　Xyris papillosa Fassett var. exserta Fassett = Xyris montana Ries ■☆

416126　Xyris parvula Malme;较小黄眼草■☆

416127　Xyris pauciflora Willd. ;少花黄眼草(葱草,红头草,黄谷精,苔花黄眼草,芴草);Chivegrass,Fewflower Yelloweyegrass ■

416128　Xyris pauciflora Willd. var. oryzetorum Miq. = Xyris pauciflora Willd. ■

416129　Xyris perroteti Steud. = Xyris anceps Lam. var. minima (Steud.) Lock ■☆

416130　Xyris peteri Poelln. ;彼得黄眼草■☆

416131　Xyris platicaulis Poir. = Xyris anceps Lam. ■☆

416132　Xyris platylepis Chapm. ;宽鳞黄眼草■☆

416133　Xyris pocockii Malme = Xyris friesii Malme ■☆

416134　Xyris popeanus Lisowski;波普黄眼草■☆

416135　Xyris porphyrea Lock;紫色黄眼草■☆

416136　Xyris pumila Rendle;微小黄眼草■☆

416137　Xyris rehmannii L. A. Nilsson;拉赫曼黄眼草■☆

416138　Xyris reptans Rendle = Xyris capensis Thunb. ■

416139　Xyris rhodesiana Malme;罗得西亚黄眼草■☆

416140　Xyris rhodolepis (Malme) Malme ex Lock;红鳞黄眼草■☆

416141　Xyris ridleyi Rendle = Xyris bancana Miq. ■☆

416142　Xyris rigidescens Welw. ex Rendle = Xyris rehmannii L. A. Nilsson ■☆

416143　Xyris robusta Mart. = Xyris indica L. ■

416144　Xyris rubella Malme;微红黄眼草■☆

416145　Xyris sanguinea Vermoesen ex Malme;血红黄眼草■☆

416146　Xyris scabridula Rendle;微糙黄眼草■☆

416147 Xyris scabrifolia R. M. Harper;糙叶黄眼草■☆

416148 Xyris schliebenii Poelln. ;施利本黄眼草■☆

416149 Xyris schoenoides Mart. = Xyris capensis Thunb. var. schoenoides（Mart.）L. A. Nilsson ■

416150 Xyris schoenoides Mart. = Xyris capensis Thunb. ■

416151 Xyris semifuscata Bojer ex Baker = Xyris capensis Thunb. var. schoenoides（Mart.）L. A. Nilsson ■

416152 Xyris semifuscata Bojer ex Baker = Xyris capensis Thunb. ■

416153 Xyris serotina Chapm. ;迟黄眼草■☆

416154 Xyris serotina Chapm. var. curtissii（Malme）Král = Xyris difformis Chapm. var. curtissii（Malme）Král ■☆

416155 Xyris setacea Chapm. = Xyris baldwiniana Schult. ■☆

416156 Xyris smalliana Nash;斯莫尔黄眼草■☆

416157 Xyris smalliana Nash var. congdonii（Small）Malme = Xyris smalliana Nash ■☆

416158 Xyris smalliana Nash var. olneyi（A. W. Wood）Gleason ex Malme = Xyris smalliana Nash ■☆

416159 Xyris sphaerocephala Malme;球头黄眼草■☆

416160 Xyris straminea L. A. Nilsson;禾色黄眼草■☆

416161 Xyris stricta Chapm. ;直立黄眼草■☆

416162 Xyris stricta Chapm. var. obscura Král;路易斯安娜黄眼草■☆

416163 Xyris subaristata Malme = Xyris aristata N. E. Br. ■☆

416164 Xyris subrubella Malme ex Hutch. = Xyris capensis Thunb. ■

416165 Xyris subtilis Lock;细黄眼草■☆

416166 Xyris sumatrana Malme = Xyris capensis Thunb. var. schoenoides（Mart.）L. A. Nilsson ■

416167 Xyris symoensii Brylska et Lisowski;西莫黄眼草■☆

416168 Xyris tennesseensis Král;田纳西黄眼草■☆

416169 Xyris theodori Malme;特奥多尔黄眼草■☆

416170 Xyris thompsonii Rendle = Xyris anceps Lam. ■☆

416171 Xyris torta Kunth = Xyris caroliniana Walter ■☆

416172 Xyris torta Kunth var. pallescens C. Mohr = Xyris caroliniana Walter ■☆

416173 Xyris torta Sm. ;纤细黄眼草;Slender Yellow-eyed-grass, Yellow-eyed Grass ■☆

416174 Xyris torta Sm. var. occidentalis Malme = Xyris torta Sm. ■☆

416175 Xyris umbilonis L. A. Nilsson = Xyris congensis Büttner ■☆

416176 Xyris unistriata Malme;单纹黄眼草■☆

416177 Xyris valida Malme;刚直黄眼草■☆

416178 Xyris vanderystii Malme = Xyris angularis N. E. Br. ■☆

416179 Xyris walkeri Kunth = Xyris complanata R. Br. ■

416180 Xyris welwitschii Rendle;韦尔黄眼草■☆

416181 Xyris zombana N. E. Br. = Xyris capensis Thunb. ■

416182 Xyroides Thouars = Xyris L. ■

416183 Xysmalobium R. Br. （1810）;止泻萝藦属■☆

416184 Xysmalobium acerateoides（Schltr.）N. E. Br. ;无角止泻萝藦■☆

416185 Xysmalobium albens（E. Mey.）D. Dietr. = Asclepias albens（E. Mey.）Schltr. ■☆

416186 Xysmalobium ambiguum N. E. Br. = Xysmalobium undulatum（L.）W. T. Aiton ■☆

416187 Xysmalobium andongense Hiern;安东止泻萝藦■☆

416188 Xysmalobium angolense N. E. Br. = Glossostelma angolensis Schltr. ■☆

416189 Xysmalobium angolense Scott-Elliot = Xysmalobium undulatum（L.）W. T. Aiton ■☆

416190 Xysmalobium appendiculatum（E. Mey.）D. Dietr. = Pachycarpus appendiculatus E. Mey. ■☆

416191 Xysmalobium asperum N. E. Br. ;粗糙止泻萝藦■☆

416192 Xysmalobium barbigerum N. E. Br. = Xysmalobium undulatum（L.）W. T. Aiton ■☆

416193 Xysmalobium baurii N. E. Br. ;鲍利止泻萝藦■☆

416194 Xysmalobium bellum N. E. Br. = Glossostelma spathulatum（K. Schum.）Bullock ■☆

416195 Xysmalobium brownianum S. Moore;布朗止泻萝藦■☆

416196 Xysmalobium carinatum（Schltr.）N. E. Br. ;龙骨止泻萝藦■☆

416197 Xysmalobium carsonii N. E. Br. = Glossostelma carsonii（N. E. Br.）Bullock ■☆

416198 Xysmalobium ceciliae N. E. Br. = Glossostelma ceciliae（N. E. Br.）Goyder ■☆

416199 Xysmalobium clavatum S. Moore;棍棒止泻萝藦■☆

416200 Xysmalobium concolor（E. Mey.）D. Dietr. = Pachycarpus concolor E. Mey. ●☆

416201 Xysmalobium confusum Scott-Elliot;混乱止泻萝藦■☆

416202 Xysmalobium congoense S. Moore;刚果止泻萝藦■☆

416203 Xysmalobium coronarium（E. Mey.）D. Dietr. = Pachycarpus coronarius E. Mey. ●☆

416204 Xysmalobium crispum（P. J. Bergius）D. Dietr. = Asclepias crispa P. J. Bergius ■☆

416205 Xysmalobium dealbatum（E. Mey.）D. Dietr. = Pachycarpus dealbatus E. Mey. ■☆

416206 Xysmalobium decipiens N. E. Br. ;迷惑止泻萝藦■☆

416207 Xysmalobium dilatatum Weim. = Xysmalobium undulatum（L.）W. T. Aiton ■☆

416208 Xysmalobium dispar N. E. Br. = Xysmalobium undulatum（L.）W. T. Aiton ■☆

416209 Xysmalobium dissolutum K. Schum. = Glossostelma lisianthoides（Decne.）Bullock ■☆

416210 Xysmalobium dolichoglossum K. Schum. = Pachycarpus spurius（N. E. Br.）Bullock ■☆

416211 Xysmalobium ensifolium（Burch. ex Scott-Elliot）N. E. Br. = Xysmalobium undulatum（L.）W. T. Aiton var. ensifolium Burch. ex Scott-Elliot ■☆

416212 Xysmalobium fluviale Bruyns;河边止泻萝藦■☆

416213 Xysmalobium fraternum N. E. Br. ;兄弟止泻萝藦■☆

416214 Xysmalobium fritillarioides Rendle = Glossostelma lisianthoides（Decne.）Bullock ■☆

416215 Xysmalobium gerrardii Scott-Elliot;杰勒德止泻萝藦■☆

416216 Xysmalobium gomphocarpoides（E. Mey.）D. Dietr. ;棒果止泻萝藦■☆

416217 Xysmalobium gomphocarpoides（E. Mey.）D. Dietr. var. parvilobum Bruyns;小裂片止泻萝藦■☆

416218 Xysmalobium gossweileri S. Moore;戈斯止泻萝藦■☆

416219 Xysmalobium gramineum S. Moore;禾状止泻萝藦■☆

416220 Xysmalobium grande N. E. Br. = Glossostelma spathulatum（K. Schum.）Bullock ■☆

416221 Xysmalobium grandiflorum（L. f.）R. Br. = Pachycarpus grandiflorus（L. f.）E. Mey. ■☆

416222 Xysmalobium graniticola A. Chev. = Dalzielia oblanceolata Turrill ■☆

416223 Xysmalobium heudelotianum Decne. ;霍氏止泻萝藦■☆

416224 Xysmalobium heudelotianum Scott-Elliot = Xysmalobium membraniferum N. E. Br. ■☆

416225 Xysmalobium holubii Scott-Elliot;霍勒布止泻萝藦■☆

416226 Xysmalobium holubyi Schltr. = Xysmalobium decipiens N. E.

Br. ■☆

416227　Xysmalobium humile（E. Mey.）D. Dietr. = Asclepias humilis（E. Mey.）Schltr. ■☆

416228　Xysmalobium involucratum（E. Mey.）Decne. ;总苞止泻萝藦■☆

416229　Xysmalobium kaessneri S. Moore;卡斯纳止泻萝藦■☆

416230　Xysmalobium lapathifolium K. Schum. = Xysmalobium undulatum（L.）W. T. Aiton ■☆

416231　Xysmalobium leucotrichum（Schltr.）N. E. Br. = Xysmalobium undulatum（L.）W. T. Aiton ■☆

416232　Xysmalobium ligulatum D. Dietr. = Pachycarpus dealbatus E. Mey. ■☆

416233　Xysmalobium linguaeforme Harv. ex Weale = Woodia mucronata（Thunb.）N. E. Br. ■☆

416234　Xysmalobium marginatum（E. Mey.）D. Dietr. = Woodia mucronata（Thunb.）N. E. Br. ■☆

416235　Xysmalobium membraniferum N. E. Br. ;膜质止泻萝藦■☆

416236　Xysmalobium mildbraedii Schltr. = Glossostelma spathulatum（K. Schum.）Bullock ■☆

416237　Xysmalobium obscurum N. E. Br. ;隐匿止泻萝藦■☆

416238　Xysmalobium orbiculare（E. Mey.）D. Dietr. ;圆形止泻萝藦■☆

416239　Xysmalobium padifolium（Baker）Scott-Elliot = Xysmalobium orbiculare（E. Mey.）D. Dietr. ■☆

416240　Xysmalobium parviflorum Harv. ex Scott-Elliot;小花止泻萝藦●☆

416241　Xysmalobium patulum S. Moore;张开止泻萝藦■☆

416242　Xysmalobium pearsonii L. Bolus;皮尔逊止泻萝藦■☆

416243　Xysmalobium pedunculatum Harv. = Xysmalobium prunelloides Turcz. ■☆

416244　Xysmalobium prismatostigma K. Schum. = Xysmalobium undulatum（L.）W. T. Aiton ■☆

416245　Xysmalobium prunelloides Harv. = Xysmalobium prunelloides Turcz. ■☆

416246　Xysmalobium prunelloides Turcz. ;夏枯草止泻萝藦■☆

416247　Xysmalobium reflectens（E. Mey.）D. Dietr. = Pachycarpus reflectens E. Mey. ■☆

416248　Xysmalobium reticulatum N. E. Br. ;网状止泻萝藦■☆

416249　Xysmalobium rhodesianum S. Moore;罗得西亚止泻萝藦■☆

416250　Xysmalobium rhomboideum N. E. Br. ;菱形止泻萝藦■☆

416251　Xysmalobium rigidum（E. Mey.）D. Dietr. = Pachycarpus rigidus E. Mey. ■☆

416252　Xysmalobium schumannianum S. Moore = Xysmalobium reticulatum N. E. Br. ■☆

416253　Xysmalobium sessile（Decne.）Decne. ;无柄止泻萝藦■☆

416254　Xysmalobium spathulatum（K. Schum.）N. E. Br. = Glossostelma spathulatum（K. Schum.）Bullock ■☆

416255　Xysmalobium speciosum S. Moore = Glossostelma cabrae（De Wild.）Goyder ■☆

416256　Xysmalobium spurium N. E. Br. = Pachycarpus spurius（N. E. Br.）Bullock ■☆

416257　Xysmalobium stellatum ?;星芒止泻萝藦■☆

416258　Xysmalobium stockenstromense Scott-Elliot;斯托肯止泻萝藦■☆

416259　Xysmalobium stocksii N. E. Br. ;斯托克斯止泻萝藦■☆

416260　Xysmalobium taschdjiani Chiov. = Xysmalobium undulatum（L.）W. T. Aiton ■☆

416261　Xysmalobium tenue S. Moore;纤细止泻萝藦■☆

416262　Xysmalobium trauseldii R. A. Dyer = Xysmalobium woodii N. E. Br. ■☆

416263　Xysmalobium trilobatum（Schltr.）N. E. Br. = Xysmalobium

undulatum（L.）W. T. Aiton ■☆

416264　Xysmalobium tysonianum（Schltr.）N. E. Br. ;泰森止泻萝藦■☆

416265　Xysmalobium undulatum（L.）W. T. Aiton;尾状止泻萝藦（乌扎拉藤）■☆

416266　Xysmalobium undulatum（L.）W. T. Aiton var. ensifolium Burch. ex Scott-Elliot;剑叶尾状止泻萝藦■☆

416267　Xysmalobium undulatum R. Br. = Xysmalobium undulatum（L.）W. T. Aiton ■☆

416268　Xysmalobium vexillare（E. Mey.）D. Dietr. = Pachycarpus vexillaris E. Mey. ■☆

416269　Xysmalobium viridiflorum（E. Mey.）D. Dietr. = Asclepias dregeana Schltr. ■☆

416270　Xysmalobium woodii N. E. Br. ;伍德止泻萝藦■☆

416271　Xysmalobium zeyheri N. E. Br. ;泽赫止泻萝藦■☆

416272　Xystidium Trin. = Perotis Aiton ■

416273　Xystrolobos Gagnep. = Ottelia Pers. ■

416274　Xystrolobos yunnanensis Gagnep. = Ottelia acuminata（Gagnep.）Dandy ■

416275　Xystrolobus Willis = Xystrolobos Gagnep. ■

416276　Yabea Koso-Pol.（1916）;亚白草属■☆

416277　Yabea Koso-Pol. = Caucalis L. ■☆

416278　Yabea microcarpa Koso-Pol. ;亚白草■☆

416279　Yadakea Makino = Pseudosasa Makino ex Nakai ●

416280　Yadakeya Makino = Pseudosasa Makino ex Nakai ●

416281　Yadakeya Makino = Yadakea Makino ●

416282　Yadakeya japonica（Siebold et Zucc. ex Steud.）Makino = Pseudosasa japonica（Siebold et Zucc. ex Steud.）Makino ex Nakai ●

416283　Yadakeya japonica（Siebold et Zucc.）Makino = Pseudosasa japonica（Siebold et Zucc. ex Steud.）Makino ex Nakai ●

416284　Yakirra Lazarides et R. D. Webster(1985);雅克黍属■☆

416285　Yakirra australiensis（Domin）Lazarides et R. D. Webster;雅克黍■☆

416286　Yakirra pauciflora（R. Br.）Lazarides et R. D. Webster;少花雅克黍■☆

416287　Yamala Raf. = Heuchera L. ■☆

416288　Yangapa Raf. = Gardenia Ellis(保留属名)●

416289　Yangua Spruce = Cybistax Mart. ex Meisn. ●☆

416290　Yanomamua J. R. Grant,Maas et Struwe(2006);亚马孙龙胆属■☆

416291　Yanomamua araca J. R. Grant,Maas et Struwe;亚马孙龙胆■☆

416292　Yarima Burret = Yarina O. F. Gook ●☆

416293　Yarina O. F. Cook = Phytelephas Ruiz et Pav. ●☆

416294　Yatabea Maxim ex Yatabe = Ranzania T. Ito ■☆

416295　Yaundea G. Schellenb. = Yaundea G. Schellenb. ex De Wild. ●

416296　Yaundea G. Schellenb. ex De Wild. = Jaundea Gilg ●

416297　Yaundea G. Schellenb. ex De Wild. = Rourea Aubl.(保留属名)●

416298　Yavia R. Kiesling et Piltz(2001);隐果掌属■☆

416299　Yavia cryptocarpa R. Kiesling et Piltz;隐果掌■☆

416300　Yeatesia Small(1896);西南美爵床属■☆

416301　Yeatesia laete-virens Small;西南美爵床■☆

416302　Yeatesia viridiflora Small;绿花西南美爵床■☆

416303　Yermo Dorn(1991);沙黄头菊属■☆

416304　Yermo xanthocephalus Dorn;沙黄头菊;Desert Yellow-head ■☆

416305　Yermoloffia Bél. = Lagochilus Bunge ex Benth. ●■

416306　Yervamora Kuntze = Bosea L. ●☆

416307　Yervamora Ludw. = Bosea L. ●☆

416308　Yervamora Ludw. ex Kuntze = Bosea L. ●☆

416309　Ygcamela Raf. = Limosella L. ■

416310 Ygramelta Raf. = Ygramela Raf. ■

416311 Yinquania Z. Y. Zhu = Cornus L. ●

416312 Yinquania Z. Y. Zhu(1984);阴茎属;Yinquania ●

416313 Yinquania muchuanensis Z. Y. Zhu;阴茎(团圆果);Yinquania ●

416314 Yinquania muchuanensis Z. Y. Zhu = Cornus oblonga Wall. ex Roxb. ●☆

416315 Yinquania oblonga (Wall.) Z. Y. Zhu = Cornus oblonga Wall. ex Roxb. ●☆

416316 Yinshania Ma et Y. Z. Zhao(1979);阴山荠属;Yinshancress, Yinshania ■★

416317 Yinshania acutangula (O. E. Schulz) Y. H. Zhang;阴山荠(锐棱岩荠, 锐棱阴山荠); Whiteflower Yinshania, Yinshancress, Yinshania

416318 Yinshania acutangula (O. E. Schulz) Y. H. Zhang subsp. microcarpa (K. C. Kuan) Al-Shehbaz, L. L. Lu et T. Y. Cheo = Yinshania microcarpa (K. C. Kuan) Y. H. Zhang ■

416319 Yinshania acutangula (O. E. Schulz) Y. H. Zhang subsp. wilsonii (O. E. Schulz) Al-Shehbaz,G. Yang,L. L. Lu et T. Y. Cheo = Yinshania qianningensis Y. H. Zhang ■

416320 Yinshania acutangula (O. E. Schulz) Y. H. Zhang subsp. wilsonii (O. E. Schulz) Al-Shehbaz, G. Yang, L. L. Lu et T. Y. Cheo;小果阴山荠（威氏阴山荠, 小果岩荠); Smallfruit Yinshancress, Smallfruit Yinshania ■

416321 Yinshania acutangula (O. E. Schulz) Y. H. Zhang var. albiflora (Ma et Y. Z. Zhao) Y. H. Zhang = Yinshania albiflora Ma et Y. Z. Zhao ■

416322 Yinshania acutangula (O. E. Schulz) Y. H. Zhang var. albiflora (Ma et Y. Z. Zhao) Y. H. Zhang = Yinshania acutangula (O. E. Schulz) Y. H. Zhang ■

416323 Yinshania acutangula (O. E. Schulz) Y. H. Zhang var. microcarpa (K. C. Kuan) Al-Shehbaz, G. Yang, L. L. Lu et T. Y. Cheo = Yinshania microcarpa (K. C. Kuan) Y. H. Zhang ■

416324 Yinshania alatipes (Hand. -Mazz.) Y. Z. Zhao;阿拉泰阴山荠(翅柄岩荠)■

416325 Yinshania alatipes (Hand. -Mazz.) Y. Z. Zhao = Cardamine fragarifolia O. E. Schulz ■

416326 Yinshania albiflora Ma et Y. Z. Zhao = Yinshania acutangula (O. E. Schulz) Y. H. Zhang ■

416327 Yinshania albiflora Ma et Y. Z. Zhao var. gobica C. H. An = Yinshania acutangula (O. E. Schulz) Y. H. Zhang ■

416328 Yinshania albiflora Ma et Y. Z. Zhao var. gobica C. H. An = Yinshania zayuensis Y. H. Zhang var. gobica (C. H. An) Y. H. Zhang ■

416329 Yinshania albiflorra Ma et Y. Z. Zhao var. gobica C. H. An = Yinshania acutangula (O. E. Schulz) Y. H. Zhang ■

416330 Yinshania exiensis Y. H. Zhang;鄂西阴山荠; W. Hubei Yinshancress, W. Hubei Yinshania ■

416331 Yinshania exiensis Y. H. Zhang = Yinshania zayuensis Y. H. Zhang ■

416332 Yinshania formosana (Hayata) Y. Z. Zhao = Yinshania rivulorum (Dunn) Al-Shehbaz et al. ■

416333 Yinshania fumarioides (Dunn) Y. Z. Zhao;紫堇叶阴山荠■

416334 Yinshania fumarioides (Dunn) Y. Z. Zhao = Hilliella fumarioides (Dunn) Y. H. Zhang et H. W. Li ■

416335 Yinshania furcatopilosa (K. C. Kuan) Y. H. Zhang;叉毛阴山荠(叉毛岩荠)■

416336 Yinshania ganluoensis Y. H. Zhang;甘洛阴山荠; Ganluo Yinshancress,Ganluo Yinshania ■

416337 Yinshania ganluoensis Y. H. Zhang = Yinshania zayuensis Y. H. Zhang ■

416338 Yinshania henryi (Oliv.) Y. H. Zhang;柔毛阴山荠(乾岩腔, 柔 毛 岩 荠); Henry Cragcress, Henry Scurvyweed, Henry Yinshancress,Henry Yinshania ■

416339 Yinshania hui (O. E. Schulz) Y. Z. Zhao;武功山阴山荠■

416340 Yinshania hui (O. E. Schulz) Y. Z. Zhao = Hilliella hui (O. E. Schulz) Y. H. Zhang et H. W. Li ■

416341 Yinshania hunanensis (Y. H. Zhang) Al-Shehbaz, G. Yang, L. L. Lu et T. Y. Cheo;湖南阴山荠■

416342 Yinshania hunanensis (Y. H. Zhang) Al-Shehbaz, G. Yang, L. L. Lu et T. Y. Cheo = Hilliella hunanensis Y. H. Zhang ■

416343 Yinshania lichuanensis (Y. H. Zhang) Al-Shehbaz, G. Yang, L. L. Lu et T. Y. Cheo;利川阴山荠■

416344 Yinshania microcarpa (K. C. Kuan) Y. H. Zhang = Yinshania acutangula (O. E. Schulz) Y. H. Zhang subsp. wilsonii (O. E. Schulz) Al-Shehbaz,G. Yang,L. L. Lu et T. Y. Cheo ■

416345 Yinshania paradoxa (Hance) Y. Z. Zhao;卵叶阴山荠■

416346 Yinshania paradoxa (Hance) Y. Z. Zhao = Hilliella paradoxa (Hance) Y. H. Zhang et H. W. Li ■

416347 Yinshania qianningensis Y. H. Zhang;乾宁阴山荠; Qianning Yinshancress, Qianning Yinshania ■

416348 Yinshania qianningensis Y. H. Zhang = Yinshania acutangula (O. E. Schulz) Y. H. Zhang subsp. wilsonii (O. E. Schulz) Al-Shehbaz,G. Yang,L. L. Lu et T. Y. Cheo ■

416349 Yinshania qianningensis Y. H. Zhang = Yinshania acutangula (O. E. Schulz) Y. H. Zhang var. albiflora (Ma et Y. Z. Zhao) Y. H. Zhang ■

416350 Yinshania qianningensis Y. H. Zhang = Yinshania albiflora Ma et Y. Z. Zhao ■

416351 Yinshania qianningensis Y. H. Zhang var. brachybotrys Y. H. Zhang;短序阴山荠; Shortspike Yinshancress, Shortspike Yinshania ■

416352 Yinshania qianningensis Y. H. Zhang var. brachybotrys Y. H. Zhang = Yinshania acutangula (O. E. Schulz) Y. H. Zhang subsp. wilsonii (O. E. Schulz) Al-Shehbaz,G. Yang,L. L. Lu et T. Y. Cheo ■

416353 Yinshania rupicola (D. C. Zhang et J. Z. Shao) Al-Shehbaz et al. ;石生阴山荠(河岸阴山荠)■

416354 Yinshania rupicola (D. C. Zhang et J. Z. Shao) Al-Shehbaz et al. = Cochlearia formosana Hayata ■

416355 Yinshania rupicola (D. C. Zhang et J. Z. Shao) Al-Shehbaz et al. = Cochlearia rupicola D. C. Zhang et J. Z. Shao ■

416356 Yinshania rupicola (D. C. Zhang et J. Z. Shao) Al-Shehbaz et al. subsp. shuangpaiensis (Z. Y. Li) Al-Shehbaz et al. ;双牌阴山荠■

416357 Yinshania rupicola (D. C. Zhang et J. Z. Shao) Al-Shehbaz et al. subsp. shuangpaiensis (Z. Y. Li) Al-Shehbaz et al. = Hilliella shuangpaiensis Z. Y. Li ■

416358 Yinshania sinuata (K. C. Kuan) Al-Shehbaz et al. ;弯缺阴山荠■

416359 Yinshania sinuata (K. C. Kuan) Al-Shehbaz et al. = Hilliella sinuata (K. C. Kuan) Y. H. Zhang et H. W. Li ■

416360 Yinshania sinuata (K. C. Kuan) Al-Shehbaz et al. subsp. qianwuensis (Y. H. Zhang) Al-Shehbaz et al. = Hilliella sinuata (K. C. Kuan) Y. H. Zhang var. qianwuensis Y. H. Zhang ■

416361 Yinshania sinuata (K. C. Kuan) Al-Shehbaz et al. subsp. qianwuensis (Y. H. Zhang) Al-Shehbaz et al. ;寻乌阴山荠■

416362 Yinshania warburgii (O. E. Schulz) Y. Z. Zhao = Hilliella fumarioides（Dunn）Y. H. Zhang et H. W. Li ■

416363　Yinshania warburgii（O. E. Schulz）Y. Z. Zhao = Hilliella warburgii（O. E. Schulz）Y. H. Zhang et H. W. Li ■

416364　Yinshania warburgii（O. E. Schulz）Y. Z. Zhao = Yinshania fumarioides（Dunn）Y. Z. Zhao ■

416365　Yinshania wenxianensis Y. H. Zhang；文县阴山荠；Wenxian Yinshancress，Wenxian Yinshania ■

416366　Yinshania wenxianensis Y. H. Zhang = Yinshania acutangula（O. E. Schulz）Y. H. Zhang ■

416367　Yinshania wenxianensis Y. H. Zhang var. songpanensis Y. H. Zhang；松潘阴山荠；Songpan Yinshancress，Songpan Yinshania ■

416368　Yinshania wenxianensis Y. H. Zhang var. songpanensis Y. H. Zhang = Yinshania acutangula（O. E. Schulz）Y. H. Zhang ■

416369　Yinshania yixianensis（Y. H. Zhang）Al-Shehbaz, G. Yang, L. L. Lu et T. Y. Cheo = Hilliella yixianensis Y. H. Zhang ■

416370　Yinshania yixianensis（Y. H. Zhang）Al-Shehbaz, G. Yang, L. L. Lu et T. Y. Cheo；黟县阴山荠■

416371　Yinshania zayuensis Y. H. Zhang；察隅阴山荠；Chayu Yinshancress，Chayu Yinshania ■

416372　Yinshania zayuensis Y. H. Zhang var. gobica（C. H. An）Y. H. Zhang；戈壁阴山荠■

416373　Yinshania zhejiangensis（Y. H. Zhang）Y. Z. Zhao = Hilliella fumarioides（Dunn）Y. H. Zhang et H. W. Li ■

416374　Yinshania zhejiangensis（Y. H. Zhang）Y. Z. Zhao = Yinshania fumarioides（Dunn）Y. Z. Zhao ■

416375　Ymnostema Neck. = Lobelia L. ●■

416376　Ymnostemma Steud. = Lobelia L. ●■

416377　Ynesa O. F. Cook = Attalea Kunth ●☆

416378　Yoania Maxim.（1872）；宽距兰属（长花柄兰属）；Yoania ■

416379　Yoania amagiensis Nakai et F. Maek.；天城山宽距兰■☆

416380　Yoania amagiensis Nakai et F. Maek. var. squamipes（Fukuy.）C. L. Yeh et C. S. Leou = Yoania japonica Maxim. ■

416381　Yoania flava K. Inoue et T. Yukawa；黄宽距兰■☆

416382　Yoania japonica Maxim.；宽距兰（长花柄兰，兰天麻，钟馗兰）；Japan Yoania ■

416383　Yoania japonica Maxim. var. squamipes Fukuy.；密鳞宽距兰（密鳞姚氏兰）■

416384　Yoania japonica Maxim. var. squamipes Fukuy. = Yoania japonica Maxim. ■

416385　Yoania squamipes（Fukuy.）Masam. = Yoania japonica Maxim. var. squamipes Fukuy. ■

416386　Yoania squamipes（Fukuy.）Masam. = Yoania japonica Maxim. ■

416387　Yodes Kurz = Iodes Blume ●

416388　Yolanda Hoehne = Brachionidium Lindl. ■☆

416389　Yongsonia Young = Fothergilla L. ●☆

416390　Youngia Cass.（1831）；黄鹌菜属；Youngia ■

416391　Youngia acaulis（Roxb.）DC. = Launaea acaulis（Roxb.）Babc. ex Kerr ■

416392　Youngia akagii（Kitag.）Kitag. = Youngia tenuicaulis（Babc. et Stebbins）De Moor ■

416393　Youngia alashanica H. C. Fu；阿拉善黄鹌菜；Alashan Youngia ■

416394　Youngia altaica（Babc. et Stebbins）De Moor；阿尔泰黄鹌菜■

416395　Youngia aspera（Schrad. ex Willd.）Steud. = Chondrilla aspera（Schrad. ex Willd.）Poir. ■

416396　Youngia bifurcata Babc. et Stebbins；黑褐黄鹌菜（顶凹黄鹌菜，二叉黄鹌菜）●■

416397　Youngia blinii（H. Lév.）Lauener；刚毛黄鹌菜（昭通黄鹌菜）■

416398　Youngia chelidoniifolia（Makino）Kitam. = Crepidiastrum chelidoniifolium（Makino）J. H. Pak et Kawano ■☆

416399　Youngia chelidoniifolia（Makino）Kitam. = Paraixeris chelidonifolia（Makino）Nakai ■

416400　Youngia chinensis（Thunb.）DC. = Ixeridium chinense（Thunb.）Tzvelev ■

416401　Youngia chrysantha Maxim. = Paraixeris denticulata（Houtt.）Nakai ■

416402　Youngia cineripappa（Babc.）Babc. et Stebbins；鼠冠黄鹌菜（灰毛黄鹌菜）；Greyhair Youngia ■

416403　Youngia conjunctiva Babc. et Stebbins；甘肃黄鹌菜■

416404　Youngia cristata C. Shih et C. Q. Cai；角冠黄鹌菜；Cristate Youngia ■

416405　Youngia cyanea S. W. Liu et T. N. Ho；蓝花黄鹌菜；Blueflower Youngia ■

416406　Youngia debilis（Poir.）DC. = Ixeris japonica（Burm. f.）Nakai ■

416407　Youngia debilis（Thunb.）DC. = Ixeris japonica（Burm. f.）Nakai ■

416408　Youngia dentata（Thunb.）DC. = Ixeridium dentatum（Thunb.）Tzvelev ■

416409　Youngia denticulata（Houtt.）Kitam. = Crepidiastrum denticulatum（Houtt.）J. H. Pak et Kawano ■☆

416410　Youngia denticulata（Houtt.）Kitam. = Ixeris denticulata（Houtt.）Stebbins ■

416411　Youngia denticulata（Houtt.）Kitam. = Paraixeris denticulata（Houtt.）Nakai ■

416412　Youngia denticulata（Houtt.）Kitam. f. pallescens（Momiy. et Tuyama）Kitam. = Crepidiastrum denticulatum（Houtt.）J. H. Pak et Kawano f. pallescens（Momiy. et Tuyama）Yonek. ■☆

416413　Youngia denticulata（Houtt.）Kitam. f. pinnatipartita（Makino）Kitam. = Crepidiastrum denticulatum（Houtt.）J. H. Pak et Kawano f. pinnatipartitum（Makino）Sennikov ■

416414　Youngia denticulata（Houtt.）Kitam. f. pinnatipartita（Makino）Kitam. = Paraixeris pinnatipartita（Makino）Tzvelev ■

416415　Youngia depressa（Hook. f. et Thomson）Babc. et Stebbins；矮生黄鹌菜；Low Youngia ■

416416　Youngia diversifolia（Ledeb. ex Spreng.）Ledeb.；细裂黄鹌菜（异叶黄鹌菜）；Diverseleaf Youngia ■

416417　Youngia diversifolia（Ledeb. ex Spreng.）Ledeb. = Youngia tenuifolia（Willd.）Babc. et Stebbins subsp. diversifolia（Ledeb. ex Spreng.）Babc. et Stebbins ■☆

416418　Youngia erythrocarpa（Vaniot）Babc. et Stebbins；红果黄鹌菜；Redfruit Youngia ■

416419　Youngia flexuosa（Ledeb.）Ledeb. = Crepis flexuosa（Ledeb.）C. B. Clarke ■

416420　Youngia flexuosa（Ledeb.）Ledeb. var. gigantea C. Winkl. ex O. Fedtsch. = Crepis flexuosa（Ledeb.）C. B. Clarke ■

416421　Youngia formosana（Hayata）H. Hara = Youngia japonica（L.）DC. ■

416422　Youngia formosana（Hayata）Yamam. = Youngia japonica（L.）DC. subsp. formosana（Hayata）Kitam. ■

416423　Youngia fusca（Babc.）Babc. et Stebbins；厚绒黄鹌菜（褐黄鹌菜）■

416424　Youngia glauca Edgew. = Crepis flexuosa（Ledeb.）C. B. Clarke ■

416425　Youngia gracilipes（Hook. f.）Babc. et Stebbins；细梗黄鹌菜；

Thinstalk Youngia ■

416426 Youngia gracilis（Hook. f. et Thomson ex C. B. Clarke）Babc. et Stebbins = Youngia stebbinsiana S. Y. Hu ■

416427 Youngia gracilis Hook. f. et Thomson = Youngia stebbinsiana S. Y. Hu ■

416428 Youngia gracilis Hook. f. ex Benth. et Hook. f. = Youngia stebbinsiana S. Y. Hu ■

416429 Youngia gracilis Miq. = Youngia japonica（L.）DC. ■

416430 Youngia hastata（Thunb.）DC. = Paraixeris denticulata（Houtt.）Nakai ■

416431 Youngia hastiformis C. Shih；戟裂黄鹌菜（顶戟黄鹌菜）■

416432 Youngia henryi（Diels）Babc. et Stebbins；长裂黄鹌菜（巴东黄鹌菜）；Henry Youngia ■

416433 Youngia heterophylla（Hemsl.）Babc. et Stebbins；异叶黄鹌菜（黄狗头）；Diversifolious Youngia ■

416434 Youngia humilis（Thunb.）DC. = Lapsana humilis（Thunb.）Makino ■

416435 Youngia humilis（Thunb.）DC. = Lapsanastrum humile（Thunb.）J.-H. Pak et K. Bremer ■

416436 Youngia humilis DC. = Lapsana humilis（Thunb.）Makino ■

416437 Youngia japonica（L.）DC.；黄鹌菜（臭头苦苴，黄瓜菜，黄花菜，黄花枝草，黄花枝香草，芥菜仔，苦菜药，毛连连，牛石花，牛屎花，雀雀包，雀雀菜，雀雀草，雀雀台，三枝香，山芥菜，台湾黄鹌菜，天葛菜，土芥菜，土山芥，鸭屎条，野菜菜，野芥兰，野青菜）；Asiatic False Hawksbeard，Japanese Hawk's Beard，Japanese Youngia，Oriental False Hawksbeard，Oriental False Hawks-beard，Oriental Hawksbeard，Taiwan Youngia ■

416438 Youngia japonica（L.）DC. subsp. ellstonii（Hochr.）Babc. et Stebbins = Youngia pseudosenecio（Vaniot）C. Shih ■

416439 Youngia japonica（L.）DC. subsp. elstonii（Hochr.）Babc. et Stebbins = Youngia japonica（L.）DC. ■

416440 Youngia japonica（L.）DC. subsp. formosana（Hayata）Kitam.；台湾黄鹌菜■

416441 Youngia japonica（L.）DC. subsp. formosana（Hayata）Kitam. = Youngia japonica（L.）DC. ■

416442 Youngia japonica（L.）DC. subsp. genuina（Hochr.）Babc. et Stebbins = Youngia japonica（L.）DC. ■

416443 Youngia japonica（L.）DC. subsp. longiflora Babc. et Stebbins；大花黄鹌菜；Longflower Japanese Youngia ■

416444 Youngia japonica（L.）DC. subsp. longiflora Babc. et Stebbins = Youngia longiflora（Babc. et Stebbins）C. Shih ■

416445 Youngia japonica（L.）DC. var. formosana（Hayata）H. L. Li = Youngia japonica（L.）DC. subsp. formosana（Hayata）Kitam. ■

416446 Youngia japonica（L.）DC. var. formosana（Hayata）H. L. Li = Youngia japonica（L.）DC. ■

416447 Youngia kangdingensis C. Shih；康定黄鹌菜；Kangding Youngia ■

416448 Youngia lanata Babc. et Stebbins；黑红黄鹌菜（绵毛黄鹌菜，绒毛黄鹌菜）■

416449 Youngia lanceolata（Houtt.）DC. = Crepidiastrum lanceolatum（Houtt.）Nakai ■

416450 Youngia longiflora（Babc. et Stebbins）C. Shih；长花黄鹌菜；Longflower Youngia ■

416451 Youngia longipes（Hemsl.）Babc. et Stebbins；戟叶黄鹌菜；Longstalk Youngia ■

416452 Youngia lyrata（Poir.）Cass. = Youngia japonica（L.）DC. ■

416453 Youngia lyrata Cass. = Youngia japonica（L.）DC. ■

416454 Youngia mairei（H. Lév.）Babc. et Stebbins；紫红黄鹌菜（东

川黄鹌菜）；Maire Youngia ■

416455 Youngia mauritiana DC. = Youngia japonica（L.）DC. ■

416456 Youngia multiflora（Thunb.）DC. = Youngia japonica（L.）DC. ■

416457 Youngia nansiensis Y. Z. Zhao et L. Ma；南寺黄鹌菜；Nansi Youngia ■

416458 Youngia napifera DC. ex Wight = Youngia japonica（L.）DC. ■

416459 Youngia nujiangensis C. Shih；怒江黄鹌菜；Nujiang Youngia ■

416460 Youngia ordosica Y. Z. Zhao et L. Ma；鄂尔多斯黄鹌菜；Eerduosi Youngia ■

416461 Youngia paleacea（Diels）Babc. et Stebbins；羽裂黄鹌菜（稃苞黄鹌菜，具苞黄鹌菜）；Pinnatifid Youngia ■

416462 Youngia paleacea（Diels）Babc. et Stebbins subsp. smithii Babc. et Stebbins = Youngia paleacea（Diels）Babc. et Stebbins ■

416463 Youngia paleacea（Diels）Babc. et Stebbins subsp. yunnanensis（Babc.）Babc. et Stebbins = Youngia paleacea（Diels）Babc. et Stebbins ■

416464 Youngia paosa Steud. = Youngia japonica（L.）DC. ■

416465 Youngia parva Babc. et Stebbins；白冠黄鹌菜（川北黄鹌菜）■

416466 Youngia pilifera C. Shih；糙毛黄鹌菜■

416467 Youngia prattii（Babc.）Babc. et Stebbins；川西黄鹌菜（黄苦麻草）；Pratt Youngia ■

416468 Youngia pseudosenecio（Vaniot）C. Shih；卵裂黄鹌菜■

416469 Youngia pygmaea（Ledeb.）Ledeb. var. purpurea C. Winkl. ex O. Fedtsch. = Crepis lactea Lipsch. ■

416470 Youngia pygmaea Ledeb. var. purpurea C. Winkl. ex O. Fedtsch. = Crepis lactea Lipsch. ■

416471 Youngia racemifera（Hook. f.）Babc. et Stebbins；总序黄鹌菜（高山黄鹌菜，旌节黄鹌菜）■

416472 Youngia rosthornii（Diels）Babc. et Stebbins；多裂黄鹌菜；Multifid Youngia ■

416473 Youngia rubida Babc. et Stebbins；川黔黄鹌菜■

416474 Youngia scaposa（C. C. Chang）Babc. et Stebbins = Youngia szechuanica（Soderb.）S. Y. Hu ■

416475 Youngia serawschanica（B. Fedtsch.）Babc. et Stebbins；长果黄鹌菜；Longfruited Youngia ■

416476 Youngia sericea C. Shih；绢毛黄鹌菜；Sericeous Youngia ■

416477 Youngia serotina Maxim. = Paraixeris serotina（Maxim.）Tzvelev ■

416478 Youngia setigera（Scott ex W. W. Sm.）Babc. et Stebbins = Youngia blinii（H. Lév.）Lauener ■

416479 Youngia simulatrix（Babc.）Babc. et Stebbins；无茎黄鹌菜；Stemless Youngia ■

416480 Youngia smithiana Hand. -Mazz. = Youngia simulatrix（Babc.）Babc. et Stebbins ■

416481 Youngia sonchifolia（Bunge）Maxim. = Crepidiastrum sonchifolium（Bunge）J. H. Pak et Kawano ■☆

416482 Youngia sonchifolia（Bunge）Maxim. = Ixeridium sonchifolium（Maxim.）C. Shih ■

416483 Youngia sonchifolia Maxim. = Ixeridium sonchifolium（Maxim.）C. Shih ■

416484 Youngia stebbinsiana S. Y. Hu；纤细黄鹌菜（细黄鹌菜）；Slender Youngia ■

416485 Youngia stenoma（Turcz.）Ledeb.；碱黄鹌菜■

416486 Youngia szechuanica（Soderb.）S. Y. Hu；少花黄鹌菜；Fewflower Youngia ■

416487 Youngia taiwaniana S. S. Ying = Youngia japonica（L.）DC.

subsp. longiflora Babc. et Stebbins ■

416488 Youngia tenuicaulis（Babc. et Stebbins）Czerep. = Youngia tenuicaulis（Babc. et Stebbins）De Moor ■

416489 Youngia tenuicaulis（Babc. et Stebbins）De Moor；叉枝黄鹤菜（细茎黄鹤菜）；Thin-branched Youngia ■

416490 Youngia tenuifolia（Willd.）Babc. et Stebbins；细叶黄鹤菜（蒲公幌）；Thinleaf Youngia ■

416491 Youngia tenuifolia（Willd.）Babc. et Stebbins subsp. diversifolia（Ledeb. ex Spreng.）Babc. et Stebbins = Youngia diversifolia（Ledeb. ex Spreng.）Ledeb. ■

416492 Youngia tenuifolia（Willd.）Babc. et Stebbins subsp. tenuicaulis Babc. et Stebbins = Youngia tenuicaulis（Babc. et Stebbins）De Moor ■

416493 Youngia terminalis Babc. et Stebbins；内毛黄鹤菜（大头黄鹤菜,光叶黄鹤菜）■

416494 Youngia thunbergiana DC. = Youngia japonica（L.）DC. ■

416495 Youngia wilsonii（Babc.）Babc. et Stebbins；栉齿黄鹤菜；E. H. Wilson Youngia ■

416496 Youngia yilingii C. Shih；艺林黄鹤菜（矮小黄鹤菜）；Yilin Youngia ■

416497 Youngia yoshinoi（Makino）Kitam. = Crepidiastrum yoshinoi（Makino）J. H. Pak et Kawano ■☆

416498 Youngia yunnanensis Babc. = Youngia paleacea（Diels）Babc. et Stebbins ■

416499 Youngia zhenduoi S. W. Liu et T. N. Ho；振铎黄鹤菜；Zhenduo Youngia ■

416500 Ypomaea Robin = Ipomoea L.（保留属名）●■

416501 Ypsilandra Franch.（1888）；丫蕊花属；Forkstamenflower ■

416502 Ypsilandra Franch. = Helonias L. ■☆

416503 Ypsilandra alpina F. T. Wang et Ts. Tang；高山丫蕊花；Alpine Forkstamenflower ■

416504 Ypsilandra cavaleriei H. Lév. et Vaniot；小果丫蕊花；Littlefruit Forkstamenflower ■

416505 Ypsilandra jinpingensis W. H. Chen, Y. M. Shui et Z. Y. Yu；金平丫蕊花；Jinping Forkstamenflower ■

416506 Ypsilandra kansuensis R. N. Zhao et Z. X. Peng；甘肃丫蕊花；Gansu Forkstamenflower ■

416507 Ypsilandra parviflora F. T. Wang et Ts. Tang = Ypsilandra cavaleriei H. Lév. et Vaniot ■

416508 Ypsilandra thibetica Franch.；丫蕊花（百合三七,峨眉石凤丹,随身丹,小飘儿菜,丫药花,一枝花）；Common Forkstamenflower, Forkstamenflower ■

416509 Ypsilandra thibetica Franch. var. angustifolia F. T. Wang et Ts. Tang = Ypsilandra thibetica Franch. ■

416510 Ypsilandra yunnanensis W. W. Sm. et Jeffrey；云南丫蕊花；Yunnan Forkstamenflower ■

416511 Ypsilandra yunnanensis W. W. Sm. et Jeffrey var. himalaica H. Hara = Ypsilandra yunnanensis W. W. Sm. et Jeffrey ■

416512 Ypsilandra yunnanensis W. W. Sm. et Jeffrey var. micrantha Hand.-Mazz. = Ypsilandra yunnanensis W. W. Sm. et Jeffrey ■

416513 Ypsilopus Summerh.（1949）；叉足兰属■☆

416514 Ypsilopus erectus（P. J. Cribb）P. J. Cribb et J. L. Stewart；直立叉足兰■☆

416515 Ypsilopus graminifolius（Kraenzl.）Summerh. = Ypsilopus longifolius（Kraenzl.）Summerh. ■☆

416516 Ypsilopus leedalii P. J. Cribb；利达尔叉足兰■☆

416517 Ypsilopus longifolius（Kraenzl.）Summerh.；长叶叉足兰■☆

416518 Ypsilopus longifolius（Kraenzl.）Summerh. subsp. erectus P. J.

Cribb = Ypsilopus erectus（P. J. Cribb）P. J. Cribb et J. L. Stewart ■☆

416519 Ypsilopus viridiflorus P. J. Cribb et J. L. Stewart；绿花叉足兰■☆

416520 Ypsilorchis Z. J. Liu, S. C. Chen et L. J. Chen（2008）；丫瓣兰属■

416521 Ypsilorchis fissipetala（Finet）Z. J. Liu, S. C. Chen et L. J. Chen；丫瓣兰（裂瓣羊耳蒜）；Splitpetal Liparis, Splitpetal Twayblade ■

416522 Ystia Compare = Schizachyrium Nees ■

416523 Ystia stagnina（Vanderyst）Compère = Schizachyrium kwiluense Vanderyst ■☆

416524 Yua C. L. Li = Parthenocissus Planch.（保留属名）●

416525 Yua C. L. Li（1990）；俞藤属；Yua ●

416526 Yua austro-orientalis（F. P. Metcalf）C. L. Li；大果俞藤（东南爬山虎）；Bigfruit Yua, SE. China Creeper, Southeastern China Creeper ●

416527 Yua chinensis C. L. Li；绿芽俞藤；Chinese Yua ●

416528 Yua chinensis C. L. Li = Yua thomsonii（M. A. Lawson）C. L. Li var. glaucescens（Diels et Gilg）C. L. Li ●

416529 Yua thomsonii（M. A. Lawson）C. L. Li；俞藤（白背乌蔹莓,粉叶地锦,粉叶爬山虎,红葡萄藤,细母猪藤）；Thomson Creeper, Thomson Yua ●

416530 Yua thomsonii（M. A. Lawson）C. L. Li var. glaucescens（Diels et Gilg）C. L. Li = Yua chinensis C. L. Li ●

416531 Yua thomsonii（M. A. Lawson）C. L. Li var. glaucescens（Diels et Gilg）C. L. Li；华西俞藤；Glaucescent Thomson Yua ●

416532 Yua thomsonii（M. A. Lawson）Planch. = Yua thomsonii（M. A. Lawson）C. L. Li ●

416533 Yuca Raf. = Yucca L. ●■

416534 Yucaratonia Burkart = Gliricidia Kunth ●☆

416535 Yucca L.（1753）；丝兰属（金棒兰属）；Adam's Needle, Adam's-needle, Bear's Grass, Beargrass, Joshua Tree, Spanish Bayonet, Spanish Dagger, Spanish Daggers, Spanish-bayonet, Yucca ●■

416536 Yucca 'Jewel'；富贵王兰■☆

416537 Yucca aletriformis Haw. = Dracaena aletriformis（Haw.）Bos ●☆

416538 Yucca aloifolia L.；千手丝兰（金棒兰,芦荟叶丝兰,千寿兰,王兰）；Adam's Needle, Aloe Yucca, Dagger Plant, Spanish Bayonet, Spanish Bayonet Yucca, Spanish Dagger, Spanish-dagger, Yucca ●

416539 Yucca aloifolia L. 'Marginata'；金边丝兰●

416540 Yucca angustifolia Engelm.；狭叶丝兰；Plains Yucca, Spanish Dagger, Yucca ●☆

416541 Yucca angustifolia Engelm. var. elata Engelm. = Yucca elata Engelm. ●

416542 Yucca angustifolia Pursh = Yucca glauca Nutt. ●

416543 Yucca angustifolia Pursh var. mollis Engelm. = Yucca arkansana Trel. ●☆

416544 Yucca angustifolia Pursh var. radiosa Engelm. = Yucca elata Engelm. ●

416545 Yucca angustissima Engelm. ex Trel.；细丝兰●☆

416546 Yucca angustissima Engelm. ex Trel. var. avia Reveal；犹他细丝兰●☆

416547 Yucca angustissima Engelm. ex Trel. var. kanabensis（McKelvey）Reveal；沙地细丝兰●☆

416548 Yucca angustissima Engelm. ex Trel. var. toftiae（S. L. Welsh）Reveal；托夫丝兰●☆

416549 Yucca arborescens（Torr.）Trel. = Yucca brevifolia Engelm. ●☆

416550 Yucca argospatha Verl. = Yucca treculeana Carrière ●☆

416551 Yucca arizonica McKelvey = Yucca baccata Torr. var. brevifolia L. D. Benson et Darrow ■☆

416552 Yucca arkansana Trel. ;阿肯色丝兰;Arkansas Yucca,Soapweed ●☆

416553 Yucca arkansana Trel. var. paniculata McKelvey = Yucca arkansana Trel. ●☆

416554 Yucca aspera Engelm. = Yucca treculeana Carrière ●☆

416555 Yucca australis Trel. = Yucca faxoniana Sarg. ●☆

416556 Yucca baccata Torr. ;长瓣丝兰;Banana Yucca, Blue Yucca, Broad-leaved Yucca, Datil, Datil Yucca, Fleshy-fruited Yucca, Longpetal Yucca,Spanish Bayonet,Spanish Dagger,Wild Date ■☆

416557 Yucca baccata Torr. var. australis Engelm. = Yucca treculeana Carrière ●☆

416558 Yucca baccata Torr. var. brevifolia L. D. Benson et Darrow;短叶长瓣丝兰;Thornber Yucca ■☆

416559 Yucca baccata Torr. var. macrocarpa Torr. = Yucca treculeana Carrière ●☆

416560 Yucca baileyi Wooton et Standl. ;贝利丝兰;Alpine Yucca ●☆

416561 Yucca baileyi Wooton et Standl. var. intermedia (McKelvey) Reveal = Yucca intermedia McKelvey ●☆

416562 Yucca baileyi Wooton et Standl. var. navajoa (J. M. Webber) J. M. Webber = Yucca baileyi Wooton et Standl. ●☆

416563 Yucca brevifolia Engelm. ;短叶丝兰;Joshua Tree,Joshua-tree, Tree Yucca ●☆

416564 Yucca brevifolia Engelm. var. herbertii (J. M. Webber) Munz = Yucca brevifolia Engelm. ●☆

416565 Yucca brevifolia Engelm. var. jaegeriana McKelvey = Yucca brevifolia Engelm. ●☆

416566 Yucca brevifolia Schott ex Trel. = Yucca baccata Torr. var. brevifolia L. D. Benson et Darrow ■☆

416567 Yucca californica Groenl. = Hesperoyucca whipplei (Torr.) Trel. ■☆

416568 Yucca californica Nutt. ex Baker = Yucca schidigera Roezl ex Ortgies ●☆

416569 Yucca campestris McKelvey;平原丝兰;Plains Yucca ●☆

416570 Yucca canaliculata Hook. = Yucca treculeana Carrière ●☆

416571 Yucca canaliculata Hook. var. pendula K. Koch = Yucca treculeana Carrière ●☆

416572 Yucca carnerosana (Trel.) McKelvey;巨丝兰; Giant Spanish Dagger,Palma Istle ●☆

416573 Yucca concava Haw. = Yucca flaccida Haw. ●

416574 Yucca confinis McKelvey = Yucca baccata Torr. var. brevifolia L. D. Benson et Darrow ■☆

416575 Yucca constricta Buckley;巴氏丝兰;Buckley's Yucca, White Rim Yucca ●☆

416576 Yucca crassifila Engelm. = Yucca treculeana Carrière ●☆

416577 Yucca draconis L. var. arborescens Torr. = Yucca brevifolia Engelm. ●☆

416578 Yucca elata Engelm. ;高丝兰;Palmella, Soap Weed, Soaptree Yucca,Soap-tree Yucca,Soap-weed Yucca,Tall Yucca ●

416579 Yucca elata Engelm. var. utahensis (McKelvey) Reveal = Yucca utahensis McKelvey ●☆

416580 Yucca elata Engelm. var. verdiensis (McKelvey) Reveal;岩坡丝兰●☆

416581 Yucca elephantipes Regel ex Trel. ;象脚王兰(大丝兰,象脚兰);Bulbstem Yucca, Giant Yucca, Spineless Yucca ●☆

416582 Yucca elephantipes Regel ex Trel. 'Variegata';乳白边象脚兰(巨丝兰)●☆

416583 Yucca faxoniana Sarg. ;西班牙丝兰;Faxon, Spanish Bayonet,Spanish Dagger ●☆

416584 Yucca filamentosa L. ;美国波罗花;Adam's Needle, Adam's Needle-and-thread, Adam's-needle, Adam's-needle Yucca, Bear Grass,Beargrass,Needle Palm,Silkgrass,Spanish Bayonet,Yucca ●

416585 Yucca filamentosa L. 'Garland Gold';加兰金丝兰;Garland Gold Yucca ☆ 416586 Yucca filamentosa L. = Yucca smalliana Fernald ●

416587 Yucca filamentosa L. var. concava (Haw.) Baker = Yucca flaccida Haw. ●

416588 Yucca filamentosa L. var. flaccida (Haw.) Engelm. = Yucca flaccida Haw. ●

416589 Yucca filamentosa L. var. glaucescens (Haw.) Baker = Yucca flaccida Haw. ●

416590 Yucca filamentosa L. var. puberula (Haw.) Baker = Yucca flaccida Haw. ●

416591 Yucca filamentosa L. var. smalliana (Fernald) H. E. Ahles = Yucca flaccida Haw. ●

416592 Yucca filamentosa Small = Yucca smalliana Fernald ●

416593 Yucca filifera Chabaud;垂花丝兰(线叶丝兰);St. Peter's Palm,Tree Yucca ●☆

416594 Yucca filifera Chabaud 'Ivory' = Yucca flaccida Haw. 'Ivory' ●

416595 Yucca flaccida Haw. ;下垂丝兰(软叶丝兰,细叶丝兰); Adam's Needle, Beargrass, Flaccid Leaf Yucca, Flaccid-leaf Yucca, Thinleaf Yucca,Weakleak Yucca,Yucca ●

416596 Yucca flaccida Haw. 'Ivory';象牙软叶丝兰●

416597 Yucca flaccida Haw. = Yucca smalliana Fernald ●

416598 Yucca flaccida Haw. var. glaucescens (Haw.) Trel. = Yucca flaccida Haw. ●

416599 Yucca flaccida Haw. var. major (Baker) Rehder = Yucca flaccida Haw. ●

416600 Yucca freemanii Shinners = Yucca flaccida Haw. ●

416601 Yucca funifera K. Koch = Hesperaloe funifera (K. Koch) Trel. ■☆

416602 Yucca gilbertiana (Trel.) Rydb. = Yucca harrimaniae Trel. ●☆

416603 Yucca glauca Nutt. ;小丝兰; Bear Grass, Beargrass, Dagger Weed, Great Plains Yucca, Indian Cabbage, Narrow-leaved Soaproot, Narrow-leaved Soapweed, Narrow-leaved Yucca, Small Soapweed, Small Yucca, Soaproot, Soapweed, Soapweed Yucca, Spanish Bayonet ●

416604 Yucca glauca Nutt. var. gurneyi McKelvey = Yucca glauca Nutt. ●

416605 Yucca glauca Nutt. var. stricta (J. Sims) Trel. = Yucca glauca Nutt. ●

416606 Yucca glaucescens Haw. = Yucca flaccida Haw. ●

416607 Yucca gloriosa L. ;凤尾丝兰(白棕,波罗花,剌叶玉兰,凤尾兰,华丽丝兰,剑麻,美国波罗花,雀芭蕉,丝兰);Adam's-needle, Candle Yucca, Mound Lily, Mound Lily Yucca, Mound-lily, Moundlily Yucca, Mound-lily Yucca, Palm Lily, Roman Candle, Spanish Dagger,Spanish Dagger Yucca,Spanish-dagger ●

416608 Yucca gloriosa L. 'Nobilis';高贵凤尾兰●

416609 Yucca gloriosa L. 'Variegata';斑叶凤尾兰■

416610 Yucca gloriosa L. var. recurvifolia (Salisb.) Engelm. ;弯叶丝兰(垂叶玉兰,卷叶丝兰);Curved-leaf Spanish Dagger, Curveleaf Yucca,Curve-leaf Yucca,Pendulous Yucca,Weeping Yucca ●

416611 Yucca graminifolia A. W. Wood = Hesperoyucca whipplei (Torr.) Trel. ■☆

416612 Yucca harrimaniae Trel. ;哈利丝兰;New Mexico Yucca ●☆

416613 Yucca harrimaniae Trel. var. gilbertiana Trel. = Yucca harrimaniae Trel. ●☆

416614 Yucca harrimaniae Trel. var. neomexicana (Wooton et Standl.) Reveal = Yucca neomexicana Wooton et Standl. ●☆

416615　Yucca harrimaniae Trel. var. sterilis Neese et S. L. Welsh = Yucca harrimaniae Trel. ●☆

416616　Yucca intermedia McKelvey;间型丝兰●☆

416617　Yucca intermedia McKelvey var. ramosa McKelvey = Yucca intermedia McKelvey ●☆

416618　Yucca kanabensis McKelvey = Yucca angustissima Engelm. ex Trel. var. kanabensis (McKelvey) Reveal ●☆

416619　Yucca longifolia Buckley = Yucca treculeana Carrière ●☆

416620　Yucca louisianensis Trel. = Yucca flaccida Haw. ●

416621　Yucca macrocarpa Coville = Yucca faxoniana Sarg. ●☆

416622　Yucca madrensis Gentry;马德雷丝兰;Sierra Madre Yucca ●☆

416623　Yucca mohavensis Sarg. = Yucca schidigera Roezl ex Ortgies ●☆

416624　Yucca navajoa J. M. Webber = Yucca baileyi Wooton et Standl. ●☆

416625　Yucca neomexicana Wooton et Standl. ;新墨西哥丝兰●☆

416626　Yucca newberryi McKelvey = Hesperoyucca newberryi (McKelvey) Clary ■☆

416627　Yucca nitida C. Wright ex S. Watson = Hesperoyucca whipplei (Torr.) Trel. ■☆

416628　Yucca orchioides Carrière var. major Baker = Yucca flaccida Haw. ●

416629　Yucca ortgensiana Roezl ex Ortgies = Hesperoyucca whipplei (Torr.) Trel. ■☆

416630　Yucca pallida McKelvey;苍白丝兰;Pale Yucca ●☆

416631　Yucca pallida McKelvey var. edentata (Trel.) Cory = Yucca pallida McKelvey ●☆

416632　Yucca parviflora J. Torr. = Hesperaloe parviflora (Torr.) J. M. Coult. ■☆

416633　Yucca pendula Groenl. = Yucca gloriosa L. var. recurvifolia (Salisb.) Engelm. ●

416634　Yucca puberula Haw. = Yucca flaccida Haw. ●

416635　Yucca radiosa (Engelm.) Trel. = Yucca elata Engelm. ●

416636　Yucca recurva Haw. = Yucca gloriosa L. var. recurvifolia (Salisb.) Engelm. ●

416637　Yucca recurvifolia Salisb. = Yucca gloriosa L. var. recurvifolia (Salisb.) Engelm. ●

416638　Yucca reverchonii Trel. ;勒韦雄丝兰;San Angelo Yucca ●☆

416639　Yucca rigida (Engelm.) Trel. ;坚丝兰;Blue Yucca,Palmilla ●☆

416640　Yucca rostrata Engelm. ex Trel. ;厚叶丝兰;Beaked Yucca,Big Bend Yucca,Tree Yucca ●☆

416641　Yucca rostrata Engelm. ex Trel. var. linearis Trel. = Yucca rostrata Engelm. ex Trel. ●☆

416642　Yucca rupicola Scheele;扭叶丝兰;Twisted-leaf Yucca ●☆

416643　Yucca rupicola Scheele var. edentata Trel. = Yucca pallida McKelvey ●☆

416644　Yucca rupicola Scheele var. tortifolia Engelm. = Yucca rupicola Scheele ●☆

416645　Yucca schidigera Ortgies = Yucca schidigera Roezl ex Ortgies ●☆

416646　Yucca schidigera Roezl ex Ortgies;莫哈维丝兰(屑片丝兰);Mohave Yucca,Mojave Yucca,Palmilla,Spanish Dagger ●☆

416647　Yucca schottii Urbina;苏特丝兰(丝氏丝兰);Schott Yucca,Schott's Yucca ●☆

416648　Yucca serrulata Haw. = Yucca aloifolia L. ●

416649　Yucca smalliana Fernald;丝兰(美国波罗花,洋波萝);Adam's Needle, Adam's Needle-and-thread, Adam's Needle Yucca, Bear Grass,Needle Palm,Silk Grass,Spanish Bayonet ●

416650　Yucca smalliana Fernald = Yucca flaccida Haw. ●

416651　Yucca standleyi McKelvey = Yucca baileyi Wooton et Standl. ●☆

416652　Yucca stricta J. Sims = Yucca glauca Nutt. ●

416653　Yucca tenuistyla Trel. ;细柱丝兰●☆

416654　Yucca thompsoniana Trel. ; 汤氏丝兰(喙丝兰);Bayoneta, Beaked Yucca,Thompson Yucca,Thompson's Yucca ●☆

416655　Yucca thornberi McKelvey = Yucca baccata Torr. var. brevifolia L. D. Benson et Darrow ■☆

416656　Yucca toftiae S. L. Welsh = Yucca angustissima Engelm. ex Trel. var. toftiae (S. L. Welsh) Reveal ●☆

416657　Yucca torreyi Shafer;托氏丝兰;Spanish Dagger,Torrey Yucca, Torrey's Yucca ☆

416658　Yucca torreyi Shafer = Yucca treculeana Carrière ●☆

416659　Yucca treculeana Carrière; 北美丝兰; Don Quixote's Lace, Palma Pita,Spanish Dagger ●☆

416660　Yucca treculeana Carrière var. succulenta McKelvey = Yucca treculeana Carrière ●☆

416661　Yucca treleasei J. F. Macbr. = Yucca baccata Torr. var. brevifolia L. D. Benson et Darrow ■☆

416662　Yucca undulata K. Koch = Yucca treculeana Carrière ●☆

416663　Yucca utahensis McKelvey;犹他丝兰●☆

416664　Yucca valida Brandegee;刚直丝兰;Datilillo ●☆

416665　Yucca verdiensis McKelvey = Yucca elata Engelm. var. verdiensis (McKelvey) Reveal ●☆

416666　Yucca whipplei Torr. = Hesperoyucca whipplei (Torr.) Trel. ■☆

416667　Yucca whipplei Torr. subsp. caespitosa (M. E. Jones) A. L. Haines = Hesperoyucca whipplei (Torr.) Trel. ■☆

416668　Yucca whipplei Torr. subsp. intermedia A. L. Haines = Hesperoyucca whipplei (Torr.) Trel. ■☆

416669　Yucca whipplei Torr. subsp. newberryi (McKelvey) Hochstätter = Hesperoyucca newberryi (McKelvey) Clary ■☆

416670　Yucca whipplei Torr. subsp. parishii (M. E. Jones) A. L. Haines = Hesperoyucca whipplei (Torr.) Trel. ■☆

416671　Yucca whipplei Torr. subsp. parishii (M. E. Jones) A. L. Haines = Yucca whipplei Torr. var. parishii M. E. Jones ■☆

416672　Yucca whipplei Torr. subsp. percursa A. L. Haines = Hesperoyucca whipplei (Torr.) Trel. ■☆

416673　Yucca whipplei Torr. var. caespitosa M. E. Jones = Hesperoyucca whipplei (Torr.) Trel. ■☆

416674　Yucca whipplei Torr. var. intermedia (A. L. Haines) J. M. Webber = Hesperoyucca whipplei (Torr.) Trel. ■☆

416675　Yucca whipplei Torr. var. parishii M. E. Jones = Hesperoyucca whipplei (Torr.) Trel. ■☆

416676　Yucca whipplei Torr. var. percursa (A. L. Haines) J. M. Webber = Hesperoyucca whipplei (Torr.) Trel. ■☆

416677　Yuccaceae J. Agardh = Agavaceae Dumort. (保留科名)●■

416678　Yuccaceae J. Agardh;丝兰科●■

416679　Yucea Raf. = Yucca L. ●■

416680　Yulania Spach = Magnolia L. ●

416681　Yulania Spach(1839);玉兰属●

416682　Yulania × soulangeana (Soul. -Bod.) D. L. Fu;二乔玉兰(二乔木兰, 朱砂玉兰, 紫背木兰); Chinese Magnolia, Saucer Magnolia,Tulip Magnolia,Twins Magnolia ●

416683　Yulania acuminata (L.) D. L. Fu = Magnolia acuminata (L.) L. ●☆

416684　Yulania amoena (W. C. Cheng) D. L. Fu;天目玉兰(木兰,天目木兰);Tianmu Magnolia, Tianmu Mountain Magnolia,Tianmushan Magnolia ●◇

416685　Yulania axilliflora (T. B. Chao,T. X. Zhang et J. T. Gao) D. L.

Fu＝Magnolia axilliflora（T. B. Chao, T. X. Zhang et J. T. Gao）T. B. Chao ●

416686　Yulania biondii（Pamp.）D. L. Fu；望春玉兰（春花, 法氏辛夷, 房木, 侯桃, 华中木兰, 姜朴花, 毛辛夷, 木笔花, 望春花, 望春木兰, 辛夷, 辛夷桃, 辛雉, 新雉, 迎春）；Biond's Magnolia ●

416687　Yulania biondii（Pamp.）D. L. Fu var. angustitepala D. L. Fu et al. ＝Yulania biondii（Pamp.）D. L. Fu ●

416688　Yulania campbellii（Hook. f. et Thomson）D. L. Fu；滇藏玉兰（大琼, 滇藏木兰, 二叶子厚朴, 木笔花, 辛夷）；Campbell Magnolia, Campbell's Magnolia, Pink Tulip Tree, Pink Tuliptree, Pink Tulip-tree ●

416689　Yulania conspicua（Salisb.）Spach＝Yulania denudata（Desr.）D. L. Fu ●

416690　Yulania cylindrica（E. H. Wilson）D. L. Fu；黄山玉兰（黄山木兰）；Huangshan Magnolia, Huangshan Mountain Magnolia ●◇

416691　Yulania cylindrica（E. H. Wilson）D. L. Fu＝Magnolia cylindrica E. H. Wilson ●◇

416692　Yulania dawsoniana（Rehder et E. H. Wilson）D. L. Fu；光叶玉兰（光叶木兰, 康定木兰）；Dawson Magnolia ●◇

416693　Yulania dawsoniana（Rehder et E. H. Wilson）D. L. Fu＝Magnolia dawsoniana Rehder et E. H. Wilson ●◇

416694　Yulania denudata（Desr.）D. L. Fu；玉兰（白玉兰, 春花, 房木, 侯桃, 姜朴, 姜朴花, 毛辛夷, 木笔, 木笔花, 木兰, 望春花, 辛夷, 辛夷桃, 辛雉, 新雉, 应春花, 迎春, 迎春花, 玉兰花, 玉堂春）；Jade Orchid, Lily Tree, Lily-tree, White Magnolia, White Yulan, Yulan, Yulan Magnolia, Yulan Tree ●

416695　Yulania denudata（Desr.）D. L. Fu subsp. pubescens（D. L. Fu et al.）D. L. Fu et al. ＝Yulania denudata（Desr.）D. L. Fu ●

416696　Yulania denudata（Desr.）D. L. Fu var. elongata（Rehder et E. H. Wilson）D. L. Fu et T. B. Chao＝Yulania sprengeri（Pamp.）D. L. Fu ●

416697　Yulania denudata（Desr.）D. L. Fu var. flava D. L. Fu et al. ＝Yulania denudata（Desr.）D. L. Fu ●

416698　Yulania denudata（Desr.）D. L. Fu var. pubescens D. L. Fu et al. ＝Yulania denudata（Desr.）D. L. Fu ●

416699　Yulania denudata（Desr.）D. L. Fu var. purpurascens（Maxim.）D. L. Fu＝Yulania denudata（Desr.）D. L. Fu ●

416700　Yulania denudata（Desr.）D. L. Fu var. pyramidalis（T. B. Chao et Zhi X. Chen）D. L. Fu＝Yulania denudata（Desr.）D. L. Fu ●

416701　Yulania elliptigemmata（C. L. Guo et L. L. Huang）N. H. Xia；椭蕾玉兰（椭蕾木兰）；Elliptic-buds Magnolia ●

416702　Yulania japonica Spach＝Yulania liliiflora（Desr.）D. L. Fu ●

416703　Yulania japonica Spach var. globosa（Hook. f. et Thomson）Parm. ＝Magnolia globosa Hook. f. et Thomson ●

416704　Yulania japonica Spach var. globosa（Rehder et E. H. Wilson）P. Parm. ＝Oyama globosa（Hook. f. et Thomson）N. H. Xia et C. Y. Wu ●

416705　Yulania japonica Spach var. obovata（Thunb.）P. Parm. ＝Houpoea obovata（Thunb.）N. H. Xia et C. Y. Wu ●

416706　Yulania jigongshanensis（T. B. Chao et al.）D. L. Fu；鸡公山玉兰●

416707　Yulania liliiflora（Desr.）D. L. Fu；紫玉兰（杜春花, 房木, 侯桃, 华夷, 姜朴, 毛辛夷, 木笔, 木笔花, 木兰, 木兰花, 望春花, 辛刭, 辛夷, 辛夷桃, 新雉, 迎春, 紫花木兰, 紫花玉兰）；Lily Magnolia, Lily-flowered Magnolia, Purple Magnolia, Purple Yulan ●

416708　Yulania liliiflora（Desr.）D. L. Fu＝Magnolia liliiflora Desr. ●

416709　Yulania mirifolia D. L. Fu et al. ；奇叶玉兰●

416710　Yulania multiflora（M. C. Wang et C. L. Min）D. L. Fu；多花玉兰（多花木兰）；Flowery Magnolia, Manyflower Magnolia ●

416711　Yulania multiflora（M. C. Wang et C. L. Min）D. L. Fu＝Magnolia multiflora M. C. Wang et C. L. Min ●

416712　Yulania pilocarpa（Z. Z. Zhao et Z. W. Xie）D. L. Fu；罗田玉兰（罗田木兰）；Luotian Magnolia ●

416713　Yulania pilocarpa（Z. Z. Zhao et Z. W. Xie）D. L. Fu＝Magnolia pilocarpa Z. Z. Zhao et Z. W. Xie ●

416714　Yulania pilocarpa（Z. Z. Zhao et Z. W. Xie）D. L. Fu var. ellipticifolia D. L. Fu et al. ＝Yulania pilocarpa（Z. Z. Zhao et Z. W. Xie）D. L. Fu ●

416715　Yulania pyriformis（T. D. Yang et T. C. Cui）D. L. Fu＝Yulania denudata（Desr.）D. L. Fu ●

416716　Yulania salicifolia（Siebold et Zucc.）D. L. Fu＝Magnolia salicifolia（Siebold et Zucc.）Maxim. ●☆

416717　Yulania sargentiana（Rehder et E. H. Wilson）D. L. Fu；凹叶玉兰（凹叶木兰, 苞谷树, 二月花, 厚皮, 花树子, 姜朴, 应春花）；Sargent Magnolia ●

416718　Yulania sargentiana（Rehder et E. H. Wilson）D. L. Fu＝Magnolia sargentiana Rehder et E. H. Wilson ●

416719　Yulania sinostellata（P. L. Chiu et Z. H. Chen）D. L. Fu＝Magnolia sinostellata P. L. Chiu et Z. H. Chen ●

416720　Yulania sinostellata（P. L. Chiu et Z. H. Chen）D. L. Fu＝Yulania stellata（Maxim.）N. H. Xia ●

416721　Yulania sprengeri（Pamp.）D. L. Fu；武当玉兰（春花, 大花玉兰, 二月花, 房木, 红花木兰, 侯桃, 湖北木兰, 姜朴花, 金山二月花, 毛辛夷, 木笔, 木笔花, 望春花, 武当木兰, 辛夷, 辛夷桃, 辛雉, 新雉, 应春花, 迎春, 迎春树, 朱砂玉兰, 紫玉兰）；Sprenger Magnolia, Sprenger Magnolia's ●

416722　Yulania sprengeri（Pamp.）D. L. Fu＝Magnolia sprengeri Pamp. ●

416723　Yulania stellata（Maxim.）N. H. Xia；星花玉兰●

416724　Yulania tomentosa（Thunb.）D. L. Fu＝Magnolia tomentosa Thunb. ●

416725　Yulania viridula D. L. Fu et al. ；青皮玉兰●

416726　Yulania wugangensis（T. B. Chao, W. B. Sun et Zhi X. Chen）D. L. Fu＝Magnolia wugangensis T. B. Chao, W. B. Sun et Zhi X. Chen ●

416727　Yulania zenii（W. C. Cheng）D. L. Fu；宝华玉兰（宝华木兰, 椭圆叶木兰）；Baohua Magnolia, Zen Magnolia ●◇

416728　Yulania zenii（W. C. Cheng）D. L. Fu＝Magnolia zenii W. C. Cheng ●◇

416729　Yunckeria Lundell＝Ctenardisia Ducke ●☆

416730　Yunckeria Lundell（1964）；尤恩紫金牛属●☆

416731　Yunckeria amplifolia（Standl.）Lundell；尤恩紫金牛●☆

416732　Yungasocereus F. Ritter＝Haageocereus Backeb. ●☆

416733　Yunnanea Hu＝Camellia L. ●

416734　Yunnanea xylocarpa Hu＝Camellia reticulata Lindl. ●◇

416735　Yunnanea xylocarpa Hu＝Camellia xylocarpa（Hu）Hung T. Chang ●

416736　Yunnanopilia C. Y. Wu et D. Z. Li＝Champereia Griff. ●

416737　Yunnanopilia C. Y. Wu et D. Z. Li（2000）；甜菜树属；Yunnanopilia ●★

416738　Yunnanopilia longistaminea（W. Z. Li）C. Y. Wu et D. Z. Li；甜菜树；Yunnanopilia ●

416739　Yunnanopilia longistaminea（W. Z. Li）C. Y. Wu et D. Z. Li＝Champereia longistaminea（W. Z. Li）D. D. Tao ●

416740　Yunquea Skottsb. = Centaurodendron Johow ●☆

416741　Yushania P. C. Keng = Sinarundinaria Nakai ●

416742　Yushania P. C. Keng（1957）；玉山竹属（玉山箭竹属）；Bamboo，Yushanbamboo，Yushania ●

416743　Yushania actinoseta W. T. Lin et Z. M. Wu = Yushania basihirsuta（McClure）Z. P. Wang et G. H. Ye ●

416744　Yushania ailuropodina T. P. Yi；紫斑玉山竹●

416745　Yushania alpina（K. Schum.）Lin = Sinarundinaria alpina（K. Schum.）C. S. Chao et Renvoize ●☆

416746　Yushania ambositrensis（A. Camus）Ohrnb. = Arundinaria ambositrensis A. Camus ●☆

416747　Yushania anceps（Mitford）W. C. Lin；喜马玉山竹；Himalayan Bamboo，Indian Fountain Bamboo ●

416748　Yushania andropogonoides（Hand.-Mazz.）T. P. Yi；草丝竹；Andropogonoid Yushania，Bluestemlike Yushanbamboo ●

416749　Yushania angustifolia T. P. Yi et J. Y. Shi；窄叶玉山竹●

416750　Yushania auctiaurita T. P. Yi；显耳玉山竹；Bigear Yushanbamboo，Marked-auriculate Yushania ●

416751　Yushania aztecorum McClure et E. W. Sm. = Otatea aztecorum（McClure et E. W. Sm.）C. E. Calderón et Soderstr. ●☆

416752　Yushania baishanzuensis Z. P. Wang et G. H. Ye；百山祖玉山竹；Baishanzu Yushanbamboo，Baishanzu Yushania ●◇

416753　Yushania basihirsuta（McClure）Z. P. Wang et G. H. Ye；毛玉山竹（南岭箭竹）；Hair Yushanbamboo，Hairy Yushania，Hirsute Chinacane ●

416754　Yushania bojieiana T. P. Yi；金平玉山竹；Bojiei Yushania，Jinping Yushanbamboo ●

416755　Yushania brevipaniculata（Hand.-Mazz.）T. P. Yi；短锥玉山竹；Shortawl Yushanbamboo，Short-paniculate Yushania ●

416756　Yushania brevis T. P. Yi；禄春玉山竹（绿春玉山竹）；Luchun Yushanbamboo，Luchun Yushania ●

416757　Yushania canoviridis G. H. Ye et Z. P. Wang；灰绿玉山竹；Grey-green Yushanbamboo，Greyish-green Yushania ●

416758　Yushania cartilaginea T. H. Wen；硬壳玉山竹；Bristly Yushania，Hardsheath Yushanbamboo ●

416759　Yushania cava T. P. Yi；空柄玉山竹；Emptystalk Yushanbamboo，Hollow Yushania ●

416760　Yushania chingii T. P. Yi；仁昌玉山竹（秦氏玉山竹）；Ching Yushanbamboo，Ching Yushania ●

416761　Yushania chungii（Keng）Z. P. Wang = Yushania brevipaniculata（Hand.-Mazz.）T. P. Yi ●

416762　Yushania chungii（Keng）Z. P. Wang et G. H. Ye = Yushania brevipaniculata（Hand.-Mazz.）T. P. Yi ●

416763　Yushania collina T. P. Yi；德昌玉山竹；Dechang Yushanbamboo，Dechang Yushania ●

416764　Yushania complanata T. P. Yi；梵净山玉山竹；Fanjingshan Yushanbamboo，Fanjingshan Yushania ●

416765　Yushania confusa（McClure）Z. P. Wang et G. H. Ye；鄂西玉山竹；Confused Yushania，W. Hubei Yushanbamboo ●

416766　Yushania crassicollis T. P. Yi；粗柄玉山竹（弯毛玉山竹）；Thick-collared Yushania，Thickstalk Yushanbamboo ●

416767　Yushania crispata T. P. Yi；波柄玉山竹；Crisped-stalked Yushania，Wavestalk Yushanbamboo ●

416768　Yushania dafengdingensis T. P. Yi；大风顶玉山竹●

416769　Yushania donganensis（B. M. Yang）T. P. Yi；东安玉山竹●

416770　Yushania elavata T. P. Yi；腾冲玉山竹；Elevated Yushania，Tengchong Yushanbamboo ●

416771　Yushania exilis T. P. Yi；沐川玉山竹；Muchuan Yushania，Weak Yushanbamboo ●

416772　Yushania exilis T. P. Yi var. pianmaensis J. R. Xue et R. Y. He；片马山竹（山竹）；Mountain Muchuan Yushania ●

416773　Yushania falcatiaurita J. R. Xue et T. P. Yi；独龙江玉山竹（粉竹）；Dulongjiang Yushanbamboo，Falcate-auriculate Yushania ●

416774　Yushania farcticaulis T. P. Yi；粉竹（独龙江玉山竹）；Dulong Yushania，Powder Yushanbamboo ●

416775　Yushania farinosa Z. P. Wang et G. H. Ye；湖南玉山竹；Hunan Yushanbamboo，Hunan Yushania ●

416776　Yushania flexa T. P. Yi；弯曲玉山竹（弯毛山竹）；Bendhair Yushanbamboo，Recurved Yushania ●

416777　Yushania glandulosa J. R. Xue et T. P. Yi；盈江玉山竹；Yingjiang Yushanbamboo，Yingjiang Yushania ●

416778　Yushania glauca T. P. Yi et T. L. Long；白背玉山竹；Grey-blue Yushania，Whiteback Yushanbamboo ●

416779　Yushania grammata T. P. Yi；棱纹玉山竹；Riblines Yushanbamboo，Rib-striped Yushania ●

416780　Yushania hirticaulis Z. P. Wang et G. H. Ye；毛秆玉山竹（毛竿玉山竹）；Hairpole Yushanbamboo，Hirsute-poled Yushania ●

416781　Yushania lacera Q. F. Zheng et K. F. Huang；撕裂玉山竹；Lacerate Yushania，Lancinate Yushanbamboo ●

416782　Yushania laetevirens T. P. Yi；亮绿玉山竹（亮叶玉山竹）；Bright-leaved Yushania，Lightgreen Yushanbamboo ●

416783　Yushania lanshanensis T. H. Wen；蓝山玉竹；Lanshan Yushanbamboo，Yushania ●

416784　Yushania lanshanensis T. H. Wen = Pseudosasa pubiflora（Keng）P. C. Keng ex D. Z. Li et L. M. Gao ●

416785　Yushania levigata T. P. Yi；光亮玉山竹；Bright Yushanbamboo，Levigated Yushania ●

416786　Yushania lineolata T. P. Yi；石棉玉山竹；Asbestus Yushanbamboo，Lineolate Yushania ●

416787　Yushania longiaurita Q. F. Zheng et K. F. Huang；长耳玉山竹；Long-auriculate Yushania，Longear Yushanbamboo，Long-ear Yushania ●

416788　Yushania longipilosa T. H. Wen et S. C. Chen；长毛玉山竹；Longhair Yushanbamboo，Long-haired Yushania ●

416789　Yushania longipilosa T. H. Wen et S. C. Chen = Yushania basihirsuta（McClure）Z. P. Wang et G. H. Ye ●

416790　Yushania longissima（T. P. Yi）T. P. Yi = Yushania complanata T. P. Yi ●

416791　Yushania longissima K. F. Huang et Q. F. Zheng；长鞘玉山竹；Longsheath Yushanbamboo，Long-sheathing Yushania ●

416792　Yushania longissima T. P. Yi = Yushania yadongensis T. P. Yi ●

416793　Yushania longiuscula T. P. Yi；蒙自玉山竹；Mengzi Yushanbamboo，Mengzi Yushania ●

416794　Yushania mabianensis T. P. Yi；马边玉山竹；Mabian Yushanbamboo，Mabian Yushania ●

416795　Yushania maculata T. P. Yi；斑壳玉山竹；Potsheath Yushanbamboo，Spotted Yushania ●

416796　Yushania mairei（Hack. ex Hand.-Mazz.）J. J. N. Campb. = Fargesia mairei（Hack. ex Hand.-Mazz.）T. P. Yi ●

416797　Yushania megalothyrsa（Hand.-Mazz.）T. H. Wen；阔叶玉山竹；Broadleaf Yushanbamboo，Broad-leaved Yushania ●

416798　Yushania megalothyrsa（Hand.-Mazz.）T. H. Wen = Gaoligongshania megathyrsa（Hand.-Mazz.）D. Z. Li，J. R. Xue et N. H. Xia ●

416799 Yushania menghaiensis T. P. Yi;隔界竹;Menghai Yushanbamboo, Menghai Yushania ●

416800 Yushania mitis T. P. Yi;泡滑竹;Soft Yushanbamboo, Soft Yushania ●

416801 Yushania monophylla T. P. Yi et B. M. Yang;单叶玉山竹; Singleleaf Yushanbamboo,Singleleaf Yushania ●

416802 Yushania monophylla T. P. Yi et B. M. Yang = Gelidocalamus stellatus T. H. Wen ●

416803 Yushania multiramea T. P. Yi;多枝玉山竹;Branchy Yushanbamboo,Yushania ●

416804 Yushania niitakayamensis (Hayata) P. C. Keng;玉山竹(玉山箭竹,玉山矢竹);Yushan Cane,Yushanbamboo ●

416805 Yushania oblonga T. P. Yi;马鹿竹(马六竹);Oblong Yushanbamboo,Oblong Yushania ●

416806 Yushania pachyclada T. P. Yi;粗枝玉山竹;Thicktwig Yushanbamboo,Thick-twigged Yushania ●

416807 Yushania papillosa (W. T. Lin) W. T. Lin = Yushania basihirsuta (McClure) Z. P. Wang et G. H. Ye ●

416808 Yushania pauciramificans T. P. Yi;少枝玉山竹;Poortwig Yushanbamboo,Rare-branched Yushania ●

416809 Yushania polytricha J. R. Xue et T. P. Yi;滑竹;Glabrous Yushania,Smooth Yushanbamboo ●

416810 Yushania punctulata T. P. Yi;抱鸡竹;Punctulate Yushania, Spot Yushanbamboo ●

416811 Yushania qiaojiaensis J. R. Xue et T. P. Yi;海竹;Qiaojia Yushanbamboo,Qiaojia Yushania ●

416812 Yushania qiaojiaensis J. R. Xue et T. P. Yi f. nuda T. P. Yi = Yushania qiaojiaensis J. R. Xue et T. P. Yi var. nuda (T. P. Yi) D. Z. Li et Z. H. Guo ●

416813 Yushania qiaojiaensis J. R. Xue et T. P. Yi var. nuda (T. P. Yi) D. Z. Li et Z. H. Guo;裸箨海竹;Naked Qiaojia Yushanbamboo, Naked Qiaojia Yushania ●

416814 Yushania racemosa (Munro) R. B. Majumdar = Arundinaria racemosa Munro ●

416815 Yushania randaiensis (Hayata) Kamik. = Sassafras randaiense (Hayata) Rehder ●

416816 Yushania rugosa T. P. Yi;皱叶玉山竹;Wrinkled Yushania, Wrinkleleaf Yushanbamboo ●

416817 Yushania straminea T. P. Yi;黄壳竹;Stramineous Yushania, Yellow-cupule Yushania,Yellowsheath Yushanbamboo ●

416818 Yushania suijiangnensis T. P. Yi;绥江玉山竹;Suijiang Yushanbamboo,Suijiang Yushania ●

416819 Yushania uniramosa J. R. Xue et T. P. Yi;单枝玉山竹;Single Yushanbamboo,Single-ramaled Yushania,Unibranch Yushania ●

416820 Yushania varians T. P. Yi;庐山玉山竹;Lushan Yushanbamboo, Lushan Yushania ●

416821 Yushania vigens T. P. Yi;长肩毛玉山竹;Longshoulderhair Yushanbamboo,Longshoulderhair Yushania ●

416822 Yushania violascens (Keng) T. P. Yi;紫花玉山竹(紫竿玉山竹);Purpleflower Yushanbamboo,Purple-flowered Yushania ●

416823 Yushania weixiensis T. P. Yi;竹扫子;Bamboobroom,Weixi Yushania ●

416824 Yushania wuyishanensis Q. F. Zheng et K. F. Huang;武夷山玉山竹;Wuyishan Yushanbamboo,Wuyishan Yushania ●

416825 Yushania xizangensis T. P. Yi;西藏玉山竹;Xizang Tibet, Xizang Yushanbamboo,Yushania ●

416826 Yushania yadongensis T. P. Yi;亚东玉山竹(长柄箭竹);

Yadong Yushanbamboo,Yadong Yushania ●

416827 Yushania yunnanensis (J. R. Xue et T. P. Yi) P. C. Keng et T. H. Wen ex T. H. Wen = Fargesia yunnanensis J. R. Xue et T. P. Yi ●

416828 Yushania yunnanensis (J. R. Xue et T. P. Yi) P. C. Keng et T. H. Wen = Fargesia yunnanensis J. R. Xue et T. P. Yi ●

416829 Yushunia Kamik. = Sassafras J. Presl ●

416830 Yushunia randaiensis (Hayata) Kamik. = Sassafras randaiense (Hayata) Rehder ●

416831 Yutajea Steyerm. (1987);尤塔茜属 ☆

416832 Yutajea liesneri Steyerm.;尤塔茜 ☆

416833 Yuyba (Barb. Rodr.) L. H. Bailey = Bactris Jacq. ●

416834 Yuyba L. H. Bailey = Bactris Jacq. ●

416835 Yvesia A. Camus (1927);马岛臂形草属■☆

416836 Yvesia madagascariensis A. Camus;马岛臂形草■☆

416837 Zaa Baill. = Phyllarthron DC. ex Meisn. ●☆

416838 Zaa ilicifolia (Pers.) Baill. = Phyllarthron ilicifolium (Pers.) H. Perrier ●☆

416839 Zabelia (Rehder) Makino = Abelia R. Br. ●

416840 Zabelia (Rehder) Makino(1948);扎氏六道木属(六道木属)●

416841 Zabelia biflora (Turcz.) Makino;六道木(二花六道木,鸡骨头,交翅,六条木,神仙菜,神仙叶子,双花六道木);Biflorous Abelia,Twinflower Abelia ●

416842 Zabelia biflora (Turcz.) Makino = Abelia biflora Turcz. ●

416843 Zabelia brachystemon (Diels) Golubk. = Abelia dielsii (Graebn.) Rehder ●

416844 Zabelia brachystemon (Diels) Golubk. = Zabelia biflora (Turcz.) Makino ●

416845 Zabelia buddleioides (W. W. Sm.) Hisauti et H. Hara = Abelia buddleioides W. W. Sm. ●

416846 Zabelia buddleioides (W. W. Sm.) Hisauti et H. Hara = Zabelia triflora (R. Br. ex Wall.) Makino ●

416847 Zabelia buddleioides (W. W. Sm.) Hisauti et H. Hara var. divergens (W. W. Sm.) Golubk. = Abelia buddleioides W. W. Sm. ●

416848 Zabelia buddleioides (W. W. Sm.) Hisauti et H. Hara var. divergens (W. W. Sm.) Golubk. = Zabelia triflora (R. Br. ex Wall.) Makino ●

416849 Zabelia buddleioides (W. W. Sm.) Hisauti et H. Hara var. stenantha (Hand. -Mazz.) Hisauti et H. Hara = Zabelia triflora (R. Br. ex Wall.) Makino ●

416850 Zabelia buddleioides (W. W. Sm.) Hisauti et H. Hara var. stenantha (W. W. Sm.) Hisauti et Hara = Abelia buddleioides W. W. Sm. ●

416851 Zabelia dielsii (Graebn.) Makino = Abelia dielsii (Graebn.) Rehder ●

416852 Zabelia dielsii (Graebn.) Makino = Zabelia biflora (Turcz.) Makino ●

416853 Zabelia integrifolia (Koidz.) Makino ex Ikuse et S. Kuros.;全叶六道木(全缘叶六道木)●☆

416854 Zabelia stenantha (Hand. -Mazz.) Golubk. = Abelia buddleioides W. W. Sm. ●

416855 Zabelia stenantha (Hand. -Mazz.) Golubk. = Zabelia triflora (R. Br. ex Wall.) Makino ●

416856 Zabelia triflora (R. Br. ex Wall.) Makino;三花六道木(醉鱼草状六道木)●

416857 Zabelia triflora (R. Br.) Makino = Abelia triflora R. Br. ex Wall. ●

416858 Zabelia triflora (R. Br.) Makino = Zabelia triflora (R. Br. ex

Wall.) Makino ●

416859　Zacateza Bullock(1954);扎卡萝藦属●☆

416860　Zacateza pedicellata（K. Schum.）Bullock;扎卡萝藦●☆

416861　Zacintha Mill. = Crepis L. ■

416862　Zacintha Vell. (1829);肖全缘轮叶属●☆

416863　Zacintha Vell. = Clavija Ruiz et Pav. ●☆

416864　Zacintha nutans Vell. ;肖全缘轮叶●☆

416865　Zacintha nutans Vell. = Clavija nutans（Vell. ）B. Stahl ●☆

416866　Zacintha verrucaria Desf. = Crepis zacintha（L. ）Babc. ■☆

416867　Zacintha verrucosa Gaertn. = Crepis zacintha（L. ）Babc. ■☆

416868　Zacyntha Adans. = Crepis L. ■

416869　Zacyntha Adans. = Zacintha Mill. ■

416870　Zaczatea Baill. = Raphionacme Harv. ■☆

416871　Zaczatea angolensis（K. Schum. ）Baill. = Raphionacme angolensis（K. Schum. ）N. E. Br. ■☆

416872　Zaga Raf. = Adenanthera L. ●

416873　Zagrosia Speta(1998);察格罗风信子属■☆

416874　Zagrosia persica（Hausskn. ）Speta;察格罗风信子■☆

416875　Zahlbruckera Steud. = Ebermaiera Nees ■

416876　Zahlbruckera Steud. = Hygrophila R. Br. ●■

416877　Zahlbruckera Steud. = Zahlbrucknera Pohl ex Nees ■

416878　Zahlbrucknera Pohl ex Nees = Ebermaiera Nees ■

416879　Zahlbrucknera Pohl ex Nees = Hygrophila R. Br. ●■

416880　Zahlbrucknera Rchb. (1832);欧洲虎耳草属■☆

416881　Zahlbrucknera Rchb. = Saxifraga L. ■

416882　Zahlbrucknera paradoxa（Sternb. ）Rchb. = Saxifraga paradoxa Sternb. ■☆

416883　Zahlbrucknera paradoxa Rchb. ;欧洲虎耳草■☆

416884　Zahleria Luer = Masdevallia Ruiz et Pav. ■☆

416885　Zahleria Luer(1790);察尔兰属■☆

416886　Zala Lour. = Pistia L. ■

416887　Zala asiatica Lour. = Pistia stratiotes L. ■

416888　Zalacca Blume = Salacca Reinw. ●

416889　Zalacca Reinw. ex Blume = Salacca Reinw. ●

416890　Zalacca Rumph. = Salacca Reinw. ●

416891　Zalaccella Becc. = Calamus L. ●

416892　Zaleia Steud. = Zaleya Burm. f. ■☆

416893　Zaleja Burm. f. = Zaleya Burm. f. ■☆

416894　Zaleya Burm. f. (1768);裂盖海马齿属(扎利草属)■☆

416895　Zaleya camillei（Cordem. ）H. E. K. Hartmann;马岛裂盖海马齿■☆

416896　Zaleya decandra（L. ）Burm. f. ;裂盖海马齿■☆

416897　Zaleya govindia（G. Don）Nair = Zaleya pentandra（L. ）C. Jeffrey ■☆

416898　Zaleya pentandra（L. ）C. Jeffrey;五蕊裂盖海马齿(五蕊假海马齿,五蕊扎利草);Horse Purslane ■☆

416899　Zaleya redimita（Melville）H. E. K. Hartmann;缠绕裂盖海马齿■☆

416900　Zaleya sennii（Chiov. ）C. Jeffrey;森恩裂盖海马齿■☆

416901　Zalitea Raf. = Euphorbia L. ●■

416902　Zallia Roxb. = Trianthema L. ■

416903　Zallia Roxb. = Zaleya Burm. f. ■☆

416904　Zalmaria B. D. Jacks. = Rondeletia L. ●

416905　Zalmaria B. D. Jacks. = Zamaria Raf. ●

416906　Zalucania Steud. = Zaluzania Pers. ■☆

416907　Zaluzania Pers. (1807);中美菊属(黄带菊属,扎卢菊属)■☆

416908　Zaluzania Pers. = Bertiera Aubl. ■☆

416909　Zaluzania augusta Sch. Bip. ;庄严中美菊■☆

416910　Zaluzania grayana B. L. Rob. et Greenm. ;格雷中美菊;Yellow Streamers ■☆

416911　Zaluzania parthenioides（DC. ）Rzed. ;菊状中美菊■☆

416912　Zaluzania robinsonii Sharp;罗宾中美菊■☆

416913　Zaluzania trilobata Hoffmanns. ;三裂中美菊■☆

416914　Zaluziana Link = Zaluzania Pers. ■☆

416915　Zaluzianskia Benth. et Hook. f. = Zaluzianskya F. W. Schmidt（保留属名）■☆

416916　Zaluzianskia Neck. (废弃属名) = Zaluzianskya F. W. Schmidt（保留属名）■☆

416917　Zaluzianskya F. W. Schmidt(1793)（保留属名）;红蕾花属;Night Phlox,Night-phlox ■☆

416918　Zaluzianskya acutiloba Hilliard;尖浅裂红蕾花■☆

416919　Zaluzianskya affinis Hilliard;近缘红蕾花■☆

416920　Zaluzianskya africana Hiern = Zaluzianskya pumila（Benth. ）Walp. ■☆

416921　Zaluzianskya alpestris Diels = Glumicalyx nutans（Rolfe）Hilliard et B. L. Burtt ■☆

416922　Zaluzianskya angustifolia Hilliard et B. L. Burtt;窄叶红蕾花■☆

416923　Zaluzianskya aschersoniana Schinz = Zaluzianskya benthamiana Walp. ■☆

416924　Zaluzianskya bella Hilliard;雅致红蕾花■☆

416925　Zaluzianskya benthamiana Walp. ;本瑟姆红蕾花■☆

416926　Zaluzianskya bolusii Hiern = Zaluzianskya benthamiana Walp. ■☆

416927　Zaluzianskya capensis（L. ）Walp. ;红蕾花;Night Phlox ■☆

416928　Zaluzianskya capensis（L. ）Walp. var. foliosa（Benth. ）Walp. = Zaluzianskya schmitziae Hilliard et B. L. Burtt ■☆

416929　Zaluzianskya capensis（L. ）Walp. var. glabriuscula Walp. = Zaluzianskya capensis（L. ）Walp. ■☆

416930　Zaluzianskya capensis（L. ）Walp. var. hirsuta（Benth. ）Walp. = Zaluzianskya capensis（L. ）Walp. ■☆

416931　Zaluzianskya capensis（L. ）Walp. var. tenuifolia（Benth. ）Walp. = Zaluzianskya capensis（L. ）Walp. ■☆

416932　Zaluzianskya chrysops Hilliard et B. L. Burtt;金黄红蕾花■☆

416933　Zaluzianskya collina Hiern;山丘红蕾花■☆

416934　Zaluzianskya coriacea（Benth. ）Walp. = Zaluzianskya capensis（L. ）Walp. ■☆

416935　Zaluzianskya crocea Schltr. ;镉黄红蕾花■☆

416936　Zaluzianskya dentata（Benth. ）Walp. = Zaluzianskya capensis（L. ）Walp. ■☆

416937　Zaluzianskya diandra Diels;二蕊红蕾花■☆

416938　Zaluzianskya distans Hiern;远离红蕾花■☆

416939　Zaluzianskya divaricata（Thunb. ）Walp. ;叉开红蕾花■☆

416940　Zaluzianskya elgonensis Hedberg;埃尔贡红蕾花■☆

416941　Zaluzianskya elongata Hilliard et B. L. Burtt;伸长红蕾花■☆

416942　Zaluzianskya falciloba Diels = Zaluzianskya pumila（Benth. ）Walp. ■☆

416943　Zaluzianskya flanaganii Hiern = Glumicalyx flanaganii（Hiern）Hilliard et B. L. Burtt ■☆

416944　Zaluzianskya gilioides Schltr. = Zaluzianskya peduncularis（Benth. ）Walp. ■☆

416945　Zaluzianskya glandulosa Hilliard;具腺红蕾花■☆

416946　Zaluzianskya glareosa Hilliard et B. L. Burtt;石砾红蕾花■☆

416947　Zaluzianskya goseloides Diels = Glumicalyx goseloides（Diels）Hilliard et B. L. Burtt ■☆

416948　Zaluzianskya gracilis Hilliard;纤细红蕾花■☆

416949　Zaluzianskya inflata Diels；膨胀红蕾花■☆

416950　Zaluzianskya karrooica Hilliard；卡卢红蕾花■☆

416951　Zaluzianskya katharinae Hiern；卡塔琳娜红蕾花■☆

416952　Zaluzianskya lanigera Hilliard；绵毛红蕾花■☆

416953　Zaluzianskya latifolia Schinz ex O. Hoffm. et Muschl. = Zaluzianskya distans Hiern ■☆

416954　Zaluzianskya longiflora（Benth.）Walp. = Zaluzianskya capensis（L.）Walp. ■☆

416955　Zaluzianskya lychnidea（D. Don）Walp. = Zaluzianskya maritima（L. f.）Walp. ■☆

416956　Zaluzianskya maritima（L. f.）Walp. ；滨海红蕾花■☆

416957　Zaluzianskya maritima（L. f.）Walp. var. atropurpurea Hiern = Zaluzianskya microsiphon（Kuntze）K. Schum. ■☆

416958　Zaluzianskya maritima（L. f.）Walp. var. breviflora Hiern = Zaluzianskya microsiphon（Kuntze）K. Schum. ■☆

416959　Zaluzianskya maritima（L. f.）Walp. var. fragrantissima Hiern = Zaluzianskya maritima（L. f.）Walp. ■☆

416960　Zaluzianskya maritima（L. f.）Walp. var. grandiflora Hiern = Zaluzianskya microsiphon（Kuntze）K. Schum. ■☆

416961　Zaluzianskya maritima（L. f.）Walp. var. pubens Hiern = Zaluzianskya pilosa Hilliard et B. L. Burtt ■☆

416962　Zaluzianskya marlothii Hilliard；马洛斯红蕾花■☆

416963　Zaluzianskya microsiphon（Kuntze）K. Schum. ；小管红蕾花■☆

416964　Zaluzianskya minima（Hiern）Hilliard；小红蕾花■☆

416965　Zaluzianskya mirabilis Hilliard；奇异红蕾花■☆

416966　Zaluzianskya montana Hiern = Zaluzianskya ovata（Benth.）Walp. ■☆

416967　Zaluzianskya muirii Hilliard et B. L. Burtt；缪里红蕾花■☆

416968　Zaluzianskya natalensis Bernh. ；纳塔尔红蕾花■☆

416969　Zaluzianskya nemesioides Diels = Reyemia nemesioides（Diels）Hilliard ■☆

416970　Zaluzianskya oreophila Hilliard et B. L. Burtt；喜山红蕾花■☆

416971　Zaluzianskya ovata（Benth.）Walp. ；卵状红蕾花■☆

416972　Zaluzianskya ovata Walp. = Zaluzianskya ovata（Benth.）Walp. ■☆

416973　Zaluzianskya pachyrrhiza Hilliard et B. L. Burtt；粗根红蕾花■☆

416974　Zaluzianskya parviflora Hilliard；小花红蕾花■☆

416975　Zaluzianskya peduncularis（Benth.）Walp. ；梗花红蕾花■☆

416976　Zaluzianskya peduncularis（Benth.）Walp. var. glabriuscula？ = Zaluzianskya peduncularis（Benth.）Walp. ■☆

416977　Zaluzianskya peduncularis（Benth.）Walp. var. hirsuta？ = Zaluzianskya peduncularis（Benth.）Walp. ■☆

416978　Zaluzianskya peduncularis Walp. = Zaluzianskya peduncularis（Benth.）Walp. ■☆

416979　Zaluzianskya pilosa Hilliard et B. L. Burtt；疏毛红蕾花■☆

416980　Zaluzianskya pseudafricana Paclt = Zaluzianskya pumila（Benth.）Walp. ■☆

416981　Zaluzianskya pulvinata Killick；叶枕红蕾花■☆

416982　Zaluzianskya pumila（Benth.）Walp. ；偃伏红蕾花■☆

416983　Zaluzianskya pusilla（Benth.）Walp. ；微小红蕾花■☆

416984　Zaluzianskya ramosa Schinz ex Hiern = Zaluzianskya benthamiana Walp. ■☆

416985　Zaluzianskya rubrostellata Hilliard et B. L. Burtt；红星红蕾花■☆

416986　Zaluzianskya schmitziae Hilliard et B. L. Burtt；施密茨红蕾花■☆

416987　Zaluzianskya selaginoides（Thunb.）Walp. = Zaluzianskya villosa F. W. Schmidt ■☆

416988　Zaluzianskya selaginoides（Thunb.）Walp. var. glabra（Benth.）

416989　Walp. = Zaluzianskya affinis Hilliard ■☆

416989　Zaluzianskya selaginoides（Thunb.）Walp. var. parviflora（Benth.）Walp. = Polycarena silenoides Harv. ex Benth. ■☆

416990　Zaluzianskya spathacea（Benth.）Walp. ；佛焰苞红蕾花■☆

416991　Zaluzianskya sutherlandica Hilliard；萨瑟兰红蕾花■☆

416992　Zaluzianskya tropicalis Hilliard；热带红蕾花■☆

416993　Zaluzianskya venusta Hilliard；雅丽红蕾花■☆

416994　Zaluzianskya villosa F. W. Schmidt；毛红蕾花■☆

416995　Zaluzianskya villosa F. W. Schmidt var. glabra（Benth.）Hiern = Zaluzianskya affinis Hilliard ■☆

416996　Zaluzianskya villosa F. W. Schmidt var. parviflora（Benth.）Hiern = Zaluzianskya affinis Hilliard ■☆

416997　Zaluzianskya violacea Schltr. ；堇色红蕾花■☆

416998　Zamaria Raf. = Rondeletia L. ●

416999　Zameioscirpus Dhooge et Goetgh. (2003)；细藨草属■☆

417000　Zameioscirpus Dhooge et Goetgh. = Isolepis R. Br. ■

417001　Zameioscirpus atacamensis（Phil.）Dhooge et Goetgh. 阿地细藨草■☆

417002　Zamia L. (1763)；泽米苏铁属（大苏铁属，泽米铁属，泽米属）；Zamia, Coontie ●☆

417003　Zamia angustifolia Jacq. ；窄叶大苏铁；Narrowleaf Zamia ●☆

417004　Zamia encephalartoides D. W. Stev. ；安氏泽米●☆

417005　Zamia fischeri Miq. ；菲舍尔泽米（菲氏大苏铁，菱叶凤尾蕉）●☆

417006　Zamia floridana A. DC. ；大苏铁（大头苏铁，佛罗里达苏铁，佛罗里达泽米，佛州苏铁）；Comtie, Coontie, Florida Zamia ●☆

417007　Zamia floridana A. DC. = Zamia angustifolia Jacq. ●☆

417008　Zamia floridana A. DC. = Zamia integrifolia L. f. ●☆

417009　Zamia friderici-guilielmi Hort. = Dioon edule Lindl. ●☆

417010　Zamia furfuracea Aiton；鳞秕大苏铁（鳞秕泽米，鳞枇苏铁）；Bran Zamia, Cardboard Palm ●☆

417011　Zamia horrida Jacq. = Encephalartos horridus Lehm. ●☆

417012　Zamia integrifolia Aiton = Zamia integrifolia L. f. ●☆

417013　Zamia integrifolia L. f. ；全缘大苏铁（全叶泽米）；Coontie, Coonties, Florida Arrowroot, Smooth Zamia ●☆

417014　Zamia kickxii Miq. ；古巴苏铁●☆

417015　Zamia latifolia Lodd. ；宽叶大苏铁；Broadleaf Zamia ●☆

417016　Zamia loddigesii Miq. ；劳氏大苏铁●☆

417017　Zamia pumila L. ；矮泽米（美洲苏铁）；Coontie, Florida Arrowroot ●☆

417018　Zamia rigida Karw. ex J. Schust. = Dioon edule Lindl. ●☆

417019　Zamia roezlii Linden；罗兹泽米（罗氏泽米）●☆

417020　Zamia silvicola Small = Zamia integrifolia L. f. ●☆

417021　Zamia skinneri Warsz. ；斯氏大苏铁●☆

417022　Zamia splendens Schutzman；华美泽米●☆

417023　Zamia umbrosa Small = Zamia integrifolia L. f. ●☆

417024　Zamiaceae Horan. (1834)；泽米苏铁科（泽米科）；Sago-palm Family ●☆

417025　Zamiaceae Rchb. = Zamiaceae Horan. ●☆

417026　Zamioculcas Schott (1856)；金钱树属（美铁芋属，雪铁芋属，雪芋属）■☆

417027　Zamioculcas boivinii Decne. = Gonatopus boivinii（Decne.）Engl. ■☆

417028　Zamioculcas lanceolata Peter = Zamioculcas zamiifolia（Lodd.）Engl. ■☆

417029　Zamioculcas loddigesii Schott；金钱树（绿元宝）■☆

417030　Zamioculcas loddigesii Schott = Zamioculcas zamiifolia（Lodd.）Engl. ■☆

417031　Zamioculcas zamiifolia（Lodd.）Engl. ；雪铁芋■☆

417032 Zamioculcas zamiifolia（Lodd.）Engl. = Caladium zamiifolium Lodd. ■☆

417033 Zamzela Raf. = Hirtella L. ●☆

417034 Zandera D. L. Schulz(1988);赞德菊属☆

417035 Zandera andersoniae（Turner）D. L. Schulz;安氏赞德菊■☆

417036 Zandera blakei（McVaugh et Lask.）D. L. Schulz;布氏赞德菊■☆

417037 Zandera hartmanii（Turner）D. L. Schulz;赞德菊■☆

417038 Zanha Hiern(1896);赞哈木属●☆

417039 Zanha africana（Radlk.）Exell;非洲赞哈木●☆

417040 Zanha golungensis Hiern;赞哈木●☆

417041 Zanha suaveolens Capuron;甜赞哈木☆

417042 Zanichelia Gilib. = Zannichellia L. ■

417043 Zanichellia Roth = Zannichellia L. ■

417044 Zannichallia Reut. = Zannichellia L. ■

417045 Zannichellia L.（1753）;角果藻属;Horned Pondweed, Horned-pondweed, Poolmat ■

417046 Zannichellia aschersoniana Graebn. = Pseudalthenia aschersoniana（Graebn.）Hartog ■☆

417047 Zannichellia contorta（Desf.）Cham. et Schltdl.;缠扭角果藻■☆

417048 Zannichellia laevis C. Presl;平滑角果藻■☆

417049 Zannichellia macrostemon J. Gay = Zannichellia contorta（Desf.）Cham. et Schltdl. ■☆

417050 Zannichellia major（Hartm.）Boenn. ex Rchb. = Zannichellia palustris L. ■

417051 Zannichellia major Boenn.;大角果藻■☆

417052 Zannichellia obtusifolia Talavera et Garcia Mur. et Smit;钝叶角果藻■☆

417053 Zannichellia palustris L.;角果藻（角茨藻,丝葛藻）;Common Poolmat, Horned Pondweed, Horned-pondweed, Poolmat ■

417054 Zannichellia palustris L. subsp. pedicellata（Wahlenb. et Rosén）Hook. f. = Zannichellia palustris L. ■

417055 Zannichellia palustris L. subsp. pedicellata（Wahlenb. et Rosén）Hook. f. = Zannichellia palustris L. var. pedicellata Wahlenb. et Rosen ■

417056 Zannichellia palustris L. subsp. pedicellata Wahlenb. et Rosen = Zannichellia palustris L. ■

417057 Zannichellia palustris L. var. indica Graebn. = Zannichellia palustris L. ■

417058 Zannichellia palustris L. var. japonica Makino = Zannichellia palustris L. ■

417059 Zannichellia palustris L. var. major（Hartm.）W. D. J. Koch = Zannichellia palustris L. ■

417060 Zannichellia palustris L. var. pedicellata Wahlenb. et Rosén;柄果角果藻（长柄角果藻）■

417061 Zannichellia palustris L. var. pedicellata Wahlenb. et Rosén = Zannichellia pedunculata Rchb. ■

417062 Zannichellia palustris L. var. pedicellata Wahlenb. et Rosén = Zannichellia palustris L. ■

417063 Zannichellia palustris L. var. pedunculata（Rchb.）A. Gray = Zannichellia palustris L. var. pedicellata Wahlenb. et Rosen ■

417064 Zannichellia palustris L. var. pedunculata（Rchb.）A. Gray = Zannichellia palustris L. ■

417065 Zannichellia palustris L. var. radicans Wallman = Zannichellia peltata Bertol. ■☆

417066 Zannichellia palustris L. var. repens（Boenn.）Koch = Zannichellia peltata Bertol. ■☆

417067 Zannichellia palustris L. var. stenophylla Asch. et Graebn. =

Zannichellia palustris L. ■

417068 Zannichellia pedicellata（Wahlenb. et Rosen）Fr. = Zannichellia palustris L. var. pedicellata Wahlenb. et Rosen ■

417069 Zannichellia pedunculata Rchb. = Zannichellia palustris L. var. pedicellata Wahlenb. et Rosen ■

417070 Zannichellia pedunculata Rchb. = Zannichellia palustris L. ■

417071 Zannichellia peltata Bertol.;盾状角果藻■☆

417072 Zannichellia stylaris C. Presl = Zannichellia palustris L. ■

417073 Zannichelliaceae Chevall.（1827）（保留科名）;角果藻科（角茨藻科）;Horned-pondweed Family, Pondweed Family, Poolmat Family ■

417074 Zannichelliaceae Dumort. = Potamogetonaceae Bercht. et J. Presl（保留科名）■

417075 Zannichelliaceae Dumort. = Zannichelliaceae Chevall.（保留科名）■

417076 Zanonia Cram. = Campelia Rich. ■

417077 Zanonia L.（1753）;翅子瓜属;Zanonia ●■

417078 Zanonia clavigera Wall. = Neoalsomitra clavigera（Roem.）Hutch. ●

417079 Zanonia clavigera Wall. = Zanonia indica L. ■

417080 Zanonia heterosperma Wall. = Hemsleya heterosperma（Wall.）C. Jeffrey ■

417081 Zanonia indica L.;翅子瓜（百症藤,棒锤瓜,穿山龙,苦藤,罗锅底,曲莲,赛金刚,细叶罗锅底）;Common Zanonia, Entireleaf Clabgourd, Entireleaf Neoalsomitra, Zanonia ■

417082 Zanonia indica L. var. pubescens Cogn.;滇南翅子瓜;Pubescent Zanonia ●■

417083 Zanonia laxa Wall. = Gynostemma laxum（Wall.）Cogn. ■

417084 Zanonia pedata（Blume）Miq. = Gynostemma pentaphyllum（Thunb.）Makino ■

417085 Zanonia pedata Miq. = Gynostemma pentaphyllum（Thunb.）Makino ■

417086 Zanonia philippinensis Merr.;菲律宾马蹄莲■☆

417087 Zanonia wightiana Arn. = Gynostemma laxum（Wall.）Cogn. ■

417088 Zanoniaceae Dumort.;翅子瓜科●■

417089 Zanoniaceae Dumort. = Cucurbitaceae Juss.（保留科名）●■

417090 Zantedeschia C. Koch = Schismatoglottis Zoll. et Moritzi ■

417091 Zantedeschia K. Koch = Schismatoglottis Zoll. et Moritzi ■

417092 Zantedeschia Spreng.（1826）（保留属名）;马蹄莲属;Altar Lily, Calla, Calla Lilies, Calla Lily, Callalily, Trumpet Lily ■

417093 Zantedeschia aethiopica（L.）Spreng.;马蹄莲（地涌金莲,野芋）;Altar Lily, Arum Lily, Calla Lily, Callalily, Common Calla, Common Calla Lily, Common Callalily, Devil's Button, Devil's Candles, Florist's Calla, Golden Calla, Horn Flower, Lily of Nile, Lily-of-the-nile, Trumpet Lily, Water Lily, White Arum, White Arum Lily ■

417094 Zantedeschia aethiopica（L.）Spreng. 'Crowborough';克罗伯勒马蹄莲■☆

417095 Zantedeschia aethiopica（L.）Spreng. 'Green Goddess';绿仙女马蹄莲■☆

417096 Zantedeschia aethiopica（L.）Spreng. var. minor Engl. = Zantedeschia aethiopica（L.）Spreng. ■

417097 Zantedeschia albomaculata（Hook.）Baill.;白马蹄莲（星点马蹄莲,银星马蹄莲）;Black-throated Arum, Spotted Arum Lily, Spotted Calla Lily, Spotted Callalily, White Callalily ■

417098 Zantedeschia albomaculata（Hook.）Baill. subsp. macrocarpa（Engl.）Letty;大果马蹄莲■☆

417099 Zantedeschia albomaculata（Hook.）Baill. subsp. valida Letty =

Zantedeschia valida (Letty) Y. Singh ■☆

417100 Zantedeschia angustiloba (Schott) Engl. ;狭裂马蹄莲;Narrow-lobed Callalily ■☆

417101 Zantedeschia angustiloba (Schott) Engl. = Zantedeschia albomaculata (Hook.) Baill. ■

417102 Zantedeschia calyptrata K. Koch = Schismatoglottis calyptrata (Roxb.) Zoll. et Moritzi ■

417103 Zantedeschia chloroleuca Engl. et Gilg = Zantedeschia albomaculata (Hook.) Baill. ■

417104 Zantedeschia elliottiana (W. Watson) Engl. ;黄花马蹄莲; Golden Arum, Golden Arum Lily, Golden Calla, Golden Callalily, Yellow Arum Lily ■☆

417105 Zantedeschia hastata (Hook.) Engl. ;澳非马蹄莲■☆

417106 Zantedeschia jucunda Letty;愉悦马蹄莲■☆

417107 Zantedeschia macrocarpa Engl. = Zantedeschia albomaculata (Hook.) Baill. subsp. macrocarpa (Engl.) Letty ■☆

417108 Zantedeschia melanoleuca (Hook. f.) Engl. ;紫心黄马蹄莲 (黑喉马蹄莲);Blackthroat Callalily, Blackwhite Callalily ■

417109 Zantedeschia melanoleuca (Hook. f.) Engl. = Zantedeschia albomaculata (Hook.) Baill. ■

417110 Zantedeschia melanoleuca (Hook. f.) Engl. var. concolor Burtt Davy = Zantedeschia albomaculata (Hook.) Baill. subsp. macrocarpa (Engl.) Letty ■☆

417111 Zantedeschia melanoleuca (Hook. f.) Engl. var. tropicalis (N. E. Br.) Traub = Zantedeschia albomaculata (Hook.) Baill. ■

417112 Zantedeschia occulta (Lour.) Spreng. = Homalomena occulta (Lour.) Schott ■

417113 Zantedeschia occulta Spreng. = Homalomena occulta (Lour.) Schott ■

417114 Zantedeschia oculata (Lindl.) Engl. = Zantedeschia albomaculata (Hook.) Baill. ■

417115 Zantedeschia oculata (Lour.) Engl. = Homalomena occulta (Lour.) Schott ■

417116 Zantedeschia odorata P. L. Perry;芳香马蹄莲■☆

417117 Zantedeschia pentlandii (Whyte ex W. Watson) Wittm. ;加州马蹄莲;California Calla ■☆

417118 Zantedeschia rehmannii Engl. ;红花马蹄莲(红马蹄莲);Calla Lily, Pink Arum, Pink Calla, Red Calla, Red Callalily, Rose Callalily, Trumpet Lily ■

417119 Zantedeschia stehmannii Sprenger = Zantedeschia rehmannii Engl. ■

417120 Zantedeschia tropicalis (N. E. Br.) Letty = Zantedeschia albomaculata (Hook.) Baill. ■

417121 Zantedeschia valida (Letty) Y. Singh;刚直马蹄莲■☆

417122 Zanthorhiza L' Hér. = Xanthorhiza Marshall ●☆

417123 Zanthorhiza apiifolia L'Hér. = Zanthorhiza simplicissima Marshall ●☆

417124 Zanthorhiza simplicissima Marshall;黄根木(黄根,黄根树,全单黄根树);Shrub Yellow Root, Shrub Yellowroot, Shrub Yellow-root, Yellowroot ●☆

417125 Zanthoxilon Franch. et Sav. = Zanthoxylum L. ●

417126 Zanthoxylaceae Bercht. et J. Presl = Rutaceae Juss. (保留科名)●■

417127 Zanthoxylaceae Martinov = Zanthoxylaceae Nees et Mart. ●

417128 Zanthoxylaceae Nees et Mart. ;花椒科●

417129 Zanthoxylaceae Nees et Mart. = Rutaceae Juss. (保留科名)●■

417130 Zanthoxylon Walter = Zanthoxylum L. ●

417131 Zanthoxylon connaroides Wight et Arn. = Trichilia connaroides (Wight et Arn.) Bentv. ●

417132 Zanthoxylum L. (1753);花椒属;Knobthorn, Knobwood, Prickly Ash, Pricklyash, Prickly-ash, Sansho, Yellow Wood, Zanthoxylum ●

417133 Zanthoxylum acanthophyllum Hayata = Zanthoxylum simulans Hance ●

417134 Zanthoxylum acanthopodium DC. ;刺花椒(毛刺花椒,木本化血丹,岩花椒,岩椒,野花椒);Spine Pricklyash, Spiny Pricklyash, Spiny Prickly-ash ●

417135 Zanthoxylum acanthopodium DC. var. deminutum (Rehder) Reeder et S. Y. Cheo = Zanthoxylum ovalifolium Wight ●

417136 Zanthoxylum acanthopodium DC. var. oligotrichum Z. M. Tan = Zanthoxylum acanthopodium DC. var. timbor Hook. f. ●

417137 Zanthoxylum acanthopodium DC. var. timbor Hook. f. ;毛刺花椒(狗花椒,木本化血丹,岩椒,野花椒,紫色果);Hairy Pricklyash, Villose Spiny Pricklyash ●

417138 Zanthoxylum acanthopodium DC. var. villosum C. C. Huang = Zanthoxylum acanthopodium DC. var. timbor Hook. f. ●

417139 Zanthoxylum ailanthoides Siebold et Zucc. ;椿叶花椒(艾油,艾子,凹头花椒,茶萸,樗叶花椒,刺椒,刺楸,蓟江某,大叶蓟葱,檫子,红刺葱,辣米油,辣子,满天星,木满天星,食茱萸,蔱,越椒); Ailanthus Prickly-ash, Ailanthusleaf Pricklyash, Ailanthus-like Pricklyash, Ailanthus-like Prickly-ash, Hemsley Pricklyash ●

417140 Zanthoxylum ailanthoides Siebold et Zucc. f. espinosum Yonek. ;无刺椿叶花椒●☆

417141 Zanthoxylum ailanthoides Siebold et Zucc. f. inermis (Hatus.) H. Ohba = Zanthoxylum ailanthoides Siebold et Zucc. f. espinosum Yonek. ●☆

417142 Zanthoxylum ailanthoides Siebold et Zucc. f. pubescens (Hatus.) H. Ohba = Zanthoxylum ailanthoides Siebold et Zucc. var. pubescens Hatus. ●

417143 Zanthoxylum ailanthoides Siebold et Zucc. var. boninshimae (Koidz. ex H. Hara) T. Yamaz. ex H. Ohba = Zanthoxylum ailanthoides Siebold et Zucc. var. inerme Rehder et E. H. Wilson ●☆

417144 Zanthoxylum ailanthoides Siebold et Zucc. var. inerme Rehder et E. H. Wilson;日本无刺椿叶花椒●☆

417145 Zanthoxylum ailanthoides Siebold et Zucc. var. inermis (Nakai) T. B. Lee = Zanthoxylum ailanthoides Siebold et Zucc. f. espinosum Yonek. ●☆

417146 Zanthoxylum ailanthoides Siebold et Zucc. var. pubescens Hatus. ;毛椿叶花椒;Pubescent Ailanthus-like Pricklyash ●

417147 Zanthoxylum ailanthoides Siebold et Zucc. var. yakumontanum (Sugim.) Hatus. = Zanthoxylum yakumontanum (Sugim.) Nagam. ●☆

417148 Zanthoxylum alatum Roxb. = Zanthoxylum armatum DC. ●

417149 Zanthoxylum alatum Roxb. = Zanthoxylum simulans Hance ●

417150 Zanthoxylum alatum Roxb. var. planispinum (Siebold et Zucc.) Rehder et E. H. Wilson = Zanthoxylum simulans Hance ●

417151 Zanthoxylum alatum Roxb. var. planispinum (Siebold et Zucc.) Rehder et E. H. Wilson = Zanthoxylum armatum DC. ●

417152 Zanthoxylum alatum Roxb. var. planispinum (Siebold et Zucc.) Rehder et E. H. Wilson f. ferrugineum Rehder et E. H. Wilson = Zanthoxylum armatum DC. var. ferrugineum (Rehder et E. H. Wilson) C. C. Huang ●

417153 Zanthoxylum alatum Roxb. var. subtrifoliatum Franch. = Zanthoxylum simulans Hance ●

417154 Zanthoxylum alatum Roxb. var. subtrifoliolatum Franch. = Zanthoxylum armatum DC. ●

417155 Zanthoxylum alpinum C. C. Huang = Zanthoxylum oxyphyllum Edgew. ●

417156   Zanthoxylum amamiense Ohwi；天见花椒●☆

417157   Zanthoxylum americanum Mill.；美洲花椒（美国花椒）；
Common Prickly Ash, Common Pricklyash, Common Prickly-ash,
Northern Prickly Ash, Northern Pricklyash, Northern Prickly-ash,
Prickly Ash, Suterberry, Toothache Tree, Yellow-wood ●☆

417158   Zanthoxylum americanum Mill. f. impuniens Fassett = Zanthoxylum
americanum Mill. ●☆

417159   Zanthoxylum arenosum Reeder et S. Y. Cheo = Zanthoxylum
armatum DC. ●

417160   Zanthoxylum argyi H. Lév. = Zanthoxylum simulans Hance ●

417161   Zanthoxylum armatum DC.；竹叶花椒（白总管，臭花椒，刺竹
叶花椒，狗花椒，狗椒，花胡椒，花椒，花椒树，鸡椒，假胡椒，见血
飞，具椒子，秦椒，三叶花椒，散血飞，山巴椒，山胡椒，山花椒，蜀
椒，搜山虎，土花椒，万花针，崖椒，岩椒，野花椒，玉椒，竹叶椒，
竹叶总管）；Bambooleaf Pricklyash, Bamboo-leaved Prickly-ash,
Chinese Wingleaf Pricklyash, Wingleaf Pricklyash ●

417162   Zanthoxylum armatum DC. var. ferrugineum（Rehder et E. H.
Wilson）C. C. Huang；毛竹叶花椒（毛刺竹叶花椒）；Hairy
Bambooleaf Pricklyash ●

417163   Zanthoxylum armatum DC. var. subtrifoliatum（Franch.）
Kitam.；三小叶花椒●☆

417164   Zanthoxylum arnottianum Maxim. = Zanthoxylum beecheyanum
K. Koch ●☆

417165   Zanthoxylum arnottianum Maxim. subsp. alatum（Nakai）
Masam. et Yanagita = Zanthoxylum beecheyanum K. Koch var. alatum
（Nakai）H. Hara ●

417166   Zanthoxylum arnottianum Maxim. var. alatum Nakai = Zanthoxylum
beecheyanum K. Koch var. alatum（Nakai）H. Hara ●

417167   Zanthoxylum aromaticum（Blume）Miq. = Melicope lunu-
ankenda（Gaertn.）T. G. Hartley ●

417168   Zanthoxylum asperum C. C. Huang = Zanthoxylum collinsae Craib ●

417169   Zanthoxylum asperum C. C. Huang var. glabrum C. C. Huang =
Zanthoxylum nitidum（Roxb.）DC. ●

417170   Zanthoxylum austrosinense C. C. Huang；岭南花椒（狗花椒，满
山香，皮子药，山胡椒，蛇总管，搜山虎，狭叶岭南花椒，总管）；
Guangdong Pricklyash, Kwangtung Pricklyash, South China
Pricklyash, South China Prickly-ash ●

417171   Zanthoxylum austrosinense C. C. Huang var. pubescens C. C.
Huang；毛叶岭南花椒；Hairyleaf South China Pricklyash ●

417172   Zanthoxylum austrosinense C. C. Huang var. stenophyllum C. C.
Huang = Zanthoxylum austrosinense C. C. Huang ●

417173   Zanthoxylum avicennae（Lam.）DC.；簕欓花椒（刺苍，刺欓，
刺倒树，狗花椒，花椒簕，画眉架，画眉筋，画眉跳，鸡屎欓，鸡嘴
簕，笋当，笋欓，勒党，勒筒，簕欓，鸟不宿，雀笼踏，山胡椒，搜山
虎，土巴椒，土花椒，乌鸦不企树，鹰不泊，鹰不沾）；Avicenna
Pricklyash, Avicenna Prickly-ash ●

417174   Zanthoxylum avicennae（Lam.）DC. var. tonkinense Pierre =
Zanthoxylum avicennae（Lam.）DC. ●

417175   Zanthoxylum becquetii（G. C. C. Gilbert）P. G. Waterman；贝凯
花椒●☆

417176   Zanthoxylum beecheyanum K. Koch；岩山椒（阿诺梯花椒）；
Arnott Pricklyash ●☆

417177   Zanthoxylum beecheyanum K. Koch var. alatum（Nakai）H.
Hara；翼山椒●

417178   Zanthoxylum bodinieri H. Lév. = Zanthoxylum dissitum Hemsl.
ex Forbes et Hemsl. ●

417179   Zanthoxylum buesgenii（Engl.）P. G. Waterman；比斯根花椒●☆

417180   Zanthoxylum bungeanum Maxim.；花椒（巴椒，川椒，大红袍，大花
椒，大椒，点椒，汉椒，汗椒，红花椒，红椒，花椒树，檓，椒，椒红，开口
川椒，陆拔，蔓椒，南椒，秦椒，青花椒，青叶茮树，蜀椒，菽藙，香椒，香
椒子，崖胡椒）；Bunge Prickly Ash, Bunge Pricklyash, Bunge Prickly-
ash, Chinese Zanthoxylum, Flat-spine Prickly Ash ●

417181   Zanthoxylum bungeanum Maxim. var. pubescens C. C. Huang；毛
叶花椒；Hairyleaf Bunge Pricklyash ●

417182   Zanthoxylum bungeanum Maxim. var. punctatum C. C. Huang；油
叶花椒；Punctate Bunge Pricklyash ●

417183   Zanthoxylum bungei Planch. = Zanthoxylum armatum DC. ●

417184   Zanthoxylum bungei Planch. et Linden = Zanthoxylum
bungeanum Maxim. ●

417185   Zanthoxylum bungei Planch. et Linden ex Hance = Zanthoxylum
bungeanum Maxim. ●

417186   Zanthoxylum bungei Planch. et Linden var. imperforatum
Franch. = Zanthoxylum bungeanum Maxim. ●

417187   Zanthoxylum bungei Planch. et Linden var. inermis Franch. =
Zanthoxylum simulans Hance ●

417188   Zanthoxylum calcicola C. C. Huang；石山花椒（石灰山花椒，岩
椒）；Calcareus Pricklyash ●

417189   Zanthoxylum calcicola C. C. Huang var. macrocarpum C. C.
Huang = Zanthoxylum leiboicum C. C. Huang ●

417190   Zanthoxylum capense（Thunb.）Harv.；好望角花椒；Fever
Tree, Knobthorn, Knobwood ●☆

417191   Zanthoxylum caribaeum Lam.；加勒比花椒●☆

417192   Zanthoxylum chaffanjonii H. Lév. = Zanthoxylum esquirolii H.
Lév. ●

417193   Zanthoxylum chalybeum Engl.；铁色花椒●☆

417194   Zanthoxylum chalybeum Engl. var. molle Kokwaro；柔软铁色花
椒●☆

417195   Zanthoxylum chevalieri P. G. Waterman；舍瓦利耶花椒●☆

417196   Zanthoxylum chinense（Merr.）C. C. Huang = Zanthoxylum
scandens Blume ●

417197   Zanthoxylum chinense C. C. Huang = Zanthoxylum scandens
Blume ●

417198   Zanthoxylum claessensii（De Wild.）P. G. Waterman；克莱森
斯花椒●☆

417199   Zanthoxylum clave-herculis L.；刺椒；Hercules' Club,
Hercules'-club, Southern Prickly Ash, Toothache Tree ●☆

417200   Zanthoxylum collinsae Craib；糙叶花椒；Rough Pricklyash,
Roughleaf Pricklyash, Scabrousleaf Pricklyash, Scabrous-leaved
Prickly-ash ●

417201   Zanthoxylum connaroides Wight et Arn. = Heynea trijuga Roxb. ●

417202   Zanthoxylum coreanum Nakai = Zanthoxylum simulans Hance ●

417203   Zanthoxylum cuspidatum（Champ.）Engl. = Zanthoxylum scandens
Blume ●

417204   Zanthoxylum cuspidatum Champ. ex Benth. = Zanthoxylum
scandens Blume ●

417205   Zanthoxylum cyrtorhachium（Hayata）C. C. Huang =
Zanthoxylum scandens Blume ●

417206   Zanthoxylum daniellii Berm. ex Daniell = Evodia daniellii（A.
W. Benn.）Hemsl. ex Forbes et Hemsl. ●

417207   Zanthoxylum davyi（I. Verd.）P. G. Waterman；戴维花椒●☆

417208   Zanthoxylum decaryi H. Perrier；得卡瑞花椒；Decary Pricklyash ●☆

417209   Zanthoxylum delagoense P. G. Waterman；迪拉果花椒●☆

417210   Zanthoxylum dimorphophyllum Hemsl. = Zanthoxylum ovalifolium
Wight ●

417211   Zanthoxylum dimorphophyllum Hemsl. var. deminutum Rehder = Zanthoxylum ovalifolium Wight ●

417212   Zanthoxylum dimorphophyllum Hemsl. var. multifoliolatum C. C. Huang = Zanthoxylum ovalifolium Wight var. multifoliolatum （ C. C. Huang） C. C. Huang ●

417213   Zanthoxylum dimorphophyllum Hemsl. var. spinifolium Rehder et E. H. Wilson = Zanthoxylum ovalifolium Wight var. spinifolium （ Rehder et E. H. Wilson） C. C. Huang ●

417214   Zanthoxylum dinklagei （ Engl.） P. G. Waterman；丁克花椒●☆

417215   Zanthoxylum dissitoides C. C. Huang = Zanthoxylum laetum Drake ●

417216   Zanthoxylum dissitum Hemsl. ex Forbes et Hemsl.；蚬壳花椒（白皮两面针，白三百棒，蚌壳花椒，蚌壳椒，大花椒，大叶花椒，单面虎，单面针，公麒麟，过山龙，黄椒根，见血飞，九百锤，麻风针，三百棒，山椒根，山枇杷，铁杆椒，岩花椒，钻山虎）；Clamshell Pricklyash，Shellfish Pricklyash，Shellfish Prickly-ash ●

417217   Zanthoxylum dissitum Hemsl. ex Forbes et Hemsl. var. acutiserratum C. C. Huang；针边蚬壳花椒（二针边虫壳花椒）；Acutemargin Shellfish Pricklyash ●

417218   Zanthoxylum dissitum Hemsl. ex Forbes et Hemsl. var. hispidum （ Reeder et S. Y. Cheo） C. C. Huang；刺蚬壳花椒；Hispid Shellfish Pricklyash ●

417219   Zanthoxylum dissitum Hemsl. ex Forbes et Hemsl. var. lanciforme C. C. Huang；长叶蚬壳花椒；Longleaf Shellfish Pricklyash ●

417220   Zanthoxylum dissitum Hemsl. ex Forbes et Hemsl. var. spinulosum Z. M. Tan = Zanthoxylum ovalifolium Wight var. spinifolium （ Rehder et E. H. Wilson） C. C. Huang ●

417221   Zanthoxylum echinocarpum Hemsl.；刺壳花椒（刺壳椒，见血飞）；Spinyfruit Pricklyash，Spiny-fruited Prickly-ash ●

417222   Zanthoxylum echinocarpum Hemsl. var. tomentosum C. C. Huang；毛刺壳花椒；Tomentose Spinyfruit Pricklyash ●

417223   Zanthoxylum elephantiasis Macfad.；厚皮花椒●☆

417224   Zanthoxylum emarginellum Miq. = Zanthoxylum ailanthoides Siebold et Zucc. ●

417225   Zanthoxylum engleri P. G. Waterman；恩格勒花椒●☆

417226   Zanthoxylum esquirolii H. Lév.；贵州花椒（文山花椒，细柄花椒，岩椒）；Esquirol Pricklyash，Esquirol Prickly-ash，Guizhou Pricklyash，Guizhou Prickly-ash ●

417227   Zanthoxylum evoideifolium Guillaumin = Zanthoxylum ovalifolium Wight ●

417228   Zanthoxylum fagara （ L.） Sarg.；岩花椒；Lime Prickly Ash，Lime Pricklyash，Wild Lime，Wild-lime ●☆

417229   Zanthoxylum fauriei （ Nakai） Ohwi；法氏花椒●☆

417230   Zanthoxylum flavum Vahl；西印度花椒（黄心花椒，黄崖椒）；Jamaican Satin Wood，West Indian Satin Wood，Yellow-heart Prickly Ash ☆

417231   Zanthoxylum fraxineum Willd.；普通花椒；Common Prickly Ash，Prickly Ash ●☆

417232   Zanthoxylum fraxineum Willd. = Zanthoxylum americanum Mill. ●☆

417233   Zanthoxylum fraxinoides Hemsl. = Zanthoxylum bungeanum Maxim. ●

417234   Zanthoxylum giganteum （ Hand. -Mazz.） Rehder = Zanthoxylum myriacanthum Wall. ex Hook. f. ●

417235   Zanthoxylum gilletii （ De Wild.） P. G. Waterman；吉勒特花椒；African Satinwood ●☆

417236   Zanthoxylum gilletii （ De Wild.） P. G. Waterman var. cordata G. C. C. Gilbert = Zanthoxylum gilletii （ De Wild.） P. G. Waterman ●☆

417237   Zanthoxylum glomeratum C. C. Huang；密果花椒；Densefruit Pricklyash，Dense-fruited Prickly-ash，Fruitful Pricklyash ●

417238   Zanthoxylum gracilipes Hemsl. = Zanthoxylum esquirolii H. Lév. ●

417239   Zanthoxylum hamiltonianum Wall. ex Hook. f. = Zanthoxylum nitidum （ Roxb.） DC. ●

417240   Zanthoxylum heitzii （ Aubrév. et Pellegr.） P. G. Waterman；海茨花椒●☆

417241   Zanthoxylum hemsleyanum Makino = Zanthoxylum ailanthoides Siebold et Zucc. ●

417242   Zanthoxylum holtzianum （ Engl.） P. G. Waterman；索马里花椒●☆

417243   Zanthoxylum holtzianum （ Engl.） P. G. Waterman var. tenuipedicellatum Kokwaro；细花梗索马里花椒●☆

417244   Zanthoxylum horridum Welw. ex Ficalho = Zanthoxylum gilletii （ De Wild.） P. G. Waterman ●☆

417245   Zanthoxylum humile （ E. A. Bruce） P. G. Waterman；矮小花椒●☆

417246   Zanthoxylum inerme （ Rehder et E. H. Wilson） Koidz. = Zanthoxylum ailanthoides Siebold et Zucc. var. inerme Rehder et E. H. Wilson ●☆

417247   Zanthoxylum integrifolium （ Merr.） Merr.；兰屿花椒（全缘花椒，全缘叶花椒）；Entireleaf Pricklyash，Entire-leaved Prickly-ash，Leaves Entire Prickly-ash ●

417248   Zanthoxylum khasianum Hook. f.；云南花椒；Khas Pricklyash，Khasi Prickly-ash，Yunnan Pricklyash ●

417249   Zanthoxylum kwangsiense （ Hand. -Mazz.） Chun ex C. C. Huang；广西花椒；Guangxi Pricklyash，Kwangsi Pricklyash，Kwangsi Prickly-ash ●

417250   Zanthoxylum laetum Drake；拟砚壳花椒（滑叶花椒）；Bright-colored Pricklyash，Bright-colored Prickly-ash，Sham Calmshell Pricklyash，Shell-fish-like Prickly-ash ●

417251   Zanthoxylum laurentii （ De Wild.） P. G. Waterman；洛朗花椒●☆

417252   Zanthoxylum laxifoliolatum （ Hayata） C. C. Huang = Zanthoxylum scandens Blume ●

417253   Zanthoxylum laxifoliolatum C. C. Huang = Zanthoxylum scandens Blume ●

417254   Zanthoxylum leiboicum C. C. Huang；雷波花椒（大果石山花椒）；Big-fruit Calcareus Pricklyash，Leibo Pricklyash，Leibo Prickly-ash ●

417255   Zanthoxylum leiorhachium （ Hayata） C. C. Huang = Zanthoxylum scandens Blume ●

417256   Zanthoxylum lemairei （ De Wild.） P. G. Waterman；迈雷花椒●☆

417257   Zanthoxylum leprieurii Guillaumin et Perr.；安哥拉花椒●☆

417258   Zanthoxylum liboense C. C. Huang；荔波花椒；Libo Pricklyash ●

417259   Zanthoxylum lindense （ Engl.） Kokwaro；林德花椒●☆

417260   Zanthoxylum lucidum Miq. = Melicope lunu-ankenda （ Gaertn.） T. G. Hartley ●

417261   Zanthoxylum macranthum （ Hand. -Mazz.） C. C. Huang；大花花椒；Big-lowered Prickly-ash，Largeflower Pricklyash ●

417262   Zanthoxylum macrophyllum Oliv.；大叶崖椒●☆

417263   Zanthoxylum macrophyllum Oliv. = Zanthoxylum gilletii （ De Wild.） P. G. Waterman ●☆

417264   Zanthoxylum madagascariense Baker；马岛花椒●☆

417265   Zanthoxylum mananarense H. Perrier；马纳纳拉花椒●☆

417266   Zanthoxylum marambong Miq. = Melicope lunu-ankenda （ Gaertn.） T. G. Hartley ●

417267   Zanthoxylum matschuricum A. W. Benn. = Zanthoxylum schinifolium Siebold et Zucc. ●

417268   Zanthoxylum mayu Bertero;梅宇崖椒●☆

417269   Zanthoxylum melanacanthum Planch. = Zanthoxylum rubescens Hook. f. ●☆

417270   Zanthoxylum mezoneurospinosum Aké Assi = Zanthoxylum leprieurii Guillaumin et Perr. ●☆

417271   Zanthoxylum micranthum Hemsl.;小花花椒(刺辣树,见血飞,野花椒);Little Flower Prickly-ash, Littleflower Pricklyash, Smallflower Pricklyash,Small-flowered Prickly-ash ●

417272   Zanthoxylum mildbraedii (Engl.) P. G. Waterman;米尔德花椒●☆

417273   Zanthoxylum molle Rehder;朵花椒(刺椿木,刺风树,刺盐肤木,大叶臭花椒,朵椒,鼓钉皮,毛海桐皮,驱风通,天星木);Pubescent Pricklyash,Pubescent Prickly-ash,Softhair Pricklyash ●

417274   Zanthoxylum montanum Blume = Turpinia montana (Blume) Kurz ●

417275   Zanthoxylum motuoense C. C. Huang;墨脱花椒;Motuo Pricklyash, Motuo Prickly-ash ●

417276   Zanthoxylum multifoliolatum Hemsl. = Zanthoxylum multijugum Franch. ●

417277   Zanthoxylum multijugum Franch.;多叶花椒(马椒,蜈蚣刺,蜈蚣藤,小叶刺椒);Leafy Pricklyash,Multijugous Prickly-ash ●

417278   Zanthoxylum myriacanthum Wall. ex Hook. f.;大叶臭花椒(刺椿木,大叶臭椒,多刺花椒,雷公木,驱风通);Bigleaf Pricklyash, Bigleafstink Pricklyash, Big-leaved Prickly-ash, Largeleaf Stink Pricklyash ●

417279   Zanthoxylum myriacanthum Wall. ex Hook. f. var. pubescens C. C. Huang;毛大叶臭花椒(炸椒);Pubescent Big-leaved Prickly-ash,Pubescet Bigleaf Pricklyash ●

417280   Zanthoxylum naranjillo Griseb.;那兰花椒●☆

417281   Zanthoxylum natalense Hochst. = Cnestis polyphylla Lam. ●☆

417282   Zanthoxylum nitidum (Roxb.) DC.;光叶花椒(出山虎,大叶猫爪筋,大叶毛枝花,钉板刺,高山花椒,狗椒,红倒钩筋,红心刺刁根,胡椒笋,花椒,花椒刺,金椒,金牛公,樛,樛子,两背针,两边针,两面针,麻药藤,马药子,蔓椒,毛两面针,鸟塔刺,牛刁茨,菜子,入地金牛,入山虎,山椒,上山虎,豕椒,疏刺花椒,双面刺,双面针,无毛糙叶花椒,豨椒,下山虎,崖椒,叶下穿针,毳椒,猪椒);Alpine Pricklyash, Alpine Prickly-ash, Glabrous Rough Pricklyash,Glittering Prickly-ash,Hairy Needle on both Sides,Loose-spiny Shinyleaf Pricklyash,Needle on Both Sides,Shiny Leaf Prickly-ash,Shinyleaf Pricklyash,Shiny-leaved Prickly-ash ●

417283   Zanthoxylum nitidum (Roxb.) DC. f. fastuosum F. C. How ex C. C. Huang = Zanthoxylum nitidum (Roxb.) DC. ●

417284   Zanthoxylum nitidum (Roxb.) DC. var. neglectum F. C. How = Zanthoxylum nitidum (Roxb.) DC. ●

417285   Zanthoxylum nitidum (Roxb.) DC. var. tomentosum C. C. Huang;毛叶两面针;Tomentose Shinyleaf Pricklyash ●

417286   Zanthoxylum odoramm (H. Lév.) H. Lév. = Zanthoxylum myriacanthum Wall. ex Hook. f. ●

417287   Zanthoxylum okinawense (Nakai) E. H. Wilson = Zanthoxylum schinifolium Siebold et Zucc. var. okinawense (Nakai) Hatus. ex Shimabuku ●☆

417288   Zanthoxylum olitorium Engl.;菜园花椒●☆

417289   Zanthoxylum olitorium Engl. = Zanthoxylum chalybeum Engl. ●☆

417290   Zanthoxylum ovalifolium Wight;异叶花椒(苍椒,刺三加,黄连木叶花椒,卵叶花椒,三叶花椒,椭圆叶花椒,羊山刺);Diversifolious Pricklyash, Ovateleaf Pricklyash, Pistache-leaf Pricklyash,Threeleaf Prickly-ash,Variable-leaved Prickly-ash ●

417291   Zanthoxylum ovalifolium Wight var. multifoliolatum (C. C. Huang) C. C. Huang;多异叶花椒;Many Diversifolious Pricklyash ●

417292   Zanthoxylum ovalifolium Wight var. spinifolium (Rehder et E. H. Wilson) C. C. Huang;刺异叶花椒(刺三加,刺叶花椒,红三百棒,黄椒,见血飞,青椒,青皮椒,散血飞);Spiny Diversifolious Pricklyash ●

417293   Zanthoxylum ovatifoliolatum (Engl.) Finkelstein;卵叶花椒●☆

417294   Zanthoxylum oxyphyllum Edgew.;尖叶花椒(大理花椒,高山花椒,西藏花椒,野花椒);Alpine Pricklyash, Dali Pricklyash, Sharpleaf Pricklyash, Sharp-leaved Prickly-ash, Tibet Pricklyash, Xizang Pricklyash, Xizang Prickly-ash ●

417295   Zanthoxylum parvifoliolum A. Chev. ex Keay;小叶花椒●☆

417296   Zanthoxylum pashanense N. Chao = Zanthoxylum stenophyllum Hemsl. ●

417297   Zanthoxylum pentandrum (Aubl.) R. A. Howard;五蕊花椒●☆

417298   Zanthoxylum piasezkii Maxim.;川陕花椒;Chuanshan Pricklyash,Piasecki Pricklyash,Piasecki Prickly-ash ●

417299   Zanthoxylum pilosiusculum (Engl.) P. G. Waterman;疏毛花椒●☆

417300   Zanthoxylum pilosulum Rehder et Wibel;微柔毛花椒;Pilose Pricklyash,Pilose Prickly-ash ●

417301   Zanthoxylum piperitum (L.) DC.;秦椒(大金花椒,日本花椒,山花椒,山椒,蜀椒);Indian Pepper,Japan Pepper Tree,Japanese Pepper,Japanese Peppercorns,Japanese Prickly Ash,Japanese Prickly-ash,Sichuan Pepper ●

417302   Zanthoxylum piperitum (L.) DC. f. brevispinosum (Makino) Makino;短刺秦椒●☆

417303   Zanthoxylum piperitum (L.) DC. f. corticosum Kusaka = Zanthoxylum piperitum (L.) DC. ●

417304   Zanthoxylum piperitum (L.) DC. f. hispidum Hayashi;毛秦椒●☆

417305   Zanthoxylum piperitum (L.) DC. f. inerme (Makino) Makino;无刺毛秦椒●☆

417306   Zanthoxylum piperitum (L.) DC. f. ovatifoliolatum (Nakai) Makino;卵叶秦椒●☆

417307   Zanthoxylum piperitum (L.) DC. f. rotundatum Yokouchi;钝叶秦椒●☆

417308   Zanthoxylum piperitum (L.) DC. f. verrucatum Kusaka = Zanthoxylum piperitum (L.) DC. ●

417309   Zanthoxylum piperitum (L.) DC. var. hispidum (Hayashi) Konta = Zanthoxylum piperitum (L.) DC. f. hispidum Hayashi ●☆

417310   Zanthoxylum pistaciiflorum Hayata = Zanthoxylum ovalifolium Wight ●

417311   Zanthoxylum planispinum Siebold et Zucc. = Zanthoxylum armatum DC. ●

417312   Zanthoxylum planispinum Siebold et Zucc. f. ferrugineum (Rehder et E. H. Wilson) C. C. Huang = Zanthoxylum armatum DC. var. ferrugineum (Rehder et E. H. Wilson) C. C. Huang ●

417313   Zanthoxylum podocarpum Hemsl. = Zanthoxylum simulans Hance ●

417314   Zanthoxylum poggei (Engl.) P. G. Waterman;波格花椒●☆

417315   Zanthoxylum psammophilum (Aké Assi) P. G. Waterman;喜沙花椒●☆

417316   Zanthoxylum pteleifolium Champ. ex Benth. = Melicope pteleifolia (Champ. ex Benth.) T. G. Hartley ●

417317   Zanthoxylum pteracanthum Rehder et E. H. Wilson;翼叶花椒(翼刺花椒);Wingespine Pricklyash, Wing-spined Prickly-ash ●

417318   Zanthoxylum pteropodum Hayata = Zanthoxylum schinifolium Siebold et Zucc. ●

417319   Zanthoxylum pterota Kunth = Zanthoxylum fagara (L.) Sarg. ●☆

417320   Zanthoxylum renieri (G. C. C. Gilbert) P. G. Waterman;雷尼尔花椒●☆

417321　Zanthoxylum rhetsoides Drake = Zanthoxylum myriacanthum Wall. ex Hook. f. ●

417322　Zanthoxylum rhetsoides Drake var. pubescens C. C. Huang = Zanthoxylum myriacanthum Wall. ex Hook. f. var. pubescens C. C. Huang ●

417323　Zanthoxylum rhombifoliolatum C. C. Huang;菱叶花椒(黄椒); Rhombicleaf Pricklyash,Rhombic-leaved Prickly-ash ●

417324　Zanthoxylum robiginosum (Reeder et S. Y. Cheo) C. C. Huang = Zanthoxylum ovalifolium Wight ●

417325　Zanthoxylum roxburghianum Cham. = Melicope lunu-ankenda (Gaertn.) T. G. Hartley ●

417326　Zanthoxylum rubescens Hook. f.;变红花椒●☆

417327　Zanthoxylum rubescens Hook. f. var. disperma (G. C. C. Gilbert) P. G. Waterman;双籽变红花椒●☆

417328　Zanthoxylum scabrum Guillaumin = Zanthoxylum collinsae Craib ●

417329　Zanthoxylum scandens Blume;花椒簕(扁轴花椒,花椒藤,尖叶花椒,山花椒,疏叶花椒,藤花椒,藤崖椒,通墙虎,凸尖花椒,弯轴花椒,乌口簕);Bent-axis Pricklyash,Climbing Pricklyash,Climbing Prickly-ash,Flattened-axis Pricklyash,Looseleaf Pricklyash ●

417330　Zanthoxylum schinifolium Siebold et Zucc.;青花椒(巴椒,翅耳崖椒,川椒,刺搜山虎,大花椒,大椒,点椒,隔山消,狗椒,汉椒,汗椒,陆拔,南椒,秦椒,青椒,雀椒,日本崖椒,散血胆,山花椒,山甲,蜀椒,蕾蔎,天椒,王椒,香椒,香椒子,小花椒,崖椒,野椒,翼柄花椒,翼柄山椒,翼柄崖椒);Green Pricklyash,Peppe Tree,Pepper-tree Prickly-ash,Peppetree,Pricklyash,Taiwan Prickly-ash,Wing-dstalked Prickly-ash,Wingedstalk Pricklyash ●

417331　Zanthoxylum schinifolium Siebold et Zucc. f. angustifolia (H. Hara) H. Hara ex Ohwi et Kitag. = Zanthoxylum schinifolium Siebold et Zucc. ●

417332　Zanthoxylum schinifolium Siebold et Zucc. f. grandifolia (H. Hara) H. Hara ex Ohwi et Kitag. = Zanthoxylum schinifolium Siebold et Zucc. ●

417333　Zanthoxylum schinifolium Siebold et Zucc. f. inerme (Nakai) T. B. Lee ex W. T. Lee;无刺青花椒●☆

417334　Zanthoxylum schinifolium Siebold et Zucc. f. microphyllum (H. Hara) W. T. Lee = Zanthoxylum schinifolium Siebold et Zucc. ●

417335　Zanthoxylum schinifolium Siebold et Zucc. var. okinawense (Nakai) Hatus. ex Shimabuku;冲绳花椒●☆

417336　Zanthoxylum senegalense DC.;西非花椒●☆

417337　Zanthoxylum senegalense DC. = Zanthoxylum xanthoxyloides Lam. ●☆

417338　Zanthoxylum setosum Hemsl. = Zanthoxylum simulans Hance ●

417339　Zanthoxylum setosum Hemsl. ex Forbes et Hemsl. = Zanthoxylum simulans Hance ●

417340　Zanthoxylum simulans Hance;野花椒(白总管,柄果花椒,臭花椒,臭椒,川椒,刺花椒,刺椒,大花椒,高脚刺,狗花椒,红总管,花胡椒,花椒,黄椒,黄总管,鸡椒,麻口皮,麻口皮子药,麻醉根,满山香,皮子药,三叶花椒,山胡椒,山花椒,四皮麻,搜山虎,天角刺,天角椒,土花椒,万花针,香椒,岩椒,玉椒,竹叶椒,总管皮);Flat Spine Prickly-ash,Flatspine Prickly Ash,Flatspine Pricklyash,Flatspine Prickly-ash,Hairy Prickly-ash,Stalkedfruit Pricklyash,Stalk-fruited Prickly-ash,Wild Pricklyash ●

417341　Zanthoxylum simulans Hance = Zanthoxylum bungeanum Maxim. ●

417342　Zanthoxylum simulans Hance var. imperforamm (Franch.) Reeder et S. Y. Cheo = Zanthoxylum bungeanum Maxim. ●

417343　Zanthoxylum simulans Hance var. podocarpum (Hemsl.) C. C. Huang = Zanthoxylum simulans Hance ●

417344　Zanthoxylum somalense (Chiov.) P. G. Waterman = Zanthoxylum holtzianum (Engl.) P. G. Waterman ●☆

417345　Zanthoxylum stenophyllum Hemsl.;狭叶花椒(巴山花椒); Narrowleaf Pricklyash, Narrow-leaved Prickly-ash, Tapa Mountain Prickly-ash ●

417346　Zanthoxylum stipitatum C. C. Huang;梗花椒(红山椒,麻口皮子药,满山香);Stipitate Pricklyash ●

417347　Zanthoxylum subspicatum H. Perrier;穗状花椒●☆

417348　Zanthoxylum szenchuanense W. P. Fang et R. X. Meng;蜀椒(毛叶花椒);Sichuan Pricklyash ●

417349　Zanthoxylum taliense C. C. Huang = Zanthoxylum oxyphyllum Edgew. ●

417350　Zanthoxylum thomense (Engl.) A. Chev. ex P. G. Waterman;爱岛花椒●☆

417351　Zanthoxylum thorncroftii (I. Verd.) P. G. Waterman;托恩花椒●☆

417352　Zanthoxylum thouvenotii H. Perrier;图弗诺花椒●☆

417353　Zanthoxylum thunbergii DC. var. grandifolia Harv. = Zanthoxylum davyi (I. Verd.) P. G. Waterman ●☆

417354　Zanthoxylum thunbergii DC. var. obtusifolia Harv. = Zanthoxylum capense (Thunb.) Harv. ●☆

417355　Zanthoxylum tibetanum C. C. Huang = Zanthoxylum oxyphyllum Edgew. ●

417356　Zanthoxylum tomentellum Hook. f.;毡毛花椒;Tomentulose Pricklyash,Tomentulose Prickly-ash ●

417357　Zanthoxylum tragodes DC.;山羊花椒●☆

417358　Zanthoxylum trifoliatum (Sw.) Wight;三叶花椒;Three-leaves Pricklyash ●

417359　Zanthoxylum trifoliatum (Sw.) Wight = Acanthopanax trifoliatus (L.) Merr. ●

417360　Zanthoxylum trifoliatum (Sw.) Wight var. spinifolium (Rehder et E. H. Wilson) C. C. Huang;刺三叶花椒;Spine Three-leaves Pricklyash ●

417361　Zanthoxylum trifoliatum L. = Acanthopanax trifoliatus (L.) Merr. ●

417362　Zanthoxylum trifoliatum L. = Eleutherococcus trifoliatus (L.) S. Y. Hu ●

417363　Zanthoxylum trijugum (Dunkley) P. G. Waterman;三对花椒●☆

417364　Zanthoxylum triphyllum (Lam.) G. Don = Melicope triphylla (Lam.) Merr. ●

417365　Zanthoxylum tsihanimposa H. Perrier;齐汉宁花椒●☆

417366　Zanthoxylum undulatifolium Hemsl.;波叶花椒(浪叶花椒); Undulateleaf Pricklyash, Undulate-leaved Prickly-ash, Waveleaf Pricklyash ●

417367　Zanthoxylum usambarense (Engl.) Kokwaro;乌桑巴拉花椒●☆

417368　Zanthoxylum usitatum Pierre ex Lannes = Zanthoxylum bungeanum Maxim. ●

417369　Zanthoxylum utile C. C. Huang;香果花椒;Useful Pricklyash ●

417370　Zanthoxylum utile C. C. Huang = Zanthoxylum myriacanthum Wall. ex Hook. f. var. pubescens C. C. Huang ●

417371　Zanthoxylum viride (A. Chev.) P. G. Waterman;绿花椒●☆

417372　Zanthoxylum wutaiense I. S. Chen;屏东花椒;Pingdong Pricklyash,Wutai Pricklyash,Wutai Prickly-ash ●

417373　Zanthoxylum xanthoxyloides Lam.;美国崖椒(崖椒)●☆

417374　Zanthoxylum xichouense C. C. Huang;西畴花椒;Xichou Pricklyash,Xichou Prickly-ash ●

417375　Zanthoxylum yakumontanum (Sugim.) Nagam.;屋久岛花椒●☆

417376　Zanthoxylum yuanjiangense C. C. Huang;元江花椒;Yuanjiang

Pricklyash, Yuanjiang Prickly-ash ●

417377 Zanthoxylum yunnanense C. C. Huang = Zanthoxylum khasianum Hook. f. ●

417378 anthoxylum zanthoxyloides (Lam.) Zepern. et Timler = Zanthoxylum xanthoxyloides Lam. ●☆

417379 Zanthyrsis Raf. = Sophora L. ●■

417380 Zantorrhiza Steud. = Xanthorhiza Marshall ●☆

417381 Zapamia Steud. = Lippia L. ●■☆

417382 Zapamia Steud. = Zapania Lam. ●■☆

417383 Zapania Lam. = Lippia L. ●■☆

417384 Zapania Nees et Mart. = Rhaphiodon Schauer ■☆

417385 Zapania citriodora Lam. = Aloysia citriodora Paláu ☆

417386 Zapania nodiflora (L.) Lam. var. rosea D. Don = Phyla nodiflora (L.) Greene var. rosea (D. Don) Moldenke ■☆

417387 Zapateria Pau = Ballota L. ●■☆

417388 Zapoteca H. M. Hern. (1987);热美朱樱花属●☆

417389 Zapoteca formosa (Kunth) H. M. Hern.;美丽热美朱樱花●☆

417390 Zapoteca gracilis (Griseb.) Bässler;细热美朱樱花●☆

417391 Zapoteca microcephala (Britton et Killip) H. M. Hern.;小头热美朱樱花●☆

417392 Zapoteca tetragona (Willd.) H. M. Hern.;热美朱樱花●☆

417393 Zappania Scop. = Salvia L. ●■

417394 Zappania Zuccagni = Lippia L. ●■☆

417395 Zappania Zuccagni = Zapania Lam. ●■☆

417396 Zarabellia Cass. = Melampodium L. ■●

417397 Zarabellia Neck. = Berkheya Ehrh. (保留属名)●■☆

417398 Zarcoa Llanos = Glochidion J. R. Forst. et G. Forst. (保留属名)●

417399 Zatarendia Raf. = Origanum L. ●■

417400 Zatarhendi Forssk. = Plectranthus L'Hér. (保留属名)●■

417401 Zataria Boiss. (1844);扎塔尔灌属●☆

417402 Zataria multiflora Boiss.;扎塔尔灌●☆

417403 Zauscheria Steud. = Zauschneria C. Presl ●■☆

417404 Zauschneria C. Presl = Epilobium L. ■

417405 Zauschneria C. Presl(1831);加州倒挂金钟属(朱巧花属);Californian Fuchsia, Fire Chalice, Fuchsia ●■☆

417406 Zauschneria arizonica Davidson;亚利桑那倒挂金钟;Arizona Wild Fuchsia ●☆

417407 Zauschneria californica C. Presl;加州倒挂金钟(朱巧花);California Fuchsia, Californian Fuchsia, Humming Bird's Trumpet, Hummingbird's Trumpet, Mexican Balsamea ●■☆

417408 Zauschneria californica C. Presl subsp. angustifolia D. D. Keck;狭叶加州倒挂金钟●■☆

417409 Zauschneria californica C. Presl var. latifolia Hook.;宽叶加州倒挂金钟●■☆

417410 Zauschneria septentrionalis D. D. Keck;北方朱巧花■☆

417411 Zazintha Boehm. = Zacintha Vell. ●☆

417412 Zea L. (1753);玉蜀黍属(玉米属);Corn, Indian Corn, Maize ■

417413 Zea Lunell = Triticum L. ■

417414 Zea amylacea Sturtev. ex L. H. Bailey;淀粉玉蜀黍;Flour Corn, Flour Maize, Soft Corn, Soft Maize ■☆

417415 Zea indentata Sturtev. = Zea indentata Sturtev. ex L. H. Bailey ■☆

417416 Zea indentata Sturtev. ex L. H. Bailey;凹齿玉蜀黍;Dent Corn ■☆

417417 Zea mays L.;玉蜀黍(包儿米,包谷,包麦米,包米,包粟,苞谷,苞芦,苞米,番麦,红须谷,陆谷,鹿角黍,粟米,西番麦,西天麦,戊菽,薏米包,纡粟,玉高粱,玉露秫秫,玉麦,玉米,玉黍,玉蜀秫,御麦,御米,珍珠芦粟,珍珠米);Asiatic Corn, Baby Corn, Babycom, Blue Corn, Bourbon, Corn, Corn Flakes, Corn Floor, Corn On The Cob, Corn Silk, Corn-on-the-cob, Dent Corn, Hen Pea, Hen Peas, Hominy, Indian Corn, Indian Wheat, Indy Corn, Maize, Mealie, Mealies, Ornamental Corn, Ornamental Maize, Polenta, Pop Corn, Striped Corn, Sweet Corn, Turkey Corn, Turkish Corn, Turkish Millet, Welsh Corn ■

417418 Zea mays L. 'Gracillima Variegata';乳纹叶玉蜀黍■☆

417419 Zea mays L. 'Harlequin';三色叶玉蜀黍■☆

417420 Zea mays L. 'Japonica';日本玉蜀黍;Japanese Maize, Ornamental Corn ■☆

417421 Zea mays L. f. variegata (G. Nicholson) Beetle = Zea mays L. var. variegata G. Nicholson ■☆

417422 Zea mays L. subsp. mexicana (Schrad.) Iltis = Euchlaena mexicana Schrad. ■

417423 Zea mays L. var. dentiformis ?;齿玉蜀黍■☆

417424 Zea mays L. var. indentata ? = Zea indentata Sturtev. ex L. H. Bailey ■☆

417425 Zea mays L. var. praecox ? = Zea praecox Steud. ■☆

417426 Zea mays L. var. rostrata ? = Zea rostrata Bonaf. ■☆

417427 Zea mays L. var. saccharata ? = Zea saccharata Sturtev. ■☆

417428 Zea mays L. var. tunicata ? = Zea tunicata Sturtev. ex L. H. Bailey ■☆

417429 Zea mays L. var. variegata G. Nicholson;彩色玉蜀黍■☆

417430 Zea mexicana Schrad. = Zea mays L. subsp. mexicana (Schrad.) Iltis ■

417431 Zea praecox Steud.;早熟玉米;Flint Corn, Flint Maize ■☆

417432 Zea rostrata Bonaf.;喙状玉蜀黍;Popcorn ■☆

417433 Zea saccharata Sturtev.;甜玉米;Sugar Corn, Sweet Corn ■☆

417434 Zea tunicata Sturtev. ex L. H. Bailey;衣玉米;Husk Corn, Pod Corn, Tunicate Corn ■☆

417435 Zeaceae A. Kern.;玉蜀黍科■

417436 Zeaceae A. Kern. = Gramineae Juss. (保留科名)■●

417437 Zeaceae A. Kern. = Poaceae Barnhart(保留科名)■●

417438 Zebrina Schnizl. = Tradescantia L. ■

417439 Zebrina pendula Schnizl. 'Discolor';美叶吊竹梅■

417440 Zebrina pendula Schnizl. 'Minima';姬吊竹梅■

417441 Zebrina pendula Schnizl. 'Quadricolor';四色吊竹梅■

417442 Zebrina pendula Schnizl. = Tradescantia zebrina Bosse ■

417443 Zederachia Fabr. = Melia L. ●

417444 Zederachia Heist. ex Fabr. = Melia L. ●

417445 Zederbauera H. P. Fuchs = Erysimum L. ■

417446 Zederbauera H. P. Fuchs(1959);齐德芥属■☆

417447 Zederbauera echinella (Hand. -Mazz.) H. P. Fuchs;齐德芥■☆

417448 Zedoaria Raf. = Curcuma L. (保留属名)■

417449 Zeduba Ham. ex Meisn. = Calanthe R. Br. (保留属名)■

417450 Zehnderia C. Cusset(1987);策恩川苔草属■☆

417451 Zehnderia microgyna C. Cusset;策恩川苔草■☆

417452 Zehneria Endl. (1833);马㼍儿属(老鼠拉冬瓜属);Zehneria ■

417453 Zehneria angolensis Hook. f.;安哥拉马㼍儿■☆

417454 Zehneria anomala C. Jeffrey;异常马㼍儿■☆

417455 Zehneria baueriana Endl. = Zehneria liukiuensis (Nakai) C. Jeffrey ex E. Walker ■

417456 Zehneria bodinieri (H. Lév.) W. J. de Wilde et Duyfjes;波氏钮子瓜(钮子瓜)■

417457 Zehneria capillacea (Schumach.) C. Jeffrey;细毛马㼍儿■☆

417458 Zehneria cerasiformis Stocks = Ctenolepis cerasiformis (Stocks) Hook. f. ■☆

417459 Zehneria cordifolia Schweinf. ex Broun et Massey = Zehneria scabra（L. f.）Sond. ●☆

417460 Zehneriadebilis Sond. = Trochomeria debilis（Sond.）Hook. f. ■☆

417461 Zehneria emirnensis（Baker）Rabenant.；埃米尔马㝔儿■☆

417462 Zehneria fernandensis Hutch. et Dalziel = Zehneria scabra（L. f.）Sond. ●☆

417463 Zehneria formosana（Hayata）S. S. Ying = Zehneria japonica（Thunb. ex A. Murray）H. Y. Liu ■

417464 Zehneria gilletii（De Wild.）C. Jeffrey；吉勒特马㝔儿■☆

417465 Zehneria guamensis（Merr.）Fosberg；台湾马㝔儿■

417466 Zehneria hallii C. Jeffrey；霍尔马㝔儿■☆

417467 Zehneria hederacea Sond. = Kedrostis nana（Lam.）Cogn. ■☆

417468 Zehneria hookeriana Arn. = Zehneria japonica（Thunb. ex A. Murray）H. Y. Liu ■

417469 Zehneria indica（Lour.）Keraudren = Neoachmandra japonica（Thunb.）W. J. de Wilde et Duyfjes■

417470 Zehneria indica（Lour.）Rabenant.；马㝔儿（单梢瓜，狗黄瓜，耗子拉冬瓜，金丝瓜，扣子草，老鼠担夯瓜，老鼠冬瓜，老鼠瓜，老鼠黄瓜，老鼠拉冬瓜，老鼠拉金瓜，马交儿，山冬瓜，山鸡仔，山熊胆，天瓜，土白薮，土花粉，野黄瓜，野苦瓜，野梢瓜，银丝莲，玉钮子）；India Melothria，India Zehneria，Indian Zehneria■

417471 Zehneria japonica（Thunb. ex A. Murray）H. Y. Liu = Zehneria japonica（Thunb.）H. Y. Liu ■

417472 Zehneria japonica（Thunb.）H. Y. Liu；日本马㝔儿（马㝔儿，台湾马㝔儿）■

417473 Zehneria japonica（Thunb.）H. Y. Liu = Bryonia japonica Thunb. ex A. Murray ■

417474 Zehneria japonica（Thunb.）H. Y. Liu = Neoachmandra japonica（Thunb.）W. J. de Wilde et Duyfjes ■

417475 Zehneria keayana R. Fern. et A. Fern.；凯伊马㝔儿■☆

417476 Zehneria kelungensis Hayata = Zehneria guamensis（Merr.）Fosberg ■

417477 Zehneria kelungensis Hayata = Zehneria mucronata（Blume）Miq. ■

417478 Zehneria liukiuensis（Nakai）C. Jeffrey ex E. Walker = Zehneria mucronata（Blume）Miq. ■

417479 Zehneria liukiuensis（Nakai）E. Walker = Zehneria guamensis（Merr.）Fosberg ■

417480 Zehneria longepedunculata A. Rich. = Zehneria scabra（L. f.）Sond. ●☆

417481 Zehneria lucida（Naudin）Hook. f. = Zehneria maysorensis（Wight et Arn.）Arn. ■

417482 Zehneria macrocarpa Sond. = Trochomeria macrocarpa（Sond.）Hook. f. ■☆

417483 Zehneria madagascariensis Rabenant.；马岛马㝔儿■☆

417484 Zehneria mannii Cogn. = Zehneria scabra（L. f.）Sond. ●☆

417485 Zehneria marginata（Blume）Keraudren = Scopellaria marginata（Blume）W. J. de Wilde et Duyfjes ■

417486 Zehneria marginata（Blume）Rabenant. = Scopellaria marginata（Blume）W. J. de Wilde et Duyfjes ■

417487 Zehneria marlothii（Cogn.）R. Fern. et A. Fern.；马洛斯马㝔儿■☆

417488 Zehneria maysorensis（Wight et Arn.）Arn.；钮子瓜（大树献钮子，红果果，钮子瓜，天罗网，土瓜，野杜瓜，野苦瓜，争文武）；Button Zehneria，Maysor Melothria，Maysor Zehneria ■

417489 Zehneria maysorensis（Wight et Arn.）Arn. = Zehneria mucronata（Blume）Miq. ■

417490 Zehneria microsperma Hook. f.；小籽马㝔儿■☆

417491 Zehneria minutiflora（Cogn.）C. Jeffrey；微花马㝔儿■☆

417492 Zehneria mucronata（Blume）Miq.；黑果马㝔儿（秤砣子，琉球马㝔儿，山刺瓜，台湾马㝔儿）；Taiwan Melothria, Taiwan Zehneria ■

417493 Zehneria mucronata（Blume）Miq. = Zehneria maysorensis（Wight et Arn.）Arn. ■

417494 Zehneria mysorensis Wight = Zehneria liukiuensis（Nakai）C. Jeffrey ex E. Walker ■

417495 Zehneria obtusiloba E. Mey. ex Sond. = Kedrostis foetidissima（Jacq.）Cogn. ■☆

417496 Zehneria oligosperma C. Jeffrey；寡籽马㝔儿■☆

417497 Zehneria pallidinervia（Harms）C. Jeffrey；白脉马㝔儿■☆

417498 Zehneria parvifolia（Cogn.）J. H. Ross；小叶马㝔儿■☆

417499 Zehneria pectinata Sond. = Trochomeria hookeri Harv. ■☆

417500 Zehneria peneyana（Naudin）Asch. et Schweinf.；佩内马㝔儿■☆

417501 Zehneria peneyana（Naudin）Schweinf. et Asch.；帕内马㝔儿■☆

417502 Zehneria perpusilla（Blume）Bole et M. R. Almeida var. deltifrons（Ohwi）H. Ohba = Zehneria maysorensis（Wight et Arn.）Arn. ■

417503 Zehneria perpusilla（Blume）Cogn. = Zehneria mucronata（Blume）Miq. ■

417504 Zehneria perrieri Rabenant.；佩里耶马㝔儿●☆

417505 Zehneria polycarpa（Cogn.）Rabenant.；多果马㝔儿●☆

417506 Zehneria racemosa Hook. f.；总花马㝔儿■☆

417507 Zehneria rutenbergiana（Cogn.）Rabenant.；鲁滕贝格马㝔儿■☆

417508 Zehneria scabra（L. f.）Sond.；粗糙马㝔儿●☆

417509 Zehneria scabra（L. f.）Sond. var. argyrea（A. Zimm.）C. Jeffrey；银色马㝔儿●☆

417510 Zehneria scrobiculata A. Rich. = Zehneria scabra（L. f.）Sond. ●☆

417511 Zehneria somalensis Thulin；索马里马㝔儿●☆

417512 Zehneria thwaitesii（Schweinf.）C. Jeffrey；思韦茨马㝔儿■☆

417513 Zehneria umbellata（Klein ex Will.）Thwaites = Solena amplexicaulis（Lam.）Gandhi ■

417514 Zehneria umbellata Thwaites = Solena amplexicaulis（Lam.）Gandhi ■

417515 Zehneria velutina Arn. = Zehneria scabra（L. f.）Sond. ●☆

417516 Zehneria viridis（A. Zimm.）C. Jeffrey = Zehneria emirnensis（Baker）Rabenant. ■☆

417517 Zehneria wallichii（C. B. Clarke）C. Jeffrey；锤果马㝔儿；Hammerfruit Zehneria, Wallich Zehneria ■

417518 Zehneria wyleana Sond. = Trochomeria debilis（Sond.）Hook. f. ■☆

417519 Zehntnerella Britton et Rose = Facheiroa Britton et Rose ●☆

417520 Zehntnerella Britton et Rose（1920）；小花杖属（小花柱属）●☆

417521 Zehntnerella squamulosa Britton et Rose；小花杖●☆

417522 Zeia Lunell = Agropyron Gaertn. ■

417523 Zeia Lunell = Triticum L. ■

417524 Zeia canina（L.）Lunell = Elymus caninus（L.）L. ■

417525 Zeia repens（L.）Lunell = Elytrigia repens（L.）Desv. ex B. D. Jacks. ■

417526 Zeia vulgaris Lunell = Triticum aestivum L. ■

417527 Zeia vulgaris Lunell var. aestiva（L.）Lunell = Triticum aestivum L. ■

417528 Zeiba Raf. = Ceiba Mill. ●

417529 Zelea Hort. ex Ten. = Carapa Aubl. ●☆

417530 Zelenkoa M. W. Chase et N. H. Williams = Oncidium Sw.（保留

属名)■☆

417531　Zelenkoa M. W. Chase et N. H. Williams（2001）；巴拿马瘤瓣兰属■☆

417532　Zelenkoa onusta（Lindl.）M. W. Chase et N. H. Williams；巴拿马瘤瓣兰■☆

417533　Zeliauros Raf. = ? Veronica L. ■

417534　Zelkova Spach（1841）（保留属名）；榉属（榉树属）；Sawleaf Zelkova，Water Elm，Waterelm，Water-elm，Zelkova ●

417535　Zelkova abelicea Boiss. ；希腊榉●☆

417536　Zelkova acuminata Planch. = Zelkova serrata（Thunb.）Makino ●

417537　Zelkova carpinifolia（Pall.）K. Koch = Zelkova crenata Spach ●☆

417538　Zelkova carpinifolia Dippel = Zelkova carpinifolia（Pall.）K. Koch ●☆

417539　Zelkova carpinifolia Dippel = Zelkova crenata Spach ●☆

417540　Zelkova crenata Spach；高加索榉；Azed，Caucasian Elm，Caucasian Zelkova，Elm Zelkova，Siberian Elm，Water-elm ●☆

417541　Zelkova cretica Spach = Zelkova abelicea Boiss. ●☆

417542　Zelkova cuspidata ? = Zelkova serrata（Thunb.）Makino ●

417543　Zelkova davidiana（Priemer）Bean. = Hemiptelea davidii（Hance）Planch. ●

417544　Zelkova davidii（Hance）Hemsl. = Hemiptelea davidii（Hance）Planch. ●

417545　Zelkova davidii（Priemer）Bean = Hemiptelea davidii（Hance）Planch. ●

417546　Zelkova davidii Bean = Hemiptelea davidii（Hance）Planch. ●

417547　Zelkova formosana Hayata；台湾榉；Taiwan Zelkova ●

417548　Zelkova formosana Hayata = Zelkova serrata（Thunb.）Makino ●

417549　Zelkova hirta C. K. Schneid. = Zelkova serrata（Thunb.）Makino ●

417550　Zelkova keaki Dippel = Zelkova serrata（Thunb.）Makino ●

417551　Zelkova keaki Maxim. = Zelkova serrata（Thunb.）Makino ●

417552　Zelkova schneideriana Hand. -Mazz.；大叶榉（大叶榉树，大叶榆，红榉，黄榉，黄栀榆，鸡油树，榉，榉木，榉榆，血榉，血榆，硬壳榔）；Bigleaf Waterelm，C. K. Schneid. Zelkova，Chinese Zelkova，Schneider Zelkova ●

417553　Zelkova schneideriana Hand. -Mazz. 'Goblin'；妖怪榉树●

417554　Zelkova schneideriana Hand. -Mazz. 'Green Vase'；绿花瓶榉树●

417555　Zelkova schneideriana Hand. -Mazz. 'Halka'；哈尔卡榉树●

417556　Zelkova schneideriana Hand. -Mazz. 'Pulverulenta'；黄斑榉树●

417557　Zelkova schneideriana Hand. -Mazz. 'Village Green'；乡村绿榉树●

417558　Zelkova serrata（Thunb.）Makino；榉树（大果榉，光光榆，光叶榉，光叶榉树，槻，鸡油，鸡油树，尖齿榉，榉，榉榔，榉木，榉榆，马柳光树，台湾榉，太鲁阁榉）；Japanese Elm，Japanese Zelkova，Keaki，Keyaki，Pointed Zelkova，Sawleaf Zelkova，Waterelm ●

417559　Zelkova serrata（Thunb.）Makino f. stipulacea（Makino）Koji Ito；托叶状榉树●☆

417560　Zelkova serrata（Thunb.）Makino var. pendula Makino；垂枝榉●☆

417561　Zelkova serrata（Thunb.）Makino var. stipulacea Makino = Zelkova serrata（Thunb.）Makino f. stipulacea（Makino）Koji Ito ●☆

417562　Zelkova serrata（Thunb.）Makino var. tarokoensis（Hayata）H. L. Li = Zelkova serrata（Thunb.）Makino ●

417563　Zelkova sinica C. K. Schneid. ；大果榉（抱树，赤肚榆，小叶榉，叶下珠，圆齿鸡油树）；Bigfruit Waterelm，Chinese Zelkova ●

417564　Zelkova tarokoensis Hayata = Zelkova serrata（Thunb.）Makino ●

417565　Zelkova tonkinensis Gagnep. ；越南榉（北部湾榉）；Tonkin

Zelkova，Vietnam Waterelm ●☆

417566　Zelmira Raf. = Calathea G. Mey. ■

417567　Zelonops Raf. = Phoenix L. ●

417568　Zeltnera G. Mans.（2004）；策尔龙胆属■☆

417569　Zeltnera trichantha（Griseb.）G. Mans. ；策尔龙胆■☆

417570　Zemisia B. Nord.（2006）；异色千里光属●☆

417571　Zemisia discolor（Sw.）B. Nord. ；异色千里光■☆

417572　Zemisne O. Deg. et Sherff = Scalesia Arn. ●☆

417573　Zenia Chun（1946）；翅荚木属（砍头树属，任豆属）；Zenia，Zenbean ●

417574　Zenia insignis Chun；翅荚木（米杠，任豆，任木）；Common Zenia，Zenbean，Zenia ◇

417575　Zenkerella Taub.（1894）；岑克尔豆属（固氮豆属）■☆

417576　Zenkerella capparidacea（Taub.）J. Léonard subsp. grotei（Harms）Temu；格罗特岑克尔豆■☆

417577　Zenkerella citrina Taub. ；岑克尔豆■☆

417578　Zenkerella egregia J. Léonard；优秀岑克尔豆●☆

417579　Zenkerella grotei（Harms）J. Léonard = Zenkerella capparidacea（Taub.）J. Léonard subsp. grotei（Harms）Temu ■☆

417580　Zenkerella pauciflora Harms = Zenkerella citrina Taub. ■☆

417581　Zenkerella perplexa Temu；缠结岑克尔豆■☆

417582　Zenkeria Arn. = Apuleia Mart.（保留属名）●☆

417583　Zenkeria Rchb. = Parmentiera DC. ●

417584　Zenkeria Trin.（1837）；山地草原草属■☆

417585　Zenkeria elegans Trin. ；山地草原草●☆

417586　Zenkerina Engl. = Staurogyne Wall. ■

417587　Zenkerina kamerunensis Engl. = Staurogyne kamerunensis（Engl.）Benoist ■☆

417588　Zenkerodendron Gilg ex Jabl. = Cleistanthus Hook. f. ex Planch. ■

417589　Zenkerophytum Engl. ex Diels = Syrrheonema Miers ●☆

417590　Zenobia D. Don（1834）；粉姬木属（白铃木属，扎诺比木属）；Zenobia ●☆

417591　Zenobia cerasiflora H. Lév. = Enkianthus chinensis Franch. ●

417592　Zenobia pulverulenta（W. Bartram ex Willd.）Pollard；粉姬木（白铃木，蓝粉扎诺比木）；Dusty Zenobia，Honeycup ●☆

417593　Zenobia pulverulenta（Willd.）Pollard = Zenobia pulverulenta（W. Bartram ex Willd.）Pollard ●☆

417594　Zenopogon Link = Anthyllis L. ■☆

417595　Zeocriton P. Beauv. = Hordeum L. ■

417596　Zeocriton Wolf = Hordeum L. ■

417597　Zephiranthes Raf. = Zephyranthes Herb.（保留属名）■

417598　Zephyra D. Don（1832）；西蒂可花属■☆

417599　Zephyra elegans D. Don；西蒂可花■☆

417600　Zephyranthaceae Salisb. ；葱莲科■

417601　Zephyranthaceae Salisb. = Amaryllidaceae J. St. -Hil.（保留科名）●■

417602　Zephyranthaceae Salisb. = Poaceae Barnhart（保留科名）■●

417603　Zephyranthella（Pax）Pax = Habranthus Herb. ■☆

417604　Zephyranthella Pax. = Habranthus Herb. ■☆

417605　Zephyranthella tubispatha（Pax）Pax = Zephyranthes tubispatha（L'Hér.）Herb. ex Traub ■☆

417606　Zephyranthes Herb.（1821）（保留属名）；葱莲属（菖蒲莲属，葱兰属，玉帘属）；Fairy Lily，Rain Lily，Rain-lily，Swamp Lily，Windflower，Zephyr Lily，Zephyr-flower，Zephyrlily，Zephyr-lily ■

417607　Zephyranthes albiella Traub；白苞葱莲■☆

417608　Zephyranthes atamasca（L.）Herb. ；阿塔葱莲（阿塔玛斯扣葱莲，大花葱莲）；Atamasco Lily，Atamascolily，Atamasco-lily，Autumn

Zephyr Lily, Carolina-lily, Easter-lily, Fairy Lily, Naked-lady, Occidental Swamp-lily, Rain Lily, Rainlily, Virginia-lily, Virginian Daffodil ■☆

417609 Zephyranthes atamasca（L.）Herb. var. treatiae（S. Watson）Meerow = Zephyranthes treatiae S. Watson ■☆

417610 Zephyranthes aurea Baker et Hook. f.；金色葱莲；Golden Zephyrlily ■☆

417611 Zephyranthes brazosensis Traub = Zephyranthes chlorosolen（Herb.）D. Dietr. ■☆

417612 Zephyranthes candida（Lindl.）Herb.；葱莲（白菖蒲莲,白花独蒜,白玉帘,菖蒲莲,葱兰,肝风草,玉帘）；Autumn Rain-lily, Autumn Zephyrlily, Fairy Lily, Flower-of-the-western-wind, Peruvian Swamp Lily, Peruvian Swamp-lily, White Amaryllis, Zephyr Flower, Zephyr Lily, Zephyrlily ■

417613 Zephyranthes carinata Herb.；韭莲（菖蒲莲,独蒜,风雨花,旱水仙,红菖蒲莲,红花菖蒲莲,红玉帘,韭兰,空心菜菜,赛番红花,山慈姑,通心韭菜）；Mexican Fairy Lily, Rain Lily, Rose Pink Zephyr Lily, Rosepink Zephyrlily ■

417614 Zephyranthes carinata Herb. = Zephyranthes grandiflora Lindl. ■

417615 Zephyranthes chlorosolen（Herb.）D. Dietr. = Cooperia drummondii Herb. ■☆

417616 Zephyranthes citrina Baker；橙黄葱莲（黄花风雨花）；Citron Rrain-lily, Citron Zephyrlily, Yellow Crocus ■☆

417617 Zephyranthes drummondii D. Don；德拉蒙德葱莲；Cebolleta, Giant Prairie Lily ■☆

417618 Zephyranthes flava Roem. et Schult.；黄葱莲 ■☆

417619 Zephyranthes grandiflora Lindl. = Zephyranthes carinata Herb. ■

417620 Zephyranthes herbertiana D. Dietr. = Zephyranthes chlorosolen（Herb.）D. Dietr. ■☆

417621 Zephyranthes insularum H. H. Hume ex Moldenke；岛屿葱莲 ■☆

417622 Zephyranthes jonesii（Cory）Traub；琼斯葱莲 ■☆

417623 Zephyranthes lindleyana Herb.；林德利葱莲 ■☆

417624 Zephyranthes longifolia Hemsl.；长叶葱莲；Cebolleta, Copper Zyphyr-lily, Rain Lily ■☆

417625 Zephyranthes macrosiphon Baker；大管葱莲 ■☆

417626 Zephyranthes pulchella J. G. Sm.；雅致葱莲；Cebolleta ■☆

417627 Zephyranthes refugiensis F. B. Jones；避难所葱莲 ■☆

417628 Zephyranthes robusta Baker = Habranthus robustus Herb. ex Sweet ■☆

417629 Zephyranthes rosea Lindl. = Zephyranthes rosea Spreng. ■☆

417630 Zephyranthes rosea Spreng.；古巴葱莲（粉红韭莲,红风雨花,小韭兰）；Cuban Zephyrlily ■☆

417631 Zephyranthes simpsonii Chapm.；辛普森葱莲 ■☆

417632 Zephyranthes smallii（Alexander）Traub；斯莫尔葱莲 ■☆

417633 Zephyranthes texana Herb. = Zephyranthes tubispatha（L'Hér.）Herb. ex Traub ■☆

417634 Zephyranthes traubii（W. Hayw.）Moldenke；特劳布葱莲 ■☆

417635 Zephyranthes treatiae S. Watson；特里特葱莲 ■☆

417636 Zephyranthes tsouii Hu = Zephyranthes grandiflora Lindl. ■

417637 Zephyranthes tubiflora（L'Hér.）Schinz；管花葱莲；Tubularflower Zephyrlily ■☆

417638 Zephyranthes tubispatha（L'Hér.）Herb. ex Traub；委内瑞拉葱莲；Barbados Snowdrop, Venezuela Zephyrlily ■☆

417639 Zephyranthes tubispatha（L'Hér.）Herb. = Zephyranthes tubispatha（L'Hér.）Herb. ex Traub ■☆

417640 Zeravschania Korovin（1948）；柴拉芹属 ■☆

417641 Zeravschania regeliana Korovin；雷格尔柴拉芹 ■☆

417642 Zerdana Boiss.（1842）；类木果芥属 ■☆

417643 Zerdana anchonioides Boiss.；类木果芥 ■☆

417644 Zerna Panz. = Vulpia C. C. Gmel. ■

417645 Zerna angrenica（Drobow）Nevski = Bromus angrenicus Drobow ■

417646 Zerna angrenica（Drobow）Nevski = Bromus paulsenii Hack. ex Paulsen ■

417647 Zerna aspera（Murray）Panz. = Bromus ramosus Huds. ■

417648 Zerna benekenii（Lange）Lindm. = Bromus benekenii（Lange）Trimen ■

417649 Zerna distachya（L.）Panz. ex B. D. Jacks. = Brachypodium distachyon（L.）P. Beauv. ■

417650 Zerna erecta（Huds.）Panz. = Bromus erectus Huds. ■

417651 Zerna himalaica（Stapf）Henrard = Bromus himalaicus Stapf ■

417652 Zerna inermis（Leyss.）Lindm. = Bromus inermis Leyss. ■

417653 Zerna korotkiji（Drobow）Nevski = Bromus korotkiji Drobow ■

417654 Zerna madritensis（L.）Panz. ex B. D. Jacks. = Bromus madritensis L. ■

417655 Zerna mairei（Hack. ex Hand. -Mazz.）Henrard = Bromus mairei Hack. ex Hand. -Mazz. ■

417656 Zerna pamirica（Drobow）Nevski = Bromus pamiricus Drobow ■

417657 Zerna pamirica（Drobow）Nevski = Bromus paulsenii Hack. ex Paulsen ■

417658 Zerna paulsenii（Hack. ex Paulsen）Nevski = Bromus paulsenii Hack. ex Paulsen ■

417659 Zerna paulsenii（Hack. ex Paulsen）Nevski subsp. pamirica（Drobow）Tzvelev = Bromus paulsenii Hack. ex Paulsen ■

417660 Zerna pumpelliana（Scribn.）Tzvelev = Bromopsis pumpelliana Scribn. ■

417661 Zerna pumpelliana（Scribn.）Tzvelev = Bromus pumpellianus Scribn. ■

417662 Zerna ramosa（Huds.）Lindm. = Bromus ramosus Huds. ■

417663 Zerna riparia（Rehmann）Nevski = Bromus riparius Rehmann ■

417664 Zerna rubens（L.）Grossh. = Bromus rubens L. ■

417665 Zerna stenostachya（Boiss.）Nevski = Bromus stenostachyus Boiss. ■

417666 Zerna sterilis（L.）Panz. = Bromus sterilis L. ■

417667 Zerna tectorum（L.）Lindm. = Bromus tectorum L. ■

417668 Zerna tectorum（L.）Panz. ex Jacks. = Bromus tectorum L. ■

417669 Zerna turkestanica（Drobow）Nevski = Bromus paulsenii Hack. ex Paulsen ■

417670 Zerna turkestanicus（Drobow）Nevski = Bromus turkestanicus Drobow ■

417671 Zerna tyttholepis Nevski = Bromus tyttholepis Nevski ■

417672 Zerna unioloides（Kunth）Lindm. = Bromus catharticus Vahl ■

417673 Zerna variegata（M. Bieb.）Nevski = Bromus variegatus M. Bieb. ■

417674 Zerna yezoensis（Ohwi）Sugim. = Bromus canadensis Michx. ■

417675 Zerna yezoensis（Ohwi）Sugim. = Bromus ciliatus L. ■

417676 Zerumbet Garsault = Kaempferia L. ■

417677 Zerumbet J. C. Wendl.（废弃属名）= Alpinia Roxb.（保留属名）■

417678 Zerumbet T. Lestib. = Zingiber Mill.（保留属名）■

417679 Zerumbet speciosum J. C. Wendl. = Alpinia zerumbet（Pers.）B. L. Burtt et R. M. Sm. ■

417680 Zerumbet speciosum Wendl. = Alpinia zerumbet（Pers.）B. L. Burtt et R. M. Sm. ■

417681 Zerumbeth Retz. = Curcuma L.（保留属名）■

417682　Zetagyne Ridl.(1921);隐雌兰属■☆

417683　Zetagyne Ridl. = Panisea (Lindl.) Lindl.(保留属名)■

417684　Zetagyne albiflora Ridl.;隐雌兰■☆

417685　Zetocapnia Link et Otto = Coetocapnia Link et Otto ■

417686　Zetocapnia Link et Otto = Polianthes L. ■

417687　Zeugandra P. H. Davis(1950);轭蕊桔梗属■☆

417688　Zeugandra iranica P. H. Davis;轭蕊桔梗■☆

417689　Zeugites P. Browne(1756);轭草属■☆

417690　Zeugites americana Willd.;轭草■☆

417691　Zeugites latifolia Hemsl.;宽叶轭草■☆

417692　Zeugites mexicana (Kunth) Trin. ex Steud.;墨西哥轭草■☆

417693　Zeuktophyllum N. E. Br.(1927);矮樱龙属■☆

417694　Zeuktophyllum calycinum (L. Bolus) H. E. K. Hartmann;萼状矮樱龙●☆

417695　Zeuktophyllum suppositum (L. Bolus) N. E. Br.;矮樱龙■☆

417696　Zeuxanthe Ridl. = Prismatomeris Thwaites ●

417697　Zeuxina Summerh. = Zeuxine Lindl.(保留属名)■

417698　Zeuxine Lindl.(1826)('Zeuxina')(保留属名);线柱兰属（腺柱兰属）;Zeuxine ■

417699　Zeuxine abbreviata (Lindl.) Hook. f. = Rhomboda abbreviata (Lindl.) Ormerod ■

417700　Zeuxine affinis (Lindl.) Benth. ex Hook. f.;宽叶线柱兰（白花线柱兰,亲种线柱兰）;Broadleaf Zeuxine ■

417701　Zeuxine africana Rchb. f.;非洲线柱兰■☆

417702　Zeuxine agyokuana Fukuy.;绿叶线柱兰（阿玉山伴兰,阿玉线柱兰,绿叶角唇兰）;Greenleaf Zeuxine ■

417703　Zeuxine arisanensis Hayata = Zeuxine affinis (Lindl.) Benth. ex Hook. f. ■

417704　Zeuxine arisanensis Hayata = Zeuxine reflexa King et Pantl. ■

417705　Zeuxine aurandiaca Schltr. = Zeuxine affinis (Lindl.) Benth. ex Hook. f. ■

417706　Zeuxine aurantiaca Schltr. = Zeuxine flava (Wall. ex Lindl.) Benth. ■

417707　Zeuxine ballii P. J. Cribb;鲍尔线柱兰■☆

417708　Zeuxine batesii Rolfe = Zeuxine tetraptera (Rchb. f.) T. Durand et Schinz ■☆

417709　Zeuxine benguetensis (Ames) Ames = Zeuxine parviflora (Ridl.) Seidenf. ■

417710　Zeuxine biloba Ridl. = Hetaeria anomala Lindl. ■

417711　Zeuxine biloba Ridl. = Hetaeria biloba (Ridl.) Seidenf. et J. J. Wood ■

417712　Zeuxine bonii Gagnep. = Pecteilis susannae (L.) Raf. ■

417713　Zeuxine boninensis Tuyama;小笠原线柱兰■☆

417714　Zeuxine boninensis Tuyama = Zeuxine parviflora (Ridl.) Seidenf. ■

417715　Zeuxine boryi (Rchb. f.) Schltr. = Cheirostylis nuda (Thouars) Ormerod ■☆

417716　Zeuxine bracteata Wight = Pecteilis susannae (L.) Raf. ■

417717　Zeuxine brevifolia Wight = Pecteilis susannae (L.) Raf. ■

417718　Zeuxine clandestina Ts. Tang et S. C. Chen = Zeuxine parviflora (Ridl.) Seidenf. ■

417719　Zeuxine cochlearis Schltr. = Zeuxine africana Rchb. f. ■☆

417720　Zeuxine cognata Ohwi et T. Koyama = Heterozeuxine nervosa (Wall. ex Lindl.) T. Hashim. ■

417721　Zeuxine cognata Ohwi et T. Koyama = Zeuxine nervosa (Wall. ex Lindl.) Trimen ■

417722　Zeuxine commelinoides A. Chev. = Zeuxine heterosepala

(Rchb. f.) Geerinck ■☆

417723　Zeuxine cristata (Blume) Schltr. = Hetaeria cristata Blume ■

417724　Zeuxine debrajiana Sud. Chowdhury = Zeuxine membranancea Lindl. ■

417725　Zeuxine elongata Rolfe;伸长线柱兰■☆

417726　Zeuxine emarginata (Blume) Lindl. = Pecteilis susannae (L.) Raf. ■

417727　Zeuxine evrardii Gagnep. = Zeuxine membranancea Lindl. ■

417728　Zeuxine flava (Wall. ex Lindl.) Benth.;黄线柱兰■

417729　Zeuxine flava (Wall.) Benth. = Zeuxine flava (Wall. ex Lindl.) Benth. ■

417730　Zeuxine flava (Wall.) Benth. = Zeuxine nervosa (Wall. ex Lindl.) Trimen ■

417731　Zeuxine fluvida Fukuy.;黄花线柱兰;Yellowflower Zeuxine ■☆

417732　Zeuxine fluvida Fukuy. = Zeuxine nervosa (Wall. ex Lindl.) Trimen ■

417733　Zeuxine formosana Rolfe = Zeuxine nervosa (Wall. ex Lindl.) Trimen ■

417734　Zeuxine franchetiana (King et Pantl.) King et Pantl. = Myrmechis pumila (Hook. f.) Ts. Tang et F. T. Wang ■

417735　Zeuxine gengmanensis (K. Y. Lang) Ormerod;耿马齿唇兰;Gengma Forkliporchis ■

417736　Zeuxine gilgiana Kraenzl. et Schltr. = Zeuxine elongata Rolfe ■☆

417737　Zeuxine godefroyi Rchb. f. = Zeuxine membranancea Lindl. ■

417738　Zeuxine goodyeroides Lindl.;白肋线柱兰;Whitevein Zeuxine ■

417739　Zeuxine gracilis (Breda) Blume;纤细线柱兰;Slender Zeuxine ■☆

417740　Zeuxine gracilis (Breda) Blume var. sakagutii (Tuyama) T. Hashim. = Zeuxine parviflora (Ridl.) Seidenf. ■

417741　Zeuxine gracilis (Breda) Blume var. sakagutii (Tuyama) T. Hashim. = Zeuxine sakagutii Tuyama ■

417742　Zeuxine gracilis (Breda) Blume var. tenuifolia (Tuyama) T. Hashim. = Zeuxine tenuifolia Tuyama ■☆

417743　Zeuxine gracilis (Breda) Blume var. tenuifolia (Tuyama) T. Hashim. = Zeuxine parviflora (Ridl.) Seidenf. ■

417744　Zeuxine gracilis Blume = Zeuxine gracilis (Breda) Blume ■☆

417745　Zeuxine grandis Seidenf.;大花线柱兰;Bigflower Zeuxine ■

417746　Zeuxine gymnochiloides Schltr. = Cheirostylis nuda (Thouars) Ormerod ■☆

417747　Zeuxine hengchuanense S. S. Ying = Zeuxine nervosa (Wall. ex Lindl.) Trimen ■

417748　Zeuxine heterosepala (Rchb. f.) Geerinck;异萼线柱兰■☆

417749　Zeuxine integerrima (Blume) Lindl. = Pecteilis susannae (L.) Raf. ■

417750　Zeuxine integrilabella C. S. Leou;全唇线柱兰;Entirelip Zeuxine ■

417751　Zeuxine inverta W. W. Sm. = Chamaegastrodia inverta (W. W. Sm.) Seidenf. ■

417752　Zeuxine kantokeiense Tatew. et Masam.;关刀溪线柱兰（港口线柱兰）;Kantokei Zeuxine ■

417753　Zeuxine lepida (Rchb. f.) Benth. ex Rolfe = Cheirostylis lepida (Rchb. f.) Rolfe ■

417754　Zeuxine leucochila Schltr.;阿里山线柱兰■

417755　Zeuxine leucochila Schltr. = Zeuxine gracilis (Breda) Blume var. tenuifolia (Tuyama) T. Hashim. ■

417756　Zeuxine leucochila Schltr. = Zeuxine parviflora (Ridl.) Seidenf. ■

417757　Zeuxine lunulata P. J. Cribb et Bowden;新月线柱兰■☆

417758　Zeuxine madagascariensis Schltr. ;马岛线柱兰■☆

417759　Zeuxine mannii（Rchb. f.）Geerinck;曼氏线柱兰■☆

417760　Zeuxine membranancea Lindl. ;膜质线柱兰■

417761　Zeuxine membranancea Lindl. = Zeuxine strateumatica（L.）Schltr. ■

417762　Zeuxine moulmeinensis（E. C. Parish et Rchb. f.）Hook. f. = Rhomboda moulmeinensis（E. C. Parish et Rchb. f.）Ormerod ■

417763　Zeuxine moulmeinensis（Parish, Rchb. f. et Sineref.）Hook. f. = Anoectochilus moulmeinensis（Parish, Rchb. f. et Sineref.）Seidenf. et Smitinand ■

417764　Zeuxine nemorosa（Fukuy.）T. P. Lin;裂唇线柱兰（裂唇指柱兰）;Splitlip Zeuxine ■

417765　Zeuxine nemorosa（Fukuy.）T. P. Lin = Cheirostylis tabiyahanensis（Hayata）Pearce et Cribb ■

417766　Zeuxine nervosa（Lindl.）Trimen = Zeuxine nervosa（Wall. ex Lindl.）Trimen ■

417767　Zeuxine nervosa（Wall. ex Lindl.）Benth. ex C. B. Clarke = Zeuxine nervosa（Wall. ex Lindl.）Trimen ■

417768　Zeuxine nervosa（Wall. ex Lindl.）Trimen;芳线柱兰（芳香线柱兰,恒春线柱兰,黄唇线柱兰,黄花线柱兰,六龟线柱兰,台湾线柱兰）;Veined Zeuxine, Yellowlip Zeuxine ■

417769　Zeuxine nervosa（Wall. ex Lindl.）Trimen = Heterozeuxine nervosa（Wall. ex Lindl.）T. Hashim. ■

417770　Zeuxine niijimae Tatew. et Masam. ;眉原线柱兰（玉线柱兰）;Niijima Zeuxine ■

417771　Zeuxine occidentalis（Summerh.）Geerinck;西方线柱兰■☆

417772　Zeuxine odorata Fukuy. ;香线柱兰;Fragrant Zeuxine ■

417773　Zeuxine odorata Fukuy. = Heterozeuxine odorata（Fukuy.）T. Hashim. ■

417774　Zeuxine parviflora（Ridl.）Seidenf. ;白花线柱兰（阿里山线柱兰,蔽花线柱兰）;White Zeuxine ■

417775　Zeuxine parviflora（Ridl.）Seidenf. = Zeuxine gracilis（Breda）Blume var. tenuifolia（Tuyama）T. Hashim. ■

417776　Zeuxine philippinensis（Ames）Ames;菲律宾线柱兰■

417777　Zeuxine procumbens Blume = Pecteilis susanne（L.）Raf. ■

417778　Zeuxine pumila（Hook. f.）King et Pantl. = Myrmechis pumila（Hook. f.）Ts. Tang et F. T. Wang ■

417779　Zeuxine reflexa King et Pantl. ;折唇线柱兰（阿里山线柱兰）;Alishan Zeuxine ■

417780　Zeuxine regia（Lindl.）Trimen;高贵线柱兰;Royl Zeuxine ■☆

417781　Zeuxine robusta Wight = Pecteilis susanne（L.）Raf. ■

417782　Zeuxine rupicola Fukuy. ;岩线柱兰■

417783　Zeuxine rupicola Fukuy. = Pecteilis susannae（L.）Raf. ■

417784　Zeuxine rupicola Fukuy. = Zeuxine strateumatica（L.）Schltr. ■

417785　Zeuxine ruwenzoriensis Kraenzl. ;鲁文佐里线柱兰■☆

417786　Zeuxine sakagutii Tuyama;黄唇线柱兰;Yellowlip Zeuxine ■

417787　Zeuxine sakagutii Tuyama = Zeuxine parviflora（Ridl.）Seidenf. ■

417788　Zeuxine sambiranoensis Schltr. = Cheirostylis nuda（Thouars）Ormerod ■☆

417789　Zeuxine shuishiehensis S. S. Ying = Zeuxine parviflora（Ridl.）Seidenf. ■

417790　Zeuxine somae Tuyama = Zeuxine nervosa（Wall. ex Lindl.）Trimen ■

417791　Zeuxine somai Tuyama;六龟线柱兰■

417792　Zeuxine somai Tuyama = Zeuxine nervosa（Wall. ex Lindl.）Trimen ■

417793　Zeuxine stammleri Schltr. ;施塔姆勒线柱兰■☆

417794　Zeuxine stenochila Schltr. = Pecteilis susanne（L.）Raf. ■

417795　Zeuxine strateumatica（L.）Schltr. ;线柱兰（细叶线柱兰,岩线柱兰）;Lawn Orchid, Soldier Orchid, Soldier's Orchid, Zeuxine ■

417796　Zeuxine strateumatica（L.）Schltr. f. rupicola（Fukuy.）T. Hashim. = Pecteilis susannae（L.）Raf. ■

417797　Zeuxine strateumatica（L.）Schltr. f. rupicola（Fukuy.）T. Hashim. = Zeuxine strateumatica（L.）Schltr. ■

417798　Zeuxine strateumatica（L.）Schltr. var. rupicola（Fukuy.）S. S. Ying = Pecteilis susannae（L.）Raf. ■

417799　Zeuxine strateumatica（L.）Schltr. var. rupicola（Fukuy.）S. S. Ying = Zeuxine strateumatica（L.）Schltr. ■

417800　Zeuxine sulcata（Roxb.）Lindl. = Pecteilis susannae（L.）Raf. ■

417801　Zeuxine sulcata（Roxb.）Lindl. = Zeuxine strateumatica（L.）Schltr. ■

417802　Zeuxine sulcata Lindl. = Zeuxine strateumatica（L.）Schltr. ■

417803　Zeuxine sutepensis Rolfe ex Downie = Zeuxine affinis（Lindl.）Benth. ex Hook. f. ■

417804　Zeuxine tabiyahanensis（Hayata）Hayata;东部线柱兰（台湾拟线柱兰）■

417805　Zeuxine tabiyahanensis Hayata = Cheirostylis tabiyahanensis（Hayata）Pearce et Cribb ■

417806　Zeuxine taiwaniana S. S. Ying = Zeuxine affinis（Lindl.）Benth. ex Hook. f. ■

417807　Zeuxine taiwaniana S. S. Ying = Zeuxine sakagutii Tuyama ■

417808　Zeuxine tenuifolia Tuyama;细叶线柱兰（毛鞘线柱兰）■☆

417809　Zeuxine tenuifolia Tuyama = Zeuxine parviflora（Ridl.）Seidenf. ■

417810　Zeuxine tetraptera（Rchb. f.）T. Durand et Schinz;四翅线柱兰■☆

417811　Zeuxine tonkinensis Gagnep. = Zeuxine parviflora（Ridl.）Seidenf. ■

417812　Zeuxine tripleura Lindl. = Pecteilis susannae（L.）Raf. ■

417813　Zeuxine uraiensis S. S. Ying = Zeuxine affinis（Lindl.）Benth. ex Hook. f. ■

417814　Zeuxine uraiensis S. S. Ying = Zeuxine sakagutii Tuyama ■

417815　Zeuxine vietnamica Aver. = Zeuxinella vietnamica（Aver.）Aver. ■☆

417816　Zeuxine vittata Rolfe ex Downie = Zeuxine nervosa（Wall. ex Lindl.）Trimen ■

417817　Zeuxine wariana Schltr. = Pecteilis susannae（L.）Raf. ■

417818　Zeuxine yakusimensis Masam. = Hetaeria cristata Blume ■

417819　Zeuxine zamboangensis（Ames）Ames = Zeuxine nervosa（Wall. ex Lindl.）Trimen ■

417820　Zeuxinella Aver.（2003）;小线柱兰属■☆

417821　Zeuxinella vietnamica（Aver.）Aver. ;小线柱兰■☆

417822　Zexmenia La Llave et Lex. = Zexmenia La Llave ●■☆

417823　Zexmenia La Llave（1824）;须冠菊属（薄翅菊属）●■☆

417824　Zexmenia brevifolia A. Gray = Jefea brevifolia（A. Gray）Strother ■☆

417825　Zexmenia frutescens S. F. Blake;灌木须冠菊●☆

417826　Zexmenia hispida（Kunth）A. Gray;柠檬须冠菊;Devil's River, Orange Zexmenia, Rough Zexmenia ■☆

417827　Zexmenia hispida（Kunth）A. Gray = Wedelia acapulcensis Kunth var. hispida（Kunth）Strother ■☆

417828　Zexmenia pittieri Greenm. = Tuxtla pittieri（Greenm.）Villaseñor et Strother ●☆

417829 Zeydora Lour. ex Gomes = Pueraria DC. ●■

417830 Zeyhera DC. = Zeyheria Mart. ●☆

417831 Zeyhera Less. = Geigeria Griess. ■●☆

417832 Zeyhera Less. = Zeyheria A. Spreng. ■●☆

417833 Zeyhera Mart. = Zeyheria Mart. ●☆

417834 Zeyherella (Engl.) Aubrév. et Pellegr. = Bequaertiodendron De Wild. ●☆

417835 Zeyherella (Engl.) Pierre ex Aubrév. et Pellegr. = Bequaertiodendron De Wild. ●☆

417836 Zeyherella (Pierre ex Baillon) Aubrév. et Pellegr = Zeyherella Pierre ex Aubrév. et Pellegr. ●☆

417837 Zeyherella (Pierre ex Engl.) Aubrév. = Englerophytum K. Krause ●☆

417838 Zeyherella Pierre ex Aubrév. et Pellegr. (1958);泽赫山榄属● ☆

417839 Zeyherella Pierre ex Aubrév. et Pellegr. = Bequaertiodendron De Wild. ●☆

417840 Zeyherella farannensis (A. Chev.) Aubrév. et Pellegr. = Englerophytum magalismontanum (Sond.) T. D. Penn. ●☆

417841 Zeyherella gossweileri (De Wild.) Aubrév. et Pellegr. = Neoboivinella gossweileri (De Wild.) Liben ●☆

417842 Zeyherella letestui Aubrév. et Pellegr. ;莱泰斯图泽赫山榄●☆

417843 Zeyherella longepedicellata (De Wild.) Aubrév. et Pellegr. ;长梗泽赫山榄●☆

417844 Zeyherella magalismontana (Sond.) Aubrév. et Pellegr. = Englerophytum magalismontanum (Sond.) T. D. Penn. ●☆

417845 Zeyherella mayumbensis (Greves) Aubrév. et Pellegr. ;马永巴泽赫山榄●☆

417846 Zeyherella rwandensis (Troupin) Liben = Afrosersalisia rwandensis (Troupin) Liben ●☆

417847 Zeyheria A. Spreng. = Geigeria Griess. ■●☆

417848 Zeyheria Mart. (1826);泽赫紫葳属●☆

417849 Zeyheria acaulis Spreng. = Geigeria ornativa O. Hoffm. ■☆

417850 Zeyheria montana C. Mart. ;泽赫紫葳●☆

417851 Zeylanidium (Tul.) Engl. (1930);斯里兰卡川苔草属■☆

417852 Zeylanidium (Tul.) Engl. = Hydrobryum Endl. ■

417853 Zeylanidium Engl. = Zeylanidium (Tul.) Engl. ■☆

417854 Zeylanidium lichenoides Engl. ;锡兰川苔草■☆

417855 Zezyphoides Parkinson = Alphitonia Reissek ex Endl. ●

417856 Zezyphoides Parkinson = Zizyphoides Sol. ex Drake ●

417857 Zhukowskia Szlach. ,R. González et Rutk. (2000);茹考夫兰属■☆

417858 Zhukowskia Szlach. , R. González et Rutk. = Spiranthes Rich. (保留属名)■

417859 Zhumeria Rech. f. et Wendelbo(1967);茹麦灌属●☆

417860 Zhumeria majdae Rech. f. et Wendelbo;茹麦灌●☆

417861 Zichia Steud. = Kennedia Vent. ●☆

417862 Zichia Steud. = Zichya Hueg. ●☆

417863 Zichya Hueg. = Kennedia Vent. ●☆

417864 Zichya Hueg. ex Benth. = Kennedia Vent. ●☆

417865 Ziegera Raf. = Miconia Ruiz et Pav. (保留属名)●☆

417866 Zieria Sm. (1798);洋茱萸属(兹利木属);Zieria ●☆

417867 Zieria arborescens Sims;树洋茱萸;Stinkwood ●☆

417868 Zieria cytisoides Sm. ;灰毛洋茱萸(灰毛兹利木);Downy Zieria ●☆

417869 Zieria laevigata Sm. ;洋茱萸●☆

417870 Zieria laxiflora (Benth.) Domin;疏花洋茱萸●☆

417871 Zieria macrophylla Bonpl. ;大叶洋茱萸●☆

417872 Zieria microphylla Bonpl. ;小叶洋茱萸●☆

417873 Zieria smithii Andréws;史氏洋茱萸●☆

417874 Zieria trifoliata Loisel. ;三小叶洋茱萸●☆

417875 Zieridium Baill. (1872);齐里橘属●☆

417876 Zieridium Baill. = Evodia J. R. Forst. et G. Forst. ●

417877 Zieridium pseudo-obtusifolium (Guillaumin) Guillaumin;假钝叶齐里橘●☆

417878 Ziervoglia Neck. = Cynanchum L. ●■

417879 Zietenia Gled. = Stachys L. ●■

417880 Zigadenus Michx. (1803);棋盘花属;Alkali Grass, Camash, Chessboard Flower, Death Camas, Death-camus, Zigadenus ■

417881 Zigadenus acutus Rydb. = Zigadenus venenosus S. Watson var. gramineus (Rydb.) O. S. Walsh ex C. L. Hitchc. ■☆

417882 Zigadenus alpinus Blank. = Zigadenus elegans Pursh ■☆

417883 Zigadenus angustifolius (Michx.) S. Watson = Zigadenus densus (Desr.) Fernald ■☆

417884 Zigadenus brevibracteatus (M. E. Jones) H. M. Hall;短苞棋盘花;Desert Death Camas ■☆

417885 Zigadenus chloranthus Richardson = Zigadenus elegans Pursh ■☆

417886 Zigadenus coloradensis Rydb. = Zigadenus elegans Pursh ■☆

417887 Zigadenus densus (Desr.) Fernald;松林棋盘花;Black Snakeroot, Crow Poison, Osceola's Plume, Pine-barren Death Camas ■☆

417888 Zigadenus diegoensis Davidson = Zigadenus venenosus S. Watson ■☆

417889 Zigadenus dilatatus Greene = Zigadenus elegans Pursh ■☆

417890 Zigadenus elegans Pursh;雅致棋盘花(雅致线柱兰);Alkali-grass, Elegant Camas, Elegant Death-camas, Elegant Zigadenus, Mountain Death Camas, White Camas ■☆

417891 Zigadenus elegans Pursh subsp. glaucus (Nutt.) Hultén;灰雅致棋盘花;Death Camas, Mountain Death Camas, White Camas ■☆

417892 Zigadenus elegans Pursh subsp. glaucus (Nutt.) Hultén = Zigadenus elegans Pursh ■☆

417893 Zigadenus elegans Pursh var. glaucus (Nutt.) Preece ex Cronquist = Zigadenus elegans Pursh subsp. glaucus (Nutt.) Hultén ■☆

417894 Zigadenus elegans Pursh var. glaucus (Nutt.) Preece ex Cronquist = Zigadenus elegans Pursh ■☆

417895 Zigadenus exaltatus Eastw. ;巨棋盘花;Giant Death Camas ■☆

417896 Zigadenus falcatus Rydb. = Zigadenus venenosus S. Watson var. gramineus (Rydb.) O. S. Walsh ex C. L. Hitchc. ■☆

417897 Zigadenus fontanus Eastw. ;泉旁棋盘花;Small-flower Death Camas ■☆

417898 Zigadenus fremontii (Torr.) S. Watson = Zigadenus fremontii (Torr.) Torr. ex S. Watson ■☆

417899 Zigadenus fremontii (Torr.) Torr. ex S. Watson;弗氏棋盘花;False Camas, Fremont's Death Camas ■☆

417900 Zigadenus fremontii (Torr.) Torr. ex S. Watson var. brevibracteatus M. E. Jones = Zigadenus brevibracteatus (M. E. Jones) H. M. Hall ■☆

417901 Zigadenus fremontii (Torr.) Torr. ex S. Watson var. inezianus Jeps. = Zigadenus fremontii (Torr.) Torr. ex S. Watson ■☆

417902 Zigadenus fremontii (Torr.) Torr. ex S. Watson var. salsus Jeps. = Zigadenus fremontii (Torr.) Torr. ex S. Watson ■☆

417903 Zigadenus glaberrimus Michx. ;沙漠棋盘花;Sandbog Death Camas ■☆

417904 Zigadenus glaucus (Nutt.) Nutt. = Zigadenus elegans Pursh subsp. glaucus (Nutt.) Hultén ■☆

417905 Zigadenus glaucus (Nutt.) Nutt. = Zigadenus elegans Pursh ■☆

417906 Zigadenus glaucus Nutt. = Zigadenus elegans Pursh ■☆

417907 Zigadenus gracilentus Greene = Zigadenus elegans Pursh ■☆

417908　Zigadenus gramineus Rydb. ;禾叶棋盘花;Death Camas ■☆

417909　Zigadenus gramineus Rydb. = Zigadenus venenosus S. Watson var. gramineus（Rydb.）O. S. Walsh ex C. L. Hitchc. ■☆

417910　Zigadenus intermedius Rydb. = Zigadenus venenosus S. Watson var. gramineus（Rydb.）O. S. Walsh ex C. L. Hitchc. ■☆

417911　Zigadenus japonicus Miq. = Veratrum maackii Regel ■

417912　Zigadenus leimanthoides A. Gray = Zigadenus densus（Desr.）Fernald ■☆

417913　Zigadenus longus Greene = Zigadenus elegans Pursh ■☆

417914　Zigadenus mexicanus（Kunth）Hemsl. = Zigadenus virescens（Kunth）J. F. Macbr. ■☆

417915　Zigadenus micranthus Eastw. ;小花棋盘花;Small-flower Death Camas ■☆

417916　Zigadenus micranthus Eastw. var. fontanus（Eastw.）O. S. Walsh ex McNeal = Zigadenus fontanus Eastw. ■☆

417917　Zigadenus mogollonensis W. J. Hess et Sivinski;莫高伦棋盘花;Mogollon Death Camas ■☆

417918　Zigadenus mohinorensis Greenm. = Zigadenus elegans Pursh ■☆

417919　Zigadenus muscitoxicus（Walter）Regel = Amianthium muscaetoxicum（Walter）A. Gray ■☆

417920　Zigadenus nuttallii（A. Gray）S. Watson;纳托尔棋盘花;Death Camas,Death Camash,Nuttall's Death Camas,Poison Sego ■☆

417921　Zigadenus paniculatus（Nutt.）S. Watson;锥状棋盘花;Foothill Death Camas,Sand-corn ■☆

417922　Zigadenus porrifolius Greene = Zigadenus virescens（Kunth）J. F. Macbr. ■☆

417923　Zigadenus salinus A. Nelson = Zigadenus venenosus S. Watson ■☆

417924　Zigadenus sibiricus（Kunth）A. Gray = Zigadenus sibiricus（L.）A. Gray ■

417925　Zigadenus sibiricus（L.）A. Gray;棋盘花;Chessboard Flower, Siberian Zigadenus ■

417926　Zigadenus texensis（Rydb.）J. F. Macbr. = Zigadenus nuttallii（A. Gray）S. Watson ■☆

417927　Zigadenus vaginatus（Rydb.）J. F. Macbr. ;鞘棋盘花;Sheathed Death Camas ■☆

417928　Zigadenus venenosus S. Watson;毒棋盘花;Death Camas, Meadow Death Camas ■☆

417929　Zigadenus venenosus S. Watson var. ambiguus M. E. Jones = Zigadenus venenosus S. Watson ■☆

417930　Zigadenus venenosus S. Watson var. gramineus（Rydb.）O. S. Walsh ex C. L. Hitchc. ;禾状毒棋盘花;Grassy Death Camas ■☆

417931　Zigadenus venenosus S. Watson var. micranthus（Eastw.）Jeps. = Zigadenus micranthus Eastw. ■☆

417932　Zigadenus virescens（Kunth）J. F. Macbr. ;绿花棋盘花;Green Death Camas ■☆

417933　Zigadenus virescens（Kunth）J. F. Macbr. var. porrifolius（Greene）O. S. Walsh ex Espejo et López-Ferr. = Zigadenus virescens（Kunth）J. F. Macbr. ■☆

417934　Zigadenus washakie A. Nelson = Zigadenus elegans Pursh ■☆

417935　Zigara Raf. = Bupleurum L. ●■

417936　Zigmaloba Raf. = Acacia Mill.（保留属名）●■

417937　Zilla Forssk.（1775）;齐拉芥属■☆

417938　Zilla macroptera Coss. = Zilla spinosa（L.）Prantl subsp. macroptera（Coss.）Maire et Weiller ■☆

417939　Zilla myagroides Forssk. = Zilla spinosa（L.）Prantl ■☆

417940　Zilla spinosa（L.）Prantl;具刺齐拉芥■☆

417941　Zilla spinosa（L.）Prantl subsp. costata Maire et Weiller;单脉齐拉芥■☆

417942　Zilla spinosa（L.）Prantl subsp. macroptera（Coss.）Maire et Weiller;大翅齐拉芥■☆

417943　Zilla spinosa（L.）Prantl subsp. myagroides（Forssk.）Maire et Weiller = Zilla spinosa（L.）Prantl ■☆

417944　Zilla spinosa（L.）Prantl var. microcarpa（Vis.）Sickenb. = Zilla spinosa（L.）Prantl ■☆

417945　Zimapania Engl. et Pax = Jatropha L.（保留属名）●■

417946　Zimmermannia Pax = Meineckia Baill. ■☆

417947　Zimmermannia Pax(1910);齐默大戟属■☆

417948　Zimmermannia acuminata Verdc. = Meineckia acuminata（Verdc.）Brunel ex Radcl. -Sm. ■☆

417949　Zimmermannia capillipes Pax = Meineckia paxii Brunel ex Radcl. -Sm. ■☆

417950　Zimmermannia grandiflora Verdc. = Meineckia grandiflora（Verdc.）Brunel ex Radcl. -Sm. ■☆

417951　Zimmermannia nguruensis Radcl. -Sm. = Meineckia nguruensis（Radcl. -Sm.）Brunel ex Radcl. -Sm. ■☆

417952　Zimmermannia ovata E. A. Bruce = Meineckia ovata（E. A. Bruce）Brunel ex Radcl. -Sm. ■☆

417953　Zimmermannia stipularis Radcl. -Sm. = Meineckia stipularis（Radcl. -Sm.）Brunel ex Radcl. -Sm. ■☆

417954　Zimmermanniopsis Radcl. -Sm.（1990）;拟齐默大戟属■☆

417955　Zimmermanniopsis Radcl. -Sm. = Meineckia Baill. ■☆

417956　Zimmermanniopsis uzungwaensis Radcl. -Sm. = Meineckia uzungwaensis（Radcl. -Sm.）Radcl. -Sm. ■☆

417957　Zingania A. Chev. = Didelotia Baill. ●☆

417958　Zingania minutiflora A. Chev. = Didelotia minutiflora（A. Chev.）J. Léonard ●☆

417959　Zingeria P. A. Smirn.（1946）;津格草属■☆

417960　Zingeria biebersteiniana（Claus）P. A. Smirn. ;津格草■☆

417961　Zingeriopsis Prob. = Zingeria P. A. Smirn. ■☆

417962　Zingiber Adans. = Zingiber Mill.（保留属名）■

417963　Zingiber Boehm. = Zingiber Mill.（保留属名）■

417964　Zingiber Mill.（1754）（'Zinziber'）（保留属名）;姜属;Ginger,Zinger ■

417965　Zingiber aromaticum Valeton;芳香姜■☆

417966　Zingiber atrorubens Gagnep. ;川东姜;Darkred Ginger, Darkred Zinger ■

417967　Zingiber bisectum D. Fang;裂舌姜■

417968　Zingiber cammuner Roxb. ;野姜;Ginger ■

417969　Zingiber capitatum Roxb. ;头状姜（锤头姜）■☆

417970　Zingiber cassumunar Roxb. ;卡萨蒙纳姜（野姜）■☆

417971　Zingiber chinensis Roscoe = Alpinia chinensis（Retz.）Roscoe ■

417972　Zingiber cochleariforme D. Fang;匙苞姜（野阳荷）;Spoonbract Zinger,Spoonshaped Ginger ■

417973　Zingiber confine Miq. ;台湾姜■

417974　Zingiber corallinum Hance;珊瑚姜;Coral Ginger,Coral Zinger ■

417975　Zingiber densissimum S. Q. Tong et Y. M. Xia;多毛姜;Densehair Ginger ■

417976　Zingiber didymoglosa K. Schum = Zingiber striolatum Diels ■

417977　Zingiber didymoglossum K. Schum. = Zingiber striolatum Diels ■

417978　Zingiber dubium Afzel. = Costus dubius（Afzel.）K. Schum. ■☆

417979　Zingiber echuanense Y. K. Yang;鄂川姜;Echuan Ginger ■

417980　Zingiber echuanense Y. K. Yang = Zingiber mioga（Thunb.）Roscoe ■

417981　Zingiber ellipticum（S. Q. Tong et Y. M. Xia）Q. G. Wu et T.

L. Wu;侧穗姜(椭圆偏穗姜);Elliptic Ginger ■

417982　Zingiber emeiense Z. Y. Zhu = Zingiber striolatum Diels ■

417983　Zingiber flavomaculosum S. Q. Tong;黄斑姜;Yellow-spot Ginger ■

417984　Zingiber fragile S. Q. Tong;脆舌姜;Fragile Ginger ■

417985　Zingiber guangxiense D. Fang;桂姜;Guangxi Ginger, Guangxi Zinger,Kwangsi Ginger ■

417986　Zingiber gulinense Y. M. Xia;古林姜;Gulin Ginger ■

417987　Zingiber hupehense Pamp. = Zingiber striolatum Diels ■

417988　Zingiber integrilabrum Hance;全唇姜;Entirelip Ginger, Entirelip Zinger ■

417989　Zingiber integrum S. Q. Tong;全舌姜■

417990　Zingiber kawagoii Hayata;毛姜(恒春姜,三奈);Hairy Ginger, Hairy Zinger ■

417991　Zingiber koshunense Hayata ex C. T. Moo;恒春姜(高雄姜)■

417992　Zingiber koshunense Hayata ex C. T. Moo = Zingiber kawagoii Hayata ■

417993　Zingiber kwangsiense D. Fang = Zingiber guangxiense D. Fang ■

417994　Zingiber laoticum Gagnep. ;梭穗姜;Lao Ginger, Lao Zinger ■

417995　Zingiber leptorrhizum D. Fang;细根姜(软姜);Thinroot Ginger ■

417996　Zingiber liangshanense Z. Y. Zhu = Zingiber striolatum Diels ■

417997　Zingiber linyyunense D. Fang;乌姜;Black Zinger, Linyun Ginger ■

417998　Zingiber longiglande D. Fang et D. H. Qin;长腺姜■

417999　Zingiber longiligulatum S. Q. Tong;长舌姜;Longiligulate Ginger ■

418000　Zingiber longyanjiang Z. Y. Zhu;龙眼姜■

418001　Zingiber menghaiense S. Q. Tong;勐海姜;Menghai Ginger ■

418002　Zingiber mioga (Thunb.) Roscoe;襄荷(巴且,猼且,莼菹,地藕,蘘葙,覆葅,观音花,嘉草,苴蓴,莲花姜,蓴苴,襄草,山姜,山麻雀,土里开花,盐蓴,羊蘘姜,阳荷,阳藿,野姜,野老姜,野生姜,芋渠);Japan Ginger,Mioga Ginger,Mioga Zinger ■

418003　Zingiber mioga (Thunb.) Roscoe 'Variegata';斑襄荷(襄荷)■☆

418004　Zingiber monglaense S. J. Chen et Z. Y. Chen;斑蝉姜;Mengla Ginger ■

418005　Zingiber montanum (J. König) Theilade;山地姜;Cassumunar Ginger ■☆

418006　Zingiber neotruncatum T. L. Wu,K. Larsen et Turland;截形姜;Truncatum Ginger ■

418007　Zingiber nigrimaculatum S. Q. Tong;黑斑姜;Blackspot Ginger ■

418008　Zingiber nigrum Gaertn. = Alpinia nigra (Gaertn.) B. L. Burtt ■

418009　Zingiber nudicarpum D. Fang;光果姜(野姜);Nakefruit Ginger ■

418010　Zingiber officinale (Willd.) Roscoe;姜(白姜,百辣姜,川姜,大肉姜,干姜,勾妆枝,火姜,均姜,母姜,生姜,炎凉小子,因地辛,子姜);Common Ginger,Garden Ginger,Ginger,Zinger ■

418011　Zingiber oligophyllum K. Schum. ;少叶姜;Fewleaf Ginger ■

418012　Zingiber oligophyllum K. Schum. = Zingiber mioga (Thunb.) Roscoe ■

418013　Zingiber omeiensis Z. Y. Zhu;峨眉姜(地莲花);Emei Ginger, Emei Zinger ■

418014　Zingiber orbiculatum S. Q. Tong;圆瓣姜;Orbiculate Ginger ■

418015　Zingiber paucipunctatum D. Fang;少斑姜■

418016　Zingiber pleiostachyum K. Schum. ;多穗姜;Manyspike Ginger, Manyspike Zinger ■

418017　Zingiber purpureum Roscoe;紫色姜(野姜,紫姜);Cassumar, Cassumar Ginger,Purple Ginger,Purple Zinger ■

418018　Zingiber recurvatum S. Q. Tong et Y. M. Xia;弯管姜;Recurvate Ginger ■

418019　Zingiber roseum (Roxb.) Roscoe;红冠姜(红柄姜,野阳荷);Red Zinger,Redish Ginger ■

418020　Zingiber sichuanense Z. Y. Zhu et al. = Zingiber officinale (Willd.) Roscoe ■

418021　Zingiber simaoense Y. Y. Qian;思茅姜;Simao Ginger, Simao Zinger ■

418022　Zingiber stipitatum S. Q. Tong;唇柄姜;Stalklip Ginger ■

418023　Zingiber striolatum Diels;阳荷(山阳荷,野姜,野生姜,野猿猴草,阴藿);Striolate Ginger,Wild Zinger ■

418024　Zingiber teres S. Q. Tong et Y. M. Xia;柱根姜■

418025　Zingiber truncatum S. Q. Tong = Zingiber neotruncatum T. L. Wu,K. Larsen et Turland ■

418026　Zingiber tuanjuum (L.) Sm. ;团聚姜(莲花姜,团矛姜)■

418027　Zingiber wandingense S. Q. Tong;畹町姜;Wanding Ginger ■

418028　Zingiber xishuangbannaense S. Q. Tong;版纳姜;Xishuangbanna Ginger ■

418029　Zingiber yingjiangense S. Q. Tong;盈江姜;Yingjiang Ginger, Yingjiang Zinger ■

418030　Zingiber yunnanense S. Q. Tong et X. Z. Liu;云南姜;Yunnan Ginger,Yunnan Zinger ■

418031　Zingiber zerumbet (L.) Sm. ;红球姜(凤姜,姜花,球姜);Bitter Ginger,Redball Ginger,Redball Zinger,Wild Ginger ■

418032　Zingiberaceae Adans. = Zingiberaceae Martinov(保留科名)■

418033　Zingiberaceae Lindl. = Zingiberaceae Martinov(保留科名)■

418034　Zingiberaceae Martinov(1820)(保留科名);姜科(襄荷科);Ginger Family ■

418035　Zinnia L. (1759)(保留属名);百日菊属(百日草属,步步高属,对叶菊属);Youth-and-old-age,Zinnia ●■

418036　Zinnia acerosa (DC.) A. Gray;沙地百日菊;Desert Zinnia,Shrubby Zinnia,White Zinnia Southern Zinnia ●☆

418037　Zinnia acerosa A. Gray = Zinnia acerosa (DC.) A. Gray ●☆

418038　Zinnia angustifolia Kunth;小百日菊(海氏百日草,墨西哥百日菊,狭叶百日菊,小百日草);Creeping Zinnia, Haoge Zinnia,Narrowleaf Zinnia,Orange Zinnia,Zinnia ■☆

418039　Zinnia angustifolia Kunth 'Orange Star' = Zinnia haageana Regel 'Orange Star'■☆

418040　Zinnia angustifolia Kunth 'Persian Carpet' = Zinnia haageana Regel 'Persian Carpet'■☆

418041　Zinnia anomala A. Gray;异形百日菊;Shortray Zinnia ●☆

418042　Zinnia bidens Retz. = Glossocardia bidens (Retz.) Veldkamp ■

418043　Zinnia elegans Jacq. ;百日草(百日草,步步登高,步步高,步登高,火求花,火毡花,江西拉花,节节高,万寿菊,鱼尾菊);Common Zinnia,Youth-and-old-age,Zinnia ■

418044　Zinnia elegans Jacq. 'Belvedere';观景楼百日菊(观景楼百日草)■☆

418045　Zinnia elegans Jacq. = Zinnia violacea Cav. ■☆

418046　Zinnia grandiflora Nutt. ;大花百日菊;Little Golden Zinnia, Plains Zinnia,Prairie Zinnia,Rocky Mountain Zinnia ●☆

418047　Zinnia haageana Regel 'Orange Star';橙星海氏百日草(橙星小百日草);Mexican Zinnia ■☆

418048　Zinnia haageana Regel 'Persian Carpet';波斯地毯海氏百日草(波斯地毯小百日草)■☆

418049　Zinnia haageana Regel = Zinnia angustifolia Kunth ■☆

418050　Zinnia hybrida Roem. et Usteri;杂种百日菊;Zinnia ■☆

418051　Zinnia liebmanii Klatt = Sanvitaliopsis liebmannii (Klatt) K. H. Schultz ex Greenm. ■☆

418052　Zinnia linearis Benth. ;小百日草■☆

418053　Zinnia maritima Kunth;海滨百日草;Palmer's Zinnia ■☆

418054　Zinnia maritima Kunth var. palmeri（A. Gray）B. L. Turner;帕氏百日草;Palmer's Zinnia ■☆

418055　Zinnia mexicana Vilm. = Zinnia angustifolia Kunth ■☆

418056　Zinnia multiflora L. = Zinnia peruviana（L.）L. ■

418057　Zinnia pauciflora L. ;少花百日菊■☆

418058　Zinnia peruviana（L.）L. ;多花百日菊(多花百日草,秘鲁百日草,山菊花,五色梅);Flowery Zinnia, Peruvian Zinnia, Redstar Zinnia, Red-star Zinnia ■

418059　Zinnia pumila A. Gray = Zinnia acerosa（DC.）A. Gray ●☆

418060　Zinnia tenuiflora Jacq. ;小花百日菊(细花百日菊)■☆

418061　Zinnia violacea Cav. ;雅致百日菊;Elegant Zinnia, Garden Zinnia ■☆

418062　Zinowiewia Turcz.（1859）;季氏卫矛属●☆

418063　Zinowiewia australis Lundell;南方季氏卫矛●☆

418064　Zinowiewia micrantha Lundell;小花季氏卫矛●☆

418065　Zinowiewia ovata Lundell;卵形季氏卫矛●☆

418066　Zinowiewia pallida Lundell;苍白季氏卫矛●☆

418067　Zinowiewia pauciflora Lundell;少花季氏卫矛●☆

418068　Zinowiewia rubra Lundell;红季氏卫矛●☆

418069　Zinziber Mill. = Zingiber Mill.（保留属名）■

418070　Zipania Pers. = Lippia L. ●■☆

418071　Zipania Pers. = Zappania Scop. ●■

418072　Zippelia Blume(1830);齐头绒属(齐头花属);Zippelia ■

418073　Zippelia Rchb. = Rhizanthes Dumort. ■☆

418074　Zippelia Rchb. ex Endl. = Rhizanthes Dumort. ■☆

418075　Zippelia begoniifolia Blume = Zippelia begoniifolia Blume ex Schult. et Schult. f. ■

418076　Zippelia begoniifolia Blume ex Schult. et Schult. f. ;齐头绒(薄叶爬崖香);Begonialeaf Zippelia ■

418077　Zippelia lappacea Benn. = Zippelia begoniifolia Blume ex Schult. et Schult. f. ■

418078　Zizania Gronov. ex L. = Zizania L. ■

418079　Zizania L.（1753）;菰属(茭白属);Water Oat, Wild Rice, Wildrice ■

418080　Zizania aquatica L. ;野茭白(白脚笋,雕菰,雕胡,菰,菰菜,菰根,菰笋草,茲,茲蒋,茲米,蒋,茭,茭白,茭白笋,茭草,茭儿菜,茭儿菜菰,茭笋,茭筍,水生菰);Annual Wild Rice, Annual Wildrice, Canada Wild Rice, Indian Rice, Manchurian Water Rice, Southern Wild Rice, Tuscarora Rice, Wild Rice, Wild-rice ■

418081　Zizania aquatica L. subsp. angustifolia（Hitchc.）Tzvelev = Zizania palustris L. ■

418082　Zizania aquatica L. subsp. brevis（Fassett）S. L. Chen;短菰;Short Wildrice ■☆

418083　Zizania aquatica L. var. angustifolia Hitchc. = Zizania palustris L. ■

418084　Zizania aquatica L. var. interior Fassett = Zizania palustris L. var. interior（Fassett）Dore ■☆

418085　Zizania aquatica L. var. interior Fassett = Zizania palustris L. ■

418086　Zizania aquatica L. var. latifolia（Griseb.）Kom. = Zizania latifolia（Griseb.）Turcz. ex Stapf ■

418087　Zizania aristata（Retz.）Kunth = Hygroryza aristata（Retz.）Nees ex Wight et Arn. ■

418088　Zizania caduciflora（Turcz. ex Trin.）Hand. -Mazz. = Zizania latifolia（Griseb.）Turcz. ex Stapf ■

418089　Zizania caduciflora（Turcz.）Hand. -Mazz. = Zizania latifolia（Griseb.）Turcz. ex Stapf ■

418090　Zizania caduciflora Hand. -Mazz. = Zizania latifolia（Griseb.）Turcz. ex Stapf ■

418091　Zizania clavulosa Michx. = Zizania aquatica L. ■

418092　Zizania dahurica Turcz. ex Steud. = Zizania latifolia（Griseb.）Turcz. ex Stapf ■

418093　Zizania interior（Fassett）Rydb. = Zizania palustris L. var. interior（Fassett）Dore ■☆

418094　Zizania latifolia（Griseb.）Stapf = Zizania latifolia（Griseb.）Turcz. ex Stapf ■

418095　Zizania latifolia（Griseb.）Turcz. ex Stapf;茭白（出隧,大叶菰,雕菰,雕胡来,菰,菰菜,菰蒋节,菰手,菰首,菰笋,黄尾草,茭杷,茭白笋,茭儿菜,茭耳菜,茭瓜,茭首,茭笋,绿节,蓬蔬,苇茎）;Fewflower Wildrice, Indian Rice, Jiaobai Wildrice, Manchurian Wildrice, Vegetable Wild-rice, Water Oats, Water Rice, Wild Rice ■☆

418096　Zizania palustris L. ;野稻茭（水生菰,沼生菰）;Interior Wild Rice, Northern Wild Rice, River Wild Rice, Wild Rice ■

418097　Zizania palustris L. = Zizania aquatica L. ■

418098　Zizania palustris L. var. interior（Fassett）Dore;北方野稻茭;Northern Wild Rice ■☆

418099　Zizania retzii Spreng. = Hygroryza aristata（Retz.）Nees ex Wight et Arn. ■

418100　Zizania subtilis（Tratt.）Raspail = Coleanthus subtilis（Tratt.）Seidl ■

418101　Zizania terrestris L. = Scleria terrestris（L.）Fassett ■

418102　Zizaniopsis Doell et Asch.（1871）;假菰属(拟菰属)■☆

418103　Zizaniopsis miliacea（Michx.）Döll et Asch. ;大假菰;Giant Cutgrass, Southern Wild Rice, Water Millet ■☆

418104　Zizia Pfeiff. = Draba L. + Alyssum L. ■

418105　Zizia Pfeiff. = Zizzia Roth ■☆

418106　Zizia W. D. J. Koch(1824);芨芨芹属(芨芨雅属)■☆

418107　Zizia aptera（A. Gray）Fernald;芨芨芹;Golden Alexanders, Heart-leaved Alexanders, Heart-leaved Golden Alexanders, Heart-leaved Meadow Parsnip, Heart-leaved Meadow-parsnip ☆

418108　Zizia aptera（A. Gray）Fernald var. occidentalis Fernald = Zizia aptera（A. Gray）Fernald ■☆

418109　Zizia aurea（L.）W. D. J. Koch;黄芨芨芹（金黄芨芨芹）;Common Golden Alexanders, Golden Alexanders, Golden Meadow-parsnip, Golden Zizia, Meadow Parsnip ■☆

418110　Zizia aurea Koch = Zizia aurea（L.）W. D. J. Koch ■☆

418111　Zizia cordata W. D. J. Koch ex DC. = Zizia aptera（A. Gray）Fernald ■☆

418112　Zizia latifolia Small;阔叶芨芨芹■☆

418113　Zizia trifoliata（Michx.）Fernald;三叶芨芨芹;Golden Alexander, Meadow Parsnip ■☆

418114　Zizifora Adans. = Ziziphora L. ●■

418115　Ziziforum Caruel = Ziziphora L. ●■

418116　Ziziphaceae Adans. = Ziziphaceae Adans. ex T. Post et Kuntze ●

418117　Ziziphaceae Adans. ex T. Post et Kuntze = Rhamnaceae Juss.（保留科名）●

418118　Ziziphaceae Adans. ex T. Post et Kuntze;枣科●

418119　Ziziphora L.（1753）;新塔花属(唇香草属);Ziziphora ●■

418120　Ziziphora biebersteiniana Grossh. ;毕氏新塔花●☆

418121　Ziziphora brantii K. Koch;布朗新塔花●☆

418122　Ziziphora breviecalyx Juz. ;短萼新塔花●☆

418123　Ziziphora bungeana Juz. ;新塔花(小叶薄荷);Bunge Ziziphora ●

418124　Ziziphora capitata L. ;头状新塔花●☆

418125　Ziziphora capitellata Juz. ;小头状新塔花●☆

418126　Ziziphora clinopodioides Lam. ;风轮新塔花（唇香草,神香草,

小叶薄荷,续则)●

418127 Ziziphora denticulata Juz.;小齿新塔花●☆

418128 Ziziphora galinae Juz.;嘎氏新塔花●☆

418129 Ziziphora hispanica L.;西班牙新塔花●☆

418130 Ziziphora interrupta Juz.;间断新塔花●☆

418131 Ziziphora pamiroalaica Juz. ex Nevski;南疆新塔花(帕米尔新塔花,小叶薄荷,须则);S. Xinjiang Ziziphora, South Sinkiang Ziziphora,South Xinjiang Ziziphora ●

418132 Ziziphora persica Bunge;波斯新塔花●☆

418133 Ziziphora pulegioides (L.) Desf. = Hedeoma pulegioides (L.) Pers. ■☆

418134 Ziziphora pungens Bunge = Nepeta pungens (Bunge) Benth. ■

418135 Ziziphora puschkinii Adams;普什新塔花●☆

418136 Ziziphora raddei Juz.;拉德新塔花●☆

418137 Ziziphora rigida (Boiss.) Heinr. Braun;硬新塔花●☆

418138 Ziziphora serpyllacea M. Bieb.;百里香新塔花●☆

418139 Ziziphora taurica M. Bieb.;克里木新塔花●☆

418140 Ziziphora tenuior L.;小新塔花;Small Ziziphora ■●

418141 Ziziphora tomentosa Juz.;天山新塔花;Tianshan Ziziphora ●

418142 Ziziphora turcomanica Juz.;土库曼新塔花●☆

418143 Ziziphora woronowii Maleev;沃氏新塔花●☆

418144 Ziziphus Mill. (1754);枣属;Jujube, Jujube-Tree ●

418145 Ziziphus abyssinica A. Rich.;阿比西尼亚枣;Abyssinia Jujube, Catchthorn ●☆

418146 Ziziphus acidojujuba C. Y. Cheng et M. J. Liu = Ziziphus spinosa (Bunge) Hu ●

418147 Ziziphus acidojujuba C. Y. Cheng et M. J. Liu = Ziziphus zizyphus (L.) Meikle var. spinosa (Bunge) Y. L. Chen ●

418148 Ziziphus amphibia A. Chev. = Ziziphus spina-christi (L.) Desf. var. microphylla A. Rich. ●☆

418149 Ziziphus apetala Hook. f.;无瓣枣;Petalless Jujube ●

418150 Ziziphus atacorensis A. Chev. = Ziziphus abyssinica A. Rich. ●☆

418151 Ziziphus atacorensis A. Chev. var. oblongifolia ? = Ziziphus abyssinica A. Rich. ●☆

418152 Ziziphus attopensis Pierre;毛果枣(老鹰枣);Hair Jujube, Hairyfruit Jujube,Hairy-fruited Jujube ●

418153 Ziziphus baguirmiae A. Chev. = Ziziphus abyssinica A. Rich. ●☆

418154 Ziziphus chinensis Spreng. = Ziziphus jujuba Mill. ●

418155 Ziziphus chloroxylon (L.) Oliv.;绿木枣;Cogwood, Cogwood Jujube,Jamaica ●☆

418156 Ziziphus crebrivenosa C. B. Rob.;葡酒枣●☆

418157 Ziziphus esquirolii H. Lév. = Hovenia acerba Lindl. ●

418158 Ziziphus flavescens Wall. = Berchemia flavescens (Wall. ex Roxb.) Brongn. ●

418159 Ziziphus floribunda Wall. = Berchemia floribunda (Wall. ex Roxb.) Brongn. ●

418160 Ziziphus fungii Merr.;褐果枣;Brown Eelgrass, Fung Jujube ●

418161 Ziziphus fungii Merr. = Ziziphus pubinervis Rehder ●

418162 Ziziphus hamosa Wall. = Sageretia hamosa (Wall. ex Roxb.) Brongn. ●

418163 Ziziphus helvola Sond. = Ziziphus zeyheriana Sond. ●☆

418164 Ziziphus incurva Roxb.;印度枣(滇枣,褐果枣,麦抱,弯叶枣);India Jujube, Indian Jujube ●

418165 Ziziphus iroensis A. Chev. = Ziziphus spina-christi (L.) Desf. ●☆

418166 Ziziphus joazeiro C. Mart.;巴西枣树●☆

418167 Ziziphus jujuba (L.) Gaertn. = Ziziphus mauritiana Lam. ●

418168 Ziziphus jujuba (L.) Lam. = Ziziphus mauritiana Lam. ●

418169 Ziziphus jujuba Mill.;枣树(白枣,边,刺枣,大枣,樲,贯枣,红枣,红枣树,还味,樻,棘,蹶泄,苦枣,老鼠屎,良枣,马驹,美枣,蜜枣,遒,酸枣,皙,洗,羊枣,要枣,枣,枣仔,枣子,枣子树);African Lotus, China Date-plum, China Eelgrass, Chinese Date, Chinese Date-plum, Chinese Datetree, Chinese Jujube, Common Jujuba,Dunks,French Jujube,Jujube,Neutral Henna,Sedra ●

418170 Ziziphus jujuba Mill. ' Tortuosa ';龙爪枣(蟠龙枣);Dragonclaw Date,Zigzag Jujube ●

418171 Ziziphus jujuba Mill. = Ziziphus zizyphus (L.) H. Karst. ●

418172 Ziziphus jujuba Mill. f. lageniformis (Nakai) Kitag.;葫芦枣;Calabash Jujube, Lageniform Jujube ●

418173 Ziziphus jujuba Mill. var. inermis (Bunge) Rehder;无刺枣(大甜枣,大枣,红枣,枣,枣树,枣子,枣子树);Chinese Date, Chinese Jujube,Common Jujube,Spineless Common Jujube,Spineless Eelgrass ●

418174 Ziziphus jujuba Mill. var. inermis (Bunge) Rehder = Ziziphus jujuba Mill. ●

418175 Ziziphus jujuba Mill. var. spinosa (Bunge) Hu ex H. F. Chow;酸枣(白刺,赤棘,红花枣,棘,棘刺花,棘针,角针,山枣树,酸枣树,硬枣);Acid Jujube, Sour Eelgrass, Spine Date ●

418176 Ziziphus laui Merr.;球枣;Ball Eelgrass, Glabose Date, Lau Jujube ●

418177 Ziziphus lotus (L.) Lam.;莲枣(百脉根枣);African Lotus, Jujube Lotus, Lotus, Lotus Bush, Lotus Fruit, Lotus Tree ●☆

418178 Ziziphus lotus (L.) Lam. subsp. saharae (Batt. et Trab.) Maire;左原枣●☆

418179 Ziziphus lotus L. = Ziziphus lotus (L.) Lam. ●☆

418180 Ziziphus macrocarpa Feng = Ziziphus mairei Dode ●

418181 Ziziphus madecassus Perr. = Ziziphus mucronata Willd. ●☆

418182 Ziziphus mairei (H. Lév.) Browicz et Lauener = Ziziphus mauritiana Lam. ●

418183 Ziziphus mairei Dode;大果枣(鸡蛋果);Big Eelgrass, Bigfuit Date, Maire Jujube ●

418184 Ziziphus mauritiana Lam.;滇刺枣(滇山刺,滇枣刺,缅枣,酸枣,印度枣);Ber, Cottony Jujube, India Ber, Indian Jujube, Yunnan Eelgrass,Yunnan Jujube,Yunnan Spiny Jujube ●

418185 Ziziphus mauritiana Lam. var. abyssinica (A. Rich.) Fiori = Ziziphus abyssinica A. Rich. ●☆

418186 Ziziphus mauritiana Lam. var. deserticola A. Chev. = Ziziphus mauritiana Lam. ●

418187 Ziziphus mauritiana Lam. var. orthacantha (DC.) A. Chev. = Ziziphus mauritiana Lam. ●

418188 Ziziphus microphylla Roxb.;小叶枣●☆

418189 Ziziphus montana W. W. Sm.;山枣;Mountain Jujube ●

418190 Ziziphus mucronata Willd.;短尖枣;Buffalo Thorn, Cape Thorn ●☆

418191 Ziziphus mucronata Willd. subsp. rhodesica R. B. Drumm.;罗得西亚枣●☆

418192 Ziziphus muratiana Maire = Ziziphus mauritiana Lam. ●

418193 Ziziphus napeca Willd.;芫菁枣●

418194 Ziziphus nummularia (Burm. f.) Wight et Arn.;圆板枣●☆

418195 Ziziphus nummularia (Burm.) Wight et Arn. var. saharae (Batt. et Trab.) A. Chev. = Ziziphus lotus (L.) Lam. subsp. saharae (Batt. et Trab.) Maire ●☆

418196 Ziziphus obtusifolia (Hook. ex Torr. et A. Gray) A. Gray;钝叶枣;Gray Thorn, Texas Buckthorn ●☆

418197 Ziziphus oenopolia (L.) Mill.;小果枣(锈毛野枣,锈毛叶野枣);Little Eelgrass, Littlefruit Jujube, Small-fruited Jujube ●

418198 Ziziphus orthacantha DC. = Ziziphus mauritiana Lam. ●

418199　Ziziphus pubescens Oliv. ;短柔毛枣●☆

418200　Ziziphus pubescens Oliv. subsp. glabra R. B. Drumm. ;光滑枣●☆

418201　Ziziphus pubinervis Rehder; 毛脉枣（毛脉野枣）; Hairyvein Jujube, Hairy-veined Jujube ●

418202　Ziziphus pubinervis Rehder = Ziziphus incurva Roxb. ●

418203　Ziziphus ramosissima（Lour.）Spreng. = Paliurus ramosissimus（Lour.）Poir. ●

418204　Ziziphus rivularis Codd;溪边枣●☆

418205　Ziziphus robertsoniana Beentje;罗伯逊枣●☆

418206　Ziziphus rotundifolia Roth = Ziziphus nummularia（Burm. f.）Wight et Arn. ●☆

418207　Ziziphus rugosa Lam. ;皱枣（弯腰果, 弯腰树）; Wrinkled Jujube ●

418208　Ziziphus saharae Batt. et Trab. = Ziziphus lotus（L.）Lam. subsp. saharae（Batt. et Trab.）Maire ●☆

418209　Ziziphus sativa Gaertn. = Ziziphus jujuba Mill. ●

418210　Ziziphus sativa Gaertn. = Ziziphus zizyphus（L.）H. Karst. ●

418211　Ziziphus sativa Gaertn. var. inermis（Bunge）C. K. Schneid. = Ziziphus jujuba Mill. var. inermis（Bunge）Rehder ●

418212　Ziziphus sativa Gaertn. var. lageniformis Nakai = Ziziphus jujuba Mill. f. lageniformis（Nakai）Kitag. ●

418213　Ziziphus sativa Gaertn. var. spinosa（Bunge）C. K. Schneid. = Ziziphus jujuba Mill. var. spinosa（Bunge）Hu ex H. F. Chow ●

418214　Ziziphus sativa Gaertn. var. spinosa（Bunge）C. K. Schneid. = Ziziphus jujuba Mill. ●

418215　Ziziphus sinensis Lam. = Ziziphus jujuba Mill. ●

418216　Ziziphus spina-christi（L.）Desf. ;叙利亚枣（刺冠枣）; Christ's Thorn, Crown of Thorns, Ilb, Lotus Tree, Syrian Christ-thorn ●☆

418217　Ziziphus spina-christi（L.）Desf. var. microphylla A. Rich. ;小叶叙利亚枣●☆

418218　Ziziphus spina-christi（L.）Desf. var. mitissima Chiov. = Ziziphus spina-christi（L.）Desf. ●☆

418219　Ziziphus spina-christi（L.）Willd. = Ziziphus spina-christi（L.）Desf. ●☆

418220　Ziziphus spinosa（Bunge）Hu = Ziziphus jujuba Mill. var. spinosa（Bunge）Hu ex H. F. Chow ●

418221　Ziziphus spinosa（Bunge）Hu ex F. H. Chen = Ziziphus jujuba Mill. var. spinosa（Bunge）Hu ex H. F. Chow ●

418222　Ziziphus suluensis Merr. ;苏卢枣●☆

418223　Ziziphus trichocarpa Hung T. Chang = Ziziphus attopensis Pierre ●

418224　Ziziphus vulgaris Lam. = Ziziphus jujuba Mill. ●

418225　Ziziphus vulgaris Lam. = Ziziphus zizyphus（L.）H. Karst. ●

418226　Ziziphus vulgaris Lam. var. inermis Bunge = Ziziphus jujuba Mill. var. inermis（Bunge）Rehder ●

418227　Ziziphus vulgaris Lam. var. inermis Bunge = Ziziphus zizyphus（L.）Meikle var. inermis（Bunge）Y. L. Chen ●

418228　Ziziphus vulgaris Lam. var. spinosa Bunge = Ziziphus jujuba Mill. var. spinosa（Bunge）Hu ex H. F. Chow ●

418229　Ziziphus vulgaris Lam. var. spinosa Bunge = Ziziphus zizyphus（L.）Meikle var. spinosa（Bunge）Y. L. Chen ●

418230　Ziziphus vulgaris Lam. var. spinosus Bunge = Ziziphus jujuba Mill. ●

418231　Ziziphus xiangchengensis Y. L. Chen et P. K. Chou;蜀枣; Xiangcheng Jujube ●

418232　Ziziphus yunnanensis C. K. Schneid. = Ziziphus incurva Roxb. ●

418233　Ziziphus zeyheriana Sond. ;泽赫枣●☆

418234　Ziziphus zizyphus（L.）H. Karst. = Ziziphus jujuba Mill. ●

418235　Ziziphus zizyphus（L.）Meikle = Ziziphus jujuba Mill. ●

418236　Ziziphus zizyphus（L.）Meikle var. inermis（Bunge）Y. L. Chen = Ziziphus jujuba Mill. var. inermis（Bunge）Rehder ●

418237　Ziziphus zizyphus（L.）Meikle var. lageniformis（Nakai）Y. L. Chen = Ziziphus jujuba Mill. f. lageniformis（Nakai）Kitag. ●

418238　Ziziphus zizyphus（L.）Meikle var. spinosa（Bunge）Y. L. Chen = Ziziphus jujuba Mill. var. spinosa（Bunge）Hu ex H. F. Chow ●

418239　Ziziphus zizyphus（L.）Meikle var. spinosa（Bunge）Y. L. Chen = Ziziphus spinosa（Bunge）Hu ●

418240　Ziziphus zizyphus（L.）Meikle var. tortuosa Hort. = Ziziphus jujuba Mill. 'Tortuosa' ●

418241　Zizyphoides Sol. ex Drake = Alphitonia Reissek ex Endl. ●

418242　Zizyphon St. -Lag. = Ziziphus Mill. ●

418243　Zizyphora Dumort. = Ziziphora L. ●■

418244　Zizyphus Adans. = Ziziphus Mill. ●

418245　Zizzia Roth = Draba L. ＋ Alyssum L. ■●

418246　Zizzia Roth = Petrocallis W. T. Aiton ■☆

418247　Zoduba Buch. -Ham. ex D. Don = Calanthe R. Br. （保留属名）■

418248　Zoduba masuca（D. Don）Buch. -Ham. = Calanthe sylvatica（Thouars）Lindl. ■

418249　Zoegea L. （1767）;掌片菊属■☆

418250　Zoegea baldshuanica C. Winkl. ;掌片菊■☆

418251　Zoegea purpurea Fresen. ;紫掌片菊■☆

418252　Zoelleria Warb. = Trigonotis Steven ■

418253　Zoellnerallium Crosa（1975）;智利百合属■☆

418254　Zoellnerallium andinum（Poepp.）Crosa;智利百合■☆

418255　Zoisia Asch. et Graebn. = Zoysia Willd. （保留属名）■

418256　Zoisia J. M. Black = Zoysia Willd. （保留属名）■

418257　Zollernia Maximil. et Nees = Zollernia Wied-Neuw. et Nees ●☆

418258　Zollernia Wied-Neuw. et Nees（1826）;佐勒铁豆属●☆

418259　Zollernia discolor Vogel;杂色佐勒铁豆●☆

418260　Zollernia falcata Maximil. et Nees ;镰形佐勒铁豆●☆

418261　Zollernia glabra（Spreng.）Yakovlev;光佐勒铁豆●☆

418262　Zollernia grandifolia Schery;大叶佐勒铁豆●☆

418263　Zollernia parvifolis Taub. ex Glaz. ;小叶佐勒铁豆●☆

418264　Zollikoferia DC. = Launaea Cass. ■

418265　Zollikoferia Nees = Chondrilla L. ■

418266　Zollikoferia Nees = Willemetia Neck. ■☆

418267　Zollikoferia angustifolia Coss. et Durieu = Launaea angustifolia（Desf.）Kuntze ■☆

418268　Zollikoferia angustifolia Coss. et Durieu var. squarrosa（Pomel）Batt. ' = Launaea angustifolia（Desf.）Kuntze ■☆

418269　Zollikoferia anomala Batt. = Launaea pumila（Cav.）Kuntze ■☆

418270　Zollikoferia arabica Boiss. = Launaea angustifolia（Desf.）Kuntze ■☆

418271　Zollikoferia arborescens Batt. = Launaea arborescens（Batt.）Murb. ■☆

418272　Zollikoferia arborescens Batt. var. cerastina Chabert = Launaea arborescens（Batt.）Murb. ■☆

418273　Zollikoferia bornmuelleri Hausskn. ex Bornm. = Launaea bornmuelleri（Haussk. ex Bornm.）Bornm. ■☆

418274　Zollikoferia cassiana（Jaub. et Spach）Boiss. = Launaea mucronata（Forssk.）Muschl. subsp. cassiana（Jaub. et Spach）N. Kilian ■☆

418275　Zollikoferia elquiensis Phil. = Malacothrix coulteri Harv. et A. Gray ☆

418276　Zollikoferia fallax（Jaub. et Spach）Boiss. = Paramicrorhynchus procumbens（Roxb.）Kirp. ■

418277　Zollikoferia fallax（Jaubert et Spach）Boiss. = Paramicrorhynchus procumbens（Roxb.）Kirp. ■

418278　Zollikoferia glomerata（Cass.）Boiss. = Launaea capitata（Spreng.）Dandy ■☆

418279　Zollikoferia leucodon Fisch. = Paramicrorhynchus procumbens（Roxb.）Kirp. ■

418280　Zollikoferia leucodon Fisch. et Mey. ex Kar. = Paramicrorhynchus procumbens（Roxb.）Kirp. ■

418281　Zollikoferia longiloba Boiss. et Reut. = Launaea fragilis（Asso）Pau ■☆

418282　Zollikoferia mucronata（Forssk.）Boiss. = Launaea mucronata（Forssk.）Muschl. ■☆

418283　Zollikoferia nudicaulis（L.）Boiss. = Launaea nudicaulis（L.）Hook. f. ■☆

418284　Zollikoferia polydichotoma（Ostenf.）Ilijin = Hexinia polydichotoma（Ostenf.）H. L. Yang ■

418285　Zollikoferia pumila Cav. = Launaea pumila（Cav.）Kuntze ■☆

418286　Zollikoferia quercifolia（Desf.）Coss. et Kralik = Launaea quercifolia（Desf.）Pamp. ■☆

418287　Zollikoferiastrum（Kirp.）Kamelin(1993);佐里菊属■☆

418288　Zollikoferiastrum brassicifolium（Boiss.）Kamelin;芥叶佐里菊■☆

418289　Zollikoferiastrum polycladum（Boiss.）Kamelin;多枝佐里菊■☆

418290　Zollikoferiastrum takhtadzhianii（Sosn.）Kamelin;佐里菊■☆

418291　Zollingeria Kurz(1872)(保留属名);佐林格无患子属■☆

418292　Zollingeria Sch. Bip.（废弃属名）= Rhynchospermum Reinw. ex Blume ■

418293　Zollingeria Sch. Bip.（废弃属名）= Zollingeria Kurz(保留属名)■☆

418294　Zollingeria macrocarpa Kurz;佐林格无患子■☆

418295　Zollingeria scandens Sch. Bip. = Rhynchospermum verticillatum Reinw. ex Blume ■

418296　Zollingeria triptera Rolfe;三翅佐林格无患子■☆

418297　Zombia L. H. Bailey(1939);海地棕属(草裙桐属,轮刺棕属)●☆

418298　Zombia antillarum（B. D. Jacks.）L. Bailey;海地棕●☆

418299　Zombiana Baill. = Rotula Lour. ●

418300　Zombitsia Keraudren(1963);佐姆葫芦属■☆

418301　Zombitsia Rabenant. = Zombitsia Keraudren ■☆

418302　Zombitsia lucorum Rabenant. ;佐姆葫芦■☆

418303　Zomicarpa Schott(1856);巴西南星属■☆

418304　Zomicarpa pythonium Schott;巴西南星■☆

418305　Zomicarpella N. E. Br.（1881）;哥伦比亚南星属■☆

418306　Zomicarpella maculata N. E. Br. ;哥伦比亚南星■☆

418307　Zonablephis Raf. = Acanthus L. ●■

418308　Zonablephis Raf. = Cheilopsis Moq. ●■

418309　Zonanthemis Greene = Hemizonia DC. ■☆

418310　Zonanthus Griseb.（1862）;带花龙胆属■☆

418311　Zonanthus cubensis Griseb. ;带花龙胆■☆

418312　Zonaria Steud. = Zornia J. F. Gmel. ■

418313　Zonotriche（C. E. Hubb.）J. B. Phipps(1964);带毛叶舌草属(流苏毛叶舌草属)■☆

418314　Zonotriche brunnea（J. B. Phipps）Clayton;布朗带毛叶舌草■☆

418315　Zonotriche decora（Stapf）J. B. Phipps;带毛叶舌草■☆

418316　Zonotriche inamoena（K. Schum.）Clayton;非洲带毛叶舌草■☆

418317　Zoophora Bernh. = Orchis L. ■

418318　Zoophthalmum P. Browne(废弃属名)= Mucuna Adans.（保留属名)●■

418319　Zootrophion Luer = Pleurothallis R. Br. ■☆

418320　Zornia J. F. Gmel.（1792）;丁葵草属;Zornia ■

418321　Zornia Moench = Lallemantia Fisch. et C. A. Mey. ■

418322　Zornia albiflora Mohlenbr. ;白花丁葵草■☆

418323　Zornia albolutescens Mohlenbr. ;黄白丁葵草■☆

418324　Zornia angustifolia Sm. = Zornia gibbosa Span. ■

418325　Zornia apiculata Milne-Redh. ;细尖丁葵草■☆

418326　Zornia bracteata J. F. Gmel. ;苞片丁葵草■☆

418327　Zornia brevipes Milne-Redh. ;短梗丁葵草■☆

418328　Zornia cantoniensis Mohlenbr. = Zornia gibbosa Span. ■

418329　Zornia capensis Pers. ;好望角丁葵草■☆

418330　Zornia capensis Pers. subsp. tropica Milne-Redh. ;热带丁葵草■☆

418331　Zornia diphylla（L.）Pers. ;二叶丁葵草(丁葵草);Barba de Burro,Trencilla,Twinleaf Zornia,Twoleaf Zornia,Zornia ■☆

418332　Zornia diphylla（L.）Pers. = Zornia gibbosa Span. ■

418333　Zornia diphylla（L.）Pers. var. ciliaris Ohwi = Zornia intecta Mohlenbr. ■

418334　Zornia filifolia Domin;线叶丁葵草■☆

418335　Zornia gibbosa Span. ;丁葵草(苍蝇翼,苍蝇翼草,丁贵草,二叶丁葵草,二叶人字草,红骨丁地草,红骨丁地青,金线吊虾蟆,金鸳鸯,苦地枕,老鸦草,铺地草,铺地锦,人字草,沙甘里,乌龙草,乌蝇翼,乌蝇翼草,斜对叶,一条根);Cystoid Zornia ■

418336　Zornia gibbosa Span. var. cantoniensis（Mohlenbr.）Ohashi = Zornia gibbosa Span. ■

418337　Zornia glochidiata Rchb. ex DC. ;钩毛丁葵草■☆

418338　Zornia graminea Span. = Zornia gibbosa Span. ■

418339　Zornia intecta Mohlenbr. ;台东丁葵草;Complete Zornia ■

418340　Zornia latifolia DC. ;宽叶丁葵草■☆

418341　Zornia linearifolia Moench = Dracocephalum ruyschiana L. ■

418342　Zornia linearis E. Mey. ;线状丁葵草■☆

418343　Zornia microphylla Desv. ;小叶丁葵草■☆

418344　Zornia milneana Mohlenbr. ;米尔恩丁葵草■☆

418345　Zornia nutans（L.）Moench = Dracocephalum nutans L. ■

418346　Zornia nutans Moench. = Dracocephalum nutans L. ■

418347　Zornia obovata（Baker f.）Mohlenbr. = Zornia setosa Baker f. subsp. obovata（Baker f.）J. Léonard et Milne-Redh. ■☆

418348　Zornia pratensis Milne-Redh. ;草原丁葵草■☆

418349　Zornia pratensis Milne-Redh. subsp. barbata J. Léonard et Milne-Redh. ;髯毛丁葵草■☆

418350　Zornia pratensis Milne-Redh. var. glabrior Milne-Redh. ;无毛丁葵草■☆

418351　Zornia puberula Mohlenbr. ;微毛丁葵草■☆

418352　Zornia punctatissima Milne-Redh. ;多斑丁葵草■☆

418353　Zornia reptans Harms;匍匐丁葵草■☆

418354　Zornia setifera Mohlenbr. = Zornia pratensis Milne-Redh. subsp. barbata J. Léonard et Milne-Redh. ■☆

418355　Zornia setosa Baker f. ;刚毛丁葵草■☆

418356　Zornia setosa Baker f. subsp. obovata（Baker f.）J. Léonard et Milne-Redh. ;倒卵刚毛丁葵草■☆

418357　Zornia tetraphylla Micheli var. obovata Baker f. = Zornia setosa Baker f. subsp. obovata（Baker f.）J. Léonard et Milne-Redh. ■☆

418358　Zornia tetraphylla Michx. var. capensis（Pers.）Harv. = Zornia capensis Pers. ■☆

418359　Zornia tetraphylla Michx. var. linearis（E. Mey.）Harv. = Zornia linearis E. Mey. ■☆

418360　Zoroxus Raf. = Polygala L. ●■

418361　Zoroxus Raf. = Polygaloides Haller ●☆

418362　Zosima Hoffm.（1814）；艾叶芹属■

418363　Zosima Phil. = Philibertia Kunth ■

418364　Zosima absinthifolia（Vent.）Link；无味艾叶芹■☆

418365　Zosima absinthifolia Link = Zosima absinthifolia（Vent.）Link ■☆

418366　Zosima gilliana Rech. f. et Riedl；吉尔艾叶芹■☆

418367　Zosima komarovii（Manden.）M. Hiroe = Semenovia dasycarpa（Regel et Schmalh.）Korovin ex Pimenov et V. N. Tikhom. ■

418368　Zosima korovinii Pimenov；艾叶芹■

418369　Zosima orientalis Hoffm. = Zosima absinthifolia（Vent.）Link ■☆

418370　Zosima pimpinelloides（Nevski）M. Hiroe = Semenovia pimpinelloides（Nevski）Manden. ■

418371　Zosima rubtzovii（Schischk.）M. Hiroe = Semenovia rubtzovii（Schischk.）Manden. ■

418372　Zosima tordylioides Korovin；环翅艾叶芹■☆

418373　Zosimia Kom. = Zosima Hoffm. ■

418374　Zosimia M. Blob. = Zosima Hoffm. ■

418375　Zoster St. -Lag. = Zostera L. ■

418376　Zostera Cavolini = Posidonia K. D. König（保留属名）■

418377　Zostera L.（1753）；大叶藻属（甘藻属）；Eelgrass, Eel-grass, Tape-grass ■

418378　Zostera americana Hartog；美洲大叶藻■☆

418379　Zostera americana Hartog = Zostera japonica Asch. et Graebn. ■

418380　Zostera angustifolia Loser；狭叶大叶藻；Narrow-leaved Eelgrass ■☆

418381　Zostera asiatica Miki；宽叶大叶藻；Broadleaf Eelgrass ■

418382　Zostera caespitosa Miki；丛生大叶藻；Cluster Eelgrass ■

418383　Zostera capensis Setch. = Nanozostera capensis（Setch.）Toml. et Posl. ■☆

418384　Zostera caulescens Miki；具茎大叶藻；Stemmed Eelgrass ■

418385　Zostera ciliata Forssk. ；缘毛大叶藻■☆

418386　Zostera ciliata Forssk. = Thalassodendron ciliatum（Forssk.）Hartog ■☆

418387　Zostera hornemanniana Tutin；窄大叶藻；Narrow-leaved Eel-grass ■

418388　Zostera hornemannii Rouy = Nanozostera noltei（Hornem.）Toml. et Posl. ■☆

418389　Zostera japonica Asch. et Graebn. ；短大叶藻（矮大叶藻，甘藻）；Dwarf Eelgrass ■

418390　Zostera japonica Asch. et Graebn. = Zostera nana Roth ■

418391　Zostera latifolia Morong；加州宽叶大叶藻■☆

418392　Zostera marina L. ；大叶藻（海草，海带，海带草，海马蔺，蕰藻）；Alva Marina, Common Eelgrass, Eelgrass, Eel-grass, Grass Wrack, Pinjane, Sea-grass, Sea-wrack, Sweet Sea Grass, Ulva Marina, Wigeon-grass ■

418393　Zostera marina L. var. stenophylla Asch. et Graebn. = Zostera marina L. ■

418394　Zostera minor Nolte ex Rchb. ；纤细大叶藻■

418395　Zostera nana Mert. ex Roth = Nanozostera noltei（Hornem.）Toml. et Posl. ■☆

418396　Zostera nana Roth；矮大叶藻（甘藻）；Dwarf Eelgrass, Dwarf Eel-grass ■

418397　Zostera nana Roth. = Zostera japonica Asch. et Graebn. ■

418398　Zostera nodosa Ucria = Cymodocea nodosa（Ucria）Asch. ■☆

418399　Zostera noltei Hornem. = Nanozostera noltei（Hornem.）Toml. et Posl. ■☆

418400　Zostera pacifica S. Watson = Zostera asiatica Miki ■

418401　Zostera pacifica S. Watson = Zostera marina L. ■

418402　Zostera stipulacea Forssk. = Halophila stipulacea（Forssk.）Asch. ■☆

418403　Zostera tridentata Solms = Halodule uninervis（Forssk.）Asch. ■

418404　Zostera uninervis Forssk. = Halodule uninervis（Forssk.）Asch. ■

418405　Zosteraceae Dumort.（1829）（保留科名）；大叶藻科（甘藻科）；Eelgrass Family, Eel-grass Family, Zostera Family ■

418406　Zosterella Small = Heteranthera Ruiz et Pav.（保留属名）■☆

418407　Zosterella Small（1913）；异药雨久花属■☆

418408　Zosterella dubia（Jacq.）Small = Heteranthera dubia（Jacq.）MacMill. ■☆

418409　Zosterella longituba Alexander = Heteranthera dubia（Jacq.）MacMill. ■☆

418410　Zosterophyllanthos Szlach. et Marg.（2002）；带叶花属■☆

418411　Zosterospermum P. Beauv. = Rhynchospora Vahl（保留属名）■

418412　Zosterostylis Blume = Cryptostylis R. Br. ■

418413　Zosterostylis arachnites Blume = Cryptostylis arachnites（Blume）Blume ■

418414　Zosterostylis walkerae Wight = Cryptostylis arachnites（Blume）Blume ■

418415　Zosterostylis zeylanica Lindl. = Cryptostylis arachnites（Blume）Blume ■

418416　Zotovia Edgar et Connor = Ehrharta Thunb.（保留属名）■☆

418417　Zotovia Edgar et Connor = Petriella Zotov ■☆

418418　Zotovia Edgar et Connor（1998）；山地皱稃草属■☆

418419　Zotovia acicularis Edgar et Connor；山地皱稃草■☆

418420　Zouchia Raf. = Pancratium L. ■

418421　Zoutpansbergia Hutch. = Callilepis DC. ■●☆

418422　Zoutpansbergia caerulea Hutch. = Callilepis caerulea（Hutch.）Leins ●☆

418423　Zoydia Pers. = Zoysia Willd.（保留属名）■

418424　Zoysia Willd.（1801）（保留属名）；结缕草属；Lawn Grass, Lawngrass, Lawn-grass, Zoysia Grass ■

418425　Zoysia × hondana Ohwi；本田结缕草■☆

418426　Zoysia japonica Steud. ；结缕草（延地青，锥子草）；Japan Lawngrass, Japanese Lawn Grass, Japanese Lawngrass, Japanese Lawn-grass, Korea Lawngrass, Korean Eelgrass, Korean Lawn Grass ■

418427　Zoysia japonica Steud. f. macrostachya H. D. Doun et L. J. Gong；大穗日本结缕草；Bigspike Japan Lawngrass, Largespike Japan Lawngrass ■

418428　Zoysia japonica Steud. var. pallida Nakai ex Honda；青结缕草■

418429　Zoysia koreana Mez = Zoysia japonica Steud. ■

418430　Zoysia liukiuensis Honda = Zoysia sinica Hance ■

418431　Zoysia macrostachya Franch. et Sav. ；大穗结缕草；Largespike Lawngrass ■

418432　Zoysia matrella（L.）Merr. ；沟叶结缕草（马尼拉芝）；Manila Grass, Manilagrass, Manila-grass ■

418433　Zoysia matrella（L.）Merr. subsp. japonica（Steud.）Masam. et Yanagita = Zoysia japonica Steud. ■

418434　Zoysia matrella（L.）Merr. subsp. tenuifolia（Willd. ex Trin.）T. Koyama = Zoysia tenuifolia Willd. ex Thiele ■

418435　Zoysia matrella（L.）Merr. var. japonica（Steud.）Sasaki = Zoysia japonica Steud. ■

418436　Zoysia matrella（L.）Merr. var. macrantha Nakai ex Honda = Zoysia sinica Hance ■

418437　Zoysia matrella（L.）Merr. var. pacifica Goudswaard = Zoysia pacifica（Goudswaard）M. Hotta et Kuroki ■

418438　Zoysia matrella（L.）Merr. var. tenuifolia（Willd.）T. Durand et Schinz ex Makino = Zoysia tenuifolia Willd. ex Thiele ■

418439　Zoysia pacifica（Goudswaard）M. Hotta et Kuroki；太平洋结缕草（细叶结缕草）■

418440　Zoysia pungens Willd. = Zoysia japonica Steud. ■

418441　Zoysia pungens Willd. = Zoysia matrella（L.）Merr. ■

418442　Zoysia pungens Willd. var. japonica（Steud.）Hack. = Zoysia japonica Steud. ■

418443　Zoysia pungens Willd. var. tenuifolia（Willd.）T. Durand et Schinz = Zoysia tenuifolia Willd. ex Thiele ■

418444　Zoysia serrulata Mez = Zoysia matrella（L.）Merr. ■

418445　Zoysia sinica Hance；中华结缕草；China Lawngrass, Chinese Eelgrass ■

418446　Zoysia sinica Hance subsp. nipponica（Ohwi）T. Koyama = Zoysia sinica Hance ■

418447　Zoysia sinica Hance subsp. nipponica（Ohwi）T. Koyama = Zoysia sinica Hance var. nipponica Ohwi ■

418448　Zoysia sinica Hance var. macrantha（Nakai ex Honda）Ohwi = Zoysia sinica Hance ■

418449　Zoysia sinica Hance var. macrantha（Nakai）Ohwi = Zoysia sinica Hance ■

418450　Zoysia sinica Hance var. nipponica Ohwi；长花结缕草（日本结缕草）；Longflower Lawngrass ■

418451　Zoysia sinica Hance var. nipponica Ohwi = Zoysia sinica Hance ■

418452　Zoysia tenuifolia Thiele = Zoysia matrella（L.）Merr. ■

418453　Zoysia tenuifolia Willd. ex Thiele；细叶结缕草（高丽芝，天鹅绒草）；Korean Lawngrass, Korean Velvet Grass, Mascaren Grass, Mascarene Grass, Thinleaf Lawngrass, Vellvetgrass, Velvet Grass ■

418454　Zoysiaceae Link = Gramineae Juss.（保留科名）■●

418455　Zoysiaceae Link = Poaceae Barnhart（保留科名）■●

418456　Zozima DC. = Zosima Hoffm.

418457　Zozimia Boiss. = Zosima Hoffm. ■

418458　Zozimia DC. = Zozima DC. ■

418459　Zozimia anethifolia DC. = Ducrosia anethefolia（DC.）Boiss. ■☆

418460　Zozimia lasiocarpa（Boiss.）Boiss. = Platytaenia lasiocarpa（Boiss.）Rech. f. et Riedl ■☆

418461　Zschokkea Müll. Arg. = Lacmellea H. Karst. ●☆

418462　Zschokkia Benth. et Hook. f. = Zschokkea Müll. Arg. ●☆

418463　Zubiaea Gand. = Daucus L. ■

418464　Zucca Comm. ex Juss. = Momordica L. ■

418465　Zuccagnia Cav.（1799）（保留属名）；细点苏木属●☆

418466　Zuccagnia Thunb. = Dipcadi Medik. ■☆

418467　Zuccagnia punctata Cav. ；细点苏木●☆

418468　Zuccangnia Thunb.（废弃属名）= Zuccagnia Cav.（保留属名）●☆

418469　Zuccarinia Blume（1827）（保留属名）；祖卡茜属■☆

418470　Zuccarinia Maerkl.（废弃属名）= Zuccarinia Blume（保留属名）■☆

418471　Zuccarinia Spreng. = Jackia Wall. ■☆

418472　Zuccarinia macrophylla Blume；祖卡茜☆ 418473　Zucchellia Decne. = Raphionacme Harv. ■☆

418474　Zucchellia angolensis（K. Schum.）Decne. = Raphionacme angolensis（K. Schum.）N. E. Br. ■☆

418475　Zuchertia Baill. = Tragia L. ●☆

418476　Zuckia Standl.（1915）；棱苞滨藜属；Siltbush ●☆

418477　Zuckia arizonica Standl. = Zuckia brandegeei（A. Gray）S. L. Welsh et Stutz var. arizonica（Standl.）S. L. Welsh ●☆

418478　Zuckia brandegeei（A. Gray）S. L. Welsh et Stutz；棱苞滨藜；Brandegee's Siltbush ●☆

418479　Zuckia brandegeei（A. Gray）S. L. Welsh et Stutz var. arizonica （Standl.）S. L. Welsh；亚利桑那棱苞滨藜；Arizona Siltbush ●☆

418480　Zuckia brandegeei（A. Gray）S. L. Welsh et Stutz var. plummeri （Stutz et S. C. Sand.）Dorn；普卢默棱苞滨藜；Plummer's Siltbush ●☆

418481　Zuelania A. Rich.（1841）；苏兰木属●☆

418482　Zuelania guidonia Britton et Millsp. ；圭东苏兰木●☆

418483　Zugilus Raf. = Ostrya Scop.（保留属名）●☆

418484　Zulatia Neck. = Miconia Ruiz et Pav.（保留属名）●☆

418485　Zulatia Neck. ex Raf. = Rhynchanthera DC.（保留属名）●☆

418486　Zuloagaea Bess = Panicum L. ■

418487　Zuloagaea Bess（2006）；鳞苞稷属；Bulb Panicgrass ■☆

418488　Zuloagaea bulbosa（Kunth）Bess；鳞苞稷；Bulb Panicgrass ■☆

418489　Zuloagaea bulbosa（Kunth）Bess = Panicum bulbosum Kunth ●☆

418490　Zuluzania Comm. ex C. F. Gaertn. = Bertiera Aubl. ☆

418491　Zunilia Lundell = Ardisia Sw.（保留属名）●■

418492　Zurloa Ten. = Carapa Aubl. ●☆

418493　Zuvanda（Dvorák）Askerova（1985）；西南亚芥属■☆

418494　Zuvanda meyeri（Boiss.）R. K. Askerova；西南亚芥■☆

418495　Zwaardekronia Korth. = Psychotria L.（保留属名）●

418496　Zwackhia Sendtn. ex Rchb. = Halacsya Dörfl. ☆

418497　Zwaelthia Sendtn. = Halacsya Dörfl. ☆

418498　Zwingera Hofer = Atropa L. ■

418499　Zwingera Hofer = Nolana L. ex L. f. ■☆

418500　Zwingera Schreb. = Quassia L. ●☆

418501　Zwingera Schreb. = Simaba Aubl. ●☆

418502　Zwingeria Fabr. = Ziziphora L. ●■

418503　Zwingeria Heist. ex Fabr. = Ziziphora L. ●■

418504　Zycona Kuntze = Allendea La Llave ■●☆

418505　Zycona Kuntze = Schistocarpha Less. ●■☆

418506　Zygadenus Endl. = Zigadenus Michx. ■

418507　Zygadenus Michx. = Zigadenus Michx. ■

418508　Zygalchemilla Rydb. = Alchemilla L. ■

418509　Zyganthera N. E. Br. = Pseudohydrosme Engl. ■☆

418510　Zyganthera buettneri（Engl.）N. E. Br. = Pseudohydrosme buettneri Engl. ■☆

418511　Zyganthera buettneri N. E. Br. = Pseudohydrosme buettneri Engl. ■☆

418512　Zygella S. Moore = Cypella Herb. ■☆

418513　Zygia Benth. et Hook. f. = Micromeria Benth.（保留属名）■●

418514　Zygia Benth. et Hook. f. = Zygis Desv. ■●

418515　Zygia Kosterm. = Pithecellobium Mart.（保留属名）●

418516　Zygia P. Browne（废弃属名）= Paralbizzia Kosterm. ●

418517　Zygia P. Browne（废弃属名）= Pithecellobium Mart.（保留属名）●

418518　Zygia Walp. = Albizia Durazz. ●

418519　Zygia cordifolia T. L. Wu = Archidendron cordifolium（T. L. Wu）I. C. Nielsen ●

418520　Zygia dulcis（Roxb.）Lyons = Pithecellobium pruinosum Benth. ●☆

418521　Zygia fastigiata E. Mey. = Albizia adianthifolia（Schumach.）W. Wight ●☆

418522　Zygia petersiana Bolle = Albizia petersiana（Bolle）Oliv. ●☆

418523　Zygia saman（Jacq.）Benth. = Samanea saman（Jacq.）Merr. ●

418524　Zygilus Post et Kuntze = Ostrya Scop.（保留属名）●

418525　Zygilus Post et Kuntze = Zugilus Raf. ●

418526　Zygis Desv. = Micromeria Benth.（保留属名）■●

418527　Zygis Desv. ex Ham.（废弃属名）= Micromeria Benth.（保留属名）■●

418528　Zygocactus Fric et K. Kreuz. = Zygocactus K. Schum. ■

418529 Zygocactus K. Schum. （1890）；蟹爪花属（蟹爪兰属，蟹爪属）；Crab Cactus，Crabcactus ■

418530 Zygocactus K. Schum. = Schlumbergera Lem. ●

418531 Zygocactus hybridus Hort. ；杂交蟹爪兰■☆

418532 Zygocactus truncatus （Haw.） K. Schum.；蟹爪兰（锦上添花，蟹爪，蟹爪花，蟹足霸王鞭）；Christmas Cactus，Claw Cactus，Crab Cactus，False Christmas Cactus，Yoke Cactus ■

418533 Zygocactus truncatus （Haw.） K. Schum. 'Bicolor'；早花蟹爪■

418534 Zygocactus truncatus （Haw.） K. Schum. 'Salmon'；橙红蟹爪■

418535 Zygocactus truncatus （Haw.） K. Schum. 'Violacea'；堇色蟹爪■

418536 Zygocactus truncatus （Haw.） K. Schum. = Schlumbergera truncata （Haw.） Moran ■

418537 Zygocarpum Thulin et Lavin(2001)；轭果豆属●☆

418538 Zygocarpum gillettii （Thulin） Thulin et Lavin；吉莱特轭果豆●☆

418539 Zygocarpum rectangulare （Thulin） Thulin et Lavin；直角轭果豆●☆

418540 Zygocarpum somalense （J. B. Gillett） Thulin et Lavin；索马里轭果豆■☆

418541 Zygocereus Frič et Kreuz. = Schlumbergera Lem. ●

418542 Zygocereus Frič et Kreuz. = Zygocactus K. Schum. ■

418543 Zygochloa S. T. Blake(1941)；怪禾木属●☆

418544 Zygochloa paradoxa （R. Br.） S. T. Blake；怪禾木●☆

418545 Zygodia Benth. = Baissea A. DC. ●☆

418546 Zygodia axillaris Benth. = Baissea axillaris （Benth.） Hua ●☆

418547 Zygodia congensis Good = Baissea campanulata （K. Schum.） de Kruif ●☆

418548 Zygodia kidengensis K. Schum. = Baissea myrtifolia （Benth.） Pichon ●☆

418549 Zygodia melanocephala （K. Schum.） Stapf = Baissea myrtifolia （Benth.） Pichon ●☆

418550 Zygodia myrtifolia Benth. = Baissea myrtifolia （Benth.） Pichon ●☆

418551 Zygodia subsessilis Benth. = Baissea urceolata （Stapf） Pichon ●☆

418552 Zygodia urceolata Stapf = Baissea urceolata （Stapf） Pichon ●☆

418553 Zygoglossum Reinw. （废弃属名） = Bulbophyllum Thouars（保留属名）■

418554 Zygoglossum Reinw. （废弃属名） = Cirrhopetalum Lindl. （保留属名）■

418555 Zygoglossum Reinw. ex Blume = Cirrhopetalum Lindl. （保留属名）■

418556 Zygoglossum umbellatum Reinw. = Bulbophyllum lemurense Bosser et P. J. Cribb ■☆

418557 Zygogonum Hutch. = Zygogynum Baill. ●☆

418558 Zygogynum Baill. （1867）；合蕊林仙属●☆

418559 Zygogynum vieillardii Baill. ；合蕊林仙●☆

418560 Zygolepis Turcz. = Arytera Blume ●

418561 Zygomenes Salisb. = Amischophacelus R. S. Rao et Kammathy ■

418562 Zygomenes axillaris （L.） Salisb. = Amischophacelus axillaris （L.） R. S. Rao et Kammathy ■☆

418563 Zygomenes axillaris （L.） Salisb. = Cyanotis axillaris （L.） Sweet ■

418564 Zygomenes pauciflora （A. Rich.） Hassk. = Cyanotis barbata D. Don ■

418565 Zygomeris Moc. et Sessé ex DC. = Amicia Kunth ■☆

418566 Zygonerion Baill. = Strophanthus DC. ●

418567 Zygonerion welwitschii Baill. = Strophanthus welwitschii （Baill.） K. Schum. ●☆

418568 Zygoon Hiern = Tarenna Gaertn. ●

418569 Zygoon graveolens Hiern = Coptosperma zygoon （Bridson） Degreef ●☆

418570 Zygopeltis Fenzl ex Endl. = Heldreichia Boiss. ■☆

418571 Zygopetalon Hook. = Zygopetalum Hook. ■☆

418572 Zygopetalon rostratum Hook. = Zygosepalum rostratum （Hook.） Rchb. f. ■☆

418573 Zygopetalum Hook. （1827）；轭瓣兰属；Zygopetalum ■☆

418574 Zygopetalum × perrenoudi Hort. ；珀氏轭瓣兰■☆

418575 Zygopetalum crinitum Lodd. ；长刚毛轭瓣兰■☆

418576 Zygopetalum discolor Rchb. f. = Warszewiczella discolor Rchb. f. ■☆

418577 Zygopetalum intermedia Lodd. ；间型轭瓣兰（美丽轭瓣兰）■☆

418578 Zygopetalum mackaii Hook. ；马氏轭瓣兰■☆

418579 Zygopetalum maxillare Lodd. ；颚骨状轭瓣兰■☆

418580 Zygophyllaceae R. Br. （1814）（保留科名）；蒺藜科；Beancaper Family，Bean-caper Family，Caltrop Family，Creosote-bush Family ●■

418581 Zygophyllidium （Boiss.） Small = Euphorbia L. ●■

418582 Zygophyllidium Small = Euphorbia L. ●■

418583 Zygophyllidium hexagonum （Nutt.） Small = Euphorbia hexagona Nutt. ■☆

418584 Zygophyllon St. -Lag. = Zygophyllum L. ●■

418585 Zygophyllum L. （1753）；驼蹄瓣属（霸王属）；Bean Caper，Beancaper，Bean-caper，Caltrop，Overlord，Twinleaf ●■

418586 Zygophyllum aegyptium Hosny；埃及驼蹄瓣●☆

418587 Zygophyllum aegyptium Hosny = Tetraena aegyptia （Hosny） Beier et Thulin ●☆

418588 Zygophyllum album L. f. ；白驼蹄瓣■☆

418589 Zygophyllum album L. f. = Tetraena alba （L. f.） Beier et Thulin ■☆

418590 Zygophyllum album L. f. subsp. gaetulum （Emb. et Maire） Quézel = Tetraena gaetula （Emb. et Maire） Beier et Thulin ●☆

418591 Zygophyllum album L. f. subsp. geslinii （Coss.） Quézel et Santa = Tetraena geslinii （Coss.） Beier et Thulin ●☆

418592 Zygophyllum album L. f. var. amblyocarpum （Baker f. ex Oliv.） Hadidi = Tetraena alba （L. f.） Beier et Thulin ■☆

418593 Zygophyllum album L. f. var. cornutum （Coss.） Murb. ；角状驼蹄瓣●☆

418594 Zygophyllum album L. f. var. geslinii （Coss.） Le Houér. = Tetraena geslinii （Coss.） Beier et Thulin ●☆

418595 Zygophyllum amblyocarpum Baker f. ex Oliv. = Tetraena alba （L. f.） Beier et Thulin ■☆

418596 Zygophyllum applanatum Van Zyl；扁平驼蹄瓣●☆

418597 Zygophyllum applanatum Van Zyl = Tetraena applanata （Van Zyl） Beier et Thulin ●☆

418598 Zygophyllum atriplicoides Fisch. = Zygophyllum atriplicoides Fisch. et C. A. Mey. ●☆

418599 Zygophyllum atriplicoides Fisch. et C. A. Mey. ；滨藜驼蹄瓣●☆

418600 Zygophyllum atriplicoides Fisch. et C. A. Mey. subsp. eurypterum （Boiss. et Buhse） Popov = Zygophyllum eurypterum Boiss. et Buhse. ●☆

418601 Zygophyllum atriplicoides Fisch. et C. A. Mey. subsp. tetramerum Popov = Zygophyllum eurypterum Boiss. et Buhse. ●☆

418602 Zygophyllum atriplicoides Fisch. et C. A. Mey. subsp. typicum Popov = Zygophyllum atriplicoides Fisch. et C. A. Mey. ●☆

418603 Zygophyllum aureum Dinter ex Engl. = Roepera pubescens （Schinz） Beier et Thulin ●■☆

418604 Zygophyllum balchaschense Boriss. ；巴尔哈什驼蹄瓣●☆

418605 Zygophyllum berenicense （Muschl.） Hadidi；贝雷尼塞驼蹄瓣●☆

418606　Zygophyllum berenicense Schweinf. = Zygophyllum berenicense（Muschl.）Hadidi ●☆

418607　Zygophyllum brachypterum Kar. et Kir.；细茎驼蹄瓣（细茎霸王）；Thinstem Beancaper ■

418608　Zygophyllum bucharicum B. Fedtsch.；布哈尔驼蹄瓣●☆

418609　Zygophyllum campanulatum Dinter ex Range = Tetraena longicapsularis（Schinz）Beier et Thulin ●☆

418610　Zygophyllum chrysopterum Retief；金翅驼蹄瓣●☆

418611　Zygophyllum chrysopterum Retief = Tetraena chrysoptera（Retief）Beier et Thulin ●☆

418612　Zygophyllum cinereum Schinz = Tetraena longicapsularis（Schinz）Beier et Thulin ●☆

418613　Zygophyllum clavatum Schltr. et Diels；棍棒驼蹄瓣●☆

418614　Zygophyllum clavatum Schltr. et Diels = Tetraena clavata（Schltr. et Diels）Beier et Thulin ●☆

418615　Zygophyllum coccineum L.；绯红驼蹄瓣●☆

418616　Zygophyllum coccineum L. = Tetraena coccinea（L.）Beier et Thulin ●☆

418617　Zygophyllum coccineum L. var. berenicense Muschl. = Zygophyllum berenicense（Muschl.）Hadidi ●☆

418618　Zygophyllum coccineum sensu Boiss. = Zygophyllum propinquum Decne. ●☆

418619　Zygophyllum cordifolium L. f. = Roepera cordifolia（L. f.）Beier et Thulin ●☆

418620　Zygophyllum cornutum Coss. = Tetraena cornuta（Coss.）Beier et Thulin ●☆

418621　Zygophyllum crassifolium Huysst. = Roepera cuneifolia（Eckl. et Zeyh.）Beier et Thulin ●☆

418622　Zygophyllum cuneifolium Eckl. et Zeyh. = Roepera cuneifolia（Eckl. et Zeyh.）Beier et Thulin ●☆

418623　Zygophyllum cuspidatum Boriss.；骤尖驼蹄瓣●☆

418624　Zygophyllum cylindrifolium Schinz；柱叶驼蹄瓣●☆

418625　Zygophyllum cylindrifolium Schinz = Tetraena cylindrifolia（Schinz）Beier et Thulin ●☆

418626　Zygophyllum darvasicum Boriss.；达尔瓦斯驼蹄瓣●☆

418627　Zygophyllum debile Cham. et Schltdl. = Roepera debilis（Cham. et Schltdl.）Beier et Thulin ■☆

418628　Zygophyllum decumbens Delile；外倾驼蹄瓣●☆

418629　Zygophyllum decumbens Delile = Tetraena decumbens（Delile）Beier et Thulin ●☆

418630　Zygophyllum decumbens Delile var. megacarpum Hosni；大果外倾驼蹄瓣●☆

418631　Zygophyllum depauperatum Drake = Tetraena madagascariensis（Baill.）Beier et Thulin ●☆

418632　Zygophyllum dichotomum Licht. ex Cham. et Schltdl.；二歧驼蹄瓣■☆

418633　Zygophyllum dielsianum Popov = Zygophyllum jaxarticum Popov ■

418634　Zygophyllum divaricatum Eckl. et Zeyh. = Roepera divaricata（Eckl. et Zeyh.）Beier et Thulin ●☆

418635　Zygophyllum dregeanum Sond.；德雷驼蹄瓣■☆

418636　Zygophyllum dumosum Boiss.；灌丛驼蹄瓣（丰塔纳驼蹄瓣,灌丛霸王）；Bean Caper ●☆

418637　Zygophyllum dumosum Boiss. = Tetraena dumosa（Boiss.）Beier et Thulin ●☆

418638　Zygophyllum eichwaldii C. A. Mey.；爱氏驼蹄瓣●☆

418639　Zygophyllum eurypterum Boiss. et Buhse.；宽翅驼蹄瓣●☆

418640　Zygophyllum eurypterum Boiss. et Buhse. subsp. gontscharowii（Boriss.）Hadidi = Zygophyllum eurypterum Boiss. et Buhse. ●☆

418641　Zygophyllum fabago L.；驼蹄瓣（豆型霸王,豆叶霸王,骆蹄瓣,骆驼蹄板,骆驼蹄瓣,蹄瓣根）；Horsebean Caltrop, Syrian Beancaper, Syrian Bean-caper ■

418642　Zygophyllum fabago L. subsp. brachypterum Popov = Zygophyllum brachypterum Kar. et Kir. ■

418643　Zygophyllum fabago L. subsp. dolichocarpum Popov ex Hadidi；长果驼蹄瓣；Longfruited Horsebean Caltrop ■

418644　Zygophyllum fabago L. subsp. orientale Boriss. ex Hadidi；短果驼蹄瓣；Oriental Horsebean Caltrop ■

418645　Zygophyllum fabago L. subsp. typicum Popov = Zygophyllum fabago L. ■

418646　Zygophyllum fabagoides Popov；拟豆叶驼蹄瓣（拟豆叶霸王）；Sham Horsebean Caltrop ■

418647　Zygophyllum fasciculatum Licht. ex Cham. et Schltdl.；簇生驼蹄瓣■☆

418648　Zygophyllum ferganense（Drobow）Boriss.；费尔干霸王；Fergan Beancaper ●

418649　Zygophyllum ferganense（Drobow）Boriss. = Sarcozygium xanthoxylum Bunge ●

418650　Zygophyllum flexuosum Eckl. et Zeyh. = Roepera flexuosa（Eckl. et Zeyh.）Beier et Thulin ●☆

418651　Zygophyllum foetidum Schrad. et J. C. Wendl. = Roepera foetida（Schrad. et J. C. Wendl.）Beier et Thulin ●☆

418652　Zygophyllum fulvum L. = Roepera fulva（L.）Beier et Thulin ●☆

418653　Zygophyllum furcatum C. A. Mey.；叉分驼蹄瓣●☆

418654　Zygophyllum fuscatum Van Zyl = Roepera fuscata（Van Zyl）Beier et Thulin ●☆

418655　Zygophyllum gaetulum Emb. et Maire；盖图拉驼蹄瓣（盖图拉霸王）●☆

418656　Zygophyllum gaetulum Emb. et Maire = Tetraena gaetula（Emb. et Maire）Beier et Thulin ●☆

418657　Zygophyllum gaetulum Emb. et Maire subsp. waterlotii（Maire）Dobignard et Jacq. et D. Jord. = Tetraena gaetula（Emb. et Maire）Beier et Thulin subsp. waterlotii（Maire）Beier et Thulin ●☆

418658　Zygophyllum gaetulum Emb. et Maire var. dolichocarpum Maire = Tetraena gaetula（Emb. et Maire）Beier et Thulin ●☆

418659　Zygophyllum garipense E. Mey. = Tetraena microcarpa（Licht. ex Cham.）Beier et Thulin ■☆

418660　Zygophyllum geslinii Coss.；热斯兰驼蹄瓣●☆

418661　Zygophyllum geslinii Coss. = Tetraena geslinii（Coss.）Beier et Thulin ●☆

418662　Zygophyllum geslinii Coss. var. ambiguum Maire = Tetraena geslinii（Coss.）Beier et Thulin ●☆

418663　Zygophyllum giessii Merxm. et A. Schreib.；吉斯驼蹄瓣●☆

418664　Zygophyllum giessii Merxm. et A. Schreib. = Tetraena giessii（Merxm. ex A. Schreib.）Beier et Thulin ●☆

418665　Zygophyllum gilfillanii N. E. Br. = Roepera lichtensteiniana（Cham. et Schltdl.）Beier et Thulin ●☆

418666　Zygophyllum glaucum E. Mey. ex Sond. = Zygophyllum sonderi H. Eichler ●☆

418667　Zygophyllum gobicum Maxim.；戈壁驼蹄瓣（戈壁霸王）；Gobi Beancaper ●■

418668　Zygophyllum gontscharovii Boriss.；高恩恰洛夫驼蹄瓣●☆

418669　Zygophyllum gontscharowii Boriss. = Zygophyllum eurypterum Boiss. et Buhse. ●☆

418670　Zygophyllum guyotii Kneuck. et Muschl. = Zygophyllum

propinquum Decne. ●☆

418671   Zygophyllum hamiense Schweinf. ;哈米驼蹄瓣●☆

418672   Zygophyllum hamiense Schweinf. = Tetraena hamiensis（Schweinf.）Beier et Thulin ●☆

418673   Zygophyllum hildebrandtii Engl. = Melocarpum hildebrandtii（Engl.）Beier et Thulin ■☆

418674   Zygophyllum hirticaule Van Zyl = Roepera hirticaulis（Van Zyl）Beier et Thulin ■☆

418675   Zygophyllum horridum Cham. = Roepera horrida（Cham.）Beier et Thulin ■☆

418676   Zygophyllum ifniense Caball. = Tetraena gaetula（Emb. et Maire）Beier et Thulin ●☆

418677   Zygophyllum iliense Popov；伊犁驼蹄瓣（伊犁霸王）；Ili Beancaper,Yili Beancaper ■

418678   Zygophyllum incanum Schinz = Tetraena longistipulata（Schinz）Beier et Thulin ●☆

418679   Zygophyllum incrustatum E. Mey. ex Sond. = Roepera incrustata（E. Mey. ex Sond.）Beier et Thulin ●☆

418680   Zygophyllum jaxarticum Popov；长果霸王（长果驼蹄瓣）；Longfruit Beancaper ■

418681   Zygophyllum kansuense Y. X. Liou；甘肃霸王（甘肃驼蹄瓣）；Gansu Beancaper ●

418682   Zygophyllum karatavicum Boriss. ;卡拉塔夫驼蹄瓣●☆

418683   Zygophyllum kaschgaricum Boriss. ；喀什霸王；Kaschgar Beancaper,Kaschgar Overlord ●

418684   Zygophyllum kaschgaricum Boriss. = Sarcozygium kaschgaricum（Boiss.）Y. X. Liou ●

418685   Zygophyllum kegense Boriss. ;凯格驼蹄瓣●☆

418686   Zygophyllum kopalense Boriss. ;科帕尔驼蹄瓣●☆

418687   Zygophyllum lanatum Willd. = Seetzenia lanata（Willd.）Bullock ■☆

418688   Zygophyllum latialatum Engl. = Tetraena rigida（Schinz）Beier et Thulin ●☆

418689   Zygophyllum latifolium Schrenk = Zygophyllum rosovii Bunge var. latifolium（Schrenk）Popov ■

418690   Zygophyllum laxum Engl. = Roepera leptopetala（E. Mey. ex Sond.）Beier et Thulin ●☆

418691   Zygophyllum lehmannianum Bunge;莱曼驼蹄瓣●☆

418692   Zygophyllum leptopetalum E. Mey. ex Sond. = Roepera leptopetala（E. Mey. ex Sond.）Beier et Thulin ●☆

418693   Zygophyllum leucocladum Diels = Roepera leucoclada（Diels）Beier et Thulin ●☆

418694   Zygophyllum lichtensteinianum Cham. et Schltdl. = Roepera lichtensteiniana（Cham. et Schltdl.）Beier et Thulin ●☆

418695   Zygophyllum loczyi Kanitz；粗茎驼蹄瓣（粗茎霸王,翼柄霸王）；Thickstem Beancaper,Wingstipe Beancaper ●■

418696   Zygophyllum longicapsulare Schinz；长裂果驼蹄瓣●☆

418697   Zygophyllum longicapsulare Schinz = Tetraena longicapsularis（Schinz）Beier et Thulin ●☆

418698   Zygophyllum longistipulatum Schinz;长托叶驼蹄瓣●☆

418699   Zygophyllum longistipulatum Schinz = Tetraena longistipulata（Schinz）Beier et Thulin ●☆

418700   Zygophyllum macrocarpum Retief = Roepera macrocarpa（Retief）Beier et Thulin ■☆

418701   Zygophyllum macropodum Boriss. ;大叶驼蹄瓣（大叶霸王）；Largeleaf Beancaper ■

418702   Zygophyllum macropterum C. A. Mey. ;大翅驼蹄瓣（大翅霸王）；Bigwing Beancaper ■

418703   Zygophyllum macropterum C. A. Mey. var. microphyllum Boriss. ;小叶大翅驼蹄瓣；Littleleaf Beancaper,Smallleaf Beancaper ■

418704   Zygophyllum maculatum Aiton = Roepera maculata（Aiton）Beier et Thulin ●☆

418705   Zygophyllum madagascariense（Baill.）Stauffer;马岛驼蹄瓣●☆

418706   Zygophyllum madagascariense（Baill.）Stauffer = Tetraena madagascariensis（Baill.）Beier et Thulin ●☆

418707   Zygophyllum madecassum H. Perrier;马德卡萨驼蹄瓣●☆

418708   Zygophyllum madecassum H. Perrier = Tetraena madecassa（H. Perrier）Beier et Thulin ●☆

418709   Zygophyllum maritimum Eckl. et Zeyh. = Roepera maritima（Eckl. et Zeyh.）Beier et Thulin ●☆

418710   Zygophyllum marlothii Engl. = Tetraena stapfii（Schinz）Beier et Thulin ●☆

418711   Zygophyllum maximiliani Schltr. ex Huysst. ;马克西米利亚诺驼蹄瓣●☆

418712   Zygophyllum megacarpum Boriss. ;大果驼蹄瓣●☆

418713   Zygophyllum meyeri Sond. = Roepera foetida（Schrad. et J. C. Wendl.）Beier et Thulin ●☆

418714   Zygophyllum microcarpum Boriss. ;小果驼蹄瓣●☆

418715   Zygophyllum microcarpum Licht. ex Cham. et Schltdl. = Tetraena microcarpa（Licht. ex Cham.）Beier et Thulin ■☆

418716   Zygophyllum microcarpum Licht. ex Cham. et Schltdl. var. macrocarpum Loes. = Tetraena microcarpa（Licht. ex Cham.）Beier et Thulin ■☆

418717   Zygophyllum microphyllum L. f. = Roepera microphylla（L. f.）Beier et Thulin ■☆

418718   Zygophyllum migiurtinorum Chiov. ;米朱蒂驼蹄瓣●☆

418719   Zygophyllum migiurtinorum Chiov. = Tetraena migiurtinora（Chiov.）Beier et Thulin ●☆

418720   Zygophyllum miniatum Cham. et Schltdl. ;朱红驼蹄瓣●☆

418721   Zygophyllum mucronatum Maxim. ;蝎虎驼蹄瓣（草霸王,鸡大腿,念念,蝎虎霸王,蝎虎草,蟹胡草）；Crab Beancaper, Crab Zygophyllum,Gecko Beancaper ●■

418722   Zygophyllum obliquum Popov；长梗驼蹄瓣（长梗霸王）；Longstalk Beancaper ■

418723   Zygophyllum obliquum Popov = Zygophyllum macropterum C. A. Mey. ■

418724   Zygophyllum oocarpum Loes. ex Huysst. = Roepera leucoclada（Diels）Beier et Thulin ●☆

418725   Zygophyllum orbiculatum Welw. ex Oliv. = Roepera orbiculata（Welw. ex Oliv.）Beier et Thulin ●☆

418726   Zygophyllum oxyanum Boriss. ;骆驼蹄草■☆

418727   Zygophyllum oxycarpum Popov;尖果驼蹄瓣（尖果霸王,骆驼蹄草）；Sharpfruit Beancaper ■

418728   Zygophyllum paradoxum Schinz = Roepera cordifolia（L. f.）Beier et Thulin ●☆

418729   Zygophyllum pfeilii Engl. = Roepera cordifolia（L. f.）Beier et Thulin ●☆

418730   Zygophyllum portulacoides Cham. ;马齿苋驼蹄瓣●☆

418731   Zygophyllum portulacoides Forssk. = Zygophyllum simplex L. ●☆

418732   Zygophyllum potaninii Maxim. ;大花驼蹄瓣（包氏霸王,大花霸王）；Bigflower Beancaper ●■

418733   Zygophyllum prismaticum Chiov. ;棱形驼蹄瓣●☆

418734   Zygophyllum prismaticum Chiov. = Tetraena prismatica（Chiov.）Beier et Thulin ●☆

418735　Zygophyllum prismatocarpum E. Mey. ex Sond. ;棱柱果驼蹄瓣●☆

418736　Zygophyllum prismatocarpum E. Mey. ex Sond. = Tetraena prismatocarpa（E. Mey. ex Sond.）Beier et Thulin ●☆

418737　Zygophyllum procumbens Adamson = Roepera spinosa（L.）Beier et Thulin ●☆

418738　Zygophyllum propinquum Decne. ;邻近驼蹄瓣●☆

418739　Zygophyllum prostratum Thunb. = Seetzenia lanata（Willd.）Bullock ■☆

418740　Zygophyllum pterocarpum Bunge;翼果驼蹄瓣（翅果霸王,翼果霸王）;Wingfruit Beancaper ●■

418741　Zygophyllum pterocarpum Bunge var. microcarpum Y. X. Liou;小翼果驼蹄瓣（小翼果霸王）;Small-fruit Wingfruit Beancaper ●

418742　Zygophyllum pterocaule Van Zyl;翼茎驼蹄瓣●☆

418743　Zygophyllum pterocaule Van Zyl = Tetraena pterocaulis（Van Zyl）Beier et Thulin ●☆

418744　Zygophyllum pubescens Schinz = Roepera pubescens（Schinz）Beier et Thulin ●■☆

418745　Zygophyllum pygmaeum Eckl. et Zeyh. = Roepera pygmaea（Eckl. et Zeyh.）Beier et Thulin ●☆

418746　Zygophyllum rangei Engl. = Roepera pubescens（Schinz）Beier et Thulin ●■☆

418747　Zygophyllum retrofractum Thunb. ;反曲驼蹄瓣●☆

418748　Zygophyllum retrofractum Thunb. = Tetraena retrofracta（Thunb.）Beier et Thulin ●☆

418749　Zygophyllum rigidum Schinz;硬驼蹄瓣●☆

418750　Zygophyllum rigidum Schinz = Tetraena rigida（Schinz）Beier et Thulin ●☆

418751　Zygophyllum robecchii Engl. = Melocarpum robecchii（Engl.）Beier et Thulin ■☆

418752　Zygophyllum rogersii Compton = Roepera rogersii（Compton）Beier et Thulin ●☆

418753　Zygophyllum rosovii Bunge;石生驼蹄瓣（若氏霸王,石生霸王）;Rockliving Beancaper,Saxicolous Beancaper ●■

418754　Zygophyllum rosovii Bunge var. latifolium（Schrenk）Popov;宽叶石生驼蹄瓣;Broadleaf Rockliving Beancaper ■

418755　Zygophyllum schaeferi Engl. = Roepera cordifolia（L. f.）Beier et Thulin ●☆

418756　Zygophyllum sessilifolium L. = Roepera sessilifolia（L.）Beier et Thulin ●☆

418757　Zygophyllum simplex L. ;简单驼蹄瓣●☆

418758　Zygophyllum simplex L. = Tetraena simplex（L.）Beier et Thulin ●☆

418759　Zygophyllum simplex L. var. herniarioides Chiov. = Tetraena simplex（L.）Beier et Thulin ●☆

418760　Zygophyllum sinkiangense Y. X. Liou;新疆驼蹄瓣（新疆霸王）;Xinjiang Beancaper ●■

418761　Zygophyllum somalense Hadidi;索马里驼蹄瓣●☆

418762　Zygophyllum somalense Hadidi = Tetraena somalensis（Hadidi）Beier et Thulin ●☆

418763　Zygophyllum sonderi H. Eichler;森诺驼蹄瓣●☆

418764　Zygophyllum sphaerocarpum Schltr. ex Huysst. = Roepera sphaerocarpa（Schltr. ex Huysst.）Beier et Thulin ●☆

418765　Zygophyllum spinosum L. = Roepera spinosa（L.）Beier et Thulin ●☆

418766　Zygophyllum stapffii Schinz;施塔普夫驼蹄瓣●☆

418767　Zygophyllum stapffii Schinz = Tetraena stapfii（Schinz）Beier et Thulin ●☆

418768　Zygophyllum stenopterum Schrenk;狭翼驼蹄瓣●☆

418769　Zygophyllum subtrijugum C. A. Mey. ;粗茎霸王;Thickstem Beancaper ●

418770　Zygophyllum suffruticosum Schinz = Tetraena rigida（Schinz）Beier et Thulin ●☆

418771　Zygophyllum sulcatum Huysst. = Roepera leucoclada（Diels）Beier et Thulin ●☆

418772　Zygophyllum teretifolium Schltr. = Roepera teretifolia（Schltr.）Beier et Thulin ●☆

418773　Zygophyllum trothai Diels = Tetraena rigida（Schinz）Beier et Thulin ●☆

418774　Zygophyllum turcomanicum Fisch. et C. A. Mey. ;土库曼驼蹄瓣●☆

418775　Zygophyllum typicum（Popov）Regel = Zygophyllum fabago L. ■

418776　Zygophyllum uitenhagense Sond. = Zygophyllum maritimum Eckl. et Zeyh. ●☆

418777　Zygophyllum waterlotii Maire = Tetraena gaetula（Emb. et Maire）Beier et Thulin subsp. waterlotii（Maire）Beier et Thulin ●☆

418778　Zygophyllum waterlotii Maire var. abbreviatum ? = Tetraena gaetula（Emb. et Maire）Beier et Thulin subsp. waterlotii（Maire）Beier et Thulin ●☆

418779　Zygophyllum waterlotii Maire var. dolichocarpum ? = Tetraena gaetula（Emb. et Maire）Beier et Thulin subsp. waterlotii（Maire）Beier et Thulin ●☆

418780　Zygophyllum xanthoxylon（Bunge）Maxim. subsp. ferganense Popov = Sarcozygium xanthoxylum Bunge ●

418781　Zygophyllum xanthoxylon（Bunge）Maxim. var. ferganense Drobov = Sarcozygium xanthoxylum Bunge ●

418782　Zygophyllum xanthoxylum（Bunge）Maxim. = Sarcozygium xanthoxylum Bunge ●

418783　Zygophyllum xanthoxylum Baill. = Sarcozygium xanthoxylon Bunge ●

418784　Zygoruellia Baill.（1890）;异芦莉草属●☆

418785　Zygoruellia richardii Baill. ;异芦莉草●☆

418786　Zygosepalum Rchb. f.（1859）;对萼兰属（接萼兰属）■☆

418787　Zygosepalum rostratum（Hook.）Rchb. f. ;对萼兰■☆

418788　Zygosicyos Humbert(1945);对瓜属（马岛瓜属）■☆

418789　Zygosicyos hirtellus Humbert;毛对瓜■☆

418790　Zygosicyos tripartitus Humbert;对瓜■☆

418791　Zygospermum Thwaites ex Baill. = Margaritaria L. f. ●

418792　Zygospermum Thwaites ex Baill. = Prosorus Dalzell ●

418793　Zygostates Lindl.（1837）;天平兰属■☆

418794　Zygostates lunata Lindl. ;天平兰■☆

418795　Zygostelma Benth.（1876）;轭冠萝藦属☆

418796　Zygostelma E. Fourn. = Lagoa T. Durand ☆

418797　Zygostelma calcaratum E. Fourn. ;轭冠萝藦☆

418798　Zygostemma Tiegh.（1909）;轭冠续断属■☆

418799　Zygostemma Tiegh. = Scabiosa L. ●■

418800　Zygostemma creticum（L.）Tiegh. ;轭冠续断■☆

418801　Zygostigma Griseb.（1838）;轭头龙胆属■☆

418802　Zygostigma australe（Cham. et Schltdl.）Griseb. ;轭头龙胆■☆

418803　Zygotritonia Mildbr.（1923）;轭观音兰属■☆

418804　Zygotritonia bongensis（Pax）Mildbr. ;邦加轭观音兰■☆

418805　Zygotritonia bongensis（Pax）Mildbr. var. robusta Mildbr. = Zygotritonia bongensis（Pax）Mildbr. ■☆

418806　Zygotritonia crocea Stapf;镉黄轭观音兰■☆

418807　Zygotritonia giorgii De Wild. = Zygotritonia nyassana Mildbr. ■☆

418808　Zygotritoniagracillima Mildbr. = Zygotritonia nyassana Mildbr. ■☆

418809　Zygotritonia homblei De Wild. = Zygotritonia nyassana Mildbr. ■☆

418810　Zygotritonia hysterantha Goldblatt;宫花轭观音兰■☆

418811　Zygotritonia nyassana Mildbr. ;尼亚萨轭观音兰■☆

418812　Zygotritonia praecox Stapf;早轭观音兰■☆

418813　Zymum Noronha ex Thouars = Tristellateia Thouars ●

418814　Zymum Thouars = Tristellateia Thouars ●

418815　Zyrphelis Cass. (1829);毛菀属●■☆

418816　Zyrphelis Cass. = Mairia Nees ■☆

418817　Zyrphelis burchellii (DC.) Kuntze;伯切尔毛菀■☆

418818　Zyrphelis corymbosa (Harv.) Kuntze = Gymnostephium papposum
　　　G. L. Nesom ■☆

418819　Zyrphelis crenata (Thunb.) Kuntze = Mairia crenata (Thunb.)

Nees ■☆

418820　Zyrphelis ecklonis (DC.) Kuntze;埃氏毛菀●☆

418821　Zyrphelis foliosa (Harv.) Kuntze;多叶毛菀■☆

418822　Zyrphelis hirsuta (DC.) Kuntze = Mairia hirsuta DC. ■☆

418823　Zyrphelis lasiocarpa (DC.) Kuntze;毛果毛菀■☆

418824　Zyrphelis microcephala (Less.) Nees;小头毛菀■☆

418825　Zyrphelis montana (Schltr.) G. L. Nesom;山地毛菀■☆

418826　Zyrphelis perezioides (Less.) G. L. Nesom;连座钝柱菊毛菀■☆

418827　Zyrphelis taxifolia (L.) Nees;落羽杉叶毛菀■☆

418828　Zyzophyllum Salisb. = Zygophyllum L. ●■

418829　Zyzygium Brongn. = Syzygium R. Br. ex Gaertn. (保留属名)●

418830　Zyzyxia Strother(1991);北喙芒菊属●☆

418831　Zyzyxia lundellii (H. Rob.) Strother;北喙芒菊●☆

主要参考文献

安德鲁·薛瓦利埃主编.1996.药用植物百科全书.梁立新,江红兵等译.2003.南宁:广西科学技术出版社,1-334.

包志毅主译,陈俊愉译审.2004.世界园林乔灌木(Trees & Shrubs).北京:中国林业出版社.

蔡永敏主编.1996.中药药名词典.北京:中国中医药出版社,1-399.

陈达夫,王新潮,车承丕等.1996.日本地名词典.北京:商务印书馆,1-1188.

陈封怀主编.1987.广东植物志(第1卷).广州:广东科技出版社,1-600.

陈焕镛主编.1964.海南植物志(第1卷).北京:科学出版社,1-517.

陈焕镛主编.1965.海南植物志(第2卷).北京:科学出版社,1-470.

陈焕镛主编.1974.海南植物志(第3卷).北京:科学出版社,1-629.

陈焕镛主编.1977.海南植物志(第4卷).北京:科学出版社,1-644.

陈嵘.1959.中国树木分类学.科学技术出版社,1-1191.

狄维忠主编.1986.贺兰山维管植物.西安:西北大学出版社.

丁广奇,王学文.1986.植物学名解释.北京:科学出版社,1-463.

福建植物志编写组.1982.福建植物志(第1卷).福州:福建科学技术出版社,1-630.

福建植物志编写组.1987.福建植物志(第2卷).福州:福建科学技术出版社,1-556.

傅坤俊主编.1989.黄土高原植物志(第5卷).北京:科学技术文献出版社,1-557.

傅立国主编.1992.中国植物红皮书(第1册).北京:科学出版社,1-736.

傅沛云主编.1995.东北植物检索表.北京:科学出版社,1-1006.

耿以礼主编.1959.中国主要植物图说(禾本科).北京:科学出版社,1-1181.

关克俭,傅立国等.1983.拉汉英种子植物名称.北京:科学出版社,1-1036.

广西科学院广西植物研究所.1991.广西植物志(第1卷).南宁:广西科学技术出版社,1-976.

广西植物研究所主办.1985－2003.广西植物(Guihaia).

贵州植物志编辑委员会.1982.贵州植物志(第1卷).贵阳:贵州人民出版社,1-393.

贵州植物志编辑委员会.1986.贵州植物志(第2卷).贵阳:贵州人民出版社,1-700.

郭本兆主编.1987.青海经济植物志.西宁:青海人民出版社,1-859.

汉拉英中国木本植物名录委会.2003.汉拉英中国木本植物名录.北京:中国林业出版社,1-577.

汉语大字典编辑委员会.1986.汉语大字典.成都:四川辞书出版社;武汉:湖北辞书出版社,1-1543.

贺士元主编.1986.河北植物志(第1卷).石家庄:河北科学技术出版社,1-830.

贺士元主编.1988.河北植物志(第2卷).石家庄:河北科学技术出版社,1-676.

贺士元主编.1991.河北植物志(第3卷).石家庄:河北科学技术出版社,1-698.

侯宽昭编,吴德邻,高蕴璋等修订.1984.中国种子植物科属词典(修订版).北京:科学出版社,1-632.

侯宽昭主编.1956.广州植物志.北京:科学出版社,1-953.

胡世平.2003.汉英拉动植物名称.北京:商务印书馆,1-561.

湖北省植物研究所.1976.湖北植物志(第1卷).武汉:湖北人民出版社,1-503.

湖北省植物研究所.1979.湖北植物志(第2卷).武汉:湖北人民出版社,1-522.

华北树木志编写组.1984.华北树木志.北京:中国林业出版社,1-743.

黄泰康,丁志遵,赵守训等主编.2001.现代本草纲目.北京:中国医药科技出版社,1-3290.

吉林省中医中药研究所等.1982.长白山植物药志.长春:吉林人民出版社,1-1476.

江纪武主编.1990.拉汉药用植物名称和检索手册.北京:中国医药科技出版社,1-1815.

江纪武主编.2005.药用植物词典.天津:天津科学技术出版社,1-1244.

江苏省植物研究所.江苏植物志(上册).南京:江苏科学技术出版社.

江苏省植物研究所.1982.江苏植物志(下册).南京:江苏科学技术出版社,1-1010.

金春星.1989.中国树木学名诠释.北京:中国林业出版社.

靳淑英.1999.中国高等植物模式标本汇编(补编).北京:中国林业出版社,1-264.

靳淑英.1994.中国高等植物模式标本汇编.北京:科学出版社,1-708.

克里斯托弗·布里克尔主编.2005.世界园林植物与花卉百科全书.杨秋生,李镇宇主译.郑州:河南科学技术出版社,1-752.

李甯汉,刘启文主编.1997.香港中草药(第7辑).香港:商务印书馆,1-216.

李甯汉,刘启文主编.2000.香港药用植物资源录.香港中草药(第1-8辑)总索引.香港:商务印书馆,1-114.

李书心主编.1988.辽宁植物志(上册).沈阳:辽宁科学技术出版社,1-1439.

李书心主编.1992.辽宁植物志(下册).沈阳:辽宁科学技术出版社,1-1245.

李衍文主编.2004.中草药异名词典.北京:人民卫生出版社,1-998.

刘尚武主编.1997.青海植物志(第1卷).西宁:青海人民出版社.

刘尚武主编.1997.青海植物志(第2卷).西宁:青海人民出版社.

刘尚武主编.1997.青海植物志(第3卷).西宁:青海人民出版社.

刘业经.1972.台湾木本植物志.台湾:国立中兴大学农学院丛书第六号.

刘媖心主编.1985.中国沙漠植物志(第1卷).北京:科学出版社,1-546.

刘媖心主编.1987.中国沙漠植物志(第2卷).北京:科学出版社,1-464.

刘媖心主编.1992.中国沙漠植物志(第3卷).北京:科学出版社,1-564.

马德滋,刘惠兰,胡福秀.2007.宁夏植物志(第二版上卷).银川:宁夏人民出版社,1-635.

马德滋,刘惠兰,胡福秀.2007.宁夏植物志(第二版下卷).银川:宁夏人民出版社,1-642.

马德滋,刘惠兰.1986.宁夏植物志(第1卷).银川:宁夏人民出版社,1-505.

马德滋,刘惠兰.1988.宁夏植物志(第2卷).银川:宁夏人民出版社,1-555.

马其云.2003.中国蕨类植物和种子植物名称总汇.青岛:青岛出版社,1-1561.

马玉明总主编.1997.内蒙古资源大词典.呼和浩特:内蒙古人民出版社,1-1822.

马毓泉主编.1985.内蒙古植物志(第1卷).呼和浩特:内蒙古人民出版社,1-294.

马毓泉主编.1978.内蒙古植物志(第2卷).呼和浩特:内蒙古人民出版社,1-890.

马毓泉主编.1977.内蒙古植物志(第3卷).呼和浩特:内蒙古人民出版社,1-309.

马毓泉主编.1979.内蒙古植物志(第4卷).呼和浩特:内蒙古人民出版社,1-223.

马毓泉主编.1980.内蒙古植物志(第5卷).呼和浩特:内蒙古人民出版社,1-442.

马毓泉主编.1982.内蒙古植物志(第6卷).呼和浩特:内蒙古人民出版社,1-355.

马毓泉主编.1983.内蒙古植物志(第7卷).呼和浩特:内蒙古人民出版社,1-282.

马毓泉主编.1985.内蒙古植物志(第8卷).呼和浩特:内蒙古人民出版社,1-372.

马毓泉主编.1998.内蒙古植物志(第二版第1卷).呼和浩特:内蒙古人民出版社,1-408.

马毓泉主编.1990.内蒙古植物志(第二版第2卷).呼和浩特:内蒙古人民出版社,1-759.

马毓泉主编.1989.内蒙古植物志(第二版第3卷).呼和浩特:内蒙古人民出版社,1-716.

马毓泉主编.1993.内蒙古植物志(第二版第4卷).呼和浩特:内蒙古人民出版社,1-907.

马毓泉主编.1994.内蒙古植物志(第二版第5卷).呼和浩特:内蒙古人民出版社,1-634.

聂绍荃,袁晓颖,杨逢建主编.2003.黑龙江植物资源志.哈尔滨:东北林业大学出版社,1-940.

牛春山主编.1990.陕西树木志.北京:中国林业出版社,1-1261.

青海省农业资源区划办公室,中国科学院西北高原生物研究所.1997.青海植物名录.西宁:青海人民出版社.

全国中草药汇编编写组.1975.全国中草药汇编(上册).北京:人民卫生出版社,1-1008.

全国中草药汇编编写组.1978.全国中草药汇编(下册).北京:人民卫生出版社,1-1020.

山东树木志编写.1984.山东树木志.济南:山东科学技术出版社.

尚衍重.2007.药用种子植物拉汉日俄英名称.北京:中国医药科技出版社,1-1927.

沈显生.2005.植物学拉丁文.合肥:中国科学技术大学出版社,1-162.

史群.1979.日本姓名词典.北京:商务印书馆,1-665.

世界姓名译名手册编译组.1987.世界姓名译名手册.北京:化学工业出版社,1-1002.

斯特恩 W T 著.1973.植物学拉丁文(上册).秦仁昌译.1981.北京:科学出版社,1-712.

斯特恩 W T 著.1973.植物学拉丁文(下册).秦仁昌译.1984.北京:科学出版社,1-344.

四川植物志编辑委员会.1981.四川植物志(第 1 卷).成都:四川人民出版社,1-509.

四川植物志编辑委员会.1985.四川植物志(第 3 卷).成都:四川人民出版社,1-309.

四川植物志编辑委员会.1988.四川植物志(第 4 卷).成都:四川人民出版社,1-493.

四川植物志编辑委员会.1988.四川植物志(第 5 卷.1).成都:四川人民出版社.

四川植物志编辑委员会.1988.四川植物志(第 5 卷.2).成都:四川人民出版社,1-457.

四川植物志编辑委员会.1988.四川植物志(第 6 卷).成都:四川人民出版社,1-410.

滕砥平,蒋芝英译.1965.(E. C. Jaeger.1955).生物名称和生物学术语的词源.北京:科学出版社,1-577.

王宇飞,赵良成,冯广平,李承森译.(James G. Harris & Melinda Woolf Harris,1994).2001.图解植物学词典.北京:1-302.

吴征镒,路安民,汤彦承等.2003.中国被子植物科属综论.北京:科学出版社,1-1209.

吴征镒主编.1983.西藏植物志(第 1 卷).北京:科学出版社,1-791.

吴征镒主编.1985.西藏植物志(第 2 卷).北京:科学出版社,1-956.

吴征镒主编.1986.西藏植物志(第 3 卷).北京:科学出版社,1-1047.

吴征镒主编.1985.西藏植物志(第 4 卷).北京:科学出版社,1-1021.

吴征镒主编.1987.西藏植物志(第 5 卷).北京:科学出版社.

吴征镒主编.1986.云南植物志(第 4 卷).北京:科学出版社,1-823.

吴征镒主编.1991.云南植物志(第 5 卷).北京:科学出版社,1-795.

吴征镒主编.1997.云南植物志(第 8 卷).北京:科学出版社.

吴征镒主编.1984.云南种子植物名录(上册).昆明:云南人民出版社,1-1070.

武汉植物研究编辑委员会.1980 – 2004.武汉植物研究(Journal of Wuhan Botanical Research).

西北植物学报编辑委员会.1985 – 2004.西北植物学报.西安:陕西科学技术出版社.

向其柏,臧德奎,孙卫邦译.2006.国际栽培植物命名法规(第 7 版).北京:中国林业出版社,1-106.

向其柏,臧德奎等译.2004.国际栽培植物命名法规(第 6 版).北京:中国林业出版社,1-206.

萧德荣主编.1988.世界地名翻译手册.北京:知识出版社,1-1616.

肖培根,连文琰.1999.中药植物原色图鉴.北京:中国农业出版社,1-628.

谢大任主编.1988.拉丁语汉语词典.北京:商务印书馆,1-601.

辛华.1978.世界地名译名手册.北京:商务印书馆,1-892.

新华通讯社译名室编(郭国荣主编).1993.世界人名翻译大辞典.北京:中国对外翻译出版公司,1-3753.

新疆植物志编辑委员会.1994.新疆植物志(第 2 卷.1).乌鲁木齐:新疆科技卫生出版社,1-393.

新疆植物志编辑委员会.1995.新疆植物志(第 2 卷.2).乌鲁木齐:新疆科技卫生出版社,1-424.

新疆植物志编辑委员会.新疆植物志(第 3 卷).乌鲁木齐:新疆科技卫生出版社.

新疆植物志编辑委员会.2004.新疆植物志(第 4 卷).乌鲁木齐:新疆科技卫生出版社,1-573.

新疆植物志编辑委员会.1999.新疆植物志(第 5 卷).乌鲁木齐:新疆科技卫生出版社,1-534.

新疆植物志编辑委员会.1996.新疆植物志(第 6 卷).乌鲁木齐:新疆科技卫生出版社,1-669.

邢福武,余明恩,张永夏主编.2002.深圳植物物种多样性及其保育.北京:中国林业出版社.

熊文愈,汪计珠,石同岱,李又芬.1993.中国木本药用植物.上海:上海科技教育出版社,1-815.

徐柱主编.1999.世界禾草属志.北京:中国农业科技出版社,1-870.

徐柱主编.1997.中国禾草属志.呼和浩特:内蒙古人民出版社,1-428.

伊藤武夫.1976.台湾植物图说(续卷).东京:国书刊行会,1-400.

伊藤武夫.1976.台湾植物图说(正卷).东京:国书刊行会,1-1083.

应俊生,张玉龙.1994.中国种子植物特有属.北京:科学出版社,1-699.

余树勋,吴应祥主编.1993.花卉词典.北京:农业出版社,1-944.

云南省园艺博览局.1999.世界园艺博览园植物名录.昆明:云南科学技术出版社.

张贵君主编.1995.中药材及饮片原色图鉴.哈尔滨:黑龙江科学技术出版社,1-564.

张丽兵译.2007.国际植物命名法规(维也纳法规).北京:科学出版社,密苏里植物园出版社,1-295.

张美珍,赖明洲等.1993.华东五省一市植物名录.上海:上海科学普及出版社,1-491.

赵士洞译.1984.国际植物命名法规(列宁格勒法规).北京:科学出版社,1-295.

赵询,等.1986.苏联百科词典.北京,上海:中国大百科全书出版社,1-2045.

赵毓棠,吉金祥.1988.拉汉植物学名辞典.长春:吉林科学技术出版社,1-726.

浙江植物志编辑委员会.1993.浙江植物志(第1卷.卷主编张朝芳,章绍尧).杭州:浙江科学技术出版社,1-411.

浙江植物志编辑委员会.1992.浙江植物志(第2卷.卷主编王景祥).杭州:浙江科学技术出版社,1-408.

浙江植物志编辑委员会.1993.浙江植物志(第3卷.卷主编韦直,何业奇).杭州:浙江科学技术出版社,1-541.

浙江植物志编辑委员会.1993.浙江植物志(第4卷.卷主编裴宝林).杭州:浙江科学技术出版社,1-423.

浙江植物志编辑委员会.1989.浙江植物志(第5卷.卷主编方云亿).杭州:浙江科学技术出版社,1-355.

浙江植物志编辑委员会.1993.浙江植物志(第6卷.卷主编郑朝宗).杭州:浙江科学技术出版社,1-390.

浙江植物志编辑委员会.1993.浙江植物志(第7卷.卷主编林泉).杭州:浙江科学技术出版社,1-584.

浙江植物志编辑委员会.1993.浙江植物志(总论.卷主编章绍尧,丁炳扬).杭州:浙江科学技术出版社,1-343.

郑万钧主编.1983.中国树木志(第1卷).北京:中国林业出版社,1-929.

郑万钧主编.1985.中国树木志(第2卷).北京:中国林业出版社,931-1996.

郑万钧主编.1997.中国树木志(第3卷).北京:中国林业出版社,1997-3969.

郑万钧主编.2004.中国树木志(第4卷).北京:中国林业出版社,3971-5429.

植物分类学报编辑委员会.1985-2008.植物分类学报(Acta Phytotaxonomica Sinica).北京:科学出版社.

植物研究编辑委员会.1985-2004.植物研究(Bulletin of Botanical Reaserch).哈尔滨:东北林业大学出版社.

中国科学院北京植物研究所主编.1985.中国高等植物图鉴(第3册).北京:科学出版社,1-1083.

中国科学院北京植物研究所主编.1985.中国高等植物图鉴(第4册).北京:科学出版社,1-932.

中国科学院北京植物研究所主编.1976.中国高等植物图鉴(第5册).北京:科学出版社,1-1146.

中国科学院编辑出版委员会名词室.1963.英拉汉植物名称.北京:科学出版社,1-986.

中国科学院昆明植物研究所主办.1980-2007.云南植物研究(Acta Botanica Yunnanica).北京:科学出版社.

中国科学院西安分院主办.1980-2007.西北植物学报(Acta Botanica Boreali-Ocidentalia Sinica).北京:科学出版社.

中国科学院西北植物研究所.1976.秦岭植物志(第1卷.1).北京:科学出版社,1-476.

中国科学院西北植物研究所.1974.秦岭植物志(第1卷.2).北京:科学出版社,1-647.

中国科学院西北植物研究所.1981.秦岭植物志(第1卷.3).北京:科学出版社,1-500.

中国科学院西北植物研究所.1983.秦岭植物志(第1卷.4).北京:科学出版社,1-421.

中国科学院西北植物研究所.1985.秦岭植物志(第1卷.5).北京:科学出版社,1-442.

中国科学院植物研究所.1996.新编拉汉英植物名称.北京:航空工业出版社,1-1166.

中国科学院植物研究所主编.1983.中国高等植物科属检索表.北京:科学出版社,1-733.

中国科学院植物研究所主编.1985.中国高等植物图鉴(第1册).北京:科学出版社,1-1157.

中国科学院植物研究所主编.1985.中国高等植物图鉴(第2册).北京:科学出版社,1-1312.

中国科学院植物研究所主编.1961.中国经济植物志.北京:科学出版社,1-2273.

中国科学院中国植物志编辑委员会.2004.中国植物志(第1卷).北京:科学出版社,1-1044.

中国科学院中国植物志编辑委员会.1978.中国植物志(第7卷).北京:科学出版社,1-542.

中国科学院中国植物志编辑委员会.1992.中国植物志(第8卷).北京:科学出版社,1-218.

中国科学院中国植物志编辑委员会.1996.中国植物志(第9卷.1).北京:科学出版社,1-761.

中国科学院中国植物志编辑委员会.2002.中国植物志(第9卷.2).北京:科学出版社,1-450.

中国科学院中国植物志编辑委员会.1987.中国植物志(第9卷.3).北京:科学出版社,1-352.

中国科学院中国植物志编辑委员会.1990.中国植物志(第10卷.1).北京:科学出版社,1-442.

中国科学院中国植物志编辑委员会.1997.中国植物志(第10卷.2).北京:科学出版社,1-339.

中国科学院中国植物志编辑委员会.1961.中国植物志(第11卷).北京:科学出版社,1-362.

中国科学院中国植物志编辑委员会.2000.中国植物志(第12卷).北京:科学出版社,1-582.

中国科学院中国植物志编辑委员会.1991.中国植物志(第13卷.1).北京:科学出版社,1-172.

中国科学院中国植物志编辑委员会.1979.中国植物志(第13卷.2).北京:科学出版社,1-242.

中国科学院中国植物志编辑委员会.1999.中国植物志(第19卷).北京:科学出版社,1-485.

中国科学院中国植物志编辑委员会.2004.中国植物志(第1卷).北京:科学出版社,1-1044.

中国科学院中国植物志编辑委员会.1982.中国植物志(第20卷.1).北京:科学出版社,1-106.

中国科学院中国植物志编辑委员会.1984.中国植物志(第20卷.2).北京:科学出版社,1-403.

中国科学院中国植物志编辑委员会.1979.中国植物志(第21卷).北京:科学出版社,1-150.

中国科学院中国植物志编辑委员会.1998.中国植物志(第22卷).北京:科学出版社,1-456.

中国科学院中国植物志编辑委员会.1998.中国植物志(第23卷.1).北京:科学出版社,1-257.

中国科学院中国植物志编辑委员会.1995.中国植物志(第23卷.2).北京:科学出版社,1-448.

中国科学院中国植物志编辑委员会.1988.中国植物志(第24卷).北京:科学出版社,1-289.

中国科学院中国植物志编辑委员会.1998.中国植物志(第25卷.1).北京:科学出版社,1-232.

中国科学院中国植物志编辑委员会.1979.中国植物志(第25卷.2).北京:科学出版社,1-262.

中国科学院中国植物志编辑委员会.1996.中国植物志(第26卷).北京:科学出版社,1-506.

中国科学院中国植物志编辑委员会.1979.中国植物志(第27卷).北京:科学出版社,1-664.

中国科学院中国植物志编辑委员会.1980.中国植物志(第28卷).北京:科学出版社,1-390.

中国科学院中国植物志编辑委员会.2001.中国植物志(第29卷).北京:科学出版社,1-343.

中国科学院中国植物志编辑委员会.1996.中国植物志(第30卷.1).北京:科学出版社,1-305.

中国科学院中国植物志编辑委员会.1979.中国植物志(第30卷.2).北京:科学出版社,1-218.

中国科学院中国植物志编辑委员会.1982.中国植物志(第31卷).北京:科学出版社,1-509.

中国科学院中国植物志编辑委员会.1999.中国植物志(第32卷).北京:科学出版社,1-594.

中国科学院中国植物志编辑委员会.1985.中国植物志(第33卷).北京:科学出版社,1-516.

中国科学院中国植物志编辑委员会.1987.中国植物志(第33卷).北京:科学出版社,1-483.

中国科学院中国植物志编辑委员会.1984.中国植物志(第34卷.1).北京:科学出版社,1-242.

中国科学院中国植物志编辑委员会.1992.中国植物志(第34卷.2).北京:科学出版社,1-306.

中国科学院中国植物志编辑委员会.1995.中国植物志(第35卷.1).北京:科学出版社,1-406.

中国科学院中国植物志编辑委员会.1979.中国植物志(第35卷.2).北京:科学出版社,1-130.

中国科学院中国植物志编辑委员会.1974.中国植物志(第36卷).北京:科学出版社,1-443.

中国科学院中国植物志编辑委员会.1986.中国植物志(第38卷).北京:科学出版社,1-167.

中国科学院中国植物志编辑委员会.1988.中国植物志(第39卷).北京:科学出版社,1-233.

中国科学院中国植物志编辑委员会.1994.中国植物志(第40卷).北京:科学出版社,1-362.

中国科学院中国植物志编辑委员会.1994.中国植物志(第40卷).北京:科学出版社,1-362.

中国科学院中国植物志编辑委员会.1995.中国植物志(第41卷).北京:科学出版社,1-400.

中国科学院中国植物志编辑委员会.1993.中国植物志(第42卷.1).北京:科学出版社,1-384.

中国科学院中国植物志编辑委员会.1998.中国植物志(第42卷.2).北京:科学出版社,1-467.

中国科学院中国植物志编辑委员会.1998.中国植物志(第43卷.1).北京:科学出版社,1-168.

中国科学院中国植物志编辑委员会.1997.中国植物志(第43卷.2).北京:科学出版社,1-250.

中国科学院中国植物志编辑委员会.1997.中国植物志(第43卷.3).北京:科学出版社,1-239.

中国科学院中国植物志编辑委员会.1994.中国植物志(第44卷.1).北京:科学出版社,1-217.

中国科学院中国植物志编辑委员会.1996.中国植物志(第44卷.2).北京:科学出版社,1-212.

中国科学院中国植物志编辑委员会.1997.中国植物志(第44卷.3).北京:科学出版社,1-150.

中国科学院中国植物志编辑委员会.1980.中国植物志(第45卷.1).北京:科学出版社,1-152.

中国科学院中国植物志编辑委员会.1999.中国植物志(第45卷.2).北京:科学出版社,1-284.

中国科学院中国植物志编辑委员会.1999.中国植物志(第45卷.3).北京:科学出版社,1-218.

中国科学院中国植物志编辑委员会.1981.中国植物志(第46卷).北京:科学出版社,1-315.

中国科学院中国植物志编辑委员会.1985.中国植物志(第47卷.1).北京:科学出版社,1-140.

中国科学院中国植物志编辑委员会.2001.中国植物志(第47卷.2).北京:科学出版社,1-243.

中国科学院中国植物志编辑委员会.1982.中国植物志(第48卷.1).北京:科学出版社,1-169.

中国科学院中国植物志编辑委员会.1998.中国植物志(第48卷.2).北京:科学出版社,1-208.

中国科学院中国植物志编辑委员会.1989.中国植物志(第49卷.1).北京:科学出版社,1-132.

中国科学院中国植物志编辑委员会.1984.中国植物志(第49卷.2).北京:科学出版社,1-208.

中国科学院中国植物志编辑委员会.1998.中国植物志(第49卷.3).北京:科学出版社,1-276.

中国科学院中国植物志编辑委员会.1998.中国植物志(第50卷.1).北京:科学出版社,1-213.

中国科学院中国植物志编辑委员会.1991.中国植物志(第50卷.2).北京:科学出版社,1-197.

中国科学院中国植物志编辑委员会.1991.中国植物志(第51卷).北京:科学出版社,1-148.

中国科学院中国植物志编辑委员会.1999.中国植物志(第52卷.1).北京:科学出版社,1-443.

中国科学院中国植物志编辑委员会.1983.中国植物志(第52卷.2).北京:科学出版社,1-192.

中国科学院中国植物志编辑委员会.1984.中国植物志(第53卷.1).北京:科学出版社,1-314.

中国科学院中国植物志编辑委员会.2000.中国植物志(第53卷.2).北京:科学出版社,1-178.

中国科学院中国植物志编辑委员会.1978.中国植物志(第54卷).北京:科学出版社,1-210.

中国科学院中国植物志编辑委员会.1979.中国植物志(第55卷.1).北京:科学出版社,1-316.

中国科学院中国植物志编辑委员会.1985.中国植物志(第55卷.2).北京:科学出版社,1-282.

中国科学院中国植物志编辑委员会.1992.中国植物志(第55卷.3).北京:科学出版社,1-272.

中国科学院中国植物志编辑委员会.1990.中国植物志(第56卷).北京:科学出版社,1-238.

中国科学院中国植物志编辑委员会.1999.中国植物志(第57卷.1).北京:科学出版社,1-239.

中国科学院中国植物志编辑委员会.1994.中国植物志(第57卷.2).北京:科学出版社,1-472.

中国科学院中国植物志编辑委员会.1991.中国植物志(第57卷.3).北京:科学出版社,1-234.

中国科学院中国植物志编辑委员会.1979.中国植物志(第58卷).北京:科学出版社,1-147.

中国科学院中国植物志编辑委员会.1989.中国植物志(第59卷.1).北京:科学出版社,1-217.

中国科学院中国植物志编辑委员会.1990.中国植物志(第59卷.2).北京:科学出版社,1-317.

中国科学院中国植物志编辑委员会.1987.中国植物志(第60卷.1).北京:科学出版社,1-166.

中国科学院中国植物志编辑委员会.1987.中国植物志(第60卷.2).北京:科学出版社,1-161.

中国科学院中国植物志编辑委员会.1992.中国植物志(第61卷).北京:科学出版社,1-347.

中国科学院中国植物志编辑委员会.1988.中国植物志(第62卷).北京:科学出版社,1-446.

中国科学院中国植物志编辑委员会.1977.中国植物志(第63卷).北京:科学出版社,1-617.

中国科学院中国植物志编辑委员会.1979.中国植物志(第64卷.1).北京:科学出版社,1-184.

中国科学院中国植物志编辑委员会.1989.中国植物志(第64卷.2).北京:科学出版社,1-253.

中国科学院中国植物志编辑委员会.1982.中国植物志(第65卷.1).北京:科学出版社,1-229.

中国科学院中国植物志编辑委员会.1977.中国植物志(第65卷.2).北京:科学出版社,1-649.

中国科学院中国植物志编辑委员会.1977.中国植物志(第66卷).北京:科学出版社,1-647.

中国科学院中国植物志编辑委员会.1978.中国植物志(第67卷.1).北京:科学出版社,1-175.

中国科学院中国植物志编辑委员会.1979.中国植物志(第67卷.2).北京:科学出版社,1-431.

中国科学院中国植物志编辑委员会.1963.中国植物志(第68卷).北京:科学出版社,1-449.

中国科学院中国植物志编辑委员会.1990.中国植物志(第69卷).北京:科学出版社,1-648.

中国科学院中国植物志编辑委员会.2002.中国植物志(第70卷).北京:科学出版社,1-392.

中国科学院中国植物志编辑委员会.1999.中国植物志(第71卷.1).北京:科学出版社,1-432.

中国科学院中国植物志编辑委员会.1999.中国植物志(第71卷.2).北京:科学出版社,1-377.

中国科学院中国植物志编辑委员会.1999.中国植物志(第71卷.3).北京:科学出版社.

中国科学院中国植物志编辑委员会.1988.中国植物志(第72卷).北京:科学出版社,1-281.

中国科学院中国植物志编辑委员会.1986.中国植物志(第73卷.1).北京:科学出版社,1-301.

中国科学院中国植物志编辑委员会.1983.中国植物志(第73卷.2).北京:科学出版社,1-203.

中国科学院中国植物志编辑委员会.1985.中国植物志(第74卷).北京:科学出版社,1-388.

中国科学院中国植物志编辑委员会.1979.中国植物志(第75卷).北京:科学出版社.

中国科学院中国植物志编辑委员会.1983.中国植物志(第76卷.1).北京:科学出版社,1-149.

中国科学院中国植物志编辑委员会.1991.中国植物志(第76卷.2).北京:科学出版社,1-319.

中国科学院中国植物志编辑委员会.1999.中国植物志(第76卷.2).北京:科学出版社,1-243.

中国科学院中国植物志编辑委员会.1983.中国植物志(第73卷.2).北京:科学出版社,1-203.

中国科学院中国植物志编辑委员会.1985.中国植物志(第74卷).北京:科学出版社,1-388.

中国科学院中国植物志编辑委员会.1979.中国植物志(第75卷).北京:科学出版社.

中国科学院中国植物志编辑委员会.1983.中国植物志(第76卷.1).北京:科学出版社,1-149.

中国科学院中国植物志编辑委员会.1991.中国植物志(第76卷.2).北京:科学出版社,1-319.

中国科学院中国植物志编辑委员会.1999.中国植物志(第76卷.3).北京:科学出版社,1-243.

中国科学院中国植物志编辑委员会.1999.中国植物志(第77卷.1).北京:科学出版社,1-369.

中国科学院中国植物志编辑委员会.1989.中国植物志(第77卷.2).北京:科学出版社,1-188.

中国科学院中国植物志编辑委员会.1987.中国植物志(第78卷.1).北京:科学出版社,1-226.

中国科学院中国植物志编辑委员会.1996.中国植物志(第79卷).北京:科学出版社,1-113.

中国科学院中国植物志编辑委员会.1997.中国植物志(第80卷.1).北京:科学出版社,1-342.

中国科学院中国植物志编辑委员会.1999.中国植物志(第80卷.2).北京:科学出版社,1-94.

中国科学院中国植物志编辑委员会.2006.中国植物志(中名和拉丁名总索引).北京:科学出版社,1-1155.

中国民族药志编辑委员会.1984.中国民族药志.北京:人民卫生出版社.

中国生物多样性国情研究报告编写组.1998.中国生物多样性国情研究报告.北京:中国环境科学出版社.

中国药材公司编著.1994.中国中药资源志要.北京:科学出版社,1-2069.

中国药学会,天津药物研究院主办.1985-2004.中草药.天津.

中国医学科学院药物研究所等.1979.中药志(第1册).北京:人民卫生出版社,1-604.

中国医学科学院药物研究所等.1982.中药志(第2册).北京:人民卫生出版社,1-608.

中国医学科学院药物研究所等.1961.中药志(第3册).北京:人民卫生出版社,1-725.

中国医学科学院药物研究所等.1998.中药志(第6册).北京:人民卫生出版社,1-404.

中国植物学会.1994.中国植物学史.北京:科学出版社,1-376.

中国植物学会.1983.中国植物学文献目录(第1册).北京:科学出版社,1-620.

中国植物学会.1983.中国植物学文献目录(第2册).北京:科学出版社,621-1226.

中国植物学会.1983.中国植物学文献目录(第3册).北京:科学出版社,1227-1793.

中国植物学会.1995.中国植物学文献目录(第4册).北京:科学出版社,1-1463.

中华林学会森林植物编辑小组.1983.中华树木名汇.中华林学会台北,1-241.

中山大学学报编辑委员会.1960-2002.中山大学学报(Act. Sci. Nat. Univ. Sunyatseni).广州.

中药辞海编写组.1993.中药辞海(第1卷).北京:中国医药科技出版社,1-2429+57.

中药辞海编写组.1996.中药辞海(第2卷).北京:中国医药科技出版社,1-2399+49.

中药辞海编写组.1997.中药辞海(第3卷).北京:中国医药科技出版社,1-1657+40.

中药辞海编写组.1998.中药辞海(第4卷).北京:中国医药科技出版社,1-159+5+332+464+460+178+7.

朱光华译.2001.国际植物命名法规(圣路易斯法规).北京:科学出版社,密苏里植物园出版社,1-410.

朱家楠主编.2001.汉拉英种子植物名称.北京:科学出版社,1-1393.

朱亚民主编.2000.内蒙古植物药志(第1卷).呼和浩特:内蒙古人民出版社,1-671.

朱亚民主编.1989.内蒙古植物药志(第2卷).呼和浩特:内蒙古人民出版社,1-523.

朱亚民主编.1989.内蒙古植物药志(第3卷).呼和浩特:内蒙古人民出版社,1-577.

朱有昌主编.1989.东北药用植物.哈尔滨:黑龙江科学技术出版社,1-1300.

庄兆祥,李甯汉主编.1978.香港中草药(第1辑).香港:商务印书馆,1-216.

庄兆祥,李甯汉主编.1981.香港中草药(第2辑).香港:商务印书馆,1-219.

庄兆祥,李甯汉主编.1983.香港中草药(第3辑).香港:商务印书馆,1-219.

庄兆祥,李甯汉主编.1985.香港中草药(第4辑).香港:商务印书馆,1-220.

庄兆祥,李甯汉主编.1986.香港中草药(第5辑).香港:商务印书馆,1-286.

庄兆祥,李甯汉主编.1985.香港中草药(第6辑).香港:商务印书馆.

北村四郎,村田源,堀胜.1983.原色日本植物图鉴.草本编(1).改订版.东京:保育社,1-297.

北村四郎,村田源,小山铁夫.1984.原色日本植物图鉴.草本编(3).改订版.东京:保育社,1-465.

北村四郎,村田源.1984.原色日本植物图鉴.草本编(2).改订版.东京:保育社,1-390.

长田武正.1976.原色日本归化植物图鉴.东京:保育社,1-425.

初岛住彦,天野铁夫.1977.琉球植物名录.冲绳:でぃご出版社,282.

大井次三郎主编.1953.日本植物志.东京:至文堂,1-1383.

大井次三郎主编.1978.日本植物志.东京:至文堂,1-1584.

角仓一.1959.有用植物の学名解.东京:广川书店,1-234.

堀田满等.1989.世界有用植物事典.东京:平凡社,1-1499.

木村康一,木村孟淳.1954.原色日本药用植物图鉴.东京:保育社,1-184.

牧野富太郎,根本莞尔.1931.日本植物总览.东京:春阳堂,1-1935.

牧野富太郎,清水藤太郎.1935.植物学名辞典.东京:第一书房,1-302.

牧野富太郎.1955(昭和三十年).牧野日本植物图鉴(增补版).东京:北隆馆,1-1304.

牧野富太郎.1977.牧野新日本植物图鉴.东京:北隆馆,1-1060.

牧野富太郎.2000.新订牧野新日本植物图鉴.东京:北隆馆.

牧野富太郎.1986.原色牧野日本植物图鉴Ⅰ.东京:北隆馆,1-395.

牧野富太郎.1986.原色牧野日本植物图鉴Ⅱ.东京:北隆馆,1-360.

牧野富太郎.1986.原色牧野日本植物图鉴Ⅲ.东京:北隆馆,1-404.

山田常雄,前川文夫,江上不二夫等.1983.岩波生物学辞典第三版.东京:岩波书店,1-1654.

松村任三.1916(大正4年).改订植物名汇.丸善株式会社.

最新園芸大辞典编集委员会.1968.最新園芸大辞典(第1卷).东京:诚文堂新光社,1-487.

最新園芸大辞典编集委员会.1968.最新園芸大辞典(第2卷).东京:诚文堂新光社,489-1030.

最新園芸大辞典编集委员会.1968.最新園芸大辞典(第3卷).东京:诚文堂新光社,1031-1596.

最新園芸大辞典编集委员会.1968.最新園芸大辞典(第4卷).东京:诚文堂新光社,1597-2182.

最新園芸大辞典编集委员会.1968.最新園芸大辞典(第5卷).东京:诚文堂新光社,2183-2763.

最新園芸大辞典编集委员会.1968.最新園芸大辞典(第6卷).东京:诚文堂新光社,2764-3255.

最新園芸大辞典编集委员会.1968.最新園芸大辞典(第7卷.索引.文献).东京:诚文堂新光社.

最新園芸大辞典编集委员会.1968.最新園芸大辞典(第8卷.补遗编).东京:诚文堂新光社,1-380.

Committee of the Flora of Taiwan. 1975. Flora of Taiwan(台湾植物志)Vol. 1. Taipei. Epoch Publishing Co. 1-561.

Committee of the Flora of Taiwan. 1976. Flora of Taiwan(台湾植物志)Vol. 2. Taipei. Epoch Publishing Co. 1-722.

Committee of the Flora of Taiwan. 1993. Flora of Taiwan(台湾植物志)Vol. 3(second edition). Taipei. Epoch Publishing Co. 1-1084.

Committee of the Flora of Taiwan. 1977. Flora of Taiwan(台湾植物志)Vol. 3. Taipei. Epoch Publishing Co. 1-1000.

Committee of the Flora of Taiwan. 1978. Flora of Taiwan(台湾植物志)Vol. 4. Taipei. Epoch Publishing Co. 1-994.

Committee of the Flora of Taiwan. 1978. Flora of Taiwan(台湾植物志)Vol. 5. Taipei. Epoch Publishing Co. 1-1166.

Committee of the Flora of Taiwan. 1979. Flora of Taiwan(台湾植物志)Vol. 6. Taipei. Epoch Publishing Co. 1-665.

D C Watts. 2000. Elsevier's Dictionary of Plant Names and Their Origin. Elsevier. 1-1001.

D J Mabberley. 1997. The Plant-Book(second edition). Cambridge. Cambridge University Press. 1-858.

Flora of North America Editorial Committee. 1993. Flora of North America. Vol. 1. New York Oxford. Oxford University Press. 1-372.

Flora of North America Editorial Committee. 1993. Flora of North America. Vol. 2. New York Oxford. Oxford University Press. 1-475.

Flora of North America Editorial Committee. 1997. Flora of North America. Vol. 3. New York Oxford. Oxford University Press. 1-590.

Flora of North America Editorial Committee. 2000. Flora of North America. Vol. 22. New York Oxford. Oxford University Press. 1-352.

Flora of North America Editorial Committee. Flora of North America. Vol. 4.

Flora of North America Editorial Committee. Flora of North America. Vol. 5.

Flora of North America Editorial Committee. Flora of North America. Vol. 7.

Flora of North America Editorial Committee. Flora of North America. Vol. 8.

Flora of North America Editorial Committee. Flora of North America. Vol. 19.

Flora of North America Editorial Committee. Flora of North America. Vol. 20.

Flora of North America Editorial Committee. Flora of North America. Vol. 21.

Flora of North America Editorial Committee. Flora of North America. Vol. 23.

Flora of North America Editorial Committee. Flora of North America. Vol. 26.

Flora of North America Editorial Committee. Flora of North America. Vol. 27.

J C Willis. 1985. A Dictionary of the Flowering Plants and Ferns (Student Edition). Cambridge. Cambridge University Press. 1-1245.

John Laird Farrar. 1995. Trees in Canada. Fitzhenry & Whiteside Limited and the Canadian Forest Service.

K Kubitzki. 1990. The Families and Genera of Vascular Plants. Vol. 1. Springer—Verlag, Berlin, Heidelberg, New York, London, Paris, Tokyo. Hong Kong, Barcelona. 1-404.

K Kubitzki. 1993. The Families and Genera of Vascular Plants. Vol. 2. Springer—Verlag, Berlin, Heidelberg, New York, London, Paris, Tokyo, Hong Kong, Barcelona, Budapest. 1-653.

K Kubitzki. 1998. The Families and Genera of Vascular Plants. Vol. 3. Springer—Verlag, Berlin, Heidelberg, New York, Barcelona, Budapest, Hong Kong, London, Milan, Paris, Singapore, Tokyo. 1-478.

K Kubitzki. 1998. The Families and Genera of Vascular Plants. Vol. 4. Springer—Verlag, Berlin, Heidelberg, New York, Barcelona, Budapest, Hong Kong, London, Milan, Paris, Singapore, Tokyo. 1-511.

K Kubitzki. 2003. The Families and Genera of Vascular Plants. Vol. 5. Springer—Verlag, Berlin, Heidelberg, New York. 1-418.

K Kubitzki. 2004. The Families and Genera of Vascular Plants. Vol. 6. Springer—Verlag, Berlin, Heidelberg, New York. 1-489.

K Kubitzki. 2004. The Families and Genera of Vascular Plants. Vol. 7. Springer—Verlag, Berlin, Heidelberg, New York. 1-478.

K Kubitzki. The Families and Genera of Vascular Plants. Vol. Springer—Verlag, Berlin, Heidelberg, New York, Barcelona, Budapest, Hong Kong, London, Milan, Paris, Singapore, Tokyo.

K Kubitzki. The Families and Genera of Vascular Plants. Vol. Springer—Verlag, Berlin, Heidelberg, New York, Barcelona, Budapest, Hong Kong, London, Milan, Paris, Singapore, Tokyo.

Murray Wrobel and Geoffrey Creber. 1996. Elsevier's Dictionary of Plant Names (In Latin, English, Franch, German and Italian). Amsterdam, Lausanne, New York. Oxyford, Shannon, Tokyo. Elsevier. 1-623.

P Macura. 1979. Elsevier's Dictionary of Botany (Ⅰ. Plant Names. in English, Franch, German, Latin and Russian). New York. Elsevier Scientific Publishing Company. 1-580.

R K Brummitt & C E Powell. 1992. Authors of Plant Names. Kew. Royal Botnic Gardens. 1-732.

R K Brummitt. 1992. Vascular Plant Familis and Genera. Kew. Royal Botanic Gardens. 1-804.

T G Tutinet al. 1968. Flora Europaea. Cambridge. Vol. 2. Cambridge Universiry Press. 1-455.

T G Tutinet al. 1964. Flora Europaea. Vol. 1. Cambridge. Cambridge Universiry Press. 1-464.

T G Tutinet al. 1972. Flora Europaea. Vol. 3. Cambridge. Cambridge Universiry Press. 1-370.

T G Tutinet al. 1976. Flora Europaea. Vol. 4. Cambridge. Cambridge Universiry Press. 1-505.

T G. 1980. Tutinet al. Flora Europaea. Vol. 5. Cambridge. Cambridge Universiry Press. 1-452.

Tatiana Wielgorskaya. 1995. Dictionary of Generic Names of Seed Plants. New York. Columbia University Press. 1-570.

The Audubon Society. 1987. Field Guide To North American Trees. The Audubon Society.

William T. 1983. Stearn. Botanical Latin (Third Edition). David & Charles. Newton Abbot London North Pomfret (Vt).

William T. 1983. Stearn. Botnical Latin. 3rd. Newton Abbot London North Pomfret. David & Charles. 1-566.

Z Y Wu, P H Raven & D Y Hong. Flora of China Vol. 10. Beijing. Science Press; and St. Louis. Missouri Botanical Garden Press.

Z Y Wu, P H Raven & D Y Hong. 2009. Flora of China Vol. 11. Beijing. Science Press; and St. Louis. Missouri Botanical Garden Press. 1-634.

Z Y Wu, P H Raven & D Y Hong. Flora of China Vol. 12. Beijing. Science Press; and St. Louis. Missouri Botanical Garden

Press.

Z Y Wu, P H Raven & D Y Hong. 2007. Flora of China Vol. 13. Beijing. Science Press; and St. Louis. Missouri Botanical Garden Press. 1-548.

Z Y Wu, P H Raven & D Y Hong. 2007. Flora of China Vol. 14. Beijing. Science Press; and St. Louis. Missouri Botanical Garden Press.

Z Y Wu, P H Raven & D Y Hong. 2001. Flora of China Vol. 15. Beijing. Science Press; and St. Louis. Missouri Botanical Garden Press.

Z Y Wu, P H Raven & D Y Hong. 2009. Flora of China Vol. 16. Beijing. Science Press; and St. Louis. Missouri Botanical Garden Press.

Z Y Wu, P H Raven & D Y Hong. 2009. Flora of China Vol. 18. Beijing. Science Press; and St. Louis. Missouri Botanical Garden Press.

Z Y Wu, P H Raven & D Y Hong. Flora of China Vol. 19. Beijing. Science Press; and St. Louis. Missouri Botanical Garden Press.

Z Y Wu, P H Raven & D Y Hong. Flora of China Vol. 2. Beijing. Science Press; and St. Louis. Missouri Botanical Garden Press.

Z Y Wu, P H Raven & D Y Hong. Flora of China Vol. 20. Beijing. Science Press; and St. Louis. Missouri Botanical Garden Press.

Z Y Wu, P H Raven & D Y Hong. Flora of China Vol. 21. Beijing. Science Press; and St. Louis. Missouri Botanical Garden Press.

Z Y Wu, P H Raven & D Y Hong. 2008. Flora of China Vol. 22. Beijing. Science Press; and St. Louis. Missouri Botanical Garden Press. 1-937.

Z Y Wu, P H Raven & D Y Hong. Flora of China Vol. 23. Beijing. Science Press; and St. Louis. Missouri Botanical Garden Press.

Z Y Wu, P H Raven & D Y Hong. 2006. Flora of China Vol. 24. Beijing. Science Press; and St. Louis. Missouri Botanical Garden Press. 1-430.

Z Y Wu, P H Raven & D Y Hong. 2010. Flora of China Vol. 25. Beijing. Science Press; and St. Louis. Missouri Botanical Garden Press. 1-570.

Z Y Wu, P H Raven & D Y Hong. Flora of China Vol. 3. Beijing. Science Press; and St. Louis. Missouri Botanical Garden Press.

Z Y Wu, P H Raven & D Y Hong. 2001. Flora of China Vol. 4. Beijing. Science Press; and St. Louis. Missouri Botanical Garden Press.

Z Y Wu, P H Raven & D Y Hong. 2001. Flora of China Vol. 6. Beijing. Science Press; and St. Louis. Missouri Botanical Garden Press.

Z Y Wu, P H Raven & D Y Hong. 2009. Flora of China Vol. 7. Beijing. Science Press; and St. Louis. Missouri Botanical Garden Press. 1-499.

Z Y Wu, P H Raven & D Y Hong. Flora of China Vol. 8. Beijing. Science Press; and St. Louis. Missouri Botanical Garden Press.

Z Y Wu, P H Raven & D Y Hong. 2004. Flora of China Vol. 9. Beijing. Science Press; and St. Louis. Missouri Botanical Garden Press.

Z Y Wu, P H Raven & D Y Hong. 2004. Flora of China Vol. 5. Beijing. Science Press; and St. Louis. Missouri Botanical Garden Press.

Z Y Wu, P H Raven & D Y Hong. 1994. Flora of China Vol. 17. Beijing. Science Press; and St. Louis. Missouri Botanical Garden Press.

А И Введенский. 1961. Флора Узбекистана Том. 5. Издательство Академии Наук Узбекской ССР. 1-667.

В Л Комаров. 1934. Флора СССР. Том. 1. Москва. Издательство Акалемик Наук СССР. 1-298.

В Л Комаров. 1941. Флора СССР. Том. 10. Москва. Издательство Акалемик Наук СССР. 1-673.

В Л Комаров. 1945. Флора СССР. Том. 11. Москва. Издательство Акалемик Наук СССР. 1-432.

В Л Комаров. 1939. Флора СССР. Том. 9. Москва. Издательство Акалемик Наук СССР. 1-539.

В Л Комаров. 1941. Флора СССР. Том. 10. Москва. Издательство Акалемик Наук СССР. 1-673.

В Л Комаров. 1945. Флора СССР. Том. 11. Москва. Издательство Акалемик Наук СССР. 1-432.

В Л Комаров. 1946. Флора СССР. Том. 12Москва. Издательство Акалемик Наук СССР. 1-679(英译本).

В Л Комаров. 1948. Флора СССР. Том. 13. Москва. Издательство Акалемик Наук СССР. 1-586.

В Л Комаров. 1949. Флора СССР. Том. 14. Москва. Издательство Акалемик Наук СССР. 1-790.

В Л Комаров. 1949. Флора СССР. Том. 15. Москва. Издательство Акалемик Наук СССР. 1-742.

В Л Комаров. 1950. Флора СССР. Том. 16. Москва. Издательство Акалемик Наук СССР. 1-648.

В Л Комаров. 1951. Флора СССР. Том. 17. Москва. Издательство Акалемик Наук СССР. 1-390.

В Л Комаров. 1952. Флора СССР. Том. 18. Москва. Издательство Акалемик Наук СССР. 1-802.

В Л Комаров. 1953. Флора СССР. Том. 19. Москва. Издательство Акалемик Наук СССР. 1-751.

В Л Комаров. 1954. Флора СССР. Том. 20. Москва. Издательство Акалемик Наук СССР. 1-555.

В Л Комаров. 1954. Флора СССР. Том. 21. Москва. Издательство Акалемик Наук СССР. 1-703.

В Л Комаров. 1955. Флора СССР. Том. 22. Москва. Издательство Акалемик Наук СССР. 1-861.

В Л Комаров. 1958. Флора СССР. Том. 23. Москва. Издательство Акалемик Наук СССР. 1-775.

В Л Комаров. 1957. Флора СССР. Том. 24. Москва. Издательство Акалемик Наук СССР. 1-501.

В Л Комаров. 1959. Флора СССР. Том. 25. Москва. Издательство Акалемик Наук СССР. 1-630.

В Л Комаров. 1961. Флора СССР. Том. 26. Москва. Издательство Акалемик Наук СССР. 1-938.

В Л Комаров. 1962. Флора СССР. Том. 27. Москва. Издательство Акалемик Наук СССР. 1-757.

В Л Комаров. 1963. Флора СССР. Том. 28. Москва. Издательство Акалемик Наук СССР. 1-653.

В Л Комаров. 1964. Флора СССР. Том. 29. Москва. Издательство Акалемик Наук СССР. 1-796.

В Л Комаров. 1960. Флора СССР. Том. 30. Москва. Издательство Акалемик Наук СССР. 1-732.

В Н Ворошилов. 1982. Определитель Растений. Издательство. Наук.

Интродукция Растений Природной. 1979. Флоры СССР. Издательство《Наука》.

М Г Попов. 1979. Флора Средней Сибири Том. 1. Издательство Академии Наук. СССР.

М Г Попов. 1959. Флора Средней Сибири Том. 2. Издательство Академии Наук. СССР.

Н В Павлов. 1956. флора Казахстана. Том. 1. Издательство Академии Наук Казахской СССР. 1-351.

Н В Павлов. 1958. флора Казахстана. Том. 2. Издательство Академии Наук Казахской СССР. 1-290.

Н В Павлов. 1960. флора Казахстана. Том. 3. Издательство Академии Наук Казахской СССР. 1-319.

Н В Павлов. 1961. флора Казахстана. Том. 4. Издательство Академии Наук Казахской СССР. 1-545.

Н В Павлов. 1961. флора Казахстана. Том. 5. Издательство Академии Наук Казахской СССР. 1-513.

Н В Павлов. 1963. флора Казахстана. Том. 6. Издательство Академии Наук Казахской СССР. 1-463.

Н В Павлов. 1964. флора Казахстана. Том. 7. Издательство Академии Наук Казахской СССР. 1-495.

Н В Павлов. 1965. флора Казахстана. Том. 8. Издательство Академии Наук Казахской СССР. 1-445.

Н В Павлов. 1966. флора Казахстана. Том. 9. Издательство Академии Наук Казахской СССР. 1-639.

О Л Петровичева. 1957. Флора Ленинградской Области. Том. 2. Ленинград. Издательство Ленинградского Университета.

П Н Овчинников. 1957. Флора Таджикской СССР. Том. 1. Москва. Издательство Академии Наук СССР.

С М Кирова. 1956. Флора Мурманской Области. Том. 3. Москва. Издательство Академии Наук СССР. 1-449.

С М Кирова. 1966. Флора Мурманской Области. Том. 5. Москва. Издательство Академии Наук СССР.

لبه